MW00759241

Biomaterials Science

Biomaterials Science

An Introduction to Materials in Medicine

Third Edition

Edited by

Buddy D. Ratner, Ph.D.
Professor, Bioengineering and Chemical Engineering,
Director of University of Washington Engineered Biomaterials (UWEB),
University of Washington, Seattle, WA, USA

Allan S. Hoffman, Sc.D.
Professor of Bioengineering and Chemical Engineering,
UWEB Investigator, University of Washington, Seattle, WA, USA

Frederick J. Schoen, M.D., Ph.D.
Professor of Pathology and Health Sciences and Technology (HST),
Harvard Medical School; Executive Vice Chairman,
Department of Pathology, Brigham and Women's Hospital;
Boston, MA, USA

Jack E. Lemons, Ph.D.
University Professor, Schools of Dentistry, Medicine and Engineering,
University of Alabama at Birmingham, Birmingham, AL, USA

AMSTERDAM • BOSTON • HEIDELBERG • LONDON • NEW YORK
OXFORD • PARIS • SAN DIEGO • SAN FRANCISCO • SYDNEY • TOKYO
Academic Press is an imprint of Elsevier

Academic Press is an imprint of Elsevier
The Boulevard, Langford Lane, Kidlington, Oxford, OX5 1GB, UK
225 Wyman Street, Waltham, MA 02451, USA

First published 1997
Second edition 2004
Third edition 2013

British Library Cataloguing in Publication Data
A catalogue record for this book is available from the British Library

Library of Congress Cataloging-in-Publication Data
A catalog record for this book is available from the Library of Congress

ISBN: 978-0-12-374626-9

For information on all Academic Press publications
visit our website at store.elsevier.com

Printed and bound in Canada

12 13 14 15 10 9 8 7 6 5 4 3 2 1

CONTENTS

PART TWO
Biology and Medicine

SECTION II.1 Some Background Concepts

SECTION II.2 Host Reaction to Biomaterials and Their Evaluation

SECTION II.3 Biological Testing of Biomaterials

SECTION II.4 Degradation of Materials in the Biological Environment

SECTION II.5 Applications of Biomaterials

Appendices

CONTRIBUTORS

Sascha Abramson
Department of Chemistry, Rutgers, The State University of New Jersey, New Brunswick, NJ, USA

D. Michael Ackermann, Jr.
Biodesign, Stanford University, Stanford, CA, USA

Robert Akins
Tissue Engineering and Regenerative Medicine Laboratory, Alfred I. duPont Hospital for Children/ Nemours Foundation, Wilmington, DE, USA

Richard Anders
Senior Lecturer, Massachusetts Institute of Technology, Cambridge, MA, USA

Phillip J. Andersen
Andersen Metallurgical, LLC

James M. Anderson
Department of Pathology, University Cleveland, OH, USA

James A. Ankrum
Harvard-MIT Division of Health Sciences & Technology, USA; Department of Medicine, Brigham and Women's Hospital, Harvard Medical School, USA

Kristi S. Anseth
Department of Chemical and Biological Engineering, Howard Hughes Medical Institute and University of Colorado, Boulder, CO, USA

Joe Antonucci
National Institute of Standards and Technology, Gaithersburg, MD, USA

Sarah Atzet
Glycosan BioSystems, Inc., Salt Lake City, Utah, USA

Stephen F. Badylak
McGowan Institute for Regenerative Medicine, University of Pittsburgh, Pittsburgh, PA, USA

Gail D. Baura
Keck Graduate Institute of Applied Life Sciences, Claremont, CA, USA

Ravi V. Bellamkonda
Department of Biomedical Engineering, Georgia Institute of Technology, Atlanta, GA, USA

Serena M. Best
University of Cambridge, Cambridge, UK

Sarindr Bhumiratana
Department of Biomedical Engineering, Columbia University, New York, USA

Richard W. Bianco
Department of Surgery, University of Minnesota, Minneapolis, MN, USA

Jack C. Bokros
On-X Life Technologies, Inc, Austin, TX, USA

Harvey S. Borovetz
University of Pittsburgh, Departments of Surgery and Departments of Bioengineering, & Chemical Engineering, Pittsburgh, PA, USA

Adele L. Boskey
Starr Chair in Mineralized Tissue Research, Hospital for Special Surgery and Professor, Weill Medical College of Cornell University; BME Faculty, Sibley School of Mechanical Engineering, Cornell, Ithaca

Justin L. Brown
Department of Bioengineering, The Pennsylvania State University, PA, USA

Bryan N. Brown
McGowan Institute for Regenerative Medicine, University of Pittsburgh, Pittsburgh, PA, USA

Stanley A. Brown
(retired)
17704 Stoneridge Dr., Gaithersburg MD, USA

John B. Brunski
Division of Plastic & Reconstructive Surgery, Department of Surgery, School of Medicine, Standford University, Standford, CA, USA

Fred Cahn
BioMedical Strategies, La Jolla, CA, USA

Alastair Campbell Ritchie
Department of Mechanical, Materials and Manufacturing Engineering, University of Nottingham, Nottingham, UK

Arnold I. Caplan
Professor of Biology, Director Skeletal Research Center, Case Western Reserve University, Cleveland, USA

Richard L. Carpenedo
The Wallace H. Coulter Department of Biomedical Engineering, Georgia Institute of Technology, Atlanta, Georgia, USA

Ashutosh Chilkoti
Center for Biologically Inspired Materials and Material Systems and Department of Biomedical Engineering, Duke University, Durham, NC, USA

Sangwon Chung
Fiber and Polymer Science, Department of Textile Engineering, Chemistry & Science, North Carolina State University, Raleigh, NC, USA; Biomedical Engineering, Joint Department of Biomedical Engineering, North Carolina State University and University of North Carolina at Chapel Hill, NC, USA

Elisa Cimetta
Department of Biomedical Engineering, Columbia University, New York, USA

Gary Cleary
Corium Inc. Menlo Park, CA, USA

Isaac P. Clements
Department of Biomedical Engineering, Georgia Institute of Technology, Atlanta, GA, USA

André Colas
Dow Corning Europe S.A., Parc Industriel, Seneffe, Belgium

Kelly P. Coleman
Medtronic Inc., Physiological Research Laboratory, Minneapolis, MN, USA

Daniel E. Conway
Robert M. Berne Cardiovascular Research Center, University of Virginia, Charlottesville, Virginia, USA

Stuart L. Cooper
William G. Lowrie Department of Chemical and Biomolecular Engineering, The Ohio State University, Columbus, OH, USA

Bill Costerton
Center for Genomic Sciences, Allegheny-Singer Research Institute, Pittsburgh, PA, USA; Department of Orthopaedic Surgery, Allegheny General Hospital, Pittsburgh, PA, USA

Arthur J. Coury
Coury Consulting Services, Boston, MA, USA

Crystal Cunanan
Tissue Engineering at Boston Scientific, Los Gatos, CA, USA

Jim Curtis
Dow Corning Corporation, Midland, MI, USA

Antonio D'Amore
University of Pittsburgh, Pittsburgh, Pennsylvania, USA

Patrick DeMeo
Department of Orthopaedic Surgery, Allegheny General Hospital, Pittsburgh, PA, USA

Tejal A. Desai
Department of Bioengineering and Therapeutic Sciences/ Department of Physiology, University of California, San Francisco, CA, USA

Sabine Dickens
American Dental Association Foundation Pattenbarger Research Center, Gaithersburg, MD, USA

Gonzalo Domingo
Author is affiliated with PATH (www.path.org)

Elaine Duncan
Paladin Medical, Inc. Stillwater, MN, USA

Suzanne G. Eskin
Department of Biomedical Engineering, Georgia Institute of Technology and Emory University School of Medicine, Atlanta, Georgia, USA

David W. Feigal, Jr.
Adjunct Faculty, Sandra Day O'Connor School of Law, Arizona State University, Santa Rosa Valley, CA, USA

Lino Ferreira
Center of Neurosciences and Cell Biology, University of Coimbra, Portugal; Biocant-Biotechnology Innovation Center, Cantanhede, Portugal

Jason Fuller
Third Rock Ventures, Boston, MA, USA

Robert P. Gallegos
Department of Surgery, University of Minnesota, Minneapolis, MN, USA

Ellen Gawalt
Department of Chemistry and Biochemistry, Duquesne University, Pittsburgh, PA, USA

Kaustabh Ghosh
Vascular Biology Program, Departments of Pathology and Surgery, Children's Hospital, Harvard Medical School, Boston, USA

Bilal Ghosn
Department of Bioengineering, University of Washington, Seattle, WA, USA

Thomas W. Gilbert
McGowan Institute for Regenerative Medicine, University of Pittsburgh, Pittsburgh, PA, USA

Drew Elizabeth Glaser
School of Engineering, University of California at Merced, Merced, CA, USA

Amandine Godier-Furnemont
Department of Biomedical Engineering, Columbia University, New York, USA

Wayne R. Gombotz
Immune Design Corp., Seattle, WA, USA

David W. Grainger
Bioengineering, University of Utah, Salt Lake City, UT, USA

Gary L. Grunkemeier
Medical Data Research Center, Providence Health & Services, Portland, Oregon, USA

S. Adam Hacking
Laboratory for Musculoskeletal Research and Innovation, Department of Orthopaedic Surgery, Massachusetts General Hospital and Harvard Medical School, Boston, MA, USA

Nadim James Hallab
Department of Orthopedic Surgery, Rush University Medical Center, Chicago, IL, USA

Luanne Hall-Stoodley
Welcome Trust Clinical Research Facility, University Hospital, Southampton Foundation Trust, Southampton, UK; NIHR Respiratory BRU, Faculty of Medicine, University of Southampton, UK

Stephen R. Hanson
Division of Biomedical Engineering, School of Medicine, Oregon Health & Science University, Portland, Oregon, USA

Axel D. Haubold
Bed Rock Ranch, Decatur, TX, USA

Kip D. Hauch
University of Washington, Seattle, WA, USA

Kenneth R. Hawkins
Author is affiliated with PATH (www.path.org)

Daniel E. Heath
William G. Lowrie Department of Chemical and Biomolecular Engineering, The Ohio State University, Columbus, OH, USA

Douglas L. Helm
Division of Plastic and Reconstructive Surgery, Brigham and Women's Hospital, Boston, MA, USA

Larry L. Hench
Imperial College, London, UK

Arne Hensten
Department of Clinical Densitry, University of Tromsø, Tromsø, Norway

Ryan T. Hill
Center for Biologically Inspired Materials and Material Systems, Duke University, Durham, NC, USA

Christopher Hobson
University of Pittsburgh, Pittsburgh, Pennsylvania, USA

Simon P. Hoerstrup
Cardiovascular Surgery Research University and University Hospital Zurich, Zurich, Switzerland

Allan S. Hoffman
Professor of Bioengineering and Chemical Engineering, UWEB Investigator, University of Washington, Seattle, WA, USA

Thomas A. Horbett
Bioengineering and Chemical Engineering, University of Washington, Seattle, WA, USA

Jeffrey A. Hubbell
EPFL (École Polytechnique Fédérale de Lausanne), Switzerland

Mark S. Humayun
University of Southern California, Los Angeles, CA, USA

Ray Ideker
The University of Alabama at Birmingham, Birmingham, AL, USA

Donald E. Ingber
Vascular Biology Program, Departments of Pathology and Surgery, Children's Hospital and Harvard Medical School, Harvard School of Engineering and Applied Sciences, Wyss Institute for Biologically Inspired Engineering at Harvard University, Boston, MA, USA

Rakhi Jain
Implant R&D, Abbott Medical Optics, Inc., Santa Ana, CA, USA

Jean Jacob
LSU Eye Center, New Orleans, LA, USA

Joshua James Jacobs
Department of Orthopedic Surgery, Rush University Medical Center, Chicago, IL, USA

Nils Jacobsen
Nordic Institute of Dental Materials Sognsveien, Ullevål Stadion, Oslo, Norway

Ruyun Jin
Medical Data Research Center, Providence Health & Services, Portland, Oregon, USA

Richard J. Johnson
Baxter Healthcare Corporation, Round Lake, IL, USA

Jeffrey M. Karp
Harvard-MIT Division of Health Sciences & Technology, USA; Department of Medicine, Brigham and Women's Hospital, Harvard Medical School, USA; Harvard Stem Cell Institute, USA

F. Kurtis Kasper
Department of Bioengineering, Rice University, Houston, TX, USA

Sandeep Kathju
Division of Plastic Surgery, University of Pittsburgh, School of Medicine, Pittsburgh, PA, USA

Ali Khademhosseini
Department of Medicine, Center for Biomedical Engineering, Brigham and Women's Hospital, Harvard Medical School, Cambridge, MA, USA and Harvard-MIT Health Sciences and Technology, Massachusetts Institute of Technology, Cambridge, MA, USA

Martin W. King
Department of Textile Engineering, Chemistry & Science, North Carolina State University, Raleigh, NC, USA; Donghua University, Shanghai, China

Lothar W. Kleiner
Drug Delivery and Medical Device Polymer Consultant, Los Altos, CA, USA

Joachim Kohn
Department of Chemistry, Rutgers, The State University of New Jersey, New Brunswick, NJ, USA

Heidi E. Koschwanez
Department of Biomedical Engineering, Duke University, Durham, NC, USA

Sangamesh G. Kumbar
Department of Orthopaedic Surgery, Department of Chemical, Materials and Biomolecular Engineering, University of Connecticut, CT, USA

Catherine K. Kuo
Department of Biomedical Engineering, Tufts University, Medford, MA, USA

Lisa LaFleur
Department of Bioengineering, University of Washington, Seattle, WA, USA

Matthew T. Lahti
Department of Surgery, University of Minnesota, Minneapolis, MN, USA

Byron Lambert
Sterilization Science, Abbott Vascular, Temecula, CA, USA

Robert Langer
Department of Chemical Engineering, Massachusetts Institute of Technology, Cambridge, MA, USA

Cato T. Laurencin
Connecticut Institute for Clinical and Translational Science; Director, Institute for Regenerative Engineering, University of Connecticut, CT, USA

David Lee-Parritz
Department of Comparative Medicine, Genzyme Corporation, Framingham, MA, USA

Jack E. Lemons
University Professor, Schools of Dentistry, Medicine and Engineering, University of Alabama at Birmingham, Birmingham, AL, USA

Mark Levin
Third Rock Ventures, Boston, MA, USA

Robert J. Levy
Abramson Pediatric Research Center, The Children's Hospital of Philadelphia, University of Pennsylvania, Philadelphia, PA, USA

Gregory M. Lewerenz
Medtronic Inc., Physiological Research Laboratory, Minneapolis, MN, USA

Wan-Ju Li
Departments of Orthopedics and Rehabilitation, and Biomedical Engineering, University of Wisconsin-Madison, Madison, WI, USA

Chien-Chi Lin
Department of Biomedical Engineering, Indiana University-Purdue, University at Indianapolis, Indianapolis, IN, USA

Fang Liu
Departments of Pharmaceutics and Pharmaceutical Chemistry, University of Utah, Salt Lake City, UT, USA

William G. Lowrie
Department of Chemical and Biomolecular Engineering, The Ohio State University, Columbus, OH, USA

Michael J. Lysaght
Center for Biomedical Engineering, Brown University, Providence, RI, USA

Robert Maidhof
Department of Biomedical Engineering, Columbia University, New York, USA

J. N. Mansbridge
Histogen, Inc., San Diego, CA, USA

M. Cristina L. Martins
INEB - Instituto de Engenharia, Biomedica, Divisão de Biomateriais, NEWTherapies Group, Universidade do Porto, Porto, Portugal

Jeffrey Martin
Director of Corporate Quality Technology, Alcon® a Novartis Company, Fort Worth, Texas, USA

Jay P. Mayesh
Kaye Scholer, New York, USA

Todd C. McDevitt
The Wallace H. Coulter Department of Biomedical Engineering, The Parker H. Petit Institute for Bioengineering and Bioscience, Georgia Institute of Technology, Atlanta, Georgia, USA

Larry V. McIntire
Department of Biomedical Engineering, Georgia Institute of Technology and Emory University School of Medicine, Atlanta, Georgia, USA

Katharine Merrit
(retired)
17704 Stoneridge Dr., Gaithersburg MD, USA

Claudio Migliaresi
Department of Materials Engineering and Industrial Technologies and BioTech Research Center, University of Trento, Trento, Italy

Antonios G. Mikos
Department of Bioengineering, Rice University, Houston, TX, USA

Carl E. Misch
DDS, MDS, Misch International Institute, Beverly Hills, MI, USA

Richard N. Mitchell
Department of Pathology, Brigham and Women's Hospital and Harvard Medical School, Boston, MA, USA

Robert B. More
Integra Life Sciences, Austin, TX, USA

Christa W. Moss
Biomedical Engineering, Case Western Reserve University, Cleveland, Ohio, OH, USA

Jennifer M. Munson
Department of Biomedical Engineering, Georgia Institute of Technology, Atlanta, GA, USA

Melba Navarro
Institute for Bioengineering of Catalonia (IBEC) Baldiri Reixac 10-12, Barcelona

Robert M. Nerem
Georgia Institute of Technology, Atlanta, GA, USA

Rei Ogawa
Division of Plastic, Reconstructive and Aesthetic Surgery, Nippon Medical School, Tokyo, Japan

Britlyn D. Orgill
Division of Plastic and Reconstructive Surgery, Brigham and Women's Hospital, Boston, MA, USA

Dennis P. Orgill
Division of Plastic and Reconstructive Surgery, Brigham and Women's Hospital, Boston, MA, USA

Robert F. Padera, Jr.
Department of Pathology, Brigham and Women's Hospital and Harvard Medical School, Boston, MA, USA

Abhay Pandit
Network of Excellence for Functional Biomaterials, National University of Ireland, Galway

Anil S. Patel
Retired Vice President of Research for Surgical Products for Alcon, Seattle, WA, USA

Roger B. Peck
Author is affiliated with PATH (www.path.org)

P. Hunter Peckham
Biomedical Engineering, Case Western Reserve University, Cleveland, Ohio, OH, USA

Nicholas A. Peppas
Chemical Engineering Department, University of Texas, Austin, TX, USA

Maria Nunes Pereira
Harvard-MIT Division of Health Sciences & Technology, USA; Department of Medicine, Brigham and Women's Hospital, Harvard Medical School, USA; Center of Neurosciences and Cell Biology, University of Coimbra, Portugal; Biocant-Biotechnology Innovation Center, Cantanhede, Portugal

Josep Planell
Institute for Bioengineering of Catalonia (IBEC) Baldiri Reixac 10-12, Barcelona

Ketul C. Popat
Department of Mechanical Engineering, School of Biomedical Engineering, Colorado State University, Fort Collins, CO, USA

Glenn D. Prestwich
Department of Medicinal Chemistry and The Center for Therapeutic Biomaterials, The University of Utah, Salt Lake City, Utah, USA

Suzie H. Pun
Department of Bioengineering, University of Washington, Seattle, WA, USA

John Rabolt
Department of Materials Science and Engineering, University of Delaware, Newark, DE, USA

Roshni S. Rainbow
Department of Anatomy and Cellular Biology, Tufts University School of Medicine, Boston, MA, USA

Taufiek Rajab
Brigham and Women's Hospital, Boston, MA, USA

Buddy D. Ratner
Professor, Bioengineering and Chemical Engineering, Director of University of Washington Engineered Biomaterials (UWEB), Seattle, WA, USA

William M. Reichert
Department of Biomedical Engineering, Duke University, Durham, NC, USA

Andrew L. Rivard
Department of Surgery, University of Minnesota, Minneapolis, MN, USA

Adrian P. Rowley
University of Southern California, Los Angeles, CA, USA

Gang Ruan
Department of Chemical and Biomolecular Engineering, The Ohio State University, Columbus, OH, USA

Michael Sacks
Department of Biomedical Engineering, The University Texas at Austin, Austin, TX, USA

Debanjan Sarkar
Harvard-MIT Division of Health Sciences & Technology, USA; Department of Medicine, Brigham and Women's Hospital, Harvard Medical School, USA; Department of Biomedical Engineering, University at Buffalo, The State University of New York, USA

Sebastian Schaefer
Harvard-MIT Division of Health Sciences & Technology, USA; Department of Medicine, Brigham and Women's Hospital, Harvard Medical School, USA

Christine E. Schmidt
Department of Biomedical Engineering, The University of Texas at Austin, Austin, TX, USA

Frederick J. Schoen
Professor of Pathology and Health Sciences and Technology (HST), Harvard Medical School, Executive Vice Chairman, Department of Pathology, Brigham and Women's Hospital, Boston, MA, USA

Stacey C. Schutte
Emory University School of Medicine, Woodruff Memorial Building, Atlanta, GA, USA

Michael V. Sefton
Institute of Biomaterials and Biomedical Engineering, University of Toronto, Ontario, Canada

Shalaby W. Shalaby
(1938–2010) was Founder, President, and Director of R&D at Poly-Med, Inc., Anderson, SC, USA

Mark Shirtliff
Department of Microbial Pathogenesis, University of Maryland, Baltimore, MD, USA

Marc A. Simon
University of Pittsburgh, Departments of Medicine, Pittsburgh, PA, USA
McGowan Institute for Regenerative Medicine, Pittsburgh, PA, USA

Milind Singh
Department of Bioengineering, Rice University, Houston, TX, USA

Steven M. Slack
(1961–2010) was Professor in the Department of Biomedical Engineering and Associate Dean of Graduate Studies in the Herff College of Engineering, at the University of Memphis, Tennesse

Francis A. Spelman
Department of Bioengineering, University of Washington, Seattle, WA, USA

Albert Starr
Oregon Health & Science University, Portland, Oregon, USA

Patrick S. Stayton
Department of Bioengineering, University of Washington, Seattle, WA, USA

Roger Steinert
Implant R&D, Abbott Medical Optics, Inc., Santa Ana, CA, USA

Paul Stoodley
National Centre for Advanced Tribology, University of Southampton, Southampton, UK

Shalu Suri
School of Chemical and Biomolecular Engineering, Georgia Institute of Technology, Atlanta GA, USA

Thomas Ming Swi Chang
Director, Artificial Cells and Organs Research Centre, Departments of Physiology, Medicine and Biomedical Engineering, Faculty of Medicine, McGill University, Montreal, QC, Canada

Nina Tandon
Department of Biomedical Engineering, Columbia University, New York, USA

Armand R. Tanguay, Jr.
University of Southern California, Los Angeles, CA, USA

M. Scott Taylor
Poly-Med, Inc., Anderson, SC, USA

Grace S.L. Teo
Harvard-MIT Division of Health Sciences & Technology, USA; Department of Medicine, Brigham and Women's Hospital, Harvard Medical School, USA

Charles K. Thodeti
Vascular Biology Program, Departments of Pathology and Surgery, Children's Hospital, Harvard Medical School, Boston, MA, USA

Joshua Tolkoff
Accelerator Executive, CIMIT, Boston, MA, USA

Matthew Treiser
Department of Biomedical Engineering, Rutgers, The State University of New Jersey, New Brunswick, NJ, USA

Rocky S. Tuan
Department of Orthopaedic Surgery and Center for Cellular and Molecular Engineering, University of Pittsburgh School of Medicine, Pittsburgh, PA, USA

Erik I. Tucker
Division of Biomedical Engineering, School of Medicine, Oregon Health & Science University, Portland, Oregon, USA

Ramakrishna Venugopalan
Auroru Burlington Clinic, Burlington, WI, USA

Angela R. Vicari
Kaye Scholer, New York, USA

Christopher Viney
School of Engineering, University of California at Merced, Merced, CA, USA

Jessica M. Voight
Department of Surgery, University of Minnesota, Minneapolis, MN, USA

Gordana Vunjak-Novakovic
Department of Biomedical Engineering, Columbia University, New York, USA

William R. Wagner
University of Pittsburgh, Departments of Surgery and Departments of Bioengineering, & Chemical Engineering, Pittsburgh, PA, USA

Lian Wang
Medical Data Research Center, Providence Health & Services, Portland, Oregon, USA

Karen R. Wasiluk
Department of Surgery, University of Minnesota, Minneapolis, MN, USA

David Christopher Watts
Professor of Biomaterials Science, The University of Manchester, School of Dentistry & Photon Science Institute, Manchester, UK

Bernhard H. Weigl
Author is affiliated with PATH (www.path.org)

James D. Weiland
University of Southern California, Los Angeles, CA, USA

John J. Whalen, III
University of Southern California, Los Angeles, CA, USA

David F. Williams
Wake Forest Institute of Regeneration Medicine Winston-Salem, North Carolina, USA

Rachel L. Williams
Clinical Engineering, Institute of Ageing and Chronic Disease, University of Liverpool, Liverpool, UK

John T. Wilson
Department of Bioengineering, University of Washington, Seattle, WA, USA

Clive G. Wilson
Strathclyde Institute of Pharmacy and Biomedical Sciences, Glasgow, Scotland, UK

Jessica Winter
Department of Chemical and Biomolecular Engineering, The Ohio State University, Columbus, OH, USA

Michael F. Wolf
Medtronic Inc., Corporate Technology and Innovation, Minneapolis, MN, USA

Jeremy C. Wright
DURECT Corporation, Cupertino, CA, USA

Paul Yager
Department of Bioengineering, University of Washington, Seattle, WA, USA

Weian Zhao
Harvard-MIT Division of Health Sciences & Technology, USA; Department of Medicine, Brigham and Women's Hospital, Harvard Medical School, USA; Sue and Bill Gross Stem Cell Research Center, Chao Family Comprehensive Cancer Center, Department of Pharmaceutical Sciences, University of California, Irvine, USA

PREFACE

Biomaterials Science: An introduction to Materials in Medicine was launched as an educational project to provide a nascent biomaterials community with an authoritative tool for training and education. Conceptual background material and a broad overview of applications were both envisioned as being integral to the textbook. In the late 1980s, biomaterials was in transition from an emerging field to a respected discipline embracing convergence (i.e., the merging of expertise from different "silos," such as provided by engineers, biomedical scientists and physicians) – the biomaterials community was in need of an integrated, comprehensive, and definitive educational resource. This rationale for launching the textbook has been validated by the success of earlier editions of *Biomaterials Science: An introduction to Materials in Medicine* – well over 25,000 copies have contributed to the biomaterials education of students and researchers around the world. The second edition has been widely adopted for classroom use, and is a reference resource for thousands of biomaterials professionals.

Biomaterials Science: An Introduction to Materials in Medicine strives for a balanced view of the biomaterials field. When this project was first launched, monographs available at that time did articulately address biomaterials, but they tended to strongly emphasize the authors' areas of expertise, while only superficially addressing other important subjects outside of their intellectual sphere. Balanced presentation means appropriate representation of:

- hard and soft biomaterials;
- orthopedic, cardiovascular, ophthalmologic, dental, and emerging applications;
- a balance of fundamental biological and materials science concepts, contemporary medical/clinical concerns, and government/commercial/societal issues that reflect the complex environment in which biomaterials are developed and used;
- coverage of the past, present, and future of biomaterials, their applications, and key challenges that lie ahead.

In this way, the reader can embrace the broad field, absorb the unifying principles common to all materials in contact with biological systems, and gain a solid appreciation for the special significance of the word *biomaterial.*

Biomaterials Science: An Introduction to Materials in Medicine, third edition, strives for curricular cohesion. Articles from the technical literature are commonly used in the classroom setting, but their scope and quality are inconsistent, and they are difficult to weave into a cohesive curriculum. Handout materials from professors are often graphically unsophisticated, and again, slanted to the specific interests of the individual. In *Biomaterials Science: An Introduction to Materials in Medicine*, by integrating the experience of many of leaders in the biomaterials field, we endeavor to present a balanced yet comprehensive perspective on an exciting and rapidly evolving discipline – and present this information in a graphically attractive, readable format, intended to be useful as an educational resource to a broad array of students, teachers, and practicing professionals.

Over 100 biomaterials experts from academia, industry, and government have contributed to this work. Certainly, such a distinguished group of authors can provide the needed balance, scope, and perspective. However, a sufficiently diverse group of authors also introduces unique complexities in a project of this type. Do the various writing styles clash? Does the presentation of material, particularly controversial material, result in one chapter contradicting another? Even with so many authors, can all subjects relevant to biomaterials be addressed? What subjects should be included, and which left out? How should such a project be refereed to ensure scientific quality, pedagogical effectiveness, and the balance we strive for? These are some of the issues the editors grappled with in the period from the conception of a revised, updated edition (shortly after the publication of the second edition), to the third edition publication in 2012. From this complex editorial process, a unique volume has evolved that the editors hope can make an ongoing contribution to the development of the biomaterials field. An educational tool has been synthesized here leading those new to biomaterials, and those with substantial experience in biomaterials (be they engineers, physicians, materials scientists, biological scientists, entrepreneurs, regulatory professionals or corporate leaders) to an appreciation of the scope, complexity, and significance of this enterprise.

What's new in this, the third, edition of *Biomaterials Science: An Introduction to Materials in Medicine*? All chapters have been updated, "colorized," and substantially rewritten. Many new chapters on important subjects have been added. The curricular organization for teaching fundamental materials science, cell and molecular biology, physiology, and pathology, key subjects that support the modern biomaterials research endeavor, has been restructured. The section on tissue engineering has been much expanded. The total content and size of the book have been significantly increased. A living website has been coupled to the book, offering supplemental material, homework problems, and other resources. The graphic design has been upgraded. This book addresses biomaterials for the 21st century.

Also, there is a new emphasis on regulatory and translational themes – how can we facilitate the progression of "smart ideas" from the biomaterials science lab bench to medical devices that save lives, improve the quality of life, generate jobs, and decrease overall healthcare costs through more effective therapies?

Acknowledgments and thanks are in order. First, the Society For Biomaterials, sponsor and inspiration for this book, is a model of "multidisciplinary cultural diversity." Composed of engineers, physicians, scientists, veterinarians, industrialists, inventors, regulators, attorneys, educators, ethicists, and others, the SFB provides the nidus for the intellectually exciting, humanitarian, and economically important biomaterials field. As was the case for the first and second editions, the editors recognize the importance of the Society For Biomaterials by donating all royalties from sales of this volume to the Society, to directly support education and career development related to biomaterials. For information on the Society For Biomaterials, visit the SFB website at http://www.biomaterials.org/.

We offer a special thanks to those who have generously invested time, energy, experience, and intelligence to author the chapters of this textbook. The many scientists, physicians, engineers, and industry leaders who contributed their expertise and perspectives are clearly the backbone of this work, and they deserve high praise – their efforts will strongly impact the education of the next generation of biomaterials scientists. We also pay respect and homage to biomaterials pioneers who have contributed to this work, but have since passed on – Jorge Heller, Michael Lysaght, Shalaby Shalaby, and Steven Slack; their contributions and collegiality are remembered and will be missed.

The organizational skills, experience, and encouragement of the staff at Elsevier Publishers, have led this third edition from a vague vision to a valuable volume – a tangible resource for the community. Thank you, Elsevier, for this contribution to the field of biomaterials.

The biomaterials field, since its inception in the 1950s, has been ripe with opportunity, intellectual stimulation, compassion, creativity, and rich collaboration. As a field we look to the horizons where the new ideas from science, technology, and medicine arise. Importantly, we strive to improve the survival and quality of life for millions through biomaterials-based medical devices and tissue engineering. We editors hope the biomaterials textbook you now hold will stimulate you as much as it has us.

Buddy D. Ratner
Allan S. Hoffman
Frederick J. Schoen
Jack E. Lemons
January, 2012

HOW TO USE THIS BOOK

Biomaterials Science: An introduction to Materials in Medicine was conceived as a learning tool to "compatibilize" through common language and fundamental principles a number of independent communities (basic sciences, engineering, medicine, dentistry, industry, regulatory, legal, etc.). Although the book has approximately 170 chapters, there is a logic of organization and curriculum that should make *Biomaterials Science: An introduction to Materials in Medicine*, third edition, straightforward to use in an academic course or as a reference work.

A guiding principle in assembling this multi-author, multidisciplinary textbook is that fundamental and translational progress in the field of biomaterials necessitates integration of concepts and tools from the full spectrum of the physical sciences, engineering, clinical medicine, biology, and the life sciences. Indeed, the discipline of biomaterials utilizes a *convergence* of multidisciplinary elements to enable the development of specific diagnostic or therapeutic devices – i.e., using biomaterials science and technology to create and implement real medical devices and other products that solve clinical problems and improve patient outcomes. Nevertheless, the editors believe (and the book has been assembled so) that a physician should be able to pick up the textbook and glean a baseline knowledge of the science, engineering, and commercialization aspects of biomaterials. A chemist could use this book to appreciate the biology behind biomaterials, the physiology associated with medical devices, and the applications in medicine. An engineer hired by a medical device company might learn the basic science underlying the technological development and details on medical applications. Similarly, for other disciplines that interface with biomaterials, this book can guide the reader through diverse but related topics that are generally not found in one volume.

The textbook has well over 100 authors. The field of biomaterials is so diverse in subject matter that a guiding principle has been that no one author can write it all – let us use the experience and wisdom of acknowledged masters of each subject to communicate the best information to the reader. But, to prevent this book from appearing to be simply a collection of review papers, considerable editorial effort has gone into ensuring logical curricular organization, continuity of ideas, and extensive cross-referencing between chapters.

Biomaterials Science: An introduction to Materials in Medicine, third edition, is divided into four parts (Materials Science and Engineering; Biology and Medicine; Practical Aspects of Biomaterials; Perspectives and Possibilities in Biomaterials Science), sections that subdivide each of the four parts (for example, under the part called "Materials Science and Engineering" there are sections on properties of materials and classes of materials used in medicine), and finally chapters, for example, chapters on each of the types of materials that are used in medicine (hydrogels, polyurethanes, titanium, etc.). Each section begins with an introduction by one of the editors that will guide the reader through the chapters, giving cohesion to the sections and highlighting key issues that are worthy of attention. Finally, there are appendices that tabulate use data and information.

Wherever possible, problems at the end of chapters are provided for classroom use and for self-testing. To permit updating problems and adding new ones, problems, and other resource materials including a full artwork catalog and downloadable images from the text, are provided on the companion website at http://www.elsevierdirect.com/companions/9780123746269.

We hope that the textbook organization, the extensive editorial effort, and the expertly authored chapters will serve their intended purpose – to guide a reader into and through this complex field of biomaterials science. The editors always appreciate feedback and commentary – contact information is provided for them.

And now, it is time to delve into the rich world of biomaterials science.

INTRODUCTION

Biomaterials Science: An Evolving, Multidisciplinary Endeavor

Buddy D. Ratner, Allan S. Hoffman, Frederick J. Schoen, and Jack E. Lemons

BIOMATERIALS AND BIOMATERIALS SCIENCE

Biomaterials Science: An Introduction to Materials in Medicine, third edition addresses the design, fabrication, testing, applications, and performance of synthetic and natural materials that are used in a wide variety of implants, devices, and process equipment that contact biological systems. These materials are referred to as biomaterials.

> The compelling, human side to biomaterials is that millions of lives are saved, and the quality of life is improved for millions more.

The field of biomaterials is some 60–70 years old (or young) at the time of publication of this third edition. It significantly impacts human health, the economy, and many scientific fields. Biomaterials and the medical devices comprised of them are now commonly used as *prostheses* in cardiovascular, orthopedic, dental, ophthalmological, and reconstructive surgery, and in other interventions such as surgical sutures, bioadhesives, and controlled drug release devices. The compelling, human side to biomaterials is that millions of lives have been saved, and the quality of life improved for millions more, based on devices fabricated from biomaterials. The biomaterials field has seen accelerating growth since the first medical devices that were based on accepted medical and scientific principles made their way into human usage in the late 1940s and early 1950s. And the growth of the field is ensured, with the aging population, the increasing standard of living in developing countries, and the growing ability to address previously untreatable medical conditions.

Biomaterials science addresses both therapeutics and diagnostics. It encompasses basic sciences (biology, chemistry, physics), and engineering and medicine. The translation of biomaterials science to clinically important medical devices is dependent on: (1) sound engineering design; (2) testing *in vitro*, in animals and in humans; (3) clinical realities; and (4) the involvement of industry permitting product development and commercialization. Figure 1 schematically illustrates the path from scientific development to the clinic.

Biomaterials science, in its modern incarnation, is an example of an important new research model – called *convergence* – an emerging paradigm that pushes multidisciplinary collaboration among experts and multidisciplinary integration of concepts and practice (Sharp and Langer, 2011). Not only biomaterials, but also bioinfomatics, synthetic biology, computational biology, nanobiology, systems biology, and other forefront fields, depend on convergence for their continued progress. This textbook aims to introduce these diverse multidisciplinary elements, particularly focusing on interrelationships rather than disciplinary boundaries, to systematize the biomaterials subject into a cohesive curriculum – a true convergence.

The title of this textbook, *Biomaterials Science: An Introduction to Materials in Medicine*, is accurate and descriptive. The intent of this work is: (1) to focus on the scientific and engineering fundamentals behind biomaterials and their applications; (2) to provide sufficient background material to guide the reader to a clear understanding and appreciation of the applications of biomaterials; and (3) to highlight the opportunities and challenges in the field. Every chapter in this textbook can serve as a portal to an extensive contemporary literature that expands on the basic ideas presented here. The magnitude of the biomaterials endeavor, its broadly integrated multidisciplinary scope, and examples of biomaterials applications will be revealed in this introductory chapter and throughout the book.

> The common thread in biomaterials is the physical and chemical interactions between complex biological systems and synthetic or modified natural materials.

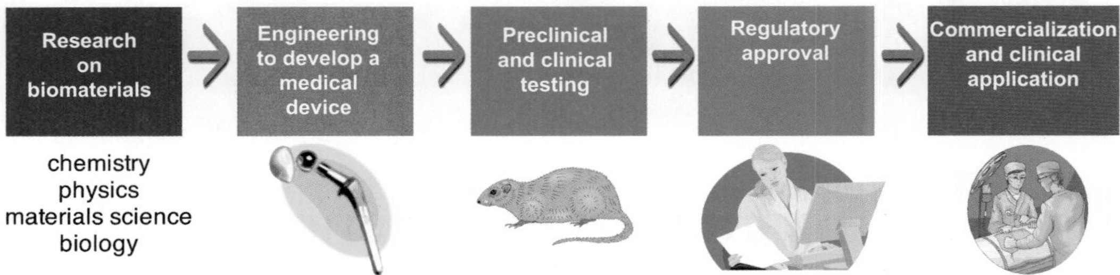

FIGURE 1 The path from the basic science of biomaterials, to a medical device, to clinical application.

Although biomaterials are primarily used for medical applications (the focus of this text), they are also used to grow cells in culture, to assay for blood proteins in the clinical laboratory, in processing equipment for biotechnological applications, for implants to regulate fertility in cattle, in diagnostic gene arrays, in the aquaculture of oysters, and for investigational cell–silicon "neuronal computers." How do we reconcile these diverse uses of materials into one field? The common thread is the physical and chemical interactions between complex biological systems and synthetic materials or modified natural materials.

In medical applications, biomaterials are rarely used as isolated materials, but are more commonly integrated into devices or implants. Chemically pure titanium can be called a biomaterial, but shaped (machined) titanium in conjunction with ultrahigh molecular weight polyethylene becomes the device, a hip prosthesis. Although this is a text on biomaterials, it will quickly become apparent that the subject cannot be explored without also considering biomedical devices and the biological response to them. Indeed, both the material and the device impact the recipient (patient) and, as we will see, the patient's host tissue impacts the device. These interactions can lead to device success or, where there is inappropriate choice of biomaterials or poor device design, device failure.

Furthermore, a biomaterial must always be considered in the context of its final fabricated, sterilized form. For example, when a polyurethane elastomer is cast from a solvent onto a mold to form the pump bladder of a heart assist device, it can elicit different blood reactions than when injection molding is used to form the same device. A hemodialysis system serving as an artificial kidney requires materials that must function in contact with a patient's blood, and also exhibit appropriate membrane permeability and mass transport characteristics. Much fabrication technology is applied to convert the biomaterials of the hemodialysis system (polysulfone, silicone rubber, polyethylene) into the complex apparatus that is used for blood purification.

Due to space limitations and the materials focus of this work, many aspects of medical device design are not addressed in this book. Consider the example of the hemodialysis system. This textbook will overview membrane materials and their biocompatibility; there will be little coverage of mass transport through membranes, the burst strength of membranes, dialysate water purification, pumps, flow systems, and monitoring electronics. Other books and articles cover these topics in detail, and chapter authors provide references useful to learn more about topics not explicitly covered.

KEY DEFINITIONS

The words "biomaterial" and "biocompatibility" have already been used in this introduction without formal definition. A few definitions and descriptions are in order, and will be expanded upon in this and subsequent chapters.

A definition of "biomaterial" endorsed by a consensus of experts in the field is:

> *A biomaterial is a nonviable material used in a medical device, intended to interact with biological systems.*
>
> ***Williams, 1987***

A biomaterial is a nonviable material used in a medical device, intended to interact with biological systems.

Although biomaterials are most often applied to meet a therapeutic or diagnostic medical need, if the word "medical" is removed, this definition becomes broader and can encompass the wide range of applications suggested above. If the word "nonviable" is removed, the definition becomes even more general, and can address many new tissue engineering and hybrid artificial organ applications where living cells are used.

"Biomaterials science" is the study (from the physical and/or biological perspective) of materials with special reference to their interaction with the biological environment. Traditionally, emphasis in the biomaterials field has been on synthesis, characterization, and the host–material interactions biology. Yet, most biomaterials (that meet the special criteria of biocompatibility – see Chapters II.3.2 and II.3.4) induce a non-specific biological reaction that we refer to as the foreign-body reaction (Chapter II.2.2). This leads us to consider a widely-used definition of biocompatibility.

"Biocompatibility" is the ability of a material to perform with an appropriate host response in a specific application.

Williams, 1987

"Biocompatibility" is the ability of a material to perform with an appropriate host response in a specific application.

Examples of "appropriate host responses" include resistance to blood clotting, resistance to bacterial colonization, and normal, uncomplicated healing. Examples of "specific applications" include a hemodialysis membrane, a urinary catheter or a hip joint replacement prosthesis. Note that the hemodialysis membrane might be in contact with the patient's blood for five hours, the catheter may be inserted for a week, and the hip joint may be in place for the life of the patient. This general concept of biocompatibility has been extended to tissue engineering, in which *in vitro* and *in vivo* processes are harnessed by careful selection of cells, materials, and metabolic and biomechanical conditions to regenerate functional tissues. Ideas central to biocompatibility are elaborated upon in Ratner (2011), and Chapter II.3.2.

In the discussion of these definitions, we are introduced to considerations that set biomaterials apart from most materials explored in materials science. Table 1 lists a few applications for synthetic materials in the body. It includes many classes of materials used as biomaterials. Note that metals, ceramics, polymers, glasses, carbons, and composite materials (combinations of different classes of materials) are listed. Such materials are used as molded or machined parts, coatings, fibers, films, membranes, foams, fabrics, and nanoparticles. Table 1 also gives estimates of the numbers of medical devices containing biomaterials that are implanted in humans each year. The human impact, and the size of the commercial market for biomaterials and medical devices, is impressive (Table 1; Table 2).

TABLE 1 Key Applications of Synthetic Materials and Modified Natural Materials in Medicine*

Application	Biomaterials Used	Number/Year – World (or World Market in US$)
Skeletal system		
Joint replacements (hip, knee, shoulder)	Titanium, stainless steel, polyethylene	2,500,000
Bone fixation plates and screws	Metals, poly(lactic acid) (PLA)	1,500,000
Spine disks and fusion hardware		800,000
Bone cement	Poly(methyl methacrylate)	($600M)
Bone defect repair	Calcium phosphates	–
Artificial tendon or ligament	Polyester fibers	–
Dental implant-tooth fixation	Titanium	($4B)
Cardiovascular system		
Blood vessel prosthesis	Dacron, expanded Teflon	200,000
Heart valve	Dacron, carbon, metal, treated natural tissue	400,000
Pacemaker	Titanium, polyurethane	600,000
Implantable defibrillator	Titanium, polyurethane	300,000
Stent	Stainless steel, other metals, PLA	1,500,000
Catheter	Teflon, silicone, polyurethane	1B ($20B)
Organs		
Heart assist device	Polyurethane, titanium, stainless steel	4000
Hemodialysis	Polysulfone, silicone	1,800,000 patients ($70B)
Blood oxygenator	silicone	1,000,000
Skin substitute	Collagen, cadaver skin, nylon, silicone	($1B)
Ophthalmologic		
Contact lens	Acrylate/methacrylate/silicone polymers	150,000,000
Intraocular lens	Acrylate/methacrylate polymers	7,000,000
Corneal bandage lens	hydrogel	–
Glaucoma drain	Silicone, polypropylene	($200M)
Other		
Cochlear prosthesis	Platinum, platinum-iridium, silicone	250,000 total users
Breast implant	Silicone	700,000
Hernia mesh	Silicone, polypropylene, Teflon	200,000 ($4B)
Sutures	PLA, polydioxanone, polypropylene, stainless steel	($2B)
Blood bags	Poly(vinyl chloride)	–
Ear tubes (Tympanostomy)	Silicone, Teflon	1,500,000
Intrauterine device (IUD)	Silicone, copper	1,000,000

*Data compiled from many sources – these numbers should be considered rough estimates that are changing with growing markets and new technologies. Where only US numbers are available, world usage is estimated at approximately 2.5× of US usage.
NOTE: M = millions, B = billions.

TABLE 2 The Biomaterials and Healthcare Market: Facts and Figures (Per Year)	
Total US healthcare expenditures (1990)	$714 billion
Total US healthcare expenditures (2009)	$2.5 trillion
Total US health research and development expenditure (2009)	$139 billion
Number of medical device companies in the US	12,000
Jobs in the US medical device industry (2008)	425,000
Sales by US medical device industry (2008)	$136 billion
World medical device market forecast for 2013*	$286 billion

*Source: Medical Market Fact Book 2008.

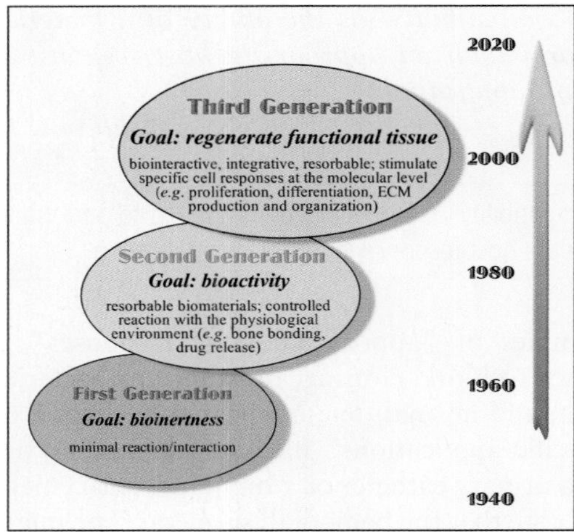

FIGURE 2 Evolution of biomaterials science and technology. (Based upon Rabkin, E. & Schoen, F. J. (2002). Cardiovascular tissue engineering. Cardiovasc Pathol, 11: 305.)

THE EVOLUTION OF THE BIOMATERIALS FIELD

Biomaterials research and development have been stimulated and guided by advances in cell and molecular biology, chemistry, materials science, and engineering. The biomaterials community has been the major contributor to the understanding of the interactions of materials with the physiological environment (often referred to as the biointerface). The development of biomaterials for medical and dental applications has evolved through three generations, each somewhat temporally overlapping, yet each with a distinct objective (Figure 2).

The goal of early biomaterials (first generation) was to achieve a suitable combination of functional properties to adequately match those of the replaced tissue without deleterious response by the host. First generation biomaterials (beginning in the 1950s and 1960s) were comprised largely of off-the-shelf, widely available industrial materials that were not developed specifically for medical use. They were selected because of the desirable combination of physical properties specific to the intended clinical use, and because they were *bioinert* (i.e., they elicited minimal response from the host tissues), and therefore they were considered *biocompatible* (see Chapter II.3.2). The widely used elastomeric polymer, silicone rubber, is prototypical. Pyrolytic carbon, originally developed in the 1960s as a coating material for nuclear fuel particles, and now widely used in modified compositions to coat components of mechanical heart valves, exemplifies one of the early biomaterials whose formulation and properties were studied, controlled according to engineering principles, and tuned specifically for medical application (Chapter I.2.8).

Second generation biomaterials evolved from those early biomaterials, and were intended to elicit a controlled reaction with the tissues into which they were implanted in order to induce a desired therapeutic effect. In the 1980s, these *bioactive* materials were in clinical use in orthopedic and dental surgeries as various compositions of bioactive glasses and ceramics (Hench and Pollak, 2002; Chapter I.2.4.A), in controlled localized drug release applications such as the Norplant hormone-loaded contraceptive formulation (Meckstroth and Darney, 2001), and in devices such as the HeartMate® left ventricular assist device for patients with congestive heart failure (Chapter II.5.3.D). This cardiac assist device has an integrally-textured polyurethane surface that fosters a controlled surface thrombotic (clotting) reaction to minimize the risk of de-adhering fragments of clotted blood into the bloodstream (Rose et al., 2001). Still another example of second generation biomaterials used in medical devices are the drug-eluting endovascular stents that have been shown to markedly limit restenosis (blood vessel closure) following balloon angioplasty (Daemen et al., 2007a and 2007b; Chapter II.5.3.B).

The second generation of biomaterials also included the development of resorbable biomaterials, with rates of degradation that could be tailored to the requirements of a desired application (Chapters I.2.6 and II.4.3). Thus, the discrete interface between the implant site and the host tissue could be eliminated in the long-term, because the foreign material would ultimately be degraded to soluble, non-toxic products by the host. A biodegradable suture composed of polyglycolic (PGA) acid has been in clinical use since the 1960s (see "History of Biomaterials"). Many groups continue to search for biodegradable polymers with needed properties such as strength, flexibility, a chemical composition conducive to tissue development, and a degradation rate consistent with the specific application. Polymeric materials with other novel properties such as shape-memory and programmable and interactive surfaces that control the cellular microenvironment are also under investigation (Langer and Tirrell, 2004; Khademhosseini et al., 2006; Lutolf et al., 2009; Chapter I.2.11).

The need for maximally effective pharmacologic dosing regimens and minimization of systemic toxicities has stimulated development of new implantable polymers and innovative systems for controlled drug delivery and gene therapy (Chapter II.5.16 and subchapters under

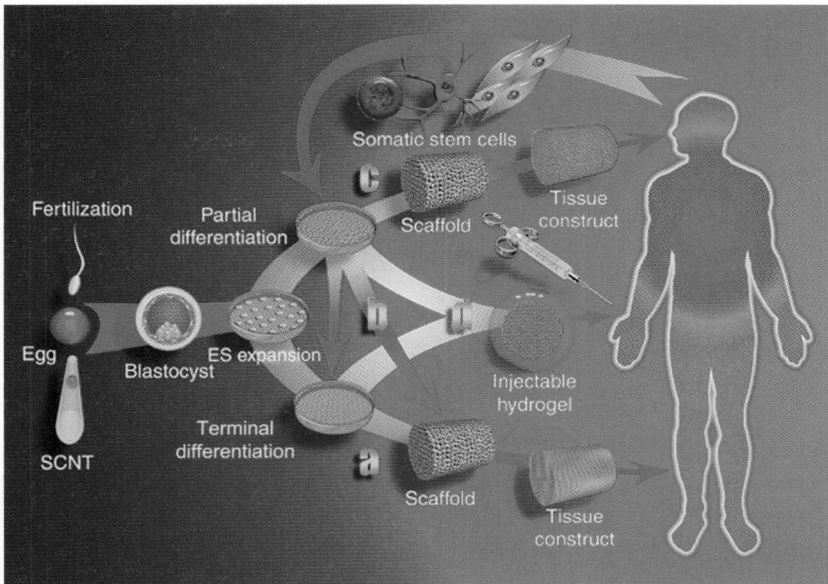

FIGURE 3 The tissue engineering paradigm – various cell types are seeded on porous scaffolds, possibly proliferated in a bioreactor, and finally implanted in various tissue sites to restore or regenerate damaged or missing tissue. (nature.com.)

this section). Also, new protein and nucleic acid-based drugs (which cannot be taken in classical pill form) created the need for new technologies to deliver these valuable pharmacologic agents. Biomaterials systems can permit delivery of drugs, active proteins, and other macromolecules localized to the site where the drug is needed. The biomaterials-intensive field of controlled drug delivery is now capable of targeting a wide range of drugs to tumors, to diseased blood vessels, to the pulmonary alveoli, etc., on a one-time or sustained basis with highly-regulated dosage (Lawson et al., 2007). Biomaterial-based slow release mechanisms have been developed for delivery of growth factors to induce vascularization and other biological responses (Kong and Mooney, 2007). Moreover, nanoparticle delivery systems (Chapters II.5.16.A and I.2.19) and the development of microelectromechanical systems (MEMS) have opened new possibilities for fine control of dosage, perhaps regulated endogenously (Sengupta et al., 2005; Goldberg et al., 2007). Other formulations can regulate cell and tissue responses through delivery of growth factors, genes or small interfering RNAs (siRNA) (Whitehead et al., 2009).

The third generation of biomaterials, the logical extension of the rapidly progressing state-of-the-art, has the goal of supporting and stimulating the regeneration of functional tissue. Through all of human history, the ability of the physician or healer to actually regenerate tissues and organs lost due to disease or trauma was essentially nonexistent; the physician's role was palliative – ease the symptoms without curing. Now, with advances in tissue engineering and regenerative medicine, it seems that true replacement with living tissue will be possible. Biomaterials play a key role in the rapidly developing field of tissue engineering and regenerative therapeutics (Section II.6). Tissue engineering is a broad term describing a set of tools at the interface of the biomedical and engineering sciences that use living cells (or attract endogenous cells) to aid tissue formation or regeneration, and thereby produce therapeutic or diagnostic benefit (Freed, 2006; Hunziker et al., 2006; Ingber, 2010; Mikos et al., 2006). In the most frequently applied paradigm, cells are seeded on a scaffold composed of synthetic polymer or natural material (collagen or chemically-treated tissue), a tissue is matured *in vitro*, and the construct is implanted in the appropriate anatomical location as a prosthesis (Figure 3). A typical scaffold is a bioresorbable polymer in a porous configuration in the desired geometry for the engineered tissue, often modified to be adhesive for cells, in some cases selective for a specific cell population; either application-specific and differentiated or undifferentiated (stem) cells are used (Stocum, 2005; Chapter II.1.7). Tissue engineering has led to the replacement in humans of damaged bladders, trachea, skin, corneal epithelium, and cartilage. Future directions in biomaterials, tissue engineering, and regenerative therapeutics are indeed exciting and priorities are under active discussion (Hacker et al., 2006; Goldberg et al., 2007; Hellman et al., 2007; Johnson, 2007; Daley et al., 2008; Kohane et al., 2008).

EXAMPLES OF TODAY'S BIOMATERIALS APPLICATIONS

Five examples of biomaterials applications now follow to illustrate important ideas. The specific devices discussed were chosen because they are widely used in humans with good success. However, key limitations with these

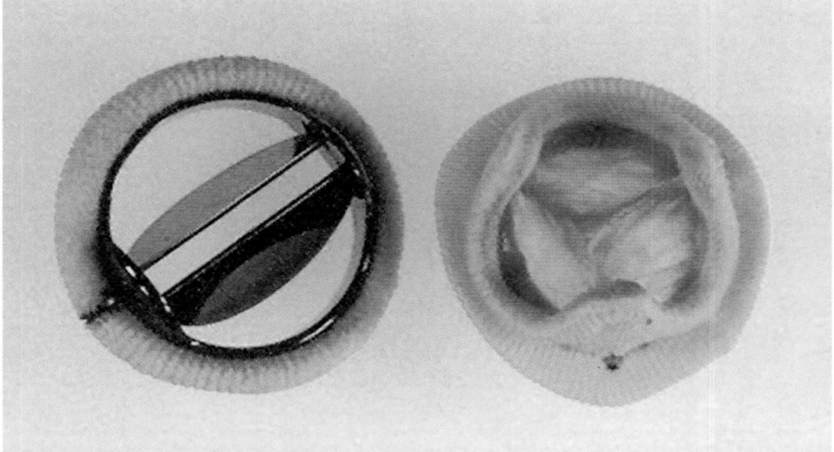

FIGURE 4 Prosthetic heart valves. Left: A bileaflet tilting disk mechanical heart valve. (St. Jude Medical Inc., St. Paul, MN.) Right: A bioprosthetic (xenograft) tissue heart valve (Hancock® valve, Medtronic Inc., MN.)

biomaterial devices are also highlighted. Each of these examples is also discussed in detail in later chapters in this textbook.

Heart Valve Prostheses

Diseases and degeneration of the heart valves often make surgical repair or replacement necessary. The natural heart valve opens and closes over 40 million times per year, and can require replacement due to disease or wear. Approximately 250,000 replacement valves are implanted each year worldwide, because of acquired damage to the natural valve and congenital heart anomalies. There are many types of heart valve prostheses, and they are fabricated from carbons, metals, elastomers, plastics, fabrics, and animal or human tissues chemically pretreated to reduce their immunologic reactivity, and to enhance durability. Figure 4 shows a bileaflet tilting disk mechanical heart valve and a bioprosthetic (xenograft) tissue heart valve, two of the most widely used designs. Generally, as soon as the valve is implanted, cardiac function is restored to near normal levels, and the patient shows rapid improvement. In spite of the overall success seen with replacement heart valves, there are problems, many of them specific to a specific type of valve; they include induction of blood clots (sometimes shed into the bloodstream as emboli), degeneration of valve tissue leaflets, mechanical failure (e.g., at weld joints), and infection. Heart valve substitutes are discussed in Chapter II.5.3.A.

Total Hip Replacement Prostheses

The human hip joint is subjected to high levels of mechanical stress, and receives considerable abuse in the course of normal and extraordinary activity. It is not surprising that after 50 or more years of cyclic mechanical stress or because of degenerative or rheumatoid disease, the natural joint wears out, leading to loss of mobility and sometimes confinement to a wheelchair. Hip joint prostheses are fabricated from a variety of materials, including titanium, stainless steel, special high-strength alloys, ceramics, composites, and ultrahigh molecular weight polyethylene (UHMWPE). Replacement hip joints (Figure 5) are implanted in more than 200,000 humans each year in the United States alone. With some types of replacement hip joints and surgical procedures which use a polymeric cement, ambulatory function is restored within days after surgery. For other types, a healing-in period is required for integration between bone and the implant before the joint can bear the full weight of the body. In most cases, good function is restored. Even athletic activities are possible, although those activities that subject the repaired joint to high stress are generally not advisable. After 10–15 years, many of these implants fail by loosening, which usually necessitates another operation (a revision procedure). Metal-on-metal implants also experience problems of corrosion and adverse responses to released metal ions. Artificial hip joint prostheses are discussed in Chapter II.5.6.

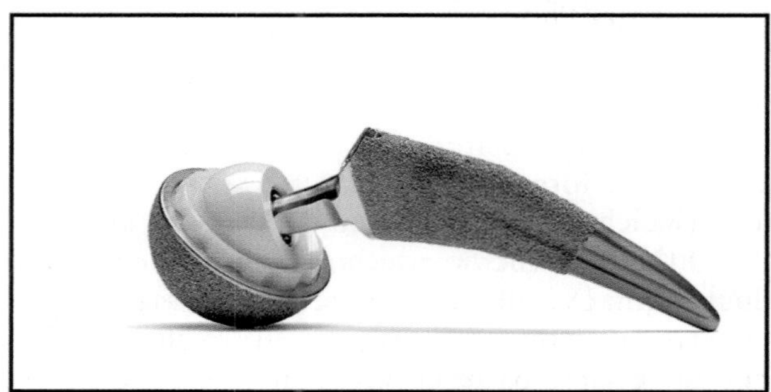

FIGURE 5 A hip prosthesis. Microplasty® titanium alloy femoral stem, Biolox® alumina-zirconia ceramic femoral head, ultra-high molecular weight polyethylene acetabular cup infused with vitamin E antioxidant. (Image courtesy of Biomet, Inc.)

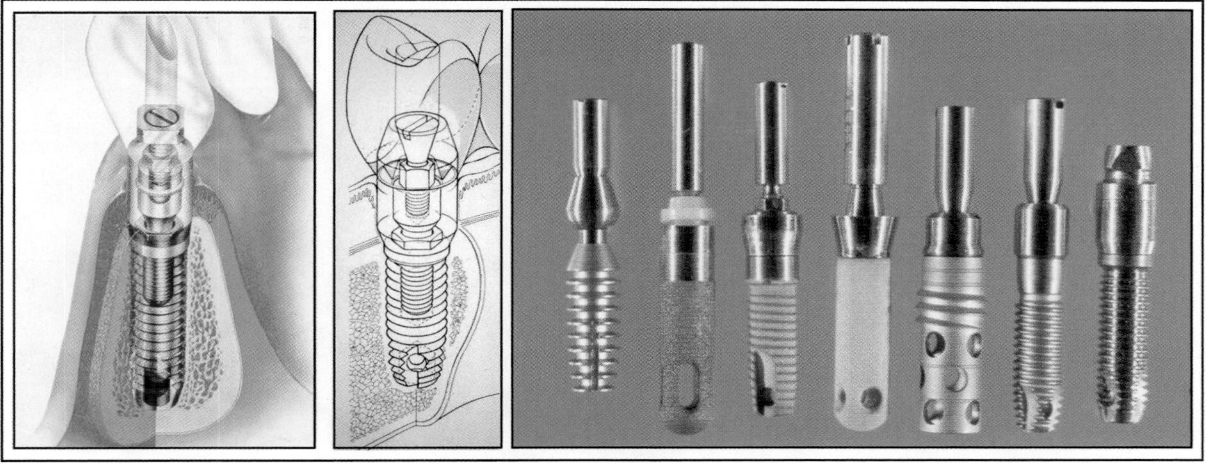

FIGURE 6 Schematic images of early dental root form implants and a photograph of several designs used in clinical practice.

Dental Implants

The development of root form designs of titanium implants (Figure 6) by Per-Ingvar Brånemark revolutionized dental implantology (Carlsson et al., 1986). These devices form an implanted artificial tooth anchor upon which a crown is affixed, and are implanted in 2,000,000 people each year in the US alone, according to the American Dental Association. A special requirement of a material in this application is the ability to form a tight seal against bacterial invasion where the implant traverses the gingiva (gum). Other issues are associated with the rapidly growing junctional epithelium inhibiting regrowth of the slower growing bone. Also, in normal physiology, the tooth is connected to the jaw by the periodontal ligament, and is not directly attached to the jawbone. One of the primary advantages originally cited for the titanium implant was its osseous integration with the bone of the jaw. In recent years, however, this attachment has been more accurately described as a tight apposition or mechanical fit, and not true bonding. Loss of tissue support leading to loosening remains an occasional problem, along with infection and issues associated with the mechanical properties of unalloyed titanium that is subjected to long-term cyclic loading. Dental implants are discussed in Chapter II.5.7.

Intraocular Lenses

Implants to replace lenses in the eye that have clouded due to cataracts are called intraocular lenses (IOLs). They have been fabricated from a variety of transparent materials including poly(methyl methacrylate), silicone elastomer, soft acrylic polymers, and hydrogels (Figure 7). By the age of 75, more than 50% of the population suffers from cataracts severe enough to warrant IOL implantation. This translates to more than three million implantations in the US alone each year, and more than double that number worldwide. Good vision is generally restored almost immediately after the lens is inserted, and the success rate with this device is high. IOL surgical procedures are well-developed, and implantation is most often performed on an outpatient basis. Observations of implanted lenses through the cornea using a microscope to directly study the implants show that inflammatory cells such as macrophages migrate to the surface of the lenses after implantation. Thus, the conventional healing pathway is seen with these devices, similar to that observed with materials implanted in other sites in the body. Outgrowth of cells onto the IOL from the posterior lens capsule, stimulated by the presence of the IOL, can cloud vision, and this is a significant complication. IOLs are discussed in Chapter II.5.9.B.

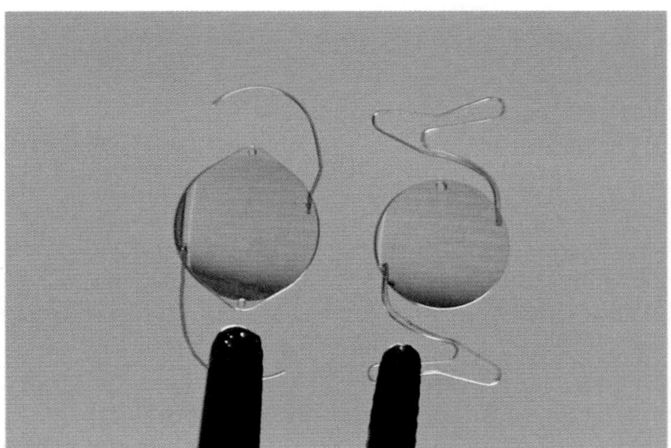

FIGURE 7 Two styles of multipiece intraocular lenses.

Left Ventricular Assist Devices

Nearly 5,000,0000 Americans are living with seriously failing hearts (congestive heart failure), and 300,000 individuals will die each year from this disease. According to the American Heart Association, 50,000–100,000 of these individuals might benefit from cardiac transplantation or mechanical assist. Since the available pool

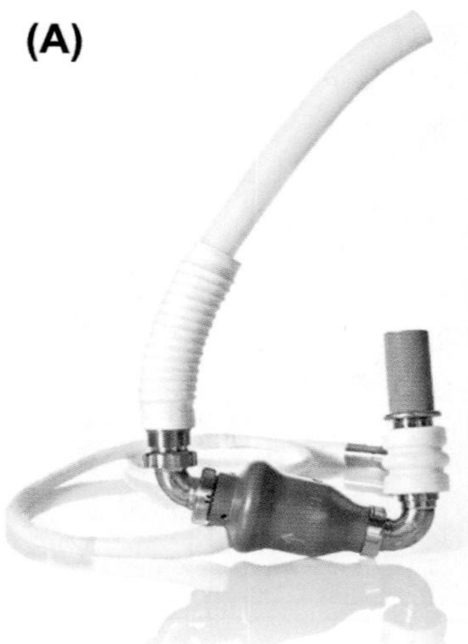

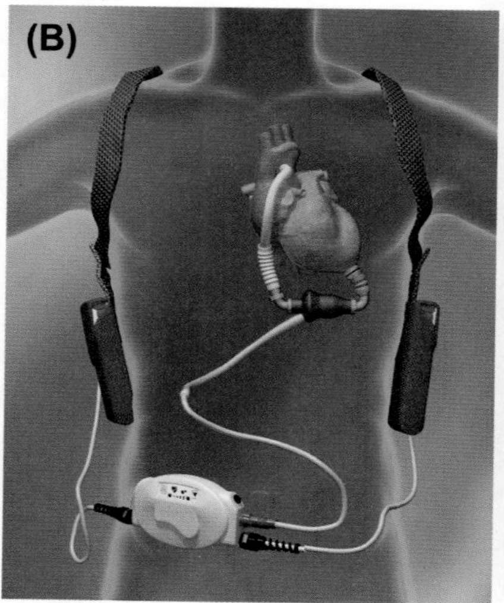

FIGURE 8 Left ventricular assist device (LVAD). (A) Continuous flow pump with associated inflow/outflow grafts and electrical drive line (Heartmate II® device). (B) Schematic of LVAD implanted as a left ventricular assist device with associated external power source. (C) Patient (human) implanted with this LVAD device illustrating freedom of activity. (Reprinted with the permission of Thoratec Corporation, Pleasantville, CA.)

of donor hearts for transplantation is only 2000–3000 per year, effective and safe mechanical cardiac assist or replacement seems like a desirable option. Left ventricular assist devices (LVADs) have evolved from a daring experimental concept, the mechanical total heart, to a life-prolonging tool. A number of devices have received FDA approval. They are now used to maintain a patient with a failing heart while the patient awaits the availability of a transplant heart, and some patients receive these LVADs as a permanent ("destination") therapy. An LVAD is illustrated in Figure 8. LVAD recipients can have considerable mobility and freedom while cardiac support is provided by the device. Patients have lived on LVAD support for more than four years. However, a patient with an LVAD is always at risk for infection and blood clots initiated within the device. These could break off (embolize) and possibly obstruct blood flow to a vital organ. Furthermore, LVADs are well over $100,000 in price, not including associated surgical, drug, and hospital expenses. Can so expensive an innovation be made available to the wide patient base that could benefit from them? LVADs are discussed in detail in Chapter II.5.3.D.

These five cases, only a small fraction of the important medical devices that could have been used as examples, spotlight central ideas and themes relevant to most medical devices interfacing with the human biology. A few generalizations are:

- Implantation in hundreds of thousands of patients with good success is noted.
- A broad range of synthetic materials of varying properties are used.
- Most anatomical sites can be interfaced with a device.
- The normal response by which the body responds to foreign bodies is observed.
- Problems, concerns, unexplained observations or unintended consequences may be noted for each device.
- Most device complications are related to biomaterials–tissue interactions.
- Companies are manufacturing devices and bringing value to shareholders (and patients).
- Regulatory agencies are carefully assessing device performance, and making policy decisions to monitor the device industry, ensure quality, and protect the patient.
- Ethical and societal issues are associated with each device.

These ideas are relevant to nearly all medical devices. As you work through this textbook, consider how these ideas impact the specific topic you are studying.

CHARACTERISTICS OF BIOMATERIALS SCIENCE

Now that we've defined key terms and reviewed specific examples highlighting successes and also complications, we can examine core characteristics of the field of biomaterials.

Multidisciplinary

More than any other field of contemporary technology, biomaterials science brings together teams of researchers from diverse academic and industrial backgrounds, who must clearly communicate and integrate complex concepts and data. Figure 9 lists some of the disciplines and key steps that are encountered in the progression from identifying the need for a biomaterial or device to its testing, regulation, manufacture, sale, and implantation.

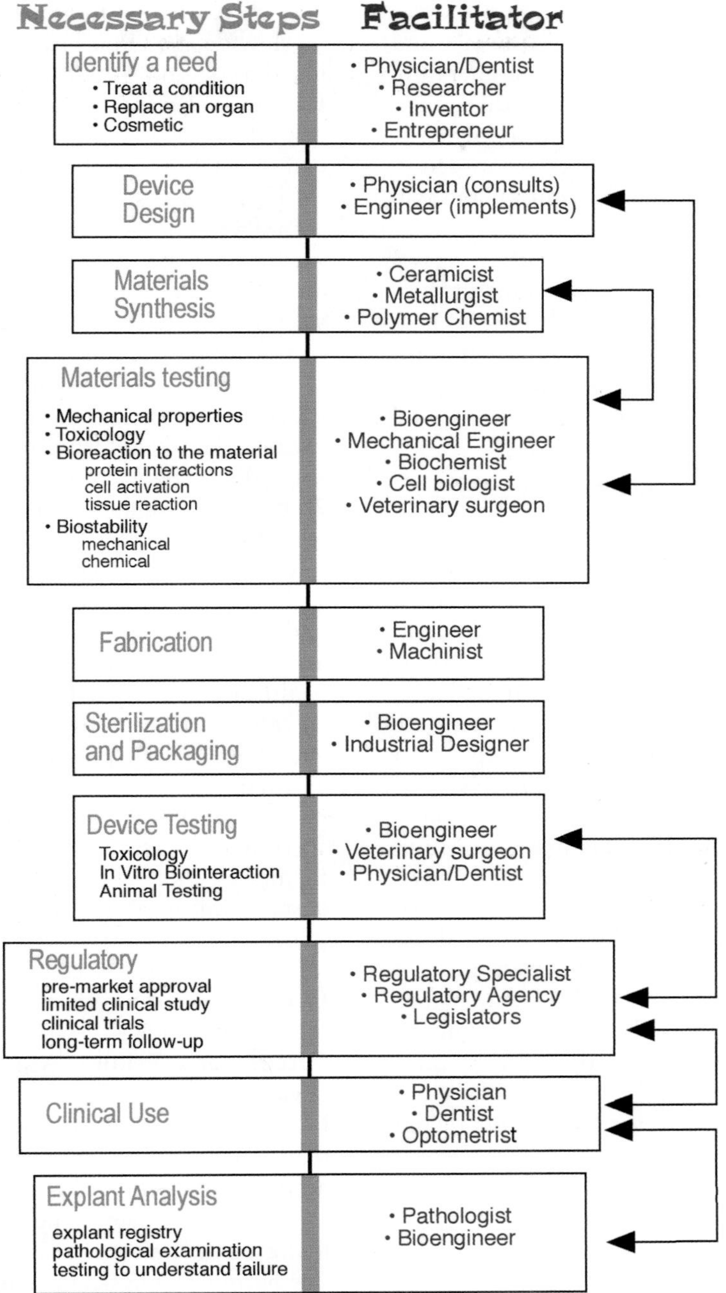

FIGURE 9 The path from an identified need to a clinical product, and some of the disciplines that facilitate this development process.

Diverse Materials are Used

The biomaterials scientist must have an appreciation of materials science including polymers, metals, ceramics, glasses, composites, and biological materials. This may range from an impressive command of the theory and practice of the field demonstrated by the professional materials scientist to a general understanding of the properties of materials that should be possessed by the physician or biologist investigator involved in biomaterials-related research.

A wide range of materials is routinely used in medical devices (Table 1), and no one researcher will be comfortable synthesizing, characterizing, designing, and fabricating with all these materials. Thus, specialization is common and appropriate. However, a broad appreciation of the properties and applications of these materials, the palette from which the biomaterials scientist "creates" medical devices, is a hallmark of professionals in the field.

There is a tendency to group biomaterials and researchers into the "hard tissue replacement" camp, typically represented by those involved in orthopedic and dental materials, and the "soft tissue replacement" camp, frequently associated with cardiovascular implants and general plastic surgery materials. Hard tissue biomaterials researchers are thought to focus on metals and ceramics, while soft tissue biomaterials researchers are considered polymer experts. In practice, this division is artificial: a heart valve may be fabricated from polymers, metals, and carbons. A hip joint will also be composed of metals and polymers (and sometimes ceramics), and will be interfaced to the body via a polymeric bone cement. There is a need for a general understanding of all classes of materials and the common conceptual theme of their interaction with the biological milieu. This book provides a background to the important classes of materials, hard and soft, and their interactions with the biological environment.

Biomaterials to Devices to Materials to Markets and Medicine

Thomas Edison once said that he would only invent things that people would buy. In an interesting way, this idea is central to biomaterials device development. The process of biomaterial/medical device innovation is driven by clinical need: a patient or a physician defines a need and then initiates an invention. However, someone must test and manufacture the device, and shepherd it though the complex, expensive regulatory process. This "someone" is a company, and a company exists (by law) to return value to its shareholders. Figure 9 illustrates multidisciplinary interactions in biomaterials, and shows the progression in the development of a biomaterial or device. It provides a perspective on how different disciplines work together, starting from the identification of a need for a biomaterial through development, manufacture, implantation, and (possibly) removal from the patient. Note that the development process for medical devices is very different from that for drugs. There are insightful reference works available to help understand this specialized device commercialization process (Zenios et al., 2010).

Magnitude of the Field

The magnitude of the medical device field expresses both the magnitude of the need and a sizeable commercial market (Table 2). Consider four commonly used biomaterial devices: a contact lens; a hip joint; a hydrocephalous drainage shunt; and a heart valve. Let's examine these devices in the contexts of human needs and commercial markets. The contact lens offers improved vision and, some will argue, a cosmetic enhancement. The hip joint offers mobility to the patient who would otherwise need a cane or crutch or be confined to a bed or wheelchair. The hydrocephalus shunt will allow an infant to survive without brain damage. The heart valve offers a longer life with improved quality of life. The contact lens may sell for \$100, and the hip joint, hydrocephalus shunt, and heart valve may sell for \$1000–\$4000 each. Each year there will be millions of contact lenses purchased worldwide, 400,000 heart valves, 160,000 hydrocephalus shunts, and 500,000 total artificial hip prostheses. Here are the issues for consideration: (1) the large number of devices (an expression of both human needs and commercial markets); (2) medical significance (cosmetic to life saving); and (3) commercial potential (who will manufacture it and why – for example, what is the market for the hydrocephalus shunt?). Always, human needs and economic issues color this field we call "biomaterials science." Medical practice, market forces, and bioethics come into play almost every day.

Lysaght and O'Laughlin (2000) have estimated that the magnitude and economic scope of the contemporary organ replacement enterprise are much larger than is generally recognized. In the year 2000, the lives of over 20 million patients were sustained, supported, or significantly improved by functional organ replacement. The impacted population grows at over 10% per year. Worldwide, first year and follow-up costs of organ replacement and prostheses exceed \$300 billion US dollars per year, and represents between 7% and 8% of total worldwide healthcare spending. In the United States, the costs of therapies enabled by organ replacement technology exceed 1% of the gross national product. The costs are also impressive when reduced to the needs of the individual patient. For example, the cost of an implanted heart valve is roughly \$5000–\$7000. The surgery to implant the device entails a hospital bill and first year follow-up costs upwards of \$40,000. Reoperation for replacing a failed valve will have these same costs. Reoperations for failed valves now exceed 10% of all valve replacements. Overall costs of healthcare and medical devices are summarized in Table 2.

Success and Failure. Most biomaterials and medical devices perform satisfactorily, improving the quality of life for the recipient or saving lives. However, no manmade construct is perfect. All manufactured devices have a failure rate. Also, all humans are different, with differing ethnicities, ages, genetics, gender, body chemistries, living environments, and degrees of physical activity. Furthermore, physicians implant or use these devices with varying degrees of skill. The other side to the medical device success story is that there are problems, compromises, complications, and unintended consequences that often occur with medical devices. Central issues for the biomaterials scientist, manufacturer, patient, physician, and attorney are: (1) is the design competent and optimal; (2) who should be responsible when devices perform "with an inappropriate host response"; and (3) what are the cost:risk or cost:benefit ratios for the implant or therapy?

Some examples may clarify these issues. Clearly, heart valve disease is a serious medical problem. Patients with diseased aortic heart valves have a 50% chance of dying within three years. Surgical replacement of the diseased valve leads to an expected survival of 10 years in 70%

of the cases. However, of these patients whose longevity and quality of life have clearly been enhanced, approximately 60% will suffer a serious valve-related complication within 10 years after the operation. Another example involves LVADs. A clinical trial called Randomized Evaluation of Mechanical Assistance for the Treatment of Congestive Heart Failure (REMATCH) led to the following important statistics (Rose et al., 2001). Patients with an implanted Heartmate® LVAD (Thoratec Laboratories) had a 52% chance of surviving for one year, compared with a 25% survival rate for patients who took medication. Survival for two years in patients with the Heartmate® was 23% versus 8% in the medication group. Also, the LVAD enhanced the quality of life for the patients – they felt better, were less depressed, and were mobile. Importantly, patients participating in the REMATCH trial were not eligible for a heart transplant. In the cases of the heart valve and the LVAD, clinical complications possibly associated with less than stellar biomaterials performance do not preclude widespread clinical acceptance.

Biomaterials science:

- multidisciplinary;
- multi-biomaterial;
- clinical need-driven;
- substantial world market;
- risk–benefit issues.

Thus, these five characteristics of biomaterials science:

- multidisciplinary;
- multi-material;
- clinical need-driven;
- substantial market; and
- risk–benefit issues.

color all aspects of the field.

In addition, there are certain unique subjects that are particularly prominent in our field and help delineate the biomaterials endeavor as a unique field of science and engineering. Let us review a few of these.

SUBJECTS INTEGRAL TO BIOMATERIALS SCIENCE

Toxicology. A biomaterial should not be toxic, unless it is specifically engineered for such requirements (e.g., a "smart" drug delivery system that targets cancer cells with a toxic drug and destroys them). Since the nontoxic requirement is the norm, toxicology for biomaterials has evolved into a sophisticated science. It deals with the substances that migrate out of biomaterials. For example, for polymers, many low molecular weight "leachables" exhibit some level of physiologic activity and cell toxicity. It is reasonable to say that a biomaterial should not give off anything from its mass unless it is specifically designed to do so. Toxicology also deals with methods to evaluate how well this design criterion is met when a new biomaterial is under development. Chapter II.3.3 provides an overview of methods in biomaterials toxicology. Implications of toxicity are addressed in Chapters II.2.5, II.3.2, and II.3.4.

Biocompatibility. The understanding and measurement of biocompatibility is unique to biomaterials science. Unfortunately, we do not have precise definitions or accurate measurements of biocompatibility. More often than not, biocompatibility is defined in terms of performance or success at a specific task. Thus, for a patient who is doing well with an implanted large diameter Dacron fabric vascular prosthesis, few would argue that this prosthesis is not "biocompatible." However, the prosthesis probably did not recellularize (although it was designed to do so), and is also embolic, although the emboli in this case usually have little clinical consequence. This operational definition of biocompatible ("the patient is alive so it must be biocompatible") offers us little insight in designing new or improved vascular prostheses. It is probable that biocompatibility may have to be specifically defined for applications in soft tissue, hard tissue, and the cardiovascular system (blood compatibility, Chapters II.2.6 and II.3.5). In fact, biocompatibility may have to be uniquely defined for each application. The problems and meanings of biocompatibility will be explored and expanded upon throughout this textbook; in particular see Chapters II.2.2, II.3.2 and II.3.4.

Inflammation and Healing. Specialized biological mechanisms are triggered when a material or device interfaces with the body. Injury to tissue will stimulate the well-defined inflammatory reaction sequence that ultimately leads to healing. Healing can be normal (physiological) or abnormal (pathological). Where a foreign body (e.g., an implant) is present in the wound site (the surgical incision), the reaction sequence is referred to as the "foreign-body reaction" (Chapters II.2.2 and II.3.4). The normal response of the body will be modulated because of the solid implant. Furthermore, this reaction will differ in intensity and duration, depending upon the anatomical site involved. An understanding of how a foreign object shifts the normal inflammatory reaction sequence is an important concern for the biomaterials scientist.

Functional Tissue Structure and Pathobiology. Biomaterials-based medical devices are implanted into almost all tissues and organs. Tissues and organs vary widely in cell composition, morphological organization, vascularization, and innervation. Implantation of a biomaterial into bone, liver, or heart will have special physiological consequences. Therefore, key principles governing the structure of normal (and abnormal) cells, tissues, and organs are important to biomaterials researchers. Also, techniques by which the structure and function of normal and abnormal tissue are studied must be mastered. In addition, fundamental mechanisms leading to abnormal cell, tissue, and organ structures (i.e., diseases and other pathologies) are critical considerations to biomaterials researchers (see Chapter II.1.5).

Dependence on Specific Anatomical Sites of Implantation. Consideration of the anatomical site of an implant is essential. An intraocular lens may be implanted into the lens capsule or the anterior chamber of the eye. A hip joint will be implanted in bone across an articulating joint space. A prosthetic heart valve will be sutured into cardiac muscle, and will contact both soft tissue and blood. A catheter may be placed in an artery, a vein or the urinary tract. Each of these sites challenges the biomedical device designer with special requirements for anatomy, physiology, geometry, size, mechanical properties, and bioresponses.

Mechanical Requirements and Physical Performance Requirements. Each biomaterial and device has mechanical and performance requirements originating from the need to perform a physiological function. These requirements can be divided into three categories: mechanical performance, mechanical durability, and physical properties.

First, consider mechanical performance. A hip prosthesis must be strong and rigid. A tendon material must be strong and flexible. A tissue heart valve leaflet must be flexible and tough. A dialysis membrane must be strong and flexible, but not elastomeric. An articular cartilage substitute must be soft and elastomeric. One significant example of a controlled micro-mechanical interface is the contact zone between a synthetic biomaterial (titanium, tantalum, alumina, zirconia, and hydroxyapatites) and oral bones. Microstrain magnitudes have been controlled through macro/micro/nano surface/topographies and construct designs to be within the microstrain limits of bone. The result has been decades of chemical stability, now called osseous integration.

Then, we must address mechanical durability. A catheter may only have to perform for three days. A bone plate may fulfill its function in six months or longer. A leaflet in a heart valve must flex 60 times per minute without tearing for the lifetime of the patient (realistically, at least for 10 or more years). A hip joint must not fail under heavy loads for 20 years or more.

Finally, the bulk physical properties impact other aspects of performance. The dialysis membrane has a specified permeability, the acetabular cup of the hip joint must have high lubricity, and the intraocular lens has transparency and refraction requirements. To meet these requirements, design principles are borrowed from physics, chemistry, mechanical engineering, chemical engineering, and materials science.

Industrial Involvement. A significant basic research effort is now under way, primarily at universities, to understand how biomaterials function and how to optimize them. At the same time, companies are producing implants for use in humans and, appropriate to the mission of a company, earning profits on the sale of medical devices. Thus, although we are now only learning about the fundamentals of biointeraction, we manufacture millions of devices for implantation in humans. How is this dichotomy explained? Basically, as a result of 50 or more years of experience we now have a set of materials that perform satisfactorily in the body. The medical practitioner can use them with reasonable confidence, and the performance in the patient is largely acceptable (generally considerably better than other alternatives). Though the devices and materials are far from perfect, the complications associated with the devices are fewer than the complications of the original diseases.

Risk–Benefit and Corporate Realities. A risk–benefit analysis must be considered in developing new devices and improving existing devices. Central to biomaterials science is the desire to alleviate suffering and death, and also the desire to improve the quality of life for patients. These considerations are convoluted with the excitement of new scientific ideas, the corporate imperative to turn a profit, and the mandate of the regulatory agencies to protect the public. For example, the acceptable risk varies with different types of medical devices. The acceptable risk of devices that sustain life (e.g., heart valve, defibrillator, cardiac assist device, hemodialysis device/access graft, hydrocephalus shunt) is greater than that of devices that alleviate pain/disability or enhance function (e.g., hip joint, artificial skin, drug delivery device, intraocular lens, intrauterine contraceptive device). Then consider the acceptable risk for devices that have only cosmetic application (e.g., collagen injections, breast implants). Obviously, ethical concerns enter into the risk–benefit picture. Remember that companies have large investments in the development, manufacture, quality control, clinical testing, regulatory clearance, and distribution of medical devices. How much of an advantage (for the company and the patient) will be realized in introducing an improved device? The improved device may indeed work better for the patient. However, the company will incur a large expense (development and regulatory costs) that will be perceived by the stockholders as reduced profits. The development of a new or improved device, as well as offering benefits, entails risks that months or years after introduction some unforeseen complication will compromise the device. Product liability issues are a major concern to manufacturers. Consider questions about the ethics of withholding improved devices from people who could benefit from them because of development costs and regulatory hurdles, the market share advantages of having a better product, and the gargantuan costs (possibly nonrecoverable) of introducing a new product into the medical marketplace. If companies did not have the profit incentive, would there be any medical devices, let alone improved ones, available for clinical application?

From the biomaterials industry, we see specialized, essential contributions to our field. Industry deals well with technologies such as packaging, sterilization, storage, distribution, quality control, and analysis. These subjects are grounded in specialized technologies, often ignored in academic communities, but having the potential to generate stimulating research questions. Also, many companies support in-house basic research laboratories, and contribute in important ways to the fundamental study of biomaterials science.

Ethics. A wide range of ethical considerations impact biomaterials science. Some key ethical questions in biomaterials science are summarized in Table 3. Typical of ethical questions, an absolute answer may be difficult to come by. Some articles have addressed ethical questions in biomaterials and debated the important points (Saha and Saha, 1987; Schiedermayer and Shapiro, 1989; Merryman, 2008). Chapter III.2.7 introduces ethics and ethical issues related to biomaterials and medical devices.

Regulation. The consumer (the patient) and the physician demand safe medical devices. To prevent inadequately tested devices and materials from coming on the market, and to screen out those clearly unqualified to produce biomaterials, the United States government has evolved a complex regulatory system administered by the US Food and Drug Administration (FDA). Most nations of the world have similar medical device regulatory bodies. The International Standards Organization (ISO) has introduced international standards for the world community. Obviously, a substantial base of biomaterials knowledge went into establishing these standards. The costs to comply with the standards and to implement materials, biological, and clinical testing are enormous. Introducing a new biomedical device to the market requires a regulatory investment of tens of millions of dollars. Are the regulations and standards truly addressing the safety issues? Is the cost of regulation inflating the cost of healthcare and preventing improved devices from reaching those who need them? Under this regulation topic, we see the intersection of all the players in the biomaterials community: government, industry, scientists, physicians, and patients. The answers are not simple, but the problems must be addressed every day. Chapters III.2.2 and III.2.3 expand on standards and regulatory concerns.

TABLE 3 Ethical Concerns Relevant to Biomaterials Science

Animals
Is the animal model relevant to human physiology? Specifically, is the experiment well-designed and outcome sufficiently important so that the data obtained will justify the suffering and sacrifice of the life of a living creature?
Human Subjects
How should human subject research be conducted to minimize negative outcomes to the patient and offer a reasonable risk-to-benefit ratio? How can we best ensure informed consent?
Industrial Involvement
Companies fund much biomaterials research and also own proprietary biomaterials. How can the needs of the patient be best balanced with the financial goals of a company? Consider that someone must manufacture devices – these would not be available if a company did not choose to manufacture them.
Researchers
Since researchers often stand to benefit financially from a successful biomedical device, and sometimes even have devices named after them, how can investigator bias be minimized in biomaterials research?
Patients
For life-sustaining devices, what is the trade-off between sustaining life and the quality of life with the device for the patient? Should the patient be permitted to "pull the plug" if the quality of life is not satisfactory?
Regulatory Agencies
With so many unanswered questions about the basic science of biomaterials, do government regulatory agencies have sufficient information to define adequate tests for materials and devices and to properly regulate biomaterials?

BIOMATERIALS LITERATURE

Over the past 50 years, the field of biomaterials has evolved from individual physicians innovating to save the lives of their patients to the science-grounded multidisciplinary endeavor we see today. In 1950, there were no biomaterials or medical device journals, and few books. Concurrent with the evolution of the discipline, a literature has also developed addressing basic science, applied science, engineering, medicine, and commercial issues. A bibliography is provided in Appendix D to highlight key reference works and technical journals in the biomaterials field. As might be expected, these journals stem from many disciplines and technical societies.

BIOMATERIALS SOCIETIES

The biomaterials field evolved from individual researchers and clinicians who intellectually associated their efforts with established disciplines such as medicine, chemistry, chemical engineering, or mechanical engineering, to a modern field called "biomaterials." This evolution is paralleled by the growing sense of professionalism and the formation of biomaterials societies as homes for the profession to develop in. Probably the first biomaterials-related society was the American Society for Artificial Internal Organs (ASAIO). Founded in 1954, this group of visionaries established a platform to consider the development of devices such as the artificial kidney and the artificial heart. A Division of Interdisciplinary Studies, the administrative home of a nascent biomaterials effort, was established at Clemson University, Clemson, South Carolina in 1969. Clemson began organizing annual International Biomaterials Symposia in 1969. The first of these symposia was titled "Use of Ceramics in Surgical Implants." About 100 scientists and physicians attended, and 17 papers were presented. The cover of the 1975 International Biomaterials Symposium program is shown in Figure 10.

In 1974–1975, these symposia evolved into the Society for Biomaterials, the world's first biomaterials

FIGURE 10 The cover of the program book for the 1975 International Biomaterials Symposium, Clemson, South Carolina.

society. Founding members, about 40, plus those who joined in 1975 and 1976, numbered less than 100, and included clinicians, engineers, chemists, and biologists. Their common interest, biomaterials, was the engaging focus for the multidisciplinary participants. The European Society for Biomaterials was founded in 1975. Shortly after that, the Canadian Biomaterials Society and the Japanese Society for Biomaterials were formed. The Controlled Release Society, a group strongly rooted in biomaterials for drug delivery, was founded in 1978. At this time there are many national biomaterials societies and specialty societies. For example, the Tissue Engineering and Regenerative Medicine International Society (TERMIS) now has about 3000 members. The development of biomaterials professionalism and a sense of identity for the field called biomaterials can be attributed to these societies and the researchers who organized and led them.

SUMMARY

This chapter provides a broad overview of the biomaterials field. It offers a vantage point from which the reader can gain a perspective to see how the sub-themes fit into the larger whole.

Biomaterials science may be the most multidisciplinary of all the sciences. Consequently, biomaterials scientists must master certain key material from many fields of science, technology, engineering, and medicine in order to be competent and conversant in this profession. The reward for mastering this volume of material is immersion in an intellectually stimulating endeavor that advances a new basic science of biointeraction and contributes to reducing human suffering.

BIBLIOGRAPHY

Carlsson, L., Rostlund, T., Albrektsson, B., Albrektsson, T., & Branemark, P. I. (1986). Osseointegration of titanium implants. *Acta. Orthopaed. Scand.*, **57**, 285–289.

Daemen, J., & Serruys, P. W. (2007a). Drug-eluting stent update 2007: Part I. A survey of current and future generation drug-eluting stents: Meaningful advances or more of the same? *Circulation*, **116**(3), 316–328.

Daemen, J., & Serruys, P. W. (2007b). Drug-eluting stent update 2007: Part II. Unsettled issues. *Circulation*, **116**(8), 961–968.

Daley, G. Q., & Scadden, D. T. (2008). Prospects for stem cell therapy. *Cell*, **132**(4), 544–548.

Freed, L. E., Guilak, F., Guo, X. E., Gray, M. L., Tranquillo, R., Holmes, J. W., Radisic, M., Sefton, M. V., Kaplan, D., & Vunjak-Novakovic, G. (2006). Advanced tools for tissue engineering: Scaffolds, bioreactors, and signaling. *Tissue Eng.*, **12**, 3285–3305.

Goldberg, M., Langer, R., & Jia, X. (2007). Nanostructured materials for applications in drug delivery and tissue engineering. *J. Biomater. Sci. Polym. Ed.*, **18**, 241–268.

Hacker, M. C., & Mikos, A. G. (2006). Trends in tissue engineering research. *Tissue Eng.*, **12**, 2049–2057.

Hellman, K. B., & Nerem, R. M. (2007). Advancing tissue engineering and regenerative medicine. *Tissue Eng.*, **13**, 2823–2824.

Hench, L. L., & Pollak, J. M. (2002). Third-generation biomedical materials. *Science*, **295**, 1014–1017.

Huebsch, N., & Mooney, D. J. (2009). Inspiration and application in the evolution of biomaterials. *Nature*, **462**, 426–432.

Hunziker, E., Spector, M., Libera, J., Gertzman, A., Woo, S. L.-Y., Ratcliffe, A., Lysaght, M., Coury, A., Kaplan, D., & Vunjak-Novakovic, G. (2006). Translation from research to applications. *Tissue Eng.*, **12**, 3341–3364.

Ingber, D. E. (2010). From cellular mechanotransduction to biologically inspired engineering. *Annals of Biomedical Engineering*, **38**(3), 1148–1161.

Johnson, P. C., Mikos, A. G., Fisher, J. P., & Jansen, J. A. (2007). Strategic directions in tissue engineering. *Tissue Eng.*, **13**, 2827–2837.

Kohane, D. S., & Langer, R. (2008). Polymeric materials in tissue engineering. *Pediatr. Res.*, **63**(5), 487–491.

Khademhosseini, A., Langer, R., Borenstein, J., & Vacanti, J. P. (2006). Microscale technologies for tissue engineering and biology. *Proc. Natl. Acad. Sci. USA*, **103**(8), 2480–2487.

Kong, H. J., & Mooney, D. J. (2007). Microenvironmental regulation of biomacromolecular therapies. *Nat. Rev. Drug Discov.*, **6**(6), 455–463.

Langer, R., & Tirrell, D. A. (2004). Designing materials for biology and medicine. *Nature*, **428**(6982), 487–492.

Lawson, H. C., Sampath, P., Bohan, E., Park, M. C., Hussain, N., Olivi, A., Weingart, J., Kleinberg, L., & Bren, H. (2007). Interstitial chemotherapy for malignant gliomas: The Johns Hopkins experience. *J. Neurooncol.*, **83**(1), 61–70.

Lutolf, M. P., Gilbert, P. M., & Blau, H. M. (2009). Designing materials to direct stem cell fate. *Nature*, **462**, 433–441.

Lysaght, M. J., & O'Laughlin, J. (2000). The demographic scope and economic magnitude of contemporary organ replacement therapies. *ASAIO J.*, **46**, 515–521.

Merryman, W. D. (2008). Development of a tissue engineered heart valve for pediatrics: A case study in bioengineering ethics. *Sci. Eng. Ethics*, **14**, 93–101.

Meckstroth, K. R., & Darney, P. D. (2001). Implant contraception. *Semin. Reprod. Med.*, **19**, 339.

Mikos, A. G., Herring, S. W., Ochareon, P., Elisseeff, J., Lu, H. H., Kandel, R., Schoen, F. J., Toner, M., Mooney, D., Atala, A., Van Dyke, M. E., Kaplan, D., & Vunjak-Novakovic, G. (2006). Engineering complex tissues. *Tissue Eng.*, **12**(12), 3307–3309.

Ratner, B. D. (2011). The biocompatibility manifesto: biocompatibility for the twenty-first century. *J. Cardiovasc. Translat. Res.*, **4**(5), 523–527.

Rose, E. A., Gelijns, A. C., Moskowitz, A. J., Heitjan, D. F., Stevenson, L. W., Dembitsky, W. Long, J. W., Ascheim, D. D., Tierney, A. R., Levitan, R. G., Watson, J. T., Ronan, N. S., Shapiro, P. A., Lazar, R. M., Miller, L. W., Gupta, L. Frazier, O. H. Desvigne-Nickens, P. Oz, M. C., Poirier, V. L., & Meier, P. (2001). Long-term use of a left ventricular assist device for end-stage heart failure. *New Engl. J. M.*, **345**, 1435–1443.

Saha, S., & Saha, P. (1987). Bioethics and applied biomaterials. *J. Biomed. Mater. Res: Appl. Biomat.*, **21**, 181–190.

Schiedermayer, D. L., & Shapiro, R. S. (1989). The artificial heart as a bridge to transplant: Ethical and legal issues at the bedside. *J. Heart Transplant.*, **8**, 471–473.

Sengupta, S., Eavarone, D., Capila, I., Zhao, G., Watson, N., Kiziltepe, T., & Sasisekharan, R. (2005). Temporal targeting of tumour cells and neovasculature with a nanoscale delivery system. *Nature*, **436**, 568–572.

Sharp, P. A., & Langer, R. (2011). Promoting convergence in biomedical science. *Science*, **333**(6042), 527.

Stocum, D. L. (2005). Stem cells in CNS and cardiac regeneration. *Adv. Biochem. Eng. Biotechnol.*, **93**, 135–159.

Whitehead, K. A., Langer, R., & Anderson, D. G. (2009). Knocking down barriers: Advances in siRNA delivery. *Nat. Rev. Drug Discov.*, **8**, 129–138.

Williams, D. F. (1987). Chester, England, March 3-5 1986, *Definitions in Biomaterials. Proceedings of a Consensus Conference of the European Society for Biomaterials* (Vol. 4). New York: Elsevier.

Zenios, S., Makower, J., & Yock, P. (Eds.), (2010). *Biodesign: The Process of Innovating Medical Technologies*. Cambridge: Cambridge University Press.

A HISTORY OF BIOMATERIALS

Buddy D. Ratner

"History consists of a series of accumulated imaginative inventions"

Voltaire

A decade into the twenty-first century, biomaterials are widely used throughout medicine, dentistry, and biotechnology. Just 60 years ago biomaterials as we think of them today did not exist. The word "biomaterial" was not used. There were no medical device manufacturers (except for external prosthetics such as limbs, fracture fixation devices, glass eyes, and dental fillings and devices), no formalized regulatory approval processes, no understanding of biocompatibility, and certainly no academic courses on biomaterials. Yet, crude biomaterials have been used, generally with poor to mixed results, throughout history. This chapter will broadly trace the history of biomaterials from the earliest days of human civilization to the dawn of the twenty-first century. It is convenient to organize the history of biomaterials into four eras: prehistory; the era of the surgeon hero; designed biomaterials/engineered devices; and the contemporary era taking us into the new millennium. The emphasis of this chapter will be on the experiments and studies that set the foundation for the field we call biomaterials, largely between 1920 and 1980.

BIOMATERIALS BEFORE WORLD WAR II

Before Civilization

The introduction of non-biological materials into the human body took place throughout history. The remains of a human found near Kennewick, Washington, USA (often referred to as the "Kennewick Man") were dated (with some controversy) to be 9000 years old. This individual, described by archeologists as a tall, healthy, active person, wandered through the region now know as southern Washington with a spear point embedded in his hip. It had apparently healed in, and did not significantly impede his activity. This unintended implant illustrates the body's capacity to deal with implanted foreign materials. The spear point has little resemblance to modern biomaterials, but it was a "tolerated" foreign material implant, just the same. Another example of the introduction of foreign material into the skin, dated to over 5000 years ago, is the tattoo. The carbon particles and other substances probably elicited a classic foreign-body reaction.

Dental Implants in Early Civilizations

Unlike the spear point described above, dental implants were devised as implants and used early in history. The Mayan people fashioned nacre teeth from sea shells in roughly 600 AD, and apparently achieved what we now refer to as osseointegration (see Chapter II.5.7), basically a seamless integration into the bone (Bobbio, 1972). Similarly, in France, a wrought iron dental implant in a corpse was dated to 200 AD (Crubezy et al., 1998). This implant, too, was described as properly osseointegrated. There was no materials science, biological understanding, or medicine behind these procedures. Still, their success (and longevity) is impressive and highlights two points: the forgiving nature of the human body and the pressing drive, even in prehistoric times, to address the loss of physiologic/anatomic function with an implant.

Sutures Dating Back Thousands of Years

There is loose evidence that sutures may have been used even in the Neolithic period. Large wounds were closed early in history primarily by one of two methods – cautery or sutures. Linen sutures were used by the early Egyptians. Catgut was used in the middle ages in Europe. In South Africa and India, the heads of large, biting ants clamped wound edges together.

Metallic sutures are first mentioned in early Greek literature. Galen of Pergamon (circa 130–200 AD) described ligatures of gold wire. In 1816, Philip Physick, University of Pennsylvania Professor of Surgery, suggested the use of lead wire sutures, noting little reaction. J. Marion Sims of Alabama had a jeweler fabricate sutures of silver wire, and in 1849 performed many successful operations with this metal.

Consider the problems that must have been experienced with sutures in eras with no knowledge of sterilization, toxicology, immunological reaction to extraneous biological materials, inflammation, and biodegradation. Yet sutures were a relatively common fabricated or manufactured biomaterial for thousands of years.

Artificial Hearts and Organ Perfusion

In the fourth century BC, Aristotle called the heart the most important organ in the body. Galen proposed that veins connected the liver to the heart to circulate "vital spirits throughout the body via the arteries." English physician William Harvey, in 1628 espoused a relatively modern view of heart function when he wrote: "the heart's one role is the transmission of the blood and its propulsion, by means of the arteries, to the extremities everywhere." With the appreciation of the heart as a pump, it was a logical idea to think of replacing the heart with an artificial pump. In 1812, the French physiologist Le Gallois expressed his idea that organs could be kept alive by pumping blood through them. A number of experiments on organ perfusion with pumps were performed from 1828 to 1868. In 1881, Étienne-Jules Marey, a brilliant scientist and thinker who published and invented in photographic technology, motion studies, and physiology, described an artificial heart device (Figure 1), primarily oriented to studying the beating of the heart.

In 1938, aviator (and engineer) Charles Lindbergh and surgeon (and Nobel Prize Winner) Alexis Carrel wrote a visionary book, *The Culture of Organs*. They addressed issues of pump design (referred to as the Lindbergh pump), sterility, blood damage, the nutritional needs of perfused organs, and mechanics. This book is a seminal document in the history of artificial organs. In the mid-1950s, Dr. Paul Winchell, better known as a ventriloquist, patented an artificial heart. In 1957, Dr. Willem Kolff and a team of scientists tested the artificial heart in animals. More modern conceptions of the artificial heart (and left ventricular assist device) will be presented below and in Chapter II.5.3.D).

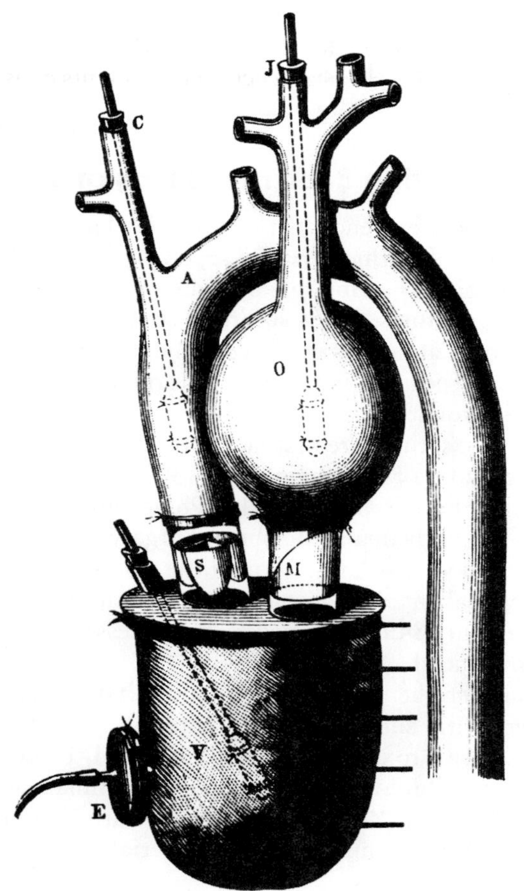

FIGURE 1 An artificial heart by Étienne-Jules Marey, Paris, 1881.

Contact Lenses

Leonardo DaVinci, in the year 1508, developed the concept of contact lenses. Rene Descartes is credited with the idea of the corneal contact lens (1632) and Sir John F. W. Herschel (1827) suggested that a glass lens could protect the eye. Adolf Gaston Eugen Fick (nephew of Adolf Eugen Fick of Fick's Law of Diffusion fame) was an optometrist by profession. One of his inventions (roughly 1860) was a glass contact lens, possibly the first contact lens offering real success. He experimented on both animals and humans with contact lenses. In the period from 1936 to 1948, plastic contact lenses were developed, primarily of poly(methyl methacrylate).

Basic Concepts of Biocompatibility

Most implants prior to 1950 had a low probability of success, because of a poor understanding of biocompatibility and sterilization. As will be elaborated upon throughout the textbook, factors that contribute to biocompatibility include the chemistry of the implant, leachables, shape, mechanics, and design. Early studies, especially with metals, focused primarily on ideas from chemistry to explain the observed bioreaction.

Possibly the first study assessing the *in vivo* bioreactivity of implant materials was performed by H. S. Levert (1829). Gold, silver, lead, and platinum specimens were studied in dogs, and platinum, in particular, was found to be well-tolerated. In 1886, bone fixation plates of nickel-plated sheet steel with nickel-plated screws were studied. In 1924, A. Zierold published a study on tissue reaction to various materials in dogs. Iron and steel were found to corrode rapidly, leading to resorption of adjacent bone. Copper, magnesium, aluminum alloy, zinc, and nickel discolored the surrounding tissue. Gold, silver, lead, and aluminum were tolerated, but inadequate mechanically. Stellite, a Co–Cr–Mo alloy, was well tolerated and strong. In 1926, M. Large noted inertness displayed by 18-8 stainless steel containing molybdenum. By 1929 Vitallium alloy (65% Co–30% Cr–5%Mo) was developed and used with success in dentistry. In 1947, J. Cotton of the UK discussed the possible use of titanium and alloys for medical implants.

The history of plastics as implantation materials does not extend as far back as metals, simply because there were few plastics prior to the 1940s. What is possibly the first paper on the implantation of a modern synthetic polymer, nylon, as a suture appeared in 1941. Papers on the implantation of cellophane, a polymer made from plant sources, were published as early as 1939, it being used as a wrapping for blood vessels. The response to this implant was described as a "marked fibrotic reaction." In the early 1940s papers appeared discussing the reaction to implanted poly(methyl methacrylate) (PMMA) and nylon. The first paper on polyethylene as a synthetic implant material was published in 1947 (Ingraham et al.). The paper pointed out that polyethylene production using a new high pressure polymerization technique began in 1936. This process enabled the production of polyethylene free of initiator fragments and other additives. Ingraham et al. demonstrated good results on implantation (i.e., a mild foreign-body reaction), and attributed these results to the high purity of the polymer they used. A 1949 paper commented on the fact that additives to many plastics had a tendency to "sweat out," and this might be responsible for the strong biological reaction to those plastics (LeVeen and Barberio, 1949). They found a vigorous foreign-body reaction to cellophane, Lucite®, and nylon, but an extremely mild reaction to "a new plastic," Teflon®. The authors incisively concluded: "Whether the tissue reaction is due to the dissolution of traces of the unpolymerized chemical used in plastics manufacture or actually to the solution of an infinitesimal amount of the plastic itself cannot be determined." The possibility that cellulose might trigger the severe reaction by activating the complement system could not have been imagined, because the complement system had not yet been discovered.

WORLD WAR II TO THE MODERN ERA: THE SURGEON/PHYSICIAN HERO

During World War I, and particularly at the end of the war, newly developed high-performance metal, ceramic, and especially polymeric materials, transitioned from wartime restriction to peacetime availability. The possibilities for using these durable, inert materials immediately intrigued surgeons with needs to replace diseased or damaged body parts. Materials, originally manufactured for airplanes, automobiles, clocks, and radios were taken "off-the-shelf" by surgeons and applied to medical problems. These early biomaterials included silicones, polyurethanes, Teflon®, nylon, methacrylates, titanium, and stainless steel.

A historical context helps us appreciate the contribution made primarily by medical and dental practitioners. Just after World War II, there was little precedent for surgeons to collaborate with scientists and engineers. Medical and dental practitioners of this era felt it was appropriate to invent (improvise) where the life or functionality of their patient was at stake. Also, there was minimal government regulatory activity, and human subject protections as we know them today were non-existent (see Chapters III.2.4 and III.2.7). The physician was implicitly entrusted with the life and health of the patient and had much more freedom than is seen today to take heroic action when other options were exhausted.[1] These medical practitioners had read about the post-World War II marvels of materials science. Looking at a patient open on the operating table, they could imagine replacements, bridges, conduits, and even organ systems based on such materials. Many materials were tried on the spur of the moment. Some fortuitously succeeded. These were high risk trials, but usually they took place where other options were not available. The term "surgeon hero" seems justified, since the surgeon often had a life (or a quality of life) at stake and was willing to take a huge technological and professional leap to repair the individual. This *laissez faire* biomaterials era quickly led to a new order characterized by scientific/engineering input, government quality controls, and a sharing of decisions prior to attempting high-risk, novel procedures. Still, a foundation of ideas and materials for the biomaterials field was built by courageous, fiercely committed, creative individuals, and it is important to look at this foundation to understand many of the attitudes, trends, and materials common today.

Intraocular Lenses

Sir Harold Ridley, MD (1906–2001) (Figure 2), inventor of the plastic intraocular lens (IOL), made early, accurate observations of biological reaction to implants consistent with currently accepted ideas of biocompatibility. After World War II, he had the opportunity to examine aviators who were unintentionally implanted in their eyes with shards of plastic from shattered canopies in Spitfire and Hurricane fighter planes (Figure 2). Most of these flyers had plastic fragments years after the war. The conventional wisdom at that time was that the human body would not tolerate implanted foreign objects, especially in the eye – the body's reaction to a splinter or a bullet was cited as examples of the difficulty of implanting materials in the body. The eye is an interesting implant site, because you can look in through a transparent window to observe the reaction. When Ridley did so, he noted that the shards had healed in place with no further reaction. They were, by his standard, tolerated

[1]The regulatory climate in the US in the 1950s was strikingly different from today. This can be appreciated in this recollection from Willem Kolff about a pump oxygenator he made and brought with him from Holland to the Cleveland Clinic (Kolff, W. J., 1998): "Before allowing Dr. Effler and Dr. Groves to apply the pump oxygenator clinically to human babies, I insisted they do 10 consecutive, successful operations in baby dogs. The chests were opened, the dogs were connected to a heart-lung machine to maintain the circulation, the right ventricles were opened, a cut was made in the interventricular septa, the septa holes were closed, the right ventricles were closed, the tubes were removed and the chests were closed. (I have a beautiful movie that shows these 10 puppies trying to crawl out of a basket).

FIGURE 2 Left: Sir Harold Ridley, inventor of the intraocular lens, knighted by Queen Elizabeth II for his achievement. Right: Shards from the canopy of the Spitfire airplane were the inspiration leading to intraocular lenses. (Image by Bryan Fury75 at fr.wikipedia [GFDL (www.gnu.org/copyleft/fdl.html), from Wikimedia Commons].)

by the eye. Today, we would describe this stable healing without significant ongoing inflammation or irritation as "biocompatible." This is an early observation of "biocompatibility" in humans, perhaps the first, using criteria similar to those accepted today. Based on this observation, Ridley traced down the source of the plastic domes, ICI Perspex® poly(methyl methacrylate). He used this material to fabricate implant lenses (intraocular lenses) that were found, after some experimentation, to function reasonably in humans as replacements for surgically removed natural lenses that had been clouded by cataracts. The first implantation in a human was on November 29, 1949. For many years, Ridley was the center of fierce controversy because he challenged the dogma that spoke against implanting foreign materials in eyes – it is hard to believe that the implantation of a biomaterial would provoke such an outcry. Because of this controversy, this industry did not instantly arise – it was the early 1980s before IOLs became a major force in the biomedical device market. Ridley's insightful observation, creativity, persistence, and surgical talent in the late 1940s evolved into an industry that presently puts more than 7,000,000 of these lenses annually in humans. Through all of human history, cataracts meant blindness or a surgical procedure that left the recipient needing thick, unaesthetic spectacle lenses that poorly corrected the vision. Ridley's concept, using a plastic material found to be "biocompatible," changed the course of history and substantially improved the quality of life for millions of individuals with cataracts. Harold Ridley's story is elaborated upon in an obituary (Apple and Trivedi, 2002).

Hip and Knee Prostheses

The first hip replacement was probably performed in 1891 by a German surgeon, Theodore Gluck, using a cemented ivory ball. This procedure was not successful. Numerous attempts were made between 1920 and 1950 to develop a hip replacement prosthesis. Surgeon M. N. Smith-Petersen, in 1925, explored the use of a glass hemisphere to fit over the ball of the hip joint. This failed due to poor durability. Chrome–cobalt alloys and stainless steel offered improvements in mechanical properties, and many variants of these were explored. In 1938, the Judet brothers of Paris, Robert and Jean, explored an acrylic surface for hip procedures, but it had a tendency to loosen. The idea of using fast setting dental acrylics to glue prosthetics to bone was developed by Dr. Edward J. Haboush in 1953. In 1956, McKee and Watson-Farrar developed a "total" hip with an acetabular cup of metal that was cemented in place. Metal-on-metal wear products probably led to high complication rates. It was John Charnley (1911–1982) (Figure 3), working at an isolated tuberculosis sanatorium in Wrightington, Manchester, England, who invented the first really successful hip joint prosthesis. The femoral stem, ball head, and plastic acetabular cup proved to be a reasonable solution to the problem of damaged joint replacement. In 1958, Dr. Charnley used a Teflon acetabular cup with poor outcomes due to wear debris. By 1961 he was using a high molecular weight polyethylene cup, and was achieving much higher success rates. Interestingly, Charnley learned of high molecular weight polyethylene from a salesman selling novel plastic gears to one of his technicians. Dr. Dennis Smith contributed in an important way to the development of the hip prosthesis by introducing Dr. Charnley to poly(methyl methacrylate) cements, developed in the dental community, and optimizing those cements for hip replacement use. Total knee replacements borrowed elements of the hip prosthesis technology, and successful results were obtained in the period 1968–1972 with surgeons Frank Gunston and John Insall leading the way.

Dental Implants

Some "prehistory" of dental implants has been described above. In 1809, Maggiolo implanted a gold post anchor into fresh extraction sockets. After allowing this to heal, he affixed to it a tooth. This has remarkable similarity to modern dental implant procedures. In 1887, this procedure was used with a platinum post. Gold and platinum gave poor long-term results, and so this procedure

FIGURE 3 Left: Sir John Charnley. Right: The original Charnley hip prosthesis. (Hip prosthesis photo courtesy of the South Australian Medical Heritage Society, Inc.)

was never widely adopted. In 1937, Venable used surgical Vitallium and Co–Cr–Mo alloy for such implants. Also around 1937, Strock at Harvard used a screw-type implant of Vitallium, and this may be the first successful dental implant. A number of developments in surgical procedure and implant design (for example, the endosteal blade implant) then took place. In 1952, a fortuitous discovery was made. Per Ingvar Brånemark, an orthopedic surgeon at the University of Lund, Sweden, was implanting an experimental cage device in rabbit bone for observing healing reactions. The cage was a titanium cylinder that screwed into the bone. After completing the experiment that lasted several months, he tried to remove the titanium device and found it tightly integrated in the bone (Brånemark et al., 1964). Dr. Brånemark named the phenomenon "osseointegration," and explored the application of titanium implants to surgical and dental procedures. He also developed low-impact surgical protocols for tooth implantation that reduced tissue necrosis and enhanced the probability of good outcomes. Most dental implants and many other orthopedic implants are now made of titanium and its alloys.

The Artificial Kidney

Kidney failure, through most of history, was a sentence to an unpleasant death lasting over a period of about a month. In 1910, at Johns Hopkins University, the first attempts to remove toxins from blood were made by John Jacob Abel. The experiments were with rabbit blood, and it was not possible to perform this procedure on humans. In 1943, in Nazi-occupied Holland, Willem Kolff (Figure 4), a physician just beginning his career at that time, built a drum dialyzer system from a 100 liter tank, wood slats, and 130 feet of cellulose sausage casing tubing as the dialysis membrane. Some successes were seen in saving lives where prior to this there was only one unpleasant outcome to kidney failure. Kolff took his ideas to the United States and in 1960, at the Cleveland Clinic, developed a "washing machine artificial kidney" (Figure 5).

Major advances in kidney dialysis were made by Dr. Belding Scribner at the University of Washington (Figure 6). Scribner devised a method to routinely access the bloodstream for dialysis treatments. Prior to this, after just a few treatments, access sites to the blood were used up and further dialysis was not possible. After seeing the potential of dialysis to help patients, but only acutely, Scribner tells the story of waking up in the middle of the night with an idea to gain easy access to the blood – a shunt implanted between an artery and vein that emerged through the skin as a "U." Through the exposed portion of the shunt, blood access could be readily achieved. When Dr. Scribner heard about the new plastic, Teflon®, he envisioned how to get the blood out of and into the blood vessels. His device, built with the assistance of Wayne Quinton (Figure 6), used Teflon tubes to access the vessels, a Dacron® sewing cuff through the skin, and a silicone rubber tube for blood flow. The Quinton–Scribner shunt made chronic dialysis possible, and is said to be responsible for more than a million patients being alive today. Interestingly, Dr. Scribner refused to patent

FIGURE 4 Dr. Willem Kolff at age 92. (Photograph by B. Ratner.)

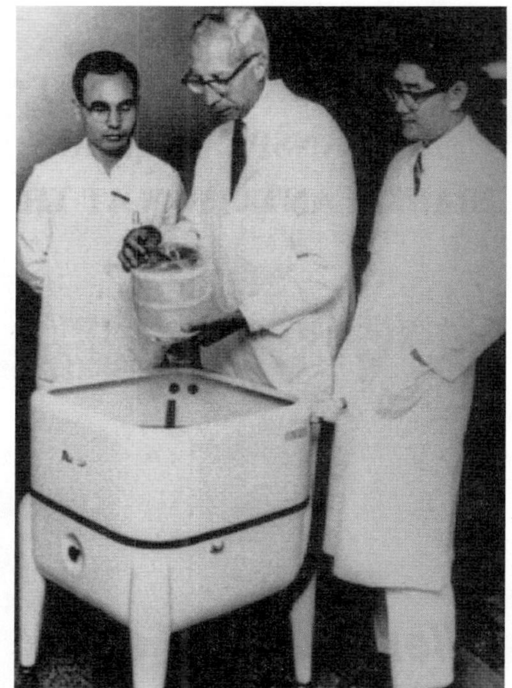

FIGURE 5 Willem Kolff (center) and the washing machine artificial kidney.

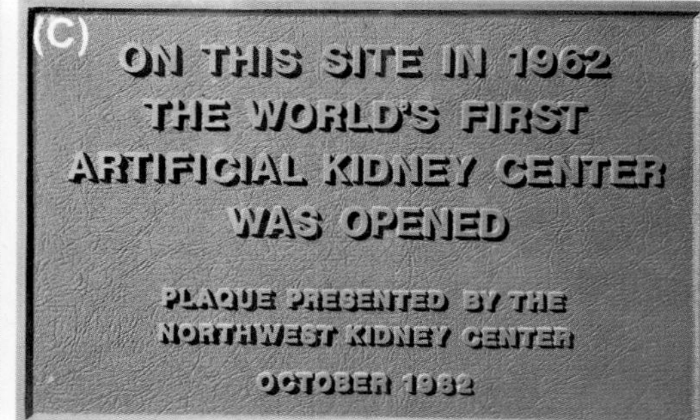

FIGURE 6 (a) Belding Scribner (Courtesy of Dr. Eli Friedman); (b) Wayne Quinton (Photo by B. Ratner); (c) Plaque commemorating the original location in Seattle of the world's first artificial kidney center.

his invention because of its importance to medical care. Additional important contributions to the artificial kidney were made by Chemical Engineering Professor Les Babb (University of Washington) who, working with Scribner, improved dialysis performance and invented a proportioning mixer for the dialysate fluid. The first dialysis center was opened in Seattle making use of these important technological advances (Figure 6). The early experience with dialyzing patients where there were not enough dialyzers to meet the demand also made important contributions to bioethics associated with medical devices (Blagg, 1998).

The Artificial Heart

Willem Kolff was also a pioneer in the development of the artificial heart. He implanted the first artificial heart in the Western hemisphere in a dog in 1957 (a Russian artificial heart was implanted in a dog in the late 1930s). The Kolff artificial heart was made of a thermosetting poly(vinyl chloride) cast inside hollow molds to prevent seams. In 1953, the heart-lung machine was invented by John Gibbon, but this was useful only for acute treatment, such as during open heart surgery. In 1964, the National Heart and Lung Institute of the NIH set a goal of a total artificial heart by 1970. Dr. Michael DeBakey implanted a left ventricular assist device in a human in 1966, and Dr. Denton Cooley and Dr. William Hall implanted a polyurethane total artificial heart in 1969. In the period 1982–1985, Dr. William DeVries implanted a number of Jarvik hearts based upon designs originated by Drs. Clifford Kwan-Gett and Donald Lyman – patients lived up to 620 days on the Jarvik 7® device.

Breast Implants

The breast implant evolved to address the poor results achieved with direct injection of substances into the breast for augmentation. In fact, in the 1960s, California and Utah classified use of silicone injections as a criminal offense. In the 1950s, poly(vinyl alcohol) sponges were implanted as breast prostheses, but results with these were also poor. University of Texas plastic surgeons Thomas Cronin and Frank Gerow invented the first silicone breast implant in the early 1960s, a silicone shell filled with silicone gel. Many variations of this device have been tried over the years, including cladding the device with polyurethane foam (the Natural Y implant). This variant of the breast implant was fraught with problems. However, the basic silicone rubber–silicone gel breast implant was generally acceptable in performance (Bondurant et al., 1999).

Vascular Grafts

Surgeons have long needed methods and materials to repair damaged and diseased blood vessels. Early in the century, Dr. Alexis Carrel developed methods to anastomose (suture) blood vessels, an achievement for which he won the Nobel Prize in medicine in 1912. In 1942 Blackmore used Vitallium metal tubes to bridge arterial defects in war wounded soldiers. Columbia University surgical intern Arthur Voorhees (1922–1992), in 1947, noticed during a post-mortem that tissue had grown around a silk suture left inside a lab animal. This observation stimulated the idea that a cloth tube might also heal by being populated by the tissues of the body. Perhaps such a healing reaction in a tube could be used to replace an artery? His first experimental vascular grafts were sewn from a silk handkerchief and then parachute fabric (Vinyon N), using his wife's sewing machine. The first human implant of a prosthetic vascular graft was in 1952. The patient lived many years after this procedure, inspiring many surgeons to copy the procedure. By 1954, another paper was published establishing the clear benefit of a porous (fabric) tube over a solid polyethylene tube (Egdahl et al., 1954). In 1958, the following technique was described in a textbook on vascular surgery (Rob, 1958): "The Terylene, Orlon or nylon cloth is bought from a draper's shop and cut with pinking shears to the required shape. It is then sewn with thread of similar material into a tube and sterilized by autoclaving before use."

Stents

Partially occluded coronary arteries lead to angina, diminished heart functionality and eventually, when the artery occludes (i.e., myocardial infarction), death of a localized portion of the heart muscle. Bypass operations take a section of vein from another part of the body and replace the occluded coronary artery with a clean conduit – this is major surgery, hard on the patient, and expensive. Synthetic vascular grafts in the 3 mm diameter size that is appropriate to the human coronary artery anatomy will thrombose, and thus cannot be used. Another option is percutaneous transluminal coronary angioplasty (PTCA). In this procedure, a balloon is threaded on a catheter into the coronary artery and then inflated to open the lumen of the occluding vessel. However, in many cases the coronary artery can spasm and close from the trauma of the procedure. The invention of the coronary artery stent, an expandable mesh structure that holds the lumen open after PTCA, was revolutionary in the treatment of coronary occlusive disease. In his own words, Dr. Julio Palmaz (Figure 7) describes the origins and history of the cardiovascular stent.

> *I was at a meeting of the Society of Cardiovascular and Interventional Radiology in February 1978 when a visiting lecturer, Doctor Andreas Gruntzig from Switzerland, was presenting his preliminary experience with coronary balloon angioplasty. As you know, in 1978 the mainstay therapy of coronary heart disease was surgical bypass. Doctor Gruntzig showed his promising new technique to open up coronary atherosclerotic blockages without the need for open chest surgery, using his own plastic balloon catheters. During his presentation, he made it clear that in a third of the cases, the treated vessel closed back after initial opening with the angioplasty balloon because of elastic recoil or delamination of the vessel wall layers. This required standby surgery facilities and personnel, in case acute closure after balloon angioplasty prompted emergency coronary bypass. Gruntzig's description of the problem of vessel reclosure elicited in my mind the idea of using some sort of support, such as used in mine tunnels or in oil well drilling. Since*

FIGURE 7 Dr. Julio Palmaz, inventor of the coronary artery stent. (Photograph by B. Ratner.)

the coronary balloon goes in small (folded like an umbrella) and is inflated to about 3–4 times its initial diameter, my idealistic support device needed to go in small and expand at the site of blockage with the balloon. I thought one way to solve this was a malleable, tubular, criss-cross mesh. I went back home in the Bay Area and started making crude prototypes with copper wire and lead solder, which I first tested in rubber tubes mimicking arteries. I called the device a BEIS or balloon-expandable intravascular graft. However, the reviewers of my first submitted paper wanted to call it a stent. When I looked the word up, I found out that it derives from Charles Stent, a British dentist who died at the turn of the century. Stent invented a wax material to make dental molds for dentures. This material was later used by plastic surgeons to keep tissues in place, while healing after surgery. The word "stent" was then generically used for any device intended to keep tissues in place while healing.

I made the early experimental device of stainless steel wire soldered with silver. These were materials I thought would be appropriate for initial laboratory animal testing. To carry on with my project I moved to the University of Texas Health Science Center in San Antonio (UTHSCSA). From 1983–1986 I performed mainly bench and animal testing that showed the promise of the technique and the potential applications it had in many areas of vascular surgery and cardiology. With a UTHSCSA pathologist, Doctor Fermin Tio, we observed our first microscopic specimen of implanted stents in awe. After weeks to months after implantation by catheterization under X-ray guidance, the stent had remained open, carrying blood flow. The metal mesh was covered with translucent, glistening tissue similar to the lining of a normal vessel. The question remained whether the same would happen in atherosclerotic vessels. We tested this question in the atherosclerotic rabbit model and to our surprise, the new tissue free of atherosclerotic plaque encapsulated the stent wires, despite the fact that the animals were still on a high cholesterol diet. Eventually, a large sponsor (Johnson & Johnson) adopted the project and clinical trials were instituted under the scrutiny of the Food and Drug Administration.

Coronary artery stenting is now performed in well over 1.5 million procedures per year.

Pacemakers

In London in 1788, Charles Kite wrote "An Essay Upon the Recovery of the Apparently Dead," where he discussed electrical discharges to the chest for heart resuscitation. In the period 1820–1880 it was already known that electric shocks could modulate the heartbeat (and, of course, consider the *Frankenstein* story from that era). The invention of the portable pacemaker, hardly portable by modern standards, may have taken place almost simultaneously in two groups in 1930–1931 – Dr. Albert S. Hyman (USA) (Figure 8), and Dr. Mark C. Lidwill (working in Australia with physicist Major Edgar Booth).

Canadian electrical engineer, John Hopps, while conducting research on hypothermia in 1949, invented an early cardiac pacemaker. Hopps' discovery was if a cooled heart stopped beating, it could be electrically restarted. This led to Hopps' invention of a vacuum tube cardiac pacemaker in 1950. Paul M. Zoll developed a pacemaker in conjunction with the Electrodyne Company in 1952. The device was about the size of a small microwave oven, was powered with external current, and stimulated the heart using electrodes placed on the chest – this therapy caused pain and burns, although it could pace the heart.

In the period 1957–1958, Earl E. Bakken, founder of Medtronics, Inc. developed the first wearable transistorized (external) pacemaker at the request of heart surgeon Dr. C. Walton Lillehei. Bakken quickly produced a prototype that Lillehei used on children with post-surgery heart block. Medtronic commercially produced this wearable, transistorized unit as the 5800 pacemaker.

In 1959 the first fully-implantable pacemaker was developed by engineer Wilson Greatbatch and cardiologist W.M. Chardack. He used two Texas Instruments transistors, a technical innovation that permitted small size and low power drain. The pacemaker was encased in epoxy to inhibit body fluids from inactivating it.

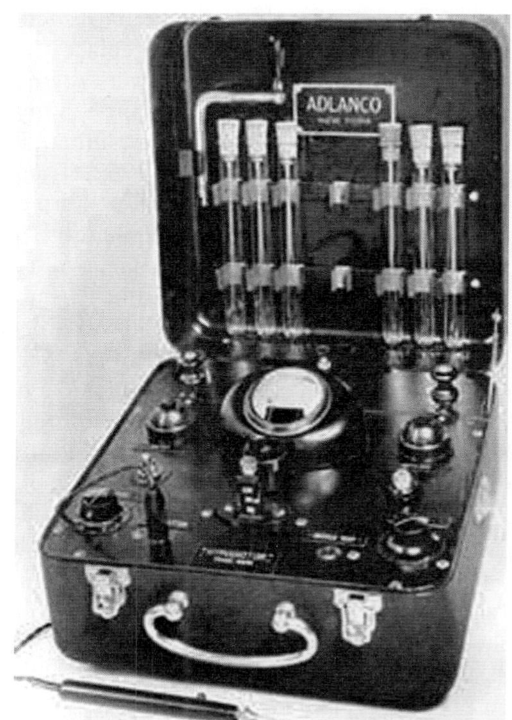

FIGURE 8 The Albert Hyman Model II portable pacemaker, circa 1932–1933. (Courtesy of the NASPE-Heart Rhythm Society History Project (www.Ep-History.org).)

Heart Valves

The development of the prosthetic heart valve paralleled developments in cardiac surgery. Until the heart could be stopped and blood flow diverted, the replacement of a valve would be challenging. Charles Hufnagel, in 1952, implanted a valve consisting of a poly(methyl methacrylate) tube and nylon ball in a beating heart (Figure 9). This was a heroic operation and basically unsuccessful, but an operation that inspired cardiac surgeons to consider that valve prostheses might be possible. The 1953 development of the heart-lung machine by Gibbon allowed the next stage in the evolution of the prosthetic heart valve to take place. In 1960, a mitral valve replacement was performed in a human by surgeon Albert Starr, using a valve design consisting of a silicone ball and poly(methyl methacrylate) cage (later replaced by a stainless steel cage). The valve was invented by engineer Lowell Edwards. The heart valve was based on a design for a bottle stopper invented in 1858. Starr was quoted as saying: "Let's make a valve that works and not worry about its looks," referring to its design that was radically different from the leaflet valve that nature evolved in mammals. Prior to the Starr–Edwards valve, no human had lived with a prosthetic heart valve longer than three months. The Starr–Edwards valve was found to permit good patient survival. The major issues in valve development in that era were thrombosis and durability. In 1969, Warren Hancock started the development of the first leaflet tissue heart valve based upon glutaraldehyde-treated pig valves, and his company and valve were acquired by Johnson & Johnson in 1979.

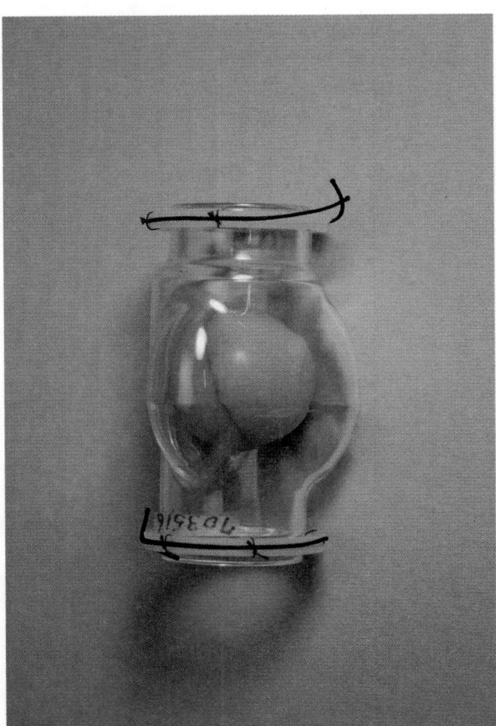

FIGURE 9 The Hufnagel heart valve consisting of a poly(methyl methacrylate) tube and nylon ball. (United States Federal Government image in the public domain.)

Drug Delivery and Controlled Release

Through most of history drugs were administered orally or by hypodermic syringe. In general, there was no effort to modulate the rate of uptake of the drug into the body. In 1949, Dale Wurster invented what is now known as the Wurster process that permitted pills and tablets to be encapsulated and therefore slow their release rate. However, modern ideas of controlled release can be traced to a medical doctor, Judah Folkman. Dr. Folkman noted that dyes penetrated deeply into silicone rubber, and he surmised from this that drugs might do the same. He sealed isoproterenol (a drug used to treat heart block) into silicone tubes, and implanted these into the hearts of dogs (Folkman and Long, 1964). He noted the delayed release and later applied this same idea to delivering a birth control steroid. He donated this development, patent-free, to the World Population Council. An entrepreneur and chemist, Alejandro Zaffaroni, heard of the Folkman work, and launched a company in 1970, Alza (originally called Pharmetrics), to develop these ideas for the pharmaceutical industry. The company developed families of new polymers for controlled release, and also novel delivery strategies. Alza was a leader in launching this new field that is so important today; further details on the field of controlled release are provided in an excellent historical overview (Hoffman, 2008).

DESIGNED BIOMATERIALS

In contrast to the biomaterials of the surgeon-hero era, when largely off-the-shelf materials were used to fabricate medical devices, the 1960s on saw the development of materials designed specifically for biomaterials applications. Here are some key classes of materials and their evolution from commodity materials to engineered/synthesized biomaterials.

Silicones

Although the class of polymers known as silicones has been explored for many years, it was not until the early 1940s that Eugene Rochow of GE pioneered the scale-up and manufacture of commercial silicones via the reaction of methyl chloride with silicon in the presence of catalysts. In Rochow's 1946 book, *The Chemistry of Silicones* (John Wiley & Sons, Publishers), he comments anecdotally on the low toxicity of silicones, but did not propose medical applications. Possibly the first report of silicones for implantation was by Lahey (1946) (see also Chapter II.5.18). The potential for medical uses of these materials was realized shortly after this. In a 1954 book on silicones, McGregor has a whole chapter titled "Physiological Response to Silicones." Toxicological studies were cited suggesting to McGregor that the quantities of silicones that humans might take into their bodies should be "entirely harmless." He mentions, without citation, the application of silicone rubber in artificial kidneys. Silicone-coated rubber grids were also used to support a dialysis membrane (Skeggs and Leonards, 1948). Many other early applications of silicones in medicine are cited in Chapter II.5.18.

Polyurethanes

Polyurethane was invented by Otto Bayer and colleagues in Germany in 1937. The chemistry of polyurethanes intrinsically offered a wide range of synthetic options leading to hard plastics, flexible films, or elastomers (Chapter I.2.2.A). Interestingly, this was the first class of polymers to exhibit rubber elasticity without covalent cross-linking. As early as 1959, polyurethanes were explored for biomedical applications, specifically heart valves (Akutsu et al., 1959). In the mid-1960s a class of segmented polyurethanes was developed that showed both good biocompatibility and outstanding flex life in biological solutions at 37°C (Boretos and Pierce, 1967). Sold under the name Biomer® by Ethicon and based upon DuPont Lycra®, these segmented polyurethanes comprised the pump diaphragms of the Jarvik 7® hearts that were implanted in seven humans.

Teflon®

DuPont chemist Roy Plunkett discovered a remarkably inert polymer, Teflon® (polytetrafluoroethylene) (PTFE), in 1938. William L. Gore and his wife, Vieve started a company in 1958 to apply Teflon® for wire insulation. In 1969, their son Bob found that Teflon®, if heated and stretched, forms a porous membrane with attractive physical and chemical properties. Bill Gore tells the story that, on a chairlift at a ski resort, he pulled from his parka pocket a piece of porous Teflon® tubing to show to his fellow ski lift passenger. The skier was a physician and asked for a specimen to try as a vascular prosthesis. Now, Goretex® porous Teflon® and similar expanded PTFEs are the leading synthetic vascular grafts, and are also used in numerous other applications in surgery and biotechnology (Chapters I.2.2.C and II.5.3.B).

Hydrogels

Hydrogels have been found in nature since life on earth evolved. Bacterial biofilms, hydrated extracellular matrix components, and plant structures are ubiquitous, water-swollen motifs in nature. Gelatin and agar were also known and used for various applictions early in human history. But the modern history of hydrogels as a class of materials designed for medical applications can be accurately traced.

In 1936, DuPont scientists published a paper on recently synthesized methacrylic polymers. In this paper, poly(2-hydroxyethyl methacrylate) (polyHEMA) was mentioned. It was briefly described as a hard, brittle, glassy polymer, and clearly was not considered of importance. After that paper, polyHEMA was essentially forgotten until 1960. Wichterle and Lim published a paper in *Nature* describing the polymerization of HEMA monomer and a cross-linking agent in the presence of water and other solvents (Wichterle and Lim, 1960). Instead of a brittle polymer, they obtained a soft, water-swollen, elastic, clear gel. Wichterle went on to develop an apparatus (built originally from a children's construction set; Figure 10) for centrifugally casting the hydrogel into contact lenses of the appropriate refractive power.

FIGURE 10 Left: Otto Wichterle (1913–1998). (Wikipedia.) Right: The centrifugal casting apparatus Wichterle used to create the first soft, hydrogel contact lenses. (Photograph by Jan Suchy, Wikipedia public domain.)

This innovation led to the soft contact lens industry, and to the modern field of biomedical hydrogels as we know them today.

Interest and applications for hydrogels have steadily grown over the years, and these are described in detail in Chapter I.2.5. Important early applications included acrylamide gels for electrophoresis, poly(vinyl alcohol) porous sponges (Ivalon) as implants, many hydrogel formulations as soft contact lenses, and alginate gels for cell encapsulation.

Poly(Ethylene Glycol)

Poly(ethylene glycol) (PEG), also called poly(ethylene oxide) (PEO) in its high molecular weight form, can be categorized as a hydrogel, especially when the chains are cross-linked. However, PEG has many other applications and implementations. It is so widely used today that its history is best discussed in its own section.

The low reactivity of PEG with living organisms has been known since at least 1944, when it was examined as a possible vehicle for intravenously administering fat-soluble hormones (Friedman, 1944). In the mid-1970s, Frank Davis and colleagues (Abuchowski et al., 1977) discovered that if PEG chains were attached to enzymes and proteins, they would a have a much longer functional residence time *in vivo* than biomolecules that were not PEGylated. Professor Edward Merrill of MIT, based upon what he called "various bits of evidence" from the literature, concluded that surface-immobilized PEG would resist protein and cell pickup. The experimental results from his research group in the early 1980s bore out this conclusion (Merrill, 1992). The application of PEGs to wide range of biomedical problems was significantly accelerated by the synthetic chemistry developments of Dr. Milton Harris while at the University of Alabama, Huntsville.

Poly(Lactic-Glycolic Acid)

Although originally discovered in 1833, the anionic polymerization from the cyclic lactide monomer in the early 1960s made creating materials with mechanical properties comparable to Dacron possible. The first publication on the application of poly(lactic acid) in medicine may be by Kulkarni et al. (1966). This group demonstrated that the polymer degraded slowly after implantation in guinea pigs or rats, and was well-tolerated by the organisms. Cutright et al. (1971) were the first to apply this polymer for orthopedic fixation. Poly(glycolic acid) and copolymers of lactic and glycolic acid were subsequently developed. Early clinical applications of polymers in this family were for sutures, based upon the work of Joe Frazza and Ed Schmitt at David & Geck, Inc. (Frazza and Schmitt, 1971). The glycolic acid/lactic acid polymers have also been widely applied for controlled release of drugs and proteins. Professor Robert Langer's group at MIT was the leader in developing these polymers in the form of porous scaffolds for tissue engineering (Langer and Vacanti, 1993).

Hydroxyapatite

Hydroxyapatite is one of the most widely studied materials for healing in bone. It is a naturally occurring mineral, a component of bone, and a synthesized material with wide application in medicine. Hydroxyapatite can be easily made as a powder. One of the first papers to describe biomedical applications of this material was by Levitt et al. (1969), in which they hot-pressed the hydroxyapatite power into useful shapes for biological experimentation. From this early appreciation of the materials science aspect of a natural biomineral, a literature of thousands of papers has evolved. In fact, the nacre implant described in the prehistory section may owe its effectiveness to hydroxyapatite – it has been shown that the calcium carbonate of nacre can transform in phosphate solutions to hydroxapatite (Ni and Ratner, 2003).

Titanium

In 1791, William Gregor, a Cornish amateur chemist, used a magnet to extract the ore that we now know as ilmenite from a local river. He then extracted the iron from this black powder with hydrochloric acid, and was left with a residue that was the impure oxide of titanium. After 1932, a process developed by William Kroll permitted the commercial extraction of titanium from mineral sources. At the end of World War II, titanium metallurgy methods and titanium materials made their way from military application to peacetime uses. By 1940, satisfactory results had already been achieved with titanium implants (Bothe et al., 1940). The major breakthrough in the use of titanium for bony tissue implants was the Brånemark discovery of osseointegration, described above in the section on dental implants.

Bioglass

Bioglass is important to biomaterials as one of the first completely synthetic materials that seamlessly bonds to bone. It was developed by Professor Larry Hench and colleagues. In 1967 Hench was Assistant Professor at the University of Florida. At that time his work focused on glass materials and their interaction with nuclear radiation. In August of that year, he shared a bus ride to an Army Materials Conference in Sagamore, New York with a US Army Colonel who had just returned from Vietnam where he was in charge of supplies to 15 MASH units. This colonel was not particularly interested in the radiation resistance of glass. Rather, he challenged Hench with the following: hundreds of limbs a week in Vietnam were being amputated because the body was found to reject the metals and polymer materials used to repair the body. "If you can make a material that will resist gamma rays why not make a material the body won't resist?"

Hench returned from the conference and wrote a proposal to the US Army Medical R and D Command. In October 1969 the project was funded to test the hypothesis that silicate-based glasses and glass-ceramics containing critical amounts of Ca and P ions would not be rejected by bone. In November 1969 Hench made small rectangles of what he called 45S5 glass (44.5 weight % SiO_2), and Ted Greenlee, Assistant Professor of Orthopaedic Surgery at the University of Florida, implanted them in rat femurs at the VA Hospital in Gainesville. Six weeks later Greenlee called: "Larry, what are those samples you gave me? They will not come out of the bone. I have pulled on them, I have pushed on them, I have cracked the bone and they are still bonded in place." Bioglass was born, and with the first composition studied! Later studies by Hench using surface analysis equipment showed that the surface of the bioglass, in biological fluids, transformed from a silicate rich composition to a phosphate rich structure, possibly hydroxyapatite (Clark et al., 1976).

THE CONTEMPORARY ERA (MODERN BIOLOGY AND MODERN MATERIALS)

It is probable that the modern era in the history of biomaterials, biomaterials engineered to control specific biological reactions, was ushered in by rapid developments in modern biology (second and third generation biomaterials; see Figure 2 in "Biomaterials Science: An Evolving, Multidisciplinary Endeavor"). In the 1960s, when the field of biomaterials was laying down its foundation principles and ideas, concepts such as cell-surface receptors, growth factors, nuclear control of protein expression and phenotype, cell attachment proteins, stem cells, and gene delivery were either controversial observations or not yet discovered. Thus, pioneers in the field could not have designed materials with these ideas in mind. It is to the credit of the biomaterials community that it has been quick to embrace and exploit new ideas from biology. Similarly, new ideas from materials science such as phase separation, anodization, self-assembly, surface modification, and surface analysis were quickly assimilated into the biomaterial scientists' toolbox and vocabulary. A few of the important ideas in biomaterials literature that set the stage for the biomaterials science we see today are useful to list:

- Protein adsorption
- Biospecific biomaterials
- Non-fouling materials
- Healing and the foreign-body reaction
- Controlled release
- Tissue engineering
- Regenerative medicine
- Nanotechnology.

Since these topics are addressed later in some detail in *Biomaterials Science: An Introduction to Materials in Medicine*, third edition, they will not be expanded upon in this history section. Still, it is important to appreciate the intellectual leadership of many researchers that promoted these ideas that comprise modern biomaterials – this is part of a recent history of biomaterials that will someday be completed. We practice biomaterials today immersed within an evolving history.

CONCLUSIONS

Biomaterials have progressed from surgeon-heroes, sometimes working with engineers, to a field dominated by engineers, chemists, and physicists, to our modern era with biologists and bioengineers as the key players. As *Biomaterials Science: An Introduction to Materials in Medicine*, third edition, is being published, many individuals who were biomaterials pioneers in the formative days of the field are well into their eighth or ninth decades of life. A number of leaders of biomaterials, pioneers who spearheaded the field with vision, creativity, and integrity, have passed away. Biomaterials is a field so new that the first-hand accounts of its roots are available. I encourage readers of the textbook to document their conversations with pioneers of the field (many of whom still attend biomaterials conferences), so that the exciting stories that led to the successful and intellectually stimulating field we see today are not lost.

BIBLIOGRAPHY

Abuchowski, A., McCoy, J. R., Palczuk, N. C., van Es, T., & Davis, F. F. (1977). Effect of covalent attachment of polyethylene glycol on immunogenicity and circulating life of bovine liver catalase. *J. Biol. Chem.*, **252**(11), 3582–3586.

Akutsu, T., Dreyer, B., & Kolff, W. J. (1959). Polyurethane artificial heart valves in animals. *J. Appl. Physiol.*, **14**, 1045–1048.

Apple, D. J., & Trivedi, R. H. (2002). Sir Nicholas Harold Ridley, Kt, MD, FRCS, FRS. *Arch. Ophthalmol.*, **120**(9), 1198–1202.

Blagg, C. (1998). Development of ethical concepts in dialysis: Seattle in the 1960s. *Nephrology*, **4**(4), 235–238.

Bobbio, A. (1972). The first endosseous alloplastic implant in the history of man. *Bull. Hist. Dent.*, **20**, 1–6.

Bondurant, S., Ernster, V., & Herdman, R. (Eds.), (1999). *Safety of Silicone Breast Implants*. Washington DC: National Academies Press.

Boretos, J. W., & Pierce, W. S. (1967). Segmented polyurethane: A new elastomer for biomedical applications. *Science*, **158**, 1481–1482.

Bothe, R. T., Beaton, L. E., & Davenport, H. A. (1940). Reaction of bone to multiple metallic implants. *Surg., Gynecol. Obstet.*, **71**, 598–602.

Branemark, P. I., Breine, U., Johansson, B., Roylance, P. J., Röckert, H., & Yoffey, J. M. (1964). Regeneration of bone marrow. *Acta. Anat.*, **59**, 1–46.

Clark, A. E., Hench, L. L., & Paschall, H. A. (1976). The influence of surface chemistry on implant interface histology: A theoretical basis for implant materials selection. *J. Biomed. Mater. Res.*, **10**, 161–177.

Crubezy, E., Murail, P., Girard, L., & Bernadou, J. -P. (1998). False teeth of the Roman world. *Nature*, **391**, 29.

Cutright, D. E., Hunsuck, E. E., & Beasley, J. D. (1971). Fracture reduction using a biodegradable material, polylactic acid. *J. Oral Surg.*, **29**, 393–397.

Egdahl, R. H., Hume, D. M., & Schlang, H. A. (1954). Plastic venous prostheses. *Surg. Forum*, **5**, 235–241.

Folkman, J., & Long, D. M. (1964). The use of silicone rubber as a carrier for prolonged drug therapy. *J. Surg. Res.*, **4**, 139–142.

Frazza, E., & Schmitt, E. (1971). A new absorbable suture. *J. Biomed. Mater. Res.*, **5**(2), 43–58.

Friedman, M. (1944). A vehicle for the intravenous administration of fat soluble hormones. *J. Lab. Clin. Med.*, **29**, 530–531.

Hoffman, A. (2008). The origins and evolution of "controlled" drug delivery systems. *Journal of Controlled Release*, **132**(3), 153–163.

Ingraham, F. D., Alexander, E., Jr., & Matson, D. D. (1947). Polyethylene, a new synthetic plastic for use in surgery. *J. Am. Med. Assoc.*, **135**(2), 82–87.

Kolff, W. J. (1998). Early years of artificial organs at the Cleveland Clinic Part II: Open heart surgery and artificial hearts. *ASAIO J.*, **44**(3), 123–128.

Kulkarni, R. K., Pani, K. C., Neuman, C., & Leonard, F. (1966). Polylactic acid for surgical implants. *Arch. Surg.*, **93**, 839–843.

Lahey, F. H. (1946). Comments (discussion) made following the speech "Results from using Vitallium tubes in biliary surgery," by Pearse H. E. before the American Surgical Association, Hot Springs, VA. *Ann. Surg.*, **124**, 1027.

Langer, R., & Vacanti, J. (1993). Tissue engineering. *Science*, **260**, 920–926.

LeVeen, H. H., & Barberio, J. R. (1949). Tissue reaction to plastics used in surgery with special reference to Teflon. *Ann. Surg.*, **129**(1), 74–84.

Levitt, S. R., Crayton, P. H., Monroe, E. A., & Condrate, R. A. (1969). Forming methods for apatite prostheses. *J. Biomed. Mater. Res.*, **3**, 683–684.

McGregor, R. R. (1954). *Silicones and their Uses*. New York: McGraw Hill Book Company, Inc.

Merrill, E. W. (1992). Poly(ethylene oxide) and blood contact. In J. M. Harris (Ed.), *Poly(ethylene glycol) chemistry: Biotechnical and biomedical applications* (pp. 199–220). New York: Plenum Press.

Ni, M., & Ratner, B. D. (2003). Nacre surface transformation to hydroxyapatite in a phosphate buffer solution. *Biomaterials*, **24**, 4323–4331.

Rob, C. (1958). Vascular surgery. In L. Gillis (Ed.), *Modern Trends in Surgical Materials* (pp. 175–185). London: Butterworth & Co.

Scales, J. T. (1958). Biological and mechanical factors in prosthetic surgery. In L. Gillis (Ed.), *Modern Trends in Surgical Materials* (pp. 70–105). London: Butterworth & Co.

Skeggs, L. T., & Leonards, J. R. (1948). Studies on an artificial kidney: Preliminary results with a new type of continuous dialyzer. *Science*, **108**, 212.

Wichterle, O., & Lim, D. (1960). Hydrophilic gels for biological use. *Nature*, **185**, 117–118.

PART I

Materials Science and Engineering

SECTION I.1

Properties of Materials

SECTION I.1

Properties of Materials

CHAPTER I.1.1 INTRODUCTION: PROPERTIES OF MATERIALS: THE PALETTE OF THE BIOMATERIALS ENGINEER

Jack E. Lemons
University Professor, Schools of Dentistry, Medicine and Engineering, University of Alabama at Birmingham, Birmingham, AL, USA

The platform, or palette, upon which the biomaterials engineer arranges information into parts for subsequent blending, let us say an art of biomaterials science, has expanded and evolved significantly in content over past decades. The depth and breadth of what is now included on this palette goes well beyond expectations expressed by founding members of the Society for Biomaterials in the late 1960s and early 1970s.

The science in biomaterials science has included fundamental aspects of physical, mechanical, chemical, electrical, and biological (compatibility) properties of the synthetic and natural origin biomaterials *per se*. Also, the methods for measuring and analyzing properties are equally applicable to the structures of the biological host. Following the recognition of the need by founders of the discipline in the 1960s, one focus has been the fundamental structure versus property relationships leading to *in vivo* biocompatibility. These relationships, and the supporting scientific information, have changed with time and experience, especially as the biological and clinical disciplines have also evolved. For example, considerations for biocompatibility are very different for biomaterials listed within biotolerant (called inert), surface bioactive (intended), and biodegradable categories. This shift of emphasis is reflected in the progression of content of the first three editions of this book. For example, initial considerations focused on materials were based primarily on substances of the metallics, ceramics, and polymerics available within various industrial applications. Thus, the emphasis in the first edition was on materials of synthetic origin and the science leading to biomaterials. The second and third editions represent the transitions from combination products to the new areas of bioactives and biodegradables. Thus, as an integrated, comprehensive, and authoritative text, this third edition reflects a broad range of biomaterials, and therefore basic properties of new generation biomaterials for regenerative medicine.

Considering relationships between biomaterial and biological systems (the interface) and the dynamics of change from nanoseconds to years, we now better understand many mechanisms of interaction at the dimensions and concentrations used to describe interactions of atoms. It is also realized that all biomaterial and host environment interactions play a role in the broader aspects of biocompatibility, especially the functionality and longevity of implant devices. In this regard, Part I on "Materials Science and Engineering" emphasizes the more basic information on the bulk and surface properties of synthetic and natural origin biomaterials. Critical aspects of constitution (chemistry) and structure (nano-, micro-, and macrodimension) relationships are presented as related to properties of implant systems. These basic considerations from the nature (I.1.2), bulk (I.1.3), and surface (I.1.5) properties are also interrelated to biomechanics (I.1.4).

In this edition, the concepts of property versus structure relationships have been expanded in terms of the role of water in biomaterials (I.1.6). The science of interactions of synthetic and natural substances with water is recognized as one of the key aspects of surgical implant biocompatibilities. This is especially important for the evolution of the discipline to include new generation biomaterials needed for future implant applications.

In summary, the content of this Part I is broadly applicable to all parts of the third edition. Therefore, students are advised to always consider the basic principles as provided in this section. This has been recognized as critical to the education of a specialist in biomaterials science leading to the selection of biomaterials for medical treatments utilizing all types of implant devices.

CHAPTER I.1.2 THE NATURE OF MATTER AND MATERIALS

Buddy D. Ratner
Professor, Bioengineering and Chemical Engineering, Director of University of Washington Engineered Biomaterials (UWEB), Seattle, WA, USA

INTRODUCTION

Biomaterials are *materials.*

Biomaterials are *materials*. What are materials and how are they structured? This is the subject of this chapter, a lead into subsequent chapters with discussions of the bulk properties of materials, mechanics of materials, surface properties, a discussion of water (since all biomaterials function in an aqueous environment, and that environment can alter both the nature of the material and the interaction that occurs with the material), and finally discussions of specific classes of materials relevant to biomaterials (polymers, metals, ceramics, etc.).

ATOMS AND MOLECULES

The key to understanding matter is to understand attractive and interactive forces between atoms.

This "Nature of Matter" section aims to communicate an understanding of the basic structure of materials that will drive their properties – both the mechanical properties important for specific applications (strong, elastic, ductile, permeable, etc.), and the surface properties that will mediate reactions with the external biological environment.

The key to understanding matter is to understand attractive and interactive forces between atoms. Argon is a gas at room temperature – it must be cooled to extremely low temperatures to transition it into liquid form. An argon atom interacts (attracts) very, very weakly with another argon atom – so at room temperature, thermal fluctuations that randomly propel the atoms exceed attractive forces that might result in the coalescence to a solid material. A titanium atom strongly interacts with another titanium atom. Extremely high temperatures are required to vaporize titanium and liberate those atoms from each other. The understanding of matter is an appreciation of interactive forces between atoms.

What holds those atoms and molecules together to make a strong nylon fiber or a cell membrane, or a hard, brittle hydroxyapatite ceramic, or a sheet of gold, or a drop of water? Even in the early 18th century, Isaac Newton was pondering this issue: "There are therefore Agents in Nature able to make the Particles of Bodies stick together by very strong Attractions."

Entropy consideration would say these molecules and atoms should "fly apart" to increase randomness. However, there is an energy term contributing to the stability of the ensemble leading to a negative Gibbs free energy, which, according to the second law of thermodynamics, should make such solids energetically favorable (of course, we intuitively know this). Thus, we must examine this energy term. We know of just four attractive forces in this universe:

- Gravitational
- Weak nuclear
- Strong nuclear
- Electromagnetic.

Gravity holds us to the surface of the planet Earth (a massive body), but the gravitational potential energy of two argon atoms is only about 10^{-52} J, 30 orders of magnitude weaker than is observed for intermolecular forces. The weak nuclear force and the strong nuclear force are only significant over 10^{-4} nm – but molecular dimensions are 5×10^{-1} nm. So these forces do not explain what holds atoms together. This leaves, by default, electromagnetic forces (positive charge attracts negative charge). Electromagnetic forces have appropriate magnitudes and distance dependencies to justify why atoms interact. Interactions can be weak, leading to liquids, or stronger, leading to solids.

Electromagnetic forces manifest themselves in a number of ways. The types of interactions usually observed between atoms (all explained by electrostatics) are summarized in Table I.1.2.1. We consider here van der Waals forces (also called induction or dispersion forces), ionic forces, hydrogen bonding, metallic forces, and covalent interactions.

Van der Waals or dispersion forces rationalize the interaction of two atoms or molecules, each without a dipole (no plus or minus faces to the molecules). For example, argon atoms can be liquefied at low temperature. Why should this happen? Why should argon atoms want to interact with each other enough to form a liquid? Figure I.1.2.1 explains the origin of dispersive forces. Such forces are important, as they dictate the properties of many materials (for example, some polymers such as polyethylene which has no obvious dipole), but also they explain why the lipids in cell membranes assemble into the bilayer structure described later in this textbook. A typical van der Waals interactive force (for example, CH_4 … CH_4) is about 9 kJ/mol. The Ar … Ar interactive force is approximately 1 kJ/mol.

Ionic forces are probably the easiest of the intermolecular forces to understand. Figure I.1.2.2 illustrates a unit cell of a sodium chloride crystal. The + and – charges are arrayed to achieve the closest interaction of opposite charges, and the furthest separation of similar charges. This unit cell can be repeated over and over in space, and the forces that hold it together are the electrostatic interaction of a permanent + charge and a permanent – charge. Typical ionic bond strengths (for example, NaCl) are about 770 kJ/mol.

Hydrogen bonding interactions are also straightforward to appreciate as electrostatic interactions. An electronegative element such as oxygen (it demands electrons) can distort the binding electron cloud from the

TABLE I.1.2.1 Forces that Hold Atoms Together

Interatomic Force	Explanation	Relative Strength	Examples
Van der Waals interactions	Transient fluctuations in the spatial localization of electron clouds surrounding atoms lead to transient positive and negative charges, and consequent interactive forces in molecules would seem to have no permanent polarity	Weak	Argon at cryogenic temperatures. Polyethylene (the forces that hold the chains together to make a solid)
Ionic	Atoms with a permanent positive (+) charge attract atoms with a permanent negative (–) charge	Very strong	NaCl $CaCl_2$
Hydrogen (H) bonding	The interaction of a covalently bound hydrogen with an electronegative atom, such as oxygen or fluorine	Medium	Water ice Nylon (the forces that hold the chains together to make a strong, high-melting point solid)
Metallic	The attractive force between a "sea" of positively charged atoms and delocalized electrons	Medium-strong	Gold Titanium metal
Covalent	A sharing of electrons between two atoms	Strong	The carbon–carbon bond Cross-links in a polyacrylamide hydrogel

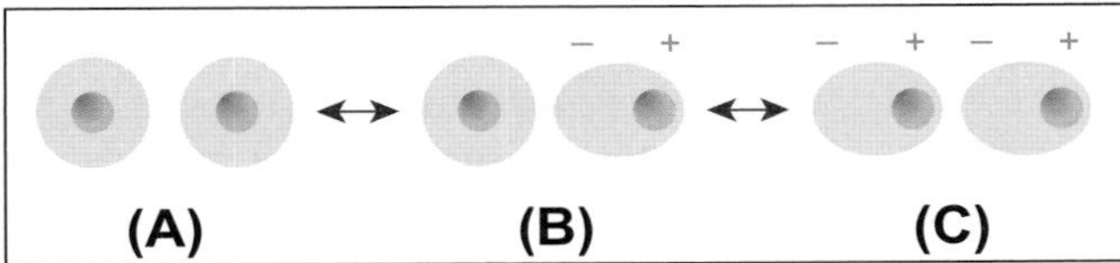

FIGURE I.1.2.1 (A) Consider the electron clouds (charge density in space) of two atoms or molecules, both without permanent dipole moments. (B) Electron clouds are continuously in motion and can shift to one side of the atom or molecule; therefore, at any moment, the atoms or molecules can create a "fluctuating instantaneous dipole." (C) The "fluctuating instantaneous dipole" in one molecule electrostatically induces such an "instantaneous dipole" in the next molecule.

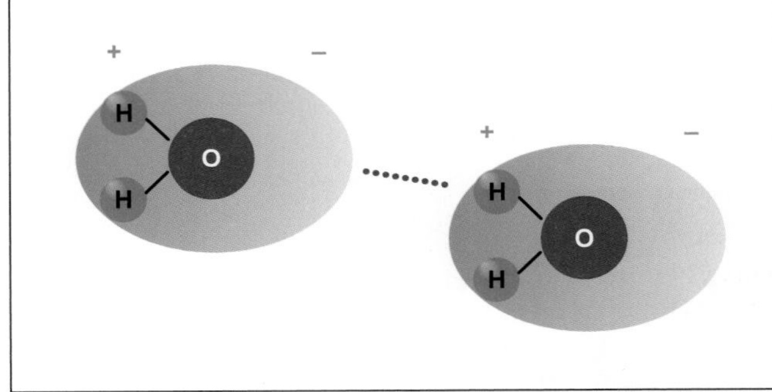

FIGURE I.1.2.3 A hydrogen bond between two water molecules.

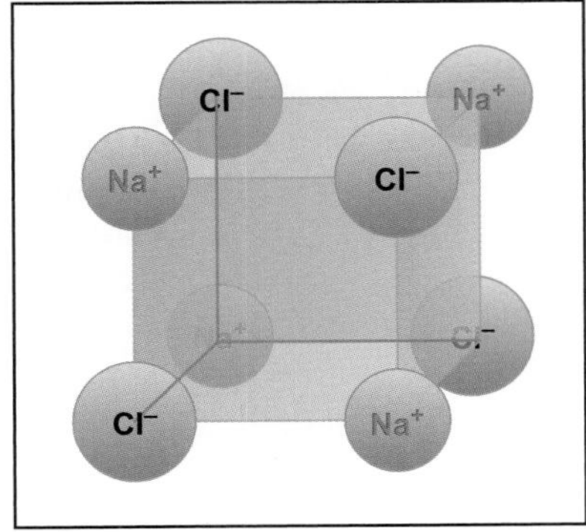

FIGURE I.1.2.2 The unit cell of a sodium chloride crystal illustrating the plus–minus electrostatic interactions.

hydrogen nucleus leaving the hydrogen (just a proton and electron), with less electron and thus more plus-charged proton. This somewhat positive charge will, in turn, then interact with an electronegative oxygen (Figure I.1.2.3). Typical hydrogen bond strengths (for example, O–H … H) are about 20 kJ/mol.

Metallic bonding is explained by a delocalized "sea" of electrons with positively charged nuclear cores dispersed within it (Figure I.1.2.4). A single metallic bond is rarely discussed. The total interactive strength is realized through the multiplicity of the plus–minus interactions. The strength of this interaction can be expressed by heats of sublimation. For example, at 25°C, aluminum will have a heat of sublimation of 325 kJ/mol, while titanium will be about 475 kJ/mol.

Covalent bonds are relatively strong bonds associated with sharing of pairs of electrons between atoms (Figure I.1.2.5). Typical covalent bond strengths (for example, C–C) are about 350 kJ/mol.

MOLECULAR ASSEMBLIES

Atoms can combine in defined ratios to form molecules (usually they combine with covalent bonds), or they can form cohesive assemblies of atoms (think of gold and metallic interactive forces, for example). Thus, materials can be made of atoms or of molecules (i.e., covalently joined atoms). The difference between the dense, lubricious plastics used in orthopedics, the soft, elastic materials of catheters, and the hard, strong metals of a hip

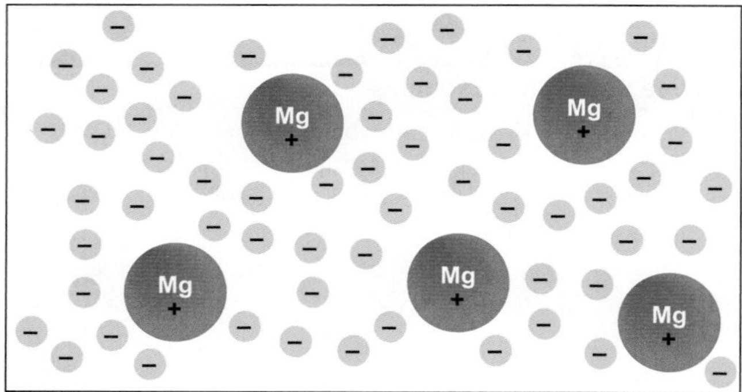

FIGURE I.1.2.4 Metallic bonding in magnesium. The 12 electrons from each Mg atom are shared among positively charged nuclear cores (the single + charge on each magnesium atom in the figure is simply intended to indicate there is some degree of positive charge on each magnesium nuclear core).

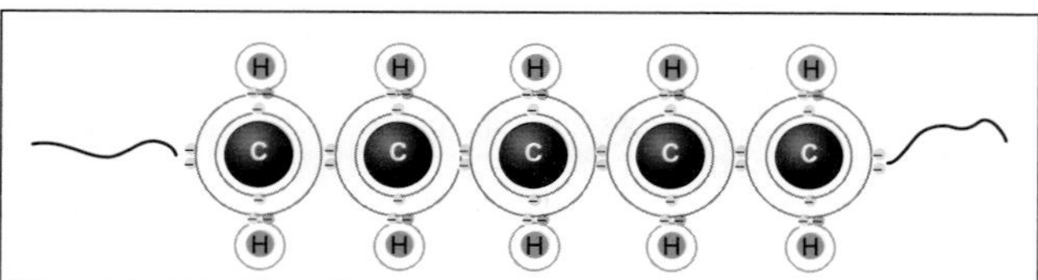

FIGURE I.1.2.5 Covalent bonding along a section of polyethylene chain. Carbons share pairs of electrons with each other, and each hydrogen shares an electron pair with carbon.

joint is associated with how those atoms and molecules are organized (due to attractive and interactive forces) in materials. Metals used in biomaterials applications can be strong, rigid and brittle, or flexible and ductile. Again, the difference is largely how the atoms making up the metal are organized, and how strongly their atoms interact.

Molecules also organize or assemble. The widely varying properties of polymers are due to molecular organization. The assembly of lipid molecules to make a cell membrane is another example of this organization.

A key concept in appreciating the properties of materials is hierarchical structures. The smallest size scale that we need consider here in materials is atoms, typically about 0.2 nm in diameter. Atoms combine to form molecules with dimensions ranging from 1 nm to 100 nm (some large macromolecules). Molecules may assemble or order to form supramolecular structures with dimensions up to 1000 nm or more. These supramolecular structures may themselves organize in bundles, fibers or larger assemblies with dimensions reaching into the range visible to the human eye. This concept of hierarchical structure is illustrated in Figure I.1.2.6, using collagen protein as the example.

The single α-chains comprising the collagen triple helix would break under tension with an application of nanograms of force. On the other hand, the collagen

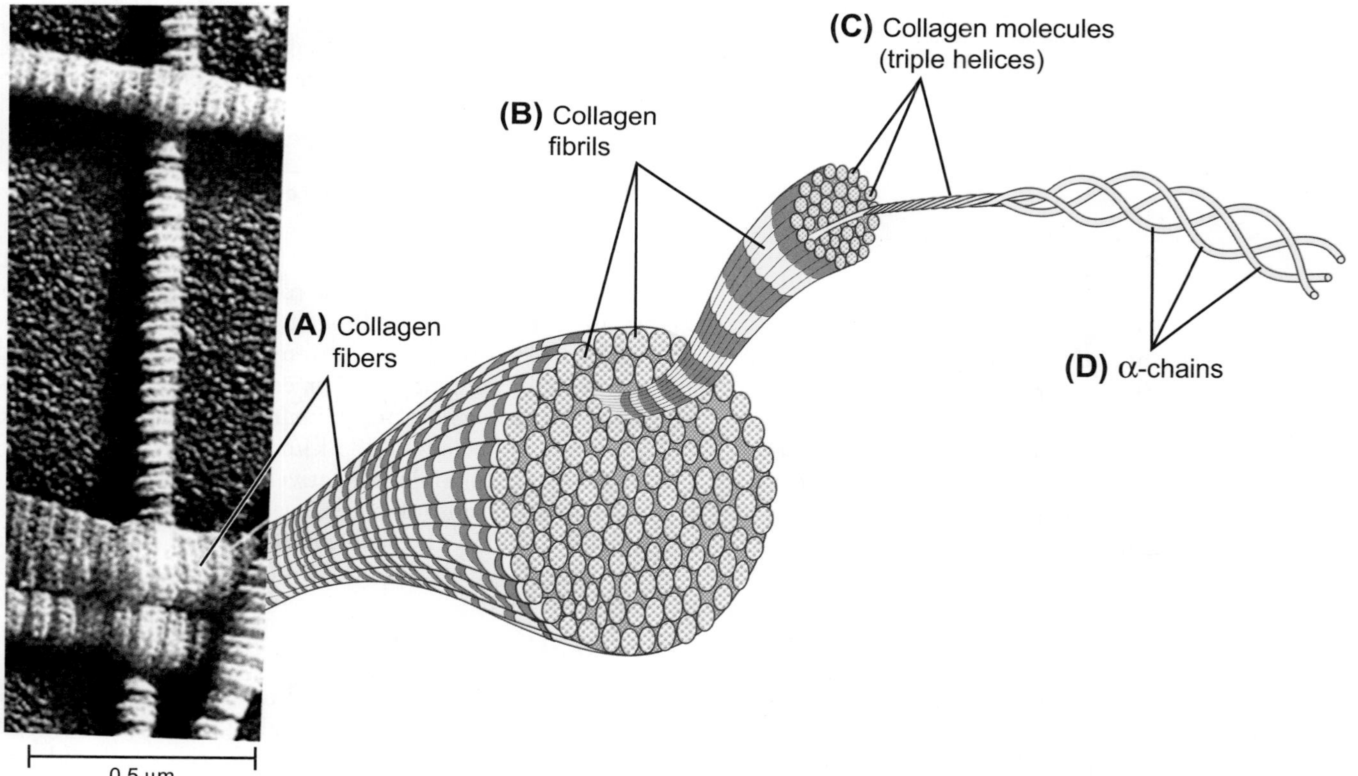

FIGURE I.1.2.6 Collagen fibers make up many structures in the body (tendons, for example). Such anatomical structures as tendons are comprised of collagen fibrils, formed of aligned bundles of collagen triple helices that are themselves made up of single collagen protein chains (α-chains). The α-chains are constructed of joined amino acid units, and the amino acids are molecules of carbon, oxygen, nitrogen, and hydrogen atoms in defined ratios and orders. (Illustration from Becker, W. (2002). *The World of the Cell*. Reprinted with permission of Pearson Education, Inc.)

fibers in a hierarchical structure such as a tendon can support many kilograms of force. Such hierarchical structures are noted frequently in both materials science and in biology.

SURFACES

As assemblies of atoms and/or molecules form, within the bulk of the material, each unit is uniformly "bathed" in a field of attractive forces of the types described in Table I.1.2.1. However, those structural units that are at the surface are pulled upon asymmetrically by just the units beneath them. This asymmetric attraction distorts the electron distributions of the surface atoms or molecules, and gives rise to the phenomenon of surface energy, an excess energy associated with this imbalance. For this reason surfaces always have unique reactivities and properties. This idea will be expanded upon in Chapter I.1.5.

CONCLUSION

In this section we have reviewed the transition from chemistry to matter. Interacting assemblies of atoms and molecules comprise matter. Without matter, we cannot have biomaterials. Matter exists because of electrostatic forces – positive and negative charges, in all cases, hold atoms together. The strength of those interactions, associated with the magnitude of the charge on each atom, and the environment the atoms are in (water, air, etc.) ultimately dictates the properties of matter (a soft gel, a hard metal, etc.). Now that we have a general idea what "matter" is, we can take these concepts from physics and chemistry and bring them to a consideration of the mechanical properties of materials, the surface properties of materials, and then into the specifics of polymers, metals, ceramics, and other types of materials.

BIBLIOGRAPHY

Barton, A. (1997). *States of Matter, States of Mind*. Bristol, UK: Institute of Physics Publishing.

Holden, A. (1992). *The Nature of Solids*. New York, NY: Dover Publications Inc.

Becker, W. M. (2003). *World of the Cell*. Kleinsmith, L. J., Hardin, J. 5th ed. Pearson Education, Inc. Upper Saddle River, NJ.

CHAPTER I.1.3 BULK PROPERTIES OF MATERIALS

Christopher Viney
School of Engineering, University of California at Merced, Merced, CA, USA

INTRODUCTION

Bulk Versus Surface

The success or failure of many biomaterials depends on the physical and chemical characteristics of their surface. It is the surface properties that dictate interactions between a material and its environment, and thus whether a permanently implanted material will be tolerated or rejected. In cases where the implanted material is required to degrade at a controlled rate (e.g., when the implant is a temporary support, such as a dissolving suture or a scaffolding for cells that regenerate tissue), then the bulk material must be capable of sustaining those properties continuously as it becomes the new surface.

Regardless of their particular surface properties, biomaterials are usually also required to exhibit certain *bulk* characteristics, especially including those attributes (mechanical properties) that relate to the ability to carry loads dependably without undue deflection or premature failure. Mechanical properties will be addressed in detail in this chapter, followed by a brief consideration of selected other bulk properties (thermal, optical) that are often significant in the context of biomaterials.

The concepts addressed in this chapter are fundamental. They can be found scattered through standard materials science or mechanical engineering textbooks, and undergraduate materials science or mechanical engineering courses. The aim in this chapter is to achieve consistency of presentation, connectivity of knowledge, and appreciation of relevance.

Microstructure and Properties

The bulk properties of a material depend not only on what types of atoms and molecules it contains (composition), but also on how those atoms and molecules are arranged (microstructure). For example, if the material is a polymer, do the individual molecules follow a random trajectory (and so occupy an approximately spherical region of space), or are they extended (so that they occupy a region of space approximated by a rectangular prism or a cylinder)? Are the molecules locally packed in a regular array? Do the molecules locally all point in the same direction? And how are these various spatial attributes controlled by processing – assuming that they can be controlled? A strong culinary analogy can be made: in addition to specifying a list of ingredients, recipes also contain instructions on how to process the ingredients to obtain the desired dish. Processing is what makes baked potatoes, mashed potatoes, fries, and shepherd's pie crust so different.

The microstructure of a material is a specification of the structural features at length scales that cannot be discerned with the naked eye. Most materials exhibit microstructural detail at several different length scales, down to the nanoscale. Therefore, the term "microstructure" encompasses structures in the size range 10^{-4} m to 10^{-9} m.

Some bulk properties, such as the stiffness of a metal, depend principally on the type (composition) of metal being characterized, and not on the microstructure. Such properties are referred to as *intrinsic properties*, and they include density and heat capacity. (Note that the stiffness of a *polymer* is not an intrinsic property, because the stiffness measured by attempting to stretch the polymer in a particular direction depends on whether or not the molecules are preferentially aligned in that direction.)

Other bulk properties, such as the yield strength of a metal, depend on attributes such as the average grain (crystal) size, as well as the number and distribution of defects in the crystal structure. Yield strength is an example of an *extrinsic property* – as is the stiffness of polymers. (Both stiffness and yield strength will be defined and discussed in detail in the section "Bulk Mechanical Properties Determined from Stress–Strain Plots.")

It is because some properties are intrinsic while others are extrinsic, and because different extrinsic properties are sensitive to changes in different microstructural attributes at different length scales, that it is possible to control and therefore optimize several bulk properties simultaneously. For example, a metal with a fine grain size will have higher yield strength than a coarser-grained metal, but both can have the same stiffness.

LOAD, NOMINAL STRESS, EXTENSION, AND NOMINAL STRAIN

When a load (force) is applied to an object, the possible consequences include translation, rotation, and deformation of the object. We are concerned here with deformation. To properly quantify the deformation caused by a load, we need to take into account the dimensions of the object to which the load is being applied. For example, we are often told that spider silk (especially the dragline) is "strong," although it is our experience that a strand of spider silk can be broken more easily than a strand of dental floss. The difference is that the dental floss is much thicker than the spider dragline; a bundle of spider dragline having a total thickness equal to that of the dental floss would be significantly more resistant to breaking than the floss. Consider a sample of material represented by the rectangular prism shown in Figure I.1.3.1.

Let a force F be applied normal to a pair of opposite faces (in this case the top and bottom faces) of the sample, which initially each have area A_1. The actual shape of the sample is not important for the analysis that follows; all that is necessary is that the cross-section perpendicular to the applied force should be uniform. A cylinder (e.g., representing a fiber) would serve just as well. The ratio:

$$\sigma_n = \frac{F}{A_1}$$

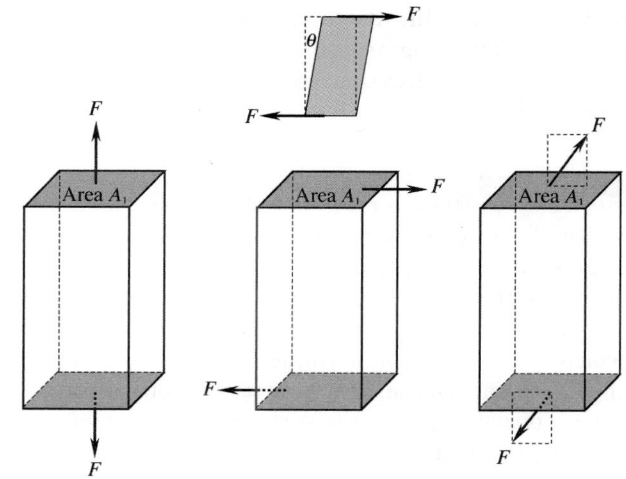

FIGURE I.1.3.1 Examples of a tensile (normal) stress, a shear stress, and a mixed stress with both normal and shear components, illustrating the relationship between load direction and surface orientation in each case. For the example of shear stress, the originally rectangular front face of the sample distorts to a parallelogram as shown, and the corresponding shear strain is defined as tan θ.

defines the *nominal stress* being applied to the sample. It is a nominal stress because the actual value of the area (cross-section) to which the load is applied will change as the sample deforms in response to the load; electing to calculate stress with reference to the initial cross-sectional area is a contrived choice. Nominal stress is often synonymously referred to as *engineering stress*. The stress is *tensile* if it is a "pull," elongating the sample, and it is *compressive* if it is a "push," shortening the sample. The SI units of stress are newtons per square meter ($N.m^{-2}$) or pascals (Pa).

The elongation of a sample in response to a tensile load will depend on the length of sample that there is to elongate. To visualize this, think of a line of atoms in the direction of the load, with the bonds between atoms being represented by springs. A tensile load causes each bond to stretch; consequently, the more bonds there are in the line, the longer the line will become in response to a given load. For purposes of comparing the responses of different types of material, it is therefore useful to quantify the effect of loads (and their related stresses) by considering the resultant *length change per unit length*, more commonly known as *strain*. If the initial length of the sample is l_1, and the applied stress causes a change to length l_2, then the *nominal* (or *engineering*) *strain* is given by:

$$\varepsilon_n = \frac{l_2 - l_1}{l_1} = \frac{\Delta l}{l_1} = \lambda_{1\rightarrow 2} - 1$$

where λ denotes the *extension ratio*, i.e., the ratio of final length to initial length.

It is a nominal strain because electing to calculate strain with reference to the initial length is again a contrived choice. The strain is *tensile* or *compressive* in accordance with whether it is caused by a tensile stress or a compressive stress. Strain, like extension ratio, is dimensionless.

The relationship between stress and resultant strain is commonly summarized in a *stress–strain plot*, where by convention the stress is plotted vertically and the strain is plotted horizontally, with data being collected while the sample is deformed at a constant strain rate.

TRUE STRESS AND TRUE STRAIN

A more realistic sense of stress imposed on a material during a stress–strain test is obtained if the applied force is scaled with reference to the *actual* cross-sectional area of the sample, A. In that case, we refer to the sample being subjected to a *true stress* defined by:

$$\sigma_t = \frac{F}{A}$$

Unfortunately, because A changes continuously during the course of a stress–strain test, it can be difficult to measure in practice. This is why the contrived, but more expedient, nominal stress is so often used instead of true stress.

The inadequacy of nominal strain as a measure of sample deformation becomes apparent when we consider the cumulative effect of successive strains. Let a sample of initial length l_1 be stretched to a length l_2, and thence to length l_3. Calculated separately, the corresponding nominal strains are:

$$\varepsilon_{n(1\to 2)} = \frac{l_2 - l_1}{l_1} \quad \text{and} \quad \varepsilon_{n(2\to 3)} = \frac{l_3 - l_2}{l_2}$$

The nominal strain for the deformation directly from length l_1 to length l_3 would be calculated as:

$$\varepsilon_{n(1\to 3)} = \frac{l_3 - l_1}{l_1} \quad \text{which is not equal to } \varepsilon_{n(1\to 2)} + \varepsilon_{n(2\to 3)}$$

To address this limitation, we consider a revised definition of strain, in which the length change of the sample is measured in very small increments dl, with each increment being scaled relative to the length l of the sample immediately prior to that increment. Thus, a small increment in *true strain* is defined as:

$$d\varepsilon_t = \frac{dl}{l}$$

The true strain corresponding to deformation from length l_1 to l_2 can then be found by integrating the above expression between limits l_1 and l_2:

$$\varepsilon_{t(1\to 2)} = \int_{l_1}^{l_2} \frac{dl}{l} = \ln\frac{l_2}{l_1}$$

If this new interpretation of strain is applied to the case of a sample of initial length l_1 being stretched to a length l_2, and thence to length l_3, the separate nominal strains are now:

$$\varepsilon_{t(1\to 2)} = \ln\frac{l_2}{l_1} \quad \text{and} \quad \varepsilon_{t(2\to 3)} = \ln\frac{l_3}{l_2}$$

The sum of these true strains is:

$$\ln\frac{l_2}{l_1} + \ln\frac{l_3}{l_2} = \ln\frac{l_3}{l_1} = \varepsilon_{t(1\to 3)}$$

In other words, true strains behave properly when we try to add them.

Stresses and strains are almost always presented without formal identification as nominal or true, in which case it can (usually) be assumed that they are nominal. True stresses and strains are (usually) identified explicitly as such. However, it is good practice to avoid any possible ambiguity by being explicit and specific when talking about stresses and strains.

It is possible to interconvert nominal and true stresses, and nominal and true strains, if we make the assumption that the sample volume does not change during deformation. The circumstances under which this assumption is valid will be addressed in the section "Condition for Incompressibility (Zero Volume Change)" below.

If volume is conserved during deformation from condition 1 (characterized by sample length l_1 and cross-sectional area A_1) to condition 2 (characterized by length l_2 and cross-section A_2), then:

$$l_2 A_2 = l_1 A_1$$

At that stage in the deformation, the true stress is:

$$\sigma_t = \frac{F}{A_2} = F\frac{l_2}{A_1 l_1} = \sigma_n\frac{l_2}{l_1} = \sigma_n(\varepsilon_n + 1)$$

The interconversion between nominal and true strains is even easier to obtain:

$$\varepsilon_{t(1\to 2)} = \ln\frac{l_2}{l_1} = \ln(\varepsilon_n + 1)$$

SHEAR STRESS AND SHEAR STRAIN

Referring again to the rectangular prism of material shown in Figure I.1.3.1, consider now the effect of applying the force F *parallel* to a pair of opposite faces (in this case again the top and bottom faces), which initially each have area A_1. For this configuration of applied load, the ratio:

$$\tau = \frac{F}{A_1}$$

defines the *shear stress* being applied to the sample.

Seen from the front of the sample, the outline becomes distorted from a rectangle to a more general parallelogram. This shape change can be used to describe the *shear strain* γ caused by the shear stress:

$$\gamma = \tan\theta$$

The use of σ to denote tensile or compressive stress, ε to denote tensile or compressive strain, τ to denote shear stress, and γ to denote shear strain is a widely adopted convention.

If a force is applied obliquely (neither perpendicular nor parallel; see Figure I.1.3.1) to a pair of opposite faces of the sample, the ensuing stress can be resolved into normal and shear components, as can the resulting strain.

BULK MECHANICAL PROPERTIES DETERMINED FROM STRESS–STRAIN PLOTS

Simple stress–strain plots can be used to define and quantify several mechanical properties of a material. Figure I.1.3.2 displays representative plots for a ductile metal (titanium), a ceramic (alumina, Al_2O_3), and a crystalline polymer (high density polyethylene, HDPE) subjected to a tensile stress.

Figure I.1.3.3 repeats the stress–strain plot for titanium schematically, emphasizing features that relate to specific mechanical properties.

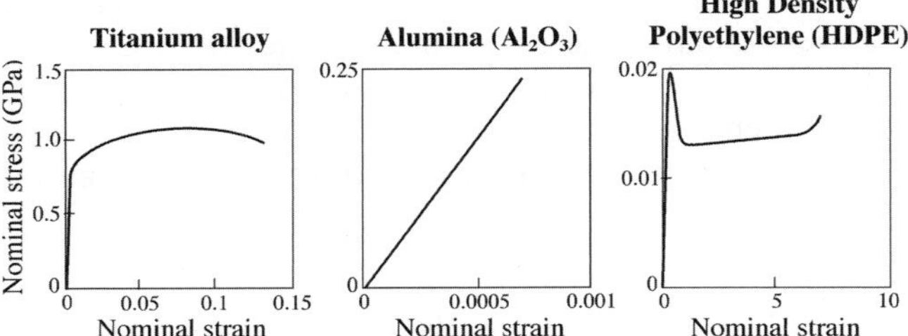

FIGURE I.1.3.2 Representative nominal stress versus nominal strain plots for three different classes of implantable material: ductile metal (titanium alloy; 6 wt% Al, 4 wt% V); ceramic (alumina); and crystalline polymer (high density polyethylene).

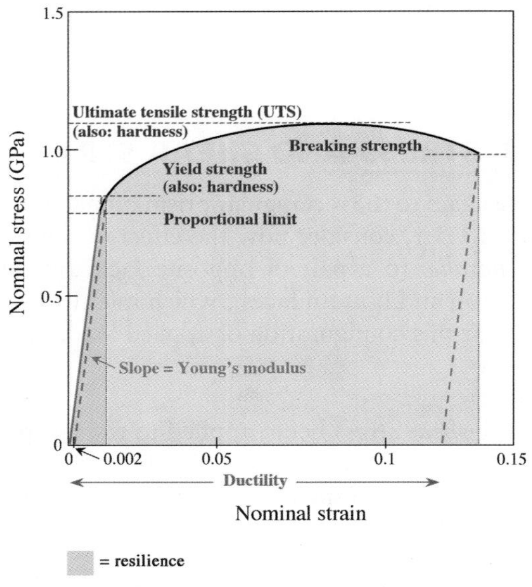

FIGURE I.1.3.3 Schematic nominal stress versus nominal strain plot for a ductile metal, emphasizing the features that relate to specific mechanical properties.

Elastic Deformation

Elastic Constants. The relationship between stress and strain is initially linear as the stress is increased from zero. This behavior is equivalent to the load-extension response of a simple spring, as first stated quantitatively by Robert Hooke and embodied in Hooke's Law. In the case of bulk material, the constant of proportionality between *tensile* (or *compressive*) stress and strain is known as the *Young's modulus, E*:

$$\sigma = E\,\varepsilon$$

The equivalent constant of proportionality in the case of a *shear* stress being applied is known as the *shear modulus*, G:

$$\tau = G\,\gamma$$

Similarly, we can define a *bulk modulus*, K, as the constant of proportionality relating pressure to the volume change caused by that pressure:

$$P = -K\frac{V_2 - V_1}{V_1} = -K\frac{\Delta V}{V_1}$$

where a pressure P causes a reduction in volume from an initial volume V_1 to a final volume V_2. The negative sign is used to obtain a convention in which the values of K are positive.

In all three cases above, the definitions of the different types of elastic modulus assume that the cause of deformation (tensile stress, shear stress or pressure) is sufficiently small to ensure that the response (tensile strain, shear strain or volume change) is *proportional* to the cause. This condition also ensures that the response is *reversible*; when the stress or pressure is removed, the material returns to its original dimensions.

There is one additional constant that is commonly encountered in the context of the elastic deformation of materials: *Poisson's ratio*, υ (Figure I.1.3.4).

For most materials subjected to a tensile stress, the resulting longitudinal strain (parallel to the direction of the applied load) is accompanied by a transverse strain (perpendicular to the direction of the applied load). In other words, when we pull on a piece of material, it usually becomes thinner as well as longer. Poisson's ratio is defined as:

$$\nu = -\frac{\varepsilon_{transverse}}{\varepsilon_{longitudinal}}$$

The negative sign is used to obtain a convention in which the values of υ are positive for most materials.

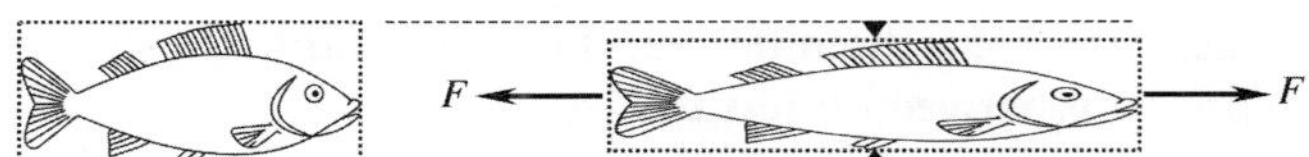

FIGURE I.1.3.4 Longitudinal elastic strain (in this case elongation) parallel to the direction of the applied load is accompanied by a transverse elastic strain (in this case contraction) perpendicular to the direction of the applied load. The negative of the ratio of transverse elastic strain to longitudinal elastic strain defines Poisson's ratio.

A relatively small number of materials – known as auxetic materials – have a negative Poisson's ratio. Typically, these are low-density, cellular materials, with cell walls that are able to hinge or buckle when loads are applied. Pulling on a piece of auxetic material causes it to become *wider* as well as longer, while attempting to squash it will cause it to become thinner as well as shorter.

For isotropic materials, i.e., those in which structure and therefore properties do not vary with direction, it can be shown that only two of the four elastic constants are independent. The relationship between the elastic constants is captured in the following two equations:

$$G = \frac{E}{2(1+\nu)} \quad \text{and} \quad K = \frac{E}{3(1-2\nu)}$$

Given any two of the constants E, G, K, and υ, these equations suffice to find the remaining two. E, G, and K all have the same units as stress, while υ is dimensionless.

In the case of many biological materials, especially those with a fibrous microstructure, E, G, and υ are *anisotropic* – in other words, the values obtained when these constants are measured depend on the direction along which load is applied to the sample. (The bulk modulus, K, relates to uniform compression of a material, and so, while its value is dependent on whether and to what extent the sample is microstructurally anisotropic, it does not depend on the orientation of the sample during tests.)

The transverse Young's modulus of fibers can be difficult to measure accurately, because of the practical challenges inherent in obtaining dependable stress versus strain data from fiber samples loaded parallel to a diameter. There are relatively few options for obtaining such data. We can attempt to use atomic force microscopy, applying and measuring force parallel to the fiber diameter on a suitably small target area of sample, and measuring the very small deflections that characterize strain in a thin (equivalent to short) sample. Alternatively, we can attempt to produce and test a significantly thicker sample of the same material while preserving the molecular alignment typical of the fiber. Better yet, we might be able to *calculate* the transverse Young's modulus from first principles, building on the methods that will be introduced in the sections "Quantitative Prediction of Elastic Behavior in an Atomic Solid" and "Quantitative Prediction of Elastic Behavior in a Molecular Solid." However, each of these methods has limitations, so that it is not at all unusual for fibers to be subjected to the simplifying – and itself incorrect – assumption that they are elastically isotropic.

Materials that are used as loadbearing implants need to have elastic properties that appropriately match the properties of the tissue that they replace. For example, if a dental filling has a Young's modulus significantly different from that of the tooth, the compressive stresses that accompany chewing will cause the filling and adjacent natural material to respond with different longitudinal strains, which in turn may cause the interface between the filling and the tooth to fail. Also, the filling and the tooth must have compatible values of Poisson's ratio. If the transverse strain induced in the filling is smaller than that which would have been induced in the material replaced by the filling, then the remaining tooth will tend to pull away from the filling.

Condition for Incompressibility (Zero Volume Change). Consider a rectangular block of material, initially having length l_1, width b_1, and thickness t_1. A force applied parallel to the length of the block stretches the sample to a new length l_1; correspondingly, the width decreases to b_1 and the thickness decreases to t_2. Initially, the volume of the block is:

$$V_1 = l_1\, b_1\, t_1$$

After stretching, the volume of the block is:

$$V_2 = l_2\, b_2\, t_2 = (l_1 + l_1\, \varepsilon_{longitudinal})\,(b_1 + b_1\, \varepsilon_{transverse})\,(t_1 + t_1\, \varepsilon_{transverse}),$$

from the definition of strain;

$$= (l_1 + l_1\, \varepsilon_{longitudinal})\,(b_1 - b_1\, \nu\varepsilon_{longitudinal})\,(t_1 - t_1\, \nu\varepsilon_{longitudinal}),$$

from the definition of υ

$$= l_1\, b_1\, t_1 + l_1\, \varepsilon_{longitudinal}\, b_1\, t_1 - l_1\, b_1\, \nu\varepsilon_{longitudinal}\, t_1 - l_1\, b_1\, t_1\, \nu\varepsilon_{longitudinal}$$

plus higher order terms in $\varepsilon_{longitudinal}$.

Because we are dealing with small strains, the higher order terms in $\varepsilon_{longitudinal}$ can be approximated as zero, so that:

$$V_2 \approx l_1\, b_1\, t_1 + l_1\, \varepsilon_{longitudinal}\, b_1\, t_1 - l_1\, b_1\, \nu\varepsilon_{longitudinal}\, t_1 - l_1\, b_1\, t_1\, \nu\varepsilon_{longitudinal}$$

Zero volume change in response to elastic deformation requires:

$$V_2 = l_1\, b_1\, t_1 + l_1\, \varepsilon_{longitudinal}\, b_1 t_1 - l_1\, b_1\, \nu\varepsilon_{longitudinal}\, t_1 - l_1\, b_1\, t_1\, \nu\varepsilon_{longitudinal} = V_1 = l_1\, b_1\, t_1$$

$$\therefore\ l_1\, \varepsilon_{longitudinal}\, b_1\, t_1 - l_1\, b_1\, \nu\varepsilon_{longitudinal}\, t_1 - l_1\, b_1\, t_1\, \nu\varepsilon_{longitudinal} = 0$$

$$\therefore\ 1 - 2\nu = 0$$

$$\therefore\ \nu = 0.5$$

This condition pertains in mammalian arteries. The walls have to "give" in response to each pulse of blood delivered by the heart, but the volume of arterial lumen occupied by fluid is necessarily constant. Therefore, the artery itself has to be incompressible.

There is a second consideration involving Poisson's ratio that is important in the context of arteries. An artery can be approximated as a (variably) pressurized cylinder. Simple elasticity theory shows that a pressurized cylinder made from incompressible material will become wider and shorter, rather than longer and thinner. Joints made to secure an arterial graft must be able to withstand the longitudinal tensile stress that develops during each pressurization cycle.

Quantitative Prediction of Elastic Behavior in an Atomic Solid. Some quite simple models can be used to predict the load versus extension, and hence stress versus strain relationship for the elastic deformation of materials, allowing us to obtain values of elastic modulus from first principles.

The *Morse potential* is often used to describe the energy of interaction, $U(r)$, which exists between two atoms, in terms of their separation, r:

$$U = U_0\,\{\exp[-2a(r - r_0)] - 2\exp[-a(r - r_0)]\}$$

where U_0, a, and r_0 are constants.

The positive and negative terms respectively represent the short-range repulsion and long-range attraction between the two atoms.

Given $U(r)$, it is possible to calculate the force $F(r)$ between the atoms, and thence the stiffness $E(r)$ of the bond between them.

Energy and force are related by the definition:

$$F = \frac{dU}{dr}$$

$$= U_0\,\{(-2a)\exp[-2a(r - r_0)] - 2(-a)\exp[-a(r - r_0)]\}$$

$$= 2aU_0\,\{-\exp[-2a(r - r_0)] + \exp[-a(r - r_0)]\}$$

Force and stiffness are related by the differential form of Hooke's Law:

$$E = \frac{dF}{dr} = \frac{d^2U}{dr^2}$$

$$= 2aU_0\,\{-(-2a)\exp[-2a(r - r_0)] + (-a)\exp[-a(r - r_0)]\}$$

$$= 2a^2U_0\,\{2\exp[-2a(r - r_0)] - \exp[-a(r - r_0)]\}$$

The initial stiffness that would be exhibited in a stress–strain test can now be predicted by setting $r = r_0$:

$$E_{initial} = 2a^2U_0\,\{2\exp[0] - \exp[0]\} = 2a^2U_0\,\{2 - 1\} = 2a^2U_0$$

This type of analysis lies at the heart of many materials modeling programs.

Quantitative Prediction of Elastic Behavior in a Molecular Solid. Here we consider materials in which the basic unit of structure is a flexible chain of covalently bonded atoms. Flexibility is afforded by the backbone bonds being able to rotate, as illustrated in Figure I.1.3.5.

Such rotations allow the chains to adopt many different conformations (shapes), and are therefore associated with increased entropy. By quantifying how stretching the material restricts the number of conformations accessible to the chains, it is possible to calculate the accompanying entropy change, and hence the change in free energy. It is then simple to proceed from the relationship between energy and strain to a relationship between force (or stress) and strain, as we saw in the previous example.

We start by fixing one end of a chain at the origin (0,0,0) of a (Cartesian) coordinate system. Let the other end be at position (x,y,z). The probability of this happening depends on the number of possible chain conformations between the ends thus positioned.

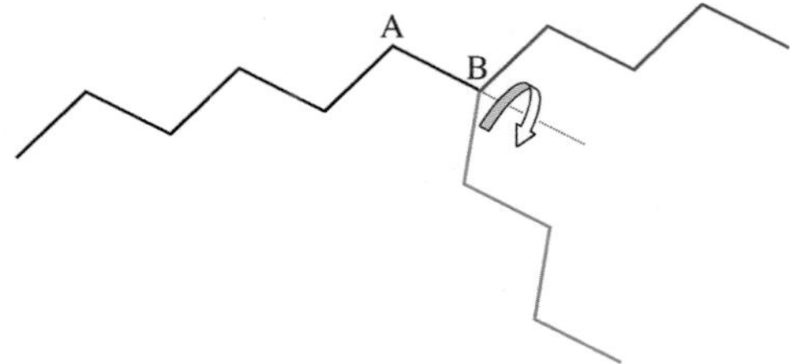

FIGURE I.1.3.5 Schematic illustration of a linear polymer chain, where vertices denote the location of backbone atoms. The black and blue lines together represent an extended (rod-like) conformation (shape) of the molecule. Rotation around backbone bonds can produce many other conformations, conferring flexibility on the molecule. For example, rotation around bond AB can move the segment denoted in blue to the position denoted in red, leading to a more compact conformation.

We make some assumptions that will simplify the subsequent mathematics without diminishing the lessons that are learned from this example. The assumptions are: (1) the chains are freely-jointed, which means that we do not impose specific valence angles on the covalently bonded atoms; (2) each chain follows a random walk between its ends, which is only strictly true while the material remains unstretched; (3) the chains have length but no width, so that self-intersection is allowed; and (4) the chains are infinitely long.

A Gaussian (statistical) analysis of random walks can be used to show that the number of conformations between the ends of a chain with ends at (0,0,0) and (x,y,z) is proportional to:

$$P(x,y,z) = \left(\frac{3}{2\pi nl^2}\right)^{\frac{3}{2}} \exp\left[-\left(\frac{3}{2nl^2}\right)(x^2 + y^2 + z^2)\right]$$

Here n is the number of links in the flexible chain, and l is the length of each link. The quantity nl^2 represents the mean square end-to-end distance of the chains $\langle r^2\rangle$, i.e., the root mean square end-to-end distance $\sqrt{\langle r^2\rangle}$ is equal to $\sqrt{n}l$, consistent with the properties of a random walk. The entropy of the chain is therefore:

$$S = k\ln W$$

where W is the number of ways in which the chain can be rearranged between its fixed end points, and k is Boltzmann's constant:

$$= k\ln[C \cdot P(x,y,z)]$$

where C is a constant:

$$= k\left[\text{constant} - \left(\frac{3}{2nl^2}\right)(x^2 + y^2 + z^2)\right]$$

Now consider stretching the chain to extension ratios λ_x, λ_y, λ_z, along axes x, y, and z respectively. After stretching:

$$S_{new} = k\left[\text{constant} - \left(\frac{3}{2nl^2}\right)(\lambda_x^2x^2 + \lambda_y^2y^2 + \lambda_z^2z^2)\right]$$

Therefore, as a result of stretching:

$$\Delta S = S_{new} - S$$
$$= -k\frac{3}{2\,nl^2}\left[(\lambda_x^2 - 1)\,x^2 + (\lambda_y^2 - 1)\,y^2 + (\lambda_z^2 - 1)\,z^2\right]$$

This change in entropy reflects the decreased number of conformations available to the chain when stretched – in other words, the stretched state is less probable than the relaxed state. Now consider all the chains in a unit volume of the material:

$$\Delta S = -k\frac{3}{2\,nl^2}\left[(\lambda_x^2 - 1)\sum x^2 + (\lambda_y^2 - 1)\sum y^2 + (\lambda_z^2 - 1)\sum z^2\right]$$

But $x^2 + y^2 + z^2 = r^2$, where r is the straight line distance between the ends of each chain. So, because the distribution of chains is spherically symmetrical in the undeformed state:

$$\sum x^2 = \sum y^2 = \sum z^2 = \frac{1}{3}\sum r^2$$

$= \frac{1}{3}N\langle r^2\rangle$ if N is the number of chains per unit volume

$= \frac{1}{3}N\,nl^2$ because freely-jointed chains are assumed.

$$\therefore \Delta S = -k\frac{3}{2\,nl^2}\left[(\lambda_x^2 - 1)\,\tfrac{1}{3}Nnl^2 + (\lambda_y^2 - 1)\,\tfrac{1}{3}Nnl^2 + (\lambda_z^2 - 1)\tfrac{1}{3}Nnl^2\right]$$
$$= -\frac{1}{2}Nk\left[\lambda_x^2 + \lambda_y^2 + \lambda_z^2 - 3\right] \text{ per unit volume.}$$

If we assume that the work done (energy needed) to stretch the chains is only used to decrease entropy, i.e., the chain conformations change, but the bonds are not stretched, then:

$$\Delta U = -T\,\Delta S = +\frac{1}{2}NkT\left[\lambda_x^2 + \lambda_y^2 + \lambda_z^2 - 3\right]$$

per unit volume,
where T denotes absolute temperature.

This result is similar in structure to the formula for stored energy per unit volume during the elastic deformation of a metal, ceramic or glass (see the section on "Resilience" below):

$$\Delta U = \frac{1}{2}(\text{Young's modulus}) \cdot (\text{strain})^2$$

We therefore refer to:
NkT as the *statistical modulus* or *rubber modulus*, and
ΔU as the *stored energy function* (SEF).
Now consider the bulk deformation of rubber.

On stretching: $l_1 \rightarrow l_1\,\lambda_x = l_2$

$$b_1 \rightarrow b_1\,\lambda_y = b_2$$
$$t_1 \rightarrow t_1\,\lambda_z = t_2$$

Conservation of volume requires that:

$$l_1\,\lambda_x \cdot b_1\,\lambda_y \cdot t_1\,\lambda_z = l_1 \cdot b_1 \cdot t_1$$
$$\therefore \lambda_x \cdot \lambda_y \cdot \lambda_z = 1$$

Assuming that the force is applied uniaxially along x, and that the transverse response is uniform ($\lambda_y = \lambda_z$) leads to:

$$\lambda_x = \frac{1}{\lambda_y^2} \equiv \frac{1}{\lambda_z^2}$$
$$\therefore \lambda_y^2 \equiv \lambda_z^2 = \frac{1}{\lambda_x}$$

Substituting this result into the SEF gives:

$$\Delta U = \frac{1}{2}NkT\left[\lambda_x^2 + \lambda_y^2 + \lambda_z^2 - 3\right] = \frac{1}{2}NkT\left[\lambda_x^2 + \frac{1}{\lambda_x} + \frac{1}{\lambda_x} - 3\right]$$
$$= \frac{1}{2}NkT\left[\lambda_x^2 + \frac{2}{\lambda_x} - 3\right]$$

The total work done in deforming the entire volume of the sample is therefore:

$$\Delta U_t = \frac{1}{2}NkT\left[\lambda_x^2 + \frac{2}{\lambda_x} - 3\right] \cdot l_1 \cdot b_1 \cdot t_1$$

This work is related to the necessary force by:

$$F = \frac{d(\Delta U_t)}{dl} = \frac{d(\Delta U_t)}{d\lambda_x}\frac{d\lambda_x}{dl} = \frac{d(\Delta U_t)}{d\lambda_x}\frac{1}{l_1}$$

$$\therefore F \cdot l_1 = \frac{d(\Delta U_t)}{d\lambda_x} = \frac{1}{2}NkT\left[2\lambda_x - \frac{2}{\lambda_x^2}\right] \cdot l_1 \cdot b_1 \cdot t_1$$

$$\therefore F = NkT\left[\lambda_x - \frac{1}{\lambda_x^2}\right] \cdot b_1 \cdot t_1$$

$$\therefore \sigma_n = \frac{F}{b_0 \cdot t_0} = NkT\left[\lambda_x - \frac{1}{\lambda_x^2}\right]$$

$$= NkT\left[(\varepsilon_n + 1) - \frac{1}{(\varepsilon_n + 1)^2}\right] \text{because } \varepsilon_n = \frac{l_2 - l_1}{l_1} = \lambda_x - 1$$

We now have a result that predicts the relationship between nominal stress and nominal strain for a molecular solid that consists of flexible chains of covalently bonded atoms. Such a material is known as an *elastomer*. The correspondence between theory and experiment is good (perhaps surprisingly, given the number of assumptions in this relatively simple model), up to an extension ratio of about six. At higher extension ratios, the chains are far from the assumed condition of a random trajectory, and the effects of finite chain length become significant. Also, at that stage of sample deformation the molecules are so well-ordered that crystallization can occur, and so the material no longer has the flexibility that is characteristic of an elastomer.

Examples of elastomers include both natural and synthetic rubber, as well as the tissue that comprises arterial walls. The relationship derived above shows that the stress required to maintain a given strain is an increasing function of temperature. We therefore see that arteries are more resistant to dilation a higher temperatures, requiring a corresponding increase in the effort expended by the heart to maintain blood circulation. This can be a contributing factor to heat stroke.

Plastic Deformation

Yield Strength and Ductility. We return now to consider additional mechanical properties that can be gleaned from nominal stress versus nominal strain curves such as those shown in Figures I.1.3.2 and I.1.3.3. As the stress on a material increases, a point may be reached where the response is no longer linear. There is a permanent (irreversible) deformation, i.e., deformation that is not recovered on removing the stress. This means that some rearrangement of the atoms or molecules must have been triggered. Irreversible deformation is known as *plastic* deformation. It occurs commonly in metals and polymers, and rarely in ceramics.

The amount of plastic deformation associated with a given stress is found by drawing a line from the point of interest on the stress–strain curve, parallel to the initial linear segment of the curve, and marking the point where this constructed line intersects the horizontal (strain) axis. The horizontal distance from the origin to the point of intersection is a measure of the plastic strain. The value of plastic strain required to break the material defines the *ductility* (Figure I.1.3.3). The stress at which departure from a linear stress–strain relationship occurs is known as the *proportional limit* (Figure I.1.3.3).

It is not always easy to ascertain the proportional limit accurately, and so a more practical determination of the condition for plastic deformation is provided by the *yield strength* (Figure I.1.3.3), which is the stress at which *noticeable* plastic strain occurs. In this context, "noticeable" is often taken to be a value of 0.002 (0.2%) for metals – although it may be chosen to be much higher in polymers, where elastic behavior persists to strains that may be ten or even a hundred times higher, and where the transition from elastic to plastic behavior may be difficult to pinpoint.

A marked *yield drop* is often displayed in the stress versus strain plot when the yield stress of a polymer is exceeded (Figure I.1.3.2). The yield drop occurs because the polymer chains become partially aligned during initial deformation, allowing easier relative motion of chains and reducing the stress needed to further deform the material. The chains then straighten and extend as deformation continues, requiring little increase in stress. Eventually, the opportunities for relative motion of chains are exhausted, along with the capacity for additional chain extension, and the stress versus strain plot then rises sharply.

Strength and Failure. Failure of a loadbearing material occurs when the material ceases to perform its loadbearing function. Different interpretations of "failure" are in common use, and so it is important to be clear about the intended meaning when we speak of materials failure.

For materials that must not undergo in situ permanent deformation, failure is synonymous with the yield stress being exceeded, and so the yield stress represents an estimate of the strength of the material in those cases.

For materials in which permanent deformation (and the attendant shape change) is acceptable, failure may be deemed to occur when a noticeable "neck" (constriction) develops in the material. The effect of the neck is to concentrate the load on a smaller area; therefore, the load that can be supported by the sample is decreased. On a nominal stress versus nominal strain plot, where stresses are referred to the *original* loadbearing areas of the sample, the stress required for additional deformation decreases. Therefore, the onset of necking corresponds to a maximum in the nominal stress versus nominal strain plot, defining the *ultimate tensile strength* (UTS) or simply the *tensile strength* of the material (Figure I.1.3.3). There is no corresponding maximum on a true stress versus true strain plot, because the stress continues to rise past the onset of necking as values of stress are obtained by dividing the applied load by the actual (small) cross-section of the neck.

On both nominal stress versus nominal strain and true stress versus true strain plots, a *breaking strength* can be defined at the point where the material actually breaks. However, to reach this point practically, at least in the case of most metals, the tensile strength of the material must first be reached, and it is tensile strength that represents the practically measurable maximum stress that a metal can survive. Ceramics, on the other hand, usually reach their breaking stress before they yield.

Thus, the perceived strength of a material depends in part on the definition of failure that is used (onset of plastic deformation, onset of necking, or actual occurrence of breaking). It also depends on the conditions that are used for testing. The *strain rate* used in a tensile test can affect the maximum stress that is reached, depending on the ability of the microstructure to undergo rearrangement to accommodate the imposed deformation.

Over long periods of time, even stresses that are below the conventionally measured yield strength of a material may be sufficient to cause gradual elongation and eventual failure via *creep*. Therefore, creep tests, in which a fixed load is applied while the strain is monitored until the sample breaks, are important for materials that are required to have a long projected service life – as is the case with many implants.

Similarly, materials that will be subjected to cyclic loading patterns during service may undergo *fatigue* failure at loads that are smaller than those needed to trigger failure in a conventional stress versus strain test. Testing to determine the relationship between average load,

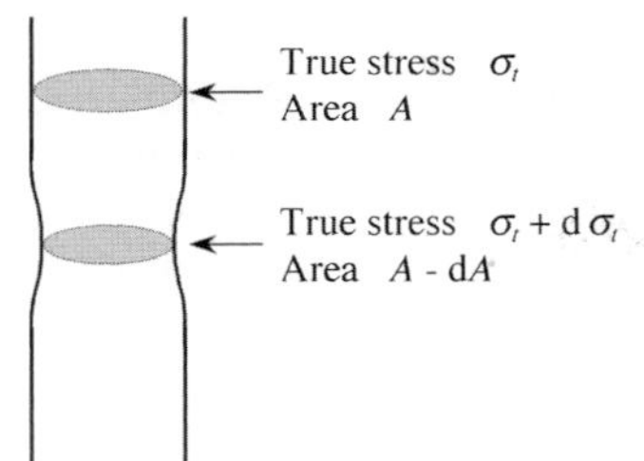

FIGURE I.1.3.6 True stress and sample geometry in the vicinity of a neck.

maximum and minimum load, and the number of loading cycles that lead to failure is required in such cases. Finally, the catastrophic failure of a material can be triggered by pre-existing flaws that locally concentrate the effect of an applied stress. We will explore this phenomenon in the section "Statistical Aspects of Failure."

Stability of Necks. A neck will be stable if its formation is accompanied by sufficient *work hardening*, i.e., by microstructural changes that increase the resistance of the material to further deformation and thus increase the load that can be carried by the necked region. Referring to Figure I.1.3.6 we see that this condition requires:

$$(\sigma_t + \mathrm{d}\sigma_t)(A - \mathrm{d}A) \geq \sigma_t A$$

$$\therefore \ \sigma_t A - \sigma_t \,\mathrm{d}A + A\mathrm{d}\sigma_t - \mathrm{d}A\mathrm{d}\sigma_t \geq \sigma_t A$$

$\therefore \ -\sigma_t \mathrm{d}A + A\mathrm{d}\sigma_t \geq 0$ if we neglect the product of infinitesimals.

$$\therefore \ \frac{\mathrm{d}\sigma_t}{\sigma_t} \geq \frac{\mathrm{d}A}{A}$$

We need to convert the change in area into a change in length, in order to write this condition in terms of stress and *strain*. Because the deformation is *plastic* at the stage where necking occurs, we can assume constant volume:

$$(A - \mathrm{d}A)(l + \mathrm{d}l) = Al$$

$$\therefore \ Al + A\mathrm{d}l - l\,\mathrm{d}A - \mathrm{d}A\mathrm{d}l = Al$$

$\therefore \ A\mathrm{d}l - l\mathrm{d}A = 0$ if we again neglect the product of infinitesimals.

$$\therefore \ \frac{\mathrm{d}A}{A} = \frac{\mathrm{d}l}{l}$$

$$\therefore \ \frac{\mathrm{d}\sigma_t}{\sigma_t} \geq \frac{\mathrm{d}l}{l}$$

$\therefore \ \dfrac{\mathrm{d}\sigma_t}{\sigma_t} \geq \mathrm{d}\varepsilon_t$ from the definition of ε_t.

But previously (see section "True Stress and True Strain") we saw that $\varepsilon_t = \ln(\varepsilon_n + 1)$. Therefore:

$$\frac{\mathrm{d}\varepsilon_t}{\mathrm{d}\varepsilon_n} = \frac{1}{\varepsilon_n + 1}$$

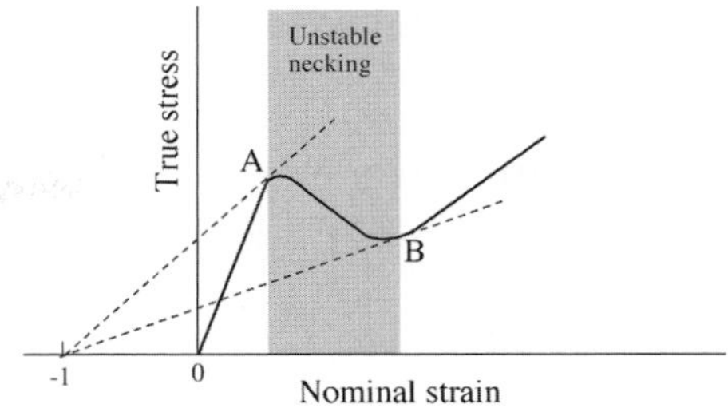

FIGURE I.1.3.7 Considère's construction for determining the stability of a neck. At all points between A and B on the true stress versus nominal strain plot shown, the local slope $\mathrm{d}\sigma_t/\mathrm{d}\varepsilon_n$ of the plot is less than the local value of $\sigma_t/(\varepsilon_n + 1)$.

$$\therefore \ \mathrm{d}\varepsilon_t = \frac{\mathrm{d}\varepsilon_n}{\varepsilon_n + 1}$$

$$\therefore \ \frac{\mathrm{d}\sigma_t}{\sigma_t} \geq \frac{\mathrm{d}\varepsilon_n}{\varepsilon_n + 1}$$

$$\therefore \ \frac{\mathrm{d}\sigma_t}{\mathrm{d}\varepsilon_n} \geq \frac{\sigma_t}{\varepsilon_n + 1} \text{ for stability}$$

This is Considère's criterion. Applied graphically (Figure I.1.3.7) it is known as Considère's construction.

Hardness

The hardness of a material is measured by applying a known load to a small indenter of known geometry (typically pyramidal or spherical) in contact with the surface of the material, for a known period of time. The dimensions of the resulting indentation are measured, and this information, together with the experimental conditions, is used to rate the hardness of the material on a relative scale.

Indents made in a soft material under a given set of conditions are larger than those made in a hard material. Therefore, hardness provides a measure of how successfully a material resists plastic deformation, which in turn is characterized by both yield strength and tensile strength. It is therefore possible to empirically develop calibration charts that can be used to convert hardness measurements into both yield strength and tensile strength values.

Hardness testing is popular for estimating the yield strength and tensile strength of materials because the equipment is relatively simple and inexpensive, and the tests are nondestructive.

Resilience

Resilience is a measure of the *elastic energy* that can be stored in a unit volume of stressed material. It corresponds to the area underneath a stress versus strain plot, extending from zero strain up to the strain at which the sample yields:

$$U_r = \int_0^{\varepsilon_y} \sigma \mathrm{d}\varepsilon$$

If the stress versus strain relationship is linear up to the yield strain (typical for ceramics, and approximate for many metals and polymers), the corresponding area under the stress versus strain plot is triangular, and the integral in the previous equation simplifies to:

$$U_r \cong \frac{1}{2}\sigma_y \varepsilon_y = \frac{\sigma_y^2}{2E}$$

Resilience is an important consideration for materials that need to be "springy," i.e., that absorb and return significant amounts of energy. Examples include natural and replacement tendons and ligaments, and materials used in foot arch reconstruction or support.

Toughness and Fracture Toughness

Although these two terms sound superficially similar, they refer to quite different properties.

Toughness is a measure of the energy required to deform a unit volume of material to its breaking point. Therefore, the definition is similar to that used for resilience, except that we now take into account the entire area under the stress versus strain plot, extending from zero strain up to the strain at which the sample breaks:

$$U_{break} = \int_0^{\varepsilon_{break}} \sigma d\varepsilon$$

Fracture toughness is a measure of how successfully a material resists the propagation of cracks. Consider a sample of material containing either an internal crack of length $2a$ or a surface crack of length a, and subjected to a stress σ, as shown in Figure I.1.3.8. We will explore the former case in detail; the results are identical in the second case.

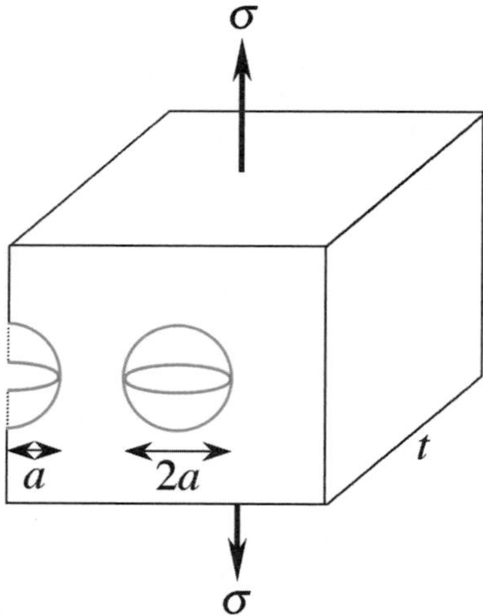

FIGURE I.1.3.8 Sample geometry used in defining the fracture toughness of a material. Examples of both a surface crack and an internal crack are shown. The cracks are assumed to be cylindrical (blue circles) for ease of mathematical description; a flattened profile (red ellipses) would be more realistic.

When the crack grows, elastic energy stored in the material is used to break bonds and create new surface. The energy released in forming a cylindrical crack is:

$$U_{volume} = \left(\frac{1}{2}\sigma\varepsilon\right)(\text{volume})$$

$$= \left(\frac{1}{2}\sigma\frac{\sigma}{E}\right)(\pi a^2 t) \text{ where } t \text{ represents sample thickness}$$

$$= \frac{\pi\, \sigma^2 a^2 t}{2E}$$

Exact calculation gives $\frac{\pi\, \sigma^2 a^2 t}{E}$ if we perform a rigorous analysis on a flattened (more realistic) crack rather than the cylindrical one. The energy required to make the surface of the crack (the new interface between the material and its environment) is:

$$U_{surface} = 2 \times 2\, at \times \Gamma_s = 4\, at\, \Gamma_s$$

taking account of the fact that *two* surfaces are created by interrupting bulk material, and using Γ_s to denote energy per unit area of surface.

Propagation (growth) of the crack requires the accompanying energy release to exceed the accompanying energy consumption:

$$\frac{\partial U_{volume}}{\partial a} > \frac{\partial U_{surface}}{\partial a}$$

$$\therefore \frac{\partial}{\partial a}\left(\frac{\pi\, \sigma^2 a^2 t}{E}\right) > \frac{\partial}{\partial a}(4\, at\, \Gamma_s)$$

$$\therefore \frac{\pi\, \sigma^2\, t}{E} 2a > 4t\, \Gamma_s$$

$$\therefore \sigma > \sqrt{\frac{2\, E\, \Gamma_s}{\pi\, a}}$$

or we can identify a critical stress σ_c for crack propagation:

$$\sigma_c = \sqrt{\frac{2\, E\, \Gamma_s}{\pi\, a}}$$

For a material in which plastic deformation occurs while cracks propagate, the factor Γ_s is increased accordingly. The previous equation can be rearranged to give:

$$2^{-1/2}\sigma_c\sqrt{\pi\, a} = \sqrt{E\, \Gamma_s}$$

or, more generally:

$$Y\sigma_c\sqrt{\pi\, a} = \sqrt{E\, \Gamma_s}$$

where Y is a constant that depends on sample and loading geometry. Since the quantities on the right-hand side of this equation are material-specific properties, it follows that the expression on the left-hand side is also a material-specific property. It is known as the *fracture toughness* of the material. It quantifies the maximum size of flaw (crack) that can be tolerated if the material is to support a particular stress. Its units are unusual: $MPa.m^{1/2}$.

The presence of the factor Γ_s in the equation for fracture toughness emphasizes that the ability of a material to resist fracture *depends on the environment* in which it is tested. In particular for biomaterials, where long in-service lifetimes are required and expected, it is important for mechanical testing to be performed under environmental conditions that are properly representative of the conditions that will be encountered by the material when in service.

Statistical Aspects of Failure

Microstructural variability between samples, especially the distribution of pre-existing flaws of different sizes, leads to variable results in tests of mechanical properties. From an engineering design point of view, materials selection should be guided by the low values of whatever property is (or properties are) relevant to the product being able to perform properly. Even if a property such as ultimate tensile strength is high on average, it is the probability of encountering a sample with a low value of tensile strength that must be used to inform the choice of safe design limits. Thus, statistical measures of spread in measured data – and hence of a material's reliability – are needed.

Consider a batch of samples, each of which is tested to determine the ultimate tensile strength. Let $\mathfrak{F}$ denote the fraction of those samples that have failed at a stress σ. $\mathfrak{F}$ therefore represents a cumulative failure probability, and it is an increasing function of σ. It can be shown that, if all the samples have the same size and shape:

$$\mathfrak{F} = 1 - \exp\left(-\frac{\sigma}{\sigma_0}\right)^m$$

where the parameters m and σ_0 are characteristic of the material, and m is known as the *Weibull modulus*. Therefore:

$$1 - \mathfrak{F} = \exp\left(-\frac{\sigma}{\sigma_0}\right)^m$$

$$\therefore \frac{1}{1-\mathfrak{F}} = \exp\left(\frac{\sigma}{\sigma_0}\right)^m$$

$$\therefore \ln\left(\frac{1}{1-\mathfrak{F}}\right) = \left(\frac{\sigma}{\sigma_0}\right)^m$$

$$\therefore \ln\left[\ln\left(\frac{1}{1-\mathfrak{F}}\right)\right] = m\ln\sigma - m\ln\sigma_0$$

So a plot of $\ln\left[\ln\left(\frac{1}{1-\mathfrak{F}}\right)\right]$ versus $\ln\sigma$ will yield a straight line with slope equal to the Weibull modulus of the material. The term "modulus" is sometimes incorrectly interpreted here as referring to a type of stiffness, because of its frequent association with elastic constants. However, in *all* the senses that it is used in this chapter, "modulus" simply means "number," in keeping with its origin in the Latin word *modulus*, a (small) measure.

Engineering materials that are required to exhibit great in-service reliability, for example the metal alloys used in critical aircraft components, exhibit Weibull moduli well over 100. In contrast, ordinary glass has a Weibull modulus of around 5, and common ceramics typically have Weibull moduli in the range $5 \le m \le 25$. Perhaps surprisingly, the Weibull modulus of natural materials can be quite low. Despite their reputation for high *average* breaking strength, stiffness, and toughness in conventional tensile tests, natural silk fibers exhibit low Weibull moduli (5.8 for silkworm silk, and 3.4 for spider dragline). The poor *reproducibility* of breaking strength constrains our ability to incorporate these materials in reliable design, and suggests that their best use might be in composites where applied loads are distributed across multiple fibers.

OTHER BULK PROPERTIES

The bulk properties of a material are of course not limited to performance in mechanical tests. Other types of bulk properties that may have to be taken into account during the selection of a biomaterial include: thermal properties; optical properties; electrical properties; and magnetic properties. Even cost – financial and environmental – can be considered to be a bulk property, given the dependence of cost on: (1) the type and volume of material that is used; and (2) the complexity of microstructural rearrangement during processing.

Thermal Properties

Thermal conductivity becomes a significant consideration if an implanted material contributes to an unnatural flow of heat through the surrounding tissue. For example, metal rods selected for their combination of stiffness, strength, fracture toughness, and biocompatibility can promote heat loss and cause the patient to feel colder than normal. Heat conduction through a metal dental filling can also be a source of discomfort, and may therefore play a role in the choice between a metal or non-metal filling.

Thermal expansion is a (nominal) strain that occurs when the temperature of a material is changed:

$$\varepsilon_{thermal} = \alpha\,(T_{final} - T_{initial})$$

where α is the thermal expansion coefficient; its units are $(\text{degrees})^{-1}$, with "degrees" measured on whatever standard scale is used to measure temperature T.

Attempts to match the properties of a dental filling to those of the surrounding tissue must therefore extend beyond achieving compatible elastic constants (see section "Elastic Constants") to include consideration of thermal expansion, to avoid premature failure of the interface.

Optical Properties

In the context of biomaterials, the most significant bulk optical properties are color, refractive index, and transparency; all three are important in the selection of materials for intraocular lenses or fluids.

The color of a transparent material is controlled by composition, and therefore demands a high degree of quality control to avoid impurities that could adversely affect color. Long-term stability of material composition is also important: components should not selectively diffuse into the surrounding tissue, and no chemical changes should occur – either by reactions between the components or in response to light.

Refractive index n of a material is defined as:

$$n = \frac{c_{vacuum}}{c_{material}}$$

where c denotes the speed of light in the subscripted medium.

When light crosses the interface between two media (materials), it is deviated from its original path by an angle that is an increasing function of the difference between the refractive indices of the media. Therefore, the effectiveness of a material as a lens is directly related to its refractive index. Refractive index increases along with the content of electron-rich atoms, which is why lead "crystal" is so useful in both lenses and decorative glassware.

Transparency is a qualitative term that describes the ability of a material to transmit light without attenuating (absorbing or scattering) it. To minimize absorption, the primary bonds in the material must be strongly covalent or ionic (and definitely not metallic). Scattering is minimized if the material is free of internal interfaces (which could reflect light) and compositional differences (which would be associated with refractive index differences that could deviate light from an uninterrupted path through the material). Thus, an optimally transparent material will either be a homogenous single crystal or it will be completely and homogeneously amorphous.

There is no commonly accepted quantitative definition of transparency. Instead, it is usual to consider the complementary property, *opacity*. This is itself a colloquial term, but it lends itself to a more formal description by way of the *mass attenuation coefficient*. If light of a given frequency (color) and initial intensity I_0 travels a distance x through a material, then the intensity is decreased to:

$$I(x) = I_0 \exp(-\mu\rho x)$$

where μ is the mass attenuation coefficient of the material, and ρ is the density.

Weibull Modulus and Non-Brittle Materials

The Weibull distribution of fracture probability for materials provides a statistical measure of brittleness, via the specification of a Weibull modulus m (Derby et al., 1992). The value of m can in theory range from zero (totally random fracture behavior, where the failure probability is the same at all stresses, equivalent to an ideally brittle material) to infinity (representing a precisely unique, reproducible fracture stress, equivalent to an ideally non-brittle material).

This quantitative approach to defining non-brittleness is more robust than qualitative description, since there is no single antonym of "brittle." Both "tough" and "ductile" are in particular senses the opposite of "brittle," but we have seen that "tough" and "ductile" are not interchangeable, and their corresponding nouns are associated with very different features on a stress–strain curve (Figure I.1.3.3).

Weibull statistics were initially adopted to quantify failure in classically brittle materials (Kelly and Macmillan, 1986) where tensile strength is dominated by the size distribution of pre-existing flaws, and the use of this method is rarely extended to materials that are regarded as non-brittle. However, there is no *a priori* reason to restrict Weibull statistics to the analysis of failure in classically brittle materials. For example, although natural silks exhibit large extensions to failure, and many of them behave like an elastomer at least during the initial stages of deformation, Weibull analysis has proved to be useful for characterizing and contextualizing the failure strength variability of these materials (Pérez-Rigueiro et al., 1998, 2001).

Why *is* a Weibull analysis appropriate for non-brittle materials? A justification can be formulated simply on phenomenological grounds: after significant deformation, silk *develops* the statistical failure *characteristics* of a brittle material, even though it initially deserves classification as an elastomer. While the existence of strength-limiting defects can be inferred from this description, their nature has yet to be confirmed; it is not clear whether they are present in as-spun material, or whether they begin to develop during the earlier stages of deformation.

The appropriateness of Weibull analysis for a non-brittle material can also be justified fundamentally (Viney, 2002), given the equation that is used as the basis for measuring m (section "Statistical Aspects of Failure"):

$$1 - \mathfrak{F} = \exp\left(-\frac{\sigma}{\sigma_0}\right)^m$$

If F is to be independent of σ, which is the hallmark of an ideally brittle material, this equation requires m to be equal to zero. However, practical Weibull analysis necessarily admits to non-zero values of m, so it is implicit that ductile contributions to failure can also be accommodated.

WORKED EXAMPLE

QUESTION:

A stainless steel rod with a circular cross-section has a length of 100 mm and a diameter of 2.0 mm. It is deformed elastically in compression by a force of 500 N applied parallel to its length. The material has a Young's modulus of 200 GPa, a shear modulus of 77 GPa, and a yield strength of 0.3 GPa. Calculate: (1) the amount by which the rod will decrease in length; and (2) the amount by which the rod will increase in diameter.

SOLUTION:

(1) Deformation is elastic, so we can use Hooke's Law:

$$\sigma = E\varepsilon$$

$$\therefore \varepsilon = \frac{\sigma}{E} = \frac{F/A_1}{E} = \frac{F}{EA_1} = \frac{F}{E\pi\left(d_1/2\right)^2} = \frac{4F}{E\pi d_1^2}$$

where the symbols are used consistently with their meaning elsewhere in this chapter; d_1 is the initial diameter of the rod. But:

$$\varepsilon = \frac{l_2 - l_1}{l_1} = \frac{\text{elongation}}{l_1}$$

$$\therefore \text{elongation} = l_1\,\varepsilon = \frac{4Fl_1}{E\pi d_1^2} = \frac{(4)(-500\text{ N})(0.1\text{ m})}{\left(200\times10^9\frac{\text{N}}{\text{m}^2}\right)\pi(0.002\text{ m})^2}$$

$$= -7.96\times10^{-5}\,m \text{ or } -0.0796\text{ mm}$$

The negative value of force is used because the rod is loaded in compression.

(2) We need a value of Poisson's ratio, which can be obtained from the two elastic constants provided if we use the following equation from the section "Elastic Constants":

$$G = \frac{E}{2(1+\nu)}$$

Rearrangement and substitution gives:

$$\nu = \frac{E}{2G} - 1 = \frac{200\times10^9\frac{\text{N}}{\text{m}^2}}{2\times77\times10^9\frac{\text{N}}{\text{m}^2}} - 1 = 0.299$$

Now we can use the definition of Poisson's ratio, υ:

$$\nu = -\frac{\varepsilon_{transverse}}{\varepsilon_{longitudinal}} = -\frac{\frac{d_2-d_1}{d_1}}{\frac{l_2-l_1}{l_1}} = -\left(\frac{d_2-d_1}{d_1}\right)\left(\frac{l_1}{l_2-l_1}\right)$$

$$\therefore d_2 - d_1 = -\nu d_1\left(\frac{l_2-l_1}{l_1}\right)$$

$$= -(0.299)(0.002\text{ m})\left(\frac{-7.96\times10^{-5}\text{ m}}{0.1\text{ m}}\right)$$

$$= 4.76\times10^{-7}\text{ m or } 4.76\times10^{-4}\text{ mm}$$

The value of the yield strength that was provided in the question is apparently superfluous. However, it is useful for checking whether the given load does indeed cause only elastic deformation as claimed. The stress on the rod is:

$$\sigma_n = \frac{F}{A_1} = \frac{-500\text{ N}}{\pi(0.001\text{ m})^2} = \frac{-3,000\text{ N}}{\pi(0.001\text{ m})^2} = 0.159\text{ GPa}$$

This result is well below the quoted yield strength of 0.3 GPa, and so is consistent with the deformation being elastic.

BIBLIOGRAPHY

Derby, B., Hills, D. A., & Ruiz, C. (1992). *Materials for Engineering: A Fundamental Design Approach*. Harlow, UK: Longman Scientific & Technical.

Kelly, A., & Macmillan, N. H. (1986). *Strong Solids* (3rd ed.). Oxford, UK: Oxford University Press.

Pérez-Rigueiro, J., Viney, C., Llorca, J., & Elices, M. (1998). Silkworm silk as an engineering material. *Journal of Applied Polymer Science*, 70, 2439–2447.

Pérez-Rigueiro, J., Elices, M., Llorca, J., & Viney, C. (2001). Tensile properties of *Argiope trifasciata* drag line silk obtained from the spider's web. *Journal of Applied Polymer Science*, 82, 2245–2251.

Viney, C. (2002). In M. Elices, & J. Llorca (Eds.), *Fibre Fracture* (pp. 303–328). Oxford, UK: Elsevier Science Ltd.

CHAPTER I.1.4 FINITE ELEMENT ANALYSIS IN BIOMECHANICS

Michael Sacks[1], Antonio D'Amore[2], and Christopher Hobson[2]

[1]Department of Biomedical Engineering, The University Texas at Austin, Austin, TX, USA

[2]University of Pittsburgh, Pittsburgh, PA, USA

INTRODUCTION

In many biomedical engineering design problems, accurate prediction of the biomaterial mechanical response is essential in the development of many prostheses and medical devices. In this process, structural analysis is a critical step in the prediction of fatigue limits and potential failure modalities. Practically, simulation methods are essential since experimental methods often require the creation of expensive, time-intensive prototypes and evaluations. Further, many experimental parameters cannot be directly measured, but can only be computed (e.g., shear stress). Moreover, analytical approaches can only provide a complete solution for simple geometries and loading conditions. In contrast, all biological and biologically derived or inspired materials exhibit pronounced mechanical anisotropy, highly nonlinear stress–strain relationships, large deformations, viscoelasticity, and strong material coupling. Taken as a whole, most biomaterials defy simple material models. Therefore, numerical approaches such as the Finite Element Method (FEM) or the Boundary Element Method (BEM) are more suitable for biomechanical problems. FEM, also known as Finite Element Analysis (FEA), represents the most popular numerical technique to perform stress and strain analysis. Although the approach was originally formulated to perform structural analyses, it has become increasingly adopted in biomechanics and biomaterial sciences to solve complex multi-physics problems. In the following, we present a brief introduction of the mathematical formulation of FEM for the biomaterials audience. We utilize the infinitesimal strain linear elasticity which, while not realistic for most biological materials, allows the basic theory to be presented with maximum clarity.

OVERVIEW OF THE FINITE ELEMENT METHOD

In FEM, a real structure is replaced by a discrete model obtained by subdivision into a number of finite elements (Figure I.1.4.1). The discretized model is composed of appropriately shaped elements defined by a series of interconnected points known as nodes. The continuum problem with infinite degrees of freedom can thus be reduced to a discrete problem with finite degrees of freedom, and solved computationally with a series of simultaneous algebraic equations. In the ordinary formulation, the displacement field within each finite element is strictly related to nodal displacement by shape functions that can be derived from the interpolation of nodal displacements. Under this assumption, the initial problem can be reduced to a discrete problem where the unknowns are the Cartesian components of the nodal displacements, in effect reducing the initial three-dimensional problem to one with only three degrees of freedom per node (Tottenham and Brebbia, 1970; Zienkiewicz, 1971; Middleton et al., 1996).

FIGURE I.1.4.1 An arbitrary area discretized with tetrahedral elements.

Problem

Consider first the simple case of a triangular plane element defined by three nodes (Figure I.1.4.2a) (Tottenham and Brebbia, 1970; Zienkiewicz, 1971). Two displacement components $u(x,y)$ and $v(x,y)$ of an arbitrary point $P(x,y)$ within the element can be expressed by the relationship (Eq. 1):

$$u(x,y) = a_1 + a_2 x + a_3 y$$
$$v(x,y) = a_4 + a_5 x + a_6 y \tag{1}$$

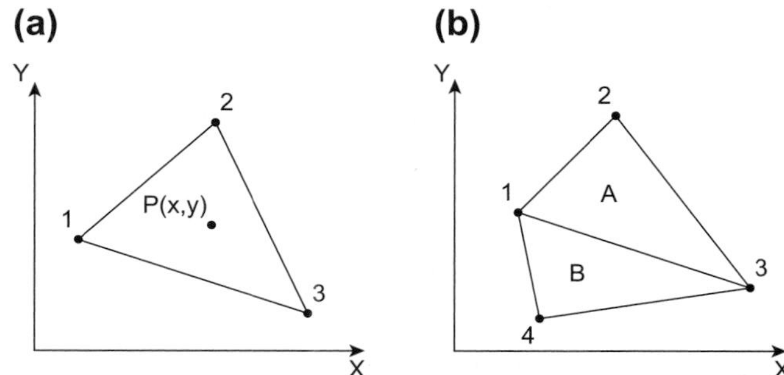

FIGURE I.1.4.2 (a) A triangular element defined by three nodes with an arbitrary point P within the element at position (x,y). (b) Two triangular elements, A and B, defined by four nodes sharing a common edge.

which can be expressed in matrix form as (Eq. 2):

$$\begin{bmatrix} u_i \\ v_i \end{bmatrix} = \begin{bmatrix} 1 & x_i & y_i & 0 & 0 & 0 \\ 0 & 0 & 0 & 1 & x_i & y_i \end{bmatrix} \cdot \begin{bmatrix} a_1 \\ a_2 \\ a_3 \\ a_4 \\ a_5 \\ a_6 \end{bmatrix} \tag{2}$$

Assembling the equations for the three nodes ($i = 1,2,3$) we can derive $\vec{d}$, the nodal displacement vector, and [C] the nodal coordinates matrix:

$$\vec{d} = [C] \cdot \vec{a} \tag{3}$$

These relationships enable the displacement of an arbitrary point P within the element to be expressed as a function of the nodal displacements. Solving Eq. 3 for $\vec{a}$ and substituting into Eq. 2.

$$\begin{bmatrix} u_i \\ v_i \end{bmatrix} = \begin{bmatrix} 1 & x_i & y_i & 0 & 0 & 0 \\ 0 & 0 & 0 & 1 & x_i & y_i \end{bmatrix} \cdot [C]^{-1} \cdot \vec{d} = [N] \cdot \vec{d} \tag{4}$$

where $[N]$ is the shape function matrix that can be written as (Eq. 4):

$$[N] = \begin{bmatrix} N_{11} & N_{12} & N_{13} & 0 & 0 & 0 \\ 0 & 0 & 0 & N_{24} & N_{25} & N_{26} \end{bmatrix} \tag{5}$$

with:

$$N_{11} = 1 - \frac{1}{2}\left[\frac{(x - x_1)(y_3 - y_2) - (x_3 - x_2)(y - y_1)}{element\ area}\right] \tag{6}$$

The other shape functions can be similarly obtained from Eq. 6. For a two-dimensional triangular element with three degrees of freedom (DoF) nodes (6), the shape functions must be linear. Given the nodal displacements from the derivative of Eq. 4, the relationship between deformation and nodal displacement can be obtained (Eq. 6).

$$\begin{bmatrix} \varepsilon_{xi} \\ \varepsilon_{yi} \\ \gamma_{xyi} \end{bmatrix} = [D] \cdot \begin{bmatrix} u_i \\ v_i \end{bmatrix} \quad \text{where } [D] = \begin{bmatrix} \frac{\partial}{\partial x} & 0 \\ 0 & \frac{\partial}{\partial y} \\ \frac{\partial}{\partial y} & \frac{\partial}{\partial x} \end{bmatrix} \tag{7}$$

$$\vec{\varepsilon} = [D] \cdot [N] \cdot \vec{d} = [B] \cdot \vec{d} \tag{8}$$

$[D]$ is a differential operator. Since the shape functions are linear, the matrix $[B]$ is defined purely by the geometry and orientation of the finite element considered. Under the hypothesis of planar tension, Hooke's law can be written as (Eq. 9):

$$\begin{bmatrix} \sigma_{xi} \\ \sigma_{yi} \\ \tau_{xyi} \end{bmatrix} = \frac{E}{(1+v)(1-2v)} \begin{bmatrix} 1-v & v & 0 \\ v & v & 0 \\ 0 & 0 & \frac{(1-2v)}{2} \end{bmatrix} \begin{bmatrix} \epsilon_{xi} \\ \epsilon_{yi} \\ \gamma_{xyi} \end{bmatrix} \tag{9}$$

$$\vec{\sigma} = [E] \cdot \vec{\varepsilon} \tag{10}$$

Nonlinear Analysis

Modern FEA packages incorporate a Newton–Raphson computational approach to solve nonlinear problems. In this approach, the load is subdivided into a series of load increments. The load increments are then applied over a series of load steps. The program performs a linear solution and checks for convergence. If convergence criteria are not met, the stiffness matrix is updated and a new solution is obtained. This iterative procedure continues until the problem converges. If convergence is not achieved, then the program attempts to solve with a smaller load increment (Bicanic and Johnson, 1979).

Substituting Eq. 8 into Eq. 10 the link between the stress field and nodal displacements (Eq. 11):

$$\vec{\sigma} = [E] \cdot [B] \cdot \vec{d} \tag{11}$$

The displacements, deformations and stress were expressed as a function of the nodal displacements. The nodal displacements can be evaluated with equilibrium equations derived using the virtual work principle, which states that for an arbitrary congruent deformation, the work (virtual) produced by external forces for the related nodal displacements will be equal to the work done by the internal stress throughout the deformation. In matrix form (Eq. 12):

$$\overrightarrow{\delta d}^{T} \cdot \vec{F} = \int_V \overrightarrow{\delta e}^{T} \cdot \vec{\sigma} dV = \int_V [B]^T \cdot [E] \cdot [B]\, dV \tag{12}$$

Where V is the element volume and $\overrightarrow{\delta d}$ and $\overrightarrow{\delta e}$ are the virtual nodal displacements and the virtual deformations vectors respectively; finally $\vec{F}$ is the nodal force vector (Eq. 13):

$$\vec{F}^T = \begin{bmatrix} F_{1x} & F_{1y} & F_{2x} & F_{2y} & F_{3x} & F_{3y} \end{bmatrix} \tag{13}$$

Fij (i = 1,2,3 and j = x,y) is the arbitrary force applied to the node i in the direction j. From Eq. 12, defining $[K]$ as the stiffness matrix (6×6) of the element given by Eq. 14 the fundamental FEM equation can be obtained representing the relationship between the forces and nodal displacements (Eq. 16).

$$[K] = \int_v [B]^T \cdot [E] \cdot [B] dv = \begin{bmatrix} k_{1x,1x} & k_{1x,1y} & k_{1x,2x} & k_{1x,2y} & k_{1x,3x} & k_{1x,3y} \\ k_{1y,1x} & \bullet & \bullet & \bullet & \bullet & \bullet \\ k_{2x,1x} & \bullet & \bullet & \bullet & \bullet & \bullet \\ k_{2y,1x} & \bullet & \bullet & \bullet & \bullet & \bullet \\ k_{3x,1x} & \bullet & \bullet & \bullet & \bullet & \bullet \\ k_{3y,1x} & \bullet & \bullet & \bullet & \bullet & \bullet \end{bmatrix} \tag{14}$$

$$\vec{F} = [K] \cdot \vec{d} \tag{15}$$

Equation 15 shows that the stiffness matrix term k_{ij} represents the force applied to the node i along the direction x that produces a unitary displacement of the node j along the direction y, assuming all of the other displacements are equal to zero. As a consequence of the Betti Theorem, the stiffness matrix must be symmetric. The stiffness matrix depends on the element geometry, orientation, and mechanical characteristic of the material described by the matrix $[E]$.

Assembly

An equation similar to Eq. 15 can be written for each of the elements within the model. These equations can be assembled into a unique system representing the entire model (Eq. 16) where the bar indicates that the vectors and the matrices refer to the global system:

$$\bar{F} = \left[\bar{K}\right] \cdot \bar{d} \tag{16}$$

In Eq. 16 $\bar{F}$ and $\bar{d}$ represent the organized force and nodal displacement vectors, whereas $\left[\bar{K}\right]$ is the system global stiffness matrix. The latter is a square symmetric matrix having size equal to the number of DoF for the entire model, which is equal to the total number of nodes multiplied by the DoF for a single node. $\left[\bar{K}\right]$ is obtained from the element stiffness matrices where $\bar{k}_{ix,jy}$ represents the force applied to the node i over the direction x producing the unitary displacement of the node j over the direction y. This force is the sum of the single forces applied to the same node that produces the same effect considering the elements that share nodes i and j separately. The arbitrary element $\bar{k}_{ix,jy}$ is, therefore, the sum of the homologous elements of the matrixes of m elements sharing the nodes i and j (Eq. 17):

$$\bar{k}_{ix,jy} = \sum_{r=1}^{m} k_{ix,jy}^{(r)} \tag{17}$$

where the index (r) indicates the element. Considering the simple case of two elements A and B sharing the nodes 1 and 3 (Figure I.1.4.2b), the global stiffness matrix is then given by Eq. 18.

$$\begin{bmatrix}
k^a_{1x,1x}+k^b_{1x,1x} & k^a_{1x,1y}+k^b_{1x,1y} & k^a_{1x,2x} & k^a_{1y,2y} & k^a_{1x,3x}+k^b_{1x,3x} & k^a_{1x,3y}+k^b_{1x,3y} & k^b_{1x,4x} & k^b_{1x,4y} \\
k^a_{1y,1x}+k^b_{1y,1x} & k^a_{1y,1y}+k^b_{1y,1y} & k^a_{1y,2x} & k^a_{1y,2y} & k^a_{1y,3x}+k^b_{1y,3x} & k^a_{1y,3y}+k^b_{1y,3y} & k^b_{1y,4x} & k^b_{1y,4y} \\
k^a_{2x,1x} & k^a_{2x,1y} & k^a_{2x,2x} & k^a_{2x,2y} & k^a_{2x,3x} & k^a_{2x,3x} & 0 & 0 \\
k^a_{2y,1x} & k^a_{2y,1y} & k^a_{2y,2x} & k^a_{2y,2y} & k^a_{2y,3x} & k^a_{2y,3y} & 0 & 0 \\
k^a_{3x,1x}+k^b_{3x,1x} & k^a_{1x,1y}+k^b_{1x,1y} & k^a_{3x,2x} & k^a_{3x,2y} & k^a_{3x,3x}+k^b_{3x,3x} & k^a_{3x,3y}+k^b_{3x,3y} & k^b_{3x,4x} & k^b_{3x,4y} \\
k^a_{3y,1x}+k^b_{3y,1x} & k^a_{3y,1y}+k^b_{3y,1y} & k^a_{3y,2x} & k^a_{3y,2y} & k^a_{3y,3x}+k^b_{3y,3x} & k^a_{3y,3y}+k^b_{3y,3y} & k^b_{3y,4x} & k^b_{3y,4y} \\
k^b_{4x,1x} & k^b_{4x,1y} & 0 & 0 & k^b_{4x,3x} & k^b_{4x,3y} & k^b_{4x,4x} & k^b_{4x,4y} \\
k^b_{4y,1x} & k^b_{4y,1y} & 0 & 0 & k^b_{4y,3x} & k^b_{4y,3y} & k^b_{4y,4x} & k^b_{4y,4y}
\end{bmatrix} \tag{18}$$

Global stiffness matrix terms referring to nodes not shared by elements are equal to zero. Due to the large number of elements composing a finite element model, this condition is frequent. Thus, the global stiffness matrix is a sparse matrix (many elements are equal to zero). Nonzero terms can be grouped along the principal diagonal by appropriately sorting nodes. The result is a band matrix which corresponds to a linear system of equations. Considering the arrangement in Figure I.1.4.3, the resulting matrix (Eq. 19) is a band matrix with a width of 5.

$$\begin{bmatrix}
k_{11} & k_{12} & k_{13} & 0 & 0 & 0 & 0 \\
k_{21} & k_{22} & k_{23} & k_{33} & 0 & 0 & 0 \\
k_{31} & k_{32} & k_{33} & k_{33} & k_{33} & 0 & 0 \\
0 & k_{42} & k_{43} & k_{44} & k_{54} & k_{46} & 0 \\
0 & 0 & * & * & * & * & * \\
0 & 0 & 0 & * & * & * & * \\
0 & 0 & 0 & 0 & * & * & *
\end{bmatrix} \tag{19}$$

The adoption of band matrices significantly reduces the computational cost to solve a system of linear equations. For this reason industrial finite element (FE) codes include advanced algorithms to optimize the global stiffness matrix. The ability to reduce the sparse global stiffness matrix to a band matrix represents one of the strengths of FEA over other numerical methods.

Isoparametric Elements

Isoparametric elements can have either straight or curvilinear boundaries, making them useful for the discretization of complex geometries with a limited number of elements. Isoparametric elements are unique in that they utilize the same shape functions to interpolate nodal coordinates and displacements. Unfortunately, switching to an isoparametric element significantly increases computational requirements (Zienkiewicz and Irons, 1969).

Problem Solution

Equation 16 represents a linear system of equations with nodal displacements and nodal forces as unknowns. At free nodes under load, nodal displacements are unknown whilst the forces are known. At constrained nodes, the forces are unknown whilst the displacements are known. It is therefore possible to partition the stiffness matrix by dividing the equation containing the unknown displacements by the equation with unknown nodal forces.

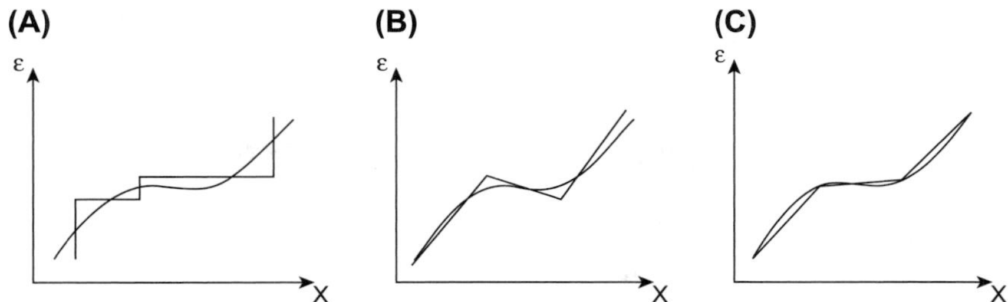

FIGURE I.1.4.3 Approximation of the deformation with higher order displacement functions. (A) Linear; (B) Quadratic; (C) Cubic.

Indicating with d_k and d_u the known and unknown nodal displacements vectors respectively, and with F_k and F_u the known and unknown nodal forces, Eq. 16 can then be rewritten as Eq. 20:

$$\begin{bmatrix} \bar{K}_{11} & \bar{K}_{12} \\ \bar{K}_{21} & \bar{K}_{22} \end{bmatrix} \cdot \frac{\vec{\bar{d}}_u}{\bar{d}_k} = \frac{\vec{\bar{F}}_k}{\bar{F}_u} \tag{20}$$

From which:

$$\bar{K}_{11} \cdot \bar{d}_u + \bar{K}_{12} \cdot \bar{d}_k = \tilde{F}_k \tag{21}$$

$$\bar{d}_u = \bar{K}_{11}^{-1} \left(\tilde{F}_k - \bar{K}_{12} \cdot \bar{d}_k \right) \tag{22}$$

and

$$\tilde{F}_u = \bar{K}_{21} \cdot \bar{d}_u + \bar{K}_{22} \cdot \bar{d}_k \tag{23}$$

Once the nodal displacements are known (Eq. 23), the global solution can be obtained by evaluating the displacement field, the deformations, and tensions on every element (Eqs. 4, 8, and 10) within the entire model.

Solution Accuracy and Convergence

The final result from FEA consists simply of nodal displacements, whereas deformations and tensions can be obtained by numerical derivation of the deformation field. Therefore, the deformations and tensions are inherently less precise than the displacements. Due to the reduction in DoFs, the solution to the elastic problem given by FEA is approximate. The number of DoFs for each finite element is a result of the reduced number of allowed deformations, which in turn are strongly related to the choice of shape function. For the triangular element example, the selection of a linear shape function produces a constant deformation and tension (Eq. 7). As a consequence the deformations and stress field within the model are approximates by step functions (Figure I.1.4.3a). The adoption of higher degree displacement functions (therefore with finite elements of higher complexity) enables for a given element size, a better approximation of the exact solution (Figure I.1.4.3b,c).

Another approach to increase the solution accuracy is to implement a finer mesh within the model, especially where high stress gradients are expected (stress concentrations, concentrated loads, etc.). However, as the number of elements is increased the size of the global stiffness matrix is also increased. The optimal number of elements within the mesh must be verified by performing a convergence test, comparing solutions from meshes of increasing density.

Model Preparation and Element Selection

Mesh generation is a fundamental step that deeply influences the quality of the result. Although general criteria cannot be provided, the following suggestion should be taken into consideration (George, 1991):

- Avoid using irregularly shaped elements. Highly deformed elements can lead to a poorly conditioned simulation, which can increase sensitivity to approximation errors;
- Appropriately refine the mesh at stress concentrations;
- Avoid the adoption of different elements types within the same model;
- Utilize both geometrical and load symmetries, when appropriate, to reduce the problem to a portion of the whole structure;
- In general, the overall precision of the results increases with the increase in the number of DoF per element.

BIOMECHANICS EXAMPLES

A Finite Element Analysis of Glenoid Replacement Prostheses in Normal and Rheumatoid Arthritic Bone

Among glenoid replacement post-operative complications, glenoid component loosening is the most reoccurring problem (Lacroix et al., 2000).

Glenoid prosthesis design can be grouped into two categories: "keeled" or "pegged". The first design is characterized by a tapered "fin" of a rectangular cross-section. The pegged design includes several circular pegs of various lengths inserted into the glenoid volume. A three-dimensional FE model including the entire scapula, cement fixation, and polyethylene keeled glenoid prostheses, was generated to investigate the hypothesis according to which keeled prostheses are performing better than pegged designs in patients affected by rheumatoid arthritis. Quantitative computed tomography (qCT) and experimental driven muscle loading were adopted to recapitulate bone geometry, material property, and a realistic mechanical loading scenario. Three-dimensional brick and wedge elements were used to generate the bone, the cement, the pegged (8148 elements) and keeled (8411 elements) prostheses meshes (Figure I.1.4.4).

Results showed that the stresses in the polyethylene components for both designs were not high enough (~2 MPa) to cause failure. In contrast, twisting of the components was shown to occur, and was more prominent in the pegged prosthesis. More importantly, the analysis demonstrated a clear distinction in the prostheses' performance depending on the patient's bone quality. The starting hypothesis derived from the clinical experience was confirmed: in normal bone keeled design (94% and 68% probability to survive for the pegged and keeled prostheses, respectively) has a larger volume of highly stressed cement, compared to the pegged design (Figure I.1.4.5A). In simulated rheumatoid arthritic bone the situation reverses. A limited fraction of cement around the keeled design (86% and 99% probability to survive for the pegged and keeled prostheses respectively) is highly stressed because the large majority of the load is carried by the cortical shell, whereas the pegged

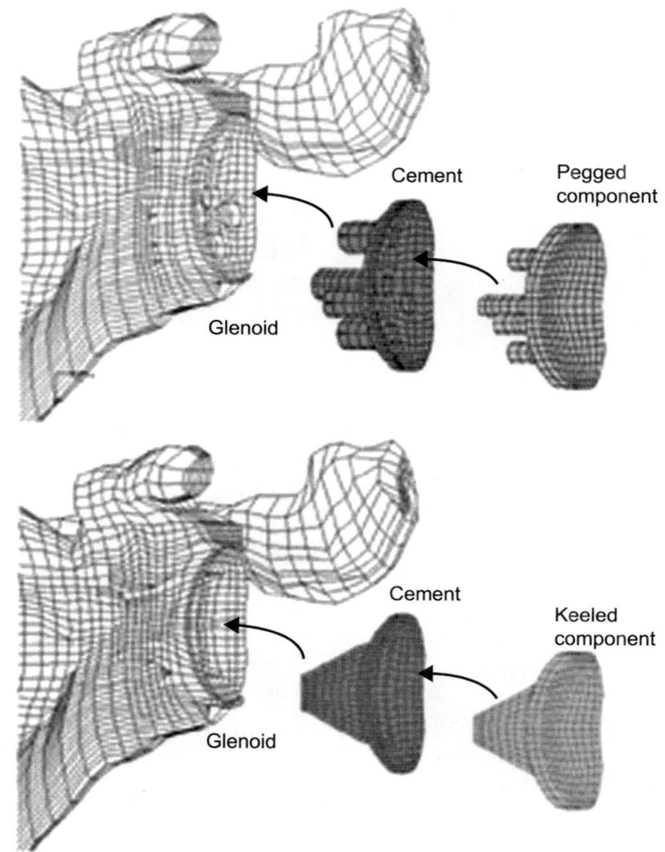

FIGURE I.1.4.4 Exploded view of the insertion of pegged and keeled prostheses.

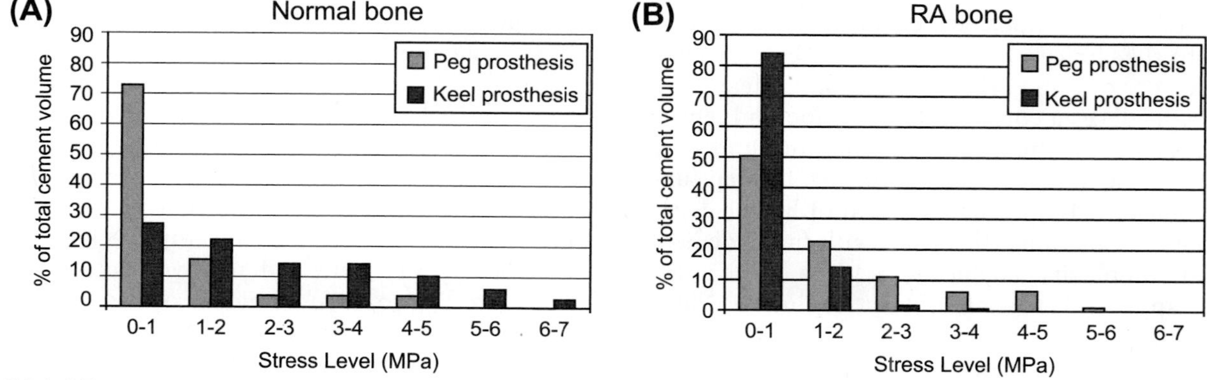

FIGURE I.1.4.5 Maximal principle stress distribution for the pegged and keeled prostheses in: (A) normal bone; and (B) rheumatoid bone.

prosthesis has a greater volume of highly stressed cement (Figure I.1.4.5B). This study demonstrates the importance of a holistic approach to the design of orthopedic prostheses in diverse patient populations.

Finite Element Analysis of a Glass Fiber Reinforced Composite Endodontic Post

In this work the mechanical behavior of a new polymeric composite endodontic post reinforced with glass fibers was analyzed through experimental and FE analysis (Pegoretti et al., 2002). Implant (Model D; Figure I.1.4.6) performance was compared with commercially available carbon fiber reinforced (Model C; Figure I.1.4.6) and gold alloy cast posts (Model B; Figure I.1.4.6). A natural tooth model (Model A; Figure I.1.4.6) restored with ideal materials, whose stiffness is equal to those of enamel and dentine, will be considered as a reference.

The main design target was to minimize the rigidity difference between the dentine and the post. The composite post consisted of unidirectional continuous E-glass fibers (Hybon® 2001 by PPG Industries) up to a volume fraction of 60% embedded in a matrix of acrylic resin, whose main component is 2,2-bis[4-(2-hydroxy-3-methacryloyloxypropoxy)-phenyl]-propane. Two-dimensional models of the vertical section of an upper central incisor were developed under the hypothesis of plane strain fields. Models were composed of 3500–4500 nodes and 3700–4400 quadrilateral and triangular elements. Elastic constants were empirically derived or selected from the literature. Von Mises equivalent stress and shear stress were identified as critical parameters for evaluation of the FE simulation results. Von Mises equivalent stress was chosen as an indicator of the average stress level. High stress values are associated with zones of higher probability of damage. Similarly, high shear stress can lead to failure of the tooth–post interface and to post detachment. Von Mises and shear stress maps were obtained for the four models under three different load configurations: a vertical load (100 N, see Figure I.1.4.7); an oblique load (50 N); and a horizontal load (10 N). FE results showed that the gold alloy cast post and core produced the greatest stress concentration at the post-dentine interface, whereas the carbon fiber-reinforced composite posts were characterized by high stresses in the cervical region due to increased flexibility and less stiff core material. Finally, the glass fiber composite, having a stiffness much closer to dentine, exhibited the lowest peak stresses within the root. The glass fiber composite post induced a stress field quite similar to that of the natural tooth. The FE analysis was crucial to identify the glass fiber composite post as the best design strategy, and to elucidate how its mechanics recapitulate the one of the natural tooth model.

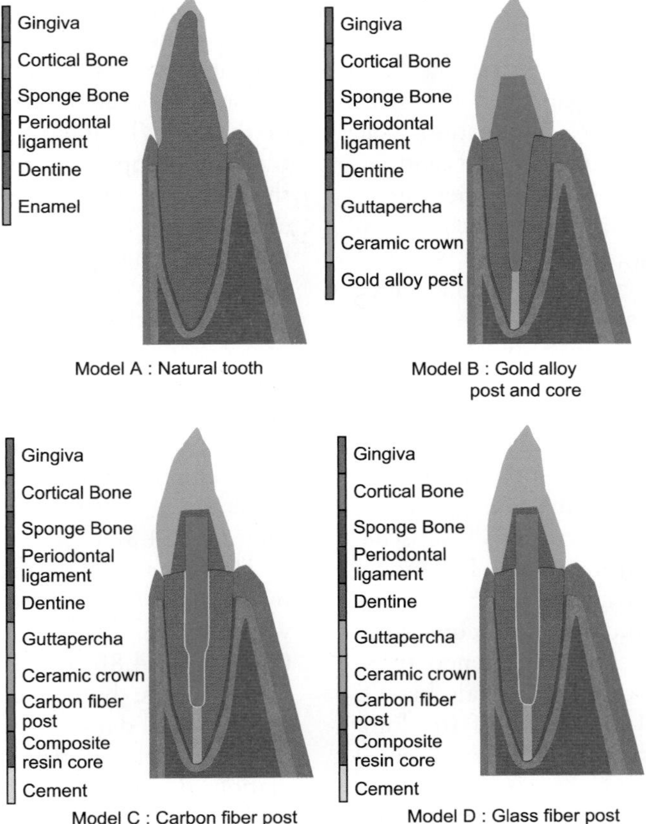

FIGURE I.1.4.6 Geometries and selected materials for the four investigated models.

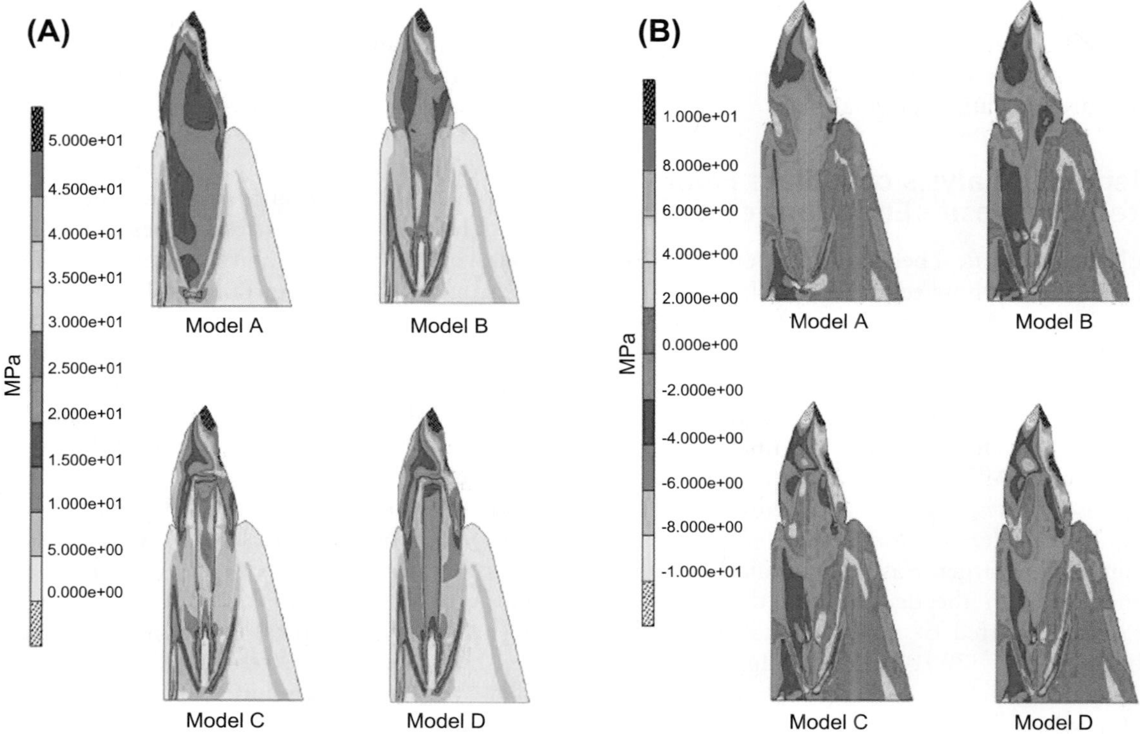

FIGURE I.1.4.7 Contour map of: (A) the von Mises stress; and (B) the shear stress in the case of a 100 N vertically applied load.

A Finite Element Multiscale Approach Methodology for Predicting Mechanical Behavior of Collagen Fibers

Most biological tissues serving a predominately mechanical function are based on collagen fiber networks (Stylianopoulos and Barocas, 2007). While the tissue characteristic functional length is of the order of centimeters, the underlying fiber network is on the micrometer scale. In a structural deterministic approach the material microstructure is described in a representative way and coupled to the macroscopic equations through a multiscale technique. A volume averaging approach replaces the material unit cell by a representative volume element (RVE). The RVE is most easily described as a heterogeneous medium where global continuum fields are locally applied and the macroscopic quantities are then derived from the volume average of the corresponding microscopic quantities. In this work a volume averaged structural model is adopted to study the inherently multiscale mechanical behavior of collagen networks. The model was formulated in two scales: the macroscopic scale that represents the functional level; and the microscopic scale that duplicates the structural level of collagen networks. The macroscopic scale was modeled by a Galerkin FE model with 720 nodes and 2160 degrees of freedom. The microscopic scale, composing the RVE, was represented by an ideal collagen network consisting of struts and nodes. Each of the 3872 RVEs contained roughly 330 cross-links, totaling almost four million degrees of freedom. Within each RVE, fibers were assumed to obey an exponential stress–strain law neglecting bending stiffness, but presumed to stretch, compress, and rotate at cross-links. The solution algorithm can be summarized in five steps. (1) The macroscopic problem is posed via suitable boundary conditions; (2) the deformation of the RVE boundary of each RVE is determined by the macroscopic deformation field (down-scaling); (3) the force balance at each RVE is solved; (4) the averaged Cauchy stress tensor is calculated (up-scaling) by volume-averaging theory; and (5) the macroscopic force balance is solved. The algorithm iterates until convergence is reached. The macroscopic problem consisted of a rectangular slab of collagen under uniaxial extension. Two different scenarios were studied: (1) a homogeneous slab and a heterogeneous slab composed of two networks of similar density but different orientation; (2) a network with a preferred alignment along the direction of extension and a nearly isotropic network (Figure I.1.4.8A).

The accuracy of the modeling approach was addressed by comparison to experimental data. The simulation was able to quantitatively assess the Poisson ratios (Figures I.1.4.8B and I.1.4.9) and engineering stress–strain curves; both were in good agreement with the experimental data.

The model was also able to predict microscale deformations. Orientation parameters equal to 1/3 for isotropic network and 1 for perfectly aligned network changed according to the local deformation field (isotropic network

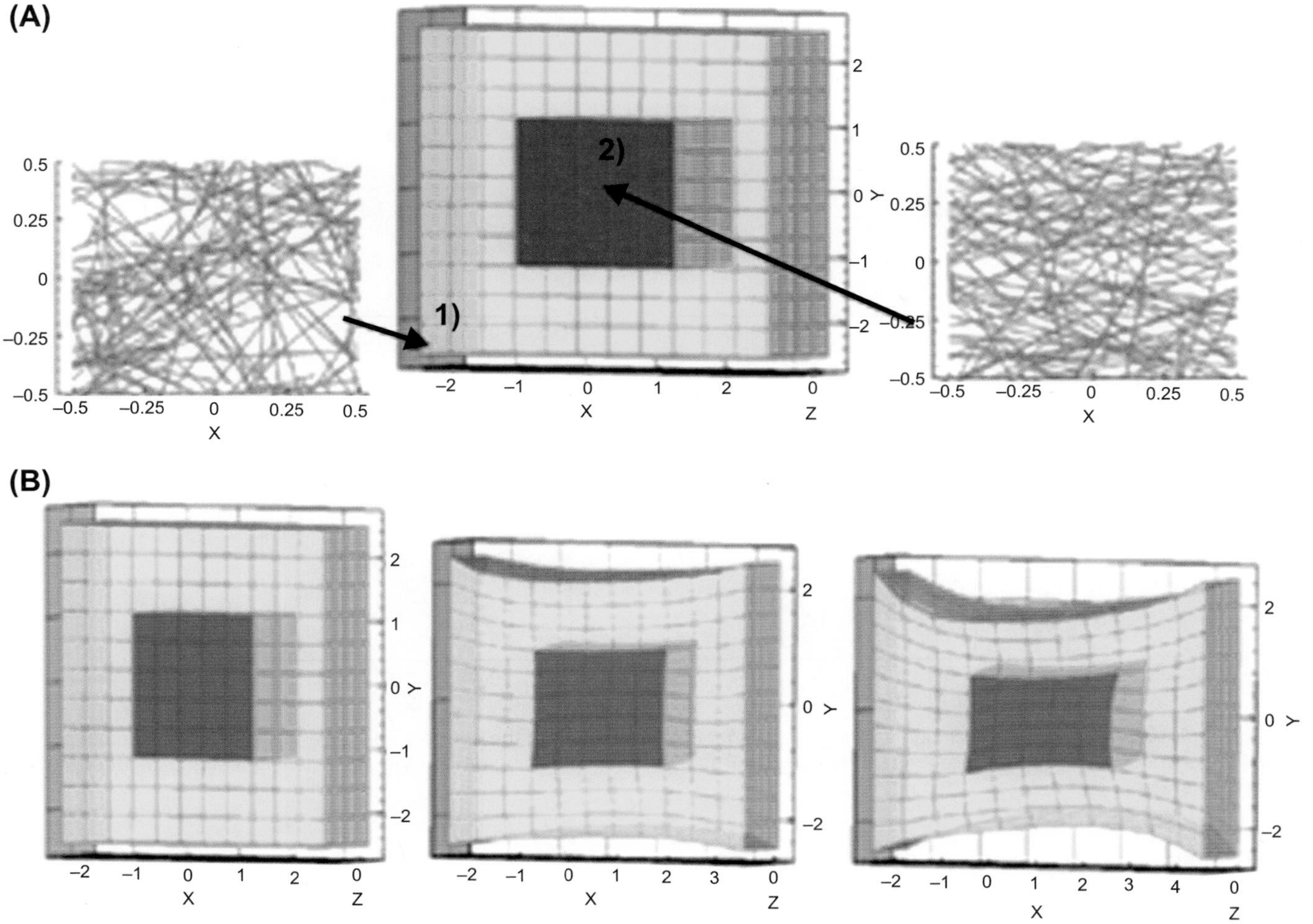

FIGURE I.1.4.8 (A) Heterogeneous collagen slab composed of two different microstructures. The central region has a preferred fiber alignment, while the surrounding region has nearly isotropic orientation. (B) Initial and deformed states of a collagen slab with heterogeneous microstructure. The picture on the left shows the initial state, the picture at the center shows the slab at 20% strain, and the picture on the right shows the slab at 40% strain.

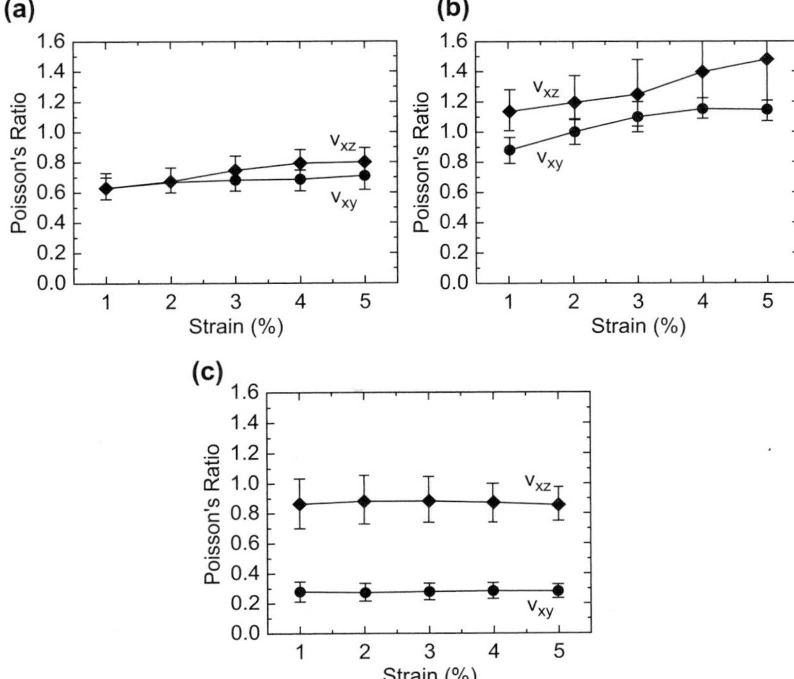

FIGURE I.1.4.9 Poisson's ratios versus strain for networks with different orientations. (a) Networks with nearly random orientation. (b) Networks with preferred alignment along the direction of stretching. (c) Networks with preferred alignment perpendicular to the direction of stretching.

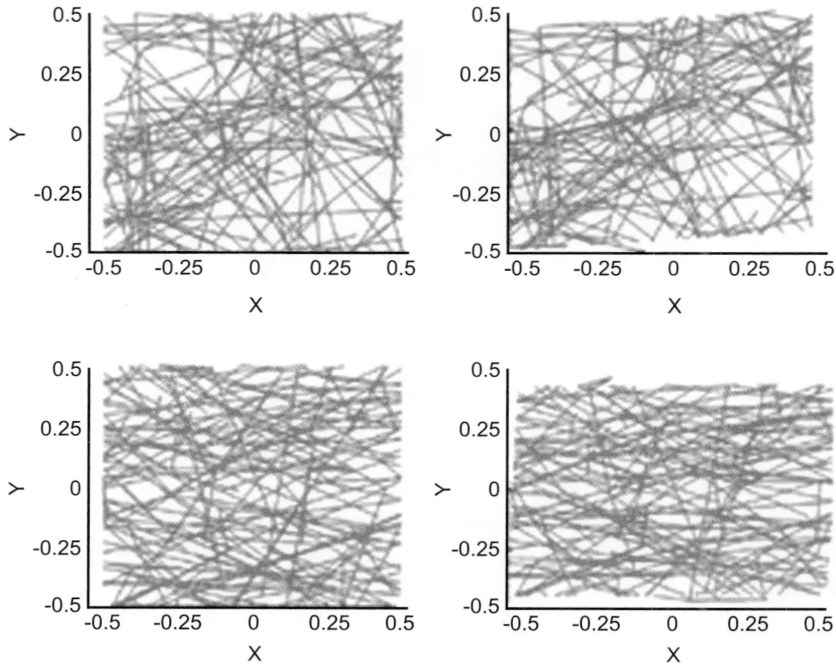

FIGURE I.1.4.10 Initial (left) and final (right) microstructures for RVEs selected from the isotropic (central) and aligned (surrounding) portions of the heterogeneous slab. Specific locations are as shown in Figure I.1.4.8A. Stretching of the RVEs and resulting rearrangement of the fiber network are evident.

from 0.35338 to 0.38189, aligned network from 0.54856 to 0.5830) indicating that the fibers realign along the direction of extension (Figure I.1.4.10). This work shows the great potential of a multiscale FE approach to both predict and link mechanical responses across different length scales.

A Three-Dimensional Numerical Simulation of Flow Through Mechanical Heart Valves

Mechanical heart valves have traditionally represented the gold standard for valvular replacement therapies (Schoen, 2005). Unfortunately, due to the possibility of thromboembolic complications, recipients require lifelong anticoagulation medication. This susceptibility is typically attributed to altered hemodynamics, specifically increased blood shear stress and turbulence. Therefore, developing an understanding of the complex and time varying mechanical forces is key to the development of a successful heart valve replacement therapy (Ge et al., 2003; Govindarajan et al., 2010). In the work presented, a three-dimensional numeric method is utilized to study a fixed bileaflet valve closely resembling the St. Jude Standard Mechanical. The model assumes a time-dependent Navier–Stokes formulation of an incompressible Newtonian fluid. The rigid aorta and leaflets are separately meshed with increasing resolution and superimposed without requiring one to one node correspondence (Figure I.1.4.11). Total node number was increased from 4×10^5 to 1.6×10^6. Steady inflow with sub-physiologic Reynolds numbers increasing from 500 to 1200 was applied.

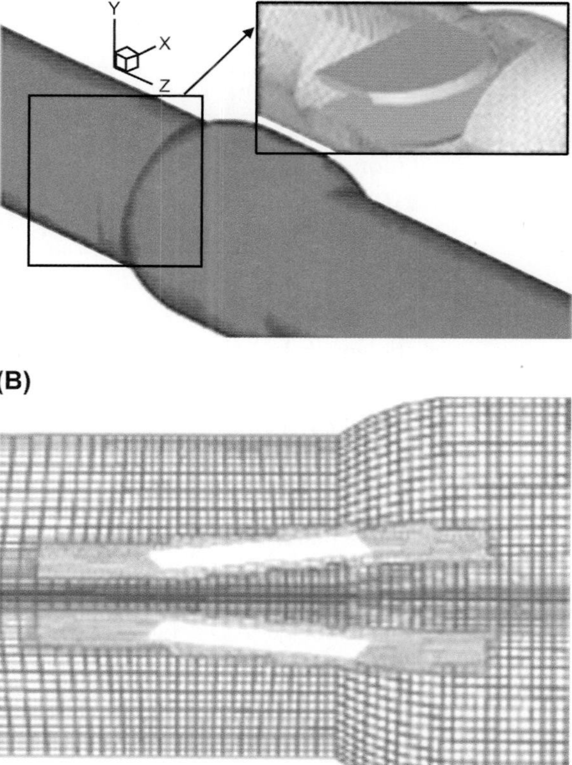

FIGURE I.1.4.11 Geometry and coordinate definitions for the finite element mesh showing: (A) the symmetric aorta, with inset of the bileaflet valve; and (B) *Y–Z* cross-section showing the overlaid leaflet meshes.

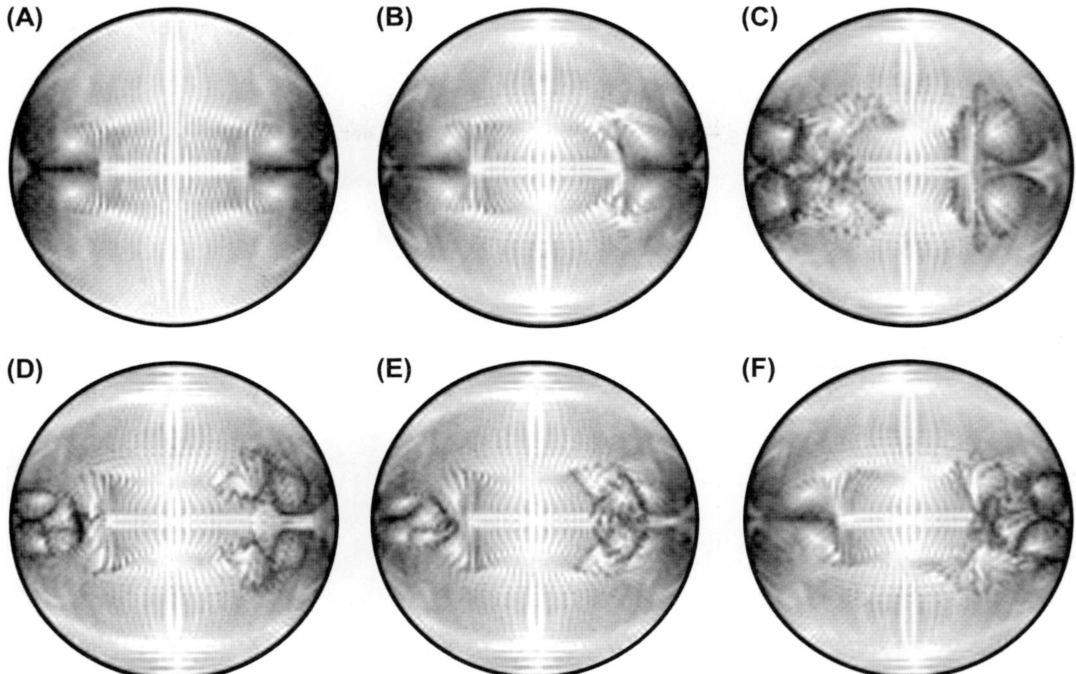

FIGURE I.1.4.12 Time course of the transverse velocity vectors at a section downstream of the valve showing the break in symmetry with impulsive acceleration from Re 1000 to Re 1200. (A) t = 0.5; (B) t = 8.4; (C) t = 8.6; (D) t = 9.2; (E) t = 9.6; and (F) t = 29.

Results demonstrated two pairs of counter-rotating vortices originating on either side of the valve orifice on the leading edge of the leaflets. These vortices increased with Reynolds number, and even at a Reynolds number of 1200 complex interactions between the vortices led to the onset of unsteady flow and a break in both planes of symmetry (Figure I.1.4.12). While it is not possible to draw any specific conclusions regarding asymmetric flow fields in a pulsatile flow regime, these results strongly suggest the importance of fully three-dimensional simulations, especially when considering higher, physiologic Reynolds numbers. During a mesh refinement study, inadequate mesh resolution was demonstrated to result in unsteady, physically unrealistic flow fields. However, the recent dramatic increase in computational power will likely enable the analysis of systems with suitably high resolution. While the techniques developed in this paper have been applied exclusively to mechanical prostheses, many of the computational developments designed to handle unsteady flows and complex geometric configurations in mechanical valves are also directly relevant to increasingly prevalent bioprosthetic valves. Essentially, a computational fluid dynamics approach can provide insight into the *in vivo* performance of prosthetic heart valves in general; however, as with any computational model of a complex system, the results must be carefully reviewed and validated.

Simulated Bioprosthetic Heart Valve Biomaterials under Quasi-Static Loading

Bioprosthetic heart valves (BHV) typically fail structurally, due to poor tissue durability and calcification (Sacks and Schoen, 2002). Clearly, an in-depth understanding of the biomechanical behavior of the BHV at both the tissue and functional prosthesis levels is essential to improving BHV design and reducing failure rates. Evaluation and simulation of the multiaxial mechanical behavior of BHV tissues is becoming more prevalent, yet testing methods have not been standardized and methodologies vary widely. As it is not possible to experimentally evaluate the effects of different biaxial test boundary conditions on test specimen internal stress distributions, numerical simulations were conducted to explore these effects (Sun et al., 2005). A nonlinear Fung-elastic constitutive model (Sun and Sacks, 2005), which fully incorporated the effects of in-plane shear, was used to simulate soft tissue mechanical behavior. Effects of boundary conditions, including varying the number of suture attachments, different gripping methods, specimen shapes, and material axes orientations were examined. Results demonstrated strong boundary effects with the clamped methods, while suture attachment methods demonstrated minimal boundary effects (Figure I.1.4.13). Suture-based methods appeared to be best suited for biaxial mechanical tests of biological materials. Moreover, the simulations demonstrated that St. Venant's effects depended significantly on the material axes orientation. While not exhaustive, these comprehensive simulations provide experimentalists with additional insight into the stress–strain fields associated with different biaxial testing boundary conditions, and may be used as a rational basis for the design of testing.

Next, these findings were applied to an actual pericardial BHV under quasi-static loading under 40, 80, and 120 mmHg pressures. Rigorous experimental validation of predicted leaflet strain field was used to validate the model

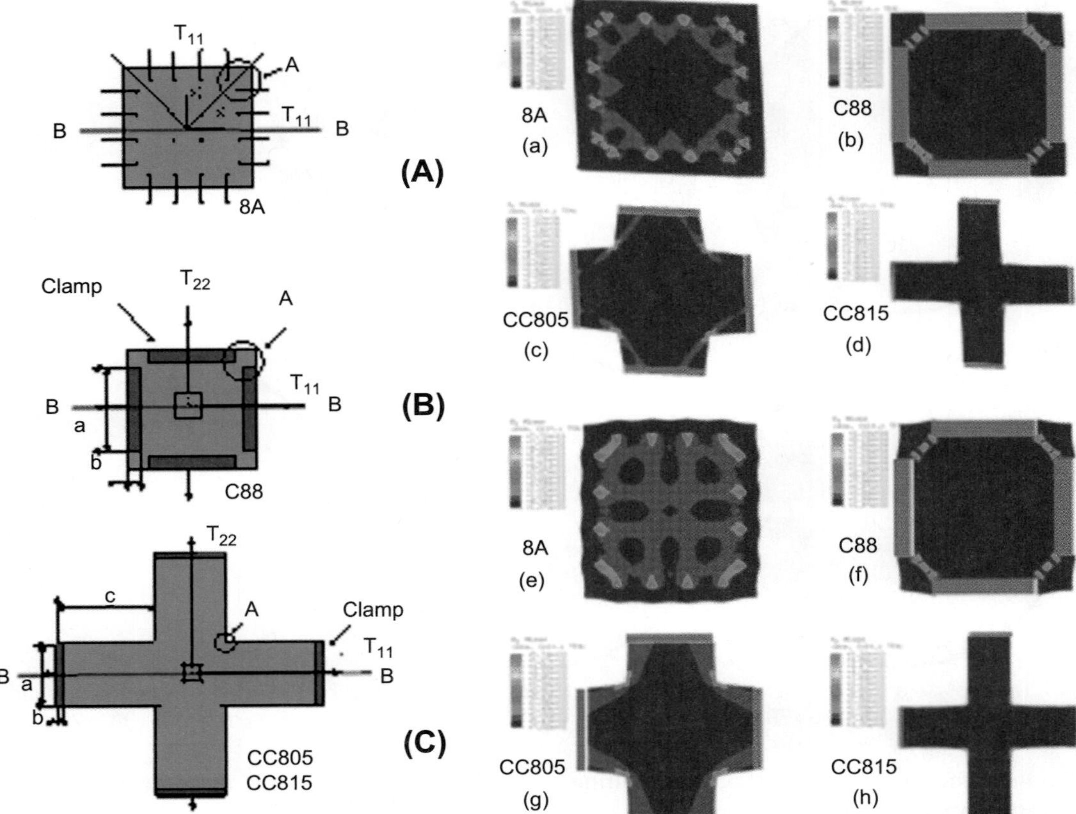

FIGURE I.1.4.13 (A) A schematic of the biaxial testing setup with four suture attachments, indicating the material axes X1–X2; (B) biaxial testing setup with gripping method as using clamp on each side of a square sample; and (C) biaxial testing setup with clamps on each side of a cruciform sample. Von Mises stresses and deformations of FE models for different gripping methods at (a–d) 45° material axes; (e–h) at 0° material axes.

results. The overall leaflet shape when loaded agreed well with experimental observations (Figure I.1.4.14). The predicted major principal strain field results demonstrated a generally smooth spatial distribution, increasing with transvalvular pressure (Figures I.1.4.14c,d).

Over most of the leaflet surface, the major principal strain values at 40 mmHg were ~8%, with focal regions of high values of ~12% located in the right and left sides just below the commissure regions. At 120 mmHg transvalvular pressure, the corresponding major principal strains increased to 12%–14%, with the focal regions' major principal strain values increasing to ~17%. Parametric studies utilizing the material parameter from one leaflet for all three pressures resulted in substantial variations in leaflet stress and strain distributions (Figure I.1.4.15) using a degree of anisotropy $DA = E_{11}/E_{22}$, where E_{11} and E_{22} are the peak strains under 1 MPa equi-biaxial stress. Note the high stresses were also observed in the belly region at all three DA values, appearing in similar locations for the aortic side. For the ventricular side of the leaflet surface, the peak stresses were slightly higher than that of the aortic side when DAs were 0.83 and 2.12; all occurring in the commissure region (Figure I.1.4.15). When DA was 1.14, a large region of high stress was concentrated in the belly region. The free edge experienced the lowest amounts of stress on the ventricular side surface, consistent with the stress levels determined at the free edge on the aortic side. This is most likely due to the contact of the leaflets that lead to the compression stress at the free edge. This result suggests that utilization of actual leaflet material properties is essential for accurate BHV FE simulations, and underscores the need for rigorous experimentation and accurate constitutive models in simulating BHV function and design.

CONCLUSION

The FE methods briefly illustrated in this chapter are applicable to a large variety of contemporary biomedical problems. As with all modeling efforts, the most important part is often focused on model development and its ability to reproduce the salient physical characteristics of interest. While established, there are many new developments in the ability of FE models to simulate increasingly complex biomedical problems, and these will increasingly be used as problems with prosthetic design and biomaterials development become increasingly sophisticated. Perhaps the most exciting aspect is the ability to reduce the often empirical nature of biomaterials development and allow novel biomaterials to be developed in a rational manner for targeted use in novel biomedical therapies.

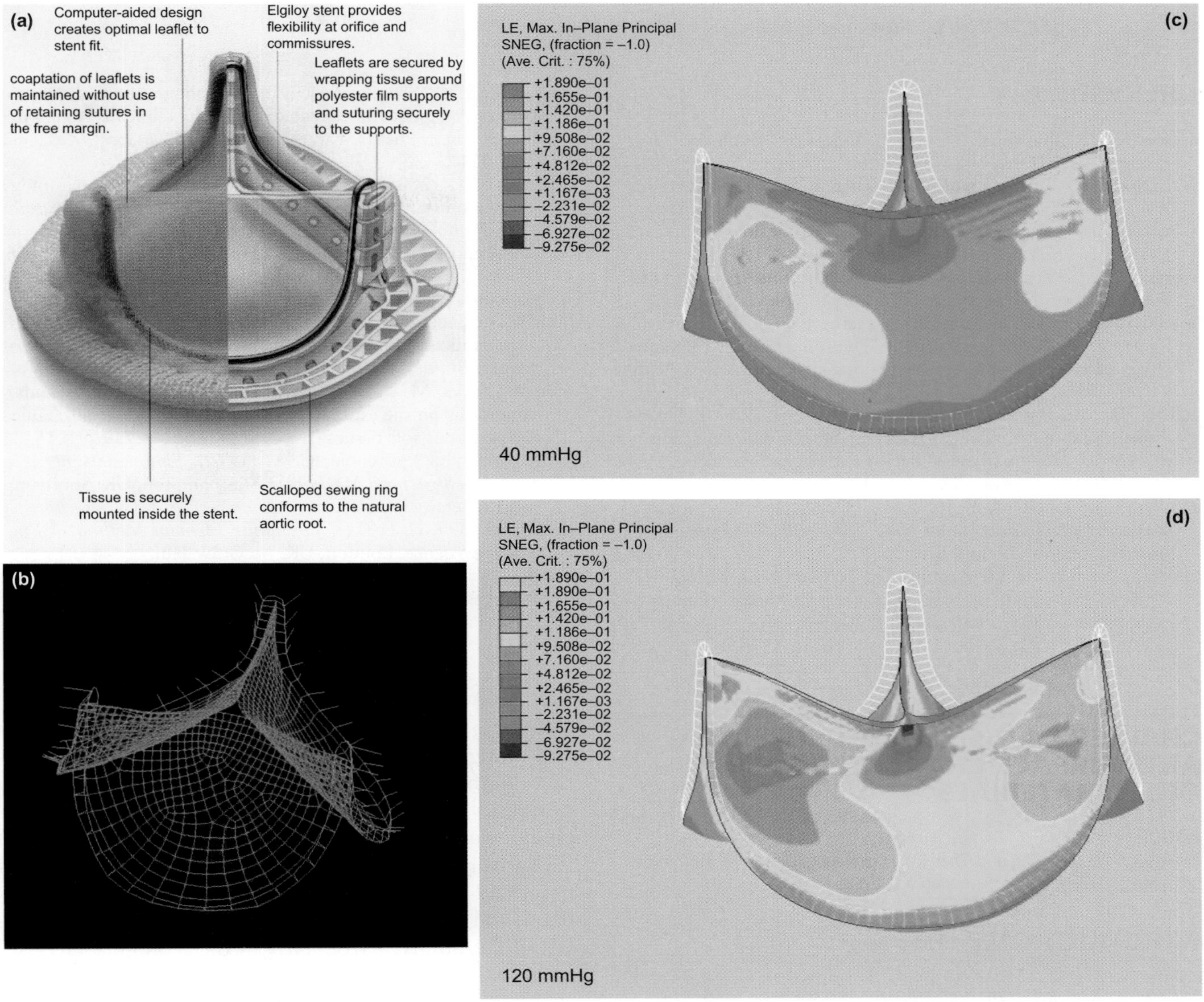

(a)

Computer-aided design creates optimal leaflet to stent fit.

Elgiloy stent provides flexibility at orifice and commissures.

coaptation of leaflets is maintained without use of retaining sutures in the free margin.

Leaflets are secured by wrapping tissue around polyester film supports and suturing securely to the supports.

Tissue is securely mounted inside the stent.

Scalloped sewing ring conforms to the natural aortic root.

(b)

(c)

LE, Max. In–Plane Principal
SNEG, (fraction = –1.0)
(Ave. Crit. : 75%)
+1.890e–01
+1.655e–01
+1.420e–01
+1.186e–01
+9.508e–02
+7.160e–02
+4.812e–02
+2.465e–02
+1.167e–03
–2.231e–02
–4.579e–02
–6.927e–02
–9.275e–02

40 mmHg

(d)

LE, Max. In–Plane Principal
SNEG, (fraction = –1.0)
(Ave. Crit. : 75%)
+1.890e–01
+1.890e–01
+1.655e–01
+1.420e–01
+1.186e–01
+9.508e–02
+7.160e–02
+4.812e–02
+2.465e–02
+1.167e–03
–2.231e–02
–4.579e–02
–6.927e–02
–9.275e–02

120 mmHg

FIGURE I.1.4.14 (a) Major components of a bioprosthetic heart valve; and (b) FE mesh (shown in simplified form). The resulting maximum in-plane principal strain magnitude (plotted using the same color fringe scale) for pressure levels of 40 mmHg and 120 mmHg (c, d), showed an asymmetrical pattern even though the valve geometry was highly symmetric.

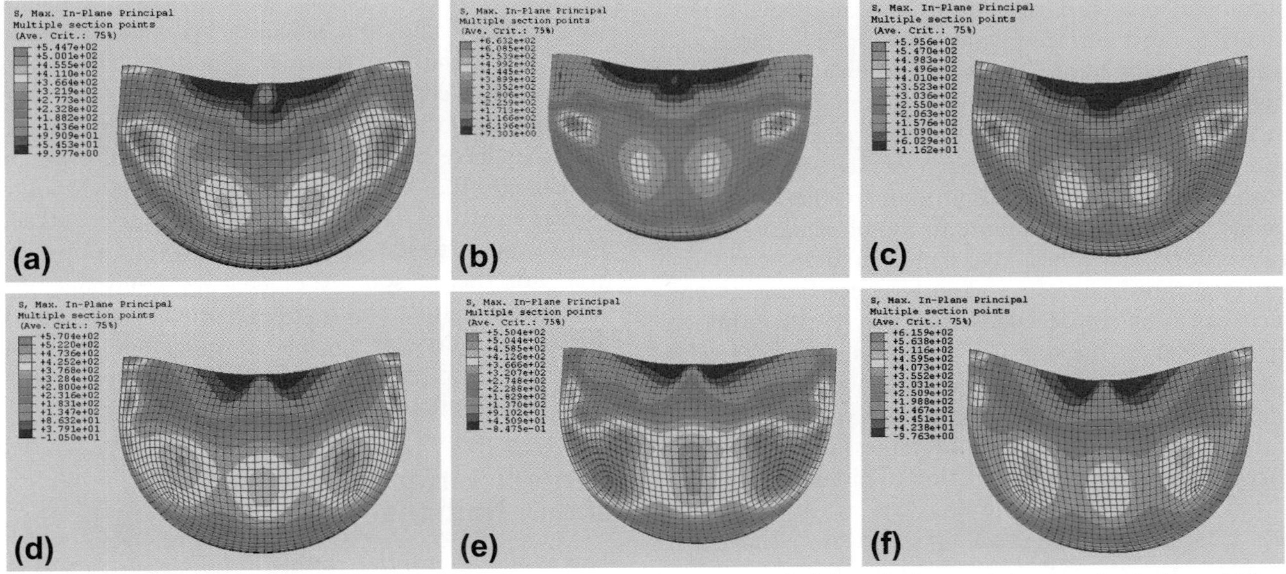

FIGURE I.1.4.15 FE results for the simulations with the same material properties of all three leaflets and degrees of anisotropy (DA) of: (a) 0.83; (b) 1.14; and (c) 2.12 on the aortic side of leaflet surface, and on the ventricular side for the same simulations with the DA of (d) 0.83; (e) 1.14; and (f) 2.12.

BIBLIOGRAPHY

Bicanic, N., & Johnson, K. W. (1979). Who was "Raphson?" *Int. J. Num. Meth. Eng.*, 148–152.

Ge, L., Jones, S. C., Sotiropoulos, F., Healy, T. M., & Yoganathan, A. P. (2003). Numerical simulation of flow in mechanical heart valves: Grid resolution and the assumption of flow symmetry. *J. Biomech. Eng.*, **125**, 709–718.

George, P. L. (1991). *Automatic Mesh Generation: Application to Finite Element Methods*. New York, NY: Wiley.

Govindarajan, V., Udaykumar, H. S., Herbertson, L. H., Deutsch, S., Manning, K. B., et al. (2010). Two-dimensional FSI simulation of closing dynamics of a tilting disc mechanical heart valve. *J. Med. Device*, **4**, 11001.

Lacroix, D., Murphy, L. A., & Prendergast, P. J. (2000). Three-dimensional finite element analysis of glenoid replacement prostheses: A comparison of keeled and pegged anchorage systems. *J. Biomech. Eng.*, **122**, 430–436.

Middleton, J., Jones, M. L., & Pande, G. N. (1996). *Computer Methods in Biomechanics and Biomedical Engineering*. Amsterdam: Gordon and Breach.

Pegoretti, A., Fambri, L., Zappini, G., & Bianchetti, M. (2002). Finite element analysis of a glass fibre reinforced composite endodontic post. *Biomaterials*, **23**, 2667–2682.

Sacks, M. S., & Schoen, F. J. (2002). Collagen fiber disruption occurs independent of calcification in clinically explanted bioprosthetic heart valves. *J. Biomed. Mater. Res.*, **62**, 359–371.

Schoen, F. J. (2005). Cardiac valves and valvular pathology: Update on function, disease, repair, and replacement. *Cardiovasc. Pathol.*, **14**, 189–194.

Stylianopoulos, T., & Barocas, V. H. (2007). Volume-averaging theory for the study of the mechanics of collagen networks. *Comput. Method. in Applied M.*, **196**, 2981–2990.

Sun, W., & Sacks, M. S. (2005). Finite element implementation of a generalized Fung-elastic constitutive model for planar soft tissues. *Biomech. Model Mechanobiol.*, **4**(2–3), 190–199.

Sun, W., Sacks, M. S., & Scott, M. J. (2005). Effects of boundary conditions on the estimation of the planar biaxial mechanical properties of soft tissues. *J. Biomech. Eng.*, **127**, 709–715.

Tottenham, H., & Brebbia, C. A. (1970). *Finite Element Techniques in Structural Mechanics*. Southampton, UK: Southampton University Press.

Zienkiewicz, O. C. (1971). *The Finite Element Method in Engineering Science*. London, UK: McGraw-Hill.

Zienkiewicz, O. C., & Irons, B. M. (1969). Finite element methods in stress analysis. In I. Holand, & K. Bell (Eds.), Trondheim Tapir Press.

CHAPTER I.1.5 SURFACE PROPERTIES AND SURFACE CHARACTERIZATION OF BIOMATERIALS

Buddy D. Ratner

Professor, Bioengineering and Chemical Engineering, Director of University of Washington Engineered Biomaterials (UWEB), Seattle, WA, USA

INTRODUCTION

> *Nothing is rich but the inexhaustible wealth of nature. She shows us only surfaces, but she is a million fathoms deep.*
>
> ***Ralph Waldo Emerson***

Biomaterials "show" to the world (and the biological environment) only surfaces. Atoms and molecules make up the outermost surface of a biomaterial (the interface between the material and the world). As we shall discuss in this section, these atoms and molecules that reside at the surface have a special organization and reactivity. They require special methods to characterize them, novel methods to tailor them, and they drive many of the biological reactions that occur in response to the biomaterial (protein adsorption, cell adhesion, cell growth, blood compatibility, etc.). The importance of surfaces for biomaterials science has been appreciated since the 1960s. Almost every biomaterials meeting will have sessions addressing surfaces and interfaces. In this chapter we focus on the special properties of surfaces, definitions of terms, methods to characterize surfaces, and some implications of surfaces as the drivers of bioreaction to biomaterials.

> Atoms and molecules that reside at the surface of a biomaterial have special reactivity and direct biological response.

During the development of biomedical implant devices and materials, we are concerned with physical properties, durability, and biocompatibility. The understanding of physical properties (e.g., mechanical strength, permeability, elasticity) has evolved over hundreds of years and is relatively well-understood – the standard tools of engineers and materials scientists are appropriate to characterize and study physical properties (see Chapters I.1.2, I.1.3). Durability, particularly in the biological environment, is less well-understood. Still, the tests we need to evaluate durability are clear (see Chapters I.1.2, II.4.3, II.4.4). Biocompatibility represents a frontier of knowledge in this field, and its study is often assigned to the biochemist, biologist, and physician (see Chapters II.3.2, II.3.3, II.3.4). However, an important question in biocompatibility is how the device or material "transduces" its structural make-up to direct or influence the response of proteins, cells, and the organism to it. For devices and materials that do not leach undesirable substances (i.e., that have passed routine toxicological evaluation; see Chapter II.3.3), this transduction occurs through the surface structure. The body "reads" the surface structure and responds to the particular chemistry and organization. For this reason we must understand the surface structure of biomaterials, and thus there is an important role for the physical scientist in understanding surface-driven biointeractions. Chapter III.1.4 elaborates on the biological implications of this idea.

General Surface Considerations and Definitions

This is the appropriate point to highlight general ideas about surfaces, especially solid surfaces. First, the surface

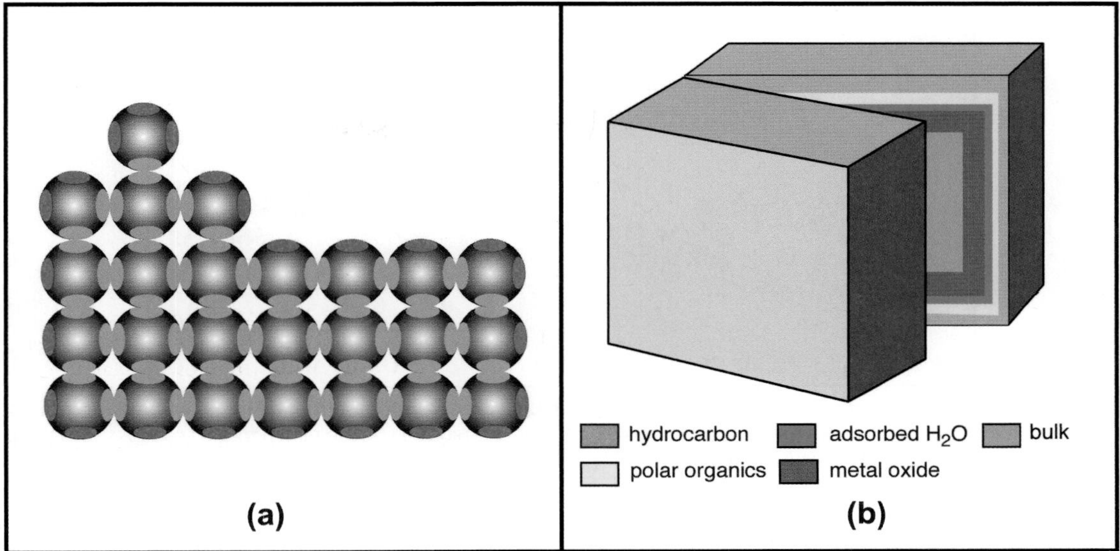

FIGURE I.1.5.1 (a) A two-dimensional representation of a crystal lattice suggesting bonding orbitals (red or pink ovals). For atoms in the center (bulk) of the crystal (pink ovals), all binding sites have associations with those of other atoms (sharing electrons). At planar exterior surfaces, one of the bonding sites is unfulfilled (red oval). At corners, two bonding sites are unfulfilled. The single atom on top of the crystal (an adatom) has three unfulfilled valencies. Energy is minimized where more of these unfulfilled valencies can interact. Where interactions do not satisfy all binding orbitals, there is an asymmetric electrical field driving a surface dipole. (b) In a "real-world" material (a block of metal from an orthopedic device, for example), if we cleave the block (under ultrahigh vacuum to prevent recontamination), we should find hydrocarbon on the outermost layer (perhaps 3 nm, surface energy ~22 ergs/cm^2), polar organic molecules (>1 nm, surface energy ~45 ergs/cm^2), adsorbed water (<1 nm, surface energy ~72 ergs/cm^2), metal oxide (approximately 5 nm, surface energy ~200 ergs/cm^2), and finally, the uniform bulk interior (surface energy ~1000 ergs/cm^2). The interface between air and material has the lowest interfacial energy (~22 ergs/cm^2). The layers are not drawn to scale.

region of a material is known to be of unique reactivity (Figure I.1.5.1a). Catalysis (for example, as used in petrochemical processing) and microelectronics both capitalize on special *surface* organization and reactivity – biology also exploits this enhanced surface reactivity to do its work. Second, the surface of a material is inevitably different from the bulk. The traditional techniques used to analyze the bulk structure of materials are not suitable for surface determination, because they typically do not have the sensitivity to observe the small amount of material comprising the unique surface chemistry/structure. Third, there is not much total mass of material at a surface. An example may help us to appreciate this – on a 1 cm^3 cube of titanium, the 100 Å oxide surrounding the cube is in the same proportion as a 5 meter-wide beach on each coast of the United States is to the roughly 5,000,000 meters distance from coast to coast. Fourth, surfaces readily contaminate with components from the vapor phase (some common examples are hydrocarbons, silicones, sulfur compounds, iodine). Under ultrahigh vacuum conditions (pressures $<10^{-7}$ Pa) we can retard this contamination. However, in view of the atmospheric pressure conditions under which all biomedical devices are used, we must learn to live with some contamination. The key questions here are whether we can make devices with controlled and acceptable levels of contamination and also avoid undesirable contaminants. This is critical so that a laboratory experiment on a biomaterial generates the same results when repeated after 1 day, 1 week or 1 year, and so that the biomedical device is dependable and has a reasonable shelf-life. Finally, the surface structure of a material is often mobile. A modern view of what might be seen at the surface of a real-world material is illustrated in Figure I.1.5.1b.

Five Points About Surfaces

1. Surfaces have unique reactivity
2. The surface is inevitably different from the bulk
3. The mass of material that makes up the surface zone is very small
4. Surfaces readily contaminate
5. Surface molecules can exhibit considerable mobility.

The movement of atoms and molecules near the surface in response to the outside environment is often highly significant. In response to a hydrophobic environment (e.g., air), more hydrophobic (lower energy) components may migrate to the surface of a material – a process that reduces interfacial energy (Figure I.1.5.1b). Responding to an aqueous environment, the surface may reverse its structure and point polar (hydrophilic) groups outward to interact with the polar water molecules. Again, energy minimization drives this process. An example of this is schematically illustrated in Figure I.1.5.2. For metal alloys, one metal tends to dominate the surface, for example, silver in a silver–gold alloy or chromium in stainless steel.

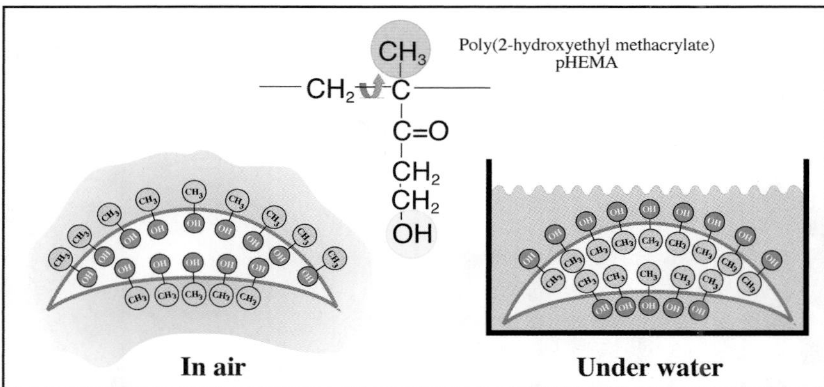

FIGURE I.1.5.2 Many materials can undergo a reversal of surface structure when transferred from air into a water environment. In this schematic illustration, a hydroxylated polymer (for example a pHEMA contact lens) exhibits a surface rich in methyl groups (from the polymer chain backbone) in air, and a surface rich in hydroxyl groups under water. This has been observed experimentally, see Ratner et al. (1978). *J Appl Polym Sci*, 22, 643; Chen et al. (1999). *J Am Chem Soc*, 121(2); 446.

The nature of surfaces is complex and is the subject of much independent investigation. The reader is referred to one of many excellent monographs on this important subject for a complete and rigorous introduction (see Somorjai, 1981; Andrade, 1985; Adamson and Gast, 1997; Garbassi et al., 1998; Somorjai and Li, 2010). Biosurfaces, with particular relevance for biomaterials, are reviewed in articles and books (Ratner, 1988; Castner and Ratner, 2002; Kasemo, 2002; Tirrell et al., 2002).

When we say "surface," a question that immediately comes to mind is: "how deep into the material does it extend?" Although formal definitions are available, for all practical purposes, the surface is the zone where the structure and composition, influenced by the interface, differs from the average (bulk) composition and structure. This value often scales with the size of the molecules making up the surface. For an "atomic" material, for example gold, after penetration of about 5 atomic layers (0.5–1 nm), the composition becomes uniform from layer to layer (i.e., you are seeing the bulk structure). At the outermost atomic layer, the organization of the gold atoms at the surface (and their reactivity) can be substantially different from the organization in the averaged bulk. The gold, in air, will always have a contaminant overlayer, largely hydrocarbon, which may be roughly 2 nm thick. There is also a difference in composition between bulk and surface, but it is not the atomic/molecular rearrangements we are discussing here. For a polymer, the unique surface zone may extend from 10 nm to 100 nm (depending on the polymeric system and the chain molecular weight). Figure I.1.5.1b addresses some of these issues about surface definitions. Two more definitions must be considered. An interface is the transition between two phases, in principle an infinitely thin separation plane. An interphase is the unique compositional zone between two phases. For the example, for gold, we might say that the interphase between gold and air is 3 nm thick (the structurally rearranged gold atoms plus the contaminant layer).

What Surface Properties are we Interested in?

A surface is fully described by many parameters (Figure I.1.5.3). The more of these parameters we measure, the more we can piece together a complete description of the surface. A complete characterization requires a cadre of techniques to examine the many facets that contribute to the surface properties. Unfortunately, we cannot yet specify which parameters are most important for understanding specific biological responses to surfaces. Studies have been published on the importance of roughness, patterns, wettability, surface mobility, chemical composition, electrical charge, crystallinity, modulus and heterogeneity to biological reaction. Since we cannot be certain which surface factors are predominant in each situation, the controlling variable or variables must be independently measured and correlated. We use surface analysis techniques to measure these surface properties.

SURFACE ANALYSIS TECHNIQUES: PRINCIPLES AND METHODS

Sample Preparation

A guiding principle of surface analysis sample preparation is that the sample should resemble, as closely as possible, the material or device in the form that it is used for biological testing or implantation. Needless to say, fingerprints on the surface of the sample will obscure surface properties of interest. If the sample is placed in a package for shipping or storage prior to surface analysis, it is critical to know whether the packaging material may deliver surface contamination. Plain paper in contact with most specimens will transfer contaminants (often metal ions) to the surface of the material. Many plastics are processed with silicone oils or other additives that can migrate to the specimen. The packaging material used should be examined by surface analysis methods

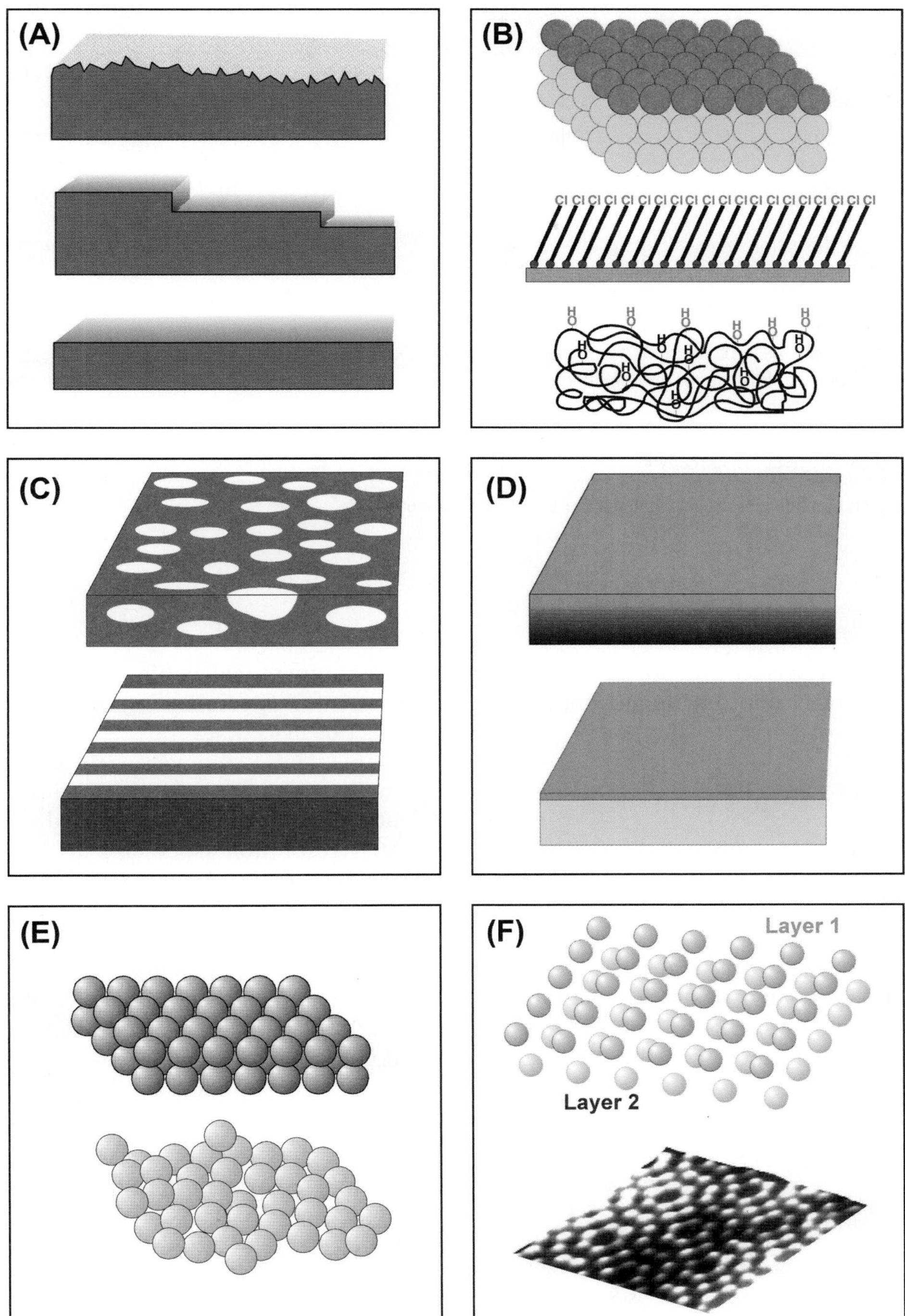

FIGURE I.1.5.3 What might be measured to define surface structure? (A) Surfaces can be rough, stepped or smooth. (B) Surfaces can be comprised of different chemistries (atomic, supramolecular, macromolecular). (C) Surfaces may be structurally or compositionally inhomogeneous in the plane of the surface such as phase-separated domains or micro-contact printed lanes. (D) Surfaces may be inhomogeneous with depth into the specimen or simply overlayered with a thin film. (E) Surfaces may be highly crystalline or disordered. (F) Crystalline surfaces are found with many organizations such as a silicon (100) unreconstructed surface or a silicon (111) (7 × 7) reconstructed surface.

to ascertain its purity. Samples can be surface analyzed prior to and after storage in containers to ensure that the surface composition measured is not due to the container. As a general rule, the polyethylene press-close bags used in electron microscopy and cell culture plasticware are clean storage containers. However, abrasive contact must be avoided, and each brand must be evaluated so that a meticulously prepared specimen is not ruined by contamination. Many brands of aluminum foil are useful for packing specimens, but some are treated with a surface layer of stearic acid that can surface-contaminate biomaterials, implants or medical devices. Aluminum

TABLE I.1.5.1 Common Methods to Characterize Biomaterial Surfaces

Method	Principle	Depth Analyzed	Spatial Resolution	Analytical Sensitivity	Cost
Contact Angles	Liquid wetting of surfaces is used to estimate the energy of surfaces	3–20 Å	1 mm	Low or high depending on the chemistry	$
ESCA (XPS)	X-rays induce the emission of electrons of characteristic energy	10–250 Å	10–150 μm	0.1 atom %	$$$
Auger Electron Spectroscopy*	A focused electron beam stimulates the emission of Auger electrons	50–100 Å	100 Å	0.1 atom %	$$$
SIMS	Ion bombardment sputters secondary ions from the surface	10 Å–1 μm**	100 Å	Very high	$$$
FTIR-ATR	IR radiation is adsorbed and excites molecular vibrations	1–5 μm	10 μm	1 mole %	$$
STM	Measurement of the quantum tunneling current between a metal tip and a conductive surface	5 Å	1 Å	single atoms	$$
SEM	Secondary electron emission induced by a focused electron beam is spatially imaged	5 Å	40 Å typically	High, but not quantitative	$$

*Auger electron spectroscopy is damaging to organic materials, and best used for inorganics.
**Static SIMS ≈ 10 Å, dynamic SIMS to 1 μm.
$: up to $5000
$$: $5000–$100,000
$$$: >$100,000

foil should be checked for surface contamination layers by surface analysis methods prior to wrapping important specimens.

Surface Analysis: General Comments

Two general principles guide sample analysis. First, all methods used to analyze surfaces also have the potential to alter the surface. The analyst must be aware of the damage potential of the method used. Second, because of the potential for artifacts and the need for corroborative information to construct a complete picture of the surface (Figure I.1.5.3), more than one method should be used whenever possible. The data derived from two or more methods should be internally consistent. When data are contradictory, be suspicious and question why. A third or fourth method may then be necessary to draw confident conclusions about the nature of a surface. These general principles are applicable to all materials.

There are properties (only a few of which will be presented here) that are specific to specific classes of materials. Compared to metals, ceramics and glasses, organic, and polymeric materials are more easily damaged by surface analysis methods. Polymeric systems also exhibit greater surface molecular mobility than inorganic systems. The surfaces of inorganic materials are contaminated more rapidly than polymeric materials because of their higher surface energy. Electrically conductive metals and carbons will often be easier to characterize than insulators using electron, X-ray, and ion interaction methods. Insulators accumulate a surface electrical charge that requires special methods (e.g., a low energy electron beam) to neutralize. To learn about other concerns in surface analysis that are specific to specific classes of materials, published papers become a valuable resource for understanding the pitfalls that can lead to artifact or inaccuracy.

Table I.1.5.1 summarizes the characteristics of many common surface analysis methods, including their depth of analysis under standard conditions and their spatial resolution (spot size analyzed). A few of the more frequently applied techniques are described in the next section. However, space limitations prevent an intensive discussion of these methods. The reader is referred to many comprehensive books on the general subject of surface analysis (Briggs and Seah, 1983; Andrade, 1985; Feldman and Mayer, 1986; Vickerman and Gilmore, 2009). References describing specific surface analysis methods will be presented in sections on each of the key methods.

Contact Angle Methods

A drop of liquid sitting on a solid surface represents a simple, but potentially powerful, method to probe surface properties. Experience tells us that a drop of water on a highly polished automobile body surface will stand up (bead up), while on an old, weathered car, the liquid will flow evenly over the surface. This observation, for those knowledgeable about contact angles, tells us that the highly polished car probably has silicones or hydrocarbons at its surface, while the unpolished car surface is oxidized material. Quantitative contact angle measurement has, in fact, been used to predict the performance of vascular grafts, and the adhesion of cells to surfaces.

The phenomenon of the contact angle can be explained as a balance between the force with which the molecules of the drop liquid are being attracted to each other (a

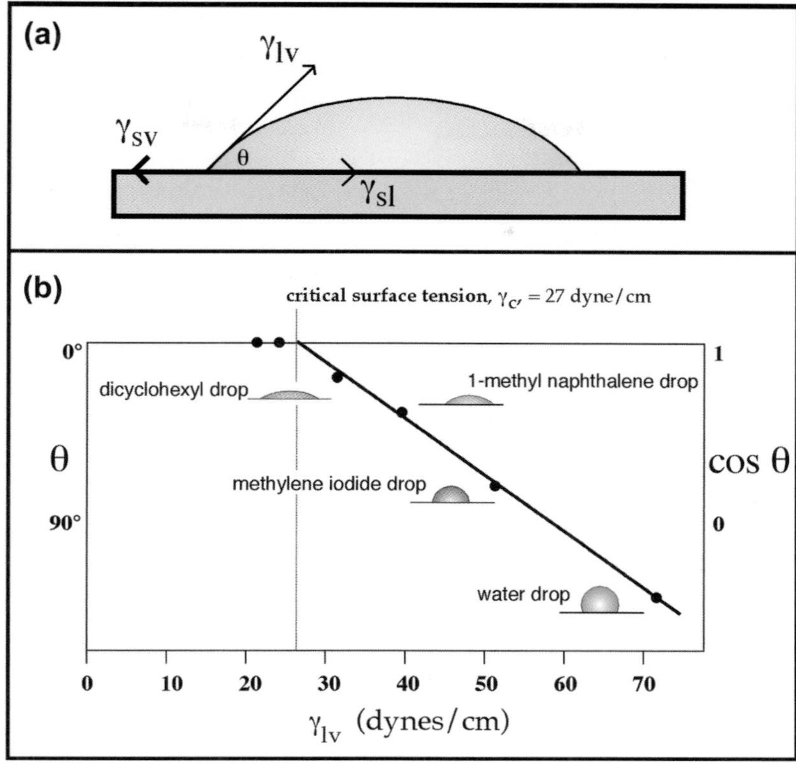

FIGURE I.1.5.4 (a) An equilibrium is established between surface tension forces contracting a liquid drop to a spherical shape and forces interacting the drop with the surface. The force balance between the liquid–vapor surface tension (γ_{lv}) of a liquid drop and the interfacial tension between a solid and the drop (γ_{sl}), manifested through the contact angle (θ) of the drop can be used to quantitatively characterize the surface–vapor interfacial tension (γ_{sv}). (b) The Zisman method permits a critical surface tension value, an approximation to the solid surface tension, to be measured. Drops of liquids of different surface tensions are placed on the solid, and the contact angles of the drops are measured. The plot of liquid surface tension versus angle is extrapolated to zero contact angle to give the critical surface tension value.

cohesive force) and the attraction of the liquid molecules for the surface (an adhesive force). An equilibrium is established between these forces. The force balance between the liquid–vapor surface tension (γ_{lv}) of a liquid drop and the interfacial tension between a solid and the drop (γ_{sl}), manifested through the contact angle (θ) of the drop with the surface, can be used to quantitatively characterize the energy of the surface (γ_{sv}) (Figure I.1.5.4a). The basic relationship describing this force balance is:

$$\gamma_{sv} = \gamma_{sl} + \gamma_{lv} \cos\theta$$

The surface energy, closely related to wettability, is a useful parameter that has often been correlated with biological interaction. Unfortunately, γ_{sv} cannot be directly obtained since this equation contains two unknowns, γ_{sl} and γ_{sv}. Therefore, the γ_{sv} is sometimes approximated by the Zisman method for obtaining the critical surface tension (Figure I.1.5.4b) or calculated by solving simultaneous equations with data from liquids of different surface tensions. Some critical surface tensions for common materials are listed in Table I.1.5.2.

Experimentally, there are a number of ways to measure the contact angle. Some of these are illustrated in Figure I.1.5.5. Contact angle methods can be inexpensive and, with some practice, easy to perform. A contact angle goniometer (a telescope to observe the drop that is equipped with a protractor eyepiece) is the least expensive method for contact angle measurement. A number of companies now offer video systems that compute the contact angle and other surface energy parameters from digitial image analysis of the liquid drop profile. Accessories to control humidity and temperature and deposit drops of liquid are available.

> Contact angles directly measure surface wettability, and indirectly probe surface energy, roughness, heterogeneity, contamination, and molecular mobility.

Contact angle measurements provide a "first line" characterization of materials and can be performed in any laboratory. Contact angle measurements provide unique insight into how the surface will interact with the external world. However, in performing such measurements, a number of concerns must be addressed to obtain meaningful data (Table I.1.5.3). Review articles are available on contact angle measurement for surface characterization (Zisman, 1964; Andrade, 1985; Ratner, 1988; Good, 1993).

TABLE I.1.5.2 Critical Surface Tension Values for Common Polymeric Materials Calculated from Contact Angle Measurements

Material	Critical Surface Tension (dynes/cm)
Polytetrafluoroethylene	19
Poly(dimethyl siloxane)	24
Poly(vinylidine fluoride)	25
Poly(vinyl fluoride)	28
Polyethylene	31
Polystyrene	33
Poly(2-hydroxyethyl methacrylate)	37
Poly(vinyl alcohol)	37
Poly(methyl methacrylate)	39
Poly(vinyl chloride)	39
Polycaproamide (nylon 6)	42
Poly(ethylene oxide)-diol	43
Poly(ethylene terephthalate)	43
Polyacrylonitrile	50

TABLE I.1.5.3 Concerns in Contact Angle Measurement

- The measurement is operator dependent (for manual, goniometer instruments)
- Surface roughness influences the results
- Surface heterogeneity influences the results
- The liquids used are easily contaminated (typically reducing their γ_{lv})
- Liquid evaporation and temperature changes can impact measurement
- The liquids used can reorient the surface structure
- The liquids used can absorb into the surface, leading to swelling
- The liquids used can dissolve the surface
- Few sample geometries are appropriate for contact angle measurement
- Information on surface structure must be inferred from the data obtained

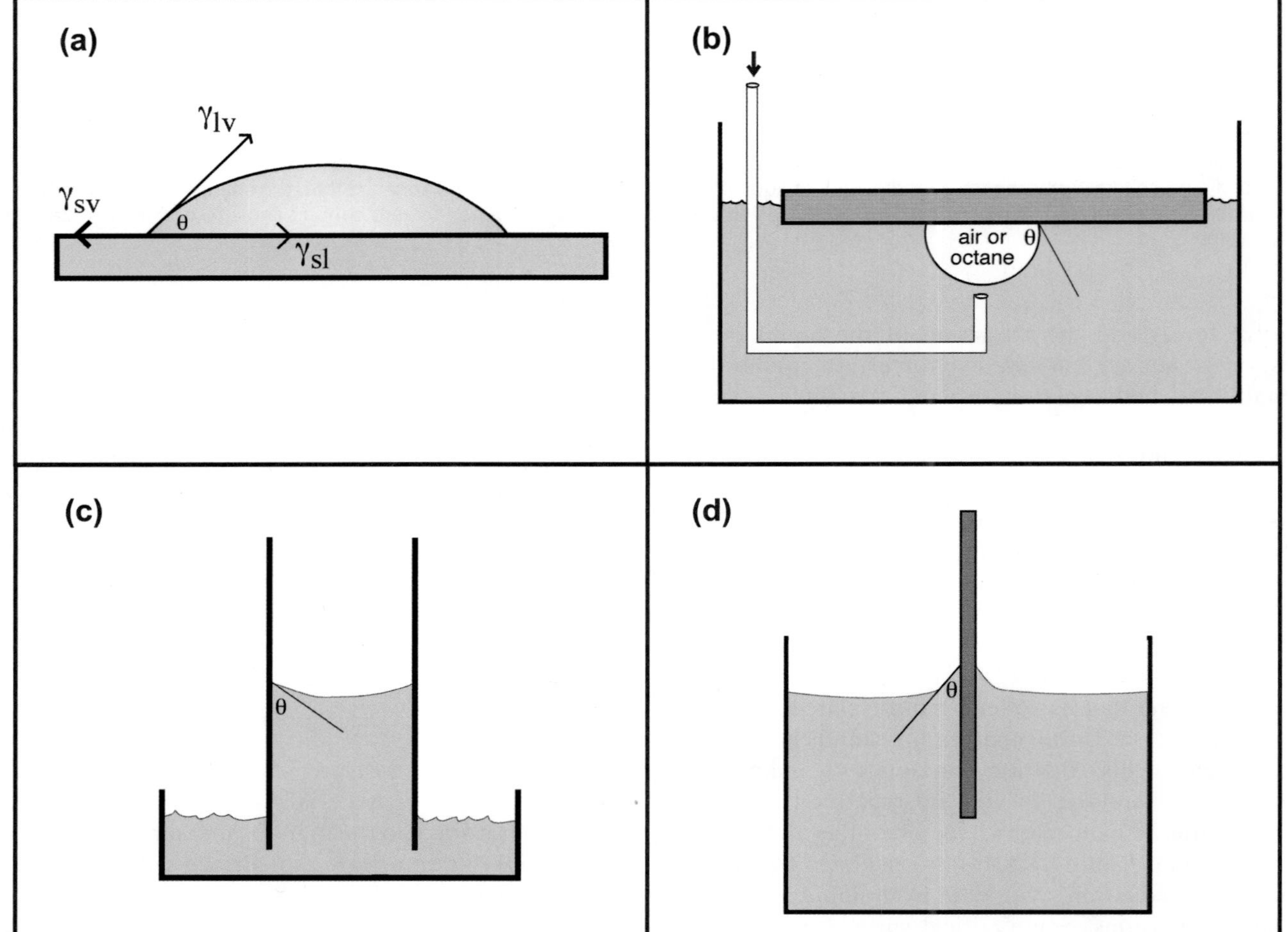

FIGURE I.1.5.5 Four possibilities for contact angle measurement: (a) sessile drop; (b) captive air bubble method; (c) capillary rise method; (d) Wilhelmy plate method.

Electron Spectroscopy for Chemical Analysis

Electron spectroscopy for chemical analysis (ESCA) provides a comprehensive qualitative and quantitative overview of a surface that would be challenging to obtain by other means (Dilks, 1981; Andrade, 1985; Ratner and McElroy, 1986; Ratner, 1988; Watts and Wolstenholme, 2003; Ratner and Castner, 2009). In contrast to the contact angle technique, ESCA requires complex, expensive apparatus (Figure I.1.5.6a) and demands considerable training to perform the measurements. However, since ESCA is available from commercial laboratories, university analytical facilities, national centers (for example, NESAC/BIO at the University of Washington), and specialized research laboratories, most biomaterials scientists can access instrumentation for sample analysis. ESCA has contributed significantly to the development of biomaterials and medical devices, and to understanding the fundamentals of biointeraction.

The ESCA method (also called X-ray photoelectron spectroscopy, XPS) is based upon the photoelectric effect, properly described by Einstein in 1905. X-rays are focused upon a specimen. The interaction of the X-rays with the atoms in the specimen causes the emission of core level (inner shell) electrons. The energy of these electrons is measured and their values provide information about the nature and environment of the atom or atoms from which they came. The basic energy balance describing this process is given by the relationship:

$$BE = h\nu - KE$$

where BE is the energy binding the electron to an atom (the value desired), KE is the kinetic energy of the emitted electron (the value measured in the ESCA spectrometer), and hv is the energy of the X-rays, a known value. A schematic diagram illustrating an ESCA instrument is shown in Figure I.1.5.6b. Table I.1.5.4 lists the types of information about the nature of a surface that can be obtained by using ESCA. The origin of the surface sensitivity of ESCA is described in Figure I.1.5.7.

ESCA has many advantages, and a few disadvantages, for studying biomaterials. The advantages include high information content, surface localization of the measurement (outermost 8–10 nm), speed of analysis, low damage potential, and the ability to analyze most samples with no special specimen preparation. The latter advantage is particularly important since biomedical devices (or parts of devices) can often be inserted, as fabricated and sterilized, directly in the analysis chamber for study. The disadvantages include the need for vacuum compatibility (i.e., no outgassing of volatile components), the vacuum environment and its impact on the specimen (particularly for hydrated specimens), the possibility of sample damage by X-rays if long analysis times are used, the need for experienced operators, and the cost associated with this complex instrumentation. The vacuum limitations can be sidestepped by using an ESCA system with a cryogenic sample stage. At liquid nitrogen temperatures, samples with volatile components, or even wet, hydrated samples, can be analyzed.

> ESCA analyzes to approximately 10 nm and gives information on elements present, their concentrations and their bonding environments.

The use of ESCA is best illustrated with a brief example. A poly(methyl methacrylate) (PMMA) ophthalmologic device is to be examined. Taking care not to touch or damage the surface of interest, the device is inserted into the ESCA instrument introduction chamber. The introduction chamber is then pumped down to 10^{-6} torr (1.33×10^{-4} Pa) pressure. A gate valve between the introduction chamber and the analytical chamber is opened and the specimen is moved into the analysis chamber. In the analysis chamber, at 10^{-9} torr (1.33×10^{-7} Pa) pressure, the specimen is positioned using a microscope or TV camera and the X-ray source is turned on. The ranges of electron energies to be observed are computer-controlled with the retardation lens on the spectrometer. First, a wide scan is made in which the energies of all emitted electrons over a 1000 eV range are detected (Figure I.1.5.8). Then, narrow scans are made in which each of the elements detected in the wide scan is examined in higher resolution (Figure I.1.5.9).

From the wide scan, we learn that the specimen contains carbon, oxygen, nitrogen, and sulfur. The presence of sulfur and nitrogen is unexpected for PMMA. We can calculate atomic percentage composition from the wide scan spectral data. The sample surface contains 58.2% carbon, 27.7% oxygen, 9.5% nitrogen, and 4.5% sulfur. The narrow scan for the carbon region (C1s spectrum) suggests four classes of compounds: hydrocarbons; carbons singly bonded to oxygen (the predominant species); carbons in amide-like molecular environments; and carbons in carboxylic acid or ester environments. This is quite different from the spectrum expected for pure PMMA. An examination of the peak position in the narrow scan of the sulfur region (S2p spectrum) suggests sulfonate-type groups. The shape of the C1s spectrum, the position of the sulfur peak, and the presence of nitrogen all suggest that heparin was immobilized to the surface of the PMMA device. Since the stoichiometry of the lens surface does not match that for pure heparin, this suggests that we are seeing either some of the PMMA substrate through a <100 Å layer of heparin or we are seeing some of the bonding links used to immobilize the heparin to the lens surface. Further ESCA analysis will permit the extraction of more detail about this surface-modified device, including an estimate of surface modification thickness, further confirmation that the coating is indeed heparin, and additional information about the nature of the immobilization chemistry.

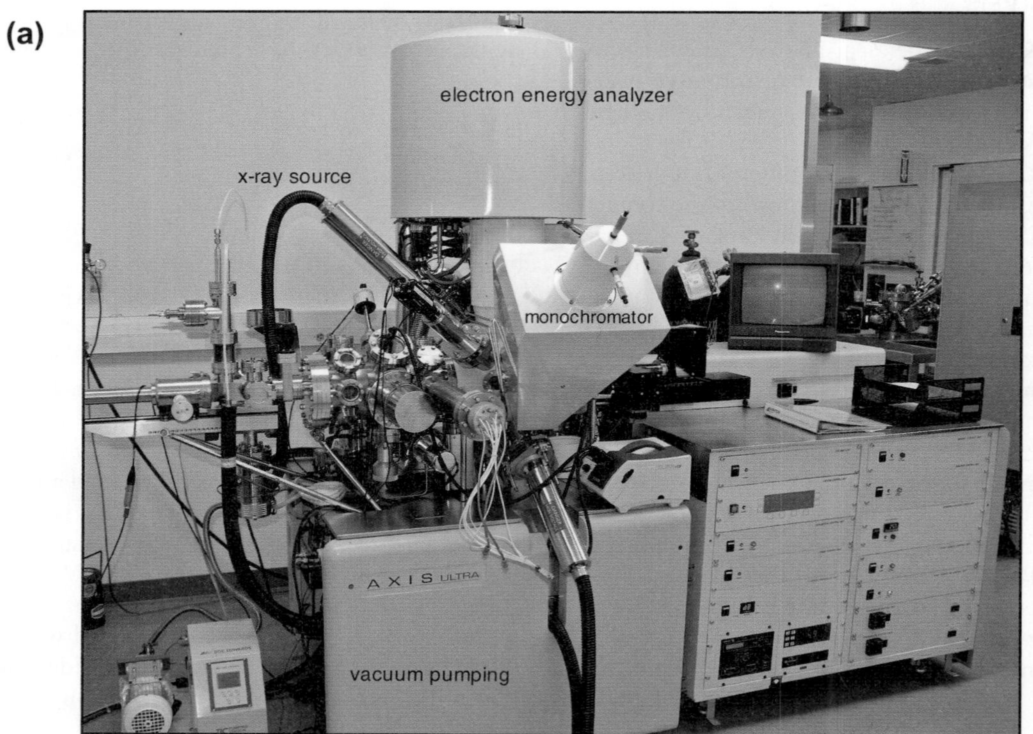

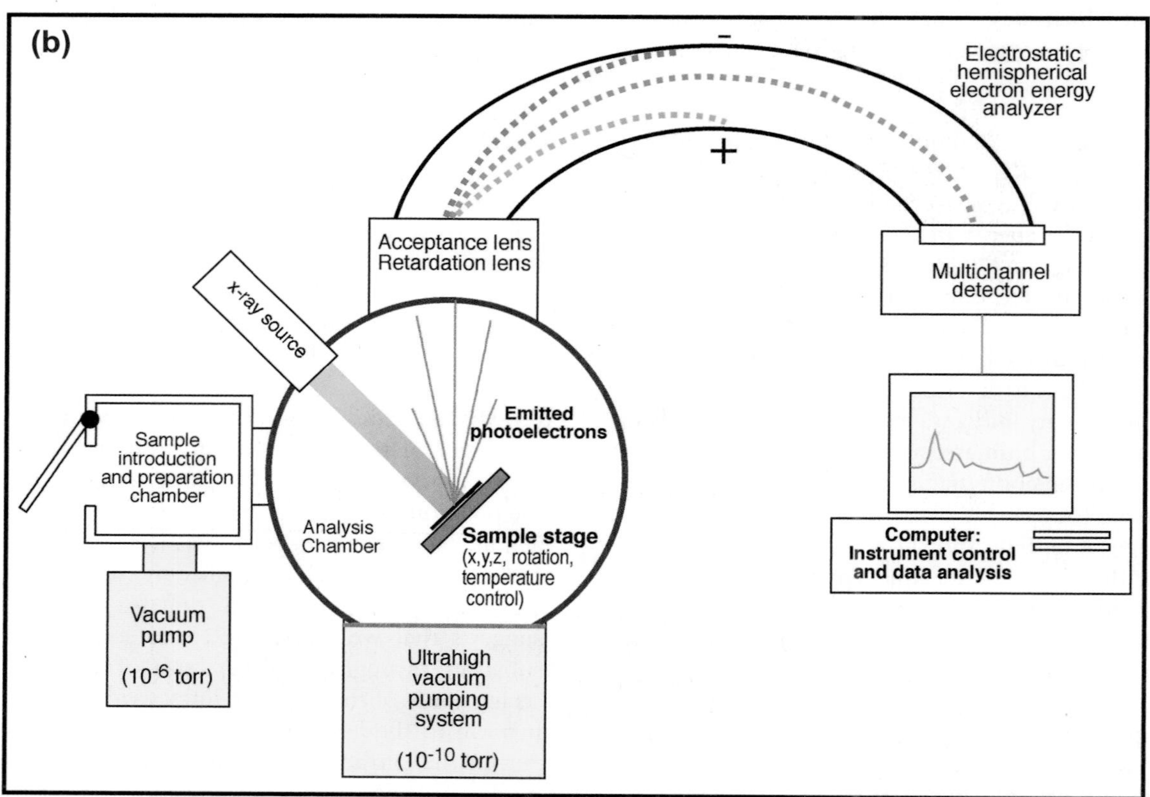

FIGURE I.1.5.6 (a) Photograph of a contemporary ESCA instrument (photo by Kratos Analytical Corp.). (b) Schematic diagram of a monochromatized ESCA instrument.

TABLE I.1.5.4 Information Derived From an ESCA Experiment
In the outermost 100 Å of a surface, ESCA can provide:
• Identification of all elements (except H and He) present at concentrations >0.1 atomic %
• Semiquantitative determination of the approximate elemental surface composition (±10%)
• Information about the molecular environment (oxidation state, bonding atoms, etc.)
• Information about aromatic or unsaturated structures from shake-up $\pi^* \leftarrow \pi$) transitions
• Identification of organic groups using derivatization reactions
• Nondestructive elemental depth profiles 100 Å into the sample and surface heterogeneity assessment using angular-dependent ESCA studies and photoelectrons with differing escape depths
• Destructive elemental depth profiles several thousand angstroms into the sample using argon etching (for inorganics)
• Lateral variations in surface composition (spatial resolution 8–150 μm, depending upon the instrument)
• "Fingerprinting" of materials using valence band spectra and identification of bonding orbitals
• Studies on hydrated (frozen) surfaces

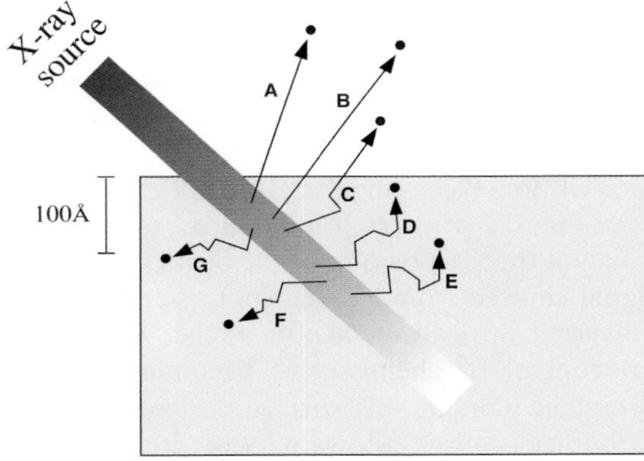

FIGURE I.1.5.7 ESCA is a surface-sensitive method. Although the X-ray beam can penetrate deeply into a specimen, electrons emitted deep in the specimen (D, E, F, G) will lose their energy in inelastic collisions and never emerge from the surface. Only those electrons emitted near the surface that lose no energy (A, B) will contribute to the ESCA signal used analytically. Electrons that lose some energy, but still have sufficient energy to emerge from the surface (C) contribute to the background signal.

Secondary Ion Mass Spectrometry

Secondary ion mass spectrometry (SIMS) is an information-rich tool that the surface analyst can bring to bear on a biomedical problem. SIMS produces a mass spectrum of the outermost 1–2 nm of a surface. Like ESCA, it requires complex instrumentation and an ultrahigh vacuum chamber for the analysis. However, it provides unique information that is complementary to ESCA, and greatly aids in understanding surface composition. Some of the analytical capabilities of SIMS are summarized in Table I.1.5.5. Review articles on SIMS are available (Benninghoven, 1983; Scheutzle et al., 1984; Ratner,

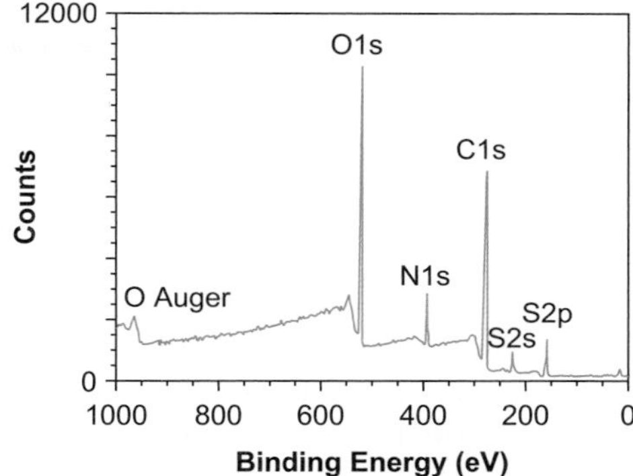

FIGURE I.1.5.8 ESCA wide scan of a surface-modified poly(methyl methacrylate) ophthalmologic device.

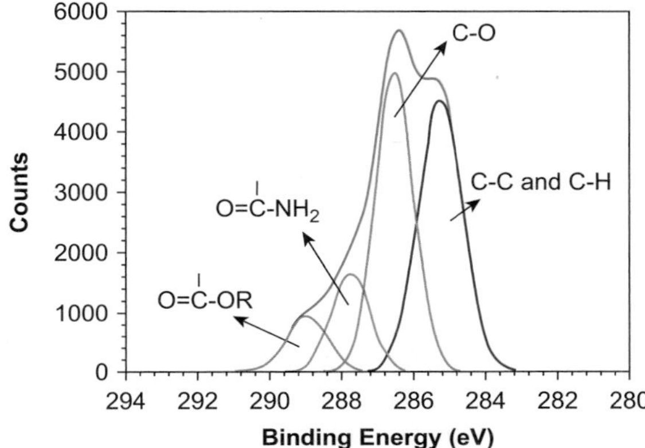

FIGURE I.1.5.9 The carbon 1s narrow scan ESCA spectrum of a surface-modified poly(methyl methacrylate) ophthalmologic device. Narrow scan spectra can be generated for each element seen in low energy resolution mode in Figure I.1.5.8.

TABLE I.1.5.5 Analytical Capabilities of SIMS

	Static SIMS	Dynamic SIMS
Identify hydrogen and deuterium	✓	✓
Identify other elements (often must be inferred from the data)	✓	✓
Suggest molecular structures (inferred from the data)	✓	—
Observe extremely high mass fragments (proteins, polymers)	✓	—
Detection of extremely low concentrations	✓	✓
Depth profile to 1 μm into the sample	*	✓
Observe the outermost 1–2 atomic layers	✓	—
High spatial resolution (features as small as approximately 400 Å)	✓	✓
Semiquantitative analysis (for limited sets of specimens)	✓	—
Useful for polymers	✓	—
Useful for inorganics (metals, ceramics, etc.)	✓	✓
Useful for powders, films, fibers, etc.	✓	✓

*Cluster ion sources allow depth profiling with static-SIMS-like information content.

1988; Briggs, 1986; Vickerman et al., 1989; Davies and Lynn, 1990; Van Vaeck et al., 1999; Belu et al., 2003; Michel and Castner, 2006).

In SIMS analysis, a surface is bombarded with a beam of accelerated ions. The collision of these ions with the atoms and molecules in the surface zone can transfer enough energy that they sputter from the surface into the vacuum phase. The process is analogous to racked pool balls that are ejected from the ball cluster by the impact of the cue ball; the harder the cue ball hits the rack of balls, the more balls are emitted from the rack. In SIMS, the "cue balls" are positive ions (cesium, gallium, bismuth, C60-buckyballs, and cluster ions are commonly used) that are accelerated at the surface to be analyzed with energies of 5000–20,000 eV. The particles ejected from the surface are positive and negative ions (secondary ions), radicals, excited states, and neutrals. Only the secondary ions are measured in SIMS. In ESCA, the energy of emitted particles (electrons) is measured. SIMS measures the mass of emitted ions (more rigorously, the ratio of mass to charge, m/z) using a time-of-flight (TOF) mass analyzer, magnetic sector analyzer or, in older instruments, a quadrupole mass analyzer.

There are two modes for SIMS analysis, dynamic and static. Dynamic SIMS uses high ion doses over a given analysis time. The primary ion beam sputters sufficient material from the surface that the surface erodes away at an appreciable rate. We can capitalize on this to depth-profile into a specimen, and this works particularly well for metallic, semiconductor, and ceramic materials. The intensity of the m/z peak of a species of interest (e.g., sodium ion, m/z = 23) might be followed as a function of time. If the ion beam is well-controlled and the sputtering rate is constant, the sodium ion signal intensity measured at any time will be directly related to its concentration at the erosion depth of the ion beam into the specimen. A concentration depth profile (sodium concentration versus depth) can be constructed from the outermost atoms to a micron or more into the specimen. Depending on the primary ion used, for example, cesium or gallium, and its energy, the high-flux ion beam can destroy organic samples and relevant organic fragments, and predominantly atomic fragments (e.g., C^-, CH^-, O^-, OH^-, Na^+, etc.) will be detected. Also, as the beam erodes deeper into the specimen, more artifacts are introduced in the data by "knock-in" and scrambling of atoms.

Static SIMS, in contrast, induces minimal surface destruction. The ion dose is adjusted so that during the period of the analysis less than 10% of one monolayer of surface atoms is sputtered. Since there are typically 10^{13}–10^{15} atoms in 1 cm^2 of surface, a total ion dose of less than 10^{13} ions/cm^2 during the analysis period is appropriate. Under these conditions, extensive degradation and rearrangement of the chemistry at the surface does not take place, and large, relatively intact molecular fragments can be ejected into the vacuum for measurement. Examples of molecular fragments are shown in Figure I.1.5.10. This figure also introduces some of the ideas behind SIMS spectral interpretation. A more complete introduction to the concepts behind static SIMS spectral interpretation can be found in Van Vaeck et al. (1999) or in standard texts on mass spectrometry.

> Static SIMS provides qualitative information on the atomic and molecular composition in the outermost 1–2 nm of surface with high analytical sensitivity and excellent *x,y* spatial resolution.

Magnetically or electrostatically focusing the primary ion beam permits the SIMS technique to have high spatial resolution in the *x,y* plane. In fact, SIMS analysis can be performed in surface regions of 10 nm diameter on optimal specimens. For static SIMS analysis, only 1–10% of the atoms in any area are sampled. Thus, as the spot size gets smaller, the challenge to achieve high analytical sensitivity increases sharply. Still, static SIMS measurements have been performed in areas as small as 40 nm. Newly-developed cluster ion sources (for example, using gold molecular clusters, Au_3, C_{60} or argon clusters as the impacting primary particles) show high secondary ion yields, and relatively low surface damage. These improve spatial resolution and also permit depth profiling of organic surfaces by sputtering down into a surface, while monitoring secondary ion emission as a function of time (Ninomiya et al., 2009).

If the focused primary ion beam is rastered over the surface, and the *x,y* position of the beam correlated with the signal emitted from a given spot, the SIMS data can be converted into an elemental image. Patterning and spatial control of chemistry is becoming increasingly important in biomaterials surface design. For example, microcontact printing allows patterned chemistry to be transferred to surfaces at the micron level using a relatively simple rubber stamp. Imaging SIMS is well-suited to studying and monitoring such spatially defined chemistry (Dubey et al., 2009; Bolles et al., 2010). An example is presented in Figure I.1.5.11. Imaging SIMS is also valuable for observing defects in thin films (pinholes), analyzing the chemistry of fine particulates or assessing causes of implant failure.

Scanning Electron Microscopy

Scanning electron microscopy (SEM) images of surfaces have great resolution and depth of field, with a three-dimensional quality that offers a visual perspective familiar to most users. SEM images are widely used, and much has been written about the technique. The comments here are primarily oriented toward SEM as a surface analysis tool.

SEM functions by focusing and rastering a relatively high-energy electron beam (typically, 5–100 keV) on a specimen that is under vacuum. Low-energy secondary electrons (1–20 eV) are emitted from each spot where the

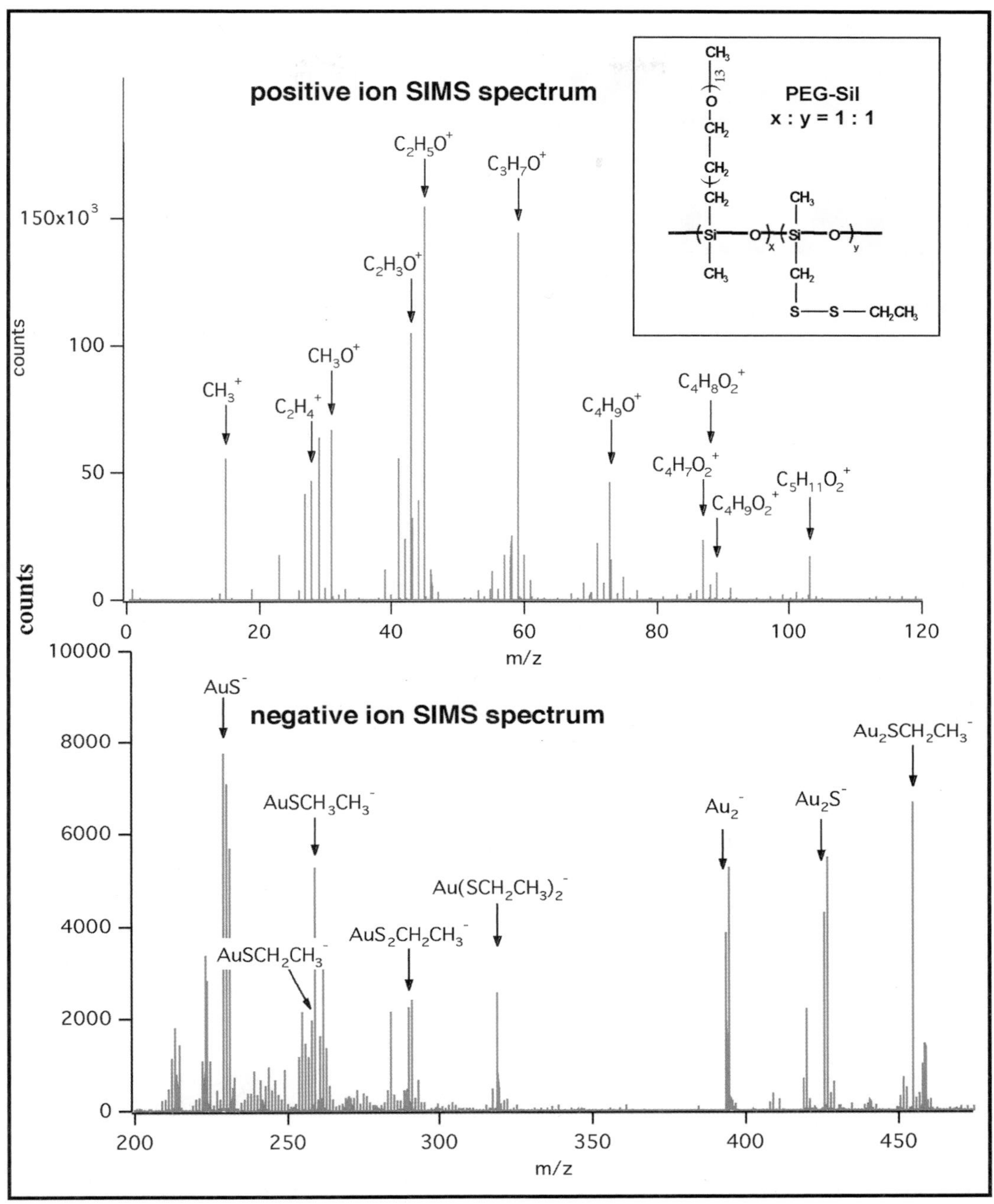

FIGURE I.1.5.10 Static positive and negative ion SIMS spectra of a poly(ethylene glycol)–poly(dimethyl siloxane) copolymer containing disulfide side groups on a gold surface. The primary peaks are identified. The low mass region of the negative ion spectrum offers little insight into the polymer structure, but the high mass region is rich in information. In this case, the low mass positive spectrum is rich in information. Further details on this class of polymers can be found in *Macromolecules*, 27, 3053 (1994). (Figure supplied by D. Castner.)

focused electron beam makes an impact. The intensity of the secondary electron emission is a function of the atomic composition of the sample and the geometry of the features under observation. The image of the surface is spatially reconstructed on a phosphor screen (or CCD detector) from the intensity of the secondary electron emission at each point. Because of the shallow penetration depth of low-energy electrons produced by the primary electron beam, only the secondary electrons generated near the surface can escape and be detected (this is analogous to the surface sensitivity described in Figure I.1.5.7). Consequently, SEM is a surface analysis method.

> SEM provides a high resolution image of the surface. On insulating materials, metallic coating is required and the image is actually of the coating surface, not the underlying material.

Nonconductive materials observed in the SEM are typically coated with a thin, electrically grounded layer

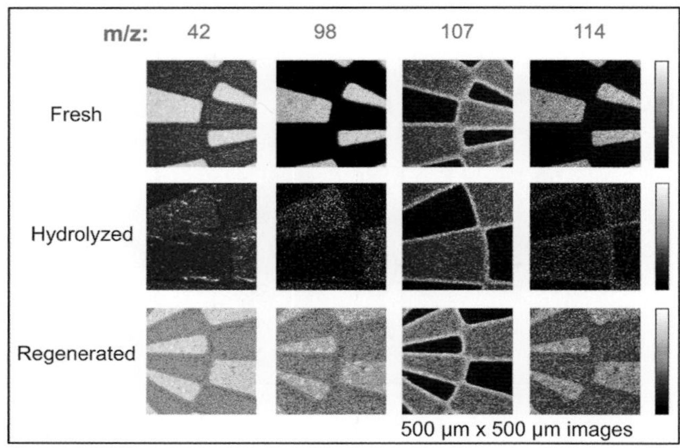

FIGURE I.1.5.11 Negative TOF-SIMS ion images exploring the N-hydroxysuccinimide (NHS) reaction used to immobilize biomolecules to surfaces. Yellow colors show regions rich in NHS groups based on peaks at m/z 42, 98, 107, and 114. The image shows a spatially defined sample with freshly prepared, hydrolyzed and regenerated photopatterned surfaces. Further details can be found in Dubey et al. (2009). (Used with permission of the publisher.)

of metal to minimize negative charge accumulation from the electron beam. However, this metal layer is always thick enough (>200 Å) so that the electrons emitted from the sample beneath cannot penetrate. Therefore, in SEM analysis of nonconductors, the surface of the metal coating is, in effect, what is being monitored. If the metal coat is truly conformal, a good representation of the surface geometry will be conveyed. However, the specimen surface chemistry no longer influences secondary electron emission. Also, at very high magnifications, the texture of the metal coat and not the surface may be under observation.

SEM, in spite of these limitations, is an important corroborative method to use in conjunction with other surface analysis methods. Surface roughness and texture can have a profound influence on data from ESCA, SIMS, and contact angle determinations. Therefore, SEM provides important information in the interpretation of data from these methods.

The development of low-voltage SEM does study the surface chemistry (and geometry) of nonconductors. With the electron accelerating voltage lowered to approximately 1 keV, charge accumulation is not as critical and metallization is not required. Low-voltage SEM has been used to study platelets and phase separation in polymers. Also, the environmental SEM (ESEM) permits wet, uncoated specimens to be studied.

The primary electron beam also stimulates the emission of X-rays. These X-rays are used to identify elements with the technique called energy-dispersive X-ray analysis (EDXA). However, the high-energy primary electron beam penetrates deeply into a specimen (a micron or more). The X-rays produced from the interaction of these electrons with atoms deep in the bulk of the specimen can penetrate through the material and be detected. Therefore, EDXA is not a surface analysis method.

The primary use of SEM is to image topography. SEM for this application is elaborated upon in the chapter on microscopy in biomaterials research (Chapter II.3.8).

Infrared Spectroscopy

Infrared spectroscopy provides information on the bond vibrations of molecular species. It is a widely used analytical method that can reveal information on specific chemistries and the orientation of structures. Fourier transform infrared (FTIR) spectrometry offers outstanding signal-to-noise ratio (S/N) and spectral accuracy. However, even with this high S/N, the small absorption signal associated with the minute mass of material in a surface region can challenge the sensitivity of the spectrometer. Also, the problem of separating the vastly larger bulk absorption signal from the surface signal must be addressed.

Surface FTIR methods couple the infrared radiation to the sample surface to increase the intensity of the surface signal and reduce the bulk signal (Allara, 1982; Leyden and Murthy, 1987; Urban, 1993; Dumas et al., 1999). Some of these sampling modes, and their characteristics, are illustrated in Figure I.1.5.12.

The attenuated total reflectance (ATR) mode of sampling has been used most often in biomaterials studies. The penetration depth into the sample is 1–5 µm. Therefore, ATR is not highly surface sensitive, but observes a broad region near the surface. However, it does offer the wealth of rich structural information common to infrared spectra. With extremely high S/N FTIR instruments, ATR studies of proteins and polymers under water have been performed. In these experiments, the water signal (which is typically 99% or more of the total signal) is subtracted from the spectrum to leave only the surface material (e.g., adsorbed protein) under observation. Micro-ATR, coupling a microscope, IR spectrometer, and micro-ATR crystal, permits ATR analysis with high spatial resolution (approximately 1 μm^2 pixel size).

ATR-IR permits detailed molecular analysis of the outermost 1–5 microns of a sample.

Another infrared method that has proven immensely valuable for observing extremely thin films on reflective surfaces is infrared reflection absorption spectroscopy (IRAS) (Figure I.1.5.12). This method has been widely applied to self assembled monolayers (SAMs), but is applicable to many surface films that are less than 10 nm in thickness. The surface upon which the thin film resides must be highly reflective and metal surfaces work best, although a silicon wafer can be used. IRAS gives information about composition, crystallinity, and molecular orientation. Infrared spectroscopy is one member of a family of methods called vibrational spectroscopies. Two other vibrational spectroscopies, sum

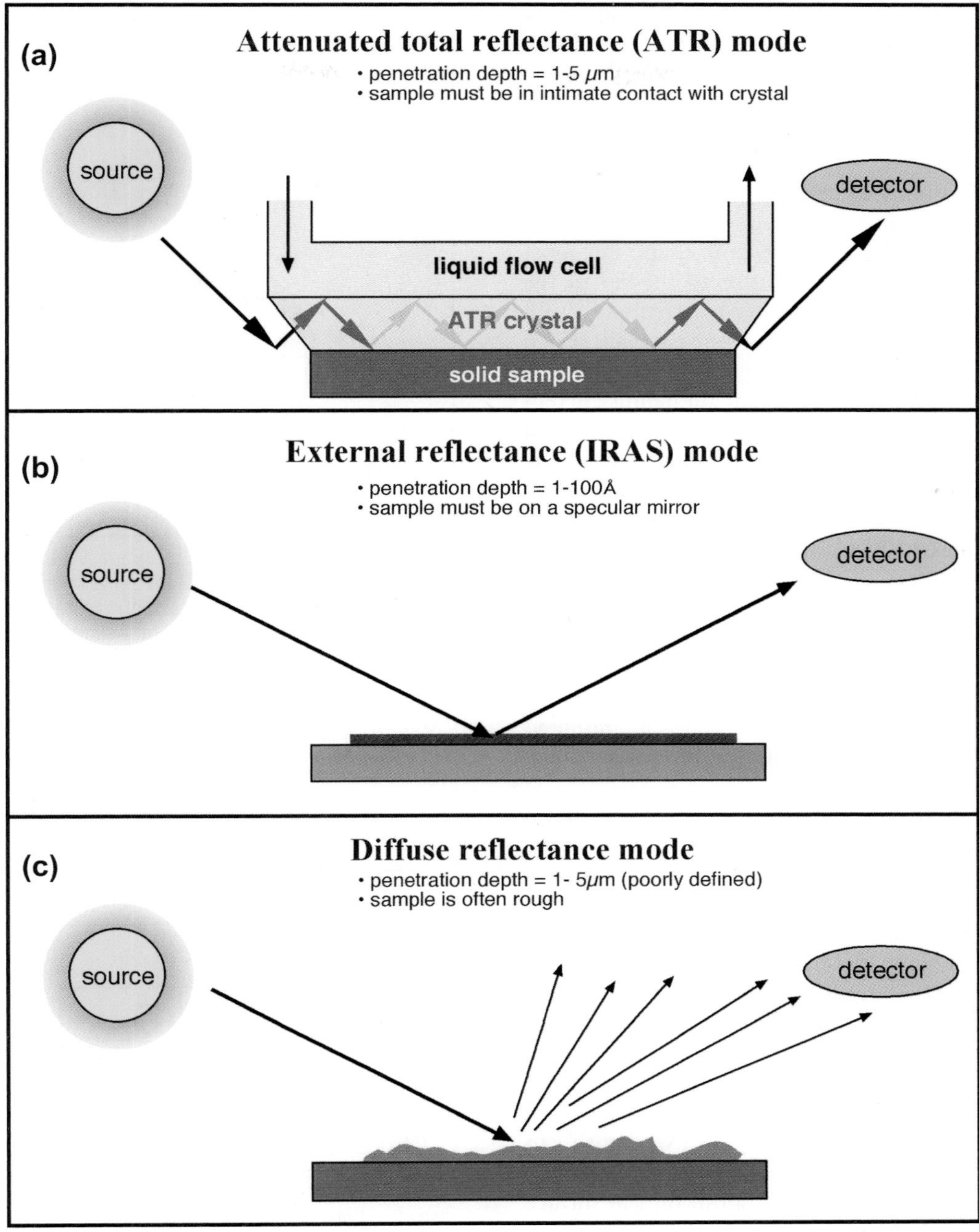

FIGURE I.1.5.12 Three surface-sensitive infrared sampling modes. (a) Attenuated total reflectance infrared (ATR-IR); (b) infrared reflection absorption spectroscopy (IRAS); (c) diffuse reflectance infrared spectroscopy.

frequency generation and Raman (in particular, surface enhanced), will be mentioned below in the section on newer methods.

Scanning Tunneling Microscopy (STM), Atomic Force Microscopy (AFM), and the Scanning Probe Microscopies (SPM). In the 16 years since the first edition of this book, scanning tunneling microscopy (STM) and atomic force microscopy (AFM) have devolved from novel research tools to key methods for biomaterials characterization. AFM is more widely used than STM because electrically conductive surfaces are not needed with AFM and quantitative force measurements can be made. General review articles are available (Binnig and Rohrer, 1986; Albrecht et al., 1988; Avouris, 1990) and articles oriented toward biological studies with these methods (Hansma et al., 1988; Miles et al., 1990; Rugar and Hansma, 1990; Jandt, 2001; Dufrêne, 2004).

The STM was invented in 1981 and led to a Nobel Prize for G. Binnig and H. Rohrer in 1986. The STM capitalizes on quantum tunneling to generate an atom-scale, electron density image of a surface. A metal tip terminating in a single atom is brought within 5–10 Å of an electrically conducting surface. At these distances, the electron cloud of the atom at the "tip of the tip" will significantly overlap the electron cloud of an atom on the surface. If a potential is applied between the tip and the surface, an electron tunneling current will be established whose magnitude, J, follows the proportionality:

$$J \propto e^{(-Ak_0S)}$$

where A is a constant, k_0 is an average inverse decay length (related to the electron affinity of the metals), and S is the separation distance in angstrom units. For most metals, a 1 Å change in the distance of the tip to the surface results in an order of magnitude change in tunneling current. Even though this current is small, it can be measured with good accuracy.

To image a surface, this quantum tunneling current is used in one of two ways. In constant current mode, a piezoelectric driver scans a tip over a surface. When the tip approaches an atom protruding above the plane of the surface, the current rapidly increases, and a feedback circuit moves the tip up to keep the current constant. Then, a plot is made of the tip height required to maintain constant current versus distance along the plane. In constant height mode, the tip is moved over the surface and the change in current with distance traveled along the plane of the surface is directly recorded. A schematic diagram of a scanning tunneling microscope is presented in Figure I.1.5.13. Two STM scanning modes are illustrated in Figure I.1.5.14.

FIGURE I.1.5.13 Schematic diagram illustrating the principle of the scanning tunneling microscope – a tip terminating in a single atom permits localized quantum tunneling current from surface features (or atoms) to tip. This tunneling current can be spatially reconstructed to form an image.

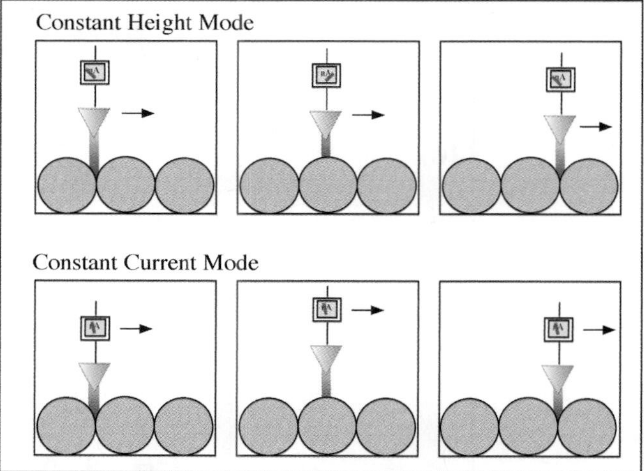

FIGURE I.1.5.14 Scanning tunneling microscopy can be performed in two modes. In constant height mode, the tip is scanned a constant distance from the surface (typically 5–10 Å) and the change in tunneling current is recorded. In constant current mode, the tip height is adjusted so that the tunneling current is always constant, and the tip distance from the surface is recorded as a function of distance traveled in the plane of the surface.

The STM measures electrical current, and therefore is well suited for conductive and semiconductor surfaces. However, biomolecules (even proteins) on conductive substrates appear amenable to imaging. STM does not "see" atoms, but rather monitors electron density. The conductive and imaging mechanism for proteins is not well-understood. Still, publications suggest that information-rich images of biomolecules on conductive surfaces can be obtained (Campbell et al., 2007).

The AFM uses a similar piezo drive mechanism. However, instead of recording tunneling current, the deflection of a tip mounted on a flexible cantilever arm due to van der Waals forces and electrostatic repulsion/attraction between an atom at the tip and an atom on the surface is measured. Atomic-dimension measurements of cantilever arm movements can be made by reflecting a laser beam off a mirror on the cantilever arm (an optical lever). A one-atom deflection of the cantilever arm can easily be measured by monitoring the position of the laser reflection on a spatially resolved photosensitive detector. Other principles are also used to measure the deflection of the tip. These include capacitance measurements and interferometry. A diagram of a typical AFM is presented in Figure I.1.5.15. Images of protein molecules recorded by AFM are shown in Figure I.1.5.16 to demonstrate the sub-nanometer resolution possible on biological specimens.

Tips are important in AFM, as the spatial resolution is significantly associated with tip terminal diameter and shape. Tips are made from microlithographically fabricated silicon or silicon nitride. Also carbon whiskers, nanotubes, and a variety of nanospherical particles have been mounted on AFM tips to increase their sharpness or improve the ability to precisely define tip geometry. Tips are also surface-modified to alter the strength and types of interactions with surfaces (static SIMS can be used to image these surface modifications). Finally, cantilevers are sold in a range of stiffnesses, so that the analysis modes can be tuned to needs of the sample and the type of data being acquired. The forces associated with the interaction of an AFM tip with a surface as it approaches

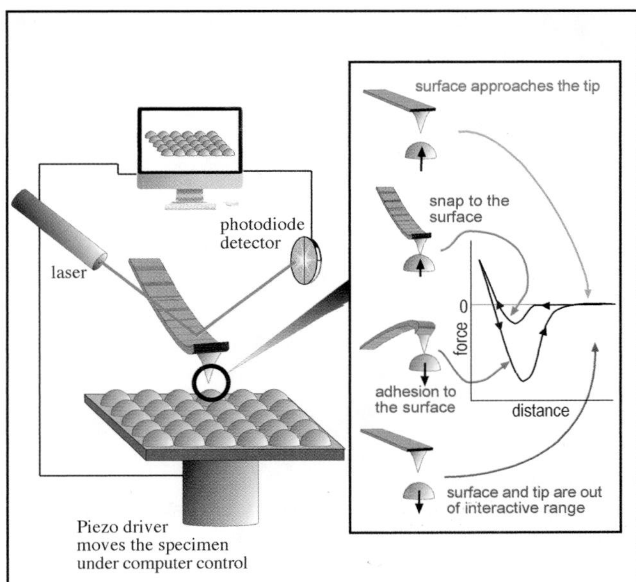

FIGURE I.1.5.15 Schematic diagram illustrating the principle of the atomic force microscope.

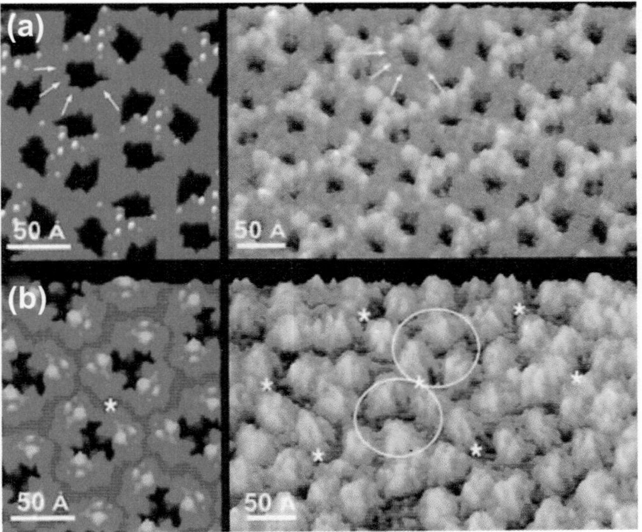

FIGURE I.1.5.16 An AFM image of porin proteins from the outer membrane of *E. coli* imaged with nanoscale resolution. Comparison of high-resolution AFM images of OmpF crystals (in brown-yellow) and the atomic model rendered at 3 Å (in blue). (From Müller, D. J. & Engel, A. (1999). Journal of Molecular Biology, **285**, 1347, used with permission of the authors and the publisher.)

and is retracted are illustrated in Figure I.1.5.15. Since force is being measured and Hooke's Law applies to the deformation of an elastic cantilever, AFM can be used to quantify the forces between surface and tip. Quantitative AFM is now widely used to measure the strength of interaction between biomolecules (Chilkoti et al., 1995; Muller and Dufrêne, 2008) and the mechanical properties of proteins (Li and Cao, 2010).

AFM instruments are commonly applied to surface problems using many possible tip interaction modes. Some AFM modes are contact, lateral force, non-contact, tapping, force modulation, and phase imaging. In contact mode, the tip is in contact with the surface (or at least the electron clouds of tip and surface essentially overlap). The pressures resulting from the force of the cantilever delivered through the extremely small surface area of the tip can be damaging to soft specimens (proteins, polymers, etc.). However, for more rigid specimens, excellent topographical imaging can be achieved in contact mode. In tapping mode, the tip is oscillated at a frequency near the resonant frequency of the cantilever. The tip barely grazes the surface. The force interaction of tip and surface can affect the amplitude of oscillation and the oscillating frequency of the tip. In standard tapping mode, the amplitude change is translated into topographic spatial information. Many variants of tapping mode have been developed allowing imaging under different conditions, and using the phase shift between the applied oscillation to the tip and the actual tip oscillation in the force field of the surface to provide information about the mechanical properties of the surface (in essence, the viscoelasticity of the surface can be appreciated).

AFM allows imaging of surfaces at sub-nanometer resolutions, and also provides detail on surface mechanics and molecular interactions.

The potential of the AFM to explore surface problems has been greatly expanded by ingenious variants of the technique. In fact, the term "atomic force microscopy" has been generalized to "scanning probe microscopy" (SPM). Table I.1.5.6 lists many of these creative applications of the AFM/STM idea.

Since the AFM measures force, it can be used with both conductive and nonconductive specimens. Force must be applied to bend a cantilever, so AFM is subject to artifacts caused by damage to fragile structures on the surface. Both AFM and STM can function well for specimens under water, in air or under vacuum. For exploring biomolecules or mobile organic surfaces, the "pushing around" of structures by the tip is a significant concern. This surface artifact can be capitalized upon to write and fabricate surface structures at the nanometer scale (Figure I.1.5.17) (Quate, 1997; Boland et al., 1998; Rosa and Liang, 2009).

Newer Methods

There are many other surface characterization methods that have the potential to become important in future years. Some of these are listed in Table I.1.5.7. A few of these evolving techniques which will be specifically mentioned here include sum frequency generation (SFG), Raman, and synchrotron methods.

SFG is a vibration spectroscopy method that looks at the outmost surface of materials in aqueous, air or vacuum. SFG uses two high intensity, pulsed laser beams, one in the visible range (frequency = $\omega_{visible}$), and one in

TABLE I.1.5.6 Scanning Probe Microscopy (SPM) Modes

Name	Acronym	Use
Contact mode	CM-AFM	Topographic imaging of harder specimens
Tapping (intermittent force) mode	IF-AFM	Imaging softer specimens
Non-contact mode	NCM-AFM	Imaging soft structures
Force modulation (allows slope of force-distance curve to be measured	FM-AFM	Enhances image contrast based on surface mechanics
Scanning surface potential microscopy (Kelvin probe microscopy)	SSPM, KPM	Measures the spatial distribution of surface potential
Magnetic force microscopy	MFM	Maps the surface magnetic forces
Scanning thermal microcopy	SThM	Maps the thermal conductivity characteristics of a surface
Recognition force microscopy	RFM	Uses a biomolecule on a tip to probe for regions of specific biorecognition on a surface
Chemical force microscopy	CFM	A tip derivatized with a given chemistry is scanned on a surface to spatially measure differences of interaction strength
Lateral force microscopy	LFM	Maps frictional force on a surface
Electrochemical force microscopy	EFM	The tip is scanned under water and the electrochemical potential between tip and surface is spatially measured
Nearfield scanning optical microscopy	NSOM	A sharp optical fiber is scanned over a surface allowing optical microscopy or spectroscopy at 100 nm resolution
Electrostatic force microscopy	EFM	Surface electrostatic potential are mapped
Scanning capacitance microscopy	SCM	Surface capacitance is mapped
Conductive atomic force microscopy	CAFM	Surface conductivity is mapped with an AFM instrument
Nanolithographic AFM	NAFM	An AFM tip etches, oxidizes or reacts a space permitting pattern fabrication at 10 nm or better resolution
Dip-pen nanolithography	DPN	An AFM tip, inked with a thiol or other molecule, writes on a surface at the nanometer scale

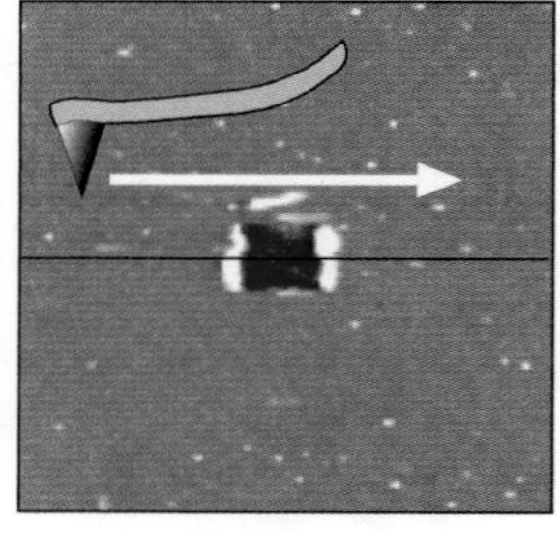

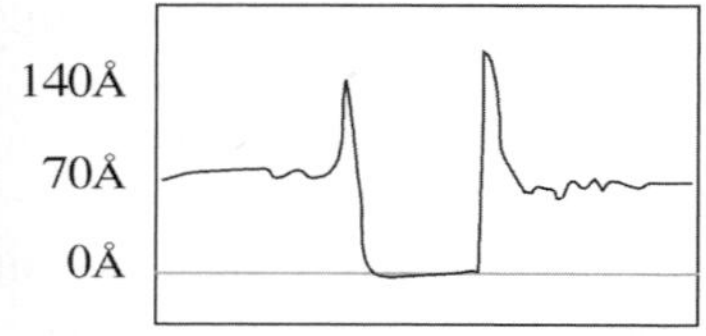

FIGURE I.1.5.17 An AFM tip, using relatively high force, was used to scratch a rectangular feature into a thin (70 Å) plasma-deposited film. The AFM also characterized the feature created.

the infrared (frequency = ω_{ir}), to illuminate a specimen. The light emitted from the specimen by a non-linear optical process, $\omega_{sum} = \omega_{visible} + \omega_{ir}$, is detected and quantified (Figure I.1.5.18). The intensity of the light at ω_{sum} is proportional to the square of the sample's second-order non-linear susceptibility (χ_2). The term susceptibility refers to the effect of the light field strength in polarizing molecules (molecular polarizability). The ω_{sum} light intensity vanishes where a material has inversion symmetry, i.e., in the bulk of the material there is compositional homogeneity so on either side of an introduced plane, there is symmetry. At an interface, the inversion symmetry is broken and an SFG signal is generated. Thus, SFG is exquisitely sensitive to the plane of the interface. In practice, ω_{ir} is scanned over a vibrational frequency range. This leads to vibrational absorptions that occur with only interface molecules. For this case, the SFG signal is resonantly enhanced and we see a vibrational spectrum. The advantages of SFG are the superb surface sensitivity, the cancellation of bulk spectral intensity (for example, this allows measurements at a water/solid interface), the richness of information from vibrational spectra, and the ability to study molecular orientation by the polarization of the laser light. SFG is not yet a routine method. The lasers and optical components are expensive and require precision alignment. However, the power of SFG for biomaterials studies has already been proven with studies on hydrated hydrogels, polyurethanes, surface active polymer additives, proteins, and non-fouling surfaces (Shen, 1989; Chen et al., 2002; Stein et al., 2009).

SFG is among the most surface-sensitive of all surface methods. It provides vibrational spectroscopic detail on surface composition, orientation, and interactions.

In Raman spectroscopy a bright light is focused on a specimen; most of the light scatters back at the same frequency as the incident beam. However, a tiny fraction of this light excites vibrations in the specimen, and thereby loses or gains energy. The frequency shift of the light corresponds to vibrational bands indicative of the molecular structure of the specimen. The Raman spectroscopic technique has been severely limited for surface studies due to its low signal level. However, in recent years

TABLE I.1.5.7 Methods Applicable for the Surface Characterization of Biomaterials

Method	Information Obtained
Second-harmonic generation (SHG)	Detect submolayer amounts of adsorbate at any light-accessible interface (air–liquid, solid–liquid, solid–gas)
Surface-enhanced Raman spectroscopy (SERS)	High-sensitivity Raman at rough metal interfaces
Ion scattering spectroscopy (ISS)	Elastically reflected ions probe only the outermost atomic layer
Laser desorption mass spectrometry (LDMS)	Mass spectra of adsorbates at surfaces
Matrix assisted laser desorption ionization (MALDI)	Though generally a bulk mass spectrometry method, MALDI has been used to analyze large adsorbed proteins
IR photoacoustic spectroscopy (IR-PAS)	IR spectra of surfaces with no sample preparation based on wavelength-dependent thermal response
High-resolution electron energy loss spectroscopy (HREELS)	Vibrational spectroscopy of a highly surface-localized region, under ultrahigh vacuum
X-ray reflection	Structural information about order at surfaces and interfaces
Neutron reflection	Thickness and refractive index information about interfaces from scattered neutrons – where H and D are used, unique information on interface organization can be obtained
Extended X-ray absorption fine structure (EXAFS)	A synchrotron method giving atomic-level chemical and nearest-neighbor (morphological) information
Near edge X-ray absorption fine structure (NEXAFS)	A synchrotron method providing information about bonding environments and chain orientation
Scanning Auger microprobe (SAM)	Spatially defined Auger analysis at the nanometer scale
Surface plasmon resonance (SPR)	Study aqueous adsorption events in real time by monitoring changes in surface refractive index
Rutherford backscattering spectroscopy (RBS)	Depth profiling of complex, multiplayer interfacial systems

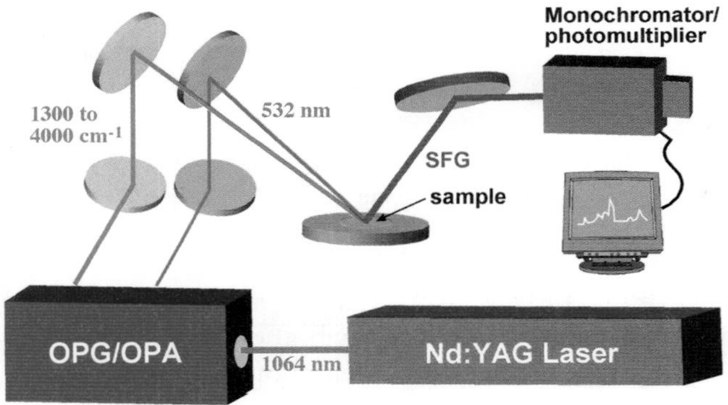

FIGURE I.1.5.18 Schematic diagram of a sum frequency generation (SFG) apparatus. (Based upon a diagram by Polymer Technology Group, Inc.)

great strides in detector sensitivity have allowed Raman to be applied for studying the minute mass of material at a surface. Also, surface enhanced Raman spectroscopy (SERS), Raman spectra taken from molecules on a roughened metal surface, can enhance Raman signal intensity by 10^6 or more. Raman spectra will be valuable for biomedical surface studies, because water, which absorbs radiation very strongly in the infrared range, has little effect on Raman spectra that are often acquired with visible light (Storey et al., 1995; Smith, 2008).

Synchrotron sources generate energetic radiation that can be used to probe matter. Such sources were originally confined to the physics community and devoted to fundamental studies. However, there are now many synchrotron sources offering access to researchers in all communities, with better instrumentation and improved data interpretation. Synchrotron sources are typically national facilities costing more than $100,000,000, and often occupying hundreds of acres (Figure I.1.5.19). By accelerating electrons to near the speed of light in a large, circular ring, energies covering a broad swathe of the electromagnetic spectrum (IR to energetic X-rays) are emitted. A synchrotron source (and ancillary equipment) permits a desired energy of the probe beam to be "dialed in" or scanned through a frequency range. Other advantages include high source intensity (bright light), and polarized light. Some of the experimental methods that can be performed with great success at synchrotron sources include crystallography, scattering, spectroscopy, microimaging, and nanofabrication. Specific surface spectroscopic methods include scanning photoemission microscopy (SPEM, 100 nm spatial resolution) (Leung et al., 2010), ultraESCA (100 μm spatial resolution, high energy resolution), and near edge X-ray absorption spectrometry (NEXAFS).

STUDIES WITH SURFACE METHODS

Hundreds of studies have appeared in the literature in which surface methods have been used to enhance the understanding of biomaterial systems. A few studies that demonstrate the power of surface analytical methods for biomaterials science are briefly described here.

Platelet Consumption and Surface Composition

Using a baboon arteriovenous shunt model of platelet interaction with surfaces, a first-order rate constant of

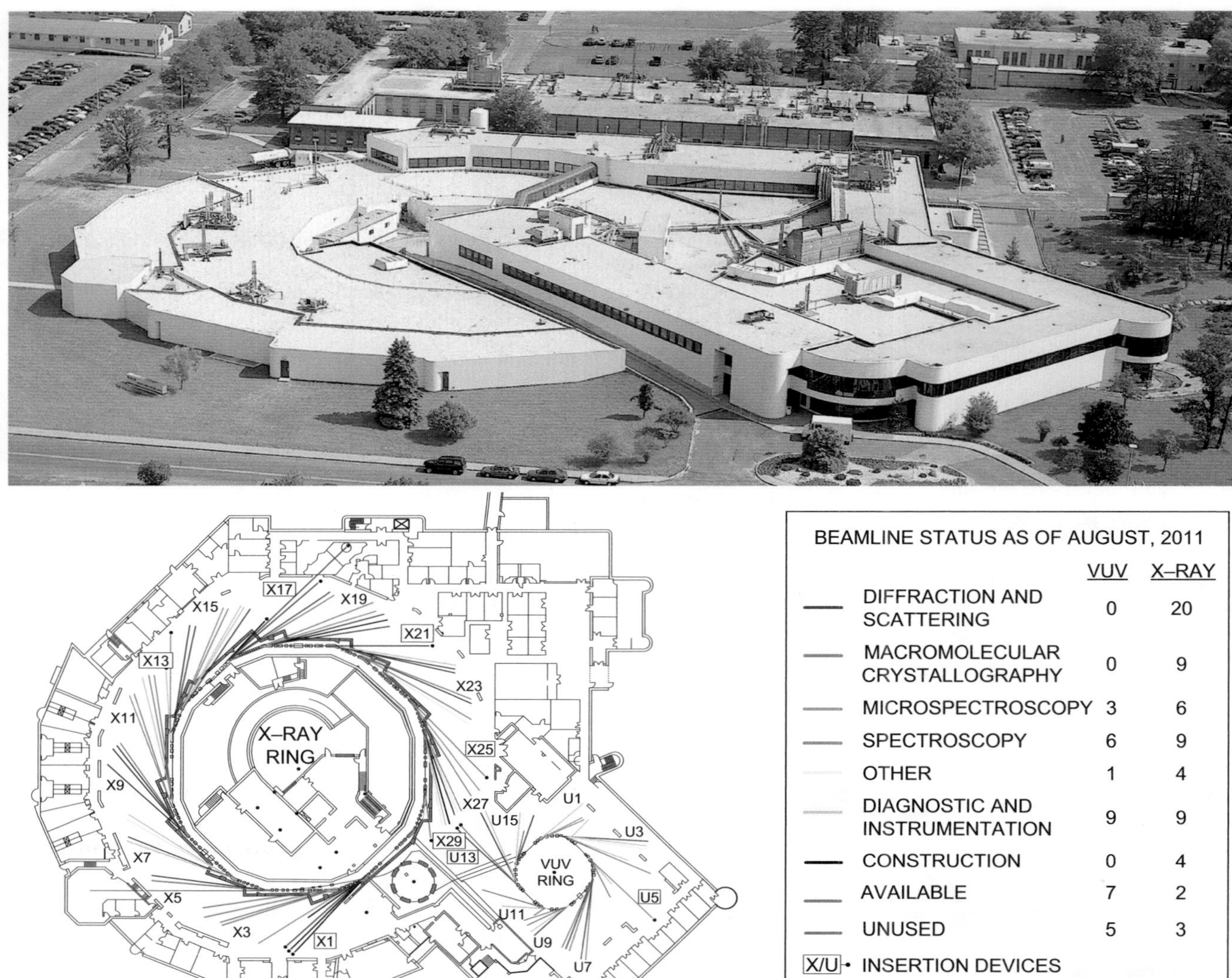

FIGURE I.1.5.19 The National Synchrotron Light Source, Brookhaven National Laboratories. (Images Courtesy of Brookhaven National Laboratory.)

reaction of platelets with a series of polyurethanes was measured. This rate constant, the platelet consumption by the material, correlated in an inverse linear fashion with the fraction of hydrocarbon-type groups in the ESCA C1s spectra of the polyurethanes (Hanson et al., 1982). Thus, surface analysis revealed a chemical parameter about the surface that could be used to predict long-term biological reactivity of materials in a complex *ex vivo* environment.

Contact-Angle Correlations

The adhesion of a number of cell types, including bacteria, granulocytes, and erythrocytes, has been shown, under certain conditions, to correlate with solid–vapor surface tension as determined from contact-angle measurements. In addition, immunoglobulin G adsorption is related to γ_{sv} (Neumann et al., 1983).

Contamination of Intraocular Lenses

Commercial intraocular lenses were examined by ESCA. The presence of sulfur, sodium, and excess hydrocarbon at their surfaces suggested contamination by sodium dodecyl sulfate (SDS) during the manufacture of the lenses (Ratner, 1983). A cleaning protocol was developed using ESCA to monitor results that produced a lens surface of clean PMMA.

Titanium

The discoloration sometimes observed on titanium implants after autoclaving was examined by ESCA and SIMS (Lausmaa et al., 1985). The discoloration was found to be related to accelerated oxide growth, with oxide thicknesses to 650 Å. The oxide contained considerable fluorine, along with alkali metals, and silicon. The source of the oxide was the cloth used to wrap the implant storage box during autoclaving. Since fluorine strongly affects oxide growth, and since the oxide layer has been associated with the biocompatibility of titanium implants, the authors advise avoiding fluorinated materials during sterilization of samples. Another paper by this group contains detailed surface characterization of titanium using a battery of surface methods, and addresses surface preparation, contamination, and cleaning (Lausmaa, 1996).

SIMS for Adsorbed Protein Identification and Quantification

All proteins are comprised of the same 20 amino acids and thus, on average, are compositionally similar. Surface analysis methods have shown the ability to detect and quantify surface-bound protein, but biological tools have, until recently, been needed to identify specific proteins. Modern static SIMS instrumentation, using a multivariate statistical analysis of the data, has demonstrated the ability to distinguish between more than 13 different proteins adsorbed on surfaces (Wagner and Castner, 2001). Also, the limits of detection for adsorbed proteins on various surfaces were compared by ESCA and SIMS (Wagner et al., 2002).

Poly(glycolic acid) Degradation Studied by SIMS

The degradation of an important polymer for tissue engineering, poly(glycolic acid), has been studied by static SIMS. As well as providing useful information on this degradation process, the study illustrates the power of SIMS for characterizing synthetic polymers and their molecular weight distributions (Chen et al., 2000).

Peptides on Polystyrene Studied by SFG and NMR. Although proteins at interfaces are critically important to biomaterials science, we have an incomplete picture of what proteins really look like and behave like on surfaces. Using the surface method, SFG, and a traditional bulk analysis method, solid-state NMR, an elegant picture of peptides on surfaces emerges. The study quantified, with site specificity and atomic resolution, the orientation and dynamics of side chains in labeled, synthetic model peptides adsorbed onto polystyrene (Weidner et al., 2010).

CONCLUSIONS

The instrumentation of surface analysis steadily advances, and newer instruments and techniques can provide invaluable information about biomaterials and medical devices. The information obtained can be used to monitor contamination, ensure surface reproducibility, and explore fundamental aspects of the interaction of biological systems with living systems. Considering that biomedical experiments, especially *in vivo* experiments, are typically expensive to perform, the costs for surface analysis are modest to ensure that the surface is as expected, stable, and identical from experiment to experiment. Surface analysis can also contribute to the understanding of medical device failure (and success). Myriad applications for surface methods are found in device optimization, manufacture, and quality control. Predicting biological reaction based on measured surface structure is a frontier area for surface analysis.

ACKNOWLEDGMENT

Support was received from the UWEB21 Engineering Research Center and the NESAC/BIO National Resource, NIH grant EB-002027, during the preparation of this chapter and for some of the studies described herein.

BIBLIOGRAPHY

Adamson, A. W., & Gast, A. (1997). *Physical Chemistry of Surfaces* (6th ed.). New York, NY: Wiley-Interscience.

Albrecht, T. R., Dovek, M. M., Lang, C. A., Grutter, P., Quate, C. F., et al. (1988). Imaging and modification of polymers by scanning tunneling and atomic force microscopy. *J. Appl. Phys.*, **64**, 1178–1184.

Allara, D. L. (1982). Analysis of surfaces and thin films by IR, Raman, and optical spectroscopy. *ACS Symp. Ser.*, **199**, 33–47.

Andrade, J. D. (1985). Surface and Interfacial Aspects of Biomedical Polymers. *Surface Chemistry and Physics* (Vol. 1). New York, NY: Plenum Publishers.

Avouris, P. (1990). Atom-resolved surface chemistry using the scanning tunneling microscope. *J. Phys. Chem.*, **94**, 2246–2256.

Belu, A. M., Graham, D. J., & Castner, D. G. (2003). Time-of-flight secondary ion mass spectrometry: Techniques and applications for the characterization of biomaterial surfaces. *Biomaterials*, **24**, 3635–3653.

Benninghoven, A. (1983). Secondary ion mass spectrometry of organic compounds (review). In A. Benninghoven (Ed.), *Springer Series of Chemical Physics: Ion Formation from Organic Solids* (Vol. 25, pp. 64–89). Berlin: Springer-Verlag.

Binnig, G., & Rohrer, H. (1986). Scanning tunneling microscopy. *IBM J. Res. Develop.*, **30**, 355–369.

Boland, T., Johnston, E. E., Huber, A., & Ratner, B. D. (1998). Recognition and nanolithography with the atomic force microscope. In B. D. Ratner, & V. V. Tsukruk (Eds.), *Scanning Probe Microscopy of Polymers Chapter 21* (Vol. 694, pp. 342–350). Washington, DC: American Chemical Society.

Bolles, K. M., Cheng, F., Burk-Rafel, J., Dubey, M., & Ratner, D. M. (2010). Imaging analysis of carbohydrate-modified surfaces using ToF-SIMS and SPRi. *Materials*, **3**, 3948–3964.

Briggs, D. (1986). SIMS for the study of polymer surfaces: A Review. *Surf. Interface Anal.*, **9**, 391–404.

Briggs, D., & Seah, M. P. (1983). *Practical Surface Analysis.* Chichester, UK: Wiley.

Campbell, S., Smith, J., Jungblut, H., & Lewerenz, H. (2007). Protein imaging on a semiconducting substrate: A scanning tunnelling microscopy investigation. *J. Electroanal. Chem.*, **599**, 313–322.

Castner, D. G., & Ratner, B. D. (2002). Biomedical surface science: Foundations to frontiers. *Surface Science*, **500**, 28–60.

Chen, J., Lee, J. -W., Hernandez, N. L., Burkhardt, C. A., Hercules, D. M., et al. (2000). Time-of-flight secondary ion mass spectrometry studies of hydrolytic degradation kinetics at the surface of poly(glycolic acid). *Macromolecules*, **33**, 4726–4732.

Chen, Z., Ward, R., Tian, Y., Malizia, F., Gracias, D. H., et al. (2002). Interaction of fibrinogen with surfaces of end-group-modified polyurethanes: A surface-specific sum-frequency-generation vibrational spectroscopy study. *J. Biomed. Mater. Res.*, **62**, 254–264.

Chilkoti, A., Boland, T., Ratner, B. D., & Stayton, P. S. (1995). The relationship between ligand-binding thermodynamics and protein-ligand interaction forces measured by atomic force microscopy. *Biophys. J.*, **69**, 2125–2130.

Dilks, A. (1981). X-ray photoelectron spectroscopy for the investigation of polymeric materials. In A. D. Baker, & C. R. Brundle (Eds.), *Electron Spectroscopy: Theory, Techniques, and Applications* (Vol. 4, pp. 277–359). London, UK: Academic Press.

Davies, M. C., & Lynn, R. A. P. (1990). Static secondary ion mass spectrometry of polymeric biomaterials. *CRC Crit. Rev. Biocompat.*, **5**, 297–341.

Dubey, M., Emoto, K., Cheng, F., Gamble, L. J., Takahashi, H., et al. (2009). Surface analysis of photolithographic patterns using ToF-SIMS and PCA. *Surf. Interface Anal.*, **41**(8), 645–652.

Dufrêne, Y. F. (2004). Using nanotechniques to explore microbial surfaces. *Nat. Rev. Microbiol.*, **2**(6), 451–460.

Dumas, P., Weldon, M. K., Chabal, Y. J., & Williams, G. P. (1999). Molecules at surfaces and interfaces studied using vibrational spectroscopies and related techniques. *Surf. Rev. Lett.*, **6**(2), 225–255.

Feldman, L. C., & Mayer, J. W. (1986). *Fundamentals of Surface and Thin Film Analysis*. New York, NY: North-Holland.

Garbassi, F., Morra, M., & Occhiello, E. (1998). *Polymer Surfaces: From Physics to Technology*. Chichester, UK: John Wiley and Sons.

Good, R. J. (1993). Contact angle, wetting, and adhesion: A critical review. In K. L. Mittal (Ed.), *Contact Angle, Wettability and Adhesion*. The Netherlands: VSP Publishers.

Hansma, P. K., Elings, V. B., Marti, O., & Bracker, C. E. (1988). Scanning tunneling microscopy and atomic force microscopy: Application to biology and technology. *Science*, **242**, 209–216.

Hanson, S. R., Harker, L. A., Ratner, B. D., & Hoffman, A. S. (1982). Evaluation of artificial surfaces using baboon arteriovenous shunt model. In G. D. Winter, D. F. Gibbons, & H. Plenk (Eds.), *Biomaterials 1980, Advances in Biomaterials* (Vol. 3, pp. 519–530). Chichester, UK: Wiley.

Jandt, K. D. (2001). Atomic force microscopy of biomaterials surfaces and interfaces. *Surface Science*, **491**, 303–332.

Kasemo, B. (2002). Biological surface science. *Surf. Sci.*, **500**, 656–677.

Lausmaa, J. (1996). Surface spectroscopic characterization of titanium implant materials. *J. Electron Spectrosc.*, **81**, 343–361.

Lausmaa, J., Kasemo, B., & Hansson, S. (1985). Accelerated oxide growth on titanium implants during autoclaving caused by fluorine contamination. *Biomaterials*, **6**, 23–27.

Leung, B. O., Brash, J. L., & Hitchcock, A. P. (2010). Characterization of biomaterials by soft X-ray spectromicroscopy. *Materials*, **3**(7), 3911–3938.

Leyden, D. E., & Murthy, R. S. S. (1987). Surface-selective sampling techniques in Fourier transform infrared spectroscopy. *Spectroscopy*, **2**, 28–36.

Li, H., & Cao, Y. (2010). Protein mechanics: From single molecules to functional biomaterials. *Accounts Chem. Res.*, **43**(10), 1331–1341.

Michel, R., & Castner, D. G. (2006). Advances in time-of-flight secondary ion mass spectrometry analysis of protein films. *Surf Interface Anal.*, **38**(11), 1386–1392.

Miles, M. J., McMaster, T., Carr, H. J., Tatham, A. S., Shewry, P. R., et al. (1990). Scanning tunneling microscopy of biomolecules. *J. Vac. Sci. Technol. A.*, **8**, 698–v702.

Muller, D. J., & Dufrêne, Y. F. (2008). Atomic force microscopy as a multifunctional molecular toolbox in nanobiotechnology. *Nat. Nanotechnol.*, 3(5), 261–269.

Neumann, A. W., Absolom, D. R., Francis, D. W., Omenyi, S. N., Spelt, J. K., et al. (1983). Measurement of surface tensions of blood cells and proteins. *Ann. New York Acad. Sci.*, **416**, 276–298.

Ninomiya, S., Ichiki, K., Yamada, H., Nakata, Y., Seki, T., et al. (2009). Precise and fast secondary ion mass spectrometry depth profiling of polymer materials with large Ar cluster ion beams. *Rapid Commun. Mass Spe.*, **23**(11), 1601–1606.

Quate, C. F. (1997). Scanning probes as a lithography tool for nanostructures. *Surf. Sci.*, **386**, 259–264.

Ratner, B. D. (1983). Analysis of surface contaminants on intraocular lenses. *Arch. Ophthal.*, **101**, 1434–1438.

Ratner, B. D. (1988). *Surface Characterization of Biomaterials*. Amsterdam: Elsevier.

Ratner, B. D., & McElroy, B. J. (1986). Electron spectroscopy for chemical analysis: Applications in the biomedical sciences. In R. M. Gendreau (Ed.), *Spectroscopy in the Biomedical Sciences* (pp. 107–140). Boca Raton, FL: CRC Press.

Ratner, B. D., & Castner, D. G. (2009). Electron spectroscopy for chemical analysis. In J. C. Vickerman, & I. S. Gilmore (Eds.), *Surface Analysis – The Principal Techniques* (2nd ed., pp. 47–112). Chichester, UK: John Wiley and Sons, Ltd.

Rosa, L. G., & Liang, J. (2009). Atomic force microscope nanolithography: Dip-pen, nanoshaving, nanografting, tapping mode, electrochemical and thermal nanolithography. *Journal of Physics: Condensed Matter*, **21**, 483001.

Rugar, D., & Hansma, P. (1990). Atomic force microscopy. *Phys. Today*, **43**, 23–30.

Shen, Y. R. (1989). Surface properties probed by second-harmonic and sum-frequency generation. *Nature*, **337**(6207), 519–525.

Scheutzle, D., Riley, T. L., deVries, J. E., & Prater, T. J. (1984). Applications of high-performance mass spectrometry to the surface analysis of materials. *Mass Spectrometry*, **3**, 527–585.

Smith, W. E. (2008). Practical understanding and use of surface enhanced Raman scattering/surface enhanced resonance Raman scattering in chemical and biological analysis. *Chem. Soc. Rev.*, **37**(5), 955.

Somorjai, G. A. (1981). *Chemistry in Two Dimensions: Surfaces*. Ithaca, NY: Cornell University Press.

Somorjai, G. A., & Li, Y. (2010). *Introduction to Surface Chemistry and Catalysis* (2nd ed.). Hoboken, NJ: John Wiley and Sons, Inc.

Stein, M., Weidner, T., McCrea, K., Castner, D., & Ratner, B. (2009). Hydration of sulphobetaine and tetra (ethylene glycol)-terminated self-assembled monolayers studied by sum frequency generation vibrational spectroscopy. *J. Physi. Chem. B*, **113**(33), 11550–11556.

Storey, J. M. E., Barber, T. E., Shelton, R. D., Wachter, E. A., Carron, K. T., et al. (1995). Applications of surface-enhanced Raman scattering (SERS) to chemical detection. *Spectroscopy*, **10**(3), 20–25.

Tirrell, M., Kokkoli, E., & Biesalski, M. (2002). The role of surface science in bioengineered materials. *Surf. Scie.*, **500**, 61–83.

Urban, M. W. (1993). *Vibrational Spectroscopy of Molecules and Macromolecules on Surfaces*. New York, NY: Wiley-Interscience.

Van Vaeck, L., Adriaens, A., & Gijbels, R. (1999). Static secondary ion mass spectrometry: (S-SIMS) Part I. Methodology and structural interpretation. *Mass Spectrom. Rev.*, **18**, 1–47.

Vickerman, J. C., & Gilmore, I. (2009). *Surface Analysis: The Principal Techniques* (2nd ed). Chichester, UK: John Wiley and Sons.

Vickerman, J. C., Brown, A., & Reed, N. M. (1989). *Secondary Ion Mass Spectrometry, Principles and Applications*. Oxford, UK: Clarendon Press.

Wagner, M. S., & Castner, D. G. (2001). Characterization of adsorbed protein films by time-of-flight secondary ion mass spectrometry with principal component analysis. *Langmuir*, **17**, 4649–4660.

Wagner, M. S., McArthur, S. L., Shen, M., Horbett, T. A., & Castner, D. G. (2002). Limits of detection for time of flight secondary ion mass spectrometry (ToF-SIMS) and X-ray photoelectron spectroscopy (XPS): Detection of low amounts of adsorbed protein. *J. Biomat. Sci. Polym. Ed.*, **13**(4), 407–428.

Watts, J. F., & Wolstenholme, J. (2003). *An Introduction to Surface Analysis by XPS and AES*. Chichester, UK: John Wiley & Sons.

Weidner, T., Breen, N. F., Li, K., Drobny, G. P., & Castner, D. G. (2010). Sum frequency generation and solid-state NMR study of the structure, orientation, and dynamics of polystyrene-adsorbed peptides. *Proc. Natl. Acad. Sci. USA*, **107**(30), 13288–13293.

Zisman, W. A. (1964). Relation of the equilibrium contact angle to liquid and solid constitution. In F. M. Fowkes (Ed.), *Contact Angle, Wettability and Adhesion ACS Advances in Chemistry Series* (Vol. 43, pp. 1–51). Washington, DC: American Chemical Society.

CHAPTER I.1.6 ROLE OF WATER IN BIOMATERIALS

Buddy D. Ratner
Professor, Bioengineering and Chemical Engineering, Director of University of Washington Engineered Biomaterials (UWEB), Seattle, WA, USA

In the world there is nothing more submissive and weak than water. Yet for attacking that which is hard and strong nothing can surpass it.

Lao Tzu

Water! Omnipresent, inert, a simple diluent – these words come to mind and prompt the question: "why is there a section of this textbook devoted to water?" The special properties of water, this substance that is critical for life as we know it, significantly impact the "materials-centric" subject of biomaterials, and the biology closely associated with biomaterials. Water is a unique, remarkable substance – in this instance, the word "unique" can be used without hyperbole. Although its chemical composition was elucidated in the 1770s by Lavoisier and others, its liquid structure continues to be explored today using state-of-the-art physical characterization methods. The importance of water for life as we know it was appreciated as early as 1913 in an interesting historical volume, *The Fitness of the Environment* (Henderson, 1913; available in reprint).

This chapter will first introduce the physical and chemical properties of water. Then water's significance for biomaterials and biology will be expanded upon.

WATER: THE SPECIAL MOLECULE

The simple "H_2O" structure of the water molecule does not immediately communicate the ability of one water molecule to interact with many other water molecules, in particular to interact not too strongly, and not too weakly. This interaction, largely via the hydrogen bond, is associated with the dipole of the water molecule (slightly more negatively charged at the oxygen and positively charged at the hydrogen, i.e., partial negative and positive charges). This dipole, combined with the bent shape (bond angle of 104.6°) of the molecule and its small size contributes to the special properties of liquid water. Figure I.1.6.1 illustrates schematically the geometry of the water molecule, and suggests some hydrogen bonding possibilities with neighboring water molecules (as well as the occasional H^+ or OH^-).

> The physical properties of water are profoundly unique compared to all other substances.

The unique physical and chemical properties of water, attributable to the 104.6° bond angle and the ability to form multiple hydrogen bonds, are discussed here. Note how special water is compared to other liquid and solid substances.

Melting Point and Boiling Point

The physical properties of H_2O that make it stand out from all other related molecules are best appreciated from the data in Table I.1.6.1.

As the molecular weight for compounds in this series of related molecules decreases, the boiling points also decrease, until you reach water, where the boiling point is (unexpectedly) up to 160°C higher. At common temperatures and pressures on earth, all the other compounds exist as gases, except water. A similar dramatic trend is seen in freezing point.

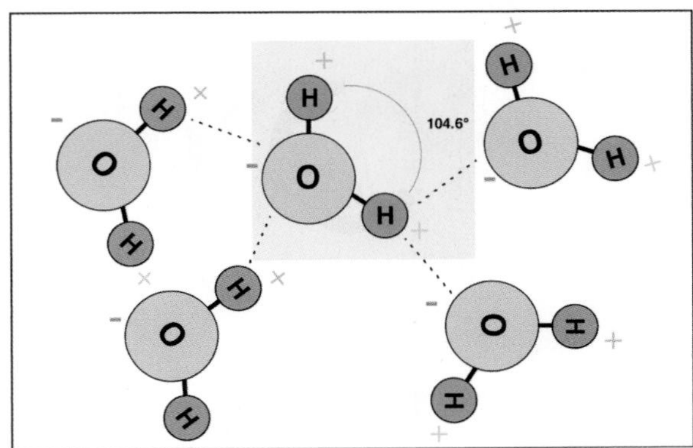

FIGURE I.1.6.1 Schematic illustration of five water molecules suggesting their geometry and dipole. A hydrogen bond is represented by a dotted line. Consider the many possibilities for these electrostatic (charge) interactions.

Density and Surface Tension

Water has a density of 0.997 g/ml at 25°C. As the temperature decreases, water, like most substances, increases in density, until 4°C where its density is 1.00 g/ml. Then, as temperature is decreased further, an unusual phenomenon is observed. The density decreases as the temperature is lowered. At 0°C, the density is 0.92 g/ml. Thus, ice floats on liquid water. There is special significance to this. If bodies of water froze from the bottom to the top (i.e., if ice were heavier than water), during cold periods in earth's geological history most aquatic life might have been destroyed.

Surface tension is a measure of the magnitude of cohesive forces holding molecules together. For substances at ambient temperature and pressure, water has the highest surface tension with one exception – mercury. The water surface tension is 72.8 dyne/cm. Compare this to, for example, ethanol at 22.3 dyne/cm (mN/m) and ethylene glycol at 47.7 dyne/cm. This surface tension permits a water strider insect (family Gerridae) to transport across lakes riding on the "surface skin" of water, a metal paper clip to float (Figure I.1.6.2), and assists with the transport of water from soil to the tops of tall trees.

TABLE I.1.6.1 Boiling Point and Freezing Point of Water Contrasted to Other H_2X Compounds

H_2X Compound		Molecular Weight	Boiling Point	Freezing Point
H_2Te	Hydrogen telluride	130	−2°C	−49°C
H_2Se	Hydrogen selenide	81	−41°C	−66°C
H_2S	Hydrogen sulfide	34	−60°C	−82°C
H_2O	Water	18	100°C	0°C

FIGURE I.1.6.2 A metal paper clip floats on the water surface "skin," a manifestation of water's high surface tension. (Photograph by Buddy Ratner.)

Specific Heat and Latent Heats of Fusion and Evaporation

Water has a specific heat of 1.0 calorie/g°C, that is, the heat per unit mass required to raise the temperature of water 1°C. For comparison, copper has a specific heat of 0.1 cal/g°C, and ethyl alcohol has a specific heat of 0.6 cal/g°C. Water has a higher specific heat than any other common substance. Related to the specific heat is the latent heat of fusion or evaporation. The latent heat is the energy in calories per gram taken up or released by matter changing phase (liquid/solid or solid/liquid) with no change in temperature. Again, water has anomalously high values compared to other common substances.

Water as a Solvent

Water is a remarkably powerful and versatile solvent. It will dissolve proteins, ions, sugars, gases, many organic liquids, and even lipids (up to the critical micelle concentration). In fact, it is the most versatile solvent that we know.

WATER: STRUCTURE

Figure I.1.6.1 suggests that water can form extended hydrogen bonding structures. These structures grow in three dimensions. The molecular crystal structure of ice is shown in Figure I.1.6.3. At temperatures of 0°C and below, thermal fluctuations of molecules (referred to as kT vibrations) are low and a continuous, ordered three-dimensional network of water molecules can form. The hydrogen bonds in water are relatively low strength, approximately 5 kcal/mol (21 kJ/mol); in contrast, a C–H bond is approximately 100 kcal/mol. At room temperature, the thermal fluctuations of water molecules can disrupt this H-bond association – if it were not so, water

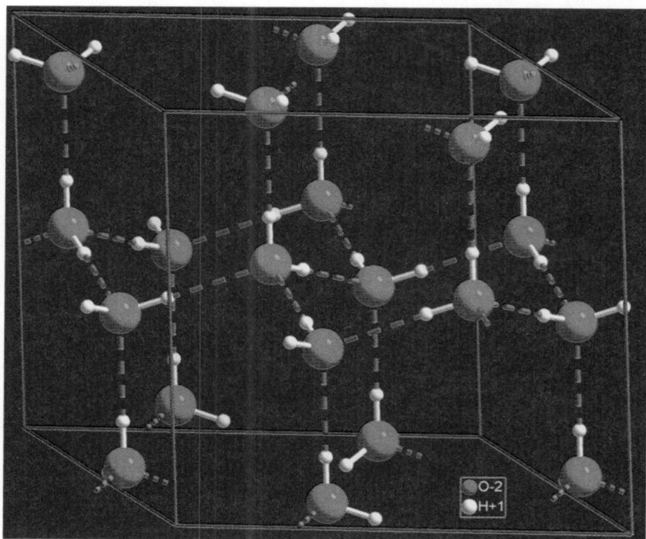

FIGURE I.1.6.3 The hexagonal ice structure. (Wikipedia public domain image.)

would coalesce into a solid at room temperature. An oft-cited model of water molecule organization suggests that disrupted groupings of water molecules rapidly form new clusters (dimers, trimers, tetramers, pentamers, hexamers, etc.) (Keutsch and Saykally, 2001). This is the "flickering cluster" model of water structure. More recently, it has been established that liquid water is closer to a hydrogen-bonded continuum with much bond angle distortion from the tetrahedral organization seen in the ice structure, and with bonds breaking and reforming so rapidly (in the order of 200 femtoseconds), that the cluster model is not an accurate description (Smith et al., 2005). The continuum model is now the most accepted model for water structure, through there is still much controversy about water structure, and new insights are frequently reported.

The water structure discussion, above, applies to bulk, pure water. The water continuum organization can be disrupted or reorganized by dissolved ions or solutes, by absorption into hydrophilic polymers (hydrogels when the polymer is cross-linked), and by solid surfaces. Each of these cases will be very briefly addressed.

Dissolved ions and solutes may have a profound impact on water organization (Marcus, 2009). Dissolved salts are dissociated to form an anion and a cation, with each ion having a shell of water of different structure from the bulk. This hydration shell perturbs the structure of water adjacent to it (Marcus, 2009). Much of the study of dissolved ions has centered around studies of the Hofmeister series of anions:

$$CO_3^{2-} > SO_4^{2-} > S_2O_3^{2-} > H_2PO_4^- > F^- > Cl^- > Br^- < NO_3^- > I^- > ClO_4^- > SCN^-$$

Ions to the left of the series are referred to as kosmotropes – they generally precipitate proteins from solution and inhibit denaturation. Ions on the right side of the series are considered chaotropes, and these increase protein solubility and are more denaturing. Note that chloride is roughly at the center of the series, and thus might be expected to induce little change in proteins and water. Though earlier theories on the Hofmeister series suggested that ions to the left enhanced water structure while those to the right destructured or disordered water, this concept is not fully supported by recent experiments. The Hofmeister effect is impacted by the degree of hydration of ions in water, which cation is involved and specific interactions of ions with solutes. Review articles that discuss the complexities of the Hofmeister effect and consider contemporary experimental and theoretical work are available (Zhang and Cremer, 2010; Paschek and Ludwig, 2011).

The "swelling" water entrained in gels (hydrogels, see Chapter I.2.5) is thought to be in three possible forms. Different nomenclatures are used to describe these forms, but basically three states have been proposed: (1) free water (similar to bulk water); (2) tightly-bound water (more structured and with limited mobility); and (3) intermediate water (with characteristics of both free and bound water) (Jhon and Andrade, 1973; Akaike et al., 1979). Below, when the hydrophobic effect is discussed, other possibilities will be suggested. Evidence for different forms of water in gels continues to accumulate (Sekine and Ikeda-Fukazawa, 2009) and implications for biomaterials and biocompatibility have been discussed (Tsuruta, 2010). The nature of the water in the gel may also impact on the diffusion of molecules through it, blood interactions, and its performance as a cell-support in tissue engineering. The water that swells a hydrogel is impacted by the polymer chains at a nanoscale, because of its close proximity to the polymer that comprises the mesh of the hydrogel. When macroscopic pores are introduced into the hydrogel, the polymer will impact the water in a different fashion, i.e., the surface of the pores will interact with water in a manner similar to surface effects discussed in the next paragraph; in the interior of the pore (away from the pore wall), the water will have an organization more similar to bulk water.

When a solid surface or biomaterial disrupts the continuum structure of water, the water near the solid surface will have to adopt a new organization to achieve energy minimization for the total system. Since all our biomaterials will first see water before proteins or cells ever diffuse or transport to the surface, the nature of this surface water layer may be, from a biomaterials science standpoint, the most important event driving biointeractions at interfaces.

> The nature of water in proximity to surfaces may be the primary driver for interactions between biomaterials and biological systems.

There is ample evidence from many analytical techniques, and also from computational methods, that water organization is altered close to a surface compared to the bulk. Most experimental data suggests this difference persists over a length of 1 to 4 water molecules, before reverting to a structure indistinguishable from the bulk water. There is much research on the adsorption of the first layer of water on materials. Clearly, the organization of this first layer will dictate the structure of subsequent layers. On many close packed metal crystal surfaces (for example, Pt, Ni, Pd) results suggest that a water bilayer exists with a structure analogous to ice (Hodgson and Haq, 2009). This may have led to the somewhat misleading term for interfacial water, "ice-like water." Although the water structure is different from the bulk liquid at all surfaces, the specific ice-like organization is predominantly seen at these close packed metal crystal surfaces. Above the water bilayer directly in contact with the metal, water structure is altered for a few molecular layers until it becomes indistinguishable from bulk water. Other studies demonstrate differences between hydrophilic and hydrophobic surfaces as to their interactions with water (Howell et al., 2010). Hydrophilic surfaces largely have a substantial effect on water organization, while at hydrophobic interfaces there appears to be a low water density zone, sometimes called a depletion zone, less than a nanometer from the surface. There are hundreds of recent studies on the water–solid interface, mostly in physical chemistry journals. A comprehensive review of this complex and still controversial subject would be impossible in this textbook. However, the take home message is relatively straightforward – the presence of an interface alters the water structure adjacent to it.

WATER: SIGNIFICANCE FOR BIOMATERIALS

When a protein or a cell approaches a biomaterial, it interacts with the surface water first. This final section will briefly review some implications of this surface water for biomaterials.

When a protein or a cell approaches a biomaterial, it interacts with the surface water first.

Hydrophobic Effect, Liposomes, and Micelles

If we place a drop of oil under water it will round up to a sphere as it floats to the water surface. The common explanation for this is that because oil molecules do not H-bond with water molecules the oil cannot strongly interact with the bulk water phase. Another way to say this is that the oil disrupts the bulk (continuum) structure of water, and the water molecules at the oil interface then have to restructure to a new (more ordered) organization. This decrease in water entropy is energetically unfavorable according to the second law of thermodynamics. So, the oil minimizes its interfacial area with the water by coalescing into a sphere (a sphere has the lowest surface area for a given volume). This is called the hydrophobic effect (Tanford, 1978; Widom et al., 2003). It is not driven by the oil–water molecular interactions, but rather by the necessity to minimize the more structured (low entropy) organization of water molecules.

If we now take a surfactant molecule comprised of an "oily" segment and a polar (water-loving) segment, by shielding the oily phase from the water with the hydrophilic head groups, contact between oil and water is minimized. Figure I.1.6.4 illustrates some hydrophobic effect-driven supramolecular aggregate structures that lead to this energy minimization. Liposomes and block copolymer micelles are widely used in drug delivery, where the hydrophobic region (micelle core or liposome bilayer) or the aqueous center (liposome) can carry a hydrophobic drug or hydrophilic drug, respectively (see Chapter II.5.18). Bilayer sheets are used in biosensors to orient and stabilize receptor proteins.

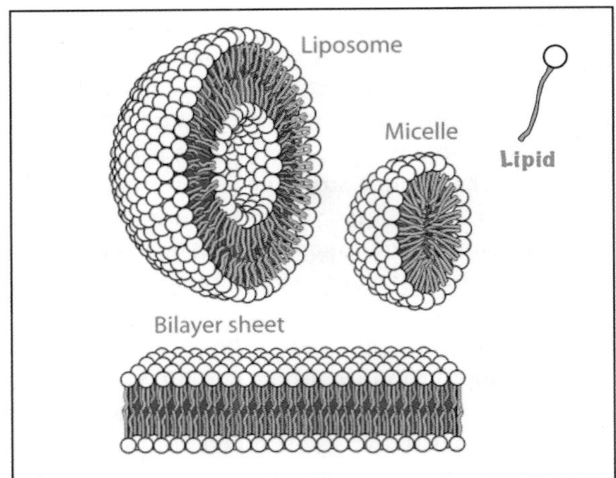

FIGURE I.1.6.4 Lipid molecules with polar head groups (white) and hydrophobic tails (brown), when placed in water, organize themselves to minimize surface area contact between the hydrophobic tail (typically comprised of methylene units, -CH2-) and water. This minimization of contact area, depending on precise conditions, can lead to micelles, liposomes or bilayer sheets. (Modified from a public domain image on Wikimedia Commons.)

Hydrogels

The structured water within hydrogels has been discussed. Water structure associated with hydrogels has been implicated in their interactions with blood (Garcia et al., 1980; Tanaka and Mochizuki, 2004). Water structure in hydrogels is also thought to be important in the performance of hydrogel contact lenses, specifically the rate of lens dehydration (Maldonado-Codina and Efron, 2005).

Protein Adsorption. Protein adsorption is discussed in detail in Chapter II.1.2, and is critically important for understanding the performance of biomaterials. A question commonly posed is: "Why do proteins bind rapidly and tenaciously to almost all surfaces?" A model that can explain this considers structured water at interfaces. All surfaces will organize water structure differently from the bulk structure; this organization almost always gives more structured (lower entropy) water. If a protein can displace the ordered water in binding to the surface, the entropy of the system will increase as the water molecules are released and gain freedom in bulk water. This is probably the driving force for protein adsorption at most interfaces. There are some specially engineered surfaces referred to as "non-fouling" or "protein-resistant" (see Chapter I.2.10). These surfaces may resist protein adsorption by binding or structuring water so strongly that the protein molecule cannot "melt" or displace the organized or tightly bound water, and thus there is no driving force for adsorption.

Life

Living systems self-assemble from smaller molecular units. For example, think about the organizational processes in going from an egg, to a fetus, to a mature creature. Much of this assembly is driven by hydrophobic interactions (i.e., entropically by water structure). Another area where water structure has a major impact on life involves DNA and its unique binding of water (Khesbak et al., 2011). Also, consider enzymes that are so essential to life. A substrate molecule enters the active site of an enzyme by displacing water molecules. The unique mechanical properties of cartilage under compression can be modeled by considering water organization. In fact, the average human is 57% by weight water, on a molar basis by far the major component in the human body. Thus we can well appreciate, as Henderson surmised in 1913, that water is essential to this phenomenon we call life. As you work through this textbook, think about how the phenomena you are reading about may be driven or controlled by water that comprises such a large fraction of all biological systems.

BIBLIOGRAPHY

Akaike, T., Sakurai, Y., Kosuge, K., Kuwana, K., Katoh, A., et al. (1979). Study on the interaction between plasma proteins and synthetic polymers by circular dichroism. *ACS Polymer Preprints*, **20**(1), 581–584.

Garcia, C., Anderson, J. M., & Barenberg, S. A. (1980). Hemocompatibility: Effect of structured water. *Transactions of the American Society for Artificial Internal Organs*, **26**, 294–298.

Henderson, L. J. (1913). *The Fitness of the Environment*. New York, NY: The Macmillan Company.

Hodgson, A., & Haq, S. (2009). Water adsorption and the wetting of metal surfaces. *Surface Science Reports*, **64**(9), 381–451.

Howell, C., Maul, R., Wenzel, W., & Koelsch, P. (2010). Interactions of hydrophobic and hydrophilic self-assembled monolayers with water as probed by sum-frequency-generation spectroscopy. *Chemical Physics Letters*, **494**(4–6), 193–197.

Jhon, M. S., & Andrade, J. D. (1973). Water and hydrogels. *Journal of Biomedical Materials Research*, **7**, 509–522.

Keutsch, F. N., & Saykally, R. J. (2001). Water clusters: Untangling the mysteries of the liquid, one molecule at a time. *Proceedings of the National Academy of Sciences*, **98**(19), 10533–10540.

Khesbak, H., Savchuk, O., Tsushima, S., & Fahmy, K. (2011). The role of water H-bond imbalances in B-DNA substrate transitions and peptide recognition revealed by time-resolved FTIR spectroscopy. *Journal of the American Chemical Society*, **133**(15), 5834–5842.

Maldonado-Codina, C., & Efron, N. (2005). An investigation of the discrete and continuum models of water behavior in hydrogel contact lenses. *Eye & Contact Lens: Science & Clinical Practice*, **31**(6), 270–278.

Marcus, Y. (2009). Effect of ions on the structure of water: Structure making and breaking. *Chemical Reviews*, **109**, 1346–1370.

Paschek, D., & Ludwig, R. (2011). Specific ion effects on water structure and dynamics beyond the first hydration shell. *Angewandte Chemie (International Edition)*, **50**(2), 352–353.

Sekine, Y., & Ikeda-Fukazawa, T. (2009). Structural changes of water in a hydrogel during dehydration. *The Journal of Chemical Physics*, **130**(3). 034501.

Smith, J. D., Cappa, C. D., Wilson, K. R., Cohen, R. C., Geissler, P. L., et al. (2005). Unified description of temperature-dependent hydrogen-bond rearrangements in liquid water. *Proceedings of the National Academy of Sciences of the United States of America*, **102**(40), 14171–14174.

Tanaka, M., & Mochizuki, A. (2004). Effect of water structure on blood compatibility: Thermal analysis of water in poly(meth) acrylate. *Journal of Biomedical Materials Research*, **68A**, 684–695.

Tanford, C. (1978). The hydrophobic effect and the organization of living matter. *Science*, **200**(4345), 1012–1018.

Tsuruta, T. (2010). On the role of water molecules in the interface between biological systems and polymers. *Journal of Biomaterials Science, Polymer Edition*, **21**(14), 1831–1848.

Widom, B., Bhimalapuram, P., & Koga, K. (2003). The hydrophobic effect. *Physical Chemistry Chemical Physics*, **5**(15), 3085.

Zhang, Y., & Cremer, P. S. (2010). Chemistry of Hofmeister anions and osmolytes. *Annual Review of Physical Chemistry*, **61**(1), 63–83.

SECTION I.2

Classes of Materials Used in Medicine

SECTION I.2

Classes of Materials Used in Medicine

CHAPTER I.2.1 INTRODUCTION: THE DIVERSITY AND VERSATILITY OF BIOMATERIALS

Allan S. Hoffman
Professor of Bioengineering and Chemical Engineering, UWEB Investigator, University of Washington, Seattle, WA, USA

In Section I.2 the reader will find chapters covering the fundamental principles of the many diverse materials used in a great variety of implants, extracorporeal medical devices, dermal and mucosal treatments and devices, drug delivery systems, sensors, and diagnostic assays. When such biomaterial devices and systems are exposed to the biologic environment, it is their surfaces that first encounter biologic species such as proteins, which usually deposit a monolayer of protein. Subsequently, cells interact with the adsorbed protein monolayers on the surfaces of the biomaterials. Therefore, it is the fundamental compositions and "nanotextures" of those biomaterials surfaces that are expected to strongly influence the overall responses of the biological environment to the materials (as well as the responses of the biomaterials to the bioenvironment), as detailed in Section II.2. Section I.2 also includes chapters that cover the key technologies used to fabricate, modify, and characterize biomaterials and their surfaces. Thus, this section is fundamental to the entire spectrum of the science and engineering of biomaterials and their surfaces for many diverse applications.

In this section, the authors often include important historical details. The wide diversity and sophistication of materials currently used in medicine and biotechnology is testimony to the significant scientific and technological advances that have occurred over the past 50 years. From World War II to the early 1960s, a few pioneering surgeons were taking commercially available polymers and metals, fabricating implants and components of medical devices from them, and applying them clinically. There was little government regulation of this activity, and yet many of those earliest implants and devices had remarkable success. However, there were also some dramatic failures, often based on a poor choice or improper fabrication of the biomaterial components (see Chapters III.1.3 and III.1.5). This led the surgeons to enlist the aid of physical, biological, and materials scientists and engineers, and the earliest multidisciplinary "bioengineering" collaborations were born some time in the 1970s–1980s.

These teams of physicians and scientists and engineers not only recognized the need to control the composition, purity, and physical properties of the materials they were using, but they also recognized the need for new materials with new and special properties. This stimulated the development of many new materials, beginning in the 1970s. New materials were designed *de novo* specifically for medical use, such as biodegradable polymers, "medical grade" silicones, pyrolytic carbon, and bioactive glasses and ceramics. Some were derived from existing materials that were then fabricated using new technologies, such as polyester fibers that were knitted or woven in the form of tubes for use as vascular grafts or cellulose acetate plastic that was processed as bundles of hollow fibers for use in artificial kidney dialysers. Some materials were "borrowed" from unexpected sources, such as pyrolytic carbons or titanium alloys that had been developed for use in space and nuclear technology. Some were the result of serendipity, such as the use of titanium in dental implants. Other materials were specifically modified to provide special biological properties, such as immobilization of heparin for anticoagulant surfaces, one of the earliest "bioengineered" biomaterials.

In this section, the reader will find chapters on the five basic classes of biomaterials: Polymers (Chapter I.2.2, focused on synthetic polymeric biomaterials, including subsections on: (A) polyurethanes, (B) silicones, (C) fluorinated biomaterials, and (D) acrylics); metals (Chapter I.2.3, including subsections on: (A) titanium and nitinol (NiTi) and (B) stainless steels); ceramics and glasses (Chapter I.2.4, including subsections on: (A) natural and synthetic hydroxyapatites and (B) alumina); engineered natural materials (Chapter I.2.7, with special focus on hyaluronic acid and collagen); and carbons (Chapter I.2.8).

There are additional chapters on the special properties and compositions of biomaterials. For example, sometimes two different classes of materials are combined together into a composite material, such as silica-reinforced silicone rubber or hydroxyapatite particle-reinforced poly(lactic acid). Such composites can be considered as a sixth class of biomaterials, and they are discussed in Composites (Chapter I.2.9). This chapter emphasizes the mechanical properties of polymers that have been reinforced with specially surface-treated particles, fibers, and fabrics.

It should be noted that some implants and devices, such as artificial heart valves, are comprised of more than one class of biomaterial. For example, the Starr-Edwards ball-in-cage heart valve includes the silicone rubber ball, the polyester fabric for fixation to the tissues, and the stainless steel struts; however, these implants should not be considered as "composites," even though one or more component may be a composite material (such as the silica-reinforced silicone rubber ball composite material in the Starr-Edwards ball-in-cage heart valve). These implants are considered in other sections and chapters in this text.

Another chapter on special forms and properties of synthetic polymers is Hydrogels (Chapter I.2.5). This chapter includes a wide variety of compositions and structures, but it does not include "smart hydrogels" or drug delivery from hydrogels, which are found in other chapters, e.g., Chapter I.2.11 in the section on "smart," environmentally-responsive polymers, and Chapter II.5.16, which is a very detailed chapter on drug delivery systems. Section I.2 also includes a chapter on bioresorbable and bioerodible materials (Chapter I.2.6) that are a very important subclass of polymeric biomaterials.

Following these chapters, there are several chapters that focus on surfaces of biomaterials, including the principles of surface modification (Chapter I.2.12), surface-immobilized biomolecules (Chapter I.2.17), and surface patterning (Chapter I.2.13). These chapters are fundamental to the bioresponses of biomaterials.

Special fabrication processes and forms of materials are also considered in chapters on electrospinning of fibers (Chapter I.2.16), medical fibers and textiles (Chapter I.2.14), textured and porous materials (Chapter I.2.15), and micro- and nanoparticles (Chapter I.2.19). A chapter on biomimetic materials (Chapter I.2.18) is included, with special focus on bio-inspired designs and compositions of biomaterials.

More recently, biomaterials scientists and engineers have developed a growing interest in natural tissues and polymers in combination with living cells. This is particularly evident in the field of tissue-engineering, which focuses on the repair or regeneration of natural tissues and organs. This interest has stimulated the isolation, purification, and application of many different natural materials, including decellularized natural tissues and spider silk. The principles and applications of these biomaterials and modified biomaterials are critically described, discussed, and reviewed in Section II.6 of this text.

CHAPTER I.2.2 POLYMERS: BASIC PRINCIPLES

Daniel E. Heath and Stuart L. Cooper
William G. Lowrie Department of Chemical and Biomolecular Engineering, The Ohio State University, Columbus, OH, USA

INTRODUCTION

Polymer materials possess an array of unique properties which make them useful in a wide variety of biomaterial applications such as orthopedics, dental, hard and soft tissue replacements, and cardiovascular devices. In fact, polymers represent the largest class of materials used in medicine. This chapter introduces the basic principles in polymer science, illustrates how polymer materials can be specifically designed to fill needs in the biomaterials field, and provides examples of how this class of materials is currently used in medical applications.

The central idea of this chapter is structure–property relationships, which means molecular characteristics such as molecular architecture, molecular weight, and chemical composition are directly related to the physical and chemical properties of the macroscopic material. For instance, polymer scientists in other fields have been able to exploit structure–property relationships to create non-stick coatings, pressure-sensitive adhesives, and the penetration-resistant materials used in bullet-proof vests. A biomaterials scientist aware of structure–property relationships can rationally engineer a polymer system for a specific need.

THE POLYMER MOLECULE

Molecular Structure of Single Polymer Molecules

The hallmark of polymer molecules is high molecular weight. A single polymer molecule could have a molecular weight of 200,000 Da compared to a water molecule, which has a molecular weight of 18 Da. Furthermore, polymer molecules are organized into very interesting architectures. Common shapes of polymer molecules are shown in Figure I.2.2.1. The simplest is the linear chain where there is a single molecular backbone. When linear chains of two different composition polymers (e.g., A and B) are linked together, the resultant polymer is called an A–B block copolymer. If another chain is added to the second chain, it may be called an A–B–C triblock copolymer, or more simply, an A–B–C block copolymer. Branched structures are also possible where a central polymer backbone has smaller side chains extending from it. Branches can occur due to undesired side reactions during synthesis, or can be purposefully incorporated into the molecular structure. So far we have discussed polymer systems in which there are discrete polymer molecules. However, what would happen if you took a linear polymer molecule and covalently bonded it to the backbone of another linear chain? You would produce a branched structure. If you repeated this act many times over you would eventually link all of the polymer chains together into one very large *network* polymer. This is also possible by using small di- or tri-functional "cross-linker" molecules to react with pendant or terminal reactive groups on linear polymer chains, eventually yielding a network polymer.

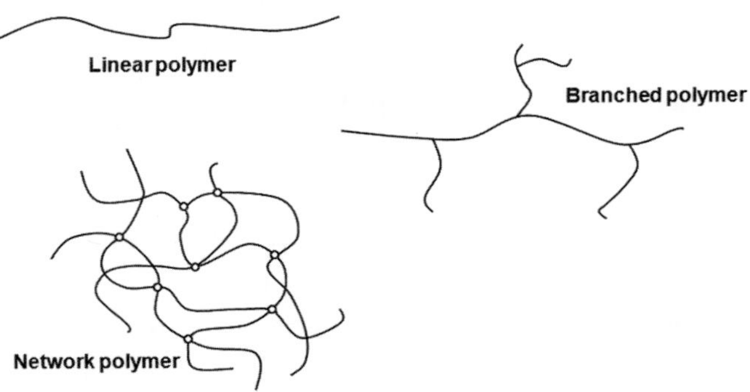

FIGURE I.2.2.1 Some of the molecular structures available for polymer molecules. In the schematic of the network, polymer open circles indicate cross-link sites.

Chemical Structure of Single Polymer Molecules

If you were able to see the individual atoms making up a polymer molecule, you would notice the same basic structure repeats over and over again. Figure I.2.2.2 shows a schematic of a linear polypropylene molecule. The polymer backbone is a series of carbon–carbon single bonds, while the hydrogen atoms and methyl group are *pendant groups*. The polymer chain is composed of many –(–CH_2–$CHCH_3$–)– groups covalently linked end-to-end. This structure is called the *repeat unit* of a polymer molecule. Etymologically, polymer comes from the Greek "poly," meaning many, and "meros," meaning part – many parts. In polymer molecules the "mer" is

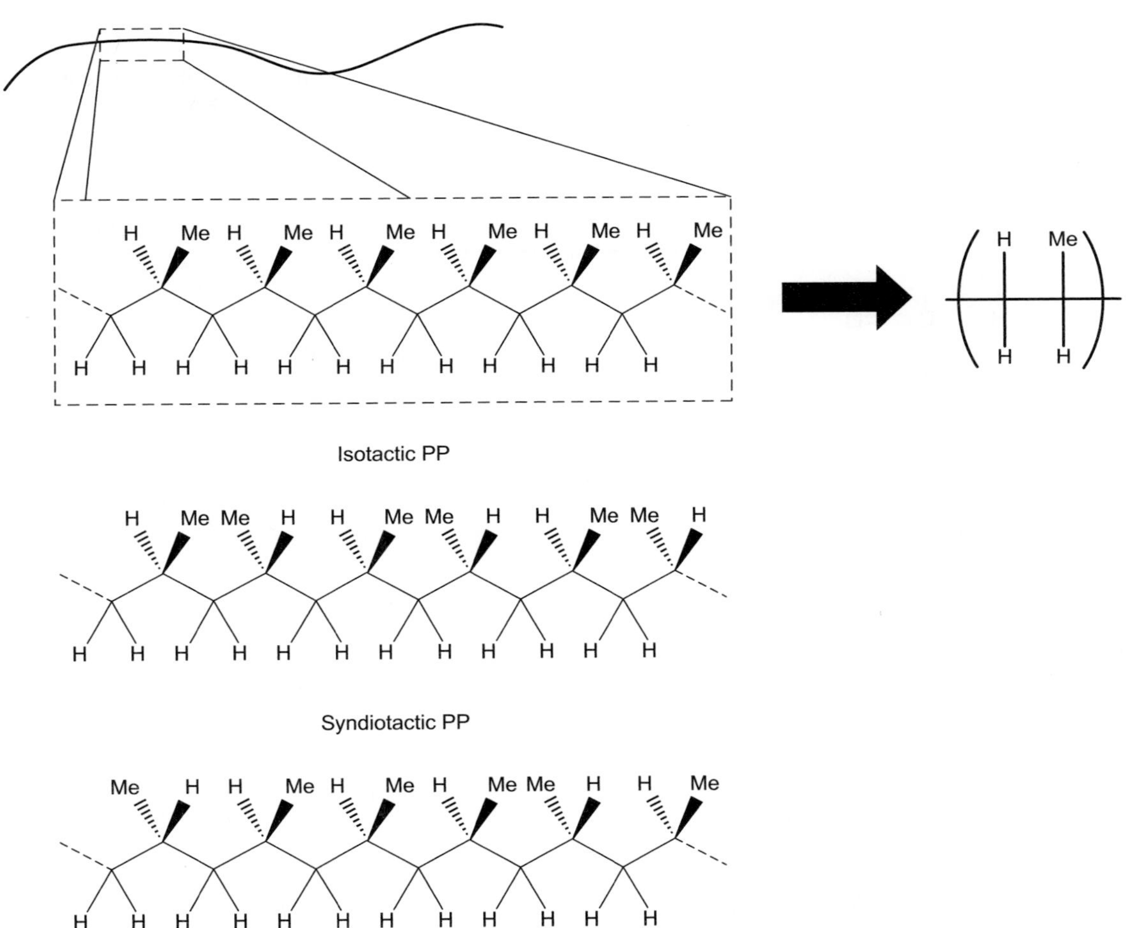

FIGURE I.2.2.2 Polypropylene repeat unit and different tactic isomers. In the schematic Me indicates a methyl group.

the repeat unit. The repeat unit can be controlled through polymer synthesis and plays a large role in the macroscopic behavior of the polymer. Figure I.2.2.3 shows the repeat units of several synthetic and natural polymers commonly used in the biomaterials field.

Copolymers

Sometimes it is advantageous to synthesize copolymers – polymers containing more than one chemically distinct repeat unit. For instance, a researcher may synthesize a

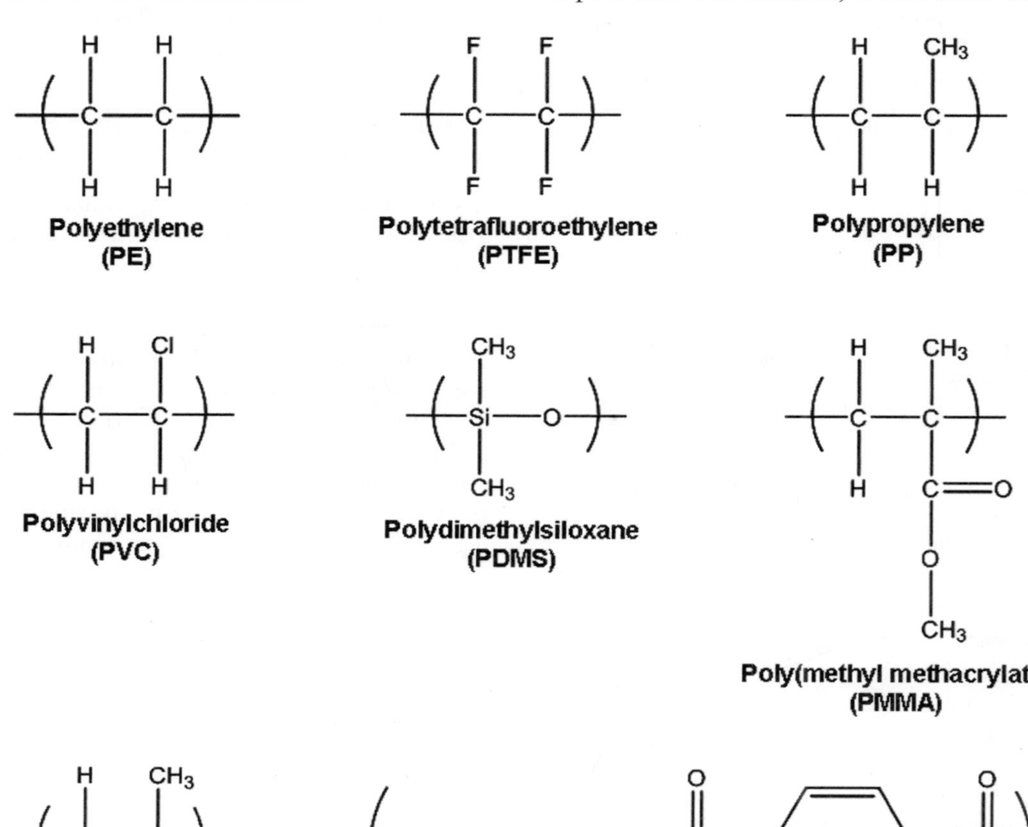

FIGURE I.2.2.3 Repeat unit of common polymer biomaterials.

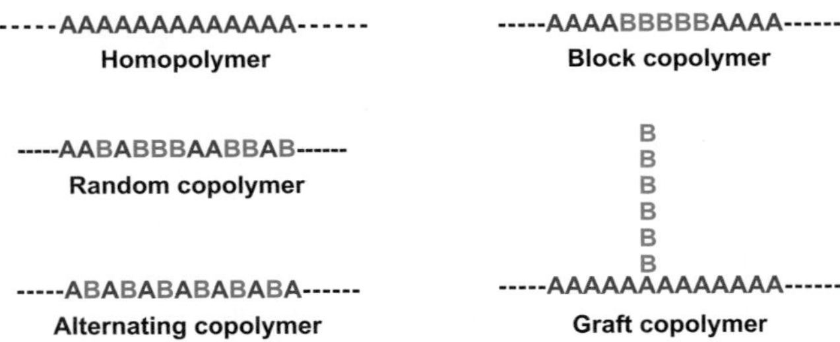

FIGURE I.2.2.4 Some of the molecular structures available for copolymers.

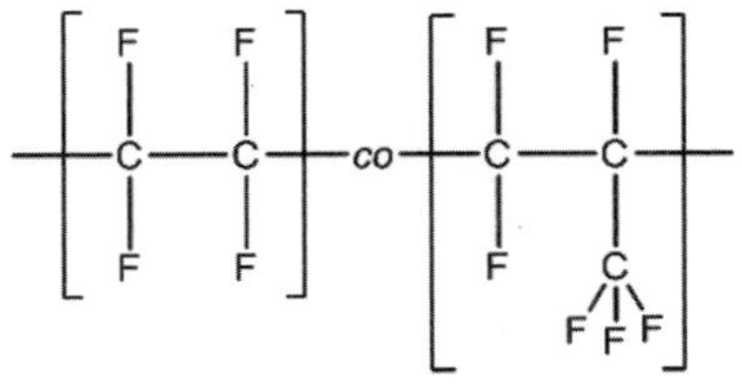

Poly(tetrafluoroethylene-co-hexafluoropropylene) random copolymer

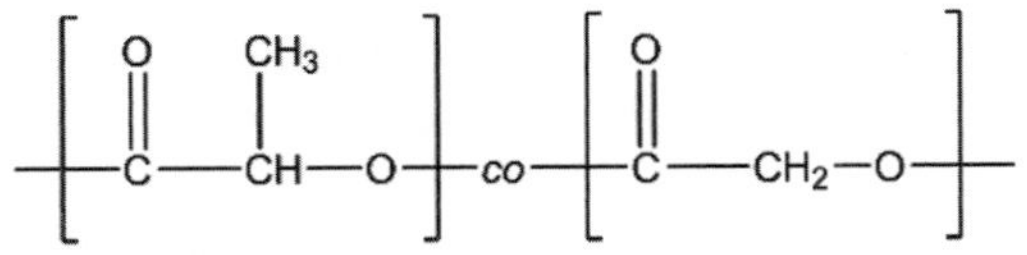

Poly (lactide-co-glycolide) random copolymer

FIGURE I.2.2.5 Repeat unit of two common copolymer biomaterials.

polymer which contains repeat units "A" and "B." As shown in Figure I.2.2.4, there are many different ways the repeat units could be organized. Random copolymers occur when the "A" and "B" repeat units have no order in the backbone; however, alternating, block, and graft copolymers are also possible and the arrangement of repeat units affects the physical behavior of the biomaterial. The repeat units of two random copolymers commonly used in the biomaterials field are shown in Figure I.2.2.5.

Determination of Chemical Composition

A researcher will often need to verify the chemical structure of polymers or determine the composition of copolymer systems. Two common techniques a scientist would use are nuclear magnetic resonance spectroscopy (NMR) and infrared spectroscopy (IR). NMR is an analytical technique which exploits the magnetic moments associated with nuclei containing an odd number of protons (most commonly the ^{1}H and ^{13}C isotopes). These protons are excited to a higher energy state through a burst of radiofrequency radiation. The nuclei relax to a lower energy state which is measured as an electric signal. Fourier transform analysis is used to convert this time domain electrical signal into the frequency domain. Due to shielding, protons attached to different structural units will display chemical shifts, meaning their peaks in the NMR spectrum will be at different frequencies. Through analysis of the peak placement and intensity, the chemical structure of molecules can be determined.

Infrared spectroscopy (IR) is also used to determine the chemical composition of polymers. The sample of interest is irradiated with infrared radiation and the sample adsorbs certain wavelengths, resulting in specific molecular motion (such as C–H stretching). The IR spectrum is created by plotting absorbance versus wavelength. Like NMR, analysis of the spectra can lead to the verification of a polymer's composition.

Often the chemical composition of the polymer surface is different from the bulk. For a medical implant the surface composition is highly important, since it will interface with the biological environment. To probe the surface composition X-ray photoelectron spectroscopy (XPS) – also known as electron spectroscopy for chemical analysis (ESCA) – is a common technique. A sample is bombarded with X-rays which results in the ejection of inner shell electrons from the atoms displayed on the material surface. The kinetic energies of the ejected electrons are measured and interpreted into information about the chemical composition of the surface (for more information on surface characterization techniques see Chapter I.1.5).

Tacticity

Tacticity describes the stereochemistry of the repeat units in polymer chains. To illustrate the discussion on tacticity let us consider a molecule of polypropylene (PP). When stretched into its *planar zigzag* form, as seen in Figure I.2.2.2, you can see that sometimes the methyl groups are all on one side of the backbone, sometimes they alternate from side to side, and sometimes they are randomly distributed. During routine synthesis, a polymer chemist would normally produce an *atactic* version of PP, one where the methyl group is randomly located in front of and behind the polymer backbone. However, when a special catalyst is used during synthesis a chemist can produce *isotactic* PP, where all the methyl groups are located on one side of the "stretched out" polymer backbone, or *syndiotactic* PP where the methyl groups regularly alternate from side to side. As will be discussed later, tacticity can drastically affect the physical behavior of the polymer system, largely by affecting the ability of the polymer molecules to crystallize. Tactic isomers occur whenever an atom in the polymer backbone has the capacity to form tetrahedral bonding and is bonded to four different ligands. Such atoms (generally carbon) are referred to as *asymmetric*.

MOLECULAR WEIGHT

The Molecular Weight Distribution and its Averages

During polymerization, polymer chains are built up from low molecular weight monomers. The number of monomer repeat units in each polymer chain is called the *degree of polymerization*. However, in a traditional free radical or condensation polymerization synthesis, each polymer chain will not have the same degree of polymerization. For instance, in the free radical polymerization of polyethylene, one polymer chain may add 3000 monomers, a second may add 4500 monomers, and a third may only add 1500. Therefore, most polymer systems have a *distribution of molecular weights*. Since polymer materials are made from molecules with a variety of molecular weights, it is incorrect to talk about *the* molecular weight of a polymer system. Instead, a polymer system is described by different averaged values of molecular weight. The two most commonly used averages are the number average molecular weight (M_n) and the weight average molecular weight (M_w). The mathematical definitions of these averages are supplied in Equations 1 and 2, where N_i is the number of molecules with "i" repeat units and M_i is the molecular weight of a polymer chain with "i" repeat units. Since M_w is calculated using the square of the molecular weight it is always greater than or equal to M_n.

$$M_n = \frac{\sum_i N_i M_i}{\sum_i N_i} \tag{1}$$

$$M_w = \frac{\sum_i N_i M_i^2}{\sum_i N_i M_i} \tag{2}$$

The ratio of M_w to M_n is the polydispersity index (PDI) (Equation 3) which is a measurement of the breadth of the molecular weight distribution. If the PDI of a polymer system is unity the number average and weight average molecular weights are identical, meaning the polymer sample is monodisperse (all chains have the same degree of polymerization). For most condensation polymers, the PDI is approximately 2.

$$PDI = \frac{M_w}{M_n} \tag{3}$$

The higher the average molecular weight, the stronger a polymer material will be. However, the melt/solution viscosity also increases with increasing average molecular weight, making the material more difficult to process. Often a molecular weight range exists within which most desired physical behaviors are achieved, yet the material is still easily processed. Generally, a number average molecular weight range from 25,000 to 50,000 Da is suitable for condensation polymers, while values from about 50,000 Da up to hundreds of thousands are preferred for addition polymers.

Characterizing the Molecular Weight Distribution

Through understanding the importance of molecular weight and its distribution, polymer scientists have developed many methods for measuring average molecular weight values. For instance, M_n can be determined through techniques such as end-group analysis, vapor pressure lowering, and freezing point depression. In the following paragraphs we will first discuss common methods historically used to measure M_n (osmotic pressure) and M_w (light scattering). Although osmotic pressure and light scattering are powerful and useful techniques in the field of polymer science, they have been replaced in recent years by gel permeation chromatography (GPC), which gives much more detail about the molecular weight distribution, and does so much quicker. From the molecular weight distribution obtained by GPC, the first and second moments of the distribution, M_n and M_w, are readily determined.

In osmotic pressure experiments, a dilute polymer solution is separated from pure solvent by a semi-permeable membrane through which solvent can freely pass but which excludes polymer. The activity of the pure solvent differs from that of the solvent molecules in the solution phase resulting in a thermodynamic driving force – the osmotic pressure (π or p) causes molecules of pure solvent to diffuse across a membrane (which is permeable only to solvent) and into a compartment containing a solute (e.g., polymer) in solution. This causes the fluid level in the solution compartment to rise, resulting in a hydrostatic pressure head. Equilibrium is reached when the pressure head exactly offsets the osmotic pressure. By measuring the rise in fluid level, the osmotic pressure can be calculated and related to the number average molecular weight. A number of polymer solutions are prepared, each with a distinct concentration. The osmotic pressure is measured for each solution and plotted against concentration, the concentration is then extrapolated to zero, and the number average molecular weight can be determined from the intercept of the regression line.

Light scattering is a common technique to determine the M_w of a polymer system. In dilute solutions, the scattering of light is directly proportional to the number of molecules. The scattered intensity is observed at a distance r and angle Q from the incident beam, and is related to the size or weight average molecular weight of the molecule. The light scattering behavior of a series of polymer solutions with varying concentrations are measured and the data are extrapolated to zero concentration to determine M_w. Once the M_n and M_w values are determined, the PDI of the polymer system can be calculated.

Gel permeation chromatography (GPC) is the most commonly used molecular weight characterization technique in modern polymer science laboratories. Unlike osmotic pressure and light scattering, which provide *average* values of molecular weight, GPC provides the entire molecular weight distribution from which all the desired average values can be calculated. In GPC experiments, a dilute polymer solution flows through a column packed with solid separation media which contains pores on the same size scale as the polymer molecules. As the polymer molecules flow though the column, smaller chains diffuse into the pores and are detained, while larger polymer chains move through the column faster since they are excluded from the pore structure, resulting in higher molecular weight material eluting from the column first, followed by lower molecular weight species. The concentration of the effluent stream is monitored and a plot of detector signal (which is proportional to polymer concentration) versus elution time can be generated. Elution time is converted to molecular weight though calibration with mono-disperse polymers of known molecular weights.

CONNECTING PHYSICAL BEHAVIOR WITH CHEMICAL CHARACTERISTICS

We have discussed the key characteristics of polymer molecules (molecular architecture, chemical composition, tacticity, and molecular weight). Now we will relate these molecular characteristics to macroscopic properties, and illustrate how these characteristics can be manipulated to create a polymer system with the desired behavior. We will focus on the tensile properties, hydrophilicity, and biodegradability of polymer systems.

Physical States of Linear Polymers. When designing a biomaterial, physical behavior is a key feature. For instance, if you were creating a cement for use in loadbearing bones (tibia or femur) you would have to ensure the material is both strong enough to act as a cement, but not so brittle that it would fail due to low fracture toughness. The physical properties of a polymer system stem from the intermolecular interactions occurring between individual polymer molecules; thus, the molecular characteristics we have discussed up to this point are of extreme importance. As you will see in the following text, the four most fundamental molecular characteristics of polymer chains which determine the physical behavior of a polymer are chain stiffness, chain composition or polarity, chain architecture or regularity, and molecular weight.

> The three fundamental molecular properties of polymer molecules are chain stiffness, chain polarity, and chain architecture. They determine two important temperatures that characterize polymer molecules: T_g, the glass transition temperature, and T_m, the crystalline melting temperature. T_m is only present when there are crystalline regions in a polymer.

In Figure I.2.2.2, we see polypropylene molecules extended into planar zigzags. Although this is a convenient way to draw polymer molecules, this type of extended structure is rarely seen in nature. More often, polymer molecules are found as unorganized and three-dimensional structures called the *random coil*. In an *amorphous* structure, each random coil is highly interpenetrated with its neighbors. Polymers in the rubbery state or the glassy state have this amorphous molecular arrangement. Under certain conditions, some polymers will arrange themselves into highly organized crystalline domains resulting in a semi-crystalline material. Each of these states will now be explored in more depth.

The Rubbery State. Rubbery polymers are amorphous. However, the random coils have enough thermal energy for rotation to occur around single bonds. If you were able look at polymer molecules in the melt state, you would see that each random coil is continuously changing shape (conformation). This molecular motion becomes more intense as the thermal energy in the system increases. Macroscopically these materials are soft, flexible, and extensible, due to the molecular motion available to the molecules.

The Glassy State. As the polymer system is cooled, rotation around single bonds becomes hindered due to energy barriers created by a segment of a chain having to move (rotate) past a neighboring segment. As the temperature drops, the rate of rotation around main chain bonds in a polymer chain becomes slower and slower, and the chain gets stiffer and stiffer. Eventually, at a low enough temperature, single bond rotation ceases and the interpenetrated random coils become frozen in space. This is called the glassy state, and the temperature where single bond rotation ceases is called the glass transition temperature (T_g). A material below its T_g is called a glass because it is hard, stiff, and brittle. Molecules in the glassy state can no longer rearrange themselves under applied stress, so deformation results in straining the secondary interactions between molecules. The opposite occurs when an amorphous polymer is heated: the amorphous region goes from hard and glassy, to "leathery," to rubbery, and if the material is not cross-linked, it will eventually flow as a viscous fluid and can be processed into shapes.

Ramen Noodles: Glassy Versus Rubbery Polymer Chains

If you're reading this you are probably a college student, which means you are very familiar with Ramen Noodles. A package of noodles can be cooked in about 10minutes. When you first remove the noodles from the package, you will notice that the individual noodles are rather random in shape and intertwined with their neighbors. Furthermore, before boiling, these noodles are very rigid and fixed in relation to one another. This food product is a pretty good example of a polymer in the glassy state, where each noodle represents one polymer molecule. As the noodles are boiled they become flexible, and can easily change their shape and slip around; however, they retain their intertwined character. These noodles are a good example of a polymer material in the rubbery state.

The Semi-Crystalline State. All polymer systems form glasses at sufficiently low temperatures. However, as a melt is cooled, certain polymers have the ability to pack into a regular lattice, leading to the formation of stable crystalline domains. In polyethylene (PE), these stable crystalline domains are formed by chains in the planar zigzag conformation, while the crystalline chains in syndiotactic poly(vinyl chloride) (PVC) have a helical conformation. Since only a portion of the long polymer chains can crystallize (some segments will not be able to pack into the crystallites) this state is called semi-crystalline. The temperature above which such crystalline regions will melt is called the crystalline melting point (T_m). Crystallites act to stiffen and reinforce the bulk material, and extend the stiffness and strength properties of a material well above the glass transition temperature until the crystallite melting point (T_m) is reached.

The Physical Behavior of Linear and Amorphous Polymers

Unlike metals which are held together by metallic bonds and atomic crystals which are held together by covalent bonds, polymer materials are held together by secondary interactions such as London forces, dipole–dipole interactions, and hydrogen bonding. For this reason, polymers are often mechanically weaker than other classes of materials; however, they can display physical behavior more similar to native tissue.

The simplest polymer system is an amorphous rubbery or glassy polymer composed of linear polymer chains. When rubbery materials are strained, the polymer molecules are able to deform and extend resulting in a material which is macroscopically soft and weak, yet highly extensible. However, as the T_g of the material approaches and exceeds the environmental conditions, the material becomes a glass and is much stronger yet stiffer. The mechanical behavior of polymer systems is often probed through stress–strain analysis. From this analysis several key material properties can be determined: modulus (a measurement of material stiffness); tensile strength (the stress at failure); and percent elongation (the amount of deformation at failure). The room temperature mechanical properties of several polymeric biomaterials are presented in Table I.2.2.1.

The Physical Behavior of Other Physical States

Cross-linking, crystallinity, and copolymerization greatly affect the physical behavior of polymer systems, and controlling these parameters gives a polymer scientist ways of specifying the physical behavior of a polymer system. Figure I.2.2.6 illustrates how each of these affects polymer modulus as a function of temperature. In Figure I.2.2.6A we see the standard modulus–temperature behavior of a linear and amorphous polymer system (solid line). Below the T_g the material is a glass and has a high modulus. However, at the glass transition temperature we see a dramatic drop in modulus due to the increased mobility within the polymer structure. At temperatures just higher than the T_g, we see a plateau on the curve where modulus declines more slowly with temperature. This is called the rubbery plateau, and in this region the polymer is still solid-like but soft, flexible, and extensible. Eventually, as the temperature is increased more, we see the modulus curve take another drop corresponding to the material beginning to flow. Notice that amorphous materials do not have a melting temperature. Melting refers to the loss of crystallinity, and since these materials are non-crystalline they never truly melt. As we will discover later, the architecture of the polymer chains, especially where there is a specific tacticity, determines if segments of a polymer chain can pack together and form crystallites.

In Figure I.2.2.6A we can also see the modulus–temperature behavior of a semi-crystalline material. In this curve we see a small drop in the modulus at the glass transition temperature, due to the onset of backbone rotation in the amorphous regions of the polymer. Since crystallites are unaffected by the T_g, the magnitude of this drop is greatly affected by the amount of crystallinity in the system. After the T_g we see that the modulus holds steady until the melting point, illustrating that crystallinity

TABLE I.2.2.1 Physical Properties and Equilibrium Water Absorption of Common Polymeric Biomaterials

Material	Tensile Modulus (GPa)	Tensile Strength (MPa)	Elongation at Break (%)	Water Absorption (%)
Polyethylene **(PE)**	0.8–2.2	30–40	130–500	0.001–0.02
Poly(methyl methacrylate) **(PMMA)**	3–4.8	38–80	2.5–6	0.1–0.4
Polytetrafluoroethylene **(PTFE)**	1–2	15–40	250–550	0.1–0.5
Polylactide **(PLA)**	3.4	53	4.1	<0.5
Poly(hydroxyethyl methacrylate)† **(PHEMA)**	0.29	0.15	71	40
Polypropylene **(PP)**	1.6–2.5	21–40	100–300	0.01–0.035
Poly(ethylene terephthalate) **(PET)**	3–4.9	42–80	50–500	0.06–0.3

†Tensile properties were measured after the polymer was equilibrated with water.

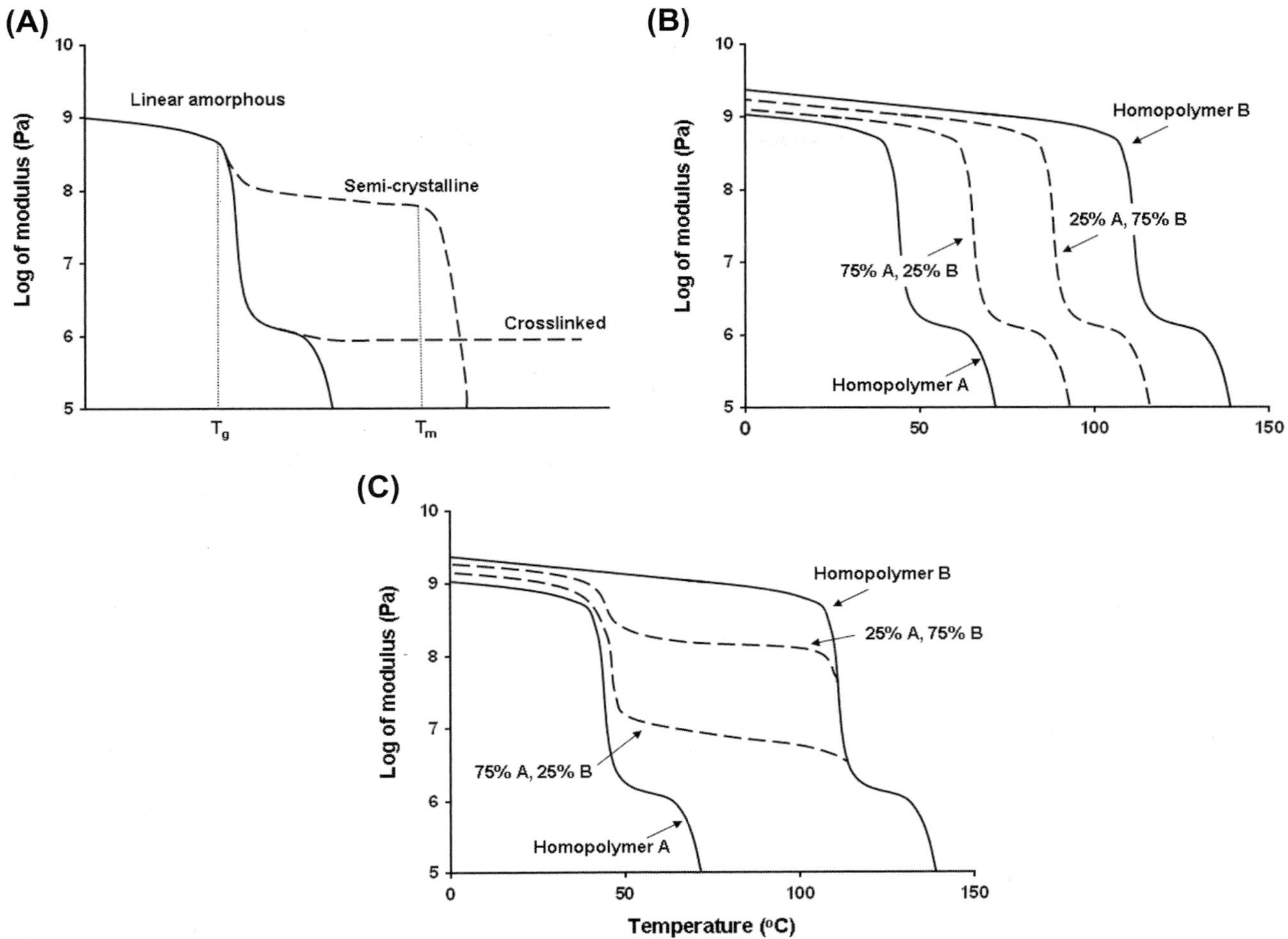

FIGURE I.2.2.6 Modulus versus temperature behavior for: (A) linear amorphous, semi-crystalline, and cross-linked polymer; (B) random copolymer; and (C) block or graft copolymer.

is a way to increase the window of temperatures in which a polymer can be employed. At T_m the crystallites melt and the polymer begins to flow. The ability of chains to crystallize is a special feature of chain architecture that is sometimes called "symmetry" or "regularity," i.e., a streamline profile in cross-section with small side groups that allows the chain segments to pack into crystallites. In special cases, a stereospecific property of some chains with asymmetric atoms is called tacticity; it can also control the ability of a polymer to crystallize. Polymers without asymmetric atoms in the backbone (and therefore having no tactic isomers) can crystallize along with isotactic and syndiotactic polymers. However, atactic materials cannot organize themselves into a lattice due to the irregularity in their chemical structure, and so are therefore permanently amorphous. Branched polymers have difficulty crystallizing, since the branch sites interfere with the ability of the polymer to organize into a lattice. In cases where branch sites are purposefully incorporated into the material they will lower the extent of crystallization (such as in the case of low density polyethylene).

Also in Figure I.2.2.6A you can see the modulus–temperature behavior of a cross-linked polymer. Again, at the glass transition temperature you will see a decrease in modulus due to the onset of single bond rotation; however, above the T_g the modulus is relatively independent of temperature because the cross-links act to tether the polymer chains in place. In fact, the cross-linked polymer will not experience a large decrease in modulus until the temperature is high enough to begin thermally degrading the bonds holding the structure together.

Copolymerization of more than one monomer unit could also be used to control the physical behavior of a material. Figure I.2.2.6B shows modulus–temperature curves for a random copolymer system where the shape of the curve is maintained, but shifted laterally. If block or graft copolymers are produced, one would see the distinct transition of both materials used to generate the copolymer, as seen in Figure I.2.2.6C.

Table I.2.2.2 provides the repeat unit, T_g, and T_m of common polymers. Chemical characteristics that facilitate backbone rotation result in lower T_g materials, while structural characteristics that hinder backbone rotation result in higher T_g materials. For instance, the –Si–O– bond in polydimethylsiloxane is unhindered, and is a freely rotating bond down to very low temperatures, resulting in a

TABLE I.2.2.2 Structural Factors Affecting the Glass Transition Temperature and the Crystalline Melting Point of Polymers

Material	T_g(°C)	T_m(°C)
$-(CH_2-CH_2)-$ Polyethylene **(PE)**	−113 to −103	125–135
$-(Si(CH_3)_2-O)-$ Polydimethylsiloxane **(PDMS)**	−125	−40
$-(CH_2-CH(CH_3))-$ Polypropylene† **(PP)**	−30 to −3	160–180
$-(CH_2-C(CH_3)_2)-$ Polyisobutylene **(PIB)**	−73	44
$-(CH_2-CH(CH_2CH_3))-$ Polybutylene† **(PB)**	−38 to −33	126–139
$-(CH_2-CHCl)-$ Poly(vinyl chloride) **(PVC)**	87	150*

†Isotactic.
*Decomposition temperature.

very low T_g. As pendant group bulkiness increases, the polymer chains along the backbone have more difficulty rotating due to steric hindrances, illustrated by the much higher T_g of polystyrene in comparison with polyethylene. However, pendant groups of four or more carbon atoms tend to create "free volume" around the polymer backbone, making main chain rotation easier, and the T_g may be lowered, as seen in the alkyl methacrylate family as the alkyl group increases in length. Increased polarity, especially where hydrogen bonds may act between side groups, increases the polar interactions between chains, slowing the rate of main chain rotations and resulting in a higher T_g. In general, the structural factors that affect the glass transition temperature (backbone flexibility, pendant group structure and polarity, main chain symmetry and polarity) similarly affect T_m, as illustrated by the data in Table I.2.2.2.

A Tragedy Caused by a Change in Temperature and its Effect on Tg and Polymer Properties

In 1986 the space shuttle *Challenger* broke apart during lift-off, killing the seven crew members aboard and halting all US space flights for the next two and a half years. This tragedy was caused by the failure of an O-ring - a polymer gasket – connecting the shuttle's external fuel tank to one of its solid rocket boosters. The O-ring failed, in part, due to the abnormally cold weather on the morning of the launch, causing stiffening of the gasket as the polymer approached its T_g, and that undermined its ability to form a tight seal. The gasket failure allowed a tongue of flame to reach both the solid rocket booster and the external fuel tank, and resulted in the structural failure of both. The shuttle then broke apart due to extreme aerodynamic loads. See NASA, 1986.

Characterizing a Polymer's Physical State and Behavior

Crystallinity plays a large role in determining polymer behavior. The *degree of crystallinity* – a measurement of how much of the polymer is incorporated into crystalline regions – can be studied with techniques such as measurement of density, X-ray diffraction (XRD), and infrared spectroscopy. In XRD experiments a polymer sample is bombarded with X-ray radiation and the intensity of the scattered X-rays is measured as a function of scattering angle. A fully amorphous material would produce a very broad peak, covering all scattering angles. However, crystalline materials will produce sharp peaks at specific angles. Furthermore, the placement of these peaks corresponds to a particular crystalline structure. By integrating the areas under the amorphous peak and the crystalline peaks, and then calculating the fractional area of the crystalline peaks relative to the total area, a researcher can determine the degree of crystallinity.

Stress–strain analysis provides information about the mechanical behavior of a polymeric biomaterial. In this test a polymer sample of known dimensions is deformed at a given rate, and the force needed to cause the deformation is recorded.

Measuring the Transition Temperatures Between States

Measuring transition temperatures (T_g and T_m) is important to understanding how a material will behave in a certain application. Two common methods for determining the transition temperatures in polymer systems are differential scanning calorimetry (DSC) and dynamic mechanical analysis (DMA). In DSC experiments a polymer sample is heated at a constant temperature rate (generally 10–20°C/min) and the amount of heat supplied to the sample to obtain the temperature increase is recorded. The output from this experiment is a plot of supplied heat versus temperature. The curves exhibit a step change at T_g due to a change in heat capacity at the onset of backbone rotation. An endothermic peak occurs in the DSC spectra at the melting point of the polymer due to the increased energy needed to melt the crystalline regions.

When performing a DMA experiment, one measures the modulus (stiffness) of the material over a temperature range, yielding the types of curves shown in Figure I.2.2.6. The glass transition temperature of the material is physically observed as the softening point of the material. In a DMA study the modulus drops by approximately three orders of magnitude at the T_g. At the T_m another drop is observed, associated with the softening due to melting of the polymer crystalline units and the onset of flow behavior.

Interactions with Water

Biomaterials are often employed in highly hydrated environments, so their interaction with water is an important design characteristic. Relatively non-polar and electrically neutral polymers such as polyethylene or poly(methylmethacrylate) are very hydrophobic and absorb <1 wt% water. However, as polarity and, in some cases, ionic character is incorporated into the polymer, it will imbibe more water due to polar (and coulombic) interactions. For instance, poly(hydroxyethyl methacrylate) (PHEMA) (the first soft contact lens) absorbs approximately 40 wt% water due to the polar hydroxyl moieties of the pendant groups. Eventually, if enough polarity or ionic character is incorporated into the material, it becomes water soluble, e.g., poly(ethylene glycol) and poly(methacrylic acid). The equilibrium water absorption of common biomaterials is presented in Table I.2.2.1.

One can tailor the interaction of a polymer with water in several ways, such as by controlling the ratio of hydrophobic and hydrophilic monomers in a copolymer. Also, crystalline regions in polymers usually resist infiltration of water molecules. If a polymer is processed in a way to control the degree of crystallization, the swelling character of the polymer can thereby be controlled. Cross-linked hydrophilic polymers generate hydrogels that can imbibe upwards of 200 wt% water. However, increasing the cross-link density can reduce the swelling of such a material, providing another method of tailoring the interactions of a polymer with water.

In addition to swelling behavior, the surface hydrophilicity of a polymer is also an important design feature. Though it is beyond the scope of this chapter, you will learn elsewhere in this text that a biomaterial surface can be physically or chemically coated with peptides or proteins to enhance cellular attachment and growth. Surfaces may also be treated to be "non-fouling," i.e., to resist protein adsorption and cell attachment.

Measuring the Hydrophilicity of Polymer Materials

Equilibrium swelling experiments are performed to study the interactions between bulk polymer and water. A dry sample of polymer of known mass is submerged in water, and the change in mass is recorded over time. Eventually the sample weight will become constant, and from this value the equilibrium water absorption can be calculated.

The surface hydrophilicity is probed through either static or dynamic contact angle experiments. In static experiments, a droplet of water is placed on a flat polymer surface and the angle the droplet makes with the material (as measured through the water phase) is measured. The degree to which water spreads on the surface is a measurement of the polarity of the interface. For instance, a material with a water contact angle of 20° has a *more polar* surface than a material with a water contact angle of 95°.

The surface composition of some materials can change depending on the external environment. For instance, a polymer may expose its hydrophobic portions to an air environment, yet it will reorganize its surface to display its hydrophilic portions when exposed to water (unless the interface has been oxidized to increase hydrophilicity). Dynamic contact angle experiments allow a researcher to probe the ability of the surface molecules to rearrange depending on the phase to which they are exposed. In such an experiment a polymer film is dipped into a bath of water and then removed. The force between the polymer sample and water is measured during immersion and retraction, and converted into contact angles. Materials with surface molecules that have the mobility to rearrange in air versus water will exhibit contact angle hysteresis, and the advancing and receding contact angles will not be the same. The advancing angle is always greater than the receding angle.

Degradation Characteristics

Depending on the application, one may desire a polymer material that is either biodegradable or biostable.

For instance, a polymer that will be used for a tissue engineered scaffold needs to be biodegradable, and to "disappear" as functioning tissue is regenerated by cells. However, some biomaterials, such as dental implants, vascular grafts, and intraocular lenses need to be biostable so that they retain their function for the lifetime of the patient. The main type of polymer degradation reaction occurring in the body is *hydrolysis*, i.e., the reaction of the polymer backbone bonds with water, which results in the hydrolysis of those bonds and loss of the polymer's mechanical properties. Eventually the polymer breaks up into small fragments that are metabolized and/or dissolve, and the residual molecules are eliminated from the body. The ultimate stability of polymers in the body depends on two key factors: water absorption and the susceptibility of main chain bonds to hydrolysis.

The carbon–carbon single bond is very stable and those materials with main chain C–C bonds, such as polyethylene, polypropylene or PMMA, do not lose their properties due to degradation of the backbone. However, the presence of other bonds in the polymer backbone (*heterochain* polymers) can lead to hydrolytic breakdown, and the rate of this hydrolysis is greatly affected by the polymer's molecular structure. For instance, the hydrolysis of amide bonds in Nylon 6,6 is so slow that it is almost non-existent at physiological conditions, making polyamide biomaterials relatively biostable. On the other hand, water can hydrolyze ester bonds (–CO–O–C–) present in some polymers. For example, poly(lactide-co-glycolide) copolymers (Figure I.2.2.5) degrade by such hydrolysis. An exception to this is poly(ethylene terephthalate) or PET, which is crystalline and hydrophobic, so that even though it has main chain ester bonds, water absorption is so low it is resistant to hydrolysis. The reader is referred to Chapter II.4.1 on degradation of biomaterial implants and to Chapter II.5.16 on drug delivery from degradable matrices for further discussion of this important topic.

POLYMER SYNTHESIS

A polymer scientist who understands structure–property relationships can design polymer molecules to fit a particular application. In this section of we discuss synthesis of polymers.

Polymerization Mechanisms

Condensation and *addition polymerization* are the two most common methods for polymer synthesis. In organic chemistry, reactions are described which occur between different functional groups. For instance, a carboxylic acid group can react with an amine group to form an amide bond and liberate a water molecule. Table I.2.2.3 illustrates some of these major organic reactions. The same reactions as those seen in Table I.2.2.3 are used to produce condensation polymers. The

TABLE I.2.2.3 Examples of Common Organic Reactions

Reactants	Product
R—OH + (epoxide ring, C–C–O) Alcohol epoxide	R—O—CH_2—CH_2—OH Ether bond
R—OH + HO—C(=O)—R' Alcohol carboxylic acid	R—O—C(=O)—R' Ester bond
R—C(=O)—OH + H_2N—R' Carboxylic acid amine	R—C(=O)—N(H)—R' Amide bond
R—NH_2 + O=C=N—R' Amine isocyanate	R—N(H)—C(=O)—N(H)—R' Urea bond
R—OH + O=C=N—R' Alcohol isocyanate	R—O—C(=O)—N(H)—R' Urethane bond

only difference is that instead of using monofunctional molecules (which contain only one functional group) difunctional monomers (which contain two functional groups) are used. Figure I.2.2.7A illustrates the formation of a poly(ethylene terephthalate) (PET) repeat unit through the reaction of ethylene glycol and dimethyl terephthalate, while Figure I.2.2.7B shows the production of a Nylon 6,6 repeat unit through the reaction of adipic acid and hexamethylenediamine. (In practice, PET is synthesized by an ester interchange reaction between dimethyl terephthalate and ethylene glycol, liberating methanol.) Notice that all of the monomers are difunctional molecules. Furthermore, the products from each reaction are also difunctional, allowing further, similar reactions to gradually produce higher molecular weight linear polymer chains.

The addition polymerization mechanism is quite different. A reactive center (most often a free radical, but sometimes an ion) is produced and reacts by adding across the C═C of vinyl monomers. The active center is not consumed during this reaction, and will continue reacting with carbon double bonds until it is terminated. High molecular weight chains are produced rapidly, and new chains are constantly being initiated. Figure I.2.2.8 is a schematic of a typical addition polymerization. First the reaction is initiated by the production of free radicals through a peroxide decomposition. The radical then propagates the polymer chain by consuming vinyl monomers. Finally the polymerization is terminated through a coupling reaction with a second radical (another method of termination called disproportionation is possible but is not discussed here) to produce a polymer molecule containing monomer repeat units.

There are newer polymerization techniques that are known as "controlled living free radical polymerizations" (or CRP) where the growth, chain transfer, and termination reactions are controlled to yield polymers with a desired molecular weight and narrow PDI. The two most popular of these techniques are known as ATRP, or Atom Transfer Radical Polymerization (Kato et al., 1995; Wang and Matyjaszewski, 1995) and RAFT,

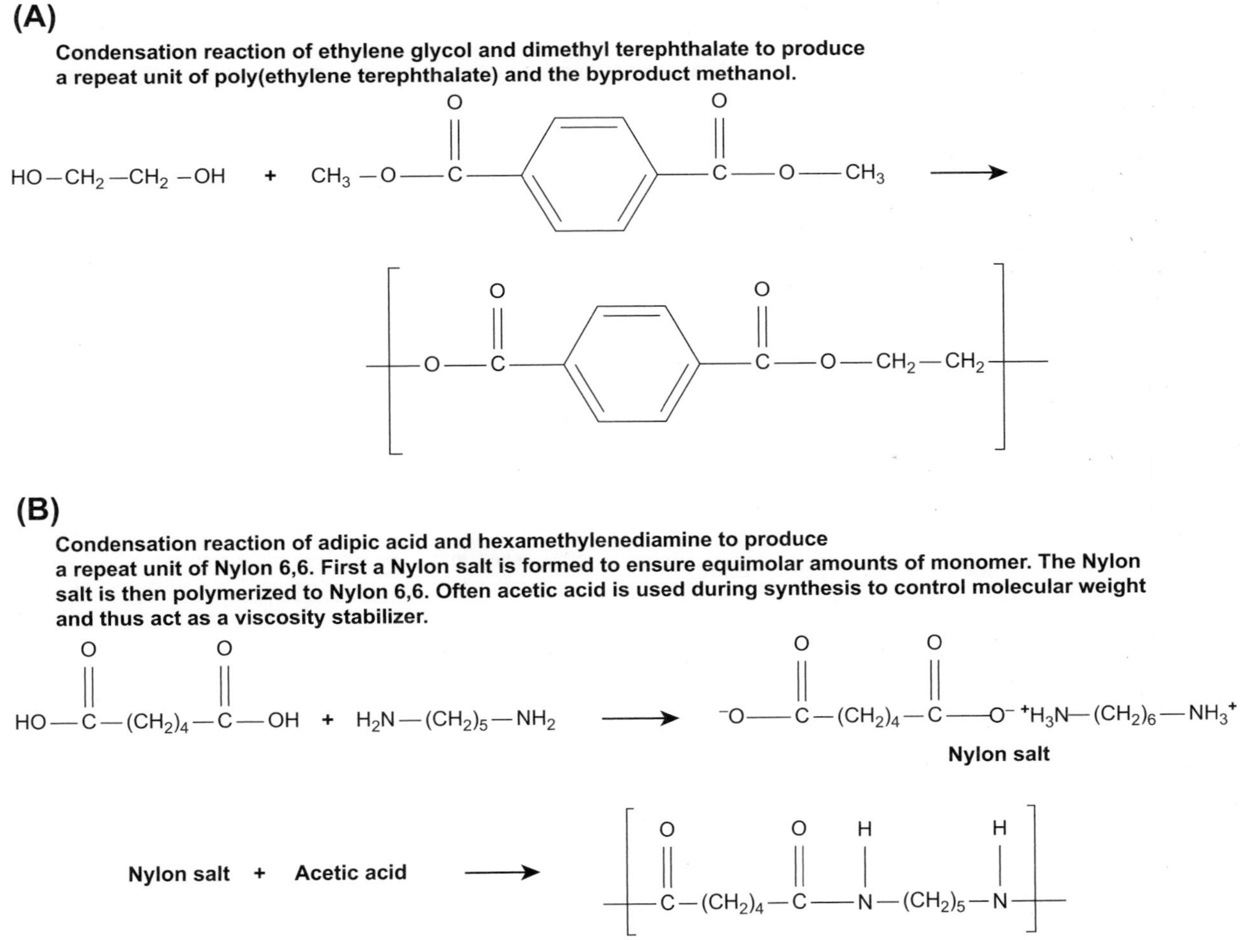

(A)

Condensation reaction of ethylene glycol and dimethyl terephthalate to produce a repeat unit of poly(ethylene terephthalate) and the byproduct methanol.

$HO-CH_2-CH_2-OH$ + $CH_3-O-C(=O)-C_6H_4-C(=O)-O-CH_3$ ⟶

$-[O-C(=O)-C_6H_4-C(=O)-O-CH_2-CH_2]-$

(B)

Condensation reaction of adipic acid and hexamethylenediamine to produce a repeat unit of Nylon 6,6. First a Nylon salt is formed to ensure equimolar amounts of monomer. The Nylon salt is then polymerized to Nylon 6,6. Often acetic acid is used during synthesis to control molecular weight and thus act as a viscosity stabilizer.

$HO-C(=O)-(CH_2)_4-C(=O)-OH$ + $H_2N-(CH_2)_5-NH_2$ ⟶ $^-O-C(=O)-(CH_2)_4-C(=O)-O^-\ {}^+H_3N-(CH_2)_6-NH_3^+$

Nylon salt

Nylon salt + **Acetic acid** ⟶ $-[C(=O)-(CH_2)_4-C(=O)-N(H)-(CH_2)_5-N(H)]-$

FIGURE I.2.2.7 Condensation reaction between: (A) ethylene glycol and dimethyl terephthalate to produce a repeat unit of poly(ethylene terephthalate); and (B) adipic acid and hexamethylenediamine first form a 1:1 Nylon salt to ensure stoichiometric balance between monomers allowing polymerization to high molecular weights. Monomers present in the Nylon salt are then polymerized to produce a repeat unit of Nylon 6,6.

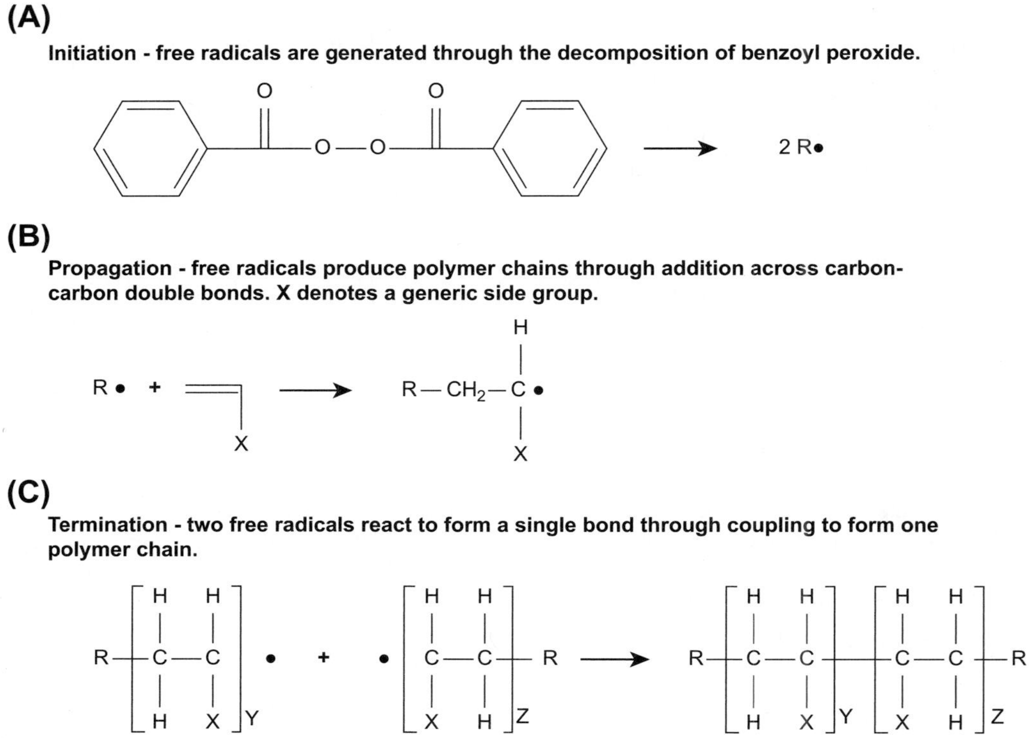

(A)

Initiation - free radicals are generated through the decomposition of benzoyl peroxide.

$\longrightarrow$ 2 R•

(B)

Propagation - free radicals produce polymer chains through addition across carbon-carbon double bonds. X denotes a generic side group.

(C)

Termination - two free radicals react to form a single bond through coupling to form one polymer chain.

FIGURE I.2.2.8 Addition polymerization of a vinyl monomer to a polymer chain: (A) the initiation step where the free radical is produced; (B) the propagation step where the free radical adds across the vinyl monomer's double bond; and (C) the termination step where two radicals react to produce dead polymer chains. In the schematic X represents a generic side group.

or Reversible Addition-Fragmentation Chain Transfer Polymerization (Chiefari et al., 1998).

The reader is referred to recent polymer textbooks (see Bibliography) for further reading on polymerization techniques.

Using Synthesis Conditions to Build the Desired Polymer

The first step towards producing a material with the desired behavior is selection of the appropriate monomer species. For instance, a researcher developing a bone cement could select a monomer with a bulky pendant group to produce a glassy polymer, while another biomaterials scientist creating a hydrogel could select a monomer with a polar pendant group to improve hydrogen bonding interactions with water.

Physical behavior can be further tailored through crystallinity and cross-linking. Addition polymers containing asymmetric carbon atoms are generally atactic, meaning the polymer will be predominantly amorphous. However, one can use special catalysts to produce isotactic or syndiotactic addition polymers that can crystallize.

In condensation polymerization, monomers containing two functional groups produce linear polymer chains. However, if one adds a small amount of a monomer with three or more functional groups, cross-linked materials are produced. Similarly, addition of vinyl type monomers containing two or more vinyl groups can produce networked materials. Linear polymers can also be cross-linked after synthesis through techniques such as vulcanization, or UV or gamma ray irradiation.

So What is a Network Polymer?

A common example of a network polymer is an automobile tire. The main polymer component of tire tread is a polystyrene/polybutadiene copolymer (SBR). The individual SBR polymer chains are covalently cross-linked into a network polymer through reaction with sulfur (a process known as vulcanization). There are other additives in car tires (carbon black, for instance, which gives the tire its black color); however, the polymer component of the tire is actually one very large, covalently cross-linked molecule.

If degradation characteristics are desired they can be achieved through control of the composition of the backbone groups. Addition polymerization produces homochain polymers that are generally biostable. Sometimes condensation polymerization mechanisms may produce a water molecule. The reverse reaction can occur between the polymer bonds being formed and the water molecule, resulting in simultaneous degradation of the polymer. In that case, water must be removed as it is formed.

CASE STUDIES

We have discussed the basic molecular characteristics that govern polymer behavior, and how to exploit chemistry in order to build polymers with the desired characteristics. Table I.2.2.4 compiles the chemical characteristics of several polymeric biomaterials, and reports their clinical applications. We will also look in more depth at the development of three polymer biomaterials that have achieved clinical success.

THE PRESENT AND THE FUTURE

In the early days of biomaterials science, materials were not designed for biomedical applications. Pre-existing materials were used and their biocompatibility was empirically assessed. As you will learn through this book, much of the current work in the biomaterials field focuses on creating materials that exhibit specific interactions with biology to achieve improved performance and better healing.

Case Study Resorbable Sutures

What problems can occur?

Sutures are often needed to hold tissue together after injury, incision, or surgery. If a non-biodegradable material is used to suture an internal wound, a secondary surgery would be necessary to remove the stitches after healing.

What properties are required of the biomaterial?

For a suture material, the most fundamental property required is appropriate tensile strength, so the wound does not reopen during healing. Also, the suture materials should degrade in the body during healing, making retrieval or removal unnecessary. Further, the degradation must occur at an appropriate rate, and the degradation products must be non-toxic.

What polymeric biomaterial is used?

Early resorbable sutures were made from sheep or beef gut. However, synthetics are more commonly used due to their ease of handling, low cost, consistent performance, and low chance of disease transfer. One commonly used material is a poly(lactide-co-glycolide) copolymer (PLGA). This is a random copolymer so the material is permanently amorphous; however, it still possesses the high tensile strength and flexibility required for this application. Also, the material biodegrades through hydrolysis of the ester bonds in the polymer backbone into lactic and glycolic acid. Both of these chemicals occur naturally in the body's metabolic pathway, and so are non-toxic in small amounts. Furthermore, the ester bonds in the lactide and glycolide repeat units hydrolyze at different rates, allowing the degradation rate to be tailored.

TABLE I.2.2.4 Common Polymeric Biomaterials with Their Applications and the Properties That Make Them Useful in the Medical Field

Material	Characteristics	Uses
Poly(methyl methacrylate) **(PMMA)**	Hydrophobic polymer which is hard, rigid, and biostable. The amorphous material is clear allowing light transmittance	Bone cement Intraocular lenses Hard contact lenses
Polyacrylamide **(PA)**	Cross-linking produces a hydrogel with molecular-sized pores and allows the gel to be used as a separation medium	Separation gel used in electrophoresis
Poly(acrylic acid) **(PAA)**	The liquid monomer can be cured with a photo-initiator. If inorganic salts are added, ionic cross-linking can occur. The material is glassy and rigid, and has the potential to bond to enamel	Glass ionomer cement used in dental restoration
Polyethylene (high density) **(PE)**	Low density PE cannot withstand sterilization temperatures; however, high density PE has good toughness and wear resistance	Tubing for drains and catheters Prosthetic joints
Poly(vinyl chloride) **(PVC)**	PVC is plasticized to make flexible materials. This material is used for short-term applications since plasticizers can be leached resulting in embrittlement of the material	Tubing Blood storage bags
Polypropylene **(PP)**	Isotactic PP is semi-crystalline, has high rigidity and tensile strength, and good biostability	Non-degradable sutures Hernia repair
Polydimethylsiloxane **(PDMS)**	Due to its silicone backbone, this material has a very low T_g making it extremely flexible and providing it with good fatigue resistance at physiological conditions	Finger joints Heart valves Breast implants Ear, chin, and nose reconstruction
Poly(ethylene terephthalate) **(PET)**	The aromatic rings in the backbone generate a polymer with a high melting point (T_m = 267°C). It is semi-crystalline and possesses excellent tensile strength	Vascular grafts Fixation of implants Hernia repair Ligament reconstruction
Cellulose acetate **(CA)**	Unique transport properties make it excellent for use in the separation of complex biological mixtures	Dialysis membranes Osmotic drug delivery devices

Case Study Soft Contact Lenses

What problem was addressed?

The first contact lens, developed in 1888, was made of glass, and could not be worn for long periods of time. A generation of PMMA lenses were developed (hard contact lenses) and experienced clinical success. However, these rigid lenses caused eye irritation, and did not allow enough oxygen to be transported to the cornea of the eye (an avascular tissue which must receive oxygen from the environment in order for normal metabolism to occur). The Czech scientists, Wichterle and Lim, wanted to make a lens which was more comfortable and would allow better transport of oxygen to the cornea. (Otto Wichterle and Drahoslav Lim, "Process for producing shaped articles from three-dimensional hydrophilic high polymers", US Patent 2,976,576, March 28, 1961.)

What properties were required of the biomaterial?

Strong mechanical properties were not a major factor in the design of soft contact lenses, since the device does not need to support large stresses. However, a successful lens needed to: (1) be transparent so that clear vision could be achieved; (2) be dimensionally stable to maintain the optical correction desired; (3) be soft and flexible to minimize eye irritation; (4) allow appropriate transport of oxygen and nutrients to the cornea; and (5) have sufficient surface wettability so as not to damage the corneal cells.

What polymeric biomaterial is used?

The major component of most modern soft contact lenses is loosely cross-linked poly(hydroxyethyl methacrylate) (PHEMA). Semi-crystalline polymers are often translucent due to light refracted by crystallites. Since PHEMA is atactic the polymer is transparent. When dry, PHEMA is glassy, allowing the lens to be fabricated through milling procedures (as is done with hard contact lenses). However, the hydroxyl-containing pendant group results in a hydrophilic polymer that imbibes 40 wt% water. The absorbed water plasticizes the polymer, producing a soft and supple material that is less irritating to the eye. Pure PHEMA lenses are not capable of providing the necessary oxygen to the cornea of the eye, so PHEMA is often copolymerized with other monomers, such as fluorinated or "siliconized" methacrylates, to improve oxygen transport either by increasing the oxygen solubility in the lens or by allowing the fabrication of a thinner lens.

Recently, much research in contact lens development has focused on the design of extended wear lenses (contacts that can be worn for up to 7–14 days without removal). A leading material for extended wear contacts is a silicone hydrogel that is composed of a silicone polymer network with an entrapped water-soluble polymer such as poly(vinyl pyrrolidone). This combination creates a material with excellent oxygen and ion permeability, and good lubricity of the lens against the cornea. (For example, see http://en.wikipedia.org/wiki/Contact_lens.)

Case Study Artificial Hip Joints

What problem was addressed?

A surgeon may perform hip arthroplasty (implantation of an artificial hip) to alleviate the pain and stiffness of severe arthritis or to repair the joint after physical damage. An artificial hip implant has two pieces: the femoral ball and stem; and the acetabular cup. The femoral ball and stem implant is a metal shaft that inserts into the femoral bone core, and is capped by a ball that articulates with the acetabular cup.

What properties were required of the biomaterial?

Sir John Charnley – a British orthopedic surgeon – conceived the initial concept that led to the modern artificial hip. Focusing on the acetabular component, tribology (the study of friction, lubrication, and wear) is of utmost importance. The ideal polymer for use in an articulating joint would be low friction and resistant to wear.

What polymeric biomaterial was used?

Initially Charnley used poly(tetrafluoroethylene) (PTFE), a very non-polar and low-friction material, for the acetabular cup. He also used stainless steel for the femoral ball and stem. Two problems arose: first, PTFE did not have the necessary wear resistance, and small pieces of PTFE debris eroded from the surface of the cup, causing severe inflammation of the surrounding tissue. Second, Charnley had to force-fit the femoral stem into the femur, and it loosened during use. Although polyethylene (PE) is a higher-friction surface than PTFE, very high molecular weight PE is a superior biomaterial for this application, since the PE did not produce excessive wear debris. Also, the PE could be simultaneously gamma radiation-sterilized and cross-linked, increasing its wear resistance. It could also be machined to a fine smoothness to further minimize wear. Charnley was advised by Dennis Smith, a biomaterials chemist working on dental implants, to use PMMA dental cement to hold the femoral stem in place during wear, and that solved the second problem.

BIBLIOGRAPHY

Allcock, H. R., Lampe, F. W., & Mark, J. E. (2004). *Contemporary Polymer Chemistry* (3rd ed.). Englewood Cliffs, NJ: Prentice Hall, Inc.

Brandrup, J., Immergut, E. H., & Grukle, E. A. (1999). *Polymer Handbook* (4th ed.). New York, NY: Wiley-Interscience.

Charnley, J., & Halley, D. K. (1975). Rate of wear in total hip replacement. *Clinical Orthopedic Related Research*, **112**, 170–179.

Chiefari, J., Chong, Y. K., Ercole, F., Krstina, J., Jeffery, J., et al. (1998). Living free-radical polymerization by reversible addition–fragmentation chain transfer: The RAFT process. *Macromolecules*, **31**, 5559–5562.

Dumitriu, S. (2002). *Polymeric Biomaterials* (2nd ed.). New York, NY: Marcel Dekker Inc.

Gilding, D. K., & Reed, A. M. (1979). Biodegradable polymers for use in surgery: Poly(glycolic)/poly(lactic acid) homo- and copolymers. *Polymer*, **20**, 1459–1464.

Flory, P. J. (1953). *Principles of Polymer Chemistry*. Ithaca, NY: Cornell University Press.

Kato, M., Kamigaito, M., Sawamoto, M., & Higashimura, T. (1995). Polymerization of methyl methacrylate with the carbon tetrachloride/dichlorotris-(triphenylphosphine)ruthenium(II)/methylaluminum bis(2,6-di-tert-butylphenoxide) initiating system: Possibility of living radical polymerization. *Macromolecules*, **28**, 1721–1723.

NASA. (1986). Challenger Accident Investigation Report, Chapter 4: The Cause of the Accident.
Painter, P. C., & Coleman, M. M. (2009). *Essentials of Polymer Science and Engineering*. Lancaster, PA: DEStech Publications.
Rodriguez, F., Cohen, C., Ober, C. K., & Archer, L. A. (2003). *Principles of Polymer Systems* (5th ed.). New York, NY: Taylor & Francis.
Vert, M. (2007). Polymeric biomaterials: Strategies of the past vs. strategies of the future. *Progress in Polymer Science*, **32**, 755–761.
Wang, J., & Matyjaszewski, K. (1995). Controlled/"living" radical polymerization: Atom transfer radical polymerization in the presence of transition-metal complexes. *J. Am. Chem. Soc.*, **117**, 5614–5615.
Wang, J., & Wu, W. (2005). Swelling behaviors, tensile properties and thermodynamic studies of water sorption of 2-hydroxyethyl methacrylate/opoxy methacrylate copolymeric hydrogels. *European Polymer Journal*, **41**, 1143–1151.
Ward, I. M., & Sweeney, J. (2004). *An Introduction to the Mechanical Properties of Solid Polymers* (2nd ed.). West Sussex, UK: John Wiley & Sons Ltd.
Wichterle, O., & Lim, D. (1960). Hydrophilic gels for biological use. *Nature*, **185**, 117–118.

A. POLYURETHANES

Daniel E. Heath and Stuart L. Cooper
William G. Lowrie Department of Chemical and Biomolecular Engineering, The Ohio State University, Columbus, OH, USA

INTRODUCTION

Polyurethanes are widely used in medical devices such as pacemakers, artificial hearts, and other blood contacting applications. Their excellent mechanical properties, stability, and good biocompatibility give them a special place in medicine. Polyurethanes were developed in the 1930s. Around this time, Nylon 6,6 – a condensation polymer of hexamethylenediamine and adipic acid – was developed and patented by Du Pont. This polymer achieved commercial success due to its good mechanical properties and fiber-producing ability. German scientists led by Otto Bayer began to explore new polymerization techniques in the hope of creating a material competitive with nylon. Initial work reacted difunctional isocyanates and amines to produce polyureas; however, these materials were too hydrophilic to be used as plastics or textiles. Diisocyanates were also reacted with diols to produce polyurethanes. One nylon-like polyurethane produced through the reaction of 1,4-butanediol and hexamethylene diisocyanate, as shown in Figure A.1, was used commercially in Germany. Chemically the polymer is the same as nylon except for the two additional oxygens per repeat unit, and the properties are similar except that the polyurethane has a lower melting point. These early polyurethanes also had lower water absorption, and better electrical and mechanical stability upon aging compared to nylon. Subsequently block polymers including polyester or polyether polyols were developed both in Germany and the United States.

Today, polyurethanes (PUs) are a class of polymer which has achieved industrial relevance due to their tough and elastomeric properties, and good fatigue resistance. PUs are used as adhesives, coatings, sealants, rigid and flexible foams, and textile fibers. Furthermore, PUs have also been employed in biomaterial applications such as artificial pacemaker lead insulation, catheters, vascular grafts, heart assist balloon pumps, artificial heart bladders, and wound dressings.

ANATOMY OF A POLYURETHANE MOLECULE

The polyurethane molecule presented in Figure A.1 has a very simple architecture. Most commercially relevant PUs today are actually block copolymers (as seen in Chapter I.2.2, Figure I.2.2.4) meaning there are alternating segments in the polymer molecules composed of solely "A" or "B" repeat units. Furthermore, the materials are designed so that one of the segments – called the *hard segment* – is glassy or crystalline at the use temperature, while the other segment – referred to as the *soft segment* – is rubbery. If the monomers are appropriately selected, the hard and soft segments of the polyurethane will be incompatible, leading to phase separation and meaning that the composition of the bulk polymer will not be homogeneous. Instead, there will be nanometer-sized regions which are rich in hard segments, and other regions rich in soft segments. The unique block copolymer structure and phase separation of PUs results in unique and useful properties. One can consider the hard phase domains as providing highly efficient reinforcing microdomains which give rise to the unusual and very attractive properties of polyurethanes, such as strength and toughness.

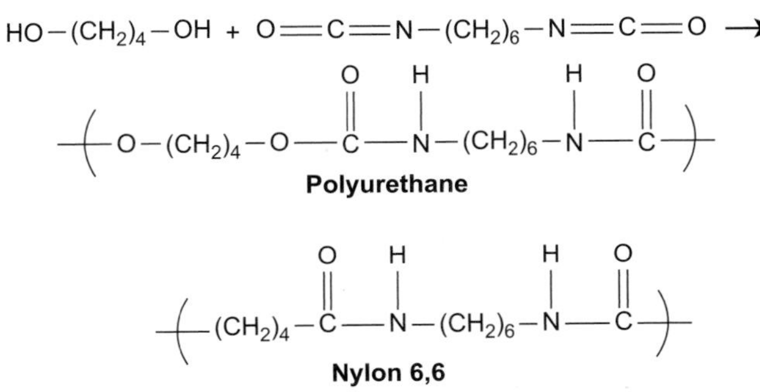

$$HO-(CH_2)_4-OH + O=C=N-(CH_2)_6-N=C=O \rightarrow$$

$$\left(O-(CH_2)_4-O-\overset{O}{\overset{\|}{C}}-\overset{H}{\overset{|}{N}}-(CH_2)_6-\overset{H}{\overset{|}{N}}-\overset{O}{\overset{\|}{C}} \right)$$

Polyurethane

$$\left((CH_2)_4-\overset{O}{\overset{\|}{C}}-\overset{H}{\overset{|}{N}}-(CH_2)_6-\overset{H}{\overset{|}{N}}-\overset{O}{\overset{\|}{C}} \right)$$

Nylon 6,6

FIGURE A.1 Synthesis scheme of a polyurethane through the condensation of 1,4-butanediol and hexamethylene diisocyanate. The repeat unit of the PU and of Nylon 6,6 are presented for comparison.

THE PHYSICAL PROPERTIES OF POLYURETHANES

Most polymer elastomers – such as rubber bands – are produced by lightly cross-linking a low T_g polymer into a loose network. When such a material is strained, the segments between cross-links can deform and elongate; however, when the load is removed the cross-links result in the material returning to its original form. Although these elastomers are useful, they are *thermosets*. Since the material is essentially one large molecule it cannot be dissolved, neither can it be made to flow through the introduction of heat and pressure. This means that once a thermoset elastomer is formed, it cannot be processed further. However, the unique segmented structure of PUs allows *thermoplastic elastomers* to be formed. When such materials are strained, the polymer segments in the soft phase will deform and elongate, while the hard phase will stabilize the structure, resulting in the recovery of the original form once the load is removed. However, unlike the covalent cross-links in thermosets, the physical cross-links present between hard segments can be undermined through the application of heat, and then reformed upon cooling, allowing the useful properties of elastomers to be combined with the simple processing procedures of thermoplastics. In addition to elastomeric properties, PUs have several useful interfacial characteristics. Foremost, PUs are abrasion- and impact-resistant, making them useful as coatings, and the materials also have good blood-contacting properties, making them useful in biomaterial applications.

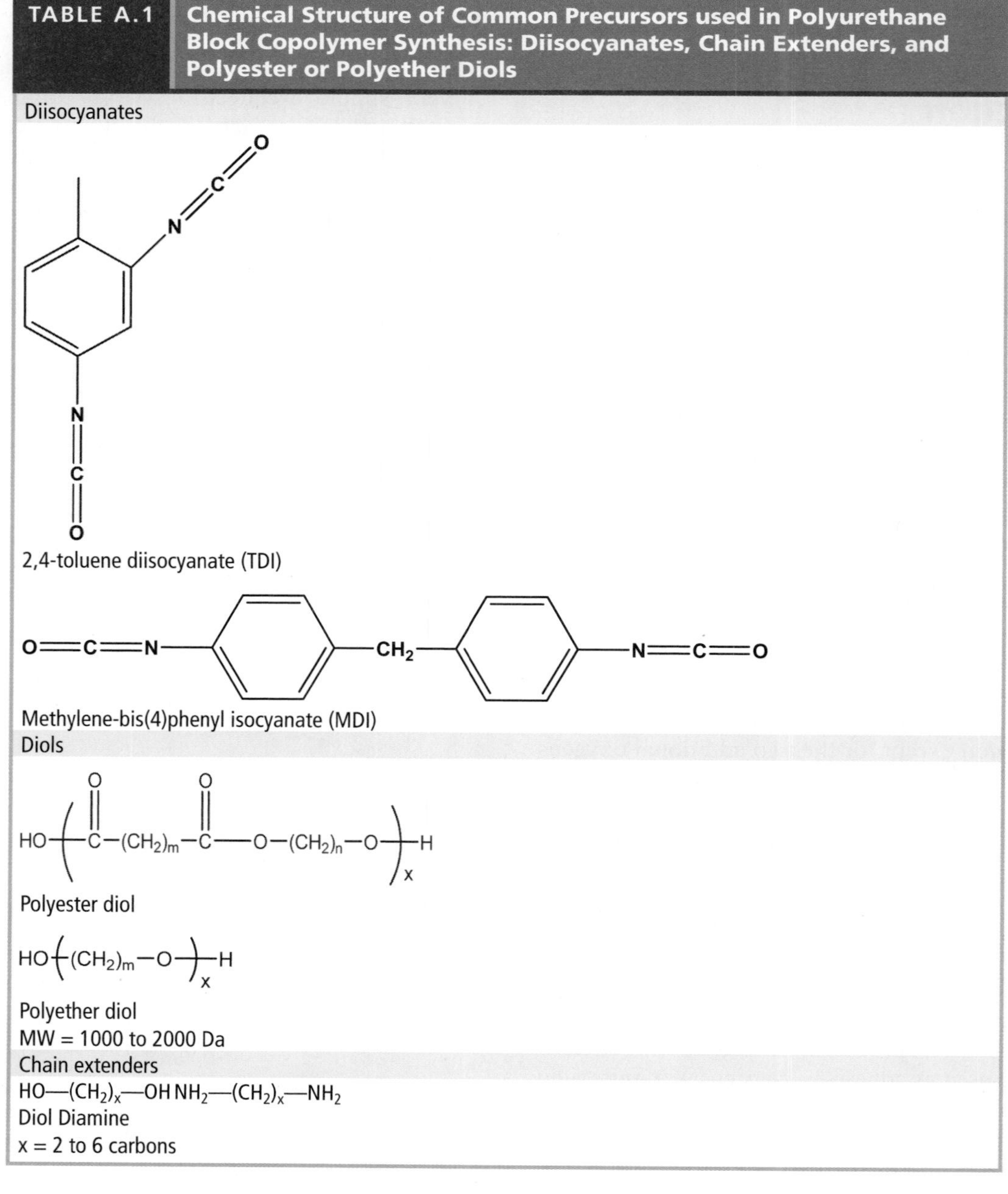

TABLE A.1 Chemical Structure of Common Precursors used in Polyurethane Block Copolymer Synthesis: Diisocyanates, Chain Extenders, and Polyester or Polyether Diols

Diisocyanates
2,4-toluene diisocyanate (TDI)
O=C=N–C_6H_4–CH_2–C_6H_4–N=C=O Methylene-bis(4)phenyl isocyanate (MDI)
Diols
$HO{-}[C(=O){-}(CH_2)_m{-}C(=O){-}O{-}(CH_2)_n{-}O]_x{-}H$ Polyester diol
$HO{-}[(CH_2)_m{-}O]_x{-}H$ Polyether diol MW = 1000 to 2000 Da
Chain extenders
$HO{-}(CH_2)_x{-}OH$ $NH_2{-}(CH_2)_x{-}NH_2$ Diol Diamine x = 2 to 6 carbons

POLYURETHANE SYNTHESIS

The Precursors

When synthesizing most PU block copolymers a two-step synthesis strategy is employed which involves three precursor molecules: diisocyanates; diols; and chain extenders. In the final polymer molecule the diisocyanates and chain extenders will form the hard segments, while the diols form the soft segments. Table A.1 illustrates the chemical structure of molecules often used to synthesize PU block copolymers.

The diisocyanate molecules used in polyurethane synthesis can be aliphatic (such as Figure A.1), but most often aromatics are used (Table A.1). The two most commonly used diisocyanates employed in PU synthesis are 2,4-toluene diisocyanate (TDI) and methylene-bis(4) phenyl isocyanate (MDI). The soft segments are formed from polyether or polyester diols (sometimes referred to as *polyols*). These molecules have a molecular weight of 1000 to 2000 Da, and are well above their T_gs and melting points at use conditions, which impart a rubbery character to the resulting polyurethane. The last precursor is the chain extender. These molecules are often short aliphatic diols or diamines containing 2 to 6 carbon atoms, and are used to build the polymer up to its final high molecular weight.

Synthesis Reactions

The isocyanate group contains two unsaturated bonds and is a highly reactive moiety. When reacted with a hydroxyl group, the result is a urethane bond, as illustrated in Figure A.1. The first step in PU synthesis reacts the diisocyanates with the polyether or polyester diols. Excess diisocyanate is used to ensure the resulting materials are end terminated with isocyanate groups, as seen in Figure A.2A. The result of this reaction is a *prepolymer* with a degree of polymerization commonly between 2 and 5. Next, the prepolymer is further reacted with the chain extenders to produce high molecular weight polyurethane molecules, as shown in Figure A.2B.

Tailoring Polyurethane Behavior

Many of the techniques described in Chapter I.2.2 to modulate polymer properties can easily be applied to polyurethanes. For instance, monomers with functionality greater than two can be used in order to produce cross-linked polymer structures. Also, as the cross-link density in thermosets controls material stiffness, the relative amounts of hard and soft segments in the PU can vary the modulus. Also, the length and chemical nature of the soft segment can be adjusted to further tune physical behavior.

For biostable biomaterial applications, MDI is generally used to form the hard segments, and polyether diols are often preferred to form the soft segments since they are more resistant to hydrolytic degradation than polyesters. However, PUs have been explored more recently as scaffolds for tissue engineering. For these applications, biodegradable PUs are required, the degradation rate must be controlled, and the degradation

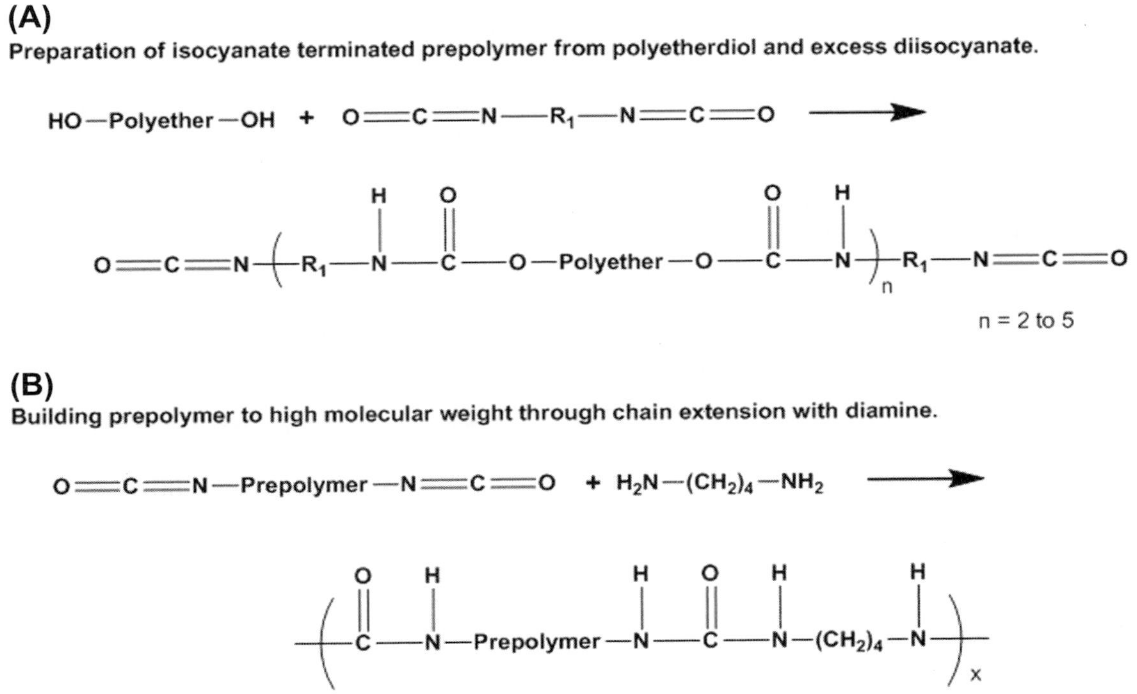

(A)
Preparation of isocyanate terminated prepolymer from polyetherdiol and excess diisocyanate.

$$\mathrm{HO{-}Polyether{-}OH} + \mathrm{O{=}C{=}N{-}R_1{-}N{=}C{=}O} \longrightarrow$$

$$\mathrm{O{=}C{=}N}\left(\mathrm{R_1{-}N(H){-}C(=O){-}O{-}Polyether{-}O{-}C(=O){-}N(H)}\right)_n\mathrm{R_1{-}N{=}C{=}O}$$

n = 2 to 5

(B)
Building prepolymer to high molecular weight through chain extension with diamine.

$$\mathrm{O{=}C{=}N{-}Prepolymer{-}N{=}C{=}O} + \mathrm{H_2N{-}(CH_2)_4{-}NH_2} \longrightarrow$$

$$\left(\mathrm{C(=O){-}N(H){-}Prepolymer{-}N(H){-}C(=O){-}N(H){-}(CH_2)_4{-}N(H)}\right)_x$$

FIGURE A.2 Two-step synthesis of a polyurethane block copolymer: (A) Isocyanate-terminated prepolymer synthesis from excess diisocyanate and polyether diol; (B) Reaction of prepolymer with diamines to build high molecular weight polyurethane chains.

Case Study — What Problem was Addressed?

An individual may require an artificial pacemaker to regulate their heartbeat. The pacemaker is an implanted device which monitors heartbeat and provides electrical stimulation to the muscles of the heart, resulting in muscle contraction when needed. The pacemaker is connected to the appropriate heart muscles through pacemaker leads, electrically conductive wires which are fed to the heart through the vasculature.

The pacemaker leads require insulation. Initially poly(dimethyl siloxane) (PDMS) or polyethylene was used as the insulator for the leads. However, both of these materials resulted in a fibrous endocardial reaction. Furthermore, PDMS has a low tensile modulus and poor tear resistance.

What properties were required of the biomaterial?

A successful insulator for pacemaker leads would not elicit a fibrous reaction from the heart, and would have high tensile strength and resistance to tearing allowing for thinner lead insulations to be produced.

What polymeric biomaterial is used?

In 1978 polyurethane was introduced as a lead insulator. Although not as flexible as PDMS, the PU had superior tensile properties and tear resistance. This allowed much thinner lead insulations to be fabricated without compromising handling properties. The thinner insulation allows multiple leads to be inserted per vein, enabling sequential pacing. Furthermore, the PU surface has lower friction in contact with blood and tissue than the PDMS surface, allowing easier insertion of the leads.

The search for the optimum lead insulation material is not over yet, however. In the 1980s it was found that metal-induced oxidation from lead metals resulted in undesired degradation of the polyurethane insulation. Lower ether content in the polyurethane was one solution to this problem, although this results in higher modulus insulation. In subsequent years there have been changes in the material used for the lead wire, and silicone rubber remains in use as well as polyurethanes for pacemaker insulation. Research is underway to find more biostable polyurethanes for this application.

products need to be non-toxic. Soft segment structure is often used to build in the appropriate degradability. For instance, soft segments containing polylactide, polyglycolide, and polycaprolactone have been investigated, all of which are susceptible to degradation and the degradation rates vary based on chemical composition. The aromatic diisocyanates can result in carcinogenic degradation products; therefore, PUs for tissue engineering applications require the use of diisocyanates other than MDI and TDI. New PU formulations have been synthesized using lysine-diisocyanate, hexamethylene diisocyanate, and 1,4 diisocyanatobutane which are believed to degrade into non-toxic products.

CONCLUDING REMARKS

Polyurethanes are a unique family of polymers with interesting properties which arise from their block copolymer nature and the resulting phase separation. These materials have achieved industrial success in everyday applications, as well as in the biomaterials field.

BIBLIOGRAPHY

Billmeyer, F. W., Jr. (1984). *Textbook of Polymer Science* (3rd ed.). New York, NY: Wiley-Interscience.

Brandrup, J., Immergut, E. H., & Grukle, E. A. (1999). *Polymer Handbook* (4th ed.). New York, NY: Wiley-Interscience.

Dumitriu, S. (2002). *Polymeric Biomaterials* (2nd ed.). New York, NY: Marcel Dekker Inc.

Lamba, N. M.K., Woodhouse, K. A., & Cooper, S. L. (1998). *Polyurethanes in Biomedical Applications*. Boca Raton, FL: CRC Press.

Mahapatro, A., & Kulshrestha, A. (2008). *Polymers for Biomedical Applications*: American Chemical Society Publication, Oxford University Press, Oxford, UK.

Randall, D., & Lee, S. (2002). *The Polyurethanes Book*. West Sussex, UK: John Wiley & Sons, Ltd.

Rodriguez, F., Cohen, C., Ober, C. K., & Archer, L. A. (2003). *Principles of Polymer Systems* (5th ed.). New York, NY: Taylor & Francis.

Santerre, J. P., Woodhouse, K., Laroche, G., & Labow, R. S. (2006). Understanding the biodegradation of polyurethanes: From classical implants to tissue engineering materials. *Biomaterials*, **26**, 7457–7470.

Saunders, J. H., & Frisch, K. C. (1965). *Polyurethanes: Chemistry and Technology, Part 1. Chemistry*. New York, NY: Interscience Publishers.

Vert, M. (2007). Polymeric biomaterials: Strategies of the past vs. strategies of the future. *Progress in Polymer Science*, **32**, 755–761.

B. SILICONES

André Colas[1] and Jim Curtis[2]

[1] Dow Corning Europe S.A., Parc Industriel, Seneffe, Belgium

[2] Dow Corning Corporation, Midland, MI, USA

Silicone materials have been widely used in medicine for over 60 years. Available in a variety of material types, they have unique chemical and physical properties that manifest in excellent biocompatibility and biodurability for many applications. Silicone elastomers have remarkably low glass-transition temperatures and maintain their flexibility over a wide temperature range, enabling them to withstand conditions from cold storage to steam autoclaving. They have high permeability to gases and many drugs, advantageous respectively in wound care or in transdermal drug delivery. They have low surface tension and remarkable chemical stability, enabling biocompatibility and biodurability in many long-term implant applications.

However, versatile as they are, present-day silicone materials still have limitations. The mechanical properties of silicone elastomers, such as tensile strength or tear resistance, are somewhat lower than for other implantable elastomers such as polyurethanes (although generally speaking, polyurethanes are less biodurable). While resistant to a wide array of chemical environments, silicone elastomers are susceptible to degradation in very strongly basic or acidic conditions, such as those found in the stomach. Like all hydrophobic implant materials, silicones are quickly coated with proteins when placed in tissue contact; and a scar tissue capsule forms to surround an implant during wound healing, walling it off from the host. Additionally, silicone elastomers are thermosetting materials, requiring different processing from conventional thermoplastics, which can on occasion be seen as a drawback.

CHEMICAL STRUCTURE AND NOMENCLATURE

Silicones are a general category of synthetic polymers whose backbone is made of repeating silicon-to-oxygen bonds. In addition to their links to oxygen to form the polymeric chain, the silicon atoms are also bonded to organic groups, typically methyl groups. This is the basis for the name "silicones," which was assigned by Kipping based on their similarity with ketones, because in most cases there is on average one silicone atom for one oxygen and two methyl groups (Kipping, 1904). Later, as these materials and their applications flourished, more specific nomenclature was developed. The basic repeating unit became known as "siloxane," and the most common silicone is polydimethylsiloxane, abbreviated as PDMS.

$$-\left(\begin{array}{c} R \\ | \\ Si-O- \\ | \\ R \end{array}\right) \quad \text{and if R is } CH_3, \quad -\left(\begin{array}{c} CH_3 \\ | \\ Si-O- \\ | \\ CH_3 \end{array}\right)_n$$

"siloxane" "polydimethylsiloxane"

Many other groups (e.g., phenyl, vinyl, and trifluoropropyl) can be substituted for the methyl groups along the chain. The simultaneous presence of organic groups attached to an inorganic backbone give silicones a combination of distinctive properties, making their use possible as fluids, emulsions, compounds, resins, and elastomers in numerous applications and diverse fields. For example, silicones are common in the aerospace industry, due principally to their low and high temperature performance. In the electronics field, silicones are used as electrical insulation, potting compounds, and other applications specific to semiconductor manufacture. Their long-term durability has made silicone sealants, adhesives, and waterproof coatings commonplace in the construction industry. Excellent biocompatibility makes many silicones well suited for use in numerous personal care, pharmaceutical, and medical device applications (see Chapter II.5.18).

Historical Milestones in Silicone Chemistry

Key milestones in the development of silicone chemistry, thoroughly described elsewhere by Lane and Burns (1996), Rochow (1945), and Noll (1968), are summarized in Table B.1.

Nomenclature

The most common silicones are the trimethylsilyloxy end-blocked polydimethylsiloxanes, with the following structure:

$$CH_3-\begin{array}{c} CH_3 \\ | \\ Si \\ | \\ CH_3 \end{array}-O-\left(\begin{array}{c} CH_3 \\ | \\ Si-O \\ | \\ CH_3 \end{array}\right)_n-\begin{array}{c} CH_3 \\ | \\ Si \\ | \\ CH_3 \end{array}-CH_3,$$

$$(n = 0,1,\ldots)$$

These are linear polymers and liquids, even for large values of n. The main chain unit, $-(Si(CH_3)_2O)-$, is often represented by the letter D for $(CH_3)_2SiO_{2/2}$, because with the silicon atom connected to two oxygen atoms this unit is capable of expanding within the polymer in two directions. M, T, and Q units are defined in a similar manner, as shown in Table B.2.

The system is sometimes modified by the use of superscript letters designating nonmethyl substituents, for example, $D^H = H(CH_3)SiO_{2/2}$ and M^ϕ or M^{Ph}

TABLE B.1	Key Milestones in the Development of Silicone Chemistry
1824	Berzelius discovers silicon by the reduction of potassium fluorosilicate with potassium: $4K + K_2SiF_6 \rightarrow Si + 6KF$. Reacting silicon with chlorine gives a volatile compound later identified as tetrachlorosilane, $SiCl_4$: $Si + 2Cl_2 \rightarrow SiCl_4$.
1863	Friedel and Crafts synthesize the first silicon organic compound, tetraethylsilane: $2Zn(C_2H_5)_2 + SiCl_4 \rightarrow Si(C_2H_5)_4 + 2ZnCl_2$.
1871	Ladenburg observes that diethyldiethoxysilane $(C_2H_5)_2Si(OC_2H_5)_2$, in the presence of a diluted acid gives an oil that decomposes only at a "very high temperature."
1901–1930s	Kipping lays the foundation of organosilicon chemistry with the preparation of various silanes by means of Grignard reactions and the hydrolysis of chlorosilanes to yield "large molecules." The polymeric nature of the silicones is confirmed by the work of Stock.
1940s	In the 1940s, silicones become commercial materials after Hyde of Dow Corning demonstrates the thermal stability and high electrical resistance of silicone resins, and Rochow of General Electric finds a direct method to prepare silicones from silicon and methylchloride.

TABLE B.2 Shorthand Notation for Siloxane Polymer Units

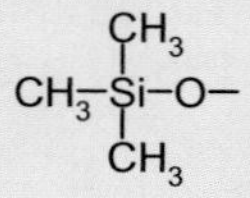 M	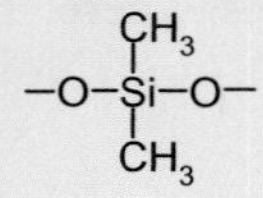 D	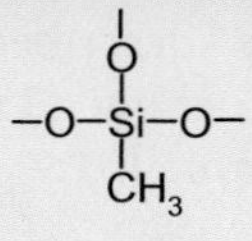 T	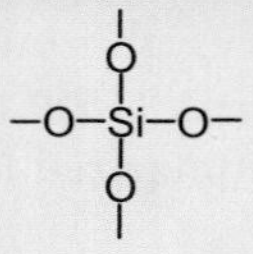 Q
$CH_3-Si(CH_3)_2-O-$	$-O-Si(CH_3)_2-O-$	$-O-Si(CH_3)(O-)-O-$	$-O-Si(O-)(O-)-O-$
$(CH_3)_3SiO_{1/2}$	$(CH_3)_2SiO_{2/2}$	$CH_3SiO_{3/2}$	$SiO_{4/2}$

= $(CH_3)_2(C_6H_5)SiO_{1/2}$ (Smith, 1991). Further examples are shown in Table B.3.

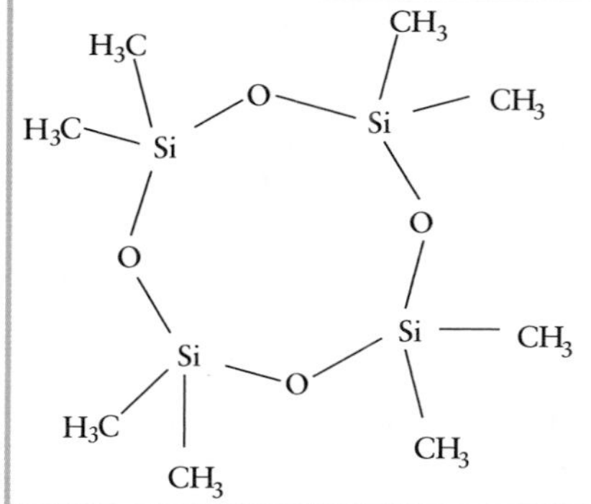

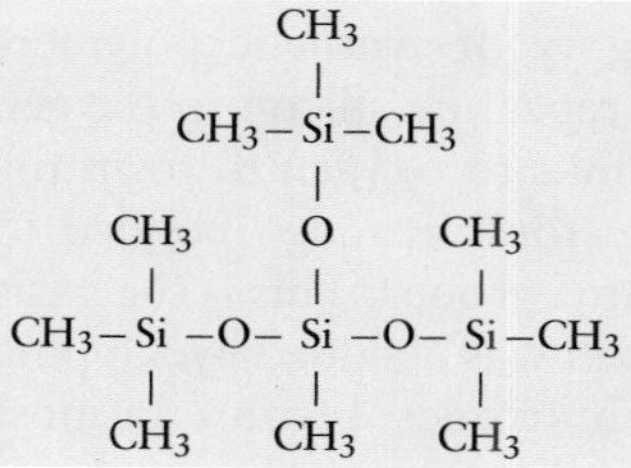

TABLE B.3 Examples of Silicone Shorthand Notation

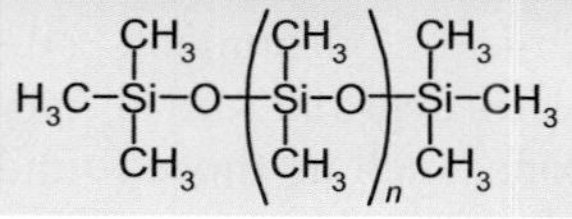

Structure	Notation
$H_3C-Si(CH_3)_2-O-(Si(CH_3)_2-O)_n-Si(CH_3)_2-CH_3$	MD_nM
Cyclic $[(CH_3)_2SiO]_4$ (octamethylcyclotetrasiloxane)	D_4
$CH_3-Si(CH_3)[O-Si(CH_3)_3]-O-Si(CH_3)_3$ ... $(CH_3)_3Si-O-Si(CH_3)(O-Si(CH_3)_3)-O-Si(CH_3)_3$	TM_3
$H-Si(CH_3)_2-O-Si[O-Si(CH_3)_3]_2-O-Si(CH_3)_2-CH_2-CH_3$	$QM_2M^HM^{C_2H_5}$ or $QM_2M^HM^{Et}$

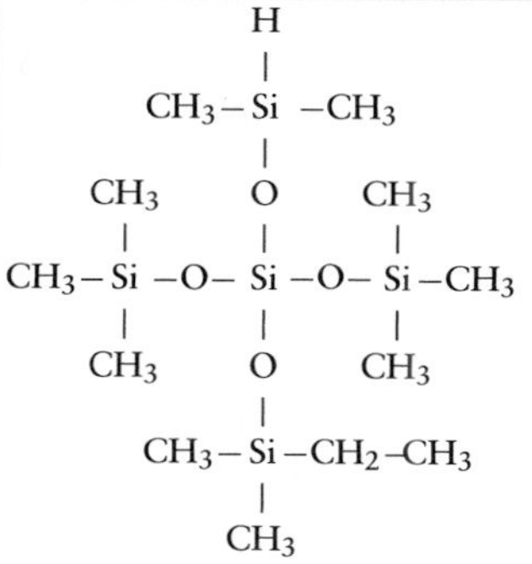

Preparation

Silicone Polymers. The modern synthesis of silicone polymers is multifaceted. It usually involves the four basic steps described in Table B.4. Only step 4 in this table will be elaborated upon here.

Polymerization and Polycondensation. The linear [4] and cyclic [5] oligomers resulting from dimethyldichlorosilane [2] hydrolysis have chain lengths too short for most applications. The cyclics must be polymerized, and the linears condensed, to give macromolecules of sufficient length (Noll, 1968).

Catalyzed by acids or bases, cyclosiloxanes $(R_2SiO)_m$ are ring-opened and polymerized to form long linear chains. At equilibrium, the reaction results in a mixture of cyclic oligomers plus a distribution of linear polymers. The proportion of cyclics depends on the substituents along the Si–O chain, the temperature, and the presence of a solvent. Polymer chain length depends on the presence and concentration of substances capable of giving chain ends. For example, in the KOH-catalyzed polymerization of the cyclic tetramer octamethylcyclotetrasiloxane $(Me_2SiO)_4$ ([5] or D_4 in shorthand notation), the average length of the polymer chains depends on the KOH concentration:

$$x(Me_2SiO)_4 + KOH \rightarrow (Me_2SiO)_y + KO(Me_2SiO)_zH$$

A stable hydroxy-terminated polymer, $HO(Me_2SiO)_zH$, can be isolated after neutralization and removal of the remaining cyclics by stripping the mixture under vacuum at elevated temperature. A distribution of chains with different lengths is obtained. The reaction can also be made in the presence of $Me_3SiOSiMe_3$, which acts as a chain end-blocker:

$$\sim\sim Me_2SiOK + Me_3SiOSiMe_3 \rightarrow \sim\sim Me_2SiOSiMe_3 + Me_3SiOK$$

where $\sim\sim$ represents the main chain.

The Me_3SiOK formed attacks another chain to reduce the average molecular weight of the linear polymer formed.

The copolymerization of $(Me_2SiO)_4$ in the presence of $Me_3SiOSiMe_3$ with Me_4NOH as catalyst displays a surprising viscosity change over time (Noll, 1968). First a peak or viscosity maximum is observed at the beginning of the reaction. The presence of two oxygen atoms on each silicon in the cyclics makes them more susceptible to a nucleophilic attack by the base catalyst than the silicon of the end-blocker, which is substituted by only one oxygen atom. The cyclics are polymerized first into very long, viscous chains that are subsequently reduced in length by the addition of terminal groups provided by the end-blocker, which is slower to react. This reaction can be described as follows:

$$Me_3SiOSiMe_3 + x(Me_2SiO)_4 \xrightarrow{cat} Me_3SiO(Me_2SiO)_nSiMe_3$$

TABLE B.4 The Basic Steps in Silicone Polymer Synthesis

Step	Reaction
1. Silica reduction to silicon	$SiO_2 + 2C \rightarrow Si + 2CO$
2. Chlorosilanes synthesis	$Si + 2CH_3Cl \rightarrow \underset{[1]}{(CH_3)_2SiCl_2} + \underset{[2]}{CH_3SiCl_3} + \underset{[3]}{(CH_3)_3SiCl} + CH_3HSiCl_2 + \cdots$
3. Chlorosilanes hydrolysis	$\underset{[1]}{Cl-Si(CH_3)_2-Cl} + 2H_2O \rightarrow \underset{\text{linears [4]}}{HO-\left(-Si(CH_3)_2-O-\right)_x-H} + \underset{\text{cyclics [5]}}{\left[Si(CH_3)_2-O\right]_{3,4,5}} + HCl$
4. Polymerization and polycondensation	$\underset{\text{cyclics [5]}}{\left[Si(CH_3)_2-O\right]_{3,4,5}} \rightarrow \underset{\text{polymer}}{-\left(-Si(CH_3)_2-O-\right)_y-}$ $\underset{\text{linears [4]}}{HO-\left(-Si(CH_3)_2-O-\right)_x-H} \rightarrow \underset{\text{polymer}}{-\left(-Si(CH_3)_2-O-\right)_z-} + z\,H_2O$

or, in shorthand notation:

$$MM + x\,D_4 \xrightarrow{cat} MD_nM$$

where $n = 4x$ (theoretically).

The ratio between D and M units defines the average molecular weight of the polymer formed.

Catalyst removal (or neutralization) is always an important step in silicone preparation. Most catalysts used to prepare silicones can also catalyze the depolymerization (attack along the chain), particularly at elevated temperatures in the presence of traces of water.

$$\sim\!\sim(Me_2SiO)_n\sim\!\sim + H_2O \xrightarrow{cat} \sim\!\sim(Me_2SiO)_yH + HO(Me_2SiO)_z\sim\!\sim$$

It is therefore essential to remove all remaining traces of the catalyst, providing the silicone optimal thermal stability. Labile catalysts have been developed. These decompose or are volatilized above the optimum polymerization temperature, and consequently can be eliminated by a brief overheating. In this way, catalyst neutralization or filtration can be avoided (Noll, 1968).

The cyclic trimer $(Me_2SiO)_3$ has internal ring tension and can be polymerized without re-equilibration of the resulting polymers. With this cyclic, polymers with narrow molecular weight distribution can be prepared, as well as polymers only carrying one terminal reactive function (living polymerization). Starting from a mixture of cyclics with different internal ring tensions also allows preparation of block or sequential polymers (Noll, 1968).

Linears can combine when catalyzed by many acids or bases to give long chains by intermolecular condensation of silanol terminal groups (Noll, 1968; Stark et al., 1982).

$$\sim\!\sim O-Si(Me)_2-OH + HO-Si(Me)_2-O\sim\!\sim \underset{+H_2O}{\overset{-H_2O}{\rightleftharpoons}} \sim\!\sim O-Si(Me)_2-O-Si(Me)_2-O\sim\!\sim$$

A distribution of chain lengths is obtained. Longer chains are favored when working under vacuum or at elevated temperatures to reduce the residual water concentration. In addition to the polymers described above, reactive polymers can also be prepared. This result can be achieved when re-equilibrating oligomers or existing polymers to obtain a polydimethylmethylhydrogenosiloxane, $MD_zD^H{}_wM$.

$$Me_3SiOSiMe_3 + x(Me_2SiO)_4 + Me_3SiO(MeHSiO)_ySiMe_3 \xrightarrow{cat} \text{cyclics} + Me_3SiO(Me_2SiO)_z(MeHSiO)_wSiMe_3 \quad [6]$$

Additional functional groups can be attached to this polymer using an addition reaction.

$$Me_3SiO(Me_2SiO)_z(MeHSiO)_wSiMe_3 + H_2C = CHR \quad [6]$$

$$\xrightarrow{\text{Pt cat}} Me_3SiO(Me_2SiO)_z(Me\underset{\substack{|\\ CH_2CH_2R}}{Si}O)_wSiMe_3$$

All the polymers heretofore shown are linear or cyclic, comprising mainly difunctional units, D. In addition, branched polymers or resins can be prepared if, during hydrolysis of the chlorosilanes, a certain amount of T or Q units are included, which allow molecular expansion in three or four directions, as opposed to just two. For example, consider the hydrolysis of methyltrichlorosilane in the presence of trimethylchlorosilane, which leads to a branched polymer:

$$x\,Me-\underset{\substack{|\\Me\\ [3]}}{\overset{\substack{Me\\|}}{Si}}-Cl + y\,Me-\underset{\substack{|\\Cl\\ [2]}}{\overset{\substack{Cl\\|}}{Si}}-Cl \xrightarrow[-HCl]{+H_2O} z$$

$$Me-\underset{\substack{|\\Me}}{\overset{\substack{Me\\|}}{Si}}-O-\underset{\substack{|\\O\\|\\Me-Si-O\sim\sim\\|\\O\\|\\Me-Si-Me\\|\\Me}}{\overset{\substack{Me\\|}}{Si}}-O-\underset{\substack{|\\OH}}{\overset{\substack{Me\\|}}{Si}}-O\sim\sim$$

The resulting polymer can be described as $(Me_3SiO_{1/2})_x(MeSiO_{3/2})_y$ or M_xT_y, using shorthand notation. The formation of three silanols on the $MeSiCl_3$ by hydrolysis yields a three-dimensional structure or resin after condensation, rather than a linear polymer. The average molecular weight depends upon the number of M units that come from the trimethylchlorosilane, which limits the growth of the resin molecule. Most of these resins are prepared in a solvent and usually contain some residual hydroxyl groups. These groups could subsequently be used to cross-link the resin and form a continuous network.

Silicone Elastomers. Silicone polymers can easily be transformed into a three-dimensional network by way of a cross-linking reaction, which allows formation of chemical bonds between adjacent chains. The majority of silicone elastomers are cross-linked according to one of the following three reactions.

1. Cross-linking with radicals. Efficient cross-linking with radicals is achieved only when some vinyl groups are present on the polymer chains. The following mechanism has been proposed for the cross-linking reaction associated with radicals generated from an organic peroxide for the initiation, propagation, and termination steps (Stark et al., 1982):

$$R^{\cdot} + CH_2 = CH-Si\equiv \rightarrow R-CH_2-CH^{\cdot}-Si\equiv$$

$$RCH_2-CH^{\cdot}-Si\equiv + CH_3-Si\equiv \rightarrow RCH_2-CH_2-Si\equiv + \equiv Si-CH_2^{\cdot}$$

$$\equiv Si-CH_2^{\cdot} + CH_2 = CH-Si\equiv \rightarrow \equiv Si-CH_2-CH_2-CH^{\cdot}-Si\equiv$$

$$\equiv Si-CH_2-CH_2-CH^{\cdot}-Si\equiv + \equiv Si-CH_3 \rightarrow \equiv Si-CH_2-CH_2-CH_2-Si\equiv + \equiv Si-CH_2^{\cdot}$$

$$2\equiv Si-CH_2^{\cdot} \rightarrow \equiv Si-CH_2-CH_2-Si\equiv$$

where ≡ represents two methyl groups and the rest of the polymer chain.

This reaction has been used for high-consistency silicone rubbers (HCRs), such as those used in extrusion or compression and injection molding, which are cross-linked at elevated temperatures. The peroxide is added before processing. During cure, some precautions are needed to avoid the formation of voids by the volatile residues of the peroxide. Post-cure may also be necessary to remove these volatiles, which can catalyze depolymerization at high temperatures.

2. Cross-linking by condensation. Although mostly used in construction sealants and caulks, condensation-cure silicone materials have also found utility in medical device manufacturing as silicone adhesives (facilitating the adherence to materials of silicone elastomers), encapsulants, and sealants.

One-part products are ready to apply and require no mixing. Cross-linking starts when the product is squeezed from the tube or cartridge and comes into contact with moisture, typically from humidity in the ambient air. These materials are formulated from a reactive polymer prepared from a hydroxy end-blocked polydimethylsiloxane and a large excess of methyltriacetoxysilane.

$$HO-(Me_2SiO)_x-H + excess\ MeSi(OAc)_3 \xrightarrow[-2AcOH]{} (AcO)_2MeSiO(Me_2SiO)_xOSiMe(OAc)_2 \quad [7]$$

$$\text{where } Ac = -\overset{\substack{CH_3\\|}}{C}=O$$

Because of this excess, the probability of two different chains reacting with the same silane molecule is remote. Consequently, all the chains are end-blocked with two acetoxy functional groups. The resulting product is still liquid and can be packaged in sealed tubes and cartridges. Upon opening and exposing the sealant to room humidity, acetoxy groups are hydrolyzed to give silanols, which allow further condensation to occur.

$$\sim\sim O-\underset{\substack{|\\OAc\\ [7]}}{\overset{\substack{Me\\|}}{Si}}-OAc \xrightarrow[-AcOH]{+H_2O} \sim\sim O-\underset{\substack{|\\OAc\\ [8]}}{\overset{\substack{Me\\|}}{Si}}-OH$$

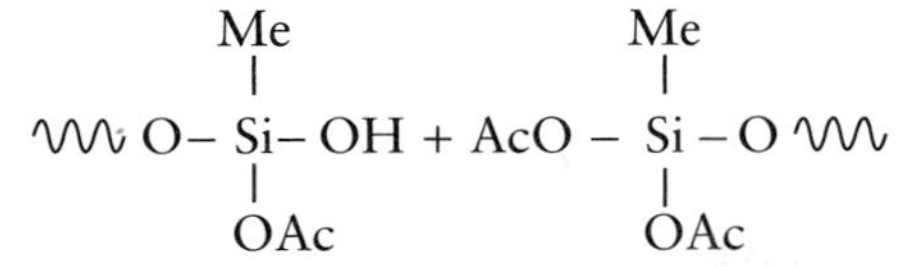

$$\begin{array}{c} \text{Me} \qquad\qquad\qquad \text{Me} \\ \sim\!\!\sim O-\underset{\text{OAc}}{\overset{|}{Si}}-OH + AcO-\underset{\text{OAc}}{\overset{|}{Si}}-O\sim\!\!\sim \\ [8] \qquad\qquad\qquad [7] \end{array}$$

$$\xrightarrow[-AcOH]{} \sim\!\!\sim O-\underset{\text{OAc}}{\overset{\text{Me}}{Si}}-O-\underset{\text{OAc}}{\overset{\text{Me}}{Si}}-O\sim\!\!\sim$$

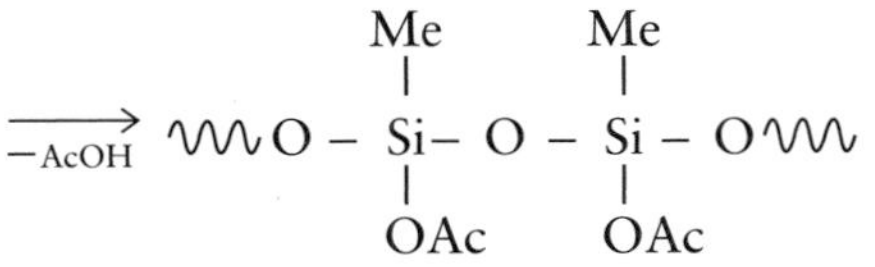

In this way, two chains have been linked, and the reaction continues from the remaining acetoxy groups. An organometallic tin catalyst is normally used, and the cross-linking reaction requires moisture to diffuse into the material. Accordingly, cure will proceed from the outside surface inward. These materials are called one-part RTV (room temperature vulcanization) sealants, but actually require moisture as a second component. Acetic acid is released as a by-product of the reaction. Problems resulting from the acid can be overcome by using other cure (cross-linking) reactions developed by replacing the methyltriacetoxysilane $MeSi(OAc)_3$ with oximosilane $RSi(ON = CR')_3$ or alkoxysilane $RSi(OR')_3$.

Condensation curing is also used in some two-part products where cross-linking starts upon mixing the two components (e.g., a hydroxy end-blocked polymer and an alkoxysilane such as tetra-*n*-propoxysilane, $Si(OnPr)_4$) (Noll, 1968):

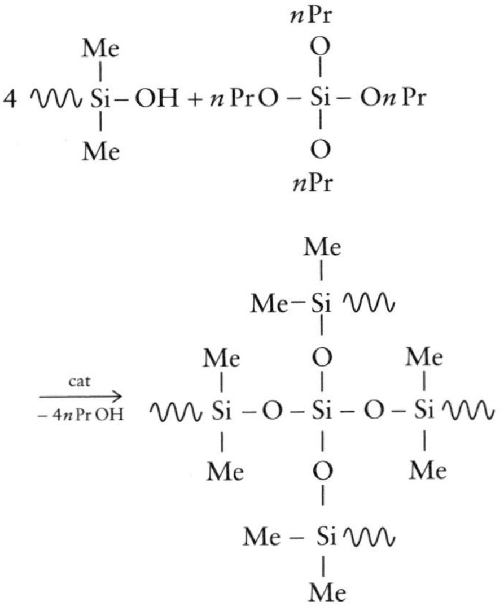

$$4 \sim\!\!\sim \underset{\text{Me}}{\overset{\text{Me}}{Si}}-OH + nPrO-\underset{\underset{nPr}{O}}{\overset{\overset{nPr}{O}}{Si}}-OnPr$$

$$\xrightarrow[-4nPrOH]{cat} \sim\!\!\sim \underset{\text{Me}}{\overset{\text{Me}}{Si}}-O-\underset{\underset{\text{Me}-\underset{\text{Me}}{Si}\sim\sim}{O}}{\overset{\overset{\text{Me}-\overset{\text{Me}}{Si}\sim\sim}{O}}{Si}}-O-\underset{\text{Me}}{\overset{\text{Me}}{Si}}\sim\!\!\sim$$

Here, no atmospheric moisture is needed. Usually an organotin salt is used as a catalyst, but it also limits the stability of the resulting elastomer at high temperatures. Alcohol is released as a by-product of the reaction, leading to some shrinkage after cure at room temperature (0.5–2% linear shrinkage). Silicones with this cure system may not be suitable for the fabrication of parts with precise tolerances.

3. Cross-linking by addition. Use of an addition-cure reaction for cross-linking can eliminate the shrinkage problem mentioned above. In addition-cure, cross-linking is achieved by reacting vinyl end-blocked polymers with ≡ Si–H groups carried by a functional oligomer such as described above [6]. A few polymers can be bonded to this functional oligomer [6] (Stark et al., 1982):

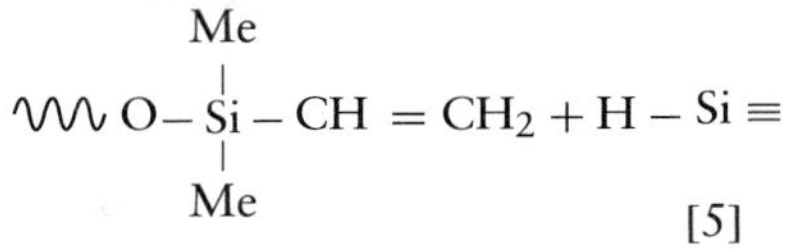

$$\sim\!\!\sim O-\underset{\text{Me}}{\overset{\text{Me}}{Si}}-CH = CH_2 + H-Si\equiv \quad [5]$$

$$\xrightarrow{cat} \sim\!\!\sim O-\underset{\text{Me}}{\overset{\text{Me}}{Si}}-CH_2-CH_2-Si\equiv$$

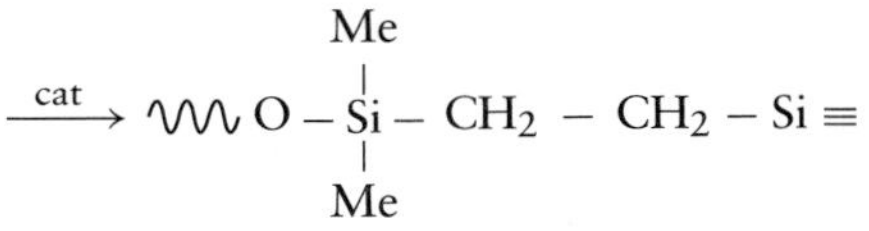

where ≡ represents the remaining valences of the Si in [6].

The addition occurs mainly on the terminal carbon and is catalyzed by Pt or Rh metal complexes, preferably as organometallic compounds to enhance their solubility. The following mechanism has been proposed (oxidative addition of the ≡ Si-H to the Pt complex, H transfer to the double bond, and reductive elimination of the product):

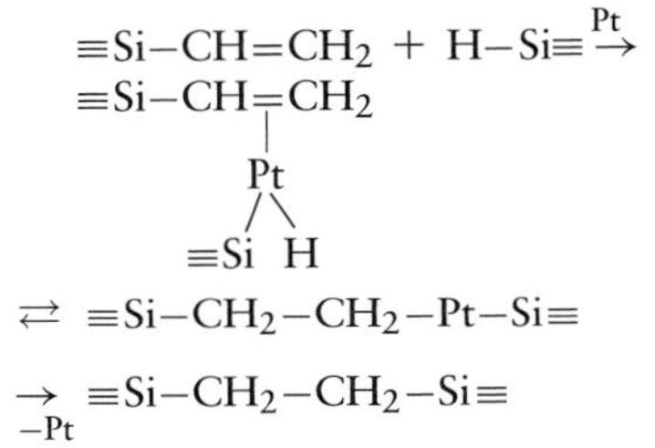

$$\equiv Si-CH=CH_2 + H-Si\equiv \xrightarrow{Pt}$$
$$\equiv Si-CH=CH_2 \text{ (coordinated to Pt, which bears } \equiv Si \text{ and H)}$$
$$\rightleftarrows \equiv Si-CH_2-CH_2-Pt-Si\equiv$$
$$\xrightarrow[-Pt]{} \equiv Si-CH_2-CH_2-Si\equiv$$

where, to simplify, other Pt ligands and other Si substituents are omitted.

There are no by-products with this reaction. Molded silicone elastomer components cured at room temperature by this addition-reaction mechanism are very accurate in terms of dimensional tolerance (i.e., there is no shrinkage). At elevated temperatures, some shrinkage occurs because of the thermal expansion during cure. However, handling these two-part products (i.e., Si–Vi polymer and Pt complex in one component, Si–Vi polymer and SiH oligomer in the other) requires some precautions. The Pt in the complex is easily bonded to electron-donating substances such as amine or organosulfur compounds to form stable complexes with these "poisons," rendering the catalyst complex inactive and inhibiting the cure.

The preferred cure system can vary by application. For example, silicone medical bonding adhesives use acetoxy cure (condensation cross-linking), while platinum cure (cross-linking by addition) is used for precise silicone parts with no by-products.

4. Elastomer filler. In addition to the silicone polymers described above, most silicone elastomers incorporate "filler." Besides acting as a material extender, as the name implies, filler acts to reinforce the cross-linked matrix. The strength of silicone polymers without filler is unsatisfactory for most applications (Noll, 1968). Like most other noncrystallized synthetic elastomers, the addition of reinforcing fillers reduces the tackiness of the silicone, increases its hardness, and enhances its mechanical strength. Fillers might also be employed to affect other properties; for example, carbon black is added for electrical conductivity, or barium sulfate to increase radiopacity. These and other materials are used to pigment the otherwise colorless elastomer; however, care must be taken to select only pigments suitable for the processing temperatures and end-use application.

Generally, the most favorable reinforcement is obtained by using fumed silica, such as Cab–O–Sil®, Aerosil®, or Wacker HDK®. Fumed silica is produced by the vapor phase hydrolysis of silicon tetrachloride vapor in a hydrogen/oxygen flame:

$$SiCl_4 + 2H_2 + O_2 \xrightarrow{1800°C} SiO_2 + 4HCl$$

Unlike many naturally occurring forms of crystalline silica, fumed silica is amorphous. The very small spheroid silica particles (in the order of 10 nm in diameter) fuse irreversibly while still semi-molten, creating aggregates. When cool, these aggregates become physically entangled to form agglomerates. Silica produced in this way possesses remarkably high surface area (100–400 m^2/g), as measured by the BET method developed by Brunauer, Emmett, and Teller (Brunauer et al., 1938; Noll, 1968; Cabot Corporation, 1990).

The incorporation of silica filler into silicone polymers is accomplished prior to cross-linking, by mixing the silica into the silicone polymers on a two-roll mill, in a twin-screw extruder, or in a Z-blade mixer capable of processing materials with this rheology.

Reinforcement occurs with polymer adsorption encouraged by the large surface area of the silica, and when hydroxyl groups on the filler surface lead to hydrogen bonds between the filler and the silicone polymer. In this way, reinforcing filler contributes to the high tensile strength and elongation capability of silicone rubber (Lynch, 1978). The addition of filler increases the already high viscosity of the polymer. Uncured silicone elastomers can have viscosities from 10,000 to well over 100,000 mPa·s. Chemical treatment of the silica filler with silanes enhances its incorporation in, and reinforcement of, the silicone elastomer, resulting in increased material strength and tear resistance (Lane and Burns, 1996) (Figure B.1).

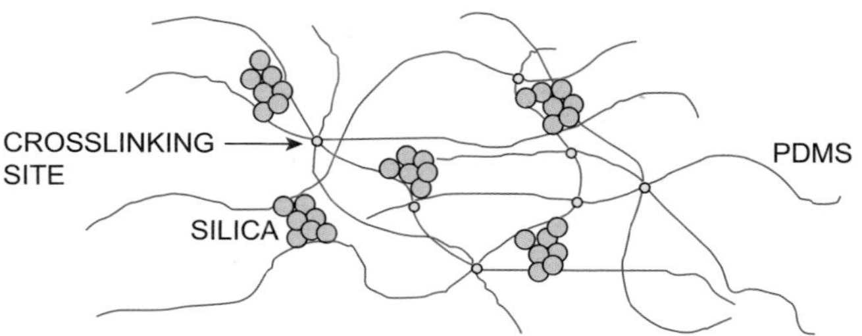

FIGURE B.1 Silicone elastomer/silica network.

> Silicone elastomers for medical applications normally use fumed silica as filler, and occasionally appropriate pigments or barium sulfate. Because of their low glass transition temperatures, these compounded and cured silicone materials are elastomeric at room and body temperatures without the use of plasticizers, unlike other medical materials such as PVC, which might contain phthalate additives.

5. Processing of silicone elastomers. In addition to the polymer blend with amorphous silica filler, other ingredients are needed: an initiator or cross-linker plus catalyst. To avoid premature cure during shipment and storage, these ingredients must be separated. Consequently, products for making silicone elastomers are generally supplied as two components or two-part kits, for example, a base and a peroxide paste, or a kit made of Part A containing polymer and catalyst, and Part B containing polymer and cross-linker. These two components are mixed at a fixed ratio at the point of use and formed into the desired shape before cure.

Silicone elastomers are thermosetting materials. They must be formed into the appropriate shape and configuration prior to cross-linking. Unlike a thermoplastic, which can be remelted and formed again, a cured silicone elastomer part cannot be reprocessed. Suitable processing methods for shaping silicone elastomers include casting, extrusion, and molding. The process selected depends on the viscosity of the feedstock elastomer material, and the shape and configuration of the desired cured elastomer product.

High Consistency Rubber (*HCR*). If very high molecular weight silicone polymers are used (silicone "gums" in the trade), the result is high consistency rubbers, which are desirable as they allow for high tear strengths and tensile elongations. Uncured HCRs are putty-like materials that require high shear equipment for processing. These are usually supplied in two parts to be mixed prior to use, either as a silicone base plus a peroxide initiator or as two-part kit using a Pt cure system. These parts are combined using high shear two-roll mills. The mixed material is then shaped into "preforms" before use in compression, transfer or injection molding at elevated temperature. *Compression molding* requires the simplest equipment: a preform is inserted in a mold and cured under high pressure at elevated temperature. The preform must be of an approximate shape that corresponds well to the mold cavity. This allows sufficient material to fill the cavity without producing excessive flash, the overflow material that remains attached to the parts from around the edge of the mold. Removal of flash requires post-processing. *Transfer molding* requires less preparation of the preform: a more precise but simply shaped preform is transferred from a receiving cavity to the mold cavity. *Injection molding* allows for more automation: an extruder system is used to inject a simple ribbon preform directly into the mold cavity. Typical considerations in this case are controlling the exact amount of material sent to the molding cavity to avoid flash, and maximizing speed but avoiding "scorching" (premature cure before the mold cavity is properly filled). HCRs are also used for *extrusion* to produce tubing, as they have enough "green strength" or mechanical integrity when leaving a cooled extruder and prior to enter a curing oven. When peroxides are used, post-cure in a well-ventilated oven is necessary to remove peroxide by-products, which could bloom at the surface and can reduce the stability of the cured elastomer. One part HCR materials, for which all ingredients have been premixed by the supplier, are also available, but have limited shelf-life depending on the cure system.

Liquid Silicone Rubber (LSR). If lower molecular weight silicone polymers are used, the silicone polymer/silica blend viscosities are lower, leading to liquid silicone rubber. These LSRs are provided as two-part materials that can be used in liquid injection molding – pumped, metered, mixed, and then directly injected in the molding cavity. Processing is eased by the shear-thinning effect that occurs during pumping and injection, reducing the viscosity of the LSR blend and the injection pressures compared to HCR processing. LSRs are particularly well-suited for long automated production runs. Mixing LSR Parts A and B from two 200-liter drums is typically automated using a static mixer prior to direct injection of the precise amount needed. In contrast with the handling of small quantities of HCR on a two-roll mill, preforming, and molding, LSR processing allows more automation. Yet liquid injection molding requires higher investment in the equipment to control the injected amount (to avoid under- or over-filling the mold) precisely. Precise mold cavity temperature control is needed, as cold material is quickly and repeatedly injected into a hot mold. With LSRs, complex molds are needed, preferably with cold runners to avoid premature cure in the feeding lines between injection cycles, with tight specifications still allowing venting (air escape from the mold cavity during injection). The molds may be equipped with complex ejectors to remove parts quickly at the end of the cure, enabling the processing of the next part without loosing thermal control of the mold. The acquisition cost of liquid injection molding machines and sophisticated molds is usually justified for large production runs, as they provide for more efficiency in terms of mold cycle time, overall processing time, and material usage. Cycle time is dependant on operator skills, equipment, and the part to be cured. Typical conditions for LSRs are 0.3–3.0 seconds injection time, 150–200°C cure temperature, and 3–5 sec/mm thickness cure time, depending on formulations (Sommer, 2003).

> Fabricators of silicone elastomer parts should be aware that these LSRs and other addition-cure products contain an inhibitor, a substance that weakly bonds to the platinum catalyst to moderate its activity, permitting sufficient pot life by avoiding premature cure. If contamination occurs with substances capable of bonding more strongly to the platinum catalyst (e.g., amino or thio compounds), the catalyst (which is present in only minute quantities, typically about 10ppm) may lose activity, resulting in inhibition of elastomer cure.

Room Temperature Vulcanizing (RTV) elastomers. In addition to HCRs and LSRs, which are designed to cure by exposure to heat, other silicone elastomers, known as RTVs, are intended to be cured at room temperature. Typically, RTV elastomers are provided in two-part systems and can be viewed as a variation of the LSRs, but with lower viscosity and less inhibitor. They can be mixed with a spatula and cast after de-airing, and are typically used for laboratory trials and commercially in medical applications for dental impression molding. RTVs are also available as one-part silicone elastomers provided ready to use, usually as adhesives. These materials rely on a condensation reaction and on moisture in the air as the second component to achieve cure.

Silicone Gels. Silicone gels are typically composed of a very lightly cross-linked silicone elastomer whose polymer network has been swollen with silicone fluids; however, these gels contain no silica or other fillers. Medical applications for silicone gels include breast, testicular, and other soft-tissue implants for tissue augmentation or to help restore one's appearance after cancer surgery. In addition, silicone gel external breast prostheses can be worn in or attached to garments, such as brassieres, for similar purposes. Silicone gel is often supplied in a two-part fluid system and cures via a platinum-catalyzed addition reaction. Parts A and B are mixed at a desired ratio and cured (usually by exposure to elevated temperature) to yield a sticky, cohesive mass of the desired consistency. The consistency of the material can be controlled by the degree of cross-linking, as well as quality and quantity of swelling fluid. After mixing but before curing, the mixture is still liquid and can be pushed through a large gauge needle, enabling the filling of a silicone elastomer implant shell or the thermoformed pouch of an external breast prosthesis made of a thin polyurethane film.

Beyond gel implants and external prostheses, silicone gels find application in skin-contacting sheet goods in wound and scar care.

In addition to gels composed entirely of silicone materials, the gel used in some gel pads for the prevention of pressure sores or for orthotic applications are comprised of a cross-linked silicone polymer network swollen with non-silicone fluids such as mineral oil.

Silicone Adhesives. Three basic types of silicone adhesives are used in medical applications: bonding, pressure-sensitive, and gel.

1. *Bonding adhesives.* Silicone bonding adhesives are used to attach components together and to seal seams and junctions. Electrical components can also be encapsulated and insulated using silicone bonding adhesive. Silicone bonding adhesives are most commonly formulated as one-part RTV elastomer systems that use a condensation cross-linking reaction, as described earlier in this chapter.
2. *Pressure-sensitive adhesives.* Silicone PSAs are typically formulated in solvent. A silanol end-blocked

POLYMER RESIN

4 HO-Si ∿∿ Si-OH + (HO-Si, Si-OH, HO-Si, Si-OH resin) $\xrightarrow[\text{Heat}]{NH_3}$ $5\ H_2O$ +

PSA

HO-Si∿∿Si-O-Si Si-O-Si∿∿Si-O-Si Si-O-Si∿∿Si-OH

HO-Si∿∿Si-O-Si Si-OH HO-Si Si-OH

PDMS undergoes a polycondensation reaction with a silicate resin in the presence of ammonia as catalyst. The ammonia is stripped with heat, and usually the solvent is exchanged.

In some applications, a hot melt silicone PSA is used ostensibly without solvent.

Silicone PSAs have properties that make them well-suited for application in the medical area. Besides their biocompatibility, the materials are highly flexible and permeable to moisture vapor, CO_2, and oxygen. Silicone PSAs can provide strong adhesion to the skin, facilitating the attachment of hairpieces, prosthetics, and other devices to the body, and they are widely used in transdermal drug delivery. Due to compatibility concerns with amine-containing drugs, an additional class of silicone PSAs have been generated by the further reaction of the PSA with hexamethyldisilazane to convert the pendant ≡ SiOH groups into ≡ $SiOSi(CH_3)_3$.

3. *Gel adhesives.* Silicone gel adhesives, also known as soft skin adhesives, are used in wound care, and are found to be gentler and less traumatic on removal than the pressure-sensitive types. Unlike PSAs, which are typically formulated in solvent, silicone gel adhesives are supplied in solventless two-part systems. In addition to wound care applications, the materials are also used in the treatment of hypertrophic and keloid scars. Evidence suggests this therapy may reduce scar height and appearance (O'Brien and Pandit, 2006).

Silicone Film-in-Place, Fast-Cure Elastomers. In addition to the cured silicone elastomers in skin contact applications, *in situ* cure materials have been developed. These materials form films when spread or sprayed on the skin, and then undergo RTV cure. Products such as spray-on wound dressings or drug-loaded lotions have been evaluated. The ability of silicone to spread and form films is related to its low surface tension as described in the Physico-Chemical Properties section below (Maxon et al., 2004).

Physico-Chemical Properties

The position of silicon just under carbon in the periodic table led to a belief in the existence of analog compounds where silicon would replace carbon. Most of these analog compounds do not exist, or behave very differently from their carbon counterparts. There are few similarities between Si–X bonds in silicones and C–X bonds (Stark et al., 1982; Corey, 1989; Hardman, 1989; Lane and Burns, 1996).

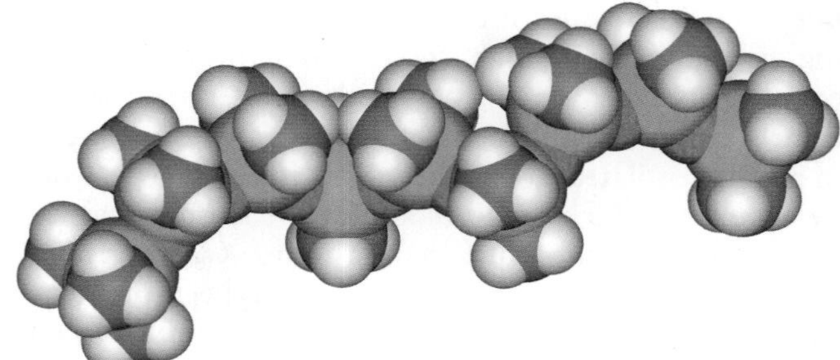

FIGURE B.2 Three-dimensional representation of eicosamethylnonasiloxane, $Me_3SiO(SiMe_2O)_7SiMe_3$ or MD_7M. *(Courtesy T. Lane, Dow Corning.)*

Between any given element and Si, bond lengths are longer than for the element and C. The lower electronegativity of silicon ($\chi_{Si} \approx 1.80$, $\chi_C \approx 2.55$) leads to a very polar Si–O bond compared to C–O. This bond polarity also contributes to strong silicon bonding; for example, the Si–O bond is highly ionic and has a high bond energy. To some extent, these values explain the stability of silicones. The Si–O bond is highly resistant to homolytic scission. On the other hand, heterolytic scissions are easy, as demonstrated by the re-equilibration reactions occurring during polymerizations catalyzed by acids or bases.

Silicones exhibit the unusual combination of an inorganic chain similar to silicates and often associated with high surface energy, but with side methyl groups that are very organic and often associated with low surface energy (Owen, 1981). The Si–O bonds are moderately polar, and without protection would lead to strong intermolecular interactions (Stark et al., 1982). Yet, the methyl groups, only weakly interacting with each other, shield the main chain (see Fig. B.2).

The surface activity of silicones is evident in many ways (Owen, 1981):

- Polydimethylsiloxanes have low surface tension (20.4mN/m) and are capable of wetting most surfaces. With the methyl groups pointing to the outside, this configuration gives very hydrophobic films and a surface with good release properties, particularly if the film is cured after application. Silicone surface tension is also in the most promising range considered for biocompatible elastomers (20–30mN/m) (Baier, 1985).
- Silicones have a critical surface tension of wetting (24mN/m), higher than their own surface tension. This means silicones are capable of wetting themselves, which promotes good film formation and surface coverage.

- Silicone organic copolymers can be prepared with surfactant properties, with the silicone as the hydrophobic part (e.g., in silicone glycol copolymers).

The low intermolecular interactions in silicones have other consequences (Owen, 1981):

- Glass transition temperatures are very low (e.g., 146°K for a polydimethylsiloxane compared to 200°K for poly-isobutylene, the analog hydrocarbon).
- The presence of a high free volume compared to hydrocarbons explains the high solubility and high diffusion coefficient of gas into silicones. Silicones have a high permeability to oxygen, nitrogen, or water vapor, even though liquid water is not capable of wetting a silicone surface. As expected, silicone compressibility is also high.
- The viscous movement activation energy is very low for silicones, and their viscosity is less dependent on temperature compared to hydrocarbon polymers. Furthermore, chain entanglements are involved at higher temperature, and contribute to limit viscosity reduction (Stark et al., 1982).

This shielding is made easier by the high flexibility of the siloxane chain. Barriers to rotation are low, and the siloxane chain can adopt many configurations. Rotation energy around a $H_2C–CH_2$ bond in polyethylene is 13.8 kJ/mol, but is only 3.3 kJ/mol around a $Me_2Si–O$ bond, corresponding to a nearly free rotation. In general, the siloxane chain adopts a configuration so that the chain exposes a maximum number of methyl groups to the outside, whereas in hydrocarbon polymers, the relative rigidity of the polymer backbone does not allow selective exposure of the most organic or hydrophobic methyl groups. Chain-to-chain interactions are low, and the distance between adjacent chains is also greater in silicones. Despite a very polar chain, silicones can be compared to paraffin, with a low critical surface tension of wetting (Owen, 1981).

CONCLUSION

Polydimethylsiloxanes are often referred to as silicones. They are used in many applications because of their stability, low surface tension, and lack of toxicity. Methyl group substitution or introduction of tri- or tetrafunctional siloxane units leads to a wide range of structures. Polymers are easily cross-linked at room or elevated temperature to form elastomers, without losing their advantageous properties.

ACKNOWLEDGMENTS

Part of this chapter (here revised) was originally published in *Chimie Nouvelle*, the journal of the Société Royale de Chimie (Belgium), Vol. 8 (30), 847 (1990) by A. Colas and is reproduced here with the permission of the editor.

BIBLIOGRAPHY

Baier, R. E. (1985). Adhesion in the biologic environment. *Biomaterials, Medical Devices, and Artificial Organs*, **12**, 133–159.

Brunauer, S., Emmett, P. H., & Teller, E. (1938). Adsorption of gases in multimolecular layers. *J. Am. Chem. Soc.*, **60**, 309.

Cabot Corporation (1990). *CAB-O-SIL Fumed Silica Properties and Functions*. Tuscola, IL: Cabot Corporation.

Corey, J. Y. (1989). Historical Overview and Comparison of Silicon with Carbon. In S. Patai, & Z. Rappoport (Eds.), *The Chemistry of Organic Silicon Compounds* (pp. 1–56). New York, NY: John Wiley & Sons. Part 1.

Dow Corning Corporation (2001). *Fabricating with Silastic High Consistency Silicone Rubber*. (Accessible at dowcorning.com/content/rubber/rubberprocess), Midland, MI: Dow Corning Corporation.

Hardman, B. (1989). Silicones. *Encyclopedia of Polymer Science and Engineering*. (Vol. 15, p. 204). New York, NY: John Wiley & Sons.

Kipping, F. S. (1904). Organic derivative of silicon. Preparation of alkylsilicon chlorides. *Proc. Chem. Soc.*, **20**, 15.

Lane, T. H., & Burns, S. A. (1996). Silica, silicon and silicones ... unraveling the mystery. *Curr. Top. Microbiol. Immunol.*, **210**, 3–12.

Lynch, W. (1978). *Handbook of Silicone Rubber Fabrication*. pp. 25–34. New York, NY: Van Nostrand Reinhold.

Maxon, B. D., Starch, M. S., & Raul, V. A. (2004). *New Silicone Film-Forming Technologies for Topical Delivery and Beyond*. Orlando, FL: Proceedings of the 23rd International Federation of the Society of Cosmetic Chemists (IFSCC) Congress.

Noll, W. (1968). *Chemistry and Technology of Silicones*. New York, NY: Academic Press.

O'Brien, L., & Pandit, A. (2006). Silicon gel sheeting for preventing and treating hypertrophic and keloid scars. *Cochrane Database of Systematic Reviews* (1), CD003826.

Owen, M. J. (1981). Why silicones behave funny. *Chemtech.*, **11**, 288.

Rochow, E. G. (1945). The direct synthesis of organosilicon compounds. *J. Am. Chem. Soc.*, **67**, 963–965.

Smith, A. L. (1991). *Introduction to Silicones. The Analytical Chemistry of Silicones*. pp. 3–19. New York, NY: John Wiley & Sons.

Sommer, J. G. (2003). *Elastomer Molding Technology: A Comprehensive and Unified Approach to Materials, Methods, and Mold Design for Elastomers*. Hudson, OH: Elastech.

Stark, F. O., Falender, J. R., & Wright, A. P. (1982). Silicones. In G. Wilkinson, F. G. A. Sone, & E. W. Ebel (Eds.), *Comprehensive Organometallic Chemistry* (Vol. 2, pp. 288–297). Oxford, UK: Pergamon Press.

C. FLUORINATED BIOMATERIALS

Fang Liu[1] *and David W. Grainger*[1,2]
[1]Departments of Pharmaceutics and Pharmaceutical Chemistry, University of Utah, Salt Lake City, UT, USA
[2]Bioengineering, University of Utah, Salt Lake City, UT, USA

INTRODUCTION

Fluorinated biomaterials have a long history of biomedical device and biomaterials applications. These materials are generally carbon-based polymers, liquids, and gas precursors to films that all contain large amounts of chemically bonded fluorine (i.e., perfluorocarbons). Fluorinated materials can be solids, liquids, thin films, coatings or gels, depending on their chemistry, processing,and mode of use. Fluoropolymers in general are thermoplastic polymers analogous to hydrocarbon-based polyethylene (PE), but highly fluorinated; some or all of the hydrogen atoms attached to the PE carbon polymer chain are replaced by fluorine (F) or fluorinated alkyl (perfluorocarbon) side groups. In some cases other halogen atoms (e.g., chlorine, Cl) are also included with F to slightly modify polymer properties (Drobny, 2005). Notably, fluoropolymers show unique propertiesnot achievable by other polymer materials, including chemical inertness, extreme hydrophobicity and solvent resistance, low coefficients of friction, high design tolerances for device fabrication, and temperature resistance (Drobny, 2005). These properties facilitate fluoropolymer use in specific technologies where most hydrocarbon-based materials do not perform well, either in device manufacturing or their applications. Nevertheless,as with all biomaterials, fluoropolymers perform best only in certain biomedical applications, and often carry a considerably greater cost than their hydrocarbon analogs into medical device designs.

Just two years after the first commercial launch of completely fluorinated polytetrafluoroethylene (PTFE), as DuPont's now well-known Teflon™ product, after World War II ended, this new polymer was implanted in animals for the first time (Leveen and Barberio, 1949). However, early PTFE materials processing difficulties limited device capabilities. In 1963, Japanese researchers reported a new process for expanding PTFE films to produce highly uniform, continuous fibrous porous structures that, after thermal processing, retained this microstructure with vastly improved mechanical strength (Oshige, 1967). In 1972, this expanded fibrous material was first used experimentally as a venous graft substitute (Soyer et al., 1972), and a year later as an arterial bypass implant (Matsumoto et al., 1973). Several years later (1976), expanded PTFE (ePTFE) was refined to production scales by Gore, allowing increased access and clinical use of this more versatile PTFE form in commercialized biomedical products (Gore, 1976). This enabled fluorinated biomaterials to be employed in biomedical applications both inside and outside the living host, facilitating entry as an important class of polymeric biomaterials that are seen in the field today.

Biomedical interest in fluorinated biomaterials focuses on several unique properties: lubricity; high sizing tolerances for device fabrication; and select aspects of reasonable biocompatibility resulting from both unique chemical and morphological properties. Understanding certain aspects of fluorinated chemistry is important to appreciate their material properties and broad utility in biomedical products.

INTERESTING FLUOROPOLYMER CHEMICAL AND PHYSICAL PROPERTIES DERIVED FROM THEIR POLYMER CHEMISTRY, MOLECULAR STRUCTURE, AND BONDING

Replacing large amounts of hydrogen in C–H and C–C bonds in organic materials with fluorine as C–F chemistry, using several different chemical means, results in dramatic changes to the fluoromaterial's physical and chemical properties. Technologically desirable characteristics of perfluorinated materials are identified from how they interact distinctly with other media (e.g., heated fabrication machines and device extruding dyes, as well as tissue, blood, proteins, and other polymers). Single fluorine bonds with carbon are the strongest carbon bonds, some 25 $kcal/mol^{-1}$ stronger than C–Cl (Smart, 1994). Fluorination also strengthens adjacent aliphatic bonds: the CF_3–CF_3 bond is 10 $kcal/mol^{-1}$ stronger than the CH_3–CH_3 bond (Smart, 1994). This makes alkyl fluorides 10^2–10^6 times more stable than the corresponding alkylchlorides in solvent and thermal reactions (Smart, 1994). Exceptional thermal and chemical stabilities observed for perfluorinated materials then result. Fluorine's high ionization potential energy and low polarizability provide relatively weak intermolecular forces, low interfacial energies, and low refractive indices in fluorinated materials. This is important to interfacial applications, (see section on Surfaces Modified by Fluorination Treatments below). Fluorine's larger atomic radius compared to hydrogen provides a rational basis for the observed structural differences between perfluorocarbons and hydrocarbons; chain movement energies (i.e., C–C chain rotational barriers) for various fluorine-substituted bonds are significantly higher than barriers in analogous hydrocarbon systems (Smart, 1994). Partially fluorinated commercial polymers, polyvinylidene fluoride (PVDF) and ethylene-trifluoroethylene copolymer (ETFE), have zig-zag polymer chain conformations with different C–F dipole alignments along the chain (Smart, 1994), while perfluorocarbons with only C–F bonds (e.g., PTFE, see below) assume helical chain orientations with C–F dipoles distributed axially around the chain helix (Doeff and Lindner, 1989; Zhang et al., 1989; Sun et al., 1994a,b; Kobayashi and Owen, 1995; Bar et al., 1997;

Stone et al., 1998). PTFE, for example, as a model for high molecular weight perfluorocarbon chains, is known to have a rich phase diagram of several distinct helical solid phases (Scheirs, 1997). This polymer helix basically encases the inner carbon–carbon polymer backbone with a tight outer shell of fluorine groups, protecting the carbon bonds from reactants, and also contributing to unique chain–chain interactions in PTFE and helical fluorinated polymers. Although most fluoropolymers share unique properties (e.g., thermal and chemical stability, lubricity), their mechanical properties are slightly different, depending on whether they are fully fluorinated or contain some hydrogen atoms. Generally, partially hydrogenated fluoropolymers exhibit higher stiffness than fully fluorinated polymers (i.e., perfluoropolymers). Perfluoropolymers exhibit greater elongation and higher maximum service temperatures that benefit device engineering and thermal processing.

DISTINGUISHING THE DIFFERENT FLUOROPOLYMERS

Fluoropolymers can be classified into homopolymers and copolymers by the monomer(s) used in their polymerization. They are divided into either "partially fluorinated" or "perfluorinated" (i.e., 100% C–F bonds) based on the amounts of fluorine in the polymer chain. Figure C.1 lists currently available commercial biomedical fluoropolymers. Polymer chemical composition affects the resulting materials' properties; only certain fluoropolymers exhibit properties attractive for biomedical products. As one selection criteria, Figure C.2 (Scheirs, 1997) compares some selected mechanical properties for different biomedical fluoropolymers (Drobny, 2005).

Polytetrafluoroethylene (PTFE)

PTFE (DuPont trade-name Teflon™) is perhaps the most commonly analyzed fluoropolymer. As described above, fluorine's large size and mutual repulsion of adjacent fluorine atoms causes PTFE macromolecule chains to exhibit a twisting helix, comprising 13 CF_2 groups per 180° turn, distinct from the classic planar zigzag chain typical for PE (Bunn and Howells, 1954). Figure C.3 shows this important conformational chain distinction that imparts certain unique physical properties to the fluoropolymer. Helical PTFE chains are often of very high molecular weight ($>10^6$) from their polymerization by reactive TFE gas. The helical polymer chains pack into solid crystallites like long, parallel, stiff molecular rods; individual PTFE polymer molecules can slip on each other as quasi-cylinders sliding under shear stress. Tight helical packing and chain–chain slip make PTFE the most lubricious polymer available, with a low coefficient of friction (0.1). This is a major selection criterion for manufacturing fluoropolymer tubing for catheters. Unfortunately, this also makes PTFE solids very susceptible to cold flow (creep) under stress – a major reason for their contraindication in mechanical applications (i.e., poor bearing and joint surfaces) (see Box 1). Mutual repulsive forces of adjacent fluorine atoms keep the PTFE chain backbone from bending. Low energy barriers to chain–chain slip events result from low chain–chain interaction energies. PTFE's very high bulk solid fractional crystallinity means that it scatters most visible light wavelengths, resulting in its characteristic opaque white color. PTFE's high molecular weight and rigid helical chain conformation also produce high melt viscosities (e.g., 10^{12} Pa/s), ~6 times higher than most thermoplastic polymers. This viscosity

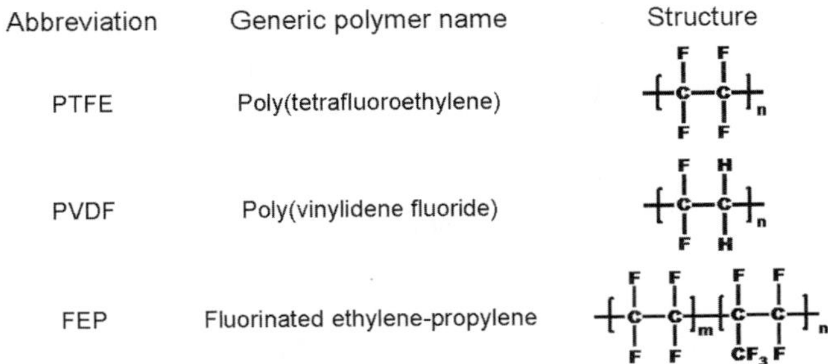

FIGURE C.1 Abbreviated names and molecular structures for some biomedical fluoropolymers.

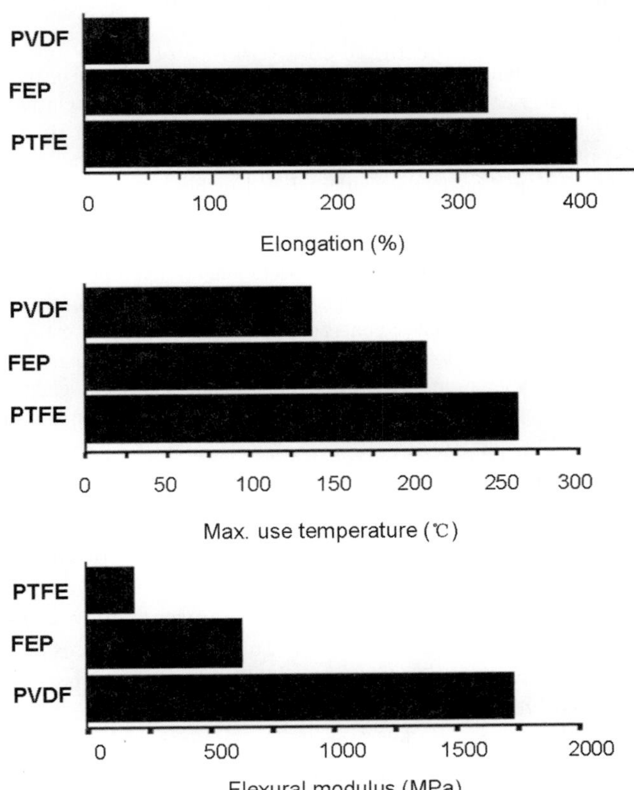

FIGURE C.2 Comparison of mechanical properties of some select biomedical fluoropolymers. (Adapted by permission from *Modern Fluoropolymers*, Scheirs, J. (Ed.). John Wiley & Sons, Ltd., copyright 1997.)

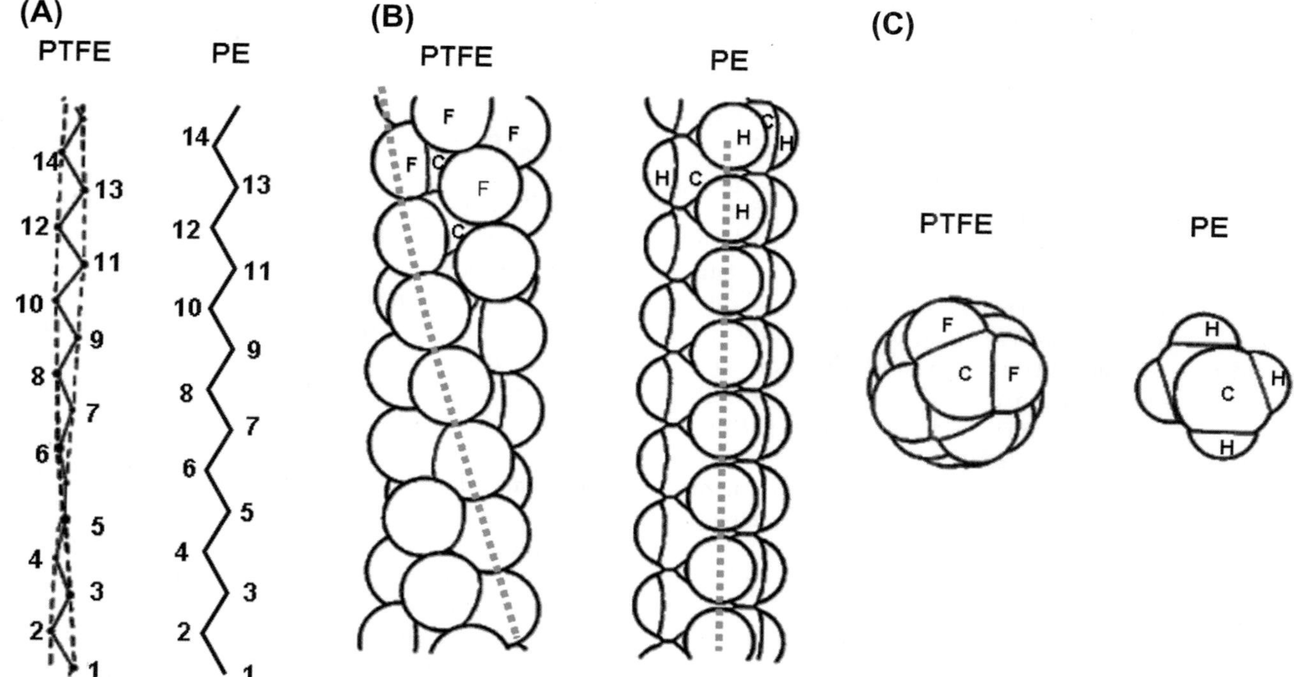

FIGURE C.3 (A) PTFE twisted zigzag chain compared to polyethylene (PE) molecular zig-zag chain; (B) Side space-filling views; and (C) top views, of PTFE chain versus hydrocarbon chain (PE). Bunn/Howells ref: "Adapted by permission from Macmillan Publishers Ltd:, Bunn, C. W. & Howells, E. R., "Structures of Molecules and Crystals of Fluoro-Carbons," Nature 174, 550, 1954. (http://www.nature.com/npg/index npg.html)

BOX 1 Sir John Charnley and the PTFE Replacement Hip Bearing

Convinced that hip joint repair could not use metal-on-metal bearings, Professor Sir John Charnley's early work on hip arthroplasty techniques in the early 1950s used a Teflon™-on-Teflon solid bearing to resurface the arthritic femoral head and acetabulum. These Teflon™-on-Teflon bearings failed within two years from mechanical creep and particulate issues. His further innovations created a femoral stem and metal head articulating against a Teflon cup-socket inserted into the acetabulum. High Teflon wear occurred with joint articulation, producing severe osteolysis, loosening in the surrounding bone and, as a result, requiring many revision surgeries. Teflon's poor performance in this articulating joint application prompted the adoption of high molecular weight polyethylene as a better bearing surface – an innovation that persists to today.

is excessive for conventional melt processing fabrication methods for thermoplastic medical devices (e.g., extrusion and injection molding). This requires a high continuous service temperature (~260°C) for processing PTFE into devices. Therefore, as PTFE does not dissolve in any solvents, PTFE processing technologies are similar to those of powder metallurgy. This involves form casting of pre-polymerized PTFE granulated powders or latex formulations, followed by mold compression and thermal sintering. PTFE's high thermal stability is key to success under the extreme processing conditions required for molded device parts (Drobny, 2005).

Fluorinated Ethylene Propylene (FEP)

FEP is a copolymer of tetrafluoroethylene (TFE) and hexafluoropropylene (HFP), first produced by DuPont in 1956 (Teflon™ FEP) to reduce PTFE's high crystallinity and melt viscosity. This improves FEP processing characteristics while maintaining high perfluorination. Bulky FEP perfluoromethyl groups produce defects in solid fluoropolymer crystallites, reducing polymer melting point, impeding chain slip, and reducing solid cold flow (Drobny, 2001). FEP combines the unique mechanical and chemical properties of PTFE with the melt-processability of more conventional polymers. FEP has a maximum service temperature of 204°C, and a slightly higher coefficient of friction than PTFE. It is used in biomedical devices in place of PTFE.

Polyvinylidene fluoride (PVDF)

PVDF is a homopolymer of the vinylidene monomer (CH_2CF_2), and is sold as Kynar™. PVDF has the highest flexural modulus of all fluoropolymers, due to the interpenetration of larger CF_2 groups crystallizing with adjacent smaller CH_2 groups on adjacent chains. Unlike other fluoropolymers, PVDF is soluble in highly polar solvents (dimethylformamide, tetrahydrofuran), acetone, and esters (Drobny, 2005). PVDF's unique high dielectric constant, high dielectric loss factor, and interesting piezoelectric behavior under certain conditions result from this chemistry and result in solid state structures. Fluorine's shielding effects to all neighboring CH_2 groups provides PVDF with good chemical resistance and thermal stability (Scheirs, 1997; Drobny, 2001). These are all valuable properties sought in specific medical device applications.

Fluoropolymer Melt Processing

Because most perfluorinated polymers do not dissolve in many solvents, their bio-medical products are often made by melt extrusion, where polymers develop flow upon melting in normal extrusion equipment. This technique, common to all polymer devices and chemistries, exposes fluoropolymers to very high temperatures in order to reduce their viscosity and improve flow characteristics for extended device production runs and product lengths. Fluoropolymers FEP and PVDF will readily melt flow when heated, typically above 260°C. This permits uninterrupted feed of fluoropolymer resin into the parts extruder to produce long continuous lengths of product (i.e., medical tubing). By contrast, PTFE extrusion is limited due to difficulties in its materials handling, size of the stock pre-forms, and tubing. However, PTFE's poor melt processing presents an important opportunity: PTFE tubing can be manufactured to very small dimensions, with wall thicknesses as small as 2.54×10^{-3} cm and tolerances of $\pm 1 \times 10^{-3}$ cm. This engineering benefit is largely due to PTFE's inability to melt flow, allowing more precise control over its use in small dimensional, high-tolerance extrusion forms. This unique PTFE property is essential for producing medical products requiring tight size tolerances, such as small diameter and multi-lumen tubing with multiple precision passages for advanced catheters.

Original Gore-Tex™ and Generic Equivalents (ePTFE)

PTFE film extrusion under anisotropic loading conditions produces expanded Teflon™ (ePTFE). Its micro-architecture exhibits pores axially aligned along the stretch direction, resulting in a unique fluoropolymer fabric material with an oriented microporous architecture. This was originally commercialized as the fabric-like Gore-Tex™ material (Gore, 1976). ePTFE's porous structure is characterized by regular PTFE nodes interconnected by PTFE fibrils (Figure C.4A), distinct from solid PTFE shown in Figure C.4B. In ePTFE internodal spacing or distance (i.e., PTFE fibril length between solid PTFE nodes) is important to control device implant behavior (McClurken et al., 1986), and this can be controlled from 1 to 100 microns (Santiago et al., 1981) while retaining some properties similar to PTFE, e.g., biological adsorption, low tensile strength, low modulus of elasticity, water penetration control, and easy sterilizability. Porous ePTFE structures also allow important mechanical modulus-matching properties in tissue sites better than other polymers for many soft biological

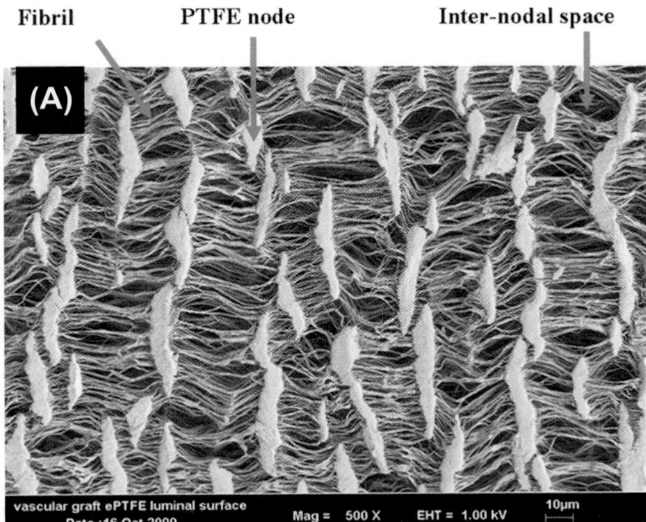

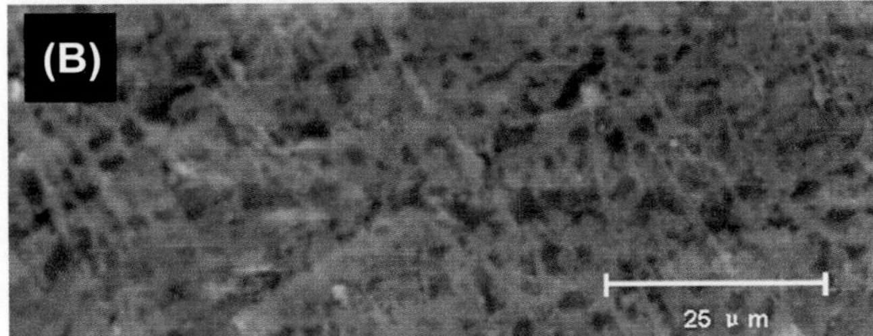

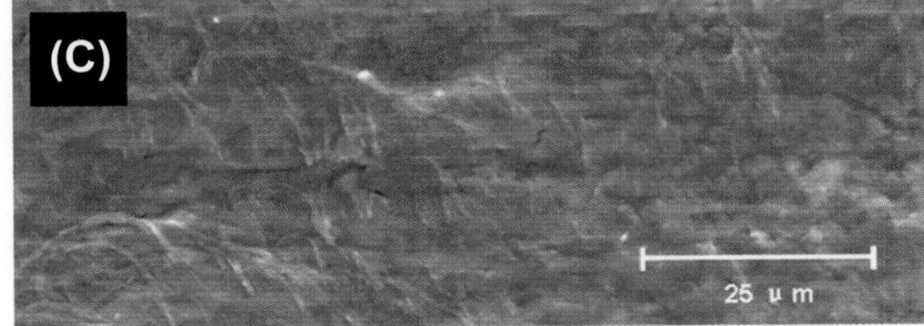

FIGURE C.4 (A) Scanning electron micrograph of ePTFE showing regular node-fiber open materials structure (scale bar, lower right corner = 10 microns). (We kindly acknowledge B. Wagner, W. L. Gore & Associates, Flagstaff, AZ, USA, for ePTFE micrographs). (B) SEM of solid Teflon™ showing the cross-sectional porous structure resulting from the compressed sintering of PTFE particles (we acknowledge P. Hogrebe for these SEM micrographs). (C) Dense continuous top surface (scale bar = 25 microns).

tissue applications (Mole, 1992). Porosity also importantly encourages in-growth of tissue, and hence moderate levels of tissue mechanical fixation. Bulk PTFE does not exhibit this property. The micropores also present active sites for stable blood clotting, an important property for conditioning implanted vascular graft surfaces to limit their chronic blood coagulation.

Surfaces Modified by Fluorination Treatments (Grainger and Stewart, 2001)

Bulk perfluorinated materials' intrinsically higher costs and often substandard mechanical properties limit their applications. When only fluorinated polymer interfacial properties are desired, then fluorinated surface layers can be chemically deposited or coated over other mechanically superior substrates to impart fluorinated properties only at the surface. This surface application limits bulk fluorocarbon costs and their performance liabilities. As developed further below, fluorinated surfaces can be produced by plasma deposition of gaseous precursor monomers, direct solid powder or solution overcoating and casting/evaporation, and solution phase fluorinated component blooming. High surface fluorocarbonation produces unique interfacial properties, imparting specific, technologically attractive features including low solid-state surface free energy (or low water wettability and permeability), surface lubricity, and chemical resistance and durability, to device surfaces. Table C.1 shows that perfluorinated polymer surfaces exhibit low interfacial energies (indicated by Zisman γ_c values), correlating directly with their utility as low adhesion, low water-wetting, and low friction surfaces. Substituting either hydrogen or another halogen for fluorine along the polymer backbone results in significant increases in γ_c values (as seen in comparisons of PTFE with high-density polyethylene (HDPE)), PVDF:poly(CH_2CHF) has a γ_c~25 mN/m, approaching that for polyethylene (HDPE, see Table C.1). Another important outcome of this low interfacial fluoromaterial energy is that fluorinated species thermodynamically prefer to coat a material's surface exposed to air. Hence, fluorinated components will migrate from deep inside fluid coating mixtures to "bloom" and enrich as a coated overlayer at the solid–air interface.

Surface-enriched films of fluorinated components reside at surfaces even if fluorinated components are doped only at trace or minority components into bulk coating materials. Therefore, fluorinated surfaces can be fabricated using minority fluorinated components added to a bulk material if they are allowed to move to that material's surface (e.g., from coating solutions), off-setting the intrinsically high costs of fluorinated materials. Emphasis on surface-fluorinated coatings and films has therefore increased as technological needs drive new, improved, and less expensive methods to put this chemistry over biomaterial surfaces. Several coating strategies have sought to create, organize, and orient $-CF_3$ and $-SF_5$ terminal groups from films over any surface to provide such properties. High surface density and organization of these groups is necessary to achieve their technological properties. Alignment of perfluorinated chains terminating in this $-CF_3$ or $-SF_5$ chemistry at the surface is required. Orienting the side chains in fluoroalkyl side-chain acrylate and methacrylate films enriches surfaces with side chain-terminating $-CF_3$ groups (Bunn and Howells, 1954; Clark and Muus, 1962; Russell et al., 1986; Naselli et al., 1989), and lowers solid interfacial energies. Perfluoroalkyl-grafted polysiloxanes also exhibit side chain orientation of their perfluorinated chemistry, sometimes with spontaneous perfluorinated group organization as a film or coating (Hare et al., 1954; Pittman, 1972; Schneider et al., 1989; Tsao et al., 1997; Clark, 1999). Gas plasma-deposited thin fluorinated coatings are also well-developed for this purpose (D'Agostino, 1990), representing a mature industry and biomedically relevant materials treatment (e.g., for vascular devices and intraocular lenses; Ratner, 1995).

TABLE C.1 Surface Energies for Perfluorocarbons versus Analogous Hydrocarbons[a]

	Solid Interfacial Energy (γ_c, mN/m)	
Substance	**Perfluorocarbon**	**Hydrocarbon**
PTFE[1]	18.5	31
PVDF	25	31
HDPE[2]	18.5	31
n-pentane	9.4	15.2
n-hexane	11.4	17.9
n-octane	13.6	21.1
Decalin	17.6	29.9
Benzene	22.6	28.5

[a]Smart, B. E. (1994). In: *Organofluorine Chemistry: Principles and Commercial Applications*; Banks, R. E., Tatlow, J. C. & Smart, B. E. (Eds.). Plenum Press: New York, NY, Chapter 3.
[1]γ_c values.
[2]$\gamma_{l/v}$ values.

BIOMEDICAL APPLICATIONS

Biomedical applications of fluoropolymers, fluorocarbon coatings, and perfluorinated fluids and gels all include clinical interventional and luminal access devices (catheters in many forms), and more permanent implants (cardiovascular (Stanley, 1982), dental (Ratner, 1993), ocular (Legeais et al., 1998), craniofacial (Valdevit et al., 2000), urological (Reid et al., 1995), and abdominal (Grannis and Wagman, 1995) applications) as well as substantial non-implanted medical tubing and biotechnology components (protein blotting and filtration membranes). Annually, millions of perfluorinated polymer components are used worldwide in biological milieu both *in vitro* and *in vivo*. PTFE (Teflon™) and ePTFE (Gore-Tex™) are widely used in medical tubing, advanced catheters, vascular grafts, meshes, sutures, and other medical implants. PVDF is used for biotechnology blotting/separation membranes. Table C.2 provides biomedical applications of more popular fluoropolymers. Clinically, ePTFE vascular grafts, including dialysis-access grafts, and Teflon™-FEP catheter components are the most widely used fluorinated material medical devices. Other biomedical applications are described below.

Fluorinated Material Biological Response

Fluoropolymers are often regarded as chemically inert under most biological conditions. As noted above, they have some mechanical shortcomings under cyclic or continuous shear (creep). This makes applications in loadbearing situations (i.e., joint replacement, wear surfaces) difficult. Additionally, fluoropolymer surfaces are not inert to host

TABLE C.2 Fluorinated Biomaterials Biomedical Applications

Decade	Biomedical Applications
1970s	Implantable vascular grafts, peripheral catheters, and catheter introducers
1980s	Guiding catheters, protein blotting membranes, tissue meshes, tubing
1990s	Endoluminal stents, blood substitutes
2000s	Drug-eluting stents

biological reactions, including protein adsorption and blood clotting, either *in vitro* or *in vivo*. In fact, extremely tight binding of serum albumin to fluoroplasma-deposited surfaces (Kiaei et al., 1992), high levels of fibronectin and hemoglobin on PVDF (Paynter and Ratner, 1985), and various serum proteins (Dekker et al., 1991) including significant fibrinogen (Chandy et al., 2000), and high levels of both fibronectin and albumin (Grainger et al., 2003) on PTFE are observed *in vitro*. Importantly, protein adsorption to fluoropolymers is also observed *in vivo* (van Wachem et al., 1985; Roald et al., 1994), including fibrinogen activation, fibrin deposition, and platelet activation from blood, often deliberately promoted to stabilize blood reactivity on ePTFE vascular graft materials *in vivo* (Hoffman et al., 1986; Callow, 1988; Roald et al., 1994). Therefore, non-specific protein adsorption is significant, often irreversible, on fluoropolymer surfaces, leading to their desired utility as efficient protein blotting membranes. But different proteins from different media (i.e., serum versus plasma versus blood) produce different interfacial reactions to fluorinated surfaces. From serum *in vitro* for example, substantial amounts of albumin adsorption hinder serum-mediated cell attachment to fluoropolymers. This albumin passivation against further biological reactivity helps render the surface bio-fouling resistant. Clinical human cell-based vascular graft endothelialization to improve their blood compatibility is also generally poor on non-porous fluoropolymer chemistries (Kempczinski et al., 1985; Callow, 1988; Dekker et al., 1991; Schmidt et al., 1991; van Kooten et al., 1992; Roald et al., 1994; Legeais et al., 1998). This poor cell–fluoropolymer attachment has often been interpreted as "biological inertness," but results instead from substantial plasma protein adsorption on fluoropolymers (Baier et al., 1984) that does not support cell attachment and growth. Serum albumin, the most abundant protein in blood, blocks most cell attachment and other protein binding (Kesler et al., 1986; Zilla et al., 1989). Fibronectin, collagens, and other trace matricellular proteins (e.g., osteopontin, laminin, vitronectin) are cell-adhesive proteins. The observed general inability of non-porous fluoropolymers to support cell attachment has been related to excessive adsorption of albumin over cell-adhesive proteins (e.g., fibronectin) from serum – a media lacking clot-forming fibrinogen (Grainger et al., 2003). Pre-adsorption of specific cell-adhesive proteins (e.g., collagen or fibronectin) to fluoropolymers is used to promote reliable cell adhesion (McClary et al., 2000; Koenig et al., 2003). Therefore, little evidence suggests that fluoropolymers can be intrinsically "biologically inert." In fact, the opposite is true; solid fluoropolymers are intrinsically reactive and adsorptive to most every protein studied, and the nature of what types of proteins are adsorbed dictates the fluoropolymer's biological reactivity. For example, porous fluoropolymers with 60–80 μm voids, including ePTFE, can facilitate rapid blood clotting, cell and bacterial in-growth in blood. These interactions occur by blood-based protein adsorption, clotting, and cell/bacterial adhesion, with additional physical integration and engagement into the ePTFE pores (Clark et al., 1974; Clowes et al., 1986).

PTFE (Teflon™) Mesh and Fabric Vascular Implants

The innovation that enabled the spinning of PTFE paste into a fiber that could be woven or knitted into fabric or mesh produced a fabric-like fluorinated material (Berry, 1951). Attracted by solid PTFE's reported "blood compatibility" claimed in early animal acute *in vivo* studies, PTFE weaves were applied early in vascular grafts (Edwards, 1959). However, these early woven vascular grafts exhibited both high early failure rates and substantial late failure rates. Graft thrombosis was the most frequent early complication, accompanied by a high mortality rate (Boyd and Midell, 1971). Gore-Tex™ (ePTFE) vascular grafts supplanted these earlier PTFE weaves in this application and are discussed below. Thrombosis is a general property of fluoropolymer meshes and weaves in blood, passivating surfaces rapidly for acute short-term use, but limiting long-term blood-contacting applications.

ePTFE and Teflon™ Soft Tissue Repair Meshes

Teflon™ fibrous mesh has been used to repair abdominal wall defects (Ludington and Woodward, 1959) and hernias (Gibson and Stafford, 1964; Snijders, 1969; Kalsbeek, 1974). However, it does not reliably integrate into body tissues, is not sufficiently infection-resistant, and exhibits wound complication rates too high for routine hernia or abdominal repair use. ePTFE mesh is clinically used to repair hernias of many types (DeBord, 1998), reducing risks of several complications (DeGuzman et al., 1995; Lo Monte et al., 2009) (see Box 2). Infection and intestinal obstructions remain an issue for ePTFE mesh. However, they are better controlled with antibiotic therapies without mesh removal. ePTFE meshes have also been used as abdominal surgery barriers against surgical adhesions (Tulandi, 1997; Morris-Stiff and Hughes, 1998). ePTFE is also FDA-approved for many different plastic surgical facial defect reconstructions and augmentations, but is contraindicated in cosmetic lip augmentation, temporomandibular joint reconstruction, cardiovascular defects, and

BOX 2 ePTFE Hernial Mesh

Nearly a million hernial repairs are performed annually in the United States. Clinically, hernial repair meshes are implanted to bulk the abdominal wall at the herniated site. Tissue in-growth and organized fibrogenesis are deliberately promoted in these meshes to stabilize them in the abdominal tissue. ePTFE meshes and their combinations with polypropylene meshes are clinically available as a hernial repair product. However, ePTFE's lack of resistance to infection, and inability to effectively engage tissue, promote on-growth and organized fibrogenesis makes these meshes inferior to more popular polypropylene and polyester synthetic polymer meshes.

dermal placement (Levine and Berman, 1995). Nonetheless, commercial sources for these plastic surgery materials are discontinued.

ePTFE Vascular Implants

Because of abundant protein adsorption, ePTFE's porous inner surface is often claimed in vascular implants (blood) to clot blood rapidly to stabilize further surface coagulation and promote formation of a host "pseudo-intimal lining" that maintains blood compatibility under high flow, high shear applications. Its porous outer surface promotes peri-graft cell and tissue infiltration that mechanically stabilizes the prosthesis in place, preventing kinking (White, 1988). Unlike most animal models, humans do not reliably form stable cellular "neo-intimal"endothelial cell linings in these grafts, limiting their success. ePTFE is currently used most widely to fabricate medium sized (4–10 micron internode) vascular grafts for use in specific clinical blood vessel replacement indications (Kannan et al., 2005), including major vessels affected by disease or trauma, and arteriovenous hemodialysis grafts (Jenkins, 1976; Konner, 2005). ePTFE vascular prostheses are most clinically successful in high blood flow, low resistance conditions (i.e., large peripheral arteries >5–6 mm diameter, such as the descending aorta). They are generally not suitable for smaller arterial reconstructions (e.g., coronary circulation and peripheral vascular placements) under high resistance, low flow conditions that allow continued clotting and loss of patency. Typically, ePTFE suffers from thrombosis, poor healing, lack of compliance, and excessive intimal and anastomotic hyperplasia leading to stenotic complications. Their routine placement in high flow vascular environments allows them to remain patent despite their thrombus generation (see Box 3).

Arteriovenous ePTFE Grafts for Dialysis Access

Hemodialysis – essential for patients with end-stage renal disease – often requires regular, repeated (weekly) access to large blood vessels capable of producing high flow rates through an external artificial kidney device. Hemodialysis patients typically undergo cannula puncture of skin, underlying tissue, and vasculature to provide this access to the external artificial kidney. Repeated trauma to patient skin, tissue, and blood vessels from 13–17 gauge access needles produces notable complications including hyperplasia, thrombosis, hematoma, occlusion, infection, and other morbidities. Vascular access complications remain the main reason for hemodialysis patient hospitalization. Synthetic ePTFE arteriovenous (A–V) grafts are surgically placed across the basilic vein and brachial artery to permit cannula access and reduce tissue trauma complications. A–V prosthetic graft failure rates are substantial (>50%), leading to increasing use of native fistulas and catheters (Li et al., 2008). However, synthetic grafts reliably provide high blood flow rates shortly after placement, as they do not require maturation before use. In A–V ePTFE grafts, stenosis occurs most commonly at the graft–venous anastomosis. Histologically, macrophages are seen in large numbers in the

BOX 3 The ePTFE Vascular Graft and a Bizarre and Costly Patent Dispute

The 1974 patent filing from W. L. Gore & Associates, Inc. for the invention of the vascular graft resulted in the eruption of one of the most expensive and notorious medical device patent disputes in history. To challenge Gore's patent validity, a competing company started from former Gore employees and consultants – International Medical Prosthetics Research Associates, Inc. (IMPRA) – filed their own competing patent based on work from a pediatric heart surgeon, Dr. Goldfarb, investigating ePTFE at Arizona State University. !MPRA also agreed to fund Goldfarb's research in exchange for exclusive rights to Goldfarb's patent. But, after an extended period without IMPRA support, Goldfarb dissolved his IMPRA ties. Eventually, IMPRA returned patent rights to the doctor, but not without a nasty struggle. Goldfarb then licensed his still-pending patent to the C. R. Bard medical device company, seeking to obtain a piece of the vascular graft market from Gore and IMPRA. In 1995, the US Patent and Trade Office Interference Board declared that the Goldfarb patent (still pending) invalidated the Gore patent on the vascular graft – a decision upheld twice in US appeals court. Goldfarb was finally awarded his patent in 2002, still licensed to C. R. Bard, invalidating the Gore patent. But Gore continued to sell its ePTFE vascular grafts until Bard filed suit in 2003. The trial, first "settled" in 2007, found Goldfarb to be the rightful graft inventor; that his patent was valid, and that Gore had willfully infringed the patent in its vascular products. But it is not over yet: in 2009, Gore's claims of inequitable conduct against Bard were denied, and damages against Gore were increased to a total of $410 million. Gore then appealed this verdict (Frankel, 2009). In 2012, citing "substantial evidence" that Gore infringed a patent for vascular grafts, the U.S Court of Appeals upheld assessed penalties on Gore for past interest, royalties and fees. Resolution of the case is estimated to yield about $1 billion for Bard from Gore, among the largest settlement in patent litigation history. Interestingly, the judgment also allows Gore to keep selling their products on the markets as it is in the public interest to allow competition in the medical device arena.

adventitial and medial layers in the anastomotic tissues from A–V ePTFE grafts (Kapadia et al., 2008).

Multi-Lumen Catheters

Fluoropolymers are important for biomedical tubing, and are central to advanced multi-lumen small gauge medical-grade tubing required in many new minimally invasive catheters. These catheters permit surgeons to perform several invasive procedures through several lumens in a single inserted catheter device without removing one entire catheter to insert another. Catheter *in vivo* exposures are usually short-term, typically using endoluminal access, and increasingly are minimally invasive. As described above, PTFE's unique properties and stable thermal processing methods allow PTFE multi-lumen tubing precision manufacture unlike any other material, and this is a very important fluoropolymer in this particular use.

Guiding Catheters

An important clinical device with a long track record, the guiding catheter helps the clinician deliver stents and other devices endo-luminally. Central to the guiding catheter is a PTFE inner liner with its superior lubricity and low friction coefficient that slides within an outer lumen. Lubricity is so critical to this device function that FEP, as the second-most lubricious material available, is insufficient as a catheter liner. During catheter construction, PTFE is chemically bonded onto the tube's outer diameter to enable slip. Bonding is accomplished by using an FEP heat-shrinkable fusing sleeve. After depositing PTFE over the liner the FEP mold is removed from the device, leaving a smooth outer PTFE jacket on the liner within the catheter outer liner.

PTFE Catheter Introducers

Now over three decades old, the PTFE "introducer" facilitates catheter insertion into a patient's vein, taking full advantage of PTFE's endoluminal lubricity and precision tubing processing. Once the catheter is inserted, the PTFE outer sheath can be removed from the patient, leaving the implanted catheter behind. The introducer exploits PTFE processing that molecularly orients the fluoropolymer material in the tubing-based sleeve over the catheter. This allows PTFE tubing to be readily split and torn longitudinally from the catheter *in situ*, enabling the surgeon to remove a PTFE introducer from a patient while the primary catheter remains in place.

Perfluorocarbon Liquids and Emulsions as Oxygen Carrying Blood Substitutes

Low molecular weight perfluoro-fluids can be aspirated into the lungs directly or injected as submicron-sized emulsion droplets into the blood to facilitate oxygen transport, exploiting their high oxygen carrying capacities (Winslow, 2005). Liquid perfluorocarbons in the lung eliminate the gas–lung interface, acting to reduce tension in the lung alveolus, and reducing mechanical work required to breathe during respiratory distress and lung surfactant deficiency. Emulsions made from dispersing perfluoro fluids into water using interfacial droplet stabilizers (emulsifiers) create sub-micron perfluoro-droplets, much smaller than red blood cells. This provides enormous oxygen carrying capacity in blood to supplement normal oxygenation in microcirculation. Several commercial products have been clinically available, most recently Oxygent™ (Alliance Pharmaceutical, USA), consisting of two different perfluorocarbon fluids stabilized as a micro-emulsion using egg phospholipids. The product is eliminated from the blood after injection by macrophage/monocyte clearance, and is exhaled eventually from the lung.

Fluorinated Liquids in the Eye as Experimental Vitreous Substitutes

Detached retinal repair and other ophthalmic surgeries require oxygen-permeable viscous liquid vitreous substitutes. Perfluorinated oils and polymers have intrinsic high oxygen permeabilities and solubilities with suitable viscosity control. Perfluoropolyethers, perfluorinated alkanes, and perfluorinated silicone oils have all been studied in ocular vitreous applications, but still lack convincing safety, toxicity or efficacy to date to produce an approved product in this context.

Fluorinated (Meth)Acrylates and (Meth) Acrylated Perfluoroalkyl Silicones as Cross-Linked Polymer Cores for Soft Contact Lenses

Exploiting the intrinsic high oxygen permeability well known for perfluorinated materials, rigid gas permeable contact lenses (RGPs) have used many variations on cross-linked perfluorinated acrylates and perfluorinated polyether silicone gels as lens cores to improve extended-wear contact lens on-eye performance. Increased ocular acuity under high-throughput inexpensive but precise lens fabrication methods, with high lens oxygen transport, is sought. Many patents describe many fluorinated polymer gels in this regard, with most major lens manufacturers developing these lens core materials.

Fluorinated Materials as Anti-Fouling Coatings for Intraocular Lenses (IOLs)

General cellular reactions to implanted polymer intraocular lenses replacing cataracts can result in cell migration onto, and adhesion to, the lens, with optical interference (clouding) requiring IOL replacement. To prevent cells from migrating onto and coating the IOL's optical lens surface (typically a thermoplastic), thin optically transparent plasma-deposited fluorocarbon films or layers of other fluorinated polymers are deposited on the IOL rear surface. Several patents describe this approach and application.

PTFE Paste Injectable Bulking Agent

Teflon™ particles formulated as an injectable paste have been reported for the treatment of vesico-ureteric reflux (VUR) (Puri, 1995), corrected by sub-ureteric injection of a PTFE paste (Puri and O'Donnell, 1984). Sub-ureteric paste injection by endoscopy has successfully addressed primary and secondary VUR in children for nearly two decades (Le Guillou et al., 1984; Dodat and Paulhac, 1987; Puri, 2000). Similarly, PTFE paste was introduced nearly 4 decades ago to treat female stress urinary incontinence (Beckingham et al., 1992; Politano, 1992; Meschia et al., 2002). However, this device does not have regulatory approval, due to high risks for PTFE microparticle migration and granuloma induction, especially in lymph nodes, kidneys, lungs, and brain (Aaronson et al., 1993). Applications of Teflon™ paste injection are still reported (Harrison et al., 1993; Lopez et al., 1993; Herschorn and Glazer, 2000), but its side-effect concerns limit clinical use.

Ligament Replacement

The Gore-Tex™ ligament prosthesis comprises a single long fiber of ePTFE woven into loops. Mechanical testing shows that the resulting ultimate tensile strength is ~3 times that of the human anterior cruciate ligament (ACL). Creep and bending fatigue testing validate this ePTFE device as a strong synthetic ACL replacement material (Bolton and Bruchman, 1985). The Gore-Tex™ ACL prosthesis is currently FDA-approved for use in patients with failed autogenous intra-articular graft procedures (Mascarenhas and Macdonald, 2008). However, while acute performance shows promising stability in the knee, extended use and implant time produce significant ligament loosening, and other knee stability problems.

Sutures

PTFE also finds limited use as a suture fiber in various forms. Both PTFE monofilament and ePTFE fibers are surgically proven, with clinically accepted surgeon handling and lubricity properties. Additionally, PTFE is blended into other common surgical sutures used for myocardial heart valve prostheses fixation. Poly(ethylene terephthalate) (PET polyester) braided sutures are impregnated with PTFE polymer to limit wrinkling of the braid and consequent swelling. PTFE hydrophobic properties likely help protect the polyester braid from water uptake and hydrolysis (Bhat, 2002).

THE MONEY JOINT

Most biomaterials used for jaw joint reconstruction were introduced to markets prior to the 1976 Medical Devices Amendment Act that required device manufacturers to prove that their devices were safe and effective. A legal loophole that required manufacturers only to prove that their devices were "substantially equivalent" to a pre-Amendment device allowed temporal mandibular joint (TMJ) implants marketed soon after 1976 to enter the market without testing. Two designs widely used as TMJ replacement surfaces were Dow Corning's Silastic® and Vitek's Proplast-Teflon product. Vitek developed and sold Proplast sheeting (Teflon™ FEP film laminated with a porous composite material of PTFE and carbon) in the 1970s. Implants modified in the 1980s comprised Teflon™ film laminated to PTFE and aluminum oxide. These implants, ~1 cm^2 in size, were cut from sheets in the operating room and sutured into the TMJ joint. In 1983, the FDA allowed Vitek to market a pre-cut disc because, under the law, the company needed only to convince the FDA that its device was "substantially equivalent" to Dow's Silastic® disc marketed years earlier. In 1986, several reports of catastrophic biomechanical failure of the PTFE implant were linked to a giant cell reaction leading to bone resorption and pain. Further analysis documented device failure rates of 10–25%. By 1992, implant success rates below 20% were reported. Animal studies performed only after failures in humans began showed complete erosion of the TMJ implant within a "few months." In early 1990, with implant failures increasing, the FDA recalled Vitek's products. Predictably, Vitek declared bankruptcy with its rising product litigation costs, but continued to market their TMJ implants. Surgeons continued to implant them until eventually the FDA seized all products from Vitek, as well as its subsidiaries.

SUMMARY

Due to their unique chemistry, fluorinated materials, primarily fluoropolymers, have attractive properties of biomaterials interest, including chemical stability, low adhesion/friction, non-wetting, high protein adsorption, high oxygen permeability, and precision tubing manufacturing. Low cell adhesion in serum and high intrinsic blood coagulation in plasma and blood both result from distinct media-dependent protein adsorption treatments and the physical form of the material. Expanded PTFE provides fabric-like properties with controlled pores and high surface area to alter device-related mechanics, processing, and tissue responses, while promoting rapid blood clotting. Solid fluoropolymers also have attractive precision engineering and device processing properties essential to producing several medical device classes where fine dimensional tolerances and biocompatibility are required. This enables fabrication of multi-lumen, high-tolerance, small dimensional tubing for advanced catheters.

GLOSSARY

Blooming: a term used to describe the spontaneous enrichment of certain chemistries at the surface of a bulk matrix, usually associated with the surface enrichment by low surface energy chemistry-like fluorinated materials.

This phenomenon is used to promote a surface enriched in trace components like expensive added fluorinated chemistry by allowing them to spontaneously diffuse from the bulk material to the surface.

Blotting membrane: a thin, porous, hydrophobic, high protein-binding capacity polymer membrane (e.g., PVDF) used to transfer proteins from a gel electrophoresis separation process for further probing with antibodies to identify the proteins as a blot or spot.

ePTFE: expanded polytetrafluoroethylene, produced from PTFE films under anisotropic stretching to yield a unique node-fibril microporous morphology in a fabric-like sheet form deemed important to biomedical utility in implanted biomaterials.

Fluorinated biomaterial: a material intended for a biomedical application made from a base material that contains significant amounts of chemically bonded fluorine.

Fluoropolymer: a fluorinated polymer, also perfluoropolymer, usually thermoplastic, with high content of fluorine atoms replacing hydrogen atoms along the carbon-based polymer chain.

Perfluorinated material: a material wherein all hydrogen atoms are replaced with fluorine, generally making C–F bonds.

PTFE: polytetrafluoroethylene, a high molecular weight, highly crystalline perfluoropolymer solid also known as Teflon™, discovered at DuPont in 1938, with unique solid properties and processing requirements.

BIBLIOGRAPHY

Aaronson, I. A., Rames, R. A., Greene, W. B., Walsh, L. G., Hasal, U. A., et al. (1993). Endoscopic treatment of reflux: Migration of Teflon to the lungs and brain. *Eur. Urol.*, **23**, 394–399.

Baier, R. E., Meyer, A. E., Natiella, J. R., Natiella, R. R., & Carter, J. M. (1984). Surface properties determine bioadhesive outcomes: Methods and results. *J. Biomed. Mater. Res.*, **18**, 327–355.

Bar, G., Thomann, Y., Brandsch, R., Cantow, H. J., Whangbo, M. H., et al. (1997). Factors affecting the height and phase images in tapping mode atomic force microscopy. Study of phase-separated polymer blends of Poly(ethene-co-styrene) and Poly(2,6-dimethyl-1,4-phenylene oxide). *Langmuir*, **13**, 3807–3812.

Beckingham, I. J., Wemyss-Holden, G., & Lawrence, W. T. (1992). Long-term follow-up of women treated with periurethral Teflon injections for stress incontinence. *Br. J. Urol.*, **69**, 580–583.

Berry, K. L. (1951). United States Patent 2559750.

Bhat, S. V. (2002). *Biomaterials*. Boston, MA: Kluwer Academic Publishers.

Bolton, C. W., & Bruchman, W. C. (1985). The GORE-TEX expanded polytetrafluoroethylene prosthetic ligament. An *in vitro* and *in vivo* evaluation. *Clin. Orthop. Relat. Res.*, **196**, 202–213.

Boyd, D. P., & Midell, A. I. (1971). Woven Teflon aortic grafts. An unsatisfactory prosthesis. *Vasc. Surg.*, **5**, 148–153.

Bunn, C. W., & Howells, E. R. (1954). Structures of molecules and crystals of fluorocarbons. *Nature*, **174**, 549–551.

Callow, A. D. (1988). Problems in the construction of a small diameter graft. *Int. Angiol.*, **7**, 246–253.

Chandy, T., Das, G. S., Wilson, R. F., & Rao, G. H. (2000). Use of plasma glow for surface-engineering biomolecules to enhance blood compatibility of Dacron and PTFE vascular prosthesis. *Biomaterials*, **21**, 699–712.

Clark, E. S. (1999). The molecular conformations of polytetrafluoroethylene: Forms II and IV. *Polymer*, **40**, 4659–4665.

Clark, E. S., & Muus, L. T. (1962). Partial disordering and crystal transitions in polytetrafluoroethylene. *Z. Kristallogr.*, **117**, 119–127.

Clark, R. E., Boyd, J. C., & Moran, J. F. (1974). New principles governing the tissue reactivity of prosthetic materials. *J. Surg. Res.*, **16**, 510–522.

Clowes, A. W., Kirkman, T. R., & Reidy, M. A. (1986). Mechanisms of arterial graft healing. Rapid transmural capillary ingrowth provides a source of intimal endothelium and smooth muscle in porous PTFE prostheses. *Am. J. Pathol.*, **123**, 220–230.

D'Agostino, R. (1990). *Plasma Deposition, Treatment, and Etching of Polymers*. Boston, MA: Academic Press.

DeBord, J. R. (1998). The historical development of prosthetics in hernia surgery. *Surg. Clin. North Am.*, **78**, 973–1006.

DeGuzman, L. J., Nyhus, L. M., Yared, G., & Schlesinger, P. K. (1995). Colocutaneous fistula formation following polypropylene mesh placement for repair of a ventral hernia: Diagnosis by colonoscopy. *Endoscopy*, **27**, 459–461.

Dekker, A., Reitsma, K., Beugeling, T., Bantjes, A., Feijen, J., et al. (1991). Adhesion of endothelial cells and adsorption of serum proteins on gas plasma-treated polytetrafluoroethylene. *Biomaterials*, **12**, 130–138.

Dodat, H., & Paulhac, J. B. (1987). Endoscopic treatment of vesico-ureteral reflux by injection of Teflon in children. Preliminary results. *Pediatrie*, **42**, 211–214.

Doeff, M. M., & Lindner, E. (1989). Structure and surface energy characteristics of a series of pseudo-perfluoroalkyl polysiloxanes. *Macromolecules*, **22**, 2951–2957.

Drobny, J. G. (2001). *Technology of Fluoropolymers*. Boca Raton, Fl: CRC Press.

Drobny, J. G. (2005). *Fluoroplastics (Rapra Review Report 184)*. Boca Raton, FL: Smithers Rapra Press.

Edwards, W. S. (1959). Progress in synthetic graft development; an improved crimped graft of Teflon. *Surgery*, **45**, 298–309.

Frankel, A. (2009). *Blood money: The American Lawyer*, November, 2.

Gibson, L. D., & Stafford, C. E. (1964). Synthetic mesh repair of abdominal wall defects: Follow up and reappraisal. *Am. Surg.*, **30**, 481–486.

Gore, R. W. (1976). United States Patent 3953566.

Grainger, D. W., & Stewart, C. W. (2001). Fluorinated Coatings and Films: Motivation and Significance. In D. G. Castner, & D. W. Grainger (Eds.), *Fluorinated Surfaces, Coatings, and Films* (pp. 1–14). Washington, DC: American Chemical Society. Chapter 1.

Grainger, D. W., Pavon-Djavid, G., Migonney, V., & Josefowicz, M. (2003). Assessment of fibronectin conformation adsorbed to polytetrafluoroethylene surfaces from serum protein mixtures and correlation to support of cell attachment in culture. *J. Biomater. Sci. Polym. Ed.*, **14**, 973–988.

Grannis, F. W., Jr., & Wagman, L. D. (1995). Repair of a complex body wall defect using polytetrafluorethylene patches. *Ann. Thorac. Surg.*, **60**, 197–199.

Hare, E. F., Shafrin, E. G., & Zisman, W. A. (1954). Properties of films of adsorbed fluorinated acids. *J. Chem. Phys.*, **58**, 236–239.

Harrison, S. C., Brown, C., & O'Boyle, P. J. (1993). Periurethral Teflon for stress urinary incontinence: Medium-term results. *Br. J. Urol.*, **71**, 25–27.

Herschorn, S., & Glazer, A. A. (2000). Early experience with small volume periurethral polytetrafluoroethylene for female stress urinary incontinence. *J. Urol.*, **163**, 1838–1842.

Hoffman, A. S., Ratner, B. D., Garfinkle, A. M., Reynolds, L. O., Horbett, T. A., & Hanson, S. R. (1986). A New Plasma Discharge Treatment for Cardiovascular Implants. In J. M. Williams, M. F. Nichols, & W. Zingg (Eds.), *Biomedical Materials* (pp. 3–17). Pittsburgh, PA: Mater. Res. Soc. Proc..

Jenkins, A. (1976). Gore-Tex: A new prosthesis for vascular access. *Br. Med. J.*, **2**, 280.

Kalsbeek, H. L. (1974). Experience with the use of Teflon mesh in the repair of incisional hernias. *Arch. Chir. Neerl.*, **26**, 71–75.

Kannan, R. Y., Salacinski, H. J., Butler, P. E., Hamilton, G., & Seifalian, A. M. (2005). Current status of prosthetic bypass grafts: A review. *J. Biomed. Mater. Res. B Appl. Biomater.*, **74**, 570–581.

Kapadia, M. R., Popowich, D. A., & Kibbe, M. R. (2008). Modified prosthetic vascular conduits. *Circulation*, **117**, 1873–1882.

Kempczinski, R. F., Rosenman, J. E., Pearce, W. H., Roedersheimer, L. R., Berlatzky, Y., et al. (1985). Endothelial cell seeding of a new PTFE vascular prosthesis. *J. Vasc. Surg.*, **2**, 424–429.

Kesler, K. A., Herring, M. B., Arnold, M. P., Glover, J. L., Park, H. M., et al. (1986). Enhanced strength of endothelial attachment on polyester elastomer and polytetrafluoroethylene graft surfaces with fibronectin substrate. *J. Vasc. Surg.*, **3**, 58–64.

Kiaei, D., Hoffman, A. S., & Horbett, T. A. (1992). Tight binding of albumin to glow discharge treated polymers. *J. Biomater. Sci. Polym. Ed.*, **4**, 35–44.

Kobayashi, H., & Owen, M. J. (1995). Surface properties of fluorosilicone. *Trends Polym. Sci.*, **3**, 330–335.

Koenig, A. L., Gambillara, V., & Grainger, D. W. (2003). Correlating fibronectin adsorption with endothelial cell adhesion and signaling on polymer substrates. *J. Biomed. Mater. Res. A.*, **64**, 20–37.

Konner, K. (2005). History of vascular access for heamodialysis. *Nephrol. Dial. Transplant.*, **20**, 2629–2635.

Le Guillou, M., Ferriere, J. M., Pourquie, J., Barthaburu, D., Amory, J. P., et al. (1984). Primary vesico-ureteral reflux in adults. 1977–1982. Department experience of 6-years. *Ann. Urol. (Paris)*, **18**, 121–123.

Legeais, J. M., Werner, L. P., Legeay, G., Briat, B., & Renard, G. (1998). *In vivo* study of a fluorocarbon polymer-coated intraocular lens in a rabbit model. *J. Cataract. Refract. Surg.*, **24**, 371–379.

Leveen, H. H., & Barberio, J. R. (1949). Tissue reaction to plastics used in surgery with special reference to Teflon. *Ann. Surg.*, **129**, 74–84.

Levine, B., & Berman, W. E. (1995). The current status of expanded polytetrafluoroethylene (Gore-Tex) in facial plastic surgery. *Ear. Nose. Throat. J.*, **74**, 681–682.

Li, L., Terry, C. M., Shiu, Y. -T. E., & Cheung, A. K. (2008). Neointimal hyperplasia associated with synthetic hemodialysis grafts. *Kidney Int.*, **74**, 1247–1261.

Lo Monte, A. I., Damiano, G., Maione, C., Gioviale, M. C., Lombardo, C., et al. (2009). Use of intraperitoneal ePTFE Gore dual-mesh plus in a giant incisional hernia after kidney transplantation: A case report. *Transpl. P.*, **41**, 1398–1401.

Lopez, A. E., Padron, O. F., Patsias, G., & Politano, V. A. (1993). Transurethral polytetrafluoroethylene injection in female patients with urinary continence. *J. Urol.*, **150**, 856–858.

Ludington, L. G., & Woodward, E. R. (1959). Use of Teflon mesh in the repair of musculofascial defects. *Surgery*, **46**, 364–373.

Mascarenhas, R., & Macdonald, P. B. (2008). Anterior cruciate ligament reconstruction: A look at prosthetics – past, present and possible future. *J. Med.*, **11**, 29–37.

Matsumoto, H., Hasegawa, T., Fuse, K., Yamamoto, M., & Saigusa, M. (1973). A new vascular prosthesis for a small caliber artery. *Surgery*, **74**, 519–523.

McClary, K. B., Ugarova, T., & Grainger, D. W. (2000). Modulating fibroblast adhesion, spreading, and proliferation using self-assembled monolayer films of alkylthiolates on gold. *J. Biomed. Mater. Res.*, **50**, 428–439.

McClurken, M. E., McHaney, J. M., & Colonel, W. M. (1986). "Physical properties and test methods for expanded polytetrafluroethylene (PTFE) grafts". In H. E. Kombic, A. Katrowitz, & P. Sung (Eds.), *Vascular Graft Update: Safety and Performance* (pp. 82–94). Philadelphia, PA: ASTM Publications.

Meschia, M., Pifarotti, P., Gattei, U., & Crosignani, P. G. (2002). Injection therapy for the treatment of stress urinary incontinence in women. *Gynecol. Obstet. Invest.*, **54**, 67–72.

Mole, B. (1992). The use of Gore-Tex implants in aesthetic surgery of the face. *Plast. Reconstr. Surg.*, **90**, 200–206.

Morris-Stiff, G. J., & Hughes, L. E. (1998). The outcomes of nonabsorbable mesh placed within the abdominal cavity: Literature review and clinical experience. *J. Am. Coll. Surg.*, **186**, 352–367.

Naselli, C., Swalen, J. D., & Rabolt, J. F. (1989). Order–disorder transitions in Langmuir–Blodgett films. IV. Structure of $[F(CF_2)_8(CH_2)_{10}COO^-]_2\ Cd^{2+}$ multilayers at ambient and elevated temperatures. *J. Chem. Phys.*, **90**, 3855.

Oshige, S. (1967). Japanese Patent No. 42–13560(67/13560).

Paynter, R. W., & Ratner, B. D. (1985). *Surface and Interfacial Aspects of Biomedical Polymers* (2nd ed.). New York, NY: Springer.

Pittman, A. G. (1972). "Surface Properties of Fluorocarbon Polymers". In L. A. Wall (Ed.), *Fluoropolymers*. New York, NY: Wiley. Chapter 13, (pp. 413–431).

Politano, V. A. (1992). Transurethral polytef injection for post-prostatectomy urinary incontinence. *Br. J. Urol.*, **69**, 26–28.

Puri, P. (1995). Ten year experience with subureteric Teflon (polytetrafluoroethylene) injection (STING) in the treatment of vesico-ureteric reflux. *Br. J. Urol.*, **75**, 126–131.

Puri, P. (2000). Endoscopic correction of vesicoureteral reflux. *Curr. Opin. Urol.*, **10**, 593–597.

Puri, P., & O'Donnell, B. (1984). Correction of experimentally produced vesicoureteric reflux in the piglet by intravesical injection of Teflon. *Br. Med. J. (Clin. Res. Ed.)*, **289**, 5–7.

Ratner, B. D. (1993). New ideas in biomaterials science: A path to engineered biomaterials. *J. Biomed. Mater. Res.*, **27**, 837–850.

Ratner, B. D. (1995). Plasma deposition of organic thin films for surface modification. *J. Photopolym. Sci. Technol.*, **8**, 481–494.

Reid, G., Busscher, H. J., Sharma, S., Mittelman, M. W., & McIntyre, S. (1995). Surface properties of catheters, stents and bacteria associated with urinary tract infections. *Surf. Sci. Rep.*, **21**, 251–273.

Roald, H. E., Barstad, R. M., Bakken, I. J., Roald, B., & Lyberg, T. (1994). Initial interactions of platelets and plasma proteins in flowing non-anticoagulated human blood with the artificial surfaces Dacron and PTFE. *Blood Coagul Fibrinolysis*, **5**, 355–363.

Russell, T. P., Rabolt, J. F., Twieg, R. J., Siemens, R. L., & Farmer, B. L. (1986). Structural characterization of semifluorinated n-alkanes. 2. Solid–solid transition behavior. *Macromolecules*, **19**, 1135–1143.

Santiago, E. J., Chatamra, K., & Taylor, D. E. (1981). Haemodynamic aspects of lower limb arterial reconstruction using Dacron and Goretex prostheses. *Ann. R. Coll. Surg. Engl.*, **63**, 253–256.

Scheirs, J. (1997). *Modern Fluoropolymers*. New York, NY: Wiley.

Schmidt, S., Decleer, W., Wagner, U., Kindermann, D., Pringle, K., et al. (1991). An approach to fetal surgery: Endoscopic use of excimer laser. *Eur. J. Obstet. Gynecol. Reprod. Biol.*, **42**(Suppl.), S84–86.

Schneider, J., Erdelen, C., Ringsdorf, H., & Rabolt, J. F. (1989). Structural studies of polymers with hydrophilic spacer groups. 2. Infrared spectroscopy of Langmuir–Blodgett multilayers of polymers with fluorocarbon side chains at ambient and elevated temperatures. *Macromolecules*, **22**, 3475–3480.

Smart, B. E. (1994). Characteristics of C–F Systems. In R. E. Banks, J. C. Tatlow, & B. E. Smart (Eds.), *Organofluorine Chemistry: Principles and Commercial Applications*. New York, NY: Plenum Press. Chapter 3.

Snijders, H. (1969). The use of Teflon gauze in the treatment of medial and recurrent inguinal hernias. *Arch. Chir. Neerl.*, **21**, 199–202.

Soyer, T., Lempinen, M., Cooper, P., Norton, L., & Eiseman, B. (1972). A new venous prosthesis. *Surgery*, **72**, 864–872.

Stanley, J. C. (1982). *Biologic and Synthetic Vascular Prostheses*. New York, NY: Grune & Stratton.

Stone, M., Nevell, T. G., & Tsibouklis, J. (1998). Surface energy characteristics of poly(perfluoroacrylate) film structures. *Materials Letters*, **37**, 102–105.

Sun, F., Grainger, D. W., Castner, D. G., & Leach-Scampavia, D. K. (1994a). Adsorption of ultrathin films of sulfur-containing siloxane oligomers on gold surfaces and their *in situ* modification. *Macromolecules*, **27**, 3053–3062.

Sun, F., Mao, G., Grainger, D. W., & Castner, D. G. (1994b). Polymer ultrathin films by self-assembly: Bound perfluorinated monolayers and their modification using *in situ* derivatization strategies. *Thin Solid Films*, **242**, 106–111.

Tsao, M. W., Hoffmann, C. L., Rabolt, J. F., Johnson, H. E., Castner, D. G., et al. (1997). Studies of molecular orientation and order in self-assembled semifluorinated *n*-alkanethiols: Single and dual component mixtures. *Langmuir*, **13**, 4317–4322.

Tulandi, T. (1997). How can we avoid adhesions after laparoscopic surgery? *Curr. Opin. Obstet. Gynecol.*, **9**, 239–243.

Valdevit, A., Turegun, M., Kambic, H., Siemionow, M., & Zins, J. (2000). Cranial defect repair using e-PTFE: Part I. Evaluation of bone stiffness. *J. Biomed. Mater. Res.*, **53**, 62–66.

van Kooten, T. G., Schakeraad, J. M., Van der Mei, H. C., & Busscher, H. J. (1992). "Influence of substractum wettability on the strength of adhesion of human fibroblasts". *Biomaterials*, **13**, 897–904.

van Wachem, P. B., Beugeling, T., Feijen, J., Bantjes, A., Detmers, J. P., et al. (1985). Interaction of cultured human endothelial cells with polymeric surfaces of different wettabilities. *Biomaterials*, **6**, 403–408.

White, R. A. (1988). The effect of porosity and biomaterial on the healing and long-term mechanical properties of vascular prostheses. *AsaioTrans.*, **34**, 95–100.

Winslow, R. W. (2005). *Blood Substitutes*. London, UK: Academic Press. p. 91.

Zhang, Y. X., Da, A. H., Hogen-Esch, T. E., & Butler, G. B. (1989). "A fluorocarbon–containing hydlophobically associating polymer". *J. Polym. Sci. Part C. Polym. Lett. Ed.*, **28**, 213–218.

Zilla, P., Fasol, R., Preiss, P., Kadletz, M., Deutsch, M., et al. (1989). Use of fibrin glue as a substrate for *in vitro* endothelialization of PTFE vascular grafts. *Surgery*, **105**, 515–522.

D. ACRYLICS

Joe Antonucci[1] and Sabine Dickens[2]

[1]National Institute of Standards and Technology, Gaithersburg, MD, USA

[2]American Dental Association Foundation Pattenbarger Research Center, Gaithersburg, MD, USA

INTRODUCTION

The modern development of tooth-colored dental polymeric composites owes much to R. L. Bowen of the American Dental Association for his pioneering studies at the National Bureau of Standards, now the National Institute of Standards and Technology. His recognition of the excellent matrix-forming potential of epoxy resins (oxiranes), as well as their poor ambient polymerization characteristics (slow under anionic catalysis and uncontrollable under the more rapid cationic catalysis then available) led him to the discovery of a unique hybrid monomer which combined the low polymerization contraction of epoxy resins with the excellent setting behavior of acrylic monomers (Bowen, 1956). His classical synthesis of the bulky, thermosetting dimethacrylate, Bis-GMA, 2,2-bis[p(2′-hydroxy-3′-methacryloxypropoxyphenyl)] propane (Figure D.1), his preparation of silica fillers that combined translucency and radiopacity while matching the refractive indices of the resin matrix, and his utilization of the technology of silane coupling agents, ushered in the modern era of esthetic dental composites (Bowen, 1963).

MONO- AND MULTI-METHACRYLATE MONOMERS

Acrylics based on the monofunctional monomer methyl methacrylate (MMA) combined with poly(methyl methacrylate) (PMMA) and a low viscosity dimethacrylate for cross-linking are mainly used in orthodontic appliances, e.g., bionator, bite plates, palatal expanders, retainers, and for removable prosthetic devices, such as partial and full dentures, temporary crowns, and bridges. Basic MMA/PMMA mixtures activated with chemical initiators have been widely used as bone cements for orthopedic applications (Shalaby et al., 2007). Further development has led to the incorporation of PMMA fibers, resulting in composites with moderately increased moduli and improved toughness (Gilbert et al., 1995). MMA and other monomethacrylates of moderate size and viscosity are only occasionally incorporated into dental restorative materials due to their relatively high molar double-bond concentration, and thus, increased polymerization shrinkage. However, functional monomethacrylates are widely used in adhesive formulations, and as coupling agents in composites, and are discussed below.

Base and Diluent Monomers

In preventive and restorative dentistry a variety of di- or multimethacrylates are photo and/or chemically cured into polymers that function as pit and fissure sealants, adhesives, veneer materials, and when combined with fillers, as esthetic or loadbearing composite restoratives. Typically, dental resins are composed of mixtures of two or more monomers that combine a relatively high viscosity dimethacrylate (base) monomer with a low viscosity dimethacrylate comonomer such as triethylene glycol dimethacrylate (TEGDMA) to obtain resins with workable rheologies. The base resins 2,2-bis[p-(2′-hydroxy-3′-methacryloxypropoxy) phenylene]propane (Bis-GMA), ethoxylated bisphenol A dimethacrylate (EBPADMA) or 1,6-bis(methacryloxy-2-ethoxycarbonylamino)-2,4,4-trimethylhexane (UDMA) (Figure D.1) are among those

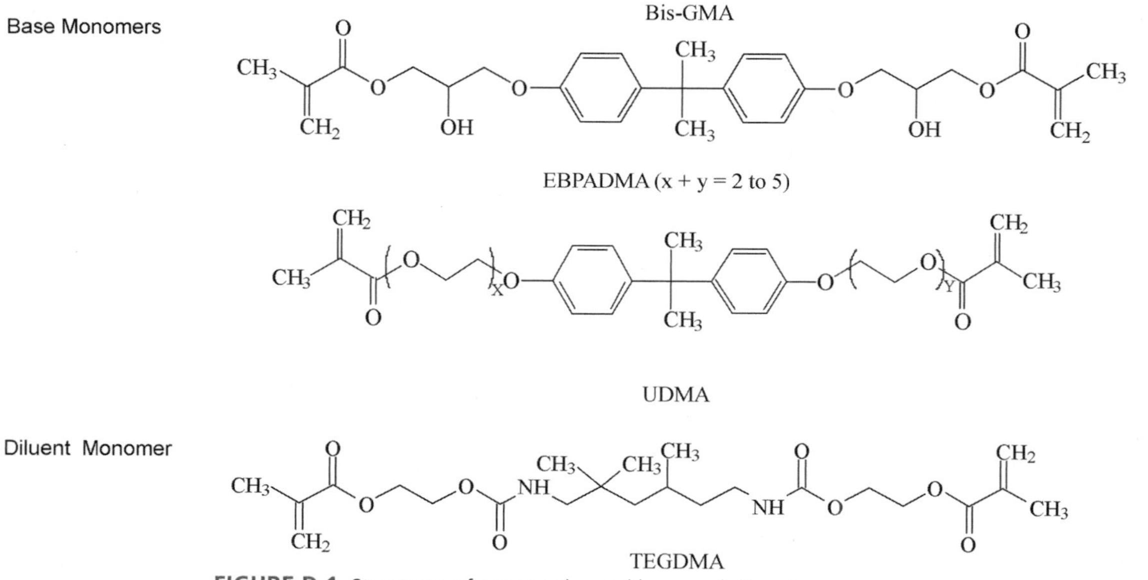

Base Monomers

Bis-GMA

EBPADMA (x + y = 2 to 5)

UDMA

Diluent Monomer

TEGDMA

FIGURE D.1 Structures of commonly used base and diluent monomers.

most commonly employed in commercial dental resin-based materials.

When cured at ambient temperature, free radicals convert the resin to a cross-linked three-dimensional polymer network. The free radicals are generated chemically from benzoyl peroxide and a tertiary aryl amine or through visible light irradiation using a photoinitiator, such as acyl-and bis-acylphosphine oxides, that yield radicals by photo cleavage or redox photoinitiator systems comprising camphorquinone or phenylpropanedione combined with aromatic or aliphatic tertiary amines that yield radicals by electron/hydrogen transfer mechanisms. For dental applications, rapid and efficient polymerization, in particular photopolymerization enabling on-command cure of dental resins, is a critical aspect of the clinical use of adhesives, sealants, and composites. Moreover, optimized cure is one of the most crucial parameters that govern the long-term performance of the polymeric material, affecting the biocompatibility by limiting the leaching of potentially cytotoxic materials from the cured polymers (Yoshii, 1997; Geurtsen et al., 1998). The extent of cure also affects the mechanical properties, e.g., fracture toughness, elastic moduli, flexural strength, and hardness (Cook, 1991; Ferracane et al., 1998). The base monomers are used in dental resin mixtures to enhance the physical and mechanical properties of the resulting polymers. Moreover, they minimize polymerization shrinkage and related stresses due to their relatively low methacrylate double-bond molar concentrations (Patel et al., 1987; Bogdal et al., 1997). While the addition of diluent monomers compromises the shrinkage, they are needed to reduce the viscosity of the base monomers to enable maximum filler loading, and to improve conversion and the physical and mechanical properties of the composite resins. When dental base monomers are combined with various proportions

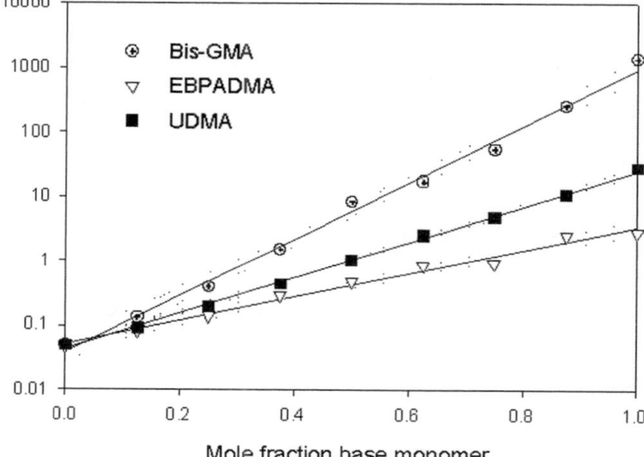

FIGURE D.2 Log viscosity of mixtures of Bis-GMA, EBPADMA, and UDMA with TEGDMA with linear regression and 95% confidence intervals (R^2 = 0.987, 0.995, and 0.998, respectively).

of the diluent monomer TEGDMA, the difference in their H-bonding potential results in compositions covering a broad range of viscosities (Figure D.2), and significantly different photopolymerization kinetics (Dickens et al., 2003). The structures of the individual monomers, and consequently the resin viscosities of the comonomer mixtures, strongly influence both the rate and extent of conversion of the photopolymerization process (Figure D.3). When compared at similar diluent concentrations, UDMA resins are significantly more reactive than Bis-GMA and EBPADMA resins. At higher diluent concentrations, EBPADMA resins provide the lowest photopolymerization reactivities. Optimum reactivities in the UDMA and EBPADMA resin systems are obtained with the addition of relatively small amounts of TEGDMA, whereas the Bis-GMA/TEGDMA resin system

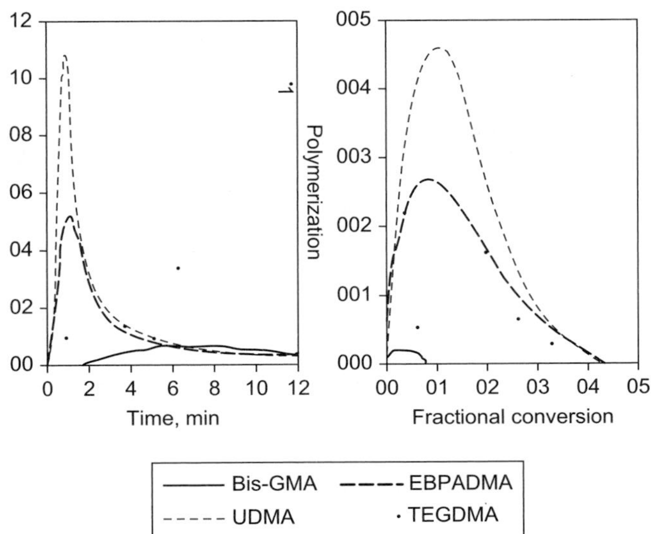

FIGURE D.3 Thermograms (a), and rate/conversion profiles (b), of the DSC homo-polymerizations of the three base resins and the diluent monomer TEGDMA (light intensity: 100 μW/cm^2). The early onset of the autoacceleration is particularly noticeable for UDMA and EBPADMA, while Bis-GMA and TEGDMA demonstrate a sluggish polymerization onset with very low conversion for Bis-GMA, and low polymerization rates.

requires near equivalent mole ratios for highest reactivity (Dickens et al., 2003).

Modified Base Monomers

Analogs and modifications of Bis-GMA and UDMA have been reported resulting in somewhat improved properties (Khatri et al., 2003; Kim et al., 2006). Derivatives of Bis-GMA having pendant n-alkyl urethane substituents exhibited lower viscosities and were more hydrophobic than Bis-GMA. Generally, the viscosity of these experimental monomers decreased with increasing chain length of the alkyl urethane substituent. Photopolymerization of the new monomers gave higher degrees of vinyl conversion compared to Bis-GMA, yet yielded lower polymerization shrinkage than Bis-GMA when compared at equivalent degrees of vinyl conversion (Antonucci and Stansbury, 1997; Peutzfeldt, 1997; Moszner and Salz, 2001, 2007). Modified urethane dimethacrylates have also been developed as alternatives for Bis-GMA; however, the improvements in physical properties were moderate, showing small changes in flexural strength and reduction in water uptake (Moszner et al., 2008).

Alternative Monomers with Improved Reactivity

In the late 1980s and early 1990s oxy bis-methacrylates were developed and evaluated as potential monomers for dental applications. It was found that these monomers undergo cyclopolymerization, resulting in polymers with high conversion yet low shrinkage (Stansbury, 1990).

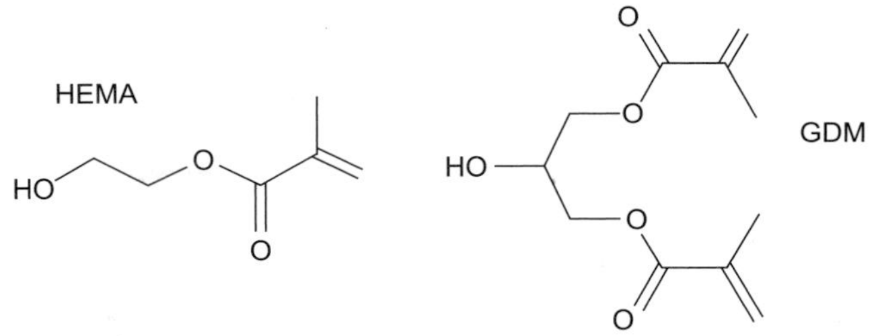

FIGURE D.4 Hydrophilic monomers: 2-hydroxyethyl methacrylate (HEMA); glycerol dimethacrylate (GDM).

Other work by Stansbury introduced cyclic carbonates as highly reactive diluent monomers for dental applications (Berchtold et al., 2008).

Functionalized Monomers with Adhesive Properties

Adhesive Monomers. Currently, there are three main classes of free radical polymerizable restoratives available: (1) resin composites, which are mainly based on the methacrylate resins described above; (2) resin-modified glass ionomer cements, which are hybrids of polyacrylic acids having pendant methacrylate groups for free radical polymerization, while the carboxylic acid groups provide the means for an acid/base setting reaction with specialty ion-leachable aluminum fluorosilicate glasses; and (3) polyacid-modified composite resins based on carboxylic acid-containing methacrylates, also known as compomers. In contrast to resin composites, groups 2 and 3 contain, in addition to methacrylate or acrylate groups, carboxylic acid groups that are attached either to a backbone (group 2) or to the center core of the monomer (group 3). The acid groups are also capable of forming ionic bonds to apatitic calcium phosphate mineral and also strong hydrogen bonds to collagen, and thus promote adhesion to both enamel and dentin. Thus, acidic monomers together with other hydrophilic monomers (Figures D.4–D.6) are significant components of adhesive primers and bonding resins. In order to form durable bonds between a resin-based restorative and tooth structure it is essential to use an adhesive system that wets enamel and dentin well. Bonding to enamel is primarily achieved by micromechanical adhesion. Formation of resin tags in patent dentin tubules, but most importantly, mechanical interlocking of dentinal substrate and adhesive resins as a result of penetration of the adhesive into the intertubular dentin is believed to be responsible for durable dentin adhesion (Nakabayashi et al., 1982). The most widely used monomers in adhesives and bonding resins are the highly hydrophilic 2-hydroxyethyl methacrylate (HEMA) and monomers that contain one or more carboxylic acid group, carboxylic acid anhydride or phosphoric acid groups (Figures D.4–D.6). The hydrophilic monomer HEMA and the acidic monomers have solubility parameters with high polar and hydrogen

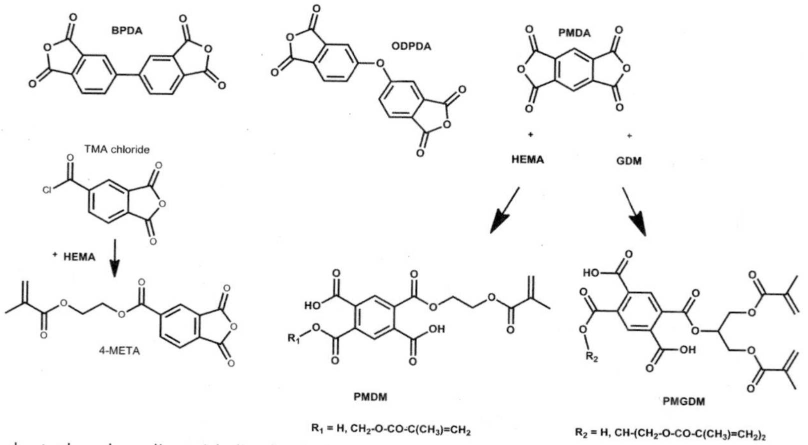

FIGURE D.5 Structures of selected carboxylic acid dianhydrides and monomers: BPDA = biphenyl dianhydride; PMDA = pyromellitic dianhydride; ODPDA = oxydiphthalic dianhydride; TMA = trimellitic anhydride. The adhesive monomers 4-META, PMDM, and PMGDM are also shown. They are formed by the reaction between TMA chloride and HEMA, PMDA, and HEMA or PMDA and GDM, respectively.

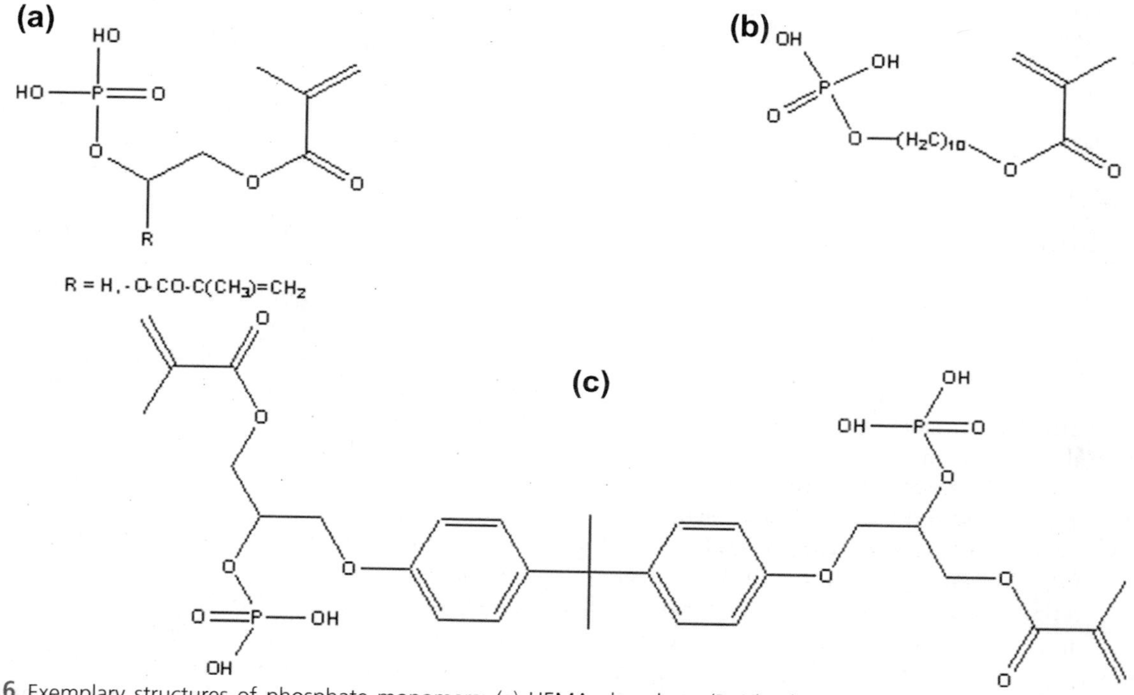

FIGURE D.6 Exemplary structures of phosphate monomers: (a) HEMA-phosphate (R=H), glycerol dimethacrylate phosphate (R= –O–CO–C(CH_3) = CH_2); (b) MDP (methacryloyldecyldihydrogen phosphate); (c) Bis-GMA-phosphate.

bonding components, resulting in high total cohesive energy density values. It has been demonstrated that bonding will be enhanced if fractional polarity and solubility parameters match those of the conditioned bonding substrate (Asmussen and Uno, 1993). Monomers with carboxylic acid groups (Figure D.5) are usually synthesized by reacting HEMA, glyceroldimethacrylate or other hydroxylated (meth)acrylate monomers with mono- or dicarboxylic acid anhydrides (Takeyama et al., 1978; Bowen et al., 1982; Venz and Dickens, 1993; Hammesfahr, 1997; Lopez-Suevos and Dickens, 2008). An adhesive monomer (4-META) with anhydride functionality is prepared from HEMA and trimellitic acid chloride (Figure D.5). To achieve improved interaction of adhesive monomers and tooth substrate beta cyclodextrines with multi-methacrylate/carboxylic-acid groups have been examined (Bowen et al., 2000; Hussain et al., 2004, 2005; Bowen et al., 2009). To improve the hydrolytic stability of monomer/polymer systems acrylamide/phosphonates have been proposed (Catel et al., 2008; Yeniad et al., 2008).

Silane Coupling Agents

Except for pure gold fillings, all dental restoratives are multiphase materials having a composite microstructure involving one or more interfaces or interphases. Polymeric dental composites are interconnected heterogeneous

γ-methacryloxypropyltrimethoxysilane (MPTMS)

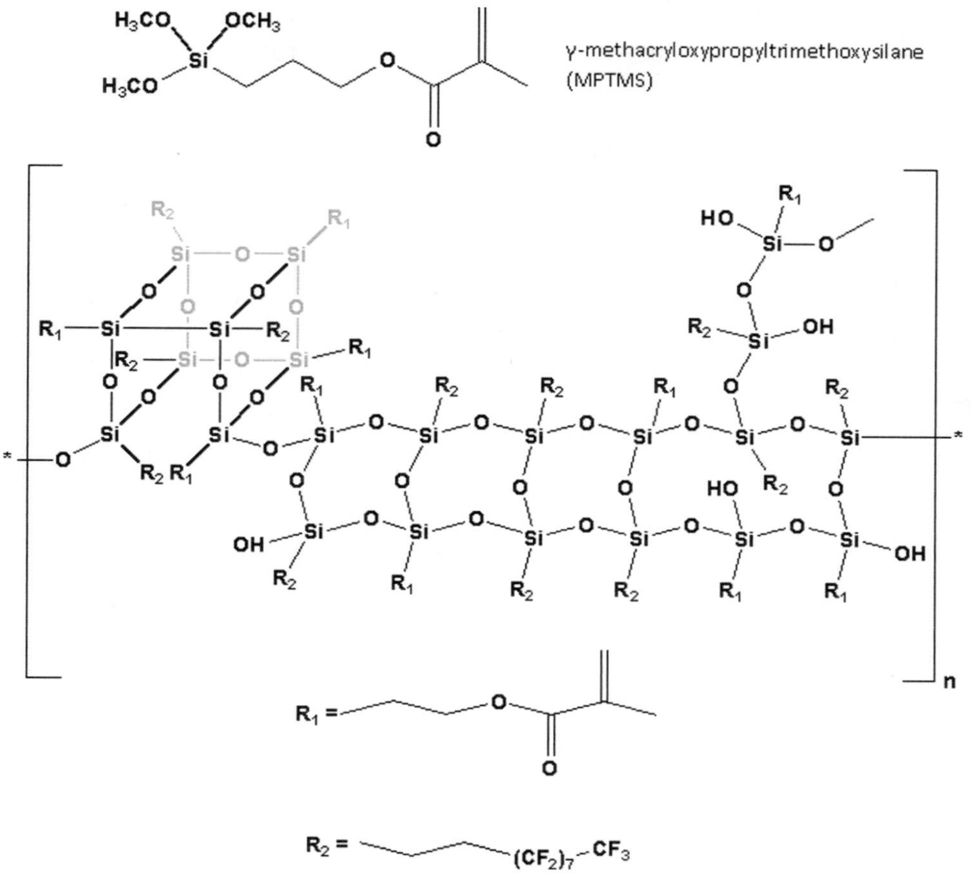

R_1 =

R_2 = $(CF_2)_7$–CF_3

FIGURE D.7 Structure of γ-methacryloxypropyltrimethoxysilane.

materials that generally have three discernable phases: (1) a polymeric matrix or continuous phase formed by polymerization of a resin system consisting of one or more monomers/oligomers activated for chemical/photochemical polymerization; (2) a higher modulus dispersed phase consisting of fillers of various types (silica, ceramic, organic, etc.), sizes, shapes, and morphologies; and (3) an interface or interphase (usually derived from silane coupling agents) that mediates bonds between the continuous soft and the hard dispersed phases, thereby enhancing the moduli and mechanical properties of the relatively low modulus polymer phase, and also facilitating stress transfer between these phases by forming a unitary material. Adhesion of lower moduli polymeric matrices to higher moduli inorganic fillers can occur as a result of van der Waals forces, ionic interactions, hydrogen bonding, ionic or covalent bonding, interpenetrating polymer network formation and, for certain types of fillers, by a micromechanical interlocking mechanism. For most mineral reinforced dental composites, the primary interphasial linkage between the polymer matrix and the filler phase is by chemical bond formation, mediated through a dual functional organosilane (Antonucci et al., 2005). The most commonly used silane coupling agent, γ-methacryloxypropyltrimethoxysilane (MPTMS), is shown in Figure D.7.

Monomers for Low Shrinking Polymers

Methacrylate Monomers: General Considerations. All methacrylate monomers undergo shrinkage when they polymerize. The volume a monomer occupies is governed by the monomer's molecular structure, and by the inherent van der Waals forces that keep monomer molecules at distinct intermolecular distances. During polymerization the double bonds of neighboring (meth) acrylate groups are converted to covalently bonded single bonds, thereby linking individual mers and forming macromers. If di- or multimethacrylates are present, a three-dimensional polymer network is formed. After the initiation of polymerization, the free radical chain growth of multimethacrylates initiated by commonly used chemical or photoinitiators is hindered by the rapid formation of the three-dimensional network, increasing the viscosity of the gel and decreasing the mobility of the monomer or short chain polymer molecules, which then limits the vinyl conversion. Thus, residual monomer and pendant double bonds are commonly present in these types of polymers to various degrees (Ruyter, 1985; Dickens et al., 2003). Monomer and short chain oligomers (a limited number of linked monomer units) not incorporated into the polymer network can potentially leach out, and both unsaturated residual and pendant methacrylate groups act as internal plasticizers that may influence the biocompatibility, physical properties, and longevity of the material (Asmussen, 1982; Ruyter, 1985). As polymerization shrinkage and remaining double bonds are directly correlated, these properties ought to be considered together when characterizing "low shrinking" materials. The polymerization process results in a reduction in free volume, measurable as linear or volumetric polymerization shrinkage. Polymerization contraction stresses occur as a consequence of polymerization shrinkage. For highly cross-linked systems, the majority of shrinkage stress develops during and after the vitrification stage (Lu et al., 2005), and is a significant concern in numerous dental applications. The most detrimental effect caused by shrinkage stress is debonding at the composite/tooth interface, which results in gap formation and microleakage that provides a route for bacterial infiltration that can contribute to secondary caries. Strategies to reduce the polymerization shrinkage in resin-based materials comprise varying the monomer structure and resin composition, increasing the number of atoms between methacrylate groups, and modifying type, size, size distribution, and amount of fillers added to the resin matrix. Most highly filled resin composites shrink between 2% and 3%.

Molecular Structure and Size Effects. In the last decade, modifications in the monomer structure for reducing the shrinkage and related stresses included hyperbranched monomers (Holter et al., 1997), liquid crystalline methacrylates (Hammesfahr, 1997; Satsangi et al., 2005), and dendritic monomers (Viljanen et al., 2005). However, these types of resins have given mixed results with respect to the physical and mechanical properties of their composites. Other recent strategies to lowering the polymerization shrinkage and stresses included use of high molecular weight bis(vinylcyclopropane)-based monomers containing a cholesteryl group (Choi et al., 2005), and dimer acid-based monomers (Trujillo-Lemon et al., 2006). It was found that these monomers contributed to significantly reduced shrinkage. The dimer acid-based monomers have found application in a commercial composite.

Thiol-ene Systems. Recently, novel photopolymerizable monomer systems have been developed which are based on thiol-ene resin chemistry involving free radical step growth rather than chain growth polymerization (Cramer and Bowman, 2001). These systems, especially oligomeric thiol-ene materials, produce polymers having high conversion and dramatically reduced polymerization shrinkage stress (Carioscia et al., 2005; Lu et al., 2005; Carioscia et al., 2007). While these systems undergo more rapid reaction with higher functional group conversion, they are mechanically inferior to current Bis-GMA/TEGDMA resins. Other work reported the application of thiol-allyl ether-methacrylate ternary systems (Kilambi et al., 2007) with an interesting approach using a two-stage polymerization process. During the first stage, highly functional oligomeric methacrylates are formed. Because this process occurs in the pre-gel stage, it does not contribute to shrinkage stress. The allyl ether conversion and shrinkage stress develop subsequently, and thus these systems have high cross-link density and improved mechanical properties, yet significantly reduced shrinkage stress when compared to previously developed thiol-ene systems.

Ring Opening Monomers: Spiroorthocarbonates. To reduce polymerization shrinkage while maintaining high conversion and acceptable material properties, non-methacrylated ring opening spiroorthocarbonate (SOC) monomers were evaluated (Thompson et al., 1979; Bailey, 1990; Byerley et al., 1992; Eick et al., 1993). Modified vinyl SOCs in combination with methacrylate monomers achieved tensile strengths that were comparable to their SOC-free controls (Stansbury, 1992). Re-examining the concept of potentially expanding resins, Stansbury et al. investigated a number of tetraoxaspirononanes either with cationic initiation alone or a combined cationic/free radical initiation (Ge et al., 2005), while others used tetraoxaspiroundecanes in combination with siloranes (Chappelow et al., 2007). Both studies confirmed the ring-opening mechanism. However, it was pointed out that even a trace amount of moisture had a significant adverse impact on both the reaction kinetics and the polymerization course of the hydrolytically vulnerable tetraoxaspirononanes (Ge et al., 2005).

Monomers Containing Silicon and/or Epoxy Functionality. Various approaches have been taken to include silicon-based monomers in addition to the silane coupling agents into dental composites. The chemistry of functional organosilanes (or silane coupling agents as they are commonly known) can be quite complex, involving hydrolytically initiated self-condensation reactions in solvents (including monomers) that can culminate in oligomeric silsesquioxane structures, exchange reactions with hydroxylated or carboxylated monomers to form silyl ethers and esters, as well as the formation of silane-derived interfaces by adhesive coupling with siliceous mineral surfaces (Antonucci et al., 2005). Examples of silsesquioxane oligomeric structures (together with MPTMS) are shown in Figure D.7. It was recently shown that the fluorinated silsesquioxane oligomers, when added to EBPADMA, led to similarly acceptable mechanical properties as the silsesquioxane-free control, while exhibiting significantly reduced polymerization shrinkage stresses (Dickens et al., 2009). It is hypothesized that the bulky nature of the oligomeric additive, possibly in combination with fluoridated pendant chains being repelled from the more hydrophilic bulk, increased the free volume of the oligomers, and thus contributed to the low shrinkage stresses in these composites. Adding 2% of the novel organic/inorganic hybrid monomer POSS-MA (polyhedral oligomeric silsesquioxane methacrylate) into Bis/TEGDMA resins showed moderate improvement of the mechanical properties. However, due to the high number of polymerizable methacrylate groups, the polymerization shrinkage was not improved (Fong et al., 2005). Cycloaliphatic epoxy (oxiranes) monomers formulated with polyol network extenders and cured with cationic onium/camphorquinone initiators have been considered as potential candidates to reduce polymerization shrinkage (Tilbrook et al., 2000). They exhibit lower polymerization shrinkage, no oxygen inhibition layer, higher strength, equivalent hardness, and acceptable glass transition temperatures as many methacrylate-based resins. However, the water uptake of these epoxy-polyol materials when immersed in water was nearly twice that of the Bis-GMA/TEGDMA control.

Even lower shrinkage and shrinkage stress values have been reported for a new class of methacrylate-free photocationically polymerizing low-shrinkage composite resins (Weinmann et al., 2005; Ilie et al., 2007). The monomers are named siloranes, indicating that their chemical structure comprises a combination of siloxanes and oxiranes. The reduction in polymerization shrinkage and resulting shrinkage stresses is based on two principles: epoxy-ring opening during setting of a large cyclic siloxane monomer core structure; and relatively slow polymerization kinetics resulting in somewhat delayed gelation and the possibility of stress relaxation. Physical properties of silorane-based materials match those of conventional composite resins. The silorane-based composites require special bonding and coupling agents based on the same chemistry to allow copolymerization between bonding agent and resin matrix, and also between the silanized reinforcing glass filler and the resin matrix.

Anticariogenic Strategies. To combat biofilm attachment to restoration surfaces and recurrent caries, a family of fluoride-releasing resins has been developed for adhesive applications, temporary crowns, and removable acrylic devices. These resins contain fluoride ions, which are covalently bound to positively charged sites within the polymer network and can be released via an exchange mechanism involving anions from oral fluids (Rawls, 1987). More recent approaches include tetrabutylammonium tetrafluoroborate incorporated into a hydrophilic monomer system (Glasspoole et al., 2001); complexing zirconium fluoride with a chelating monomer (Ling et al., 2009); or incorporating antibacterial functionality into monomers as demonstrated for methacryloyloxydodecyl-pyridinium bromide monomer (Imazato et al., 1994; Imazato, 2009).

SUMMARY

Dental resin chemistry has made significant progress in the last decade in improving clinical performance and addressing some of the most pertinent problems in restorative dentistry, i.e., polymerization shrinkage and shrinkage stress that potentially affect recurrent caries through bacterial ingress into marginal crevices around dental restorations. Thus, the most significant improvements for better dental health are the successful reduction of polymerization shrinkage and polymerization shrinkage stresses for improved marginal integrity; the development of resin chemistries that enhance the hydrolytic stability of the restoratives; and the development of monomers with antibacterial properties designed to increase the resistance of tooth structure to bacterial attack.

Disclaimer. Certain commercial materials and equipment are identified in this paper for adequate definition of the experimental procedure. In no instance does such identification imply recommendation or endorsement by the National Institute of Standards and Technology or the ADA Foundation or that the material or equipment identified is necessarily the best available for the purpose.

ACKNOWLEDGMENTS

This study was supported in part by the American Dental Association Foundation, the National Institute of Standards and Technology, and the National Institute of Dental and Craniofacial Research, Grant No. DE13298, and the NIDCR/NIST Interagency Agreement YI-DE 7005-01.

BIBLIOGRAPHY

Antonucci, J. M., & Stansbury, J. W. (1997). Molecular Designed Dental Polymers. In R. Arshady (Ed.), *Desk Reference of Functional Polymers: Synthesis and Applications* (pp. 719–738). Washington, D.C.: American Chemical Society.

Antonucci, J. M., Dickens, S. H., Fowler, B. O., Xu, H. H. K., & McDonough, W. G. (2005). Chemistry of silanes: Interfaces in dental polymers and composites. *J. Res. Natl. Inst. Stand. Technol.*, **110**(5), 541–558.

Asmussen, E. (1982). Restorative resins: Hardness and strength vs. quantity of remaining double bonds. *Scand J. Dent. Res.*, **90**(6), 484–489.

Asmussen, E., & Uno, S. (1993). Solubility parameters, fractional polarities, and bond strengths of some intermediary resins used in dentin bonding. *J. Dent. Res.*, **72**(3), 558–565.

Bailey, W. J. (1990). Matrices that expand on curing for high-strength composites and adhesives. *Mater. Sci. Eng. A.*, **126**, 271–279.

Berchtold, K. A., Nie, J., Stansbury, J. W., & Bowman, C. N. (2008). Reactivity of monovinyl (meth)acrylates containing cyclic carbonates. *Macromolecules*, **41**(23), 9035–9043.

Bogdal, D., Pielichowski, J., & Boron, A. (1997). Application of diol dimethacrylates in dental composites and their influence on polymerization shrinkage. *J. Appl. Polym. Sci.*, **66**(12), 2333–2337.

Bowen, R. L. (1956). Use of epoxy resins in restorative materials. *J. Dent. Res.*, **35**(3), 360–369.

Bowen, R. L. (1963). Properties of a silica-reinforced polymer for dental restorations. *J. Am. Dent. Assoc.*, **66**(1), 57–64.

Bowen, R. L., Cobb, E. N., & Rapson, J. E. (1982). Adhesive bonding of various materials to hard tooth tissues: Improvement in bond strength to dentin. *J. Dent. Res.*, **61**(9), 1070–1076.

Bowen, R. L., Farahani, M., Dickens, S. H., & Guttman, C. M. (2000). MALDI-TOF MS analysis of a library of polymerizable cyclodextrin derivatives. *J. Dent. Res.*, **79**(4), 905–911.

Bowen, R. L., Carey, C. M., Flynn, K. M., & Guttman, C. M. (2009). Synthesis of polymerizable cyclodextrin derivatives for use in adhesion-promoting monomer formulations. *J. Res. Natl. Inst. Stand. Technol.*, **114**(1), 1–9.

Byerley, T. J., Eick, J. D., Chen, G. P., Chappelow, C. C., & Millich, F. (1992). Synthesis and polymerization of new expanding dental monomers. *Dent. Mater.*, **8**(5–6), 345–350.

Carioscia, J. A., Lu, H., Stansbury, J. W., & Bowman, C. N. (2005). Thiol-ene oligomers as dental restorative materials. *Dent. Mater.*, **21**(12), 1137–1143.

Carioscia, J. A., Stansbury, J. W., & Bowman, C. N. (2007). Evaluation and control of thiol-ene/thiol-epoxy hybrid networks. *Polymer.*, **48**(6), 1526–1532.

Catel, Y., Degrange, M., Le Pluart, L., Madec, P. J., Pham, T. N., et al. (2008). Synthesis, photopolymerization and adhesive properties of new hydrolytically stable phosphonic acids for dental applications. *J. Polym. Sci. [A1]*, **46**(21), 7074–7090.

Chappelow, C. C., Pinzino, C. S., Chen, S. S., Jeang, L., & Eick, J. D. (2007). Photopolymerization of a novel tetraoxaspiroundecane and silicon-containing oxiranes. *J. Appl. Polym. Sci.*, **103**(1), 336–344.

Choi, S., Hessamian, A., Tabatabai, M., Fischer, U. K., Moszner, N., et al. (2005). Bis(vinylcyclopropane) and bis(methacrylate) monomers with cholesteryl group for dental composites. *E-Polymers*.

Cook, W. D. (1991). Fracture and structure of highly cross-linked polymer composites. *J. Appl. Polym. Sci.*, **42**(5), 1259–1269.

Cramer, N. B., & Bowman, C. N. (2001). Kinetics of thiol-ene and thiol-acrylate photopolymerizations with real-time Fourier transform infrared. *J. Polym. Sci. Part. A. – Polym. Chem.*, **39**(19), 3311–3319.

Dickens, S. H., Stansbury, J. W., Choi, K. M., & Floyd, C. J. E. (2003). Photopolymerization kinetics of methacrylate dental resins. *Macromolecules*, **36**(16), 6043–6053.

Dickens, S. H., Luther, D., & Antonucci, J. M. (2009). Effects of silane oligomers on composite properties. *J. Dent. Res.*, **87**(Spec. Iss. A), 1648.

Eick, J. D., Byerley, T. J., Chappell, R. P., Chen, G. R., Bowles, C. Q., et al. (1993). Properties of expanding soc/epoxy copolymers for dental use in dental composites. *Dent. Mater.*, **9**(2), 123–127.

Ferracane, J. L., Berge, H. X., & Condon, J. R. (1998). *In vitro* aging of dental composites in water: Effect of degree of conversion, filler volume, and filler/matrix coupling. *J. Biomed. Mater. Res.*, **42**(3), 465–472.

Fong, H., Dickens, S. H., & Flaim, G. M. (2005). Evaluation of dental restorative composites containing polyhedral oligomeric silsesquioxane methacrylate. *Dent. Mater.*, **21**(6), 520–529.

Ge, J. H., Trujillo, M., & Stansbury, J. (2005). Synthesis and photopolymerization of low shrinkage methacrylate monomers containing bulky substituent groups. *Dent. Mater.*, **21**(12), 1163–1169.

Geurtsen, W., Lehmann, F., Spahl, W., & Leyhausen, G. (1998). Cytotoxicity of 35 dental resin composite monomers/additives in permanent 3T3 and three human primary fibroblast cultures. *J. Biomed. Mater. Res.*, **41**(3), 474–480.

Gilbert, J. L., Ney, D. S., & Lautenschlager, E. P. (1995). Self-reinforced composite poly(methyl methacrylate), static and fatigue properties. *Biomaterials*, **16**(14), 1043–1055.

Glasspoole, E. A., Erickson, R. L., & Davidson, C. L. (2001). A fluoride-releasing composite for dental applications. *Dent. Mater.*, **17**(2), 127–133.

Hammesfahr, P. D. (1997). Compomers and hydromers for use in restorative dental procedures. *Am. Chem. Soc. Div. Polym. Chem. Polym. Prepr.*, **38**(2), 131–132.

Holter, D., Frey, H., Mulhaupt, R., & Klee, J. E. (1997). Branched bismethacrylates based on Bis-GMA: A systematic route to low shrinkage composites. *Am. Chem. Soc.*, **214**. 9–Poly.

Hussain, L. A., Dickens, S. H., & Bowen, R. L. (2004). Effects of polymerization initiator complexation in methacrylated beta-cyclodextrin formulations. *Dent. Mater.*, **20**(6), 513–521.

Hussain, L. A., Dickens, S. H., & Bowen, R. L. (2005). Properties of eight methacrylated beta-cyclodextrin composite formulations. *Dent. Mater.*, **21**(3), 210–216.

Ilie, N., Jelen, E., Clementino-Luedemann, T., & Hickel, R. (2007). Low-shrinkage composite for dental application. *Dent. Mater. J.*, **26**(2), 149–155.

Imazato, S. (2009). Bio-active restorative materials with antibacterial effects: New dimension of innovation in restorative dentistry. *Dent. Mater. J.*, **28**(1), 11–19.

Imazato, S., Torii, M., Tsuchitani, Y., McCabe, J. F., & Russell, R. R. B. (1994). Incorporation of bacterial inhibitor into resin composite. *J. Dent. Res.*, **73**(8), 1437–1443.

Khatri, C. A., Stansbury, J. W., Schultheisz, C. R., & Antonucci, J. M. (2003). Synthesis, characterization and evaluation of urethane derivatives of Bis-GMA. *Dent. Mater.*, **19**(7), 584–588.

Kilambi, H., Reddy, S. K., Schneidewind, L., Lee, T. Y., Stansbury, J. W., et al. (2007). Design, development, and evaluation of monovinyl acrylates characterized by secondary functionalities as reactive diluents to diacrylates. *Macromolecules*, **40**(17), 6112–6118.

Kim, J. W., Kim, L. U., Kim, C. K., Cho, B. H., & Kim, O. Y. (2006). Characteristics of novel dental composites containing 2,2-bis[4-(2-methoxy-3-methacryloyloxy propoxy) phenyl] propane as a base resin. *Biomacromolecules*, **7**(1), 154–160.

Ling, L., Xu, X., Choi, G. Y., Billodeaux, D., Guo, G., et al. (2009). Novel F-releasing composite with improved mechanical properties. *J. Dent. Res.*, **88**(1), 83–88.

Lopez-Suevos, F., & Dickens, S. H. (2008). Degree of cure and fracture properties of experimental acid-resin modified composites under wet and dry conditions. *Dent. Mater.*, **24**(6), 778–785.

Lu, H., Stansbury, J. W., & Bowman, C. N. (2005). Impact of curing protocol on conversion and shrinkage stress. *J. Dent. Res.*, **84**(9), 822–826.

Moszner, N., & Salz, U. (2001). New developments of polymer dental composites. *Prog. Polym. Sci.*, **26**(4), 535–576.

Moszner, N., & Salz, U. (2007). Recent developments of new components for dental adhesives and composites. *Macromol. Mater. Eng.*, **292**(3), 245–271.

Moszner, N., Fischer, U. K., Angermann, J., & Rheinberger, V. (2008). A partially aromatic urethane dimethacrylate as a new substitute for Bis-GMA in restorative composites. *Dent. Mater.*, **24**(5), 694–699.

Nakabayashi, N., Kojima, K., & Masuhara, E. (1982). The promotion of adhesion by the infiltration of monomers into tooth substrates. *J. Biomed. Mater. Res.*, **16**(3), 265–273.

Patel, M. P., Braden, M., & Davy, K. W. (1987). Polymerization shrinkage of methacrylate esters. *Biomaterials*, *8*(1), 53–56.

Peutzfeldt, A. (1997). Resin composites in dentistry: The monomer systems. *Eur. J. Oral. Sci.*, **105**(2), 97–116.

Rawls, H. R. (1987). Fluoride-releasing acrylics. *J. Biomater. Appl.*, **1**(3), 382–405.

Ruyter, I. E. (1985). Monomer Systems and Polymerization. In G. Vanherle, & D. C. Smith (Eds.), *Posterior Composite Resin Dental Restorative Materials* (pp. 109–135). The Netherlands: Peter Szulc Publishing Co.

Satsangi, N., Rawls, H. R., & Norling, B. K. (2005). Synthesis of low-shrinkage polymerizable methacrylate liquid-crystal monomers. *J. Biomed. Mater. Res. Part. B:. Appl. Biomater.*, **74B**(2), 706–711.

Shalaby, S. W., Nagatomi, S. D., & Peniston, S. J. (2007). Polymeric Biomaterials for Articulating Joint Repair and Total Joint Replacement. In S. W. Shalaby, & U. Salz (Eds.), *Polymers for Dental and Orthopedic Applications*. Boca Raton, London, New York: CRC Press.

Stansbury, J. W. (1990). Cyclopolymerizable monomers for use in dental resin composites. *J. Dent. Res.*, **69**(3), 844–848.

Stansbury, J. W. (1992). Synthesis and evaluation of new oxaspiro monomers for double-ring opening polymerization. *J. Dent. Res.*, **71**(7), 1408–1412.

Takeyama, M., Kashibuchi, N., Nakabayashi, N., & Masuhara, E. (1978). Studies on dental self-curing resins (17). Adhesion of PMMA with bovine enamel or dental alloys. *J. Jpn. Dent. Appar. Mater.*, **19**, 179–185.

Thompson, V. P., Williams, E. F., & Bailey, W. J. (1979). Dental resins with reduced shrinkage during hardening. *J. Dent. Res.*, **58**(5), 1522–1532.

Tilbrook, D. A., Clarke, R. L., Howle, N. E., & Braden, M. (2000). Photocurable epoxy-polyol matrices for use in dental composites I. *Biomaterials*, **21**(17), 1743–1753.

Trujillo-Lemon, M., Ge, J. H., Lu, H., Tanaka, J., & Stansbury, J. W. (2006). Dimethacrylate derivatives of dimer acid. *J. Polym. Sci. [A1]*, **44**(12), 3921–3929.

Venz, S., & Dickens, B. (1993). Modified surface-active monomers for adhesive bonding to dentin. *J. Dent. Res.*, **72**(3), 582–586.

Viljanen, E. K., Lassila, L. V. J., Skrifvars, M., & Vallittu, P. K. (2005). Degree of conversion and flexural properties of a dendrimer/methyl methacrylate copolymer: Design of experiments and statistical screening. *Dent. Mater.*, **21**(2), 172–177.

Weinmann, W., Thalacker, C., & Guggenberger, R. (2005). Siloranes in dental composites. *Dent. Mater.*, **21**(1), 68–74.

Yeniad, B., Albayrak, A. Z., Olcum, N. C., & Avci, D. (2008). Synthesis and photopolymerizations of new phosphonated monomers for dental applications. *J. Polym. Sci. [A1]*, **46**(6), 2290–2299.

Yoshii, E. (1997). Cytotoxic effects of acrylates and methacrylates: Relationships of monomer structures and cytotoxicity. *J. Biomed. Mater. Res.*, **37**(4), 517–524.

CHAPTER I.2.3 METALS: BASIC PRINCIPLES

John B. Brunski
Division of Plastic & Reconstructive Surgery, Department of Surgery, School of Medicine, Standford University, Standford, CA, USA

INTRODUCTION

Since large segments of the medical device industry rely on implants with one or more metallic parts – e.g., stents, heart valves, orthopedic hips and knees, plus oral/maxillofacial implants – metals have a highly significant place in the biomaterials market. In view of this wide utilization of metallic biomaterials, this chapter provides some basic principles underlying the quantification and design-related manipulation of composition, structure, and properties of metallic biomaterials. Major themes are the metallurgical principles underlying structure–property relationships, and the larger problem of design, production, and proper utilization of medical devices.

STEPS IN THE FABRICATION OF METALLIC BIOMATERIALS

Understanding the structure and properties of metallic implant materials requires an appreciation of the metallurgical significance of the material's processing history. Typically, any metallic medical device will differ in exactly how it is manufactured, so it is useful to look briefly at some generic processing steps in implant manufacture (Figure I.2.3.1A).

Metal-Containing Ore to Raw Metal Product

With the exception of noble metals such as gold (which do not represent a major fraction of implant metals), metals exist in the Earth's crust in mineral form, wherein the metal is chemically combined with other elements, as in the case of metal oxides. These mineral deposits, or "ore," must be located, mined, separated, and enriched for further processing into pure metal or various alloys. For example, in the case of titanium, certain mines in the southeastern United States yield sands containing common quartz along with mineral deposits of zircon, titanium, iron, and rare earth elements. The sandy mixture can be concentrated by using water flow and gravity to isolate titanium-containing compounds such as rutile (TiO_2) and ilmenite ($FeTiO_3$). To obtain rutile, which is particularly good for making metallic titanium, further processing typically involves electrostatic separations. Then, to extract titanium metal from the rutile, one method involves treating the ore with chlorine to make titanium tetrachloride liquid, which in turn is treated with magnesium or sodium to produce chlorides of the latter metals along with bulk titanium "sponge" according to the Kroll process. At this stage, the titanium sponge is not of controlled purity, so depending on the purity ("grade") of the final titanium product that is sought, it is necessary to refine it further by using vacuum furnaces, remelting, and additional steps. All of this is critical in producing titanium with the appropriate properties, as exemplified in the production of the most common grades of commercially pure (CP) titanium; these grades differ in oxygen content by only tenths of a percent, yet these small differences in oxygen content make major differences in mechanical properties, including yield, tensile and fatigue strength of titanium.

After extraction steps, the resulting raw metal product eventually emerges in some type of bulk form, such as ingots, which can be supplied to raw materials' vendors or metal manufacturers. For instance, in the case of multicomponent metallic implant alloys (i.e., made up of more than one element), the raw metal product will usually have to be further processed both chemically and physically. Processing steps can include remelting, addition of specific alloying elements, and controlled solidification from the melt, in order to produce an alloy that meets certain chemical and metallurgical specifications. For example, to make ASTM (American Society for Testing and Materials) F138 316L stainless steel, iron is alloyed with specific amounts of carbon, silicon, nickel, and chromium; and to make ASTM F75 or F90 alloy, cobalt is alloyed with specific amounts of chromium, molybdenum, carbon, nickel, and other elements. Table I.2.3.1 lists ASTM designations and typical properties of common metallic alloys for surgical implants.

Raw Metal Product to Stock Metal Shapes

A metal supplier will typically further process the bulk raw metal product (metal or alloy) into stock bulk shapes, such as bars, wire, sheet, rods, plates, tubes or powders. These stock shapes may then be sold to implant manufacturers, who typically want a stock shape that is closer to the final implant shape, e.g., a maker of screw-shaped dental implants would often buy rod stock of the appropriate metal as feedstock for screw-manufacturing machines.

A metal supplier might transform the raw metal product into stock shapes by a variety of processes, including

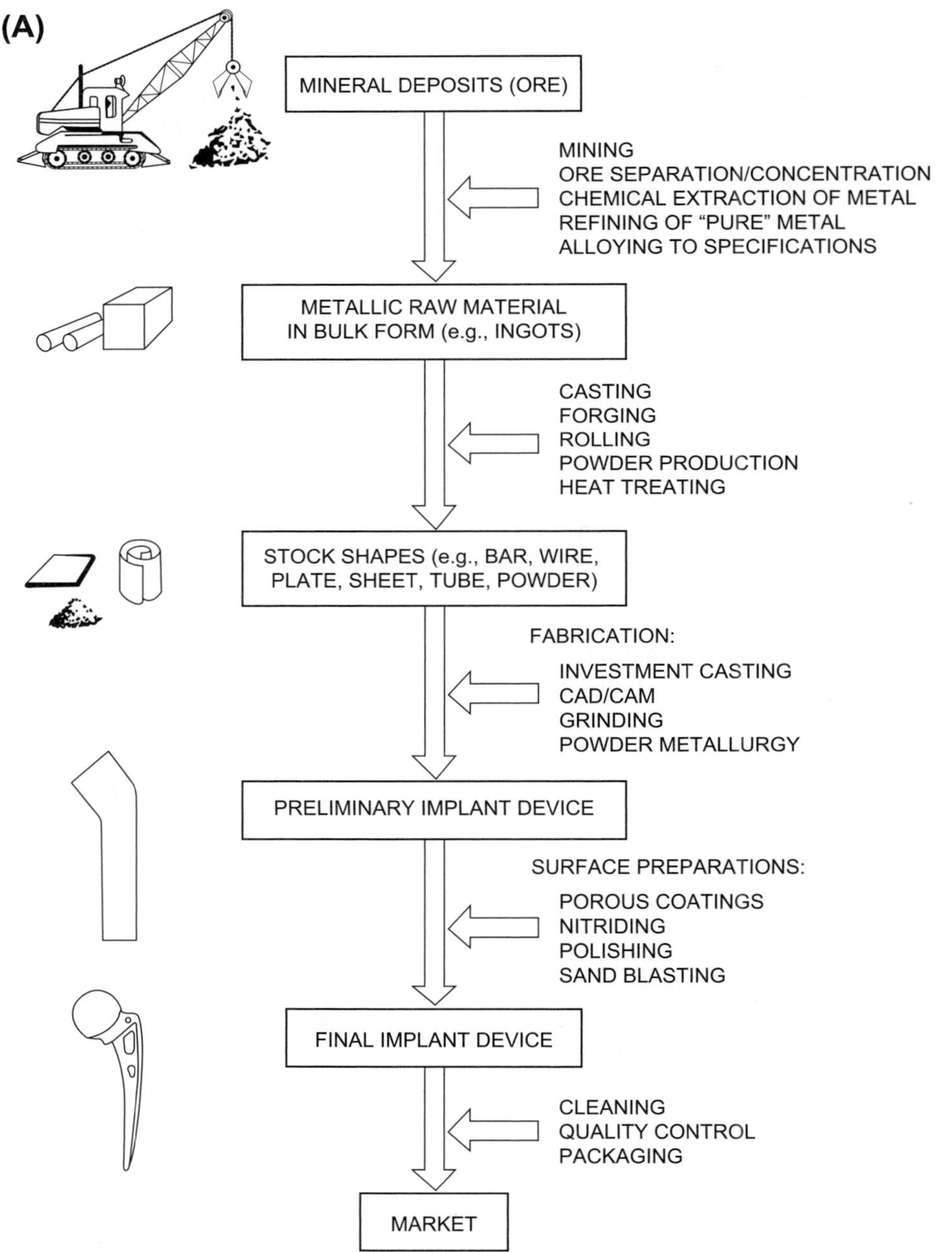

FIGURE I.2.3.1 (A) Generic processing history of a typical metallic implant device, in this case a hip implant.

FIGURE I.2.3.1 (B) Image of one step during the investment casting ("lost wax") process of manufacturing hip stems; a rack of hip stems can be seen attached to a system of sprues through which molten metal can flow. At this point, ceramic investment material composes the mold into which the molten metal will flow and solidify during casting, thereby replicating the intended shape of a hip stem.

remelting and continuous casting, hot rolling, forging, cold drawing through dies, etc. Depending on the metal there might also be heat-treating steps (i.e., carefully controlled heating and cooling cycles) designed to: facilitate further working or shaping of the stock; relieve the effects of prior plastic deformation (e.g., as in annealing) or produce a specific microstructure and properties in the stock material. Because of the high chemical reactivity of some metals at elevated temperatures, high temperature processing may require vacuum conditions or inert atmospheres to prevent unwanted uptake of oxygen by the metal, all of which can add to the cost of production. For instance, in the production of fine powders of ASTM F75 Co-Cr-Mo alloy, molten metal is ejected through a small nozzle to produce a fine spray of atomized metal droplets that solidify while cooling in an inert argon atmosphere.

Typically, stock shapes are chemically and metallurgically tested at this early stage to ensure that the chemical composition and microstructure of the metal sold to an implant company conforms to industry standards for surgical implants (e.g., ASTM Standards), as discussed later in this chapter. It makes sense that an implant manufacturer will demand quality assurance that they are buying an appropriate grade of metal from the supplier of the raw metal stock.

Stock Metal Shapes to Preliminary and Final Metal Devices

Typically, an implant manufacturer will buy stock material and then fabricate preliminary and final forms of the device. Specific steps depend on a number of factors, including the final geometry of the implant, the forming and machining properties of the metal, and the costs of alternative fabrication methods. Typical fabrication methods include investment casting (the "lost wax" process), conventional and computer-based machining (CAD/CAM), forging, powder metallurgical processes (e.g., hot isostatic pressing or HIP), and a range of grinding and polishing steps. A variety of fabrication methods are required because not all implant alloys can be feasibly or economically fabricated into a final form in the same way. As one example, cobalt-based alloys are extremely difficult to machine into the complicated shapes of some implants by conventional machining methods. Therefore, many cobalt-based alloys are frequently shaped into implant forms by investment casting (e.g., Figure I.2.3.1B) or by powder metallurgy. On the other hand, titanium is relatively difficult to cast, and is therefore often machined.

Other aspects of fabrication (which to some degree more accurately fall under the heading of surface treatment) involve the application of macro-, micro-, and/or nano-level coatings on implants, often with the intent of producing certain ranges of surface roughness. Such surface modifications have become more popular in recent years as a means to improve fixation of implants in bone or to improve interfacial bone growth. The surface coating or roughening can take various forms, and can require different fabrication technologies. In some cases, modifications to the surface have the potential to negatively affect the metallurgical properties of the underlying bulk implant metal. For example, in the case of alloy beads or "fiber metal" coatings for macro-porous implants, the manufacturer may employ high temperature sintering to apply the coating over specific regions of the implant surface (e.g., on the proximal portion of a femoral hip stem). Sintering involves heating the coating and substrate to about half (or more) of the alloy's melting temperature in order to encourage diffusion, which in turn helps form the necks that join the beads to one another and to the implant's surface. Such reasonably high temperatures can modify the underlying metallic substrate by allowing grain growth or other mechanisms, which could negatively affect the properties. An alternative to sintering is plasma or flame spraying a metal onto an implant's surface. In this process, hot, high velocity gas plasma is charged with a metallic powder and directed at appropriate regions of

TABLE I.2.3.1 Typical Mechanical Properties of Implant Metals[a]

Material	ASTM Designation	Condition	Young's Modulus (GPa)	Yield Strength (MPa)	Tensile Strength (MPa)	Fatigue Endurance Limit Strength (at 10^7 cycles, $R = -1$[c]) (MPa)
Stainless steel	F745	Annealed	190	221	483	221–280
	F55, F56, F138, F139	Annealed	190	331	586	241–276
		30% Cold-worked	190	792	930	310–448
		Cold forged	190	1213	1351	820
Co–Cr alloys	F75	As-cast/annealed	210	448–517	655–889	207–310
		P/M HIP[b]	253	841	1277	725–950
	F799	Hot forged	210	896–1200	1399–1586	600–896
	F90	Annealed	210	448–648	951–1220	Not available
		44% Cold-worked	210	1606	1896	586
	F562	Hot forged	232	965–1000	1206	500
		Cold-worked, aged	232	1500	1795	689–793 (axial tension $R = 0.05$, 30 Hz)
Ti alloys	F67	30% Cold-worked Grade 4	110	485	760	300
	F136	Forged annealed	116	896	965	620
		Forged, heat treated	116	1034	1103	620–689

[a] Data collected from references noted at the end of this chapter, especially Table 1 in Davidson and Georgette (1986).
[b] P/M HIP: Powder metallurgy product, hot-isostatically pressed.
[c] R is defined as $\sigma_{min}/\sigma_{max}$.

an implant surface. The powder particles fully or partially melt, and then fall onto the substrate surface, where they solidify rapidly to form a tough coating. This has the advantage of not excessively heating the substrate.

Still more surface treatments are available, including ion implantation (to produce better surface properties), nitriding, and coating with a thin diamond film. For example, in nitriding, a high energy beam of nitrogen ions is directed at the implant under vacuum so that nitrogen atoms penetrate the surface and come to rest at sites in the substrate. Depending on the alloy, this process can produce enhanced properties, such as improved surface hardness and wear properties. (Many examples of different levels of surface modifications – ranging from the nano-level on up to the macro level – can be seen by perusing the advertisements for oral and maxillofacial implants in any one of the many journals devoted to such implants.)

Finally, a manufacturer of a metallic implant device will normally perform a set of finishing steps. These vary with the metal and manufacturer, but typically include chemical cleaning and passivation (i.e., rendering the metal inactive if exposed to a corrosive environment). Electrolytically-controlled treatments may also be used to remove machining chips or impurities that may have become embedded in the implant's surface. As a rule, these steps are conducted according to good manufacturing practice (GMP), and ASTM specifications for cleaning and finishing implants. Notably, these finishing steps can be extremely important to the overall biological performance of the implant because they can affect the detailed surface properties of the medical device – which govern the surface of the device that comes in direct contact with the blood and other tissues at the implant site.

MICROSTRUCTURES AND PROPERTIES OF IMPLANT METALS

In order to understand the properties of each alloy system in terms of microstructure and processing history, it is essential to know: (1) the chemical and crystallographic identities of the phases present in the microstructure; (2) the relative amounts, distribution, and orientation of these phases; and (3) the effects of the phases on properties. This section of the chapter emphasizes mechanical properties of metals used in implant devices, even though other properties, such as surface properties and wear properties, must also be considered and may in some instances be more critical to control. (Surface properties of materials are reviewed in more depth in Chapter I.1.5 of this textbook.) The basic principles are illustrated here in the context as applied to the stainless steels, cobalt-based alloys, and titanium-based alloys – which are very commonly used in medical devices. (More detailed discussions follow in the separate chapters in this textbook on stainless steel (Chapter I.2.3.B) and titanium (Chapter I.2.3.A).)

Microstructure and Mechanical Properties

There are many ways to manipulate microstructures and properties of metallic biomaterials. The following sections highlight just a few of the more commonly-encountered principles that apply to the most common metallic biomaterials, including the stainless steels, Co-Based alloys, and Ti-based alloys.

316L Stainless Steel. In this alloy, two common strengthening methods are cold-working and controlling grain size. The basis of each method is the idea of increasing the difficulty of slip of dislocations. In cold-working, the idea is to introduce more and more plastic deformation such that additional plastic flow becomes even more difficult. In decreasing grain size, the idea is to have more grain boundaries to interfere with the flow of dislocations on slip systems within each grain.

With 316L stainless (ASTM F138), typically it is used in a 30% cold-worked state, because this cold-worked metal has a markedly increased yield, ultimate tensile, and fatigue strength relative to the annealed state (Table I.2.3.1). The trade-off in this case is decreased ductility of the cold-worked metal, but ordinarily this is not a major concern in implant products. In dealing with grain size in 316L, the recommended grain size is ASTM #6 or finer, in which the grain size number n is defined in the formula $N = 2^{n-1}$, where N is the number of grains counted in a 1 in^2 area at 100 × magnification (0.0645 mm^2 actual area). As an example, when $n = 6$, the grain size is about 100 micrometers. The emphasis on a fine grain size is explained by the well-known Hall–Petch-type relationship (Hall, 1951; Petch, 1953), which relates mechanical yield stress and grain diameter as follows:

$$t_y = t_i + kd^{-m}$$

where t_y and t_i are the yield and friction stress, respectively; d is the grain diameter; k is a constant associated with propagation of deformation across grain boundaries; and m is approximately 0.5. This equation indicates that a higher yield stress may be achieved by a metal with a smaller grain diameter d, all other things being equal. A key determinant of grain size is manufacturing history, including details on solidification conditions, cold-working, annealing cycles, and recrystallization.

Cobalt-Based Alloys. Cobalt-based alloys include Haynes-Stellite 21 and 25 (ASTM F75 and F90, respectively), forged Co-Cr-Mo alloy (ASTM F799), and multiphase (MP) alloy MP35N (ASTM F562). The F75 and F799 alloys are virtually identical in composition, each being about 58–70% Co and 26–30% Cr, with the key difference in their processing history. The other two alloys, F90 and F562, have slightly less Co and Cr, but more Ni (F562) or more tungsten (F90).

To consider the example of F75 alloy (which has a long history in both the aerospace and biomedical implant industries), the main attribute of this alloy is corrosion resistance in chloride environments, which is related to its bulk composition and surface oxide (nominally Cr_2O_3). When F75 is cast into shape by investment casting (e.g., the "lost wax" process; Figure I.2.3.1B), the alloy is first melted at 1350–1450°C and then poured or pressurized into ceramic molds of the desired shape (e.g., femoral stems for artificial hips, oral implants, dental partial bridgework). The sometimes intricately-shaped molds are made by fabricating a wax pattern to near-final dimensions of the implant and then coating (or "investing") the pattern with a special ceramic, which then holds its shape after the wax is burned out prior to casting – hence the name "lost wax" for the process. Molten metal is poured into the ceramic mold through sprues, or pathways, and then, once the metal has solidified into the shape of the mold, the ceramic mold is cracked away and processing of the metal continues toward the final device.

Depending on the exact details of the casting process, at least three microstructural features can come into play as strong determinants of implant properties. First, as-cast F75 alloy typically consists of a Co-rich matrix (alpha phase) plus interdendritic and grain boundary carbides (primarily $M_{23}C_6$, where M represents Co, Cr or Mo). Overall, the relative amounts of the alpha and carbide phases should be approximately 85% and 15%, respectively, but due to non-equilibrium cooling, it is possible that a "cored" microstructure may develop. In this situation, the interdendritic regions become rich in solute (Cr, Mo, C) and contain carbides, while the dendrites become depleted in Cr and richer in Co. This creates an unfavorable electrochemical situation, with the Cr-depleted regions being anodic with respect to the rest of the microstructure. (This is also an unfavorable situation if a porous coating will subsequently be applied by sintering to this bulk metal.) Subsequent solution-annealing heat treatments at 1225°C for 1 hour can help alleviate this situation.

A second issue with F75 solidification is that dendrites will form, along with a relatively large grain size. This is generally undesirable because it decreases the yield strength via a Hall–Petch relationship between yield strength and grain diameter (recall the discussion from the section on stainless steel above). An example of dendritic growth and large grain diameter (approximately 4 mm) can be easily seen in Figures I.2.3.2A and I.2.3.2B, which show metallographic cross-sectional views through a femoral hip stem, and also a dental implant (Figures I.2.3.2C, I.2.3.2D) manufactured by investment casting.

A third issue with cast F75 is that casting defects may arise, e.g., Figure I.2.3.2B shows an inclusion in the middle of the cross-section through the distal third of the femoral hip stem. The inclusion was a particle of the ceramic mold (investment) material, which presumably broke off and became trapped within the interior of the mold while the metal was solidifying. This contributed to a fatigue fracture of the implant device *in vivo*, most likely because of stress concentrations and crack initiation sites associated with the ceramic inclusion. For similar reasons, it is also desirable to avoid macro- and microporosity arising from metal shrinkage upon solidification of castings. As an example of such porosity, Figures I.2.3.2C and I.2.3.2D exemplify a markedly dendritic microstructure with large grain size and

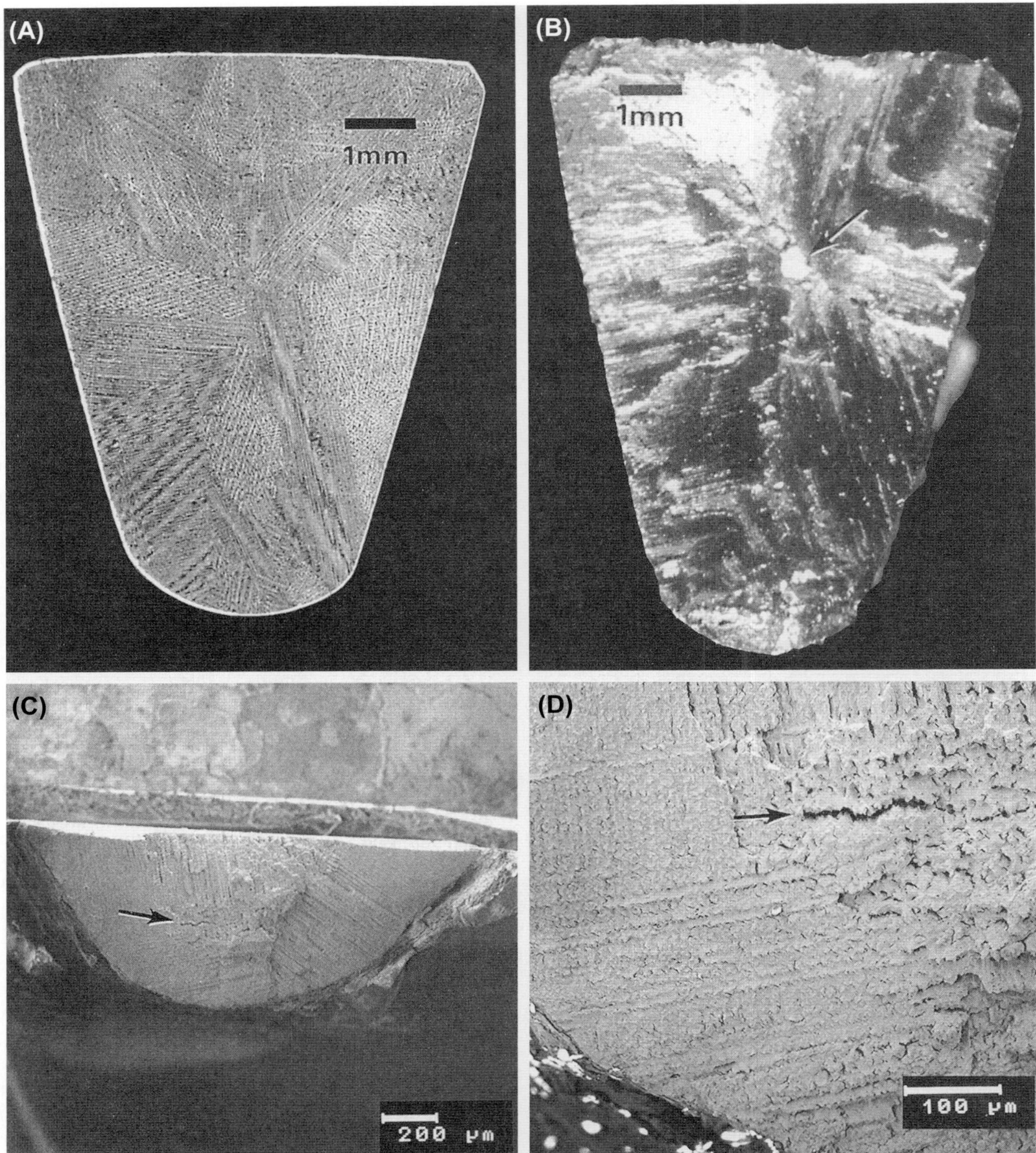

FIGURE I.2.3.2 (A) Macrophoto of a metallographically polished and etched cross-section of a cast Co–Cr–Mo ASTM F75 femoral hip stem, showing dendritic structure and large grain size. (B) Macrophoto of the fracture surface of the same Co–Cr–Mo ASTM F75 hip stem as in (A). Arrow indicates large inclusion within the central region of the cross section. Fracture of this hip stem occurred *in vivo*. (C), (D) Scanning electron micrographs of the fracture surface from a cast F75 subperiosteal dental implant. Note the large grain size, dendritic microstructure, and interdendritic microporosity (arrows).

microporosity at the fracture surface of a dental implant fabricated by investment casting.

To avoid problems such as the ones described above with cast F75, powder metallurgical techniques have been designed and used by some implant manufacturers. For example, in hot isostatic pressing (HIP), a fine powder of F75 alloy is compacted and sintered together under appropriate pressure and temperature conditions (about 100 MPa at 1100°C for 1 hour) and then forged to final shape. The typical microstructure shows a much smaller grain size than with casting (e.g., about 8 microns), which has the benefit of a much higher yield strength (via the Hall–Petch relationship), and better ultimate and fatigue properties than the as-cast alloy (Table I.2.3.1).

One final example with F75 alloy is the issue of porous-coated implants made by sintering. Here the idea is to sinter (join) the beads together, and to the underlying bulk substrate. With Co-based alloys like F75, however, sintering can be difficult, requiring temperatures near the melting point (1225°C), which in turn can decrease the fatigue strength of the substrate alloy. For example, cast/solution-treated F75 alloy has a fatigue strength of about 200–250 MPa, but this strength can decrease to about 150 MPa after porous coating treatments – evidently from further phase changes in the non-equilibrium cored microstructure in the original cast F75 alloy. On the other hand, it has been found that a modified sintering treatment can return the fatigue strength back up to about 200 MPa (Table I.2.3.1).

Titanium-Based Alloys. Commercially pure (CP) titanium (ASTM F67) and extra-low interstitial (ELI) Ti-6Al-4V alloy (ASTM F136) are the two most common titanium-based implant biomaterials (although within the category of CP Ti there are four grades). F67 CP Ti is 98.9–99.6% Ti. The oxygen content of CP Ti, as well as the content of other interstitial elements (e.g., C and N), affect its yield, tensile, and fatigue strengths significantly. With Ti-6Al-4V ELI alloy, the individual Ti-Al and Ti-V phase diagrams suggest the effects of the alloying additions in the ternary alloy; since Al is an alpha (HCP) phase stabilizer and V is a beta (BCC) phase stabilizer, Ti-6Al-4V alloy used for implants is an alpha-beta alloy.

For CP titanium implants (as exemplified by many current dental implants) typical microstructures are made up of single-phase alpha titanium (having the HCP structure), in which there is typically mild (30%) cold-work and grain diameters in the range of 10–150 microns, depending on manufacturing. The nominal mechanical properties are listed in Table I.2.3.1. Beyond cold-work, interstitial elements (O, C, N) in both CP titanium and the Ti-6Al-4V alloy strengthen the metal through interstitial solid solution strengthening mechanisms, with nitrogen having approximately twice the hardening effect (per atom) of either carbon or oxygen. As noted earlier, the oxygen content of CP Ti (and the interstitial content generally) affects its yield, tensile, and fatigue strengths significantly. For example, data available in the ASTM standard show that at 0.18% oxygen (grade 1), the yield strength is about 170 MPa, whereas at 0.40% (grade 4) the yield strength is about 485 MPa.

Likewise, the ASTM standard shows that the tensile strength increases with oxygen content. The literature also reveals that the fatigue limit of unalloyed CP Ti is typically increased by interstitial content, in particular the oxygen content. For example, Figure I.2.3.3A shows data from Beevers and Robinson (1969), who tested

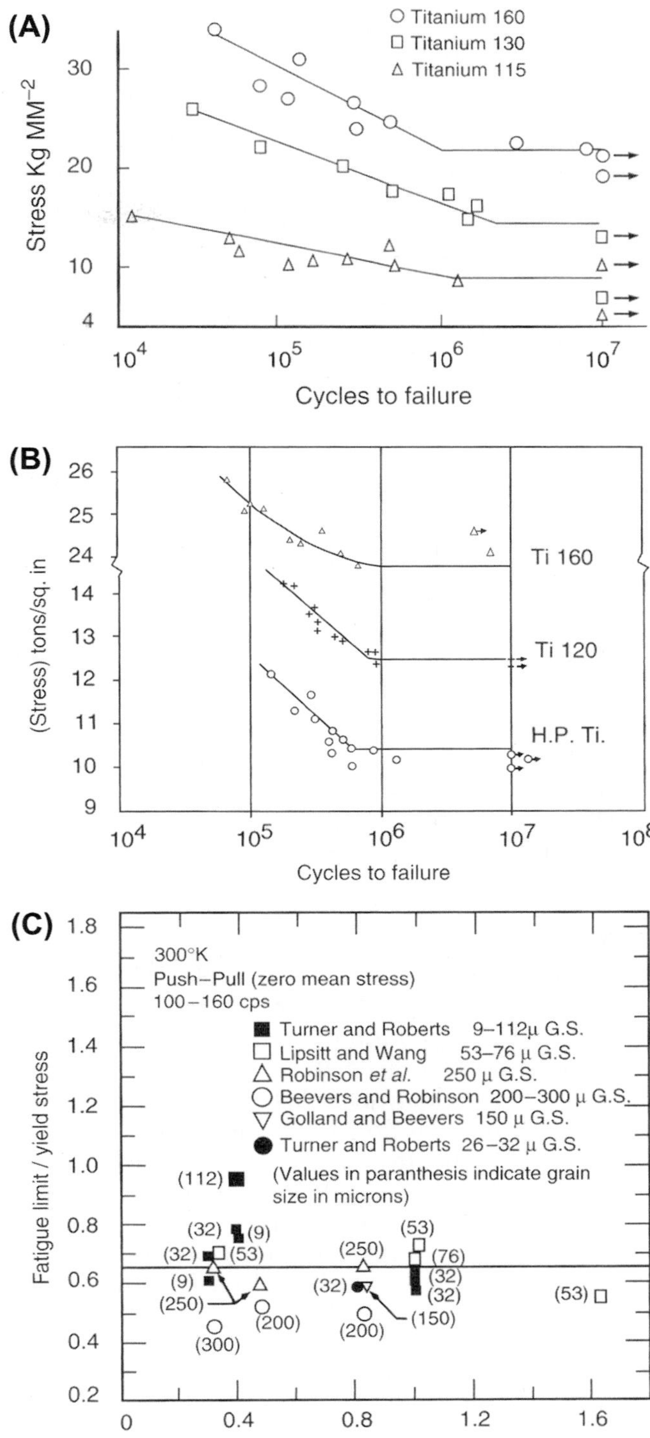

FIGURE I.2.3.3 (A) S–N curves (stress amplitude–number of cycles to failure) at room temperature for CP Ti with varying oxygen content (see text for O content of Ti 160, 130, and 115), from Beevers and Robinson (1969). (B) S–N curves at room temperature for CP Ti with varying oxygen content (see text), from Turner and Roberts (1968a). (C) Ratio of fatigue limit to yield stress in unalloyed Ti at 300°K as a function of at% oxygen and grain size, from Conrad et al. (1973). See Bibliography for details of references given within this figure.

vacuum-annealed CP Ti having a grain size in the range 200–300 microns in tension-compression at a mean stress of zero, at 100 cycles/sec. The 10^7 cycle endurance limit, or fatigue limit, for Ti 115 (0.085 wt% O, grade 1), Ti 130 (0.125 wt% O, grade 1), and Ti 160 (0.27 wt% O, grade 3) was 88.3, 142, and 216 MPa, respectively. Figure I.2.3.3B shows similar results from Turner and Roberts' (1968a) fatigue study on CP Ti (tension-compression, 160 cycles/sec, mean stress equals zero) having a grain size in the range 26–32 micrometers. Here the fatigue limit for "HP Ti" (0.072 wt% O, grade 1), Ti 120 (0.087 wt% O, grade 1), and Ti 160 (0.32 wt% O, grade 3) was 142, 172, and 295 MPa, respectively – again showing an increasing endurance limit with increasing oxygen content. Also, for grade 4 Ti in the cold-worked state, Steinemann et al. (1993) reported a 10^7 endurance limit of 430 MPa. Other workers (Conrad et al., 1973) have reported fatigue studies on CP Ti at 300°K, and noted that the ratio of fatigue limit to yield stress is relatively constant at about 0.65, independent of interstitial content and grain size (Figure I.2.3.3C). The work of Turner and Roberts also reported that the ratio *f* (defined as fatigue limit/ultimate tensile strength) – also referred to as the "fatigue ratio" in materials design textbooks (e.g., see Charles and Crane, 1989, p. 106) – was 0.43 for the high-purity Ti (0.072 wt% O), 0.5 for Ti 120 (0,087 wt% O), and 0.53 for Ti 160 (0.32 wt% O). It seems clear that interstitial content affects the yield and tensile and fatigue strengths in CP Ti.

At the same time, cold-work appears to increase the fatigue properties of CP Ti. For example, Disegi (1990) quoted bending fatigue data for annealed versus cold-worked CP Ti in the form of un-notched 1.0 mm thick sheet, and showed a moderate increase in ultimate tensile strength (UTS) and "plane bending fatigue strength," when comparing annealed versus cold-rolled Ti samples. Based on these data, the ratio of fatigue strength to ultimate tensile strength ("endurance ratio" or "fatigue ratio," see paragraph above) varied between 0.45 and 0.66. On the other hand, the *ASM Handbook Fatigue and Fracture* (Wagner et al., 1996) noted that the fatigue limit for high purity Ti only increased about 10% relative to the annealed material, while Disegi's data showed that the fatigue strength increased by about 28%, on average, with cold-work.

CONCLUDING REMARKS

Metallurgical structure–property relationships inform materials selection in medical implant design, just as they do in the design of any well-engineered product. Although this chapter's emphasis has been on mechanical properties (for the sake of specificity), other properties – including surface properties – are receiving increasing attention in relation to biological performance of implants. Examples of this latter theme are efforts to attach relevant biomolecules to metallic implant surfaces to promote certain desired interfacial activities, and attempts to develop special surface textures on implants to guide molecular and cellular reactions; these subjects can be researched further based on the following Bibliography.

Another point to remember is that the intrinsic material properties of metallic implants – e.g., Young's elastic modulus, yield strength, or fatigue strength – are not the sole determinant of implant performance and success; overall implant design is also critical. So while it is certainly true that inadequate attention to intrinsic material properties can doom a device to failure, it is also true that even with the best intrinsic properties, a device may still fail because of faulty structural properties, inappropriate use of the implant, surgical error or overall inadequate mechanical design of the implant. As an illustration of this latter point, Figure I.2.3.4 shows a plastically-deformed 316L stainless steel Harrington spinal distraction rod (*circa* 1980s) that failed *in vivo* because of metallurgical fatigue. A failure analysis of this case concluded that failure did not occur because of any shortcomings in the rod's 316L cold-worked stainless steel; rather, the fracture occurred because of a combination of other factors, namely: (1) the surgeon plastically bent the rod to make it fit a bit better in the patient's back (and the rod company sold rod-bending jigs!), but

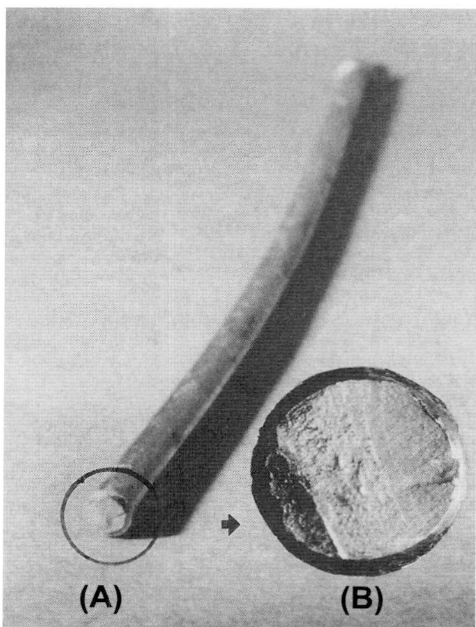

FIGURE I.2.3.4 The smooth part of a 316L stainless steel Harrington spinal distraction rod that fractured by fatigue *in vivo*. Note the bend in the rod (the rod was originally straight) and (insert) the relationship of the crack initiation zone of the fracture surface to the bend. The inserted photo shows the nature of the fatigue fracture surface, which is characterized by a region of "beach marks" and a region of sudden overload failure. (Photo courtesy of Brunski, J. B., Hill, D. C. & Moskowitz, A. (1983). Stresses in a Harrington distraction rod: Their origin and relationship to fatigue fractures *in vivo*. *J. Biomech. Eng.*, **105**, 101–107.)

the bend in the rod increased the bending moment and bending stresses on the rod at the first ratchet junction, which was a known problem area in terms of fractures *in vivo*; (2) the stress concentrations at the fracture site at the ratchet end of the rod were severe enough to significantly increase the tensile stresses at the first ratchet junction, which was the eventual site of the fatigue fracture; and (3) spinal fusion did not occur in the patient, which contributed to relatively persistent loading of the rod over several months post-implantation, thus further predisposing the implant to fatigue fracture. Here the point is that all three of these factors could have been anticipated and considered in the original design of the rod, during which both structural and material properties could have been considered in various stress analyses to forestall fatigue failure. While this example comes from spinal implants, there are many similar examples with the same theme in the orthopedic, oral, and maxillofacial areas. The take-home message here is that implant design is generally a multifaceted problem in which biomaterials selection is only a part of the overall design problem.

BIBLIOGRAPHY

American Society for Testing and Materials (1978). *ASTM Standards for Medical and Surgical Materials and Devices, Authorized Reprint from Annual Book of ASTM Standards*. Philadelphia, PA: ASTM.

Beevers, C. J., & Robinson, J. (1969). Some observations on the influence of oxygen content on the fatigue behavior of alpha-titanium. *J. Less-Common Metals*, **17**, 345–352.

Brunski, J. B., Hill, D. C., & Moskowitz, A. (1983). Stresses in a Harrington distraction rod: Their origin and relationship to fatigue fractures. *in vivo J. Biomech. Eng.*, **105**, 101–107.

Charles, J. A., & Crane, F. A. (1989). *Selection and Use of Engineering Materials* (2nd ed.). Halley Court, Oxford, UK: Butterworth-Heinemann Ltd.

Compte, P. (1984). Metallurgical observations of biomaterials. In J. W. Boretos, & M. Eden (Eds.), *Contemporary Biomaterials* (pp. 66–91). Park Ridge, NJ: Noyes Publishers.

Conrad, H., Doner, M., & de Meester, B. (1973). Critical review: Deformation and fracture. In R. I. Jaffee, & H. M. Burte (Eds.), *Titanium Science and Technology* (Vol. 2,) (pp. 969–1005). New York, NY: Plenum Press.

Davidson, J. A., & Georgette, F. S. (1986). State-of-the-art materials for orthopaedic prosthetic devices. In *Implant Manufacturing and Material Technology*. Itasca, IL: Proc. Soc of Manufacturing Engineering.

Disegi, J. (1990). AO/ASIF Unalloyed Titanium Implant Material. Technical brochure available from Synthes (USA), PO Box 1766, 1690 Russell Road, Paoli, PA, 19301.

Golland, D. I., & Beevers, C. J. (1971a). Some effects of prior deformation and annealing on the fatigue response of alpha titanium. *J. Less-Common Metals*, **23**, 174.

Golland, D. I., & Beevers, C. J. (1971b). The effect of temperature on the fatigue response of alpha-titanium. *Met. Sci. J.*, **5**, 174.

Gomez, M., Mancha, H., Salinas, A., Rodriguez, J. L., Escobedo, J., et al. (1997). Relationship between microstructure and ductility of investment cast ASTM F-75 implant alloy. *J. Biomed. Mater. Res.*, **34**, 157–163.

Hall, E. O. (1951). The deformation and ageing of mild steel: Discussion of results. *Proc. Phys. Soc., (London)*, **64B**, 747–753.

Hamman, G., & Bardos, D. I. (1980). Metallographic quality control of orthopaedic implants. In J. L. McCall, & P. M. French (Eds.), *Metallography as a Quality Control Tool* (pp. 221–245). New York, NY: Plenum Publishers.

Honeycombe, R. W.K. (1968). *The Plastic Deformation of Metals*. New York, NY: St. Martin's Press. p. 234.

Kasemo, B., & Lausmaa, J. (1988). Biomaterials from a surface science perspective. In B. D. Ratner (Ed.), *Surface Characterization of Biomaterials* (pp. 1–12). New York: Elsevier. Chapter 1.

Lipsitt, H. A., & Wang, D. Y. (1961). The effects of interstitial solute atoms on the fatigue limit behavior of titanium. *Trans. AIME*, **221**, 918.

Nanci, A., Wuest, J. D., Peru, L., Brunet, P., Sharma, V., et al. (1998). Chemical modification of titanium surfaces for covalent attachment of biological molecules. *J. Biamed. Mater. Res.*, **40**, 324–335.

Petch, N. J. (1953). The cleavage strength of polycrystals. *J. Iron. Steel. Inst., (London)*, **173**, 25.

Pilliar, R. M., & Weatherly, G. C. (1984). Developments in implant alloys. *CRC. Crit. Rev. Biocompatibility*, **1**(4), 371–403.

Richards Medical Company (1985). *Medical Metals*. Memphis, TN: Richards Medical Company Publication No 3922, Richards Medical Co. [Note: This company is now known as Smith & Nephew Richards, Inc.].

Robinson, S. L., Warren, M. R., & Beevers, C. J. (1969). The influence of internal defects on the fatigue behavior of alpha-titanium. *J. Less-Common Metals*, **19**, 73–82.

Steinemann, S. G., Mansli, P. -A., Szmuckler-Moncler, S., Semlitsch, M., Pohler, O., et al. (1993). Beta-titanium alloy for surgical implants. In F. H. Froes, & I. Caplan (Eds.), *Titanium 92 Science and Technology* (pp. 2689–2698). The Minerals, Metals & Materials Society, Warrendale, PA.

Turner, N. G., & Roberts, W. T. (1968a). Fatigue behavior of titanium. *Trans. Met. Soc. AIME*, **242**, 1223–1230.

Turner, N. G., & Roberts, W. T. (1968b). Dynamic strain ageing in titanium. *J. Less-Common Metals*, **16**, 37.

Wagner, L., et al. (1996). Fatigue life behavior. In S. Lampman, G. M. Davidson, F. Reidenbach, R. L. Boring, & A. Hammel (Eds.), *ASM Handbook Fatigue and Fracture* (Vol. 19) (pp. 837–853). ASM International, Metals Park, Ohio.

Zimmer, USA. (1984a). *Fatigue and Porous Coated Implants*. Warsaw, IN: Zimmer Technical Monograph, Zimmer USA.

Zimmer, USA. (1984b). *Metal Forming Techniques in Orthopaedics*. Warsaw, IN: Zimmer Technical Monograph, Zimmer USA.

Zimmer, USA. (1984c). *Physical and Mechanical Properties of Orthopaedic Alloys*. Warsaw, IN: Zimmer Technical Monograph, Zimmer USA.

Zimmer, USA. (1984d). *Physical Metallurgy of Titanium Alloy*. Warsaw, IN: Zimmer Technical Monograph, Zimmer USA.

A. TITANIUM AND NITINOL (NiTi)

Abhay Pandit[1], *Josep Planell*[2], *and Melba Navarro*[2]
[1]Network of Excellence for Functional Biomaterials, National University of Ireland, Galway
[2]Institute for Bioengineering of Catalonia (IBEC) Baldiri Reixac 10-12, Barcelona

Titanium and its alloys have attracted much attention in the medical implants field, not only because of their unique mechanical properties and resistance to corrosion, they do not initiate an allergenic response and have the probably the best tolerance among metallic biomaterials in the body. Currently only three alloy systems have extensive use in the industry, specifically commercially pure titanium (cpTi), Ti-6Al-4V, and Ti-6A1-7Nb. These alloy systems represent a very significant percentage of the market for titanium biomaterials.

In its elemental form, titanium has a hexagonal closed packed crystal structure (hcp), named α, up to a temperature of 882.5°C, where it transforms into a body centered cubic structure (bcc), named β, up to its high melting point temperature at 1668°C (Collings, 1994). Due to this allotropic change, and depending on the alloying elements, titanium alloys are classified as α, near-α, α + β, metastable β, or stable β, depending on the phase and microstructure stabilized at room temperature (Polmear, 1981). Alloying elements are classified as α-stabilizers, such as Al, O, N, C; β-stabilizers, such as Mo, V, Nb, Ta, Fe, W, Cr, Si, Co, Mn, H; and neutral, such as Zr. α and near α alloys exhibit superior corrosion resistance, but find limited applications as biomaterials due to their low room temperature strength. The α + β alloys exhibit higher strength (with an elastic modulus around 110 GPa) due to the combination of both phases, whereas the β alloys exhibit a unique combination of low elastic modulus (Young's modulus around 60 GPa) and superior corrosion resistance (Bania, 1993; Schutz, 1993). The final properties of the alloy depend on the chemical composition which affects the α and β phase relative proportions that can be modified by thermal treatment, and thermo-mechanical processing conditions.

FABRICATION

Ti and Ti alloys are fabricated into semi-finished bulk products by conventional plastic forming methods, such as forging, rolling, pressing, and drawing. Most Ti materials are processed to avoid an excessive intake of O, N, and H during heating and annealing, since these interstitial elements may induce embrittlement. The hot-working temperature for fabrication has to be carefully selected, since this depends on the alloy composition, and has to be controlled in order to obtain the desired mechanical properties and grain structure. Machining of Ti alloys is to some extent more difficult than that of austenitic stainless steels since they yield under the tool pressure due to their low Young's modulus. For this same reason the tools used in machining tend to stick to the material. Therefore, sharp tools, rigid setups, heavy feeds, slow speeds, and abundant coolant are required (Breme and Biehl, 1998).

Metallic macroporous structures in orthopedic implants are attracting increasing attention because they combine high strength, which is important to avoid deformation or fracture, relatively low stiffness that reduces the stress-shielding effect, high toughness that prevents brittle fracture, and porosity that allows full bone in-growth and stable long-term fixation. Some fabrication methods used to obtain both open-cell or closed-cell porous metallic structures include sintering of Ti powder, plasma spraying, combustion synthesis, replication, and rapid prototyping (Bansiddhi et al., 2008; Ryan et al., 2008).

CORROSION RESISTANCE

To some extent, interest in Ti and its alloys for biomaterials applications stems from their superior corrosion resistance when compared to other metals and alloys. This is attributed to the spontaneous formation of a few nm thick titanium dioxide film that protects the metal from further oxidation (Textor et al., 2001). This behavior depends on alloy composition and corrosive medium (Steinemann et al., 1993): anodic polarization tests indicate that Ti-12Mo-6Zr-2Fe protective oxide has a breakdown resistance equal to that of Ti-6Al-4V, whereas for Ti-5Mo-5Zr-3Al alloy, corrosion current densities are lower than for cpTi (Steinemann et al., 1993; Wang et al., 1993). In addition, electrochemical measurements for Ti-13Nb-13Zr also confirmed the potential of Ti, Nb, and Zr to develop highly protective passive layers, resulting in a better corrosion resistance compared to Ti-6Al-4V (Mishra et al., 1993; Cai et al., 2003).

BIOCOMPATIBILITY AND SURFACE MODIFICATION

The biocompatibility of a metallic alloy has to be understood in terms of both the biocompatibility of the alloy itself, which is closely associated with its corrosion resistance, and the biocompatibility of its by-products as a result of corrosion. It is generally accepted that Ti and its alloys are relatively "bioinert," and exhibit acceptable *in vitro* and *in vivo* responses for the desired application. When implanted in bone, tissue heals in close apposition to the metal, although there maybe a thin fibrous layer separating the metallic implant and the bone. In order to improve osteointegration, bioactivity, biocompatibility, and corrosion resistance, several surface treatments based on chemical and physical modifications have been developed (Liu et al., 2004). Methods include chemical and electrochemical treatment, sol-gel, chemical vapor deposition (CVD), and biochemical modification, while physical methods include machining, grinding, polishing, blasting, thermal and plasma spraying, physical vapor deposition, and ion implantation and deposition, among others. The aim of all these methods is to obtain specific surface topography and chemistry to improve either the mechanical bone/implant interlocking, or the deposition of an apatite or another calcium phosphate bioactive layer to promote direct material–bone bonding, or the activation of particular cell responses by the use of specific biomolecules, such as peptides and proteins (Endo, 1995; Sargeant et al., 2008).

MECHANICAL PROPERTIES

Ti and its alloys are very attractive for use in different biomedical devices, particularly orthopedics, due to their elastic modulus which is closer to that of bone in comparison to other alloys. The mechanical properties of cpTi and some Ti-based alloys are summarized in Table A.1. The Young's moduli of $\alpha + \beta$ and β-type titanium alloys for biomedical applications such as Ti-6Al-4V and Ti-35Nb-5Ta-7Zr are much lower than those of stainless steel and Co-based alloys. However, their Young's moduli are still significantly higher than that of cortical bone. Much effort is currently being carried out to develop new alloys with low elastic modulus that mimic that of bone tissue. Some Ti-based alloys within systems such as Ti–Zr, Ti–Mo, Ti–Ta, Ti–Ta–Zr, Ti–Nb–Hf, Ti–Nb–Zr, Ti–Nb–Sn, Ti–Nb–Ta–Zr, Ti–Fe–Ta, Ti–Mo–Zr–Sn, Ti–Sn–Nb–Ta, Ti–Mo–Zr–Fe, Ti–Mo–Nb–Si, Ti–Mo–Ga, Ti–Mo–Ge, and Ti–Mo–Al have recently been reported as potential biomaterials (Niinomi 1999; Nitta et al., 2001). Many of these alloys are β-type, and contain rather large percentages of Nb, Ta, Zr, Mo, and/or Sn, as these elements are relatively non-toxic and non-allergenic, as reported in previous studies (Ahmed et al., 1995; Steinemann 1980).

Among the alloys under investigation, the Ti–Nb–Ta–Zr system known as Gum metal (Ti-29Nb-13Ta-4.6Zr) (Saito, 2001) is gaining attention in the field because of its "super elastic" properties. This alloy has a Young's modulus of 40 GPa, and an elastic strain of 2.5% (Niinomi et al., 2002). Another alloy, Ti-35Nb-7Zr-5Ta, also known as TiOsteum® β-Ti alloy, is an alloy that has shown enhanced osseointegration, improved ductility, adequate mechanical strength, optimal hot and cold workability, and low elastic modulus (55 GPa)(Jablokov et al., 2005).

NiTi ALLOY

Nitinol (*Nickel-Titanium Naval Ordnance Laboratory*), an equiatomic alloy of nickel and titanium with shape memory properties, was discovered in the early 1960s by Buehler and his co-workers (Buehler and Wang, 1967). Since then, intensive research has been carried out in order to elucidate the mechanics of its basic behavior. Although NiTi had been proposed and used in different clinical applications, such as clamps for orthopedic and traumatological bone fixation, filters to retain emboli in vascular surgery or orthodontic wires, it was only in the mid-1990s that the first widespread commercial NiTi stents made their appearance. At present, NiTi alloys have been introduced in the medical implant field for their superelastic and shape memory alloy properties, despite related drawbacks associated with Ni ionic release (Huang et al., 2003). Ni is allergenic at high concentrations, and thus the use of Ti–Ni shape memory alloys is quite restrictive (Cederbrant et al., 2003). NiTi shape memory alloy also exhibits two different temperature-dependent crystal structures or phases, named martensite, low temperature phase, and austenite, high temperature or parent phase. When martensite NiTi is heated, it transforms into austenite. The temperature at which this phenomenon starts is called austenite start temperature (A_s). The temperature at which this phenomenon is completed is called austenite finish temperature (A_f). When austenite NiTi is cooled, it transforms into martensite. The temperature at which this

TABLE A.1 Titanium Alloys Developed for Orthopedic Applications and Their Mechanical Properties (Adapted from Long and Rack, 1998)

Alloy	Microstructure	Elastic Modulus E (GPa)	Yield Strength YS (MPa)	Ultimate Strength UTS (MPa)
cpTi	α	105	692	785
Ti-6Al-4V	α/β	110	850–900	960–970
Ti-6Al-7Nb	α/β	105	921	1024
Ti-5Al-2.5Fe	Metastable β	110	914	1033
Ti-15Mo-5Zr-3Al	Metastable β	82	771	812
Ti-Zr	Cast α'/β	N/A	N/A	900
Ti-13Nb-13Zr	α'/β	79	900	1030
Ti-15Mo-3Nb-0.30	Metastable β + silicides	82	1020	1020
Ti-35Nb-5Ta-7Zr	Metastable β	55	530	590
Ti-35Nb-5Ta-7Zr-0.40	Metastable β	66	976	1010
Stainless steel 316L	–	205–210	170–750	465–950
Co–Cr–Mo	–	220–230	275–1585	600–1785
Bone	–	10–40		90–140

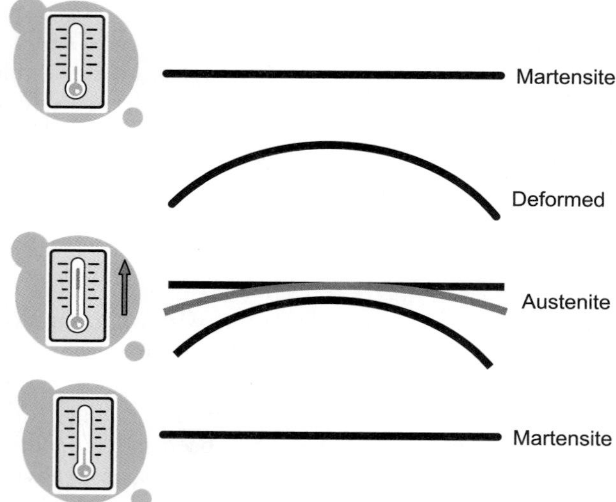

FIGURE A.1 Schematic of the shape memory effect created when the surrounding temperature changes; NiTi alloy modifies its shape to a preprogrammed structure due to the austenite-to-martensite phase transformations.

phenomenon starts is called martensite start temperature (M_s). The temperature at which martensite is again completely reverted is called martensite finish temperature (M_f) (Buehler and Wang, 1967).

A_s, A_f, M_s, and M_f depend on the chemical composition of the alloy, and can be adjusted by introducing small variations in the almost equiatomic composition of these alloys. The unique behavior of NiTi is based on the temperature-dependent austenite-to-martensite phase transformations on an atomic scale, which is also called thermoelastic martensitic transformation, and causes the shape recovery as a result of the need for the crystal lattice structure to accommodate to the minimum energy state for a given temperature (Otsuka and Wayman, 1998). As a consequence, the shape memory effect is created when the surrounding temperature changes; NiTi alloy modifies its shape to a preprogrammed structure (see Figure A.1).

Superelasticity (or pseudoelasticity) refers to the ability of NiTi to return to its original shape upon unloading after substantial deformation. This is based on stress-induced martensite formation. The application of an external stress causes macroscopic deformation which is accommodated by the formation of martensite at temperatures higher than M_s. When the stress is released, the martensite transforms back into austenite, and the specimen returns back to its original shape. Superelastic NiTi can be strained several times, i.e., cycled, without being plastically deformed, reflecting a rubber-like behavior. While most metals sustain plastic deformation by slip or dislocation movement, NiTi sustains large elastic deformations by simply changing the orientation of its crystal structure through the movement of twin boundaries. This phenomenon is only observed over a specific range of temperatures. The highest temperature at which martensite can no longer be stress induced is called M_d. Above M_d, NiTi alloy deforms in the austenite phase by slipping, like ordinary metals. Below A_s, the material is martensitic and does not recover. Superelasticity appears in a temperature range around A_f up to M_d. The ability to recover is maximal when temperature is close to A_f (Duerig and Pelton, 1994).

TABLE A.2 Mechanical Properties of NiTi Alloy (Duerig and Pelton, 1994)

	NiTi	
	Austenitic	**Martensitic**
Ultimate tensile strength (MPa)	800–1500	103–1100
Tensile yield strength (MPa)	100–800	50–300
Modulus of elasticity (GPa)	70–110	21–69
Elongation at failure (%)	1–20	up to 60

The mechanical properties of NiTi depend on the phase which is present at a certain temperature (Buehler and Wang, 1967; Van Humbeeck et al., 1998). Fully austenitic NiTi material generally has suitable properties for surgical implantation. The common mechanical properties of martensitic and austenitic NiTi are presented in Table A.2 (Buehler and Wang, 1967; Honma, 1987; Breme and Biehl, 1998; Van Humbeeck et al., 1998).

NiTi has unique high fatigue and ductile properties, which are also related to its martensitic transformation. These properties are favorable for their use in orthopedic implants. However, wear resistance has been reported to be lower when compared to the Co–Cr–Mo alloys (Sekiguchi, 1987). Moreover, NiTi is a non-magnetic alloy and consequently, magnetic resonance imaging is possible.

SURFACE MODIFICATIONS OF NiTi

It is well known that titanium is well tolerated when implanted in the human body; however, it is also known that certain nickel levels are toxic and may trigger allergic responses. NiTi alloys form a passive titanium oxide layer that improves the metal resistance to corrosion, and acts as a barrier to Ni oxidation and release. The original NiTi surface has shown a tendency towards preferential oxidation of titanium. This behavior is in agreement with the fact that free enthalpy formation of titanium oxides is negative, and exceeds in absolute value the enthalpy formation of nickel oxides by at least two to three times (Chan et al., 1990). As a result, Ni is mostly present in an ionic state on the NiTi surfaces. Various approaches are currently under development to prevent undesirable Ni release and improve NiTi tolerance in the body. Laser and plasma treatment, ion implantation, hydroxyapatite, as well as TiN and TiCN chemical vapor deposits, have been used to modify the surface of NiTi alloys (Averback and Kirk, 1985; Filip et al., 1997; Villermaux et al., 1997). Induction or development of bioactive surfaces by immersion of the material in either simulated body fluid (SBF) or Hank's solution, and protein coatings are also strategies for improving NiTi biocompatibility and reducing Ni release. In addition, techniques such as oxidation in air/oxygen atmosphere, oxidation in boiling water and steam, and chemical/electrochemical passivation have been successfully used (Michiardi et al., 2005).

APPLICATIONS

Superelasticity is a unique property of great interest in the biomedical field, and this is the reason why NiTi alloy has been proposed for different applications, such as orthopedic plates, screws and staples; cardiovascular devices, mainly stents; surgical instruments, such as surgical endoscopic tools and orthodontic devices (Duerig et al., 1999).

Since superelasticity and shape memory effect are highly advantageous, but the risk of metallic allergy is still high, investigations in developing Ni-free superelastic and shape memory titanium alloys containing non-toxic elements are of significant interest to researchers in the field. Some candidates that are currently investigated include the Ti–Nb–Sn system alloy. The martensite transformation temperature (M_s) of this system decreases by increasing the amount of Nb or Sn, and the shape memory effect is recognized when the alloy is deformed below austenite transformation temperature (A_f) similar to the case of Ti–Ni. It has been reported that superelastic strain of 3.5% is obtained at the composition of Ti-18Nb-4Sn (Nitta et al., 2001). Other systems being investigated also include Ti–Mo–Ga, Ti–Mo–Ge or Ti–Mo–Al, Ti–Ta, Ti–Ta–Zr, and Ti–Sc–Mo (Niinomi, 2003).

BIBLIOGRAPHY

Ahmed, T., Long, M., Silvestri, J., Ruiz, C. & Rack, H. J. (1995). A new low modulus, biocompatible titanium alloy. Presented at the 8th World Titanium Conference. Birmingham, UK.

Averback, R., & Kirk, M. (1985). Atomic displacement process in ion-irradiated materials. In L. Rehn, S. Piceaux, & H. Wiedersieln (Eds.), *Surface Alloying by Ion, Electron, and Laser Beams* (pp. 91–132). Metal Park, OH: ASM.

Bania, P. J. (1993). Beta titanium alloys and their role in the titanium industry. In D. Eylon, R. R. Boyer, & D. A. Koss (Eds.), *Titanium Alloys in the 1990s* (pp. 3–14). Warrendale, PA: Metals & Materials Society. The Mineral.

Bansiddhi, A., Sargeant, T. D., Stupp, S. I., & Dunand, D. C. (2008). Porous NiTi for bone implants: A review. *Acta Biomaterialia*, **4**, 773–782.

Breme, J., & Biehl, V. (1998). Metallic biomaterials. In J. Black, & G. Hastings (Eds.), *Handbook of Biomaterial Properties* (pp. 135–213). Chapman & Hall, London, UK.

Buehler, W. J., & Wang, F. E. (1967). A summary of recent research on the nitinol alloys and their potential application in ocean engineering. *Ocean Engineering*, **1**, 105–120.

Cai, Z., Shafer, T., Watanabe, I., Nunn, M. E., & Okabe, T. (2003). Electrochemical characterization of cast titanium alloys. *Biomaterials*, **24**, 213–218.

Cederbrant, K., Anderson, C., Andersson, T., Marcusson-Stahl, M., & Hultman, P. (2003). Cytokine production, lymphocyte proliferation and T-cell receptor V beta expression in primary peripheral blood mononuclear cell cultures from nickel-allergic individuals. *International Archives of Allergy and Immunology*, **132**, 373–379.

Chan, C., Trigwell, C., & Duerig, T. (1990). Oxidation of an NiTi alloy. *Surface and Interface Analysis*, **15**, 349–354.

Collings, E. W. (1994). Classification of Titanium Alloys. In R. Boyer, G. Welsch, & E. W. Collings (Eds.), *Materials Properties Handbook: Titanium Alloys* (pp. 5–11). ASM International.

Duerig, T. W., & Pelton, A. R. (1994). Ti-Ni shape memory alloys. In R. Boyer, G. Welsch, & E. W. Collings (Eds.), *Materials properties Handbook: Titanium Alloys* (pp. 1035–1048). Materials Park, OH: ASM International.

Duerig, T., Pelton, A., & Stöckel, D. (1999). An overview of nitinol medical applications. *Materials Science and Engineering A*, **273–275**, 149–160.

Endo, K. (1995). Chemical modification of metallic implant surface with biofunctional proteins. Molecular structure and biological activity of a modified NiTi alloy surface. *Dental Materials Journal*, **14**(2), 185–188.

Filip, P., Kneissl, K., & Mazanec, K. (1997). Physics of hydroxyapatite plasma coatings on NiTi shape memory materials. *Materials Science and Engineering A*, **234–236**, 422–425.

Honma, T. (1987). Types and mechanical characteristics of shape memory alloys. In H. Funakubo (Ed.), *Shape Memory Alloys* (pp. 61–172). New York, NY: Gordon and Breach Science Publishers.

Huang, H., Chiu, Y.-H., Lee, T.-H., Wu, S.-C., Yang, H.-W., et al. (2003). Ion release from NiTi orthodontic wires in artificial saliva with various acidities. *Biomaterials*, **24**, 3585–3592.

Jablokov, V., Murray, N., Rack, H., & Freese, H. (2005). Influence of oxygen content on the mechanical properties of titanium-35 Niobium-7 Zirconium-5 Tantalum Beta Titanium alloy. *Journal of ASTM International (JAI)*, **2**(8), 12.

Liu, X., Chu, P. K., & Ding, C. (2004). Surface modification of titanium, titanium alloys, and related materials for biomedical applications. *Materials Science and Engineering R*, **47**, 49–121.

Long, M., & Rack, H. J. (1998). Titanium alloys in total joint replacement: A materials science perspective. *Biomaterials*, **19**, 1621–1639.

Michiardi, A., Aparicio, C., Planell, J. A., & Gil, J. (2005). New oxidation treatment of NiTi shape memory alloys to obtain Ni-free surfaces and to improve biocompatibility. *Journal Biomedical Materials Research*, **77B**(2), 249–256.

Mishra, A. K., Davidson, J. A., Kovacs, P., & Poggie, R. A. (1993). Ti-13Nb-13Zr: A new low modulus, high strength, corrosion resistant near-beta alloy for orthopaedic implants. In *Beta Titanium in the 1990s. The Minerals* (pp. 61–72). Warrendale, PA: Metals & Materials Society.

Niinomi, M. (1999). Recent research and development on titanium for biomedical applications in Japan. *JOM*, **51**, 32–34.

Niinomi, M. (2003). Recent research and development in titanium alloys for biomedical applications and healthcare goods. *Science and Technology of Advanced Materials*, **4**, 445–454.

Niinomi, M., Fukui, H., Hattori, T., Kyo, K., & Suzuki, A. (2002). Development of high biocompatible Ti alloy, Ti-29Nb-13Ta-4.6Zr. *Mater. Jpn.*, **41**, 221–223.

Nitta, K., Watanabe, S., Masahashi, N., Hosoda, H., & Hanawa, S. (2001). Ni-free Ti–Nb–Sn shape memory alloys. In M. Niinomi, T. Okabe, E. M. Taleff, D. R. Lesuer, & H. F. Lippard (Eds.), *Structural Biomaterials for the 21st Century* (pp. 25–34). TMS, Warrendale, PA.

Otsuka, K., & Wayman, C. M. (1998). *Mechanism of Shape Memory Effect and Superelasticity, Shape Memory Materials.* Cambridge: Cambridge University Press. pp. 27–48.

Polmear, I. J. (1981). Titanium Alloys. In R. Honeycombe, & P. Hancock (Eds.), *Light Alloys* (pp. 248–315). Arnold, London, UK.

Ryan, G. E., Pandit, A. S., & Apatsidis, D. P. (2008). Porous titanium scaffolds fabricated using a rapid prototyping and powder metallurgy technique. *Biomaterials*, **29**(27), 3625–3635.

Saito, S. (2001). Super elastic titanium alloy [Gum metal]. *Report of 26th Workshop of JSPS 156 Committee*, 7–8.

Sargeant, T., Rao, M., Koh, C.-Y., & Stupp, S. I. (2008). Covalent functionalization of NiTi surfaces with bioactive peptide amphiphile nanofibers. *Biomaterials*, **29**(8), 1085–1098.

Schutz, R. W. (1993). An overview of Beta-Titanium alloy environmental behavior. In D. Eylon, R. Boyer, K. Koss (Eds.), *Beta Titanium Alloys in the 1990s* (pp. 75–91). Warrendale, PA.

Sekiguchi, Y. (1987). Applications of Shape Memory Alloys: Medical Applications. In H. Funakubo (Ed.), *Shape Memory Alloys* (Vol. 1, pp. 226–269). New York, NY: Gordon and Breach Science Publishers.

Steinemann, S. G. (1980). Corrosion of surgical implants: *In vivo* and *in vitro* tests. In G. D. Winter, J. L. Leray, & K. de Groot (Eds.), *Evaluation of Biomaterials* (pp. 1). New York, NY: Wiley.

Steinemann, S. G., Mäusli, P. A., Szmucler-Moncler, S., Semlitsch, M., Pohler, O., et al. (1993). Beta-titanium alloy for surgical implants. In *Titanium'92 Science and Technology. The Minerals* (pp. 2689–2696). Warrendale, PA: Metals & Materials Society.

Textor, M., Sittig, C., Frauchiger, V., Tosatti, S., & Brunette, D. M. (2001). Properties and biological significance of natural oxide films on Ti and its alloys. In D. M. Brunette, P. Tengvall, M. Textor, & P. Thompsen (Eds.), *Titanium in Medicine* (pp. 171–230). Berlin: Springer.

Van Humbeeck, J., Stalmans, R., & Besselink, P. A. (1998). Shape memory alloys. In Jef A. Helsen, & H. Jurgen Breme (Eds.), *Metals as Biomaterials* (pp. 73–101). Chichester, UK: Wiley.

Villermaux, F., Tabrizian, M., Yahia, L., Meunier, M., & Piron, D. (1997). Exeimer laser treatment of NiTi shape memory alloy biomaterials. *Applied Surface Science*, **109/110**, 62–66.

Wang, K., Gustavson, L., & Dumbleton, J. (1993). Low modulus, high strength, biocompatible titanium alloy for medical implants. In *Titanium'92 Science and Technology. The Minerals* (pp. 2697–2704). Warrendale, PA: Metals & Materials Society.

B. STAINLESS STEELS

Phillip J. Andersen
Andersen Metallurgical, LLC

INTRODUCTION

There are a large number of stainless steels available commercially but only a few of these alloys are used as biomaterials for implantable devices. Use of stainless steel implants began in the 1920s and 1930s; stainless steel applications have expanded as new medical procedures and improved materials were developed.

The performance of these alloys depends on their chemical composition and processing history. While 316L is the most common stainless steel, alloys with enhanced corrosion resistance and mechanical properties are available.

METALLURGICAL AND CHEMICAL CONSIDERATIONS

Implantable stainless steels all have a crystal structure known as face centered cubic (models of metallic crystal structures are shown in Figure I.3.1.1, Chapter I.1.3); in steels this crystal structure is called austenite. Austenitic stainless steels are essentially non-magnetic; this is crucial since patients may be exposed to high intensity magnetic fields during MRI examinations. It should be noted that modest amounts of heating and displacement can occur for these alloys under MRI exposure. Under some processing conditions, it is possible to form different phases within the austenitic alloys such as delta ferrite (which is magnetic); the presence of these additional phases is deleterious and is not allowed under the standard specifications that control these materials for implant use.

Table B.1 shows the chemical composition ranges specified for several common implantable stainless steels. Chromium is present in these alloys, primarily to form a protective Cr_2O_3 surface layer (passive film) that is crucial to their corrosion resistance. Since chromium stabilizes the magnetic ferrite (body centered cubic) structure, other alloying elements must be added to stabilize the desired austenite phase. This role is filled primarily by additions of nickel, manganese, and nitrogen. Nitrogen additions also increase mechanical strength and corrosion resistance. Molybdenum additions have a beneficial impact on the pitting corrosion resistance of stainless steels.

Carbon is controlled to be at low levels to prevent the formation of chromium carbides (the "L" in 316L designates low carbon). Formation of these carbides can result in a phenomenon known as sensitization. If enough carbon is available, chromium carbides can form when austenitic stainless steels are held at temperatures in the range of 450°C to 815°C; the time required to form these carbides depends on the temperature. Under these conditions, carbides tend to form preferentially along grain boundaries, leaving the adjacent areas with depleted chromium levels which are prone to attack by corrosion. Slow cooling after welding is a classic process that can cause sensitization if time and cooling rates are not controlled.

Examination of the chemical compositions in Table B.1 reveals some trends that have changed over time in the chemical composition of implantable stainless steels. While 316 and 316L stainless steels have been used successfully for implants for more than 50 years, stronger and more corrosion-resistant alloys have been developed. The alloys known as Rex 734 (also known as Ortron 90) and 22-13-5 were developed in the 1980s. These alloys contain higher levels of chromium, manganese,

TABLE B.1 Compositions of Common Implantable Stainless Steels (Weight Percent)

Alloy	Cr	Ni	Mn	Mo	C	N	Nb	V	Si	Cu	P	S
316L ASTM F138, ISO 5832-1	17–19	13–15	<2 max	2.25–3	<0.030	<0.10	–	–	<0.75	<0.5	<0.025	<0.010
22-13-5 ASTM F1314	20.5–23.5	11.5–13.5	4–6	2–3	<0.030	0.2–0.4	0.1–0.3	0.1–0.3	<0.75	<0.5	<0.025	<0.010
Rex 734, Ortron 90 ASTM F1586 ISO 5832-9	19.5–22	9–11	2–4.25	2–3	<0.08	0.25–0.5	0.25–0.8	–	<0.75	<0.25	<0.25	<0.010
BioDur® 108 ASTM F 2229	19–23	<0.050	21–24	0.5–1.5	<0.08	0.85–1.10	–	–	<0.75	<0.25	<0.03	<0.010

and nitrogen than 316L, leading to improved mechanical properties and corrosion resistance. Recent concern about patients with an allergic response to nickel (see Orthopedic Applications, Chapter II.5.6 for additional discussion of this topic) has led to development of stainless steels which are essentially nickel free. One example is an alloy known as BioDur®108, which uses additions of manganese and nitrogen instead of nickel to stabilize the austenite phase. The nitrogen level in this alloy (1%) is much higher than the level found in the prior implantable stainless steels.

Empirical formulas have been developed to describe the contribution of alloying elements to pitting corrosion resistance of austenitic stainless steels; this is known as the Pitting Resistance Equivalent number or PRE. The formula for PRE depends on several factors, including the amount of nitrogen in the alloy. At low nitrogen levels an appropriate approximation of the PRE is: PRE = %Cr + 3.3 × %Mo + 16 × %N; at nitrogen levels nearing 1% a more appropriate approximation is: PRE = %Cr + 3.2 × %Mo + 8 × %N (Gebeau and Brown, 2001). Higher PRE numbers indicate improved resistance to pitting corrosion.

MECHANICAL PROPERTIES

The mechanical properties of metals and alloys depend on their chemical composition and processing history. To understand how these factors influence mechanical properties it is important to consider the mechanisms involved when permanent deformation takes place (plastic deformation). Plastic deformation occurs primarily by movement of dislocations within the material ("slip"). Dislocations are defects or disturbed regions in the crystal lattice. If the resistance to dislocation motion is increased, strength properties such as yield strength increase, while deformation capacity (ductility) decreases.

Chemical composition also influences strength by a process known as solid solution strengthening. In stainless steels the metallic alloying elements (Cr, Ni, etc.) replace iron atoms at random locations within the crystal structure. Since the various atoms are not the same size, additions of alloying elements lead to distortion of the crystal lattice. This distortion makes deformation of the material more difficult (dislocation movements), thus increasing strength (and generally decreasing ductility). Smaller atoms, such as N, fit within the gaps between the larger metallic atoms; for this reason they are often referred to as interstitials. Additions of these atoms lead to large strength increases due to the interactions between the solute atom and dislocations.

Another strengthening mechanism available for these materials involves work hardening (also known as "cold-working"). Processes commonly used to change the mechanical properties of other alloys such as aging reactions during heat treatment are not applicable to austenitic stainless steels.

Work hardening occurs due to the increased difficulty of dislocation motion, as more and more dislocations interact within the material. As the material is deformed, increasing quantities of dislocations develop within the grains of the alloy. If the deformation process takes place at high temperatures (which is typical of the large-scale processes used to produce metal products), or if the material is heated above its recrystallization temperature after deformation, the dislocations are removed by formation of new, annealed grains (recrystallization). Deformation at lower temperatures does not eliminate dislocations. As the density of dislocations within the material increases, they begin to interact making further dislocation motion more difficult. This leads to increases in strength and reductions in ductility. More information on solid solution strengthening and work hardening can be found in a number of materials science textbooks (Hosford, 2005; Meyers & Chawla, 2008).

Evidence of deformation within the grains of a moderately cold-worked 316L sample is shown in Figure B.1, which compares the structure of annealed and ~30% cold-worked 316L.The amount of cold-work is routinely specified by the percentage reduction of area; this is calculated by $(A_i–A_f/A_i) \times 100$, where A_i is the initial cross-sectional area of the product and A_f is the final cross-sectional area. As the amount of cold-work

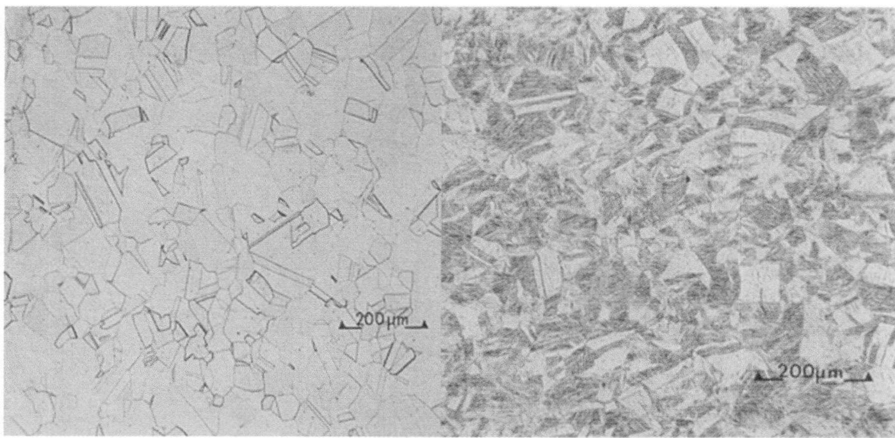

FIGURE B.1 Microstructures of annealed 316L (left), and ~30% cold-worked 316L (right). Deformation within many of the grains of the cold-worked material is evident.

TABLE B.2 Approximate Mechanical Properties of Stainless Steels

Alloy	Material Condition	Ultimate Tensile Strength (MPa)	Yield Strength (MPa)	%Elongation	!0⁷ Cycle Endurance Limit (MPa)	Reference
316L	Annealed	550	240	55	180	Shetty & Ottersberg, 1995
316L	30% Cold-worked	896	827	20	380	Shetty & Ottersberg, 1995
316L	60% Cold-worked	1240	1000	12	450	Shetty & Ottersberg, 1995
Rex 734	Hot forged	1140–1230	1050–1179	15–19	585	Windler & Steger, 2003
22-13-5	Annealed	965	760	35	380	Shetty & Ottersberg, 1995
22-13-5	30% Cold-worked	1240	1170	15	530	Shetty & Ottersberg, 1995
22-13-5	60% Cold-worked	1585	1480	9	670	Shetty & Ottersberg, 1995
BioDur®108	Annealed	827–930	517–605	30–50	380	ASTM F 2229 and Technical Data Sheet BioDur®108
BioDur®108	35% Cold-worked	1580	1350	15		Technical Data Sheet BioDur®108
BioDur®108	65% Cold-worked	2000	1790	5		Technical Data Sheet BioDur®108

increases, strength parameters (yield strength and ultimate strength) increase (along with hardness), while the ductility decreases. This behavior gives designers the ability to specify a wide range of mechanical properties by proper selection of the alloy composition, processing temperature, and the amount of cold-work. Table B.2 lists approximate tensile and fatigue strength properties for some implantable stainless steels produced under different conditions. The influence of chemical composition and amount of cold-work on mechanical properties is evident in these data.

It should be noted that there are limits to the uniformity of the cold-working response which is a function of the cross-sectional area of the material. In practical terms, this means that a large cross-section may have a gradient in the amount of cold-work; thus the outer regions of the material may be more heavily worked than the interior. Another effect of cold-working is that heavily worked stainless steels, such as 304 and 305, can become magnetic due to a stress-induced transformation of crystal structure from austenite (non-magnetic) to martensite (magnetic). This can result in negative clinical results if fragments of these alloys, which are routinely used as guide wires, break off during a procedure (FDA Public Health Notification, 2008). It is also important to realize that elevated temperature processes such as welding will result in softening of a cold-worked material, and reduction of mechanical properties to annealed levels.

A high level of fatigue strength is necessary for many stainless steel medical devices (see Chapter I.1.3 of this volume for more information on fatigue). The fatigue properties of metallic materials depend on a number of factors including: composition; grain size; processing history; surface finish; and test method. The fatigue data shown in Table B.2 provide approximate values; these data should not be used for implant design purposes since the properties of specific devices are strongly influenced by manufacturing processes, product design features, the inherent variation from batch to batch of raw material, and variability in the processes used to produce the device.

CORROSION BEHAVIOR

With the exception of noble metals such as gold or platinum, the common implantable metals and alloys rely on the formation of protective oxide films to control corrosion of the material to acceptable levels. These oxides are commonly referred to as passive films. They are very thin (typically less than 100 microns), dense, and adhere strongly to the underlying substrate. They limit transport of metallic ions to the implant surface. The alloys that contain substantial amounts of chromium (austenitic stainless steels and implantable cobalt base alloys) form a Cr_2O_3 layer while alloys rich in titanium (titanium alloys and Nitinol) form a TiO_2 layer. More detailed information on passive layers is presented in Chapter II.5.6 Orthopedic Applications. Chapter II.4.4 Degradative Effects of the Biological Environment on Metals and Ceramics includes a great deal of information on corrosion of implant metals.

In general terms, the austenitic stainless steels are not considered to be quite as corrosion resistant as either cobalt–chromium alloys or titanium alloys. The more recent alloys (22-13-5, Rex 734, and BioDur®108) exhibit improved corrosion resistance when compared to 316L, due to their higher levels of Cr and N. One measure of relative corrosion resistance is the pitting potential, as determined in an anodic polarization test; pitting potential varies somewhat depending on the testing conditions, but one author found it was 346 mV for 316L, 1030 mV for 22-13-5, and 1120 mV for BioDur®108 (Zardiackas et al., 2003). Another factor which influences the corrosion of stainless steels is the presence of foreign particles, known as inclusions. These are typically oxide particles such as alumina or silicates which are formed during the initial melting of the alloy and become trapped within the material during subsequent processing. Since these inclusions have different corrosion behavior than the alloy, they can act as corrosion initiation sites if they are found at the surface of a component. Careful control of alloy melting practice and subsequent processing is required to minimize the number of inclusions within these alloys. The ASTM specifications for implantable stainless steels include sections that limit the number and size of inclusions.

SUMMARY

A limited group of austenitic stainless steels are routinely implanted because they offer sufficient corrosion resistance and mechanical strength for long-term applications in the human body. Although 316L is one of the oldest and most common alloys in use, more recent developments have led to stainless steels with improved mechanical strength and corrosion resistance. The corrosion resistance and mechanical properties of these materials are determined by chemical composition, and the processing used to produce the component.

The ability to choose from a range of properties is advantageous. A soft, ductile, annealed condition can be specified for a component that must be contoured in the operating room, while a higher strength, less ductile, work-hardened material may be used for devices that must withstand higher fatigue loads. A long history of success in an application may also be a factor in the use of stainless steels in some devices.

BIBLIOGRAPHY

BioDur. (2008). *Technical Data Sheet BioDur®108*. http://www.cartech.com/techcenter.aspx?id=1692. Date Accessed 22/1/2008.

FDA. (2008). *FDA Public Health Notification: Unretrieved Device Fragments*. http://www.fda.gov/cdrh/safety/011508-udf.html. Date Accessed 22/1/2008.

Gebeau, R. C., & Brown, R. S. (2001). Biomedical implant alloy. *Advanced Materials & Processes*, 46–48. Sept.

Hosford, W. F. (2005). *Mechanical Behavior of Materials*. Cambridge University Press.

Meyers, M. A., & Chawla, K. K. (2008). *Mechanical Behavior of Materials*. Cambridge University Press.

Shetty, R. H., & Ottersberg, W. H. (1995). Metals in Orthopedic Surgery. In *Encyclopedic Handbook of Biomaterials and Bioengineering – Part B Applications* (pp. 509–540). New York, NY: Marcel Decker.

Windler, M., & Steger, R. (2003). Mechanical and Corrosion Properties of Forged Hip Stems made of High Nitrogen Stainless Steel. In G. L. Winters, & M. J. Nutt (Eds.), *Stainless Steels for Medical and Surgical Applications ASTM STP 1438* (pp. 39–49). ASTM International.

Zardiackas, L. D., et al. (2003). Comparison of Anodic Polarization and Galvanic Corrosion of a Low-Nickel Stainless Steel to 316LS and 22Cr-13Ni-5Mn Stainless Steels. In G. L. Winters, & M. J. Nutt (Eds.), *Stainless Steels for Medical and Surgical Applications ASTM STP 1438* (pp. 107–118). ASTM International.

CHAPTER I.2.4 CERAMICS, GLASSES, AND GLASS-CERAMICS: BASIC PRINCIPLES

Larry L. Hench[1], *Serena M. Best*[2]
[1]Imperial College, London, UK
[2]University of Cambridge, Cambridge, UK

Ceramics, glasses, and glass-ceramics include a broad range of inorganic/nonmetallic compositions. In the medical industry, these materials have been essential for eyeglasses, diagnostic instruments, chemical ware, thermometers, tissue culture flasks, and fiber optics for endoscopy. Insoluble porous glasses have been used as carriers for enzymes, antibodies, and antigens, offering the advantages of resistance to microbial attack, pH changes, solvent conditions, temperature, and packing under high pressure required for rapid flow (Hench and Ethridge, 1982). Insoluble glasses have also been developed as a microinjectable delivery system for radioactive isotopes for *in situ* treatment of tumors. The glass microspheres go to the site of the tumor by way of the blood supply, and the radiation kills the cancer cells with very little damage to the other tissues, saving thousands of patients (Hench et al., 2010). Ceramics are also widely used in dentistry as restorative materials, such as in gold–porcelain crowns, glass-filled ionomer cements, and dentures. These dental ceramics are discussed by Phillips (1991). Glass-ceramics are also widely used for dental restorations, including inlays, onlays, crowns, and multi-unit bridges. This is one of the most rapidly growing applications of this class of biomaterials (Hench et al., 2010).

This chapter focuses on ceramics, glasses, and glass-ceramics used as implants. Although dozens of compositions have been explored in the past, relatively few have achieved clinical success. This chapter examines differences in processing and structure, describes the chemical and microstructural basis for their differences in physical properties, and relates properties and tissue response to particular clinical applications. For a historical review of these biomaterials, see Hulbert et al. (1987). For a summary of clinical applications see Hench (1998) and Hench et al. (2004).

TYPES OF BIOCERAMICS: TISSUE ATTACHMENT

It is essential to recognize that no one material is suitable for all biomaterial applications. As a class of biomaterials, ceramics, glasses, and glass-ceramics are generally used to repair or replace skeletal hard connective tissues. Their success depends upon: (1) achieving a stable attachment to connective tissue when used as bulk implants; or (2) stimulating repair and regeneration of bone when used as particulates for bone grafting.

The mechanism of tissue attachment is directly related to the type of tissue response at the implant–tissue interface. No material implanted in living tissue is inert because all materials elicit a response from living tissues. There are four types of tissue response (Table I.2.4.1) and four different means of attaching prostheses to the skeletal system (Table I.2.4.2).

A comparison of the relative chemical activity of the different types of bioceramics, glasses, and glass-ceramics is shown in Figure I.2.4.1. The relative reactivity that is shown in Figure I.2.4.1A correlates very closely with the rate of formation of an interfacial bond of ceramic, glass or glass-ceramic implants with bone (Figure I.2.4.1B). Figure I.2.4.1B is discussed in more detail in the section on bioactive glasses and glass-ceramics in this chapter.

The relative level of reactivity of an implant influences the thickness of the interfacial zone or layer between the material and tissue. Analyses of implant material failures often show failure originating at the biomaterial–tissue interface. When biomaterials are nearly inert (type 1 in Table I.2.4.2 and Figure I.2.4.1) and the interface is not chemically or biologically bonded, there is relative movement and progressive development of a fibrous capsule in soft and hard tissues. The presence of movement at the biomaterial–tissue interface eventually leads to deterioration in function of the implant or the tissue at the interface, or both. Wear particles can accelerate the deterioration of the tissue–implant interface. The thickness of the non-adherent capsule varies, depending upon both material (Figure I.2.4.2) and extent of relative motion.

The fibrous tissue at the interface of dense Al_2O_3 (alumina) implants is very thin. Consequently, if alumina devices are implanted with a very tight mechanical fit and are loaded primarily in compression, they are very successful. In contrast, if a type 1 nearly inert implant is loaded so that interfacial movement can occur, the fibrous capsule can become several hundred micrometers thick, and the implant can loosen very quickly.

The mechanism behind the use of nearly inert microporous materials (type 2 in Table I.2.4.2 and Figure I.2.4.1) is the ingrowth of tissue into pores on the surface or throughout the implant. The increased interfacial area between the implant and the tissues results in an increased resistance to movement of the device in the tissue. The interface is established by the living tissue in the pores. Consequently, this method of attachment is often termed "biological fixation." It is capable of withstanding more complex stress states than type 1 implants

TABLE I.2.4.1 Types of Implant–Tissue Response
If the material is toxic, the surrounding tissue dies.
If the material is nontoxic and biologically inactive (nearly inert), a fibrous tissue of variable thickness forms.
If the material is nontoxic and biologically active (bioactive), an interfacial bond forms.
If the material is nontoxic and dissolves, the surrounding tissue replaces it.

TABLE I.2.4.2 Types of Bioceramic Tissue Attachment and Their Classification

Type of attachment	Example
1. Dense, nonporous, nearly inert ceramics attach by bone growth into surface irregularities by cementing the device into the tissues or by press-fitting into a defect (termed "morphological fixation").	Al_2O_3 (single crystal and polycrystalline)
2. For porous inert implants, bone ingrowth occurs that mechanically attaches the bone to the material (termed "biological fixation").	Al_2O_3 (polycrystalline) Hydroxyapatite-coated porous metals
3. Dense, nonporous surface-reactive ceramics, glasses, and glass-ceramics attach directly by chemical bonding with the bone (termed "bioactive fixation").	Bioactive glasses Bioactive glass-ceramics Hydroxyapatite
4. Dense, nonporous (or porous) resorbable ceramics are designed to be slowly replaced by bone.	Calcium sulfate (Plaster of Paris) Tricalcium phosphate Calcium–phosphate salts

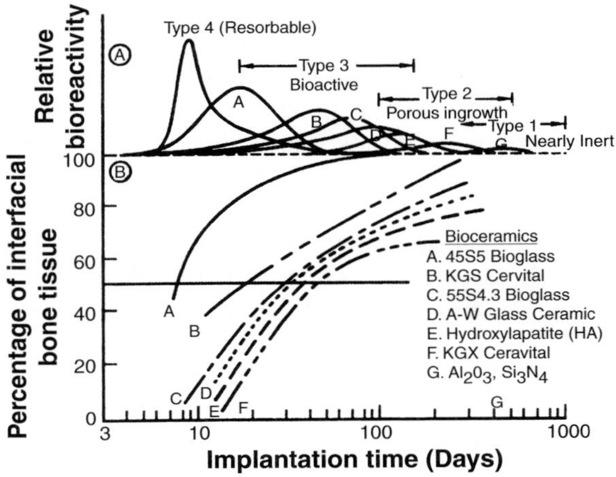

FIGURE I.2.4.1 Bioactivity spectra for various bioceramic implants: (A) Relative rate of bioreactivity; (B) Time-dependence of formation of bone bonding at an implant interface.

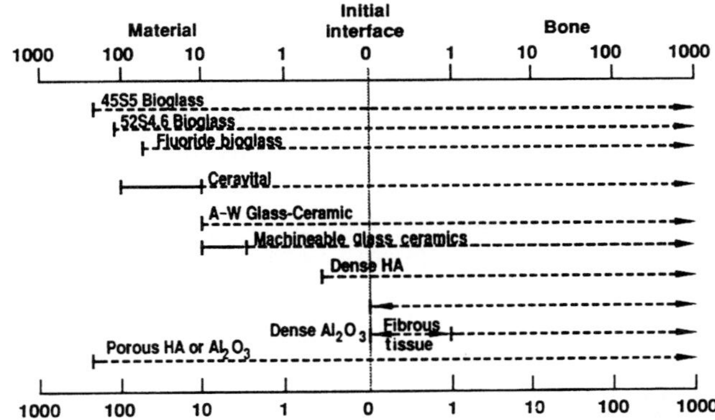

FIGURE I.2.4.2 Comparison of interfacial thickness (μm) of reaction layer of bioactive implants of fibrous tissue of inactive bioceramics in bone.

with "morphological fixation." The limitation with type 2 porous implants, however, is that for the tissue to remain viable and healthy, it is necessary for the pores to be greater than 50 to 150 μm (Figure I.2.4.2). The large interfacial area required for porosity is due to the need to provide a blood supply to the ingrown connective tissue (vascular tissue does not appear in pore sizes less than 100 μm). Also, if micromovement occurs at the interface of a porous implant and tissue is damaged, the blood supply may be cut off, the tissues will die, inflammation will ensue, and the interfacial stability will be destroyed. When the material is a porous metal, the large increase in surface area can provide a focus for corrosion of the implant and loss of metal ions into the tissues. This can be mediated by using a bioactive ceramic material such as hydroxyapatite (HA) as a coating on the metal. The fraction of large porosity in any material also degrades the strength of the material proportional to the volume fraction of porosity. Consequently, this approach to solving interfacial stability works best when materials are used as coatings or as unloaded space fillers in tissues.

Resorbable biomaterials (type 4 in Table I.2.4.2 and Figure I.2.4.1) are designed to degrade gradually over a period of time, and to be replaced by the natural host tissue. This leads to a very thin or nonexistent interfacial thickness (Figure I.2.4.2). This is the optimal biomaterial solution, if the requirements of strength and short-term performance can be met, since natural tissues can repair and replace themselves throughout life. Thus, resorbable biomaterials are based on biological principles of repair that have evolved over millions of years. Complications in the development of resorbable bioceramics are: (1) maintenance of strength and the stability of the interface during the degradation period and replacement by the natural host tissue; and (2) matching resorption rates to the repair rates of body tissues (Figure I.2.4.1A) (e.g., some materials dissolve too rapidly and some too slowly). Because large quantities of material may be replaced, it is also essential that a resorbable biomaterial consist only of metabolically acceptable substances. This criterion imposes considerable limitations on the compositional design of resorbable biomaterials. Successful examples of resorbable polymers include poly(lactic acid) and poly(glycolic acid) used for sutures, which are metabolized to CO_2 and H_2O and therefore are able to function for an appropriate time and then dissolve and disappear. Porous or particulate calcium phosphate ceramic materials such as tricalcium phosphate (TCP) have proved successful for resorbable hard tissue replacements when low loads are applied to the material.

Another approach to solving problems of interfacial attachment is the use of bioactive materials (type 3 in Table I.2.4.2 and Figure I.2.4.1). Bioactive materials are intermediate between resorbable and bioinert. A bioactive material is one that elicits a specific biological response at the interface of the material, resulting in the formation of a bond between the tissues and the material. This concept has now been expanded to include a large number of bioactive materials with a wide range of rates of bonding and thicknesses of interfacial bonding layers (Figures I.2.4.1 and I.2.4.2). They include bioactive glasses such as 45S5 Bioglass; bioactive glass-ceramics such as A-W glass-ceramic; dense HA and bioactive composites such as HA-polyethylene. All of these materials form an interfacial bond with adjacent tissue. However, the time dependence of bonding, the strength of bond, the mechanism of bonding, and the thickness of the bonding zone differ for the various materials. Relatively small changes in the composition of a biomaterial can dramatically affect whether it is bioinert, resorbable or bioactive. These compositional effects on surface reactions are discussed in the section on bioactive glasses and glass-ceramics.

CHARACTERISTICS AND PROCESSING OF BIOCERAMICS

The types of implants listed in Table I.2.4.2 are made using different processing methods. The characteristics and properties of the materials, summarized in Table I.2.4.3, differ greatly, depending upon the processing method used.

The primary methods of processing ceramics, glasses, and glass-ceramics are summarized in Figure I.2.4.3. These methods yield five categories of microstructures:

1. Glass;
2. Cast or plasma-sprayed polycrystalline ceramic;
3. Liquid-phase sintered (vitrified) ceramic;
4. Solid-state sintered ceramic;
5. Polycrystalline glass-ceramic.

Differences in the microstructures of the five categories are primarily a result of the different thermal processing steps required to produce them. Alumina and calcium phosphate bioceramics are made by fabricating the product from fine grained particulate solids. For example, a desired shape may be obtained by mixing the particulates with water and an organic binder, then pressing them in a mold. This is termed "forming." The formed piece is called green ware. Subsequently, the temperature is raised to evaporate the water (i.e., drying) and the binder is burned out, resulting in bisque ware. At a very much higher temperature, the part is densified during firing. After cooling to ambient temperature, one or more finishing steps may be applied, such as polishing. Porous ceramics are produced by adding a second phase that decomposes prior to densification, leaving behind holes or pores (Schors and Holmes, 1993) or transforming natural porous organisms, such as coral, to porous HA by hydrothermal processing (Roy and Linnehan, 1974).

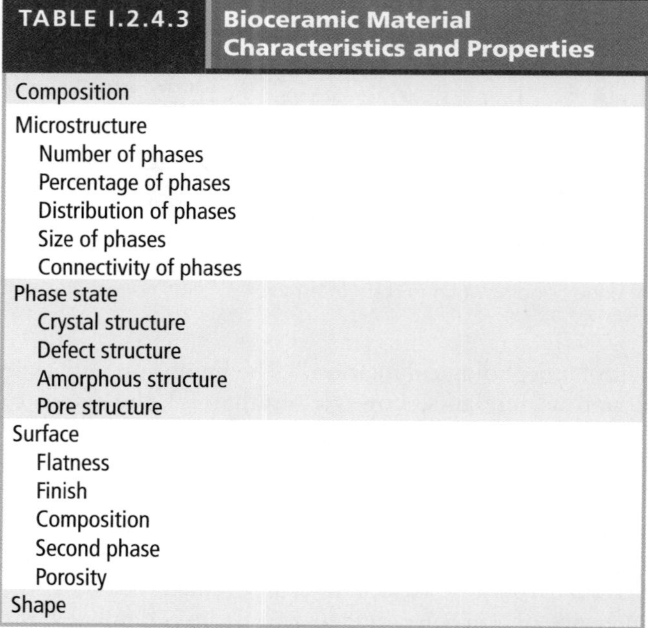

TABLE I.2.4.3 Bioceramic Material Characteristics and Properties
Composition
Microstructure
Number of phases
Percentage of phases
Distribution of phases
Size of phases
Connectivity of phases
Phase state
Crystal structure
Defect structure
Amorphous structure
Pore structure
Surface
Flatness
Finish
Composition
Second phase
Porosity
Shape

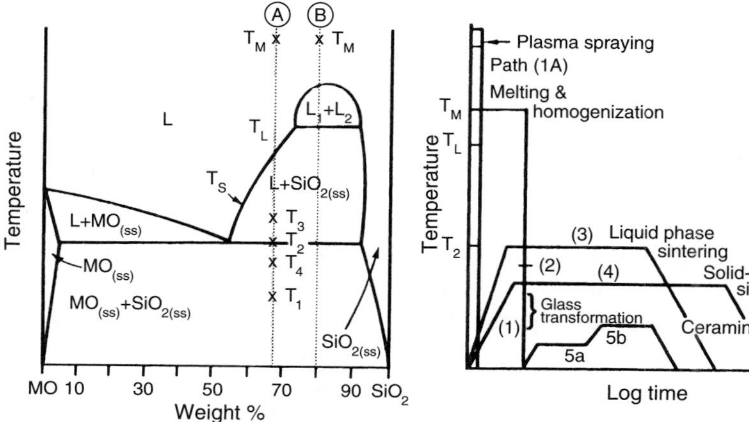

FIGURE I.2.4.3 Relation of thermal processing schedules of various bioceramics to equilibrium phase diagram.

The interrelation between microstructure and thermal processing of various bioceramics is shown in Figure I.2.4.3, which is a binary phase diagram consisting of a network-forming oxide such as SiO_2 (silica), and some arbitrary network modifier oxide (MO) such as CaO. When a powdered mixture of MO and SiO_2 is heated to the melting temperature T_m, the entire mass will become liquid (L). The liquid will become homogeneous when held at this temperature for a sufficient length of time. When the liquid is cast (paths 1B, 2, 5), forming the shape of the object during the casting, either a glass or a polycrystalline microstructure will result. Plasma spray coating follows path 1A. However, a network-forming oxide is not necessary to produce plasma-sprayed coatings such as hydroxyapatites, which are polycrystalline (Lacefield, 1993).

If the starting composition contains a sufficient quantity of network former (SiO_2), and the casting rate is sufficiently slow, a glass will result (path 1B). The viscosity of the melt increases greatly as it is cooled, until at approximately T_1, the glass transition point, the material is transformed into a solid.

If either of these conditions is not met, a polycrystalline microstructure will result. The crystals begin growing at T_L and complete growth at T_2. The final material consists of the equilibrium crystalline phases predicted by the phase diagram. This type of cast object is not often used commercially, because the large shrinkage cavity and large grains produced during cooling make the material weak and subject to environmental attack.

If the MO and SiO_2 powders are first formed into the shape of the desired object and fired at a temperature T_3, a liquid-phase sintered structure will result (path 3). Before firing, the composition will contain approximately 10–40% porosity, depending upon the forming process used. A liquid will be formed first at grain boundaries at the eutectic temperature, T_2. The liquid will penetrate between the grains, filling the pores, and will draw the grains together by capillary attraction. These effects decrease the volume of the powdered compact. Since the mass remains unchanged and is only rearranged, an increased density results. Should the compact be heated for a sufficient length of time, the liquid content can be predicted from the phase equilibrium diagram. However, in most ceramic processes, liquid formation does not usually proceed to equilibrium owing to the slowness of the reaction and the expense of long-term heat treatments.

The microstructure resulting from liquid-phase sintering (or vitrification as it is commonly called) will consist of small grains from the original powder compact surrounded by a liquid phase. As the compact is cooled from T_3 to T_2, the liquid phase will crystallize into a fine-grained matrix surrounding the original grains. If the liquid contains a sufficient concentration of network formers, it can be quenched into a glassy matrix surrounding the original grains.

A powder compact can be densified without the presence of a liquid phase by a process called solid-state sintering. This is the process usually used for manufacturing alumina and dense HA bioceramics. Under the driving force of surface energy gradients, atoms diffuse to areas of contact between particles. The material may be transported by grain boundary diffusion, volume diffusion, creep, or any combination of these, depending upon the temperature or material involved. Because long-range migration of atoms is necessary, sintering temperatures are usually in excess of one-half of the melting point of the material: $T > T_L/2$ (path 4).

The atoms move to fill up the pores and open channels between the grains of the powder. As the pores and open channels are closed during heat treatment, the crystals become tightly bonded together, and the density, strength, and fatigue resistance of the object improve greatly. The microstructure of a material that is prepared by sintering consists of crystals bonded together by ionic-covalent bonds with a very small amount of remaining porosity.

The relative rate of densification during solid-state sintering is slower than that of liquid-phase sintering, because material transport is slower in a solid than in a liquid. However, it is possible to solid-state sinter individual component materials such as pure oxides, since liquid development is not necessary. Consequently, when high purity and uniform fine-grained microstructures are required (e.g., for bioceramics) solid-state sintering is essential.

The fifth class of microstructures is called glass-ceramics, because the object starts as a glass and ends up as a polycrystalline ceramic. This is accomplished by first quenching a melt to form the glass object. The glass is transformed into a glass-ceramic in two steps. First, the glass is heat treated at a temperature range of 500–700°C (path 5a) to produce a large concentration of nuclei from which crystals can grow. When sufficient nuclei are present to ensure that a fine-grained structure will be obtained, the temperature of the object is raised to a range of 600–900°C, which promotes crystal growth (path 5b). Crystals grow from the nuclei until they impinge and up to 100% crystallization is achieved. The resulting microstructure is nonporous and contains fine-grained, randomly oriented crystals that may or may not correspond to the equilibrium crystal phases predicted by the phase diagram. There may also be a residual glassy matrix, depending on the duration of the heat treatment, called ceraming. When phase separation occurs (indicated by composition B in Figure I.2.4.3), a nonporous, phase-separated microstructure can be produced that consists of glass phases suspended in a glass matrix. Crystallization of phase-separated glasses results in very complex microstructures, as reviewed in Höland and Beal (2005). Glass-ceramics can also be made by pressing powders and a grain boundary glassy phase (Kokubo, 1993). For additional details on the processing of ceramics, see Reed (1988) or Onoda and Hench (1978), and for processing of glass-ceramics, see Höland and Beal (2005).

NEARLY INERT CRYSTALLINE CERAMICS

High-density, high-purity (>99.5%) alumina is used in the articulating surfaces of total joint prostheses because of its excellent corrosion resistance, good biocompatibility, high wear resistance, and high strength (Hulbert et al., 1987; Christel et al., 1988; Hulbert, 1993; Miller et al., 1996). Most Al_2O_3 devices are very fine-grained polycrystalline α-Al_2O_3 ceramics produced by pressing and sintering at $T = 1600$–$1700°C$. A very small amount of MgO (< 0.5%) is used to aid sintering and limit grain growth during sintering.

Strength, fatigue resistance, and fracture toughness of polycrystalline α-Al_2O_3 ceramics are a function of grain size and percentage of sintering aid (i.e., purity). Al_2O_3 ceramics with an average grain size of <4 μm and >99.7% purity exhibit good flexural strength and excellent compressive strength. These and other physical properties are summarized in Table I.2.4.4, along with the International Standards Organization (ISO) requirements for alumina implants. Extensive testing has shown that alumina implants that meet or exceed ISO standards have excellent resistance to dynamic and impact fatigue, and also resist subcritical crack growth (Dörre and Dawihl, 1980). An increase in average grain size to >17 μm can decrease mechanical properties by about 20%. High concentrations of sintering aids must be avoided, because they remain in the grain boundaries and degrade fatigue resistance.

Methods exist for lifetime predictions and statistical design of proof tests for loadbearing ceramics. Applications of these techniques show that load limits for specific prostheses can be set for an Al_2O_3 device based upon the flexural strength of the material and its use environment (Ritter et al., 1979). Loadbearing lifetimes of 30 years at 12,000 N loads have been predicted (Christel et al., 1988). Results from aging and fatigue studies show that it is essential that Al_2O_3 implants be produced at the highest possible standards of quality assurance, especially as they are often used as orthopedic prostheses in younger patients. Alumina has been used in orthopedic surgery for >30 years (Miller et al., 1996). Its use has been motivated largely by its exceptionally low coefficient of friction and low wear rates.

The superb tribiologic properties (friction and wear) of alumina occur only when the grains are very small (<4 μm) and have a very narrow size distribution. These conditions lead to very low surface roughness values Ra <0.02 μm (Table I.2.4.4). If large grains are present, they can pull out and lead to very rapid wear of bearing surfaces owing to local dry friction.

Alumina on alumina loadbearing, wearing surfaces, such as in hip prostheses, must have a very high degree of sphericity, which is produced by grinding and polishing the two mating surfaces together. For example, the alumina ball and socket in a hip prosthesis are polished together and used as a pair. The long-term coefficient of friction of an alumina–alumina joint decreases with time and approaches the values of a normal joint. This leads to wear on alumina-articulating surfaces being nearly 10 times lower than metal–polyethylene surfaces (Figure I.2.4.4), and eliminates formation of polyethylene wear particles that are associated with aspetic loosening of total joint prostheses, as discussed elsewhere in this textbook.

Low wear rates have led to widespread use in Europe of alumina noncemented cups press-fitted into the acetabulum of the hip. The cups are stabilized by the growth of bone into grooves or around pegs. The mating femoral ball surface is also made of alumina, which is bonded to a metallic stem. Long-term results in general are good, especially for younger patients. However, Christel et al. (1988) caution that stress shielding, owing to the high elastic modulus of alumina, may be responsible for cancellous bone atrophy and loosening of the acetabular cup in old patients with senile osteoporosis or rheumatoid arthritis. Consequently, it is essential that the age of the patient, nature of the disease of the joint, and biomechanics of the repair be considered carefully before any prosthesis is used, including alumina ceramics.

TABLE I.2.4.4 Physical Characteristics of Al_2O_3 Bioceramics

	High Alumina Ceramics	ISO Standard 6474
Alumina content (% by weight)	>99.8	≥99.50
Density (g/cm3)	>3.93	≥3.90
Average grain size (μm)	3–6	<7
Ra (μm)[a]	0.02	
Hardness(Vickers hardness number, VHN)	2300	>2000
Compressive strength (MPa)	4500	
Bending strength (MPa) (after testing in Ringer's solution)	550	400
Young's modulus (GPa)	380	
Fracture toughness (K_1C) (MPa^{12})	5–6	
Slow crack growth	10–52	

[a]Surface roughness value.

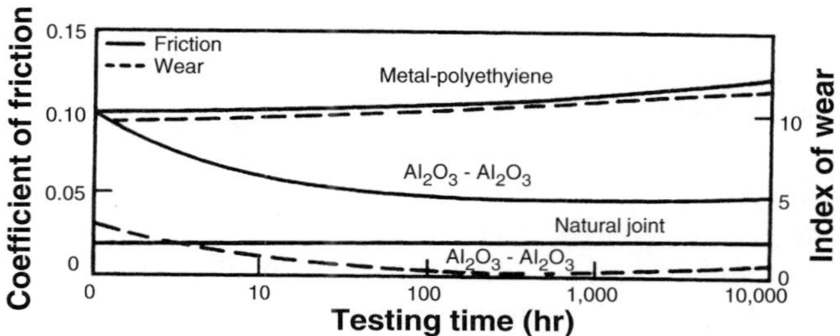

FIGURE I.2.4.4 Time dependence of coefficient of friction and wear of alumina–alumina versus metal–polyethylene hip joint (*in vitro* testing).

Yttria-stabilised zirconia was also used in a similar way to alumina for articulating applications in hip and knee replacement (Hench and Wilson, 1993). However a series of implant failures around the year 2000, resulted in the withdrawal of zirconia for these applications. Today, there is significant interest in the use of zirconia toughened alumina implants due to the potential to enhance strength and toughness properties over those of alumina. Zirconia is used widely for dental applications due to aesthetic and mechanical benefits the material offers.

Other clinical applications of alumina prostheses reviewed by Hulbert et al. (1987) include knee prostheses; bone screws; alveolar ridge and maxillofacial reconstruction; ossicular bone substitutes; keratoprostheses (corneal replacements); segmental bone replacements; and blade, screw, and post-dental implants.

POROUS CERAMICS

The potential advantage offered by a porous ceramic implant (type 2, Table I.2.4.2, Figures I.2.4.1 and I.2.4.2) is its inertness combined with the mechanical stability of the highly-convoluted interface that develops when bone grows into the pores of the ceramic. The mechanical requirements of prostheses, however, severely restrict the use of low-strength porous ceramics to nonloadbearing applications. Studies reviewed by Hench and Ethridge (1982), Hulbert et al. (1987), and Schors and Holmes (1993) have shown that when loadbearing is not a primary requirement, porous ceramics can provide a functional implant. When pore sizes exceed 100 μm, bone will grow within the interconnecting pore channels near the surface and maintain its vascularity and long-term viability. In this manner, the implant serves as a structural bridge or scaffold for bone formation.

Commercially available porous products originate from two sources: hydroxyapatite converted from coral or animal bone. The optimal type of porosity is still uncertain. The degree of interconnectivity of pores may be more critical than the pore size. Eggli et al. (1988) demonstrated improved integration in interconnected 50–100 μm pores compared with less connected pores with a size of 200–400 μm. Similarly Kühne et al. (1994) compared two grades of 25–35% porous coralline apatite with average pore sizes of 200 and 500 μm, and reported bone ingrowth to be improved in the 500 μm pore sized ceramic. Holmes (1979) suggests that porous coralline apatite when implanted in cortical bone requires interconnections of osteonic diameter for transport of nutrients to maintain bone ingrowth. The findings clearly indicate the importance of thorough characterization of porous materials before implantation, and Hing (1999) has recommended a range of techniques that should be employed.

Porous materials are weaker than the equivalent bulk form in proportion to the percentage of porosity, so that as the porosity increases, the strength of the material decreases rapidly. Much surface area is also exposed, so that the effects of the environment on decreasing the strength become much more important than for dense, nonporous materials. The aging of porous ceramics, with their subsequent decrease in strength, requires bone ingrowth to stabilize the structure of the implant. Clinical results for nonloadbearing implants are good (Schors and Holmes, 1993).

BIOACTIVE GLASSES AND GLASS-CERAMICS

Certain compositions of glasses, ceramics, glass-ceramics, and composites have been shown to bond to bone (Hench and Ethridge, 1982; Gross et al., 1988; Yamamuro et al., 1990; Hench, 1991; Hench and Wilson, 1993). These materials have become known as bioactive ceramics. Some specialized compositions of bioactive glasses will bond to soft tissues as well as bone (Wilson et al., 1981). A common characteristic of bioactive glasses and bioactive ceramics is a time-dependent, kinetic modification of the surface that occurs upon implantation. The surface forms a biologically-active carbonated HA layer (HCA) that provides the bonding interface with tissues. Materials that are bioactive develop an adherent interface with tissues that resist substantial mechanical forces. In many cases, the interfacial strength of adhesion is equivalent to or greater than the cohesive strength of the implant material or the tissue bonded to the bioactive implant.

Bonding to bone was first demonstrated for a compositional range of bioactive glasses that contained SiO_2, Na_2O, CaO, and P_2O_5 in specific proportions (Hench et al., 1971) (Table I.2.4.5). There are three key compositional features to these bioactive glasses that distinguish them from traditional soda–lime–silica glasses: (1) less than 60 mol% SiO_2; (2) high Na_2O and CaO content; and (3) a high CaO/P_2O_5 ratio. These features make the surface highly reactive when it is exposed to an aqueous medium.

Many bioactive silica glasses are based upon the formula called 45S5, signifying 45 wt.% SiO_2 (S = the network former) and 5:1 ratio of CaO to P_2O_5. Glasses with lower ratios of CaO to P_2O_5 do not bond to bone. However, substitutions in the 45S5 formula of 5–15 wt.% B_2O_3 for SiO_2 or 12.5 wt.% CaF_2 for CaO or heat treating the bioactive glass compositions to form glass-ceramics has no measurable effect on the ability of the material to form a bone bond. However, adding as little as 3 wt.% Al_2O_3 to the 45S5 formula prevents bonding to bone.

The compositional dependence of bone and soft tissue bonding on the $Na_2O–CaO–P_2O_5–SiO_2$ glasses is illustrated in Figure I.2.4.5. All the glasses in Figure I.2.4.5 contain a constant 6 wt.% of P_2O_5. Compositions in the middle of the diagram (region A) form a bond with bone. Consequently, region A is termed the bioactive bone-bonding boundary. Silicate glasses within region B (e.g., window or bottle glass or microscope slides) behave as nearly inert materials and elicit a fibrous capsule at

TABLE I.2.4.5 Composition of Bioactive Glasses and Glass-Ceramics (in Weight Percent)

	45S5 Bioglass	45S5F Bioglass	45S5.4F Bioglass	40S5B5 Bioglass	52S4.6 Bioglass	55S4.3 Bioglass	KGC Ceravital	KGS Ceravital	KGy213 Ceravital	A-W GC	MB GC
SiO_2	45	45	45	40	52	55	46.2	46	38	34.2	19–52
P_2O_5	6	6	6	6	6	6				16.3	4–24
CaO	24.5	12.25	14.7	24.5	21	19.5	20.2	33	31	44.9	9–3
$Ca(PO_3)_2$							25.5	16	13.5		
CaF_2		12.25	9.8							0.5	
MgO							2.9			4.6	5–15
MgF_2											
Na_2O	24.5	24.5	24.5	24.5	21	19.5	4.8	5	4		3–5
K_2O							0.4				3–5
Al_2O_3									7		12–33
B_2O_3				5							
Ta_2O_5/TiO_2									6.5		
Structure	Glass	Glass	Glass	Glass	Glass		Glass-ceramic	Glass-ceramic		Glass-ceramic	Glass-ceramic
Reference	Hench et al. (1982)	Hench et al. (1982)	Hench et al. (1982)	Hench et al. (1982)	Hench et al. (1982)	Hench et al. (1982)	Gross et al. (1988)	Gross et al. (1988)		Nakamura et al. (1985)	Höeland and Vogel (1993)

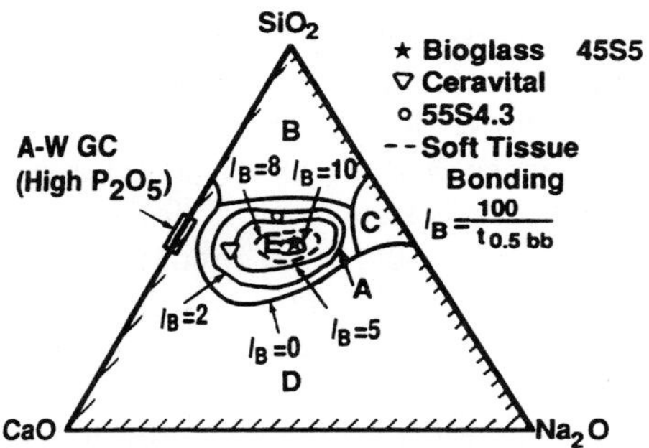

FIGURE I.2.4.5 Compositional dependence (in wt.%) of bone bonding and soft tissue bonding of bioactive glasses and glass-ceramics. All compositions in region A have a constant 6 wt.% of P_2O_5. A-W glass ceramic has higher P_2O_5 content (see Table I.2.4.5 for details). I_B, Index of bioactivity.

the implant–tissue interface. Glasses within region C are resorbable and disappear within 10 to 30 days of implantation. Glasses within region D are not technically practical, and therefore have not been tested as implants.

The collagenous constituent of soft tissues can strongly adhere to the bioactive silicate glasses that lie within the dashed line region in Figure I.2.4.5. The interfacial thicknesses of the hard tissue–bioactive glasses are shown in Figure I.2.4.2 for several compositions. The thickness decreases as the bone-bonding boundary is approached.

Gross et al. (1988) and Gross and Strunz (1985) have shown that a range of low-alkali (0 to 5 wt.%) bioactive silica glass-ceramics also bond to bone. They found that small additions of Al_2O_3, Ta_2O_5, TiO_2, Sb_2O_3 or ZrO_2 inhibit bone bonding (Table I.2.4.5, Figure I.2.4.1). A two-phase silica–phosphate glass-ceramic composed of apatite $[Ca_{10}(PO_4)_6(OH_1F_2)]$ and wollastonite [CaO, SiO_2] crystals and a residual silica glassy matrix, termed A-W glass-ceramic (A-WGC) (Nakamura et al., 1985; Yamamuro et al., 1990; Kokubo, 1993), also bonds with bone. The addition of Al_2O_3 or TiO_2 to the A-W glass-ceramic also inhibits bone bonding, whereas incorporation of a second phosphate phase, B-whitlockite (3CaO, P_2O_5), does not.

Another multiphase bioactive phosphosilicate containing phlogopite (Na, K) $Mg_3[AlSi_3O_{10}]F_2$ and apatite crystals bonds to bone even though Al is present in the composition (Höhland and Vogel, 1993). However, the Al^{3+} ions are incorporated within the crystal phase, and do not alter the surface reaction kinetics of the material. The compositions of these various bioactive glasses and glass-ceramics are compared in Table I.2.4.5. The important compositions that are used in the clinic are 45S5 Bioglass and A/W glass-ceramic.

The surface chemistry of bioactive glass and glass-ceramic implants is best understood in terms of six possible types of surface reactions (Hench and Clark, 1978). A high-silica glass may react with its environment by developing only a surface hydration layer. This is called a type I response. Vitreous silica (SiO2) and some inert glasses at the apex of region B (Figure I.2.4.5) behave in this manner when exposed to a physiological environment.

When sufficient SiO_2 is present in the glass network, the surface layer that forms from alkali–proton exchange can repolymerize into a dense SiO_2-rich film that protects the glass from further attack. This type II surface is characteristic of most commercial silicate glasses, and their biological response of fibrous capsule formation is typical of many within region B in Figure I.2.4.5.

At the other extreme of the reactivity range, a silicate glass may undergo rapid, selective ion exchange of alkali ions, with protons or hydronium ions leaving a thick but highly porous and nonprotective SiO_2-rich film on the surface (a type IV surface). Under static or slow flow conditions, the local pH becomes sufficiently alkaline ($pH > 9$) that the surface silica layer is dissolved at a high rate, leading to uniform bulk network or stoichiometric

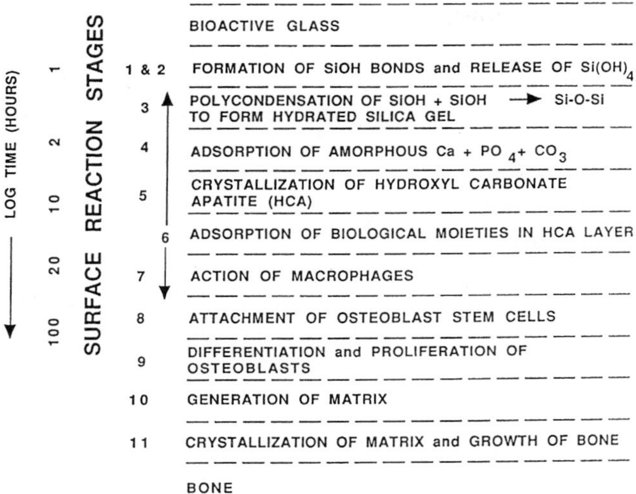

FIGURE I.2.4.6 Types of silicate glass interfaces with aqueous or physiological solutions.

dissolution (a type V surface). Both type IV and V surfaces fall into region C of Figure I.2.4.5.

Type IIIA surfaces are characteristic of bioactive glasses. The first five reaction stages depicted in Figure I.2.4.6 are characteristic of Type IIIA bioactive glasses such as 45S5 Bioglass. A $CaO\text{-}P_2O_5$ layer forms on top of the alkali-depleted SiO_2-rich film. When multivalent cations such as Al^{3+}, Fe^{3+}, and Ti^{4+} are present in the glass or solution, multiple layers form on the glass as the saturation of each cationic complex is exceeded, resulting in a type IIIB surface, which does not bond to tissues.

A general equation describes the overall rate of change of glass surfaces and gives rise to the interfacial reaction profiles shown in Figure I.2.4.1. The reaction rate (R) depends on at least four terms (for a single-phase glass). For glass-ceramics, which have several phases in their microstructures, each phase will have a characteristic reaction rate, R_i

$$R = -k_1 t^{0.5} - k_2 t^{1.0} - k_3 t^{1.0} + k_4 t^y + k_n t^z \tag{1}$$

The first term describes the rate of alkali extraction from the glass, and is called a stage 1 reaction. A type II nonbonding glass surface (region B in Figure I.2.4.5) is primarily undergoing stage 1 attack. Stage 1, the initial or primary stage of attack shown in Figure I.2.4.6, is a process that involves an exchange between alkali ions from the glass and hydrogen ions from the solution, during which the remaining constituents of the glass are not altered. During stage 1, the rate of alkali extraction from the glass is parabolic ($t^{1/2}$) in character.

The second term describes the rate of interfacial network dissolution that is associated with a stage 2 reaction. A type IV surface is a resorbable glass (region C in Figure I.2.4.5), and experiences a combination of stage 1 and stage 2 reactions. A type V surface is dominated by a stage 2 reaction. Stage 2, the second stage of attack, is a process by which the silica structure breaks down, and the glass totally dissolves at the interface. Stage 2 kinetics are linear ($t^{1.0}$).

A glass surface with a dual protective film is designated type IIIA. The thickness of the secondary films can vary considerably – from as little as 0.01 μm for Al_2O_3–SiO_2-rich layers on inactive glasses, to as much as 30 μm for CaO–P_2O_5-rich layers on bioactive glasses (Figure I.2.4.2). A type III surface forms as a result of the repolymerization of SiO_2 on the glass surface by the condensation of the silanols (Si-OH) formed from the stage 1 reaction. For example:

$$Si-OH+OH-Si \rightarrow Si-O-Si+H_2O \tag{2}$$

Stage 3 protects the glass surface. The SiO_2 polymerization reaction contributes to the enrichment of surface SiO_2-rich gel layer, Figure I.2.4.6. It is described by the third term in Eq. (1). This reaction is interface controlled with a time dependence of $+k_3t^{1.0}$. The interfacial thickness of the most reactive bioactive glasses shown in Figure I.2.4.2 is largely due to this reaction.

The fourth term in Eq. (1), $+k_4t^y$ (stage 4), describes the precipitation reactions that result in the multiple films characteristic of type III glasses. When only one secondary film forms, the surface is type IIIA. When several additional films form, the surface is type IIIB.

In stage 4, an amorphous calcium phosphate film precipitates on the silica-rich layer and is followed by crystallization to form carbonated HA crystals, Figure I.2.4.6. The calcium and phosphate ions in the glass or glass-ceramic provide the nucleation sites for crystallization. The most bioactive glasses reach completion of this stage of surface reaction very rapidly. Thus, the biological response of living cells at the interface with the bioactive glass is to the presence of the growing HCA layer, and not a foreign-body. Carbonate anions (CO_3^{2-}) substitute for OH^- in the apatite crystal structure to form a carbonate hydroxyapatite (HCA) similar to that found in living bone (see the following section for details). Incorporation of CaF_2 in the glass results in incorporation of fluoride ions in the apatite crystal lattice. Crystallization of HCA occurs around collagen fibrils present at the implant interface, and results in interfacial bonding.

In order for the material to be bioactive and form an interfacial bond, the kinetics of reaction in Eq. (1), and especially the rate of stage 4, must match the rate of biomineralization that normally occurs *in vivo*. If the rates in Eq. (1) are too rapid, the implant is resorbable, and if the rates are too slow, the implant is not bioactive. The interfacial reaction rates for 45S5 Bioglass are summarized in Figures I.2.4.1 and I.2.4.6.

By changing the compositionally-controlled reaction kinetics (Eq. (1)), the rates of formation of hard tissue at an implant interface can be altered, as shown in Figure I.2.4.1. Thus, the level of bioactivity of a material can be related to the time for more than 50% of the interface to be bonded ($t_{0.5bb}$) (e.g., I_B index of bioactivity: = $(100/t_{0.5bb})$) (Hench, 1988). It is necessary to impose a 50% bonding criterion for an I_B, since the interface between an implant and bone is irregular (Gross et al., 1988). The initial concentration of macrophages, osteoblasts, chondroblasts, and fibroblasts at an implant site varies as a function of the fit of the implant or particle size if the implant is a bone graft particulate, and especially the nature of the bony defect. Consequently, all bioactive implants require an incubation period before bone proliferates and bonds. The length of this incubation period varies widely, depending on the composition of the implant.

The compositional dependence of I_B indicates that there are isoI_B contours within the bioactivity boundary, as shown in Figure I.2.4.5 (Hench, 1988). The change of I_B with the $SiO_2/(Na_2O + CaO)$ ratio is very large as the bioactivity boundary is approached. The addition of multivalent ions to a bioactive glass or glass-ceramic shrinks the isoI_B contours, which will contract to zero as the percentage of Al_2O_3, Ta_2O_5, ZrO_2 or other multivalent (M^{n+}) increases in the material. Consequently, the isoI_B boundary shown in Figure I.2.4.5 indicates the contamination limit for bioactive glasses and glass-ceramics. If the composition of a starting implant is near the I_B boundary, it may take only a few parts per million of multivalent cations to shrink the I_B boundary to zero and eliminate bioactivity. Also, the sensitivity of fit of a bioactive implant and length of time of immobilization postoperatively depends on the I_B value and closeness to the $I_B = 0$ boundary. Implants near the I_B boundary require more precise surgical fit and longer fixation times before they can bear loads. In contrast, increasing the surface area of a bioactive implant by using them in particulate form for bone augmentation expands the bioactive boundary. Small (<200 μm) bioactive glass granules behave as a partially resorbable implant and stimulate new bone formation (Hench, 1994). Slow dissolution of the bioactive glass particles provides ionic stimuli to osteoprogenitor cells and leads to enhanced osteogenesis, as described below.

Bioactive implants with intermediate I_B values do not develop a stable soft tissue bond; instead, the fibrous interface progressively mineralizes to form bone. Bioactive materials with this level of activity are referred to as Class B bioactive. Consequently, there appears to be a critical isoI_B boundary beyond which bioactivity is restricted to stable bone bonding and growth of bone along the bioactive material interface, termed osteoconduction. Inside the critical isoI_B boundary, the bioactivity includes both stable bone and soft-tissue bonding, depending on the progenitor stem cells in contact with the implant. Such bioactive materials are considered to exhibit Class A bioactivity and possess properties of both osteoconduction and osteostimulation, discussed below. This soft tissue-critical isoI_B limit for Class A bioactivity is shown by the dashed contour in Figure I.2.4.5.

The thickness of the bonding zone between a bioactive implant and bone is proportional to its I_B (compare Figure I.2.4.1 with Figure I.2.4.2). The failure strength of a bioactive bond is inversely dependent on the thickness of the bonding zone. For example, 45S5 Bioglass with a very high I_B develops a gel bonding layer of 200 μm, which has relatively low shear strength. In contrast, A-W glass-ceramic, with an intermediate I_B value, has a bonding interface in the range of 10–20 μm and a very high resistance to shear. Thus, the interfacial bonding strength appears to be optimum for I_B values of ~4. However, it is important to recognize that the interfacial area for bonding is time dependent, as shown in Figure I.2.4.1. Therefore, interfacial strength is time-dependent and is a function of such morphological factors as the change in interfacial area with time, progressive mineralization of the interfacial tissues, and resulting increase of elastic modulus of the interfacial bond, as well as shear strength per unit of bonded area. A comparison of the increase in interfacial bond strength of bioactive fixation of implants bonded to bone with other types of fixation is given in Figure I.2.4.7 (Hench, 1987).

Clinical applications of bioactive glasses and glass-ceramics are reviewed by Yamamuro et al. (1990), Hench and Wilson (1993), (Hench 1998), and Hench et al. (2004). The high level of clinical success of 20-year use of A-W glass-ceramic in vertebral surgery (Yamamuro et al., 1990), 45S5 Bioglass in endosseous ridge maintenance (Stanley et al., 1996), and middle-ear replacement are important accomplishments of this field. Table I.2.4.6 lists the clinical applications of 45S5 bioactive glass. Use of 45S5 Bioglass in repair of periodontal defects (Wilson and Low, 1992; Wilson, 1994; Hench and Wilson, 1996; Lovelace et al., 1998; Mengle et al., 2003) and in a variety of dental, maxillo-facial, and orthopedic applications is summarized in Hench et al. (2004). This 2004 review includes an extensive bibliography of 800 references listed chronologically since the first paper by Hench et al. (1971) described bone bonding to bioactive glasses. The year 2009 heralded the 40th anniversary of the discovery

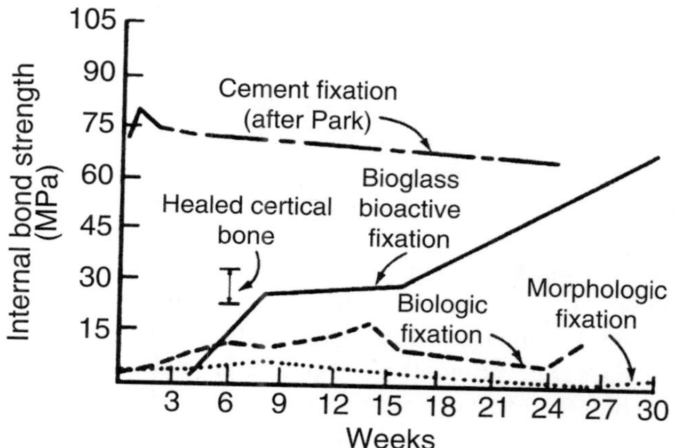

FIGURE I.2.4.7 Time dependence of interfacial bond strength of various fixation systems in bone. (After Hench, 1987.)

TABLE I.2.4.6 Typical Clinical Applications of Bioactive Ceramics, Glasses and Glass Ceramics (highlighting 45S5 and Hydroxyapatite as Examples)

Orthopedics
Trauma:
Long bone fracture (acute and/or comminuted); alone and with internal fixation
Femoral non-union repair
Tibial plateau fracture
Arthroplasty
Filler around implants (acetabular reconstruction)
Impaction grafting
General
Filling of bone after cyst/tumor removal
Spine Fusion
Interbody fusion (cervical, thoracolumbar, lumbar)
Posterolateral fusion
Adolescent idiopathic scoliosis
Cranial-Facial
Cranioplasty
Facial reconstruction
General oral/dental defects
Extraction sites
Ridge Augmentation
Sinus elevation
Cystectomies
Osteotomies
Periodontal Repair
Dental- Maxillofacial- ENT
Toothpaste and treatments for dentinal hypersensitivity and inhibition of gingivitis
Pulp capping
Sinus obliteration
Repair of orbital floor fracture
Endosseous ridge maintenance implants
Middle ear ossicular replacements (Douek MED)

of 45S5 Bioglass, and the landmark sales of the one millionth dose of the bone graft product (NovaBone) and the one millionth tube of toothpaste containing 445S5 particulate (NovaMin) designed to occlude dentinal tubules and remineralize the surface of teeth, thereby eliminating the cause of dentinal hypersensitivity. The excellent clinical success of this bioactive material is due to the stimulation of osteogenesis, as summarized above, and a clinical response equivalent to use of autogeneous bone grafts without second site morbidity (Ilharreborde et al., 2008; Ameri et al., 2009).

BIOACTIVITY REACTION STAGES

There is a sequence of 11 reaction stages that occur at the surface of a Class A bioactive glass, as summarized in Figure I.2.4.6. The log time axis shown in Figure I.2.4.6 is based upon studies of surface reaction rates, as discussed above. The first five stages of surface reactions occur very rapidly and go to completion within 24 hours for the Class A bioactive glasses with highest levels of bioactivity, e.g., 45S5 Bioglass. The effect of the surface reactions is rapid release of soluble ionic species from the glass into the interfacial solution. A high surface area hydrated silica and polycrystalline hydroxyl-carbonate apatite (HCA) bilayer is formed on the glass surface within hours (Figure I.2.4.6, Stages 1–5). The reaction layers enhance adsorption and desorption of growth factors produced by the cells (Figure I.2.4.6, Stage 6), and decrease greatly the length of time macrophages are required to prepare the implant site for tissue repair (Figure I.2.4.6, Stage 7).

Attachment of osteoprogenitor stem cells (Figure I.2.4.6, Stage 8) and synchronized proliferation and differentiation of the cells towards the mature osteoblast phenotype (Figure I.2.4.6, Stage 9) rapidly occurs on the surface of Class A bioactive materials. Several weeks are required for similar cellular events to occur on the surface of Class B bioactive materials. Differentiation of progenitor cells into a mature osteoblast phenotype does not occur on bioinert materials, because of the lack of ionic stimuli. In contrast, osteoprogenitor cells colonize the surface of Class A bioactive materials within 24–48 hours, and begin production of various growth factors which stimulate cell division, mitosis, and production of extracellular matrix proteins (Figure I.2.4.6, Stage 10). Mineralization of the matrix follows soon thereafter and mature osteocytes, encased in a collagen-HCA matrix, are the final product by 6–12 days *in vitro* and *in vivo* (Figure I.2.4.6, Stage 11).

A rapid completion of all 11 reaction stages shown in Figure I.2.4.6 is the key to regenerative repair of bone, which involves: (1) control of the population of cells that are capable of entering into active phases of the cell cycle; (2) complete mitosis of cells with accurate replication of genes (cell proliferation); and (3) cellular differentiation into a phenotype capable of synthesizing a full complement of extracellular proteins that constitute mature osteocytes.

Such osteoblast cell cycle control is achieved *in vivo* by the controlled release of critical concentrations of Ca and Si ionic dissolution products from 45S5 bioactive glass particulate (Buttery et al., 2001; Xynos et. al., 2000, 2001) or 58S bioactive gel-glass (Christodoulou et al., 2005). Osteoprogenitor cells colonize the surface of the bioactive glass (Figure I.2.4.6), and slow release of the critical concentration of soluble Si and Ca ions at the cell–solution interface activates seven families of genes responsible for osteogenesis in the cells. Controlled rates of dissolution of the glass provide the critical concentration of the biologically active ions to the cells via the interfacial solution. The families of genes that are upregulated and/or activated are shown in Table I.2.4.7.

The upregulated genes encode nuclear transcription factors and potent growth factors, especially IGF-II, along with IGF-binding proteins and proteases that cleave IGF-II from their binding proteins. The growth

TABLE I.2.4.7 Families of Genes in Human Osteoblasts Activated or Up-Regulated by Ionic Dissolution Products of Bioactive Glasses
(1) Transcription Factors and Cell Cycle Regulators
(2) DNA Synthesis, Repair and Recombination
(3) Apoptosis Regulators
(4) Growth Factors and Cytokines
(5) Cell Surface Antigens and Receptors
(6) Signal Transduction Molecules
(7) Extracellular Matrix Compounds

factors are present in a biologically active state, as confirmed by analysis of the upregulation of IGF-II mRNA with quantitative real-time PCR (Buttery, 2001; Hench, 2003). Similar bioactive induction of the transcription of extracellular matrix components and their secretion and self-organization into a mineralized matrix is responsible for the rapid formation and growth of bone nodules and differentiation of the mature osteocyte phenotype. Details are given in Xynos et al. (2000, 2001), and reviewed in Hench (2003).

CALCIUM PHOSPHATE CERAMICS

Bone typically consists by weight of 25% water, 15% organic materials, and 60% mineral phases. The mineral phase consists primarily of calcium and phosphate ions, with traces of magnesium, carbonate, hydroxyl, chloride, fluoride, and citrate ions. Hence, calcium phosphates occur naturally in the body, but they also occur within nature as mineral rocks, and certain compounds can be synthesized in the laboratory. Table I.2.4.8 summarizes the mineral name, chemical name, and composition of various phases of calcium phosphates.

Within the past 20–30 years interest has intensified in the use of calcium phosphates as biomaterials, but only certain compounds are useful for implantation in the body, since both their solubility and speed of hydrolysis increase with a decreasing calcium-to-phosphorus ratio. Driessens (1983) stated that those compounds with a Ca/P ratio of less than 1:1 are not suitable for biological implantation.

The main crystalline component of the mineral phase of bone is a calcium deficient carbonate hydroxyapatite, and a wide variety of methods have been investigated to produce synthetic hydroxyapatite. The most commercially popular routes are based on aqueous precipitation or conversion from other calcium compounds. The first reports appeared in the literature in the 1960s, and since then there have been many approaches based on those original methods. Aqueous precipitation is most often performed in one of two ways: a reaction between a calcium salt and an alkaline phosphate (Collin et al., 1963; Eanes et al., 1965; Jarcho et al., 1976; Kijima and Tsutsumi, 1979; Denissen et al., 1980; Young and Holcomb, 1982; Bonel et al., 1987) or a reaction between calcium hydroxide or calcium carbonate and phosphoric acid (Mooney, 1961; Rao and Boehm, 1974; McDowell et al., 1977; Akao et al., 1981; Irvine, 1981; Nagai et al., 1985). Since these papers were published there have been many reports of alternative wet chemical routes, but most of these are, in fact, based on similar reactants to those first described in the 1960s, 1970s, and 1980s.

Other routes include those based on solid-state processing (originally described by Lehr et al., 1967; Fowler, 1974; Rootare et al., 1978; Monma et al., 1981; Young and Holcomb, 1982), hydrolysis (Schleede et al., 1932; Perloff et al., 1956; Morancho et al., 1981; Young and Holcomb, 1982) or hydrothermal synthesis (Perloff and Posner, 1960; Roy, 1971; Skinner, 1973; Fowler, 1974; Arends et al., 1979; Young and Holcomb, 1982).

The route and conditions under which synthetic HA is produced will greatly influence its physical and chemical characteristics. Factors that affect the rate of resorption of the implant include physical factors, such as the physical features of the material (e.g., surface area, crystallite size), chemical factors, such as atomic and ionic substitutions in the lattice, and biological factors, such as the types of cells surrounding the implant and location, age, species, sex, and hormone levels.

The thermodynamic stability of the various calcium phosphates is summarized in the phase diagram shown in Figure I.2.4.8. The binary equilibrium phase diagram between CaO and P_2O_5 gives an indication of the compounds formed between the two oxides, and by comparing this with Table I.2.4.8 it is possible to identify

TABLE I.2.4.8 Calcium Phosphates

Ca:P	Mineral Name	Formula	Chemical Name
1.0	Monetite	$CaHPO_4$	Dicalcium phosphate (DCP)
1.0	Brushite	$CaHPO_4 \cdot 2H_2O$	Dihydrate (DCPD) Dicalcium phosphate
1.33	—	$Ca_8(HPO_4)_2(PO_4)_4 \cdot 5H_2O$	Octocalcium phosphate (OCP)
1.43	Whitlockite	$Ca_{10}(HPO_4)(PO_4)_6$	
1.5	—	$Ca_3(PO_4)_2$	Tricalcium phosphate (TCP)
1.67	Hydroxyapatite	$Ca_{10}(PO_4)_6(OH)_2$	
2.0		$Ca_4P_2O_9$	Tetracalcium phosphate

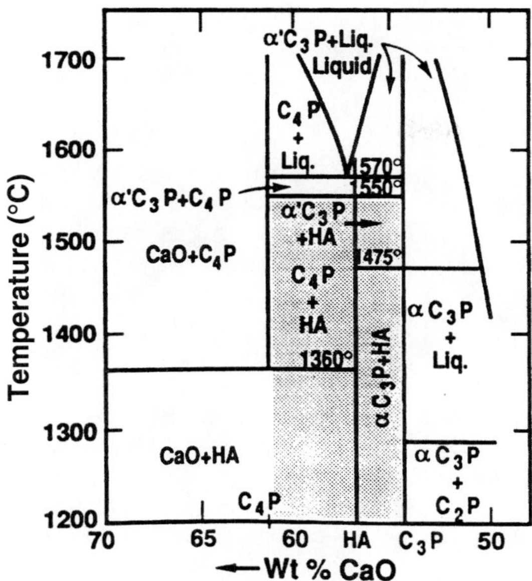

FIGURE I.2.4.8 Phase equilibrium diagram of calcium phosphates in a water atmosphere. Shaded area is processing range to yield HA-containing implants. (After K. de Groot (1988). Ann. N. Y. Acad. Sci. 523: 227.)

the naturally occurring calcium phosphate minerals. The diagram does not indicate the phase boundaries of apatite due to the absence of hydroxyl groups. However, from the binary diagram an indication may be obtained of the stability of other calcium phosphates with temperature.

The stoichiometry of HA is highly significant where thermal processing of the material is required. Slight imbalances in the stoichiometric ratio of calcium and phosphorus in HA (from the standard molar ratio of 1.67) can lead to the appearance of either α- or β-tricalcium phosphate on heat treatment. Many early papers concerning the production and processing of HA powders reported problems in avoiding the formation of these extraneous phases (Jarcho et al., 1976; Peelen et al., 1978; De With et al., 1981a,b). However, using stoichiometric hydroxyapatite it should be possible to sinter, without phase purity problems, at temperatures in excess of 1300°C.

X-ray diffraction and infrared spectroscopy should be used to reveal the phase purity and level of hydroxylation of hydroxyapatite (Figures I.2.4.9 and I.2.4.10). Kijima and Tsutsumi (1979) used these techniques to study hydroxyapatite sintered at different temperatures, and reported that after sintering at 9000°C the material was fully hydroxylated, but after sintering at temperatures higher than this, dehydroxylation occurred. In fact, the dehydration of hydroxyapatite, produced by processes such as high temperature solid-state reactions, result in the formation of oxyhydroxyapatite: $Ca_{10}(PO_4)_6(OH)_{2-2x}O_xV_x$, (where V is a hydroxyl vacancy). Hydroxyapatite has a $P6_3/_m$ space group: this signifies that the lattice is primitive Bravais,

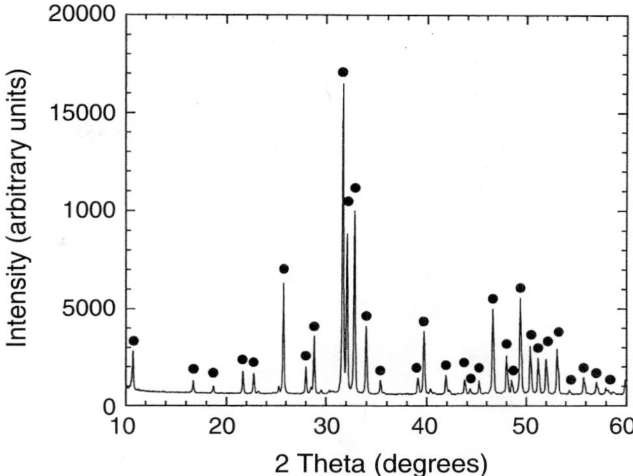

FIGURE I.2.4.9 X-ray diffraction of hydroxyapatite.

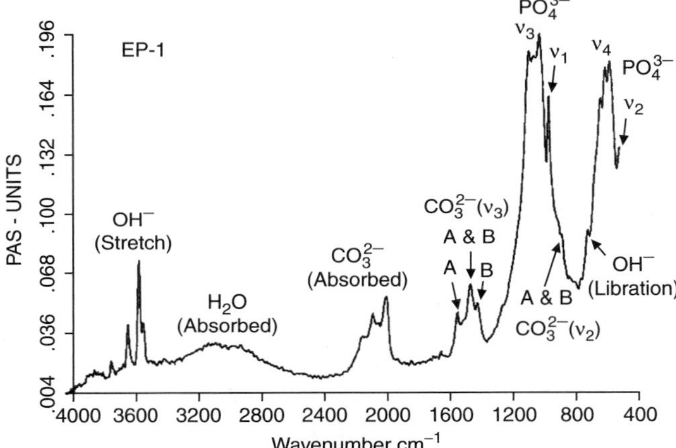

FIGURE I.2.4.10 Typical FT-IR spectrum for a bone mineral-derived hydroxyapatite.

there is a six-fold axis parallel to the c-axis and a 1/2 (3/6) translation along the length of the c-axis (a screw axis) with a mirror plane situated perpendicular to the screw axis and the c-axis. The lattice parameters for hydroxyapatite have been determined to be 0.9418 nm and 0.6884 nm for the a and c parameters, respectively (JCPDS #09-0432).

The structure assumed by any solid is such that, on an atomic level, the configuration of the constituents is of the lowest possible energy. In phosphates, this energy requirement results in the formation of discrete subunits within the structure, and the $PO_4{}^{3-}$ group forms a regular tetrahedron with a central P^{5+} ion and O^{2-} ions at the four corners. In a similar manner, the $(OH)^-$ groups are also ionically bonded. In terms of the volume occupied, oxygen ions exceed all other elements in phosphates. Any other elements present may therefore be considered as filling the interstices, with the exact position being determined by atomic radius and charge (see Figure I.2.4.11).

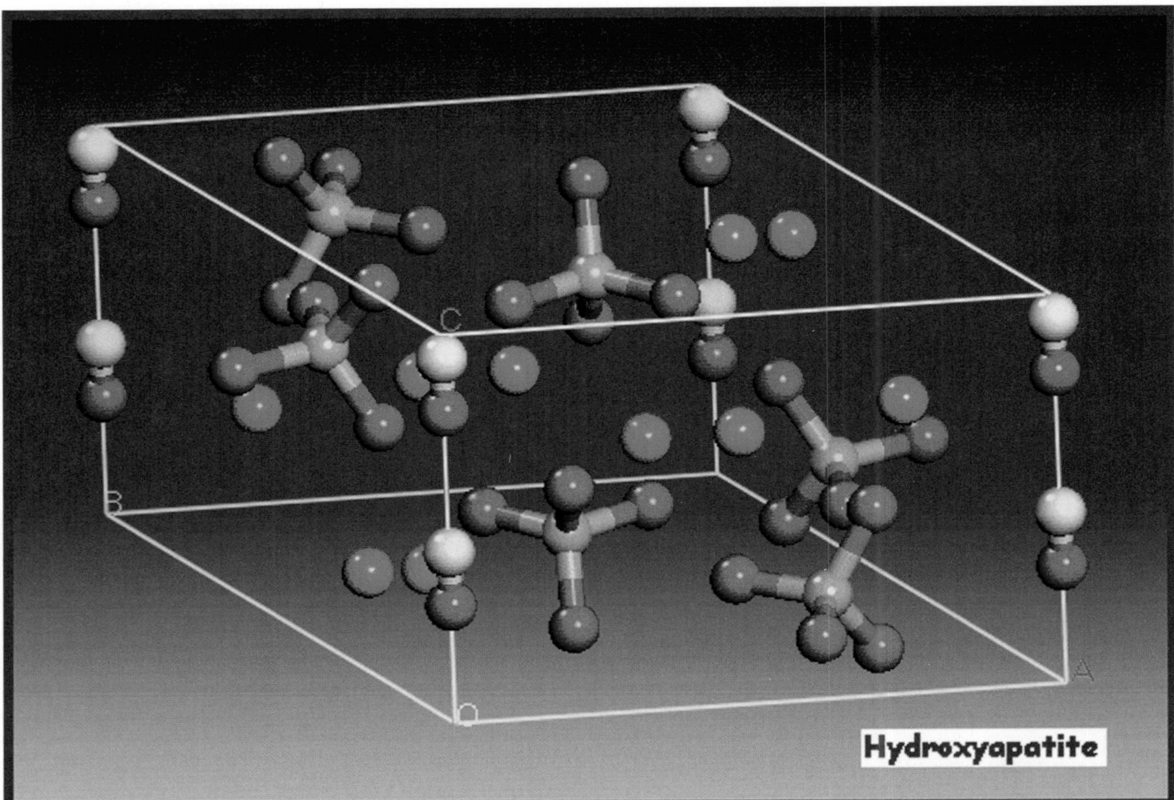

FIGURE I.2.4.11 Unit cell of hydroxyapatite (from PhD thesis, University of Cambridge, H. Chappell 2008).

The hydroxyapatite lattice contains two kinds of calcium positions: columnar and hexagonal. There is a net total of four "columnar calcium" ions that occupy the [1/3, 2/3, 0] and [1/3, 2/3, 1/2] lattice points. The "hexagonal calcium" ions are located on planes parallel to the basal plane at $c = 1/4$ and $c = 3/4$, and the six PO_4^{3-} tetrahedra are also located on these planes. The $(OH)^-$ groups are located in columns parallel to the *c*-axis, at the corners of the unit cell, which may be viewed as passing through the centers of the triangles formed by the "hexagonal calcium" ions. Successive hexagonal calcium triangles are rotated through 60° (see Figure I.2.4.12).

Defects and impurities in hydroxyapatite may be identified as either substitutional or discrete extraneous crystalline phases (as discussed above). Methods of detection of impurities include X-ray diffraction, infrared spectroscopy, and spectrochemical analysis. It is important to make a full spectrochemical analysis of hydroxyapatite, since contact with any metal ions during production can lead to high levels of impurties in the product. Typical data for one commercial hydroxyapatite powder are shown in Table I.2.4.9.

Other ions which may be incorporated into the HA structure, either intentionally or unintentionally, include carbonate ions (substituting for hydroxyl or phosphate groups), fluoride ions (substituting for hydroxyl groups), silicon or silicate ions (substituting for phosphorus or phosphate groups), and magnesium ions substituting for calcium (e.g., Newsley, 1963; Le Geros, 1965; Barralet, 1997; Jha et al., 1998; Gibson et al., 1999).

Based on the research work performed by Carlisle in the 1970s, in which the nutritional effects of silicon were studied, Bonfield, Best, and co-workers later reported the development of silicate-substituted hydroxyapatites (Si-HA) (Gibson et al., 1999; Patel et al., 2002; Porter et al., 2003). *In vivo* studies comparing the rates of bone apposition to HA and Si-HA ceramic implants demonstrated a significant increase in the amount of bone apposition and organization around silicon-substituted HA (Si-HA) implants, illustrating their potential as bone graft materials (Patel et al., 2002). Several other groups have also investigated the production and characterization of Si-HA and Si-TCP (Ruys, 1993; Balas et al., 2003; Gasqueres et al., 2008).

Silicate-substituted HA (Si-HA) may be prepared via a variety of synthesis routes. Gibson et al. (1999) produced phase-pure Si-HA via an aqueous precipitation route, and proposed the substitution mechanism given the equation below:

$$10Ca^{2+} + (6-x)PO_4^{3-} + xSiO_4^{4-} + (2-x)OH^- \rightarrow Ca_{10}(PO_4)_{6-x}(SiO_4)_x(OH)_{2-x}$$

Structural analysis has confirmed that silicate (SiO_4^{4-}) ions can substitute for the PO_4 sites in HA.

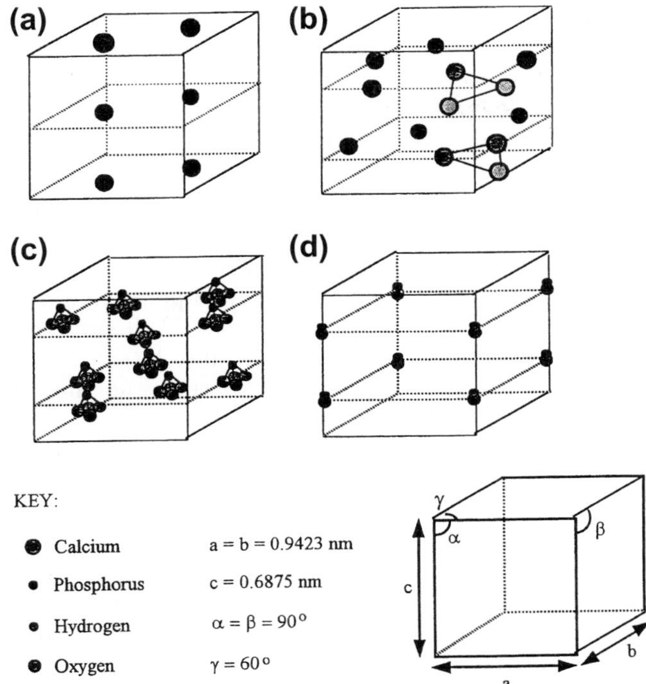

FIGURE I.2.4.12 Theoretical positions of the ionic species within the unit cell of hydroxyapatite. (Hing, 1995).

TABLE I.2.4.9 Trace Elements in a Commercial Hydroxyapatite

Trace Element	PPM
Al	600
Cu	1
Fe	1000
Ge	100
Mg	2000
Mn	300
Na	3000
Pb	4
Si	500
Ti	30

This substitution results in changes in the crystal lattice dimension, with a decrease in the a-axis and an increase in the c-axis (Gibson et al., 1999; Patel et al., 2002). This work led to a significant level of interest in silicate-substituted apatite-based bone grafts, and these materials are now used in a range of orthopedic applications (Waked and Grauer, 2008).

The presence of carbonate may be observed directly, using infrared spectroscopy, in the form of weak peaks at between 870 and 880 cm^{-1}, and a stronger doublet between 1460 and 1530 cm^{-1}, and also through alterations in the hydroxyapatite lattice parameters from X-ray diffraction (LeGeros, 1965; Rootare and Craig, 1978; Barralet et al., 1997). The substitution of electronegative anions such as fluorine and chlorine for $(OH)^-$ have also been reported to alter the lattice parameters

TABLE I.2.4.10 Typical Mechanical Properties of Dense Hydroxyapatite Ceramics

Property	Value
Theoretical density	3.156 g cm^3
Hardness	500–800 HV, 2000–3500 Knoop
Tensile strength	40–100 MPa
Bend strength	20–80 MPa
Compressive strength	100–900 MPa
Fracture toughness	approx. 1 Mpa $m^{0.5}$
Young's modulus	70–120 GPa

of the material (e.g., Kay et al., 1964; Young and Elliot, 1966; see also Elliot, 1994).

Hydroxyapatite may be processed as a ceramic using compaction (die pressing, isostatic pressing, slip casting, etc.) followed by solid-state sintering (discussed earlier in this chapter). When reporting methods for the production and sintering of hydroxyapatite powders, it is very important to adequately characterize the morphology of the product including the surface area, particle size distribution, mean particle size, and physical appearance of the powders, since this will greatly influence the handling and processing characteristics of the material (Best and Bonfield, 1994). There is a great deal of variation in the reported mechanical performance of dense hydroxyapatite ceramics, dependent on phase purity, density and grain size, but the properties cited generally fall in the range shown in Table I.2.4.10.

The bonding mechanism of dense HA implants appear to be very different from that described above for bioactive glasses. The bonding process for HA implants has been described by Jarcho (1981). A cellular bone matrix from differentiated osteoblasts appears at the surface, producing a narrow amorphous electron-dense band only 3 to 5 μm wide. Between this area and the cells, collagen bundles are seen. Bone mineral crystals have been identified in this amorphous area. As the site matures, the bonding zone shrinks to a depth of only 0.05 to 0.2 μm (Figure I.2.4.2). The result is normal bone attached through a thin epitaxial bonding layer to the bulk implant. TEM image analysis of dense HA bone interfaces shows an almost perfect epitaxial alignment of some of the growing bone crystallites with the apatite crystals in the implant. A consequence of this ultra-thin bonding zone is a very high gradient in elastic modulus at the bonding interface between HA and bone. This is one of the major differences between the bioactive apatites and the bioactive glasses and glass-ceramics. The implications of this difference for the implant interfacial response to Wolff's law are discussed in Hench and Ethridge (1982), Chapter 14.

CALCIUM PHOSPHATE COATINGS

The clinical application of calcium phosphate ceramics is largely limited to bone grafting applications or, if in the form of a monolithic or porous component, to non-major loadbearing parts of the skeleton. This is because of relatively poor mechanical strength and toughness, and inferior mechanical properties, and it was partly for this reason that interest was directed toward the use of calcium phosphate coatings on metallic implant subtrates. Good reviews of the production of calcium phosphate coatings have been published by de Groot et al. (1998), Sun et al. (2001), and Heimann (2006). Many techniques are available for the deposition of hydroxyapatite coatings, including electrophoresis, sol–gel routes, electrochemical routes, biomimetic routes, and sputter techniques, but the most popular commercial routes are those based on plasma spraying. In plasma spraying, an electric arc is struck between two electrodes and a stream of gases is passed through the arc. The arc converts the gases into plasma with a speed of up to 400 m/sec and a temperature within the arc of 20,000°K. The ceramic powder is suspended in the carrier gas and fed into the plasma, where it can be fired at a substrate. There are many variables in the process including the gases used, the electrical settings, the nozzle/substrate separation, and the morphology, particle size, and particle size distribution of the powder. Because of the very high temperatures but very short times involved, the behavior of the hydroxyapatite powder particle is somewhat different than might be predicted in an equilibrium phase diagram. However, according to the particular conditions used, it is likely that at least a thin outer layer of the powder particle will be in a molten state and will undergo some form of phase transformation, but by careful control of the operating variables the transformed material should represent a relatively small volume fraction of the coating, and the product should maintain the required phase purity and crystallinity (Mayer et al., 1986; Cook et al., 1988; Wolke et al., 1992).

Early work on plasma sprayed coatings established that there are a number of factors that influence the properties of the resulting coating, including coating thickness (this will influence coating adhesion and fixation – the agreed optimum now seems to be 50–100 μm) (de Groot et al., 1987; Soballe et al., 1993), crystallinity (this affects the dissolution and biological behavior) (Le Geros et al., 1992; Klein et al., 1994a,b; Clemens, 1995), biodegradation (affected by phase purity, chemical purity, porosity, crystallinity) (Tang et al., 2010), and adhesion strength (these may range between 5 and 65 MPa (de Groot, 1998).

Plasma sprayed coatings have been found to be highly successful, and are now widely used in hip joint replacement, particularly for younger patients (Sun et al., 2001). However, the mechanical mismatch between the coating and the substrate can lead to high levels of residual interfacial stress. For this reason, a number of other "low temperature" and thin film deposition techniques been investigated, including electrophoresis, sol–gel routes, electrochemical routes, biomimetic routes, electrohyhdrodynamic spray deposition sputter techniques, and bone-like apatite coatings through simulated body fluid treatments.

Sputtering results in thin (<1 micron) coatings on metallic, ceramic or polymeric substrates (Hulshoff et al., 1996, 1997; Lo et al., 2000; Ong, 2002; Boyd et al., 2003; Cairns et al., 2008). Pure calcium phosphate coatings need to be deposited using magnetron sputtering. Using a combination of two or more targets allows the co-deposition of Si with the calcium phosphate, and the resulting films have been found to exhibit enhanced biological performance (Thian et al., 2005, 2006). However, further investigation of sputtered coatings is required to provide better understanding of their potential long-term clinical performance.

The preparation of SiHA films via sol–gel routes (Hijon et al., 2006) and pulsed laser deposition techniques (Garcia-Sanz et al., 1997) has also been investigated. Fluoride-substituted HA coatings on metallic surfaces using a sol–gel method (Wang et al., 2007). Biomimetic approaches have also been investigated. Substrates are immersed in a supersaturated or metastable solution at 37°C with ion concentrations and pH value similar to human blood plasma. After several days, the formation and growth occurs of a thin "bone-like" apatite layer (Ohtsuki et al., 1991). Biomimetic coatings are generally produced using simple, low cost, low temperature processes. However, long induction and growth periods are required to form the coatings, and adhesion strength between the coating and substrate still requires optimization.

Electrostatic spray deposition allows the fabrication of dense, porous or nanostructured CaP coatings (Leeuwenburgh et al., 2003; Huang et al., 2005; Lim et al., 2005; Ahmad et al., 2006; Huang et al., 2008). Research indicates that this route offers the potential for using CaP coatings as carriers for a variety of drugs (Siebers et al., 2004, 2006).

CALCIUM PHOSPHATE IMPLANTS: MECHANICAL PROPERTIES AND POROSITY

The mechanical behavior of calcium phosphate ceramics strongly influences their application as implants. Tensile and compressive strength and fatigue resistance depend on the total volume of porosity. Porosity can be in the form of micropores (<1 μm diameter, due to incomplete sintering) or macropores (>100 μm diameter, created to permit bone growth). The dependence of compressive strength (σ_c) and total pore volume (V_p) is described in de Groot et al. (1990) by:

$$\sigma_c = 700\exp - 5V_p(\text{in MPa}) \quad (3)$$

where (V_p is in the range of 0–0.5.

Tensile strength depends greatly on the volume fraction of microporosity ((V_m):

$$\sigma_t = 220\exp - 20\ V_m (\text{in MPa}) \quad (4)$$

The Weibull factor (n) of HA implants is low in physiological solutions (n = 12), which indicates low reliability under tensile loads. Consequently, in clinical practice, calcium phosphate bioceramics should be used: (1) as powders; (2) in small, unloaded implants such as in the middle ear; (3) with reinforcing metal posts, as in dental implants; (4) as coatings (e.g., composites) or (5) in porous implants where bone growth acts as a reinforcing phase.

There are a number of commercially available porous products on the market, and these originate from three main sources: hydroxyapatite that has been chemically synthesized (e.g., Apapore®); material has been converted from synthetic sources; coral (e.g., Pro Osteon®, Interpore®, Bio Eye®) or animal bone (e.g., Endobon®). There has been some discussion in the literature regarding the optimum "type" of porosity, and a number of authors have suggested that the degree of interconnectivity is more critical than the pore size. Eggli et al. (1988) demonstrated improved integration in interconnected 50–100 μm pores compared with less connected pores with a size of 200–400 μm. Similarly, Kühne et al. (1994) compared two grades of 25–35% porous coralline apatite with average pore sizes of 200 and 500 μm and reported bone ingrowth to be improved in the 500 μm pore sized ceramic. Holmes (1979), who also studied a porous coralline apatite, suggested that when implanted in cortical bone interconnections of osteonic diameter were required for transport of nutrients to the bone ingrowth. The findings clearly indicate the importance of thorough characterization of the porous materials before implantation, and Hing (1999) has recommended a range of techniques which should be employed.

RESORBABLE CALCIUM PHOSPHATES

Resorption or biodegradation of calcium phosphate ceramics is caused by three factors:

1. Physiochemical dissolution, which depends on the solubility product of the material and local pH of its environment. New surface phases may be formed, such as amorphous calcium phosphate, dicalcium phosphate dihydrate, octacalcium phosphate, and anionic-substituted HA.
2. Physical disintegration into small particles as a result of preferential chemical attack of grain boundaries.
3. Biological factors, such as phagocytosis, which causes a decrease in local pH concentrations (de Groot and Le Geros, 1988).

Ideally, one might wish for a replacement material to be slowly resorbed by the body once its task of acting as a scaffold for new bone has been completed. Degradation or resorption of calcium phosphates *in vivo* occurs by a combination of phagocytosis of particles and the production of acids. However, when selecting a resorbable material for implantation, care must be taken to match the rate of resorption with that of the expected bone tissue regeneration. Where the solubility of calcium phosphates is higher than the rate of tissue regeneration, they will not be of use to fill bone defects. As mentioned previously, the rate of dissolution increases with decreasing calcium-to-phosphorus ratio and, consequently, tricalcium phosphate, with a Ca:P ratio of 1.5, is more rapidly resorbed than hydroxyapatite. Tricalcium phosphate has four polymorphs: α; β; γ; and super α. The γ polymorph is a high pressure phase, and the super α polymorph is observed at temperatures above approximately 1500°C (Nurse et al., 1959). Therefore, the most frequently observed polymorphs in bioceramics are α- and β-TCP. X-ray diffraction studies indicate that the β polymorph transforms to the α polymorph at temperatures between 1120°C and 1290°C (Gibson et al., 1996).

All calcium phosphate ceramics biodegrade to varying degrees; the rate of biodegradation increases as:

1. Surface area increases (powders > porous solid > dense solid);
2. Crystallinity decreases;
3. Crystal perfection decreases;
4. Crystal and grain size decrease;
5. There are ionic substitutions of CO^{-3}_2, Mg^{2+}, and Sr^{2+} in HA.

Factors that tend to decrease the rate of biodegradation include: (1) F^- substitution in HA; (2) Mg^{2+} substitution in β-TCP; and (3) lower β-TCP/HA ratios in biphasic calcium phosphates.

CALCIUM PHOSPHATE BONE CEMENTS

In the 1980s, calcium phosphate bone cements were developed. These materials offer the potential for *in situ* molding and injectability. There are a variety of different combinations of calcium compounds (e.g., α-tricalcium phosphate and dicalcium phosphate) which are used in the formulation of these bone cements, but the end product is normally based either on a calcium deficient hydroxyapatite (Fernandez et al., 1998, 1999a,b) or on brushite (although *in vivo* these tend to convert to hydroxyapatite) (Lemaitre et al., 1987; Constantz et al., 1998). In the development and production of the bone cements a number of factors need to be considered, including the processing parameters (such as solid and liquid component composition, particle size, and liquid-to-powder ratio), setting properties, cohesion time, and the injectability of the paste (Doroszhkin, 2009). These will in turn significantly influence the microstructure and porosity, and hence mechanical behavior of the cement.

The cements have been found to perform well *in vivo*, and are now commercially available (Bohner, 2000). There is considerable interest in the future potential for these materials to be used for drug delivery (Ginebra et al., 2001).

CLINICAL APPLICATIONS OF HA

Calcium phosphate-based bioceramics have been used in medicine and dentistry for over 20 years (Jarcho, 1981; de Groot, 1983; Hulbert et al., 1987; Le Geros, 1988; de Groot, 1988; de Groot et al., 1990; Le Geros and Le Geros, 1993). Applications include dental implants, periodontal treatment, alveolar ridge augmentation, orthopedics, maxillofacial surgery, and otolaryngology (Table I.2.4.6).

Most authors agree that HA is bioactive, and it is generally agreed that the material is osseoconductive, where osseoconduction is the ability of a material to encourage bone growth along its surface when placed in the vicinity of viable bone or differentiated bone-forming cells. Good reviews of porous hydroxyapatite from different sources for eventual clinical application have been prepared by Hing et al. (1998, 2005), and Damien and Revell (2004). Hing et al. (1998) observed that there are a large number of "experimental parameters," including specimen, host, and test parameters, which need to be carefully controlled in order to allow adequate interpretation of data.

Hydroxyapatite has been investigated and used in a range of different forms for clinical applications. It has been utilized as a dense, sintered ceramic for middle ear implant applications, (van Blitterswijk, 1990) and alveolar ridge reconstruction and augmentation (Quin and Kent, 1984; Cranin et al., 1987), in porous form for orbital implants (Suter et al., 2002), as granules for filling bony defects in dental and orthopaedic surgery, such as for spinal fusion and impaction grafting (Froum et al., 1986; Galgut et al., 1990; Oonishi et al., 1990; Wilson and Low, 1992; Aoki, 1994; Fujishiro, 1997; Bolder et al., 2002; Hing et al., 2007; Wheeler al., 2007; Coathup et al., 2008), and as a coating on metal implants (de Groot, 1987; Cook et al., 1992a,b; Camazzola et al., 2009). There are now numerous reports of the successful use of hydroxyapatite coated femoral prostheses and acetabular cups, and clinical results indicate that these implants are especially successful for younger patients (Stack et al., 2006; Shah, 2009).

Another successful clinical application of hydroxyapatite has been in polymer composites. The original concept of a bioceramic filler in a polymer composite was introduced by Bonfield et al. (1981) with the aim of producing an analog of the mineral-reinforced organic matrix in cortical bone. The material developed by Bonfield and co-workers contains up to 50 vol% hydroxyapatite in a polyethylene matrix, has a stiffness similar to cortical bone, has high toughness, and has been found to exhibit bone-bonding *in vivo*. The material has been used as an orbit implant for orbital floor fractures and volume augmentation (Tanner et al., 1994), and is now in used in middle ear implants, commercialized under the trade name, HAPEX® (Bonfield, 1996).

Other approaches have been made to produce bioactive and either bioresorbable or biodurable composites. The incorporation of glass ceramics A-W and Bioglass® have been investigated within a wide variety of polymer matrices (Thompson et al., 1998; Rea et al., 2004). Another recent advance has been in the development of a compositionally-graded mineralized collagen-GAG scaffold for cartilage and ligament repair (Lynn et al., 2005).

REFERENCES

Adachi, N., Ochi, M., Deie, M., & Ito, Y. (2005). Transplant of mesenchymal stem cells and hydroxyapatite ceramics to treat severe osteochondral damage after septic arthritis of the knee. *J. Rheumatol.*, **32**, 1615–1618.

Ahmad, Z., Huang, J., Edirisinghe, M. J., Jayasinghe, S. N., Best, S. M., Bonfield, W., Brooks, R. A., & Rushton, N. (2006). Electrohydrodynamic print-patterning of nanohydroxyapatite. *J. Biomed. Nanotechnol.*, **2**, 201–207.

Akao, M., Aoki, H., & Kato, K. (1981). Mechanical properties of sintered hydroxyapatite for prosthetic applications. *J. Mat. Sci.*, **16**, 809.

Ameri, E., Behtash, H., Mobini, B., Omidi-Kashani, F., & Nojomi, M. (2009). Bioactive glass versus autogenous iliac crest bone graft in adolescent idiopathic scoliosis surgery. *Acta Medica Iranica*, **47**(1), 41–45.

Andreiotelli, M., Wenz, H. J., & Kohal, R. J. (2009). Are ceramic implants a viable alternative to titanium implants? A systematic literature review. *Clin. Oral Implants Res.*, **20**(Suppl 4), 32–47.

ASTM F1185-03. (2003). *Standard Specification for Composition of Hydroxylapatite for Surgical Implants, Book of Standards, 13.01*.

Baikie, T., Mercier, P. H., Elcombe, M. M., Kim, J. Y., Le Page, Y., Mitchell, L. D., White, T. J., & Whitfield, P. S. (2007). Triclinic apatites. *Acta. Crystallogr. B.*, **63**, 251–256.

Balas, F., Perez-Pariente, J., & Vallet-Regi, F. (2003). *In vitro* bioactivity of silicon substituted hydroxyapatites. *J. Biomed. Mater. Res. A*, **66**, 364–375.

Barralet, J. E., Best, S. M., & Bonfield, W. (1998). Carbonate substitution in precipitated hydroxyapatite: An investigation into the effects of reaction temperature and bicarbonate Ion concentration. *J. Biomed. Mater. Res.*, **41**, 79–86.

Barralet, J. E., Gaunt, T., Wright, A. J., Gibson, I. R., & Knowles, J. C. (2002). Effect of porosity reduction by compaction on compressive strength and microstructure of calcium phosphate cement. *J. Biomed. Mater. Res.*, **63**, 1–9.

Bertoni, E., Bigi, A., Cojazzi, G., Gandolfi, M., Panzavolta, S., & Roveri, N. (1998). Nano–crystals of magnesium and fluoride substituted hydroxyapatite. *J. Inorg. Biochem.*, **72**, 29–35.

Best, S. M., & Bonfield, W. (1994). Processing behaviour of hydroxyapatite powders of contrasting morphology. *J. Mater. Sci. Mat. Med.*, **5**, 516.

Blumenthal, N. C., Betts, F., & Posner, A. S. (1975). Effect of carbonate and biological macromolecules on formation and properties of hydroxyapatite. *Calcif. Tissue Res.*, **18**, 81–90.

Bocholz, R. W., Carlton, A., & Holme, R. E. (1987). Hydroxyapatite and tricalcium phosphate bone graft substitute. *Orthop. Clin. North Am.*, **18**, 323–334.

Bohner, M. (2000). Calcium orthophosphates in medicine: From ceramics to calcium phosphate cements. *Injury Int J. Care Injured*. 31 S-D37–47.

Bolder, S. B., Verdonschot, N., Schruers, B. W., & Buma, P. (2002). Acetabular defect reconstruction with impacted morcellized bone grafts or TCP/HA particles. A study on the mechanical stability of cemented cups in an artificial acetabulum model. *Biomaterials*, **23**(3), 659–666.

Bonel, G., Heughebeart, J-C., Heughebaert, M., Lacout, J. L., & Lebugle, A. (1987). Apatitic calcium orthophosphates and related compounds for biomaterials preparation. *Annals of the New York Academy of Science*, **523**, 115–130.

Bonfield, W. (1996). Composite biomaterials. In T. Kukubo, T. Nakamura, & F. Miyaji (Eds.), *Bioceramics 9*. Pergamon: Proceedings of the 9th International Symposium on Ceramics in Medicine.

Bonfield, W., Grynpas, M. D., Tully, A. E., Bowman, J., & Abram, J. (1989). Hydroxyapatite reinforced polyethylene: A mechanically compatible implant. *Biomaterials*, **2**, 185–186.

Boskey, A. (2007). Mineralization of Bones and Teeth. In Nita Sahai (Ed.), *Elements Magazine* (3, pp. 385–392). : .

Boskey, A. L. (1997). Amorphous calcium phosphate: The contention of bone. *J. Dent. Res.*, **76**, 1433–1436.

Boskey, A. L., & Mendelsohn, R. (2005). Infrared spectroscopic characterization of mineralized tissues. *Vib. Spectrosc.*, **38**, 107–114.

Boyd, A., Akay, M., & Meenan, B. J. (2003). Influence of target surfqof r.f. magnetron sputtered calcium phosphate coatings. *Surface and Interface Analysis*, **35**(2), 188–198.

Bucholz, R. W. (2002). Nonallograft osteoconductive bone graft substitutes. *Clin. Orthop. Relat. Res.*, **395**, 44–52.

Buttery, L. D. K., Bourne, S., Xynos, I. D., Wood, H., Hughes, F. J., Hughes, S. P. F., Episkopou, V., & Polak, J. M. (2001). Differentiation of osteoblasts and *in vitro* bone formation from murine embryonic stem cells. *Tissue Eng.*, **7**, 89–99.

Cairns, M. L., Meenan, B. J., Burke, G. A., & Boyd, A. (2008). Effect of nanoscale topography on fibronectin adsorption to sputter deposited calcium phosphate thin films. Int. J. *Nano and Biomaterials*, **1**(3), 280–298.

Camazzola, D., Hammond, T., Gandhi, R., & Davey, J. R. (2009). A randomized trial of hydroxyapatite-coated femoral stems in total hip arthroplasty: A 13 year follow-up. *J. Arthroplasty*, **24**(1), 33–37.

Carden, A., & Morris, M. D. (2000). Application of vibrational spectroscopy to the study of mineralized tissues. *J. Biomed. Opt.*, **5**, 259–268.

Carlisle, E. M. (1970). Silicon: A possible factor in bone calcification. *Science*, **167**, 279–280.

Cavalcanti, A. N., Foxton, R. M., Watson, T. F., Oliveira, M. T., Giannini, M., & Marchi, G. M. (2009). Y-TZP ceramics: Key concepts for clinical application. *Oper. Dent*, **34**, 344–351.

Chai, C., Ben-Nissan, B., Pyke, S., & Evans, L. (1995). Sol gel derived hydroxyapatite coatings for biomedical applications. *Mater. Manuf. Processes.*, **10**, 205–216.

Chiroff, R. T., White, E. W., Weber, K. N., & Roy, D. M. (1975). Tissue ingrowth of Replamineform implants. *J. Biomed. Mater. Res.*, **9**, 29–45.

Chowdhury, S., Thomas, V., Dean, D., Catledge, S. A., & Vohra, Y. K. (2005). Nanoindentation on porous bioceramic scaffolds for bone tissue engineering. *J. Nanosci. Nanotechnol.*, **5**, 1816–1820.

Christel, P. S. (1992). Biocompatibility of surgical-grade dense polycrystalline alumina. *Clin. Orthop. Relat. Res*, 10–18.

Christodoulou, I., Buttery, L. D., Saravanapavan, P., Tai, G. P., Hench, L. L., & Polak, J. M. (2005). Characterization of human foetal osteoblasts by microarray analysis following stimulation with 58S bioactive gel-glass ionic dissolution products. *Journal of Biomedical Materials Research Part B-Applied Biomaterials*, **77B**, 431–446.

Ciocca, D. R., Frayssinet, P., & Cuello-Carrion, F. D. (2007). A pilot study with a therapeutic vaccine based on hydroxyapatite ceramic particles and self-antigens in cancer patients. *Cell. Stress Chaperones.*, **12**, 33–43.

Coathup, M., Smith, N., Kinglsley, C., Buckland, T., Dattani, R., Ashcroft, G. P., & Blunn, G. (2008). Impaction grafting with a bone-graft substitute in a sheep model of revision hip replacement. *J. Bone and Joint Surg*, **90-B**(2), 246–253.

Collin, R. L. (1959). Strontium–calcium hydroxyapatite solid solutions: Preparations and lattice constant measurements. *J. Am. Chem. Soc.*, **81**, 5275.

Constantz, B. R., Barr, B. M., Ison, I. C., Fulmer, M. T., Baker, J., McKinney, L. A., Goodman, Ss. B, Gunasekaren, S., Delaney, D. C., Ross, J., & Poser, R. D. (1998). Histological, chemical and crystallographic analysis of four calcium phosphate cements in different rabbit osseous sites. *J. Biomed. Mater. Res.: Appl. Biomats.*, **43**, 451–461.

Cook, S. D., Thomas, K. A., Dalton, J. E., Volkman, R. K., White-cloud, T. S., & Kay, J. E. (1999). Hydroxylapatite coating of porous implants improves bone ingrowth and interface attachment strength. *J. Biomed. Mater. Res.*, **26**, 989–1001.

Cook, S. D., Thomas, K. A., Kay, J. F., & Jarcho, M. (1998). Hydroxylapatite coated titanium for orthopaedic implant applications. *Clin. Orthop. Rel. Res.*, **232**, 225–243.

Costantino, P. D., Friedman, C. D., Jones, K., Chow, L. C., Pelzer, H. J., & Sisson Sr, G. A. (1991). Hydroxyapatite cement: Basic chemistry and histologic properties. *Arch Otolaryngol. Head Neck Surg*, 117,379–384.

Covani, U., Giacomelli, L., Krajewski, A., Ravaglioli, A., Spotorno, L., Loria, P., Das, S., & Nicolini, C. (2007). Biomaterials for orthopedics: A roughness analysis by atomic force microscopy. *J. Biomed. Mater. Res. A.*, **82**, 723–730.

Cranin, A. N., Tobin, G. P., & Gelbman, J. (1987). Applications of hydroxyapatite in oral and maxillofacial surgery, part II: Ridge augmentation and repair of major oral defects. *Compend. Contin. Educ. Dent.*, **8**, 334–345.

Damien, E., & Revell, P. A. (2004). Coralline hydroxyapatite bone graft substitute: A review of experimental studies and biomedical applications. *J. Appl. Biomechanics*, **2**, 65–73.

de Groot, K. (1983). *Bioceramics of Calcium-Phosphate*. Boca Raton, FL: CRC Press.

de Groot, K. (1987). Hydroxylapatite coatings for implants in surgery. In P. Vincenzini (Ed.), *High Tech Ceramics*. Amsterdam: Elsevier.

de Groot, K. (1988). Effect of porosity and physicochemical properties on the stability, resorption, and strength of calcium phosphate ceramics. In Bioceramics: Material Characteristics versus In-Vivo Behavior *Ann. N. Y. Acad. Sci.* (523, pp. 227).

de Groot, K., & Le Geros, R. (1988). In P. Ducheyne, & J. Lemons (Eds.), *Position Papers in Bioceramics: Materials Characteristics versus In-Vivo Behavior. Ann. N. Y. Acad. Sci.*. (523). : 227, 268, 272, .

de Groot, K., Geesink, R., Klein, C. P., & Serekian, P. (1987). Plasma sprayed coatings of hydroxylapatite. *J. Biomed. Mater. Res.*, **21**, 1375–1381.

de Groot, K., Klein, C. P. A. T., Wolke, J. G. C., & de Blieck-Hogervorst, J. (1990). Chemistry of calcium phosphate bioceramics. In T. Yamamuro, L. L. Hench, & J. Wilson (Eds.), *Handbook on Bioactive Ceramics*. (Vol. II). Boca Raton, FL: CRC Press. Ch. 1.

De Maeyer, E. A., Verbeeck, R. M., & Pieters, I. Y. (1996). Effect of K(+) on the stoichiometry of carbonated hydroxyapatite obtained by the hydrolysis of monetite. *Inorg. Chem.*, **35**, 857–863.

Della Bona, A., & Kelly, J. R. (2008). The clinical success of all-ceramic restorations. *J. Am. Dent. Assoc.*, **139**(Suppl), 8S–13S.

Denissen, H. W., de Groot, K., Driessen, A. A., Wolke, J. G.C., Peelen, J. G.J., van Dijk, H. J.A., Gehring, A. P., & Klopper, P. J. (1980). Hydroxylapatite implants: Preparation, properties and use in alveolar ridge preservation. *Sci. Ceram.*, **10**, 63.

Detsch, R., Mayr, H., & Ziegler, G. (2008). Formation of osteoclast-like cells on HA and TCP ceramics. *Acta. Biomater.*, **4**, 139–148.

Dickens, B., Schroeder, L. W., & Brown, W. E. (1974). Crystallographic studies of the role of Mg as a stabilising impurity in β-tricalcium phosphate. *J. Solid State Chem.*, **10**, 232.

Dorozhkin, S. V. (2008). Calcium othophosphate cements for biomedical application. *J. Mater. Sci.*, **43**, 3028–3057.

Driessen, A. A., Klein, C. P.A.T., & de Groot, K. (1982). Preparation and some properties of sintered ß-Whitlockite. *Biomaterials*, **3**, 113–116.

Ellies, L. G., Nelson, D. G., & Featherstone, J. D. (1988). Crystallographic structure and surface morphology of sintered carbonated apatites. *J. Biomed. Mater. Res.*, **22**, 541–553.

Elliot, J. R. (1994). *Structure and Chemistry of Apatites and Other Calcium Orthophosphate*. Amsterdam: Elsevier.

Ellis, D. E., Terra, J., Warschkow, O., Jiang, M., González, G. B., Okasinski, J. S., Bedzyk, M. J., Ross, i. A.M., & Eon, J. G. (2006). A theoretical and experimental study of lead substitution in calcium hydroxyapatite. *Phys. Chem. Chem. Phys.*, **8**, 967–976.

Fabbri, M., Celotti, G. C., & Ravaglioli, A. (1995). Hydroxyapatite-based porous aggregates: Physico-chemical nature, structure, texture and architecture. *Biomaterials*, **16**, 225–228.

Fernandez, E., Gil, F. X., Ginebra, M. P., Driessens, F. C.M., Planell, J. A., & Best, S. M. (1999a). Calcium phosphate bone cements for clinical applications. Part I: Solution chemistry. *J. Mater. Sci. Mater. in Med.*, **10**, 169–176.

Fernandez, E., Gil, F. X., Ginebra, M. P., Driessens, F. C.M., Planell, J. A., & Best, S. M. (1999b). Calcium phosphate bone cements for clinical applications. Part II: Precipitate formation during setting reactions. *J. Mater. Sci. Mater. Med.*, **10**, 177–184.

Fernandez, E., Planell, J. A., Best, S. M., & Bonfield, W. (1998). Synthesis of dahllite through a cement setting reaction. *J. Mater. Sci. Mater. in Med.*, **9**, 789–792.

Fowler, B. O. (1974). Infrared studies of apatites. Part II: Preparation of normal and isotopically substituted calcium, strontium and barium hydroxyapatites and spectra-structure correlations. *Inorg. Chem.*, **13**, 207.

Garcia-Sanz, F. J., Mayor, M. B., Arias, J. L., Pou, J., Leon, B., & Perez-Amor, M. (1997). Hydroxyapatite coatings: A comparative study between plasma-sprayed and pulsed laser deposition techniques. *J. Mater Sci: Mater Med.*, **8**, 861–865.

Gasqueres, G., Bonhomme, P., Maquet, J., Babonneau, F., Hayakawa, S., Kanaya, T., & Osaka, A. (2008). Revisiting silicate substituted hydroxyapatite by solid state NMR. *Magn. Reson. Chem.*, **46**, 342–346.

Gibson, I. R., Best, S. M., & Bonfield, W. (1996). Phase transformations of tricalcium phosphates using high temperature x-ray diffraction. In Nakamura Kokubo, & Miyaji (Eds.), *Bioceramics 9* (pp. 173–176). Oxford: Publ. Pergamon Press.

Gibson, I. R., Best, S. M., & Bonfield, W. (1999). Chemical characterization of silicon-substituted hydroxyapatite. *J. Biomed. Mater. Res.*, **44**, 422–428.

Gibson, I. R., Best, S. M., & Bonfield, W. (1999). Chemical characterisation of silicon-substituted hydroxyapatite. *J. Biomed. Mater. Res.*, **44**, 422–428.

Ginebra, M. P., Rilliard, A., Fernandez, E., Elvira, C., San Roman, J., & Planell, J. A. (2001). Mechanical and rheological improvement of a calcium phosphate cement by the addition of a polymeric drug. *J. Biomed. Mater. Res.*, **57**, 113–118.

Gross, V., & Strunz, V. (1985). The interface of various glasses and glass-ceramics with a bony implantation bed. *J. Biomed. Mater. Res.*, **19**, 251.

Gross, V., Kinne, R., Schmitz, H. J., & Strunz, V. (1988). The response of bone to surface active glass/glass-ceramics. *CRC Crit. Rev. Biocompatibility*, **4**, 2.

Guo, Y., Shi, D., Lian, J., Dong, Z., Wang, W., Cho, H., Liu, G., Wang, L., & Ewing, R. C. (2008). Quantum dot conjugated hydroxylapatite nanoparticles for *in vivo* imaging. *Nanotechnology*, **19**, 175102–175108.

Habibovic, P., Yuan, H., van der Valk, C. M., Meijer, G., van Blitterswijk, C. A., & de Groot, K. (2005). 3D microenvironment as essential element for osteoinduction by biomaterials. *Biomaterials*, **26**, 3565–3575.

Harding, I. S., Rashid, N., & Hing, K. A. (2005). Surface charge and the effect of excess calcium ions on the hydroxyapatite surface. *Biomaterials*, **26**, 6818–6826.

Heimann, R. B. (2006). Thermal spraying of biomaterials. *Surf. Coating Tech.*, **201**, 2012–2019.

Hench, L. L. (1991). Bioceramics: From concept to clinic. *J. Am. Ceram. Soc.*, **74**, 1487–1510.

Hench, L. L. (1998). Bioceramics. *J. Am. Ceram. Soc.*, **81**(7), 1705–1728.

Hench, L. L. (2003). Glass and Genes: The 2001 W. E. S. Turner Memorial Lecture. *Glass Technology*, **44**, 1–10.

Hench, L. L., & Clark, D. E. (1978). Physical chemistry of glass surfaces. *J. Non-Cryst. Solids*, **28**(1), 83–105.

Hench, L. L., & Ethridge, E. C. (1982). *Biomaterials: An Interfacial Approach.*. New York: Academic Press.

Hench, L. L., & Wilson, J. W. (1993). *An Introduction to Bioceramics*. Singapore: World Scientific.

Hench, L. L., & Wilson, J. W. (1996). *Clinical Performance of Skeletal Prostheses*. London: Chapman and Hall.

Hench, L. L., Day, D., & Hohland, W. (2010). Int. Journal of Applied GlassTechnology. *Amer. Ceram. Soc.*, **1**, 101.

Hench, L. L., Splinter, R. J., Allen, W. C., & Greenlee, T. K., Jr. (1971). Bonding mechanisms at the interface of ceramic prosthetic materials. *J. Biomed. Res. Symp.*, 117–141. No. 2. Interscience, New York.

Hench, L. L., Wilson, J. W., & Greenspan, D. C. (2004). *J. Australian Ceram. Soc.*, **40**(1), 1–42.

Heymann, D., Pradal, G., & Benahmed, M. (1999). Cellular mechanisms of calcium phosphate ceramic degradation. *Histol. Histopathol.*, **14**, 871–877.

Hijon, N., Cabanas, M. V., Pena, J., & Vallet-Regi, M. (2006). Dip coated silicon-substituted hydroxyapatite films. *Acta. Biomater*, **2**, 567–574.

Hing, K. A. (2005). Bioceramic bone graft substitutes: Influence of porosity and chemistry. *Int. J. Appl Ceram. Tech.*, **2**(3), 184–199.

Hing, K. A., Best, S. M., & Bonfield, W. (1999). Characterisation of porous hydroxyapatite. *J. Mater. Sci. Mater. Med.*, **10**, 135–160.

Hing, K. A., Best, S. M., Tanner, K. E., Revell, P. A., & Bonfield, W. (1998). Histomorphological and biomechanical characterisation of calcium phosphates in the osseous environment. *Proc. Inst. Mech. Engrs.*, **212**(part H), 437.

Hing, K. A., Wilson, L. F., & Buckland, T. (2007). Comparative performance of three ceramic bone graft substitutes. *Spine J*, **7**(4), 475–490.

Holmes, R. E. (1979). Bone regeneration within a coralline hydroxyapatite implant. *Plast. Reconstr. Surg.*, **63**, 626–633.

Holt, G. E., Halpern, J. L., Dovan, T. T., Hamming, D., & Schwartz, H. S. (2005). Evolution of an *in vivo* bioreactor. *J. Orthop. Res.*, **23**, 916–923.

Hsieh, M. F., Perng, L. H., Chin, T. S., & Perng, H. G. (2001). Phase purity of sol–gel-derived hydroxyapatite ceramic. *Biomaterials*, **22**, 2601–2607.

Huang, J., Jayasinghe, S. N., Best, S. M., Edirisinghe, M. J., Brooks, R. A., & Bonfield, W. (2004). Electrospraying of a nanohydroxyapatite suspension. *J. Mater. Sci.*, **39**, 1029–1032.

Huang, J., Jayasinghe, S. N., Best, S. M., Edirisinghe, M. J., Brooks, R. A., Rushton, N., & Bonfield, W. (2005). Novel deposition of nano-sized silicon substituted hydroxyapatite by electrostatic spraying. *J. Mater. Sci. Mater. Med.*, **16**, 1137–1142.

Huber, F. X., McArthur, N., Hillmeier, J., Kock, H. J., Baier, M., Diwo, M., Berger, I., & Meeder, P. J. (2006). Void filling of tibia compression fracture zones using a novel resorbable nanocrystalline hydroxyapatite paste in combination with a hydroxyapatite ceramic core: First clinical results. *Arch Orthop. Trauma Surg.*, **126**, 533–540.

Huckstep, R. L. (1987). Stabilization and prosthetic replacement in difficult fractures and bone tumors. *Clin. Orthop. Relat. Res*, 12–25.

Hulbert, S. (1993). The use of alumina and zirconia in surgical implants. In L. L. Hench, & J. Wilson (Eds.), *An Introduction to Bioceramics* (pp. 25–40). Singapore: World Scientific.

Hulbert, S. F., Young, F. A., Mathews, R. S., Klawitter, J. J., Talbert, C. D., & Stelling, F. H. (1970). Potential of ceramic materials as permanently implantable skeletal prosthesis. *J. Biomed. Mater. Res.*, **4**, 433–456.

Hulshoff, J. E., Hayakawa, T., van Dijk, K., Leijdekkers-Govers, A. F., van der Waerden, J. P., & Jansen, J. A. (1997). Mechanical and histologic evaluation of Ca-P plasma-sprayed and magnetron sputter-coated implants in trabecular bone of the goat. *J. Biomed. Mater. Res.*, **36**, 75–83.

Hulshoff, J. E., van Dijk, K., van der Waerden, J. P., Wolke, J. G., Kalk, W., & Jansen, J. A. (1996). Evaluation of plasma-sprayed and magnetron-sputter Ca-P-coated implants: An *in vivo* experiment using rabbits. *J. Biomed. Mater. Res.*, **31**, 329–337.

Ilharreborde, B., Morel, E., Fitoussi, F., Presedo, A., Souchet, P., Pennecot, G., & Mazda, K. (2008). Bioactive glass as a bone substitute for spinal fusion in adolescent idiopathic scoliosis: A comparative study with iliac crest autograft. *J. Pediatr. Orthop.*, **28**, 347–351.

Isago, T., Nozaki, M., Kikuchi, Y., Honda, T., & Nakazawa, H. (2004). Sinking skin flap syndrome: A case of improved cerebral blood flow after cranioplasty. *Ann. Plast. Surg.*, **53**, 288–292.

Jäger, C., Welzel, T., Meyer-Zaika, W., & Epple, M. (2006). A solid-state NMR investigation of the structure of nanocrystalline hydroxyapatite. *Magn. Reson. Chem.*, **44**, 573–580.

Jarcho, M. (1981). Calcium phosphate ceramics as hard tissue prosthetics. *Clin. Orthop. Rel. Res.*, **157**, 281–288.

Jarcho, M. (1981). Calcium phosphate ceramics as hard tissue prosthetics. *Clin. Orthop. Relat. Res.*, **157**, 259.

Jarcho, M., Bolen, C. H., Thomas, M. B., Bobick, J., Kay, J. F., & Doremus, R. H. (1976). Hydroxylapatite synthesis in dense polycrystalline form. *J. Mat. Sci.*, **11**, 2027.

Jha, L., Best, S. M., Knowles, J., Rehman, I., Santos, J., & Bonfield, W. (1997). Preparation and characterisation of fluoride-substituted apatites. *J. Mater. Sci. Mater. in Med.*, **8**, 185–191.

Jones, A. C., Arns, C. H., Hutmacher, D. W., Milthorpe, B. K., Sheppard, A. P., & Knackstedt, M. A. (2009). The correlation of pore morphology, interconnectivity and physical properties of 3D ceramic scaffolds with bone ingrowth. *Biomaterials*, **30**(7), 1440–1451.

Jones, C. B., Sabatino, C. T., Badura, J. M., Sietsema, D. L., & Marotta, J. S. (2008). Improved healing efficacy in canine ulnar segmental defects with increasing recombinant human bone morphogenetic protein-2/allograft ratios. *J. Orthop. Trauma.*, **22**, 550–559.

Jones, J. R., Lee, P. D., & Hench, L. L. (2006). Hierarchical porous materials for tissue engineering. *Philosophical Transactions of the Royal Society A*, **364**, 263–281.

Joschek, S., Nies, B., Krotz, R., & Göferich, A. (2000). Chemical and physicochemical characterization of porous hydroxyapatite ceramics made of natural bone. *Biomaterials*, **21**, 1645–1658.

Jun, Y. K., Kim, W. H., Kweon, O. K., & Hong, S. H. (2003). The fabrication and biochemical evaluation of alumina reinforced calcium phosphate porous implants. *Biomaterials*, **24**, 3731–3739.

Kawata, M., Uchida, H., Itatani, K., Okada, I., Koda, S., & Aizawa, M. (2004). Development of porous ceramics with well-controlled porosities and pore sizes from apatite fibers and their evaluations. *J. Mater. Sci. Mater. Med.*, **15**, 817–823.

Kay, M. I., Young, R. A., & Posner, A. S. (1964). Crystal structure of hydroxyapatite. *Nature*, **204**, 1050–1052.

Kay, M. I., Young, R. A., & Posner, A. S. (1964). Crystal structure of hydroxyapatite. *Nature*, **12**, 1050.

Khan, S. N., Fraser, J. F., Sandhu, H. S., Cammisa, F. P., Jr., Girardi, F. P., & Lane, J. M. (2005). Use of osteopromotive growth factors, demineralized bone matrix, and ceramics to enhance spinal fusion. *J. Am. Acad. Orthop. Surg.*, **13**, 129–137.

Kijima, T., & Tsutsumi, M. (1979). Preparation and thermal properties of dense polycrystalline oxyhydroxyapatite. *J. Am. Ceram. Soc.*, **62**(9), 455.

Kim, H. W., Knowles, J. C., Li, L. H., & Kim, H. E. (2005). Mechanical performance and osteoblast-like cell responses of fluorine-substituted hydroxyapatite and zirconia dense composite. *J. Biomed. Mater. Res. A.*, **72**, 258–268.

Klein, C. P., Driessen, A. A., de Groot, K., & van den Hooff, A. (1983). Biodegration behaviour of various calcium phosphate materials in bone tissue. *J. Biomed. Mater. Res.*, **17**, 769–784.

Klein, C. P., Wolke, J. G., de Blieck-Hogervorst, J. M., & de Groot, K. (1994). Features of calcium phosphate coatings: An *in vivo* study. *J. Biomed. Mater. Res.*, **28**, 909–917.

Kobayashi, S., Hara, H., Okudera, H., Takemae, T., & Sugita, K. (1987). Usefulness of ceramic implants in neurosurgery. *Neurosurgery*, **21**, 751–755.

Kokubo, T. (1993). A/W glass-ceramics: Processing and properties. In L. L. Hench, & J. Wilson (Eds.), *An Introduction to Bioceramics* (pp. 75–88). Singapore: World Scientific.

Kokubo, T., Kim, H. M., & Kawashita, M. (2003). Novel bioactive materials with different mechanical properties. *Biomaterials*, **24**, 2161–2175.

Kothapalli, C. R., Wei, M., Legeros, R. Z., & Shaw, M. T. (2005). Influence of temperature and aging time on HA synthesized by the hydrothermal method. *J. Mater. Sci. Mater. Med.*, **16**, 441–446.

Kubo, M., Kuwayama, N., Hirashima, Y., Takaku, A., Ogawa, T., & Endo, S. (2003). Hydroxyapatite ceramics as a particulate embolic material: Report of the clinical experience. *A.J.N.R. Am. J. Neuroradiol.*, **24**, 1545–1547.

Kühne, J. H., Bartl, R., Frish, B., Hanmer, C., Jansson, V., & Zimmer, M. (1994). Bone formation in coralline hydroxyapatite: Effects of pore size studied in rabbits. *Acta. Orthop. Scand*, **65**(3), 246–252.

Kundu, B., Sinha, M. K., Mitra, S., & Basu, D. (2005). Synthetic hydroxyapatite-based integrated orbital implants: A human pilot trial. Indian J. *Ophthalmol.*, **53**, 235–241.

Lacefield, W. R. (1993). Hydroxylapatite coatings. In L. L. Hench, & J. Wilson (Eds.), *An Introduction to Bioceramics* (pp. 223–238). Singapore: World Scientific.

Landi, E., Tampieri, A., Celotti, G., Langenati, R., Sandri, M., & Sprio, S. (2005). Nucleation of biomimetic apatite in synthetic body fluids: Dense and porous scaffold development. *Biomaterials*, **26**, 2835–2845.

Landis, W. J., Moradian-Oldak, J., & Weiner, S. (1991). Topographic imaging of mineral and collagen in the calcifying turkey tendon. *Connect. Tissue Res.*, **25**, 181–196.

Lang, J. E., Whiddon, D. R., Smith, E. L., & Salyapongse, A. K. (2008). Use of ceramics in total hip replacement. *J. Surg. Orthop. Adv.*, **17**, 51–57.

Geros, Le, R.Z., & Le Geros, J. P. (1993). Dense hydroxyapatite. In L. L. Hench, & J. Wilson (Eds.), *An Introduction to Bioceramics* (pp. 139–180). Singapore: World Scientific.

Le Huec, J. C., Schaeverbeke, T., Clement, D., Faber, J., & Le Rebeller, A. (1995). Influence of porosity on the mechanical resistance of hydroxyapatite ceramics under compressive stress. *Biomaterials*, **10**, 113–118.

Lee, S. J., Yoon, Y. S., Lee, M. H., & Oh, N. S. (2007). Nanosized hydroxyapatite powder synthesized from eggshell and phosphoric acid. *J. Nanosci. Nanotechnol.*, **7**, 4061–4064.

Leeuwenburgh, S., Wolke, J., Schoonman, J., & Jansen, J. A. (2003). Electrostatic spray deposition (ESD) of calcium phosphate coatings. *J. Biomed. Mater. Res.*, **66A**, 330–334.

LeGeros, R. Z. (1965). Effect of carbonate on the lattice parameters of apatite. *Nature*, **206**, 403.

LeGeros, R. Z. (1988). Calcium phosphate materials in restorative dentistry: A review. *Adv. Dent. Res.*, **2**(1), 164.

LeGeros, R. Z. (2002). Properties of osteoconductive biomaterials: Calcium phosphates. *Clin. Orthop. Relat. Res.*, **395**, 81–98.

LeGeros, R. Z., & LeGeros, J. P. (2008). Hydroxyapatite. In T. Kokubo (Ed.), *Handbook of Bioceramics and their Applications* (pp. 367–394). London: Woodhead Publishing Ltd.

LeGeros, R. Z., Daculsi, G., Orly, I., Gregoire, M., Heughebeart, M., Gineste, M., & Kijkowska, R. (1992). Formation of carbonate apatite on calcium phosphate materials: Dissolution/precipitation processes. In P. Ducheynes, T. Kukubo, & C. A. van Blitterswijk (Eds.), *Bone Bonding Biomaterials* (pp. 78088). The Netherlands: Reed Healthcare Communications, Leiderdorp.

Lemaitre, J., Mirtchi, A., & Mortier, A. (1987). Calcium phosphate cements for medical use: State of the art and perspectives of development. *Silicate Industries*, **9-10**, 141–146.

Leventouri, T., Bunaciu, C. E., & Perdikatsis, V. (2003). Neutron powder diffraction studies of silicon-substituted hydroxyapatite. *Biomaterials*, **24**, 4205–4211.

Levin, E. M., Robbins, C. R., & McMurdie, H. F. (Eds.), (1964). *American Ceramic Society, Phase Diagrams for Ceramists* (pp. 107). Ohio: American Ceramics Society.

Li, B., Chen, X., Guo, B., Wang, X., Fan, H., & Zhang, X. (2009). Fabrication and cellular biocompatibility of porous carbonated biphasic calcium phosphate ceramics with a nanostructure. *Acta. Biomater.*, **5**, 134–143.

Lieberman, J. R., Daluiski, A., & Einhorn, T. A. (2002). The role of growth factors in the repair of bone. Biology and clinical applications. *J. Bone Joint. Surg. Am.*, **84A**, 1032–1044.

Lim, Y. M., Kim, B. H., Jeon, Y. S., Jeon, K. O., & Hwang, KS. (2005). Calcium phosphate films deposited by electrostatic spray deposition and an evaluation of their bioactivity. *J. Ceram. Proc. Res.*, **6**, 255–258.

Lin, F. H., Liao, C. J., Chen, K. S., Sun, J. S., & Lin, C. Y. (2000). Preparation of betaTCP/HAP biphasic ceramics with natural bone structure by heating bovine cancellous bone with the addition of $(NH_{(4)})(2)HPO_{(4)}$. *J. Biomed. Mater. Res.*, **51**, 157–163.

Liu, D. M. (1997). Fabrication of hydroxyapatite ceramic with controlled porosity. *J. Mater. Sci. Mater. Med.*, **8**, 227–232.

Lo, W. J., Grant, D. M., Ball, M. D., Welsh, B. S., Howdle, S. M., Antonov, E. N., Bagratasshvili, V. N., & Popov, V. K. (2000). Physical, chemical and biological characterisation of pulsed laser deposited and plasma sputtered hydroxyapatite thin films on titanium alloy. *J. Biomed. Mater. Res. Part A*, **50**(4), 536–545.

Lovelace, T. B., Mellonig, J. T., Meffert, R. M., Jones, A. A., Nummikoski, P. V., & Cochran, D. L. (1998). Clinical evaluation of bioactive glass in the treatment of periodontal osseous defects in humans. *J. Periodontol.*, **69**, 1027–1035.

Lu, J., Blary, M. C., Vavasseur, S., Descamps, M., Anselme, K., & Hardouin, P. (2004). Relationship between bioceramics sintering and micro-particles-induced cellular damages. *J. Mater. Sci. Mater. Med.*, **15**, 361–365.

Lynn, A. K., Nakamura, T., Patel, N., Porter, A. E., Renouf, A. C., Laity, P. R., Best, S. M., Cameron, R. E., Shimizu, Y., & Bonfield, W. (2005). Composition-controlled nanocomposites of apatite and collagen incorporating silicon as an osseopromotive agent. *J. Biomed Mater Res A*, **1**, **74**, **3**, 447–453.

Maxian, S. H., Zawaddsky, J. P., & Dunn, M. G. (1994). Effect of calcium phosphate coating resorption and surgical fit on the bone/implant interface. *J. Biomed. Mater. Res.*, **28**, 1311–1319.

McDowell, H., Gregory, T. M., & Brown, W. E. (1977). Solubility of $Ca_5(PO_4)_3OH$ in the system $Ca(OH)_2 - H_3PO_4 - H_2O$ at 5, 15, 25 and 37°C. *J. Res. Natl. Bureau Standards*, **81A**, 273.

Medina Ledo, H., Thackray, A. C., Jones, I. P., Marquis, P. M., Macaskie, L. E., & Sammons, R. L. (2008). Micro structure and composition of biosynthetically synthesised hydroxyapatite. *J. Mater. Sci. Mater. Med.*, **19**, 3419–3427.

Mengel, R., Soffner, M., & Flores-de-Jacoby, L. (2003). Bioabsorbable membrane and bioactive glass in the treatment of intrabony defects in patients with generalized aggressive periodontitis: Results of a 12-month clinical and radiological study. *J. Periodontol*, **74**(6), 899–908.

Moioli, E. K., Clark, P. A., Chen, M., Dennis, J. E., Erickson, H. P., Gerson, S. L., & Ma, J. J. (2008). Synergistic actions of hematopoietic and mesenchymal stem/progenitor cells in vascularizing bioengineered tissues. *PloS One*, **3**, 1–11.

Nakamura, T., Yamumuro, T., Higashi, S., Kokubo, T., & Itoo, S. (1985). A new glass-ceramic for bone replacement: Evaluation of its bonding to bone tissue. *J. Biomed. Mater. Res.*, **19**, 685.

Newsley, H. (1963). Crystallographic and morphological study of carbonate-apatite, M.h.f. *Chem.*, **95**, 270.

O'Donnell, M. D., Fredholm, Y., de Rouffignac, A., & Hill, R. G. (2008). Structural analysis of a series of strontium-substituted apatites. *Acta. Biomater.*, **4**, 1455–1464.

Ohtsuki, C., Kushitani, H., Kokubo, T., Kotani, S., & Yamamuro, T. (1991). Apatite formation on the surface of ceravital-type glass-ceramic in the body. *J. Biomed. Mater. Res.*, **25**, 1363–1370.

Okazaki, M., & Takahashi, J. (1999). Synthesis of functionally graded CO_3 apatite as surface biodegradable crystals. *Biomaterials*, **20**, 1073–1078.

Okumura, T., Oda, Y., & Mori, K. (1984). Tissue compatibility of newly developed bioceramics. *Neurol. Med. Chir. (Tokyo)*, **24**, 909–914.

Ong, J. L., Bessho, K., Cavin, R., & Carnes, D. L. (2002). Bone response to radio frequency sputtered calcium phosphate implants and titanium implants *in vivo*. *J. Biomed. Mater. Res.*, **59A**, 184–190.

Ono, I., Tateshita, T., Satou, M., Sasaki, T., Matsumoto, M., & Kodama, N. (1999). Treatment of large complex cranial bone defects by using hydroxyapatite ceramic implants. *Plast. Reconstr. Surg.*, **104**, 339–349.

Onoda, G., & Hench, L. L. (1978). *Ceramic Processing before Firing*. New York: Wiley.

Otto, B., & Ogilvie, A. (1998). The response of peritoneal macrophages after implantation of several ceramics as measured by the change of ectoenzyme activity. *Biomaterials*, **19**, 1049–1055.

Patel, N., Gibson, I. R., Hing, K. A., Best, S. M., Revell, P. A., & Bonfield, W. (2002). A comparative study on the *in vivo* behaviour of hydroxyapatite and silicon substituted hydroxyapatite granules. *Journal of Materials Science: Materials in Medicine*, **13**, 1199–1206.

Patel, N., Gibson, I. R., Ke, S., Best, S. M., & Bonfield, W. (2001). Calcining influence on the powder properties of hydroxyapatite. *J. Mater. Sci. Mater. Med.*, **12**, 181–188.

Paul, W., Nesamony, J., & Sharma, C. P. (2002). Delivery of insulin from hydroxyapatite ceramic microspheres: Preliminary *in vivo* studies. *J. Biomed. Mater. Res.*, **61**, 660–662.

Perloff, A., & Posner, A. S. (1956). Preparation of pure hydroxyapatite crystals. *Science*, **124**, 583.

Peters, F., Schwarz, K., & Epple, M. (2000). The structure of bone studied with synchrotron X-ray diffraction, X-ray absorption spectroscopy and thermal analysis. *Thermochim. Acta.*, **361**, 131–138.

Phillips, R. W. (1991). Skinners Science of Dental Materials. In Ralph W. Phillips (Ed.), (9th ed.). Philadelphia: Saunders.

Pietak, A. M., Reid, J. W., Stott, M. J., & Sayer, M. (2007). Silicon substitution in the calcium phosphate bioceramics. *Biomaterials*, **2**, 84023–84032.

Polack, F. M., & Heimke, G. (1980). Ceramic keratoprostheses. *Ophthalmology*, **87**, 693–698.

Popat, K. C., Leary Swan, E. E., Mukhatyar, V., Chatvanichkul, K. I., Mor, G. K., Grimes, C. A., & Desai, T. A. (2005). Influence of nanoporous alumina membranes on long-term osteoblast response. *Biomaterials*, **26**, 4516–4522.

Porter, A. E., Best, S. M., & Bonfield, W. (2003). Comparison of the ultrastructure of hydroxyapatite and silicon-substituted hydroxyapatite for biomedical applications. *Journal of Biomedical Materials Research*, **68A**(1), 133–141.

Porter, A. E., Patel, N., Skepper, J. N., Best, S. M., & Bonfield, W. (2004). Effect of sintered silicate-substituted hydroxyapatite on remodelling processes at the bone-implant interface. *Biomaterials*, **25**(16), 3303–3314.

Qian, J., Kang, Y., Zhang, W., & Li, Z. (2008). Fabrication, chemical composition change and phase evolution of biomorphic hydroxyapatite. *J. Mater. Sci. Mater. Med.*, **19**, 3373–3383.

Ramanathan, C., & Ackerman, J. L. (1999). Quantitative solid-state NMR imaging of synthetic calcium phosphate implants. *Magn. Reson. Med.*, **41**, 1214–1220.

Rea, S. M., & Bonfield, W. (2004). Biocomposites for medical applications. *J. Aust. Ceram. Soc.*, **40**(1), 43–57.

Reed, J. S. (1988). *Introduction to Ceramic Processing*. New York: Wiley.

Reynolds, M. A., Aichelmann-Reidy, M. E., Branch-Mays, G. L., & Gunsolley, J. C. (2003). The efficacy of bone replacement grafts in the treatment of periodontal osseous defects. A systematic review. *Ann. Periodontol.*, **8**, 227–265.

Ritter, J. E., Jr., Greenspan, D. C., Palmer, R. A., & Hench, L. L. (1979). Use of fracture mechanics theory in lifetime predictions for alumina and bioglass-coated alumina. *J. Biomed. Mater. Res.*, **13**, 251–263.

Roy, D. M. (1971). Crystal growth of hydroxyapatite. *Mater. Res. Bull.*, **6**, 1337.

Roy, D. M., & Linnehan, S. K. (1974). Hydroxyapatite formed from coral skeletal carbonate by hydrothermal exchange. *Nature*, **247**, 220–222.

Ruys, A. (1993). J., Silicon-doped hydroxyapatite. *J. Aust. Ceram Soc.*, **29**, 71–80.

Ruys, A. J., Wei, M., Sorrell, C. C., Dickson, M. R., Brandwood, A., & Milthorpe, B. K. (1995). Sintering effects on the strength of hydroxyapatite. *Biomaterials*, **16**, 409–415.

Sakamoto, M., Nakasu, M., Matsumoto, T., & Okihana, H. (2007). Development of superporous hydroxyapatites and their examination with a culture of primary rat osteoblasts. *J. Biomed. Mater. Res. A.*, **82**, 238–242.

Salyer, K. E., & Hall, C. D. (1989). Porous hydroxyapatite as an onlay bone-graft substitute for maxillofacial surgery. *Plast Reconstr. Surg.*, **84**, 236–244.

Sato, K., Kogure, T., Iwai, H., & Tanaka, J. (2008). Atomic-scale (101-0) interfacial structure in hydroxyapatite determined by high-resolution transmission electron microscopy. *J. Am. Ceramic Soc.*, **85**, 3054–3058.

Schmitz, J. P., Hollinger, J. O., & Milam, S. B. (1999). Reconstruction of bone using calcium phosphate bone cements: A critical review. *J. Oral. Maxillofac. Surg.*, **57**, 1122–1126.

Schors, E. C., & Holmes, R. E. (1993). Porous hydroxyapatite. In L. L. Hench, & J. Wilson (Eds.), *An Introduction to Bioceramics* (pp. 181–198). Singapore: World Scientific.

Schuessele, A., Mayr, H., Tessmar, J., & Goepferich, A. (2008). Enhanced bone morphogenetic protein-2 performance on hydroxyapatite ceramic surfaces. *J. Biomed. Mater. Res. A*, **90**(4), 959–971.

Sebers, M. C., Walboomers, X. F., Leeuwenburgh, SC. G., Wolke, JC. G., Boerman, O. C., & Jansen, J. A. (2006). Transforming growth factor – beta 1 release from a porous electrostatic spray deposition-derived calcium phosphate coating. *Tissue Eng.*, **12**, 2449–2456.

Shah, N. N., Edge, A. J., & Clark, D. W. (2009). Hydroxyapatite ceramic-coated components in young patients followed up for 16 to 19 years. *J. Bone and Joint Surgery*, **91-B**(7), 865–869.

Shapoff, C. A., Alexander, D. C., & Clark, A. E. (1997). Clinical use of a bioactive glass particulate in the treatment of human osseous defects. *Compendium Contin. Educ. Dent*, **18**(4), 352–363.

Siebers, M. C., Walboomers, X. F., Leeuwenburgh, S. C., & Wolke, J. G. (2004). Jansen J. A., Electrostatic spray deposition (ESD) of calcium phosphate coatings, an *in vitro* study with osteoblast-like cells. *Biomater.*, **25**, 2019–2027.

Siebers, M. C., Walboomers, X. F., Leeuwenburgh, S. C., Wolke, JG., & Jansen, J. A. (2006). The influence of the crystallinity of electrostatic spray deposition-derived coatings on osteoblast-liked cell behavior, *in vitro*. *J. Biomed. Mater. Res.*, **78A**, 258–267.

Skinner, H. C.W. (1973). Phase relations in the CaO-P_2O_5–H_2O system from 300 to 600°C at 2kb H_2O pressure. *J. Am. Sci.*, **273**, 545.

Sngh, S., Trikha, S. P., & Edge, A. J. (2004). Hydroxyapatite ceramic-coated femoral stems in young patients. A prospective ten-year study. *J. Bone Joint. Surg. Br.*, **86**, 1118–1123.

Soballe, K., Hansen, E. S., Brockstedt-Rasmussen, H. B., & Bunger, C. (1993). Hydroxyapatite coating converts fibrous tissue to bone around loaded implants. *J. Bone Jt. Surgery*, **75B**, 270–278.

Sogo, Y., Ito, A., Onoguchi, M., Oyane, A., Tsurushima, H., & Ichinose, N. (2007). Formation of a FGF-2 and calcium phosphate composite layer on a hydroxyapatite ceramic for promoting bone formation. *Biomed. Mater.*, **2**, S175–S180.

Spence, G., Patel, N., Brooks, R., & Rushton, N. (2009). Carbonate substituted hydroxy–apatite: Resorption by osteoclasts modifies the osteoblastic response. *J. Biomed. Mater. Res. A*, **90**(1), 217–224.

Stack, R., Tindall, A., Shetty, A. A., James, K. D., & Rand, C. (2006). 15-year follow-up results of the hydroxyapatite ceramic-coated femoral stem. *J. Orthop.Surg*, **14**(4), 11–154.

Stanley, H. R., Clark, A. E., & Hench, L. L. (1996). Alveolar ridge maintenance implants. In *Clinical Performance of Skeletal Prostheses* (pp. 237–254). London: Chapman and Hall.

Stephenson, P. K., Freeman, M. A. R. F., Revell, P. A., German, J., Tuke, M., & Pirie, C. J. (1991). The effect of hydroxyapatite coating on ingrowth of bone into cavities in an implant. *J. Arthroplasty*, **6**(1), 51–58.

Sudo, A., Hasegawa, M., Fukuda, A., & Uchida, A. (2008). Treatment of infected hip arthroplasty with antibiotic-impregnated calcium hydroxyapatite. *J. Arthroplasty*, **23**, 145–150.

Sun, L., Berndt, C. C., Gross, K. A., & Kucuk, A. (2001). Materials fundamentals and clinical performance of plasma sprayed hydroxyapatite coatings: A review. *J. Biomed. Mater. Res. Appl. Biomater*, **58**, 570–592.

Suter, A. J., Molteno, A. C., Bevin, T. H., Fulton, J. D., & Herbison, P. (2002). Long term follow up of bone derived hydroxyapatite orbital implants. *Br. J. Ophthalmol*, **86**(11), 1287–1292.

Swan, E. E., Popat, K. C., & Desai, T. A. (2005a). Peptide-immobilized nanoporous alumina membranes for enhanced osteoblast adhesion. *Biomaterials*, **26**, 1969–1976.

Swan, E. E., Popat, K. C., Grimes, C. A., & Desai, T. A. (2005b). Fabrication and evaluation of nanoporous alumina membranes for osteoblast culture. *J. Biomed. Mater. Res. A*, **72**, 288–295.

Tadic, D., & Epple, M. (2004). A thorough physicochemical characterization of 14 calcium phosphate-based bone substitution materials in comparison to natural bone. *Biomaterials*, **25**, 987–994.

Tadic, D., Beckmann, F., Schwarz, K., & Epple, M. (2004). A novel method to produce hydroxyapatite objects with interconnecting porosity that avoids sintering. *Biomaterials*, **25**, 3335–3340.

Tampieri, A., Celotti, G., & Landi, E. (2005). From biomimetic apatites to biologically inspired composites. *Anal. Bioanal. Chem.*, **381**, 568–576.

Tang, Q., Brooks, R. A., Rushton, N., & Best, S. (2010). Production and characterization of HA and SiHA coatings. *J. Mater Sci. Mater Med.*, **21**, 173–181.

Tanner, K. E., Downes, R. N., & Bonfield, W. (1994). Clinical applications of hydroxyapatite reinforced materials. *Brit. Ceram. Trans.*, **4**(93), 104–107.

Tateiwa, T., Clarke, I. C., Williams, P. A., Garino, J., Manaka, M., Shishido, T., Yamamoto, K., & Imakiire, A. (2008). Ceramic total hip arthroplasty in the United States: safety and risk issues revisited. *Am. J. Orthop. (Belle Mead NJ)*, **37**, E26–E31.

Teixeira, S., Monteiro, F. J., Ferraz, M. P., Vilar, R., & Eugénio, S. (2007). Laser surface treatment of hydroxyapatite for enhanced tissue integration: Surface characterization and osteoblastic interaction studies. *J. Biomed. Mater. Res. A.*, **81**, 920–929.

TenHuisen, K. S., & Brown, P. W. (1998). Formation of calcium-deficient hydroxyapatite from alpha-tricalcium phosphate. *Biomaterials*, **19**, 2209–2217.

Thian, E. S., Huang, J., Best, S. M., & Barber, Z. H. (2006). Bonfield W., Novel silicon-doped hydroxyapatite (Si-HA) for biomedical coatings: An *in vitro* study using acellular simulated body fluid. *J. Biomed. Mater. Res. Appl. Biomater*, **76B**, 326–333.

Thian, E. S., Huang, J., Best, S. M., Barber, Z. H., & Bonfield, W. (2005). Magnetron co-sputtered silicon-containing hydroxyapatite thin films – an *in vitro* study. *Biomater.*, **26**, 2947–2956.

Thian, E. S., Huang, J., Best, S. M., Barber, Z. H., & Bonfield, W. (2005). A novel way of incorporating silicon in hydroxyapatite (Si-HA) thin films. *J. Mater. Sci. Mater. Med.*, **16**, 411–415.

Thian, E. S., Huang, J., Best, S. M., Barber, Z. H., Brooks, R. A., Rushton, N., & Bonfield, W. (2006). The response of osteoblasts to nanocrystalline silicon-substituted hydroxyapatite thin films. *Biomater*, **27**, 2692–2698.

Thian, E. S., Huang, J., Vickers, M. E., Best, S. M., Barber, Z. H., & Bonfield, W. (2006). Silicon-substituted hydroxyapatite (SiHA): A novel calcium phosphate coating for biomedical applications. *J. Mater. Sci.*, **41**, 709–717.

Thompson, I., & Hench, L. L. (1998). Mechanical properties of bioactive glasses, glass ceramics and composites. *Proc. I.Mech.E., Part H: Journal of Engineering in Medicine*, **212**(2). 127-126.

Tiselius, A. S., Hjerten, S., & Levin, O. (1956). Protein chromatography on calcium phosphate columns. *Arch. Biochem.*, **65**, 132–155.

Trikha, S. P., Singh, S., Raynham, O. W., Lewis, J. C., Mitchell, P. A., & Edge, A. J. (2005). Hydroxyapatite-ceramic-coated femoral stems in revision hip surgery. *J. Bone Joint. Surg. Br.*, **87**, 1055–1060.

Turhani, D., Watzinger, E., Weissenböck, M., Yerit, K., Cvikl, B., Thurnher, D., & Ewers, R. (2005). Three-dimensional composites manufactured with human mesenchymal cambial layer precursor cells as an alternative for sinus floor augmentation: An *in vitro* study. *Clin. Oral Implants Res.*, **16**, 417–424.

van der Meulen, J., & Koerten, H. K. (1994). Inflammatory response and degradation of three types of calcium phosphate ceramic in a non-osseous environment. *J. Biomed. Mater. Res.*, **28**, 1455–1463.

van Dijk, S., Dean, D. D., Liu, Y., Zhao, Y., Chirgwin, J. M., Schwartz, Z., & Boyan, B. D. (1998). Purification, amino acid sequence, and cDNA sequence of a novel calcium-precipitating proteolipid involved in calcification of *corynebacterium matruchotii. Calcif. Tissue Int.*, **62**, 350–358.

Vaz, L., Lopes, A. B., & Almeida, M. (1999). Porosity control of hydroxyapatite implants. *J. Mater. Sci. Mater. Med.*, **10**, 239–242.

Waked, W., & Grauer, J. (2008). Silicates and bone fusion. *Orthopaedics*, **21**, 591–598.

Wang, C., Duan, Y., Markovic, B., Barbara, J., Rolfe Howlett, C., Zhang, X., & Zreiqat, H. (2004). Proliferation and bone-related gene expression of osteoblasts grown on hydroxyapatite ceramics sintered at different temperature. *Biomaterials*, **25**, 2949–2956.

Wang, Y., Zhang, S., Zeng, X., Ma, L. L., Weng, W., Yan, W., & Qian, M. (2007). Osteoblastic cell response on fluoridated hydroxyapatite coatings. *Acta. Biomater.*, **3**, 191–197.

Webster, T. J., Ergun, C., Doremus, R. H., Siegel, R. W., & Bizios, R. (2000). Enhanced functions of osteoblasts on nanophase ceramics. *Biomaterials*, **21**, 1803–1810.

Wenisch, S., Stahl, J. P., Horas, U., Heiss, C., Kilian, O., Trinkaus, K., Hild, A., & Schnettler, R. (2003). *In vivo* mechanisms of hydroxyapatite ceramic degradation by osteoclasts: Fine structural microscopy. *J. Biomed. Mater. Res. A.*, **67**, 713–718.

Wheeler, D. L., Jenis, L., Kovach, M. E., Marini, J., & Turner, AS. (2007). Efficacy of silicated calcium phosphate graft in posterolateral lumbar fusion in sheep. *Spine J*, **7**(3), 308–317.

White, A. A., Best, S. M., & Kinloch, I. A. (2007). Hydroxyapatite-carbon nanotube composites for biomedical applications: A review. *International Journal of Applied Ceramic Technology*, **4**(1), 1–13.

White, E., & Schors, E. C. (1986). Biomaterials aspects of interpore-200 porous hydroxyapatite. *Dent. Clin. North Am.*, **30**, 49–67.

Wilson, J. (1994). Clinical applications of bioglass implants. In O. H.Andersson (Ed.), *Bioceramics-7*. Oxford, England: Butterworth–Heinemann.

Wilson, J., & Low, S. B. (1992). Bioactive ceramics for periodontal treatment: Comparative study in Patus monkey. *J. App. Biomat.*, **2**, 123–129.

Wilson, J., & Low, S. B. (1992). Bioactive ceramics for periodontal treatment: Comparative studies. *J. Appl. Biomaterials*, **3**, 123–169.

Wilson, J., Pigott, G. H., Schoen, F. J., & Hench, L. L. (1981). Toxicology and biocompatibility of bioglass. *J. Biomed. Mater. Res.*, **15**, 805.

Wilson, R. M., Elliott, J. C., Dowker, S. E., & Smith, R. I. (2004). Rietveld structure refinement of precipitated carbonate apatite using neutron diffraction data. *Biomaterials*, **25**, 2205–2213.

Wolke, J. G., de Blieck-Hogervorst, J. M., Dhert, W. J., Klein, C. P., & de Groot, K. (1992). Studies on thermal spraying of apatite bioceramics. *J. Thermal Spray Technology*, **1**, 75–82.

Xynos, I. D., Edgar, A. J., Buttery, L. D., Hench, L. L., & Polak, J. M. (2000). Ionic products of bioactive glass dissolution increase proliferation of human osteoblasts and induce insulin-like growth factor II mRNA expression and protein synthesis. *Biochem. Biophys. Res. Comm*, **276**, 461–465.

Xynos, I. D., Edgar, A. J., Buttery, L. D., Hench, L. L., & Polak, J. M. (2001). Gene-expression profiling of human osteoblasts following treatment with the ionic products of Bioglass 45S5 dissolution. *J. Biomed. Mater. Res.*, **55**(2), 151–157.

Xynos, I. D., Hukkanen, M. V., Batten, J. J., Buttery, I. D., Hench, L. L., & Polak, J. M. (2000). Bioglass 45S5 stimulates osteoblast turnover and enhances bone formation *in vitro*: Implications and applications for bone tissue engineering. *Calcif. Tiss. Int.*, **67**, 321–329.

Yamamuro, T., Hench, L. L., & Wilson, J. (1990). Handbook on Bioactive Ceramics. *Bioactive Glasses and Glass-Ceramics.* (Vol. I). Boca Raton, FL: CRC Press. Vol. II: Calcium-Phosphate Ceramics.

Yao, X., Tan, S., & Jiang, D. (2005). Fabrication of hydroxyapatite ceramics with controlled pore characteristics by slip casting. *J. Mater. Sci. Mater. Med.*, **16**, 161–165.

Yoshikawa, H., & Myoui, A. (2005). Bone tissue engineering with porous hydroxyapatite ceramics. *J. Artif. Organs*, **8**, 131–136.

Yoshikawa, H., Tamai, N., Murase, T., & Myoui, A. (2009). Interconnected porous hydroxyapatite ceramics for bone tissue engineering. *J. R. Soc. Interface*, **6**(Suppl 3), S341–S348.

Young, R. A., & Elliot, J. C. (1966). Atomic scale bases for several properties of apatites. *Arch. Oral Biol.*, **11**, 699.

A. NATURAL AND SYNTHETIC HYDROXYAPATITES

Adele L. Boskey
Starr Chair in Mineralized Tissue Research, Hospital for Special Surgery and Professor, Weill Medical College of Cornell University; BME Faculty, Sibley School of Mechanical Engineering, Cornell, Ithaca

INTRODUCTION

The mineral in bone and dentin and some pathologic deposits is a poorly-crystalline analog of the geologic mineral hydroxylapatite ($Ca_{10}(PO_4)_6(OH)_2$), HAP. The analogy was recognized from X-ray diffraction analyses performed almost a century ago. The crystal structure of a single crystal of stoichiometric hydroxylapatite was solved in 1964 (Kay et al., 1964). More recent refinements have been made on powder diffraction patterns based on neutron diffraction data (Leventouri et al., 2003) and by Reitfeld refinement (Wilson et al., 2004). The salient features of the structure have not changed and are summarized in Figure A.1 as a projection on the x,y plane (i.e., the a and b axes are in the plane of the paper). The elevations shown in each symbol relate to the positions along the c-(needle) axis. Transmission electron microscopy (TEM) shows that the crystals (depending on how they are viewed) are needle- or plate-like (Landis et al., 1991); however, a plate-like crystal on its side resembles a needle, and *vice versa*. Nanocrystalline HAP crystals that are synthesized in the laboratory have structural properties similar to bone HAP, and range in size from 30 to 50 nm in length, with a diameter, based on TEM, of about 10 nm (Jäger et al., 2006).

As can be seen from the wide angle X-ray powder diffraction patterns (Figure A.2), the peaks for bone are broader than those for the highly crystalline hydroxyapatite, but are more similar to that of a synthetic apatite made at room temperature. The CaP ceramic shown in the figure has peaks that are distinct from both bone apatite and hydroxylapatite. The broadening of the peaks is caused by the imperfections in the crystal lattice and the small size of the crystals. The crystallite size (and perfection) can be estimated by the Scherrer equation (Eq. 1):

$$\beta 1/2 = K\lambda/(d \cos \Theta) \qquad (1)$$

where $\beta 1/2$ is the peak width at half maximum in degrees 2 Θ, K is a constant (usually 0.9 for HAP), λ is the wavelength of the X-ray source used (in Angstroms), d is the average domain size (assumed to be crystallite size), and Θ is the position of the reflection peak in radians. For HAP examined with Cu Kα radiation, the wavelength is 1.541, the 002 reflection (used to measure c-axis length) is 25.85 degrees 2 Θ.

In general, bone apatite is hydroxide deficient (Boskey, 2007), and hence is called hydroxyapatite or "apatite." Detailed analyses of the structure of synthetic and bone mimetic-apatites have shown them to be calcium deficient, with magnesium, sodium, carbonate, and fluoride substitutions. They may also be phosphate deficient, and lattice vacancies are also found (Legeros, 2002; Tampieri et al., 2005). The lack of stoichiometry of bone apatite has been confirmed by chemical analyses, which finds Ca/P molar ratio distinct from the predicted 1.67, ranging from 1.4:1 to 2.0:1.

Substitution of foreign ions and vacancies into the hydroxyapatite crystal structure are common, especially in biologic apatites, and adsorption of foreign ions and molecules onto the surfaces of the crystals is equally common, because of the relatively small size and large surface area of the biologic apatites. The most common vacancies are in the calcium positions; carbonate can substitute for either phosphate or hydroxyl; fluoride substitutes for hydroxyl; and the cations substitute in the calcium positions. Substitution of calcium by lead (Ellis et al., 2006), sodium and potassium (De Maeyer et al., 1996), zinc (Tampieri et al., 2005), magnesium (Bertoni et al., 1998), silicon (Gibson et al., 1999; Pietak et al., 2007), and strontium (O'Donnell et al., 2008) destabilizes the structure, increasing the solubility of the substituted apatites. Since the clinical application of HAP ceramics is bone and tooth repair, it is important to note that some of the substituted apatites have been shown to enhance bone cell growth in

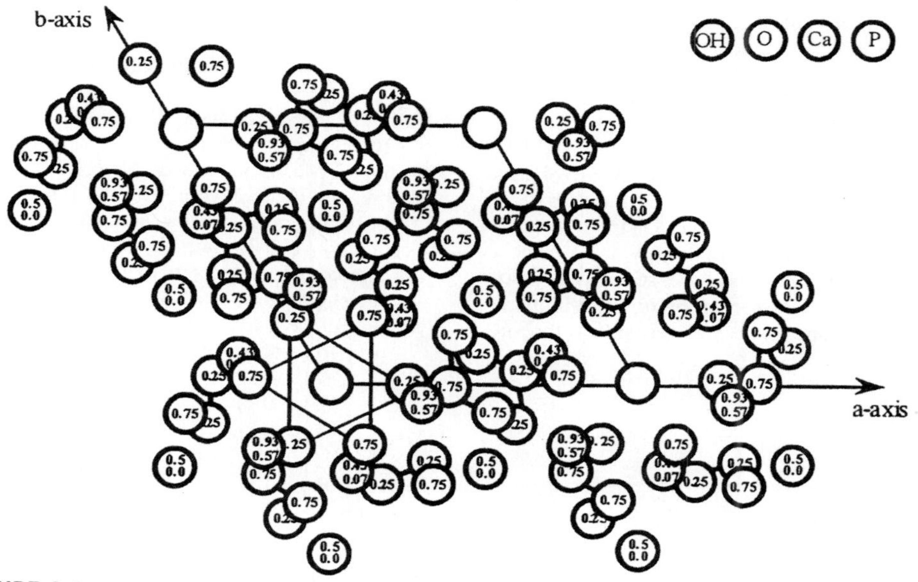

FIGURE A.1 Structure of hydroxyapatite projected on the x, y plane. (*Adapted from* Kay et al., 1964.)

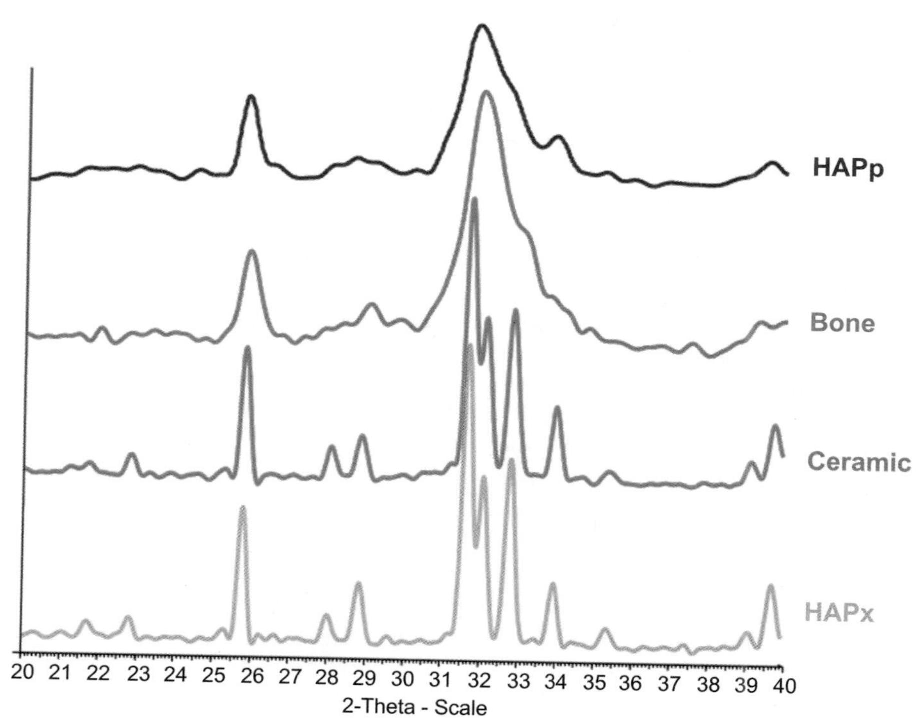

FIGURE A.2 X-ray diffraction powder patterns for (top to bottom) - poorly crystalline HAP (HAPp), bovine cortical bone (Bone), HAP ceramic, and highly crystalline HAP (HAPx).

culture and *in situ* (Table A.1). Substitution of hydroxyl and/or phosphate groups by carbonate increases the apatite solubility (Okazaki and Takahashi, 1999). Carbonate substitution is said to enhance osteoconduction (Spence et al., 2008), but the mechanism of this effect is unknown. The most common hydroxyl ion substituents are chloride and fluoride (LeGeros et al., 2002), both of which significantly change the HAP crystal structure. In the case of chloride, the symmetry of the structure changes and hydroxyapatite becomes closer to the monoclinic chloroapatite structure (Baikie et al., 2007). Fluoride substitution shortens the HAP crystals' c-axis length and, due to hydrogen bonding, decreases the hydroxyapatite solubility (Bertoni et al., 1998). As discussed below, each of these ionic substitutions has been included alone or in combination, in

TABLE A.1 Effect of Ionic Substitution on Synthetic HA Properties

Ion	Effect on Unit Cell	Effect on Bone Growth	Effect on Bone Remodeling
Si	Shorter a-axis; longer c-axis; decreased crystallinity[a,b]	Enhanced osteoblast growth	
Sr	Increased a and c axis[c]; increased crystallinity[c]	Enhanced osteoblast gene expression[d]	Decreased osteoclast activity[d]
CO_3	Decreased c-axis[e]	Enhanced osteogenesis[f]	Enhanced resorption[f]
Mg		Decreased osteogenesis[g]	
Fe		Inhibits growth[h]	

[a]Gibson, I.R., Best, S.M. & Bonfield, W. (1999). Chemical characterization of silicon-substituted hydroxyapatite. *J.Biomed. Mater. Res.*, 44; 422–428.
[b]Balamurugan, A., Rebelo, A.H., Lemos, A.F., Rocha, J.H., Ventura, J.M. & Ferreira JM. (2008). Suitability evaluation of sol–gel derived Si-substituted hydroxyapatite for dental and maxillofacial applications through *in vitro* osteoblast response. *Dent. Mater*. 24; 1374–1380.
[c]O'Donnell, M.D., Fredholm, Y., de Rouffignac, A. & Hill, R.G. (2008). Structural analysis of a series of strontium-substituted apatites. *Acta Biomater*. 4; 1455–1464.
[d]Capuccini, C., Torricelli, P., Sima, F., Boanini, E., Ristoscu, C., Bracci, B., Socol, G., Fini, M., Mihailescu, I.N. & Bigi, A. (2008). Strontium-substituted hydroxyapatite coatings synthesized by pulsed-laser deposition: *In vitro* osteoblast and osteoclast response. *Acta Biomater*. 4; 1885–1893.
[e]Müller, L., Conforto, E, Caillard, D. & Müller, F.A. (2007). Biomimetic apatite coatings – carbonate substitution and preferred growth orientation. *Biomol. Eng*. 24; 462–466.
[f]Spence, G., Patel, N., Brooks, R. & Rushton, N. (2009). Carbonate substituted hydroxyapatite: Resorption by osteoclasts modifies the osteoblastic response. *J. Biomed. Mater. Res. A*. 90(1); 217–224.
[g]Serre, C.M., Papillard, M., Chavassieux, P., Voegel, J.C. & Boivin, G. (1998). Influence of magnesium substitution on a collagen-apatite biomaterial on the production of a calcifying matrix by human osteoblasts. *J. Biomed. Mater. Res*. 42; 626–633.
[h]Guggenbuho, P.K., Filmon, R., Mabilleau, G., Basle, M.F. & Chappard, D. (2008). Iron inhibits Hydroxyapatite crystal growth *in vitro*. *Metabolism*, 57; 903–910.

HAP ceramics in an effort to optimize the ceramics' functional properties.

APPLICATIONS OF HAP CERAMICS

The HAP ceramics are designed to mimic the mineral in bone or other mineralized tissues. The ideal HAP mimetic should be biocompatible, easy to work with, and preferably osteoinductive. Most of the HAP ceramics, however, are osteoconductive at best. Osteoinductivity refers to the ability to generate bone. In contrast, osteoconductivity indicates that bone will grow on the surface of the material in question, but will not penetrate the structure. Biocompatibility refers to the ability of the material to respond appropriately in its desired *in situ* function, of which the HAP ceramics have many.

Clinical applications of HAP ceramics are quite varied, and range from prosthetic grafts in orthopedic surgery (Bucholz, 2002; Sing et al., 2004; Adachi et al., 2005; Trinkha et al., 2005; Huber et al., 2006), including spine (Khan et al., 2005), plastic (Isago et al., 2004), and reconstructive surgery (Ono et al., 1999; Schmitz et al, 1999; Reynolds et al., 2003), orbital and cranial reconstruction (Kundu et al., 2005), and oral surgery (Turhani et al., 2005), to their use as an embolic material (Kubo et al., 2003) in which capillary beds in tumors are occluded. Along with these clinical applications, HAP ceramics have also been tried as and drug delivery devices, for example for delivery of insulin (Paul et al., 2002), antibiotics (Sudo et al., 2008), and anti-tumor vaccines (Ciocca et al., 2007). Preclinical models have also examined the release of growth factors and cytokines from HAP ceramics (Sogo et al., 2007; Jones et al., 2008), but few of these are in clinical use (Yoshikawa et al., 2009). The HAP ceramics are used in a variety of forms (beads, pastes, chips, pre-formed structures), depending on application. In some cases the HAP ceramics are reported to be too brittle unless porosity is introduced (Yoshikawa et al., 2009). HAP ceramics are used to coat metallic prostheses to improve mechanical strength, while encouraging osteointegration. Addition of pores of sufficient size that penetrate the ceramic has been a challenge in the synthesis of HAP ceramics. A detailed review of the history of HAP clinical applications was recently published (LeGeros and LeGeros, 2008). HAP ceramics are also used in chromatographic columns, and as abrasives (LeGeros and LeGeros, 2008), conjugated with quantum dots for *in vivo* imaging (Guo et al., 2008).

SYNTHESIS OF HAP CERAMICS

There are two categories of HAP ceramics, those with minimal porosity (dense ceramics) and those in to which porosity is introduced (porous ceramics). In general, both types of HAP ceramics are prepared by sintering, and because the strength of the ceramics decreases as the sintering temperature increases (Ruys et al., 1995) a variety of techniques, discussed below, have been introduced to improve the ceramics' mechanical strengths. Sintered HAPs have higher crystallinity and lack carbonate substitution in contrast with bone apatites (Peters et al., 2000). Often the synthesis can involve two heating processes: calcining, which refers to the heating of the material turning it into ash (usually done at a lower temperature); and sintering, derived from the word for "cinder" which by definition is a method for making objects from powders by heating them below their melting point until they adhere to one another. Sintering can be done with pressure (referred to as hot pressing or hot isostatic pressuring) and without added pressure, which is usually done for nanoparticles.

Several distinct processes can be used to make HAP ceramics (LeGeros and LeGeros, 2008). These include solid-state reactions in which calcium- and phosphate-containing compounds are mixed, compressed, and sintered above 950°C; hydrothermal reactions in which the reactants of the solid-state reactions are heated under pressure; sol–gel reactions in which solutions (sol) or colloids are mixed to form a colloid solution (gel), which is then aged, dried, and sintered (Chai et al., 1995); and variants thereof. A number of calcium phosphate phases can serve as starting materials for the preparation of HAP ceramics. The list includes alpha-tricalcium phosphate (TenHuisen and Brown, 1998) which can be hydrolyzed to form a calcium deficient HAP; carbonate-containing hydroxyapatite that loses its carbonate on heating (Ellies et al., 1988), and hydroxyapatite itself. These starting phases are listed in Table A.2. While a number of phases are listed, the most common are HA and ACP, but only HA, whitlokite, and brushite ($CaHPO_4.2H_2O$) persist in the body, the other phases are remodeled or dissolved. It is important to note that not every calcium phosphate ceramic is HAP, requiring analysis of the phases present before clinical use (ASTM F1185-03). For example, because the ceramics are prepared by sintering at high temperatures, depending on starting material they may contain calcium oxide or calcium carbonate (Qian et al., 2008) or some of the other phases listed in Table A.2.

To make the HAP starting material, variations of the Tiselius direct precipitation method (Tiselius et al., 1956), in which a calcium source and a phosphate source are mixed in solution at pH >7.4 and heated above 60°C, are used. A hydrolysis method wherein non-apatitic calcium phosphates (Table A.2) are hydrolyzed in base is also used frequently. Amorphous calcium phosphate (ACP) prepared from 0.4 M calcium chloride and 0.3 M ammonium chloride mixed in pH 7.4 buffer converts to HAP at room temperature in a few hours (Boskey, 1997). ACP, which can be recognized from the total lack of features in its X-ray diffraction pattern, can be flame sprayed or otherwise heated to generate the ceramic. HAP has been made by addition of phosphoric acid into a calcium hydroxide suspension followed by spray drying the suspension. The powder thus formed is calcined at 700°C and jet-milled for conditioning (Kubo et al., 2003). Recently, calcium nitrate tetrahydrate was mixed with triethyl phosphate following the sol gel method, after which the gel was dried slowly and calcined at 600°C, yielding a CaO-free HAP (Hsieh et al., 2001). Preparing the HAP particles in an autoclave with stirring results in variation in particle size and shape (Kothanpalli et al., 2005). Specifically, the aspect ratio of the particles increased with the reaction temperature. The length of the HA particles increased with increasing temperature when the temperature was below 170°C, and decreased above that temperature.

HAP ceramics can be made by sintering bovine bone (Joschek et al., 2000), coral (Chiroff et al., 1975), red algae (Turhani et al., 2005) or egg shells (calcium carbonate) mixed with phosphoric acid (Lee et al., 2007). In some cases bone is soaked in sodium pyrophosphate before calcining to provide a mixture of HAP and the more soluble (biodegradable) beta-TCP (Jun et al., 2003). Recently, bacterial synthesis has been used to produce HAP (Medina Ledo et al, 2008). While oral bacteria are known to produce HAP (van Dijk et al., 1998), in this novel approach the bacterium *Serratia* forms a biofilm on polyurethane, and then is incubated in calcium chloride and 2-glycerophosphate to produce HAP in bulk. The bacteria hydrolyze the phosphate and the initial precipitate formed is a calcium deficient HAP. Sintering increases the crystallite length.

The sintered bovine bone product, which retains the shape and porosity of the starting material is marketed as Bio-Oss® (Osteohealth, Luitpold Pharmaceuticals, Inc, Luitpold, NY); the coral product as Pro-Osteon and Interpore 200 (Interpore Cross Int); and the algae based product as C GRAFT/Algipore (Clinician's Preference LLC, Golden, CO). This is just a sample of the commercially-available HAP ceramics (see Box 1); in addition there are several that are mixtures of HAP and beta-tricalcium phosphate, the so-called biphasic calcium phosphates (BCPs) (e.g., KGBONE (Kasios, La Croix, France) or TRICOSS (Baxter, US)). The physicochemical characteristics of 14 CaP bone substitutes available in

TABLE A.2 Calcium Phosphate Phases Used to Prepare HAP Ceramics

Compound	Formula	Acronym	Geologic Name
Alpha-tricalcium phosphate	$Ca_3(PO_4)_2$	α-TCP	
Amorphous calcium phosphate	$(Ca_3(PO_4)_2)$	ACP	
Anhydrous monocalcium phosphate	$Ca(H_2PO_4)_2$	MCPA	
Beta-tricalcium phosphate	$Ca_3(PO_4)_2$	β-TCP	B-Whitlokite
Dicalcium phosphate anhydrous	$CaHPO_4$	DCPA	Monetite
Dicalcium phosphate dihydrate	$CaHPO_4.2H_2O$	DCPD	Brushite
Hydroxyapatite	$Ca_5(PO_4)_6OH$	HAP, HA	Hydroxylapatite
Monocalcium phosphate monohydrate	$(Ca(H_2PO4)_2.H_2O$	MCPM	
Octacalcium phosphate	$Ca_8(HPO_4)_2(PO_4)_4.5H_2O$	OCP	
Tetracalciumphosphate	$Ca_4(PO_4)_2O$	TTCP	

BOX 1 Commercially Available HAP Ceramics

Product	Made From
Aperceram	Synthetic
Bio-Oss	Bovine bone (not sintered)
Bioroc	Synthetic
BONECERAM	Synthetic
Calcibon	Synthetic
Calcititek	Synthetic
C GRAFT/Algipore	Algae and starfish
Cerament	HAP + Ca sulfate
Collapat	Synthetic on bovine type I collagen fibrils
Cortoss	Synthetic
Endobon	Bovine bone (sintered)
FRIOS	Marine fluoroapatite
HAPEX	HAP + polymer
Hapset	HA + Ca sulfate
Healos	HAP + collagen composite
Interpore 200	Coral
Perossal	Synthetic
Nano-XIM	Synthetic
NuOSS	Bovine cortical bone
Ostogen	Synthetic (unsintered)
Osteograft	Bovine bone (sintered)
Ostim	Synthetic (unsintered)
Osprovit	Synthetic
Pro-Osteon	Coral (incomplete conversion)
Radiesse	Synthetic
Truebone	Bovine bone (sintered)

2004 were described in detail by Tadic and Epple (2004), and those available in Japan in 2008 were reviewed by Yoshikawa et al. (2009).

Ideally, a HAP ceramic that is clinically useful must be made from a material that is mechanically strong enough to bear the load on bones and teeth, while being sufficiently porous to allow cell penetration. In addition, the pores should be interconnected so that the cells can invade the entire construct and also receive nutrition. To improve the mechanical properties, as sintered HAP tends to be very brittle (Barralet et al., 2002), reinforcement with magnesium (Landi et al., 2005), zirconium (Kim et al., 2005), aluminum (Jun et al., 2003), lithium (Vaz et al., 1999), and a variety of trace metals (Kokubo et al., 2003) has been applied. Addition of PMMA (Yao et al., 2005) has been used to increase the strength of the HAP ceramics. Coating HAP ceramics on metallic surfaces or on pliable scaffolds (biotemplates) has also been used to increase mechanical toughness. There are also composite materials consisting of both HAP ceramics and synthetic or naturally-occurring polymers (such as bovine collagen, silk or tree bark) as biotemplates (Qian et al., 2008), and multilayered ceramics with various porosities and mechanical properties within the different layers (Vaz et al., 1999).

To allow cells to penetrate the HAP ceramic, pores are introduced into the sintered material. Pore diameters must be >100 um (Jones et al., 2009). Smaller connecting pores are satisfactory for nutrient flow into and fluid flow out of the cells, but not for cell penetration or new matrix synthesis. A variety of methods for increasing porosity have been described including, but not limited to, microwave processing at 1150 to 1200°C for short periods of time (1 to 5 minutes); heating in the presence of activated charcoal (Li et al., 2009); using starting materials with different morphology, adjusting density, changing sintering time and temperature; and mixing ammonium carbonate in the HAP powder during the consolidation process. The ammonia and carbon dioxide gases that are released on heating generate the pores. Pore-forming agents such as polyvinyl butyl and polyvinyl polyacrylate, which are also removed on heating, have been used to generate more uniform pore sizes (Liu, 1997; Vaz et al., 1999). HAP slurries have been impregnated into cellulose sponges which were then calcined at 1280°C (Fabbri et al., 1995). These processes are reviewed in detail elsewhere (Tadic et al., 2004). Other organic materials such as carbon beads (Kawata et al., 2004) have used to produce uniform porosities. "Super porous" HAP ceramics with interconnecting pores were made by including surfactants during the heating process (Sakamoto et al., 2007). The sol–gel can also be infiltrated into "biotemplates," and subsequently sintered at elevated temperatures to obtain porous HA ceramics. A "foam–gel" method that involves preparing an HAP gel in a cross-linking slurry (made from polyethyleneimine), casting the foamy slurry, and after removal from the mold drying and sintering, yields uniformed size interconnecting pores and a compressive strength of 12 Mpa, is currently in clinical trial in Japan (Yoshikawa et al., 2009).

Porosity alters the microstructure (Barralet et al., 2002), and the mechanical strength of HAP ceramics is dependent both on the number and size of the pores (Le Huec et al., 1995). The surface area and porosity of HAP ceramics decrease with increasing preparation temperature (Salver and Hall, 1989), while crystal size increases (Habibovic et al., 2005). The surface area of HAP ceramics is ~5–7 m^2/g at 1000°C and ~70–80 m^2/g at 400°C (Patel et al., 2001). The HAP structure degrades if sintered at temperatures higher than 1300°C (Lin et al., 2000), while an inflammatory response develops if sintering is performed at too low a temperature (Lu et al., 2004). With this in mind, HAP ceramics sintered at 1100–1200°C are reported to be the most osteogenic (Wang et al., 2004; Habibovic et al., 2005). Table A.3 compares some mechanical properties of cortical and cancellous bone to that of dense and porous HAP ceramics. Increasing porosity decreases compressive strength (Le Huec et al., 1995); hence, both dense and porous ceramics have advantages, which may be addressed by composites of the two.

Recently, emphasis has been placed on non-sintered HAP materials that are porous, but more similar in crystal

TABLE A.3 Mechanical Properties of Adult Human Cortical and Cancellous Bones and HAP Ceramics*

Property	Cortical Bone	Cancellous Bone	HAP Ceramics*
Elastic modulus GPa	3–20	8	34–117 (p)
Mechanical strength Compressive MPa	~140	41–62	7–69 (p) 207–897 (d)
Strength – tensile MPa	~70	3.4	2.5 (p) 69–193 (d)
Ultimate strength MPa	130	~10	13 (p)
Ultimate stress – compression MPa	~200 MPa	~10 MPa	~6
Ultimate strain	~0.02%		~0.20% (d)
Ultimate stress – tension	~130	~10	~125
Yield Strain	~0.008%	~0.80%	

(p) = porous (d) = dense

*Range of values from multiple sources

Literature Sources:

Bayraktar, H.H., Morgan, E.F., Niebur, G.L., Morris, G.E., Wong, E.K. & Keaveny, T.M. (2004). Comparison of the elastic and yield properties of human femoral trabecular and cortical bone tissue. *J. Biomech.* 37; 27–35.

Currey, J. (2004). Tensile yield in compact bone is determined by strain, post-yield behavior by mineral content. *J. Biomech.* 37; 549–556.

Haddock, S.M., Debes, J.C., Nauman, E.A., Fong, K.E., Arramon, Y.P. & Keaveny, T.M. (1990). Structure–function relationships for coralline hydroxyapatite bone substitute. *J. Biomed. Mater. Res.* 47; 71–78.

Hsu, Y.H., Turner, I.G. & Miles, A.W. (2007). Mechanical characterization of dense calcium phosphate bioceramics with interconnected porosity. *J. Mater. Sci. Mater. Med.* 18; 2319–2329.

Hsu, Y.H., Turner, I.G. & Miles, A.W. (2007). Fabrication of porous bioceramics with porosity gradients similar to the bimodal structure of cortical and cancellous bone. *J. Mater. Sci. Mater. Med.* 18; 2251–2256.

Kopperdahl, D.L. & Keaveny, T.M. (1998). Yield strain behavior of trabecular bone. *J. Biomech.* 31; 601–608.

Le Huec, J.C., Schaeverbeke, T., Clement, D., Faber, J. & Le Rebeller, A. (1995), Influence of porosity on the mechanical resistance of hydroxyapatite ceramics under compressive stress. *Biomaterials*, 16; 113–118.

size and composition to bone apatite. Tadic et al. (2004) produced such a nanomaterial ceramic by synthesizing HAP from mixtures of 5:1 $(NH_4)_2PO_4$, $(NH_4)_2CO_3$, and $Ca(NO_3)_2$, such that the Ca/P molar ratio was 1.67:1. Following ball milling and sieving of the powder that was formed to 250–400 μm, it was mixed with polyvinylalcohol fibers and NaCl crystals and cold isostatically pressed at room temperature. The NaCl and polyvinylalcohol were extracted in warm water, but the mechanical stability was markedly decreased, with compressive strength about 1/10th to 1/20th of that of the material without the pores.

The porous and nanomaterial HAP ceramics can be prepared in a variety of forms, depending on the desired use. These forms (Table A.4) are made by molding, grinding, three-dimensional printing or mixing with other materials. Some, like the injectable Radiesse, which is a dermal filler in a carrier gel made up of water, glycerin, and carboxymethylcellulose, combine multiple forms. To enhance cellular ingrowth, cell proliferation, and even vascular ingrowth, growth factors and cytokines, e.g., bone morphogenetic proteins, endothelial elements, and vascular endothelial growth factor (VEGF) have been included within the porous ceramics (Lieberman et al., 2002; Holt et al., 2005; Moioli et al., 2008; Schussele et al., 2009). Modification of surface texture, which increases surface area, has also been used to enhance cell interactions with implants (Harding et al., 2005).

TABLE A.4 Forms of HAP Ceramics

• Granular (granule size 300 to 3000 μm)
• Microspheres
• Injectable (Radiesse)
• Beads
• Pastes
• Moldable
• Gels
• Porous blocks
• Cements (Biobon)

STRUCTURE CHARACTERIZATION OF HAP CERAMICS

The American Society for Testing of Materials (ASTM) mandates analyses of HAP ceramics and related materials. Analyses can be done by X-ray and electron diffraction, vibrational spectroscopy, electron microscopy, and NMR. Changes in the composition of bone and synthetic apatites and apatite ceramics can be detected by infrared and Raman spectroscopy (Carden and Morris, 2000; Tadic and Epple, 2004; Boskey and Mendelsohn, 2005). The FTIR (Figure A.3) and Raman (Figure A.4) spectra for the same materials analyzed by X-ray diffraction in Figure A.2 are presented. The peak assignments for these spectra are summarized in Table A.5. As in the X-ray diffraction patterns, the "gold standard" method for analyses of these materials, the peaks for the synthetic and

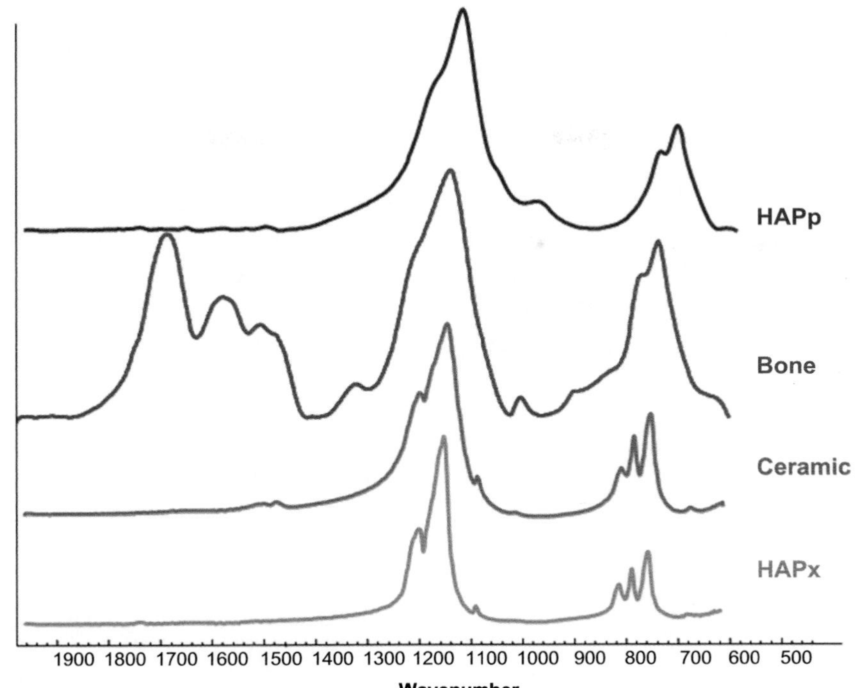

FIGURE A.3 Infrared spectra of (top to bottom): - poorly crystalline HAP (HAPp), bovine cortical bone (Bone), HAP ceramic, and highly crystalline HAP (HAPx).

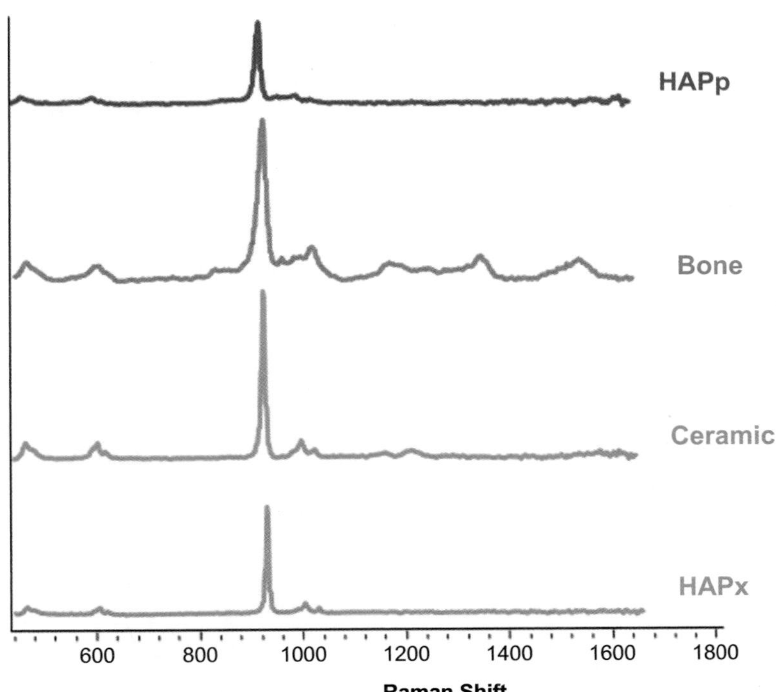

FIGURE A.4 Raman spectra of (top to bottom): - poorly crystalline HAP (HAPp), bovine cortical bone (Bone), HAP ceramic, and highly crystalline HAP (HAPx).

bone apatites are broadened relative to the geologic apatite and the HAP ceramic. The broadening of the peaks in the vibrational spectra indicate changes in the environment of the bond being analyzed. A more ordered environment leads to a sharper line. Both the sharpening of the band at 631 cm^{-1} in the infrared and the peak area and intensity ratios of sub-bands at 1030 cm^{-1} and 1020 cm^{-1} (Gadaleta et al., 1996) have been used as a measure of crystal size and perfection (Blumenthal et al., 1975). Other peaks can be used to indicate the presence of additional phases or impurities in the material. When the template (be it natural or a biotemplate) is included, the vibrations from organic moieties can reveal the relative content of proteins, proteoglycans, lipids, and nucleotides.

NMR spectroscopy can also be used to analyze the composition of the HAP containing tissues and synthetic materials (Figure A.5). As in the other methods, the peaks characteristic of highly crystalline apatites are broadened due to the change in the environment of the processing nuclide. One advantage of this technology is that it can be used in real-time to monitor the mass of the implant independent of the presence of native bone (Ramanathan and Ackerman, 1999).

TABLE A.5 Peak Assignments for FTIR and Raman Spectra of HAP Ceramics and Bone

Vibration	*FTIR Peak cm^{-1}	**Raman Peak cm^{-1}
PO_4 v2		422
PO_4 v4	560–580	550–590
Proline ring		855
OHProline ring		880
PO_4 v_1 symmetric stretch	960	960–962 (vs)
Phenylalanine ring breathing		1002 (s)
PO_4 in HAP	982,984	
PO_4 v3 asymmetric stretch	999–1004	
PO_4 v3 asymmetric stretch in non-stoichiometric HAP	1020	
PO_4 v3 asymmetric stretch stoichiometric HAP	1030	1031 (m)
PO_4 v3 asymmetric stretch – CO_3 substitution	1045	
PO_4 v3 asymmetric stretch	1055–1060	
CO_3-2 v1 in plane stretch		1064 (s)
PO_4 v3 asymmetric stretch	1076	
v3 PO_4 asymmetric stretch	1096	
v3 PO_4 asymmetric stretch, Immature HA (HPO_4)	1112	
v3 PO_4 asymmetric stretch, mature HA	1116	
v3 PO_4 asymmetric stretch, HPO_4	1123–1126	
HPO_4	1145	
C-N		1130–1150
Amide III, adenine C-N-H stretch		1245–1250
Adenine, thymine, guanine		1339–1375
C-H, CH_2 wag		1445–1450
Amide I C-C-N stretch	1580–1720	1650–1660
Amide II	1500–1575	

*FTIR from Rey et al., 1991 and Fowler et al., 1996
**Raman from Carden and Morris, 2000

TEM and other high resolution microscopic tools, such as scanning electron microscopy (SEM) and atomic force microscopy (AFM) are also used to characterize bones, teeth, and the HAP ceramics. AFM is used to determine the average sizes of ceramic crystals, pore morphology, surface topology, and nanoindentation modulus (Chowdhury et al., 2005; Covani et al., 2007; Teixeira et al., 2007). TEM and SEM are used to determine the crystallite size, interaction of cells with the cements, and other morphometric features (LeGeros and LeGeros, 2008; Qian et al., 2008; Sato et al., 2008; Li et al., 2009).

STABILITY, BIOCOMPATIBILITY, AND OSTEOINTEGRATION OF HAP CERAMICS

It is highly desirable that the HAP ceramic is replaced by native tissue (bone or dentin), and the implant dissolves slowly as the new tissue forms. The relatively large crystal size, small specific surface, and low porosity of the HAP ceramics makes them relatively insoluble. Modification of any of these properties tends to increase the solubility. In general, of the calcium phosphates, at physiologic pH based on solubility products, brushite, tetracalcium phosphate, and amorphous calcium phosphate are the most soluble, followed by alpha- and beta-tricalcium phosphate and then HAP. Solubility is dependent on the solution properties (pH, buffer concentration, degree of saturation, HAP/solution ratio, temperature, etc.), as well as on the crystallite size, nature of substituents, and porosity of the cement (LeGeros and LeGeros, 2008).

The HAP ceramics dissolve when there is mineral–cell contact, due to the phagocytic capability of the bone resorbing cells, the osteoclasts (Wenisch et al., 2003), and the low pH they produce in their environment. Thus, when new bone forms within the implant, there is remodeling and replacement of the implant ceramic

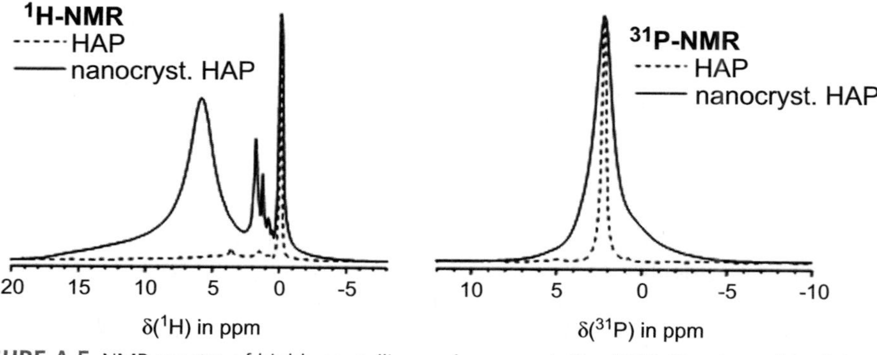

FIGURE A.5 NMR spectra of highly crystalline and nanocrystalline HAP. (Courtesy of Prof. Jaeger.)

by the patient's own tissue. Factors that control implant solubility and osteoclastic activity were reviewed by Heymann et al. (1999) and Detsch et al. (2008). The problem is that most of the sintered HAP ceramics, even the porous ones, have such large crystals that they are difficult to remodel. The biphasic calcium phosphates tend to be more soluble, but are less osteoinductive. The nanocrystalline HAP materials are more soluble, and hence may be more osteogenic (Webster et al., 2000).

Excellent biocompatibility of HAP ceramics has been consistently reported in bone, soft tissues, and other organs (Okumura et al, 1984; Costantino et al., 1991; Yoshikawa and Myoui, 2005). There are, however, reports of immune response, perhaps to proteins that adhere to the ceramic particles (van der Meulen and Koerten, 1994; Otto and Ogilvie 1998; Lu et al., 2005). The immune response usually takes the form of inflammation, which has not been reported to be severe, and which subsides rapidly. The optimal ceramic implants will dissolve slowly and be replaced by native bone, thus maximizing the mechanical properties of the tissues. Osteogenic activity of the bone and tooth implant materials is their most important property. This activity is dependent on the pore size, pore structure, pore volume, and amount of pore interconnectivity. Thus, selection of the optimal HAP ceramic may involve compromise as those that are stronger are unlikely to be those that are the most porous or the most osteoconductive. The osteoconductive properties are most frequently examined with bone forming or bone resorbing cells in culture, followed by analyses in preclinical animal models.

REFERENCES

Adachi, N., Ochi, M., Deie, M., & Ito, Y. (2005). Transplant of mesenchymal stem cells and hydroxyapatite ceramics to treat severe osteochondral damage after septic arthritis of the knee. *J. Rheumatol.*, **32**, 1615–1618.

ASTM F1185-03. (2003). *Standard Specification for Composition of Hydroxylapatite for Surgical Implants.* Book of Standards, 13.01.

Baikie, T., Mercier, P. H., Elcombe, M. M., Kim, J. Y., Le Page, Y., Mitchell, L. D., White, T. J., & Whitfield, P. S. (2007). Triclinic apatites. *Acta Crystallogr. B.*, **63**, 251–256.

Barralet, J. E., Gaunt, T., Wright, A. J., Gibson, I. R., & Knowles, J. C. (2002). Effect of porosity reduction by compaction on compressive strength and microstructure of calcium phosphate cement. *J. Biomed. Mater. Res.*, **63**, 1–9.

Bertoni, E., Bigi, A., Cojazzi, G., Gandolfi, M., Panzavolta, S., & Roveri, N. (1998). Nanocrystals of magnesium and fluoride substituted hydroxyapatite. *J. Inorg. Biochem.*, **72**, 29–35.

Blumenthal, N. C., Betts, F., & Posner, A. S. (1975). Effect of carbonate and biological macromolecules on formation and properties of hydroxyapatite. *Calcif. Tissue Res.*, **18**, 81–90.

Boskey, A. L. (1997). Amorphous calcium phosphate: the contention of bone. *J. Dent. Res.*, **76**, 1433–1436.

Boskey, A. (2007). Mineralization of Bones and Teeth. In Nita Sahai (Ed.), *Elements Magazine* (3, pp. 385–392).

Boskey, A. L., & Mendelsohn, R. (2005). Infrared spectroscopic characterization of mineralized tissues. *Vib. Spectrosc.*, **38**, 107–114.

Bucholz, R. W. (2002). Nonallograft osteoconductive bone graft substitutes. *Clin. Orthop. Relat. Res.*, **395**, 44–52.

Carden, A., & Morris, M. D. (2000). Application of vibrational spectroscopy to the study of mineralized tissues. *J. Biomed. Opt.*, **5**, 259–268.

Chai, C., Ben-Nissan, B., Pyke, S., & Evans, L. (1995). Sol gel derived hydroxyapatite coatings for biomedical applications. *Mater. Manuf. Processes.*, **10**, 205–216.

Chiroff, R. T., White, E. W., Weber, K. N., & Roy, D. M. (1975). Tissue ingrowth of Replamineform implants. *J. Biomed. Mater. Res.*, **9**, 29–45.

Chowdhury, S., Thomas, V., Dean, D., Catledge, S. A., & Vohra, Y. K. (2005). Nanoindentation on porous bioceramic scaffolds for bone tissue engineering. *J. Nanosci. Nanotechnol.*, **5**, 1816–1820.

Ciocca, D. R., Frayssinet, P., & Cuello-Carrion, F. D. (2007). A pilot study with a therapeutic vaccine based on hydroxyapatite ceramic particles and self-antigens in cancer patients. *Cell. Stress Chaperones.*, **12**, 33–43.

Costantino, P. D., Friedman, C. D., Jones, K., Chow, L. C., Pelzer, H. J., & Sisson Sr, G. A. (1991). Hydroxyapatite cement: I. basic chemistry and histologic properties. *Arch Otolaryngol. Head Neck Surg.*, 117, 379–384.

Covani, U., Giacomelli, L., Krajewski, A., Ravaglioli, A., Spotorno, L., Loria, P., Das, S., & Nicolini, C. (2007). Biomaterials for orthopedics: a roughness analysis by atomic force microscopy. *J. Biomed. Mater. Res. A.*, **82**, 723–730.

De Maeyer, E. A., Verbeeck, R. M., & Pieters, I. Y. (1996). Effect of K(+) on the Stoichiometry of Carbonated Hydroxyapatite Obtained by the Hydrolysis of Monetite. *Inorg. Chem.*, **35**, 857–863.

Detsch, R., Mayr, H., & Ziegler, G. (2008). Formation of osteoclast-like cells on HA and TCP ceramics. *Acta Biomater.*, **4**, 139–148.

Ellies, L. G., Nelson, D. G., & Featherstone, J. D. (1988). Crystallographic structure and surface morphology of sintered carbonated apatites. *J. Biomed. Mater. Res.*, **22**, 541–553.

Ellis, D. E., Terra, J., Warschkow, O., Jiang, M., González, G. B., Okasinski, J. S., Bedzyk, M. J., Ross, i A. M., & Eon, J. G. (2006). A theoretical and experimental study of lead substitution in calcium hydroxyapatite. *Phys. Chem. Chem. Phys.*, **8**. 876–967.

Fabbri, M., Celotti, G. C., & Ravaglioli, A. (1995). Hydroxyapatite-based porous aggregates: physico-chemical nature, structure, texture and architecture. *Biomaterials.*, **16**, 225–228.

Fowler, B. O., Moreno, E. C., & Brown, E. (1966). Infra-red spectra of hydroxyapatite, octacalcium phosphate and pyrolysed octacalcium phosphate. *Arch. oral Biol.*, **11**, 477–492.

Gadaleta, S. J., Paschalis, E. P., Betts, F., Mendelsohn, R., & Boskey, A. L. (1996). Fourier transform infrared spectroscopy of the solution-mediated conversion of amorphous calcium phosphate to hydroxyapatite: new correlations between X-ray diffraction and infrared data. *Calcif. Tissue Int.*, **58**, 9–16.

Gibson, I. R., Best, S. M., & Bonfield, W. (1999). Chemical characterization of silicon-substituted hydroxyapatite. *J. Biomed. Mater. Res.*, **44**, 422–428.

Guo, Y., Shi, D., Lian, J., Dong, Z., Wang, W., Cho, H., Liu, G., Wang, L., & Ewing, R. C. (2008). Quantum dot conjugated hydroxylapatite nanoparticles for in vivo imaging. *Nanotechnology*, **19**, 175102–175108.

Habibovic, P., Yuan, H., van der Valk, C. M., Meijer, G., van Blitterswijk, C. A., & de Groot, K. (2005). 3D microenvironment as essential element for osteoinduction by biomaterials. *Biomaterials.*, **26**, 3565–3575.

Harding, I. S., Rashid, N., & Hing, K. A. (2005). Surface charge and the effect of excess calcium ions on the hydroxyapatite surface. *Biomaterials.*, **26**, 6818–6826.

Heymann, D., Pradal, G., & Benahmed, M. (1999). Cellular mechanisms of calcium phosphate ceramic degradation. *Histol. Histopathol.*, **14**, 871–877.

Holt, G. E., Halpern, J. L., Dovan, T. T., Hamming, D., & Schwartz, H. S. (2005). Evolution of an in vivo bioreactor. *J. Orthop. Res.*, **23**, 916–923.

Hsieh, M. F., Perng, L. H., Chin, T. S., & Perng, H. G. (2001). Phase purity of sol-gel-derived hydroxyapatite ceramic. *Biomaterials.*, **22**, 2601–2607.

Huber, F. X., McArthur, N., Hillmeier, J., Kock, H. J., Baier, M., Diwo, M., Berger, I., & Meeder, P. J . (2006). Void filling of tibia compression fracture zones using a novel resorbable nanocrystalline hydroxyapatite paste in combination with a hydroxyapatite ceramic core: first clinical results. *Arch Orthop. Trauma Surg.*, **126**, 533–540.

Isago, T., Nozaki, M., Kikuchi, Y., Honda, T., & Nakazawa, H. (2004). Sinking skin flap syndrome: a case of improved cerebral blood flow after cranioplasty. *Ann. Plast. Surg.*, **53**, 288–292.

Jäger, C., Welzel, T., Meyer-Zaika, W., & Epple, M. (2006). A solid-state NMR investigation of the structure of nanocrystalline hydroxyapatite. *Magn. Reson. Chem.*, **44**, 573–580.

Jones, A. C., Arns, C. H., Hutmacher, D. W., Milthorpe, B. K., Sheppard, A. P., & Knackstedt, M. A. (2009). The correlation of pore morphology, interconnectivity and physical properties of 3D ceramic scaffolds with bone ingrowth. *Biomaterials.*, **30**, 1–12.

Jones, C. B., Sabatino, C. T., Badura, J. M., Sietsema, D. L., & Marotta, J. S. (2008). Improved healing efficacy in canine ulnar segmental defects with increasing recombinant human bone morphogenetic protein-2/allograft ratios. *J. Orthop. Trauma.*, **22**, 550–559.

Joschek, S., Nies, B., Krotz, R., & Göferich, A. (2000). Chemical and physicochemical characterization of porous hydroxyapatite ceramics made of natural bone. *Biomaterials.*, **21**, 1645–1658.

Jun, Y. K., Kim, W. H., Kweon, O. K., & Hong, S. H. (2003). The fabrication and biochemical evaluation of alumina reinforced calcium phosphate porous implants. *Biomaterials.*, **24**, 3731–3739.

Kawata, M., Uchida, H., Itatani, K., Okada, I., Koda, S., & Aizawa, M. (2004). Development of porous ceramics with well-controlled porosities and pore sizes from apatite fibers and their evaluations. *J. Mater. Sci. Mater. Med.*, **15**, 817–823.

Kay, M. I., Young, R. A., & Posner, A. S. (1964). Crystal Structure of Hydroxyapatite. *Nature.*, **204**, 1050–1052.

Khan, S. N., Fraser, J. F., Sandhu, H. S., Cammisa, F. P., Jr., Girardi, F. P., & Lane, J. M. (2005). Use of osteopromotive growth factors, demineralized bone matrix, and ceramics to enhance spinal fusion. *J. Am. Acad. Orthop. Surg.*, **13**, 129–137.

Kim, H. W., Knowles, J. C., Li, L. H., & Kim, H. E. (2005). Mechanical performance and osteoblast-like cell responses of fluorine-substituted hydroxyapatite and zirconia dense composite. *J. Biomed. Mater. Res. A.*, **72**, 258–268.

Kokubo, T., Kim, H. M., & Kawashita, M. (2003). Novel bioactive materials with different mechanical properties. *Biomaterials.*, **24**, 2161–2175.

Kothapalli, C. R., Wei, M., Legeros, R. Z., & Shaw, M. T. (2005). Influence of temperature and aging time on HA synthesized by the hydrothermal method. *J. Mater. Sci. Mater. Med.*, **16**, 441–446.

Kubo, M., Kuwayama, N., Hirashima, Y., Takaku, A., Ogawa, T., & Endo, S. (2003). Hydroxyapatite ceramics as a particulate embolic material: report of the clinical experience. *A.J.N.R. Am. J. Neuroradiol.*, **24**, 1545–1547.

Kundu, B., Sinha, M. K., Mitra, S., & Basu, D. (2005). Synthetic hydroxyapatite-based integrated orbital implants: a human pilot trial. *Indian J. Ophthalmol.*, **53**, 235–241.

Landi, E., Tampieri, A., Celotti, G., Langenati, R., Sandri, M., & Sprio, S. (2005). Nucleation of biomimetic apatite in synthetic body fluids: dense and porous scaffold development. *Biomaterials*, **26**, 2835–2845.

Landis, W. J., Moradian-Oldak, J., & Weiner, S. (1991). Topographic imaging of mineral and collagen in the calcifying turkey tendon. *Connect. Tissue Res*, **25**, 181–196.

Lee, S. J., Yoon, Y. S., Lee, M. H., & Oh, N. S. (2007). Nanosized hydroxyapatite powder synthesized from eggshell and phosphoric acid. *J. Nanosci. Nanotechnol.*, **7**, 4061–4064.

LeGeros, R. Z. (2002). Properties of osteoconductive biomaterials: calcium phosphates. *Clin. Orthop. Relat. Res.*, **395**, 81–98.

LeGeros, R. Z., & LeGeros, J. P. (2008). Hydroxyapatite. In T. Kokubo (Ed.), *Handbook of Bioceramics and their Applications* (pp. 367–394). London: Woodhead Publishing Ltd.

Le Huec, J. C., Schaeverbeke, T., Clement, D., Faber, J., & Le Rebeller., A. (1995). Influence of porosity on the mechanical resistance of hydroxyapatite ceramics under compressive stress. *Biomaterials*, **10**, 113–118.

Leventouri, T., Bunaciu, C. E., & Perdikatsis, V. (2003). Neutron powder diffraction studies of silicon-substituted hydroxyapatite. *Biomaterials.*, **24**, 4205–4211.

Li, B., Chen, X., Guo, B., Wang, X., Fan, H., & Zhang, X. (2009). Fabrication and cellular biocompatibility of porous carbonated biphasic calcium phosphate ceramics with a nanostructure. *Acta Biomater.*, **5**, 134–143.

Lieberman, J. R., Daluiski, A., & Einhorn, T. A. (2002). The role of growth factors in the repair of bone. Biology and Clinical Applications. *J. Bone Joint. Surg. Am.*, **84A**, 1032–1044.

Lin, F. H., Liao, C. J., Chen, K. S., Sun, J. S., & Lin, C. Y. (2000). Preparation of betaTCP/HAP biphasic ceramics with natural bone structure by heating bovine cancellous bone with the addition of (NH(4))(2)HPO(4). *J. Biomed. Mater. Res.*, **51**, 157–163.

Liu, D. M. (1997). Fabrication of hydroxyapatite ceramic with controlled porosity. *J. Mater. Sci. Mater. Med.*, **8**, 227–232.

Lu, J., Blary, M. C., Vavasseur, S., Descamps, M., Anselme, K., & Hardouin, P. (2004). Relationship between bioceramics sintering and micro-particles-induced cellular damages. *J. Mater. Sci. Mater. Med.*, **15**, 361–365.

Medina Ledo, H., Thackray, A. C., Jones, I. P., Marquis, P. M., Macaskie, L. E., & Sammons, R. L. (2008). Micro structure and composition of biosynthetically synthesised hydroxyapatite. *J. Mater. Sci. Mater. Med.*, **19**, 3419–3427.

Moioli, E. K., Clark, P. A., Chen, M., Dennis, J. E., Erickson, H. P., Gerson, S. L., & Ma, J. J. (2008). Synergistic actions of hematopoietic and mesenchymal stem/progenitor cells in vascularizing bioengineered tissues. *PloS One*, **3**, 1–11.

O'Donnell, M. D., Fredholm, Y., de Rouffignac, A., & Hill, R. G. (2008). Structural analysis of a series of strontium-substituted apatites. *Acta Biomater.*, **4**, 1455–1464.

Okazaki, M., & Takahashi, J. (1999). Synthesis of functionally graded CO3 apatite as surface biodegradable crystals. *Biomaterials.*, **20**, 1073–1078.

Okumura, T., Oda, Y., & Mori, K. (1984). Tissue compatibility of newly developed bioceramics. *Neurol. Med. Chir. (Tokyo)*, **24**, 909–914.

Ono, I., Tateshita, T., Satou, M., Sasaki, T., Matsumoto, M., & Kodama, N. (1999). Treatment of large complex cranial bone defects by using hydroxyapatite ceramic implants. *Plast. Reconstr. Surg.*, **104**, 339–349.

Otto, B., & Ogilvie, A. (1998). The response of peritoneal macrophages after implantation of several ceramics as measured by the change of ectoenzyme activity. *Biomaterials*, **19**, 1049–1055.

Patel, N., Gibson, I. R., Ke, S., Best, S. M., & Bonfield, W. (2001). Calcining influence on the powder properties of hydroxyapatite. *J. Mater. Sci. Mater. Med.*, **12**, 181–188.

Paul, W., Nesamony, J., & Sharma, C. P. (2002). Delivery of insulin from hydroxyapatite ceramic microspheres: preliminary in vivo studies. *J. Biomed. Mater. Res.*, **61**, 660–662.

Peters, F., Schwarz, K., & Epple, M. (2000). The structure of bone studied with synchrotron X-ray diffraction, X-ray absorption spectroscopy and thermal analysis. *Thermochim. Acta*, **361**, 131–138.

Pietak, A. M., Reid, J. W., Stott, M. J., & Sayer, M. (2007). Silicon substitution in the calcium phosphate bioceramics. *Biomaterials.*, **2**, 84023–84032.

Qian, J., Kang, Y., Zhang, W., & Li, Z. (2008). Fabrication, chemical composition change and phase evolution of biomorphic hydroxyapatite. *J. Mater. Sci. Mater. Med.*, **19**, 3373–3383.

Ramanathan, C., & Ackerman, J. L. (1999). Quantitative solid-state NMR imaging of synthetic calcium phosphate implants. *Magn. Reson. Med.*, **41**, 1214–1220.

Rey, C., Shimizu, M., Collins, B., & Glimcher, M. J. (1991). Resolution-enhanced Fourier transform infrared spectroscopy study of the environment of phosphate ion in the early deposits of a solid phase of calcium phosphate in bone and enamel and their evolution with age: 2. Investigations in the nu3PO4 domain. *Calcif Tissue Int.*, **49**, 383–388.

Reynolds, M. A., Aichelmann-Reidy, M. E., Branch-Mays, G. L., & Gunsolley, J. C. (2003). The efficacy of bone replacement grafts in the treatment of periodontal osseous defects. A systematic review. *Ann. Periodontol.*, **8**, 227–265.

Ruys, A. J., Wei, M., Sorrell, C. C., Dickson, M. R., Brandwood, A., & Milthorpe, B. K. (1995). Sintering effects on the strength of hydroxyapatite. *Biomaterials.*, **16**, 409–415.

Sakamoto, M., Nakasu, M., Matsumoto, T., & Okihana, H. (2007). Development of superporous hydroxyapatites and their examination with a culture of primary rat osteoblasts. *J. Biomed. Mater. Res. A.*, **82**, 238–242.

Salyer, K. E., & Hall, C. D. (1989). Porous hydroxyapatite as an onlay bone-graft substitute for maxillofacial surgery. *Plast Reconstr Surg*, **84**, 236–244.

Sato, K., Kogure, T., Iwai, H., & Tanaka, J. (2008). Atomic-scale {101-0} Interfacial structure in Hydroxyapatite determined by high-resolution transmission electron microscopy. *J. Am. Ceramic Soc.*, **85**, 3054–3058.

Schmitz, J. P., Hollinger, J. O., & Milam, S. B. (1999). Reconstruction of bone using calcium phosphate bone cements: a critical review. *J. Oral. Maxillofac. Surg.*, **57**, 1122–1126.

Schuessele, A., Mayr, H., Tessmar, J., & Goepferich, A. (2009). Enhanced bone morphogenetic protein-2 performance on hydroxyapatite ceramic surfaces. *J. Biomed. Mater. Res. A.*, **90**, 959–971.

Singh, S., Trikha, S. P., & Edge, A. J. (2004). Hydroxyapatite ceramic-coated femoral stems in young patients. A prospective ten-year study. *J. Bone Joint. Surg.Br.*, **86**, 1118–1123.

Sogo, Y., Ito, A., Onoguchi, M., Oyane, A., Tsurushima, H., & Ichinose, N. (2007). Formation of a FGF-2 and calcium phosphate composite layer on a hydroxyapatite ceramic for promoting bone formation. *Biomed. Mater.*, **2**, S175–S180.

Spence, G., Patel, N., Brooks, R., & Rushton, N. (2008). Carbonate substituted hydroxyapatite: Resorption by osteoclasts modifies the osteoblastic response. *J. Biomed. Mater. Res. A.*, **90**, 1635–1640.

Sudo, A., Hasegawa, M., Fukuda, A., & Uchida, A. (2008). Treatment of infected hip arthroplasty with antibiotic-impregnated calcium hydroxyapatite. *J. Arthroplasty.*, **23**, 145–150.

Tadic, D., Beckmann, F., Schwarz, K., & Epple, M. (2004). A novel method to produce hydroxyapatite objects with interconnecting porosity that avoids sintering. *Biomaterials.*, **25**, 3335–3340.

Tadic, D., & Epple, M. (2004). A thorough physicochemical characterization of 14 calcium phosphate-based bone substitution materials in comparison to natural bone. *Biomaterials.*, **25**, 987–994.

Tampieri, A., Celotti, G., & Landi, E. (2005). From biomimetic apatites to biologically inspired composites. *Anal. Bioanal. Chem.*, **381**, 568–576.

Teixeira, S., Monteiro, F. J., Ferraz, M. P., Vilar, R., & Eugénio, S. (2007). Laser surface treatment of hydroxyapatite for enhanced tissue integration: surface characterization and osteoblastic interaction studies. *J. Biomed. Mater. Res. A.*, **81**, 920–929.

TenHuisen, K. S., & Brown, P. W. (1998). Formation of calcium-deficient hydroxyapatite from alpha-tricalcium phosphate. *Biomaterials.*, **19**, 2209–2217.

Tiselius, A. S., Hjerten, S., & Levin, O. (1956). Protein chromatography on calcium phosphate columns. *Arch. Biochem*, **65**, 132–155.

Trikha, S. P., Singh, S., Raynham, O. W., Lewis, J. C., Mitchell, P. A., & Edge, A. J. (2005). Hydroxyapatite-ceramic-coated femoral stems in revision hip surgery. *J. Bone Joint. Surg. Br.*, **87**, 1055–1060.

Turhani, D., Watzinger, E., Weissenböck, M., Yerit, K., Cvikl, B., Thurnher, D., & Ewers, R. (2005). Three-dimensional composites manufactured with human mesenchymal cambial layer precursor cells as an alternative for sinus floor augmentation: an in vitro study. *Clin. Oral Implants Res*, **16**, 417–424.

van der Meulen, J., & Koerten, H. K. (1994). Inflammatory response and degradation of three types of calcium phosphate ceramic in a non-osseous environment. *J. Biomed. Mater. Res.*, **28**, 1455–1463.

van Dijk, S., Dean, D. D., Liu, Y., Zhao, Y., Chirgwin, J. M., Schwartz, Z., & Boyan, B. D. (1998). Purification, amino acid sequence, and cDNA sequence of a novel calcium-precipitating proteolipid involved in calcification of corynebacterium matruchotii. *Calcif. Tissue Int.*, **62**, 350–358.

Vaz, L., Lopes, A. B., & Almeida, M. (1999). Porosity control of hydroxyapatite implants. *J. Mater. Sci. Mater. Med*, **10**, 239–242.

Wang, C., Duan, Y., Markovic, B., Barbara, J., Rolfe Howlett, C., Zhang, X., & Zreiqat, H. (2004). Proliferation and bone-related gene expression of osteoblasts grown on hydroxyapatite ceramics sintered at different temperature. *Biomaterials.*, **25**, 2949–2956.

Webster, T. J., Ergun, C., Doremus, R. H., Siegel, R. W., & Bizios, R. (2000). Enhanced functions of osteoblasts on nanophase ceramics. *Biomaterials.*, **21**, 1803–1810.

Wenisch, S., Stahl, J. P., Horas, U., Heiss, C., Kilian, O., Trinkaus, K., Hild, A., & Schnettler, R. (2003). In vivo mechanisms of hydroxyapatite ceramic degradation by osteoclasts: fine structural microscopy. *J. Biomed. Mater. Res. A.*, **67**, 713–718.

Wilson, R. M., Elliott, J. C., Dowker, S. E., & Smith, R. I. (2004). Rietveld structure refinement of precipitated carbonate apatite using neutron diffraction data. *Biomaterials.*, **25**, 2205–2213.

Yao, X., Tan, S., & Jiang, D. (2005). Fabrication of hydroxyapatite ceramics with controlled pore characteristics by slip casting. *J. Mater. Sci. Mater. Med.*, **16**, 161–165.

Yoshikawa, H., & Myoui, A. (2005). Bone tissue engineering with porous hydroxyapatite ceramics. *J. Artif. Organs*, **8**, 131–136.

Yoshikawa, H., Tamai, N., Murase, T., & Myoui, A. (2009). Interconnected porous hydroxyapatite ceramics for bone tissue engineering. *J. R. Soc. Interface.*, **53**, 5341–5348.

B. ALUMINA

Ketul C. Popat[1] *and Tejal A. Desai*[2]
[1]Department of Mechanical Engineering, School of Biomedical Engineering, Colorado State University, Fort Collins, CO, USA
[2]Department of Bioengineering and Therapeutic Sciences/ Department of Physiology, University of California, San Francisco, CA, USA

INTRODUCTION

Aluminum oxide or alumina has been used since 1970 as a material for manufacture of components of prostheses and surgical devices. It is very inert and resistant to corrosion in an *in vivo* environment. It elicits minimal response from the tissues and remains stable for many years of service. It is found in nature as corundum in emery, topaz, and emerald, and as the precious gemstones ruby and sapphire. Commercially it is extracted from ores such as bauxite and cryolite using the Bayer process. Its most significant use is in the production of aluminum metal. Other than applications in prostheses and surgical devices, it is commonly used as an abrasi·· due to its hardness, and as a refractory material due to its high melting point.

PRODUCTION OF ALUMINA

Commercially, alumina is extracted from bauxite using the Bayer process. The Bayer process can be considered in three stages:

1. Extraction: The aluminum-bearing minerals in bauxite are selectively extracted from the insoluble components by dissolving them in a solution of sodium hydroxide.

$$Al(OH)_3 + Na^+ + OH^- \rightarrow Al(OH)^{4-} + Na^+$$

2. Precipitation: Crystalline aluminum trihydroxide is precipitated. This is the reverse of the extraction process, except the chemistries are well-controlled.

$$Al(OH)^{4-} + Na^+ \rightarrow Al(OH)^3 + Na^+ + OH^-$$

3. Calcination: Aluminum trihydroxyde is calcined to form alumina. The water is driven off to form alumina. This process dictates the properties of the final product.

$$2Al(OH)_3 \rightarrow Al_2O_3 + 3H_2O$$

STRUCTURE OF ALUMINA

The alumina obtained by the Bayer process is recrystallized, based on the application for which it will be used. Alumina exists in many crystal phases: α, δ, γ, η, θ, ϱ, and χ, depending on the heat-treatment conditions. However, α-alumina or corundum is most thermodynamically stable form, and hence is used as a biomaterial. It has has a rhombohedral structure (a = 4.758 Å and c = 12.991 Å) (Figure B.1). The structure of can be viewed as a hexagonal close-packed array of O^- atoms with 2/3 of the octahedral sites occupied by Al^{3+} ions. Thus, each Al^{3+} ion is bonded to six O^- atoms in a distorted octahedron. Each such octahedron shares a face with one on the upper and one on the lower layers. The distortion is caused by repulsion between Al^{3+} ions in octahedra sharing the faces. Such close packing of the aluminum and oxygen atoms within this structure leads to its good mechanical and thermal properties.

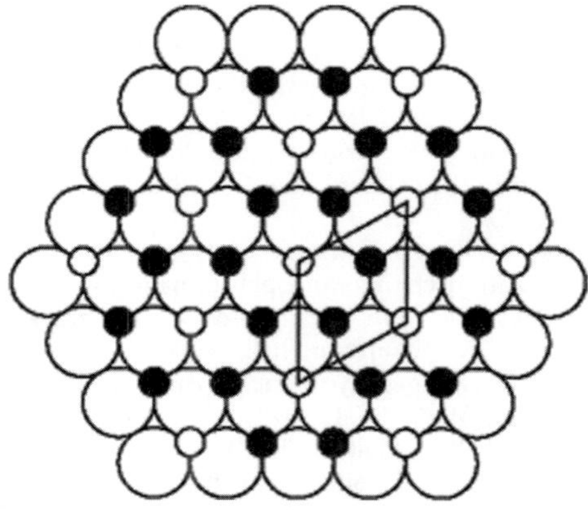

FIGURE B.1 Crystal structure of alumina (○: aluminum; ●: oxygen).

PROPERTIES OF ALUMINA

α-alumina is dense (specific gravity of 3.97), nonporous, and nearly inert. It is extremely hard and scratch-resistant (9 on the Mohs scale, next only to diamond). It has excellent corrosion resistance in *in vivo* environments. The physical properties of alumina such as strength, fatigue resistance, and fracture resistance depend on the grain size, porosity, and purity. Further, alumina exhibits high surface wettability, resulting in a low coefficient of friction. These excellent wetting characteristics are due to an outer surface of alumina that adsorbs a molecular film of water and biological molecules. The low coefficient of friction and wear rates of alumina occur only when the surface roughness is less than 0.02 μm, i.e., grain size is smaller than 4 μm with a very narrow size distribution. If the surface roughness is higher than 0.02 μm, large grains can pull out and lead to very rapid wear of loadbearing surfaces. Table B.1 lists some of the mechanical properties of 99.5% pure alumina.

The porosity of the alumina can be changed by how it is processed

Two primary synthetic processes are used for the production of alumina:

1. The traditional ceramic powder process: This processing involves three basic steps: powder-processing, shape-forming, and densification, often with a final mechanical finishing step. The ceramic produced by this technique is very dense, with virtually 0% porosity.
2. The solution–gelation or "sol–gel" process: This process is primarily used for the production of porous alumina or alumina coatings with porosity up to 40%. It involves four basic steps: dispersion, gelation, drying, and firing.

TABLE B.1	Mechanical Properties of 99.5% Alumina
Density	3.97 gm/cm^3
Flexural strength	345 MPa
Elastic modulus	300 GPa
Shear modulus	124 GPa
Bulk modulus	172 GPa
Poisson's ratio	0.21
Compressive strength	2100 MPa
Hardness	1000 kg/mm^2
Fracture toughness	3.5 MPa.m$^{1/2}$

The environmental impact of alumina and alumina-based ceramics is in general negligible; however, this is not true for the processing methods. Both of the above processes can have a significant environmental impact. For the traditional ceramic powder process, the most direct environmental impact arises from the shape-forming process where various binders, solvents, and other potentially toxic agents are used. For the sol–gel process, strong acids, plasticizers, binders, and solvents are used. Depending on the firing conditions, variable amounts of organic materials such as binders and plasticizers may be released as combustion products, causing significant environmental impact.

ALUMINA AS A BIOMATERIAL

Alumina is very inert and resistant to corrosion in an *in vivo* environment (Christel, 1992). It elicits minimal response from the tissues, and remains stable for many years of service. It is not bioresorbable; hence, the body does recognize it as a foreign material and attempts to isolate it by forming a fibrous capsule around the implant. However, as soon as a device made from alumina is implanted, within minutes proteins and other biomolecules adsorb on the surface, thus protecting the implant from the body's immune system. Further, the interface between alumina and the tissue can be engineered, preventing formation of a fibrous capsule around the implant. Although alumina is biocompatible, alumina particles formed by wear of the implant can induce a considerable foreign-body reaction (Tateiwa et al., 2008).

Nanoporous alumina (Popat et al., 2005; Swan et al., 2005a,b)

Cylindrical and planar anodized alumina nanoporous templates with pore diameters ranging from 10 to 130 nm can be fabricated using dilute solutions of either sulfuric or oxalic acid with anodizing voltages of approximately 8 V to 100 V. Pore length is determined by the thickness of the resultant alumina layer, and can vary from nanometers to several tens of microns. A two-step anodization process is used to ensure high pore uniformity. The aluminum substrate is first anodized in an oxalic or sulfuric acid solution. The anodization is stopped after consuming a few microns of aluminum, and the porous alumina film is removed through etching. The etchant, a mixture of chromic acid and phosphoric acid, is highly selective attacking alumina much faster than aluminum. The remaining aluminum is dimpled, with the dimples serving as a uniform seed layer upon which a highly uniform porous layer can then be achieved through a second anodization step. The exact pore size and pore structure is determined in this second anodization through control of the anodizing voltage. The pore diameter increases with anodization voltage, at ≈1.29 nm/volt. Constant anodizing voltages result in straight uniform pores, while step-wise changes in the anodizing voltage can be used to introduce branching channels with smaller diameters, enabling fabrication of a nanoscale tree-like structure. Such nanoporous templates can either be used as coatings for orthopedic implants to enhance osseointegration or as membranes for protein or virus separation (Figure B.2).

ALUMINA IN JOINT REPLACEMENTS

In the 1970s it was first realized that the properties of alumina ceramics could be exploited to provide better implants for orthopedic applications. Since then, alumina has often been used for wear surfaces in joint replacement prostheses due to its excellent wear resistance and ability to create smooth and polished surfaces. Typically, alumina is used to fabricate femoral heads for hip replacement implants and wear plates for knee replacement implants. To date more than 2.5 million alumina femoral heads have been implanted (Lang et al., 2008). Figure B.3 shows a schematic of total hip replacement and knee replacement implants. In hip replacements, the alumina femoral head is used in conjunction with a metallic femoral stem and an acetabular cup made from ultra-high molecular weight polyethylene (UHMWPE) or alumina for the opposing articulating surface. The success of such joint replacement implants depends on two factors:

1. The frictional and wear behavior of materials;
2. The reliability of the anchorage of the implant to the natural tissue.

Studies have shown that wear rates for alumina on UHMWPE are almost 20 times lower than those for metal on UHMWPE, resulting in less debris formation (Andreiotelli et al., 2009). Wear debris leads to complicated conditions such as osteolysis, resulting in long-term failure of the implants. An alumina–alumina ceramic bearing couple is superior to alumina–UHMWPE or metal, resulting in better wear resistance and preventing the inflammatory response elicited by polyethylene particles. Further, studies have also shown that ceramic debris is less inflammatory than that from UHMWPE. Fracture rates of alumina heads range between 0% for ceramics manufactured after 1990 and 13.4% for ceramics manufactured before 1990. The high fracture rates for alumina heads prior to 1990 were caused as low-density alumina with a very coarse microstructure was used. Over the years, material scientists have substantially improved the fabrication strategies of alumina, thus resulting in improved mechanical strength. The current generation is an alumina that is hot isostatic pressed,

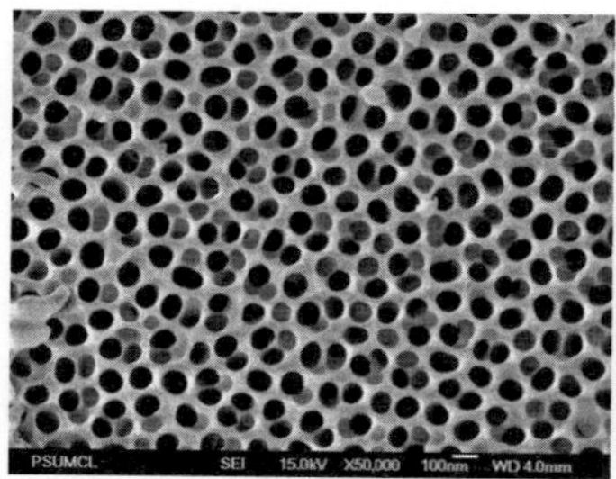

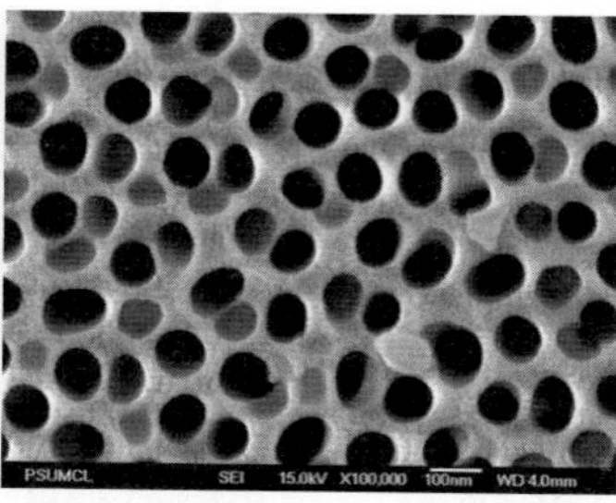

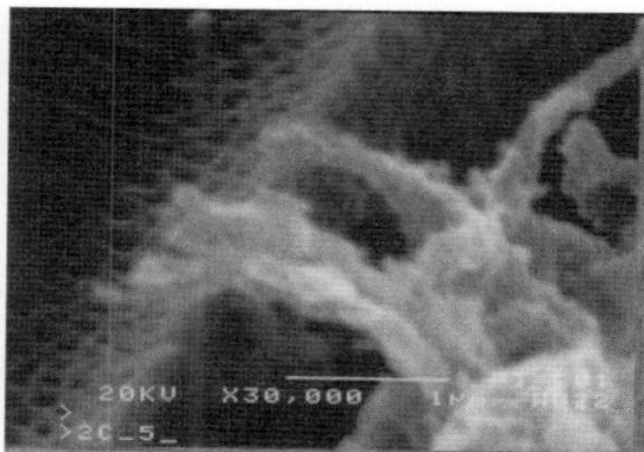

FIGURE B.2 Nanoporous alumina fabricated using the anodization process (left and center). Osteoblast interaction with the nanoporous architecture (right).

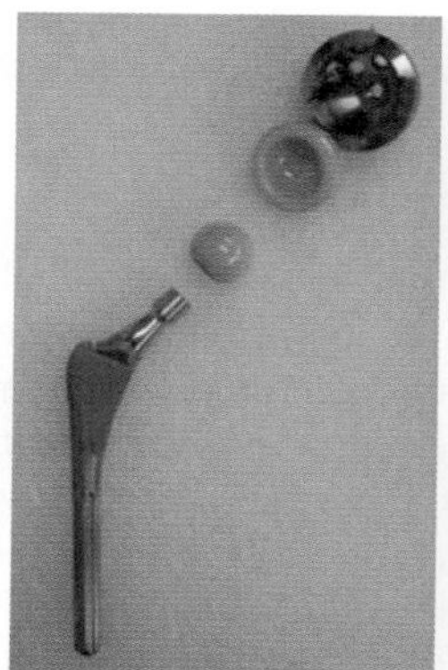

FIGURE B.3 Total hip replacement (left) and total knee replacement (right) implants.

laser marked, and proof tested. This material has been on the market since 1994. The fracture rate for femoral heads is 0.004%. In total knee replacements, alumina is used in portions that come in contact with bone, and a combination of alumina and UHMWPE is used in sliding portions. In such implants, problems caused by polyethylene debris are much more severe than in hip implants (Figure B.3).

Alumina is not the optimal material for orthopedic applications. The Young's modulus for medical grade alumina is 380–420 GPa. This is much higher than that of cancellous bone (0.05–0.5 GPa) or cortical bone (7–25 GPa) (Young's modulus also depends on the age of the person, and the location of the bone tissue in the body). Thus, there is a mismatch in the mechanical properties of bone and alumina. Alumina implant will shield the bone from any mechanical loading, and the entire load will be carried by the implant. This mechanical shielding will result in compressive stresses on the bone, leading to resorption and weakening of the bone which will eventually result in failure of the implant.

ALUMINA IN BONE SPACERS

Alumina with more than 30% porosity can be used as bone spacers to replace sections of bone that have been removed due to cancer or traumatic injury (Huckstep, 1987). Bone spacers are inserted in natural tissue using metallic pins. The interconnected porous nature of the implant allows for bone cell infiltration into the implant, eventually resulting in new bone formation. Typical pore sizes are greater than 100 μm. This will not only allow the new bone to grow, but will also encourage vascularization. Porous alumina is fabricated using either the sol–gel process or a hydrothermal exchange process in which microstructures of certain corals are used to cast the interconnected porous network. Porous alumina is weaker than the dense form of alumina, with its strength decreasing with increase in porosity.

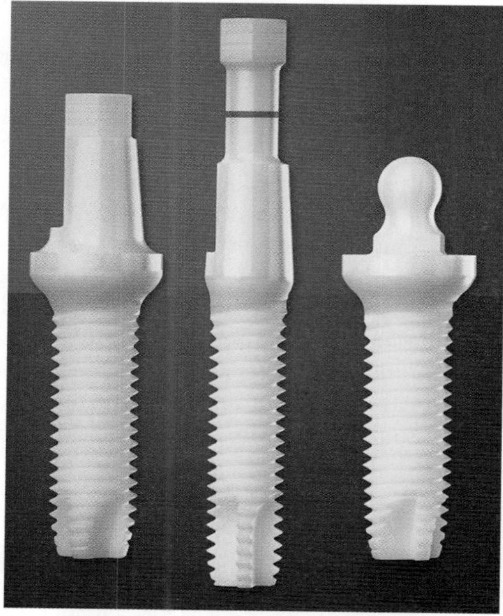

FIGURE B.4 Ceramic dental implants.

ALUMINA IN DENTAL APPLICATIONS

Dense alumina has been used in dental applications, specifically for replacement of teeth (Figure B.4). Single crystal alumina is used for dental implants, since polycrystalline alumina can be fractured while inserting the implant in the dental root. Single crystal alumina has a higher bending strength (13,000 kg/cm^2) compared to that of polycrystalline alumina (3500 kg/cm^2). Typically, dental implants are made of single crystal alumina cylindrical core around which polycrystalline alumina is fused. There are several disadvantages of alumina dental implants. They have a high modulus of elasticity compared to the native tissue, and cannot bend like metal implants. Thus, alumina dental implants are currently declining in popularity and being replaced by material such as dental porcelain (Della Bona and Kelly, 2008).

OTHER APPLICATIONS OF ALUMINA

Apart from joint replacements, bone spacers, and dental implants, alumina has been used for ENT applications and in maxillo-facial surgery. Alumina implants have also been used for neurosurgical operations such as cranioplasty (Kobayashi et al., 1987). Alumina has also been used in keratoprostheses (corneal replacement) (Polack and Heimke, 1980). Porous alumina has also been used for controlled delivery of hormones, vaccines, and drugs over long periods of time.

ALUMINA MATRIX COMPOSITES

Alumina matrix composites are composed of alumina, which represents 82% of the volume of the overall material. Nanoparticles of zirconium oxide (zirconia) are added to the alumina matrix and represent 17% of the volume (Cavalcanti et al., 2009). These particles are stabilized in the tetragonal phase, since it provides the best mechanical performance. The composites are reinforced in three different forms:

1. *In situ* formation of elongated crystals of strontium oxide in the alumina matrix, deflecting any subcritical cracks;
2. The addition of homogeneously dispersed small zirconia particles in the alumina matrix, creating a transformation toughening;
3. The creation of a solid solution of alumina with chromium oxide, resulting in increased hardness.

The US FDA has approved alumina matrix composites for use in ball heads of implant as a replacement for UHMWPE. The first clinical trials were performed in 2000, since then more than 65,000 ball heads have been implanted all over the world.

CONCLUSION

Alumina is very inert and resistant to corrosion in an *in vivo* environment. It elicits minimal response from the tissues, and remains stable for many years of service. Most of the alumina used for implant applications is polycrystalline, although some dental implants are single crystal. It is widely used in loadbearing applications due to its excellent biocompatibility, corrosion resistance, high wear resistance, and high compressive strength. Further, alumina has the tribological properties most suitable for articulating surfaces in orthopedic implants. However, high elastic modulus, slow crack growth, cyclic failure, low toughness, and sensitivity to tensile strength are some of the concerns for high loadbearing applications. The recent development of alumina matrix ceramics may extend the implant life beyond the current 10–12 years. It is expected that the use of alumina will continue to grow in orthopedic, dental, and maxillo-facial applications in the near future.

FIGURE B.5 BIOLOX®* delta ceramic femoral head.

BIOLOX®* Delta Ceramic Femoral Head

The new alumina matrix composite, BIOLOX® delta developed by CeramTec AG, is used by Zimmer Inc. to fabricate femoral heads. This composite has excellent biocompatibility, low wear rates, high hardness, and is very stable in *in vivo*-like environments. The implant made from this composite meets the increasing demands of development of a durable implant for young and active patients. The heads fabricated from this ceramic have improved mechanical properties compared to standard alumina heads. The composite consists of about 75% alumina, 24% zirconia, and 1% chromium oxide. Alumina provides the hardness and the wear resistance, while zirconia along with chromium oxide provides improved mechanical properties. The pink color of the implant is also derived from chromium oxide (Figure B.5).

BIBLIOGRAPHY

Andreiotelli, M., Wenz, H. J., & Kohal, R. J. (2009). Are ceramic implants a viable alternative to titanium implants? A systematic literature review. *Clin. Oral Implants Res.*, **20**(Suppl 4), 32–47.

Cavalcanti, A. N., Foxton, R. M., Watson, T. F., Oliveira, M. T., Giannini, M., & Marchi, G. M. (2009). Y-TZP ceramics: key concepts for clinical application. *Oper. Dent.*, **34**, 344–351.

Christel, P. S. (1992). Biocompatibility of surgical-grade dense polycrystalline alumina. *Clin. Orthop. Relat. Res.*, 10–18.

Della Bona, A., & Kelly, J. R. (2008). The clinical success of all-ceramic restorations. *J. Am. Dent. Assoc.*, **139**(Suppl), 8S–13S.

Huckstep, R. L. (1987). Stabilization and prosthetic replacement in difficult fractures and bone tumors. *Clin. Orthop. Relat. Res.*, 12–25.

Kobayashi, S., Hara, H., Okudera, H., Takemae, T., & Sugita, K. (1987). Usefulness of ceramic implants in neurosurgery. *Neurosurgery*, **21**, 751–755.

Lang, J. E., Whiddon, D. R., Smith, E. L., & Salyapongse, A. K. (2008). Use of ceramics in total hip replacement. *J. Surg. Orthop. Adv.*, **17**, 51–57.

Polack, F. M., & Heimke, G. (1980). Ceramic keratoprostheses. *Ophthalmology*, **87**, 693–698.

Popat, K. C., Leary Swan, E. E., Mukhatyar, V., Chatvanichkul, K. I., Mor, G. K., Grimes, C. A., & Desai, T. A. (2005). Influence of nanoporous alumina membranes on long-term osteoblast response. *Biomaterials*, **26**, 4516–4522.

Swan, E. E., Popat, K. C., & Desai, T. A. (2005a). Peptide-immobilized nanoporous alumina membranes for enhanced osteoblast adhesion. *Biomaterials*, **26**, 1969–1976.

Swan, E. E., Popat, K. C., Grimes, C. A., & Desai, T. A. (2005b). Fabrication and evaluation of nanoporous alumina membranes for osteoblast culture. *J. Biomed. Mater. Res. A*, **72**, 288–295.

Tateiwa, T., Clarke, I. C., Williams, P. A., Garino, J., Manaka, M., Shishido, T., Yamamoto, K., & Imakiire, A. (2008). Ceramic total hip arthroplasty in the United States: safety and risk issues revisited. *Am. J. Orthop. (Belle Mead NJ)*, **37**, E26–E31.

CHAPTER I.2.5 HYDROGELS

Nicholas A. Peppas[1] *and Allan S. Hoffman*[2]

[1]Chemical Engineering Department, University of Texas, Austin, TX, USA

[2]Professor of Bioengineering and Chemical Engineering, UWEB Investigator, University of Washington, Seattle, WA, USA

INTRODUCTION

Hydrogels have received significant attention because of their high water contents and related potential for many biomedical applications. Hydrogels are polymeric structures held together as water-swollen gels by: (1) primary covalent cross-links; (2) ionic forces; (3) hydrogen bonds; (4) affinity or "bio-recognition" interactions; (5) hydrophobic interactions; (6) polymer crystallites; (7) physical entanglements of individual polymer chains; or (8) a combination of two or more of the above interactions. The classic book by Andrade (1976) offers some of the best work that was available prior to 1975 (e.g., Ratner and Hoffman, 1976). Since then, numerous reviews and several books have addressed the preparation, structure, characterization, and applications of hydrogels (e.g., Peppas 1987, 2001; Brannon-Peppas and Harland, 1990; Lee and Mooney, 2001; Qiu and Park, 2001; Hennink and van Nostrum, 2002; Jeong et al., 2002; Miyata et al., 2002; Drury and Mooney, 2003; Patterson et al., 2010). In this chapter we focus on the preparation and characterization of the structure, and chemical and physical properties of *synthetic* hydrogels. Many natural polymers such as collagen, gelatin, fibrin, hyaluronic acid, heparin, alginates, pectins, chitosan, and others can be used to form hydrogels, and some of these gels have been used in biomedical applications. Details on these types of materials can be found throughout this textbook in other chapters.

CLASSIFICATION AND BASIC STRUCTURES OF HYDROGELS

Depending on their method of preparation, ionic charge, or physical structure features, hydrogels may be classified in several categories. Based on the method of preparation, they may be: (1) *homopolymer hydrogels*; (2) *copolymer hydrogels*; (3) *multi-polymer hydrogels*; and (4) *interpenetrating network (IPN) hydrogels*. *Homopolymer hydrogels* are cross-linked networks of one type of hydrophilic monomer unit, whereas *copolymer hydrogels* are produced by cross-linking of chains composed of two comonomer units, at least one of which must be hydrophilic to render them water swellable. *Multi-polymer hydrogels* are produced from three or more comonomers reacting together (e.g., Lowman and Peppas, 1997 and 1999). Finally, *interpenetrating network (IPN) hydrogels* are produced by two methods, one within a pre-formed network and the other in solution; the most common method is to polymerize one monomer within a different cross-linked hydrogel network. The monomer polymerizes to form a polymer or a second cross-linked network that is intermeshed with the first network.

There are several other ways that hydrogels may be classified. *Ionic hydrogels*, with ionic charges on the backbone polymers, may be classified as: (1) *neutral hydrogels* (uncharged); (2) *anionic hydrogels* (having negative charges only); (3) *cationic hydrogels* (having positive charges only); or (4) *ampholytic hydrogels* (having both positive and negative charges). These last gels may end up with a net negative, positive or neutral charge.

Based on physico-chemical structural features of the network, hydrogels may also be classified as: (1) *amorphous hydrogels* (having covalent cross-links); or (2) *semi-crystalline hydrogels* (may or may not have covalent cross-links). In amorphous hydrogels, the macromolecular chains are arranged randomly. Semi-crystalline hydrogels are characterized by self-assembled regions of ordered macromolecular chains (crystallites).

Another type of classification of hydrogels includes the "*complexation*" *hydrogels*, which are held together by specific types of secondary forces. These include hydrogen bonds, hydrophobic group associations, and affinity "complexes" (e.g., (1) hetero-dimers (peptide/peptide interactions called "coil–coil"); (2) biotin/streptavidin; (3) antibody/antigen; (4) conA/glucose; (5) poly(D-lactic acid)/poly(L-lactic acid) (PDLA/PLLA) stereo-complexes; and (6) cyclodextrin (CD) inclusion complexes). The physical properties of the gels held together by such secondary associations are critically dependent on the network density of these interactions, as well as on the many environmental conditions that can affect them. Hydrogels may also be classified as stable or degradable, with the latter further categorized as hydrolytically or enzymatically degradable.

Structural evaluation of hydrogels reveals that ideal networks are only rarely observed. Figure I.2.5.1a shows an ideal macromolecular network (hydrogel) indicating tetra-functional cross-links (junctions) produced by covalent bonds. However, in real networks it is possible to encounter multi-functional junctions (Figure I.2.5.1b) or physical molecular entanglements (Figure I.2.5.1c) playing the role of semi-permanent junctions. Hydrogels with molecular defects are always possible. Figures I.2.5.1d and I.2.5.1e indicate two such effects: unreacted functionalities with partial entanglements (Figure I.2.5.1d) and chain loops (Figure I.2.5.1e).

The terms "cross-link," "junction" or "tie-point" (shown by an open circle symbol in Figure I.2.5.1d) indicate the covalent or secondary connection points of several chains. In the case of covalent linkages, these junctions may be carbon atoms, but they are usually small chemical bridges (e.g., an acetal bridge in cross-linked poly(vinyl alcohol), or an ethylene glycol diester bridge in the polyHEMA contact lens gel) with molecular weights much smaller than those of the cross-linked polymer chains. In other situations, a junction may be crystallites or other secondary interactions, such as described above, of a permanent or semi-permanent nature. Thus, in reality, the junctions should never be considered as points without volume, which is the usual assumption made when developing theoretical models for prediction of the properties of cross-linked hydrogels (Flory, 1953). Instead, they have a finite size and contribute to the physical properties during biomedical applications.

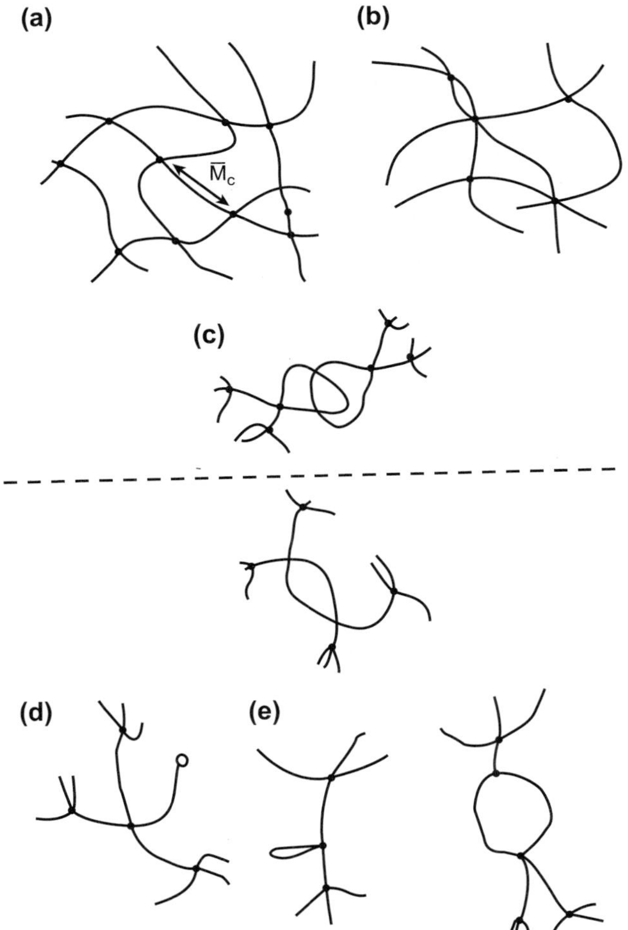

FIGURE I.2.5.1 (a) Ideal macromolecular network of a hydrogel; (b) Network with multifunctional junctions; (c) Physical entanglements in a hydrogel; (d) Unreacted functionality in a hydrogel; (e) Chain loops in a hydrogel.

SYNTHESIS OF HYDROGELS

Covalently cross-linked hydrogels are usually prepared by bringing together small multi-functional molecules such as monomers and oligomers, and reacting to form a network structure. Sometimes large polymer molecules may be cross-linked with the same small multi-functional molecules. Such cross-linking may be achieved by reaction of two chemical groups on two different molecules, which can be initiated by catalysts, by photo-polymerization or by radiation cross-linking (see reviews by Peppas et al. 2000; Hoffman, 2002; Ottenbrite et al., 2010; Jin & Dijkstra, 2010).

Several methods for forming cross-linked hydrogels are based on free radical reactions. The first involves a copolymerization-cross-linking reaction between one or more monomers and one multi-functional monomer that are present in relatively small quantities. In a related method, two water-soluble polymers may be cross-linked together by formation of free radicals on both polymer molecules, which combine to form the cross-link (Figure I.2.5.2). These reactions are free radical polymerization or cross-linking reactions, and such processes for synthesis of cross-linked hydrogels can be initiated by decomposition of peroxides or azo compounds, or by using ionizing radiation or UV light. Ionizing radiation methods utilize electron beams, gamma rays or X-rays to excite a polymer and produce a cross-linked structure via free radical reactions (e.g., Chapiro, 1962). Such free radical reactions can lead to rapid formation of a three-dimensional network, and are usually carried out

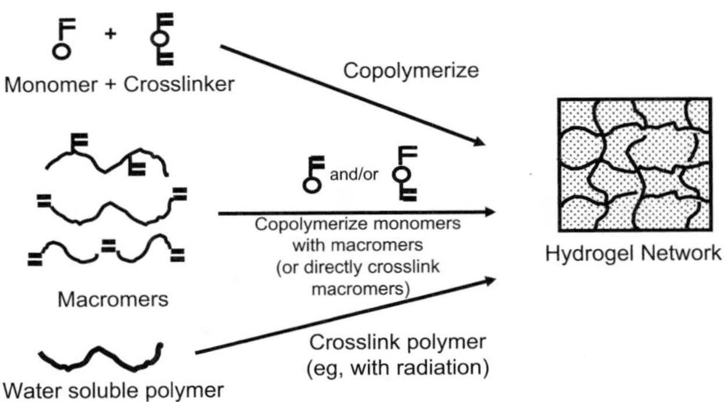

FIGURE I.2.5.2 Synthesis of hydrogels by free radical polymerization reactions and cross-linking reactions. (Hoffman, 2002.)

in the absence of oxygen or air (note: some polymers are degraded by radiation, especially in air).

In another method, chemical cross-linking calls for direct reaction of a linear or branched polymer with a di-functional or multi-functional, small molecular weight, cross-linking agent. This agent usually links two longer molecular weight chains through its di- or multi-functional groups (Figure I.2.5.3a). There are a number of well-known reactions that can be used for linking hydrophilic polymers together with each other or with cross-linkers to form hydrogels, including some recent methods with growing popularity, such as the Michael addition of dithiol compounds with divinyl compounds (e.g., see Schoenmakers et al., 2004; van de Wetering et al., 2005) and the reaction of alkynes plus azides to form triazoles (Click reaction; see Kolb et al., 2001). The reader is referred to the excellent, comprehensive book detailing such chemistries, *Bioconjugate Techniques* (Hermanson, 2008).

A similar method involves the reaction of a small bi-functional molecule and linear polymeric chains having pendant or terminal reactive groups such as $-OH$, $-NH_2$, NCO, or $-COOH$ that are cross-linked by the bi-functional molecule (Figure I.2.5.3b). Natural polymers such as proteins can also be cross-linked in a similar way by using enzymes. For example, transglutaminase catalyses the reaction of protein glutamine amide groups with lysine amino groups (or with pendant or terminal amine groups on a synthetic polymer backbone) (Sperinde and Griffith, 1997 and 2000). This reaction is:

$$-(CH_2)_3-\mathbf{CONH_2}+-(CH_2)_4-NH_2+\text{transglutaminase}\rightarrow$$
$$-(CH_2)_3-\mathbf{CONH}-(CH_2)_4-\ +NH_3$$

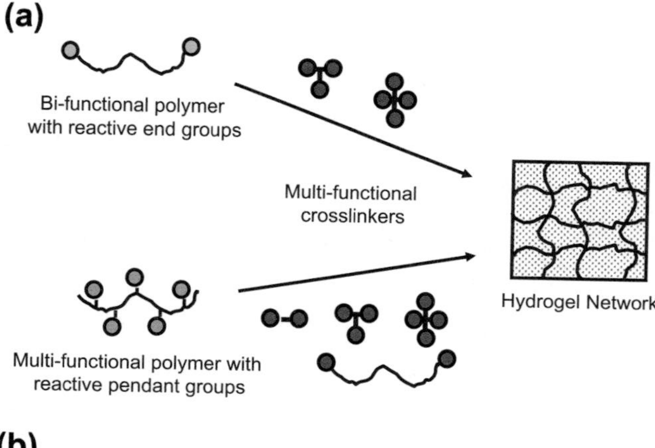

(b)

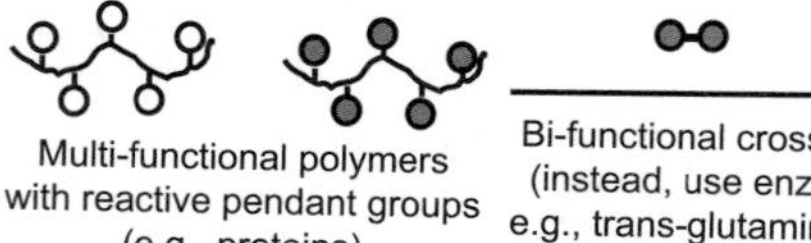

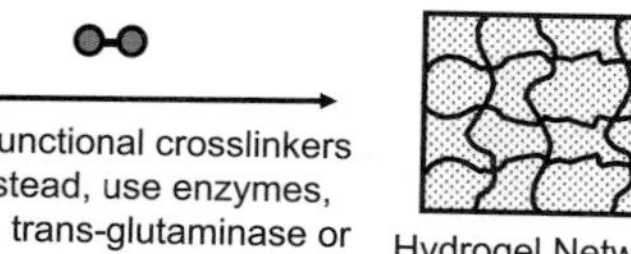

Hydrogel Network

FIGURE I.2.5.3 (a) Synthesis of hydrogels by cross-linking reactive polymers with multi-functional cross-linkers. (Hoffman, 2002.) (b) Synthesis of hydrogels by cross-linking of multi-functional polymers with small bi-functional molecules. (Hoffman, 2002.)

In another enzyme-catalysed cross-linking reaction, hydroxyphenylpropionic acid (tyramine) was conjugated to gelatin and cross-linked to form a hydrogel by the horseradish peroxidase (HRP)-catalysed oxidation reaction with hydrogen peroxide (H_2O_2) (Wang et al., 2010; Jin and Dijkstra, 2010).

Formation of hydrogel networks may also result from physico-chemical interactions, and some examples are highlighted below.

Poly(ethylene glycol) (PEG) molecules or block polymers containing PEG can "thread" through cyclodextrins (CDs), and sequences of CDs threaded on different PEG molecules can then self-associate, helping to hold together the hydrogel that is formed (Harada et al., 1992; Li et al., 2001, 2003a,b, 2006). Hetero-dimer peptide sequences that are conjugated as pendant groups on different polymer chains can complex together in "coil–coil" associations, and thereby "tie" chains together to form hydrogels (Yang et al., 2006; Xu and Kopecek, 2008). Stereo-complexes can form between D and L forms of poly(lactic acid) (known as PDLA/PLLA stereo-complexes); this complexation can lead to hydrogel formation if block copolymers that contain both PDLA and PLLA blocks, along with hydrophilic blocks as PEG are mixed together (Jin and Dijkstra, 2010).

Figure I.2.5.4 shows the formation of interpenetrating network (IPN) hydrogels. Figure I.2.5.5 shows how hydrogels can be formed by ionic interactions. Figures I.2.5.6a and I.2.5.6b show how hydrogels can be formed by affinity recognition reactions (see also Miyata, 2010, for more on formation of affinity recognition hydrogels).

SWELLING BEHAVIOR OF HYDROGELS

The physical behavior of biomedical hydrogels is dependent on their dynamic swelling and equilibrium in water and in aqueous solutions. Much of the water within swollen hydrogels may be bound to the polymer chains by either polar or hydrophobic interactions (e.g., Ilavsky, 1982). Many solutes can diffuse into and through hydrogels only within unbound or "free" water channels. Solutes that are chaotropic may diffuse into and through hydrogels by destructuring such bound water layers around the polymer chains.

The Flory–Huggins theory is an ideal thermodynamic description of polymer solutions, and it does not consider network imperfections or the real, finite volumes of network chains and cross-links, and in the case of aqueous solutions it does not consider the presence of "bound" (versus "free") water around the network chains. It can be used to calculate thermodynamic quantities related to that mixing process. Flory (1953) developed the initial theory of the swelling of cross-linked polymer gels using a Gaussian distribution of the polymer chains. His model describing the equilibrium degree of cross-linked polymers postulated that the degree to which a polymer network swelled was governed by the elastic retractive

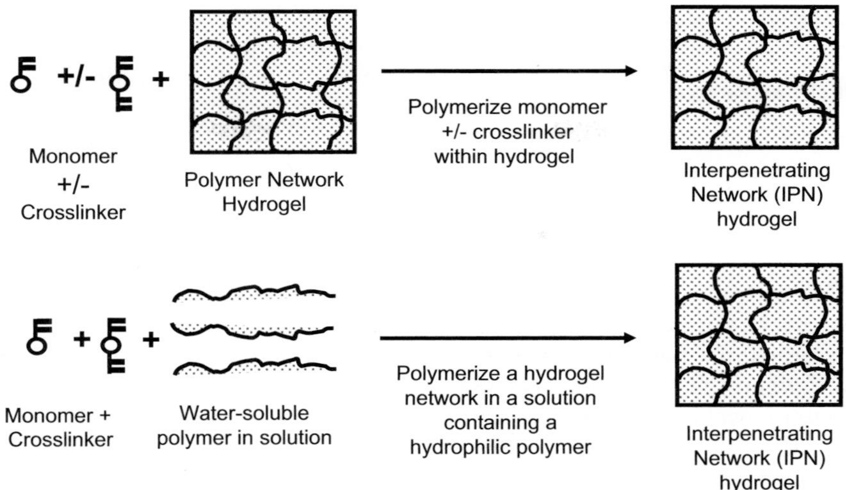

FIGURE I.2.5.4 Two methods for formation of an interpenetrating network (IPN) hydrogel. (Hoffman, 2002.)

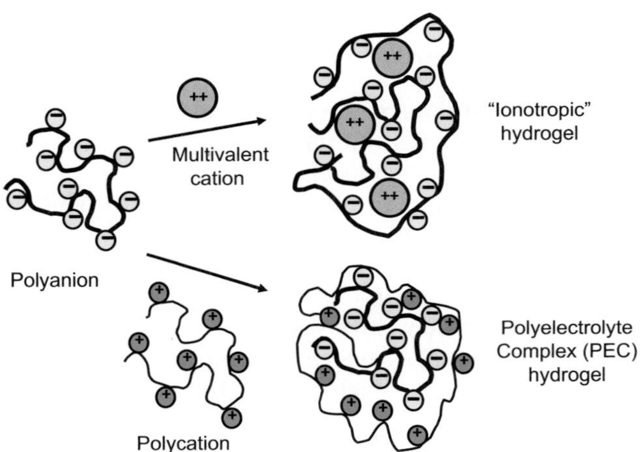

FIGURE I.2.5.5 Formation of ionic hydrogels. (Hoffman, 2002.)

forces of the polymer chains and the thermodynamic compatibility of the polymer and the solvent molecules. In terms of the free energy of the system, the total free energy change upon swelling was written as:

$$\Delta G = \Delta G_{\text{elastic}} + \Delta G_{\text{mix}} \quad (1)$$

Here, $\Delta G_{\text{elastic}}$ is the contribution due to the elastic retractive forces and ΔG_{mix} represents the thermodynamic compatibility of the polymer and the swelling agent (water).

Upon differentiation of Equation (1) with respect to the water molecules in the system, an expression can be derived for the chemical potential change of water in terms of the elastic and mixing contributions due to swelling.

$$\mu_1 - \mu_{1,0} = \Delta\,\mu_{\text{elastic}} + \Delta\,\mu_{\text{mix}} \quad (2)$$

Here, μ_1 is the chemical potential of water within the gel and $\mu_{1,0}$ is the chemical potential of pure water.

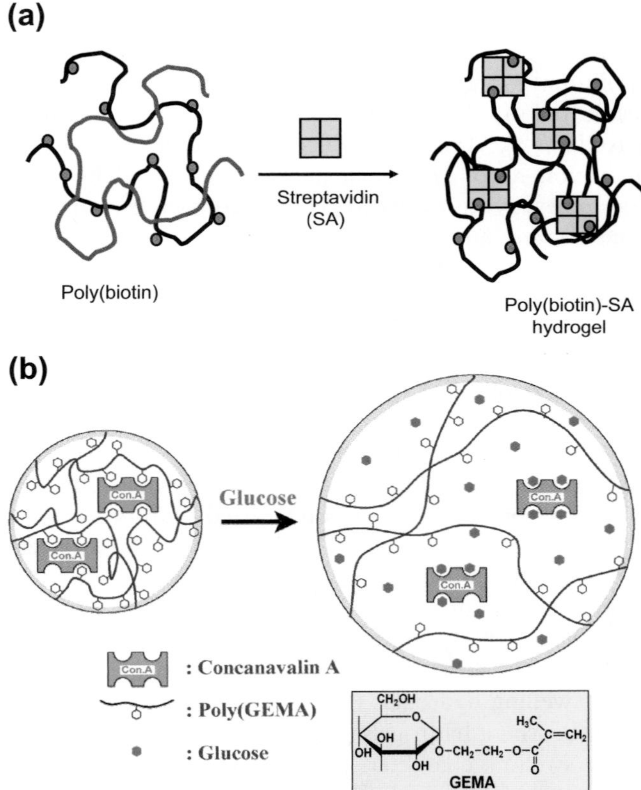

FIGURE I.2.5.6 (a) Formation of an affinity hydrogel between polybiotin and streptavidin. (Morris et al., 1993.) (b) Glucose-responsive hydrogel swells when free glucose competes with polymeric glucose groups in a ConA-cross-linked GEMA hydrogel. (Miyata et al., 1996.)

At equilibrium, the chemical potentials of water inside and outside of the gel must be equal. Therefore, the elastic and mixing contributions to the chemical potential will balance one another at equilibrium. The chemical potential change upon mixing can be determined from the heat of mixing and the entropy of mixing. Using the

Flory–Huggins theory, the chemical potential of mixing can be expressed as:

$$\Delta \mu_{mix} = RT\left(\ln(1 - \upsilon_{2,s}) + \upsilon_{2,s} + \chi_1 \upsilon_{2,s}^2\right) \quad (3)$$

where χ_1 is the polymer–water interaction parameter, $\upsilon_{2,s}$ is the polymer volume fraction of the gel, T is absolute temperature, and R is the gas constant.

This thermodynamic swelling contribution is counterbalanced by the retractive elastic contribution of the cross-linked structure. The latter is usually described by the rubber elasticity theory and its variations (Peppas, 1987). Equilibrium is attained in a particular solvent at a particular temperature when the two forces become equal. The volume degree of swelling, Q (i.e., the ratio of the actual volume of a sample in the swollen state divided by its volume in the dry state) can then be determined from Equation (4).

$$\upsilon_{2,s} = \frac{\text{volume of polymer}}{\text{volume of swollen gel}} = \frac{V_p}{V_{gel}} = 1/Q \quad (4)$$

Researchers working with hydrogels for biomedical applications prefer to use other parameters in order to define the equilibrium-swelling behavior. For example, Yasuda et al. (1969) introduced the use of the so-called hydration ratio, H, which has been accepted by those researchers who use hydrogels for contact lens applications (Peppas and Yang, 1981). Another definition is that of the weight degree of swelling, q, which is the ratio of the weight of the swollen sample over that of the dry sample.

In general, highly water-swollen hydrogels include those of cellulose derivatives, poly(vinyl alcohol), poly(N-vinyl 2-pyrrolidone) (PNVP), and poly(ethylene glycol), among others. Moderately and poorly swollen hydrogels are those of poly(hydroxyethyl methacrylate) (PHEMA) and many of its copolymers. In general, a basic hydrophilic monomer can be copolymerized with other more or less hydrophilic monomers to achieve desired swelling properties. Such processes have led to a wide range of swellable hydrogels, as Gregonis et al. (1976), Peppas (1987, 1997), and others have noted. Park and coworkers have developed a family of high water content, rapid swelling hydrogels (Omidian and Park, 2010) and superabsorbent hydrogels (Mun et al., 2010). Knowledge of the swelling characteristics of a polymer is of utmost importance in biomedical and pharmaceutical applications since the equilibrium degree of swelling influences: (1) the solute diffusion coefficient through these hydrogels; (2) the surface properties and surface molecule mobility; (3) the optical properties, especially in relation to contact lens applications; and (4) the mechanical properties.

DETERMINATION OF STRUCTURAL CHARACTERISTICS

The parameter that describes the basic structure of the hydrogel is the molecular weight between cross-links, $\overline{M}_c$ (as shown in Figure I.2.5.1a). This parameter defines the average molecular size between two consecutive junctions regardless of the nature of those junctions and can be calculated by Equation (5).

$$\frac{1}{\overline{M}_c} = \frac{2}{\overline{M}_n} - \frac{(\upsilon/V_1)\left[\ln(1 - \upsilon_{2,s}) + \upsilon_{2,s} + \chi_1 \upsilon_{2,s}^2\right]}{\left(\upsilon_{2,s}^{1/3} - \frac{\upsilon_{2,s}}{2}\right)} \quad (5)$$

An additional parameter of importance in structural analysis of hydrogels is the cross-linking density, ρ_x, which is defined by Equation (6).

$$\rho_x = \frac{1}{\upsilon \overline{M}_c} \quad (6)$$

In these equations, υ is the specific volume of the polymer (i.e., the reciprocal of the amorphous density of the polymer), and $\overline{M}_n$ is the initial molecular weight of the uncross-linked polymer.

BIOMEDICAL HYDROGELS

Acrylic Hydrogels

Hydrogels with desired physical or chemical properties for a specific biomedical application may be "molecularly engineered" by choosing among the many types of acrylic monomers and cross-linkers available. This has led to many publications, describing a large family of acrylic hydrogels (e.g., Peppas et al., 2000; Peppas, 2001; Ottenbrite et al., 2010).

The most widely used hydrogel is water-swollen, cross-linked PHEMA, which was introduced as a biological material by Wichterle and Lim (1960). The hydrogel is inert to normal biological processes, shows resistance to degradation, is permeable to metabolites, is not absorbed by the body, is biocompatible, withstands heat sterilization without damage, and can be prepared in a variety of shapes and forms. The swelling, mechanical, diffusional, and biomedical characteristics of PHEMA gels have been studied extensively. The properties of these hydrogels are dependent upon their method of preparation, polymer volume fraction, degree of cross-linking, temperature, and swelling agent (Michalek et al., 2010).

Other acrylic hydrogels of biomedical interest include polyacrylamides and their derivatives. Tanaka (1979) has done extensive studies on the abrupt swelling and de-swelling of partially hydrolyzed acrylamide gels with changes in swelling agent composition, curing time, degree of cross-linking, degree of hydrolysis, and temperature. These studies have shown that the ionic groups produced in an acrylamide gel upon hydrolysis give the gel a structure that shows a discrete transition in equilibrium-swollen volume with environmental changes.

Discontinuous swelling in partially hydrolyzed polyacrylamide gels has been studied by Gehrke et al. (1986). Copolymers of HEMA and acrylamides with methacrylic acid (MAA) and methyl methacrylate (MMA) have proven useful as hydrogels in biomedical applications (see below).

Small amounts of MAA as a comonomer have been shown to dramatically increase the swelling of PHEMA polymers. Owing to the hydrophobic nature of MMA, copolymers of MMA and HEMA have a lower degree of swelling than pure PHEMA (Brannon-Peppas and Peppas, 1991a). One particularly interesting IPN is the double network (DN) hydrogel of Gong and Murosaki (Murosaki and Gong, 2010). These DN hydrogels are composed of two interpenetrating cross-linked networks of PAAm and PAMPS, and exhibit the unusual combination of exceptionally strong mechanical properties and high water contents. All of these materials have potential uses in advanced technology applications, including biomedical separations, drug delivery devices, and as scaffolds for tissue engineering.

Poly(vinyl alcohol) (PVA) Hydrogels

Another hydrophilic polymer that has received much attention is poly(vinyl alcohol). This material holds great promise as a biological drug delivery matrix because it is non-toxic. Two methods exist for the preparation of PVA gels. In the first method, linear PVA chains are cross-linked using glyoxal, glutaraldehyde or borate. In the second method, semi-crystalline gels are prepared by exposing aqueous solutions of PVA to repeated freezing and thawing (Peppas and Hassan, 2000). The freezing and thawing induces crystal formation in the materials and allows for the formation of a network structure cross-linked with the quasi-permanent crystallites. The latter method is the preferred method for preparation as it allows for the formation of a "pure" network without the need to add cross-linking agents. Ficek and Peppas (1993) used PVA gels for the release of bovine serum albumin using novel PVA microparticles.

Poly(ethylene glycol) (PEG) Hydrogels

Hydrogels of poly(ethylene oxide) (PEO) and poly(ethylene glycol) (PEG) have received increasing attention recently for biomedical applications because of the non-toxic behavior of PEG, and its wide use in PEGylation of nanoscale drug carriers (e.g., Graham, 1992; Harris, 1992; Griffith and Lopina, 1995; Kofinas et al., 1996; Lee and He, 2010; Oishi and Nagasaki, 2010).

Three major techniques exist for the preparation of PEG networks: (1) chemical cross-linking between PEG chains, such as reaction of di-functional PEGs and multi-functional cross-linking agents; (2) radiation cross-linking of PEG chains to each other; and (3) physical interactions of hydrophobic blocks of triblock copolymers containing both hydrophobic blocks and PEG blocks (e.g., see Jeong et al., 2002, and Lee and He, 2010 for detailed discussion of pioneering work by S.W. Kim and co-workers on such block copolymer hydrogels. See also discussions of PEG hydrogels in the sections on Degradable Hydrogels and Temperature-Sensitive Hydrogels in this chapter.)

The advantage of using radiation-cross-linked PEG networks is that no toxic cross-linking agents are required. However, it is difficult to control the network structure of these materials. Stringer and Peppas (1996) prepared PEG hydrogels by radiation cross-linking. In this work, they analyzed the network structure in detail. Additionally, they investigated the diffusional behavior of smaller molecular weight drugs, such as theophylline, in these gels. Kofinas et al. (1996) have prepared PEG hydrogels by a similar technique. In this work, they studied the diffusional behavior of various macromolecules in these gels. They noted an interesting, yet previously unreported, dependence between the cross-link density and protein diffusion coefficient, and the initial molecular weight of the linear PEGs.

Lowman et al. (1997) described a method for the preparation of PEG gels with controllable structures. In this work, highly cross-linked and tethered PEG gels were prepared from PEG-dimethacrylates and PEG-monomethacrylates. The diffusional behavior of diltiazem and theophylline in these networks was studied. The technique described in this work has been used for the development of a new class of functionalized PEG-containing gels that are used for a variety of drug release applications.

Degradable Hydrogels

Hydrogels may degrade and dissolve by either of two mechanisms: hydrolysis or enzymolysis of main chain, side chain, or cross-linker bonds (e.g., Gombotz and Pettit, 1995). Degradable hydrogels have mainly been designed and synthesized for applications in drug delivery and, more recently, as tissue engineering scaffolds (e.g., Park, 1993; Atzet et al., 2008; Garcia et al., 2010).

Hydrolytically degradable hydrogels have been synthesized from triblock copolymers of A–B–A structure that form hydrogels held together by hydrophobic forces, where A (or B) may be PLA, PLGA or other hydrophobic polyesters that form hydrophobic blocks, and B (or A) is PEG, a hydrophilic block. These polyester hydrogels degrade into natural, endogenous metabolites such as lactic or glycolic acids, and the PEG block is then excreted through the kidneys (e.g., Lee and He, 2010; Jeong et al., 2002). A variation of this type of degradable hydrogel is formed by an A–B–A triblock copolymer composed of PEG-degradable polyester-PEG blocks mixed with cyclodextrin (CD) molecules, which thread onto the PEG blocks, after which the CDs self-assemble, forming the hydrogel (e.g., see the work of

Harada et al., 1992; Li et al., 2001, 2003(a,b), 2006; Li, 2009).

Polymerizable, cross-linked and degradable PEG gels have been prepared from acrylate- or methacrylate-terminated block copolymers that include PEG as a hydrophilic block (see the work of Hubbell and co-workers, e.g., Sawhney et al., 1993; Schoenmakers et al., 2004; van de Wetering et al., 2005; Lutolf and Hubbell, 2005; Raeber et al., 2005; Patterson et al., 2010; Patterson and Hubbell, 2010). These gels may have the simple A–B–A triblock structure of (methacrylate)–PEG–methacrylate which is photo-polymerized and later degrades by hydrolysis of the ester bonds linking PEG to the methacrylate cross-links. Another gel was formed with a more elaborate structure, (methacrylate–oligolactide–PEG–oligolactide–methacrylate), which is photo-polymerized, and later degrades and dissolves mainly by hydrolysis of the PLA in the main chains (see also Atzet et al., 2008; Kloxin et al., 2009). A third type of PEG gel may include a fibrin peptide cross-linking block, where the peptide is sequenced from fibrin that is a substrate for a naturally-occuring, fibrinolytic enzyme. The peptide block reacts by thiol addition of HS–peptide–SH to the acrylate vinyl groups, cross-linking the (acrylate)–PEG–acrylate triblock, to form: {-peptide–S–acrylate–PEG–acrylate–S–peptide–}, which then degrades and dissolves by proteolysis of the peptide by a natural, endogenous fibrinolytic enzyme.

Star Polymer and Dendrimer Hydrogels

Dendrimers and star polymers (Dvornik and Tomalia, 1996; Oral and Peppas, 2004) are exciting new materials because of the large number of functional groups available in a very small molecular volume. Such systems could have great promise in drug targeting applications. In 1993 Merrill published an exceptional review of PEO star polymers and applications of such systems in the biomedical and pharmaceutical fields. Griffith and Lopina (1995) prepared gels of controlled structure and large biological functionality by irradiation of PEO star polymers. Such structures could have particularly promising drug delivery applications when combined with emerging new technologies such as molecular imprinting.

Self-Assembled Hydrogel Structures

Recently there have been new, creative methods of preparation of novel hydrophilic polymers and hydrogels that may have significant drug delivery applications in the future (e.g., Li et al., 2001, 2003(a,b); Yang et al., 2006; Jin and Dijkstra, 2010; Wang et al., 2010; Miyata, 2010). In one unusual example, Stupp et al. (1997) synthesized self-assembled triblock copolymer nanogels having well-defined molecular architectures.

> Hydrogels usually exhibit swelling behavior dependent on the external environment. Over the last 30 years there has been a significant interest in the development and analysis of environmentally-responsive hydrogels. These types of gels show large and significant changes in their swelling ratio due to small changes in environmental conditions, such as pH, temperature, ionic strength, nature and composition of the swelling agent (including affinity solutes), light (visible versus UV), electrical, and magnetic stimuli (e.g., Peppas, 1991, 1993; Hoffman, 1997; Hoffman et al. 2000; Yoshida and Okano, 2010). In most responsive networks, a critical point exists at which this transition occurs. These gels are sometimes referred to as "smart" or "intelligent" hydrogels.

"SMART" OR "INTELLIGENT," STIMULI-RESPONSIVE HYDROGELS AND THEIR APPLICATIONS

An interesting characteristic of numerous stimuli-responsive "smart" gels is that the mechanism causing the network structural changes can be entirely reversible in nature. The ability of pH- or temperature-responsive gels to exhibit rapid changes in their swelling behavior and pore structure in response to changes in environmental conditions lends these materials favorable characteristics as carriers for delivery of drugs, including peptides and proteins. This type of behavior may also allow these materials to serve as self-regulated, pulsatile or oscillating drug delivery systems (Yoshida and Okano, 2010).

pH-Sensitive Hydrogels

One of the most widely studied types of physiologically-responsive hydrogels is pH-responsive hydrogels. These hydrogels are swollen ionic networks containing either acidic or basic pendant groups. In aqueous media of appropriate pH and ionic strength, the pendant groups can ionize and develop fixed charges on the gel, leading to rapid swelling. All ionic hydrogels exhibit both pH and ionic strength sensitivity, especially around the pK of the pH-sensitive group. These gels typically contain ionizable pendant groups such as carboxylic acids or amine groups. The most commonly studied ionic polymers include poly(acrylic acid) (PAA), poly(methacrylic acid) (PMAA), poly(diethylaminoethyl methacrylate) (PDEAEMA), and poly(dimethylaminoethyl methacrylate) (PDMAEMA). The swelling and drug delivery characteristics of anionic copolymers of PMAA and PHEMA (PHEMA–co–MAA) have been investigated. In acidic media, the gels did not swell significantly; however, in neutral or basic media, the gels swelled to a high degree due to ionization of the pendant acid group (this is similar behavior to that of polymers used as enteric coatings). Brannon-Peppas and Peppas (1991b) have also studied the oscillatory swelling behavior of these gels.

One interesting example of pH-responsive "smart" polymers with great sensitivity to pK is the behavior of two poly(alkylacrylic acids): poly(ethylacrylic acid) (PEAA) (Tirrell et al., 1985) and poly(propylacrylic acid) (PPAA) (Cheung et al., 2001). These polymers phase separate sharply as pH is lowered below their pK, and this can lead to lipid bilayer membrane disruption in acidic liposome solutions or within the acidic environments of endosomes and lysosomes of cells, where the membranes of those vesicles contain proton pumps. This behavior makes such polymers very useful for endosomal escape and cytosolic delivery of biomolecular drugs such as protein and nucleic acid drugs (Stayton and Hoffman, 2008).

The swelling forces developed in pH-responsive gels are significantly increased over non-ionic hydrogels. This increase in swelling is due to the presence of mobile counter-ions (such as Na^+) that electrostatically balance the fixed charges on the polymer backbone. The concentration of such counter-ions will be dependent on the concentration of the fixed polymer charges, which in turn will be dependent on the composition of the network polymer and the pH. As a result, the water content and mesh size of an ionic polymeric network can change significantly with small changes in pH, as the osmotic pressure of the counter-ions within the gel changes.

pH-Responsive Complexation Hydrogels

Another promising class of hydrogels that exhibit responsive behavior is complexing hydrogels. Bell and Peppas (1995) have discussed a class of graft copolymer gels of PMAAc grafted with PEG: poly(MAAc–g–EG). These gels exhibited pH-dependent swelling behavior due to the presence of acidic pendant groups and the formation of interpolymer H-bonded complexes at low pH between the ether groups on the graft chains and protonated pendant groups. In these covalently cross-linked, complexing poly(MAA–g–EG) hydrogels, complexation resulted in the formation of temporary physical cross-links due to hydrogen bonding between the PEG grafts and the pendant and protonated -COOH groups in PMAAc. The physical cross-links were reversible in nature and dependent on the pH and ionic strength of the environment. As a result, these complexing hydrogels exhibit drastic changes in their mesh size in response to small changes of pH, which could be useful for drug delivery in varying pH environments in the body, such as in the GI tract, mouth and vagina, and on the skin.

In another study of complexation hydrogels, Hayashi et al. (2007) formed gels from PEGylated papain and PAAc at low pH. They showed how the molecular weight of the PEG and the addition of free PEG significantly affected the release rate of the PEGylated protein from the gel. A complexation gel formed as a result of H-bonding between PAAc and PAAm chains at low pH. It was unusual in that it exhibited temperature-responsive behavior, and went from a gel to a solution state as temperature rose above 30°C (Katono et al., 1991).

Temperature-Sensitive Hydrogels

Another class of environmentally-sensitive gels exhibits sharp temperature-sensitive swelling-de-swelling behavior due to a change in the polymer/swelling agent compatibility over the temperature range of interest. Temperature-sensitive polymers typically exhibit a lower critical solution temperature (LCST), below which the polymer is soluble. Above this temperature, the polymers may lose their hydrophobically-bound water, and phase separate, causing the gel to collapse. Below the LCST, the cross-linked gel re-swells to significantly higher degrees because of the increased hydrophobic bonding with water. Poly(N-isopropyl acrylamide) (PNIPAAm) has been the most widely studied temperature-responsive polymer and hydrogel, with an LCST around 32–34°C (Dong and Hoffman, 1986; Park and Hoffman, 1990, 1992; Kim, 1996; Hoffman et al., 2000; Yoshida and Okano, 2010).

Some of the earliest work with PNIPAAm hydrogels was carried out by Dong and Hoffman (1986). They immobilized an enzyme in copolymer hydrogels of NIPAAm and AAm, and observed a maximum in the specific activity of the enzyme in each hydrogel as the temperature was raised. They concluded that above the maximum the gel collapsed as the copolymer LCST was surpassed, and the collapse blocked substrate diffusion into, and product out of, the gel. As the ratio of AAm/NIPAAm increased, the maximum shifted to higher temperatures due to the increasing hydrophilicity, which caused the increase in LCST of the copolymer hydrogel. Most interesting was the fact that the curves were reversible up to the maximum reached for the highest AAm/NIPAAm ratio; above that LCST the enzyme was denatured due to the high temperature.

In other early work, Hirotsu et al. (1987) synthesized cross-linked PNIPAAm gels and determined that the LCST of the gels was 34.3°C; below this transition temperature significant gel swelling occurred. They noted that the de-swelling-swelling above and below the LCST was reversible. Similar to Dong and Hoffman (1986), they also noted that the transition temperature was raised by copolymerizing PNIPAAm with small amounts of hydrophilic ionic monomers. Dong and Hoffman (1991) prepared heterogeneous PNIPAAm gels containing silicone polymer regions; these unusual gels collapsed at significantly faster rates than homopolymers of PNIPAAm. Park and Hoffman (1990, 1992) studied the effect of temperature cycling on the efficiency of enzyme turnover in a temperature-controlled packed bed of PNIPAAm hydrogel microparticles containing the enzyme. They noted a significant increase in productivity of the reactor with thermal cycling of temperatures from above to below the LCST, where the increased efficiency was due to the collapse of the gel particles above the LCST, "squeezing" out the product followed by the re-swelling of the gel below the LCST, enhancing uptake of substrate.

Yoshida et al. (1995) and Kaneko et al. (1996) developed an ingenious method to prepare comb-type graft hydrogels of PNIPAAm chains grafted to a PNIPAAm hydrogel network. Under conditions of gel collapse (above the LCST), hydrophobic regions were developed in the pores of the gel by the collapse of the grafted chains, drawing the network chains together with the collapsing grafted chains, resulting in a very rapid collapse of the gel. These materials had the ability to collapse from a fully swollen conformation in less than 20 minutes, while comparable gels that did not contain graft chains required up to a month to fully collapse. Such systems show promise for rapid or oscillatory release of drugs such as peptides and proteins.

There is a whole class of thermally-sensitive hydrogels based on physical interactions of hydrophobic blocks of triblock copolymers that also contain PEG blocks. Pluronic® block polymers (e.g., PPO–PEO–PPO) form such gels, but they are not degradable. The most interesting block copolymers for biomedical applications are hydrolytically degradable since they contain blocks of PLA or PLGA (e.g., PLGA–PEG–PLGA or PEG–PLA–PEG), which hydrolyze and release PEG chains that can be eliminated through the kidneys. These thermally-gelling block copolymers form *in situ* gels when injected subcutaneously, and act as drug delivery depots, releasing entrapped drugs as they degrade (e.g., see Jeong et al., 2002, and Lee and He, 2010 for detailed discussion of pioneering work by S. W. Kim and co-workers on such block copolymer hydrogels). One very interesting class of such degradable block copolymer hydrogels is formed by stereocomplexation of the two stereoisomers of PLA in PLLA–PEG and PDLA–PEG block copolymers (Fujiwara et al., 2010).

Affinity Hydrogels

Some hydrogels may exhibit environmental sensitivity due to the formation of complexes between chains that hold them together as a gel. Polymer complexes are macromolecular structures formed by the non-covalent association of groups on multi-functional molecules or on polymer chains that exhibit affinity for different groups on another polymer molecule. Sometimes this complexation is due to affinity recognition interactions, such as between streptavidin, with four binding sites for biotin, and a polymer with multiple pendant biotins (see Figure I.2.5.7a and Morris et al., 1993), or Concanavalin A with four binding sites for glucose and a polymer with multiple pendant glucose units (see Figure I.2.5.7b and Miyata et al., 1996), or an antibody with two binding sites for its antigens (Miyata, 2010). The complexes may form by association of repeating units on different chains (interpolymer complexes) or on separate regions of the same chain (intrapolymer complexes). The stability of these affinity hydrogels is dependent on such factors as the affinity constant of the association, the concentration of a competing, mobile affinity agent, temperature, pH, ionic strength, network composition, and structure, especially the length of the network polymer chains between association points. In these types of hydrogel, complex formation results in the formation of physical cross-links in the gel. As the degree of such physical cross-linking is increased, the network mesh size and degree of swelling will be significantly reduced. As a result, if such hydrogels are used as drug carriers, the rate of drug release will decrease dramatically upon the formation of inter-polymer complexes.

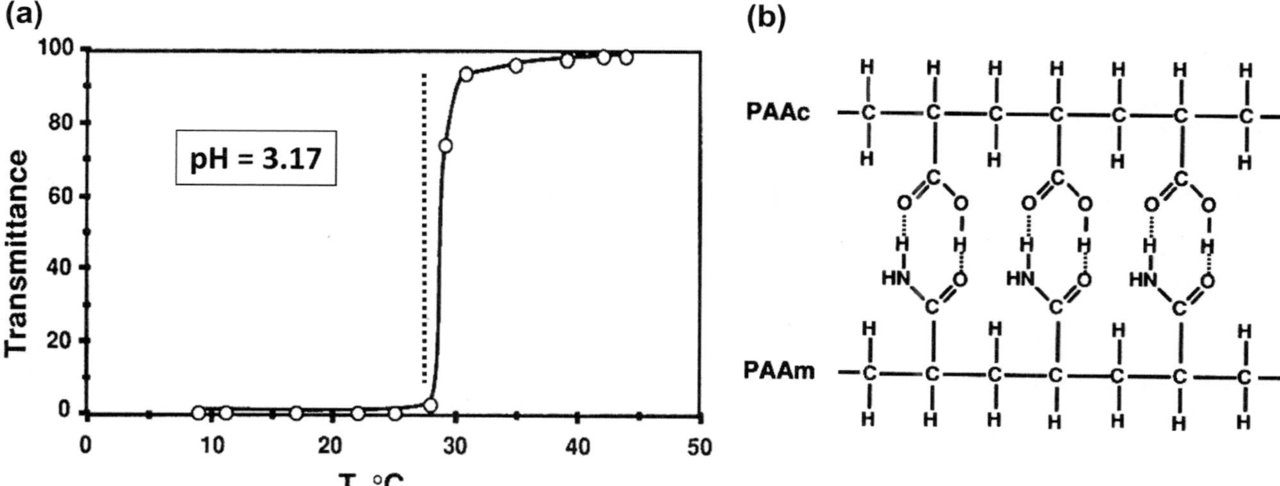

FIGURE I.2.5.7 Temperature dependence of light transmission for two H-bonded polymers, PAAc (Polyacrylic acid) and PAAm (Polyacrylamide) at pH 3.17; (a) shows the temperature dependence of light transmission and (b) shows the hypothetical H-bonded structure that would exist at low pH and at temperatures below 30°C, where the COOH groups are protonated and the polymer chains are complexed. The H-bonding is disrupted as temperature rises above 30°C. Data are for an aqueous solution at pH 3.17 (adjusted by HCl). Polymer concentration (wt. %): PAAc, 0.5%; PAAm, 0.5%. (Katono et al., (1991).

The hydrophilic character of hydrogels makes them attractive for a variety of biomedical and pharmaceutical applications. Because of their normally high water contents, hydrogels have been useful for delivering drugs from ingested tablets and osmotic pumps. Further, they have been successful as contact lenses applied to the eye, or as drug-releasing coatings on mucosal, skin or open wound surfaces. They have also been applied as non-fouling coatings on implants and devices that may contact blood, such as catheters. More recently they are being developed as scaffolds for tissue engineering implants.

BIOMEDICAL APPLICATIONS OF HYDROGELS

Contact Lenses

One of the earliest biomedical applications of hydrogels was the use of PHEMA hydrogels in contact lenses (Wichterle and Lim, 1960). Hydrogels are particularly useful as contact lenses because of their relatively good mechanical stability and favorable refractive index (see also Tighe 1976; Peppas and Yang, 1981; Michalek et al., 2010). More recently, extended wear contact lenses have been fabricated from an IPN composed of PNVP chains entrapped within a silicone hydrogel network. In this system silicone monomers and cross-linkers are polymerized in a solution containing PNVP, and an IPN hydrogel is formed. The PNVP acts to lubricate the surface of the lens against the cornea, and the silicone hydrogel provides high oxygen transport to the cornea, as well as enhanced permeability of small nutrient molecules and ions (McCabe et al., 2003).

Blood-Contacting Hydrogels

Hydrogels also exhibit properties that make them desirable candidates for blood-contacting biomaterials (Merrill et al., 1987). Nonionic hydrogels have been prepared from poly(vinyl alcohol), polyacrylamides, PNVP, PHEMA, and poly(ethylene oxide), (PEO, sometimes also referred to as PEG) (Peppas et al., 1999). Heparinized polymer hydrogels (Sefton, 1987) and heparin-based hydrogels (Tae et al., 2007) also show promise as materials for blood-contacting applications.

Drug Delivery from Hydrogels

Applications of hydrogels in controlled drug delivery systems (DDS) have become very popular in recent years. They include equilibrium-swollen hydrogels, i.e., matrices that have a drug incorporated in them and are swollen to equilibrium, releasing the drug. This category of *solvent-activated*, matrix-type, controlled-release devices comprises two important types of systems: *(1) rapidly swelling, diffusion-controlled devices*; and *(2) slowly swelling, swelling-controlled devices*. In general, a drug-loaded hydrogel may be prepared by swelling the hydrogel to equilibrium in a drug solution, and carefully drying it. In the dry state it becomes a glassy polymer that can be swollen when brought in contact with water or simulated biological fluids. This swelling process may or may not be the controlling mechanism for diffusional release, depending on the relative rates of the macromolecular relaxation of the polymer and drug diffusion from the gel.

In swelling-controlled release systems, the bioactive agent is dispersed into the polymer to form nonporous films, disks, or spheres. Upon contact with an aqueous dissolution medium, a distinct front (interface) is observed that corresponds to the water penetration front into the polymer and separates the glassy from the rubbery (gel-like) state of the material. Under these conditions, the macromolecular relaxation of the polymer influences the diffusion mechanism of the drug through the rubbery state. This water uptake can lead to considerable swelling of the polymer, with a thickness that depends on time. The swelling process proceeds toward equilibrium at a rate determined by the water activity in the system and the structure of the polymer. If the polymer is cross-linked, or if it is of sufficiently high molecular weight (so that chain entanglements can maintain structural integrity), the equilibrium state is a water-swollen gel. The equilibrium water content of such hydrogels can vary from ~30% to over 90%. If the dry hydrogel contains a water-soluble drug, the drug is essentially immobile in the glassy matrix, but begins to diffuse out as the polymer swells with water. Drug release thus depends on the simultaneous rate processes of water migration into the device, polymer chain hydration and relaxation, followed by drug dissolution and diffusion outward through the swollen gel. An initial burst effect is frequently observed in matrix devices, especially if the drying process brings a higher concentration of drug to the surface. The continued swelling of the matrix causes the drug to diffuse increasingly easily, mitigating the slow tailing off of the release curve. The net effect of the swelling process is to prolong and "linearize" the release curve. Details of the process of drug delivery from hydrogels have been presented by Korsmeyer and Peppas (1981) for poly(vinyl alcohol) systems, and by Reinhart et al. (1981) for PHEMA systems and their copolymers. One of numerous examples of such swelling-controlled systems was reported by Franson and Peppas (1983) who prepared cross-linked copolymer gels of poly(HEMA–co–MAA) of varying compositions. Theophylline release was studied and it was found that near zero-order release could be achieved using copolymers containing 90% PHEMA.

Targeted Drug Delivery from Hydrogels

Promising new methods for the delivery of chemotherapeutic agents using hydrogels have been recently reported. Novel biorecognizable sugar-containing copolymers have been investigated for use in targeted delivery of anti-cancer drugs. Peterson et al. (1996) have used poly(N-2-hydroxypropyl methacrylamide) carriers for the treatment of ovarian cancer.

Tissue Engineering Scaffolds from Hydrogels

This is an application area that continues to expand. It is driven by the same attractive properties that drive the use of hydrogels for drug delivery applications: high water content gels that may be synthesized with degradable backbone polymers, with an added advantage of being able to attach cell adhesion ligands to the network polymer chains. There are a number of natural polymer-based hydrogel scaffolds that have been studied (e.g., collagen, gelatin, alginates, hyaluronic acid, chitosan, etc.) and the reader is referred to two chapters in this text (see Chapters I.2.7 and II.6.3) and some excellent review articles (Lee and Mooney, 2001; Lutolf and Hubbell, 2005; Jin and Dijkstra, 2010).

One very interesting observation with hydrogels that has recently been reported is that they may stimulate stem cell differentiation; that is, when stem cells are deposited on some hydrogel surfaces, depending on the composition and/or mechanical stiffness of the surface, differentiation of the stem cells into certain phenotypes may occur (e.g., Liu et al., 2010; Nguyen et al., 2011).

Miscellaneous Biomedical Applications of Hydrogels

Other potential applications of hydrogels mentioned in the literature include artificial tendon materials, wound-healing bioadhesives, artificial kidney membranes, articular cartilage, artificial skin, maxillofacial and sexual organ reconstruction materials, and vocal cord replacement materials (Byrne et al., 2002a,b).

BIBLIOGRAPHY

Andrade, J. D. (1976). Hydrogels for Medical and Related Applications. *ACS Symposium Series*. (Vol. 31). Washington, DC: American Chemical Society.

Atzet, S., Curtin, S., Trinh, P., Bryant, S., & Ratner, B. (2008). Degradable poly(2-hydroxyethyl methacrylate)-co-polycaprolactone hydrogels for tissue engineering scaffolds. *Biomacromol.*, **9**, 3370–3377.

Bell, C. L., & Peppas, N. A. (1995). Biomedical membranes from hydrogels and interpolymer complexes. *Adv. Polym. Sci.*, **122**, 125–175.

Brannon-Peppas, L., & Harland, R. S. (1990). *Absorbent Polymer Technology*. Amsterdam: Elsevier.

Brannon-Peppas, L., & Peppas, N. A. (1991a). Equilibrium swelling behavior of dilute ionic hydrogels in electrolytic solutions. *J. Controlled Release*, **16**, 319–330.

Brannon-Peppas, L., & Peppas, N. A. (1991b). Time-dependent response of ionic polymer networks to pH and ionic strength changes. *Inter. J. Pharm.*, **70**, 53–57.

Byrne, M. E., Henthorn, D. B., Huang, Y., & Peppas, N. A. (2002a). Micropatterning Biomimetic Materials for Bioadhesion and Drug Delivery. In A. K. Dillow, & A. M. Lowman (Eds.), *Biomimetic Materials and Design: Biointerfacial Strategies, Tissue Engineering and Targeted Drug Delivery* (pp. 443–470). New York, NY: Dekker.

Byrne, M. E., Park, K., & Peppas, N. A. (2002b). Molecular imprinting within hydrogels. *Adv. Drug. Deliv. Revs.*, **54**, 149–161.

Chapiro, A. (1962). *Radiation Chemistry of Polymeric Systems*. New York, NY: Interscience.

Cheung, C. Y., Murthy, N., Stayton, P. S., & Hoffman, A. S. (2001). A pH-sensitive polymer that enhances cationic lipid-mediated gene transfer. *Bioconj. Chem.*, **12**, 906–910.

Dong, L. C., & Hoffman, A. S. (1986). Thermally reversible hydrogels: III. Immobilization of enzymes for feedback reaction control. *J. Contr. Rel.*, **4**, 223–227.

Dong, L. C., & Hoffman, A. S. (1991). A novel approach for preparation of pH-sensitive hydrogels for enteric drug delivery. *J. Controlled Release*, **15**, 141–152.

Drury, J. L., & Mooney, D. J. (2003). Hydrogels for tissue engineering: Scaffold design variables and applications. *Biomaterials*, **24**, 4337–4351.

Dvornik, P. R., & Tomalia, D. A. (1996). Recent advances in dendritic polymers. *Current Opinion in Colloid & Interface Science*, **1**, 221–235.

Ficek, B. J., & Peppas, N. A. (1993). Novel preparation of poly(vinyl alcohol) microparticles without cross-linking agent. *J. Controlled Rel.*, **27**, 259–264.

Flory, P. J. (1953). *Principles of Polymer Chemistry*. Ithaca, NY: Cornell University Press.

Franson, N. M., & Peppas, N. A. (1983). Influence of copolymer composition on water transport through glassy copolymers. *J. Appl. Polym. Sci.*, **28**, 1299–1310.

Fujiwara, T., Yamaoka, T., & Kimura, Y. (2010). Thermo-responsive Biodegradable Hydrogels from Stereocomplexed Polylactides. In R. M. Ottenbrite, K. Park, & T. Okano (Eds.), *Biomedical Applications of Hydrogels Handbook* (pp. 157–178). New York, NY: Springer.

Garcia, L., Aguilar, M. R., & San Román, J. (2010). Biodegradable Hydrogels for Controlled Drug Delivery. In R. M. Ottenbrite, K. Park, & T. Okano (Eds.), *Biomedical Applications of Hydrogels Handbook* (pp. 147–155). New York, NY: Springer.

Gehrke, S. H., Andrews, G. P., & Cussler, E. L. (1986). Chemical aspects of gel extraction. *Chem. Eng. Sci.*, **41**, 2153–2160.

Gombotz, W. R., & Pettit, D. K. (1995). Biodegradable polymers for protein and peptide drug delivery. *Bioconjugate Chemistry*, **6**, 332–351.

Graham, N. B. (1992). Poly(ethylene glycol) Gels and Drug Delivery. In J. M. Harris (Ed.), *Poly(Ethylene Glycol) Chemistry, Biotechnical and Biomedical Applications* (pp. 263–281). New York, NY: Plenum Press.

Gregonis, D. E., Chen, C. M., & Andrade, J. D. (1976). The Chemistry of Some Selected Methacrylate Hydrogels. In J. D. Andrade (Ed.), *ACS Symposium Series: Hydrogels for Medical and Related Applications Vol. 31*, (pp. 88–104). Washington, DC: American Chemical Society.

Griffith, L., & Lopina, S. T. (1995). Network structures of radiation cross-linked star polymer gels. *Macromolecules*, **28**, 6787–6794.

Harada, A., Li, J., & Kamachi, M. (1992). The molecular necklace: A rotaxane containing many threaded α-cyclodextrins. *Nature*, **356**, 325–327.

Harris, J. M. (Ed.), (1992). *Poly(Ethylene Glycol) Chemistry, Biotechnical and Biomedical Applications*. New York, NY: Plenum Press.

Hassan, C. M., & Peppas, N. A. (2000). Cellular Freeze/Thawed PVA hydrogels. *J. Appl. Polym. Sci.*, **76**, 2075–2079.

Hayashi, Y., Harris, J. M., & Hoffman, A. S. (2007). Delivery of PEGylated drugs from mucoadhesive formulations by pH-induced disruption of H-bonded complexes of PEG-drug with poly(acrylic acid). *Reactive and Functional Polymers*, **67**, 1330–1337.

Hennink, W. E., & van Nostrum, C. F. (2002). Novel cross-linking methods to design hydrogels. *Adv. Drug. Del. Revs.*, **54**, 13–36.

Hermanson, G. T. (2008). *Bioconjugate Techniques* (2nd ed.). New York, NY: Elsevier.

Hickey, A. S., & Peppas, N. A. (1995). Mesh size and diffusive characteristics of semicrystalline poly(vinyl alcohol) membranes. *J. Membr. Sci.*, **107**, 229–237.

Hirotsu, S., Hirokawa, Y., & Tanaka, T. (1987). Swelling of gels. *J. Chem. Phys.*, **87**, 1392–1395.

Hoffman, A. S. (1997). Intelligent Polymers. In K. Park (Ed.), *Controlled Drug Delivery*. Washington, DC: ACS Publications, ACS.

Hoffman, A. S. (2002). Hydrogels for biomedical applications. *Adv. Drug. Del. Revs.*, **43**, 3–12.

Hoffman, A. S., Stayton, P. S., Bulmus, V., Chen, J., Cheung, C., et al. (2000). Really smart bioconjugates of smart polymers and receptor proteins. *J. Biomed. Mater. Res.*, **52**, 577–586.

Ilavsky, M. (1982). Phase transition in swollen gels. *Macromolecules*, **15**, 782–788.

Jeong, B., Kim, S. W., & Bae, Y. H. (2002). Thermosensitive sol-gel reversible hydrogels. *Adv. Drug. Del. Revs.*, **54**, 37–51.

Jin, R., & Dijkstra, P. J. (2010). Hydrogels for Tissue Engineering Applications. In R. M. Ottenbrite, K. Park, & T. Okano (Eds.), *Biomedical Applications of Hydrogels Handbook* (pp. 203–226). New York, NY: Springer.

Kaneko, Y., Saki, K., Kikuchi, A., Sakurai, Y., & Okano, T. (1996). Fast swelling/deswelling kinetics of comb-type grafted poly(N-isopropyl acrylamide) hydrogels. *Macromol. Symp.*, **109**, 41–53.

Katono, H., Maruyama, A., Sanui, K., Ogata, N., Okano, T., & Sakurai, Y. (1991). Thermo-responsive swelling and drug release switching of interpenetrating polymer networks composed of poly(acrylamide-co-butyl methacrylate) and poly (acrylic acid). *J. Contr. Rel.*, **16**, 215–227.

Kim, S. W. (1996). Temperature Sensitive Polymers for Delivery of Macromolecular Drugs. In N. Ogata, S. W. Kim, J. Feijen, & T. Okano (Eds.), *Advanced Biomaterials in Biomedical Engineering and Drug Delivery Systems* (pp. 125–133). Tokyo: Springer.

Kloxin, A. M., Kasko, A., Salinas, C. N., & Anseth, K. S. (2009). Photodegradable hydrogels for dynamic tuning of physical and chemical properties. *Science*, **324**, 59–63.

Kofinas, P., Athanassiou, V., & Merrill, E. W. (1996). Hydrogels prepared by electron beam irradiation of poly(ethylene oxide) in water solution: Unexpected dependence of cross-link density and protein diffusion coefficients on initial PEO molecular weight. *Biomaterials*, **17**, 1547–1550.

Kolb, H. C., Finn, M. G., & Sharpless, K. B. (2001). Click chemistry: Diverse chemical function from a few good reactions. *Angewandte Chemie International Edition*, **40**(11), 2004–2021.

Korsmeyer, R. W., & Peppas, N. A. (1981). Effects of the morphology of hydrophilic polymeric matrices on the diffusion and release of water soluble drugs. *J. Membr. Sci.*, **9**, 211–227.

Lee, D. S., & He, C. (2010). *In-Situ* Gelling Stimuli-Sensitive PEG-Based Amphiphilic Copolymer Hydrogels. In R. M. Ottenbrite, K. Park, & T. Okano (Eds.), *Biomedical Applications of Hydrogels Handbook* (pp. 123–146). New York, NY: Springer.

Lee, K. Y., & Mooney, D. J. (2001). Hydrogels for tissue engineering. *Chem. Revs.*, **101**, 1869–1879.

Li, J. (2009). Cyclodextrin inclusion polymers forming hydrogels. *Adv. Polym. Sci.*, **222**, 79–113.

Li, J., Li, X., Zhou, Z., Ni, H., & Leong, K. W. (2001). Formation of supramolecular hydrogels induced by inclusion complexation between Pluronics and cyclodextrin. *Macromolecules*, **34**, 7236–7237.

Li, J., Ni, X., Zhou, Z., & Leong, K. W. (2003a). Preparation and characterization of polypseudorotaxanes based on block-selected inclusion complexation between poly(propyleneoxide)-poly(ethylene oxide)-poly(propylene oxide) triblock copolymers and a-cyclodextrin. *J. Am. Chem. Soc.*, **125**, 1788–1795.

Li, J., Ni, X., & Leong, K. W. (2003b). Injectable drug-delivery systems based on supramolecular hydrogels formed by poly(ethylene oxide)s and cyclodextrin. *J. Biomed. Mater. Res.*, **65A**, 196–202.

Li, J., Li, X., Ni, X., Wang, X., Li, H., & Leong, K. W. (2006). Self-assembled supramolecular hydrogels formed by biodegradable PEO-PHB-PEO triblock copolymers and a-cyclodextrin for controlled drug delivery. *Biomaterials*, **27**, 4132–4140.

Liu, S. Q., Tay, R., Khan, M., Lai, P., Ee, R., et al. (2010). Synthetic hydrogels for controlled stem cell differentiation. *Soft Matter.*, **6**, 67–81.

Lowman, A. M., & Peppas, N. A. (1997). Analysis of the complexation/decomplexation phenomena in graft copolymer networks. *Macromolecules*, **30**, 4959–4965.

Lowman, A. M., & Peppas, N. A. (1999). Hydrogels. In E. Mathiowitz (Ed.), *Encyclopedia of Controlled Drug Delivery* (pp. 397–418). New York, NY: Wiley.

Lowman, A. M., Dziubla, T. D., & Peppas, N. A. (1997). Novel networks and gels containing increased amounts of grafted and cross-linked poly(ethylene glycol). *Polymer Preprints*, **38**, 622–623.

Lutolf, M. P., & Hubbell, J. A. (2005). Synthetic biomaterials as instructive extracellular microenvironments for morphogenesis in tissue engineering. *Nature Biotechnology*, **23**, 47–55.

McCabe, K., et al. (2003). Biomedical devices containing internal wetting agents. Johnson & Johnson, US Patent # 20030162862.

Merrill, E. W. (1993). Poly(ethylene oxide) star molecules: Synthesis, characterization, and applications in medicine and biology. *J. Biomater. Sci. Polym. Edn*, **5**, 1–11.

Merrill, E. W., Pekala, P. W., & Mahmud, N. A. (1987). Hydrogels for Blood Contact. In N. A. Peppas (Ed.), *Hydrogels in Medicine and Pharmacy* (Vol. 3, pp. 1–16). Boca Raton, FL: CRC Press.

Michalek, J., Hobzova, R., Pradny, M., & Duskova, M. (2010). Hydrogel Contact Lenses. In R. M. Ottenbrite, K. Park, & T. Okano (Eds.), *Biomedical Applications of Hydrogels Handbook* (pp. 303–316). New York, NY: Springer.

Miyata, T. (2010). Biomolecule-Responsive Hydrogels. In R. M. Ottenbrite, K. Park, & T. Okano (Eds.), *Biomedical Applications of Hydrogels Handbook* (pp. 65–86). New York, NY: Springer.

Miyata, T., Jikihara, A., Nakamae, K., Uragami, T., Hoffman, A. S., Kinomura, K., & Okumura, M. (1996). Preparation of glucose-sensitive hydrogels by entrapment or copolymerization of concanavalin A in a glucosyloxyethyl mathacrylate hydrogel. In N. Ogata, S. W. Kim, J. Feijen, & T. Okano (Eds.), *Advanced Biomaterials in Biomedical Engineering and Drug Delivery Systems* (pp. 237–238). Springer.

Miyata, T., Uragami, T., & Nakamae, K. (2002). Biomolecule-sensitive hydrogels. *Adv. Drug. Del. Revs.*, **54**, 79–98.

Morris, J. E., Fischer, R., & Hoffman, A. S. (1993). Affinity Precipitation of Proteins with Polyligands. *J. Anal. Biochem.*, **41**, 991–997.

Mun, G., Suleimenov, I., Park, K., & Omidian, H. (2010). Superabsorbant Hydrogels. In R. M. Ottenbrite, K. Park, & T. Okano (Eds.), *Biomedical Applications of Hydrogels Handbook* (pp. 375–392). New York, NY: Springer.

Murosaki, T., & Gong, J. P. (2010). Double Network Hydrogels as Tough, Durable Tissue Substitutes. In R. M. Ottenbrite, K. Park, & T. Okano (Eds.), *Biomedical Applications of Hydrogels Handbook* (pp. 285–302). New York, NY: Springer.

Nguyen, L. H., Kudva, A. K., Guckert, N. L., Linse, K. D., & Roy, K. (2011). Unique biomaterial compositions direct bone marrow stem cells into specific chondrocyte phenotypes corresponding to the various zones of articular cartilage. *Biomaterials.*, **32**, 1327–1338.

Oishi, M., & Nagasaki, Y. (2010). Stimuli-responsive PEGylated Nanogels for Smart Nanomedicine. In R. M. Ottenbrite, K. Park, & T. Okano (Eds.), *Biomedical Applications of Hydrogels Handbook* (pp. 87–106). New York, NY: Springer.

Okano, T. (1993). Molecular Design of Temperature-Responsive Responsive Polymers as Intelligent Materials. In K. Dusek (Ed.), *Gels: Volume Transitions II*. New York, NY: Springer-Verlag.

Omidian, H., & Park, K. (2010). Engineered High Swelling Hydrogels. In R. M. Ottenbrite, K. Park, & T. Okano (Eds.), *Biomedical Applications of Hydrogels Handbook* (pp. 351–374). New York, NY: Springer.

Oral, E., & Peppas, N. A. (2004). Responsive and recognitive hydrogels using star polymers. *J. Biomed. Mater. Res.*, **68A**, 439–447.

Ottenbrite, R. M., Park, K., & Okano, T. (Eds.), (2010). *Biomedical Applications of Hydrogels Handbook*. New York, NY: Springer.

Park, K. (1993). *Biodegradable Hydrogels for Drug Delivery*. Lancaster, PA: Technomic Publishing Co., Inc.

Park, T. G., & Hoffman, A. S. (1990). Immobilized biocatalysts in reversible hydrogels. *NY. Acad. Sci.*, **613**, 588–593.

Park, T. G., & Hoffman, A. S. (1992). Synthesis and characterization of pH and temperature-sensitive hydrogels. *J. Appl. Poly. Sci.*, **46**, 659–671.

Patterson, J., & Hubbell, J. A. (2010). Enhanced proteolytic degradation of molecularly-engineered PEG hydrogels in response to MMP-1 and MMP-2. *Biomaterials*, **31**, 7836–7845.

Patterson, J., et al. (2010). Biomimetic materials in tissue engineering. *Mater. Today*, **13**, 14–22.

Peppas, N. A. (1987). *Hydrogels in Medicine and Pharmacy*. Boca Raton, FL: CRC Press.

Peppas, N. A. (1991). Physiologically-responsive hydrogels. *J Bioact. Compat. Polym.*, **6**, 241–246.

Peppas, N. A. (1993). Fundamentals of pH- and Temperature-sensitive Delivery Systems. In R. Gurny, H. E. Juninger, & N. A. Peppas (Eds.), *Pulsatile Drug Delivery* (pp. 41–56). Stuttgart: Wissenschaftliche Verlagsgesellschaft.

Peppas, N. A. (1997). Hydrogels and drug delivery. *Critical Opinion in Colloid and Interface Science*, **2**, 531–537.

Peppas, N. A. (2001). Gels for Drug Delivery. In *Encyclopedia of Materials: Science and Technology* (pp. 3492–3495). Amsterdam: Elsevier.

Peppas, N. A., & Yang, W. H. M. (1981). Properties-based optimization of the structure of polymers for contact lens applications. *Contact Intraocular Lens. Med. J.*, **7**, 300–321.

Peppas, N. A., Keys, K. B., Torres-Lugo, M., & Lowman, A. M. (1999). Poly(ethylene glycol)-containing hydrogels in drug delivery. *J. Controlled Release*, **62**, 81–87.

Peppas, N. A., Huang, Y., Torres-Lugo, M., Ward, J. H., & Zhang, J. (2000). Physicochemical foundations and structural design of hydrogels in medicine and biology. *Ann. Revs. Biomed. Eng.*, **2**, 9–29.

Peppas, N. A., Wood, K. M., & Blanchette, J. O. (2004). Hydrogels for oral delivery of therapeutic proteins. *Expert Opin. Biol. Ther.*, **4**, 881–887.

Peppas, N. A., Hilt, J. Z., Khademhosseini, A., & Langer, R. (2006). Hydrogels in biology and medicine: From fundamentals to bionanotechnology. *Adv. Mater.*, **18**, 1345–1360.

Peterson, C. M., Lu, J. M., Sun, Y., Peterson, C. A., Shiah, J. G., Straight, R. C., & Kopecek, J. (1996). Combination chemotherapy and photodynamic therapy with N-(2-hydroxypropyl) methacrylamide copolymer-bound anticancer drugs inhibit human ovarian carcinoma heterotransplanted in nude mice. *Cancer Research*, **56**, 3980–3985.

Qiu, Y., & Park, K. (2001). Environment-sensitive hydrogels for drug delivery. *Adv. Drug. Del. Revs.*, **53**, 321–339.

Raeber, G. P., Lutolf, M., & Hubbell, J. A. (2005). Molecularly engineered PEG hydrogels: A novel model system for proteolytically mediated cell migration. *Biophys. J.*, **89**, 1374–1388.

Ratner, B. D., & Hoffman, A. S. (1976). Synthetic Hydrogels for Biomedical Applications. In J. D. Andrade (Ed.), *Hydrogels for Medical and Related Applications* (Vol. 31, pp. 1–36). Washington, DC: ACS Symposium Series, American Chemical Society.

Reinhart, C. T., Korsmeyer, R. W., & Peppas, N. A. (1981). Macromolecular Network Structure and its Effects on Drug and Protein Diffusion. *Intern. J. Pharm. Techn.*, **2**(2), 9–16.

Sawhney, A. S., Pathak, C. P., & Hubbell, J. A. (1993). Bioerodible hydrogels based on photopolymerized poly(ethylene glycol)-co-poly(alpha-hydroxy acid) diacrylate macromers. *Macromolecules*, **26**, 581–587.

Schoenmakers, R. G., van de Wetering, P. Elbert, D. L., & Hubbell, J. A. (2004). The effect of the linker on the hydrolysis rate of drug-linked ester bonds. *J. Contr. Rel.*, **95**, 291–300.

Sefton, M. V. (1987). Heparinized Hydrogels. In N. A. Peppas (Ed.), *Hydrogels in Medicine and Pharmacy* (Vol. 3, pp. 17–52). Boca Raton, FL: CRC Press.

Sperinde, J. J., & Griffith, L. G. (1997). Synthesis and characterization of enzymatically-cross-linked poly(ethylene glycol) hydrogels. *Macromolecules*, **30**, 5255–5264.

Sperinde, J. J., & Griffith, L. G. (2000). Control and prediction of gelation kinetics in enzymatically cross-linked poly(ethylene glycol) hydrogels. *Macromolecules*, **33**, 5476–5480.

Stayton, P. S., & Hoffman, A. S. (2008). Smart pH-Responsive Carriers for Intracellular Delivery of Biomolecular Drugs. In V. Torchilin (Ed.), *Multifunctional Pharmaceutical Nanocarriers*. New York, NY: Springer.

Stringer, J. L., & Peppas, N. A. (1996). Diffusion in radiation-cross-linked poly(ethylene oxide) hydrogels. *J. Controlled Rel.*, **42**, 195–202.

Stupp, S. I., LeBonheur, V., Walker, K., Li, L. S., Huggins, K. E., et al. (1997). Supramolecular materials: Self-organized nanostructures. *Science*, **276**, 384–389.

Tae, G., Kim, Y. J., Choi, W. I., Kim, M., Stayton, P. S., et al. (2007). Formation of a novel heparin-based hydrogel in the presence of heparin-binding biomolecules. *Biomacromol.*, **8**, 1979–1986.

Tanaka, T. (1979). Phase transitions in gels and a single polymer. *Polymer*, **20**, 1404–1412.

Tighe, B. J. (1976). The design of polymers for contact lens applications. *Brit. Polym. J.*, **8**, 71–90.

Tirrell, D. A., Takigawa, D. Y., & Seki, K. (1985). pH sensitization of phospholipid vesicles via complexation with synthetic poly(carboxylic acid)s. *Ann. NY. Acad. Sci.*, **446**, 237–248.

van de Wetering, P., Metters, A. T., Schoenmakers, R. G., & Hubbell, J. A. (2005). Poly(ethylene glycol) hydrogels formed by conjugate addition with controllable swelling, degradation, and release of pharmaceutically active proteins. *J. Contr. Rel.*, **102**, 619–627.

Wang, L. S., Boulaire, J., Chan, P. P.Y., Chung, J. E., & Kurisawa, M. (2010). The role of stiffness of gelatin-hydroxyphenylpropionic acid hydrogels formed by enzyme-mediated cross-linking on the differentiation of human mesenchymal stem cell. *Biomaterials*, **31**, 1148–1157 and 8608–8616.

Wichterle, O., & Lim, D. (1960). Hydrophilic gels for biological use. *Nature*, **185**, 117–118.

Xu, C., & Kopecek, J. (2008). Genetically engineered block copolymers: Influence of the length and structure of the coiled-coil block on hydrogel self-assembly. *Pharm. Res.*, **25**, 674–682.

Yang, J., Xu, C., Wang, C., & Kopecek, J. (2006). Refolding hydrogels self-assembled from HPMA graft copolymers by antiparallel coiled-coil formation. *Biomacromol.*, 7, 1187–1195.

Yasuda, H., Peterlin, A., Colton, C. K., Smith, K. A., & Merrill, E. W. (1969). Permeability of solutes through hydrated polymer membranes. III. Theoretical background for the selectivity of dialysis membranes. *Makromol. Chemie.*, **126**, 177–186.

Yoshida, R., & Okano, T. (2010). Stimuli-Responsive Hydrogels and their Application to Functional Materials. In R. M. Ottenbrite, K. Park, & T. Okano (Eds.), *Biomedical Applications of Hydrogels Handbook* (pp. 19–44). New York, NY: Springer.

Yoshida, R., Uchida, K., Kaneko, Y., Sakai, K., Kikcuhi, A., et al. (1995). Comb-type grafted hydrogels with rapid deswelling response to temperature changes. *Nature*, **374**, 240–242.

CHAPTER I.2.6 DEGRADABLE AND RESORBABLE BIOMATERIALS

Matthew Treiser,[1] Sascha Abramson,[2] Robert Langer,[3] and Joachim Kohn[2]

[1]Department of Biomedical Engineering, Rutgers, The State University of New Jersey, New Brunswick, NJ, USA

[2]Department of Chemistry, Rutgers, The State University of New Jersey, New Brunswick, NJ, USA

[3]Department of Chemical Engineering, Massachusetts Institute of Technology, Cambridge, MA, USA

INTRODUCTION

Clinical circumstances often require the application of implants that must serve a temporary rather than a permanent purpose (Table I.2.6.1). In these situations degradable polymers are of interest because the implants fabricated from these materials do not need to be surgically removed. The surgical removal of an implant with a temporary purpose is undesirable, as the process creates another wound with the possibility of surgical complication and infection. Additionally, the use of degradable implants can sometimes circumvent problems related to the long-term safety of permanent implants, such as long-term immune rejection, chronic inflammation at the implant–tissue interface, and failure of the device. However, degradable implants are not without their own safety concerns, such as the toxicity of their degradation products, and the degradation-related, premature failure of the implant. Therefore, designing a degradable implant requires careful testing for potential toxicity of its degradation products and careful consideration of the implant's mechanical integrity during the required service life of the implant. To facilitate a better understanding of the complex decisions that have to be made during the design of a degradable implant, this chapter covers basic definitions relating to the process of degradation and/or erosion, the most prominent types of *synthetic*, degradable polymers available today, a classification of degradable medical implants, and considerations specific for the design and use of degradable medical polymers.

DEFINITIONS RELATING TO THE PROCESSES OF DEGRADATION VERSUS BIODEGRADATION, AND EROSION VERSUS BIOEROSION

The term "degradation" refers to a chemical process resulting in the cleavage of covalent bonds. Hydrolysis is the most common chemical process by which polymers degrade, but degradation can also occur via oxidative, photodegradative, and enzymatic mechanisms. In contrast, the term "erosion" refers to physical changes in size, shape or mass of a device, which could be the consequence of degradation, dissolution, ablation or mechanical wear. Thus, it is important to realize that erosion can occur in the absence of degradation, and degradation can occur in the absence of erosion. For example, erosion without degradation occurs when a sugar cube is placed in water: the sugar cube loses its mass through dissolution, e.g., it erodes, but the sugar does not degrade as its chemical structure remains unchanged. Degradation without erosion is observed when an implant made of poly(lactic acid) (PLA) is first placed into the body of a patient: PLA clearly starts to degrade, as evidenced by the reduction of its molecular weight, but the initial degradation of the polymer is not associated with erosion, as evidenced by the lack of mass loss of the implant. Only when the degradation of PLA is well advanced do the first signs of implant erosion become evident.

It is important to distinguish between the many different terms that describe the conversion of a solid biomaterial into a solution of the biomaterial or its molecular fragments. The list of terms includes *degradation*, *biodegradation*, *bioabsorption*, *bioresorption*, *erosion*, *bioerosion*, *surface erosion*, and *bulk erosion*. These terms are often used in an inconsistent fashion in the biomaterials literature; the reader is encouraged to read this section carefully and also the Glossary of Key Terms at the end of the chapter.

All polymers undergo some degree of degradation. Therefore, the practice of designating certain polymers as "degradable" and others as "non-degradable" represents a certain amount of arbitrariness. For the purposes

TABLE I.2.6.1 Some "Short Term" Medical Applications of Degradable Polymeric Biomaterials

Application	Comments
Sutures	The earliest successful application of synthetic, degradable polymers in human medicine
Drug delivery devices	One of the most widely investigated medical applications for degradable polymers. Commonly used in clinical devices for the delivery of chemotherapeutics for the treatment of cancer
Orthopedic fixation devices	Requires polymers of exceptionally high mechanical strength and stiffness
Adhesion prevention	Requires polymers that can form soft membranes or films
Temporary vascular grafts and stents made of degradable polymers	Only investigational devices are presently available. Blood compatibility and fouling, as well as device fragmentation upon degradation, are major concerns
Tissue engineering or guided tissue regeneration scaffold	Attempts to recreate or improve native tissue function using degradable scaffolds. Cells may be seeded with the scaffolds before implantation. Bioactives (e.g., growth factors, short selective peptide sequences, etc.) are sometimes included in the scaffolds to modulate cellular responses

of this chapter, we will use the criteria as described by Göpferich that "degradable polymers" are those that degrade within the time scales of their expected service life or shortly thereafter (Göpferich, 1996). Conversely, "non-degradable" polymers have degradation times that are substantially longer than their service life.

Currently, at least four different terms (biodegradation, bioerosion, bioabsorption, and bioresorption) are used to indicate that a given material or device will eventually disappear after having been introduced into a living organism. However, within the literature, no consistent distinctions in the meaning of these four terms are evident. Likewise, the meaning of the prefix "bio" is not well established, often leading to the interchangeable use of the terms "degradation" and "biodegradation", or "erosion" and "bioerosion." Although efforts have been made to establish generally applicable and widely accepted definitions for all aspects of biomaterials research (Williams, 1987), there is still confusion even among experienced researchers in the field as to the correct terminology of various degradation processes.

In the context of this chapter, we follow the usage suggested by the Consensus Conference of the European Society for Biomaterials (Williams, 1987), and refer to "biodegradation" when we wish to emphasize that a biological agent (enzyme, cell or microorganism) is causing the chemical breakdown of the implanted device. After extensive discussion in the literature, it is now widely believed that the chemical degradation of the polymeric backbone of poly(lactic acid) (PLA) is predominantly controlled by simple hydrolysis and occurs independently of any biological agent (Li, 1999). Consequently, in these instances, the degradation of PLA to lactic acid should not be described as "biodegradation" but rather as "degradation." (Only under rare conditions may the enzyme proteinase K catalyze the hydrolytic degradation of PLA (Williams, 1981; Tsuji and Miyauchi, 2001); that situation is so rare that we will not consider it further here.)

In agreement with Heller's suggestion (Heller, 1987), we define a "bioerodible polymer" as a water-insoluble polymer that is converted under physiological conditions into water-soluble material(s) without regard to the specific mechanism involved in the erosion process. "Bioerosion" includes, therefore, both physical processes (such as dissolution) and chemical processes (such as backbone cleavage). Here the prefix "bio" indicates that the erosion occurs under physiological conditions, as opposed to other erosion processes caused, for example, by high temperature, strong acids or bases, UV light, mechanical stresses or weather conditions. The terms "bioresorption" and "bioabsorption" are used interchangeably and often imply that the polymer or its degradation products are removed by cellular activity (e.g., phagocytosis) in a biological environment. While commonly found within the literature, these terms have not been consistently defined.

An understanding of the distinct definitions of degradation and bioerosion are necessary to facilitate the subsequent discussion of degradable polymers and devices. The mechanisms and of degradation and bioerosion are discussed more thoroughly on pages 189–190: The Process of Bioerosion, Mechanisms of Chemical Degradation, and Factors that Influence the Rate of Bioerosion.

OVERVIEW OF CURRENTLY AVAILABLE DEGRADABLE POLYMERS

The development of highly stable materials has always been a major research challenge. Today, many polymers are available for clinical purposes that are virtually non-destructible in biological systems, e.g., Teflon™ (poly(tetrafluoroethylene)), Kevlar™ (poly(paraphenylene terephthalamide)), poly(ether-ether-ketone) (PEEK), or poly(ethylene terephthalate) (PET). On the other hand, the development of degradable biomaterials is a relatively new area of research. The variety of available, degradable biomaterials is still too limited to cover a wide enough range of diverse material properties. Thus, the design and synthesis of new, degradable biomaterials is currently an important research challenge. Within the context of tissue engineering, new biomaterials that provide predetermined and controlled cellular responses are a critically needed component of most practical applications (Anderson et al., 2004).

Degradable materials must fulfill more stringent requirements, in terms of their biocompatibility, than non-degradable materials. In addition to the potential problem of toxic contaminants leaching from the implant (residual monomers, stabilizers, polymerization initiators, emulsifiers, sterilization byproducts), one must also consider the potential toxicity of the degradation products and subsequent metabolites. The practical consequence of this consideration is that only a limited number of non-toxic, monomeric starting materials have been successfully applied to the preparation of degradable biomaterials.

Over the last decade a number of hydrolytically unstable polymers have been suggested as degradable biomaterials. However, in most cases, no attempts have been made to develop these new materials for specific medical applications. Thus, detailed toxicological studies *in vivo*, investigations of degradation rate and mechanism, and careful evaluations of the physico-mechanical properties have so far been published for only a small fraction of those polymers. An even smaller number of synthetic, degradable polymers have so far been used in medical implants and devices that have gained approval by the Food and Drug Administration (FDA) for use in patients. The FDA does not approve polymers or materials *per se*, but only specific medical devices and drug delivery formulations. As of 2006, only seven distinct synthetic, degradable polymer classes have been approved for use in a narrow range of clinical applications (Kohn et al., 2007). These polymers are polyesters containing lactic acid, polyesters containing glycolic acid, polyesters containing dioxanone, polyesters containing caprolactone, poly(trimethylene carbonate)s, polyanhydrides containing sebacic acid, and tyrosine-derived polyarylates. Various other synthetic, degradable biomaterials currently in clinical use are blends or copolymers of these base materials, such as a wide range of copolymers of lactic and glycolic acid. This listing does not include polymers derived from animal sources such as collagen, gelatin or hyaluronic acid.

> It is important to note that no polymer will degrade or erode by either hydrolytic or enzymatic processes unless water molecules can reach and "interact with" the molecular groups of the polymer chains. Thus, the "degree of hydrophobicity," the presence of crystallites, and the resultant absorption of water by the polymer are critical factors for all degradation/resorption processes.

Recent research has led to several well-established investigational polymers that may find practical applications as degradable implants within the next decade. This chapter will concern itself mostly with *synthetic* degradable polymers, since *natural* polymers (e.g., polymers derived from animal or plant sources) are described elsewhere in this book. Furthermore, the chapter on Hydrogels (Chapter I.2.5) contains a section on degradable hydrogels, and therefore, that subject won't be covered in this chapter.

Table I.2.6.2 provides an overview of some representative degradable polymers. For completeness, some of the natural polymers have also been included here. Structural formulas of commonly investigated synthetic degradable polymers are provided in Figure I.2.6.1. A large proportion of the currently investigated, *synthetic*, degradable polymers are polyesters. It remains to be seen whether some of the alternative backbone structures such as polyanhydrides, polyphosphazenes, polyphosphonates, polyphosphoesters, polyamides, or polycarbonates will be able to challenge the predominant position of the polyesters in the future.

Poly(glycolic acid) and poly(lactic acid) and their copolymers are currently the most widely investigated, and most commonly used, synthetic, bioerodible polymers. In view of their importance in the field of biomaterials, their properties and applications will be described in more detail.

Poly(glycolic acid) (PGA) is the simplest linear, aliphatic polyester (Figure I.2.6.1). Since PGA is highly crystalline, it has a high melting point and low solubility in organic solvents. PGA was used in the development of the first totally synthetic, absorbable suture (Herrmann et al., 1970). PGA sutures have been commercially available under the trade name "Dexon™" since 1970. A practical limitation of Dexon™ sutures is that they tend to lose their mechanical strength rapidly, typically over a period of two to four weeks after implantation. PGA has also been used in the design of internal bone fixation devices (bone pins). These pins have become commercially available under the trade name "Biofix™."

To adapt the material properties of PGA to a wider range of possible applications, copolymers of glycolic acid with the more hydrophobic lactic acid were intensively investigated (Gilding and Reed, 1979). The hydrophobicity of lactic acid limits the water uptake of poly(lactic acid) (PLA) thin films to about 2%, and reduces the rate of backbone hydrolysis as compared to PGA. Sutures composed of copolymers of lactic acid and glycolic acid, poly(lactide-co-glycolide) (PLGA), were developed as an alternative to PGA sutures. The suture and the 90:10 PLGA copolymer it is composed of are distributed under the trade names "Vicryl™" and "Polyglactin 910™."

It is noteworthy that there is no linear relationship between the ratio of glycolic acid to lactic acid and the physico-mechanical properties of the corresponding copolymers. Whereas PGA is highly crystalline, crystallinity is rapidly lost in copolymers of glycolic acid and lactic acid. These morphological changes lead to an increase in the rates of hydration and hydrolysis. Thus, 50:50 PLGA copolymers degrade more rapidly than either PGA or PLA.

Since lactic acid is a chiral molecule, it exists in two stereoisomeric forms which give rise to four morphologically distinct polymers: the two-stereoregular polymers, D-PLA and L-PLA; and the racemic form D,L-PLA. A fourth morphological form, meso-PLA, can be obtained from D,L lactide but is rarely used in practice.

TABLE I.2.6.2 Degradable Polymers and Applications Under Investigation

Synthetic Degradable Polyesters	Current Major Research Applications
Poly(glycolic acid) (PGA), poly(lactic acid) (PLA) and copolymers	Barrier membranes, drug delivery, hormone delivery, guided tissue regeneration (in dental applications), orthopedic applications, vascular/urological stents, staples, sutures, injectable fillers, dura mater substitutes, skin replacement materials, tissue engineering
Poly(hydroxybutyrate) (PHB), poly(hydroxyvalerate) (PHV), and copolymers thereof	Long-term drug delivery, orthopedic applications, stents, artificial skin, surgical patching materials for congenital heart defects, sutures
Polycaprolactone (PCL)	Long-term drug delivery, implantable contraceptive drug devices, orthopedic applications, staples, stents
Polydioxanone (PDS)	Fracture fixation in non-load-bearing bones, sutures, wound clips
Other Synthetic Degradable Polymers	
Polyanhydrides	Drug delivery
Polycyanoacrylates	Adhesives, drug delivery
Poly(amino acid)s and "pseudo"-poly(amino acid)s	Drug delivery, tissue engineering, orthopedic applications, stents, anti-adhesion barriers
Poly(ortho ester) (POE)	Drug delivery, and stents
Polyphosphazenes	Blood contacting devices, drug delivery, skeletal reconstruction, vaccine adjuvants
Poly(propylene fumarate) (PPF)	Orthopedic applications
Poly(trimethylene carbonate) (PTMC)	Sutures, orthopedic applications
Some Natural Resorbable Polymers	
Collagen	Drug delivery, gene delivery, artificial skin, coatings to improve cellular adhesion, guided tissue regeneration in dental applications, spinal dural repair, orthopedic applications, soft tissue augmentation, tissue engineering, scaffold for reconstruction of blood vessels, wound closure, hemostatic agents
Fibrinogen and fibrin	Tissue sealant, cell delivery
Elastin-like peptides (ELP)	Drug delivery, coating of vascular grafts
Gelatin	Capsule coating for oral drug delivery, hemorrhage arrester
Hyaluronic acid	Wound dressing applications, drug delivery, tissue engineering, synthetic bone grafts, synovial fluid substitutes
Cellulose	Adhesion barrier, hemostat
Various polysaccharides such as chitosan, alginate	Drug/vaccine delivery, encapsulation of cells, sutures, wound dressings/healing
Starch and amylose	Oral drug delivery

The polymers derived from the optically active D and L monomers are semicrystalline materials, while the optically inactive D,L-PLA is always amorphous. Generally, L-PLA is more frequently employed than D-PLA, since the hydrolysis of L-PLA yields L(+) lactic acid, which is the naturally occurring stereoisomer of lactic acid.

The differences in the crystallinity of D,L-PLA and L-PLA have important practical ramifications. Since D,L-PLA is an amorphous polymer, it is usually considered for applications such as drug delivery, where it is important to have a homogeneous dispersion of the active species within the carrier matrix. On the other hand, the semicrystalline L-PLA is preferred in applications where high mechanical strength and toughness are required, such as sutures and orthopedic devices.

PLA and PGA and their co-polymers have been investigated for more applications than any other degradable polymer. The high interest in these materials is not based on their superior material's properties, but mostly on the fact that these polymers have already been used successfully in a number of approved medical implants and are considered safe, non-toxic, and biocompatible by regulatory agencies in virtually all developed countries. Therefore, implantable devices prepared from PLA, PGA or their copolymers can be brought to market in less time and for a lower cost than similar devices prepared from novel polymers whose biocompatibility is still unproven.

Currently available and approved products include: sutures; suture reinforcements; GTR membranes for dentistry; orthopedic fixation devices; injectable fillers for restoration of facial volume; skin replacement materials; dura mater substitutes; hormone delivery systems; and implantable drug delivery systems. The polymers are also being widely investigated in the design of vascular and urological stents, and as scaffolds for tissue engineering and tissue reconstruction. In many of these applications, PLA, PGA and their copolymers have performed with moderate to high degrees of success. However, there are still unresolved issues: first, in tissue culture experiments, most cells do not attach to PLA or PGA surfaces and do

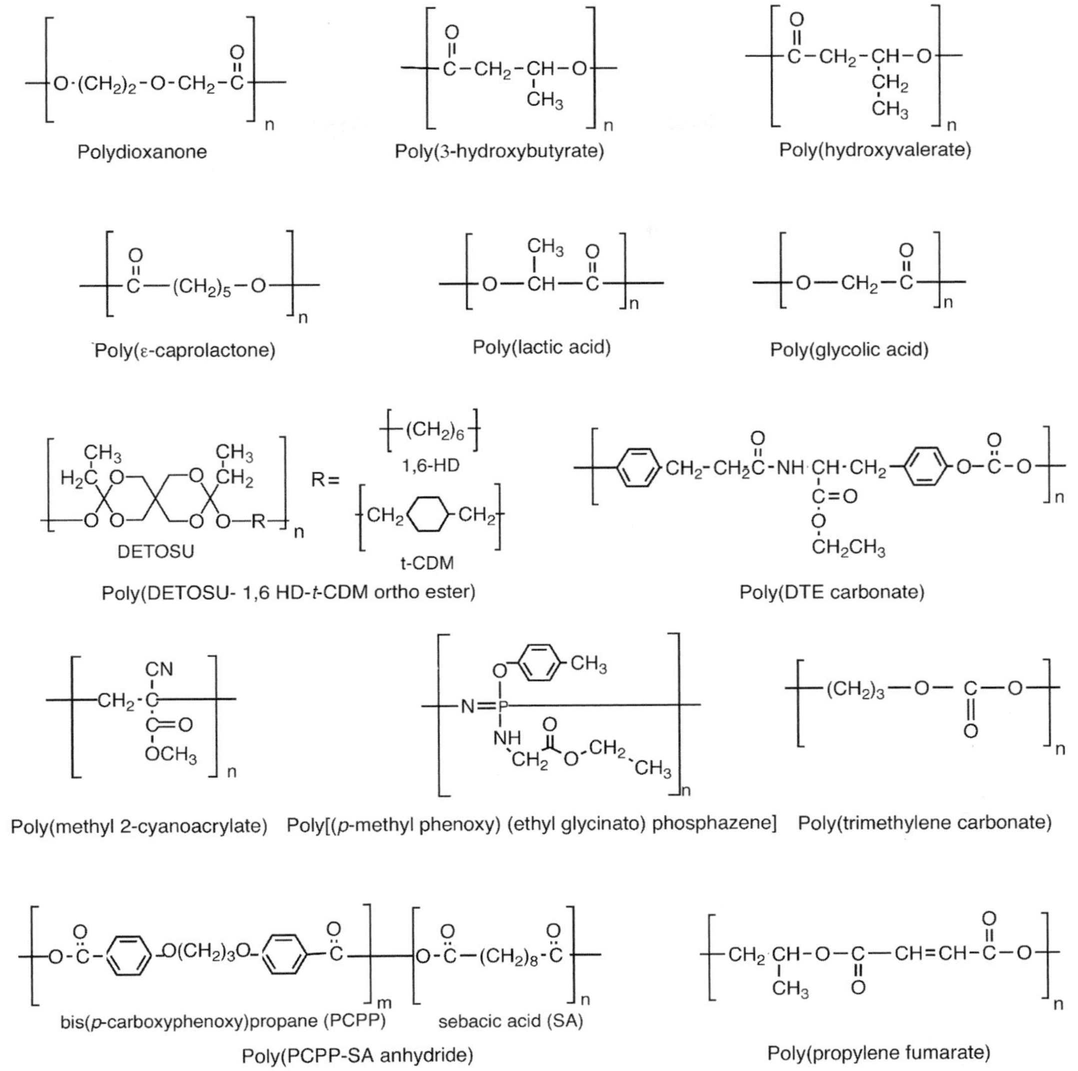

Polydioxanone
Poly(3-hydroxybutyrate)
Poly(hydroxyvalerate)
Poly(ε-caprolactone)
Poly(lactic acid)
Poly(glycolic acid)
DETOSU
1,6-HD
t-CDM
Poly(DETOSU- 1,6 HD-*t*-CDM ortho ester)
Poly(DTE carbonate)
Poly(methyl 2-cyanoacrylate)
Poly[(*p*-methyl phenoxy) (ethyl glycinato) phosphazene]
Poly(trimethylene carbonate)
bis(*p*-carboxyphenoxy)propane (PCPP)
sebacic acid (SA)
Poly(PCPP-SA anhydride)
Poly(propylene fumarate)

FIGURE I.2.6.1 Chemical structures of widely investigated degradable polymers.

not grow as vigorously as on the surface of other materials, indicating that these polymers are actually poor substrates for cell growth *in vitro*. Second, the degradation products of PLA and PGA are relatively strong acids (lactic acid and glycolic acid). When these degradation products accumulate at the implant site, a delayed inflammatory response is often observed months to years after implantation (Bergsma et al., 1995; Ignatius and Claes, 1996; Törmälä et al., 1998).

Poly(ethylene glycol) (PEG) has been used to initate lactide and/or glycolide ring opening, producing A–B–A triblock copolymers, which are degradable, producing LA, GA, and PEG as byproducts. They have been made into a variety of PEG-based hydrogels and used in drug delivery, and as tissue engineering scaffolds. They are discussed in detail in the chapter on Hydrogels (Chapter I.2.5).

Polydioxanone (PDS) is a poly(ether-ester) made by a ring opening polymerization of p-dioxanone monomer. PDS was first introduced in the early 1980s and has gained increasing interest in the medical and pharmaceutical fields due to the low toxicity of its degradation products *in vivo*. The degradation time of PDS *in vivo* is approximately 60 days, with complete mass loss within 9 to 12 months (Maurus and Kaeding, 2004). This degradation time is comparable to PGA. PDS has a lower modulus and glass transition temperature than PLA or PGA. PDS became the first degradable polymer to be used to make a monofilament suture (Ray et al., 1981). PDS has also been introduced to the market as a suture clip, a degradable ligating device, as well as a bone pin marketed under the name OrthoSorb® in the USA and Ethipin® in Europe.

Poly(hydroxybutyrate) (PHB), poly(hydroxyvalerate) (PHV), and their copolymers represent examples of resorbable polyesters that are derived from microorganisms. PHB and PHV are intracellular storage polymers providing a reserve of carbon and energy to microorganisms similar to the role of starch in plant metabolism. Therefore, although these classes of polymers are examples of *natural* materials (as opposed to *synthetic* materials), they are included here because they have similar properties and similar areas of application as the widely investigated poly(lactic acid). The polymers can be degraded by soil bacteria (Senior et al., 1972), but are relatively stable under physiological conditions (pH 7.4, 37°C). PHB is currently available as either poly(3-hydroxybutyrate) (P3HB) or poly(4-hydroxybutyrate) (P4HB). P3HB and its copolymers with up to 30% of 3-hydroxyvaleric acid are commercially available under the trade name "Biopol®" (Miller and Williams, 1987). P4HB is likewise available for clinical applications under the trade name "PHA4400." Within a relatively narrow window, the rate of degradation can be modified slightly by varying the copolymer composition between PHB and PHV. Members of the P3HB family of polymers require several years for complete resorption *in vivo*, while P4HB members require 8–52 weeks. *In vivo*, P3HB and P4HB degrade to D-3-hydroxybutyric acid and D-4-hydroxybutyric acid, respectively, which are natural metabolites present in the brain, heart, lung, liver, kidney, and muscle (Nelson et al., 1981). The low toxicity of PHBs may at least in part be due to this fact.

The mechanical properties of PHBs are dependent on the morphology considered. P3HB homopolymer is a highly crystalline and brittle material, while the copolymers of PHB with hydroxyvaleric acid are less crystalline, more flexible, and more readily processible (Barham et al., 1984). In contrast, P4HB is a strong, pliable thermoplastic material that is more flexible than PLA and PGA with an elongation break around 1000% (Martin and Williams, 2003). These polymers have been considered in several biomedical applications, such as: controlled drug release devices; sutures; artificial skin; surgical patching materials for congenital cardiovascular defects; heart valves; vascular grafts; and also in industrial applications such as medical disposables (Martin and Williams, 2003). P3HB is especially attractive for orthopedic applications due its slow degradation time. The polymer typically retains 80% of its original stiffness over 500 days upon *in vivo* degradation (Knowles, 1993).

Polycaprolactone (PCL) was first synthesized via ring-opening polymerization in the 1930s (Van Natta et al., 1934). However, its degradability was identified in the 1970s and it became available commercially following efforts at Union Carbide to identify synthetic polymers that could be degraded by microorganisms (Huang, 1985). Based on a large number of tests, ε-caprolactone and polycaprolactone are currently regarded as non-toxic and tissue-compatible materials. It is a semicrystalline polymer having a low glass transition temperature. The high solubility of polycaprolactone, its low melting point (59–64°C), and exceptional ability to form blends has stimulated research on its application as a biomaterial. Polycaprolactone degrades at a slower rate than PLA and has a high permeability to many drugs, therefore prompting its use in drug delivery devices that remain active for over one year (Sinha et al., 2004). The release characteristics of polycaprolactone have been investigated in detail by Pitt and his co-workers (Pitt et al., 1979). The Capronor™ system, a one-year implantable contraceptive device, has been commercially available in Europe and the USA. The toxicology of polycaprolactone has been extensively studied as part of the evaluation of Capronor™. In Europe, polycaprolactone is already in clinical use as a degradable staple (for wound closure). A monofilament suture derived from copolymers of polycaprolactone and PGA is widely used clinically and is available under the brand name Monocryl™ (Bezwada et al., 1995). Caprolactones have also been used in block copolymers with glycolide, lactide, and poly(ethylene glycol) for a commercially available drug delivery device under the brand name SynBiosys™.

Polyanhydrides were explored as possible substitutes for polyesters in textile applications, but failed due to their pronounced hydrolytic instability. It was this property that prompted Langer and his co-workers to explore polyanhydrides as degradable implant materials (Tamada and Langer, 1993). The main advantages of polyanhydrides are that: their production requires a single-step synthesis scheme with inexpensive reagents; they have controllable molecular weights and predictable degradation/drug release profiles; they are easily processable; and they are relatively non-toxic upon degradation (Kumar et al., 2002).

The polyanhydride library includes a large number of synthetic materials that all contain degradable anhydride bonds within the backbone of the polymer chain. To date, polyanhydrides have been synthesized with aliphatic monomers, unsaturated monomers, aromatic monomers, and linear fatty acid monomers (Nair and Laurencin, 2007). Aliphatic polyanhydrides degrade within days, whereas some aromatic polyanhydrides degrade over several years. Thus, aliphatic–aromatic copolymers are usually employed to achieve intermediate rates of degradation which are dependant on the monomer composition (Göpferich and Tessmar, 2002).

Polyanhydrides are among the most reactive and hydrolytically unstable polymers currently used as biomaterials. The high chemical reactivity is both an advantage and a limitation of polyanhydrides. Many polyanhydrides degrade by surface erosion without the need to incorporate various catalysts or excipients into the device formulation. While this property makes them good candidates for drug delivery applications, their hydrolytic instability limits shelf life. To maintain their integrity, polyanhydride materials must be stored under anhydrous and low temperature conditions. Additionally, polyanhydrides can react with drugs containing free amino groups or other nucleophilic functional groups, especially during high temperature processing (Leong et al., 1986). The potential reactivity of the polymer matrix toward nucleophiles limits the type of drugs that can be successfully incorporated into a polyanhydride matrix by melt processing techniques.

A comprehensive evaluation of the toxicity of polyanhydrides, as reviewed by Katti et al. demonstrates that, in general, the polyanhydrides possess excellent *in vivo* biocompatibility (Katti et al., 2002). The most immediate applications for polyanhydrides have been in the field of drug delivery. Drug loaded devices made of polyanhydrides can be prepared by compression, injection or melt molding, solvent casting or microencapsulation (Kumar et al., 2002). Depending on the monomer composition, polyanhydrides are able to release drugs with zero order kinetics over time periods ranging from days to years (Nair and Laurencin, 2007). A wide variety of drugs and proteins including insulin, bovine growth factors, angiogenesis inhibitors (e.g., heparin and cortisone), enzymes (e.g., alkaline phosphatase and β-galactosidase), and anesthetics have been incorporated into polyanhydride matrices with their *in vitro* and *in vivo* release characteristics evaluated (Katti et al., 2002; Kumar et al., 2002). The first polyanhydride-based drug delivery system to enter clinical use was for the delivery of chemotherapeutic agents. An example of this application is the delivery of BCNU (bis-chloroethylnitrosourea) to the brain for the treatment of glioblastoma multiforme, a highly fatal brain cancer (Brem et al., 1993). For this application, BCNU-loaded implants made of the polyanhydride derived from bis-p-carboxyphenoxypropane and sebacic acid received FDA regulatory approval in the fall of 1996 and are currently being marketed under the name Gliadel™. Septacin™, a drug delivery vehicle for gentamicin sulfate has been developed for the treatment of osteomyelitis (Li et al., 2002).

Poly(ortho ester)s (POE) are a family of synthetic, degradable polymers that have been under development for a number of years (Heller and Gurny, 1999). Currently, there are four major types of poly(ortho ester)s: first generation (POE I); second generation (POE II); third generation (POE III); and fourth generation (POE IV) (Heller et al., 2002). (See also the chapter on Drug Delivery, Chapter II.5.16.)

POE I were first prepared by Choi and Heller via the transesterification of 2,2′-dimethoxyfuran with a diol (Heller and Gurny, 1999). Unfortunately, the hydrolysis of POE I is an acid-sensitive process that itself results in the release of acidic degradation products. This leads to autocatalysis of the polymer, limiting its utility for clinical applications. The second generation of poly(ortho ester)s (POE II) were based on an acid catalyzed addition reaction of diols with diketeneacetals (Heller, 1990). Unlike POE I, the degradation products of this polymer are neutral in pH, therefore preventing the autocatalysis of the polymer. The mechanical properties of the POE II can be controlled to a large extent by the choice of the diols used in the synthesis. Materials ranging from stiff materials to those that are quite soft at room temperature can be achieved. Since the ortho ester linkage in POE I and II is far more stable in base than in acid, Heller and his coworkers controlled the rate of polymer degradation by incorporating acidic or basic excipients into the polymer matrix. The third generation of poly(ortho ester)s (POE III) were very soft and were even viscous liquids at room temperature (Merkli et al., 1993). POE III can be used in the formulation of drug delivery systems that are injected rather than implanted into the body. Release profiles of the chemotherapeutic agent 5-fluorouracil demonstrated zero order kinetics and were investigated for ocular implants (Heller, 2005). Unfortunately, difficulties in synthesis and reproducibility limited the practical applications of this material, and currently it is no longer under development. The final generation of poly(ortho ester)s (POE IV) was synthesized by Heller and his co-workers to address the need for excipients in POE I and II (Ng et al., 1997). This polymer achieves consistent degradation profiles via the incorporation of glycolic or lactic acid to the polymer backbone. Upon degradation, the acid segments are released, therefore catalyzing the hydrolysis of the ortho ester linkages.

POE IV is naturally "surface eroding", while POE I and II can erode by "surface erosion" if appropriate excipients are incorporated into the polymeric matrix. One concern about the "surface erodability" of poly(ortho ester)s is that the incorporation of highly water-soluble drugs into the polymeric matrix can result in swelling of the polymer matrix. The increased amount of water imbibed into the matrix can then cause the polymeric device to exhibit "bulk erosion" instead of "surface erosion" (see the section The Process of Bioerosion, pp. 189–190, for a more detailed explanation of these erosion mechanisms) (Okada and Toguchi, 1995). Since surface eroding, slab-like devices tend to release drugs embedded within the polymer at a constant rate, poly(ortho ester)s appear to be particularly suited for controlled release drug delivery applications. For example, poly(ortho ester)s have been studied for the controlled delivery of cyclobenzaprine and steroids, and a significant number of publications describe the use of poly(ortho ester)s for various drug delivery applications (Heller and Gurny, 1999) (see Chapter II.5.16).

Poly(amino acid)s and "pseudo"-poly(amino acid)s: since proteins are composed of amino acids, researchers explored the possible use of poly(amino acid)s in biomedical applications (Anderson et al., 1985). Poly(amino acid)s were regarded as promising candidates since the amino acid side chains offer sites for the attachment of drugs, cross-linking agents, or pendent groups that can be used to modify the physico-mechanical properties of the polymer. Additionally, they initially held potential as degradable polymer materials because they possessed long range order, defined conformations in solution, and their amino acid-derived building blocks lacked toxicity (Kohn, 1993). Poly(amino acid)s have been investigated for use as suture materials (Williams, 1982), as artificial skin substitutes (Spira et al., 1969), and as drug delivery systems (Matsumura, 2008). Various drugs have been attached to the side chains of poly(amino acid)s, usually via a spacer unit that distances the drug from the backbone. Poly(amino acid)–drug combinations investigated for chemotherapy include poly(L-lysine) with methotrexate and pepstatin

(Campbell et al., 1980), poly(glutamic acid) with adriamycin and cisplatin (Van Heeswijk et al., 1985), and PEG-polyaspartate with paclitaxel (Matsumura, 2008).

Despite their apparent potential as biomaterials, poly(amino acid)s have actually found few practical applications due to several synthetic disadvantages: the expense of production; insolubility in common organic solvents; and thermal degradation upon melting limiting processibility. Poly(amino acid)s have a pronounced tendency to swell in aqueous media, therefore making the prediction of drug release rates difficult. Furthermore, the antigenicity of polymers containing three or more amino acids limits their use in biomedical applications (Kohn, 1993). Due to these difficulties, only a few poly(amino acid)s have been investigated for clinical uses. While some poly(amino acid) derivatives of poly(glutamic acid) carrying various pendent chains at the γ-carboxylic acid group have been investigated as implant materials (Lescure et al., 1989), the majority of their clinical success has been in the drug delivery of chemotherapeutics. As of 2008, PEG-poly(glutamic acid) nanocarrier formulations are in Phase 1 clinical trials for the delivery of paclitaxel and camptothecin, while Phase 2 trials are underway for cisplatin delivery against stomach cancer (Matsumura, 2008). So far, no implantable devices made of a poly(amino acid) have been approved for clinical use in the USA.

In an attempt to circumvent the problems associated with conventional poly(amino acid)s, backbone-modified "pseudo"-poly(amino acid)s were introduced in 1984 (Kohn and Langer, 1984, 1987). The first "pseudo"-poly(amino acid)s investigated were a polyester from N-protected trans-4-hydroxy-L-proline, and a polyiminocarbonate derived from tyrosine dipeptide. The tyrosine-derived "pseudo"-poly(amino acid)s are processable by solvent or heat methods and exhibit a high degree of biocompatibility.

The reason for the improved physico-mechanical properties of "pseudo"-poly(amino acid)s relative to conventional poly(amino acid)s can be traced to the reduction in the number of interchain hydrogen bonds. In conventional poly(amino acid)s, individual amino acids are polymerized via repeated amide bonds, leading to strong interchain hydrogen bonding. In natural peptides, hydrogen bonding is one of the interactions leading to the spontaneous formation of secondary structures such as α-helices or β-pleated sheets. Strong hydrogen bonding also results in high processing temperatures and low solubility in organic solvents, which tends to lead to intractable polymers with limited applications. In "pseudo"-poly(amino acid)s, half of the amide bonds are replaced by other linkages (such as, for example, carbonate, ester, or iminocarbonate bonds) which have a much lower tendency to form interchain hydrogen bonds, leading to better processibility and, generally, a loss of crystallinity.

For example, tyrosine-derived polycarbonates (Pulapura and Kohn, 1992; Bourke and Kohn, 2003) are high strength materials that may be useful in the formulation of degradable orthopedic implants. One of the tyrosine-derived pseudo-poly(amino acid)s, poly(DTE carbonate) exhibits a high degree of bone conductivity (e.g., bone tissue will grow directly along the polymeric implant) (Choueka et al., 1996; James et al., 1999; Asikainen et al., 2006). As of 2012, tyrosine derived "pseudo"-poly(amino acid)s have been used in two FDA approved devices (a hernia repair device and an antimicrobial pacemaker pouch from TyRx Pharma, Inc.) and a third is in clinical trials (a resorbable cardiovascular stent from REVA Medical, Inc.).

Polycyanoacrylates are used as bioadhesives and have been investigated for drug delivery. These adhesives are known for their high rates of curing, high adhesive strength, and good adhesion to both tissue and inorganic materials. Cyanoacrylates undergo spontaneous polymerization at room temperature in the presence of water. Their toxicity and erosion rate after polymerization differs with the length of their alkyl chain (Vezin and Florence, 1980). Methyl cyanoacrylates are more commonly used as general-purpose glues and are commercially available as "Crazy Glue." Methyl cyanoacrylate was used during the Vietnam War as an emergency tissue adhesive, but is no longer used today. Butyl cyanoacrylate is a slowly degrading polymer approved for use in Canada and Europe as a dental adhesive. n-Butyl-cyanoacrylate is used in veterinary glues, while 2-octyl cyanoacrylate (Dermabond™) is used in liquid bandages. Polycyanoacrylates have several limiting properties: first, the monomers (cyanoacrylates) are very reactive compounds that often have significant toxicity; second, upon degradation polycyanoacrylates release formaldehyde resulting in intense inflammation in the surrounding tissue. In spite of these inherent limitations, polycyanoacrylates have been investigated as potential drug delivery matrices, have been suggested for use in ocular drug delivery (Deshpande et al., 1998) and the delivery of nucleic acids (Fattal and Couvreur, 2004), and are in late stage clinical trials for cancer therapy (Vauthier et al., 2003).

Polyphosphazenes are polymers whose backbone consists of nitrogen–phosphorous bonds. These polymers are at the interface between inorganic and organic polymers, and have unusual material properties. Polyphosphazenes have found industrial applications, mainly because of their high thermal stability. They have also been used in investigations for the formulation of controlled drug delivery systems (Allcock, 1990). Polyphosphazenes are interesting biomaterials in many respects. Most polyphosphazenes are not degradable, but the incorporation of certain chemical groups into the polymer structure may render these materials degradable (Allcock, 1999). They have been claimed to be biocompatible, and their chemical structure provides a readily accessible "pendent chain" to which various drugs, peptides or other biological compounds can be attached and later released via hydrolysis. Polyphosphazenes have been examined for use in skeletal tissue regeneration and have been shown to be osteocompatible (Qiu and Zhu, 2000; Nair et al., 2006). Another novel use of polyphosphazenes is in vaccine design, where these materials were used as immunological adjuvants (Andrianov et al., 1998).

Poly(propylene fumarate)s (PPF) and Poly(trimethylene carbonate) (PTMC) are polyester based materials, like PGA and PLA. PPF is most commonly synthesized using a two-step process (Shung et al., 2002) that results in the formation of either a viscous liquid or a yellow solid. The formation of the liquid or the solid depends on the molecular weight of the synthesized polymer, with high molecular weights yielding the solid, and low molecular weights yielding the liquid. PPF, unlike PGA and PLA, is an unsaturated polyester. Since a double bond is present within the backbone, various chemical modifications, including cross-linking, can be used to alter the physicomechanical properties of the material. Cross-linking of the liquid polymer allows the material to be processed and molded into any number of shapes both pre-implantation and *in situ*. The ability to harden the materials *in situ* via thermal or photo initiated processes coupled with their osteoconductivity *in vitro* (Yaszemski et al., 1995) and *in vivo* (Lewandrowski et al., 2000; Fisher et al., 2002) has prompted exploration of these materials as cements and scaffolding materials for bone. PFF is degraded via hydrolytic cleavage of its ester bonds yielding propylene glycol and fumarate upon complete degradation. Both propylene glycol and fumarate are non-toxic *in vivo*. PPF, like PGA and PLA, releases acidic products upon degradation. However, PPF seems to lack susceptibility to autocatalyzed degradation as cross-linking density governs degradation rates to a far greater extent than local pH (Timmer et al., 2003).

PTMC is an elastomeric polyester with high flexibility but poor mechanical properties. It is synthesized via the ring opening polymerization of trimethyl carbonate and has a degradation rate that is 20 times slower than poly (ε-caprolactone) in phosphate buffered saline *in vitro* (Zhu et al., 1991). Interestingly, the *in vivo* degradation time of PTMC is significantly faster than its *in vitro* degradation, thus strongly suggesting a role for enzymatic degradation (Zhu et al., 1991; Zhang et al., 2006). This makes PTMC a polymer that truly undergoes "biodegradation." Copolymers of trimethylene carbonate and glycolide have been used as sutures and orthopedic fixation devices under the brand names Maxon™ and Acufex™ respectively. Additionally, a terpolymer of trimethylene carbonate, glycolide, and dioxane has been used to create a degradable suture under the brand name of BioSyn™ (Suzuki and Ikada, 2005).

Polyurethanes represent a class of elastomeric materials that have been used in the fabrication of a number of medical devices. While considered to be resistant to hydrolysis, polyurethane pacemaker leads were found to degrade through oxidative, enzymatic, and cellular mechanisms (Mcmillin, 1991). Recent work has focused on the incorporation of degradable linkages to impart predictable degradation kinetics. Linkages based on chemicals such as lysine diisocyanate have allowed the creation of degradable polyurethanes that are non-toxic. Polyurethanes are of particular interest in implantables because they have robust mechanical properties and good biocompatibility (Gunatillake and Adhikari, 2003). Non-degradable polyurethanes are currently used in a number of clinical devices, but have gained particular interest as vascular grafts. This stems from the observation that protein adsorption, and therefore clotting, is less pronounced on polyurethanes as compared to other materials. Degradable polyurethanes have also been investigated as artificial skin (Bruin et al., 1990), orthopedic injectables, and tissue engineering scaffolds (Degrapol™) (Saad et al., 1997; Zhang et al., 2000).

APPLICATIONS OF SYNTHETIC, DEGRADABLE POLYMERS AS BIOMATERIALS

Classification of Degradable Medical Implants

Some typical short-term applications of biodegradable polymers are listed in Table I.2.6.1. From a practical perspective, it is convenient to distinguish between five main types of degradable implants: the temporary support device; the temporary barrier; the drug delivery device; the tissue-engineering scaffold; and the multi-functional implant.

A temporary support device is used in those circumstances in which the natural tissue bed has been weakened by disease, injury or surgery, and requires some artificial support. A healing wound, a broken bone or a damaged blood vessel are examples of such situations. Sutures, bone fixation devices (e.g., bone nails, screws or plates), and vascular grafts would be examples of the corresponding support devices. In all of these instances, the degradable implant would provide temporary, mechanical support until the natural tissue heals and regains its strength. In order for a temporary support device to work properly, a gradual stress transfer should occur: as the natural tissue heals, the degradable implant should gradually weaken. The need to adjust the degradation rate of the temporary support device to the healing of the surrounding tissue represents one of the major challenges in the design of such devices.

A successful example of a temporary support device, the first synthetic, degradable sutures were made of PGA and became available under the trade name "Dexon™" in 1970. This represented the first routine use of a degradable polymer in a major clinical application (Frazza and Schmitt, 1971). Later, copolymers of lactic acid and glycolic acid were developed. The widely used "Vicryl™" suture, for example, is a 90:10 copolymer of lactic and glycolic acid, introduced into the market in 1974. Sutures made of polydioxanone (PDS) became available in the USA in 1981. Sutures of ε-caprolactone and trimethylene carbonate with glycolide have been developed under the names "MonoCryl™" and "Maxon™," respectively. "Biosyn™" is a terpolymer suture composed of trimethylene carbonate, glycolide, and dioxane. In spite of extensive research efforts in many laboratories, no other degradable polymers are currently used to any significant extent in the formulation of degradable sutures.

A temporary barrier has its major medical use in the prevention of postsurgical adhesions. During surgical closure, barrier devices are interposed between tissues that must remain separate following surgery. Adhesions are formed between two tissue sections by clotting of blood in the extravascular tissue space, followed by inflammation and fibrosis. If this natural healing process occurs between surfaces that were not meant to bond together, the resulting adhesion can cause pain, functional impairment, and problems during subsequent surgery. Surgical adhesions are a significant cause of morbidity, and represent one of the most significant complications of a wide range of surgical procedures such as abdominal, cardiac, spinal, and tendon surgery. A temporary barrier could take the form of a thin polymeric film or a mesh-like device that would be placed between adhesion-prone tissues at the time of surgery. To be useful, such a temporary barrier would have to prevent the formation of scar tissue connecting adjacent tissue sections, followed by the slow resorption of the barrier material. This sort of barrier has also been investigated for sealing of breaches of the lung tissue that cause air leakage.

Another important example of a temporary barrier is in the field of skin reconstruction. Several products have become available that are generally referred to as "artificial skin." The first such product consists of an artificial, degradable collagen/glycosaminoglycan matrix that is placed on top of the skin lesion to stimulate the regrowth of a functional dermis. Another product consists of a degradable collagen matrix with pre-seeded human fibroblasts. Synthetic degradable polymers have also been used to create "artificial skin." Dermagraft™ is an artificial skin product that uses a PLGA-based mesh scaffold to provide structural support. Again, the goal is to stimulate the regrowth of a functional dermis. These products are used in the treatment of burns and other deep skin lesions, and represent an important application for temporary barrier type devices.

Implantable drug delivery devices and **injectable polymer-drug depot delivery systems** are capable of delivering drugs to specific locations over an extended period of time. They are, by necessity, temporary devices, for the drug reservoir will eventually be depleted or the need for delivery of a specific drug will be eliminated once the disease is adequately treated. The development of implantable drug delivery systems is probably the most widely investigated application of degradable polymers (Nair and Laurencin, 2007). One can expect that the future acceptance of implantable and injectable drug delivery devices by physicians and patients alike will depend on the availability of degradable systems that do not have to be surgically explanted.

Since PLA and PGA have an extensive safety profile based on their use as sutures, these polymers have been very widely investigated in the formulation of implantable controlled release devices. Several implantable, controlled release formulations based on copolymers of lactic and glycolic acid have already become available. However, a wide range of other degradable polymers has also been investigated. Particularly noteworthy is the use of a new type of polyanhydride in the formulation of an intracranial, implantable device for the administration of BCNU (a chemotherapeutic agent) to patients suffering from glioblastoma multiforme, a usually lethal form of brain cancer (Gliadel™). A clinical trial by Westphal et al. in 2003 demonstrated a 29% reduction in the risk of death in patients treated with Gliadel™ versus those who were not (Westphal et al., 2003). The topic of drug delivery systems is discussed in more detail in Chapter II.5.16.

The term **tissue engineering scaffold** will be used in this chapter to describe a degradable implant that is designed to act as an artificial extracellular matrix by providing space for cells to grow into and reorganize into functional tissue (Hutmacher, 2001).

It has become increasingly obvious that man-made implantable prostheses do not function as well as the native tissue, or maintain the functionality of native tissue over long periods of time. Therefore, tissue engineering has emerged as an interdisciplinary field that utilizes degradable polymers, among other substrates and biologics, to develop treatments that enhance the body's potential to heal itself without the need for permanently implanted, artificial prosthetic devices. In the ideal case, a tissue engineering scaffold is implanted to restore lost tissue function, maintain tissue function or enhance existing tissue function (Langer and Vacanti, 1993). These scaffolds can take the form of a felt-like material obtained from knitted or woven fibers or from fiber meshes. Alternatively, various processing techniques can be used to obtain foams or sponges. For all tissue engineering scaffolds, pore volume and pore interconnectivity are key properties, as cells need to be able to migrate and grow throughout the entire scaffold. Thus, industrial foaming techniques used, for example, in the fabrication of furniture cushions, are not applicable to the fabrication of tissue engineering scaffolds, as these industrial foams are designed to contain "closed pores," while tissue engineering scaffolds require an "open pore" structure to attain the requisite interconnectivity. Tissue engineering scaffolds may be pre-seeded with cells *in vitro* prior to implantation. Alternatively, tissue-engineering scaffolds may consist of a cell-free structure that is either invaded and "colonized" by cells after implantation, or a surgeon may seed the scaffold with autologous cells intraoperatively (e.g., combining bone marrow aspirate with demineralized bone matrix to treat bone defects). In either case, the tissue engineering scaffold must allow the formation of functional tissue *in vivo*, followed by the safe resorption of the scaffold material.

There has been some debate in the literature as to the exact definition of the related term "guided tissue regeneration" (GTR). Guided tissue regeneration is a term traditionally used in dentistry. This term sometimes implies that the scaffold encourages the growth of specific types of tissue. For example, in the treatment of periodontal disease, periodontists use the term "guided tissue regeneration" when using implants that favor new bone growth in the periodontal pocket over soft tissue ingrowth (scar formation) (Asikainen et al., 2006).

One of the major challenges in the design of tissue engineering scaffolds is the need to adjust the rate of scaffold degradation to the rate of tissue healing. Depending upon the application of the scaffold, the polymer may need to maintain integrity in the order of days to months. Scaffolds intended for the reconstruction of bone illustrate this point: in most applications, the scaffold must maintain some mechanical strength to support the bone structure while new bone is formed. Premature degradation of the scaffold material can be as detrimental to the healing process as a scaffold that remains intact for excessive periods of time. The future use of tissue engineering scaffolds has the potential to revolutionize the way aging, trauma, and disease related loss of tissue function could be treated.

Multifunctional devices, as the name implies, combine several of the above-mentioned functions within a single device. Over the last few years, there has been a trend toward increasingly sophisticated applications for degradable biomaterials. Usually, these applications envision the combination of several functions within the same device and require the design of custom-made materials with a narrow range of predetermined material properties. For example, the availability of biodegradable bone nails and bone screws made of ultra high strength poly(lactic acid) opens the possibility of combining the "mechanical support" function of the device with a "site-specific drug delivery" function: a biodegradable bone nail that holds the fractured bone in place and can simultaneously stimulate the growth of new bone tissue at the fracture site by slowly releasing bone growth factors (e.g., bone morphogenetic protein or transforming growth factor β) throughout its degradation process.

Likewise, biodegradable stents for implantation into coronary arteries are currently being investigated (Ramcharitar and Serruys, 2008). The stents are designed to mechanically prevent the collapse and restenosis (reblocking) of arteries that have been opened by balloon angioplasty. Ultimately, the stents could deliver an anti-inflammatory or anti-thrombogenic agent directly to the site of vascular injury. Again, it would potentially be possible to combine a mechanical support function with site-specific drug delivery.

Various functional combinations involve the tissue engineering scaffold. Perhaps the most important multi-functional device for future applications is a tissue engineering scaffold that also serves as a drug delivery system for cytokines, growth hormones or other agents that directly affect cells and tissue in the vicinity of the implanted scaffold.

The Process of Bioerosion

One of the most important prerequisites for the successful use of a degradable polymer for any medical application is a thorough understanding of the way the device will degrade/erode and ultimately resorb from the implant site. Within the context of this chapter, we are limiting our discussion to the case of a solid, polymeric implant. The transformation of such an implant into water-soluble material(s) is best described by the term "bioerosion." The bioerosion process of a solid, polymeric implant is associated with macroscopic changes in the appearance of the device, changes in its physico-mechanical properties and in physical processes such as swelling, deformation or structural disintegration, weight loss and the eventual depletion of drug or loss-of-function. It is important to note that the bioerosion of a solid device is not necessarily due to the chemical cleavage of the polymer backbone, cross-links or side chains. Rather, simple solubilization of the intact polymer chains, for instance, due to changes in pH, may also lead to the erosion of a solid device.

Two distinct modes of bioerosion have been described in the literature: bulk erosion and surface erosion. In bulk erosion, the rate of water penetration into the solid device exceeds the rate at which the polymer is transformed into water-soluble material(s). Consequently, an erosion process that occurs throughout the entire volume of the solid device follows the uptake of water. Due to the rapid penetration of water into the matrix of the polymer, most of the currently available polymers will give rise to bulk eroding devices. In a typical bulk erosion process, cracks and crevices will form throughout the device, which may crumble into pieces. A good illustration for a typical bulk erosion process is the disintegration of an aspirin tablet that has been placed into water. Depending on the specific application, the uncontrollable tendency of bulk eroding devices to crumble into little pieces can be a disadvantage.

Alternatively, in surface erosion, the bioerosion process is limited to the surface of the device. Therefore, the device will become thinner with time, while maintaining its structural integrity throughout much of the erosion process. In surface erosion, the polymer must impede the rapid transport of water into the interior of the device. In addition, the rate at which the polymer is transformed into water-soluble material(s) has to be fast relative to the rate of water penetration into the device. Under these conditions, scanning electron microscopic evaluation of surface eroding devices has sometimes shown a sharp border between the eroding surface layer and the intact polymer in the core of the device (Mathiowitz et al., 1990).

The determination of whether a material is surface eroding or bulk eroding is dependent on several factors. Since the transport of water into the interior of a device is necessary for erosion, polymer hydrophobicity plays an important role. However, polymer chemistry is not the sole mediator of whether a material undergoes surface or bulk erosion. Von Burkersroda et al. have proposed that the physical architecture of the device may determine if it is surface or bulk eroding (Von Burkersroda et al., 2002). In this publication, a critical thickness is calculated for a selection of polymeric materials. Above this critical thickness, a material is surface eroding, while below this critical thickness it is bulk eroding. This critical thickness is so large for some polymer classes (in the order of meters) that with regard to implantable devices they cannot practically be surface eroding. Therefore, surface eroding devices have so far been obtained only from a small number of hydrophobic polymers containing hydrolytically highly reactive linkages in the backbone. Currently, polyanhydrides, and poly(ortho ester)s are the best-known examples of polymers that can be fabricated into surface eroding devices.

A possible exception to this general rule is enzymatic surface erosion. The inability of enzymes to penetrate into the interior of a solid, polymeric device may result in an enzyme-mediated surface erosion mechanism. In the case of enzymatic surface erosion, the term biodegradation may be correctly applied, for enzymatic degradation of the polymer backbone dominates the rate of hydrolysis under physiological conditions. It is possible for a polymeric device to be classified as bulk eroding and biodegrading if the polymer imbibes sufficient water to allow enzymes to penetrate the swollen network of polymer chains.

Mechanisms of Chemical Degradation

Although bioerosion can be caused by the solubilization of an intact polymer, chemical degradation of the polymer is usually the underlying cause for the bioerosion of a solid, polymeric device. Several distinct types of chemical degradation mechanisms have been identified (Figure I.2.6.2) (Rosen et al., 1988). Chemical reactions can lead to cleavage of cross-links between water-soluble polymer chains (Mechanism I), to cleavage of polymer side chains resulting in the formation of polar or charged groups (Mechanism II), or to cleavage of the polymer backbone (Mechanism III). Obviously, combinations of these mechanisms are possible: for instance, a cross-linked polymer may first be partially solubilized by the cleavage of cross-links (Mechanism I), followed by the cleavage of the backbone itself (Mechanism III). It should be noted that water is key to all of these degradation schemes. Even enzymatic degradation occurs in an aqueous environment.

Since the chemical cleavage reactions described above can be mediated by water or by biological agents, such as enzymes and microorganisms, it is possible to distinguish between hydrolytic degradation and biodegradation. It has often been stated that the availability of water is virtually constant in all soft tissues and varies little from person to person. On the other hand, the levels of enzymatic activity may vary widely not only from person to person, but also between different tissue sites in the same person. Thus polymers that undergo hydrolytic cleavage tend to have more predictable *in vivo* erosion rates than polymers whose degradation is mediated predominantly by enzymes. The latter polymers tend to be generally less useful as degradable medical implants.

Factors that Influence the Rate of Bioerosion

Although the solubilization of intact polymer, as well as several distinct mechanisms of chemical degradation have been recognized as possible causes for the observed bioerosion of a solid, polymeric implant, virtually all currently available implant materials erode due to the hydrolytic cleavage of the polymer backbone (Mechanism III in Figure I.2.6.2). We therefore limit the following discussion to solid devices that bioerode in this manner.

The main factors that determine the overall rate of the erosion process are the chemical stability of the hydrolytically-susceptible groups in the polymer backbone, the hydrophilic–hydrophobic character of the repeat units, the morphology of the polymer, the initial molecular weight and molecular weight distribution of the polymer, the fabrication process used to prepare the device, the presence of catalysts, additives or plasticizers, and the geometry (specifically the surface area to volume ratio) of the implanted device.

The susceptibility of the polymer backbone to hydrolytic cleavage is probably the most fundamental parameter. Anhydride bonds are the most susceptible to hydrolysis, followed by carbonate, ester, urethane, ortho ester, and amide bonds (Figure I.2.6.3) (Gombotz and Pettit, 1995). Thus, polyanhydrides will tend to degrade faster than polyesters, which in turn will have a higher tendency to bioerode than polyamides. Based on the known susceptibility of the polymer backbone structure to hydrolysis, it is possible to make predictions about the bioerosion of a given polymer.

However, the actual erosion rate of a solid polymer cannot be predicted on the basis of the polymer backbone structure alone. The observed erosion rate is strongly dependent on the ability of water molecules to penetrate into the polymeric matrix. The hydrophilic versus hydrophobic character of the polymer, which is a function of the structure of the monomeric starting materials, can therefore have an overwhelming influence on the observed bioerosion rate. For instance, the erosion rate of polyanhydrides can be slowed by about three orders of magnitude when the less hydrophobic sebacic acid is

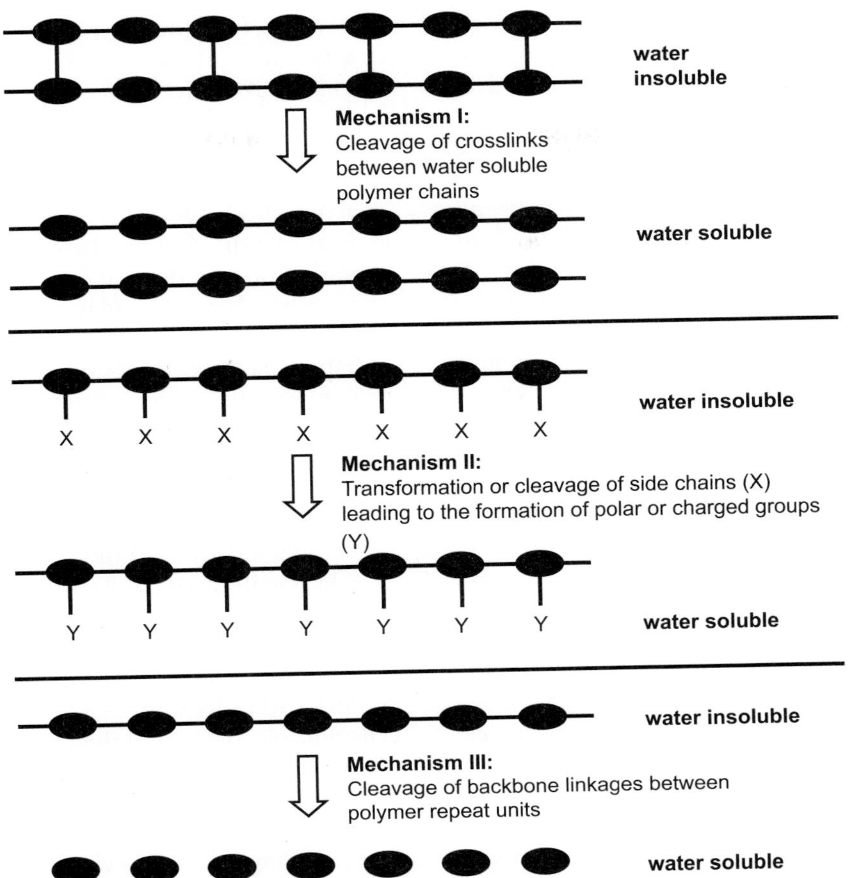

FIGURE I.2.6.2 Mechanisms of chemical degradation, adapted from Gombotz and Pettit (1995). Mechanism I involves the cleavage of degradable cross-links between water-soluble polymer chains. Mechanism II involves the cleavage or chemical transformation of hydrophobic polymer side chains, resulting in the formation of charged or polar groups. The presence of charged or polar groups leads then to the solubilization of the intact polymer chain. Mechanism III involves the cleavage of unstable linkages in the polymer backbone, followed by solubilization of the low molecular weight fragments.

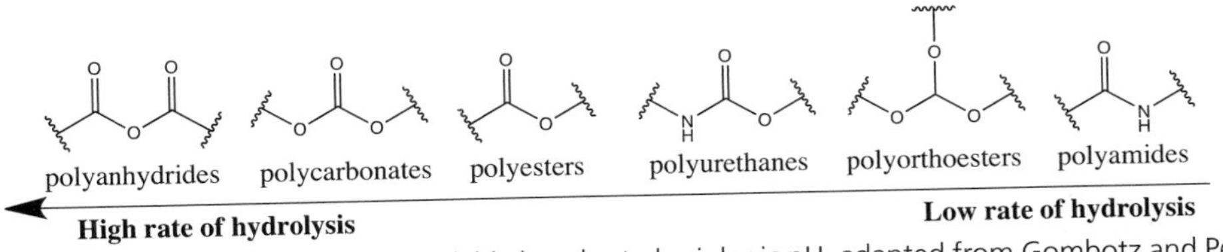

polyanhydrides polycarbonates polyesters polyurethanes polyorthoesters polyamides

High rate of hydrolysis — Low rate of hydrolysis

FIGURE I.2.6.3 Relative hydrolytic susceptibility of degradable bonds at physiologic pH, adapted from Gombotz and Pettit (1995). In general, polymers with high rates of main chain bond hydrolysis will degrade more quickly than polymers with low rates of main chain bond hydrolysis, but many other factors, such as pH and autocatalysis at low pHs, copolymer composition and structure, percent crystallinity, and related equilibrium water content, and geometry of the degradable polymer, will influence the polymer/device degradation rate. For example, at acidic pHs, the order in this table would change, because poly(ortho ester)s will have a higher rate of hydrolysis, perhaps somewhere between polyanhydrides and polycarbonates. In a related example, with thicker samples of polyesters such as PLGA, the acidic byproducts (lactic acid and glycolic acid) may accumulate within the bulk, lowering the internal bulk pH and leading to more rapid degradation within the bulk of the sample than at the surface.

replaced by the more hydrophobic bis(carboxy phenoxy) propane as the monomeric starting material (Tamada and Langer, 1993). Likewise, devices made of PGA erode faster than identical devices made of the more hydrophobic PLA, although the ester bonds have about the same chemical reactivity toward water in both polymers.

The observed bioerosion rate is further influenced by the morphology of the polymer. Polymers can be classified as either semicrystalline or amorphous. At body temperature (37°C) amorphous polymers with Tg (glass transition temperature) above 37°C will be in a glassy state, and polymers with a Tg below 37°C will be in a rubbery state. In this discussion it is therefore necessary to consider three distinct morphological states: semicrystalline; amorphous-glassy; and amorphous-rubbery.

In the crystalline state, the polymer chains are densely packed and organized into crystalline domains that resist the penetration of water. Consequently, backbone hydrolysis tends to occur in the amorphous regions of a semi-crystalline polymer and at the surface of the crystalline regions. This phenomenon is of particular importance to the erosion of devices made of poly(L-lactic acid) and poly(glycolic acid), which tend to have high degrees of crystallinity, around 50%.

Another good illustration of the influence of the polymer morphology on the rate of bioerosion is provided by a comparison of poly(L-lactic acid) and poly(D,L-lactic acid); although these two polymers have chemically identical backbone structures and an identical degree of hydrophobicity, devices made of poly(L-lactic acid) tend to degrade much more slowly than identical devices made of poly(D,L-lactic acid). The slower rate of bioerosion of poly(L-lactic acid) is due to the fact that this stereoregular polymer is semi-crystalline, while the racemic poly(D,L-lactic acid) is an amorphous polymer.

Likewise, a polymer in its glassy state is less permeable to water than the same polymer when it is in its rubbery state. This observation could be of importance in cases where an amorphous polymer has a glass transition temperature that is not far above body temperature (37°C). In this situation, water absorption into the polymer could lower its Tg below 37°C, resulting in abrupt changes in the bioerosion rate.

The manufacturing process may also have a significant effect on the erosion profile. For example, Mathiowitz and co-workers (Mathiowitz et al., 1990) showed that polyanhydride microspheres produced by melt encapsulation were very dense and eroded slowly, whereas when the same polymers were formed into microspheres by solvent evaporation, the microspheres were very porous (and therefore more water permeable) and eroded more rapidly.

The above examples illustrate an important technological principle in the design of bioeroding devices; the bioerosion rate of a given polymer is not an unchangeable property, but depends to a very large degree on readily controllable factors such as the presence of plasticizers or additives, the manufacturing process, the initial molecular weight of the polymer, and the geometry of the device.

Storage Stability, Sterilization, and Packaging

It is important to minimize premature polymer degradation during fabrication and storage. Traces of moisture can seriously degrade even relatively stable polymers, such as poly(bisphenol A carbonate), during injection molding or extrusion. Degradable polymers are particularly sensitive to hydrolytic degradation during high temperature processing. The industrial production of degradable implants therefore often requires the construction of facilities with controlled atmospheres where the moisture content of the polymer and the ambient humidity can be strictly controlled.

After fabrication, γ-irradiation or exposure to ethylene oxide may be used for the sterilization of degradable implants. Both methods have disadvantages, and as a general rule the choice is between the lesser of two evils. γ-Irradiation at a dose of 2 to 3 Mrad can result in significant backbone degradation. Since the aliphatic polyesters PLA, PGA, and PDS are particularly sensitive to radiation damage, these materials are usually sterilized by exposure to ethylene oxide, and not by γ-irradiation. Unfortunately, the use of the highly dangerous ethylene oxide gas represents a serious safety hazard by potentially leaving residual traces in the polymeric device. Polymers sterilized with ethylene oxide must be degassed for extended periods of time.

Additionally, for applications in tissue engineering, biodegradable scaffolds may be pre-seeded with viable cells or may be impregnated with growth factors or other biologics. There is currently no method that could be used to sterilize scaffolds that contain viable cells without damaging the cells. Therefore, such products must be manufactured under sterile conditions, and must be used within a very short time after manufacture. Currently, a small number of products containing pre-seeded, living cells are in clinical use. These products are expensive, are shipped in special containers, and have a short shelf life.

Likewise, it has been shown that sterilization of scaffolds containing osteoinductive or chondroinductive agents leads to significant losses in bioactivity, depending on the sterilization method used (Athanasiou et al., 1998). The challenge of producing tissue engineering scaffolds that are pre-seeded with viable cells or that contain sensitive biological agents has not yet been fully solved.

After sterilization, degradable implants are usually packaged in air-tight aluminum-backed plastic-foil pouches. In some cases, refrigeration may also be required to prevent backbone degradation during storage.

Evaluating Toxicity of Bioerodible Materials

A polymeric medical device is not completely pure. Device toxicity could originate from many sources, such as those listed below:

Polymer impurities: Initiators, residual monomers and oligomers, byproducts.

Solvents: Contamination from solvents used in synthesis, to clean equipment or in fabrication (e.g., solvent casting processes to produce films, microspheres or fibers).

Additives: Stabilizers, emulsifiers, antioxidants, plasticizers.

Processing byproducts: Byproducts of high-temperature processing and sterilization.

Degradation products: Products of hydrolytic or enzymatic degradation and their subsequent metabolites.

Testing for potential toxicity is governed by ISO 10993 and the FDA G95-1 Memorandum. Cytotoxicity, sensitization, irritation, carcinogenicity, genotoxicity, hemocompatibility, and systemic toxicity are evaluated using cell culture methods and animal testing:

Mammalian cell culture: Indirect (agar overlay) or direct sample contact or exposure to device extract.

Animal testing: Mouse, rat, guinea pig, and dog models are typically used with evaluation by histopathology, gross observation of redness or swelling, lymph node extraction, micronucleous assay, and blood thrombogenicity, lysis, and complement activation.

GLOSSARY OF KEY TERMS

Degradation: A chemical process by which covalent bonds are cleaved.

Erosion: A process that results in physical changes to the size, shape, or mass of a material.

Biodegradation: Degradation as a consequence of a biological agent, such as an enzyme, cell, or microorganism.

Bioerosion: Conversion of a water-insoluble material into a water-soluble material(s) under physiological conditions, regardless of the mechanism of conversion.

Surface erosion: A process of polymer erosion where the rate of conversion to water-soluble material(s) exceeds the rate of water infiltration into the interior of the device. As a consequence, material erodes from the surface of the device, while the core remains structurally consistent.

Bulk erosion: A process of polymer erosion where the rate of water penetration into the interior of the device exceeds the rate of conversion into water-soluble material(s). As a consequence, material erosion is equally likely to happen throughout the bulk of the device.

BIBLIOGRAPHY

Allcock, H. R. (1990). Polyphosphazenes as New Biomedical and Bioactive Materials. In M. Chasin, & R. Langer (Eds.), *Biodegradable Polymers as Drug Delivery Systems* (pp. 163–193). New York, NY: Marcel Dekker.

Allcock, H. R. (1999). Inorganic–organic polymers as route to biodegradable materials. *Macromol. Symp.*, **144**, 33–46.

Anderson, D. G., Burdick, J. A., & Langer, R. (2004). Smart biomaterials. *Science*, **305**(5692), 1923–1924.

Anderson, J. M., Spilizewski, K. L., & Hiltner, A. (1985). Poly(α-amino acids as biomedical polymers. In D. F. Williams (Ed.), *Biocompatability of Tissue Analogs* (pp. 67–88). Boca Raton, FL: CRC Press.

Andrianov, A. K., Sargent, J. R., Sule, S. S., Le Golvan, M. P., Woods, A. L., et al. (1998). Synthesis, physicochemical properties and immunoadjuvant activity of water-soluble phosphazene polyacids. *J. Bioact. Compat. Polym.*, **13**(4), 243–256.

Asikainen, A. J., Noponen, J., Lindqvist, C., Pelto, M., Kellomaki, M., et al. (2006). Tyrosine-derived polycarbonate membrane in treating mandibular bone defects An experimental study. *J. R. Soc. Interface*, **3**(10), 629–635.

Athanasiou, K. A., Agrawal, C. M., Barber, F. A., & Burkhart, S. S. (1998). Orthopaedic applications for PLA–PGA biodegradable polymers. *Arthroscopy*, **14**(7), 726–737.

Barham, P. J., Keller, A., Otun, E. L., & Holmes, P. A. (1984). Crystallization and morphology of a bacterial thermoplastic: poly-3-hydroxybutyrate. *J. Mater. Sci.*, **19**(9), 2781–2794.

Bergsma, J. E., De Bruijn, W. C., Rozema, F. R., Bos, R. R., & Boering, G. (1995). Late degradation tissue response to poly(L-lactide) bone plates and screws. *Biomaterials*, **16**(1), 25–31.

Bezwada, R. S., Jamiolkowski, D. D., Lee, I. Y., Agarwal, V., Persivale, J., et al. (1995). Monocryl suture, a new ultra-pliable absorbable monofilament suture. *Biomaterials*, **16**(15), 1141–1148.

Bourke, S. L., & Kohn, J. (2003). Polymers derived from the amino acid L-tyrosine: Polycarbonates, polyarylates and copoly-mers with poly(ethylene glycol). *Adv. Drug. Deliv. Rev.*, **55**(4), 447–466.

Brem, H., Tamargo, R. J., & Olivi, A. (1993). Delivery of drugs to the brain by use of a sustained-release polyanhydride polymer system. In H. Salem, & S. I. Baskin (Eds.), *New Technologies & Concepts for Reducing Drug Toxicities* (pp. 33–39). Boca Raton, FL: CRC Press.

Bruin, P., Smedinga, J., Pennings, A. J., & Jonkman, M. F. (1990). Biodegradable lysine diisocyanate-based poly(glycolide-co-epsilon-caprolactone)-urethane network in artificial skin. *Biomaterials*, **11**(4), 291–295.

Campbell, P., Glover, G. I., & Gunn, J. M. (1980). Inhibition of intracellular protein degradation by pepstatinyl, poly(L-lysine), and pepstatinyl-poly(L-lysine). *Arch. Biochem. Biophys.*, **203**(2), 676–680.

Choueka, J., Charvet, J. L., Koval, K. J., Alexander, H., James, K. S., et al. (1996). Canine bone response to tyrosine-derived polycarbonates and poly(L-lactic acid). *J. Biomed. Mater. Res.*, **31**(1), 35–41.

Deshpande, A. A., Heller, J., & Gurny, R. (1998). Bioerodible polymers for ocular drug delivery. *Crit. Rev. Ther. Drug. Carrier Syst.*, **15**(4), 381–420.

Fattal, E., & Couvreur, P. (2004). Poly(alkylcyanoacrylate) Nanoparticles for Nucleic Acid Delivery. In M. M. Amiji (Ed.), *Polymeric Gene Delivery* (pp. 387–396). Boca Raton, FL: CRC Press.

Fisher, J. P., Vehof, J. W., Dean, D., Van Der Waerden, J. P., Holland, T. A., et al. (2002). Soft and hard tissue response to photocrosslinked poly(propylene fumarate) scaffolds in a rabbit model. *J. Biomed. Mater. Res.*, **59**(3), 547–556.

Frazza, E. J., & Schmitt, E. E. (1971). A new absorbable suture. *J. Biomed. Mater. Res.*, **5**(2), 43–58.

Gilding, D. K., & Reed, A. M. (1979). Biodegradable polymers for use in surgery. Poly(ethylene oxide)-poly(ethylene terephthalate) (PEO/PET) copolymers: 1. *Polymer*, **20**(12), 1454–1458.

Gombotz, W. R., & Pettit, D. K. (1995). Biodegradable polymers for protein and peptide drug delivery. *Bioconjugate Chemistry*, **6**(4), 332–351.

Göpferich, A. (1996). Mechanisms of polymer degradation and erosion. *Biomaterials*, **17**(2), 103–114.

Göpferich, A., & Tessmar, J. (2002). Polyanhydride degradation and erosion. *Adv. Drug Deliv. Rev.*, **54**(7), 911–931.

Gunatillake, P. A., & Adhikari, R. (2003). Biodegradable synthetic polymers for tissue engineering. *Eur. Cell Mater*, **5**, 1–16. discussion 16.

Heller, J. (1987). Controlled Drug Delivery, Fundamentals and Applications. In J. R. Robinson, & V. H. L. Lee (Eds.), (2nd edn. pp. 180–210). New York, NY: Marcel Dekker.

Heller, J. (1990). Development of poly(ortho esters): A historical overview. *Biomaterials*, **11**(9), 659–665.

Heller, J. (2005). Ocular delivery using poly(ortho esters). *Adv. Drug Deliv. Rev.*, **57**(14), 2053–2062.

Heller, J., & Gurny, R. (1999). Poly(ortho esters). In E. Mathiowitz (Ed.), *Encylopedia of Controlled Drug Delivery* (pp. 852–874). New York, NY: John Wiley.

Heller, J., Barr, J., Ng, S. Y., Abdellauoi, K. S., & Gurny, R. (2002). Poly(ortho esters): Synthesis, characterization, properties and uses. *Adv. Drug Deliv. Rev.*, **54**(7), 1015–1039.

Herrmann, J. B., Kelly, R. J., & Higgins, G. A. (1970). Polyglycolic acid sutures. Laboratory and clinical evaluation of a new absorbable suture material. *Arch. Surg.*, **100**(4), 486–490.

Huang, S. (1985). Biodegradable Polymers. In F. H. Mark, N. M. Bikales, C. G. Overberger, G. Menges, & J. I. Kroshwits (Eds.), *Encyclopedia of Polymer Science & Engineering* (pp. 220–243). New York, NY: John Wiley.

Hutmacher, D. W. (2001). Scaffold design and fabrication technologies for engineering tissues: State of the art and future perspectives. *J. Biomater Sci. Polym. Ed.*, **12**(1), 107–124.

Ignatius, A. A., & Claes, L. E. (1996). *In vitro* biocompatibility of bioresorbable polymers: Poly(L, DL-lactide) and poly(L-lactide-co-glycolide). *Biomaterials*, **17**(8), 831–839.

James, K., Levene, H., Parsons, J. R., & Kohn, J. (1999). Small changes in polymer chemistry have a large effect on the bone-implant interface: Evaluation of a series of degradable tyrosine-derived polycarbonates in bone defects. *Biomaterials*, **20**(23–24), 2203–2212.

Katti, D. S., Lakshmi, S., Langer, R., & Laurencin, C. T. (2002). Toxicity, biodegradation and elimination of polyanhydrides. *Adv. Drug Deliv. Rev.*, **54**(7), 933–961.

Knowles, J. C. (1993). Development of a natural degradable polymer for orthopaedic use. *J. Med. Eng. Technol.*, **17**(4), 129–137.

Kohn, J. (1993). Design, synthesis and possible applications of pseudo-poly(amino acids). *Trends in Polymer Science, (Cambridge, United Kingdom)*, **1**(7), 206–212.

Kohn, J., & Langer, R. (1984). A new approach to the development of bioerodible polymers for controlled release applications employing naturally occurring amino acids. *Polymeric Materials Science & Engineering*, **51**, 119–121.

Kohn, J., & Langer, R. (1987). Polymerization reactions involving the side chains of the alpha-L-amino acids. *Journal of the American Chemical Society*, **109**(3), 817–820.

Kohn, J., Welsh, W. J., & Knight, D. (2007). A new approach to the rationale discovery of polymeric biomaterials. *Biomaterials*, **28**(29), 4171–4177.

Kumar, N., Langer, R. S., & Domb, A. J. (2002). Polyanhydrides: An overview. *Advanced Drug Delivery Reviews*, **54**(7), 889–910.

Langer, R., & Vacanti, J. P. (1993). Tissue engineering. *Science*, **260**(5110), 920–926.

Leong, K. W., D'amore, P. D., Marletta, M., & Langer, R. (1986). Bioerodible polyanhydrides as drug-carrier matrices. II. Biocompatibility and chemical reactivity. *J. Biomed. Mater. Res.*, **20**(1), 51–64.

Lescure, F., Gurny, R., Doelker, E., Pelaprat, M. L., Bichon, D., et al. (1989). Acute histopathological response to a new biodegradable polypeptidic polymer for implantable drug delivery system. *J. Biomed. Mater. Res.*, **23**(11), 1299–1313.

Lewandrowski, K.-U., Gresser, J. D., Wise, D. L., White, R. L., & Trantolo, D. J. (2000). Osteoconductivity of an injectable and bioresorbable poly(propylene glycol-co-fumaric acid) bone cement. *Biomaterials*, **21**(3), 293–298.

Li, L. C., Deng, J., & Stephens, D. (2002). Polyanhydride implant for antibiotic delivery: From the bench to the clinic. *Adv. Drug. Deliv. Rev.*, **54**(7), 963–986.

Li, S. (1999). Hydrolytic degradation characteristics of aliphatic polyesters derived from lactic and glycolic acids. *J. Biomed. Mater. Res.*, **48**(3), 342–353.

Martin, D. P., & Williams, S. F. (2003). Medical applications of poly-4-hydroxybutyrate: a strong flexible absorbable biomaterial. *Biochemical Engineering Journal*, **16**(2), 97–105.

Mathiowitz, E., Kline, D., & Langer, R. (1990). Morphology of polyanhydride microsphere delivery systems. *Scanning Microsc.*, **4**(2), 329–340.

Matsumura, Y. (2008). Poly (amino acid) micelle nanocarriers in preclinical and clinical studies. *Adv. Drug. Deliv. Rev.*, **60**(8), 899–914.

Maurus, P. B., & Kaeding, C. C. (2004). Bioabsorbable implant material review. *Operative Techniques in Sports Medicine*, **12**(3), 158–160.

Mcmillin, C. R. (1991). An assessment of elastomers for biomedical applications. In M. Szycher (Ed.), *High Performance Biomaterials: A Comprehensive Guide to Medical & Pharmaceutical Applications*. Lancaster, PA: Technomic Publishing Company, Inc.

Merkli, A., Heller, J., Tabatabay, C., & Gurny, R. (1993). Synthesis and characterization of a new biodegradable semi-solid poly(ortho ester) for drug delivery systems. *J. Biomater. Sci. Polym. Ed.*, **4**(5), 505–516.

Miller, N. D., & Williams, D. F. (1987). On the biodegradation of poly-beta-hydroxyburate (PHB) homopolymer and poly-beta-hydroxyburate-hydroxyvalerate copolymers. *Biomaterials*, **8**(2), 129–137.

Nair, L. S., & Laurencin, C. T. (2007). Biodegradable polymers as biomaterials. *Progress in Polymer Science*, **32**(8–9), 762–798.

Nair, L. S., Lee, D. A., Bender, J. D., Barrett, E. W., Greish, Y. E., et al. (2006). Synthesis, characterization, and osteocompatibility evaluation of novel alanine-based polyphosphazenes. *J. Biomed. Mater. Res. A*, **76**(1), 206–213.

Nelson, T., Kaufman, E., Kline, J., & Sokoloff, L. (1981). The extraneural distribution of gamma-hydroxybutyrate. *J. Neurochem.*, **37**(5), 1345–1348.

Ng, S. Y., Vandamme, T., Taylor, M. S., & Heller, J. (1997). Controlled drug release from self-catalyzed poly(ortho esters). *Ann NY Acad. Sci.*, **831**, 168–178.

Okada, H., & Toguchi, H. (1995). Biodegradable microspheres in drug delivery. *Crit. Rev. Ther. Drug Carrier Syst.*, **12**(1), 1–99.

Pitt, C. G., Gratzl, M. M., Jeffcoat, A. R., Zweidinger, R., & Schindler, A. (1979). Sustained drug delivery systems II: Factors affecting release rates from poly(epsilon-caprolactone) and related biodegradable polyesters. *J. Pharm. Sci.*, **68**(12), 1534–1538.

Pulapura, S., & Kohn, J. (1992). Tyrosine-derived polycarbonates: Backbone-modified "pseudo"-poly (amino acids) designed for biomedical applications. *Biopolymers*, **32**(4), 411–417.

Qiu, L. Y., & Zhu, K. J. (2000). Novel biodegradable polyphosphazenes containing glycine ethyl ester and benzyl ester of amino acethydroxamic acid as cosustituents. *Journal of Applied Polymer Science*, **77**(13), 2987–2995.

Ramcharitar, S., & Serruys, P. W. (2008). Fully biodegradable coronary stents: Progress to date. *Am. J. Cardiovasc Drugs*, **8**(5), 305–314.

Ray, J. A., Doddi, N., Regula, D., Williams, J. A., & Melveger, A. (1981). Polydioxanone (PDS), a novel monofilament synthetic absorbable suture. *Surg. Gynecol. Obstet.*, **153**(4), 497–507.

Rosen, H., Kohn, J., Leong, K., & Langer, R. (1988). Bioerodible polymers for controlled release systems. In D. Hsieh (Ed.), *Controlled Release Systems: Fabrication Technology* (pp. 83–110). Boca Raton, FL: CRC Press.

Saad, B., Hirt, T. D., Welti, M., Uhlschmid, G. K., Neuenschwander, P., et al. (1997). Development of degradable polyesterurethanes for medical applications: *In vitro* and *in vivo* evaluations. *J. Biomed. Mater. Res.*, **36**(1), 65–74.

Senior, P. J., Beech, G. A., Ritchie, G. A., & Dawes, E. A. (1972). The role of oxygen limitation in the formation of poly-hydroxybutyrate during batch and continuous culture of *Azotobacter beijerinckii*. *Biochem. J.*, **128**(5), 1193–1201.

Shung, A. K., Timmer, M. D., Jo, S., Engel, P. S., & Mikos, A. G. (2002). Kinetics of poly(propylene fumarate) synthesis by step polymerization of diethyl fumarate and propylene glycol using zinc chloride as a catalyst. *J. Biomater. Sci. Polym. Ed.*, **13**(1), 95–108.

Sinha, V. R., Bansal, K., Kaushik, R., Kumria, R., & Trehan, A. (2004). Poly-epsilon-caprolactone microspheres and nanospheres: An overview. *Int. J. Pharm.*, **278**(1), 1–23.

Spira, M., Fissette, J., Hall, W., Hardy, S. B., & Gerow, F. J. (1969). Evaluation of synthetic fabrics as artificial skin grafts to experimental burn wounds. *J. Biomed. Mater. Res.*, **3**(2), 213–234.

Suzuki, M., & Ikada, Y. (2005). Biodegradable polymers in medicine. In R. L. Reis, & J. S. Roman (Eds.), *Biodegradable Systems in Tissue Engineering & Regenerative Medicine* (pp. 2–13). Boca Raton, FL: CRC Press.

Tamada, J. A., & Langer, R. (1993). Erosion kinetics of hydrolytically degradable polymers. *Proc. Natl. Acad. Sci., USA*, **90**(2), 552–556.

Timmer, M. D., Ambrose, C. G., & Mikos, A. G. (2003). *In vitro* degradation of polymeric networks of poly(propylene fumarate) and the crosslinking macromer poly(propylene fumarate)-diacrylate. *Biomaterials*, **24**(4), 571–577.

Törmälä, P., Pohjonen, T., & Rokkanen, P. (1998). Bioabsorbable polymers: Materials technology and surgical applications. *Proc. Inst. Mech. Eng. [H]*, **212**(2), 101–111.

Tsuji, H., & Miyauchi, S. (2001). Poly(-lacitde): 7. Enzymatic hydrolysis of free and resticted amorphous regions in poly(L-lactide) films with different crystallinities and a fixed crystaline thickness. *Polymer*, **42**(9), 4463–4467.

Van Heeswijk, W. A. R., Hoes, C. J. T., Stoffer, T., Eenink, M. J. D., Potman, W., et al. (1985). The synthesis and characterization of polypeptide-adriamycin conjugates and its complexes with adriamycin. Part I. *J. Controlled Release*, **1**(4), 301–315.

Van Natta, F. J., Hill, J. W., & Carruthers, W. H. (1934). Polymerization and ring formation , ε-caprolactone and its polymers. *J. Am. Chem.*, **56**, 455–459.

Vauthier, C., Dubernet, C., Chauvierre, C., Brigger, I., & Couvreur, P. (2003). Drug delivery to resistant tumors: The potential of poly(alkyl cyanoacrylate) nanoparticles. *J. Control Release*, **93**(2), 151–160.

Vezin, W. R., & Florence, A. T. (1980). In vitro heterogeneous degradation of poly(n-alkyl alpha-cyanoacrylates). *J. Biomed. Mater. Res.*, **14**(2), 93–106.

Von Burkersroda, F., Schedl, L., & Gopferich, A. (2002). Why degradable polymers undergo surface erosion or bulk erosion. *Biomaterials*, **23**(21), 4221–4231.

Westphal, M., Hilt, D. C., Bortey, E., Delavault, P., Olivares, R., et al. (2003). A phase 3 trial of local chemotherapy with biodegradable carmustine (BCNU) wafers (Gliadel wafers) in patients with primary malignant glioma. *Neuro Oncol.*, **5**(2), 79–88.

Williams, D. F. (1981). Enzymatic Hydrolysis of Poly(Lactic Acid). *Eng. Med.*, **10**(1), 5–7.

Williams, D. F. (1982). Biodegradation of surgical polymers. *J. Mater. Sci.*, **17**(5), 1233–1246.

Williams, D. F. (1987). *Definitions in Biomaterials: Proceedings of a Consensus Conference of the European Society for Biomaterials, Vol. 4 in the series: Progress in Biomaterials*. New York, NY: Elsevier.

Yaszemski, M. J., Payne, R. G., Hayes, W. C., Langer, R., & Mikos, A. G. (1995). The in vitro mechanical strength and in vivo bone ingrowth of a degrading polymeric composite biomaterial. *Mater. Res. Soc. Symp. Proc.*, **394**(Polymers in Medicine & Pharmacy), 21–24.

Zhang, J. Y., Beckman, E. J., Piesco, N. P., & Agarwal, S. (2000). A new peptide-based urethane polymer: synthesis, biodegradation, and potential to support cell growth *in vitro*. *Biomaterials*, **21**(12), 1247–1258.

Zhang, Z., Kuijer, R., Bulstra, S. K., Grijpma, D. W., & Feijen, J. (2006). The *in vivo* and *in vitro* degradation behavior of poly(trimethylene carbonate). *Biomaterials*, **27**(9), 1741–1748.

Zhu, K. J., Hendren, R. W., Jensen, K., & Pitt, C. G. (1991). Synthesis, properties, and biodegradation of poly(1,3-trimethylene carbonate). *Macromolecules*, **24**(8), 1736–1740.

CHAPTER I.2.7 ENGINEERED NATURAL MATERIALS

Glenn D. Prestwich[1] and Sarah Atzet[2]

[1]Department of Medicinal Chemistry and The Center for Therapeutic Biomaterials, The University of Utah, Salt Lake City, Utah, USA

[2]Glycosan BioSystems, Inc., Salt Lake City, Utah, USA

INTRODUCTION TO COMMONLY USED NATURAL MATERIALS

Naturally occurring materials offer several advantages, since the surrounding biological environment can recognize and metabolically process implanted substances through established pathways. However, natural materials have historically been plagued with several drawbacks, including immunogenicity, lot-to-lot variability, structural complexity, and inadequate biomechanical properties. Recent developments in the field have led to a reduction of these unfavorable properties, and a substantial number of successful biomaterials and biomaterial applications based on the unique properties of naturally occurring polymers and their chemically-modified derivatives have been developed (Shin et al., 2003; Khademhosseini et al., 2009).

A unique advantage of naturally derived materials is that through either enzymatic or hydrolytic degradation the host can successfully clear and metabolize the implanted substance. While this may not be desirable in permanent and long-term implants such as hip replacements, it is extremely advantageous when timed biological resorption is desired, i.e., in delivery vehicles for drugs or therapeutic cells. Natural materials can be chemically modified and cross-linked to match the degradation rate to a particular biological application (Augst et al., 2006; Badylak and Gilbert, 2008).

One frequent concern about natural materials is the immunogenic response that can occur following implantation. This response results from the fact that the implanted materials, although similar to the endogenous

TABLE I.2.7.1 General Properties and Sources of Select Natural Materials			
Alginate	Algae (Kelp)	Anionic polysaccharide, limited degradation unless modified, forms hydrogels	Wound dressing (Phytacare™) (see ref. FDA PhytaCare)
Chitosan	Crustacean exoskeletons	Positively charged, enzymatic degradation, can form hydrogels	Hemostats/wound dressings (HemCon™) (see ref. FDA HemCon)
Silk	Synthesized by arthropods (spider or silk worm)	Physiologic function is as a protective cocoon, strong and can be woven, slow or minimal degradation, reported biocompatibility issues	Sutures (Ethicon) (see ref. Ethicon Silk Sutures)
Elastin	Animal tissues	Structural ECM protein found in connective tissues, low solubility, reversible deformation, soluable precursor is tropoelastin	Surgical Mesh (Bard) (see ref. FDA Collamend)
Elastin-Like Peptides	Synthetically produced	Synthetic mimic of elastin based on highly repetative amino acid motifs, reversible thermal phase transition	N/A
Collagen	Animal tissues/ cell culture/ bacterial fermentation	Abundant ECM protein, triple helix structure, provide cell attachment sites, well documented	Dermal filler (CosmoDerm™) (see ref. FDA CosmoDerm)
Gelatin	Denatured collagen	Inexpensive form of collagen, used for cell attachment in culture	Gelfoam™ Sterile Sponge (see ref. FDA GelFoam)
Fibrin/ Fibrinogen	Animal tissues or plasma	Fibrin results from polymerization of fibrinogen with thrombin, crucial component for clot formation	Fibrin Sealent (Evicel™) (see ref. FDA Evicel)
Hyaluronic Acid	Animal tissues/ bacterial fermentation	Lubricating polymer only non-sulfated GAG, negatively charged, can form hydrogels	Dermal fillers (Restylane™) / Wound dressings (see ref. FDA Restylane)
Heparin	Animal tissues/plasma	Strongly negatively charged GAG, binds to numerous GFs, anti-coagulant activity, used for coating stents	Stent and catheter coatings (see ref. FDA TyCo)
Chondroitin Sulfate	Animal tissues (commonly shark cartilage)	Negatively charged GAG, major component of cartilage	Nutritional supplements/ Wound Dressings (gels) (Applied Nutritionals) (see ref. FDA Applied Nutritionals)
Decellularized Tissue	Animal tissue	Complex mixture of proteins/GAGs that retains tissue's ECM composition and structure, residual cellular components can cause immune reactions	Decellularized pulmonary artery patch (Allograft™) (see ref. FDA Allograft)

host extracellular matrix component, may not be identical, and thus contain antigenic determinants. This issue is more common among protein-based materials, but is typically significantly lower among polysaccharides including glycosaminoglycans. Additionally, this immunogenic effect can be reduced by either chemical modification or through sourcing of the natural material (Badylak and Gilbert, 2008). Chemical modification and cross-linking are common methods for altering mechanical properties such as elasticity and tensile strength of natural materials that are often not appropriate for dynamic physiologic environments (i.e., the beating myocardium). However, the complex structures of natural materials can complicate technical modifications that are simple to perform in synthetic polymers. Nonetheless, numerous research studies have demonstrated the successful modification and purification of natural-based materials. Another common issue arising with natural materials is the lot-to-lot variability in molecular structure that occurs due to animal sourcing. Variability arises not only due to species-to-species variation, but also from tissue-to-tissue variation. This can complicate processing and quantification of the materials. Recently bacterial recombinant techniques have been use to produce several natural materials including hyaluronic acid (HA) and collagen. This production technique reduces both the variability and immunogenicity issues with natural materials. The following section will discuss common types of natural based biomaterials, their origins, and uses. Table I.2.7.1 details several of these natural materials, their common sources, characteristics, and notable FDA approved products. Figure I.2.7.1 illustrates the structures of natural polymers commonly used in biomaterials and medical devises. The remainder of this chapter will focus on biomaterials engineered from one specific biopolymer, HA.

Alginate

Alginates constitute a family of linear anionic polysaccharides isolated from coastal brown algae species (commonly known as kelp) such as *Laminaria hyperborean* (Augst et al., 2006). Alginates are composed of (1,4)-linked β-D-mannuronic acid (M) and α-L-guluronic acid (G) residues in homopolymer blocks of either similar or strictly alternating sections (i.e., MMMMM, GGGGG, MGMGMG) (Rowley et al., 1999). The relative amounts of the blocks are dependent on the origin of the alginate. Due to axial links, G blocks are stiffer than M blocks and alternating sequences in turn increase the solubility of the alginate. The molecular weight of alginates

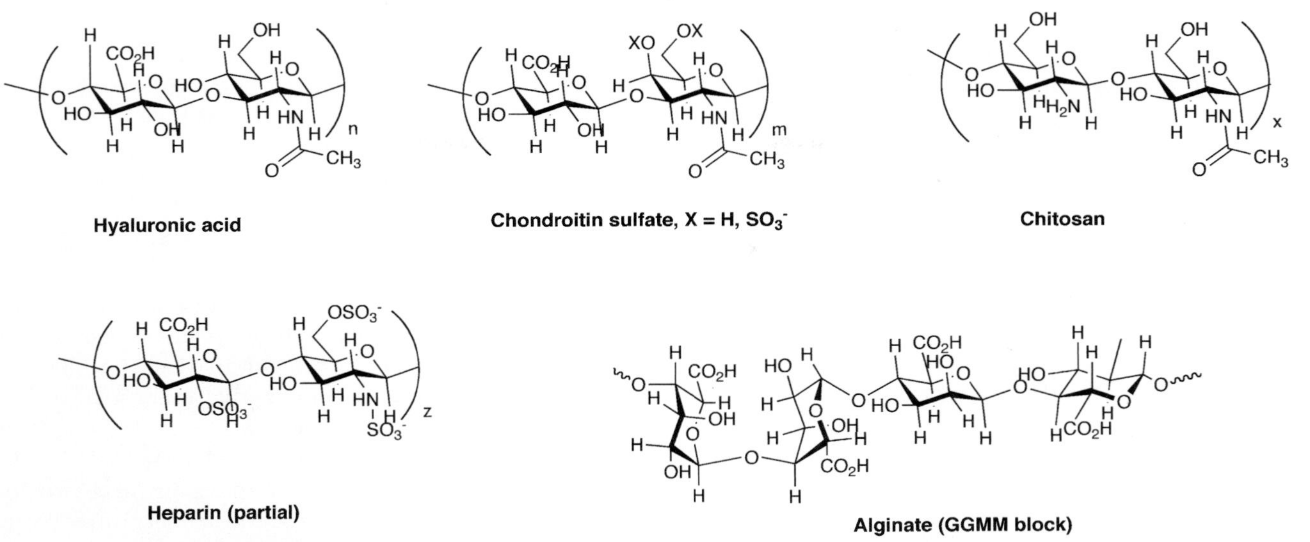

FIGURE I.2.7.1 Structures of natural polymers found in biomaterials and medical devices.

can vary between 50 kDa and 100,000 kDa. Hydrogels can be formed by various cross-linking methods. A common technique is to use a divalent cation (Ca^{2+} or Ba^{2+}) to form ionic bonds between the carboxylic groups of neighboring strands, thereby inducing chain–chain associations. Alginate can also be cross-linked by partial oxidation of the poly(guluronate), followed by reaction with adipic dihydrazide to form covalent cross-links. A third cross-linking strategy uses polyethylene glycol diamines to covalently link alginate chains. Gel stiffness and elasticity vary with composition and cross-linking method. The stiffness of divalent cation cross-linked alginate increases with the following trends in composition: MG < MM < GG, whereas elasticity increases following GG < MM < MG. Thus the M:G ratio, cross-linking method, and molecular weight can all be used to tune mechanical properties of alginate gels.

Because alginate is foreign to mammalian cells, alginate materials are not naturally broken down enzymatically *in vivo*, and as a result have poorly controlled degradation *in vivo*. When the divalent cations are exchanged with Na^+ ions, ionically cross-linked alginate typically undergoes slow uncontrolled dissolution at neutral pH (i.e., greater than three months) (Mi et al., 2002a; FDA PhytaCare 510(k), 2010; FDA HemCon 510(k), 2010). Degradation kinetics of covalently cross-linked alginate generally depends on the type and density of cross-linker (Mi et al., 2002a). However, there are strategies to adjust degradation kinetics by modifying polymer molecular weight, composition (M:G), and by partial oxidation.

Another unique property of alginate is that it is relatively biologically inert. Cells do not naturally adhere to alginate, which has fostered investigations in the attachment of extracellular matrix (ECM) proteins and peptide sequences to hydrogels. Adhesion proteins such as laminin, fibronectin, and collagen have all been coupled to alginate, as well as the integrin-binding, fibronectin-derived, cell adhesive peptide arginine–glycine–aspartic acid (RGD) and its derivatives (FDA PhytaCare 510(k), 2010; FDA Collamend 510(k), 2010). These modifications have led to significant changes in cellular attachment to alginate materials.

Specifically tailored alginate hydrogels have numerous proposed uses, including as bulking agents, drug delivery vehicles, and scaffolds for regenerative medicine. Alginate hydrogels have been clinically investigated as a carrier for transplanted chondrocytes to treat intrinsic sphincter disease, and as a delivery vehicle for mineralization promoting proteins. Alginate-based biomaterials have been approved by the FDA for topical use in wound healing. The PhytaCare™ Alginate Hydrogel and the Silverlon™ Calcium Alginate Dressing products both have 510(k) clearance as barrier wound dressings that provide a microbial barrier for partial to full thickness wounds including ulcers, graft sites, surgical wounds, and burns.

Chitosan

Chitosan is a polycationic copolymer of *N*-acetyl-glucosamine and *N*-glucosamine, and is generally obtained by alkaline deacetylation of chitin, a linear homopolymer of *N*-acetyl-glucosamine. Chitin is structurally related to glycosaminoglycans, but lacks the uronic acid groups. It is the primary component in fungi cell walls, and the exoskeleton of crustaceans such as shrimp, crab, and lobster (Khor and Lim, 2003). When the degree of deacetylation (DD) (conversion to *N*-glucosamine) is greater than 50%, the biopolymer is termed chitosan. The biopolymer is soluble in dilute acidic solutions, such as acetic acid or hydrochloric acid, and as a result the amino groups become protonated (Jiang et al., 2008). Chitosan hydrogels can be formed by entangled networks, ionic cross-links/polyelectrolyte complexes or by covalent cross-linking (Mi et al., 2002b; Berger et al., 2004). Once dissolved, entangled networks of chitosan can gel by increasing the pH or extruding the solution into a nonsolvent. Entangled networks have limited use due to a lack of mechanical strength and a tendency to dissolve. Due to the cationic nature of chitosan, combination with a negatively charged polymer, such as chondroitin sulfate or hyaluronan, will result in an ionic bound network. Covalently cross-linked hydrogels can be prepared by reaction with various di-functional molecules such as glutaraldehyde, poly(ethylene glycol) dialdehyde diethyl acetal, or genipin (Dal Pozzo et al., 2000). Chitosan has been shown to be metabolized by human enzymes such as lysozyme, and thus is considered biodegradable. Degradation rates depend on the type of hydrogel network, and specifically for covalent networks, the type and density of cross-linker (Freier et al., 2005a). As with alginate, much research has explored the conjugation of biomolecules to chitosan, either by ionic, adsorption, or covalent modification of ECM proteins and peptide sequences to modify cellular response (Suh and Matthew, 2000; Ho et al., 2005). Chitosan provides multiple hydroxyl and amino functional sites for chemical modification opportunities (Berger et al., 2005). Chitosan-based hydrogels have been researched and developed for applications in drug and growth factor delivery, tissue regenerative scaffolds, and wound dressings (Rao and Sharma, 1997; Mi et al., 2002c; VandeVord et al., 2002; Mi et al., 2003; Freier et al., 2005b). Currently, chitosan based biomaterials have been approved by the FDA for topical use in wound healing and as hemostats. Scion Cardio-Vascular, Inc. developed and markets a soft absorbent pad called the Clo-Sur P.A.D™ as a noninvasive topical approach to hemostasis based on chitosan and its absorbent properties. This product has been cleared as a Class III medical device by the FDA. Other devices with FDA 510(k) clearance are Aquanova's super absorbent dressing for wounds such as ulcers, and Celox's granulated chitosan for hemostasis.

Collagen

Collagen is the most prevalent protein in the connective tissue of animals, and constitutes approximately 25% of total body protein in vertebrates. The molecular subunit of collagen, called tropocollagen, is a rigid rod with a molecular weight ~300 kDa (Ratner and Hoffman, 2004). Tropocollagen self assembles into larger structural units called fibrils or fibers. While there are several types of collagen, these larger units, or fibers, all share the characteristic triple helical structure, but have variations in the length of the non-helical and helical sections, and variations in the number of carbohydrates attached to the helical fraction (Stenzel et al., 1974). There are four commonly used types of collagen, and here we will follow the nomenclature proposed by Kauzmann (Kauzmann, 1959). Type I collagen is the type commonly found in skin, bone, and tendons, while type II is common in cartilage. Type III collagen is prevalent in the vasculature structures (i.e., blood vessel walls), and can also be found in small amounts in skin. Type IV collagen is unique in that it is largely non-helical and does not form fibrils. Collagen Type IV is found in the basement membrane that separates epithelial and mesodermal tissues (Kauzmann, 1959; Kuntz and Kauzmann, 1974). Gelatin is an irreversibly denatured, hydrolyzed form of collagen that is commonly used in pharmaceuticals, cosmetics, and foodstuffs. Most medical-grade collagen and gelatin are obtained from closely regulated bovine and porcine tissue. However, human collagen can be obtained by recombinant methodology, isolation from tissues such as lung and placenta, or from human cell culture (Lee et al., 2001).

Collagen is used as a biomaterial because of its widespread abundance in nature and ease of modification (Stenzel et al., 1974; Stone et al., 1997; Friess, 1998). Collagen is composed of some 20 amino acids, each providing pendant side groups available for chemical modification (White et al., 1973; Lin et al., 2003). The prevalence of hydroxyl, amine, and carboxylic acid groups allows simple cross-linking between and within collagen units. Cross-linking collagen can decelerate the degradation of implanted materials, and is thought to reduce immunogenicity in xenografted collagen materials (Nimni et al., 1987). Self-cross-linking of collagen via interchain peptide bonds can be induced by dehydration when exposed to high temperatures (i.e., >105°C). Dialdehydes are also commonly used to cross-link collagen, and have long been used in the leather industry as tanning agents. Exposure to glutaraldehyde produces collagen with firm covalent cross-links and elastic behavior. Numerous commercial products have been developed from collagen and collagen-based technologies; a few examples are dermal fillers CosmoDerm™ and Zyderm™ (Allergan), and the sponge-like Gelfoam™ (Pfizer). CosmoDerm™ is a human collagen-based implant generated and purified from human fibroblast cell culture. Type I collagen is the major component of this material while Type III provides minor contributions, as determined by Western blot analysis. Alternatively, Zyderm™ is based on bovine-derived Type I collagen, although both products are chemically cross-linked by treatment with glutaraldehyde. Gelfoam™ is an absorbable, sterile compressed sponge produced from denatured collagen I from porcine skin. It is water-soluble, non-elastic and porous, and is used for compression, and in combination with human thrombin as a hemostat.

Fibrin

Fibrin is a naturally occurring biodegradable matrix that can be produced via polymerization of fibrinogen and thrombin to form a mesh network (Ahmed et al., 2008). Fibrinogen is a glycoprotein synthesized by the liver and soluble in the bloodstream (Linnes et al., 2007). Thrombin, also produced in the liver and found in the bloodstream, initiates end-to-end polymerization of the soluble fibrinogen to form long strands which precipitate to form a fibrin mesh. This polymerization occurs naturally at a wound site and contributes, in conjunction with platelets, to form a clot. *In vivo*, fibrin gels are degraded by the proteolytic enzyme plasmin. Fibrin gels often do not possess the required mechanical strength for some physically dynamic environments. To overcome this, researchers have combined various natural or synthetic polymers, such as collagen or polyglactin, to provide stronger gels (Eyrich et al., 2007; Dare et al., 2009). Commercially available fibrin sealants are used to control bleeding and speed wound healing. Fibrinogen can be isolated from animal plasma, including human plasma. Thus, autologous scaffolds for tissue engineering applications can be generated from a patient's own blood.

Elastin and Elastin-Like-Peptides

Elastin is an extracellular matrix protein that provides elasticity to tissues and organs. It is abundant in blood vessels, ligament, lung tissue, and skin; all organs that commonly stretch and relax (Geutjes et al., 2006; Daamen et al., 2007). Tropoelastin, the soluble protein precursor to elastin, can align and form intermolecular cross-links resulting in extremely stable insoluble elastin. As a biomaterial, elastin can have various forms. Insoluble elastin can be used in autografts or allografts, it can be hydrolyzed to be used in soluble elastin preparations, or tropoelastin can provide components of biomaterials. Elastin-like peptides (ELPs) have been synthetically produced (Urry et al., 2002; McHale et al., 2005). These polypeptides are composed of the highly repetitive amino acid sequences found in elastin. Physical properties of the ELPs can be manipulated based on amino acid sequence and resulting hydrophobicity. The polypeptides can coacervate and then be cross-linked by gamma irradiation. Researchers have produced non-woven fabrics and fibers by electrospinning ELPs (Huang et al., 2000).

Decellularized Tissues

Tissues are composed of both cellular material and extracellular matrix proteins in various densities. Decellularized tissues are natural materials, usually from an animal or human source, that have been purified in such a way as to remove all cellular components and debris but maintain the three-dimensional structural and functional proteins of the ECM (Gilbert et al., 2006). The tissue site, species of origin, decellularization process methods, and end point sterilization techniques can vary greatly for these materials. It is necessary to remove the cellular components and antigens as they are recognized by host tissue as foreign and irrevocably induce an inflammatory response or immune-mediated rejection of the decellularized tissue (Badylak et al., 1995). On the other hand, the structure of ECM components is generally well-conserved between species, and typically well-tolerated by host recipients. Ideally, a decellularization protocol, which often includes multiple stages of physical, chemical, and enzymatic treatments, should gently remove all cellular and nuclear components while minimizing effects on the ECM. Reported batch-to-batch variations are large due to tissue source, composition, and density, even when decellularization protocols are predominately similar (Daamen et al., 2007). A number of ECM-based devices have received regulatory approval, including porcine heart valves (Synergraft™, CryoLife™), porcine small intestinal submucosa (SurgiSIS™, Cook Biotech), and a pulmonary artery patch (Allograft™).

Glycosaminoglycans

Glycosaminoglycans (GAGs) are a family of linear polysaccharides that can generally be described as an alternating copolymer, and are found in all animal tissues. The repeat units of GAGs consist of a hexamine unit (glucosamine or galactosamine) and a sugar (galactose, glucuronic acid, or iduronic acid). GAGs are found on cell surfaces in the extracellular matrix and are well-preserved among species (Hildebrandt, 2002; Prestwich 2008a; Prestwich and Kuo, 2008). When GAGs are bound to a protein core the resulting macromolecule is termed a proteoglycan and is often described as having a "bottlebrush-like" configuration. The molecular weights of GAGs generally range from 5 to 5000 kDa. There are several specific GAGs used as biomaterials including heparin, heparan sulfate, keratan sulfate, dermatan sulfate, chondroitin sulfate, and HA. GAGs are naturally degraded by enzymes such as hyaluronidase, chondroitinase, and heparanase.

Heparin is a highly sulfated GAG and is widely used as an anticoagulant (Breckwoldt et al., 1991; van der Giessen et al., 1998; Hildebrandt, 2002). Due to the sulfation levels it is very strongly negatively charged. Pharmaceutical heparin is commonly derived from bovine or porcine mucosal tissues. Heparin is chemically related to heparan sulfate and is composed of variably sulfated disaccharides. The predominant saccharide units are D-glucosamine and uronic acid (either D-glucuronic acid or L-iduronic acid) with sulfation of some hydroxyl groups. Due to the strong negative charge, heparin has been found to bind to growth factors, such as vascular endothelial growth factor (VEGF) and basic fibroblast growth factor (bFGF), and stabilize them *in vivo*, thus regulating their release (Peattie et al., 2008). Growth factor release is dependent on the strength of association with heparin, and can vary depending on the growth factor and environment from days to weeks. Heparin was one of the first biological molecules to be covalently immobilized to biomaterials (Hoffman et al., 1972; Schmer, 1972; Basmadjian and Sefton, 1983), and has since been used extensively for coating in catheters, stents, and other medical devices (Breckwoldt et al., 1991; van der Giessen et al., 1998; Cauda et al., 2008; Jiang et al., 2009; FDA TyCo, 2010).

Chondroitin sulfate (CS) is also a negatively charged GAG, but it is composed of repeating glucuronic acid and

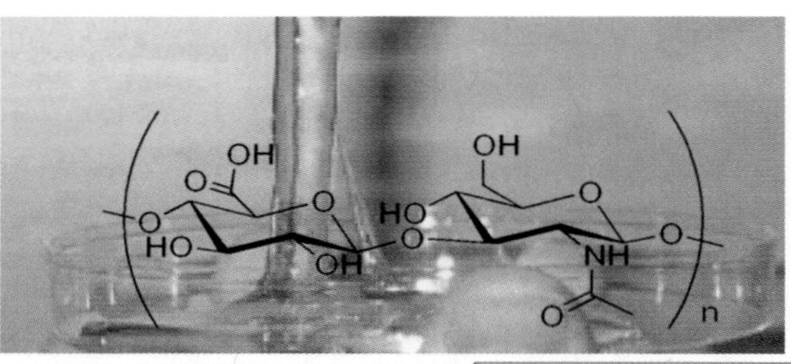

Biology:	**Commercial Products:**
– Tissue hydration and permeability	– Eye surgery and hydration
– Lubrication of joints	– Reduce OA pain
– Enhance wound repair	– Dermal fillers and anti-ageing
	– Prevent adhesions

FIGURE I.2.7.2 The non-sulfated glycosaminoglycan hyaluronan (HA) forms viscoelastic solutions that are important in human biology.

N-acetylgalactosamine units. Most disaccharide units contain a single sulfate at C-4 or C-6 hydroxyl groups, as well as a carboxyl group that can be chemically modified (Yuan et al., 2008; Baldwin and Kiick, 2010). CS is a major component of cartilage tissue, and being capable of readily absorbing water is credited for the compressive strength of that tissue. It is generally found attached to proteoglycans as a structural protein or on cell surfaces or basement membranes where it functions as a receptor. Commercial chondroitin sulfate is often sourced and purified from shark or whale cartilage, or from bovine or porcine tissues (Baldwin and Kiick, 2010).

Hyaluronic acid (HA) is the only non-sulfated glycosaminoglycan, and consists of repeating disaccharide units of N-acetylglucosamine and glucuronic acid (Figure I.2.7.2). In addition to being unsulfated, HA is unique among GAGs in that it exists in much higher molecular weights (50–5000 kDa), is secreted by cells, and (with a few exceptions) does not occur as a proteoglycan, that is, bound to a protein core (Prestwich and Kuo, 2008). HA has been extensively studied as a biomaterial, in part due to these and other unique properties (Shu et al., 2002, 2004, 2006). The remainder of this chapter will now focus on HA as a unique glycosaminoglycan, and discuss its chemistry, biology, and its future as the basis of clinical products for reparative and regenerative medicine.

HA IN MEDICINE: THE OLD AND THE NEW

The molecular weight can vary from over 6000 kDa in umbilical cords and rooster combs, to 100 kDa and below in serum. HA is found in the ECM of all tissues of the body, and is particularly abundant in umbilical cords, embryonic hearts, the eye, synovial fluid, and cartilage (Laurent, 1989). HA occurs primarily as the sodium salt under physiological conditions; collectively, HA, sodium HA, and other HA salts are often referred to as hyaluronan. The two major commercial sources of HA are rooster combs for >2000 kDa HA and bacterial fermentation for 50–100 kDa HA, and these are polydisperse. Recently, a third option, unique to Hyalose,

Monolithic HA Products	Living HA Products
◆ Fixed chemical entity during use	◆ Variable composition during use
◆ May be fabricated into different physical forms (mesh, sheet, particles)	◆ May be cast into different physical forms before crosslinking
◆ Cannot form new covalent bonds in the presence of cells or tissue	◆ *In situ* crosslinkable, forming new covalent bonds *in vivo*
◆ Cannot separately adjust viscoelasticity and biogradability	◆ Separate engineering of viscoelasticity and degradability
◆ Difficult to add cells or therapeutic agents	◆ Facile addition of cells or drugs prior to crosslinking
e.g., Hylan®, Restylane®, HYAFF®	*e.g., HyStem™, Extracel™*

FIGURE I.2.7.3 Chemically-modified HA can occur in monolithic or living forms, and this table summarizes the principal differences with examples of specific products.

is HA and labeled HA oligosaccharides produced enzymatically to give monodisperse polymers ranging from only a few disaccharide units up to 20 kDa (DeAngelis et al., 2003). HA forms lubricious and viscoelastic solutions that can protect eye tissues during ophthalmic surgeries, provide viscosupplementation in joints, reduce post surgical adhesions, and deliver drugs (Kuo, 2006). However, the short residence time *in vivo* of HA and its poor mechanical properties led to many chemical modifications of HA (Shu and Prestwich, 2004; Allison and Grande-Allen, 2006). In the last two decades, HA-based medical devices and coatings have addressed clinical needs with different levels of success.

Chemically-modified HA derivatives can be grouped into two types: monolithic and living (Figure I.2.7.3) (Prestwich and Kuo, 2008). The chemistry and biology of HA is exhaustively reviewed on the Glycoforum website (http://glycoforum.gr.jp/science/hyaluronan/). Monolithic derivatives arise when the chemical modification gives a final form that cannot form new chemical bonds in the presence of cells or tissues, and can only be physically fabricated into different forms. In contrast, living derivatives contain modifications that permit the formation of new covalent bonds in the presence of cells or tissues, enabling a change in physical form during *in vivo* or *in vitro* biological use. For example, an *in situ* cross-linkable material for adhesion prevention or cell encapsulation would be a living modification.

The Old: Monolithic Ha Derivatives

Carbodiimide-mediated reaction products. In water at pH <5, the carboxylic acid groups of HA can react with an ethyl dimethylaminopropylcarbodiimide (EDCI) to form an intermediate O-acylurea that can be trapped or rearranged to the more stable *N*-acylurea adduct (Pouyani et al., 1994). Interaction of the cationic aminopropyl moiety with remaining carboxylates can lead to physical gelation. Seprafilm™ (Genzyme) is the product of EDCI plus HA and carboxymethylcelluose, yielding a bioabsorbable material that is used to prevent post-surgical adhesions in abdominal and gynecological indications. Seprafilm™ is brittle when dry, and sticky when moist, and is therefore difficult for surgeons to handle. The use of biscarbodiimides affords chemically cross-linked HA

(Kuo et al., 1991), a technology found in the post-surgical adhesion product Incert™ (Anika Therapeutics).

Divinylsulfone cross-linking. Hylans™ (Biomatrix, Genzyme) are HA derivatives that are cross-linked through the hydroxyl groups, leaving the majority of the carboxylic acids unmodified (Band, 1998). Hylan A™ develops HA–protein cross-links with formaldehyde, giving HA products with increased molecular weight. In contrast, Hylan B™ is a highly swollen gel with an infinite network of divinylsulfone cross-links. Hylans™ are more viscoelastic than HA and have longer half-lives *in vivo*. Hylans™ are used as injectable materials (Synvisc) to treat osteoarthritic joints (Arrich et al., 2005), as dermal fillers (Beasley et al., 2009) to treat facial wrinkles (Hylaform™), and as space-occupying stents in nasal/sinus surgery (Hylasine™, Sepragel Sinus) (Kuo, 2006).

Esterification. HA esters are prepared by the reaction of the quaternary ammonium salt of HA with an alcohol that imparts hydrophobicity, using a non-aqueous solvent such as dimethylformamide for the reaction (Campoccia et al., 1998). For example, the benzyl ester HYAFF-11 (Fidia Advanced Biopolymers) is less susceptible to enzymatic degradation than native HA. HYAFF-11 scaffolds are cell adhesive and are stable in aqueous solution for over three weeks. A three-dimensional HYAFF scaffold supports proliferation of keratinocytes, fibroblasts, chondrocytes, and mesenchymal stem cells. The dermal replacement (Hyalograft 3D™) consists of HYAFF-11 with autologous fibroblasts combined with an epidermal replacement (Laserskin) made of a microperforated HYAFF-11 membrane containing autologous keratinocytes. A clinical study showed 59% of treated pressure ulcers reached complete closure in less than four months with this product. Hyalograft C™, which consists of autologous chondrocytes grown on a HYAFF-11 scaffold, is used to treat cartilage defects of the knee, and patients improved continuously up to three years.

HA carboxyl groups can be activated with 2-chloro-1-methylpyridinium iodide (CMPI) to form internal ester bonds. The so-called autocross-linked polymer (ACP; Fidia) which can be used as a gel (Hyalobarrier™, Hyaloglide™) was effective in the prevention of adhesions in gynecological and joint surgeries in preclinical model studies (Acunzo et al., 2003; Brunelli et al., 2005).

Bis-epoxide cross-linked HA. HA that can be cross-linked with a diglycidyl ether under strongly alkaline conditions to give a biocompatible material with *in vivo* residence times up to 4 weeks (Beasley et al., 2009). For example, Restylane™ (Q-Med) and Juvederm™ (Allergan) are used as cosmetic dermal fillers. HA can also be cross-linked twice with the same bis-epoxide, first under basic conditions and then under acidic conditions. Double cross-linking reduced hyaluronidase degradation compared to the single cross-linked intermediate. Puragen™ (Mentor) is a doubly cross-linked HA gel, and is used as dermal filler in the same manner as Restylane™ and Juvederm™.

The New: Living HA Derivatives

Reversible cross-linking using carbodiimide-mediated hydrazide chemistry. As described above, HA-derived O-acylureas can undergo further coupling reactions with compounds that are nucleophilic at pH 4–6, e.g., N-hydroxysuccinimide, aminooxy compounds, and hydrazides (Shu and Prestwich, 2004). When hydrazides containing disulfide bonds are used, living HA derivatives can be produced (Shu et al., 2002). The resulting thiol-modified HA can be cross-linked to itself or to other synthetic and natural polymers bearing thiol groups. Moreover, a variety of physical properties and rates of biodegradation can be obtained by controlling several parameters for the biocompatible HA hydrogels, including: (1) molecular weight of starting HA employed; (2) percentage thiol modification; (3) concentration of thiol-modified HA; (4) choice of chemistry for polyfunctional cross-linker; (5) molecular weight of polyfunctional cross-linker; and (6) ratio of thiols to electrophiles (Prestwich, 2008b).

The cross-linked hydrogel can act as a barrier to cell infiltration and migration. Such a material promotes scar-free wound healing and prevents post-operative adhesion formation. This formulation has been employed for scar-free mucosal healing following endoscopic sinus surgery (Proctor et al., 2006), for prevention of post-surgical abdominal adhesions (Liu et al., 2007a), and to minimize subglottic stenosis in airway stents (Sondrup et al., 2006). Veterinary products from Sentrx Animal Care have been successfully employed for equine and small animal wound care (Figure I.2.7.4).

Electrophilic HA. HA derivatives equipped with thiol-reactive electrophilic esters can react with thiol-modified macromolecules to give "cross-linker-free" hydrogels. HA was converted to the bromoacetate and iodoacetate derivatives (Serban and Prestwich, 2007). Cross-linker-free semi-synthetic ECM (sECM) hydrogels were prepared by combining the polyfunctional electrophilic HA haloacetates with thiol-modified HA (CMHA-S). The resulting hydrogel was cytocompatible, but did not support the attachment of primary fibroblasts. Thus, the HA haloacetates offer an alternative living macromonomer for producing cross-linker-free sECM biomaterials that can function as anti-adhesive barriers or for culture of non-adherent cell types.

Photo-cross-linked HA. Methacrylation of HA by modification of the hydroxyl groups produces macromonomers that can be seeded with cells and photocross-linked into hydrogels with either UV or visible light plus an initiator (Ifkovits and Burdick, 2007). Potential uses for *in situ* photopolymerized HA hydrogels include prevention of adhesions and tissue regeneration. For example, chondrocytes encapsulated in photocross-linkable HA methacrylate retained the chondrocytic phenotype and synthesized cartilage matrix *in vitro*, and accelerated healing in an osteochondral defect model *in vivo* (Chung and Burdick, 2009). Importantly, photocross-linked HA methacrylates have been used for self-renewal of human embryonic stem cells without differentiation (Gerecht et al., 2007).

Summary

The first generation of chemically-modified HA derivatives were "monolithic" materials that could be injection molded, shaped, spun, or woven for a variety of clinical uses. While these monolithic materials continue to enjoy commercial and clinical success, an important second generation of "living" HA derivatives have emerged. The living polymers offer many opportunities for three-dimensional cell culture, injectable tissue engineering, wound repair, scar-free healing, drug delivery, and toxicology testing. The applications of the living HA derivatives to produce a synthetic ECM, and the use of sECMs to produce clinical biomaterials, will be described below.

RECREATING THE EXTRACELLULAR MATRIX

Design Criteria

What is the problem to be solved? The native ECM is a heterogeneous network of soluble and fibrous proteins and glycosaminoglycans, which is maintained by both covalent and non-covalent interactions (Figure I.2.7.5). Recapitulating the composition and structure of the ECM is a daunting yet irresistible challenge to engineers and chemists.

Scientific Criteria

A biomaterial ECM-mimetic needs to recapitulate the principal functions of the natural ECM in orchestrating cell proliferation, migration, differentiation, angiogenesis, and invasion. Several objective design criteria

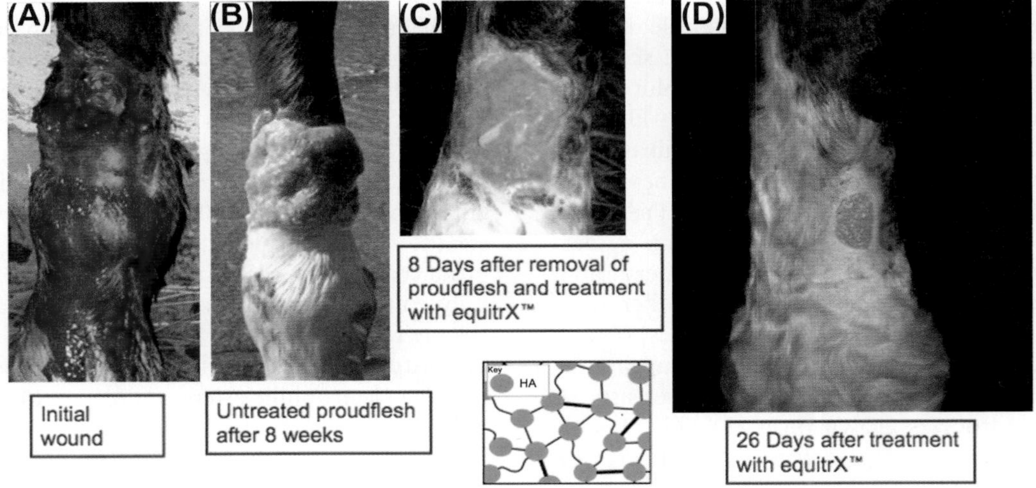

FIGURE I.2.7.4 The initial injury to this horse's leg (A) went untreated for two months, resulting in the formation of exuberant granulation tissue, or proud flesh (B). The wound was then debrided by a veterinarian, treated with the cross-linked HA product equitrX™. After eight days, the proud flesh had not recurred (C) and 26 days after an additional equitrX™ treatment, the wound was over 80% healed. Courtesy of Sentrx Animal Care (www.sentrxanimalcare.com).

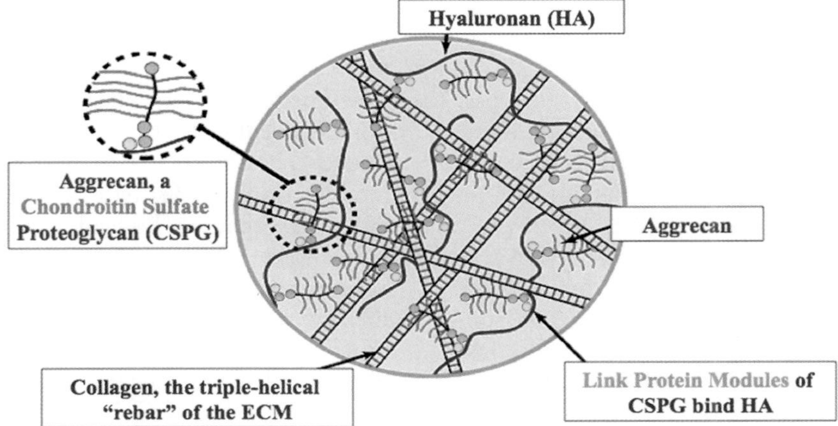

FIGURE I.2.7.5 The native ECM is a heterogeneous network of glycosaminoglycans, proteoglycans, and proteins that consists of both covalent and non-covalent interactions.

must be met to recreate the intertwined biochemical and mechanical cues in the cellular microenvironment, and allow experimental versatility in fabrication (Prestwich, 2007). These performance criteria should include: (1) experimental control over composition; (2) experimental control over compliance; (3) controllable biodegradability *in vitro* and *in vivo*; (4) moldable *in situ* cross-linking during cell encapsulation and fabrication; (5) batch-to-batch consistency; (6) ease of use at physiological temperature and pH; (7) transparency for ease of visualization; and (8) seamless translation from lab to clinic. Modularized sECMs that meet these criteria would allow end users to add soluble factors, attachment peptides, matricellular proteins, cross-linkers, and macromomers to customize them for research or treatment.

Market Criteria

The needs of the end-user must also be included as design criteria (Prestwich, 2008a), whether for cell therapy or for drug evaluation and research tools. These additional design criteria (see Box 1) are derived from the realization that commercialization is *sine qua non* for translating technology from the academic lab to the private sector. This requires significant financial commitment to developing the technology and to recruiting a team capable of reducing the technology to practice. Investor interest is highest for products for which a large market can be readily penetrated. Specifically, for biomaterials to be used in regenerative medicine, several key parameters must be satisfied.

Building a Semi-synthetic ECM

The solution: A modular sECM. We selected the thiol-modified HA described in the section The Old: Monolithic HA Derivatives as the starting point to make an *in situ* cross-linkable synthetic mimic of the extracellular matrix (sECM). We deconstructed the ECM into modular units containing one, two, or several chemically modified macromonomers that could be covalently cross-linked into a hydrogel network (Figure I.2.7.6) (Serban and Prestwich, 2007). The resulting sECM offers user-controllable composition and compliance, to which a variety of growth factors or matricellular proteins may be added. Moreover, the elastic moduli of hydrogels could be tailored to match the cell type to achieve optimal regeneration of a new tissue. Importantly, it is also easy to use and meets all of the scientific criteria in the section on Divinylsulfone cross-linking (Figure I.2.7.7).

BOX 1 Commercial Constraints for a Clinical Biomaterial

It's all about the difference between sales and marketing. In developing a technology, we sell our concepts and results to review committees, journals, students, and our peers. Converting a technology to a product is different. Creating a product requires identifying and meeting the needs of your customers, the end-users. While technological advances are driven by curiosity, asking questions, and conducting experiments with many variables, products must be simple, consistent, robust, reproducible, and must meet a need. Consciously including these real-world design criteria is crucial for developing high-cost biomaterials for use in a competitive and highly-regulated environment.

Design Criteria for a Clinical Biomaterial

- **Meet needs of end-users: patients & physicians**
- **Simple, effective composition**
- **Simple to use**
- **Easy to manufacture under GMP**
- **Clear path to regulatory approval**
- **Reimbursable by health care payers**

Interest by investors and industry partners

Clinically useful product

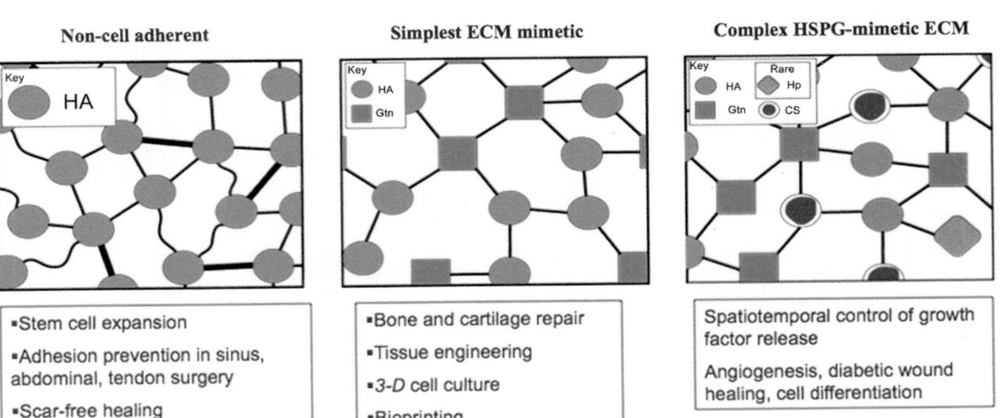

FIGURE I.2.7.6 Living sECMs are based on deconstruction of the native ECM and modular reassembly into one, two, or multi-component biomaterials.

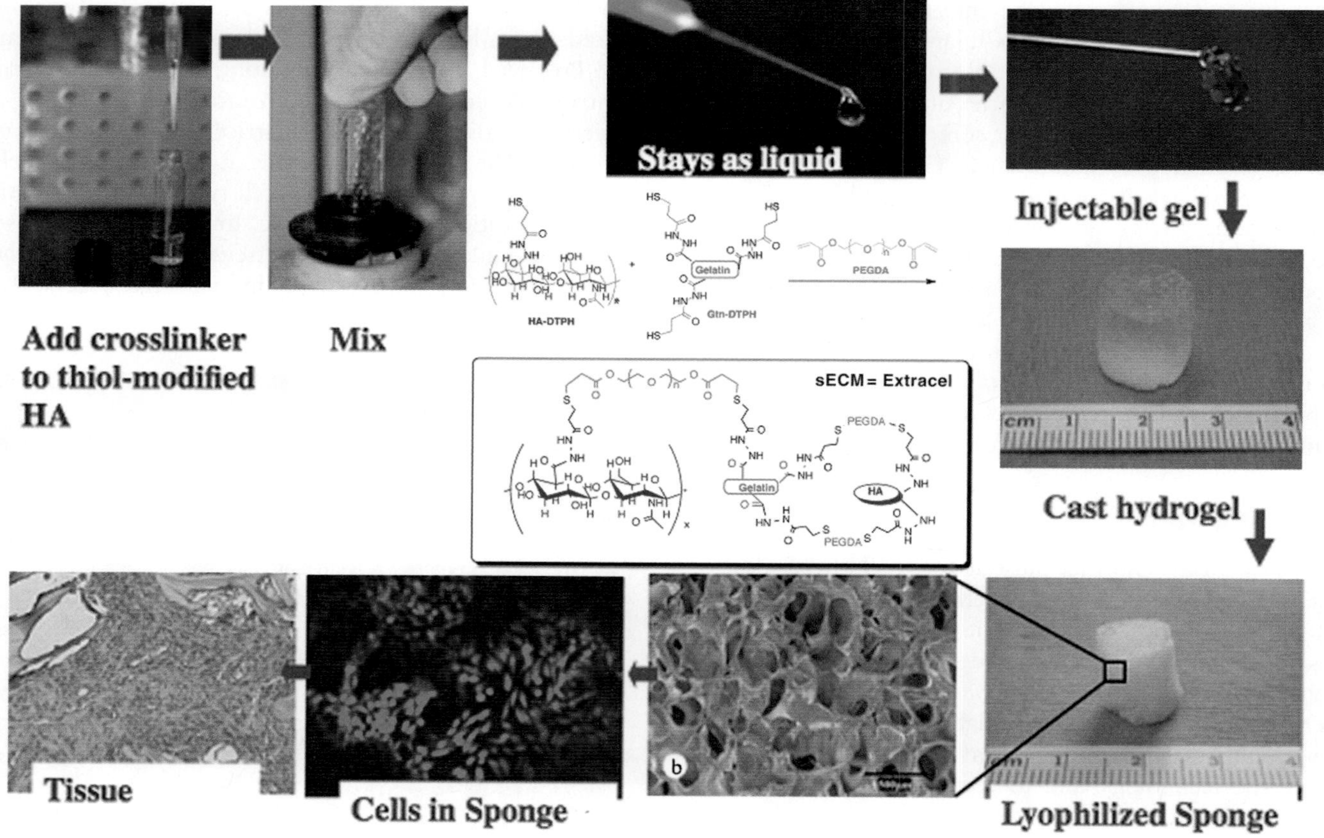

FIGURE I.2.7.7 Easy protocol for using modular sECMs.

The sECMs may be cross-linked *in situ* in the presence of cells to provide an injectable cell-seeded hydrogel for tissue repair or without cells for the controlled, localized delivery of drugs or growth factors. Thiol-modified glycosaminoglycans and proteins are also living macromonomers that can be cross-linked with biocompatible polyvalent electrophiles to form biocompatible, bioerodable hydrogels and porous sponges. This new technology offers a powerful set of new tools for culturing primary cells in three dimensions (3-D). Indeed, the applications of this technology include *in vitro* and *in vivo* growth of 3-D cellularized, vascularized tissues using films, sponges, and hydrogels.

The thiol modified carboxymethylated HA known as CMHA-S was readily cross-linkable and yielded biocompatible hydrogels that did not support cell attachment and proliferation. Cross-linking was accomplished slowly via disulfide bond formation in air, or with better control of cross-linking time and hydrogel stiffness using cytocompatible bivalent cross-linkers such as polyethylene glycol diacrylate (PEGDA) or other PEG-based biselectrophiles (Vanderhooft et al., 2007). As indicated above in the section The Old: Monolithic HA Derivatives, the one-component gel has proven utility in scar-free wound healing, post-surgical adhesion prevention, and stem cell expansion.

Incorporation of attachment factors provides an sECM that supports cell attachment, growth, and proliferation in 3-D. By co-cross-linking thiol-modified gelatin and CMHA-S, we obtained materials to which cells readily attached and spread. Gelatin can be replaced by an RGD peptide or selected domains of fibronectin (Ghosh et al., 2006). These two-component sECMs have been employed in repair of osteochondral defects (Liu and Prestwich, 2006), centrifugal casting of endothelial cells (Mironov et al., 2005), repair of vocal fold defects (Duflo et al., 2006), growth of tumor xenografts (Liu et al., 2007b), and *ex vivo* culture of cells in 2-D and 3-D (Shu et al., 2006). Both *in vivo* and *in vitro*, cells invade the compliant hydrogel, remodel it with matrix metalloproteinases, and secrete a new tissue-like cell-specific ECM. The fundamental role of the sECM hydrogel is to enhance the natural biological repair processes mediated by endogenous cells.

Engineered tissues require a blood supply to survive, and developing an internal vasculature in the neotissue and connecting that network to the host are formidable challenges to overcome. Thiol-modified heparin (HP-DTPH) and chondroitin sulfate (CS-DTPH) macromonomers provide the biomimetic equivalents of heparan sulfate (HS) and CS proteoglycans (Cai et al., 2005). Other heparin–HA bioconjugates have also been produced for controlled growth factor release (Liu et al., 2002). These hydrogels provided spatiotemporal control of growth factor release over a time course of weeks to months. In addition, they allowed dual growth factor release, and simultaneously enhanced the rate of neovascularization *in vivo* (Pike et al., 2006). These materials can also be used to deliver bFGF to accelerate excisional wound repair in diabetic mice (Liu et al., 2007c).

MEETING THE TRANSLATIONAL CHALLENGE

No technology will reach patients and physicians unless it is commercialized. The translational challenge implicit in this statement is three-fold. First, the inventors and origin of the technology must be willing and effective in identifying licensing or start-up options and following through on their intentions. Second, the technology must be robust and fulfill a market need. Third, the investors and management team must identify key funding and market opportunities, and then guide the company unerringly to the successful fulfillment of the plan.

Too often, scientists have underestimated the importance of the financial, production, and regulatory constraints during the development of a biomaterial for reparative or regenerative medicine. In contrast, a marketing approach driven by the needs of the end-users places these crucial and limiting pragmatic factors as part of the initial design criteria.

The native ECM evolved to co-optimize biological and structural features. Rather than rebuilding the entire biological complexity *de novo*, the modular sECM approach uses the minimum number of components necessary to enable cells to rebuild tissues. Biology is best at doing the "heavy lifting" of tissue engineering.

The message of the translational imperative is summarized in Figure I.2.7.8 (Prestwich, 2008a). As scientists, we must **embrace the complexity** of biology, but we cannot let this complexity dominate the design of a clinical material. Since no one formulation can fulfill every need, we should **engineer flexibility/versatility** into our products, so that a single suite of approvable materials can be exploited in a variety of clinical niches. Finally, for regenerative medicine to succeed as a reality, we must **deliver simplicity**. The penetration of new technology into any market is not determined only by improvement in patient outcome. The true limiting factors are cost, familiarity, and ease of use. The living sECM technology addresses a balance between the physiological and the practical requirements, and is currently used in clinical and veterinary medicine, and for translational laboratory research.

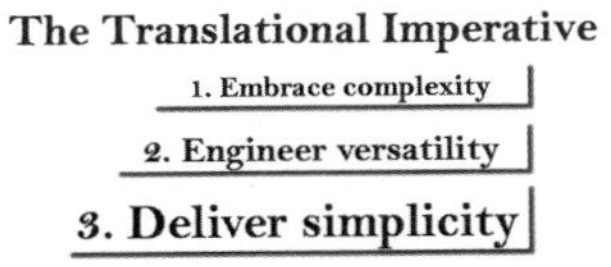

The Translational Imperative

1. Embrace complexity
2. Engineer versatility
3. **Deliver simplicity**

FIGURE I.2.7.8 The translational imperative.

GLOSSARY OF ACRONYMS

bFGF: Basic fibroblast growth factor
CMHA-S: Carboxymethylated hyaluronic acid
CMPI: 2-chloro-1-methylpyridinium iodide
CS: Chondroitin sulfate
CS-DTPH: Thiol-modified chondroitin sulfate
CTB: Center for Therapeutic Biomaterials
DD: Degree of deacetylation
ECM: Extracellular matrix
EDCI: Ethyl dimethylaminopropylcarbodiimide
ELP: Elastin-like peptide
ENT: Ear-nose-throat
G: a-L-guluronic acid
GAG: Glycosaminoglycan
GMP: Good manufacturing practice
HA: Hyaluronic acid
HP-DTPH: Thiol-modified heparin
HS: Heparan sulfate
IP: Intellectual property
M: b-D-mannuronic acid
PEG: Poly (ethylene glycol)
PEGDA: Poly (ethylene glycol) diacrylate
RGD: Arginine-glycine-aspartic acid peptide sequence
SBIR: Small business innovative research
sECM: Semi-synthetic extracellular matrix
VC: Venture capital
VEGF: Vascular endothelial growth factor

DISCLOSURE STATEMENT

Glenn Prestwich holds equity in Glycosan, Sentrx Animal Care, and Carbylan Biosurgery which uses the sECM technology. Sarah Atzet is an employee of Glycosan.

Case Study — Fields of Use

How Sentrx Surgical became three separate companies

Figure I.2.7.9 shows how a university-based technology can be effectively commercialized. The original discovery of the chemical modification of HA to give monolithic and later living polymers occurred over 15 years with industry, state, and federal support. As inventions were reduced to practice, patent applications were filed to protect the intellectual property (IP), and papers were published on the chemical and biological results. Technology transfer offices at two universities assisted with the patenting process and with interactions with potential licensees. In the end, negotiations reached a standstill. However, at this juncture, the State of Utah provided financial support to create a Center of Excellence, the Center for Therapeutic Biomaterials (CTB), to facilitate commercialization of technologies in this area. The CTB then spun out Sentrx Surgical, so named because two of the co-founders were otolargynologists (ear, nose, and throat (ENT) surgeons). The name incorporated our mission – ENT plus Rx – biomaterials for ENT treatments. Sentrx applied for, and received, four SBIR grants, using non-dilutive federal funding as well as state funding to the CTB, to develop its IP portfolio and compelling preclinical results to show to potential investors.

During the "road shows" to raise money, a Utah venture capital (VC) group offered to lead the investment if we could raise additional funding to match their terms. In identifying additional investors, a VC firm expressed interest, but only wanted exclusive control human medical device and drug device combination uses of the technology, and also wanted the university to have less equity. This was accomplished by removing other fields from the license to Carbylan Biosurgery, with the exception of non-exclusive rights for cell therapy uses. Within two months, we started two additional companies: Glycosan Biosystems, with an exclusive license for research uses and a co-exclusive for cell therapy, and Sentrx Animal Care, with exclusive rights to veterinary uses. All three companies use the same original IP, which they each customize for use in their respective market segments.

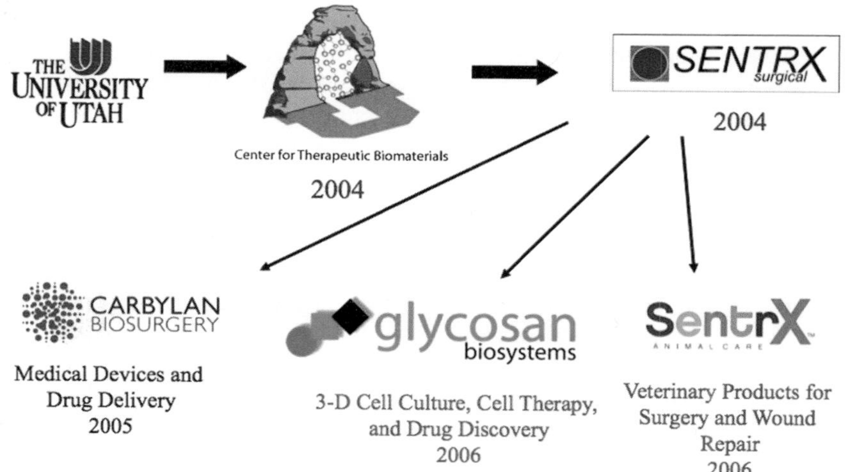

FIGURE I.2.7.9 Paradigm for commercialization of a university technology with state and federal support followed by venture investments and division into distinct product-oriented fields of use.

BIBLIOGRAPHY

Acunzo, G., Guida, M., Pellicano, M., Tommaselli, G. A., Sardo, A. D., et al. (2003). Effectiveness of auto-cross-linked hyaluronic acid gel in the prevention of intrauterine adhesions after hysteroscopic adhesiolysis: A prospective, randomized, controlled study. *Hum. Reprod.*, **18**(9), 1918–1921.

Ahmed, T. A., Dare, E. V., & Hincke, M. (2008). Fibrin: A versatile scaffold for tissue engineering applications. *Tissue Eng. Part B. Rev.*, **14**(2), 199–215.

Allison, D., & Grande-Allen, K. (2006). Hyaluronan: A powerful tissue engineering tool. *Tissue Eng.*, **12**(8), 2131–2140.

Arrich, J., Piribauer, F., Mad, P., Schmid, D., Klaushofer, K., et al. (2005). Intra-articular hyaluronic acid for the treatment of osteoarthritis of the knee: Systematic review and meta-analysis. *Cmaj.*, **172**(8), 1039–1043.

Augst, A. D., Kong, H. J., & Mooney, D. J. (2006). Alginate hydrogels as biomaterials. *Macromol Biosci.*, **6**(8), 623–633.

Badylak, S. F., & Gilbert, T. W. (2008). Immune response to biologic scaffold materials. *Semin Immunol*, **20**(2), 109–116.

Badylak, S. F., Tullius, R., Kokini, K., Shelbourne, K. D., Klootwyk, T., et al. (1995). The use of xenogeneic small intestinal submucosa as a biomaterial for Achilles tendon repair in a dog model. *J. Biomed. Mater. Res.*, **29**(8), 977–985.

Baldwin, A. D., & Kiick, K. L. (2010). Polysaccharide-modified synthetic polymeric biomaterials. *Biopolymers*, **94**(1), 128–140.

Band, P. A. (1998). Hyaluronan Derivatives: Chemistry and Clinical Applications. In T. C. Laurent (Ed.), *The Chemistry, Biology and Medical Applications of Hyaluronan and its Derivatives* (pp. 33–42). London: Portland Press.

Basmadjian, D., & Sefton, M. V. (1983). Relationship between release rate and surface concentration for heparinized materials. *J. Biomed. Mater. Res.*, **17**(3), 509–518.

Beasley, K. L., Weiss, M. A., & Weiss, R. A. (2009). Hyaluronic acid fillers: A comprehensive review. *Facial Plast. Surg.*, **25**(2), 86–94.

Berger, J., Reist, M., Mayer, J. M., Felt, O., Peppas, N. A., et al. (2004). Structure and interactions in covalently and ionically cross-linked chitosan hydrogels for biomedical applications. *Eur. J. Pharm. Biopharm.*, **57**(1), 19–34.

Berger, J., Reist, M., Chenite, A., Felt-Baeyens, O., Mayer, J. M., et al. (2005). Pseudo-thermosetting chitosan hydrogels for biomedical application. *Int. J. Pharm.*, **288**(1), 17–25.

Breckwoldt, W. L., Belkin, M., Gould, K., Allen, M., Connolly, R. J., et al. (1991). Modification of the thrombogenicity of a self-expanding vascular stent. *J. Invest. Surg.*, **4**(3), 269–278.

Brunelli, G., Longinotti, C., Bertazzo, C., Pavesio, A., & Pressato, D. (2005). Adhesion reduction after knee surgery in a rabbit model by Hyaloglide(R), a hyaluronan derivative gel. *J. Orthopaed. Res.*, **23**(6), 1377–1382.

Cai, S., Liu, Y., Zheng Shu, X., & Prestwich, G. D. (2005). Injectable glycosaminoglycan hydrogels for controlled release of human basic fibroblast growth factor. *Biomaterials*, **26**(30), 6054–6067.

Campoccia, D., Doherty, P., Radice, M., Brun, P., Abatangelo, G., et al. (1998). Semisynthetic resorbable materials from hyaluronan esterification. *Biomaterials*, **19**(23), 2101–2127.

Cauda, F., Cauda, V., Fiori, C., Onida, B., & Garrone, E. (2008). Heparin coating on ureteral Double J stents prevents encrustations: An *in vivo* case study. *J. Endourol*, **22**(3), 465–472.

Chung, C., & Burdick, J. (2009). Influence of 3D hyaluronic acid microenvironments on mesenchymal stem cell chondrogenesis. *Tissue Eng. A.*, **15**, 243–254.

Daamen, W. F., Veerkamp, J. H., van Hest, J. C., & van Kuppevelt, T. H. (2007). Elastin as a biomaterial for tissue engineering. *Biomaterials*, **28**(30), 4378–4398.

Dal Pozzo, A., Vanini, L., Fagnoni, M., Guerrini, M., De Benedittis, A., et al. (2000). Preparation and characterization of poly(ethylene glycol)-cross-linked reacetylated chitosans. *Carbohydr. Polym.*, **42**, 201–206.

Dare, E. V., Griffith, M., Poitras, P., Kaupp, J. A., Waldman, S. D., et al. (2009). Genipin cross-linked fibrin hydrogels for *in vitro* human articular cartilage tissue-engineered regeneration. *Cells Tissues Organs*, **190**(6), 313–325.

DeAngelis, P. L., Oatman, L. C., & Gay, D. F. (2003). Rapid chemoenzymatic synthesis of monodisperse hyaluronan oligosaccharides with immobilized enzyme reactors. *J. Biol. Chem.*, **278**(37), 35199–35203.

Duflo, S., Thibeault, S. L., Li, W., Shu, X. Z., & Prestwich, G. D. (2006). Vocal fold tissue repair *in vivo* using a synthetic extracellular matrix. *Tissue Eng.*, **12**, 2171–2180.

Ethicon Silk Sutures. http://www.ecatalog.ethicon.com/sutures-non-absorbable/view/perma-hand-suture (3/24/10).

Eyrich, D., Brandl, F., Appel, B., Wiese, H., Maier, G., et al. (2007). Long-term stable fibrin gels for cartilage engineering. *Biomaterials*, **28**(1), 55–65.

FDA Allograft Artery Patch PMA. http://www.accessdata.fda.gov/cdrh_docs/pdf8/K081438.pdf (3/24/10).

FDA Applied Nutritionals Chondroitin Wound Dressing. http://www.accessdata.fda.gov/cdrh_docs/pdf8/K081724.pdf (3/24/10).

FDA Collamend 510(k). http://www.accessdata.fda.gov/cdrh_docs/pdf8/K082687.pdf (3/24/10).

FDA CosmoDerm PMA. http://www.accessdata.fda.gov/cdrh_docs/pdf/P800022S050b.pdf (3/24/10).

FDA Evicel PMA. http://www.accessdata.fda.gov/cdrh_docs/pdf7/K070575.pdf (3/24/10).

FDA GelFoam PMA. http://www.accessdata.fda.gov/scripts/cdrh/cfdocs/cfPMA/pma.cfm?id=9023 (3/24/10).

FDA HemCon 510(k). http://www.accessdata.fda.gov/cdrh_docs/pdf8/K080818.pdf (3/24/10).

FDA PhytaCare 510(k). http://www.accessdata.fda.gov/cdrh_docs/pdf5/K053538.pdf (3/24/10).

FDA Restylane PMA. http://www.accessdata.fda.gov/cdrh_docs/pdf2/P020023a.pdf (3/24/10).

FDA TyCo. (Healthcare Heparin Coated Catheter). http://www.accessdata.fda.gov/cdrh_docs/pdf6/K062671.pdf (3/24/10).

Freier, T., Koh, H. S., Kazazian, K., & Shoichet, M. S. (2005a). Controlling cell adhesion and degradation of chitosan films by N-acetylation. *Biomaterials*, **26**(29), 5872–5878.

Freier, T., Montenegro, R., Shan Koh, H., & Shoichet, M. S. (2005b). Chitin-based tubes for tissue engineering in the nervous system. *Biomaterials*, **26**(22), 4624–4632.

Friess, W. (1998). Collagen: Biomaterial for drug delivery. *Eur. J. Pharm. Biopharm.*, **45**(2), 113–136.

Gerecht, S., Burdick, J., Ferreira, L., Townsend, S., Langer, R., et al. (2007). Hyaluronic acid hydrogel for controlled self-renewal and differentiation of human embryonic stem cells. *Proc. Natl. Acad. Sci., USA*, **104**, 11298–11303.

Geutjes, P. J., Daamen, W. F., Buma, P., Feitz, W. F., Faraj, K. A., et al. (2006). From molecules to matrix: Construction and evaluation of molecularly defined bioscaffolds. *Adv. Exp. Med. Biol.*, **585**, 279–295.

Ghosh, K., Ren, X. D., Shu, X. Z., Prestwich, G. D., & Clark, R. A. (2006). Fibronectin functional domains coupled to hyaluronan stimulate adult human dermal fibroblast responses critical for wound healing. *Tissue Eng.*, **12**(3), 601–613.

Gilbert, T. W., Sellaro, T. L., & Badylak, S. F. (2006). Decellularization of tissues and organs. *Biomaterials*, **27**(19), 3675–3683.

Hildebrandt, P. (2002). Glycosaminoglycans: All round talents in coating technology. *Biomed. Tech. (Berl.)*, **47**(Suppl. 1 Pt. 1), 476–478.

Ho, M. H., Wang, D. M., Hsieh, H. J., Liu, H. C., Hsien, T. Y., et al. (2005). Preparation and characterization of RGD-immobilized chitosan scaffolds. *Biomaterials*, **26**(16), 3197–3206.

Hoffman, A. S., Schmer, G., Harris, C., & Kraft, W. G. (1972). Covalent binding of biomolecules to radiation-grafted hydrogels on inert polymer surfaces. *Trans. Am. Soc. Artif. Intern. Organs.*, **18**(1), 10–18.

Huang, L., McMillan, R. A., Apkarian, R. P., Pourdeyhimi, B., et al. (2000). Generation of synthetic elastin-mimic small diameter fibers and fiber networks. *Macromolecules*, **33**, 2989–2997.

Ifkovits, J., & Burdick, J. (2007). Photopolymerizable and degradable biomaterials for tissue engineering applications. *Tissue Eng.*, **13**, 2369–2385.

Jiang, T., Kumbar, S. G., Nair, L. S., & Laurencin, C. T. (2008). Biologically active chitosan systems for tissue engineering and regenerative medicine. *Curr. Top. Med. Chem.*, 8(4), 354–364.

Jiang, T., Wang, G., Qiu, J., Luo, L., & Zhang, G. (2009). Heparinized poly(vinyl alcohol): Small intestinal submucosa composite membrane for coronary covered stents. *Biomed. Mater.*, **4**(2). 025012.

Kauzmann, W. (1959). Some factors in the interpretation of protein denaturation. *Adv. Protein. Chem.*, **14**, 1–63.

Khademhosseini, A., Vacanti, J. P., & Langer, R. (2009). Progress in tissue engineering. *Sci. Am.*, **300**(5), 64–71.

Khor, E., & Lim, L. Y. (2003). Implantable applications of chitin and chitosan. *Biomaterials*, **24**(13), 2339–2349.

Kneser, U., Schaefer, D. J., Polykandriotis, E., Horch, R. E. (2006). Tissue engineering of bone: The reconstructive surgeon's point of view. *J. Cell Mol. Med.*, **10**(1), 7–19.

Knothe, Tate, M. L., (2003). "Whither flows the fluid in bone?" an osteocyte's perspective. *J. Biomech.*, **36**(10), 1409–1424.

Kuntz, I. D., Jr., & Kauzmann, W. (1974). Hydration of proteins and polypeptides. *Adv. Protein Chem.*, **28**, 239–345.

Kuo, J. W. (2006). *Practical Aspects Of Hyaluronan Based Medical Products*. Boca Raton: CRC/Taylor & Francis.

Kuo, J. W., Swann, D. A., & Prestwich, G. D. (1991). Chemical modification of hyaluronic acid by carbodiimides. *Bioconjugate Chem.*, **2**, 232–241.

Laurent, T. (1989). The biology of hyaluronan. Introduction. *Ciba Found Symp.*, **143**, 1–20.

Lee, C. H., Singla, A., & Lee, Y. (2001). Biomedical applications of collagen. *Int. J. Pharm.*, **221**(1–2), 1–22.

Lin, C., Lee, T., Lee, H., Sung, H. (2003). Chemical modification of biomedical materials with genipin, US Patent US6608040, Filed Sep 27 2001, Issued Aug 19, Application # 09/297,808.

Linnes, M. P., Ratner, B. D., & Giachelli, C. M. (2007). A fibrinogen-based precision microporous scaffold for tissue engineering. *Biomaterials*, **28**(35), 5298–5306.

Liu, L. S., Ng, C. -K., Thompson, A. Y., Poser, J. W., & Spiro, R. C. (2002). Hyaluronate-heparin conjugate gels for the delivery of basic fibroblast growth factor (FGF-2). *J. Biomed. Mater. Res.*, **62**, 128–135.

Liu, Y., Shu, X. Z., & Prestwich, G. D. (2006). Osteochondral defect repair with autologous bone-marrow derived mesenchymal stem cells in an injectable, *in situ* cross-linked synthetic extracellular matrix. *Tissue Eng.*, **12**, 3405–3416.

Liu, Y., Shu, X. Z., & Prestwich, G. D. (2007a). Reduced postoperative intra-abdominal adhesions using carbylanTM-SX, a semi-synthetic glycosaminoglycan hydrogel. *Fertil & Steril*, **87**, 940–948.

Liu, Y., Shu, X. Z., & Prestwich, G. D. (2007b). Tumor engineering: Orthotopic cancer models in mice using cell-loaded, injectable, cross-linked hyaluronan derived hydrogels. *Tissue Eng.*, **13**, 1091–1101.

Liu, Y., Cai, S., Shu, X. Z., Shelby, J., & Prestwich, G. D. (2007c). Sustained release of basic fibroblast growth factor from a cross-linked glycosaminoglycan hydrogel promotes wound healing in genetically diabetic mice. *Wound Repair Regen.*, **15**, 245–251.

McHale, M. K., Setton, L. A., & Chilkoti, A. (2005). Synthesis and *in vitro* evaluation of enzymatically cross-linked elastin-like polypeptide gels for cartilaginous tissue repair. *Tissue Eng.*, **11**(11–12), 1768–1779.

Mi, F. L., Sung, H. W., & Shyu, S. S. (2002a). Drug release from chitosan-alginate complex beads reinforced by a naturally occurring cross-linking agent. *Carbohydr. Polym.*, **48**, 61–72.

Mi, F. L., Tan, Y. C., Liang, H. F., & Sung, H. W. (2002b). *In vivo* biocompatibility and degradability of a novel injectable-chitosan-based implant. *Biomaterials*, **23**(1), 181–191.

Mi, F. L., Lin, Y. M., Wu, Y. B., Shyu, S. S., & Tsai, Y. H. (2002c). Chitin/PLGA blend microspheres as a biodegradable drug-delivery system: Phase-separation, degradation and release behavior. *Biomaterials*, **23**(15), 3257–3267.

Mi, F. L., Shyu, S. S., Lin, Y. M., Wu, Y. B., et al. (2003). Chitin/PLGA blend microspheres as a biodegradable drug delivery system: A new delivery system for protein. *Biomaterials*, **24**(27), 5023–5036.

Mironov, V., Kasyanov, V., Zheng Shu, X., Eisenberg, C., Eisenberg, L., et al. (2005). Fabrication of tubular tissue constructs by centrifugal casting of cells suspended in an *in situ* cross-linkable hyaluronan-gelatin hydrogel. *Biomaterials*, **26**(36), 7628–7635.

Nimni, M. E., Cheung, D., Strates, B., Kodama, M., & Sheikh, K. (1987). Chemically modified collagen: A natural biomaterial for tissue replacement. *J. Biomed. Mater. Res.*, **21**(6), 741–771.

Peattie, R. A., Pike, D. B., Yu, B., Cai, S., Shu, X. Z., et al. (2008). Effect of gelatin on heparin regulation of cytokine release from hyaluronan-based hydrogels. *Drug Deliv.*, **15**(6), 389–397.

Pike, D. B., Cai, S., Pomraning, K. R., Firpo, M. A., Fisher, R. J., et al. (2006). Heparin-regulated release of growth factors *in vitro* and angiogenic response *in vivo* to implanted hyaluronan hydrogels containing VEGF and bFGF. *Biomaterials*, **27**, 5242–5251.

Pouyani, T., Harbison, G. S., & Prestwich, G. D. (1994). Novel hydrogels of hyaluronic acid: Synthesis, surface morphology and solid-state NMR. *J. Am. Chem. Soc.*, **116**, 7515–7522.

Prestwich, G. D. (2007). Simplifying the extracellular matrix for 3-D cell culture and tissue engineering: A pragmatic approach. *J. Cell. Biochem.*, **101**, 1370–1383.

Prestwich, G. D. (2008a). Engineering a clinically-useful matrix for cell therapy. *Organogenesis*, **4**(1), 42–47.

Prestwich, G. D. (2008b). Evaluating drug efficacy and toxicology in three dimensions: Using synthetic extracellular matrices in drug discovery. *Acc. Chem. Res.*, **41**(1), 139–148.

Prestwich, G. D., & Kuo, J. W. (2008). Chemically-modified HA for therapy and regenerative medicine. *Curr. Pharm. Biotechnol.*, **9**(4), 242–245.

Proctor, M., Proctor, K., Shu, X. Z., McGill, L. D., Prestwich, G. D., et al. (2006). Composition of hyaluronan affects wound healing in the rabbit maxillary sinus. *Am. J. Rhinology*, **20**, 206–211.

Rao, S. B., & Sharma, C. P. (1997). Use of chitosan as a biomaterial: studies on its safety and hemostatic potential. *J. Biomed. Mater. Res.*, **34**(1), 21–28.

Ratner, B. D. (2004). *Biomaterials Science: An Introduction to Materials in Medicine* (2nd ed.). In A. S. Hoffman, F. J. Schoen & J. E. Lemons (Eds.), San Diego: Elsevier Academic Press.

Rowley, J. A., Madlambayan, G., & Mooney, D. J. (1999). Alginate hydrogels as synthetic extracellular matrix materials. *Biomaterials*, **20**(1), 45–53.

Schmer, G. (1972). The biological activity of covalently immobilized heparin. *Trans. Am. Soc. Artif. Intern. Organs*, **18**(1), 321–324. 333.

Serban, M. A., & Prestwich, G. D. (2007). Synthesis of hyaluronan haloacetates and biology of novel cross-linker-free synthetic extracellular matrix hydrogels. *Biomacromolecules*, **8**, 2821–2828.

Serban, M. A., & Prestwich, G. D. (2008). Making modular extracellular matrices: Solutions for the puzzle. *Methods*, **45**, 93–98.

Shu, X. Z., Liu, Y., Luo, Y., Roberts, M. C., & Prestwich, G. D. (2002). Disulfide cross-linked hyaluronan hydrogels. *Biomacromolecules*, 3(6), 1304–1311.

Shin, H., Jo, S., & Mikos, A. G. (2003). Biomimetic materials for tissue engineering. *Biomaterials*, **24**(24), 4353–4364.

Shu, X. Z., Ghosh, K., Liu, Y., Palumbo, F. S., Luo, Y., et al. (2004). Attachment and spreading of fibroblasts on an RGD peptide-modified injectable hyaluronan hydrogel. *J. Biomed. Mater. Res. A*, **68**(2), 365–375.

Shu, X. Z., & Prestwich, G. D. (2004). Therapeutic Biomaterials from Chemically Modified Hyaluronan. In H. G. Garg, & C. A. Hales (Eds.), *Chemistry and Biology of Hyaluronan* (pp. 475–504). Amsterdam: Elsevier.

Shu, X. Z., Ahmad, S., Liu, Y., & Prestwich, G. D. (2006). Synthesis and evaluation of injectable, *in situ* cross-linkable synthetic extracellular matrices for tissue engineering. *J. Biomed. Mater. Res. A*, **79**(4), 902–912.

Sondrup, C., Liu, Y., Shu, X. Z., Prestwich, G. D., & Smith, M. E. (2006). Cross-linked hyaluronan-coated stents in the prevention of airway stenosis. *Otolaryngol Head Neck Surg.*, **135**(1), 28–35.

Stenzel, K. H., Miyata, T., & Rubin, A. L. (1974). Collagen as a biomaterial. *Annu. Rev. Biophys. Bioeng.*, **3**(0), 231–253.

Stone, K. R., Steadman, J. R., Rodkey, W. G., & Li, S. T. (1997). Regeneration of meniscal cartilage with use of a collagen scaffold. Analysis of preliminary data. *J. Bone. Joint Surg. Am.*, **79**(12), 1770–1777.

Suh, J. K., & Matthew, H. W. (2000). Application of chitosan-based polysaccharide biomaterials in cartilage tissue engineering: A review. *Biomaterials*, **21**(24), 2589–2598.

Urry, D. W., Hugel, T., Seitz, M., Gaub, H. E., Sheiba, L., et al. (2002). Elastin: A representative ideal protein elastomer. *Philos. Trans. R. Soc. Lond. B Biol. Sci.*, **357**(1418), 169–184.

van der Giessen, W. J., van Beusekom, H. M., Eijgelshoven, M. H., Morel, M. A., & Serruys, P. W. (1998). Heparin-coating of coronary stents. *Semin. Interv. Cardiol.*, 3(3–4), 173–176.

Vanderhooft, J. L., Mann, B. K., & Prestwich, G. D. (2007). Synthesis and characterization of novel thiol-reactive poly(ethylene glycol) cross-linkers for biomaterials. *Biomacromolecules*, **8**, 2883–2889.

VandeVord, P. J., Matthew, H. W., DeSilva, S. P., Mayton, L., Wu, B., et al. (2002). Evaluation of the biocompatibility of a chitosan scaffold in mice. *J. Biomed. Mater. Res.*, **59**(3), 585–590.

White, M. J., Kohno, I., Rubin, A. L., Stenzel, K. H., & Miyata, T. (1973). Collagen films: Effect of cross-linking on physical and biological properties. *Biomater. Med. Devices Artif. Organs*, **1**(4), 703–715.

Yuan, N., Tsai, R., Ho, M., Wang, D., Lai, J., et al. (2008). Fabrication and characterization of chondroitin sulfate-modified chitosan membranes for biomedical applications. *Desalination*, **23**(1–3), 166–174.

CHAPTER I.2.8 PYROLYTIC CARBON FOR LONG-TERM MEDICAL IMPLANTS

Robert B. More,[1] *Axel D. Haubold,*[2] *and Jack C. Bokros*[3]

[1]Integra Life Sciences, Austin, TX, USA

[2]Bed Rock Ranch, Decatur, TX, USA

[3]On-X Life Technologies, Inc, Austin, TX, USA

INTRODUCTION

Carbon materials are ubiquitous and of great interest because the majority of substances that make up living organisms are carbon compounds. Although many engineering materials and biomaterials are based on carbon or contain carbon in some form, elemental carbon itself is also an important and very successful biomaterial. Furthermore, there exists enough diversity in their structure and properties for elemental carbons to be considered as a unique class of materials beyond the traditional molecular carbon focus of organic chemistry, polymer chemistry, and biochemistry. Through a serendipitous interaction between researchers during the late 1960s the outstanding blood compatibility of a special form of elemental pyrolytic carbon deposited at high temperature in a fluidized bed was discovered. The material was found to have not only remarkable blood compatibility, but also the structural properties needed for long-term use in artificial heart valves (LaGrange et al., 1969). The blood compatibility of pyrolytic carbon was recognized empirically using the Gott vena cava ring test. This test involved implanting a small tube made of a candidate material in a canine vena cava and observing the development of thrombosis within the tube in time. Prior to pyrolytic carbon, only surfaces coated with graphite, benzylalkonium chloride, and heparin would resist thrombus formation when exposed to blood for long periods. The incorporation of pyrolytic carbon in mechanical heart valve implants was declared "an exceptional event" (Sadeghi, 1987) because it added the durability and stability needed for heart valve prostheses to endure for a patient's lifetime. The objective of this chapter is to present the elemental pyrolytic carbon materials currently used in the fabrication of medical devices, and to describe their manufacture, characterization, and properties.

ELEMENTAL CARBON

Elemental carbon is found in nature as two crystalline allotrophic forms: graphite and diamond. Elemental carbon also occurs as a spectrum of imperfect, turbostratic crystalline forms that range in degree of crystallinity from amorphous to the perfectly crystalline allotropes. Recently a third crystalline form of elemental carbon, the fullerene structure, has been discovered. The crystalline polymorphs of elemental carbon are shown in Figure I.2.8.1.

The properties of the elemental carbon crystalline forms vary widely according to their structure. Diamond with its tetrahedral sp^3 covalent bonding is one of the hardest materials known. In the diamond crystal structure, covalent bonds of length 1.54 Å connect each carbon atom with its four nearest neighbors. This tetrahedral symmetry repeats in three dimensions throughout

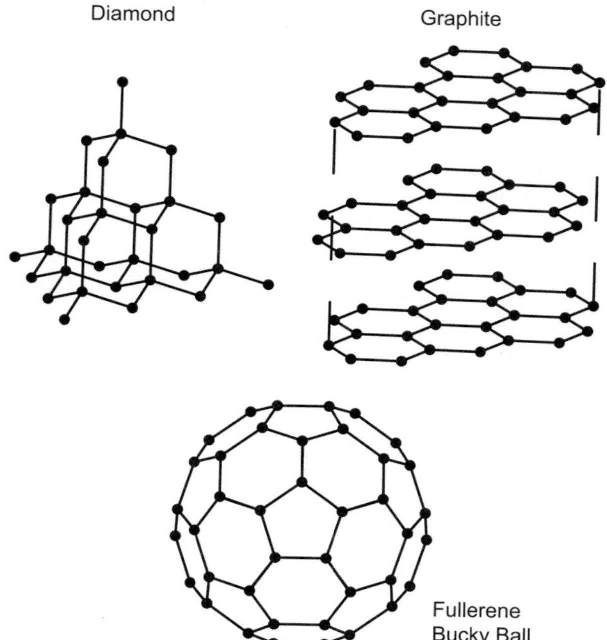

FIGURE I.2.8.1 Allotropic crystalline forms of carbon: diamond, graphite, and fullerene.

the crystal (Pauling, 1964). In effect, the crystal is a giant isotropic covalently-bonded molecule; therefore, diamond is very hard.

Graphite, with its anisotropic layered in-plane hexagonal covalent bonding and interplane van der Waals bonding structure, is a soft material. Within each planar layer, each carbon atom forms two single bonds and one double bond with its three nearest neighbors. This bonding repeats in-plane to form a giant molecular (graphene) sheet. The in-plane atomic bond length is 1.42 Å, which is a resonant intermediate (Pauling, 1964) between the single bond length of 1.54 Å and the double bond length of 1.33 Å. The planer layers are held together by relatively weak van der Waals bonding at a distance of 3.4 Å, which is more than twice the 1.42 Å bond length (Pauling, 1964). Graphite has low hardness and a lubricating property because the giant molecular sheets can readily slip past one another against the van der Waals bonding. Nevertheless, although large-crystallite-size natural graphite is used as a lubricant, some artificially produced graphites can be very abrasive if the crystallite sizes are small and randomly oriented.

Fullerenes have yet to be produced in bulk, but their properties on a microscale are entirely different from those of their crystalline counterparts. Fullerenes and nanotubes consist of a graphene layer that is rolled up or folded (Sattler, 1995) to form a tube or ball. These large molecules, C_{60} and C_{70} fullerenes and (C_{60} + 18*j*) nanotubes, are often mentioned in the literature (Sattler, 1995) along with more complex multilayer "onion skin" structures.

There exist many possible forms of elemental carbon that are intermediate in structure and properties between those of the allotropes diamond and graphite. Such "turbostratic" carbons occur as a spectrum of amorphous, through mixed amorphous, graphite-like and diamond-like, to the perfectly crystalline allotropes (Bokros, 1969). Because of the dependence of properties upon structure, there can be considerable variability in properties for the turbostratic carbons. Glassy carbons and pyrolytic carbons, for example, are two turbostratic carbons with considerable differences in structure and properties. Consequently, it is not surprising that carbon materials are often misunderstood through oversimplification. Properties found in one type of carbon structure can be totally different in another type of structure. Therefore, it is very important to specify the exact nature and structure when discussing carbon.

PYROLYTIC CARBON (PyC)

The biomaterial known as pyrolytic carbon is not found in nature; it is manmade. The successful pyrolytic carbon biomaterial was developed at General Atomic during the late 1960s using a fluidized-bed reactor (Bokros, 1969). In the original terminology, this material was considered a low-temperature isotropic carbon (LTI carbon). Since the initial clinical implant of a pyrolytic carbon component in the DeBakey–Surgitool mechanical valve in 1968, 95% of the mechanical heart valves implanted worldwide have at least one structural component made of pyrolytic carbon. On an annual basis this translates into approximately 500,000 components (Haubold, 1994). Pyrolytic carbon components have been used in more than 25 different prosthetic heart valve designs since the late 1960s, and have accumulated a clinical experience in the order of 16 million patient-years. Clearly, pyrolytic carbon is one of the most successful, critical biomaterials both in function and application. Among the materials available for mechanical heart valve prostheses, pyrolytic carbon has the best combination of blood compatibility, physical and mechanical properties, and durability. However, the blood compatibility of pyrolytic carbon in heart valve applications is not perfect; chronic anticoagulant therapy is needed for patients with mechanical heart valves. Whether the need for anticoagulant therapy arises from the biocompatible properties of the material itself or from the particular hydrodynamic interaction of a given device and the blood remains to be resolved.

The term "pyrolytic" is derived from "pyrolysis," which is thermal decomposition. Pyrolytic carbon is formed from the thermal decomposition of hydrocarbons such as propane, propylene, acetylene, and methane, in the absence of oxygen. Without oxygen the typical decomposition of the hydrocarbon to carbon dioxide and water cannot take place; instead a more complex cascade of decomposition products occurs that ultimately results in a "polymerization" of the individual carbon atoms into large macroatomic arrays.

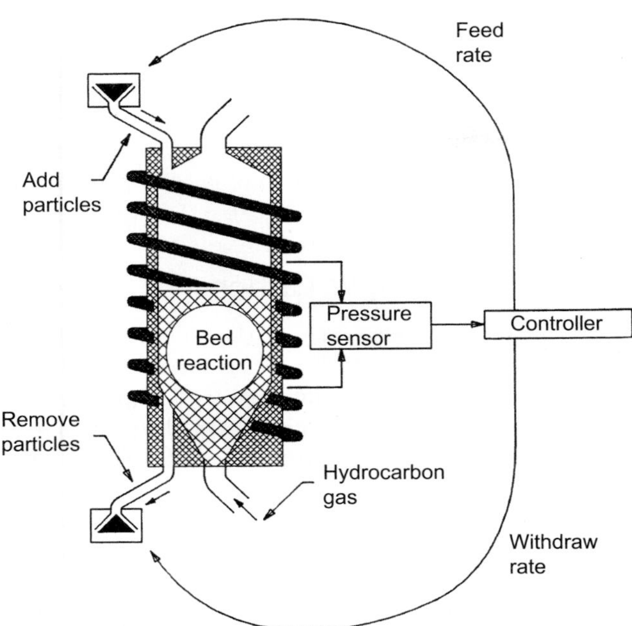

FIGURE I.2.8.2 Fluidized-bed reactor schematic.

Pyrolysis of the hydrocarbon is normally carried out in a fluidized-bed reactor such as the one shown in Figure I.2.8.2. A fluidized-bed reactor typically consists of a vertical tube furnace that may be induction or resistance heated to temperatures of 1000 to 2000°C (Bokros, 1969). Reactor diameters ranging from 2 cm to 25 cm have been used; however, the most common size used for medical devices has a diameter of about 10 cm. These high-temperature reactors are expensive to operate, and the reactor size limits the size of device components able to be produced.

Small refractory ceramic particles are placed into the vertical tube furnace. When a gas is introduced into the bottom of the tube furnace, the gas causes the particle bed to expand. Interparticle spacing increases to allow for the flow of the gas. Particle mixing occurs and the bed of particles begins to "flow" like a fluid. Hence the term "fluidized bed." Depending upon the gas flow rate and volume, this expansion and mixing can be varied from a gentle bubbling bed to a violent spouting bed. An oxygen-free, inert gas, such as nitrogen or helium, is used to fluidize the bed, and the source hydrocarbon is added to the gas stream when needed.

At a sufficiently high temperature, pyrolysis or thermal decomposition of the hydrocarbon can take place. Pyrolysis products range from free carbon and gaseous hydrogen to a mixture of $C_x H_y$ decomposition species. The pyrolysis reaction is complex and is affected by the gas flow rate, composition, temperature, and bed surface area. Decomposition products, under the appropriate conditions, can form gas-phase nucleated droplets of carbon/hydrogen, which condense and deposit on the surfaces of the wall and bed particles within the reactor (Bokros, 1969). Indeed, the fluidized-bed process was originally developed to coat small (200–500 micrometer) diameter spherical particles of uranium/thorium carbide or oxide with pyrolytic carbon. These coated particles were used as the fuel in the high temperature gas-cooled nuclear reactor (Bokros, 1969).

Pyrolytic carbon coatings produced in vertical tube reactors can have a variety of structures, such as laminar or isotropic, granular or columnar (Bokros, 1969). The structure of the coating is controlled by the gas flow rate (residence time in the bed), hydrocarbon species, temperature, and bed surface area. For example, an inadequately fluidized or static bed will produce a highly anisotropic, laminar pyrolytic carbon (Bokros, 1969).

Control of the first three parameters (gas flow rate, hydrocarbon species, and temperature) is relatively easy. However, until recently, it was not possible to measure the bed surface area while the reactions were taking place. As carbon deposits on the particles in the fluidized bed, the diameter of the particles increases. Hence the surface area of the bed changes, which in turn influences the subsequent rate of carbon deposition. As surface area increases, the coating rate decreases, since a larger surface area now has to be coated with the same amount of carbon available. Thus, the process is not in equilibrium. The static-bed process was adequate to coat nuclear fuel particles without attempting to control the bed surface area because such thin coatings (25–50 μm thick) did not appreciably affect the bed surface area.

It was later found that larger objects could be suspended within the fluidized bed of small ceramic particles and also become uniformly coated with carbon. This finding led to the demand for thicker, structural coatings, an order of magnitude thicker (250–500 μm). Bed surface area control and stabilization became an important factor (Akins and Bokros, 1974) in achieving the goal of thicker, structural coatings. In particular, with the discovery of the blood-compatible properties of pyrolytic carbon (LaGrange et al., 1969), thicker structural coatings with consistent and uniform mechanical properties were needed to realize the application to mechanical heart valve components. Quasi-steady-state conditions as needed to prolong the coating reaction were achieved empirically by removing coated particles and adding uncoated particles to the bed while the pyrolysis reaction was taking place (Akins and Bokros, 1974). However, the rates of particle addition and removal were based upon little more than good guesses.

Three of the four parameters that control carbon deposition could be accurately measured and controlled, but a method to measure and control bed surface area was lacking. Thus, the quasi-steady-state process was more of an art than a science. If too many coated particles were removed, the bed became too small to support the larger components within it and the bed collapsed. If too few particles were removed, the rate of deposition decreased, and the desired amount of coating was not achieved in the anticipated time. Furthermore, there were considerable variations in the mechanical properties of the coating from batch to batch. It was found that in order to consistently achieve the hardness needed for wear resistance in prosthetic heart valve applications, it was necessary to add a small amount of β-silicon carbide to the carbon coating. The dispersed silicon carbide particles within the pyrolytic carbon matrix added sufficient hardness to compensate for potential variations in the properties of the pyrolytic carbon matrix. The β-silicon carbide was obtained from the pyrolysis of methyl-trichlorosilane, CH_3SiCl_3. For each mole of silicon carbide produced, the pyrolysis of methyltrichlorosilane also produces three moles of hydrogen chloride gas. Handling and neutralization of this corrosive gas added substantial complexity and cost to an already complex process. Nevertheless, this process allowed consistency for the successful production of several million components for use in mechanical heart valves.

A process has been developed and patented that allows precise measurement and control of the bed surface area. A description of this process is given in the patent literature and elsewhere (Emken et al., 1993, 1994; Ely et al., 1998). With precise control of the bed surface area, it is no longer necessary to include the silicon carbide. Elimination of the silicon carbide has produced a stronger, tougher, and more deformable pure pyrolytic carbon. Historically, pure carbon was the original objective of the development program because of the potential for superior biocompatibility (LaGrange et al., 1969). Furthermore, the enhanced mechanical and physical properties of the pure pyrolytic carbon now possible with the improved process control allow prosthesis design improvements in the hemodynamic contribution to thromboresistance (Ely et al., 1998).

Structure of Pyrolytic Carbons

X-ray diffraction patterns of the biomedical-grade fluidized-bed pyrolytic carbons are broad and diffuse because of the small crystallite size and imperfections. In silicon-alloyed pyrolytic carbon, a diffraction pattern characteristic of the β form of silicon carbide also appears in the diffraction pattern along with the carbon bands. The carbon diffraction pattern indicates a turbostratic structure (Kaae and Wall, 1996) in which there is order within carbon layer planes, as in graphite; but, unlike graphite, there is no order between planes. This type of turbostratic structure is shown in Figure I.2.8.3 compared to that of graphite. In the disordered crystalline structure, there may be lattice vacancies and the layer planes are curved or kinked. The ability of the graphite layer planes to slip is inhibited, which greatly increases the strength and hardness of the pyrolytic carbon relative to that of graphite. From the Bragg

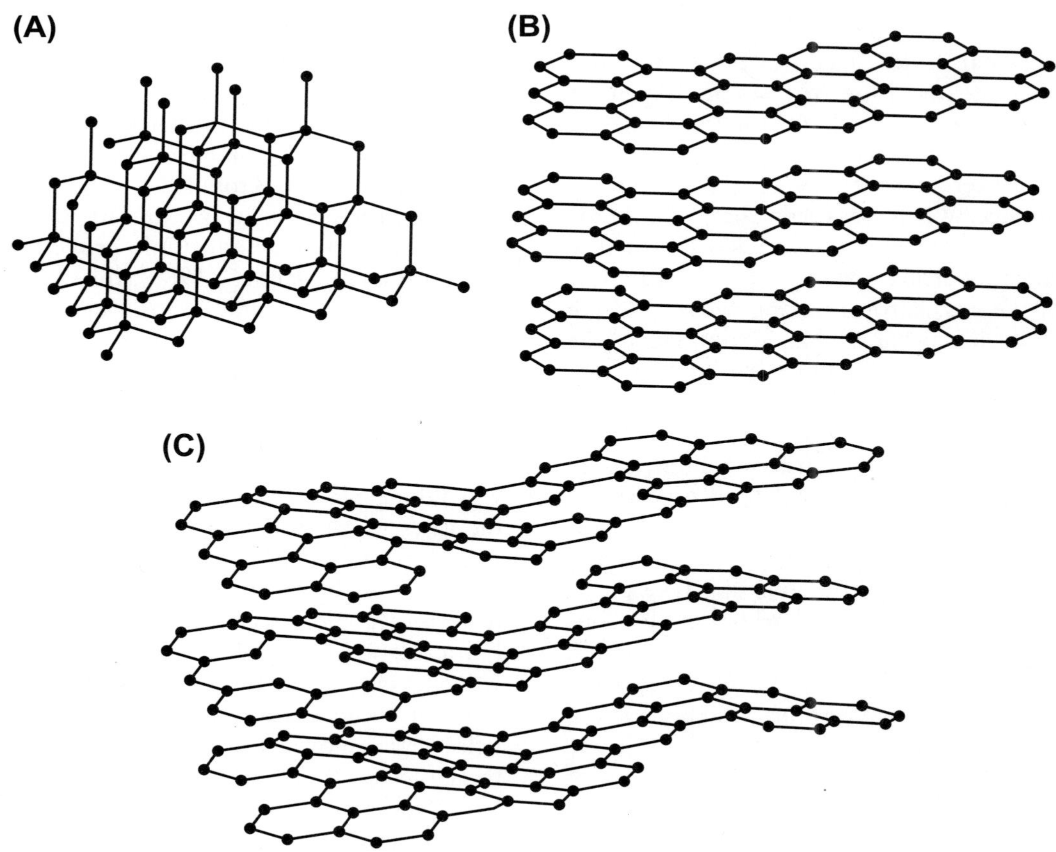

FIGURE I.2.8.3 Structures of: (A) diamond; (B) graphite; and (C) turbostratic pyrolytic carbon.

equation, the pyrolytic carbon layer spacing is reported to be 3.48 Å, which is larger than the 3.35 Å graphite layer spacing (Kaae and Wall, 1996). The increase in layer spacing relative to graphite is due to both the layer distortion and the small crystallite size, and is a common feature for turbostratic carbons. From the Scherrer equation the crystallite size is typically 25–40 Å (Kaae and Wall, 1996).

During the coating reaction, gas-phase nucleated droplets of carbon/hydrogen form that condense and deposit on the surfaces of the reactor wall and bed particles within the reactor. These droplets aggregate, grow, and form the coating. When viewed with high-resolution transmission electron microscopy, a multitude of near-spherical polycrystalline growth features are evident, as shown in Figure I.2.8.4 (Kaae and Wall, 1996). These growth features are considered to be the basic building blocks of the material, and the shape and size are related to the deposition mechanism. In the silicon-alloyed carbon, small silicon carbide particles are present within the growth features, as shown in Figure I.2.8.5. Based on a crystallite size of 33 Å, each growth feature contains about 3×10^9 crystallites. Although the material is quasi-crystalline on a fine level, the crystallites are very small and randomly oriented in the fluidized bed pyrolytic carbons so that overall the material exhibits isotropic behavior.

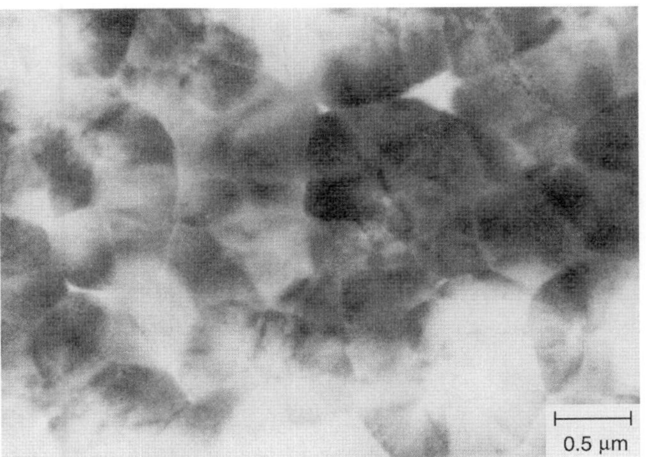

FIGURE I.2.8.4 Electron micrograph of pure pyrolytic carbon microstructure showing near-spherical polycrystalline growth features formed during deposition (Kaae and Wall, 1996).

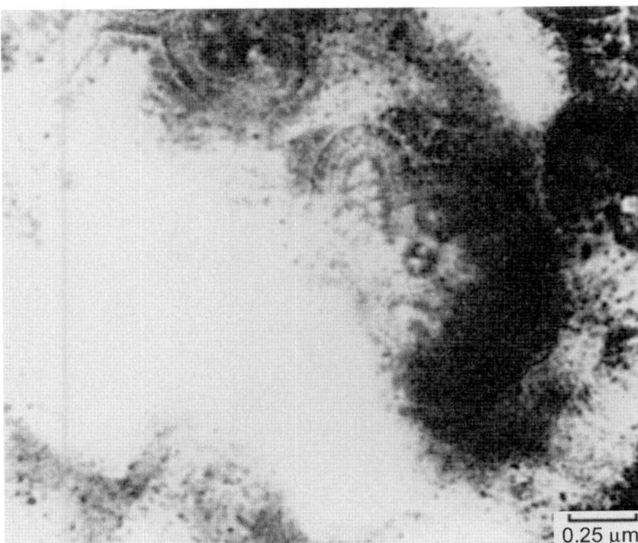

FIGURE I.2.8.5 Electron micrograph of silicon-alloyed pyrolytic carbon microstructure showing near-spherical polycrystalline growth features formed during deposition (Kaae and Wall, 1996). Small silicon carbide particles are shown in concentric rings in the growth features.

Glassy carbon, also known as vitreous carbon or polymeric carbon, is another turbostratic carbon form that has been proposed for use in long-term implants. However, its strength is low and the wear resistance is poor. Glassy carbons are quasi-crystalline in structure, and are named "glassy" because the fracture surfaces closely resemble those of glass (Haubold et al., 1981).

Vapor-deposited carbons are also used in heart valve applications. Typically, the coatings are thin (<1 μm) and may be applied to a variety of materials in order to confer the biochemical characteristics of turbostratic carbon. Some examples are vapor-deposited carbon coatings on heart valve sewing cuffs and metallic orifice components (Haubold et al., 1981).

Mechanical Properties

Mechanical properties of pure pyrolytic carbon, silicon-alloyed pyrolytic carbons, and glassy carbon are given in Table I.2.8.1. Pyrolytic carbon flexural strength is

TABLE I.2.8.1 Mechanical Properties of Biomedical Carbons

Property	Pure PyC	Typical Si-alloyed PyC	Typical Glassy Carbon
Flexural strength (MPa)	493.7 ± 12	407.7 ± 14.1	175
Young's modulus (GPa)	29.4 ± 0.4	30.5 ± 0.65	21
Strain-to-failure (%)	1.58 ± 0.03	1.28 ± 0.03	-
Fracture toughness (MPa $\sqrt{m}$)	1.68 ± 0.05	1.17 ± 0.17	0.5–0.7
Hardness (DPH, 500 g load)	235.9 ± 3.3	287 ± 10	150
Density (g/cm^3)	1.93 ± 0.01	2.12 ± 0.01	<1.54
CTE (10^{-6} cm/cm°C)	6.5	6.1	-
Silicon content (%)	0	6.58 ± 0.32	0
Wear resistance	Excellent	Excellent	Poor

high enough to provide the necessary structural stability for a variety of implant applications, and the density is low enough to allow for components to move easily under the applied forces of circulating blood. With respect to orthopedic applications, Young's modulus is in the range reported for bone (Reilly and Burstein, 1974; Reilly et al., 1974), which allows for compliance matching. Relative to metals and polymers, the pyrolytic carbon strain-to-failure rate is low; it is a nearly ideal linear elastic material and requires consideration of brittle material principles in component design. Strength levels vary with the effective stressed volume or stressed area, as predicted by classical Weibull statistics (De Salvo, 1970). The flexural strengths cited in Table I.2.8.1 are for specimens tested in four-point bending, third-point loading (More et al., 1993) with effective stressed volumes of 1.93 mm^3. The pyrolytic carbon material Weibull modulus is approximately 10 (More et al., 1993).

Fracture toughness levels reflect the brittle nature of the material, but the fluidized-bed isotropic pyrolytic carbons are remarkably fatigue resistant. Recent fatigue studies indicate the existence of a fatigue threshold that is very nearly the single-cycle fracture strength (Gilpin et al., 1993; Ma and Sines, 1996, 1999, 2000). Fatigue-crack propagation studies indicate very high Paris-law fatigue exponents, on the order of 80, and display clear evidence of a fatigue-crack propagation threshold (Ritchie et al., 1990; Beavan et al., 1993; Cao, 1996).

Crystallographic mechanisms for fatigue-crack initiation, as occur in metals, do not exist in the pyrolytic carbons (Haubold et al., 1981). In properly designed and manufactured components, and in the absence of externally induced damage, fatigue does not occur in pyrolytic-carbon mechanical heart valve components. In the 30 years of clinical experience, there have been no clear instances of fatigue failure. Few pyrolytic carbon component fractures have occurred, less than 60 out of more than four million implanted components (Haubold, 1994), and most are attributable to induced damage from handling or cavitation (Kelpetko et al., 1989; Kafesjian et al., 1994; Richard and Cao, 1996).

Wear resistance of the fluidized-bed pyrolytic carbons is excellent. The strength, stability, and durability of pyrolytic carbon are responsible for the extension of mechanical valve lifetimes from less than 20 years to more than the recipient's expected lifetime (Schoen et al., 1982; Schoen, 1983; More and Silver, 1990; Wieting, 1996).

Pyrolytic carbon in heart valve prostheses is often used in contact with metals, either as a carbon disk in a metallic valve orifice or as a carbon orifice stiffened with a metallic ring. Carbon falls with the noble metals in the galvanic series (Haubold et al., 1981), the sequence being silver, titanium, graphite, gold, and platinum. Carbon can accelerate corrosion when coupled to less noble metals *in vivo*. However, testing using mixed potential corrosion theory and potentiostatic polarization has determined that no detrimental effects occur when carbon is coupled with titanium or cobalt–chrome alloys (Thompson et al., 1979; Griffin et al., 1983). Carbon couples with stainless steel alloys are not recommended.

STEPS IN THE FABRICATION OF PYROLYTIC CARBON COMPONENTS

To convert a gaseous hydrocarbon into a shiny, polished black component for use in the biological environment is not a trivial undertaking. Furthermore, because of the critical importance of long-term implants to a recipient's health, all manufacturing operations are performed to stringent levels of quality assurance under the auspices of US Food and Drug Administration Good Manufacturing Practices and International Standards Organization ISO-9000 regulations. As in the case of fabrication of metallic implants, numerous steps are involved. Pyrolytic carbon is not machined from a block of material, as is the case with most metallic implants, nor is it injection or reactive molded, as are many polymeric devices. An overview of the processing steps leading to a finished pyrolytic carbon coated component for use in a medical device is shown in Figure I.2.8.6, and is further described in the following sections.

Substrate Material

Since pyrolytic carbon is a coating, it must be deposited on an appropriately shaped, preformed substrate (preform). Because the pyrolysis process takes place at high temperatures, the choice of substrates is severely

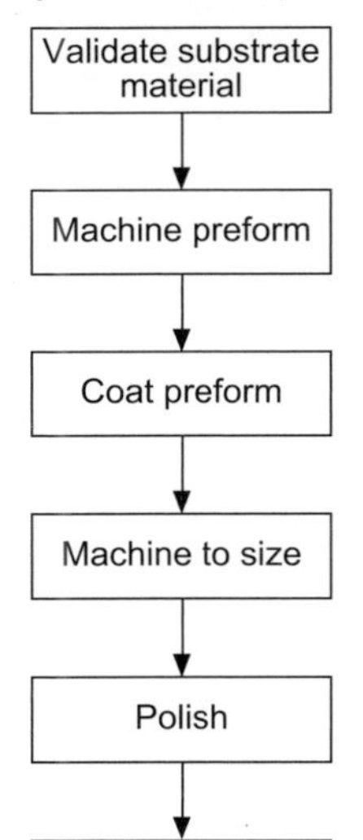

FIGURE I.2.8.6 Schematic of manufacturing processing steps.

limited. Only a few of the refractory materials, such as tantalum or molybdenum/rhenium alloys and graphite, can withstand the conditions at which the pyrolytic carbon coating is produced. Some refractory metals have been used in heart valve components; for example, Mo/Re preforms were coated to make the struts for the Beall–Surgitool mitral valve. It is important for the thermal expansion characteristics of the substrate to closely match those of the applied coating. Otherwise, upon cooling of the coated part to room temperature the coating will be highly stressed and can spontaneously crack. For contemporary heart valve applications, fine-grained isotropic graphite is the most commonly used substrate. This substrate graphite can be doped with tungsten in order to provide radioopacity for X-ray visualizations of the implants. The graphite substrate does not impart structural strength. Rather, it provides a dimensionally stable platform for the pyrolytic carbon coating, both at the reaction temperature and at room temperature.

Preform

Once the appropriate substrate material has been selected and prior to making a preform, it must be inspected to ensure that the material meets the desired specifications. Typically, the strength and density of the starting material are measured. Thermal expansion is ordinarily validated and monitored through process control. The preform, which is an undersized replica of the finished component, is normally machined using conventional machining methods. Because the fine-grained isotropic graphite is very abrasive, standard machine tools have given way to diamond-plated or single-point diamond tools. In the case of heart valves, numerical control machining methods are often required to maintain critical component dimensional tolerances. After the preform is completed, it is inspected to ensure that its dimensions fall within the specified tolerances and that it contains no visible flaws or voids.

Coating

Generally numerous preforms are coated in one furnace run. A batch to be coated is made up of substrates from a single lot of preforms. Such batch processing by lot is required in order to maintain "forward and backward" traceability. In other words, ultimately it is necessary to know all of the components that were prepared using a specific material lot, given either the starting material lot number (forward) or given the specific component serial number (backward). The number of parts that can be coated in one furnace run is dictated by the size of the furnace and the size and weight of the parts to be coated. The batch of substrates is placed within the fluidized bed in the vertical tube furnace and is coated to the desired thickness. Coating times are generally on the order of a few hours, but the entire cycle (heat-up, coating, and cool-down) may take as long as a full day.

FIGURE I.2.8.7 Metallographic mount cross-section of heat valve component. The light-colored pyrolytic carbon layer is coated over the interior, darker-colored granular-appearing graphite substrate.

A statistical sample from each coating lot is taken for analysis. At this point, typical measurements include coating thickness, microhardness, and microstructure. The microhardness and microstructure are determined from a metallographically prepared cross-section of the coated component taken perpendicular to the plane of deposition. Thus, this test is destructive. An example of a metallographically prepared cross-section of a pyrolytic carbon component is shown in Figure I.2.8.7.

Machine to Size

The components used to manufacture medical devices have strict dimensional requirements. Because of the inability, until recently, to precisely measure and control bed size, and indirectly coating thickness, the preforms were generally coated more thickly than necessary to ensure adequate pyrolytic carbon coating thickness on the finished part. The strict dimensional requirements were then achieved through precision grinding or other machining operations. Because pyrolytic carbon is very hard, conventional machine tools again cannot be used. Diamond-plated grinding wheels and other diamond tooling are required. The dimensions of final machined parts are again verified.

Polish

The surface of as-deposited, machined, and polished components is shown in Figure I.2.8.8. It was found early on in experiments (LaGrange et al., 1969; Sawyer et al., 1975; Haubold et al., 1981) that clean polished pyrolytic carbon surfaces of tubes when placed within the vasculature of experimental animals accumulated minimal if any thrombus; and certainly less than

FIGURE I.2.8.8 Scanning electron microscope micrographs of: (A) as-coated; and (B) as-polished surfaces.

TABLE I.2.8.2 Surface Finish (R_a, Average, and R_q, Root Mean Square) of Pyrolytic Carbon Heat Valve Components[a]

Specimen	R_a (nm)	R_q (nm)	Comments
Glass microscope slide	17.14	26.80	
On-X leaflet	33.95	42.12	Clinical
Sorin Bicarbon leaflet	40.12	50.63	Nonclinical
SJM leaflet	49.71	62.74	Clinical
CMI (SJM) leaflet	67.98	85.56	Nonclinical
Sorin Monoleaflet	99.59	128.10	Clinical
DeBakey–Surgitool ball	129.78	157.93	Nonclinical
As-coated slab	389.07	503.72	

[a]Components/prepared by: On-X/Medical Carbon Research Institute, Austin, TX, USA; Sorin/Sorin Biomedica, Saluggia, Italy; SJM/Saint Jude Medical, Saint Paul, MN, USA; CMI (SJM)/CarboMedics, Austin, TX, USA; DeBakey-S/CarboMedics, San Diego, USA (circa 1968). "Clinical" was from as-packaged valve; "nonclinical" lacks component traceability.

pyrolytic carbon tubes with the as-deposited surface. Consequently, the surfaces of pyrolytic carbon have historically been polished, either manually or mechanically, using fine diamond or aluminum oxide pastes and slurries. The surface finish achieved has roughness measured on the scale of nanometers. As can be seen from Table I.2.8.2 (More and Haubold, 1996), the surfaces of polished pyrolytic carbon (30–50 nm) are an order of magnitude smoother than the as-deposited surfaces (300–500 nm).

Once the desired surface quality is achieved, components are again inspected. The final component inspection may include measurement of dimensions, X-ray inspection in two orientations to verify coating thickness, and visual inspection for surface quality and flaws. In many cases, automated inspection methods with computer-controlled coordinate measurement machines are used. X-ray inspection can be used to ensure that minimum coating thickness requirements are met. Two orthogonal views ensure that machining and grinding of the coating was achieved uniformly, and that the coating is symmetrical. The machining and grinding operation after coating is not without the risk of inducing cracks

FIGURE I.2.8.9 Components for On-X bileaflet heart valve.

or flaws in the coating, which may subsequently affect the service life of the component. Such surface flaws are detected visually or with the aid of dye-penetrant techniques. Components may also be proof-tested to detect and eliminate components with subsurface flaws. With the advent of bed size control, which allows coating to exact final dimensions, the concerns about flaws introduced during the machining and grinding operation have been eliminated.

The polished and inspected components, thus prepared, are now ready for assembly into devices, or are packaged and sterilized in the case of standalone devices. Shown in Figure I.2.8.9 are the three pyrolytic carbon components for a bileaflet mechanical heart valve. The components were selected and matched for assembly using the data generated from the final dimensional inspection to achieve the dimensional requirements specified in the device design. In Figure I.2.8.10, the pyrolytic carbon components for a

FIGURE I.2.8.10 Replacement metacarpophalangeal total joint prosthesis components, Ascension Orthopedics, Austin TX, USA.

replacement metacarpophalangeal total joint prosthesis are shown.

Assembly

The multiple components of a mechanical heart must be assembled. The brittleness of pyrolytic carbon poses a significant assembly problem. Because the strain-to-failure is on the order of 1.28% to 1.58%, there is a limited range of deformation that can be applied in order to achieve a proper fit. Relative fit between the components defines the capture and the range of motion for components that move to actuate valve opening and closing. Furthermore, component obstructive bulk and tolerance gaps are critical to hemodynamic performance.

In designs that use a metallic orifice, the metallic components are typically deformed in order to insert the pyrolytic carbon occluder disk. For the all-carbon bileaflet designs, the carbon orifice must be deflected in order to insert the leaflets. As the valve diameter decreases, and as the section modulus of the orifice design increases, the orifice stiffness increases. The possibility of damage or fracture during assembly was a limiting factor in early orifice design. For this reason, the orifices in valve designs using silicon-alloyed pyrolytic carbon were simple cylindrical geometries, and the smallest sizes were limited to the equivalent of a 19 mm diameter tissue annulus. The simple cylindrical orifice designs are often reinforced with a metallic stiffening ring that is shrunk on after assembly. The stiffening ring ensures that physiological loading will not produce deflections that can inhibit valve action or result in leaflet escape.

The increased strain-to-failure of pure pyrolytic carbon, relative to the silicon-alloyed carbon, allows designs with more complex orifice section moduli. This allows designers to utilize hydrodynamically efficient shapes such as flared inlets and to incorporate external stiffening bands that eliminate the need for a metallic stiffening ring. The increased strain-to-failure of On-X carbon has been used to advantage in the On-X mechanical heart valve design (Ely et al., 1998).

Cleaning and Surface Chemistry

Pyrolytic carbon surface chemistry is important because the manufacturing and cleaning operations to which a component is subjected can change and redefine the surface that is presented to the blood. Oxidation of carbon surfaces can produce surface contamination that detracts from blood compatibility (LaGrange et al., 1969; Bokros et al., 1969). Historically, the initial examinations of pyrolytic carbon biocompatibility assumed *de facto* that the surface needed to be treated with a thromboresistant agent such as heparin (Bokros et al., 1969). It was found, however, that the non-heparin-coated surface was actually more blood compatible than the treated surface. Hence, the efforts toward surface coating with heparin were abandoned.

In general it is desired to minimize the surface oxygen and any other non-carbon surface contaminants. From X-ray photoelectron spectroscopy (XPS) analyses, a typical heart valve component surface has 76–86% C, 12–21% O, 0–2% Si, and 1–2% Al (King et al., 1981; Smith and Black, 1984; More and Haubold, 1996). Polishing compounds tend to contain alumina, and some alumina particles may become imbedded in the carbon surface. Other contaminants that may be introduced at low levels (<2% each) are Na, B, Cl, S, Mg, Ca, Zn, and N. The XPS carbon 1s peak when scanned at high resolution can be deconvoluted to determine carbon oxidation states. The carbon 1s peak will typically consist primarily of hydrocarbon-like carbon (60–81%), ether alcohol/ester-like carbon (10–24%), ketone-like carbon (0–6%), and ester/acid-like carbon (1–12%) (More and Haubold, 1996). Each manufacturing, cleaning, and sterilizing operation potentially redefines the surface. The effect of modified surface chemistry on blood compatibility is not well characterized, so this adds a level of uncontrolled variability when considering the blood compatibility of pyrolytic-carbon heart valve materials from different manufacturing sources and different investigators. In general, the presence of oxygen and surface contaminants should be eliminated.

BIOCOMPATIBILITY OF PYROLYTIC CARBON

The suitability of a material for use in an implant is a complex issue. Biocompatibility testing is the focus of other chapters. In the case of pyrolytic carbon, its successful history interfacing with blood in mechanical heart valves attests to its suitability for this application. A note of caution, however, is in order. Until about a decade ago, the pyrolytic carbon used so successfully in mechanical heart valves was produced by a single manufacturer; the material, many applications in the biological environment, and the processes for producing the material, were all patented. Since the expiration of the last of these patents in 1989, other sources for pyrolytic carbon have appeared that are copies of the original General Atomic material. When considering alternative carbon materials, it is important to recognize that the proper combination of physical, mechanical, and blood-compatible properties is required for the success of the implant application. Furthermore, because there are a number of different possible pyrolysis processes, it should be recognized that each can result in different microstructures with different properties. Just because a material is carbon, a turbostratic carbon or a pyrolytic carbon does not qualify its use in a long-term human implant (Haubold and Ely, 1995). For example, pyrolytic carbons prepared by chemical vapor deposition processes, other than the fluidized-bed process, are known to exhibit anisotropy, nonhomogeneity, and considerable variability in mechanical properties (Agafonov et al., 1999). Although these materials may exhibit biocompatibility, the potential for variability in structural stability and durability may lead to valve dysfunction.

The original General Atomic-type fluidized-bed pyrolytic carbons all demonstrate negligible reactions in the standard Tripartite and ISO 10993-1 type biocompatibility tests. Results from such tests are given in Table I.2.8.3 (Ely et al., 1998). Pure pyrolytic carbon is so non-reactive that it can serve as a negative control for these tests.

It is believed that pyrolytic carbon owes its demonstrated blood compatibility to its inertness and to its ability to quickly absorb proteins from blood without triggering a protein denaturing reaction. Ultimately, the blood compatibility is thought to be a result of the protein layer formed upon the carbon surface. Baier observed that pyrolytic carbon surfaces have a relatively high critical surface tension of 50 dyn/cm, which immediately drops to 28–30 dyn/cm following exposure to blood (Baier et al., 1970). The quantity of sorbed protein was thought to be an important factor for blood compatibility. Lee and Kim (1974) quantified the amount of radiolabeled proteins sorbed from solutions of mixture proteins (albumin, fibrinogen, and gamma-globulin). While pyrolytic carbon does adsorb albumin, it also adsorbs a considerable quantity of fibrinogen, as shown in Figure I.2.8.11. As can be seen in Figure I.2.8.11, the amount of fibrinogen adsorbed on pyrolytic carbon surfaces is far greater than the amount of albumin on these surfaces, and is comparable to the amount of fibrinogen that is sorbed on silicone rubber. The mode of albumin adsorption, however, appears to be drastically different for these two materials. Albumin sorbs immediately on the pyrolytic carbon surfaces, whereas the build-up of fibrinogen is much slower. In the case of silicone rubber, both proteins sorb at a much slower rate. It appears that

TABLE I.2.8.3 Biological Testing of Pure PyC

Test Description	Protocol	Results
Klingman maximization	ISO/CD 10993–10	Grade 1; not significant
Rabbit pyrogen	ISO/DIS 10993–11	Nonpyrogenic
Intracutaneous injection	ISO 10993–10	Negligible irritant
Systemic injection	ANSI/AAMI/ISO 10993–11	Negative – same as controls
Salmonella typhimurium reverse mutation assay	ISO 10993–3	Nonmutagenic
Physico-chemical	USP XXIII, 1995	Exceeds standards
Hemolysis – rabbit blood	ISO 10993–4/NIH 77–1294	Nonhemolytic
Elution test (L929 mammalian cell culture)	ISO 10993–5, USP XXIII, 1995	Noncytotoxic

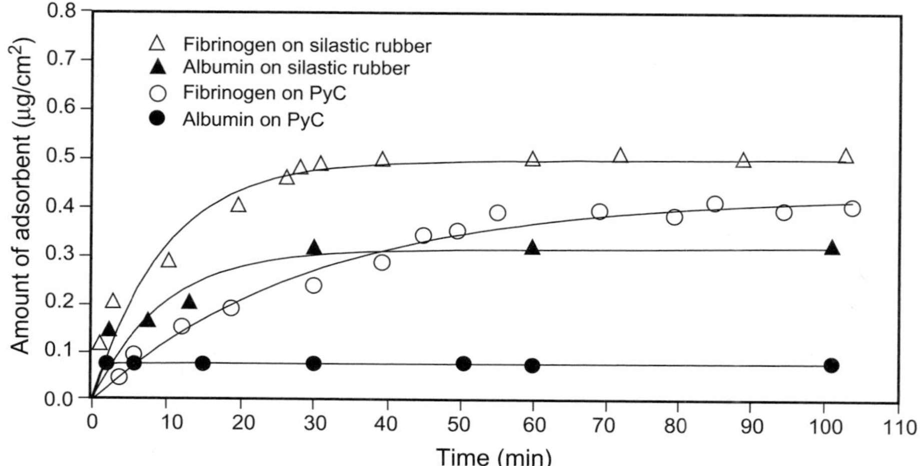

FIGURE I.2.8.11 Fibrinogen and albumin adsorption on pyrolytic carbon (PyC) and Silastic silicone rubber.

the mode of protein adsorption is important, and not the total amount sorbed.

Nyilas and Chiu (1978) studied the interaction of plasma proteins with foreign surfaces by measuring directly the heats of adsorption of selected proteins onto such surfaces using microcalorimetric techniques. They found that the heats of adsorption of fibrinogen, up to the completion of first monolayer coverage, are a factor of eight smaller on pyrolytic carbon surfaces than on the known thrombogenic control (glass) surface, as shown in Figure I.2.8.12. Furthermore, the measured net heats of adsorption of gamma globulin on pyrolytic carbon were about 15 times smaller than those on glass. They concluded that low heats of adsorption onto a foreign surface imply small interaction forces with no conformational changes of the proteins that might trigger the clotting cascade. It appears that a layer of continuously exchanging blood proteins in their unaltered state "masks" the pyrolytic carbon surfaces from appearing as a foreign body.

There is further evidence that the minimally altered sorbed protein layers on pyrolytic carbon condition blood compatibility. Salzman et al. (1977), for example, observed a significant difference in platelet reaction with pyrolytic carbon beads in packed columns prior to and after pretreatment with albumin. With no albumin preconditioning treatment, platelet retention by the columns was high, but the release of platelet constituents was low. However, with albumin pretreatment, platelet retention and the release of constituents was minimal.

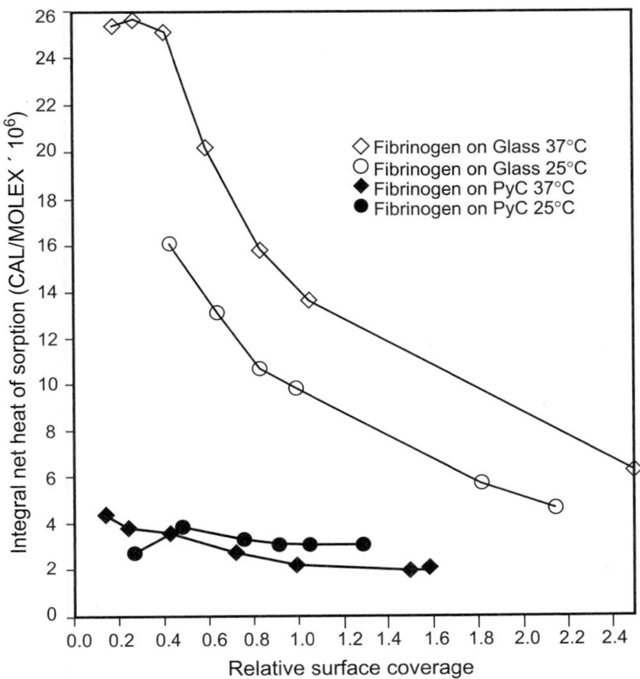

FIGURE I.2.8.12 Integral heat of sorption for fibrinogen on glass and fibrinogen on PyC at two different temperatures (Nyilas and Chiu, 1978).

The foregoing observations led to the view that pyrolytic carbon owes its demonstrated blood compatibility to its inertness, and to its ability to quickly adsorb proteins from blood without triggering a protein-denaturing reaction (Nyilas and Chiu, 1978; Haubold et al., 1981). However, the assertion that pyrolytic carbon is an inert material and induces minimal conformational changes in adsorbed protein was re-examined by Feng and Andrade (1994). Using differential scanning calorimetry and a variety of proteins and buffers, they found that pyrolytic carbon surfaces denatured all of the proteins studied. They concluded that whether or not a surface denatures protein cannot be the sole criteria for blood compatibility. Their suggestion was that the specific proteins and the sequence in which they are denatured may be important. For example, it was suggested that pyrolytic carbon may first adsorb and denature albumin, which forms a layer that subsequently passivates the surface and inhibits thrombosis.

Chinn et al. (1994) re-examined the adsorption of albumin and fibrinogen on pyrolytic carbon surfaces and noted that relatively large amounts of fibrinogen were adsorbed, and speculated that the adsorbed fibrinogen was rapidly converted to a non-elutable form. If the elutable form is more reactive to platelets than the non-elutable form, then the non-elutable protein layer may contribute to the passivating effect.

Work on visualizing the carbon surface and platelet adhesion done by Goodman et al. (1995) using low accelerating-voltage scanning electron microscopy, along with critical-point drying techniques, has discovered that the platelet spreading on pyrolytic carbon surfaces is more extensive than previously observed (Haubold et al., 1981). However, platelet loading was in a static flow situation that does not model the physiological flow that a heart valve is subjected to. Hence, this approach cannot resolve kinetic effects on platelet adhesion. However, Okazaki, Tweden, and co-workers observed adherent platelets on valves following implantation in sheep that were not treated with anticoagulants (Okazaki et al., 1997). There were no instances of valve thrombosis, even though platelets were present on some of the valve surfaces. But the relevance of this observation to clinical valve thromboses is not clear, because human patients with mechanical heart valves undergo chronic anticoagulant therapy (Edmunds, 1987), and have a hemostatic system different from that of sheep.

A more contemporary version of the mechanism of pyrolytic carbon blood compatibility might be to reject the assumption that the surface is inert, as it is now thought by some that no material is totally inert in the body (Williams, 1998), and to accept that the blood–material interaction is preceded by a complex, interdependent, and time-dependent series of interactions between the plasma proteins and the surface (Hanson, 1998) that is as yet poorly understood. To add to the confusion, it must also be recognized that much of the aforementioned conjecture depends on the assumption that all of the carbon surfaces studied were in fact pure and comparable to one another.

CLINICAL APPLICATIONS

Widespread clinical use of pyrolytic carbon components for heart valve replacement began in October of 1968 when Dr. Michael DeBakey implanted an aortic valve with a hollow Pyrolite® carbon ball occluder. Following this first implant, several million PyC mechanical valve prostheses have been implanted worldwide, generating an experience on the order of 20 million patient-years. Use of PyC to replace polymers in valve prostheses was declared an "exceptional event" (Sadeghi, 1987) because the superior durability, stability, and biocompatibility of PyC enabled valves to endure for the patient's lifetime.

However, patients with mechanical valve prostheses require chronic anticoagulation therapy because of the risk of valve-related hemostatic complications. In efforts to reduce the risk of hemostatic complications and the need for chronic anticoagulation, it was hypothesized that valve-related hemostatic complications were in part due to flow-induced mechanical trauma to the blood, and to the presence of PyC itself because of the potential thrombogenicity of the silicon-carbide alloy constituent (LaGrange et al., 1969).

In 1992, advances in pyrolytic carbon manufacturing technology enabled precise control of processing parameters (Bokros et al., 1994). With this precise control it was possible to produce pure isotropic pyrolytic carbons having significantly improved properties relative to silicon-alloyed PyC, thus eliminating the need for the silicon-carbide alloy (see Table I.2.8.1). Precise process control also enabled a coat-to-size capability needed to eliminate surface blemishes caused by post-coating machining processes such as grinding.

The On-X valve shown in Figure I.2.8.9 was specifically designed to exploit the improved pure PyC which enables features to reduce flow-induced trauma (Bokros et al., 1994, 1997, 1998, 2003), and introduced into clinical practice (FDA, 2001, 2002). The success of the On-X valve in general clinical experience, particularly in noncompliant anticoagulant therapy patient populations (Williams and van Riet, 2006) and animal studies (Flameng and Meuris, 2002) strongly indicated improvements in mechanical valve-related hemostatic complications. Data from this experience was used to justify the first and only FDA approved non-warfarin and reduced warfarin prospective randomized trial for a mechanical heart valve. This trail for the On-X Prosthetic Heart Value, with the objective to reduce anticoagulation bleeding risk, was initiated at Emory Crawford Long Hospital, Atlanta in 2006. The trial is currently ongoing at 40 institutions to include 1200 patients. The high risk aortic group has been fully enrolled, with more than 500 pt-yr experience, maintained with a reduced dose of Coumadin (INR of 1.5 to 2.0) and aspirin (81 mg/dy). The low risk aortic group is maintained with platelet inhibitors, (75 mg/dy plavix) and aspirin. The mitral group enrollment was extended to 2012; this group is maintained at an INR of 2.0 to 2.5 and aspirin (81 mg/dy). A successful outcome for this trial would offer further improvements in quality of life for a significant number of On-X mechanical heart valve recipients.

During the past 28 years PyC has also been used as a loadbearing material for small orthopedic joint replacement implants. Such prostheses relieve pain, correct deformity, and improve the appearance of joints damaged by disease such as rheumatoid arthritis, osteoarthritis, and post-traumatic conditions. However, all PyC joint replacements require careful patient selection for good quality bone and soft tissue. As is true with all implants, prosthesis sizing, alignment, and interactions with soft tissue are critical considerations during implantation surgery and rehabilitation.

Successful applications for upper limb total joint prostheses include the metacarpophalangeal (MCP) joint (Figure I.2.8.10) and the proximal interphalangeal (PIP) joint (Cook et al., 1999; Bravo et al., 2007). Pure isotropic PyC is a nearly ideal material for orthopedic application, with demonstrated advantages over traditional materials such as polymers, ceramics, and metals (Stanley et al., 2008) which include:

- Elimination of wear-related failures
- Absence of osteolytic adverse tissue reactions
- Excellent fatigue resistance
- Non-cemented fixation via bone apposition
- Minimization of stress shielding effects and bone resorption
- Excellent compatibility with joint cartilage and bone tissues.

This excellent compatibility with cartilage and bone tissue enables a number of hemiarthroplasty applications, in which only one component of the joint is replaced leaving the PyC device bearing and articulating against native synovial surfaces. Successful devices for hemiarthroplasty include the MCP and PIP joints, carpometacarpal (CMC) joints, radial head, lunate, and inter-positional articulating surface spacers for use in the CMC joint. Currently, approximately 18,000 PyC small joint and hemiarthroplasty devices have been implanted worldwide.

Given the clinical success of the small joint implants, enhanced compatibility with joint tissue, superior durability, and potential significantly extended device lifetimes, efforts are currently underway to use PyC as a platform for larger joint implants such as the shoulder, knee, and hip. In the larger loadbearing joints, a viable strategy is to use PyC as the bearing surface in conservative resurfacing devices and in total joint modular devices. The mechanical valve experience has demonstrated excellent PyC compatibility with traditional implant material metals, ceramics, and polymers. This PyC materials compatibility lends great versatility in design for modular devices. We fully expect that PyC devices will prove more functional, aesthetic, durable, and complication free than implants with traditional materials only.

CONCLUSION

Because the blood compatibility of pyrolytic carbon in mechanical heart valves is not perfect, anticoagulant therapy is required for mechanical heart valve patients. However, pyrolytic carbon has been the most successful material in heart valve applications because it offers excellent blood and tissue compatibility which, combined with the appropriate set of physical and mechanical properties and durability, allows for practical implant device design and manufacture. Improvements in biocompatibility are desired, of course, because when heart valves and other implants are used, a deadly or disabling disease is often treated by replacing it with a less pathological, more manageable chronic condition. Ideally, an implant should not lead to a chronic condition.

It is important to recognize that the mechanism for the blood compatibility of pyrolytic carbons is not fully understood, nor is the interplay between the biomaterial itself, design-related hemodynamic stresses, and the ultimate biological reaction. The elucidation of the mechanism for blood and tissue compatibility of pyrolytic carbon remains a challenge.

It is also worth restating that the suitability of carbon materials from new sources for long-term implants is not assured simply because the material is carbon. Elemental carbon encompasses a broad spectrum of possible structures and mechanical properties. Each new candidate carbon material requires a specific assessment of biocompatibility based on its own merits, and not by reference to the historically successful General Atomic-type pyrolytic carbons.

BIBLIOGRAPHY

Agafonov, A., Kouznetsova, E., Kouznetsova, V., & Reif, T. (1999). TRI carbon strength and macroscopic isotropy of boron carbide alloyed pyrolytic carbon. *Artif. Organs*, **23**(7), 80.

Akins, R. J., & Bokros, J. C. (1974). The deposition of pure and alloyed isotropic carbons and steady state fluidized beds. *Carbon*, **12**, 439–452.

Baier, R. E., Gott, V. L., & Feruse, A. (1970). Surface chemical evaluation of thrombo-resistant materials before and after venous implantation. *Trans. Am. Soc. Artif. Intern. Organs*, **16**, 50–57.

Beavan, L. A., James, D. W., & Kepner, J. L. (1993). Evaluation of Fatigue in Pyrolite Carbon. In P. Ducheyne, & D. Christiansen (Eds.), *Bioceramics* Vol. 6, (pp. 205–210). Oxford, UK: Butterworth-Heinemann.

Bokros, J. C. (1969). Deposition, Structure and Properties of Pyrolytic Carbon. In P. L. Walker (Ed.), *Chemistry and Physics of Carbon* Vol. 5, (pp. 1–118). New York, NY: Marcel Dekker.

Bokros, J. C., Gott, V. L., LaGrange, L. D., Fadall, A. M., Vos, K. D., et al. (1969). Correlations between blood compatibility and heparin adsorptivity for an impermeable isotropic pyrolytic carbon. *J. Biomed. Mater. Res.*, **3**, 497–528.

Bokros, J. C., Emken, M. R., Haubold, A. D., et al. (1994). *Prosthetic heart valve*. US Patent No. 5,308,361: Issued May 3, 1994.

Bokros, J. C., Ely, J. L., Emken, M. R., et al. (1997). *Prosthetic heart valve*. US Patent No. 5,642,324: Issued June 24, 1997.

Bokros, J. C., Ely, J. L., Emken, M. R., et al. (1998). *Prosthetic heart valve with improved blood flow*. US Patent No. 5,772,694: Issued June 30, 1998.

Bokros, J. C., Stupka, J. C., More, R. B., et al. (2003). *Prosthetic heart valve*. US Patent No. 6,096,075: Issued August 1, 2003.

Bravo, C. J., Rizzo, R., Hormel, K. B., & Beckenbaugh, R. D. (2007). Pyrolytic carbon proximal interphalangeal joint arthroplasty: Results with minimum two-year follow-up evaluation. *J. Hand Surg.*, **32A**, 1–11.

Cao, H. (1996). Mechanical performance of pyrolytic carbon in prosthetic heart valve applications. *J. Heart Valve Dis.*, **5**(Suppl. I), S32–S49.

Chinn, J. A., Phillips, R. E., Lew, K. R., & Horbett, T. A. (1994). Fibrinogen and albumin adsorption to pyrolite carbon. *Trans. Soc. Biomater.*, **17**, 250.

Cook, S. D., Beckenbaugh, R. D., Redondo, J., Popich, L. S., Klawitter, J. J., et al. (1999). Long term follow-up of pyrolytic carbon metacarpophalangeal implants. *J. Bone and Joint Surg.*, **81A**(5), 635–648.

De Salvo, G. (1970). *Theory and Structural Design Applications of Weibull Statistics, Report WANL-TME-2688,* Westinghouse Electric Corporation.

Dillard, J. G. (1995). X-ray Photoelectron Spectroscopy (XPS) and Electron Spectroscopy for Chemical Analysis (ESCA). In H. Ishida, & L. E. Fitzpatrick (Eds.), *Characterization of Composite Materials* Vol. 1, (pp. 22). Boston, MA: Butterworth-Heinemann.

Edmunds, L. H. (1987). Thrombotic and bleeding complications of prosthetic heart valves. *Ann. Thorac. Surg.*, **44**, 430–445.

Ely, J. L., Emken, M. R., Accuntius, J. A., Wilde, D. S., Haubold, A. D., et al. (1998). Pure pyrolytic carbon: Preparation and properties of a new material, on-X carbon for mechanical heart valve prostheses. *J. Heart Valve Dis.*, **7**, 626–632.

Emken, M. R., Bokros, J. C., Accuntius, J. A., & Wilde, D. S. (1993). *Precise control of pyrolytic carbon coating*. Buffalo, New York: Presented at the 21st Biennial Conference on Carbon. June 13–18, 1993, Extended Abstracts and Program Proceedings, pp. 531–532.

Emken, M. R., Bokros, J. C., Accuntius, J. A., & Wilde, D. S. (1994). *U.S. Patent No. 5,284,676, Pyrolytic deposition in a fluidized bed*. Feb. 8, 1994.

Feng, L., & Andrade, J. D. (1994). Protein adsorption on low-temperature isotropic carbon: I. Protein conformational change probed by differential scanning calorimetry. *J. Biomed. Mater. Res.*, **28**, 735–743.

Flameng, W., & Meuris, B. (2002). *Performance of bileaflet heart valve prostheses in an animal model of heart valve thrombosis*. 6th Annual Hilton Head Workshop on Prosthetic Heart Valves. South Carolina: Hilton Head. March 610.

Gilpin, C. B., Haubold, A. D., & Ely, J. L. (1993). Fatigue Crack Growth and Fracture of Pyrolytic Carbon Composites. In P. Ducheyne, & D. Christiansen (Eds.), *Bioceramics* Vol. 6, (pp. 217–223). Oxford, UK: Butterworth-Heinemann.

Goodman, S. L., Tweden, K. S., & Albrecht, R. M. (1995). Three-dimensional morphology and platelet adhesion on pyrolytic carbon heart valve materials. *Cells Mater.*, **5**(1), 15–30.

Griffin, C. D., Buchanan, R. A., & Lemons, J. E. (1983). *In vitro* electrochemical corrosion study of coupled surgical implant materials. *J. Biomed. Mater. Res.*, **17**, 489–500.

Hanson, S. R. (1998). Blood–Material Interactions. In J. Black, & G. Hastings (Eds.), *Handbook of Biomaterial Properties* (pp. 545–555). London, UK: Chapman and Hall.

Haubold, A. D. (1994). On the durability of pyrolytic carbon *in vivo*. *Medi. Prog. Technol.*, **20**, 201–208.

Haubold, A. D., & Ely, J. L. (1995). *Carbons used in mechanical heart valves. Transactions Society for Biomaterials*. San Francisco: 21st Annual Meeting. 275.

Haubold, A. D., Shim, H. S., & Bokros, J. C. (1981). Carbon in Medical Devices. In D. P. Williams (Ed.), *Biocompatibility of Clinical Implant Materials* (pp. 3–42). Boca Raton, Florida: CRC Press.

Kaae, J. L., & Wall, D. R. (1996). Microstructural characterization of pyrolytic carbon for heart valves. *Cells Mater.*, **4**, 281–290.

Kafesjian, R., Howanec, M., Ward, G. D., Diep, L., Wagstaff, L., et al. (1994). Cavitation damage of pyrolytic carbon in mechanical heart valves. *J. Heart Valve Dis.*, **3**(Suppl. I), S2–S7.

Kelpetko, V., Moritz, A., Mlczoch, J., Schurawitzki, H., Domanig, E., et al. (1989). Leaflet fracture in Edwards-Duromedics bileaflet valves. *J. Thorac. Cardiovasc. Surg.*, **97**, 90–94.

King, R. N., Andrade, J. D., Haubold, A. D., & Shim, H. S. (1981). Surface Analysis of Silicon: Alloyed and Unalloyed LTI Pyrolytic Carbon. In D. W. Dwight, T. J. Fabish, & H. R. Thomas (Eds.), *Photon, Electron and Ion Probes of Polymer Structure and Properties, ACS Symposium Series* **162** (pp. 383–404). Washington, DC: American Chemical Society.

LaGrange, L. D., Gott, V. L., Bokros, J. C., & Ramos, M. D. (1969). Compatibility of Carbon and Blood. In R. J. Hegyeli (Ed.), *Artificial Heart Program Conference Proceedings* (pp. 47–58). Washington, DC: US Government Printing Office.

Lee, R. G., & Kim, S. W. (1974). Adsorption of proteins onto hydrophobic polymer surfaces: Adsorption isotherms and kinetics. *J. Biomed. Mater. Res.*, **8**, 251.

Ma, L., & Sines, G. (1996). Fatigue of isotropic pyrolytic carbon used in mechanical heart valves. *J. Heart Valve Dis.*, **5**(Suppl. I), S59–S64.

Ma, L., & Sines, G. (1999). Unalloyed pyrolytic carbon for implanted heart valves. *J. Heart Valve Dis.*, **8**(5), 578–585.

Ma, L., & Sines, G. (2000). Fatigue behavior of pyrolytic carbon. *J. Biomed. Mater. Res.*, **51**, 61–68.

More, R. B., & Haubold, A. D. (1996). Surface chemistry and surface roughness of clinical pyrocarbons. *Cells Mater.*, **6**, 273–279.

More, R. B., & Silver, M. D. (1990). Pyrolytic carbon prosthetic heart valve occluder wear: *in vivo* vs. *in vitro* results for the Björk–Shiley prosthesis. *J. Appl. Biomater.*, **1**, 267–278.

More, R. B., Kepner, J. L., & Strzepa, P. (1993). Hertzian Fracture in Pyrolite Carbon. In P. Ducheyne, & D. Christiansen (Eds.), *Bioceramics* Vol. 6, (pp. 225–228). Oxford, UK: Butterworth-Heinemann.

Nyilas, E., & Chiu, T. H. (1978). Artificial surface/sorbed protein structure/hemocompatibility correlations. *Artif. Organs*, **2**(Suppl.), 56–62.

Okazaki, Y., Wika, K. E., Matsuyoshi, T., Fukamachi, K., Kunitomo, R., et al. (1997). Platelets were early postoperative depositions on the leaflet of a mechanical heart valve in sheep without postoperative anticoagulants or antiplatelet agents. *ASAIO J.*, **42**, M750–M754.

Pauling, L. (1964). *College Chemistry* (3rd ed). San Francisco, CA: W. H. Freeman and Company.

Reilly, D. T., & Burstein, A. H. (1974). The mechanical properties of bone. *J. Bone Joint Surg. Am.*, **56**, 1001.

Reilly, D. T., Burstein, A. H., & Frankel, V. H. (1974). The elastic modulus for bone. *J. Biomech.*, **7**, 271.

Richard, G., & Cao, H. (1996). Structural failure of pyrolytic carbon heart valves. *J. Heart Valve Dis.*, **5**(Suppl. I), S79–S85.

Ritchie, R. O., Dauskardt, R. H., Yu, W., & Brendzel, A. M. (1990). Cyclic fatigue-crack propagation, stress corrosion and fracture toughness behavior in pyrolite carbon coated graphite for prosthetic heart valve applications. *J. Biomed. Mat. Res.*, **24**, 189–206.

Sadeghi, H. (1987). Dysfonctions des prostheses valvulaires cardaques et leur traitment chirgical. *Schwiez. Med. Wochenschr.*, **117**, 1665–1670.

Salzman, E. W., Lindon, J., Baier, D., & Merril, E. W. (1977). Surface-induced platelet adhesion, aggregation and release. *Ann. NY. Acad. Sci.*, **283**, 114.

Sattler, K. (1995). Scanning tunneling microscopy of carbon nanotubes and nanocones. *Carbon*, **7**, 915–920.

Sawyer, P. N., Lucas, L., Stanczewski, B., Ramasamy, N., Kammlott, G. W., et al. (1975). Evaluation techniques for potential cardiovascular prosthetic alloys experience with titanium aluminum 6–4 ELI tubes. *Proceedings of the San Diego Biomedical Symposium*, Vol. 14, 423–427.

Schoen, F. J. (1983). Carbons in Heart Valve Prostheses: Foundations and Clinical Performance. In M. Zycher (Ed.), *Biocompatible Polymers, Metals and Composites* (pp. 240–261). Lancaster, PA: Technomic.

Schoen, F. J., Titus, J. L., & Lawrie, G. M. (1982). Durability of pyrolytic carbon-containing heart valve prostheses. *J. Biomed. Mater. Res.*, **16**, 559–570.

Smith, K. L., & Black, K. M. (1984). Characterization of the treated surfaces of silicon alloyed pyrolytic carbon and SiC. *J. Vac. Sci. Technol.*, **A2**, 744–747.

Stanley, J., Klawitter, J., & More, R. (2008). Replacing Joints with Pyrolytic Carbon. In P. A. Revell (Ed.), *Joint Replacement Technology: New Developments* (pp. 631–656). Cambridge, UK: Woodhead Pub and CRC Press LLC.

FDA (US Food and Drug Administration) (2001). *Summary of Safety and Effectiveness, On-X® Prosthetic Heart Valve*. FDA PMA P000037 May 30.

FDA (US Food and Drug Administration) (2002). *Summary of Safety and Effectiveness, On-X® Prosthetic Heart Valve*. FDA P000037/S1, March 6, 2002, European Primary Trial, updated May 31, 2003.

Thompson, N. G., Buchanan, R. A., & Lemons, J. E. (1979). *In vitro* corrosion of Ti-6Al-4V and Type 316L stainless steel when galvanically coupled with carbon. *J. Biomed. Mater. Res.*, **13**, 35–44.

Wieting, D. W. (1996). The Björk–Shiley Delrin tilting disc heart valve: Historical perspective, design and need for scientific analyses after 25 years. *J. Heart Valve Dis.*, **5**(Suppl. I), S157–S168.

Williams, D. F. (1998). General Concepts of Biocompatibility. In J. Black, & G. Hastings (Eds.), *Handbook of Biomaterial Properties* (pp. 481–489). London, UK: Chapman and Hall.

Williams, M. A., & van Riet, S. (2006). The On-X® prosthetic heart valve: Mid-term results in a poorly anticoagulated population. *J. Heart Valve Dis.*, **15**, 80–86.

CHAPTER I.2.9 COMPOSITES

Claudio Migliaresi
Department of Materials Engineering and Industrial Technologies and BioTech Research Center, University of Trento, Trento, Italy

INTRODUCTION

The word *composite* means "consisting of two or more distinct parts." At the atomic level, materials such as metal alloys and polymeric materials could be called composite materials in that they consist of different and distinct atomic groupings. At the microstructural level (about 1 to 10 microns), a metal alloy such as a plain carbon steel containing ferrite and pearlite could be called a composite material since the ferrite and pearlite are distinctly visible constituents as observed in the optical microscope.

In engineering, a composite material usually refers to a material consisting of constituents in the nano- to micro- to macrosize range, each having a distinct interface separating them. Such composites usually consist of one or more discontinuous phases embedded within a continuous phase. The discontinuous phase is usually harder and stronger than the continuous phase, and is called the *reinforcement* or *reinforcing material*, whereas the continuous phase is termed the *matrix*. In some cases, tough fillers, e.g., rubber particles, are combined with brittle matrices in order to produce higher toughness materials with better impact strength. In other cases, the "reinforcement" could be aimed at achieving specific functional properties, such as bioactivity in the case of biomedical composites. Many body tissues are composites, such as extracellular matrix (ECM), tendons, ligaments, skin, bone, and so on, with an additional complexity due to their hierarchical structure.

The addition to a matrix of harder, stronger or tougher fillers may improve to different extents the resulting material stiffness, strength or toughness, depending on the filler type, content, and filler–matrix adhesion. Properties of composites are strongly influenced by the properties of their constituent materials, their distribution and content, and the interface and interphase interactions between them. The interface is particularly important for short fiber- or particle-reinforced composites. External applied stresses are in fact transferred from the matrix to the filler through the interface, and properties of composites with the same constituents but strong or weak interfaces can be better or worse than those of the pure matrix.

The composite properties may be the volume fraction sum of the properties of the constituents, or the constituents may interact in a synergistic way (e.g., due to geometrical orientation) so as to provide properties in the composite that are not accounted for by a simple volume fraction sum. Thus, in describing a composite material, besides specifying the constituent materials and their properties, one needs to specify the geometry of the reinforcement, their concentration, distribution, and orientation. At increasing filler surface-to-volume ratio, i.e., at decreasing filler dimensions, the role of interface/interphase becomes more and more important, and, for this reason, properties of nanocomposites (matrices loaded with fillers of nanometric sizes) are difficult to predict.

Most composite materials are fabricated to provide desired mechanical properties such as strength, stiffness, toughness, and fatigue resistance. Therefore, it is natural to study together the composites that have a common strengthening mechanism. The strengthening mechanism strongly depends upon the geometry of the reinforcement. This approach leads to a classification of composites on the basis of the geometry of a representative unit of reinforcement more than on the type of matrix, e.g., composites with long fibers, short fibers or particles (see Figure I.2.9.1). A fiber is characterized as having a length that is much greater than its cross-sectional dimensions. Particle-reinforced composites are sometimes referred to as *particulate composites*. Short or long fiber-reinforced composites are, understandably, called *fibrous composites*. Matrices loaded with nanofibers or nanoparticles are referred as *nanocomposites*. *Laminates* are composite structures made by stacking laminae of oriented fiber composites. Such composites are characterized by the number and orientation of the laminae.

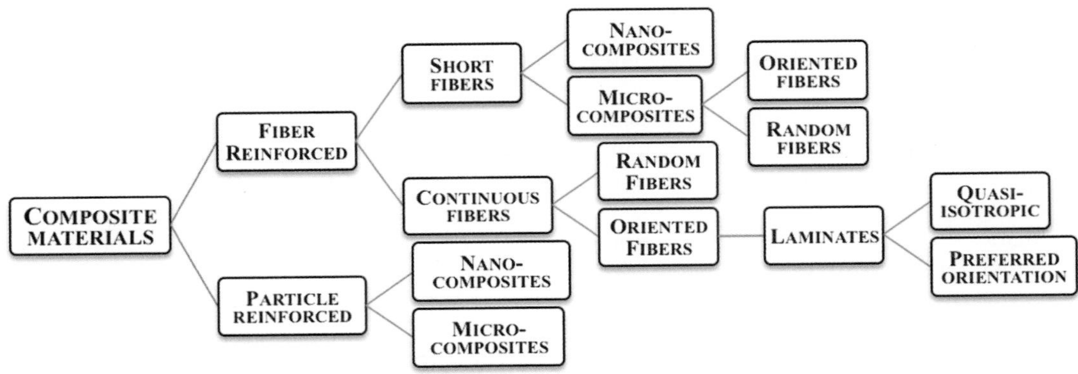

FIGURE I.2.9.1 Classification of composite materials.

Fibers are much more mechanically effective than particles, and polymer–fiber composites can reach stiffness and strengths per unit area that are comparable to metals and even higher. Moreover, while particle-reinforced composites are isotropic, fiber-reinforced composites are anisotropic. Properties in different directions can, in most cases, be designed to match specific requirements. The ability to design anisotropic properties into a composite material is without doubt one of the most important advantages offered by composites with respect to more traditional monolithic materials. Most of the composite body tissues have anisotropic properties, and the match between the properties of the prosthesis and the properties of the substituted tissues is, when possible, the first design criterion.

Failure of a composite material implant can expose fibers or particles to the surrounding biological environment. In many cases failure in composites is preceded by the failure of the interface between filler and matrix, this being due to hydrothermal aging or stresses exceeding the interface strength (Pegoretti and Migliaresi, 2002). Sterilization methods and conditions can play an important role, due to the potential sensitivity of the matrix, and sometimes the filler and its interface, to temperature, humidity or radiation (Godara et al., 2007).

Proper arrangement of fibers in a soft polymer matrix can generate materials mimicking structure and mechanical properties of soft tissues, such as tendons, ligaments, arteries, and so on. These advantages can be used to construct arterial prostheses (Gershon et al., 1990, 1992), or intervertebral discs mimicking the natural structure (Ambrosio et al., 1996; Gloria et al., 2007), or fixation plates and nails with controlled stiffness (Veerabagu et al., 2003), or composite scaffolds for cartilage tissue engineering (Moutos and Guilak, 2008). The addition of hydroxyapatite particles into a stiff biodegradable polymer may improve the compression properties of the matrix while enhancing osteoblast proliferation and improving bone mineralization (Wahl and Czernuszka, 2006). For some applications, moreover, radiolucency is considered to be a further potential advantage. An example is when external or internal fracture fixation devices do not shield the bone fracture site in the X-ray radiograph (Migliaresi et al., 2004).

Design flexibility, strength, and lightweight properties have helped to make carbon fiber-reinforced materials into ideal materials for orthotic aids that can return walking and even athletic abilities to impaired people (Dawson, 2000). Composite material prostheses can be tailored to meet the specific requirements for amputee athletes aiming to compete. Recently, C-shaped carbon fiber composite blades that simulated the spring action of a normal foot allowed a South African double-amputee sprinter, Oscar Pistorius, to run and to compete with the strongest unimpaired world-class athletes. Examples of composite prosthetics for athletes can be found in McCarvill (2005) and Nolan (2008).

As with all biomaterials, the question of biocompatibility (tissue response to the material) is paramount. Being composed of two or more materials, composites may have enhanced probability of causing adverse tissue reactions. Also, the fact that one constituent (the reinforcement) usually has dimensions on or even below the cellular scale always leaves open the question of the effect of cellular ingestion of the particulate which can result in either the expression and secretion of tissue-lysing enzymes by the cell, or transport into the lymph system.

REINFORCING SYSTEMS

The main reinforcing materials that have been used in biomedical composites are carbon fibers, polymer fibers, ceramic particles, and glass fibers and particles. Depending upon the application, the reinforcements have either been inert or absorbable. Several nanocomposites have also been studied in the last few years that contain carbon nanotubes, nanoclays, silica, and hydroxyapatite nanoparticles.

Carbon Fiber

Carbon fiber is a lightweight, flexible, high strength, and high tensile modulus material produced by the pyrolysis of organic precursor fibers, such as rayon, polyacrylonitrile (PAN), and pitch in an inert environment (Matsumoto, 1985).

The term carbon is often indifferently interchanged with the term graphite, but carbon and graphite fibers differ in the temperature of fabrication, thermal treatment, and the content of carbon (93–95% for carbon fibers; more than 99% for graphite fibers). Due to their low density (depending on the precursor, density ranges from 1.7 to 2.1 g/cm^3) and high mechanical properties (elastic moduli up to 900 GPa and strengths up to 4.5 GPa, depending on the precursor and on the fabrication process – hence they can be much stiffer and stronger than steel!) these fibers are used in composites in a variety of applications that demand light weight and high mechanical properties. Their disadvantage is that carbon fibers have poor shear strength and can be brittle.

In medicine, several products have used carbon fibers. Some of them reached the market, some others clinical studies in humans. Examples of carbon fiber composites that are already in the market are the Zimmer BAK® Vista® interbody fusion implants poly(ether-ether-ketone) (PEEK) and carbon fibers), the OCELOT™ Stackable Cage System made by DePuy (PEEK and carbon fibers), the QUANTUM™ Humeral Composite Nailing System produced by Invibio (PEEK and carbon fibers), and the different prosthetics (spine, limb, knee, foot, etc.) produced by Ossur (UK).

Studies on the use of carbon fiber composites for biomedical applications date back to 1990 (Kwarteng, 1990). Examples are spine cages (Brooke et al., 1997) or hip prosthesis stems with tailored rigidity where long carbon fibers have been coupled with epoxy resins, polysulfone or poly(ether-ether-ketone) (PEEK). A follow-up of 51 carbon fiber/epoxy resin composite hip joint stem prostheses implanted in humans did show failure at the bone–implant interface without osteointegration, but never showed failure or wear of the composites (Adam et al., 2002). Some early devices, however, have experienced severe negative effects and have been recalled from the market. Two examples are:

- Short carbon fiber-reinforced ultra-high molecular weight polyethylene (UHMWPE) was developed for orthopedic applications. The assumption was that increase of strength and decrease of creep would increase the longevity of the articular surfaces of the hip implant. The favorable indications of reduced wear in laboratory tests contrasted with the *in vivo* results, where many patients were presented with osteolysis and failure of the tibial inserts (Kurtz et al., 1999).
- In the 1980s carbon fibers were used to develop a scaffolding device to induce tendon or ligament repair. The low shear strength of fibers caused fiber breakage and the formation of harmful debris. A resorbable polymeric coating was somewhat successful in preventing carbon fiber breakage and localizing debris. However, due to poor performance and permanent wear debris in the joint, the carbon fiber device was not approved by the US Food and Drug Administration (FDA) for anterior cruciate ligament (ACL) reconstruction (Dunn, 1998).

In spite of these early failures, however, carbon fibers display unique properties for the fabrication of loadbearing medical devices. Most successful applications are for prosthetics where carbon fiber composites exhibit unique properties in term of lightness, stiffness, and strength.

Polymer Fibers

Polymer fibers are not comparable to carbon fibers in strength or stiffness when used to reinforce other polymers, with the possible exceptions of aramid fibers or UHMWPE fibers. For biomedical applications, biocompatibility, of course, and high strength and fatigue resistance are compulsory, while stiffness is a design parameter to be adapted to the specific conditions. This is why poly(ethylene terephthalate) (PET) fibers have been used for some applications. In addition, thanks to their absorbability, not to their mechanical superiority, certain absorbable fibers have been employed (see below).

- *Aramid* is the generic name for aromatic polyamide fibers. The most well-known aramids are Kevlar™ and Nomex™ (DuPont trademarks), and Twaron™ (made by Teijin/Twaron of Japan). Kevlar is produced by spinning a sulfuric acid/poly(p-phenylene terephthalamide) solution through an air layer into a coagulating water bath. Aramid fibers are relatively light (density = 1.44 g/cm^3), stiff (the modulus can go up to 190 GPa), and strong (tensile strength about 3.6 GPa). Moreover, they resist impact and abrasion damage, due to the high energy dissipation mechanism involved in their failure (they break into microfibrils). In particular, aramide fibers are less stiff and strong than carbon, but being lighter, they possess comparable specific strength (i.e., strength-to-weight ratio). A negative point, that can be relevant for biomedical applications, is that aramid fibers absorb moisture, and worth noting is their poor compressive strength, about 1/8 of the tensile strength. Aramid fiber composites are used commercially where high tensile strength and stiffness, resistance to impact damage, and resistance to fatigue and stress rupture are important. In medicine, these composites have not seen extensive use, due perhaps to some concerns about their biocompatibility or long-term fate. Main applications have been in dentistry and for ligament prostheses (Wening et al., 1994). The use of partially degradable Kevlar-reinforced composites for bone fracture fixation plates has also been proposed (Jayabalan, 2009). In this paper the author refers to publications assessing the *in vivo* biocompatibility of Kevlar; these papers, while assessing the absence of short-term negative reaction to a PMMA–Kevlar composite and to Kevlar bundles implanted subcutaneously in rats, did not fully elucidate the long-term reaction to free Kevlar fibers in the body (Henderson et al., 1987; Wening et al., 1989).
- Commercially available high-strength, high-modulus polyethylene fibers include Spectra® from Honeywell Performance Fibers (Colonial Heights, VA), Dyneema® from DSM (Heerlen, The Netherlands), and Toyobo fibers from Toyobo (Shiga, Japan). Ultra-high molecular weight polyethylene (UHMWPE) fibers are produced by a gel-spinning technique starting from an approximately 2–8 weight percent solution of the ultra-high molecular weight polymer ($Mw > 10^6$) in a common solvent, such as decalin. Spinning at 130–140°C and hot drawing at very high draw ratios produces fibers with the highest specific strength of all commercial fibers available to date. UHMWPE fibers possess high modulus and strength, besides displaying light weight (density about 0.97 g/cm^3) and high-energy dissipation ability, compared to other fibers. In addition, polyethylene fibers resist abrasion and do not absorb water. However, the chemical properties of UHMWPE fibers are such that few resins bond well to the fiber surfaces, and so the structural properties expected from the fiber properties are often not fully realized in a composite. The low melting point of the fibers (about 147°C) impedes high temperature fabrication. Bulk UHMWPE has extensive applications in medicine for the fabrication of bearings for joint prostheses, displaying excellent biocompatibility, but with lifetime restricted by its wear resistance. Bulk UHMW polyethylene demonstrates excellent biocompatibility.

Polyethylene fibers have been used to reinforce acrylic resins for application in dentistry or to make intervertebral disc prostheses (Kotani et al., 2002). They have also been used for the fabrication of ligament augmentation devices (Guidoin et al., 2000).

- Dacron™ is the DuPont trade name commonly used for poly(ethylene terephthalate) (PET) fibers. These fibers have several biomedical uses, mostly in cardiovascular surgery for arterial grafts or as fabrics for implant fixation, as sewing rings on artificial heart valves. PET fibers have also been proposed in orthopedics for the fabrication of artificial tendons or ligaments (Kolarik et al., 1981) and ligament augmentation devices, as fibers or fabrics alone or as a reinforcing fabric embedded in a polymer matrix, such as in the casing for ventricles in the total artificial heart. Other proposed applications include soft tissue prostheses, intervertebral discs (Ambrosio et al., 1996), and plastic surgery applications.
- Poly(lactic acid) (PLA), poly(glycolic acid) (PGA) and their copolymers (poly(lactic-co-glycolic acid), PLGA) are the principal biodegradable polymers used for the fabrication of biodegradable fibers. These fibers have been used for a number of years in absorbable sutures. Properties of these fibers depend upon several factors, such as LA/GA ratio, water absorption, degree of crystallinity, molecular weight, purity, etc. (Migliaresi and Fambri, 1997). Fibers and tissues of these polymers have been proposed for ligament reconstruction or as scaffolds for tissue engineering applications (Lu and Mikos, 1996; Guarino et al., 2008). PLA and PGA fibers have been employed in composites, e.g., in combination with PLA biodegradable matrices. Examples are the intramedullary biodegradable pins and plates (Middleton and Tipton, 2000), and biodegradable scaffolds for bone regeneration. Electrospun nonwoven nets of PLA and PGA have been recently used in combination with several natural and synthetic polymer matrices to construct degradable scaffolds for tissue engineering applications.
- Polycaprolactone (PCL) is a relatively hydrophobic, slowly degrading, semicrystalline polymer used as a homopolymer or as a component of copolymers with PLA and PGA in implant applications for drug release systems, adhesion barrier film, plastic surgery, and, more recently, for the fabrication of tissue engineering scaffolds. PCL fibers for biomedical applications have been produced by electrospinning, loaded by hydroxyapatite powders (Erisken et al., 2008; Gupta et al., 2009) or used as reinforcement for natural or synthetic polymer matrices.

Ceramics

A number of different ceramic materials have been used to reinforce biomedical composites. Since most biocompatible ceramics, when loaded in tension or shear, are relatively weak and brittle materials compared to metals, the preferred form for this reinforcement has usually been particulate. These reinforcements have included various calcium phosphates, aluminum- and zinc-based phosphates, glass and glass–ceramics, and bone mineral. Minerals in bone are numerous. In the past, bone has been defatted, ground, and calcined or heated to yield a relatively pure mix of the naturally occurring bone minerals. It was recognized early that this mixture of natural bone mineral was poorly defined and extremely variable. Consequently, its use as an implant material was limited.

Tricalcium phosphates and hydroxyapatite are commonly referred as bioceramics, i.e., bioactive ceramics. The definition refers to their ability to elicit a specific biological response that results in the formation of a bond between the tissues and material (Hench et al., 1971).

The calcium phosphate ceramic systems have been the most intensely studied ceramic systems. Of particular interest are the calcium phosphates having calcium to phosphorous ratios of 1.5–1.67. Tricalcium phosphate and hydroxyapatite form the boundaries of this compositional range. At present, these two materials are used clinically for dental and orthopedic applications. Tricalcium phosphate has a nominal composition of $Ca_3(PO_4)_2$. The common mineral name for this material is *whitlockite*. It exists in two crystographic forms, α- and β-whitlockite. In general, it has been used in the β-form. Moreau and Xu (2009) investigated the properties of calcium phosphate–chitosan composites to be used as injectable bone repair materials. The ceramic hydroxyapatite has received a great deal of attention. Hydroxyapatite is, of course, the major mineral component of bone. The nominal composition of this material is $Ca_{10}(PO_4)_6(OH)_2$.

Hydroxyapatite (HA) ceramic and tricalcium phosphates are used in orthopedics and dentistry, alone or in combination with other substances or also as coating of metal implants. The rationale behind the use of bioceramics in combination with polymeric matrices for composites is in their potential ability to enhance the integration in bone, while improving the device mechanical properties. One example is the HA–Polyethylene composite developed by Bonfield (Bonfield, 1988; Bonfield et al., 1998), and today commercialized with the name of HAPEX (Smith & Nephew ENT, Memphis, TN, USA).

Many papers have been published on HA-reinforced polymers. A recent example is the use of HA to reinforce PLGA for the fabrication of nanofibrous scaffolds for tissue engineering applications (Jose et al., 2009).

Glasses

Glass fibers, namely C glasses and S glasses, are used to reinforce plastic matrices to form structural composites and molding compounds. Commercial glass fiber plastic composite materials have the following favorable characteristics: high strength-to-weight ratio; good dimensional stability; good resistance to heat, cold, moisture, and corrosion; good electrical insulation properties; ease of fabrication; and relatively low cost. The following are examples of glass composites:

- De Santis et al. (2000) have stacked glass and carbon/PEI (poly(ether-imide)) laminae to manufacture a hip prosthesis with constant tensile modulus, but with bending modulus increasing in the tip-head direction.
- An isoelastic intramedullary nail made of PEEK and chopped glass fibers has been evaluated by Lin et al. (1997). Zhao et al. (2009) compared properties and *in vivo* behavior of intramedullary nails that were implanted in rabbits; the nails were either made of titanium or a Bis-GMA–TEGDMA/E-glass fiber composite that was HA-coated. They observed that the composite materials implants induced proper healing of the fixed femurs with no signs of adverse tissue reactions or implant failure.

The term isoelastic has often been improperly used. It means "having the same elastic modulus as." For bone fracture fixation plates, the fact that the plate has an elastic modulus similar to bone (i.e., isoelastic) has been claimed to be a major requirement. The concept is misleading, because the deformation of a plate would not only depend on the elastic modulus of the material that it is composed of, but also on the size and shape of the plate (for loading mode other than tension or compression). For example, a toothpick and the tree from which it was made are isoelastic, but no one could imagine supporting a tree with a toothpick!

Glass fibers have also been used to increase the mechanical properties of acrylic resins for applications in dentistry (Chen et al., 2001). Zimmerman et al. (1991) and Lin (1986) introduced an absorbable polymer composite reinforced with an absorbable calcium phosphate glass fiber. This allowed for the fabrication of a completely absorbable composite implant material. Commercial glass fiber produced from a lime–aluminum–borosilicate glass typically has a tensile strength of about 3 GPa and a modulus of elasticity of 72 GPa. Lin (1986) estimates the absorbable glass fiber to have a modulus of 48 GPa, comparing favorably with the commercial fiber. The tensile strength, however, was significantly lower, approximately 500 MPa.

Nanoparticles and Nanofibers: "Nanofillers"

In recent years, polymers filled with nanoparticles or nanofibers have been proposed for several applications in medicine. Due to the huge amount of interface or interphase being created between fillers and matrix by nanofillers, small amounts of such fillers can cause large modifications in both mechanical and functional properties of the matrix. The list of nano-reinforcements that have been investigated for biomedical use includes silica nanoparticles, nanoclays, carbon nanotubes, and POSS (polyhedral oligomeric silsesquioxane, silicon and oxygen atoms linked together in a cubic form, with silicon atoms occupying the corners). Here are some examples of nanocomposites:

- Nanoclays, such as montomorillonite, have been used as reinforcement in poly(L-lactic acid) polymer (PLLA) electrospun fibers (Lee et al., 2005).
- A recent work suggested the use of poly(carbonate)–urethane–POSS nanocomposites for blood contacting applications (Sarkar et al., 2009). These materials were found to be thrombo-resistant, biostable, and more compliant than PTFE vascular grafts *in vitro*.
- Carbon nanotubes are geometrically nanofibers, having diameters of the order of 10 nm (multi-walled nanotubes) or 1 nm (single-walled nanotubes), and lengths equal to 1000–2000 times their diameter. Their exceptionally high mechanical properties (Elastic modulus = 1 TeraPascal; Strength (calculated) = 50 GPa) have stimulated numerous researchers to study their application for composites, also for biomedical applications. For instance, carbon nanotubes have been used to produce electrospun silk fibroin nanocomposites for potential tissue engineering applications (Gandhi et al., 2009). However, carbon nanotubes can be cytotoxic (Tian et al., 2006), and this should be clearly kept in mind when investigating their use for biomedical composites.

The use of nanoparticles to reinforce polymer matrices has brought novelty and excitement to the field of composites, and as often occurs, this trend has also become very popular for biomedical applications. However, the biomaterials scientist should be aware of the fact that different sized objects of the same material can induce very different biological responses, especially in this case, where the size scale of at least one dimension of a nanoparticle or "nanorod" (e.g., nanofiber or nanotube) is much smaller than a cell.

MATRIX SYSTEMS

Ceramic matrix or metal matrix composites have important technological applications, but their use is mostly in non-biomedical applications (e.g., cutting tools, power generation equipment, process industries, aerospace), with just a few examples for biomedical applications (e.g., calcium phosphate bone cements). Most biomedical composites have polymeric matrices, mostly thermoplastic, that can be bioabsorbable or not.

The most common matrices in biomedical products are synthetic non-absorbable polymers, such as polysulfone, poly(ether-ether-ketone) (PEEK), ultra-high molecular weight polyethylene (UHMWPE), polytetrafluoroethylene (PTFE) and poly(methylmethacrylate)

(PMMA). These matrices may be reinforced with carbon fibers, polyethylene fibers, and ceramic particles. They have been used as components of prosthetic hip stems, fracture fixation devices, artificial joint bearing surfaces, artificial tooth roots, and bone cements. A comprehensive review on PEEK biomaterials in medicine is found in Kurtz and Devine (2007). Epoxy composite materials have also been used. However, due to concerns about the toxicity of monomers (Morrison et al., 1995) the research activity on epoxy composites for implantable devices gradually decreased. Hydrogel matrix composites have been also studied for soft tissue applications (Ambrosio et al., 1996; Kolarik et al., 1981).

The materials used and some proposed applications are reported in Table I.2.9.1. Not all the proposed systems have undergone clinical trials, and only some of them are commercialized today. A review on biomedical applications of composites is found in Ramakrishna et al. (2001).

Absorbable composite implants can be produced from absorbable α-polyester materials such as poly(lactic acid) (PLA) and poly(glycolic acid) (PGA) polymers. Previous work has demonstrated that, for most applications, it is necessary to reinforce these polymers to obtain adequate mechanical strength. PGA was the first biodegradable polymer synthesized (Frazza and Schmitt, 1971). It was followed by PLA and copolymers of the two (Gilding and Reed, 1979). These α-polyesters have been investigated for use as sutures, and as implant materials for the repair of a variety of

TABLE I.2.9.1 Some Examples of Biomedical Composite Systems

Applications	Matrix/Reinforcement*	Reference
External fixator	Epoxy resin/CF	Baidya et al., 2001; Migliaresi et al., 2004
Bone fracture fixation plates, pins, screws	Epoxy resins/CF	Ali et al.,1990; Veerabagu et al., 2003; Pemberton et al., 1994
	PMMA/CF	Woo et al., 1974
	PSU/CF	Claes et al., 1997
	PP/CF	Christel et al., 1980
	PE/CF	Rushton and Rae, 1984
	PBT/CF	Gillett et al., 1986
	PEEK/CF	Fujihara et al., 2001
	PEEK/GF	Lin et al., 1997
	PLLA/HA	Furukawa et al., 2000 a,b
	PLLA/PLLA fibers	Tormala,1992; Rokkanen et al., 2000
	PGA/PGA fibers	Tormala,1992; Rokkanen et al., 2000
Spine surgery	PU/Bioglass	Claes et al., 1999
	PSU/Bioglass	Marcolongo et al., 1998
	PEEK/CF	Ciappetta et al., 1997
	Hydrogels/PET fibers	Ambrosio et al., 1996
	PLA/PLA fibers/CP	Huttunen et al., 2006
Bone cement	PMMA/HA particles	Morita et al., 1998
	PMMA/Glass beads	Shinzato et al., 2000
	Calcium phosphate/aramid fibers,CF,GF,PLGA fibers	Xu et al., 2000
	PMMA/UHMWPE fibers	Yang et al., 1997
Dental cements and other dental applications	Bis-GMA/inorganic particles	Moszner and Salz , 2001
	PMMA/KF	Pourdeyhimi et al., 1986; Vallittu, 1996
Acetabular cups	PEEK/CF	Wang et al., 1998
Hip prostheses stem	PEI/CF-GF	De Santis et al., 2000
	PEEK/CF	Akay and Aslan, 1996; Kwarteng, 1990
	CF/PA12	Campbell et al., 2008
Bone replacement, substitute	PE/HA particles	Bonfield, 1988; Bonfield et al., 1998
Bone filling, regeneration	Poly(propylene fumarate)/TCP	Yaszemski et al., 1996
	PEG-PBT/HA	Qing et al., 1997
	PLGA/HA fibers	Thomson et al., 1998
	P(DLLA-CL)/HA particles	Ural et al., 2000
	Starch/HA particles	Reis and Cunha, 2000; Leonor et al., 2003
Tendons and ligaments	Hydrogels/PET	Kolarik et al., 1981; Iannace et al., 1995
	Polyolefins/UHMWPE fibers	Kazanci et al., 2002
Vascular grafts	PELA/Polyurethane fibers	Gershon et al., 1990; Gershon et al., 1992
Prosthetic limbs	Epoxy resins/CF, GF, KF	Dawson, 2000

*See Glossary of Terms.

osseous and soft tissues. Other biodegradable polymers that have been developed include poly(ortho esters), synthesized by Heller and co-workers (Heller et al., 1980), and a class of bioerodable dimethyl-trimethylene carbonates (DMTMCs) (Tang et al., 1990). A review of absorbable polymers by Nair and Laurencin (2007) included PGA, PLA, PLGA polydioxanone, polycaprolactone (PCL), poly(trimethylenecarbonate), bacterial polyesters, polyurethanes, poly(esteramide), poly(orthoesters), polyanhydrides, poly(anhydride-co-imide), cross-linked polyanhydrides, poly(propylene fumarate), poly(pseudo amino acids), poly(alkyl cyanoacrylates), polyphosphazenes, and polyphosphoesters. Another review by Middleton and Tipton (2000) focused on biodegradable polymers suited for orthopedic applications, mainly PGA, PLA, and PLGA. The authors examined the chemistry, fabrication, degradation mechanisms, and biocompatibility of different polymers and their devices. Degradable polymers are also discussed in other chapters of this text.

In addition to synthetic biodegradable polymers, natural polymers have been also investigated as matrices for biomedical composites (see also Chapter I.2.7). Examples are collagen (Wahl and Czernuszka, 2006), silk fibroin (Wang et al., 2010), and chitosan (Kong et al., 2005). Reis et al. (1998) have studied blends of starch with various thermoplastic polymers. They were proposed for a large range of applications such as temporary hard tissue replacement, bone fracture fixation, drug delivery devices or tissue engineering scaffolds.

FABRICATION OF COMPOSITES

Composite materials can be fabricated with different technologies. Some of them are particular for the type of filler (particle, short or long fiber) and matrix (thermoplastic or thermosetting). Some methods, such as compression molding, extrusion or injection molding, are the same that are used for polymers. Some make use of solvents that are unacceptable, even in trace amounts, and are therefore not suitable for the fabrication of biomedical composites. Some biomedical composites, moreover, are fabricated *in situ*. This is the case for composite bone cements.

The selection of the most appropriate manufacturing technology is also influenced by the relatively low volumes of the materials used in the product compared to other applications, and by the relatively low importance of the manufacturing cost compared to the overall cost of the device.

Properties of composites mainly depend on the properties and geometry of the constituents, their volume fraction, and their interfacial interactions with the matrix. For example, properties such as strength for particulate or short fiber composites and stiffness/strength for off-axis loaded fiber composites are influenced by the filler–matrix adhesion. Poor adhesion between reinforcing filler and matrix can be also responsible for undesirable hydrolytic breakdown of composites in the body environment (Pegoretti and Migliaresi, 2002). Hydrolytic instability of the fillers can also lead to degradation of composites in the body (Brown et al., 2005). In order to improve the strength of the interface, fillers can be surface treated in different ways to increase their surface energy, i.e., wettability or by binding to the filler surface molecules that will react with the matrix (e.g., silanes for glassy fillers).

Fabrication of Particle-Reinforced Composites

Injection molding, compression molding, and extrusion are the most common fabrication technologies for biomedical particulate composites. In some applications composites are manufactured *in situ*. This is the case of dental restorative composites and particle-reinforced bone cements.

Fabrication of Fiber-Reinforced Composites

Hand Lay-Up. This technique is mostly used for the preparation of thermosetting matrix, long fiber composite laminates, with relatively high fiber content (i.e., 50–60% by volume) and well-controlled in-plane mechanical properties. The basic structure is a thin sheet of matrix embedded with uniaxially or 0–90° oriented reinforcing fibers named lamina. Unidirectional or 0–90° oriented fiber sheets are cut in the desired shape and deposited with different orientation on a mold, then impregnated with the proper amount of resin. Impregnation can be performed after the deposition of each layer or previously resin-impregnated laminae ("prepregs") may be used. After lamination, the system is usually enclosed in a plastic vacuum bag, vacuum is applied to extract air (which can inhibit free radical curing processes) and compact the laminae, and finally the assembly will undergo a thermal program ("cure") to achieve resin cross-linking. For better quality composites, cure is often made in an autoclave at pressures of 6–8 atmospheres. (Composite materials produced by this method are currently used in aircraft and aerospace applications.) The conditions for curing vary depending upon the material. After being removed from the autoclave, the composite part is stripped from its tooling and is ready for further finishing operations.

The stacking sequence of the laminae and their number will determine the mechanical properties of the resulting composite (see Appendix). Composites with the same number and type of laminae but with different stacking sequences will have the same tensile modulus but different bending moduli.

The hand lay-up processing technique is potentially useful for the production of fracture fixation devices and total hip stems. The technique has been applied to the fabrication of various types of biomedical prostheses, such as hip prostheses or intervertebral discs.

Filament-Winding Process. The filament-winding process is an open-mold process used to produce high-strength hollow cylinders. In this process, the fiber reinforcement is fed through a resin bath and then wound on a suitable mandrel. When sufficient layers have been applied, the fiber–polymer wound mandrel is cured. The molded part is then stripped from the mandrel. The high degree of fiber orientation and high fiber loading with this method can produce extremely high tensile strengths. Biomedical applications for this process include intramedullary rods for fracture fixation, prosthetic hip stems, ligament prostheses, and interverterbral discs.

Other Processes. There are many closed-mold methods used for producing fiber-reinforced plastic materials. The methods of most importance to biomedical composites are compression and injection molding, and continuous "pultrusion."

In compression molding, the previously-fabricated prepregs are arranged in a two-piece mold that is then heated under pressure to produce the laminated part. This method is particularly useful for use with thermoplastic matrices.

In injection molding the fiber–matrix mix is injected into a mold at elevated temperature and pressure. The finished part is removed after cooling. This is an extremely fast and inexpensive technique that has application to chopped fiber-reinforced thermoplastic composites. It offers the possibility of producing composite devices, such as bone plates and screws, at much lower costs than comparable metallic devices. In resin transfer molding (RTM), the resin is injected in a closed mold that contains the fiber mat or fabric already deposited in the desired geometrical sequence. This procedure has been used for the production of composite hip prostheses.

Continuous pultrusion is a process used for the manufacture of fiber-reinforced plastics of constant cross-section, such as structural shapes, beams, channels, pipe, and tubing. In this process, continuous-strand fibers are impregnated in a resin bath and then are drawn through a heated die, which determines the shape of the finished stock. Highly oriented parts cut from this stock can then be used in other structures or they can be used alone in such applications as intramedullary rods or pin fixation of bone fragments.

ABSORBABLE MATRIX COMPOSITES

Absorbable matrix composites have been used in situations where absorption of the matrix is desired. Matrix absorption may be desired to expose surfaces to tissue or to release admixed materials such as antibiotics or growth factors (drug release) (Yasko et al., 1992). However, the most common reasons for the use of this class of matrices for composites has been to accomplish time-varying mechanical properties and ensure complete dissolution of the implant, eliminating long-term biocompatibility concerns. This type of composite would contain an absorbable matrix as well as absorbable reinforcing fillers. A typical clinical example is fracture fixation (Daniels et al., 1990; Tormala, 1992).

Fracture Fixation

Rigid internal fixation of fractures has conventionally been accomplished with metallic plates, screws, and rods. During the early stages of fracture healing, rigid internal fixation maintains alignment and promotes primary osseous union by stabilization and compression. Unfortunately, as healing progresses, or after healing is complete, rigid fixation may cause bone to undergo stress protection atrophy. This can result in significant loss of bone mass and osteoporosis. Additionally, there may be a basic mechanical incompatibility between the metal implants and bone. The elastic modulus of cortical bone ranges from 17 to 24 GPa, depending upon the age and location of the specimen, while the commonly used alloys have moduli ranging from 110 GPa (titanium alloys) to 210 GPa (316L steel). This large difference in stiffness can result in disproportionate load sharing, which can lead to relative motion between the implant and bone upon loading, as well as to high stress concentrations at bone–implant junctions.

Another potential problem is that the metal alloys currently used for plates corrode to some degree. Metal ions are released and they have been reported to cause adverse local tissue reactions, which in turn raises questions of adverse effects on bone mineralization, as well as adverse systemic responses such as local tumor formation (Bauer et al., 1987). Consequently, it is usually recommended that a second operation be performed to remove the metal hardware after healing.

The advantages of absorbable devices are thus twofold. First, the devices degrade mechanically with time, reducing stress protection and the accompanying osteoporosis. Second, there is no need for secondary surgical procedures to remove absorbable devices. The state of stress at the fracture site gradually returns to normal, allowing normal bone remodeling.

Absorbable fracture fixation devices have been produced from poly(L-lactic acid) polymer (PLLA), PGA polymer, and polydioxanone. An excellent review of the mechanical properties of biodegradable polymers was prepared by Daniels and co-workers (Daniels et al., 1990). Their review revealed that unreinforced biodegradable polymers are initially 36% as strong in tension as annealed stainless steel, and 54% in bending, but only 3% as stiff in either test mode. With fiber reinforcement, highest initial strengths exceeded those of stainless steel. Stiffness reached 62% of stainless steel with nondegradable carbon fibers, 15% with degradable inorganic fibers, but only 5% with degradable polymeric fibers.

Most previous work on absorbable composite fracture fixation has been performed with PLLA polymer. PLLA possesses three major characteristics that make it a potentially attractive biomaterial:

- It degrades in the body. The degradation rate depends on the polymer purity, crystallinity, and molecular weight.
- The degradation product of PLLA is mainly lactic acid, which is nontoxic, biocompatible, easily absorbed into and eliminated from the body. Lactic acid enters the lactic acid cycle of metabolites within cells. Ultimately it is metabolized to carbon dioxide and water.
- The PLLA rate of degradation can be controlled by copolymerizing lactic acid with glycolic acid to form PLGA or blending it with PGA or PLGA.

Poly(L-lactic acid) polymer reinforced with randomly oriented chopped carbon fiber was used to produce partially degradable bone plates (Corcoran et al., 1981). It was demonstrated that the plates, by virtue of the fiber reinforcement, exhibited mechanical properties superior to those of pure polymer plates. *In vivo*, the matrix degraded and the plates lost rigidity, gradually transferring load to the healing bone. However, the mechanical properties of such chopped fiber plates were relatively low; consequently, the plates were only adequate for low-load situations. Zimmerman et al. (1987) used composite theory to determine an optimum fiber lay-up for a long fiber composite bone plate. Composite analysis predicted the mechanical superiority of a 0°/±45° laminae lay-up. Although the 0°/±45° carbon/poly(lactic acid) composite possessed adequate initial mechanical properties, water absorption and subsequent delamination degraded the properties rapidly in an aqueous environment. The fibers did not chemically bond to the matrix.

In an attempt to develop a totally absorbable composite material, a calcium–phosphate-based glass fiber has been used to reinforce poly(lactic acid). Experiments were pursued to determine the biocompatibility and *in vitro* degradation properties of the composite (Zimmerman et al., 1991). These studies showed that the glass fiber–PLA composite was biocompatible, but its degradation rate was too high for use as an orthopedic implant.

Shikinami and Okuno (2001) have used compression molding to produce miniplates, rods, and screws made of PLLA containing 30% and 40% by weight of hydroxyapatite particles. These composites were suggested for indications such as repair of bone fracture in osteosynthesis, and fixation of bony fragments in bone grafting and osteotomy, exhibiting total resorbability and osteological bioactivity while retaining sufficiently high stiffness for a long period of time to achieve bony union. Furukawa et al. (2000a,b) have investigated the *in vivo* biodegradation behavior of hydroxyapatite/poly (L-lactide) composite rods implanted subcutaneously and in the intramedullary cavities of rabbits, showing that after 25 weeks of implantation, rods maintained a bending strength higher than 200 MPa. Their conclusion was that such strength was sufficient for application of the rods for the fixation of human bone fractures. In another *in vivo* study, Hasegawa et al. (2006) marked the excellent biodegradability and osteoconductivity of the HA/PLLA composites with no apparent late aseptic reactions after over seven years of implantation. HA/PLLA composite plates have been commercialized with the name of Fixsorb-MX and are presently commercialized in the USA and Europe with the name of OSTEOTRANS™ for several bone fixation applications.

By using a sintering technique, Tormala et al. (1988) produced self-reinforced PGA rods that have been used in the treatment of fractures and osteotomies. Afterwards, by using the same technique, self-reinforced PLLA pins and screws have been produced. The self-reinforced PLGA had higher initial mechanical properties, but degraded faster with respect to the parent PLLA material, which is reabsorbed in 12–16 months. These products are commercially available.

NON-ABSORBABLE MATRIX COMPOSITES

Non-absorbable matrix composites are generally used to provide specific mechanical properties unattainable with homogeneous materials. Particulate and chopped-fiber reinforcement have been used in bone cements and bearing surfaces to stiffen and strengthen these structures. Because the matrices and fillers are non-absorbable, they are used for "lifelong" implants such as orthopedic appliances and total joint replacements.

Reduced-stiffness carbon fiber-reinforced epoxy bone plates have been studied for fracture fixation with the aim of reducing "stress protection," which can cause osteoporosis. Many authors have demonstrated the negative stress protecting effect exerted by high rigidity fixation plates. Composites can offer unique advantages, since it is possible to produce plates equal in size, shape, and materials, but with different mechanical properties, such as with different ratios of tensile stiffness to bending or torsion stiffness (e.g., this is achieved by changing fiber orientation or fiber orientation sequence within the plate thickness).

Fujihara et al. (2003) investigated the influence of braiding angles and plate thickness on the bending performance of braided carbon/PEEK composite bone plates compared to metal plates. Different braiding angles produced composite plates with different stiffnesses (see Figure I.2.9.2) that could be adapted better than metals to different bone fixation and stiffness requirements.

Epoxy/carbon fiber fracture plates, produced by Orthodynamics Ltd, UK, are currently commercially available. As reported by the company, these plates have been implanted in more than 1000 cases since 1981,

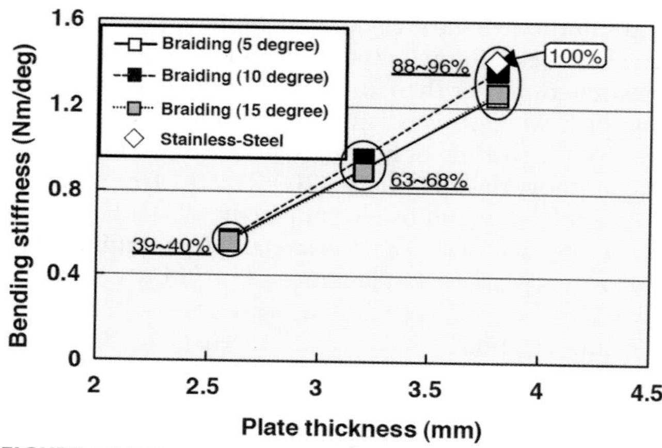

FIGURE I.2.9.2 Bending stiffness of carbon/PEEK composite bone fracture fixation plate compared to metal. (Reprinted from *Biomaterials,* Vol. 24, Fujihara et al., (2003). Performance study of braided carbon/PEEK composite compression bone plates, 2661-2667, with permission from Elsevier.)

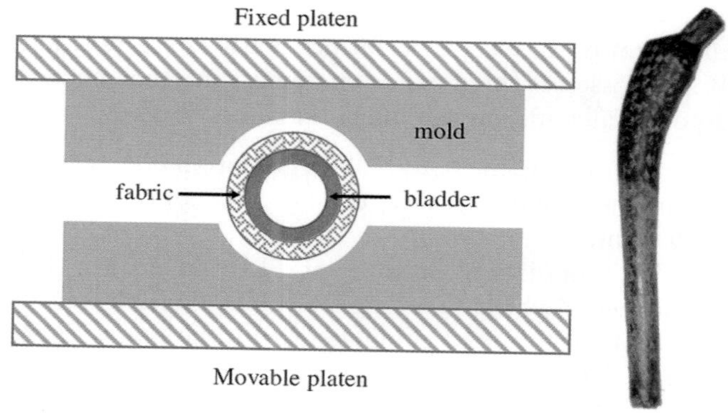

FIGURE I.2.9.3 An example of technology to produce composite hip prosthesis stems. (Reprinted from Composites Part A: Applied Science and Manufacturing. Vol. 39, Campbell et al., (2008). CF/PA12 composite femoral stems: Manufacturing and properties, 796-804, with permission from Elsevier.)

showing significant advantages over the more rigid metal plates (Pemberton et al., 1994).

Total Joint Replacement

By far the most studied, and potentially the most valuable, use of non-absorbable composites has been in total joint replacement. Bone resorption in the femur leading to aseptic loosening is an all-too-common occurrence associated with the implantation of metallic femoral hip replacement components. Several publications have shown that a relationship exists between bone remodeling and the state of stress and strain of the cortical femur, and that bone loss can occur after total hip replacement (Rosenthall et al., 2000).

It has long been recognized that bone adapts to functional stress by remodeling to re-establish a stable mechanical environment. When applied to the phenomenon of bone loss around implants, one can postulate that the relative stiffness of the metallic component is depriving bone of its accustomed load. Clinical and experimental results have shown the significant role that implant elastic characteristics play in allowing the femur to attain a physiologically acceptable stress state. Femoral stem stiffness has been indicated as an important determinant of cortical bone remodeling (Cheal et al., 1992). Composite materials technology offers the ability to alter the elastic characteristics of an implant and provide a better mechanical match with the host bone, potentially leading to a more favorable bone remodeling response (Bougherara et al., 2007).

Using different polymer matrices reinforced with carbon fiber, a large range of mechanical properties is possible. Campbell et al. (2008) have recently proposed carbon fibers/polyamide 12 (CF/PA12) composites for the fabrication of isoelastic stems of hip protheses. In their process, overlaid braided sleeves of CF/PA12 were placed around an inflatable bladder, compressed against a mold, and cured at temperature higher than the melting temperature of the matrix (see Figure I.2.9.3).

St. John (1983) reported properties for ±15° laminated test specimens) with moduli ranging from 18 to 76 GPa. However, the best reported study involved a press-fit device constructed of carbon fiber/polysulfone composite implanted in dogs (Magee et al., 1988). The femoral component was designed to have strength and elastic properties that matched well with the proximal canine femur. The component contained a core of unidirectional carbon/polysulfone composite enveloped with a bidirectional-braided layer composed of carbon/polysulfone composite covering the core. These regions were encased in an outer coating of pure polysulfone. Finite-element stress analysis predicted that this construction would cause minimal disruption of the normal stresses in the intact cortical bone. Canine animal studies carried out over four years showed a favorable bone remodeling response. The authors proposed that implants fabricated from carbon/polysulfone composites should have the potential for use in loadbearing applications. An implant with appropriate elastic properties provides the opportunity for the natural bone remodeling response to enhance implant stability.

Poly(ether-ether-ketone) (PEEK) matrix has been one of the most studied matrices for bone interfacing prosthesis. An *in vitro* study of Scotchford et al. (2003) showed a similar osteoblast attachment and proliferation on a PEEK/carbon fiber stem component as referred to Ti6Al4V stems. A reduced stiffness acetabular cup prosthesis made by a ceramic bearing cup mounted on a PEEK/carbon fiber composite back that was investigated by Mathias and Tabeshfar (2006) surpassed five million cycles in a fatigue test at a load three times higher than the one specified by ISO standards for femoral stems.

According to information available from the manufacturer (Orthodynamics Ltd, Dorset, UK), a hip protheses with a stem made with metal coated with PEEK/carbon fiber composite was implanted in 65 patients between 1992 and 1998, with favorable early clinical results.

This concept was also pursued by Simoes and Marques (2005), who developed a prosthetic stem made with a Cr–Co core and an external, controlled stiffness layer of a hybrid braided carbon/glass fiber composite. The stems were produced by a molding technology (see Figure I.2.9.4) that could be easily applied to the manufacture of other long and short fiber composite prostheses. The technology is an example of a manufacturing process that can be applied to the fabrication of laminated composites with or without an internal core.

Adam et al. (2002) reported on the revision of 51 press-fit hip prostheses implanted in humans that were made of epoxy resin/carbon fiber composites. Their results showed that within six years 92% of the prostheses displayed aseptic loosening, i.e., did not induce bone ingrowth. Authors attributed the failure to the smoothness of the stem surface. No osteolysis or wear or inflammatory reactions were, however, observed.

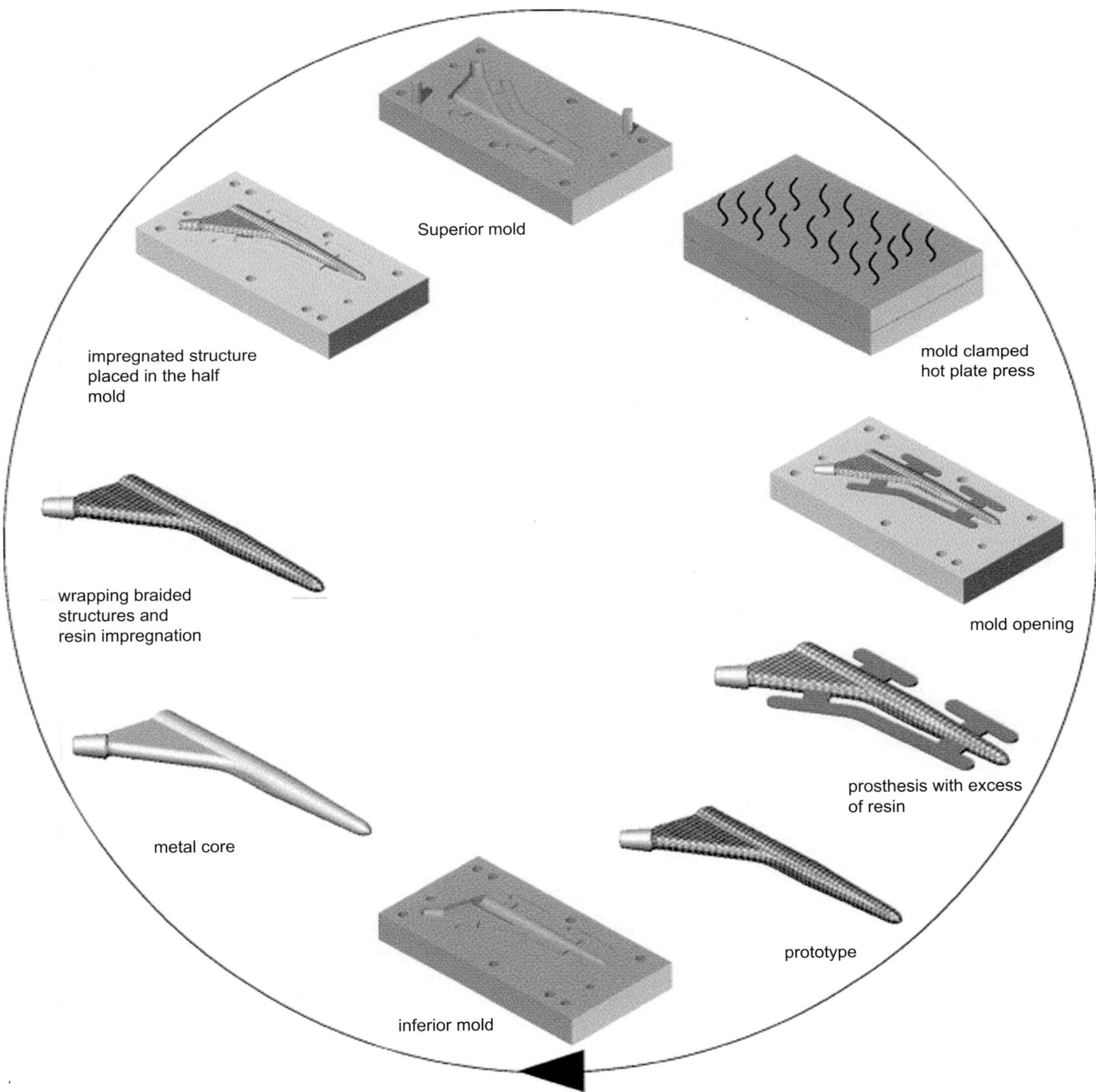

FIGURE I.2.9.4 Fabrication of a composite hip prosthesis. (Reprinted from *Materials & Design*, Vol. 26, Simões and Marques Design of a composite hip femoral prothesis, 391-401, (2005) with permission from Elsevier.)

Different fiber matrices and fabrication technologies have been proposed for the fabrication of hip prostheses. A review of materials and methods is found in the papers of Ramakrishna et al. (2001) and de Oliveira, Simões and Marques (2001).

The EPOCH® hip prosthesis developed by Zimmer was approved by the FDA to begin clinical trials in 2002. The stem of the prosthesis is made by an inner core of Cr–Co, an intermediate layer of PEEK (the configuration was modified in 2006), and an outer layer of titanium fibers. This is not a composite, as strictly defined, but nevertheless is a composite structure where different materials make a stem less stiff with respect to the more traditional metal stems.

Tissue Engineering Scaffolds

Composites have been widely proposed as tissue engineering (TE) scaffolds. The aim of the addition of fillers to a biodegradable polymer matrix is not only to reinforce the matrix, but also to impart to the scaffold specific bioactive properties, such as drug delivery. The role of the scaffold is to provide a support to the newly forming tissue through a favorable interaction with cells, but also fulfilling specific requirements of the implant site (Hollister, 2005; Agrawal et al., 2006).

Sometimes, polymeric particles have been added to polymeric matrices to promote sustained release of specific molecules. For instance, gelatin microparticles have been added to injectable cross-linked oligo(ethylene glycol fumarate) matrices to release drugs or growth factors for regenerating tissue in an osteochondral defect (Guo et al., 2009).

In a similar type of study, the addition of PLGA particles to silk fibroin porous sponges (Wenk et al., 2009), or PLGA particles to a poly(propylene fumarate) matrix (Kempen et al., 2006) was investigated. A three-dimensional PGA structure impregnated with gels of agarose and fibroin showed mechanical properties similar to the native articular cartilage; it was also seeded with chondrocytes (Moutos et al., 2007).

The reader should also refer to Section II.6 on tissue engineering in this volume.

SUMMARY

Biomedical composites have demanding properties that allow few, if any, "off-the-shelf" materials to be used. Few biomedical composites are yet in general clinical use. In many cases, those that have been developed to date have not been designed for a specific clinical use. They are simple laminates, or chopped fiber- or particulate-reinforced systems, sometimes without attempts made to react or bond the phases together. Such bonding may be accomplished by altering the surface chemistry, surface texture of the filler or by the introduction of coupling agents, i.e., molecules that can react with both filler and matrix. However, concerns about the biocompatibility of coupling agents and the high development costs of surface texture alteration procedures have curtailed major developments in this area. It is also possible to provide three-dimensional reinforcement with complex fiber weaving and impregnation procedures that are now regularly used in high-performance aerospace composites. Unfortunately, the high development costs associated with these techniques have restricted their application in biomedical composites.

Biomedical composites, because of their unique requirements, are probably the first general class of materials developed exclusively for implantation purposes. In recent years, nano-composite prostheses and composite scaffolds for tissue engineering applications have been proposed. In both cases, specific functional properties were provided by the filler addition. For nanocomposites, often the proposed solutions did not consider the biocompatibility implications behind the use of nanoparticles that are so small they can easily cross the cell membrane.

APPENDIX 1: MECHANICAL AND PHYSICAL PROPERTIES OF COMPOSITES

CONTINUOUS FIBER COMPOSITES

Laminated continuous fiber-reinforced composites are described from either a micro- or macro-mechanical point of view. Micromechanics is the study of composite material behavior wherein the interaction of the constituent materials is examined on a local basis. Macro-mechanics is the study of composite material behavior wherein the material is presumed homogeneous and the effects of the constituent materials are detected only as averaged apparent properties of the composite.

Micro-mechanics

There are two basic approaches to understanding the micro-mechanics of composite materials: the "mechanics-of-materials" approach and the elasticity approach. The mechanics-of-materials approach embodies the concept of simplifying assumptions regarding the hypothesized behavior of the mechanical system. It is the simpler of the two and the traditional choice for micro-mechanical evaluation.

The most prominent assumption made in the mechanics-of-materials approach is that strains in the fiber direction of a unidirectional fibrous composite are the same in the fibers and the matrix. This assumption allows the planes to remain parallel to the fiber direction. It also allows the longitudinal normal strain to vary linearly throughout the member with the distance from the neutral axis. Accordingly, the stress will also have a linear distribution.

Some other important assumptions are as follows:

1. The lamina is macroscopically homogeneous, linearly elastic, orthotropic, and initially stress-free.
2. The fibers are homogeneous, linearly elastic, isotropic, regularly spaced, and perfectly aligned.
3. The matrix is homogeneous, linearly elastic, and isotropic.

In addition, no voids are modeled in the fibers, the matrix or between them.

The mechanical properties of a lamina are determined by fiber orientation. The most often used laminate coordinate system has the length of the laminate in the x direction and the width in the y direction. The principal fiber direction is the 1 direction, and the 2 direction is normal to that. The angle between the x and 1 directions is Φ. A counterclockwise rotation of the 1-2 system yields a positive Φ.

The mechanical properties of the lamina are dependent on the material properties and the volume content of the constituent materials. The equations for the mechanical properties of a lamina in the 1-2 directions are:

$$E_1 = E_f v_f + E_m V_m \tag{1}$$

$$E_2 = E_m \frac{1+\xi\eta V_f}{1-\eta V_f} \tag{2}$$

$$G_{12} = G_m \frac{1+\xi\eta V_f}{1-\eta V_f} \tag{3}$$

$$\nu_{12} = V_m \nu_m + V_f \nu_f \tag{4}$$

$$\nu_{21} E_1 = \nu_{12} E_2 \tag{5}$$

where E = modulus, G = shear modulus, V = volume fraction, ν = Poisson's ratio, and subscripts f and m = fiber and matrix properties, respectively. The equations 2 and 3 are known as the Halpin-Tsai equation. This equation can be used to predict the stiffness of the lamina in the direction orthogonal to the fibers, but also the lamina shear modulus or the stiffness of short fibers laminae, or other properties, by adjusting the numerical parameter ξ. The factor η depends on ξ and on the fibers/matrix properties.

Macromechanics of a Lamina

The generalized Hooke's law relating stresses to strains is

$$\sigma_i = C_{ij}\varepsilon_j \qquad ij = 1, 2, \ldots, 6 \tag{6}$$

where σ_i = stress components, C_{ij} = stiffness matrix, ε_j = strain components. An alternative form of the stress-strain relationship is

$$\varepsilon_j = S_{ij}\sigma_i \qquad ij = 1, 2, \ldots, 6 \tag{7}$$

where S_{ij} = compliance matrix.

Given that $C_{ij} = C_{ji}$, the stiffness matrix is symmetric, thus reducing its population of 36 elements to 21 independent constants. We can further reduce the matrix size by assuming the laminae are orthotropic. There are 9 independent constants for orthotropic laminae. In order to reduce this three-dimensional situation to a two-dimensional situation for plane stress, we have

$$\tau_3 = 0 = \sigma_{23} = \sigma_{13} \tag{8}$$

thus reducing the stress-strain relationship to

$$\begin{vmatrix} \varepsilon_1 \\ \varepsilon_2 \\ \gamma_{12} \end{vmatrix} = \begin{vmatrix} S_{11} & S_{12} & 0 \\ S_{21} & S_{22} & 0 \\ 0 & 0 & S_{66} \end{vmatrix} \begin{vmatrix} \sigma_1 \\ \sigma_2 \\ \tau_{12} \end{vmatrix} \tag{9}$$

The stress-strain relation can be inverted to obtain

$$\begin{vmatrix} \sigma_1 \\ \sigma_2 \\ \tau_{12} \end{vmatrix} = \begin{vmatrix} Q_{11} & Q_{12} & 0 \\ Q_{21} & Q_{22} & 0 \\ 0 & 0 & Q_{66} \end{vmatrix} \begin{vmatrix} \varepsilon_1 \\ \varepsilon_2 \\ \gamma_{12} \end{vmatrix} \tag{10}$$

where Q_{ij} are the reduced stiffnesses. The equations for these stiffnesses are

$$Q_{11} = \frac{E_1}{1-\nu_{21}\nu_{12}} \tag{11}$$

$$Q_{12} = \frac{\nu_{12}E_2}{1-\nu_{12}\nu_{21}} = \frac{\nu_{21}E_1}{1-\nu_{12}\nu_{21}} = Q_{21} \tag{12}$$

$$Q_{22} = \frac{E_2}{1-\nu_{12}\nu_{21}} \tag{13}$$

$$Q_{66} = G_{12} \tag{14}$$

The material directions of the lamina may not coincide with the body coordinates. The equations for the transformation of stresses in the 1-2 direction to the x-y direction are

$$\begin{vmatrix} \sigma_x \\ \sigma_y \\ \tau_{xy} \end{vmatrix} = \begin{vmatrix} T^{-1} \end{vmatrix} \begin{vmatrix} \sigma_1 \\ \sigma_2 \\ \tau_{12} \end{vmatrix} \tag{15}$$

where $[T^{-1}]$ is

$$\begin{vmatrix} T^{-1} \end{vmatrix} = \begin{vmatrix} \cos^2\varphi & \sin^2\varphi & -2\sin\varphi\cos\varphi \\ \sin^2\varphi & \cos^2\varphi & 2\sin\varphi\cos\varphi \\ \sin\varphi\cos\varphi & \sin\varphi\cos\varphi & \cos^2\varphi - \sin^2\varphi \end{vmatrix} \tag{16}$$

The x and 1 axes form angle φ. This matrix is also valid for the transformation of strains,

$$\begin{vmatrix} \varepsilon_x \\ \varepsilon_y \\ \gamma_{xy} \end{vmatrix} = \begin{vmatrix} T^{-1} \end{vmatrix} \begin{vmatrix} \varepsilon_1 \\ \varepsilon_2 \\ \gamma_{12} \end{vmatrix} \tag{17}$$

Finally, it can be demonstrated that

$$\begin{vmatrix} \sigma_x \\ \sigma_y \\ \tau_{xy} \end{vmatrix} = \begin{vmatrix} \overline{Q}_{ij} \end{vmatrix} \begin{vmatrix} \varepsilon_x \\ \varepsilon_y \\ \gamma_{xy} \end{vmatrix} \tag{18}$$

where $[\overline{Q}_{ij}]$ is the transformed reduced stiffness. The transformed reduced stiffness matrix is

$$[\overline{Q}_{ij}] = \begin{vmatrix} \overline{Q}_{11} & \overline{Q}_{12} & \overline{Q}_{16} \\ \overline{Q}_{21} & \overline{Q}_{22} & \overline{Q}_{26} \\ \overline{Q}_{16} & \overline{Q}_{26} & \overline{Q}_{66} \end{vmatrix} \tag{19}$$

where

$$\begin{aligned}
\overline{Q}_{11} &= Q_{11}\cos^4\Phi + Q_{22}\,\mathrm{sen}^4\Phi + 2(Q_{12}+2Q_{66})\mathrm{sen}^2\Phi\cos^2\Phi \\
\overline{Q}_{22} &= Q_{11}\,\mathrm{sen}^4\Phi + Q_{22}\cos^4\Phi + 2(Q_{12}+2Q_{66})\mathrm{sen}^2\Phi\cos^2\Phi \\
\overline{Q}_{12} &= (Q_{11}+Q_{22}-4Q_{66})\mathrm{sen}^2\Phi\cos^2\Phi + Q_{12}(\cos^4+\mathrm{sen}^4\Phi) \\
\overline{Q}_{66} &= (Q_{11}+Q_{22}-2Q_{12}-2Q_{66})\mathrm{sen}^2\Phi\cos^2\theta \\
&\quad + Q_{66}(\cos^4\Phi+\mathrm{sen}^4\Phi) \\
\overline{Q}_{16} &= (Q_{11}-Q_{12}-2Q_{66})\cos^3\Phi\,\mathrm{sen}\,\Phi \\
&\quad -(Q_{22}-Q_{12}-2Q_{66})\cos\Phi\,\mathrm{sen}^3\Phi \\
\overline{Q}_{26} &= (Q_{11}-Q_{12}-2Q_{66})\cos\Phi\,\mathrm{sen}^3\Phi \\
&\quad -(Q_{22}-Q_{12}-2Q_{66})\cos^3\Phi\,\mathrm{sen}\,\Phi
\end{aligned} \tag{20}$$

The transformation matrix $[T^{-1}]$ and the transformed reduced stiffness matrix $[\overline{Q}_{ij}]$ are very important matrices in the macromechanical analysis of both laminae and laminates. These matrices play a key role in determining the effective in-plane and bending properties and how a laminate will perform when subjected to different combinations of forces and moments.

Macromechanics of a Laminate

The stiffness matrices of the laminate can be calculated through the following equations, where the $\overline{Q}_{ij}$ coefficients of the laminae have been defined before.

$$\begin{aligned}
|A_{ij}| &= \sum_{k=1}^{n}(\overline{Q}_{ij})(h_k - h_{k-1}) \\
|B_{ij}| &= \frac{1}{2}\sum_{k=1}^{n}(\overline{Q}_{ij})(h_k^2 - h_{k-1}^2) \\
|D_{ij}| &= \frac{1}{3}\sum_{k=1}^{n}(\overline{Q}_{ij})(h_k^3 - h_{k-1}^3)
\end{aligned} \tag{21}$$

The matrix [A] is called the *extensional stiffness matrix* because it relates the resultant forces to the midplaine strains, while matrix [D] is called the *bending stiffness matrix* because it relates the resultant moments to the laminate curvature. The so called *coupling stiffness matrix*, [B], accounts for coupling between bending and extension, which means that normal and shear forces acting at the laminate midplane are causing laminate curvature or that bending and twisting moments are accompanied by midplane strain.

The letter k denotes the number of laminae in the laminate with a maximum number (N). The letter h represents the distances from the neutral axis to the edges (k and k-1) of the respective laminae. A standard procedure for numbering laminae is used where the 0 lamina is at the bottom of a plate and the Kth lamina is at the top.

It is possible to demonstrate that the deformation in each point of the laminate at a distance z from the midplane can be correlated to the deformation, ε, and the curvature, k, of the midplane.

The integration of eq. 18 to the thickness of the laminate results in the calculation of the components of the forces and of the momentum per unit of width acting on the laminate:

$$\begin{vmatrix} N_x \\ N_y \\ N_{xy} \end{vmatrix} = |A| \begin{vmatrix} \varepsilon_x \\ \varepsilon_y \\ \gamma_{xy} \end{vmatrix} + |B| \begin{vmatrix} k_x \\ k_y \\ k_{xy} \end{vmatrix} \tag{22}$$

$$\begin{vmatrix} M_x \\ M_y \\ M_{xy} \end{vmatrix} = |B| \begin{vmatrix} \varepsilon_x \\ \varepsilon_y \\ \gamma_{xy} \end{vmatrix} + |D| \begin{vmatrix} k_x \\ k_y \\ k_{xy} \end{vmatrix} \tag{23}$$

As told, the above equations, which hold for generic laminates, couple extensional forces with curvatures and momentum with warping. In other words, if extensional forces and momentum are applied to a generic laminate, it bends and warps.

The coupling between bending and extension can be eliminated if the coefficient of the B_{ij} matrix are equal to zero, that means if the laminate is fabricated symmetric with respect to its midplane.

$$\begin{vmatrix} N_x \\ N_y \\ N_{xy} \end{vmatrix} = |A| \begin{vmatrix} \varepsilon_x \\ \varepsilon_y \\ \gamma_{xy} \end{vmatrix} \tag{24}$$

In a generic laminate, normal stresses N_x and/or N_y (or thermal stresses or liquid sorption) will cause deformations in the directions x and/or y, but also shear strains, unless A16 and A26 of the extensional stiffness matrix are equal to 0. These coefficient become 0 if the laminate is balanced, i.e. has the same number of laminae oriented at Φ and -Φ. Once the laminate strains are determined, the stresses in the *xy* direction for each lamina can be calculated. The most useful information gained from the *ABD* matrices involves the determination of generalized in-plane and bending properties of the laminate.

The equivalent elastic constants (E_x, E_y, G_{xy}, ν_{xy}) of a symmetric and balanced laminate can be easily evaluated from the A_{ij} coefficients (Barbero, 1998)

$$E_x = \frac{1}{h}\frac{A_{11}A_{22} - A_{12}^2}{A_{22}} \tag{25}$$

$$E_y = \frac{1}{h}\frac{A_{11}A_{22} - A_{12}^2}{A_{11}} \tag{26}$$

$$G_{xy} = \frac{1}{h}A_{66} \tag{27}$$

$$\nu_{xy} = \frac{A_{12}}{A_{22}} \tag{28}$$

SHORT-FIBER COMPOSITES

A distinguishing feature of the unidirectional laminated composites discussed above is that they have higher strength and modulus in the fiber direction, and thus their properties are amenable to alteration to produce specialized laminates. However, in some applications, unidirectional multiple-ply laminates may not be required. It may be advantageous to have isotropic lamina. An effective way of producing an isotropic lamina is to use randomly oriented short fibers as the reinforcement. Of course, molding compounds consisting of short fibers that can be easily molded by injection or compression molding may be used to produce generally isotropic composites. The theory of stress transfer between fibers and matrix in short-fiber composites goes beyond this text; it is covered in detail by Agarwal and Broutman (1980). However, from the theory, the longitudinal and transverse moduli (E_L and E_T, respectively) for an aligned shortfiber lamina can be found from

$$\frac{E_L}{E_m} = \frac{1+(2l/d))\eta V_f}{1-\eta_L V_f} \tag{29}$$

$$\frac{E_T}{E_m} = \frac{1+2\eta_T V_f}{1-\eta_T V_f} \tag{30}$$

$$\eta_L = \frac{E_f/E_m - 1}{E_f/E_m + 2(l/d))} \qquad \eta_r = \frac{E_f/E_m - 1}{E_f/E_m + 2} \tag{31}$$

The problem of predicting properties of *randomly oriented* short-fiber composites is more complex. The following empirical equation can be used to predict the modulus of composites containing fibers that are randomly oriented in a plane:

$$E_{random} = \frac{3}{8}E_L + \frac{5}{8}E_T \tag{32}$$

where E_L and E_T are respectively the longitudinal and transverse moduli of an aligned short-fiber composite having the same fiber aspect ratio and fiber volume fraction as the composite under consideration. Moduli E_L and E_T can either be determined experimentally or calculated using Eqs. 34 and 35.

PARTICULATE COMPOSITES

The reinforcing effect of particles on polymers was first recognized for rubbery matrices while studying the effect of carbon black on the properties of natural rubber.

Several models have been introduced to predict the effect of the addition of particles to a polymeric matrix, starting from the equation developed by Einstein in 1956 to predict the viscosity of suspensions of rigid spherical inclusions. The paper by Ahmed and Jones, 1990, well reviews theories developed to predict strength and modulus of particulate composites.

One of the most versatile equations predicting the shear modulus of composites of polymers and spherical fillers is due to Kerner, 1956:

$$G_c = G_m\left(1+\frac{V_f 15(1+v_m)}{V_m(8-10v_m)}\right) \tag{33}$$

Nielsen developed a more generalized form,

$$M = M_m\frac{1+ABV_f}{1-B\delta V_f} \tag{34}$$

where the function Bδ depends on the particle packing fraction and M is the shear, elastic or bulk modulus.

By using a finite element analysis method Guild and Bonfield, 1993, predicted the elastic modulus of hydroxyapatitepolyethylene reinforced composites for various filler content. Their result, indicated a good agreement between theoretical and experimental data, except at higher hydroxyapatite volume fraction.

While elastic modulus of a particulate composites increases with the filler content, strength decreases in tension and increases in compression. Size and shape of the inclusion play an important role, with a higher stress concentration cause by non-regularly shaped inclusions. For spherical particles, the tensile strength σ_{cu} of the particulate composite can be correlated to the one of the matrix, σ_{mu}, through the equation (Nicolais and Narkis, 1971):

$$\sigma_{cu} = \sigma_{mu}(1-1.21V_f^{2/3}) \tag{35}$$

GLOSSARY OF TERMS

Bis-GMA: Bis-glycidyl dimethacrylate
CF: carbon fibers
CP: calcium phosphate
DMTMC: dimethyl-trimethylene carbonate
GF: glass fibers
HA: hydroxyapatite
KF: Kevlar fibers
PA12: polyamide 12
PBT: poly(butylene terephthalate)
PCL: polycaprolactone
PDLA: poly(D-lactic acid)
PDLLA: poly(D,L-lactic acid)
P(DLLA-CL): poly(D,L-lactide/-caprolactone)
PE: polyethylene
PEEK: poly(ether-ether-ketone)
PEG: poly(ethyleneglycol)
PEI: poly(ether-imide)
PELA: poly(ethylene oxide)-b-poly(lactic acid) block copolymer
PET: poly(ethylene terephthalate)
PGA: poly(glycolic acid)
PLA: poly(lactic acid)
PLGA: poly(lactic-co-glycolic acid)
PLLA: poly(L-lactic acid)
PMMA: poly(methylmethacrylate)
POSS: Polyhedral Oligometric Silsequioxane
PP: polypropylene
PSU: polysulfone
PTFE: polytetrafluoroethylene
PU: polyurethane
TCP: tricalciumphosphate
TEGDMA: Tetraethyleneglycol Dimethacrylate
UHMWPE: ultra-high molecular weight polyethylene

BIBLIOGRAPHY

Adam, F., Hammer, D. S., Pfautsch, S., & Westermann, K. (2002). Early failure of a press-fit carbon fiber hip prosthesis with a smooth surface. *J. Arthroplasty*, **17**, 217–223.

Agarwal, B. D., & Broutman, L. J. (1980). *Analysis and Performance of Fiber Composites*. New York, NY: Wiley-Interscience.

Agrawal, C. M., Carter, J., & Ong, J. L. (2006). Basics of polymeric scaffolds for tissue engineering. *Journal of ASTM International*, (3), 9.

Ahmed, S., & Jones, F. R. (1990). A review of particulate reinforcement theories for polymer composites. *J. Mater. Sci.*, **25**, 4933–4942.

Akay, M., & Aslan, N. (1996). Numerical and experimental stress analysis of a polymeric composite hip joint prosthesis. *J. Biomed. Mater. Res.*, **31**, 167–182.

Ali, M. S., Hastings, G. W., Rushton, N., Ross, E. R. S., & Wynn-Jones, C. H. (1990). Carbon fiber composite plates. *J. Bone Joint Surg.*, **72-B**, 586–591.

Ambrosio, L., Netti, P., Iannace, S., Huang, S. J., & Nicolais, L. (1996). Composite hydrogels for intervertebral disc prostheses. *J. Mater. Sci.: Mater. Med.*, **7**, 251–254.

Baidya, K. P., Ramakrishna, S., Rahman, M., & Ritchie, A. (2001). Advanced textile composite ring for Ilizarov external fixator system. *Proc. Inst. Mech. Eng. Part H, J. Eng. in Medicine*, **215**, 11–23.

Barbero, E. J. (1998). *Introduction to Composite Materials Design*. Philadelphia (PA), USA: Taylor and Francis, Inc.

Bauer, T. W., Manley, M. T., Stern, L. S., et al. (1987). Osteosarcoma at the site of total hip replacement. *Trans. Soc. Biomater.*, **10**, 36.

Bonfield, W. (1988). Composites for bone replacement. *J. Biomed. Eng.*, **10**, 522.

Bonfield, W., Wang, M., & Tanner, K. E. (1998). Interfaces in analogue biomaterials. *Acta Mater.*, **7**, 2509–2518.

Bougherara, H., Bureau, M., Campbell, M., Vadean, A., & Yahia, L. (2007). Design of a biomimetic polymer-composite hip prosthesis. *J. Biomed. Mater. Res.*, **82**(1), 27.

Brooke, N. S., Rorke, A. W., King, A. T., & Gullan, R. W. (1997). Preliminary experience of carbon fibre cage prostheses for treatment of cervical spine disorders. *Brit. J. Neurosurg.*, **11**, 221.

Brown, E. N., Davis, A. K., Jonnalagadda, K. D., & Sottos, N. R. (2005). Effect of surface treatment on the hydrolytic stability of E-glass fiber bundle tensile strength. *Compos. Sci. Technol.*, **65**, 129.

Campbell, M., Denault, J., Yahia, L., & Bureau, M. N. (2008). CF/PA12 composite femoral stems: Manufacturing and properties. *Compos. Part A: Appl. Sci. Manuf.*, **39**(5), 796.

Cheal, E. J., Spector, M., & Hayes, W. C. (1992). Role of loads and prosthesis material properties on the mechanics of the proximal femur after total hip arthroplasty. *J. Orthop. Res.*, **10**, 405–422.

Chen, S. Y., Liang, W. M., & Yen, P. S. (2001). Reinforcement of acrylic denture base resin by incorporation of various fibers. *J. Biomed. Mater. Res.*, **58**(2), 203–208.

Christel, P., Leray, J., Sedel, L., & Morel, E. (1980). Mechanical Evaluation and Tissue Compatibility of Materials for Composite Bone Plates. In G. Hasting, & D. F. Williams (Eds.), *Mechanical Properties of Biomaterials* (pp. 367–377).

Ciappetta, P., Boriani, S., & Fava, G. P. (1997). A carbon fiber reinforced polymer cage for vertebral body replacement: Technical note. *Neurosurgery*, **4**(5), 1203–1206.

Claes, L., Hutter, W., & Weiss, R. (1997). Mechanical Properties of Carbon Reinforced Polysulfone Plates for Internal Fixation. In P. Christel, A. Meunier, & A. J. C. Lee (Eds.), *Biological and Biomechanical Performance of Biomaterials* (pp. 81–86). Elsevier.

Claes, L., Schultheiss, M., Wolf, S., Wilke, H. J., Arand, M., et al. (1999). A new radiolucent system for vertebral body replacement: Its stability in comparison to other systems. *J. Biomed. Mater. Res., Appl. Biomater.*, **48**(1), 82–89.

Corcoran, S., Koroluk, J., Parsons, J. R., Alexander, H., & Weiss, A. B. (1981). The Development of a Variable Stiffness, Absorbable Composite Bone Plate. In H. K. Uhthoff (Ed.), *Current Concepts for Internal Fixation of Fractures* (p. 136). New York, NY: Springer-Verlag.

Daniels, A. U., Melissa, K. O., & Andriano, K. P. (1990). Mechanical properites of biodegradable polymers and composites proposed for internal fixation of bone. *J. Appl. Biomater.*, **1**(1), 57–78.

Dawson, D. K. (2000). Medical Devices. In A. Kelly, & C. Zweben (Eds.), *Comprehensive Composite Materials* (Vol. 6, pp. 755–786). Oxford: Elsevier.

de Oliveira Simoes, J. A., & Marques, A. T. (2001). Determination of stiffness properties of braided composites for the design of hip prostheses. *Compos. Part A: Appl. Sci.*, **32**, 655–662.

De Santis, R., Ambrosio, L., & Nicolais, L. (2000). Polymer-based composite hip prosthesis. *J. Inorg. Biochem.*, **79**, 97–102.

Dunn, M. G. (1998). Anterior Cruciate Ligament Prostheses. In T. D. Fahey (Ed.), *Encyclopedia of Sports Medicine and Science*. Internet Society for Sport Science, http://sportsci.org.

Einstein, A. (1956). In: *Investigation of Theory of Brownian Motion*. NY: Dover.

Erisken, C., Kalyon, D. M., & Hongjun Wang, H. (2008). Functionally graded electrospun polycaprolactone and β-tricalcium phosphate nanocomposites for tissue engineering applications. *Biomaterials*, **29**, 4065–4073.

Frazza, E. J., & Schmitt, E. E. (1971). A new absorbable suture. *J. Biomed. Mater. Res.*, **10**, 43.

Fujihara, K., Huang, Z. M., Ramakrishna, S., Satkunanantham, K., & Hamada, H. (2001). Development of braided carbon/PEEK composite bone plates. *Adv. Compos. Lett.*, **10**, 449–456.

Fujihara, K., Huang, Z. M., Ramakrishna, S., Satknanantham, K., & Hamada, H. (2003). Performance study of braided carbon/PEEK composite compression bone plates. *Biomaterials*, **24**(15), 2661.

Furukawa, T., Matsusue, Y., Yasunaga, T., Nakagawa, Y., Shikinami, Y., et al. (2000a). Bone bonding ability of a new biodegradable composite for internal fixation of bone fractures. *Clin. Orthop.*, **379**, 247–258.

Furukawa, T., Matsusue, Y., Yasunaga, T., Shikinami, Y., Okuno, M., et al. (2000b). Biodegradation behavior of ultra-high-strength hydroxyapatite/poly(L-lactide) composite rods for internal fixation of bone fractures. *Biomaterials*, **21**, 889–898.

Gandhi, M., Yang, H., Shor, L., & Ko, F. (2009). Post-spinning modification of electrospun nanofiber nanocomposite from *Bombyx mori* silk and carbon nanotubes. *Polymer*, **50**(8), 1918–1924.

Gershon, B., Cohn, D., & Marom, G. (1990). The utilization of composite laminate theory in the design of synthetic soft tissue for biomedical prostheses. *Biomaterials*, **11**, 548–552.

Gershon, B., Cohn, D., & Marom, G. (1992). Compliance and ultimate strength of composite arterial prostheses. *Biomaterials*, **13**, 38–43.

Gilding, D. K., & Reed, A. M. (1979). Biodegradable polymers for use in surgery: PGA/PLA homo- and copolymers: 1. *Polymer*, **20**, 1459.

Gillett, N., Brown, S. A., Dumbleton, J. H., & Pool, R. P. (1986). The use of short carbon fiber reinforced thermoplastic plates for fracture fixation. *Biomaterials*, **6**, 113–121.

Gloria, A., Causa, F., De Santis, R., Netti, P. A., & Ambrosio, L. (2007). Dynamic-mechanical properties of a novel composite intervertebral disc prosthesis. *J. Mater. Sci.: Mater. Med.*, **18**(11), 2159–2165.

Godara, A., Raabe, D., & Green, S. (2007). The influence of sterilization processes on the micromechanical properties of carbon fiber-reinforced PEEK composites for bone implant applications. *Acta Biomaterialia*, 3(2), 209–220.

Gong, J. P., (2003). Double-network hydrogels with extremely high mechanical strength. *Advanced Materials*, **15**(4), 1155–1158.

Guarino, V., Causa, F., Taddei, P., di Foggia, M., Ciapetti, G., et al. (2008). Polylactic acid fiber-reinforced polycaprolactone scaffolds for bone tissue engineering. *Biomaterials*, **29**(27), 3662–3670.

Guidoin, M. F., Marois, Y., Bejui, J., Poddevin, N., King, M. W., et al. (2000). Analysis of retrieved polymer fiber based replacements for the ACL. *Biomaterials*, **21**, 2461–2474.

Guild, F. J., & Bonfield, W. (1993). Predictive modeling of hydroxyapatite-polyethylene composite. *Biomaterials*, **14**(13), 985–993.

Guo, X., Park, H., Liu, G., Liu, W., Cao, Y., et al. (2009). *In vitro* generation of an osteochondral construct using injectable hydrogel composites encapsulating rabbit marrow mesenchymal stem cells. *Biomaterials*, **30**(14), 2741–2752.

Gupta, D., Venugopal, J., Mitra, S., Giri Dev, V. R., & Ramakrishna, S. (2009). Nanostructured biocomposite substrates by electrospinning and electrospraying for the mineralization of osteoblasts. *Biomaterials*, **30**(11), 2085–2094.

Hasegawa, S., Ishii, S., Tamura, J., Furukawa, T., Neo, M., et al. (2006). A 5–7 year *in vivo* study of high-strength hydroxyapatite/poly(L-lactide) composite rods for the internal fixation of bone fractures. *Biomaterials*, **27**, 1327.

Heller, J., Penhale, D. W. H., & Helwing, R. F. (1980). Preparation of poly(ortho esters) by the reaction of diketene acetals and polyols. *J. Polym. Sci. Polym. Lett. Ed.*, **18**, 619.

Hench, L. L., Splinter, R. J., Allen, W. C., & Greenlee, T. K. (1971). Bonding mechanisms at the interface of ceramic prosthetic materials. *J. Biomed. Mater. Res.*, **2**(72), 117–141.

Henderson, J. D., Jr., Mullarky, R. H., & Ryan, D. E. (1987). Tissue biocompatibility of kevlar aramid fibers and polymethylmethacrylate composites in rabbits. *J. Biomed. Mater. Res.*, **21**(1), 59–64.

Hollister, S. J. (2005). Porous scaffold design for tissue engineering. *Nature Materials*, **4**, 518.

Huttunen, M., Ashammaki, N., Tormala, P., & Kellomaki, M. (2006). Fiber reinforced bioresorbable composites for spinal surgery. *Acta Biomaterialia*, 2(5), 575.

Iannace, S., Sabatini, G., Ambrosio, L., & Nicolais, L. (1995). Mechanical behavior of composite artificial tendons and ligaments. *Biomaterials*, **16**, 675–680.

Jayabalan, M. (2009). Studies on poly(propylene fumarate-co-caprolactone diol) thermoset composites towards the development of biodegradable bone fixation devices. *Int. J. Biomater.*, 1–10.

Jose, M. V., Thomas, V. T., Johnson, K. T., Dean, D. R., & Nyairo, E. (2009). Aligned PLGA/HA nanofibrous nanocomposite scaffolds for bone tissue engineering. *Acta Biomaterialia*, **5**(1), 305–315.

Kazanci, M., Cohn, D., Marom, G., Migliaresi, C., & Pegoretti, A. (2002). Fatigue characterization of polyethylene fiber reinforced polyolefin biomedical composites. *Compos. Part A: Appl. Sci.*, **33**, 453–458.

Kempen, D. H. R., Lu, L., Kim, C., Zhu, X., Dhert, W. J. A., et al. (2006). Controlled drug release from a novel injectable biodegradable microsphere/scaffold composite based on poly(propylene fumarate). *J. Biomed. Mater. Res. Part A*, **77**(1), 103–111.

Kerner, E. H. (1956). The elastic and thermo-elastic properties of composite media. *Proceedings of the Physics Society, B*, **69**, 808–813.

Kolarik, J., Migliaresi, C., Stol, M., & Nicolais, L. (1981). Mechanical properties of model synthetic tendons. *J. Biomed. Mater. Res.*, **15**, 147.

Kong, L., Gao, Y., Cao, W., Gong, Y., Zhao, N., et al. (2005). Preparation and characterization of nano-hydroxyapatite/chitosan composite scaffolds. *J. Biomed. Mater. Res., Part A*, **75**(2), 275.

Kotani, Y., Abumi, K., Shikinami, Y., Takada, T., Kadoya, K., et al. (2002). Artificial intervertebral disc replacement using bioactive three-dimensional fabric: Design, development and preliminary animal study. *Spine*, **27**(9), 929–935.

Kurtz, M., & Devine, J. N. (2007). PEEK biomaterials in trauma, orthopedic, and spinal implants. *Biomaterials*, **28**(32), 4845–4869.

Kurtz, S. M., Muratoglu, O. K., Evans, M., & Edidin, A. A. (1999). Advances in the processing, sterilization, and crosslinking of ultra-high molecular weight polyethylene for total joint arthroplasty. *Biomaterials*, **20**, 1659–1688.

Kwarteng, K. B. (1990). Carbon fiber reinforced PEEK (APC-2/AS-4) composites for orthopaedic implants. *SAMPE Quarterly*, 10–14.

Lee, Y. H., Lee, J. H., An, I. -G., Kim, C., Lee, D. S., et al. (2005). Electrospun dual-porosity structure and biodegradation morphology of Montmorillonite reinforced PLLA nanocomposite scaffolds. *Biomaterials*, **26**(16), 3165–3172.

Leonor, I. B., Ito, A., Kanzaki, N., & Reis, R. L. (2003). *In vitro* bioactivity of starch thermoplastic/hydroxyapatite composite biomaterials: An *in situ* study using atomic force microscopy. *Biomaterials*, **24**, 579–585.

Lin, T. C. (1986). Totally absorbable fiber reinforced composite from internal fracture fixation devices. *Trans. Soc. Biomater.*, **9**, 166.

Lin, T. W., Corvelli, A. A., Frondoza, C. G., Roberts, J. C., & Hungerford, D. S. (1997). Glass peek composite promotes proliferation and osteocalcin production of human osteoblastic cells. *J. Biomed. Mater. Res.*, **36**(2), 137–144.

Lu, L., & Mikos, A. G. (1996). The importance of new processing techniques in tissue engineering. *MRS Bull.*, **21**(11), 28–32.

Magee, F. P., Weinstein, A. M., Longo, J. A., Koeneman, J. B., & Yapp, R. A. (1988). A canine composite femoral stem. *Clin. Orthop. Rel. Res.*, **235**, 237.

Marcolongo, M., Ducheyne, P., Garino, J., & Schepers, E. (1998). Bioactive glass fiber/polymeric composites bond to bone tissue. *J. Biomed. Mater. Res.*, **39**(1), 161–170.

Mathias, M. J., & Tabeshfar, K. (2006). Design and development of a new acetabular cup prosthesis. *Mater. Sci. Eng.: C*, **26**(8), 1428–1433.

Matsumoto, T. (1985). Mesophase pitch and its carbon fiber. *Pure Appl. Chem.*, **57**(11), 15.

McCarvill, S. (2005). Essay: Prosthetics for athletes. *The Lancet*, **366**(1), S10.

Middleton, J. C., & Tipton, A. J. (2000). Syntetic biodegradable polymers as orthopedic devices. *Biomaterials*, **21**, 2335–2346.

Migliaresi, C., & Fambri, L. (1997). Processing and degradation of poly(L-lactic acid) fibers. *Macromol. Symp.*, **123**, 155–161.

Migliaresi, C., Nicoli, F., Rossi, S., & Pegoretti, A. (2004). Novel uses of carbon composites for the fabrication of external fixators. *Compos. Sci. Technol.*, **64**, 873–883.

Moreau, J. L., & Xu, H. K. (2009). Mesenchymal stem cell proliferation and differentiation on an injectable calcium phosphate–chitosan composite scaffold. *Biomaterials*, **30**(14), 2675–2682.

Morita, S., Furuya, K., Ishihara, K., & Nakabayashi, N. (1998). Performance of adhesive bone cement containing hydroxyapatite particles. *Biomaterials*, **19**(17), 1601–1606.

Morrison, C., Macnair, R., MacDonald, C., Wykman, A., Goldie, I., et al. (1995). *In vitro* biocompatibility testing of polymers for orthopaedic implants using cultured fibroblasts and osteoblasts. *Biomaterials*, **16**(3), 987–992.

Moszner, N., & Salz, U. (2001). New developments of polymeric dental composites. *Progress in Polymer Science*, **26**(4), 535–576.

Moutos, F. T., & Guilak, F. (2008). Composite scaffolds for cartilage tissue engineering. *Biorheology*, **45**, 501–512.

Moutos, F. T., Freed, L. E., & Guilak, F. (2007). A biomimetic three-dimensional woven composite scaffold for functional tissue engineering of cartilage. *Nature Materials*, **6**, 162–167.

Nair, L. S., & Laurencin, C. T. (2007). Biodegradable polymers as biomateriale. *Prog. Polym. Sci.*, **32**, 762.

Nicolais, L., & Narkis, M. (1971). *Polym. Eng. Sci.*, **194**.

Nielsen, L. E. (1970). Generalized equation for the elastic moduli of composite materials. *J. Appl. Phys.*, **41**, 4626–4627.

Nolan, L. (2008). Carbon fiber prostheses and running in amputees: A review. *Foot Ankle Surg.*, **14**(3), 125.

Pegoretti, A., & Migliaresi, C. (2002). Effect of hydrothermal aging on the thermo-mechanical properties of a composite dental prosthetic material. *Polym. Compos.*, **23**(3), 342–351.

Pemberton, D. J., Evans, P. D., Grant, A., & McKibbin, B. (1994). Fractures of the distal femur in the elderly treated with a carbon fiber supracondylar plate. *Injury*, **25**(5), 317.

Pourdeyhimi, B., Wagner, H. D., & Schwartz, P. (1986). A Comparison of the Mechanical Properties of Kevlar 29 Reinforced Bone and Dental Cements. *J. Mat. Sci.*, **21**, 4468–4474.

Qing, L., de Wijn, J. R., & van Blitterswijk, C. A. (1997). Nano-apatite/polymer composites: Mechanical and physicochemical characteristics. *Biomaterials*, **18**, 1263–1270.

Ramakrishna, S., Mayer, J., Wintermantel, E., & Leong, K. W. (2001). Biomedical applications of polymer composite materials: A review. *Compos. Sci. Technol.*, **61**, 1189–1224.

Reis, R. L., & Cunha, A. M. (2000). New degradable load-bearing biomaterials based on reinforced thermoplastic starci incorporatine blends. *J. Appl. Med. Polym.*, **4**, 1–5.

Reis, R. L., Cunha, A. M., & Bevis, M. J. (1998). Shear controlled orientation injection molding of polymeric composites with enhanced properties. Atlanta, USA: *SPE Proceedings,* 57th Annual Technical Conference, 487–493.

Rokkanen, P. U., Bostman, O., Hirvensalo, E., Makela, E. A., Partio, E. K., et al. (2000). Bioabsorbable fixation in orthopaedic surgery and traumatology. *Biomaterials*, **21**, 2607–2613.

Rosenthall, L., Bobyns, D. J., & Tanzer, M. (2000). Periprosthetic bone densitometry of the hip: Influence of prosthetic design and hydroxyapatite coating on regional bone remodelling. *J. Musculoskel. Neuron. Interact.*, **1**, 57.

Rushton, N., & Rae, T. (1984). The intra-articular response to particulate carbon fiber reinforced high density polyethylene and its constituents: An experimental study in mice. *Biomaterials*, **5**, 352–356.

Sarkar, S., Burriesci, B., Wojcik, A., Aresti, N., Hamilton, G., et al. (2009). Manufacture of small calibre quadruple lamina vascular bypass grafts using a novel automated extrusion-phase-inversion method and nanocomposite polymer. *J. Biomech.*, **42**(6), 722–730.

Scotchford, C. A., Garle, M. J., Batchelor, J., John Bradley, J., & Grant, D. M. (2003). Use of a novel carbon fiber composite material for the femoral stem component of a THR system: *In vitro* biological assessment. *Biomaterials*, **24**(26), 4871–4879.

Shikinami, Y., & Okuno, M. (2001). Bioresorbable devices made of forged composites of hydroxyapatite (HA) particles and poly l-lactide (PLLA). Part II: Practical properties of miniscrews and miniplates. *Biomaterials*, **22**, 3197–3211.

Shinzato, S., Kobayashi, M., Farid, Mousa W., Kamimura, M., Neo, M., et al. (2000). Bioactive polymethyl methacrylate-based bone cement: Comparison of glass beads, apatite- and wollastonite-containing glass-ceramic, and hydroxyapatite fillers on mechanical and biological properties. *J. Biomed. Mater. Res.*, **51**(2), 258–272.

Simoes, J. A., & Marques, A. T. (2005). Design of a composite hip femoral prosthesis. *Mater. Design*, **26**, 391–401.

St. John, K. R. (1983). Applications of Advanced Composites in Orthopaedic Implants. In M. Szycher (Ed.), *Biocompatible Polymers, Metals, and Composites* (p. 861). Lancaster, PA: Technomic.

Tang, R., Boyle, W. J., Jr., Mares, F., & Chiu, T. -H. (1990). Novel bioresorbable polymers and medical devices. *Trans. 16th Ann. Mtg. Soc. Biomater.*, **13**, 191.

Tian, F., Cui, D., Schwarz, H., Estrada, G. G., & Kobayashi, H. (2006). Cytotoxicity of single-wall carbon nanotubes on human fibroblasts. *Toxicol. In Vitro*, **20**(7), 1202–1212.

Thomson, R. C., Yaszemski, M. J., Powers, J. M., & Mikos, A. G. (1998). Hydroxyapatite fiber reinforced poly(α-hydroxy ester) foams for bone regeneration. *Biomaterials*, **19**, 1935–1943.

Tormala, P. (1992). Biodegradable self-reinforced composite materials: Manufacturing structure and mechanical properties. *Clin. Mater.*, **10**, 29–34.

Tormala, P., Rokkanen, P., Laiho, J., Tamminmaki, M., & Vainionpaa, S. (1988). Material for osteosynthesis devices. *US Patent*, **4**(734), 257.

Ural, E., Kesenci, K., Fambri, L., Migliaresi, C., & Piskin, E. (2000). Poly(D,L-lactide/ϵ-caprolactone)/hydroxyapatite composites. *Biomaterials*, **21**, 2147–2154.

Vallittu, P. K., (1996). A review of fibre-reinforced denture based resins. *J. Prosthodont.*, **5**, 270–276.

Veerabagu, S., Fujihara, K., Dasari, G. R., & Ramakrishna, S. (2003). Strain distribution analysis of braided composite bone plates. *Compos. Sci. Technol.*, **61**, 427–435.

Wagner, J. D., & Mosby, E. L. (1990). Assessment of proplast-teflon disc replacements. *J. Oral. Max. Surg.*, **48**, 1140.

Wahl, D. A., & Czernuszka, J. T. (2006). Collagen-hydroxyapatite composites for hard tissue repair. *Euro. Cells Mater.*, **11**, 43.

Wang, A., Lin, R., Polineni, V. K., Essner, A., Stark, C., et al. (1998). Carbon fiber reinforced polyether ether ketone composite as a bearing surface for total hip replacement. *Tribol. Int.*, **31**, 661–667.

Wang, G. H., Yang, H., Li, M., Lu, S., Chen, X., et al. (2010). The use of silk fibroin/hydroxyapatite composite co-cultured with rabbit bone-marrow stromal cells in the healing of a segmental bone defect. *J. Bone Joint Surg. Br.*, **92**(2), 320.

Wening, J. V., Langendorff, U., Delling, G., Marquardt, H., Hoffmann, M., et al. (1989). Initial results of the biocompatibility, cytotoxicity and genotoxicity of Aramid. *Unfallchirurgie*, **15**(5), 215–220.

Wening, J. V., Katzer, A., Nicolas, V., Hahn, M., Jungbluth, K. H., et al. (1994). Imaging of alloplastic ligament implant. An *in vivo* and *in vitro* study exemplified by Kevlar. *Unfallchirurgie*, **20**(2), 61–65.

Wenk, E., Meinela, A. J., Wildya, S., Merklea, H. P., & Meinel, L. (2009). Microporous silk fibroin scaffolds embedding PLGA microparticles for controlled growth factor delivery in tissue engineering. *Biomaterials*, **30**(13), 2571–2581.

Woo, S. L. Y., Akeson, W. H., Levenetz, B., Coutts, R. D., Matthews, J. V., et al. (1974). Potential application of graphite fiber and methylmethacrylate resin composites as internal fixation plates. *J. Biomed. Mater. Res.*, **8**, 321–328.

Xu, H. K., Eichmiller, F. C., & Giuseppetti, A. A. (2000). Reinforcement of a self-setting calcium phosphate cement with different fibers. *J. Biomed. Mater. Res.*, **52**(1), 107–114.

Yang, J. M., Huang, P. Y., Yang, M. C., & Lo, S. K. (1997). Effect of MMA-g-UHMWPE grafted fiber on mechanical properties of acrylic bone cement. *J. Biomed. Mater. Res.*, **38**(4), 361–369.

Yasko, A., Fellinger, E., Waller, S., Tomin, A., Peterson, M., et al. (1992). Comparison of biological and synthetic carriers for recombinant human BMP induced bone formation. *Trans. Orth. Res. Soc.*, **17**, 71.

Yaszemski, M. J., Paune, R. G., Hayes, W. C., Langer, R., & Mikos, A. G. (1996). *In vitro* degradation of a poly(propylene fumarate)-based composite material. *Biomaterials*, **17**, 2127–2130.

Zhao, D. S., Moritz, N., Laurila, P., Mattila, R., Lassila, L. V. J., et al. (2009). Development of a multi-component fiber-reinforced composite implant for load-sharing conditions. *Med. Eng. Phys.*, **31**(4), 461–469.

Zimmerman, M. C., Parsons, J. R., & Alexander, H. (1987). The design and analysis of a laminated partially degradable composite bone plate for fracture fixation. *J. Biomed. Mater. Res. Appl. Biomater.*, **21A**(3), 345.

Zimmerman, M. C., Alexander, H., Parsons, J. R., & Bajpai, P. K. (1991). The Design and Analysis of Laminated Degradable Composite Bone Plates for Fracture Fixation. In T. Vigo (Ed.), *Hi-Tech Textiles*. Washington, DC: ACS Publications.

CHAPTER I.2.10 NON-FOULING SURFACES

Buddy D. Ratner[1] and Allan S. Hoffman[2]

[1]Professor, Bioengineering and Chemical Engineering, Director of University of Washington Engineered Biomaterials (UWEB), Seattle, WA, USA

[2]Professor of Bioengineering and Chemical Engineering, UWEB Investigator, University of Washington, Seattle, WA, USA

INTRODUCTION

"Non-fouling" surfaces (NFSs) refer to surfaces that resist the adsorption of proteins and/or adhesion of cells. They are also loosely referred to as protein-resistant surfaces and "stealth" surfaces. It is generally acknowledged that surfaces that strongly adsorb proteins will generally bind cells, and that surfaces that resist protein adsorption will also resist cell adhesion. It is also generally recognized that hydrogel surfaces are more likely to resist protein adsorption, and that hydrophobic surfaces will usually adsorb a monolayer of tightly adsorbed protein. Exceptions to these generalizations exist but, overall, they are accurate statements. Further details on protein adsorption can be found in Chapter II.1.2.

Given that proteins have a strong tendency to adsorb to almost all surfaces, how might NFSs work to inhibit such adsorption? Although there are many potential mechanisms for action, most non-fouling surfaces seem to have strong interactions with water. This water–polymer interaction highly hydrates and expands surface hydrophilic polymer chains. Also, the NFSs bind water tightly, and this water shield separates the proteins from the material of the surface. These mechanisms will be elaborated on below.

An important area for NFSs focuses on bacterial biofilms (see Chapter II.2.8). Bacteria are believed to adhere to surfaces via a "conditioning film" of organic molecules that adsorbs first to the surface. The bacteria stick to this conditioning film and begin to exude a gelatinous slime layer (the biofilm) that aids in their protection from external agents (for example, antibiotics). Such layers are particularly troublesome in devices such as urinary catheters and endotracheal tubes. However, they also form on vascular grafts, hip joint prostheses, heart valves, and other long-term implants, where they can stimulate significant inflammatory reaction to the infected device. If the conditioning film can be inhibited, bacterial adhesion and biofilm formation can also be reduced. NFSs offer this possibility.

> Non-fouling surfaces are important in medical devices where they may inhibit bacterial colonization and blood cell adhesion. They are also valuable for microfluidic devices, surface-based diagnostic assays, preventing marine organism build-up on ships, and reducing biofilm formation on heat exchangers.

NFSs have medical and biotechnology uses as blood-compatible materials (where they may resist fibrinogen adsorption and platelet attachment), implanted devices (where they reduce the foreign-body reaction; simplify device removal), urinary catheters, biosensors, affinity separation processes, microchannel flow devices, intravenous syringes and tubing, and non-medical uses as biofouling-resistant heat exchangers and ship hulls. It is important to note that many of these applications involve *in vivo* implants or extra-corporeal devices, and others involve *in vitro* diagnostic assays, sensors, and affinity separation processes. As well as having considerable medical and economic importance, they offer significant experimental and theoretical insights into an important phenomenon in biomaterials science: protein adsorption. Hence, NFSs have been the subject of many investigations. Aspects of non-fouling surfaces are addressed in many other chapters of this textbook, including the chapters on water at interfaces (Chapter I.1.6), surface modification (Chapter I.2.12), and protein adsorption (Chapter II.1.2).

The significant portion of the literature on non-fouling surfaces focuses on surfaces containing the relatively simple polymer, poly(ethylene glycol) or PEG:

$$HO(-CH_2CH_2O-)_n\ H$$

When n is in the range of 15–3500 (molecular weights of approximately 400–100,000), the PEG designation is used. Where molecular weights are greater than 100,000, the molecule is commonly referred to as poly(ethylene oxide) (PEO). Where n is in the range of 2–15, the term oligo(ethylene glycol) (OEG) is often used. Other natural and synthetic polymers besides PEG show non-fouling behavior, and they will also be discussed in this chapter. Figure I.2.10.1 schematically illustrates types of surfaces that show non-fouling properties.

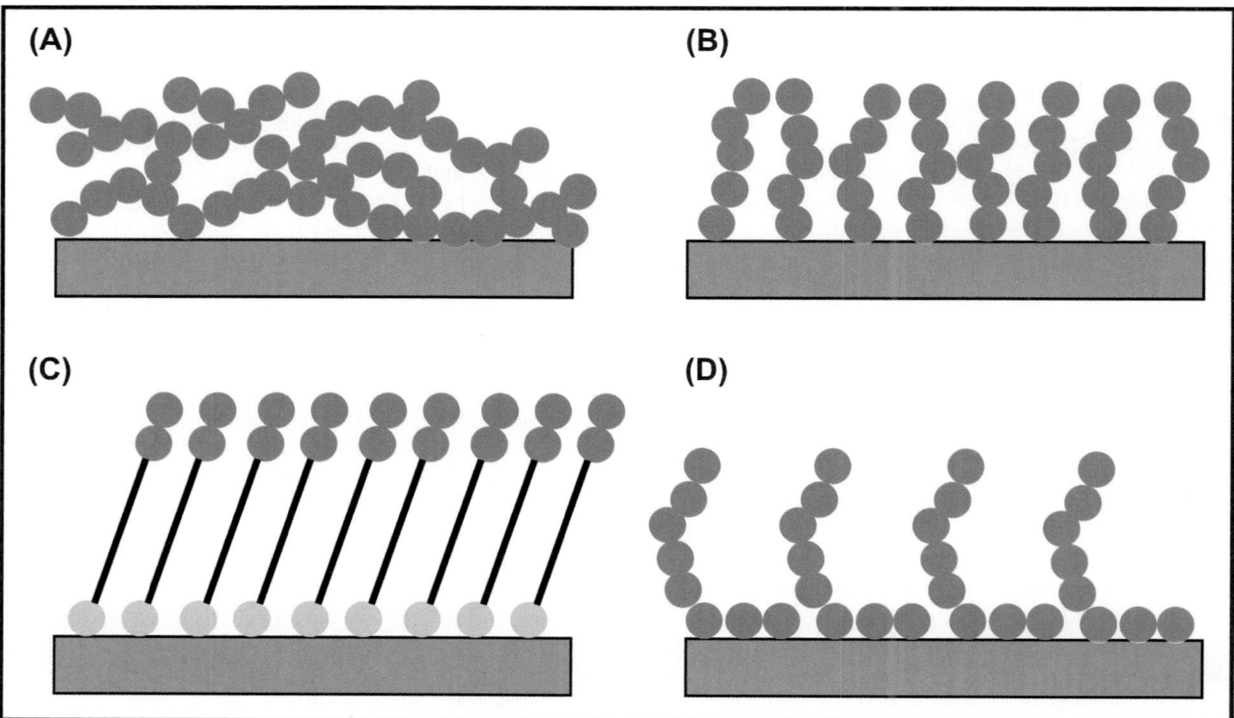

FIGURE I.2.10.1 Four surfaces that can show non-fouling behavior. The blue dot represents a non-fouling chemical moiety such as $(CH_2CH_2O)_n$. (A) Cross-linked network of long, polymeric chains; (B) polymer brush grown off the surface; (C) oligo-non-fouling headgroups on a self-assembled monolayer (the yellow dot is the anchor group such as a thiol); (D) surfactant adsorbed to the surface. The hydrophobic tails are represented by turquoise dots.

BACKGROUND

The published literature on protein and cell interactions with biomaterial surfaces has grown significantly in the past 40 years, and the following concepts have emerged:

- It is well-established that hydrophobic surfaces have a strong tendency to adsorb proteins, often irreversibly (Hoffman, 1986; Horbett and Brash, 1987, 1995). The driving force for this action is most likely the unfolding of the protein on the surface, accompanied by release of many bound water molecules from the solid surface at the interface (see Chapter I.1.6), leading to a large entropy gain for the system (Hoffman, 1999). Note that adsorbed proteins can be displaced (or "exchanged") from the surface by solution phase proteins (Brash et al., 1974).
- It is also well-known that, at low ionic strengths, cationic proteins bind to anionic surfaces and anionic proteins bind to cationic surfaces (Horbett and Hoffman, 1975; Bohnert et al., 1988; Hoffman, 1999). The major thermodynamic driving force for these actions is a combination of ion–ion coulombic interactions, accompanied by an entropy gain due to the release of counter-ions along with their waters of hydration. However, these interactions are diminished at physiologic conditions by shielding of the protein ionic groups at 0.15 N ionic strength (Horbett and Hoffman, 1975). Still, lysozyme, a highly charged cationic protein at physiologic pH, strongly binds to hydrogel contact lenses containing anionic monomers (see Bohnert et al., 1988, and Chapter II.5.9A).
- It has been a common observation that proteins tend to adsorb in monolayers, i.e., proteins do not adsorb non-specifically onto their own monolayers (Horbett, 1993). This is probably due to retention of hydration water by the adsorbed protein molecules, preventing close interactions of the protein molecules in solution with the adsorbed protein molecules. In fact, adsorbed protein films are, in themselves, reasonable non-fouling surfaces with regard to other proteins (but not necessarily to cells).
- Early studies were carried out on surfaces coated with physically- or chemically-immobilized PEG, and a conclusion was reached that a minimum PEG molecular weight (MW) was needed in order to provide good protein repulsion (Mori et al., 1983; Gombotz et al., 1991; Merrill, 1992). This seemed to be the case whether PEG was chemically bound: (1) as a side-chain of a polymer that was grafted to the surface (Mori et al., 1983); (2) by one end to the surface (Gombotz et al., 1991; Merrill, 1992); or was incorporated (3) as segments in a cross-linked network (Merrill, 1992). The minimum MW to achieve an NFS with these low molecular weight polymeric PEGs was found to be ca. 500–2000, depending on packing density (Mori et al., 1983; Gombotz et al., 1991; Merrill, 1992).

- The mechanism of protein resistance by the polymeric PEG surfaces may be due to a combination of factors, including the resistance of the polymer coil to compression due to its desire to retain the more expanded volume of a random coil (called "entropic repulsion" or "elastic network" resistance) plus the resistance of the PEG molecule to release both bound and free water from within the hydrated coil (called "osmotic repulsion") (Gombotz et al., 1991; Antonsen and Hoffman, 1992; Heuberger et al., 2005). It is interesting, as noted above, that the non-fouling character of PEG surfaces begins when the MWs of PEG reach around 500–2000. As that size is reached, Gombotz et al. postulated that the chain segments will begin to form random coils, which could retain water more efficiently between the chain segments. The retention of water by shorter chains would not be as efficient as by the higher MW coils, especially at relatively low surface OEG/PEG densities. The size of the adsorbing protein and its resistance to unfolding may also be important factors determining the extent of adsorption on any surface (Lim and Herron, 1992).

The thermodynamic principles governing the adsorption of proteins onto surfaces involve a number of enthalpic and entropic terms favoring or resisting adsorption. These terms are summarized in Table I.2.10.1. The major factors favoring adsorption will be the entropic gain of released water and the enthalpy loss due to cation–anion attractive interactions between ionic protein groups and surface groups. The major factors favoring resistance to protein adsorption will be the retention of bound water plus, in the case of an immobilized hydrophilic polymer, entropic and osmotic repulsion of the polymer coils.

Although evidence was discussed above for a PEG molecular weight effect, excellent protein resistance can be achieved with very short chain PEGs (OEGs) and PEG-like surfaces (Lopez et al., 1992; Prime and Whitesides, 1993; Sheu et al., 1993). When these surfaces with only short OEG units have sufficiently high (CH_2CH_2O) unit molar densities, they exhibit excellent non-fouling behavior under conditions of low protein concentration. For example, Prime and Whitesides (1993) have noted that non-fouling behavior begins when around 50–60% of the surface groups in their self-assembled monolayer (SAM) are conjugated with OEGs having six EG units. This may be due to formation of a "contiguous" surface film of bound water that is extensive in area and a nanometer or so thick. Another way to view this phenomenon is that, if the OEG units are spaced isotropically on the surface, the remaining "bare" or exposed surface areas on average are less than the diameter of a typical globular protein. (Note: At higher protein solution concentrations, these oligo surfaces foul with adsorbed protein. This solution concentration effect will be discussed shortly.)

Observations on self-assembled monolayers (SAMs) of lipid-oligoEG molecules have been particularly important in the understanding of NFSs (SAMs are described in Chapter I.2.12). SAMs prepared to explore NFSs were formed on gold with a set of molecules based on this alkyl thiol structure: $HS(CH_2)_{11}(OCH_2CH_2)_nOH$. The oligo(ethylene glycol) chain length, n, was 0, 1, 2, 4, 6 or 17. OEGs with larger n (more EG units) were more protein-resistant. The surface concentration of these OEG-thiol units needed to achieve non-fouling behavior decreased with increasing n (Prime and Whitesides, 1993). Note that when $n = 17$, the OEG chain has a molecular weight over 700. Thus, the excellent non-fouling behavior of this surface can be attributed to both chain entropy arguments, and the ability of EG units to interact with water. Thus, protein resistance by OEG-coated surfaces may be related to a "cooperativity" between the hydrated, short OEG chains in the "plane of the surface," wherein the OEG chains interact together to bind water to the surface in a manner analogous to the hydrated coil and its osmotic repulsion, as described above. It was observed that a minimum of three EG units are needed for highly effective protein repulsion (Harder et al., 1998), although Prime and Whitesides (1993) demonstrated that good protein resistance could be achieved with $n = 2$, if the surface density was high enough. Based on all of these observations, one may describe the mechanism as being related to the conformation of the individual oligoEG chains, along with their packing density in the SAM. It has been proposed that helical or amorphous OEG conformations lead to stronger water–OEG interactions than an all-trans OEG conformation (Harder et al., 1998).

Packing density of the non-fouling groups on the surface is clearly an important factor in preparing non-fouling surfaces. Nevertheless, one may conclude that the one common factor connecting all NFSs is their resistance to release of bound water molecules from

TABLE I.2.10.1 Thermodynamics of Protein Adsorption

Favoring Adsorption	
ΔH_{ads}	- van der Waals interactions (short range) - ion–ion interactions (long range)
ΔS_{ads}	+ desorption of many H_2O's + unfolding of protein
Opposing Adsorption	
ΔH_{ads}	+ dehydration (interface between surface and protein) + unfolding of protein + chain compression (PEO)
ΔS_{ads}	- adsorption of protein - protein hydrophobic exposure - chain compression (PEO) - osmotic repulsion (PEO)

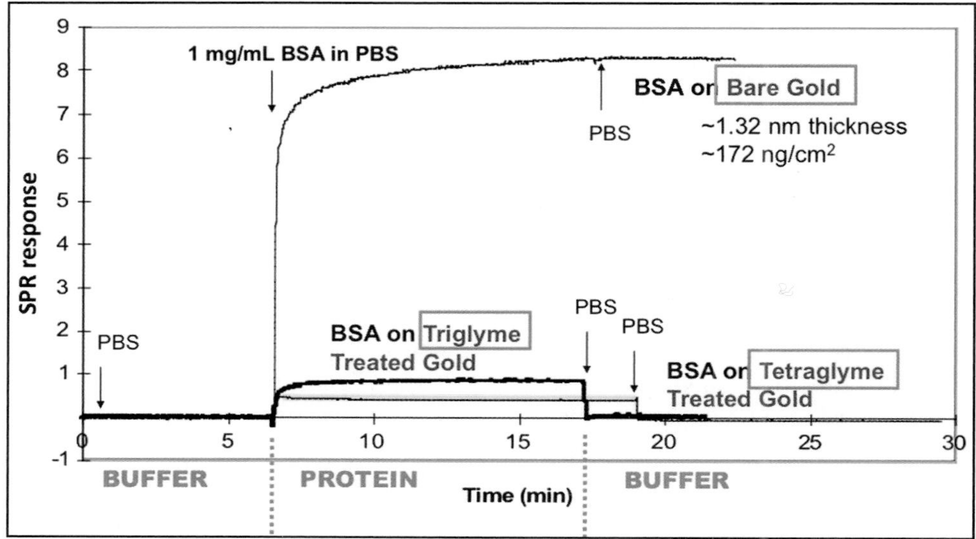

FIGURE I.2.10.2 A surface plasmon resonance (SPR) study of protein adsorption to plasma-deposited triglyme and tetraglyme surfaces. The SPR response (*y* axis) is proportional to protein concentration. At minute 7, bovine serum albumin (BSA) solution (1 mg/mL) is introduced into the buffer flow over the surfaces being examined. A gold surface adsorbs significant levels of BSA. Both the triglyme and the tetraglyme surfaces show much lower SPR responses. At 17 minutes, the protein flow is switched to a buffer flow. No protein desorbs from the gold. Essentially all the protein on the triglyme and tetraglyme surfaces desorbs, bringing the SPR signal for those surfaces back to baseline. This experiment is described in Mar et al., 1999.

the surface. This argument is supported by neutron reflectivity measurements by Skoda et al. (2009), where they found a depletion of protein concentration around 4–5 nm above an OEG surface that might be attributable to a tightly-bound water layer interfering with the protein approach to the actual surface. Also, direct measurements by D_2O exchange and sum frequency generation (SFG, see Chapter I.1.5) of water organization at the interface of non-fouling and fouling SAMs demonstrate that NFSs lead to a more bound, organized water (Stein et al., 2009). Interfacial water may be reorganized both by hydrophobic groups and by hydrophilic (primarily via hydrogen bonds) interactions. In the latter case, the water may be H-bonded to neutral polar groups, such as hydroxyl (-OH) or ether (-C-O-C-) groups, or it may be polarized by ionic groups, such as $-COO^-$ or $-NH_3^+$. The more profound molecular reorganizations of water seen in NFSs are associated with certain hydrophilic surfaces. The overall conclusion from all of the above observations is that resistance to protein adsorption at biomaterial interfaces is directly related to retention by interfacial groups of their bound waters of hydration.

> Resistance to protein adsorption at biomaterial interfaces is directly related to resistance of interfacial groups to the release of their bound waters of hydration.

Based on these conclusions, it is obvious why the most common approaches to reducing protein and cell binding to biomaterial surfaces have been to form surfaces that strongly interact with water. This has been accomplished by a variety of strategies:

- Chemical immobilization of a hydrophilic polymer (e.g., PEG or OEG) on the biomaterial surface using: (1) UV or ionizing radiation to graft copolymerize a hydrophilic monomer onto surface groups; (2) deposition of such a polymer from the vapor of a precursor monomer in a gas discharge process; (3) covalent immobilization of a pre-formed hydrophilic polymer or surfactant to the surface using radiation, gas discharge or wet-chemical coupling processes; or (4) growth of a non-fouling polymer brush off the surface by controlled radical polymerization methods.

Examples of some of these strategies follow.

- Gas discharge has been used to covalently bind non-fouling surfactants such as Pluronic® polyols to polymer surfaces (Sheu et al., 1993), and it has also been used to deposit an "oligoEG-like" coating from vapors of triglyme or tetraglyme (Lopez et al., 1992; Mar et al., 1999) (Figure I.2.10.2). Highly fouling-resistant brushes of PEG chains have been grown from surfaces using atom transfer radical polymerization (ATRP) methods (Ma et al., 2004).
- Chemical derivitization of surface groups with neutral groups, such as hydroxyls, or with negatively-charged groups (especially since most proteins and cells are negatively charged) such as carboxylic acids or their salts, or sulfonates.
- Physical adsorption of surfactants or other surface-active molecules that are coupled to non-fouling moieties. Examples include: adsorption of poly(ethylene

gycol-propylene glycol) block copolymers (Lee et al., 1989); surface attachment of biomimetic analogs of muscle adhesive peptides coupled to PEG chains (Dalsin et al., 2003); and immobilization of PEG brushes to metal oxides through interaction with a positively charged polylysine backbone (Kenausis et al., 2000).

- Hydrophilic polymers containing zwitterionic groups along their backbones have been extensively studied for their non-fouling properties. Examples include: poly[2-(methacryloyloxy)ethyl phosphorylcholine] (Ishihara et al., 1998; Iwasaki et al., 1999; Ishihara and Takai, 2009); sulfbetaines (Stein et al., 2009); and carboxybetaines (Zhang et al., 2008).
- Coatings of many hydrogels, including poly(2-hydroxyethyl methacrylate) and polyacrylamide, show reasonable non-fouling behavior (Lopez et al., 1993).

A number of naturally-occurring biomolecules such as albumin, casein, hyaluronic acid, and mucin have been coated on surfaces and exhibited resistance to non-specific adsorption of proteins (Vogt et al., 1987; Van Beek et al., 2008). Albumin and non-fat milk solids (mainly containing casein) are often used as "blocking" agents to prevent non-specific protein adsorption in ELISA assays. Naturally-occurring ganglioside lipid surfactants having saccharide head groups have been used to make "stealth" liposomes (Lasic and Needham, 1995). One paper even suggested that the protein resistance of PEGylated surfaces is related to the "partitioning" of albumin into the PEG layers, causing those surfaces to "look like native albumin" (Vert and Domurado, 2000).

SAMs presenting an interesting series of head group molecules that can act as H-bond acceptors, but not as H-bond donors, have been shown to yield surfaces with unexpected protein resistance (Chapman et al., 2000; Ostuni et al., 2001; Kane et al., 2003). Interestingly, PEG fits in this category of H-bond acceptors but not donors; Merrill first noted this in 1983 (Merrill and Salzman, 1983). However, this generalization does not explain all non-fouling surfaces; consider that mannitol groups with H-bond donor -OH groups were found to be non-fouling (Luk et al., 2000). Also, adsorbed protein films are non-fouling, and these have H-bond donor and acceptor groups. Another hypothesis proposes that the functional groups that impart a non-fouling property are kosmotropes, order-inducing molecules (Kane et al., 2003). Perhaps because of the ordered water surrounding these molecules, they cannot penetrate the ordered water shell surrounding proteins, therefore strong intermolecular interactions between surface group and protein cannot occur. An interesting kosmotrope molecule with good non-fouling ability described in this paper is taurine $H_3N^{+}(CH_2)2SO^3$. Table I.2.10.2 summarizes some of the different compositions that have been applied as non-fouling surfaces.

It is worthwhile mentioning computational papers (supported by some experiments) that offer new insights

TABLE I.2.10.2 "Non-Fouling" Surface Compositions

Synthetic Hydrophilic Surfaces
- PEG polymers and surfactants
- Neutral polymers
  poly(2-hydroxyethyl methacrylate)
  polyacrylamide and poly(N-methyl acrylamide)
  poly(N-vinyl 2-pyrrolidone)
  poly(N-isopropyl acrylamide) (below 31°C)
- Anionic polymers
- Phosphoryl choline polymers
- Sulfobetaines, carboxybetaines, taurine-functionalized materials
- Poly(2-methyl-2-oxazoline)
- Gas discharge-deposited coatings (especially from PEG-like monomers)
- Self-assembled n-alkyl molecules with oligo-PEG headgroups

Natural Hydrophilic Surfaces
- Passivating proteins (e.g., albumin and casein)
- Polysaccharides (e.g., hyaluronic acid)
- Liposaccharides
- Phospholipid bilayers
- Glycoproteins (e.g., mucin)

Other Molecules (see Ostuni et al., 2001).

and ideas on NFSs (Lim and Herron, 1992; Pertsin and Grunze, 2000; Pertsin et al., 2002; Xu et al., 2011). Also, many new experimental methods have been applied to study the mechanism of non-fouling surfaces, including neutron reflectivity to measure the water density in the interfacial region (Schwendel et al., 2003; Skoda et al., 2009), scanning force microscopy (Feldman et al., 1999; Heuberger et al., 2005), and sum frequency generation (Zolk et al., 2000; Stein et al., 2009).

The significance of protein solution concentration for surface protein resistance requires further discussion. In relatively dilute protein solutions, such as are used in diagnostic assays and sometimes in cell culture, all of the surfaces described so far will demonstrate non-fouling behavior. However, at protein concentrations similar to that in the bloodstream or in body fluids, many of these surfaces no longer appear non-fouling. The surfaces that exhibit non-fouling behavior at higher protein concentration are comprised of longer hydrophilic chains with high conformational freedom (Ma et al., 2004; Zhang et al., 2008). Surfaces comprised of short oligo units of non-fouling chemistries (Figure I.2.10.1C or the glyme surfaces described in Figure I.2.10.2) probably function by tightly binding water. It may be that at high solution protein content, the protein molecules compete for the bound surface water, and the non-fouling effect vanishes. On the other hand, when longer, flexible chains are present, the entropic repulsion effect described earlier permits these surfaces to remain free of adsorbed protein, as long as the hydrophilic polymer surface density is high enough.

Finally, it should be noted that bacteria tend to adhere and colonize almost any type of surface, perhaps even many protein-resistant NFSs. However, the best NFSs can provide acute resistance to bacteria and biofilm build-up better than most surfaces (Johnston et al., 1997). Resistance to bacterial adhesion remains an unsolved problem in surface science. Also, it has been pointed out that susceptibility of PEGs to oxidative damage may reduce their utility as non-fouling surfaces in real-world situations (Kane et al., 2003).

CONCLUSIONS AND PERSPECTIVES

It is remarkable how many different surface compositions appear to be non-fouling (although the majority of surfaces used in medicine and biology adsorb protein). Although the unifying mechanisms are not fully clarified, it appears that the major factor favoring resistance to protein adsorption will be the retention of bound water by the surface molecules, plus, in the case of an immobilized hydrophilic polymer, entropic and osmotic repulsion of the polymer coils. Little is known about how long a non-fouling surface will remain non-fouling *in vivo*. Longevity and stability for non-fouling biomaterials remains an uncharted frontier. Defects (e.g., pits, uncoated areas) in NFSs may provide "footholds" for bacteria and cells to begin colonization. Enhanced understanding of how to optimize the surface density and composition of NFSs will lead to improvements in quality and fewer micro-defects. Finally, it is important to note that a clean, "non-fouled" surface may not always be desirable. In the case of cardiovascular implants or devices, emboli may be shed when such a surface is exposed to flowing blood (Hoffman et al., 1982). This can lead to undesirable consequences, even though (or perhaps especially because) the surface is an effective, non-fouling surface. In the case of contact lenses, a protein-free lens may seem desirable, but there are concerns that such a lens will not be comfortable (mucin adsorption, for example, may contribute to comfort). Although biomaterials scientists can presently create surfaces that are non-fouling for a period of time, applying such surfaces must take into account the specific application, the biological environment, and the intended service life.

BIBLIOGRAPHY

Antonsen, K. P., & Hoffman, A. S. (1992). Water Structure of PEG Solutions by DSC Measurements. In J. M. Harris (Ed.), *Polyethylene Glycol Chemistry: Biotechnical and Biomedical Applications* (pp. 15–28). New York, NY: Plenum Press.

Brash, J. L., Uniyal, S., & Samak, Q. (1974). Exchange of albumin adsorbed on polymer surfaces. *Trans. Am. Soc. Artif. Int. Org.*, **20**, 69–76.

Bohnert, J. L., Horbett, T. A., Ratner, B. D., & Royce, F. H. (1988). Adsorption of proteins from artificial tear solutions to contact lens materials. *Invest. Ophth. Vis. Sci.*, **29**(3), 362–373.

Chapman, R. G., Ostuni, E., Takayama, S., Holmlin, R. E., Yan, L., et al. (2000). Surveying for surfaces that resist the adsorption of proteins. *JACS*, **122**, 8303–8304.

Dalsin, J. L., Hu, B.-H., Lee, B. P., & Messersmith, P. B. (2003). Mussel adhesive protein mimetic polymers for the preparation of nonfouling surfaces. *J. Am. Chem. Soc.*, **125**(14), 4253–4258.

Feldman, K., Hahner, G., Spencer, N. D., Harder, P., & Grunze, M. (1999). Probing resistance to protein adsorption of oligo(ethylene glycol)-terminated self-assembled monolayers by scanning force microscopy. *J. Am. Chem. Soc.*, **121**(43), 10134–10141.

Gombotz, W. R., Wang, G. H., Horbett, T. A., & Hoffman, A. S. (1991). Protein adsorption to PEO surfaces. *J. Biomed. Mater. Res.*, **25**, 1547–1562.

Harder, P., Grunze, M., Dahint, R., Whitesides, G., & Laibinis, P. (1998). Molecular conformation in oligo (ethylene glycol)-terminated self-assembled monolayers on gold and silver surfaces determines their ability to resist protein adsorption. *J. Phys. Chem. B*, **102**, 426–436.

Heuberger, M., Drobek, T., & Spencer, N. D. (2005). Interaction forces and morphology of a protein-resistant poly(ethylene glycol) layer. *Biophys. J.*, **88**(1), 495–504.

Hoffman, A. S. (1986). A general classification scheme for hydrophilic and hydrophobic biomaterial surfaces. *J. Biomed. Mater. Res.*, **20**(9). ix–xi.

Hoffman, A. S. (1999). Non-fouling surface technologies. *J. Biomater. Sci., Polymer Ed.*, **10**, 1011–1014.

Hoffman, A. S., Horbett, T. A., Ratner, B. D., Hanson, S. R., Harker, L. A., et al. (1982). Thrombotic events on grafted polyacrylamide-Silastic surfaces as studied in a baboon. *ACS Adv. Chem. Ser.*, **199**, 59–80.

Horbett, T. A. (1993). Principles underlying the role of adsorbed plasma proteins in blood interactions with foreign materials. *Cardiovasc. Pathol.*, **2**, 137S–148S.

Horbett, T. A., & Brash, J. L. (1987). Proteins at Interfaces: Current Issues and Future Prospects. In T. A. Horbett, & J. L. Brash (Eds.), *Proteins at Interfaces, Physicochemical and Biochemical Studies ACS Symposium Series* (Vol. 343, pp. 1–33). Washington, DC: American Chemical Society.

Horbett, T. A., & Brash, J. L. (1995). Proteins at Interfaces: An Overview. In T. A. Horbett, & J. L. Brash (Eds.), *Proteins at Interfaces II: Fundamentals and Applications ACS Symposium Series* (Vol. 602, pp. 1–25). Washington, DC: American Chemical Society.

Horbett, T. A., & Hoffman, A. S. (1975). Bovine Plasma Protein Adsorption to Radiation Grafted Hydrogels Based Hydroxyethylmethacrylate and N-Vinyl-Pyrrolidone. In R. Baier (Ed.), *Advances in Chemistry Series, 145, Applied Chemistry at Protein Interfaces* (pp. 230–254). Washington DC: American Chemical Society.

Ishihara, K., & Takai, M. (2009). Bioinspired interface for nanobiodevices based on phospholipid polymer chemistry. *J. Roy. Soc. Interface*, **6**(Suppl. 3), S279–S291.

Ishihara, K., Nomura, H., Mihara, T., Kurita, K., Iwasaki, Y., et al. (1998). Why do phospholipid polymers reduce protein adsorption? *J. Biomed. Mater. Res.*, **39**, 323–330.

Iwasaki, Y., Nakabayashi, N., Nakatani, M., Mihara, T., Kurita, K., et al. (1999). Competitive adsorption between phospholipid and plasma protein on a phospholipid polymer surface. *J. Biomater. Sci. Polym. Edn.*, **10**, 513–529.

Johnston, E. E., Ratner, B. D., & Bryers, J. D. (1997). RF Plasma Deposited PEO-Like Films: Surface Characterization and Inhibition of *Pseudomonas aeruginosa* Accumulation. In R. d'Agostino, P. Favia, & F. Fracassi (Eds.), *Plasma Processing of Polymers* (pp. 465–476). Dordrecht, Netherlands: Kluwer Academic Publishers.

Kane, R. S., Deschatelets, P., & Whitesides, G. M. (2003). Kosmotropes form the basis of protein-resistant surfaces. *Langmuir*, **19**, 2388–2391.

Kenausis, G. L., Vörös, J., Elbert, D. L., Huang, N., Hofer, R., et al. (2000). Poly (L-lysine)-g-poly (ethylene glycol) layers on metal oxide surfaces: Attachment mechanism and effects of polymer architecture on resistance to protein adsorption. *J. Phys. Chem. B*, **104**(14), 3298–3309.

Lasic, D. D., & Needham, D. (1995). The "stealth" liposome: A prototypical biomaterial. *Chem. Rev.*, **95**(8), 2601–2628.

Lee, J. H., Kopecek, J., & Andrade, J. D. (1989). Protein-resistant surfaces prepared by PEO-containing block copolymer surfactants. *J. Biomed. Mater. Res.*, **23**, 351–368.

Lim, K., & Herron, J. N. (1992). Molecular Simulation of Protein-PEG Interaction. In J. M. Harris (Ed.), *Polyethylene Glycol Chemistry: Biotechnical and Biomedical Applications* (pp. 29). New York, NY: Plenum Press.

Lopez, G. P., Ratner, B. D., Tidwell, C. D., Haycox, C. L., Rapoza, R. J., et al. (1992). Glow discharge plasma deposition of tetraethylene glycol dimethylether for fouling-resistant biomaterial surfaces. *J. Biomed. Mater. Res.*, **26**, 415–439.

Lopez, G. P., Ratner, B. D., Rapoza, R. J., & Horbett, T. A. (1993). Plasma deposition of ultrathin films of poly(2-hydroxyethyl methacrylate): Surface analysis and protein adsorption measurements. *Macromolecules*, **26**, 3247–3253.

Luk, Y. Y., Kato, M., & Mrksich, M. (2000). Self-assembled monolayers of alkanethiolates presenting mannitol groups are inert to protein adsorption and cell attachment. *Langmuir*, **16**, 9605.

Ma, H., Hyun, J., Stiller, P., & Chilkoti, A. (2004). "Non-fouling" oligo(ethylene glycol)- functionalized polymer brushes synthesized by surface-initiated atom transfer radical polymerization. *Adv. Mater.*, **16**(4), 338–341.

Mar, M. N., Ratner, B. D., & Yee, S. S. (1999). An intrinsically protein-resistant surface plasmon resonance biosensor based upon a RF-plasma-deposited thin film. *Sensor. Actuator. B*, **54**, 125–131.

Merrill, E. W. (1992). Poly(ethylene oxide) and Blood Contact: A Chronicle of One Laboratory. In J. M. Harris (Ed.), *Polyethylene Glycol Chemistry: Biotechnical and Biomedical Applications* (pp. 199–220). New York, NY: Plenum Press.

Merrill, E. W., & Salzman, E. W. (1983). Polyethylene oxide as a biomaterial. *ASAIO Journal*, **6**, 60.

Mori, Y., et al. (1983). Interactions between hydrogels containing peo chains and platelets. *Biomaterials*, **4**, 825–830.

Ostuni, E., Chapman, R. G., Holmlin, R. E., Takayama, S., & Whitesides, G. M. (2001). A survey of structure–property relationships of surfaces that resist the adsorption of protein. *Langmuir*, **17**, 5605–5620.

Pertsin, A. J., & Grunze, M. (2000). Computer simulation of water near the surface of oligo(ethylene glycol)-terminated alkanethiol self-assembled monolayers. *Langmuir*, **16**(23), 8829–8841.

Pertsin, A. J., Hayashi, T., & Grunze, M. (2002). Grand canonical Monte Carlo simulations of the hydration interaction between oligo(ethylene glycol)-terminated alkanethiol self-assembled monolayers. *J. Phys. Chem. B.*, **106**(47), 12274–12281.

Prime, K. L., & Whitesides, G. M. (1993). Adsorption of proteins onto surfaces containing end-attached oligo(ethylene oxide): A model system using self-assembled monolayers. *J. Am. Chem. Soc.*, **115**, 10714–10721.

Schwendel, D., Hayashi, T., Dahint, R., Pertsin, A., Grunze, M., et al. (2003). Interaction of water with self-assembled monolayers: Neutron reflectivity measurements of the water density in the interface region. *Langmuir*, **19**(6), 2284–2293.

Sheu, M.-S., Hoffman, A. S., Terlingen, J. G.A., & Feijen, J. (1993). A new gas discharge process for preparation of non-fouling surfaces on biomaterials. *Clin. Mater.*, **13**, 41–45.

Skoda, M., Schreiber, F., Jacobs, R., Webster, J., Wolff, M., et al. (2009). Protein density profile at the interface of water with oligo (ethylene glycol) self-assembled monolayers. *Langmuir*, **25**(7), 4056–4064.

Stein, M., Weidner, T., McCrea, K., Castner, D., & Ratner, B. D. (2009). Hydration of sulphobetaine and tetra(ethylene glycol)-terminated self-assembled monolayers studied by sum frequency generation vibrational spectroscopy. *J. Phys. Chem. B.*, **113**(33), 11550–11556.

Van Beek, M., Jones, L., & Sheardown, H. (2008). Hyaluronic acid containing hydrogels for the reduction of protein adsorption. *Biomaterials*, **29**(7), 780–789.

Vert, M., & Domurado, D. (2000). PEG: Protein-repulsive or albumin-compatible? *J. Biomater. Sci. Pol. Ed.*, **11**, 1307–1317.

Vogt, R. V., Jr., Phillips, D. L., Omar Henderson, L., Whitfield, W., & Spierto, F. W. (1987). Quantitative differences among various proteins as blocking agents for ELISA microtiter plates. *J. Immunol. Methods*, **101**(1), 43–50.

Xu, X., Cao, D., & Wu, J. (2011). Density functional theory for predicting polymeric forces against surface fouling. *Soft Matter*, **6**(19), 4631–4646.

Zhang, Z., Zhang, M., Chen, S., Horbett, T. A., Ratner, B. D., et al. (2008). Blood compatibility of surfaces with superlow protein adsorption. *Biomaterials*, **29**, 4285–4291.

Zolk, M., Eisert, F., Pipper, J., Herrwerth, S., Eck, W., et al. (2000). Solvation of oligo(ethylene glycol)-terminated self-assembled monolayers studied by vibrational sum frequency spectroscopy. *Langmuir*, **16**(14), 5849–5852.

CHAPTER I.2.11 APPLICATIONS OF "SMART POLYMERS" AS BIOMATERIALS

Allan S. Hoffman
Professor of Bioengineering and Chemical Engineering, UWEB Investigator, University of Washington, Seattle, WA, USA

INTRODUCTION

Stimulus-responsive, "intelligent" polymers are polymers that respond to small changes in physical or chemical conditions near a critical condition with sharp and relatively large phase or property changes. They are also known as "smart," stimuli-responsive or "environmentally-sensitive" polymers. These polymers can take many forms; for example, they may be dissolved in aqueous solutions, adsorbed or grafted on aqueous–solid interfaces, or cross-linked in the form of hydrogels. They may also be combined chemically or physically with other molecules, especially a variety of bioactive molecules. A number of reviews over the past 20–25 years have highlighted applications of smart polymers in the biomedical field (Hoffman, 1987, 1995, 1997; Hoffman et al., 2000; Okano et al., 2000; Roy and Gupta, 2003; Roy et al., 2010).

Many different stimuli-responsive polymers have been investigated, and the various stimuli that researchers have used are listed in Table I.2.11.1. Typically, when the polymer's smart response is stimulated, the resultant behavior will depend on the initial state of the polymer

(Figure I.2.11.1). Three examples of this are described below:

- When a smart polymer is dissolved in an aqueous solution, it will exhibit a sudden onset of turbidity as it phase separates, and if its concentration is high enough it may convert from a viscous solution to a gel.
- When a smart polymer is chemically grafted to an aqueous–solid interface and is stimulated to phase separate, it will collapse, converting that interface from a hydrophilic to a hydrophobic interface. If the smart polymer is dissolved in solution and is stimulated to phase separate in the presence of a solid–aqueous interface, it may physically adsorb at the interface, especially when the surface composition has a balance of hydrophobic and polar groups similar to the phase-separating smart polymer.
- A smart polymer may be chemically cross-linked into a network. When it is swollen in an aqueous solution below its critical condition, it may be called a "smart hydrogel." When the polymer network chains are stimulated to phase separate at their critical condition, the hydrogel will collapse and release much of its swelling solution as a burst.

TABLE I.2.11.1 Environmental Stimuli
Physical
Temperature
Ionic strength
Solvents
Radiation (UV, visible)
Electric fields
Mechanical stress
High pressure
Sonic radiation
Magnetic fields
Chemical
pH
Specific ions
Chemical agents
Biochemical
Enzyme substrates
Affinity ligands
Other biological agents

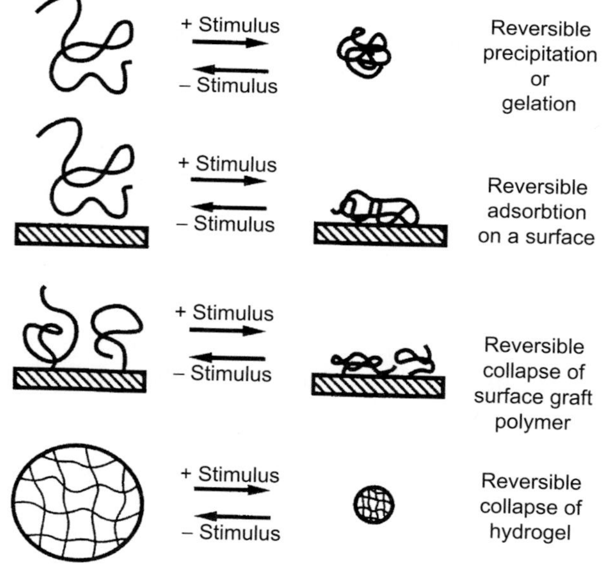

FIGURE I.2.11.1 Schematic illustration showing the different types of responses of "intelligent" polymer systems to environmental stimuli. Note that all systems are reversible when the stimulus is reversed. (Hoffman et al., (2000) *Journal of Biomedical Materials Research* (JBMR): Reproduced with permission of *JBMR*, J. Wiley, Publisher.)

These phenomena all reverse when the stimulus is reversed. The phase separation of a water-solvated smart polymer chain is driven by the release of the hydrophobically bound water molecules on the polymer backbone, which creates a large gain in entropy in the system. Often the rate of reversion back to the hydrated state is slower than the collapse, because in the reverse process the hydrophobic groups of the polymer have to be rehydrated, and that process is thermodynamically opposed by the resultant decline in entropy of the system (the reswelling process is favored by the exothermic hydration and expansion of the polymer chains). The rate of collapse of smart polymer systems and its reversal can also depend on the dimensions of the smart polymer system. Rates will be more rapid for systems with smaller dimensions, e.g., nanoscale versus microscale dimensions.

The rate of collapse of smart polymer systems and its reversal can depend on the dimensions of the smart polymer system. Rates will be more rapid for systems with smaller dimensions, e.g., nanoscale versus microscale dimensions.

Smart polymers may be physically mixed with, complexed with, or chemically conjugated to biomolecules to yield a large and diverse family of polymer–biomolecule "biohybrid" systems that can respond to biological, as well as to physical and chemical, stimuli. Sometimes such biohybrids are called "doubly smart." Biomolecules that may be combined with smart polymer systems include proteins and peptides, sugars and polysaccharides, single and double-stranded oligonucleotides, RNA and DNA, simple lipids and phospholipids, and a wide spectrum of recognition ligands and synthetic drug molecules. In addition, polyethylene glycol (PEG) may be conjugated to the smart polymer backbone to provide it with "stealth" properties (Figure I.2.11.2 schematically shows many different molecules, one or several together, which may be conjugated to a single smart polymer chain). The reader is referred to recent reviews of the synthesis and applications of smart polymer-protein conjugates (e.g.,

Hoffman, et al., 2000; Heredia and Maynard, 2007; Hoffman and Stayton, 2010; Grover and Maynard, 2010; Broyer et al., 2011).

Among the most studied of these smart polymer–biomolecule systems are conjugates of the smart polymer with drugs, enzymes, and antibodies (see below). Such smart bioconjugates may be utilized as free conjugates in solution, or they may be physically adsorbed or chemically immobilized at solid interfaces. A free enzyme, antibody or drug may also be physically or chemically entrapped within smart hydrogels or smart polymer phase-separated masses (phase-separated polymer–drug masses *in vivo* are known as "depot" drug delivery systems). There have been a number of successful applications in both medicine and biotechnology for such smart polymer–biomolecule systems and, as such, they represent an important and exciting extension of polymeric biomaterials beyond their well-known uses in implants, medical devices, and drug delivery (see Chapter II.5.16 on Drug Delivery Systems for more details).

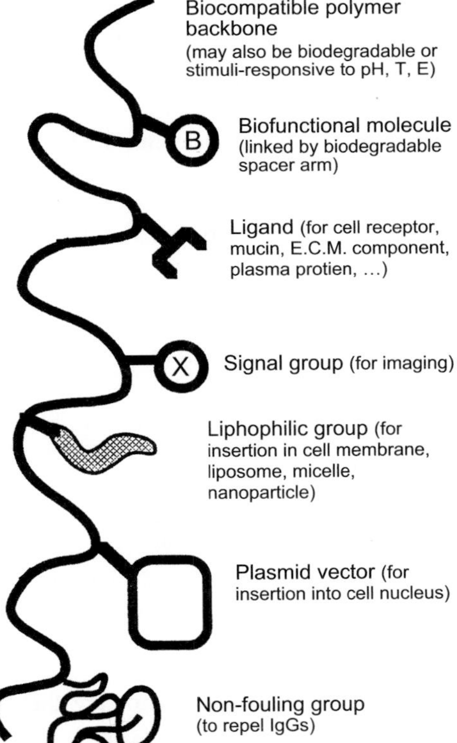

FIGURE I.2.11.2 Schematic illustration showing the variety of natural or synthetic biomolecules that may be conjugated to a smart polymer. In some cases, only one molecule may be conjugated, such as a recognition protein, which may be linked to the protein at a reactive terminal group of the polymer, or it may be linked at a reactive pendant group along the polymer backbone. In other cases more than one molecule may be conjugated along the polymer backbone, such as a targeting ligand along with many drug molecules. (Hoffman et al., (2000) *Journal of Biomedical Materials Research* (JBMR): Reproduced with permission of *JBMR*, J. Wiley, Publisher.)

Smart polymer–biomolecule conjugates in solution, on surfaces, and within hydrogels represent an important and exciting extension of polymeric biomaterials beyond their well-known uses in implants, medical devices, and drug delivery.

SMART POLYMERS IN SOLUTION

There are many polymers that exhibit thermally-induced precipitation out of aqueous solutions (Table I.2.11.2), and the polymer that has been the most extensively studied is poly(*N*-isopropyl acrylamide), or PNIPAAm. This polymer is soluble in water below 32°C, and it precipitates sharply as temperature is raised above 32°C (Heskins and Guillet, 1968; Schild, 1992). The precipitation temperature is called the lower critical solution temperature, or LCST. If the solution contains buffer and salts the LCST may be reduced several degrees. If NIPAAm monomer is copolymerized with more hydrophilic monomers, such as acrylamide, the LCST will increase and may even disappear. If NIPAAm monomer is copolymerized with more hydrophobic monomers, such as *n*-butylacrylamide, the LCST decreases (Figure I.2.11.3) (Priest et al., 1987). NIPAAm may also be copolymerized with pH-sensitive monomers, leading to random copolymers with both temperature- and pH-responsive behavior (Murthy, et al., 1999; Zareie et al., 2000; Murthy et al., 2003a,b; Yin, et al., 2006). Tirrell, and more recently, Stayton, Hoffman, and co-workers have studied the behavior of pH-sensitive alpha-alkylacrylic acid polymers

TABLE I.2.11.2 Some Polymers and Surfactants that Exhibit Thermally-Induced Phase Separation in Aqueous Solutions
Polymers/Surfactants with Ether Groups
Poly(ethylene oxide) (PEO)
Poly(ethylene oxide/propylene oxide) random copolymers [poly(EO/PO)]
PEO–PPO–PEO triblock surfactants (Polyoxamers or Pluronics)
PLGA–PEO–PLGA triblock polymers
Alkyl–PEO block surfactants (Brij)
Poly(vinyl methyl ether)
Polymers with Alcohol Groups
Poly(hydropropyl acrylate)
Hydroxypropyl cellulose
Methylcellulose
Hydroxypropyl methylcellulose
Poly(vinyl alcohol) derivatives
Polymers with Substituted Amide Groups
Poly(N-substituted acrylamides)
Poly(N-acryloyl pyrrolidine)
Poly(N-acryloyl piperidine)
Poly(acryl-l-amino acid amides)
Others
Poly(methacrylic acid)

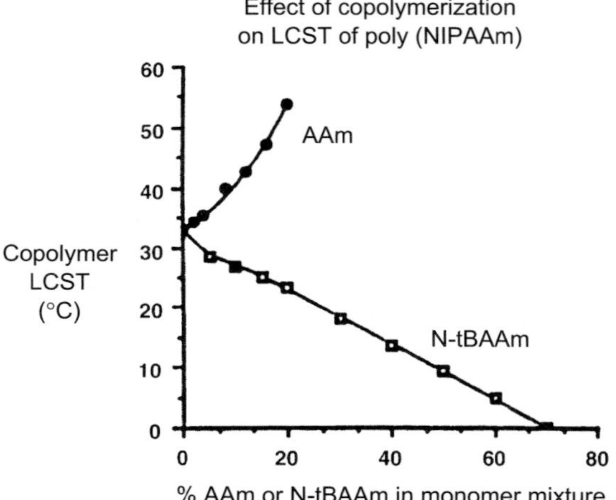

FIGURE I.2.11.3 Copolymerization of a thermally sensitive polymer, PNIPAAm, with a more hydrophilic comonomer, AAm, raises the LCST of the copolymer, whereas copolymerization with a more hydrophobic comonomer, N-tBAAm, lowers the LCST. (Hoffman et al., (2000) *Journal of Biomedical Materials Research* (JBMR): Reproduced with permission of *JBMR*, J. Wiley, Publisher.)

in solution (Tirrell, 1987; Lackey et al., 1999; Murthy et al., 1999; Stayton et al., 2000; Murthy et al., 2003a,b; Yin et al., 2006; Stayton and Hoffman, 2008). As pH is lowered, these polymers become increasingly protonated and hydrophobic, and eventually phase separate; this transition can be sharp, resembling the phase transition at the LCST. If a polymer such as poly(ethylacrylic acid) (PEAA) or poly(propylacrylic acid) (PPAA) is in the vicinity of a lipid bilayer membrane as pH is lowered, the polymer will interact with the membrane and disrupt it. These polymers have been applied by Stayton and Hoffman and co-workers to stimulate endosomal escape of the drug (with or without its nanocarrier) to the cytosol. This action is a consequence of the drop in pH in the endosome, which causes PPAA to become hydrophobic and disrupt the endosomal membrane (e.g., Stayton and Hoffman, 2008; also see detailed discussion and referencing of this work in Chapter II.5.16 on Drug Delivery Systems). Bae and co-workers have also developed interesting temperature- and pH-sensitive polymers for endosomal release of drug formulations (e.g., Na et al., 2006; Kang and Bae, 2007).

NIPAAm has been copolymerized with pH-responsive macromonomers of acrylic acid (AAc), which yielded graft copolymers that independently exhibit two distinct stimuli responses, one for the backbone PNIPAAm and one for the side chain PAAc (Chen and Hoffman, 1995). Block copolymers of two such smart polymers also exhibit two distinct stimuli responses if the two individual blocks are each long enough.

A family of thermally gelling and biodegradable triblock copolymers has been developed by Sung Wan Kim and co-workers for injectable formulations that form degradable, drug depot masses at body conditions (Vernon et al., 2000; Lee et al., 2001; Jeong et al., 2002). The formulation is a medium viscosity, injectable solution at room temperature, and a phase-separated hydrogel mass at 37°C. These polymers are based on blocks of hydrophobic, degradable polyesters combined with blocks of PEO. The copolymers are triblocks with varying MW segments of PLGA and PEO. Typical compositions are PEO–PLGA–PEO and PLGA–PEO–PLGA. They are discussed further in Chapter II.5.16 on Drug Delivery Systems.

Doo Sung Lee, You Han Bae, and co-workers have developed similar biocompatible and degradable A–B–A triblock copolymers having both pH and temperature sensitivity; they are based on two blocks of poly(caprolactone-co-lactic acid) copolymers (PCLA), combined with a central PEG block. The pH sensitivity is derived from short blocks containing the sulfonamide group (OSM) that are attached at each end of the triblock. Typical formulae of the copolymers are PCLA–PEG–PCLA and, when modified with oligomers containing the sulfonamide group at each end, OSM–PCLA–PEG–PCLA–OSM (Shim et al., 2005, 2006).

SMART POLYMER–PROTEIN BIOCONJUGATES IN SOLUTION

Smart polymers may be conjugated randomly to proteins by binding the reactive end of the polymer or reactive pendant groups along the polymer backbone to reactive sites on the protein. One may utilize functionalized free radical initiator or chain transfer agents to synthesize oligomers with one functional end group, which can then be derivatized to form a reactive group that can be conjugated to the protein. For example, NIPAAm has been copolymerized with reactive monomers such as *N*-hydroxysuccinimide acrylate (NHS acrylate) to yield a random copolymer with pendant NHS groups, which have then been conjugated to the protein. (These synthesis methods are described in Cole et al., 1987; Monji and Hoffman, 1987; Chen et al., 1990; Chen and Hoffman, 1990; Yang et al., 1990; Takei et al., 1993a; Chen and Hoffman, 1994; Ding et al., 1996, 1998.) Vinyl monomer groups may also be conjugated to proteins to provide "comonomer" vinyl groups for copolymerization with free monomers such as NIPAAm (Shoemaker et al., 1987).

Normally the lysine amino groups are the most reactive protein sites for random polymer conjugation to proteins, and *N*-hydroxysuccinimide (NHS) attachment chemistry is often utilized. Other possible sites include –COOH groups of aspartic or glutamic acid residues, –OH groups of serine or tyrosine residues, –SH groups of cysteine residues, and terminal amino groups. The most likely attachment site will be determined by the reactive group on the polymer and the reaction conditions, especially the pH. Because these conjugations are generally carried out in a nonspecific way, the conjugated polymer can interfere sterically with the protein's active site or

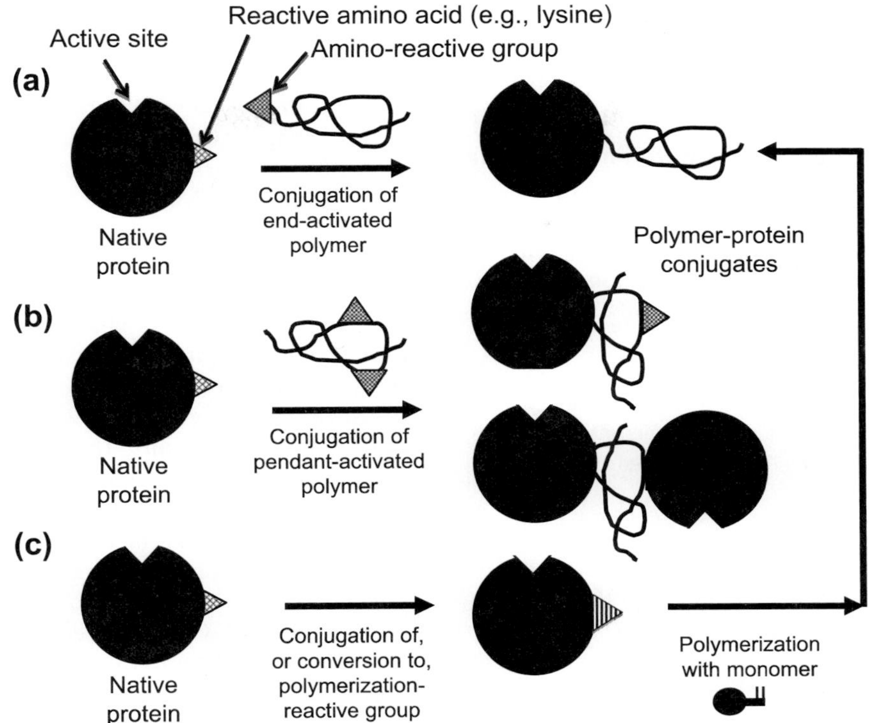

FIGURE I.2.11.4 Various methods for synthesizing polymer–protein bioconjugates.

modify its microenvironment, typically reducing the bioactivity of the protein. Although PEG is not a smart polymer like PNIPAAm, in the special case of PEGylation of the new recombinant protein drugs, it is critical to attach the PEG molecule in a way so as to minimize such steric hindrance. In this situation, the pH may be lowered to favor conjugation of PEG–NHS to the terminal amino group, rather than to randomly located lysine amino groups. On rare occasions the conjugation of a polymer may increase the activity of the protein (e.g., Ding et al., 1998). Figure I.2.11.4 shows various ways that polymers may be directly conjugated to proteins or formed by growth from the protein (Hoffman and Stayton, 2010).

Random conjugation of temperature-sensitive (mainly) and pH-sensitive (occasionally) polymers to proteins has been extensively investigated, and applications of these conjugates have been focused on immunoassays, affinity separations, enzyme recovery, and drug delivery. One of the earliest applications of a smart polymer–biomolecule conjugate was an immunoassay developed by Hoffman, along with co-workers at a Seattle company called Genetic Systems Corp. They conjugated PNIPAAm to an antibody and used it to capture an antigen, then added a second, labeled antibody to the antigen, and warmed to phase separate the labeled immune complex sandwich. The assay was similar to an ELISA assay, but was much faster and was as accurate as ELISA. This concept was first presented by Hoffman at a PIMS meeting in 1986 (Hoffman et al., 1986), published in 1987 (Monji and Hoffman, 1987), and patented in 1988 (Monji et al., 1988). Other interesting work on soluble–insoluble polymer–biomolecule conjugates can be found in the following references: Schneider et al., 1981; Okamura et al., 1984; Nguyen and Luong, 1989; Taniguchi et al., 1989; Chen and Hoffman, 1990; Monji et al., 1990; Pecs et al., 1991; Taniguchi et al., 1992; Takei et al., 1993b; Galaev and Mattiasson, 1993; Takei et al., 1994a; Fong et al., 1999; Anastase-Ravion et al., 2001.

In some cases the "smart" polymer may be a polyligand, such as polybiotin or poly(glycosyl methacrylate), which is used to phase separate target molecules by complexation to multiple binding sites on target proteins, such as streptavidin and Concanavalin A, respectively (Larsson and Mosbach, 1979; Morris et al., 1993; Nakamae et al., 1994).

Wu et al. have synthesized PNIPAAm–phospholipid conjugates for use in drug delivery formulations as components of thermally sensitive composites and liposomes (Wu et al., 1992, 1993).

There has been some concern about the toxicity of PNIPAAm, related to the toxic nature of the acrylamide monomer which is similar to NIPAAm. If any NIPAAm monomer residues are present in the polymer product, it is most likely they can be washed out. Although there are no significant reports of PNIPAAm toxicity, this issue has not been adequately investigated to date.

Although there are no significant reports of PNIPAAm toxicity, that issue has not been adequately investigated to date.

SMART POLYMERS AND THEIR BIOMOLECULE CONJUGATES ON SURFACES

Smart polymers, especially PNIPAAm, have been coated on various surfaces by physical adsorption, chemical conjugation, affinity complexation, radiation-induced polymerization, direct or chain-transfer initiated polymerization from the surface, and plasma discharge deposition. These methods and various applications of them are discussed in this section.

> Smart polymers, especially PNIPAAm, have been coated on various surfaces by physical adsorption, chemical conjugation, affinity complexation, radiation-induced polymerization, direct or chain-transfer initiated polymerization from the surface, and plasma discharge deposition.

One may covalently graft PNIPAAm to a surface by exposing the surface to ionizing radiation in the presence of the monomer (and in the absence of air) or by pre-irradiating the polymer surface in air, and later contacting the surface with the monomer solution and heating in the absence of air to expose trapped radicals. These surfaces exhibit stimulus-responsive changes in wettability (e.g., Uenoyama and Hoffman, 1988; Takei et al., 1994b; Kidoaki et al., 2001). Okano and Yamato and co-workers have been pioneers in the area of smart polymer cell culture surfaces. They applied the radiation grafting technique to form surfaces having a surface layer of grafted PNIPAAm (Yamato and Okano, 2001; Shimizu et al., 2003). They cultured cells to confluent sheets on these surfaces at 37°C, which is above the LCST of the polymer. When the PNIPAAm collapses, the interface becomes hydrophobic and leads to physical adsorption of cell adhesion proteins from the culture medium, enhancing the cell culture process. Then, when the temperature is lowered, the interface becomes hydrophilic as the PNIPAAm chains rehydrate, and the cell sheets are released from the surface along with the cell adhesion proteins, which remain bound to the cell surfaces. The cell sheet can be recovered and used in tissue engineering, e.g., for artificial corneas and heart patches, and in periodontal applications. These exciting new applications of smart surface cell sheets are being pursued in the clinic. Patterned PNIPAAm surfaces have also been prepared for cell culture (Yamato et al., 2001).

Ratner and co-workers have used a gas plasma discharge technique to deposit temperature-responsive coatings from a NIPAAm monomer vapor phase plasma discharge. When cells are cultured on these surfaces, they also form reversible cell sheets in a manner similar to Okano and Yamato's PNIPAAm-grafted surfaces (Pan et al., 2001).

Smart polymers may also be grafted to surfaces by polymerizing the monomer directly from the surface. This may be done by first attaching to the surface either a radical initiator, such as an azo or peroxide initiator, or a chain transfer agent (CTA) for controlled radical polymerizations such as reversible addition-fragmentation chain transfer polymerization (RAFT) (e.g., Chiefari et al., 1998; McCormick and Lowe, 2004; Moad et al., 2008) or atom transfer radical polymerization (ATRP) (e.g., Wang and Matyjaszewski, 1995; Matyjaszewski and Xia, 2001; Matyjaszewski and Tsarevsky, 2009). (The ATRP of the smart polymer from a biomolecule such as a protein to form the bioconjugate directly was discussed above in this chapter.) Then the polymerization is initiated at the surface or in the surrounding solution, and the result is a smart polymer grafted to the surface. In the case of RAFT or ATRP controlled radical polymerizations, the molecular weights of the grafted polymer can be controlled to a narrow range. In a recent example, Golden and co-workers conjugated a RAFT CTA to the surface of a microporous, hydroxylated nylon membrane, and graft polymerized NIPAAm directly from the surface, forming a graft polymer of around 9000 MW (Golden et al., 2010). Then this surface was heated, and the collapsed PNIPAAm-coated porous membrane was used in a microfluidic channel to sequester dilute PNIPAAm-antibody conjugates that had earlier captured a biomarker antigen in a blood test sample. In this way, the captured biomarker was concentrated on the membrane surface and later released by cooling to provide a concentrated pulse flowing downstream in the channel for this smart polymer microfluidic immunoassay.

The research team of Stayton and Hoffman, and co-workers have also developed a variety of nanoparticles (nps) coated with PNIPAAm for various applications in diagnostic immunoassays. They include polystyrene latex nanoparticles, magnetic nanoparticles, and gold nanoparticles. In the case of the latex nps, Malmstadt and co-workers functionalized aminated polystyrene latex nps with PNIPAAm by conjugating an NHS-terminated PNIPAAm chain to the surface amino groups. The aminated beads were also functionalized with an NHS–PEG–biotin polymer, and streptavidin was bound to the biotin groups. The beads were localized in a microfluidic channel by raising the temperature in one section of the channel. The PNIPAAm- and antibody-functionalized beads were then used to capture labeled, biotinylated antigens in a test sample, to produce a smart polymer competition immunoassay carried out in a microfluidic device (Malmstadt et al., 2003a,b, 2004).

The research team of Lai and Nash and co-workers developed PNIPAAm and PNIPAAm-antibody coatings on magnetic and gold nps, which were used to capture dilute antigen biomarkers in a test sample, then later were captured on the PNIPAAm-grafted porous membrane surface described above (Golden et al., 2010). After that they were released by cooling, forming a concentrated pulse of labeled nps for assay downstream in a microfluidic device. PNIPAAm-coated magnetic and gold nps were also applied in lateral flow, "dipstick" assay by the Stayton–Hoffman group.

In a process for coating magnetic nps with PNIPAAm, Lai and co-workers formed magnetic nps in the presence of a dodecyl-terminated PNIPAAm. The dodecyl-terminated PNIPAAm polymer formed micelles that concentrated the pentacarbonyl iron precursor of magnetic iron oxides in the dodecyl core of the micelle before it was converted at high temperature to magnetic iron oxide (Lai et al., 2007, 2009). To coat the gold nps, Nash bound RAFT-polymerized PNIPAAm chains to the gold surface via sulfur's affinity to gold, utilizing the sulfur-containing groups in the dithioester or trithio carbonate terminal groups from the original CTA of the RAFT polymer chain (Nash et al., 2010a,b,c). The coated magnetic particles were mainly used for separating and concentrating the captured antigen, and the coated gold particles were mainly used as color (red) labels to indicate the approximate concentration of the captured antigen.

PNIPAAm may also be physically adsorbed to surfaces. Work by Miura and Hoffman and co-workers established the importance of matching the smart polymer composition with the surface composition in order to enhance the stimulus-driven adsorption of the smart polymer on the surface (Miura et al., 1994).

Smart polymers may also be grafted in gradients to surfaces to provide surfaces of gradually varying hydrophilicity and hydrophobicity as a function of the polymer composition and conditions. This phenomenon has been applied by Okano, Kikuchi, and co-workers to prepare chromatographic column packing for use in eluate-free ("green") chromatographic separations (Kobayashi et al., 2001; Kikuchi and Okano, 2002). Ishihara et al. (1982, 1984) developed photo-responsive coatings and membranes that reversibly change surface wettability or swelling due to the photo-induced isomerization of an azobenzene-containing polymer.

SITE-SPECIFIC SMART POLYMER BIOCONJUGATES

Conjugation of a responsive polymer to a specific site near the "activity" pocket of a genetically engineered protein is a powerful new concept initially demonstrated by Chilkoti et al. (1994) (see also later work by Stayton and Hoffman et al., 2000). Such site-specific protein–smart polymer conjugates can permit sensitive environmental control of the protein's recognition process, which controls all living systems. They conjugated the smart polymer to a reactive –SH thiol group from a cysteine that had been inserted by site-specific mutagenesis at a site nearby the active site of the protein (Figure I.2.11.5). To do this, they first used cassette mutagenesis to insert a site-specific mutation into the DNA sequence of the protein, and then cloned the mutant in cell culture. This method is applicable only to proteins whose complete peptide sequence is known. The preparation of the reactive smart polymer is similar to methods described above, but in this case, the reactive end or pendant groups and the reaction conditions were specifically designed to

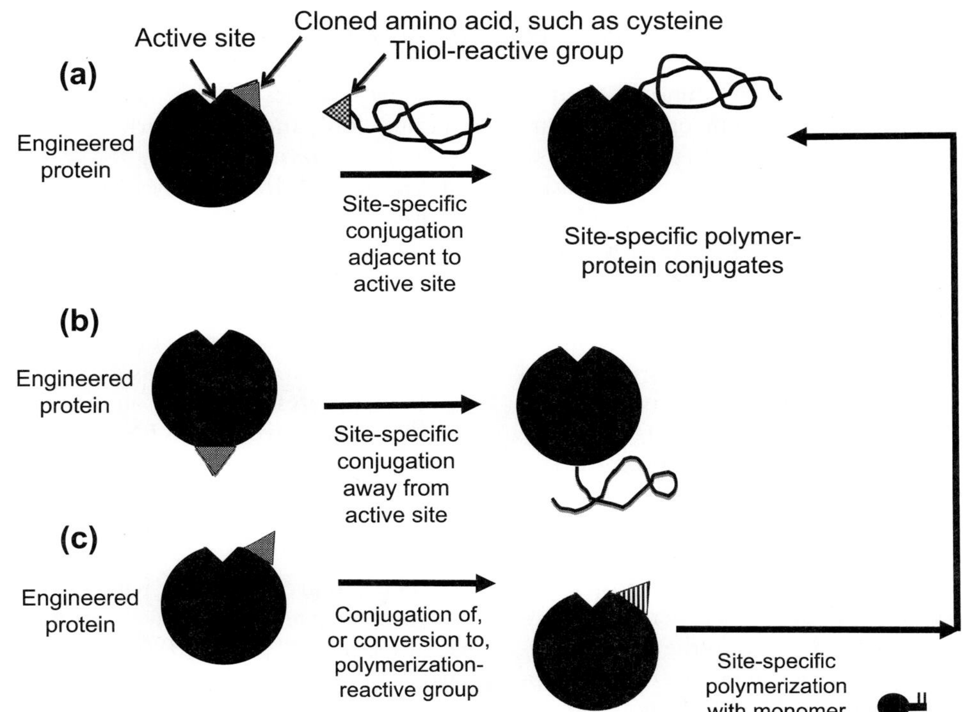

FIGURE I.2.11.5 Schematic illustration of various processes for preparing a site-specific conjugate of a polymer with a genetically-engineered, mutant protein.

favor conjugation to –SH groups rather than to $–NH_2$ groups. Typical SH-reactive polymer end groups include maleimide and vinyl sulfone groups.

The specific site for polymer conjugation can also be located far away from the active site (Chilkoti et al., 1994), in order to avoid interference with the biological functioning of the protein, or nearby or even within the active site, in order to control the ligand–protein recognition process, and the resulting biological activity of the protein (Figure I.2.11.5), (see also Hoffman and Stayton, 2010). Temperature-, pH-, and light-sensitive smart polymers have been used to form such novel, "really smart" bioconjugates (Ding et al., 1999; Bulmus et al., 1999; Stayton et al., 2000; Ding et al., 2001; Shimoboji et al., 2001, 2002a,b, 2003). Since the objective is to control the activity of the protein, and not to phase separate it, in practice these smart polymer-engineered protein bioconjugates would be immobilized on the surfaces of microarray chips, microbeads or nanobeads.

Site-specific conjugates of PNIPAAm to protein mutants of streptavidin and the enzyme endocellulase have been extensively studied by the Stayton–Hoffman group. Site-specific bioconjugates of PNIPAAm with streptavidin have been used to control access of biotin to its binding site on streptavidin (Stayton et al., 1995), and have enabled separation of biotinylated proteins according to the size of the protein (Ding et al., 2001). Ding, Stayton, and Hoffman et al. (1999) also found that raising the temperature to induce the collapse of the PNIPAAm chain "triggered" the release of the bound biotin molecules (perhaps due to a reduced biotin "off" rate compared to the "on" rate) (Ding et al., 1999). For the site-specific enzyme conjugates, a combined temperature- and light-sensitive polymer was conjugated to specific sites on endocellulase, which provided "on-off" control of the enzyme activity with either light or temperature (Shimoboji et al., 2001, 2002a,b, 2003). Bulmus et al. (1999) conjugated a pH- and temperature-responsive PNIPAAm copolymer to a specific site on streptavidin.

Triggered release of bound ligands by the smart polymer-engineered protein bioconjugates could be used to release therapeutics, such as for topical drug delivery to the skin or mucosal surfaces of the body, and also for localized delivery of drugs within the body by stimulated release at pretargeted sites using noninvasive, focused stimuli such as ultrasound or delivery of light stimuli from catheters. Triggered release could also be used to release and recover affinity-bound ligands from chromatographic and other supports in eluate-free conditions, including capture and release of specific cell populations to be used in stem cell and bone marrow transplantation. These processes could involve two different stimuli-responsive polymers with sensitivities to the same or different stimuli. For delicate target ligands such as peptides and proteins, recovery could be affected without the need for time-consuming and harsh elution conditions. Triggered release could also be used to remove inhibitors, toxins or fouling agents from the recognition sites of enzymes and affinity molecules, such as those used in biosensors, diagnostic assays or affinity separations. This could be used to "regenerate" such recognition proteins for extended process use. Light-controlled binding and release of site-specific protein conjugates may be utilized as a molecular switch for various applications in biotechnology, medicine, and bioelectronics, including hand-held diagnostic devices, biochips, and lab-on-a-chip devices.

Fong, Stayton, and Hoffman (Fong et al., 1999) have developed an interesting construct to control the distance of the PNIPAAm from the active site. For this purpose, they conjugated one sequence of complementary nucleotides to a specific site near the binding pocket of streptavidin, and a second sequence to the end of a PNIPAAm chain. Then, by controlling the location and length of the complementary sequence, the self-assembly via hybridization of the two single-chain DNA sequences could be used to control the distance of the polymer from the streptavidin binding site.

SMART POLYMER HYDROGELS

When a smart polymer is cross-linked to form a gel, it will collapse and reswell in water as a stimulus raises or lowers it through its critical condition. PNIPAAm hydrogels have been extensively studied, starting with the pioneering work of Toyoichi Tanaka in 1981 (Tanaka, 1981). In the mid- to late-1980s, Hoffman and co-workers were among the first to recognize the potential of PNIPAAm hydrogels as biomaterials; they showed that smart gels could be used to entrap enzymes and cells, and then by inducing cyclic collapse and swelling of the gel, the enzymes (or enzymes within the cells) could be turned "on" and "off." Park and Hoffman also showed enhanced enzyme efficiency for an entrapped enzyme in a packed bed, hydrogel bead reactor, since the collapse delivered product faster than it could diffuse out, and the reswelling imbibed substrate faster than it could diffuse in. They also could deliver drugs or remove toxic biomolecules, by stimulus-induced collapse or swelling. Dong and Hoffman developed an interesting pH- and temperature-responsive gel that swelled linearly at 37°C and pH 7.4, at rates depending on the AAc monomer content in the gel. This work with PNIPAAm gels in drug delivery systems (DDS) is discussed in detail in Chapter II.5.16 on Drug Delivery Systems (Dong and Hoffman, 1986, 1987, 1990, 1991; Park and Hoffman, 1988, 1990a,b,c, 1994) (Figure I.2.11.6).

The research team of Kim, Okano, Bae, and their co-workers were also actively studying smart hydrogels in the late 1980s and 1990s. For example, they investigated smart gels containing entrapped cells that could be used as artificial organs (Vernon et al., 2000). Their drug delivery work with smart hydrogels is discussed in detail in Chapter II.5.16 on Drug Delivery Systems.

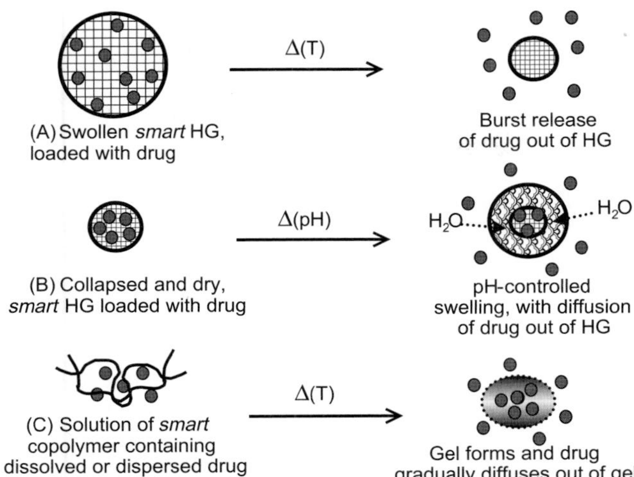

FIGURE I.2.11.6 Schematic illustration showing three ways that smart gel formulations may be stimulated to release bioactive agents: (A) thermally-induced collapse, which is relevant to skin or mucosal drug delivery; (B) pH-induced swelling, which is relevant to oral drug delivery, where the swelling is induced by the increase in pH in going from the gastric to enteric regions; and (C) sol-to-gel formation, which is relevant to injectable or topical formulations of a triblock copolymer solution that are thermally gelled at body temperature. For *in vivo* uses, the block copolymer is designed to be degradable. The first two apply to cross-linked gels applied topically or orally, and the third is relevant to thermally-induced formation of gels from polymer solutions that are delivered topically or by injection. (See also Chapter II.5.16 Drug Delivery Systems.)

Since then, the properties of PNIPAAm hydrogels have been widely investigated in the form of beads, slabs or discs, and multilamellar laminates (Park and Hoffman, 1992a,b; Hu et al., 1995, 1998; Mitsumata et al., 2001; Gao and Hu, 2002; Kaneko et al., 2002).

Okano and co-workers designed and synthesized PNIPAAm gels that collapse much more rapidly than PNIPAAm gels that are simply cross-linked with a divinyl comonomer. Yoshida, Okano, and co-workers copolymerized NIPAAm monomer with a PNIPAAm macromer and a cross-linker, which yielded a gel with PNIPAAm chains grafted to the PNIPAAm network. The grafted chains collapsed first at the LCST, enhancing the rate of collapse of the network, along with the related rate of expulsion of water from the gel (Yoshida et al., 1995; Masahiko et al., 2003).

Peppas and co-workers (e.g., Peppas et al., 2000; Robinson and Peppas, 2002) have extensively studied pH-sensitive acrylic acid-acrylate copolymer smart gels. This work is discussed in more detail in Chapter I.2.5 on Hydrogels (see also Peppas, 1997, 2001; Peppas et al., 1999, 2000).

Smart hydrogel compositions have been developed that are both thermally gelling and biodegradable (Zhong et al., 2002; Yoshida et al., 2003). These sol–gel systems have been used to deliver drugs by *in vivo* injections, and are discussed in the section on Smart Polymers in Solution (see also Chapter I.2.5 on Hydrogels, and Chapter II.5.16 on Drug Delivery Systems).

Matsuda and co-workers incorporated PNIPAAm into physical mixtures with natural polymers such as hyaluronic acid and gelatin, for use as tissue engineering scaffolds (Ohya et al., 2001a,b). Park and Hoffman developed a unique process for synthesizing mm-size spherical beads of PNIPAAm (Park and Hoffman, 1992c).

Nakamae, Hoffman, and co-workers developed novel compositions of smart gels containing phosphate groups that were used to bind cationic proteins as model drugs, and then to release them by a combination of thermal stimuli and ion exchange (Nakamae et al., 1992, 1997; Miyata et al., 1994).

A number of smart hydrogel drug delivery systems have been designed to respond to biologic signals in a feedback manner. Many of these gels contain an immobilized enzyme. They are discussed in detail in the Chapter II.5.16 on Drug Delivery Systems.

CONCLUSIONS

Smart polymers in solution, on surfaces, and as hydrogels have been utilized in many interesting ways, especially in combination with biomolecules such as proteins and drugs. Important applications include affinity separations, enzyme processes, immunoassays, drug delivery, and toxin removal. These smart polymer–biomolecule systems represent an important extension of polymeric biomaterials beyond their well-known uses in implants and medical devices.

BIBLIOGRAPHY

Anastase-Ravion, S., Ding, Z., Pelle, A., Hoffman, A. S., & Letourneur, D. (2001). New antibody purification procedure using a thermally-responsive polyNIPAAm-dextran derivative conjugate. *J. Chromatogr. B.*, **761**, 247–254.

Broyer, R. M., Grover, G. N., & Maynard, H. D. (2011). Emerging synthetic techniques for protein–polymer conjugations. *Chem. Commun.*, **47**, 2212–2226.

Bulmus, V., Ding, Z., Long, C. J., Stayton, P. S., & Hoffman, A. S. (1999). Design, synthesis and site-specific conjugation of a pH- and temperature-sensitive polymer to streptavidin for pH-controlled binding and triggered release of biotin. *Bioconj. Chem.*, **11**, 78–83.

Chen, G., & Hoffman, A. S. (1994). Synthesis of carboxylated poly(NIPAAm) oligomers and their application to form thermo-reversible polymer-enzyme conjugates. *J. Biomater. Sci. Polymer. Edn.*, **5**, 371–382.

Chen, G., & Hoffman, A. S. (1995). Graft copolymer compositions that exhibit temperature-induced transitions over a wide range of pH. *Nature*, **373**, 49–52.

Chen, J. P., & Hoffman, A. S. (1990). Polymer-protein conjugates. II. Affinity precipitation of human IgG by poly(N-isopropyl acrylamide)-protein A conjugates. *Biomaterials*, **11**, 631–634.

Chen, J. P., Yang, H. J., & Hoffman, A. S. (1990). Polymer–protein conjugates. I. Effect of protein conjugation on the cloud point of poly(N-isopropyl acrylamide). *Biomaterials*, **11**, 625–630.

Chiefari, J., Chong, Y. K., Ercole, F., Krstina, J., Jeffery, J., et al. (1998). Living free-radical polymerization by reversible addition-fragmentation chain transfer: The RAFT process. *Macromolecules*, **31**(16), 5559–5562. doi:10.1021/ma9804951.

Chilkoti, A., Chen, G., Stayton, P. S., & Hoffman, A. S. (1994). Site-specific conjugation of a temperature-sensitive polymer to a genetically-engineered protein. *Bioconj. Chem.*, **5**, 504–507.

Cole, C. A., Schreiner, S. M., Priest, J. H., Monji, N., & Hoffman, A. S. (1987). N-isopropyl acrylamide and N-acryl succinimide copolymers: A thermally reversible water soluble activated polymer for protein conjugation. In P. Russo (Ed.), *Reversible Polymeric Gels and Related Systems, ACS Symposium Series* (Vol. 350, pp. 245–254). Washington, DC: ACS.

Ding, Z. L., Chen, G., & Hoffman, A. S. (1996). Synthesis and purification of thermally-sensitive oligomer–enzyme conjugates of poly(NIPAAm)–trypsin. *Bioconj. Chem.*, **7**, 121–125.

Ding, Z. L., Chen, G., & Hoffman, A. S. (1998). Properties of polyNIPAAm–trypsin conjugates. *J. Biomed. Mater. Res.*, **39**, 498–505.

Ding, Z. L., Long, C. J., Hayashi, Y., Bulmus, E. V., Hoffman, A. S., et al. (1999). Temperature control of biotin binding and release with a streptavidin–polyNIPAAm site-specific conjugate. *J. Bioconj. Chem.*, **10**, 395–400.

Ding, Z. L., Shimoboji, T., Stayton, P. S., & Hoffman, A. S. (2001). A smart polymer shield that controls the binding of different size biotinylated proteins to streptavidin. *Nature*, **411**, 59–62.

Dong, L. C., & Hoffman, A. S. (1986). Thermally reversible hydrogels: III. Immobilization of enzymes for feedback reaction control. *J. Contr. Rel.*, **4**, 223–227.

Dong, L. C., & Hoffman, A. S. (1987). Thermally reversible hydrogels: Swelling characteristics and activities of copoly(NIPAAm-AAm) gels containing immobilized asparaginase. In P. Russo (Ed.), *Reversible Polymeric Gels and Related Systems, ACS Symposium Series* (Vol 350, pp. 236–244). Washington, DC: ACS.

Dong, L. C., & Hoffman, A. S. (1990). Synthesis and application of thermally-reversible heterogels for drug delivery. *J. Contr. Release*, **13**, 21–32.

Dong, L. C., & Hoffman, A. S. (1991). A novel approach for preparation of pH- and temperature-sensitive hydrogels for enteric drug delivery. *J. Contr. Release*, **15**, 141–152.

Fong, R. B., Ding, Z. L., Long, C. J., Hoffman, A. S., & Stayton, P. S. (1999). Thermoprecipitation of streptavidin via oligonucleotide-mediated self-assembly with poly(NIPAAm). *Bioconj. Chem.*, **10**, 720–725.

Galaev, I. Y., & Mattiasson, B. (1993). Affinity thermoprecipitation: Contribution of the efficiency and access of the ligand. *Biotechnol. Bioeng.*, **41**, 1101–1106.

Gao, J., & Hu, Z. B. (2002). Optical properties of N-isopropylacrylamide microgel spheres in water. *Langmuir*, **18**, 1360–1367.

Golden, A. L., Battrell, C. F., Pennell, S., Hoffman, A. S., Lai, J. J., et al. (2010). Simple fluidic system for purifying and concentrating diagnostic biomarkers using stimuli-responsive antibody conjugates and membranes. *Bioconj. Chem.*, **21**, 1820–1826.

Grover, G. N., & Maynard, H. D. (2010). Protein–polymer conjugates: Synthetic approaches by controlled radical polymerizations and interesting applications. *Current Opin. Chem. Biol.*, **14**, 818–827.

Heredia, K. L., & Maynard, H. D. (2007). Synthesis of protein–polymer conjugates. *Organic & Biomol. Chem.*, **5**, 45–53.

Heskins, H., & Guillet, J. E. (1968). Solution properties or poly (*N*-isopropyl acrylamide. *J. Macromol. Sci. Chem.*, **A2**(6), 1209.

Hoffman, A. S. (1987). Applications of thermally reversible polymers and hydrogels in therapeutics and diagnostics. *J. Contr. Rel.*, **6**, 297–305.

Hoffman, A. S. (1995). Intelligent polymers in medicine and biotechnology. *Macromol. Symp.*, **98**, 645–664.

Hoffman, A. S. (1997). Intelligent polymers in medicine and biotechnology. In K. Park (Ed.), *Controlled Drug Delivery*. Washington, DC: ACS Publications.

Hoffman, A. S., & Stayton, P. S. (2010). Conjugates of stimuli-responsive polymers and proteins. *Prog. Polym. Sci.*, **32**, 922–932.

Hoffman, A. S., et al. (1986). Novel Application of Polymers in Bioseparations and Diagnostics. *Proceedings of Polymers in Medicine and Surgery (PIMS)*. V: Holland. **37**(1), 2.

Hoffman, A. S., Stayton, P. S., Bulmus, V., Chen, G., Chen, J., et al. (2000). Really smart bioconjugates of smart polymers and receptor proteins. *J. Biomed. Mater. Res.*, **52**, 577–586.

Hu, Z. B., Zhang, X. M., & Li, Y. (1995). Synthesis and application of modulated polymer gels. *Science*, **269**, 525.

Hu, Z. B., Chen, Y. Y., Wang, C. J., Zheng, Y. Y., & Li, Y. (1998). Polymer gels with engineered environmentally responsive surface patterns. *Nature*, **393**, 149.

Ishihara, K., Okazaki, A., Negishi, N., Shinohara, I., Okano, T., et al. (1982). Photo-induced change in wettability and binding ability of azoaromatic polymer. *J. Appl. Polymer. Sci.*, **27**, 239–245.

Ishihara, K., Hamada, N., Kato, S., & Shinohara, I. (1984). Photo-induced swelling control of amphiphilic azoaromatic polymer membrane. *Polymer. Sci. (Polymer Chem. Ed.)*, **22**, 21–128.

Jeong, B., Kim, S. W., & Bae, Y. H. (2002). Thermosensitive sol–gel reversible hydrogels. *Adv. Drug. Delivery. Rev.*, **54**, 37–51.

Kaneko, D., Gong, J. P., & Osada, Y. (2002). Polymer gels as soft and wet chemomechanical systems: An approach to artifical muscles. *J. Mater. Chem.*, **12**, 2169–2177.

Kang, H. C., & Bae, Y. H. (2007). pH-Tunable endosomolytic oligomers for enhanced nucleic acid delivery. *Advanced Functional Materials*, **17**, 1263–1272.

Kidoaki, S., Ohya, S., Nakayama, Y., & Matsuda, T. (2001). Thermoresponsive structural change of a PNIPAAm graft layer measured with AFM. *Langmuir*, **17**, 2402–2407.

Kikuchi, A., & Okano, T. (2002). Intelligent thermoresponsive polymeric stationary phases for aqueous chromatography of biological compounds. *Progr. Polymer. Sci.*, **27**, 1165–1193.

Kobayashi, J., Kikuchi, A., Sakai, K., & Okano, T. (2001). Aqueous chromatography utilizing pH-/temperature-responsive polymers as column matrix surfaces for separation of ionic bioactive compounds. *Anal. Chem.*, **73**, 2027–2033.

Lackey, C. A., Murthy, N., Press, O. W., Tirrell, D. A., Hoffman, A. S., et al. (1999). Hemolytic activity of pH-responsive polymer–streptavidin bioconjugates. *Bioconj. Chem.*, **10**, 401–405.

Lai, J., Hoffman, A. S., & Stayton, P. S. (2007). Dual magnetic-temperature responsive nanoparticles for microfluidic separations and assays. *Langmuir*, **23**, 7385–7391.

Lai, J. J., Nelson, K. E., Nash, M. A., Hoffman, A. S., Yager, P., et al. (2009). Dynamic bioprocessing and microfluidic transport control with smart magnetic nanoparticles in laminar-flow devices. *Lab. Chip.*, **9**, 1997–2002.

Larsson, P. O., & Mosbach, K. (1979). Affinity precipitation of enzymes. *FEBS Lett*, **98**, 333–338.

Lee, D. S., Shim, M. S., Kim, S. W., Lee, H., Park, I., et al. (2001). Novel thermoreversible gelation of biodegradable PLGA-block-PEO-block-PLGA triblock copolymers in aqueous solution. *Macromol. Rapid. Commun.*, **22**, 587–592.

Masahiko, A., Matsuura, T., Kasai, M., Nakahira, T., Hara, Y., et al. (2003). Preparation of comb-type N-isopropylacrylamide hydrogel beads and their application for size-selective separation media. *Biomacromolecules*, **4**, 395–403.

Malmstadt, N., Yager, P., Hoffman, A. S., & Stayton, P. S. (2003a). A Smart Microfluidic Affinity Chromatography Matrix Composed of Poly(N-isopropylacrylamide)-Coated Beads. *Anal. Chem.*, **75**, 2943–2949.

Malmstadt, N., Hyre, D., Ding, Z., Hoffman, A. S., & Stayton, P. S. (2003b). Affinity Thermoprecipitation and Recovery of Biotinylated Biomolecules via a Mutant Streptavidin-Smart Polymer Conjugate. *Bioconj. Chem.*, **14**, 575–580.

Malmstadt, N., Hoffman, A. S., & Stayton, P. S. (2004). Smart Mobile Affinity Matrix for Microfluidic Immunoassays. *Lab. Chip.*, **4**, 412–415.

Matyjaszewski, K., & Tsarevsky, N. V. (2009). Nanostructured functional materials prepared by atom transfer radical polymerization. *Nature Chemistry*, **1**, 276–288. doi:10.1038/NCHEM.257.

Matyjaszewski, K., & Xia, J. (2001). Atom transfer radical polymerization. *Chem. Rev.*, **101**, 2921–2990. doi:10.1021/cr940534g. ISSN 0009-2665. PMID 11749397.

McCormick, C., & Lowe, A. B. (2004). Aqueous RAFT polymerization: Recent developments in synthesis of functional water-soluble (co)polymers with controlled structures. *Accounts of Chemical Research*, **37**, 312–325. doi:10.1021/ar0302484.

Mitsumata, T., Gong, J. P., & Osada, Y. (2001). Shape memory functions and motility of amphiphilic polymer gels. *Polymer. Adv. Technol.*, **12**, 136–150.

Miura, M., Cole, C. A., Monji, N., & Hoffman, A. S. (1994). Temperature-dependent adsorption/desorption behavior of lcst polymers on various substrates. *J. Biomtls. Sci. (Polymer Ed)*, **5**, 555–568.

Miyata, T., Nakamae, K., Hoffman, A. S., & Kanzaki, Y. (1994). Stimuli-sensitivities of hydrogels containing phosphate groups. *Macromol. Chem. Phys.*, **195**, 1111–1120.

Moad, G., Rizzardo, E., & Thang, S. H. (2008). Radical addition-fragmentation chemistry in polymer synthesis. *Polymer*, **49**, 1079–1131. doi:10.1016/j.polymer.2007.11.020.

Monji, N., & Hoffman, A. S. (1987). A novel immunoassay system and bioseparation process based on thermal phase separating polymers. *Appl. Biochem. Biotechnol.*, **14**, 107–120.

Monji, N., Hoffman, A. S., Priest, J. H. & Houghton, R. L. (1988). Thermally-Induced Phase Separation Immunoassay. U.S. Patent 4,780,409 (10/25/88).

Monji, N., Cole, C. A., Tam, M., Goldstein, L., Nowinski, R. C., et al. (1990). Application of a thermally-reversible polymer–antibody conjugate in a novel membrane-based immunoassay. *Biochem. Biophys. Res. Commun.*, **172**, 652–660.

Morris, J. E., Hoffman, A. S., & Fisher, R. R. (1993). Affinity precipitation of proteins by polyligands. *Biotechnol. Bioeng.*, **41**, 991–997.

Murthy, N., Stayton, P. S., & Hoffman, A. S. (1999). The design and synthesis of polymers for eukaryotic membrane disruption. *J Controlled Release*, **61**, 137–143.

Murthy, N., Campbell, J., Fausto, N., Hoffman, A. S., & Stayton, P. S. (2003a). Bioinspired polymeric carriers that enhance intracellular delivery of biomolecular therapeutics. *Bioconj. Chem.*, **14**, 412–419.

Murthy, N., Campbell, J., Fausto, N., Hoffman, A. S., & Stayton, P. S. (2003b). Design and synthesis of pH-responsive polymeric carriers that target uptake and enhance the intracellular delivery of oligonucleotides to hepatocytes. *J. Contr. Rel.*, **89**(3), 365–374.

Na, K., Lee, D. H., Hwang, D. J., Lee, K. H., & Bae, Y. H. (2006). pH-Sensitivity and pH-dependent structural change in polymeric nanoparticles of poly(vinyl sulfadimethoxine)-deoxycholic acid conjugate. *Eur. Polym. J.*, **42**, 2581–2588.

Nakamae, K., Miyata, T., & Hoffman, A. S. (1992). Swelling behavior of hydrogels containing phosphate groups. *Macromol. Chem.*, **193**, 983–990.

Nakamae, K., Miyata, T., Jikihara, A., & Hoffman, A. S. (1994). Formation of poly(glucosyloxyethyl methacrylate)-concanavalin a complex and its glucose sensitivity. *J. Biomtls. Sci. (Polymer Ed)*, **6**, 79–90.

Nakamae, K., Nizuka, T., Miyata, T., Furukawa, M., Nishino, T., et al. (1997). Lysozyme loading and release from hydrogels carrying pendant phosphate groups. *J. Biomater. Sci. (Polymer Ed)*, **9**, 43–53.

Nash, M. A., Yager, P., Hoffman, A. S., & Stayton, P. S. (2010a). Mixed stimuli-responsive magnetic and gold nanoparticle system for rapid purification, enrichment and detection of biomarkers. *Bioconj. Chem.*, **21**, 2197–2204.

Nash, M. A., Lai, J. J., Hoffman, A. S., Yager, P., & Stayton, P. S. (2010b). "Smart" diblock copolymers as templates for magnetic-core gold-shell nanoparticle synthesis. *Nano. Lett.*, **10**, 85–91.

Nash, M. A., Hoffman, J. M., Stevens, D. Y., Hoffman, A. S., Stayton, P. S., et al. (2010c). Laboratory-scale protein striping system for patterning biomolecules onto paper-based immunochromatographic test strips. *Lab. Chip.*, **10**, 2279–2282.

Nguyen, A. L., & Luong, J. H.T. (1989). Syntheses and application of water soluble reactive polymers for purification and immobilization of biomolecules. *Biotechnol. Bioeng.*, **34**, 1186–1190.

Ohya, S., Nakayama, Y., & Matsuda, T. (2001a). Thermoresponsive artificial extracellular matrix for tissue engineering: Hyaluronic acid bioconjugated with poly(N-isopropylacrylamide) grafts. *Biomacromolecules*, **2**, 856–863.

Ohya, S., Nakayama, Y., & Matsuda, T. (2001b). Material design for an artificial extracellular matrix: Cell entrapment in poly(N-isopropylacrylamide) (PNIPAM)-grafted gelatin hydrogel. *J Artif Organs*, **4**, 308–314.

Okamura, K., Ikura, K., Yoshikawa, M., Sakaki, R., & Chiba, H. (1984). Soluble–insoluble interconvertible enzymes. *Agric. Biol. Chem.*, **48**, 2435–2440.

Okano, T., Kikuchi, A., & Yamato, M. (2000). Intelligent hydrogels and new biomedical applications. In: *Biomaterials and Drug Delivery toward the New Millennium* (pp. 77–86). Seoul, Korea: Han Rim Won Publishing Co.

Pan, Y. V., Wesley, R. A., Luginbuhl, R., Denton, D. D., & Ratner, B. D. (2001). Plasma-polymerized N-isopropylacylamide: Synthesis and characterization of a smart thermally responsive coating. *Biomacromolecules*, **2**, 32–36.

Park, T. G., & Hoffman, A. S. (1988). Effect of temperature cycling on the activity and productivity of immobilized β-galactosidase in a thermally reversible hydrogel bead reactor. *Appl. Biochem. Biotechnol.*, **19**, 1–9.

Park, T. G., & Hoffman, A. S. (1990a). Immobilization and characterization of β-galactosidase in thermally reversible hydrogel beads. *J. Biomed. Mater. Res.*, **24**, 21–38.

Park, T. G., & Hoffman, A. S. (1990b). Immobilization of *A. simplex* cells in a thermally-reversible hydrogel: Effect of temperature cycling on steroid conversion. *Biotech. Bioeng.*, **35**, 52–159.

Park, T. G., & Hoffman, A. S. (1990c). Immobilized biocatalysts in reversible hydrogels. In A. Tanaka (Ed.), *Enzyme Engineering X Ann NY Acad Sci.*, **613**, 588–593.

Park, T. G., & Hoffman, A. S. (1992a). Preparation of large, uniform size temperature-sensitive hydrogel beads. *J. Polymer. Sci. A. Polymer. Chem.*, **30**, 505–507.

Park, T. G., & Hoffman, A. S. (1992b). Synthesis and characterization of pH- and/or temperature-sensitive hydrogels. *J. Appl. Polymer. Sci.*, **46**, 659–671.

Park, T. G., & Hoffman, A. S. (1992c). Preparation of large, uniform size temperature-sensitive hydrogel beads. *J. Poly. Sci. A., Poly. Chem.*, **30**, 505–507.

Park, T. G., & Hoffman, A. S. (1994). Estimation of temperature-dependent pore sizes in poly(NIPAAm) hydrogel beads. *Biotechnol. Progr.*, **10**, 82–86.

Pecs, M., Eggert, M., & Schügerl, K. (1991). Affinity precipitation of extracellular microbial enzymes. *J. Biotechnol.*, **21**, 137–142.

Peppas, N. A. (1997). Hydrogels and drug delivery. *Critical Opinion in Colloid and Interface Science*, **2**, 531–537.

Peppas, N. A. (2001). Gels for drug delivery. In: *Encyclopedia of Materials: Science and Technology* (pp. 3492–3495). Amsterdam: Elsevier.

Peppas, N. A., Keys, K. B., Torres-Lugo, M., & Lowman, A. M. (1999). Poly(ethylene glycol)-containing hydrogels in drug delivery. *J. Controlled Release*, **62**, 81–87.

Peppas, N. A., Huang, Y., Torres-Lugo, M., Ward, J. H., & Zhang, J. (2000). Physicochemical foundations and structural design of hydrogels in medicine and biology. *Ann. Revs. Biomed. Eng.*, **2**, 9–29.

Priest, J. H., Murray, S., Nelson, R. G., & Hoffman, A. S. (1987). LCSTs of aqueous copolymers of N-isopropyl acrylamide and other N-substituted acrylamides. In P. Russo (Ed.), *Reversible Polymeric Gels and Related Systems* (Vol. 350, pp. 255–264). Washington, DC: ACS. ACS Symposium Series.

Robinson, D. N., & Peppas, N. A. (2002). Preparation and characterization of pH-responsive poly(methacrylic acid-g-ethylene glycol) nanospheres. *Macromolecules*, **35**, 3668–3674.

Roy, D., Cambre, J. N., & Sumerlin, B. S. (2010). Future perspectives and recent advances in stimuli-responsive materials. *Prog. in Polymer. Sci.*, **35**, 278–301.

Roy, I., & Gupta, M. N. (2003). Smart polymeric materials: Emerging biochemical applications. *Chemistry & Biology*, **10**, 1161–1171.

Schild, H. G. (1992). Poly(N-isopropylacrylamide): Experiment, theory and application. *Prog. Polym. Sci.*, **17**, 163–249.

Schneider, M., Guillot, C., & Lamy, B. (1981). The affinity precipitation technique: Application to the isolation and purification of trypsin from bovine pancreas. *Ann. NY Acad. Sci.*, **369**, 257–263.

Shim, W. S., Yoo, J. S., Bae, Y. H., & Lee, D. S. (2005). Novel injectable pH and temperature sensitive block copolymer hydrogel. *Biomacromolecules*, **6**, 2930–2934.

Shim, W. S., Kim, J. H., Park, H., Kim, K., Kwon, I. C., et al. (2006). Biodegradability and biocompatibility of a pH- and thermo-sensitive hydrogel formed from a sulfonamide-modified poly(e-caprolactone-co-lactide)–poly(ethylene glycol)–poly (e-caprolactone-co-lactide) block copolymer. *Biomaterials*, **27**, 5178–5185.

Shimizu, T., Yamato, M., Kikuchi, A., & Okano, T. (2003). Cell sheet engineering for myocardial tissue reconstruction. *Biomaterials*, **24**, 2309–2316.

Shimoboji, T., Ding, Z., Stayton, P. S., & Hoffman, A. S. (2001). Mechanistic investigation of smart polymer–protein conjugates. *Bioconj. Chem.*, **12**, 314–319.

Shimoboji, T., Ding, Z. L., Stayton, P. S., & Hoffman, A. S. (2002a). Photoswitching of ligand association with a photoresponsive polymer–protein conjugate. *Bioconj. Chem.*, **13**, 915–919.

Shimoboji, T., Larenas, E., Fowler, T., Kulkarni, S., Hoffman, A. S., et al. (2002b). Photoresponsive polymer–enzyme switches. *Proc. Natl. Acad. Sci. USA*, **99**, 16592–16596.

Shimoboji, T., Larenas, E., Fowler, T., Hoffman, A. S., & Stayton, P. S. (2003). Temperature-induced switching of enzyme activity with smart polymer–enzyme conjugates. *Bioconj. Chem.*, **14**, 517–525.

Shoemaker, S., Hoffman, A. S., & Priest, J. H. (1987). Synthesis of vinyl monomer-enzyme conjugates. *Appl. Biochem. and Biotechnology*, **15**, 11.

Stayton, P. S., & Hoffman, A. S. (2008). Smart pH-responsive carriers for intracellular delivery of Biomolecular Drugs. In V. Torchilin (Ed.), *Multifunctional Pharmaceutical Nanocarriers*. New York, NY: Springer Publishers.

Stayton, P. S., Shimoboji, T., Long, C., Chilkoti, A., Chen, G., et al. (1995). Control of protein–ligand recognition using a stimuli-responsive polymer. *Nature*, **378**, 472–474.

Stayton, P. S., Hoffman, A. S., Murthy, N., Lackey, C., Cheung, C., et al. (2000). Molecular engineering of proteins and polymers for targeting and intracellular delivery of therapeutics. *J. Contr. Rel.*, **65**, 203–220.

Takei, Y. G., Aoki, T., Sanui, K., Ogata, N., Okano, T., et al. (1993a). Temperature-responsive bioconjugates. 1. Synthesis of temperature-responsive oligomers with reactive end groups and their coupling to biomolecules. *Bioconj. Chem.*, **4**, 42–46.

Takei, Y. G., Aoki, T., Sanui, K., Ogata, N., Okano, T., et al. (1993b). Temperature-responsive bioconjugates. 2. Molecular design for temperature-modulated bioseparations. *Bioconj. Chem.*, **4**, 341–346.

Takei, Y. G., Matsukata, M., Aoki, T., Sanui, K., Ogata, N., et al. (1994a). Temperature-responsive bioconjugates. 3. Antibody-poly(N-isopropylacrylamide) conjugates for temperature-modulated precipitations and affinity bioseparations. *Bioconj. Chem.*, **5**, 577–582.

Takei, Y. G., Aoki, T., Sanui, K., Ogata, N., Sakurai, Y., et al. (1994b). Dynamic contact angle measurements of temperature-responsive properties for PNIPAAm grafted surfaces. *Macromolecules*, **27**, 6163–6166.

Tanaka, T. (1981). Gels. *Scientific American*, **244**, 124.

Taniguchi, M., Kobayashi, M., & Fujii, M. (1989). Properties of a reversible soluble–insoluble cellulase and its application to repeated hydrolysis of crystalline cellulose. *Biotechnol. Bioeng.*, **34**, 1092–1097.

Taniguchi, M., Hoshino, K., Watanabe, K., Sugai, K., & Fujii, M. (1992). Production of soluble sugar from cellulosic materials by repeated use of a reversibly soluble-autoprecipitating cellulase. *Biotechnol. Bioeng.*, **39**, 287–292.

Tirrell, D. (1987). Macromolecular switches for bilayer membranes. *J. Contr. Rel.*, **6**, 15–21.

Uenoyama, S., & Hoffman, A. S. (1988). Synthesis and characterization of AAm/NIPAAm grafts on silicone rubber substrates. *Radiat. Phys. Chem.*, **32**, 605–608.

Vernon, B., Kim, S. W., & Bae, Y. H. (2000). Thermoreversible copolymer gels for extracellular matrix. *J. Biomed. Mater. Res.*, **51**, 69–79.

Wang, J., & Matyjaszewski, K. (1995). Controlled/"living" radical polymerization. Atom transfer radical polymerization in the presence of transition-metal complexes. *J. Am. Chem. Soc.*, **117**, 5614–5615. doi:10.1021/ja00125a035.

Wu, X. S., Hoffman, A. S., & Yager, P. (1992). Conjugation of phosphatidylethanolamine to poly(NIPAAm) for potential use in liposomal drug delivery systems. *Polymer*, **33**, 4659–4662.

Wu, X. S., Hoffman, A. S., & Yager, P. (1993). Synthesis of and insulin release from erodible polyNIPAAm-phospholipid composites. *J. Intell. Mater. Syst. Struct.*, **4**, 202–209.

Yamato, M., & Okano, T. (2001). Cell sheet engineering for regenerative medicine. *Macromol. Chem. Symp.*, **14**(2), 21–29.

Yamato, M., Kwon, O. H., Hirose, M., Kikuchi, A., & Okano, T. (2001). Novel patterned cell co-culture utilizing thermally responsive grafted polymer surfaces. *J. Biomed. Mater. Res.*, **55**, 137–140.

Yang, H. J., Cole, C. A., Monji, N., & Hoffman, A. S. (1990). Preparation of a thermally phase-separating copolymer with a controlled number of active ester groups per polymer chain. *J. Polymer. Sci. A. Polymer. Chem.*, **28**, 219–226.

Yin, X., Stayton, P. S., & Hoffman, A. S. (2006). Temperature- and pH-responsiveness of poly(n-isopropylacrylamide-co-propylacrylic acid) copolymers prepared by RAFT polymerization. *Biomacromol.*, **7**, 1381–1385.

Yoshida, R., Uchida, K., Kaneko, Y., Sakai, K., Kikuchi, A., et al. (1995). Comb-type grafted hydrogels with rapid de-swelling response to temperature changes. *Nature*, **374**, 240–242.

Yoshida, T., Aoyagi, T., Kokufuta, E., & Okano, T. (2003). Newly designed hydrogel with both sensitive thermoresponse and biodegradability. *J. Polymer. Sci. A Polymer. Chem.*, **41**, 779–787.

Zareie, H. M., Bulmus, V., Gunning, P. A., Hoffman, A. S., Piskin, E., et al. (2000). Investigation of a pH- and temperature-sensitive polymer by AFM. *Polymer*, **41**, 6723–6727.

Zhong, Z., Dijkstra, P. J., Feijen, J., Kwon, Y. -Mi, Bae, Y. H., et al. (2002). Synthesis and aqueous phase behavior of thermoresponsive biodegradable poly(d, l-3-methyl glycolide)-b-poly(ethylene glycol)-b-poly(d, l-3-methyl glycolide) triblock copolymers. *Macromol. Chem. Phys.*, **203**, 1797–1803.

CHAPTER I.2.12 PHYSICOCHEMICAL SURFACE MODIFICATION OF MATERIALS USED IN MEDICINE

Buddy D. Ratner[1] *and Allan S. Hoffman*[2]
[1]Professor, Bioengineering and Chemical Engineering, Director of University of Washington Engineered Biomaterials (UWEB), Seattle, WA, USA
[2]Professor of Bioengineering and Chemical Engineering, UWEB Investigator, University of Washington, Seattle, WA, USA

INTRODUCTION

Much effort goes into the design, synthesis, and fabrication of biomaterials and devices to ensure that they have the appropriate mechanical properties, durability, and functionality. To cite a few examples, a hip joint should withstand the high stresses associated with walking and running, a hemodialyzer should have the requisite permeability characteristics, and the pumping bladder in an artificial heart should flex for millions of cycles without failure. The bulk composition and organization of materials govern these properties.

The biological response to biomaterials and devices, on the other hand, is controlled largely by their surface chemistry and structure (see Chapters I.1.5, II.1.3, and II.1.4). The rationale for the surface modification of biomaterials is therefore straightforward: to retain the key physical properties of a biomaterial while modifying only the outermost surface to influence the biointeraction. If such surface modification is properly effected, the mechanical properties and functionality of the device will be unaffected, but the bioresponse related to the tissue–device interface will be improved or modulated.

Materials can be surface modified by using biological, mechanical or physicochemical methods. Many biological surface modification schemes are covered in Chapter I.2.17. Surface modification for creating non-fouling surfaces is reviewed in Chapter I.2.12. Generalized examples of physicochemical surface modifications, the focus of this chapter, are illustrated schematically in Figure I.2.12.1. Surface modification with Langmuir–Blodgett (LB) films has elements of both biological modification and physicochemical modification. LB films will be discussed later in this chapter. Some applications for surface modified biomaterials are listed in Table I.2.12.1. Physical and chemical surface modification methods, and the types of materials to which they can be applied, are listed in Table I.2.12.2. Methods to modify or create surface texture or roughness will not be explicitly covered here (see Chapter I.2.15).

GENERAL PRINCIPLES

Surface modifications fall into three categories: (1) chemically or physically altering the atoms, compounds or molecules in the existing surface (chemical modification, etching, mechanical roughening); (2) overcoating the existing surface with a material having a different composition (coating, grafting, thin film deposition); (3) creating surface textures or patterns (Figure I.2.12.1). A few general principles provide guidance when undertaking surface modifications.

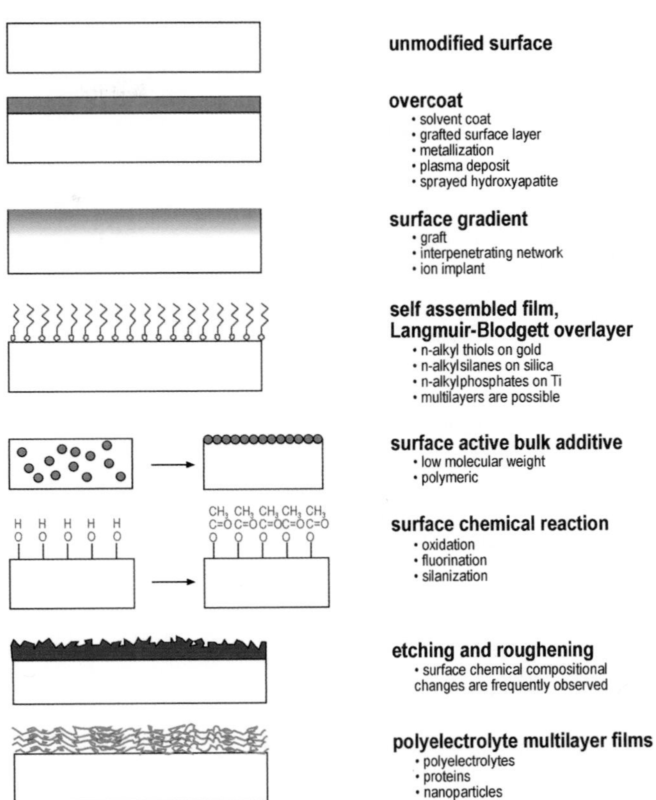

FIGURE I.2.12.1 Schematic representations of methods to modify surfaces.

Thin Surface Modifications

Thin surface modifications are desirable. Modified surface layers that are too thick can change the mechanical and functional properties of the material. Thick coatings are also more subject to delamination and cracking. How thin should a surface modification be? Ideally, alteration of only the outermost few molecular layers (3–10 nm) should be sufficient. In practice, thicker films than this will be necessary, since it is difficult to ensure that the original surface is uniformly covered when coatings and treatments are molecularly thin. This is because of intrinsic non-uniformities in many coating processes, and also because of surface roughness that exists on almost all surfaces at the nanometer scale. Also, extremely thin layers may be more subject to surface reversal (see below) and mechanical erosion. Some coatings intrinsically have a specific thickness. For example, the thickness of LB films is related to the length of the amphiphilic molecules that comprise them (2–5 nm). Other coatings, such as poly(ethylene glycol) (PEG) protein-resistant layers, may require a minimum thickness (a dimension

TABLE I.2.12.1 Examples of Surface Modified Biomaterials by Physicochemical Methods

To Modify Blood Compatibility
Octadecyl group attachment to surfaces (albumin affinity)
Silicone-containing block copolymer additive
Plasma fluoropolymer deposition
Plasma siloxane polymer deposition
Grafted poly(ethylene glycol) (PEG) or PEG-containing polymers
Chemically-modified polystyrene for heparin-like activity
To Modulate Cell Adhesion and Growth
Oxidized polystyrene surface
Ammonia plasma-treated surface
Plasma-deposited acetone or methanol film
Plasma fluoropolymer deposition (reduce corneal endothelial adhesion and modify blood interactions)
Varying surface modulus
Anti-bacterial treatments?
To Control Protein Adsorption
Surface with immobilized poly(ethylene glycol) (reduce adsorption)
Treated ELISA dish surface (increase adsorption)
Affinity chromatography column
Surface cross-linked contact lens (reduce adsorption)
To Improve Lubricity
Plasma treatment?
Radiation grafting (hydrogels)
Interpenetrating polymeric networks
To Improve Wear Resistance and Corrosion Resistance
Ion implantation
Diamond deposition
Anodization
To Alter Transport Properties
Polyelectrolyte grafting
Surface self-assembled film barrier
Plasma-deposited barrier layer
To Modify Electrical Characteristics
Polyelectrolyte grafting
Magnetron sputtering of titanium
Surface fluoropolymer insulation

related to the molecular weight of chains) to function (see Chapter I.2.10). In general, surface modifications should be the minimum thickness needed for uniformity, durability, and functionality, but no thicker. This is often experimentally defined for each system.

Delamination Resistance

The surface modified layer should be resistant to delamination and cracking. Resistance to delamination is achieved by covalently bonding the modified region to the substrate, intermixing the components of the substrate and the surface film at an interfacial zone (for example, an interpenetrating network or IPN), applying a compatibilizing ("primer") layer at the interface or incorporating appropriate functional groups for strong intermolecular adhesion between a substrate and an overlayer (Wu, 1982). Where the mechanical properties of the substrate and overlayer are significantly mismatched, delamination becomes increasingly of concern.

Surface Rearrangement

Surface rearrangement can readily occur. It is driven by a thermodynamic minimization of interfacial energy and enhanced by molecular mobility. Surface chemistries and structures can "switch" due to diffusion or translation of surface atoms or molecules in response to the external environment (see Chapter I.1.5 and Figure I.1.5.2 in that chapter). A newly formed surface chemistry can migrate from the surface into the bulk or molecules from the bulk can diffuse to cover the surface. Such reversals occur in metallic and other inorganic systems, as well as in polymeric systems. Terms such as "reconstruction," "relaxation," and "surface segregation" are often used to describe mobility-related alterations in surface structure and chemistry (Ratner and Yoon, 1988; Garbassi et al., 1989; Somorjai, 1990, 1991; Cometa et al., 2010). The driving force for these surface changes is a minimization of the interfacial energy. However, sufficient atomic or molecular mobility must exist for the surface changes to occur in reasonable periods of time. For a modified surface to remain as intended, surface reversal must be prevented or inhibited. This can be done by cross-linking, by sterically blocking the ability of surface structures to move, by incorporating a rigid, impermeable layer between the substrate material and the surface modification or by reducing the hydrophilic character of the modified surface, which will reduce water uptake and the mobility of surface atoms.

Surface Analysis

Surface modification and surface analysis are complementary and sequential technologies. The surface modified region is usually thin, and consists of only minute amounts of material. Undesirable contamination can readily be introduced during modification reactions. The potential for surface reversal to occur during surface modification is also high. The surface reaction should be monitored to ensure that the intended surface is indeed being formed. Since conventional analytical methods are often insufficiently sensitive to detect surface modifications, special surface analytical tools are called for (Chapter I.1.5).

Commercializability

The end products of biomaterials research are devices and materials that are manufactured to exacting specifications for use in humans. A surface modification that is too complex will be challenging and expensive to reproducibly manufacture and commercialize. It is best to minimize the number of steps in a surface modification process, and to design each step to be relatively insensitive to small changes in reaction conditions.

TABLE I.2.12.2 Physical and Chemical Surface Modification Methods

	Polymer	Metal	Ceramic	Glass
Noncovalent Coatings				
Solvent coating	✓	✓	✓	✓
Langmuir–Blodgett film deposition	✓	✓	✓	✓
Surface active additives	✓	✓	✓	✓
Vapor deposition of carbons and metals*	✓	✓	✓	✓
Vapor deposition of Parylene (p-xylylene)	✓	✓	✓	✓
Covalently Attached Coatings				
Radiation grafting (electron accelerator and gamma)	✓	–	–	–
Photografting (UV and visible sources)	✓	–	–	✓
Plasma (gas discharge) (RF, microwave, acoustic)	✓	✓	✓	✓
Gas phase deposition:				
• Ion beam sputtering	✓	✓	✓	✓
• Chemical vapor deposition (CVD)	–	✓	✓	✓
• Flame spray deposition	–	✓	✓	✓
Chemical grafting (e.g., ozonation + grafting)	✓	✓	✓	✓
Silanization	✓	✓	✓	✓
Biological modification (biomolecule immobilization)	✓	✓	✓	✓
Modifications of the Original Surface				
Ion beam etching (e.g., argon, xenon)	✓	✓	✓	✓
Ion beam implantation (e.g., nitrogen)	–	✓	✓	✓
Plasma etching (e.g., nitrogen, argon, oxygen, water vapor)	✓	✓	✓	✓
Corona discharge (in air)	✓	✓	✓	✓
Ion exchange	✓**	✓	✓	✓
UV irradiation	✓	✓	✓	✓
Chemical Reaction				
• Nonspecific oxidation (e.g., ozone)	✓	✓	✓	✓
• Functional group modifications (oxidation, reduction)	✓	–	–	–
• Addition reactions (e.g., acetylation, chlorination)	✓	–	–	–
Conversion coatings (phosphating, anodization)	–	✓	–	–
Mechanical roughening and polishing	✓	✓	✓	✓

*Some covalent reaction may occur.
**For polymers with ionic groups.

METHODS FOR MODIFYING THE SURFACES OF MATERIALS

General methods to modify the surfaces of materials are illustrated in Figure I.2.12.1, with many examples listed in Table I.2.12.2. A few of the more widely used of these methods will be briefly described. Some of the conceptually simpler methods, such as solution coating of a polymer onto a substrate or metallization by sputtering or thermal evaporation, will not be elaborated upon here.

Chemical Reaction

There are hundreds of chemical reactions that can be used to modify the chemistry of a surface. Chemical reactions, in the context of this article, are those performed with reagents that react with atoms or molecules at the surface, but do not overcoat those atoms or molecules with a new layer. Chemical reactions can be classified as nonspecific and specific.

Nonspecific reactions leave a distribution of different functional groups at the surface. An example of a nonspecific surface chemical modification is the chromic acid oxidation of polyethylene surfaces leading to numerous surface oxygen species including carboxylic acid, carbonyl, hydroxyl, etc. Other examples include the corona discharge modification of materials in air; radio frequency glow discharge (RFGD) (plasma) treatment of materials in oxygen, argon, nitrogen, carbon dioxide, fluorinated gases or water vapor; and the oxidation of metal surfaces to a mixture of metallic suboxides.

Specific chemical surface reactions change only one functional group into another with a high yield and few side reactions. A few examples of specific chemical surface modifications for polymers are presented in Figure I.2.12.2. Certain specific reactions can be used for inorganic (silicon) surfaces as well as polymeric surfaces (Filler and Bent, 2003). Still another example (expanded upon in the section, below, titled "Specific Chemical Reactions for Forming Surface Grafts") uses atomic transfer radical polymerization (ATRP) chain growth on surfaces (Hucknall et al., 2009). Others have also used radical addition chain transfer polymerization (RAFT) to graft polymer chains to a surface with high precision.

(a)
Alkylation of poly(chlorotrifluoroethylene)

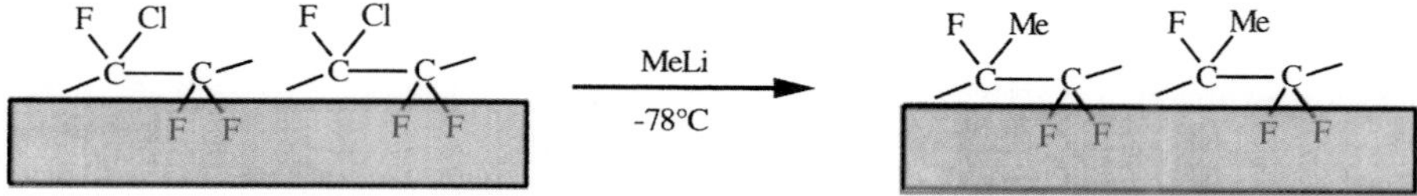

(b)
Trifluoroacetic anhydride reaction of a hydroxylated surface

$(CF_3C{=}O)_2O$

(c)
Glycidyl group introduction into a polysiloxane

Pt catalyst

FIGURE I.2.12.2 Some specific chemical reactions to modify surfaces.

This may be done by conjugating the chain transfer agent (CTA) to the surface and then initiating graft polymerization of a monomer to the CTA using a typical free radical initiator (Boyer et al., 2009; Golden et al., 2010).

Protein immobilization is typically (although not always) a surface modification reaction performed with high chemical specificity. Detailed chemistries for biomolecule immobilization are described in Chapter I.2.17.

Surface Grafting: Radiation Grafting, Photografting, and Newer Methods

Surface grafting methods have been widely applied for the surface modification of biomaterials since the early days of the field (Hoffman et al., 1972). Comprehensive review articles are available (Ratner, 1980; Hoffman, 1981; Hoffman et al., 1983; Stannett, 1990; Safrany, 1997; Ma et al., 2007). Surface grafting methods can be divided into two general categories: (1) using energetic or highly reactive species to activate chemically inert surface permitting the attachment of desired surface species; and (2) taking advantage of well-defined, reactive surface chemical groups to covalently attach the surface modifying species (a polymer, a low molecular weight compound, a drug, a protein, etc.). These category (2) specific reactions for forming surface grafts will be addressed in a separate section below.

The earliest biomedical surface modification studies focused on attaching chemically reactable groups (-OH, -COOH, $-NH_2$, etc.) to the surfaces of relatively inert hydrophobic polymers using methods from category (1), above. Four types of reactions can be distinguished in category (1): grafting using ionizing radiation sources (most commonly, a cobalt 60 or cesium 137 gamma radiation source) (Dargaville et al., 2003); grafting using UV radiation (photografting) (Matsuda and Inoue, 1990; Dunkirk et al., 1991; Swanson, 1996; Sebra et al., 2007); grafting using high-energy electron beams (Singh and Silverman, 1992); and grafting using a reactive, surface-activating species such as ozone (Chiang et al., 2009). In all cases, similar processes occur. The radiation or reactive chemical breaks chemical bonds in the material to be grafted, forming free radicals, peroxides or other reactive species. These reactive surface groups are then exposed to a monomer. The monomer reacts with the free radicals at the surface and propagates as a free radical chain reaction incorporating other monomers into a surface grafted polymer. Electron beams and gamma radiation sources are also used for biomedical device sterilization (see Chapter III.1.2). Note that glow discharge plasma reactions to surfaces also have similarities to these non-specific surface modification approaches. Plasma reaction will be discussed in its own section, below.

High energy surface modification technologies are strongly dependent on the source energy, the radiation dose rate, and the amount of the dose absorbed. The energy of the radiation produced by gamma sources is roughly 1 MeV (1 eV = 23.06 kcal/mol; for comparison, a carbon–carbon bond has a strength of approximately 85 kcal/mol). Typical energies for electron beam processing are in the range 5–10 MeV. UV radiation sources are of much lower energy (<6 eV). Radiation dose rates are low for UV and gamma, and very high for electron beams. The amount of energy absorbed is measured in units of Grays (Gy), where 1 kiloGray (Kgy) = 1000 joules/kilogram. Units of megarads (MR) are often used for gamma sources, 1 MR = 1 × 10^6 ergs/gram.

Three distinct reaction modes can be described: (1) in the mutual irradiation method, the substrate material is immersed in an oxygen-free solution (monomer ± solvent) that is then exposed to the radiation source; (2) the substrate materials can also be exposed to the radiation under an inert atmosphere or at low temperatures (to stabilize free radicals) (in this case, the materials are later contacted with a monomer solution to initiate the graft process); (3) finally, the exposure to the radiation can take place in air or oxygen, leading to the formation of peroxide groups on the surface. Heating the material to be grafted in the presence of monomer or addition of a redox reactant (e.g., Fe^{++}) which will decompose the peroxide groups to form free radicals can initiate the graft polymerization (in O_2-free conditions).

Graft layers formed by energetic irradiation of the substrate are often relatively thick (>1 μm) and are comprised of high molecular weight polymer chains. However, they are typically well-bonded to the substrate material, and thus resist delamination. Since many polymerizable monomers are available, a wide range of surface chemistries can be created. Mixtures of monomers can form unique graft copolymers (Ratner and Hoffman, 1980). For example, the hydrophilic/hydrophobic ratio of surfaces can be controlled by varying the ratio of a hydrophilic and a hydrophobic monomer in the grafting mixture (Ratner et al., 1979; Ratner and Hoffman, 1980).

Photoinitiated grafting (usually with visible or UV light) represents a unique subcategory of surface modifications for which there is growing interest (Deng et al., 2009; Park et al., 2009). There are many approaches to effect this photoinitiated covalent coupling. For example, a phenyl azide group can be converted to a highly reactive nitrene upon UV exposure. This nitrene will quickly react with many organic groups. If a synthetic polymer is prepared with phenyl azide side groups, and this polymer is exposed simultaneously to UV light and a substrate polymer or polymeric medical device, the polymer containing the phenyl azide side groups will be immobilized to the substrate (Matsuda and Inoue, 1990). Another method involves the coupling of a benzophenone molecule to a hydrophilic polymer (Dunkirk et al., 1991). In the presence of UV irradiation, the benzophenone is excited to a reactive triplet state that can covalently couple with many polymers. As still another example, a dithiocarbamate-functionalized polyurethane was used to photo-pattern an extracellular-matrix-like coating (Sebra et al., 2007).

Radiation, electron beam, and photografting have frequently been used to bond hydrogels to the surfaces of hydrophobic polymers (Matsuda and Inoue, 1990; Dunkirk et al., 1991; Sebra et al., 2007). Electron beam grafting of N-isopropyl acrylamide to polystyrene has been used to create a new class of temperature-dependent surfaces for cell growth (Kwon et al., 2000) (also see Chapter I.2.17). The protein interactions (Horbett and Hoffman, 1975), cell interactions (Ratner et al., 1975; Matsuda and Inoue, 1990), blood compatibility (Chapiro, 1983; Hoffman et al., 1983), and tissue reactions (Greer et al., 1979) of hydrogel graft surfaces have been investigated.

RFGD Plasma Deposition and Other Plasma Gas Processes

RFGD plasmas, as used for surface modification, are low-pressure ionized gas environments typically at ambient (or slightly above ambient) temperature. They are also referred to as glow discharge or gas discharge depositions or treatments. Plasmas can be used to modify existing surfaces by ablation or etching reactions or, in a deposition mode, to overcoat surfaces (Figure I.2.12.1). Good review articles on plasma deposition and its application to biomaterials are available (Yasuda and Gazicki, 1982; Hoffman, 1988; Ratner et al., 1990; Chu et al., 2002; Kitching et al., 2003; Desmet et al., 2009). Some biomedical applications of plasma-modified biomaterials are listed in Table I.2.12.3.

RFGD plasma surface modifications are widely used in biomaterials research and development. Because such coatings and treatments are now used in commercialized biomaterials products and have special promise for improved biomaterials, they will be emphasized in this chapter. The specific advantages of plasma-deposited films (and to some extent, plasma-treated surfaces) for biomedical applications are:

1. They are conformal. Because of the penetrating nature of a low-pressure gaseous environment in which mass transport is governed by both molecular (line-of-sight) diffusion and convective diffusion, complex geometric shapes can be treated.
2. They are free of voids and pinholes. This continuous barrier structure is suggested by transport studies and electrical property studies (Charlson et al., 1984).
3. Plasma-deposited films can coat almost any clean solid, including polymers, metals, ceramics, and semiconductors. Other surface grafting or surface modification technologies are highly dependent upon the chemical nature of the substrate.

TABLE I.2.12.3 Biomedical Applications of Glow Discharge Plasma-Induced Surface Modification Processes

A. Plasma Treatment (Etching)
   1. Clean
   2. Sterilize
   3. Cross-link surface molecules

B. Plasma Treatment (Etching) and Plasma Deposition
   1. Form barrier films
      a. Protective coating
      b. Electrically insulating coating
      c. Reduce absorption of material from the environment
      d. Inhibit release of leachables
      e. Control drug delivery rate
   2. Modify cell and protein reactions
      a. Modulate biointeractions
      b. Promote selective protein adsorption
      c. Enhance cell adhesion
      d. Improve cell growth
      e. Form nonfouling surfaces
      f. Increase lubricity

Anti-Bacterial Properties?
   3. Provide reactive sites
      a. For grafting or polymerizing polymers
      b. For immobilizing biomolecules

4. They exhibit good adhesion to the substrate. The energetic nature of the gas phase species in the plasma reaction environment can induce mixing, implantation, penetration, and reaction between the overlayer film and the substrate.
5. Unique film chemistries can be produced. The chemical structure of the polymeric overlayer films generated from the plasma environment usually cannot be synthesized by conventional chemical methods.
6. They can serve as excellent barrier films because of their pinhole-free and dense, cross-linked nature.
7. Plasma-deposited layers generally show low levels of leachables. Due to their highly cross-linked nature, plasma-deposited films contain negligible amounts of low molecular weight components that might lead to an adverse biological reaction. They can also inhibit leaching of low molecular weight material from the substrate and, in the case of drug delivery implants or inserts, they can reduce the rate of drug diffusion out of the device (Hendricks et al., 2000).
8. These films are easily prepared. Once the apparatus is set up and optimized for a specific deposition, treatment of additional substrates is rapid and simple.
9. Plasma deposition is a mature technology. The microelectronics industry has made extensive use of inorganic plasma-deposited films for many years (Sawin and Reif, 1983; Nguyen, 1986).
10. Plasma surface modifications, although chemically complex, can be characterized by infrared (IR) (Inagaki et al., 1983; Haque and Ratner, 1988; Krishnamurthy et al., 1989), nuclear magnetic resonance (NMR) (Kaplan and Dilks, 1981), electron spectroscopy for chemical analysis (ESCA) (Chilkoti et al., 1991a), chemical derivatization methods (Everhart and Reilley, 1981; Gombotz and Hoffman, 1988; Griesser and Chatelier, 1990; Chilkoti et al., 1991a), and static secondary ion mass spectrometry (SIMS) (Chilkoti et al., 1991b, 1992; Johnston and Ratner, 1996).
11. Plasma-treated surfaces are sterile when removed from the reactor, offering an additional advantage for cost-efficient production of medical devices.
12. New developments in the generation of atmospheric pressure plasmas may permit plasma processing without the need for a vacuum environment (Pappas, 2011).

Although plasma treatments have many advantages, there are concerns with plasma surface modification that should be discussed. First, the chemistry produced on a surface is often ill-defined. For example, if tetrafluoroethylene gas is introduced into the reactor, polytetrafluoroethylene (PTFE) will not be deposited on the surface. Rather, a complex, branched fluorocarbon polymer will be produced with a stoichiometry different from PTFE. This scrambling of monomer structure has been addressed in studies dealing with retention of monomer structure in the final film (Lopez and Ratner, 1991; Lopez et al., 1993; Panchalingam et al., 1993). Second, the apparatus used to produce plasma depositions can be expensive. A good laboratory-scale reactor will cost \$10,000–\$30,000, and a production reactor can cost \$100,000 or more. Third, uniform reaction within tubes and narrow pores can be difficult to achieve. Finally, contamination can be a problem, and care must be exercised to prevent extraneous gases and pump oils from entering the reaction zone. However, the advantages of plasma reactions outweigh these potential disadvantages for many types of modifications that cannot be accomplished by other methods.

The Nature of the Plasma Environment

Plasmas are atomically and molecularly dissociated gaseous environments. A plasma environment contains positive ions, negative ions, free radicals, electrons, atoms, molecules, and photons (visible and near UV). Typical conditions within the plasma include an electron energy of 1–10 eV, a gas temperature of 25–60°C, an electron density of 10^{-9} to $10^{-12}/cm^2$, and an operating pressure of 0.025–1.0 torr.

Many processes can occur on the substrate surface that lead to the observed surface modification or deposition. First, a competition takes place between deposition and etching by the high-energy gaseous species (ablation) (Yasuda, 1979). When ablation is more rapid than deposition, no deposition will be observed. Because of its energetic nature, the ablation or etching process can

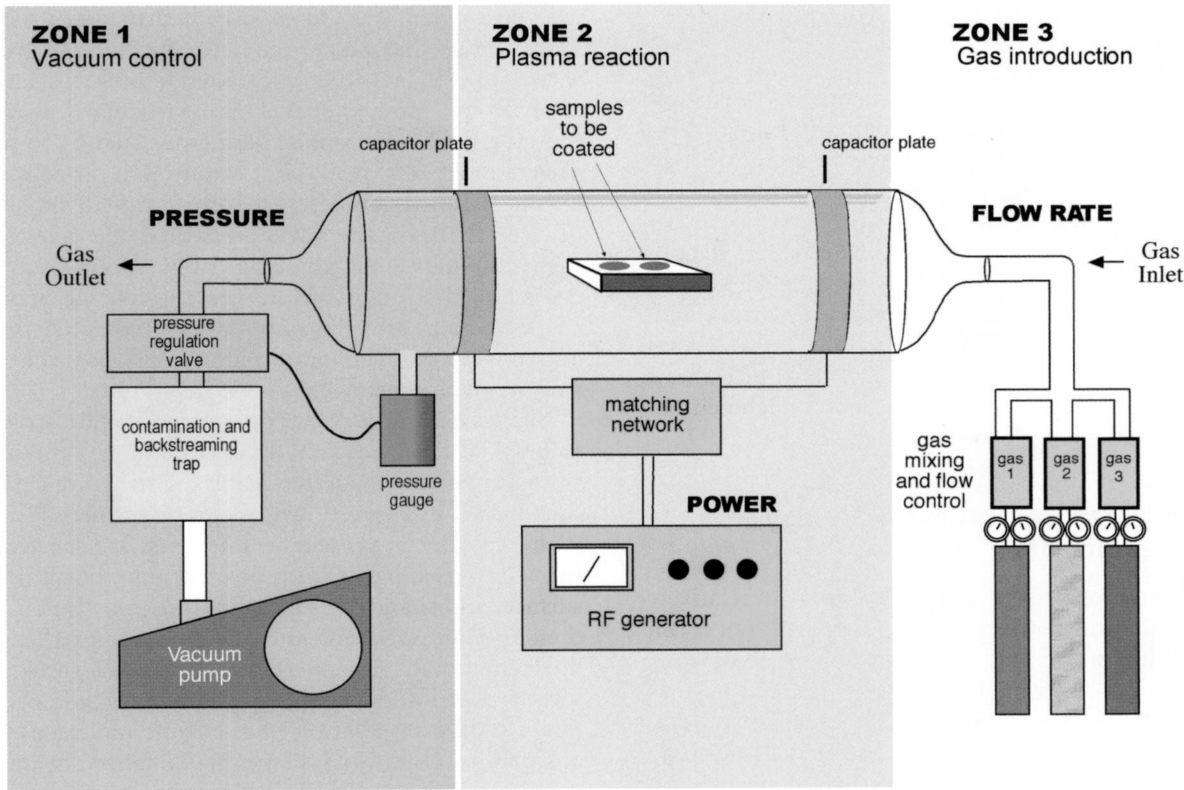

FIGURE I.2.12.3 A capacitively-coupled RF plasma reactor. Redish colors indicate gas storage and mixing. Yellow colors are components that power the reactor. Zone 1: Vacuum system (pressure measurement and control); Zone 2: Plasma generation and sample placement; Zone 3: Gas introduction and flow control.

result in substantial chemical and morphological changes to the substrate.

A number of mechanisms have been postulated for the deposition process. The reactive gaseous environment may create free radical and other reactive species on the substrate surface that react with and polymerize molecules from the gas phase. Alternatively, reactive small molecules in the gas phase could combine to form higher molecular weight units or particulates that may settle or precipitate onto the surface. Most likely, the depositions observed are formed by some combination of these two processes.

The Apparatus to Generate Plasmas for Surface Modification

Many experimental variables relating both to reaction conditions and to the substrate onto which the deposition is placed affect the final outcome of the plasma deposition process. A diagram of a typical capacitively-coupled radio frequency plasma reactor is presented in Figure I.2.12.3. The major subsystems that comprise this apparatus are a gas introduction system (control of gas mixing, flow rate, and mass of gas entering the reactor), a vacuum system (measurement and control of reactor pressure and inhibition of backstreaming of molecules from the pumps), an energizing system to efficiently couple energy into the gas phase within the reactor, and a reactor zone in which the samples are treated. Radio frequency, acoustic or microwave energy can be coupled to the gas phase. Devices for monitoring the molecular weight of the gas phase species (mass spectrometers), the optical emission from the glowing plasma (spectrophotometers), and the deposited film thickness (ellipsometers, vibrating quartz crystal microbalances) are also commonly found on plasma reactors. Technology has been developed permitting atmospheric pressure plasma deposition (Klages et al., 2000; Massines et al., 2000; Pappas, 2011). Another important development is "reel-to-reel" (continuous) plasma processing opening the way to low cost, high-throughput treatment of films, fibers, and tubes. Finally, methods have been developed to assist in the retention of polymer chemical structure during plasma deposition. These methods include using pulsed plasmas (Panchalingam et al., 1993), condensing vapors on a cold stage during the plasma process (Lopez and Ratner, 1991), and using very low plasma powers (Pan et al., 2001). Such controlled chemistry plasma deposition might be thought of as at some intermediate location on a continuum between the scrambled surface chemistries achieved with conventional plasma deposition of polymers, and the precise graft structures described in the section below titled "Specific Chemical Reactions for Forming Surface Grafts."

RFGD Plasmas for the Immobilization of Molecules

Plasmas have often been used to introduce reactive organic functional groups (e.g., amine, hydroxyl) on a surface that can be activated to attach biomolecules (see Chapter I.2.17). Certain reactive gas environments can also be used for directly immobilizing organic molecules such as surfactants. For example, a poly(ethylene glycol)-n-alkyl surfactant will adsorb to polyethylene (PE) via the hydrophobic alkyl block. If the polyethylene surface with the adsorbed surfactant is briefly exposed to an argon plasma, the n-alkyl chains will cross-link to surface atoms of PE, thereby leading to the covalent attachment of pendant poly(ethylene glycol) chains (Sheu et al., 1992).

High Temperature and High-Energy Plasma Treatments

The plasma environments described above are of relatively low energy and low temperature. Consequently, they can be used to deposit organic layers on polymeric or inorganic substrates. Under higher energy conditions, plasmas can effect unique and important inorganic surface modifications on inorganic substrates. For example, flame-spray deposition involves injecting a high-purity, relatively finely divided (~100 mesh) metal powder into a high-velocity plasma or flame. The melted or partially melted particles impact the surface and rapidly solidify (see Chapter I.2.3). Examples of plasma thermal spray coating on titanium are seen in Gruner (2001), and Vidigal et al. (2009).

Specific Chemical Reactions for Forming Surface Grafts

Techniques such as radiation grafting and plasma deposition create polymer surface grafts, although the polymeric species are generally highly cross-linked, chemically rearranged, and difficult to characterize. A new generation of polymer reactions permits surface grafts with precision in chain length, control of chain architectures (for example, diblock and triblock units), and minimization of unwanted branching and other side reactions. These polymer reactions applied to surface grafting include ATRP (Hucknall et al., 2009), reversible addition-fragmentation chain transfer (RAFT) polymerization (Boyer et al., 2009; Golden et al., 2010), and CLICK chemistry (Fleischmann et al., 2008).

Silanization

Silane treatments of surfaces involve a liquid phase or vapor phase chemical reaction, and are straightforward to perform and of low cost. A typical silane surface modification reaction is illustrated in Figure I.2.12.4 (although some details of this reaction have been questioned, see Schlecht and Maurer, 2011). Silane reactions are most often used to modify hydroxylated surfaces. Since glass, silicon, alumina, titania, and quartz surfaces, as well as other metal oxide surfaces, are rich in hydroxyl groups, silanes are particularly useful for modifying these materials. Numerous silane compounds are commercially available, permitting a broad range of chemical functionalities to be incorporated on surfaces (Table I.2.12.4). The advantages of silane reactions are their simplicity and stability, attributed to their covalent, cross-linked structure. However, the linkage between a silane and a hydroxyl group is also subject to hydrolysis, and film breakdown under some conditions must be considered (Wasserman et al., 1989).

Silanes can form two types of surface film structures. If only surface reaction occurs (perhaps catalyzed by traces of adsorbed surface water), a structure similar to that shown in Figure I.2.12.4 can be formed. However, if more water is present, a thicker silane layer can be formed consisting of both Si–O groups bonded to the surface, and silane units participating in a "bulk," three-dimensional, polymerized network. The initial stages in the formation of a thicker silane film are suggested by the further reaction of the group at the right side of Figure I.2.12.4(d) with solution phase silane molecules. Without careful control of silane liquid purity, water, and reaction conditions, thicker silane films can form and be rough and inhomogeneous.

A new class of silane-modified surfaces based upon monolayer silane films and yielding self-assembled, highly ordered structures is of particular interest in the precision engineering of surfaces (Pomerantz et al., 1985; Maoz et al., 1988; Heid et al., 1996; Haensch et al., 2010). These self-assembled monolayers are described in more detail later in this chapter.

Many general reviews and basic science studies on surface silanization are available (Arkles, 1977; Plueddemann, 1980; Rye et al., 1997). Applications for silanized surface modified bomaterials are on the increase, and include cell attachment (Hickman and Stenger, 1994; Matsuzawa et al., 1997), biomolecule and polymer immobilization (Mao et al., 1997; Xiao et al., 1997), non-fouling surfaces (Lee and Laibinis, 1998), surfaces for DNA studies (Hu et al., 1996), biomineralization (Archibald et al., 1996), and model surfaces for biointeraction studies (Jenney and Anderson, 1999).

Ion Beam Implantation

The ion beam method injects accelerated ions with energies ranging from 10^1 to 10^6 eV (1 eV = 1.6×10^{-19} joules) into the surface zone of a material to alter its properties. It is largely, but not exclusively, used with metals and other inorganics such as ceramics, glasses, and semiconductors. Ions formed from most of the atoms in the periodic table can be implanted, but not all provide useful modifications to the surface properties. Important potential applications for biomaterial surfaces include

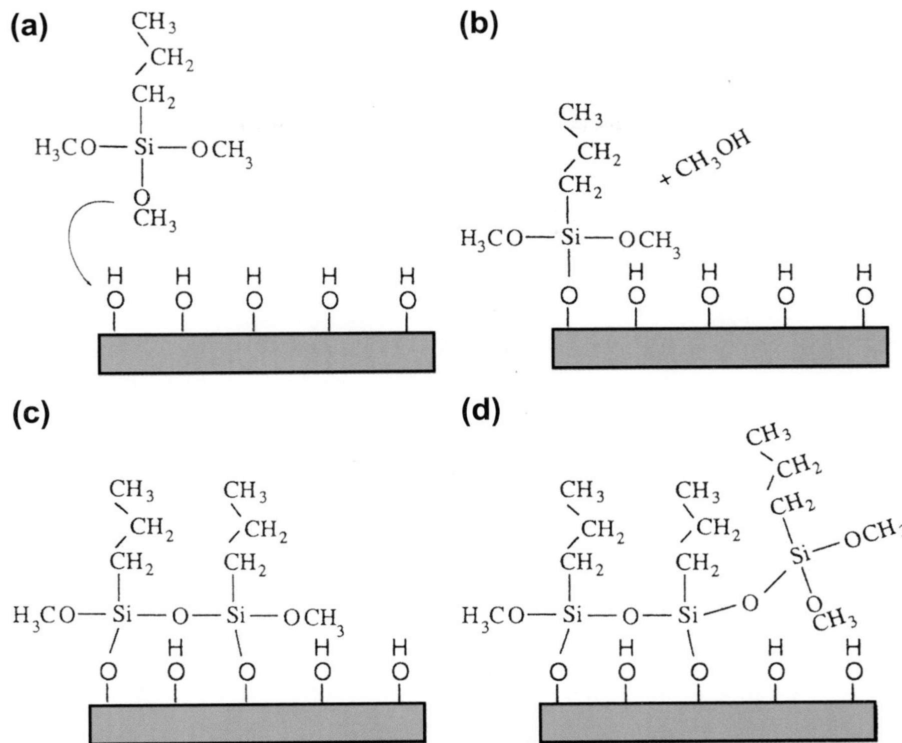

FIGURE I.2.12.4 The chemistry of a typical silane surface modification reaction: (a) a hydroxylated surface is immersed in a solution containing n-propyl trimethoxysilane (nPTMS); (b) one of the methoxy groups of the nPTMS couples with a hydroxyl group releasing methanol; (c) two of the methoxy groups on another molecule of the nPTMS have reacted, one with a hydroxyl group and the other with a methoxy group from the first nPTMS molecule; (d) a third nPTMS molecule has reacted only with a methoxy group. This molecule is tied into the silane film network, but is not directly bound to the surface.

TABLE I.2.12.4 Silanes for Surface Modification of Biomaterials

$X_3\text{–Si–}R$ (X – Si – R, with two further X groups on Si)

X = Leaving Group	R = Functional Group
$-Cl$	$-(CH_2)_nCH_3$
$-OCH_3$	$-(CH_2)_3NH_2$
$-OCH_2CH_3$	$-(CH_2)_2(CF_2)_5CF_3$
	$-(CH_2)_3O\text{–}C(=O)\text{–}C(CH_3)=CH_2$
	$-CH_2CH_2\text{–}C_6H_5$

modification of hardness (wear), lubricity, toughness, corrosion, conductivity, and bioreactivity.

If an ion with kinetic energy greater than a few electron volts impacts a surface, the probability that it will enter the surface is high. The impact transfers much energy to a localized surface zone in a very short time interval. Some considerations for the ion implantation process are illustrated in Figure I.2.12.5. These surface changes must be understood quantitatively for engineering of modified surface characteristics. Many review articles and books are available on ion implantation processes and their application for tailoring surface properties (Picraux and Pope, 1984; Colligon, 1986; Sioshansi, 1987; Nastasi et al., 1996; Rautray et al., 2010).

Specific examples of biomaterials that have been surface altered by ion implantation processes are plentiful. Iridium was ion implanted in a Ti-6Al-4V alloy to improve corrosion resistance (Buchanan et al., 1990). Nitrogen implanted into titanium greatly reduces wear (Sioshansi, 1987). The ion implantation of boron and carbon into type 316L stainless steel improves the high cycle fatigue life of these alloys (Sioshansi, 1987). Silver ions implanted into polystyrene permit cell attachment (Tsuji et al., 1998).

Langmuir–Blodgett Deposition

The Langmuir–Blodgett (LB) deposition method overcoats a surface with one or more highly ordered layers of surfactant molecules. Each of the molecules that assemble into this layer contains a polar "head" group and a non-polar "tail" group. The deposition of an LB film using an LB trough is illustrated schematically in Figure I.2.12.6. By withdrawing the vertical plate out of the aqueous phase and through the air–water interface (keeping the surface film at the air–water interface compressed at all times,

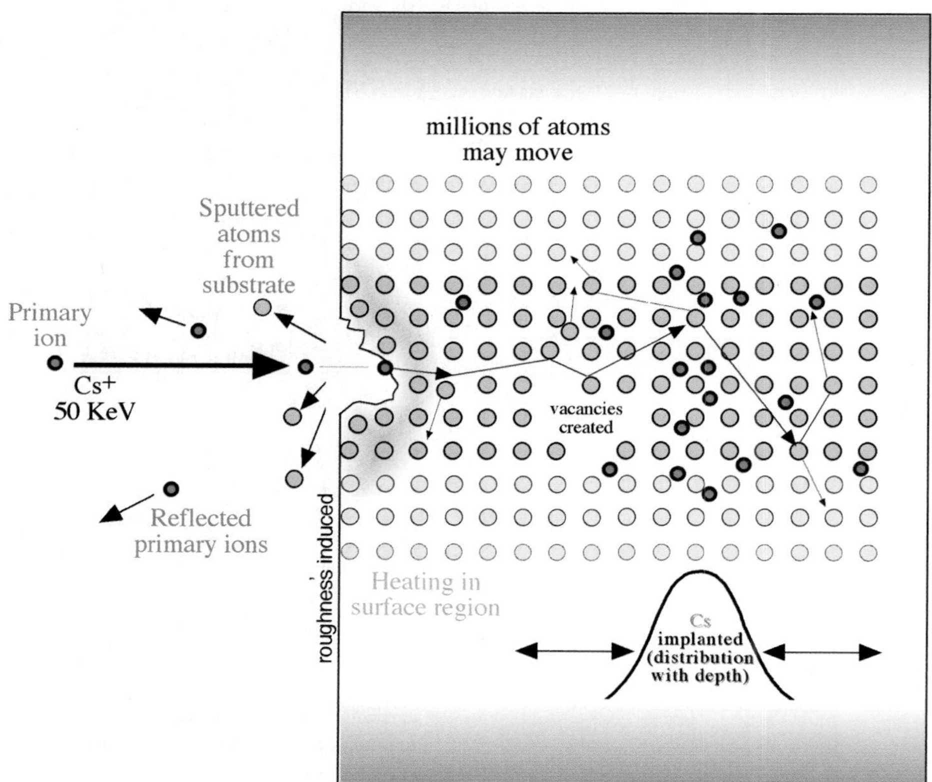

FIGURE I.2.12.5 Ion impact and implantation considerations with a 50 KeV cesium ion accelerated into a surface.

as illustrated in Figure I.2.12.6), an assembled structure coats the glass with the hydrophilic tail group contacting the hydrophilic glass surface. By next pushing the plate back down through the air–water interface, another layer can be deposited, this time with the hydrophobic tails of the two layers in contact; dipping in and out with the film at the water surface always compressed permits multilayer structures to be created. Some compounds that form organized LB layers are shown in Figure I.2.12.7. Other techniques to create such ordered lipd structures at surfaces include Langmuir–Schaefer transfer and vesicle fusion (Li et al., 2008). The advantages of films deposited on surfaces by this method are their high degree of order and uniformity, and also their resemblance to the lipid bilayer membranes surrounding living cells. Also, since a wide range of chemical structures can form LB films, there are many options for incorporating new chemistries at surfaces. The stability of LB films can be improved by cross-linking or internally polymerizing the molecules after film formation, often through double bonds in the alkyl portion of the chains (Meller et al., 1989). A number of research groups have investigated LB films for biomedical applications (Hayward and Chapman, 1984; Bird et al., 1989; Cho et al., 1990; Heens et al., 1991; Knoll et al., 2008). A cross between silane thin films and LB layers has been developed for biomedical surface modification (Takahara et al., 2000). Many general reviews on these surface structures are available (Knobler, 1990; Ulman, 1991; Park and Advincula, 2011).

Self-Assembled Monolayers

Self-assembled monolayers (SAMs) are surface films that spontaneously form as highly ordered structures (two-dimensional crystals) on specific substrates (Maoz et al., 1988; Ulman, 1990, 1991; Whitesides et al., 1991; Knoll, 1996; Raynor et al., 2009). In some ways SAMs resemble LB films, but there are important differences, in particular their ease of formation. Examples of SAM films include n-alkyl silanes on hydroxylated surfaces (silica, glass, alumina), alkane thiols (e.g., $CH_3(CH_2)_nSH$) and disulfides on coinage metals (gold, silver, copper), amines and alcohols on platinum, carboxylic acids on aluminum oxide, and silver and phosphates (phosphoric acid or phosphonate groups) on titanium or tantalum surfaces. Silane SAMs and thiols on gold are the most commonly used types. Most molecules that form SAMs have the general characteristics illustrated in Figure I.2.12.8. Two processes are particularly important for the formation of SAMs (Ulman, 1991): a moderate-to-strong adsorption of an anchoring chemical group to the surface (typically 30–100 kcal/mol), and van der Waals interaction of the alkyl chains. The bonding to the substrate (chemisorption) provides a driving force to fill every site on the surface, and to displace contaminants from the surface. This process is analogous to the compression to the LB film by the movable barrier in the trough. Once adsorption sites are filled on the surface, the chains will be in sufficiently close proximity so that the weaker van der Waals interactive forces between

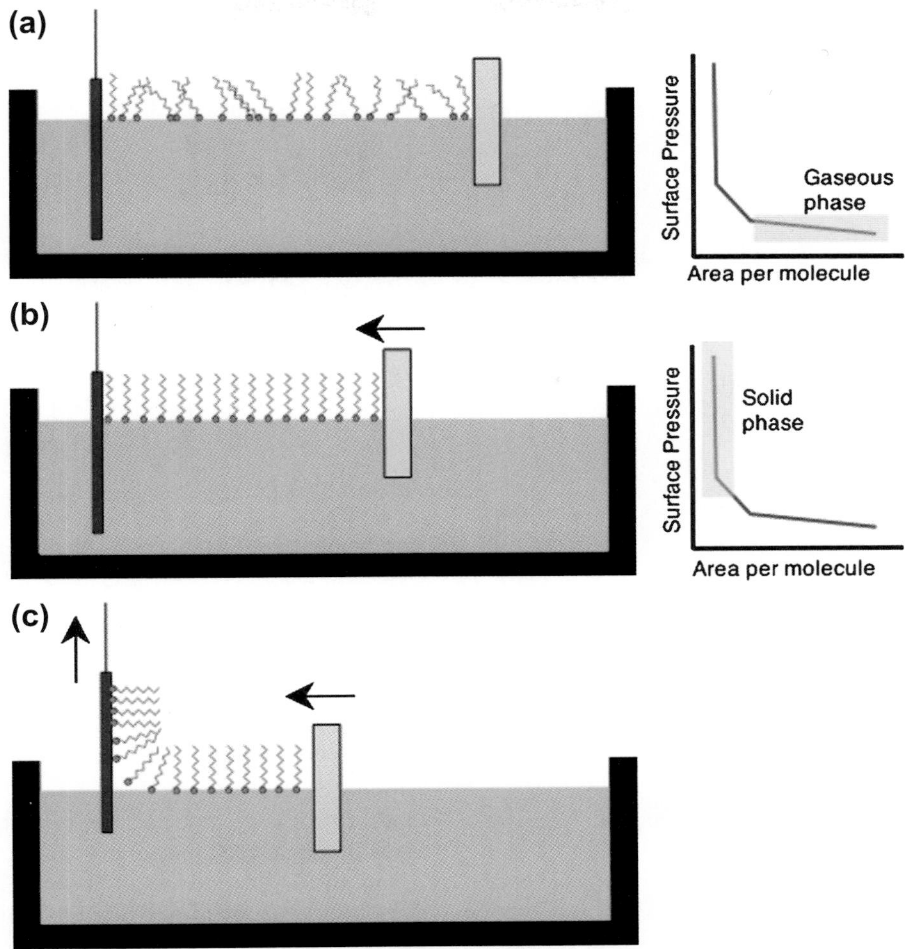

FIGURE I.2.12.6 Deposition of a lipid film onto a glass slide by the Langmuir–Blodgett method: (a) the lipid film is floated on the aqueous layer; (b) the lipid film is compressed by a moveable barrier; (c) the vertical glass slide is withdrawn while pressure is maintained on the floating lipid film with the moveable barrier.

chains can exert their influence and lead to a crystallization of the alkyl groups. Too few CH_2 groups do not provide sufficient interactive force to stabilize the two-dimensional quasi-crystal and are difficult to assemble. More than 24 CH_2 groups have too many options for defects in the crystal, and are also difficult to assemble. Molecules with lengths between 9 and 24 methylene groups will assemble well. Molecular mobility is an important consideration in this surface crystal formation process, so that: (1) the molecules have sufficient time to maneuver into position for tight packing of the binding end groups at the surface; and (2) the chains can enter the quasi-crystal.

The advantages of SAMs are their ease of formation, their chemical stability (often considerably higher than chemically-related LB films), and the many options for changing the outermost group that interfaces with the external environment. Many biomaterials applications have already been suggested for SAMs (Lewandowska et al., 1989; Mrksich and Whitesides, 1996; Ferretti et al., 2000; Raynor et al., 2009). Useful SAMs for creating molecularly-engineered functional surfaces include headgroups of ethylene glycol oligomers, biotin, free-radical initiators, N-hydroxysuccinimide esters, ATRP or RAFT chain transfer agents, anhydrides, perfluoro groups, and amines, just to list a small sample of the many possibilities. Although most SAMs are based on n-alkyl chain assembly, SAMs can form from other classes of molecules including peptides, proteins (Sara and Sleytr, 1996), porphyrins, nucleotide bases, and aromatic ring hydrocarbons.

LAYER-BY-LAYER DEPOSITION AND MULTILAYER POLYELECTROLYTE DEPOSITION

Layer-by-layer (L-b-L) deposition of thin films, and the most common variant of this method, multilayer polyelectrolyte adsorption, has become an important strategy for the surface modification of biomaterials (Decher, 1996; Ariga et al., 2007). For L-b-L surface treatment, a molecule (or polymer or protein) is chosen that adsorbs strong to the substrate (adsorbate 1). Then the substrate is rinsed (retaining adsorbate 1), and dipped in a solution of a molecule that interacts strongly with adsorbate 1

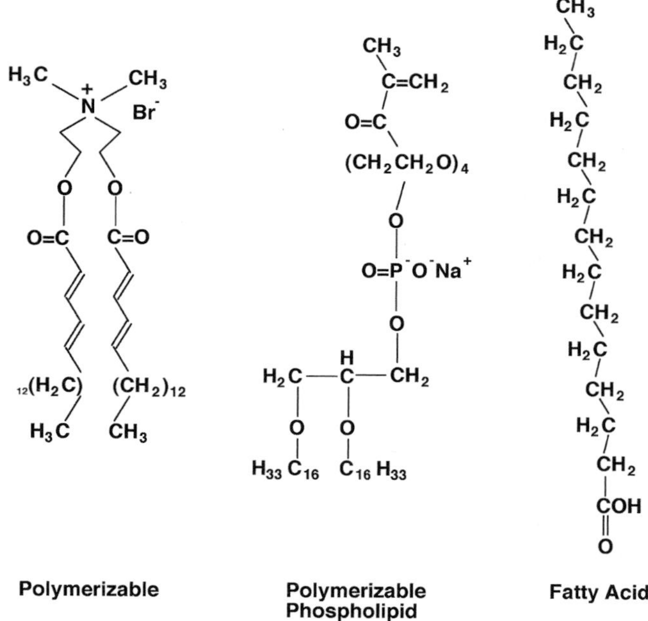

FIGURE I.2.12.7 Three examples of molecules that form organized Langmuir–Blodgett Films.

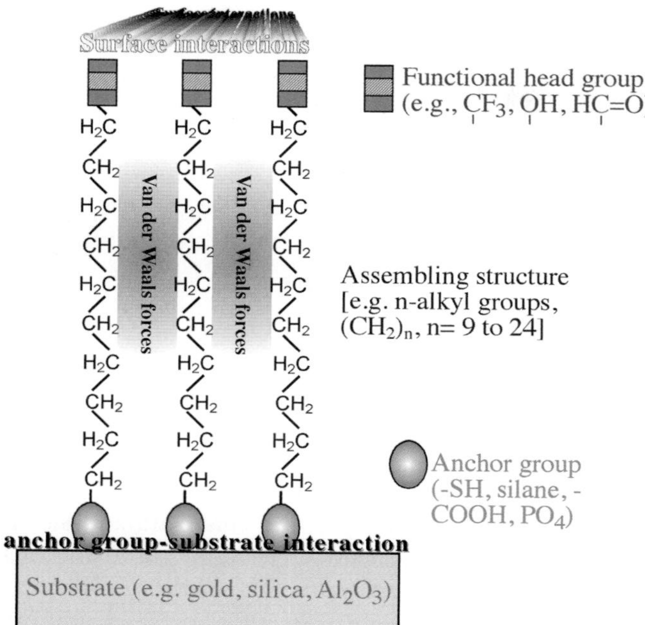

FIGURE I.2.12.8 General characteristics of molecules that form self-assembled monolayers.

(adsorbate 2). After rinsing, the surface layer is again dipped in a solution of adsorbate 1. Absorbate 1, adsorbate 2, and the rinsing step are alternated until the coating has the desired multilayer thickness (Figure I.2.12.9). Commonly, the substrate has a fixed positive or negative charge, permitting multilayers of polyanions and polycations. The rinse step is important for the removal of any unadsorbed polyions, as well as for removal of the counter-ions released upon formation of each layer of the polyion complex. Some surfaces are intrinsically charged (for example, mica), others can be modified with methods already described in this chapter. Once a thin layer of a charged component absorbs, it will repel additional adsorption of the component of the same charge, thus tightly controlling the layer thickness and uniformity. The outermost layer can be the positively or negatively charged component. This strategy works well with charged biomolecules, for example hyaluronic acid (–) and chitosan (+). Layers formed are durable, and assembly of these multilayer structures is simple. The pH and ionic strength of polyelectrolyte solutions are important process variables. The L-b-L process has been used with interacting layers of proteins, nucleotides, saccharides, nanoparticles, acid–base pairs, and H-bonding pairs. L-b-L has been used to coat surfaces, microspheres, nanoparticles, and even living cells. L-b-L layers can be used as reservoirs to contain drug for release or as rate-limiting diffusion barriers. L-b-L films have been used to treat contact lenses, to improve comfort and retard biofilm build-up.

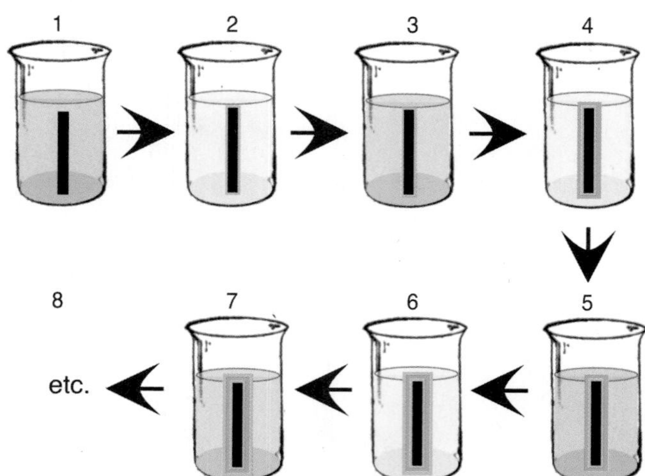

FIGURE I.2.12.9 Layer by layer deposition of a multilayer surface coating. A series of adsorption-rinse steps lead to a multilayer surface deposition: (1) the "blue" compound is adsorbed to the substrate; (2) the adsorbed layer is rinsed in solvent (frequently water) to remove weakly adherent material; (3) the "pink" compound is adsorbed to the blue layer on the substrate; (4) the adsorbed layer is rinsed in solvent to remove weakly adherent material; (5) the "blue" compound is adsorbed to the "pink" adsorbed layer; (6) the adsorbed layer is rinsed in solvent to remove weakly adherent material; (7) the "pink" compound is adsorbed to the blue layer; (8) this process continues to build as many layers as desired.

Surface Modifying Additives

Specifically designed and synthesized surface-active compositions can be added in low concentrations to a material during fabrication, and will spontaneously rise to and dominate the surface (Ward, 1989; Wen et al., 1997). These surface modifying additives (SMAs) are well-known for both organic and inorganic systems. A driving force to minimize the interfacial energy leads to the concentration of the SMA at the surface after blending homogeneously with a material. For efficient surface concentration, two factors must be considered. First, the magnitude of interfacial energy difference between the system without the additive and the same system with the SMA at the surface will determine the driving force leading to an SMA-dominated surface. Second, the molecular mobility of the bulk material and the SMA additive molecules within the bulk will determine the rate at which the SMA reaches the surface or if it will get there at all. An additional concern is the durability and stability of the SMA at the surface.

A typical SMA designed to alter the surface properties of a polymeric material will be a relatively low molecular weight diblock or triblock copolymer (see Chapter I.2.2.). The "A" block will be soluble in, or compatible with, the bulk material into which the SMA is being added. The "B" block will be incompatible with the bulk material and will have lower surface energy. Thus, the A block will anchor the B block into the material to be modified at the interface. This is suggested schematically in Figure I.2.12.10. During initial fabrication, the SMA might be distributed uniformly throughout the bulk. After a period for curing or an annealing step, the SMA will migrate to the surface. Low molecular weight endgroups on polymer chains can also provide the driving force to bring the endgroup to the surface.

As an example, an SMA for a polyurethane might have a low molecular weight polyurethane "A" block and a poly(dimethyl siloxane) (PDMS) "B" block. The PDMS component on the surface may confer improved blood compatibility to the polyurethane. The "A" block will anchor the SMA in the polyurethane bulk (the polyurethane "A" block should be reasonably compatible with the bulk polyurethane), while the low surface energy, highly flexible, silicone "B" block will be exposed at the air surface to lower the interfacial energy (note that air is "hydrophobic"). The "A" block anchor should confer stability to this system. However, consider that if the system is placed in an aqueous environment, a low surface energy polymer (the "B" block) is now in contact with water – a high interfacial energy situation. If the system, after fabrication, still exhibits sufficient chain mobility, it might phase-invert to bring the bulk polyurethane or the "A" block to the surface. Unless the system is specifically engineered to do such a surface phase reversal, this inversion is undesirable. Proper choice of the bulk polymer and the "A" block can impede surface phase inversion.

An example of a polymer additive that was developed by 3M specifically to take advantage of this surface chemical inversion phenomena is a stain inhibitor for fabric. Although not intended as a biomaterial, it illustrates design principles for this type of system. The compound has three "arms." A fluoropolymer arm, the lowest energy component, resides at the fabric surface in air. Fluoropolymers and hydrocarbons (typical stains) do not mix, so hydrocarbons are repelled. A second arm of hydrophilic poly(ethylene oxide) will come to the surface in water, and assist with the washing out of any material on the surface. Finally, a third arm of hydrocarbon anchors this additive into the fabric.

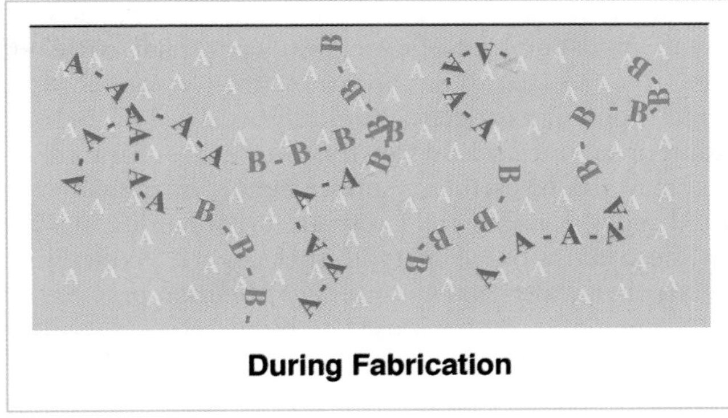

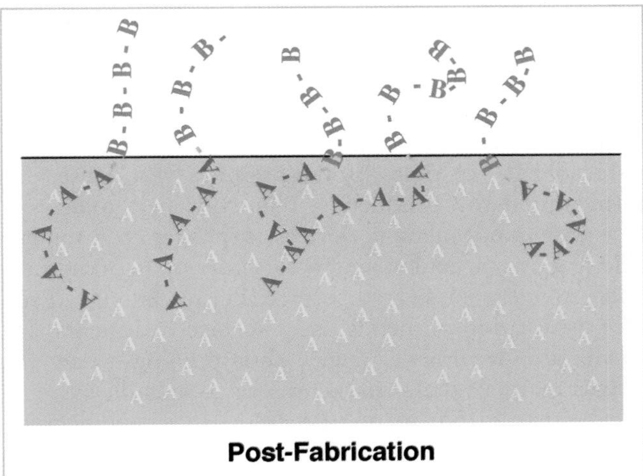

FIGURE I.2.12.10 A block copolymer surface-modifying additive comprised of an "A" block and a "B" block is blended into a support polymer (the bulk) with a composition similar to "A" block. During fabrication, the block copolymer is randomly distributed throughout the support polymer. After curing or annealing, the "A" block anchors the surface-modifying additive into the support, while the low-energy "B" block migrates to the air–polymer interface.

Many SMAs for inorganic systems are known. For example, very small quantities of nickel will completely alter the structure of a silicon (111) surface (Wilson and Chiang, 1987). Copper or silver will accumulate at the surface of gold alloys (Tanaka et al., 1988). Also, in stainless steels, chromium will concentrate (as the oxide) at the surface, imparting corrosion resistance.

There are a number of additives that spontaneously surface-concentrate, but are not necessarily designed as SMAs. A few examples for polymers include PDMS, some extrusion lubricants (Ratner, 1983), and some UV stabilizers (Tyler et al., 1992). The presence of such additives at the surface of a polymer may be unplanned, and they will not necessarily form stable, durable surface layers. However, they can significantly contribute (either positively or negatively) to the bioresponse to the surface.

Conversion Coatings

Conversion coatings modify the surface of a metal into a dense oxide-rich layer that imparts corrosion protection (Sørensen et al., 2009), enhanced adhesivity, altered appearance (e.g., color), and sometimes lubricity to the metal. For example, steel is frequently phosphated (treated with phosphoric acid) or chromated (treated with chromic acid). Aluminum is electrochemically anodized in chromic, oxalic or sulfuric acid electrolytes. Electrochemical anodization may also be useful for surface modifying titanium and Ti-Al alloys (Kasemo and Lausmaa, 1985; Bardos, 1990).

The conversion of metallic surfaces to "oxide-like," electrochemically passive states is a common practice for base metal alloy systems used as biomaterials. Standard and recommended techniques have been published (e.g., ASTM F4-86), and are relevant for most musculoskeletal loadbearing surgical implant devices. The background literature supporting these surface passivation technologies has been summarized (von Recum, 1986).

Base metal alloy systems, in general, are subject to electrochemical corrosion ($M * M^+ + e^-$) within saline environments. The rate of this corrosion process is reduced 10^3–10^6 times by the presence of a dense, uniform, minimally conductive, relatively inert oxide surface. For many metallic devices, exposure to a mineral acid (e.g., nitric acid in water) for times up to 30 minutes will provide a passivated surface. Plasma-enhanced surface passivation of metals, laser surface treatments, and mechanical treatments (e.g., "shot peening") can also impart many of these characteristics to metallic systems.

The reason that many of these surface modifications are called "oxide-like" is that the structure is complex, including OH, H, and subgroups that may or may not be crystalline. Since most passive surfaces are thin films (5–500 nm), and are transparent or metallic in color, the surface appears similar before and after passivation. Further details on surfaces of this type can be found in Chapters I.1.5, I.2.3, and II.4.4.

Parylene Coating

Parylene (para-xylylene) coatings occupy a unique niche in the surface modification literature, because of their wide application to biomedical devices and the good quality of the thin film coatings formed (Loeb et al., 1977a; Nichols et al., 1984). The deposition method is also unique and involves the simultaneous evaporation, pyrolysis, deposition, and polymerization of the monomer, di-para-xylylene (DPX), according to the following reaction:

Di-para-xylylene → Para-xylylene → Poly (para-xylylene)

1) vaporize 2) pyrolyze 3) deposit

The DPX monomer is vaporized at 175°C and 1 torr, pyrolyzed at 700°C and 0.5 torr, and finally deposited on a substrate at 25°C and 0.1 torr. The coating has excellent electrical insulation and moisture barrier properties, and has been used for protection of implant electrodes (Loeb et al., 1977b; Nichols et al., 1984) and implanted electronic circuitry (Spivack and Ferrante, 1969). A parylene coating has been used on stainless steel cardiovascular stents between the metal and a drug-eluting polymer layer (see Chapters II.5.3.B and II.5.18).

Laser Methods

Lasers can rapidly and specifically induce surface changes (roughness, crystallinity, chemistry) in organic and inorganic materials (Picraux and Pope, 1984; Dekumbis, 1987; Chrisey et al., 2003). The advantages of using lasers for such modifications are the precise control of the frequency of the light, the wide range of frequencies available, the high energy density, the ability to focus and raster the light, the possibilities for using both heat and specific excitation to effect change, and the ability to pulse the source and control reaction time. Lasers commonly used for surface modification include ruby, neodymium:yttrium aluminum garnet (Nd:YAG), argon, argon-fluoride eximer, and CO_2. Treatments are pulsed (nanosecond to picoseconds pulse times) and continuous wave (CW), with interaction times often less than 1 msec. Laser-induced surface alterations include annealing, etching, deposition, and polymerization. Polymers, metals, ceramics, and even tooth dentin have been effectively surface modified using laser energy. The major considerations in designing a laser surface treatment include the absorption (coupling) between the laser energy and the material, the penetration depth of the laser energy into the material, the interfacial reflection and scattering, and heating induced by the laser.

Hundreds of papers have been published on laser surface modification. Examples include: (1) a Nd:YAG laser was used to treat titanium implants permitting improved hydroxyapatite nucleation and more rapid integration into bone (Faeda et al., 2009); (2) laser irradiation at 810 nm of silicone rubber induced a "lotus leaf-like" texture that imparted superhydrophobic properties to the silicone polymer (Yoon et al., 2008); (3) eximer laser irradiation at 193 nm permitted fine spatial patterning of polystyrene for improved protein adsorption and cell interactions (Pfleging et al., 2009).

Patterning

Essentially all of the surface modification methods described in this chapter can be applied to biomaterial surfaces, either as a uniform surface treatment or as patterns on the surface with length scales of millimeters, microns or even nanometers. There are many options to pattern biomaterial surfaces. These include ion beam etching, electron beam lithography, laser methods, inkjet printers, and stochastic patterns made by phase separation of two components (Takahara et al., 2000). Chapter I.2.13 elaborates on methods to pattern biomaterials.

CONCLUSIONS

Surface modifications are widely applied to modulate or enhance the biointeractions of biomedical devices and improve other aspects of performance. Surface modification is used in the laboratory setting, and also for the treatment of medical devices used in the clinic. Since a medical device may already have appropriate physical properties and be well-understood in the clinic, surface modification provides a means to alter only the biointeractions of the device without the need for redesign, retooling for manufacture, and retraining of medical personnel.

BIBLIOGRAPHY

Archibald, D. D., Qadri, S. B., & Gaber, B. P. (1996). Modified calcite deposition due to ultrathin organic films on silicon substrates. *Langmuir*, **12**, 538–546.

Ariga, K., Hill, J. P., & Ji, Q. (2007). Layer-by-layer assembly as a versatile bottom-up nanofabrication technique for exploratory research and realistic application. *Phys. Chem. Chem. Phys.*, **9**(19), 2319–2340.

Arkles, B. (1977). Tailoring surfaces with silanes. *Chemtech*, **7**, 766–778.

Bardos, D. I. (1990). Titanium and titanium alloys. In E. Williams, R. W. Cahn, & M. B. Bever (Eds.), *Concise Encyclopedia of Medical and Dental Materials* (1st ed., pp. 360–365). Oxford, UK: Pergamon Press.

Bird, R. R., Hall, B., Hobbs, K. E.F., & Chapman, D. (1989). New haemocompatible polymers assessed by thrombelastography. *J. Biomed. Eng.*, **11**, 231–234.

Boyer, C., Bulmus, V., Davis, T., Ladmiral, V., Liu, J., et al. (2009). Bioapplications of RAFT polymerization. *Chem. Rev.*, **109**, 5402–5436.

Buchanan, R. A., Lee, I. S., & Williams, J. M. (1990). Surface modification of biomaterials through noble metal ion implantation. *J. Biomed. Mater. Res.*, **24**, 309–318.

Chapiro, A. (1983). Radiation grafting of hydrogels to improve the thrombo-resistance of polymers. *Eur. Polym. J.*, **19**, 859–861.

Charlson, E. J., Charlson, E. M., Sharma, A. K., & Yasuda, H. K. (1984). Electrical properties of glow-discharge polymers, parylenes, and composite films. *J. Appl. Polym. Sci. Appl. Polym. Symp.*, **38**, 137–148.

Chiang, Y. C., Chang, Y., Higuchi, A., Chen, W. Y., & Ruaan, R. C. (2009). Sulfobetaine-grafted poly (vinylidene fluoride) ultrafiltration membranes exhibit excellent antifouling property. *J. Membrane Sci.*, **339**(1), 151–159.

Chilkoti, A., Ratner, B. D., & Briggs, D. (1991a). Plasma-deposited polymeric films prepared from carbonyl-containing volatile precursors: XPS chemical derivatization and static SIMS surface characterization. *Chem. Mater.*, **3**, 51–61.

Chilkoti, A., Ratner, B. D., & Briggs, D. (1991b). A static secondary ion mass spectrometric investigation of the surface structure of organic plasma-deposited films prepared from stable isotope-labeled precursors. Part I. Carbonyl precursors. *Anal. Chem.*, **63**, 1612–1620.

Chilkoti, A., Ratner, B. D., Briggs, D., & Reich, F. (1992). Static secondary ion mass spectrometry of organic plasma deposited films created from stable isotope-labeled precursors. Part II. Mixtures of acetone and oxygen. *J. Polym. Sci., Polym. Chem. Ed.*, **30**, 1261–1278.

Cho, C. S., Takayama, T., Kunou, M., & Akaike, T. (1990). Platelet adhesion onto the Langmiur–Blodgett film of poly(gamma-benzyl L-glutamate)-poly(ethylene oxide)-poly(gamma-benzyl L-glutamate) block copolymer. *J. Biomed. Mater. Res.*, **24**, 1369–1375.

Chrisey, D. B., Piqué, A., McGill, R. A., Horowitz, J. S., Ringeisen, B. R., et al. (2003). Laser deposition of polymer and biomaterial films. *Chem. Rev.*, **103**, 553–576.

Chu, P. K., Chen, J. Y., Wang, L. P., & Huang, N. (2002). Plasma surface modification of biomaterials. *Mater. Sci. Eng. Reports*, **36**, 143–206.

Colligon, J. S. (1986). Surface modification by ion beams. *Vacuum*, **36**, 413–418.

Cometa, S., Chiellini, F., Bartolozzi, I., Chiellini, E., De Giglio, E., et al. (2010). Surface segregation assessment in poly(ε-caprolactone)-poly(ethylene glycol) multiblock copolymer films. *Macromol. Biosci.*, **10**(3), 317–327.

Dargaville, T. R., George, G. A., Hill, D. J.T., & Whittaker, A. K. (2003). High energy radiation grafting of fluoropolymers. *Prog. Polym. Sci.*, **28**, 1355–1376.

Decher, G. (1996). Layered nanoarchitectures via directed assembly of anionic and cationic molecules. In J. -P. Sauvage, & M. W. Hosseini (Eds.), *Comprehensive Supramolecular Chemistry Templating, Self-Assembly and Self-Organization* (Vol. 9, pp. 507–528). Oxford, UK: Pergamon Press.

Dekumbis, R. (1987). Surface treatment of materials by lasers. *Chem. Eng. Prog.*, **83**, 23–29.

Deng, J., Wang, L., Liu, L., & Yang, W. (2009). Developments and new applications of UV-induced surface graft polymerizations. *Prog. Polym. Sci.*, **34**(2), 156–193.

Desmet, T., Morent, R., Geyter, N. D., Leys, C., Schacht, E., et al. (2009). Nonthermal plasma technology as a versatile strategy for polymeric biomaterials surface modification: A review. *Biomacromolecules*, **10**(9), 2351–2378.

Dunkirk, S. G., Gregg, S. L., Duran, L. W., Monfils, J. D., Haapala, J. E., et al. (1991). Photochemical coatings for the prevention of bacterial colonization. *J. Biomater. Appl.*, **6**(2), 131–156.

Everhart, D. S., & Reilley, C. N. (1981). Chemical derivatization in electron spectroscopy for chemical analysis of surface functional groups introduced on low-density polyethylene film. *Anal. Chem.*, **53**, 665–676.

Faeda, R. S., Tavares, H. S., Sartori, R., Guastaldi, A. C., & Marcantonio, E. (2009). Biological performance of chemical hydroxyapatite coating associated with implant surface modification by laser beam: Biomechanical study in rabbit tibias. *J. Oral Maxil. Surg.*, **67**(8), 1706–1715.

Ferretti, S., Paynter, S., Russell, D. A., & Sapsford, K. E. (2000). Self-assembled monolayers: A verstile tool for the formulation of bio-surfaces. *Trends Analyt. Chem.*, **19**(9), 530–540.

Filler, M. A., & Bent, S. F. (2003). The surface as molecular reagent: Organic chemistry at the semiconductor interface. *Prog. Surf. Sci.*, **73**, 1–56.

Fleischmann, S., Hinrichs, K., Oertel, U., Reichelt, S., Eichhorn, K. J., et al. (2008). Modification of polymer surfaces by click chemistry. *Macromol. Rapid Comm.*, **29**(12–13), 1177–1185.

Folch, A., & Toner, M. (2000). Microengineering of cellular interactions. *Ann. Rev. Bioeng.*, **2**, 227–256.

Garbassi, F., Morra, M., Occhiello, E., Barino, L., & Scordamaglia, R. (1989). Dynamics of macromolecules: A challenge for surface analysis. *Surf. Interface Anal.*, **14**, 585–589.

Goessl, A., Garrison, M. D., Lhoest, J., & Hoffman, A. S. (2001). Plasma lithography: Thin-film patterning of polymeric biomaterials by RF plasma polymerization I: Surface preparation and analysis. *J. Biomater. Sci. Polymer*, **12**(7), 721–738.

Golden, A. L., Battrell, C. F., Pennell, S., Hoffman, A. S., Lai, J. J., et al. (2010). Simple fluidic system for purifying and concentrating diagnostic biomarkers using stimuli-responsive antibody conjugates and membranes. *Bioconj. Chem.*, **21**, 1820–1826.

Gombotz, W. R., & Hoffman, A. S. (1988). Functionalization of polymeric films by plasma polymerization of allyl alcohol and allylamine. *J. Appl. Polym. Sci. Appl. Polym. Symp.*, **42**, 285–303.

Greer, R. T., Knoll, R. L., & Vale, B. H. (1979). Evaluation of tissue-response to hydrogel composite materials. *SEM*, **2**, 871–878.

Griesser, H. J., & Chatelier, R. C. (1990). Surface characterization of plasma polymers from amine, amide and alcohol monomers. *J. Appl. Polym. Sci. Appl. Polym. Symp.*, **46**, 361–384.

Gruner, H. (2001). Thermal spray coating on titanium. In D. M. Brunette, P. Tengvall, M. Textor, & P. Thomsen (Eds.), *Titanium in Medicine*. Berlin: Springer-Verlag.

Haensch, C., Hoeppener, S., & Schubert, U. S. (2010). Chemical modification of self-assembled silane based monolayers by surface reactions. *Chem. Soc. Rev.*, **39**(6), 2323–2334.

Haque, Y., & Ratner, B. D. (1988). Role of negative ions in the RF plasma deposition of fluoropolymer films from perfluoropropane. *J. Polym. Sci., Polym. Phys. Ed.*, **26**, 1237–1249.

Hayward, J. A., & Chapman, D. (1984). Biomembrane surfaces as models for polymer design: The potential for haemocompatibility. *Biomaterials*, **5**, 135–142.

Heens, B., Gregoire, C., Pireaux, J. J., Cornelio, P. A., Gardella, J. A. Jr. (1991). On the stability and homogeneity of Langmuir–Blodgett films as models of polymers and biological materials for surface studies: An XPS study. *Appl. Surf. Sci.*, **47**, 163–172.

Heid, S., Effenberger, F., Bierbaum, K., & Grunze, M. (1996). Self-assembled mono- and multilayers of terminally functionalized organosilyl compounds on silicon substrates. *Langmuir*, **12**(8), 2118–2120.

Hendricks, S. K., Kwok, C., Shen, M., Horbett, T. A., Ratner, B. D., et al. (2000). Plasma-deposited membranes for controlled release of antibiotic to prevent bacterial adhesion and biofilm formation. *J. Biomed. Mater. Res.*, **50**(2), 160–170.

Hickman, J. J., & Stenger, D. A. (1994). Interactions of cultured neurons with defined surfaces. In D. A. Stenger, & T. N. Mckenna (Eds.), *Enabling Technologies for Cultured Neural Networks* (pp. 51–76). SanDiego, CA: Academic Press.

Hoffman, A. S. (1981). A review of the use of radiation plus chemical and biochemical processing treatments to prepare novel biomaterials. *Radiat. Phys. Chem.*, **18**, 323–342.

Hoffman, A. S. (1988). Biomedical applications of plasma gas discharge processes. *J. Appl. Polym. Sci. Appl. Polym. Symp.*, **42**, 251–267.

Hoffman, A. S., Schmer, G., Harris, C., & Kraft, W. G. (1972). Covalent binding of biomolecules to radiation-grafted hydrogels on inert polymer surfaces. *Trans. Am. Soc. Artif. Int. Organs*, **18**, 10–17.

Hoffman, A. S., Cohn, D. C., Hanson, S. R., Harker, L. A., Horbett, T. A., et al. (1983). Application of radiation-grafted hydrogels as blood-contacting biomaterials. *Radiat. Phys. Chem.*, **22**, 267–283.

Horbett, T. A., & Hoffman, A. S. (1975). Bovine plasma protein adsorption on radiation-grafted hydrogels based on hydroxyethyl methacrylate and N-vinyl-pyrrolidone. In R. E. Baier (Ed.), *Applied Chemistry at Protein Interfaces, Advances in Chemistry Series* (pp. 230–254). Washington, DC: American Chemical Society.

Hu, J., Wang, M., Weier, U. G., Frantz, P., Kolbe, W., et al. (1996). Imaging of single extended DNA molecules on flat (aminopropyl)triethozysilane-mica by atomic force microscopy. *Langmuir*, **12**(7), 1697–1700.

Hucknall, A., Simick, A. J., Hill, R. T., Chilkoti, A., Garcia, A., et al. (2009). Versatile synthesis and micropatterning of nonfouling polymer brushes on the wafer scale. *Biointerphases*, **4**(2), FA50–FA57.

Inagaki, N., Nakanishi, T., & Katsuura, K. (1983). Glow discharge polymerizations of tetrafluoroethylene, perfluoromethylcyclohexane and perfluorotoluene investigated by infrared spectroscopy and ESCA. *Polym. Bull.*, **9**, 502–506.

Jenney, C. R., & Anderson, J. M. (1999). Alkylsilane-modified surfaces: Inhibition of human macrophage adhesion and foreign body giant cell formation. *J. Biomed. Mater. Res.*, **46**, 11–21.

Johnston, E. E., & Ratner, B. D. (1996). XPS and SSIMS characterization of surfaces modified by plasma deposited oligo(glyme) films. In B. D. Ratner, & D. G. Castner (Eds.), *Surface Modification of Polymeric Biomaterials* (pp. 35–44). New York, NY: Plenum Press.

Kaplan, S., & Dilks, A. (1981). A solid state nuclear magnetic resonance investigation of plasma-polymerized hydrocarbons. *Thin Solid Films*, **84**, 419–424.

Kasemo, B., & Lausmaa, J. (1985). Metal selection and surface characteristics. In P. I. Branemark, G. A. Zarb, & T. Albrektsson (Eds.), *Tissue-Intergrated Prostheses* (pp. 99–116). Chicago, Il: Quintessence Publishing.

Kitching, K. J., Pan, V., & Ratner, B. D. (2003). Biomedical applications of plasma-deposited thin films. In H. Biederman (Ed.), *Plasma Polymer Films*. London, UK: Imperial College Press.

Klages, C. -P., Höpfner, K., & Kläke, N. (2000). Surface functionalization at atmospheric pressure by dbd-based pulsed plasma polymerization. *Plasmas Polym.*, **5**, 79–89.

Knobler, C. M. (1990). Recent developments in the study of monolayers at the air–water interface. *Adv. Chem. Phys.*, **77**, 397–449.

Knoll, W. (1996). Self-assembled microstructures at interfaces. *Curr. Opin. Coll. & Interface Sci.*, **1**, 137–143.

Knoll, W., Naumann, R., Friedrich, M., Robertson, J., Lösche, M., et al. (2008). Solid supported lipid membranes: New concepts for the biomimetic functionalization of solid surfaces. *Biointerphases*, **3**(2), FA125–FA135.

Krishnamurthy, V., Kamel, I. L., & Wei, Y. (1989). Analysis of plasma polymerization of allylamine by FTIR. *J. Polym. Sci., Polym. Chem. Ed.*, **27**, 1211–1224.

Kwon, O. H., Kikuchi, A., Yamato, M., Sakuri, Y., & Okano, T. (2000). Rapid cell sheet detachment from poly(n-isopropylacrylamide)-grafted porous cell culture membranes. *J. Biomed. Mater. Res.*, **50**, 82–89.

Lee, S. -W., & Laibinis, P. E. (1998). Protein-resistant coatings for glass and metal oxide surfaces derived from oligo(ethylene glycol)-terminated alkyltrichlorosilane. *Biomaterials*, **19**, 1669–1675.

Lewandowska, K., Balachander, N., Sukenik, C. N., & Culp, L. A. (1989). Modulation of fibronectin adhesive functions for fibroblasts and neural cells by chemically derivatized substrata. *J. Cell. Physiol.*, **141**, 334–345.

Li, M., Chen, M., Sheepwash, E., Brosseau, C. L., Li, H., et al. (2008). AFM studies of solid-supported lipid bilayers formed at an Au (111) electrode surface using vesicle fusion and a combination of Langmuir–Blodgett and Langmuir–Schaefer techniques. *Langmuir*, **24**(18), 10313–10323.

Loeb, G. E., Bak, M. J., Salcman, M., & Schmidt, E. M. (1977a). Parylene as a chronically stable, reproducible microelectrode insulator. *IEEE Trans. Biomed. Eng., BME-24*, 121–128.

Loeb, G. E., Walker, A. E., Uematsu, S., & Konigsmark, B. W. (1977b). Histological reaction to various conductive and dielectric films chronically implanted in the subdural space. *J. Biomed. Mater. Res.*, **11**, 195–210.

Lopez, G. P., & Ratner, B. D. (1991). Substrate temperature effects of film chemistry in plasma deposition of organics. I. Nonpolymerizable precursors. *Langmuir*, **7**, 766–773.

Lopez, G. P., Ratner, B. D., Rapoza, R. J., & Horbett, T. A. (1993). Plasma deposition of ultrathin films of poly(2-hydroxyethyl methacrylate): Surface analysis and protein adsorption measurements. *Macromolecules*, **26**, 3247–3253.

Ma, Z., Mao, Z., & Gao, C. (2007). Surface modification and property analysis of biomedical polymers used for tissue engineering. *Colloids Surf. B: Biointerfaces*, **60**(2), 137–157.

Mao, G., Castner, D. G., & Grainger, D. W. (1997). Polymer immobilization to alkylchlorosilane organic monolayer films using sequential derivatization reactions. *Chem. Mater.*, **9**(8), 1741–1750.

Maoz, R., Netzer, L., Gun, J., & Sagiv, J. (1988). Self-assembling monolayers in the construction of planned supramolecular structures and as modifiers of surface properties. *J. Chim. Phys.*, **85**, 1059–1064.

Massines, F., Gherardi, N., & Sommer, F. (2000). Silane-based coatings of polypropylene, deposited by atmospheric pressure glow discharge plasmas. *Plasmas Polym.*, **5**, 151–172.

Matsuda, T., & Inoue, K. (1990). Novel photoreactive surface modification technology for fabricated devices. *Trans. Am. Soc. Artif. Int. Organs*, **36**, M161–M164.

Matsuzawa, M., Umemura, K., Beyer, D., Sugioka, K., & Knoll, W. (1997). Micropatterning of neurons using organic substrates in culture. *Thin Solid Films*, **305**, 74–79.

Meller, P., Peters, R., & Ringsdorf, H. (1989). Microstructure and lateral diffusion in monolayers of polymerizable amphiphiles. *Coll. Polym. Sci.*, **267**, 97–107.

Mrksich, M., & Whitesides, G. M. (1996). Using self-assembled monolayers to understand the interactions of manmade surfaces with proteins and cells. *Annu. Rev. Biophys. Biomol. Struct.*, **25**, 55–78.

Nastasi, M., Mayer, J., & Hirvonen, J. K. (1996). *Ion–Solid Interactions: Fundamentals and Applications*. Cambridge, UK: Cambridge University Press.

Nguyen, S. V. (1986). Plasma assisted chemical vapor deposited thin films for microelectronic applications. *J. Vac. Sci. Technol. B.*, **4**, 1159–1167.

Nichols, M. F., Hahn, A. W., James, W. J., Sharma, A. K., & Yasuda, H. K. (1984). Evaluating the adhesion characteristics of glow-discharge plasma-polymerized films by a novel voltage cycling technique. *J. Appl. Polym. Sci. Appl. Polym. Symp.*, **38**, 21–33.

Pan, Y. V., Wesley, R. A., Luginbuhl, R., Denton, D. D., & Ratner, B. D. (2001). Plasma polymerized n-isopropylacylamide: Synthesis and characterization of a smart thermally responsive coating. *Biomacromolecules*, **2**(1), 32–36.

Panchalingam, V., Poon, B., Huo, H. H., Savage, C. R., Timmons, R. B., et al. (1993). Molecular surface tailoring of biomaterials via pulsed RF plasma discharges. *J. Biomater. Sci. Polymer. Edn.*, **5**(1/2), 131–145.

Pappas, D. (2011). Status and potential of atmospheric plasma processing of materials. *J. Vac. Sci. Technol. A: Vacuum, Surfaces, and Films*, **29**. 020801-1-17.

Park, E. J., Carroll, G. T., Turro, N. J., & Koberstein, J. T. (2009). Shedding light on surfaces: Using photons to transform and pattern material surfaces. *Soft Matter*, **5**(1), 36.

Park, J. Y., & Advincula, R. C. (2011). Nanostructuring polymers, colloids, and nanomaterials at the air–water interface through Langmuir and Langmuir–Blodgett techniques. *Soft Matter*, **7**(21), 9829–9843.

Pfleging, W., Torge, M., Bruns, M., Trouillet, V., Welle, A., et al. (2009). Laser-and UV-assisted modification of polystyrene surfaces for control of protein adsorption and cell adhesion. *Appl. Surf. Sci.*, **255**(10), 5453–5457.

Picraux, S. T., & Pope, L. E. (1984). Tailored surface modification by ion implantation and laser treatment. *Science*, **226**, 615–622.

Plueddemann, E. P. (1980). Chemistry of silane coupling agents. In D. E. Leyden (Ed.), *Silylated Surfaces* (pp. 31–53). New York, NY: Gordon and Breach Science Publishers.

Pomerantz, M., Segmuller, A., Netzer, L., & Sagiv, J. (1985). Coverage of Si substrates by self-assembling monolayers and multilayers as measured by IR, wettability and x-ray diffraction. *Thin Solid Films*, **132**, 153–162.

Ratner, B. D. (1980). Characterization of graft polymers for biomedical applications. *J. Biomed. Mater. Res.*, **14**, 665–687.

Ratner, B. D. (1983). ESCA studies of extracted polyurethanes and polyurethane extracts: Biomedical implications. In K. L. Mittal (Ed.), *Physicochemical Aspects of Polymer Surfaces* (pp. 969–983). New York, NY: Plenum Publishing.

Ratner, B. D., & Hoffman, A. S. (1980). Surface grafted polymers for biomedical applications. In M. Szycher, & W. J. Robinson (Eds.), *Synthetic Biomedical Polymers. Concepts and Applications* (pp. 133–151). Westport, CT: Technomic Publishing.

Ratner, B. D., & Yoon, S. C. (1988). Polyurethane surfaces: Solvent and temperature induced structural rearrangements. In J. D. Andrade (Ed.), *Polymer Surface Dynamics* (pp. 137–152). New York, NY: Plenum Press.

Ratner, B. D., Horbett, T. A., Hoffman, A. S., & Hauschka, S. D. (1975). Cell adhesion to polymeric materials: Implications with respect to biocompatibility. *J. Biomed. Mater. Res.*, **9**, 407–422.

Ratner, B. D., Hoffman, A. S., Hanson, S. R., Harker, L. A., & Whiffen, J. D. (1979). Blood compatibility-water content relationships for radiation grafted hydrogels. *J. Polym. Sci., Polym. Symp.*, **66**, 363–375.

Ratner, B. D., Chilkoti, A., & Lopez, G. P. (1990). Plasma deposition and treatment for biomaterial applications. In R. D'Agostino (Ed.), *Plasma Deposition, Treatment and Etching of Polymers* (pp. 463–516). San Diego, CA: Academic Press.

Rautray, T. R., Narayanan, R., Kwon, T. -Y., & Kim, K. -H. (2010). Surface modification of titanium and titanium alloys by ion implantation: A review. *J. Biomed. Mater. Res.*, **93B**, 581–591.

Raynor, J. E., Capadona, J. R., Collard, D. M., Petrie, T. A., & García, A. J. (2009). Polymer brushes and self-assembled monolayers: Versatile platforms to control cell adhesion to biomaterials (Review). *Biointerphases*, **4**(2), FA3–FA16.

Rye, R. R., Nelson, G. C., & Dugger, M. T. (1997). Mechanistic aspects of alkylchlorosilane coupling reactions. *Langmuir*, **13**(11), 2965–2972.

Safrany, A. (1997). Radiation processing: Synthesis and modification of biomaterials for medical use. *Nucl. Instrum. Methods Phys. Res., Sect. B.*, **131**, 376–381.

Sara, M., & Sleytr, U. B. (1996). Crystalline bacterial cell surface layers (S-layers): From cell structure to biomimetics. *Prog. Biophys. Mol. Biol.*, **65**(1/2), 83–111.

Sawin, H. H., & Reif, R. (1983). A course on plasma processing in integrated circuit fabrication. *Chem. Eng. Ed.*, **17**, 148–152.

Schlecht, C. A., & Maurer, J. A. (2011). Functionalization of glass substrates: Mechanistic insights into the surface reaction of trialkoxysilanes. *RSC Adv.*, **1**(8), 1446–1448.

Sebra, R. P., Reddy, S. K., Masters, K. S., Bowman, C. N., & Anseth, K. S. (2007). Controlled polymerization chemistry to graft architectures that influence cell–material interactions. *Acta Biomater.*, **3**(2), 151–161.

Sheu, M. -S., Hoffman, A. S., & Feijen, J. (1992). A glow discharge process to immobilize PEO/PPO surfactants for wettable and non-fouling biomaterials. *J. Adhes. Sci. Technol.*, **6**, 995–1101.

Singh, A., & Silverman, J. (Eds.), (1992). *Radiation Processing of Polymers*. New York, NY: Oxford University Press.

Sioshansi, P. (1987). Surface modification of industrial components by ion implantation. *Mater. Sci. Eng.*, **90**, 373–383.

Somorjai, G. A. (1990). Modern concepts in surface science and heterogeneous catalysis. *J. Phys. Chem.*, **94**, 1013–1023.

Somorjai, G. A. (1991). The flexible surface. Correlation between reactivity and restructuring ability. *Langmuir*, **7**, 3176–3182.

Sørensen, P. A., Kiil, S., Dam-Johansen, K., & Weinell, C. E. (2009). Anticorrosive coatings: A review. *J. Coat. Technol. Res.*, **6**(2), 135–176.

Spivack, M. A., & Ferrante, G. (1969). Determination of the water vapor permeability and continuity of ultrathin parylene membranes. *J. Electrochem. Soc.*, **116**, 1592–1594.

Stannett, V. T. (1990). Radiation grafting: State-of-the-art. *Radiat. Phys. Chem.*, **35**, 82–87.

Swanson, M. J. (1996). A unique photochemical approach for polymer surface modification. In K. L. Mittal, & K. W. Lee (Eds.), *Polymer Surfaces and Interfaces: Characterization, Modification and Application*. The Netherlands: VSP.

Takahara, A., Ge, S., Kojio, K., & Kajiyama, T. (2000). *In situ* atomic force mircroscopic observation of albumin adsorption onto phase-separated organosilane monolayer surface. *J. Biomater. Sci. Polymer. Edn.*, **11**(1), 111–120.

Tanaka, T., Atsuta, M., Nakabayashi, N., & Masuhara, E. (1988). Surface treatment of gold alloys for adhesion. *J. Prosthet. Dent.*, **60**, 271–279.

Tsuji, H., Satoh, H., Ikeda, S., Ikemoto, N., Gotoh, Y., et al. (1998). Surface modification by silver-negative-ion implantation for controlling cell-adhesion properties of polystyrene. *Surf. Coat. Technol.*, **103-104**, 124–128.

Tyler, B. J., Ratner, B. D., Castner, D. G., & Briggs, D. (1992). Variations between Biomer™ lots. 1. Significant differences in the surface chemistry of two lots of a commercial polyetherurethane. *J. Biomed. Mater. Res.*, **26**, 273–289.

Ulman, A. (1990). Self-assembled monolayers of alkyltrichlorosilanes: Building blocks for future organic materials. *Adv. Mater.*, **2**, 573–582.

Ulman, A. (1991). *An Introduction to Ultrathin Organic Films*. Boston, MA: Academic Press.

Vidigal, G. M., Groisman, M., De Sena, L. Á, & De Almeida Soares, G. (2009). Surface characterization of dental implants coated with hydroxyapatite by plasma spray and biomimetic process. *Implant Dent.*, **18**(4), 353–361.

von Recum, A. F. (1986). *Handbook of Biomaterials Evaluation* (1st ed.). New York, NY: Macmillan Publishing Company.

Ward, R. S. (1989). Surface modifying additives for biomedical polymers. *IEEE Eng. Med. Biol.*, 22–25. June.

Wasserman, S. R., Tao, Y. -T., & Whitesides, G. M. (1989). Structure and reactivity of alkylsiloxane monolayers formed by reaction of alkyltrichlorosilanes on silicon substrates. *Langmuir*, **5**, 1074–1087.

Wen, J. M., Gabor, S., Lim, F., & Ward, R. (1997). XPS study of surface composition of a segmented polyurethane block copolymer modified by PDMS end groups and its blends with phenoxy. *Macromolecules*, **30**, 7206–7213.

Whitesides, G. M., Mathias, J. P., & Seto, C. T. (1991). Molecular self-assembly and nanochemistry: A chemical strategy for the synthesis of nanostructures. *Science*, **254**, 1312–1319.

Wilson, R. J., & Chiang, S. (1987). Surface modifications induced by adsorbates at low coverage: A scanning-tunneling-microscopy study of the Ni/Si(111) $\sqrt{19}$ surface. *Phys. Rev. Lett.*, **58**, 2575–2578.

Wu, S. (1982). *Polymer Interface and Adhesion*. New York, NY: Marcel Dekker.

Xiao, S. J., Textor, M., Spencer, N. D., Wieland, M., Keller, B., et al. (1997). Immobilization of the cell-adhesive peptide arg-gly-asp-cys (RGDC) on titanium surfaces by covalent chemical attachment. *J. Mat. Sci.: Mat. Med.*, **8**, 867–872.

Yasuda, H. K. (1979). Competitive ablation and polymerization (CAP) mechanisms of glow discharge polymerization. In M. Shen, & A. T. Bell (Eds.), *ACS Symposium Series 108: Plasma Polymerization* (pp. 37–52). Washington DC: American Chemical Society.

Yasuda, H. K., & Gazicki, M. (1982). Biomedical applications of plasma polymerization and plasma treatment of polymer surfaces. *Biomaterials*, **3**, 68–77.

Yoon, T. O., Shin, H. J., Jeoung, S. C., & Park, Y. I. (2008). Formation of superhydrophobic poly (dimethysiloxane) by ultrafast laser-induced surface modification. *Opt. Express*, **16**(17), 12715–12725.

CHAPTER I.2.13 SURFACE PATTERNING

Ryan T. Hill[1] *and Ashutosh Chilkoti*[2]

[1]Center for Biologically Inspired Materials and Material Systems, Duke University, Durham, NC, USA

[2]Center for Biologically Inspired Materials and Material Systems and Department of Biomedical Engineering, Duke University, Durham, NC, USA

INTRODUCTION

Surface patterning is the use of surface modification methods to create chemically or physically demarcated regions on a surface. The art of patterning surfaces has a long history; the recent antecedent methods to pattern biomolecules are derived, in part, from the microelectronics industry (Xia and Whitesides, 1998; Whitesides et al., 2001; Geissler and Xia, 2004). As hand-wiring small, complex electronic devices became increasingly difficult, new methods were developed to pattern substrates such that many electrical components and interconnects between the different components of these devices could be created on the surface of a relatively large substrate in a batch process. These processes that have been instrumental to the development of all modern electronic devices have been driven by a set of technologies to create micro- and nanoscale features on the surface of semiconductors that are collectively and somewhat loosely termed "photolithography."

Many of these surface patterning techniques, especially photolithography, have become important tools for biomedical research. The unique demands of biomedical research (for example soft hydrated materials such as proteins and hydrogels in an aqueous environment), impose a special set of demands and constraints on patterning methods. Seen from this broader perspective, it might be argued that the conceptual origins of many of the techniques used to pattern biomolecules on surfaces extend back to the pre-industrial era, to techniques that were developed to print on paper and textiles, such as block printing, screen printing, lithography, and batik.

This chapter focuses on surface patterning in the realm of biomedical research. The goal of this chapter is to familiarize the reader with various surface patterning techniques, and how these techniques can be used in biomedical research. First we discuss important figures-of-merit of patterning methods that allow different methods to be compared with each other, and allow a user to decide which patterning method is most appropriate for a specific application. Then, we discuss various types of patterning techniques, and follow up with a brief conclusion.

Some chapters in this textbook that provide background information for this chapter include I.1.5 (Surface Properties and Surface Characterization), I.2.10 (Non-Fouling Surfaces), and I.2.12 (Physico-Chemical Surface Modification of Materials Used in Medicine). However, as will quickly become apparent, this chapter has implications for many other chapters in the book.

COMMON CONCERNS IN BIOMOLECULAR SURFACE PATTERNING

Surface patterning is carried out for a wide variety of reasons and, in turn, the ultimate goal of the patterned substrate defines a set of constraints or concerns that dictate how the fabrication process should be carried out. Surface patterning within the scope of biomaterials research is largely focused on the patterning of biomolecules or the patterning of cells. Listed below are important figures-of-merit (FOMs) and concerns that typically need to be addressed in patterning biomolecules or cells on solid substrates for a specific application. We also briefly discuss how the selection of an appropriate patterning methodology for a particular application is not usually influenced by a single FOM, but is usually dictated by the technique that offers the best combination of several FOMs.

Resolution

Resolution refers to the smallest feature size that can be reliably created by a patterning technique. Although the resolution of new techniques is constantly pushed towards smaller feature sizes and finer control, it is also useful to know what *range* of feature sizes a particular technique is capable of producing. Figure I.2.13.1 shows an example of different size patterns created with a high-resolution technique, as well as patterns created with a lower-resolution technique but one that is capable of a wider range of feature sizes. The highest resolution is not necessarily better, as the desired resolution depends upon the application. Thus, to pattern cells whose dimension are at least several micrometers, a technique with a relatively low resolution in the micrometers range is likely adequate, as long as it can pattern a large enough area, i.e., have a high enough throughput, which is discussed next.

Throughput

While resolution refers to the smallest possible feature size that can be created by a patterning technique, *throughput* is a measure of the surface area that can be patterned in a given time. The throughput of a patterning technique is especially important for its use in manufacturing of devices where it is imperative that the patterned substrates can be produced on large length scales and on a timescale that is useful for manufacturing – minutes to hours, rather than days to weeks. Although the different types of patterning techniques are discussed later in this chapter, here we briefly discuss a few patterning methods to illustrate throughput considerations. Direct-write or "on-the-fly" types of patterning methods are those that use a stylus and ink to pattern a surface. These techniques usually provide precise control over the patterning process and finer feature sizes – often sub-100 nm – than some other printing methods such as microcontact printing, and can often create arbitrary feature shapes. Patterning large areas (i.e., greater than hundreds of square micrometers) rapidly, however, is challenging for most variants of this type of patterning. On the other hand, techniques based on printing using a stamp can readily pattern large areas of a surface quickly (in the order of seconds to minutes), but may only be able to access certain shapes of features (e.g., circles, lines or squares), and may have limited resolution (i.e., on the micrometer scale or greater).

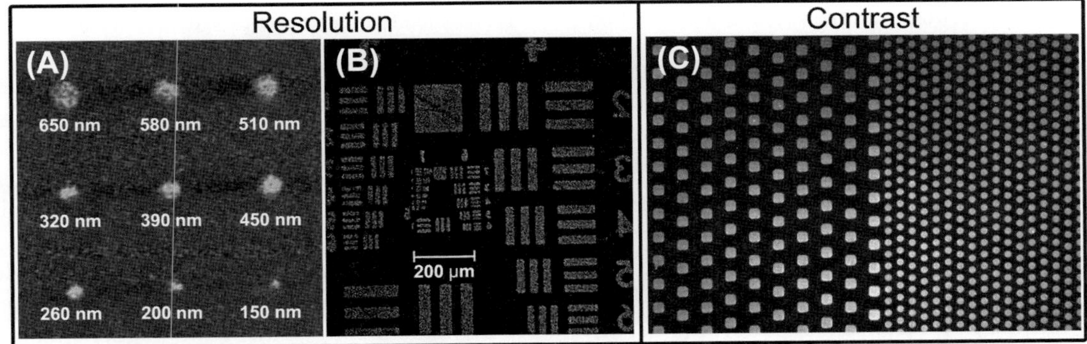

FIGURE I.2.13.1 Examples of Resolution and Contrast in Surface Patterning. (A, B) (Two "Resolution" images): features of different sizes created with a high-resolution patterning technique: (A) Lee et al., 2006a; and relatively lower-resolution technique: (B) Yang et al., 2000. (C) (Two "Contrast" images): the example in (C) represents a higher throughput process, since a large range of feature sizes and a larger surface area can be patterned in a one-patterning process. This is an example of high contrast due to patterning on a non-fouling surface that minimizes background signal (10 μm square features and 5 μm circular features) (Hucknall et al., 2009a). (A) is adapted with permission from Wiley-Blackwell. (B) is adapted with permission from (Yang, Z., Frey, W., Oliver, T., & Chilkoti, A. (2000). Light-activated affinity micropatterning of proteins on self-assembled monolayers on gold. Langmuir, 16(4), 1751–1758). Copyright (2000) American Chemical Society. (C) is adapted and reprinted under the Open Access Creative Commons License with permission from the authors.

Contrast

Contrast is the degree to which the components in the pattern are distinguishable from the background substrate (Figure I.2.13.1). In many surface patterning methodologies, contrast is achieved by patterning one material on the surface of another material. In patterning biomolecules, however, contrast is usually defined as the surface density of the deliberately patterned biomolecule of interest relative to other biomolecules that might adventitiously be bound to the background regions of the same surface, as a consequence of the patterning method. Contrast in patterns of active molecules, such as enzymes or antibodies, can be assessed by measuring the activity of the patterned molecules after the patterning process. For example, enzyme patterns can be visualized by monitoring the concentration of insoluble fluorescent products created by the enzyme (usually via their fluorescence intensity) from the patterned regions relative to the unpatterned background regions when incubated with a soluble, non-fluorescent enzyme substrate. Another example of biomolecular patterning in which the activity of the patterned molecule can be used to measure the contrast is when the patterned molecule has an affinity for another molecule, as is the case for antibody/antigen and many protein–ligand interactions (e.g., avidin-biotin). Here, contrast can be visualized by exposing the pattern to a complementary target molecule containing a fluorescent, radioactive or plasmonic tag that enables visualization of the pattern.

In rare cases, biomolecular patterning involves molecules with a functional activity or activity that is difficult to assess in a purely *in vitro* setting. In these cases, pattern contrast is often visualized by exposure to a secondary processing step for pattern visualization. For example, contrast in patterned peptides that act as cues for cell adhesion can be visualized by exposing the pattern to cells in culture, and the degree of cell patterning is an indirect measure of the fidelity and contrast of the pattern.

Contrast in biomolecular patterning is heavily dependent on surface properties of not only the pattern itself, but also of the background. Biomolecular patterns are often exposed to solutions that can contain complex mixtures of proteins and other molecules in the course of their use, and most surfaces are particularly prone to *nonspecific* and *irreversible adsorption* of proteins. Removing nonspecifically adsorbed proteins requires the use of detergents or other harsh conditions that are typically not suitable for many biological applications. Thus, maintaining contrast in biomolecular patterning usually involves some sort of passivation step to prevent two sets of undesirable events: (1) adsorption of the molecule to be patterned on the unpatterned background regions; (2) adsorption of biomolecules – that are to be subsequently exposed to the patterned surface – onto the patterned features or the background regions of the surface (Figure I.2.13.2).

Non-specific adsorption can be reduced by methods of varying degrees of complexity (also see Chapter I.2.10). Perhaps the most common and simplest way to reduce non-specific adsorption is to use proteins that adsorb to surfaces, such as bovine serum albumin (BSA), under the assumption that controlled deposition of a surface-active protein will prevent the immobilization of most other molecules to the substrate. However, adsorption of BSA often leads to sub-monolayer coverage of the surface, and it has been shown that proteins can adsorb to a BSA layer (Hyun et al., 2001), so that this strategy, despite its widespread use, is of limited utility.

A more effective way of reducing non-specific adsorption is to create a *"non-fouling"* surface. Non-fouling surfaces are typically defined as surfaces that have a molecular or polymer layer that prevents the adsorption of proteins on a surface. A common method to create a non-fouling surface is to form a dense layer of polyethylene glycol (PEG) on the surface. Here, the network of ethylene glycol molecules effectively creates a hydrated barrier between the bulk solution and the solid surface. Proteins in solutions can diffuse into and out of the network without becoming immobilized to the surface. The degree of protein resistance is related to the density of PEG molecules on the surface (Ma et al., 2006a; Hucknall et al., 2009a; Zauscher and Chilkoti, 2009). Unfortunately, most methods to adsorb PEG or chemically attach it to the surface (Lee et al., 1989; Amiji and Park, 1993; Llanos and Sefton, 1993; Prime and Whitesides, 1993; Elbert and Hubbell, 1998; Liu et al., 2002; Xia et al., 2002; Bearinger et al., 2003) do not provide a high enough density of oligoethylene glycol moieties on the surface. In these cases the PEG will decrease, but will not eliminate the adsorption of proteins, especially from highly complex and concentrated protein mixtures such as serum and plasma.

In situ polymerization methods, also termed surface-initiated polymerization, in which an initiator is tethered to a surface and a PEG-like coating is directly grown from the surface, are a powerful method to grow a dense layer of a polymer that presents a high density of oligoethylene glycol moieties at the surface. Such surfaces are highly resistant to protein adsorption (Ma et al., 2004, 2006a,b; Hucknall et al., 2009a,b,c; Zauscher and Chilkoti, 2009), and patterns made on these non-fouling surfaces remain functional for extended periods of time in cell culture (Ma et al., 2004).

Other methods that are also useful to create non-fouling interfaces include self-assembled monolayers of alkanethiols that present an oligoethylene glycol moiety at the solid–water interface (Prime and Whitesides, 1993; Herrwerth et al., 2003), inter-penetrating polymer networks that present PEG at their solvent interface (Bearinger et al., 2003), plasma-deposited "PEG-like" films deposited from oligoglyme precursors (López et al., 1992), and the deposition of PEG-poly(lysine) (PEG-PLL) by electrostatic adsorption (Kenausis et al., 2000;

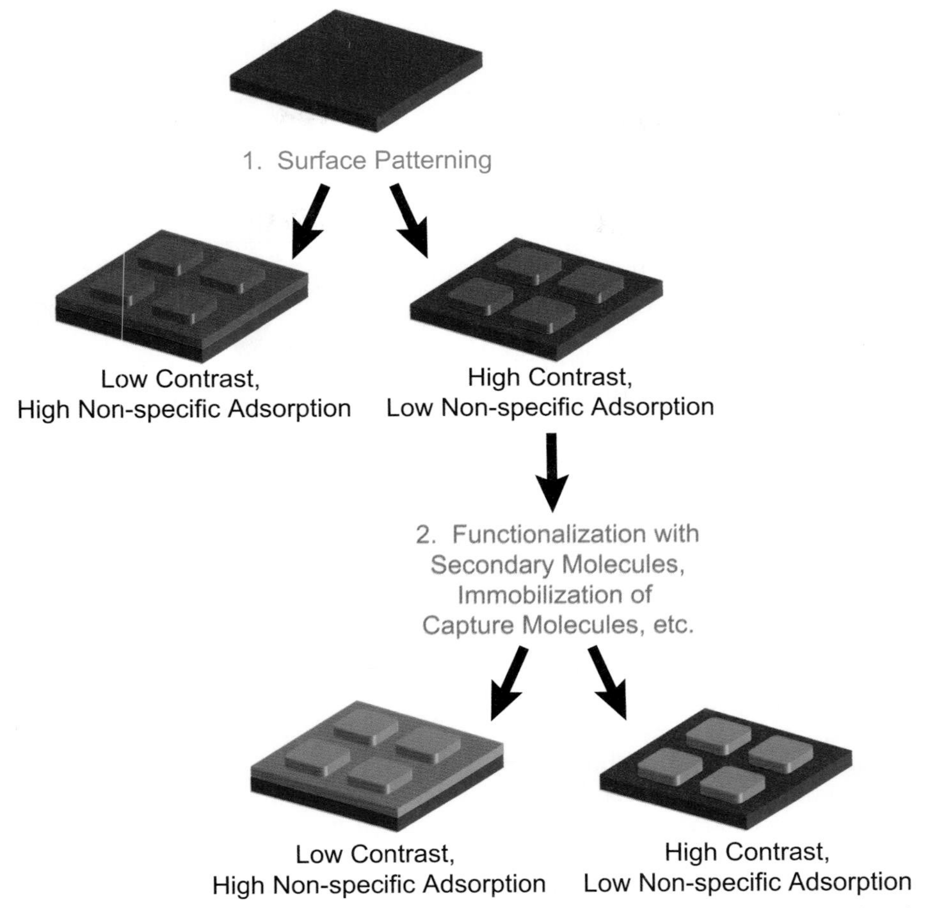

FIGURE I.2.13.2 Contrast and Non-Specific Adsorption. Maintaining contrast during patterning of biomolecules often requires that: (1) the patterned materials have low non-specific adsorption to the background substrate; and (2) that any molecules used in subsequent steps (functionalization of pattern with secondary, reporter molecules; immobilization of antigens or capture molecules from solutions; etc.) also have low non-specific adsorption to the background.

Huang et al., 2001). However, some of these methods are somewhat limited in their utility across a range of polymer substrates. For example, PEG-PLL is only useful for negatively-charged surfaces, and has typically been most successful with metal oxide surfaces that have a negatively-charged surface, while oligoethylene glycol-terminated self-assembled monolayers (SAMs) are only applicable to gold and silver.

Bioactivity

Preserving the bioactivity of a biomolecule during the patterning process is a significant and non-trivial concern. Many patterning processes, especially those that are adapted from the semiconductor processing industry, can involve exposure of the surface to extreme environments, such as high vacuum, extreme temperatures, and non-aqueous conditions, which can denature or degrade biomolecules. Methods that are compatible with aqueous conditions and ambient temperature are hence preferable when deciding upon a method to pattern a biomolecule – especially proteins – on surfaces, as they are especially prone to denaturation compared to other biomolecules such as peptides and oligonucleotides.

Shelf-Life and Durability

The shelf-life and durability of the pattern over time is another consideration when choosing a patterning methodology. This is especially important in the case where patterns need to be generated for use in diagnostic or point-of-care devices. The ultimate convenience in such a situation would be one where a fully functional pattern can be manufactured, stored in ambient conditions for an unspecified amount of time, and then used by the end-user. These conditions can easily be met when patterning "hard" materials, such as those used in microelectronics, because they are typically not very sensitive to their environment. However, when "soft-wet" materials – hydrated polymers, proteins, and cells – are patterned, the exposure of these materials to vacuum, high temperature or extremes of pH becomes a critical and often limiting issue, because many of these materials are easily degraded, destroyed or irreversibly denatured under these conditions. For this reason, special care is needed to ensure that all biological components involved in the patterning process are maintained in a functional state. Ensuring this may involve strategies such as sealing the devices in a hydrated environment so that they are protected from the environment until they are ready for use, using preconditioning steps to prepare the surface prior to analysis or integration of rinsing steps into the analysis scheme. In some instances it may be necessary to create the patterned substrates that act as a template for patterning biomolecules, and then functionalizing the pattern with the biomolecule of interest, on demand, immediately prior to its use. An example of this approach would be the fabrication of a pattern of a non-fouling polymer on a substrate, which can be stored for an extended period of time under ambient conditions. Immediately prior to use, hydration of the surface and incubation in a protein solution drives adsorption of the protein only to the unpatterned regions, leading to the spontaneous formation of a protein pattern driven by the thermodynamic template on the patterned surface. This approach has the advantage of obviating the need to store the surface in a hydrated environment, which would have been necessary if the surface had been patterned with a protein at the outset, rather than on-demand.

In addition, it is also useful to know how long a pattern will stay patterned under the end-use conditions. This is especially important for substrates patterned for cell growth or for implantation into the body that need to retain functionality for extended periods of time. For this reason, researchers will often report an FOM that alludes to the life of the pattern by stating, for example, how long the cells remain confined to the pattern in cell culture.

PATTERNING TECHNIQUES

The number of different surface patterning techniques increases daily, with new publications that present novel patterning schemes or variants of existing methodologies. This chapter seeks to provide a short discussion of the different categories of patterning techniques, and within each category points out some interesting examples; this chapter is not an exhaustive compendium of all patterning techniques.

Much of the overall organization or structure of this section owes an intellectual debt to the excellent and comprehensive review by Geissler and Xia (2004). However, the content of this section places greater emphasis on patterning for biomedical applications. We first present direct-write patterning techniques; these techniques typically provide high resolution at the expense of speed and throughput. Presented next are mask- and master-based patterning techniques, which provide high throughput and can pattern large surface areas, but often have lower resolution compared to direct-write techniques. Next, patterning by self-assembly is described, which is a powerful technique for patterning large surface areas with high-resolution features by exploiting intermolecular interactions. We also present a brief section on the use of dynamic patterning for interrogation of biological environments. Finally, we conclude with a few examples of three-dimensional patterning techniques that can be used to control the three-dimensional microenvironment of cells.

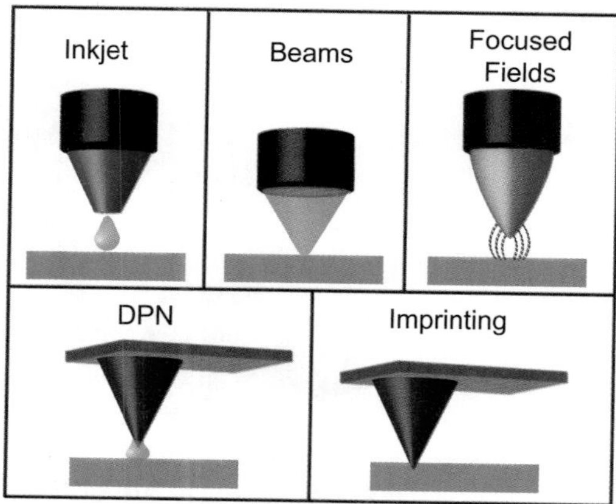

FIGURE I.2.13.3 Types of Direct-Write Patterning. Inkjet, beams, focused fields, dip-pen nanolithography (DPN), and imprinting.

Direct-Write Patterning

In direct-write patterning techniques, patterns are fabricated by serially scanning a patterning element across a substrate. Direct-write patterning is a useful method because patterns of arbitrary feature shape and size – within the resolution limit of the technique – can be fabricated on-the-fly, in a process analogous to writing with a pen. In addition, direct-write patterning can generate extremely high-resolution features with great spatial accuracy. The drawbacks of direct-write patterning techniques are that they are typically slow, low-throughput, and not particularly suitable for large area patterning, because of the need to serially write the pattern. This limitation can be addressed, at the cost of greatly increased complexity of the instrumentation, by using independently actuated pens. Several direct-write techniques are described next and are grouped by the type of "pen" that is used to write the pattern (Figure I.2.13.3).

Writing with a Stylus. This section describes direct-write patterning techniques that use a rigid stylus to write the pattern. Dip-pen nanolithography (DPN), nanoimprinting, and nanoengraving (also called nanoshaving) are typical direct-write methods that use tips under control of an atomic force microscope (AFM) to write the pattern. Very fine AFM tips are available, and AFMs have nanometer positional control, thus, DPN, nanoengraving, and nanoshaving can all produce very high-resolution, sub-100 nm features. In contrast, inkjet printing produces coarser features, with the best lateral resolution of ~10 μm, and more often in the order of 100 μm. Inkjet systems, however, can be easily multiplexed, making them very attractive for patterning biomolecules in diagnostic devices, such as protein and DNA microarrays.

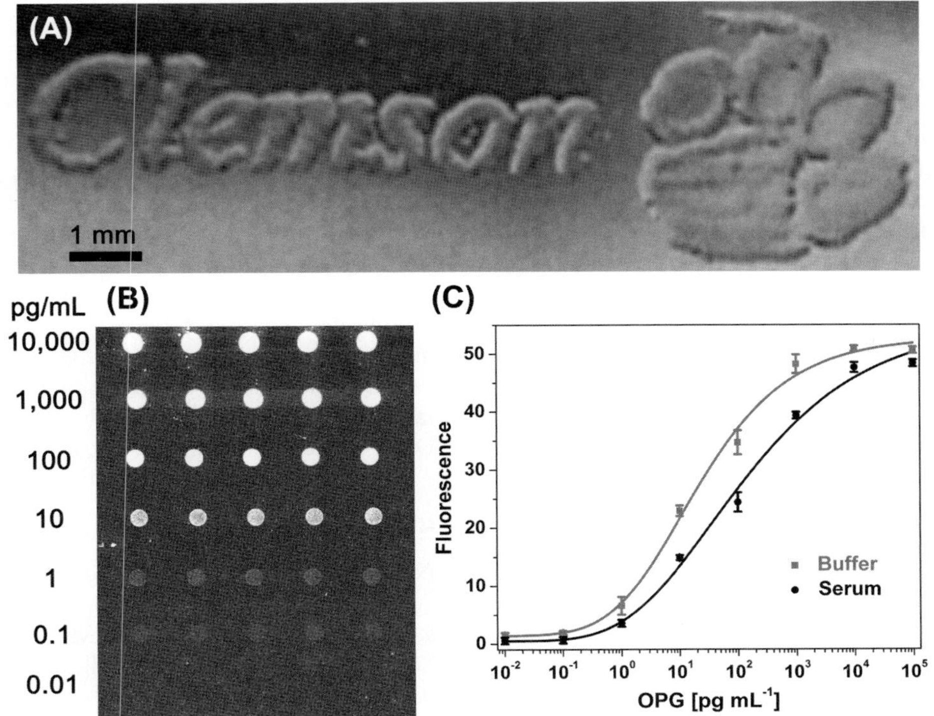

FIGURE I.2.13.4 Inkjet Printing. A colony of bacteria printed in the shape of a university logo by a modified consumer-grade inkjet printer (A) (Xu et al., 2004). An inkjet printed antibody microarray on a non-fouling surface (B) is used for ultrasensitive detection in complex mixtures such as blood and serum (C) (Hucknall et al., 2009c). (A–C) are adapted/reprinted with permission from Wiley-Blackwell.

Printing with Inkjets, Quills, and Pins. Inkjet printing refers to any method where liquid is pumped through a nozzle and onto a substrate in a defined pattern. Inkjet printers deliver liquid solutions as small droplets – typically 10–20 pL in volume (Geissler and Xia, 2004) – that are either continuously pumped through the nozzle head, electrostatically charged, and then guided via electric fields or generated on demand by thermally or mechanically triggered pressure pulses (Derby, 2008). The liquid inks are typically stored in reservoirs and pumped into the printer as needed. Inkjet nozzles are usually very small (20–30 μm in diameter) (Geissler and Xia, 2004), which enables them to be mounted on nozzle heads that can be interfaced with computers to control the fabrication of patterns. The resolution limit of inkjet printing is in the range of tens of micrometers, and is limited by the droplet volume, droplet spreading on the substrate (which varies depending on the ink and the surface), and other mechanical factors. Some useful features of inkjet printing are: (1) patterning is contact-free and can be performed under ambient conditions on a range of solid substrates; (2) the method has high positional accuracy; (3) multiplexed printing of analytes is easily accomplished by use of multiple nozzles and/or nozzle heads; and (4) high-throughput patterning over large surface areas can be accomplished by robotic automation.

Inkjet printing of biomolecules has gained popularity in recent years, in part due to the simplicity of the technique and the widespread availability of cheap printers. Thomas Boland's research group demonstrated that low-cost consumer inkjet printers could be adapted to print "biological inks" with their demonstration of the inkjet printing of proteins (Pardo et al., 2003), live cells (Xu et al., 2005, 2006a), and bacterial colonies (Xu et al., 2004). Figure I.2.13.4 shows a "living advertisement" – a colony of bacteria printed in the shape of a university logo – printed by Boland and co-workers.

Automated robotic inkjet printing systems designed to print "biological inks" such as DNA or proteins are typically called microarray printers, and are much more sophisticated and expensive than consumer-grade inkjet printers. These systems are often self-contained in an environmentally controlled chamber, and can run complex protocols involving delivery of multiple solutions as well as rinsing steps. Such systems are used to create spatially addressable arrays of biomolecules, such as DNA microarrays (Marshall and Hodgson, 1998; Bier et al., 2008). However, much of this work used arrays fabricated not by inkjet printers, but by pin-based arrayers in which pin arrays are dipped into solutions and then contact printed onto a surface (DeRisi et al., 1997), or else arrays that were fabricated by on-chip, photolithographic synthesis of spatially addressable arrays of oligonucleotide probes, such as the Affymetrix oligonucleotide array technology (Fodor et al., 1991). Since the advent of the human genome project and the rise of genomics, microarrays have migrated into other fields because they provide a convenient, high-throughput, and multiplexed methodology to characterize biomolecules. For example, protein arrays are used in proteomics (Caiazzo et al., 2009), and have been used to identify protein markers for disease (Orchekowski et al., 2005). Pin-based arrayers have also largely been used to fabricate protein arrays (MacBeath and Schreiber, 2000; Ray et al., 2007), because it is significantly cheaper and less complex to manufacture arrayers with a large number of pins or quills, compared with scaling-up the number of independently actuated inkjets beyond a dozen. Similarly, both pin-based (Pilobello et al., 2005, 2007; Pilobello and Mahal, 2007) and inkjet microarrays (Nagaraj et al., 2008) have been used to fabricate lectin arrays for glycomics. Microarrays have also been used as clinical diagnostics to detect specific analytes (Delehanty and Ligler, 2002; Hucknall et al., 2009c). For example, the Chilkoti group has used inkjet printed microarrays of antibodies on a non-fouling polymer brush for the detection of multiple protein analytes from serum and blood with a femtomolar limit-of-detection (Hucknall et al., 2009c) (Figure I.2.13.4).

Dip-Pen Nanolithography (DPN)

DPN is the nanoscale equivalent of writing with a quill. In DPN, AFM probes are inked with a solution of molecules and then scanned along a surface to write a pattern of the ink. The ink solution forms a small meniscus between the tip of the probe and the surface, which transports biomolecules from the tip to the surface. DPN has been used to pattern alkanethiols (Hong et al., 1999; Piner et al., 1999), as well as a range of biomolecules including DNA (Demers et al., 2002), proteins (Lee et al., 2002, 2006a), functional enzymes (Hyun et al., 2004a), biopolymers (Hyun et al., 2004b), and viruses (Vega et al., 2005) on many different materials. The resolution limit of DPN is on the order of tens of nanometers, and this method can create patterns that extend over several hundred micrometers, which is the limit of most AFM scanners. Single probe DPN patterning is also relatively low-throughput, as it is a somewhat slow and serial patterning technique. For example, in an early report on DPN (Piner et al., 1999), 10 minutes were required to create a 1 μm × 1 μm feature.

High-throughput DPN can be performed by mounting multiple probes on a single scan head. Perhaps the most impressive work to date in this regard is that of Chad Mirkin's group, where 55,000 DPN probes were inked with a thiol solution, and used simultaneously to generate massively high-throughput, high-resolution patterns (Figure I.2.13.5A) (Salaita et al., 2006). In this demonstration, the face of Thomas Jefferson from a US five-cent coin was generated 55,000 times, with each face containing 470 *million* nanoscale features, in less than 30 minutes (Figure I.2.13.5B). Commercial DPN systems are now available that are capable of multi-probe patterning of nanoscale-to-microscale patterns over areas as large as 40 mm^2 (NanoInk, Inc.). For good reviews of DPN, see Krämer et al. (2003), and Mendes et al. (2007).

Imprinting/Engraving

Nanoscale imprinting and engraving of surfaces can be accomplished by indenting or scraping a surface with a hard stylus, such as an AFM probe. Such techniques have similar advantages and limitations as DPN, as the patterning mechanism is very similar. *Nanoshaving* is a term used to describe the process of scratching patterns in thin molecular layers, such as SAMs, on a bulk substrate (Liu et al., 2000). Here, arbitrary patterns in the molecular layer are engraved by dragging the probe on the surface. A two-step process can be used to functionalize the exposed, underlying substrate with a secondary molecule. For example, nanoshaving has been used to selectively degrade a protein-resistant SAM on gold, which was then exposed to a solution of an antibody (Kenseth et al., 2001). Nanopatterns of antibodies were formed on the surface as the antibodies adsorbed to the exposed regions of the substrate that were no longer protein resistant. *Nanografting* – a variant of nanoshaving – involves scratching the surface in the presence of a secondary molecule with high affinity to the underlying substrate (Xu and Liu, 1997; Liu et al., 2000) so that the secondary molecule fills in the regions that are cleared away by the scanning probe. Nanografting has also been used to create nanopatterns of enzymes, antibodies (Wadu-Mesthrige et al., 1999, 2008; Hu et al., 2005), and DNA (Liu et al., 2002) on a surface.

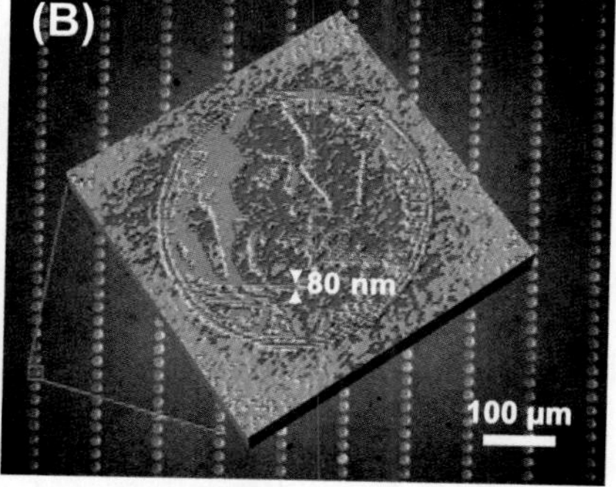

FIGURE I.2.13.5 High-throughput, high-resolution dip-pen nanolithography. For details of figures (A) and (B) please see the text. (Salaita et al., 2006). Adapted with permission from Wiley-Blackwell.

Writing with Beams

This section describes patterning methods that scan an energetic beam over a surface to create a pattern. The spatial resolution of this class of patterning techniques is generally limited by the spot size of the beam. Beams of photons are used for patterning in direct-write photolithography, and thus the spot size (and hence, resolution limit) is set by the optical diffraction limit, which is ~200 nm. However, the diffraction limit can be overcome by using near-field methods, such as scanning near-field lithography, to create features in which the smallest dimension is ~tens of nanometers. Direct-write photolithography is particularly conducive to patterning biomolecules, as these methods are compatible with aqueous conditions. Electron beam lithography and focused ion beam lithography represent two other methods in this class of patterning techniques that have higher spatial resolution than conventional far-field photolithography. In these methods, shorter wavelength radiation is used than in photolithography, which leads to smaller spot sizes, and hence higher lateral resolution. Electron beam lithography and focused ion beam lithography, however, usually require high vacuum and dry samples, which can be a serious impediment to patterning biomolecules.

Direct-Write Photolithography

In *direct-write photolithography*, a beam of focused light is used to pattern a substrate directly. Here, light is usually focused into small spots using conventional light optics, and patterns are formed by photochemical or physical modification of the surface. The resolution of features patterned with conventionally focused beams of light is governed by the optical diffraction limit, which is roughly half of the wavelength of the light used for excitation (Dunn, 1999). Thus, patterns formed by focused light typically have a minimum feature size of several hundred nanometers. Feature sizes can be scaled upwards using larger spot sizes; however, intensity requirements of the patterning process or the amount of light-per-area necessary to form patterns, can in some cases be a limiting factor in producing larger features, as light intensity decreases with larger spot sizes. Although broadband light sources can be used for direct-write photopatterning, lasers are more commonly used for this application, because they provide high-intensity beams of a single wavelength which allows precise control of specific photochemical reactions.

Patterning with light is generally performed using a photochemical mechanism or a mechanism that induces a spatially localized physical change in the underlying substrate. In direct-write patterning via photochemistry, light is used to locally activate photoactive molecules that subsequently either form a pattern directly or react with another species to form a pattern. In some cases, the activated species is created in one step, and then another step is necessary to expose the activated species to the biomolecule to be patterned. However, in other cases the photoactivated species is generated in the presence of the secondary species to be patterned, and thus the pattern is created in one step. Another one-step patterning scenario is the case where the photoactive molecule is covalently attached to the molecule to be patterned. Photochemical patterning with light has been widely used to create patterns of proteins on a substrate. For example, Holden and Cremer used a laser to pattern enzymes directly within flow channels (Holden et al., 2004) (Figure I.2.13.6A), a methodology that can be applicable to lab-on-a-chip devices.

In the second methodology to pattern biomolecules using light, a laser is used to pattern substrates by physically

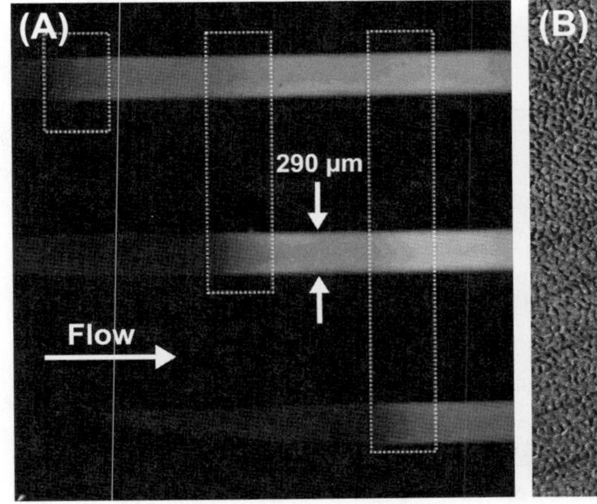

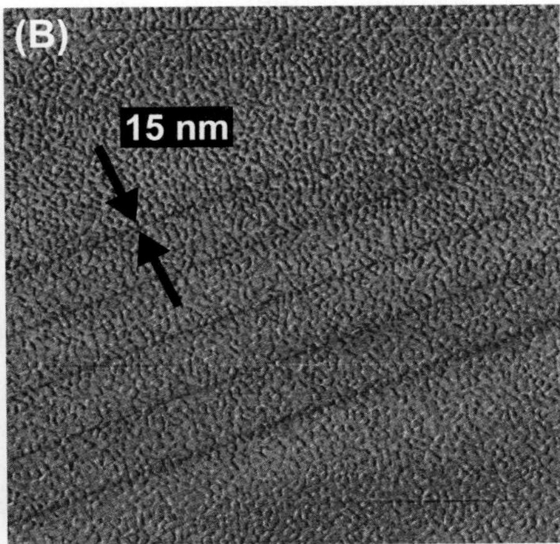

FIGURE I.2.13.6 Examples of Direct-Write Protein Patterning with Light. (A) Patches of enzymes (locations indicated by dashed boxes in the figure) patterned within flow channels produce fluorescent products when exposed to fluorogenic reagents; (B) Sub-diffraction limit features patterned with light using near-field excitation. Here, 15 nm lines were written in a self-assembled monolayer on gold (Ducker et al., 2007b). (A) is adapted with permission from (Holden, M. A., Jung, S. -Y., & Cremer, P. S. (2004). Patterning enzymes inside microfluidic channels via photoattachment chemistry. Anal. Chem., 76(7), 1838–1843). Copyright (2004) American Chemical Society. (B) is adapted with permission from the authors and SPIE.

altering a spatially localized region of a substrate. This type of patterning is conceptually similar to engraving with AFM tips, with the only difference being that the substrate material – polymers or resins – are ablated, melted or deformed due to the high energy of the focused laser spot, instead of being physically excavated (Joglekar et al., 2003). For example, Nielson and Shear used a high-energy pulsed laser to ablate pores in a membrane separating adjacent flow cells (Nielson and Shear, 2006). They showed that this technique could be used to dose cells with precisely located chemical gradients. In another example, the Yasuda group used a focused laser to photothermally etch microchambers in agar for use in single cell experiments (Kojima et al., 2003; Sugio et al., 2004; Suzuki et al., 2004).

Patterning with light can be used to create feature sizes that are much smaller than those created using diffraction-limited optics. An example of ultra-small direct-write photolithography is patterning by scanning near-field photolithography (SNP) that uses a near-field scanning optical microscope (NSOM) to create nanoscale patterns using light in the near field. In SNP, light is passed through a metal-coated optical fiber that is tapered to a ~50 nm aperture at its end that acts as a sub-wavelength light source. When in ~10 nm (Betzig and Trautman, 1992; Wegscheider et al., 1995) proximity to a photoactive substrate, the light source can excite chromophores in the near-field via the non-diffracting evanescent field that emerges from the tip, and thus can create patterns that are well below the diffraction limit. Leggett and co-workers have extensively used this technique to create nanoscale patterns of proteins and self-assembled monolayers (Sun et al., 2002; Sun and Leggett, 2004; Ducker and Leggett, 2006; Ducker et al., 2007a,b) (Figure I.2.13.6B).

Electron Beam Lithography

Direct-write *electron beam lithography* (EBL) uses similar patterning principles as that of patterning with focused light. However, in EBL, a focused beam of electrons (e-beam) is used as the stylus to write patterns in an e-beam sensitive material. The focused energy from the e-beam can initiate cross-linking (Brough et al., 2007; Christman et al., 2009) or functionalization (Eck et al., 2000; Golzhauser et al., 2001; Schmelmer et al., 2007; Steenackers et al., 2007) of surface moieties on the substrate. Since the e-beam radiation is of a much shorter wavelength than that used in direct-write photolithography, the spatial resolution of EBL is greater than direct-write photolithography. For this reason, EBL is a very powerful technique for creating patterns at the nanoscale, with feature sizes of 10–100 nm easily fabricated by EBL (Ducker et al., 2008). EBL, however, has several limitations: first, although existing scanning electron microscopy (SEM) systems can be easily configured for EBL, they are expensive, so EBL is not a routine technique that can be set up in a laboratory, and hence requires access to a dedicated instrument that is typically located in a shared instrumentation facility. Second, EBL requires high vacuum, which limits its utility for patterning biological molecules. Third, EBL is slow and can only pattern relatively small areas, although larger areas can be "stitched together" by moving the substrate using a motorized stage. The high resolution, low throughput, and high cost make EBL useful largely as a method for high-resolution fabrication of nanoscale patterns over a small area. As biological molecules generally do not remain functional in high vacuum, and are easily damaged by e-beam radiation, substrates are typically first patterned by EBL and then exposed to a biological molecule of interest for selective attachment of the biomolecule to the patterned features. An elegant example of patterning by EBL is the work done by Heather Maynard's group, where multi-component protein patterns were fabricated by e-beam cross-linking of functionalized PEG molecules. These functional PEG patterns were then used to immobilize proteins from solutions via specific binding interactions between the proteins and the ligands presented by the PEG patterns (Christman et al., 2009) (Figure I.2.13.7, left).

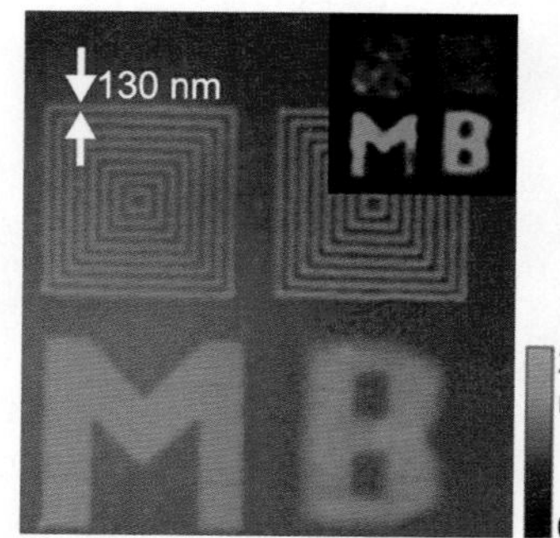

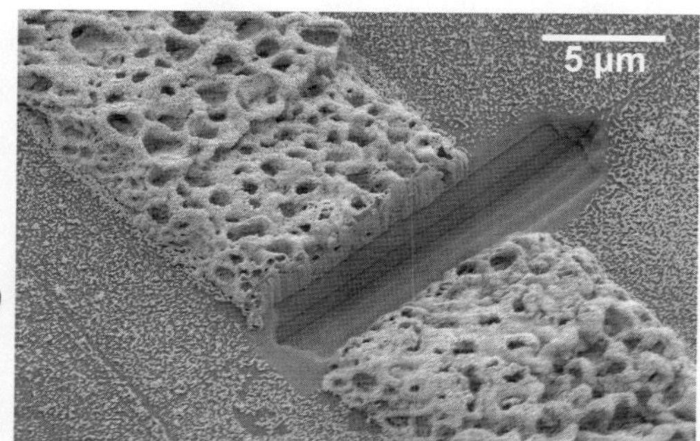

FIGURE I.2.13.7 Examples of EBL (left) and FIB (right) Patterning. Shown on the left is an example of EBL where multi-component protein patterns were formed by electron beam cross-linking and subsequent functionalization of poly(ethylene glycol) molecules. Shown on the right is an example where FIB milling was used to open a conductive circuit by milling a gap in a metallized protein-based wire (scale bar 5 μm). (Left) is adapted with permission from (Christman, K. L., Schopf, E., Broyer, R. M., Li, R. C., Chen, Y., et al. (2009). Positioning multiple proteins at the nanoscale with electron beam cross-linked functional polymers. J. Am. Chem. Soc., 131(2), 521–527.). Copyright (2009) American Chemical Society. (Right) is adapted with permission from (Hill, R. T., Lyon, J., Allen, R., Stevenson, K., & Shear, J. B. (2005). Microfabrication of three-dimensional bioelectronic architectures. J. Am. Chem. Soc., 127(30), 10707–10711). Copyright (2005) American Chemical Society.

Focused Ion Beam

A *focused ion beam* (FIB) can also be used for direct-write patterning. FIB patterning is similar to EBL in its advantages and limitations; however, FIB patterning makes use of high-mass ions, such as gallium ions, as the energetic particles instead of electrons as in EBL. FIB patterning is an inherently destructive process, as the bombardment of a surface with ions causes atomic sputtering of atoms from the surface. For this reason, FIB is typically used as a milling technique to engrave a surface with submicrometer size features. For example, Hill et al. used an FIB to open a conductive circuit by milling a gap in a metallized protein-based wire (Hill et al., 2005) (Figure I.2.13.7, right). FIB can also be used to sputter metals (Reetz et al., 1997) onto a surface, which makes FIB useful in repairing integrated circuits (Geissler and Xia, 2004).

Writing with Fields

Direct-write patterning with fields includes scanning electric and magnetic fields across a substrate to form a pattern. Features generated by direct-write electric field scanning typically have a lower spatial resolution than that attainable with other direct-write techniques. Similar to other scanning techniques, writing with fields is slow and is unsuitable for large area patterning. However, scanning electrodes allow one to perform extremely localized electrochemical reactions. Direct-write magnetic patterning is mostly geared towards data storage and, as it has not been used for sputtering biomolecules, only the underlying principle will be briefly described in this section.

Electric Field

Direct-write patterning with electric fields is accomplished by scanning an electrode near a surface to locally modify the surface using charge or current. Patterning is mediated through a variety of electrochemical processes, which include localized charging, ohmic heating, and redox reactions. Scanning electrochemical microscopy (SECM), which was developed by Allan Bard and co-workers, can be used in a patterning mode to perform localized redox reactions on a surface, resulting in the formation of microstructures (Bard et al., 1990). SECM patterning is performed with ultramicroelectrodes in electrolyte solutions, and has been used to pattern enzymes (Shiku et al., 1997; Wittstock and Schuhmann, 1997) and SAMs on a surface (Wittstock et al., 1997). The feature sizes of patterns generated by SECM are typically larger than those created with other types of scanning probe techniques (Geissler and Xia, 2004). Scanning tunneling microscopy (STM) and AFM can be used to generate higher resolution features. STM has the highest resolution of all proximal-probe lithography methods, as it can manipulate single atoms (Eigler and Schweizer, 1990; Manoharan et al., 2000), but as it must be used under high vacuum to achieve atomic resolution, it is of limited utility for biomolecular patterning. *Anodization lithography* is a methodology in which biased AFM tips are used to selectively grow oxide nanostructures on a silicon oxide substrate (Ducker et al., 2008). Stefan Zauscher's group has used anodization lithography to localize polymerization precursors, which were subsequently used to create nanopatterns of polymer brushes (Lee et al., 2006b).

Magnetic Field

Direct-write patterning with magnetic fields has mostly been geared towards data storage applications to date, and thus this topic will not be covered in detail in this section other than to provide the reader a basic idea of its underling physical principle. In direct-write magnetic patterning, an inductive element is scanned above a magnetizable material, such that localized regions of the surface become magnetized. These local magnetized areas allow data to be recorded as bits, and can read the stored data using the same type of inductive element that is used to generate the patterns. For more information on magnetic direct-write patterning, see the review papers by Chou (1997) and Geissler and Xia (2004). Magnetic patterning by self-assembly is discussed in a later section of this chapter.

Patterning with Masks

Patterning with masks is a highly developed form of patterning, as it has been the core technology of the microelectronics industry. Because of its sophistication and maturity, it has also been adapted for the patterning of biomolecules and cells. A mask is broadly defined as a template that can spatially modulate a field or radiation that passes through it, or a template that physically masks spatially defined regions of the underlying substrate from exposure to inks or etchants. Masking light to create patterns on substrates is the most common form of patterning with masks. Patterning with light and masks is generally called photolithography, even though photolithography can be performed in a mask-less, direct-write manner, as described previously. Mask-based patterning can also be performed with radiation (other than light), liquid inks or with chemical or biological etchants. Mask-based patterning is powerful and versatile for the following reasons: (1) large areas – typically 4–6 inch wafers – can be rapidly patterned in batch mode, which makes it a high-throughput patterning technique; (2) spatially intricate patterns can be fabricated that are only limited by the design of the mask; and (3) layered features can be fabricated by performing multiple patterning steps in succession, each with a different etching or deposition step. With advances in light sources and optics, the

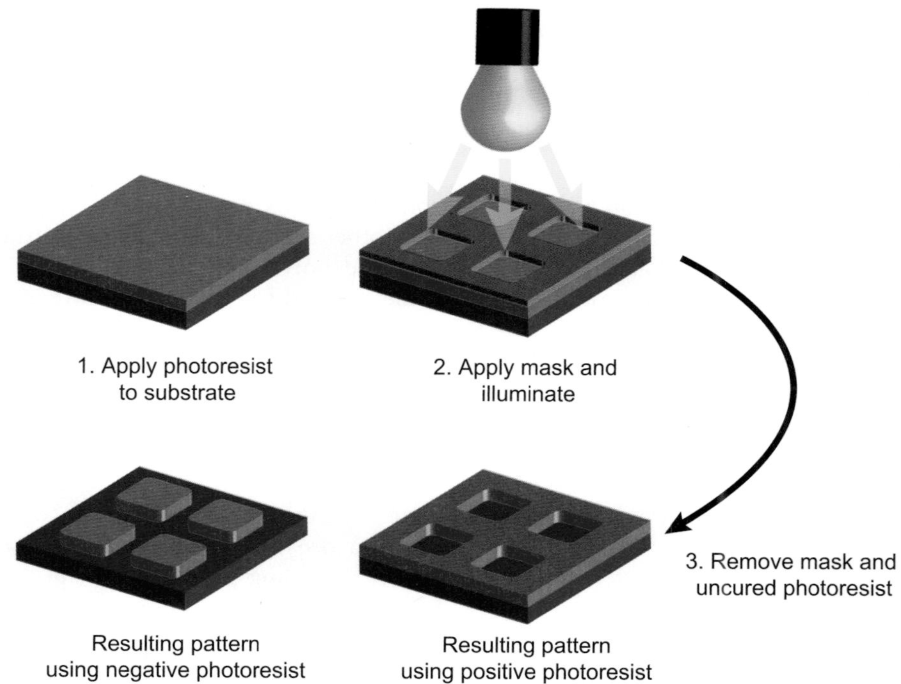

FIGURE I.2.13.8 Mask-Based Photolithography.

spatial resolution of mask-based approaches is now in the sub-200 nm regime (Totzeck et al., 2007).

Photolithography with Masks

In *photolithography*, a mask can be any material that has optically transparent and opaque regions. Masks can be rigid, such as those fabricated out of metal or printed on quartz glass, or they can be fabricated on flexible substrates such as transparencies. The critical requirement of a mask-based photolithographic patterning process is that light must pass through spatially defined regions of the mask and lead to the formation of a pattern that conforms to or is dictated by the pattern of the mask, on a substrate that is located below the mask. However, historically, the term *photolithography* encompasses a set of processes that derived from the semiconductor industry, which involves more than simply using light and masks to fabricate patterns.

In photolithography, patterning is facilitated by a polymer *photoresist* that is spin-coated onto a silicon wafer, activated by light, and developed in a developing solution (Figure I.2.13.8) to create a pattern of the photoresist, which is then subsequently processed to create a pattern in the underlying substrate. There are two types of photoresists. (1) A *positive photoresist* becomes soluble in the developing solution in areas that are exposed to light, while the unexposed regions of the photoresist are insoluble in the developing solution. After exposure and development, the areas of the positive photoresist that were exposed to light are dissolved and rinsed away to expose the underlying substrate. (2) A *negative photoresist* works on the opposite principle; areas of the photoresist that were exposed to the light will become less soluble in the developing solution, so that incubation in the developing solution will dissolve the unexposed regions of the photoresist, and expose the underlying substrate. The photolithography process is completed by removing parts of the exposed or unexposed photoresist. Removal is usually followed by an etching step in which the exposed thin metal/oxide films coating a bulk substrate are dissolved or a deposition step in which metal or other materials are preferentially deposited on to exposed regions.

A mask can be placed relative to the light source and substrate in three modes: contact mode; proximity mode or projection mode. In contact mode, the mask is placed directly on the substrate. Contact mode photolithography produces features at a 1:1 size relative to the feature size in the photomask, and thus resolution is limited by the feature size of the photomask. In contact mode, it is essential that the substrate and mask are free of particulate contaminants, as any trapped debris between the mask and substrate will contaminate the pattern and/or damage the mask or substrate. Proximity mode patterning is a variant of contact mode, in that the mask is slightly separated from the substrate along the propagating light axis. Proximity mode photolithography also produces features in the substrate that are the same size as are patterned on the mask, but the presence of foreign contaminants is not as devastating as in contact mode, as the mask and the substrate are not in contact. In projection mode photolithograpy, the mask is placed at an image plane that is well separated from the surface of the substrate along the optical axis that is conjugate to the image plane at the surface to be patterned. Projection mode photolithography can produce a smaller feature size than is embedded in the mask, as optics can be used to project the mask pattern at very small physical dimensions. For example, a microscope can be used to reduce the mask pattern 100 times by using a 100 × microscope objective. Projection mode photolithography also enables higher-throughput processing, as mask alignment can be done quickly and efficiently due to reduced risk of damaging the mask or the substrate because they are well separated.

Typical photolithography with masks, as developed in the microelectronics industry, involves specialized photoresists composed of polymers that are designed to polymerize or depolymerize upon exposure to UV light. However, light-based patterning with masks can be accomplished with any material that is light responsive. An extreme example is mask-based patterning of bacteria in plates of agar by Levskaya et al., (2005). Here, bacteria were genetically engineered so that lacZ gene expression, which encodes β-galactosidase, could be repressed by light exposure. Photomasks were used to spatially localize lacZ gene expression, and hence β-galactosidase, to areas within the bacterial colony that were masked from light. Patterns were visualized by testing for β-galactosidase activity within the bacterial colony using media that formed black precipitates in the presence of β-galactosidase (Figure I.2.13.9).

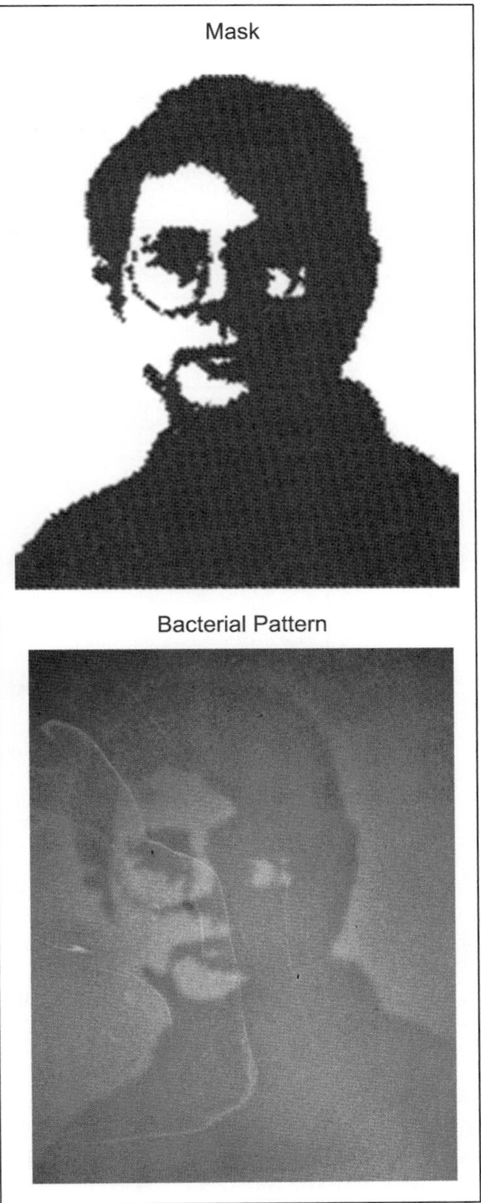

FIGURE I.2.13.9 Projection Mask Photolithography of Bacterial Cultures. Patterns are formed via light-responsive gene expression within the bacterial culture (Levskaya et al., 2005). Adapted with permission from Nature Publishing Group.

Deposition/Etching with Masks

As mentioned above, mask-based deposition and etching is often a finishing step in a typical photolithography process. In fact, the actual photolithographic component of the process, as derived from the semiconductor/microelectronics industry, is the creation of a polymer-based mask on a substrate that is meant to protect specific parts of the underlying substrate from a subsequent deposition or etching process. However, it is beneficial to think of deposition and etching with masks as standalone patterning techniques, because these techniques extend beyond the context of photolithography. Simply put, mask-based deposition is just like using a stencil to paint letters or graphics on paper, clothing or any other material. The stencil, or mask, can be made of any material that conforms to the constraints of the experiments at hand. Examples of some experimental constraints are the following: the nature of the material to be deposited (i.e., is it hard, soft, deformable, etc.?); the nature of the substrate; the nature of etchant; etc. Masks can be made of metal or metal/glass as in traditional photolithography; such masks are rigid and are easy to clean. However, other masks can be made of elastomeric polymers, such as poly(dimethyl siloxane) (PDMS), which provide the benefits of being able to form watertight seals with the substrate, compatibility with curved surfaces, and flexible peel-off (see Chapter I.2.2.B). In a deposition process, the mask is applied to a substrate, the combination of which is then exposed to the material to be deposited. Once the mask is removed, areas protected by the mask remain free of the deposited material. In an etching process, the mask/substrate is exposed to a form of radiation or an etching chemical, and the areas exposed by the mask are selectively etched.

There is a seemingly endless array of photolithography and mask-based deposition/etching examples. Mask-based photolithographic patterning of biomolecules most likely started with works such as Stephen Fodor's, where arrays of photoactive chemicals were patterned to create

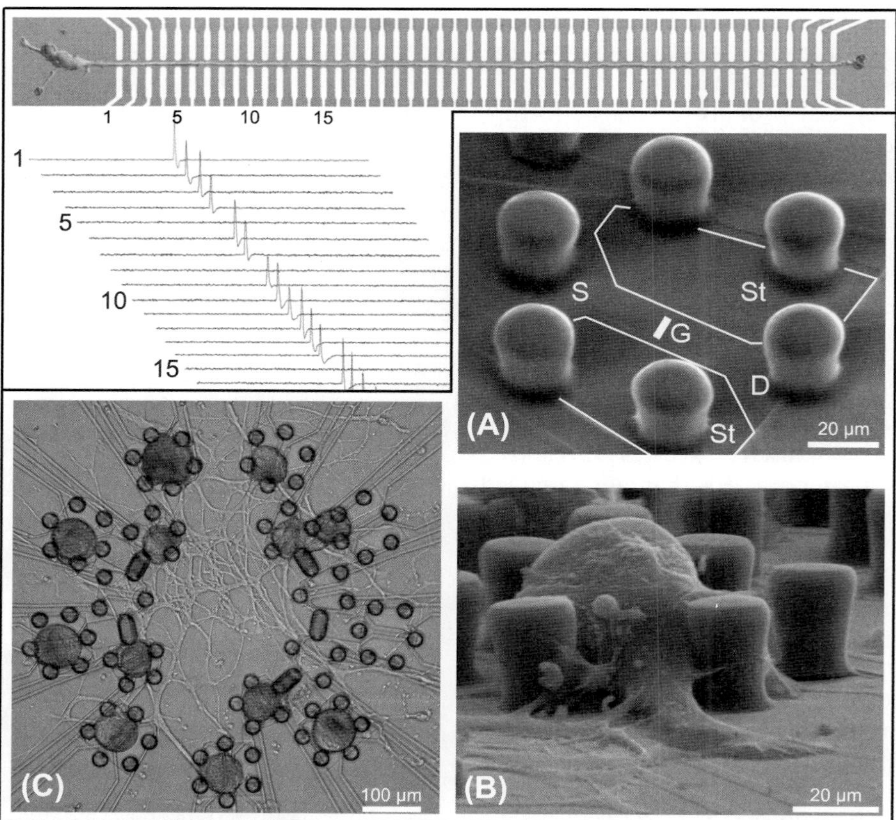

FIGURE I.2.13.10 Photolithography and Masked Deposition/Etching Used to Create Microelectronics for Interfacing with Neurons. Upper left: a massive array of electrodes (top image) is used to record electrical signals from a single neuron (Patolsky et al., 2006). The signals are shown in the plots below the image. Lower right: polymeric posts are used to confine neurons atop underlying transistors (Zeck and Fromherz, 2001). The device is diagrammed in (A) indicating "Stimulator wings" (St) and the transistor components (S, source; D, drain; G, gate). The device is shown with a single cell in (B). An array of devices is used to interface with a neuronal network (C). Upper left is from [Patolsky, F., Timko, B. P., Yu, G., Fang, Y., Greytak, A. B., et al. (2006). Detection, stimulation, and inhibition of neuronal signals with high-density nanowire transistor arrays. Science, **313**(5790), 1100–1104.]. Adapted with permission from AAAS and from the authors. Lower Right is adapted with permission from (Zeck, G., & Fromherz, P. (2001). Noninvasive neuroelectronic interfacing with synaptically connected snail neurons immobilized on a semiconductor chip. Proc. Natl. Acad. Sci. USA, **98**(18), 10457–10462.). Copyright (2001) National Academy of Sciences, U.S.A.

microarrays for peptide synthesis (Fodor et al., 1991). Also, around the same time (the early 1990s) photolithographic processes were being used to create patterned surface chemistries to direct cell growth on substrates (Lom et al., 1993; Healy et al., 1994). This work was initiated by collaborative efforts from Kevin Healy's and Philip Hockberger's groups at Northwestern University. They combined photolithography and organosilane chemistry to produce materials with precise control over the regional distribution of surface chemistry. They went on to use these surfaces to study the mechanisms by which mammalian cells were distributed spatially when exposed to these micropatterned surfaces. These seminal studies established the utility of masked photolithography for biomolecular patterning as an important research tool in biomaterials.

An exciting and relatively more modern biological application of mask-based photolithography combined with deposition and etching is the development of microelectronic interfaces for cells. Much work has been done creating electrode arrays and transistor arrays that can stimulate and record from cultures of neuronal cells in an attempt to learn more about brain activity on a single-cell level, as well as from controlled populations of networks of neurons (Gross et al., 1985; Gross and Schwalm, 1994; Zeck and Fromherz, 2001; Patolsky et al., 2006). Figure I.2.13.10 details two separate applications in which electronic devices were created by mask-based photolithography and deposition/etching.

Patterning with Masters

A *master* is a template that is used to replicate patterns, usually in a batch-type process. Rigid masters can be used directly to imprint patterns on substrates. However, the most widespread use of masters for patterning is to create a *mold*, typically of an elastomeric polymer, which is subsequently used to mold the patterns. Molds are usually fabricated by curing polymeric materials in the presence of the master, which embeds the features of the master into the mold. Once separated from the master, molds can be used for printing features on substrates using molecular inks that are first applied to ink the mold, and then are transferred by contact with the protruding features of the mold with a surface. Alternatively, they can be brought into contact with a surface to create channels of the void features in the mold on a substrate, which can be used for patterning by a liquid ink flowing through the channel. Patterning with masters is a relatively cheap methodology to mass-produce features over large surface areas. Also, the use of conformal molds allows patterning on curved surfaces (Jackman et al., 1995). Cylindrical molds with patterned features also offer the possibility of a direct roll-on patterning process of a planar surface (Xia et al., 2004). Typically, patterning with masters does not offer as high spatial resolution as some direct-write patterning techniques.

Imprinting with a Master

Imprinting with a master is essentially the same concept as imprinting with a stylus, as described above, except that a master is used to imprint many features at once, instead of using a stylus and serially imprinting or writing one feature at a time. The master is fabricated from a relatively hard material such that it resists deformation when it is pressed against the substrate. The master contains a rigid relief pattern that molds the substrate into the shape which it is pressed into. Imprinting with a master is used often to pattern polymer films; however, other substrates can be used with the main requirement being that substrate must be soft enough to be deformed by the master. The resolution of features obtained from imprinting with a master is dictated by the durability of the master, the physical properties of the substrate, and the mechanical pressure applied during embossing. Feature sizes as small as 10 nm are attainable by imprinting with masters (Geissler and Xia, 2004).

Printing (Use of Protruding Features of a Mold)

Printing of patterns using molds is most widely done by *soft lithography*, a suite of related techniques that generally refers to the use of soft, elastomeric polymer molds to create patterns on substrates. Here, liquid monomer is poured onto a rigid master and then cured; when separated from the master the polymer mold retains the topography of the original master. These topographical features are then transferred to another substrate for patterning. Masters can be created using photolithographic techniques (described above), direct-write patterning or even relatively low-tech methods such as gluing rigid objects together so that they create necessary voids in the polymer (Allen et al., 2005; Hill and Shear, 2006). The most widely used polymer in soft lithography is PDMS. PDMS has numerous advantageous qualities (Whitesides et al., 2001) that contribute to its popularity, some of which include that it is low cost, available commercially, easily prepared, flexible, optically flat (good for optical microscopy), insulating, resistant to harsh chemicals (including hydrofluoric acid, a popular glass etchant), and relatively permeable to biological gases. PDMS has tunable surface properties, and can be reversibly or irreversibly bound to surfaces such as glass, creating liquid-tight seals for controlled flows. The elasticity and adherent nature of PDMS make it an extremely versatile mold, as these properties enable the mold to remain in close conformation with curved or otherwise non-uniform surfaces, which contributes to the ability to pattern large surface areas (up to hundreds of square centimeters) (Xia and Whitesides, 1998) with a single mold.

Printing in the context of soft lithography is commonly referred to as microcontact printing (Figure I.2.13.11), and was first demonstrated by Whitesides and co-workers at Harvard University in the early 1990s (Kumar and Whitesides, 1993). In microcontact printing, an elastomeric stamp created from a rigid master is used to print molecular inks (dried) on substrates with patterned feature sizes in the micrometer to nanometer size range (Xia and Whitesides, 1998; Whitesides et al., 2001; Geissler and Xia, 2004; Ducker et al., 2008). This technology was first demonstrated with printing of alkanethiol SAMs on gold surfaces (Kumar and Whitesides, 1993), and was benefitted by the fact that alkanethiols have a high affinity to gold and form dense monolayers on gold surfaces (Love et al., 2005). Thus, when a PDMS stamp containing SAM-forming molecules contacts the gold surface, the molecules dissociate from the PDMS stamp and assemble on the surface. The feature sizes of patterns generated by microcontact printing are determined by the mechanical properties of the stamp, the deposition conditions (i.e., pressure and duration of application), and the leakiness or spreading properties of the molecular ink during the stamping process.

Two papers from the Whitesides group from the 1990s stand out as being particularly relevant to the field of biomaterials. The first by Lopez et al. (1993) demonstrated the power of surface engineering to spatially control the adhesion of cells in adherent cultures. Although the patterns in this study were not fabricated using soft lithography, but were instead created in a direct-write process, many of the conceptual ideas in this paper reappeared in subsequent papers from the Whitesides group and others that used soft lithography to pattern alkanethiols on surfaces. A second seminal paper, by Singhvi et al. (1994), followed from this study and was one of the earliest works demonstrating the important role that geometrical patterns on a surface could play in confining cell shape, and hence their function and/or behavior. This work revealed the power of micropatterning on the length scale of the cell.

Since the initial demonstration of the printing of SAMs, other researchers have extended the use of soft

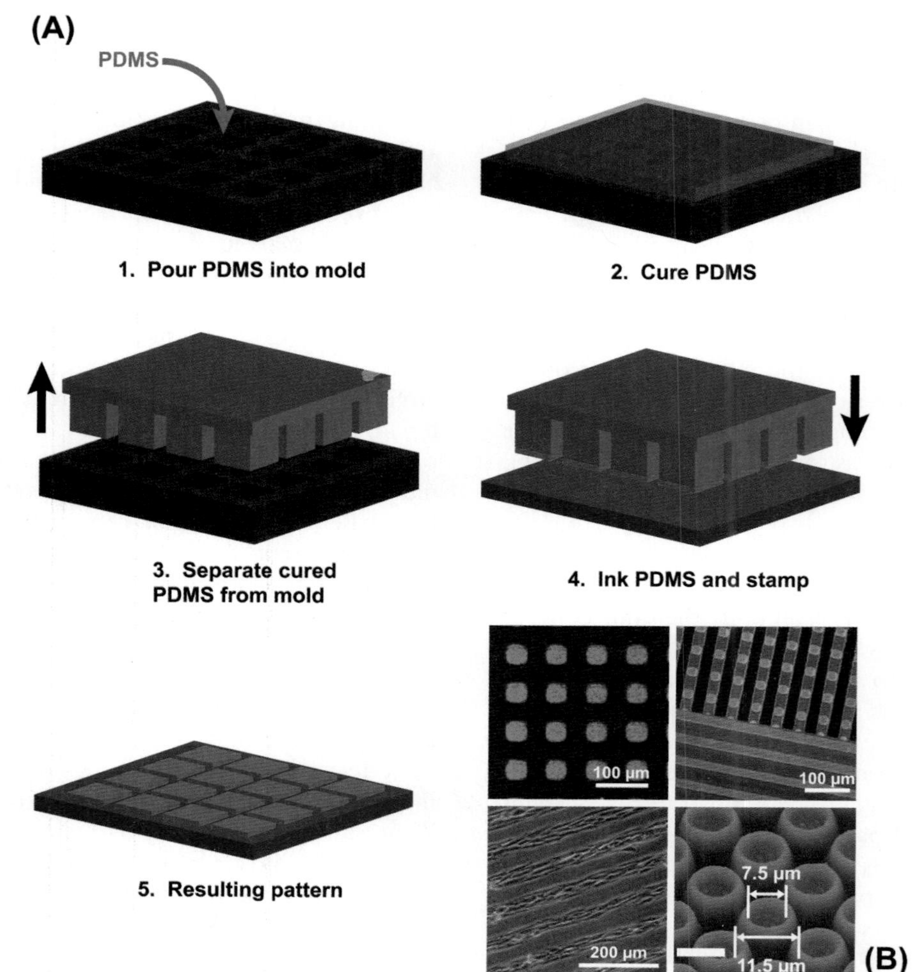

FIGURE I.2.13.11 Microcontact Printing. (A) The typical steps involved in microcontact printing. (B) Clockwise from upper left: checkered pattern of fluorescent protein formed by microcontact printing on a polymer; multicomponent polymer brush patterns formed by microcontact printing; anemone-shaped patterns of stimulus-responsive polymer patterned by microcontact printing (Chen et al., 2009); cell patterning on poly(ethylene terephthalate) by microcontact printing of a protein-resistant comb polymer and backfilling with fibronectin (Hyun et al., 2003). Upper left is adapted with permission from (Ma H., Li, D Sheng, X., Zhao B., & Chilkoti, A. (2006b). Protein-resistant polymer coatings on silicon oxide by surface-initiated atom transfer radical polymerization. Langmuir, 22(8), 3751–3756.). Copyright (2006) American Chemical Society. Upper right is adapted with permission from (Zhou, F., Zheng, Z., Yu, B., Liu, W., & Huck, W. T. S. (2006). Multicomponent polymer brushes. J. Am. Chem. Soc., 128(50), 16253–16258.). Copyright (2006) American Chemical Society. Lower right and left are adapted with permission from Wiley–Blackwell.

lithography to include patterning of polymers (James et al., 1998; Zhou et al., 2006; Chen et al., 2009), proteins (including functional antibodies and enzymes) (Bernard et al., 1998, 2000, 2001a; Kung et al., 2000; Renault et al., 2002; Tan et al., 2002; Ma et al., 2006b), and cells (Hyun et al., 2003) on various surfaces, including glass, silicon, and metals. In one example of diversifying beyond printing alkanethiol SAMs on gold, Chilkoti and co-workers showed that a class of random oligoethylene glycol-functionalized methacrylate terpolymers that are protein resistant could be spin-coated on to a micropatterned PDMS stamp (Ma et al., 2005). Bringing the stamp into contact with a substrate resulted in transfer of the polymer only in regions of the stamp that were in contact with surface. This selective delamination process worked with a surprisingly large number of materials, and provided a simple method to stamp protein-resistant polymer features with controlled lateral dimensions and height (controlled by the thickness of the polymer layer spin-coated on to the stamp) on a surface.

Use of Void Features of a Mold

While printing makes use of the protruding features of a mold, a complementary set of patterning techniques utilize the void features in the mold (Xia et al., 1996a, 1997). When a mold with a striped pattern is affixed to a substrate, the voids in the mold form channels that can be filled with liquid to create patterns. In *microfluidic patterning*, a network of channels is used to deliver liquids to specific locations on a substrate. Here, liquids can be used to deliver reagents locally to microdevices

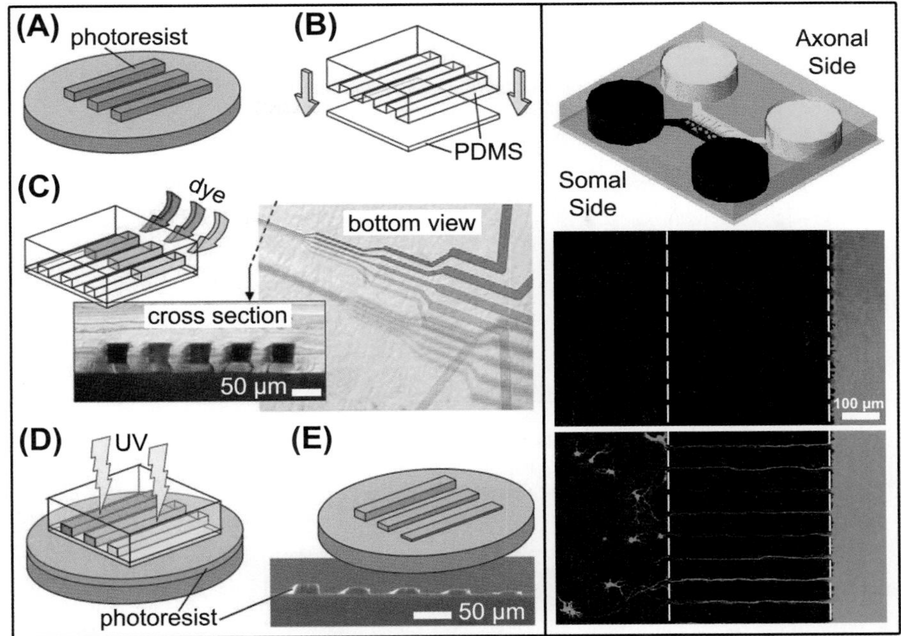

FIGURE I.2.13.12 Microfluidics. Left: microfluidic channels containing fluids of varying opacity are used as a mask for photolithography (Chen et al., 2003). Flow channels are created using an elastomertic mold (B) developed from a master pattern (A). The channels are filled with fluids of varying opacity (C), which modulates the light exposure to the underlying photoresist (D) and hence creates patterns that vary in height (E). Right: a microfluidic cell culture platform (top) is used to isolate axons of neurons, which can be targeted with effectors located in either of the fluid chambers (Taylor et al., 2005). Fluidic isolation of a dye on the axonal side of the chamber is shown in the middle panel. The bottom panel shows neurons that became stained with a dye that is located in the axonal side of the chamber. Left is adapted with permission from (Chen, C., Hirdes, D., & Folch, A. (2003). Gray-scale photolithography using microfluidic photomasks. Proc. Natl. Acad. Sci., 100(4), 1499–1504.). Copyright (2003) National Academy of Sciences, U.S.A. Right is adapted with permission from Nature Publishing Group.

on microfluidic chips (Figure I.2.13.12, right) (Juncker et al., 2002; Taylor et al., 2005), to deposit biomolecules onto the surface through adsorption or attachment chemistries (Bernard et al., 2001b; Dertinger et al., 2002; Fosser and Nuzzo, 2003), to locally dissolve parts of the underlying substrate (Rodriguez et al., 2003) or to serve as a mask for photolithography (Figure I.2.13.12, left) (Chen et al., 2003). See McDonald et al. (2000) for a good review of microfluidic systems.

Channels in a mold can also be used to deliver a polymer solution to a surface in a spatially delimited manner, which are then hardened or cured (Kim et al., 1995). Subsequent removal of the elastomeric mold leaves behind deposits of cured polymer in the regions of the surface in contact with the channels of the original mold. Several variants of this type of patterning have been developed, such as capillary force lithography (CFL) (Suh et al., 2005; Liu et al., 2006), micromolding in capillaries (MIMIC) (Xia et al., 1996b; Beh et al., 1999), and solvent-assisted micromolding (SAMIM) (Kim et al., 1997).

Patterning by Self-Assembly of Polymers and Colloids

Patterning by self-assembly of polymers and colloids is a powerful technique because, when properly executed, it requires minimal effort to produce the patterns. Self-assembled patterns are created by using experimental conditions where intermolecular or inter-particle forces cause polymers or colloidal particles to phase separate or aggregate in a spatially defined pattern driven by minimization of the free energy of the system. Self-assembled patterning can be used to pattern large surface areas with features ranging in resolution from the nanoscale to the microscale. Three examples of patterning by self-assembly are highlighted in this section: block copolymer self-assembly; nanosphere lithography; and magnetic self-assembly (Figure I.2.13.13).

Block Copolymer Self-Assembly

Block copolymer self-assembly is an example of self-assembled patterning at the molecular level. Patterning with block copolymers is accomplished by using a block copolymer that has two or more blocks that are immiscible with each other. A solution of the block copolymer is deposited on to a surface where it undergoes microphase segregation as the solvent evaporates. This segregation then leads to the formation of highly ordered surface patterns on the nanometer length scale that can readily cover large surface areas, although defects are a common problem on larger coverage areas. The nanometer-scale lateral resolution over a macroscopic surface area afforded by block copolymer self-assembly makes it a high-resolution, high-throughput patterning technique (Li and Ober, 2006). Patterns can

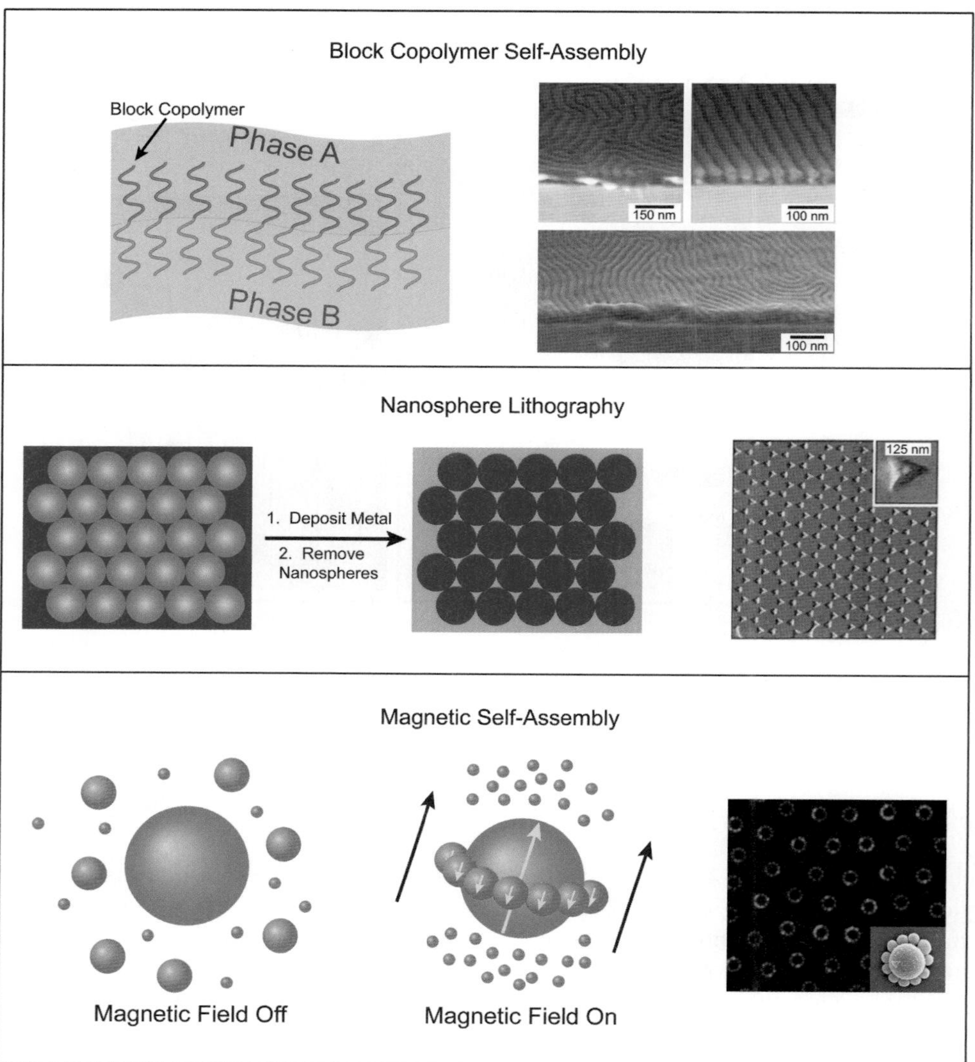

FIGURE I.2.13.13 Patterning by Self-Assembly. Top: block copolymer patterning on surfaces due to microphase separation between the polymer blocks (Ouk Kim et al., 2003). Middle: in nanosphere lithography, self-assembled, close packed nanospheres are used as a mask for creating metallic patterns on surfaces. Bottom: magnetic fields can be used to pattern magnetically susceptible objects. Here, rings of non-magnetic spheres are localized around larger paramagnetic spheres in solutions of magnetic ferrofluid (Erb et al., 2009). We note that the nature of this work was to position and align objects in three dimensions; however, this technique should be readily adaptable to surface patterning of three-dimensional objects. Data panels on the right: Top and Bottom are adapted with permission from Nature Publishing Group. Middle is adapted with permission from (Haynes, C., & Van Duyne, R. (2001). Nanosphere lithography: A versatile nanofabrication tool for studies of size-dependent nanoparticle optics. J. Phys. Chem. B., 105(24), 5599–5611.). Copyright (2001) American Chemical Society.

be controlled by altering a number of variables: the sequence and length of the blocks; the degree of immiscibility of the blocks; the physico-chemical properties of the solvent; the ambient conditions during thin film deposition; film thickness; and surface chemistry (Nie and Kumacheva, 2008). As described in Nie and Kumacheva's review (2008), patterned block copolymer films have been used as photonic crystals (Valkama et al., 2004), optical waveguides (Kim et al., 2005), and as templates for the growth of inorganic materials (Shin et al., 2002). They have also been used to create surfaces with tunable wettability (Xu et al., 2006b), and they can also be used as masks for patterning. In this approach, one phase of the block copolymer pattern is selectively removed, which leaves a nanoporous film behind that can serve as a mask (Nie and Kumacheva, 2008). For examples of block copolymer patterning, as well as reviews, see references Hamley, 2003; Park et al., 2003; Lie and Ober, 2006; Kim and Hinsberg, 2008; Nie and Kumacheva, 2008.

Nanosphere Lithography

Nanosphere lithography exploits the self-assembly of nanoscale objects to create a periodic mask or pattern. In nanosphere lithography, a solution of relatively monodisperse nanospheres – usually composed of a polymer – are delivered to a flat substrate from solution. Under specific conditions, the nanospheres self-assemble into a hexagonal, close-packed monolayer as the solvent evaporates, which leaves voids between the nanospheres where the underlying substrate is exposed. The nanosphere monolayer is then used as a mask to pattern the underlying substrate. Typically, nanosphere masks are used to mask the deposition of metals onto a substrate. After metal deposition, the nanospheres are removed from the surface, and leave behind metal pyramids in the regions corresponding to the void volumes between the nanospheres. For a detailed review on creating metal nanoparticles by nanosphere lithography, see Haynes and Van Duyne (2001). Template stripping has been used in combination with nanosphere lithography to create ultra-flat metal triangles (instead of pyramids) on surfaces using a similar deposition scheme (Frey et al., 2000). However, not all forms of nanosphere lithography involve metal deposition. For example, a metal microgrid template was used to direct dewetting of solutions on surfaces containing colloidal particles (including nanospheres and microspheres) (Celio et al., 2006). Once dewetting is complete, highly ordered patterns of the particles are formed over a large surface area on the underlying substrate.

Magnetic Self-Assembly

In *magnetic self-assembly*, nano-to-microscale particles with a magnetic susceptibility align or orient themselves in a magnetic field. In one implementation of this approach, magnetic nanoparticles can be directed to surfaces in a pattern, which can then serve as a mask for deposition of proteins or other biological materials (Bardea and Naaman, 2009). Magnetic particles have also been attached to cells to enable magnetism-directed cell patterning (Alsberg et al., 2006; Ino et al., 2007; Lin et al., 2008). Magnetic ferrofluid can also be used to template the self-assembly of non-magnetic materials, such as microspheres and cells (Yellen et al., 2006; Erb et al., 2009; Krebs et al., 2009). In this case, non-magnetic objects are positioned in three-dimensional space, due to the self-assembly of the magnetic medium surrounding them as the medium aligns with the external magnetic field.

Dynamic Patterning

The discussion of patterning thus far has mostly related to the formation of static patterns, which offer little control to the designer after formation, and often are subject to erosion or contamination over time. An emerging field of surface patterning is focused on the creation of dynamic or "smart" patterns. These patterns are designed to respond to external cues in the local environment. One motivation for the creation of dynamic patterns or materials is to create micro-mechanical components for use in microfluidic devices, diagnostics, theraputics or other sensing devices (Beebe et al., 2000; Jager et al., 2000; Terray et al., 2002; Alexander and Shakesheff, 2006; Dong et al., 2006; Peppas et al., 2006; Sidorenko et al., 2007; Kaehr and Shear, 2008). A motivation for dynamic patterning more relevant to the focus of this chapter, however, is to enable spatio-temporal control over interfaces used to study the cellular micro-environment, as well as those used in chip-based sensing devices. Dynamic patterning is particularly important, because most biological processes are triggered by temporal cues in the local environment. For example, intracellular signal cascades that are driven by ligand binding to receptors, clustering of receptors in membranes, and remodeling of the extracellular matrix due to mechanochemical stimuli, are all dynamic processes in biology. Engineered dynamic interfaces present a biologically inspired – although abiological – methodology to provide temporal control of biological interactions at artificial surfaces. For example, Milan Mrksich's group at the University of Chicago has created SAMs on gold that present ligands at the SAM–water interface that can be switched between an active or inactive state via electrochemical signals from an underlying gold electrode (Mrksich, 2002, 2009). They have used these surfaces to study the interactions between ligands and cell membranes during cell attachment to micropatterned surfaces (Yeo et al., 2003) and heterogeneous (Yousaf et al., 2001) cell patterning. Furthermore, the Chilkoti group has used elastin-like biopolymers, which actuate with a thermal or saline stimulus, to enable dynamic and reversible spatial localization of proteins to patterned templates (Frey et al., 2003a,b; Nath and Chilkoti, 2003; Hyun et al., 2004b) (Figure I.2.13.14) with application for the fabrication of rewriteable surfaces for cell studies and biosensors.

Three-Dimensional Patterning

Three-dimensional patterning is a new and exciting development in the field of surface patterning. A major motivation for the development of three-dimensional patterning is the fact that cells exist in a three-dimensional environment in nature. However, cell-based experimentation in laboratories typically involves cell cultures growing on planar substrates, where they interact only in two dimensions and are often irrelevant because cells exist in three dimensions in the body. The benefits of three-dimensional patterning are also relevant for the design of biosensor and diagnostics, as control of an extra dimension in device fabrication will enable the design of new sensing architectures and transduction modalities with potentially improved figure-of-merit. In addition, three-dimensional control of patterning will also yield better control over the design of microfluidic lab-on-a-chip devices.

Many of the techniques described in the preceding sections can be used for three-dimensional patterning. For

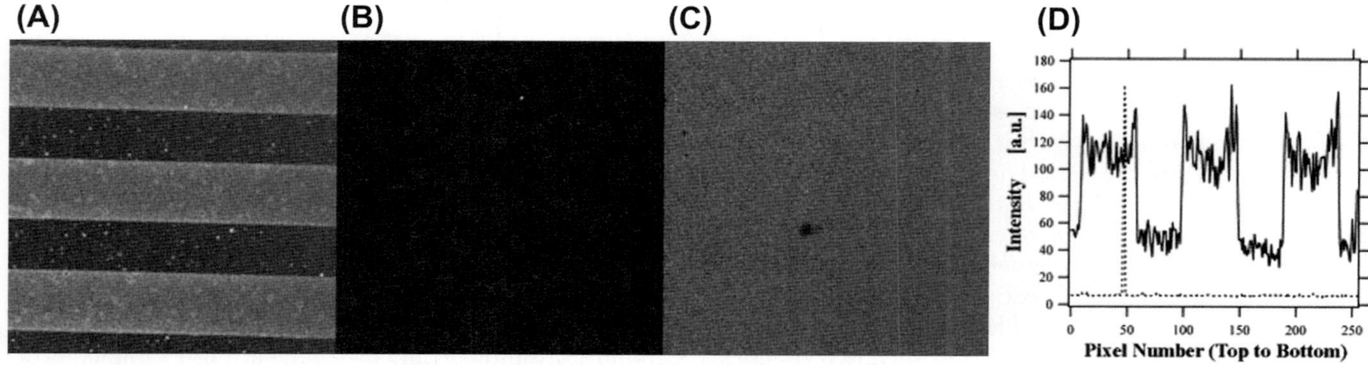

FIGURE I.2.13.14 Dynamic Patterning (Frey et al., 2003b). Shown in (A) is a confocal fluorescence image of a pattern of fluorescent antibodies localized to regions containing antigen coupled to a stimulus-responsive polymer (light regions of the image). Once triggered, the pattern is completely erased, as shown in (B) – an image using the same gain settings as in (A) – and again in (C), an image with enhanced gain settings to show uniformity of the background. (D) shows vertical intensity line plots drawn through (A) (solid line) and (B) (dashed line), which further verify complete elimination of the initial pattern shown in (A). Reprinted with permission from Wiley-Blackwell.

example, *mask-based photolithography* in combination with masked deposition or etching can be used to fabricate three-dimensional structures using multiple steps to build up features in the third dimension layer-by-layer (Zeck and Fromherz, 2001). This technique works well for patterning large surface areas; however, the processing involved can be quite laborious and time-consuming, as each layer requires a standalone photolithographic process. Three-dimensional features can also be deposited from voids in *elastomeric molds*, as discussed in the section on soft lithography. In addition, soft lithography (Hammond and Whitesides, 1995; Clark et al., 1997; Clark and Hammond, 1998; Jiang and Hammond, 2000) and scanning probe techniques (Lee et al., 2009) can be used to confine precursors in two dimensions on a substrate that can initiate *self-assembly* of features in a third dimension via molecularly thin, layer-by-layer (LBL) deposition of oppositely charged polyelectrolytes. This approach provides nanometer spatial resolution in the third dimension; however, assembly of polyelectrolyte layers via LBL deposition can be time-consuming when substantial feature heights are desired.

Another direct-write example of three-dimensional patterning is photothermal needle etching, in which an agarose gel is photothermally ablated by scanning a needle, with a laser that is continually focused on it, throughout the gel (Moriguchi and Yasuda, 2006). In a final example of three-dimensional patterning, robotic *inkjet printing* of viscoelastic polymers on surfaces has been used to create very complex three-dimensional architectures in a direct-write manner (Gratson et al., 2004), and some of these patterned three-dimensional structures have been used in tissue engineering (Woodfield et al., 2004), and for microfluidic (Therriault et al., 2003) applications. This technique suffers the drawbacks of any direct-write method, in that it is laborious to pattern large surface areas via serial scanning. However, inkjet printing has the attractive features that the spatial resolution is in range of the hundreds of nanometers, and complete three-dimensional control is possible as arbitrary features that can be fabricated by this methodology.

The rest of this section will detail some other techniques for three-dimensional patterning that use a different set of principles than those described so far.

Multiphoton Lithography

Multiphoton lithography (MPL) is a laser-based, direct-write approach to three-dimensional patterning in which multiphoton excitation of photoinitiators or photosensitizers is used to photocross-link or photopolymerize materials at the focus of a high instantaneous-energy, pulsed laser beam. As multiphoton excitation varies nonlinearly with the intensity of the excitation source, these photochemical reactions are confined to a small volume, or voxel, within the laser focus where intensity is large enough to excite a photochemical process of interest. This voxel can be translated in three dimensions within a solution of photoactive materials to locally cure photoactive material from solution. The remaining uncured solution can be rinsed away, leaving behind hardened, freestanding three-dimensional features. The resolution limit of this technique is ~250 nm, and can be scaled to the microscale; however, fabrication of larger features usually requires higher laser power, which can be a limiting factor in some applications. The use of a motorized mechanical stage allows automated patterning in the X–Y dimensions, while auto-focusing mechanisms can provide automated control over the Z dimension.

MPL has been widely used to create three-dimensional features from UV-curable synthetic resins. Complex serial scanning algorithms have enabled the fabrication of micronscale replicas of livestock, sculptures (Serbin et al., 2003), houses (Lafratta et al., 2007), and the fabrication of three-dimensional tattoos on biological specimens (Baldacchini et al., 2004), free-standing metallized wires and other electrical microcomponents (Lafratta et al., 2007; Li and Fourkas, 2007), micro-mechanical parts (Maruo

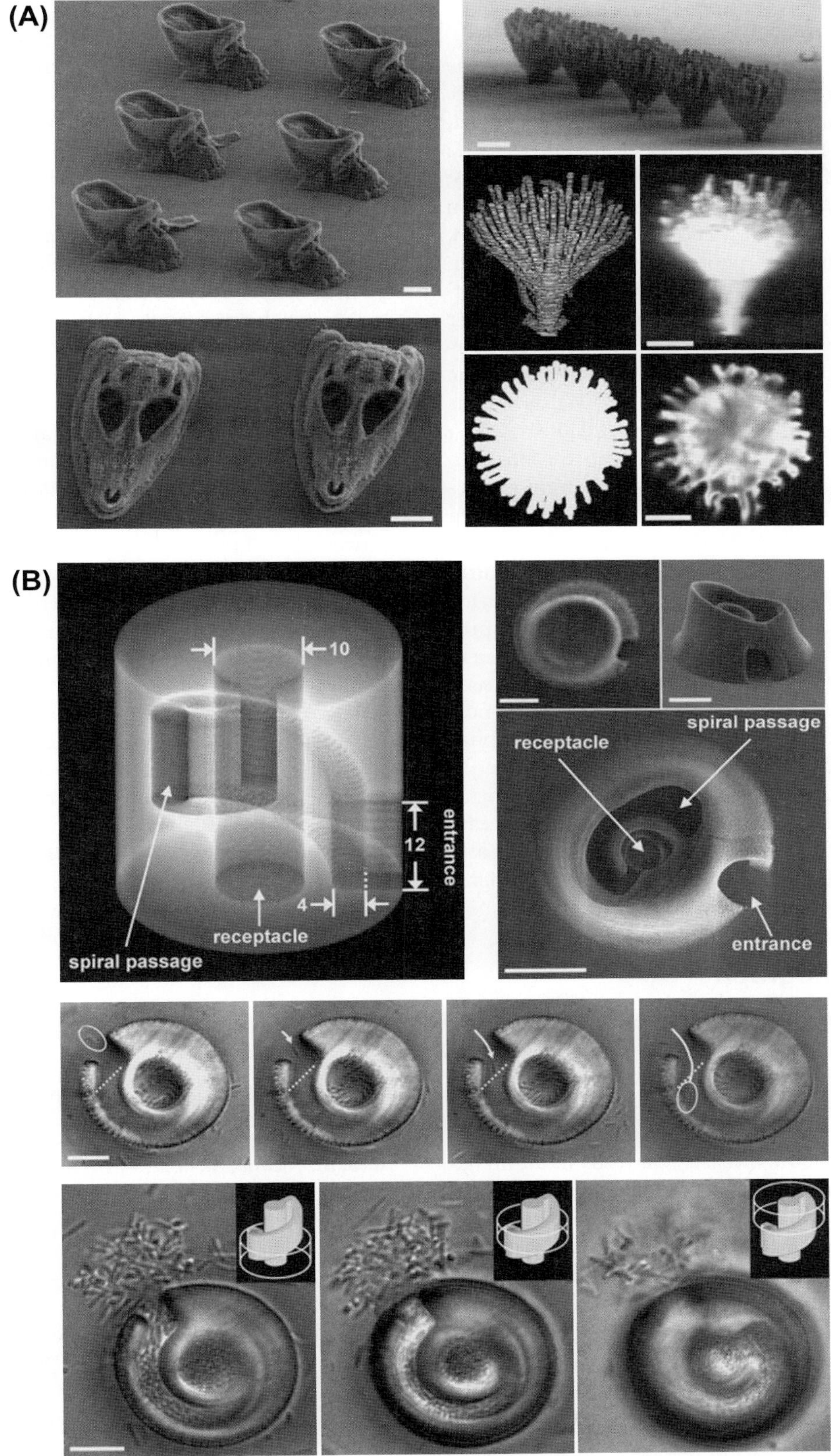

FIGURE I.2.13.15 Three-dimensional protein microstructures fabricated by multiphoton lithography (Neilson et al., 2009). (A) Complex protein microstructures are created by laser scanning with a digital micromirror device. (B) Houses for bacteria made from three-dimensional protein microstructures. All Scale bars, 10 μm. Adapted with permission from Wiley-Blackwell.

et al., 2003), and photonic crystals (Seet et al., 2006; Tetreault et al., 2006; Lafratta et al., 2007; Li and Fourkas, 2007). Perhaps more biologically relevant, however, is the use of MPL to pattern protein-based materials in three dimensions (Pitts et al., 2000, 2002; Basu and Campagnola, 2004; Basu et al., 2004, 2005; Kaehr et al., 2004, 2006; Allen et al., 2005; Hill et al., 2005; Hill and Shear, 2006; Kaehr and Shear, 2007, 2008; Lyon et al., 2007; Nielson et al., 2009; Seidlits et al., 2009). In this approach, solutions of proteins and photosensitizers are used in place of synthetic resins to create biologically compatible three-dimensional protein microstructures. Shear and co-workers have used this methodology to generate three-dimensional protein structures on-the-fly in the presence of living neurons (Kaehr et al., 2004), as well as protein-based houses for bacteria (Kaehr and Shear, 2007; Nielson et al., 2009). This work opens up the possibility of directing single cell and cell–cell interaction experiments in real time, as opposed to relying on patterning a substrate before cell seeding. They have also created three-dimensional protein–hydrogel structures for cell cultures (Seidlits et al., 2009), three-dimensional multi-enzyme reactors (Allen et al., 2005; Hill and Shear, 2006), protein-based conductive wires (Hill et al., 2005), and protein hydrogel microactuators (Kaehr and Shear, 2008). Laser scanning of masks combined with scanning in the vertical dimension was demonstrated as a mask-based, multiphoton variant of projection lithography in three dimensions (Kaehr and Shear, 2007). More recently, laser scanning of a dynamic digital micro-mirror device was used to create protein-based tissue micro-replicas of monkey skulls and insects using high-resolution X-ray computed tomography data (Nielson et al., 2009) (Figure I.2.13.15).

Holographic Lithography

Holographic lithography is a three-dimensional patterning technique based on the projection of three-dimensional images using interference patterns from intersecting laser beams. In this method, multiple coherent lasers are aligned, such that they interfere with one another to create periodic intensity patterns in an image plane. By controlling the number of interfering beams and their relative alignment, phase, amplitude, wave vector, and polarization, one can vary the complexity of the projected images. With the appropriate number of beams and alignment (Del Campo and Greiner, 2007), three-dimensional periodic lattices can be projected on to an image plane. This projection can be used to polymerize photoresists and create three-dimensional objects from the projected images. Holographic patterning is typically used to create three-dimensional templates for fabrication of photonic crystals (Geissler and Xia, 2004; Del Campo and Greiner, 2007). A major advantage of holographic patterning over other methods of three-dimensional patterning is that, once the necessary optics are aligned and configured, it is relatively simple to generate three-dimensional patterns quickly without the use of masks or serial scanning. The drawbacks of this method, however, are the complexity of the alignment procedure, as well as the optical components and sources needed. A holographic approach that is somewhat less technically challenging and more user-friendly uses a spatial light modulator that is illuminated by a single beam and projects three-dimensional holograms from computer generated bitmap images (Jenness et al., 2008). Simultaneous generation of multiple three-dimensional polymeric and protein microstructures has been accomplished using this technique (Jenness et al., 2010).

CONCLUSIONS

Surface patterning has become an extremely important tool in biomedical research. The ability to precisely define the spatial location of biomolecules and/or cells on a surface or in three dimensions provides researchers with a powerful tool to examine the interplay between biomolecules, cells, and artificial materials in a highly controlled spatial environment. This chapter has reviewed a diverse set of patterning techniques, common concerns that need to be addressed in patterning biomolecules and cells, and important figures-of-merit of different patterning methods that should provide the reader with an intelligent stepping stone to sources in the primary literature for a deeper, in-depth investigation of this area of research. This chapter should also help the reader in making a rational decision about the choice of the patterning methodology that is suitable for a particular application. In conclusion, we hope that this chapter will serve as a beginner's guide for those who wish to delve deeper into the field of biomolecular and cellular patterning in two and three dimensions.

BIBLIOGRAPHY

Alexander, C., & Shakesheff, K. (2006). Responsive polymers at the biology/materials science interface. *Adv. Mater.*, **18**(24), 3321–3328.

Allen, R., Nielson, R., Wise, D. D., & Shear, J. B. (2005). Catalytic three-dimensional protein architectures. *Anal. Chem.*, **77**(16), 5089–5095.

Alsberg, E., Feinstein, E., Joy, M., Prentiss, M., & Ingber, D. (2006). Magnetically-guided self-assembly of fibrin matrices with ordered nano-scale structure for tissue engineering. *Tissue Eng.*, **12**(11), 3247–3256.

Amiji, M., & Park, K. (1993). Surface modification of polymeric biomaterials with poly (ethylene oxide), albumin, and heparin for reduced thrombogenicity. *J. Biomater. Sci., Polym. Ed.*, **4**(3), 217–234.

Baldacchini, T., LaFratta, C., Farrer, R. A., Teich, M., Saleh, B., et al. (2004). Acrylic-based resin with favorable properties for three-dimensional two-photon polymerization. *J. Appl. Phys.*, **95**, 6072.

Bard, A., Denuault, G., Lee, C., Mandler, D., & Wipf, D. (1990). Scanning electrochemical microscopy: A new technique for the characterization and modification of surfaces. *Accounts Chem. Res.*, **23**(11), 357–363.

Bardea, A., & Naaman, R. (2009). Magnetolithography: From bottom-up route to high throughput. *Small*, **5**(3), 316–319.

Basu, S., & Campagnola, P. (2004). Enzymatic activity of alkaline phosphatase inside protein and polymer structures fabricated via multiphoton excitation. *Biomacromolecules*, **5**(2), 572–579.

Basu, S., Wolgemuth, C., & Campagnola, P. (2004). Measurement of normal and anomalous diffusion of dyes within protein structures fabricated via multiphoton excited cross-linking. *Biomacromolecules*, **5**(6), 2347–2357.

Basu, S., Cunningham, L., Pins, G., Bush, K., Taboada, R., et al. (2005). Multiphoton excited fabrication of collagen matrices cross-linked by a modified benzophenone dimer: Bioactivity and enzymatic degradation. *Biomacromolecules*, **6**(3), 1465–1474.

Bearinger, J. P., Terrettaz, S., Michel, R., Tirelli, N., Vogel, H., et al. (2003). Chemisorbed poly(propylene sulphide)-based copolymers resist biomolecular interactions. *Nat. Mater.*, **2**(4), 259–264.

Beebe, D., Moore, J., Bauer, J., Yu, Q., Liu, R., et al. (2000). Functional hydrogel structures for autonomous flow control inside microfluidic channels. *Nature*, **404**(6778), 588–590.

Beh, W., Kim, I., Qin, D., Xia, Y., & Whitesides, G. (1999). Formation of patterned microstructures of conducting polymers by soft lithography, and applications in microelectronic device fabrication. *Adv. Mater.*, **11**(12), 1038–1041.

Bernard, A., Delamarche, E., Schmid, H., Michel, B., Bosshard, H., et al. (1998). Printing patterns of proteins. *Langmuir*, **14**(9), 2225–2229.

Bernard, A., Renault, J., Michel, B., Bosshard, H., & Delamarche, E. (2000). Microcontact printing of proteins. *Adv. Mater.*, **12**(14), 1067–1070.

Bernard, A., Fitzli, D., Sonderegger, P., Delamarche, E., Michel, B., et al. (2001a). Affinity capture of proteins from solution and their dissociation by contact printing. *Nat. Biotechnol.*, **19**(9), 866–869.

Bernard, A., Michel, B., & Delamarche, E. (2001b). Micromosaic immunoassays. *Anal. Chem.*, **73**(1), 8–12.

Betzig, E., & Trautman, J. K. (1992). Near-field optics: Microscopy, spectroscopy, and surface modification beyond the diffraction limit. *Science*, **257**(5067), 189–195.

Bier, F. F., von Nickisch-Rosenegk, M., Ehrentreich-Förster, E., Reiss, E., Henkel, J., et al. (2008). DNA microarrays. *Adv. Biochem. Eng. Biotechnol.*, **109**, 433–453.

Brough, B., Christman, K., Wong, T., Kolodziej, C., Forbes, J., et al. (2007). Surface initiated actin polymerization from top-down manufactured nanopatterns. *Soft Matter*, **3**(5), 541–546.

Caiazzo, R. J., Maher, A. J., Drummond, M. P., Lander, C. I., Tassinari, O. W., et al. (2009). Protein microarrays as an application for disease biomarkers. *Prot. Clin. Appl.*, **3**(2), 138–147.

Celio, H., Barton, E., & Stevenson, K. J. (2006). Patterned assembly of colloidal particles by confined dewetting lithography. *Langmuir*, **22**(26), 11426–11435.

Chen, C., Hirdes, D., & Folch, A. (2003). Gray-scale photolithography using microfluidic photomasks. *Proc. Natl. Acad. Sci.*, **100**(4), 1499–1504.

Chen, T., Zhang, J., Chang, D. P., Garcia, A., & Zauscher, S. (2009). Fabrication of micropatterned stimulus-responsive polymer-brush "Anemone.". *Adv. Mater.*, **21**(18), 1825–1829.

Chou, S. (1997). Patterned magnetic nanostructures and quantized magnetic disks. *Proc. IEEE*, **85**(4), 652–671.

Christman, K. L., Schopf, E., Broyer, R. M., Li, R. C., Chen, Y., et al. (2009). Positioning multiple proteins at the nanoscale with electron beam cross-linked functional polymers. *J. Am. Chem. Soc.*, **131**(2), 521–527.

Clark, S., & Hammond, P. T. (1998). Engineering the microfabrication of layer-by-layer thin films. *Adv. Mater.*, **10**(18), 1515–1519.

Clark, S., Montague, M., & Hammond, P. T. (1997). Selective deposition in multilayer assembly: SAMs as molecular templates. *Supramole. Sci.*, **4**(1–2), 141–146.

Del Campo, A., & Greiner, C. (2007). SU-8: A photoresist for high-aspect-ratio and 3D submicron lithography. *J. Micromechanics and Microengineering*, **17**(6), 81–95.

Delehanty, J. B., & Ligler, F. S. (2002). A microarray immunoassay for simultaneous detection of proteins and bacteria. *Anal. Chem.*, **74**(21), 5681–5687.

Demers, L. M., Ginger, D. S., Park, S.-J., Li, Z., Chung, S.-W., et al. (2002). Direct patterning of modified oligonucleotides on metals and insulators by dip-pen nanolithography. *Science*, **296**(5574), 1836–1838.

Derby, B. (2008). Bioprinting: Inkjet printing proteins and hybrid cell-containing materials and structures. *J. Mater. Chem.*, **18**(47), 5717–5721.

DeRisi, J. L., Iyer, V. R., & Brown, P. O. (1997). Exploring the metabolic and genetic control of gene expression on a genomic scale. *Science*, **278**(5338), 680–686.

Dertinger, S. K.W., Jiang, X., Li, Z., Murthy, V. N., & Whitesides, G. M. (2002). Gradients of substrate-bound laminin orient axonal specification of neurons. *Proc. Natl. Acad. Sci. USA*, **99**(20), 12542–12547.

Dong, L., Agarwal, A., Beebe, D., & Jiang, H. (2006). Adaptive liquid microlenses activated by stimuli-responsive hydrogels. *Nature*, **442**(7102), 551–554.

Ducker, R. E., & Leggett, G. J. (2006). A mild etch for the fabrication of three-dimensional nanostructures in gold. *J. Am. Chem. Soc.*, **128**(2), 392–393.

Ducker, R. E., Janusz, S., Sun, S., & Leggett, G. J. (2007a). One-step photochemical introduction of nanopatterned protein-binding functionalities to oligo(ethylene glycol)-terminated self-assembled monolayers. *J. Am. Chem. Soc.*, **129**(48), 14842–14843.

Ducker, R. E., Montague, M., Sun, S., & Leggett, G. J. (2007b). Fabrication of sub-diffraction-limit molecular structures by scanning near-field photolithography. *Proc. SPIE*, **6645**, 664513.

Ducker, R. E., Garcia, A., Zhang, J., Chen, T., & Zauscher, S. (2008). Polymeric and biomacromolecular brush nanostructures: Progress in synthesis, patterning and characterization. *Soft Matter*, **4**(9), 1774–1786.

Dunn, R. (1999). Near-field scanning optical microscopy. *Chem. Rev.*, **99**, 2891–2928.

Eck, W., Stadler, V., Geyer, W., Zharnikov, M., Golzhauser, A., et al. (2000). Generation of surface amino groups on aromatic self-assembled monolayers by low energy electron beams: A first step towards chemical lithography. *Adv. Mater.*, **12**(11), 805–808.

Eigler, D., & Schweizer, E. (1990). Positioning single atoms with a scanning tunnelling microscope. *Nature*, **344**(6266), 524–526.

Elbert, D. L., & Hubbell, J. A. (1998). Reduction of fibrous adhesion formation by a copolymer possessing an affinity for anionic surfaces. *J. Biomed. Mat. Research.*, **42**(1), 55–65.

Erb, R. M., Son, H. S., Samanta, B., Rotello, V. M., & Yellen, B. B. (2009). Magnetic assembly of colloidal superstructures with multipole symmetry. *Nature*, **457**(7232), 999–1002.

Fodor, S. P., Read, J. L., Pirrung, M. C., Stryer, L., Lu, A. T., et al. (1991). Light-directed, spatially addressable parallel chemical synthesis. *Science*, **251**(4995), 767–773.

Fosser, K. A., & Nuzzo, R. G. (2003). Fabrication of patterned multicomponent protein gradients and gradient arrays using microfluidic depletion. *Anal. Chem.*, **75**(21), 5775–5782.

Frey, W., Woods, C., & Chilkoti, A. (2000). Ultraflat nanosphere lithography: A new method to fabricate flat nanostructures. *Adv. Mater.*, **12**(20), 1515–1519.

Frey, W., Meyer, D., & Chilkoti, A. (2003a). Thermodynamically reversible addressing of a stimuli responsive fusion protein onto a patterned surface template. *Langmuir*, **19**(5), 1641–1653.

Frey, W., Meyer, D., & Chilkoti, A. (2003b). Dynamic addressing of a surface pattern by a stimuli-responsive fusion protein. *Adv. Mater.*, **15**(3), 248–251.

Geissler, M., & Xia, Y. (2004). Patterning: Principles and some new developments. *Adv. Mater.*, **16**(15), 1249–1269.

Golzhauser, A., Eck, W., Geyer, W., Stadler, V., Weimann, T., et al. (2001). Chemical nanolithography with electron beams. *Adv. Mater.*, **13**(11), 803–806.

Gratson, G. M., Xu, M., & Lewis, J. A. (2004). Microperiodic structures: Direct writing of three-dimensional webs. *Nature*, **428**(6981), 386.

Gross, G. W., & Schwalm, F. U. (1994). A closed flow chamber for long-term multichannel recording and optical monitoring. *J. Neurosci. Methods*, **52**, 73–85.

Gross, G. W., Wen, W. Y., & Lin, J. W. (1985). Transparent indium-tin oxide electrode patterns for extracellular, multisite recording in neuronal cultures. *J. Neurosci. Methods*, **15**(3), 243–252.

Hamley, I. (2003). Nanostructure fabrication using block copolymers. *Nanotechnology*, **14**(10), 39–39.

Hammond, P. T., & Whitesides, G. (1995). Formation of polymer microstructures by selective deposition of polyion multilayers using patterned self-assembled monolayers as a template. *Macromolecules*, **28**(22), 7569–7571.

Haynes, C., & Van Duyne, R. (2001). Nanosphere lithography: A versatile nanofabrication tool for studies of size-dependent nanoparticle optics. *J. Phys. Chem. B.*, **105**(24), 5599–5611.

Healy, K. E., Lom, B., & Hockberger, P. E. (1994). Spatial distribution of mammalian cells dictated by material surface chemistry. *Biotechnol. Bioeng.*, **43**(8), 792–800.

Herrwerth, S., Eck, W., Reinhardt, S., & Grunze, M. (2003). Factors that determine the protein resistance of oligoether self-assembled monolayers: Internal hydrophilicity, terminal hydrophilicity, and lateral packing density. *J. Am. Chem. Soc.*, **125**(31), 9359–9366.

Hill, R. T., & Shear, J. B. (2006). Enzyme-nanoparticle functionalization of three-dimensional protein scaffolds. *Anal. Chem.*, **78**(19), 7022–7026.

Hill, R. T., Lyon, J., Allen, R., Stevenson, K., & Shear, J. B. (2005). Microfabrication of three-dimensional bioelectronic architectures. *J. Am. Chem. Soc.*, **127**(30), 10707–10711.

Holden, M. A., Jung, S.-Y., & Cremer, P.S. (2004). Patterning enzymes inside microfluidic channels via photoattachment chemistry. *Anal. Chem.*, **76**(7), 1838–1843.

Hong, S., Zhu, J., & Mirkin, C. (1999). Multiple ink nanolithography: Toward a multiple-pen nano-plotter. *Science*, **286**(5439), 523–525.

Hu, Y., Das, A., Hecht, M. H., & Scoles, G. (2005). Nanografting de novo proteins onto gold surfaces. *Langmuir*, **21**(20), 9103–9109.

Huang, N., Michel, R., Voros, J., Textor, M., Hofer, R., et al. (2001). Poly (L-lysine)-g-poly (ethylene glycol) layers on metal oxide surfaces: Surface-analytical characterization and resistance to serum and fibrinogen adsorption. *Langmuir*, **17**(2), 489–498.

Hucknall, A., Simnick, A., Hill, R. T., Chilkoti, A., Garcia, A., et al. (2009a). Versatile synthesis and micropatterning of nonfouling polymer brushes on the wafer scale. *Biointerphases*, **4**(2), FA50–FA57.

Hucknall, A., Rangarajan, S., & Chilkoti, A. (2009b). In pursuit of zero: Polymer brushes that resist the adsorption of proteins. *Adv. Mater.*, **21**(23), 2441–2446.

Hucknall, A., Kim, D.-H., Rangarajan, S., Hill, R. T., Reichert, W. M., et al. (2009c). Simple fabrication of antibody microarrays on nonfouling polymer brushes with femtomolar sensitivity for protein analytes in serum and blood. *Adv. Mater.*, **21**(19), 1968–1971.

Hyun, J., Zhu, Y., Liebmann-Vinson, A., Beebe, T., Jr., & Chilkoti, A. (2001). Microstamping on an activated polymer surface: Patterning biotin and streptavidin onto common polymeric biomaterials. *Langmuir*, **17**(20), 6358–6367.

Hyun, J., Ma, H., Zhang, Z., Beebe, T. P., Jr., & Chilkoti, A. (2003). Universal route to cell micropatterning using an amphiphilic comb polymer. *Adv. Mater.*, **15**(78), 576–579.

Hyun, J., Kim, J., Craig, S. L., & Chilkoti, A. (2004a). Enzymatic nanolithography of a self-assembled oligonucleotide monolayer on gold. *J. Am. Chem. Soc.*, **126**(15), 4770–4771.

Hyun, J., Lee, W.-K., Nath, N., Chilkoti, A., & Zauscher, S. (2004b). Capture and release of proteins on the nanoscale by stimuli-responsive elastin-like polypeptide "switches." *J. Am. Chem. Soc.*, **126**(23), 7330–7335.

Ino, K., Ito, A., & Honda, H. (2007). Cell patterning using magnetite nanoparticles and magnetic force. *Biotechnol. Bioeng.*, **97**(5), 1309–1317.

Jackman, R. J., Wilbur, J. L., & Whitesides, G. M. (1995). Fabrication of submicrometer features on curved substrates by microcontact printing. *Science*, **269**(5224), 664–666.

Jager, E., Smela, E., & Inganas, O. (2000). Microfabricating conjugated polymer actuators. *Science*, **290**(5496), 1540.

James, C., Davis, R., Kam, L., Craighead, H., Isaacson, M., et al. (1998). Patterned protein layers on solid substrates by thin stamp microcontact printing. *Langmuir*, **14**(4), 741–744.

Jenness, N. J., Wulff, K. D., Johannes, M. S., Padgett, M. J., Cole, D. G., et al. (2008). Three-dimensional parallel holographic micropatterning using a spatial light modulator. *Opt. Express*, **16**(20), 15942–15948.

Jenness, N. J., Hill, R. T., Hucknall, A., Chilkoti, A., & Clark, R. L. (2010). A versatile diffractive maskless lithography for single-shot and serial microfabrication. *Opt. Express*, **18**(11), 11754–11762.

Jiang, X., & Hammond, P. T. (2000). Selective deposition in layer-by-layer assembly: Functional graft copolymers as molecular templates. *Langmuir*, **16**(22), 8501–8509.

Joglekar, A., Liu, H., Spooner, G., Meyhöfer, E., Mourou, G., et al. (2003). A study of the deterministic character of optical damage by femtosecond laser pulses and applications to nanomachining. *Appl. Phys. B: Lasers Opt.*, **77**(1), 25–30.

Juncker, D., Schmid, H., Drechsler, U., Wolf, H., Wolf, M., et al. (2002). Autonomous microfluidic capillary system. *Anal. Chem.*, **74**(24), 6139–6144.

Kaehr, B., & Shear, J. B. (2007). Mask-directed multiphoton lithography. *J. Am. Chem. Soc.*, **129**(7), 1904–1905.

Kaehr, B., & Shear, J. B. (2008). Multiphoton fabrication of chemically responsive protein hydrogels for microactuation. *Proc. Natl. Acad. Sci. USA*, **105**(26), 8850–8854.

Kaehr, B., Allen, R., Javier, D. J., Currie, J., & Shear, J. B. (2004). Guiding neuronal development with *in situ* microfabrication. *Proc. Natl. Acad. Sci. USA*, **101**(46), 16104–16108.

Kaehr, B., Ertas, N., Nielson, R., Allen, R., Hill, R. T., et al. (2006). Direct-write fabrication of functional protein matrixes using a low-cost Q-switched laser. *Anal. Chem.*, **78**(9), 3198–3202.

Kenausis, G., Voros, J., Elbert, D., Huang, N., Hofer, R., et al. (2000). Poly (l-lysine)-g-poly (ethylene glycol) layers on metal oxide surfaces: Attachment mechanism and effects of polymer architecture on resistance to protein adsorption. *J. Phys. Chem. B.*, **104**(14), 3298–3309.

Kenseth, J., Harnisch, J., Jones, V., & Porter, M. (2001). Investigation of approaches for the fabrication of protein patterns by scanning probe lithography. *Langmuir*, **17**(13), 4105–4112.

Kim, D. H., Lau, K. H. A., Robertson, J. W. F., Lee, O.-J., Jeong, U., et al. (2005). Thin films of block copolymers as planar optical waveguides. *Adv. Mater.*, **17**(20), 2442–2445.

Kim, E., Xia, Y., & Whitesides, G. (1995). Polymer microstructures formed by moulding in capillaries. *Nature*, **376**(17), 581–584.

Kim, E., Xia, Y., Zhao, X., & Whitesides, G. (1997). Solvent-assisted microcontact molding: A convenient method for fabricating three-dimensional structures on surfaces of polymers. *Adv. Mater. – Including CVD – Chemical Vapor Deposition*, **9**(8), 651–653.

Kim, H., & Hinsberg, W. (2008). Surface patterns from block copolymer self-assembly. *J. Vac. Sci. & Technol. A: Vacuum, Surfaces, and Films*, **26**, 1369.

Kojima, K., Moriguchi, H., Hattori, A., Kaneko, T., & Yasuda, K. (2003). Two-dimensional network formation of cardiac myocytes in agar microculture chip with 1480 nm infrared laser photo-thermal etching. *Lab on a Chip*, **3**(4), 292–296.

Krämer, S., Fuierer, R. R., & Gorman, C. B. (2003). Scanning probe lithography using self-assembled monolayers. *Chem. Rev.*, **103**(11), 4367–4418.

Krebs, M. D., Erb, R. M., Yellen, B. B., Samanta, B., Bajaj, A., et al. (2009). Formation of ordered cellular structures in suspension via label-free negative magnetophoresis. *Nano Lett.*, **9**(5), 1812–1817.

Kumar, A., & Whitesides, G. (1993). Features of gold having micrometer to centimeter dimensions can be formed through a combination of stamping with an elastomeric stamp and an alkanethiol "ink" followed by chemical etching. *Appl. Phys. Lett.*, **63**, 2002.

Kung, L., Kam, L., Hovis, J., & Boxer, S. (2000). Patterning hybrid surfaces of proteins and supported lipid bilayers. *Langmuir*, **16**(17), 6773–6776.

Lafratta, C. N., Fourkas, J. T., Baldacchini, T., & Farrer, R. A. (2007). Multiphoton fabrication. *Angew. Chem. Int. Ed.*, **46**(33), 6238–6258.

Lee, G., Shin, Y.-H., & Son, J. Y. (2009). Formation of self-assembled polyelectrolyte multilayer nanodots by scanning probe microscopy. *J. Am. Chem. Soc.*, **131**(5), 1634–1635.

Lee, J. H., Kopecek, J., & Andrade, J. D. (1989). Protein-resistant surfaces prepared by PEO-containing block copolymer surfactants. *J. Biomed. Mat. Research*, **23**(3), 351–368.

Lee, K.-B., Park, S.-J., Mirkin, C. A., Smith, J. C., & Mrksich, M. (2002). Protein nanoarrays generated by dip-pen nanolithography. *Science*, **295**(5560), 1702–1705.

Lee, S. W., Oh, B.-K., Sanedrin, R. G., Salaita, K., Fujigaya, T., et al. (2006a). Biologically active protein nanoarrays generated using parallel dip-pen nanolithography. *Adv. Mater.*, **18**(9), 1133–1136.

Lee, W.-K., Caster, K. C., Kim, J., & Zauscher, S. (2006b). Nanopatterned polymer brushes by combining AFM anodization lithography with ring-opening metathesis polymerization in the liquid and vapor phase. *Small*, **2**(7), 848–853.

Levskaya, A., Chevalier, A. A., Tabor, J. J., Simpson, Z. B., Lavery, L. A., et al. (2005). Synthetic biology: Engineering *Escherichia coli* to see light. *Nature*, **438**(7067), 441–442.

Li, L., & Fourkas, J. T. (2007). Multiphoton polymerization. *Mater. Today*, **10**(6), 30–37.

Li, M., & Ober, C. (2006). Block copolymer patterns and templates. *Mater. Today*, **9**(9), 30–39.

Lin, R.-Z., Chu, W.-C., Chiang, C.-C., Lai, C.-H., & Chang, H.-Y. (2008). Magnetic reconstruction of three-dimensional tissues from multicellular spheroids. *Tissue Eng. Part C: Methods*, **14**(3), 197–205.

Liu, G., Xu, S., & Qian, Y. (2000). Nanofabrication of self-assembled monolayers using scanning probe lithography. *Acc. Chem. Res.*, **33**(7), 457–466.

Liu, M., Amro, N., Chow, C., & Liu, G. (2002). Production of nanostructures of DNA on surfaces. *Nano Lett.*, **2**(8), 863–867.

Liu, V., Jastromb, W., & Bhatia, S. (2002). Engineering protein and cell adhesivity using PEO-terminated triblock polymers. *J. Biomed. Mat. Research*, **60**(1), 126–134.

Liu, Y., Klep, V., & Luzinov, I. (2006). To patterned binary polymer brushes via capillary force lithography and surface-initiated polymerization. *J. Am. Chem. Soc.*, **128**(25), 8106–8107.

Llanos, G., & Sefton, M. (1993). Review: Does polyethylene oxide possess a low thrombogenicity? *J. Biomat. Sci., Polymer Edition*, **4**(3), 381–400.

Lom, B., Healy, K. E., & Hockberger, P. (1993). A versatile technique for patterning biomolecules onto glass coverslips. *J. Neurosci. Methods*, **50**(3), 385–397.

Lopez, G., Albers, M., Schreiber, S., Carroll, R., Peralta, E., et al. (1993). Convenient methods for patterning the adhesion of mammalian cells to surfaces using self-assembled monolayers of alkanethiolates on gold. *J. Am. Chem. Soc.*, **115**(13), 5877–5878.

López, G. P., Ratner, B. D., Tidwell, C. D., Haycox, C. L., Rapoza, R. J., et al. (1992). Glow discharge plasma deposition of tetraethylene glycol dimethyl ether for fouling-resistant biomaterial surfaces. *J. Biomed. Mater. Res.*, **26**(4), 415–439.

Love, J. C., Estroff, L., Kriebel, J., Nuzzo, R., & Whitesides, G. M. (2005). Self-assembled monolayers of thiolates on metals as a form of nanotechnology. *Chem. Rev.*, **105**(4), 1103–1169.

Lyon, J. L., Hill, R. T., Shear, J. B., & Stevenson, K. J. (2007). Direct electrochemical and spectroscopic assessment of heme integrity in multiphoton photo-cross-linked cytochrome C structures. *Anal. Chem.*, **79**(6), 2303–2311.

Ma, H., Hyun, J., Stiller, P., & Chilkoti, A. (2004). "Non-fouling" oligo(ethylene glycol)- functionalized polymer brushes synthesized by surface-initiated atom transfer radical polymerization. *Adv. Mater.*, **16**(4), 338–341.

Ma, H., Hyun, J., Zhang, Z., Beebe, T., Jr., & Chilkoti, A. (2005). Fabrication of biofunctionalized quasi-three-dimensional microstructures of a nonfouling comb polymer using soft lithography. *Adv. Funct. Mats.*, **15**(4), 529–540.

Ma, H., Wells, M., Beebe, T. P., & Chilkoti, A. (2006a). Surface-initiated atom transfer radical polymerization of oligo(ethylene glycol) methyl methacrylate from a mixed self-assembled monolayer on gold. *Adv. Funct. Mater.*, **16**(5), 640–648.

Ma, H., Li, D., Sheng, X., Zhao, B., & Chilkoti, A. (2006b). Protein-resistant polymer coatings on silicon oxide by surface-initiated atom transfer radical polymerization. *Langmuir*, **22**(8), 3751–3756.

MacBeath, G., & Schreiber, S. L. (2000). Printing proteins as microarrays for high-throughput function determination. *Science*, **289**(5485), 1760–1763.

Manoharan, H., Lutz, C., & Eigler, D. (2000). Quantum mirages formed by coherent projection of electronic structure. *Nature*, **403**(6769), 512–515.

Marshall, A., & Hodgson, J. (1998). DNA chips: An array of possibilities. *Nat. Biotechnol.*, **16**(1), 27–31.

Maruo, S., Ikuta, K., & Korogi, H. (2003). Force-controllable, optically driven micromachines fabricated by single-step two-photon microstereolithography. *J. Microelectromechanical Sys.*, **12**(5), 533–539.

McDonald, J., Duffy, D., Anderson, J., Chiu, D., Wu, H., et al. (2000). Fabrication of microfluidic systems in poly (dimethylsiloxane). *Electrophoresis*, **21**(1), 27–40.

Mendes, P. M., Yeung, C. L., & Preece, J. A. (2007). Bio-nanopatterning of surfaces. *Nanoscale Res. Lett.*, **2**(8), 373–384.

Moriguchi, H., & Yasuda, K. (2006). Photothermal microneedle etching: Improved three-dimensional microfabrication method for agarose gel for topographical control of cultured cell communities. *Japanese J. Appl. Phys.*, **45**(30), L796–L799.

Mrksich, M. (2002). What can surface chemistry do for cell biology? *Curr. Opin. Chem. Biol.*, **6**(6), 794–797.

Mrksich, M. (2009). Using self-assembled monolayers to model the extracellular matrix. *Acta Biomater.*, **5**(3), 832–841.

Nagaraj, V. J., Eaton, S., Thirstrup, D., & Wiktor, P. (2008). Piezoelectric printing and probing of Lectin NanoProbeArrays for glycosylation analysis. *Biochem. Biophys. Res. Commun.*, **375**(4), 526–530.

Nath, N., & Chilkoti, A. (2003). Fabrication of a reversible protein array directly from cell lysate using a stimuli-responsive polypeptide. *Anal. Chem.*, 75(4), 709–715.

Nie, Z., & Kumacheva, E. (2008). Patterning surfaces with functional polymers. *Nat. Mater.*, 7(4), 277–290.

Nielson, R., & Shear, J. B. (2006). Parallel chemical dosing of subcellular targets. *Anal. Chem.*, **78**(17), 5987–5993.

Nielson, R., Kaehr, B., & Shear, J. B. (2009). Microreplication and design of biological architectures using dynamic-mask multiphoton lithography. *Small*, 5(1), 120–125.

Orchekowski, R., Hamelink, D., Li, L., Gliwa, E., vanBrocklin, M., et al. (2005). Antibody microarray profiling reveals individual and combined serum proteins associated with pancreatic cancer. *Cancer Res.*, **65**(23), 11193–11202.

Ouk Kim, S., Solak, H. H., Stoykovich, M. P., Ferrier, N. J., De Pablo, J. J., et al. (2003). Epitaxial self-assembly of block copolymers on lithographically defined nanopatterned substrates. *Nature*, **424**(6947), 411–414.

Pardo, L., Wilson, W. C., & Boland, T. (2003). Characterization of patterned self-assembled monolayers and protein arrays generated by the ink-jet method. *Langmuir*, **19**(5), 1462–1466.

Park, C., Yoon, J., & Thomas, E. (2003). Enabling nanotechnology with self assembled block copolymer patterns. *Polymer*, **44**(22), 6725–6760.

Patolsky, F., Timko, B. P., Yu, G., Fang, Y., Greytak, A. B., et al. (2006). Detection, stimulation, and inhibition of neuronal signals with high-density nanowire transistor arrays. *Science*, **313**(5790), 1100–1104.

Peppas, N., Hilt, J., Khademhosseini, A., & Langer, R. (2006). Hydrogels in biology and medicine: From molecular principles to bionanotechnology. *Adv. Mater.*, **18**(11), 1345–1360.

Pilobello, K. T., & Mahal, L. K. (2007). Deciphering the glycocode: The complexity and analytical challenge of glycomics. *Curr. Opin. Chem. Biol.*, **11**(3), 300–335.

Pilobello, K. T., Krishnamoorthy, L., Slawek, D., & Mahal, L. K. (2005). Development of a lectin microarray for the rapid analysis of protein glycopatterns. *Chembiochem.*, **6**(6), 985–989.

Pilobello, K. T., Slawek, D. E., & Mahal, L. K. (2007). A ratiometric lectin microarray approach to analysis of the dynamic mammalian glycome. *Proc. Natl. Acad. Sci. USA*, **104**(28), 11534–11539.

Piner, R., Zhu, J., Xu, F., Hong, S., & Mirkin, C. (1999). "Dip-pen" nanolithography. *Science*, **283**(5402), 661.

Pitts, J., Campagnola, P., Epling, G., & Goodman, S. (2000). Submicron multiphoton free-form fabrication of proteins and polymers: Studies of reaction efficiencies and applications in sustained release. *Macromolecules*, **33**(5), 1514–1523.

Pitts, J., Howell, A., Taboada, R., Banerjee, I., Wang, J., et al. (2002). New photoactivators for multiphoton excited three-dimensional submicron cross-linking of proteins: Bovine serum albumin and Type 1 collagen. *Photochemistry and Photobiology*, **76**(2), 135–144.

Prime, K., & Whitesides, G. (1993). Adsorption of proteins onto surfaces containing end-attached oligo (ethylene oxide): A model system using self-assembled monolayers. *J. Am. Chem. Soc.*, **115**(23), 10714–10721.

Ray, S., Britschgi, M., Herbert, C., Takeda-Uchimura, Y., Boxer, A., et al. (2007). Classification and prediction of clinical Alzheimer's diagnosis based on plasma signaling proteins. *Nat. Med.*, **13**(11), 1359–1362.

Reetz, M., Winter, M., Dumpich, G., Lohau, J., & Friedrichowski, S. (1997). Fabrication of metallic and bimetallic nanostructures by electron beam induced metallization of surfactant stabilized Pd and Pd/Pt clusters. *J. Am. Chem. Soc.*, **119**(19), 4539–4540.

Renault, J., Bernard, A., Juncker, D., Michel, B., Bosshard, H., et al. (2002). Fabricating microarrays of functional proteins using affinity contact printing. *Angewandte Chemie (International ed. Print)*, **41**(13), 2320–2323.

Rodriguez, I., Spicar-Mihalic, P., Kuyper, C., Fiorini, G., & Chiu, D. (2003). Rapid prototyping of glass microchannels. *Anal. Chim. Acta*, **496**(1–2), 205–215.

Salaita, K., Wang, Y., Fragala, J., Vega, R. A., Liu, C., et al. (2006). Massively parallel dip-pen nanolithography with 55,000-pen two-dimensional arrays. *Angew. Chem. Int. Ed.*, **45**(43), 7220–7223.

Schmelmer, U., Paul, A., Küller, A., Steenackers, M., Ulman, A., et al. (2007). Nanostructured polymer brushes. *Small*, **3**(3), 459–465.

Seet, K., Mizeikis, V., Juodkazis, S., & Misawa, H. (2006). Three-dimensional horizontal circular spiral photonic crystals with stop gaps below 1 μm. *Appl. Phys. Lett.*, **88**, 221101.

Seidlits, S. K., Schmidt, C. E., & Shear, J. B. (2009). High-resolution patterning of hydrogels in three dimensions using direct-write photofabrication for cell guidance. *Adv. Funct. Mater.*, **19**(22), 3543–3551.

Serbin, J., Egbert, A., Ostendorf, A., Chichkov, B. N., Houbertz, R., et al. (2003). Femtosecond laser-induced two-photon polymerization of inorganic organic hybrid materials for applications in photonics. *Opt. Lett.*, **28**(5), 301–303.

Shiku, H., Uchida, I., & Matsue, T. (1997). Microfabrication of alkylsilanized glass substrate by electrogenerated hydroxyl radical using scanning electrochemical microscopy. *Langmuir*, **13**(26), 7239–7244.

Shin, K., Leach, K., Goldbach, J., Kim, D., Jho, J., et al. (2002). A simple route to metal nanodots and nanoporous metal films. *Nano. Lett.*, **2**(9), 933–936.

Sidorenko, A., Krupenkin, T., Taylor, A., Fratzl, P., & Aizenberg, J. (2007). Reversible switching of hydrogel-actuated nanostructures into complex micropatterns. *Science*, **315**(5811), 487.

Singhvi, R., Kumar, A., Lopez, G., Stephanopoulos, G., Wang, D., et al. (1994). Engineering cell shape and function. *Science*, **264**(5159), 696–698.

Steenackers, M., Küller, A., Ballav, N., Zharnikov, M., Grunze, M., et al. (2007). Morphology control of structured polymer brushes. *Small*, **3**(10), 1764–1773.

Sugio, Y., Kojima, K., Moriguchi, H., Takahashi, K., Kaneko, T., et al. (2004). An agar-based on-chip neural-cell-cultivation system for stepwise control of network pattern generation during cultivation. *Sensor. Actuat.: B Chem.*, **1**, 156–162.

Suh, K., Choi, S., Baek, S., Kim, T., & Langer, R. (2005). Observation of high-aspect-ratio nanostructures using capillary lithography. *Adv. Mater.*, **17**(5), 560–563.

Sun, S., & Leggett, G. J. (2004). Matching the resolution of electron beam lithography by scanning near-field photolithography. *Nano Lett.*, 4(8), 1381–1384.

Sun, S., Chong, K. S. L., & Leggett, G. J. (2002). Nanoscale molecular patterns fabricated by using scanning near-field optical lithography. *J. Am. Chem. Soc.*, **124**(11), 2414–2415.

Suzuki, I., Sugio, Y., Moriguchi, H., Jimbo, Y., & Yasuda, K. (2004). Modification of a neuronal network direction using stepwise photo-thermal etching of an agarose architecture. *J. Nanobiotechnology*, **2**(1), 7.

Tan, J., Tien, J., & Chen, C. (2002). Microcontact printing of proteins on mixed self-assembled monolayers. *Langmuir*, **18**(2), 519–523.

Taylor, A. M., Blurton-Jones, M., Rhee, S. W., Cribbs, D. H., Cotman, C. W., et al. (2005). A microfluidic culture platform for CNS axonal injury, regeneration and transport. *Nat. Methods*, **2**(8), 599–605.

Terray, A., Oakey, J., & Marr, D. (2002). Microfluidic control using colloidal devices. *Science*, **296**(5574), 1841.

Tetreault, N., von Freymann, G., Deubel, M., Hermatschweiler, M., Pérez-Willard, F., et al. (2006). New route to three-dimensional photonic bandgap materials: Silicon double inversion of polymer templates. *Adv. Mater.*, **18**(4), 457–460.

Therriault, D., White, S. R., & Lewis, J. A. (2003). Chaotic mixing in three-dimensional microvascular networks fabricated by direct-write assembly. *Nat. Mater.*, **2**(4), 265–271.

Totzeck, M., Ulrich, W., Göhnermeier, A., & Kaiser, W. (2007). Semiconductor fabrication: Pushing deep ultraviolet lithography to its limits. *Nat. Photonics*, **1**, 629–631.

Valkama, S., Kosonen, H., Ruokolainen, J., Haatainen, T., Torkkeli, M., et al. (2004). Self-assembled polymeric solid films with temperature-induced large and reversible photonic-bandgap switching. *Nat. Mater.*, **3**(12), 872–876.

Vega, R. A., Maspoch, D., Salaita, K., & Mirkin, C. A. (2005). Nanoarrays of single virus particles. *Angew. Chem. Int. Ed. Engl.*, **44**(37), 6013–6015.

Wadu-Mesthrige, K., Xu, S., Amro, N., & Liu, G. (1999). Fabrication and imaging of nanometer-sized protein patterns. *Langmuir*, **15**(25), 8580–8583.

Wadu-Mesthrige, K., Amro, N. A., Garno, J. C., Xu, S., & Liu, G.-Y. (2008). Fabrication of nanometer-sized protein patterns using atomic force microscopy and selective immobilization. *Biophys. J.*, **80**(4), 1891–1899.

Wegscheider, S., Kirsch, A., Mlynek, J., & Krausch, G. (1995). Scanning near-field optical lithography. *Thin Solid Films*, **264**(2), 264–267.

Whitesides, G. M., Ostuni, E., Takayama, S., Jiang, X., & Ingber, D. (2001). Soft lithography in biology and biochemistry. *Ann. Rev. Biomed. Eng.*, **3**(1), 335–373.

Wittstock, G., & Schuhmann, W. (1997). Formation and imaging of microscopic enzymatically active spots on an alkanethiolate-covered gold electrode by scanning electrochemical microscopy. *Anal. Chem.*, **69**(24), 5059–5066.

Wittstock, G., Hesse, R., & Schuhmann, W. (1997). Patterned self-assembled alkanethiolate monolayers on gold. Patterning and imaging by means of scanning electrochemical microscopy. *Electroanalysis (Weinheim)*, **9**(10), 746–750.

Woodfield, T. B. F., Malda, J., de Wijn, J., Péters, F., Riesle, J., et al. (2004). Design of porous scaffolds for cartilage tissue engineering using a three-dimensional fiber-deposition technique. *Biomaterials*, **25**(18), 4149–4161.

Xia, N., Hu, Y., Grainger, D., & Castner, D. (2002). Functionalized poly (ethylene glycol)-grafted polysiloxane monolayers for control of protein binding. *Langmuir*, **18**(8), 3255–3262.

Xia, Y., & Whitesides, G. M. (1998). Soft lithography. *Angew. Chem. Int. Ed.*, **37**, 550–575.

Xia, Y., Kim, E., Zhao, X., Rogers, J., Prentiss, M., et al. (1996a). Complex optical surfaces formed by replica molding against elastomeric masters. *Science*, **273**(5273), 347–349.

Xia, Y., Kim, E., & Whitesides, G. (1996b). Micromolding of polymers in capillaries: Applications in microfabrication. *Chem. Mater.*, **8**(7), 1558–1567.

Xia, Y., McClelland, J., Gupta, R., Qin, D., Zhao, X., et al. (1997). Replica molding using polymeric materials: A practical step toward nanomanufacturing. *Adv. Mater. (Weinheim)*, **9**(2), 147–149.

Xia, Y., Qin, D., & Whitesides, G. (2004). Microcontact printing with a cylindrical rolling stamp: A practical step toward automatic manufacturing of patterns with submicrometer-sized features. *Adv. Mater.*, **8**(12), 1015–1017.

Xu, C., Wayland, B., Fryd, M., Winey, K., & Composto, R. (2006b). pH-responsive nanostructures assembled from amphiphilic block copolymers. *Macromolecules*, **39**(18), 6063–6070.

Xu, S., & Liu, G. (1997). Nanometer-scale fabrication by simultaneous nanoshaving and molecular self-assembly. *Langmuir*, **13**(2), 127–129.

Xu, T., Petridou, S., Lee, E. H., Roth, E. A., Vyavahare, N. R., et al. (2004). Construction of high-density bacterial colony arrays and patterns by the ink-jet method. *Biotechnol. Bioeng.*, **85**(1), 29–33.

Xu, T., Jin, J., Gregory, C. A., Hickman, J. J., & Boland, T. (2005). Inkjet printing of viable mammalian cells. *Biomaterials*, **26**(1), 93–99.

Xu, T., Gregory, C. A., Molnar, P., Cui, X., Jalota, S., et al. (2006a). Viability and electrophysiology of neural cell structures generated by the inkjet printing method. *Biomaterials*, **27**(19), 3580–3588.

Yang, Z., Frey, W., Oliver, T., & Chilkoti, A. (2000). Light-activated affinity micropatterning of proteins on self-assembled monolayers on gold. *Langmuir*, **16**(4), 1751–1758.

Yellen, B., Erb, R. M., Halverson, D., Hovorka, O., & Friedman, G. (2006). Arraying nonmagnetic colloids by magnetic nanoparticle assemblers. *IEEE Trans. Magn.*, **42**(10), 3548–3553.

Yeo, W., Yousaf, M., & Mrksich, M. (2003). Dynamic interfaces between cells and surfaces: Electroactive substrates that sequentially release and attach cells. *J. Am. Chem. Soc.*, **125**(49), 14994–14995.

Yousaf, M. N., Houseman, B. T., & Mrksich, M. (2001). Using electroactive substrates to pattern the attachment of two different cell populations. *Proc. Natl. Acad. Sci. USA*, **98**(11), 5992–5996.

Zauscher, S., & Chilkoti, A. (2009). Biological applications of polymer brushes. *Biointerphases*, **4**. FA1.

Zeck, G., & Fromherz, P. (2001). Noninvasive neuroelectronic interfacing with synaptically connected snail neurons immobilized on a semiconductor chip. *Proc. Natl. Acad. Sci. USA*, **98**(18), 10457–10462.

Zhou, F., Zheng, Z., Yu, B., Liu, W., & Huck, W. T.S. (2006). Multicomponent polymer brushes. *J. Am. Chem. Soc.*, **128**(50), 16253–16258.

CHAPTER I.2.14 MEDICAL FIBERS AND BIOTEXTILES

Martin W. King[1,2] *and Sangwon Chung*[3,4]

[1]Department of Textile Engineering, Chemistry & Science, North Carolina State University, Raleigh, NC, USA

[2]Donghua University, Shanghai, China

[3]Fiber and Polymer Science, Department of Textile Engineering, Chemistry & Science, North Carolina State University, Raleigh, NC, USA

[4]Biomedical Engineering, Joint Department of Biomedical Engineering, North Carolina State University and University of North Carolina at Chapel Hill, NC, USA

The use of textiles in medicine is not new. It goes back to the early Egyptians who used linen threads to ligate blood vessels and approximate skin tissue after injury (Shalaby, 1985). Over the past several decades the use of fibers and textiles in medicine has grown dramatically as new and innovative fibers, structures, and therapies have been developed. Advances in fiber spinning methods, fabric assembly technologies, and surface modification treatments have led to numerous new concepts for both products and therapies, some of which are still in development or in clinical trials.

The generic term **biomedical textile** encompasses a wide range of medical products and devices for both internal and external use. Their role is to improve health and wellness, maintain comfort and hygiene, prevent or treat an injury, avoid infection or control disease, assist in repairing an internal injury or surgical operation, or replacing a diseased or injured internal organ or tissue.

BOX 1 What are Biotextiles?

The following definition of **biotextiles** has been derived from a parallel definition of **biomaterials** from Professor D. F. Williams (Williams, 1987).

"Biotextiles are non-viable, permanent or temporary, fibrous textile structures created from synthetic or natural materials that are used in an *in vivo*, *ex vivo* or *in vitro* biological environment as a medical device for the prevention, treatment or diagnosis of an injury or disease, and as such, serve to improve the health, medical condition, comfort and wellness of the patient" (King, 2001).

The types of external biomedical textile products can vary from the simple bandage, external casting or support, feminine hygiene product, diaper or incontinence pad to surgical drapes, protective gowns, and masks worn in the clinic and operating room. Since the field of biomaterials science deals with implantable devices and artificial organs, the focus of this chapter will be limited to those biomedical textile products that are called **biotextiles** and are used for **medical applications** only (see Box 1).

As can be seen from the contents of this textbook, today's surgeon has an impressive armamentarium of artificial devices as well as new viable living tissue products at his or her disposal, many of which are textile-based products that contain textile fibers. These biotextiles are implanted for a number of different reasons, including the replacement or bypass of blood vessels, the replacement of heart valves, knee ligaments, and the repair of hernias and prolapses using open and endoscopic surgical techniques. These and other current commercial biotextile products are listed in Table I.2.14.1 to illustrate the scope and range of biotextiles in surgical applications.

The field of biotextiles lies at the intersection of the disciplines of polymer and fiber science, textile technology, biomedical engineering, surface science, biomechanics, cell biology, human anatomy, and physiology. Obviously, collaboration between these different fields is necessary in order to undertake meaningful research and to develop innovative implantable biomedical devices that improve the health, well-being, and quality-of-life of individuals. For example, fiber-based textiles can provide thin, strong, flexible, lightweight, porous structures that have excellent fatigue properties, a very large surface area desirable for drug delivery and cell attachment, the ability to be folded or compressed into a small volume for less invasive delivery through a catheter, the ability to tolerate needle and other iatrogenic damage, and to encourage the infiltration of cells into the structure for acceptable biocompatibility. Iatrogenic refers to any adverse condition induced in a patient by a physician's activity, manner or therapy. The process of developing a new biotextile product involves an understanding of the materials or polymers, the elements or fibers, the components or yarns, and the hierarchical architecture associated with the textile fabrication (e.g., weaving, knitting or braiding), as well as the surface modification or finishing.

In this chapter, we present this series of concepts in a logical order so that the reader can appreciate how the selection of the polymer, the fiber, and yarn, the fabrication process and the surface finish can all impact the final structure, properties, and performance of the biomaterial device.

FIBER FORMING POLYMERS

While natural and manufactured textile fibers are made from a wide range of materials including metals (e.g., stainless steel) and ceramic materials (e.g., glass), the majority of fibers of interest to biomaterial scientists are derived from natural and synthetic fiber-forming polymers. Not all polymer materials lend themselves to forming fibers. In order to be able to spin or convert a polymer into a manufactured fiber, its chemistry needs to meet certain structural requirements. The following features are preferred:

1. Intermediate to high molecular weight (range approximately 20,000 to 250,000 Da).
2. Linear polymer chain with absence of bulky side groups, cross-links or side chains.
3. Ability to rapidly form ordered or crystalline structures when solidified from the melt or when precipitated from solution. This depends on the regularity of the repeat unit to allow close packing between chains.
4. The level of intermolecular bonding and the level of chain entanglement should not be too high so as to prevent the polymer chains from forming an aligned and preferentially ordered structure on stretching or drawing.

An appropriate molecular weight range of 20,000 to 30,000 Da is typical for condensation polymers such as nylon and PET, whereas addition polymers such as polypropylene, and aliphatic polyesters such as PGA and PLA typically have molecular weights in the upper end of the range. These characteristics are common to all of the permanent and resorbable polymer types listed in Table I.2.14.2.

Because of their unique monomer repeat unit size and shape, different polymers have a different inherent ability to crystallize. Some like silk, PTFE, PVDF, polyethylene, and polydioxanone have a high degree of crystallinity that influences their mechanical properties, their ability to adsorb moisture, and their biostability. Table I.2.14.2 provides examples of natural and synthetic permanent polymers, as well as resorbable polymers that have been designed or engineered to resorb after implantation within a biological environment. These resorbable polymers were originally designed as suture materials to extend the range of resorbable collagen (catgut) sutures but, as seen in Table I.2.14.2, they can now be applied to a wide range of biomedical applications ranging from

TABLE I.2.14.1 Commercial Biotextile Structures and their Clinical Applications

Clinical Application	Devices with Biotextile Component	Polymer Type*	Fiber Structure	Fabric Structure	Resorption Time	Examples of Commercial Products
General Surgery	Esophageal stent	Polyester (PET)	Monofilament	Braided	Permanent	Polyflex®
	Hernia repair mesh	Polypropylene	Microfiber	Nonwoven	Permanent	Surgimesh®
		Polypropylene	Monofilament	Warp knit	Permanent	Atrium®, Sepramesh®
		PTFE	Expanded film	ePTFE membrane	Permanent	Dualmesh®, Dulex™
		PGCL and polypropylene	Monofilaments	Warp knit	Partially resorbable	Ultrapro®
		Polylactide (PLA)	Monofilament	Self fixation, pile component	Partially resorbable	Paritex ProGrip™
	Patch	Polyester (PET)	Multifilament	Weft knit, double velour, nonwoven	Permanent	Bard®, Ultramax™
		PTFE	Expanded film	ePTFE membrane	Permanent	Impra®, iVena™
	Prolapse repair mesh	Polypropylene	Monofilament	Warp knit	Permanent	Prolene®, Marlex®, Avaulta®
		PTFE	Expanded film	ePTFE membrane	Permanent	Gore-Tex®
	Permanent suture	Nylon 6	Monofilament	—	Permanent	Ethilon™
			Multifilament	Braided	Permanent	Nurolon™
		Polyester (PET)	Monofilament	—	Permanent	Surgidac™
			Multifilament	Braided	Permanent	Mersilene™, Ethibond™,Ticron™
		Polypropylene	Monofilament	—	Permanent	Prolene™, Surgipro™
		PBT-TMEG	Monofilament	—	Permanent	Vascufil™, Novafil™
		PVDF-HFP	Monofilament	—	Permanent	Pronova™
		Silk	Multifilament	Braided	Permanent	Perma-hand™, Sofsilk™
		Stainless steel	Monofilament	—	Permanent	Steel
	Resorbable suture	Collagen (Catgut)	Monofilament	—	70–90 days	Chromic/plain
		Polydioxanone (PDO)	Monofilament	—	180–250 days	PDS™
		Polyglycolide (PGA)	Monofilament	—	60–90 days	Dexon™
			Multifilament	Braided		Visob™
		PGCL	Monofilament	—	90–120 days	Monocryl™
		PLGA	Monofilament	—	40–70 days	Vicryl™
			Multifilament	Braided		Vicryl™, Polysorb™
		PGA-PCL-TMC-PLA	Monofilament	—	20–60 days	Caprosyn™
		PGA-PDO-TMC	Monofilament	—	90–110 days	Biosyn™
		PGA-TMC	Monofilament	—	60–180 days	Maxon™

(Continued)

TABLE I.2.14.1 Commercial Biotextile Structures and their Clinical Applications *(Continued)*

Clinical Application	Devices with Biotextile Component	Polymer Type*	Fiber Structure	Fabric Structure	Resorption Time	Examples of Commercial Products
Cardiovascular	Annuloplasty ring	Polyester (PET)	Multifilament	Weft knit	Permanent	Carpentier-Edwards®, Duran Ancore®
		PTFE	Multifilament	Weft knit	Permanent	Koehler MRS
	Arteriovenous shunt	PTFE	Expanded film	ePTFE membrane	Permanent	Vectra®, Flixene™
	Blood filter *(ex vivo)*	Polyester (PET)	Multifilament	Nonwoven	Permanent	MacoPharma
	Cardiac support device	Polyester (PET)	Multifilament	Knit	Permanent	Acorn CorCap™
	Embolic vena cava filter	PTFE-FEP	Monofilament	Knotted	Permanent	Crux Biomedical
	Endovascular stent-graft	Polyester (PET)	Multifilament	Woven	Permanent	Zenith®, Endurant®
		PTFE	Expanded film	ePTFE membrane	Permanent	Excluder®, PowerLink®
	Heart valve sewing ring	Polyester (PET)	Multifilament	Weft knit	Permanent	St.Jude Medical, Sorin
		PTFE	Multifilament	Weft knit	Permanent	Edwards Lifesciences
	Septal defect repair device	Polyester (PET)	Multifilament	Knit	Permanent	Amplatzer®, CardioSEAL®
		PTFE	Expanded film	ePTFE membrane	Permanent	Helex®
	Vascular prosthesis	Polyester (PET)	Multifilament	Warp knit, woven	Permanent	DeBakey®, Gelweave®, Barone®
		PTFE	Expanded film	ePTFE membrane	Permanent	Gore-Tex®, Flixene™
Dental	Dental reinforcing ribbon	UHMWPE	Multifilament	Leno weave	Permanent	Ribbond®
Neural	Nerve guide prosthesis	Polyglycolide (PGA)	Multifilament	Woven	60–90 days	NeuraGen®
Orthopedic	Bone graft/cement	Carbon fiber	Staple fiber/ Multifilament	Reinforced PMMA composite	Permanent	DePuy
	Ligament and tendon prosthesis	Polyester (PET)	Multifilament	Braided/Woven/Knitted	Permanent	Stryker-Meadox, Leeds-Keio
		PTFE	Expanded film	ePTFE membrane	Permanent	Gore-Tex®
		UHMWPE	Multifilament	Braided	Permanent	Richards Polytex
	Spinal disc nucleus prosthesis	UHMWPE	Multifilament	Woven	Permanent	Raymedica PDN®
	Spinal support	Carbon fiber	Staple fiber/ Multifilament	Reinforced PEEK composite	Permanent	Ocelot™

Skin and Wound Dressing	Skin graft	PLGA	Multifilament	Knitted	40–70 days	Dermagraft®
		Nylon	Multifilament	Knitted	Temporary	Biobrane®
		Nylon/Collagen	Multifilament	Knitted velour	Temporary	TransCyte®
	Wound dressing	Cotton	Staple yarn	Woven	Temporary	Kendall Curity®
		UHMWPE/Viscose rayon and PET	Multifilament/ Staple fiber	Knitted/Nonwoven composite	Temporary	Acticoat™
		Carboxymethylcellulose	Staple fiber	Nonwoven	Temporary	Acquacel™
		Cotton/Viscose rayon	Staple yarn	Leno weave	Temporary	Jelonet™
		Viscose rayon/Nylon	Multifilament	Knitted	Temporary	Kband™
		Polyester (PET)/Viscose rayon	Staple fiber	Nonwoven composite	Temporary	Mepore™
		Polyproplyene/Cellulose	Staple fiber/Pulp fiber	Nonwoven composite	Temporary	Mesorb™
		Nylon	Multifilament	Leno weave	Temporary	Tegapore™
		Calcium or sodium alginate	Staple fiber	Nonwoven	Temporary	Kaltostat™

*Abbreviations:
ePTFE: Expanded polytetrafluoroethylene
FEP: Fluorinated ethylene propylene polymer
PBT-TMEG: Poly(butylene terephthalate)-co-poly(tetramethylene ether glycol)
PCL: Polycaprolactone
PDO: Polydioxanone
PEEK: Polyethyletherketone
PET: Poly(ethylene terephthalate)
PGA: Polyglycolide
PGCL: Poly(glycolide-co-caprolactone)
PLA: Polylactide
PLGA: Poly(lactide-co-glycolide)
PMMA: Polymethylmethacrylate
PTFE: Poly(tetrafluoroethylene)
PVDF-HFP: Poly(vinylidene fluoride-co-hexafluoropropylene)
TMC: Trimethylene carbonate
UHMWPE: Ultra-high molecular weight polyethylene

TABLE I.2.14.2 Examples of Permanent and Resorbable Fiber-Forming Polymers

Type	Chemical and Physical Aspects	Construction/ Useful Forms	Comments/Applications	Resorption Time
Nylon 6	Thermoplastic; Hydrophobic $T_m = 220°C; T_g = 45°C$ Crystallinity 50% as spun	Melt spun into monofilaments and multifilament yarns	Sutures (Nurolon™; Ethilon™)	Permanent
Nylon 66	Thermoplastic; Hydrophobic $T_m = 265°C; T_g = 50°C$ Crystallinity 50% as spun	Melt spun into monofilaments and multifilament yarns	Sutures (Monosof™; Dermalon™; Surglion™)	Permanent
Polyethylene (PE)	High-density PE (HDPE) Thermoplastic; Hydrophobic $T_m = 125°C$ Crystallinity up to 85%	Melt spun into fibers for polymer composite reinforcement	Orthopedic and craniofacial implants	Permanent
	Ultra high molecular weight PE (UHMWPE) $T_m = 140–150°C$ Crystallinity 100% High tensile strength and modulus	Converted to very high tenacity yarn by gel spinning	Ligament prostheses and loadbearing orthopedic composites	
Poly(ethylene terephthalate) (PET)	Thermoplastic; Hydrophobic $T_m = 265°C; T_g = 65–105°C$ 100% amorphous as spun so need to be drawn and annealed to have crystallinity	Melt spun into monofilaments and multifilament yarns for weaving, knitting, and braiding	Sutures (Mersilene™; Surgidac™; Ticron™); hernia repair meshes; ACL prostheses; heart valve sewing rings; vascular and endovascular grafts	Permanent
Polypropylene (PP)	Predominantly isotactic Thermoplastic; Hydrophobic $T_m = 165–175°C$ Crystallinity 40–46% Higher fracture toughness than HDPE Susceptible to degradation from heat and radiation	Melt spun into monofilaments, hollow fibers or spunbonded and melt blown to nonwoven fabrics	Sutures (Prolene™; Surgipro™); hernia repair meshes; blood filters (e.g., renal dialysis machines)	Permanent
Poly(tetrafluoroethylene) (PTFE)	Thermoplastic; Hydrophobic $T_m = 325°C$ Crystallinity 50–75% for processed material	Melt extruded into films and often thermo-mechanically expanded into ePTFE	Sutures; vascular and endovascular grafts; heart valve sewing rings; embolic vena cava filter; ACL ligaments	Permanent
Poly(vinylidene fluoride) (PVDF)	Thermoplastic; Hydrophobic $T_m = 165–175°C$ Crystallinity 52–66%	Melt spun into monofilaments and multifilaments	Sutures (Pronova™)	Permanent
Silk	Natural bicomponent fiber from fibroin protein with sericin sheath; Hydrophilic Crystallinity 70% Low elastic recovery	Spun by *Bombyx mori* silk worm with sericin coating; sericin is removed before processing	Sutures (Perma-Hand™; Sofsilk™)	Classified as permanent but known to degrade in 1–2 years

Viscose Rayon	First manufactured fiber in 1892; Hydrophilic Crystallinity 33% Decomposition begins at 250°C	Wet spun from alkali sodium cellulose xanthate solution into continuous multifilaments	Wound dressing	Permanent
Polycaprolactone (PCL)	Thermoplastic; Rubbery at room temperature $T_m = 58–63°C; T_g = -65°C$	Melt spun into monofilament and multifilament yarns; can be electrospun into nanofibers; can be copolymerized with other resorbable polymers	Used as scaffolds in tissue engineering	>24 months
Polydioxanone (PDO)	Thermoplastic $T_m = 110–115°C, T_g = -10–0°C$ Crystallinity 55%	Melt spun into monofilament and multifilament yarns; can be electrospun into nanofibers; can be copolymerized with other resorbable polymers	Sutures (PDS™); intramedullar pins; ligating clips	6–9 months
Polyglycolide (PGA)	Thermoplastic $T_m = 225°C; T_g = 40–45°C$; Crystallinity 45–55% Simplest linear aliphatic polyester Rigid mechanical properties	Melt spun into monofilament and multifilament yarns; can be electrospun into nanofibers; can be copolymerized with other resorbable polymers	Sutures (Dexon™); meshes (for defect repairs and periodontal inserts); used as scaffolds in tissue engineering	6–12 months
Polylactide (PLA)	Thermoplastic $T_m = 173–178°C; T_g = 60–65°C$		Stents; drug delivery devices; used as scaffolds in tissue engineering	>24 months
Polyglactin 910 10/90 Poly(l-lactide-co-glycolide)	Thermoplastic $T_m = 205°C; T_g = 43°C$ Crystallinity 40%	Multifilament yarns, for weaving, knitting and braiding	Sutures (Vicryl™; Polysorb™); meshes	3 months
Poly(glycolide-co-caprolactone) (PGCL)	Thermoplastic $T_g = -43–18°C$ depending on the copolymer ratio and molecular weight	Monofilament	Sutures (Monocryl™)	3–4 months
Poly(glycolide-trimethylene carbonate)	$T_m = 206°C; T_g = 20°C$	Monofilament	Sutures (Maxon™)	2–6 months

implanted drug delivery systems and degradable stents to tissue engineering scaffolds. When in contact with the body, these polymers degrade either by hydrolysis or by enzymatic degradation into nontoxic byproducts. As biotextiles, they break down or degrade through an erosion process, which starts at the exterior surface of the fiber and continues until the fiber has been totally resorbed. Sometimes the process is accelerated in the center of the fiber where the acid byproducts act as catalysts for further degradation.

It is important to note that most permanent synthetic polymers currently used in medicine were originally developed as commercial polymers for non-medical applications and usually contain additives such as delustrants, stabilizers, antioxidants, antistatic agents, and dye linkage sites. Some of these chemicals may be cytotoxic for biomedical applications and, as leachable contaminants, they must be removed prior to use. To illustrate this point, poly(ethylene terephthalate) (PET), which at present is the material of choice for large caliber textile vascular grafts and other applications as seen in Table I.2.14.1, was originally developed for apparel use. A rigorous cleaning or scouring process is required before the material can be used as an implantable device (King, 1991; Fages et al., 1998).

MEDICAL FIBERS

All textile-based medical devices are composed of structures fabricated from textile fibers.

What are Textile Fibers?

Textile fiber is: "a generic term for the various types of matter that form the basic elements of fabrics and other textile structures. More specifically, a **textile fiber** is a unit of matter that is characterized by having an aspect ratio (length divided by width) of at least 100, and which can be spun into yarn or made into a fabric" (ASTM, 1996).

Textile fibers can be spun in several different forms, such as continuous monofilaments and multifilament yarns, which can vary in length from 500 meters of natural silk reeled or unwound from a cocoon to several hundred kilometers of a manufactured yarn on a single bobbin. Usually the units of fiber size, thickness or yarn linear density are expressed in **decitex (dtex)**, which is the mass in grams of 10,000 meters of fiber. In North America, the unit of linear density is usually expressed in **denier**, which is the mass in grams of 9000 meters of fiber. For example, a typical polyester yarn used to knit a vascular prosthesis has a linear density of 100 dtex, as shown in Table I.2.14.3. If this were a single round monofilament then the filament diameter would correspond to about 100 μm. However, as indicated in Table I.2.14.3, this is a multifilament yarn consisting of 54 round filaments. Therefore, the linear density of each filament is 1.9 dtex and the diameter of each filament is 13.2 μm, which makes the multifilament yarn significantly more flexible, with a much lower bending rigidity than a 100 dtex monofilament yarn.

TABLE I.2.14.3 Examples of Structure and Properties of Knitted Polyester Vascular Prostheses

Structure/Properties	DeBakey® Vascular II, C.R. Bard, USA		Dialine II, Cardial, France		Barone Microvelour®, Argentina	VP1200K™ (Unsealed Gelsoft®) Vascutek, UK
Type of yarn	2 single textured multifilament		2 ply textured multifilament	Single flat multifilament	3 ply textured multifilament	Single textured multifilament
Yarn nominal linear density (dtex)	76	110	130	74	340	100
Approx. filament count	40	54	88	44	66	54
Filament nominal linear density (dtex)	1.9	2.0	1.5	1.7	5.2	1.9
Filament cross-section	Round	Round	Round	Round	Round	Round
Filament diameter (μm)	13.3	13.6	11.7	12.5	22.6	13.2
Type of knit	Pile warp knit		Warp knit		Weft knit	Warp knit
Type of stitch	Locknit		Locknit		Single jersey	Locknit
Wale count (cm^{-1})	14		17		16	14
Course count (cm^{-1})	31		28		21	15
Stitch density (cm^{-2})	430		492		324	210
Mass per unit area (g/cm^2)	224		265		278	182
Thickness (mm)	0.57		0.55		0.63	0.57
Fabric density (g/cm^3)	0.99		0.89		0.94	1.06
Total porosity (%)	72		65		68	77
Water permeability ($ml/cm^2/min$)	2150		1050		3660	2330
Bursting strength (N)	166		186		215	145
Dilation at 120 mmHg (%)	1.6		4.1		4.1	9.6
Suture retention strength (N)	22.8		25.1		28.4	31.5

In addition to synthetic polymers, a class of fibers exists which is composed of natural biopolymer-based materials. In contrast to synthetic fibers, which have been adapted for medical use, natural fibers have evolved naturally and so can be particularly suited for medical applications. Cellulose, which is obtained from processed cotton or wood pulp, is one of the most common fiber-forming biopolymers. Because of their high thrombogenicity, cellulose fibers and their derivatives are used for applications such as wound dressings and hemostats for open wounds. Also of growing interest are fibers wet spun into multifilament fibers from modified polysaccharides including alginates from algae, chitosan from crustacean shells, dextran and reticulated cellulose from bacterial fermentation which are used to fabricate surgical sutures and meshes, and tissue engineering scaffold materials (Shalaby and Shah, 1991; Ohkawa et al., 2004; El-Mekawy et al., 2010).

Traditionally, natural textile fibers of vegetable and animal origin had a diameter range of 10–50 μm and manufactured fibers until recently have also targeted the same size range. However, with recent advances in spinning technology for converting polymers into fibers, the size range has been extended down to below one micron as illustrated in Figure I.2.14.1. In the following sections, the different fiber spinning techniques are discussed in greater detail.

Melt Spinning

In melt spinning, the polymer resin is heated above its melting temperature and extruded through a spinneret. The number of holes in the spinneret defines the number of filaments in the mono- or multifilament yarn being produced. For example, a spinneret for a monofilament fiber contains one hole, whereas 54 holes are required to produce the multifilament yarn which is commonly used in vascular graft construction. In this process, the molten resin is extruded through the spinning head, and cold air is typically used to cool and solidify the continuous threadline, as shown in Figure I.2.14.2A. It is then drawn, lubricated, twisted or entangled prior to winding on a bobbin. Spin finish is applied as a lubricant to reduce friction and to improve texturing, knitting, and weaving efficiencies. Melt spinning is typically used with thermoplastic polymers that are not affected by the elevated temperatures required in the melt spinning process as listed in Table I.2.14.2. Depending on the shape of the spinneret hole, melt spun fibers can exhibit various cross-sectional shapes such as round and trilobal hollow (Figure I.2.14.3).

Wet/Gel Spinning

If the polymer system is not thermoplastic or experiences degradation at elevated temperatures, a low temperature wet solution or gel spinning process can be used, as is the case with cellulose (viscose rayon), chitosan, alginate or cyclodextrin. The same spinning process applies when converting these polymers into a fiber-based drug delivery system so as to maintain the activity of the drug component (Tuzlakoglu et al., 2004; Malheiro et al., 2009). In this process, the polymer is dissolved in a solvent and

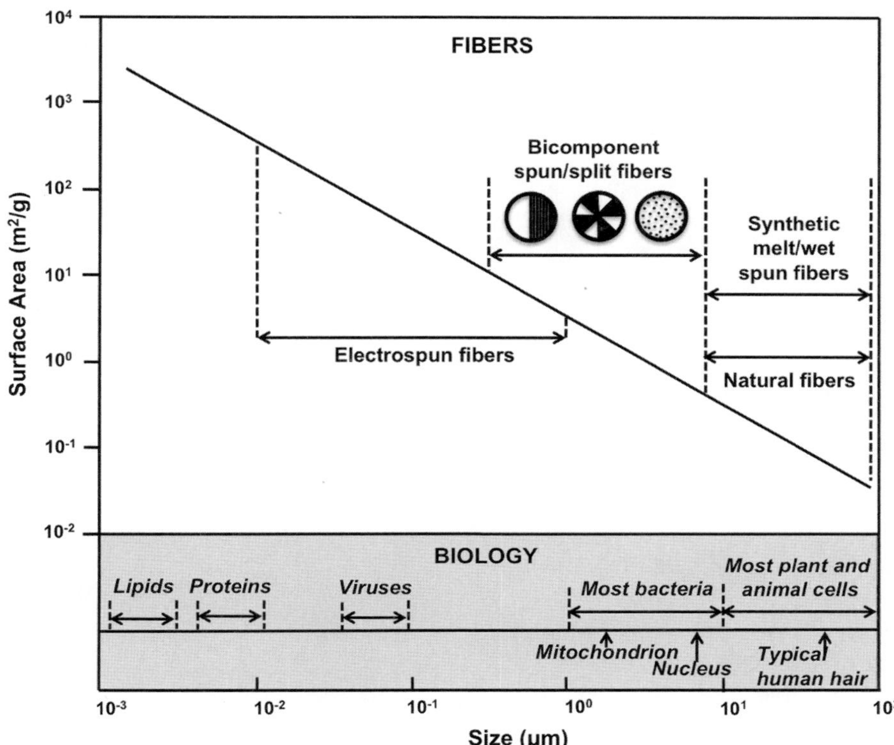

FIGURE I.2.14.1 Relationship between fiber dimensions and surface area compared to a biological scale.

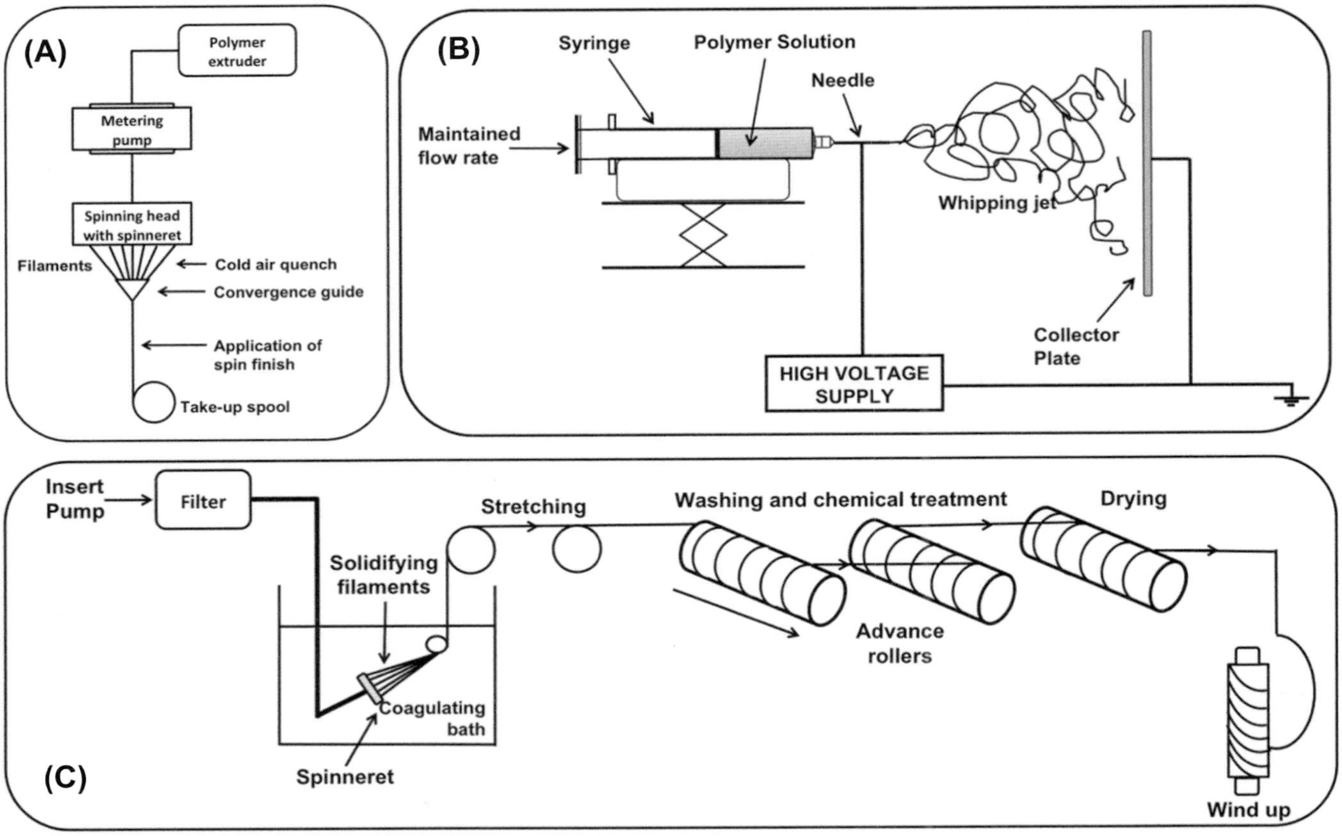

(A)
Polymer extruder
Metering pump
Spinning head with spinneret
Filaments
Cold air quench
Convergence guide
Application of spin finish
Take-up spool

(B)
Syringe
Polymer Solution
Needle
Maintained flow rate
Whipping jet
Collector Plate
HIGH VOLTAGE SUPPLY

(C)
Insert Pump
Filter
Solidifying filaments
Coagulating bath
Spinneret
Stretching
Washing and chemical treatment
Advance rollers
Drying
Wind up

FIGURE I.2.14.2 Spinning processes: (A) Melt spinning; (B) Wet/Gel spinning; (C) Electrospinning.

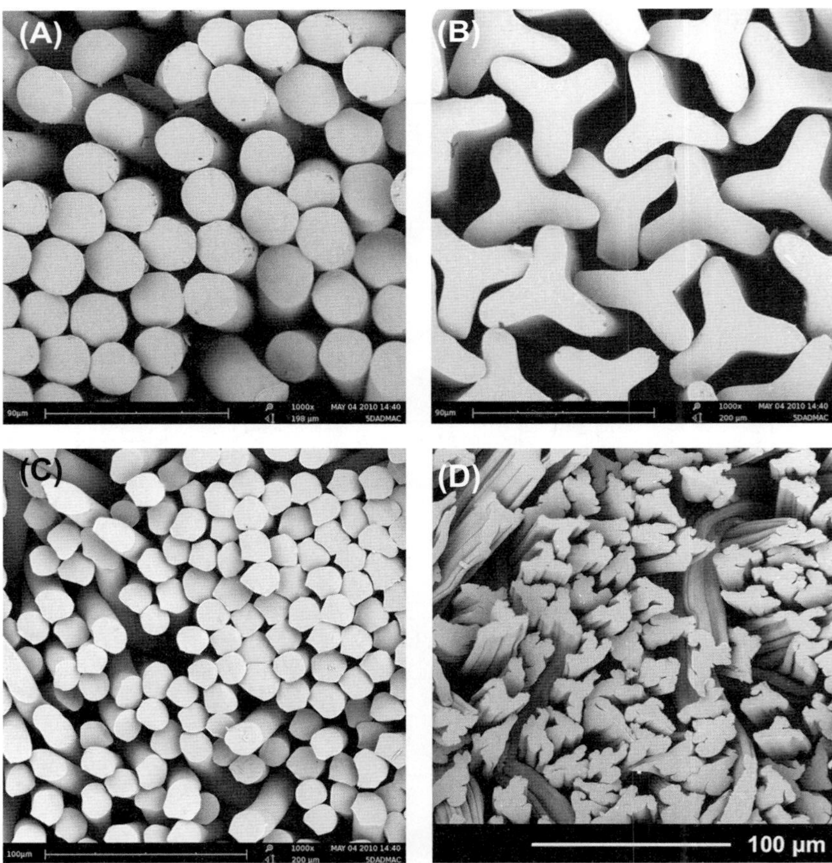

FIGURE I.2.14.3 Cross-sections of: (A) Round nylon; (B) Trilobal nylon; (C) Round PET; (D) Crenulated viscose rayon (common scale bar: 100 μm).

then extruded through a spinneret into a non-solvent in a spin bath, as shown in Figure I.2.14.2B. For example, for spinning chitosan, an acetic acid solvent is used and an ethyl alcohol or phosphate phthalate salt solution serves as the coagulation bath (Knaul et al., 1999). In the case of spinning alginate fibers, aqueous solutions of sodium carbonate and caustic soda are used as the solvent, and the fibers are precipitated by acidification with hydrochloric acid. Because the polymer is soluble in the solvent but not in the spin bath, the continuous polymer solution stream coagulates into continuous solid filaments, which are then washed to remove all solvents and non-solvents, drawn and dried before winding (Adanur, 1995). The cross-sectional shape of the resulting fibers is often crenulated with multiple grooves due to the slow diffusion of the solvent out of the fiber during the coagulation process, as shown in Figure I.2.14.3D.

Electrospinning

The diameters of fibers spun by melt spinning and wet solution spinning are controlled by the size of the hole in the spinneret and the amount of draw or stretch applied to the filament prior to winding up. So the diameters of conventional spun fibers fall in the range from about 10 μm for multifilament yarns to 500 μm or thicker for monofilaments. To obtain finer fiber diameters it is necessary to employ alternative spinning technologies, such as the bicomponent fiber spinning approach (see next section), or an electrospinning technique as illustrated in Figure I.2.14.2B. This method of manufacturing microfibers and nanofibers has been known since 1934 when the first patent was filed (Formhals, 1934). Start-ups such as eSpin Technologies, NanoTechnics, and KATO Tech are some of the companies seeking to reap the unique advantages offered by electrospinning, while companies such as Donaldson Company and Freudenberg have been using electrospinning processes to manufacture air filtration products for the last two decades (Ramakrishna et al., 2006).

Electrospinning occurs when a polymer solution or melt is exposed to an electrostatic field by the application of a high voltage, which overcomes the surface tension of the polymer and accelerates fine jets of the liquid polymer towards a grounded target (Doshi and Reneker, 1995; Ma et al., 2005; Lannutti et al., 2007). As the polymer jets cool and/or lose solvent they are drawn in a series of unstable loops, solidified, and collected as an interconnected nonwoven web of fine fibers on a grounded rotating drum or other specially shaped target, as illustrated in Figure I.2.14.2B. There are many variables that influence the spinning efficiency/stability and the fineness of the fibers produced, such as the type of polymer and solvent system, the solution concentration, melt viscosity and conductivity, the strength and uniformity of the applied electric field, and the geometry and operating conditions of the spinning system. Fiber diameters in the range of submicron to 100 nm or less have been reported. As well as using electrospinning techniques to fabricate ultra-thin filtration membranes, these techniques have also been applied to the production of nonwoven mats for wound dressings and scaffolds for tissue engineering applications (Kumbar et al., 2008). Nonwoven scaffolds electrospun from Type I collagen and synthetic polymers such as polylactide, poly(lactide-co-glycolide), polyvinyl alcohol, poly(ethylene-co-vinyl acetate), polyethylene oxide, poly(lactide-co-ε-caprolactone), polyurethanes, and polycarbonates have been reported (Huang et al., 2000; Boland et al., 2001; Huang et al., 2006; Pham et al., 2006; Barnes et al., 2007; Chung et al., 2009; Leong et al., 2009; Chung et al., 2010a).

Bicomponent Spinning

Bicomponent, hybrid or multicomponent fiber spinning refers to the technology of bringing two or more different polymers together at the spinneret hole so that each spun filament contains all the polymer components in separate parts of the cross-section (Durany et al., 2009). Over the past few years, a number of different fiber cross-sectional configurations have been developed, as illustrated in Figures I.2.14.1 and I.2.14.3B. The main motivations for these developments have been: (1) to take advantage of the properties of more than one polymer component within the same fiber; and (2) to achieve smaller fiber dimensions by splitting or separating components after co-spinning. For example, one of the configurations of a bicomponent fiber is to spin a resorbable polymer sheath around an inner core of a second non-absorbable polymer. The advantage of this fiber configuration for vascular applications is that the sheath modulates a faster inflammatory or foreign body response and a more complete healing process, while the core component maintains the mechanical integrity of the device (King et al., 1999). The composition and molecular weight of the polymer and the thickness of the sheath regulates its absorption rate. As shown in Figure I.2.14.4, within 28 days the PGA sheath component is in the process of resorbing into fragments while provoking acute inflammatory response with the recruitment of many macrophages and collagen matrix deposition. Additionally, drugs can be incorporated into the outer resorbable sheath and delivered at predefined rates, depending on the choice and thickness of the outer polymer. By using this bicomponent spinning technology, both the material strength profile and the biological properties can be engineered into the fiber to meet specific medical requirements. In addition, polymers with different thermal properties can be spun into sheath–core bicomponent configurations to be thermally bonded as a nonwoven web for use as protective gown fabrics, where the sheath component acts as an adhesive having a lower softening temperature than the core component.

As illustrated in Figure I.2.14.1, the eight-component segmented-pie configuration contains two different polymers, and, since these two have no attraction or adhesion between them, they can be split into finer triangular segmented fibers by applying external forces

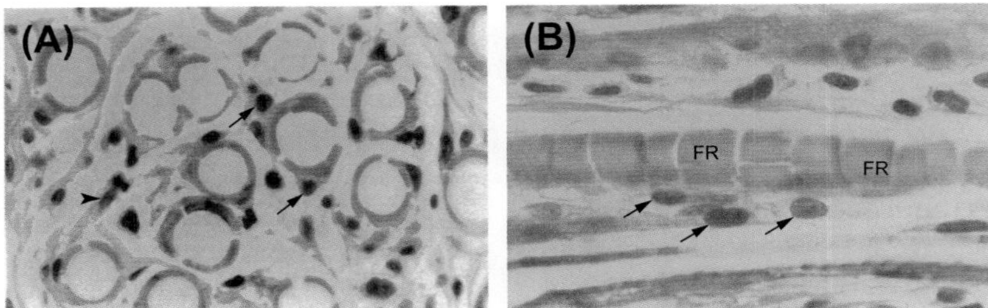

FIGURE I.2.14.4 Histological images of PGA/PP sheath/core bicomponent fibers showing the foreign body response after 28 days in a rat subcutaneous model with hematoxylin, phloxine and safran stain. (A) Cross-section; (B) Longitudinal section (arrows identify activated macrophages) (Common scale bar: 10 μm).

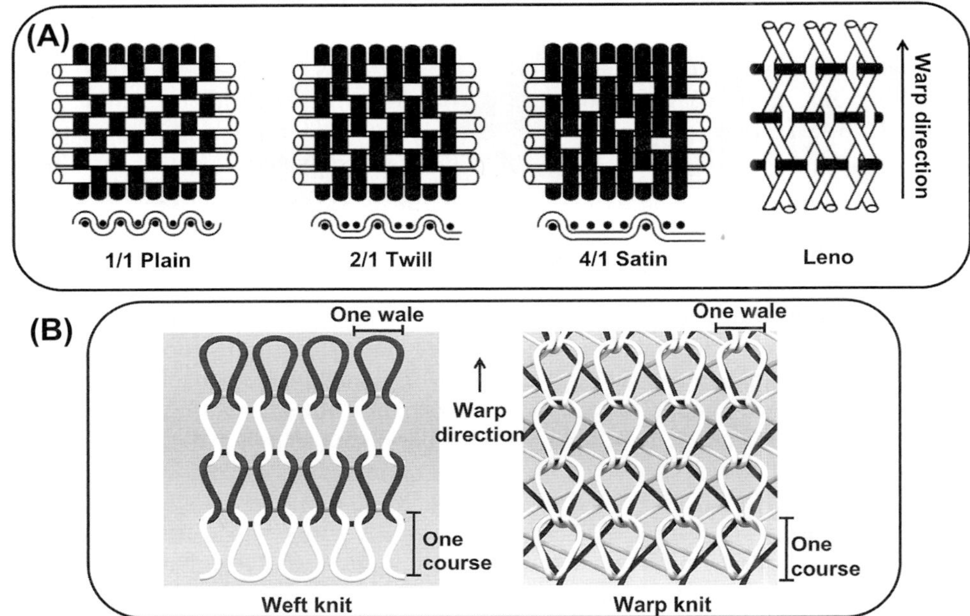

FIGURE I.2.14.5 Schematic diagram of textile fabrics (A) Woven structures; (B) Knitted structures.

(e.g., mechanical or hydraulic) (Gong and Nikoukhesal, 2009). Alternative configurations, such as "islands-in-the-sea" and a 16-component segmented pie with a core are spun as precursor fibers to create grooved or ultra fine fibers by removing the white sacrificial polymer component in a subsequent post-spinning process (Chung et al., 2010b).

TEXTILE STRUCTURES

After a fiber or yarn is produced, it is then fabricated into a textile structure in order to obtain the desired mechanical and biological properties. Typical biotextile structures used for medical applications include wovens, knits, braids and nonwovens. Within each of these configurations, many variations exist. Each type of construction has positive and negative attributes and, in most cases, the final choice represents a compromise between desired and actual fabricated properties. For example, woven fabrics typically are stronger and more dimensionally stable, and can be fabricated with lower porosities or water/blood permeability compared to knits, but are stiffer, less flexible, and more difficult to handle and suture. Knits have higher permeability and flexibility than woven designs and are easier to suture, but may dilate after implantation. Braids have high longitudinal tensile properties, but can be unstable when subject to torsional loads. Each construction is a compromise.

Woven

What is a Woven Structure?

The term **woven** is used to describe a textile configuration where the primary structural yarns are oriented at 90° to each other. The machine direction is called the warp direction and the cross direction is identified as the weft or filling direction. Because of the orthogonal relationship between the warp and weft yarns, woven structures display low elongation and high breaking strength in both directions. There are many types of woven constructions including plain, twill, satin, and leno weaves, as illustrated in Figure I.2.14.5A, which are commonly used in vascular graft and dental ribbon fabrications.

TABLE I.2.14.4 Examples of Structure and Properties of Woven Polyester Vascular Prostheses

Structure/ Properties	DeBakey Soft Woven®, C.R. Bard, USA		Twill Weave® (Unsealed Gelweave™), Vascutek, UK		Woven Double Velour (Unsealed Hemashield®), Boston Scientific, USA	
Fabric direction	Warp	Weft	Warp	Weft	Warp	Weft
Type of yarn	Single flat multifilament	Single flat multifilament	Textured multifilament	Textured multifilament	Flat + textured multifilament	Flat + textured multifilament
Yarn nominal linear density (dtex)	190	110	102	118	105 + 120	105 + 120
Approx. filament count	108	54	54	54	54 + 54	54 + 54
Filament nominal linear density (dtex)	1.8	1.9	1.9	2.2	1.9 + 2.2	1.9 + 2.2
Filament cross-section	Round	Round	Round	Round	Round	Round
Filament diameter (μm)	13.0	13.3	13.2	14.2	13.5 + 14.4	13.5 + 14.4
Type of weave	Plain		Plain + twill		Plain + satin	
Woven fabric count (ends/cm)	58		42p + 22t		36p + 36s	
Woven fabric count (picks/cm)	35		48		38	
Mass per unit area (g/cm^2)	152		172		184	
Thickness (mm)	0.27		0.35		0.32	
Fabric density (g/cm^3)	0.81		0.88		0.80	
Total porosity (%)	59		64		58	
Water permeability (ml/cm^2/min)	180		330		310	
Bursting strength (N)	211		280		310	
Dilation at 120 mmHg (%)	0.2		0.0		0.7	
Suture retention strength (N)	26.9		22.0		19.0	

Table I.2.14.4 lists a number of commercial woven vascular graft designs and presents their textile structure and performance properties. Water permeability is one of the critical parameters used in the assessment of textile structures for vascular implants, which is a measure of the water flux through a fabric under a fixed pressure of 120 mmHg (ANSI/AAMI/ISO (R), 2004). Surgeons use this parameter as a guide to determine if "pre-clotting" of a graft material is required prior to implantation. Fabric grafts with water permeability values less than 300 ml/cm^2/min usually do not require pre-clotting prior to implantation, and so are safe to use for emergency surgery when there is no time for pre-clotting (also when a patient's blood clotting ability is unknown). Water permeability is controlled by the linear density of the yarns, as well as the woven fabric count that is measured in terms of the frequency of warp yarns (ends/cm) and the frequency of weft yarns (picks/cm). Plain weaves, in contrast to twill and satin weaves, can be made very thin (e.g., 0.27 mm) and have thus become the material of choice for many endovascular graft designs. However, plain weaves woven from flat multifilament yarns do not permit good cellular attachment. Therefore, the choice of textured multifilament yarns in twill or satin weave provides a more open, porous, and thicker structure that permits greater tissue infiltration, which is desirable for long-term healing of vascular prostheses. Figures I.2.14.6A and I.2.14.6B show the external and internal (luminal) surfaces of a DeBakey Soft Woven® prosthesis with a smooth tightly woven 1/1 plain weave compared to a Twill Weave® (unsealed Gelweave™) prosthesis with textured yarns providing bulkier floats on the surface. The structure and properties of these devices are listed in Table I.2.14.4.

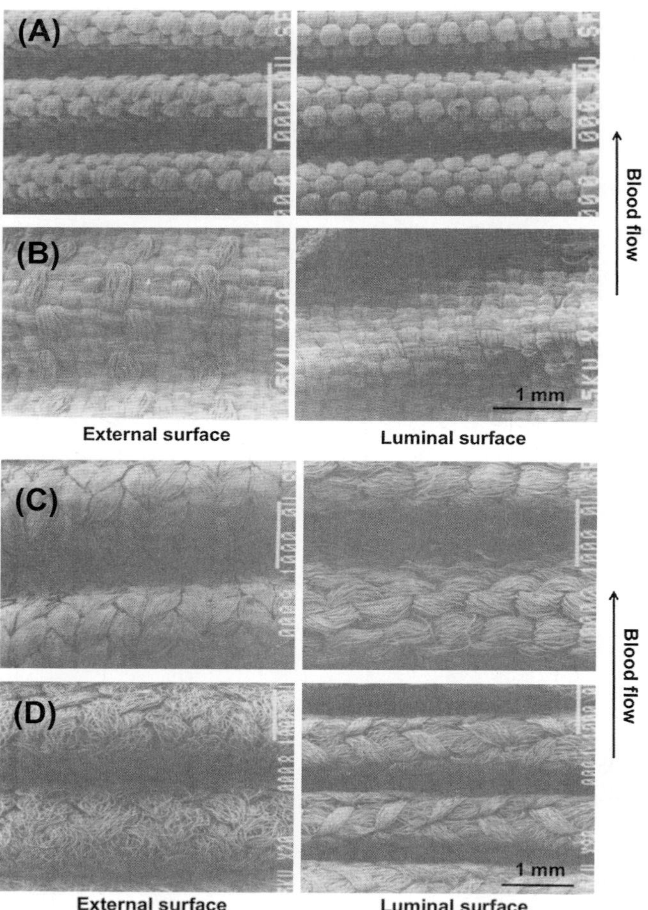

FIGURE I.2.14.6 SEM photomicrographs of external and luminal surfaces of woven and knitted vascular prostheses: (A) DeBakey Soft Woven®; (B) Twill Weave®; (C) Barone Microvelour® weft knit; (D) Vascutek VP1200K™ warp knit.

Knitted

What is a Knitted Structure?

Knitted constructions are made by interlooping yarns in horizontal rows (wales) and vertical columns (courses) of stitches. They are softer, more flexible and easily conformable, and have better handling characteristics than woven structures. As is the case with woven structures, there are several variations in knits; the most common are the weft knit and warp knit constructions as seen in Figure I.2.14.5B. When used for vascular grafts, knitted structures can have water permeability values as high as 5000 ml/cm^2/min, and therefore they need to be coated or impregnated with collagen or gelatin so that the surgeon does not have to perform the time-consuming preclotting process at the time of surgery.

Table I.2.14.3 lists a number of commercial knitted vascular graft designs and presents their textile structure and performance properties. Most yarns used for these knitted vascular grafts are textured so as to impart thickness to the fabric for improved softness, flexibility, ease of suturing, and to provide larger pores for tissue ingrowth. Knitted structures have an inherently open porous structure with total porosity values greater of at least 65%, which requires special processing to shrink the yarns and tighten the looped structure by heat setting. Because of their open structure, knits are typically easier to suture and have better handling characteristics. Warp knitted structures have less stretch than weft knits, and therefore are inherently more dimensionally stable, being associated with less dilation *in vivo*. In Figures I.2.14.6C and I.2.14.6D, the external and internal (luminal) surfaces of the weft knitted Barone Microvelour® are compared to those of the warp knitted VP1200K™ (unsealed Gelsoft®) prosthesis. The structure and properties of these devices are listed in Table I.2.14.3.

Warp knits do not run or ravel when cut at an angle, and can be further modified by the addition of an extra yarn in the structure, which adds thickness, bulk, and surface roughness to the fabric (King, 1991). This structure is commonly known as a velour or pile knit, as indicated in Table I.2.14.3. The addition of the pile yarn, while making the fabric feel softer, results in a thicker thrombotic deposit on initial blood contact, which leads to an increased amount of tissue ingrowth into the fabric during healing.

Braided

Braids have found their way into medical use primarily in the manufacture of suture materials and anterior cruciate ligament (ACL) prostheses. Common braided structures involve the diagonal intersection of an even number of yarns that are interlaced at different angles and frequencies, as shown in Figure I.2.14.7. The variables in constructing a braid include the horizontal repeat distance l, called a line, the vertical repeat distance s, called a stitch, w the width of the yarn, and θ the braid angle between the yarn and machine directions. A myriad of structural forms can be achieved with three-dimensional braiding, such as "I" beams, channels, and solid tubes. The challenge for using a braided structure as an implantable device is in securing the many yarns at both ends of the device. Some progress has been made in fabricating a braided structure from a single wire for an esophageal stent (Polyflex®) and for thoracic aortic aneurysm repair (Murgo et al., 1998).

Nonwoven

By definition, a nonwoven is a textile structure produced directly from fibers without the intermediate step of yarn production. The short staple fibers or long continuous fibers are either bonded or interlocked together by means of mechanical, hydraulic or thermal action, or by using an adhesive or solvent, or a combination of these approaches. The fibers may be oriented randomly or preferentially in one or more directions, and by combining multiple layers one can engineer the mechanical properties independently in the machine (lengthwise) and cross directions.

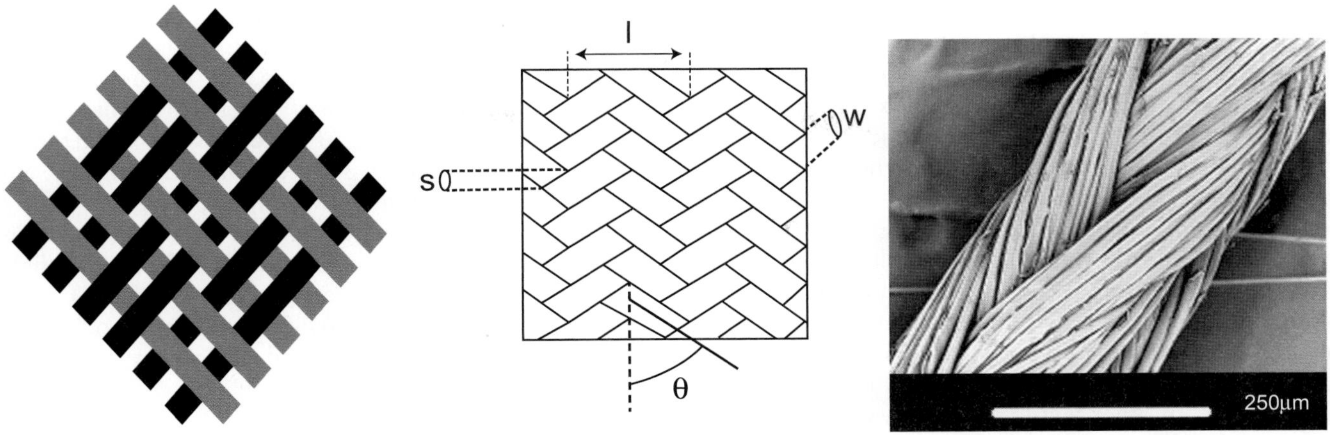

FIGURE I.2.14.7 Schematic diagrams of braided structures and SEM photomicrograph of a braided silk suture.

The total porosity, average pore size, and pore size distribution of a nonwoven web can be controlled by changing certain variables, such as the density of fibers, fiber diameter and length, orientation of the web, and the method of bonding. Nonwoven structures have been used as *ex vivo* blood filters and as a composite multilayered wound dressing as listed in Table I.2.14.1, as well as for scaffold structures for tissue engineering applications (Mao and Russell, 2004; Unger et al., 2004).

FINISHING

Once a textile structure has been fabricated from yarns or fibers, the subsequent processing steps are known as finishing. As mentioned previously, the starting yarn may contain additives, which can result in cytotoxicity and adverse reactions when in contact with tissue. Some of these additives, such as titanium dioxide that is used as a delusterant to increase the amount of light reflected, are inside the spun fiber and therefore cannot be removed during finishing. On the other hand other surface finishes, such as yarn lubricants, can be removed with the proper cleaning and scouring operations. Typically such surface additives are mineral oil or silicone-based and demand specially designed aqueous-based washing procedures or dry-cleaning techniques with organic solvents to ensure complete removal. Since each polymer and fabrication process is different, the finishing operation must be material and device specific. Finishing includes such steps as cleaning, heat setting, bleaching, shrinking, inspection, packaging and sterilization, and will influence the ultimate properties of the biotextile fabric. Figure I.2.14.8 represents a schematic of a typical vascular graft manufacturing operation including the finishing steps. If the cleaning process is properly designed with the use of pyrogen-free water, all extractables are removed and no additional pyrogens are added during the finishing process. Testing of the finished product for cytotoxicity and residual extractables is typically used to ensure that all surface additives are removed from the product prior to packaging and sterilization.

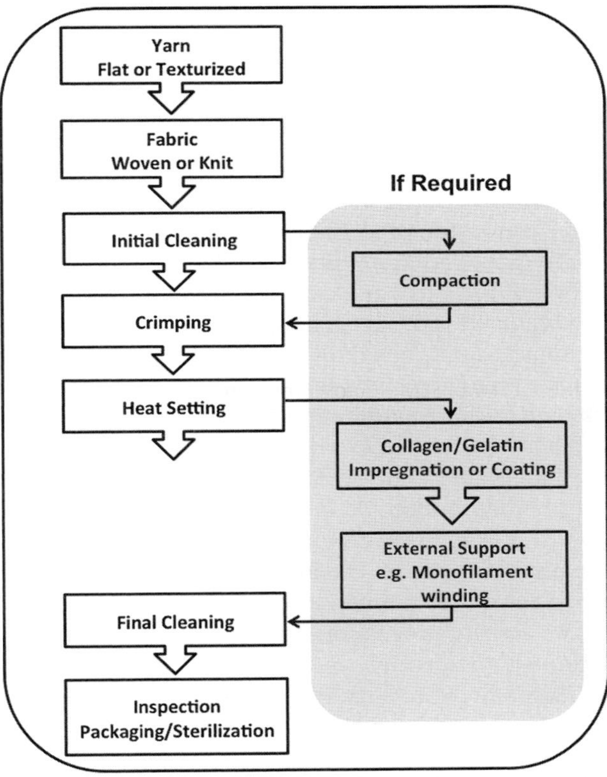

FIGURE I.2.14.8 Typical vascular graft manufacturing operations.

BIOTEXTILE PRODUCTS

General Surgery

Two main applications of biotextiles in general surgery are tissue repair meshes and surgical sutures. Various forms of porous mesh fabrics are used as a support material in hernia and prolapse repair or as tissue patches, as listed in Table I.2.14.1. Traditional constructions are warp knitted from polypropylene or polyester monofilaments or ePTFE membranes, which are preferred since they are more resistant to infection compared to multifilament yarns. More recently, three-dimensional warp knits using polyester multifilament yarns have been found to be more flexible, and therefore can be implanted endoscopically. While most of these meshes are non-degradable and therefore permanent, recent developments have included a resorbable component to promote a more acute inflammatory response and faster healing. Some designs include a protein coating (collagen and/or gelatin) or a microporous PTFE layer on one side only, which reduces the risk of unwanted adhesions *in vivo*. As with other textile structures, various properties can be engineered into the mesh to meet design goals that may include added flexibility, increased strength, reduced thickness, improved handling, and better suture holding strength.

Commercial sutures are either monofilaments or braided multifilament structures, and can be constructed from natural materials such as silk, collagen (catgut), or synthetic materials such as nylon, polyester, polypropylene, and stainless steel, as listed in Table I.2.14.1. Further discussion of the structure and properties of surgical sutures can be found in Chapter II.5.17.

Among the recent innovations related to sutures is the creation of the barbed suture, which can anastomose tissue without the need for a knot. The concept mimics the porcupine's quill, whereby the oriented barbs protrude from the monofilament's surface and mechanically interlock with the surrounding tissue. The advantages of such self-anchoring barbed sutures include the shorter time for suturing because there are no knots to tie, and the more uniform distribution of holding forces which reduces suture slippage, tissue distortion and necrosis, and leads to improved cosmesis (Leung et al., 2002). Resorbable barbed sutures are preferred clinically for cosmetic and plastic face-lift procedures and eyelid ptosis. Work is continuing to identify the optimum barb dimensions, frequency, and geometry for use with different types of tissues (Ingle and King, 2009; Ingle et al., 2010).

Cardiovascular

Biotextiles developed for cardiovascular uses include applications such as heart valve sewing rings, annuloplasty rings that provide dimensional stability to incompetent cardiac heart valves, vascular grafts and endovascular stent grafts, as listed in Table I.2.14.1. One of the key uses of biotextiles in surgery is in the fabrication of large diameter vascular grafts (10–40 mm in diameter). As previously noted, polyester (PET) is the principal polymer used to fabricate vascular grafts. These grafts can either be knitted or woven and are produced in straight or bifurcated configurations. Within each type of construction, various properties can be incorporated into the product, as illustrated in Tables I.2.14.3 and I.2.14.4.

The ideal criteria for a successful vascular prosthesis include the following requirements: available in different shapes and sizes; sterilizable; ease of handling and suturability; cut edges will not fray, ravel, or run; hemostatic; biocompatible; low rate of infection; encourages rapid healing; dimensionally stable; non-thrombogenic; and compliant (Edwards and Tapp, 1957; King, 1991).

No single biomaterial can meet all of these requirements, but textile structures being flexible, porous, lightweight, and compliant satisfy most of these criteria. Woven structures can provide hemostasis and dimensional stability whereas knitted devices are more compliant, easier to handle and suture, and encourage more rapid healing (Tables I.2.14.3 and I.2.14.4). Warp knits are more dimensionally stable and do not ravel compared to weft knits, which are more compliant. Therefore a compromise is necessary, and one needs to identify the key properties for a particular implantation site or disease state before choosing the material and selecting the structure. In order to avoid kinking and facilitate handling, the concept of crimping or imparting radial folds prior to finishing was introduced by Dr. Sterling Edwards in the 1950s (Edwards and Tapp, 1957).

Today a substantial amount of research activity is being directed towards the development of a small vessel prosthesis with diameters less than 6 mm for coronary artery bypass and tibial/popliteal artery replacement. Currently due to high thrombogenicity and compliance mismatch of existing materials and complications due to intimal hyperplasia, no successful commercial product exists to meet this market need. The question still remains as to whether a biotextile will work as a small vessel prosthesis if it is fabricated to have the required compliance and mechanical properties, and its surface is modified with a surface coating, growth factors and other bioactive agents to prevent thrombosis and thrombo-embolic events.

Current research activities are directed towards developing biological-based grafts, surface-modified materials, and tissue-engineering constructs (Sipehia et al., 1996; Niklason et al., 1999; Chung et al., 2010a). The first clinical application of an artificial vessel based on a synthetic scaffold system was reported to successfully reconstruct a low-pressure pulmonary outflow tract in pediatric patients with a cyanotic congenital defect (Shin'oka et al., 2001). In this approach, autologous bone marrow cells were seeded into tubes made from a copolymer of lactide and caprolactone and reinforced with a woven PGA sleeve. While the grafts were not suitable for high pressure arterial implantation, this study still demonstrated the feasibility of a tissue-engineered approach. So far, these grafts have been implanted in at least 22 patients as venous conduits for reconstructive cardiovascular surgery, and >95% patency at 1 year has been reported without evidence of stenosis, thrombogenic complications or aneurysm formation (Matsumura et al., 2003).

During the past 20 years, large investments of resources have been focused on the development of endovascular stent grafts, which are used for minimally invasive aortic aneurysm repair, occlusive disease, and vascular trauma (Hinchliffe and Hopkinson, 2007). The advantages over open surgery are that there is less patient trauma, blood loss, post-operative complications, risk of infection and exposure time to anesthesia. Also the costs associated with hospitalization, patient care, and the time for recovery and rehabilitation are significantly lower than for open surgery. Endovascular prostheses or stent grafts are tubular biotextiles with either an internal or external stent, which are typically made from nitinol or stainless steel wire. Since the stent graft needs to be collapsed and folded onto a balloon catheter and inserted through a distal artery, the structure needs to be thin, flexible, and hemostatic. These requirements can be met with either an ePTFE membrane or a polyester tube tightly woven from fine (45 dtex or less) untextured multifilament yarns so as to minimize the overall wall thickness (Cartes-Zumelzu et al., 2002; Guidoin et al., 2004). However, the long-term durability and fatigue resistance of such structures needs to be evaluated by *in vitro* testing and predictive modeling (Zhao et al., 2009, 2010). (Also see Chapter II.5.18 with a section on drug-eluting stents.)

Traditionally knitted textile structures have been used exclusively as sewing rings for aortic and mitral valves, due to their high suture retention strength and their ability to tolerate needle and iatrogenic damage during surgical procedures. More recently, researchers have reported developing heart valve leaflets from textile-based structures because of their superior flexural fatigue properties using either permanent polyester or resorbable PCL multifilament yarns for tissue engineering (Lieshout et al., 2006; Heim et al., 2008).

Orthopedic

Attempts have been made to construct replacement ligaments and tendons using woven and braided fabrications. One design using a high tenacity polyester woven web inside a pre-stretched knitted graft was evaluated for the repair of separated shoulder joints and anterior cruciate ligaments (ACL) with limited success. In general, biotextiles have had premature clinical failures in orthopedic ligament and tendon applications as a result of abrasion wear problems, inadequate strength, and poor bone attachment (Guidoin et al., 2000). An attempt was made to use a braided PTFE structure for ACL repair, but early failures occurred as a result of creep of the PTFE polymer, resulting in patients experiencing increased knee instability over time (Roolker et al., 2000). Three-dimensional braided and woven tissue engineering scaffolds using resorbable fibers spun from PGA, PLA, and their copolymers have been reported for ligament and cartilage replacement (Cooper et al., 2005; Lu et al., 2005a; Freeman et al., 2007; Moutos et al., 2007). A successful ACL ligament replacement would be a significant advance for orthopedic surgery, but at present no biotextile or other type of prosthesis has shown clinical promise.

Traditionally, loadbearing prostheses used in orthopedic applications have been fabricated primarily from metals and ceramics due to their high strength and modulus. With recent advances in fiber spinning technologies, it is now possible to produce fibers with superior mechanical performance, such as ultra-high molecular weight polyethylene (UHMWPE), carbon, and arylamide (Kevlar®) fibers. These fibers are mostly used as a reinforced composite structure for bone graft cement or for spinal support devices.

Wound Dressings and Skin Grafts

Another common application of biotextiles and fiber technology is in wound dressings and hemostats. Their primary functions are to serve as a physical barrier to prevent infection and to promote moisture absorption and blood coagulation. Woven cotton gauze has been used traditionally as a wound dressing because of its superior moisture absorption and blood clotting ability. However, since it is composed of short staple fibers of cellulose, it adheres to the wound and causes infection and trauma upon removal. Other materials such as polyester, polypropylene, nylon, and viscose rayon, and the addition of antibiotics, growth factors, and a barrier layer have been combined with absorbent cellulose in order to fabricate a more effective wound dressing. A typical example of a layered laminate structure includes a protective outer barrier of polyester or polypropylene with a cellulose-based absorbent inner layer. Examples of commercial products are listed in Table I.2.14.1. Collagen, chitosan, and cellulose derivatives can also serve as hemostatic devices in layered fibril, foam, and powdered forms.

Patients who have suffered severe burn injuries or chronic diabetic ulcers, need a temporary skin graft for protection against infection and maintenance of hydration prior to an autologous dermal transplant. For such patients, the first commercial skin graft product available was Epicel® made with sheets of epidermal keratinocytes (Guerra et al., 2009). While being biocompatible, this type of graft with only biological components lacked mechanical integrity and was challenging to handle and transport. To overcome these limitations, alternative skin graft products have been developed to include both biological components such as porcine collagen and allogenic fibroblasts with a synthetic textile layer as listed in Table I.2.14.1 (Marston et al., 2003). (See also Chapter II.5.14 on Burn Dressings and Skin Substitutes.)

FUTURE DIRECTIONS

In the future, innovations and new applications for medical fibers and biotextiles will be driven by both the direction of clinical practice, which is moving towards more arthroscopic and less invasive surgical techniques, as well as the move towards regenerative medicine whereby resorbable scaffolds serve as temporary templates for the proliferation and regeneration of new viable tissues. Biotextiles are thinner, stronger, more flexible, and lighter weight structures than other biomaterials such as metals and ceramics, and they have already demonstrated their superiority to withstand the compression, tensile, shear, and bending forces that accompany folding, compaction, and delivery through a catheter. Clearly, less invasive approaches to implantation which are currently used for endovascular stent grafts, aortic heart valves, hernia and prolapse meshes, embolic protection devices, spinal disc prostheses, and septal defect repair, will continue to expand as surgeons and biomedical engineers rise to meet new and currently unforeseen clinical challenges.

These are five areas within the field of medical fibers and biotextiles where we anticipate further advances to be made in the near future.

1. In the past the regulatory environment has delayed getting new polymer materials to market. We note that there are now a plethora of novel fiber-forming polymers in development which have unique properties, such as shape memory, that are electroactive, elastomeric, rapidly resorbable, responsive, and "smart" materials. When spun into fibers or as biotextiles, they can serve as biosensors, actuators, and drug delivery systems by responding to changes in temperature, pH, moisture level or drug concentration (Wang and McCord, 2007; Shi et al., 2008; Shankar et al., 2009). In addition, through advances in genetic engineering, we are able to capitalize on the unique properties of biopolymers, as found in high tenacity spider silk, and use a biomimetic approach to design and synthesize such novel fiber-forming protein biopolymers with unique strength and mechanical performance for use as ophthalmic sutures (Vollrath and Knight, 2001).
2. Figure I.2.14.1 has already shown how new fiber spinning techniques have extended the range of manufactured fiber diameters from about 10 μm down to less than 50 nm, and have also facilitated the spinning of fibers with a range of different cross-sectional shapes and bicomponent cross-sections. These finer noncircular and bicomponent fibers appear to provoke a different cellular response that as yet is not fully understood, but lends itself to fabricating implantable devices and tissue engineering constructs that have uniquely tunable cellular responses, rates of resorption, and drug delivery performance (Vaughn and Carman, 2001; Lu et al., 2005b).
3. While textile fabrics have traditionally been thin, flexible, two-dimensional structures, there is increasing use of pile weaving and knitting techniques, as well as other technologies such as flocking, tufting, and embroidery to create a thick three-dimensional structure that contains fibers and yarns oriented in the thickness (z-axis) direction (Wollenweber et al., 2006; Pereira et al., 2007; Walther et al., 2007; Rentsch et al., 2009). There is a need to apply these velvet, velour, and toweling fabrication techniques to the development of the thicker tissue engineered scaffolds so as to improve cell migration and tissue regeneration through the thickness of the construct.
4. Surface modification is a well-established technique for ensuring that biomaterials, regardless of their bulk chemical structure, have appropriate biocompatibility in the acute period after implantation for their particular clinical application. While textiles have traditionally always relied on the appropriate final finishing operations to ensure cleanliness, uniformity of color, structural stability, and surface functionality, such as coating, water repellency or hydrophilicity, it is only recently that more advanced surface modification techniques, such as the immobilization of biomolecules, growth factors, and antimicrobial agents, and the use of radio frequency plasma for bioactivation, sterilization or surface grafting have been applied to biotextile structures (Michielsen and Lee, 2007; El-Mekawy et al., 2010; Gorensek et al., 2010; Patterson et al., 2010). We anticipate seeing surface modification becoming increasingly important in the future as our understanding of the potential benefits of these techniques being applied to medical fibers and biotextiles with such large surface areas continues to expand. (See also Chapter I.2.12 on surface modification of biomaterials.)
5. Given that the diameters for some medical fibers lie in the micro-range, which is close to the dimensions of cells and microorganisms, this is one of the reasons why biotextiles can be engineered to elicit precise and reproducible cellular responses. Further, some fibers with diameters in the nano-range may have special effects on living cells, as noted above. Not only are biotextile structures versatile in their elemental design and hierarchical components, but they are also fabricated by well-established textile manufacturing technologies that are designed to make large quantities of defined structures with identical and precisely specified property requirements under good quality and manufacturing controls. As such, biotextiles and fiber-based tissue engineering scaffolds will soon be considered as standardized "off-the-shelf" products which can be scaled up to be fabricated in large quantities. Biotextiles are therefore well suited to commercialization, and their production can be readily scaled up to meet growing clinical demand.

In summary, the use of medical fibers and biotextiles in medicine will continue to grow as new synthetic polymers and genetically-engineered biopolymers, fiber spinning technologies, three-dimensional constructions, coatings, and surface modification processes are introduced. This will meet the future needs of less-invasive interventional procedures and surgical delivery, the long-term cyclic fatigue resistance of permanent implants, and the previously unobtainable goal of regenerating tissue-engineered organs.

ACKNOWLEDGMENTS

The authors wish to thank Mike Ferguson and Judy Elson for their technical assistance in preparing the figures.

REFERENCES

Adanur, S. (1995). *Wellington Sears Handbook of Industrial Textiles*. Lancaster, PA: Technomic Publishing Company.

ANSI/AAMI/ISO (R). (2004). *Cardiovascular implants: Tubular vascular prostheses. 8.2.2. Determination of Water Permeability*.

ASTM (1996). *Standard Terminology Relating to Textiles. D*. 123–196.

Barnes, C. P., Sell, S. A., Boland, E. D., Simpson, D. G., & Bowlin, G. L. (2007). Nanofiber technology: Designing the next generation of tissue engineering scaffolds. *Advanced Drug Delivery Reviews*, **59**, 1413.

Boland, E. D., Wnek, G. E., Simpson, D. G., Pawlowski, K. J., & Bowlin, G. L. (2001). Tailoring tissue engineering scaffolds using electrostatic processing techniques: A study of poly(glycolic acid) electrospinning. *Journal of Macromolecular Science – Pure Applied Chemistry*, **A38**, 1231–1243.

Cartes-Zumelzu, F., Lammer, J., Hoelzenbein, T., Cejna, M., Schoder, M., et al. (2002). Endovascular placement of a nitinol-ePTFE stent-graft for abdominal aortic aneurysms: Initial and midterm results. *Journal of Vascular and Interventional Radiology*, **13**, 465–473.

Chung, S., Moghe, A. K., Montero, G. A., Kim, S. -H., & King, M. W. (2009). Nanofibrous scaffolds electrospun from elastomeric biodegradable poly(L-lactide-co-e-caprolactone) copolymer. *Biomedical Materials*, **4**, 015019.

Chung, S., Ingle, N. P., Montero, G. A., Kim, S. H., & King, M. W. (2010a). Bioresorbable elastomeric vascular tissue engineering scaffolds via melt spinning and electrospinning. *Acta Biomaterialia*, **6**, 1958–1967.

Chung, S., Gamcsik, M., & King, M. W. (2010b). High surface area tissue engineering scaffolds with multigrooved fibers. *34th Annual Meeting Society of Biomaterials*, **32**, 461.

Cooper, J. A., Lu, H. H., Ko, F. K., Freeman, J. W., & Laurencin, C. T. (2005). Fiber-based tissue-engineered scaffold for ligament replacement: Design considerations and *in vitro* evaluation. *Biomaterials*, **26**, 1523–1532.

Doshi, J., & Reneker, D. H. (1995). Electrospinning process and applications of electrospun fibers. *Journal of Electrostatics*, **35**, 151–160.

Durany, A., Anantharamaiah, N., & Pourdeyhimi, B. (2009). High surface area nonwovens via fibrillating spunbonded nonwovens comprising Islands-in-the-Sea bicomponent filaments: Structure–process–property relationships. *Journal of Materials Science*, **44**, 5926–5934.

Edwards, W. S., & Tapp, J. S. (1957). Two and a half years experience with crimped nylon grafts. *ASAIO Journal*, **3**, 70–72.

El-Mekawy, A., Hudson, S., El-Baz, A., Hamza, H., & El-Halafawy, K. (2010). Preparation of chitosan films mixed with superabsorbent polymer and evaluation of its haemostatic and antibacterial activities. *Journal of Applied Polymer Science*, **116**, 3489–3496.

Fages, J., Poddevin, N., King, M. W., Marois, Y., Bronner, J., et al. (1998). Use of supercritical fluid extraction as a method of cleaning anterior cruciate ligament prostheses. *ASAIO Journal*, **44**, 278–288.

Formhals, A. (1934). *Process and apparatus for preparing artificial threads*. US Patent 1975504.

Freeman, J. W., Woods, M. D., & Laurencin, C. T. (2007). Tissue engineering of the anterior cruciate ligament using a braid-twist scaffold design. *Journal of Biomechanics*, **40**, 2029.

Gong, R. H., & Nikoukhesal, A. (2009). Hydro-entangled bicomponent microfiber nonwovens. *Polymer Engineering and Science*, **49**, 1703–1707.

Gorensek, M., Gorjanc, M., Bukosek, V., Kovac, J., Jovancic, P., et al. (2010). Functionalization of PET fabrics by corona and nano silver. *Textile Research Journal*, **80**, 253–262.

Guerra, L., Dellambra, E., Panacchia, L., & Paionni, E. (2009). Tissue engineering for damaged surface and lining epithelia: Stem cells, current clinical applications, and available engineered tissues. *Tissue Engineering Part B, Reviews*, **15**, 91–112.

Guidoin, M. -F., Marois, Y., Beijui, J., Poddevin, N., King, M. W., et al. (2000). Analysis of retrieved polymer fiber based replacements for the ACL. *Biomaterials*, **21**, 2461–2474.

Guidoin, R., Douville, Y., Baslé, M. F., King, M., Marinov, G. R., et al. (2004). Biocompatibility studies of the Anaconda stent-graft and observations of Nitinol corrosion resistance. *Journal of Endovascular Therapy*, **11**, 385–403.

Heim, F., Durand, B., & Chakfe, N. (2008). Textile heart valve prosthesis: Manufacturing process and prototype performances. *Textile Research Journal*, **78**, 1124.

Hinchliffe, R. J., & Hopkinson, B. R. (2007). Development of endovascular stent-grafts. *Proceedings of the Institution of Mechanical Engineers – Part H – Journal of Engineering in Medicine*, **221**, 547–560.

Huang, C., Chen, S., Lai, C., Reneker, D. H., Qiu, H., et al. (2006). Electrospun polymer nanofibres with small diameters. *Nanotechnology*, **17**, 1558–1563.

Huang, L., Mcmillan, R. A., Apkarian, R. P., Pourdeyhimi, B., Conticello, V. P., et al. (2000). Generation of synthetic elastin-mimetic small diameter fibers and fiber networks. *Macromolecules*, **33**, 2989–2997.

Ingle, N. P., & King, M. W. (2009). Optimizing the tissue anchoring performance of barbed sutures in skin and tendon tissues. *Journal of Biomechanics*, **43**, 303–309.

Ingle, N. P., King, M. W., & Zikry, M. A. (2010). Finite element analysis of barbed sutures in skin and tendon tissues. *Journal of Biomechanics*, **43**, 879–886.

King, M. W. (1991). Designing fabrics for blood vessel replacement. *Canadian Textile Journal*, **108**, 24–30.

King, M. W. (2001). Overview of opportunities in medical textiles. *Canadian Textile Journal*, **118**, 28–36.

King, M. W., Ornberg, R., Marois, Y., Marinov, G., & Cadi, R. (1999). Healing responses of partially bioresorbable bicomponent fibers: A subcutaneous rat study. *25th Annual Meeting Society for Biomaterials*, **22**, 60.

Knaul, J. Z., Hudson, S. M., & Creber, K. A.M. (1999). Improved mechanical properties of chitosan fibers. *Journal of Applied Polymer Science*, **72**, 1721–1732.

Kumbar, S., Nukavarapu, S., Roshan, J., & Hogan, M. (2008). Recent patents on electrospun biomedical nanostructures: An overview. *Recent Patents Biomed. Eng*, **1**, 68–78.

Lannutti, J., Reneker, D., Ma, T., Tomasko, D., & Farson, D. (2007). Electrospinning for tissue engineering scaffolds. *Materials Science and Engineering: C*, **27**, 504.

Leong, M. F., Chian, K. S., Mhaisalkar, P. S., Ong, W. F., & Ratner, B. D. (2009). Effect of electrospun poly(D, L-lactide) fibrous scaffold with nanoporous surface on attachment of porcine esophageal epithelial cells and protein adsorption. *Journal of Biomedical Materials Research Part A*, **89**, 1040–1048.

Leung, J. C., Ruff, G. L., & Megaro, M. A. (2002). Barbed, bi-directional medical sutures: Biomechanical properties and wound closure efficacy study. *28th Anuual Meeting Society for Biomaterials*, **25**, 724.

Lieshout, M. V., Peters, G., Rutten, M., & Baaijens, F. (2006). A knitted, fibrin-covered polycaprolactone scaffold for tissue engineering of the aortic valve. *Tissue Engineering*, **12**, 481–488.

Lu, H., Cooper, J. A., Manuel, S., Freeman, J., Attawia, M., et al. (2005a). Anterior cruciate ligament regeneration using braided biodegradable scaffolds: *In vitro* optimization studies. *Biomaterials*, **26**, 4805–4886.

Lu, Q., Simionescu, A., & Vyavahare, N. (2005b). Novel capillary channel fiber scaffolds for guided tissue engineering. *Acta Biomaterialia*, **1**, 607–614.

Ma, Z., Kotaki, M., Inai, R., & Ramakrishna, S. (2005). Potential of nanofiber matrix as tissue-engineering scaffolds. *Tissue Engineering*, **11**, 101–109.

Malheiro, V., Caridade, S., Alves, N., & Mano, J. (2009). New poly(epsilon-caprolactone)/chitosan blend fibers for tissue engineering applications. *Acta Biomaterialia*, **6**(2), 418–428.

Mao, N., & Russell, S. J. (2004). Nonwoven wound dressings. *Textile Progress*, **36**, 1–57.

Marston, W. A., Hanft, J., Norwood, P., & Pollak, R. (2003). The efficacy and safety of Dermagraft in improving the healing of chronic diabetic foot ulcers. *Diabetes Care*, **26**, 1701–1705.

Matsumura, G., Hibino, N., Ikada, Y., Kurosawa, H., & Shin'oka, T. (2003). Successful application of tissue engineered vascular autografts: Clinical experience. *Biomaterials*, **24**, 2303–2308.

Michielsen, S., & Lee, H. J. (2007). Design of a superhydrophobic surface using woven structures. *Langmuir*, **23**, 6004–6010.

Moutos, F. T., Freed, L. E., & Guilak, F. (2007). A biomimetic three-dimensional woven composite scaffold for functional tissue engineering of cartilage. *Nature Materials*, **6**, 162–167.

Murgo, S., Dussaussois, L., Golzarian, J., Cavenaile, J., Abada, H., et al. (1998). Penetrating atherosclerotic ulcer of the descending thoracic aorta: Treatment by endovascular stent-graft. *CardioVascular and Interventional Radiology*, **21**, 454–458.

Niklason, L. E., Gao, J., Abbott, W. H., Hirschi, K. K., Houser, S., et al. (1999). Functional arteries grown *in vitro*. *Science*, **284**, 489–493.

Ohkawa, K., Cha, D., Kim, H., Nishida, A., & Yamamoto, H. (2004). Electrospinning of chitosan. *Macromolecular Rapid Communications*, **25**, 1600–1605.

Patterson, J., Martino, M., & Hubbell, J. (2010). Biomimetic materials in tissue engineering. *Materials Today*, **13**, 14–22.

Pereira, S., Anand, S., Rajendran, S., & Wood, C. (2007). A study of the structure and properties of novel fabrics for knee braces. *Journal of Industrial Textiles*, **36**, 279–300.

Pham, Q. P., Sharma, U., & Mikos, A. G. (2006). Electrospinning of polymeric nanofibers for tissue engineering applications: A review. *Tissue Engineering*, **12**, 1197–1211.

Ramakrishna, S., Fujihara, K., Teo, W. -E., Yong, T., Ma, Z., et al. (2006). Electrospun nanofibers: Solving global issues. *Materials Today*, **9**, 40–50.

Rentsch, B., Hofmann, A., Breier, A., Rentsch, C., & Scharnweber, D. (2009). Embroidered and surface modified polycaprolactone-co-lactide scaffolds as bone substitute: *In vitro* characterization. *Annals of Biomedical Engineering*, **37**, 2118–2128.

Roolker, W., Patt, T. W., van Dijk, C. N., Vegter, M., & Marti, R. K. (2000). The Gore-Tex prosthetic ligament as a salvage procedure in deficient knees. *Knee Surgery, Sports Traumatology, Arthroscopy*, **8**, 20–25.

Shalaby, S. W. (1985). Fibrous Materials for Biomedical Applications. In M. Lewin, & J. Preson (Eds.), *High Technology Fibers*. New York, NY: Marcel Dekker.

Shalaby, S. W., & Shah, K. R. (1991). Chemical Modification of Natural Polymers and their Technological Relevance. In S. W. Shalaby, G. B. Butler, & C. L. McCormick (Eds.), *Water Soluble Polymers: Chemistry and Applications*. Washington, DC: American Chemical Society.

Shankar, R., Ghosh, T. K., & Spontak, R. J. (2009). Mechanical and actuation behavior of electroactive nanostructured polymers. *Sensors and Actuators A: Physical*, **151**, 46–52.

Shi, G., Zhang, Z., & Rouabhia, M. (2008). The regulation of cell functions electrically using biodegradable polypyrrole-polylactide conductors. *Biomaterials*, **29**, 3792–3798.

Shin'oka, T., Imai, Y., & Ikada, Y. (2001). Transplantation of a tissue-engineered pulmonary artery. *New England Journal of Medicine*, **344**, 532–533.

Sipehia, R., Martucci, G., & Lipscombe, J. (1996). Transplantation of human endothelial cell monolayer on artificial vascular prosthesis: The effect of growth-support surface chemistry, cell seeding density, ECM protein coating, and growth factors. *Artificial Cells, Blood Substitutes and Biotechnology*, **24**, 51–63.

Tuzlakoglu, K., Alves, C. M., Mano, J. F., & Reis, R. L. (2004). Production and characterization of chitosan fibers and 3-D fiber mesh scaffolds for tissue engineering applications. *Macromolecular Bioscience*, **4**, 811–819.

Unger, R. E., Wolf, M., Peters, K., Motta, A., Migliaresi, C., et al. (2004). Growth of human cells on a non-woven silk fibroin net: A potential for use in tissue engineering. *Biomaterials*, **25**, 1069–1075.

Vaughn, E., & Carman, B. (2001). Expanded surface area fibers: A means for medical product enhancement. *Journal of Industrial Textiles*, **30**, 303–310.

Vollrath, F., & Knight, D. P. (2001). Liquid crystalline spinning of spider silk. *Nature*, **410**, 541–548.

Walther, A., Bernhardt, A., Pompe, W., Gelinsky, M., Mrozik, B., et al. (2007). Development of novel scaffolds for tissue engineering by flock technology. *Textile Research Journal*, **77**, 892–899.

Wang, X., & McCord, M. G. (2007). Grafting of poly(N-isopropylacrylamide) onto nylon and polystyrene surfaces by atmospheric plasma treatment followed with free radical graft copolymerization. *Journal of Applied Polymer Science*, **104**, 3614–3621.

Williams, D. F. (Ed.) (1987). Definitions in Biomaterials. Amsterdam: Elsevier.

Wollenweber, M., Domaschke, H., Hanke, T., Boxberger, S., Schmack, G., et al. (2006). Mimicked bioartificial matrix containing chondroitin sulphate on a textile scaffold of poly(3-hydroxybutyrate) alters the differentiation of adult human mesenchymal stem cells. *Tissue Engineering*, **12**, 345–359.

Zhao, H., Wang, L., Li, Y., Liu, X., & King, M. W. (2009). *In vitro* fatigue properties of prototype textile components of endovascular devices. *Fibers and Polymers*, **10**, 91–97.

Zhao, H., Wang, L., Li, Y., Liu, X., & King, M. W. (2010). The mathematical model for evaluating the fatigue resistance of stent-graft tubular fabric: Relationships between textile parameters and fatigue performance. *Journal of Biomaterials Applications*, **24**, 579–590.

CHAPTER I.2.15 TEXTURED AND POROUS MATERIALS

Heidi E. Koschwanez and William M. Reichert
Department of Biomedical Engineering, Duke University, Durham, NC, USA

INTRODUCTION

The use of porous and textured implants to stimulate tissue ingrowth, disrupt fibrosis and promote angiogenesis dates back as far as the 1940s, and is an early example of a tissue engineering-like approach to biomaterials. Device function and placement in the body dictates what material would be most suitable for a particular application. Numerous materials (metals, ceramics, natural and synthetic polymers) can be textured or rendered porous using a wide variety of techniques (Tables I.2.15.1 and I.2.15.2).

The extent to which a textured or porous material can influence the host tissue response is related to pore size and porosity. While no one pore size fits all applications, the majority of porous materials employed possess an open architecture with interconnecting pores that are sufficiently large to support cell and tissue infiltration, but small enough to disrupt fibrous tissue deposition (Sharkawy et al., 1998). Textured implants with open architecture pore structures also appear to form thinner foreign-body capsules (Salzmann et al., 1997; Sharkawy et al., 1997; Updike et al., 2000; Ward et al., 2002) (also see Chapter II.3.2). Porous surfaces promote tissue ingrowth that minimizes interfacial cell necrosis from mechanical shear forces, that in turn results in less inflammation and reduced capsule thickness (Rosengren et al., 1999).

Tissue ingrowth strongly influences the long-term success of orthopedic, dental, ocular, percutaneous, and cardiovascular implants. In most instances, texturing is used to promote tissue ingrowth for improved tissue–implant stabilization. Other important effects incited by implant texturing include disruption of fibrous encapsulation,

TABLE I.2.15.1 Common Materials used in Orthopedic and Dental Applications for Fabricating Porous Coatings or Porous Scaffolds

Material		Common Methods to Render Porous	Advantages	Disadvantages
Ceramics (Sun et al., 2001; Karageorgiou and Kaplan, 2005; Simske et al., 1997)	• Hydroxyapatite (HA) • Natural coral • Calcium phosphate based material • Amorphous glasses	• Plasma-spray (coating) • Sinter (scaffold)	• Similar composition and structure to natural bone • Enhanced osseointegration using apatite-like materials • Biocompatible • Easy to render porous • Low corrosion	• Coatings: Long-term stability (i.e. resportion, delamination) • Scaffolds: Brittle, slow degradation • Not amenable to machining • Difficult to control pore sizes of naturally occurring ceramics
Metals (Sun et al., 2001; Karageorgiou and Kaplan, 2005; Simske et al., 1997; Ryan et al., 2006; Bobyn et al., 1999)	• Titanium (Ti) and Ti alloys • Cobalt-chromium (Co-Cr) and Co-Cr alloys • Tantalum	• Plasma spray • Sinter • Grit-blast • Wire mesh • Combustion synthesis • Vapor deposition	• Bioinert • Excellent mechanical properties (i.e. less susceptible to fatigue than ceramics, polymers)	• Corrosion related complications • Stress shielding complications from significantly higher Young's modulus than bone
Natural Polymers (Karageorgiou and Kaplan, 2005)	• Collagen • Hyaluronic acid • Silk fibroin	• Freeze-drying • Salt leaching • Gas foaming • Crosslinking	• Biocompatible • Naturally biodegradable • Amicable for growth factor incorporation, supporting cells	• Lack mechanical strength • Rapid degradation rates • Chemical modifications can reduce biocompatibility
Synthetic Polymers (Karageorgiou and Kaplan, 2005; Simske et al., 1997; Ryan et al., 2006)	• Poly lacticide (PLA) • Poly glycolide (PGA) • Poly ethylene glycol (PEG) • Polyvinyl alcohol (PVA) • Polyethylene terephthalate (PET) • polyurethane	• Salt leaching • Gas foaming • Electrospinning • Sintering	• Scaffolds can easily be tailored to meet application • Amicable for growth factor incorporation, supporting cells	• Lack mechanical strength

Note: Stainless steel is not commonly used in porous form since it is not as corrosive resistant as Ti, Co-Cr and their alloys (Simske et al., 1997).

TABLE I.2.15.2 Summary of Methods Developed for Fabricating Porous Three-Dimensional Biodegradable Scaffolds for Tissue Engineering (Chen et al., 2002; Hutmacher, 2000; Yang et al., 2001)

Method	Description	Processing Time	Merits of Method	Challenges
Gas foaming * (Mooney et al., 1996)	High pressure CO_2 gas dissolved in polymer to create pores	~ 2 days	• Tuneable porosity and pore structure • No organic solvents	• Closed pore architecture • Non-porous surface
Fiber bonding (Mikos et al., 1993)	Physical bonding non-woven polymer fibers to create interconnecting mesh	~ 4 days	• Interconnecting, open pore architecture • Large surface area for cell attachment	• Lack of mechanical strength • Difficulty controlling porosity • Amenable to only certain polymers and solvents
Three-dimensional printing (Wu et al., 1996)	Computer controlled solvent application to sequential layers of powdered polymer, creating a porous scaffold	> 7 days	• Open pore architecture • Complex 3D scaffolds possible • Incorporation of bioactive compounds	• Lack of mechanical strength • User and technique sensitive • Instrument requirements
Phase separation (Schugens et al., 1996a,b)	Sublimation of solvent in polymer to form macropores	>2 days	• Open pore architecture • Can incorporate bioactive molecules	• Difficult to control scaffold morphology
Emulsion freeze-drying (Whang et al., 1995)	Polymer, solvent and water (all immiscible) homogenized, quenched in liquid N_2, then freeze-dried to form pores	>7 days	• Open pore architecture • Can incorporate bioactive molecules	• Difficult to control scaffold morphology • User and technique sensitive
Porogen leaching ‡ (Mikos et al., 1994)	Dissolution of salt in polymer to form pores	>7 days	• Open pore architecture • Tuneable pore size and porosity	• Lack of mechanical strength • Possible solvent residue • Can not incorporate bioactive molecules • Visible surface skin

*Gas foaming with CO_2 has also been combined with salt leaching to create open pore architecture with a processing time of approximately 4 days (Harris et al., 1998).
‡Porogen leaching has also been combined with gas foaming using ammonium bicarbonate as salt leaching/gas foaming agent, reducing processing time to 24 hours with no surface skin layer (Nam et al., 2000).

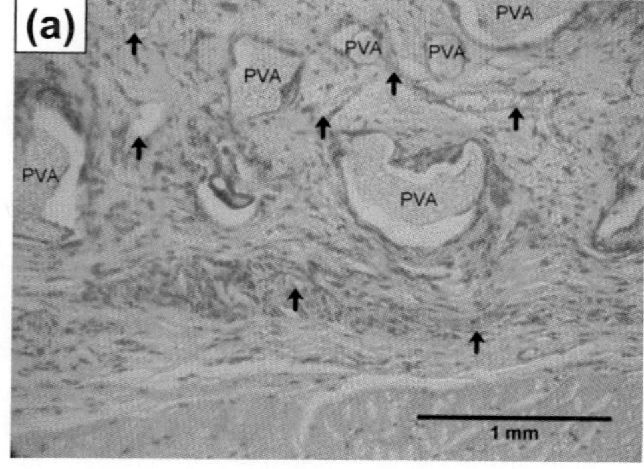

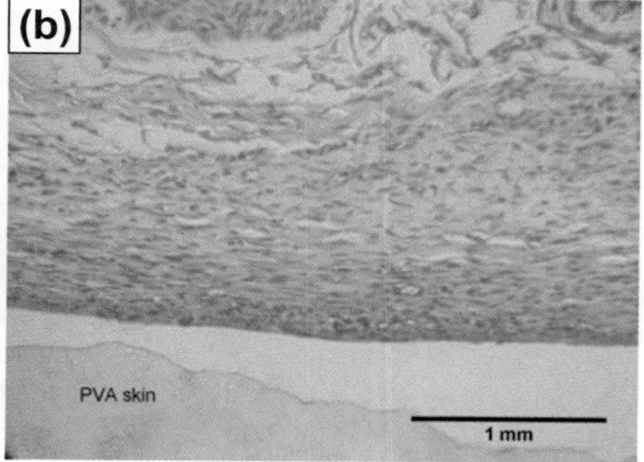

FIGURE I.2.15.1 Difference in cellular and tissue response around polyvinyl alcohol (PVA) implanted in rat subcutaneous tissue after 3 weeks: (a) porous PVA (300 μm pores) promoted fibrovascular ingrowth (arrows denote capillaries); and (b) smooth PVA developed dense avascular capsule adjacent to material surface. (From Koschwanez, H. E. (2004). Unpublished data.)

improved tissue healing, and increased vascularization of the tissue surrounding the implant (Figure I.2.15.1).

These recurring concepts appear throughout this chapter in several medical devices, i.e., breast implants, bone regenerative implants and orthopedic implant coatings, ocular implants and glaucoma drainage devices, synthetic vascular grafts, and percutaneous devices, with a separate section for percutaneous glucose sensors. For each device, the benefits and challenges of using textured surfaces will be discussed, in addition to optimal pore size, choice of implant material, and theorized mechanisms of how these surface topologies influence host response.

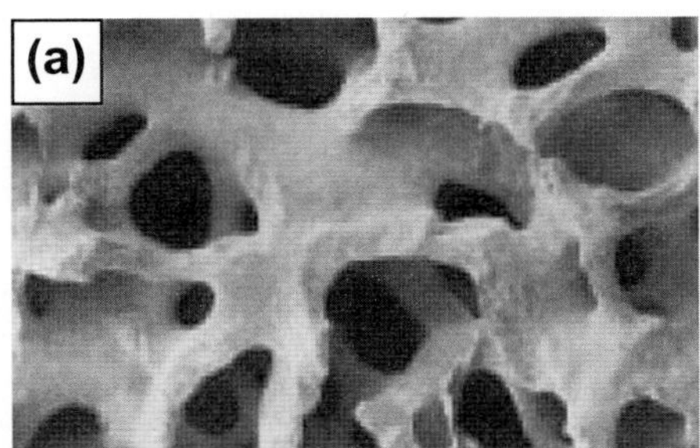

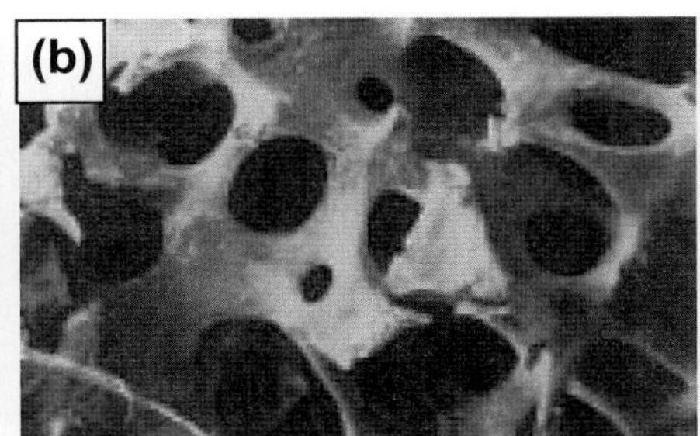

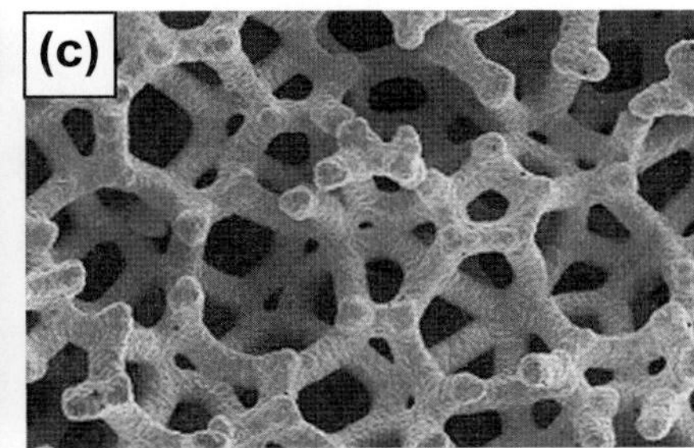

FIGURE I.2.15.2 Similar porous structure and composition between: (a) coralline hydroxyapatite; and (b) human cancellous bone. (c) Chemical vapor deposition of tantalum on a vitreous carbon skeleton creates a porous metal construct that also mimics native bone structure. (a,b: Courtesy of Interpore).

STIMULATING TISSUE INGROWTH

Porosity to Promote Bone Regeneration and Implant Fixation

Porous materials are often used to repair severely fractured or malformed bone. These bone regeneration scaffolds must provide structural support to the newly forming bone, share mechanical properties similar to the host bone, be biocompatible, and biodegrade at a rate synergistic with bone remodeling (Karageorgiou and Kaplan, 2005). Controlling pore size and porosity is essential in creating a scaffold that meets the needs of both the specific repair site (mechanical, mineral, and cellular properties) and the patient (age, activity level, nutrition, disease) (Karageorgiou and Kaplan, 2005).

Of the over 2,000,000 total joint replacements performed annually worldwide (Kuster, 2002; de Pina et al., 2011), an increasing number of implants are employing textured fixation layers to minimize latent complications of prostheses loosening (Chang et al., 1996). These fixation layers mostly consist of metal beads sintered to the implant surface, thin coatings of hydroxyapatite or grooves or pitted surfaces (Simske et al., 1997; Sun et al., 2001). Candidates for textured implants are typically individuals with replacement joints (total knee, acetabular component) implanted in cancellous bone, as this tissue type consistently achieves a stable implant–host bone attachment (Bloebaum et al., 2007). Improved implant fixation through osteointegration into porous implants also reduces stress shielding (a condition that occurs from uneven load distribution between the host bone and implant; see Chapter II.5.6). As the stiffer implant material bears more load than the bone, the bone becomes weakened and susceptible to fracture and implant loosening.

Rate of bone growth into a porous material is highly dependent on pore size (Klawitter et al., 1976) and pore connectivity (Hing et al., 2004). Bone is naturally porous, with pore sizes ranging from 1 to 100 μm (Simske et al., 1997) (Figure I.2.15.2). The optimum pore size for bone ingrowth into porous-coated prostheses, regardless of material, ranges from 100–400 μm (Simske et al., 1997; Kienapfel et al., 1999; Ryan et al., 2006); however, interconnecting pores as small as 50 μm have been shown to be effective primarily for encouraging blood vessel growth within the implant (Klawitter et al., 1976; Bobyn et al., 1980; Simske et al., 1997; Ryan et al., 2006). To support continuous tissue ingrowth, a minimum pore size of 100 μm is required. Although increased fixation strength occurs with increased pore size when using pores less than 100 μm (Kienapfel et al., 1999), no such relationship has been observed for pore size in the 150–400 μm range. Interestingly, while the size of the interconnecting pores is critical for osteointegration, pore shape does not appear to influence biological response (Turner et al., 1986; Ryan et al., 2006).

Osteointegration will not occur if fibrous tissue ingrowth precedes first. Fibrous ingrowth has been shown to occur with pores smaller than 15 μm (Simske et al., 1997) or larger than 1000 μm (Ryan et al., 2006), suggesting different cell types have preferential pore sizes. Increased micromotion between the implant and host bone will also promote fibrous connective tissue ingrowth instead of bony ingrowth (Kienapfel et al., 1999; Ryan et al., 2006). This fibrotic tissue may inhibit bone formation or prevent bony ingrowth from the host tissue, ultimately reducing implant fixation strength (Ryan et al., 2006).

Several porous ceramics, metals, and polymers have been investigated for orthopedic implants, with porous metals being the most widely employed for loadbearing applications (Ryan et al., 2006) (Table I.2.15.1). However, porous materials may have substantially reduced dynamic fatigue strength. Additionally, the pores may create stress concentration sites for microfractures, compounding implant fatigue (Simske et al., 1997). Additionally, porous metals may exhibit higher corrosion rates, due to increased surface area, raising long-term safety concerns (Ryan et al., 2006). Composite implants blending two or more materials, such as plasma-sprayed hydroxyapatite coatings on titanium, could provide the best opportunity to match implant and bone properties (Simske et al., 1997; Sun et al., 2001). Another interesting possibility is creating scaffolds with gradient pore size and porosity to promote osteogenesis at one end of the scaffold and osteochondrial ossification at the other,

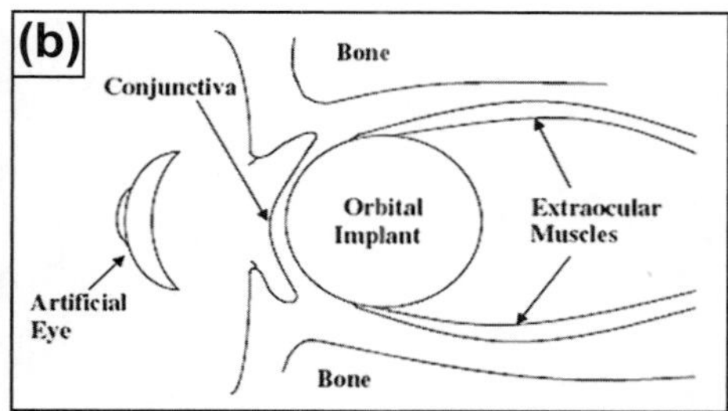

FIGURE I.2.15.3 (a) Examples of porous hydroxyapatite (left) and polyethylene (right) orbital implants; (b) sagittal view of human orbit showing placement of orbital implant.

ideal for bone-cartilage repair. Investigations are also underway to integrate biomolecules into scaffold materials to further promote bone repair (Karageorgiou and Kaplan, 2005).

Texturing to Improve Healing and Restore Motion in Eye Implants

When the eye is enucleated due to severe trauma, intraocular cancer or removal of a blind and painful eye, the lost tissue may be compensated with an orbital implant (Chalasani et al., 2007; Sami et al., 2007). Important considerations with orbital implants include fit, optimized prosthesis motility, and minimized long-term complications, such as implant exposure, extrusion, migration, and infection (Goldberg et al., 1994).

Numerous materials have been used in orbital implant design, including glasses, silicones, and acrylics (Chalasani et al., 2007). Porous materials such as hydroxyapatite (HA), porous polyethylene (PPE), and aluminum oxide (AO) are showing increased popularity (Chalasani et al., 2007; Sami et al., 2007) (Figure I.2.15.3). HA, the mineral component of bone, has good biocompatibility; however, suturing this material to the extraocular muscles is difficult. Abrasion from roughened HA implants may exacerbate inflammation and lead to latent implant exposure. The roughness of HA also makes surgical insertion more challenging (Chalasani et al., 2007). PPE is flexible, easily molded, relatively inexpensive, and can be sutured to surrounding tissue (Chalasani et al., 2007). PPE appears to incite less inflammation and fibrosis than HA (Goldberg et al., 1994). AO is robust, biocompatible, easy to manufacture, and less expensive than HA (Chalasani et al., 2007). AO is the newest porous material used in orbital implants, with only one commercially available implant (Chalasani et al., 2007). More long-term studies are required to ensure clinical efficacy of this material as an orbital implant (Sami et al., 2007).

HA and PPE roughly resemble the structure of trabecular bone (Chalasani et al., 2007; Sami et al., 2007), and encourage fibrovascularization (tissue ingrowth) of the implant within weeks of implantation (Chalasani et al., 2007). Optimal pore size for orbital implants ranges from 150 μm to 400 μm, with preference closer to 400 μm for more complete fibrovascularization (Goldberg et al., 1994; Rubin et al., 1994). Highly porous materials with interconnecting pores of uniform diameter are more favourable for facilitating cell movement and nutrient flow within the implant, encouraging tissue ingrowth (Chalasani et al., 2007).

Porous orbital materials are stiffer in relation to the conjunctiva and surrounding tissue (Chalasani et al., 2007). This compliance mismatch coupled with constant implant movement by extraocular muscles may exacerbate inflammation, leading to tissue necrosis and eventual implant exposure (Chalasani et al., 2007). Porous hydrogels may be a promising alternative to the current implant materials, as these water-based materials possess physical properties similar to native tissue (Chalasani et al., 2007).

DISRUPTING FIBROSIS

Texturing to Disrupt Capsular Contracture of Breast Prostheses

One of the most common cosmetic surgical procedures in the United States is breast augmentation (Gampper et al., 2007). All breast implants are constructed with a smoothed or textured shell of silicone elastomer, a material pervasively used in medical devices. The majority of implants are filled with saline rather than silicone gel, after the 1992 FDA voluntary moratorium on silicone gel implants (Gampper et al., 2007); however, gel-filled implants are making a comeback owing to more lifelike mechanical properties.

Fibrous capsular contraction is the most common complication in breast augmentation surgery, with long-term contracture incidence reported at 15–25% (Malata et al., 1997; Benediktsson and Perbeck, 2006). As with any foreign object implanted in the body, the host response to a breast implant is to construct a fibrous capsule around the foreign-body, "walling off" the implant from the rest of the body. During the first several months post-implantation the fibrous capsule begins to contract, resulting in breast firmness that may lead to patient discomfort and disfigurement.

Several possible factors contribute to breast implant capsular contracture, including implant surface (smooth versus porous), implant placement (submuscular versus subglandular), implant shape (i.e., fill volume), bacterial infection, bleeding, surgical technique, post-operative care (i.e., breast massage), and patient health (Burkhardt et al., 1986; Gabriel et al., 1997; Becker and Springer, 1999; Embrey et al., 1999). Contracture has been found to be an independent breast-based, not patient-based, phenomenon (i.e., one breast may experience contracture while the other breast may not) (Burkhardt, 1984; Burkhardt et al., 1986).

Surface texturing has been investigated as a means to reduce capsular contracture for decades. Texturing has been reported to reduce early capsular contracture in subglandular breast augmentation (Wong et al., 2006); however, it is not conclusive whether texturing reduces contracture incidence (Bucky et al., 1994; Tarpila et al., 1997; Fagrell et al., 2001) or simply delays contracture onset (Wong et al., 2006).

The type of texturing, pore size or implant material does not appear to be as important as simply disrupting surface smoothness (McCurdy, 1990; Caffee, 1994; Danino et al., 2001). One clinical study comparing saline-filled silicone implants textured with either 75–150 μm or 600–800 μm pore size concluded that both morphologies were effective in reducing capsular contracture (Danino et al., 2001). The mechanisms contributing to reduced capsular contracture with textured implants still remain poorly understood (Pollock, 1993; Caffee, 1994; Wong et al., 2006). Disrupting the smooth surface with texturing influences collagen arrangement and structure, macrophage population, micromotion, and risk of infection (Taylor and Gibbons, 1983; McCurdy, 1990). The collagen arrangement hypothesis is that irregularly arranged collagen fibers at textured surfaces are less able to generate cooperative contractile forces typical of maturing scar tissue (McCurdy, 1990; Pennisi, 1990), while another hypothesis is that irregularly aligned collagen fibers are more susceptible to collagenase degradation (Raykhlin and Kogan, 1961; Taylor and Gibbons, 1983). Textured implants have been shown to have more macrophages and less collagen surrounding them compared to smooth implants, indicating that macrophages may be degrading the capsule as it forms, minimizing capsule thickness (Taylor and Gibbons, 1983; McCurdy, 1990). More functional hypthotheses are that the tissue integration promoted by the textured surface may be either reducing micromotion, thus reducing the extent of fibrosis or eliminating the exudate-filled periprosthetic space around the implant that is prone to infection and/or chronic inflammation (McCurdy, 1990).

While capsular contracture is a challenge from a clinical standpoint (McCurdy, 1990), patient preference suggests that breast firmness is not a definitive factor in deciding which type of breast implant is preferred. Textured implants often have an unnatural "feel," because textured implant shells lack flexibility (Caffee, 1994). A higher incidence of skin wrinkling over the breast also occurs with textured implants (McCurdy, 1990; Wong et al., 2006), presumably from tissue integration that immobilizes the implant and allows implant surface wrinkling to be more visible. This may be because smooth implants reside in a fluid-like periprosthetic pocket, reducing skin wrinkling as the implant is able to move within this space. Since implant feel and appearance are important patient considerations, smooth surface implants are often preferred by patients in the absence of significant contracture (Wong et al., 2006).

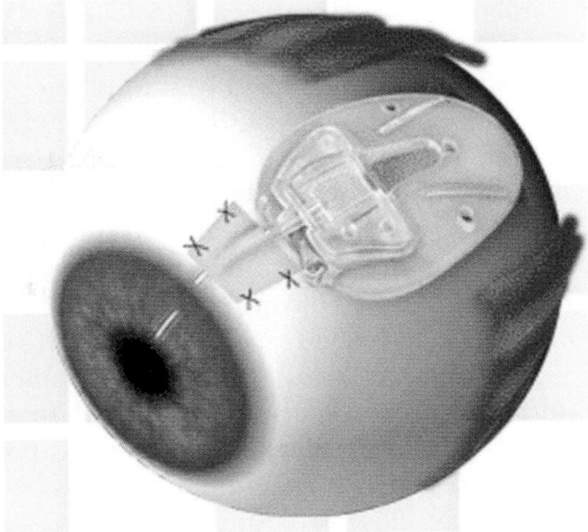

FIGURE I.2.15.4 Schematic of glaucoma drainage device placement in eye. (Ahmed™ Glaucoma Valve glaucoma drain, New World Medical, Inc.)

Texturing to Minimize Fibrotic Encapsulation of Glaucoma Drainage Devices

Glaucoma is the obstruction of aqueous drainage that causes the intraocular pressure (IOP) to increase to the point of nerve damage (Adatia and Damji, 2005). Treatment methods to decrease IOP include topical eye drops, surgery, and glaucoma drainage devices in refractory cases (Boswell et al., 1999; Ayyala et al., 2006). Glaucoma drainage devices (GDDs) are silicone tubes inserted into the anterior chamber of the eye. Attached to the tube is a plate that is sutured beneath the conjunctiva (Jacob et al., 1998) (Figure I.2.15.4). The plate becomes enclosed by a fibrous capsule 3–6 weeks post-operative, creating a "filtering bleb" (a small bubble) through which aqueous humor drains and is reabsorbed by the surrounding tissue (Jacob et al., 1998). Over two to four years, a thick fibrous capsule forms around the device, causing the filtering bleb to fail at least 20% of the time (Jacob et al., 1998). Encapsulation decreases absorption of the drained aqueous humor, leading to increased pressure in the anterior chamber (Boswell et al., 1999).

Implant material, rigidity, flexibility, and shape (Ayyala et al., 2006) are potential contributors to long-term device failure. Micromotion of a smooth, rigid plate against the scleral surface has also been attributed to exacerbating scar tissue formation by causing constant, low-grade inflammation (Jacob et al., 1998; Ayyala et al., 2006). To reduce micromotion, porous plate designs (50 μm pore size silicone attached to smooth plate) were investigated to encourage fibrotic tissue anchorage of the implant (Jacob et al., 1998). However, this particular porous design has been linked with higher incidences of diplopia (double vision), possibly due to implant adhesion to the recti muscles and limiting bleb elevation compared with smooth designs (Ayyala et al., 2006).

Preliminary experiments in rabbits with ePTFE porous plates (60 μm pore size) found thinner, less dense, and more vascular capsules forming than around commercial, non-porous plates. This capsule provides less restriction to aqueous draining into the subconjuntival space, and the close proximity of microvessels also facilitates aqueous reabsorption (Boswell et al., 1999). Commercial GDDs enclosed in a two-layer ePTFE membrane have been reported to have minimal capsule thickness, with highly vascular tissue encasing the device after six weeks in rabbits (Ahmad et al., 2004). The smaller pores within the inner layer prevent cells from infiltrating the device lumen and valve or restricting aqueous humor outflow. The larger outer pore layer stimulates tissue integration and blood vessel formation adjacent to the device, as well as reducing capsule thickness (Ahmad et al., 2004).

PROMOTING ANGIOGENESIS

Improving Long-Term Performance of Vascular Grafts

Synthetic vascular grafts are used to repair damaged or occluded blood vessels. For decades, researchers and manufacturers have been challenged to create synthetic grafts that mimic physiological function (Davids et al., 1999). Graft material porosity is related to graft healing (Wesolowski et al., 1961; Kuzuya et al., 2004), and is critical as part of the strategy to form a stable, endothelium-lined lumen to provide an anti-thrombotic surface that is similar to native vessels (Wesolowski et al., 1961; White et al., 1983; Clowes et al., 1986; Hirabayashi et al., 1992; Nagae et al., 1995). ePTFE and Dacron® (knitted and woven) are common synthetic graft materials (Nagae et al., 1995; Davids et al., 1999).

Numerous studies conclude that high porosity (≥60 μm pores) ePTFE small diameter grafts have superior graft healing compared to low porosity (≤30 μm pores) grafts (Hess et al., 1984; Branson et al., 1985; Golden et al., 1990; Hirabayashi et al., 1992; Nagae et al., 1995) in terms of patency as well as the rate, stability, and completeness of luminal endothelialization. Optimal pore size for achieving stable luminal endothelium coverage in small diameter grafts is 60 μm (Hess et al., 1989; Golden et al., 1990; Hirabayashi et al., 1992; Nagae et al., 1995; Miura et al., 2002). This pore size provides adequate porosity for transmural fibrovascular tissue ingrowth. Infiltrating capillaries from the surrounding tissue facilitate endothelialization and smooth muscle cell formation along the graft surface, forming the stable neointima (Davids et al., 1999). Dacron® has a greater prominence in large diameter grafts due to excessive graft surface thrombus. Typically, Dacron® graft lumens become coated in a thick, stable layer of fibrin which limits endothelialization and transmural tissue ingrowth (Davids et al., 1999) (see Chapter II.5.3.B for more discussion on vascular grafts).

Minimizing Infection and Epithelial Downgrowth

Percutaneous devices penetrate the body through a surgically created defect in the skin to provide a conduit between an implanted medical device or artificial organ and the extracorporeal space (von Recum, 1984; Yu et al., 1999). Percutaneous devices include catheters (i.e., peritoneal dialysis, intravascular), prosthetic attachments, dental implants, feeding and tracheal tubes, and needle-type glucose sensors (Fukano et al., 2006; Isenhath et al., 2007). However, breaking the skin barrier provides a route to infection and increases complications associated with wound non-closure (Yu et al., 1999). The incidence of catheter related bloodstream infections ranges between 80,000 and 250,000 annually in the USA, with health related expenses costing several billions of dollars (Isenhath et al., 2007).

Strategies to reduce percutaneous device-related infection involve antibiotics applied topically, administered prophylactically or incorporated into the device (Dasgupta, 2000; Fukano et al., 2006; Isenhath et al., 2007). Antibiotics, while clinically effective, raise concerns about the development of antibiotic-resistant strains (Dasgupta, 2000).

Percutaneous devices also fail from mechanical irritation (avulsion) or epithelial downgrowth that forms a pocket around the implant (marsupialization). Acute mechanical interfacial stresses tear the device from its implantation bed (von Recum, 1984), while chronic, small mechanical stresses cause localized injury, resulting in inflammation and increased susceptibility to infection (von Recum, 1984). Mechanical forces also prevent an epidermal seal from forming, increasing risk of infection (von Recum, 1984). Marsupialization occurs primarily around smooth implants, when epidermis surrounding the implant grows parallel to the implant surface and unites under the implant, surrounding the implant in an epidermal pocket. The percutaneous device has now become extracutaneous, and the tissue pocket becomes a source for infection since an epidermal seal has not formed between the implant and skin (von Recum, 1984).

Perimigration is the migration of epidermal cells into a porous material (von Recum, 1984). Competition between macrophages, giant cells, and fibroblasts within the pores prevents connective tissue maturation and scar tissue formation. The immature connective tissue soon becomes displaced by proliferating and maturing epidermal cells, possibly with assistance from enzymatic lysis. As the epidermal cells reach the underlying connective tissue bed, they begin to proliferate and form an epidermis-lined pocket around the percutaneous device, causing extrusion (von Recum, 1984).

Dermal integration is necessary to prevent infection, avulsion, and marsupialization, and to ensure the long-term performance of percutaneous devices (von Recum, 1984; Chehroudi and Brunette, 2002; Isenhath et al., 2007). Several porous and textured biomaterials have been investigated to improve percutaneous device performance and longevity. To reduce catheter exit-site infection, "cuffs" are attached to the catheter at the location where the external segment of catheter exits the body. These cuffs are implanted beneath the skin.

Dacron® velour cuffs are commonly used in commercial catheters (Dasgupta, 2000). These cuffs become fibrotically encapsulated, serving both to anchor the catheter and to provide a barrier for bacterial entry (Dasgupta, 2000). However, avascularity of the fibrotic tissue has been reported to encourage bacterial attachment in the tissue surrounding the catheter (Dasgupta, 2000). Microporous silicone cuffs have been shown to reduce exit site infection by 60% in canine models (Moncrief et al., 1995). The microporous structure encourages vascular tissue integration to facilitate immune cell surveillance at the exit site, as well as to reduce scar formation (Moncrief et al., 1995).

Microtexture and porous biomaterials have been shown to be effective in reducing infection, avulsion, and epithelial downgrowth in several animal models (Chehroudi and Brunette, 2002). Highly textured surfaces are reported to have significantly more tissue attachment, thinner capsules, and less tissue–implant separation compared to smooth surface implants (Kim et al., 2006). Grooved surfaces encourage random fibroblast orientation, lessening the effects of contractile forces that contribute to tissue–implant separation (Kim et al., 2006). In contrast, parallel collagen fibers against smooth surfaces create poorly integrated, avascular capsules prone to separation from the implant during wound contracture (Kim et al., 2006). Stable tissue attachment also reduces the effects of micromotion (Kim et al., 2006), minimizing avulsion risk. Capsule thickness appears to relate to the amount of mechanical stresses imparted on the implant (von Recum, 1984). Stability may also promote healing, and encourage thinner capsule formation (Kim et al., 2006).

Optimal pore size for tissue ingrowth is 40 μm or larger (Winter, 1974; von Recum, 1984; Isenhath et al., 2007; Fukano et al., 2010); however, pores as small as 3 μm have been observed with fibrous tissue ingrowth (Squier and Collins, 1981). Most rapid epithelium migration occurs during the first two weeks post-implantation, regardless of pore size (Squier and Collins, 1981). Following the initial two weeks, epithelial migration appears to be inversely related to pore size: increased endothelium downgrowth with decreased pore size (Squier and Collins, 1981). Speed of epithelial migration differs between species (Chehroudi and Brunette, 2002), for example migration in rodents is considerably faster than in humans (Chehroudi and Brunette, 2002). Porosity is also thought to be a key factor affecting the rate of migration (Squier and Collins, 1981), and is required for connective tissue and possibly epithelial ingrowth into percutaneous implants (Isenhath et al., 2007; Fukano et al., 2010).

While textured and porous surfaces encourage dermal integration, there are concerns that surfaces with irregular topologies may be more prone to harboring bacteria (Chehroudi and Brunette, 2002). Bacteria may outcompete tissue cells in adhering to an implant surface, thus preventing tissue integration and promoting infection. Future investigations to optimize topological properties to minimize bacterial contamination (Chehroudi and Brunette, 2002) are recommended.

Porous Coatings to Improve Glucose and Oxygen Transport to Implanted Sensors

Percutaneous glucose sensors must be removed after three to seven days to prevent host inflammation, wound healing, and subsequent foreign-body encapsulation from jeopardizing sensor reliability. The foreign-body response ultimately causes impedance of glucose and oxygen transport to the sensor, resulting in sensor signal deterioration, and frequently sensor failure. Methods that improve analyte (glucose, oxygen) transport to indwelling sensors could allow sensors to reliably measure interstitial glucose concentrations for several weeks, as opposed to days.

Attempts to improve long-term sensor performance have included surface chemical modification, various coatings (Gifford et al., 2005; Nablo et al., 2005; Shin and Schoenfisch, 2006), release of molecular mediators (Friedl, 2004; Ward et al., 2004; Norton et al., 2005), and surface topography (Wisniewski and Reichert, 2000; Wisniewski et al., 2000). The effect of surface texturing on the tissue that surrounds implanted biomaterials is well-documented for devices such as total joint arthroplasty (Bauer and Schils, 1999; Ryan et al., 2006) and percutaneous devices (Tagusari et al., 1998; Walboomers and Jansen, 2005; Kim et al., 2006).

Topographical approaches for improving long-term sensor performance were first proposed by Woodward (1982), who suggested that the best coating for an implanted glucose sensor was a sponge that encourages tissue ingrowth and disrupts fibrosis (Figure I.2.15.5). Efforts to create tissue-modifying textured coatings for implantable sensors are attractive, because their impact is not dependent on a depletable drug reservoir, unlike drug eluting techniques.

A significant range of materials and pore sizes are capable of promoting angiogenesis and reducing capsule

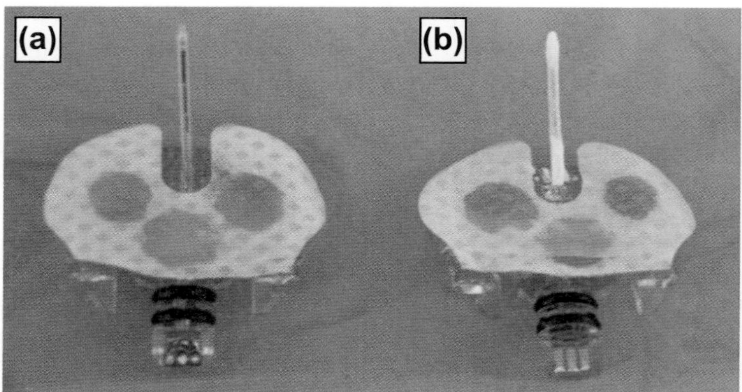

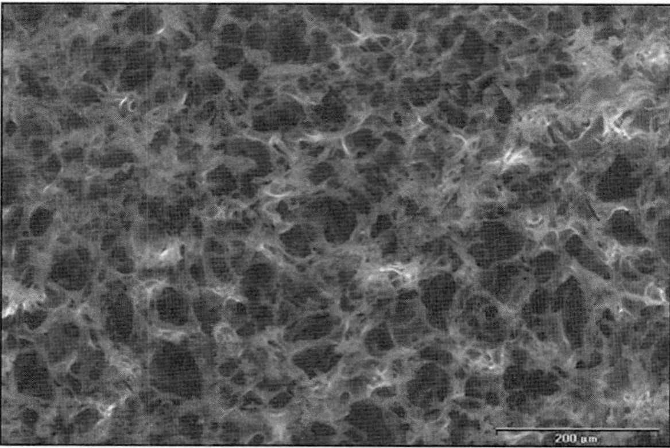

FIGURE I.2.15.5 Example of a: (a) Medtronic MiniMed SOF-SENSOR^{TM} glucose sensor; and (b) an experimental porous poly-L-lactic acid (PLLA) coating applied to the sensor tip for investigational purposes. Inset: environmental scanning electron microscope image of porous PLLA coating fabricated using salt-leaching/gas foaming with ammonium bicarbonate (NH_4HCO_3). (Koschwanez, H. E. (2006). Unpublished data.) Courtesy of John Wiley and Sons.

TABLE I.2.15.3 Summary of Pore Sizes that Yielded Optimal Tissue Response (Promoted Angiogenesis and/or Reduced Capsule Thickness) Around Biomaterials or Glucose Sensors

Investigator	Porous Material	Optimal Pore Size	Application and Test Subject	Duration of Investigation
Brauker et al., 1995	PTFE	5 μm	Membrane implanted in rat subcutis	3 weeks
Sharkawy et al., 1998	PVA	60 μm	Membrane implanted in rat subcutis	12-16 weeks
Ward et al., 2002	ePTFE and PVA	ePTFE: 1 μm PVA: 60 μm	Membrane implanted in rat subcutis	7 weeks
Marshall et al., 2004	HEMA hydrogels	35 μm	Hydrogel implanted in mouse subcutis	4 weeks
Updike et al., 2000	ePTFE	5-10 μm (Shults et al., 2006)	Glucose sensor implanted in dog subcutis	162 days* (best of 6 sensors)
Yu et al., 2006	Epoxy-enhanced polyurethane	Not specified	Glucose sensor implanted in rat subcutis	56 days* (best of 9 sensors)
Gilligan et al., 2004	ePTFE	Not specified	Glucose sensor implanted in human subcutis	185 days* (best of 5 sensors)

NOTE: PTFE (polytetrafluoroethylene), ePTFE (expanded polytetrafluoroethylene), PVA (polyvinyl alcohol), HEMA (hydroxyethyl methacrylate).
*Maximum time sensor remained functional *in vivo.*

thickness (Ward et al., 2002). Geometry, rather than chemical composition, of the material appears to determine biomaterial–microvasculature interactions (Brauker et al., 1995; Sieminski and Gooch, 2000). Table I.2.15.3 summarizes leading research in the area of porosity and porous coatings for glucose sensors, including the pore size reported to yield the greatest vascularization and/or the least capsule formation around the implant. Variations in pore size and pore structure in implanted biomaterials may, however, limit the conclusions that can be drawn about how pore size influences tissue response (Marshall et al., 2004).

While porous biomaterials seemingly create the ideal environment for an indwelling glucose sensor, porous coatings applied to sensors have had less than ideal results. A critical factor in sensor failure *in vivo* is the fibrotic capsule that forms around glucose sensors (Dungel et al., 2008). Despite porous coatings stimulating the formation of vascular networks around glucose sensors in rats, newly formed vessels within porous coatings have been unable to overcome the diffusion barriers imparted by the collagen capsule (Dungel et al., 2008). Failing sensor sensitivity was found to correlate with increasing collagen deposition within the sponge implant (Dungel et al., 2008). Additionally, other factors, such as mechanical stresses imposed by the percutaneously implanted sensor, may have overshadowed the angiogenic-inducing, collagen-reducing properties of porous coatings (Koschwanez et al., 2008).

Recently, human studies (Gilligan et al., 2004) were performed using sensors covered with a porous angiogenic and bioprotective ePTFE membrane (Updike et al., 2000; Shults et al., 2006). Unfortunately, inflammation within the angiogenic layer in 80% of sensors, in addition to packaging failure in 60% of sensors, resulted in only 20% sensor survival after six months (Gilligan et al., 2004).

CONCLUSION

The use of textured and porous materials is widespread in medical device applications. Research and development on porous structures continues because of good clinical outcomes, important research findings, and a straightforward regulatory pathway. While a single pore size or textured morphology does not fit all applications, the majority of porous and textured biocompatible materials used in medical devices share the common characteristic of open architectures with interconnecting pores that support nutrient transfer, and promote cell migration and proliferation. Stable integration of the implant with the surrounding host tissue reduces irritation caused by micromotion, and promotes stable fibrovascular tissue ingrowth that promotes healing and minimizes infection.

BIBLIOGRAPHY

Adatia, F. A., & Damji, K. F. (2005). Chronic open-angle glaucoma: Review for primary care physicians. *Canadian Family Physician*, **51**, 1229–1237.

Ahmad, S., Lee, M., Klitzman, B., Olbrich, K., Mordes, D., et al. (2004). *Comparison of Conventional Ahmed Implant With Newly Designed Valved Implant: A Pilot Study*. Ft Lauderdale, FL: Association for Research in Vision and Ophthalmology.

Ayyala, R. S., Duarte, J. L., & Sahiner, N. (2006). Glaucoma drainage devices: State of the art. *Expert Review of Medical Devices*, **3**, 509–521.

Bauer, T. W., & Schils, J. (1999). The pathology of total joint arthroplasty I. Mechanisms of implant fixation. *Skeletal Radiology*, **28**, 423–432.

Becker, H., & Springer, R. (1999). Prevention of capsular contracture. *Plastic and Reconstructive Surgery*, **103**, 1766–1768.

Benediktsson, K., & Perbeck, L. (2006). Capsular contracture around saline-filled and textured subcutaneously-placed implants in irradiated and non-irradiated breast cancer patients: Five years of monitoring of a prospective trial. *Journal of Plastic Reconstructive and Aesthetic Surgery*, **59**, 27–34.

Bloebaum, R. D., Willie, B. M., Mitchell, B. S., & Hofmann, A. A. (2007). Relationship between bone ingrowth, mineral apposition rate, and osteoblast activity. *Journal of Biomedical Materials Research Part A*, **81A**, 505–514.

Bobyn, J. D., Pilliar, R. M., Cameron, H. U., & Weatherly, G. C. (1980). The optimum pore-size for the fixation of porous-surfaced metal implants by the ingrowth of bone. *Clinical Orthopaedics and Related Research*, 263–270.

Bobyn, J. D., Stackpool, G. J., Hacking, S. A., Tanzer, M., & Krygier, J. J. (1999). Characteristics of bone ingrowth and interface mechanics of a new porous tantalum biomaterial. *Journal of Bone and Joint Surgery-British Volume*, **81B**, 907–914.

Boswell, C. A., Noecker, R. J., Mac, M., Snyder, R. W., & Williams, S. K. (1999). Evaluation of an aqueous drainage glaucoma device constructed of ePTFE. *Journal of Biomedical Materials Research*, **48**, 591–595.

Branson, D. F., Picha, G. J., & Desprez, J. (1985). Expanded polytetrafluoroethylene as a microvascular graft: A study of 4 fibril lengths. *Plastic and Reconstructive Surgery*, **76**, 754–763.

Brauker, J. H., Carrbrendel, V. E., Martinson, L. A., Crudele, J., Johnston, W. D., et al. (1995). Neovascularization of synthetic membranes directed by membrane microarchitecture. *Journal of Biomedical Materials Research*, **29**, 1517–1524.

Bucky, L. P., Ehrlich, H. P., Sohoni, S., & May, J. W. (1994). The capsule quality of saline-filled smooth silicone, textured silicone, and polyurethane implants in rabbits: A long-term study. *Plastic and Reconstructive Surgery*, **93**, 1123–1131.

Burkhardt, B. R. (1984). Comparing contracture rates: Probability-theory and the unilateral contracture. *Plastic and Reconstructive Surgery*, **74**, 527–529.

Burkhardt, B. R., Dempsey, P. D., Schnur, P. L., & Tofield, J. J. (1986). Capsular contracture: A prospective-study of the effect of local antibacterial agents. *Plastic and Reconstructive Surgery*, **77**, 919–930.

Caffee, H. H. (1994). The effect of siltex texturing and povidone-iodine irrigation on capsular contracture around saline inflatable breast implants: Discussion. *Plastic and Reconstructive Surgery*, **93**, 129–130.

Chalasani, R., Poole-Warren, L., Conway, R. M., & Ben-Nissan, B. (2007). Porous orbital implants in enucleation: A systematic review. *Survey of Ophthalmology*, **52**, 145–155.

Chang, Y. S., Oka, M., Kobayashi, M., Gu, H. O., Li, Z. L., et al. (1996). Significance of interstitial bone ingrowth under load-bearing conditions: A comparison between solid and porous implant materials. *Biomaterials*, **17**, 1141–1148.

Chehroudi, B., & Brunette, D. M. (2002). Subcutaneous microfabricated surfaces inhibit epithelial recession and promote long-term survival of percutaneous implants. *Biomaterials*, **23**, 229–237.

Chen, G. P., Ushida, T., & Tateishi, T. (2002). Scaffold design for tissue engineering. *Macromolecular Bioscience*, **2**, 67–77.

Clowes, A. W., Kirkman, T. R., & Reidy, M. A. (1986). Mechanisms of arterial graft healing: Rapid transmural capillary ingrowth provides a source of intimal endothelium and smooth-muscle in porous PTFE prostheses. *American Journal of Pathology*, **123**, 220–230.

Danino, A. M., Basmacioglu, P., Saito, S., Rocher, F., Blanchet-Bardon, C., et al. (2001). Comparison of the capsular response to the Biocell RTV and Mentor 1600 Siltex breast implant surface texturing: A scanning electron microscopic study. *Plastic and Reconstructive Surgery*, **108**, 2047–2052.

Dasgupta, M. K. (2000). Exit-site and catheter related infections in peritoneal dialysis: Problems and progress. *Nephrology*, **5**, 17–25.

Davids, L., Dower, T., & Zilla, P. (1999). The lack of healing in conventional vascular grafts. In P. Zilla, & H. P. Greisler (Eds.), *Tissue Engineering of Vascular Prosthetic Grafts*. Georgetown, TX: Landes Company.

de Pina, M. F., Ribeiro, A. I., & Santos, C. (2011). Epidemiology and Variability of Orthopaedic Procedures Worldwide. In G. Bentley (Ed.), *European Instructional Lectures* (pp. 9–19). Berlin, Heidelberg: Springer. doi:10.1007/978-3-642-18321-8_2.

Dungel, P., Long, N., Yu, B., Moussy, Y., & Moussy, F. (2008). Study of the effects of tissue reactions on the function of implanted glucose sensors. *Journal of Biomedical Materials Research Part A*, **85A**, 699–706.

Embrey, M., Adams, E. E., Cunningham, B., Peters, W., Young, V. L., et al. (1999). A review of the literature on the etiology of capsular contracture and a pilot study to determine the outcome of capsular contracture interventions. *Aesthetic Plastic Surgery*, **23**, 197–206.

Fagrell, D., Berggren, A., & Tarpila, E. (2001). Capsular contracture around saline-filled fine textured and smooth mammary implants: A prospective 7.5-year follow-up. *Plastic and Reconstructive Surgery*, **108**, 2108–2112.

Friedl, K. E. (2004). Corticosteroid modulation of tissue responses to implanted sensors. *Diabetes Technology and Therapeutics*, **6**, 898–901.

Fukano, Y., Knowles, N. G., Usui, M. L., Underwood, R. A., Hauch, K. D., et al. (2006). Characterization of an *in vitro* model for evaluating the interface between skin and percutaneous biomaterials. *Wound Repair and Regeneration*, **14**, 484–491.

Fukano, Y., Usui, M. L., Underwood, R. A., Isenhath, S., Marshall, A. J., et al. (2010). Epidermal and dermal integration into sphere-templated porous poly(2-hydroxyethyl methacrylate) implants in mice. *Journal of Biomedical Materials Research Part B: Applied Biomaterials*, **94**(4), 1172–1186.

Gabriel, S. E., Woods, J. E., Ofallon, M., Beard, C. M., Kurland, L. T., et al. (1997). Complications leading to surgery after breast implantation. *New England Journal of Medicine*, **336**, 677–682.

Gampper, T. J., Khoury, H., Gottlieb, W., & Morgan, R. F. (2007). Silicone gel implants in breast augmentation and reconstruction. *Annals of Plastic Surgery*, **59**, 581–590.

Gifford, R., Batchelor, M. M., Lee, Y., Gokulrangan, G., Meyerhoff, M. E., et al. (2005). Mediation of *in vivo* glucose sensor inflammatory response via nitric oxide release. *Journal of Biomedical Materials Research Part A*, **75A**, 755–766.

Gilligan, B., Shults, M., Rhodes, R. K., Jacobs, P. G., Brauker, J. H., et al. (2004). Feasibility of continuous long-term glucose monitoring from a subcutaneous glucose sensor in humans. *Diabetes Technology and Therapeutics*, **6**, 378–386.

Goldberg, R. A., Dresner, S. C., Braslow, R. A., Kossovsky, N., & Legmann, A. (1994). Animal-model of porous polyethylene orbital implants. *Ophthalmic Plastic and Reconstructive Surgery*, **10**, 104–109.

Golden, M. A., Hanson, S. R., Kirkman, T. R., Schneider, P. A., & Clowes, A. W. (1990). Healing of polytetrafluoroethylene arterial grafts is influenced by graft porosity. *Journal of Vascular Surgery*, **11**, 838–845.

Harris, L. D., Kim, B. -S., & Mooney, D. J. (1998). Open pore biodegradable matrices formed with gas foaming. *Journal of Biomedical Materials Research*, **42**, 396–402.

Hess, F., Jerusalem, C., Grande, P., & Braun, B. (1984). Significance of the inner-surface structure of small-caliber prosthetic blood-vessels in relation to the development, presence, and fate of a neo-intima: A morphological evaluation. *Journal of Biomedical Materials Research*, **18**, 745–755.

Hess, F., Steeghs, S., & Jerusalem, C. (1989). Neointima formation in expanded polytetrafluoroethylene vascular grafts with different fibril lengths following implantation in the rat aorta. *Microsurgery*, **10**, 47–52.

Hing, K. A., Best, S. M., Tanner, K. E., Bonfield, W., & Revell, P. A. (2004). Mediation of bone ingrowth in porous hydroxyapatite bone graft substitutes. *Journal of Biomedical Materials Research Part A*, **68A**, 187–200.

Hirabayashi, K., Saitoh, E., Ijima, H., Takenawa, T., Kodama, M., et al. (1992). Influence of fibril length upon ETFE graft healing and host modification of the implant. *Journal of Biomedical Materials Research*, **26**, 1433–1447.

Hutmacher, D. W. (2000). Scaffolds in tissue engineering bone and cartilage. *Biomaterials*, **21**, 2529–2543.

Isenhath, S. N., Fukano, Y., Usui, M. L., Underwood, R. A., Irvin, C. A., et al. (2007). A mouse model to evaluate the interface between skin and a percutaneous device. *Journal of Biomedical Materials Research Part A*, **83A**, 915–922.

Jacob, J. T., Burgoyne, C. F., Mckinnon, S. J., Tanji, T. M., Lafleur, P. K., et al. (1998). Biocompatibility response to modified Baerveldt glaucoma drains. *Journal of Biomedical Materials Research*, **43**, 99–107.

Karageorgiou, V., & Kaplan, D. (2005). Porosity of 3D biomaterial scaffolds and osteogenesis. *Biomaterials*, **26**, 5474–5491.

Kienapfel, H., Sprey, C., Wilke, A., & Griss, P. (1999). Implant fixation by bone ingrowth. *The Journal of Arthroplasty*, **14**, 355–368.

Kim, H., Murakami, H., Chehroudi, B., Textor, M., & Brunette, D. M. (2006). Effects of surface topography on the connective tissue attachment to subcutaneous implants. *International Journal of Oral & Maxillofacial Implants*, **21**, 354–365.

Klawitter, J. J., Bagwell, J. G., Weinstein, A. M., Sauer, B. W., & Pruitt, J. R. (1976). Evaluation of bone-growth into porous high-density polyethylene. *Journal of Biomedical Materials Research*, **10**, 311–323.

Koschwanez, H. E., Yap, F. Y., Klitzman, B., & Reichert, W. M. (2008). *In vitro* and *in vivo* characterization of porous poly-L-lactic acid coatings for subcutaneously implanted glucose sensors. *Journal of Biomedical Materials Research*, **87A**, 792–803.

Kuster, M. (2002). Exercise recommendations after total joint replacement: A review of the current literature and proposal of scientifically based guidelines. *Sports Medicine*, **32**, 433–445.

Kuzuya, A., Matsushita, M., Oda, K., Kobayashi, M., Nishikimi, N., et al. (2004). Healing of implanted expanded polytetrafluoroethylene vascular access grafts with different internodal distances: A histologic study in dogs. *European Journal of Vascular and Endovascular Surgery*, **28**, 404–409.

Malata, C. M., Feldberg, L., Coleman, D. J., Foo, I. T.H., & Sharpe, D. T. (1997). Textured or smooth implants for breast augmentation? Three year follow-up of a prospective randomised controlled trial. *British Journal of Plastic Surgery*, **50**, 99–105.

Marshall, A. J., Irvin, C. A., Barker, T., Sage, E. H., Hauch, K. D., et al. (2004). Biomaterials with tightly controlled pore size that promote vascular in-growth. *Abstracts of Papers of the American Chemical Society*, **228**. U386–U386.

McCurdy, J. A. (1990). Relationships between spherical fibrous capsular contracture and mammary prosthesis type: A comparison of smooth and textured implants. *American Journal of Cosmetic Surgery*, **7**, 235–238.

Mikos, A. G., Bao, Y., Cima, L. G., Ingber, D. E., Vacanti, J. P., et al. (1993). Preparation of poly(glycolic acid) bonded fiber structures for cell attachment and transplantation. *Journal of Biomedical Materials Research*, **27**, 183–189.

Mikos, A. G., Thorsen, A. J., Czerwonka, L. A., Bao, Y., Langer, R., et al. (1994). Preparation and characterization of poly(l-lactic acid) foams. *Polymer*, **35**, 1068–1077.

Miura, H., Nishibe, T., Yasuda, K., Shimada, T., Hazama, K., et al. (2002). The influence of node-fibril morphology on healing of high-porosity expanded polytetrafluoroethylene grafts. *European Surgical Research*, **34**, 224–231.

Moncrief, J., Popovich, R., Seare, W., Moncrief, D., Simmons, V., et al. (1995). A new porous surface modification technology for peritoneal dialysis catheters as an exit-site cuff to reduce exit-site infections. *Advances in Peritoneal Dialysis*, **11**, 197–199.

Mooney, D. J., Baldwin, D. F., Suh, N. P., Vacanti, J. P., & Langer, R. (1996). Novel approach to fabricate porous sponges of poly(-lactic-co-glycolic acid) without the use of organic solvents. *Biomaterials*, **17**, 1417–1422.

Nablo, B. J., Prichard, H. L., Butler, R. D., Klitzman, B., & Schoenfisch, M. H. (2005). Inhibition of implant-associated infections via nitric oxide. *Biomaterials*, **26**, 6984–6990.

Nagae, T., Tsuchida, H., Peng, S. K., Furukawa, K., & Wilson, S. E. (1995). Composite porosity of expanded polytetrafluoroethylene vascular prosthesis. *Cardiovascular Surgery*, **3**, 479–484.

Nam, Y. S., Yoon, J. J., & Park, T. G. (2000). A novel fabrication method of macroporous biodegradable polymer scaffolds using gas foaming salt as a porogen additive. *Journal of Biomedical Materials Research*, **53**, 1–7.

Norton, L. W., Tegnell, E., Toporek, S. S., & Reichert, W. M. (2005). *In vitro* characterization of vascular endothelial growth factor and dexamethasone releasing hydrogels for implantable probe coatings. *Biomaterials*, **26**, 3285–3297.

Pennisi, V. R. (1990). Long-term use of polyurethane breast prostheses: A 14-year experience. *Plastic and Reconstructive Surgery*, **86**, 368–371.

Pollock, H. (1993). Breast capsular contracture: A retrospective study of textured versus smooth silicone implants. *Plastic and Reconstructive Surgery*, **91**, 404–407.

Raykhlin, N. T., & Kogan, A. K. (1961). The development and malignant degeneration of the connective tissue capsules around plastic implants. *Prob. Oncol.*, **7**, 11–14.

Rosengren, A., Danielsen, N., & Bjursten, L. M. (1999). Reactive capsule formation around soft-tissue implants is related to cell necrosis. *Journal of Biomedical Materials Research*, **46**, 458–464.

Rubin, P. A.D., Popham, J. K., Bilyk, J. R., & Shore, J. W. (1994). Comparison of fibrovascular ingrowth into hydroxyapatite and porous polyethylene orbital implants. *Ophthalmic Plastic and Reconstructive Surgery*, **10**, 96–103.

Ryan, G., Pandit, A., & Apatsidis, D. P. (2006). Fabrication methods of porous metals for use in orthopaedic applications. *Biomaterials*, **27**, 2651–2670.

Salzmann, D. L., Kleinert, L. B., Berman, S. S., & Williams, S. K. (1997). The effects of porosity on endothelialization of ePTFE implanted in subcutaneous and adipose tissue. *Journal of Biomedical Materials Research*, **34**, 463–476.

Sami, D., Young, S., & Peterson, R. (2007). Perspective on orbital enucleation implants. *Survey of Ophthalmology*, **52**, 244–265.

Schugens, C., Maquet, V., Grandfils, C., Jerome, R., & Teyssie, J. (1996a). Polylactide macroporous biodegradable implants for cell transplantation. II. Preparation of polylactide foams by liquid–liquid phase separation. *Journal of Biomedical Materials Research*, **30**, 449–461.

Schugens, C., Maquet, V., Grandfils, C., Jerome, R., & Teyssie, P. (1996b). Biodegradable and macroporous polylactide implants for cell transplantation. 1. Preparation of macroporous polylactide supports by solid–liquid phase separation. *Polymer*, **37**, 1027–1038.

Sharkawy, A. A., Klitzman, B., Truskey, G. A., & Reichert, W. M. (1997). Engineering the tissue which encapsulates subcutaneous implants. 1. Diffusion properties. *Journal of Biomedical Materials Research*, **37**, 401–412.

Sharkawy, A. A., Klitzman, B., Truskey, G. A., & Reichert, W. M. (1998). Engineering the tissue which encapsulates subcutaneous implants. II. Plasma-tissue exchange properties. *Journal of Biomedical Materials Research*, **40**, 586–597.

Shin, J. H., & Schoenfisch, M. H. (2006). Improving the biocompatibility of *in vivo* sensors via nitric oxide release. *Analyst*, **131**, 609–615.

Shults, M. C., Updike, S. J., & Rhodes, R. K. (2006). Device and method for determining analyte levels. *US Patent No 7110803*.

Sieminski, A. L., & Gooch, K. J. (2000). Biomaterial-microvasculature interactions. *Biomaterials*, **21**, 2233–2241.

Simske, S. J., Ayers, R. A., & Bateman, T. A. (1997). Porous materials for bone engineering. *Materials Science Forum*, **250**, 151.

Squier, C. A., & Collins, P. (1981). The relationship between soft-tissue attachment, epithelial downgrowth and surface porosity. *Journal of Periodontal Research*, **16**, 434–440.

Sun, L. M., Berndt, C. C., Gross, K. A., & Kucuk, A. (2001). Material fundamentals and clinical performance of plasma-sprayed hydroxyapatite coatings: A review. *Journal of Biomedical Materials Research*, **58**, 570–592.

Tagusari, O., Yamazaki, K., Litwak, P., Kojima, A., Klein, E. C., et al. (1998). Fine trabecularized carbon: Ideal material and texture for percutaneous device system of permanent left ventricular assist device. *Artificial Organs*, **22**, 481–487.

Tarpila, E., Ghassemifar, R., Fagrell, D., & Berggren, A. (1997). Capsular contracture with textured versus smooth saline-filled implants for breast augmentation: A prospective clinical study. *Plastic and Reconstructive Surgery*, **99**, 1934–1939.

Taylor, S. R., & Gibbons, D. F. (1983). Effect of surface texture on the soft-tissue response to polymer implants. *Journal of Biomedical Materials Research*, **17**, 205–227.

Turner, T. M., Sumner, D. R., Urban, R. M., Rivero, D. P., & Galante, J. O. (1986). A comparative-study of porous coatings in a weight-bearing total hip-arthroplasty model. *Journal of Bone and Joint Surgery (American Volume)*, **68A**, 1396–1409.

Updike, S. J., Shults, M. C., Gilligan, B. J., & Rhodes, R. K. (2000). A subcutaneous glucose sensor with improved longevity, dynamic range, and stability of calibration. *Diabetes Care*, **23**, 208–214.

von Recum, A. F. (1984). Applications and failure modes of percutaneous devices: A review. *Journal of Biomedical Materials Research*, **18**, 323–336.

Walboomers, X. F., & Jansen, J. A. (2005). Effect of microtextured surfaces on the performance of percutaneous devices. *Journal of Biomedical Materials Research Part A*, **74A**, 381–387.

Ward, W. K., Slobodzian, E. P., Tiekotter, K. L., & Wood, M. D. (2002). The effect of microgeometry, implant thickness and polyurethane chemistry on the foreign body response to subcutaneous implants. *Biomaterials*, **23**, 4185–4192.

Ward, W. K., Wood, M. D., Casey, H. M., Quinn, M. J., & Federiuk, I. F. (2004). The effect of local subcutaneous delivery of vascular endothelial growth factor on the function of a chronically implanted amperometric glucose sensor. *Diabetes Technology and Therapeutics*, **6**, 137–145.

Wesolowski, S. A., Fries, C. C., Karlson, K. E., Debakey, M., & Sawyer, P. N. (1961). Porosity: Primary determinant of ultimate fate of synthetic vascular grafts. *Surgery*, **50**, 91–96.

Whang, K., Thomas, C. H., Healy, K. E., & Nuber, G. (1995). A novel method to fabricate bioabsorbable scaffolds. *Polymer*, **36**, 837–842.

White, R., Goldberg, L., Hirose, F., Klein, S., Bosco, P., et al. (1983). Effect of healing on small internal diameter arterial graft compliance. *Biomaterials Medical Devices and Artificial Organs*, **11**, 21–29.

Winter, G. D. (1974). Transcutaneous implants: Reactions of skin–implant interface. *Journal of Biomedical Materials Research*, **8**, 99–113.

Wisniewski, N., & Reichert, M. (2000). Methods for reducing biosensor membrane biofouling. *Colloids and Surfaces B-Biointerfaces*, **18**, 197–219.

Wisniewski, N., Moussy, F., & Reichert, W. M. (2000). Characterization of implantable biosensor membrane biofouling. *Fresenius Journal of Analytical Chemistry*, **366**, 611–621.

Wong, C. H., Samuel, M., Tan, B. K., & Song, C. (2006). Capsular contracture in subglandular breast augmentation with textured versus smooth breast implants: A systematic review. *Plastic and Reconstructive Surgery*, **118**, 1224–1236.

Woodward, S. C. (1982). How fibroblasts and giant-cells encapsulate implants: Considerations in design of glucose sensors. *Diabetes Care*, **5**, 278–281.

Wu, B. M., Borland, S. W., Giordano, R. A., Cima, L. G., Sachs, E. M., et al. (1996). Solid free-form fabrication of drug delivery devices. *Journal of Controlled Release*, **40**, 77–87.

Yang, S. F., Leong, K. F., Du, Z. H., & Chua, C. K. (2001). The design of scaffolds for use in tissue engineering. Part 1. Traditional factors. *Tissue Engineering*, **7**, 679–689.

Yu, B. Z., Long, N., Moussy, Y., & Moussy, F. (2006). A long-term flexible minimally-invasive implantable glucose biosensor based on an epoxy-enhanced polyurethane membrane. *Biosensors and Bioelectronics*, **21**, 2275–2282.

Yu, C., Sun, Y., Bradfield, J., Fiordalisi, I., & Harris, G. D. (1999). A novel percutaneous barrier device that permits safe subcutaneous access. *Asaio Journal*, **45**, 531–534.

CHAPTER I.2.16 ELECTROSPINNING FUNDAMENTALS AND APPLICATIONS

Robert Akins[1] *and John Rabolt*[2]

[1]Tissue Engineering and Regenerative Medicine Laboratory, Alfred I. duPont Hospital for Children/Nemours Foundation, Wilmington, DE, USA

[2]Department of Materials Science and Engineering, University of Delaware, Newark, DE, USA

MOTIVATION FOR USING ELECTROSPUN MEMBRANES

Electrospinning is an excellent candidate process for producing tissue engineering and vascular graft constructs, since the resulting electrospun materials possess many of the desired properties appropriate for successful cell and tissue growth, such as a high surface-to-volume ratio and an interconnected three-dimensional porous network (Table I.2.16.1). It is a simple and robust method to produce micro- and nanometer diameter fibers, and can be thought of as a biomimetic process since the materials produced resemble the nanoscale architecture of extracellular matrix (ECM) (Figure I.2.16.1), and can promote and enhance biological activities. The experimental arrangement generally consists of a pipette or syringe with a blunt needle, which is filled with a polymer solution or melt. An electrode from a high voltage power supply (0–30 kV) is placed in contact with the solution and an electric field is applied.

The electric charge overcomes the surface tension of the solution droplet, and a single fiber in the form of a jet is emitted. The jet is collected on a counter electrode in the form of a nonwoven fibrous mat (Figure I.2.16.1). The fibers produced are generally 100–400 nm in diameter, but the size of the fibers can easily be changed through variations in the processing parameters, such as applied electric field strength and solute concentrations (Doshi and Reneker, 1995). In addition, a wide variety of polymers that are desirable for tissue engineering applications have been electrospun. These include collagen, poly(lactic acid)/poly(glycolic acid), and their copolymers, fibrinogen, elastin, and spider silk (Matthews et al., 2002; Stephens et al., 2003; Boland et al., 2004; Shields et al., 2004). Coupling the ability to control the fiber size within a matrix with the ability to choose a variety of polymers with different physical/chemical/mechanical properties will make it possible to create a construct that has comparable properties to a specific tissue. Another desirable aspect of electrospinning is that only a small amount of starting material (<50 mg) is required compared to more traditional fiber formation methods (10–20 pounds). This becomes critically important for next generation materials, such as biopolymers like spider-silk and those synthesized in small-scale laboratory preparations, because they are generally produced in small quantities (milligrams). Recent attempts to scale-up the electrospinning process using a multiple jet system have shown considerable promise (see for example Elmarco at www.elmarco.com), and in the near future useful materials for transfer to the clinical environment will be developed.

TABLE I.2.16.1 List of Desired Properties of Tissue Engineering Constructs

Desired Properties	Electrospinning
Interconnected porous network	Yes
Large void volume (cell seeding and penetration)	Yes
Large surface-to-volume ratio	Yes
Specific surface chemistry and surface microstructure	Yes
Mechanical strength (comparable to physiological stresses)	Yes
Controlled degradation rate (matches tissue regeneration)	Yes
Incorporation of growth factors	Yes

HISTORICAL PERSPECTIVE

Electrospinning has been around for more than a century, first invented by Formhals in 1934 and improved by different research groups over the years. Many polymers have been routinely electrospun, and several review articles summarizing these techniques have appeared (for example, see Greiner and Wendorff, 2007). Unique micro- and nanoscale features on the surface of polymer fibers have been produced at the University of Delaware using the electrospinning process. These features are advantageous for tissue engineering and filtration applications because they provide an increase in surface area and roughness. Surface textures, in the form of micro- and nanopores, can be formed on the surface of fibers electrospun from volatile solvents. Pores ranging in size from 20–1000 nm (having depths ranging from 50–70 nm) have been observed on the surface of several amorphous and semicrystalline polymer fibers, such as polystyrene (PS), poly(methylmethacrylate) (PMMA), PC, and poy(ethylene oxide) (PEO) (Megelski et al., 2002; Casper et al., 2004) (see Figure I.2.16.2). As we further optimized the processing protocols for electrospinning fibers we observed that, with a judicious choice of electrospinning parameters and solvent, nanoweb structures (Stephens et al., 2003) could also be produced. Nanowebs are composed of small fibrils (~10–15 nm) that interconnect the larger fibers (~70–100 nm) of the electrospun mat to produce a web-like structure that is similar in appearance to that produced by the spider. These fibrillar nanowebs have been observed in several different types of polymers and bioderived materials (collagen, denatured collagen, spider silk, nylon) two examples of which are shown in Figure I.2.16.3. We have identified the processing parameters and the mechanisms that cause the surface texturing and nanowebs to form, and this has provided us with the ability to produce fibers and fiber mats with the desired features for different applications. For example, in Figure I.2.16.3, the electrospinning conditions to promote (Stephens et al., 2003) the formation of

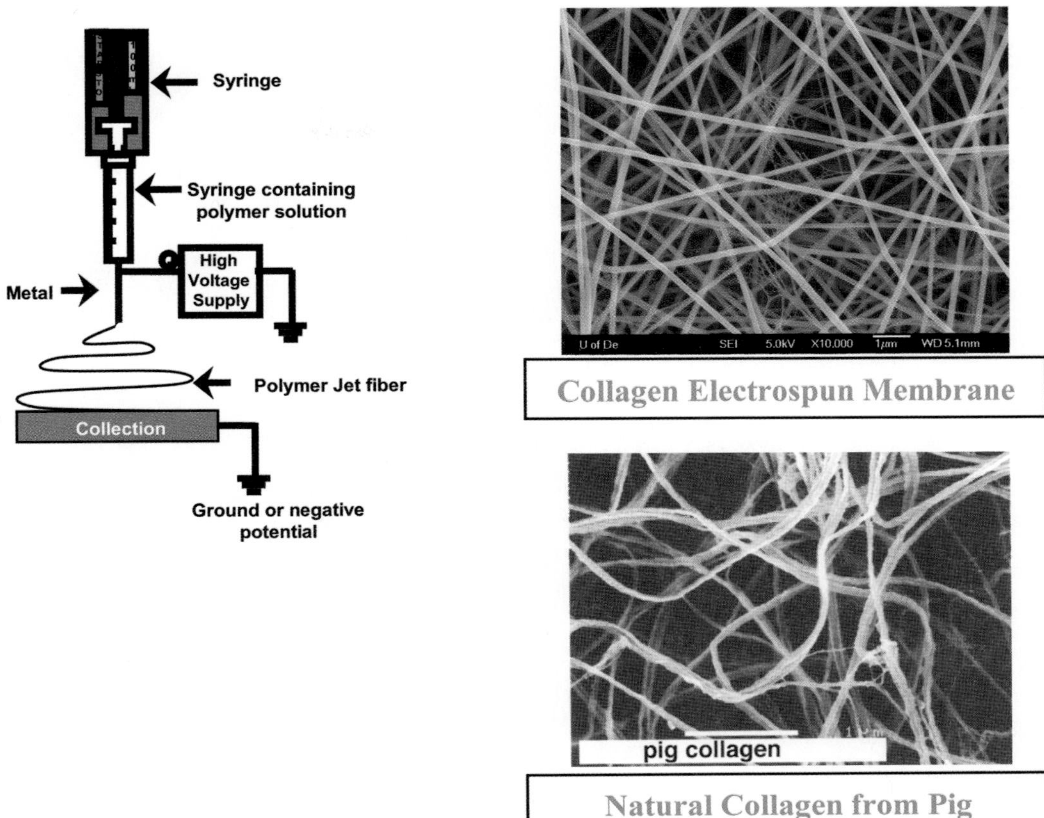

FIGURE I.2.16.1 Schematic of electrospinning set up and field emission scanning electron microscope (FE-SEM) micrographs comparing an electrospun collagen fibrous membrane with natural collagen fibrils obtained from a pig ECM. (Gemmell et al., 1995)

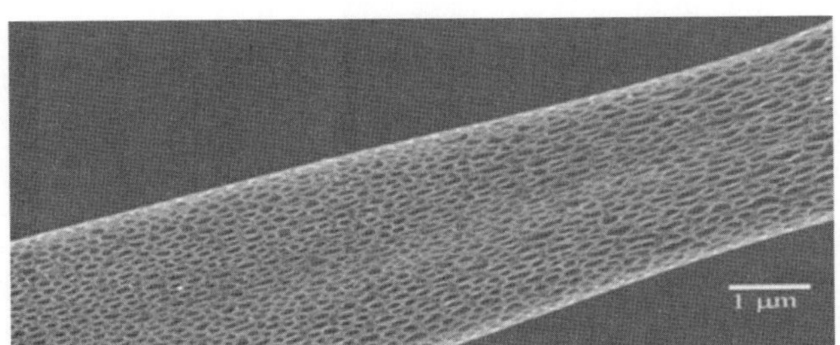

FIGURE I.2.16.2 Micro- and nanotextured electrospun fibers; PS electrospun from tetrahydrofuran (THF).

nanowebs were the following: Figure I.2.16.3(a) collagen (type I) at 30 wt% from formic acid, 23 gauge needle, 10 cm syringe to counter-electrode gap, 7 kV; and Figure I.2.16.3(b) synthetic spider silk at 20 wt%, 23 gauge needle, 10 cm gap, 7 kV.

In addition, we have also demonstrated that electrospun fibers can be functionalized with macromolecules that may improve the patency of vascular graft materials. Specifically, four-arm star-shaped poly(ethylene glycol) PEG polymers with termini derivatized with low molecular weight heparin (PEG-LMWH) have been incorporated into electrospun fibers comprised of either poly(ethylene oxide) (PEO) or poly(lactic-glycolic acid) PLGA. The synthesis of the PEG-LMWH star bioconjugate and its incorporation into biopolymeric fibers offers opportunities for the production of fibrous materials capable of local and sustained release of therapeutically relevant proteins.

The PEG-LMWH can be easily incorporated into fibrous matrices via electrospinning protocols, which has been demonstrated (Casper et al., 2005) with PEO and PLGA fibers. The concentration of the polymeric solutions used during electrospinning varied depending on the carrier polymer employed, but in both cases, PEG-LMWH was easily incorporated into fibers without altering the fiber size or shape (Figure I.2.16.4). PEO (M_w = 300,000 g/mol) was used at a 10 wt% concentration in water, and a flow rate of 0.07 mL/min was employed. PLGA (75:25, M_w = 90,000–126,000 g/mol) solutions were made at a 45 wt% concentration in dimethylformamide (DMF) and required a flow rate of 0.26 mL/hr.

Scanning electron microscopy confirmed that the incorporation of the PEG-LMWH did not alter fiber morphology (Figure I.2.16.4). Multiphoton microscopy of fibers electrospun with fluorescently labeled PEG-LMWH confirmed the presence of LMWH throughout the depth of the electrospun matrix. Toluidine blue spectrophotometric assays that detect heparin were used to determine that the amount of LMWH per mg of fibers was in the range of 3.5–85 μg, depending on the sample examined. Although both LMWH and PEG-LMWH can be processed into fibers, the incorporation of PEG-LMWH resulted in functional advantages such as improved growth factor binding, which likely results from improvements in retention of the PEG-LMWH in the fibers over the LMWH alone. The PEG-LMWH is

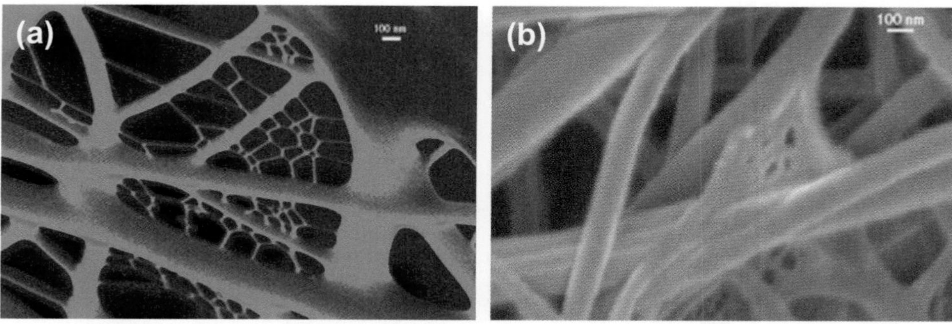

FIGURE I.2.16.3 Nanowebs formed by electrospinning: (a) collagen; (b) spider silk.

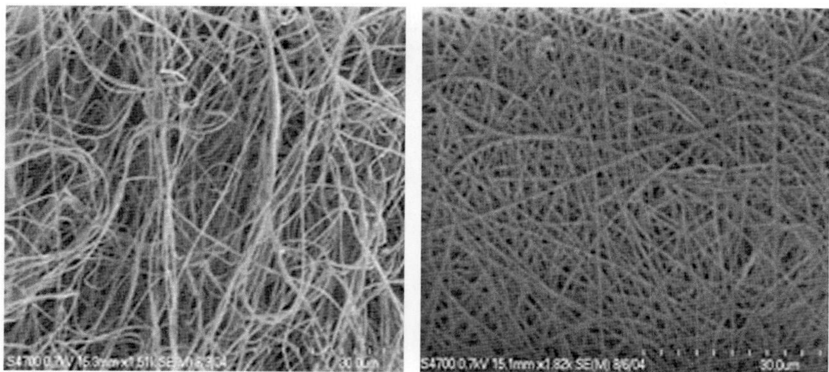

FIGURE I.2.16.4 FE-SEM micrographs of electrospun PLGA (left) and PEG-LMWH/PLGA (right). Fiber diameters range from 200 nm to 1 micron. (Casper et al., 2005.)

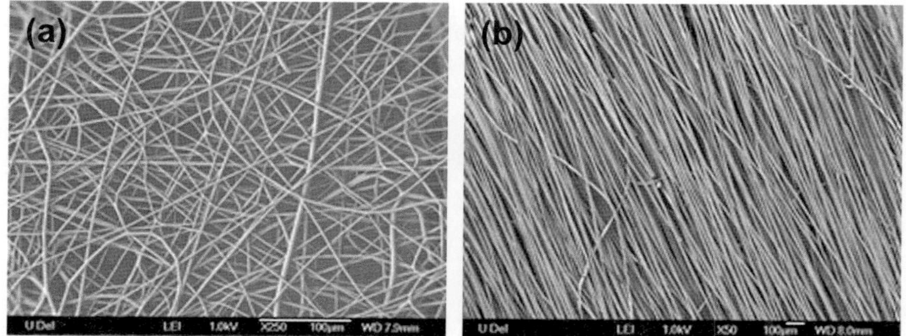

FIGURE I.2.16.5 FE-SEM images of two sets of electrospun fibers: (a) isotropically oriented nylon 6 fibers (Lee et al., 2008); (b) aligned polystyrene fibers. (Mandrel assembly courtesy of Vahik Krikorian.)

retained in the fibers for at least 14 days, in contrast to LMWH, which is almost completely released from the fibers after 24 hours. A slower release of PEG-LMWH, and any bound growth factors, would allow for drug delivery over a period of time that would more closely match the timescales needed for cell proliferation, for example, along the inner surface of a vascular graft.

It is sometimes advantageous to macroscopically align the polymer fibers, since some mechanical and optical properties of the material can be improved. Shown in Figure I.2.16.5 are FE-SEM images of an isotropic polymer fibrous mat collected on a metallic plate (Figure I.2.16.5(a)) and fibers that have been wound up on a grounded rotating mandrel (Figure I.2.16.5(b)). In addition to the obvious fiber alignment that may improve properties, it has been demonstrated that fiber alignment can influence cell phenotype, an observation that will be illustrated in more detail later in this chapter.

CHARACTERIZATION METHODS

The application of standard confocal fluorescence and electron microscopic characterization of electrospun membranes provides facile determination of the extent of fiber orientation, diameter, and surface/interior functionalization. However, the functionality of polymer nanofibers is critically dependent on the macromolecular structure at the primary, secondary, and tertiary levels. Also, backbone orientation and potential changes in structure during dynamic deformation in the electrospinning process impact nanofiber functionality. The various forms of vibrational spectroscopy and wide angle X-ray

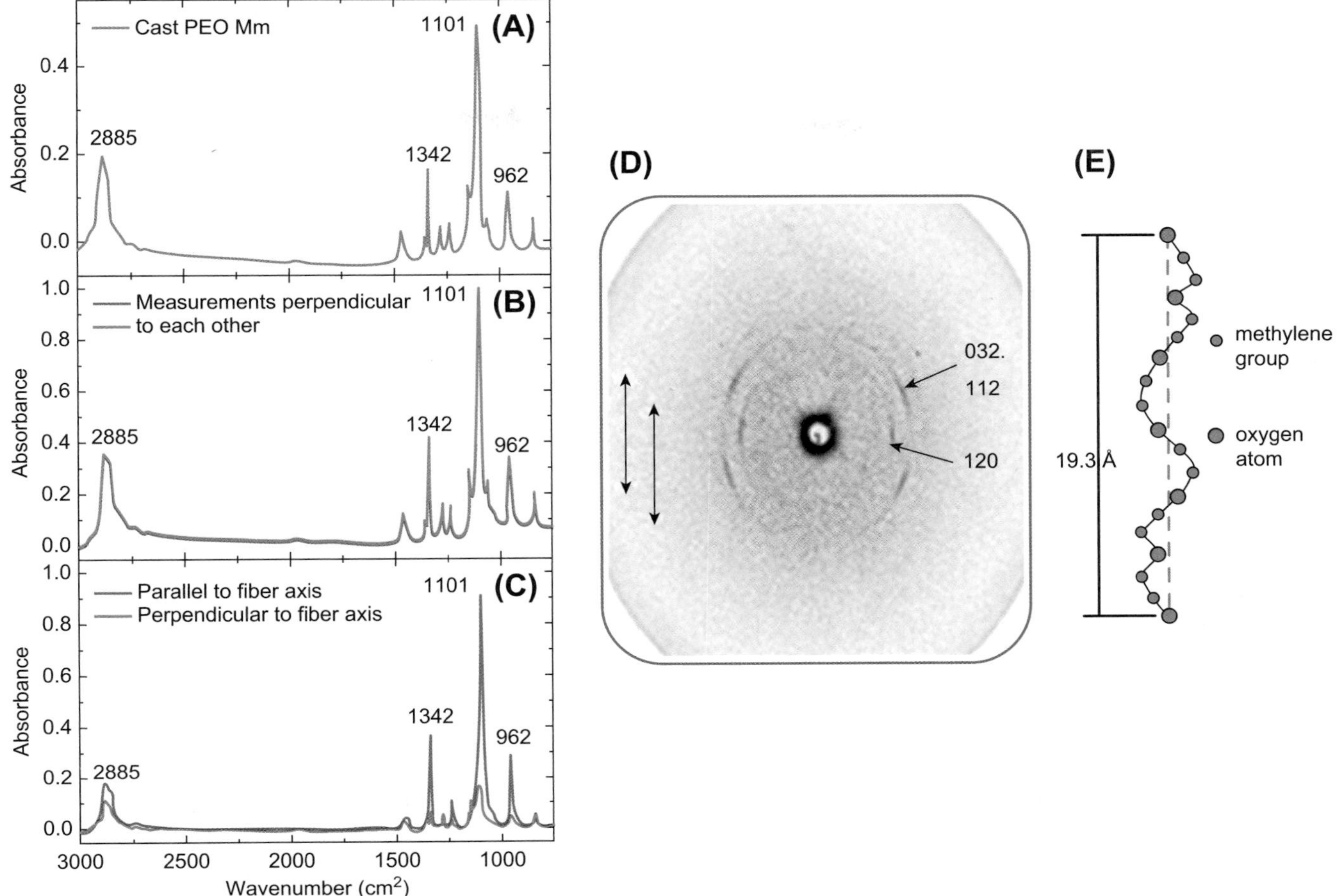

FIGURE I.2.16.6 (A) A typical FTIR spectra of pure PEO cast film; (B) polarized FTIR spectra for isotropic PEO fibrous mat show that the peak intensities for measurements parallel to fiber axis were very similar to the perpendicular measurements; (C) polarized FTIR spectra for uniaxially oriented PEO nanofibers show that the peak intensities for measurements parallel to fiber axis were greater compared to the perpendicular measurements; (D) wide-angle X-ray diffraction pattern of PEO nanofibers with the arrows indicating the direction of the fiber axis and the numbers indicating crystal planes in the reciprocal lattice; (E) a schematic of the PEO 7/2 helix indicating its orientation along the fiber axis.

diffraction employed in routine characterization studies of polymers also play a key role in understanding the structure/property relationships in polymer nanofibers.

The extent of macroscopic orientation of electrospun micro- and nanoscale diameter fibers can be designed into the fibrous mat by electric field assisted fiber collection, by controlling the wind-up speed of the mandrel relative to the fiber spinning speed, by using a specially designed "banded" rod for collection, and by translating the rotating mandrel at various speeds during the collection process. If, in addition, the molecular orientation of the polymer backbone with respect to the fiber axis is controlled simultaneously, then it should be possible to use fiber orientation to optimize mechanical properties (compliance, modulus, tenacity, etc.). In order to assess the amount of molecular orientation within fibers, polarized Raman scattering (Frisk et al., 2004) can also be used in addition to the fourier transform infrared (FTIR) and X-ray diffraction methods illustrated in Figure I.2.16.6. It is possible to estimate the extent of polymer backbone alignment in the case of uniaxially oriented samples, such as fibers, using experimental quantities obtained by employing a combination of different polarization directions and sample orientation. These procedures can be used to determine an orientation distribution function (Kakade et al., 2007). We have successfully carried out such studies (Kakade et al., 2007) on a number of synthetic polymers (e.g., polyethylene, poly(ethylene terephthalate), and poly(trimethylene terephthalate)) and the experimental and theoretical protocols can easily be adapted to the study of biopolymers, such as collagen, PLGA/PEG-LMWH, and elastin electrospun fibers, in order to correlate processing conditions with extent of backbone orientation, which in turn will correlate with mechanical properties.

As shown also in Figure I.2.16.6(D), wide angle X-ray diffraction (WAXD) is also useful in determining molecular orientation in macroscopic polymer nanofibers. Instead of the ring pattern indicative of randomly oriented crystallites, intense arcs are observed in the PEO electrospun fibers collected on a charged plate. By indexing the crystal planes in the reciprocal lattice, it was possible to ascertain that the polymer chains are oriented parallel to the fiber axis and that the PEO backbone maintains its

7/2 helical conformation, confirming the results obtained by polarized FTIR and Raman spectroscopy. Studies on Nylon 6 nanofibers produced by electrospinning using an atomic force microscopy (AFM) tip and collected on a charged plate also showed molecular orientation relative to the fiber axis (Gururajan et al., 2011). This molecular alignment will produce improved mechanical, optical, and electrical properties.

BIOMEDICAL APPLICATIONS FOR ELECTROSPUN MATERIALS

Tissue Engineering and Regenerative Medicine

Tissue engineering (TE) is an emerging strategy in the broader field of regenerative medicine (RM) (see Section II.6, "Applications of Biomaterials in Functional Tissue Engineering"). In TE/RM, living cells are employed in the construction of *de novo* tissues for use as implants, diagnostic testing devices, and translational research models. Biomaterials of different types are used to support cellularized constructs. The development and implementation of TE/RM strategies, therefore, involve engineering cells and cell scaffolds to direct the organization and function of biological–material composites to desired ends. Electrospun materials have several attributes that recommend them as platforms for TE/RM approaches:

- Fibers can be prepared from a variety of natural and synthetic polymers that are biocompatible, amenable to surface modification with bioactive molecules, and capable of directing cell attachment and growth.
- The small diameter filaments within electrospun scaffolds mimic the fibrous architecture of native extracellular matrix.
- Composite materials can be prepared wherein biologically-active molecules are enmeshed in the fibrous matrix.
- Scaffolds can be engineered with desirable physical characteristics, such as stiffness.
- Materials can be designed to biodegrade over desired timescales.
- Scaffolds can be prepared with extensive networks of inter-connected pores that allow cell invasion into three-dimensional constructs and the subsequent perfusion of neo-tissues.

The physical and chemical characteristics of electrospun materials can be manipulated during production in order to control subsequent interactions between the produced scaffold and cells of different types. Materials can be designed to act as barriers to block cell penetration or as scaffolds to promote cell adhesion and migration, and to enable the diffusion of nutrients and wastes (for examples, see references Lu and Mikos, 1996; Riboldi et al., 2005; Zong et al., 2005). The specifics of cell–material interactions are of critical concern, and the adhesion, growth, organization, and biomolecular characteristics of cells on scaffolds are important biological parameters to consider in developing TE/RM strategies.

Electrospinning has been explored as a strategy to engineer a number of tissues including vasculature (Xu et al., 2004; Inoguchi et al., 2007), nerve (Prabhakaran et al., 2009), bone (Liu et al., 2009), and ligament/tendon (Ouyang et al., 2003; Bashur et al., 2009; Inanc et al., 2009). Many types of polymers have been employed in these early TE/RM efforts including poly(l-lactide), PLGA, PEO, polycaprolactone, polyurethane, polyethylene glycol, different types of collagen, elastin, hyaluronic acid, chitosan, fibroin, and combinations of these, to name a few. The selection of polymer and the specific conditions for electrospinning are motivated by the design characteristics of the finished product, and there are many choices and variations that can be employed. In general, the chemical nature of electrospun fibers can be varied across a broad range by altering polymer selection and production conditions.

In conjunction with the chemical characteristics of electrospun fibers, the physical characteristics of electrospun fiber scaffolds are also important. The porosity of scaffolds, which impacts cell penetration (Heydarkhan-Hagvall et al., 2008), fiber diameter, which affects cell morphology and phenotype (Yang et al., 2005), and the lateral alignment of fibers, which impacts cell morphology and local cell–cell interactions (Chew et al., 2008; Rockwood et al., 2008; Bashur et al., 2009), are all important determinants of tissue organization and function. These three parameters are inter-related, and in combination they may have complex effects on cultured cells. For example, suitable scaffold porosity in combination with appropriate polymer composition improves the three-dimensional migration of adipocyte-derived stem cells into composite gelatin/PCL scaffolds (Heydarkhan-Hagvall et al., 2008). In experiments to study the effects of both fiber alignment and fiber diameter on neural stem cells (NSCs), it was found that culture on anisotropic materials with aligned poly(L-lactic acid) (PLLA) fibers resulted in directed neurite outgrowth of significantly greater length than those found on cells cultured using isotropic PLLA scaffolds (Yang et al., 2005). In addition, the NSCs differentiated more readily on nanofibers when compared to the same cells grown on microfibers; this effect was independent of fiber alignment (Yang et al., 2005). In general, structural anisotropy and the diameter of filaments in fibrous scaffold materials may profoundly affect the activity of cells from other tissues as well.

In other work looking at aligned fibers, experiments using human ligament fibroblasts have shown that cells synthesize significantly more collagen when grown on aligned scaffolds versus randomly oriented nanofibers (Lee et al., 2005). Studies with Schwann cells grown on aligned electrospun polycaprolactone (ES-PCL) scaffolds indicated that culture on nanofibrous supports can improve cell maturation, with aligned scaffolds providing the greatest degree of maturation (Chew et al., 2008). Similarly, electrospun polyurethane (ES-PU) of different types has also been used to study the engineering of ligament-like tissues (Bashur et al., 2009) and myocardial-like tissues (Rockwood et al., 2008); in both these cases, culture on the aligned ES-PU substrata resulted in a more mature cellular phenotype.

In our laboratories, we have investigated the effects of ES-PU anisotropy on the phenotype of heart cells. One of the critical but sometimes underappreciated functions of the heart is to release neurohormones such as atrial natriuretic peptide (ANP). ANP production and secretion by heart muscle cells are highly controlled, and an increased level of ANP in the blood has dramatic physiologic impact, resulting in a decrease in circulating blood volume. We found that the organization of ES-PU substrata had a profound effect on ANP production and release (Rockwood et al., 2008). As shown in Figure I.2.16.7, when identical cells were grown on anisotropic versus isotropic ES-PU, ANP production was significantly attenuated in the aligned ES-PU group. This attenuation is similar to that seen in three-dimensional culture (Akins et al., 2009), and occurred in cells that maintained the levels of other molecular markers for cardiac muscle. These results indicate a specific and significant change in cell phenotype associated with substrate alignment.

The specific mechanisms by which ES fiber orientation impacts cell organization and function may differ from system to system, but cellular effects associated with electrospun support materials are likely related to differences in cell morphology and the altered interactions among neighboring cells. In particular, cells grown on oriented fibers tend to be more elongated with a spindle-like shape than cells grown on isotropic supports. In addition, anisotropic materials tend to be laterally aligned with significant cell–cell contacts along the length of neighboring cells. The ability to impact cell organization and function by engineering the physical and chemical characteristics of electrospun scaffolds is a powerful technology in nascent TE/RM approaches.

Drug Delivery

Another potent use for electrospun materials is in the area of drug and biomolecule delivery. Electrospun materials are amenable to surface modification with bioactive molecules, and it is possible to enmesh compounds into matrices during production (see discussion earlier in this chapter). This flexibility allows engineers to incorporate therapeutic compounds into implant materials to benefit patients or to include biomolecules in electrospun scaffolds to influence cells in tissue engineered constructs. Drug delivery strategies employing electrospun materials are being developed for both these applications. Examples of drug delivery approaches using electrospun materials include the introduction of antibiotics (Bolgen et al., 2007), anti-tumor drugs (Xie and Wang, 2006), bioactive proteins (Kim et al., 2007), and DNA (Nie and Wang, 2007). Applications employing antibiotics or DNA illustrate some key concepts.

Several groups have investigated the potential use of electrospun mats for the local delivery of antibiotics. The high surface area-to-volume ratios of these materials allows for extensive drug loading. Different approaches can be used to incorporate drugs into electrospun mats; mats can be prepared and then coated with the desired agent or the

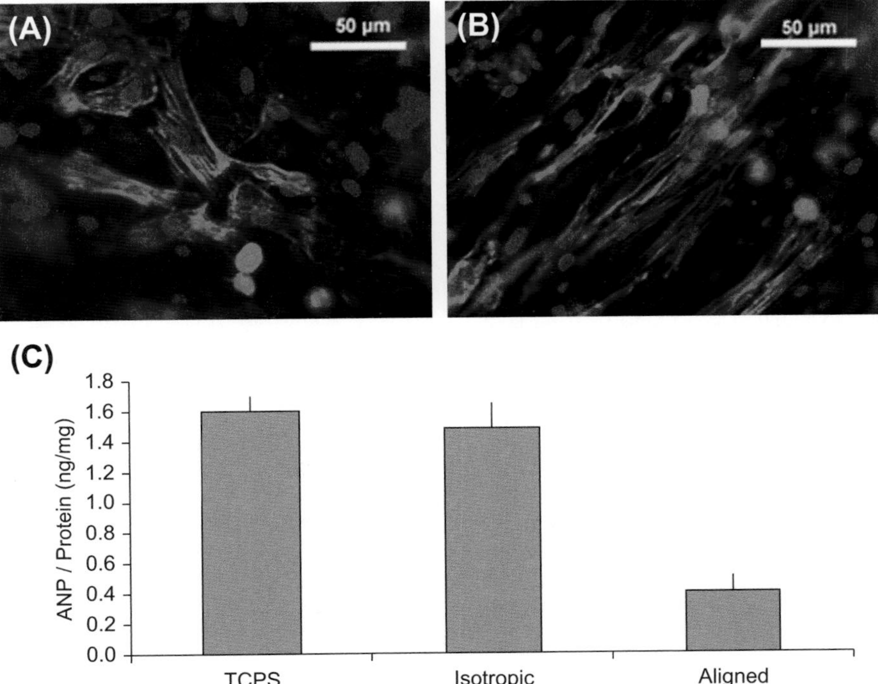

FIGURE I.2.16.7 (A) Primary heart cells grown on isotropic ES-PU. Cardiac muscle cells appear red (anti-sarcomeric myosin immunofluorescence) and nuclei appear blue (Hoechst 33258 dye). (B) Primary heart cells grown on aligned ES-PU. (C) Cellular ANP levels for cells grown on tissue culture polystyrene (TCPS), isotropic ES-PU, and aligned ES-PU. Data are mean ± S.D., n = 3. (Akins et al., 2009).

agent can be included with the polymer solution prior to electrospinning. The former approach has been applied to absorb polycaprolactone (PCL) membranes with a commercial antibiotic for use in preventing post-surgical abdominal adhesions in an animal model (Bolgen et al., 2007). The investigators found that approximately 80% of the antibiotic was released from the PCL within 3 hours, with the remaining amount released within 18 hours; however, the PCL scaffold remained in place substantially longer. The antibiotic-soaked PCL was effective in limiting adhesions, and exhibited improved healing compared to the control and to PCL mats without antibiotic.

An alternative approach using PLGA-based electrospun mats in which a different antibiotic was added to the pre-spun polymer solution found that incorporation of the drug yielded more consistent fibers with a decreased fiber diameter (Kim et al., 2004). The change in fiber diameter and morphology was likely a result of changes in the charge characteristics of the polymer/drug solution. The activity of the antibiotic was retained throughout the solvation and spinning processes. Interestingly, the inclusion of PEG-PLA block copolymer into the PLGA backbone allowed the amount of antibiotic embedded in the nanofibers to be increased, due to the altered hydrophilicity of the material.

The inclusion of DNA in composite scaffolds containing PLGA and hydroxyapatite has been used as a strategy to deliver plasmids containing bone morphogenetic protein gene sequences in efforts to augment bone tissue regeneration (Nie and Wang, 2007). The DNA is taken up by cells and the encoded protein expressed. It was found that higher relative amounts of hydroxyapatite in the scaffold resulted in faster DNA release from the material; perhaps due to the hydrophilicity of the components. These results further emphasize the need to understand the interactions between electrospun materials and the molecule to be delivered in the development of drug delivery approaches.

SUMMARY

Electrospinning provides a facile method to fabricate porous and fibrous biomaterials from many different starting polymers. The fiber size can be comparable to that found in ECM. The ability to control fiber spacing and orientation permits material to be optimized for specific applications. Applications in tissue engineering and drug delivery are envisioned.

BIBLIOGRAPHY

Akins, R. E., Rockwood, D., Robinson, K. G., Sandusky, D., Rabolt, J., et al. (2009). Three-dimensional culture alters primary cardiac cell phenotype. *Tissue Engineering Part A*, **16**(2), 629–641.

Bashur, C. A., Shaffer, R. D., Dahlgren, L. A., Guelcher, S. A., & Goldstein, A. S. (2009). Effect of fiber diameter and alignment of electrospun polyurethane meshes on mesenchymal progenitor cells. *Tissue Engineering*, **15**, 2435.

Boland, E. D., Matthews, J. A., Pawlowski, K. J., Simpson, D. G., Wnek, G. E., et al. (2004). Electrospinning collagen and elastin: Preliminary vascular tissue engineering. *Frontiers in Bioscence*, **9**, 1422.

Bolgen, N., Vargel, I., Korkusuz, P., Menceloglu, Y. Z., & Piskin, E. (2007). *In vivo* performance of antibiotic embedded electrospun PCL membranes for prevention of abdominal adhesions. *Journal of Biomedical Materials Research Part B: Applied Biomaterials*, **81**, 530.

Casper, C., Stephens, J., Tassi, N., Chase, D. B., & Rabolt, J. F. (2004). Controlling surface morphology of electrospun polystyrene fibers: Effect of humidity and molecular weight in the electrospinning process. *Macromolecules*, **37**, 573.

Casper, C., Yamaguchi, N., Kiick, K. L., & Rabolt, J. F. (2005). Functionalizing electrospun fibers with biologically relevant macromolecules. *Biomacromolecules*, **6**, 1998.

Chew, S. Y., Mi, R., Hoke, A., & Leong, K. W. (2008). The effect of the alignment of electrospun fibrous scaffolds on Schwann cell maturation. *Biomaterials*, **29**, 653.

Doshi, J., & Reneker, D. H. (1995). Electrospinning process and application of electrospun fibers. *Journal of Electrostatics*, **35**, 151.

Frisk, S., Ikeda, R. M., Chase, D. B., & Rabolt, J. F. (2004). Determination of the molecular orientation of poly(propyleneterephthalate) fibers using polarized raman spectroscopy: A comparison of methods. *Applied Spectroscopy*, **58**, 279.

Gemmell, C. H., et al. (1995). Platelet activation in whole blood by artificial surfaces: Identification of platelet-derived microparticles and activated platelet binding to leukocytes as material-induced activation events. *J.Lab. and Clinic. Med.*, **125**, 276–287.

Greiner, A., & Wendorff, J. H. (2007). Electrospinning: A fascinating method for the preparation of ultrathin fibers. *Angewandte Chemie (International edn.)*, **46**, 5670.

Gururajan, G. S., Beebe, S. P., Chase, T. P., & Rabolt, D. B. (2011). Continuous electrospinning of Nylon 6 nanofibers using an atomic force microscopy probe tip. *Nanoscale*, **3**, 3300–3308.

Heydarkhan-Hagvall, S., Schenke-Layland, K., Dhanasopon, A. P., Rofail, F., Smith, H., et al. (2008). Three-dimensional electrospun ECM-based hybrid scaffolds for cardiovascular tissue engineering. *Biomaterials*, **29**, 2907.

Inanc, B., Arslan, Y. E., Seker, S., Elcin, A. E., & Elcin, Y. M. (2009). Periodontal ligament cellular structures engineered with electrospun poly(DL-lactide-co-glycolide) nanofibrous membrane scaffolds. *Journal of Biomedical Materials Research*, **90**, 186.

Inoguchi, H., Tanaka, T., Maehara, Y., & Matsuda, T. (2007). The effect of gradually graded shear stress on the morphological integrity of a huvec-seeded compliant small-diameter vascular graft. *Biomaterials*, **28**, 486.

Kakade, M. V., Givens, S., Gardner, K., Lee, K.-Y., Chase, D. B., et al. (2007). Electric field induced orientation of polymer chains in macroscopically aligned electrospun polymer nanofibers. *Journal of the American Chemical Society*, **10**, 2777.

Kim, K., Luu, Y. K., Chang, C., Fang, D., Hsiao, B. S., et al. (2004). Incorporation and controlled release of a hydrophilic antibiotic using poly(lactide-co-glycolide)-based electrospun nanofibrous scaffolds. *Journal of Control Release*, **98**, 47.

Kim, T. G., Lee, D. S., & Park, T. G. (2007). Controlled protein release from electrospun biodegradable fiber mesh composed of poly(epsilon-caprolactone) and poly(ethylene oxide). *International Journal of Pharmaceutics*, **338**, 276.

Lee, C. H., Shin, H. J., Cho, I. H., Kang, Y. M., Kim, I. A., et al. (2005). Nanofiber alignment and direction of mechanical strain affect the ECM production of human ACL fibroblast. *Biomaterials*, **26**, 1261.

Lee, K.-H., Kim, K.-W., Pesapane, A., Kim, H.-Y., & Rabolt, J. F. (2008). Polarized FT-IR study of macroscopically oriented electrospun Nylon 6 nanofibers. *Macromolecules*, **41**, 1494.

Liu, X., Smith, L. A., Hu, J., & Ma, P. X. (2009). Biomimetic nanofibrous gelatin/apatite composite scaffolds for bone tissue engineering. *Biomaterials*, **30**, 2252.

Lu, L., & Mikos, A. G. (1996). The importance of new processing in tissue engineering. *MRS Bulletin*, **31**, 28.

Matthews, J. A., Wnek, G. E., Simpson, D. G., & Bowlin, G. L. (2002). Electrospinning of collagen nanofibers. *Biomacromolecules*, **3**, 232.

Megelski, S., Stephens, J. S., Chase, D. B., & Rabolt, J. F. (2002). Arrays of micro- and nanopores on electrospun polymer fibers. *Macromolecules*, **35**, 8456.

Nie, H., & Wang, C. H. (2007). Fabrication and characterization of PLGA/HAp composite scaffolds for delivery of BMP-2 plasmid DNA. *Journal of Control Release*, **120**, 111.

Ouyang, H. W., Goh, J. C., Thambyah, A., Teoh, S. H., & Lee, E. H. (2003). Knitted poly-lactide-co-glycolide scaffold loaded with bone marrow stromal cells in repair and regeneration of rabbit Achilles tendon. *Tissue Engineering*, **9**, 431.

Prabhakaran, M. P., Venugopal, J. R., & Ramakrishna, S. (2009). Mesenchymal stem cell differentiation to neuronal cells on electrospun nanofibrous substrates for nerve tissue engineering. *Biomaterials*, **30**, 4996.

Riboldi, S. A., Sampaolesi, M., Neuenschwander, P., Cossu, G., & Mantero, S. (2005). Electrospun degradable polyesterurethane membranes: Potential scaffolds for skeletal muscle tissue engineering. *Biomaterials*, **26**, 4606.

Rockwood, D. N., Akins, R. E., Jr., Parrag, I. C., Woodhouse, K. A., & Rabolt, J. F. (2008). Culture on electrospun polyurethane scaffolds decreases atrial natriuretic peptide expression by cardiomyocytes *in vitro*. *Biomaterials*, **29**, 4783.

Shields, K. J., Beckman, M. J., Bowlin, G. L., & Wayne, J. S. (2004). Mechanical properties and cellular proliferation of electrospun collagen type II. *Tissue Engineering*, **10**, 1510.

Stephens, J., Rabolt, J. F., Fahnestock, S., & Chase, D. (2003). From the spider to the web: Biomimetic processing of protein polymers and collagen. *MRS Proceedings*, **774**, 31.

Xie, J., & Wang, C. H. (2006). Electrospun micro- and nanofibers for sustained delivery of paclitaxel to treat C6 glioma *in vitro*. *Pharmaceutical Research*, **23**, 1817.

Xu, C. Y., Inai, R., Kotaki, M., & Ramakrishna, S. (2004). Aligned biodegradable nanofibrous structure: A potential scaffold for blood vessel engineering. *Biomaterials*, **25**, 877.

Yang, F., Murugan, R., Wang, S., & Ramakrishna, S. (2005). Electrospinning of nano/micro scale poly(L-lactic acid) aligned fibers and their potential in neural tissue engineering. *Biomaterials*, **26**, 2603.

Zong, X., Bien, H., Chung, C.-Y., Yin, L., Fang, D., et al. (2005). Electrospun fine-textured scaffolds for heart tissue constructs. *Biomaterials*, **26**, 5330.

CHAPTER I.2.17 SURFACE-IMMOBILIZED BIOMOLECULES

Allan S. Hoffman[1] *and Jeffrey A. Hubbell*[2]

[1]Professor of Bioengineering and Chemical Engineering, UWEB Investigator, University of Washington, Seattle, WA, USA

[2]EPFL (École Polytechnique Fédérale de Lausanne), Switzerland

INTRODUCTION

Biomolecules such as enzymes, antibodies, affinity proteins, cell receptor ligands, and drugs of all kinds have been immobilized on and within biomaterial surfaces for a wide range of therapeutic, diagnostic, tissue regeneration, separation, and bioprocess applications. Immobilization of heparin on polymer surfaces is one of the earliest examples of a surface-modified, biologically functional biomaterial (Gott et al., 1963). Living cells may also be combined with biomaterials, especially when their surfaces contain cell adhesion peptides or proteins, and the fields of cell culture, artificial organs, and tissue engineering include important examples of cell–surface interactions. These "hybrid" combinations of natural and synthetic materials confer "biological functionality" to the synthetic biomaterial. Many sections and chapters in this textbook cover various aspects of this topic, including adsorption of proteins and adhesion of cells and bacteria on biomaterial surfaces, non-fouling surfaces, cell culture, tissue engineering, artificial organs, drug delivery, and others; this chapter will focus on the methodology involving physical adsorption and chemical immobilization of biomolecules on biomaterial surfaces, especially for applications requiring bioactivity of the immobilized biomolecule.

Among the different classes of biomaterials that could be biologically modified, synthetic polymers are especially interesting because their surfaces may contain reactive groups such as –OH, –COOH or NH_2 groups, or they may be readily modified with other reactive groups such as azide, alkyne, and SH groups. All of these groups can be used to covalently link biomolecules.

Another advantage of polymers as supports for biomolecules is that the polymers may be fabricated in many forms, including films, membranes, tubes, fibers, fabrics, particles, capsules, and porous structures. Furthermore, macromolecular structures can also vary substantially. The latter can include homopolymers, random, alternating, block, and graft copolymers, hyperbranched (comb-like) and star-shaped structures (see Chapter I.2.2 on Polymers).

Living anionic polymerization techniques, along with newer methods of living free-radical polymerizations, now provide fine control of molecular weights with narrow distributions. The molecular forms of solid polymers include non-cross-linked chains that are insoluble at physiologic conditions, cross-linked networks, physical blends, and interpenetrating networks (IPNs) (e.g., Piskin and Hoffman, 1986; see also Chapter I.2.2). "Smart" polymers are sharply responsive in solubility behavior to stimuli, such as temperature, pH, and salt concentration (see Chapter I.2.11 on "Smart" Polymers).

For surfaces of metals, metal oxides, inorganic glasses or ceramics, biological functionality can sometimes be added via a chemically immobilized or physically adsorbed polymeric or surfactant adlayer, or by use of techniques such as plasma gas discharge, corona discharge in air or ozone to modify polymer surface compositions with functional groups (see also Chapter I.2.12). Several researchers have applied mussel adhesive chemistry based on self-condensation of dopamine to form tight bonding layers of polydopamine on a variety of surfaces, including metals, metal oxides, and glasses (Lee et al., 2007; Ku et al., 2010). The amine groups in these polymers may be further functionalized with biomolecules.

PATTERNED SURFACE COMPOSITIONS

(See also Chapter I.2.13.)

Biomaterial surfaces may also be functionalized in geometric patterns (Bernard et al., 1998; Blawas and Reichert, 1998; James et al., 1998; Ito, 1999; Kane et al., 1999; Folch and Toner, 2000). Sometimes the patterned surfaces will have regions that are non-binding to proteins (so-called "non-fouling" compositions) while others may contain covalently-linked cell receptor ligands (Neff et al., 1999; Alsberg et al., 2002; Csucs et al., 2003; VandeVondele et al., 2003), or may have physically adsorbed cell adhesion proteins (McDevitt et al., 2002; Ostuni et al., 2003). A huge industry has also evolved based on "biochips" that contain microarrays of immobilized, single-stranded DNA (for genomic assays) or peptides or proteins (for proteomic assays) (Houseman and Mrksich, 2002; Lee and Mrksich, 2002). The majority of these microarrays utilize inorganic silica chips rather than polymer substrates directly, but it is possible to incorporate functionality through chemical modification with silane chemistries (Puleo, 1997) or adsorption of a polymeric adlayer (Scotchford et al., 2003; Winkelmann et al., 2003).

A variety of methods have been used for the production of these patterned biochips, including photo-initiated synthesis through patterned masks (Ellman and Gallop, 1998; Folch and Toner, 2000), microfluidic fluid exposure (Ismagilov et al., 2001), and protection with adhesive organic protecting layers that are lifted off after exposure to the biomolecular treatment (Jackman et al., 1999).

IMMOBILIZED BIOMOLECULES AND THEIR USES

Many different biologically functional molecules can be chemically or physically immobilized on polymeric supports (Table I.2.17.1) (Laskin, 1985; Tomlinson and Davis, 1986). Examples of applications of these immobilized biological species are listed in Table I.2.17.2. When hydrophilic, molecularly cross-linked or entangled solids are water-swollen above about 15–25% water content, they become hydrogels and biomolecules may be immobilized on the outer gel surface, as well as within the water-containing regions ("meshes") of the swollen polymer gel network. It can be seen that there are many diverse uses of such biofunctional systems in both the medical and biotechnology fields. For example, a number of immobilized enzyme supports and reactor systems (Table I.2.17.3) have been developed for therapeutic uses in the clinic (Table I.2.17.4) (De Myttenaere et al., 1967; Kolff, 1979; Sparks et al., 1969; Chang, 1972; Nose et al., 1983; Schmer et al., 1981; Callegaro and Denri, 1983; Nose et al., 1984; Lavin et al., 1985; Sung et al., 1986). Advantages and disadvantages of immobilized enzymes are listed in Table I.2.17.5.

TABLE I.2.17.1 Examples of Biologically Active Molecules that may be Immobilized on or within Polymeric Biomaterials
Proteins/Peptides
Enzymes
Antibodies
Antigens
Cell adhesion molecules
"Blocking" proteins
Saccharides
Sugars
Oligosaccharides
Polysaccharides
Lipids
Fatty acids
Phospholipids
Glycolipids
Other
Conjugates or mixtures of the above
Drugs
Antithrombogenic agents
Anticancer agents
Antibiotics
Contraceptives
Drug antagonists
Peptide, protein drugs
Ligands
Hormone receptors
Cell surface receptor ligands (peptides, saccharides)
Avidin, biotin
Nucleic Acids, Nucleotides
Single or double-stranded
DNA, RNA (e.g., antisense oliogonucleotides)

IMMOBILIZED CELL LIGANDS AND CELLS

Cell interactions with foreign materials are usually mediated by a biological intermediate, such as adsorbed proteins, as described in Chapter II.1.2. An approach using biologically-functionalized materials can be much more direct, by adsorbing or covalently grafting ligands for cell–surface adhesion receptors, such as integrins, to the material surface (Lutolf and Hubbell, 2005; Patterson et al., 2010) (Table I.2.17.6).

This has been accomplished with peptides grafted randomly over a material surface (Massia and Hubbell,

TABLE I.2.17.2 Application of Immobilized Biomolecules and Cells

Enzymes	Bioreactors (industrial, biomedical) Bioseparations Biosensors Diagnostic assays Biocompatible surfaces
Antibodies, peptides, and other affinity molecules	Biosensors Diagnostic assays Affinity separations Targeted drug delivery Cell culture
Drugs	Thrombo-resistant surfaces Drug delivery systems
Lipids	Thrombo-resistant surfaces Albuminated surfaces
Nucleic acid derivatives and nucleotides	DNA probes Gene therapy
Cells	Bioreactors (industrial) Bioartificial organs Biosensors

TABLE I.2.17.3 Bioreactors, Supports and Designs

"Artificial cell" suspensions
(microcapsules, RBC ghosts, liposomes, reverse micelles [w/o] microspheres)

Biologic Supports
(membranes and tubes of collagen, fibrin ± glycosaminoglycans)

Synthetic Supports
(porous or asymmetric hollow fibres, particulates, parallel plate devices)

TABLE I.2.17.4 Examples of Immobilized Enzymes in Therapeutic Bioreactors

Medical Application	Substrate	Substrate Action
Cancer Treatment		
l-Asparaginase	Asparagine	Cancer cell nutrient
l-Glutaminase	Glutamine	Cancer cell nutrient
l-Arginase	Arginine	Cancer cell nutrient
l-Phenylalanine	Phenylalanine	Toxin lyase
Indole-3-alkane α hydroxylase	Tryptophan	Cancer cell nutrient
Cytosine deaminase	5-Fluorocytosine	Toxin
Liver Failure (Detoxification)		
Bilirubin oxidase	Bilirubin	Toxin
UDP-Gluceronyl transferase	Phenolics	Toxin
Other		
Heparinase	Heparin	Anticoagulant
Urease	Urea	Toxin

TABLE I.2.17.5 Some Advantages and Disadvantages of Immobilized Enzymes

Advantages
Enhanced stability
Can modify enzyme microenvironment
Can separate and reuse enzyme
Enzyme-free product
Lower cost, higher purity product
No immunogenic response (therapeutics)
Disadvantages
Difficult to sterilize
Fouling by other biomolecules
Mass transfer resistances (substrate in and product out)
Adverse biological responses of enzyme support surfaces (*in vivo* or *ex vivo*)
Greater potential for product inhibition

TABLE I.2.17.6 Selected* Peptide Adhesion Domains in Cell Adhesion Proteins

Fibronectin	RGDS LDV REDV
Vitronectin	RGDS
Laminin A chain	LRGDN IKVAV
Laminin B1 chain	YIGSR
Laminin B2 chain	RNIAEIIKDA
Collagen I	RGDT DGEA
Thrombospondin	RGD

*A large number of receptor-binding peptide domains have been identified in numerous adhesion proteins that are able to recapitulate some of the binding character of the entire protein. Only a small selection of widely studied peptides is listed here.

1991), as well as with peptides presented in a pre-clustered manner (Irvine et al., 2001). The latter has important advantages. Cells normally cluster their adhesion receptors into nanoscopic assemblies referred to as "focal contacts" (Geiger et al., 2009), in which both adhesion ligands and receptors are co-clustered. This clustering plays an important role in both cell adhesion mechanics (Ward and Hammer, 1993), and cell signaling (Maheshwari et al., 2000; Geiger et al., 2009). In addition to peptides, saccharides have also been grafted to polymer surfaces to confer biological functionality (Griffith and Lopina, 1998; Chang and Hammer, 2000).

Nanotechnology methods of ligand immobilization have been used extensively to achieve very precise control of ligand immobilization density, ligand clustering, and arrangement, and even ligand exposure from a previously hidden state or hiding after an exposed state. For example, nano-patterning methods have been used to show that cells are capable of sensing and responding to clustering on the length scale of 70 nm of the RGD peptide, which binds to integrins such as $\alpha_v\beta_3$ (Huang et al., 2009). At longer length scales, the cell is unable to sense clustering. Methods of formation of self-assembling monolayers, including stimulus-sensitive self-assembling monolayers, have been used to conduct investigations on the migration- and differentiation-inducing influences of different cell adhesion molecules, and also to identify new receptor–ligand biomolecular pairs (Mrksich, 2009).

Specific biomolecules can be immobilized in order to control cellular interactions; one important class of such functionalizations is the polypeptide growth factor. Such molecules can be immobilized and retain their ability to provide biological cues that signal specific cellular behavior, such as support of liver-specific function of hepatocytes (Kuhl and Griffith-Cima, 1996), induction of neurite extension in neurons (Sakiyama-Elbert et al., 2001), induction of angiogenesis (Zisch et al., 2001) or the differentiation of mesenchymal stem cells into bone-forming osteoblasts (Lutolf et al., 2003a,b; Martino et al., 2009). It is important to understand that these effects are not merely induced by the mechanics of adhesion, i.e., what matters are the details of which adhesion ligand is displayed, and thus which adhesion receptor is ligated and what downstream signaling is activated. For example, stimulation of mesenchymal stem cells with a ligand for integrin $\alpha_5\beta_1$ stimulated osteogenesis, whereas stimulation with a ligand for integrin $\alpha_v\beta_3$ did not. Other molecules may be immobilized that can take part in enzymatic reactions at the surface. McClung et al. (2001, 2003) have immobilized lysines, whose ε-amino groups may interact with pre-adsorbed tissue plasminogen activator (tPA) during coagulation, to enhance fibrin clot dissolution at that surface.

The above paragraphs deal with biomaterial surfaces as though they are two-dimensional. In some devices, this is indeed the case; however, biomaterials have been developed to display their surfaces in three-dimensional situations, even with triggering by cellular remodeling. For example, biomaterial gels have been developed, consisting of cross-linked PEG chains, where the cross-linker is sensitive to proteases, such as plasmin or matrix metalloproteinases, that are activated by cells as they migrate (Lutolf et al., 2003a,b). In this way, as cells migrate in a material and remodel it, they can expose new surfaces and be stimulated by those new surfaces. Using such materials as tools, a number of regenerative medical applications have been targeted, where the biomaterial surface displays both adhesion ligands and polypeptide growth factors (Lutolf and Hubbell, 2005). Such materials, through careful selection of adhesion ligand composition, can control very specific biological processes, such as embryonic stem cell self-renewal (Lee et al., 2010). Materials have been developed that allow for very precise control of the adhesion ligand display and cross-linked nature of the three-dimensional material, e.g., through locally-controlled photochemical manipulation (Kloxin et al., 2009, 2010).

IMMOBILIZATION METHODS

There are three major methods for immobilizing biomolecules: physical adsorption; physical "entrapment;" and covalent attachment (Stark, 1971; Zaborsky, 1973; Dunlap, 1974). Physical adsorption includes: (1) van der Waals interactions; (2) electrostatic interactions; and (3) affinity recognition. Once adsorbed, the molecules may be further cross-linked to each other. Physical "entrapment" systems include: (1) microcapsules; (2) hydrogels; and (3) physical mixtures such as matrix drug delivery systems. Covalent attachment includes: (1) soluble polymer conjugates; (2) conjugates on solid surfaces; or (3) conjugates within hydrogels.

It is clear that the first two are physically based, while the third is based on covalent or "chemical" attachment to the support molecules. However, sometimes the physical attachment process may involve pairs of molecules with very strong affinity interactions, verging on covalent force levels, such as biotin with streptavidin. Thus, it is important to note that the term "immobilization" can refer to a short-term, long-term or "permanent" localization of the biomolecule on or within a support. In the case of a drug delivery system, the immobilized drug is supposed to be released from the support, either over a short period or over a longer-term, while an immobilized enzyme or cell adhesion peptide or protein in an artificial organ is designed to remain attached to or entrapped within the support over the duration of use. Either physical or chemical immobilization can lead to relatively long-term or "permanent" retention on or within a solid support, especially if the immobilized biomolecule is large.

If the polymer support is biodegradable, then the chemically immobilized biomolecule will be released as the matrix erodes or degrades away. Many researchers have chemically immobilized cell adhesion peptides such as RGD onto biodegradable matrices such as PLGA for use as tissue engineering scaffolds. The cells will have the time to bind and regenerate tissue if the support is slowly degrading. If the support degrades more rapidly than the cells' ability to reach and bind to the adhesion peptides, then the matrix will not function in the intended way. The immobilized biomolecule may also be susceptible to enzymatic degradation *in vivo*, and this remains an interesting aspect that has received relatively little attention.

A large and diverse group of methods have been developed for covalent binding of biomolecules to soluble or solid polymeric supports (Weetall, 1975; Carr and Bowers, 1980; Dean et al., 1985; Gombotz and Hoffman, 1986; Shoemaker et al., 1987; Park and Hoffman, 1990; Yang et al., 1990; Schense and Hubbell, 1999; Lutolf et al., 2003b). Many of these methods are schematically illustrated in Figure I.2.17.1. The same biomolecule may be immobilized by many different methods; specific

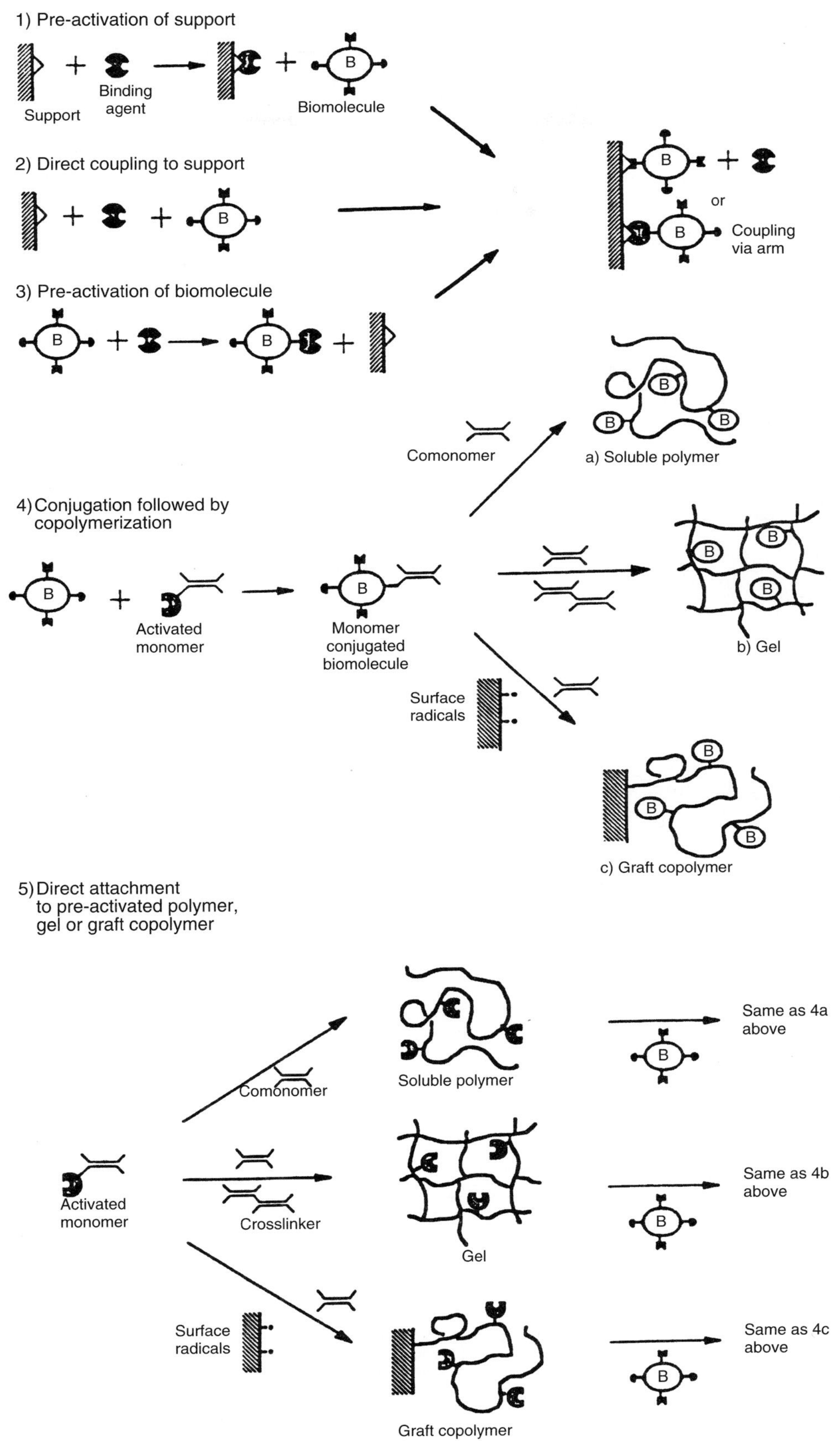

FIGURE I.2.17.1 Schematic cartoons showing various methods for covalent biomolecule immobilization.

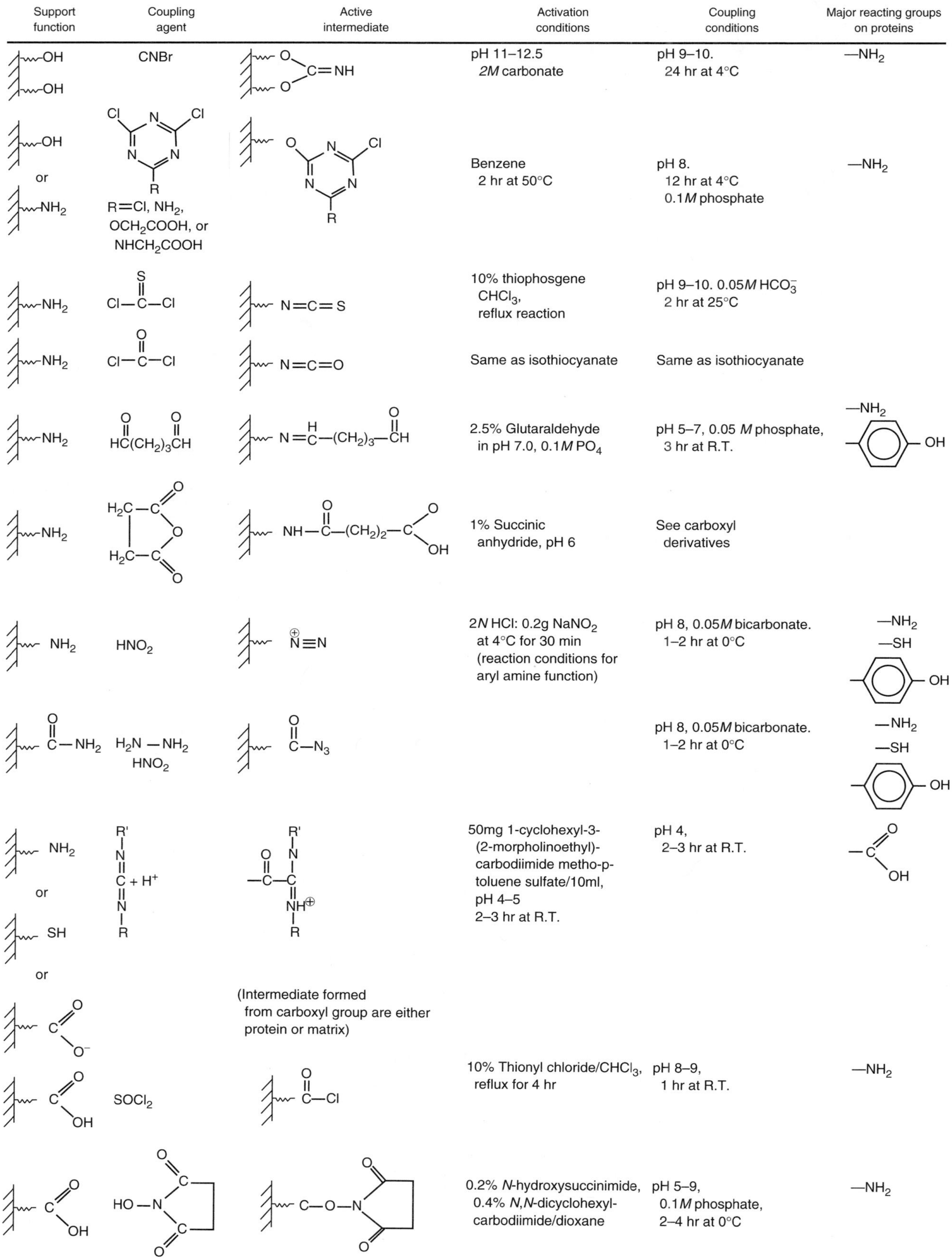

Support function	Coupling agent	Active intermediate	Activation conditions	Coupling conditions	Major reacting groups on proteins
~OH, ~OH	CNBr	~O–C(=NH)–O~	pH 11–12.5, 2*M* carbonate	pH 9–10. 24 hr at 4°C	$—NH_2$
~OH or ~NH_2	Cl, Cl, R-triazine; R=Cl, NH_2, OCH_2COOH, or $NHCH_2COOH$	~O–triazine(Cl)(R)	Benzene 2 hr at 50°C	pH 8. 12 hr at 4°C 0.1*M* phosphate	$—NH_2$
~NH_2	Cl—C(=S)—Cl	~N=C=S	10% thiophosgene $CHCl_3$, reflux reaction	pH 9–10. 0.05*M* HCO_3^- 2 hr at 25°C	
~NH_2	Cl—C(=O)—Cl	~N=C=O	Same as isothiocyanate	Same as isothiocyanate	
~NH_2	HC(=O)$(CH_2)_3$CH(=O)	~N=CH—$(CH_2)_3$—CH(=O)	2.5% Glutaraldehyde in pH 7.0, 0.1*M* PO_4	pH 5–7, 0.05 *M* phosphate, 3 hr at R.T.	$—NH_2$; —C_6H_4—OH
~NH_2	Succinic anhydride (H_2C—C(=O), H_2C—C(=O), O)	~NH—C(=O)—$(CH_2)_2$—C(=O)OH	1% Succinic anhydride, pH 6	See carboxyl derivatives	
~ NH_2	HNO_2	~ $\overset{\oplus}{N}$≡N	2*N* HCl: 0.2g $NaNO_2$ at 4°C for 30 min (reaction conditions for aryl amine function)	pH 8, 0.05*M* bicarbonate. 1–2 hr at 0°C	$—NH_2$; —SH; —C_6H_4—OH
~ C(=O)—NH_2	$H_2N—NH_2$, HNO_2	~ C(=O)—N_3		pH 8, 0.05*M* bicarbonate. 1–2 hr at 0°C	$—NH_2$; —SH; —C_6H_4—OH
~ NH_2 or ~ SH or ~ C(=O)O^-	R'—N=C=N—R + H^+	—C(=O)—C(=N$H^{\oplus}$—R)—NH—R' (Intermediate formed from carboxyl group are either protein or matrix)	50mg 1-cyclohexyl-3-(2-morpholinoethyl)-carbodiimide metho-p-toluene sulfate/10ml, pH 4–5 2–3 hr at R.T.	pH 4, 2–3 hr at R.T.	—C(=O)OH
~ C(=O)OH	$SOCl_2$	~ C(=O)—Cl	10% Thionyl chloride/$CHCl_3$, reflux for 4 hr	pH 8–9, 1 hr at R.T.	$—NH_2$
~ C(=O)OH	HO—N(succinimide)	~ C(=O)—O—N(succinimide)	0.2% *N*-hydroxysuccinimide, 0.4% *N,N*-dicyclohexyl-carbodiimide/dioxane	pH 5–9, 0.1*M* phosphate, 2–4 hr at 0°C	$—NH_2$

FIGURE I.2.17.2 Examples of various chemical methods used to bond biomolecules directly to reactive supports. (Carr and Bowers, 1980.)

examples of the most common chemical reactions utilized are shown in Figure I.2.17.2. This figure does not include some important reactions, such as thiol-maleimide, thiol-vinyl sulfone, amino-vinyl (Michael additions), and azide-alkyne (click) conjugation reactions. The reader is referred to Hermansson (2008) for many useful details on numerous conjugation chemistries.

For covalent binding to an inert solid polymer surface such as polyethylene or silicone rubber, the surface must first be chemically modified to provide reactive groups for the subsequent immobilization step. If the "inert" polymer support does not contain such groups, then it is necessary to modify it in order to permit covalent immobilization of biomolecules to the surface. A wide number of solid surface modification techniques have been used, including ionizing radiation-initiated graft copolymerization, plasma gas discharge, photochemical grafting, chemical modification (e.g., ozone grafting), and chemical derivatization (Hoffman et al., 1972, 1986; Gombotz and Hoffman, 1986, 1987; Hoffman, 1987, 1988) (see also Chapter I.2.12 on surface modification techniques).

A covalently-immobilized biomolecule may also be attached via a spacer group, sometimes called an "arm" or a "tether" (Cuatrecasas and Anfinsen, 1971; Hoffman et al., 1972; Hoffman, 1987). One of the most popular tethers is a PEG molecule that has been derivatized with different reactive end groups (e.g., Kim and Feijen, 1985), and several companies currently offer a variety of homo- or hetero-bifunctional PEGs with end group chemistries such as N-hydroxysuccinimide (NHS), maleimide, pyridyl disulfide, thiol, vinyl sulfone, alkyne, and azide (the last two are for click chemical attachments). Such spacers can provide greater steric freedom, and thus greater specific activity for the immobilized biomolecule. The spacer arm may also be either hydrolytically or enzymatically degradable, and therefore will release the immobilized biomolecule as it degrades (Kopecek, 1977; Hern and Hubbell, 1998).

Inert surfaces, whether polymeric, metal or ceramic, can also be functionalized through modification of a polymeric adlayer. Such physically adsorbed or chemisorbed polymers can be bound to the surface via electrostatic interactions (VandeVondele et al., 2003), hydrophobic interactions (Neff et al., 1999) or specific chemical interactions, such as that between gold and sulfur atoms (Harder et al., 1998; Bearinger et al., 2003). Metal or ceramic surfaces may also be derivatized with functional groups using silane chemistry, such as with functionalized triethoxysilanes (Massia and Hubbell, 1991; Puleo, 1997) or dopamine polymer chemistry, as described above. Plasma gas discharge has been used to deposit polymeric amino groups for conjugation of hyaluronic acid to a metal surface (Verheye et al., 2000).

As noted earlier, hydrophobic interactions have been used to functionalize hydrophobic surfaces, utilizing

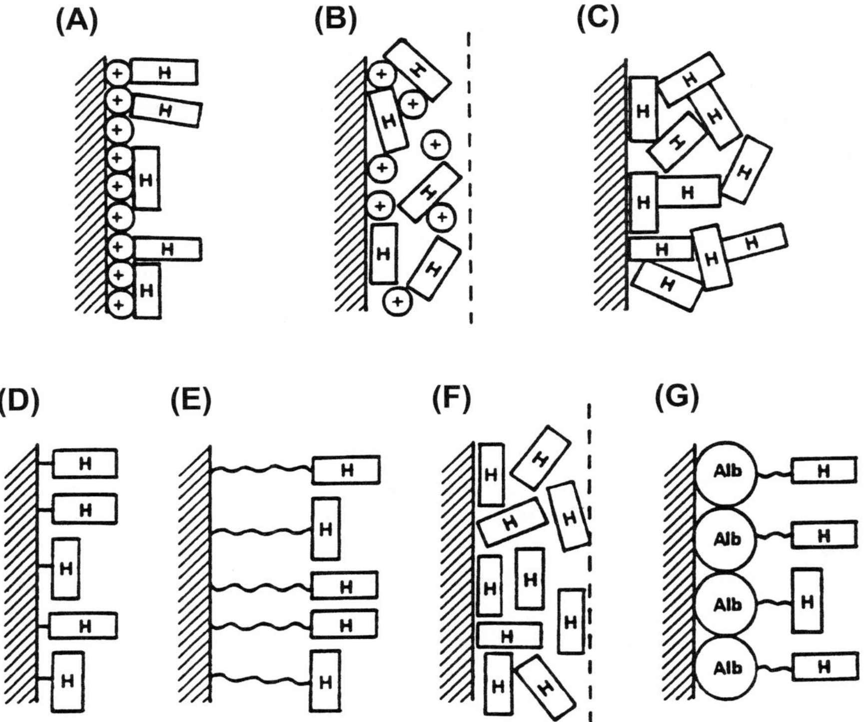

FIGURE I.2.17.3 Various methods for heparinization of surfaces: (A) heparin bound ionically on a positively charged surface; (B) heparin ionically complexed to a cationic polymer, physically coated on a surface; (C) heparin physically coated and self-cross-linked on a surface; (D) heparin covalently linked to a surface; (E) heparin covalently immobilized via spacer arms; (F) heparin dispersed into a hydrophobic polymer; (G) heparin–albumin conjugate immobilized on a surface. (Kim and Feijen, 1985.)

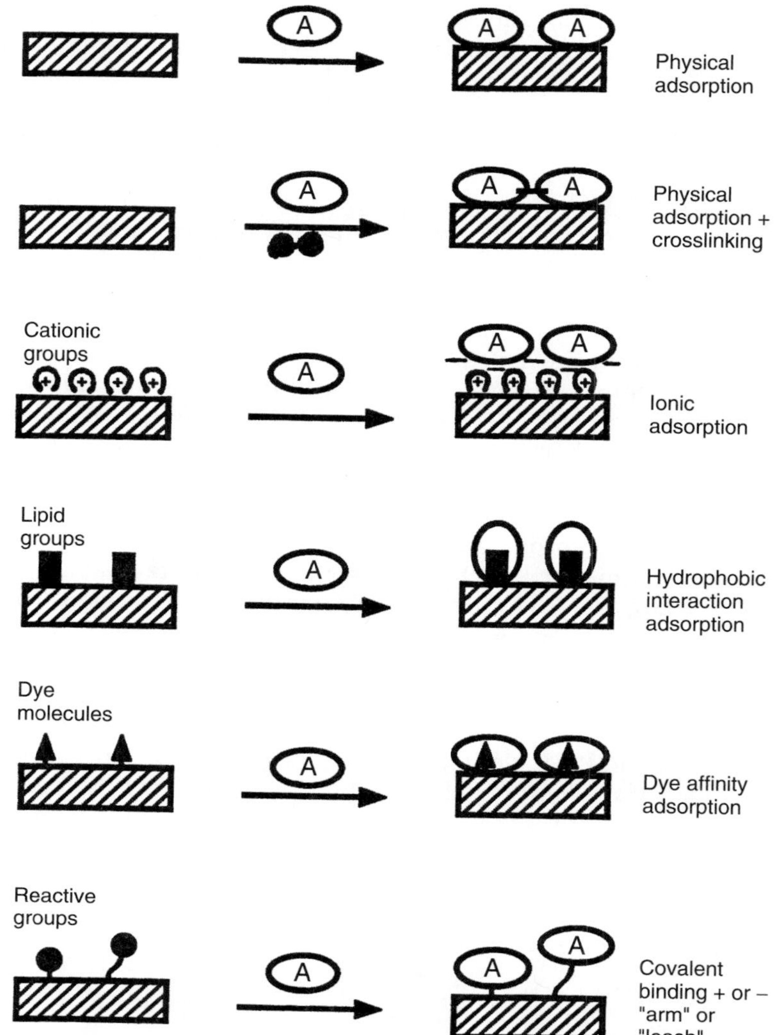

FIGURE I.2.17.4 Schematic of various ways that albumin may be immobilized on a surface. Albumin is often used as a "passivating" protein, to minimize adsorption of other proteins to a surface.

biomolecules such as ligands attached to hydrophobic sequences (e.g., Ista et al., 1999; Nath and Chilkoti, 2003). Surfaces with hydrophobic gradients have also been prepared for this purpose (Detrait et al., 1999). An interesting surface-active product was developed several years ago that was designed to convert a hydrophobic surface to a cell adhesion surface by hydrophobic adsorption; it had an RGD cell adhesion peptide coupled at one end to a hydrophobic peptide sequence.

Sometimes more than one biomolecule may be immobilized to the same support. For example, a soluble polymer designed to "target" a drug molecule may have separately conjugated to it a targeting moiety such as an antibody, along with the drug molecule, which may be attached to the polymer backbone via a biodegradable spacer group (Ringsdorf, 1975; Kopecek, 1977; Goldberg, 1983). For some nucleic acid drugs, sometimes a "nuclear localization signal" or NLS is added to enhance intracellular delivery to the nucleus (e.g., Nair et al., 2003). In another example, the wells in an immunodiagnostic microtiter plate will usually be coated first with an antibody, and then with albumin or casein (to block nonspecific adsorption during the assay); each is physically adsorbed to the well surface. In the case of affinity chromatography supports, the affinity ligand may be covalently coupled to the solid packing, and a "blocking" protein such as albumin or casein may be similarly added to block nonspecific adsorption to the support.

It is evident that there are many different ways in which the same biomolecule can be immobilized to a polymeric support. Heparin and albumin are two common biomolecules that have been immobilized by a number of widely differing methods. These are illustrated schematically in Figures I.2.17.3 and I.2.17.4.

Some of the major features of the different immobilization techniques are compared and contrasted in Table I.2.17.7. The important molecular criteria for successful immobilization of a biomolecule are that a large fraction of the available biomolecules should be immobilized, and a large fraction of those immobilized biomolecules should retain an acceptable level of bioactivity over an economically and/or clinically appropriate time period.

TABLE I.2.17.7 Biomolecule Immobilization Methods

Method	Physical and Electrostatic Adsorption	Cross-linking (After Physical Adsorption)	Entrapment	Covalent Binding
Ease:	High	Moderate	Moderate to low	Low
Loading level possible:	Low (unless high S/V)	Low (unless high S/V)	High	(Depends on S/V and site density)
Leakage (loss):	Relatively high (sens. to ΔpH salts)	Relatively low	Low to none[a]	Low to none
Cost:	Low	Low to moderate	Moderate	High

[a]Except for drug delivery systems.

CONCLUSIONS

It can be seen that there is a wide and diverse range of materials and methods available for immobilization of biomolecules and cells on or within biomaterial supports. Combined with the great variety of possible biomedical and biotechnological applications, this represents a very exciting and fertile field for applied research in biomaterials.

BIBLIOGRAPHY

Alsberg, E., Anderson, K. W., Albeiruti, A., Rowley, J. A., & Mooney, D. J. (2002). Engineering growing tissues. *Proc. Natl. Acad. Sci.*, **99**, 12025.

Bearinger, J. P., Terrettaz, S., Michel, R., Tirelli, N., Vogel, H., et al. (2003). Chemisorbed poly(propylene sulphide)-based copolymers resist biomolecular interactions. *Nat. Mater.*, **2**, 259–264.

Bernard, A., Delamarche, E., Schmid, H., Michel, B., Bosshard, H. R., et al. (1998). Printing patterns of proteins. *Langmuir*, **14**, 2225–2229.

Blawas, A. S., & Reichert, W. M. (1998). Protein patterning. *Biomaterials*, **19**, 595–609.

Callegaro, L., & Denri, E. (1983). Applications of bioreactors in medicine. *Int. J. Artif. Organs*, **6**(Suppl. 1), 107.

Carr, P. W., & Bowers, L. D. (1980). *Immobilized Enzymes in Analytical and Clinical Chemistry: Fundamentals and Applications*. New York, NY: Wiley.

Chang, K. C., & Hammer, D. A. (2000). Adhesive dynamics simulations of sialyl-Lewis(x)/E-selectin-mediated rolling in a cell-free system. *Biophys. J.*, **79**, 1891–1902.

Chang, T. M. S. (1972). *Artificial Cells*. Springfield, IL: C. C. Thomas.

Csucs, G., Michel, R., Lussi, J. W., Textor, M., & Danuser, G. (2003). Microcontact printing of novel co-polymers in combination with proteins for cell-biological applications. *Biomaterials*, **24**, 1713–1720.

Cuatrecasas, P., & Anfinsen, C. B. (1971). Affinity chromatography. *Ann. Rev. Biochem.*, **40**, 259.

Dean, P. D. G., Johnson, W. S., & Middle, F. A. (Eds.), (1985). *Affinity Chromatography*. Oxford, UK: IRL Press.

De Myttenaere, M. H., Maher, J., & Schreiner, G. (1967). Hemoperfusion through a charcoal column for glutethimide poisoning. *Trans. ASAIO*, **13**, 190.

Detrait, E., Lhoest, J. B., Bertrand, P., & de Aguilar, V. B. (1999). Fibronectin–pluronic coadsorption on a polystyrene surface with increasing hydrophobicity: Relationship to cell adhesion. *J. Biomed. Mater. Res.*, **45**, 404–413.

Dunlap, B. R. (Ed.), (1974). *Immobilized Biochemicals and Affinity Chromatography*. New York, NY: Plenum.

Ellman, J. A., & Gallop, M. A. (1998). Combinatorial chemistry. *Curr. Opin. Chem. Biol.*, **2**, 17–319.

Folch, A., & Toner, M. (2000). Microengineering of cellular interactions. *Annu. Rev. Biomed. Eng.*, **2**, 227–256.

Geiger, B., Spatz, J. P., & Bershadsky, A. D. (2009). Environmental sensing through focal adhesions. *Nat. Rev. Mol. Cell. Biol.*, **10**, 21–33.

Goldberg, E. (Ed.), (1983). *Targeted Drugs*. New York, NY: Wiley-Interscience.

Gombotz, W. R., & Hoffman, A. S. (1986). Immobilization of biomolecules and cells on and within synthetic polymeric hydrogels. In N. A. Peppas (Ed.), *Hydrogels in Medicine and Pharmacy* (Vol. 1, pp. 95–126). Boca Raton, FL: CRC Press.

Gombotz, W. R., & Hoffman, A. S. (1987). Gas discharge techniques for modification of biomaterials. In D. Williams (Ed.), *Critical Reviews in Biocompatibility* (Vol. 4, pp. 1–42). Boca Raton, FL: CRC Press.

Gott, V. L., Whiffen, J. D., & Dutton, R. C. (1963). Heparin bonding on colloidal graphite surfaces. *Science*, **142**, 1297–1298.

Griffith, L. G., & Lopina, S. (1998). Microdistribution of substratum-bound ligands affects cell function: Hepatocyte spreading on PEO-tethered galactose. *Biomaterials*, **19**, 979–986.

Harder, P., Grunze, M., Dahint, R., Whitesides, G. M., & Laibinis, P. E. (1998). Molecular conformation and defect density in oligo (ethylene glycol)-terminated self-assembled monolayers on gold and silver surfaces determine their ability to resist protein adsoption. *J. Phys. Chem. B*, **102**, 426–436.

Hermansson, G. (2008). *Bioconjugate Techniques* (2nd ed.). New York, NY: Academic Press.

Hern, D. L., & Hubbell, J. A. (1998). Incorporation of adhesion peptides into nonadhesive hydrogels useful for tissue resurfacing. *J. Biomed. Mater. Res.*, **39**, 266–276.

Hoffman, A. S. (1987). Modification of material surfaces to affect how they interact with blood. *Ann. NY Acad. Sci.*, **516**, 96–101.

Hoffman, A. S. (1988). Applications of plasma gas discharge treatments for modification of biomaterial surfaces. *J. Appl. Polymer Sci. Symp.*, **42**, 251.

Hoffman, A. S., Schmer, G., Harris, C., & Kraft, W. G. (1972). Covalent binding of biomolecules to radiation-grafted hydrogels on inert polymer surfaces. *Trans. Am. Soc. Artif. Internal. Organs*, **18**, 10.

Hoffman, A. S., Gombotz, W. R., Uenoyama, S., Dong, L. C., & Schmer, G. (1986). Immobilization of enzymes and antibodies to radiation grafted polymers for therapeutic and diagnostic applications. *Radiat. Phys. Chem.*, **27**, 265–273.

Houseman, B. T., & Mrksich, M. (2002). Towards quantitative assays with peptide chips: A surface engineering approach. *Trends. Biotechnol.*, **20**, 279–281.

Huang, J., Grater, S. V., Corbellini, F., Rinck, S., Bock, E., et al. (2009). Impact of order and disorder in RGD nanopatterns on cell adhesion. *Nano. Lett.*, **9**, 1111–1116.

Irvine, D. J., Mayes, A. M., & Griffith, L. G. (2001). Nanoscale clustering of RGD peptides at surfaces using comb polymers. 1. Synthesis and characterization of comb thin films. *Biomacromolecules*, **2**, 85–94.

Ismagilov, R. F., Ng, J. M.K., Kenis, P. J.A., & Whitesides, G. M. (2001). Microfluidic arrays of fluid–fluid diffusional contacts as detection elements and combinatorial tools. *Anal. Chem.*, 73, 5207–5213.

Ista, L. K., Pérez-Luna, V. H., & López, G. P. (1999). Surface-grafted, environmentally sensitive polymers for biofilm release. *Appl. Environ. Microbiol.*, **65**, 1603–1609.

Ito, Y. (1999). Surface micropatterning to regulate cell functions. *Biomaterials*, **20**, 2333–2342.

Jackman, R. J., Duffy, D. C., Cherniavskaya, O., & Whitesides, G. M. (1999). Using elastomeric membranes as dry resists and for dry lift-off. *Langmuir*, **15**, 2973–2984.

James, C. D., Davis, R. C., Kam, L., Craighead, H. G., Isaacson, M., et al. (1998). Patterned protein layers on solid substrates by thin stamp microcontact printing. *Langmuir*, **14**, 741–744.

Kane, R. S., Takayama, S., Ostuni, E., Ingber, D. E., & Whitesides, G. M. (1999). Patterning proteins and cells using soft lithography. *Biomaterials*, **20**, 2363–2376.

Kim, S. W., & Feijen, J. (1985). Methods for immobilization of heparin. In D. Williams (Ed.), *Critical Reviews in Biocompatibility* (pp. 229–260). Boca Raton, FL: CRC Press.

Kloxin, A. M., Kasko, A. M., Silans, C. N., & Anseth, K. S. (2009). Photodegradable hydrogels for dynamic tuning of physical and chemical properties. *Science*, **324**, 59–63.

Kloxin, A. M., Tibbitt, M. W., Kasko, A. M., Fairbaim, J. A., & Anseth, K. S. (2010). Tunable hydrogels for external manipulation of cellular microenvironments through controlled photodegradation. *Adv. Mater.*, **22**, 61–66.

Kolff, W. J. (1979). Artificial organs in the seventies. *Trans. ASAIO*, **16**, 534.

Kopecek, J. (1977). Soluble biomedical polymers. *Polymer. Med.*, 7, 191.

Ku, S. H., Ryu, K., Hong, S. K., Lee, H., & Park, C. B. (2010). General functionalization route for cell adhesion on non-wetting surfaces. *Biomaterials*, **31**, 2535–2541.

Kuhl, P. R., & Griffith-Cima, L. G. (1996). Tethered epidermal growth factor as a paradigm for growth factor-induced stimulation from the solid phase. *Nat. Med.*, **2**, 1022–1027.

Laskin, A. I. (Ed.), (1985). *Enzymes and Immobilized Cells in Biotechnology*. Menlo Park, CA: Benjamin/Cummings.

Lavin, A., Sung, C., Klibanov, A. M., & Langer, R. (1985). Enzymatic removal of bilirubin from blood: A potential treatment for neonatal jaundice. *Science*, **230**, 543.

Lee, H., Dellatore, S. M., Miller, W. M., & Messersmith, P. S. (2007). Mussel inspired surface chemistry for multifunctional coatings. *Science*, **318**, 426–430.

Lee, S. T., Yun, J. I., Jo, Y. S., Mochizuki, M., van der Vlies, A. J., et al. (2010). Engineering integrin signaling for promoting embryonic stem cell self-renewal in a precisely defined niche. *Biomaterials*, **31**, 1219–1226.

Lee, Y. S., & Mrksich, M. (2002). Protein chips: From concept to practice. *Trends Biotechnol.*, **20**, S14–S18.

Lutolf, M. P., & Hubbell, J. A. (2005). Synthetic biomaterials as instructive extracellular microenvironments for morphogenesis in tissue engineering. *Nat. Biotechnol.*, **23**, 47–55.

Lutolf, M. R., Weber, F. E., Schmoekel, H. G., Schense, J. C., Kohler, T., et al. (2003a). Repair of bone defects using synthetic mimetics of collagenous extracellular matrices. *Nat. Biotechnol.*, **21**, 513–518.

Lutolf, M. P., Raeber, G. P., Zisch, A. H., Tirelli, N., & Hubbell, J. A. (2003b). Cell-responsive synthetic hydrogels. *Adv. Mater.*, **15**, 888–892.

McClung, W. G., Clapper, D. L., Hu, S. P., & Brash, J. L. (2001). Lysine-derivatized polyurethane as a clot lysing surface: Conversion of plasminogen to plasmin and clot lysis *in vitro*. *Biomaterials*, **22**, 1919–1924.

McClung, W. G., Clapper, D. L., Anderson, A. B., Babcock, D. E., & Brash, J. L. (2003). Interactions of fibrinolytic system proteins with lysine-containing surfaces. *J. Biomed. Mater. Res.*, **66A**, 795–801.

McDevitt, T. C., Angelo, J. C., Whitney, M. L., Reinecke, H., Hauschka, S. D., et al. (2002). *In vitro* generation of differentiated cardiac myofibers on micropatterned laminin surfaces. *J. Biomed. Mater. Res.*, **60**, 472–479.

Maheshwari, G., Brown, G., Lauffenburger, D. A., Wells, A., & Griffith, L. G. (2000). Cell adhesion and motility depend on nanoscale RGD clustering. *J. Cell. Sci.*, **113**, 1677–1686.

Martino, M. M., Mochizuki, M., Rothenfluh, D. A., Rempel, S. A., Hubbell, J. A., et al. (2009). Controlling integrin specificity and stem cell differentiation in 2D and 3D environments through regulation of fibronectin domain stability. *Biomaterials*, **30**, 1089–1097.

Massia, S. P., & Hubbell, J. A. (1991). An RGD spacing of 440 nm is sufficient for integrin $\alpha v\beta 3$-mediated fibroblast spreading and 140 nm for focal contact and stress fiber formation. *J. Cell Biol.*, **114**, 1089–1100.

Mrksich, M. (2009). Using self-assembled monolayers to model the extracellular matrix. *Acta. Biomater.*, **5**, 823–841.

Nair, R., Carter, P., & Burkhard, R. (2003). NLSdb: Database of nuclear localization signals. *Nucleic. Acids. Res.*, **31**, 397–399.

Nath, N., & Chilkoti, A. (2003). Fabrication of reversible functional arrays of proteins directly from cells using a stimuli responsive polypeptide. *Anal. Chem.*, **75**, 709–715.

Neff, J. A., Tresco, P. A., & Caldwell, K. D. (1999). Surface modification for controlled studies of cell–ligand interactions. *Biomaterials*, **20**, 2377–2393.

Nose, Y., Malchesky, P. S., & Smith, J. W. (Eds.), (1983). *Plasmapheresis: New Trends in Therapeutic Applications*. Cleveland, OH: ISAO Press.

Nose, Y., Malchesky, P. S., & Smith, J. W. (Eds.), (1984). *Therapeutic Apheresis: A Critical Look*. Cleveland, OH: ISAO Press.

Ostuni, E., Grzybowski, B. A., Mrksich, M., Roberts, C. S., & Whitesides, G. M. (2003). Adsorption of proteins to hydrophobic sites on mixed self-assembled monolayers. *Langmuir*, **19**, 1861–1872.

Park, T. G., & Hoffman, A. S. (Eds.). (1990). Immobilizaiton of *Arthrobacter simplex* in a thermally reversible hydrogel: Effect of temperature cycling on steroid conversion. *Biotech. Bioeng.*, **35**, 152–159.

Patterson, J., Martino, M. M., & Hubbell, J. A. (2010). Biomimetic materials in tissue engineering. *Materials Today*, **13**, 14–22.

Piskin, E., & Hoffman, A. S. (Eds.), (1986). *Polymeric Biomaterials*. Dordrecht, The Netherlands: M. Nijhoff.

Puleo, D. A. (1997). Retention of enzymatic activity immobilized on silanized Co–Cr–Mo and Ti-6Al-4V. *J. Biomed. Mater. Res.*, **37**, 222–228.

Ringsdorf, H. (1975). Structure and properties of pharmacologically active polymers. *J. Polymer. Sci.*, **51**, 135.

Sakiyama-Elbert, S. E., Panitch, A., & Hubbell, J. A. (2001). Development of growth factor fusion proteins for cell-triggered drug delivery. *FASEB J*, **15**, 1300–1302.

Schense, J. C., & Hubbell, J. A. (1999). Cross-linking exogenous bifunctional peptides into fibrin gels with factor XIIIa. *Bioconjugate Chem.*, **10**, 75–81.

Schmer, G., Rastelli, L., Newman, M. O., Dennis, M. B., & Holcenberg, J. S. (1981). The bioartificial organ: Review and progress report. *Internat. J. Artif. Organs*, **4**, 96.

Scotchford, C. A., Ball, M., Winkelmann, M., Voros, J., Csucs, C., et al. (2003). Chemically patterned, metal-oxide-based surfaces produced by photolithographic techniques for studying protein- and cell-interactions. II: Protein adsorption and early cell interactions. *Biomaterials*, **24**, 1147–1158.

Shoemaker, S., Hoffman, A. S., & Priest, J. H. (1987). Synthesis and properties of vinyl monomer/enzyme conjugates: Conjugation of l-asparaginase with N-succinimidyl acrylate. *Appl. Biochem. Biotechnol.*, **15**, 11.

Sparks, R. E., Solemme, R. M., Meier, P. M., Litt, M. H., & Lindan, O. (1969). Removal of waste metabolites in uremia by microencapsulated reactants. *Trans. ASAIO*, **15**, 353.

Stark, G. R. (Ed.), (1971). *Biochemical Aspects of Reactions on Solid Supports*. New York, NY: Academic Press.

Sung, C., Lavin, A., Klibanov, A., & Langer, R. (1986). An immobilized enzyme reactor for the detoxification of bilirubin. *Biotech. Bioeng.*, **28**, 1531.

Tomlinson, E., & Davis, S. S. (1986). *Site-Specific Drug Delivery: Cell Biology, Medical and Pharmaceutical Aspects*. New York, NY: Wiley.

VandeVondele, S., Voros, J., & Hubbell, J. A. (2003). RGD-grafted poly-l-lysine-graft-(polyethylene glycol) copolymers block non-specific protein adsorption while promoting cell adhesion. *Biotechnol. Bioeng.*, **82**, 784–790.

Verheye, S., Markou, C. P., Salame, M. Y., Wan, B., King, S. B., III, et al. (2000). Reduced thrombus formation by hyaluronic acid coating of endovascular devices. *Arterioscler. Thromb. Vasc. Biol.*, **20**, 1168–1172.

Ward, M. D., & Hammer, D. A. (1993). A theoretical analysis for the effect of focal contact formation on cell–substrate attachment strength. *Biophys. J.*, **64**, 936–959.

Weetall, H. H. (Ed.), (1975). *Immobilized Enzymes, Antigens, Antibodies and Peptides: Preparation and Characterization*. New York, NY: Dekker.

Winkelmann, M., Gold, J., Hauert, R., Kasemo, B., Spencer, N. D., et al. (2003). Chemically patterned, metal oxide based surfaces produced by photolithographic techniques for studying protein– and cell–surface interactions I: Microfabrication and surface characterization. *Biomaterials*, **24**, 1133–1145.

Yang, H. J., Cole, C. A., Monji, N., & Hoffman, A. S. (1990). Preparation of a thermally phase-separating copolymer, poly(N- isopropylacrylamide-co-N-acryloxysuccinimide) with a controlled number of active esters per polymer chain. *J. Polymer Sci. A Polymer Chem.*, **28**, 219–220.

Zaborsky, O. (1973). *Immobilized Enzymes*. Cleveland, OH: CRC Press.

Zisch, A. H., Schenk, U., Schense, J. C., Sakiyama-Elbert, S. E., & Hubbell, J. A. (2001). Covalently conjugated VEGF-fibrin matrices for endothelialization. *J. Controlled Release*, **72**, 101–113.

CHAPTER I.2.18 BIOMIMETIC MATERIALS

Drew Elizabeth Glaser and Christopher Viney
School of Engineering, University of California at Merced, Merced, CA, USA

INTRODUCTION: WHAT ARE BIOMIMETIC MATERIALS?

Biomimetic materials are designed to replicate one or more attributes of a material produced by a living organism. This attempt at a definition highlights a shared characteristic of *biomimetic materials* and *biomaterials*, since successful biomaterials serve to either: (1) restore a natural function where the original material is absent or unable to perform properly; or (2) sustain an environment that is optimally conducive to processes such as cell culture, tissue growth, biomolecular assays, and biotechnology-based manufacturing.

How Can Biomaterials Science Benefit From Biomimicry?

In both biomaterials science and the science of biomimetic materials, there is an underlying premise that the desirable characteristics of a natural material can be replicated successfully. However, biomimicry is not just about replicating the useful *properties* of a natural material (human or other animal tissue) in a surrogate product. Biomimicry also draws upon Nature's lessons in the chemistry, processing, and structure of materials. Biomimicry therefore can provide guidance for every step in the procedure by which molecules are selected and manipulated during the design and fabrication of a biomaterial. This chapter highlights biomimicry as a source of inspiration for design and modification of biomaterials and biomedical devices.

Biomimicry recognizes that the functional materials produced by living organisms (and the natural processes by which such materials are made) have evolved over a very long period of time and that, by definition, are biocompatible. If a natural material does an excellent job (and a better job than an existing synthetic counterpart), then mimicry of that material should be considered. In this context, "excellent job" and "better job" are qualified with reference to a product having superior properties, which may include such attributes as being inexpensive or robust or environmentally friendly, or simply being unobtrusive.

It is necessary to take care about what we understand by a *natural material*. Since humans are themselves an integral part of Nature, we have to recognize that human technology and the materials that it creates can be regarded in principle as just further examples of what Nature can achieve. To avoid possible ambiguity, we will consider the term "natural material" to refer to materials that are produced *metabolically* by living organisms.

A CLASSIFICATION OF BIOMIMETIC MATERIALS

Within the context of (bio)materials, it can be helpful to recognize – and convenient to distinguish between – various classes of biomimicry, based on what type of natural attribute is being mimicked. The use of parentheses in "(bio)materials" here signifies that Nature's lessons for materials science can be widely applicable, and are not restricted specifically to biomaterials.

Functional biomimicry occurs when Nature provides the inspiration for one or more physical and/or chemical *properties* of a (bio)material. According to our preceding definition of a biomimetic material, there is some element of functional biomimicry in every successful biomaterial.

In its most direct sense, functional biomimicry is achieved when material is recovered from a donor organism, sterilized, and remodeled into a replacement for damaged tissue in a host. In such cases, Nature's contribution extends beyond that of muse to delivery of a ready-made, working solution. A rudimentary example is provided by false teeth made from ivory. A more sophisticated example is the use of pig mucosa to repair rotator cuff tears (Derwin et al., 2010).

Molecular biomimicry is the result of Nature inspiring the *types of atoms and molecules* that are selected (and that usually are available from natural sources) in the production of a (bio)material. An example is filamentous phage, which is versatile, tailorable and robust as a promoter of liquid crystalline phase formation (Reed et al., 2009), and therefore can drive self-assembly of a material. If the tendency to self-assemble is sufficiently strong, filamentous phage – or, significantly, related genetically engineered (fusion) phagemids – can serve as a host that guides the self-assembly of complex structures, including scaffolds that promote the patterned alignment and growth of cells (Rong et al., 2008). In this context, the fusion phagemid accomplishes molecular biomimicry by stabilizing liquid crystalline order in the same way that a native filamentous phage would. It also fulfills a functional biomimicry role in being biodegradable, and thus a disposable component of the repaired tissue.

Process biomimicry arises when Nature provides inspiration for the *pathway* or *mechanism* by which atoms and molecules acquire their final position and orientation in the (bio)material. For example, consider the fact that the natural silk produced by spiders and some insect larvae is spun from aqueous solution, although the final product is water-insoluble. This change in solubility is accompanied by changes in the folded shape (conformation) of the constituent protein chains, but the *chemical* nature (configuration) of the chains remains unaltered. The water-soluble protein molecules secreted by the silk glands are globular; they aggregate into a processable liquid crystalline fluid within the ducts leading from the glands to the spinnerets, and thence undergo shear-nucleated reorganization into a solid fiber stabilized by water-insoluble crystals (Viney, 1997). Nature thus provides a blueprint for the environmentally benign processing of polymer fiber. Silk will feature as a recurring example throughout this chapter, as a versatile source of principles to illustrate various facets of biomimicry. It will also provide the topic for a case study (see Case Study 2).

Structure biomimicry happens when Nature's inspiration guides the pattern of atomic and/or molecular organization – i.e., the *nanostructure* and *microstructure* – that is obtained in a (bio)material. Nature is particularly adept at achieving *multifunctionality*, i.e., the simultaneous optimization of more than one useful property in its materials, by the use of hierarchical structure to allow the independent control of properties that are a consequence of structure at different length scales (Dunlop and Fratzl, 2010).

An example of multifunctionality arising from structure is provided by the red "sweat" (which is really an oily secretion, as distinct from the watery excretion that is usually implied by the term "sweat") of the hippopotamus (Reed et al., 2009). The fluid sweat contains two types of liquid crystalline spherulite. Some of the spherulites have a fine substructure on a scale that is effective at scattering light, so that the sweat performs well as a sunblock; the other spherulites, which are microstructurally coarser, reduce the viscosity of the sweat and thus promote its ability to spread on the hippo's skin. Both of these considerations could be useful in the design of new skincare products – especially if combined with molecular biomimicry of the red sweat, which contains molecules that are effective as sunscreens (Saikawa et al., 2004) in addition to their ability to self-assemble a sun-blocking microstructure.

A second opportunity for structure biomimicry is obtained by again referring to silk – in this case the fibers spun by spiders from the secretion of their major ampullate glands to form the principal component of the drag line and the radial threads of a web. The nanostructure of this silk contains a bimodal size distribution of crystalline regions (Trancik et al., 2006). The larger crystals promote the high initial stiffness of the fiber when it is deformed in tension; however, their imperfect internal structure prevents them from acting prematurely as stress concentrators that could accelerate failure. The smaller crystals do not contribute significantly to initial stiffness, but they provide stability to the material during plastic deformation. Attempts to achieve a similarly optimal combination of mechanical properties in other materials can be usefully guided by this type of microstructure.

The different types of biomimicry identified above are collated in Figure I.2.18.1, in the context of a single biological material: actin. Particular features of actin that merit attention for possible biomimicry are noted in green. Possible benefits of such biomimicry are noted in blue. While actin is not used on its own as a biomaterial, it offers broadly applicable lessons for materials and biomaterials design.

THE ORIGINS OF BIOMIMICRY

The classification scheme outlined above emphasizes that Nature can provide useful guidance regarding optimization at all levels of the complex relationship between the composition, synthesis, assembly, structure, properties, and performance of a (bio)material.

In the broader context of engineering, biomimicry can extend beyond materials to entire devices, as exemplified by Leonardo da Vinci sketching birds with the objective of studying and eventually emulating their ability to fly (Zanon, 2009). Attempts at device or systems biomimicry continue today in a biomedical context, with the development of artificial tissues and organs which may incorporate several different biomaterials. However, for the sake of tractability, the present discussion will focus

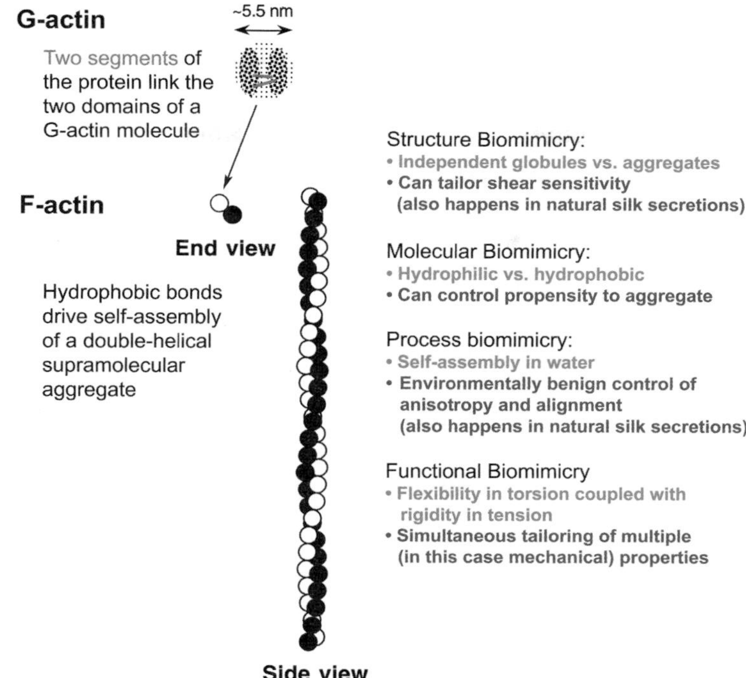

FIGURE I.2.18.1 Actin as a versatile source of lessons in biomimicry. The structural source of torsional flexibility is highlighted in red. Particular features of actin that merit attention for possible biomimicry are noted in green. Possible benefits of such biomimicry are noted in blue. All four classes of (bio)materials biomimicry are represented.

on Nature's lessons that we may wish to mimic in the production of a single biomaterial.

Biomimicry Involves Transferring Nature's Technology to Solve Engineering Problems

In biomimicry, we recognize, adopt and adapt Nature's tried-and-tested principles to solve engineering problems (Bar-Cohen, 2011). Biomimicry can be invoked broadly as a proof of concept, for example. Nature demonstrates that powered flight is possible (birds and insects can fly), that fibers can be spun from organic feed-stock (silkworms and spiders do it), and that nanomachines can be built (flagellar motors exist). Biomimicry is increasingly used to guide design in areas as diverse as architecture (e.g., to optimize airflow and minimize heat loss in buildings), telecommunications (e.g., to plan networks that maximize information flow while minimizing infrastructure), and transportation (e.g., to design vehicle profiles that minimize drag).

In the 17th century, Robert Hooke speculated in his book *Micrographia* as to the possibility of producing fibers artificially, based on what could be learned from studying silkworms:

"And I have often thought, that probably there might be a way found out, to make an artificial glutinous composition, much resembling, if not full as good, nay better, than that Excrement, or whatever other substance it be out of which, the Silk-worm wire-draws his clew. If such a composition were found, it were certainly an easier matter to find very quick ways of drawing it out into small wires for use. I need not mention the use of such an Invention, nor the benefit that is likely to accrue to the finder, they being sufficiently obvious. This hint therefore may, I hope, give some Ingenious inquisitive Person an occasion of making some trials, which if successful, I have my aim, and I suppose he will have no occasion to be displeas'd."

At the time when these thoughts were first being expressed, it was supposed possible for artificial silk fiber to be spun from plant matter, since silkworms feed on mulberry leaves. Motivated by the goal of inexpensive functional biomimicry of silk, catalyzed by misconceptions of the process biomimicry of silk, but lacking any correct knowledge of the molecular biomimicry of silk, experimenters spent the next 200 years trying to draw fibers from solutions or suspensions prepared from wood pulp and similar plant-sourced raw materials. Thus the first manufactured organic polymer fiber was rayon, reconstituted from solutions of cellulose (initially mulberry bark), born from attempts to emulate what we now know to be a protein-based natural fiber.

As a concept, biomimicry has been around for a long time. As an academic discipline it is sufficiently new to not yet have been codified (or limited) by minutely defined rules and principles. It is destined to remain very much a work in progress, as we become more adept at *recognizing* what Nature does as well as *applying* what Nature does. Thanks to our burgeoning insights into the molecular basis of life, the molecular biomimicry needed to underpin rapid progress in biomimetic biomaterials is within our grasp.

SOME ATTRACTIONS OF BIOMIMICRY

The Big Picture

Some aspects of biomimicry have a broad appeal throughout the context of applied biomaterials. These universally desirable attributes include:

Biocompatibility. A biomaterial that is produced with significant input from molecular and structure biomimicry will necessarily achieve a high degree of functional biomimicry. In particular, we can expect the surface properties of a "closely" biomimetic material to be comparable to those of the corresponding natural material. Thus, a closely biomimetic material is also likely to be biocompatible, if it is used within an environment that is comparable to the native environment of the material being mimicked. *These related consequences are perhaps the most attractive feature of biomimicry in the context of applied biomaterials.*

Integration with Biotechnology. An additional benefit of biocompatibility is that the metabolic pathways of living organisms can be harnessed via the techniques of biotechnology to obtain controlled purity, economic scale up and simple convenience in molecular biomimicry. We have already encountered this advantage implicitly (see section "A Classification of Biomimetic Materials") as it pertains to fusion phagemids. Our recurring example of silk is also relevant here, as various transgenic hosts – including bacteria, goats, and potatoes – are being explored as possible sources of silk-based biomaterials (Scheller et al., 2001; Vendrely and Scheibel, 2007).

Environmental Impact. The opportunities for biocompatibility that go hand-in-hand with biomimicry extend beyond integration with biotechnology. Nature offers lessons in how to select raw materials, adopt processes, and even limit the generation of waste products in ways that use molecules and energy efficiently, and that are environmentally benign. The case study titled "Echinoderm Collagens and Fiber-Reinforced Composites as Examples for Biomimicry" (see Case Study 1) explores a detailed example of Nature's superior efficiency in materials use.

As materials engineers become increasingly concerned about whole lifecycle design (the sourcing, processing, deployment, decommissioning, and disposition of materials), so consideration of the options for recycling become more pressing, and materials that are compatible with the recycling that occurs in Nature become especially desirable. Again, silk provides an instructive example. Spiders have to be able to depend on (and from!) their dragline as a component of webs, and as a means for making controlled vertical descents. Yet, spiders can eat silk that has served its mechanical purpose; they produce digestive enzymes that allow them to recycle the silk proteins into their constituent amino acids. In this way, the spiders cut down on the amount of new dietary protein that they have to trap in order to maintain their webs and their mobility. The reader is referred to Case Study 1: Echinoderm Collagens and Fiber-Reinforced Composites as Examples for Biomimicry.

A Specific Snapshot: Proteins as Versatile Fiber-Forming Materials

A biomaterial may or may not have an immediate counterpart in Nature to serve as a template for biomimicry. However, if it is desirable for the biomaterial to have a fibrous structure (for example, in an engineered biomaterial scaffold that supports cells during tissue reconstruction), then there are various ways in which biomimicry can guide the design and manufacture of the fibers. Nature is highly adept at producing fibers by self-assembly from solutions of protein (Renuart and Viney, 2000). We have already noted (see section "A Classification of Biomimetic Materials") how some natural fibers can capitalize on a hierarchical microstructure (i.e., microstructure at different length scales) to simultaneously optimize more than one mechanical property. Another illustration of this capability is afforded by actin (Viney, 2001; Figure I.2.18.1), which exhibits both rigidity in tension and flexibility in torsion: the tensile rigidity is maintained by the intermolecular forces that bind G-actin globular protein chains into rodlike F-actin aggregates, while the torsional flexibility is conferred by the internal structure of the G-actin globules.

Successful fiber self-assembly usually requires molecules that can interact with each other as "appropriate" building blocks, an environment that facilitates processing, a structural anisotropy, and intermolecular interactions (usually secondary bonds, but in some cases covalent cross-links) that stabilize the assembled fiber. All of these requirements can be met by proteins, and are demonstrated in the diversity of fibrous structures that Nature self-assembles. The versatility and ubiquity of proteins in providing stable, self-assembling, anisotropic structures is illustrated in Figure I.2.18.2.

Variety. The unique properties of a given protein depend on which amino acids are present, the relative amounts of those amino acids, the specific sequence in which they are arranged, and the overall length of the molecule. In synthetic (non-natural) protein materials, the amino acids do not have to be restricted to ones that are found in naturally-occurring protein (for example, under contrived conditions, *p*-fluorophenylalanine, trifluoroleucine, norleucine or selenomethionine can be incorporated into protein chains in place of phenylalanine, leucine, isoleucine or methionine respectively). Also, following protein synthesis, some of the amino acid side groups may undergo derivatization, such as occurs naturally in the conversion of proline to 4-hydroxyproline and lysine to 5-hydroxylysine during the biosynthesis of collagen. The variety of possible natural proteins to use as building blocks is therefore extremely large, and the variety of possible synthetic proteins is even larger, limited only by the practicalities (ultimately, kinetic considerations) of getting long molecules to fold into a conformation that ensures the desired properties.

It could be reasonable to question the value of incorporating non-natural amino acids into biomimetic

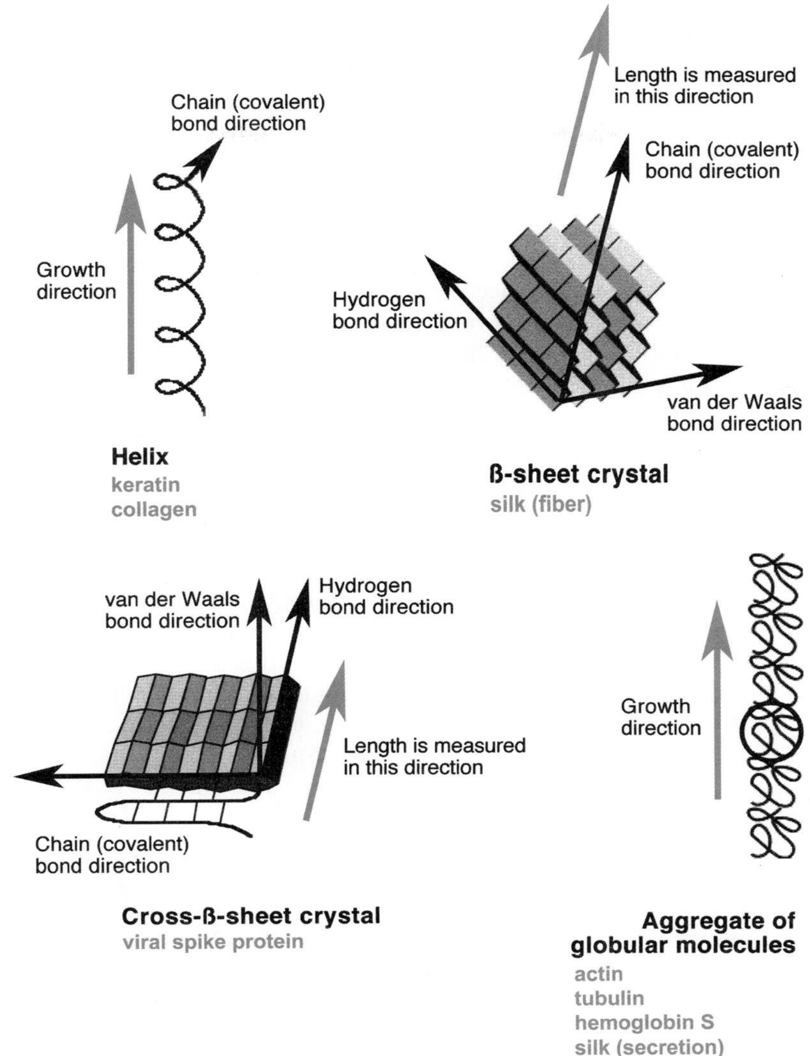

FIGURE I.2.18.2 Examples of stable, self-assembling, anisotropic structures formed by proteins. Red arrows emphasize the direction in which anisotropy is propagated.

synthetic proteins, on the grounds that the non-natural amino acids would surely detract from the biomimetic properties of the product. However, it is important to remember that biomimicry is a process of being *guided* by lessons from nature; it is not merely a process of duplicating outcomes. We will return to this thought in the section "Limitations of Biomimicry."

Processability. Two characteristics of protein molecules underlie their versatile fiber-forming ability:

(1) The backbones are intrinsically flexible, so proteins have the potential to fold into a large variety of possible conformations.

(2) The side groups of the constituent amino acids confer a locally hydrophilic (acidic, basic or neutral) or hydrophobic character to the molecules. In aqueous environments, the charge distribution along the molecules will therefore be sensitive to the presence of other charged species, and so will depend on pH and dissolved ions. The flexible protein chains will tend to fold in a way that allows hydrophobic segments to be screened from the water by the more hydrophilic segments, while also tending to maximize the extent to which charged and polar sites participate in hydrogen bonding.

Structural Anisotropy. The intrinsic flexibility of proteins allows them to form *geometrically anisotropic* structures such as helices or β-sheet crystals. The stability of these structures is sensitive to the amino acid sequence of the protein, and to the pattern of intramolecular and intermolecular bonding favored by interactions between particular residues or sequences along the protein chain as well as between the protein and the surrounding medium.

Even protein molecules that have a *globular* conformation in water can also form geometrically anisotropic structures, in this case by *aggregating*. This is the process by which F-actin assembles from G-actin, and by which microtubules assemble from α- and β-tubulin. It is also (as noted in the section "A Classification of Biomimetic Materials" above) complicit in the formation of a processable fluid from solutions of globular fibroin during the natural spinning of silk fibers. The aggregation of globular molecules into anisotropic structures is promoted by liquid crystalline order in the protein solution: synergism develops between alignment-enhanced assembly and shape-enhanced alignment of the aggregates.

In an aqueous environment, helices can form superhelices (coiled coils) stabilized by hydrophobic bonding across the area of helix–helix contact. This is how procollagen, the intracellular precursor of extracellular collagen, is assembled into triple helices. Likewise, hydrophobic bonding stabilizes cross-β-sheets into hollow trimeric box beams (resembling the packaging of Toblerone® chocolate), which serve as the spikes on human adenovirus, and are noted for their high stiffness and yield strength (Hoess et al., 1992).

Every fibrous product found in Nature, and the process by which that product is made, has the capacity to inspire biomimicry in a fibrous biomaterial. The lessons from Nature represent a transferable technology that may potentially be capitalized on – in whole or in part, depending on the particular application being considered. The reader is referred to Case Study 2: Silk as an Example of a Biomaterial for Biomimicry.

A Blueprint for Progress in Design

Biomimicry offers a blueprint for the engineering design process itself. In Nature, evolution drives innovation (and innovation drives evolution) via a cycle of events that can be summarized as follows:

- One or more problems, needs or opportunities arise in response to continuously and/or suddenly changing circumstances.
- Multiple possible solutions are spontaneously and simultaneously explored.
- Appropriate solutions are selected.

New and more pressing problems, needs or opportunities can arise at any stage. Any existing design embodies a history of successful response to changing circumstances.

Thus, in Nature, the emergence of problems and the generation of solutions occur side-by-side, allowing a seamless response. In traditional engineering, solutions follow problems sequentially and therefore often discontinuously. To the biomaterials engineer, Nature offers a palette of ready-made solutions, waiting for the right problem or need to be expressed. Biomimicry, most simply, is a process of asking Nature the right question, to capitalize on the diversity and optimization of solutions that have evolved over time.

Validation of Non-Biomimetic Technology

The terms "biomimetic" and "biomimicry" conventionally refer to lessons that are *proactively* learned from studying Nature's materials, processes, and devices. In other words, Nature provides the impetus for discovered knowledge and adapted technology. However, it can be reassuring to discover that a specific product developed without recourse to active biomimicry is, with *hindsight*, biomimetic in one or more of its features. We can in such cases be confident that the underlying design really is sound.

In the 1960s, the idea of using a polymer gel network to absorb several hundred times its own weight in water was developed by the United States Department of Agriculture (USDA) into a product for boosting water conservation in agriculture, and subsequently adopted for use in diapers, surgical pads, and wound dressings. The fact that slugs and snails have evolved a highly effective, multifunctional surface coating (mucus) based on the properties of polymer gel networks was not known in such molecular detail at the time of the USDA work, but the albeit unintentional biomimicry does suggest that polymer gel formation is indeed a particularly efficient way to retain absorbed water. Also, there is the opportunity to build on this serendipitous biomimicry to enhance the biocompatibility and biodegradability of personal care products that contain superabsorbent polymers. The ability of secreted mucin granules to swell rapidly on contact with water (Verdugo et al., 1992) is felicitously emulated in superporous hydrogels (Omidiana et al., 2005) that have applications in drug delivery and other technologies where a combination of high surface area and fast mass transport is desirable.

LIMITATIONS OF BIOMIMICRY

Although biomimicry should be a tool that all biomaterials engineers keep close to hand, it should never be the only tool in the box. As is also the case with any other tool used by engineers, an understanding of its limitations is a necessary prerequisite if the tool is to be used appropriately and effectively. Biomimicry is not a biomaterials design panacea. We will therefore explore some of the most important misconstrued and real limitations of biomimicry.

Toxicity of Many Natural Materials. The section "The Big Picture" identifies biocompatibility as one of the broadly appealing consequences of biomimicry. Yet, many naturally produced materials and processing byproducts are toxic. There is no inconsistency here, because biomimicry properly involves recognizing *appropriate* lessons that Nature offers, and applying them *in the proper context*. Also, when nature does produce toxins, these are usually selective in their action, and they rarely persist in the environment.

Copycats have Fewer than Nine Lives. "Biomimicry" (denoting a learning and design process) and "biomimetics" (denoting a field of study in which the lessons and designs are applied) are perhaps unfortunate terms in that they may foster an impression that Nature is being copied – an overly literal interpretation of mimicked – without thought being given to the value or limitations or consequences of such copying. However, just like inter-species mimicry in nature, biomimicry is properly a process of *selective* learning and copying. Biomimicry is about being inspired and guided by Nature's processes and products; it is not a synonym for bioduplication.

We have previously noted (see section "The Origins of Biomimicry") the debt that powered flight owes to observations made on birds, dating back at least as far as the work of Leonardo da Vinci. This is an excellent example of what biomimicry can lead to by alerting engineers to a *possibility*. Nevertheless, people did sometimes get lost along the path to our present-day success in aviation (and our ability to fly heavier loads much faster than birds can), by straying too far away from bioinspiration and too close to bioduplication. Useful flying machines do not work by flapping their wings up and down, but by capitalizing on Nature's lesson that a wing with variable curvature is the key to a workable trade-off between lift and drag over a range of speeds. Similarly, successful biomimetic materials are those where "what if" questions are informed by knowledge of biology and the answers are open to refinement from any pertinent field of expertise.

The inspirational component of biomimicry can be highly significant for the (bio)materials engineer. The idea that you *can* make adhesives capable of setting in an aqueous environment (inspired by mussel and barnacle "glue") or biodegradable lubricants (inspired by slug mucus) or superhydrophobic surfaces (inspired by lotus leaves) or photonic materials with non-fading and perhaps cosmetically useful color (inspired by butterfly scales), is what matters.

Nature's Best May Not Be Good Enough. Natural materials are impressive in many ways. However, natural materials are not always or even necessarily superior to artificial ones; indeed there is no intrinsic reason why they *should* always be superior. Nature and technology serve different goals, usually under very different constraints (Vogel, 2003).

- In Nature, the chemical building blocks of materials are synthesized from relatively few types of atom – most of the periodic table does not feature in the ingredients of natural materials, and, as far as we know, natural organisms do not make metallic materials. If the objective of a biomaterial is to replace like with like, then the absence of metals in Nature's materials toolbox isn't an issue. However, if a biomaterial is expected to do a better job than the material that it replaces, it may be necessary to look beyond Nature for a solution. A good example here is Charnley's hip (Charnley, 1976), which combines metal and synthetic polymers to produce a highly successful implant. The result certainly is functionally biomimetic – it behaves like a hip – although it is not a product of molecular, process or structure biomimicry.
- Nature's materials processing strategies operate at near-ambient temperature, and under near-equilibrium conditions. Thus, the same attributes that are attractive because of their low environmental impact will limit processing (production) rates compared to what might be achieved at elevated temperature and under significantly off-equilibrium conditions. Limited production rates are, however, less likely to be a critical factor in the small volume, high value context of a specialty biomaterial than they are in the high volume world of more traditional engineering materials.
- The ubiquitous presence of organic matter in natural materials precludes the use of these materials at high temperatures or in chemically aggressive environments. While this limitation can be a deal-breaker concerning the use of biomimetic materials in many high-tech applications, it is not a major consideration in the context of biomaterials.

Natural Materials Do Fail. Indeed, this is why the need for biomaterials arises in the first place! Here we encounter another divergence between the goals of Nature and the goals of technology. In Nature, where survival of the species is more important than the survival of any individual, failures are tolerated as being part of (indeed necessary for) the process of evolutionary improvement. It is acceptable to lose a few or even many individuals, as long as there are enough survivors. A related approach was adopted in medieval battles – not just in treating soldiers as arrow fodder (provided that one had *enough* soldiers), but also in the practice of using decoys to confuse the enemy ("I think there be six Richmonds in the field: Five have I slain today instead of him" – William Shakespeare, *Richard III*, Act 5, Scene 4).

Although failure can also be regarded as essential to successful engineering design (Petroski, 1992), engineers are also deeply concerned about survival of the individual. The goal of engineering is that *no* machines will break, *no* bridge will fall down, *no-one* will be injured or killed. On that basis, an engineer may be reluctant to replace a failed piece of human tissue or a worn-out body part with something that suffers from the same weakness, wishing instead to implant something more durable. However, the engineer would also be wise to exercise *selective* biomimicry, learning from those molecular, processing, and structural aspects of the original material that are successful, and attempting improvements only where necessary.

Wonders and Woes of Water as a Processing Medium. Nature's elegant solution for spinning water-insoluble fiber from aqueous solutions of silk protein (see section "A Classification of Biomimetic Materials") has a downside with respect to the properties of the fiber. The hydrophilic amino acid sequences responsible for the aqueous solubility of the protein (in its globular conformation) are retained in the product. Thus, although the fiber microstructure is stabilized by water-insoluble β-sheet crystals (formed by hydrophobic sequences that are exposed to water when the globular molecules are denatured by shear), the fibers still absorb water from their environment. The resultant plasticization (increased molecular mobility, due to hydrogen bonds between protein chains being replaced by hydrogen bonds between the chains and absorbed water molecules) leads to the fibers having poor creep resistance. While much attention has been garnered by the tensile strength and stiffness of silk, the long-term performance under load is unspectacular. The use of silk in biomaterial scaffolds, capitalizing on biocompatibility, biodegradability, ease of processing, and not requiring high loads, is therefore more successful than deployment in critical loadbearing situations.

Hydrophobic bonding between cross-β-sheets stabilizes the trimeric box beam structure adopted by adenovirus capsid spikes (see section "A Specific Snapshot: Proteins as Versatile Fiber-Forming Materials"), contributing to the ability of those structures to resist deformation under axial compression. Hence the spikes function well, as a nanoscale battering ram in broaching the outer wall or membranes of a target cell, and as a nanoscale syringe in delivering virus DNA into the target. Because high compressive strength and stiffness (singly or in combination) have been a long-term but elusive goal of polymer science, attempts were made to spin fiber from solutions of recombinant spike protein. For ease of dissolving the protein and convenience in spinning the fibers, 1,1,1,3,3,3-hexafluoro-2-propanol was used as the solvent (O'Brien et al., 1994). However, the mechanical properties of the resulting fiber were disappointing, and the trimeric box beams were not found in the hierarchical microstructure of the fibers (Gillespie et al., 1994). There are several possible reasons for this failure, illustrating the pitfalls of attempting biomimicry outside of an appropriate context: (1) spinning fibers artificially under non-equilibrium conditions will not necessarily allow time for the assembly of structures resembling those formed naturally under near-equilibrium conditions; (2) the driving force to form hydrophobic bonds will be lower in 1,1,1,3,3,3-hexafluoro-2-propanol than in water; and (3) even if the fibers had been processed from an aqueous environment under near-equilibrium conditions, there is no *a priori* reason to suppose that a structure stabilized by hydrophobic bonds would persist in samples that are being evaluated for their mechanical properties in air. Simply put, the natural material is designed to be assembled and to perform in an aqueous medium, and attempts to mimic its properties must take this reality into account.

THE FUTURE OF BIOMIMICRY IN BIOMATERIALS SCIENCE

The examples presented in this chapter have focused mainly on fibrous and fiber-reinforced materials. On Nature's scale of achievements, the object lessons chosen here have been relatively simple. More challenging examples await detailed study by (bio)materials scientists. A case in point: insect antennae are self-assembled fibrous structures too. They are also small, mechanically robust, self-repairing, responsive to chemical and thermal information, and able to undergo controlled and rapid changes in shape and orientation – all of which are attributes that could find application in biomaterial sensors and actuators.

Nature's lessons do not stop there. Self-assembly is used by Nature to produce *all* its materials for structures and devices. The principles explored in this chapter are also reference points from which to venture into the emerging subject of Nature's nanomotors (Goel and Vogel, 2008) to consider their assembly and repair, their ability to transduce energy, and their connectivity to functional structures such as cilia, flagella, and antennae. We can imagine a future in which some biomaterials are assembled *in situ* without the need for surgery, and in which parallel advances in nanotoxicology (Shvedova et al., 2010) permit functional, process, and structure biomimicry of natural nanomotors with synthetic nanomaterials.

We can also look forward to learning (bio)materials lessons from species that remain to be characterized in detail – or are yet undiscovered. Although we have been able to summarize some biomimicry principles, we must recognize that these are subject to continuous refinement as we gain further insights into the opportunities and constraints of biomimicry. The principles gleaned thus far should not be regarded as immutable rules, but as guidelines that are often useful but not invariably right. We are still a long way from unlocking the full potential of biomimicry or from reaching its limits.

ACKNOWLEDGMENT

DEG's contributions to this chapter are adapted from a graduate class assignment at the University of California, Merced. The authors are grateful to Prof. Lilian Davila for many helpful discussions that accompanied the assignment.

WORKED EXAMPLE

Question

The petiole (waist) of a wasp is a narrow constriction occurring between the first and second abdominal segments. In some species, for example the mud dauber *Sceliphron caementarium* (Hymenoptera: Sphecidae), the petiole is approximately cylindrical. What biomaterials-related lesson(s) might one hope to learn from studying the microstructure of a wasp petiole?

Solution

A number of tissues and functions critical to the wasp's ability to live must pass through this effective structural bottleneck. The petiole has to accommodate and protect the alimentary canal, nerve tissue, an aorta, and open circulatory functions that communicate between the front and rear of the wasp.

From these requirements, we can deduce that the cuticular material making up the petiole wall must be crush-resistant, puncture-resistant, and fracture-resistant. Such characteristics are indicators that the cuticle has a high fracture toughness, in turn suggesting that the material is multilayered (lots of interfaces to prevent or deter crack propagation), and that there may also be some sort of composite structure *within* layers. The cuticular material must also confer stiffness and resilience in bending, especially when the load on the petiole increases as a consequence of the wasp using its stinger to immobilize prey. All these properties must be achieved without the material contributing unduly to the weight of the insect. Thus, we might expect to learn particular lessons pertaining to the hierarchical structure and self-assembly of biocompatible tubes that are optimized in regard to multiple mechanical properties. More broadly, the petiole presents a model system from which to learn lessons about: (1) the relationship between efficient use of material, molecular organization, and mechanical property optimization in the context of a simple object geometry; and (2) the factors that limit molecular and energy transport when such an object is self-assembled.

Studies performed by scanning electron microscopy and transmission electron microscopy (Reed et al., 2007) have revealed several details about the structure of the mud dauber wasp petiole, including:

- The cuticle is indeed multilayered, consisting of a large number (43) of concentric layers.
- The inside surface of the petiole contains ribs of tissue that run either longitudinally (on the dorsal half) or circumferentially (on the ventral half). These different orientations of structural reinforcement help to resist buckling when the ventral surface is in compression, while increasing the restoring force that opposes longitudinal extension of the dorsal surface – a useful response to the loading conditions that would be experienced while the wasp stings its prey.
- There are spiral strands of fibrous, crystalline chitin within the individual concentric layers of the petiole. The diameter of the spiral occupies the perpendicular distance between layers, so that the petiole has a structure related to that of multi-layer corrugated cardboard. However, whereas the corrugated "medium" in cardboard confers a highly anisotropic resistance to bending (sheets are stiff if bent around an axis running normal to the corrugations, but not if bent around an axis running parallel to the corrugations), the coiled medium within the layers of petiole will impart a resilient response to bending in any direction.

Collectively, the above observations are consistent with the exceptional ability of cuticle to undergo large, multiaxial, reversible deformations without catastrophic failure and while maintaining a "respectable" stiffness-to-weight ratio.

Case Study 1 Echinoderm Collagens and Fiber-Reinforced Composites as Examples for Biomimicry

We will consider here one of Nature's lessons for optimizing the amount of material (fiber) used to achieve a particular property (reinforcement of a fibrous composite). This particular lesson has been gleaned from studies of both sea cucumber dermis and the "catch" ligament associated with the ball-and-socket joint at the base of sea urchin spines. Both materials are among the living world's oldest fibrous composites, so their survival is in itself a testimony to their successful design. Interest in these materials also stems from the fact that their mechanical properties are adaptive: their stiffness can be quickly and reversibly changed in response to external stimuli. Therefore, they are serving as models for new biomaterials that can be used as adaptive substrates for intracortical microelectrodes (Capadona et al., 2008).

The collagen fibers in sea cucumber dermis and sea urchin ligament are *tapered* rather than cylindrical. The shape is ensured by the nucleation and growth mechanism of supramolecular self-assembly (Trotter et al., 2000) by which the fibers are formed (Figure I.2.18.3).

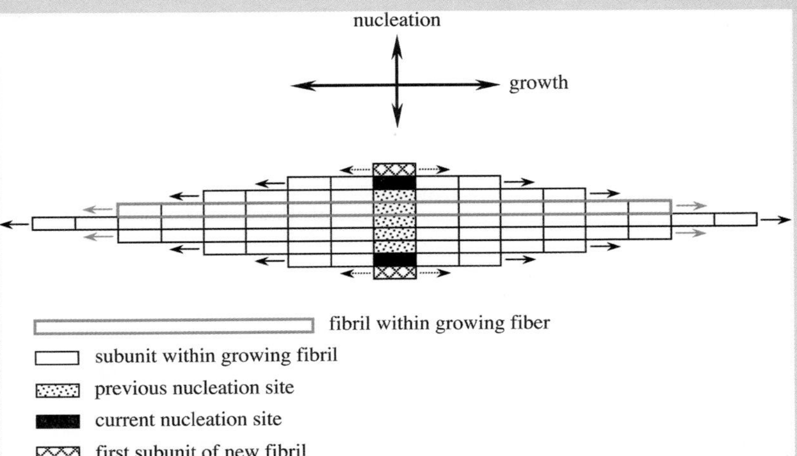

FIGURE I.2.18.3 Mechanism of collagen fiber growth in sea cucumber dermis and sea urchin ligament, based on literature descriptions (Trotter et al., 2000). The mechanism ensures both a tapered shape and (for a fiber consisting of a sufficiently large number of parallel fibrils) a consistent axial ratio.

Case Study 1 Echinoderm Collagens and Fiber-Reinforced Composites as Examples for Biomimicry—*cont'd*

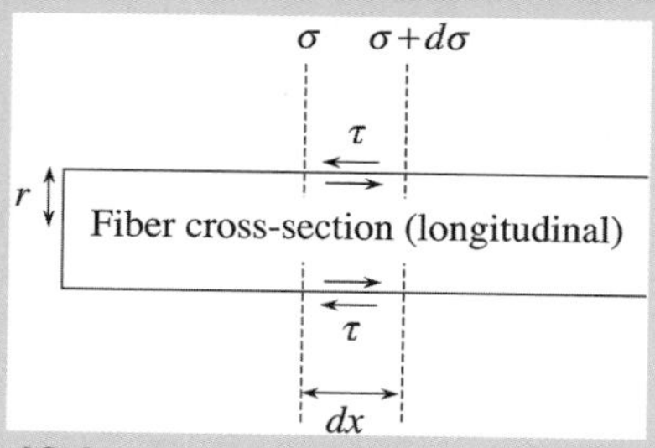

FIGURE I.2.18.4 Schematic description of the geometry and variables needed to set up Equation (1).

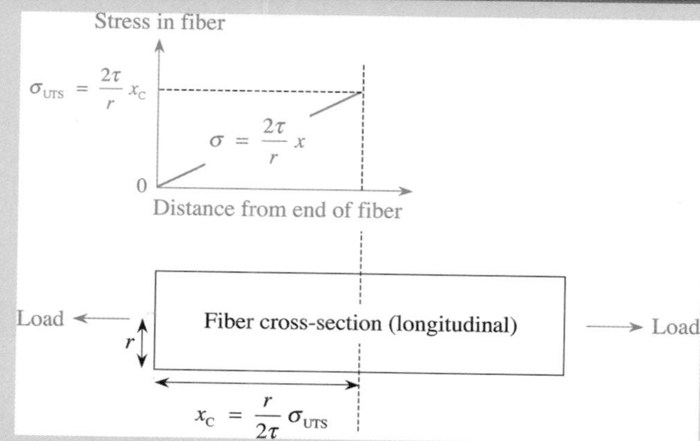

FIGURE I.2.18.5 Schematic representation of a critical length of fiber in longitudinal cross-section. A plot of internal stress versus distance from the fiber end, obtained from Equations (3) and (4), is superimposed.

We will explore how this shape reduces the volume of fiber that is not exploited at close-to-maximum loadbearing capacity, and that therefore would have been a waste of resources to produce.

In the case of conventional reinforcing fibers in a composite, the fiber cross-section is approximately constant at all points along the length of any given fiber. If, as is expected, the fibers deform less readily than the matrix, then load is transferred from the matrix to the fibers through the action of the resulting shear stress that develops at the fiber–matrix interfaces. Ideally, discontinuous fibers will be long enough for the stress at their midpoint to approach the fiber failure strength.

To quantify the nominal tensile stress in the fibers as a function of increasing distance from either end, we will proceed with some simplifying assumptions: (1) the fibers are discontinuous and perfectly aligned; (2) a tensile load is applied to the composite along a direction parallel to the length of the fibers; and (3) the fibers are cylindrical and have radius r.

Figure I.2.18.4 designates a small reference length dx near to one end of a fiber. The tensile stress in the fiber increases by $d\sigma$ over this distance. The increase in tensile stress is achieved by the interfacial shear stress τ acting on the area of fiber–matrix interface accommodated within the distance dx. A simple force balance (Martin, 2002) now gives:

$$d\sigma(\pi r^2) = \tau(2\pi r dx) \tag{1}$$

$$\therefore d\sigma = \frac{2\tau}{r} dx \tag{2}$$

Overall, the tensile stress increases from $\sigma = 0$ (at $x = 0$, i.e., the end of the fiber) to $\sigma = \sigma_1$ (at $x = x_1$, measured from the end of the fiber). We can integrate Equation (2) and substitute these limits:

$$\int_0^{\sigma_1} d\sigma = \frac{2\tau}{r}\int_0^{x_1} dx$$

$$\therefore \sigma_1 = \frac{2\tau}{r} x_1 \tag{3}$$

Thus, the tensile *stress in the fiber increases linearly as a function of distance* from the end of the fiber. The distance from the end of the fiber to the point where the stress has attained its maximum is called the *transfer length*. Given a sufficiently long fiber, the maximum stress can reach the ultimate tensile strength σ_{UTS} of the fiber. In that case, the transfer length is known as the *critical transfer length* (which we will denote as x_C), and the corresponding overall length of the fiber (twice the critical transfer length, since the fiber has two ends) is known as the *critical length*. If the fiber is shorter than the critical length, then it cannot live up to its full potential to carry load transferred from the matrix. If it is longer than the critical length, then it will break, and so the additional material does not contribute an ability to carry additional transferred load.

We can substitute $\sigma_1 = \sigma_{UTS}$ and $x_1 = x_C$ into Equation (3), and rearrange to obtain a useful expression for the critical transfer length:

$$\sigma_{UTS} = \frac{2\tau}{r} x_C$$

$$\therefore x_C = \frac{r}{2\tau}\sigma_{UTS} \tag{4}$$

Figure I.2.18.5 schematically shows a critical length of stressed fiber in longitudinal cross-section, together with a superimposed plot of the internal (tensile) stress versus distance from the fiber end. It is apparent that much of the material in the fiber is wasted, in that its tensile strength is not being properly exploited along almost the whole length of the fiber! The load near the ends of the fiber could be carried adequately by a smaller fiber cross-section, compared to the cross-section needed to carry the load near the middle of the fiber. A less wasteful use of material, and a more efficient exploitation of the fiber properties, is therefore achieved with a fiber that *tapers* from the middle towards the ends.

Case Study 2 Silk as an Example of a Biomaterial for Biomimicry

Functional Biomimicry

Silk has multiple biomimetic uses. Aqueous solutions of silkworm fibroin (silk protein) have been processed artificially into materials that mimic the properties of native silk (Jin and Kaplan, 2003). Aqueous processing has also been used to produce prototypical silk lenses and diffraction gratings from solutions of silk fibroin; their transparency is ensured by controlling the crystallization of β-sheets, and the products benefit from the mechanical properties and biodegradability of silk (Lawrence et al., 2008).

Decorating reconstituted silk networks with peptides and hormones can create bioactive surfaces with numerous possible biomaterials

Case Study 2 Silk as an Example of a Biomaterial for Biomimicry—*cont'd*

applications that include use as a scaffold on which to culture tissues such as bone (Sofia et al., 2001). Because silk is biodegradable, its use as a cell and tissue scaffold can avoid any need for subsequent surgical removal. The biodegradability of silk fibroin scaffolds has been tested (Wang et al., 2008), with degradation times being controlled by the processing environments. Biodegradation also makes silk suitable for use as a controlled release drug delivery vector (Hardy et al., 2008).

Processing

Sericin (the glue-like protein that binds fiber to maintain the integrity of silkworm cocoons) can be removed by heating the cocoons in an aqueous solution of at least 0.1% (w/v) Na_2CO_3 for an hour at a temperature ranging from 60°C to 90°C, followed by rinsing in distilled water and drying at room temperature. There are many variants of this degumming procedure.

The degummed silk is then dissolved in a concentrated aqueous solution of LiBr (9.3–9.5 M) at 60°C and dialyzed against pure water. Subsequent lyophilization yields a powder that can be manipulated into a variety of forms that include electrospun fibers, films, microparticles, and three-dimensional scaffolds.

Example 1: Microparticles

Microparticles and nanoparticles are attractive vehicles for drug and growth factor delivery. Silk microparticles are formed by dropwise addition of either DMSO or ethanol to reconstituted silk fibroin solutions. The resulting particles are then collected by centrifugation, rinsed with deionized water, and lyophilized once more to obtain the product (Bessa et al., 2010).

Example 2: Scaffolds

It is possible to form scaffolds by various techniques that include salt leaching, gas foaming, and freeze-drying (Nazarov et al., 2004). Scaffolds formed by gas foaming have the highest compressive strength and modulus, and they display the desired pore size for bone and cartilage tissue grafts.

Due to the ultimate strength, elasticity, and long-term biodegradability displayed by silk fibroin scaffolds, their use has been focused on bone matrix and cartilage generation, on replacements for tendons such as the anterior cruciate ligament (ACL), and on blood vessel grafts:

- Scaffold seeded with both endothelial cells and osteoblast progenitor cells can form vascularized networks surrounded by osteoblasts within four weeks (Fuchs et al., 2009).
- Human chondrocytes seeded into silk scaffolds deposit cartilage-like extracellular matrix, raising realistic prospects of cartilage replacement therapies (Wang et al., 2006).
- Cell-seeded twisted silk cord has shown promise as a replacement for damaged ACL (Altman et al., 2003): two weeks after cell seeding, silk fiber is completely encased in extracellular matrix deposited by bone marrow stromal cells.
- A silk blood vessel seeded with bone marrow cells and transplanted into the aorta of a rat showed long-term patency, without any indication of thrombosis or aneurism; over time the silk fibroin vessel walls were replaced by collagen (Enomoto et al., 2010).

BIBLIOGRAPHY

Altman, G. H., Diaz, F., Jakuba, C., Calabro, T., Horan, R. L., et al. (2003). Silk-based biomaterials. *Biomaterials*, **24**, 401–416.

Bar-Cohen, Y. (2011). *Biomimetics: Nature-Based Innovation.* Boca Raton, FL: CRC Press.

Bessa, P. C., Balmayor, E. R., Azevedo, H. S., Nürnberger, S., Casal, M., et al. (2010). Silk fibroin microparticles as carriers for delivery of human recombinant BMPs. Physical characterization and drug release. *Journal of Tissue Engineering and Regenerative Medicine*, **4**, 349–355.

Capadona, J. R., Shanmuganathan, K., Tyler, D. J., Rowan, S. J., & Weder, C. (2008). Stimuli-responsive polymer nanocomposites inspired by the sea cucumber dermis. *Science*, **319**, 1370–1374.

Charnley, J. (1976). *Principles and Practice in Hip Replacement. Proceedings of the Royal Society.* London B192, pp. 191–198.

Derwin, K. A., Badylak, S. F., Steinmann, S. P., & Iannotti, J. P. (2010). Extracellular matrix scaffold devices for rotator cuff repair. *Journal of Shoulder and Elbow Surgery*, **19**, 467–476.

Dunlop, J. W. C., & Fratzl, P. (2010). Biological composites. *Annual Review of Materials Research*, **40**, 1–24.

Enomoto, S., Sumi, M., Kajimoto, K., Nakazawa, Y., Takahashi, R., et al. (2010). Long-term patency of small-diameter vascular graft made from fibroin, a silk-based biodegradable material. *Journal of Vascular Surgery*, **51**, 155–164.

Fuchs, S., Jiang, X., Schmidt, H., Dohle, E., Ghanaati, S., et al. (2009). Dynamic processes involved in the pre-vascularization of silk fibroin constructs for bone regeneration using outgrowth endothelial cells. *Biomaterials*, **30**, 1329–1338.

Gillespie, D. B., Thiel, B. L., Trabbic, K. A., Viney, C., & Yager, P. (1994). Structural Investigation of (Ad 11)$_{26}$ Fiber, a Novel Bioengineered Material Based on a Viral Spike Protein. *Macromolecules*, **27**, 6177–6182.

Goel, A., & Vogel, V. (2008). Harnessing biological motors to engineer systems for nanoscale transport and assembly. *Nature Nanotechnology*, **3**, 465–475.

Hardy, J. G., Romer, L. M., & Scheibel, T. R. (2008). Polymeric materials based on silk proteins. *Polymer*, **49**, 4309–4327.

Hoess, R. H., O'Brien, J. P., & Salemme, F. R. (1992). *World Patent Application WO 92/09695.*

Jin, H.-J., & Kaplan, D. L. (2003). Mechanism of silk processing in insects and spiders. *Nature*, **424**, 1057–1061.

Lawrence, B. D., Cronin-Golomb, M., Georgakoudi, I., Kaplan, D. L., & Omenetto, F. G. (2008). Bioactive silk protein biomaterial systems for optical devices. *Biomacromolecules*, **9**, 1214–1220.

Martin, J. W. (2002). *Materials for Engineering* (2nd ed.). Leeds: UK. Maney Publishing (for the Institute of Materials).

Nazarov, R., Jin, H.-J., & Kaplan, D. L. (2004). Porous 3-D scaffolds from regenerated silk fibroin. *Biomacromolecules*, **5**, 718–726.

O'Brien, J. P., Hoess, R. H., Gardner, K. H., Lock, R. L., Wasserman, Z. R., et al. (1994). Design, Synthesis, and Fabrication of a Novel Self-Assembling Fibrillar Protein. In D. L. Kaplan, W. W. Adams, B. L. Farmer, & C. Viney (Eds.), *Silk Polymers – Materials Science and Biotechnology* (pp. 104–117). Washington, DC: American Chemical Society.

Omidiana, H., Rocca, J. G., & Park, K. (2005). Advances in superporous hydrogels. *Journal of Controlled Release*, **102**, 3–12.

Petroski, H. (1992). *To Engineer is Human. The Role of Failure in Successful Design.* New York, NY: Vintage Books.

Reed, E. J., Dunlap, M. R., Jasinski, J., & Viney, C. (Eds.), (2007). Microstructure, Nanostructure and Properties of the Wasp Petiole. *Materials Research Society Symposium Proceedings 975E*. Warrendale, PA, paper 0975-DD07-04.

Reed, E. J., Klumb, L., Koobatian, M., & Viney, C. (2009). Biomimicry as a route to new materials: What kinds of lessons are useful? *Philosophical Transactions of the Royal Society of London Series A*, **367**, 1571–1585.

Renuart, E., & Viney, C. (2000). Biological Fibrous Materials: Self-Assembled Structures and Optimised Properties. In M. Elices (Ed.), Chapter 8, (pp. 221–267). *Structural Biological Materials* Oxford, UK: Pergamon/Elsevier Science.

Rong, J., Lee, L. A., Li, K., Harp, B., Mello, C. M., et al. (2008). Oriented cell growth on self-assembled bacteriophage M13 thin films. *Chemical Communications*, **41**, 5185–5187.

Saikawa, Y., Hashimoto, K., Nakata, M., Yoshihara, M., Nagai, K., et al. (2004). Pigment chemistry: The red sweat of the hippopotamus. *Nature*, **429**, 363.

Scheller, J., Gührs, K.-H., Grosse, F., & Conrad, U. (2001). Production of spider silk proteins in tobacco and potato. *Nature Biotechnology*, **19**, 573–577.

Shvedova, A. A., Kagan, V. E., & Fadeel, B. (2010). Close encounters of the small kind: Adverse effects of man-made materials interfacing with the nano-cosmos of biological systems. *Annual Review of Pharmacology and Toxicology*, **50**, 63–88.

Sofia, S., McCarthy, M. B., Gronowicz, G., & Kaplan, D. L. (2001). Functionalized silk-based biomaterials for bone formation. *Journal of Biomedical Materials Research*, **54**, 139–148.

Trancik, J. E., Czernuszka, J. T., Bell, F. I., & Viney, C. (2006). Nanostructural features of a spider dragline silk as revealed by electron and X-ray diffraction studies. *Polymer*, **47**, 5633–5642.

Trotter, J. A., Kadler, K. E., & Holmes, D. F. (2000). Echinoderm collagen fibrils grow by surface-nucleation-and-propagation from both centers and ends. *Journal of Molecular Biology*, **300**, 531–540.

Vendrely, C., & Scheibel, T. (2007). Biotechnological production of spider-silk proteins enables new applications. *Macromolecular Bioscience*, **7**, 401–409.

Verdugo, P., Deyrup-Olsen, I., Martin, A. W., & Luchtel, D. L. (1992). Polymer gel phase transition: The molecular mechanism of product release in mucin secretion. In E. Karalis (Ed.), *Swelling of Polymer Networks* (pp. 671–681). Heidelberg: Springer-Verlag.

Viney, C. (1997). Natural silks: Archetypal supramolecular assembly of polymer fibres. *Supramolecular Science*, **4**, 75–81.

Viney, C. (2001). Natural Protein Fibers. In K. H. J. Buschow, R. W. Cahn, M. C. Flemings, B. Ilschner, E. J. Kramer, & S. Mahajan (Eds.), *Encyclopedia of Materials: Science and Technology* Vol. 6, (pp. 5948–5957). Oxford, UK: Pergamon/Elsevier Science.

Vogel, S. (2003). Nature's swell, but is it worth copying? *MRS Bulletin*, **28**, 404–408.

Wang, Y., Blasioli, D. J., Kim, H.-J., Kim, H. S., & Kaplan, D. L. (2006). Cartilage tissue engineering with silk scaffolds and human articular chondrocytes. *Biomaterials*, **27**, 4434–4442.

Wang, Y., Rudym, D. D., Walsh, A., Abrahamsen, L., Kim, H.-J., Kim, H. S., et al. (2008). *In vivo* degradation of three-dimensional silk fibroin scaffolds. *Biomaterials*, **29**, 3415–3428.

Zanon, E. (2009). *The Book of the Codex on Flight: From the Study of Bird Flight to the Flying Machine*. Milan: Italy: Leonardo3.

CHAPTER I.2.19 MICROPARTICLES AND NANOPARTICLES

Shalu Suri[1], Gang Ruan[2], Jessica Winter[2], Christine E. Schmidt[3].*

[1]School of Chemical and Biomolecular Engineering, Georgia Institute of Technology, Atlanta, GA, USA

[2]Department of Chemical and Biomolecular Engineering, The Ohio State University, Columbus, OH, USA

[3]Department of Biomedical Engineering, The University of Texas at Austin, Austin, TX, USA.

**Note: will have new address effective Jan. 1, 2013:* J. Crayton Pruitt Family Department of Biomedical Engineering, Biomedical Sciences Building, Gainesville, FL, USA

INTRODUCTION

Microparticles and nanoparticles have had an enormous impact on a wide-range of biomedical applications including drug delivery, imaging, and basic research. Miniaturization of therapeutic devices to the micron (1–1000 μm), sub-micron (100–1000 nm), and nanometer (1–100 nm) scales has facilitated the integration of biomedical devices with therapeutic biomolecules for improved clinical efficacy (George et al., 2005). Despite an extensive database on microparticles and nanoparticles, a clear universal boundary between nano- and microsize does not exist in the literature. Expert opinions in the micro- and nano-sciences have emphasized that 1–100 nm is the optimum nanoscale range; however, in biotechnology and medicine, the definition of "nano" is less stringent (Ferrari, 2005). Design of any miniaturized system is dependent on the endpoint application. For example, systemic (intravascular) application requires use of particles less than 500 nm in diameter, whereas for intramuscular application or in some cases oral delivery applications, particles greater than 1 micron and less than 125 μm can be easily administered (Jain, 2000). Other than size, another critical parameter that can significantly modulate the function of these particles is shape (Champion et al., 2007); shape can impact cellular uptake by immune cells, release behavior of biomolecules, and cell targeting. A variety of materials have been synthesized as micro- and nanoparticles, mostly for imaging or targeted delivery of therapeutic biomolecules such as hormones, vaccine antigens and adjuvants, peptides, and anti-inflammatory agents (Aukunuru et al., 2003; Kim et al., 2004). In addition, these particles have been used as biosensors and in affinity bioseparations, immunological assays, cell labeling, and cell sorting. Also see Chapters II.5.16.B1 to II.5.16.B9 for applications of various particulate systems in drug delivery. The following discussion focuses on preparation, characterization, and applications of microparticles and nanoparticles.

MICROPARTICLES

Microparticles have been an active area of research in the biomedical field because of their unique advantages: their small sizes (few microns) render them excellent candidates for injectable drug delivery as compared to surgical implants; high surface area allows for surface functionalization to target specific cells, tissues or organs; and the ability to carry fluorescent molecules enables their use in microscale detection of cells and cellular components (e.g., proteins, DNA). Further, magnetic microparticles have been used for cell sorting and hyperthermia applications.

Materials Used for Microparticle Synthesis

The choice of biomaterial for microparticle synthesis is dependent on the intended use, duration of therapy, desired nature of the polymer (for example, hydrophobic or hydrophilic, neutral or charged), bioactive agents to be delivered (for drug delivery applications), and the chemistry needed for further functionalization and modification. The vast majority of current microparticle research has focused on site-specific drug delivery, controlled drug release, and minimizing toxicity and degradation of drugs (Dumitriu, 2002). Table I.2.19.1 lists some of the natural and synthetic materials used in microparticle drug delivery applications. Only a few commonly used polymers are discussed here.

Natural Materials

Chitosan. Chitosan is a cationic polysaccharide exhibiting β(1→4) linkages, and is obtained by alkaline deacetylation of chitin, a polymer present in the exoskeleton of many insects and marine crustaceans. Its biocompatibility, complete elimination from the body post-degradation, and abundance of primary amines make chitosan a popular material for drug delivery applications (Agnihotri et al., 2004). Primary amines render chitosan soluble in aqueous acidic solutions, thus eliminating the use of organic chemicals during synthesis, and also offer numerous possibilities for ionic attachment of negatively charged biomolecules (e.g., DNA, RNA). Further, chitosan microparticles exhibit enhanced mucosal residence time, increasing muscosal permeability of encapsulated proteins and drugs for nasal drug delivery (van der Lubben et al., 2001). Chitosan microparticles are also taken up by the Peyer's patch of gut-associated lymphoid tissue (GALT), making them appropriate candidates for oral drug delivery (van der Lubben et al., 2001c; Ahire et al., 2007). Chitosan has been used to orally deliver proteins and antigens such as insulin (Ubaidulla et al., 2007), diphtheria (van der Lubben et al., 2003) or tetanus toxoid (Ahire et al., 2007). Chitosan microparticles have also been successfully used for DNA and vaccine delivery, facilitating transfection and preventing DNA degradation by exogenous Dnase (Guliyeva et al., 2006).

Alginate. Alginate is an unbranched copolymer of D-mannuronic and L-guluronic acid linked by β(1→4) linkages, and is derived from brown seaweed. This polymer has been actively explored in the pharmaceutical sciences because of its biocompatibility, low immunogenicity, and unique property of sol–gel transition in the presence of multivalent cations (e.g., Ca^{++}, Mg^{++}). The preparation of alginate microparticles involves mild reaction conditions, which minimizes or eliminates the use of organic solvents and high temperatures that could otherwise reduce the bioactivity of encapsulated proteins, polypeptides or nucleic acids (Dumitriu, 2002). Alginate microparticles have been used extensively for protein, drug, and cell microencapsulation (Joki et al., 2001; Wang et al., 1997). Protein drugs have been encapsulated in calcium chloride cross-linked alginate microspheres, with release fine-tuned by varying the cross-linking factors (Wheatley et al., 1991). For example, applying polycations to the outer microcapsule surface can reduce the burst effect and produce more sustained release (Wheatley et al., 1991). Drug-loaded alginate microparticles have been primarily studied for nasal and oral delivery (Bowersock et al., 1996; Hari et al., 1996). Cell encapsulation has been used to obtain high concentrations of hybridoma monoclonal antibodies for mammalian cell engineering applications (King et al., 1987), and to encapsulate rat islet cells for diabetes treatment (Lim and Moss, 1981; De Vos et al., 1997). Alginate microspheres have also been prepared with other polymers such as chitosan. The addition of chitosan alters the microparticle characteristics and the release profile for potential gastrointestinal tract delivery via an oral route (Hari et al., 1996).

Gelatin. Gelatin is the denatured form of animal-derived collagen, and has excellent biocompatibility and biodegradability. Gelatin microparticles have been studied for sustained release applications (Ratcliffe et al., 1984; Tanaka et al., 1963). Micropellets of gelatin were first prepared in the early 1960s by Tanaka et al. (Tanaka et al., 1963). Since then, gelatin microparticles have been explored for the delivery of mitomycin C, 5-fluorouracil or adriamycin (Oner and Groves, 1993), and used commercially in agglutination tests for diagnostic detection of HIV-1, syphilis, and measles (Hesketh et al., 2006; Barbara et al., 1989; Sato et al., 1997). Mikos and co-workers encapsulated different growth factors inside gelatin microparticles for applications in cartilage tissue engineering. Growth factors such as transforming growth factor-beta 1 (TGF-β1) (Holland et al., 2005, 2007; Park et al., 2007a, insulin-like growth factor-1 (IGF-1) (Holland et al., 2007; Pham et al., 2008), bone morphogenetic protein-2 (BMP-2) (Patel et al., 2008b,c), and vascular endothelial growth factor (VEGF) (Patel et al., 2008a,c) have been successfully encapsulated. Dual growth factor delivery systems devised to deliver two growth factors simultaneously result in improved cell growth and bone matrix regeneration, in comparison to single growth factor delivery systems (Holland et al.,

TABLE I.2.19.1 Examples of Materials Used in Microsphere Drug Delivery Applications

Polymer	Product Name	Active Ingredient	Application	Synthesis Method	Reference
PLGA	Sandostatin LAR® Depot	Octreotide acetate	Acromegaly	Emulsion solvent evaporation/ extraction	Mundargi et al., 2008
	Nutropin Depot®	Growth hormone	Growth hormone deficiency	Spray freeze-drying	Mundargi et al., 2008
	Lupron® Depot	Leuprolide acetate	Prostate cancer	Emulsion solvent evaporation/ extraction	Mundargi et al., 2008
	–	DNA, protein antigen, siRNA, CpG oligos, monophosphoryl lipid (MPL)	Delivery of vaccines and immuno-modulatory agents	Emulsion solvent evaporation/extraction, spray drying	Bates et al., 2006 Singh et al., 2008 Malyala et al., 2009
PHEMA	–	Vasopressin	Oral delivery	–	Lee et al., 2007
Alginate		Protein	Drug delivery	Ionic gelation	Wheatley et al., 1991
		Pancreatic islets	Diabetes	Ionic gelation	De Vos et al., 1997
		Recombinant fibroblasts	Non-autologous somatic gene therapy	Ionic gelation	Chang et al., 1994
SPCL	–	Dexamethasone (DEX)	Drug delivery and tissue engineering	Emulsion solvent evaporation/ extraction	Balmayor et al., 2009
Poly sebacic anhydrides	–	Rhodamine B	Controlled drug release	Spraying with acoustic excitation	Berkland et al., 2004
Poly orthoesters	–	Plasmid antigens	DNA delivery	Emulsion solvent evaporation/extraction	Nguyen et al., 2008
Poly caprolactones	–	Antigen, anti-hypertensive drugs, Taxol, antibiotics, ribozymes, nerve growth factor (NGF), insulin, heparin	Drug delivery	Emulsion solvent evaporation/extraction, spray drying, solution enhanced dispersion method Hot-melt process	Heller et al., 2002
Collagen		Hydroxyapatite	Osteoblast-based grafting material	Water-in-oil emulsion	Wu et al., 2004
Chitosan		Anti-cancer drugs, anti-inflammatory drugs, cardiac agents, calcium channel blockers, antibiotics, antithrombotic agent, steroids, proteins, antigens, antidiabetic agents, growth factors, DNA encapsulation, cytokines	Drug delivery, gene therapy	Ionic gelation, emulsification and ionotropic gelation, coacervation and complex-coacervation, precipitation–chemical cross-linking, thermal cross-linking, solvent evaporation method, spray drying	Sinha et al., 2004b
Hyaluronic acid		Hydrocortisone	Drug delivery	Solvent evaporation	Benedetti et al., 1990
Hyaluronic acid/ chitosan		Gentamicin sulphate	Mucoadhesive nasal drug delivery	Emulsion solvent evaporation/ extraction	Lim et al., 2000

PHEMA: poly(2-hydroxyethyl methacrylate)
PLGA: poly(lactic-co-glycolic acid)
SPCL: starch-poly-e-caprolactone

2005, 2007; Patel et al., 2008c; Park et al., 2008). Glutaraldehyde is a common cross-linker used to form microparticles, and its concentration can control the release rate of biomolecules. To slow degradation, enhance sustained release of drug, and enhance *in vivo* biocompatibility, some researchers have utilized the naturally-occurring cross-linker genipin to cross-link gelatin microparticles. The resulting particles had a reduced degradation rate and inflammation *in vivo*, in comparison to glutaraldehyde-cross-linked microparticles (Liang et al., 2003).

Dextran. Dextran is a branched homopolysaccharide of glucose, and is synthesized naturally from sucrose by some bacterial strains. Numerous biomedical studies have employed dextran as a biomaterial since it is biocompatible, biodegradable, available in different molecular weights, and can be easily derivatized. Microparticles of dextran have been studied for biomedical applications ranging from drug delivery (Chen et al., 2006; Cheung et al., 2005; Diwan et al., 2001) and cell therapy (Demetriou et al., 1986) to diagnostics and biosensors (Zhang et al., 2008). In a number of studies, dextran has been employed in the preparation of beads that were later functionalized with antibodies or biomolecules for diagnostic applications (Zhang et al., 2008; Tansey and Cammer, 1998). For example, hydroxyl ethyl methacrylated dextran microspheres have been studied for the delivery of human growth factor (Vlugt-Wensink et al., 2006, 2007). Adding different excipients changed protein loading efficiency and release. Collagen-coated dextran microparticles have also been used for liver tissue engineering. Hepatocytes were attached to their surfaces and injected intraperitoneally into rats, in an attempt to devise therapies for hepatic failures (Demetriou et al., 1986).

Other Natural Materials. In addition to the above mentioned materials, other natural materials have also been employed for microparticle synthesis. For example, collagen microparticles have been used for delivery of glucocorticsteroids (Berthold et al., 1998) and lipophilic drugs (Rossler et al., 1995; Swatschek et al., 2002). Similarly, hyaluronic acid microparticles have been utilized to deliver biomolecules (i.e., recombinant human growth hormone) (Kim et al., 2005), and antibiotics (Lee et al., 2002). Natural materials are biodegradable and well-tolerated in the body, which renders them suitable for delivery applications. See Chapter I.2.7 for details on natural polymers used in medicine.

Synthetic Polymers

Poly(lactic acid) (PLA), Poly(glycolic acid) (PGA), and Copolymers (PLGA). Bioresorbable aliphatic polyesters of lactic acid and glycolic acid are used clinically for controlled drug delivery (Lewis, 1990). Structural details of homopolymers (PLA, PGA) and copolymers (PLGA) have been discussed (Chapter I.2.6). Optically inactive poly(D,L)lactic acid is amorphous in nature, and thus is mechanically weaker than L-PLA or D-PLA, making it a suitable candidate for drug delivery applications. PLGA microparticles were first developed as drug delivery carriers in the late 1970s (Tice and Lewis, 1980). Since then, these polymers have been synthesized in various morphologies, including microparticles, microneedles, pellets, nanocapsules, films, and implants (Park et al., 2007b; Siegel et al., 2006; Teixeira et al., 2005; Marchais et al., 1998; Gumusderelioglu and Deniz, 2000; Webber et al., 1998). As discussed (Chapter I.2.6), there exists no linear relationship between homopolymer ratios of PLGA and degradation behavior of the copolymer. The degradation rate depends on several factors that facilitate water accessibility to the ester linkages in the microspheres (Park, 1995; Shive and Anderson, 1997). The crystallinity of the homopolymer or copolymer plays a critical role; hydrolytically-cleavable ester linkages in PLGA are more accessible to penetrating water molecules compared to linkages in PLA, thus resulting in faster degradation kinetics. In a classic study, microspheres formed using PLGA copolymers with homopolymer ratios PLGA 50:50 and 70:30 degraded much faster than PLGA 80:20 and 90:10 (Figure I.2.19.1) (Park, 1995). This study illustrated that PLGA microparticles degrade in a heterogeneous fashion (Figure I.2.19.1A). During the degradation process the microspheres retain their shape until they completely disintegrate (Figure I.2.19.1A). Because the end products of PLA or PLGA microparticles are primarily lactic or/and glycolic acids, L-lactic acid content in the degradation medium was quantified. Figure I.2.19.1C shows an increase in L-lactic acid content with an increase in PGA homopolymer content, because the PGA linkage is the primary hydrolytically cleavable site in these polyesters (Park, 1995). Nonetheless, degradation is highly tunable. For example, in separate *in vivo* studies, PLA microspheres injected in rat gastrocnemius muscles degraded between 360–480 days, whereas PLGA (50:50) microspheres degraded in 63 days (Visscher et al., 1985, 1986, 1987, 1988). Because of their excellent biocompatibility and the ability to fine-tune monomer ratio to obtain varying degradation, this polymer has been studied extensively for delivery of a range of biomolecules, including nucleic acids, proteins and peptides, and hydrophilic and hydrophobic drugs (Singh et al., 2008).

Poly(ε-caprolactone) (PCL). PCL is a biodegradable, semicrystalline polymer with a glass transition temperature of ~60°C. The ring-opening polymerization of PCL was first studied in 1934 (Van Natta et al., 1934), but the biodegradable nature of the polymer was not recognized until the 1970s. Since then, various groups have explored microparticles of PCL for drug delivery applications because of its degradation characteristics, biocompatibility, high permeability with drugs, and lack of toxicity (Murthy, 1997). Further, their slower degradation rates and non-acidic degradation environment, compared to PLA/PLGA, make PCL microparticles suitable for prolonged drug release. PCL blends well with other polymers, which results in altered permeability and degradation rates, and therefore, altered drug release profiles (Pitt et al., 1979a,b). For example, release kinetics of fertility-regulating agents from PCL microparticles mixed with cholesterol, which serves as a viscosity thickener and enhances the drug solubility or by PCL microparticles made from polymer pretreated with formic acid, have been compared to PLA/PGA/PLGA microparticles (Benagiano and Gabelnick, 1979). PCL microparticles have also been studied for encapsulation and release of antibacterial agents (Dubernet et al., 1987), and for combined delivery of antigens and adjuvants for enhanced oral immunization (Youan et al., 1999; Murillo et al., 2002). Lipophilic and hydrophilic drugs have been co-encapsulated in PCL microparticles using a double emulsion technique (Hombreiro Pérez

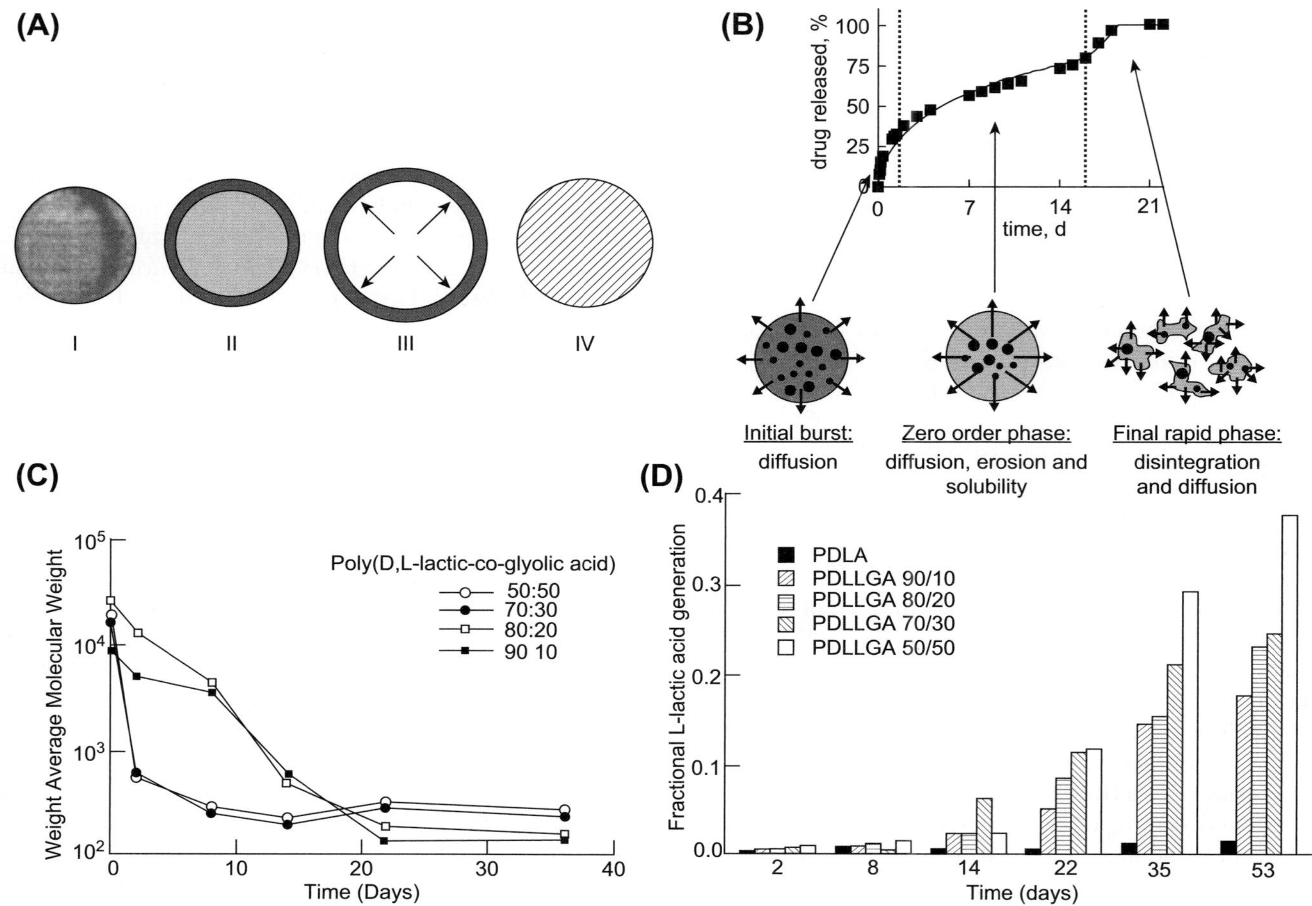

FIGURE I.2.19.1 (A) Schematic diagram of microsphere degradation mechanism (Park, 1995); (B) drug release from and drug release mechanisms in 5-fluorouracil (5-FU)-loaded, poly(lactic-co-glycolic acid) (PLGA)-based microparticles (Siepmann and Siepmann, 2006); (C) decrease in molecular weight of microspheres with varied PLA/PGA ratio, as a function of time (Park, 1995); (D) fractional L-lactic acid concentration in the medium as a function of time, with varied PLA/PGA ratio (Park, 1995).

et al., 2000). For a detailed review on PCL microparticles, refer to the review by Sinha et al. (Sinha et al., 2004a).

Poly(ortho ester) (POE). Another class of biodegradable materials developed specifically for controlled drug delivery is the poly(ortho esters) (POEs). The biocompatibility and biodegradability of POEs have been studied extensively. There are four different families of POEs, which have been reviewed in detail by Heller et al. (Heller et al., 1983, 2002). POEs degrade by surface erosion, as opposed to the bulk erosion of PLGA polymers, resulting in slower degradation and rendering devices made from this polymer highly stable. Degradation time can be shortened by modifying the polymer backbone with short segments of lactic acid or glycolic acid. Varying the number of glycolic acid or lactic acid segments produces a range of degradation rates suitable for a wide range of delivery applications (Heller et al., 2002). POE microparticles have been examined for protein and nucleic acid delivery (Yang et al., 2001; Wan et al., 2001). For example, the release of plasmid DNA vaccines yielded a potent immune response. Slower surface erosion produced few degradation products, which may have otherwise caused DNA degradation, diminishing the immune response. Microparticles have also been synthesized from the triblock copolymer POE–PEG–POE employing the double emulsion process. The incorporation of PEG in the POE chain resulted in increased hydrophilicity, leading to higher bovine serum albumin (BSA) encapsulation and faster protein release compared to unmodified POE particles.

Other Synthetic Materials. A number of synthetic materials have been examined as microparticles. For example, polyacrylamide microparticles were prepared almost three decades ago. Their distribution and elimination have been extensively studied in rodent models (Sjoholm and Edman, 1979), and they have been used in drug delivery applications (Kriwet et al., 1998). Similarly, poly(vinyl alcohol) microparticles (Ficek and Peppas, 1993) and poly(2-hydroxyethyl methacrylate) (PHEMA) microparticles (Lehr et al., 1992) have been used for controlled protein delivery and for the release of peptide drugs, respectively. Polyketal microparticles have been examined for the release of therapeutic agents after microparticle phagocytosis, in which release is mediated by microparticle degradation in the acidic phagosome environment (Lee et al., 2007). These materials have also been mixed with other natural or synthetic materials to achieve desired characteristics, such as varying degradation rate, release profile, and selectivity for target tissue.

MICROPARTICLE PREPARATION

Polymer microparticles are prepared from synthetic monomers, natural polymers, and sometimes semi-synthetic polymers, using techniques that create polymer matrices of desirable sizes and structures. State-of-the-art techniques include single and double emulsion solvent evaporation/extraction methods, coacervation, spray drying, ionic gelation, and suspension cross-linking.

Single and Double Emulsion Solvent Evaporation

The emulsification solvent evaporation technique is one of the most commonly used in microparticle preparation, particularly for drug encapsulation. The two-step technique calls for initial emulsification of polymer solution in a volatile organic solvent, followed by internal phase solvent evaporation or extraction that results in hardening and precipitation of microparticles. Solvent evaporation is generally performed at atmospheric pressure (or sometimes under reduced pressure) to promote evaporation of the volatile solvent. The solvent extraction process, however, involves quenching in excess water or relevant quench medium, allowing for solvent diffusion through oil droplets. Commonly used solvents include methylene chloride (also called dichloromethane or DCM) (Singh et al., 2008), ethyl acetate (Cho and Sah, 2005), and acetone/methanol mixtures. The single emulsion technique is generally employed for synthesis of microparticles encapsulating hydrophobic drugs. The polymer is dissolved in one phase (e.g., PLGA in oil phase or chitosan in aqueous phase), followed by either addition of water to oil (W/O emulsion) or oil to water (O/W emulsion). To prevent coalescence of emulsion droplets, surfactants such as polyvinyl alcohol (PVA) are used. The two phase system is then homogenized at speeds of 8,000–20,000 rpm for 2–5 minutes. Emulsifying agents (i.e., surfactants) play an important role in emulsion selection for microparticle synthesis. If the emulsifier is oil soluble then it favors W/O type emulsion, and *vice versa*. Since the stability of the emulsion also depends on the difference in specific gravities between oil and water phases, W/O emulsions are sometimes difficult to create. However, stable systems can be created (Okochi and Nakano, 2000). For an O/W system, final microparticles are obtained as the volatile organic solvent diffuses outward through the solvent-saturated aqueous phase from the internal organic phase, eventually evaporating and producing hardened microparticles. Microparticle porosity is controlled by rate of evaporation, ratio of internal aqueous phase, and viscosity of the polymer. Rapid solvent evaporation, as in the solvent extraction process, often results in porous microparticles as compared to the slow solvent evaporation process.

Unlike the O/W single emulsion, which is suitable for encapsulating lipophilic molecules like steroids, the double emulsion process or water-in-oil-in-water (W/O/W) method is commonly used to synthesize microparticles encapsulating hydrophilic biomolecules and drugs such as peptides, proteins, and antigens (vaccines). The W/O/W double emulsion is a three-phase system in which polymer is dissolved in the oil phase with an internal aqueous phase, and immersed in an external aqueous phase containing surfactant or emulsifying agent. Although various techniques to create W/O/W double emulsions exist (e.g., one-stage emulsification method, two-stage emulsification method, phase inversion method), microparticle synthesis for biomedical applications has primarily focused on the two-stage method, because of its superior drug loading capability (Okochi and Nakano, 2000). In a typical W/O/W double emulsion process, polymer (e.g., PLGA, PCL, PEG–PLA copolymer) is dissolved in an organic (oil phase) solvent followed by addition of a small volume (100–500 μl) of aqueous phase. This solution is subjected to vigorous homogenization or sonication to yield the primary emulsion (Dorati et al., 2008). This primary emulsion is then poured into an excess of aqueous solution containing emulsifier (e.g., PVA), and further homogenized. The volatile organic solvent is eliminated from the resultant W/O/W double emulsion through solvent evaporation or extraction. Similarly, chitosan microparticles are often prepared by an extended W/O emulsification process in which chitosan in the aqueous phase is emulsified in oil with surfactant, and then cross-linked using glutaraldehyde or ethyleneglycol diglycidyl ether to form hard droplets (Agnihotri et al., 2004).

The formation mechanism of initial emulsion droplets and resultant microparticles in an oil-in-water (O/W) single emulsion and W/O/W double emulsions has been examined (Figure I.2.19.2) (Rosca et al., 2004). During the initial stage of the O/W emulsion, emulsion droplets are large because the aqueous phase is initially saturated with solvent (Figure I.2.19.2B). As solvent evaporates, solvent concentration in the droplet decreases, causing a rapid shrinkage of droplet size and the formation of hardened microparticles (Figure I.2.19.2C). In contrast, in the W/O/W emulsion, different types of droplets are formed (Figure I.2.19.2D): (1) microcapsule with no inner aqueous phase entrapped; (2) microcapsule with only one aqueous phase; and (3) a single droplet containing multiple microcapsules. The inner and outer structures of these microcapsules are further clarified by the scanning electron micrograph (SEM) image (Figure I.2.19.2E). Droplets without any inner aqueous phase form plain polymer microparticles, indicated as "p," thin-walled open microcapsules "c1" are formed when the emulsion droplet is similar in size to the inner droplet, and microcapsules "c2" are formed when a small difference exists between the microdroplet and emulsion droplet size.

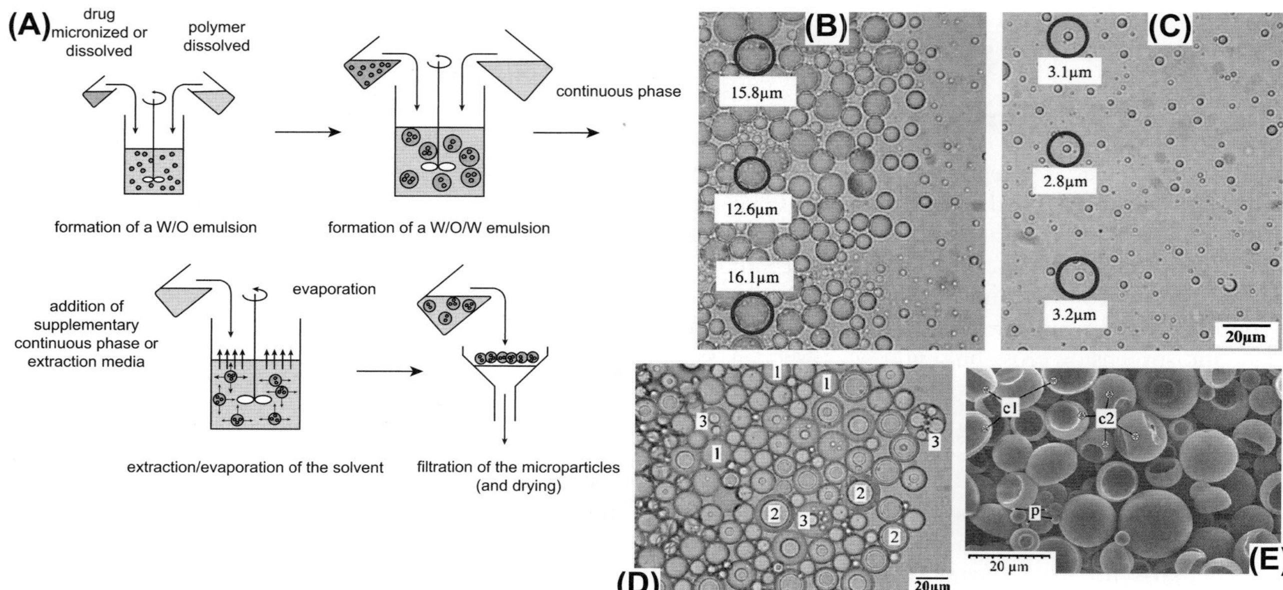

FIGURE I.2.19.2 (A) Preparation of microparticles using a water-in-oil-in-water (W/O/W) technique (Siepmann and Siepmann, 2006); (B,C) the first and last images, respectively, of the emulsion microdroplets transformation into the final microparticles; (D) initial W/O/W microdroplets; and (E) final microparticle SEM (scanning electron micrograph) image (Rosca et al., 2004). See text.

Precipitation and Coacervation

Coacervation uses the physiochemical characteristics of polymers to form microparticles. Coacervation is a three-step process in which a W/O emulsion is formed with polymer dissolved in the organic phase and drug dispersed in the aqueous phase. The solubility of the polymer is altered (e.g., by mixing with another polymer or organic non-solvent or by changing pH or ionic strength). The non-solvent used in the initial steps should be miscible with the organic solvent used to dissolve the polymer; however, the polymer and/or drug should not be soluble in the non-solvent (e.g., vegetable oil, paraffin oils, silicone oils). This results in phase separation to form a polymer-rich coacervate phase surrounded by a dilute supernatant phase, essentially producing a coating of coacervate around the dispersed drug molecules. Stabilization of the coating occurs with solidification of microparticles in another organic non-solvent followed by extensive washing, centrifugation, and freeze drying (Jain, 2000).

Spray Drying

Conventional microparticle fabrication methods like double emulsion and coacervation often have limitations, including particle aggregation as a result of exceptionally high surface energies and mutual adhesion between particles, inefficient encapsulation, multi-step processing, and residual organic solvents (Jain, 2000; Jalil and Nixon, 1990). Spray drying is a faster, more efficient process for encapsulating drugs in polymeric microparticles (Jain, 2000). This technique involves dispersion (emulsification) of aqueous drugs or direct addition of hydrophobic drugs in a polymer solution, which is then sprayed through a fine nozzle into a chamber (generally heated) where the solvent evaporates and particles are collected (Figure I.2.19.3A). This technique generally yields microparticles from 1–100 μm in diameter and drug encapsulation of up to 99%, depending on drug–polymer interaction, solution viscosity, and nozzle characteristics (Jain, 2000; Wagenaar and Muller, 1994). Spray drying has been used for synthetic polymers such as PLGA, PLA, and PCL, and also for natural polymers including gelatin and polysaccharides (Lorenzo-Lamosa et al., 1998). For polysaccharides (e.g., chitosan), the aqueous acetic acid-based solution containing a mixture of polysaccharide, drug, and cross-linker is atomized similarly in a heated chamber. Cross-linkers such as glutaraldehyde form an aldehyde–amine bond with the polysaccharide that results in hardened microparticles; however, cross-linker addition is not essential for microparticle formation. Various parameters influence particle size, including nozzle radius, chamber temperature, flow rate and pressure, and cross-linker concentration (Agnihotri et al., 2004). Microparticles may be lost because of agglomeration and adhesion to the chamber walls, and some research groups have developed a novel dual-nozzle system that simultaneously delivers both the polymer–drug mixture and an anti-adherent solution (e.g., mannitol) to prevent adhesion and aggregation (Takada et al., 1995).

The low temperature spray drying method reported by Alkermes, Inc. (ProLease® technology) was specifically developed to maintain integrity and activity of encapsulated biomolecules during the synthesis process. In this technique, powdered biomolecules and stabilizing

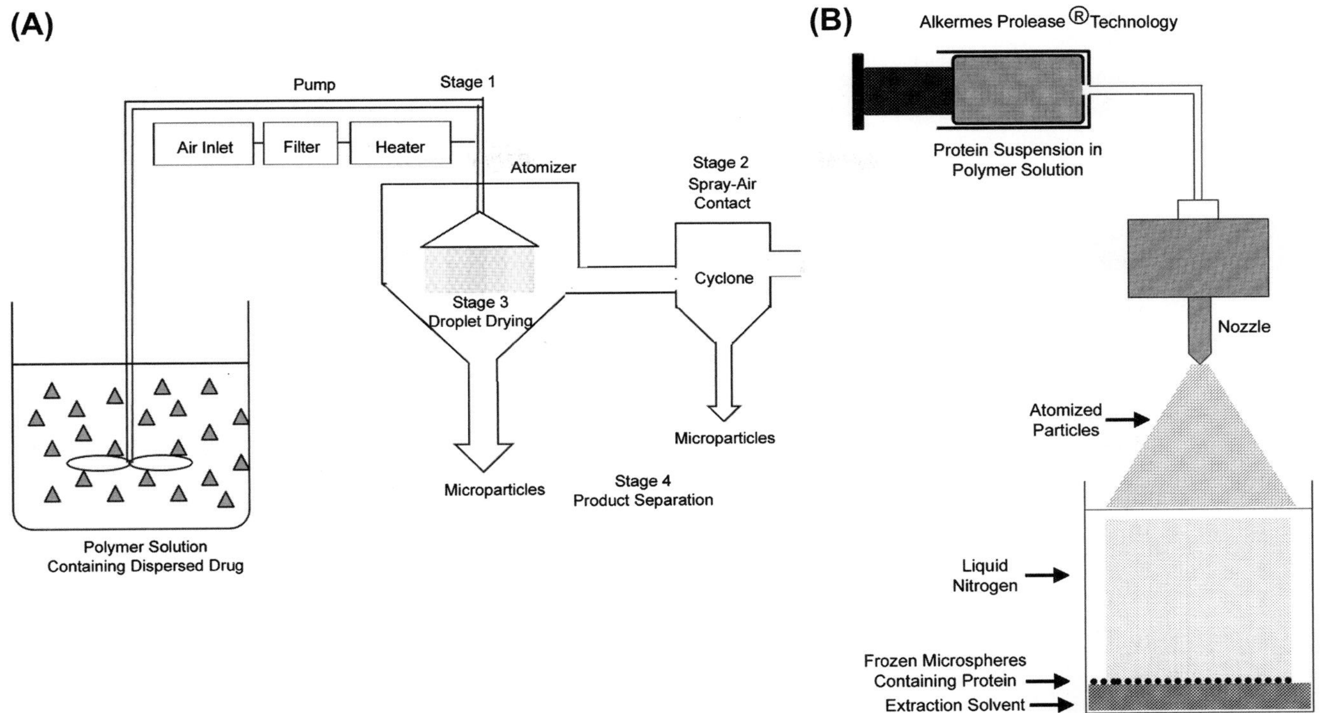

FIGURE I.2.19.3 Preparation of microparticles using spray drying techniques: (A) conventional spray drying method (adapted from Re, 2006); (B) low temperature spray drying method (adapted from Johnson et al., 1997).

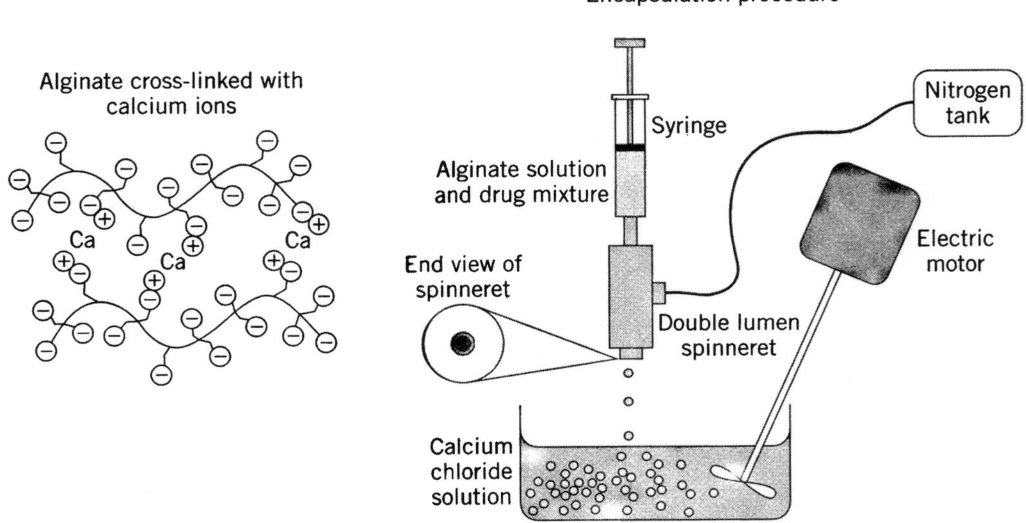

FIGURE I.2.19.4 Schematic showing the synthesis of microparticles by the ionic gelation method.

excipients are suspended in a polymer solution in an organic solvent (e.g., DCM, acetone, ethyl acetate) and sprayed into a vessel containing liquid nitrogen followed by evaporation of the liquid nitrogen (Figure I.2.19.3B). The vessel also contains a frozen solvent such as ethanol for extraction of the polymer solvent from liquid nitrogen frozen droplets. With this technique, microparticles with high encapsulation efficiency (~95%) and sizes ranging from 50–60 μm can be readily obtained (Johnson et al., 1997).

Ionic Gelation

Ionic gelation is a physiochemical technique in which ionic polyelectrolytes (or charged polymers) are chelated with multivalent counterions, producing cross-links (Figure I.2.19.4). Ionic gelation is widely used with polysaccharides such as chitosan and the sodium salt of alginic acid, because of its simplicity and mild conditions (Bodmeier et al., 1989; Li et al., 2008; Park et al., 2004; Shu and Zhu, 2001). Chitosan is a positively charged polysaccharide which forms cross-links in anionic sodium triphosphate or tripolyphosphate (TPP) or $[Fe(CN)_6]^{4-}$ or $[Fe(CN)_6]^{3-}$ (Li et al., 2008). The ionic gelation process is fairly simple and involves drop-wise addition of polymer solution to the counter ion solution with dispersed drug, under continuous stirring. The ionization degree of TPP is dependent on the solution pH. At slightly acidic pH, only $(P_3O_{10})^{5-}$ anions are available, favoring ionic gelation with the acetic acid-based chitosan solution (Ko et al., 2002). The ionic gelation process can result in both nano- and microparticles ranging from several nanometers to 700 μm. The process has been extensively applied for encapsulating drugs (e.g., insulin, cyclosporine A, felodipine) with encapsulation efficiencies ranging from 55% to 90%, depending on deacetylation of chitosan and reaction conditions (Agnihotri et al., 2004; Ko et al., 2002). Ionic gelation performed at high concentrations of TPP results in slower drug release and *vice versa*. Drug release is further influenced by the extent of cross-linking (Agnihotri et al., 2004).

SUBMICRON-SIZED PARTICLES

Liposomes and Micelles

Liposomes (Figure I.2.19.5) are submicron-sized spheres with an aqueous core and a bilayer membrane (e.g., lamella), whereas micelles consist of an aqueous core surrounded by a single layer of lipids. Liposomes are formed by one or several lipids, including sphingomyelin, egg phosphatidylcholine (PC), cholesterol, monosialoganglioside (GM1), distearoylphosphatidylethanolamine (DSPE) or dioleylphosphatidylethanolamine (DOPE). Liposomes can be classified as multilamellar (0.5–5 μm diameter), small unilamellar (~0.1 μm), or large unilamellar (0.2–0.8 μm) vesicles. Liposomes have excellent biocompatibility, and can encapsulate hydrophobic drug between the lipophilic bilayers and hydrophilic drug in the core. Liposomal delivery systems have been used for therapeutics, including rifampicin, budesonide, diclofenac, methotrexate, vaccine antigens, nucleic acids, peptides, and immunomodulatory agents (Torchilin, 2005).

Conventional liposomes have a short blood circulation time and are quickly removed by the reticuloendothelial system (RES), as was demonstrated in early studies with immunoliposomes (i.e., liposomes modified with antibodies) that resulted in mass accumulation in the liver. Extensive research has been performed to develop long-circulating liposomes. Surface immobilization of PEG imparts "stealth" properties to liposomes (Figure I.2.19.5C), which minimizes RES uptake. Liposomes modified with PEG have been used in delivering anticancer drugs such as doxorubicin and paclitaxel (Skubitz, 2003), and have demonstrated improved drug bioavailability and pharmacokinetics (Allen and Hansen, 1991). Long-circulating liposomes have also been used in infectious disease treatment and diagnostic applications (Bakker-Woudenberg, 2002). PEG can also be combined with other targeting methodologies (i.e., antibodies) through covalent modifications or hydrophobic interactions to produce long-circulating immunoliposomes (Figure I.2.19.5D,E) (Torchilin, 2005; Blume and Cevc, 1990; Abra et al., 2002).

NANOPARTICLES

Nanoparticles are generally acknowledged as particles with dimensions less than ~100 nm. Nanoparticles may be composed of virtually any material, but unlike those used in microparticles, some of the most common materials are semiconductors and metals. Nanoparticles have generated significant interest as biomaterials, primarily because of their unique size. From a purely physical perspective, nanoparticles are small enough to interface with objects that control some of the most basic cellular functions (e.g., ~2 nm diameter DNA). In comparison,

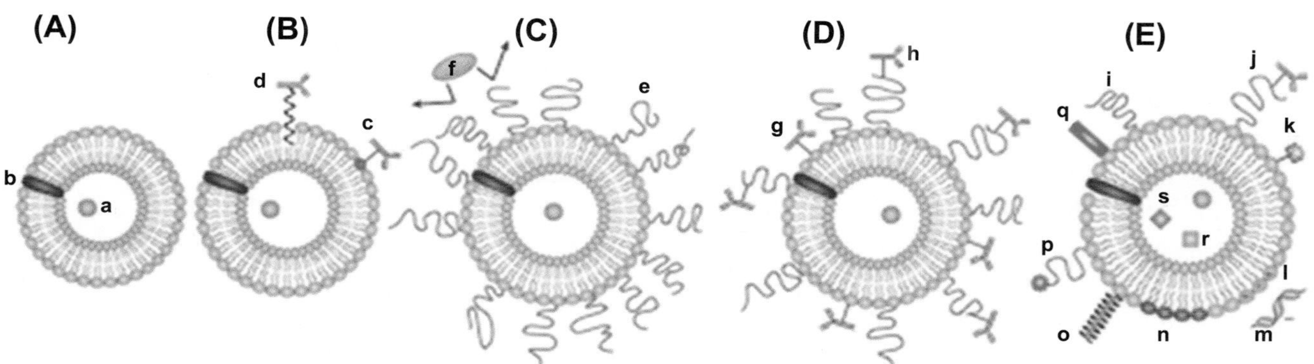

FIGURE I.2.19.5 Evolution of liposomes: (A) "Plain" phospholipid liposomes. Water soluble drug (a) is entrapped in the aqueous interior, and water-insoluble drug (b) is incorporated in the liposomal membrane (these designations are not repeated on other figures); (B) antibody-targeted immunoliposome. Antibody is covalently coupled (c) or hydrophobically anchored (d) to the liposomal membrane; (C) long-circulating liposome. Protective polymer (e.g., PEG) (e) shields the surface from opsonizing antibodies (f), which may otherwise lead to liposome phagocytosis; (D) long-circulating immunoliposome. Antibody is attached to the surface (g) or, preferably, to the grafted polymeric chain (h); (E) new-generation liposome. Multivalent liposomal systems can contain surface-bound protective polymer (i), targeting ligands (e.g., antibodies) (j), diagnostic labels (k), positively charged lipids (l) that permit complexation with DNA (m), stimuli-sensitive lipids (n) or polymers (o), cell-penetrating peptides (p) or viral components (q). The internal cavity can contain drug, magnetic particles (r) for magnetic targeting, and/or colloidal gold or silver particles (s) for electron microscopy (Torchilin, 2005).

cells are much larger (average diameter ~10 μm). Also, their small size allows them to interface with existing microelectronic components, such as field effect transistors (FETs), permitting the development of diagnostic biosensors.

In addition to these direct benefits of size, nanoparticles are also interesting biomaterials because of their unique, size-dependent physical properties. For example, semiconductor quantum dots and gold nanoparticles produce optical responses that are not evident in bulk materials. Iron oxide particles display a different type of magnetism when presented as nanoparticles than when presented as microparticles or in bulk. These properties originate from different principles, but are all derived from the size of the material, which lies between bulk and atomic composition. Because of these interesting properties, nanoparticles have found application as contrast agents in biomedical imaging, signaling molecules that enhance diagnostic ability, carriers for drug delivery, and as therapeutic elements.

MATERIALS USED FOR NANOPARTICLE SYNTHESIS

Noble Metals

Noble metal nanoparticles have a long history. They were first discovered by the Romans (Wagner et al., 2000), who used them to create colored glass. More recently, they were extensively studied by Michael Faraday (Faraday, 1857), whose samples are still on display at the Faraday Museum in London. In biomedical applications, the mostly commonly investigated materials have been gold and silver, which are created primarily through chemical reduction. For example, gold nanoparticles (Figure I.2.19.6A) can be created by the reduction of chloroauric acid ($HAuCl_4$) by sodium citrate (Turkevich, 1951) and silver nanoparticles by the reduction of silver nitrate ($AgNO_3$) by sodium borohydride ($NaBH_4$). In both cases, the reducing agent serves also as a passivating ligand, coating the surface of the nanoparticles.

Noble metals are widely regarded as non-toxic to humans; in fact silver nanoparticles are noted for their anti-bacterial properties (Sondi and Salopek-Sondi, 2004). As a result, noble metals have been applied as carriers for drug and gene delivery, and in anti-fouling coatings. In addition, they are attractive for imaging because they exhibit surface plasmon resonance (SPR) (Ritchie, 1957). Metals have many free electrons that typically reside in the conduction band. These electrons form a "cloud" surrounding the surface of the material that can deflect incoming light, producing a reflective surface. However, as we know from quantum mechanics, electrons can also behave as waves. If light of the same frequency as the oscilliations of the electron cloud is absorbed, the cloud vibrates or resonates, producing an absorbance signal. This signal is altered by the thickness of the material and, in the case of nanomaterials, is enhanced by the large fraction of atoms at the material surface. This signal is also altered by adsorption of

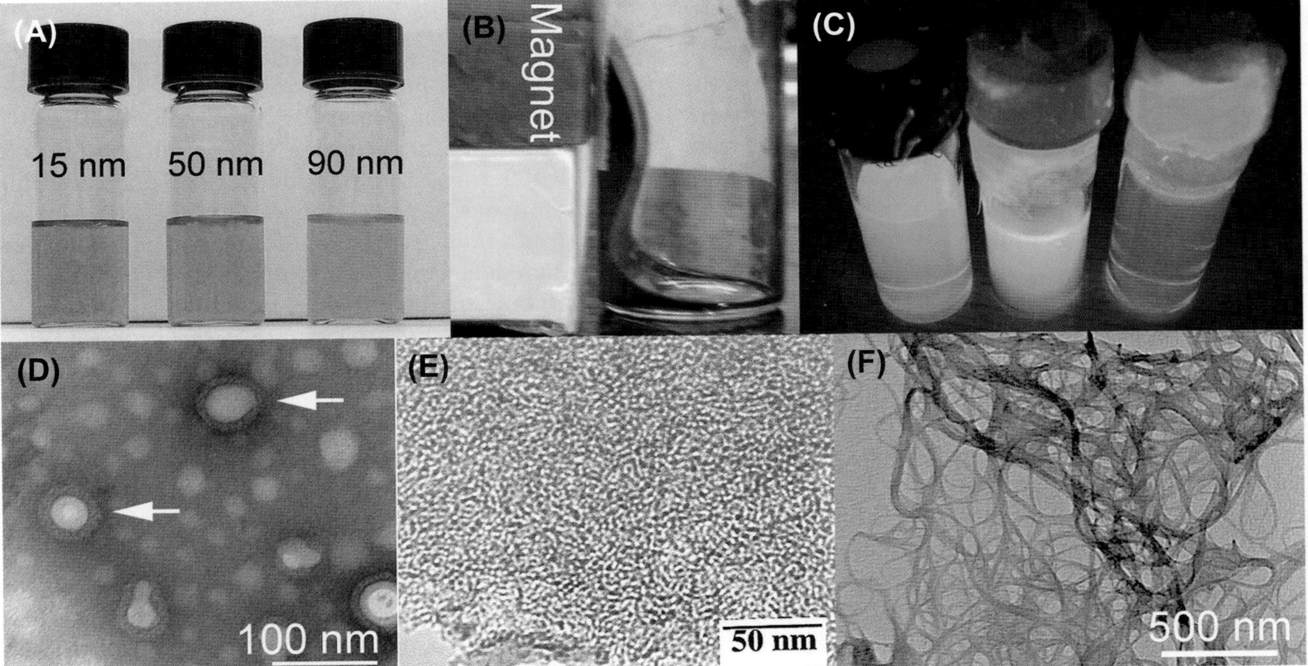

FIGURE I.2.19.6 Nanoparticles and their properties: (A) gold nanoparticles experience a color change with increasing size; (B) magnetic nanoparticles are attracted to magnets; (C) quantum dots fluoresce under UV excitation; (D) antibody-conjugated micelles (arrows) are shown in negative TEM (transmission electron microscopy) staining; (E) mesoporous silica contains multiple nanosized pores; (F) TEM image of single-walled carbon nanotubes. (Figures A–D courtesy of Dr. Jessica Winter, The Ohio State University; Figure E reproduced from Huo et al., 1997; Figure F courtesy of Dr. Maurizio Prato, University of Trieste, Trieste, Italy.)

other molecules to the material surface and nanoparticle aggregation; a property that has been exploited for the development of noble metal biosensors (He et al., 2000). Similarly, the strong absorbance produced by SPR has been used in photoacoustic imaging (Wang et al., 2003), which uses light to produce a signal detected via ultrasound.

Magnetic Materials

Although early research on magnetic nanoparticles dates back several decades (Crick and Hughes, 1949), interest in their use as a biomaterial is more recent (e.g., 1980s) (Sun et al., 2008). Superparamagnetic iron oxide nanoparticles (SPIONs), which are usually <15 nm in diameter, are the most commonly employed magnetic materials for these applications (Figure I.2.19.6B). SPION synthesis was originally performed through precipitation and microemulsion routes (Gobe et al., 1983), but more recently a high temperature precursor decomposition method (Hyeon et al., 2001) has been developed that yields small, uniform nanoparticles. One of the most important features of SPIONs is their unique magnetic response. Unlike larger nano- and microsized particles (e.g., >~15 nm), which are ferromagnetic, SPIONs display superparamagnetic behavior. This arises from their small size, which is below that of magnetic domain boundaries. Large particles consist of several magnetic domains that can be aligned in an external field, producing permanent magnetization. However, SPIONs have a single magnetic domain. They respond to a magnetic field, but do not display permanent magnetization, as evidenced by their lack of hysteresis on a magnetization (M) versus magnetic field strength (H) plot. This is because thermal fluctuations prevent permanent magnetic alignment of the single domain. Thus, SPIONs have the appearance of paramagnetic behavior (i.e., no permanent magnetization), while still responding to a magnetic field. This behavior is called superparamagnetism.

These magnetic properties have been used for a number of biomedical applications. The force generated on SPIONs in an external magnetic field has been harnessed for cell separation, drug delivery, and biomolecule manipulation (Pankhurst et al., 2003). Additionally, SPIONs have been used as contrast agents for magnetic resonance imaging (MRI) (Weissleder et al., 1990). MR imaging is accomplished by examining either the time that it takes for the spin magnetization of an individual proton to return to its original state (T1) or the time for the net magnetic moment for all protons to return to zero (T2). The magnetization produced by SPIONs interacts with water molecules in the body to shorten their T2 relaxation time. This produces an area of negative contrast in an image. In addition to MRI, SPIONs have been used for magnetic hyperthermia. When SPIONs are placed in an alternating magnetic field, they attempt to align their magnetic moments with that field. This produces internal friction that manifests itself as heat. If magnetic nanoparticles are placed in a tumor and a temperature above 45°C is reached, the tumor cells can be killed in a process known as hyperthermia (Pankhurst et al., 2003). Alternatively, if the temperature is raised to ~42°C, the efficiency of chemotherapy can be improved.

Quantum Dots

Quantum dots are small, fluorescent nanocrystals made from semiconductor materials, usually of spherical shape and <10 nm in diameter (Figure I.2.19.6C). The most commonly studied quantum dots are composed of a cadmium selenide (CdSe) core surrounded by a zinc sulfide (ZnS) shell, and are commercially available. Quantum dots are synthesized primarily through high temperature precursor decomposition (Peng and Peng, 2001); however, some aqueous arrested precipitation routes are also available (Winter et al., 2005). Quantum dots were first identified by Louis Brus at Bell Labs (Rossetti et al., 1983), and first employed for biomedical applications in 1998 (Chan & Nie, 1998; Bruchez Jr. et al., 1998). Quantum dots are so named because they experience quantum confinement, which means that the electron wavefunction is limited by the size of the particle. Because of this, quantum dots have a band gap energy (i.e., energy difference between the ground and excited states of an electron) that is dependent on particle size. Optical properties of quantum dots can thus be tuned by adjusting their size (Alivisatos, 1996).

Quantum dots have been used in a broad range of biological applications, including single molecule biophysics (Courty et al., 2006), optical barcoding (Han et al., 2001), and molecular (Courty et al., 2006), cellular (Jaiswal et al., 2003), and *in vivo* imaging (Gao et al., 2004). Compared to organic dyes and fluorescent proteins, quantum dots are about 10–100 times brighter, mainly because of their large absorption cross-sections, 100–1000 times more stable against photobleaching, and show narrower and more symmetric emission spectra. In addition, a single light source can be used to excite quantum dots with different emission wavelengths. CdSe has become the prototypical quantum dot material for biological applications, because it can produce fluorescence at wavelengths from green (500 nm wavelength) to red (650 nm wavelength) simply by adjusting the nanocrystal diameter from 2 to 8 nm. Other materials may be used to span other spectral regions, such as the ultraviolet region (CdS, ZnS, ZnSe), the near-infrared (CdTe, $CdSe_xTe_{1-x}$, PbS), and the mid-infrared (PbSe). A major limitation of quantum dots is their relatively large size compared with organic dyes, which may lead to steric hindrance in some applications (Alivisatos et al., 2005).

Polymers and Lipids

Polymers (Marty et al., 1978) and lipids (Gregoriadis et al., 1971) were among the first nanomaterials employed in biological applications. Polymer nanoparticles can consist of solid, porous polymer spheres or of nanocapsules consisting of an external polymer layer surrounding a liquid, aqueous core. Polymers can also form vesicles or micelles, sometimes called polymersomes. Lipid particles can consist of liposomes, lipid micelles or microemulsions (Figure I.2.19.6D). Polymeric nanoparticles can be fabricated via emulsion routes, in which case particle size is determined by the size of the suspended droplets, or via interfacial interactions with another solution, in which case size is determined by droplet size and mixing (Pinto Reis et al., 2006). Lipid-based microemulsions can be created by sonication of lipids in water, where size is determined by the extent of sonication. Vesicles/micelles of either lipids or polymers can be formed by self-assembly, in which case the size is determined by the thermodynamics of the system.

Polymer and lipid nanoparticles have been primarily used as carriers for therapeutic and imaging agents, and offer several advantages in this application. Polymer and lipid biomaterials can be used to encapsulate and protect therapeutic agents while they are transported to desired locations by passive or active targeting mechanisms. Furthermore, therapeutic agents (e.g., small molecule drugs, antibodies, small interfering RNAs) can be released from the nanocarriers in a controlled manner, for optimum therapeutic efficacy with minimal side effects. Additionally, these nanoparticles can incorporate a variety of factors, such as targeting, controlled release, imaging, and therapy into a single device.

Silica

The most common type of silica employed as a biomaterial is mesoporous silica (Figure I.2.19.6E). These particles, which have been created in both micro- and nanoparticle form, consist of silica (SiO_2) containing pores ranging from 2–10 nm in diameter. Mesoporous silica particles were first created in 1990 (Yanagisawa et al., 1990), and were improved by the development of controlled templating (Kresge et al., 1992) and synthetic advances to enhance particle monodispersity (Huo et al., 1997). In addition to mesoporous silica particles, silica is also widely-used as a nanoparticle coating. For example, silica has been employed to render quantum dots and iron oxide nanoparticles biocompatible. Most commonly, silica is deposited using the Stöber method (Stöber et al., 1968), which is a condensation reaction occurring in an alcohol–water system.

The primary advantages of silica as a biomaterial are its biocompatibility, easy surface modification, and porous structure. These properties have led to applications in imaging, drug delivery, and gene therapy. The surface of silica can be readily altered using siloxanes (R-$SiOCH_3$), a number of which are commercially available (e.g., terminal –SH or –NH_2 modification). This property has permitted silica nanoparticles to be conjugated to biomolecules that can specifically target certain cells or tissues. Additionally, the highly porous nature of silica materials has been used to encapsulate other materials, including imaging agents, drugs, DNA, and even other nanoparticles (Tallury et al., 2008). These materials can be slowly released at the target site to provide therapeutic benefit or a method of biodegradability.

Carbon

Several carbonaceous nanomaterials have been investigated as biomaterials including fullerenes, carbon nanotubes, and carbon nanodiamonds. The first of these to be discovered were carbon nanodiamonds, which were identified as an explosion by-product in 1961 (DeCarli and Jamieson, 1961); however, these materials were not investigated as a biomaterial until much later (Kossovsky et al., 1995). Carbon nanodiamonds are usually <10 nm diameter and display the diamond crystal structure. The next carbonaceous materials to be identified were fullerenes (Kroto et al., 1985), also known as buckyballs because of their structural similarity to geodesic domes designed by architect Richard Buckminster Fuller. Their discovery earned the Nobel Prize in Chemistry in 1996 for co-inventors Richard Smalley, Robert Curl, and Harold Kroto. Work on fullerenes led to the discovery of carbon nanotubes (CNTs) (Iijima, 1991), which can be viewed as rolled-up sheets of graphene (Figure I.2.19.6F). CNTs can consist of a single sheet (i.e., single-walled carbon nanotubes (SWCNTs)) or multiple sheets (i.e., multi-walled carbon nanotubes (MWCNTs)), and can be either semiconducting or metallic depending on the orientation of the rolled sheet. Carbonaceous nanomaterials can be produced through detonation (Danilenko, 2004), arc discharge (Kroto et al., 1985), laser ablation (Yang et al., 1998; Guo et al., 1995), and chemical vapor deposition (Feldman et al., 1995). Native carbonaceous materials are generally extremely hydrophobic; however, acid treatment can produce –COOH functionalization that renders particles water soluble (Chen et al., 1998). Because of their extreme hydrophobicity there have been substantial concerns regarding the biocompatibility of carbonaceous materials (Cui et al., 2005; Schrand et al., 2007). However, toxicity appears to be related to a number of controllable factors including dimensions, surface functionalization, manufacturing method, and processing (Helland et al., 2007).

Carbonaceous nanomaterials exhibit a number of unique, size-dependent properties, including high strength (e.g., CNTs are 100× as strong as steel (Treacy et al., 1996)), and electrical and optical characteristics, which have been exploited in biomedical applications ranging from molecular separation, biomolecule delivery and therapeutic treatment to imaging. For example, CNTs have been used for gene (Pantarotto et al., 2004) and drug (Kam et al., 2005) delivery, because they can penetrate cells through a mechanism believed to be independent of endocytosis (i.e., penetration of the lipid membrane) (Pantarotto et al., 2004). Because of their conductive and semiconductive properties, CNTs have been examined as scaffolds for nerve regeneration (Mattson et al., 2000), and as electrodes for neural prostheses (Lovat et al., 2005; Gheith et al., 2005). Conductivity has also been used to create biosensors (Kong et al., 2000) with detection down to the single molecule level (Besteman et al., 2003). However, two of the more interesting properties of carbonaceous materials are their ability to generate reactive oxygen species (ROS) and to produce multiple imaging signals.

ROS are oxygen or oxygen-containing molecules that have a highly reactive unpaired electron (e.g., $O_2{}^1$, singlet oxygen). ROS can cause significant damage to cells by interacting with the mitochondrial respiration system or DNA, and can be used to kill targeted cells (e.g., photodynamic therapy). Fullerenes generate ROS by light-induced excitation of a free electron to the singlet excited state, which is quickly followed by decay to the triplet state (e.g., the spin of the electron is reversed) (Yamakoshi et al., 2003). Conversely, fullerenes not exposed to light can neutralize ROS. Despite numerous π-bonded rings, fullerenes are electron acceptors, rather than donors (Dugan et al., 1996), and easily react with the ROS-excited electron. This property can be used to reduce damage to cells, for example, neutralizing ROS generated by UV irradiation.

In addition to ROS generation, carbonaceous materials produce a number of optical signals, which, when coupled with a low toxicity, make them attractive candidates for *in vivo* imaging. Carbon nanodiamonds exhibit fluorescence that is believed to arise from excitation of nitrogen vacancies (known as N-V centers) (Holt, 2007), and can also produce electron spin resonance (ESR/EPR) signals (Balasubramanian et al., 2008). Similarly, semiconducting CNTs can exhibit bandgap fluorescence, as well as fluorescent emission resulting from surface states. CNTs also produce strong Raman scattering signals, which result when light striking an object loses or gains energy from the collision. In the case of CNTs, Raman scattering occurs from interactions of light with the electron cloud surrounding the CNT structure. Both of these properties have been used for cellular imaging (Heller et al., 2005).

NANOPARTICLE PREPARATION

Arrested Precipitation

This method has been applied for the synthesis of semiconductor quantum dots (e.g., CdS), metals (e.g., Au), and iron oxide nanoparticles. In the absence of stabilizing agents, particles produced using these reactions exhibit very low solubility in water, hence they precipitate. However, when another agent is added to the reaction system that is water-soluble and can also bind to the nanoparticle surface, stable nanoparticles are produced. For example, in the presence of a Cd-binding ligand, mixing Cd^{2+} and S^{2-} salts produces CdS nanocrystals with size-dependent fluorescence (Rossetti et al., 1984). Similarly, gold nanoparticles can be produced using the Turkevich procedure (Scheme 1) (Ji et al., 2007):

$$HAuCl_4 + Na_3C_3H_5O(CO_2)_3 \xrightarrow{boiling}$$
$$[AuCl_2(OH)_2]^- + [AuCl(OH)_3]^- + [HC_3H_5O(CO_2)_3]^{-2} + [C_3H_5O(CO_2)_3]^{-3} \rightarrow$$
$$\text{AuNPs - Citrate ions} \quad (1)$$

In this process, a sodium citrate solution is added to a $HAuCl_4$ solution with heating and vigorous stirring. Sodium citrate first acts as a reducing agent, and then the negatively-charged citrate ions bind to the gold nanoparticle surface, preventing aggregation (Turkevich, 1951). Similarly, iron oxide nanoparticles can be synthesized by reduction in base (Gupta and Gupta, 2005):

$$Fe^{2+} + 2Fe^{3+} + 8OH^- \rightarrow Fe_3O_4 + 4H_2O \quad (2)$$

Iron ions are obtained from iron chloride salts (e.g., $FeCl_2$, $FeCl_3$), and sodium hydroxide (NaOH) serves as the base. The unifying feature of all these procedures is that particles are produced in the presence of a solubilizing agent, or ligand, that prevents aggregation and controls growth through steric hindrance and occupation of potential reaction sites on the particle surface.

High Temperature Precursor Decomposition

In 1993, the high temperature precursor decomposition method was developed to synthesize high quality quantum dots (Murray et al., 1993). This method relies on the use of a solid precursor, which decomposes at high temperature to yield reactive ions. As an example, to produce CdSe quantum dots, dimethylcadmium precursor and Se shot are mixed in solvents (e.g., trioctylphosphine, trioctylphosphine oxide, hexadecylamine) at high temperature. These solvents also serve as ligands whose basic functional groups (e.g., phosphines, phosphine oxides, amines) attach to the quantum dot surface, displaying their hydrophobic alkyl chains to the exterior. Slow growth and annealing results in uniform particle size and regularity in core structure. However, fluorescence quantum yield of quantum dots produced by this procedure decays over time, due to surface defects. In 1997, a higher band gap energy material (e.g., CdS) was used to coat CdSe quantum dots; the CdSe/CdS core shell structure resulted in a dramatic improvement in the fluorescence quantum yield (as much as 10 times) (Peng et al., 1997). Recently, the high temperature synthesis method has been adapted to produce other types of nanoparticles (e.g., iron oxide nanoparticles) (Hyeon et al., 2001).

Self-Assembly

The self-assembly method is driven primarily by thermodynamic considerations. For example, amphiphilic molecules, which possess both hydrophobic and hydrophilic segments, will attempt to minimize exposure of their hydrophobic segments in aqueous solution, spontaneously forming micelles, vesicles, rods, lamellae, tubules, and other nanoparticulates. The shape and size of these particles are mainly determined by the molecular structure of the amphiphiles employed (Zhang & Eisenberg, 1995). In practice, the self-assembly process can be conducted using different methods, such as direct dissolution, dialysis, and film hydration. The direct dissolution process simply involves adding amphiphile to water, but can only be used for water-soluble amphiphiles. Dialysis can be used when the amphiphile is not easily soluble in water, but can be readily dissolved in a water-miscible organic solvent such as dimethylformamide. In this case, a mixture of the amphiphile and water-miscible solvent is dialyzed against water, driving the organic solvent to the aqueous phase and inducing nanoparticle formation. If the amphiphile is only soluble in non-polar solvents, film hydration can be used. In this process, a thin film is formed by dissolving amphiphile in a low boiling point, non-polar solvent (e.g., chloroform), which is removed by evaporation. Water is then added, hydrating the film and driving spontaneous nanoparticle formation (Allen et al., 1999; Lasic, 1993).

Stöber Method

The Stöber method is a classic technique for producing silica nanoparticles, films, and coatings discovered in 1968 by Werner Stöber and colleagues (Stöber et al., 1968). The technique is a sol–gel method based on the hydrolysis and subsequent condensation of silicon-containing precursors (Scheme 3).

$$\begin{aligned} &\equiv \text{SiOR} + \text{H}_2\text{O} \leftrightarrow \equiv \text{SiOH} + \text{ROH} \\ &\equiv \text{SiOH} + \text{RO-Si} \leftrightarrow \equiv \text{Si-O-Si} \equiv + \text{ROH} \\ &\equiv \text{SiOH} + \text{HO-Si} \equiv \leftrightarrow \equiv \text{Si-O-Si} + \text{H}_2\text{O} \end{aligned} \quad (3)$$

Typically, tetramethoxysilane (TMOS) or tetraethylorthoxysilane (TEOS) are used as the silicon source. Both of these compounds contain alkoxide groups (e.g., $-OCH_3$ for TMOS, $-OCH_2CH_3$ for TEOS) that react with water to yield Si-OH intermediates and R-OH by-products. Si-OH can then react with other Si-OH or Si-OR groups to yield silica and ROH and H_2O by-products. The reaction is usually performed in an alcohol–water mixture, with alcohols being methanol, ethanol, *n*-propanol or *n*-butanol. Ammonia is used as a catalyst for colloidal growth. This reaction produces nearly monodisperse particles in sizes ranging from the micron to nanometer scale. This method was further modified by Unger in 1997 (Grun et al., 1997) to create monodisperse, porous silica spheres. This was accomplished by adding cetyltrimethylammonium bromide (CTAB), a cationic surfactant, which forms micelles in the reaction solution. Micelles are encapsulated during sphere formation. This surfactant is subsequently removed during particle calcination (i.e., heating), producing pores within the silica spheres. This method of particle formation is known as the modified Stöber method.

Methods to Produce Carbonaceous Nanoparticles

Carbonaceous materials can be produced via explosion, arc discharge, laser ablation, and chemical vapor deposition. A unifying feature of these methods is the heating of carbon precursors to high temperatures. Explosion is primarily used to create carbon nanodiamonds (DeCarli and Jamieson, 1961; Danilenko, 2004), and produces high temperatures and pressures that force carbon into the diamond region of its phase diagram. The carbon source may consist of a separate substrate or originate from the explosives themselves (Shenderova et al., 2002). Nanodiamond is manufactured in several commercial operations, primarily in Russia, its former republics, and Japan.

Arc discharge (Iijima, 1991; Krätschmer et al., 1990) and laser ablation (Kroto et al., 1985; Guo et al., 1995) are used to produce fullerenes and CNTs. Each method produces a carbon gas cloud, either by passing current through graphitic electrodes (i.e., arc discharge) or by using lasers to heat solid, carbon substrates (i.e., laser ablation) (Shenderova et al., 2002), reaching temperatures of 3000–4000°C. Gaseous carbon then condenses, producing carbonaceous nanostructures. This method is often performed in the presence of a catalyst, for example cobalt, yttrium or nickel, to control the shape of the resultant nanostructures. Critical parameters for arc discharge are the identity and pressure of the gas separating the electrodes, and the strength of the current. High gas pressure and low current tend to favor CNT production, whereas lower gas pressures favor fullerene formation. Laser ablation often produces a mixture of desired particles and undesired products. Separation processes, such as reflux in nitric acid, have been developed to isolate desired materials (Liu et al., 1998).

Chemical vapor deposition (CVD) has been used to produce carbon nanodiamonds (Eto et al., 1992) and CNTs (Kong et al., 1998). For nanodiamond, the CVD method primarily yields films, rather than freestanding particles (Shenderova et al., 2002). In this method, a hydrocarbon gas is deposited on a metal catalyst (e.g., iron, cobalt, nickel), which breaks the carbon-containing molecules into smaller components. These molecules serve as nucleation centers for additional particle growth. Reaction takes place at temperatures substantially below those of arc discharge or laser ablation (e.g., near ~1000°C). Acid treatment can be used to purify the materials produced.

SURFACE MODIFICATION OF MICRO/NANOPARTICLES

Surface properties of particles are governed by the charge and hydrophilic or hydrophobic behavior of the starting materials. For example, the surface charge of PLGA microparticles is slightly negative because of the presence of carboxylic groups, and the surface of polystyrene-based microparticles is essentially hydrophobic because of hydrophobic monomers. Surface properties can modulate particle interactions with the body permitting passive targeting to specific biological elements (e.g., through charge or hydrophobic interactions) (Singh et al., 2000). Particles can also be modified with biomolecules (e.g., antibodies, peptides) for active targeting to specific organs, cells or cellular components. Passive adsorption is one of the oldest surface modification methods and has been extensively employed to attach proteins, antibodies, antigens, nucleic acids, and polymers to particles (Bates et al., 2006). Loading is initiated by the interaction of biomolecule functional groups with the micro/nanoparticle surface (e.g., interaction of nucleic acid amines with carboxylic groups on PLGA). To obtain high loading efficiency and retain bioactivity, adsorption should be performed at a pH near the isoelectric point of the biomolecule. In addition to passive loading, numerous bioconjugation strategies exist (see Chapter I.2.17). One class of molecules that have been particularly important in particle surface modification are poly(ethylene oxides) (PEOs) and their shorter chain polyethylene glycol (PEG) derivatives. PEO/PEG modification reduces protein adsorption to the particle surface, prolonging circulation time (Cui et al., 2001; Coombes et al., 1997), and reducing uptake by clearance systems (Akerman et al., 2002).

CHARACTERIZATION OF MICRO/NANOPARTICLES

Particle behavior depends on a number of factors. Surface charge and morphology, size, chemical composition, and aggregation tendency all play a crucial role, with minor alterations in any of these properties potentially resulting in significant changes. The most common methods for physiochemical characterization of micro/nanoparticles are electron microscopy (i.e., SEM, TEM), dynamic light scattering (DLS), zeta-potential measurement, X-ray photoelectron spectroscopy (XPS) analysis, atomic force microscopy (AFM), and optical light or fluorescence microscopy. Microscopy techniques, including optical, fluorescence, and transmission and scanning electron microscopy, are often utilized to study particle surface morphology, size, and shape (Champion et al., 2007). It should be noted, however, that EM techniques may involve sample preparation and pre-handling steps, including dehydration and coating, which might alter particle properties. Further, high vacuum renders imaging of aqueous samples problematic and may cause structural damage. To obtain size information on a distribution or large number of particles, light scattering is preferred. Light scattering measures particle size based on the scattering intensity, which changes with variation in scattering angle, size of particles, and refractive indices of particle and the medium. However, there are some limitations associated with this method, such as the difficulty of evaluating particles with a large size distribution. Also, measurements are dependent on algorithms that require particles to be spherical, and to have a large difference between particle refractive index and dispersion medium. Therefore, light scattering is often combined with other techniques such as microscopy methods. AFM is used less frequently and can provide similar information on particle shape, size, and morphology. In addition, AFM can gather quantitative data on elastic modulus, and particle adhesion and deformation (Vakarelski et al., 2001; Tagit et al., 2008). XPS can provide information on the particle surface chemical composition (Xie et al., 2006). In particular, XPS has been exploited to detect the presence of biomolecules/drugs on the particle surface (Xie et al., 2006; Chesko et al., 2008). Zeta potential measures net surface charge, and is used to determine particle stability in suspension. A large absolute value of zeta potential indicates more stable particles that are less likely to form aggregates. Zeta potentials can also be used to determine whether a surface is positively or negatively charged, which is important for particle modification. For example, cationic microparticles can be prepared by modifying PLGA microparticles with surfactants, with the surface positive charge confirmed by zeta potential studies. Positively charged surfaces permit tethering of negatively-charged DNA for DNA delivery applications (Singh et al., 2000). Similarly, anionic microparticles of PLGA have been created with similar techniques (Singh et al., 2004). Apart from these commonly exercised characterization techniques, differential scanning calorimetry (DSC) has been used to assess changes in polymer thermal properties, infrared spectra (IR or FTIR) has been used to analyze interactions between multiple materials, and surface tension measurements have been used to evaluate particle aggregation. Chapters I.1.5 (surface characterization) and II.3.8 (microscopy) will be useful in understanding the characterization of microparticles and nanoparticles.

APPLICATIONS OF MICRO/NANOPARTICLES

Detection and Imaging

Microparticles have been used for detection and imaging applications, however, to a significantly lesser extent than nanoparticles. Fluorescent microparticles have been employed to detect diseased tissues. For example, fluorescent microparticles (1–10 μm diameter) were used for the detection of myocardium at risk of infarction in dogs (Fieno et al., 2000). Furthermore, because of the presence of multiple fluorescent dyes, microparticles with different fluorescent tags can be imaged in the same specimen. Fluorescent particles have also been used to image patterned scaffolds (Lu et al., 2006). Patterned scaffolds were fabricated with fluorescent microparticles as a prototype to demonstrate the feasibility of patterning scaffolds with multiple regions with varying properties (i.e., regions with

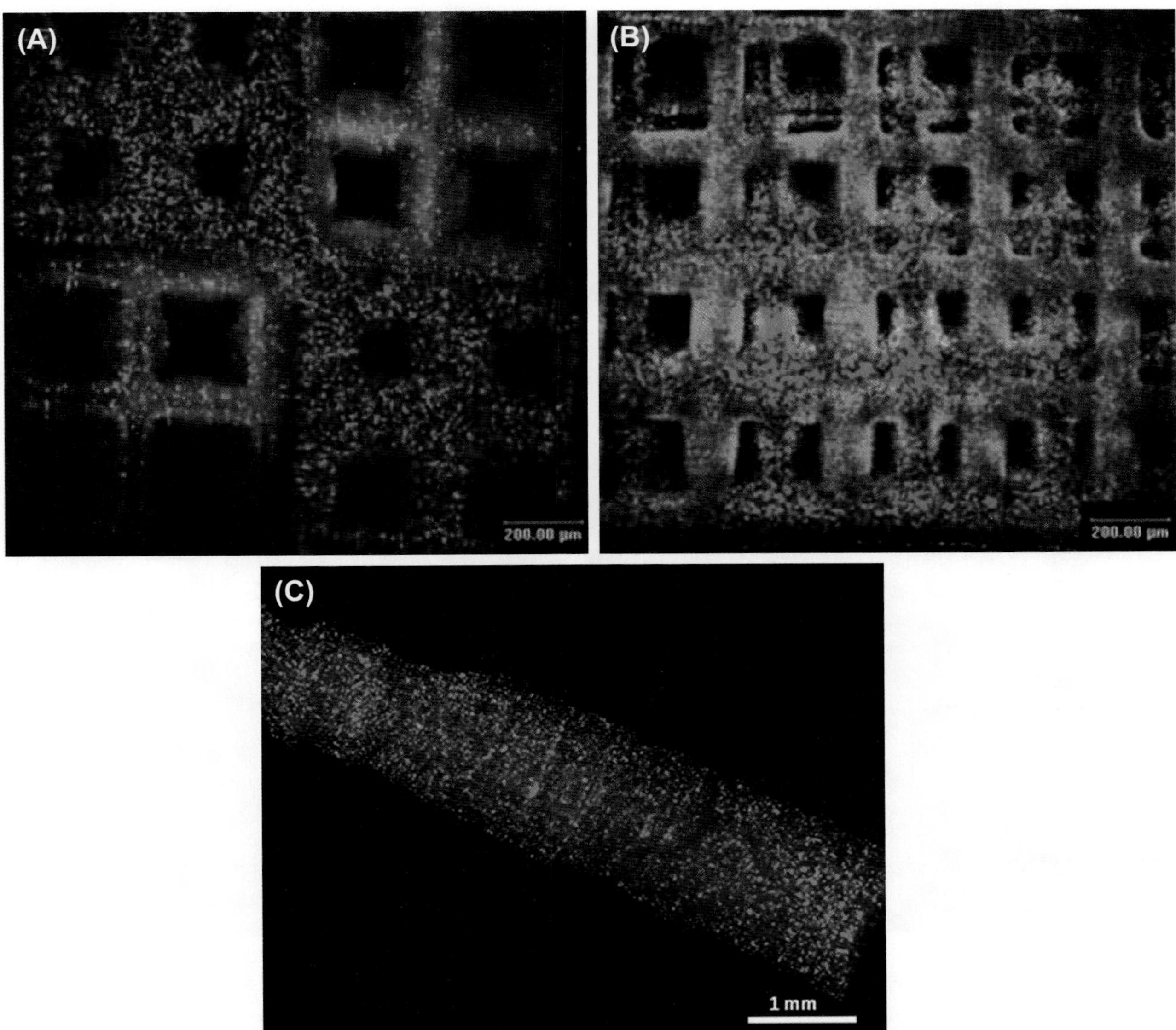

FIGURE I.2.19.7 Spatial patterning in PEG scaffolds using fluorescent microparticles (A and B) (Lu et al., 2006); (A) single-layered scaffold with spatially-patterned regions; (B) multi-layered scaffold with each layer fabricated with different fluorescent microparticles; (C) scaffold exhibiting the gradient of two different fluorescent microparticles running in opposite directions. (Figure C courtesy of Shalu Suri, Li Hsin Han, Shaochen Chen, and Christine Schmidt from The University of Texas at Austin, TX, USA.)

different biomolecules, growth factors, and cell types). For example, Cyt5- and FITC-labeled polystyrene microparticles were utilized to create spatially patterned multilayered PEG scaffolds (Figure I.2.19.7). Microparticles have also been used to create growth factor gradients as a proof of concept to create biomolecule gradients (Figure I.2.19.7C).

In addition to fluorescent microparticles, metal iron oxide microparticles (IOMPs) have also had an impact on *in vivo* cellular imaging and tracking (McAteer et al., 2007; Shapiro et al., 2004). Because of their larger size, in comparison to iron oxide nanoparticles, IOMPs are not suitable for molecular detection applications. However, IOMPs have successfully been used for imaging of single cells, mouse embryos, and tissue samples such as the brain. These microparticles are advantageous since the iron oxide content is orders of magnitude higher than that in nanoparticles, which results in improved MRI contrast. Antibody-conjugated IMOPs are endocytosed by cells and provide quantifiable contrast even at very low concentration in cells, unlike nanoparticles. In addition, IOMPs are less susceptible to nonspecific cellular uptake than nanoparticles. Furthermore, fluorescent IOMPs offer the advantage of being able to perform MRI and fluorescence microscopy in the same specimen (Shapiro et al., 2004).

Nanoparticles in particular can play a critical role in molecular detection by serving as probes of biomolecular processes. Their size similarity to biomolecules and their unique size-dependent properties make them particularly attractive agents for imaging, detection, and treatment. For example, quantum dots (Chan & Nie, 1998; Bruchez Jr. et al., 1998) have achieved tremendous

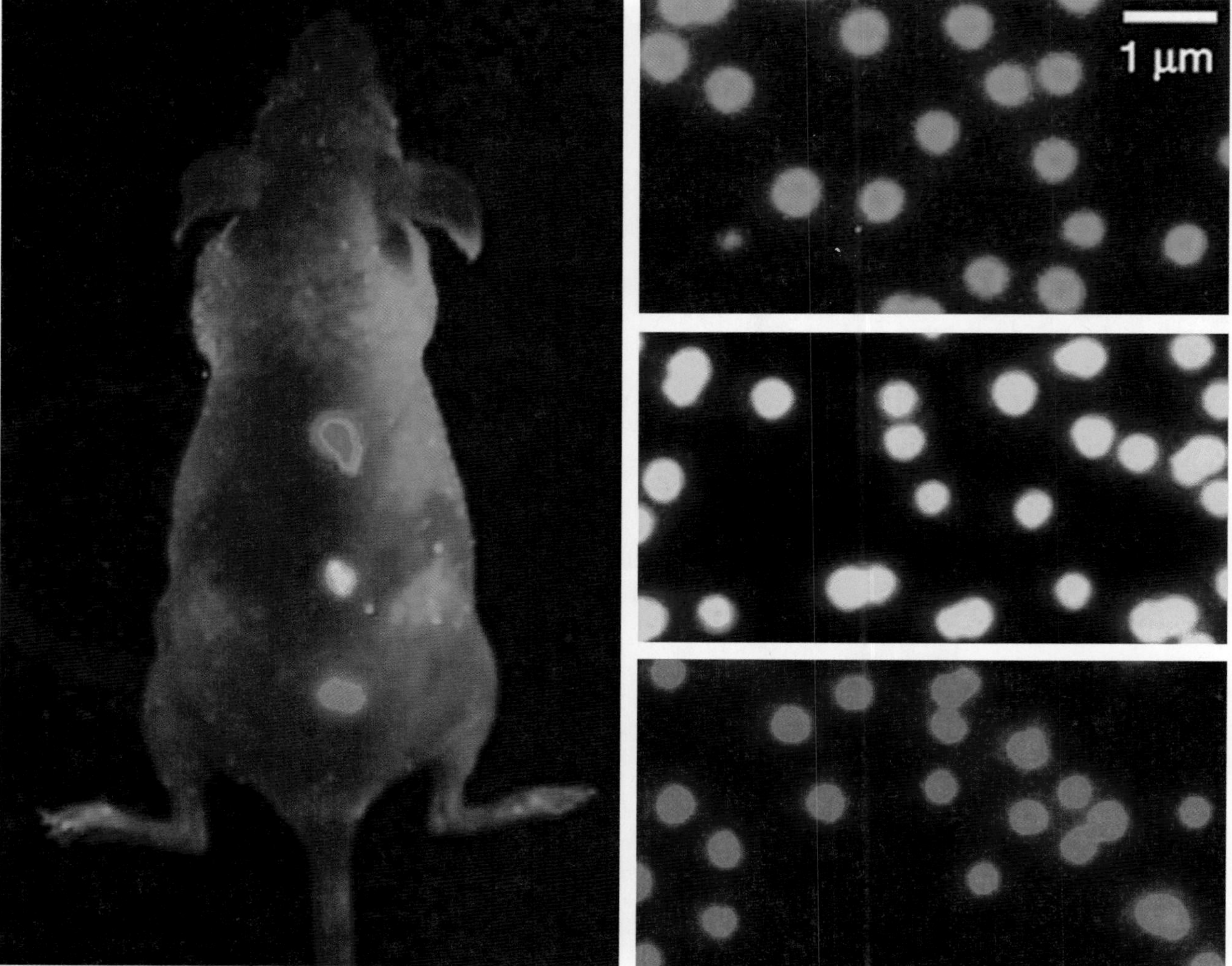

FIGURE I.2.19.8 Multiple color imaging. Three different colors of quantum dot microbeads (right) were imaged simultaneously (left). Approximately 1–2 million beads in each color were injected subcutaneously at three adjacent locations on a host animal. (Image from Gao et al., 2004.)

success in biomedical imaging because of their unique fluorescence properties. They exhibit a bright signal that has led to dramatically improved signal-to-noise ratios (compared with organic fluorescent dyes). Additionally, because of their narrow fluorescence emission peaks, they can be used to image multiple biomolecules simultaneously (i.e., multiplexed imaging). This is important because many diseases are often best identified by multiple, rather than single, biomarkers. The first clinically relevant biomolecular investigation using quantum dots was the detection and imaging of an important breast cancer marker, her-2 (Wu et al., 2003), the presence of which is used to direct treatment (e.g., eligible to receive herceptin) and can impact patient survival. Quantum dots of different emission wavelengths have been used to detect and quantify as many as four biomarkers simultaneously in tissue sections (Xing et al., 2007; Nie et al., 2007). *In vivo*, simultaneous targeting and imaging of three distinct tumors has been demonstrated (Gao et al., 2004) (Figure I.2.19.8). Further, quantum dots have extraordinary stability against photobleaching, making them ideal for tracking dynamic biological processes. Quantum dots have been used to observe development of *Xenopus* frog embryos (Dubertret et al., 2002), trafficking of glycine receptors in the neural synapse (Dahan et al., 2003), motion of individual kinesin motor proteins (Courty et al., 2006), and intracellular dynamics of the Tat peptide, which is responsible for the cellular invasion of HIV virus (Ruan et al., 2007).

Whereas quantum dots have made a significant impact in imaging, gold nanoparticles have been instrumental in molecular detection. The SPR effect of gold nanoparticles has been extensively employed for detection of nucleic acids (e.g., DNA) (Elghanian et al., 1997). Gold nanoparticles are easily conjugated to oligonucleotides which, upon exposure to complementary target sequences, cross-link, reducing particle separation distance. This change in distance leads to an SPR effect, resulting in a solution color change detectable by the naked eye (Figure I.2.19.6A). Another property which has been exploited for molecular detection is surface-enhanced Raman scattering (SERS), which is the enhancement of the Raman scattering signal upon binding of a reporter molecule to the gold nanoparticle surface. The enhancement factor can be as large as 10^{14}–10^{15}, which has permitted *in vivo* tumor imaging (Qian et al., 2008). Compared with fluorescence detection, an important advantage of SERS-based detection is its potential for greater multiplexing. Each SERS spectrum is unique, and is composed of multiple bands that are much narrower than fluorescence spectra of dyes or quantum dots permitting many more signals to be simultaneously discriminated. It has been reported that up to three respiratory viruses in the same sample were detected by SERS-based gold nanoparticles (Boisselier and Astruc, 2009), and in theory multiplexing capacity could be greatly expanded.

In addition to these two nanoparticles, several others have made an impact in molecular medicine. SPIONs have been used as a contrast agent for MR imaging because of their ability to shorten T2 relaxation times (Pankhurst et al., 2003; Yang et al., 2009). SWCNTs have been used as tips in AFM imaging to selectively detect DNA from single-base mismatch sequences (Woolley et al., 2000). CNTs have also been used to create sensors. For example, the electrochemical signal of guanine bases was amplified for label-free electrochemical detection of DNA (Wang et al., 2004).

Drug and Gene Delivery

Microparticles are of immense interest in the field of drug and gene delivery, and have been researched predominantly for delivery applications. Microparticles offer versatile advantages including: multiple routes of administration; encapsulation of a range of drugs including hydrophobic, hydrophilic or neutral drugs; protection of the biomolecule from degradation in the stringent environment and from body clearance; and provision of site-specific controlled release. The drug or biomolecule can either be encapsulated in the bulk of the particle or conjugated outside using surface modification techniques. Multiple biomolecules can also be delivered from the same microparticles to target more than one application. When encapsulated, the drug release occurs either via diffusion through swelling or via bulk or surface degradation. Therefore, varying release profiles can be obtained simply by modulating the microparticle degradation rate. For example, PLA/PGA microparticles were used for 5-fluorouracil delivery; different PLA/PGA ratios resulted in altered release profiles (Park, 1995) (Figure I.2.19.1).

One of the most active and thoroughly investigated areas of microparticle drug delivery is for vaccine applications. Particularly, polymeric microparticles in the size range of 1–5 μm are easily recognized by immune cells as foreign and easily phagocytosed, thus making microparticles a popular system for vaccine delivery. Microparticles of natural and synthetic origin have been utilized to deliver nucleic acids such as DNA, short interfering RNA (siRNA), proteins, and peptides. Immunogenic DNA antigens have been extensively encapsulated inside or conjugated outside PLGA microparticles for preclinical studies with Hepatitis B and various cancer models (Luo et al., 2003). Advancement in the synthesis/fabrication techniques has opened the field towards the delivery of multiple biomolecules. For example, modification of conventional PLGA microparticles has facilitated delivery of DNA with siRNA by encapsulating siRNA in the core of the particles and conjugating DNA to the surface (Singh et al., 2008). Further, hybrids of PLGA and poly β amino esters (PBAE, pH sensitive polyesters) have resulted in microparticles with superior intracellular delivery of DNA antigens for tumor rejection, as compared to conventional DNA encapsulated PLGA microparticles (Little et al., 2004).

Microparticles have also been extensively used for oral delivery of proteins and peptides. Mucoadhesive chitosan-based microparticles have been used for delivery of proteins to the gastrointestinal tract, where they can be taken up by the M cells of the Peyer's patch (van der Lubben et al., 2001c; Ahire et al., 2007). These particles exhibit prolonged retention in the mucous layer, thus increasing the availability of microparticles for cellular uptake. Further examples of microparticle applications based on material properties have been summarized in the section "Materials Used for Microparticle Synthesis" and in Table I.2.19.1.

Nanoparticles are attractive biomolecule carriers because of their small size, which provides several benefits. First, the dosing potential of nanoparticles is significant. Nearly 1 million, 5 nm particles could fit within the cytoplasm of a 10 μm cell versus ~1–10 micron-sized particles, with typical numbers in the range of tens of thousands. Second, nanoparticles consist primarily of surface (e.g., surface-to-volume ratio is very high), which is easily modified with biomolecules to be delivered. Thus, the loading capability of nanoparticle carriers is high on a volume basis. Additionally, easily modified surfaces permit conjugation to targeting molecules that can direct nanoparticles to the cell/organ of interest. Because nanoparticles are much smaller than cells, they can be directed not only to specific cell types, but even to specific regions of a cell (i.e., nucleus for gene delivery). Finally, the small size of nanoparticles has proven an advantage in evading the body's normal uptake and clearance system, the RES, which permits nanoparticle carriers to circulate for much longer than drug alone.

One of the most active areas of drug delivery has been in the field of cancer treatment, primarily because of the ability of nanoparticles to take advantage of the enhanced permeation and retention (EPR) effect (Matsumura and Maeda, 1986). Tumor vasculature is commonly poorly organized and "leaky" compared to native vasculature. As a result, particles with diameters <~200 nm can pass through the blood vessels and accumulate at the tumor site. Polymer and lipid nanoparticles were

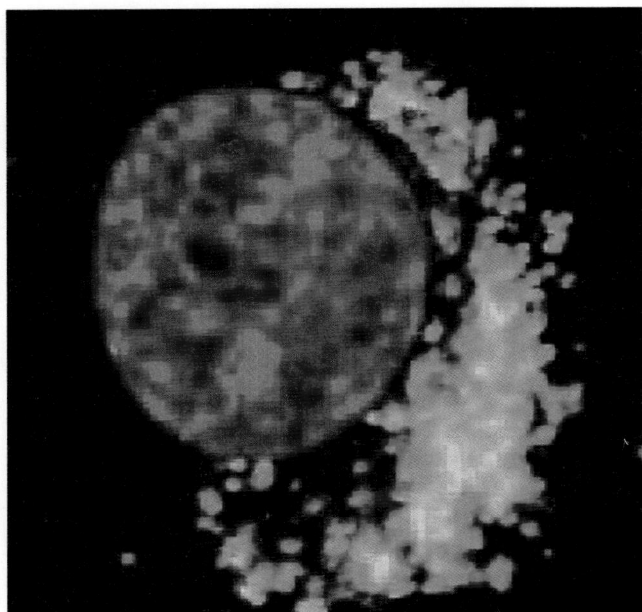

FIGURE I.2.19.9 Dual detection of fluorescent-DNA-CNT (green) internalized into a HeLa cell with the nucleus stained by DRAQ5 fluroescent dye (red). (Photo from Kam et al., 2005).

among the first studied as drug delivery carriers (Marty et al., 1978), primarily for delivery of doxorubicin, a classic chemotherapy drug. Silica (Roy et al., 2003), gold (Paciotti et al., 2004), quantum dots (Gao et al., 2004), and CNTs (Figure I.2.19.9) (Kam et al., 2005) have all been used as simple drug carriers, whereas magnetic nanoparticles and gold nanoshells have demonstrated enhanced delivery effects as a result of their unique properties. Accumulation of magnetic nanoparticles at the target site can be improved by the application of magnetic fields (Lubbe and Bergemann, 1998). Gold nanoshells, which are grown on an inert core, generate heat when exposed to near-infrared light that can be used to expand temperature-sensitive polymers and release drug at selected intervals (Sershen et al., 2000).

Nanoparticles have also been actively studied for gene delivery, and have several advantages as gene carriers. Despite the fact that their transfection efficiency can be considerably lower than their viral counterparts, they are often much less toxic. Additionally, it has been shown that the increased bulk of nanoparticles versus free DNA can enhance *ex vivo* transfection (Luo & Saltzman, 2000). Finally, many nanoparticles possess additional functionalities that can enhance delivery or permit simultaneous monitoring of transfection. As with drug delivery, the first use of nanoparticles included liposomes (Fraley et al., 1980) and polymers (Cohen et al., 2000), although silica (Kneuer et al., 2000), fullerenes (Nakamura et al., 2000), gold (McIntosh et al., 2001), magnetic nanoparticles (Scherer et al., 2002), CNTs (Pantarotto et al., 2004; Kam et al., 2005), and quantum dots (Srinivasan et al., 2006) have also been employed. One interesting application of gene carriers specific to nanoparticles is magnetotransfection (Scherer et al., 2002), where magnetic nanoparticle gene carriers are exposed to a magnetic field that enhances their uptake at the target site.

Therapeutic Treatment

In addition to serving as carriers of drugs and other therapeutic compounds, occasionally the unique properties of nanoparticles themselves can be applied for treatment. Two excellent examples of this are the use of nanoparticles for hyperthermia and interactions with ROS for photodynamic therapy and antioxidant activity. Hyperthermia is the death of tissue caused by elevated temperature (e.g., >~41°C), and is primarily used for cancer treatment. It was originally induced by direct application of heat or by circulating heated fluids through the vasculature; however, both of these methods produced significant collateral damage. Nanoparticles, because of their small size and ability to be precisely targeted, offer the possibility of localized hyperthermia. Two types of nanoparticles have been targeted in this application. The first are gold nanoshells (O'Neal et al., 2004), which strongly absorb light in the near-IR and convert it to thermal vibrations (e.g., heat) (Figure I.2.19.10). Light absorption in the near-IR is advantageous because the near-IR and IR offer greatest tissue penetration. The second type of nanoparticles used for hyperthermia are iron oxide magnetic nanoparticles (Chan et al., 1993), which produce heat when placed in an alternating magnetic field. In both cases, heat is localized primarily to the tumor site, reducing collateral damage and producing near microscale targeting of cancer treatment.

The second application of nanoparticles as a direct therapy is their interaction with the ROS pathway. Usually nanoparticles are examined for photodynamic therapy, which is the light-induced generation of ROS for producing cell death, but they have also been applied as anti-oxidants that neutralize ROS. In photodynamic therapy, light excites an electron from a sensitizer molecule to an excited state. The energy in this excited electron can be transferred to oxygen or oxygen-containing species to produce ROS. Traditionally, photodynamic therapy has been accomplished using molecular sensitizers; however, nanoparticles offer a significant advantage in that their excitation wavelengths can be precisely tuned by adjusting nanoparticle size. Both quantum dots (Samia et al., 2003) and fullerenes (Tabata et al., 1997) have been used to produce ROS.

CHALLENGES FACING NANOPARTICLES

Despite their success as biomaterials, there are several challenges facing nanoparticles. Greatest among these are questions regarding their fate in the body. Although most of the materials used have been widely characterized in bulk form, there is evidence that nanoparticles

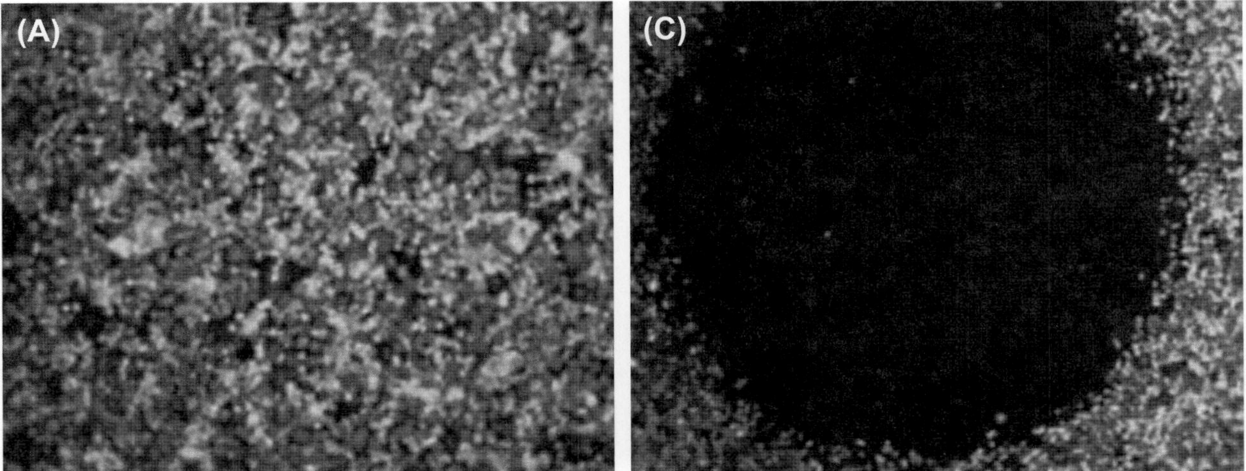

FIGURE I.2.19.10 Breast carcinoma cells were treated with nanoshells conjugated with either: (A) control; or (C) targeting biomolecule. Upon NIR (near infrared) laser treatment, cells exposed to targeted nanoshells were effectively destroyed (C), as demonstrated by the circular region corresponding to the laser spot that lacks staining with a fluorescent viability marker, while the nanoshells bound to the control antibody did not produce this effect (A). Silver staining was then performed to visualize nanoshell binding to cells. (Photo from Hirsch et al., 2006).

of the same material can exhibit substantially different behavior. For one, the surface-to-volume ratio is significantly enhanced in nanomaterials, altering the chemical interface with the body. Also, many nanoparticles exhibit unique properties not found in the bulk that can influence biological response. Finally, because of their small size they can be internalized by cells and organs, especially those of the RES, yet they are often too large to be cleared by the kidneys. Understanding the biological behavior of nanosized versions of well-known biomaterials is crucial to their continued application.

Toxicity

A major concern for nanoparticles is their potential toxicity. Although some materials (e.g., gold, silica, many polymer and lipid formulations, magnetic nanoparticles) have demonstrated exceptional biocompatibility, there are lingering questions regarding the toxicity of carbon nanostructures and quantum dots. Carbon nanostructures offer several potential modes of toxicity. They are often synthesized in the presence of metallic catalysts, such as nickel and iron, which can provoke cellular responses (Kagan et al., 2006). Careful isolation of the final product can minimize these effects. Also, they are usually extremely hydrophobic. Unmodified carbon nanostructures can penetrate the cell membrane (Monteiro-Riviere et al., 2005) and form granulomas in tissue (Warheit et al., 2004). This effect can be mitigated by acidic oxidation to yield hydrophilic surfaces. Finally, carbon nanostructures have been shown to cause direct toxicity, most likely through the generation of ROS (Oberdorster, 2004). Surface coatings are especially important in controlling these responses.

Quantum dots have evoked similar toxicity concerns (Kotov et al., 2009), most notably because a primary constituent material is often cadmium. Cadmium can cause acute toxicity, resulting in death in large doses, and in small doses is a carcinogen disrupting DNA mismatch repair. However, cadmium in nanoparticle form is presented as a complex with another material. Primary modes of quantum dot toxicity are believed to be their ability to generate ROS, and also given the structural similarity of Cd and Zn, to bind metallothenein, disrupting zinc metabolic pathways. Toxicity of quantum dot nanostructures can be minimized by judicious use of surface coatings (Derfus et al., 2004), but biocompatibility remains a significant concern hindering clinical application in patients (Figure I.2.19.11).

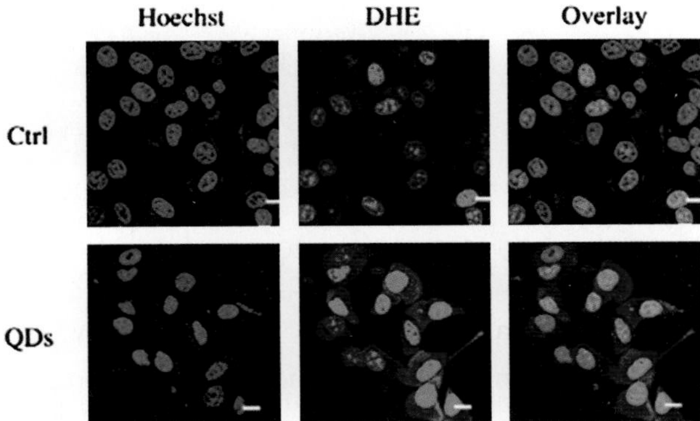

FIGURE I.2.19.11 Quantum dots induce generation of ROS. MCF-7 cells were stained with a marker for ROS (DHE, red) and for the cell nucleus (Hoechst, blue). Four hours after quantum dot exposure, quantum dot-treated cells (bottom) produce more ROS (red) than control cells. The scale bar represents 10 μm. (Figure from Lovric et al., 2005).

Targeting

Another significant challenge for nanoparticles is their controlled targeting both *in vitro* and *in vivo*. In theory,

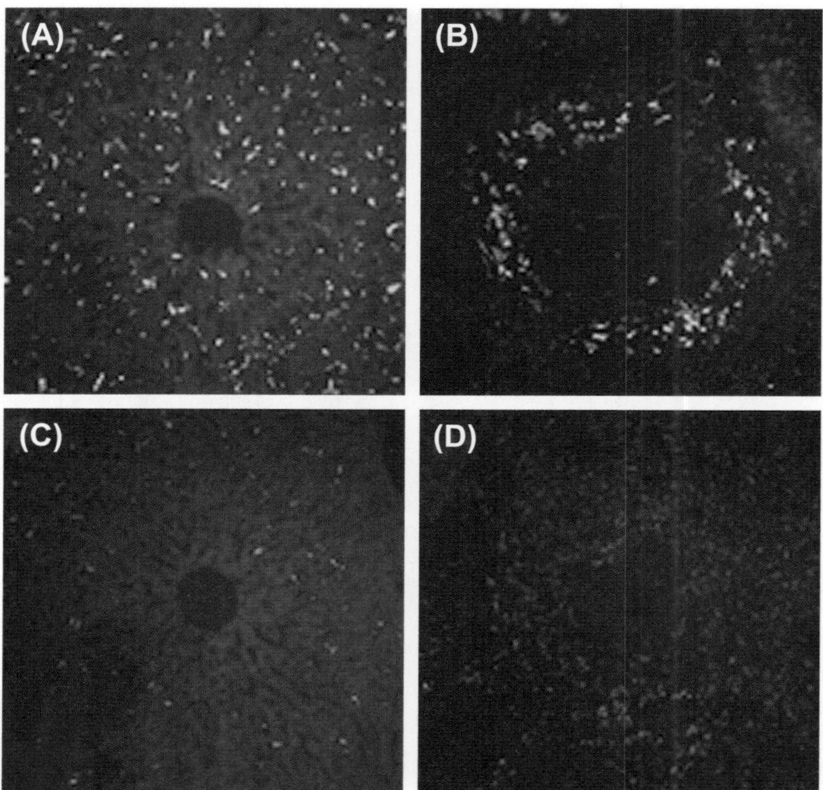

FIGURE I.2.19.12 Nanoparticle uptake by the RES is reduced by PEG coating. Green quantum dots without (A,B) or with (C,D) an added PEG coating were injected into the tail vein of mice. Quantum dot localization was observed by epifluorescent microscopy of the liver (A,C) and spleen (B,D). Original magnifications: 200×. (Figure from Akerman et al., 2002.)

nanoparticles can be precisely directed to a specific location using antibodies, peptides or other biorecognition molecules. However, in practice, nanoparticles are subject to non-specific binding and endocytosis *in vitro* (Gomez et al., 2005), and uptake by the RES *in vivo* (Akerman et al., 2002). Non-specific binding usually results from the nanoparticle charge, and can be reduced by using appropriate blocking agents or by encapsulating nanoparticles in another protective material. Endocytosis, however, is more problematic. Free nanoparticles in solution and nanoparticles targeted to cell surface receptors are readily internalized by cells through endocytotic pathways. It has been shown that nanoparticles accumulate in the perinuclear region (Shukla et al., 2005), where they presumably remain indefinitely or until metabolic dissolution. Endocytosis is related to nanoparticle size (Osaki et al., 2004), with optimal uptake for particles ~50 nm. *In vitro*, endocytosis can be prevented by performing experiments or analyses at low temperature (e.g., 4°C). The *in vivo* picture is more complicated. Since most of the accumulated data on cellular uptake is from *in vitro* systems, it is not clear that particles would be internalized *in vivo* at the same rate. Nonetheless, the possibility of nanoparticle internalization and permanent residency in the cell is troubling. Finally, many nanoparticles never make it to their cellular destinations *in vivo*, having been filtered from the blood by the RES. Evasion of RES uptake is possible, usually using PEG as a coating (Figure I.2.19.12) (Akerman et al., 2002). PEG increases nanoparticle circulation time because it resists protein adsorption (Eugene, 2004), preventing recognition by immune cells of the RES.

Clearance

Nanoparticle biodistribution is a key issue for clinical application. Unlike bulk materials that remain in the body indefinitely (e.g., hip implant, pacemaker), nanoparticles have the potential to interact with the body's native clearance system, the RES. The RES consists of the lymph nodes, spleen, and liver, and is comprised of immune cells responsible for identifying and removing foreign matter from the blood. Alternatively, very small molecules may be filtered by the kidneys. Since many applications of nanoparticles utilize blood circulation to promote particle delivery, nanoparticles have ample opportunity to interact with the RES. Polymer, lipid, and magnetic nanostructures degrade *in vivo*, and their ultimate biological fate is less of a concern; however, there is little evidence that quantum dots, metallic nanoparticles, silica or carbon nanostructures are eliminated through natural degradation processes. The fate of particles is controlled not only by their composition, but also by their surface coating, further

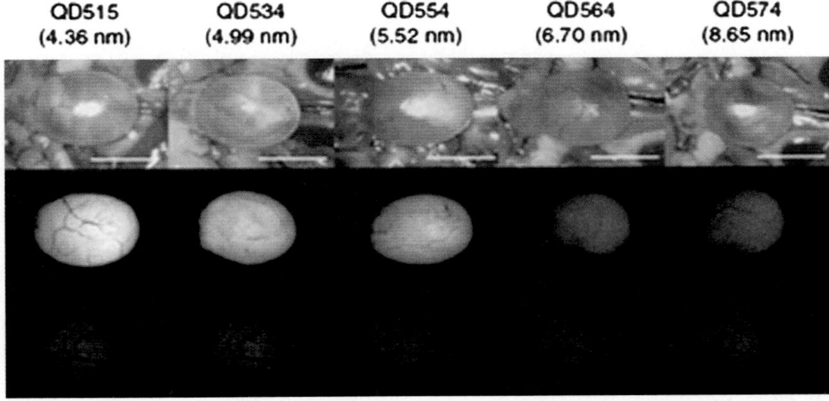

FIGURE I.2.19.13 Mouse bladders after intravenous injection of quantum dots of defined diameter. Shown are frames from a color video (top) and fluorescence images for bladder 4 hours after injection (middle) and for uninjected control bladder (bottom). Scale bar: 1 cm. (Figure from Choi et al., 2007.)

complicating the issue; and as this field is just emerging, current understanding of the fate of nanoparticles in the body is incomplete.

Gold nanoshells (James et al., 2007) and nanoparticles (Kim et al., 2007), quantum dots (Fischer et al., 2006), silica nanoparticles (He et al., 2008), nanodiamond (Yuan et al., 2009), and fullerenes (Yamago et al., 1995) have all been found to accumulate in RES organs. It has been shown that particles with diameters <~6 nm (Choi et al., 2007; Burns et al., 2009), can be cleared through renal pathways (Figure I.2.19.13), but most nanoparticles intended for clinical use are >30 nm. There is some evidence that quantum dots (Chen et al., 2008) and silica (He et al., 2008) can be at least partially metabolized by the liver, and excreted through feces or urine. Carbon nanotube distribution has been more controversial, with several conflicting reports claiming elimination via kidney (Singh et al., 2006) versus uptake by the RES with little clearance (McDevitt et al., 2007). A recent report provides some clarity, demonstrating that individual CNTs can cross renal barriers for excretion, whereas CNTs that are aggregated are taken up by the RES (Lacerda et al., 2008).

CONCLUDING REMARKS

Despite their identical composition to bulk materials, microparticles and nanoparticles represent a new class of biomaterials with distinct properties, interactions with biological components, and distribution within the body. Microparticles and nanoparticles have been used for a range of applications in the biomedical field. Their properties can be easily modulated by varying their composition, shape, size, and surface chemistry. Microparticles were first used in the 1950s, and are one of most extensively studied systems in the biomedical field. Microparticles are most commonly employed in drug and vaccine delivery. Nanomaterials represent an emerging field; careful study of their biological activity will be required before there is large-scale clinical application. Because of their small size scale and unique physical properties, nanoparticles offer great hope for the future, particularly in the areas of imaging and drug delivery. They have already been successfully applied in the clinic as MRI contrast agents and drug delivery vehicles. These applications will undoubtedly continue to expand as more materials are moved from the bench to the bedside.

BIBLIOGRAPHY

Abra, R. M., Bankert, R. B., Chen, F., Egilmez, N. K., Huang, K., et al. (2002). The next generation of liposome delivery systems: Recent experience with tumor-targeted, sterically-stabilized immunoliposomes and active-loading gradients. *J. Liposome Res.*, **12**(1–2), 1–3.

Agnihotri, S. A., Mallikarjuna, N. N., & Aminabhavi, T. M. (2004). Recent advances on chitosan-based micro- and nanoparticles in drug delivery. *J. Control. Release.* **100**(1), 5–28.

Ahire, V. J., Sawant, K. K., Doshi, J. B., & Ravetkar, S. D. (2007). Chitosan microparticles as oral delivery system for tetanus toxoid. *Drug Dev. Ind. Pharm.*, **33**(10), 1112–1124.

Akerman, M. E., Chan, W. C., Laakkonen, P., Bhatia, S. N., & Ruoslahti, E. (2002). Nanocrystal targeting *in vivo*. *Proc. Natl. Acad. Sci. USA*, **99**(20), 12617–12621.

Alivisatos, A. P. (1996). Semiconductor clusters, nanocrystals, and quantum dots. *Science*, **271**(5251), 933–937.

Alivisatos, A. P., Gu, W., & Larabell, C. (2005). Quantum dots as cellular probes. *Annu. Rev. Biomed. Eng.*, **7**, 55–76.

Allen, C., Maysinger, D., & Eisenberg, A. (1999). Nano-engineering block copolymer aggregates for drug delivery. *Colloid. Surface. B – Biointerfaces*, **16**(1-4), 3–27.

Allen, T. M., & Hansen, C. (1991). Pharmacokinetics of stealth versus conventional liposomes: Effect of dose. *Biochim. Biophys. Acta*, **1068**(2), 133–141.

Aukunuru, J. V., Ayalasomayajula, S. P., & Kompella, U. B. (2003). Nanoparticle formulation enhances the delivery and activity of a vascular endothelial growth factor antisense oligonucleotide in human retinal pigment epithelial cells. *J. Pharm. Pharmacol.*, **55**(9), 1199–1206.

Bakker-Woudenberg, I. A. (2002). Long-circulating sterically stabilized liposomes as carriers of agents for treatment of infection or for imaging infectious foci. *Int. J. Antimicrob. Ag.*, **19**, 299–311.

Balasubramanian, G., Chan, I. Y., Kolesov, R., Al-Hmoud, M., & Tisler, J. (2008). Nanoscale imaging magnetometry with diamond spins under ambient conditions. *Nature*, **455**(7213), 648–651.

Balmayor, E. R., Tzulakoglu, K., Azevedo, H. S., & Reis, R. L. (2009). Preparation and characterization of starch-poly-epsilon-caprolactone microparticles incorporating bioactive agents for drug delivery and tissue engineering applications. *Acta Biomater.*, **5**(4), 1035–1045.

Barbara, J. A., Salker, R., Challis, P., & Contreras, M. (1989). Gelatin particle agglutination assay for HIV antibodies: A rapid, economical modification with increased sensitivity. *Med. Lab. Sci.*, **46**(2), 135–140.

Bates, I. R., Hébert, B., Luo, Y., Liao, J., Bachir, A. I., et al. (2006). Membrane lateral diffusion and capture of CFTR within transient confinement zones. *Biophys. J.*, **91**(3), 1046–1058.

Benagiano, G., & Gabelnick, H. L. (1979). Biodegradable systems for the sustained release of fertility-regulating agents. *J. Steroid Biochem.*, **11**(1, Part 2), 449–455.

Benedetti, L. M., Topp, E. M., & Stella, V. J. (1990). Microspheres of hyaluronic acid esters: Fabrication methods and *in vitro* hydrocortisone release. *J. Control. Release*, **13**(1), 33–41.

Berkland, C., Kipper, M. J., Narasimhan, B., Kim, K. K., & Pack, D. W. (2004). Microsphere size, precipitation kinetics and drug distribution control drug release from biodegradable polyanhydride microspheres. *J. Control. Release*, **94**(1), 129–141.

Berthold, A., Cremer, K., & Kreuter, J. (1998). Collagen microparticles: Carriers for glucocorticosteroids. *Eur. J. Pharm. Biopharm.*, **45**(1), 23–29.

Besteman, K., Lee, J. O., Wiertz, F. G.M., Heering, H. A., & Dekker, C. (2003). Enzyme-coated carbon nanotubes as single-molecule biosensors. *Nano Lett.*, **3**(6), 727–730.

Blume, G., & Cevc, G. (1990). Liposomes for the sustained drug release *in vivo*. *Biochim. Biophys. Acta*, **1029**(1), 91–97.

Bodmeier, R., Chen, H. G., & Paeratakul, O. (1989). A novel approach to the oral delivery of micro- or nanoparticles. *Pharm. Res.*, **6**(5), 413–417.

Boisselier, E., & Astruc, D. (2009). Gold nanoparticles in nanomedicine: Preparations, imaging, diagnostics, therapies and toxicity. *Chem. Soc. Rev.*, **38**(6), 1759–1782.

Bowersock, T. L., Hogenesch, H., Suckow, M., Porter, R. E., Jackson, R., et al. (1996). Oral vaccination with alginate microsphere systems. *J. Control. Release*, **39**, 209–220.

Bruchez, M., Jr., Moronne, M., Gin, P., Weiss, S., & Alivisatos, A. P. (1998). Semiconductor nanocrystals as fluorescent biological labels. *Science*, **281**(5385), 2013–2016.

Burns, A. A., Vider, J., Ow, H., Herz, E., Penate-Medina, O., et al. (2009). Fluorescent silica nanoparticles with efficient urinary excretion for nanomedicine. *Nano Lett.*, **9**(1), 442–448.

Champion, J. A., Katare, Y. K., & Mitragotri, S. (2007). Making polymeric micro- and nanoparticles of complex shapes. *Proc. Natl. Acad. Sci. USA*, **104**(29), 11901–11904.

Chan, D. C. F., Kirpotin, D. B., & Bunn, P. A. (1993). Synthesis and evaluation of colloidal magnetic iron-oxides for the site-specific radiofrequency-induced hyperthermia of cancer. *J. Magn. and Magn. Mater.*, **122**(1–3), 374–378.

Chan, W. C., & Nie, S. (1998). Quantum dot bioconjugates for ultrasensitive nonisotopic detection. *Science*, **281**(5385), 2016–2018.

Chang, P. L., Hortelano, G., Awrey, D. E., & Tse, M. (1994). Growth of recombinant fibroblasts in alginate microcapsules. *Biotechnol. Bioeng.*, **43**(10), 925–933.

Chen, J., Hamon, M. A., Hu, H., Chen, Y., Rao, A. M., et al. (1998). Solution properties of single-walled carbon nanotubes. *Science*, **282**(5386), 95–98.

Chen, F. M., Wu, Z. F., Sun, H. H., Wu, H., Xin, S. N., et al. (2006). Release of bioactive BMP from dextran-derived microspheres: A novel delivery concept. *Int. J. Pharm.*, **307**(1), 23–32.

Chen, Z., Chen, H., Meng, H., Xing, G., Gao, X., et al. (2008). Bio-distribution and metabolic paths of silica coated CdSeS quantum dots. *Toxicol. Appl. Pharm.*, **230**(3), 364–371.

Chesko, J., Kazzaz, J., Ugozzoli, M., Singh, M., O'Hagan, D. T., et al. (2008). Characterization of antigens adsorbed to anionic PLG microparticles by XPS and TOF-SIMS. *J. Pharm. Sci.*, **97**(4), 1443–1453.

Cheung, R. Y., Ying, Y., Rauth, A. M., Marcon, N., & Yu Wu, X. (2005). Biodegradable dextran-based microspheres for delivery of anticancer drug mitomycin C. *Biomaterials*, **26**(26), 5375–5385.

Cho, M., & Sah, H. (2005). Formulation and process parameters affecting protein encapsulation into PLGA microspheres during ethyl acetate-based microencapsulation process. *J. Microencapsul.*, **22**(1), 1–12.

Choi, H. S., Liu, W., Misra, P., Tanaka, E., Zimmer, J. P., et al. (2007). Renal clearance of quantum dots. *Nat. Biotechnol.*, **25**(10), 1165–1170.

Cohen, H., Levy, R. J., Gao, J., Fishbein, I., Kousaev, V., et al. (2000). Sustained delivery and expression of DNA encapsulated in polymeric nanoparticles. *Gene Ther.*, **7**(22), 1896–1905.

Coombes, A. G., Tasker, S., Lindblad, M., Holmgren, J., Hoste, K., et al. (1997). Biodegradable polymeric microparticles for drug delivery and vaccine formulation: The surface attachment of hydrophilic species using the concept of poly(ethylene glycol) anchoring segments. *Biomaterials*, **18**(17), 1153–1161.

Courty, S., Bouziques, C., Luccardini, C., Ehrensperger, M. V., Bonneau, S., et al. (2006). Tracking individual proteins in living cells using single quantum dot imaging. *Method. Enzymol.*, **414**, 211–228.

Crick, F., & Hughes, A. (1949). The physical properties of the cytoplasm: A study by means of the magnetic particle method. *Exp. Cell Res.*, **1**, 37–80.

Cui, X., Lee, V. A., Raphael, Y., Wiler, J. A., Hetke, J. F., et al. (2001). Surface modification of neural recording electrodes with conducting polymer/biomolecule blends. *J. Biomed. Mater. Res.*, **56**(2), 261–272.

Cui, D., Tian, F., Ozkan, C. S., Wang, M., & Gao, H. (2005). Effect of single wall carbon nanotubes on human HEK293 cells. *Toxicol. Lett.*, **155**(1), 73–85.

Dahan, M., Lévi, S., Luccardini, C., Rostaing, P., Riveau, B., et al. (2003). Diffusion dynamics of glycine receptors revealed by single-quantum dot tracking. *Science*, **302**(5644), 442–445.

Danilenko, V. V. (2004). On the history of the discovery of nanodiamond synthesis. *Phys. Solid State*, **46**(4), 595–599.

De Vos, P., De Haan, B. J., Wolters, G. H., Strubbe, J. H., & Van Schilfgaarde, R. (1997). Improved biocompatibility but limited graft survival after purification of alginate for microencapsulation of pancreatic islets. *Diabetologia*, **40**(3), 262–270.

DeCarli, P. S., & Jamieson, J. C. (1961). Formation of diamond by explosive shock. *Science*, **133**(3467), 1821–1822.

Demetriou, A. A., Whiting, A. F., Feldman, D., Levenson, S. M., Chowdhury, N. R., et al. (1986). Replacement of liver function in rats by transplantation of microcarrier-attached hepatocytes. *Science*, **233**(4769), 1190–1192.

Derfus, A. M., Chan, W. C.W., & Bhatia, S. N. (2004). Probing the cytotoxicity of semiconductor quantum dots. *Nano Lett.*, **4**(1), 11–18.

Diwan, M., Khar, R. K., & Talwar, G. P. (2001). Tetanus toxoid loaded "preformed microspheres" of cross-linked dextran. *Vaccine*, **19**(28-29), 3853–3859.

Dorati, R., Genta, I., Tomasi, C., Modena, T., Colonna, C., et al. (2008). Polyethylenglycol-co-poly-D, L-lactide copolymer based microspheres: Preparation, characterization and delivery of a model protein. *J. Microencapsul.*, **25**(5), 330–338.

Dubernet, C., Benoit, J. P., Couarraze, G., & Duchene, D. (1987). Microencapsulation of nitrofurantoin in poly([epsilon]-caprolactone): tableting and *in vitro* release studies. *Int. J. Pharm.*, **35**(1–2), 145–156.

Dubertret, B., Skourides, P., Norris, D. J., Noireaux, V., Brivanlou, A. H., et al. (2002). *In vivo* imaging of quantum dots encapsulated in phospholipid micelles. *Science*, **298**(5599), 1759–1762.

Dugan, L. L., Gabrielsen, J. K., Yu, S. P., Lin, T. S., & Choi, D. W. (1996). Buckminsterfullerenol free radical scavengers reduce excitotoxic and apoptotic death of cultured cortical neurons. *Neurobiol. Dis.*, **3**(2), 129–135.

Dumitriu, S. (2002). *Polymeric Biomaterials* (2nd ed). , New York: Marcel Dekker. xiv, 1168 p.

Elghanian, R., Storhoff, J. J., Mucic, R. C., Letsinger, R. L., & Mirkin, C. A. (1997). Selective colorimetric detection of polynucleotides based on the distance-dependent optical properties of gold nanoparticles. *Science*, **277**(5329), 1078–1081.

Eto, H., Tamou, Y., Ohsawa, Y., & Kikuchi, N. (1992). TEM observations of diamond films prepared by microwave plasma CVD. *Diam. Rel. Mater.*, **1**(2-4), 373–379.

Eugene, M. (2004). Polyethyleneglycols and immunocamouflage of the cells tissues and organs for transplantation. *Cell Mol. Biol. (Noisy-le-Grand)*, **50**(3), 209–215.

Faraday, M. (1857). Experimental relations of gold (and other metals) to light. *Philosophical Transactions*, **147**, 145–181.

Feldman, Y., Wasserman, E., Srolovitz, D. J., & Tenne, R. (1995). High-rate, gas-phase growth of Mos2 nested inorganic fullerenes and nanotubes. *Science*, **267**(5195), 222–225.

Ferrari, M. (2005). Cancer nanotechnology: Opportunities and challenges. *Nat. Rev. Cancer*, **5**(3), 161–171.

Ficek, B. J., & Peppas, N. A. (1993). Novel preparation of poly(vinyl alcohol) microparticles without cross-linking agent for controlled drug-delivery of proteins. *J. Control. Release*, **27**(3), 259–264.

Fieno, D. S., Kim, R. J., Chen, E. L., Lomasney, J. W., Klocke, F. J., et al. (2000). Contrast-enhanced magnetic resonance imaging of myocardium at risk: Distinction between reversible and irreversible injury throughout infarct healing. *J. Am. Coll. Cardiol.*, **36**(6), 1985–1991.

Fischer, H. C., Liu, L., Pang, K. S., & Chan, W. C.W. (2006). Pharmacokinetics of nanoscale quantum dots: *In vivo* distribution, sequestration, and clearance in the rat. *Adv. Funct. Mater.*, **16**(10), 1299–1305.

Fraley, R., Subramani, S., Berg, P., & Papahadjopoulos, D. (1980). Introduction of liposome-encapsulated Sv40 DNA into cells. *J. Biol. Chem.*, **255**(21), 431–435.

Gao, X., Cui, Y., Levenson, R. M., Chung, L. W., & Nie, S. (2004). *In vivo* cancer targeting and imaging with semiconductor quantum dots. *Nat. Biotechnol.*, **22**(8), 969–976.

George, P. M., Lyckman, A. W., LaVan, D. A., Hegde, A., Leung, Y., et al. (2005). Fabrication and biocompatibility of polypyrrole implants suitable for neural prosthetics. *Biomaterials*, **26**(17), 3511–3519.

Gheith, M. K., Sinani, V. A., Wicksted, J. P., Matts, R. L., & Kotov, N. A. (2005). Single-walled carbon nanotube polyelectrolyte multilayers and freestanding films as a biocompatible platform for neuroprosthetic implants. *Adv. Mater.*, **17**(22), 2663–2670.

Gobe, M., Konno, K., Kyori, K., & Kitahara, A. (1983). Preparation and characterization of monodisperse magnetite sols in W/O microemulsion. *J. Colloid Interf. Sci.*, **93**(1), 293–295.

Gomez, N., Winter, J. O., Shieh, F., Saunders, A. E., Korgel, B. A., et al. (2005). Challenges in quantum dot-neuron active interfacing. *Talanta*, **67**(3), 462–471.

Gregoriadis, G., Leathwood, P. D., & Ryman, B. E. (1971). Enzyme entrapment in liposomes. *FEBS Lett.*, **14**(2), 95–99.

Grun, M., Lauer, I., & Unger, K. K. (1997). The synthesis of micrometer- and submicrometer-size spheres of ordered mesoporous oxide MCM-41. *Adv. Mater.*, **9**(3), 254–257.

Guliyeva, U., Oner, F., Ozsoy, S., & Hasiroğlu, R. (2006). Chitosan microparticles containing plasmid DNA as potential oral gene delivery system. *Eur. J. Pharm. Biopharm.*, **62**(1), 17–25.

Gumusderelioglu, M., & Deniz, G. (2000). Sustained release of mitomycin-C from poly(DL-lactide)/poly(DL-lactide-co-glycolide) films. *J. Biomater. Sci. Polym. Ed.*, **11**(10), 1039–1050.

Guo, T., Nikolaev, P., Thess, A., Colbert, D. T., & Smalley, R. E. (1995). Catalytic growth of single-walled nanotubes by laser vaporization. *Chem. Phys. Lett.*, **243**(1-2), 49–54.

Gupta, A. K., & Gupta, M. (2005). Synthesis and surface engineering of iron oxide nanoparticles for biomedical applications. *Biomaterials*, **26**(18), 3995–4021.

Han, M., Gao, X., Su, J. Z., & Nie, S. (2001). Quantum-dot-tagged microbeads for multiplexed optical coding of biomolecules. *Nat. Biotechnol.*, **19**(7), 631–635.

Hari, P. R., Chandy, T., & Sharma, C. P. (1996). Chitosan/calcium-alginate beads for oral delivery of insulin. *J. Appl. Polym. Sci.*, **59**(11), 1795–1801.

He, L., Musick, M. D., Nicewarner, S. R., Salinas, F. G., Benkovic, S. J., et al. (2000). Colloidal Au-enhanced surface plasmon resonance for ultrasensitive detection of DNA hybridization. *J. Am. Chem. Soc.*, **122**(38), 9071–9077.

He, X. X., Nie, H., Wang, K., Tan, W., Wu, X., et al. (2008). *In vivo* study of biodistribution and urinary excretion of surface-modified silica nanoparticles. *Anal. Chem.*, **80**(24), 9597–9603.

Helland, A., Wick, P., Koehler, A., Schmid, K., & Som, C. (2007). Reviewing the environmental and human health knowledge base of carbon nanotubes. *Environ. Health Perspect.*, **115**(8), 1125–1131.

Heller, J., Hellwing, R. F., Baker, R. W., & Tuttle, M. E. (1983). Controlled release of water-soluble macromolecules from bioerodible hydrogels. *Biomaterials*, **4**(4), 262–266.

Heller, J., Barr, J., Ng, S. Y., Abdellauoi, K. S., & Gurny, R. (2002). Poly(ortho esters): Synthesis, characterization, properties and uses. *Adv. Drug Deliv. Rev.*, **54**(7), 1015–1039.

Heller, D. A., Baik, S., Eurell, T. E., & Strano, M. S. (2005). Single-walled carbon nanotube spectroscopy in live cells: Towards long-term labels and optical sensors. *Adv. Mater.*, **17**(23), 2793–2799.

Hesketh, T., Li, L., Ye, X., Wang, H., Jiang, M., et al. (2006). HIV and syphilis in migrant workers in eastern China. *Sex Transm. Infect.*, **82**(1), 11–14.

Hirsch, L. R., Gobin, A. M., Lowery, A. R., Tam, F., Drezek, R. A., et al. (2006). Metal nanoshells. *Ann. Biomed. Eng.*, **34**(1), 15–22.

Holland, T. A., Tabata, Y., & Mikos, A. G. (2005). Dual growth factor delivery from degradable oligo(poly(ethylene glycol) fumarate) hydrogel scaffolds for cartilage tissue engineering. *J. Control. Release*, **101**(1-3), 111–125.

Holland, T. A., Bodde, E. W., Cuijpers, V. M., Baggett, L. S., Tabata, Y., et al. (2007). Degradable hydrogel scaffolds for *in vivo* delivery of single and dual growth factors in cartilage repair. *Osteoarthr. Cartilage*, **15**(2), 187–197.

Holt, K. B. (2007). Diamond at the nanoscale: Applications of diamond nanoparticles from cellular biomarkers to quantum computing. *Philos. Transact. A Math Phys. Eng. Sci.*, **365**(1861), 2845–2861.

Hombreiro Pérez, M., Zinutti, C., Lamprecht, A., Ubrich, N., Astier, A., et al. (2000). The preparation and evaluation of poly([epsilon]-caprolactone) microparticles containing both a lipophilic and a hydrophilic drug. *J. Control. Release*, **65**(3), 429–438.

Huo, Q. S., Feng, J., Schuth, F., & Stucky, G. D. (1997). Preparation of hard mesoporous silica spheres. *Chem. of Mater.*, **9**(1), 14–17.

Hyeon, T., Lee, S. S., Park, J., Chung, Y., & Na, H. B. (2001). Synthesis of highly crystalline and monodisperse maghemite nanocrystallites without a size-selection process. *J. Am. Chem. Soc.*, **123**(51), 12798–12801.

Iijima, S. (1991). Helical microtubules of graphitic carbon. *Nature*, **354**(6348), 56–58.

Jain, R. A. (2000). The manufacturing techniques of various drug loaded biodegradable poly(lactide-co-glycolide) (PLGA) devices. *Biomaterials*, **21**(23), 2475–2490.

Jaiswal, J. K., Mattoussi, H., Mauro, J. M., & Simon, S. M. (2003). Long-term multiple color imaging of live cells using quantum dot bioconjugates. *Nat. Biotechnol.*, **21**(1), 47–51.

Jalil, R., & Nixon, J. R. (1990). Biodegradable poly(lactic acid) and poly(lactide-co-glycolide) microcapsules: Problems associated with preparative techniques and release properties. *J. Microencapsul.*, **7**(3), 297–325.

James, W. D., Hirsch, L. R., West, J. L., O'Neal, P. D., & Payne, J. D. (2007). Application of INAA to the build-up and clearance of gold nanoshells in clinical studies in mice. *J. Radioanaly. Nucl. Chem.*, **271**(2), 455–459.

Ji, X., Song, X., Li, J., Bai, Y., Yang, W., et al. (2007). Size control of gold nanocrystals in citrate reduction: The third role of citrate. *J. Am. Chem. Soc.*, **129**(45), 13939–13948.

Johnson, O. L., Jaworowicz, W., Cleland, J. L., Bailey, L., Charnis, M., et al. (1997). The stabilization and encapsulation of human growth hormone into biodegradable microspheres. *Pharm. Res.*, **14**(6), 730–735.

Joki, T., Machluf, M., Atala, A., Zhu, J., Seyfried, N. T., et al. (2001). Continuous release of endostatin from microencapsulated engineered cells for tumor therapy. *Nat. Biotechnol.*, **19**(1), 35–39.

Kagan, V. E., Tyurina, Y. Y., Tyurin, V. A., Konduru, N. V., Potapovich, A. I., et al. (2006). Direct and indirect effects of single walled carbon nanotubes on RAW 264.7 macrophages: Role of iron. *Toxicol. Lett.*, **165**(1), 88–100.

Kam, N. W., O'Connell, M., Wisdom, J. A., & Dai, H. (2005). Carbon nanotubes as multifunctional biological transporters and near-infrared agents for selective cancer cell destruction. *Proc. Natl. Acad. Sci. USA*, **102**(33), 11600–11605.

Kim, S., Lim, Y. T., Soltesz, E. G., De Grand, A. M., Lee, J., et al. (2004). Near-infrared fluorescent type II quantum dots for sentinel lymph node mapping. *Nat. Biotechnol.*, **22**(1), 93–97.

Kim, S. J., Hahn, S. K., Kim, M. J., Kim, D. H., & Lee, Y. P. (2005). Development of a novel sustained release formulation of recombinant human growth hormone using sodium hyaluronate microparticles. *J. Control. Release*, **104**(2), 323–335.

Kim, D., Park, S., Lee, J. H., Jeong, Y. Y., & Jon, S. (2007). Antibiofouling polymer-coated gold nanoparticles as a contrast agent for *in vivo* X-ray computed tomography imaging. *J. Am. Chem. Soc.*, **129**(24), 7661–7665.

King, G. A., Daugulis, P., Faulkner, M., & Goosen, F. A. (1987). Alginate-polylysine microcapsules of controlled membrane molecular weight cutoff for mammalian cell culture engineering. *Biotechnol. Bioeng.*, **3**(4), 231–240.

Kneuer, C., Sameti, M., Haltner, E. G., Schiestel, T., Schirra, H., et al. (2000). Silica nanoparticles modified with aminosilanes as carriers for plasmid DNA. *Int. J. Pharm.*, **196**(2), 257–261.

Ko, J. A., Park, H. J., Hwang, S. J., Park, J. B., & Lee, J. S. (2002). Preparation and characterization of chitosan microparticles intended for controlled drug delivery. *Int. J. Pharm.*, **249**(1-2), 165–174.

Kong, J., Soh, H. T., Cassell, A. M., Quate, C. F., & Dai, H. (1998). Synthesis of individual single-walled carbon nanotubes on patterned silicon wafers. *Nature*, **395**(6705), 878–881.

Kong, J., Franklin, N. R., Zhou, C., Chapline, M. G., Peng, S., et al. (2000). Nanotube molecular wires as chemical sensors. *Science*, **287**(5453), 622–625.

Kossovsky, N., Gelman, A., Hnatyszyn, H. J., Rajguru, S., Garrell, R. L., et al. (1995). Surface-modified diamond nanoparticles as antigen delivery vehicles. *Bioconjug. Chem.*, **6**(5), 507–511.

Kotov, N. A., Winter, J. O., Clements, I. P., Jan, E., Timko, B. P., et al. (2009). Nanomaterials for neural interfaces. *Adv. Mater.*, **21**(40), 3970–4004.

Krätschmer, W., Lamb, L. D., Fostiropoulous, K., & Huffman, D. R. (1990). Solid C-60: A new form of carbon. *Nature*, **347**(6291), 354–358.

Kresge, C. T., Leonowicz, M. E., Roth, W. J., Vartuli, J. C., & Beck, J. S. (1992). Ordered mesoporous molecular-sieves synthesized by a liquid-crystal template mechanism. *Nature*, **359**(6397), 710–712.

Kriwet, B., Walter, E., & Kissel, T. (1998). Synthesis of bioadhesive poly(acrylic acid) nano- and microparticles using an inverse emulsion polymerization method for the entrapment of hydrophilic drug candidates. *J. Controll. Release*, **56**(1–3), 149–158.

Kroto, H. W., Heath, J. R., O'Brien, S. C., Curl, R. F., & Smalley, R. E. (1985). C-60 –Buckminsterfullerene. *Nature*, **318**(6042), 162–163.

Lacerda, L., Herrero, M. A., Venner, K., Bianco, A., Prato, M., et al. (2008). Carbon-nanotube shape and individualization critical for renal excretion. *Small*, **4**(8), 1130–1132.

Lasic, D. D. (1993). *Liposomes: From Physics to Applications*. Amsterdam, Netherlands: Elsevier.

Lee, J. E., Park, J. C., Lee, K. H., Oh, S. H., Kim, J. G., et al. (2002). An infection-preventing bilayered collagen membrane containing antibiotic-loaded hyaluronan microparticles: Physical and biological properties. *Artif. Organs*, **26**(7), 636–646.

Lee, S., Yang, S. C., Heffernan, M. J., Taylor, W. R., & Murthy, N. (2007). Polyketal microparticles: A new delivery vehicle for superoxide dismutase. *Bioconjug. Chem.*, **18**(1), 4–7.

Lehr, C. M., Bouwstra, J. A., Kok, W., De Boer, A. G., Tukker, J. J., et al. (1992). Effects of the mucoadhesive polymer polycarbophil on the intestinal absorption of a peptide drug in the rat. *J. Pharm. Pharmacol.*, **44**(5), 402–407.

Lewis, D. H. (1990). Controlled release of bioactive agents from lactide/glycolide polymers. In R. Langer, & M. C. Hasin (Eds.), *Biodegradable Polymers as Drug Delivery Systems* (pp. 1–41). New York, NY: Marcel Dekker.

Li, X., Kong, X., Shi, S., Zheng, X., Guo, G., et al. (2008). Preparation of alginate coated chitosan microparticles for vaccine delivery. *BMC Biotechnol.*, **8**, 89.

Liang, H. C., Chang, W. H., Lin, K. J., & Sung, H. W. (2003). Genipin-cross-linked gelatin microspheres as a drug carrier for intramuscular administration: *In vitro* and *in vivo* studies. *J. Biomed. Mater. Res. A.*, **65**(2), 271–282.

Lim, F., & Moss, R. D. (1981). Microencapsulation of living cells and tissues. *J. Pharm. Sci.*, **70**(4), 351–354.

Lim, S. T., Martin, G. P., Berry, D. J., & Brown, M. B. (2000). Preparation and evaluation of the *in vitro* drug release properties and mucoadhesion of novel microspheres of hyaluronic acid and chitosan. *J. Control. Release*, **66**(2–3), 281–292.

Little, S. R., Lynn, D. M., Ge, Q., Anderson, D. G., Puram, S. V., et al. (2004). Poly-beta amino ester-containing microparticles enhance the activity of nonviral genetic vaccines. *Proc. Natl. Acad. Sci. USA*, **101**(26), 9534–9539.

Liu, J., Rinzler, A. G., Dai, H., Hafner, J. H., Bradley, R. K., et al. (1998). Fullerene pipes. *Science*, **280**(5367), 1253–1256.

Lorenzo-Lamosa, M. L., Remuñán-Lopez, C., Vila-Jato, J. L., & Alonso, M. J. (1998). Design of microencapsulated chitosan microspheres for colonic drug delivery. *J. Control. Release*, **52**(1–2), 109–118.

Lovat, V., Pantarotto, D., Lagostena, L., Cacciari, B., Grandolfo, M., et al. (2005). Carbon nanotube substrates boost neuronal electrical signaling. *Nano Lett.*, **5**(6), 1107–1110.

Lovric, J., Cho, S. J., Winnick, F. M., & Maysinger, D. (2005). Unmodified cadmium telluride quantum dots induce reactive oxygen species formation leading to multiple organelle damage and cell death. *Chem. Biol.*, **12**(11), 1227–1234.

Lu, Y., Mapili, G., Suhali, G., Chen, S., & Roy, Kl (2006). A digital micro-mirror device-based system for the microfabrication of complex, spatially patterned tissue engineering scaffolds. *J. Biomed. Mater. Res. A.*, 77(2), 396–405.

Lubbe, A. S., & Bergemann, C. (1998). Magnetically-controlled drug targeting. *Cancer J.*, **11**(3), 104–105.

Luo, D., & Saltzman, W. M. (2000). Enhancement of transfection by physical concentration of DNA at the cell surface. *Nat. Biotechnol.*, **18**(8), 893–895.

Luo, Y., O'Hagan, D., Zhou, H., Singh, M., Ulmer, J., et al. (2003). Plasmid DNA encoding human carcinoembryonic antigen (CEA) adsorbed onto cationic microparticles induces protective immunity against colon cancer in CEA-transgenic mice. *Vaccine*, **21**(17–18), 1938–1947.

Malyala, P., O'Hagan, D. T., & Singh, M. (2009). Enhancing the therapeutic efficacy of CpG oligonucleotides using biodegradable microparticles. *Adv. Drug Deliv. Rev.*, **61**(3), 218–225.

Marchais, H., Benali, S., Irache, J. M., Tharasse-Bloch, C., Lafont, O., et al. (1998). Entrapment efficiency and initial release of phenylbutazone from nanocapsules prepared from different polyesters. *Drug Dev. Ind. Pharm.*, **24**(9), 883–888.

Marty, J. J., Oppenheim, R. C., & Speiser, P. (1978). Nanoparticles - New colloidal drug delivery system. *Pharm. Acta Helv.*, **53**(1), 17–23.

Matsumura, Y., & Maeda, H. (1986). A new concept for macromolecular therapeutics in cancer-chemotherapy – Mechanism of tumoritropic accumulation of proteins and the antitumor agent Smancs. *Cancer Res.*, **46**(12), 6387–6392.

Mattson, M. P., Haddon, R. C., & Rao, A. M. (2000). Molecular functionalization of carbon nanotubes and use as substrates for neuronal growth. *J. Mol. Neurosci.*, **14**(3), 175–182.

McAteer, M. A., von Zur Muhlen, C., Anthony, D. C., Sibson, N. R., & Choudhury, R. P. (2007). *In vivo* magnetic resonance imaging of acute brain inflammation using microparticles of iron oxide. *Nat. Med.*, **13**(10), 1253–1258.

McDevitt, M. R., Chattopadhyay, D., Kappel, B. J., Jaggi, J. S., Schiffman, S. R., et al. (2007). Tumor targeting with antibody-functionalized, radiolabeled carbon nanotubes. *J. Nucl. Med.*, **48**(7), 1180–1189.

McIntosh, C. M., Esposito, E. A., 3rd, Boal, A. K., Simard, J. M., Martin, C. T., et al. (2001). Inhibition of DNA transcription using cationic mixed monolayer protected gold clusters. *J. Am. Chem. Soc.*, **123**(31), 7626–7629.

Monteiro-Riviere, N. A., Nemanich, R. J., Inman, A. Q., Wang, Y. Y., & Riviere, J. E. (2005). Multi-walled carbon nanotube interactions with human epidermal keratinocytes. *Toxicol. Lett.*, **155**(3), 377–384.

Mundargi, R. C., Babu, V. R., Rangaswamy, V., Patel, P., & Aminabhavi, T. M. (2008). Nano/micro technologies for delivering macromolecular therapeutics using poly(D, L-lactide-co-glycolide) and its derivatives. *J. Control. Release*, **125**(3), 193–209.

Murillo, M., Goñi, M. M., Irache, J. M., Arangoa, M. A., Blasco, J. M., et al. (2002). Modulation of the cellular immune response after oral or subcutaneous immunization with microparticles containing Brucella ovis antigens. *J. Control. Release*, **85**(1-3), 237–246.

Murray, C. B., Norris, D. J., & Bawendi, M. G. (1993). Synthesis and characterization of nearly monodisperse Cde (E = S, Se, Te) semiconductor nanocrystallites. *J. Am. Chem. Soc.*, **115**(19), 8706–8715.

Murthy, R. S.R. (1997). Biodegradable polymers. In N. K. Jain (Ed.), *Controlled and Novel Drug Delivery* (pp. 27–51). New Delhi, India: CBS Publisher.

Nakamura, E., Isobe, H., Tomita, N., Sawamura, M., Jinno, S., et al. (2000). Functionalized fullerene as an artificial vector for transfection. *Angew. Chem. Int. Edit.*, **39**(23), 4254–4257.

Nguyen, D. N., Raghavan, S. S., Tashima, L. M., Lin, E. C., Fredette, S. J., et al. (2008). Enhancement of poly(orthoester) microspheres for DNA vaccine delivery by blending with poly(ethylenimine). *Biomaterials*, **29**(18), 2783–2793.

Nie, S. M., Xing, Y., Kim, G. J., & Simons, J. W. (2007). Nanotechnology applications in cancer. *Ann. Rev. Biomed. Eng.*, **9**, 257–288.

O'Neal, D. P., Hirsch, L. R., Halas, N. J., Payne, J. D., & West, J. L. (2004). Photo-thermal tumor ablation in mice using near infrared-absorbing nanoparticles. *Cancer Lett.*, **209**(2), 171–176.

Oberdorster, E. (2004). Manufactured nanomaterials (fullerenes, C60) induce oxidative stress in the brain of juvenile largemouth bass. *Environ. Health Perspect.*, **112**(10), 1058–1062.

Okochi, H., & Nakano, M. (2000). Preparation and evaluation of w/o/w type emulsions containing vancomycin. *Adv. Drug Deliv. Rev.*, **45**(1), 5–26.

Oner, L., & Groves, M. J. (1993). Optimization of conditions for preparing 2- to 5-micron-range gelatin microparticles by using chilled dehydration agents. *Pharm. Res.*, **10**(4), 621–626.

Osaki, F., Kanamori, T., Sando, S., Sera, T., & Aoyama, Y. (2004). A quantum dot conjugated sugar ball and its cellular uptake on the size effects of endocytosis in the subviral region. *J. Am. Chem. Soc.*, **126**(21), 6520–6521.

Paciotti, G. F., Myer, L., Weinreich, D., Goia, D., Pavel, N., et al. (2004). Colloidal gold: A novel nanoparticle vector for tumor directed drug delivery. *Drug Deliv.*, **11**(3), 169–183.

Pankhurst, Q. A., Connolly, J., Jones, S. K., & Dobson, J. (2003). Applications of magnetic nanoparticles in biomedicine. *J. Phys. D – Appl. Phys.*, **36**(13), R167–R181.

Pantarotto, D., Singh, R., McCarthy, D., Erhardt, M., Briand, J. P., et al. (2004). Functionalized carbon nanotubes for plasmid DNA gene delivery. *Angew. Chem. Int. Edit.*, **43**(39), 5242–5246.

Park, S. B., Kang, H. W., Haam, S., Park, H. Y., & Kim, W. S. (2004). Ca-alginate microspheres encapsulated in chitosan beads. *J. Microencapsul.*, **21**(5), 485–497.

Park, H., Temenoff, J. S., Tabata, Y., Caplan, A. I., & Mikos, A. G. (2007a). Injectable biodegradable hydrogel composites for rabbit marrow mesenchymal stem cell and growth factor delivery for cartilage tissue engineering. *Biomaterials*, **28**(21), 3217–3227.

Park, J. H., Choi, S. O., Kamath, R., Yoon, Y. K., Allen, M. G., et al. (2007b). Polymer particle-based micromolding to fabricate novel microstructures. *Biomed. Microdevices*, **9**(2), 223–234.

Park, H., Temenoff, J. S., Tabata, Y., Caplan, A. I., Raphael, R. M., et al. (2008). Effect of dual growth factor delivery on chondrogenic differentiation of rabbit marrow mesenchymal stem cells encapsulated in injectable hydrogel composites. *J. Biomed. Mater. Res. A.*, **88**(4), 889–897.

Park, T. G. (1995). Degradation of poly(lactic-co-glycolic acid) microspheres: Effect of copolymer composition. *Biomaterials*, **16**(15), 1123–1130.

Patel, Z. S., Ueda, H., Yamamoto, M., Tabata, Y., & Mikos, A. G. (2008a). *In vitro* and *in vivo* release of vascular endothelial growth factor from gelatin microparticles and biodegradable composite scaffolds. *Pharm. Res.*, **25**(10), 2370–2378.

Patel, Z. S., Yamamoto, M., Ueda, H., Tabata, Y., & Mikos, A. G. (2008b). Biodegradable gelatin microparticles as delivery systems for the controlled release of bone morphogenetic protein-2. *Acta Biomater.*, **4**(5), 1126–1138.

Patel, Z. S., Young, S., Tabata, Y., Jansen, J. A., Wong, M. E., et al. (2008c). Dual delivery of an angiogenic and an osteogenic growth factor for bone regeneration in a critical size defect model. *Bone*, **43**(5), 931–940.

Peng, Z. A., & Peng, X. G. (2001). Formation of high-quality CdTe, CdSe, and CdS nanocrystals using CdO as precursor. *J. Am. Chem. Soc.*, **123**(1), 183–184.

Peng, X. G., Schlamp, M. C., Kadavanich, A. V., & Alivisatos, A. P. (1997). Epitaxial growth of highly luminescent CdSe/CdS core/shell nanocrystals with photostability and electronic accessibility. *J. Am. Chem. Soc.*, **119**(30), 7019–7029.

Pham, Q. P., Kasper, F. K., Scott Baggett, L., Raphael, R. M., Jansen, J. A., et al. (2008). The influence of an *in vitro* generated bone-like extracellular matrix on osteoblastic gene expression of marrow stromal cells. *Biomaterials*, **29**(18), 2729–2739.

Pinto Reis, C., Neufeld, R. J., Ribeiro, A. J., & Veiga, F. (2006). Nanoencapsulation I. Methods for preparation of drug-loaded polymeric nanoparticles. *Nanomedicine: The official journal of the American Academy of Nanomedicine*, **2**(1), 8–21.

Pitt, C. G., Gratzl, M. M., Jeffcoat, A. R., Zweidinger, R. A., & Schindler, A. (1979a). Sustained drug delivery systems II: Factors affecting release rates from poly(epsilon-caprolactone) and related biodegradable polyesters. *J. Pharm. Sci.*, **68**(12), 1534–1538.

Pitt, C. G., Jeffcoat, A. R., Zweidinger, R. A., & Schindler, A. (1979b). Sustained drug delivery systems. I. The permeability of poly(epsilon-caprolactone), poly(DL-lactic acid), and their copolymers. *J. Biomed. Mater. Res.*, **13**(3), 497–507.

Qian, X., Peng, X. H., Ansari, D. O., Yin-Goen, Q., Chen, G. Z., et al. (2008). *In vivo* tumor targeting and spectroscopic detection with surface-enhanced Raman nanoparticle tags. *Nat. Biotechnol.*, **26**(1), 83–90.

Ratcliffe, J. H., Hunneyball, I. M., Smith, A., Wilson, C. G., & Davis, S. S. (1984). Preparation and evaluation of biodegradable polymeric systems for the intra-articular delivery of drugs. *J. Pharm. Pharmacol.*, **36**(7), 431–436.

Re, M. (2006). Formulating drug delivery systems by spray drying. *Dry. Technol.*, **23**, 433–446.

Ritchie, R. H. (1957). Plasma losses by fast electrons in thin films. *Phys. Rev.*, **106**(5), 874.

Rosca, I. D., Watari, F., & Uo, M. (2004). Microparticle formation and its mechanism in single and double emulsion solvent evaporation. *J. Control. Release*, **99**(2), 271–280.

Rossetti, R., Nakahara, S., & Brus, L. E. (1983). Quantum size effects in the redox potentials, resonance raman-spectra, and electronic-spectra of CdS crystallites in aqueous-solution. *J. Chem. Phys.*, **79**(2), 1086–1088.

Rossetti, R., Ellison, J. L., Gibson, J. M., & Brus, L. E. (1984). Size effects in the excited electronic states of small colloidal CdS crystallites. *J. Chem. Phys.*, **80**(9), 4464–4469.

Rossler, B., Kreuter, J., & Scherer, D. (1995). Collagen microparticles: Preparation and properties. *J. Microencapsul.*, **12**(1), 49–57.

Roy, I., Ohulchanskyy, T. Y., Pudavar, H. E., Bergey, E. J., Oseroff, A. R., et al. (2003). Ceramic-based nanoparticles entrapping water-insoluble photosensitizing anticancer drugs: A novel drug-carrier system for photodynamic therapy. *J. Am. Chem. Soc.*, **125**(26), 7860–7865.

Ruan, G., Agrawal, A., Marcus, A. I., & Nie, S. (2007). Imaging and tracking of tat peptide-conjugated quantum dots in living cells: New insights into nanoparticle uptake, intracellular transport, and vesicle shedding. *J. Am. Chem. Soc.*, **129**(47), 14759–14766.

Samia, A. C., Chen, X., & Burda, C. (2003). Semiconductor quantum dots for photodynamic therapy. *J. Am. Chem. Soc.*, **125**(51), 15736–15737.

Sato, T. A., Miyamura, K., Sakae, K., Kobune, F., Inouye, S., et al. (1997). Development of a gelatin particle agglutination reagent for measles antibody assay. *Arch. Virol.*, **142**(10), 1971–1977.

Scherer, F., Anton, M., Schillinger, U., Henke, J., Bergemann, C., et al. (2002). Magnetofection: Enhancing and targeting gene delivery by magnetic force *in vitro* and *in vivo*. *Gene Ther.*, **9**(2), 102–109.

Schrand, A. M., Huang, H., Carlson, C., Schlager, J. J., Omacr Sawa, E., et al. (2007). Are diamond nanoparticles cytotoxic? *J. Phys. Chem. B*, **111**(1), 2–7.

Sershen, S. R., Westcott, S. L., Halas, N. J., & West, J. L. (2000). Temperature-sensitive polymer-nanoshell composites for photothermally modulated drug delivery. *J. Biomed. Mater. Res.*, **51**(3), 293–298.

Shapiro, E. M., Skrtic, S., Sharer, K., Hill, J. M., Dunbar, C. E., et al. (2004). MRI detection of single particles for cellular imaging. *Proc. Natl. Acad. Sci. USA*, **101**(30), 10901–10906.

Shenderova, O. A., Zhirnov, V. V., & Brenner, D. W. (2002). Carbon nanostructures. *Cr. Rev. Sol. State Mater. Sci.*, **27**(3–4), 227–356.

Shive, M. S., & Anderson, J. M. (1997). Biodegradation and biocompatibility of PLA and PLGA microspheres. *Adv. Drug Deliv. Rev.*, **28**(1), 5–24.

Shu, X. Z., & Zhu, K. J. (2001). Chitosan/gelatin microspheres prepared by modified emulsification and ionotropic gelation. *J. Microencapsul.*, **18**(2), 237–245.

Shukla, R., Bansal, V., Chaudhary, M., Basu, A., Bhonde, R. R., et al. (2005). Biocompatibility of gold nanoparticles and their endocytotic fate inside the cellular compartment: a microscopic overview. *Langmuir*, **21**(23), 10644–10654.

Siegel, S. J., Kahn, J. B., Metzger, K., Winey, K. I., Werner, K., et al. (2006). Effect of drug type on the degradation rate of PLGA matrices. *Eur. J. Pharm. Biopharm.*, **64**(3), 287–293.

Siepmann, J., & Siepmann, F. (2006). Microparticles used as drug delivery systems. In F. Kremer, & W. Richtering (Eds.), *Prog. Coll. Pol. Sci.* (Vol. 133, pp. 15–21). Germany: Springer.

Singh, M., Briones, M., Ott, G., & O'Hagan, D. (2000). Cationic microparticles: A potent delivery system for DNA vaccines. *Proc. Natl. Acad. Sci. USA*, **97**(2), 811–816.

Singh, M., Kazzaz, J., Chesko, J., Soenawan, E., Ugozzoli, M., et al. (2004). Anionic microparticles are a potent delivery system for recombinant antigens from *Neisseria meningitidis* serotype B. *J. Pharm. Sci.*, **93**(2), 273–282.

Singh, R., Pantarotto, D., Lacerda, L., Pastorin, G., Klumpp, C., et al. (2006). Tissue biodistribution and blood clearance rates of intravenously administered carbon nanotube radiotracers. *Proc. Natl. Acad. Sci. USA*, **103**(9), 3357–3362.

Singh, A., Nie, H., Ghosn, B., Qin, H., Kwak, L. W., et al. (2008). Efficient modulation of T-cell response by dual-mode, single-carrier delivery of cytokine-targeted siRNA and DNA vaccine to antigen-presenting cells. *Mol. Ther.*, **16**(12), 2011–2021.

Sinha, V. R., Bansal, K., Kaushik, R., Kumria, R., & Trehan, A. (2004a). Poly-epsilon-caprolactone microspheres and nanospheres: An overview. *Int. J. Pharm.*, **278**(1), 1–23.

Sinha, V. R., Singla, A. K., Wadhawan, S., Kaushik, R., Kumria, R., et al. (2004b). Chitosan microspheres as a potential carrier for drugs. *Int. J. Pharm.*, **274**(1–2), 1–33.

Sjoholm, I., & Edman, P. (1979). Acrylic microspheres *in vivo*. I. Distribution and elimination of polyacrylamide microparticles after intravenous and intraperitoneal injection in mouse and rat. *J. Pharmacol. Exp. Ther.*, **211**(3), 656–662.

Skubitz, K. (2003). Phase I trial of pegylated-liposomal doxorubicin (Doxil) in sarcoma. *Cancer Invest.*, **21**(2), 167–176.

Sondi, I., & Salopek-Sondi, B. (2004). Silver nanoparticles as antimicrobial agent: A case study on *E-coli* as a model for Gram-negative bacteria. *J. Coll. Interf. Sci.*, **275**(1), 177–182.

Srinivasan, C., Lee, J., Papadimitrakopoulos, F., Silbart, L. K., Zhao, M., et al. (2006). Labeling and intracellular tracking of functionally active plasmid DNA with semiconductor quantum dots. *Mol. Ther.*, **14**(2), 192–201.

Stöber, W., Fink, A., & Bohn, E. (1968). Controlled growth of monodisperse silica spheres in the micron size range. *J. Colloid Interf. Sci.*, **26**, 62–69.

Sun, C., Lee, J. S.H., & Zhang, M. Q. (2008). Magnetic nanoparticles in MR imaging and drug delivery. *Adv. Drug Deliver. Rev.*, **60**(11), 1252–1265.

Swatschek, D., Schatton, W., Müller, W., & Kreuter, J. (2002). Microparticles derived from marine sponge collagen (SCMPs): preparation, characterization and suitability for dermal delivery of all-trans retinol. *Eur. J. Pharm. Biopharm.*, **54**(2), 125–133.

Tabata, Y., Murakami, Y., & Ikada, Y. (1997). Photodynamic effect of polyethylene glycol-modified fullerene on tumor. *Jpn. J. Cancer Res.*, **88**(11), 1108–1116.

Tagit, O., Tomczak, N., & Vancso, G. J. (2008). Probing the morphology and nanoscale mechanics of single poly(N-isopropylacrylamide) microgels across the lower-critical-solution temperature by atomic force microscopy. *Small*, **4**(1), 119–126.

Takada, S., Uda, Y., Toguchi, H., & Ogawa, Y. (1995). Application of a spray drying technique in the production of TRH-containing injectable sustained-release microparticles of biodegradable polymers. *PDA J. Pharm. Sci. Technol.*, **49**(4), 180–184.

Tallury, P., Payton, K., & Santra, S. (2008). Silica-based multimodal/multifunctional nanoparticles for bioimaging and biosensing applications. *Nanomedicine*, **3**(4), 579–592.

Tanaka, N., Takino, S., & Utsumi, I. (1963). A new oral gelatinized sustained-release dosage form. *J. Pharm. Sci.*, **52**, 664–667.

Tansey, F. A., & Cammer, W. (1998). Differential uptake of dextran beads by astrocytes, macrophages and oligodendrocytes in mixed glial-cell cultures from brains of neonatal rats. *Neurosci. Lett.*, **248**(3), 159–162.

Teixeira, M., Alonso, M. J., Pinto, M. M., & Barbosa, C. M. (2005). Development and characterization of PLGA nanospheres and nanocapsules containing xanthone and 3-methoxyxanthone. *Eur. J. Pharm. Biopharm.*, **59**(3), 491–500.

Tice, T. R., & Lewis, D. H. (1980). *Microencapsulation Process*. Cincinnati, OH, USA: Stolle Research and Development Corporation.

Torchilin, V. P. (2005). Recent advances with liposomes as pharmaceutical carriers. *Nat. Rev. Drug Discov.*, **4**(2), 145–160.

Treacy, M. M.J., Ebbesen, T. W., & Gibson, J. M. (1996). Exceptionally high Young's modulus observed for individual carbon nanotubes. *Nature*, **381**(6584), 678–680.

Turkevich, J. (1951). A study of the nucleation and growth processes in the synthesis of colloidal gold. *Discuss. Faraday Soc.*, **11**, 55–75.

Ubaidulla, U., Khar, A. K., Ahmed, F. J., & Panda, A. K. (2007). Development and *in vivo* evaluation of insulin-loaded chitosan phthalate microspheres for oral delivery. *J. Pharm. Pharmacol.*, **59**(10), 1345–1351.

Vakarelski, I. U., Toritani, A., Nakayama, M., & Higashitani, K. (2001). Deformation and adhesion of elastomer microparticles evaluated by AFM. *Langmuir*, **17**(16), 4739–4745.

van der Lubben, I. M., Konings, F. A., Borchard, G., Verhoef, J. C., & Junginger, H. E. (2001a). *In vivo* uptake of chitosan microparticles by murine Peyer's patches: Visualization studies using confocal laser scanning microscopy and immunohistochemistry. *J. Drug Target.*, **9**(1), 39–47.

van der Lubben, I. M., Verhoef, J. C., Borchard, G., & Junginger, H. E. (2001b). Chitosan for mucosal vaccination. *Adv. Drug Deliv. Rev.*, **52**(2), 139–144.

van der Lubben, I. M., Verhoef, J. C., van Aelst, A. C., Borchard, G., & Junginger, H. E. (2001c). Chitosan microparticles for oral vaccination: Preparation, characterization and preliminary *in vivo* uptake studies in murine Peyer's patches. *Biomaterials*, **22**(7), 687–694.

van der Lubben, I. M., Kersten, G., Fretz, M. M., Beuvery, C., Coos Verhoef, J., et al. (2003). Chitosan microparticles for mucosal vaccination against diphtheria: Oral and nasal efficacy studies in mice. *Vaccine*, **21**(13–14), 1400–1408.

Van Natta, F. J., Hill, J. W., & Carothers, W. H. (1934). Studies of polymerization and ring formation, ε-caprolactone and its polymers. *J. Am. Chem. Soc.*, **56**, 455–459.

Visscher, G. E., Robison, R. L., Maulding, H. V., Fong, J. W., Pearson, J. E., et al. (1985). Biodegradation of and tissue reaction to 50:50 poly(DL-lactide-co-glycolide) microcapsules. *J. Biomed. Mater. Res.*, **19**(3), 349–365.

Visscher, G. E., Robison, R. L., Maulding, H. V., Fong, J. W., Pearson, J. E., et al. (1986). Biodegradation of and tissue reaction to poly(DL-lactide) microcapsules. *J. Biomed. Mater. Res.*, **20**(5), 667–676.

Visscher, G. E., Robison, M. A., & Argentieri, G. J. (1987). Tissue response to biodegradable injectable microcapsules. *J. Biomater. Appl.*, **2**(1), 118–131.

Visscher, G. E., Pearson, J. E., Fong, J. W., Argentieri, G. J., Robison, R. L., et al. (1988). Effect of particle size on the *in vitro* and *in vivo* degradation rates of poly(DL-lactide-co-glycolide) microcapsules. *J. Biomed. Mater. Res.*, **22**(8), 733–746.

Vlugt-Wensink, K. D., Jiang, X., Schotman, G., Kruijtzer, G., Vredenberg, A., et al. (2006). *In vitro* degradation behavior of microspheres based on cross-linked dextran. *Biomacromolecules*, **7**(11), 2983–2990.

Vlugt-Wensink, K. D., Meijer, Y. J., van Steenbergen, M. J., Verrijk, R., Jiskoot, W., et al. (2007). Effect of excipients on the encapsulation efficiency and release of human growth hormone from dextran microspheres. *Eur. J. Pharm. Biopharm.*, **67**(3), 589–596.

Wagenaar, B. W., & Muller, B. W. (1994). Piroxicam release from spray-dried biodegradable microspheres. *Biomaterials*, **15**(1), 49–54.

Wagner, F. E., Haslbeck, S., Stievano, L., Calogero, S., Pankhurst, Q. A., et al. (2000). Before striking gold in gold-ruby glass. *Nature*, **407**(6805), 691–692.

Wan, J. P., Yang, Y. Y., Chung, T. S., Tan, D., Ng, S., et al. (2001). POE-PEG-POE triblock copolymeric microspheres containing protein. II. Polymer erosion and protein release mechanism. *J. Control. Release*, **75**(1–2), 129–141.

Wang, T., Lacík, I., Brissová, M., Anilkumar, A. V., Prokop, A., et al. (1997). An encapsulation system for the immunoisolation of pancreatic islets. *Nat. Biotechnol.*, **15**(4), 358–362.

Wang, X. D., Pang, Y., Ku, G., Xie, X., Stoica, G., et al. (2003). Noninvasive laser-induced photoacoustic tomography for structural and functional *in vivo* imaging of the brain. *Nat. Biotechnol.*, **21**(7), 803–806.

Wang, J., Liu, G. D., & Jan, M. R. (2004). Ultrasensitive electrical biosensing of proteins and DNA: Carbon-nanotube derived amplification of the recognition and transduction events. *J. Am. Chem. Soc.*, **126**(10), 3010–3011.

Warheit, D. B., Laurence, B. R., Reed, K. L., Roach, D. H., Reynolds, G. A., et al. (2004). Comparative pulmonary toxicity assessment of single-wall carbon nanotubes in rats. *Toxicol. Sci.*, **77**(1), 117–125.

Webber, W. L., Lago, F., Thanos, C., & Matiowitz, E. (1998). Characterization of soluble, salt-loaded, degradable PLGA films and their release of tetracycline. *J. Biomed. Mater. Res.*, **41**(1), 18–29.

Weissleder, R., Elizondo, G., Wittenberg, J., Lee, A. S., Josephson, L., et al. (1990). Ultrasmall superparamagnetic iron-oxide: An intravenous contrast agent for assessing lymph-nodes with MR imaging. *Radiology*, **175**(2), 494–498.

Wheatley, M. A., Chang, M., Park, E., & Langer, R. (1991). Coated alginate microspheres - factors influencing the controlled delivery of macromolecules. *J. Appl. Polym. Sci.*, **43**(11), 2123–2135.

Winter, J. O., Gomez, N., Gatzert, S., Schmidt, C. E., & Korgel, B. A. (2005). Variation of cadmium sulfide nanoparticle size and photoluminescence intensity with altered aqueous synthesis conditions. *Colloid. Surface. A-Physicochemical and Engineering Aspects*, **254**(1–3), 147–157.

Woolley, A. T., Guillemette, C., Li Cheung, C., Housman, D. E., & Lieber, C. M. (2000). Direct haplotyping of kilobase-size DNA using carbon nanotube probes. *Nat. Biotechnol.*, **18**(7), 760–763.

Wu, X., Liu, H., Liu, J., Haley, K. N., Treadaway, J. A., et al. (2003). Immunofluorescent labeling of cancer marker Her2 and other cellular targets with semiconductor quantum dots. *Nat. Biotechnol.*, **21**(1), 41–46.

Wu, T. J., Huang, H. H., Lan, C. W., Lin, C. H., Hsu, F. Y., et al. (2004). Studies on the microspheres comprised of reconstituted collagen and hydroxyapatite. *Biomaterials*, **25**(4), 651–658.

Xie, J., Marijnissen, J. C., & Wang, C. H. (2006). Microparticles developed by electrohydrodynamic atomization for the local delivery of anticancer drug to treat C6 glioma *in vitro*. *Biomaterials*, **27**(17), 3321–3332.

Xing, Y., Chaudry, Q., Shen, C., Kong, K. Y., Zhau, H. E., et al. (2007). Bioconjugated quantum dots for multiplexed and quantitative immunohistochemistry. *Nat. Protoc.*, **2**(5), 1152–1165.

Yamago, S., Tokuyama, H., Nakamura, E., Kikuchi, K., Kananishi, S., et al. (1995). *In vivo* biological behavior of a water-miscible fullerene – C-14 labeling, absorption, distribution, excretion and acute toxicity. *Chem. Biol.*, **2**(6), 385–389.

Yamakoshi, Y., Umezawa, N., Ryu, A., Arakane, K., Miyata, N., et al. (2003). Active oxygen species generated from photoexcited fullerene (C-60) as potential medicines: O-2(-center dot) versus O-1(2). *J. Am. Chem. Soc.*, **125**(42), 12803–12809.

Yanagisawa, T., Shimizu, T., Kuroda, K., & Kato, C. (1990). The preparation of alkyltrimethylammonium-kanemite complexes and their conversion to microporous materials. *Bull. Chem. Soc. Jpn.*, **63**(4), 988–992.

Yang, G. W., Wang, J. B., & Liu, Q. X. (1998). Preparation of nano-crystalline diamonds using pulsed laser induced reactive quenching. *J. Phys.-Condens. Mat.*, **10**(35), 7923–7927.

Yang, Y. Y., Wan, J. P., Chung, T. S., Pallathadka, P. K., Ng, S., et al. (2001). POE-PEG-POE triblock copolymeric microspheres containing protein. I. Preparation and characterization. *J. Control. Release*, **75**(1–2), 115–128.

Yang, L., Mao, H., Wang, Y. A., Cao, Z., Peng, X., et al. (2009). Single chain epidermal growth factor receptor antibody conjugated nanoparticles for *in vivo* tumor targeting and imaging. *Small*, **5**(2), 235–243.

Youan, B. B., Benoit, M. A., Rollmann, B., Riveau, G., & Gillard, J. (1999). Protein-loaded poly(epsilon-caprolactone) microparticles. II. Muramyl dipeptide for oral controlled release of adjuvant. *J. Microencapsul.*, **16**(5), 601–612.

Yuan, Y., Chen, Y., Liu, J. H., Wang, H., & Liu, Y. (2009). Biodistribution and fate of nanodiamonds *in vivo*. *Diam. Rel. Mater.*, **18**(1), 95–100.

Zhang, H. L., Lai, G. S., Han, D. Y., & Yu, A. M. (2008). An amperometric hydrogen peroxide biosensor based on immobilization of horseradish peroxidase on an electrode modified with magnetic dextran microspheres. *Anal. Bioanal. Chem.*, **390**(3), 971–977.

Zhang, L. F., & Eisenberg, A. (1995). Multiple morphologies of crew-cut aggregates of polystyrene-B-poly(acrylic acid) blockcopolymers. *Science*, **268**(5218), 1728–1731.

PART II

Biology and Medicine

SECTION II.1

Some Background Concepts

SECTION II.1

Some Background Concepts

CHAPTER II.1.1 INTRODUCTION: BIOLOGY AND MEDICINE – KEY CONCEPTS IN THE USE OF BIOMATERIALS IN SURGERY AND MEDICAL DEVICES

Buddy D. Ratner
Professor, Bioengineering and Chemical Engineering, Director of University of Washington Engineered Biomaterials (UWEB), Seattle, WA, USA

Biomaterials Science: An Introduction to Materials in Medicine directly addresses the multidisciplinary nature of the biomaterials field by providing concise tutorials in the key concepts essential for practitioners of biomaterials science. The word "biomaterials" implies biology and materials. This section addresses the fundamental "bio" science relevant to biomaterials (Section I.2 serves a similar function for the fundamental materials science relevant to biomaterials). The background biology is important, because this scientific foundation will help us to understand, predict, and engineer the *in vivo* bioresponses observed by the biomaterials scientist, the physician, and the patient.

When a synthetic material is placed in a biological milieu (implantation in a living organism, for example), a series of reactions is initiated almost instantaneously. Water molecules (and H^+ and OH^-), being the lowest molecular weight species, should reach the biomaterial surface first (in milliseconds). Their considerable impact on subsequent reactions is elaborated upon in Chapters I.1.6 and I.2.10. Although water is not considered a biomolecule, it sets the stage for the biological reactions to follow.

Proteins, dissolved in serum at a concentration of approximately 7 g/dL, arrive next at the surface by diffusion or in some cases by convection. There are 700 or more proteins in blood. Some are relatively non-bioreactive, such as albumin, and others have numerous biological activities, such as fibrinogen or fibronectin. When proteins diffuse to a biomaterial interface, they will arrive in order of molecular weight (small proteins like thrombin, 35 kDa, arrive before heavier proteins like IgG, at 160 kDa). Once at the surface, proteins can adsorb, desorb, exchange with other proteins, denature, increase their biological activity or decrease their biological activity. This complex landscape, comprised of many proteins in many states of organization and elaborated upon in Chapter II.1.2, becomes the environment to which cells will respond (and cells do indeed respond to proteins).

The fundamentals of *in vitro* cell interaction with surfaces (or more accurately, interactions with the protein layer on biomaterial interfaces in water) are described in Chapter II.1.3. This chapter is particularly relevant for surface diagnostic devices (ELISA assays, DNA microarrays, protein microarrays), cell culture, and for biosensors, but also pertains to many reactions observed *in vivo*.

Chapter II.1.4 discusses the biology of cells, with an emphasis on what the cells do under various conditions and what happens when they are injured (you can't have a surgical site without cell injury; also rubbing and mechanical forces associated with medical implants can induce cell injury).

Single cell types rarely exist alone. They aggregate or interact to form tissues (Chapter II.1.5). Some principles that govern tissue formation from individual cell types are relatively simple and have been elaborated upon by researchers such as Malcolm S. Steinberg, who demonstrated that a cell membrane surface tension associated with cell surface cadhedrin molecules can drive cell aggregation into organized multicellular structures. Other aspects of tissue formation are governed by the complex mechanisms observed in studying the developmental biology of organisms. Biomaterials and medical devices are interfaced against tissues, so understanding tissues, their organization, and their properties is important. Furthermore, the cells in tissues make a "grout" that holds them together; this material is called extracellular matrix (ECM). ECM serves a mechanical function in stabilizing the tissues but is, in its own right, highly biologically active. For example, it has been shown that triple-helical collagen I, the primary component of ECM, can bind in a biospecific manner to more than 50 other biomolecules, including some with high biological activity such as fibronectin. Thus, the ECM presents to the cells bioactive biomolecules in optimal conformation and orientation, and thus directs their functions and actions. This concept is particularly important to tissue engineering (Section II.6). Also, the degradation products of ECM can be highly bioactive. For example, the work of Judah Folkman and colleagues demonstrated that degradation fragments of some collagens are inhibitory to angiogenesis. Since ECM is synthesized by tissue cells and is broken down by cells such as the macrophage, the bioactive fragments from the breakdown can impact tissue formation and biological reaction.

Chapter II.1.6 addresses mechanical forces, their effects on cells and tissues, and their importance for biomaterials (Chapter I.1.4 on biomechanics provides some of the concepts needed to appreciate Chapter II.1.6). Living organisms move, and blood flows. Thus, all implants in living organisms are subjected to mechanical forces. Although Chapter II.1.6 focuses on the liquid–biomaterial interface (e.g., blood interactions), the concepts of force communication from the environment external to the cell, to the cell membrane, particularly from an engineering standpoint, are elaborated upon. The work of Donald Ingber and others helps us to understand how this mechanical trigger at the cell membrane can signal to the cell nucleus, leading to downregulation or upregulation of specific genes, which then leads to expression of new cytokines and/or other signaling molecules.

Mechanical forces can impact cells macroscopically, but also impact at a microscopic level. In 1965, S. B. Carter published a paper in *Nature* in which he showed that cells follow micron-dimension scratches in a plastic film, a phenomenon he named contact guidance. We use that term to this day to describe this mechanically-induced response of cells to micron scale features. Chapters I.2.13 and I.2.15 give us patterning, texture, and porosity tools to manipulate cell–mechanical interactions at the micro- and even the nanolevel.

Section II.1 ends with Chapter II.1.7, which is on key concepts relating to stem cells. When the second edition of this textbook was being assembled, one could hardly imagine the importance that stem cells would assume for biomaterials and tissue engineering. Now, some 10 years later, stem cells are seen to be of critical importance to tissue engineering, but are also important for wound healing, tissue development, and even for the evolution of cancerous tumors. Furthermore, stem cell differentiation to specific tissue cells seems to be significantly impacted by mechanical forces communicated to the cells by surface modulus and surface topographies (textures and porosities) at the micron scale (surface chemical composition can also impact differentiation). Thus, we see relationships between this chapter and Chapter II.1.6 (effects of mechanical forces on cells) and Section II.6 (tissue engineering).

CHAPTER II.1.2 ADSORBED PROTEINS ON BIOMATERIALS

Thomas A. Horbett
Bioengineering and Chemical Engineering, University of Washington, Seattle, WA, USA

INTRODUCTION

The replacement of injured or diseased tissues with devices made from materials that are not of biologic origin is the central approach in current biomaterials science and clinical practice. The prevalence of this approach is due largely to the fact that these materials are not attacked by the immune system, unlike donor tissues or organs. This fundamental difference arises from the presence of immunologically recognizable biologic motifs on donor tissue, and their absence on synthetic materials.

Nonetheless, there are other types of biological responses to implanted biomaterials that often impair their usefulness, including the clotting of blood and the foreign-body reaction. Clearly, the body does recognize and respond to biomaterials. The basis for these reactions is the adsorption of adhesion proteins to the surface of the biomaterials that are recognized by the integrin receptors present on most cells. The adsorption of adhesion proteins to the biomaterial converts it into a biologically recognizable material, as illustrated in Figure II.1.2.1. The protein adsorption event is rapid (seconds) and happens on all materials implanted into biological systems. Note: familiarity with protein structure is essential to appreciate the content of this chapter. All introductory biochemistry textbooks can provide this critical background material.

The interaction of adhesion receptors with adhesion proteins thus constitutes a major cellular recognition system for biomaterials. Therefore, the role of adsorbed adhesion proteins in mediating cellular interactions with biomaterials will be the primary focus of this chapter. Examples illustrating the ability of adsorbed adhesion proteins to influence cellular interactions with foreign materials are presented first. Then, some of the major physicochemical characteristics of protein adsorption are illustrated and discussed, including rapid kinetics, monolayer adsorption, and competitive adsorption. Molecular spreading events related to the conformational stability of the protein are presented at some length, as background for a section on how the biological activity of adsorbed adhesion proteins is affected by the substrate. The final section summarizes the principles underlying the role of adsorbed proteins in mediating platelet response to biomaterials, as an illustrative case study representative of many other types of cellular responses.

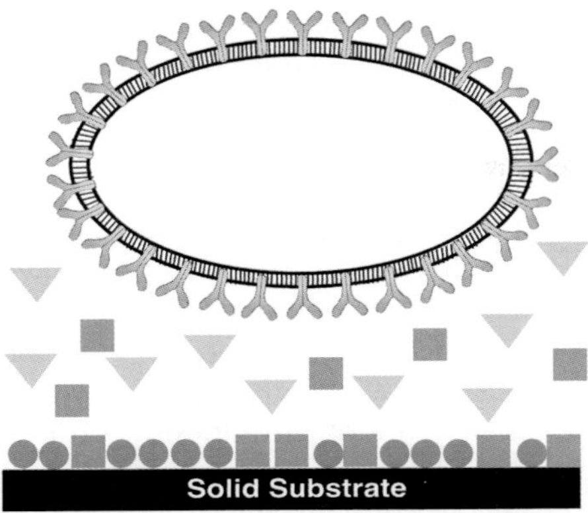

FIGURE II.1.2.1 Cell interactions with foreign surfaces are mediated by integrin receptors with adsorbed adhesion proteins that sometimes change their biological activity when they adsorb. The cell is shown as a circular space with a bilayer membrane in which the adhesion receptor protein molecules (the slingshot-shaped objects) are partly embedded. The proteins in the extracellular fluid are represented by circles, squares, and triangles. The receptor proteins recognize and cause the cell to adhere to only the surface bound form of one protein, the one represented by a solid blue circle. The bulk phase of this same adhesion protein is represented by a blue triangle, indicating that the solution and solid phase forms of this same protein have a different biological activity. The figure is schematic and not to scale.

This chapter emphasizes the biological effects of adsorbed adhesion proteins, and includes material that is discussed in greater detail in several previous review articles by the author (Horbett 1993, 1994, 1996, 1999, 2003; Horbett and Brash 1995). Those articles also give citations to all the work discussed here. A useful recent review that emphasizes thermodynamic and other physicochemical aspects of protein adsorption has been provided by Latour (2008).

Most studies of protein interactions with biomaterials and their effects on cells have been done *in vitro* after relatively short contact periods, leading to the focus here on effects in the shorter-term that involve undegraded adhesion proteins that mediate cell interactions. However, the adsorbed protein layer on biomaterials implanted in humans or animals for longer times includes smaller size fragments (observed on breast implants (Backovic et al., 2007), glucose sensors (Gifford et al., 2006), and kidney dialyzers (Cornelius and Brash, 1993)) that are presumably due to proteolytic attack. The functional role of any of these changes to the adsorbed proteins on the interaction of biomaterials with the body remains to be elucidated.

EXAMPLES OF THE EFFECTS OF ADHESION PROTEINS ON CELLULAR INTERACTIONS WITH MATERIALS

Protein adsorption to materials can be performed with a single protein, typically in a buffer solution, or from complex, multi-protein solutions, such as blood plasma which can contain hundreds of proteins. Experiments done with single proteins in buffer can be used to study fundamental aspects of protein adsorption, such as adsorption rates or conformational changes, and to study biological responses such as cell adhesion to each protein. Adsorption from complex media approximates the adsorption conditions encountered by implanted biomaterials, and thus provides a more realistic insight into the functional role of adsorbed proteins. Examples of the effects on cellular responses of protein adsorption done under both conditions are presented.

The Effects of Preadsorption with Purified Adhesion Proteins

Preadsorption of certain proteins onto a solid substrate, such as tissue culture polystyrene, greatly increases its adhesiveness to many kinds of cells, and such proteins are called adhesion proteins. The increased adhesiveness is due to the fact that many cells have receptors in their cell membrane that bind specifically to these specialized proteins. The adhesion receptors involved in cell adhesion to biomaterials and extracellular matrices such as collagen are called integrins. For example, fibronectin preadsorption greatly increases the adhesion of fibroblasts, while albumin preadsorption prevents it. Experiments of this type have been done with a wide variety of cells and adhesion proteins, with basically similar results. Adhesion proteins also promote the flattening out or spreading of the cell onto the surface. A specific example is provided by measuring the percentage of attached cells that spread on polystyrene surfaces pretreated with increasing concentrations of fibronectin. Spreading is only about 5% on albumin coated surfaces, but increases to nearly 100% as the fibronectin concentrations in the preadsorption solution are increased from 0.03 to 3 μg/ml.

In studies of protein-mediated cell adhesion, it is necessary to eliminate or reduce the role of non-specific adhesion directly to the uncoated surfaces, as this is often high, so that the special ability of adhesion proteins compared to other proteins would be masked. Thus, when testing for the ability of a protein to specifically cause cell adhesion, any residual uncoated sites left after preadsorption with the putative adhesion protein are filled by a second adsorption step with a protein such as albumin, which is commonly used to block surfaces because it greatly reduces most cell adhesion. The results for this surface are compared to a control surface that is only adsorbed with albumin. Under these conditions, any elevation in cell adhesion compared to the albumin control can be ascribed to the test protein, rather than non-specific adhesion due to direct contact with the surface.

An example of the effect of fibronectin adsorption is shown in Figure II.1.2.2, which also contrasts it with the effects of the non-adhesive protein immunoglobulin G. As shown in the figure, the adhesion of

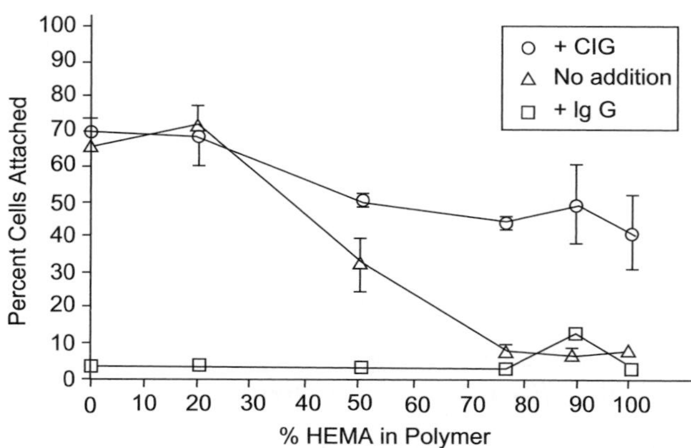

FIGURE II.1.2.2 3T3 cell adhesion to HEMA–EMA copolymers varying from hydrophilic (HEMA-rich) to hydrophobic (EMA-rich): effect of no adsorbed protein, preadsorption with fibronectin (designated CIG in the figure) or preadsorption with immunoglobulin G (IgG). The data is from the author's laboratory.

the fibroblast-like 3T3 cells to a series of polymers and copolymers of 2-hydroxyethyl methacrylate (HEMA) and ethyl methacrylate (EMA) not previously adsorbed with protein (and without proteins in the cell suspension) varies, being much less on the hydrophilic polyHEMA-rich surfaces than on the hydrophobic polyEMA-rich surfaces. These data are an example of direct or non-specific cell adhesion in which the cells adhere directly to the surfaces, rather than to adsorbed proteins. The preadsorption of the surfaces with immunoglobulin G (IgG) greatly reduces the adhesion of the cells to all the surfaces, because this protein effectively blocks the non-specific adhesion of the cells to the surfaces. In the presence of IgG, the high non-specific adhesion to EMA rich surfaces cannot occur, as the cells do not have direct contact with the surfaces. In contrast, surfaces preadsorbed with fibronectin (designated CIG in the figure) are much more adhesive than the same surface adsorbed with IgG. After fibronectin coating, even surfaces rich in HEMA that are not adhesive become adhesive, as the cells now interact not with HEMA, but with the adhesive protein.

Preadsorption of adhesion proteins also affects cell interactions with surfaces studied under *in vivo* conditions. For example, blood platelet deposition onto polymeric arteriovenous shunts in dogs is greatly increased when fibrinogen or fibronectin are preadsorbed to the surfaces in comparison to albumin coated surfaces (see Chapters II.2.6 and II.3.5).

Depletion Studies

Although adsorption of purified adhesion proteins to a surface is one way to see their effect on cell adhesion, as presented in the section "The Effects of Preadsorption with Purified Adhesion Proteins," it does not mimic very well what occurs with implanted biomaterials. This is because implants are exposed to complex mixtures of proteins such as plasma or serum, so the adhesion protein must compete with many others for adsorption to the biomaterial. Under those conditions, a given adhesion protein present in the bulk phase protein mixture may really play little or no role. It is possible that very little of the adhesion protein may adsorb to the surface, as it is "out-competed" by other proteins for the limited surface sites. Thus, a more biologically relevant way to understand the role of an adhesion protein in reactions to implants is to study the effect of their selective depletion from the complex mixture. The observations presented in this section, and the articles they are based on, are presented in greater detail in a review article (Horbett, 1999).

Selective depletion means that only one of the proteins is removed from the mixture at a time, either by immunoadsorption chromatography, by use of plasma from mutant individuals lacking the adhesion protein of interest, or by selective enzymatic degradation. Thus, the more important role of adsorbed vitronectin, as opposed to fibronectin, in mediating attachment and spreading of cells on many surfaces has emerged from immunoadsorption studies. Several studies illustrate the important role that adsorbed fibrinogen plays in the adhesion of platelets, neutrophils, and macrophages.

The effects of removal of fibronectin or vitronectin or both proteins from serum on the adhesion of endothelial cells depend on surface chemistry. On tissue culture polystyrene (TCPS) fibronectin removal has little effect, while vitronectin removal greatly reduces adhesion. The results clearly show the primary role of adsorbed vitronectin in supporting endothelial cell adhesion to TCPS. In contrast, adhesion to a surface modified by ammonia in a glow discharge (see Chapter I.2.12) requires fibronectin, since removal of that protein greatly reduces adhesion to this surface, while vitronectin removal has little effect. However, the results for TCPS are more typical, i.e., on most surfaces studied by this method it appears that vitronectin, not fibronectin, is the primary agent responsible for cell adhesion.

Platelet adhesion to surfaces preadsorbed with plasma deficient in fibrinogen is much less than to the same surface preadsorbed with normal plasma, as illustrated in Figure II.1.2.3. Most of the adhesion supporting activity can be restored to fibrinogen deficient plasma by addition of exogenous fibrinogen at concentrations well below that in normal plasma. Platelets are very sensitive to fibrinogen as they adhere to some surfaces at fibrinogen adsorption levels of only 5 ng/cm^2, in comparison with adsorption values in the range of 100–200 ng/cm^2 in the presence of normal fibrinogen concentrations. In contrast, removal of fibronectin or vitronectin or von Willebrand's factor from plasma has little effect on platelet adhesion (data not shown), even though these other plasma proteins act as adhesion proteins when adsorbed

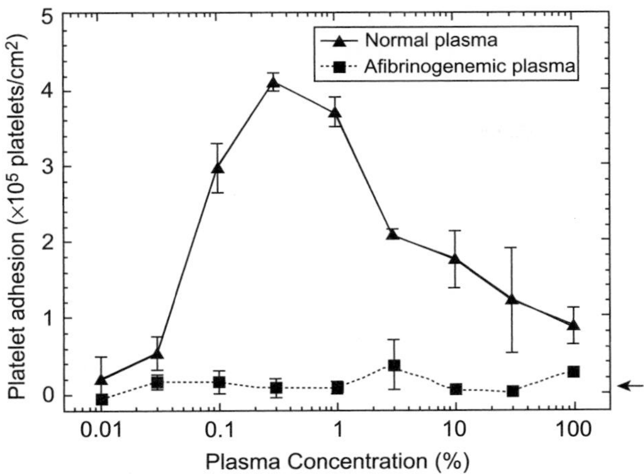

FIGURE II.1.2.3 Platelet adhesion to Immulon I® preadsorbed with normal plasma (triangles) or afibrinogenemic plasma (squares). The solid line represents the platelet adhesion to Immulon I® preadsorbed with a series of dilutions of normal plasma, whereas the dotted line represents the platelet adhesion to Immulon I® preadsorbed with a series of dilutions of afibrinogenemic plasma. The arrow at the lower right corner indicates platelet adhesion to Immulon I® preadsorbed with 2% BSA only (BSA is bovine serum albumin). (Source: Figure 4 in Tsai and Horbett, 1999. Copyright permission received.)

as single proteins to surfaces. It appears that too little of these other proteins adsorb from plasma to make much difference, i.e., competition from fibrinogen and other proteins keeps their surface concentration too low, and so their removal has no effect.

The apparently negligible role of von Willebrand's factor in platelet adhesion discussed above only applies when platelet adhesion from static suspensions is measured. At higher wall shear rates (e.g., 500 sec^{-1}), platelet adhesion to biomaterials is highly dependent on the presence of adsorbed von Willebrand's factor. This dependence is most clearly illustrated by the effects of increasing the shear rate on platelet adhesion. On surfaces preadsorbed with normal plasma (i.e., containing von Willebrand's factor, fibrinogen, and the other proteins), platelet adhesion is much higher at higher shear rates than low shear rates, because of the enhanced transport of the platelets to the surface, i.e., the greater number of platelet–surface encounters at high shear rate allows for an increase in the number of adhesive outcomes. On surfaces preadsorbed with plasma deficient in von Willebrand's factor only, raising the shear rate not only does not result in an increase in adhesion, but the adhesion at high shear is actually lower than at low shear, exactly the opposite of what happens in the presence of von Willebrand's factor. Thus, von Willebrand's factor adsorption plays a major role in platelet adhesion at higher shear rates. The special role of von Willebrand's factor in high shear adhesion is thought to be due to the ability of the platelet receptor GPIb-IX to form and break bonds very rapidly with immobilized von Willebrand's factor, causing the platelets to roll along the surface until they are activated and can form a permanent bond that stops the rolling. The role of von Willebrand's factor in shear-mediated adhesion was documented in several recent studies (summarized in Zhang et al., 2008). Interestingly, the adsorbed von Willebrand's factor molecule is itself subject to distortion by both mechanical and fluid shear (Siedlecki et al., 1996). That may play a role in the surface activation of von Willebrand factor by exposing platelet binding sites on the molecule.

When mice are depleted of fibrinogen by treatment with an enzyme that degrades it, the adhesion of neutrophils and macrophages to a polymer implanted in their peritoneal cavity is greatly reduced. The fibrinogen depleted animals exhibited near normal neutrophil and macrophage adhesion to the implants, if the implants are preadsorbed with fibrinogen. These studies clearly illustrate the power of the depletion method. Previously, it had been thought that either complement or IgG would be the main adhesion proteins for neutrophils and monocytes, due to the presence of receptors on these cells that bind these proteins. Instead, it appears that an integrin receptor for fibrinogen (CD11b/CD18, also known as Mac-1) plays a major role, at least during the initial or acute phase of the foreign-body response in the mouse peritoneal cavity.

Inhibition of Receptor Activity with Antibodies

Another way to show the role of adhesion proteins in cell interactions with biomaterials is to add specific inhibitors of their function. Adhesion proteins cause cell adhesion by binding to integrin receptors that specifically recognize the adhesion protein. One way to inhibit this reaction is to add an antibody that binds to the adhesion receptor or adhesion protein. Examples of these approaches are now presented.

Platelet receptor-mediated interactions appear to be the primary mechanism of platelet interaction *in vivo* with certain vascular grafts, because platelet deposition is largely inhibited by antibodies to the glycopro IIb/IIIa receptor, the major integrin on the surface of platelets. *In vitro* platelet adhesion to surfaces preadsorbed with blood plasma is also inhibited by anti-glycoprotein IIb/IIIa in a dose-dependent manner, as illustrated in Figure II.1.2.4. In this study, samples of the polyurethane Biomer were preadsorbed with plasma and then exposed to platelets in an albumin containing buffered saline suspension that had increasing amounts of the antibody. As shown in the figure, adhesion declined to very low values when high concentrations of the anti-integrin antibody were present.

There is currently much interest in the ability of biomaterials to regulate stem cell differentiation, including the role of proteins adsorbed to the biomaterials. Thus, antibodies to vitronectin and fibronectin were used to study the role of these two proteins in osteogenic differentiation of mesenchymal stem cells on polycaprolactone. The vitronectin antibody selectively blocked adhesion and differentiation, but the fibronectin antibody did not, showing the importance of vitronectin-mediated adhesion in the behavior of this type of stem cell (Chastain et al., 2006). Combinatorial screening methods to find biomaterials that promote neural stem cell differentiation found a role for adsorbed fibronectin and other adhesion factors in combination with immobilized growth factors (Nakajima et al., 2007). A much broader combinatorial screen (3456 assays) for biomaterials conducive to human mesenchymal stem cell adhesion has also been done (Anderson et al., 2005). Modification of biomaterials with various adhesion promoting peptides to promote stem cell adhesion and differentiation has also come under active study, because it is thought to be a good way to "provide a microenvironment where adhesive moieties are expressed in a spatial and temporal manner to control cellular behaviors" (Hwang et al., 2008).

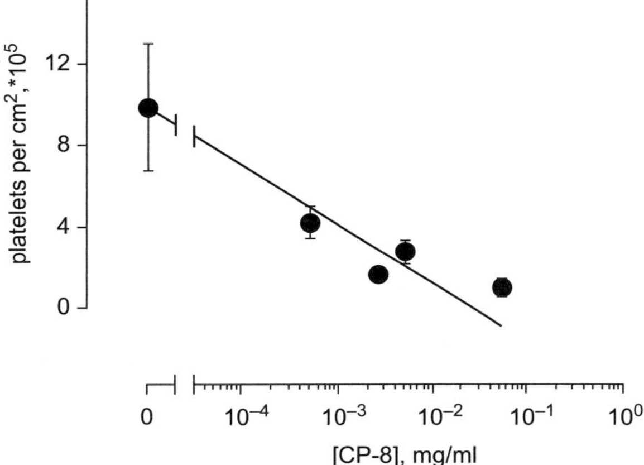

FIGURE II.1.2.4 Effect of anti-IIb/IIIa antibody on platelet adhesion to Biomer preadsorbed with plasma. Adhesion of platelets incubated in monoclonal antibody CP-8 (monovalent Fab′ fragment directed against the glycoprotein (GP) IIb/IIIa complex) to Biomer. Substrates were contacted with 1.0% plasma for 2 hours, then with washed, antibody-treated platelets for 2 hours. (From: Chinn J.A., Horbett T.A., Ratner B.D. (1991). Baboon Fibrinogen Adsorption and Platelet Adhesion to Polymeric Materials. *Thromb Haemost*, **65**, 608–17.)

THE ADSORPTION BEHAVIOR OF PROTEINS AT SOLID–LIQUID INTERFACES

Adsorption Transforms the Interface

Figure II.1.2.5 illustrates an experiment that is performed by the author to demonstrate the adsorption of proteins to surfaces. As illustrated in part (A), water droplets sprayed on the surface of an unused polystyrene cell culture dish are easily visible because they bead up, i.e., the contact angle between the droplet and the polystyrene surface is very high due to the water repellent, hydrophobic nature of polystyrene (Chapter I.1.5). If a cell were placed on a polystyrene dish instead of the water droplet, it would also encounter a very non-wettable surface. Part (B) of the figure illustrates the results of spraying water droplets on the surface of a polystyrene dish that had first been exposed to a protein solution for a short time, and then rinsed extensively with water. As illustrated, no water droplets can be seen on this surface, reflecting the fact that in this case the added drop of water completely spread out over the surface of the preadsorbed dish. This happens because the water in part (B) was not able to interact with the polystyrene surface, because the surface had become coated with a layer of the hydrophilic protein adsorbate. Similarly, cells that come into contact with surfaces adsorbed with proteins do not directly "see" the substrate, but instead they interact with the intervening protein adsorbate.

Rapid Adsorption Kinetics and Irreversibility

The kinetics of adsorption of proteins to solid surfaces typically consists of a very rapid initial phase, followed by a slower phase upon approach to the steady-state value. Initially, proteins adsorb as quickly as they arrive at the largely empty surface. In this phase, adsorption is linear when plotted against time$^{1/2}$, characteristic of a diffusion-controlled process. In the later, slower phase, it is more difficult for the arriving proteins to find and fit into an empty spot on the surface.

Figure II.1.2.6 shows the time course of adsorption of lysozyme on silica measured with a high speed,

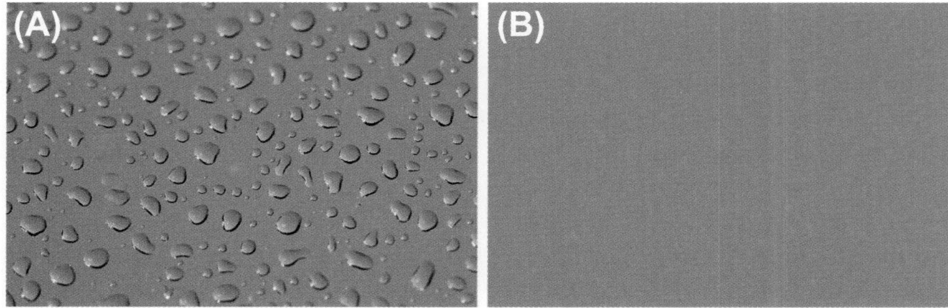

FIGURE II.1.2.5 The conversion of non-wettable polystyrene surface (A) into one completely wettable by water (B) is due to the adsorption of proteins (simulated images based on actual observations).

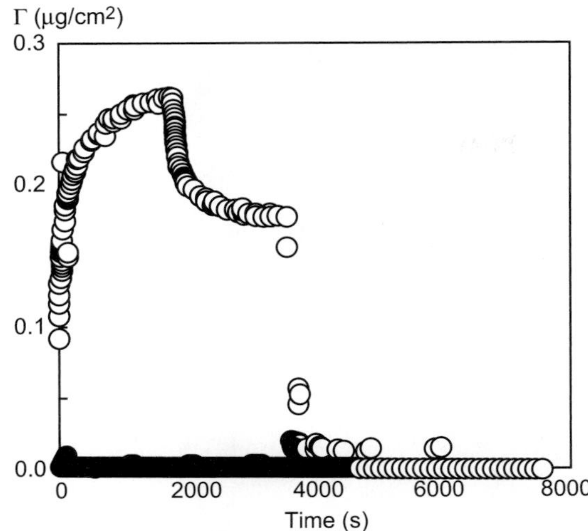

FIGURE II.1.2.6 The adsorption kinetics of lysozyme to a silica surface as studied with ellipsometry. The adsorbed amount versus time for adsorption of lysozyme to silica followed by buffer rinsing after 1800 seconds, addition of surfactant (sodium dodecyl sulfate) after 3600 seconds, and a final rinse with buffer after 5400 seconds (open circles). Adsorption from a mixture of the protein and surfactant for 1800 seconds followed by rinsing is also included (closed circles). The experiments were carried out at 25°C in 0.01 M phosphate buffer, 0.15 M NaCl, pH 7. (Reprinted with permission from: Arnebrandt and Wahlgen, 1995. Copyright © 1995 American Chemical Society.)

automated ellipsometer capable of very rapid measurements (Arnebrant and Wahlgen, 1995). At the earliest measurement time, less than a second into the study, the adsorption has reached almost half of the steady-state value. At 2000 seconds, the protein solution was replaced with a buffer, resulting in some removal of loosely bound protein, but the adsorption stabilizes and would have remained at this value for much longer than shown, due to the tight, irreversible binding. At 4000 seconds, a solution of the detergent sodium dodecyl sulfate (SDS) was infused, leading to complete removal of the protein. Thus, this experiment illustrates the rapid adsorption of proteins. It also illustrates that most of the adsorbed protein is irreversibly bound, as indicated by the fact that washing the surface with buffer does not remove the protein. The adsorbed protein is only removed when a strong surfactant (SDS in this example) is used. All these features are characteristic of protein adsorption to solid surfaces.

Figure II.1.2.7 provides an example from modeling studies illustrating the complex array of interactions possible for a protein interacting with a surface (Agashe et al., 2005). The illustrations are for the gamma chain of fibrinogen interacting with either a CH_3 or a COOH terminated self-assembled monolayer (SAM) (Chapter I.2.12), and shows that many atoms in the protein's amino acid side chains come near the surface, and could thus contribute to bonding. These modeling studies start with high resolution protein structures, and apply molecular dynamic calculations that allow the protein molecule and its atoms to move in space, in order to find favorable interactions with surfaces. It is found that the protein rolling over the surface eventually oriented itself with several residues near the surface, as shown.

For the CH_3-terminated SAM, the position of the protein illustrated is a stable orientation that has hydrophobic groups, namely the side chains of alanine and leucine, and the CH_2 of a lysine side chain, lying close to the surface forming hydrophobic interactions with the methyl surface. In this orientation, the hydrophilic functional groups of the protein residues remained hydrated and well separated from the surface.

For the COOH surface, the position of the molecule that is a preferred orientation is quite different than for the CH_3 surface, as comparison of the overall shapes of the molecule against the two surfaces indicates. A clearer indication of these differences is the types of amino acid side chains found near the COOH interface. As shown, the near neighbors are three lysines, a glutamine, and an arginine, providing favorable interactions with the COOH. The COOH-terminated SAM was modeled as 50% deprotonated/50% protonated to represent pH 7.4.

The Monolayer Model

The existence of a close packed monolayer of adsorbed protein is suggested by studies with single protein solutions, in which a saturation effect can often be observed in the adsorption isotherm (Figure II.1.2.8). Adsorption to surfaces exposed to different concentrations of protein until steady-state adsorption is achieved (2 hours or more) increases steeply at low bulk phase concentrations, but typically reaches a plateau or saturation value at higher bulk concentrations. This behavior is called a Langmuir isotherm, and has the form shown in the following equation:

$$\theta = \frac{[P]}{K_D + [P]}$$

where θ is the fractional occupancy of the surface (with a range of 0 to 1), i.e., Γ/Γ_{max} where Γ is the adsorption of a particular bulk protein concentration ($[P]$) and Γ_{max} is the maximum adsorption in the saturation or plateau region. The constant K_D is the dissociation constant for the protein–surface interaction:

$$P + S \rightleftarrows P - S$$

$$K_D = \frac{[P] \cdot [S]}{[P - S]}$$

where S is an absorption site on the surface.

Despite the fact that protein adsorption to biomaterials is an irreversible process, the concentration dependence of adsorption for pure protein solutions often follows the Langmuir isotherm that is only applicable for reversible adsorption processes such as gas adsorption. One explanation for this is that the space occupied per protein molecule in a close packed layer

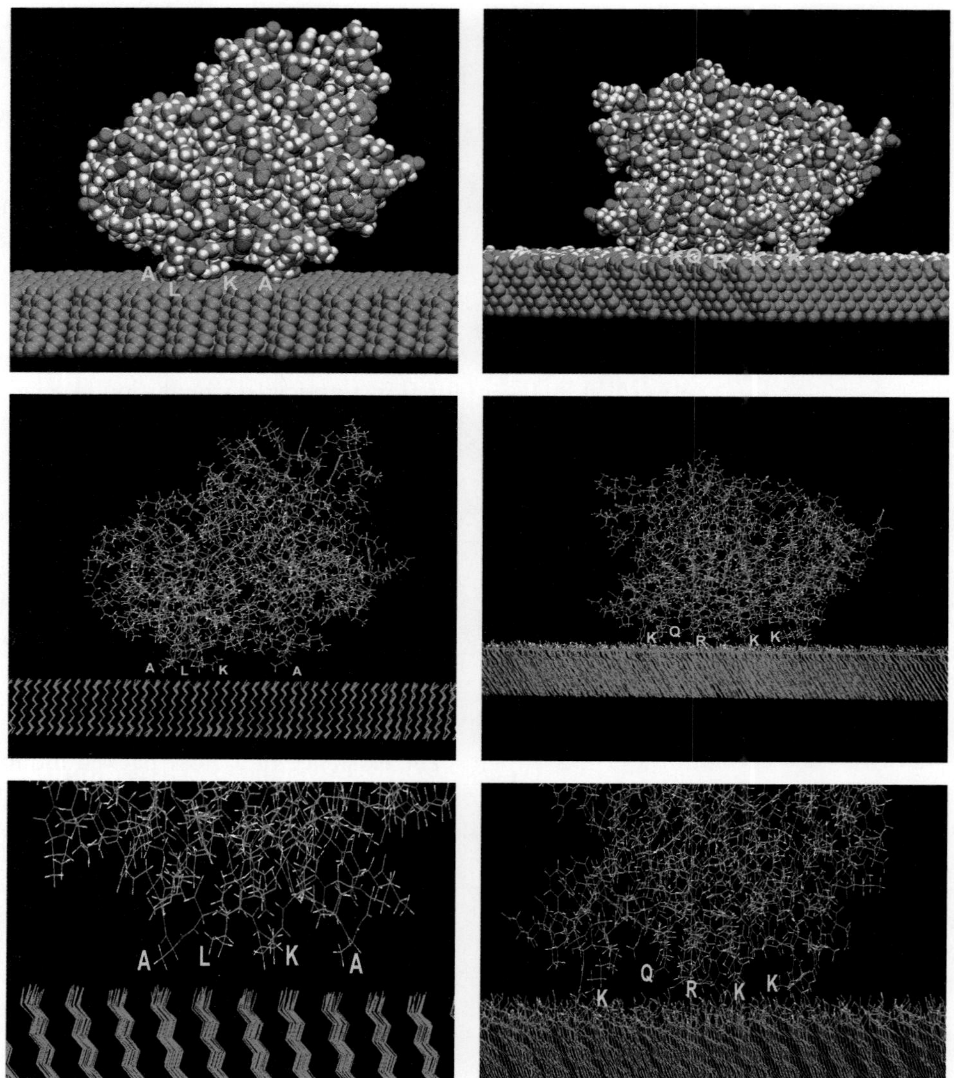

FIGURE II.1.2.7 Molecular interactions of the 30 kDa C-terminus γ-chain fragment of fibrinogen with self-assembled monolayers with terminal CH_3 (left) and COOH (right) groups. Spacefill and wireframe models were newly created by R. Latour from the trajectory data for simulations described by Agashe et al., (2005). The models for the CH_3 surface show the interactions of the side chains of two alanines residues (A), one leucine (L), and the CH_2 groups of a lysine residue (K). The models for the COOH surface show the interactions of the side chains of three lysines (K), a glutamine (Q), and an arginine (R). Color code: Green = C; Red = O; Blue = N; White = H; Yellow = S. For clarity, interactions not easily seen in the full molecule views are not labeled.

(its "footprint") varies depending on the adsorption conditions. Thus, at low concentrations, the proteins have time to molecularly spread onto the surface and form a larger footprint, i.e., a packed layer of more spread protein molecules is formed. In contrast, when adsorption is done at higher concentrations, the adjacent sites are rapidly filled with other proteins, and so the spreading of each molecule into these sites on the surface is inhibited. In this case, a packed layer of protein is also formed, but the molecules are less spread so that there can be more of them per unit area than when adsorption occurs at lower concentration (Wertz and Santore, 2001). Support for the existence of variations in molecular spreading and protein footprints is provided by the fact that proteins adsorbed at lower concentrations tend to be more tightly bound, as judged by lower desorption upon exposure to buffer or other proteins or detergents.

Usually, the plateau adsorption value observed in measured adsorption isotherms falls within the range expected for a close-packed monolayer of protein (about 0.1–0.5 μg /cm^2, depending on the diameter and orientation assumed for the protein). The adsorption values from complex protein mixtures are also typically in the monolayer range. For example, the sum total of the amount of adsorption of the three major proteins in plasma (albumin, IgG, and fibrinogen) on the HEMA-EMA series of surfaces is also in the range 0.1–0.5 μg/cm^2. In addition, the fact that competition exists between proteins for adsorption to a surface (see section

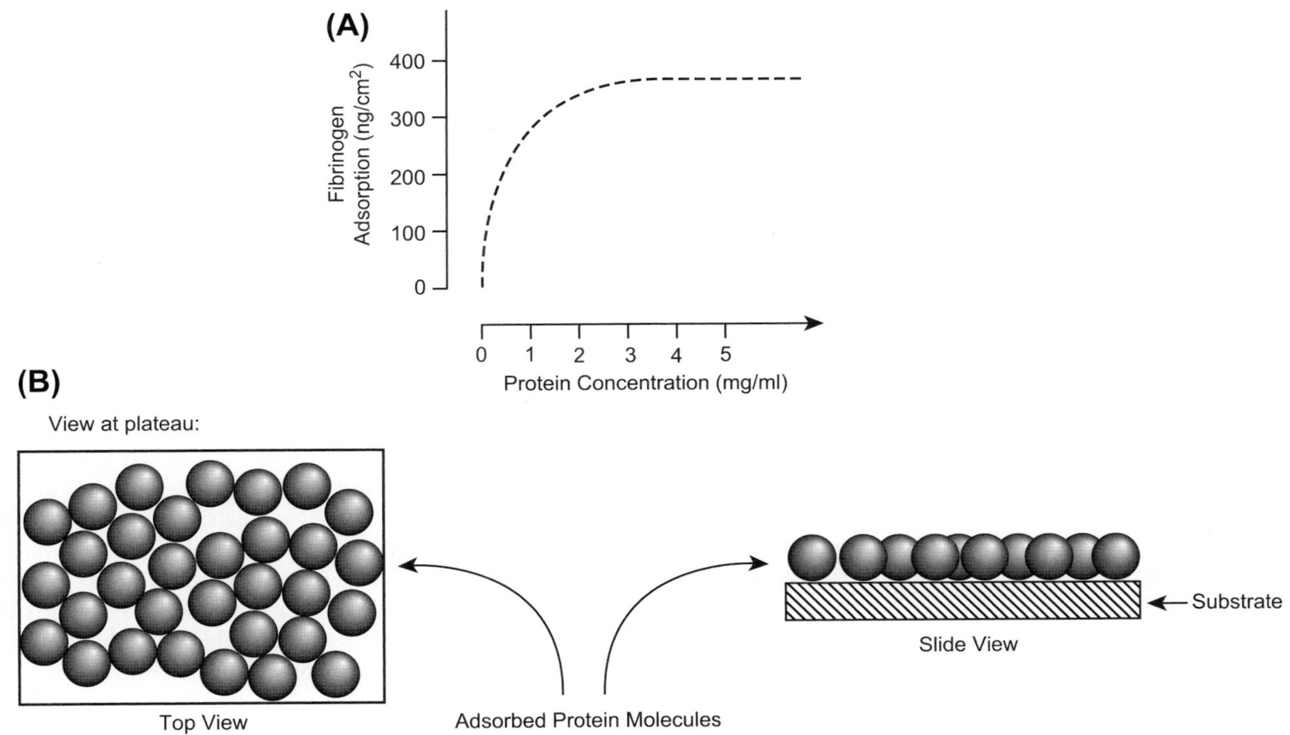

FIGURE II.1.2.8 Adsorption isotherms (A) and the monolayer concept (B).

"Competitive Adsorption of Proteins to Surfaces from Protein Mixtures," below) also indicates the existence of limited sites. Thus, when a monolayer is the limit, there must be some selection for which proteins are present in the adsorbed film. It should be noted that well-defined plateaus are not always observed, but instead adsorption rises much more slowly at higher bulk phase concentrations than at low concentrations, i.e., Freundlich isotherms do occur.

Competitive Adsorption of Proteins to Surfaces from Protein Mixtures

Adsorption from mixtures of proteins is selective, leading to enrichment of the surface phase in certain proteins. In this context, enrichment means that the fraction of the total mass of the adsorbed protein layer corresponding to a given protein is often much higher than the fraction of this protein in the bulk phase mixture from which it adsorbed. Since the solid can accommodate only a small fraction of the total protein present in the bulk phase, and proteins vary greatly in their affinity for surfaces, some adsorbed proteins are present in greater amounts than others. Studies of surfaces exposed to plasma have shown that many different proteins are present in the adsorbed film.

The competitive phenomena underlying differential enrichment from multi-protein mixtures are most clearly illustrated in binary mixtures of proteins. Figure II.1.2.9 has three separate curves in it, which overlap at the high and low ends. These curves represent the typical outcome of binary mixture studies, but for three different conditions. For example, when a radiolabeled protein such as fibrinogen ("A" in the figure) is mixed with various amounts of an unlabeled protein such as albumin ("B" in the figure), the adsorption of fibrinogen ("A") always declines when sufficiently high amounts of albumin ("B") are present. However, the amount of competing protein needed to inhibit the adsorption of the labeled protein is different in each curve. This is meant to illustrate that, for a given pair of competing proteins, the competition curves will be different if the surfaces they are competing for are different. In addition, if the surfaces are kept the same, but the competition of different pairs of proteins are studied, the curves will differ because the ability of proteins to compete for surface sites is quite different for different proteins. A convenient way to characterize

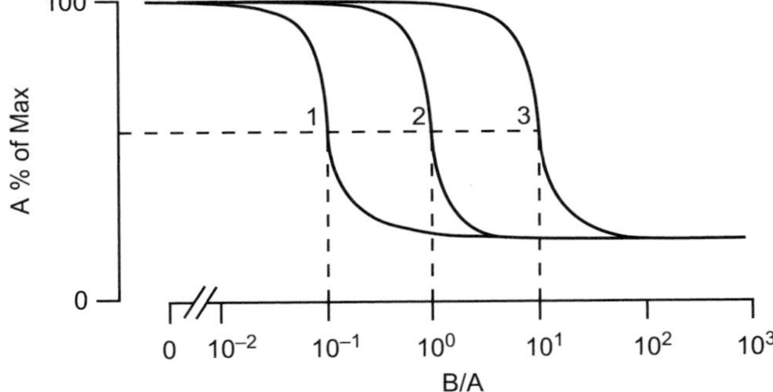

FIGURE II.1.2.9 Competitive adsorption of two proteins from a mixture. (From: Horbett, 1993.)

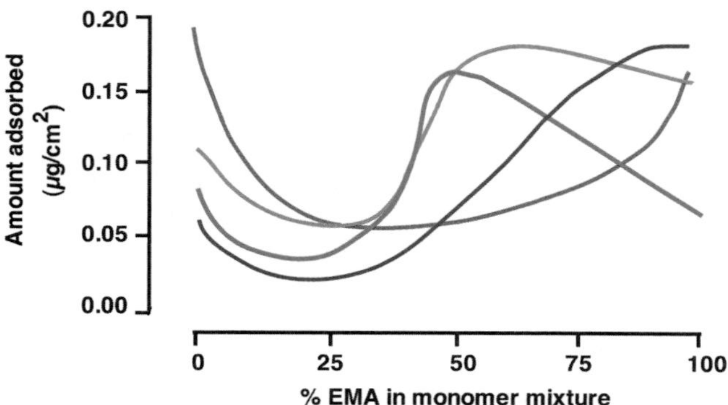

FIGURE II.1.2.10 Differential affinity of proteins to a series of polymers: adsorption of Fg (green line); IgG (purple line); Alb (red line); and Hb (blue line) from plasma to hydroxyethyl methacrylate – ethyl methacrylate copolymers. (Redrawn version of a figure from: Horbett, 1981.)

these binary competition curves is the ratio of competing protein it takes to cause 50% inhibition of the adsorption of the labeled protein. If this ratio is 1, it reflects the fact that the proteins are equal in affinity for the surface; if the ratio is much less (or much greater) than one, it signifies that much less (or much more) of the competing protein is needed to reduce the adsorption, and so the competitor has a much higher (or much lower) affinity for the surface.

For example, inhibition of fibrinogen adsorption to polyethylene requires roughly 10-fold excess by weight of lower affinity competing proteins such as albumin, but is effectively inhibited by the higher affinity protein hemoglobin, even when hemoglobin is present at only 10% of the mass of fibrinogen. However, the amounts needed for this inhibition will depend on the surface chemistry, because the affinity of proteins changes with surface chemistry. Affinity variation is thus a major principle determining the outcome of the competitive adsorption processes.

An experimental example of surface chemistry-dependent selective adsorption of proteins from a complex protein mixture is shown in Figure II.1.2.10. In these studies, a radiolabeled protein (e.g., 125I-fibrinogen) added to blood plasma enabled measurement of the protein's adsorption to a series of polymers and copolymers of 2-hydroxyethyl methacrylate (HEMA) and ethyl methacrylate (EMA). This series varies greatly in its surface properties, from water swollen hydrophilic HEMA to water repelling hydrophobic EMA. The studies were repeated four times, once each with 125I labeled fibrinogen, albumin, IgG or hemoglobin. The dependence of the adsorption of each protein on the changes in surface chemistry in this series is different for each protein, as reflected in the very different shape of lines for each protein. For example, albumin displayed preferential adsorption to both HEMA and EMA, in that adsorption to these polymers was much greater than to any of the HEMA–EMA copolymers. Conversely, fibrinogen adsorption was maximal at intermediate HEMA–EMA compositions. Another behavior is shown by the adsorption of hemoglobin and IgG, both of which display maximal adsorption to EMA, with minimal adsorption on surfaces made with about 20% EMA. Another way to look at this data is to consider that the very different trends in adsorption across the series for each protein means that the relative amount of each protein is unique to each surface, and varies quite a bit among the surfaces. On HEMA, the adsorption is seen to be in the order albumin>hemoglobin>fibrinogen>IgG, while on EMA the order is quite different: IgG, hemoglobin, and albumin are all about the same, but much greater than fibrinogen adsorption. On the 50–50 HEMA–EMA copolymer, the order is also different: fibrinogen and hemoglobin adsorption are about the same, but much greater than albumin or IgG. Examination of the amounts of adsorption for each protein on each surface shows that the overall composition of the adsorbed layer is different on each and every polymer in the series. The differences seen in these studies are typical for many other polymers that have been studied. Namely, the relative enrichment of a given protein when adsorbed from a complex mixture such as plasma is unique to each polymer studied. The effect of surface chemistry in causing differences in the relative enrichment of proteins adsorbed from the complex mixture of proteins in plasma is due to differences in the relative competitive affinity of the proteins for each surface that were discussed above in regard to binary competitive adsorption studies.

Other examples of surface enrichment from complex protein mixtures are readily available. The examples presented in this paragraph emphasize protein affinity differences, in contrast to the surface affinity examples presented in the previous paragraph. Although fibrinogen is only the third most concentrated protein in plasma, after IgG and albumin, biomaterials exposed to plasma are enriched in fibrinogen in the adsorbed phase. Hemoglobin is present in very low concentrations in plasma (0.01 mg/mL or less), but it is still adsorbed in amounts similar to the more predominant proteins because of its high surface affinity. Albumin, a lower affinity protein, presents a counter example. Albumin concentration in plasma is much higher than fibrinogen, yet the surface concentration of albumin adsorbed from plasma is typically about the same as fibrinogen. The high concentration of albumin in the plasma drives it onto the surface according to the law of mass action. Similarly, fibrinogen adsorption is higher from plasmas that contain higher concentrations of fibrinogen. Thus, mass concentration in the bulk phase is the second major factor determining competitive adsorption behavior.

The adsorption of fibrinogen from plasma exhibits some unusual behavior. On some surfaces, fibrinogen adsorption is maximal at intermediate dilutions of plasma (see example in Figure II.1.2.11A). In addition, fibrinogen adsorption from full-strength or moderately diluted plasma is higher at very early adsorption times than at

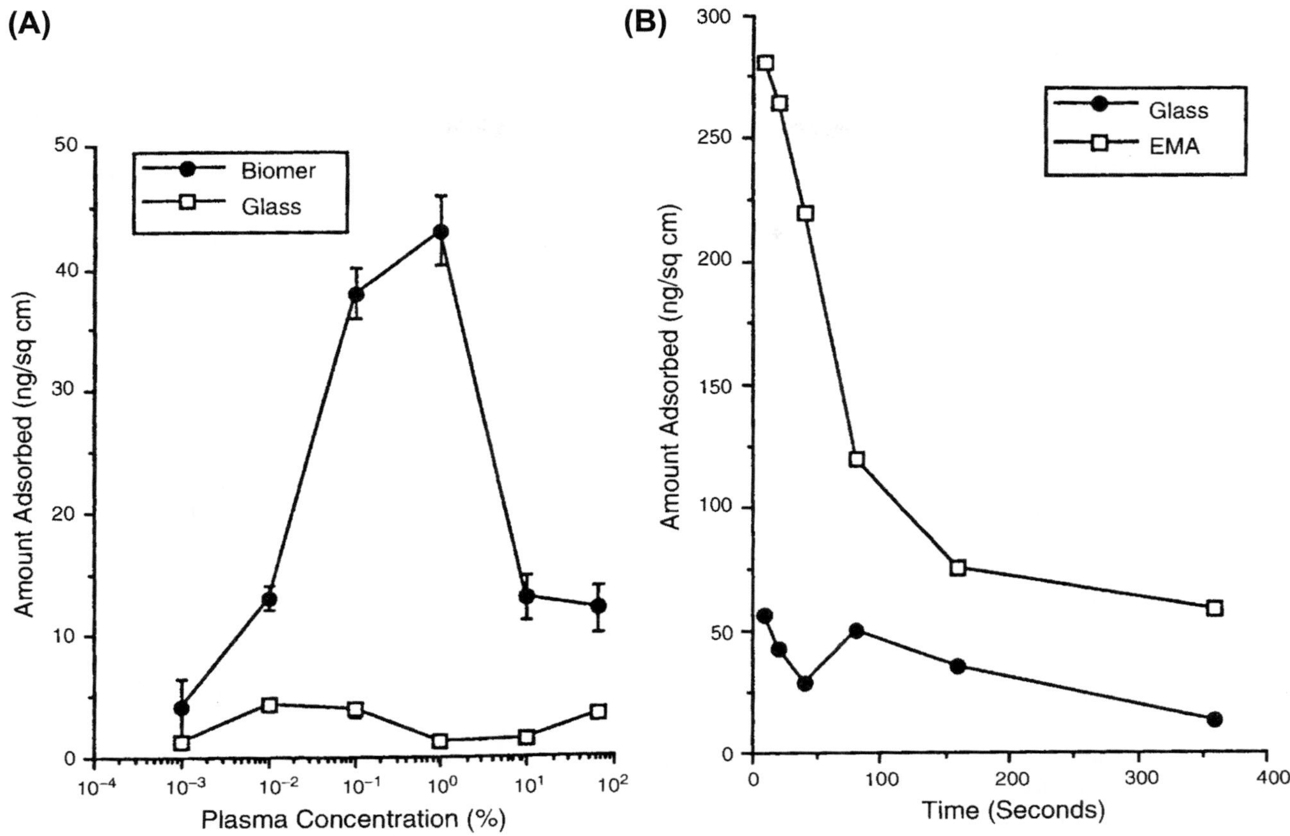

FIGURE II.1.2.11 The Vroman effect: (A) fibrinogen adsorption to Biomer and glass from various concentrations of blood plasma; (B) time course of fibrinogen adsorption to glass and poly(ethyl methacrylate) (PEMA) from undiluted plasma. (Reprinted with permission from Slack and Horbett, 1995. Copyright © 1995 American Chemical Society.)

later times (example shown in Figure II.1.2.11B). These are examples of the Vroman effect for fibrinogen. This phenomenon is a clear example of the unique effects of competitive adsorption on both the steady-state and the transient composition of the adsorbed layer that forms from plasma. The Vroman effect appears to involve displacement of initially adsorbed fibrinogen by later arriving, more surface-active plasma proteins, especially high molecular weight kininogen, and transitions in the adsorbed fibrinogen which make it less displaceable with adsorption time (reviewed in Slack and Horbett, 1995). Vroman effects, i.e., peaks in adsorption at intermediate dilution, are also observed for fibronectin adsorption from serum, and even for binary protein mixtures under certain conditions.

MOLECULAR SPREADING EVENTS: CONFORMATIONAL AND BIOLOGICAL CHANGES IN ADSORBED PROTEINS

Proteins that adsorb to solid surfaces can undergo conformational changes, because of the relatively low structural stability of proteins and the tendency to unfold to allow further bond formation with the surface. Conformational changes can be detected with many types of physicochemical methods, and also by measuring changes in the biological activity of the adsorbed proteins.

Physicochemical Studies of Conformational Changes

"Soft" proteins are found to adsorb more readily and more tenaciously than "hard" proteins. In this context, a "soft" protein is one with a low thermodynamic stability, while a "hard" protein is more stable to unfolding in solution in response to denaturing conditions, such as elevated temperature. This concept and the articles supporting the following discussion are presented in detail elsewhere (Horbett and Brash, 1995).

Comparison of the adsorptive behavior of different proteins to their molecular properties indicates that less stable proteins are more adsorptive. The important role of structural stability in adsorption is also supported by recent studies with engineered mutant proteins with single amino acid substitutions that vary in stability. Lysozyme adsorption at the solid–liquid interface, and tryptophan synthase occupation of the air–water interface are greater for less stable mutants, and more stable lysozyme variants tend to be displaced by less stable mutants (Lee et al., 2004).

Several studies with differential scanning calorimetry (DSC) methods seemed to indicate that adsorbed proteins may lose much of their structure, depending on how "soft" they are. Heat is taken up at a certain elevated temperature for proteins in solution, due to unfolding of the native protein at the transition temperature. An absence or reduction of this effect for an adsorbed protein suggests that the adsorbed protein has already undergone the transition, i.e., that it has already unfolded upon adsorption. The transition enthalpy of lysozyme adsorbed to negatively charged polystyrene was much less than for the protein in solution (0–170 kJ/mol for the adsorbed protein versus about 600 kJ/mol for the native protein depending on the pH). However, for lysozyme adsorbed on hematite, the unfolding enthalpy is only about 20% less than for the native protein, indicating that changes in the enthalpy of unfolding depend on the adsorbing surface. Furthermore, for lactalbumin the heat released is nearly zero when adsorbed on either the polystyrene or the hematite surface, suggesting complete unfolding of lactalbumin on both surfaces. These observations are consistent with the lower stability of lactalbumin in comparison to lysozyme. Several proteins adsorbed to pyrolytic carbon do not show any release of heat at the expected transition temperature, suggesting that pyrolytic carbon induces complete unfolding; a result that is consistent with the tenacious binding of proteins to this surface. It has also been shown that albumin and lysozyme adsorbed to polystyrene exhibit no unfolding enthalpy, while lysozyme adsorbed to a hydrophilic contact lens still exhibits about 50% of the heat released by the native protein. Streptavidin adsorbed to polystyrene displays an unfolding enthalpy that is very similar to that for the native protein in solution, probably because of the greater stability of streptavidin in comparison to lysozyme or albumin.

However, more recent studies of adsorbed proteins by the DSC method in conjunction with other, more direct conformational measurements such as circular dichroism (CD), show that at least some adsorbed proteins that appear to be completely denatured, as judged by DSC, still have considerable amounts of their native structure, as measured by CD. It thus appears that some proteins become somewhat more stable after adsorption, and thus do not show heat release at the normal melting temperature.

The concept of molecular spreading of the adsorbed protein suggested by these observations has been used to explain differences in the amount of IgG adsorbed during stepwise adsorption. When the final concentration of bulk protein is achieved in a series of smaller concentration steps, as opposed to bringing the bulk concentration to its final value in one step, adsorption is smaller. This result is thought to be due to spreading and an enlarged footprint of a protein molecule that can occur at lower bulk concentrations, and is similar to the explanation of Langmuir isotherms given earlier. Conformational changes upon adsorption of fibronectin to polystyrene beads and Cytodex® microcarrier beads have been detected using electron spin resonance spectroscopy. Many other physicochemical studies are consistent with partial unfolding of the adsorbed proteins (Andrade, 1985; Lundstrom, 1985; Horbett and Brash, 1995).

Changes in Biological Properties of Adsorbed Proteins

While physicochemical studies sometimes suggest complete denaturation of adsorbed proteins, most probes for biological activity suggest the changes are more limited (reviewed with citations in Horbett, 1993). Thus, enzymes retain at least some of their activity in the adsorbed state, especially when the surfaces are more fully loaded with enzyme. Measurements of enzyme activity or retention times during passage over hydrophobic chromatography matrices have shown that the degree of denaturation is highly dependent on the protein, the length of time the protein has spent on the matrix, the solvent, and other conditions, and is not necessarily complete.

Changes in the binding of a monoclonal antibody to fragment D of fibrinogen upon fibrinogen adsorption to polystyrene have been attributed to changes in the conformation of fibrinogen after adsorption. Thus, solution-phase fibrinogen does not bind the antibody raised to fragment D, but the surface-adsorbed fibrinogen does. Furthermore, bulk fibrinogen does not compete for the binding of the antibody to the surface-adsorbed fibrinogen. The RIBS (receptor-induced binding site) antibodies are similar: they bind to fibrinogen only after the fibrinogen has bound to either a solid surface or to the platelet IIb/IIIa receptor. The binding of the RIBS antibodies and others that bind to the platelet binding regions of fibrinogen varies with the length of time after adsorption of the fibrinogen. Platelet adhesion to polymethacrylates has been correlated with the amount of antifibrinogen binding, suggesting that the adsorbed fibrinogen is in different conformations on the various polymethacrylates.

Fibrinogen undergoes a time-dependent transition after its adsorption to a surface that results in reduced platelet and antibody binding to the adsorbed fibrinogen, as well as reduction in the SDS and plasma displaceability, and changes in amide II frequency of the adsorbed fibrinogen. The losses in platelet binding, antibody binding, and SDS elutability are prevented if albumin is included in the storage buffer. An example showing time dependent losses in antibody binding to fibrinogen and its prevention by albumin is shown in Figure II.1.2.12. Vitronectin also appears to undergo conformational changes upon adsorption that affect its ability to bind heparin and its infrared spectra.

Modulation of the biologic activity of fibronectin has been shown in several studies in which the ability of fibronectin adsorbed to various surfaces to support cell attachment or spreading was found to differ. For

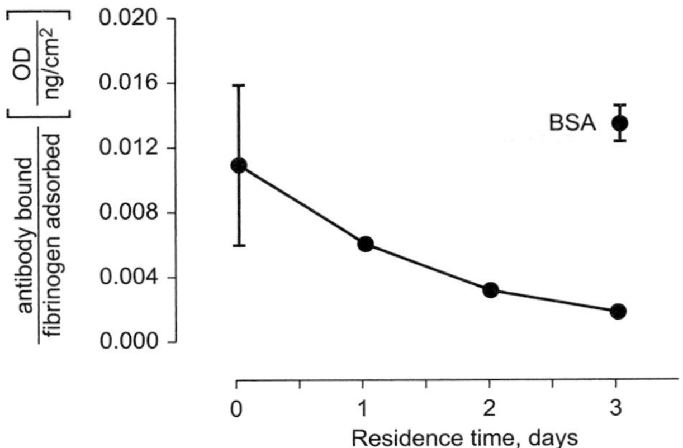

FIGURE II.1.2.12 Transitions in adsorbed fibrinogen. The effect of three-day residence in buffer or buffered albumin solution upon anti-fibrinogen binding to fibrinogen adsorbed from dilute plasma to Biomer polyurethane is shown. BSA: bovine serum albumin. (From: Fig. 3A in Chinn et al., 1992.)

example, fibronectin adsorbed to tissue culture grade polystyrene was able to support cell attachment and spreading, whereas fibronectin adsorbed to ordinary polystyrene does not support spreading very well, unless some albumin is added to the fibronectin solution. Fibronectin adsorbed to self-assembled monolayer (SAM) films with various functional end groups also varies in its ability to interact with cells. On hydrophobic SAMs there is poor cell spreading unless albumin is coadsorbed (albumin "rescuing"). The albumin "rescuing" phenomenon observed for SAMs is similar to the albumin effect on fibronectin's ability to promote cell spreading on polystyrene, and to the effect of albumin addition in preventing losses in platelet adhesion to fibrinogen adsorbed surfaces discussed above. The ability of fibronectin adsorbed to a series of polymers to support the outgrowth of corneal epithelial cells has also been found to vary a great deal, despite the presence of similar amounts of fibronectin on the surfaces.

The effect of albumin addition in enhancing the adhesivity of fibronectin-coated surfaces is opposite to what one might expect, because the added albumin should reduce the amount of adsorbed fibronectin as albumin competes for sites on the surfaces. The explanation for the albumin effect is thought to be that by adsorbing along with the fibronectin to the surface, the albumin molecules occupy surface sites near the fibronectin molecules. The adsorbed albumin molecules thus keep the adsorbed fibronectin molecules from undergoing structural changes that they would otherwise do in trying to spread into formerly empty surface sites, but they cannot do so if albumin molecules fill those sites.

Highly uniform SAM surfaces with different terminal functional groups have proved useful in showing how surface chemistry modulates the biologic properties of adsorbed fibronectin. When increasing amounts of fibronectin were adsorbed to SAMs with terminal OH, COOH, NH_2 or CH_3 functional groups, it was found that the amount of adsorbed fibronectin required to cause cell adhesion varied widely. Thus, on the CH_3 surface it took far higher amounts of fibronectin to obtain equivalent maximal adhesion than on the OH surfaces (Keselowsky et al., 2003). Similarly, the binding affinity of antibodies that bind to the cell binding domain and of the integrin $\alpha_5\beta_1$ was much higher for fibronectin on the OH surface than on the CH_3 surface. Thus, there are significant differences in the functional presentation of the central cell-binding domain of fibronectin upon adsorption to different surface chemistries. The theoretical predictions of differences in preferred orientation of adsorbed molecules on surfaces with different chemistry (see Figure II.1.2.7) suggest that these biological effects are also due to orientational effects. With an osteogenic cell line, genes that mark osteoblastic differentiation and actual mineralization were both elevated in the cells on the fibronectin-coated OH and NH_2 surfaces compared with the COOH and CH_3 SAMs (Keselowsky et al., 2005). Myoblast differentiation on fibronectin adsorbed to this same series of SAMs also varied greatly, in the order OH> CH_3> NH_2 = COOH (Lan et al., 2005). The ability of myoblasts to rearrange adsorbed fibronectin and deposit new fibronectin was also much greater on the CH_3 surface than on the OH surface.

Figure II.1.2.13 shows an experimental example of the modulation of the biologic activity of adsorbed fibrinogen, namely its ability to mediate platelet adhesion. In these studies, both fibrinogen adsorption and binding of a fibrinogen specific antibody were measured from plasma diluted to various degrees on a series of chemically modified polystyrene surfaces. Platelet adhesion to the plasma preadsorbed surfaces was also measured. The left part of Figure II.1.2.13 shows that there was no obvious correlation between platelet adhesion and fibrinogen adsorption, despite the fact that we know that fibrinogen adsorption is required for adhesion, as shown in Figure II.1.2.3. However, when the availability of the platelet binding region in fibrinogen adsorbed to these surfaces was measured using a monoclonal antibody (M1) that binds to the C-terminal region of the gamma chain of fibrinogen, it was found to vary greatly. Platelet adhesion was well correlated with the antibody binding, as shown in the right panel in Figure II.1.2.13. The poor correlation with total fibrinogen adsorption, and the greatly improved correlation with platelet binding site availability, is strong support for the idea that the functional activity of adsorbed fibrinogen in supporting platelet adhesion depends on the surface to which it is adsorbed. Thus, fibrinogen's bioadhesive activity is modulated by the surface chemistry.

The studies with platelets, fibroblasts, epithelial cells, osteocytic cells, and myoblasts show that substrate properties somehow modulate the ability of adsorbed proteins to interact with cells. These differences may arise, at least in part, from differences in the availability of epitopes on adhesive proteins for the cell surface receptor. That

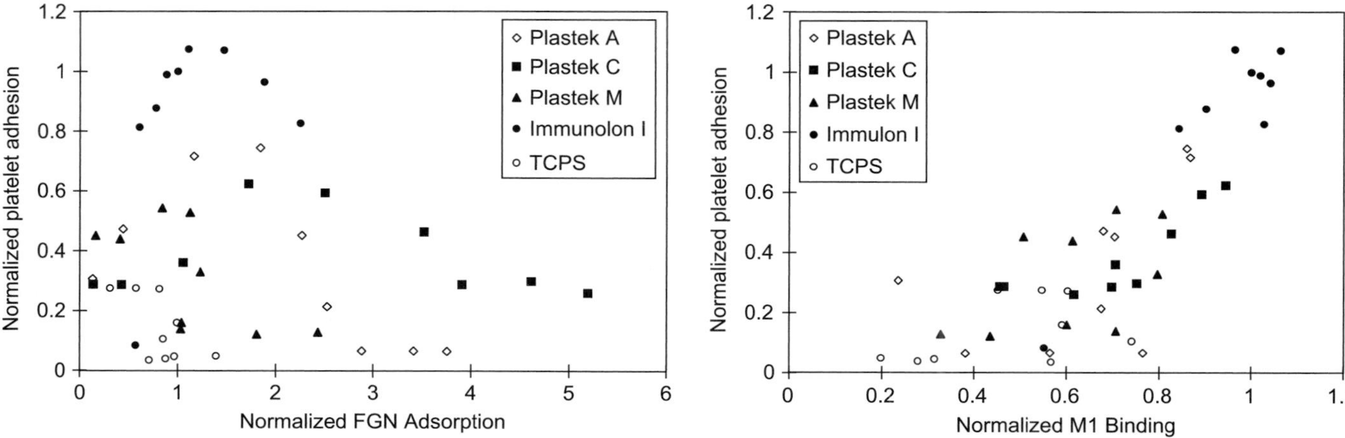

FIGURE II.1.2.13 Cross plots of platelet adhesion against the amount of fibrinogen (FGN) adsorption or M1 antibody binding on five different surfaces. (From: Tsai et al., 2003.)

is, both the amount of the adsorbed adhesive protein, as well as its "bioreactivity," are actively influenced by the properties of the surface to which it is adsorbed.

THE IMPORTANCE OF ADSORBED PROTEINS IN BIOMATERIALS

Table II.1.2.1 summarizes the principles underlying the influence of adsorbed proteins in biomaterials used in contact with the blood. All of the principles listed also apply in other environments, such as the extravascular spaces, albeit with other proteins and other cell types (e.g., macrophages in the peritoneum adhere via other receptors and other adhesion proteins). The blood platelets therefore provide a "case study," and we close this chapter by considering this case.

The high sensitivity of platelet–surface interactions to adsorbed proteins is fundamentally due to the presence of adhesion receptors in the platelet membrane that bind to certain plasma proteins. There are only a few types of proteins in plasma that are bound by the adhesion receptors. The selective adsorption of these proteins to synthetic surfaces, in competition with the many non-adhesive proteins that also tend to adsorb, is thought to mediate platelet adhesion to these surfaces, which in turn leads to platelet activation events such as their ability to catalyze the formation of thrombin. However, since the dissolved plasma-phase adhesion proteins do not bind to adhesion receptors unless the platelets are appropriately stimulated, while unstimulated platelets can adhere to adsorbed adhesion proteins, it appears that adsorption of proteins to surfaces accentuates and modulates the adhesion receptor–adhesion protein interaction. The type of surface to which the adhesion protein is adsorbed affects the ability of the protein to support platelet adhesion, due to variations in the amount of adsorbed fibrinogen (Horbett, 1993), and in the availability of the platelet binding regions in the fibrinogen molecule (Tsai et al., 2003). The principles that determine protein adsorption to biomaterials include monolayer adsorption, the intrinsic surface activity and bulk concentration of the protein, and the effect of different surfaces on the selectivity of adsorption and biologic activity of the adsorbed protein.

TABLE II.1.2.1 Principles Underlying the Influence of Adsorbed Plasma Proteins on Blood Platelet Interactions with Biomaterials

1. Synthetic foreign materials acquire bioreactivity only after first interacting with dissolved proteins. The principal means by which the transformation from an inert, nonthrombogenic polymer to a biologically active surface takes place is the interaction of the proteins with the surface that then mediates cell adhesion.
2. Platelets are an important example of why and how adsorbed proteins are influential in cell–biomaterials interactions.
3. Sensitivity of platelets to adsorbed proteins is due to:
   a. Some proteins in plasma are strongly adhesive for platelets: fibrinogen, fibronectin, vitronectin, and von Willebrand factor.
   b. Concentrating, localizing, immobilizing effects of adsorbing the proteins at the interface accentuates the receptor–adhesion protein interaction.
   c. Platelets have receptors (IIb/IIIa and Ib/IX) that bind specifically to a few of the plasma proteins, mediating adhesion.
4. Principles of protein adsorption to biomaterials:
   a. Monolayer adsorption and consequent competition for available adsorption sites means that not all proteins in the plasma phase can be equally represented on the surface.
   b. Driving forces for adsorption are the intrinsic surface activity and bulk phase concentration of the proteins.
   c. Surfaces vary in selectivity of adsorption.
   d. Biological activity of the adsorbed protein varies on different surfaces.

More generally, all proteins are known to have an inherent tendency to deposit very rapidly on surfaces as a tightly bound adsorbate that strongly influences subsequent interactions of many different types of cells with the surfaces. It is therefore thought that the particular properties of surfaces, as well as the specific properties of individual proteins, together determine the organization of the adsorbed protein layer, and that the nature of this layer in turn determines the cellular response to the adsorbed surfaces.

The importance of adsorbed proteins in biomaterials science is also well-illustrated by the amount of effort that has been expended to develop protein "non-fouling" or protein repellent materials in recent years (see Chapter I.2.10 and Horbett, 2003). Since fibrinogen adsorption is required for platelet adhesion, an obvious approach to improve blood compatibility is to produce biomaterials that reduce fibrinogen adsorption, and thus much recent research has focused on finding protein repellent materials. For example, many studies show that fibrinogen adsorption can be reduced with poly(ethylene oxide) (PEO) coatings of various types (e.g., PEO/polylactide copolymers, PEO grafted polyurethanes (PUs), PEO grafted PU/polystyrene, ethylene glycol terminated SAMs or glow discharge deposited polymers of the low molecular weight polyether tetraglyme to make PEO-like coatings), and that platelet adhesion is reduced, both *in vitro* and *in vivo*. However, since even low amounts of adsorbed fibrinogen (ca. 5 ng/cm^2; Tsai and Horbett, 1999) are sufficient to support nearly full-scale platelet adhesion, and many of the surfaces tested fail to reduce fibrinogen adsorption to that degree, it has proved difficult to make biomaterials that fully achieve this design criterion.

BIBLIOGRAPHY

Agashe, M., Raut, V., Stuart, S. J., & Latour, R. A. (2005). Molecular simulation to characterize the adsorption behavior of a fibrinogen γ-chain fragment. *Langmuir*, **21**, 1103–1117.

Anderson, D. G., Putnam, D., Lavik, E. B., Mahmood, T. A., & Langer, R. (2005). Biomaterial microarrays: Rapid, microscale screening of polymer–cell interaction. *Biomaterials*, **26**, 4892–4897.

Andrade, J. D. (1985). Principles of protein adsorption. In J. D. Andrade (Ed.), *Surface and Interfacial Aspects of Biomedical Polymers* (pp. 1–80). New York, NY: Plenum Press.

Arnebrandt, T., & Wahlgen, M. (1995). Protein–surfactant interactions at solid surfaces. In T. A. Horbett, & J. Brash (Eds.), *Proteins at Interfaces II: Fundamentals and Applications, ACS Symposium Series* (**602**, pp. 239–254). Washington, DC: American Chemical Society.

Backovic, A., Huang, H. L., Del Frari, B., Piza, H., Huber, L. A., et al. (2007). Identification and dynamics of proteins adhering to the surface of medical silicones *in vivo* and *in vitro*. *J. Proteome Res.*, **6**, 376–381.

Chastain, S. R., Kundu, A. K., Dhar, S., Calvert, J. W., & Putnam, A. J. (2006). Adhesion of mesenchymal stem cells to polymer scaffolds occurs via distinct ECM ligands and controls their osteogenic differentiation. *J. Biomed. Mater. Res. A.*, **78**, 73–85.

Chinn, J. A., Horbett, T. A., & Ratner, B. D. (1991). Baboon fibrinogen adsorption and platelet adhesion to polymeric materials. *Thromb. Haemostasis*, **65**, 608–617.

Chinn, J. A., Posso, S. E., Horbett, T. A., & Ratner, B. D. (1992). Post-adsorptive transitions in fibrinogen adsorbed to polyurethanes: Changes in antibody binding and sodium dodecylsulfate elutability. *J. Biomed. Mater. Res.*, **26**, 757–778.

Cornelius, R. M., & Brash, J. L. (1993). Identification of proteins absorbed to hemodialyser membranes from heparinized plasma. *J. Biomater. Sci. Polym. Ed.*, **4**, 291–304.

Gifford, R., Kehoe, J. J., Barnes, S. L., Kornilayev, B. A., Alterman, M. A., et al. (2006). Protein interactions with subcutaneously implanted biosensors. *Biomaterials*, **27**, 2587–2598.

Horbett, T. A. (1981). Adsorption of Proteins from Plasma to a Series of Hydrophilic-Hydrophobic Copolymers. II. Compositional Analysis with the Prelabelled Protein Technique. *J. Biomed. Mater. Res.*, **15**, 673–695.

Horbett, T. A. (1993). Principles underlying the role of adsorbed plasma proteins in blood interactions with foreign materials. *Cardiovasc. Pathol.*, **2**, 137S–148S.

Horbett, T. A. (1994). The role of adsorbed proteins in animal cell adhesion. *Coll. Surf. B: Biointerfaces*, **2**, 225–240.

Horbett, T. A. (1996). Proteins: Structure, properties, and adsorption to surfaces. In B. D. Ratner, A. S. Hoffman, F. Schoen, & J. E. Lemons (Eds.), *Biomaterials Science* (pp. 133–141). San Diego, CA: Academic Press.

Horbett, T. A. (1999). The role of adsorbed adhesion proteins in cellular recognition of biomaterials. *BMES Bulletin*, **23**, 5–9.

Horbett, T. A. (2003). Biological activity of adsorbed proteins. In M. Malmsten (Ed.), *Biopolymers at Interfaces,* 2nd ed. (pp. 393–413). New York, NY: Marcel Dekker, Inc. Chapter 15.

Horbett, T. A., & Brash, J. L. (1995). Proteins at interfaces: An overview. In T. A. Horbett, & J. Brash (Eds.), *Proteins at Interfaces II: Fundamentals and Applications, ACS Symposium Series* (**602**, pp. 1–25). Washington, DC: American Chemical Society.

Hwang, N. S., Varghese, S., & Elisseeff, J. (2008). Controlled differentiation of stem cells. *Adv. Drug Deliv. Rev.*, **60**, 199–214.

Keselowsky, B. G., Collard, D. M., & García, A. J. (2003). Surface chemistry modulates fibronectin conformation and directs integrin binding and specificity to control cell adhesion. *J. Biomed. Mater. Res.*, **66A**, 247–259.

Keselowsky, B. G., Collard, D. M., & García, A. J. (2005). Integrin binding specificity regulates biomaterial surface chemistry effects on cell differentiation. *Proc. Natl. Acad. Sci. USA*, **102**, 5953–5957.

Lan, M. A., Gersbach, C. A., Michael, K. E., Keselowsky, B. G., & García, A. J. (2005). Myoblast proliferation and differentiation on fibronectin-coated self assembled monolayers presenting different surface chemistries. *Biomaterials*, **26**, 4523–4531.

Latour, R. A. (2008). Biomaterials: Protein–surface interactions. In G. E. Wnek, & G. L. Bowlin (Eds.), *Encyclopedia of Biomaterials and Biomedical Engineering*, 2nd ed., (Vol. 1, pp. 270–284). New York, NY: Informa Healthcare.

Lee, W. -K., McGuire, J., & Bothwell, M. K. (2004). Competitive adsorption of bacteriophage T4 lysozyme stability variants at hydrophilic glass surfaces. *J. Colloid Interf. Sci.*, **269**, 251–254.

Lundstrom, I. (1985). Models of protein adsorption on solid surfaces. *Prog. Coll. Polym. Sci.*, **70**, 76–82.

Nakajima, M., Ishimuro, T., Kato, V., Ko, I. -K., Hirata, I., et al. (2007). Combinatorial protein display for the cell-based screening of biomaterials that direct neural stem cell differentiation. *Biomaterials*, **28**, 1048–1060.

Siedlecki, C. A., Lestini, B. J., Kottke-Marchant, K. K., Eppell, S. J., Wilson, D. L., et al. (1996). Shear-dependent changes in the three-dimensional structure of human von Willebrand factor. *Blood*, **88**, 2939–2950.

Slack, S. M., & Horbett, T. A. (1995). The Vroman effect: A critical review. In T. A. Horbett, & J. Brash (Eds.), *Proteins at Interfaces II: Fundamentals and Applications*, *ACS Symposium Series* (**602**, pp. 112–128). Washington, DC: American Chemical Society.

Tsai, W. -B., & Horbett, T. A. (1999). Human plasma fibrinogen adsorption and platelet adhesion to polystyrene. *J. Biomed. Mater. Res.*, **44**, 130–139.

Tsai, W. -B., Grunkemeier, J. M., & Horbett, T. A. (2003). Variations in the ability of adsorbed fibrinogen to mediate platelet adhesion to polystyrene-based materials: A multivariate statistical analysis of antibody binding to the platelet binding sites of fibrinogen. *J. Biomed. Mater. Res.*, **67A**, 1255–1268.

Wertz, C. F., & Santore, M. M. (2001). Effect of surface hydrophobicity on adsorption and relaxation kinetics of albumin and fibrinogen: Single-species and competitive behavior. *Langmuir*, **17**, 3006–3016.

Zhang, M., Wu, Y., Hauch, K., & Horbett, T. A. (2008). Fibrinogen and von Willebrand's factor mediated platelet adhesion to polystyrene under flow conditions. *J. Biomater. Sci., Polymer Edition*, **19**(10), 1383–1410.

CHAPTER II.1.3 CELLS AND SURFACES *IN VITRO*

S. Adam Hacking[1] *and Ali Khademhosseini*[2]

[1]Laboratory for Musculoskeletal Research and Innovation, Department of Orthopaedic Surgery, Massachusetts General Hospital and Harvard Medical School, Boston, MA, USA

[2]Department of Medicine, Center for Biomedical Engineering, Brigham and Women's Hospital, Harvard Medical School, Cambridge, MA, USA and Harvard-MIT Health Sciences and Technology, Massachusetts Institute of Technology, Cambridge, MA, USA

INTRODUCTION

There is no doubt that *in vitro* methods have been and will continue to be of great value to the progression of medical science. The culture of mammalian cells and the study of their behavior *in vitro* has led to new understandings of cell physiology (Neher and Sakmann, 1976), molecular biology (Mello, 2007), and many disease processes (Schwarz et al., 1985). In many experimental settings, the ability to extract and grow cells from host tissues is a powerful tool since experimental question(s) can be evaluated under standardized conditions without the many complicating influences from the host physiology.

As science has progressed, new *in vitro* techniques have developed beyond the culture of cells in glass or plastic dishes. Cell–substrate interactions are of particular interest since cells interact physically with the extracellular matrix (ECM) and often modify it in response to various stimuli. As a result, efforts to examine and recapitulate the *in vivo* environment *in vitro* often require modification of the cell culture substrates. Cell–substrate interactions at the micro- and nanoscales have been facilitated by new techniques, applications, and approaches that have emerged from the combination of engineering and biology. In practice, the chemistry, topography, and elasticity of the culture surface may be altered, and mechanical forces can also be applied to cells by physical deformation of the culture substrate.

With respect to tissue engineering and organ regeneration, advances in our understanding of cell–substrate interactions are essential to controlling cell behavior. For example, the ability to pattern cells with micron scale precision on a variety of substrates presents new opportunities to study and recreate cell–cell interactions required to generate functional tissues *in vitro* for subsequent implantation *in vivo*. This is of primary importance, since organs contain multiple cell types that are precisely arranged in close proximity to form complex structures.

This chapter provides a basic understanding of *in vitro* principles, beginning with a fundamental overview of the culture of cells and their interactions with surfaces. We then describe a number of techniques employed to manipulate the *in vitro* environment at the micro- and nanoscale and then discuss how these fundamental techniques can be applied to investigate and influence cell behavior.

A BASIC OVERVIEW OF CELL CULTURE

Tissue culture is a general term for the harvest of cells, tissues or organs, and their subsequent growth or maintenance in an artificial environment. Mammalian cells cultured *in vitro* have the same basic requirements as cells growing within an organism. Specialized approaches have been developed to replace a number of mammalian systems essential for life. *In vitro*, blood is replaced by the culture media, which bathes the cells and provides an energy source (glucose), essential nutrients (salts and amino acids), proteins and hormones (from added serum), and a buffer (to maintain pH balance). During culture, the byproducts of cellular metabolism are released into the media as its constituents are depleted. Since the culture media is not continually circulated and purified, it must be changed regularly to maintain optimal conditions for cell function. Not surprisingly, these culture conditions are also favorable for the growth of unwanted organisms such as fungi and bacteria. Since there is no immune system to control infection *in vitro*, anti-fungal and anti-bacterial agents can be added to the media on a prophylactic basis. To further reduce the likelihood of contamination by microorganisms, cells are manipulated under a laminar flow hood using strict aseptic (sterile) techniques. Laminar hoods control and direct filtered air to facilitate a sterile working environment by reducing the contact of air-borne bacteria and particulates with the culture dish and hood surfaces.

Cells in media are contained in culture dishes or flasks and incubated in a temperature-controlled (37°C) and humidified (95%) chamber. While the lids of the culture dishes cover and protect the media, they are not sealed and enable gas exchange within the incubator. The exchange of gas at the media surface acts like the lungs to maintain

TABLE II.1.3.1 Examples of Some Commonly Used Cell Lines

Cell Line	Organism	Tissue of Origin	Further Cell Information
3T3	Mouse	Embryonic fibroblast	Fibroblasts
AML-12	Mouse	Liver	Liver cells
HeLa	Human	Cervical cancer	Epithelium (first cell line reported)
HUVEC	Human	Umbilical cord vein	Endothelium (stem cells)
MC3T3-E1	Mouse	Calvarial fibroblast	Differentiate to osteoblast

the gas balance necessary for metabolism. CO_2 is usually added to the incubator at a low concentration (5%). The CO_2 interacts with the bicarbonate buffer in the media to help maintain a pH of 7.0–7.4. Buffering counteracts pH changes in the media as it accumulates waste from cellular activity. Phenol red is commonly added as an indicator to monitor pH, and a change in the color of the media is often a sign of poor culture conditions. Since the culture dishes are not sealed they are prone to evaporation. To reduce evaporation, and prevent a subsequent change in media concentration, a water dish (with an anti-fungal/anti-bacterial agent) is usually placed in the bottom of the incubator to maintain a high level of ambient humidity.

Primary Culture

Cells for *in vitro* work may be obtained from tissue (primary culture) or from cell lines. Cells for primary culture are obtained by surgical dissection of living tissues. Cells can be obtained by passing media through the marrow cavities of the long bones, by collecting cells that grow out from a piece of tissue (*explant culture*) or by enzymatic digestion of tissue that contains cells. Enzymatic digestion of small pieces of tissue immersed in collagenase at 37°C breaks down the surrounding ECM and releases entrapped cells (*enzymatic dissociation*). After harvest, cells may be cultured in suspension or on culture substrates. Anchorage dependent cells are plated for a few days to remove dead or unwanted (non-adherent) cell types. Cells anchored to the surface of the culture plate are "released" by adding a mixture of trypsin, a protease found in the digestive system, and ethylenediaminetetraacetic acid (EDTA). Trypsin degrades proteins, including important cell adhesion proteins (CAPs) that anchor the cell to the culture surface. Over short treatment times, trypsin is generally not harmful to cells. EDTA is a chelating agent that binds calcium and magnesium ions. EDTA increases the effectiveness of trypsin and prevents cell aggregation after release from the culture dish.

Harvested cells are often filtered to produce a suspension of single cells. The single cell suspension is counted, and portions of the initial cell population are transferred to new culture dishes for expansion. Expansion is also referred to as *sub-culturing* or *passaging* the cells. Passaging cells ensures that cells are plated at a low density so they have space around them to proliferate. Under appropriate conditions, most cells proliferate and form a monolayer on the culture surface. The point at which cells completely cover the culture surface is referred to as *confluence*. At confluence, proliferation will generally cease because cell–cell contact results in *contact inhibition*. Unlike cell lines, primary cells are often used without passaging or passaged only a few times, since passaging can result in loss of cell characteristics (phenotype).

Cell Lines

Generating primary cultures by continually harvesting and isolating cells is a time-consuming task. As a result, it is often preferable to utilize cells that are obtained from stock maintained in the laboratory. Cell lines refer to cells that can be passaged many times without loss of their phenotype. Physiologically, these cells can divide repeatedly, without shortening of their telomeres (Hayflick, 1965). Cell lines differ from other cells in that they have escaped the Hayflick limit and are *immortalized* (Hayflick, 1985). Examples of a handful of commonly used cell lines, from a large number of existing cell lines, are presented in Table II.1.3.1.

Cell lines also offer an experimental advantage. In terms of strict experimental uniformity, primary cultures may contain unwanted cell types unintentionally included during harvest. Primary cultures may also contain slight genetic variations due to genetic differences arising from the harvest of cells from different animals. Since cell lines are derived from the same cell they are all genetic clones of the parent cell.* As a result, experiments using cell lines can be reproduced by different laboratories using the same cell, with obvious advantages for collaboration, verification, and reproducibility.

A large number of cell lines are derived from tumors or cells that have undergone spontaneous mutation or some form of manipulation. The HeLa cell line is an example that has been in use since the early 1950s and was derived from cervical cancer cells taken from Henrietta Lacks (Scherer et al., 1953). It is helpful to recognize that while cell lines may behave similarly to normal cells, they have been obtained by repeated culture and selection (Puck and Marcus, 1955; Browne and Al-Rubeai, 2007),

* If the cell line has not been contaminated or has not undergone spontaneous genetic change.

by treating or transforming normal cells (using viruses, oncogenes, radiation, drugs or chemicals) (Prasad et al., 1994; Eiges et al., 2001; Groger et al., 2008), from transgenic mice (Connelly et al., 1989; Wu et al., 1994) or isolated from tumors (Riches et al., 2001; Yasuda et al., 2009). For example, transformed cells may present different morphologies and altered metabolisms (Priori et al., 1975; Wittelsberger et al., 1981), and often do not exhibit robust contact inhibition (Bell, 1977; Erickson, 1978). These behaviors strongly suggested that cell lines have the inherent potential to demonstrate responses that may not be typical of primary cells (Steele et al., 1991; Heckman, 2009). Therefore, when establishing new experimental protocols using cell lines it is recommended that the results of the protocol be validated with primary cells.

Cell lines can be obtained from nonprofit organizations such as the American Type Culture Collection (ATCC), the European Collection of Cell Cultures (ECACC), and the Coriell Institute for Medical Research (CIMR). The ATCC (www.atcc.org), ECACC (www.hpacultures.org.uk), and the CIMR (www.coriell.org) provide high-quality cell lines which are subject to careful testing and validation. The testing process ensures that cells provided are free of unwanted organisms such as mycoplasma or that other unwanted cell types are not present. Cell lines obtained from tissue banks are usually meticulously characterized and preserved to maintain these materials in a manner that permits reproducibility of results across time and among laboratories around the world. In some cases, cell lines can be obtained from other research laboratories; however, caution should be exercised since the possibility exists that the cell lines may be contaminated with microorganisms or unwanted cell types.

Characteristics of Cultured Cells

Cultured cells often display characteristic behaviors and growth patterns. A well-known characteristic is *contact inhibition*. In 1954, Abercrombie and Heaysman proposed that cells do not form multilayers because they do not move over or under their neighbors (Abercrombie and Heaysman, 1954). Cells exhibiting contact inhibition will preferably migrate to unoccupied areas of the culture substrate, and will not use the upper surface of a confluent cell monolayer as substrate. Contact inhibition is closely related to the cessation of cell proliferation (Timpe et al., 1978), and is one reason why most cells cultured *in vitro* produce monolayers of cells (Garrod and Steinberg, 1973). When cultured cells are in complete contact with each other, plasma membrane ruffling and cell motility are inhibited, resulting in an overall cessation of cell growth (Heckman, 2009). For this reason, cells maintained in culture are frequently replated at lower densities so that growth and proliferation may continually occur.

Detecting Cancerous Cells *In Vitro*

The phenomenon of contact inhibition has diagnostic implications for assessing the malignancy of cells (Amitani and Nakata, 1977; Abercrombie, 1979). A predominant behavior of malignant cells is uncontrolled cell growth. The natural phenomenon of contact inhibition has been used as a test to distinguish cancerous from noncancerous cells (Dehner, 1998). Interestingly, unlike normal cells, malignant cells will also proliferate in agar (a three-dimensional matrix) where no suitable binding sites for cell attachment exist (Whang-Peng et al., 1986).

Cultured cells also often display characteristic shapes that can be used as visual cues to assess the health and differentiation of a culture population. For example, preosteoblastic cells are fibroblastic in shape, possessing a long "drawn out" morphology. Upon subsequent differentiation into mature osteoblasts, cells display a cuboidal appearance. Dead or non-adherent cells are often spherical in shape and float in the culture media.

UNDERSTANDING CELL–SUBSTRATE INTERACTIONS

In vivo, cell–surface interactions are often explored as a means to enhance or control peri-implant tissue formation. Surface chemistry, topography, and elastic modulus (stiffness) of the substrate are all means to control and guide cell activity, and ultimately modulate tissue formation. This triad of properties is important for tissue engineering and organ regeneration where multiple cell types must be directed to form complex structures all in close proximity, and often with micron scale precision. In terms of three-dimensional scaffolds, tuning of scaffold chemistry, topography, and stiffness are all means to selectively guide cell development and differentiation within the bulk material.

Since most cells interact with a substrate (e.g., within the ECM) it is often desirable to control the interaction of cells with substrates *in vitro*. Modulation of the culture surface by fabrication of "engineered surfaces" comprised of micro- and nanosized chemical and topographical patterns or coatings have been used to investigate cell behavior such as adhesion, morphology, migration, proliferation, cell–cell communication, gene expression, production of ECM, differentiation, and responsiveness to extracellular signaling.

Surfaces for Cell Culture

It has long been established that most mammalian cells respond favorably to culture conditions that promote cell attachment. Cell attachment is often characterized by a change in cell morphology. Adherent cells possess a "flattened" appearance, often with an irregular cell shape and the extension of cellular processes (Figure II.1.3.1). Regular observation of cells *in vitro* is essential for successful culture as cell density, state, and contamination can all be readily assessed visually. In this regard, optically clear materials are practical as they permit simple assessment of cell growth and morphology during culture, and are well suited to many histological techniques that involve staining for quantification of cell activity. Glass, once a common form of culture ware, has been replaced by tissue culture polystyrene (TCP). While TCP is most widely used as a culture substrate, other polymers such as polyvinylchloride, polycarbonate, and polytetrafluoroethylene have been used. Advantages of TCP are that it is inexpensive, clear, easily formed and treated to modify surface groups, and unlike glass, it can be embedded and sectioned.

Process of Cell Attachment *In Vitro*

Cell attachment is the initial step in a cascade of cell–biomaterial interactions, and is important to cellular processes such as cell guidance, proliferation, and differentiation (Ruoslahti, 1996). For many adherent cells, cell proliferation can only occur on a substrate. As a result, cell–surface interactions are fundamental to understanding cell behavior, and provide an important means to control and quantify cell activity.

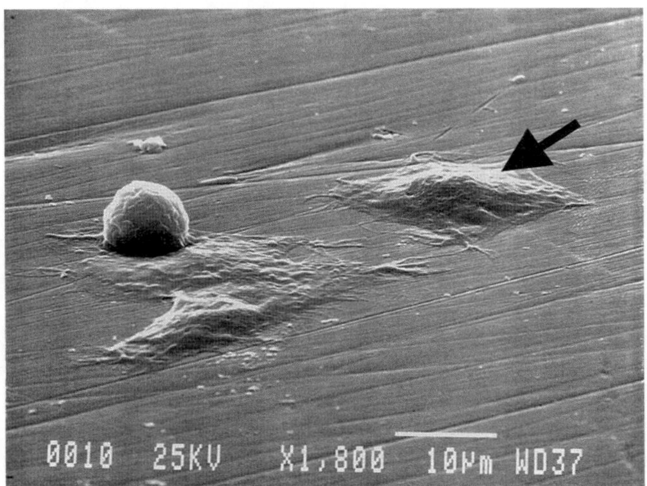

FIGURE II.1.3.1 Scanning electron microscopy image of MC3T3-E1 cells at various stages of adherence on polished titanium substrate demonstrating changes in cell morphology during attachment. Attached cells have a flattened morphology and form intimate contact with the substrate. Detached cells are round. SEM 1800X.

The process of cell attachment is largely understood (Ruoslahti and Pierschbacher, 1987; Ruoslahti and Reed, 1994; Ruoslahti, 1996; Garcia, 2005). *In vitro*, when hydrophilic surfaces like TCP are exposed to culture media containing serum, they are rapidly coated by a thin (~20 nm) layer comprised mainly of proteins (Andrade and Hlady, 1987; Kasemo and Lausmaa, 1994; Nakamura et al., 2007) that adsorb to the culture surface in a monolayer. The process of cell attachment to TCP (and many other materials) is indirect, since cells do not bind directly to TCP but instead bind to the adsorbed protein monolayer (Fredriksson et al., 1998). Cells make contact with, and anchor to, the adsorbed proteins at discrete peptide regions referred to as focal contacts (Gallant and Garcia, 2007). Cells possess heterodimeric transmembrane proteins composed of α and β subunits called *integrins*. Integrins are receptors that recognize and bind to specific anchoring proteins present on the conditioned TCP surface (Peter and Ma, 2006). The adsorbed proteins present short but specific peptide sequences (Table II.1.3.2) that are recognized by integrins as binding sites. Integrins recognize and bind to specific ligands such as fibronectin, vitronectin, collagen, and laminin. Over 18 α and 8 β receptor subunits have been identified, and these subunits combine to produce a wide variety of receptor types (Hynes, 2002). Of these combinations, the $\alpha_v\beta_3$ integrin is one of the most abundant and binds to vitronectin, fibronectin, von Willebrand factor, osteopontin, tenascin, bone sialoprotein, and thrombospondin (Hynes, 2002). A common peptide receptor for $\alpha_v\beta_3$ integrin is the RGD (arginine–glycine–aspartic acid) sequence (Ruoslahti and Pierschbacher, 1987). For the majority of cells cultured *in vitro*, fibronectin and vitronectin are important for cell attachment to TCP (Massia and Hubbell, 1991a; Steele et al., 1993).

In their capacity as adhesion receptors, integrins play an important role in controlling various steps in the

TABLE II.1.3.2 Specific Peptide Sequences on Cell Anchoring Proteins

Peptide Sequence	Derived from ECM Protein	Conjugate Receptor	References
RGD	Fibronectin, laminin α-chain, collagen, vitronectin	Multiple integrins	(Ruoslahti and Pierschbacher, 1987; Massia and Hubbell, 1990a,b; Massia and Hubbell, 1991a,b; Ruoslahti, 1996)
YIGSR	Laminin β1-chain	β_1 integrins	(Boateng et al., 2005; Weber et al., 2007; Weber and Anseth, 2008)
IKVAV	Laminin α-chain	LBP110	(Tashiro et al., 1989; Weber et al., 2007; Weber and Anseth, 2008)
REDV	Fibronectin	$\alpha_4\beta_1$ integrin	(Hubbell et al., 1991)
DGEA	Collagen type I	$\alpha_2\beta_1$ integrin	(Staatz et al., 1991; Weber et al., 2007; Weber and Anseth, 2008)
KQAGDV	Fibronectin γ-chain	β_3 integrins	(Mann and West, 2002; Gobin and West, 2003)
VAPG	Elastin	Elastase receptor, $\alpha_5\beta_3$ integrin	(Mann and West, 2002; Gobin and West, 2003)

signaling pathways that regulate processes as diverse as cytoskeletal organization, cell proliferation, differentiation, apoptosis, and migration. In most cell types certain biochemical signals necessary for cell growth, function, and survival are triggered by integrins upon attachment. Without attachment, the cell eventually undergoes *apoptosis*, which is also known as programmed cell death (Ruoslahti and Reed, 1994).

Commercial and Experimental Modifications of Culture Surfaces

Polystyrene is a hydrophobic material that is not well suited to the attachment of most mammalian cells (Curtis et al., 1983). For the culture of non-adherent cells, untreated polystyrene dishes are routinely used. As previously described, most cells require modified surfaces that are suitable for protein adsorption before attachment can occur.

A variety of treatment methods have been employed to improve cell attachment and growth on the TCP surface (Table II.1.3.3). TCP treatments used to enhance cell adhesion are often proprietary and vendor specific, but some methods such as chemical treatment in sulfuric acid (Curtis et al., 1983), chemisorption (Shen and Horbett, 2001; Bain and Hoffman, 2003a), ionizing radiation (Callen et al., 1993) or exposure to plasma-based ionizing processes such as glow discharge (Amstein and Hartman, 1975; Koller et al., 1998) have been described in the literature. It is generally understood that these treatments impart a charge to the TCP surface that enhances its ability to interact with and bind specific CAPs present in the serum added to the culture media (Jacobson and Ryan, 1982; Lee et al., 1997).

It is important to recognize that the preparation of culture surfaces can have profound effects on the behavior of cultured cells (Shen and Horbett, 2001; Bain and Hoffman, 2003b). For some cell types, commercially available surfaces may be far from optimal, and identification of optimized substrates can be a laborious task, especially for cells that are difficult to culture *in vitro*. One approach has been the generation of culture surfaces comprised of mixtures of protein-adhesive and "protein-repellant" molecules or groups. Bain and Hoffman used a combination of protein adhesive diamine groups (N_2) and hydrophobic trifluoropropyl groups (F3) to generate a large number of culture surfaces by varying the ratio of monomers in a silanization bath (Bain and Hoffman, 2003a). The different surfaces produced possessed varying affinities for protein adhesion and, as a result, cell adhesion. A surprising finding of the work was that the growth and activity of insulin-secreting cells was best supported by the most hydrophobic (F3) surfaces in the study (Bain and Hoffman, 2003b).

More direct approaches to tune substrate chemistry can be achieved by anchoring specific adhesion molecules to the culture substrate. In such investigations, it is often advantageous to retain only the molecule(s) of interest, while reducing or excluding non-specific protein adhesion as might occur with cells grown in normal, serum-enriched media (Koepsel and Murphy, 2009). Self-assembled monolayers (SAMs) are a versatile and valuable tool that has been used to study protein adhesion (Prime and Whitesides, 1991; Li et al., 2007). SAMs are comprised of long alkane thiols anchored at the thiol end to the gold-coated substrate, while the other end is derivatized with small functional groups such as cell adhesion peptides or "protein-repellant" oligo(ethylene oxide) sequences. The functional groups can also be conjugated to specific proteins, such as monoclonal antibodies, or can interact with proteins in the media. SAMs can be tailored to exhibit specific amounts of protein adsorption, by controlling the composition of the monolayer (mixture of functional groups).

The control of ligand density and type provides another means to modify substrate chemistry and cell behavior. Gradients of immobilized ECM components and growth factors have also been shown to direct cell migration and shape (Liu et al., 2007; Inoue et al., 2009). Such experimental aspects can also be explored efficiently with microfluidic systems where the facile and reproducible generation of gradients utilizing small reagent volumes has been applied to the study of cell behavior *in vitro* (Du et al., 2009).

Dynamic Control of Cell Culture Surfaces

Controllable surfaces for cell culture have been generated that enable a predictable change in the adhesive or morphological characteristics of the surface. Changes in surface characteristics may be achieved by mechanical or physico-chemical methods, and the latter methods

TABLE II.1.3.3 Treatment Methods Employed to Improve Cell Attachment and Growth on the TCP Surface

Approach	Technique	References
Surface modification	Corona discharge, Glow discharge	(Amstein and Hartman, 1975; Ertel et al., 1990; Chinn et al., 1994; Lee et al., 1994; Lee et al., 2003)
	Oxidation	(Curtis et al., 1983; Chang and Sretavan, 2008; Frimat et al., 2009)
	Plasma etching	(Claase et al., 2003; Rhee et al., 2005)
	Surface chemical reaction	(Maroudas, 1977; Curtis et al., 1983; Kowalczynska and Kaminski, 1991)
	Ultraviolet irradiation	(Nakayama et al., 1993; Matsuda and Chung, 1994; Welle et al., 2002)
	Radiation grafting (thermo-sensitive polymers)	(Yamato et al., 2001; Yamato et al., 2002)

are the focus of this section. A wide variety of dynamic surfaces have been fabricated from SAMs with specific functional groups that are responsive to thermal, chemical, electrical or electromagentic (UV or visible light) stimuli. Dynamic culture surfaces can be used as a means to control cell adhesion. More sophisticated applications have been used to selectively pattern surfaces to generate a culture environment with multiple cell types specifically arranged in groups in close proximity to each other.

Dynamic culture surfaces have also been used to detach cells from substrates. Cell sheets have been spontaneously detached by cooling a thermally-sensitive, radiation- or plasma-polymerized poly(N-isopropylacrylamide) (poly[NIPAAm]) polymer (Cheng et al., 2004). Okano's group in Japan has pioneered in this method, using electron beam radiation grafting of NIPAAm monomer on the surfaces of cell culture dishes. (Yamato et al., 2001, 2002) (see Table II.1.3.3). At temperatures suitable for cell culture (above 32°C) the poly[NIPAAm] surface is hydrophobic and readily adsorbs cell adhesive proteins (CAPs), leading to the growth of a cell monolayer to confluence. However, when the temperature is reduced below a critical temperature (32°C) the poly[NIPAAm] surface becomes hydrophilic, and the cell sheet detaches as a contiguous, confluent monolayer, carrying with it the CAP monolayer. Such dynamic thermal control of cell adhesion can be particularly useful for the intact removal of a sheet of cells by avoiding the use of proteolytic enzymes such as trypsin that can disrupt natural cell–cell networks and compromise cellular integrity. Cell sheets have a variety of applications in tissue engineering (Elloumi-Hannachi et al., 2010; Gauvin et al., 2010), and the Okano group has applied them clinically as monolayers of corneal cells for repairing damaged or diseased eye surfaces.

Methods that switch surfaces from cell-adhesive to non-adhesive have also been reported using photo-responsive SAMs. These approaches generally modify the presentation of CAPs making them less accessible to the cell. Upon exposure to UV light, surfaces based on a combination of azobenzene groups and RGD ligands will "hide" the RGD groups by UV-induced conversion of the azobenzene functional group from polar to non-polar character (Auernheimer et al., 2005). Exposure to UV light reduced the exposed length of the azobenzene-RGD segments, and reduced cell adhesiveness by 17%. Nitro-spiropyran and methyl methacrylate polymer coatings have also been employed as photo-responsive surfaces (Higuchi et al., 2004). Nitro-spiropyran is a photo-responsive polymer that switches from a cell adhesive state to a non-adhesive (hydrophilic) state upon exposure to UV light.

Investigating Cell–Substrate Interactions

The cellular microenvironment consists of a myriad of signals arising from surrounding cells, the ECM, and soluble molecules. The ability to control and modulate cell behavior, through the cell environment, is important for understanding cell behavior as well as controlling a wide range of biomaterial interactions. In general, cell behavior can be modified by controlling substrate chemistry, topography, and elasticity, and investigations have traditionally focused on broad changes to the cell culture surface, such as those obtained by application of a coating. However, cell–cell and cell–ECM interactions in nature usually occur over very small, nanoscale distances.

New insights into cell behavior and physiology have been gained by utilizing and extending a variety of techniques and materials that can be used to influence the cell environment at the microscale. The following sections examine some of the fundamental concepts learned from examining cell behavior in response to substrate chemistry, topography, elasticity, and strain, alone and in combination at macro-, micro-, and nanoscales.

Limitations of Biomaterial Evaluation *In Vitro*

It is important to recognize that the *in vitro* evaluation of biomaterials is subject to specific constraints arising from the reduced complexity of the *in vitro* environment. First and foremost, there is no immune or inflammatory response *in vitro*. Nor is there the same cascade of events resulting from implantation, e.g., the interaction with components of blood, formation of a clot, vascularization, and recruitment of a variety of cells that participate in the wound healing response.

CELL RESPONSE TO SUBSTRATE CHEMISTRY

Surface chemistry is perhaps the most obvious, and in some cases, the most accessible parameter to control when modifying *in vitro* conditions to assess their effects on cell behavior. At the macroscale, a variety of strategies can be employed to modify culture surface chemistry. Most coating strategies aim to increase cell adhesion or to preferentially select for certain cell types. Uniform coatings may be achieved by either spin coating a solution of it on the substrate surface, or by immersion of the substrate in a solution of it, followed by draining and drying. Substrates have been coated in these ways with a variety of organic and inorganic compounds, such as collagen (Chlapowski et al., 1983; Kataropoulou et al., 2005), fibronectin (Klein-Soyer et al., 1989), gelatin (Aframian et al., 2000), and poly-L-lysine (Jensen and Koren, 1979). Culture surfaces have also been broadly coated with specific adhesion-related peptides. The peptide sequence arginine–glycine–aspartate (RGD) has been immobilized on a number of materials as a means of enhancing two-dimensional cell attachment (Massia and Hubbell, 1990a, 1991b; Cutler and Garcia, 2003; Petrie et al., 2008). Studies suggest that 1 pmol/mm^2 of RGD may be required as a minimal surface density (Chollet et al., 2009), and that an increase in RGD density is generally beneficial to cell attachment (Massia and Hubbell, 1991b). However, it is worth noting that there may be an optimum ligand density for maximum cell attachment. The spatial organization of various cell ligands may also be achieved on some surfaces. This approach has been used to enhance the initial number of attached cells, to increase their proliferation or to preferentially select for specific cell types (Zemljic Jokhadar et al., 2007).

There are, however, instances where substrate coatings are of limited benefit, and examples of such include cultures where the applied coating may rapidly degrade or detach from the culture substrate. A common strategy to enhance cell adhesion is to modify the substrate chemistry to encourage the adsorption and presentation of certain CAPs or ligands (Keselowsky et al., 2003). Ligands are present on CAPs in the serum added to culture media. As previously discussed with TCP, a number of processes have been developed to enable modification of large culture surface areas to enhance CAP protein adsorption.

Micrometer-Scale Chemical Patterns

Patterning at the microscale is associated with the control of cell shape, position, and density, and presents a powerful tool to examine and control the cell microenvironment. A common approach is to generate a pattern of CAPs on a two-dimensional surface and then to plate with cells that adhere preferentially to the patterned, adhesive regions. Usually the non-adhesive regions are PEGylated. A multitude of patterns in various configurations have been attained using printing methods, as well as methods including dynamic surfaces as previously discussed.

Cells and proteins have been previously patterned on various substrates using SAMs (Mrksich and Whitesides, 1996), metal templates (O'Neill et al., 1990), stamped proteins, peptides (Hyun et al., 2001), and biopolymers (Bhatia et al., 1997), microfluidic channels (Takayama et al., 1999), membranes (Folch et al., 2000; Ostuni et al., 2000), polysaccharides (Khademhosseini et al., 2004; Suh et al., 2004), and cross-linked poly(ethylene glycol) (PEG) (Khademhosseini et al., 2003). Non-contact methods can also be used to generate protein patterns and include films that are cross-linked by photolithography that selectively promote or exclude cell adhesion (Chien et al., 2009). In this manner, non-specific biological molecules can be used to create patterns of adhesive and/or non-adhesive regions. Short chains of PEG have been assembled on surfaces to form a brush-like layer that resists protein adsorption and, as a consequence, cell attachment.

Protein patterning has also been used to produce regions of preferential adhesion for specific cell types, to alter cell shape, to cluster cells together, to restrict their proliferation or to position different cell types in close proximity. Components of the ECM which participate in cell adhesion *in vivo* are candidates for studies involving chemical patterning *in vitro*. Fibronectin, laminin, vitronectin, and collagen are examples of ECM molecules commonly used for patterning cell adhesive regions. Such proteins may also be classified as CAPs.

Direct transfer of proteins or cells from a stamp to the substrate can also be used to generate micro- and nanoscale patterns. Microstamping or microcontact printing (μCP) is a technique whereby polydimethylsiloxane (PDMS) molds are used to pattern substrates by the selective transfer of materials of interest (Mrksich and Whitesides, 1996). In one approach, the relief patterns on a PDMS stamp can be used to form patterns of SAMs on the surface of a substrate through conformal contact. Alkane thiols are used to pattern gold-coated surfaces by forming densely packed SAMs. The thiol end groups strongly interact with the gold surfaces, and the rest of the molecule can then be used to control surface chemistry, wetting, and protein and cell adhesion. For example, alkane thiols can be PEGylated at the opposite end from the –SH group to render gold-coated surfaces "non-fouling," or protein and cell resistant. To micropattern surfaces using alkane thiols, a variety of methods have been employed such as μCP. In this approach, PDMS molds are "inked" with the alkanethiol solution, and transferred to the gold surface by conformal contact between the relief pattern and the substrate. Using μCP, micropatterns of SAMs terminated with PEG chains have been generated. These micropatterns have been used to immobilize proteins and cells on specific regions of a surface, and inhibit their contact with other regions by selectively modifying those regions with the non-fouling PEGs.

Using micropatterned fibronectin islands, Chen et al. reported that the area available for cell adhesion and the resultant cell shape governed whether individual cells grow or die, regardless of the type of matrix protein or antibody to integrin used to mediate adhesion (Chen et al., 1997, 1998) (Figure II.1.3.2). For bovine capillary cells, apoptosis significantly increased as fibronectin island size was reduced below 400 μm^2. Using micropatterned fibronectin McBeath et al. demonstrated that cell shape regulated the differentiation of human mesenchymal stem cells (hMSCs) (McBeath et al., 2004); hMSCs that were permitted to adhere, spread, and flatten, differentiated into osteoblasts, while those that were not permitted to attach formed spherical shapes and resulted in adipocytes.

Non-Fouling Surfaces in Cell Culture

As previously described, non-fouling surfaces may be created with PEGylated alkane thiols formed into SAMs. Other approaches include the generation and deposition of non-fouling coatings or films, such as plasma-deposited triglyme and tetraglyme films of the Ratner group (Johnston et al., 2005), plasma-deposited tetraglyme, and fluorocarbon-based films (Goessl et al., 2001; Cao et al., 2007).

PEG has been used in many chemical variations to generate non-fouling surfaces. Polymerization of PEG monomers to form brush-like non-fouling coatings has been investigated. Beginning with surface-initiated atom

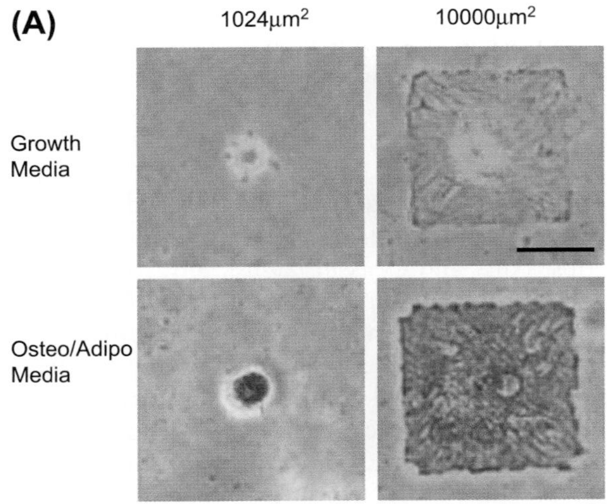

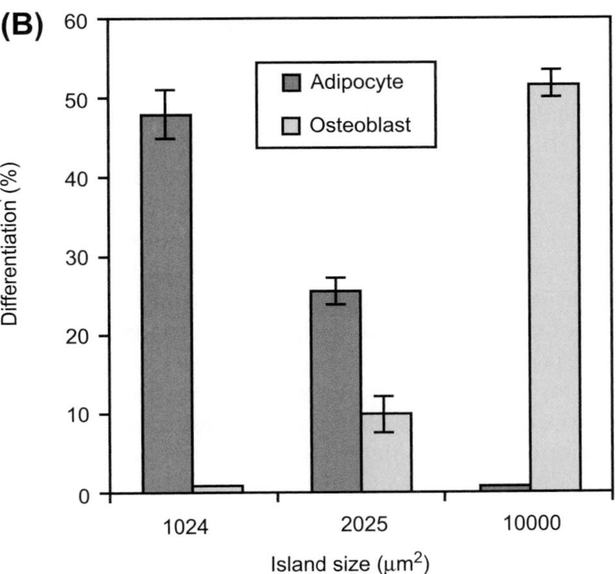

FIGURE II.1.3.2 Cell shape modulates hMSC differentiation: (A) Brightfield images of hMSCs plated onto small (1024 μm²) or large (10,000 μm²) fibronectin spots after 1 week. Fibronectin spots were patterned on a mixed SAM substrate by micro-contact printing. Large fibronectin spots supported osteogenesis (blue) while small fibronectin spots supported adipogenesis (red). Scale bar 50 μm. (B) Differentiation of hMSCs on 1024, 2025, or 10,000 μm² islands after 1 week of culture. Small fibronectin spots resulted in adipogenesis, medium fibronectin spots supported both adipogenesis and osteogenesis, and large fibronectin spots resulted in osteogenesis. (Reprinted from McBeath et al. (2004). Cell shape, cytoskeletal tension, and RhoA regulate stem cell lineage commitment, *Developmental Cell,* **6**(4), 483–495. Copyright (2004), with permission from Elsevier.)

transfer radical polymerization, Lavanant et al. grew robust brush-like polymers of poly(ethylene glycol) methacrylate (PEGMA) on low density PE by photobromination followed by polymerization (Lavanant et al., 2010). Functionalization of the peptide brushes with the RGD containing peptide ligand, GGGRGDS, increased cell adhesion compared to the polymerized PEGMA surface, as expected.

Changes in the structure of the component molecules of SAMs can also reduce cell adhesion. For example, SAMs comprised of oligo(ethylene glycol) (OEG) with EG side chain lengths of 4, 9, and 23 were grafted to titanium as a non-fouling surface. Surfaces were exposed to 3T3 fibroblast cells in media with 10% FBS, and uncoated titanium was used as a control. At four hours, cells were visible on the Ti surface, but there was no cell adhesion and no visible difference between the SAMs with different side chain lengths. At 35 days, cell adhesion on surfaces was related to side chain length, and longer side chains resulted in fewer adhered cells. Eventually cells covered all the surfaces; however, complete cell coverage was observed at 7, 10, and 11 weeks for samples with side chain lengths of 4, 9, and 23 (Fan et al., 2006).

For many patterning applications, the ability to isolate or sequester cell adhesive regions by surrounding them with surfaces that resist cell attachment is of great interest. The practical applications include the fabrication of arrays of cells for high throughput testing, as well as patterned co-cultures for investigating complex interactions between different groups of homogenous cell populations. Microarrays of captured molecules surrounded by non-fouling backgrounds are one approach that has been used to identify and screen combinations of cell adhesion peptides to develop bioactive surfaces. Monchaux and Vermette grafted RGD and RGE (a non-adhesive control) as well as other known endothelial ligands (REDV and SVVYGLR) on a low-fouling carboxymethyl dextran background (Monchaux and Vermette, 2007). They concluded that RGD was necessary for cell adhesion, and that endothelial cells could discriminate against and would not adhere to RGE. Co-immobilization of the RGD peptide with VEGF or SVVYGLR significantly enhanced endothelial cell adhesion, while co-immobilization of RGD with either REDV or SVVYGLR induced a reduction in endothelial cell spreading.

The development of non-fouling surfaces from biodegradable or biologically relevant compounds is of great interest. Biodegradable ultra-low fouling peptide surfaces have been generated from natural amino acids that offer new and innovative means to control cell adhesion (Chen et al., 2009). Amino acids (H2N-CH-R-COOH) are molecules that possess an amine group (H2N), a carboxylic acid group (COOH), and an organic side chain (R) of varying length and composition. Amino acids are the building blocks of proteins, and are the ubiquitous components of many organisms. Ultra-low fouling surfaces based on net neutral zwitterionic amino acids have been developed that can "catch and release" molecules of interest (Mi et al., 2010), potentially enabling a new generation of *in vitro* surfaces for diagnostic and assay-based testing.

Chemical Patterning for the Co-Culture of Cells

Cell–cell interactions are important for tissue development and function. During embryogenesis, the temporal and spatial signaling between cells of different origins is part of the cascade of events that support tissue growth and development (Yu et al., 2008). However, the process of harvesting tissue and isolating cells, as well as the purification of specific cell types, disrupts many of the cell–cell interactions found *in vivo*, resulting in a loss of cell–cell contact, signaling, structure, and general heterogeneity.

The co-culture of different cell types is one approach to enhance the *in vitro* environment and maintain cell phenotype *in vitro* (Nandkumar et al., 2002; Chang Liu and Chang, 2006). Co-cultures can be established by simply mixing two or more cell populations prior to plating; however, this provides little control over cell–cell contact, spacing or interaction. Greater control can be obtained through the use of patterned co-cultures which primarily control homotypic and heterotypic cell–cell contact. Using patterning techniques, co-cultures can be established where cell–cell contact, spacing, and interaction can be controlled with high resolution. Both cells and cell adhesive proteins may be patterned using selective adhesion to micropatterned substrates (Folch and Toner, 2000; Hui and Bhatia, 2007), by flow through a microfluidic channel (Chiu et al., 2000; Takayama et al., 2001), by stamping with a stencil-based approach (Chen et al., 1998; Folch and Toner, 2000) or by seeding on surfaces that dynamically switch from cell-adhesive to cell-repulsive (Edahiro et al., 2005; Kikuchi et al., 2009). A common requirement of most co-culture (and patterning) applications is the ability to selectively control and limit cell adhesion to specific areas of the substrate. As a result, non-fouling surfaces can be of considerable benefit.

A practical application of co-culturing is the study of cells from the liver (hepatocytes). Hepatocytes rapidly lose their ability to generate a characteristic protein (albumin) when cultured *in vitro*. Using mask-to-mask registration and photolithography, Hui and Bhatia (2007) generated surfaces with three distinct chemistries for the co-cultures of hepatocytes and 3T3-J2 fibroblasts. Hepatocytes preferentially adhered to collagen coated regions, and fibroblasts to serum coated glass regions (Figure II.1.3.3). A non-fouling PEG region was included that separated the two cell types and controlled the amount of contact or interaction between co-cultured populations for up to seven days (Figure II.1.3.3). At eight days, hepatocytes retained the ability to produce albumin.

High-Throughput Screening

Microarrays represent a specific type of chemical patterning. The arrays consist of a large number of small spots, each with a defined location and a unique chemical composition. This approach is suitable for testing biomaterials and microenvironments to enable the screening of large numbers of chemical combinations in a high-throughput manner. A high-throughput microarray that probes the cell–cell, cell–ECM, and cell–biomaterial interactions can be useful for studying cell adhesion, proliferation, and differentiation (Khademhosseini, 2005; Khademhosseini et al., 2006).

Typically, cell–ECM interactions are studied by using purified matrix proteins adsorbed to cell culture substrates. This approach requires large amounts of protein per 96- or 384-well plate, and can be prohibitively expensive. To address these challenges, new methods have been developed using pin tools (Cleveland and Koutz, 2005), piezo tips (Niles and Coassin, 2005), ultrasound (Strobl et al., 2004), and microarray (Flaim et al., 2005) technology for nanoliter liquid handling. Recently, robotic spotters capable of dispensing and immobilizing nanoliters of materials have been used to fabricate microarrays, where cell–matrix interactions can be tested and optimized in a high-throughput manner. For example, synthetic biomaterial arrays have been fabricated to test the interaction of stem cells with various extracellular signals (Anderson et al., 2004). Using this approach, thousands of polymeric materials were synthesized and their effect on the differentiation of human embryonicstem (ES) cells (Anderson et al., 2004) and human mesenchymal stem cells (hMSC) (Anderson et al., 2005) were evaluated. These interactions have led to unexpected and novel cell–material interactions. Although the molecular mechanisms associated with the biological responses have yet to be clarified, such technology may be widely applicable in cell–microenvironment studies, and in the identification of cues that induce desired cell responses. Also, the materials which yield desired responses could be used as templates for tissue engineering scaffolds. Such an approach is a radical change from traditional methods of developing new biomaterials, where polymers have been individually developed and tested for their effect on cells. In addition to the screening of libraries of synthetic materials, the effect of natural ECM molecules on cell fate can be similarly evaluated in a high-throughput manner. In one example, combinatorial matrices of various natural ECM proteins were tested for their ability to maintain the function of differentiated hepatocytes, and to induce hepatic differentiation from murine ES cells (Flaim et al., 2005). A peptide and small molecule microarray made by a DNA microarray spotter was also used to study high-throughput cell adhesion (Falsey et al., 2001). A microarray of immobilized ligands was analyzed with three different biological assays such as protein-binding assay, functional phosphorylation assay, and cell adhesion assay. This array can be used to rapidly screen and analyze the functional properties of various ligands.

Nanometer-Scale Chemical Patterning

Nanosized chemical patterns generally do not direct cell shape or orientation as a result of their extremely small size relative to the cell. However, surfaces with chemical features on the nanoscale do modulate cell functions such as adhesion, proliferation, migration, differentiation, and gene expression.

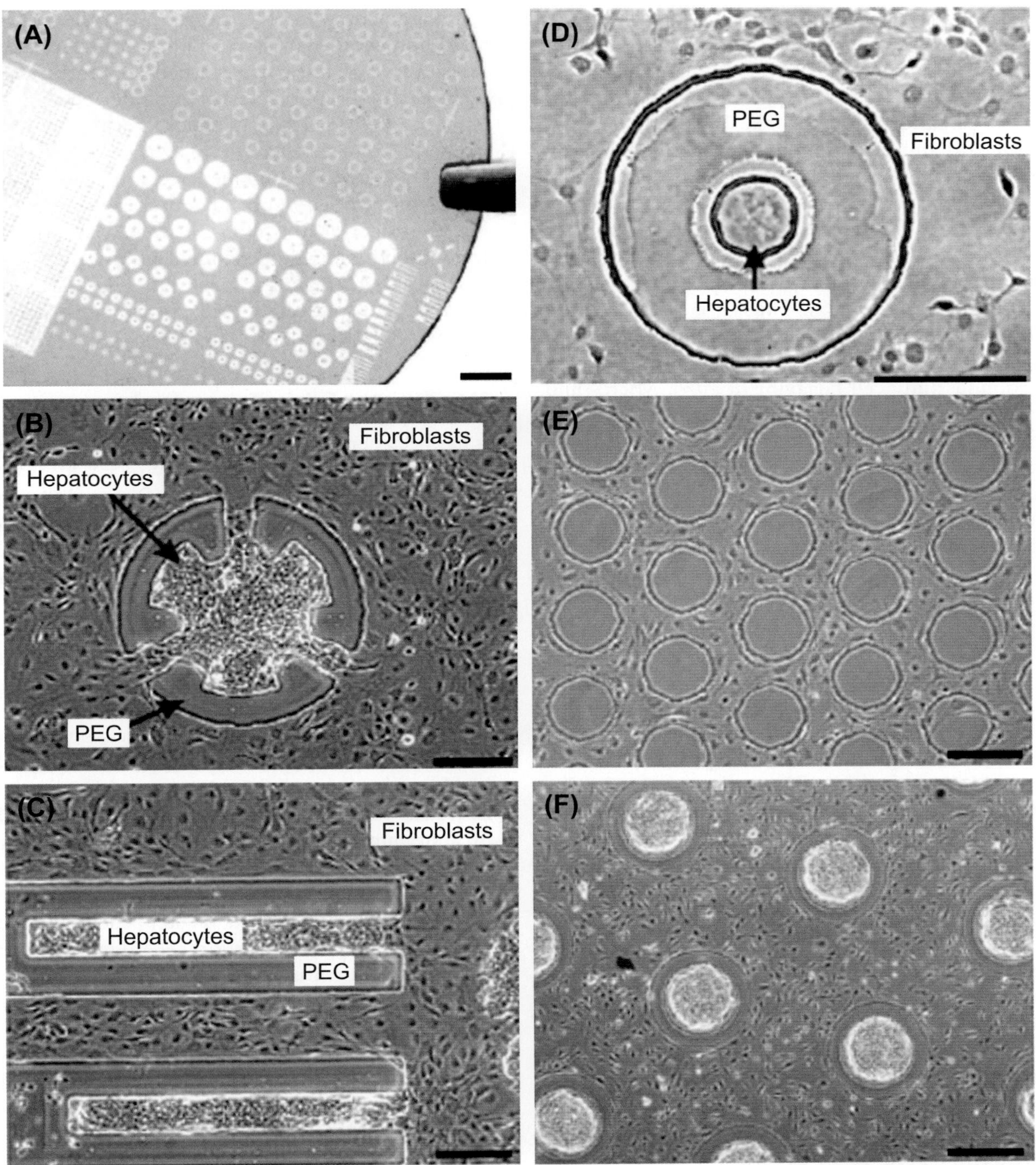

FIGURE II.1.3.3 Examples of patterned co-cultures: (A) Glass coverslip with patterned photoresist. Scale bar is 2 mm. (B,C) Spatially restricted contact between hepatocytes and fibroblasts (day 6). (D) Phase contrast and fluorescence overlay showing expression of intracellular albumin in green after 8 days of culture, indicating retained hepatocyte function. (B–D) Scale bars are 200 μm. Control of cell organization via patterned surface chemistry: (E) 3T3-J2 fibroblasts excluded from islands of PEG-disilane (day 5). (F) Hepatocytes and 3T3-J2 fibroblasts patterned on a combination of collagen and PEG-disilane (day 2). (E,F) Scale bars 500 μm. (Reprinted in part with permission from Hui and Bhatia. (April 2007). Langmuir, Vol **23**(8), 4103–4107. Copyright 2007 American Chemical Society.)

CELL RESPONSE TO SUBSTRATE TOPOGRAPHY

All cells are exposed to some type of physical environment *in vivo*. Cell–cell interactions, the ECM, and biomaterials all present a surface to the cell that is quantifiable in terms of its surface features (Abrams et al., 2003; Liliensiek et al., 2009). As a field of study, surface topography defines the specific morphological characteristics of a surface. Surface topography may be generally described as isotropic (uniformity in all directions) or anisotropic (uniformity in one direction).

Surface topography may be quantified by a number of techniques such as atomic force microscopy (AFM) or white light interferometery (WLI).

Unlike periodic surfaces typically fabricated for *in vitro* investigation (Clark et al., 1987; Chou et al., 1998; Brunette and Chehroudi, 1999), most biological surfaces present non-ordered features at the macro- and microscale which generally become more ordered at the nanoscale (Fratzl, 2008). Fibers in the ECM and pores in basement membranes are just a few examples of prevalent and quantifiable topographical features (Abrams et al., 2003; Liliensiek et al., 2009). Thus, it is not surprising that cells respond to topographical cues, and that surface topography of the culture surface presents another means to control and study cell behavior *in vitro*. Micro- and nanosized features are known to influence cell behavior, and both generally coexist in many biological surfaces. Accurate quantification of these surfaces is challenging, since most tools are optimized for quantifying micro- or nano-features individually, but not both at the same time (Guruprasad Sosale and Vengallatore, 2008).

In terms of cell activity, surface topography has been reported to affect proliferation, gene expression (Carinci et al., 2003), cell adhesion (Hamilton et al., 2006), motility (Clark et al., 1987), alignment (Clark et al., 1991), differentiation (Martinez et al., 2009), and matrix production (Hacking et al., 2008). In sensing and interacting with the topographical environment, evidence suggests that cells extend fine processes termed *filopodia* (Morris et al., 1985; Dalby et al., 2003a, 2004; Choi et al., 2007). While the specific mechanisms are poorly understood, changes in cell activity resulting from interaction with surface features have been linked to changes in cytoskeletal arrangement (Wojciak-Stothard et al., 1995) including actin filaments (Gerecht et al., 2007), nuclear shape (Dalby et al., 2003b), and ion channels (Tobasnick and Curtis, 2001). It is likely that these effects represent a small portion of the cellular events relating to interactions with surface topography, and different mechanisms or combinations of mechanisms may be activated by different feature sizes and shapes (Ball et al., 2007).

Micrometer-Scale Topography

Investigations concerning the response of cells to precisely patterned, microfabricated surfaces began decades ago. Fibroblasts (Brunette, 1986a), osteoblasts (Hamilton and Brunette, 2007), macrophages (Wojciak-Stothard et al., 1996), neutrophils (Tan and Saltzman, 2002), epithelial (Brunette, 1986b), endothelial (Bettinger et al., 2006), cardiomyocytes (Kim et al., 2005), and neuronal cells (Clark et al., 1991), have all demonstrated a reproducible response to surface texture on the microscale (Ireland et al., 1987; Britland et al., 1992; Curtis and Wilkinson, 1998; Fitton et al., 1998; Dalby et al., 2002b; Berry et al., 2004). Polarization or cell alignment along the direction of ridged surfaces is a phenomenon that is commonly observed when cells are cultured on linearly patterned substrates. Cell alignment to topographical features is referred to as contact guidance (Wood 1988; Wojciak-Stothard et al., 1996; Brunette and Chehroudi, 1999). With microscale-sized features, cell alignment generally increases with increasing groove depth and decreasing groove spacing (Bettinger et al., 2006); however, cell response to topography is highly dependent upon cell type (Figure II.1.3.4). For example, epithelial cells and fibroblasts are differently affected by surfaces with micron-sized grooves (Clark et al., 1987, 1990).

Nanometer-Scale Topography

It has been demonstrated that cells can sense and respond to features as small as 10–30 nm (Wojciak-Stothard et al., 1996; Dalby et al., 2002b). Compared to the microscale, where cells may be profoundly affected by a single feature, at the nanoscale a repetition of similar features has the greatest effect and provides the most predictable results regarding cell behavior (Curtis et al., 2004; Ball et al., 2007). Once again, cell response to topography is highly dependent upon cell type. Using surfaces of aligned ridges possessing a constant ridge spacing of 260 nm and depths varying from 100–400 nm, Clark demonstrated that neuronal cells were highly oriented in the ridge direction for all depths, epithelial-like cells were highly oriented for all depths, whereas the orientation of fibroblast-like cells increased with increasing depth from 50% orientation at 100 nm to nearly 95% orientation at 400 nm (Clark et al., 1991).

At the microscale, increasing the depth and decreasing the width of grooved or stepped surface features increases the probability of cell alignment with the topographical feature(s). At the nanoscale, it appears that a finite limit in feature sizes exists with respect to cell alignment to surface topography. By culturing fibroblasts on linearly patterned surfaces fabricated by e-beam lithography, it was determined that groove depths below 35 nm or ridge widths smaller than 100 nm did not result in cell alignment, and that 35 nm may be the threshold for whole cell fibroblast alignment (Loesberg et al., 2007). Similarly Teixeira et al. demonstrated a finite limit in the response of human corneal epithelial cells to surfaces with linearly patterned surfaces with groove spacing ranging from 400 to 4000 nm (groove width 330 to 2100 nm) (Teixeira et al., 2006). At groove spacings greater than 400 nm, cells were oriented in an increasingly parallel manner to the linear pattern (Figure II.1.3.4); however, at groove spacings less than 400 nm, cells were perpendicularly arranged. These findings further demonstrate a cell-specific sensitivity with regard to cell alignment to surface topography, and also imply that a limit exists in feature sizes that cells are able to recognize.

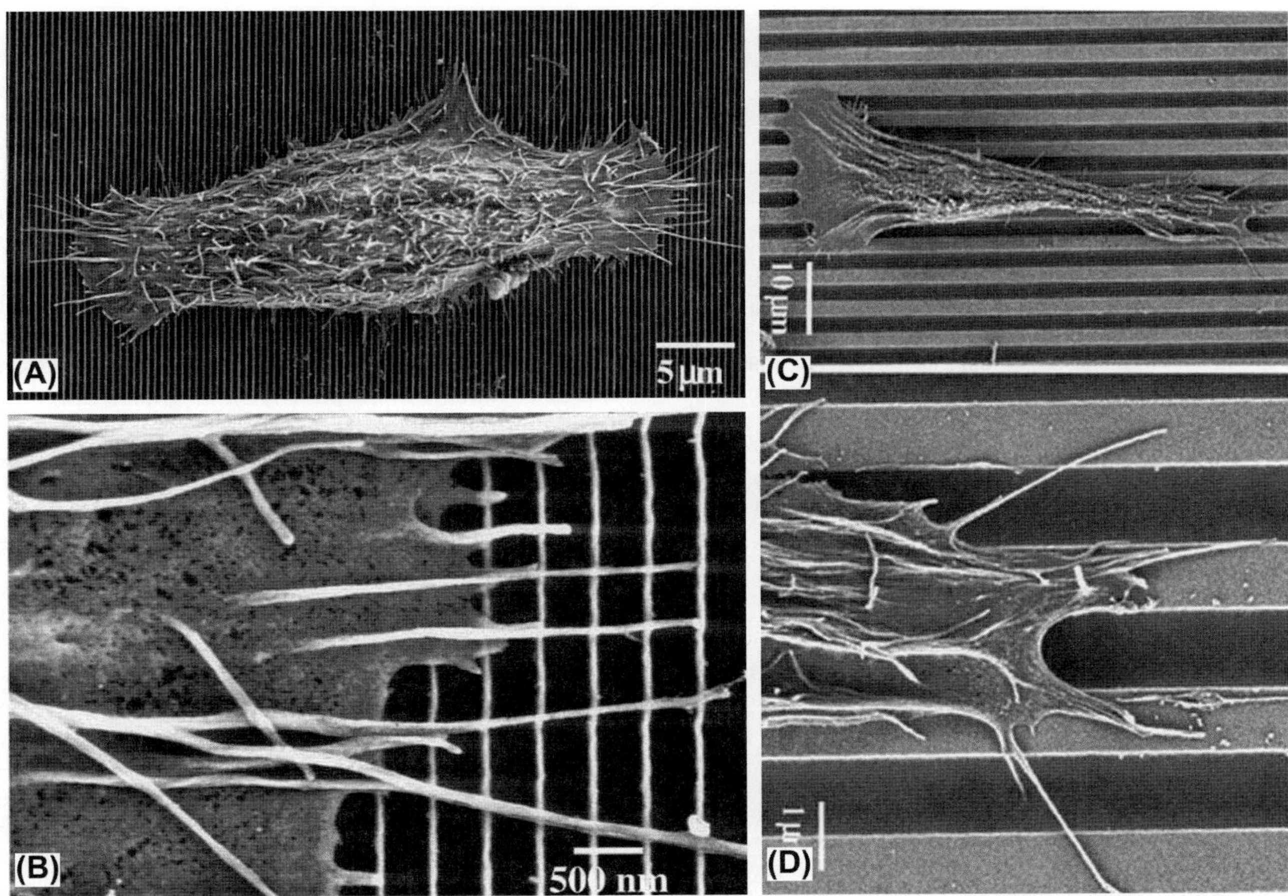

FIGURE II.1.3.4 Effect of micro and nano patterns on corneal epithelial cell alignment. (A) Cells were aligned perpendicularly to ridges that were 70 nm wide and 330 nm apart. (B) Expanded view of (A) showing filopodia also aligned perpendicularly to the patterns. (C) With increasing width and spacing (1900 nm ridge width, 2100 nm spacing), cells were aligned with the ridges; and (D) filopodia were guided by the topographic pattern. (Reprinted from Teixeira et al. (2006). The effect of environmental factors on the response of human corneal epithelial cells to nanoscale substrate topography. *Biomaterials*, **27**(21), 3945–3954. Copyright (2006), with permission from Elsevier.)

The effects of nano-topography on cell differentiation have also been reported. Proteomic-based studies have also shown that cells respond to nm-sized pits and pores in irregular patterns, leading to increased differentiation and matrix production by human osteoprogenitor cells (Kantawong et al., 2008). Likewise, studies with surfaces fabricated by the arrangement of titanium nanotubes have demonstrated that the nanotube diameter affects both hMSC differentiation and adhesion (Oh et al., 2009). Smaller diameter nanotubes (30 nm) increased adhesion, while larger diameters (100 nm) increased differentiation. Other authors using different cell lines and culture conditions have reported optimal nanotube diameters and spacing as low as 15 nm (Park et al., 2007). Topography at the nanoscale can also be used to effectively reduce cell adhesion. Kunzler et al. generated a constant gradient of nano-features (65 nm diameter and height) with the spacing as the only changing parameter along the gradient (Kunzler et al., 2007). This study demonstrated that the spacing or density of non-adhesive nanoscale features can disrupt cell adhesion, likely by restricting receptor–ligand interaction.

High-Throughput Screening of Surface Topography

Like microarrays used to test many chemical compounds at one time in a high-throughput manner, topographical arrays have been developed to assess the effects of systematic variances in surface topography. Lovmand et al. developed a topographical array consisting of 504 topographically distinct surface structures fabricated on silicon wafers that were coated with tantalum prior to cell seeding (Lovmand et al., 2009). Each structure consisted of a topographical pattern comprised of a series of circles, squares or rectangles that varied systematically in height from 0.6 to 2.4 µm, and feature size and spacing from 1.1 to 6.6 µm. Murine osteoblast cells (MC3T3-E1) were cultured on the array for up to 28 days to assess proliferation, cell area, and mineralization. It was determined that the height of the features had the greatest effect on mineralization, followed by feature size, then the gap size. The combined effects of feature size and gap size became important as feature height decreased.

Clinical Applications of Surface Morphology

Osteoblasts are substrate-dependent cells that have been widely used for investigating responses to surface topography. A considerable amount of work has been produced concerning the response of osteoblast cells to surface topography, and the end result of this work has been widely applied clinically to enhance the long-term fixation of dental and orthopedic implants by bone (Hacking et al., 1999, 2003). Mechanical fixation of an implant by the direct apposition of bone is referred to as osseointegration (Branemark et al., 1977; Albrektsson et al., 1981). Irregularly textured surfaces on titanium implants have been created by blasting the surface with small hard particles (Al_2O_3) that deform the relatively ductile titanium surface and produce a distinctly unordered micron-sized surface texture. It is common with dental implants that these surfaces are further processed by immersion in acid that superimposes a finer secondary nanosized structure upon the textured surface, and increases osseointegration.

A Note about Cell Lines in the Study of Surface Topography

To enhance optical clarity, the TCP surface is very smooth (R_a~1.2 nm) (Chang et al., 2007), and this has certain ramifications for the growth and development of cells *in vitro*. Adherent cell lines are selected preferentially for their ability to proliferate and differentiate on glass or tissue culture plastic. As a result, many cell lines have been selected based on their behavior on smooth, flat surfaces that are rarely found *in vivo*. Studies comparing the response of primary cells to cell lines derived from similar tissue sources have reported significant differences in cell activity (Fisher and Tickle, 1981; McCartney and Buck, 1981). This point becomes especially salient when working with surfaces where the effects of surface topography are the subject of primary interest (Clark et al., 1991; Hacking et al., 2008).

CELL RESPONSE TO SUBSTRATE ELASTICITY

While surface chemistry and topography modulate cell function, it is increasingly evident that cells respond to the physical or mechanical properties (stiffness) of the substrate. Stiffness can be described as the resistance of a solid material to deformation, and is commonly defined by elastic modulus (*E*) and reported in units referred to as a "pascal" (Pa). Excluding bone ($10–20 \times 10^9$ Pa), most tissues have an elastic moduli in the range of 10^1 to 10^6 Pa, substantially lower than the elastic moduli of most culture surfaces like TCP ($3–3.5 \times 10^9$ Pa).

Hydrogels are well-suited to the study of cell substrate interactions since their stiffness can be varied by controlling the water content of the gel, which is controlled by modifying the polymer concentration or the relative extent of cross-linking. Hydrogels can also be made from a very diverse variety of natural and synthetic materials, such as hyaluronic acid (HA), fibrin, alginate, agarose, chitosan, polyacrylamide, and PEG. In terms of collagen hydrogels, gel stiffness may be increased by increasing the collagen concentration (Brigham et al., 2009) or reducing the thickness of the gel and immobilizing it on a glass substrate (Arora et al., 1999).

The stiffness of the substrate affects cell adhesion, spreading, and migration, especially the last two. In two-dimensional cultures, cells preferentially migrate towards surfaces of greater stiffness, a phenomenon referred to as mechanotaxis (Lo et al., 2000). In two-dimensional systems endothelial and fibroblast cells cultured on collagen-coated substrates that were classified as compliant ($\sim 5 \times 10^3$ Pa) or stiff ($\sim 70 \times 10^3$ Pa) demonstrated remarkably different behaviors (Opas, 1994). Cells cultured on compliant substrates had reduced spreading, increased lamellipodia activity, and greater migration. Cells cultured on stiffer substrates generally increased in proliferation; however, the specific effect seen is cell-type dependent (Pelham and Wang, 1997). There is substantial evidence that substrate stiffness also affects cell proliferation in two-dimensional as well as three-dimensional culture systems (Wong et al., 2003).

Substrate elasticity can also be used to direct the differentiation of certain cell types. MSCs cultured on collagen-coated substrates of varying stiffness can be directed toward neuronal, myogenic, and osteogenic lineages (Engler et al., 2006). For example, adult neural stem cells will selectively differentiate to neurons on softer substrates ($1 \times 10^2 – 5 \times 10^2$ Pa) or glial cell types on stiffer substrates ($1 \times 10^3 – 1 \times 10^4$ Pa) (Saha et al., 2008). These findings suggest that stem cells are able to recognize and differentiate according to physiologically relevant substrate stiffness.

CELL RESPONSE TO MECHANICAL DEFORMATION (STRAIN)

There has been considerable work assessing the response of cells to mechanical strain or deformation (Buckwalter and Cooper, 1987). Forces can be applied directly to adherent cells *in vitro* by either elongation or compression of their substrate (two-dimensional) or matrix (three-dimensional). With respect to cyclic tensile loads, in both two-dimensional and three-dimensional cultures, cells predominantly align in the direction of the applied load and assume an elongated morphology (Toyoda et al., 1998; Henshaw et al., 2006; Haghighipour et al., 2007).

Because of the loadbearing function and well-known adaptation to load of the musculoskeletal system, cells of this system have been evaluated on surfaces under conditions of tensile and compressive strain. Toyoda et al. cultured cells harvested from the anterior cruciate ligament (ACL) and synovium (Toyoda et al., 1998). Cells were subjected to cyclic tensile load for 24 hours on culture plates with flexible rubber bases. For both cell types, tensile load increased cell alignment and elongation; however, tensile load only increased the production of collagen type I in cells derived from the ACL (Toyoda et al., 1998). These results have been supported by similar studies where fibroblasts subjected to cyclic tensile strain increased the formation of organized ECM, increased collagen production, and increased metabolic activity compared to unloaded controls (Hannafin et al., 2006; Joshi and Webb, 2008).

Bone is another well-known loadbearing tissue, and the extent of mineralization and bone mineral density is load dependent (Rodionova and Oganov, 2003). *In vitro*, cyclic strain in both tensile (Hanson et al., 2009) and compressive (Goldstein, 2002) loading enhances osteoblast mineralization. Thus, strain appears to be an important stimulus for the generation of mineralized and tendon-like tissues *in vitro*. These findings have led to the fabrication of bioreactors that can apply physiologic loads to developing tissue to enhance the production and alignment of ECM in a number of tissues including cardiac (Kortsmit et al., 2009), bone (van Griensven et al., 2009), cartilage (Preiss-Bloom et al., 2009), and tendon (Riboh et al., 2008).

COMPARISON AND EVALUATION OF SUBSTRATE CUES

As previously discussed, cells respond to a number of surface cues such as chemistry, topography, elasticity, and strain (sometimes cyclic strain). Evaluating cell response to multiple cues is inherently difficult, since substrate properties are often interrelated. In many *in vitro* investigations of biomaterials it may be difficult to clearly separate the effects of surface chemistry from those of morphology and elasticity. This is particularly problematic in cases where the unique properties of a material may arise from the manufacturing process (Bobyn et al., 1999; Hacking et al., 2002). In such investigations, it is important to recognize that subtle changes in biomaterial properties can have profound effects on cellular response, and ultimately *in vivo* function. This is especially true, for example, of implant coatings where responses to surface morphology need to be dissociated from other factors, such as surface chemistry. Experimental controls must not just approximate surface morphology, but must match it precisely, since subtle changes in surface morphology have the potential to confound experimental findings. In many cases, however, it may not be practical or even possible to produce an exact morphological control from a different biomaterial. A morphological control may be generated by coating the surface of interest with a thin-film that is dense, durable, homogenous, and does not alter the surface morphology. There are a variety of thin-film deposition techniques collectively referred to as physical vapor deposition (PVD) that have been employed by several research groups (Hacking et al., 2002, 2007; Meredith et al., 2007). While other coating methods exist, PVD is advantageous since samples can be processed at lower temperatures in a relatively inert environment.

Control of substrate elasticity can be achieved by varying the composition and curing conditions. PDMS, for example, can be prepared with a range of elasticity (2.8–1882 kPa) by varying the ratio of base to curing agent (Cheng et al., 2009). Since PDMS can be cast into many shapes, patterned surfaces with varying elasticity can be created. These surfaces can be subsequently coated to achieve uniform and comparable surface chemistry. Similarly, by progressively reducing the thickness of a gel to a thin film, chemical composition can be maintained while substrate elasticity changes.

While the contributions of topography, chemistry, elasticity, and strain to cell behavior have been evaluated and documented for a wide variety of cell types, much of the literature has evaluated the effects of only one stimulus in each experimental setting. *In vivo*, however, it is reasonable to expect that cells experience a wide range of stimuli simultaneously. Since physical cues such as chemistry, topography, elasticity, and strain can produce both synergistic and opposing effects on cell behavior, it is of interest to evaluate their simultaneous effect on cell behavior, and if possible describe a hierarchy of cellular cues. However, since the cellular response to physical stimuli is often dependent upon cell phenotype, and is further influenced by soluble factors, broad exploration of these phenomena presents a challenging combinatorial experimental environment.

Chemistry and Topography

Using a combination of patterning techniques including photolithography, Curtis established that fibroblasts respond to both chemical and topographic patterns (Curtis and Wilkinson, 1999). However, when both chemical and topographical stimuli were combined and orthogonally opposed on the same surface, cells aligned with nanosized topographical patterns. Curtis coined the term "topographic reaction" to describe these events (Curtis and Wilkinson, 1999). Similarly, a preference of cellular alignment on chemical patterns versus response to topographic cues in the presence of both topographic and chemical patterns has been demonstrated in a number of studies (Charest et al., 2006; Gomez et al., 2007).

Chemistry and Strain

Hyun et al. developed a system to evaluate the combined effects of chemical patterning and strain on cell alignment (Hyun et al., 2006). Thin paraffin films were patterned with adhesive regions of fibronectin to which NIH 3T3 fibroblasts adhered and aligned. After cells aligned in the direction of the chemical patterning, the paraffin films with the adherent cells were subjected to cyclical stretch. Mechanical force was applied perpendicular to the cell alignment resulting from chemical patterning, and the actin cytoskeleton realigned along the axis of applied mechanical stress. Stretched cells showed altered gene expression of cytoskeletal and matrix proteins in response to mechanical deformation.

Topography and Strain

Cheng et al. (2009) fabricated patterned PDMS substrates consisting of parallel microchannels (2 μm wide, 2 μm separation, and 10 μm deep) that were subjected to cyclic compressive strain applied perpendicular to the direction of the channels. After six hours of compression, fibroblasts cultured on the patterned PDMS substrate aligned in the direction of the microchannels (topographical cue) and perpendicular to the direction of compression. Elastic microgrooved surfaces have also been utilized to maintain the cellular orientation of mesenchymal stem cells (MSCs) while subjected to cyclic tensile strain (Kurpinski et al., 2006).

SUMMARY

In vivo, cells are responsive to a myriad of substrate-related signals within the environment. At the micro- and nanoscale, interactions with the ECM expose the cell to a variety of chemical moieties and topographical features. Micro- and nanoscale technologies have provided many new tools to manipulate the *in vitro* environment. Cell behavior may be modified by patterns of cell adhesive ligands and topographical features. Further modification of cell behavior can be achieved by altering substrate mechanical properties, such as elasticity, or by deformation of the cell shape. With respect to generating complex tissue structures, substrate cues like topography, chemistry, and matrix elasticity are promising candidates for controlling cell differentiation. Unlike soluble factors, substrate cues can provide specific information for directing and controlling cell behavior. The combination of multiple physical cues presents new and potentially useful methods to direct and control cell development and behavior.

GLOSSARY OF TERMS

ACL: anterior cruciate ligament
AFM: atomic force microscopy
Al_2O_3: alumina oxide
ATCC: American Type Culture Collection
CAPs: Cell adhesive proteins
CIMR: Coriell Institute for Medical Research
DNA: deoxyribonucleic acid
***E*:** elastic modulus
ECM: extracellular matrix
EDTA: ethylenediaminetetraacetic acid
HA: hyaluronic acid
hMSCs: human mesenchymal stem cells
MSCs: mesenchymal stem cells
NIPAAm: N-isopropylacrylamide
Pa: pascal
PE: polyethylene
PDMS: polydimethylsiloxane
PEG: poly(ethylene glycol)
PVD: physical vapor deposition
REDV: Arginine-Glutamic Acid-Aspastic Acid Valine
RGD: arginin–glycine–aspartate, adhesion-related peptide sequence
RGE: Arginine-Glycine-Glutamic Acid
SAMs: self-assembled monolayers
SVVTGLR: Serie-Valine-Valine-Tyrosine-Glycine-Leucine-Arginine
Ta: Tantalum
TCP: tissue culture polystyrene
Ti: Titanium
VEGF: Vascular Endotthelial Growth Factor
WLI: white light interferometery

BIBLIOGRAPHY

Abercrombie, M. (1979). Contact inhibition and malignancy. *Nature*, **281**(5729), 259–262.

Abercrombie, M., & Heaysman, J. E. (1954). Observations on the social behaviour of cells in tissue culture. II. Monolayering of fibroblasts. *Exp. Cell Res.*, **6**(2), 293–306.

Abrams, G. A., Murphy, C. J., Wang, Z. Y., Nealey, P. F., & Bjorling, D. E. (2003). Ultrastructural basement membrane topography of the bladder epithelium. *Urol. Res.*, **31**(5), 341–346.

Aframian, D. J., Cukierman, E., Nikolovski, J., Mooney, D. J., Yamada, K. M., et al. (2000). The growth and morphological behavior of salivary epithelial cells on matrix protein-coated biodegradable substrata. *Tissue Eng.*, **6**(3), 209–216.

Albrektsson, T., Branemark, P. I., Hansson, H, A., & Lindstrom, J. (1981). Osseointegrated titanium implants. Requirements for ensuring a long-lasting, direct bone-to-implant anchorage in man. *Acta. Orthop. Scand.*, **52**(2), 155–170.

Amitani, K., & Nakata, Y. (1977). Characteristics of osteosarcoma cells in culture. *Clin. Orthop. Relat. Res.*, **122**, 315–324.

Amstein, C. F., & Hartman, P. A. (1975). Adaptation of plastic surfaces for tissue culture by glow discharge. *J. Clin. Microbiol.*, **2**(1), 46–54.

Anderson, D. G., Levenberg, S., & Langer, R. (2004). Nanoliter-scale synthesis of arrayed biomaterials and application to human embryonic stem cells. *Nat. Biotechnol.*, **22**(7), 863–866.

Anderson, D. G., Putnam, D., Lavik, E. B., Mahmood, T. A., & Langer, R. (2005). Biomaterial microarrays: Rapid, microscale screening of polymer-cell interaction. *Biomaterials*, **26**(23), 4892–4897.

Andrade, J. D., & Hlady, V. (1987). Plasma protein adsorption: The big twelve. *Ann. NY. Acad. Sci.*, **516**, 158–172.

Arora, P. D., Narani, N., & McCulloch, C. A. (1999). The compliance of collagen gels regulates transforming growth factor-beta induction of alpha-smooth muscle actin in fibroblasts. *Am. J. Pathol.*, **154**(3), 871–882.

Auernheimer, J., Dahmen, C., Hersel, U., Bausch, A., & Kessler, H. (2005). Photoswitched cell adhesion on surfaces with RGD peptides. *J. Am. Chem. Soc.*, **127**(46), 16107–16110.

Bain, J. R., & Hoffman, A. S. (2003a). Tissue-culture surfaces with mixtures of aminated and fluorinated functional groups. Part 1. Synthesis and characterization. *J. Biomater. Sci. Polym. Ed.*, **14**(4), 325–339.

Bain, J. R., & Hoffman, A. S. (2003b). Tissue-culture surfaces with mixtures of aminated and fluorinated functional groups. Part 2. Growth and function of transgenic rat insulinoma cells (betaG I/17). *J. Biomater. Sci. Polym. Ed.*, **14**(4), 341–367.

Ball, M. D., Prendergast, U., O'Connell, C., & Sherlock, R. (2007). Comparison of cell interactions with laser machined micron- and nanoscale features in polymer. *Exp. Mol. Pathol.*, **82**(2), 130–134.

Bell, P. B., Jr. (1977). Locomotory behavior, contact inhibition and pattern formation of 3T3 and polyoma virus-transformed 3T3 cells in culture. *J. Cell Biol.*, **74**(3), 963–982.

Berry, C. C., Campbell, G., Spadiccino, A., Robertson, M., & Curtis, A. S. (2004). The influence of microscale topography on fibroblast attachment and motility. *Biomaterials*, **25**(26), 5781–5788.

Bettinger, C. J., Orrick, B., Misra, A., Langer, R., & Borenstein, J. T. (2006). Microfabrication of poly (glycerol-sebacate) for contact guidance applications. *Biomaterials*, **27**(12), 2558–2565.

Bhatia, S. N., Yarmush, M. L., & Toner, M. (1997). Controlling cell interactions by micropatterning in co-cultures: Hepatocytes and 3T3 fibroblasts. *J. Biomed. Mater. Res.*, **34**(2), 189–199.

Bobyn, J. D., Stackpool, G. J., Hacking, S. A., Tanzer, M., & Krygier, J. J. (1999). Characteristics of bone ingrowth and interface mechanics of a new porous tantalum biomaterial. *J. Bone Joint Surg. Br.*, **81**(5), 907–914.

Branemark, P. I., Hansson, B. O., Adell, R., Breine, U., Lindstrom, J., et al. (1977). Osseointegrated implants in the treatment of the edentulous jaw. Experience from a 10-year period. *Scand. J. Plast Reconstr. Surg. Suppl.*, **16**, 1–132.

Brigham, M. D., Bick, A., Lo, E., Bendali, A., Burdick, J. A., et al. (2009). Mechanically robust and bioadhesive collagen and photocrosslinkable hyaluronic acid semi-interpenetrating networks. *Tissue Eng. Part A*, **15**(7), 1645–1653.

Britland, S., Clark, P., Connolly, P., & Moores, G. (1992). Micropatterned substratum adhesiveness: A model for morphogenetic cues controlling cell behavior. *Exp. Cell Res.*, **198**(1), 124–129.

Browne, S. M., & Al-Rubeai, M. (2007). Selection methods for high-producing mammalian cell lines. *Trends Biotechnol.*, **25**(9), 425–432.

Brunette, D. M. (1986a). Fibroblasts on micromachined substrata orient hierarchically to grooves of different dimensions. *Exp. Cell Res.*, **164**(1), 11–26.

Brunette, D. M. (1986b). Spreading and orientation of epithelial cells on grooved substrata. *Exp. Cell Res.*, **167**(1), 203–217.

Brunette, D. M., & Chehroudi, B. (1999). The effects of the surface topography of micromachined titanium substrata on cell behavior *in vitro* and *in vivo*. *J. Biomech. Eng.*, **121**(1), 49–57.

Buckwalter, J. A., & Cooper, R. R. (1987). Bone structure and function. *Instr. Course Lect.*, **36**, 27–48.

Callen, B. W., Sodhi, R. N., Shelton, R. M., & Davies, J. E. (1993). Behavior of primary bone cells on characterized polystyrene surfaces. *J. Biomed. Mater. Res.*, **27**(7), 851–859.

Cao, L., Chang, M., Lee, C. Y., Castner, D. G., Sukavaneshvar, S., et al. (2007). Plasma-deposited tetraglyme surfaces greatly reduce total blood protein adsorption, contact activation, platelet adhesion, platelet procoagulant activity, and *in vitro* thrombus deposition. *J. Biomed. Mater. Res. A.*, **81**(4), 827–837.

Carinci, F., Pezzetti, F., Volinia, S., Francioso, F., Arcelli, D., et al. (2003). Analysis of osteoblast-like MG63 cells' response to a rough implant surface by means of DNA microarray. *J. Oral. Implantol.*, **29**(5), 215–220.

Chang, T. Y., Yadav, V. G., De Leo, S., Mohedas, R., Rajalingam, B., et al. (2007). Cell and protein compatibility of parylene-C surfaces. *Langmuir*, **23**(23), 11718–11725.

Chang Liu, Z., & Chang, T. M. (2006). Coencapsulation of hepatocytes and bone marrow cells: *In vitro* and *in vivo* studies. *Biotechnol. Annu. Rev.*, **12**, 137–151.

Charest, J. L., Eliason, M. T., Garcia, A. J., & King, W. P. (2006). Combined microscale mechanical topography and chemical patterns on polymer cell culture substrates. *Biomaterials*, **27**(11), 2487–2494.

Chen, C. S., Mrksich, M., Huang, S., Whitesides, G. M., & Ingber, D. E. (1997). Geometric control of cell life and death. *Science*, **276**(5317), 1425–1428.

Chen, C. S., Mrksich, M., Huang, S., Whitesides, G. M., & Ingber, D. E. (1998). Micropatterned surfaces for control of cell shape, position, and function. *Biotechnol. Prog.*, **14**(3), 356–363.

Chen, S., Cao, Z., & Jiang, S. (2009). Ultra-low fouling peptide surfaces derived from natural amino acids. *Biomaterials*, **30**(29), 5892–5896.

Cheng, C. M., Steward, R. L., Jr., & LeDuc, P. R. (2009). Probing cell structure by controlling the mechanical environment with cell-substrate interactions. *J. Biomech.*, **42**(2), 187–192.

Cheng, X., Wang, Y., Hanein, Y., Bohringer, J. F., & Ratner, B. D. (2004). Novel cell patterning using microheater-controlled thermoresponsive plasma films. *J. Biomed. Mater. Res. A.*, **70**(2), 159–168.

Chien, H. W., Chang, T. Y., & Tsai, W. B. (2009). Spatial control of cellular adhesion using photo-crosslinked micropatterned polyelectrolyte multilayer films. *Biomaterials*, **30**(12), 2209–2218.

Chiu, D. T., Jeon, N. L., Huang, S., Kane, R. S., Wargo, C. J., et al. (2000). Patterned deposition of cells and proteins onto surfaces by using three-dimensional microfluidic systems. *Proc. Natl. Acad. Sci. U.S.A.*, **97**(6), 2408–2413.

Chlapowski, F. J., Minsky, B. D., Jacobs, J. B., & Cohen, S. M. (1983). Effect of a collagen substrate on the growth and development of normal and tumorigenic rat urothelial cells. *J. Urol.*, **130**(6), 1211–1216.

Choi, C. H., Hagvall, S. H., Wu, B. M., Dunn, J. C., Beygui, R. E., et al. (2007). Cell interaction with three-dimensional sharp-tip nanotopography. *Biomaterials*, **28**(9), 1672–1679.

Chollet, C., Chanseau, C., Remy, M., Guignandon, A., Bareille, R., et al. (2009). The effect of RGD density on osteoblast and endothelial cell behavior on RGD-grafted polyethylene terephthalate surfaces. *Biomaterials*, **30**(5), 711–720.

Chou, L., Firth, J. D., Uitto, V. D., & Brunette, D. M. (1998). Effects of titanium substratum and grooved surface topography on metalloproteinase-2 expression in human fibroblasts. *J. Biomed. Mater. Res.*, **39**(3), 437–445.

Clark, P., Connolly, P., Curtis, A. S., Dow, J. A., & Wilkinson, C. D. (1987). Topographical control of cell behaviour. I. Simple step cues. *Development*, **99**(3), 439–448.

Clark, P., Connolly, P., Curtis, A. S., Dow, J. A., & Wilkinson, C. D. (1990). Topographical control of cell behaviour: II. Multiple grooved substrata. *Development*, **108**(4), 635–644.

Clark, P., Connolly, P., Curtis, A. S., Dow, J. A., & Wilkinson, C. D. (1991). Cell guidance by ultrafine topography. *In vitro. J. Cell Sci.*, **99**(Pt 1), 73–77.

Cleveland, P. H., & Koutz, P. J. (2005). Nanoliter dispensing for uHTS using pin tools. *Assay Drug Dev. Technol.*, **3**(2), 213–225.

Connelly, C. S., Fahl, W. E., & Iannaccone, P. M. (1989). The role of transgenic animals in the analysis of various biological aspects of normal and pathologic states. *Exp. Cell Res.*, **183**(2), 257–276.

Curtis, A. S., & Wilkinson, C. D. (1998). Reactions of cells to topography. *J. Biomater. Sci. Polym. Ed.*, **9**(12), 1313–1329.

Curtis, A., & Wilkinson, C. (1999). New depths in cell behaviour: Reactions of cells to nanotopography. *Biochem. Soc. Symp.*, **65**, 15–26.

Curtis, A. S., Forrester, J. V., McInnes, C., & Lawrie, F. (1983). Adhesion of cells to polystyrene surfaces. *J. Cell Biol.*, **97**(5 Pt 1), 1500–1506.

Curtis, A. S., Gadegaard, N., Dalby, M. J., Riehle, M. O., Wilkinson, C. D., et al. (2004). Cells react to nanoscale order and symmetry in their surroundings. *IEEE Trans. Nanobioscience*, **3**(1), 61–65.

Cutler, S. M., & Garcia, A. J. (2003). Engineering cell adhesive surfaces that direct integrin alpha5beta1 binding using a recombinant fragment of fibronectin. *Biomaterials*, **24**(10), 1759–1770.

Dalby, M. J., Riehle, M. O., Johnstone, H. J., Affrossman, S., & Curtis, A. S. (2002a). *In vitro* reaction of endothelial cells to polymer demixed nanotopography. *Biomaterials*, **23**(14), 2945–2954.

Dalby, M. J., Riehle, M. O., Johnstone, H. J., Affrossman, S., & Curtis, A. S. (2002b). Polymer-demixed nanotopography: Control of fibroblast spreading and proliferation. *Tissue Eng.*, **8**(6), 1099–1108.

Dalby, M. J., Childs, S., Riehle, M. O., Johnstone, H. J., Affrossman, S., et al. (2003a). Fibroblast reaction to island topography: Changes in cytoskeleton and morphology with time. *Biomaterials*, **24**(6), 927–935.

Dalby, M. J., Riehle, M. O., Yarwood, S. J., Wilkinson, C. D., & Curtis, A. S. (2003b). Nucleus alignment and cell signaling in fibroblasts: Response to a micro-grooved topography. *Exp. Cell Res.*, **284**(2), 274–282.

Dalby, M. J., Riehle, M. O., Sutherland, D. S., Aghelis, H., & Curtis, A. S. (2004). Changes in fibroblast morphology in response to nano-columns produced by colloidal lithography. *Biomaterials*, **25**(23), 5415–5422.

Dehner, L. P. (1998). The evolution of the diagnosis and understanding of primitive and embryonic neoplasms in children: Living through an epoch. *Mod. Pathol.*, **11**(7), 669–685.

Du, Y., Shim, J., Vidula, M., Hancock, M. J., Lo, E., et al. (2009). Rapid generation of spatially and temporally controllable long-range concentration gradients in a microfluidic device. *Lab. Chip*, **9**(6), 761–767.

Edahiro, J., Sumaru, K., Tada, Y., Ohi, K., Takagi, T., et al. (2005). *In situ* control of cell adhesion using photoresponsive culture surface. *Biomacromolecules*, **6**(2), 970–974.

Eiges, R., Schuldiner, M., Drukker, M., Yanuka, O., Itskovitz-Eldor, J., et al. (2001). Establishment of human embryonic stem cell-transfected clones carrying a marker for undifferentiated cells. *Curr. Biol.*, **11**(7), 514–518.

Elloumi-Hannachi, I., Yamato, M., & Okano, T. (2010). Cell sheet engineering: A unique nanotechnology for scaffold-free tissue reconstruction with clinical applications in regenerative medicine. *J. Intern. Med.*, **267**(1), 54–70.

Engler, A. J., Sen, S., Sweeney, H. L., & Discher, D. E. (2006). Matrix elasticity directs stem cell lineage specification. *Cell*, **126**(4), 677–689.

Erickson, C. A. (1978). Contact behaviour and pattern formation of BHK and polyoma virus-transformed BHK fibroblasts in culture. *J. Cell Sci.*, **33**, 53–84.

Falsey, J. R., Renil, M., Park, S., Li, S., & Lam, K. S. (2001). Peptide and small molecule microarray for high throughput cell adhesion and functional assays. *Bioconjug. Chem.*, **12**(3), 346–353.

Fan, X., Lin, L., & Messersmith, P. B. (2006). Cell fouling resistance of polymer brushes grafted from Ti substrates by surface-initiated polymerization: Effect of ethylene glycol side chain length. *Biomacromolecules*, **7**(8), 2443–2448.

Fisher, P. E., & Tickle, C. (1981). Differences in alignment of normal and transformed cells on glass fibres. *Exp. Cell Res.*, **131**(2), 407–410.

Fitton, J. H., Dalton, B. A., Beumer, G., Johnson, G., Griesser, H. J., et al. (1998). Surface topography can interfere with epithelial tissue migration. *J. Biomed. Mater. Res.*, **42**(2), 245–257.

Flaim, C. J., Chien, S., & Bhatia, S. N. (2005). An extracellular matrix microarray for probing cellular differentiation. *Nature Methods*, **2**(2), 119–125.

Folch, A., & Toner, M. (2000). Microengineering of cellular interactions. *Annu. Rev. Biomed. Eng.*, **2**, 227–256.

Folch, A., Jo, B. H., Hurtado, O., Beebe, D. J., & Toner, M. (2000). Microfabricated elastomeric stencils for micropatterning cell cultures. *J. Biomed. Mater. Res.*, **52**(2), 346–353.

Fratzl, P. (2008). Bone fracture: When the cracks begin to show. *Nat. Mater.*, **7**(8), 610–612.

Fredriksson, C., Khilman, S., Kasemo, B., & Steel, D. M. (1998). *In vitro* real-time characterization of cell attachment and spreading. *J. Mater. Sci. Mater. Med.*, **9**(12), 785–788.

Gallant, N. D., & Garcia, A. J. (2007). Model of integrin-mediated cell adhesion strengthening. *J. Biomech.*, **40**(6), 1301–1309.

Garcia, A. J. (2005). Get a grip: Integrins in cell-biomaterial interactions. *Biomaterials*, **26**(36), 7525–7529.

Garrod, D. R., & Steinberg, M. S. (1973). Tissue-specific sorting-out in two dimensions in relation to contact inhibition of cell movement. *Nature*, **244**(5418), 568–569.

Gauvin, R., Ahsan, T., Larouche, D., Lévesque, P., Dubé, J., et al. (2010). A novel single-step self-assembly approach for the fabrication of tissue-engineered vascular constructs. *Tissue Eng. Part A*, **16**(5). 1737–1347.

Gerecht, S., Bettinger, C. J., Zhang, Z., Borenstein, J. T., Vunjak-Novakovic, G., et al. (2007). The effect of actin disrupting agents on contact guidance of human embryonic stem cells. *Biomaterials*, **28**(28), 4068–4077.

Goessl, A., Golledge, S. L., & Hoffman, A. S. (2001). Plasma lithography: Thin-film patterning of polymers by RF plasma polymerization II: Study of differential binding using adsorption probes. *J. Biomater. Sci. Polym. Ed.*, **12**(7), 739–753.

Goldstein, S. A. (2002). Tissue engineering: Functional assessment and clinical outcome. *Ann. NY. Acad. Sci.*, **961**, 183–192.

Gomez, N., Chen, S., & Schmidt, C. E. (2007). Polarization of hippocampal neurons with competitive surface stimuli: Contact guidance cues are preferred over chemical ligands. *J.R. Soc. Interface*, **4**(13), 223–233.

Groger, S., Michel, J., & Meyle, J. (2008). Establishment and characterization of immortalized human gingival keratinocyte cell lines. *J. Periodontal. Res.*, **43**(6), 604–614.

Guruprasad Sosale, S., & Vengallatore, S. T. (2008). Topography analysis of grit-blasted and grit-blasted-acid-etched titanium implant surfaces using multi-scale measurements and multi-parameter statistics. *J. Materials Research*, **23**(10), 2704–2713.

Hacking, S. A., Bobyn, J. D., Tanzer, M., & Krygier, J. J. (1999). The osseous response to corundum blasted implant surfaces in a canine hip model. *Clin. Orthop. Relat. Res.*, **364**, 240–253.

Hacking, S. A., Tanzer, M., Harvey, E. J., Krygier, J. J., & Bobyn, J. D. (2002). Relative contributions of chemistry and topography to the osseointegration of hydroxyapatite coatings. *Clin. Orthop. Relat. Res.*, **405**, 24–38.

Hacking, S. A., Harvey, E. J., Tanzer, M., Krygier, J. J., & Bobyn, J. D. (2003). Acid-etched microtexture for enhancement of bone growth into porous-coated implants. *J. Bone Joint Surg. Br.*, **85**(8), 1182–1189.

Hacking, S. A., Zuraw, M., Harvey, E. J., Tanzer, M., Krygier, J. J., et al. (2007). A physical vapor deposition method for controlled evaluation of biological response to biomaterial chemistry and topography. *J. Biomed. Mater. Res. A.*, **82**(1), 179–187.

Hacking, S. A., Harvey, E., Roughley, P., Tanzer, M., & Bobyn, J. (2008). The response of mineralizing culture systems to microtextured and polished titanium surfaces. *J. Orthop. Res.*, **26**(10), 1347–1354.

Hacking, S. A., Harvey, E., Roughley, P., Tanzer, M., & Bobyn, J. (2008). The response of mineralizing culture systems to microtextured and polished titanium surfaces. *J. Orthop. Res.*, **26**(10), 1347–1354.

Haghighipour, N., Tafazzoli-Shadpour, M., Shokrgozar, M. A., Amini, S., Amanzadeh, A., et al. (2007). Topological remodeling of cultured endothelial cells by characterized cyclic strains. *Mol. Cell Biomech.*, **4**(4), 189–199.

Hamilton, D. W., & Brunette, D. M. (2007). The effect of substratum topography on osteoblast adhesion mediated signal transduction and phosphorylation. *Biomaterials*, **28**(10), 1806–1819.

Hamilton, D. W., Wong, K. S., & Brunette, D. M. (2006). Microfabricated discontinuous-edge surface topographies influence osteoblast adhesion, migration, cytoskeletal organization, and proliferation and enhance matrix and mineral deposition *in vitro*. *Calcif. Tissue Int.*, **78**(5), 314–325.

Hannafin, J. A., Attia, E. A., Henshaw, R., Warren, R. F., & Bhargava, M. M. (2006). Effect of cyclic strain and plating matrix on cell proliferation and integrin expression by ligament fibroblasts. *J. Orthop. Res.*, **24**(2), 149–158.

Hanson, A. D., Marvel, S. W., Bernacki, S. H., Banes, A. J., van Aalst, J., et al. (2009). Osteogenic effects of rest inserted and continuous cyclic tensile strain on hASC lines with disparate osteodifferentiation capabilities. *Ann. Biomed. Eng.*, **37**(5), 955–965.

Hayflick, L. (1965). The limited *in vitro* lifetime of human diploid cell strains. *Exp. Cell Res.*, **37**, 614–636.

Hayflick, L. (1985). The cell biology of aging. *Clin. Geriatr. Med.*, **1**(1), 15–27.

Heckman, C. A. (2009). Contact inhibition revisited. *J. Cell Physiol.*, **220**(3), 574–575.

Henshaw, D. R., Attia, E., Bhargava, M., & Hannafin, J. A. (2006). Canine ACL fibroblast integrin expression and cell alignment in response to cyclic tensile strain in three-dimensional collagen gels. *J. Orthop. Res.*, **24**(3), 481–490.

Higuchi, A., Hamamura, A., Shindo, Y., Kitamura, H., Yoon, B. O., et al. (2004). Photon-modulated changes of cell attachments on poly(spiropyran-co-methyl methacrylate) membranes. *Biomacromolecules*, **5**(5), 1770–1774.

Hui, E. E., & Bhatia, S. N. (2007). Microscale control of cell contact and spacing via three-component surface patterning. *Langmuir*, **23**(8), 4103–4107.

Hynes, R. O. (2002). Integrins: Bidirectional, allosteric signaling machines. *Cell*, **110**(6), 673–687.

Hyun, J., Zhu, Y. J., et al. (2001). Microstamping on an activated polymer surface: Patterning biotin and streptavidin onto common polymeric biomaterials. *Langmuir*, **17**(20), 6358–6367.

Hyun, J., Chen, J., Setton, L. A., & Chilkoti, A. (2006). Patterning cells in highly deformable microstructures: Effect of plastic deformation of substrate on cellular phenotype and gene expression. *Biomaterials*, **27**(8), 1444–1451.

Inoue, S., Iida, Y., Otani, Y., Hirano, Y., & Tabata, Y. (2009). Adhesion behavior of human adipo-stromal cells on self-assembled monolayers with different surface densities or gradients of RGD peptide. *J. Biomater. Sci. Polym. Ed.*, **20**(4), 495–510.

Ireland, G. W., Dopping-Hepenstal, P., Jordan, P., & O'Neill, C. (1987). Effect of patterned surfaces of adhesive islands on the shape, cytoskeleton, adhesion and behaviour of Swiss mouse 3T3 fibroblasts. *J. Cell Sci. Suppl.*, **8**, 19–33.

Jacobson, B. S., & Ryan, U. S. (1982). Growth of endothelial and HeLa cells on a new multipurpose microcarrier that is positive, negative or collagen coated. *Tissue Cell*, **14**(1), 69–83.

Jensen, P. J., & Koren, H. S. (1979). Depletion of NK by cellular immunoadsorption. *J. Immunol.*, **123**(3), 1127–1132.

Johnston, E. E., Bryers, J. D., & Ratner, B. D. (2005). Plasma deposition and surface characterization of oligoglyme, dioxane, and crown ether nonfouling films. *Langmuir*, **21**(3), 870–881.

Joshi, S. D., & Webb, K. (2008). Variation of cyclic strain parameters regulates development of elastic modulus in fibroblast/substrate constructs. *J. Orthop. Res.*, **26**(8), 1105–1113.

Kantawong, F., Burchmore, R., Gadegaard, N., Oreffo, R. O., & Dalby, M. J. (2008). Proteomic analysis of human osteoprogenitor response to disordered nanotopography. *J.R. Soc. Interface*, **6**(40), 1075–1086.

Kasemo, B., & Lausmaa, J. (1994). Material-tissue interfaces: The role of surface properties and processes. *Environ. Health Perspect*, **102**(Suppl. 5), 41–45.

Kataropoulou, M., Henderson, C., & Grant, M. H. (2005). Metabolic studies of hepatocytes cultured on collagen substrata modified to contain glycosaminoglycans. *Tissue Eng.*, **11**(7–8), 1263–1273.

Keselowsky, B. G., Collard, D. M., & Garcia, A. J. (2003). Surface chemistry modulates fibronectin conformation and directs integrin binding and specificity to control cell adhesion. *J. Biomed. Mater. Res. A.*, **66**(2), 247–259.

Khademhosseini, A. (2005). Chips to Hits: Microarray and microfluidic technologies for high-throughput analysis and drug discovery. *Expert Rev. Mol. Diagn.*, **5**(6), 843–846.

Khademhosseini, A., Jon, S., Suh, K. Y., Eng, G., Yeh, J., Tran, T. T., Langer, R., (2003). Direct patterning of protein and cell resistant polymeric monolayers and microstructures. *Adv. Mat.*, **15**(23), 1995–2000.

Khademhosseini, A., Suh, K. Y., Yang, J. M., Eng, G., Yeh, J., et al. (2004). Layer-by-layer deposition of hyaluronic acid and poly-L-lysine for patterned cell co-cultures. *Biomaterials*, **25**(17), 3583–3592.

Khademhosseini, A., Langer, R., Borenstein, J., & Vacanti, J. P. (2006). Microscale technologies for tissue engineering and biology. *Proc. Natl. Acad. Sci. U.S.A.*, **103**(8), 2480–2487.

Kikuchi, K., Sumaru, K., Edahiro, J., Ooshima, Y., Sugiura, S., et al. (2009). Stepwise assembly of micropatterned co-cultures using photoresponsive culture surfaces and its application to hepatic tissue arrays. *Biotechnol. Bioeng.*, **103**(3), 552–561.

Kim, D. H., Kim, P., Suh, K., Kyu Choi, S., Ho Lee, S., et al. (2005). Modulation of adhesion and growth of cardiac myocytes by surface nanotopography. *Conf. Proc. IEEE Eng. Med. Biol. Soc.*, **4**, 4091–4094.

Klein-Soyer, C., Hemmendinger, S., & Cazenave, J. P. (1989). Culture of human vascular endothelial cells on a positively charged polystyrene surface, primaria: Comparison with fibronectin-coated tissue culture grade polystyrene. *Biomaterials*, **10**(2), 85–90.

Koepsel, J. T., & Murphy, W. L. (2009). Patterning discrete stem cell culture environments via localized self-assembled monolayer replacement. *Langmuir*, **25**(21), 12825–12834.

Koller, M. R., Palsson, M. A., Manchel, I., Maher, R. J., & Palsson, B. O. (1998). Tissue culture surface characteristics influence the expansion of human bone marrow cells. *Biomaterials*, **19**(21), 1963–1972.

Kortsmit, J., Rutten, M., Wijlaars, M. W., & Baaijens, F. P. (2009). Deformation controlled load application in heart valve tissue-engineering. *Tissue Eng. Part C Methods*, **15**(4), 707–716.

Kunzler, T. P., Huwiler, C., Drobek, T., Vörös, J., & Spencer, N. D. (2007). Systematic study of osteoblast response to nanotopography by means of nanoparticle-density gradients. *Biomaterials*, **28**(33), 5000–5006.

Kurpinski, K., Chu, J., Hashi, C., & Li, S. (2006). Anisotropic mechanosensing by mesenchymal stem cells. *Proc. Natl. Acad. Sci. U.S.A.*, **103**(44), 16095–16100.

Lavanant, L., Pullin, B., Hubbell, J. A., & Klok, H. A. (2010). A facile strategy for the modification of polyethylene substrates with non-fouling, bioactive poly(poly(ethylene glycol) methacrylate) brushes. *Macromol. Biosci.*, **10**(1), 101–108.

Lee, J. H., Lee, J. W., Khang, G., & Lee, H. B. (1997). Interaction of cells on chargeable functional group gradient surfaces. *Biomaterials*, **18**(4), 351–358.

Li, L., Chen, S., & Jiang, S. (2007). Protein interactions with oligo(ethylene glycol) (OEG) self-assembled monolayers: OEG stability, surface packing density and protein adsorption. *J. Biomater. Sci. Polym. Ed.*, **18**(11), 1415–1427.

Liliensiek, S. J., Nealey, P., & Murphy, C. J. (2009). Characterization of endothelial basement membrane nanotopography in rhesus macaque as a guide for vessel tissue engineering. *Tissue Eng. Part A*, **15**(9), 2643–2651.

Liu, L., Ratner, B. D., Sage, E. H., & Jiang, S. (2007). Endothelial cell migration on surface-density gradients of fibronectin, VEGF, or both proteins. *Langmuir*, **23**(22), 11168–11173.

Lo, C. M., Wang, H. B., Dembo, M., & Wang, Y. L. (2000). Cell movement is guided by the rigidity of the substrate. *Biophys. J.*, **79**(1), 144–152.

Loesberg, W. A., te Riet, J., van Delft, F. C., Schön, P., Figdor, C. G., et al. (2007). The threshold at which substrate nanogroove dimensions may influence fibroblast alignment and adhesion. *Biomaterials*, **28**(27), 3944–3951.

Lovmand, J., Justesen, J., Foss, M., Lauridsen, R. H., Lovmand, M., et al. (2009). The use of combinatorial topographical libraries for the screening of enhanced osteogenic expression and mineralization. *Biomaterials*, **30**(11), 2015–2022.

Martinez, E., Lagunas, A., MillsRodríguez-Seguí, C. A., Estévez, M., et al. (2009). Stem cell differentiation by functionalized micro- and nanostructured surfaces. *Nanomed.*, **4**(1), 65–82.

Massia, S. P., & Hubbell, J. A. (1990a). Covalent surface immobilization of Arg-Gly-Asp- and Tyr-Ile-Gly-Ser-Arg-containing peptides to obtain well-defined cell-adhesive substrates. *Anal. Biochem.*, **187**(2), 292–301.

Massia, S. P., & Hubbell, J. A. (1990b). Covalently attached GRGD on polymer surfaces promotes biospecific adhesion of mammalian cells. *Ann. N.Y. Acad. Sci.*, **589**, 261–270.

Massia, S. P., & Hubbell, J. A. (1991a). Human endothelial cell interactions with surface-coupled adhesion peptides on a nonadhesive glass substrate and two polymeric biomaterials. *J. Biomed. Mater. Res.*, **25**(2), 223–242.

Massia, S. P., & Hubbell, J. A. (1991b). An RGD spacing of 440 nm is sufficient for integrin alpha V beta 3-mediated fibroblast spreading and 140 nm for focal contact and stress fiber formation. *J. Cell Biol.*, **114**(5), 1089–1100.

McBeath, R., Pirone, D. M., Nelson, C. M., Bhadriraju, K., & Chen, C. S. (2004). Cell shape, cytoskeletal tension, and RhoA regulate stem cell lineage commitment. *Dev. Cell*, **6**(4), 483–495.

McCartney, M. D., & Buck, R. C. (1981). Comparison of the degree of contact guidance between tumor cells and normal cells *in vitro*. *Cancer Res.*, **41**(8), 3046–3051.

Mello, C. C. (2007). Return to the RNAi world: Rethinking gene expression and evolution. *Cell Death Differ.*, **14**(12), 2013–2020.

Meredith, D. O., Riehle, M. O., Curtis, A. S., & Richards, R. G. (2007). Is surface chemical composition important for orthopaedic implant materials? *J. Mater. Sci. Mater. Med.*, **18**(2), 405–413.

Mi, L., Bernards, M. T., Cheng, G., Yu, Q., & Jiang, S. (2010). pH responsive properties of non-fouling mixed-charge polymer brushes based on quaternary amine and carboxylic acid monomers. *Biomaterials*, **31**(10), 2919–2925.

Monchaux, E., & Vermette, P. (2007). Bioactive microarrays immobilized on low-fouling surfaces to study specific endothelial cell adhesion. *Biomacromolecules*, **8**(11), 3668–3673.

Morris, R. J., Beech, J. N., Barber, P. C., & Raisman, G. (1985). Early stages of Purkinje cell maturation demonstrated by Thy-1 immunohistochemistry on postnatal rat cerebellum. *J. Neurocytol.*, **14**(3), 427–452.

Mrksich, M., & Whitesides, G. M. (1996). Using self-assembled monolayers to understand the interactions of man-made surfaces with proteins and cells. *Annu. Rev. Biophys. Biomol. Struct.*, **25**, 55–78.

Nakamura, H. K., Butz, F., Saruwatari, L., & Ogawa, T. (2007). A role for proteoglycans in mineralized tissue-titanium adhesion. *J. Dent. Res.*, **86**(2), 147–152.

Nandkumar, M. A., Yamato, M., Kushida, A., Konno, C., Hirose, M., et al. (2002). Two-dimensional cell sheet manipulation of heterotypically co-cultured lung cells utilizing temperature-responsive culture dishes results in long-term maintenance of differentiated epithelial cell functions. *Biomaterials*, **23**(4), 1121–1130.

Neher, E., & Sakmann, B. (1976). Single-channel currents recorded from membrane of denervated frog muscle fibres. *Nature*, **260**(5554), 799–802.

Niles, W. D., & Coassin, P. J. (2005). Piezo- and solenoid valve-based liquid dispensing for miniaturized assays. *Assay Drug Dev. Technol.*, **3**(2), 189–202.

Oh, S., Brammer, K. S., Li, Y. S., Teng, D., Engler, A. J., et al. (2009). Stem cell fate dictated solely by altered nanotube dimension. *Proc. Natl. Acad. Sci. U.S.A.*, **106**(7), 2130–2135.

O'Neill, C., Jordan, P., Riddle, P., & Ireland, G. (1990). Narrow linear strips of adhesive substratum are powerful inducers of both growth and total focal contact area. *J. Cell Sci.*, **95**(Pt 4), 577–586.

Opas, M. (1994). Substratum mechanics and cell differentiation. *Int. Rev. Cytol.*, **150**, 119–137.

Ostuni, E., Kane, R., et al. (2000). Patterning mammalian cells using elastomeric membranes. *Langmuir*, **16**(20), 7811–7819.

Park, J., Bauer, S., von der Mark, K., & Schmuki, P. (2007). Nanosize and vitality: TiO_2 nanotube diameter directs cell fate. *Nano Lett.*, **7**(6), 1686–1691.

Pelham, R. J., Jr., & Wang, Y. (1997). Cell locomotion and focal adhesions are regulated by substrate flexibility. *Proc. Natl. Acad. Sci. U.S.A.*, **94**(25), 13661–13665.

Peter, X., & Ma, J. H.E. (2006). *Scaffolding in Tissue Engineering*. Boca Raton, FL: CRC Press.

Petrie, T. A., Raynor, J. E., Reyes, C. D., Burns, K. L., Collard, D. M., et al. (2008). The effect of integrin-specific bioactive coatings on tissue healing and implant osseointegration. *Biomaterials*, **29**(19), 2849–2857.

Prasad, K. N., Carvalho, E., Kentroti, S., Edwards-Prasad, J., Freed, C., et al. (1994). Establishment and characterization of immortalized clonal cell lines from fetal rat mesencephalic tissue. *In Vitro Cell. Dev. Biol. Anim.*, **30A**(9), 596–603.

Preiss-Bloom, O., Mizrahi, J., Elisseeff, J., & Seliktar, D. (2009). Real-time monitoring of force response measured in mechanically stimulated tissue-engineered cartilage. *Artif. Organs*, 33(4), 318–327.

Prime, K., & Whitesides, G. (1991). Self-assembled organic monolayers: Model systems for studying adsorption of proteins at surfaces. *Science*, **252**(5009), 1164–1167.

Priori, E. S., Wilbur, J. R., Allen, P. T., East, J. L., & Dmochowski, L. (1975). Transformation of cells in human bone tumor cultures. *Bibl. Haematol.*, **40**, 185–196.

Puck, T. T., & Marcus, P. I. (1955). A rapid method for viable cell titration and clone production with hela cells in tissue culture: The use of x-irradiated cells to supply conditioning factors. *Proc. Natl. Acad. Sci. U.S.A.*, **41**(7), 432–437.

Riboh, J., Chong, A. K., Pham, H., Longaker, M., Jacobs, C., et al. (2008). Optimization of flexor tendon tissue engineering with a cyclic strain bioreactor. *J. Hand Surg. [Am.]*, **33**(8), 1388–1396.

Riches, A., Peddie, C., Rendell, S., Bryant, P., Zitzelsberger, H., et al. (2001). Neoplastic transformation and cytogenetic changes after gamma irradiation of human epithelial cells expressing telomerase. *Radiat. Res.*, **155**(1 Pt 2), 222–229.

Rodionova, N. V., & Oganov, V. S. (2003). Changes of cell-vascular complex in zones of adaptive remodeling of the bone tissue under microgravity conditions. *Adv. Space Res.*, **32**(8), 1477–1481.

Ruoslahti, E. (1996). RGD and other recognition sequences for integrins. *Annu. Rev. Cell and Develop Biol.*, **12**, 697–715.

Ruoslahti, E., & Pierschbacher, M. D. (1987). New perspectives in cell adhesion: RGD and integrins. *Science*, **238**(4826), 491–497.

Ruoslahti, E., & Reed, J. C. (1994). Anchorage dependence, integrins, and apoptosis. *Cell*, **77**(4), 477–478.

Saha, K., Keung, A. J., Irwin, E. F., Li, Y., Little, L., et al. (2008). Substrate modulus directs neural stem cell behavior. *Biophys. J.*, **95**(9), 4426–4438.

Scherer, W. F., Syverton, J. T., & Gey, G. O. (1953). Studies on the propagation *in vitro* of poliomyelitis viruses. IV. Viral multiplication in a stable strain of human malignant epithelial cells (strain HeLa) derived from an epidermoid carcinoma of the cervix. *J. Exp. Med.*, **97**(5), 695–710.

Schwarz, E., Freese, U. K., Gissmann, L., Mayer, W., Roggenbuck, B., et al. (1985). Structure and transcription of human papillomavirus sequences in cervical carcinoma cells. *Nature*, **314**(6006), 111–114.

Shen, M., & Horbett, T. A. (2001). The effects of surface chemistry and adsorbed proteins on monocyte/macrophage adhesion to chemically modified polystyrene surfaces. *J. Biomed. Mater. Res.*, **57**(3), 336–345.

Steele, J. G., Dalton, B. A., Underwood, P. A., & Smith, G. J. (1991). Differences in adhesion to tissue culture plastic of clonally related transformed and control sublines from an epithelial cell strain. *J. Cell Sci.*, **100**(Pt 1), 195–203.

Steele, J. G., Dalton, B. A., Johnson, G., & Underwood, P. A. (1993). Polystyrene chemistry affects vitronectin activity: An explanation for cell attachment to tissue culture polystyrene but not to unmodified polystyrene. *J. Biomed. Mater. Res.*, **27**(7), 927–940.

Strobl, C. J., von Guttenberg, Z., & Wixforth, A. (2004). Nano- and pico-dispensing of fluids on planar substrates using SAW. *IEEE Trans. Ultrason Ferroelectr. Freq. Control*, **51**(11), 1432–1436.

Suh, K. Y., Khademhosseini, A.Yang, J. M., Tran, T. T, Eng, G. & Langer, R. (2004). Soft lithographic patterning of hyaluronic acid on solid substrates wing molding and printing, Adv. Mat., **16**(7), 584–588.

Takayama, S., McDonald, J. C., Ostuni, E., Liang, M. N., Kenis, P. J., et al. (1999). Patterning cells and their environments using multiple laminar fluid flows in capillary networks. *Proc. Natl. Acad. Sci. U.S.A.*, **96**(10), 5545–5548.

Takayama, S., Ostuni, E., LeDuc, P., Naruse, K., Ingber, D. E., et al. (2001). Subcellular positioning of small molecules. *Nature*, **411**(6841), 1016.

Tan, J., & Saltzman, W. M. (2002). Topographical control of human neutrophil motility on micropatterned materials with various surface chemistry. *Biomaterials*, **23**(15), 3215–3225.

Teixeira, A. I., McKie, G. A., Foley, J. D., Bertics, P. J., Nealey, P. F., et al. (2006). The effect of environmental factors on the response of human corneal epithelial cells to nanoscale substrate topography. *Biomaterials*, **27**(21), 3945–3954.

Timpe, L., Martz, E., & Steinberg, M. S. (1978). Cell movements in a confluent monolayer are not caused by gaps: Evidence for direct contact inhibition of overlapping. *J. Cell Sci.*, **30**, 293–304.

Tobasnick, G., & Curtis, A. S. (2001). Chloride channels and the reactions of cells to topography. *Eur. Cell Mater.*, **2**, 49–61.

Toyoda, T., Matsumoto, H., Fujikawa, K., Saito, S., & Inoue, K. (1998). Tensile load and the metabolism of anterior cruciate ligament cells. *Clin. Orthop. Relat. Res.*, **353**, 247–255.

van Griensven, M., Diederichs, S., Roeker, S., Boehm, S., Peterbauer, A., et al. (2009). Mechanical strain using 2D and 3D bioreactors induces osteogenesis: Implications for bone tissue engineering. *Adv. Biochem. Eng. Biotechnol.*, **112**, 95–123.

Whang-Peng, J., Triche, T. J., Knutsen, T., Miser, J., Kao-Shan, S., et al. (1986). Cytogenetic characterization of selected small round cell tumors of childhood. *Cancer Genet. Cytogenet.*, **21**(3), 185–208.

Wittelsberger, S. C., Kleene, K., & Penman, S. (1981). Progressive loss of shape-responsive metabolic controls in cells with increasingly transformed phenotype. *Cell*, **24**(3), 859–866.

Wojciak-Stothard, B., Curtis, A. S., Monaghan, W., McGrath, M., Sommer, I., et al. (1995). Role of the cytoskeleton in the reaction of fibroblasts to multiple grooved substrata. *Cell Motil. Cytoskeleton*, **31**(2), 147–158.

Wojciak-Stothard, B., Curtis, A., Monaghan, W., MacDonald, K., & Wilkinson, C. (1996). Guidance and activation of murine macrophages by nanometric scale topography. *Exp. Cell Res.*, **223**(2), 426–435.

Wong, J. Y., Velasco, A., et al. (2003). Directed movement of vascular smooth muscle cells on gradient-compliant hydrogels. *Langmuir*, **19**(5), 1908–1913.

Wood, A. (1988). Contact guidance on microfabricated substrata: The response of teleost fin mesenchyme cells to repeating topographical patterns. *J. Cell Sci.*, **90**(Pt 4), 667–681.

Wu, J. C., Merlino, G., & Fausto, N. (1994). Establishment and characterization of differentiated, nontransformed hepatocyte cell lines derived from mice transgenic for transforming growth factor alpha. *Proc. Natl. Acad. Sci. U.S.A.*, **91**(2), 674–678.

Yamato, M., Utsumi, M., Kushida, A., Konno, C., Kikuchi, A., et al. (2001). Thermo-responsive culture dishes allow the intact harvest of multilayered keratinocyte sheets without dispase by reducing temperature. *Tissue Eng.*, **7**(4), 473–480.

Yamato, M., Konno, C., Utsumi, M., Kikuchi, A., & Okano, T. (2002). Thermally responsive polymer-grafted surfaces facilitate patterned cell seeding and co-culture. *Biomaterials*, **23**(2), 561–567.

Yasuda, T., Kanamori, M., Nogami, S., Hori, T., Oya, T., et al. (2009). Establishment of a new human osteosarcoma cell line, UTOS-1: Cytogenetic characterization by array comparative genomic hybridization. *J. Exp. Clin. Cancer Res.*, **28**(1), 26.

Yu, J., Jin, F., Deng, Z., Li, Y., Tang, L., et al. (2008). Epithelial-mesenchymal cell ratios can determine the crown morphogenesis of dental pulp stem cells. *Stem Cells Dev.*, **17**(3), 475–482.

Zemljic Jokhadar, S., Znidarcic, T., Svetina, S., & Batista, U. (2007). The effect of substrate and adsorbed proteins on adhesion, growth and shape of CaCo-2 cells. *Cell Biol. Int.*, **31**(10), 1097–1108.

CHAPTER II.1.4 CELL FUNCTION AND RESPONSE TO INJURY

Richard N. Mitchell and Frederick J. Schoen
Department of Pathology, Brigham and Women's Hospital and Harvard Medical School, Boston, MA, USA

Composed of nucleic acids, proteins, and other large and small molecules, *cells* constitute the basic structural building blocks of all living matter. Except for the circulating elements in blood and lymph, cells are joined via cell–cell junctions to form *tissues* comprising a handful of general types: *epithelium*; *connective tissue*; *muscle*; and *nerve*. *Hematopoietic* and *lymphoid* tissues – responsible for producing red and white blood cells and platelets – also exhibit similar features as they generate their final circulating progeny. Organs are assembled from the basic tissue types, and are "glued" together within an *extracellular matrix* (ECM) composed of myriad proteins, glycoproteins, mucopolysaccharides (sugar polymers) and proteoglycans (polymeric sugars attached to a protein core) synthesized by the individual cells. The organs, in turn, perform the various functions required by the intact living organism, including circulation, respiration, digestion, excretion, movement, and reproduction.

A major goal in this and the following chapter is to describe how biological structure is adapted to perform physiologic function. This general and overarching concept extends from cells (and their *subcellular* constituents) to the organization of tissues, and of organs. Beginning with the smallest subunits of cellular organization, we will build to progressively more complex systems. In the following chapter, we will extend the concepts of structure–function correlation beyond cells to include extracellular matrix and complex tissues, and will also describe the technologies by which histologists and pathologists examine normal and abnormal tissues. In these chapters, we also provide an introduction to the physiologic responses to environmental stimuli, the mechanisms of cell injury, cell–materials interactions, and the methodologies by which these are all studied.

In this chapter on function and response to injury, and in the following chapter on tissues and the extracellular matrix, we will highlight the following fundamental concepts of cells and the tissues they form.

CELLS: FUNCTION AND RESPONSE TO INJURY

- General characteristics and functions of cells
- Compartmentalization of regionally specialized function by membranes
- Cellular specialization to facilitate unique functions
- Regulation and coordination of cell function
- Cellular differentiation
- Response of cells to injury, including mechanisms of cell death.

TISSUES AND THE EXTRACELLULAR MATRIX

- Structure and function of the extracellular matrix (ECM)
- Organization of cells into tissues
- Integration of tissues into organs
- Remodeling of the extracellular matrix
- Interaction of cells, tissues, and foreign materials
- Basic methods used to study cells and tissues.

NORMAL CELL HOUSEKEEPING

In very broad strokes, we will first outline the general organization of a prototypical cell, using it to identify the functional considerations required for maintaining viability. We will then revisit each of these structural features to illustrate important concepts in *homeostatic maintenance* (keeping the cell within certain functional boundary conditions) and response to the environment.

Conceptually, cells may be viewed as independent collections of self-replicating enzymes and structural proteins that carry out certain general functions. The most essential cell attributes are:

- protection from the environment
- acquisition of nutrients
- movement
- communication
- catabolism of extrinsic molecules
- degradation and renewal of senescent intrinsic molecules
- energy generation
- self-replication.

Intracellular constituents exist in a microcosm of water, ions, sugars, and small molecular weight molecules called the *cytosol* or *cytoplasm*. Within the cytosol is also a source of energy, typically *adenosine triphosphate* (ATP). Although long conceptualized as a randomly diffusing bag of soluble molecules, the cell is, in fact, a structurally highly-ordered and functionally integrated assembly of organelles, cytoskeletal elements, and enzymes.

The cytosol is delimited and protected from the environment by a phospholipid bilayer, the *plasma membrane*, which permits the cell to maintain cytosolic constituents at concentrations different than those in the surrounding environment. Due to its hydrophobic inner core, the plasma membrane is impermeable to charged and/or large polar molecules; however, it is rendered selectively permeable (i.e., permits specific passage) to incoming or outgoing material (ions, sugars, amino acids, etc.) by pore or transport proteins inserted through it. Most nutrient acquisition is accomplished by the movement of desired substrates either through *protein channels* (driven by electrochemical gradients) or by energy-driven transport through *carrier proteins*. Cells also have the capacity to internalize material by capturing solute or even particles from the extracellular environment in invaginated folds of the plasma membrane called *vesicles*. Depending on the volume and size of the ingested material, the process is denoted *phagocytosis* ("cell eating") or *pinocytosis* ("cell drinking"). *Transcytosis* denotes the movement of pinocytic vesicles from one side of a cell to the other (typically across endothelial cells lining blood vessels); this process clearly permits bulk fluid transport and can play an important role in mediating the increased vascular permeability ("leaky vessels") that occurs around tumors or at sites of inflammation. Plasma membranes also characteristically express a variety of specific surface molecules that facilitate interactions with other cells, soluble ligands (e.g., insulin), and/or with the extracellular matrix (communication).

Many of a cell's normal "housekeeping" functions are compartmentalized within membrane-bounded intracellular organelles (Figure II.1.4.1), thus permitting adjacent regions of the cell to have vastly different chemistries. By isolating certain cellular functions within distinct compartments, potentially injurious degradative enzymes or toxic metabolites can be kept at usefully high concentrations locally, without causing damage to more delicate intracellular constituents. Moreover, compartmentalization also allows the creation of unique intracellular environments (e.g., low pH, high calcium

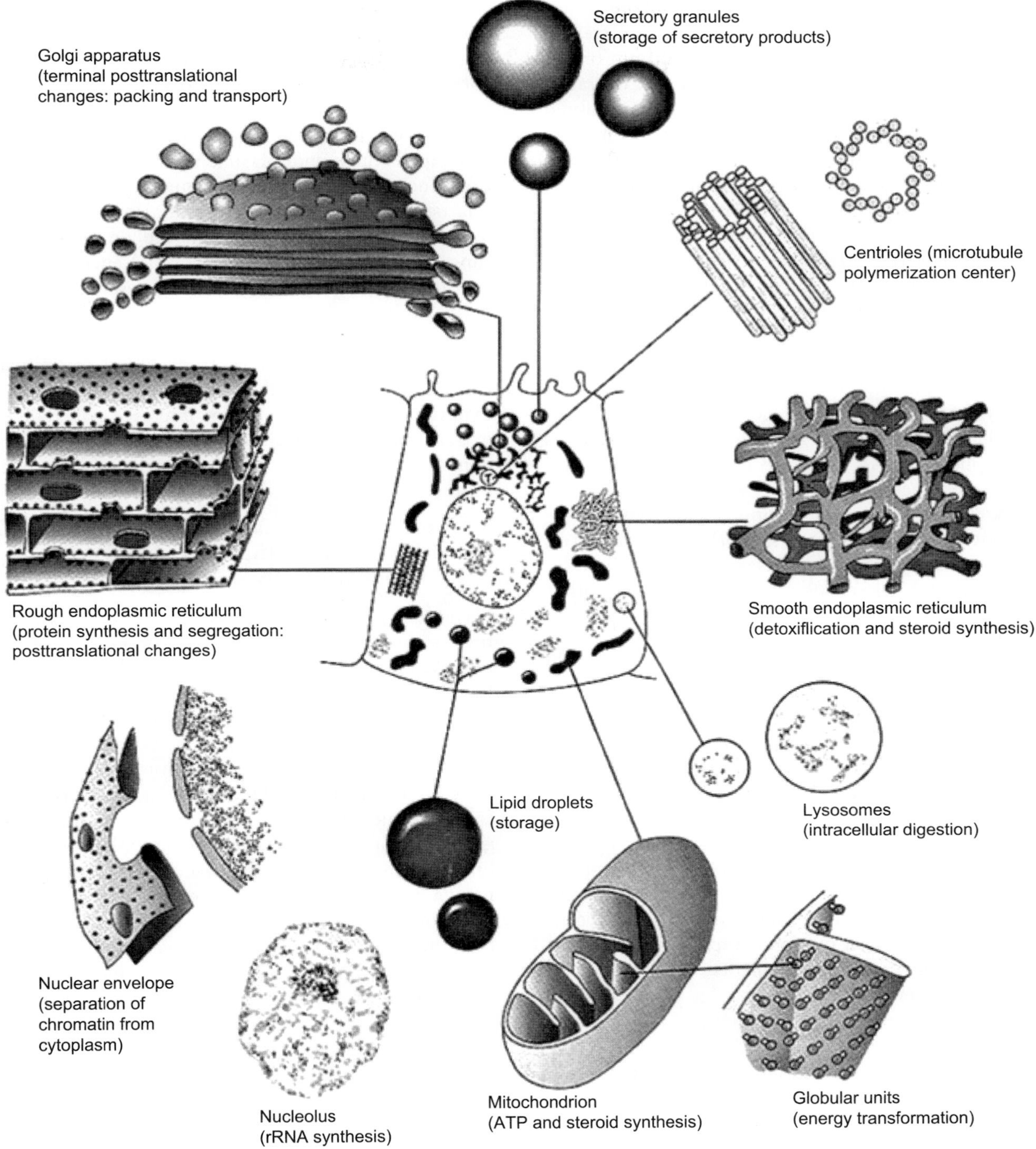

FIGURE II.1.4.1 General schematic of a typical mammalian cell, demonstrating the general organization and major organelles. Note that each compartment has distinct functions made possible by selectively permeable membranes. (Reproduced by permission from Bergman, A. R., Afifi, A. K. & Heideger, P. M. (Eds.). (1996). *Histology*. Saunders: Philadelphia, PA.)

or containing reactive oxygen species or proteases) that permit more efficient functioning of certain chemical processes, enzymes or metabolic pathways.

The enzymes and structural proteins of the cell are constantly being renewed by ongoing synthesis tightly balanced with intracellular degradation. Oversight for the new synthesis of macromolecules, including *deoxyribonucleic acid* (DNA) and *ribonucleic acid* (RNA) is provided by the *nucleus*. New proteins destined for the plasma membrane or for secretion into the extracellular environment are synthesized and packaged in the *rough endoplasmic reticulum* (RER) and *Golgi apparatus*; proteins intended for remaining in the cytosol are synthesized on free *ribosomes*.

Smooth endoplasmic reticulum (SER) may be abundant in certain cell types where it is used for steroid hormone and lipoprotein synthesis, as well as for the modification of hydrophobic compounds into water-soluble molecules for export, or in some cases adapted to store intracellular calcium. Degradation of internalized molecules or senescent self-molecules into their constituent amino acids, sugars, and lipids (catabolism) is the primary responsibility of the lysosomes and proteasomes. Peroxisomes play a specialized role in the catabolism of fatty acids, generating hydrogen peroxide in the process. Intracellular vesicles busily shuttle internalized material to appropriate intracellular site(s) for catabolism or direct newly-synthesized materials to the plasma membrane or relevant target organelle. The three-dimensional architecture of the cell is maintained by a scaffolding of intracellular proteins collectively called the cytoskeleton, analogous in some ways to the support provided by the bones of our bodies. Directed movement of organelles, and even selected proteins, within the cell is accomplished by using the cytoskeleton as a connected network of guide wires; molecular motors (called *dyneins* and *kinesins*) literally "walk" attached organelles from one site to another along these tightropes. These structural proteins also provide basic cellular shape and intracellular organization, which are necessary for the maintenance of *cell polarity* (differences in cell structure and function at the top of a cell versus its side or base). For example, in many cell types, and particularly in epithelial tissues, it is critical for cells to distinguish – and keep separated – the top (*apical*) versus the bottom and side (*basolateral*) surfaces.

Cell movement within its environment is accomplished through rearrangement of the cytoskeleton; the various protein networks have to be disassembled and re-established in a coordinated fashion to allow directed motion. The direction of motion is typically determined by external concentration gradients of chemotactic agents that differentially signal one edge of a cell more than another. The major energy currency for macromolecular synthesis, metabolite degradation, and intra- and extracellular transport is ATP, which is generated in the *mitochondrion* using oxidative phosphorylation. Finally, all of these organelles must be replicated (organellar biogenesis) and correctly apportioned in daughter cells following mitosis.

The Specific Function(s) of a Given Cell are Reflected by the Relative Amounts and Types of Organelles it Contains. The relative predominance of specific types of organelles can be inferred by examination of tissue sections prepared by standard histologic techniques, and can be confirmed by transmission electron microscopy. For example, cells with substantial energy requirements can be expected to have a significantly greater capacity to generate energy. Thus, kidney tubular epithelial cells (which reabsorb sodium and chloride against concentration gradients), and cardiac myocytes (which rhythmically contract 50–100 times per minute) have a generous complement of mitochondria. In comparison, cells specifically adapted to synthesize and export selected proteins (e.g., insulin in a pancreatic islet cell or antibody produced by a plasma cell) have a well-developed rough endoplasmic reticulum. A corollary to this general concept is that the nutritional and extracellular structural needs of a cell can also be inferred from the organellar constituents and the presumed cellular function. Thus, cells with substantial energy demands, but little in the way of intracellular reserves (e.g., myocardial cells), will need a continuous source of oxygen and glucose – provided by a dense capillary supply with minimal diffusion distance from blood to mitochondria. Conversely, cells that absorb from one surface but export products at a different interface will require an extracellular basement membrane and specific cell–cell connections to provide the proper orientation.

THE PLASMA MEMBRANE: PROTECTION, NUTRIENT ACQUISITION, AND COMMUNICATION

Plasma membranes, as well as all other organellar membranes, are dynamic, fluid, and inhomogeneous lipid bilayers containing a variety of embedded proteins. Biologic membranes are composed of phospholipids that are *amphipathic*; i.e., they have a polar head-group that prefers to interact with water (*hydrophilic*), and non-polar fatty acid chains that resist interaction with water (*hydrophobic*). These phospholipids spontaneously assemble to form two-dimensional sheets with their hydrophilic head groups facing towards the aqueous cytoplasm or extracellular fluid, and their hydrophobic lipid tails interacting to form a central core that largely excludes water. Since the lipid core of plasma membrane is intrinsically resistant to the movement of large or polar molecules, continued cell function requires that specific protein channels or transport mechanisms be in place to facilitate uptake of ions and metabolites (Figure II.1.4.2). In addition, the plasma membrane is also the interface between the extracellular environment and the inner cellular domains; cell surface proteins can link extracellular matrix with the intracellular cytoskeleton or can act as receptors for extracellular ligands that can trigger intracellular signaling events.

The lipid confers structural integrity and barrier function to the membrane, and the inserted proteins provide specialized membrane functions (see below). In addition, the specific composition of the lipid bilayer, e.g., relative amounts of various phospholipids, cholesterol, and glycolipids, will alter the physico-chemical properties of the membrane, as well as the functioning of associated proteins. Although the various lipid and protein constituents can move about easily in the plane of the membrane, certain components have a predilection for each other, and different domains with distinct lipid compositions

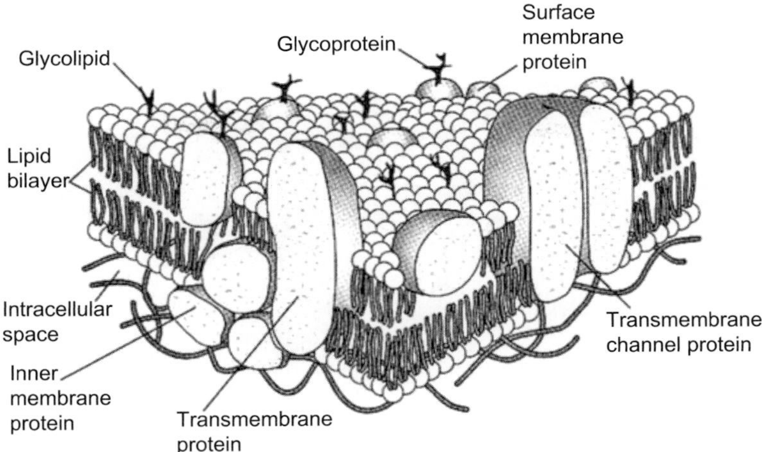

FIGURE II.1.4.2 Model of a prototypical cell membrane. Note the lipid bilayer with outer, hydrophilic headgroups (exposed to an aqueous environment) and inner, hydrophobic tailgroups (maintaining a barrier to solute movement). Inserted through the membrane or attached to either inner or outer planes are various proteins that permit transport, cell–cell and signal molecule interactions, and linkage of the membrane to the intracellular cytoskeleton or extracellular matrix. (Reproduced by permission from Bergman, A. R., Afifi, A. K. & Heideger, P. M. (Eds.). (1996). *Histology*. Saunders: Philadelphia, PA.)

are thereby created. Since inserted membrane proteins have different solubilities in these various lipid domains, the membrane lipid inhomogeneities result in functionally distinct islands and patches of protein molecules. This has significance in terms of cell–cell and cell–matrix interactions, as well as in intracellular signaling. The lipid component of the plasma membrane is also asymmetric, that is, the inner and outer leaflets have different general compositions. The asymmetry has functional significance in that gangliosides, conferring a net negative charge, and glycolipids – both of which are on the outer face of the bilayer – are important for cell–cell and cell–matrix interactions, local electrostatic effects, and creation of barriers to infection. In contrast, inositol-containing phospholipids are primarily arrayed on the inner face, presumably because they play an important role in intracellular signaling.

How Proteins Associate with Membranes Reflects their Function. Most proteins inserted into membranes are *integral* or *transmembrane* proteins, having one or more relatively hydrophobic α-helical segments that traverse the lipid bilayer and securely anchor the protein. Proteins involved in forming pores or transporting other molecules will typically be transmembrane. In other cases, proteins are attached to the membrane via weaker lipid–protein or protein–protein interactions. Non-integral proteins can potentially interact with a variety of membrane molecules, can mediate cytoskeletal interactions that permit movement or can be used to transduce transient signals.

Plasma Membrane Proteins Frequently Function Together as Large Complexes. These complexes may be primarily assembled as the proteins are synthesized in the RER or may arise by lateral diffusion in the plasma membrane. Such *de novo* complex formation, followed by signal cascade inside the cell, is a typical paradigm employed to translate ligand–surface receptor binding into an intracellular response. Large complexes also form the basis for intercellular connections such as *tight junctions* (see below). Similar interactions between like proteins on adjacent cells (*homotypic interactions*) create a zone that separates the apical versus basolateral aspects of cells in epithelial layers, and thereby establish cell polarity.

The Hydrophobic Lipid Core of Plasma Membranes is an Effective Barrier to the Passage of Most Polar Molecules. Small, nonpolar molecules, such as O_2 and CO_2, readily dissolve in lipid bilayers, and therefore rapidly diffuse across them. Large hydrophobic molecules, such as steroid hormones (testosterone and estrogen, for example), also readily cross lipid bilayers. Even polar molecules, if sufficiently small (e.g., water, ethanol, and urea at molecular weights of 18, 46, and 60 daltons, respectively) rapidly cross membranes. In contrast, glucose, at a molecular weight of only 180 daltons is effectively excluded, and lipid bilayers are completely impermeable to ions, regardless of size, due to their charge and high degree of hydration. Therefore, to facilitate the entrance or egress of most molecules, specific transport proteins are required. For small molecular weight molecules (ions, glucose, nucleotides, and amino acids up to approximately 1000 daltons) there are two main categories: *carrier proteins* and *channel proteins*. For larger molecules or even particles, uptake is mediated by specific receptors, internalized via *endocytosis*. Large molecules destined for export are packaged in *secretory vesicles* that fuse with the plasma membrane and expel their contents in a process called *exocytosis*.

Each type of transported molecule (ion, sugar, nucleotide, etc.) requires a unique membrane transport protein to facilitate passage. These transporters typically exhibit strong specificity. Thus, a particular transporter

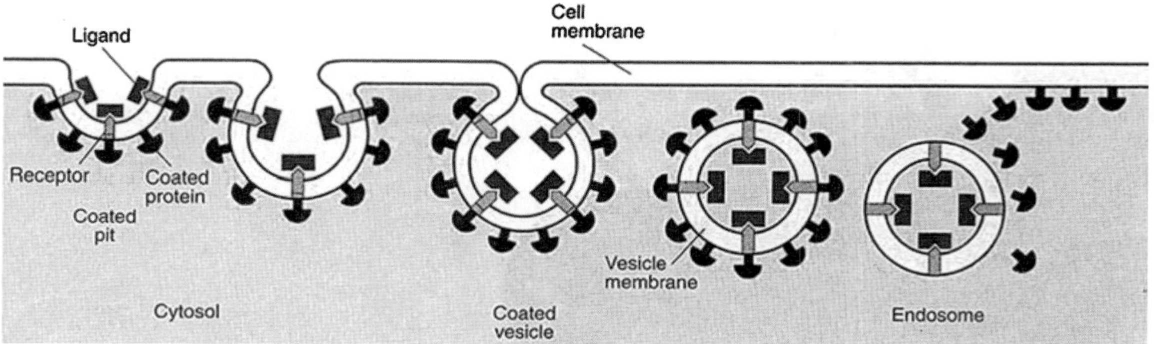

FIGURE II.1.4.3 Schematic of receptor-mediated endocytosis, enabling large molecules (e.g., transferrin with bound iron) and extracellular solutes to enter the cell. After binding to specific receptors, proteins are internalized on *clathrin-coated pits*, which pinch off to form *clathrin-coated vesicles*. The clathrin coat is then removed and the vesicle fuses with *endosomes* delivering its bound contents. Acidification of the endosome releases ligands; the receptors can be reutilized by vesicles pinching off from the endosome and recycling to the plasma membrane (*exocytosis*, not shown). (Reproduced by permission from Bergman, A. R., Afifi, A. K. & Heideger, P. M. (Eds.). (1996). *Histology*. Saunders: Philadelphia, PA.)

will move glucose but not galactose; another transporter will move potassium but not sodium. Carrier proteins bind their specific ligand and undergo a series of conformational changes to transfer it across the membrane; their transport is relatively slow. In comparison, channel proteins create hydrophilic pores; when open, these permit rapid movement of selected solutes. In most cases, a concentration and/or electrical gradient between the inside and outside of the cell drives solute movement by passive transport (virtually all plasma membranes have an electrical potential difference across them, with the inside of a cell negative relative to the outside). In some cases, active transport against a concentration gradient is accomplished by carrier molecules (never channels) and requires energy expenditure (provided by the breakdown of ATP). Transporters also include the multidrug resistance (MDR) protein, which pumps polar compounds (for example, chemotherapeutic drugs) out of cells, and may render cancer cells more resistant to treatment. Similar transport mechanisms also regulate intracellular and intraorganellar pH; human beings and most of their cytosolic enzymes prefer to work at pH 7.2, while lysosomes function best at pH 5 or less.

Because the plasma membrane is freely permeable to water, water will move into or out of cells along its concentration gradient (by osmosis). Extracellular salt in excess of that seen in the cytoplasm (hypertonicity) will cause a net movement of water out of cells; conversely, hypotonicity will cause a net movement of water into cells. Since the intracellular environment is rich in charged molecules that attract a large number of charged counter-ions (tending to increase osmolarity), cells need to constantly and actively regulate their intracellular osmolarity by pumping out small inorganic ions (typically sodium and chloride). The loss of the ability to generate energy (e.g., in a cell injured by toxins or lacking oxygen), therefore results in an osmotically swollen cell which may eventually rupture if normal homeostatic mechanisms do not soon supervene.

Uptake and Metabolism of Large Extracellular Molecules Requires Vesicle Targeting and Membrane Recycling. Proteins, large carbohydrates, and other macromolecules cannot enter cells by either channels or carriers. Instead, they are internalized by endocytosis (see Figure II.1.4.3). Endocytosis typically begins at a specialized region of the plasma membrane called the *clathrin-coated pit*, which rapidly invaginates and pinches off to form a *clathrin-coated vesicle* (about 2500 per minute in the average fibroblast). *Clathrin* is a hexamer of proteins that spontaneously assemble into a basket-like lattice to drive the budding process of endocytosis. Trapped within the vesicle will be a minute gulp of the extracellular milieu, as well as any molecules specifically bound to receptors on the internalized bit of plasma membrane (*receptor-mediated endocytosis*). This is the pathway, for example, by which cells internalize iron from the circulation: ionized iron, bound to a protein called *transferrin*, interacts with cell surface *transferrin receptors* which are then internalized via receptor-mediated endocytosis. The same pathway – but using a different receptor – is responsible for the binding and uptake of low-density lipoproteins (LDL) or any of a variety of other extracellular macromolecules. Notably, the same clathrin-coated pit or vesicle can contain many different receptors; all it takes to localize receptors to the clathrin endocytic pathway are specific amino acid motifs expressed on the cytosolic tails of the relevant proteins.

The vesicles rapidly lose their clathrin coat and then fuse with an acidic intracellular structure called the *endosome* where the lower pH causes dissociation of ligands from their receptors. This allows released ligands to be shuttled to the lysosome (or other locations) while receptors can either return to the plasma membrane for another cycle (e.g., the transferrin receptor) or can be degraded themselves (e.g., the LDL receptor) (Figure II.1.4.4 and below). Degradation of a receptor after internalization (*receptor down-regulation*) provides an important control for receptor expression and

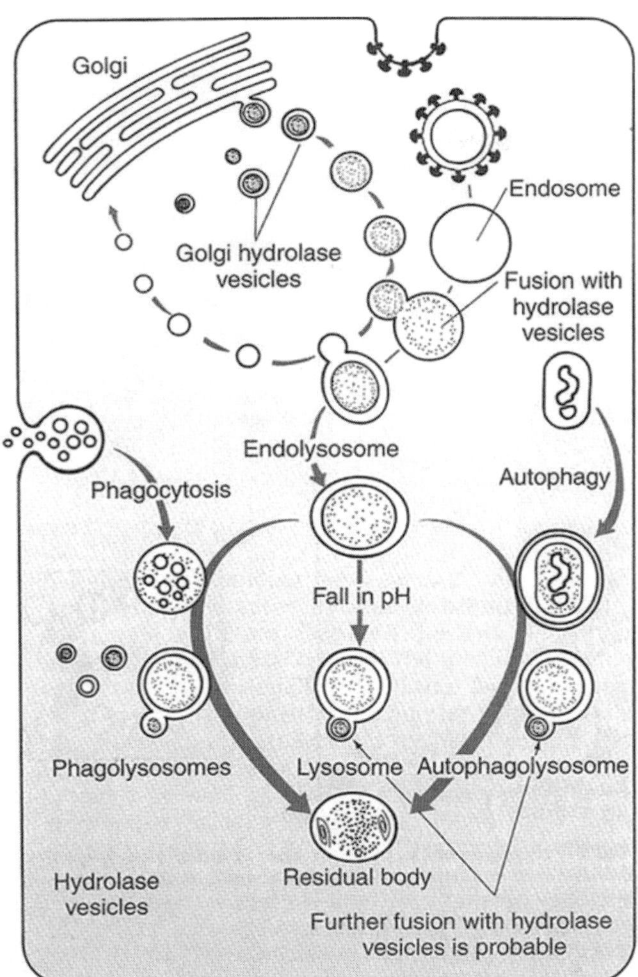

FIGURE II.1.4.4 Schematic demonstrating the pathways by which internalized material is degraded. There is convergence of *endocytic*, *phagocytic*, and *autophagic* vesicles as they fuse with newly synthesized catabolic enzyme-containing vesicles or with pre-existing *lysosomes*. The contents are nearly completely degraded to constituent amino acids, sugars, and lipids; non-degradable material will accumulate in residual bodies. (Reproduced by permission from Bergman, A. R., Afifi, A. K. & Heideger, P. M. (Eds.). (1996). *Histology*. Saunders: Philadelphia, PA.)

receptor-mediated signaling. Endocytosis can also deliver material completely across a cell, i.e., from the apical surface to the basolateral face (*transcytosis*), forming, for example, the basis for transport of nutrients from the gastrointestinal tract to the bloodstream. *Endocytosis* is an ongoing process, with constant recycling of vesicles back to the plasma membrane (*exocytosis*). Endocytosis and exocytosis must be tightly coupled, since a cell can internalize 10–20% of its own cell volume each hour or about 1–2% of its plasma membrane each minute.

Cell Communication is Critical in Multicellular Organisms. At the most basic level, extracellular signals may determine whether a cell lives or dies, whether it remains quiescent, proliferates, migrates or otherwise becomes active to perform its specific function(s). Intercellular signaling is critical in the developing embryo in order for cells to appear in the correct quantity and location, and to maintain tissue organization. Intercellular signaling is also important in the intact organism, ensuring that all tissues act in appropriate concert in response to stimuli as divergent as food or threat to life. Loss of communication may be reflected in a congenital structural defect, in unregulated cell growth (cancer) or an ineffective response to an extrinsic stress.

Cells Communicate Over Short and Long Distances. Adjacent cells can communicate via *gap junctions* – hydrophilic cylindrical pores that effectively connect the two cells' cytoplasm, formed by hexamers of proteins called *connexins*. These pores permit movement of small ions (e.g., Ca^{++}), various metabolites, and potential second messenger molecules, but not larger macromolecules. *Extracellular signaling* by soluble mediators occurs in three different forms: (1) *paracrine*, meaning that it affects cells only in the immediate vicinity. To accomplish this, there can be only minimal diffusion, with the signal rapidly degraded, taken up by other cells or trapped in the ECM; (2) *synaptic*, classically where activated neural tissue secretes *neurotransmitters* at a specialized cell junction (*synapse*) onto target cells; (3) *endocrine*, where a regulatory substance, such as a hormone, is released into the bloodstream and acts on target cells at a distance.

Since most signaling molecules are present at extremely low concentrations ($\leq 10^{-8}$ M), binding to the appropriate target cell receptor is typically a high affinity and exquisitely specific interaction. Receptor proteins may be on the cell surface or they may be intracellular; in the latter case, ligands must be sufficiently hydrophobic to enter the cell (e.g., vitamin D or steroid and thyroid hormones). For intracellular receptors, ligand binding leads to formation of a receptor–ligand complex that directly associates with nuclear DNA, and subsequently either activates or turns off gene transcription. For cell-surface receptors, the binding of ligand can: (1) open ion channels (e.g., at the synapse between electrically-excitable cells); (2) activate an associated GTP-binding regulatory protein (*G protein*); or (3) activate an associated enzyme. Secondary intracellular downstream events frequently involve phosphorylation (via *kinases*) or dephosphorylation (via *phosphatases*) of target molecules, with subsequent changes in enzymatic activity, intracellular localization or the ability to bind to DNA and induce transcription (see Figure II.1.4.5).

Besides *soluble signals* that arrive through cell surface (and intracellular) receptor–ligand interactions (as in Figure II.1.4.5), cell behavior and differentiation are also regulated by mechanical forces – so-called *mechanotransduction*. Fluid shear, pressure or tension on a tissue modulates cellular activity through transmembrane proteins called *integrins* that link extracellular matrix molecules with the intracellular actin cytoskeleton. Tugging

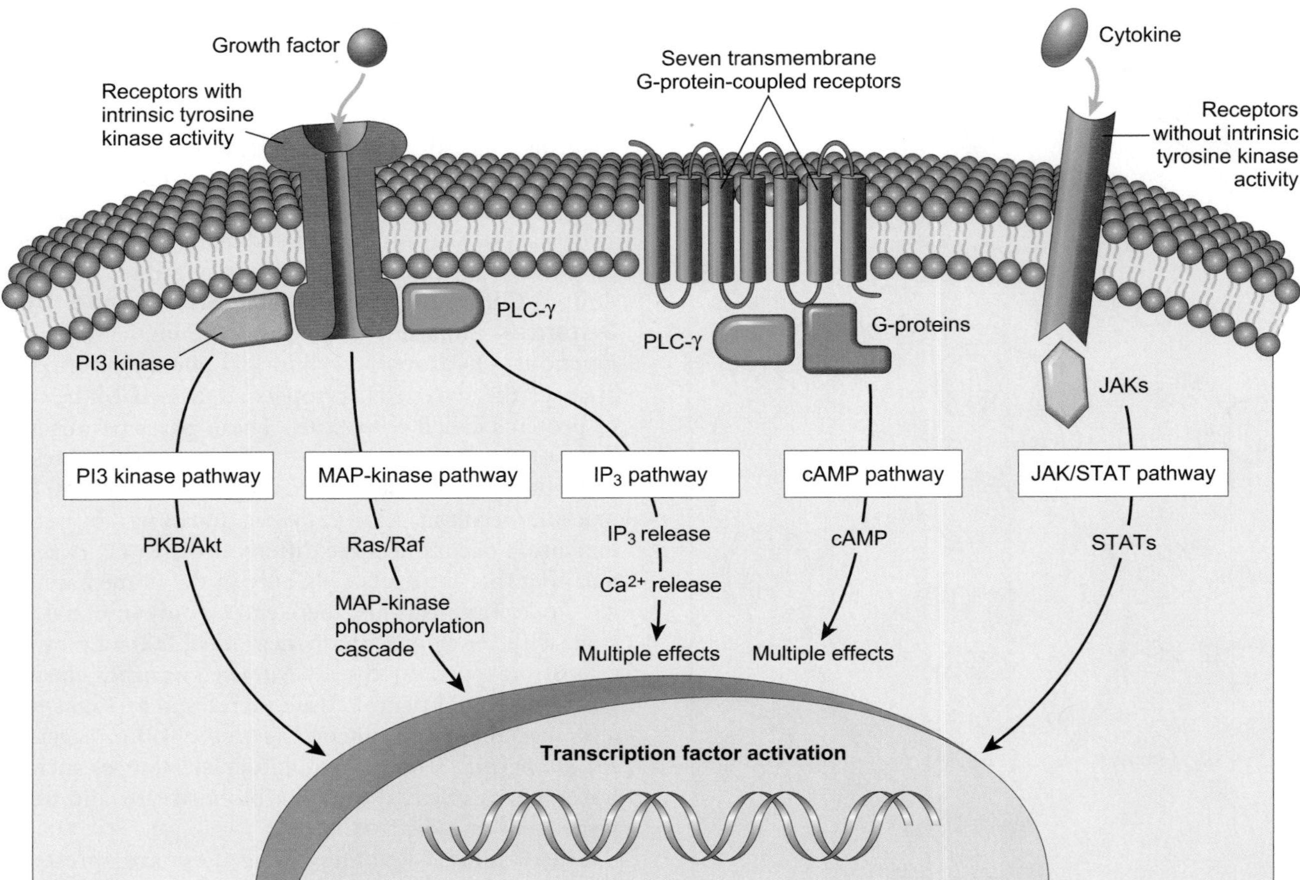

FIGURE II.1.4.5 Overview of intracellular and cell surface receptor signaling pathways. Binding of ligands to cell surface receptors can open ion channels, activate associated G proteins or activate associated enzymes, many of which have kinase activity (the ability to phosphorylate other protein substrates). Kinase activity may activate additional kinases that may in turn phosphorylate other proteins (e.g., the JAK-STAT pathway). Subsequent intracellular downstream mediators (e.g., calcium, IP_3, cAMP) or events (e.g., the translocation of a phosphorylated protein to the nucleus) eventually induce changes in the transcriptional activity of specific genes. Ligand binding to intracellular receptors (e.g., estrogen and estrogen receptor; not shown) leads to formation of complexes that can directly associate with nuclear DNA to activate or turn off transcription. PI3 kinase, phosphoinositide-3 kinase; PLC-γ, phospholipase C- γ; JAK, Janus kinase; MAP kinase, mitogen activated protein kinase; IP_3, inositol trisphosphate; cAMP, cyclic adenosine monophosphate; STAT, signal transducers and activators of transcription. (Reproduced by permission from Kumar, V., Abbas, A. K. & Fausto, N. (Eds.). (2005). *Robbins and Cotran Pathologic Basis of Disease,* 7th edn. Saunders: Philadelphia, PA.)

on the extracellular matrix will thereby elicit cytoskeletal and nuclear conformation changes that can alter genetic expression. A cell membrane complex of proteins associated with the integrins (the *focal adhesion complex*) can also transduce extracellular forces through the mechanical activation of cell membrane kinases and ion channels (Figure II.1.4.6).

An individual cell is exposed to a remarkable cacophony of signals which it must sort through and integrate into a rational response. Some ligands or forces induce a given cell type to differentiate, others signal proliferation, and yet others direct the cell to perform a specialized function. Multiple signals at once, in a certain ratio, may induce yet another totally unique response. Many cells require certain signals just to continue living; in the absence of appropriate exogenous ligand – or even simple connection to the extracellular matrix – they may undergo a form of cellular suicide (programmed cell death) called apoptosis (discussed in detail later in this chapter).

THE CYTOSKELETON: CELLULAR INTEGRITY AND MOVEMENT

The ability of cells to adopt a particular shape, maintain cell polarity, organize the relationship of intracellular organelles, and move depends on the intracellular scaffolding of proteins called the cytoskeleton. In eukaryotic cells there are three major classes of cytoskeletal proteins: 6–8 nm diameter actin microfilaments; 10 nm diameter intermediate filaments; and 25 nm diameter microtubules. Although these proteins can impart intracellular structure and cell shape (especially the intermediate filaments) it should also be emphasized that they are all dynamic. In particular, actin and microtubules enable movement and cellular locomotion.

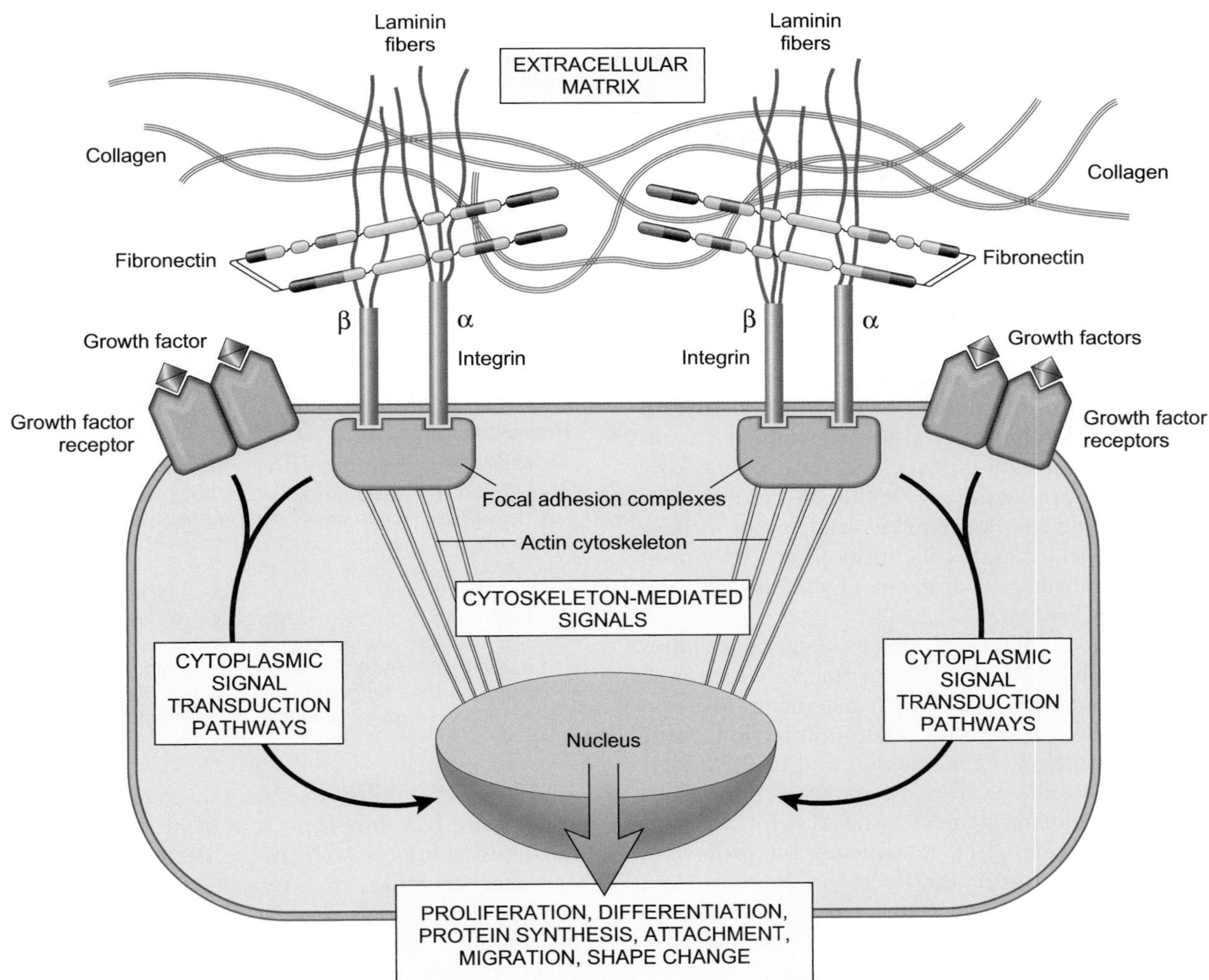

FIGURE II.1.4.6 Soluble and mechanotransduction pathways by which a variety of cell behaviors – e.g., proliferation, differentiation, protein synthesis, locomotion or shape change – can be modified. Note that soluble mediators can act through plasma membrane receptors (or intracellular receptors – not shown) to affect changes in nuclear transcription, as depicted in greater detail in Figure II.1.4.5. Extrinsic mechanical forces can also lead to changes in cellular phenotypes by protein–protein interactions linking matrix proteins (e.g., fibronectin, and laminin) with cell surface integrins that are in turn connected to the nucleus through cytoskeletal elements such as actin microfilaments. Integrins are also important components of focal adhesion complexes, which can provide soluble secondary signals (e.g., through protein phosphorylation) when external mechanical forces are applied to proteins connected to them. PDGF, platelet-derived growth factor; EGF, epidermal growth factor; IL-1/TNF, interleukin-1/tumor necrosis factor; TGF-β, transforming growth factor-β; TIMPs, tissue inhibitors of metalloproteinases (Reproduced by permission from Kumar, V., Abbas, A. K. & Fausto, N. (Eds.). (2005). *Robbins and Cotran Pathologic Basis of Disease*, 7th edn. Saunders: Philadelphia, PA.)

Actin microfilaments: the globular protein actin (G-actin) is the major subunit of microfilaments, and is the most abundant cytosolic protein in cells. The monomers polymerize into long double-stranded helical filaments (F-actin), with a defined polarity (one end is stable; the other end grows or shrinks). In muscle cells, the filamentous protein *myosin* binds to actin, and moves along it – driven by ATP hydrolysis – forming the basis of muscle contraction. In non-muscle cells, F-actin and an assortment of *actin-binding proteins* form coordinated bundles and networks that control cell shape and surface movements.

Intermediate filaments comprise a large and heterogeneous family of closely related structural proteins. Although they generally perform similar functions, each member of the family has a relatively restricted expression in specific cell types. These rope-like fibers are found predominantly in a stable polymerized form within cells; they are not usually actively reorganizing like actin and microtubules. Intermediate filaments impart strength and carry mechanical stress, e.g., in epithelia where they connect *spot desmosomes* (see below). They also form the major structural proteins of skin and hair (i.e., keratin).

Microtubules consist of polymerized dimers of α- and β-*tubulin* arrayed in constantly elongating or shrinking hollow tubes. These have a defined polarity with ends designated "+" or "–." In most cells, the "–" end is embedded in a microtubule organizing center (MTOC or centrosome) which lies near the nucleus; the "+" end elongates or recedes in response to various stimuli by the addition or subtraction of the tubulin dimers. Within cells, microtubules may serve as mooring lines for protein "molecular motors" that hydrolyze ATP to move vesicles, organelles or other molecules around cells; the polarity of the microtubules allows cells to direct whether attached structures are "coming" or "going" relative to the nucleus. In neurons, microtubules are critical for the delivery of molecules synthesized in the nuclear area to the far-flung axonal cytosol, which may be as far away as 10,000 times the width of the cell body (indeed, the nucleus of some motor neurons in the spinal cord have axons extending to the muscles of the great toe over 3 feet away!). Microtubules are also used to pull chromosome pairs apart during mitosis, and thus play a basic role in cellular proliferation. Finally, in certain cells, microtubules and their associated molecular motors have been harnessed to facilitate cellular motility – they form cilia to move mucus and dust out of the airways, and flagella to propel sperm.

Maintaining cellular and tissue integrity requires cell–cell and cell–ECM interactions, mediated through the proteins embedded in the plasma membrane and connected to the cytoskeleton. The organization of tissues requires attaching cells together and to the underlying ECM scaffolding. The mechanical connections between extracellular elements and the intracellular cytoskeleton – mediated by transmembrane proteins – also means that extracellular forces on a tissue (e.g., torsion or tension) can be translated into intracellular events. Indeed, we have learned that "normal" cell behavior can critically depend on external physical forces, such as laminar shear stress on endothelial cells that line blood vessels.

As discussed above, the external face of a cell membrane is diffusely studded with carbohydrate-modified (*glycosylated*) proteins and lipids. This cell coat (or *glycocalyx*) functions primarily as a chemical and mechanical barrier, but also serves an important role in cell–cell and cell–matrix interactions, including sperm–egg attachment, blood clotting, lymphocyte recirculation, and inflammatory responses. For example, neutrophils (cells involved in acute inflammation) are recruited to sites of infection or injury by localized endothelial cell expression of lectin-like molecules (*selectins*) that bind the sugar motifs expressed on the neutrophil surface glycoproteins.

Cell–cell connections include *occluding junctions* (*tight junctions*) and *anchoring junctions* (*desmosomes*) (Figure II.1.4.7) (gap junctions (described above) function primarily in cell-to-cell communication and do not materially contribute to cellular adhesiveness). Tight junctions seal cells together in a continuous sheet, preventing even small molecules (but not water) from leaking from one side to the other. These junctions are the basis of the high electrical resistance of many epithelia,

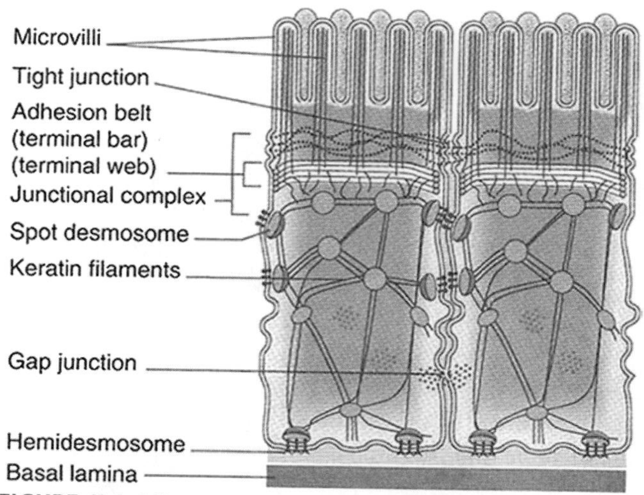

FIGURE II.1.4.7 Diagram of two prototypical epithelial cells (as part of a larger planar sheet) demonstrating attachments to the underlying basal lamina (*hemidesmosomes*), and to each other (*tight junction*, *adhesion belt*, and *spot desmosomes*). Note the cytoskeletal elements underlying each attachment point giving structural integrity to the individual cells as well as to the larger epithelial structure. *Gap junctions* do not confer intercellular adhesion, but are responsible for direct cell–cell signaling. (Reproduced by permission from Bergman, A. R., Afifi, A. K. & Heideger, P. M. (Eds.). (1996). *Histology*. Saunders: Philadelphia, PA.)

as well as the ability to segregate apical and basolateral spaces. It is important to remember, however, that these junctions (as well as the desmosomes) are also dynamic structures that intermittently dissociate and reform. This allows processes such as the healing of an epithelial wound or passage of inflammatory cells across vascular endothelium to sites of infection. Intracellular actin microfilaments can stretch across cells, connecting tight junctions on different faces of the same cell; such spanning creates a circumferential band (*adhesion belt*) of cytoskeleton that provides structural integrity and shear strength to sheets of interconnected cells (Figure II.1.4.7).

Desmosomes mechanically attach cells (and their cytoskeletons) to other cells or to the ECM. When they occur in broad belts or bands between cells they are referred to as *belt desmosomes*; when they are small and rivet-like they are denoted as *spot desmosomes*. Desmosomes are formed by a homotypic association (two proteins of the same type) of transmembrane glycoproteins called *cadherins*. The cytosolic ends of cadherins are associated with cytoskeletal actin microfilaments and intermediate filaments. *Focal adhesion complexes* or *hemidesmosomes* (literally, half a desmosome) are "spot welds" that connect cells to the extracellular matrix; in the case of epithelial tissues, the connections are to a dense ECM meshwork called the *basal lamina* or *basement membrane* (see Chapter II.1.5). The plasma membrane proteins forming the basis of these interactions are called *integrins*; like cadherins, they attach to intracellular actin microfilaments. These focal adhesion complexes, connecting cells to the ECM, also act to generate intracellular signals when cells are subjected to abnormal shear stress (such as endothelium in a turbulent area of the bloodstream).

THE NUCLEUS: CENTRAL CONTROL

With the exception of the terminally-differentiated hematopoietic cells (erythrocytes and platelets), every human cell has a central regulatory nucleus containing nucleic acids (DNA and RNA) and proteins that determine the sequence and rate of macromolecular synthesis. The full complement of DNA in a cell is called its *genome*.

The Nucleus is not a Uniform Static Repository of Molecules. At any given time, and depending on the functional state of the cell, there is *replication* (duplication of DNA) or *transcription* (conversion of DNA into messenger RNA) of selected subsets of the genome. Clearly, proliferating cells need to generate an entire copy of all nuclear material so that daughter cells can each be afforded a complete set of chromosomes. In mitotically-quiescent cells, however (those in G_0 of the cell cycle), there is no DNA replication; nuclear activity is essentially all devoted to gene transcription. While certain housekeeping functions must be constitutively active to maintain normal cellular viability, in differentiated cells only a small fraction of all possible genes is transcribed.

Nuclear Proteins and Nucleic Acids are Organized into Clumps and Clusters Called Chromatin. Two basic forms of chromatin are recognizable by light microscopy and routine staining correlating with the activity of the gene. When genes are transcriptionally inactive, they are tightly coiled in compact aggregates wrapped around protein *histones*, and are not accessible to transcription machinery; this results in a cytochemically dense appearance called *heterochromatin*. In actively transcribing genes, the nuclear material uncoils into a more extended linear form, which is cytochemically disperse and called *euchromatin*. The degree of cellular activation or gene transcription may thus be inferred from the general staining characteristics of nuclei (Figure II.1.4.8). The nucleolus is a subcompartment of the nucleus dedicated to ribosomal RNA synthesis and ribosome assembly. Its size may also reflect the translational activity of the cell; the greater the protein synthesis, the more prominent the nucleolus.

Movement of Molecules into and Out of the Nucleus is Restricted and Tightly Regulated. The nucleus is surrounded by a *nuclear envelope* formed by two concentric membranes supported by networks of intermediate filaments. The outer membrane is continuous with the endoplasmic reticulum, and is joined to the inner membrane at numerous *nuclear pores* that punctuate the envelope. The pores are elaborate, gated structures which permit the active transport of molecules to and from the cytosol. Since the pores are readily permeable

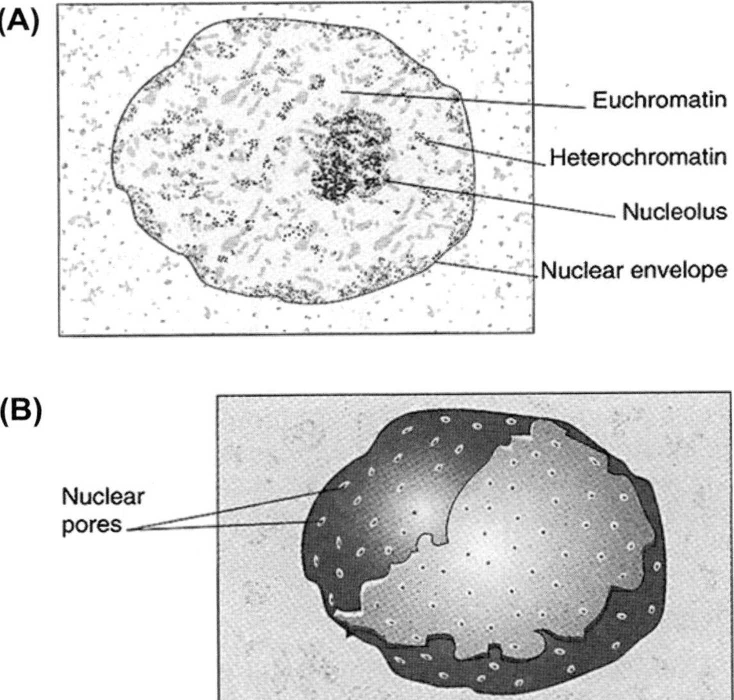

FIGURE II.1.4.8 Diagram of the nucleus and nuclear pore system. (A) Illustration of *heterochromatin* (clumped non-transcribing nuclear material), *euchromatin* (dispersed, transcribing nuclear material), and *nucleolus* (site of ribosome synthesis); (B) Schematic representation of nuclear pores; note that they are not passive openings in the nuclear envelope, but rather represent selective active transporters. (Reproduced by permission from Bergman, A. R., Afifi, A. K. & Heideger, P. M. (Eds.). (1996). *Histology*. Saunders: Philadelphia, PA.)

only to globular proteins ≤50,000 daltons, the process of moving large molecules and complexes (e.g., ribosomes out of the nucleus or histones and polymerase complexes into the nucleus) is accomplished by specific receptor proteins. Macromolecules destined for the nucleus are identified by specific nuclear localization signals (typically certain amino acid sequences) that permit binding to the pore receptor proteins. Of note, these localization signals may be cryptic, as in some inactive steroid receptor proteins. Subsequent binding of the steroid ligand causes a conformational change that uncovers the localization signal; the receptor can then be translocated to the nucleus.

ROUGH AND SMOOTH ENDOPLASMIC RETICULUM, AND GOLGI APPARATUS: BIOSYNTHETIC MACHINERY

The endoplasmic reticulum (ER) is the site for the synthesis of all the transmembrane proteins and lipids for most of a cell's organelles, including the ER itself. It is also the initial site for the synthesis of the majority of molecules destined for residence in the lumens of ER, Golgi, and lysosomes or for export out of the cell. The ER is organized into a mesh-like maze of interconnecting branching tubes and flattened vesicles (see Figure II.1.4.1) forming a continuous sheet around a single highly convoluted space topologically equivalent to the extracellular environment. The ER is composed of contiguous but distinct domains, distinguished by the presence (rough ER or RER) or absence (smooth ER or SER) of ribosomes.

Membrane-bound ribosomes on the cytosolic face of RER are actively translating mRNA into proteins, which are folded and edited in the lumen of the ER (Figure II.1.4.9). This process, called translation, is directed by amino acid signal peptides generally present on the N-termini of nascent proteins; if a new protein with a signal peptide is produced on a free ribosome, the signal peptide directs the entire complex to attach to the ER membrane. Proteins synthesized in this way are directly inserted into the ER as they are being made. Within the ER lumen, the proteins fold and form multi-subunit complexes under the scrutiny of ER chaperone molecules. These chaperones interact with a variety of proteins within the ER, ensuring that they are completely assembled and functional. Failure to appropriately fold results in retention and eventually degradation within the ER; thus the ER has an editing function for safeguarding the fidelity of the transcriptional apparatus. In some cases where unfolded proteins overwhelm the editing capacity, a so-called ER "stress response" (also called the unfolded protein response) triggers the affected cell to undergo apoptotic cell death.

For proteins lacking a signal sequence, the translation apparatus remains on free ribosomes in the cytosol, frequently forming *polyribosomes* as multiple translation units attach to the mRNA; the resulting protein remains within the cytoplasm and is not exported from the cell.

After leaving the RER, proteins and lipids are modified in the Golgi apparatus and sorted for intracellular delivery (Figure II.1.4.10). In the Golgi proteins are glycosylated – modified by a step-wise addition of various sugars. These modifications are important in subsequent sorting of molecules to various intracellular sites, and also because glycosylation is critical for surface molecules

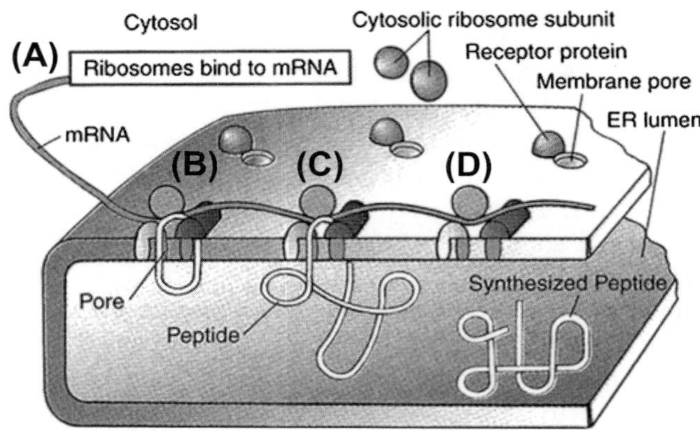

FIGURE II.1.4.9 Schematic demonstrating the general steps in the synthesis of proteins on *rough endoplasmic reticulum* (RER). (A) Peptide synthesis from mRNA begins on free ribosomes in the cytosol; (B) Signal peptide sequences on the nascent proteins direct the entire complex to attach to the ER membrane with insertion of the synthesizing protein into the RER lumen; (C) The signal sequence is cleaved and the protein is completely synthesized, eventually detaching from the ribosome; (D) The protein is folded and assembled (if necessary) into multi-subunit complexes; the ribosome detaches from the ER surface and returns to the cytoplasm. (Reproduced by permission from Bergman, A. R., Afifi, A. K. & Heideger, P. M. (Eds.). (1996). *Histology.* Saunders: Philadelphia, PA.)

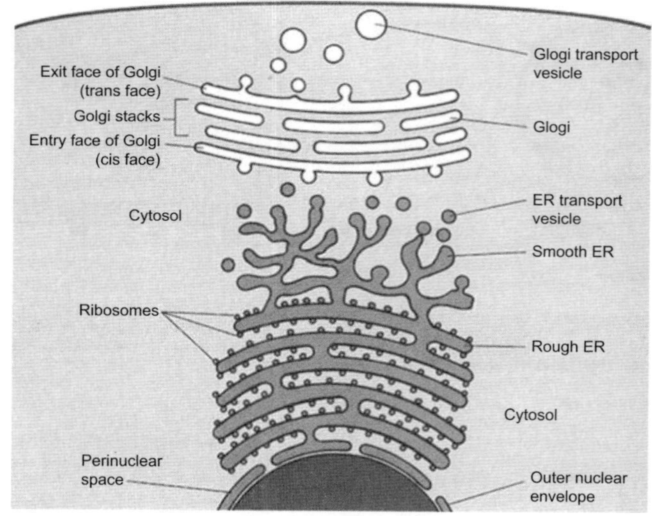

FIGURE II.1.4.10 Functional, schematic diagram of the relationship between the *rough endoplasmic reticulum* (RER), *smooth endoplasmic reticulum* (SER), and the *Golgi apparatus*. Proteins destined for export or for other intracellular organelles pass from the RER to the SER where they form vesicles for transport to the Golgi; there they are progressively modified and sorted, eventually leaving in transport vesicles for their appropriate final destination. (Reproduced by permission from Bergman, A. R., Afifi, A. K. & Heideger, P. M. (Eds.). (1996). *Histology.* Saunders: Philadelphia, PA.)

involved in cell–cell or cell–matrix interactions. In addition to the step-wise glycosylation of lipids and proteins, the trans-Golgi network sorts molecules for dispatch to other organelles (including the plasma membrane) or secretory vesicles destined for extracellular release. The Golgi complex is especially prominent in cells specialized for secretion, including goblet cells of the intestine or bronchial epithelium (making large amounts of polysaccharide-rich mucus), and plasma cells (secreting antibodies).

The SER has a Role in Steroid Hormone Synthesis, Modification of Certain Metabolites, and Intracellular Calcium Regulation. The *SER* in most cells is relatively sparse. However, in cells that synthesize steroid hormones (e.g., the adrenal cortex or gonads) or that catabolize lipid-soluble molecules (e.g., liver cells make certain drugs more water-soluble so that they may be excreted), the SER may be particularly conspicuous. The SER is also responsible for sequestering intracellular Ca++; release of Ca++ from the SER is a mechanism by which cells can rapidly respond to extracellular signals. Finally, in muscle cells, specialized SER called *sarcoplasmic reticulum* regulates the successive cycles of myofiber contraction (Ca++ released into the cytosol) and relaxation (Ca++ pumped back into the SER).

LYSOSOMES, PROTEASOMES AND PEROXISOMES: WASTE DISPOSAL

To digest internalized macromolecules or senescent organelles, cells primarily rely upon lysosomes. *Lysosomes* are membrane-bounded organelles containing a large assortment (>40) of acid hydrolase enzymes including proteases, nucleases, lipases, glycosidases, phosphatases, and sulfatases. Each have an optimal activity at pH 5, which is a protective feature since these enzymes will do less damage should they leak into the pH 7.2 cytosol. Materials destined for catabolism arrive in the lysosomes by one of three pathways:

(1) internalized by *fluid-phase* or *receptor-mediated endocytosis*. Material passes from plasma membrane to early endosome to late endosome, and ultimately into the lysosome (see Figure II.1.4.4).
(2) obsolete organelles within cells (the average mitochondrion, for example, only lives 10 days) are shuttled into lysosomes by a process called *autophagy*. The resultant *autophagosome* then fuses with lysosomes and the organelle is catabolized. Besides normal organellar turnover, this pathway may be important in pathologic settings leading to cellular and tissue atrophy (e.g., disuse atrophy of skeletal muscle); it has also been increasingly recognized as another mechanism of programmed cell death distinct from apoptosis.
(3) *phagocytosis* of microorganisms or large fragments of matrix or debris occurs primarily in professional phagocytes (*macrophages* or *neutrophils*). The material is engulfed to form a *phagosome* that subsequently fuses with a lysosome.

Proteasomes degrade cytosolic molecules that are senescent or require constant turnover to regulate their activity. The cytosol also needs a mechanism by which misfolded proteins can either be refolded or degraded, and to regulate the longevity of certain other proteins that turn over at discrete rates. Chaperones (also known as heat shock proteins) are intracellular proteins that are upregulated following cell stress, and have the capacity to refold partially denatured molecules (Figure II.1.4.11). Failing successful reconstitution of normal structure or in cases where a protein has outlived its usefulness, molecules are marked for destruction by covalently binding one or more 76-amino acid

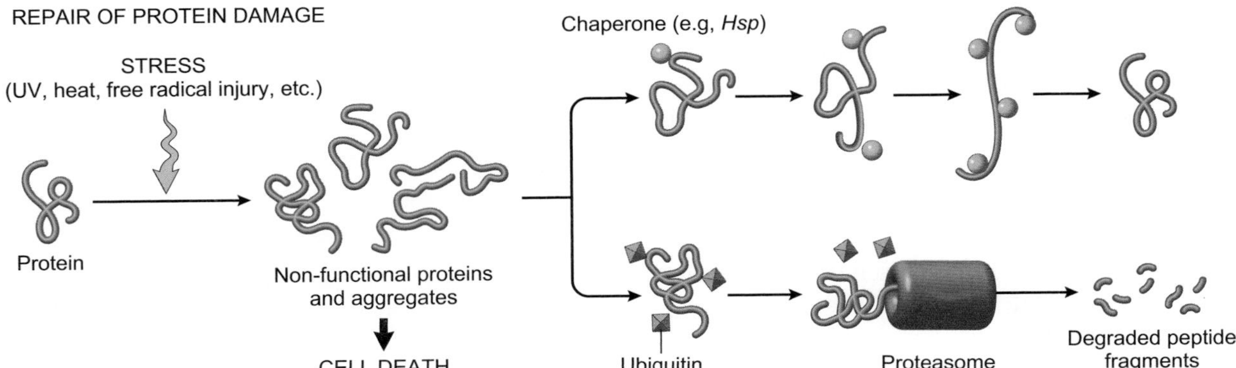

FIGURE II.1.4.11 Mechanism of intracellular protein repair and degradation. The normal activity of damaged and denatured proteins can be recovered through the action of chaperones (also called *heat shock proteins*). If the proteins cannot be successfully refolded, proteasome degradation will occur. This is also a mechanism by which cytoplasmic proteins that are no longer required (e.g., signaling molecules) can be regulated. Senescent or damaged cytoplasmic proteins are marked for destruction by the proteasome by ubiquitination. The proteasome is a cylinder composed of several different proteases, each with its active site pointed at the hollow core; proteins are degraded into small (6–12 amino acids) fragments. (Reproduced by permission from Kumar, V., Abbas, A. K. & Fausto, N. (Eds.). (2005). *Robbins and Cotran Pathologic Basis of Disease,* 7th edn. Saunders: Philadelphia, PA.)

proteins called ubiquitin. Ubiquitin-tagged molecules are then targeted to polymeric protein complexes called proteasomes; hundreds to thousands of proteasomes are present in any given cell. Each proteasome is a cylinder composed of several different proteases, each with its active site pointed at the hollow core; proteins are degraded into small (6–12 amino acids) fragments. This degradative mechanism is thought to be a holdover from prokaryotes (i.e., primitive cellular organisms) that lack lysosomes (Figure II.1.4.11).

Peroxisomes are probably the vestiges of ancient eukaryotic structures that catabolized oxygen when it would have been a highly toxic molecule, and not necessarily useful in generating energy. These organelles derive their name from the numerous enzymes they contain to remove hydrogen ions from various substrates and transfer them to oxygen, forming *peroxide* in the bargain. In addition, they contain large quantities of oxidative enzymes such as catalase (which breaks down hydrogen peroxide) and urate oxidase. A major function of peroxisomes in mammalian cells (shared by mitochondria) is the *β-oxidation* of fatty acids, whereby two-carbon blocks are successively removed, converted to acetyl CoA, and shuttled to the cytoplasm to undergo additional biosynthetic reactions. Peroxisomes are also the only organelle capable of catabolizing *very long chain fatty acids* (≥20 carbons). An interesting sidelight is that no ATP is generated by peroxisomes in the course of their fatty acid breakdown; the energy is released in the form of heat. Interestingly, these are self-replicating organelles; new peroxisomes can only derive from pre-existing peroxisomes. They self-assemble from selected proteins (marked by a serine-lysine-leucine triple amino acid motif) synthesized in the cytosol and imported through binding to specific peroxisome membrane receptors.

MITOCHONDRIA: ENERGY GENERATION

Energy to run intracellular processes is (mostly) provided by adenosine triphosphate (ATP) generated by mitochondria. ATP hydrolysis, the chemical reaction that removes a terminal phosphate from ATP to form adenosine diphosphate (ADP), is accompanied by the release of a large amount of energy. The energy derived from this hydrolysis is the essential currency used for such activities as ion and macromolecular transport across membranes, synthesis of molecules for cell housekeeping and for export, and specialized cell functions such as contraction of muscle.

ATP production begins with cytosolic glycolysis (e.g., metabolizing glucose to pyruvate or lactate), followed by the mitochondrial conversion of pyruvate to carbon dioxide and water through the Krebs cycle, with subsequent oxidative phosphorylation. These reactions – particularly oxidative phosphorylation (the process of generating ATP from substrate oxidation) – are critically dependent on the availability of oxygen. When oxygen is present, 38 ATP are generated from metabolism of one glucose molecule; in the absence of oxygen, only two molecules of ATP are generated. Although carbohydrate metabolism is the primary source of ATP (in the brain, glucose is essentially the *only* energy source), fat and protein breakdown can also contribute to ATP production.

Each mitochondrion has two separate and specialized membranes. There is a core *matrix space* (containing most of the enzymes to catabolize glucose and its primary metabolites) surrounded by an *inner membrane* (containing the enzymes to transfer electrons to oxygen) folded into *cristae*; these constitute the major working parts of the organelle. The inner membrane is enclosed by the *intermembrane space* (site of ADP-to-ATP phosphorylation) which is in turn surrounded by the *outer membrane*; the latter is studded with a transport protein called *porin*, which forms aqueous channels permeable to low molecular weight substrates (Figure II.1.4.12). Interestingly, mitochondria are also central to the pathways leading to *apoptosis* (see later).

Mitochondria probably evolved from ancestral prokaryotes engulfed by primitive eukaryotes about 1–2 billion years ago. This explains why mitochondria contain their own DNA (circularized, about 1% of the total cellular DNA) encoding for approximately one-fifth of the proteins involved in oxidative phosphorylation. Although the mitochondrial DNA codes for only a very small number of proteins, mitochondria have the machinery necessary to carry out all the steps of DNA replication, transcription, and translation. Consistent with its evolutionary origin, mitochondrial translational machinery is similar to present-day bacteria. For example, protein synthesis is initiated with the same modified amino acid

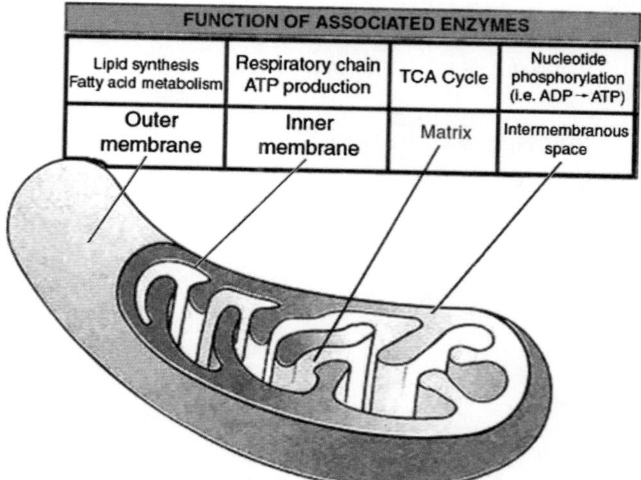

FIGURE II.1.4.12 Schematic diagram of the mitochondrion demonstrating the functional segregation of the enzymatic machinery required to generate ATP. TCA: tricarboxylic acid. (Reproduced by permission from Bergman, A. R., Afifi, A. K. & Heideger, P. M. (Eds.). (1996). *Histology*. Saunders: Philadelphia, PA.)

as bacteria (N-formyl methionine), and is sensitive to the same antibiotics. It is noteworthy that – like peroxisomes – new mitochondria can only derive from pre-existing mitochondria. Thus, since only the ovum contributes the cytoplasmic organelles in the fertilized zygote, mitochondrial DNA is *maternally inherited*.

CELL SPECIALIZATION AND DIFFERENTIATION

As discussed above, basic functional attributes of cells include protection, signaling, nutrient absorption and assimilation, energy generation, macromolecule synthesis, growth, and reproduction. Without these basic activities, cells cannot live. However, most cells also exhibit specialization – that is, they have additional capabilities, such as irritability, conductivity, absorption or secretion of molecules. Multicellular organisms are thus composed of individual cells with marked specialization of structure and function. These differentiated cells allow a division of labor in the performance and coordination of complex functions carried out in architecturally distinct and organized tissues.

Determining a Cellular Phenotype

In many cases, the nature of well-differentiated tissues can be readily identified by routine histologic staining – that is putting thin sections on a glass slide and dipping them in *hematoxylin* (which stains negatively charged molecules such as DNA blue) and *eosin* (a counter-stain that stains virtually everything else pink), so-called *H&E* stain. Thus, skeletal muscle can be easily distinguished from liver; and brain from fat (just as examples). Histologists have also developed a variety of other stains to highlight various tissue components (e.g., calcium, elastin, collagen, etc.) that can further help to identify tissue differentiation. However, immature or poorly differentiated tissues – and especially individual cells – can be extremely difficult to identify by observation alone. The following techniques take advantage of some characteristic feature of targeted cells and tissues, most commonly a specific protein or mRNA made only by a specific cell to establish the phenotype; occasionally, a functional characteristic (e.g., excitability of a neuron or secretion of insulin by β-islet cells) can also be used to determine cellular identity. Note that techniques can typically distinguish the *location* of a particular cell or tissue within an organ or can *quantify* how much of a particular product is being made, but *cannot do both.*

- *Immunofluorescence or immunohistochemistry*: antibodies to a specific cell surface or intracellular molecule are applied to tissue slices or to individual cells. Presence of bound antibody is assessed by fluorescent or enzymatic tags on the antibodies. This demonstrates which cell in a tissue (or even which tissue) is making a particular constituent, but is generally not quantitative.
- *Western blot or enzyme-linked immunosorbent assay* (ELISA): tissues are ground up or otherwise solubilized by detergent extraction or other means. The protein constituents are separated by electrophoretic techniques on the basis of molecular weight and/or charge (Western blot) or are simply incubated in the presence of antibodies adsorbed onto a plastic plate (ELISA). In the Western blot, the presence of specific molecules is detected by the binding of antibodies to the separated protein bands, as assessed by enzymatic or radioactive tags on the antibodies. In the ELISA, protein bound to the plate-adsorbed antibodies is detected by a second set of antibodies linked to an enzyme. These techniques are quantitative, but will not typically identify which cell out of a mixture is making the molecule of interest.
- *Polymerase chain reaction* (PCR): messenger RNA is extracted from a cell population or tissue of interest and specific sequences representing the transcripts that are translated into a particular protein are subjected to enzymatic amplification; amplified fragments are detected by electrophoretic techniques that separate the mRNA based on size. The presence of a selected mRNA band is therefore taken as evidence of a particular cell phenotype. Although the technique is typically quite sensitive and can be quantitative for the mRNA, it does not necessarily equate to the quantity of a protein produced. Also, note that the technique will not determine which cell in a mixture is making the mRNA of interest.
- In-situ *hybridization*: complementary DNA sequences (labeled with fluorescent or other tags) that bind to a particular mRNA are applied to tissue sections. Binding is detected by fluorescence or enzymatic assays. This technique allows identification of which cell in a tissue is making a particular message, but is generally not quantitative.
- *Gene array technology*: mRNA from cells or tissues of interest is converted to complementary DNA (cDNA) fragments, typically using fluorescently-tagged nucleotides of a particular color (e.g., green or red). These are then bound to a panel of different DNA transcripts, where each sequence in the panel – representing a particular gene – is adsorbed to a discrete location on a plastic grid or "chip" (thousands of genes can be represented on each chip). Binding of a cDNA is detected by fluorescence imaging, and the *pattern* of mRNA expression (i.e., multiple mRNA species can be discerned) by the original cell or tissue can be inferred. A particular strength of the technique is that such labeled cDNAs can also be prepared from cells or tissues treated with different reagents, and then compared on the same chip. For example, a stem cell population can be treated with one kind of drug (drug A) with the cDNAs prepared from it being labeled with green nucleotides. The same stem cell population can also be treated with a different drug (drug B) with the cDNAs being labeled with red nucleotides. Equal amounts of the red and green cDNAs are then put on the same chip and the relative binding of red versus green cDNA can be quantitated. If a particular gene on the chip glows red it indicates that drug B induced that message; if green, then drug A induces it. If both drugs induce a message comparably, the result is yellow fluorescence, and if neither induces a particular mRNA, the result is black.

Differentiated cells have developed well-defined structural and/or functional characteristics associated with increasing specialization. For example, striated muscle cells have well-organized actin and myosin filaments that slide over one another, facilitating cellular contraction. Gastric (stomach) epithelial cells have large numbers of mitochondria to generate the ATP necessary to pump hydrogen ions out of the cell against a concentration gradient and acidify the stomach contents. Skin keratinocytes function as a protective barrier by undergoing autophagic programmed cell death and becoming scale-like structures filled with durable, non-living keratin (an intermediate filament). Specialized phagocytic cells of the immune system must detect infectious microorganisms

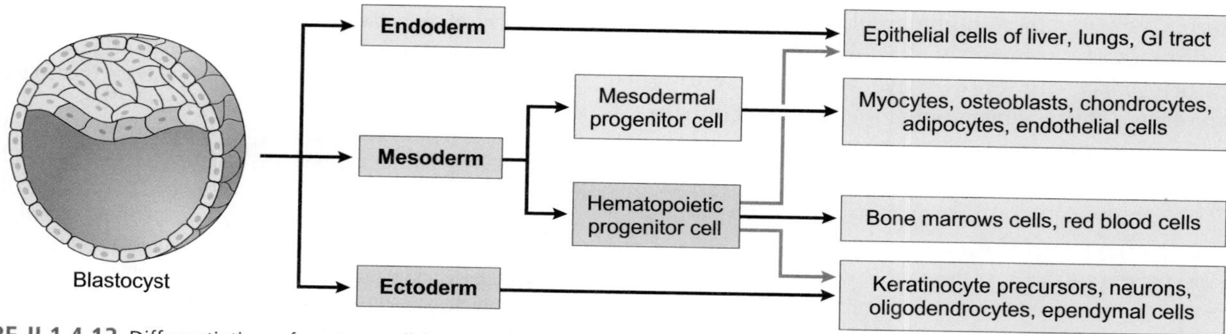

FIGURE II.1.4.13 Differentiation of mature cell lineages from the totipotential blastocyst. Cells in the early embryo have the capacity to proliferate and differentiate into all adult tissues. As the embryo acquires more cells, distinct layers form from the original sphere (endoderm, mesoderm, and ectoderm – based on general location of the cells within the developing fetus) that will become three general pluripotential lineages with more limited cellular differentiation repertoires. Eventually, after multiple rounds of replication and additional differentiation, the final adult tissues develop. Adult stem cells derived from the original three lineages typically persist within the organs they formed, with the capacity to regenerate cells of that tissue. Moreover, adult hematopoietic progenitors from the original mesoderm have the relatively unique capacity to generate tissues of the endoderm and ectoderm in addition to making marrow-derived blood cells. (Reproduced by permission from Kumar, V., Abbas, A. K. & Fausto, N. (Eds.). (2005). *Robbins and Cotran Pathologic Basis of Disease*, 7th edn. Saunders: Philadelphia, PA.)

(e.g., bacteria, parasites, and viruses), actively migrate to them, and then ingest and destroy them. *Polymorphonuclear leukocytes* (also called *PMNs* or *neutrophils*) are particularly active against bacteria, and *macrophages* react to other types of organisms and foreign material. B-lymphocytes are not phagocytic, but contribute to immunity by producing antibodies that can bind and neutralize infectious agents.

All specialized cells derive originally from the totipotential cells that comprise the early blastocyst. As the fetus grows and develops, subsets of these totipotential cells differentiate and assume specific functions that will eventually lead to the formation of mature tissues, e.g., bone, muscle, liver, skin, and brain (Figure II.1.4.13). To some extent, even in fully developed organisms, there persist populations of multipotential cells variably able to repopulate adult tissues where there is an ongoing balance between cellular proliferation, programmed cell death (*apoptosis*), and differentiation into end-stage cells with special "talents" but no further ability to reproduce (Figure II.1.4.14). This next section will briefly describe the concepts of stem cells and cellular differentiation.

Cellular Differentiation Involves an Alteration in Gene Expression. Every cell in the body has the same complement of genes (called the *genotype*). With progressive differentiation, selected subsets of genes are preferentially expressed, yielding a distinct biological profile (called the *phenotype*). As cells progressively specialize, more and more of the "unnecessary" genes in the differentiating cell are turned off (usually irreversibly). Some genes are active at all times (*constitutively expressed*); others may be selectively activated or *modulated* depending on external influences (e.g., injury). Thus, from a pluripotent cell, differential activation of specific genes can engender the development of endothelial cells, skeletal muscle, fat, bone or cartilage (Figure II.1.4.15).

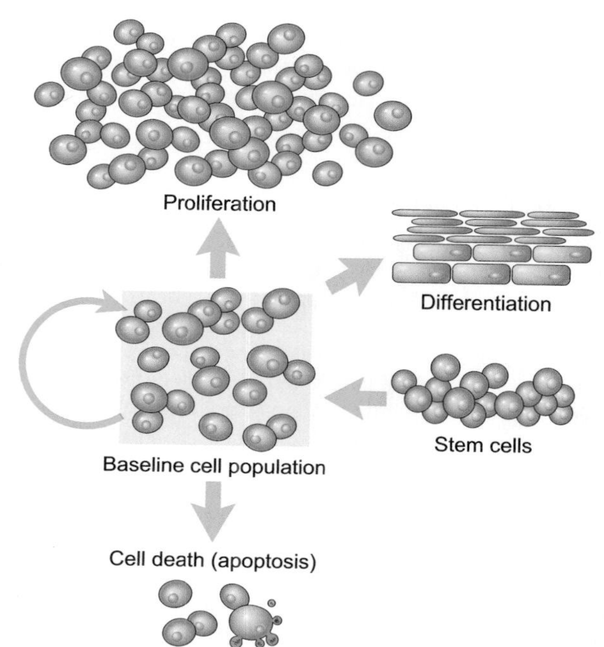

FIGURE II.1.4.14 Achieving steady-state populations of cells. Total cell numbers in a baseline tissue population can be regulated by changes in stem cell input, loss of cells through programmed cell death (apoptosis), cellular proliferation or differentiation. (Reproduced by permission from Kumar, V., Abbas, A. K. & Fausto, N. (Eds.). (2005). *Robbins and Cotran Pathologic Basis of Disease*, 7th edn. Saunders: Philadelphia, PA.)

The structural and functional changes that occur during cellular differentiation are usually irreversible. Moreover, increasing specialization results in a loss of cell potentiality, as well as a loss in the capacity for cell division. For example, the newly fertilized ovum is absolutely undifferentiated and has the capacity to divide extensively, ultimately giving rise to progeny that comprise all the cells of the body. Cells with the capacity to both divide and yield differentiated cells of one or more types are called stem cells. "Stemness" can be

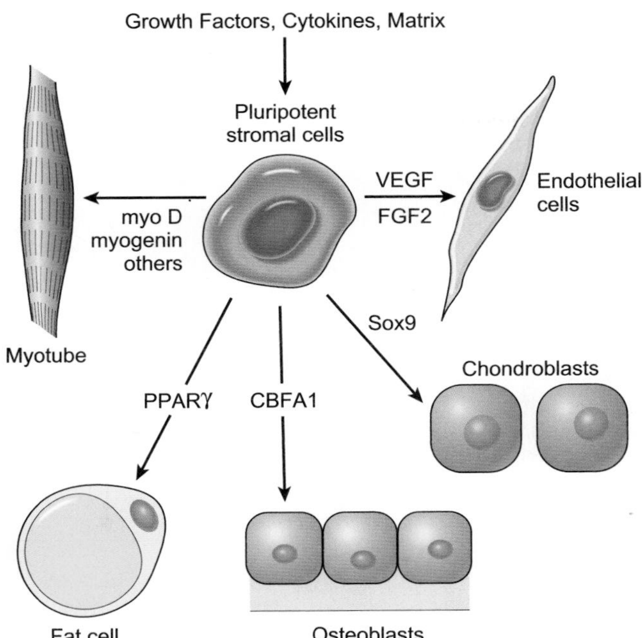

FIGURE II.1.4.15 Differentiation of pluripotent mesodermal progenitor cells. The same stromal cell can differentiate down different developmental pathways depending on the activation of specific genes; this can occur through exogenous soluble factors or cell–matrix interactions. Thus, vascular endothelial growth factor (VEGF) and fibroblast growth factor-2 (FGF2) are extrinsic signals that induce endothelial differentiation. Sox9, CBFA1, PPARγ, and myo D are intracellular transcription factors whose activities will induce the specific lineages shown. (Reproduced by permission from Kumar, V., Abbas, A. K. & Fausto, N. (Eds.). (2005). *Robbins and Cotran Pathologic Basis of Disease,* 7th edn. Saunders: Philadelphia, PA.)

functionally understood as the capacity for asymmetric division: one progeny cell maintains totipotentiality (it can still potentially become any cell type) while the other daughter cell transcribes selected subsets of genes to generate more differentiated tissues.

The stem cells in early embryos (found only rarely in adult cell populations) are *totipotential* and have virtually limitless replicative capacity. As cells differentiate along particular specialization or tissue pathways, they lose the ability to interconvert and develop into all cell types, although they may be able to form most or all the cells of a particular tissue. Thus, gastrointestinal stem cells in the small bowel can differentiate into absorptive epithelial cells, mucus-producing epithelial cells, and Paneth crypt cells that produce bacteriocidal proteins; however, they cannot differentiate into smooth muscle cells or even other types of epithelium. Stem cells in these lineages are said to be *pluri-* or *multipotential*; they may also have slightly limited replicative capacity. With further terminal differentiation, cells typically lose the ability to replicate at all. In the example of the small bowel, the terminally-differentiated absorptive cells actually divide no further and undergo programmed cell death within 3–4 days of their genesis. Classically, cardiac muscle cells and neurons have also been considered as terminally-differentiated cells without replicative capacity. Obviously, this has clinical significance when these very important terminally-differentiated cells are injured (e.g., by heart attack or stroke). Nevertheless, there is increasing evidence that there may be small populations of cardiac and neuronal stem cells that we can tap into to regenerate these tissues; moreover, we are increasingly adept at isolating and cultivating totipotential stem cells – and coaxing them to differentiate into desirable cell lineages. Finally, there is also accumulating evidence to suggest that cells of end-stage, highly differentiated tissues can, under very limited conditions (e.g., some forms of injury), de-differentiate into multipotent stem cells.

CELL REGENERATION AND PROLIFERATION

At a tissue level, the ability to regenerate is a function of the level of differentiation of the stem cells within that tissue and their overall replicative potential; these fall into three broad general categories: labile; stable; and permanent tissues.

Labile tissues have pluripotent stem cells that are constantly in cell cycle, and thus are constantly generating new cells that will differentiate into the final population comprising the parenchyma of that particular organ. As those final differentiated cells senesce, they typically undergo programmed cell death (apoptosis, discussed later); nevertheless, they are continuously being replaced by the "next generation," and thus the integrity of the tissue is maintained. These tissues include many epithelia such as the stratified squamous surfaces of the skin, oral cavity, vagina, and cervix; the cuboidal epithelia of the ducts draining exocrine organs (e.g., salivary glands, pancreas, biliary tract); the columnar epithelium of the gastrointestinal tract, uterus, and fallopian tubes; and the transitional epithelium of the urinary tract. The hematopoietic cells of the bone marrow are also continuously replenishing the circulating peripheral blood cell elements. As a corollary of their ongoing proliferation, labile tissues typically repopulate sites of injury relatively quickly. However, they are also most sensitive to extrinsic insults such as radiation or chemotherapy for tumors – which is why patients being treated for cancer lose their hair (hair follicle epithelium), malabsorb food (gastrointestinal epithelium), and are anemic and prone to infections and bleeding (bone marrow).

Stable tissues have pluripotent stem cells that typically rest in the G_0 stage of the cell cycle and are not actively turning over. However, these cells can be kick-started into cell cycle, and have substantial regenerative capacity. They include the parenchymal cells of organs such as liver, kidney, and pancreas, as well as endothelial cells, and mesenchymal cells such as fibroblasts and smooth muscle.

Permanent tissues include neurons and cardiac myocytes. Classically, these cells cannot replicate, and irreversible injury results only in scarring. Skeletal muscle is also generally placed in this category, although this tissue does have a limited regenerative capacity, largely due to the proliferation and transformation of a small population of stem cells associated with endomysial sheaths. Moreover, we are also learning that terminally differentiated permanent cells – previously thought to be irreplaceable once destroyed (e.g., nerve and muscle) – may be potentially repopulated by bone marrow-derived stem cells or even small populations of tissue-endogenous stem cells. While exciting, the extent to which we can utilize this replacement process appears to be very limited; for all practical purposes (for now), dead permanent cells become scar.

These concepts have important ramifications not only for embryogenesis, but also for the budding tissue engineer. For applications that require large numbers of differentiated cells, expansion of terminally-differentiated cell populations may work for some cell types (e.g., skin). However, cells such as heart or nerve will probably require directed differentiation from expanded stem cells.

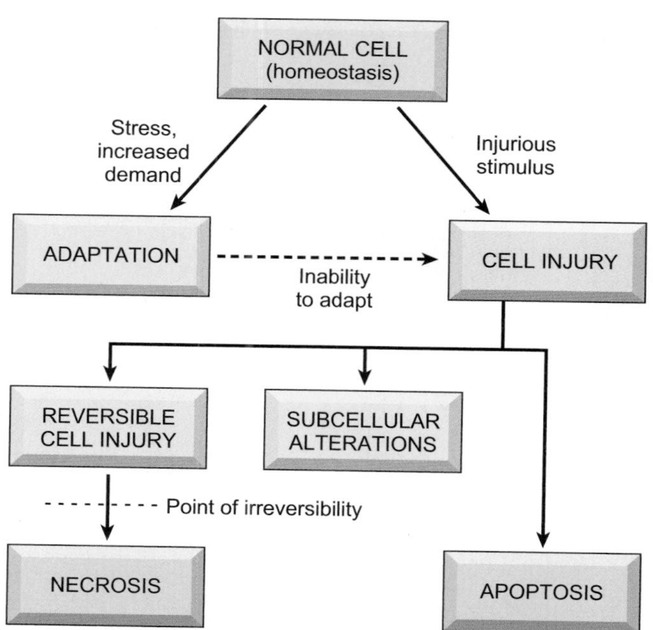

FIGURE II.1.4.16 General overview of potential cellular responses to stressors and injurious stimuli. (Reproduced by permission from Kumar, V., Abbas, A. K., Fausto, N. & Aster, J. C. (Eds.). (2010). *Robbins and Cotran Pathologic Basis of Disease,* 8th edn. Saunders: Philadelphia, PA.)

CELL INJURY AND REGENERATION

Cells constantly adjust their structure and function to accommodate alterations in their environment, particularly responding to chemical and mechanical stressors; inability to successfully adapt results in cell injury and ultimately cell death (Figure II.1.4.16).

Cells attempt to maintain their intracellular milieu within relatively narrow physiologic parameters; that is, they maintain *normal homeostasis*. For relatively small changes in the environment, subtle modifications of cell behavior will likely permit normal "business as usual." However, as cells and tissues encounter more challenging physiologic stresses or pathologic stimuli, they undergo more extensive changes (*adaptation*) designed to achieve a new steady-state, but in all cases to *preserve viability*. The principal significant adaptive responses are:

hypertrophy: an increase in size of individual cells;
hyperplasia: an increase in cell number;
atrophy: a decrease in size, without appreciable change in cell number;
metaplasia: transformation from one mature cell type to another.

There may also be more subtle changes in the expression of selected genes that are functionally beneficial but are not necessarily reflected in alterations of morphology.

Usually, if the extracellular stressors recede, the cells and tissues will revert to their prestressed state. However, if the stressors persist and a cell's adaptive capability is exceeded, *cell injury* develops. Up to a point, cell injury itself is *reversible*, and with normalization of the stimulus the cell returns to its baseline state, usually no worse for wear. However, with severe or persistent stress, the cell suffers *irreversible* injury and dies.

For example, when heart muscle cells are subjected to persistent increased load (e.g., high blood pressure), the cells adapt by undergoing *hypertrophy* (enlargement of the individual myocytes, and eventually the entire heart) to compensate for the higher pressures they must pump against. Conversely, in periods of prolonged starvation (as can happen in prolonged illness or with malignant tumors), all myocytes (and thus the heart) will undergo *atrophy*. The same myocytes, subjected to an imbalance between blood supply and energy demand due to an occluded coronary artery (*ischemia*), may be *reversibly injured* if the occlusion is incomplete or sufficiently brief; alternatively, they may undergo *irreversible injury* (i.e., cell death, also called *necrosis*, as in *myocardial infarction*) following complete or prolonged occlusion (Figure II.1.4.17).

Beyond adaptive responses, environmental stimuli can also induce genetic changes which can trigger abnormal proliferation and differentiation. Such behavior is uncoordinated relative to normal tissues, has lost its responsiveness to normal growth controls, and persists after cessation of the stimuli that initiated it. This condition is called "*neoplasia*;" in its malignant form, it is more commonly called *cancer*.

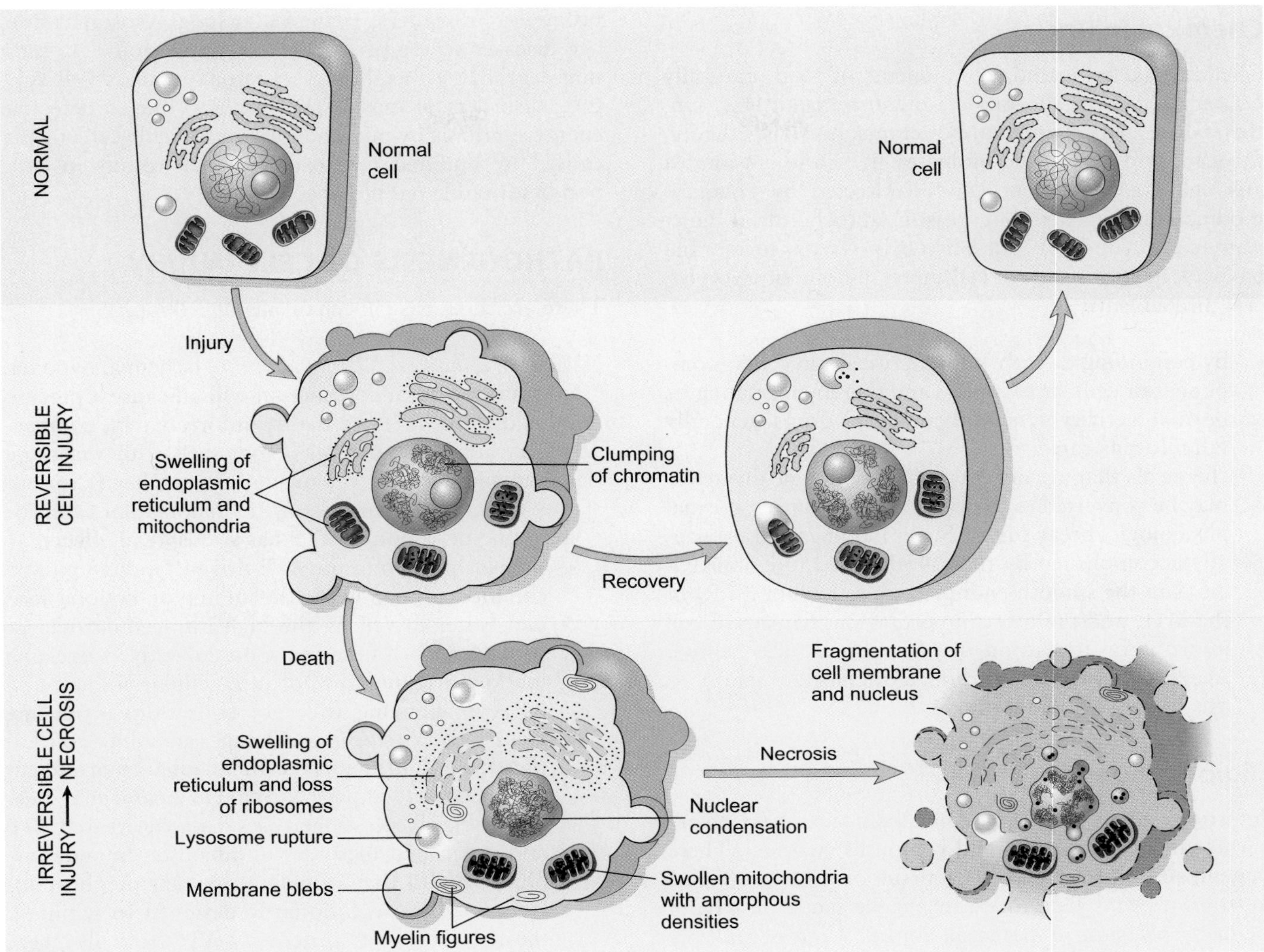

FIGURE II.1.4.17 Schematic of the cellular responses and morphologic changes to reversible and irreversible injury, the latter eventually culminating in necrotic cell death. Reversible injury is characterized by a generalized cell and organellar swelling due to a loss of ATP-driven mechanisms to maintain normal ionic gradients; there will also be plasma membrane blebbing (as cytoskeletal elements degrade), ribosome detachment from the ER with diminished protein synthesis, and nuclear chromatin clumping with reduced transcription. As the cells transition to irreversible injury, the changes become more severe, and eventually the cell cannot recover normal function – even after oxygen and glucose are restored to generate ATP. In that case, necrosis supervenes with lysosomal rupture, swollen mitochondria with amorphous densities within them, membrane fragmentation, and condensation of the nucleus (*pyknosis*), fragmentation of the nucleus (*karyorrhexis*), and dissolution of the nucleus (*karyolysis*). (Reproduced by permission from Kumar, V., Abbas, A. K. & Fausto, N. (Eds.). (2005). *Robbins and Cotran Pathologic Basis of Disease,* 7th edn. Saunders: Philadelphia, PA.)

CAUSES OF CELL INJURY

Hypoxia and Ischemia

Recall that the efficient production of ATP by mitochondria – and therefore the energy needed to run all cell activities – is critically dependent on oxygen. *Hypoxia* is decreased O_2 supply relative to the needs of a particular cell or tissue. *Anoxia* is the complete absence of oxygen. Tissue hypoxia can occur in any setting where there is reduced oxygen tension including causes as diverse as high altitude, anemia, and carbon monoxide poisoning (the latter diminishes the oxygen carrying capacity of hemoglobin). However, far and away the most common cause of tissue hypoxia is diminished blood flow, called *ischemia*; irreversible tissue injury (*necrosis*) due to ischemia is called *infarction*. It bears repeating that although diminished blood flow will invariably lead to hypoxia, oxygen deficiency can occur in the setting of adequate tissue perfusion. It is also important to note that diminished blood flow will also affect the ability to remove waste products from a metabolizing tissue so that, all things considered, *ischemia is far worse than simple hypoxia.*

Chemical Injury

Chemical agents include components of food, naturally occurring toxins, hormones, neurotransmitters, synthetic drugs, environmental pollutants, poisons, ethanol, tobacco, and even toxic biomaterials. Although almost any cellular constituent can be affected by chemical modification, agents that poison mitochondrial function (e.g., cyandide) will obviously wreak substantial havoc. Chemicals induce cell injury by one of two general mechanisms:

- by combining directly with a critical molecular component or cellular organelle and thereby inhibiting its normal activity (chemotherapeutic drugs generally fall into this category).
- chemicals that are not intrinsically biologically active may be converted to toxic metabolites during normal physiologic breakdown. Such modification is usually accomplished by the P-450 mixed function oxidases in the smooth endoplasmic reticulum (SER) of the liver, and the most important mechanism of cell injury is by formation of *free radicals* (see below). Acetaminophen and carbon tetrachloride belong to this category of compounds.

Biologic Agents

Infectious agents run the gamut from virus and bacteria to fungi, protozoans, and helminths (worms). There is generally a preferred cell or tissue of invasion (called a *tropism*), and therefore each agent tends to have a defined spectrum of potential injury. *Viruses* multiply intracellularly, appropriating host biosynthetic machinery in the process; cell lysis may occur directly or as a result of the immune system's recognition and destruction of infected cells. More insidiously, viruses may compromise the ability of a cell or tissue to perform its normal functions; worse still, viruses may play a role in transformation to malignant neoplasms. *Bacteria* have toxic cell wall constituents (e.g., endotoxin), and can release a variety of exotoxins. Moreover, the very process of eradicating infections by the host immune system can also cause injury – and in many cases constitutes the bulk of the pathology.

Physical Injury

Injury can result by direct mechanical force (trauma, pressure), temperature extremes (burn, frostbite), electric shock or ionizing radiation.

Genetic Defects

Mutations in a variety of cellular proteins can lead to cellular dysfunction and eventually irreparable cellular injury. Congenital defects generally manifest as progressive disorders; examples include: lysosomal storage diseases where progressive accumulation of certain non-degradable metabolites eventually causes cell rupture; disorders of muscle (*myopathies*) due to defective energy synthesis by mitochondria; and sickle cell anemia caused by mutated hemoglobin which results in stiff, non-deformable red blood cells.

PATHOGENESIS OF CELL INJURY

There are three basic mechanisms of cell injury:

(1) *Loss of adequate ATP production.* Ischemia, hypoxia, or mitochondrial dysfunction will all cause a precipitous drop in ATP synthesis; unfortunately, compensatory anaerobic pathways that suffice for yeast and other single-celled organisms cannot begin to satisfy the energy demands of the mammalian cell. The resulting depletion of ATP has widespread effects:
   - the plasma membrane ATP-driven "sodium pump" cannot keep up with the influx of counter-ions and water driven by the high intracellular charge density and osmolarity. Consequently, there is a marked accumulation of intracellular sodium and its accompanying water of hydration, producing *acute cellular swelling*. Changes in solute concentrations and in the intracellular ionic environment will negatively impact normal enzymatic activities.
   - anaerobic glycolysis increases due to decreased ATP and associated increases in adenosine monophosphate (AMP) that stimulate the enzyme phosphofructokinase. Evolutionarily designed to maintain the cell's energy by generating ATP from glycogen, activation of this pathway leads to rapid *depletion of glycogen stores*. Increased glycolysis also results in accumulation of lactic acid and inorganic phosphates from hydrolysis of phosphate esters, and lowers the intracellular pH, in addition to exacerbating the already increased osmotic load.
   - dropping pH and ATP levels cause ribosomes to detach from the rough endoplasmic reticulum (RER) and polysomes to dissociate into monosomes, with a resultant *reduction in protein synthesis*.
   - if ATP generation is not restored the cytoskeleton disperses, resulting in loss of ultrastructural features such as microvilli, and the formation of cell surface "blebs." Mitochondria, endoplasmic reticulum, and indeed whole cells usually appear swollen owing to loss of osmotic regulation. If ATP synthesis is recovered (usually by restoration of oxygen), all the above disturbances are reversible; if not, irreversible injury (cell death) follows (Figure II.1.4.17).

(2) *Oxygen and oxygen-derived free radicals.* Lack of oxygen (and loss of ATP production) obviously underlies the pathogenesis of ischemic cell injury. In

addition, *partially-reduced, activated oxygen* species are important mediators of cell death. *Free radicals* are chemical species with a single unpaired electron in an outer orbital; they are extremely unstable and readily react with inorganic or organic chemicals. The most important free radicals in biological systems are oxygen-derived and include: *hydroxyl* (OH•) radicals (from the hydrolysis of water, e.g., by ionizing radiation); *superoxide radicals* (O_2–•); and nitric oxide radicals (NO•). Free radicals initiate autocatalytic reactions; molecules that react with free radicals are in turn converted into free radicals, further propagating the chain of damage. When generated in cells, free radicals cause single-strand breaks in DNA, fragment lipids in membranes via lipid peroxidation, and fragment or cross-link proteins, leading to accelerated degradation or loss of enzymatic activity. Free radical damage is a pathogenic mechanism in such varied processes as chemical and radiation injury, oxygen and other gaseous toxicity, cellular aging, microbial killing by phagocytic cells, inflammatory damage, and tumor destruction by macrophages, among others.

It is important to note that besides being a consequence of chemical and radiation injury, free radical generation is also a normal part of respiration and other routine cellular activities, including microbial defense. It therefore makes sense that cells and tissues have developed mechanisms to degrade free radicals and thereby minimize any injury. Fortunately, free radicals are inherently unstable, and generally spontaneously decay; superoxide, for example, rapidly breaks down in the presence of water into oxygen and hydrogen peroxide. The rate of such decay is significantly increased by the action of superoxide dismutases (SOD) found in many cell types. Other enzymes, e.g., glutathione (GHS) peroxidase, also protect against injury by catalyzing free radical breakdown, and catalase directs the degradation of hydrogen peroxide. In addition, endogenous or exogenous antioxidants, e.g., vitamin E, may either block free radical formation or scavenge them once they have formed.

(3) *Failure of intracellular calcium homeostatic mechanisms.* Cytosolic free calcium is normally maintained by ATP-dependent calcium transporters at extremely low concentrations (less than 0.1 μM), while extracellular calcium and sequestered mitochondrial and endoplasmic reticulum calcium stores are typically at roughly 10^4-fold higher concentrations. Consequently, any cell ischemia or injury potentially looses a flood-gate of calcium. Increased cytosolic calcium in turn activates a variety of *phospholipases* (promoting membrane damage), *proteases* (catabolizing structural and membrane proteins), *ATPases* (accelerating ATP depletion), and *endonucleases* (fragmenting genetic material).

Ischemia-Reperfusion Injury

If cells are reversibly injured in ischemic circumstances, restoration of blood flow should theoretically result in cell recovery. However, in many cases, restoration of blood flow to ischemic but otherwise viable tissues results in paradoxically exacerbated and accelerated injury. As a result, tissues experience additional cell loss and dysfunction in excess of that which occurs in just the irreversibly damaged areas. This so-called *ischemia/ reperfusion injury* is a clinically important process that significantly contributes to tissue damage in myocardial and cerebral infarctions. Although the exact mechanisms are unclear, reperfusion into ischemic tissues may cause further damage by the following:

- Restoration of blood flow bathes compromised cells in high concentrations of calcium when they are not able to fully regulate their ionic environment; increased intracellular calcium activates the pathways described above and causes loss of cellular integrity.
- Restoration of blood flow into an area that is already irreversibly injured results in activation of innate immunity pathways (including complement) and a locally augmented inflammatory cell recruitment. These cells release high levels of potentially deleterious oxygen-derived reactive species and cytokine inflammatory mediators.
- Damaged mitochondria yield incomplete oxygen reduction and therefore increased production of free radical species; in addition, ischemically-injured cells have compromised antioxidant defense mechanisms.

RESPONSES TO CELL INJURY

Whether a specific form of stress induces adaptation or causes reversible or irreversible injury depends not only on the nature and severity of the stress, but also on several other cell-specific variables including vulnerability, differentiation, blood supply, nutrition, and previous state of the cell.

- *Cellular response to injurious stimuli depends on the type of injury, its duration, and its severity.* Thus, low doses of toxins or a brief period of ischemia may only lead to reversible cell injury, whereas larger toxin doses or longer ischemic intervals may result in irreversible injury and cell death.
- *Consequences of an injurious stimulus are dependent on the type of cell being injured*, its *current status* (nutritional, hormonal, etc.), and its *adaptability*. For example, striated skeletal muscle in the leg can tolerate complete ischemia for 2–3 hours without suffering irreversible injury, while cardiac muscle will die after only 20–30 minutes, and central nervous system (CNS) neurons are dead after 2–3 minutes. Similarly, a well-nourished liver (with reserves of intracellular glycogen) can withstand an ischemic or anaerobic

challenge far better than a liver without any potential energy stockpile.

REVERSIBLE VERSUS IRREVERSIBLE INJURY

Regardless of the nature of a particular cellular insult, four cell systems are particularly vulnerable: (1) cell membrane integrity; (2) ATP generation, largely via mitochondrial aerobic respiration; (3) protein synthesis; and (4) the integrity of the genetic apparatus.

Within limits, cells can compensate for disturbance of any of these, and if the injurious stimulus abates will return to normalcy. Persistent or excessive injury, however, causes cells to pass the threshold into *irreversible injury*. In other words, they die; pathologically, the cells become *necrotic* or are said to undergo *necrosis* (Figure II.1.4.17).

An important concept on the spectrum from normal to reversible injury, and thence onto irreversible injury, relates to cellular and tissue function. The specialized activities relating to cell function may shut down relatively early in the setting of injury, and in most cases long before there is any cell death. For example, myocardial cells become non-contractile after 1–2 minutes of ischemia, although they will not die until 20–30 minutes of ischemia have elapsed. Consequently, if injury can be reversed before cells become irretrievably damaged, then restoration of function can be anticipated. Of course, lack of cardiac function may lead to death of other important structures (e.g., the brain) even before the heart cells themselves reach the point of no return.

NECROSIS

Two phenomena consistently characterize irreversible injury. The first is the *inability to reverse mitochondrial dysfunction* (lack of oxidative phosphorylation and ATP generation) even upon restoration of oxygen; the second is the development of *profound disturbances in membrane function*. Massive calcium influx into the cell occurs, particularly if ischemic tissue is reperfused after the point of irreversible injury, with broad activation of calcium-dependent catabolic enzymes. In addition, proteins, essential coenzymes, and ribonucleic acids seep out through the newly permeable membranes, and the cells also lose metabolites vital for the reconstitution of ATP. Injury to the lysosomal membranes results in leakage of their enzymes into the cytoplasm; the catabolic enzymes are activated in the reduced intracellular pH of the ischemic cell, and will further degrade cytoplasmic and nuclear components.

Since necrotic tissue is a potential nidus for secondary infections, the body rapidly mobilizes inflammatory cells to clean up the resulting debris, and initiates the process of either rebuilding the dead tissue or laying down a fibrous scar (see Chapter II.1.5). As mentioned previously, the recruited inflammation can in itself be a cause of further local injury. Moreover, having to replace a specialized tissue with a non-functional matrix scar is also a less than optimal outcome.

APOPTOSIS

The cell death that has been thus far described is the consequence of irreversible injury; in a way, this can be thought of as cellular "homicide." However, there is also an important form of controlled or *programmed* cell death that can be conceptualized as cellular "suicide." *Apoptosis* (from root words meaning "a falling away from") is a distinctive and important mode of cell death that must be differentiated from standard *necrosis* (Figure II.1.4.18 and Table II.1.4.1). Apoptosis is responsible for programmed cell death in several important physiologic (as well as pathologic) processes, including:

- the programmed destruction of cells during embryogenesis, including implantation, organogenesis, and developmental involution.
- hormone-dependent physiologic involution, such as the endometrium during the menstrual cycle or the lactating breast after weaning, or pathologic atrophy, as in the prostate after castration.
- cell deletion in proliferating populations such as intestinal crypt epithelium or cell death in tumors.
- deletion of autoreactive T cells in the thymus, cell death of cytokine-starved lymphocytes or cell death induced by cytotoxic T cells.
- a variety of mild injurious stimuli (heat, radiation, cytotoxic cancer drugs, etc.) that cause irreparable DNA damage that in turn triggers cell suicide pathways (e.g., via the tumor-suppressor protein *p53*).

Indeed, failure of cells to undergo physiologic apoptosis may result in unimpeded tumor proliferation, autoimmune diseases or aberrant development.

Apoptosis usually involves single cells or clusters of cells with condensed nuclear chromatin or chromatin fragments. The cells rapidly shrink, form *cytoplasmic buds*, and fragment into *apoptotic bodies* composed of membrane-bound vesicles of cytosol and organelles (see Figure II.1.4.18). These fragments are quickly extruded, and then phagocytosed and degraded by neighboring cells, and *do not elicit an inflammatory response*. The nuclear changes are due to fragmentation of DNA into histone-sized pieces through the activation of endonucleases.

The mechanisms underlying apoptosis are roughly schematized in Figure II.1.4.19. The basic mechanisms comprise four separable but overlapping components:

(1) *Signaling pathways*. The process may be triggered by a variety of signals ranging from an intrinsic

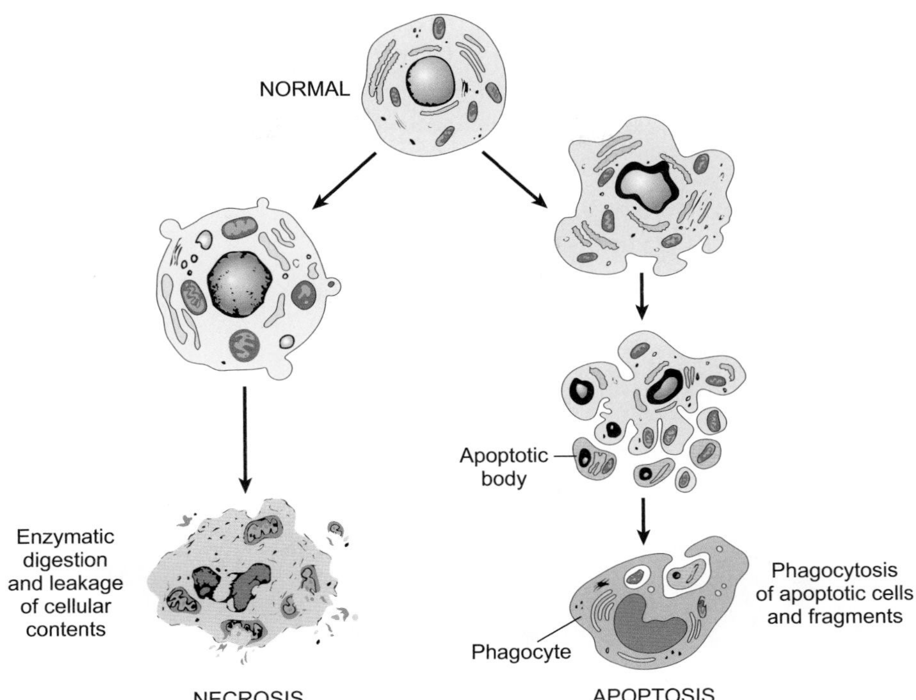

FIGURE II.1.4.18 The cellular structural changes of *necrosis* (left) and *apoptosis* (right). In necrosis, there is chromatin clumping, organelle swelling, and eventual membrane damage; dead cells are typically degraded and digested by recruited inflammatory cells. In apoptosis, there is nuclear chromatin condensation and fragmentation, followed by cytoplasmic budding of *apoptotic bodies*, which are phagocytized by adjacent cells. (Reproduced by permission from Kumar, V., Abbas, A. K. & Fausto, N. (Eds.). (2005). *Robbins and Cotran Pathologic Basis of Disease,* 7th edn. Saunders: Philadelphia, PA.)

TABLE II.1.4.1 Cellular Necrosis *Versus* Apoptosis

Feature	Necrosis	Apoptosis
Cell size	Enlarged (swelling)	Reduced (shrinkage)
Nucleus	Pyknosis, karyorrhexis, karyolysis	Fragmentation into nucleosome-sized fragments
Plasma membrane	Disrupted	Intact but altered, especially lipid orientation
Cellular contents	Enzymatic digestion; may leak out of cell	Intact; released in apoptotic bodies
Enzymatic mediator	Calpains	Caspases
Inflammation	Frequent	No
Physiologic or Pathologic role	Pathologic	*Pathologic* after some forms of injury *Physiologic* means of eliminating unwanted cells

Adapted by permission from Kumar, V., Abbas, A. K. & Fausto, N. (Eds.). (2005). *Robbins and Cotran Pathologic Basis of Disease*, 7th edn. Saunders: Philadelphia, PA.

programmed event (e.g., in development), a lack of growth factor, specific receptor–ligand interactions, release of granzymes from cytotoxic T cells or selected injurious agents (e.g., radiation). Transmembrane signals may either suppress pre-existing death programs (and are thus survival stimuli) or may initiate a death cascade. The most important in this latter group are those that belong to the tumor necrosis factor receptor (TNFR) superfamily of plasma membrane molecules (includes the FAS surface molecule). These plasma membrane receptors share an intracellular "death domain" protein sequence which, when oligomerized (typically trimerized) leads to a cascade of enzyme activation culminating in cell death.

(2) *Control and integration.* This is accomplished by specific proteins that connect the original death signals to the final execution program. These proteins are important because their actions may result in either "commitment" or abortion of potentially lethal signals. There are two broad schemes for this stage:

- direct transmission of death signals *adapter proteins* to the execution mechanism
- members of the *Bcl-2 family of proteins* that *regulate mitochondrial permeability.*

Various intracellular agonists (Ca^{+2}, free radicals, etc.) can affect mitochondria by causing *mitochondrial permeability transitions*. These are pores within the inner mitochondrial membrane that

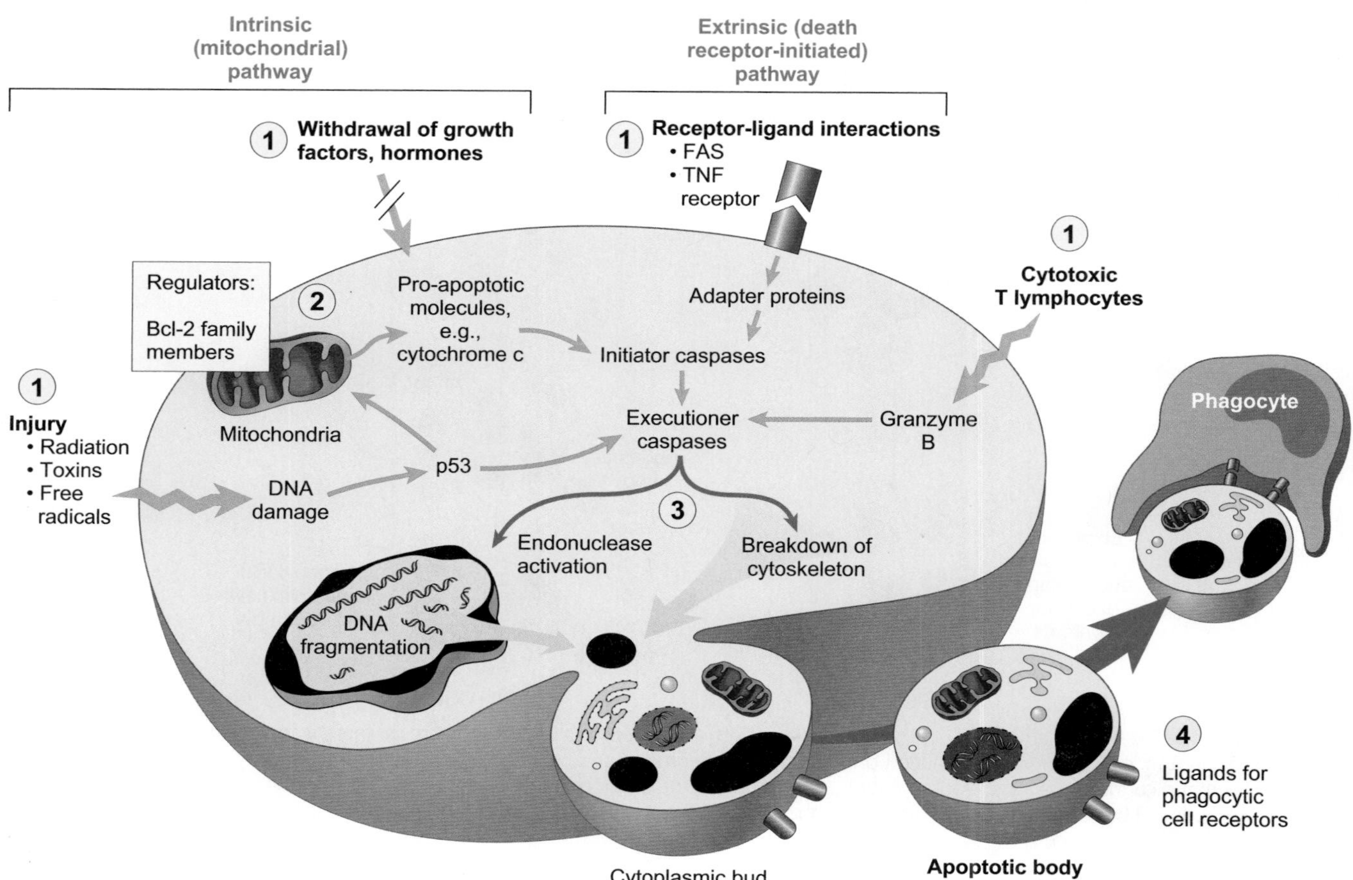

FIGURE II.1.4.19 Schematic of events occurring in apoptosis. Items labeled (1) are various stimuli for apoptosis; some involve direct activation of caspases (cytotoxic T cells) while others act via adaptor proteins (e.g., surface receptors such as FAS or the tumor necrosis factor (TNF) receptor) or via mitochondrial release of cytochrome c. Items labeled (2) are inhibitors (e.g., Bcl-2) or promoters that modulate whether or not mitochondria will develop membrane permeability transitions to release cytochrome c; cytochrome c interacts with additional cytoplasmic proteins (e.g., *Apaf-1*) to induce caspase activation. Executioner caspases (3) activate endonucleases that degrade nuclear chromatin, and intracellular proteases that degrade the cytoskeleton. The end result (4) is extruded apoptotic bodies containing various organelles and cytosolic components that express surface phosphatidylserine residues which induce their uptake by adjacent phagocytic cells. (Reproduced by permission from Kumar, V., Abbas, A. K. & Fausto, N. (Eds.). (2005). *Robbins and Cotran Pathologic Basis of Disease,* 7th edn. Saunders: Philadelphia, PA.)

result in reduced membrane potential and diminished ATP production; increased permeability of outer mitochondrial membranes also releases an apoptotic trigger, *cytochrome c*, into the cytosol. Released cytochrome c binds certain cytosolic proteins (e.g., *pro-apoptotic protease activating factor* or *Apaf-1*), and activates them, triggering in turn caspase activation (see below) and setting in motion proteolytic events that will eventually allow the cell to kill itself. *Bcl-2* (found in the mitochondrial membrane) suppresses apoptosis by preventing increased mitochondrial permeability, and by stabilizing proteins to prevent caspase activation. Other members of the Bcl-2 family bind to Bcl-2 and modulate its antiapoptotic effect; thus Bcl-XL inhibits apoptosis while Bax and Bad promote programmed cell death.

(3) *Execution.* The final pathway of apoptosis, resulting from the synthesis and activation of a number of cytosolic catabolic enzymes, and culminating in the morphologic changes described above. Although there are subtle variations, the final execution pathways exhibit common themes generally applicable to all forms of apoptosis:

- *Protein cleavage* by a class of proteases named *caspases*, so-called because they have an active site cysteine, and cleave after aspartic acid residues. Activation of one or more such caspase enzymes putatively leads to a cascade of activation of other proteases, inexorably culminating in cell suicide. Downstream *endonuclease* activation results in the characteristic DNA fragmentation of apoptosis, while cell volume and shape changes may in part be attributable to cleavage of components of the cytoskeleton.
- *Extensive protein cross-linking* via *transglutaminase* activation converts soluble cytoplasmic proteins, and particularly cytoskeletal proteins, into a covalently-linked condensed shell that readily fragments into apoptotic bodies.

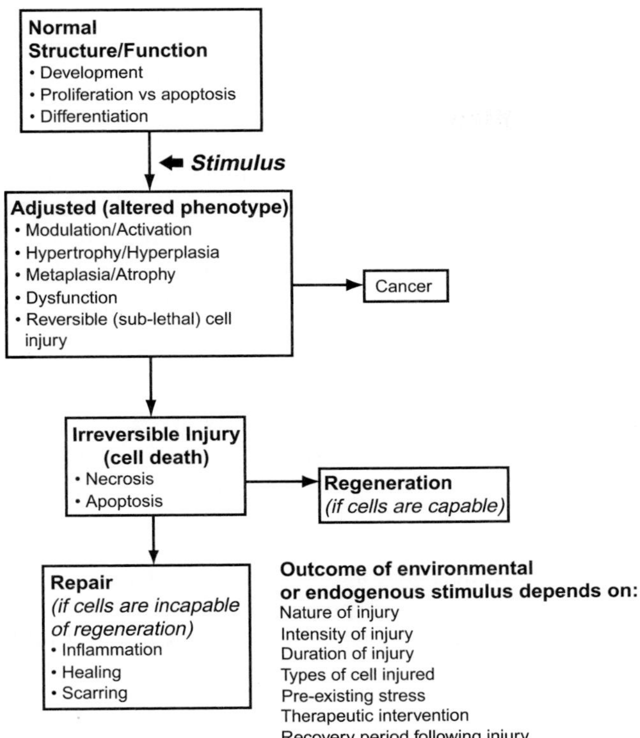

FIGURE II.1.4.20 Cellular mechanisms of disease, emphasizing the general concepts of activation and other phenotypic alterations, reversible and irreversible cell injury, and the various possible outcomes of cell injury.

- *DNA breakdown* into 180–200 base pair fragments (the distance between nucleosomes) occurs through the action of Ca^{+2} and Mg^{+2} dependent endonucleases. This may be visualized as a distinctive "laddering" of DNA into discrete sized pieces on agarose gel electrophoresis; this pattern is different from the random DNA fragmentation (forming a "smear" on agarose gels) typically seen in necrotic cells.

(4) *Removal of dead cells.* Apoptotic cells and their fragments have marker molecules on their surfaces that signal uptake and disposal by adjacent cells or phagocytes. This occurs by the flipping of phosphatidylserine from the inner cytoplasmic face of the apoptotic cells to the extracellular face. This (and other) alterations allow the early recognition and phagocytosis of apoptotic cells without release of pro-inflammatory mediators. The process is so efficient that dead cells typically disappear without leaving a trace, and inflammation is virtually absent.

The general framework of cell injury is summarized in Figure II.1.4.20. In the following chapter, we will extend the concepts of structure–function correlation beyond cells to include the extracellular matrix and complex tissues, examine what happens following cell and tissue injury, and describe how histologists and pathologists examine normal and abnormal tissues.

BIBLIOGRAPHY

Cell Biology:

Bergman, A. R., Afifi, A. K., & Heideger, P. M. (Eds.). (1996). *Histology*. Philadelphia, PA: Saunders.

Cooper, G. M., & Hausman, R. E. (2007). *The Cell: A Molecular Approach* (4th ed.). Sunderland, MA: Sinauer Associates, Inc.

Lodish, H., Berk, A., Kaiser, C. A., Krieger, M., Scott, M. P., et al. (2007). *Molecular Cell Biology* (6th ed.). New York, NY: W.H. Freeman and Co.

Stem Cells:

Kørbling, M., & Estrov, Z. (2003). Adult stem cells for tissue repair – a new therapeutic concept? *N. Engl. J. Med.*, **349**, 570–582.

Murry, C. E., & Keller, G. (2008). Differentiation of embryonic stem cells to clinically relevant populations: Lessons from embryonic development. *Cell*, **132**, 661–680.

Rossi, D. J., Jamieson, C. H., & Weissman, I. L. (2008). Stems cells and the pathways to aging and cancer. *Cell*, **132**, 681–696.

Cell Injury:

Anaya-Prado, R., Toledo-Pereyra, L. H., Lentsch, A. B., & Ward, P. A. (2002). Ischemia/reperfusion injury. *J. Surg. Res.*, **105**, 248–258.

Danial, N. N., & Korsmeyer, S. J. (2004). Cell death: Critical control points. *Cell*, **116**, 205–219.

Dong, Z., Saikumar, P., Weinberg, J. M., & Venkatachalam, M. A. (2006). Calcium in cell injury and death. *Annu. Rev. Pathol.*, **1**, 405–434.

Giorgi, C., Romagnoli, A., Pinton, P., & Rizzuto, R. (2008). Ca^{2+} signaling, mitochondria and cell death. *Curr. Mol. Med.*, **8**, 119–130.

Kroemer, G., Galluzzi, L., & Brenner, C. (2007). Mitochondrial membrane permeabilization in cell death. *Physiol. Rev.*, **87**, 99–163.

Kumar, V., Abbas, A. K., & Fausto, N. (Eds.). (2005). *Robbins and Cotran Pathologic Basis of Disease* (7th ed.). Philadelphia, PA: Elsevier.

Levine, B., & Kroemer, G. (2008). Autophagy in the pathogenesis of disease. *Cell*, **132**, 27–42.

Lin, J. H., Walter, P., & Yen, T. S. (2008). Endoplasmic reticulum stress in disease pathogenesis. *Annu. Rev. Pathol.*, **3**, 399–425.

Ryter, S. W., Kim, H. P., Hoetzel, A., Park, J. W., Nakahira, K., et al. (2007). Mechanisms of cell death in oxidative stress. *Antioxid. Redox. Signal.*, **9**, 49–89.

Szabo, C. (2005). Mechanisms of cell necrosis. *Crit. Care. Med.*, **33**, S530–S534.

Tsujimoto, Y., & Shimizu, S. (2007). Role of the mitochondrial membrane permeability transition in cell death. *Apoptosis*, **12**, 835–840.

Valko, M., Leibfritz, D., Moncol, J., Cronin, M. T., Mazur, M., et al. (2007). Free radicals and antioxidants in normal physiological functions and human disease. *Int. J. Biochem. Cell. Biol.*, **39**, 44–84.

Vanlangenakker, N., Berghe, T. V., Krysko, D. V., Festjens, N., & Vandenabeele, P. (2008). Molecular mechanisms and pathophysiology of necrotic cell death. *Curr. Mol. Med.*, **8**, 207–220.

CHAPTER II.1.5 TISSUES, THE EXTRACELLULAR MATRIX, AND CELL–BIOMATERIAL INTERACTIONS

Frederick J. Schoen and Richard N. Mitchell
Department of Pathology, Brigham and Women's Hospital and Harvard Medical School, Boston, MA, USA

This chapter will extend the concepts discussed previously in Chapter II.1.4 (Cells, Cell Function and Cell Injury) to describe how cells are organized to form specialized tissues and organs, how environmental stimuli affect tissue structure and function, how tissues respond to various insults, including those related to the insertion of biomaterials, and how cells interact with extracellular matrix and biomaterials. Four general areas will be covered:

1. Principles governing the structure and function of normal tissue, organs, and systems;
2. Processes leading to and resulting from abnormal (injured or diseased) tissues and organs;
3. Concepts in cell–biomaterials interactions;
4. Approaches to study the structure and function of tissues.

Germane to this discussion are two basic definitions: *histology* is the microscopic study of tissue structure; *pathology* is the study of the molecular, biochemical, and structural alterations and their consequences in diseased tissues and organs, and the underlying mechanisms that cause these changes. The mainstay of both histology and pathology is microscopic examination of tissues, aided by stains and other methods that enhance tissue contrast, indicate specific chemical and molecular moieties present, and convey structural and functional information.

Cells and *extracellular matrix* comprise the structural elements of the tissues of the body. The structure of cells, tissues, and organs comprise the morphologic expression of the complex (and dynamic) activities that comprise body function. The underlying theme is that *structure is adapted to perform specific functions (conversely, changes in function may alter structure)*. Some excellent general references on tissue biology and pathology are available (Rubin and Strayer 2012; Kumar et al., 2010).

Cells and *extracellular matrix* are the structural elements of the tissues of the body. The structure of cells, tissues, and organs comprise the morphologic expression of the complex (and dynamic) activities that comprise body function. The underlying theme is that *structure is adapted to perform specific functions (conversely, changes in function may change and alter structure)*.

STRUCTURE AND FUNCTION OF NORMAL TISSUES

Cells, the living component of the body (as discussed in detail in the Chapter II.1.4), are surrounded by and obtain their nutrients and oxygen, and discharge wastes via blood, tissue fluid (also known as extracellular fluid), and lymph. *Blood* consists of blood cells suspended in a slightly viscous fluid called *plasma*; *serum* results when the coagulation proteins are removed from plasma, Capillaries exude a clear watery liquid called tissue fluid that permeates the amorphous intercellular substances lying between capillaries and cells. More tissue fluid is produced than can be absorbed back into the capillaries; the excess is carried away as lymph by a series of vessels called lymphatics, which ultimately empty the lymph into the bloodstream.

In all tissues and organs, cells are assembled during embryonic development into coherent groupings by virtue of specific cell–cell and cell–matrix interactions (Ingber, 2010a). Each type of tissue has a distinctive and genetically-determined pattern of structural organization adapted to its particular function; each pattern is strongly influenced by both metabolic (Carmeliet, 2003) and/or mechanical factors (Wozniak and Chen, 2009). *Homeostasis*, a normal process by which cells adjust their function to adapt to minor changes in day-to-day physiological demands, is regulated by:

1. Genetic programs of metabolism, differentiation, and specialization;
2. Influences of neighboring cells, the extracellular matrix, and signals that trigger specific cellular responses;
3. Environmental conditions, such as mechanical forces, temperature, and ionic content;
4. Availability of oxygen and metabolic substrates.

When normal limits of homeostasis are exceeded, pathological changes may result.

The Need for Tissue Perfusion

Since all mammalian cells require perfusion (i.e., blood flow bringing oxygen and nutrients, and carrying away wastes) for their survival, most tissues have a rich vascular network. Thus, the circulatory (also known as the cardiovascular) system is a key feature of tissue and organ structure and function (Figure II.1.5.1). Perfusion (i.e., delivery of blood) to a tissue or an organ is provided by the cardiovascular system, composed of a pump (the *heart*), a series of distributing and collecting tubes (*arteries* and *veins*), and an extensive system of thin walled vessels (*capillaries*) that permit exchange of substances between the blood and the tissues. Circulation of blood transports and distributes essential substances to the tissues, and removes byproducts of metabolism. Implicit in these functions are the intrinsic capabilities of the cardiovascular system to buffer pulsatile flow in order to ensure steady flow in the capillaries, regulate blood pressure and volume at all levels of the vasculature, maintain circulatory continuity while permitting free exchange between capillaries/venules and the extravascular compartments,

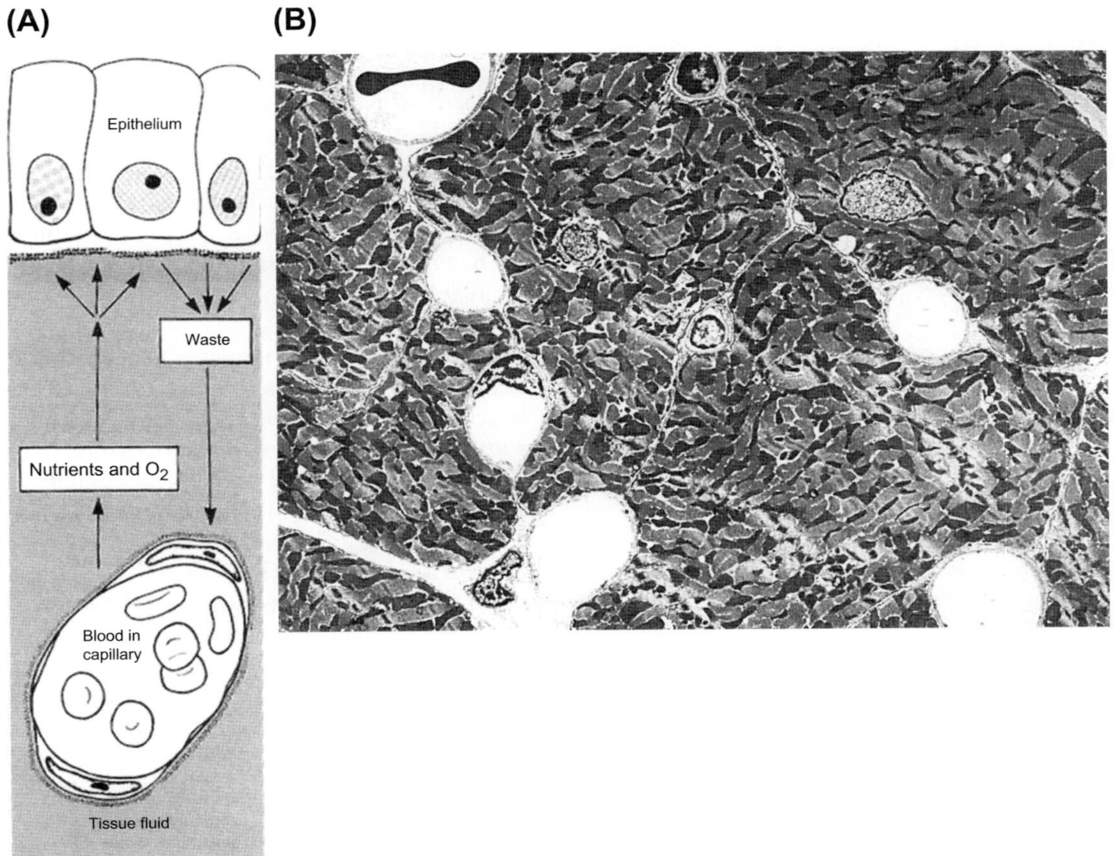

FIGURE II.1.5.1 Role of the vasculature in tissue function. (A) Schematic diagram of the route by which the cells in a tissue obtain their nutrients and oxygen from underlying capillaries. Metabolic waste products pass in the reverse direction and are carried away in the bloodstream. In each case, diffusion occurs through the tissue fluid that permeates the amorphous intercellular substances lying between the capillaries and the tissue cells. (B) Myocardium, a highly metabolic tissue, has a rich vascular/capillary network, as demonstrated by transmission electron microscopy. The six open round spaces are capillaries. A red blood cell is noted in the capillary at upper left. ((A): Reproduced by permission from Cormack, D. H. (1987). *Ham's Histology,* 9th Edn. Lippincott: Philadelphia, PA.)

and control hemostasis (managing hemorrhage by a coordinated response of vasoconstriction and plugging of vascular defects by coagulation and platelet clumps). Other functions of the circulation include such homeostatic (control) mechanisms as regulation of body temperature, and distribution of various regulating substances (e.g., hormones, cytokines and other inflammatory mediators, growth factors). Moreover, the circulatory system distributes immune and inflammatory cells to their sites of action, and the endothelial cells that line blood vessels have important immunological and inflammatory functions (Mitchell and Schoen, 2010).

Although the cardiac output is intermittent owing to the cyclical nature of the pumping of the heart, continuous flow to the periphery occurs by virtue of distention of the aorta and its branches during ventricular contraction (*systole*), and elastic recoil of the walls of the aorta and other large arteries with forward propulsion of the blood during ventricular relaxation (*diastole*). Blood moves rapidly through the aorta and its arterial branches, which become narrower and whose walls become thinner and change histologically toward the peripheral tissues. By adjusting the degree of contraction of their circular muscle coats, the distal arteries (*arterioles*) control the distribution of tissue blood flow to various tissues, and also permit regulation of local and systemic blood pressure. Blood returns to the heart from the *capillaries*, the smallest and thinnest walled vessels, by passing through *venules* and then through veins of increasing size.

Blood entering the right ventricle of the heart via the right atrium is pumped through the pulmonary arterial system at mean pressures about 1/6 of those developed in the systemic arteries. The blood then passes through the lung capillaries in the alveolar walls, where carbon dioxide is released across the pulmonary alveolar septa to, and oxygen taken up from, the alveoli. The oxygen-rich blood returns through the pulmonary veins to the left atrium and ventricle to complete the cycle.

Large diameter blood vessels are effective in delivering blood. Smaller vessels are most effective in regulating blood flow, and the smallest (the capillaries) regulate diffusional transport to and from the surrounding tissues. Thus, owing to their very thin walls and slow velocity of blood flow, which falls to approximately 0.1 cm/sec from 50 cm/sec in the aorta, capillaries are the sites of most exchange of oxygen, nutrients, and cellular wastes.

Capillary density is determined by the diffusion limit of oxygen; approximately 100–200 μm in most highly metabolic tissues (recall Figure II.1.5.1). Thus, highly metabolic tissues (e.g., heart muscle) have a dense network of blood vessels, and three-dimensional tissue formation and growth requires the formation of new blood vessels, a process called *angiogenesis*. It also follows that tissues that require less nutrition (e.g., cartilage and resting skeletal muscle) and those that are relatively thin (e.g., heart valve leaflets) may either require a sparse vascular network or none at all.

In the following sections we will cover the general functional principles of tissue organization and response to various types of injury, highlighted by specific illustrative examples.

Extracellular Matrix

Extracellular matrix (ECM) comprises the biological material produced by, residing in-between, and supporting cells. ECM, cells, and capillaries are physically and functionally integrated in tissues and organs (Figure II.1.5.2). The ECM provides physical support and a matrix in which cells can adhere, signal each other, and interact. However, the ECM is more than a scaffold that maintains tissue and organ structure. Indeed, the ECM regulates many aspects of cell behavior, including cell proliferation and growth, survival, change in cell shape, migration, and differentiation (the so-called "cell fate decisions"). (See Figures II.1.5.3 and II.1.5.4.)

The principal functions of the ECM are:

- Mechanical support for cell anchorage
- Determination of cell orientation
- Control of cell growth
- Maintenance of cell differentiation
- Scaffolding for orderly tissue renewal
- Establishment of tissue microenvironments
- Sequestration, storage, and presentation of soluble regulatory molecules.

Some extracellular matrices are specialized for a particular function, such as strength (tendon), filtration (the basement membranes in the kidney glomerulus) or adhesion (basement membranes supporting most epithelia). To produce additional mechanical strength in bones and teeth, the ECM is calcified. Even in a tissue as "simple" as a heart valve leaflet, the coordinated interplay of several ECM components, and the spatial and temporal dynamics of these interactions, are critical to function (Schoen, 2008). Moreover, the scaffolds used in tissue engineering applications often replicate the several functions of, and are intended to stimulate the production of, natural ECM (Lutolf and Hubbell, 2005; Lutolf et al, 2009).

ECM components are synthesized, secreted, and remodeled by cells in response to environmental cues. Virtually all cells secrete and degrade ECM to some extent. Certain cell types (e.g., fibroblasts and smooth muscle cells) are particularly active in production of interstitial ECM (i.e., the ECM between cells). Epithelial cells also

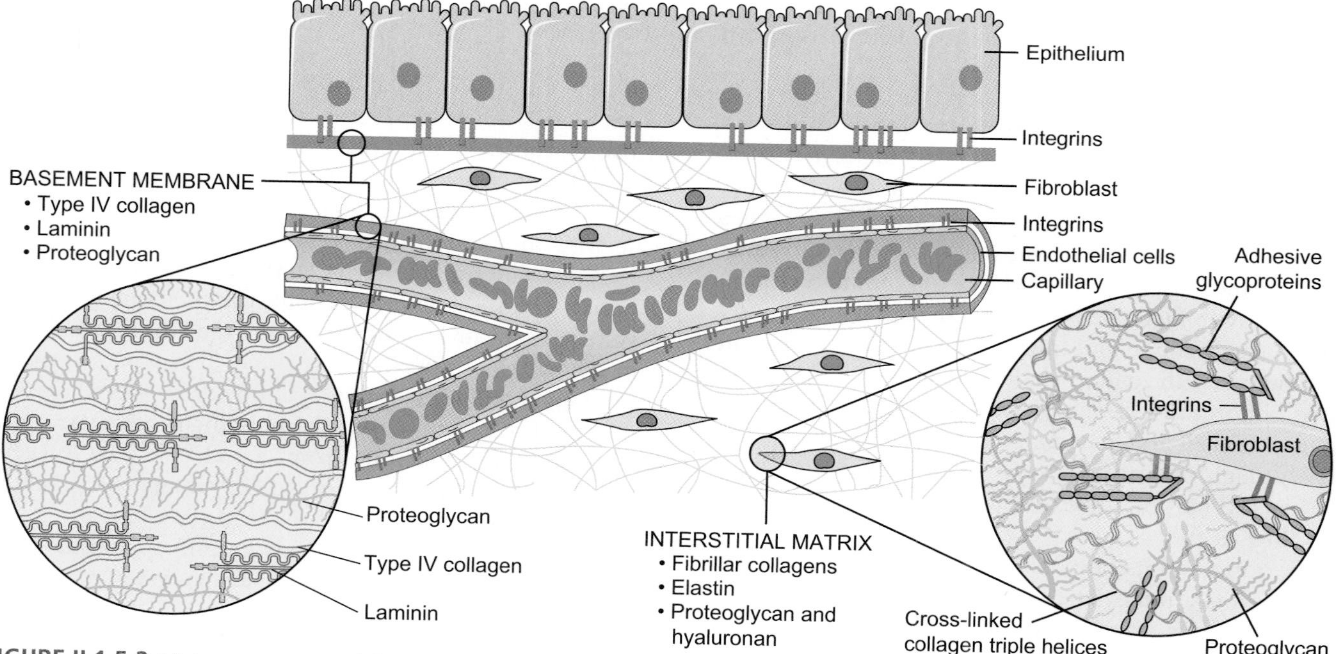

FIGURE II.1.5.2 Main components of the extracellular matrix, including collagen, proteoglycans, and adhesive glycoproteins. Both epithelial cells and fibroblasts interact with ECM via integrins. Basement membranes and interstitial ECM have differing architecture and general composition, although there is some overlap. Some ECM components (e.g., elastin) are not included for the sake of simplification. (Reproduced by permission from Kumar, V., Abbas, A. K., Fausto, N. & Aster, J. C. (eds.). (2010). *Robbins and Cotran Pathologic Basis of Disease*, 8th Edn. Saunders: Philadelphia, PA.)

synthesize the ECM of their basement membranes (see Figure II.1.5.5D).

Determinants of Tissue Form and Function

The mechanical forces that cells experience from their surroundings markedly influence the maintenance of cellular phenotypes, and affect cell shape, orientation, and differentiated function through interaction with receptors for specific ECM molecules on cell surfaces (such as *integrins*) (Chen, 2008; Mammoto and Ingber, 2010; Reilly and Engler, 2010). The resultant changes in cytoskeletal organization and in the production of second messengers can modify gene expression. ECM plays a critical role in cell differentiation, organogenesis, and as a scaffold allowing orderly repair following injury. The reciprocal instructions between cells and ECM are termed *dynamic reciprocity*.

> The ECM provides physical support and a matrix in which cells can adhere, signal each other, and interact. However, the ECM is more than a scaffold that maintains tissue and organ structure. Indeed, the ECM regulates many aspects of cell behavior, including cell proliferation and growth, survival, change in cell shape, migration and differentiation (the so-called "cell fate decisions").

ECM consists of large molecules synthesized by cells, exported to the intercellular space, and structurally linked. Present to some degree in all tissues, and particularly abundant as an intercellular substance in *connective tissues*, ECM is composed of: (1) fibrous structural and adhesive proteins (e.g., collagen, laminins, fibronectin, vitronectin, and elastin); (2) specialized proteins (including growth factors); and (3) a largely amorphous interfibrillary matrix (mainly proteoglycans, solutes, and water). The precise composition varies from tissue to tissue. The large fibrous molecules are interlinked within an expansile glycoprotein–water gel, resembling a fiber-reinforced composite. The specific components are described below.

Fibrillar Proteins: Collagens and Elastin

Collagen comprises a family of closely related but genetically, biochemically, and functionally distinct molecules, which are responsible for tissue tensile strength. The most common protein in the animal world, collagen provides the extracellular framework for all multicellular organisms. The collagens are composed of a triple helix of three polypeptide α chains; about 30 different α chains form approximately 20 distinct collagen types. Types I, II, and III are the *interstitial* or *fibrillar collagens*; they are the most abundant and the most important for the present discussion. Type I is ubiquitous in hard and soft tissues; Type II is rich in cartilage, intervertebral disk, and the vitreous of the eye; and Type III is prevalent in soft tissues, especially those regions that are healing following injury, and in the walls of hollow organs. Types IV and V are nonfibrillar and are present in basement membranes, and soft tissues and blood vessels, respectively. Collagens are: (1) synthesized in cells as soluble procollagen precursors in discrete protein subunits; (2) secreted

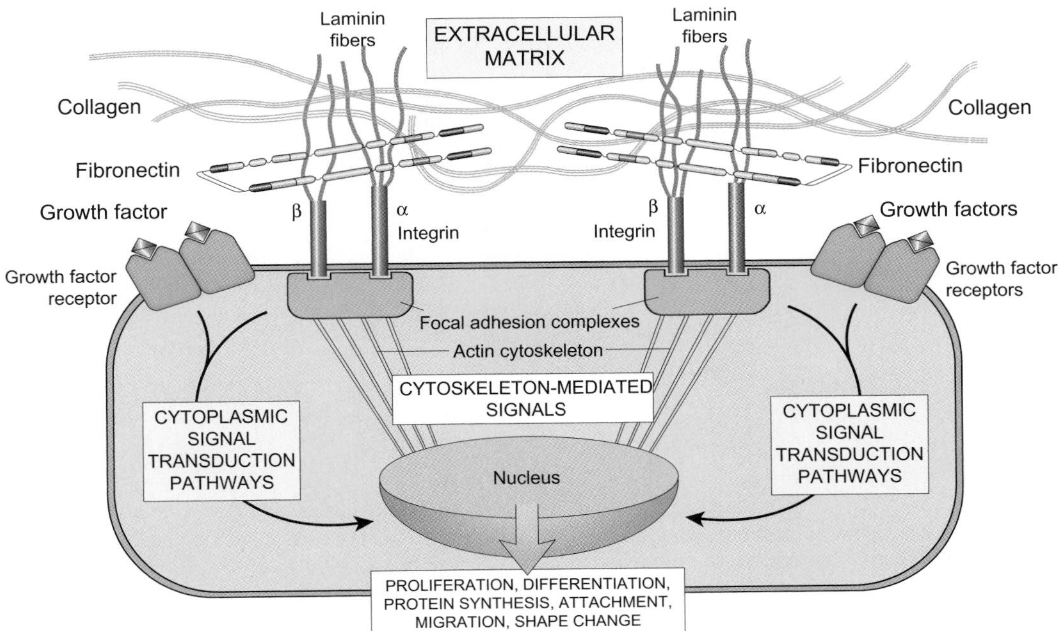

FIGURE II.1.5.3 Integrin-ECM interaction. Schematic showing the mechanisms by which ECM (e.g., collagen, fibronectin, and laminin) and growth factors interact with cells, activate signaling pathways, and can influence gene expression, growth, motility, differentiation, and protein synthesis. Integrins bind ECM and interact with the cytoskeleton at focal adhesion complexes. This can initiate the production of intracellular messengers, or can directly mediate nuclear signals. Cell surface receptors for growth factors also initiate second signals. (Reproduced by permission from Kumar, V., Abbas, A. K., Fausto, N. & Aster, J. C. (eds.). (2010). *Robbins and Cotran Pathologic Basis of Disease*, 8th Edn. Saunders: Philadelphia, PA.)

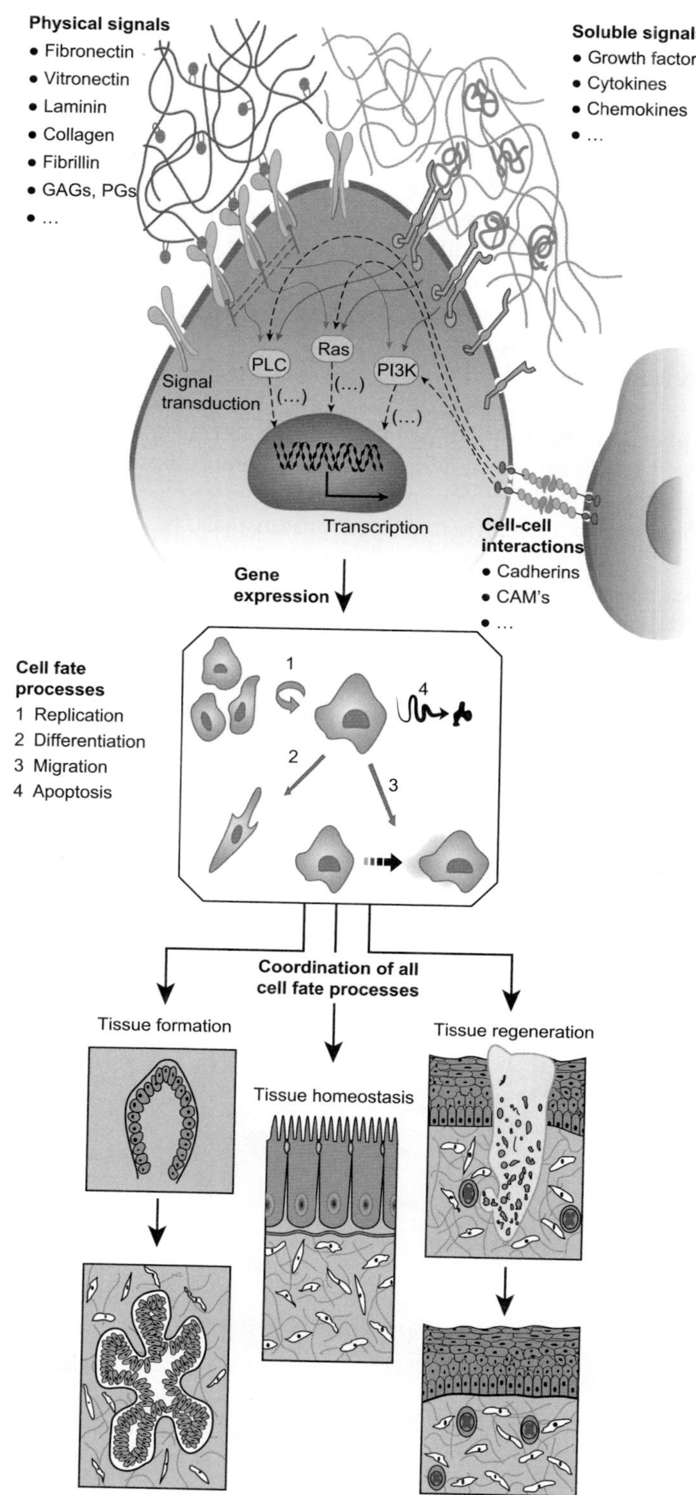

FIGURE II.1.5.4 Individual cell behavior and the dynamic state of multicellular tissues is regulated by reciprocal molecular interactions between cells and their surroundings. (Reproduced by permission from Lutolf, M. P. & Hubbell, J. A. (2005). Synthetic biomaterials as instructive extracellular microenvironments for morphogenesis in tissue engineering. *Nature Biotechnology*, **23**, 47–55.)

into the extracellular environment self-assembled; and (3) matured insoluble collagen molecules in the extracellular space (called "post-translational modification").

During collagen synthesis, the α chains are subjected to a number of enzymatic modifications, including hydroxylation of proline and lysine residues, providing collagen with a high content of hydroxyproline (10%). Vitamin C is required for hydroxylation of the collagen propeptide; a requirement that explains inadequate wound healing in vitamin C deficiency (scurvy). After the modifications, but still inside the cell, the procollagen chains align and form the triple helix. However, the procollagen molecule is still soluble and contains N-terminal and C-terminal propeptides. During or shortly after secretion from the cell, procollagen peptidases clip the terminal propeptide chains, promoting formation of fibrils, often called *tropocollagen*, and oxidation of specific lysine and hydroxylysine residues occurs by the extracellular enzyme lysyl oxidase. This results in cross-linkages between α chains of adjacent molecules stabilizing the array that is characteristic of collagen. The mechanical properties of collagen reflect this structure; collagen fibrils are both non-extensible even at very high loads, and incompressible.

Elastic fibers are composed of the protein elastin. They confer passive recoil to various tissues, are critical components of vascular tissues (especially the aorta where repeated pulsatile flow would cause unacceptable shears on non-compliant tissue and elastic recoil is important), and of intervertebral disks (where the repetitive forces of ambulation along the spine are dissipated). Unlike collagen, elastin can be stretched). The stretching of an artery every time the heart pumps blood into an artery is followed by the recoil of elastin which restores the artery's former diameter between heartbeats.

Amorphous Matrix: Glycosaminoglycans (GAGs), Proteoglycans, and Hyaluronan

Amorphous intercellular substances contain carbohydrate bound to protein. The carbohydrate is in the form of long chain polysaccharides called *glycosaminoglycans* (GAGs). When GAGs are covalently bound to proteins, the molecules are called *proteoglycans*. GAGs are highly charged (usually sulfated) polysaccharide chains up to 200 sugars long, composed of repeating unbranched disaccharide units (one of which is always an amino sugar – hence the name glycosaminoglycan). GAGs are divided into four major groups on the basis of their sugar residues:

- Hyaluronic acid: a component of loose connective tissue and of joint fluid, where it acts as a lubricant;
- Chrondroitin sulfate and dermatan sulfate;
- Heparan sulfate and heparin;
- Keratin sulfate.

With the exception of hyaluronic acid (not sulfated and thereby unique among the GAGs), all GAGs are covalently attached to a protein backbone to form proteoglycans, with a structure that schematically resembles a bottle brush. Proteoglycans are diverse, owing to different core proteins and different glycosaminoglycans. Proteoglycans are named according to the structure of their principal repeating disaccharide. Some of the most common are heparan sulfate, chondroitin sulfate, and dermatan sulfate. Proteoglycans can also be integral membrane proteins, and are thus modulators of cell growth and differentiation. The syndecan family has a core protein that spans the plasma membrane and contains a short cytosolic domain, as well as a long external domain to which a small number of heparan sulfate chains are attached. Syndecan binds collagen, fibronectin, and thrombospondin in the ECM, and can modulate the activity of growth factors. Hyaluronan consists of many repeats of a simple disaccharide stretched end-to-end and binds a large amount of water, forming a viscous hydrate gel, which gives connective tissue turgor pressure and an ability to resist compression forces. This ECM component helps provide resilience, as well as a lubricating feature to many types of connective tissue, notably that found in the cartilage in joints.

Adhesive Molecules

Adhesive proteins, including fibronectin, laminin, and entactin permit the attachment to, and movement of, cells within the ECM.

- *Fibronectin* is a ubiquitous, multi-domain glycoprotein possessing binding sites for a wide variety of other ECM components. It is synthesized by many different cell types, with the circulating form produced mainly by hepatocytes. Fibronectin is important for linking cells to the ECM via cell surface integrins. Fibronectin's adhesive character also makes it a crucial component of blood clots, and of pathways followed by migrating cells. Thus, fibronectin-rich pathways guide and promote the migration of many kinds of cells during embryonic development and wound healing.
- *Laminin* is an extremely abundant component of the basement membrane; it is a tough, thin, sheet-like layer on which epithelial cells sit, and is important for cell differentiation, adhesion to the substrate, and tissue remodeling. Laminin polypeptides are arranged in the form of an elongated cross, with individual chains held together by disulfide bonds. Like fibronectin, laminin has a distinct domain structure; different regions of the molecule bind to Type IV collagen (an important component of the basement membrane), heparin sulfate, entactin (a short protein that cross-links each laminin molecule to Type IV collagen), and cell surface integrins.

Extracellular Matrix Remodeling

The maintenance of the extracellular matrix requires ongoing remodeling of collagen, itself dependent on ongoing collagen synthesis and catabolism. Connective tissue remodeling, either physiological or pathological, is in most cases a highly organized and regulated process that involves the selective action of a group of related proteases that collectively can degrade most, if not all, components of the extracellular matrix. These proteases are known as the matrix metalloproteinases (MMPs). Subclasses include the interstitial collagenases, stromelysins, and gelatinases.

In many tissues, cells degrade and reassemble the ECM in response to external signals as part of physiological homeostasis in mature, healthy tissues. Although matrix turnover is generally quite low in normal mature tissues, rapid and extensive remodeling characterizes embryological development (which involves branching morphogenesis, and stem cell homing, proliferation and differentiation), and wound repair, as well as various adaptive and pathologic states, such as tumor cell invasion and metastasis. ECM remodeling is mediated (and regulated) by signaling through ECM receptors, including the integrins and the ECM-modifying proteins, such as matrix metalloproteinases (MMPs), serene proteases (e.g., plasmin, plasminogen activator), and cysteine protease (e.g., cathepsins) (Page-McCaw et al., 2007; Daley et al., 2008). An understanding of cell–substrate interactions and matrix remodeling are central to the application of next-generation biomaterials technology, tissue engineering, and regenerative therapeutics (Gurtner et al., 2008).

Enzymes that degrade collagen are synthesized by macrophages, fibroblasts, and epithelial cells. Collagenases are specific for particular types of collagens, and many cells contain two or more different such enzymes. For example, fibroblasts synthesize a host of matrix components, as well as enzymes involved in matrix degradation, such as MMPs and serine proteases. Particularly important in tissue remodeling are myofibroblasts, a particular phenotype of cells that are characterized by features of both smooth muscle cells (e.g., contractile proteins such as α-actin) and features of fibroblasts (e.g., rough endoplasmic reticulum in which proteins are synthesized) (Hinz et al., 2007). These cells may also be responsible for the production of (and likely respond to) tissue forces during remodeling, thereby regulating the evolution of tissue structure according to mechanical requirements.

Evidence suggests that growth factors and hormones (autocrine, paracrine, and endocrine) are pivotal in orchestrating both synthesis and degradation of ECM components. Cytokines such as TGF-β, PDGF, and IL-1 clearly play an important role in the modulation of collagenase and tissue inhibitors of metalloproteinases (TIMPs) expression. MMP enzymatic activities are regulated by TIMPs, which are especially important during wound repair. As natural inhibitors of the MMPs, TIMPS are multifunctional proteins with both MMP inhibitor activity and cell growth modulating properties. Degradation of the extracellular matrix is mediated by an excess of MMP activity over that of TIMPs. Distortion of the balance between matrix synthesis and turnover may result in altered matrix composition and amounts.

Cell–Cell and Cell–Matrix Interactions. Like cell–cell interactions, cell–matrix interactions have a high degree of specificity, requiring recognition, adhesion, electrical and chemical communication, cytoskeletal reorganization, and/or cell migration. Moreover, adhesion receptors may also act as transmembrane signaling molecules that transmit information about the environment to the inside of cells, in some cases all the way to the nucleus, and mediate the effects of signals initiated by growth factors or compounds controlling gene expression, phenotypic modulation and cell replication, differentiation and apoptosis (Figure II.1.5.3). Although many of the components of the extracellular matrix (*ligands*) with which cells interact are immobilized and not in solution (e.g., the integrin adhesion receptors and the vascular selectins (the latter modulate interaction of circulating inflammatory cells with the endothelium)), some soluble (secreted) factors also modulate cell–cell communication in the normal and pathologic regulation of tissue growth and maturation. Indeed, the array of specific binding of specialized cues with cell-surface receptors induces complex interactions that regulate tissue formation, homeostasis, and regeneration (Figure II.1.5.4). Moreover, the scaffolds used in tissue engineering applications often replicate the several functions of, and are intended to stimulate the production of, natural ECM (Lutolf et al., 2009).

The *integrins* comprise a family of cell receptors with diverse specificity that bind ECM proteins, other cell surface proteins and plasma proteins, and control cell growth, differentiation, gene expression, and motility (Bokel and Brown, 2002). Some integrins bind only a single component of the ECM, e.g., fibronectin, collagen, or laminin (see above). Other integrins can interact with several of these polypeptides. In contrast to hormone receptors, which have high affinity and low abundance, the integrins exhibit low affinity and high abundance, so that they can bind weakly to several different but related matrix molecules. This property allows the integrins to promote cell–cell interactions as well as cell–matrix binding.

BASIC TISSUES

Humans have over a hundred distinctly different types of cells allocated to four types of basic tissues (Table II.1.5.1 and Figure II.1.5.5): (1) epithelium; (2) connective tissue;

TABLE II.1.5.1 The Basic Tissues: Classification and Examples

Basic Tissues	Examples
Epithelial tissue	
Surface	Skin epidermis, gut mucosa
Glandular	Thyroid follicles, pancreatic acini
Special	Retinal or olfactory epithelium
Connective tissue	
Connective tissue proper	
Loose	Skin dermis
Dense (regular, irregular)	Pericardium, tendon
Special	Adipose tissue
Hemopoietic tissue, blood and lymph	Bone marrow, blood cells
Supportive tissue	Cartilage, bone
Muscle tissue	
Smooth	Arterial or gut smooth muscle
Skeletal	Limb musculature, diaphragm
Cardiac muscle	Heart
Nerve tissue	Brain cells, peripheral nerve

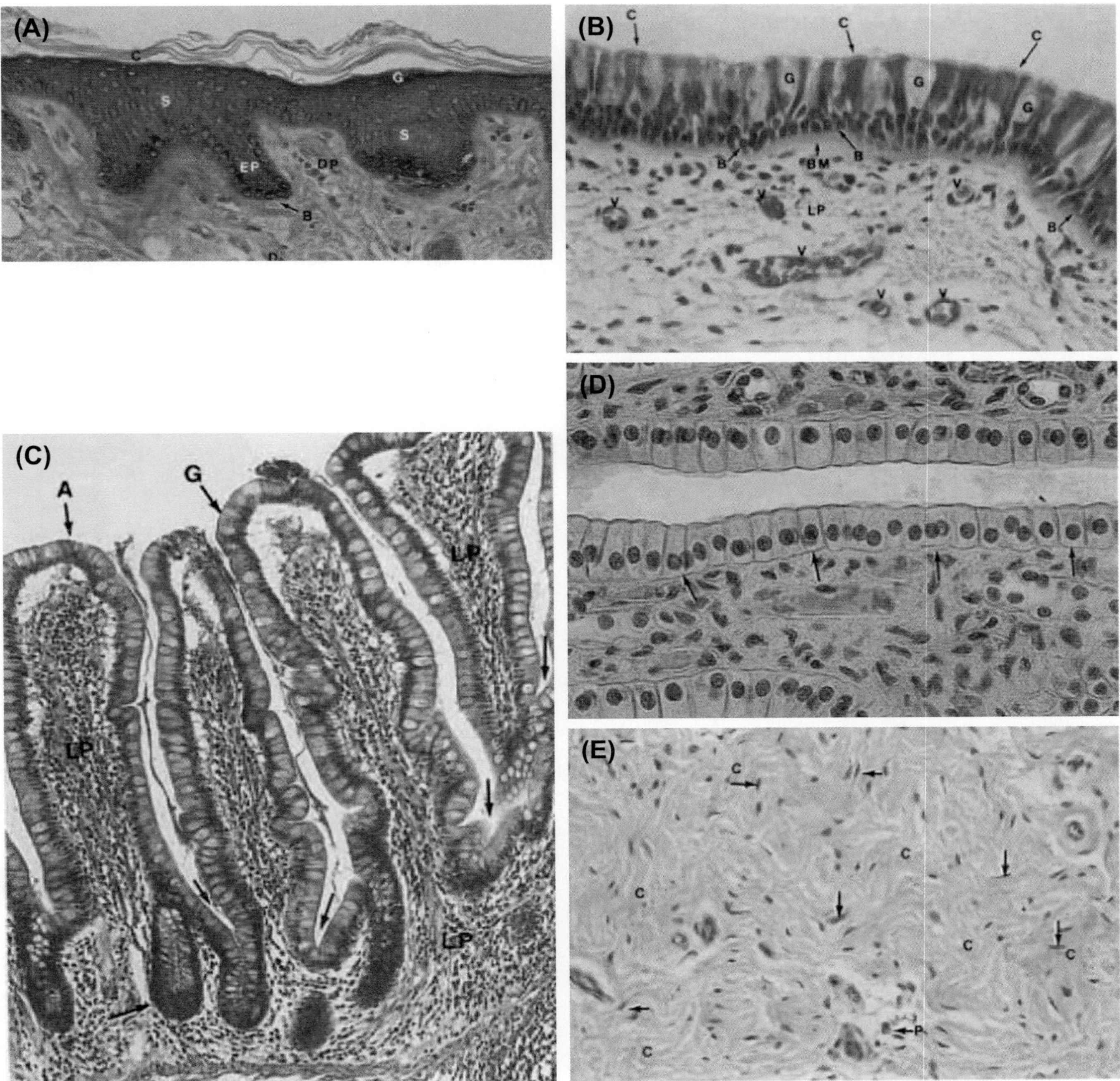

FIGURE II.1.5.5 Photomicrographs of basic tissues, emphasizing key structural features. (A)–(D) Epithelium; (E), (F) connective tissue; (G) muscle; and (H) nervous tissue. (A) Skin. Note the thin stratum corneum (c) and stratum granulosum (g). Also shown are the stratum spinosum (*s*), stratum basale (b), epidermal pegs (ep), dermal papilla (dp), and dermis (d). (B) Trachea, showing goblet cells (g), ciliated columnar cells (*c*), and basal cells (*b*). Note the thick basement membrane (bm) and numerous blood vessels (v) in the lamina propria (lp). (C) Mucosa of the small intestine (ileum). Note the goblet (g) and columnar absorbing (a) cells, the lamina propria (lp), and crypts (arrows). (D) Epithelium of a kidney collecting duct resting on a thin basement membrane (arrows). (E) Dense irregular connective tissue. Note the wavy unorientated collagen bundles (c) and fibroblasts (arrows), (p), plasma cells. (F) Cancellous bone clearly illustrating the morphologic difference between inactive bone lining (endosteal, osteoprogenitor) cells (bl) and active osteoblasts (ob). The clear area between the osteoblasts and calcified bone represents unmineralized matrix or osteoid. (cl), Cement lines; (o), osteocycles. (G) Myocardium (cardiac muscle). The key features are centrally placed nuclei, intercalated discs (representing end-to-end junctions of adjoining cells) and the sarcomere structure visible as cross-striations in the cells. (H) Small nerve fascicles (n) with perineurium (p) separating it from two other fascicles (n). ((A)–(F) and (H) reproduced by permission from Berman, I. (1993). Color Atlas of Basic Histology, Appleton and Lange; (G) reproduced by permission from Schoen, F. J. (2004). The Heart. In: Cotran, R. S., Kumar, V. & Collins, T. (eds.). *Robbins Pathologic Basis of Disease*, 7th Edn. Saunders: Philadelphia, PA.)

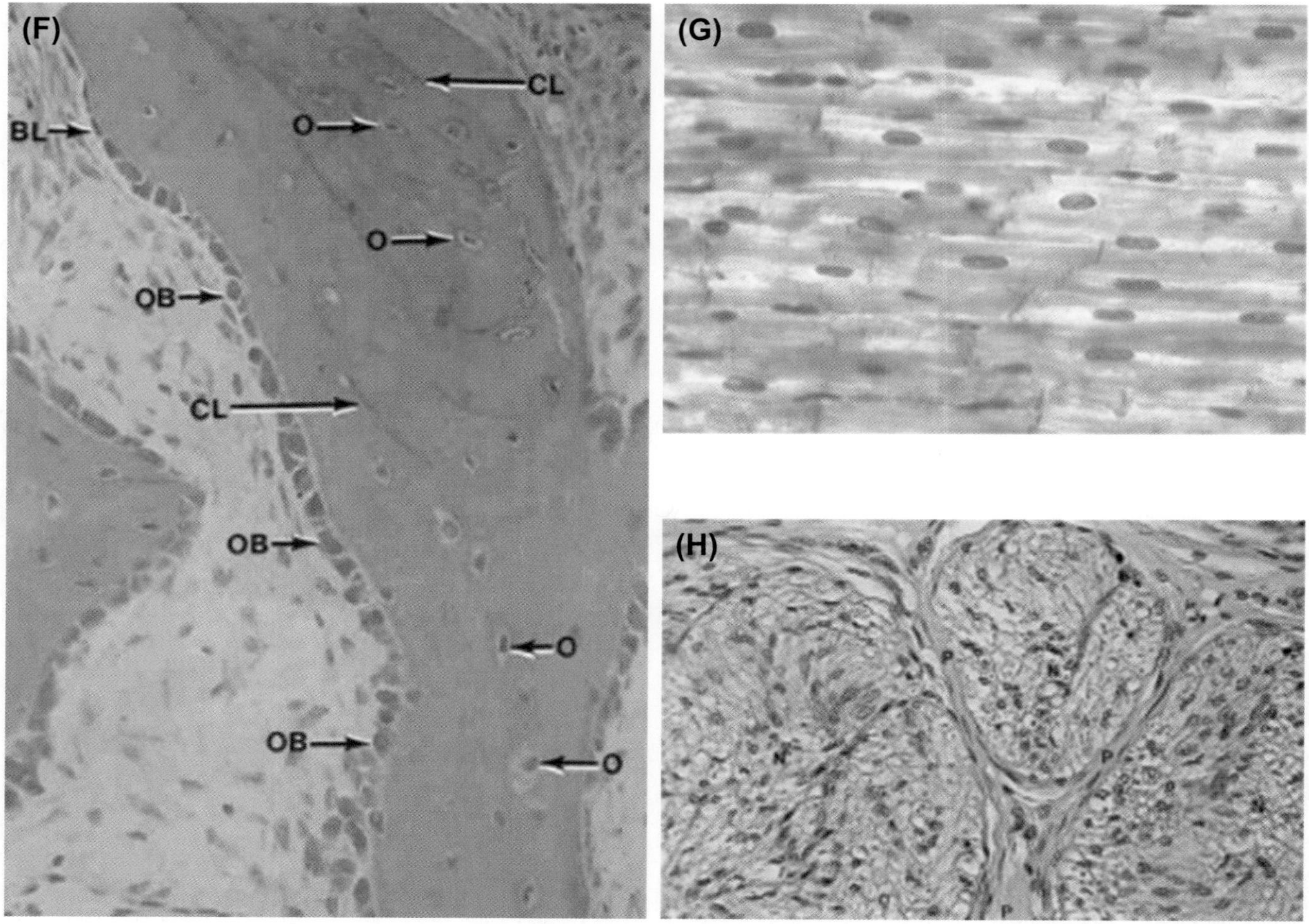

FIGURE II.1.5.5 cont'd

(3) muscle; and (4) nervous tissue. The different basic tissues have distinctive microscopic appearances that belie their specific functional roles. The basic tissues have their origins in embryological development; early events include the formation of a tube with three *germ layers* in its wall: (1) an outer layer of ectoderm (which gives rise to skin and the nervous system); (2) an inner layer of endoderm (from which the lining membranes of the gut, respiratory tract, and genitourinary system, and the liver and pancreas derive); and (3) a middle layer of mesoderm (which is the origin of muscle, bone, blood cells, and the blood forming organs and the heart).

Epithelia cover the internal and external body surfaces and accommodate diverse functions. An epithelial surface can be: (1) a protective dry, cutaneous membrane barrier (as in skin); (2) a moist, mucous membrane, lubricated by glandular secretions and variably absorptive (digestive and respiratory tracts); (3) a moist membrane lined by mesothelium, lubricated by fluid that derives from blood plasma (peritoneum, pleura, pericardium); (4) the inner lining of the circulatory system, called endothelium; and (5) the source of internal and external secretions (e.g., endocrine and sweat glands, respectively). Epithelium derives mostly from ectoderm and endoderm, but also from mesoderm.

Epithelial cells play fundamental roles in the directional movement of ions, water, and macromolecules between biological compartments, including absorption, secretion, and exchange. Therefore, the architectural and functional organization of epithelial cells includes structurally, biochemically, and physiologically distinct plasma membrane domains that contain region-specific ion channels, transport proteins, enzymes and lipids, and cell–cell junctional complexes. These components help integrate multiple cells to form an interface between biological compartments in organs. Subcellular epithelial specializations are not apparent to the naked eye – or even necessarily to light microscopy; they are best studied by transmission electron microscopy (TEM) and by functional assays (e.g., assessing synthetic products, permeability, and transport).

Supporting the other tissues of the body, connective tissue arises from embryonic mesenchyme, a derivative of mesoderm. Connective tissue also serves as a scaffold and conduit, through which course the nerves and blood vessels that support the various epithelial tissues. Other types of tissue with varying functions are also of mesenchymal origin. These include dense connective tissue, adipose (fat) tissue, cartilage and bone, circulating cells (blood cells and their precursors in bone marrow), as well as inflammatory cells that defend the body against infectious organisms and other foreign agents.

Muscle cells develop from mesoderm and are highly specialized for contraction. They have the contractile proteins *actin* and *myosin* in varying amounts and configuration, depending on cell function. Muscle cells are of three types: smooth muscle; skeletal muscle; and cardiac muscle. The latter two have a striated microscopic appearance, owing to their discrete bundles of actin and myosin organized into sarcomeres. Smooth muscle cells, which have a less compact arrangement of myofilaments, are prevalent in the walls of blood vessels and the gastrointestinal tract. Their slow, non-voluntary contraction regulates blood vessel caliber, and proper movement of food and solid waste, respectively.

Nerve tissue, which derives from ectoderm, is highly specialized with respect to irritability and conduction. Nerve cells not only have cell membranes that generate electrical signals called action potentials, but also secrete neurotransmitters, molecules that trigger adjacent nerve or muscle cells to either transmit an impulse or to contract.

Organs

Several different types of tissues arranged into a functional unit constitute an organ. These have a composite structure in which epithelial cells typically perform the specialized work of the organ, while connective tissue and blood vessels support and provide nourishment to the epithelium. There are two basic organ patterns: tubular (or hollow) and compact (or solid) organs. Tubular organs include the blood vessels and the digestive, urogenital, and respiratory tracts; they have similar architectures in that each is composed of layers of tissue arranged in a specific sequence. For example, each has an inner coat consisting of a lining of epithelium, a middle coat consisting of layers of muscle (usually smooth muscle) and connective tissue, and an external coat consisting of connective tissue, often covered by epithelium (for example, the intestines or vascular walls) (Figure II.1.5.6). Specific variations in architecture reflect organ-specific functional requirements. While the outer coat of an organ that blends into surrounding structures is called the *adventitia*, the outside epithelial lining of an organ suspended in a body cavity (e.g., thoracic or abdominal) is called a *serosa*.

The histologic composition and organization, as well as the thickness, of these three layers of tubular systems varies characteristically with the physiologic functions performed by specific regions. The structure/function correlations are particularly well-demonstrated in the cardiovascular system (Figure II.1.5.7). Blood vessels have three layers: an *intima* (primarily endothelium); a *media* (primarily smooth muscle and elastin); and an adventitia (primarily collagen). The amounts and relative proportions of layers of the blood vessels are influenced by mechanical factors (especially blood pressure, which determines the amount and arrangement of muscular tissue), and metabolic factors (reflecting the local nutritional needs of the tissues). Three features will illustrate the variation in site-specific structure–function correlations:

- As discussed above, capillaries have a structural reduction of the vascular wall to only endothelium and minimal supporting structures to facilitate exchange (i.e., the metabolic function dominates). Thus, capillaries are part of the tissues they supply and, unlike larger vessels, do not appear as a separate anatomical unit.
- Arteries and veins have distinctive structures. The regional structural changes are primarily controlled by mechanical factors. The arterial wall is generally thicker than the venous wall, in order to withstand the higher blood pressures that prevail within arteries compared with veins. The thickness of the arterial wall gradually diminishes as the vessels themselves become smaller, but the wall–to–lumen ratio becomes greater in the periphery. Veins have a larger overall diameter, a larger lumen, and a narrower wall than the corresponding arteries with which they course.
- In essence, the heart is a blood vessel specialized for rhythmic contraction; its media is the *myocardium*, containing muscle cells (*cardiac myocytes*).

The blood supply of an organ comes from its outer aspect. In tubular organs, large vessels penetrate the outer coat, perpendicular to it, and give off branches that run parallel to the tissue layers (Figure II.1.5.8). These vessels divide yet again to give off penetrating branches that course through the muscular layer, and branch again in the connective tissue parallel to the layers. The small blood vessels have junctions (*anastomoses*) with one another in the connective tissue. These junctions may provide collateral pathways that can allow blood to bypass obstructions. Compact, solid organs have an extensive connective tissue framework, surrounded by a dense, connective tissue capsule (Figure II.1.5.9). Such organs have an area of thicker connective tissue where blood vessels and other conduits (e.g., airways in the lungs) enter the organ; this region is called the *hilus*. From the hilus, strands of connective tissue extend into the organ and may divide it into lobules.

In both tubular and compact organs, the dominant cells comprising specialized tissues are the *parenchyma* (e.g., epithelial cells lining the gut, thyroglobulin-hormone producing epithelial cells in the thyroid or cardiac muscle

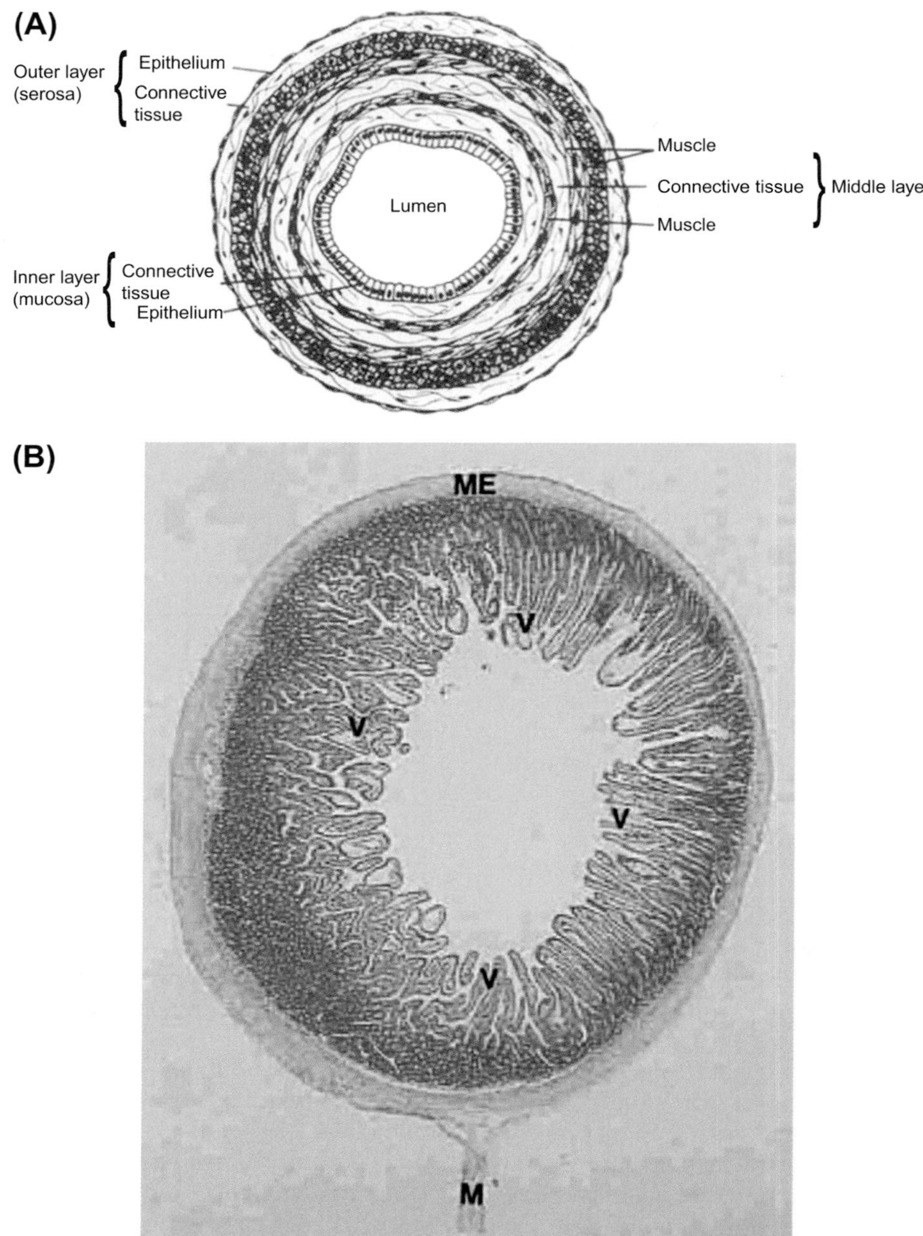

FIGURE II.1.5.6 General structure of a tubular organ system, as demonstrated by the gastrointestinal tract. (A) Organization of tissue layers in the digestive tract (e.g., stomach or intestines). (B) Photomicrograph of the dog jejunum illustrating villi (v), the muscularis external (me), and mesentery (M). In this organ the epithelium is organized into folds (the villi) in order to increase the surface area for absorption. ((A) Reproduced by permission from Borysenko, M. & Beringer, T. (1989). Functional Histology, 3rd Edn. Copyright 1989 Little, Brown, and Co.; (B) Reproduced by permission from Berman, I. (1993). *Color Atlas of Basic Histology*. Appleton and Lange.)

cells in the heart). Parenchyma occurs in masses (e.g., endocrine glands), cords (e.g., liver), or tubules (e.g., kidney). Parenchymal cells can be arranged uniformly in an organ, or they may be segregated into a subcapsular region (*cortex*) and a deeper region (*medulla*), each performing a distinct functional role. The remainder of the organ has a delicate structural framework, including supporting cells, extracellular matrix, and vasculature (essentially the "service core"), which constitutes the *stroma*. After entering an organ, the blood supply branches repeatedly to small arteries, and ultimately capillaries in the parenchyma; veins and nerves generally follow the course of the arteries.

Parenchymal cells are generally more sensitive to chemical, physical, or ischemic (i.e., low blood flow) injury than stroma. Moreover, when an organ is injured, orderly repair and regrowth of parenchymal cells requires an intact underlying stroma.

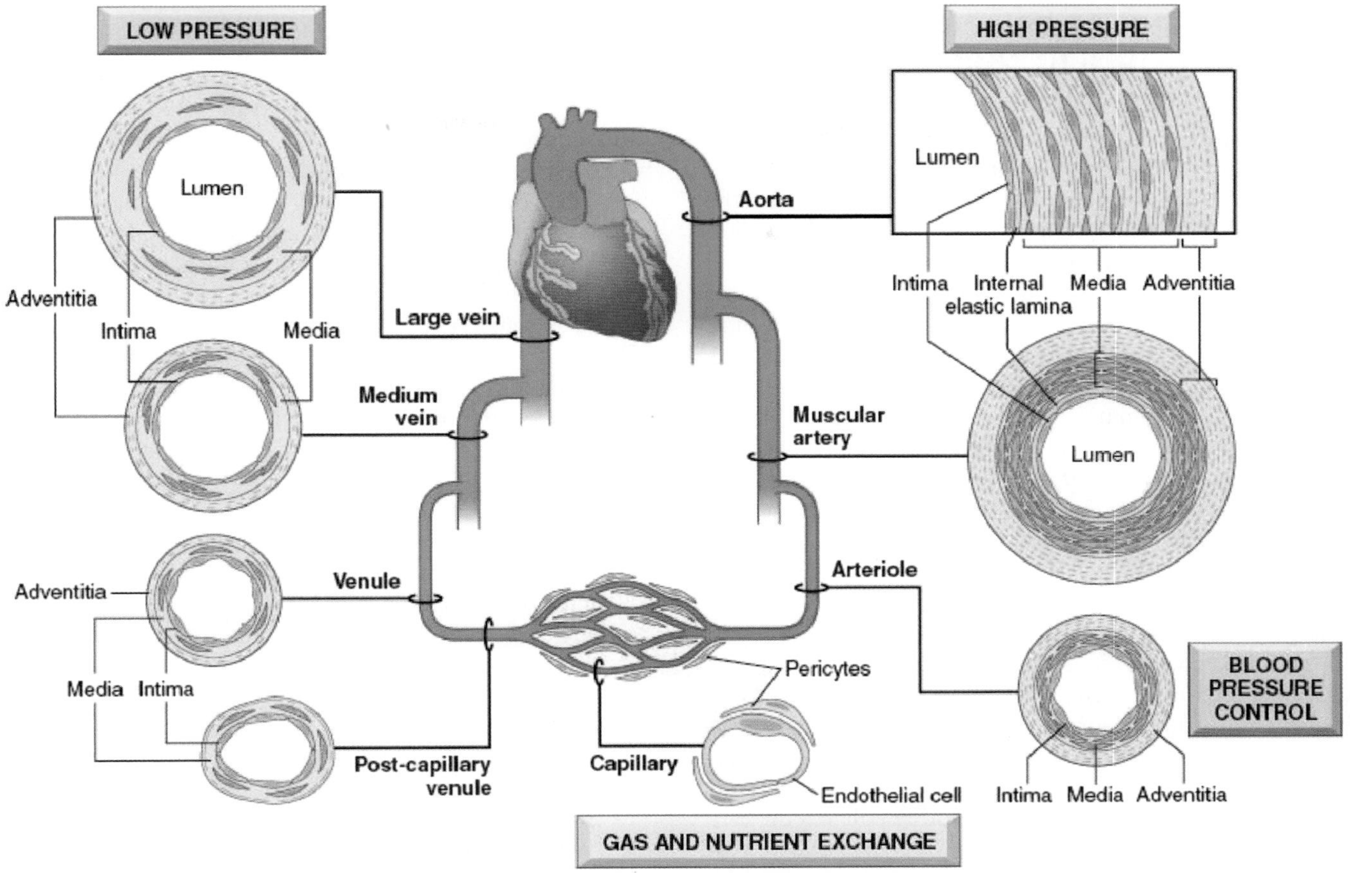

FIGURE II.1.5.7 Regional structural variations in the cardiovascular system. Although the basic organization is constant, the thickness and composition of the various layers differ, as a functional response to differential hemodynamic forces and tissue metabolic requirements. (Reproduced by permission from Mitchell, R. N. & Schoen, F. J. (2010). Blood Vessels. In: Kumar, V., Abbas, A. K., Fausto, N. & Aster, J. C. *Robbins and Cotran Pathologic Basis of Disease*, 8th Edn. Saunders: Philadelphia, PA.)

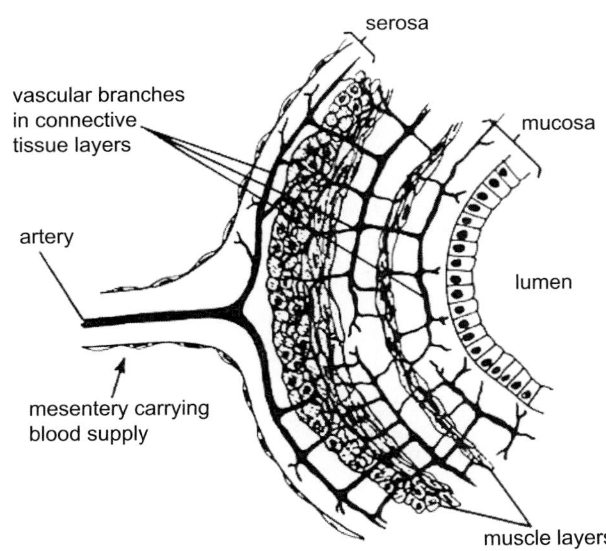

FIGURE II.1.5.8 Vascularization of hollow organs. (Reproduced by permission from Borysenko, M. & Beringer, T. (1989). *Functional Histology*, 3rd Edn. Copyright, 1989 Little, Brown, and Co.)

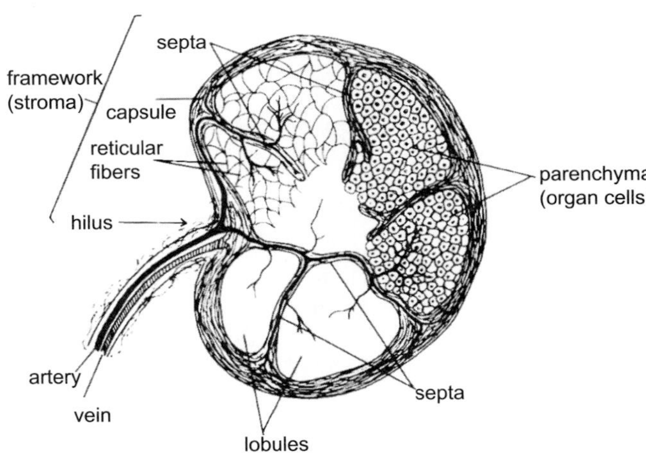

FIGURE II.1.5.9 Organization of compact organs. (Reproduced by permission from Borysenko, M. & Beringer, T. (1989). *Functional Histology*, 3rd Edn. Copyright, 1989 Little, Brown, and Co.)

TISSUE RESPONSE TO INJURY: INFLAMMATION, REPAIR, AND REGENERATION

A key protective response of an organism is its ability to eliminate damaged tissues and foreign invaders, such as microbes or exogenous non-biological materials (the latter including therapeutic biomaterials). Elimination of the intruder rids the organism of both the cause and the consequences of cell and tissue injury. Without this protective process, tissue wounds would not heal and infections would go unchecked. However, inappropriately triggered or uncontrolled inflammation can also be deleterious. Inflammation is usually a highly coordinated response involving proteins, leukocytes, and phagocytic cells (macrophages) that are derived from circulating cells, all of which travel in the blood and are recruited to a site where they are needed.

The outcome of tissue injury depends primarily on the tissue type affected, the extent of the injury, and whether it is persistent (Figure II.1.5.10). When an environmental perturbation or injury is transient or short-lived, tissue destruction is small, and the tissue is capable of homeostatic adjustment or regeneration; the outcome is restoration of normal structure and function. However, when tissue injury is extensive or occurs in tissues that do not regenerate, inflammation and repair occur, and the result is scarring (Singer and Clark, 1999). When an infection or foreign material cannot be eliminated, the body "controls" the foreign body or infection by creating a wall around it (and thereby isolating it as much as possible from healthy tissue). Thus, most inert biomaterials elicit an early inflammatory response, followed by a late tissue reaction characterized by encapsulation by a relatively thin fibrous tissue capsule (composed of collagen and fibroblasts, and similar to a scar). The nature of the reaction is largely dependent on the chemical and physical characteristics of the implant.

As described above, inflammation and repair follow cell and tissue injury induced by various exogenous and endogenous stimuli, both eliminating (i.e., diluting, destroying or isolating) the cause of the injury (e.g., microbes or toxins), and disposing of necrotic cells and tissues that occur as a result of the injury. In doing so, the inflammatory response initiates the process that heals and reconstitutes the tissue, by replacing the wound with native parenchymal cells or by filling up the defect with fibroblastic scar tissue, or some combination of both (such as skin reconstituting the epidermis and scarring the dermis) (Figure II.1.5.11). Inflammation and repair constitute an overlapping sequence of several processes (Figure II.1.5.12):

- Acute inflammation: The immediate and early response to injury, of relatively short duration, characterized by fluid and plasma protein exudation into the tissue, and by accumulation of neutrophils (polymorphonuclear leukocytes). In the case of an infection that cannot be easily eliminated, an abscess (i.e., a localized collection of acute inflammation and infectious organisms) is the outcome.
- Chronic inflammation: This phase is manifested histologically as lymphocytes and macrophages, often with concurrent tissue destruction, and can evolve into repair involving fibrosis and new blood vessel proliferation. A special type of chronic inflammation characterized by activated macrophages and often multinucleated giant cells, organized around an irritating focus, is called a *granuloma* or *granulomatous inflammation*. The pattern also occurs in pathologic states where the inciting agent is not removable, including the foreign body reaction (see below).

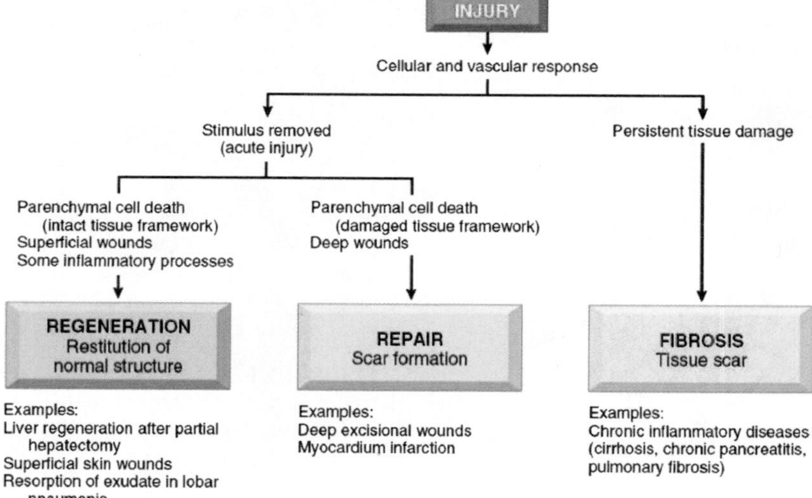

FIGURE II.1.5.10 Regeneration, repair, and fibrosis after injury and inflammation. (Reproduced by permission from Kumar, V., Abbas, A. K., Fausto, N. & Aster, J. C. (2010). *Robbins and Cotran Pathologic Basis of Disease*, 8th Edn. Saunders: Philadelphia, PA.)

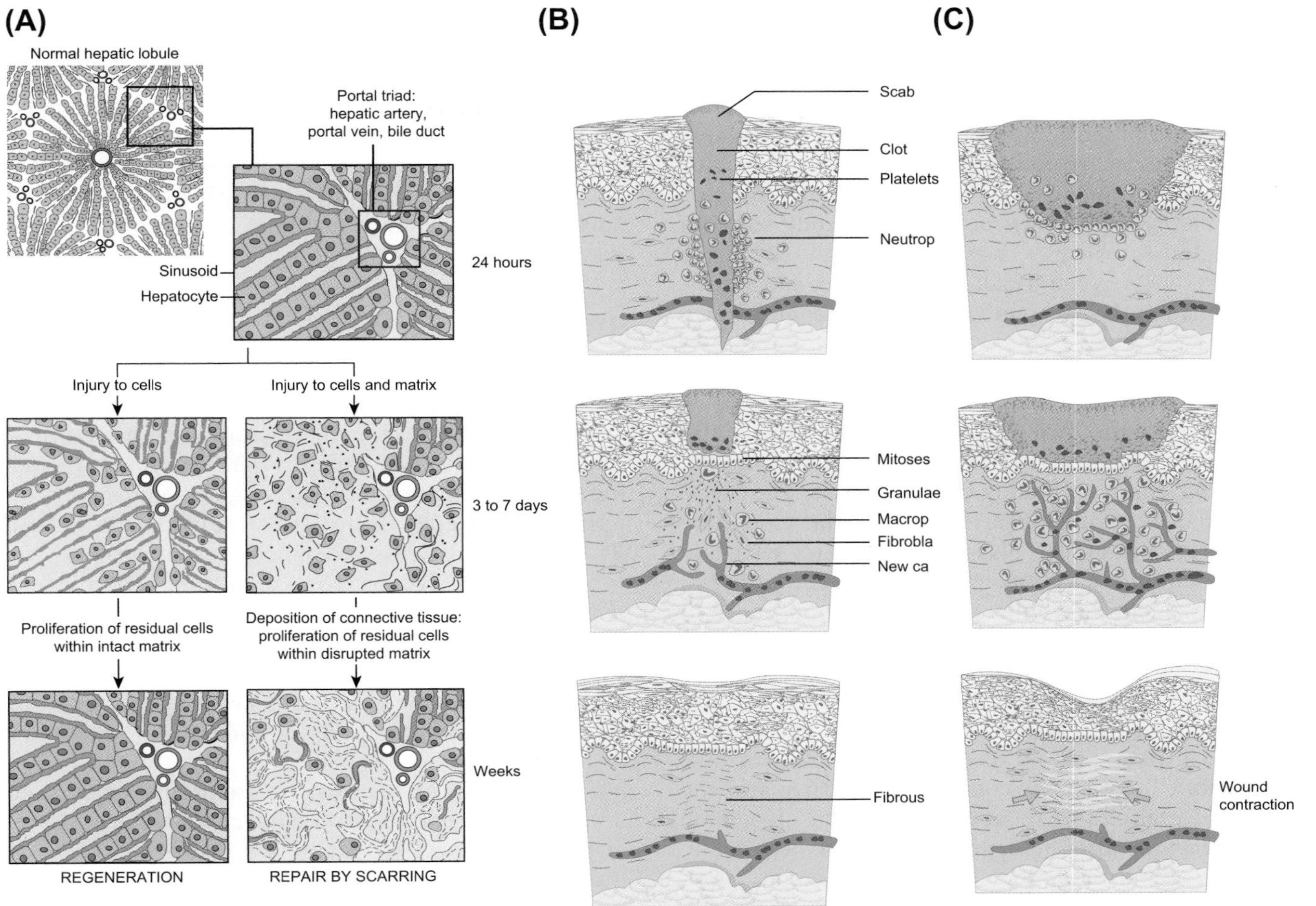

(A)
Normal hepatic lobule
Portal triad: hepatic artery, portal vein, bile duct
Sinusoid
Hepatocyte
24 hours
Injury to cells
Injury to cells and matrix
3 to 7 days
Proliferation of residual cells within intact matrix
Deposition of connective tissue: proliferation of residual cells within disrupted matrix
Weeks
REGENERATION
REPAIR BY SCARRING

(B)
Scab
Clot
Platelets
Neutrop
Mitoses
Granulae
Macrop
Fibrobla
New ca
Fibrous

(C)
Wound contraction

FIGURE II.1.5.11 Key concepts in tissue responses following injury. (A) Role of the extracellular matrix in regeneration versus repair, illustrated for liver injury. Regeneration requires an intact matrix and persistence of cells capable of proliferation. If both cells and matrix are damaged, repair occurs by fibrous tissue deposition and scar formation. (B) and (C) Wound healing and scar formation in skin. (B) Healing of a wound that caused little injury leads to a thin scar with minimal contraction. (C) Healing of a large wound causes substantial scar and more contraction and distortion of the tissue. (Reproduced by permission from Kumar, V., Abbas, A. K., Fausto, N. & Aster, J. C. (2010). *Robbins and Cotran Pathologic Basis of Disease*, 8th Edn. Saunders: Philadelphia, PA.)

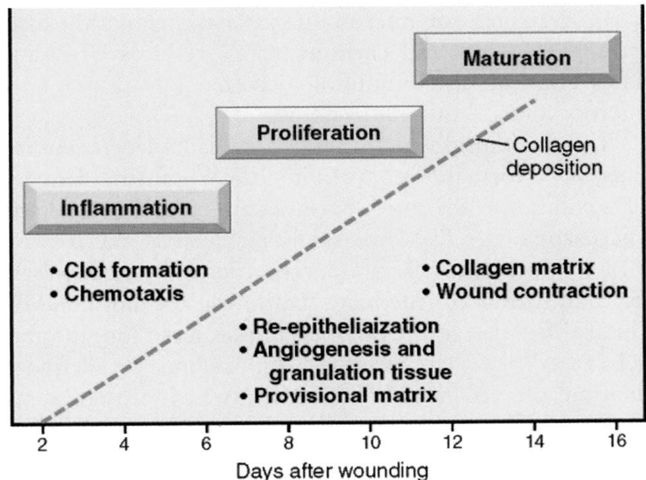

FIGURE II.1.5.12 Phases of cutaneous wound healing: Inflammation, proliferation, and maturation. (Reproduced by permission from Kumar, V., Abbas, A. K., Fausto, N. & Aster, J. C. (2010) *Robbins and Cotran Pathologic Basis of Disease*, 8th Edn. Saunders: Philadelphia, PA.)

Synthetic biomaterials are generally not immunogenic, and are therefore generally not "rejected" like a transplanted organ. However, they typically elicit the *foreign body reaction* (*FBR*), a special form of non-immune inflammation. The most prominent cells in the FBR are macrophages, which presumably attempt to phagocytose the material, but degradation is difficult. Depending on the local milieu and the resultant chemical signals, microphages will differentiate (activate) in different ways, either pro-inflammatory or pro-healing and fibrosis (Mosser and Edwards, 2008; Brown et al, 2012). Multinucleated *giant cells* in the vicinity of a foreign body are generally considered evidence of a more severe FBR in which the material is particularly irritating. This reaction is frequently called a *foreign body granuloma*. The more "biocompatible" the implant, the more quiescent is the ultimate response.

- Scarring: In situations where repair cannot be accomplished by regeneration, scarring occurs as a composite of three sequential processes: (1) formation of new blood vessels (*angiogenesis*); (2) deposition of collagen (*fibrosis*); and (3) maturation and remodeling of the scar (*remodeling*). The early healing tissue rich in new capillaries and proliferation of fibroblasts is called *granulation tissue* (not to be confused with granuloma, above). The essential features of the healing process are usually advanced by 4–6 weeks, although full scar remodeling may require much longer.

Inflammation is also associated with the release of chemical mediators from plasma, cells or extracellular matrix, which regulates the subsequent vascular and cellular events, and may modify their evolution. The chemical mediators of inflammation include the vasoactive amines (e.g., histamine), plasma proteases (of the coagulation, fibrinolytic, kinin, and complement systems), arachidonic acid metabolites (eicosinoids) produced in the cyclooxygenase pathway (the prostaglandins) and the lipoxygenase pathway (the leukotrienes), platelet-activating factor, cytokines (e.g., interleukin1 [IL-1], tumor necrosis factor-α [TNF-α] and interferon-γ [IFN-γ]), nitric oxide and oxygen-derived free radicals, and various intracellular constituents, particularly the lysosomal granules of inflammatory cells. Polypeptide growth factors also influence repair and healing by affecting cell growth, locomotion, contractility, and differentiation. Growth factors may act by endocrine (systemic), paracrine (stimulating adjacent cells) or autocrine (same cell carrying receptors for their own endogenously produced factors) mechanisms. Growth factors involved in mediating angiogenesis, fibroblast migration, proliferation, and collagen deposition in wounds include epidermal growth factor (EGF, important in proliferation of epithelial cells and fibroblasts), platelet-derived growth factor (PDGF, involved in fibroblast and smooth muscle cell migration), fibroblast growth factors (FGFs), transforming growth factor-β (TGF-β, with a central role in fibrosis), and vascular endothelial growth factor (VEGF, with a central role in angiogenesis).

TECHNIQUES FOR ANALYSIS OF CELLS AND TISSUES

Cells in culture (*in vitro*) often continue to perform many of the normal functions they have in the body (*in vivo*). Through measurement of changes in secreted products under different conditions, for example, culture methods can be used to study how cells respond to certain stimuli. However, since cells in culture do not have the usual chemical and physical environment that they have in tissues, normal physiological function may not always be present.

Techniques commonly used to study the structure of either normal or abnormal tissues, and the purpose of each mode of analysis, are summarized in Table II.1.5.2 and Figure II.1.5.13. The most widely used technique (historically and presently) for examining tissues is light microscopy, described below. Several advanced tissue analysis techniques are also increasingly being used to answer more detailed questions concerning cell and tissue function.

Light Microscopy

The conventional light microscopy technique involves obtaining the tissue sample, followed by fixation, paraffin embedding, sectioning, mounting on a glass slide, staining, and examination. Photographs of conventional tissue sections taken through a light microscope (photomicrographs) were illustrated in Figure II.1.5.5. Photographs of a tissue sample, paraffin block, and resulting tissue section on a glass slide are shown in Figure II.1.5.14. The tissue is obtained by either surgical excision (removal), biopsy (sampling) or autopsy (postmortem examination). A sharp instrument is used to dissect the tissue to avoid distortion from crushing. The key processing steps are summarized in the following paragraphs.

To preserve the structural relationships among cells, their environment, and subcellular structures in tissues, it is necessary to cross-link and preserve the tissue in a permanent, non-viable state called *fixation*. Specimens should be placed in fixatives as soon as possible after removal. Fixative solutions prevent degradation of the tissue when it is separated from its source of oxygen and nutrition (i.e., autolysis) by coagulating (i.e., cross-linking, denaturing, and precipitating) proteins. This prevents cellular hydrolytic enzymes (which are released when cells die) from degrading tissue components and spoiling tissues for microscopic analysis. Fixation also immobilizes fats and carbohydrates, reduces or eliminates enzymic and immunological reactivity, and kills microorganisms present in tissues.

A 37% solution of formaldehyde is called *formalin*; thus, 10% formalin is approximately 4% formaldehyde. This solution is the routine fixative in pathology for light microscopy. For TEM and scanning electron microscopy (SEM), glutaraldehyde preserves structural elements better than formalin. Adequate fixation in formalin and/or glutaraldehyde requires tissue samples less than 1.0 and 0.1 cm, respectively, in largest dimension. For adequate fixation, the volume of fixative into which a tissue sample is placed should generally be at least 5 to 10 times that of the tissue volume.

Dehydration and Embedding

In order to support the specimen during sectioning, specimen water (approximately 70% of tissue mass) must be

TABLE II.1.5.2 Techniques Used to Study Tissue

Technique	Purpose
Gross examination	Determine overall specimen configuration; many diseases and processes can be diagnosed at this level
Light microscopy (LM)	Study overall microscopic tissue architecture and cellular structure
Transmission electron microscopy (TEM)	Study ultrastructure and identify cells and their and environment
Enzyme histochemistry	Demonstrate the presence and location of enzymes in gross or microscopic sections
Immunohistochemistry	Identify and locate specific molecules, usually proteins, for which a specific antibody is available as a probe
Microbiologic cultures	Diagnose the presence of infectious organisms
Morphometric studies (at gross, LM or TEM levels)	Quantitate the amounts, configuration, and distribution of specific structures
Chemical, biochemical and spectroscopic analysis	Assess bulk concentration of molecular or elemental constituents
Energy dispersive x-ray analysis (EDXA)	Perform site-specific elemental analysis in sections
In-situ hybridization	Identify presence and location of mRNA or DNA within cells and tissues in histological sections
Northern blot	Detect the presence, amount, and size of specific mRNA molecules
Western blot	Identify the presence and amount of specific proteins
Southern blot	Detect and identify specific DNA sequences
Polymerase chain reaction (PCR)	Cyclical amplification of DNA to produce sufficient material for analysis
Tissue microarray	Permits study of multiple lesions simultaneously using tissue from multiple patients or blocks on the same slide; produced using a needle to biopsy conventional paraffin blocks and placing the cores into an array on a recipient paraffin block

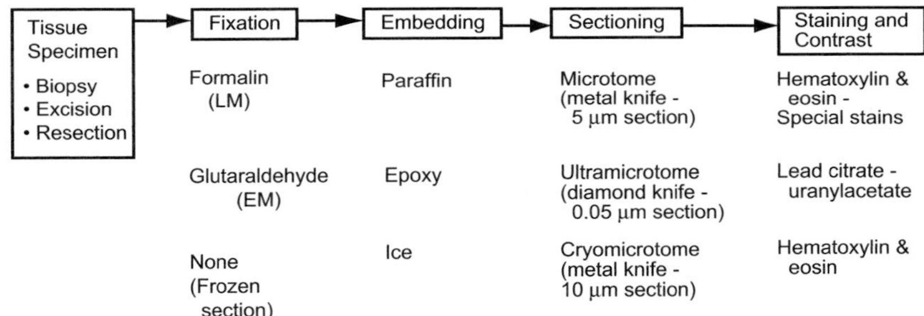

FIGURE II.1.5.13 Key features of tissue processing for examination by light and electron microscopy.

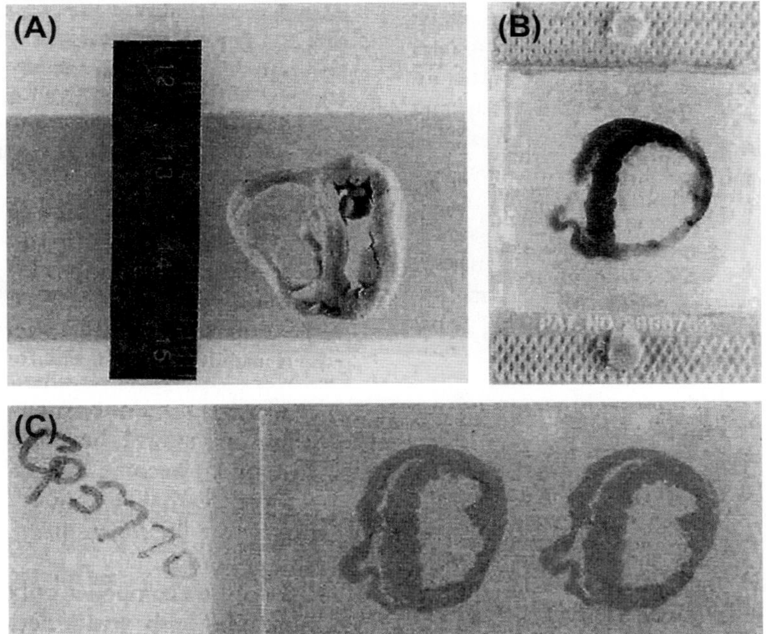

FIGURE II.1.5.14 Tissue processing steps for light microscopy. (A) Tissue section. (B) Paraffin block. (C) Resulting histologic section.

replaced by paraffin wax or other embedding medium, such as glycolmethacrylate. This is done through several steps, beginning with dehydration of the specimen through increasing concentrations of ethanol (eventually to absolute ethanol). However, since alcohol is not miscible with paraffin (the final embedding medium), xylol (an organic solvent) is used as an intermediate solution.

Following dehydration, the specimen is soaked in molten paraffin and placed in a mold larger than the specimen, so that tissue spaces originally containing water, as well as a surrounding cube, are filled with wax. The mold is cooled, and the resultant solid block containing the specimen (see Figure II.1.5.14B) can then be easily handled.

Sectioning

Tissue specimens are sectioned on a microtome (which has a blade similar to a single-edged razor blade), that is progressively advanced through the specimen block. The shavings are picked up on glass slides. Sections for light microscopic analysis must be thin enough to both transmit light and avoid superimposition of various tissue components. Typically sections are approximately 5 μm thick – slightly thicker than a human hair, but thinner than the diameter of most cells. If thinner sections are required (e.g., approximately 0.06 μm thick ultrathin sections are necessary) for TEM analysis, a harder supporting (embedding) medium (usually epoxy plastic) and a correspondingly harder knife (usually diamond) are used. Sections for TEM analysis are cut on an ultramicrotome. Because the conventional paraffin technique requires overnight processing, frozen sections are often used to render an immediate diagnosis (e.g., during a surgical procedure that might be modified according to the diagnosis). In this method, the specimen itself is frozen, so that the solidified internal water acts as a support medium, and sections are then cut in a cryostat (i.e., a microtome in a cold chamber). Although frozen sections are extremely useful for immediate tissue examination, the quality of the appearance is inferior to that obtained by conventional fixation and embedding methods. Frozen sections may also be used to section specimens that contain a component of interest that would be dissolved out or otherwise altered by processing with the organic solvents required by paraffin embedding (e.g., some polymers).

Staining

Tissue components have no intrinsic contrast, and are of fairly uniform optical density. Therefore, in order for tissue to be visible by light microscopy, tissue elements must be distinguished by selective adsorption of dyes (Luna, 1968). Since most stains are aqueous solutions of dyes, staining requires the paraffin in the tissue section to be removed and replaced by water (rehydration). The stain used routinely in histology involves sequential incubation in the dyes hematoxylin and eosin (H&E). Hematoxylin has an alkaline (basic) pH that stains blue–purple; substances stained with hematoxylin typically have a net negative charge and are said to be "basophilic" (e.g., cell nuclei containing DNA). In contrast, substances that stain with eosin, an acidic pigment that colors positively-charged tissue components pink–red, are said to be "acidophilic" or "eosinophilic" (e.g., cell cytoplasm, collagen). The tissue sections shown in Figure II.1.5.5 were stained with hematoxylin and eosin.

Special Staining and Immunohistochemistry

There are special staining methods for highlighting components that do not stain well with routine stains (e.g., microorganisms) or for indicating the chemical nature or the location of a specific tissue component (e.g., collagen, elastin; Table II.1.5.3). There are also special techniques for demonstrating the specific chemical activity of a compound in tissues (e.g., enzyme histochemistry). In this case, the specific substrate for the enzyme of interest is reacted with the tissue; a colored product precipitates in the tissue section at the site of the enzyme. In contrast, immunohistochemical staining takes advantage of the immunological properties (antigenicity) of a tissue component to demonstrate its nature and location by identifying sites of antibody binding. Antibodies to the particular tissue constituent are attached to a dye, usually a compound activated by a peroxidase enzyme, and reacted with a tissue

TABLE II.1.5.3 Stains for Light Microscopic Histology[a]

To Demonstrate	Stain
Overall morphology	Hematoxylin and eosin (H&E)
Collagen	Masson's trichrome
Elastin	Verhoeff-van Gieson
Glycosoaminoglycans (GAGs)	Alcian blue
Collagen–elastic–GAGs	Movat
Bacteria	Gram
Fungi	Methenamine silver or periodic acid-Schiff (PAS)
Iron	Prussian blue
Calcium phosphates (or calcium)	von Kossa (or alizarin red)
Fibrin	Lendrum or phosphotungstic acid hematoxylin (PTAH)
Amyloid	Congo red
Inflammatory cell types	Esterases (e.g., chloroacetate esterase for neutrophils, nonspecific esterase for macrophages)

[a]Reproduced by permission from F. J. Schoen, *Interventional and Surgical Cardiovascular Pathology: Clinical Correlations and Basic Principles*, Saunders, 1989.

section (immunoperoxidase technique) or the antibody is attached to a compound that is excited by a specific wavelength of light (immunofluorescence). Although some antigens and enzymatic activity can survive the conventional histological processing technique, both enzyme activity and immunological reactivity are often largely eliminated by routine fixation and embedding. Therefore, histochemistry and immunohistochemistry are frequently done on frozen sections. Nevertheless special preservation and embedding techniques now available often allow immunological methods to be carried out on carefully fixed tissue.

Electron Microscopy

Contrast in the electron microscope depends on relative electron densities of tissue components. Sections are stained with salts of heavy metals (osmium, lead, and uranium), which react differentially with different structures, creating patterns of electron density that reflect tissue and cellular architecture. An example of an electron photomicrograph is shown in Figure II.1.5.1B.

It is often possible to derive quantitative information from routine tissue sections using various manual or computer-aided methods. Morphometric or stereologic methodology, as these techniques are called, can be extremely useful in providing an objective basis for otherwise subjective measurements (Loud and Anversa, 1984). Techniques for sampling information from small amounts of tissue include laser microdissection, confocal microscopy, two-photon microscopy, etc.

Three-Dimensional Interpretation

Interpretation of tissue sections depends on the reconstruction of three-dimensional information from two-dimensional observations on tissue sections that are usually thinner than a single cell. Therefore, a single section may yield an unrepresentative view of the whole. A particular structure (even a very simple one) can look very different, depending on the plane of section. Figure II.1.5.15 shows how multiple sections must be examined to appreciate the actual configuration of an object or a collection of cells.

Artifacts

Artifacts are unwanted or confusing features in tissue sections that result from errors or technical difficulties in obtaining, processing, sectioning or staining the specimen. Recognition of artifacts avoids misinterpretation. The most frequent and important artifacts are autolysis, tissue shrinkage, separation of adjacent structures, precipitates formed by poor buffering or by degradation of fixatives or stains, folds or wrinkles in the tissue sections, knife nicks or rough handling (e.g., crushing) of the specimen.

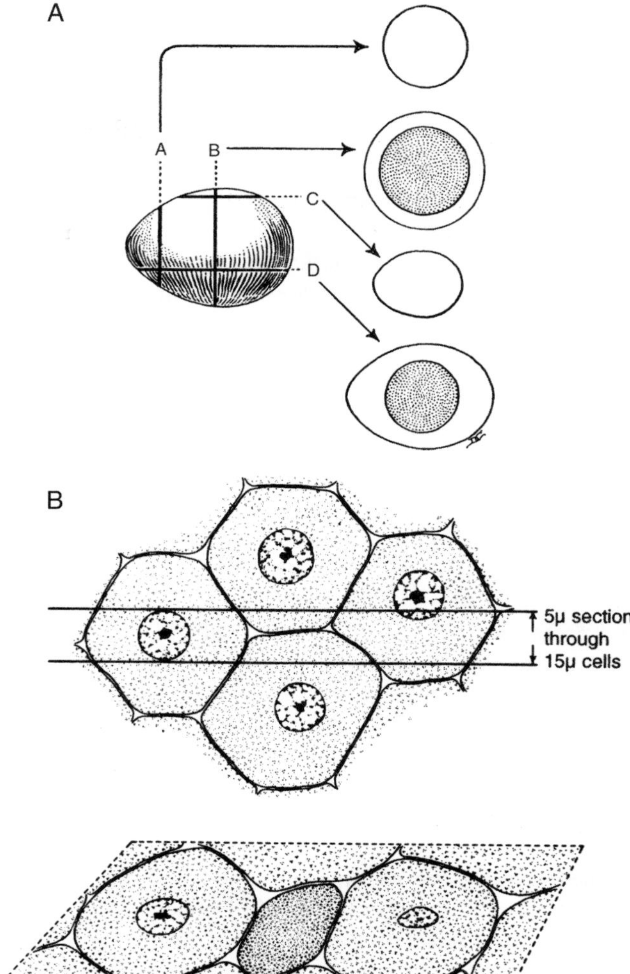

FIGURE II.1.5.15 Considerations for three-dimensional interpretation of two-dimensional information. Sections through a subject in different levels and orientations can give different impressions about its structure, here illustrated (A) for a hard-boiled egg, and (B) for a section through cells of a uniform size that suggests that the cells are heterogenous. (Modified by permission from Cormack, D. H. (1993). *Essential Histology*. Copyright 1993, Lippincott.)

Identification, Genotyping, and Functional Assessment of Cells, Including Synthetic Products, in Cells or Tissue Sections

It is frequently necessary to accurately ascertain or verify the identity of a cell, or to determine some aspect of its function, including the production of synthetic molecules. For such an assay, either isolated cells or whole tissues are used, depending on the objective of the study. Isolated cells or minced tissue have the advantage of allowing molecular and/or biochemical analyses on the cells and/or products, and often allow the acquisition of quantitative data. Nevertheless, the major advantage of whole tissue preparations is the ability to spatially localize molecules of interest in the context of architectural features of the tissue. Methods for identification and determination of the function of cells and identification of extracellular components in situ are summarized in Table II.1.5.4. New techniques are available. Cellular apoptosis and proliferation can be quantified (Watanabe et al., 2002). Immunohistochemical markers allow detection of proteins that are highly expressed in a tissue section. However, the relevant antibodies to proteins expressed in high concentration must be available, and the expense of such studies limits their usefulness. In-situ hybridization permits similar investigation of gene expression but, as with immunohistochemistry, only a discrete panel of previously predicted genes can be probed.

Advanced Tissue and Cell Analysis

Several very exciting new and evolving techniques are available to assay the anatomy and function of tissues and cells with extraordinary precision and/or spatial localization. These can be used individually or coupled. Very exciting are various forms of functional genomic and proteomic analyses (Chen et al., 2009). In particular, gene expression profiling shows the complete array of genes expressed in cells or tissues; the technology may identify pathogenetically distinct subtypes of any lesion and search for fundamental mechanisms even when candidate genes are unknown (Duggal et al., 2009). Confocal microscopy helps localize a particular component in a living cell by observing a series of optical sections (planes) which are reconstructed into a three-dimensional image (Watson et al., 2000; Howell et al., 2002). Tissue microassays permit the comparative examination of potentially hundreds of individual specimens in a single paraffin block (Hewitt, 2012). In addition, laser-assisted microdissection techniques permit isolation of an individual or a homogenous population of cells on selected cell populations under direct visualization from a routine histological section of complex, heterogeneous tissue (Edwards, 2007; Kurkalli et al., 2010). Very exciting new imaging technologies may permit analysis of viable tissues *in vivo* (Pinaud et al., 2006; Georgakoudi et al., 2008; Weissleder and Pittet, 2008).

TABLE II.1.5.4 Determination of Cell Function, Gene Expression, and Synthetic Products

Special histologic stains
Electron microscopy
Histochemistry/cytochemistry
Autoradiography
Antibody methods for light microscopy
Immunofluorescence
Immunoperoxidase
Immuno-electron microscopy
Flow cytometry
Molecular methods
In-situ hybridization
In-situ polymerase chain reaction (PCR)
TUNEL staining (for apoptosis)
Gene expression profiling (with or without laser capture microdissection)
Molecular imaging
Confocal microscopy

REGENERATIVE CAPACITY OF CELLS AND TISSUES

Organs are theoretically capable of renewing via three routes: (1) proliferation and differentiation of resident stem cells; (2) dedifferentiation of resident cells followed by proliferation/differentiation; and (3) homing and differentiation of circulating stem cells. Most types of cell populations can undergo turnover, but the process is highly regulated, and the production of cells of a particular kind generally ceases until some are damaged or another need arises. Rates of proliferation are different among various cell populations, and are frequently divided into three categories: (1) renewing (also called labile) cells have continuous turnover, with proliferation balancing cell loss that accrues by death or physiological depletion; (2) expanding (also called stable) cells, normally having a low rate of death and replication, retain the capacity to divide following stimulation; and (3) static (also called permanent) cells, not only have no normal proliferation, but they have also lost their capacity to divide. The relative proliferative (and regenerative) capacity of various cell types is summarized in Table II.1.5.5.

In normally renewing (labile) cell populations (e.g., skin, intestinal epithelium, bone marrow), tissue-specific (often called "adult") stem cells proliferate to form daughter cells that can become differentiated and repopulate the damaged cells. A particular stem cell produces many such daughter cells, and, in some cases, several different kinds of cells can arise from a common multipotential ancestor cell (also called multipotency, e.g., bone marrow stem cells lead to several different types of blood cells). In epithelia (such as the

TABLE II.1.5.5 Regenerative Capacity of Cells Following Injury

Category	Normal rate of Replication	Response to Stimulus/ Injury	Examples
Renewing/ labile	High	Modest increase	Skin, intestinal mucosa, bone marrow
Expanding/ stable	Low	Marked increase	Endothelium, fibroblasts, liver cells
Static/ permanent	None	No replication; replacement by scar	Heart muscle cells, nerves

epidermis layer of skin), the stem cells reside at the base of the tissue layer, away from the surface; differentiation and maturation occur as the cells move toward the surface. Regenerative tissues, especially those with high proliferative capacity, such as the hematopoietic system (e.g., bone marrow) and the epithelia of the skin and the gastrointestinal tract, renew themselves continually and can regenerate following injury, providing that the stem cells of these tissues have not been destroyed. Expanding (stable) populations can increase their rate of replication in response to suitable stimuli. Stable cell populations include glandular epithelial cells, liver, fibroblasts, vascular smooth muscle cells, osteoblasts, and endothelial cells. Cells that die are generally replaced by new ones of the same kind. In contrast, permanent (static) cells have minimal, if any, normal mitotic capacity and, in general, cannot be induced to regenerate. Following injury, they are replaced by scar. The inability to regenerate large, functional amounts of certain tissue types following injury can result in a clinically important deficit, since the function of the damaged tissue is irretrievably lost. For example, an area of heart muscle or brain which is damaged by ischemic injury (e.g., myocardial infarction) cannot be effectively replaced by viable cells; the necrotic area is repaired by scar has no functional potential. Therefore, in heart the remainder of the heart muscle must assume the workload of the lost tissue. In brain, the majority of function is lost in the cells damaged by a stroke.

While the classical concepts enumerated above continue to hold true from a practical standpoint, recent evidence suggests that some regeneration of neural tissue and heart muscle cells can occur under certain circumstances following injury. Both the extent to which this can occur and effective strategies to harness this potential and exciting source of new tissue are as yet unknown (Orive et al., 2009; Hosoda et al., 2010; Yi et al., 2010).

CELL/TISSUE–BIOMATERIALS INTERACTIONS

For most applications, biomaterials are in contact with cells and tissues (either hard tissue, including bone; soft tissue, including cardiovascular tissues; and blood in the case of cardiovascular implants or extracorporeal devices), often for prolonged periods. Thus, rational and sophisticated use of biomaterials and design of medical devices requires some knowledge of the general concepts concerning the interaction of cells with non-physiological surfaces. This discussion complements that described in Chapter II.1.2. and II.1.3.

Most tissue-derived cells require attachment to a solid surface for viability, growth, migration, and differentiation. The nature of that attachment is an important regulator of those functions. Moreover, the behavior and function of adherent cells (e.g., shape, proliferation, synthetic function) depend on the characteristics of the substrate, particularly its adhesiveness and mechanical properties (Reilly & Engler, 2010).

Following contact with tissue or blood, a bare surface of a biomaterial is covered rapidly (usually in seconds) with proteins that are adsorbed from the surrounding body fluids. The chemistry of the underlying substrate (particularly as it affects wettability and surface charge) controls the nature of the adherent protein layer. For example, macrophage fusion and platelet adhesion/aggregation are strongly dependent on surface chemistry. Moreover, although cells are able to adhere, spread, and grow on bare biomaterial surfaces *in vitro*, proteins absorbed from the adjacent tissue environment or blood and/or secreted by the adherent cells themselves markedly enhance cell attachment, migration, and growth.

Cell adhesion to biomaterials and the ECM is mediated by cytoskeletally-associated receptors in the cell membrane, which interact with cell adhesion proteins that adsorb to the material surface from the surrounding plasma and other fluids (Figure II.1.5.16). Cell adhesion triggers multiple functional biochemical signaling

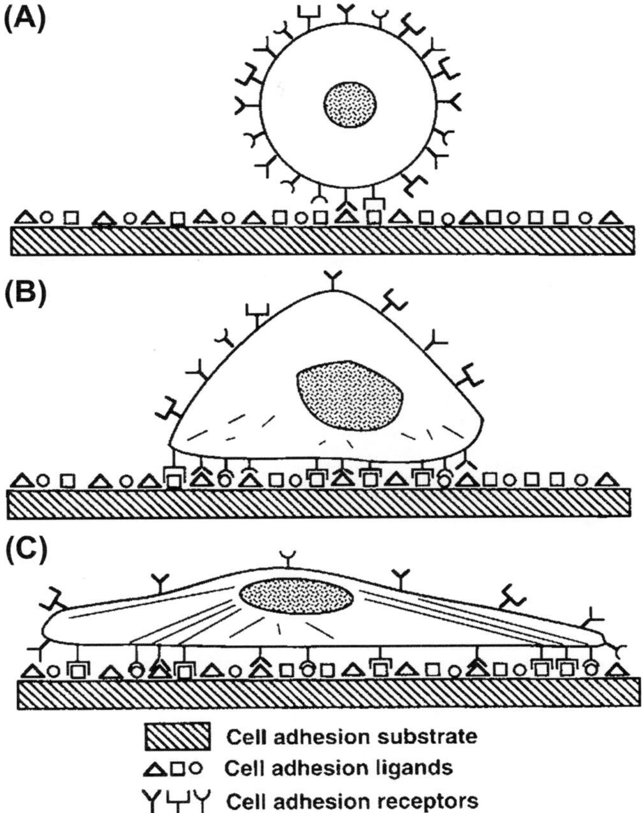

FIGURE II.1.5.16 Progression of anchorage-dependent cell adhesion. (A) Initial contact of cell with solid substrate. (B) Formation of bonds between cell surface receptors and cell adhesion ligands. (C) Cytoskeletal reorganization with progressive spreading of the cell on the substrate for increased attachment strength. (Reproduced by permission from Massia, S. P. (1999). Cell–Extracellular Matrix Interactions Relevant to Vascular Tissue Engineering. In: *Tissue Engineering of Prosthetic Vascular Grafts*, Zilla, P. & Greisler, H. P. (eds.). RG Landes Co., 1999.)

pathways within the cell (Chen, 2008). The interactions are critical to cell-growth control responsive to mechanical forces mediated through associated changes in cell shape and cytoskeletal tension (Ingber, 2010b). Focal adhesions are considered to represent the strongest such interactions. They comprise a complex assembly of intra- and extracellular proteins, coupled to each other through transmembrane integrins. Cell-surface integrin receptors promote cell attachment to substrates, and especially those covered with the extracellular proteins fibronectin and vibronectin. These receptors transduce biochemical signals to the nucleus by activating the same intracellular signaling pathways that are used by soluble growth factor receptors. The more cells spread, the higher their rate of proliferation. The importance of cell spreading on their proliferation has been emphasized by classical experiments that used endothelial cells cultured on microfabricated substrates containing fibronectin-coated islands of various defined shapes and sizes of a micrometer scale (Figure II.1.5.17) (Chen et al., 1997). Cells spread to the limits of the islands containing a fibronectin substrate; cells on circular islands were circular while cells on square islands became square in shape and had 90° corners. When the spreading of the cells was restricted by small adhesive islands (10–30 μm), proliferation was arrested, while larger islands (80 μm) permitted proliferation. When the cells were grown on micropatterned substrates with 3–5 μm dots, comprising multiple adhesive islands that permitted the cells to extend over multiple islands while maintaining the total ECM contact similar to that of one small island (that was associated with inhibited growth), they proliferated. This confirmed that the ability of cells to proliferate depended directly on the degree to which the cells were allowed to distend physically, and not on the actual surface area of substrate binding. Thus, cell distortion is a critical determinant of cell behavior.

Interactions of cells with ECM differ from those with soluble regulatory factors owing to the reciprocal interactions between the ECM and the cell's actin cytoskeleton (Ingber, 2010b). For example, rigid substrates promote cell spreading and growth in the presence of soluble mitogens; in contrast, flexible scaffolds that cannot resist cytoskeletal forces promote cell retraction, inhibit growth, and promote differentiation. Thus, the properties of the nature and configuration of the surface-bound ECM on a substrate, and the properties of the substrate itself, can regulate cell–biomaterials interactions. The key

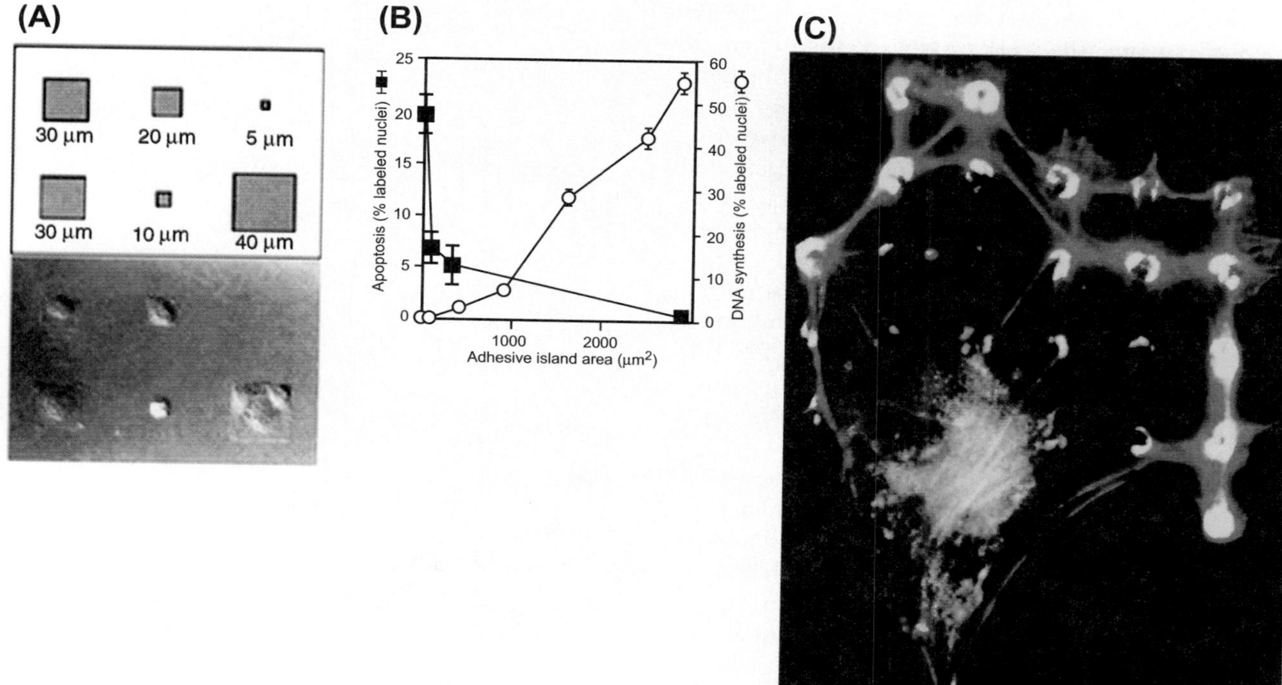

FIGURE II.1.5.17 Effect of spreading on cell growth and apoptosis. (A) Schematic diagram showing the initial pattern design containing different-sized square adhesive islands and Nomarski views of the final shapes of bovine adrenal capillary endothelial cells adherent to the fabricated substrate. Distances indicate lengths of the square's sides. (B) Apoptotic index (percentage of cells exhibiting positive TUNEL staining) and DNA synthesis index (percentage of nuclei labeled with 5-bromodeoxyuridine) after 24 hours, plotted as a function of the projected cell area. Data were obtained only from islands that contained single adherent cells; similar results were obtained with circular or square islands, and with human or bovine endothelial cells. (C) Fluorescence micrograph of an endothelial cell spread over a substrate containing a regular array of small (5 μm diameter) circular ECM islands separated by nonadhesive regions created with a microcontact printing technique. Yellow rings and crescents indicate colocalization of vinculin (green) and F-actin (red) within focal adhesions that form only on the regulatory spaced circular ECM islands. ((A), (B) Reproduced by permission from Chen, C. S. et al. (1997) Geometric control of cell life and death. *Science*, **276**, 1425; (C) Reproduced by permission from Ingber, D. E. (2003). Mechanosensation through integrins: Cells act locally but think globally. *Proc Natl Acad. Sci.*, **100**, 1472.)

concept is that a biomaterial surface can contain specific chemical and structural information that controls tissue formation, in a manner analogous to cell–cell communication and patterning during embryological development.

The exciting potential of this strategy is exemplified by tissue engineering approaches that employ biomaterials with surfaces designed to stimulate biophysical and biochemical microenviromental use of highly precise spatial and temporal reactions with proteins and cells at the molecular level, to trigger cell fate decisions such as adhesion, proliferation, migration, and specific gene-expression profiles and differentiation patterns (Lutolf et al, 2009). Such materials provide the scientific foundation for molecular design of scaffolds that could be seeded with cells *in vitro* for subsequent implantation or specifically attract endogenous functional cells *in vivo*. A key challenge in tissue engineering is to understand quantitatively how cells respond to various molecular signals and integrate multiple inputs to generate a given response, and to control non-specific interactions between cells and a biomaterial, so that cell responses specifically follow desired receptor–ligand interactions. Understanding the principles regulating these responses, the means to chart them, and the ability to fabricate elaborate interfaces with carefully controlled ligand type, density, clustering, and spatial distribution (potentially in three dimensions) will ultimately permit novel biosensors, smart biomaterials, advanced medical devices, microarrays, and tissue regenerative approaches.

BIBLIOGRAPHY

Berman, I. (1993). *Color Atlas of Basic Histology*. Appleton and Lanse.

Bokel, C., & Brown, N. H. (2002). Integrins in development: Moving on, responding to, and sticking to the extracellular matrix. *Develop. Cell*, **3**, 311–321.

Borysenko, M., & Bringer, T. (1989). *Functional Histology* (3rd ed.). Little, Brown, and Co.

Bozsenko, M., & Beringer., T. (1989). *Functional Histology* (3rd ed.). : Little Brown and co.

Brown, B. N., Ratner, B. D., Goodman, s. B., Amar, S., & Badylak, S. F. (2012). Macrophage polarization: An opportunity for improved outcomes in biomaterials and regenerative medicine. *Biomaterials*, **33**, 3792–3802.

Carmeliet, P. (2003). Angiogenesis in health and disease. *Nature Med.*, **9**, 653–660.

Chen, C. S. (2008). Mechanotransduction: A field pulling together? *J. Cell Sci.*, **121**, 3285–3292.

Chen, C. S., Mrksich, M., Huang, S., Whitesides, G., & Ingber, D. E. (1997). Geometric control of cell life and death. *Science*, **276**, 1425–1428.

Chen, X., Jorgenson, E., & Cheung, S. T. (2009). New tools for functional genomic analysis. *Drug Discov. Today*, **14**, 754–760.

Cormack, D. H. (1993). *Essential Histology*. (2nd ed.) Philadelphia PA: Lippincott.

Cormack, D. H. (1987). *Han's Histology* (9th ed.). Philadelphia, PA: Lippincott.

Daley, W. P., Peters, S. B., & Larsen, M. (2008). Extracellular matrix dynamics in development and regenerative medicine. *J. Cell Sci*, **121**, 255–264.

Duggal, S., Fronsdal, K. B., Szöke, K., Shahdadfar, A., Melvik, J. E., et al. (2009). Phenotype and gene expression of human mesenchymal stem cells in alginate scaffolds. *Tissue Eng. Part A*, **15**, 1763–1773.

Edwards, R. A. (2007). Laser capture microdissection of mammalian tissue. *J. Vis. Exp.*, **8**, 309.

Georgakoudi, I., Rice, W. L., Hronik-Tupaj, M., & Kaplan, D. L. (2008). Optical spectroscopy and imaging for the noninvasive evaluation of engineered tissues. *Tissue Eng. Part B*, **14**, 321–340.

Gurtner, G. C., Werner, S., Barrandon, Y., & Longaker, M. T. (2008). Wound repair and regeneration. *Nature*, **453**, 314–321.

Hewitt, S. M. (2012). Tissue microarrays as a tool in the discovery and validation of predictive biomarkers. *Methods Mol Biol.*, **823**, 201–214.

Hinz, B., Phan, S. H., Thannickal, V. J., Galli, A., Bochaton-Piallat, M. L., et al. (2007). The myofibroblast: One function, multiple origins. *Am. J. Pathol.*, **170**, 1807–1816.

Hosoda, T., Kajstura, J., Leri, A., & Anversa, P. (2010). Mechanisms of myocardial regeneration. *Circ. J.*, **74**, 13–17.

Howell, K., Hopkins, N., & Mcloughlin, P. (2002). Combined confocal microscopy and stereology: A highly efficient and unbiased approach to quantitative structural measurement of tissues. *Exp. Physiol.*, **87**, 747–756.

Ingber, D. E. (2003). Mechanosensation through integrins: Cells act locally but think globally. *Proc Natl Acad Sci*, **100**, 1472.

Ingber, D. E. (2010a). Mechanical control of tissue and organ development. *Development*, **137**, 1407–1420.

Ingber, D. E. (2010b). From cellular mechanotransduction to biologically inspired engineering. *Ann. Biomed. Eng.*, **38**, 1148–1161.

Kumar, V., Abbas, A. K., Fausto, N., & Aster, J. C., (2010). *Robbins and Cotran Pathologic Basis of Disease* (8th ed.). Philadelphia, PA: Saunders, pp. 487–528.

Kurkalli, B. G., Gurevitch, O., Sosnik, A., Cohn, D., & Slavin, S. (2010). Repair of bone defect using bone marrow cells and demineralized bone matrix supplemented with polymeric materials. *Curr. Stem Cell Res. Ther.*, **5**, 49–56.

Loud, A. V., & Anversa, P. (1984). Morphometric analysis of biological processes. *Lab. Invest.*, **50**, 250–261.

Luna, M. G. (1968). *Manual of Histologic Staining Methods of the Armed Forces Institute of Pathology* (3rd ed.). New York, NY: McGraw-Hill.

Lutolf, M. P., & Hubbell, J. A. (2005). Synthetic biomaterials as instructive extracellular microenvironments for morphogenesis in tissue engineering. *Nat. Biotechnol.*, **23**, 47–55.

Lutolf, M. P., Gilbert, P. M., & Blau, H. M. (2009). Designing materials to direct stem-cell fate. *Nature*, **26**(462), 433–441.

Mammoto, T., & Ingber, D. E. (2010). Mechanical control of tissue and organ development. *Development*, **137**, 1407–1420.

Massia, S. P. (1999). Cell–Extracellular Matrix Interactions Relevant to Vascular Tissue Engineering. In P. Zilla, & H. P. Greisler (Eds.), *Tissue Engineering of Prosthetic Vascular Grafts.* : RG Landes Co., 1999.

Mitchell, R. N., & Schoen, F. J. (2010). Blood Vessels. In V. Kumar, N. Fausto, J. C. Aster, & A. Abbas (Eds.), *Robbins and Cotran Pathologic Basis of Disease* (8th ed.). Philadelphia, PA: Saunders, pp. 487–528.

Mosser, D. M., & Edwards, J. P. (2008). Exploring the full spectrum of macrophage activation. *Nature*, **8**, 958–969.

Orive, G., Anitua, E., Pedraz, J. L., & Emerich, D. F. (2009). Biomaterials for promoting brain protection, repair and regeneration. *Nat. Rev. Neurosci.*, **10**, 682–692.

Page-McCaw, A., Ewald, A. J., & Werb, Z. (2007). Matrix metalloproteinases and the regulation of tissue remodeling. *Nat. Rev. Mol. Cell Biol.*, **8**, 221–233.

Pinaud, F., Michalet, X., Bentolila, L. A., Tsay, J. M., Doose, S., et al. (2006). Advances in fluorescence imaging with quantum dot bio-probes. *Biomaterials*, **27**, 1679–1687.

Reilly, G. C., & Engler, A. J. (2010). Intrinsic extracellular matrix properties regulate stem cell differentiation. *J. Biomech.*, **43**, 55–62.

Rubin, R., & Strayer, D. S. (2012). *Rubin's Pathology* (6th ed.). Lippincott, Williams & Wilkins Philadelphia.

Schoen, F. J. (1989). *Interventional and Surgical Cardiovascular Pathology: Clinical Correlations and Basic Principles*: W.B. Saunders.

Schoen, F. J. (2004). The heart. In R. S. Cotran., V. Kumar., & T. Collins (Eds.), *Robbins Pathologic Basis of Diseases* (7th ed.). Philadelphia, PA: W.B. Saunders.

Schoen, F. J. (2008). Evolving concepts of heart valve dynamics. The continuum of development, functional structure, pathology and tissue engineering. *Circulation*, **118**, 864–1880.

Singer, A. J., & Clark, R. A. (1999). Cutaneous wound healing. *N. Engl. J. Med.*, **341**, 738–746.

Watanabe, M., Hitomi, M., van der Wee, K., Rothenberg, F., Fisher, S. A., et al. (2002). The pros and cons of apoptosis assays for use in the study of cells, tissues and organs. *Microsc. Microanal.*, **8**, 375–391.

Watson, T. F., Azzopardi, A., Etman, M., Cheng, P. C., & Sidhu, S. K. (2000). Confocal and multi-photon microscopy of dental hard tissues and biomaterials. *Am. J. Dent.*, **13**, 19D–24D.

Weissleder, R., & Pittet, M. J. (2008). Imaging in the era of molecular oncology. *Nature*, **452**, 580–589.

Wozniak, M. A., & Chen, C. S. (2009). Mechanotransduction in development: A growing role for contractility. *Nat. Rev. Cell Mol. Biol.*, **10**, 34–43.

Yi, B. A., Wernet, O., & Chien, K. R. (2010). Pregenerative medicine: Developmental paradigms in the biology of cardiovascular regeneration. *J. Clin. Invest.*, **120**, 20–28.

CHAPTER II.1.6 EFFECTS OF MECHANICAL FORCES ON CELLS AND TISSUES (THE LIQUID–CELL INTERFACE)

Daniel E. Conway[1], Suzanne G. Eskin[2], and Larry V. McIntire[2]

[1]Robert M. Berne Cardiovascular Research Center, University of Virginia, Charlottesville, Virginia, USA

[2]Department of Biomedical Engineering, Georgia Institute of Technology and Emory University School of Medicine, Atlanta, Georgia, USA

Mechanical properties of materials have already been covered in Chapter I.1.3 on "Bulk Properties of Materials." In this chapter, we will cover the effects of mechanical forces on cells within tissues, on the surfaces of biomaterials or within polymer scaffolds, focusing primarily on the effects of fluid forces. Because host cells interact with implanted materials or because cells are implanted as therapeutic entities in themselves, often within a biomaterial scaffold, the response of cells to mechanical forces is important to consider in order to predict the success of an implant. Cells that are particularly adapted for functioning in concert with physical forces are those of the cardiovascular and musculoskeletal systems, but all cells experience some mechanical forces. In fact, the appropriate local mechanical environment appears to be crucial for maintenance of proper cell phenotype. This chapter will focus on physiological and pathological flow-induced shear stress and fluid-induced strains that cells experience with emphasis on the effects of these forces on cell function. The key role of mechanical forces in tissue formation and maintenance of cell–extracellular matrix and cell–cell interactions will be discussed in Chapter II.6.5 in the context of micromechanical design criteria for tissue-engineered biomaterials.

CELLULAR DETECTION OF MECHANICAL FORCES

Mechanotransduction is the transduction of mechanical forces applied to a cell or tissue into chemical and biological responses. Sensors of mechanical forces include ion channels, G-protein-coupled receptors, intercellular junctions, and focal adhesions (Davies, 1995; Papadaki and Eskin, 1997). Most of these sensors have direct connections to the cytoskeleton, and often are unable to transduce physical forces when the attachment to the cytoskeleton is lost, suggesting that coupling of surface mechano-receptors to the cytoskeleton is required for mechanical transduction (Alenghat and Ingber, 2002). An additional indirect effect of mechanical force is in the redistribution of extracellular signaling molecules. Mechano-sensing cells, such as hair cells in the inner ear, are specialized for their role in sensing mechanical forces. However, most other cells with primary functions that do not include sensing mechanical forces are responsive to such forces if exposed to them.

> Cells that are particularly adapted for functioning in concert with physical forces are those of the cardiovascular and musculoskeletal systems, but all cells experience some mechanical forces. In fact, the appropriate local mechanical environment appears to be crucial for maintenance of proper cell phenotype.
>
> Mechano-sensing cells, such as hair cells in the inner ear, are specialized for their role in sensing mechanical forces. However, most other cells with primary functions that do not include sensing mechanical forces are responsive to such forces if exposed to them.

BLOOD VESSELS

Mechanical forces resulting from blood flow directly affect cellular functions and morphology and thus, the physiology of the cardiovascular system (Davies, 2009). The forces to which cells of the blood vessels are subjected directly include wall shear stress and circumferential strain (Figure II.1.6.1). Disruption or distortion of these forces affects local cellular homeostasis, and may prime the area for development of pathology. Sites of atherosclerosis in humans have been shown to be non-random, and are strongly correlated with curved or branched sites in the vasculature that, as a result of their geometry, experience low time-averaged shear stress, shear stress reversal, and spatial and temporal gradients in shear stress (Glagov et al., 1988; Davies, 2008). Disturbances in mechanical forces due to implantation of medical devices can also alter cellular function. Small diameter arterial grafts (both synthetic and autologous vein) frequently fail because of compliance mismatch with the native vessel and resulting flow and wall strain disturbances at the anastomoses (see Box 1) (Singh et al., 2007)

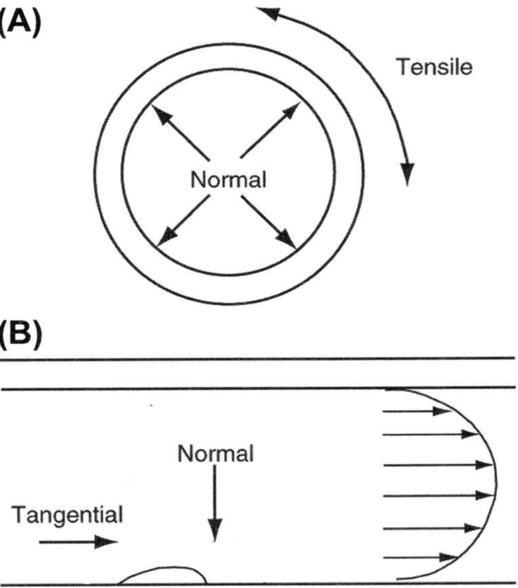

FIGURE II.1.6.1 Schematic representation of the hemodynamic forces acting on the artery wall. (A) Cross-section through the artery, in which all layers are constantly subjected to tensile stress (cyclic strain) which arises from normal forces generated by the pusatility of blood flow. (B) Longitudinal section through the artery. Fluid flow profile (illustrated by the parabola) imparts tangential forces (shear stress) to the blood vessel wall. Endothelial cells lining the blood vessel wall respond first to this force.

Blood vessels are comprised of three major cell types: endothelial cells; smooth muscle cells; and fibroblasts. A one-cell thick layer of endothelial cells forms the endothelium that lines the lumen of the entire vasculature. Large arteries also include a middle layer, the media, which consists of smooth muscle cells with secreted extracellular matrix, and an outermost layer, the adventitia, consisting primarily of fibroblasts and extracellular matrix, which also contains blood vessels and nerves supplying the artery itself.

Extensive studies that apply shear stress or cyclic strain to endothelial cells *in vitro* confirm that these cells actively participate in vascular physiology. Although only one cell thick, the endothelium provides a permeability barrier; controls thrombosis and hemostasis by maintaining an active thromboresistant surface, which no biomaterial has yet been able to match, and acts as a mechanosensor for the underlying tissue. Smooth muscle cells in the medial layer contract, relax, proliferate, synthesize matrix or migrate in response to shear stress, cyclic strain, and paracrine factors (biochemical signals) from endothelial cells. Vessels denuded of the endothelium expose the smooth muscle cells and matrix to blood flow, resulting in binding of platelets and, subsequently, thrombosis (Wagner and Frenette, 2008).

Effect of Shear Stress on Blood Vessels

The flow of blood over the endothelium generates viscous drag forces in the direction of flow. The resulting tangential force exerted per unit area of vessel surface at the blood–endothelium surface defines shear stress. Mathematically, the product between the viscosity (μ) and the

Box 1 Hemodialysis Access Fistulas

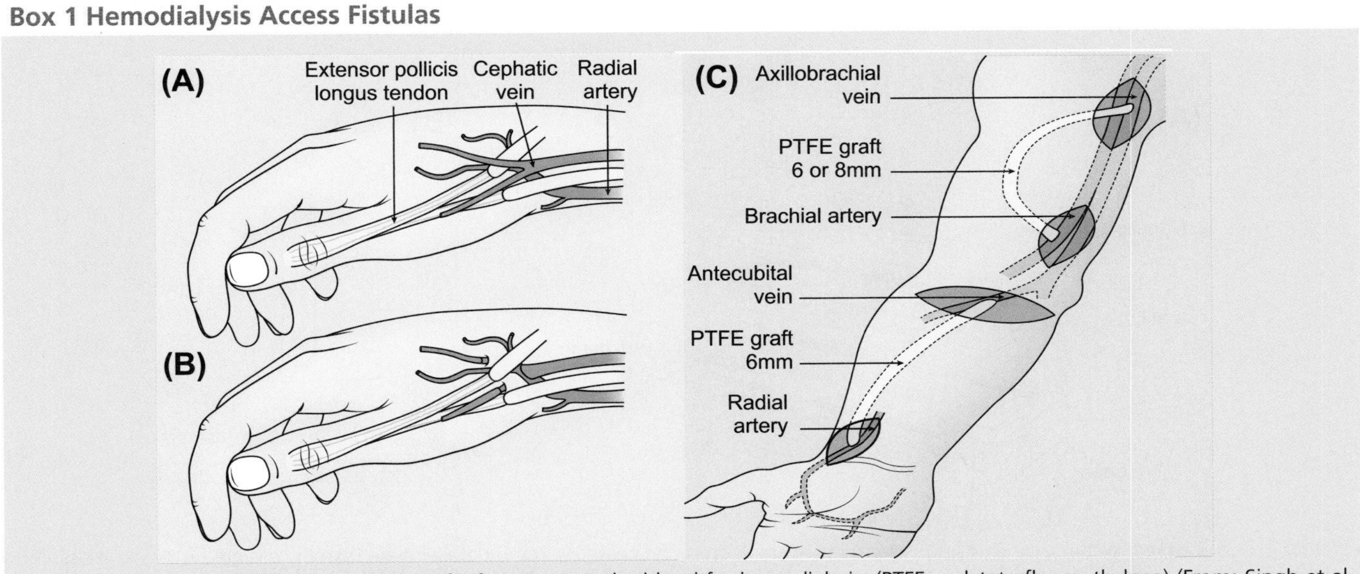

Placement of (A) and (B) fistulas, and (C) grafts for access to the blood for hemodialysis. (PTFE, polytetrafluoroethylene) (From: Singh et al. (2007). Successful angioaccess. *Surg. Clin. North Am.*, **87**, p. 1213. Elsevier.)

velocity gradient at the wall, also known as the wall shear rate (γ), equates to wall shear stress (τ_w) (Equation 1):

$$\tau_w = \mu\gamma \quad (1)$$

With ventricular contraction, momentum propagates as waves down the aorta, but diminishes in amplitude on the arterial side of the circulation, with distance from the heart. Pulsatility is generated by the pumping action of the heart, and gives rise to pulsatile shear stress and cyclic strain. Typical mean arterial values of shear stress range from 6 to 40 dyn/cm^2, but can vary from 0 to well over 100 dyn/cm^2 elsewhere in the vasculature (Goldsmith and Turitto, 1986; Dobrin et al., 1989). While pulsing down the arterial tree, blood flow remains mostly laminar; however, it often becomes complex and/or disturbed (reversing and/or recirculating) at areas of arterial branching, triggering spatial and/or temporal gradients in shear stress (e.g., Figure II.1.6.2).

Endothelial cells have been studied extensively for their ability to respond to changes in shear stress. Two *in vitro* systems have been developed to characterize the response of endothelial cells to a variety of shear stresses: the parallel plate flow chamber; and the cone and plate system (Figure II.1.6.3). In the parallel plate model, cells are grown on a glass slide, which sandwiches a silicone gasket onto a polycarbonate parallel plate (Frangos

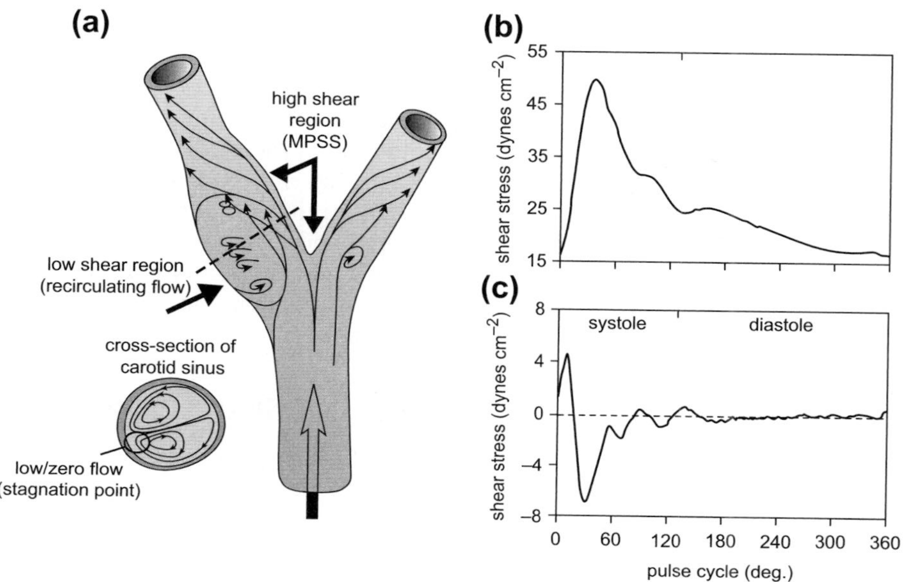

FIGURE II.1.6.2 Representation of flow features at the carotid bifurcation. (a) Change in wall shear stress throughout the course of the cardiac cycle at two locations within the carotid bifurcation, low shear (including recirculating flow) and high shear regions (MPSS, mean positive shear stress). (b) waveform of shear stress during one cardiac cycle (1 sec.) measured at the high shear stress region; (c) waveform of shear stress at the low shear stress region, showing reversing (recirculating) flow. (Reprinted with permission from the Royal Society; White and Frangos, (2007). *Philos. Trans. R. Soc. Lond. B. Biol. Sci.* **362**(1484), 1459–1467).

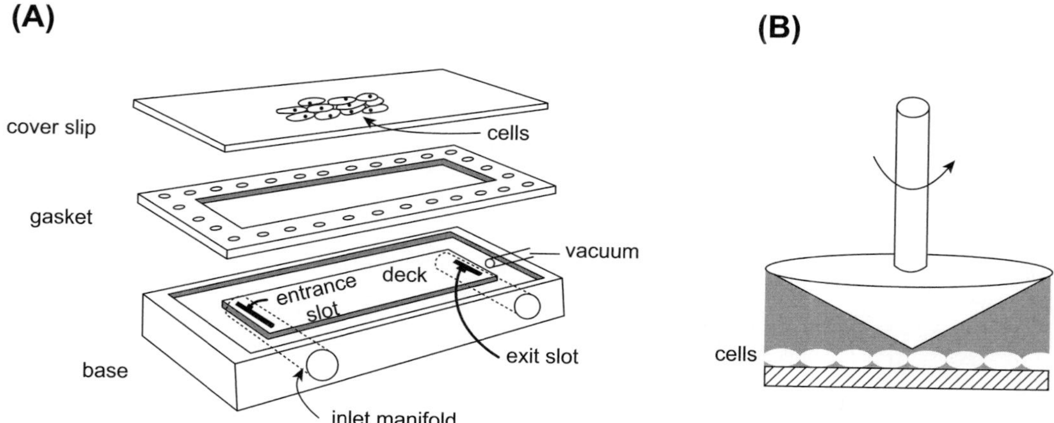

FIGURE II.1.6.3 *In vitro* systems for studying the effects of shear stress on cells. (A) In the parallel plate system cells are grown on glass slides and mounted onto a parallel plate chamber. The gasket provides a known gap height. Fluid flows in the inlet, across the deck, and out the exit slot (B). In the cone and plate system cells are grown on a circular dish, which forms a viscometer at the bottom. A small fixed angle cone is rotated at a constant angular velocity providing a constant shear stress across the cells.

et al., 1987). This creates a very small gap width through which tissue culture medium is circulated. The geometric dimensions are known, allowing for the calculation of the fluid shear stress from the flow rate. For a Newtonian fluid flowing through a parallel plate flow chamber with a rectangular geometry, the steady, laminar shear stress at the wall is:

$$\tau_w = \frac{6\mu Q}{bh^2} \tag{2}$$

in which τ_w = wall shear stress, μ = viscosity, Q = flow rate, b = channel width, and h = channel height. By varying the chamber geometry, the flow rate or the viscosity, the entire physiological range of wall shear stresses can be investigated. The flow rate is varied using a syringe pump or gravity driven system to provide the desired wall shear stress across the cells. The cone and plate system consists of a tissue culture dish as the plate, and a fixed angle cone mounted to a motor that imparts an angular velocity. For small cone angles, the shear rate (and therefore the shear stress) is essentially constant throughout the flow field. Since the dimensions of this system are also characterized, the fluid shear stress can be calculated for a given angular velocity.

If the cell property studied responds to shear stress in a dose-dependent fashion (i.e., varies directly with the level of shear stress imposed), it is referred to as "shear stress-dependent." If the property under study responds to changes in shear rate, but not directly with the level of shear stress imposed, it is referred to as "transport-dependent." "Flow-dependent" can include both shear stress- and shear rate-dependent phenomena. The difference arises from inclusion of mass-transport phenomena (convection and diffusion) in flow-dependent processes, whereas the term "shear stress-dependent" refers to the mechanical force only. Differences between effects due to shear stress alone, and those due to mass and momentum transfer, can be examined by circulating media of different viscosities (e.g., with the addition of high molecular weight dextran).

Endothelial cells, in either of these *in vitro* shear stress systems, exposed to long-term (24 hours or greater) steady shear stress at arterial levels (10–25 dynes/cm^2), as well as pulsatile non-reversing shear stress (Figure II.1.6.2b), have been shown to produce a more anti-inflammatory and anti-proliferative phenotype than when exposed to pulsatile reversing conditions or low shear stress or to static culture (Figure II.1.6.2c) (McCormick et al., 2001; Dai et al., 2004; Chien, 2008; Yee et al., 2008). Acute responses to arterial shear stress include release of signaling molecules such as nitric oxide and prostacyclin, phosphorylation of membrane proteins, and activation of GTPases and tyrosine kinases (Frangos et al., 1985; White and Frangos, 2007). Released signaling molecules, such as prostacyclin and nitric oxide, can in turn act on smooth muscle cells to mediate vasorelaxation. Response of endothelial cells to arterial level shear stresses of 24 hours or longer include alignment of the actin cytoskeleton, movement of the microtubule organizing center toward the direction of flow, and cell elongation and alignment in the direction of flow (Eskin and McIntire, 1988; Orr et al., 2006).

Endothelial cells respond differently to reversing or oscillatory shear stress as compared to non-reversing shear stress. Steady or non-reversing shear stress transiently induces proinflammatory and proliferative pathways, which are subsequently downregulated by long-term exposure to shear stress. However, oscillatory or reversing shear stress results in sustained activation of these pathways, leading to the hypothesis that under steady shear stress cells adapt and downregulate these pathways, whereas under disturbed flow the continued changes in flow magnitude and direction lead to sustained activation (Dai et al., 2004; Orr et al., 2006; Chien, 2008). A summary of the changes induced by reversing shear stress as compared to non-reversing shear stress are summarized in Table II.1.6.1. The increases in cell proliferation, lipid metabolism, and inflammation observed in endothelial cells exposed to reversing shear stress *in vitro* suggest that *in vivo* disturbed hemodynamics may prime local sites for atherosclerosis. Consistent with biochemical differences, cells do not elongate or align in the direction of flow at sites of atherosclerosis (Figure II.1.6.4).

TABLE II.1.6.1 Comparison of the effects of reversing flow to non-reversing flow on endothelial gene expression and function

	Reversing Flow	Non-reversing Flow
Flow Pattern	Fluid reversal, small net flow	No reversal, large net flow
Location in Arteries	Bifurcations and high curvature	Straight regions
Endothelial Cell Alignment	Not aligned	Aligned parallel to flow
Inflammatory Genes	Up-regulated	Down-regulated
Monocyte Adhesion	Increased	Decreased
Cell Proliferation	Increased	Decreased
Anti-oxidant Genes	Decreased	Increased
Vasodilating molecules	Decreased	Increased
Effect on Atherogenesis	Pro-atherogenic	Anti-atherogenic

Adapted from Chien, S. (2008) Effects of disturbed flow on endothelial cells. *Ann Biomed Eng*, 36, 554–562.

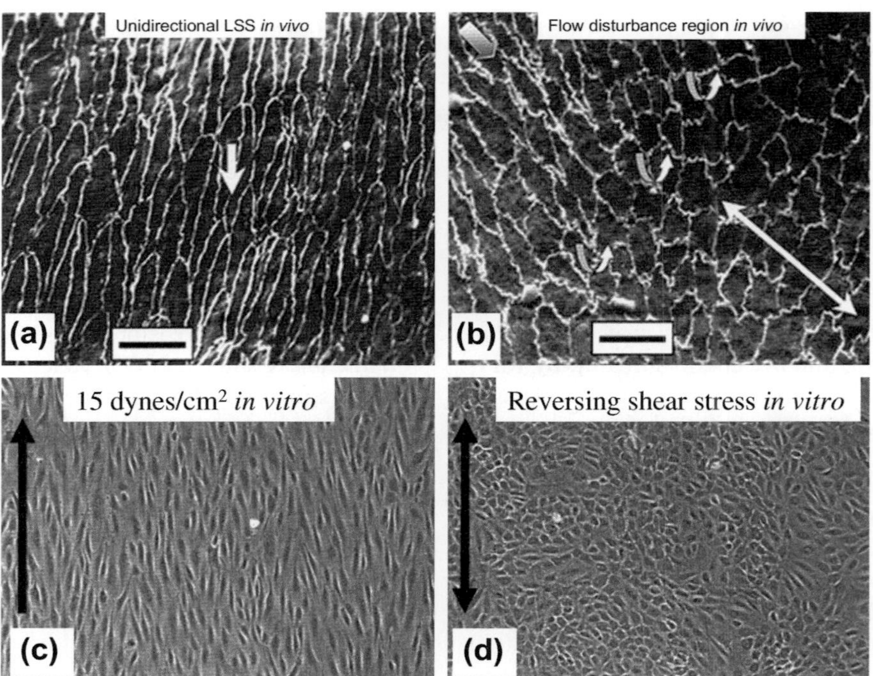

FIGURE II.1.6.4 Arterial endothelial cell alignment *in vivo* (porcine) and *in vitro* (human). (a) Endothelial cell alignment in undisturbed unidirectional laminar flow in the descending thoracic aorta (LSS, laminar shear stress), and (b) absence of transition from alignment in disturbed flow adjacent to a branch of the aorta. In (b), a region of cell alignment changes abruptly to polygonal cell morphology beyond a line of flow separation (curved arrows) that marks the boundary of a disturbed flow region in which oscillating, multidirectional flow typically occurs (double-headed arrow). Scale bar in each panel = 15 μm. (c) Endothelial cell alignment under *in vitro* steady shear stress (15 dynes/cm^2), and (d) reversing shear stress (modeled after the wall shear stress at the carotid sinus) is similar to *in vivo* alignment. ((a) and (b) reprinted with permission from Springer: Davies, P. F. (2008). Endothelial transcriptome profiles *in vivo* in complex arterial flow fields, *Annals of Biomedical Engineering*, **36**, 563–570, 2008; (c) and (d) unpublished data (D. E. Conway).)

There is evidence that endothelial cells respond to shear stress within seconds. Rapid changes occur through the activation of ion channels, G-proteins, and stimulation of protein kinases (Davies et al., 2003; White and Frangos, 2007). Shear stress may act directly on the cell membrane to deform the cell surface and activate unknown sensor proteins. However, recent evidence has suggested that shear stress forces are transmitted by the cytoskeleton to intracellular locations where signaling can occur, such as intercellular junctions, focal adhesions, the nuclear membrane, and lipid rich regions of the cell membrane known as caveolae (Stamatas and McIntire, 2001; Davies et al., 2003; Boyd et al., 2003). In addition to remodeling of the cytoskeleton in response to shear stress there is also evidence of focal adhesion remodeling and activation of proteins at these adhesion sites (Davies et al., 2003), supporting the hypothesis that the majority of shear stress sensing is directly coupled to the cytoskeleton.

More recently, the glycocalyx has re-emerged as an important mechanosensor of shear stress in endothelial cells. The glycocalyx is an extracellular membrane-bound layer of glycoproteins and plasma proteins that has been shown to extend up to 500 μm from the endothelial cell membrane (Figure II.1.6.5). *In vivo*, this structure is large enough to prevent cells and even large plasma proteins from reaching the surface of the cell. Removal of key glycocalyx molecules results in a reduced response to shear stress (reduction in nitric oxide production, absence of cell elongation in response to flow, and loss of shear stress-induced suppression of cell proliferation), suggesting that the glycocalyx is an important mechano-sensor of fluid shear stress in endothelial cells (Tarbell and Pahakis, 2006). A reduction in the glycocalyx has also been observed in a number of disease states, including hypertension, inflammation, ischemia and reperfusion, hyperglycemia, and at known sites of atherosclerosis (internal carotid sinus). Furthermore, reduced glycocalyx expression has been correlated with increased leukocyte adhesion, suggesting that the glycocalyx may provide protection against atherogenesis (Van Teeffelen et al., 2007). Recently it has been suggested that *in vitro* endothelial cells do not form a hemodynamically relevant glycocalyx, highlighting the need for a better understanding of the formation and regulation of this potentially important mechano-sensitive extracellular structure (Potter and Damiano, 2008).

Animal models have further validated data generated from *in vitro* systems. Suo et al. used computational fluid dynamics to model the wall shear stress in the mouse aortic arch (Suo et al., 2007). Although the mouse vasculature has higher overall shear stresses than the human, inflammatory markers, such as intracellular adhesion molecule-1 (ICAM-1) and vascular cell adhesion molecule-1

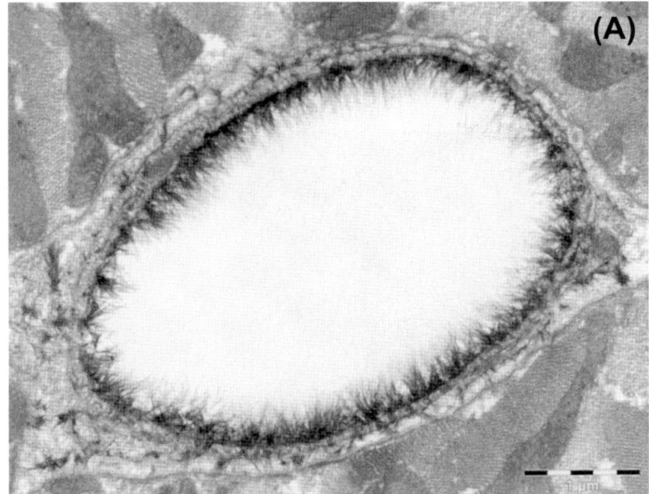

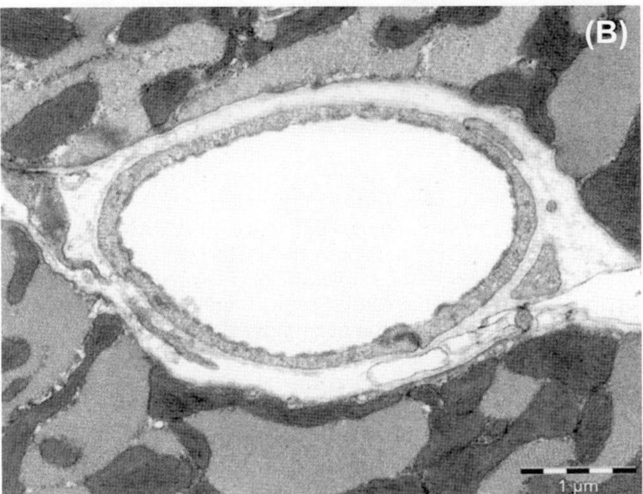

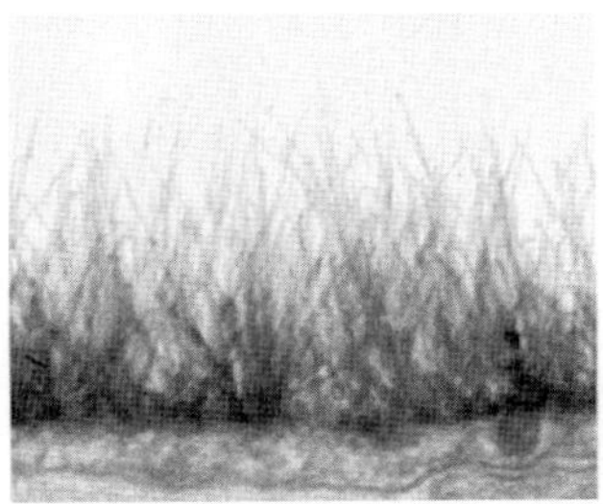

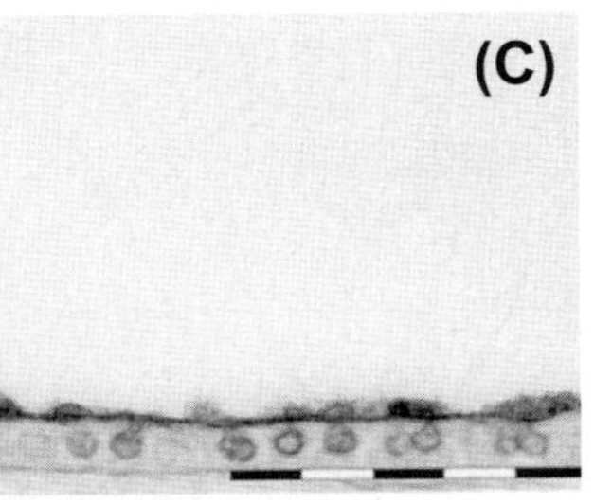

FIGURE II.1.6.5 The endothelial glycocalyx can extend up to 0.5 μm into the vessel lumen. (A) Electron microscopic overview of an Alcian blue 8GX-stained rat left ventricular myocardial capillary (bar = 1 μm). (B) After hyaluronidase treatment, before Alcian blue staining (bar = 1 μm). (C) Detailed pictures of glycocalyx on normal (left) and hyaluronidase-treated (right) capillaries (bar = 0.5 μm). (Source: van den Berg, B. M. et al. (2003). *Circ. Res.*, **92**, 592–594.)

(VCAM-1), were upregulated at sites with low time-averaged shear stress and changes in shear stress direction (Figure II.1.6.6) (Suo et al., 2007). Microarray analysis of disturbed flow and undisturbed flow regions in the porcine aorta showed increased inflammatory cytokines and receptors in regions of disturbed flow (Passerini et al., 2004). These *in vivo* studies suggest that shear stress at sites of atherosclerosis may prime the endothelium toward an inflammatory state, which then may be exacerbated by additional risk factors (e.g., cholesterol, smoking, exercise).

Effect of Cyclic Strain on Blood Vessels

Along with momentum, pressure propagates as waves down the arterial tree, leading to a periodic circumferential tensile stress in the vessel lumen (Figure II.1.6.1). *In vivo*, blood pressure measures the variation in force against the blood vessel wall as the blood is ejected from the heart during each heartbeat (each cardiac cycle). In measuring blood pressure in humans, the brachial artery is occluded by a pressure cuff (attached to a sphygmomanometer), while the sound of the blood flow is measured downstream with a stethoscope. The pressure cuff is slowly loosened, allowing the pressure to decrease and the blood to flow downstream. The pressure at which the blood flow begins is the systolic pressure. As the pressure continues to decrease, the value at which the sound disappears is the diastolic pressure. Since the arterial wall is compliant, this periodic pressure difference gives rise to a cyclic wall strain. Since arteries and some synthetic substrates on which cells are cultured are elastic, cyclic strain can be measured as the percentage change in diameter between the systolic and diastolic pressures. In normal circulation, the internal diameter and, thus, circumference of large mammalian arteries increases cyclically between 2% and 18% over the cardiac cycle at a frequency of approximately 1 Hz (60 cycles/min) (Dobrin, 1978). The arterial cyclic strain can increase by 15% in hypertension (Gupta and Grande-Allen, 2006). Typical systole/diastole values in large human arteries range from 90/60 mmHg to 120/80 mmHg.

To study cyclic strain effects *in vitro*, cells must be cultured on a deformable substrate, usually silicone rubber or segmented polyurethane coated with extracellular matrix proteins, then subjected to cyclic deformation at a rate similar to the heart rate (1 Hz). Most frequently, custom built uniaxial strain devices are driven mechanically by an eccentric cam, which imposes a nearly uniform strain along the substrate at a frequency simulating the heart rate (1 Hz) (Frye et al., 2005; Yung et al., 2008). Custom built biaxial strain units have also been used (Kim et al., 1999). The mechanical forces generated in *in vitro* studies on uniaxial cyclic strain optimally include a "motion control" condition. "Motion control" controls for the reversing shear stress (less than 0.5 dyne/cm^2) from fluid motion accompanying cyclic strain of the membrane. Some endothelial cells are more responsive to motion control than to cyclic strain (Sung et al., 2007). Cultured cells have been exposed to cyclic strain in commercially available Flexercell Strain Units (Flexcell International, Corp.), which can deform in uniaxial or biaxial modes (Matheson et al., 2007). These units are driven by vacuum pressure beneath 6-well plates with flexible elastomeric bottoms, on which cells are cultured, thus deforming substrate and cells (Haseneen et al., 2003). It

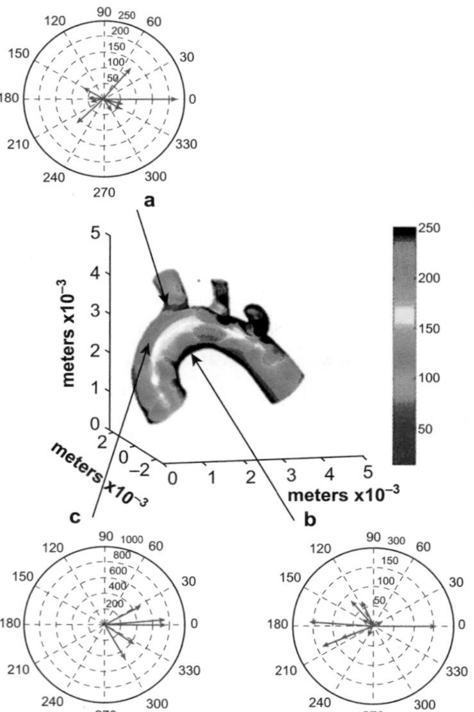

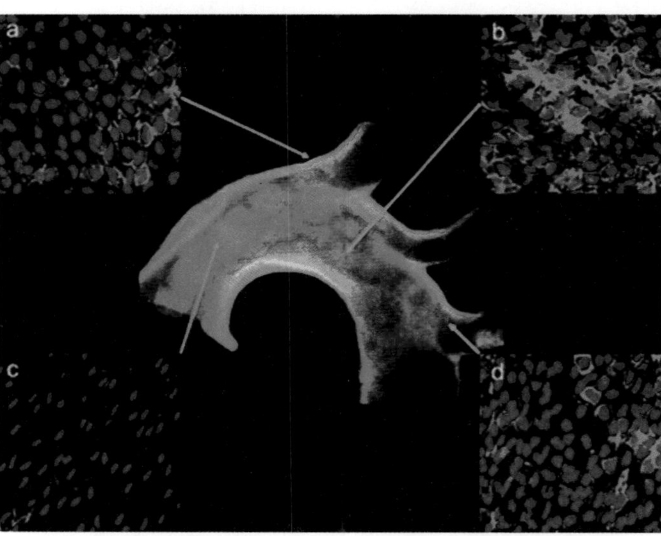

FIGURE II.1.6.6 Regional differences in wall shear stress in the mouse aortic arch correlates with regional differences in VCAM-1 expression. Left: The computed, time-varying wall shear stress vectors are depicted at areas (a), (b), and (c) in one mouse aorta. The color coding indicates the mean wall shear stress distribution (defined as the wall shear stress magnitude averaged over the cardiac cycle). The vector diagrams indicate the time-varying changes in the direction and magnitude of wall shear stress throughout the cardiac cycle. There is a greater variation in the direction of the instantaneous wall shear stress vectors in areas (a) and (b), as can be seen in the polar plots of wall shear stress. Note that the relatively low mean wall shear stress zone at the inner curvature of the aortic arch (b), has instantaneous magnitudes of wall shear stress up to approximately 150 dynes/cm^2, whereas values exceeding 600 dynes/cm^2 were found along the lateral wall of the ascending aorta (c). Right: High expression of VCAM-1 protein correlates with areas of the aortic arch with low and time-varying shear stresses (a, b, and d) whereas VCAM-1 expression in the ascending aorta (c) is lower, consistent with unidirectional flow. (Source: Suo, J. et al. (2007). *Arteriosclerosis, Thrombosis, and Vascular Biology,* **27**, 346–351.)

should be noted that not only do the mechanical forces that impinge on a population of cultured cells alter their function, but also the substrate on which the cells are cultured may affect cell response. Cyclic strain has been shown to alter phenotype of cultured smooth muscle cells on polymeric scaffolds (Kim et al., 1999), and to cause differentiation of embryonic stem cells into smooth muscle cells (Shimizu et al., 2008).

Although the effects of shear stress and cyclic strain are most frequently studied independently of each other, recent work has suggested that the synchronization of the two forces when applied simultaneously can affect gene expression and morphology (Owatverot et al., 2005). The stress phase angle has been defined as the temporal phase angle between cyclic strain and wall shear stress (Figure II.1.6.7) (Dancu et al., 2004). The stress phase angle has been shown to be most negative at sites prone to atherosclerosis when compared to other regions of the vasculature, and can be made even more negative with hypertension (Dancu et al., 2004). Cells exposed to identical wall shear stress and cyclic strain, but at a stress phase angle of −180° instead of 0°, had reduced endothelial nitric oxide production and increased endothelin-1 production (Dancu et al., 2004), suggesting that asynchronous wall shear stress and cyclic strain could lead to endothelial cell dysfunction.

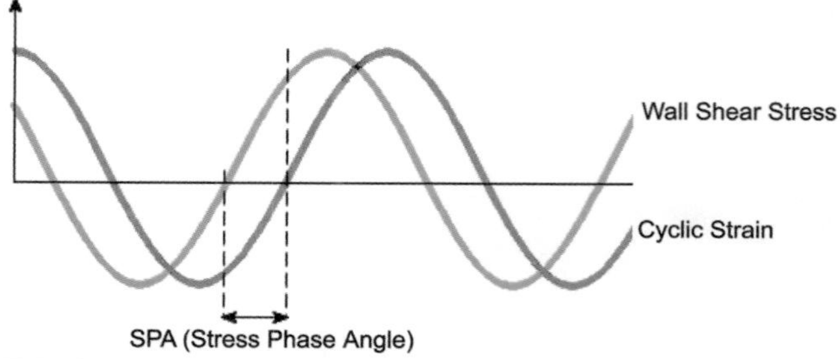

FIGURE II.1.6.7 Stress phase angle (SPA) is the phase shift between cyclic strain and wall shear stress. (D. E. Conway drawing.)

INTERSTITIAL FLUID

Twenty percent of the mass of the human body consists of interstitial fluid, the fluid that surrounds most cells (Swartz and Fleury, 2007). This fluid is constantly in motion, driven by lymphatic drainage, which returns plasma that has leaked out of capillaries to the blood

Case Study: Mechanical Forces in the Embryonic-Cardiovascular System

The embryonic circulatory system is highly dynamic, forming and remodeling while simultaneously providing blood flow to the developing organs. Until recently, the role of hemodynamic forces in the development of the heart and blood vessels was unknown. This case study shows the effects of hemodynamic forces on the vascular remodeling of the mouse embryo yolk sac (Lucitti et al., 2007). Endothelial cells in the yolk sac first assemble into a primitive vascular mesh network, but begin to remodel in the presence of flow into a network of branched, hierarchical network of large and small vessels that are surrounded by smooth muscle cells. Using a mouse knockout model (Mlc2a-null) that had impaired cardiac contractility (resulting in impaired flow and red blood cell movement), the authors showed that reduced blood flow led to impaired vessel remodeling. Reductions in the red blood cells in normal mice also impaired the vessel remodeling. To determine if the impaired vessel remodeling is due to lowered oxygen transport or reduced blood viscosity, the plasma in the mice with reduced red blood cells was supplemented with synthetic sugars to restore the viscosity to normal levels, which resulted in the development of a normal vascular system. Their results show that mechanical forces are essential for proper blood vessel development.

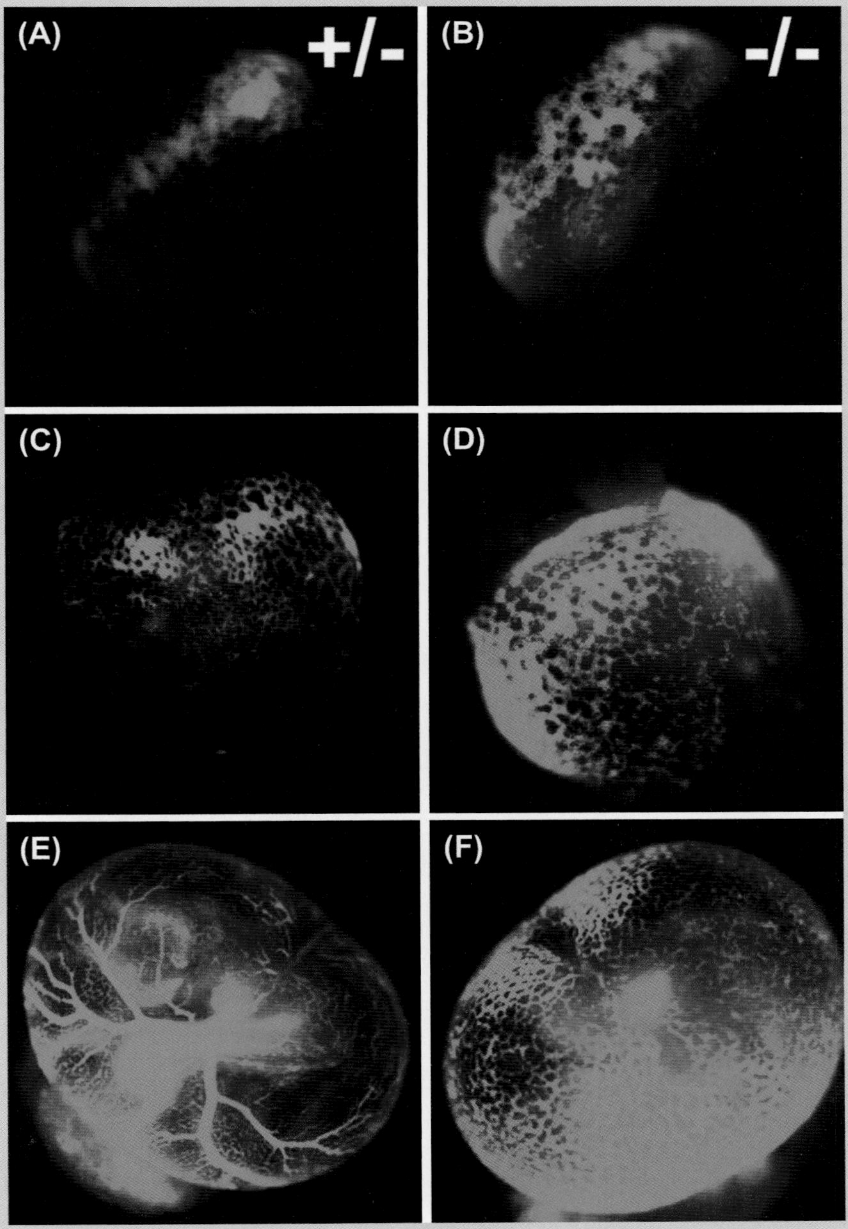

Heterozygous (A,C,E) and knockout (B,D,F) mouse embryos with fluorescently labeled red blood cells. At embryonic day 8.5, blood islands have formed in both wild-type (A) and *Mlc2a*-null (B) embryos, and red blood cells begin to circulate in both wild-type (C) and *Mlc2a* mutant (D) embryos. However, the vascular network fails to remodel by embryonic day 9.5 in the mutant embryos (F), as compared with wild-type (E). (Reprinted with permission from The Company of Biologists, Lucitti, J. L., Jones, E. A., Huang, C., Chen, J., Fraser, S. E. & Dickinson, M. E. (2007). Vascular remodeling of the mouse yolk sac requires hemodynamic force. *Development,* **134**, 3317–3326.)

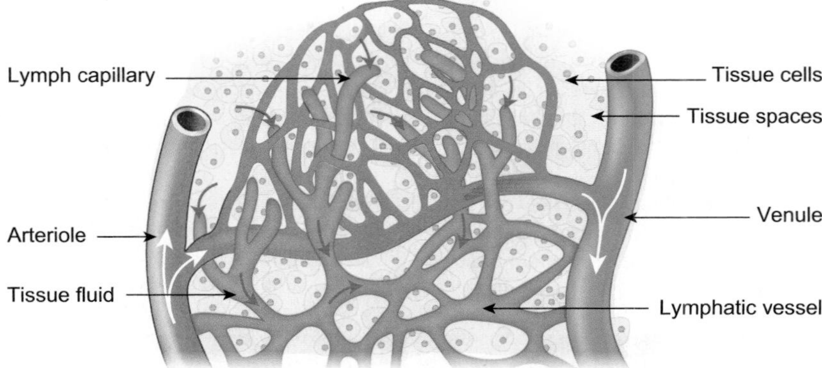

FIGURE II.1.6.8 Lymphatic vessels carry fluid away from the tissues. Lymphatic vessels start as blunt-ended capillaries and come together to form larger lymphatic vessels. (Source: National Cancer Institute http://training.seer.cancer.gov/module_anatomy/unit8_2_lymph_compo.html.)

through the lymphatic system (Figure II.1.6.8). The smallest lymphatic vessels are the lymph capillaries, which are found in most organs of the body, excluding the bone marrow and central nervous system. These capillaries start as blunt ends and come together to form larger contractile lymphatic vessels that have valves that prevent backflow, and smooth muscle cells that can provide contractility. The forces responsible for interstitial flow are hydrostatic and osmotic pressure differences (also known as Starling forces) between the blood, interstitium, and lymphatic system. Contractility by larger lymphatic vessels can further increase pressure gradients. The velocity of interstitial fluid is estimated to be 0.1 to 2 μm/s, but can be increased during inflammation (Swartz and Fleury, 2007). The resistance to interstitial fluid flow is affected by the specific composition of the extracellular matrix (ECM), cell density, and integrity of the lymphatic system; however, interstitial fluid pressure is also affected by external factors, such as movement, exercise, blood pressure, and hydration (Swartz and Fleury, 2007).

In addition to providing exchange of cell nutrients and wastes, interstitial flow has been shown to affect blood and lymphatic capillary morphogenesis, chondrocyte and osteocyte function, fibroblast differentiation, smooth muscle cell cytokine production, ocular function, morphogenesis in perfused three-dimensional cultures, and patterning in embryo development (see Box 2) (Nonaka, 2005; Swartz and Fleury, 2007). Interstitial flow can drive both mechanical and biochemical cellular responses by imparting mechanical forces, such as shear stress and cyclic strain, both on the cell surface and the extracellular matrix that is directly connected to the cellular cytoskeleton, and biochemical changes through redistribution of extracellular signaling molecules (Figure II.1.6.9) (Rutkowski and Swartz, 2007). In three-dimensional matrices with interstitial fluid flow of 1 μm/s, the shear stress at the cell surface has been estimated to average 6×10^{-3} dynes/cm^2, with a peak shear stress of 1.5×10^{-2} dynes/cm^2, over three orders of magnitude lower than the wall shear stress in a large artery (Rutkowski and Swartz, 2007). It is unclear if cells can directly sense such a low shear stress, and therefore may rely on the surrounding extracellular matrix to amplify the signals through shear-induced strain or alternatively the effects may be due to convective mass transfer of signaling molecules.

Lymphatic flow has recently been shown to enable autologous chemotaxis, a mechanism whereby a cell can receive directional cues for cellular migration, but simultaneously can be the source of such cues (Fleury et al., 2006). Previously it was assumed that cells migrating towards a signaling concentration gradient were following a gradient secreted by an upstream cell or tissue. However, through autologous chemotaxis, subtle flows can create assymetrical pericellular protein concentrations and trancellular gradients that increase in the direction of fluid flow (Fleury et al., 2006). This mechanism could explain how leukocytes and tumor cells use gradients in interstitial fluid to locate and migrate to draining lymphatics (Rutkowski and Swartz, 2007).

In large arteries, the endothelial cells are the cells exposed directly to shear stress, while all the cell types in the vessel wall are exposed to cyclic strain. But when the endothelial cells are damaged, the hydraulic conductivity of an artery is dramatically increased. This increase in hydraulic conductivity leads to elevated interstitial flow through the artery wall. Models of interstitial flow in blood vessels with an intact endothelium have shown that medial smooth muscle cells are on average exposed to 1 dyne/cm^2 shear stress, but the smooth muscle cells closest to the endothelial cells can experience a shear stress of 10 dynes/cm^2, similar to the fluid shear stress directly imparted to the endothelial cells by blood flow (Tada and Tarbell, 2002). A study by Garanich et al. showed that increased shear stress enhanced the migratory properties of adventitial fibroblasts (Garanich et al., 2007). These fibroblasts have been reported to participate in the arterialization of capillaries, and also contribute to intimal hyperplasia (vessel thickening) that occurs in response to vessel injury. Changes in

Box 2 Interstitial Fluid Flow Affects Vertebrate Embryo Patterning

Interstitial fluid flow has also been shown to be important in development and tissue morphogenesis. Nonaka et al. showed that in the mouse embryo the nodal cilia generate a unidirectional leftward flow of extraembryonic fluid which is the first step in the generation left–right asymmetry (Nonaka et al., 1998). Recent work has shown that convection, not just diffusion alone, of such signaling molecules can drive left–right asymmetry of internal organs (e.g., determining that the heart is located on the left side of the body). Reversal of this flow direction resulted in reversal of the asymmetry (Nonaka et al., 2002).

It is currently unclear if nodal flow affects transport of signaling molecules or acts directly as a mechanical stress on the cells (Shiratori and Hamada, 2006). Morphological analysis of the nodal cilia has shown that the cilia are not completely perpendicular, but are instead tilted in the posterior direction (A and B) (Nonaka et al., 2005). This tilt, combined with the clockwise rotation of the cilia results in a more efficient stroke in the leftward direction, and a less efficient stroke in the rightward direction, resulting in a net flow in the left direction (C).

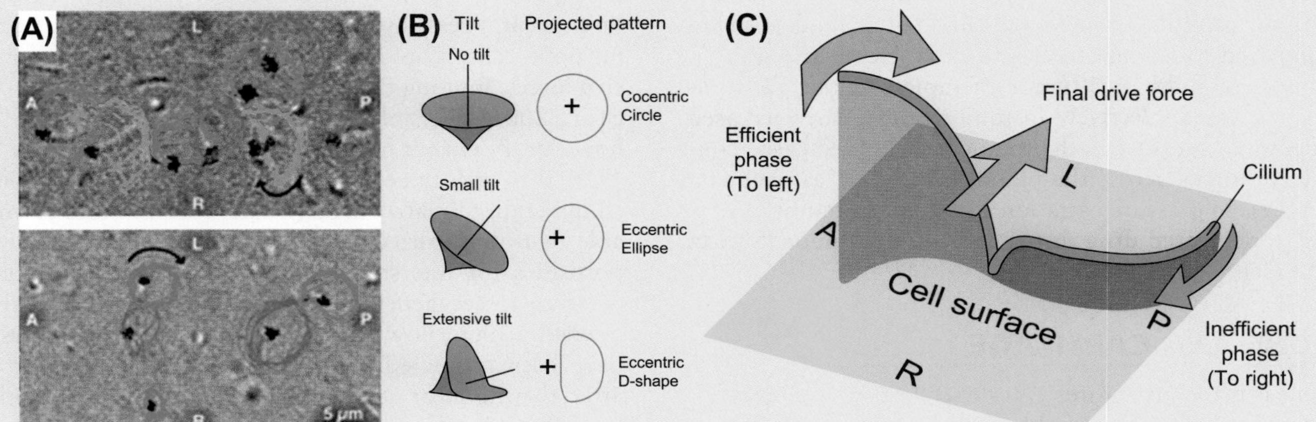

Left/right symmetry of vertebrate embryo. (A) Most cilia have a pattern consistent with the projection of a tilted cone (blue and green, see text) whereas some cilia move in a D-shape (orange). A, P, L, and R refer to anterior, posterior, left, and right sides of the node, respectively. The direction of cilia rotation was clockwise (arrows). (B) Relationship between movement of cilia and their projected images at various tilt angles. (C) Circular clockwise motion of a cilium can generate directional leftward flow if its axis is not perpendicular to the cell surface but tilted posteriorly. Due to distance from the cell surface, a cilium in the leftward phase (red arrow) drags surrounding water more efficiently than the rightward phase (blue arrow), resulting in a net leftward force (pink arrow). (Images are from Nonaka, S., Yoshiba, S., Watanabe, D., Ikeuchi, S., Goto, T., et al. (2005). *De novo* formation of left–right asymmetry by posterior tilt of nodal cilia. *PLoS Biol.* 3(8): e268. This journal releases its work to the public domain).

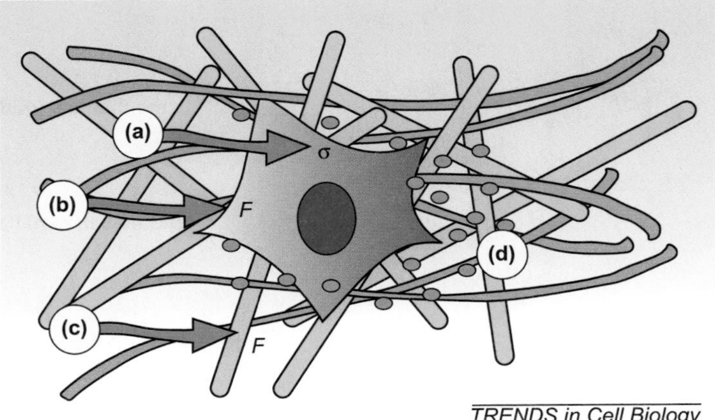

FIGURE II.1.6.9 The direct effects of interstitial flow on cells. Interstitial flow can induce (a) fluid shear stress, σ, on the cell surface; (b) forces normal to the cell surface (*F*); (c) shear and normal forces to the pericellular matrix that is mechanically coupled to the cytoskeleton; and (d) redistribution of pericellular proteins (autocrine and paracrine signals) that bind cell receptors. (Source: Rutkowski, J. M. & Swartz, M. A. (2007). A driving force for change: Interstitial flow as a morphoregulator. *Trends in Cell Biology,* **17**(1), 44–50. Elsevier.)

interstitial flow through the arterial wall may be important in mediating blood vessel wall remodeling.

Interstitial flow is important in the uptake of pharmaceuticals in local tissues or tumor delivery. Traditionally, therapeutics are given systemically, and therefore have been limited to small molecules with high diffusivities. However, the development of protein and antibody therapies and synthetic drug carriers, which are designed for selective uptake by a specific tissue, require a better understanding of the local interstitial transport in the target tissue. Tumors frequently have leaky blood vessels, in combination with a nonfunctional lymphatic system, resulting in increased interstitial fluid pressure compared to normal tissues (Swartz and Fleury, 2007). This can hinder anti-tumor therapies seeking to accumulate drugs selectively in tumor tissue. The increased interstitial pressure in the tumor leads to decreased convective forces across the blood vessel wall, as well as a decrease in pressure variation within in the tumor, resulting in decreased drug distribution in the tumor (Swartz and Fleury, 2007).

BONE AND CARTILAGE

Bone is made up of three types of cells: osteocytes, which maintain function; osteoblasts, which build bone; and osteoclasts which degrade bone. These cells reside in a matrix of collagen and hydroxyapatite, a complex tetracalcium phosphate, which is produced by osteoblasts. The osteocytes are mechanosensors, which release prostaglandins in response to mechanical forces, principally shear stress (Jiang et al., 2007). They transmit signals to other bone cells and the matrix through thin channels in the matrix to influence bone remodeling.

Bone is a tissue that is constantly undergoing remodeling in response to mechanical loading, forming new bone parallel to the loading direction, and losing density in unstressed regions. Bone loss from sedentary lifestyle, limb paralysis or microgravity during space flight has provided conclusive *in vivo* evidence for the positive effects of physiological loading of skeletal tissues (Orr et al., 2006). Mechanical loading generates fluid shear stress, hydrostatic compression, and stretch on bone cells. The dense bone matrix resists compressive forces, limiting the transfer of force to the cells (Orr et al., 2006). Physiological strains are only 0.04–0.3%; however, *in vitro* it has been shown that strains must be 1–10% to induce cellular responses (Orr et al., 2006). Unmineralized matrix around osteocytes is more permeable than mineralized bone, creating lacuno-canalicular porosities for interstitial fluid flow. As a result, compressive forces generate pressure gradients that drive interstitial fluid flow in the matrix; these shear stresses have been estimated to be 8–30 dynes/cm^2, similar to arterial wall shear stress (Orr et al., 2006). Interstitial fluid flow due to mechanical loading is thought to be the primary mechanism for action of mechanical forces on bone (Figure II.1.6.10) (Klein-Nulend et al., 2005). Although *in vitro* both osteoclasts and osteoblasts have been shown to respond to shear stress, the geometry of

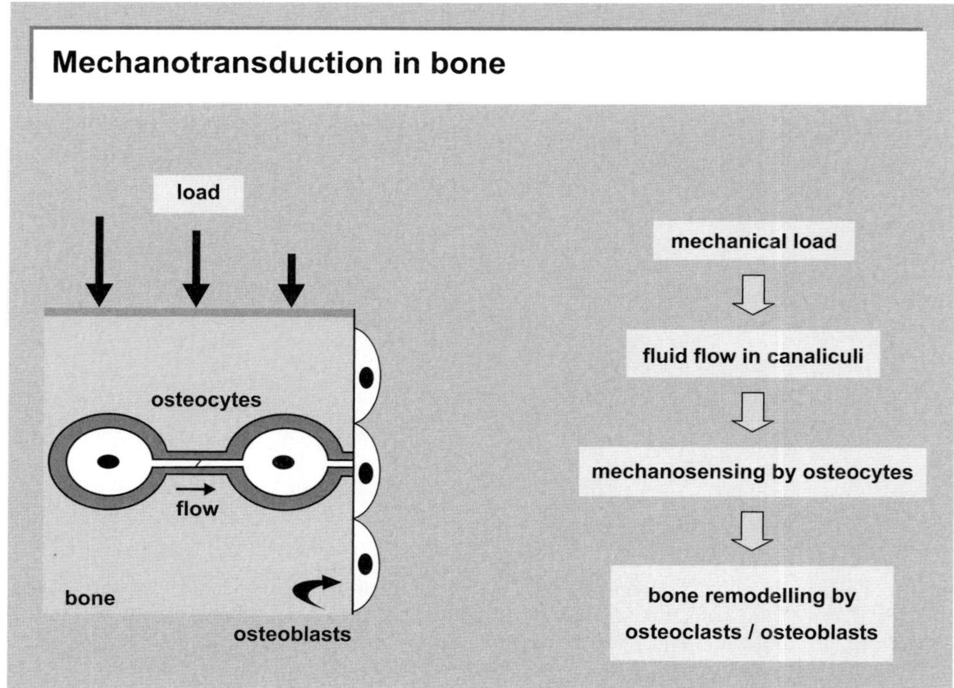

FIGURE II.1.6.10 Model for the transduction of mechanical strain to osteocytes in bone. Left: The osteocyte-lining cell network of bone tissue under stress (large arrows). Loading results in flow of interstitial fluid in the canalicular non-mineralized matrix (horizontal arrow). (Source: Klein-Nulend, J., Bacabac, R. G. & Mullender, M. G. (2005). Mechanobiology of bone tissue. *Pathologie Biologie,* **53**(10), 576–580. Elsevier.)

the lacuno-canalicular porosity suggests that interstitial fluid flow is preferentially sensed by the osteoclasts (Orr et al., 2006). Osteocytes respond to mechanical stimuli by secreting factors that can then, in turn, modulate both osteoblasts and osteoclasts (Klein-Nulend et al., 2005). Recent models, however, have suggested that the interstitial fluid flow applies drag forces to the pericellular matrix of the osteocyte that are 20 times larger than the fluid shear forces applied directly to the cell surface (Han et al., 2004). The fluid flow through the pericellular matrix induces strain on the actin filaments of the cytoskeleton, which results in strain amplification at the cellular level an order of magnitude greater than tissue-level strains. These recent models question the hypothesis that the regulatory mechanical force on osteocytes is fluid shear stress, and instead suggest that shear-induced strain may be the principal mechanotransducing force (Orr et al., 2006; Han et al., 2004).

> Bone is a tissue that is constantly undergoing remodeling in response to mechanical loading, forming new bone parallel to the loading direction, and losing density in unstressed regions. Bone loss from sedentary lifestyle, limb paralysis or microgravity during space flight has provided conclusive *in vivo* evidence for the positive effects of physiological loading of skeletal tissues.

Exercise-stimulated bone remodeling may be the response of osteoblasts to interstitial fluid flow (Reich et al., 1990; Hillsley and Frangos, 1994). Furthermore, temporal gradients in interstitial fluid flow, imposed by pulsatile shear stress (designed to simulate mechanical loading) stimulate osteoblast proliferation *in vitro* (Jiang et al., 2002), whereas steady shear stress does not stimulate osteoblast signal transduction or proliferation. Exposure of mouse osteocytes and mixed-population human bone cells to pulsating shear stress *in vitro* increased prostaglandin and nitric oxide release, similar to the effects of shear stress on vascular endothelial cells, suggesting that there may be some similar mechanisms for sensing fluid flow (Klein-Nulend et al., 2005). Strain rate has been shown to be more important than strain amplitude in inducing bone formation in response to loading, supporting the hypothesis that bone formation is stimulated by dynamic rather than static loads (Klein-Nulend et al., 2005).

Cartilage, located on the articulating surfaces of joints, provides low friction for freely moving joints. In the growing embryo, cartilage is the precursor to bone. Cartilage cells (chondrocytes) secrete a matrix of collagen fibers embedded in mucopolysaccharide (e.g., chondroitin sulfate). Although cartilage is relatively asvascular compared with bone, cartilage responds to mechanical loading similarly to bone. Increased load leads to increased matrix production, resulting in stronger tissue. Cartilage must withstand tensile and shear forces in addition to compression (Kim et al., 1994). Most of the work on cartilage loading focuses on compression and hydrostatic pressure. Cyclic compression of explants (0.1 Hz, 2–3% compression) stimulated matrix synthesis by chondrocytes (Shieh and Athanasiou, 2003).

SUMMARY

Nearly all mammalian cells and tissues are subject to mechanical forces. In the preceding pages we have highlighted the cardiovascular and skeletal systems adapted specifically to mechanical forces, to variations in blood flow and pressure, and to loadbearing, respectively. *In vitro* models have been employed to study how isolated endothelial and smooth muscle cells from arteries and veins respond to conditions of controlled shear stress and cyclic strain. Gene expression is sensitively regulated by alterations in shear stress and cyclic strain, which translates to functional changes in the cells. The elucidation of roles of different bone cell types has been clarified using *in vitro* models. We are just beginning to understand interstitial fluid flow in the lymphatic system, and the influence mechanical forces may have on cell migration and tumor growth.

In order to properly design biomaterials which successfully simulate native tissues, understanding the mechanical environment in which the biomaterial is implanted is important. Cells exposed to disturbed mechanical environments (e.g., reversing flow in blood vessels, compliance mismatch of blood vessel grafts, lack of weight-bearing on bone) typically have reduced function, and may promote the development of a pathological condition. In designing a biomaterial, we must attempt to recreate the mechanical environment that is to receive the implant for maximum success and long-term viability.

BIBLIOGRAPHY

Alenghat, F. J., & Ingber, D. E. (2002). Mechanotransduction: All signals point to cytoskeleton, matrix, and integrins. *Sci. STKE*, **2002**, PE6.

van den Berg, B. M., Vink, H., & Spaan, J. A. (2003). The endothelial glycocalyx protects against myocardial edema. *Circ. Res.*, **92**(6), 592–594.

Boyd, N. L., Park, H., Yi, H., Boo, Y. C., Sorescu, G. P., Sykes, M., & Jo, H. (2003). Chronic shear induces caveolae formation and alters ERK and Akt responses in endothelial cells. *Am. J. Physiol. Heart Circ. Physiol.*, **285**, H1113–H1122.

Chien, S. (2008). Effects of disturbed flow on endothelial cells. *Ann. Biomed. Eng.*, **36**, 554–562.

Dai, G., Kaazempur-Mofrad, M. R., Natarajan, S., Zhang, Y., Vaughn, S., Blackman, B. R., Kamm, R. D., Garcia-Cardena, G., & Gimbrone, M. A., Jr. (2004). Distinct endothelial phenotypes evoked by arterial waveforms derived from atherosclerosis-susceptible and -resistant regions of human vasculature. 41, *Proc. Natl. Acad. Sci. USA*, **101**, 14871–14876.

Dancu, M. B., Berardi, D. E., Vanden Heuvel, J. P., & Tarbell, J. M. (2004). Asynchronous shear stress and circumferential strain reduces endothelial NO synthase and cyclooxygenase-2 but induces endothelin-1 gene expression in endothelial cells. *Arterioscler. Thromb. Vasc. Biol.*, **24**, 2088–2094.

Davies, P. F. (1995). Flow-mediated endothelial mechanotransduction. *Physiol. Rev.*, **75**, 519–560.

Davies, P. F. (2008). Endothelial transcriptome profiles *in vivo* in complex arterial flow fields. *Ann. Biomed. Eng.*, **36**, 563–570.

Davies, P. F. (2009). Hemodynamic shear stress and the endothelium in cardiovascular pathophysiology. *Nat. Clin. Pract. Cardiovasc. Med.*, **6**, 16–26.

Davies, P. F., Zilberberg, J., & Helmke, B. P. (2003). Spatial microstimuli in endothelial mechanosignaling. *Circ. Res.*, **92**, 359–370.

Dobrin, P. B. (1978). Mechanical properties of arteries. *Physiol. Rev.*, **58**, 397–460.

Dobrin, P. B., Littooy, F. N., & Endean, E. D. (1989). Mechanical factors predisposing to intimal hyperplasia and medial thickening in autogenous vein grafts. *Surgery*, **105**, 393–400.

Eskin, S. G., & McIntire, L. V. (1988). Hemodynamic effects on atherosclerosis and thrombosis. *Semin. Thromb. Hemost.*, **14**, 170–174.

Fleury, M. E., Boardman, K. C., & Swartz, M. A. (2006). Autologous morphogen gradients by subtle interstitial flow and matrix interactions. *Biophys. J.*, **91**, 113–121.

Frangos, J. A., Eskin, S. G., McIntire, L. V., & Ives, C. L. (1985). Flow effects on prostacyclin production by cultured human endothelial cells. *Science*, **227**, 1477–1479.

Frangos, J. A., McIntire, L. V., & Eskin, S. G. (1987). Shear stress induced stimulation of mammalian cell metabolism. *Biotech. Bioeng.*, **32**, 1053–1060.

Frye, S. R., Yee, A., Eskin, S. G., Guerra, R., Cong, X., & McIntire, L. V. (2005). cDNA microarray analysis of endothelial cells subjected to cyclic mechanical strain: Importance of motion control. *Physiol. Genomics*, **21**, 124–130.

Garanich, J. S., Mathura, R. A., Shi, Z. D., & Tarbell, J. M. (2007). Effects of fluid shear stress on adventitial fibroblast migration: Implications for flow-mediated mechanisms of arterialization and intimal hyperplasia. *Am. J. Physiol. Heart Circ. Physiol.*, **292**, H3128–H3135.

Glagov, S., Zarins, C., Giddens, D. P., & Ku, D. N. (1988). Hemodynamics and atherosclerosis. Insights and perspectives gained from studies of human arteries. *Arch. Pathol. Lab. Med.*, **112**, 1018–1031.

Goldsmith, H. L., & Turitto, V. T. (1986). Rheological aspects of thrombosis and haemostasis: basic principles and applications. ICTH-Report – Subcommittee on Rheology of the International Committee on Thrombosis and Haemostasis. *Thromb. Haemost.*, **55**, 415–435.

Gupta, V., & Grande-Allen, K. J. (2006). Effects of static and cyclic loading in regulating extracellular matrix synthesis by cardiovascular cells. *Cardiovasc. Res.*, **72**, 375–383.

Han, Y., Cowin, S. C., Schaffler, M. B., & Weinbaum, S. (2004). Mechanotransduction and strain amplification in osteocyte cell processes. *Proc. Natl. Acad. Sci. USA.*, **101**, 16689–16694.

Haseneen, N. A., Vaday, G. G., Zucker, S., & Foda, H. D. (2003). Mechanical stretch induces MMP-2 release and activation in lung endothelium: Role of EMMPRIN. *Am. J. Physiol. Lung. Cell. Mol. Physiol.*, **284**, L541–L547.

Hillsley, M. V., & Frangos, J. A. (1994). Bone tissue engineering: The role of interstitial fluid flow. *Biotechnol. Bioeng.*, **43**, 573–581.

Jiang, G. L., White, C. R., Stevens, H. Y., & Frangos, J. A. (2002). Temporal gradients in shear stimulate osteoblastic proliferation via ERK1/2 and retinoblastoma protein. *Am. J. Physiol. Endocrinol. Metab.*, **283**, E383–E389.

Jiang, J. X., Siller-Jackson, A. J., & Burra, S. (2007). Roles of gap junctions and hemichannels in bone cell functions and in signal transmission of mechanical stress. *Front. Biosci.*, **12**, 1450–1462.

Kim, B. S., Nikolovski, J., Bonadio, J., & Mooney, D. J. (1999). Cyclic mechanical strain regulates the development of engineered smooth muscle tissue. *Nat. Biotechnol.*, **17**, 979–983.

Kim, Y. J., Sah, R. L., Grodzinsky, A. J., Plaas, A. H., & Sandy, J. D. (1994). Mechanical regulation of cartilage biosynthetic behavior: physical stimuli. *Arch. Biochem. Biophys.*, **311**(1), 1–12.

Klein-Nulend, J., Bacabac, R. G., & Mullender, M. G. (2005). Mechanobiology of bone tissue. *Pathol. Biol. (Paris)*, **53**(10), 576–580.

Lucitti, J. L., Jones, E. A., Huang, C., Chen, J., Fraser, S. E., & Dickinson, M. (2007). Vascular remodeling of the mouse yolk sac requires hemodynamic force. *Development*, **134**, 3317–3326.

Matheson, L. A., Maksym, G. N., Santerre, J. P., & Labow, R. S. (2007). Differential effects of uniaxial and biaxial strain on U937 macrophage-like cell morphology: Influence of extracellular matrix type proteins. *J. Biomed. Mater Res. A.*, **81**, 971–981.

McCormick, S. M., Eskin, S. G., McIntire, L. V., Teng, C. L., Lu, C. M., Russell, C. G., & Chittur, K. K. (2001). DNA microarray reveals changes in gene expression of shear stressed human umbilical vein endothelial cells. *Proc. Natl. Acad. Sci. USA*, **98**, 8955–8960.

Nonaka, S., Tanaka, Y., et al. (1998). Randomization of left-right asymmetry due to loss of nodal cilia generating leftward flow of extraembryonic fluid in mice lacking KIF3B motor protein. *Cell*, **95**(6), 829–837.

Nonaka, S., Shiratori, H., Y., Saijoh, Y., & Hamada, H. (2002). Determination of left–right patterning of the mouse embryo by artificial nodal flow. *Nature*, **418**(6893), 96–99.

Nonaka, S., Yoshiba, S., Watanabe, D., Ikeuchi, S., Goto, T., Marshall, W. F., & Hamada, H. (2005). *De novo* formation of left–right asymmetry by posterior tilt of nodal cilia. *PLoS Biol.*, 3(8), e268.

Orr, A. W., Helmke, B. P., Blackman, B. R., & Schwartz, M. A. (2006). Mechanisms of mechanotransduction. *Dev. Cell.*, **10**, 11–20.

Owatverot, T. B., Oswald, S. J., Chen, Y., Wille, J. J., & Yin, F. C. (2005). Effect of combined cyclic stretch and fluid shear stress on endothelial cell morphological responses. *J. Biomech. Eng.*, **127**, 374–382.

Papadaki, M., & Eskin, S. G. (1997). Effects of fluid shear stress on gene regulation of vascular cells. *Biotech. Prog.*, **13**, 209–221.

Passerini, A. G., Polacek, D. C., Shi, C., Francesco, N. M., Manduchi, E., Grant, G. R., Pritchard, W. F., Powell, S., Chang, G. Y., Stoeckert, C. J., Jr., & Davies, P. F. (2004). Coexisting proinflammatory and antioxidative endothelial transcription profiles in a disturbed flow region of the adult porcine aorta. *Proc. Natl. Acad. Sci. USA*, **101**, 2482–2487.

Potter, D. R., & Damiano, E. R. (2008). The hydrodynamically relevant endothelial cell glycocalyx observed *in vivo* is absent *in vitro*. *Circ. Res.*, **102**, 770–776.

Reich, K. M., Gay, C. V., & Frangos, J. A. (1990). Fluid shear stress as a mediator of osteoblast cyclic adenosine monophosphate production. *J. Cell. Physiol.*, **143**, 100–104.

Rutkowski, J. M., & Swartz, M. A. (2007). A driving force for change: Interstitial flow as a morphoregulator. *Trends Cell. Biol.*, **17**(1), 44–50.

Shieh, A. C., & Athanasiou, K. A. (2003). Principles of cell mechanics for cartilage tissue engineering. *Ann. Biomed. Eng.*, **31**(1), 1–11.

Shimizu, N., Yamamoto, K., Obi, S., Kumagaya, S., Masumura, T., Shimano, Y., Naruse, K., Yamashita, J. K., Igarashi, T., & Ando, J. (2008). Cyclic strain induces mouse embryonic stem cell differentiation into vascular smooth muscle cells by activating PDGF receptor beta. *J. Appl. Physiol.*, **104**(3), 766–772.

Shiratori, H., & Hamada, H. (2006). The left–right axis in the mouse: From origin to morphology. *Development*, **133**(11), 2095–2104.

Singh, N., Starnes, B. W., & Andersen, C. (2007). Successful angioaccess. *Surg. Clin. North. Am.*, **87**, 1213–1228. xi.

Stamatas, G. N., & McIntire, L. V. (2001). Rapid flow-induced responses in endothelial cells. *Biotechnol. Prog.*, **17**, 383–402.

Sung, H. J., Yee, A., Eskin, S. G., & McIntire, L. V. (2007). Cyclic strain and motion control produce opposite oxidative responses in two human endothelial cell types. *Am. J. Physiol. Cell. Physiol.*, **293**, C87–C94.

Suo, J., Ferrara, D. E., Sorescu, D., Guldberg, R. E., Taylor, W. R., & Giddens, D. P. (2007). Hemodynamic shear stresses in mouse aortas: Implications for atherogenesis. *Arterioscler. Thromb. Vasc. Biol.*, **27**, 346–351.

Swartz, M. A., & Fleury, M. E. (2007). Interstitial flow and its effects in soft tissues. *Annu. Rev. Biomed. Eng.*, **9**, 229–256.

Tada, S., & Tarbell, J. M. (2002). Flow through internal elastic lamina affects shear stress on smooth muscle cells (3D simulations). *Am. J. Physiol. Heart. Circ. Physiol.*, **282**, H576–H584.

Tarbell, J. M., & Pahakis, M. Y. (2006). Mechanotransduction and the glycocalyx. *J. Intern. Med.*, **259**, 339–350.

Van Teeffelen, J. W., Brands, J., Stroes, E. S., & Vink, H. (2007). Endothelial glycocalyx: Sweet shield of blood vessels. *Trends Cardiovasc. Med.*, **17**, 101–105.

Wagner, D. D., & Frenette, P. S. (2008). The vessel wall and its interactions. *Blood*, **111**, 5271–5281.

White, C. R., & Frangos, J. A. (2007). The shear stress of it all: The cell membrane and mechanochemical transduction. *Philos. Trans. R. Soc. Lond. B. Biol. Sci.*, **362**(1484), 1459–1467.

Yee, A., Bosworth, K. A., Conway, D. E., Eskin, S. G., & McIntire, L. V. (2008). Gene expression of endothelial cells under pulsatile non-reversing vs. steady shear stress; comparison of nitric oxide production. *Ann. Biomed. Eng.*, **36**, 571–579.

Yung, Y. C., Vandenburgh, H., & Mooney, D. J. (2008). Cellular strain assessment tool (CSAT): Precision-controlled cyclic uniaxial tensile loading. *J. Biomech.*, **42**, 178–182.

CHAPTER II.1.7 STEM CELLS: KEY CONCEPTS

Richard L. Carpenedo[1] and Todd C. McDevitt[1,2]

[1]The Wallace H. Coulter Department of Biomedical Engineering,

[2]The Parker H. Petit Institute for Bioengineering and Bioscience, Georgia Institute of Technology, Atlanta, Georgia, USA

INTRODUCTION

Stem cells have become of increasing interest and relevance to the scientific community over the past 10–20 years as a result of the discovery and derivation of numerous types of stem cells that are now known to exist. Within the field of biomaterials research, stem cells have rightfully garnered more attention as it has been recognized that stem cell–material interactions can be critical to regulating cell fate, and that biomaterials may significantly improve the efficacy of therapeutic applications of stem cells currently being developed. Although stem cells are often broadly referred to as a single class of cells, in reality a wide variety of stem cells exist that exhibit a range of different molecular, cellular, and functional properties. Thus, it is imperative to the understanding of stem cell biology and the performance of stem cell research to be aware of the common nomenclature used to denote the similarities and differences among various types of stem cells. In this chapter, different classes of stem cells will be described and contrasted to one another based on functional definitions of their potential ("potency") to develop into mature cell types. This is followed by a discussion of the different physical factors comprising the environments ("niches") that regulate stem cell self-renewal and differentiation.

STEM CELL POTENCY

Historically, different types of stem cells have been classified according to the stage of development and/or tissue of origin from which they were originally isolated. This has resulted in frequent distinction between stem cells, without paying heed to the functional attributes of different stem cell populations. The misnomers "embryonic" and "adult" have been frequently used to convey different types of stem cells, whereas the reality is that a continuum of stem cells exists at all stages of mammalian development spanning varying levels of differentiation potential. It is generally accepted that stem cells derived from early stages of development (i.e., embryonic) can differentiate into a broader range of cells than stem cells isolated from more mature tissues formed during fetal and post-natal development. This hierarchical arrangement of stem cells has been classically depicted as immature cells giving rise to increasingly mature cells with more restricted differentiation potential.

In the simplest terms, stem cells are functionally defined by their dual capacity for self-renewal and differentiation (Figure II.1.7.1). Self-renewal is the process whereby cells stably maintain their phenotype and ability to differentiate as they undergo cell growth and division. Differentiation, on the other hand, results in an altered cell phenotype that typically has specialized functions, and that slows or loses ability for proliferation and undergoes a reduction in subsequent differentiation potential. The strictest functional test for a *bona fide* stem cell is for it to be clonogenic; that is for a single cell to demonstrate that it can give rise to phenotypically identical cells and differentiated progeny. Although most stem cells have traditionally been identified and referred to by their stage of development and tissue of origin, different classes of stem cells are more accurately categorized by the breadth of their capacity for differentiation – commonly referred to as "potency" (Figure II.1.7.2). Pluripotent stem cells have the ability to differentiate into cells from all three germ lineages (endoderm, mesoderm, and ectoderm), as well as germ cells that can give rise to viable offspring (i.e., sperm, ova). Pluripotent stem cells are thought to be normally found only during the earliest stages of embryogenesis, and become quickly depleted as tissue development proceeds. Embryonic stem cells (ESCs) are the classic example of a pluripotent cell, but more recently fully differentiated somatic cells from a variety of tissue sources have been successfully reverted to a pluripotent state and are referred to as "*induced*" pluripotent

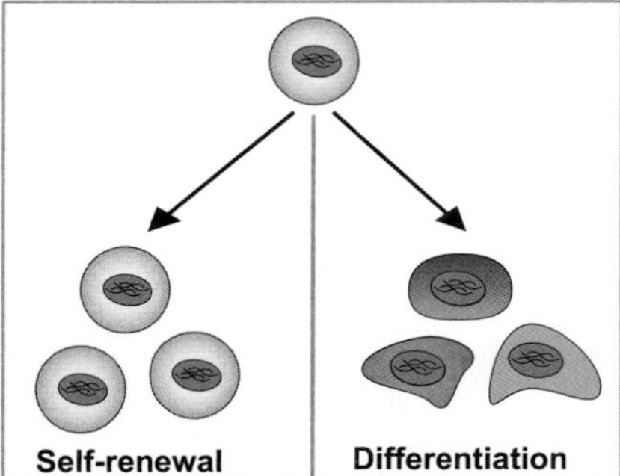

FIGURE II.1.7.1 Stem Cell Definition. Stem cells are defined functionally by their ability to both self-renew and differentiate.

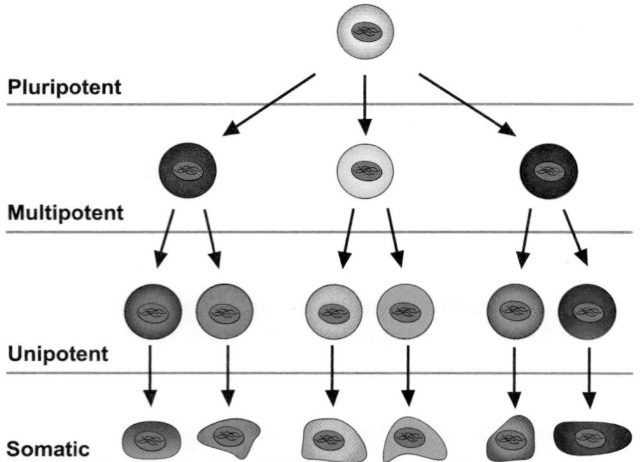

FIGURE II.1.7.2 Stem Cell Potency. Stem cells can be classified according to the breadth of their potential to differentiate into various progenitor and somatic cells.

stem cells (iPS cells or iPSCs). Multipotent stem cells, in contrast, are capable of giving rise to multiple cell types, yet cannot differentiate into all somatic cell types (like pluripotent stem cells). Generally speaking, differentiation of multipotent stem cells is restricted to a single germ lineage from which the cells initially originated. Multipotent stem cells can be found widely distributed throughout the body in different physiological systems. Hematopoietic and mesenchymal stem cells are common examples of multipotent cells that exhibit the capacity to differentiate into various blood cells and connective tissue, respectively. Finally, unipotent stem cells, sometimes also referred to as "progenitor" cells, are generally restricted to differentiate to only a single cell type, yet also retain the ability to self-renew. Unipotent stem cells are commonly found in tissues of adult organisms that exhibit frequent turnover, such as the skin and the intestines. Situated within the respective tissues, unipotent stem cells locally serve to replenish cells in order to maintain tissue homeostasis.

Tissue-specific (also often called "adult") stem cells are also contained in many organs that do not turn over at a high rate, such as skeletal muscle, lungs, liver, kidney, etc. These cells are induced by specific injury to proliferate and provide more differentiated cells of one or several very restricted subtypes of an organ. Recent studies have also suggested that cardiac and neural stem cells also exist in heart, but their capability for proliferation is yet uncertain. Although the plasticity of stem cells can be generally categorized as pluri-, multi- or unipotent, the full spectrum of differentiation potential exhibited by the various types of stem cells that have been identified and characterized is actually a continuum that can't be fully delineated presently by these general classifications. However, the notion of "potency" is important, in that it provides a general framework whereupon the functional attributes of different stem cells can be compared. As a result, it also conveys that the diversity of stem cells broadens as the potency of different types of stem cells decreases (Figure II.1.7.3). However, as stem cell research progresses the hierarchy among stem cells and distinguishing characteristics between different types of stem cells continue to become better defined.

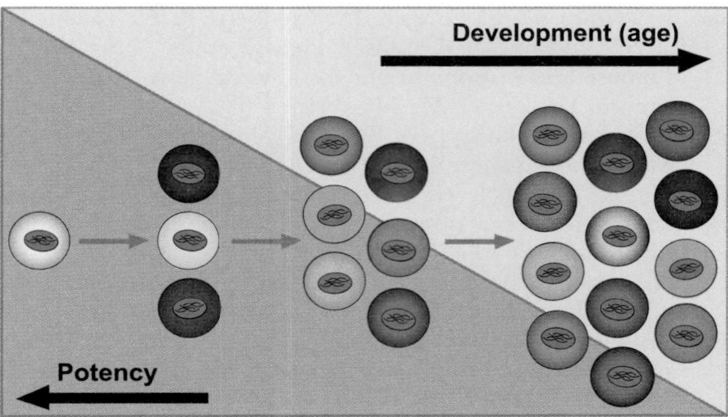

FIGURE II.1.7.3 Stem Cells and Development. An inverse relationship exists between the potency of stem cells and the relative stage of development with which they are associated. Increasing numbers of different stem cells emerge over time that individually exhibit more restricted differentiation potential.

Pluripotent Stem Cells

Pluripotent stem cells are distinguished from all other stem cells by their unique ability to differentiate into all 200-plus somatic cell types of mature mammalian organisms. Pluripotent stem cells are considered to have infinite self-renewal capacity when consistently maintained under optimal growth conditions *in vitro*, and normally grow at a faster rate than other stem or somatic cell types. Pluripotent stem cells are similar to epithelial cells, in that they express the cell adhesion marker E-cadherin, and remain closely associated with neighboring stem cells as they grow. As pluripotent stem cells differentiate, many of the cells undergo an epithelial-to-mesenchymal phenotypic transition whereby the cells decrease E-cadherin expression, become more spindle-shaped, and migrate away from one another coincident with the epigenetic changes that occur within the cells during differentiation.

Embryonic carcinoma cells (ECs) were the first class of pluripotent cells to be isolated and studied *in vitro* (Martin

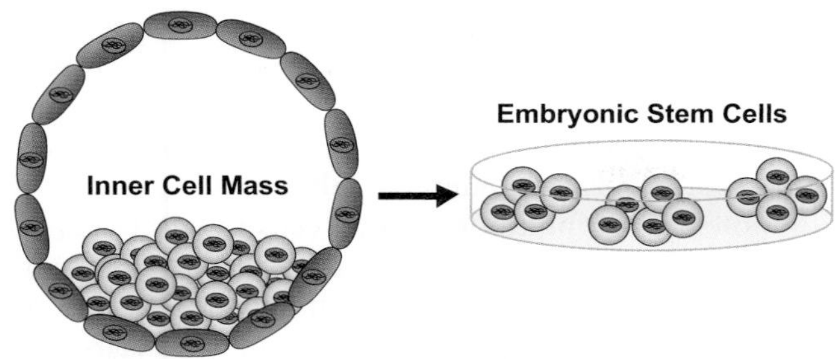

FIGURE II.1.7.4 Derivation of Embryonic Stem Cells. Embryonic stem cells are derived from the inner cell mass of blastocysts and propagate indefinitely in culture.

and Evans, 1975). ECs are derived from teratocarcinomas, a malignant tumor that arises in the gonads from germ cells, and are capable of giving rise to cells from all three germ lineages both within the tumor, and following isolation and *in vitro* culture. Traditionally, ECs have been utilized as a model system for various aspects of tumor biology, including growth kinetics, drug resistance, and vascularization (Yuhas et al., 1977; Casciari et al., 1988; Shweiki et al., 1995; Helmlinger et al., 1997). Many of the methods and assays originally developed for EC culture were adopted for the subsequent cultivation and differentiation of the second class of pluripotent cells to be defined, *embryonic stem cells* (ESCs), discussed below.

ESCs, derived from the inner cell mass (ICM) during the blastocyst stage of embryonic development (Figure II.1.7.4), are the best known example of pluripotent stem cells (Evans and Kaufman, 1981; Martin, 1981; Doetschman et al., 1985). ESCs naturally give rise to all types of cells and tissues of the body. ESCs were originally derived from mice in the early 1980s, and largely used for the purpose of creating transgenic mice and to serve as an accessible *in vitro* model system for mammalian developmental biology studies (Baribault and Kemler, 1989). Nearly two decades later, ESCs were also successfully derived from primates; first rhesus monkeys (Thomson et al., 1995) and shortly thereafter, humans (Thomson et al., 1998). The establishment of human ESC lines heralded the prospect of not only developmental studies with human cells, but also the potential use of human pluripotent cells for regenerative cell therapies. Embryonic germ-like cells (EGCs), derived around the same time as human ESCs, are another example of a pluripotent cell type (Shamblott et al., 1998). EGCs are derived from primordial germ cells found within gonadal ridges and mesenteries. Human EGCs display similar phenotypic markers to those of ESCs, and when cultured as multicellular spheroids known as "embryoid bodies," give rise to cells from all three germ lineages (Shamblott et al., 2001). Although human ESC and EGC cell lines were originally derived at nearly identical times, researchers have more commonly worked with and published on ESCs, which have therefore served as the basis for most of what is currently known about human pluripotent biology.

The characterization of the genomic signature of pluripotent stem cell populations has recently led to the reprogramming of somatic cell types to induce pluripotency in terminally differentiated cells (iPS cells or iPSCs). The introduction of transcription factors found to be enriched in ESCs and other stem cells (i.e., Oct4, Sox2, c-Myc, and Klf4) into somatic cells has been shown to transform fully-differentiated cells into pluripotent cells (Takahashi and Yamanaka, 2006; Takahashi et al., 2007; Wernig et al., 2007; Yu et al., 2007; Okita and Yamanaka, 2011). Initially this novel proof-of-concept was demonstrated with different types of fibroblasts, but soon thereafter also performed with other somatic cells from various tissues, including peripheral blood, stomach, and liver (Aoi et al., 2008; Loh et al., 2009). Subsequent work has further demonstrated that different combinations of reprogramming factors in diverse cell types, or in conjunction with various co-factors, can be used to successfully produce iPS cells. Since the original creation of iPS cells, multiple approaches have been developed to introduce the necessary exogenous factors in the form of DNA, RNA or protein into cells to induce reprogramming (Okita et al., 2008; Zhou et al., 2009; Warren et al., 2010). Overall, independent of the particular method used, pluripotency appears to be re-established by the reactivation of endogenous genes (i.e., Oct4, Sox2, Nanog) that are normally permanently silenced by epigenetic mechanisms in somatic cells. The precise molecular mechanisms, signaling pathways, and transition points by which differentiated cells acquire a pluripotent state have yet to be defined, and remain the focus of many ongoing studies. The production of iPS cells is perceived to be advantageous over ESC isolation in that it does not require donated ova or blastocysts as a source of cells, but rather adult somatic cells. Furthermore, the ability to revert somatic cells to a pluripotent state provides a feasible route to the production of patient-specific iPS cells that could circumvent immunological issues associated with transplantation of ESC derivatives. Moreover, iPS cells provide a unique opportunity to examine developmental and mechanistic aspects of human cells for which viable sources are not readily accessible (i.e., heart muscle, neurons, islets). iPS cells derived from individuals with various congenital and familial diseases may also enable the study of progressive human disorders, such as Parkinson's, Huntington's, Alzheimer's, and diabetes (Park et al., 2008). However, many challenges remain in the realm of iPS cells; for instance, the efficiency of reprogramming remains quite low (typically <1%), although different somatic cells and methods display varying levels of reprogramming efficiency. It also remains unclear how similar/different iPSCs are compared to ESCs or how much heterogeneity exists amongst iPS cell lines created from different individuals or using different methods. In addition to creating iPS cells, as with other pluripotent stem cells, developing methods to efficiently direct iPSCs to specific differentiated cell types is a significant hurdle that needs to be overcome in order to realize the envisioned clinical applicability of iPS cells to human diseases and regenerative therapies.

Multipotent Stem Cells

Multipotent stem cells include many cell types traditionally classified as adult stem cells, including hematopoietic, mesenchymal, and neural stem cells. Multipotent stem cells exhibit a variety of different types of morphologies that in many cases can be readily distinguished from one another, but are not necessarily uniquely identifiable from other types of differentiated cells. Multipotent stem cells tend to proliferate more slowly than pluripotent stem cells, and have a limited capacity for expansion, particularly *in vitro*. In culture, multipotent stem cells tend to eventually senesce after a finite number of population doublings. Most types of multipotent stem cells have been successfully derived from pluripotent stem cells, in addition to being isolated from mature tissue sources. The proliferative ability and differentiation potential of multipotent stem cells in tissues generally declines with age; however, the exact reasons why remain unclear. The two most commonly held beliefs are that it is due to the gradual depletion of stem cells over time and/or the reduced potency of individual cells due to aging and disease.

"Creating" Stem Cells

Prior to the discovery of iPS cells, stem cells could only be obtained by isolating, purifying/enriching and propagating cells from different tissue sources to ultimately arrive at either primary isolates or established cell lines. Reverting fully-differentiated (somatic) cells to a less mature state is a paradigm-shifting notion for cell biology and all related fields (Figure II.1.7.5). The origins of cellular reprogramming date back to the 1950s, when somatic frog nuclei were transplanted into enucleated eggs (Gurdon et al., 1958); a process that later became known as "cloning." The demonstration that cells can be reprogrammed indicates that differentiation should no longer be considered a uni-directional, irreversible path whereby cells progressively adopt more mature and restricted cell fates during the course of development. The results of reprogramming studies demonstrate that the traditional arrows depicting stem cell hierarchy can go in the opposite direction as well. Although the focus of most initial reprogramming studies was on reverting terminally differentiated cells to the most primitive (pluripotent) state, the concept of reprogramming has been extended to generate multipotent and progenitor cells from somatic cell sources, most often from the tissue the stem cells are most commonly associated with. It is anticipated that as the mechanisms of reprogramming become better understood, new technologies will be developed and new insights will be gained that will broadly impact the fields of biology and the biomedical sciences, including cancer, regenerative medicine, and disease modeling.

Hematopoietic stem cells (HSCs) are rare populations of cells that give rise to all cell types of the blood that were initially identified by their colony-forming capacity in the spleens of irradiated mice after bone marrow transfer (Till and McCulloch, 1961; Becker et al., 1963). HSCs can differentiate into both lymphoid (B- and T-lymphocytes, natural killer cells) and myeloid cells (neutrophils, basophils, eosinophils, monocytes, macrophages, erythrocytes, and platelets). HSCs reside primarily in the bone marrow, but can also be found in the thymus, and the liver or spleen during fetal development. HSCs can exit the bone marrow compartment and transit in the circulatory system to distant sites within the body. The inherent ability of HSCs to be mobilized to the peripheral circulation allows for HSCs to be harvested directly from the blood by a process referred to as apheresis. HSC mobilization from the bone marrow into the circulation is stimulated by cytokines, such as G-CSF and/or GM-CSF. HSCs were the earliest stem cells to be discovered, and as a result more is known about the hierarchical distinctions and associated molecular markers for HSCs than any other stem cells. The low prevalence of HSCs in the blood makes it hard to recognize the cells accurately on a morphological basis, therefore isolation of HSCs is dependent on markers, particularly cell surface antigens, to identify and select HSCs away from other cells contained in the non-adherent fraction of blood or bone marrow aspirate. Of tremendous importance clinically, HSCs have the capacity to repopulate the entire hematopoietic system when transplanted into compromised recipients, such as following bone marrow

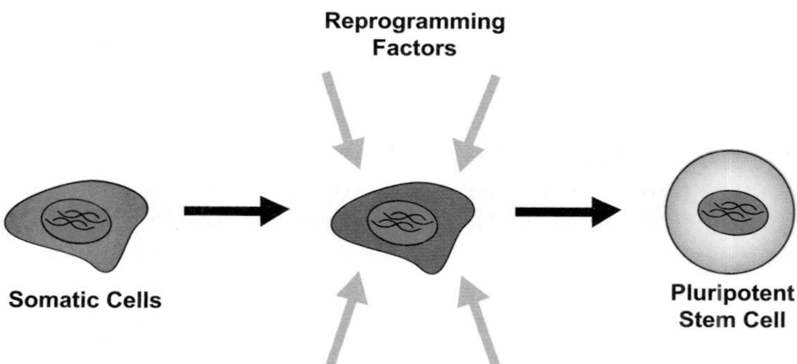

FIGURE II.1.7.5 Reprogramming of Somatic Cells. Somatic cells can be reverted to a pluripotent state by the influence of various types of reprogramming factors.

ablation. The transplantation of HSCs has been successfully employed as a therapeutic strategy for a variety of diseases, but is most common for the treatment of hematologic and lymphoid cancers. Hematopoietic stem cell transplantation has been performed clinically since 1957 (Appelbaum, 2007). According to the Center for International Blood and Marrow Transplant Research, in 2009 more than 30,000 autologous and 25,000 allogenic HSC transplants were performed worldwide, with the numbers steadily increasing each year. The therapeutic demand for HSCs prompted the need to explore alternative tissues sources, one of which, umbilical cord blood, has proven to be enriched with HSCs (Lu et al., 1993). The use of cord blood for HSC therapies has led to the creation of public cord blood banks and to the common practice of parents opting to cryopreserve their children's cord blood at the time of birth with private commercial entities. Current challenges associated with HSC transplantation include immune response to allogenic sources (termed graft versus host disease), variability in HSC sources (i.e., age, gender, race), and the often extensive and exhaustive preparative regimens that are necessary for many transplants.

Mesenchymal stem cells (MSCs) were first isolated from bone marrow based on their ability to adhere to culture substrates versus the rest of the non-adherent cell fraction, and largely resemble fibroblasts based on their morphology (Friedenstein et al., 1974). MSCs have subsequently been derived from a variety of other tissues of mesenchymal origin, including fat, umbilical cords, and dental pulp, to name just a few. MSCs have the proven capacity to differentiate into connective tissue cell types such as osteo-, adipo-, and chondrogenic cell lineages, making them of interest for many musculoskeletal studies and orthopedic tissue engineering efforts. It has been commonly reported that MSCs exhibit myogenic potential, although less frequently than other aforementioned phenotypes. In some cases, MSCs or sub-sets of cells within MSC populations have even been reported to differentiate into neuronal or cardiomyogenic cell fates; however, subsequent studies have failed to confirm such findings, thus consensus agreement on these latter, more distant phenotypes has not been reached. Some of the controversy over MSC differentiation potential could be due to the fact that MSC cell populations are known to be heterogeneous, consisting of morphologically similar but distinct sub-sets of cells expressing subtly different types or levels of phenotypic markers.

One of the more recently appreciated and functional characteristics of MSCs, independent of their differentiation potential, is their ability to secrete an array of trophic factors that can affect other cells in a paracrine manner. MSCs can produce molecules that impact tissue repair and remodeling processes, such as vasculogenesis and fibrosis, as well as immunomodulatory molecules that can act locally and systemically. Other aspects of MSCs are described in Chapter II.6.4 "Cell Sources for Tissue Engineering"). Neural stem cells (NSCs) are found in the proliferating regions of fetal or adult brain tissues, typically in the subventricular zone (SVZ) or the hippocampus (Palmer et al., 1997; Doetsch et al., 1999; Martino et al., 2011). NSCs are therefore most commonly extracted from one of these two regions. NSCs have the capacity to self-renew and differentiate into neurons as well as glial cells, including astrocytes and oligodendrocytes both *in vitro* and *in vivo* (Gage, 2000). The use of NSCs and their functional derivatives has been of great clinical interest due to the incidence of neurodegenerative disorders that impair cognitive and motor function, including Parkinson's, Alzheimer's, and Huntington's diseases, Multiple Sclerosis, and Amyotrophic Lateral Sclerosis (ALS), as well as traumatic brain and spinal cord injuries (Kim and de Vellis, 2009).

The differentiation of various subtypes of motor and sensory neurons from NSCs, including those responsible for producing specific neurotransmitters, such as dopamine, has been achieved (Studer et al., 2000). In addition to molecular markers, the functional characterization of electrically excitable neurons created from NSCs *in vitro* has been confirmed using electrophysiological techniques, such as patch clamping and extracellular recordings. Transplantation of NSCs and neurons or glia derived from NSCs has been attempted for the treatment of Parkinson's disease, as well as ALS and spinal cord injury, with promising results.

Unipotent Stem Cells

Progenitor cells capable of self-renewal and differentiation, albeit it into only a single cell lineage, are referred to as unipotent stem cells. Unipotent stem cells, similar to multipotent stem cells in many respects, exhibit limited proliferative capacity and constitute a minority of the cells within any given tissue. In many cases, the similarities in morphology and marker expression make it technically difficult to readily discern stem cells from somatic cells in the surrounding tissue, but their spatial location can often be identified based on previous characterization studies. Depending on the tissue, some unipotent stem cells remain quiescent until activated by the need for tissue repair, such as skeletal myoblasts, whereas others are almost continuously generating progeny, such as epithelial stem cells of the intestines, to keep up with the high cell turnover of the tissues. As a result of their inherent specification for particular tissues, the heterogeneity and number of unipotent stem cells is much greater than for pluripotent or even multipotent stem cells.

Satellite cells are the precursors to skeletal myoblasts that reside in skeletal muscle located under the basal lamina immediately adjacent to mature muscle fibers. Upon traumatic injury or muscle degeneration, satellite cells are activated to proliferate, differentiate, and restore contractile muscle. Activation of the myogenic regulatory factors (MRFs) – MyoD, Myf5, myogenin, and MRF4

(Myf6) – initiates the myogenic differentiation of satellite cells into skeletal myoblasts, along with the myocyte enhancer factor 2 (MEF2) family of transcription factors that act cooperatively with MRFs. Myogenin and MRF4 expression are required for fusion of individual myoblasts to form multi-nucleated myotubes as terminal differentiation proceeds. During the course of differentiation, a sub-set of cells commit to replenishing the satellite cell compartment to permit continual repair of mucle tissue, whereas the majority of cells terminally differentiate into mature striated muscle fibers. Myoblasts have been investigated as a cell source for muscular dystrophy and cardiac regeneration, due to the fact that they are capable of producing contractile muscle.

Epithelial stem cells are among the most common types of stem cells, due to the abundance of epithelial tissues found throughout the body. Epithelial stem cells, such as those of the epidermis and intestine, are relatively long-lived compared to most other "adult" stem cells, due to the rapid turnover of these tissues. Epidermal stem cells and intestinal stem cells reside below the surface in the basal layers of their respective tissues, and although they are typically classified as unipotent under homeostatic conditions, they may exhibit multipotent differentiation potential, especially following tissue injury. Epithelial stem cells typically reside in a specific location, such as the "bulge" in skin or the "crypt" in the intestine, and proceed to transiently amplify and terminally differentiate as they move away from these sites of their normal residence.

STEM CELL NICHES

Controlling the proliferation, migration, and differentiation of stem cells is essential for proper tissue homeostasis and repair. The dynamic and complex microenvironment in which stem cells reside *in vivo* is a critical regulator of cell fate decisions, including quiescence, self-renewal, and differentiation. The components that make up this microenvironment are collectively referred to as the stem cell "niche" (Scadden, 2006). Neighboring cells, which provide direct cell–cell contacts, secreted paracrine factors, and extracellular matrix, are the key elements that comprise the niche. Stem cell niches are distinctly located within most tissues, and in many cases are distributed at a regular intervals or spacing. The specific role of different niche components has been studied for a variety of stem cell types, including MSCs, HSCs, and NSCs; however, much uncertainty remains as to how stem cell populations are precisely controlled within complex niche microenvironments throughout the lifespan of an organism.

Soluble Factors

Signaling molecules secreted by cells located within the niche play a critical role in directing the fate of stem cells. Unlike cell adhesions, secreted molecules can diffuse over large distances within the niche, allowing interaction with many stem cells. The secretion of morphogens by cells comprising the niche also allows for concentration gradients to be established, which contribute to the precise control of stem cell fate. Although a wide range of secreted factors are important in regulation of specific stem cell niches, the Wnt/β-Catenin and TGF-β signaling families appear to be involved in regulating the action of stem cells in multiple niches. The canonical Wnt/β-Catenin signaling pathway plays a role in dictating cellular processes, including differentiation and proliferation, in a variety of cell types at different stages of development. The Wnt/β-Catenin pathway is conserved across species, yet plays different roles in controlling stem cell behavior in different niches. For example, Wnt signaling has been shown to promote self-renewal in HSCs (Reya et al., 2003), MSCs (Boland et al., 2004; Baksh and Tuan, 2007), and stem cells in the intestinal epithelium (van Es et al., 2005), yet induce differentiation in others, such as hair follicle precursors (Huelsken et al., 2001), skeletal myoblasts (Rochat et al., 2004), and NSCs (Muroyama et al., 2004). Bone morphogenetic proteins (BMPs), a sub-set of the transforming growth factor-β (TGF-β) superfamily, are another class of signaling molecules involved in regulation of multiple stem cell types. BMPs are large, dimeric, secreted proteins that contribute to various aspects of embryological development, as well as the maintenance of adult tissue. BMPs are involved in the regulation of the HSC niche; interestingly, studies suggest that rather than acting directly on HSCs, BMPs interact with osteoblasts within the HSC niche, which in turn signal to regulate HSC behavior. BMPs have also been implicated in the maintenance of the MSC niche (Zhang et al., 2003). Treatment of MSCs with different BMP isoforms *in vitro* has been shown to result in both osteogenesis (Friedman et al., 2006) and chondrogenesis (Gooch et al., 2002). BMP signaling is likewise important in neuro-induction, as well as differentiation of NSCs. Inhibition of BMP signaling via antagonists Noggin and Chordin is required for neuroectoderm formation during embryological development. In the late embryo and adult subventricular zone (SVZ), BMP signaling is required for glial cell differentiation, while Noggin potently induces neuronal differentiation (Lim et al., 2000).

Extracellular Matrix

The stem cell microenvironment consists not only of cells, but also of extracellular matrix, including proteins, glycoproteins, and glycosaminoglycans. The interaction of cells with the ECM is generally mediated by a class of membrane proteins called integrins. Integrins are heterodimeric proteins, consisting of an α and a β subunit. Integrins function to attach cells to the ECM, as well as in signal transduction from the ECM to the cells. The interaction of integrins with ECM components has been studied in the context of many stem cell niches. Osteoblasts within the HSC bone marrow niche secrete the matricellular sialoprotein osteopontin (OPN), and the amount of OPN, in turn, can control the numbers of stem cells within the marrow in a dose-dependent manner (Nilsson et al., 2005; Stier et al., 2005). Tenascin C (TenC), an ECM glycoprotein, appears to be important in regulating NSC function in the SVZ. Mice lacking TenC exhibit increased numbers of NSCs, and a delayed progression of NSC differentiation from primarily neuronal to glial cells, suggesting that TenC is involved in coordinating NSC development (Garcion et al., 2004). Mice deficient in TenC also display reduced colony forming capacity of bone marrow cells, suggesting a role for TenC in the hematopoietic niche (Ohta et al., 1998). Laminins are a class of ECM proteins expressed in nearly all tissues and predominantly found as a component of basement membranes. Laminin-5 appears to play a role in MSC differentiation, as MSCs plated on laminin-5 displayed enhanced osteogenic differentiation (Klees et al., 2005, 2007). Laminins, as well as other common ECM molecules, such as collagens and fibronectin, are expressed by bone marrow stromal cells (Zuckerman and Wicha, 1983) and many other cell types, thus such molecules likely play a role in regulating not only the HSC niche, but many other stem cell niches as well.

Additionally, ECM stiffness may contribute to the regulation of stem cell niches. *In vitro* studies in which substrate stiffness on which MSCs were cultured was systematically varied, revealed that differentiation was a function of substrate modulus. Cells cultured on hard substrates displayed preferential osteogenesis; cells on soft substrates displayed neurogenesis, and cells on intermediate substrates were myogenic (Engler et al., 2006). Studies performed with adult neural stem cells cultured on interpenetrating polymer networks with variable moduli revealed that softer substrates supported neuronal differentiation, while substrates with higher moduli promoted differentiation to glial cells (Saha et al., 2008). The physical properties of substrates may also play a role in ESC differentiation, as stiffer matrices have been shown to enhance ESC spreading and proliferation, as well as specifically promote mesoderm induction and osteogenic differentiation (Evans et al., 2009). Altogether, these studies suggest that physiological changes in the mechanical properties of ECM, in addition to its biochemical composition, may play an important role in regulating stem cell behavior in various niches.

Cell–cell Interactions. The interaction of stem cells with neighboring cell types is critical for proper regulation of stem cell fate. While the role of secreted factors has been extensively studied, the role of direct adhesions of stem cells with neighboring cells is less well-understood. Direct cell–cell contacts are generally mediated by a class of calcium-dependant transmembrane proteins called cadherins. There are multiple subclasses of cadherins, including E-, P-, N-, and VE-cadherin, that typically interact with each other in a homophilic manner. Cadherins have been implicated in regulation of the HSC niche, as N-cad+ osteoblasts have been observed to play a role in supporting a subset of HSCs known as long-term HSCs (Zhang et al., 2003). Cadherins may play other roles in HSC niches, including the balance between proliferation and quiescence (Wilson et al., 2004), as well as influencing asymmetric division (Muguruma et al., 2006). Cadherins also play a role in embryonic development and in ESC differentiation. Proper embryo development requires an epithelial-to-mesenchymal transition (EMT), in which E-cadherin is downregulated, while N-cadherin expression increases (Thiery et al., 2009). Interestingly, the reverse process, mesenchymal-to-epithelial transition, also occurs during development, particularly in the kidneys and heart. ESCs, which characteristically express E-cadherin, undergo EMT during the course of differentiation in order to form mesenchymal cell lineages that express N- and VE-cadherin.

Additionally, the Notch signaling pathway is involved in the regulation of stem cell niches. Notch and its ligands, including Jagged and Delta-like, are each transmembrane proteins that require cell–cell contact in order to communicate. Osteoblasts in the bone marrow niche can express high levels of the Notch ligand jagged 1, which in turn can activate the Notch receptor on HSCs, suggesting a functional role for Notch signaling in HSC maintenance (Calvi et al., 2003). The role of Notch in NSC self-renewal and differentiation is not well-understood, as studies show that Notch is necessary for NSC maintenance (Chambers et al., 2001; Hitoshi et al., 2002), while others indicate that Notch activation accelerates glial cell differentiation of neural crest stem cells while inhibiting neuronal differentiation (Morrison et al., 2000). The Notch pathway is also a regulatory component of ESC fate, as mouse ESCs constitutively expressing Notch display enhanced neural differentiation efficiency upon removal of self-renewal stimuli, and Notch signaling activation in human ESCs is required to differentiate to all three germ layers (Lowell et al., 2006; Yu et al., 2008).

CONCLUSIONS

Stem cells are a unique class of cells that exist in most tissues of the body at all stages of development, from early embryogenesis all the way throughout adult life. The unique ability of stem cells to self-renew and differentiate distinguishes them functionally from other cell types, and the varying potency of stem cells to differentiate into different phenotypes establishes a hierarchical organization of different classes of stem cells. Multiple factors which comprise the physical environments, or "niches," within which stem cells reside can significantly influence cell fate decisions. Thus, the engineering of biomaterial properties to control and direct stem cell fate *in vitro* and *in vivo* has tremendous potential to aid in the advancement of stem cell research, and translation of stem cells into regenerative medicine therapies and cell-based diagnostics.

BIBLIOGRAPHY

Aoi, T., Yae, K., Nakagawa, M., Ichisaka, T., Okita, K., et al. (2008). Generation of pluripotent stem cells from adult mouse liver and stomach cells. *Science*, **321**, 699–702.

Appelbaum, F. R. (2007). Hematopoietic-cell transplantation at 50. *N. Engl. J. Med.*, **357**, 1472–1475.

Baksh, D., & Tuan, R. S. (2007). Canonical and non-canonical Wnts differentially affect the development potential of primary isolate of human bone marrow mesenchymal stem cells. *J. Cell Physiol.*, **212**, 817–826.

Baribault, H., & Kemler, R. (1989). Embryonic stem cell culture and gene targeting in transgenic mice. *Mol. Biol. Med.*, **6**, 481–492.

Becker, A. J., McCulloch, E. A., & Till, J. E. (1963). Cytological demonstration of the clonal nature of spleen colonies derived from transplanted mouse marrow cells. *Nature*, **197**, 452–454.

Boland, G. M., Perkins, G., Hall, D. J., & Tuan, R. S. (2004). Wnt 3a promotes proliferation and suppresses osteogenic differentiation of adult human mesenchymal stem cells. *J. Cell Biochem.*, **93**, 1210–1230.

Calvi, L. M., Adams, G. B., Weibrecht, K. W., Weber, J. M., Olson, D. P., et al. (2003). Osteoblastic cells regulate the haematopoietic stem cell niche. *Nature*, **425**, 841–846.

Casciari, J. J., Sotirchos, S. V., & Sutherland, R. M. (1988). Glucose diffusivity in multicellular tumor spheroids. *Cancer Res.*, **48**, 3905–3909.

Chambers, C. B., Peng, Y., Nguyen, H., Gaiano, N., Fishell, G., et al. (2001). Spatiotemporal selectivity of response to Notch1 signals in mammalian forebrain precursors. *Development*, **128**, 689–702.

Doetsch, F., Caille, I., Lim, D. A., Garcia-Verdugo, J. M., & Alvarez-Buylla, A. (1999). Subventricular zone astrocytes are neural stem cells in the adult mammalian brain. *Cell*, **97**, 703–716.

Doetschman, T. C., Eistetter, H., Katz, M., Schmidt, W., & Kemler, R. (1985). The *in vitro* development of blastocyst-derived embryonic stem cell lines: Formation of visceral yolk sac, blood islands and myocardium. *J. Embryol. Exp. Morphol.*, **87**, 27–45.

Engler, A. J., Sen, S., Sweeney, H. L., & Discher, D. E. (2006). Matrix elasticity directs stem cell lineage specification. *Cell*, **126**, 677–689.

Evans, M. J., & Kaufman, M. H. (1981). Establishment in culture of pluripotential cells from mouse embryos. *Nature*, **292**, 154–156.

Evans, N. D., Minelli, C., Gentleman, E., LaPointe, V., Patankar, S. N., et al. (2009). Substrate stiffness affects early differentiation events in embryonic stem cells. *Eur. Cell Mater.*, **18**, 1–13, discussion 13–14.

Friedenstein, A. J., Deriglasova, U. F., Kulagina, N. N., Panasuk, A. F., Rudakowa, S. F., et al. (1974). Precursors for fibroblasts in different populations of hematopoietic cells as detected by the *in vitro* colony assay method. *Exp. Hematol.*, **2**, 83–92.

Friedman, M. S., Long, M. W., & Hankenson, K. D. (2006). Osteogenic differentiation of human mesenchymal stem cells is regulated by bone morphogenetic protein-6. *J. Cell Biochem.*, **98**, 538–554.

Gage, F. H. (2000). Mammalian neural stem cells. *Science*, **287**, 1433–1438.

Garcion, E., Halilagic, A., Faissner, A., & ffrench-Constant, C. (2004). Generation of an environmental niche for neural stem cell development by the extracellular matrix molecule tenascin C. *Development*, **131**, 3423–3432.

Gooch, K. J., Blunk, T., Courter, D. L., Sieminski, A. L., Vunjak-Novakovic, G., et al. (2002). Bone morphogenetic proteins-2, -12, and -13 modulate *in vitro* development of engineered cartilage. *Tissue Eng.*, **8**, 591–601.

Gurdon, J. B., Elsdale, T. R., & Fischberg, M. (1958). Sexually mature individuals of *Xenopus laevis* from the transplantation of single somatic nuclei. *Nature*, **182**, 64–65.

Helmlinger, G., Netti, P. A., Lichtenbeld, H. C., Melder, R. J., & Jain, R. K. (1997). Solid stress inhibits the growth of multicellular tumor spheroids. *Nat. Biotechnol.*, **15**, 778–783.

Hitoshi, S., Alexson, T., Tropepe, V., Donoviel, D., Elia, A. J., et al. (2002). Notch pathway molecules are essential for the maintenance, but not the generation, of mammalian neural stem cells. *Genes Dev.*, **16**, 846–858.

Huelsken, J., Vogel, R., Erdmann, B., Cotsarelis, G., & W. Birchmeier, W. (2001). beta-Catenin controls hair follicle morphogenesis and stem cell differentiation in the skin. *Cell*, **105**, 533–545.

Kim, S. U., & de Vellis, J. (2009). Stem cell-based cell therapy in neurological diseases: A review. *J. Neurosci. Res.*, **87**, 2183–2200.

Klees, R. F., Salasznyk, R. M., Kingsley, K., Williams, W. A., Boskey, A., et al. (2005). Laminin-5 induces osteogenic gene expression in human mesenchymal stem cells through an ERK-dependent pathway. *Mol. Biol. Cell*, **16**, 881–890.

Klees, R. F., Salasznyk, R. M., Vandenberg, S., Bennett, K., & Plopper, G. E. (2007). Laminin-5 activates extracellular matrix production and osteogenic gene focusing in human mesenchymal stem cells. *Matrix Biol.*, **26**, 106–114.

Lim, D. A., Tramontin, A. D., Trevejo, J. M., Herrera, D. G., Garcia-Verdugo, J. M., et al. (2000). Noggin antagonizes BMP signaling to create a niche for adult neurogenesis. *Neuron*, **28**, 713–726.

Loh, Y. H., Agarwal, S., Park, I. H., Urbach, A., Huo, H., et al. (2009). Generation of induced pluripotent stem cells from human blood. *Blood*, **113**, 5476–5479.

Lowell, S., Benchoua, A., Heavey, B., & Smith, A. G. (2006). Notch promotes neural lineage entry by pluripotent embryonic stem cells. *PLoS. Biol.*, **4**, e121.

Lu, L., Xiao, M., Shen, R. N., Grigsby, S., & Broxmeyer, H. E. (1993). Enrichment, characterization, and responsiveness of single primitive CD34 human umbilical cord blood hematopoietic progenitors with high proliferative and replating potential. *Blood*, **81**, 41–48.

Martin, G. R. (1981). Isolation of a pluripotent cell line from early mouse embryos cultured in medium conditioned by teratocarcinoma stem cells. *Proc. Natl. Acad. Sci. USA*, **78**, 7634–7638.

Martin, G. R., & Evans, M. J. (1975). Differentiation of clonal lines of teratocarcinoma cells: Formation of embryoid bodies *in vitro*. *Proc. Natl. Acad. Sci. USA*, **72**, 1441–1445.

Martino, G., Pluchino, S., Bonfanti, L., & Schwartz, M. (2011). Brain regeneration in physiology and pathology: The immune signature driving therapeutic plasticity of neural stem cells. *Physiol. Rev.*, **91**, 1281–1304.

Morrison, S. J., Perez, S. E., Qiao, Z., Verdi, J. M., Hicks, C., et al. (2000). Transient Notch activation initiates an irreversible switch from neurogenesis to gliogenesis by neural crest stem cells. *Cell*, **101**, 499–510.

Muguruma, Y., Yahata, T., Miyatake, H., Sato, T., Uno, T., et al. (2006). Reconstitution of the functional human hematopoietic microenvironment derived from human mesenchymal stem cells in the murine bone marrow compartment. *Blood*, **107**, 1878–1887.

Muroyama, Y., Kondoh, H., & Takada, S. (2004). Wnt proteins promote neuronal differentiation in neural stem cell culture. *Biochem. Biophys. Res. Commun.*, **313**, 915–921.

Nilsson, S. K., Johnston, H. M., Whitty, G. A., Williams, B., Webb, R. J., et al. (2005). Osteopontin, a key component of the hematopoietic stem cell niche and regulator of primitive hematopoietic progenitor cells. *Blood*, **106**, 1232–1239.

Ohta, M., Sakai, T., Saga, Y., Aizawa, S., & Saito, M. (1998). Suppression of hematopoietic activity in tenascin-C-deficient mice. *Blood*, **91**, 4074–4083.

Okita, K., & Yamanaka, S. (2011). Induced pluripotent stem cells: Opportunities and challenges. *Philos. Trans. R. Soc. Lond. B. Biol. Sci.*, **366**, 2198–2207.

Okita, K., Nakagawa, M., Hyenjong, H., Ichisaka, T., & Yamanaka, S. (2008). Generation of mouse induced pluripotent stem cells without viral vectors. *Science*, **322**, 949–953.

Palmer, T. D., Takahashi, J., & Gage, F. H. (1997). The adult rat hippocampus contains primordial neural stem cells. *Mol. Cell Neurosci.*, **8**, 389–404.

Park, I. H., Arora, N., Huo, H., Maherali, N., Ahfeldt, T., et al. (2008). Disease-specific induced pluripotent stem cells. *Cell*, **134**, 877–886.

Reya, T., Duncan, A. W., Ailles, L., Domen, J., Scherer, D. C., et al. (2003). A role for Wnt signalling in self-renewal of haematopoietic stem cells. *Nature*, **423**, 409–414.

Rochat, A., Fernandez, A., Vandromme, M., Moles, J. P., Bouschet, T., et al. (2004). Insulin and wnt1 pathways cooperate to induce reserve cell activation in differentiation and myotube hypertrophy. *Mol. Biol. Cell*, **15**, 4544–4555.

Saha, K., Keung, A. J., Irwin, E. F., Li, Y., Little, L., et al. (2008). Substrate modulus directs neural stem cell behavior. *Biophys. J.*, **95**, 4426–4438.

Scadden, D. T. (2006). The stem cell niche as an entity of action. *Nature*, **441**, 1075–1079.

Shamblott, M. J., Axelman, J., Wang, S., Bugg, E. M., Littlefield, J. W., et al. (1998). Derivation of pluripotent stem cells from cultured human primordial germ cells. *Proc. Natl. Acad. Sci. USA*, **95**, 13726–13731.

Shamblott, M. J., Axelman, J., Littlefield, J. W., Blumenthal, P. D., Huggins, G. R., et al. (2001). Human embryonic germ cell derivatives express a broad range of developmentally distinct markers and proliferate extensively *in vitro*. *Proc. Natl. Acad. Sci. USA*, **98**, 113–118.

Shweiki, D., Neeman, M., Itin, A., & Keshet, E. (1995). Induction of vascular endothelial growth factor expression by hypoxia and by glucose deficiency in multicell spheroids: Implications for tumor angiogenesis. *Proc. Natl. Acad. Sci. USA*, **92**, 768–772.

Stier, S., Ko, Y., Forkert, R., Lutz, C., Neuhaus, T., et al. (2005). Osteopontin is a hematopoietic stem cell niche component that negatively regulates stem cell pool size. *J. Exp.Med.*, **201**, 1781–1791.

Studer, L., Csete, M., Lee, S. H., Kabbani, N., Walikonis, J., et al. (2000). Enhanced proliferation, survival, and dopaminergic differentiation of CNS precursors in lowered oxygen. *J. Neurosci.*, **20**, 7377–7383.

Takahashi, K., & Yamanaka, S. (2006). Induction of pluripotent stem cells from mouse embryonic and adult fibroblast cultures by defined factors. *Cell*, **126**, 663–676.

Takahashi, K., Tanabe, K., Ohnuki, M., Narita, M., Ichisaka, T., et al. (2007). Induction of pluripotent stem cells from adult human fibroblasts by defined factors. *Cell*, **131**, 861–872.

Thiery, J. P., Acloque, H., Huang, R. Y., & Nieto, M. A. (2009). Epithelial-mesenchymal transitions in development and disease. *Cell*, **139**, 871–890.

Thomson, J. A., Kalishman, J., Golos, T. G., Durning, M., Harris, C. P., et al. (1995). Isolation of a primate embryonic stem cell line. *Proc. Natl. Acad. Sci. USA*, **92**, 7844–7848.

Thomson, J. A., Itskovitz-Eldor, J., Shapiro, S. S., Waknitz, M. A., Swiergiel, J. J., et al. (1998). Embryonic stem cell lines derived from human blastocysts. *Science*, **282**, 1145–1147.

Till, J. E., & McCulloch, E. A. (1961). A direct measurement of the radiation sensitivity of normal mouse bone marrow cells. *Radiat. Res.*, **14**, 213–222.

van Es, J. H., Jay, P., Gregorieff, A., van Gijn, M. E., Jonkheer, S., et al. (2005). Wnt signalling induces maturation of Paneth cells in intestinal crypts. *Nat. Cell. Biol.*, **7**, 381–386.

Warren, L., Manos, P. D., Ahfeldt, T., Loh, Y. H., Li, H., et al. (2010). Highly efficient reprogramming to pluripotency and directed differentiation of human cells with synthetic modified mRNA. *Cell Stem Cell*, **7**, 618–630.

Wernig, M., Meissner, A., Foreman, R., Brambrink, T., Ku, M., et al. (2007). *In vitro* reprogramming of fibroblasts into a pluripotent ES-cell-like state. *Nature*, **448**, 318–324.

Wilson, A., Murphy, M. J., Oskarsson, T., Kaloulis, K., Bettess, M. D., et al. (2004). c-Myc controls the balance between hematopoietic stem cell self-renewal and differentiation. *Genes Dev.*, **18**, 2747–2763.

Yu, J., Vodyanik, M. A., Smuga-Otto, K., Antosiewicz-Bourget, J., Frane, J. L., et al. (2007). Induced pluripotent stem cell lines derived from human somatic cells. *Science*, **318**, 1917–1920.

Yu, X., Zou, J., Ye, Z., Hammond, H., Chen, G., et al. (2008). Notch signaling activation in human embryonic stem cells is required for embryonic, but not trophoblastic, lineage commitment. *Cell Stem Cell*, **2**, 461–471.

Yuhas, J. M., Li, A. P., Martinez, A. O., & Ladman, A. J. (1977). A simplified method for production and growth of multicellular tumor spheroids. *Cancer Res.*, **37**, 3639–3643.

Zhang, J., Niu, C., Ye, L., Huang, H., He, X., et al. (2003). Identification of the haematopoietic stem cell niche and control of the niche size. *Nature*, **425**, 836–841.

Zhou, H., Wu, S., Joo, J. Y., Zhu, S., Han, D. W., et al. (2009). Generation of induced pluripotent stem cells using recombinant proteins. *Cell Stem Cell*, **4**, 381–384.

Zuckerman, K. S., & Wicha, M. S. (1983). Extracellular matrix production by the adherent cells of long-term murine bone marrow cultures. *Blood*, **61**, 540–547.

SECTION II.2

Host Reaction to Biomaterials and Their Evaluation

SECTION

II.2

Host Reaction to Biomaterials and Their Evaluation

CHAPTER II.2.1 INTRODUCTION: "BIOLOGICAL RESPONSES TO BIOMATERIALS"

Frederick J. Schoen
Professor of Pathology and Health Sciences (HST), Harvard Medical School, Executive Vice Chairman, Department of Pathology, Brigham and Women's Hospital, Boston, MA, USA

Biomaterials are now commonly used as implants and other tissue-contacting medical devices over a wide range of applications, including, for example, as *prostheses* in cardiovascular, orthopedic, dental, ophthalmological, and reconstructive surgery, in minimally invasive interventions such as stent placement in the biliary tree or in the blood vessels, and in extracorporeal devices as hemodialysis membranes, in surgical sutures or bioadhesives, and as controlled drug-release devices. Most implants serve their recipients well for extended periods by alleviating the conditions for which they were implanted. However, some implants and extracorporeal devices ultimately develop complications – adverse interactions of the patient with the device or *vice versa* – which constitute device failure, and thereby may cause harm to or death of the patient. Complications are costly, both to patient outcomes (mortality and morbidity) and to the healthcare budget.

Complications of biomaterials and medical devices result largely as a consequence of biomaterial–tissue interactions, which all implants have with the environment into which they are placed. Effects of both the implant on the host tissues and the host on the implant are important in mediating complications and device failure (Figure II.2.1.1). Moreover, these processes can be related (Anderson et al., 2008). The chapters in Section II.2 contain overview discussions of the most important host reactions to biomaterials and their evaluation, including non-specific inflammation and specific immunological reactions, systemic effects, blood–materials interactions, tumor formation, and infection. To a great extent, these interactions arise from alterations of physiological (normal) processes (e.g., immunity, inflammation, blood coagulation) comprising host defense mechanisms that function to protect an organism from deleterious external threats (such as bacteria and other microbiologic organisms, injury, and foreign materials). The chapters in Section II.2 address degradation mechanisms in biomaterials (i.e., the effect of the host on biomaterials). Several key concepts of biomaterials–tissue interactions are emphasized below, in an effort to guide the reader and facilitate the use of this section.

THE INFLAMMATORY REACTION TO BIOMATERIALS

In their respective chapters, Anderson (II.2.2), Mitchell (II.2.3), and Johnson (II.2.4) describe the inflammatory and potential immunological interactions that occur with biomaterials. Most biomaterials of potential clinical interest typically elicit the *foreign-body reaction* (FBR), a special form of non-specific inflammation (see Chapter II.2.2 of this book). The most prominent cells in the FBR are macrophages, which attempt to phagocytose the material and are variably successful, but complete engulfment and degradation are often difficult. The macrophages, activated in the process of interacting with a biomaterial, may elaborate cytokines which stimulate inflammation or fibrosis. Macrophages are also the first line of defense against pathogens, and the mode of their activation will determine the success or failure of the host response to pathogens (Mège et al., 2011). Based on limited numbers of markers, activated macrophages can be classified along a phenotypic spectrum as classically activated (M1) macrophages that support microbicidal activity, are proinflammatory, and elicit dense connective tissue and/or scarring, or alternatively activated (M2) macrophages that are not competent to eliminate pathogens that are regulatory/homeostatic. Thus, a more constructive remodeling outcome may be due to the recruitment and survival of M2 cell populations to the sites of remodeling associated with materials (Brown et al., 2012). Multinucleated *giant cells* in the vicinity of a foreign-body are generally considered evidence of a more severe FBR (Brodbeck and Anderson, 2009). The more "biocompatible" the implant, the more quiescent (less inflammation) is the ultimate response. When the implant is a source of particles not easily controlled, such as wear debris from articulating joint surfaces (Pandit et al., 2008; Catelas et al., 2011), the inability of inflammatory cells to adhere

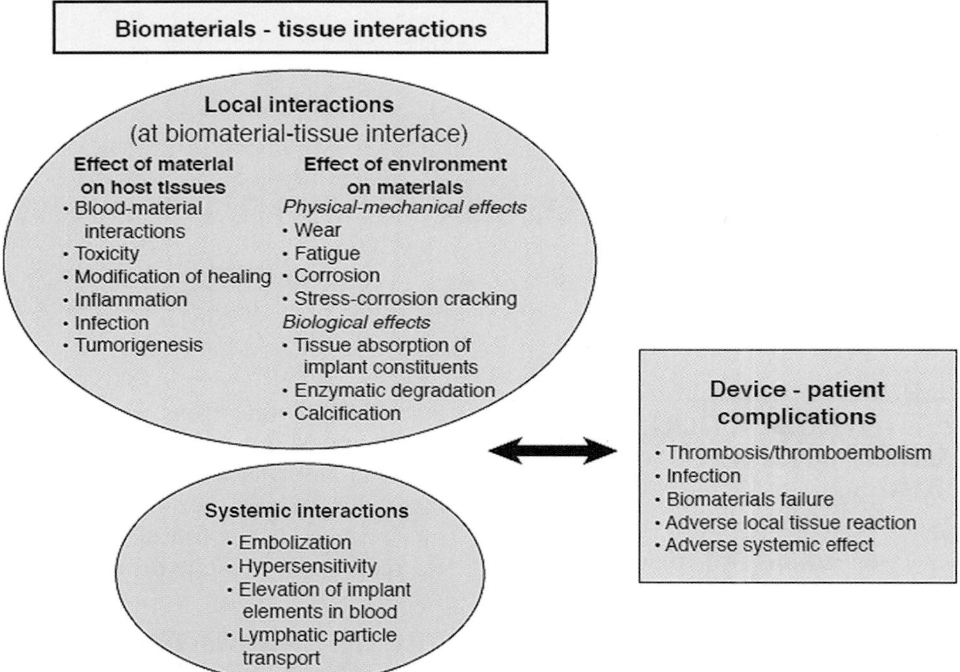

FIGURE II.2.1.1 Biomaterials–tissue interactions. (Reproduced from Schoen, F. J. (2002). Prosthetic-materials: Past, present and future. In: *Advances in Cardiovascular Medicine* (Harvey 1602–2002 Symposium, on the 4th Centenary of William Harvey's Graduation at the University of Padua), Thiene, G., & Pessina, A. C. (Eds.), Universita degli Studi di Padova, pp. 289–307 and Schoen, F. J. & Padera, R. (2012). Cardiovascular Pathology. In: *Cardiac Surgery in the Adult*, 4th edn., Cohn, L. H. & Edmunds, L. H. (Eds.). McGraw–Hill, pp. 95–148.)

to but not phagocytose particles larger than a critical size ("frustrated phagocytosis") can lead to release of enzymes (exocytosis) and chemical mediators (e.g., prostaglandin, tumor necrosis factor-alpha, and interleukin-1, etc.), and cause harm to the extracellular environment. Thus, inflammatory cell products that are critical in killing microorganisms in typical inflammation can damage tissue adjacent to foreign-bodies.

In contrast to living organ transplants, biomaterials are not generally "rejected." The process of tissue or whole organ rejection denotes an inflammatory process that results from a specific immune response (i.e., a response to specific antigens having certain characteristics that are described by Mitchell in Chapter II.2.3), and which synthetic biomaterials typically do not generate. Indeed, the usual response to biomaterials comprises a form of non-specific inflammation known as the "foreign body response" (see Chapter II.2.2).

However, as summarized by Mitchell (2001), tissue-derived biomaterials (such as bioprosthetic heart valves) may express foreign histocompatibility antigens, and be antigenic and capable of eliciting an immune response mediated by antibodies and antigen-specific T cells. Nevertheless, as explained in more detail below, it is important to understand that:

(1) the capacity for tissue immunogenicity does not necessarily induce immunologically mediated device dysfunction;
(2) specific immunological responses can be either a cause of or result from device failure;
(3) although mononuclear inflammatory cell infiltrates (containing macrophages and lymphocytes) are characteristically in organ/tissue rejection on histological examination, mononuclear inflammatory infiltrates are themselves non-specific, and comprise a largely stereotyped and generic response to tissue injury. Therefore, the presence of mononuclear cells does not necessarily denote a rejection pathogenesis.

In order to invoke an immunological reaction to a biomaterial as the cause of a device failure, an immunological variant of the classical Koch's Postulates that are the objective criteria for concluding that a disease is infectious and caused by a specific microbiologic agent would be appropriate. The classic Koch's Postulates state that (Inglis, 2007):

(1) a suspected infectious agent should be recoverable from the pathologic lesions of the human host;
(2) the agent should cause the pathologic lesions when inoculated into an animal host;
(3) the agent should then be recoverable from the pathologic lesions in the animal.

Mitchell (2001) describes an immunological variant of Koch's Postulates to test an immunological hypothesis for biomaterials failure, as exemplified by the widely quoted but non-proven immunological calcific and non-calcific bioprosthetic valve failure (Mitchell, 2001; Schoen and Levy, 2005) as follows:

(1) antigen-specific elements (antibodies or cells) should be directly associated with failing valves. Moreover, control experiments should be performed to demonstrate that any antibodies or cells on implanted valves are not simply present because of surgical manipulation or aberrant flow conditions;
(2) antibodies or cells from experimental animals that have dysfunctional implanted valves transferred into an appropriate second host (immunologically-matched to the original valve donor) should cause valve failure;
(3) adoptively transferred cells or antibodies should be detectable on a failed valve in the second recipient.

Although some evidence for these criteria can be obtained in humans, carefully designed animal investigations provide the only rigorous way to satisfy them. With respect to tissue heart valves, although some investigators have demonstrated that such tissues can be immunogenic, *there exists no evidence that clinical valve destruction or loss-of-function is mediated by immune elements or that blockade of immune mechanisms by immunosuppression prevents that outcome.*

The nature of the reaction is largely dependent on the chemical and physical characteristics of the implant. For most inert biomaterials, the late tissue reaction is *encapsulation* by a fibrous tissue capsule (composed of collagen and fibroblasts). Tissue interactions can be modified by changing the chemistry of the surface (e.g., by adding specific chemical groupings to stimulate adhesion or bone formation in orthopedic implants), inducing roughness or porosity to enhance physical binding to the surrounding tissues, incorporating a surface–active agent to chemically bond the tissue, using a bioresorbable component to allow slow replacement by tissue to simulate natural healing properties, or by systemic or localized administration of corticosteroid drugs.

SYSTEMIC AND REMOTE EFFECTS

Hensten and Jacobsen (Chapter II.2.5) summarize biomaterials-related systemic toxicity and hypersensitivity reaction (through lymphatics and the bloodstream) in animals and patients with either stainless steel or cobalt–based orthopedic total joint replacement components, where elevations of metallic content occur in tissue (at both local and remote sites) and in serum and urine. Transport of particulates over large distances by macrophages to regional lymph nodes and to the lungs and liver has been considered a systemic and remote effect. As consequence, an enlarged, hard axillary lymph node in a woman who received a silicone-gel breast prosthesis for reconstruction following mastectomy for a carcinoma can be misdiagnosed as tumor (Hausner et al., 1978) and prosthetic hip joint failure may present as liver problems (Peoc'h, 1996).

"Metal allergy" is well–recognized and is frequently associated in women with the use of cheap, high–nickel alloy costume jewelry or earrings, and can also occur in association with metallic implants (Hallab et al., 2001). By themselves, metal ions lack the structural complexity required to challenge the immune system. However, when combined with proteins, such as those available in the skin, connective tissues, and blood, a wide variety of metals induce immune responses, and this can have clinical effects. Cobalt, chromium, and nickel are included in this category, with nickel perhaps the most potent; at least 10% of a normal population will be sensitive by skin test to one or more of these metals, at some threshold level.

THROMBOEMBOLIC COMPLICATIONS

Hanson and Tucker (Chapter II.2.6) emphasize that exposure of blood to an artificial surface can induce thrombosis, embolization, and consumption of platelets and plasma coagulation factors, as well as the systemic effects of activated coagulation and complement products, and platelet activation. It is clear that no synthetic or modified biological surface generated by man is as resistant to thrombosis (*thromboresistant*) as normal unperturbed endothelium (the cellular lining of the circulatory system). However, it is important to understand that under some circumstances endothelial cells can be "dysfunctional," and although physically intact can express prothrombotic molecules that can induce thrombosis (Bonetti et al., 2003).

Thromboembolic complications are a major cause of mortality and morbidity with cardiovascular devices. Both fibrin (red) thrombus and platelet (white) thrombus form in association with valves and other cardiovascular devices. As in the cardiovascular system in general, Virchow's triad (surface thrombogenicity, hypercoagulability, and locally static blood flow) largely predicts the relative propensity toward thrombus formation and location of thrombotic deposits with cardiovascular prostheses (Bennett et al., 2009). However, despite over a quarter century of intense research, the physical and chemical characteristics of materials that control the outcome of blood–surface interaction are incompletely understood.

Considerable evidence implicates a primary regulatory role for blood platelets in the thrombogenic response to artificial surfaces (see Chapter II.2.6). Platelet adhesion to artificial surfaces strongly resembles that of adhesion to the vascular subendothelium that has been exposed by injury. Nevertheless, the major clinical approach to controlling thrombosis in cardiovascular devices is the use of systemic anticoagulants, particularly Coumadin® (warfarin), which inhibits thrombin formation but does not inhibit platelet–mediated thrombosis.

TUMORIGENESIS

Schoen (Chapter II.2.7) emphasizes that although animals frequently have sarcomas at the site of an experimental biomaterial implant, neoplasms occurring at the site of clinically implanted medical devices are rare, despite the large numbers of implants used clinically in humans over an extended duration. Moreover, a neoplasm at an implant site does not prove that the implant had a causal role. Cancers associated with foreign-bodies can appear at any postoperative interval, but tend to occur many years postoperatively. The pathogenesis of implant–induced tumors is not well understood; most experimental data indicate that the physical rather than chemical characteristics of the foreign-body primarily determine tumorigenicity. The possibility that implants may be causal to tumor formation is an ever present problem, with contemporary questions related to metal-on-metal hip joints and breast prostheses (see Chapter II.2.7).

INFECTION

As emphasized by Stoodley et al. (Chapter II.2.8), infection occurs in as many as 5–10% of patients with implanted prosthetic devices, and is a major source of morbidity and mortality (Tanner, 1997; Vlessis, 1997; Schierholz, 2001). *Infections associated with medical devices are often resistant to antibiotics and host defenses, often persisting until the devices are removed.* Early implant infections (less than approximately 1–2 months postoperatively) are most likely due to intraoperative contamination from airborne sources or nonsterile surgical technique, or to early postoperative complications such as wound infection. In contrast, late infections likely occur by a hematogenous (blood-borne) route, and are often initiated by bacteremia induced by therapeutic dental or genitourinary procedures. Perioperative prophylactic antibiotics and periodic antibiotic prophylaxis given shortly before diagnostic and therapeutic procedures protect against implant infection. Infections associated with foreign bodies are characterized microbiologically by a high prevalence of *Staphylococcus epidermidis* and other staphylococci, especially *S. aureus*. Ordinarily, *S. epidermidis* is an organism with low virulence, and is thus an infrequent cause of non–prosthesis–associated deep infections. This emphasizes the unique environment in the vicinity of a foreign-body.

The presence of a foreign-body per se *potentiates infection*. A classic experiment indicated that the staphylococcal bacterial inoculum required to cause infection in the presence of a foreign implant was 10,000 less than that when no foreign-body was present (Elek and Conen, 1957). Devices could facilitate infection in several ways. Microorganisms are provided access to the circulation and to deeper tissue by damage to natural barriers against infection during implantation or subsequent function of a prosthetic device. Moreover, an implanted foreign-body could: (1) limit phagocyte migration into infected tissue or (2) interfere with inflammatory cell phagocytic mechanisms, through release of soluble implant components or surface–mediated interactions, thus allowing bacteria to survive adjacent to the implant. As Stoodley et al. emphasize, adhesion of bacteria to the prosthetic surface and the formation of microcolonies within an adherent biofilm are fundamental steps in the pathogenesis of clinical and experimental infections associated with foreign bodies.

BIBLIOGRAPHY

Anderson, J. M., Rodriguez, A., & Chang, D. T. (2008). Foreign body reaction to biomaterials. *Semin. Immunol.*, **20**, 86–100.

Bennett, P. C., Silverman, S. H., Gill, P. S., & Lip, G. Y. (2009). Peripheral arterial disease and Virchow's triad. *Thromb. Haemost.*, **101**, 1032–1040.

Bonetti, P. O., Lerman, L. O., & Lerman, A. (2003). Endothelial dysfunction: A marker of atherosclerotic risk. *Arteriosci. Thromb. Vasc. Biol.*, **23**, 168–175.

Brodbeck, W. G., & Anderson, J. M. (2009). Giant cell formation and function. *Curr. Opin. Hematol.*, **16**, 53–57.

Brown, B. N., Londono, R., Tottey, S., Zhang, L., Kukla, K. A., et al. (2012). Macrophage phenotype as a predictor of constructive remodeling following the implantation of biologically derived surgical mesh materials. *Acta Biomater.*, **8**(3), 978–987.

Catelas, I., Wimmer, M. A., & Utzschneider, S. (2011). Polyethylene and metal wear particles: Characteristics and biological effects. *Semin. Immunopathol.*, **33**, 257–271.

Elek, S. D., & Conen, P. E. (1957). The virulence of *Staphylococcus pyogenes* for man: A study of the problems of wound infection. *Br. J. Exp. Pathol.*, **38**, 573–586.

Hallab, N., Merritt, K., & Jacobs, J. J. (2001). Metal sensitivity in patients with orthopaedic implants. *J. Bone Joint Surg.*, **83**, 428–436.

Hausner, R. J., Schoen, F. J., & Pierson, K. K. (1978). Foreign body reaction to silicone in axillary lymph nodes after prosthetic augmentation mammoplasty. *Plast. Reconstr. Surg.*, **62**, 381–384.

Inglis, T. J. (2007). Principia aetiologica: Taking causality beyond Koch's postulates. *J. Med. Microbiol.*, **56**, 1419–1422.

Mège, J.-L., Mehraj, V., & Capo, C. (2011). Macrophage polarization and bacterial infections. *Curr. Opin. Infect. Dis.*, **24**, 230–234.

Mitchell, R. N. (2001). Don't blame the lymphocyte: Immunologic processes are NOT important in tissue valve failure. *J. Heart Valve Dis.*, **10**, 467–470.

Pandit, H., Glyn-Jones, S., McLardy-Smith, P., Gundle, R., Whitwell, D., et al. (2008). Pseudotumours associated with metal-on-metal hip resurfacings. *J. Bone Joint Surg. (Br).*, **90**, 847–851.

Peoc'h, M. (1996). Case records of the Massachusetts General Hospital. *New Engl. J. Med.*, **335**, 133.

Schierholz, J. M., & Beuth, J. (2001). Implant infections: A haven for opportunistic bacteria. *J. Hosp. Infect.*, **49**, 87–93.

Schoen, F. J. (2002). Biomaterials–Tissue Interactions. In G.Thiene, & A.C. Pessina (Eds.), *Advances in Cardiovascular Medicine* (Harvey 1602–2002 Symposium, on the 4th Centenary of William Harvey's Graduation at the University of Padua) (pp. 289–307). Universita degli Studi di Padova.

Schoen, F. J., & Levy, R. J. (2005). Calcification of tissue heart valve substitutes: Progress toward understanding and prevention. *Ann. Thorac. Surg.*, 79, 1072–1080.

Schoen, F. J., & Padera, R. (2012). Cardiovascular Pathology. In L. H. Cohn, & L. H. Edmunds (Eds.), *Cardiac Surgery in the Adult* (4th ed., pp. 95–148). McGraw–Hill.

Tanner, A., Maiden, M. F., Lee, K., Shulman, L. B., & Weber, H. P. (1997). Dental implant infections. *Clin. Infect. Dis.*, 25, S213–S217.

Vlessis, A. A., Khaki, A., Grunkemeier, G. L., Li, H. H., & Starr, A. (1997). Risk, diagnosis and management of prosthetic valve endocarditis: A review. *J. Heart Valve Dis.*, 6, 443–465.

CHAPTER II.2.2 INFLAMMATION, WOUND HEALING, AND THE FOREIGN-BODY RESPONSE

James M. Anderson
Department of Pathology, University Cleveland, OH, USA

Inflammation, wound healing, and foreign-body reaction are generally considered as parts of the tissue or cellular host responses to injury. Table II.2.2.1 lists the sequence/continuum of these events following injury. Overlap and simultaneous occurrence of these events should be considered; e.g., the foreign-body reaction at the implant interface may be initiated with the onset of acute and chronic inflammation. From a biomaterials perspective, placing a biomaterial in the *in vivo* environment requires injection, insertion or surgical implantation, all of which injure the tissues or organs involved.

Biocompatibility and Implantation

Implantation of a biomaterial, medical device, or prosthesis results in tissue injury that initiates host defense systems, e.g., inflammatory, wound healing, and foreign-body responses. The extent and time-dependent nature of these responses, in the context of the characteristics and properties of the biomaterial, form the basis for determining the biocompatibility or safety of the biomaterial. In addition to defining the biocompatibility of a biomaterial, a fundamental understanding of these responses permits their use as biological design criteria.

The placement procedure initiates a response to injury by the tissue, organ or body, and mechanisms are activated to maintain homeostasis. The degrees to which the homeostatic mechanisms are perturbed, and the extent to which pathophysiologic conditions are created and undergo resolution, are a measure of the host reactions to the biomaterial and may ultimately determine its biocompatibility. Although it is convenient to separate homeostatic mechanisms into blood–material or tissue–material interactions, it must be remembered that the various components or mechanisms involved in homeostasis are present in both blood and tissue, and are a part of the physiologic continuum. Furthermore, it must be noted that host reactions may be tissue-dependent, organ-dependent, and species-dependent. Obviously, the extent of injury varies with the implantation procedure.

TABLE II.2.2.1 Sequence/Continuum of Host Reactions Following Implantation of Medical Devices
Injury
Blood–material interactions
Provisional matrix formation
Acute inflammation
Chronic inflammation
Granulation tissue
Foreign-body reaction
Fibrosis/fibrous capsule development

OVERVIEW

Inflammation is generally defined as the reaction of vascularized living tissue to local injury. Inflammation serves to contain, neutralize, dilute, or wall off the injurious agent or process. In addition, it sets in motion a series of events that may heal and reconstitute the implant site through replacement of the injured tissue by regeneration of native parenchymal cells, formation of fibroblastic scar tissue, or a combination of these two processes.

Immediately following injury, there are changes in vascular flow, caliber, and permeability. Fluid, proteins, and blood cells escape from the vascular system into the injured tissue in a process called exudation. Following changes in the vascular system, which also include changes induced in blood and its components, cellular events occur and characterize the inflammatory response. The effect of the injury and/or biomaterial *in situ* on plasma or cells can produce chemical factors that mediate many of the vascular and cellular responses of inflammation.

Blood–material interactions and the inflammatory response are intimately linked, and in fact early responses to injury involve mainly blood and vasculature. Regardless of the tissue or organ into which a biomaterial is implanted, the initial inflammatory response is activated by injury to vascularized connective tissue (Table II.2.2.2). Since blood and its components are involved in the initial inflammatory responses, blood clot formation and/or thrombosis also occur. Blood coagulation and thrombosis are generally considered humoral responses that may be influenced by other homeostatic mechanisms, such as the extrinsic and intrinsic coagulation systems, the complement system, the fibrinolytic system, the kinin-generating system, and platelets. Thrombus or blood clot formation on the surface of a biomaterial is related to the well-known Vroman effect (see Chapter II.3.2), in which a hierarchical and dynamic series of collision, absorption, and exchange processes, determined by protein mobility and concentration, regulate early time-dependent changes in blood protein adsorption. From a wound-healing perspective, blood protein deposition on a biomaterial surface is described as provisional matrix formation. Blood interactions with biomaterials are generally considered under the category of hematocompatibility and are discussed elsewhere in this book.

Injury to vascularized tissue in the implantation procedure leads to immediate development of the provisional matrix at the implant site. This provisional matrix consists of fibrin, produced by activation of the coagulation and thrombosis systems, and inflammatory products released by the complement system, activated platelets, inflammatory cells, and endothelial cells. These events occur early, within minutes to hours following implantation of a medical device. Components within or released from the provisional matrix, i.e., fibrin network (thrombosis or clot), initiate the resolution, reorganization, and repair processes such as inflammatory cell and fibroblast recruitment. The provisional matrix appears to provide both structural and biochemical components to the process of wound healing. The complex three-dimensional structure of the fibrin network with attached adhesive proteins provides a substrate for cell adhesion and migration. The presence of mitogens, chemoattractants, cytokines, and growth factors within the provisional matrix provides for a rich milieu of activating and inhibiting substances for various cellular proliferative and synthetic processes. The provisional matrix may be viewed as a naturally derived, biodegradable, sustained release system in which mitogens, chemoattractants, cytokines, and growth factors are released to control subsequent wound-healing processes. In spite of the increase in our knowledge of the provisional matrix and its capabilities, our knowledge of the control of the formation of the provisional matrix and its effect on subsequent wound healing events is poor. In part, this lack of knowledge is due to the fact that much of our knowledge regarding the provisional matrix has been derived from *in vitro* studies, and there is a paucity of *in vivo* studies that provide for a more complex perspective. Little is known regarding the provisional matrix which forms at biomaterial and medical device interfaces *in vivo*, although attractive hypotheses have been presented regarding the presumed ability of materials and protein adsorbed materials to modulate cellular interactions through their interactions with adhesive molecules and cells.

The predominant cell type present in the inflammatory response varies with the age of the inflammatory injury (Figure II.2.2.1). In general, neutrophils predominate during the first several days following injury and then are replaced by monocytes as the predominant cell type. Three factors account for this change in cell type: neutrophils are short lived and disintegrate and disappear after

TABLE II.2.2.2 Cells and Components of Vascularized Connective Tissue

Intravascular (Blood) Cells
Erythrocytes (RBC)
Neutrophils (PMNs, polymorphonuclear leukocytes)
Monocytes
Eosinophils
Lymphocytes
Plasma cells
Basophils
Platelets
Connective Tissue Cells
Mast cells
Fibroblasts
Macrophages
Lymphocytes
Extracellular Matrix Components
Collagens
Elastin
Proteoglycans
Fibronectin
Laminin

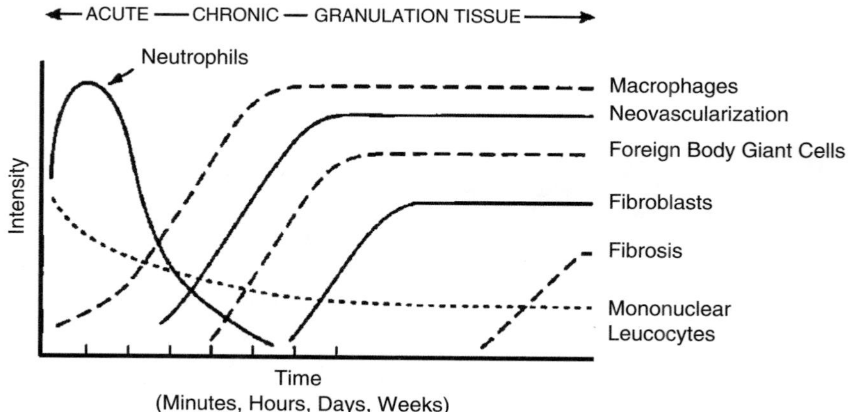

FIGURE II.2.2.1 The temporal variation in the acute inflammatory response, chronic inflammatory response, granulation tissue development, and foreign-body reaction to implanted biomaterials. The intensity and time variables are dependent upon the extent of injury created in the implantation and the size, shape, topography, and chemical and physical properties of the biomaterial.

24–48 hours; neutrophil emigration from the vasculature to the tissues is of short duration; and chemotactic factors for neutrophil migration are activated early in the inflammatory response. Following emigration from the vasculature, monocytes differentiate into macrophages and these cells are very long-lived (up to months). Monocyte emigration may continue for days to weeks, depending on the injury and implanted biomaterial, and chemotactic factors for monocytes are activated over longer periods of time. In short-term (24 hour) implants in humans, administration of both H1 and H2 histamine receptor antagonists greatly reduced the recruitment of macrophages/monocytes and neutrophils on polyethylene terephthalate surfaces. These studies also demonstrated that plasma-coated implants accumulated significantly more phagocytes than did serum-coated implants.

The temporal sequence of events following implantation of a biomaterial is illustrated in Figure II.2.2.1. The size, shape, and chemical and physical properties of the biomaterial may be responsible for variations in the intensity and duration of the inflammatory or wound-healing process. Thus, intensity and/or time duration of the inflammatory reaction may characterize the biocompatibility of a biomaterial.

While injury initiates the inflammatory response, the chemicals released from plasma, cells or injured tissue mediate the inflammatory response. Important classes of chemical mediators of inflammation are presented in Table II.2.2.3. Several points must be noted in order to understand the inflammatory response and how it relates to biomaterials. First, although chemical mediators are classified on a structural or functional basis, different mediator systems interact and provide a system of checks and balances regarding their respective activities and functions. Second, chemical mediators are quickly inactivated or destroyed, suggesting that their action is predominantly local (i.e., at the implant site). Third, generally the lysosomal proteases and the oxygen-derived free radicals produce the most significant damage or injury. These chemical mediators are also important in the degradation of biomaterials.

TABLE II.2.2.3 Important Chemical Mediators of Inflammation Derived from Plasma, Cells or Injured Tissue

Mediators	Examples
Vasoactive Agents	Histamines, serotonin, adenosine, endothelial-derived relaxing factor (EDRF), prostacyclin, endothelin, thromboxane α_2
Plasma Proteases	
Kinin system	Bradykinin, kallikrein
Complement system	C3a, C5a, C3b, C5b–C9
Coagulation/fibrinolytic system	Fibrin degradation products, activated Hageman factor (FXIIA), tissue plasminogen activator (tPA)
Leukotrienes	Leukotriene B_4 (LTB_4), hydroxyeicosatetranoic acid (HETE)
Lysosomal proteases	Collagenase, elastase
Oxygen-derived free radicals	H_2O_2, superoxide anion
Platelet activating factors	Cell membrane lipids
Cytokines	Interleukin 1 (IL-1), tumor necrosis factor (TNF)
Growth factors	Platelet-derived growth factor (PDGF), fibroblast growth factor (FGF), transforming growth factor TGF-α or (TGF-β), epithelial growth factor (EGF)

ACUTE INFLAMMATION

Acute inflammation is of relatively short duration, lasting for minutes to hours to days, depending on the extent of injury. Its main characteristics are the exudation of fluid and plasma proteins (edema) and the emigration of leukocytes (predominantly neutrophils). Neutrophils (polymorphonuclear leukocytes, PMNs) and other motile white cells emigrate or move from the blood vessels to the perivascular tissues and the injury (implant) site. Leukocyte emigration is assisted by "adhesion molecules" present on leukocyte and endothelial surfaces. The surface expression of these adhesion molecules can be induced, enhanced or altered by inflammatory agents and chemical mediators. White cell emigration is controlled, in part, by chemotaxis, which is the unidirectional migration of cells along a chemical gradient. A wide variety of exogenous and endogenous substances have been identified as chemotactic agents. Specific receptors for chemotactic agents on the cell membranes of leukocytes are important in the emigration or movement of leukocytes. These and other receptors also play a role in the transmigration of white cells across the endothelial lining of vessels, and the activation of leukocytes. Following localization of leukocytes at the injury (implant) site, phagocytosis and the release of enzymes occur following activation of neutrophils and macrophages. The major role of the neutrophil in acute inflammation is to phagocytose microorganisms and foreign materials. Phagocytosis is seen as a three-step process in which the injurious agent undergoes recognition and neutrophil attachment, engulfment, and killing or degradation. In regard to biomaterials, engulfment and degradation may or may not occur, depending on the properties of the biomaterial.

Although biomaterials are not generally phagocytosed by neutrophils or macrophages because of the disparity in size (i.e., the surface of the biomaterial is greater than the size of the cell), certain events in phagocytosis may occur. The process of recognition and attachment is expedited when the injurious agent is coated by naturally occurring serum factors called "opsonins." The two major opsonins are immunoglobulin G (IgG) and the complement-activated fragment, C3b. Both of these plasma-derived proteins are known to adsorb to biomaterials, and neutrophils and macrophages have corresponding cell-membrane receptors for these opsonization proteins. These receptors may also play a role in the activation of the attached neutrophil or macrophage. Other blood proteins such as fibrinogen, fibronectin, and vitronectin may also facilitate cell adhesion to biomaterial surfaces. Owing to the disparity in size between the biomaterial surface and the attached cell, frustrated phagocytosis may occur. This process does not involve engulfment of the biomaterial, but does cause the extracellular release of leukocyte products in an attempt to degrade the biomaterial.

Henson (1971) has shown that neutrophils adherent to complement-coated and immunoglobulin-coated nonphagocytosable surfaces may release enzymes by direct extrusion or exocytosis from the cell. The amount of enzyme released during this process depends on the size of the polymer particle, with larger particles inducing greater amounts of enzyme release. This suggests that the specific mode of cell activation in the inflammatory response in tissue depends upon the size of the implant, and that a material in a phagocytosable form (i.e., powder, particulate or nanomaterial) may provoke a different degree of inflammatory response than the same material in a nonphagocytosable form (i.e., film). In general, materials greater than 5 microns are not phagocytosed, while materials less than 5 microns, i.e., nanomaterials, can be phagocytosed by inflammatory cells.

Acute inflammation normally resolves quickly, usually less than 1 week, depending on the extent of injury at the implant site. However, the presence of acute inflammation (i.e., PMNs) at the tissue/implant interface at time periods beyond 1 week (i.e., weeks, months, or years) suggests the presence of an infection (Figure II.2.2.2).

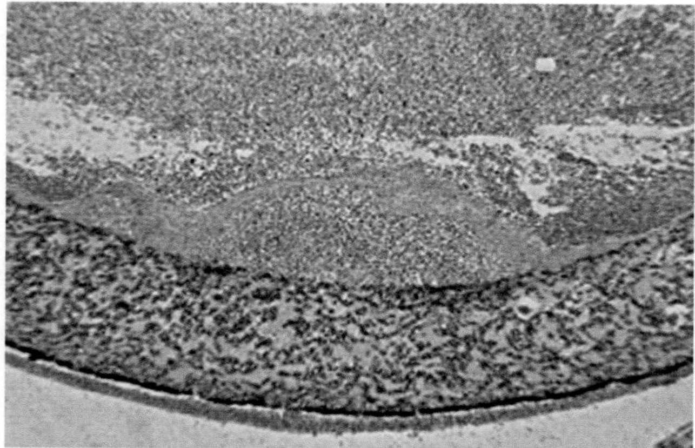

FIGURE II.2.2.2 Acute inflammation, secondary to infection, of an ePTFE (expanded poly tetrafluoroethylene) vascular graft. A focal zone of polymorphonuclear leukocytes is present at the lumenal surface of the vascular graft, surrounded by a fibrin cap, on the blood-contacting surface of the ePTFE vascular graft. Hematoxylin and eosin stain. Original magnification 4 ×.

CHRONIC INFLAMMATION

Chronic inflammation is less uniform histologically than acute inflammation. In general, chronic inflammation is characterized by the presence of macrophages, monocytes, and lymphocytes, with the proliferation of blood vessels and connective tissue. Many factors can modify the course and histologic appearance of chronic inflammation.

Persistent inflammatory stimuli lead to chronic inflammation. While the chemical and physical properties of the biomaterial in themselves may lead to chronic inflammation, motion in the implant site by the biomaterial or infection may also produce chronic inflammation. The chronic inflammatory response to biomaterials is usually of short duration, and is confined to the implant site. The presence of mononuclear cells, including lymphocytes and plasma cells, is considered chronic inflammation, whereas the foreign-body reaction with the development of granulation tissue is considered the normal wound healing response to implanted biomaterials (i.e., the normal foreign-body reaction). Chronic inflammation with the presence of collections of lymphocytes and monocytes at extended implant times (weeks, months, years) may also suggest the presence of a long-standing infection (Figures II.2.2.3A and II.2.2.3B). The prolonged presence of acute and/or chronic inflammation also may be due to toxic leachables from a biomaterial.

Lymphocytes and plasma cells are involved principally in immune reactions, and are key mediators of antibody production and delayed hypersensitivity responses. Although they may be present in nonimmunologic injuries and inflammation, their roles in such circumstances are largely unknown. Little is known regarding humoral immune responses and cell-mediated immunity to synthetic biomaterials. The role of macrophages must be considered in the possible development of immune responses to synthetic biomaterials. Macrophages and dendritic cells process and present the antigen to immunocompetent cells, and thus are key mediators in the development of immune reactions.

Monocytes and macrophages belong to the mononuclear phagocytic system (MPS), also known as the reticuloendothelial system (RES). These systems consist of cells in the bone marrow, peripheral blood, and specialized tissues. Table II.2.2.4 lists the tissues that contain cells belonging to the MPS or RES. The specialized cells in these tissues may be responsible for systemic effects in organs or tissues secondary to the release of components or products from implants through various tissue–material interactions (e.g., corrosion products, wear debris, degradation products) or the presence of implants (e.g., microcapsule or nanoparticle drug-delivery systems).

The macrophage is probably the most important cell in chronic inflammation, because of the great number of biologically active products it can produce. Important

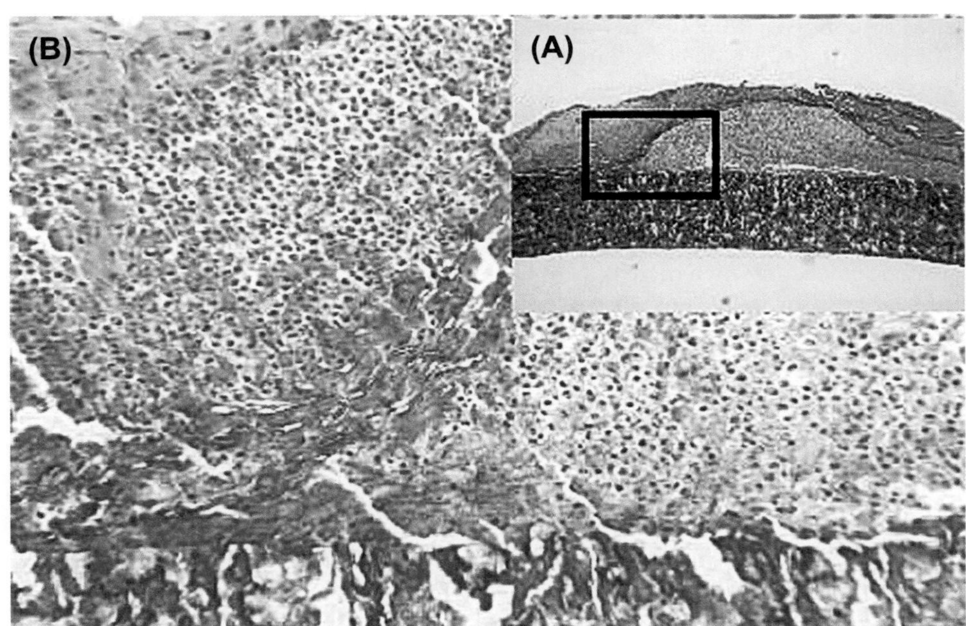

FIGURE II.2.2.3 Chronic inflammation, secondary to infection, of an ePTFE arteriovenous shunt for renal dialysis. (A) Low-magnification view of a focal zone of chronic inflammation. (B) High-magnification view of the outer surface with the presence of monocytes and lymphocytes at an area where the outer PTFE wrap had peeled away from the vascular graft. Hematoxylin and eosin stain. Original magnification (A) 4 ×, (B) 20 ×.

TABLE II.2.2.4 Tissues and Cells of MPS and RES

Tissues	Cells
Implant sites	Inflammatory macrophages
Liver	Kupffer cells
Lung	Alveolar macrophages
Connective tissue	Histiocytes
Bone marrow	Macrophages
Spleen and lymph nodes	Fixed and free macrophages
Serous cavities	Pleural and peritoneal macrophages
Nervous system	Microglial cells
Bone	Osteoclasts
Skin	Langerhans cells
Lymphoid tissue	Dendritic cells

classes of products produced and secreted by macrophages include neutral proteases, chemotactic factors, arachidonic acid metabolites, reactive oxygen metabolites, complement components, coagulation factors, growth-promoting factors, and cytokines.

Growth factors such as platelet-derived growth factor (PDGF), fibroblast growth factor (FGF), transforming growth factor-β (TGF-β), TGF-α/epidermal growth factor (EGF), and interleukin-1 (IL-1) or tumor necrosis factor (TNF-α) are important to the growth of fibroblasts and blood vessels, and the regeneration of epithelial cells. Growth factors released by activated cells can stimulate production of a wide variety of cells; initiate cell migration, differentiation, and tissue remodeling; and may be involved in various stages of wound healing.

GRANULATION TISSUE

Within one day following implantation of a biomaterial (i.e., injury), the healing response is initiated by the action of monocytes and macrophages. Fibroblasts and vascular endothelial cells in the implant site proliferate and begin to form granulation tissue, which is the specialized type of tissue that is the hallmark of healing inflammation. Granulation tissue derives its name from the pink, soft granular appearance on the surface of healing wounds, and its characteristic histologic features include the proliferation of new small blood vessels and fibroblasts (Figure II.2.2.4). Depending on the extent of injury, granulation tissue may be seen as early as 3–5 days following implantation of a biomaterial.

The new small blood vessels are formed by budding or sprouting of pre-existing vessels in a process known as neovascularization or angiogenesis. This process involves proliferation, maturation, and organization of endothelial cells into capillary vessels. Fibroblasts also proliferate in developing granulation tissue, and are active in synthesizing collagen and proteoglycans. In the early stages of granulation tissue development, proteoglycans predominate but later collagen, especially type III collagen, predominates and forms the fibrous capsule. Some fibroblasts in developing granulation tissue may have the features of smooth muscle cells, i.e., actin microfilaments. These cells are called myofibroblasts, and are considered to be responsible for the wound contraction seen during the development of granulation tissue. Macrophages are almost always present in granulation tissue. Other cells may also be present if chemotactic stimuli are generated.

The wound-healing response is generally dependent on the extent or degree of injury or defect created by the implantation procedure. Wound healing by primary union or first intention is the healing of clean, surgical incisions in which the wound edges have been approximated by surgical sutures. Healing under these conditions occurs without significant bacterial contamination, and with a minimal loss of tissue. Wound healing by secondary union or second intention occurs when there is a large tissue defect that must be filled or there is extensive loss of cells and tissue. In wound healing by secondary intention, regeneration of parenchymal cells cannot completely reconstitute the original architecture, and much larger amounts of granulation tissue are formed that result in larger areas of fibrosis or scar formation. Under these conditions, different regions of tissue may show different stages of the wound-healing process simultaneously.

Granulation tissue is distinctly different from granulomas, which are small collections of modified macrophages called epithelioid cells. Langhans' or foreign-body-type giant cells may surround nonphagocytosable particulate materials in granulomas. Foreign-body giant cells (FBGCs) are formed by the fusion of monocytes and macrophages in an attempt to phagocytose the material (Figure II.2.2.5).

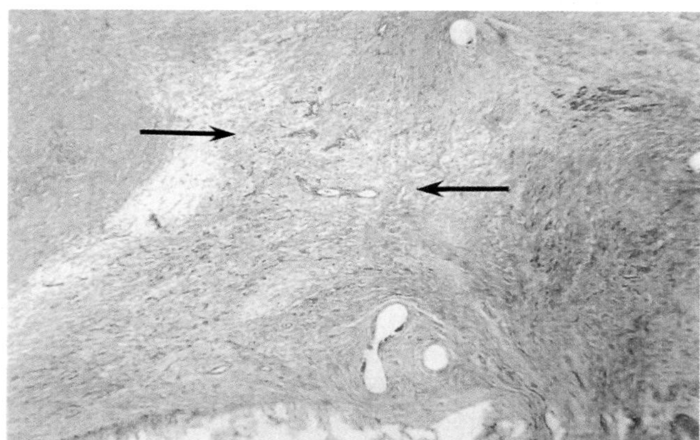

FIGURE II.2.2.4 Granulation tissue in the anastomotic hyperplasia at the anastomosis of an ePTFE vascular graft. Capillary development (red slits) and fibroblast infiltration with collagen deposition (blue) from the artery form the granulation tissue (arrows). Masson's Trichrome stain. Original magnification 4 ×.

FOREIGN-BODY REACTION

The foreign-body reaction to biomaterials is composed of foreign-body giant cells and the components of granulation tissue (e.g., macrophages, fibroblasts, and capillaries in varying amounts), depending upon the form and topography of the implanted material (Figure II.2.2.6). Relatively flat and smooth surfaces such as those found on breast prostheses have a foreign-body reaction that is composed of a layer of macrophages one-to-two cells in thickness. Relatively rough surfaces such as those found on the outer surfaces of expanded poly tetrafluoroethylene (ePTFE) or Dacron vascular prostheses have a foreign-body reaction composed of macrophages and foreign-body giant cells at the surface. Fabric materials generally have a surface response composed of macrophages and foreign-body giant cells, with varying degrees of granulation tissue subjacent to the surface response (Figure II.2.2.7).

As previously discussed, the form and topography of the surface of the biomaterial determine the composition of the foreign-body reaction. With biocompatible materials, the composition of the foreign-body reaction in the implant site may be controlled by the surface properties of the biomaterial, the form of the implant, and the relationship between the surface area of the biomaterial and the volume of the implant. For example, high surface-to-volume implants, such as fabrics, porous materials, particulate or microspheres will have higher ratios of macrophages and foreign-body giant cells in the implant site than smooth-surface implants, which will have fibrosis as a significant component of the implant site.

The foreign-body reaction consisting mainly of macrophages and/or foreign-body giant cells may persist at the tissue–implant interface for the lifetime of the implant (Figure II.2.2.1). Generally, fibrosis (i.e., fibrous encapsulation) surrounds the biomaterial or implant with its interfacial foreign-body reaction, isolating the implant and foreign-body reaction from the local tissue environment. Early in the inflammatory and wound-healing response, the macrophages are activated upon adherence to the material surface.

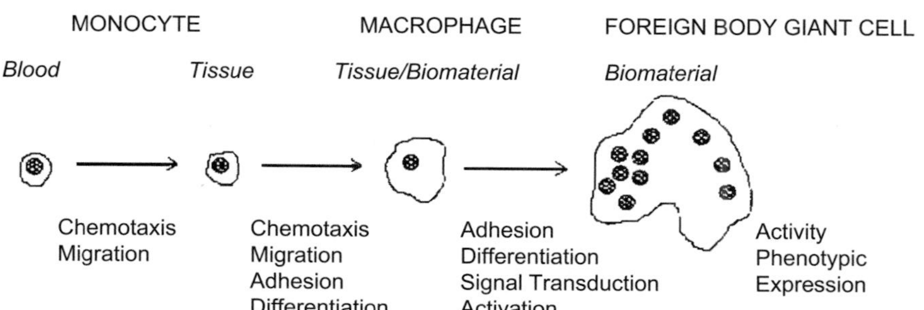

FIGURE II.2.2.5 *In vivo* transition from blood-borne monocyte to biomaterial adherent monocyte/macrophage to foreign-body giant cell at the tissue–biomaterial interface. Little is known regarding the indicated biological responses, which are considered to play important roles in the transition to FBGC development.

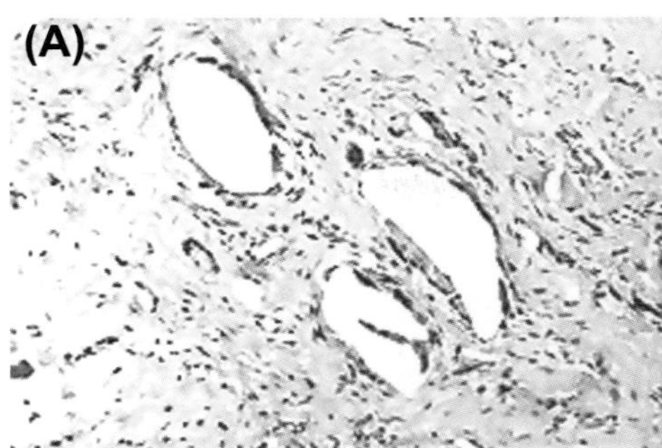

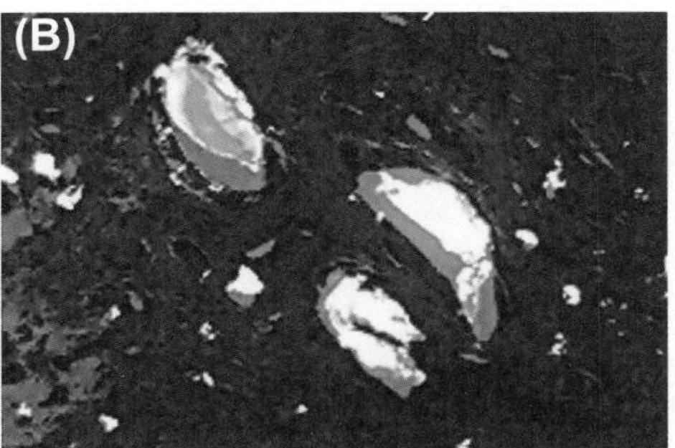

FIGURE II.2.2.6 (A) Focal foreign-body reaction to polyethylene wear particulate from a total knee prosthesis. Macrophages and foreign-body giant cells are identified within the tissue and lining the apparent void spaces indicative of polyethylene particulate. Hematoxylin and eosin stain. Original magnification 20 ×. (B) Partial polarized light view. Polyethylene particulate is identified within the void spaces commonly seen under normal light microscopy. Hematoxylin and eosin stain. Original magnification 20 ×.

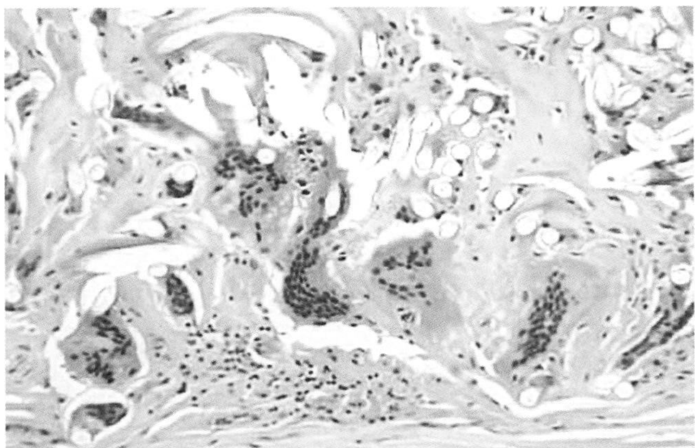

FIGURE II.2.2.7 Foreign-body reaction with multinucleated foreign body giant cells and macrophages at the periadventitial (outer) surface of a Dacron vascular graft. Fibers from the Dacron vascular graft are identified as clear oval voids. Hematoxylin and eosin stain. Original magnification 20 ×.

Although it is generally considered that the chemical and physical properties of the biomaterial are responsible for macrophage activation, the subsequent events regarding the activity of macrophages at the surface are not clear. Tissue macrophages, derived from circulating blood monocytes, may coalesce to form multinucleated foreign-body giant cells. It is not uncommon to see very large foreign-body giant cells containing large numbers of nuclei on the surface of biomaterials. While these foreign-body giant cells may persist for the lifetime of the implant, it is not known if they remain activated, releasing their lysosomal constituents, or become quiescent.

Figure II.2.2.5 demonstrates the progression from circulating blood monocyte, to tissue macrophage, to foreign-body giant cell development that is most commonly observed. Indicated in the figure are important biological responses that are considered to play an important role in FBGC development. Material surface chemistry may control adherent macrophage apoptosis (i.e., programmed cell death) (see Chapter II.3.3) that renders potentially harmful macrophages nonfunctional, while the surrounding environment of the implant remains unaffected. The level of adherent macrophage apoptosis appears to be inversely related to the surface's ability to promote fusion of macrophages into FBGCs, suggesting a mechanism for macrophages to escape apoptosis.

Figure II.2.2.8 demonstrates the sequence of events involved in inflammation and wound healing when medical devices are implanted. In general, the PMN predominant acute inflammatory response and the lymphocyte/monocyte predominant chronic inflammatory response resolve quickly (i.e., within 2 weeks) depending on the type and location of the implant. Studies using IL-4 or IL-13, respectively, demonstrate the role for Th2 helper lymphocytes and/or mast cells in the development of the foreign-body reaction at the tissue/material interface. Integrin receptors of IL-4-induced FBGC are characterized by the early constitutive expression of $\alpha V\beta 1$, and the later induced expression of $\alpha 5\beta 1$ and $\alpha X\beta 2$, which indicate potential interactions with adsorbed complement C3, fibrin(ogen), fibronectin, Factor X, and vitronectin. Interactions through indirect (paracrine) cytokine and chemokine signaling have shown a significant effect in enhancing adherent macrophage/FBGC activation at early times, whereas interactions via direct (juxtacrine) cell–cell mechanisms dominate at later times. Th2 helper lymphocytes have been described as "anti-inflammatory" based on their cytokine profile, of which IL-4 is a significant component.

FIBROSIS/FIBROUS ENCAPSULATION

The end-stage healing response to biomaterials is generally fibrosis or fibrous encapsulation (Figure II.2.2.9). However, there may be exceptions to this general statement (e.g., porous materials inoculated with parenchymal cells or porous materials implanted into bone) (Figure II.2.2.10). As previously stated, the tissue response to implants is, in part, dependent upon the extent of injury or defect created in the implantation procedure and the amount of provisional matrix.

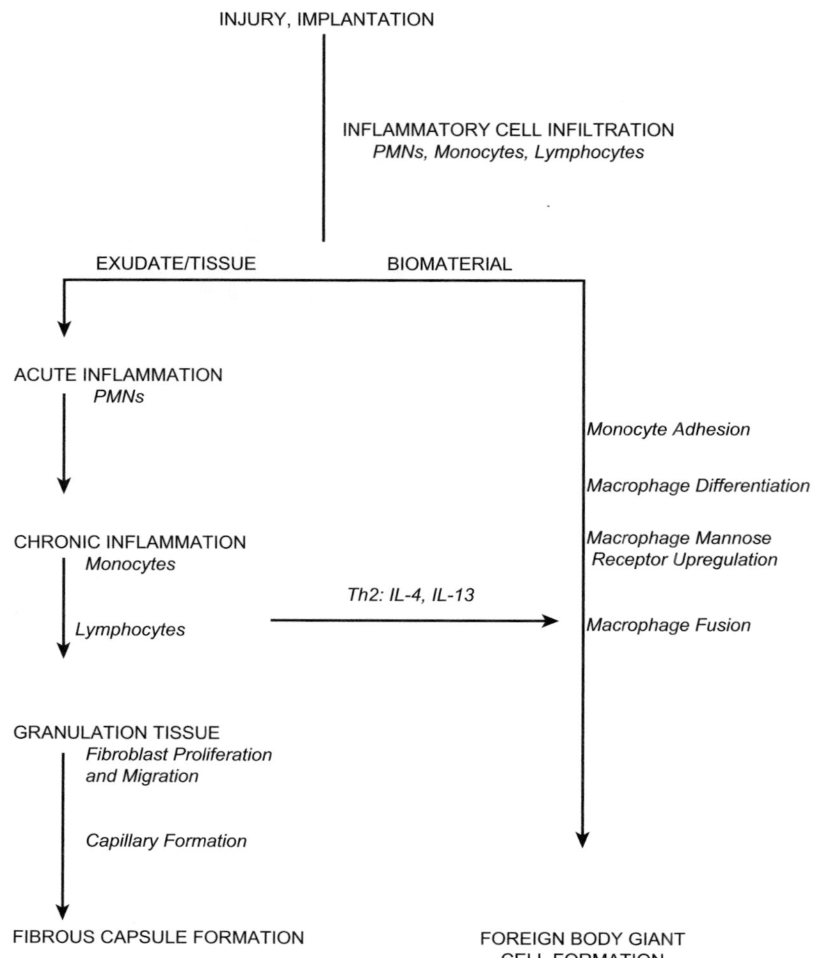

FIGURE II.2.2.8 Sequence of events involved in inflammatory and wound-healing responses leading to foreign-body giant cell formation.

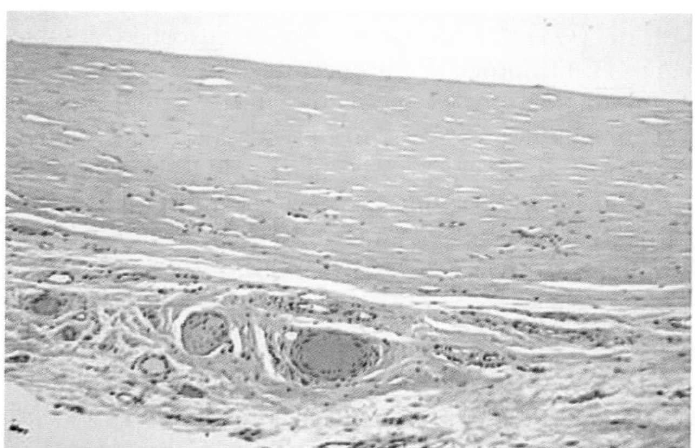

FIGURE II.2.2.9 Fibrous capsule composed of dense, compacted collagen. This fibrous capsule had formed around a Mediport catheter reservoir. Loose connective tissue with small arteries, veins, and a nerve is identified below the acellular fibrous capsule.

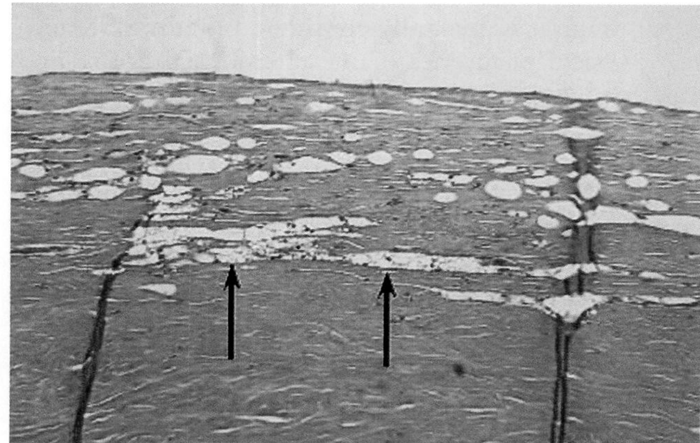

FIGURE II.2.2.10 Fibrous capsule with a focal foreign-body reaction to silicone gel from a silicone gel–filled silicone-rubber breast prosthesis. The breast prosthesis–tissue interface is at the top of the photomicrograph. Oval void spaces lined by macrophages and a few giant cells are identified and a focal area of foamy macrophages (arrows) indicating macrophage phagocytosis of silicone gel is identified. Hematoxylin and eosin stain. Original magnification 10 ×.

Injury and Repair at Implant Sites

The ultimate goal of tissue engineering and regenerative medicine is replacement of injured tissue by cells that reconstitute normal tissue and organ structures. Numerous approaches, including stem cells, scaffolds, growth factors, etc., are currently being investigated. However, the relatively rapid responses of inflammation, wound healing, and the foreign-body reaction, as well as other significant factors in tissue regeneration, present major challenges to the successful achievement of this goal.

Repair of implant sites can involve two distinct processes: regeneration, which is the replacement of injured tissue by parenchymal cells of the same type or replacement by connective tissue that constitutes the fibrous capsule. These processes are generally controlled by either: (1) the proliferative capacity of the cells in the tissue or organ receiving the implant and the extent of injury as it relates to the destruction; or (2) persistence of the tissue framework of the implant site. See Chapter II.3.4 for a more complete discussion of the types of cells present in the organ parenchyma and stroma, respectively.

The regenerative capacity of cells allows them to be classified into three groups: labile; stable (or expanding); and permanent (or static) cells (see Chapter II.3.3). Labile cells continue to proliferate throughout life; stable cells retain this capacity but do not normally replicate; and permanent cells cannot reproduce themselves after birth. Perfect repair with restitution of normal structure can theoretically occur only in tissues consisting of stable and labile cells, whereas all injuries to tissues composed of permanent cells may give rise to fibrosis and fibrous capsule formation with very little restitution of the normal tissue or organ structure. Tissues composed of permanent cells (e.g., nerve cells and cardiac muscle cells) most commonly undergo an organization of the inflammatory exudate, leading to fibrosis. Tissues of stable cells (e.g., parenchymal cells of the liver, kidney, and pancreas); mesenchymal cells (e.g., fibroblasts, smooth muscle cells, osteoblasts, and chondroblasts); and vascular endothelial and labile cells (e.g., epithelial cells, lymphoid and hematopoietic cells) may also follow this pathway to fibrosis or may undergo resolution of the inflammatory exudate, leading to restitution of the normal tissue structure.

The condition of the underlying framework or supporting stroma of the parenchymal cells following an injury plays an important role in the restoration of normal tissue structure. Retention of the framework with injury may lead to restitution of the normal tissue structure, whereas destruction of the framework most commonly leads to fibrosis. It is important to consider the species-dependent nature of the regenerative capacity of cells. For example, cells from the same organ or tissue, but from different species, may exhibit different regenerative capacities and/or connective tissue repair.

Following injury, cells may undergo adaptations of growth and differentiation. Important cellular adaptations are atrophy (decrease in cell size or function), hypertrophy (increase in cell size), hyperplasia (increase in cell number), and metaplasia (change in cell type). Other adaptations include a change by cells from producing one family of proteins to another (phenotypic change) or marked overproduction of protein. This may be the case in cells producing various types of collagens and extracellular matrix proteins in chronic inflammation and fibrosis. Causes of atrophy may include decreased workload (e.g., stress-shielding by implants), diminished blood supply, and inadequate nutrition (e.g., fibrous capsules surrounding implants).

Local and systemic factors may play a role in the wound-healing response to biomaterials or implants. Local factors include the site (tissue or organ) of implantation, the adequacy of blood supply, and the potential for infection. Systemic factors may include nutrition, hematologic derangements, glucocortical steroids, and pre-existing diseases such as atherosclerosis, diabetes, and infection.

Finally, the implantation of biomaterials or medical devices may be best viewed at present from the perspective that the implant provides an impediment or hindrance to appropriate tissue or organ regeneration and healing. Given our current inability to control the sequence of events following injury in the implantation procedure, restitution of normal tissue structures with function is rare. Current studies directed toward developing a better understanding of the modification of the inflammatory response, stimuli providing for appropriate proliferation of permanent and stable cells, and the appropriate application of growth factors may provide keys to the control of inflammation, wound healing, and fibrous encapsulation of biomaterials.

Case Study **Implant Toxicity**

In vivo implantation studies were carried out subcutaneously in rats and rabbits with naltrexone sustained release preparations that included placebo (polymer only) beads and naltrexone containing beads. Histopathological tissue reactions utilizing standard procedures and light microscopic evaluation to the respective preparations were determined at days 3, 7, 14, 21, and 28. The only significant histological finding in both rats and rabbits at any time period was the inflammation that occurred focally around the naltrexone containing beads. The focal inflammatory cell density in both rats and rabbits was higher for the naltrexone beads than for the placebo beads at days 14, 21, and 28, respectively. This difference in inflammatory response between naltrexone beads and placebo beads increased with increasing time of implantation. Considering the resolution of the inflammatory response for the placebo beads with implantation time in both rats and rabbits, it is suggested that the naltrexone drug itself is identified as the causative agent of the focal inflammation present surrounding the naltrexone beads in the implant sites (Yamaguchi & Anderson, 1992).

The important lesson from this case study is the necessary use of appropriate control materials. If no negative control, i.e., placebo polymer-only material, had been used, the polymer in the naltrexone containing beads also would have been considered as a causative agent of the extended inflammatory response. Similar inflammatory responses have been identified with drugs, polymer plasticizers and other additives, fabrication and manufacturing aids, and sterilization residuals. Each case presents its own unique factors in a risk assessment process necessary for determining safety (biocompatibility) and benefit versus risk in clinical application.

BIBLIOGRAPHY

Anderson, J. M. (2001). Biological responses to materials. *Ann. Rev. Mater. Res.*, **31**, 81–110.

Anderson, J. M., & Jones, J. A. (2007). Phenotypic dichotomies in the foreign-body reaction. *Biomaterials*, **28**, 5114–5120.

Anderson, J. M., Rodriguez, A., & Chang, D. T. (2008). Foreign-body reaction to biomaterials. *Seminars in Immunology*, **20**, 86–100.

Brodbeck, W. G., & Anderson, J. M. (2009). Giant cell formation and function. *Curr. Opin. In Hematology*, **16**, 53–57.

Browder, T., Folkman, J., & Pirie-Shepherd, S. (2000). The hemostatic system as a regulator of angiogenesis. *J. Biol. Chem.*, **275**, 1521–1524.

Broughton, G., Janis, J. E., & Attinger, C. E. (2006). The basic science of wound healing. *Plast. Reconstr. Surg.*, **117**(Suppl.), 12S–34S.

Chang, D. T., Colton, E., & Anderson, J. M. (2009). Paracrine and juxtacrine lymphocyte enhancement of adherent macrophage and foreign-body giant cell activation. *J. Biomed. Mater. Res.*, **89A**(2), 490–498.

Clark, R. A.F. (Ed.), (1996). *The Molecular and Cellular Biology of Wound Repair*. New York, NY: Plenum Publishing.

Gallin, J. I., & Snyderman, R. (Eds.), (1999). *Inflammation: Basic Principles and Clinical Correlates* (3rd ed.). New York, NY: Raven Press.

Henson, P. M. (1971). The immunologic release of constituents from neutrophil leukocytes: II. Mechanisms of release during phagocytosis, and adherence to nonphagocytosable surfaces. *J. Immunol.*, **107**, 1547.

Hunt, T. K., Heppenstall, R. B., Pines, E., & Rovee, D. (Eds.), (1984). *Soft and Hard Tissue Repair*. New York, NY: Praeger Scientific.

Hynes, R. O. (2002). Integrins: Bidirectional, allosteric signaling machines. *Cell*, **110**, 673–687.

Hynes, R. O., & Zhao, Q. (2000). The evolution of cell adhesion. *J. Cell Biol.*, **150**, F89–F96.

Jones, J. A., Chang, D. T., Meyerson, H., Colton, E., Kwon, I. K., et al. (2007). Proteomic analysis and quantification of cytokines and chemokines from biomaterial surface-adherent macrophages and foreign-body giant cells. *J. Biomed. Mater. Res.*, **83A**, 585–596.

Jones, K. S. (2008a). Assays on the influence of biomaterial on allogeneic rejection in tissue engineering. *Tissue Engineering: Part B*, **14**(4), 407–417.

Jones, K. S. (2008b). Effects of biomaterial-induced inflammation on fibrosis and rejection. *Seminars in Immunology*, **20**, 130–136.

Kumar, V., Abbas, A. K., & Fausto, N. (Eds.), (2005). *Robbins and Cotran Pathologic Basis of Disease* (7th ed.), (pp. 47–118). Philadelphia, PA: W. B. Saunders.

Marchant, R. E., Anderson, J. M., & Dillingham, E. O. (1986). *In vivo* biocompatibility studies. VII. Inflammatory response to polyethylene and to a cytotoxic polyvinylchloride. *J. Biomed. Mater. Res.*, **20**, 37–50.

McNally, A. K., & Anderson, J. M. (1995). Interleukin-4 induces foreign-body giant cells from human monocytes/macrophages. Differential lymphokine regulation of macrophage fusion leads to morphological variants of multinucleated giant cells. *Am. J. Pathol.*, **147**, 1487–1499.

McNally, A. K., MacEwan, S. R., & Anderson, J. M. (2007). Alpha subunit partners to Beta1 and Beta2 integrins during IL-4-induced foreign-body giant cell formation. *J. Biomed. Mater. Res.*, **82A**, 568–574.

McNally, A. K., Jones, J. A., MacEwan, S. R., Colton, E., & Anderson, J. M. (2008). Vitronectin is a critical protein adhesion substrate for IL-4-induced foreign-body giant cell formation. *J. Biomed. Mater. Res.*, **86A**(2), 535–543.

Nel, A., Xia, T., Mädler, L., & Li, N. (2006). Toxic potential of materials at the nano level. *Science*, **311**, 622–627.

Nguyen, L. L., & D'Amore, P. A. (2001). Cellular interactions in vascular growth and differentiation. *Int. Rev. Cytol.*, **204**, 1–48.

Pierce, G. F. (2001). Inflammation in nonhealing diabetic wounds: The space–time continuum does matter. *Am. J. Pathol.*, **159**(2), 399–403.

Sefton, M. V., Babensee, J. E., & Woodhouse, K. A. (Eds.), (2008). Special issue on innate and adaptive immune responses in tissue engineering. *Seminars in Immunology*, **20**(2).

Yamaguchi, K., & Anderson, J. M. (1992). Biocompatibility studies of naltrexone sustained release formulations. *J. Controlled Rel.*, **19**, 299–314.

Zdolsek, J., Eaton, J. W., & Tang, L. (2007). Histamine release and fibrinogen adsorption mediate acute inflammatory responses to biomaterial implants in humans. *Journal of Translational Medicine*, **5**, 31–37.

CHAPTER II.2.3 INNATE AND ADAPTIVE IMMUNITY: THE IMMUNE RESPONSE TO FOREIGN MATERIALS

Richard N. Mitchell

Department of Pathology, Brigham and Women's Hospital and Harvard Medical School, Boston, MA, USA

The goal of this chapter is to understand the general organization of the immune system (both specific and non-specific components), how the multiple different elements recognize apparent "invaders," and what effector responses are generated to eliminate perceived threats. "Immunology," as this subject is called, typically encompasses an entire course (with its own introductory text). Thus, an overview chapter can only paint with very broad brushstrokes what is a beautifully complex and intricate canvas of concepts regarding *innate* and *adaptive* immunity. The intent here is to provide a fundamental understanding of the body's response to the insertion of a foreign device, so that the reader can anticipate the issues that arise at the biomaterials–tissue interface. More extensive discussion of any aspect of the immune system can be found in a number of excellent basic immunology texts (Coico et al., 2003; Abbas et al., 2007; Murphy et al., 2007), and from there to the primary literature (citations sprinkled throughout).

OVERVIEW

The immune system exists primarily to *defend the host against infectious organisms*. The general strategy begins with distinguishing "self" from "non-self." Once the immune system determines that something does not belong, the responding population proliferates, recruits additional cells, and generates mediators to neutralize and destroy the invader. In most cases, the immune response is so exquisitely specific that "self" is not perceived as foreign, and so well-regulated that host tissues do not sustain any "innocent bystander" injury in the process.

Unfortunately, this is not always the case. *Autoimmunity* (inappropriate immune responses to self) can be the source of chronic, debilitating, and life-threatening disease. Moreover, infections can occasionally lead to substantially greater tissue injury, and even death, occurring as a direct consequence of host immune responses. Finally, one of the most powerful triggers of the immune system is tissue injury. This makes good functional sense, since evolutionarily most cases of injury are either caused by microbial agents, or tissues damaged by another cause often becoming secondarily infected. Thus, although the system evolved primarily to identify and eliminate infectious agents, *noninfectious foreign materials* (including biomaterials in medical devices or the process by which they are implanted) can also elicit immune responses. Occasionally, these can result – even if uninfected – in severe tissue damage.

Consequently, a more inclusive definition of **immunity** is the body's *reaction to any substance* (microbes, proteins, polysaccharides, self, silastic implants, etc.) *regardless of the pathologic consequences*. In this chapter, we will initially focus on the physiologic pathways of immune activation in response to infectious agents, and describe the subsequent effector responses. We will then show how these same pathways can have pathologic outcomes, with special emphasis being given to the mechanisms underlying a response to a foreign biomaterial and/or implant.

An abbreviated glossary of terms is provided here to help the reader navigate the subsequent pages; other terms will be explained as they arise:

Adaptive immunity: Immune responses conferred by T lymphocytes or antibodies. The complete repertoire of T cells and antibodies in any given host has the capacity to bind uniquely to a huge number of antigens or distinct molecular configurations (a feature called *specificity*). Subsequent exposure to the same molecular configuration leads to more robust, faster responses (a feature called *memory*).

Allogeneic: Different than oneself, usually referring to the same molecule that has different configurations in different members of the same species (such as Major Histocompatibility Complex (MHC) molecules).

Allograft: Transplanted tissue or organ from a different donor, usually of the same species. A heart transplant from one human to another is called an *allograft*.

Antigen: A molecular configuration that is recognizable by the adaptive immune system. This molecular configuration can be a linear sequence of amino acids or a folded three-dimensional conformation; it can also be sugars, nucleotides or small molecules, or modified versions of any of these. Thus, a phosphorylated tyrosine in a short peptide can be *antigenic*. It is important to understand that antigens need not necessarily be foreign and can also be self-molecules. Moreover, just because a molecule is *antigenic* does not necessarily mean that it will elicit an immune response; issues related to the mode of antigen presentation, access, recognition, and tolerance induction all impact on host reactivity.

Autograft: Transplanted tissue or organ from the same individual. Thus explanted skin that is expanded in culture *ex vivo* and returned to the same donor is an *autograft*.

Immunogenic: Capable of eliciting a specific (adaptive) immune response.

Innate immunity: Immune responses conferred by macrophages and neutrophils, as well as complement and other proteins that recognize a discrete set of fixed motifs. There is a limited repertoire of *pathogen recognition receptors*, and responding cells have the same response with each contact (i.e., there is no *memory*).

Isograft: Transplanted tissue or organ from a different individual with identical genetic make-up. This is more typical in experimental situations involving *inbred animal strains*, but can occur in the setting of identical human twin donor and recipient pairs.

Syngeneic: The same as oneself, usually referring to genetic make-up, and more specifically in transplantation, identity for Major Histocompatibility Complex (MHC) molecule expression.

Xenogeneic: Differences across species. Thus, a porcine renal transplant into a baboon is a *xenograft*.

INNATE AND ADAPTIVE IMMUNITY

Defense against microbes is a two-stage process, beginning with a relatively non-specific *innate* response to "injury," followed by a targeted *adaptive* response more uniquely focused on the specific causal agent (Figure II.2.3.1; Table II.2.3.1).

COMPONENTS OF INNATE IMMUNITY

Also called "natural" or "native" immunity, this is the initial, rapid (hours) response to infection or injury; it is mediated by both cellular and protein constituents

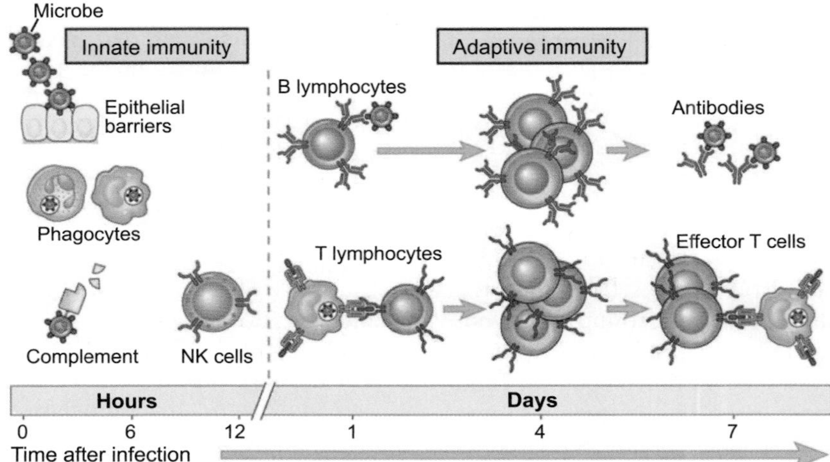

FIGURE II.2.3.1 Innate and adaptive immunity. Although with a limited ability to recognize invading organisms, the relatively primitive (evolutionarily-speaking) components of innate immunity provide a very effective first line of defense against the vast majority of microbial infections. Adaptive immunity, with exquisite specificity to any particular infectious agent, develops sometime later after innate components have responded. Notably, the elements of innate immunity not only respond first, but direct subsequent adaptive immunity; in turn, elements of the adaptive immune responses orchestrate a more efficient and vigorous response by the components of innate immunity. The specific kinetics of the responses shown are approximations, and will vary depending on the inciting agent. (Figure reprinted with permission from Abbas et al. (2007), *Cellular and Molecular Immunology,* 6th ed.)

TABLE II.2.3.1 Components of Innate and Adaptive Immunity

	Innate	Adaptive
Cellular and chemical barriers	Skin, mucosal epithelium, antimicrobial proteins (e.g., defensins, cathelicidins)	Lymphocytes in epithelia, secreted antibodies
Blood proteins	Complement, mannose-binding protein, C-reactive protein	Antibodies
Cells	Phagocytes (macrophages, neutrophils) natural killer cells	Lymphocytes

Adapted from Abbas et al., 2007.

(Figure II.2.3.2; Table II.2.3.2) (Medzhitov and Janeway, 2000; Medzhitov, 2007):

- physical and chemical barriers such as epithelia and anti-microbial proteins (e.g., *defensins*)
- phagocytic cells (neutrophils and macrophages) that ingest (*phagocytize*) and destroy microbes (see later)
- natural killer (NK) cells that kill virally-infected cells
- circulating proteins (complement, coagulation factors, C-reactive protein, etc.) that either directly insert pore-forming proteins in microbes that lead to cell death (e.g., complement, see below and also Chapter II.2.4), or that *opsonize* microbes (rendering them more "attractive" and readily phagocytized)
- cytokines: proteins secreted by cells of innate or adaptive immunity that regulate and coordinate the cellular response.

There is a bewildering plethora of cytokines with a wide array of activities influencing everything from endothelial cell function, to fibroblast synthetic activity, to lymphocyte differentiation (and more). Different cells will make different cohorts of cytokines, and any given cytokine can affect target cells differently. Multiple cytokines are in play in a particular location at any given time and can have synergistic or antagonistic effects on one another. Clearly, this is a fairly complex topic that warrants its own extensive treatment; for the purposes of this chapter, only selected cytokines will be discussed fleetingly, and the interested reader is encouraged to at least look at more comprehensive texts (Coico et al., 2003; Abbas et al., 2007; Murphy et al., 2007).

Innate immunity is an evolutionarily primitive system found even in invertebrates, and to some extent in plants (Litman et al., 2005); in most cases it can quite capably dispatch infections without the benefit of the lymphocytes or antibodies of adaptive immunity. Nevertheless, in cases where additional (and more specific) ammunition is required, the components of innate immunity are critical in mobilizing the subsequent lymphocyte and antibody effectors to clear invading microorganisms (see later).

RECOGNITION IN INNATE IMMUNITY

Innate immunity is triggered by molecular structures – called *pathogen-associated molecular patterns* (PAMPs) – that are common to groups of related microbes (Akira et al., 2006; Meylan et al., 2006; Palm and Medzhitov, 2009). These structures are characteristic of microbial pathogens and are not present on mammalian tissues;

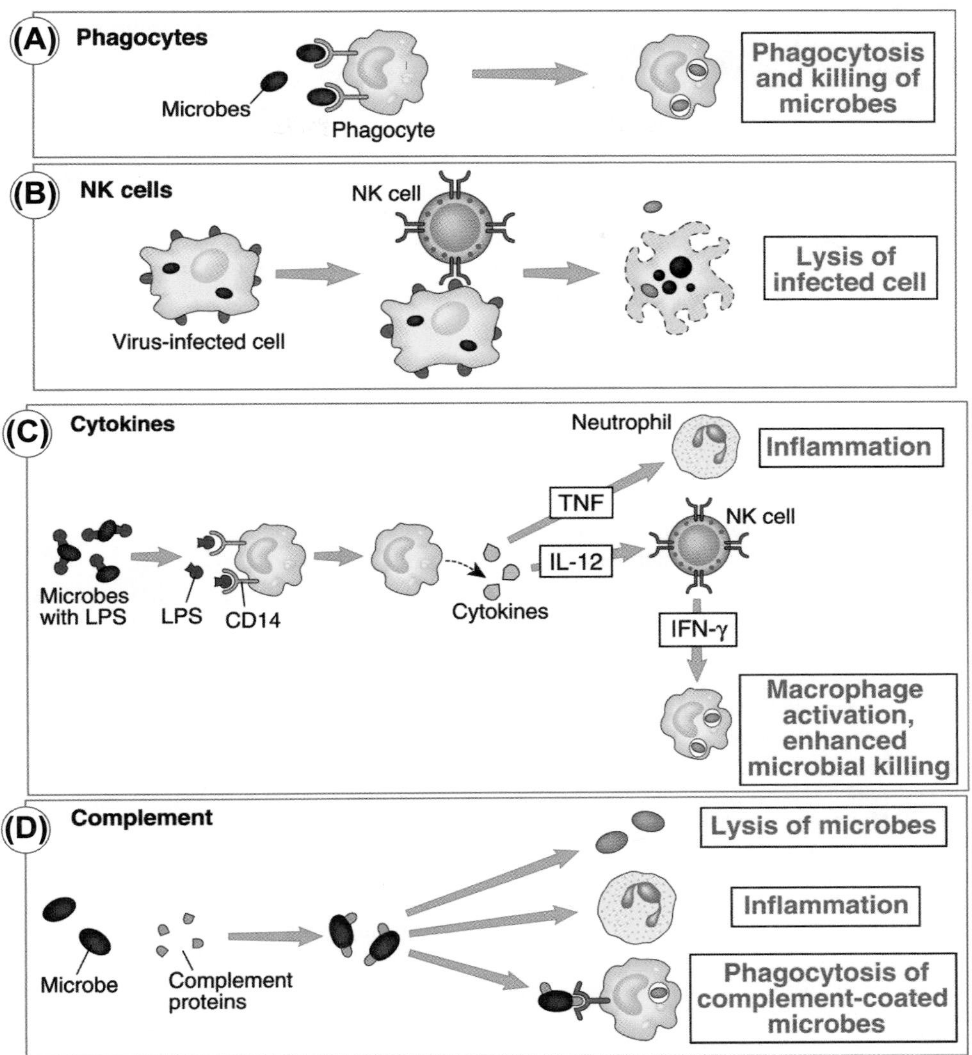

FIGURE II.2.3.2 Basic mechanisms of innate immunity. The principal cellular components of innate immunity in defense against microbial infection include phagocytes and natural killer (NK) cells; the principal protein constituents are cytokines and complement (a proteolytic cascade of related proteins, see also Chapter II.2.4). (A) Phagocytes (*neutrophils* and *macrophages*) will directly bind, ingest, and intracellularly degrade various microbes; in addition, macrophages can secrete *cytokines* to recruit and activate additional inflammatory cell types (e.g., *neutrophils*) to sites of infection, and will help drive the activation of the T lymphocytes of the adaptive immune response. Note that macrophages may also require the production of cytokines (e.g., interferon-γ) by other cell types in order to be most active. (B) *Natural killer* (NK) cells directly lyse virally-infected cells, as well as cells coated with bound antibody; in addition, they are a source of cytokines (e.g., interferon-γ) that can activate macrophages and T lymphocytes. (C) *Cytokines* are proteins produced by cells of innate and adaptive immunity that locally and systemically affect inflammation. In the figure, macrophages have been activated by lipopolysaccharide (LPS, a constituent of bacterial cell walls) to secrete tumor necrosis factor (TNF) and interleukin-12 (IL-12). TNF results (among other activities) in the production, recruitment, and activation of neutrophils. IL-12 drives NK cell activation, resulting in the production of interferon-γ that in turn activates macrophages. (D) *Complement* proteins form pores in the membranes of microbes to cause direct cell lysis; in addition, complement components will incite inflammatory cell recruitment, and augment phagocytosis (*opsonize* microbes). (Figure reprinted with permission from Abbas et al. (2000), *Cellular and Molecular Immunology,* 4th ed.)

consequently, recognition via this pathway will distinguish self and non-self. Moreover, because the microbial products that are recognized are usually essential for the survival of the microorganism, they cannot be discarded or mutated. The *pattern recognition receptors* for pathogen-associated structures have a fairly limited diversity (numbering about 20 different types of molecules), and have no capacity to make fine distinctions between different substances; each cell of the innate system also expresses essentially the same cohort of receptors (Figure II.2.3.3). The components of innate immunity have no functional memory; they react in essentially the same way each time they encounter the same infectious agent, so that there is no mechanism to allow more rapid or specific responses upon a second encounter with the same agent.

TABLE II.2.3.2 Components and Functions of Innate Immunity

Component	Function
Barriers	
Epithelium	Prevent microbial entry
Defensins, cathelicidins	Microbial killing
Intraepithelial leukocytes	Microbial killing
Circulating proteins	
Complement	Microbial killing, opsonization, leukocyte recruitment and activation
Mannose-binding protein	Opsonization, complement activation
C-reactive protein	Opsonization, complement activation
Coagulation factors	Thrombosis, leukocyte activation
Circulating cells	
Neutrophils	Early phagocytosis and killing
Macrophages	Subsequent, definitive phagocytosis and killing, cytokine production
Natural killer (NK) cells	Killing of infected cells, macrophage activation
Cytokines	
Chemokines	Leukocyte recruitment and activation
Interleukin-1 (IL-1) and tumor necrosis factor (TNF)	Leukocyte recruitment and activation, systemic inflammatory responses, endothelial activation, fibroblast activation
Interferon-α and -β	Anti-viral activity
Interferon-γ	Macrophage activation
Interleukin-12 (IL-12)	NK cell activation
Transforming growth factor-β (TGF-β)	Modulation of inflammation

Adapted from Abbas et al., 2000.

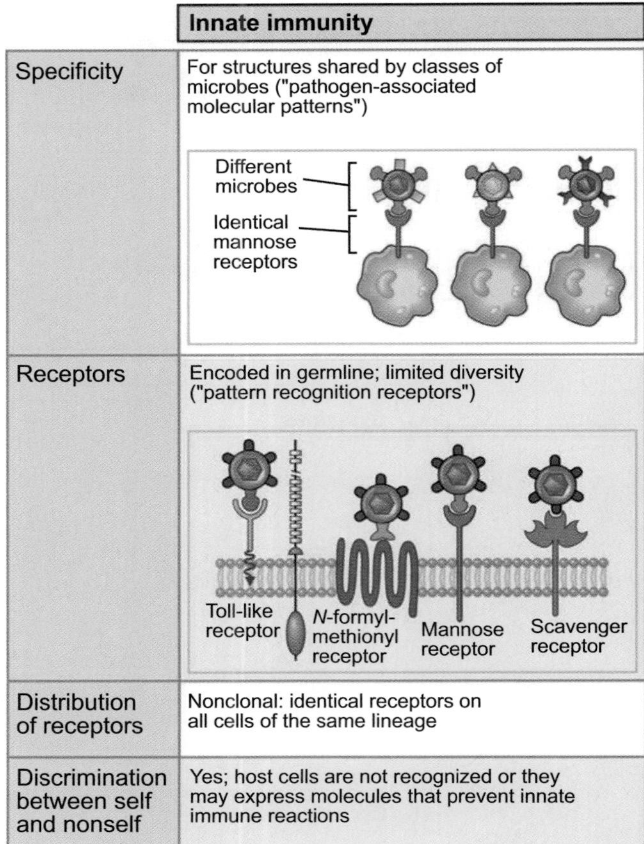

FIGURE II.2.3.3 Pattern recognition receptors. Cells of innate immunity have a limited repertoire of receptors for foreign molecular structures; the same receptors are present in all cells, and they can be on the cell surface (e.g., the mannose receptor or scavenger receptor) or cytosolic (e.g., Nod-like receptors). Circulating molecules such as C-reactive protein and mannose-binding protein are also pattern recognition receptors; their binding to microbes can trigger subsequent complement activation or phagocytosis. Because the number of different receptors is relatively small, they are all encoded in the germline. In comparison, the recognition of antigens by the adaptive immune system is specific and unique for each potential antigen (see Figure II.2.3.10); to achieve the potential diversity of roughly 10^{10} different specificities requires somatic recombination of different gene segments. (Figure reprinted with permission from Abbas et al. (2007), *Cellular and Molecular Immunology,* 6th ed.)

The PAMPs include:

- Double-stranded RNA found in cells containing replicating viruses. This induces cytokine production by infected cells leading to the destruction of the intracellular virus.
- Unmethylated CpG DNA sequences characteristic of bacterial infections. These induce autocrine macrophage activation and more effective intracellular killing of phagocytosed organisms.
- N-formylmethionine peptides from bacterial protein synthesis. Binding to receptors on neutrophils and macrophages causes *chemotaxis* (movement up a concentration gradient) and activation. Similar chemotaxis can be engendered by protein fragments released during complement activation, lipid mediators of inflammation, and chemokine proteins released by "stressed" cells.
- Mannose-rich oligosaccharides from bacterial or fungal cell walls. Engagement of receptors on macrophages induces phagocytosis; soluble mannose-binding protein in the plasma *opsonizes* or enhances phagocytosis of microbes bearing mannose.
- Bacterial or fungal wall oligosaccharides directly activate complement and induce either direct microbial lysis or microbial coating with complement that markedly enhances phagocytosis.
- Phosphorylcholine in bacterial cell walls binds to circulating *C-reactive protein* (CRP; also called *pentraxin*); CRP induces opsonization and also activates complement.
- *Lipopolysaccharide* (LPS) from bacterial cell walls binds to circulating LPS-binding protein which in turn binds to CD14 surface molecules on macrophages, activating it through an associated *toll-like receptor* (TLR). The macrophages respond by producing a host of cytokines including tumor necrosis factor (TNF) and interleukin-12 that recruit and activate neutrophils and NK cells, respectively. By similar

pathways, LPS induces severe systemic responses that can culminate in *septic shock* (Hotchkiss and Karl, 2003).

Components of the innate system also recognize sites of injury, anticipating either that these may be primarily caused by infection or may be subject to subsequent infection. Heat shock proteins, altered membrane phospholipids, and urates released from the degradation of nucleic acids are among the danger signals released by necrotic tissues that can lead to macrophage activation (Kono and Rock, 2008). Moreover, components of the coagulation cascade, denatured connective tissue elements (such as might occur at sites of trauma), or denatured circulating proteins can also bind to macrophage cell surface receptors and induce activation. In particular, denatured proteins expressing previously cryptic RGD (arginine-glycine-aspartic acid) motifs that bind to cells via integrins can participate in the recruitment and activation of the cells of innate immunity. These become especially important in the context of the implantation of foreign bodies where otherwise minor trauma, and the presence of denatured proteins non-specifically adhering to "inert" surfaces, leads to macrophage activation (Tang and Eaton, 1999; Anderson et al., 2008).

EFFECTOR MECHANISMS OF INNATE IMMUNITY

As we will discuss in greater detail with the effector functions of adaptive immunity, the immune system needs to deal with pathogens lurking in two general environments – those that live in the extracellular space, and those that are intracellular. Extracellular pathogens are dispatched by a combination of direct killing (NK cells and complement) and ingestion (macrophages and neutrophils). Intracellular microbes (e.g., viruses) require either that the infected cell marshal its own defense mechanisms (such as interferon-α to reduce viral replication), or – failing that – be killed by NK cells (presumably taking down the virus at the same time), or even commit cellular suicide (*apoptosis*, see Chapter II.2.4).

Although the cells and proteins of innate immunity are primarily intended to clear infections, inappropriate, over-exuberant or poorly controlled effector activity can also result in host tissue injury. Once set into motion, activated proteases, macrophages and neutrophils, cytokines and other mediators will execute their functions somewhat blindly, and without regard to whether the "victim" is a microbe or self. This will continue until the initial stimulus abates or regulatory mechanisms supervene. Even business-as-usual for innate immunity can cause host cell death (i.e., the action of NK cells killing virally-infected cells), or loss of normal function (i.e., scarring of a tissue).

COMPLEMENT

Complement is an important protein constituent of the recognition and effector arms of innate immunity (Roozendaal and Carroll, 2006), as well as an important mediator of adaptive immune function (Carroll, 2004) (see also Chapter II.2.4). It is comprised of a number of related proteins (called C1–C9, plus other co-factors B, D, and properdin) that form a self-amplifying proteolytic cascade – in other words, complement components earlier in the cascade enzymatically cleave and activate subsequent components. Late products in the cascade have the ability to form pores in cell membranes and cause cell lysis (a complex of C5b, C6, C7, C8, and C9 called the membrane attack complex or MAC), while many of the proteolytic fragments generated along the way (e.g., C3b, C5a, C5b) have additional useful activities (Figure II.2.3.4 and below) (Walport, 2001a,b). Complement in the circulation is synthesized by the liver; in tissues, macrophages constitute the major source. The complement cascade can be activated (albeit somewhat inefficiently) by binding directly to microbial surfaces or it can be activated by circulating proteins of the innate system (e.g., mannose binding protein) that bind to microbial PAMPs. However, it can be efficiently activated (10^3–10^4 better than direct activation) by antibody first binding to a particular antigen (discussed later).

The complement cascade can induce microbial death and/or cellular injury by:

- direct *cytolysis* via C5b–9, the membrane-attack complex (MAC) punching holes in a cell's plasma membrane
- *opsonization* (via the C3b fragment), enhancing phagocytosis by macrophages and neutrophils
- C3a and C5a (so-called *anaphylotoxins*) mediate *increased vascular permeability* and smooth muscle relaxation (*vasodilation*), mainly by inducing mast cells to degranulate, releasing histamine from mast cells
- C5a also activates the lipoxygenase pathway in arachidonic acid catabolism, resulting in *increased leukotriene synthesis*
- C5a mediates *chemotaxis* of polymorphonuclear cells (PMN; neutrophils) and monocytes.

NEUTROPHILS AND MACROPHAGES: CELLS OF INNATE IMMUNITY

The primary function of all the various recruiting and activating factors is to attract phagocytes (literally, *eating cells*) into a site of infection to ingest and destroy microbes (Figure II.2.3.5). The primary responding cell in the earliest stages of injury or infection is the *neutrophil*, a short-lived (hours) phagocytic cell capable of ingesting and destroying microbes, as well as releasing a

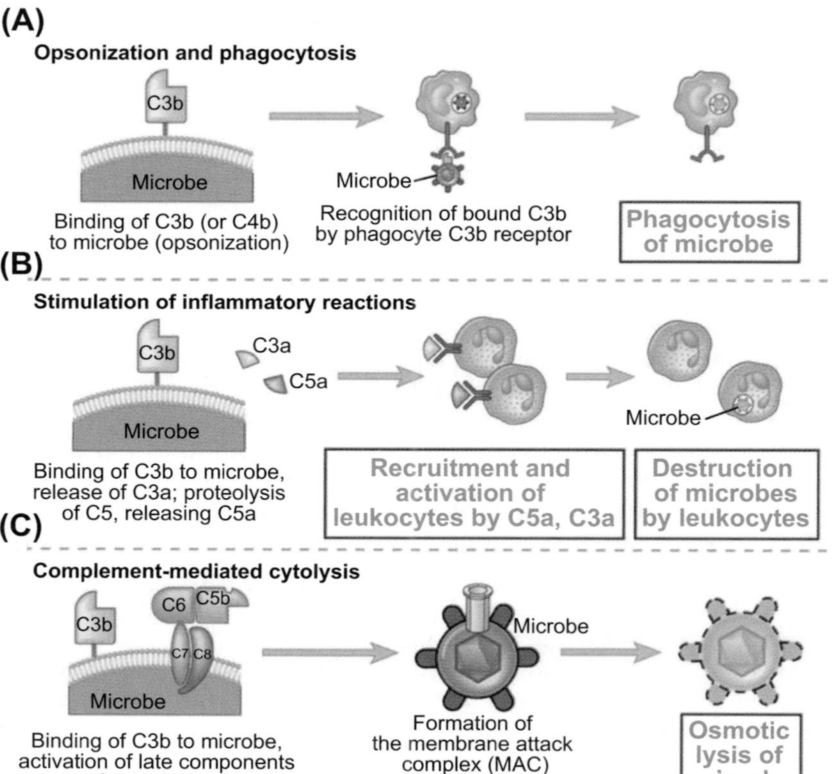

FIGURE II.2.3.4 The major functions of complement. (A) Complement fragment C3b on the surface of microbes (or other cells) promotes phagocytosis (also called *opsonization*). (B) C3a and C5a proteolytic fragments increase vascular permeability and cause vasodilation by releasing histamine from mast cells; C5a is also chemotactic and enhances leukocyte binding to endothelium (shown here is a neutrophil), while stimulating leukotriene synthesis and the production of reactive oxygen species. (C) The C5b–C9 complex forms a membrane attack complex (MAC) that punches holes in microbes (and other cells) leading to osmotic rupture. (Figure reprinted with permission from Abbas et al. (2007), *Cellular and Molecular Immunology,* 6th ed.)

variety of potent proteases. Neutrophils are the characteristic cellular feature of *acute inflammation*; these cells are typically first on the scene following injury (within 6–12 hours), and this phase of the host response to damage lasts for roughly 24–48 hours. Thereafter, the host response enters a phase of *chronic inflammation*, characterized by *macrophage* infiltration. Macrophages generally constitute the second wave of inflammatory cells recruited to sites of injury; they are substantially longer-lived and can persist at sites of inflammation, making them the dominant effector cell type in late stage innate immunity.

Both neutrophils and macrophages are recruited to sites of injury by changes in *adhesion molecule* expression on endothelial cells in the vicinity, and by *chemotactic signals* (acting e.g., through G-protein coupled receptors on neutrophils) delivered by injured cells (i.e., chemokines), by complement components (generated during complement activation), and by microorganisms themselves. The successive waves of inflammatory cell recruitment (neutrophils followed by macrophages) are regulated by distinct subsets of adhesion molecules and chemokines with different specificities. The recruited phagocytic cells clear *opsonized* microorganisms, kill them with *reactive oxygen intermediates* (superoxide, oxyhalide molecules, nitric oxide, and hydrogen peroxide), and degrade them with proteases (Figure II.2.3.5) (Segal, 2005). However, release of such cytotoxic and degradative molecules into the neighboring environment will also cause local tissue injury. Severe local injury due to excessive neutrophil activation results in an *abscess* with total destruction of parenchyma and stroma. C3b opsonization can also lead *indirectly* to tissue injury; large, non-ingestible cells or tissue can result in "frustrated phagocytosis" by neutrophils or macrophages. The attempted intracellular lysis results instead in the extracellular release of proteases and toxic oxygen metabolites (neutrophils), and/or cytokines (macrophages).

In addition, activated macrophages (and neutrophils to a more limited extent) release a variety of cytokines and other factors that can have both local and systemic effects (Figure II.2.3.6):

- *tumor necrosis factor* (TNF) recruits and activates neutrophils
- *interleukin-12* (IL-12) activates T cells and NK cells
- activates coagulation pathways (*tissue factor* elaboration)
- secretes angiogenic factors (new blood vessel formation)
- fibroblast activating factors (e.g., *platelet-derived growth factor*; PDGF) that induce fibroblast proliferation
- *transforming growth factor-β* (TGF-β) expression increases extracellular matrix (ECM) synthesis and down-regulates many aspects of inflammation
- matrix metalloproteinases that remodel the ECM.

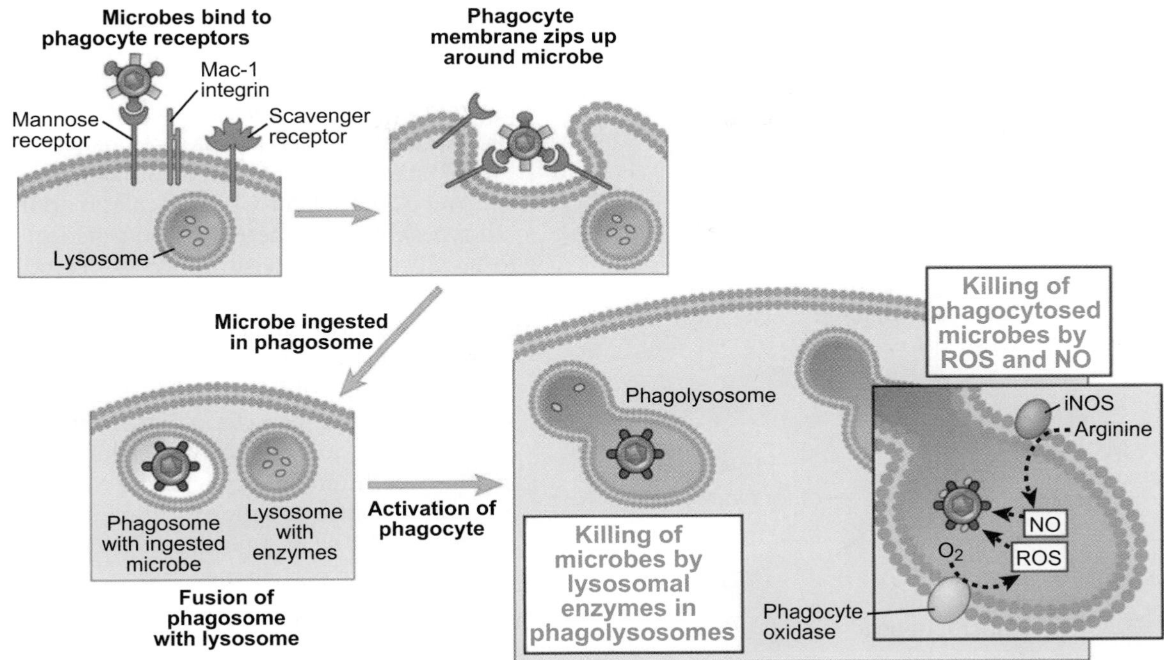

FIGURE II.2.3.5 Phagocytosis and intracellular destruction of microorganisms. Surface receptors on phagocytes can either bind microbes directly, or may bind opsonized microbes (for example, Mac-1 integrin binds microorganisms after they have been coated with complement proteins). After binding to one (or more) of the variety of surface receptors, microbes are internalized into phagosomes, which subsequently fuse with intracellular lysosomes to form *phagolysosomes*. Following fusion, *reactive oxygen species* (ROS, e.g., superoxide and hydrogen peroxide) and *nitric oxide* (NO) are selectively generated within the phagolysosomes; these reactive oxygen species kill the microbes largely via free radical injury; the lysosomes also contain a wealth of catabolic enzymes that digest the microbes. (Figure reprinted with permission from Abbas et al. (2007), *Cellular and Molecular Immunology,* 6th ed.)

Thus (as can be inferred from the preceding descriptions), in the setting of prolonged activation, innate immunity ultimately drives tissue fibrosis and scarring (Figure II.2.3.6) (Henry and Garner, 2003; Martin and Leibovich, 2005). This clearly has ramifications in the setting of inserted foreign bodies intended to persist for many years.

ADAPTIVE IMMUNITY

Adaptive immunity (also called "specific" or "acquired" immunity) is comprised of cellular (*lymphocyte*) and humoral circulating protein (*antibody*) mediators; temporally, these will follow innate immunity in recognizing and resolving infections.

Adaptive immunity is more evolutionarily advanced than its innate counterpart, and is first seen in phylogenic development with the jawed vertebrates. Lymphocytes have the cardinal features of: (1) exquisite *specificity* for distinct macromolecules; and (2) *memory*, the latter being the ability to respond more vigorously to subsequent exposures to the same microorganism. The substances that induce specific immune responses are called *antigens*. These can be proteins, carbohydrates, lipids or other small molecules; they can also be self or non-self, microbial or non-infectious. The cells and antibodies of adaptive immunity each have the theoretical capacity to recognize 10^9–10^{11} distinct antigenic determinants. Since each cell of the adaptive immune system can only recognize and respond to a single antigenic determinant, the impressive diversity of the system requires an equally large number of different cells. Given that the human body only has 10^{13}–10^{14} cells in *total*, there obviously cannot be more than a few tens to hundreds of cells of any one specificity at baseline. This has important ramifications in eliminating an infection; the small numbers of cells that can initially recognize an invader will not be up to the task on their own. Rather, following recognition of a pathogen, the responding cell(s) must clonally expand, as well as recruit an additional army of additional effectors (see later).

The other important concept is that the elements of adaptive immunity (like innate immunity) need to be able to respond to two basic infectious challenges – pathogens that are extracellular and those that reside within cells. As we will see when we look at effector mechanisms, the humoral and cellular elements complement each other in this enterprise – and also take advantage of the innate immune system. Thus, not only does the initial innate response influence the strength and nature of adaptive immune responses (e.g., whether antibodies or cellular mediators are produced), but the same effectors of innate immunity are also conscripted by adaptive immunity to clear infectious agents.

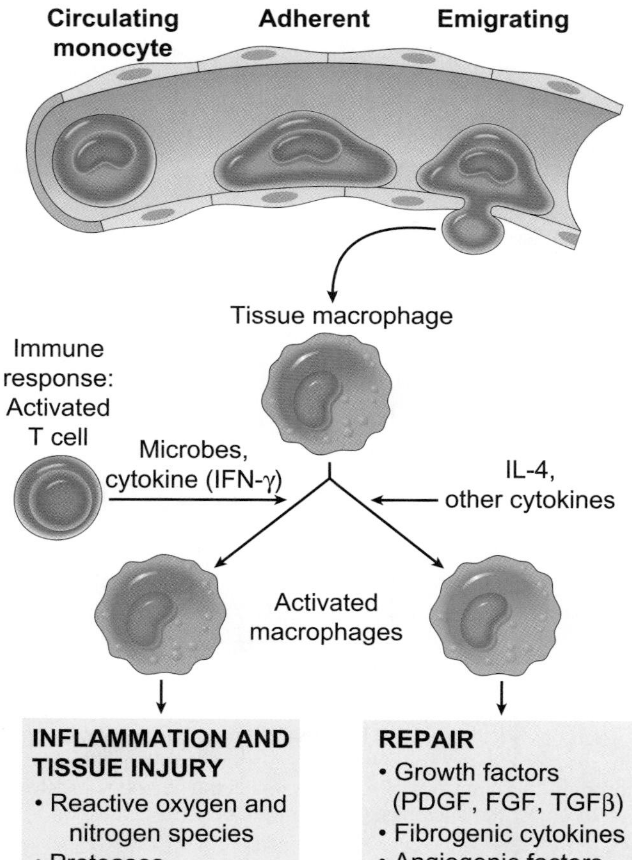

FIGURE II.2.3.6 Macrophage recruitment and local tissue effects after activation. Circulating monocytes are recruited to sites of tissue injury by changes in adhesion molecule expression on endothelial cells in the vicinity of the injury, as well as by *chemotactic signals* (chemokines) delivered by injured cells or neutrophils, by complement components (generated during complement activation), and by microorganisms themselves. Once monocytes emigrate from the vasculature, they become tissue *macrophages*. Macrophages are activated by interferon-γ (IFN-γ) from various sources – including activated NK cells or T cells – or by non-immunologic stimuli such as endotoxin. Activated macrophages aggressively phagocytize microorganisms and necrotic debris, and also secrete a number of eicosanoids (arachadonic acid or AA metabolites such as prostglandins and leukotrienes), reactive oxygen intermediates, and proteases that can sterilize the local environment, but will also incite tissue injury. Overall, the cytokine mediators produced by activated macrophages will tend to induce fibrosis (scar). (Figure reprinted with permission from Kumar et al. (2007), *Robbins Basic Pathology,* 8th ed.)

COMPONENTS OF ADAPTIVE IMMUNITY

The principal components of adaptive immunity are:

- *B lymphocytes* (also called B cells), responsible for making antibodies
- Antibodies: proteins secreted by B lymphocytes with specificity for a specific antigen
- *T lymphocytes* (also called T cells) are functionally divided into *helper T cells* (*Th cells*) that produce *cytokines* to orchestrate the activity of other cell types, and *cytotoxic T cells* (*Tc cells*) that kill selected target cells. Th cells express a cell surface protein designated CD4 ("CD" stands for "cluster of differentiation;" there are over 300 different CD molecules marking various cell types), and are therefore also generically referred to as $CD4^+$ ("CD4-positive") cells. Tc cells express CD8 and may be referred to as $CD8^+$ cells. All T cells also express CD3 (actually a complex of γ, δ, and ε protein chains) that is involved in T cell activation.
- Cytokines: many of the cytokines produced by T cells are also secreted by cells of innate immunity; several, however, are relatively T-cell specific, and are involved in modulating immune responses or in effector cell differentiation. Although typically synthesized by Th cells, Tc cells and even B cells can be sources of these protein mediators.

Not all the possible adaptive responses are elicited concurrently in response to a particular pathogen. In some cases, it may be more advantageous to induce primarily a B cell antibody-mediated response; in other circumstances, a cytotoxic T cell response may be most warranted. Moreover, the adaptive immune response needs to be tightly regulated to prevent ongoing tissue injury, and therefore a negative-regulatory feedback must exist. The regulation of which of these potential outcomes occurs derives from the helper T cells, and more specifically the nature of the cytokines that they produce.

ANTIBODIES

The generic antibody (also called an immunoglobulin or Ig) is a vaguely "Y"-shaped structure composed of four disulfide-linked proteins – two smaller (25 kD) light chains and two larger (50 kD) heavy chains (Figure II.2.3.7); for any given Ig the light chains on each arm of the "Y" are identical, as are the heavy chains. The N-termini of both chains have genetically (germline and somatic) variable domains (designated V_L and V_H in Figure II.2.3.7) that are the source of the tremendous antigen-binding diversity of the total antibody repertoire. The C-termini of each chain are genetically homogeneous for any given Ig, and are designated as constant domains (C_L or C_H). The two "arms" of the Ig are each composed of heavy and light chains, and are called the Fab or antigen-binding *fragment* of the molecule; the "body" of the Ig is composed of paired heavy chain constant regions, and is called the Fc portion of the molecule (for historical reasons, the "c" stands for "crystallizable" but "constant" is equally applicable).

Once antigen is bound to the Fab portion of an Ig (more on this recognition step later), the protein undergoes a conformational change in the Fc region that allows the antibody to activate complement, or to attach to one

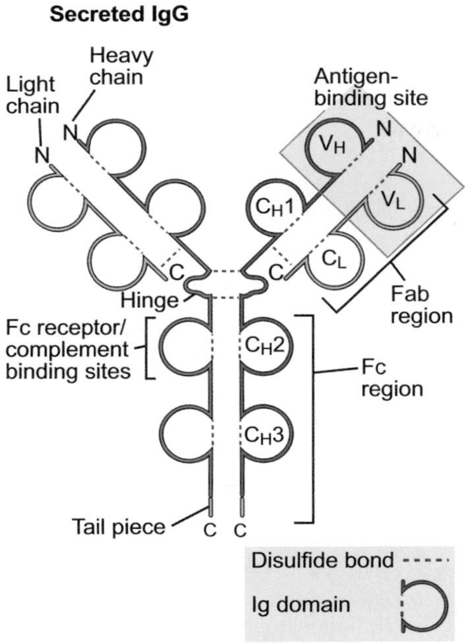

FIGURE II.2.3.7 Generic antibody structure. Antibodies typically adopt a "Y"-shaped structure composed of two light chains and two heavy chains. The N-termini are the antigen binding fragments (Fab) and contain variable regions that allow the necessary diversity. The C-termini are constant for any particular immunoglobulin isotype (called Fc fragment). Antigen binding to the Fab area causes a conformational change in the Fc region that allows the antibody to activate complement, or to attach to Fc receptors. (Figure reprinted with permission from Abbas et al. (2007), *Cellular and Molecular Immunology,* 6th ed.)

of a variety of surface-bound Fc receptors on different cell types (e.g., macrophages, neutrophils, B cells, NK cells, etc.). Based on the structure of the heavy chain constant regions, antibodies occur as five major isotypes, IgA, IgD, IgE, IgG, and IgM; IgA has two subtypes and IgG has four. IgG is the most common of the antibody isotypes. All except IgD (which is only found in the membrane-bound form on naïve B cells) exist as circulating *and* B cell membrane-bound forms; the membrane-bound form of Ig is the antigen receptor for B cells and engagement drives B cell activation and differentiation. The different antibody isotypes are specialized to activate specific effector mechanisms (e.g., phagocytosis, complement activation or NK cell killing); we will also specifically revisit one (IgE) in detail later when we discuss allergy.

T LYMPHOCYTES

As we will see, T cells can participate in the adaptive immune response and elimination of extracellular pathogens. However, they are also the main mechanism by which *intracellular pathogens* (e.g., viruses and certain bacteria) – not accessible to phagocytes and circulating antibodies – can be targeted (see later).

In broad strokes, the helper *versus* cytotoxic T cell dichotomy addresses the requirements of defending against extracellular (Th) and intracellular (Tc) pathogens; having said that, Th cells are also involved in immunity to intracellular microbes, and help drive the Tc response.

Th CELLS

$CD4^+$ T helper cells achieve their function by secreting specific cytokines to stimulate the proliferation and differentiation of other cells. Two basic subsets of Th cells are classically described, called Th1 and Th2; each produce a handful of specific cytokines that will induce a different general immunologic response (Moss et al., 2004; Pulendran, 2004). Once a Th1- or a Th2-specific reaction is established against a particular pathogen, the responding population tends to maintain that pattern of differentiation. This is accomplished because Th1 cell cytokines promotes further Th1 differentiation and inhibits Th2 cells, and *vice versa*.

Generally speaking, Th1 cells are induced by intracellular pathogens that infect or activate macrophages and NK cells. The Th1 cytokines include interferon-γ (IFN-γ, the signature cytokine of Th1 cells), lymphotoxin, and tumor necrosis factor (TNF); these promote phagocyte-mediated defenses, especially against intracellular microbes (Figure II.2.3.8).

Th2 cells are induced by extracellular pathogens (especially worms) and certain polymeric antigens (*allergens*). The cardinal cytokines of this subset include interleukin (IL)-4, -5, and -10; these promote IgE and mast cell/eosinophil activation (and inhibit macrophage activation); these pathways putatively evolved to deal with helminthic infections that are not effectively eradicated by macrophages and NK cells (Figure II.2.3.9); however, Th2 differentiation also underlies many allergic responses.

More recently, a unique third Th subset has been described, generated in the context of TGF-β and IL-6 or IL-1 cytokine production; IFNγ and IL-4 actually inhibit the expansion of this subset. These are called Th17 cells because they characteristically produce IL-17, a cytokine that, in association with IL-22 and other mediators, induces neutrophil-rich inflammatory responses (Korn et al., 2007, 2009). Since Th17 cells can potentially significantly contribute to tissue damage in the setting of local inflammation (due to the neutrophil-rich infiltrates), and because they are lineage-related to *anti-inflammatory regulatory T cells*, the pathways controlling their differentiation and activation are a focus of intense research.

Tc CELLS

$CD8^+$ cytotoxic T lymphocytes (also called CTL) are effector cells that function to eliminate intracellular infections. The strategy is similar to NK cells, in that infected cells are induced to undergo apoptotic cell death. However, there is greater specificity of these cells *versus* NK cells, and $CD8^+$ T cells also have the attribute of memory – subsequent exposure to the same antigen results in a faster, more robust secondary response.

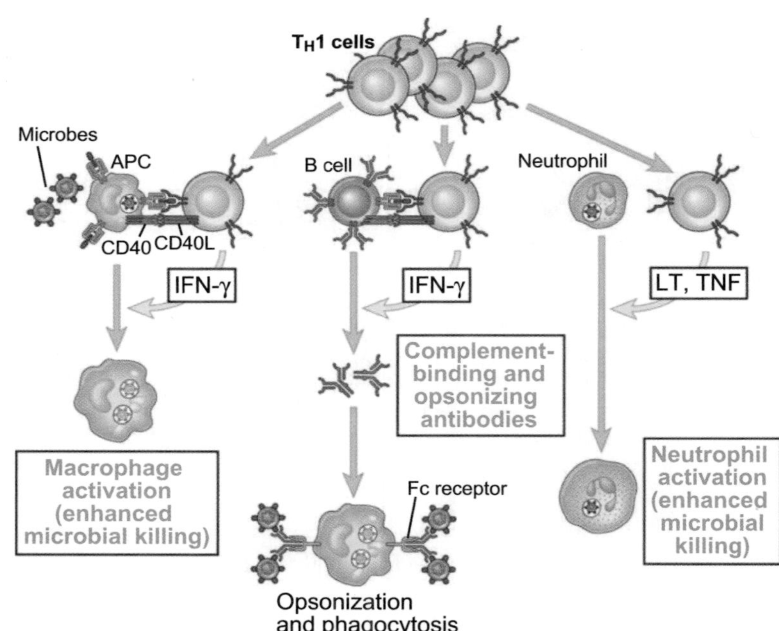

FIGURE II.2.3.8 Th1 cell functions. Differentiation of CD4+ T cells into Th1 cells results in the production of interferon-γ (IFN-γ), lymphotoxin (LT), and tumor necrosis factor (TNF). IFN-γ drives macrophage activation increasing their capacity to kill intracellular microbes; IFN-γ also induces the differentiation of B cells to produce antibodies that are better at opsonization and complement activation (e.g., IgG). LT and TNF function to recruit and activate additional inflammatory cells at a site of injury; shown here is a neutrophil, but macrophages are also involved. The net result of the Th1 cytokines is the enhancement of phagocyte-mediated defenses against infection; this involves both extracellular pathogens and those that can persist within non-activated macrophages. (Figure reprinted with permission from Abbas et al. (2007), *Cellular and Molecular Immunology,* 6th ed.)

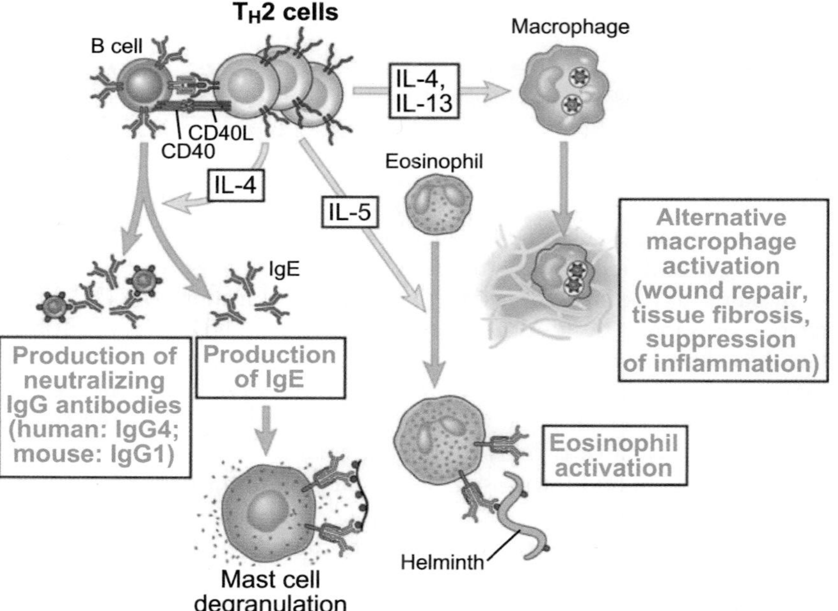

FIGURE II.2.3.9 Th2 cell functions. Differentiation of CD4+ T cells into Th2 cells results in the production of interleukin (IL)-4, IL-5, and Il-10. IL-4 drives the differentiation of B cells to produce antibody isotypes that either bind well to mast cell Fc receptors (IgE) or bind antigen but do not activate complement or induce opsonization (e.g., IgG_4); IL-4 (along with IL-10) also modulates macrophage activation status, antagonizing the actions of IFN-γ. IL-5 is involved in the recruitment and activation of eosinophils that are potent mediators, particularly against helminths. The net effect is to augment responses that will be effective against large extracellular pathogens such as worms. (Figure reprinted with permission from Abbas et al. (2007), *Cellular and Molecular Immunology,* 6th ed.)

RECOGNITION IN ADAPTIVE IMMUNITY

B cells, antibodies, and T cells can potentially recognize 10^{11} different antigenic specificities; this is achieved by somatic mutation and rearrangement of limited numbers of germline sequences, but the genetic details are beyond the scope of this discussion. However, *what* these different components of adaptive immunity recognize *is* important. Thus, B cell receptor and secretory Ig binds to *intact* antigens, although the antigen need not necessarily be in its native conformation (Figure II.2.3.10). In comparison, T cells cannot recognize antigens until they have been degraded into smaller fragments and associated with self *histocompatibility molecules* on antigen presenting cells (Figure II.2.3.11). The following sections will elaborate on these basic tenets.

B CELL AND ANTIBODY RECOGNITION

As shown in Figure II.2.3.10, antibodies can bind to conformations that depend on some combination of primary, secondary, and tertiary structure. Antigens can be linear sequences of amino acids or sugars, or can be composed of elements that are juxtaposed by virtue of

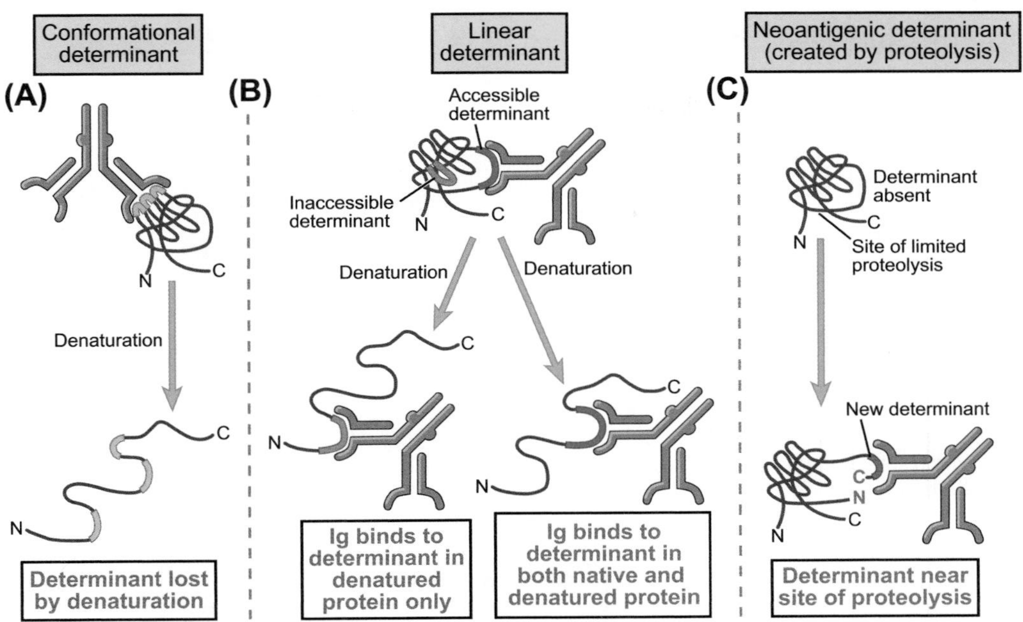

FIGURE II.2.3.10 Antigen–antibody binding. (A) Antibodies can bind to three-dimensional conformations of proteins with determinants (or epitopes) comprised of multiple amino acids (or other small molecules) from different loops; denaturation of the protein will result in loss of antibody binding to this particular kind of antigen. (B) Antibodies can bind to primary sequences of amino acids (or other small molecules) that can be present in the native molecule or be occult – that is, hidden within the folded tertiary structure. Denaturation of the protein will not affect antibody binding to this antigenic structure, and indeed, may expose previous inaccessible determinants. (C) Antibodies can bind to determinants that are only exposed or created by proteolytic degradation. (Figure reprinted with permission from Abbas et al. (2007), *Cellular and Molecular Immunology,* 6th ed.)

three-dimensional folding; the same molecule can have multiple different potential antigenic *epitopes* (also called *determinants*). The nature of the epitopes is significant because in native, intact molecules, certain determinants may be internally folded and inaccessible to antibody. However, denaturation or proteolytic cleavage (e.g., at sites of tissue injury) can conceivably expose previously occult epitopes, and allow antibody binding and subsequent effector activity.

T CELL RECOGNITION

T cells have surface receptors that cannot "see" intact antigens, but rather only recognize proteolytically digested antigen fragments ("processed antigen") presented on the surface of certain host cells in association with *major histocompatibility complex* molecules (Figure II.2.3.11). For Th cells, these accessory or *antigen-presenting cells* (APC) include macrophages, one of the major cell types of the innate immune response; in this manner, the innate system can influence adaptive immune responses.

Histocompatibility molecules are grouped together on chromosomes into clusters generically called *Major Histocompatibility Complexes* or MHC. Proteins of this complex are denoted as "histocompatibility" molecules because they were first recognized as the major determining element in tissue ("histo") compatibility in organ transplantation. When inbred strains of animals shared the same MHC determinants, tissue grafts could be transplanted with relative impunity; if the donor and host were MHC-disparate, grafts were said to be histo-incompatible, and the organs ultimately failed by a process called *rejection.* This is due to the host "interpreting" foreign MHC as a potential pathogen, and then engaging the full fury of the various immune effectors. Tissue incompatibility clearly has ramifications in the assembly and implantation of devices prepared from non-autologous (*autologous* = self) viable human cells; in most cases, these cells will express host-incompatible MHC and trigger specific immune responses (see discussion at the end of this chapter).

In humans, this MHC cluster occurs on chromosome 6, and the molecules are called *Human Leukocyte Antigens* or HLA. There are two general categories of MHC molecules, called *class I* and *class II*. In humans, MHC class I (MHC I) molecules are called HLA-A, -B, and -C; MHC class II (MHC II) molecules are called HLA-DP, -DQ, and -DR. In the general population, each type of HLA molecule (e.g., HLA-A, HLA-B, etc.) has hundreds of possible different alleles. Thus, outside of haplo-identical twins, it is virtually impossible to have a perfect match among all the possible class I and class II MHC combinations.

The MHC class I molecules present peptide fragments derived from intracellular antigens to $CD8^+$ Tc cells; in this manner, intracellular infections can be detected and the infected cell can be killed. In comparison, MHC class II molecules present peptide fragments from extracellular antigens to $CD4^+$ Th cells. In this pathway, the APC is not killed; rather, extracellular infections are detected by the Th cells that subsequently release cytokines to recruit

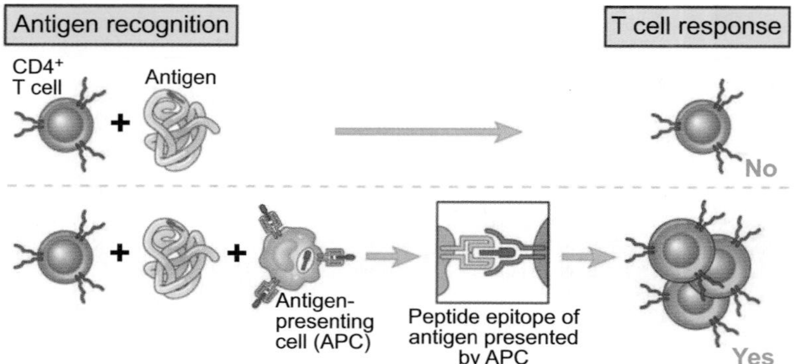

FIGURE II.2.3.11 T cells cannot recognize unprocessed antigen. Intact protein antigens will not bind the T cell receptor (TCR). Instead, the TCR can only bind to smaller (10–15 amino acids in length) polypeptides, and then only when complexed to self major histocompatibility (MHC) molecules. Thus, exogenous antigen must be taken up by antigen presenting cells (APC), delivered to intracellular endosomes for proteolytic processing, and the resulting fragments joined to newly synthesized MHC before being displayed on the cell surface. (Figure reprinted with permission from Abbas et al. (2007), *Cellular and Molecular Immunology,* 6th ed.)

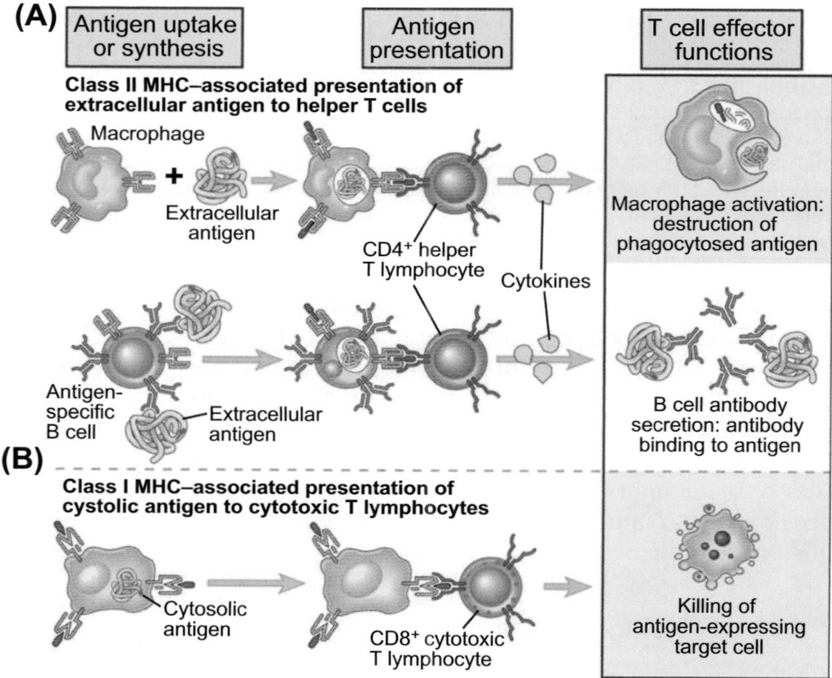

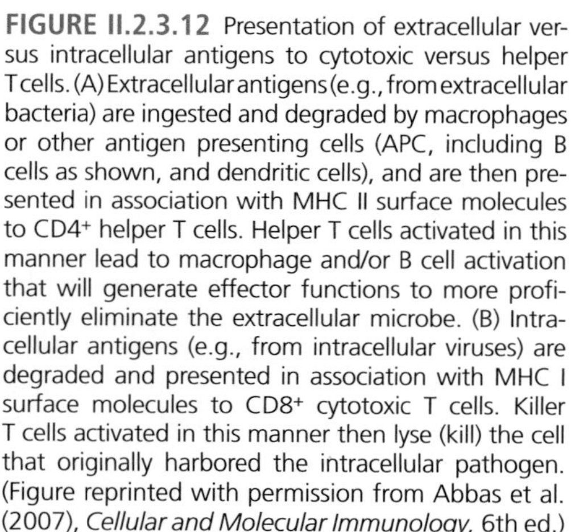

FIGURE II.2.3.12 Presentation of extracellular versus intracellular antigens to cytotoxic versus helper T cells. (A) Extracellular antigens (e.g., from extracellular bacteria) are ingested and degraded by macrophages or other antigen presenting cells (APC, including B cells as shown, and dendritic cells), and are then presented in association with MHC II surface molecules to $CD4^+$ helper T cells. Helper T cells activated in this manner lead to macrophage and/or B cell activation that will generate effector functions to more proficiently eliminate the extracellular microbe. (B) Intracellular antigens (e.g., from intracellular viruses) are degraded and presented in association with MHC I surface molecules to $CD8^+$ cytotoxic T cells. Killer T cells activated in this manner then lyse (kill) the cell that originally harbored the intracellular pathogen. (Figure reprinted with permission from Abbas et al. (2007), *Cellular and Molecular Immunology,* 6th ed.)

additional effectors (Figure II.2.3.12) (Klein and Sato, 2000a,b).

T cell recognition of antigen fragments bound to MHC molecules results in *T cell activation*. This recognition step is accomplished by T cell receptors (*TCR*) on the surface of T lymphocytes interacting with the CD3 complex and additional molecules to transduce a signal to the nucleus, resulting in selected gene transcription (Kuhns et al., 2006). Complete activation of T cells also requires additional interplay between other molecules (called *co-stimulator molecules*) on the surface of T cells and antigen presenting cells of the innate immune system. *Incomplete activation* of T cells (i.e., without the co-stimulators) may result in *anergy* (no response) to the antigen (Peter and Warnatz, 2005) (Figure II.2.3.13).

EFFECTOR PATHWAYS IN ADAPTIVE IMMUNITY

Depending on the nature of the inciting pathogen and the preceding innate response, different elements of adaptive immunity will be engaged; as a result, different effector functions will also obviously be involved.

Figure II.2.3.14 depicts schematically how antibodies can be marshaled to defend against pathogens. Binding to microbial surface receptor molecules can prevent access to target cells; similarly, binding to circulating toxins can serve to block their activity. Antibody opsonization will increase the phagocytosis (and intracellular destruction) of pathogens, as well as facilitate NK cell killing; both of these happen through the interaction

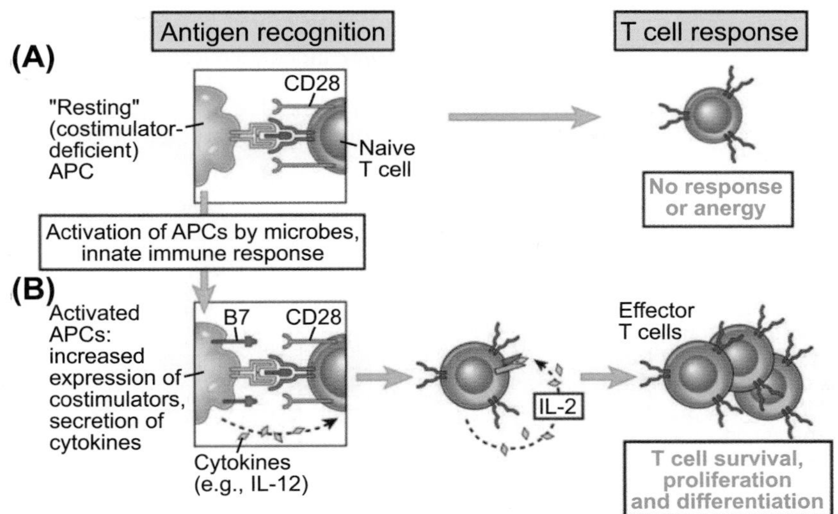

FIGURE II.2.3.13 Role of co-stimulation in T cell activation. (A) Antigen presenting cells (APC) that are not activated will express few or no co-stimulator molecules. In that setting, even though the APC display processed antigen in the appropriate MHC context, the T cells will fail to respond. Indeed, such co-stimulator-poor APC presentation may result in a long-term *anergy* (inability to respond) to particular antigens. (B) Microbes and cytokines produced during innate immune responses activate APC to make co-stimulator molecules (such as B7, shown here) that will result in "complete" activation of the T cells. Activated APC also produce additional cytokines such as interleukin-12 (IL-12) that also participate in stimulating T cell activation and differentiation. (Figure reprinted with permission from Abbas et al. (2007), *Cellular and Molecular Immunology,* 6th ed.)

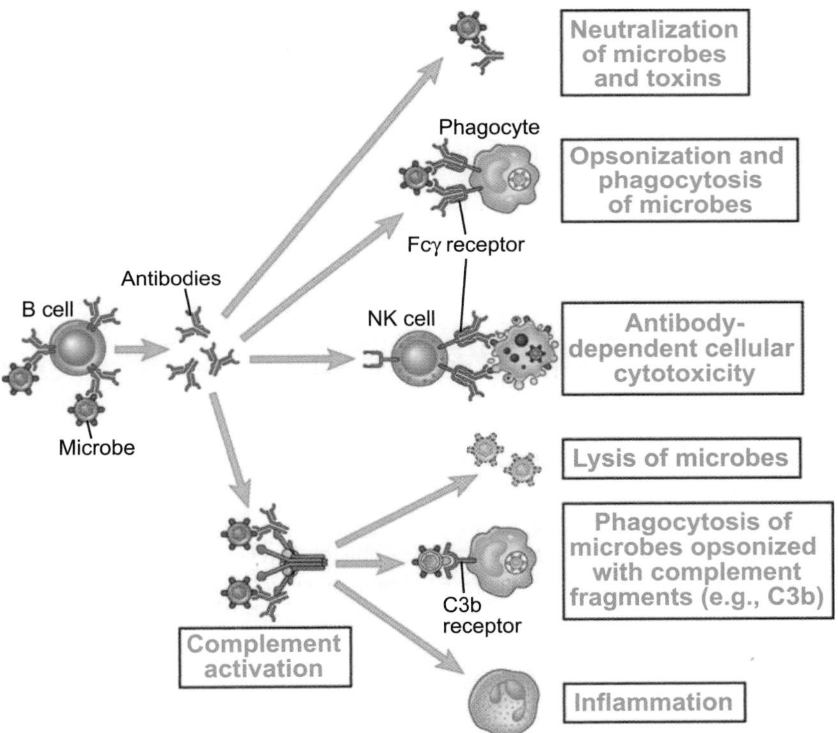

FIGURE II.2.3.14 Effector functions of antibodies. From top to bottom, antibodies can: bind pathogens (or their toxins) and neutralize them (i.e., prevent their attachment or entrance into host cells); bind to microbes and induce opsonization (ingestion by Fc-receptor-bearing phagocytes); bind to microbes and direct natural killer (NK) cell cytolysis; and activate complement to cause direct lysis, opsonization, or additional inflammatory cell recruitment. (Figure reprinted with permission from Abbas et al. (2007), *Cellular and Molecular Immunology,* 6th ed.)

of bound Ig and cells bearing specific Fc receptors. Finally, antibody binding also efficiently activates complement with subsequent MAC formation, additional opsonization, and the augmented recruitment and activation of innate inflammatory cells. The downside to these activities occurs when antibody is bound – either specifically or non-specifically – to self tissues: the full array of antibody-mediated effector functions are then brought to bear and can cause substantial injury (see later).

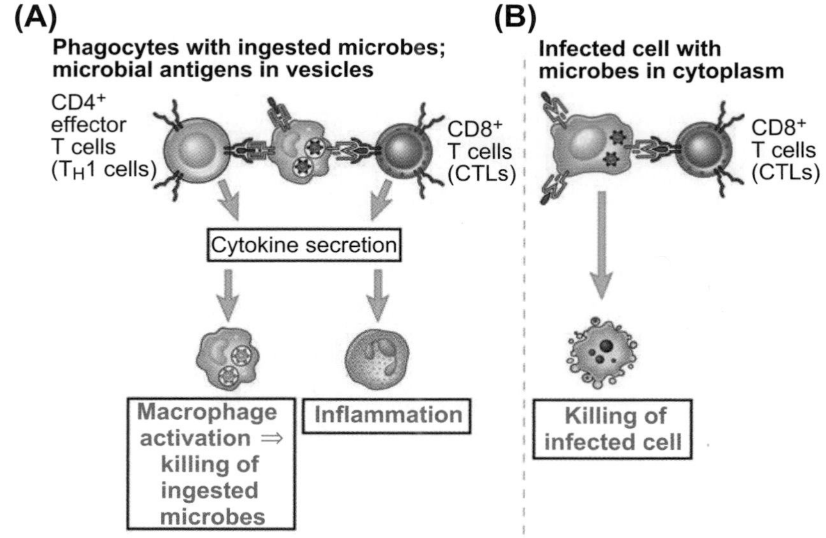

FIGURE II.2.3.15 Effector functions of T lymphocytes. (A) When $CD4^+$ T cells encounter antigenic fragments presented appropriately by antigen presenting cells (APC) they respond by the production of cytokines that will drive subsequent effector cell recruitment and activation (see also Figures II.2.3.8 and II.2.3.9); shown here is the response of a Th1 cell producing cytokines that increase macrophage activation and inflammatory cell recruitment. $CD8^+$ T cells can also be a source of cytokines that affect the local inflammatory response. (B) Recognition of processed antigen on target cells by $CD8^+$ cytotoxic T lymphocytes (CTL) typically results in the directed CTL killing of the target. (Figure reprinted with permission from Abbas et al. (2007), *Cellular and Molecular Immunology,* 6th ed.)

In comparison, Figure II.2.3.15 shows the general effector mechanism of cell-mediated immunity. Recognition of infected cells by $CD8^+$ CTL leads to their cytolysis. Antigen recognition by $CD4^+$ Th cells leads to cytokine production that recruits and activates additional effector cells to more efficiently carry out their microbicidal activities. The down-side to CTL is that inappropriate targeting of host cells (*autoimmunity*) will cause their destruction; even the appropriate targeting of infected cells can wreak substantial havoc if the infection is widespread. If Th activity is not well-regulated or if Th cells begin recognizing self-antigens, then exuberant innate immunity activation will also cause significant tissue damage and scarring (see also below).

OVERVIEW: IMMUNITY TO PATHOGENS

Recall that following pathogen recognition – and depending on the location of the invader – effector functions need to be appropriately coordinated to neutralize and/or destroy either intracellular or extracellular microbes.

Extracellular pathogens: Innate immunity deals with an extracellular challenge largely through the activation of complement and phagocytosis, as well as by being effective at recruiting additional cells that can do more of the same. Antibodies are the major components of adaptive immunity that contribute to the battle against extracellular pathogens; they do so by binding and neutralizing microbes or their toxins, and by highly effectively inducing complement activation and phagocytosis. Th cells can also participate by secreting cytokines that recruit neutrophils, and by recruiting and activating macrophages, augmenting their phagocytic capacity to kill.

Intracellular pathogens: Intracellular pathogens trigger the involvement of innate immunity through interferon-α and -β production, by the activation of macrophages to more effectively kill ingested pathogens, and by the activation of NK cells to kill infected cells. On the adaptive immune side, CTL kill virally infected cells, and Th cells mediate highly effective macrophage activation. As it turns out, the most important immune response to intracellular bacteria – as well as large extracellular pathogens such as fungi – involves T cell-driven activation of macrophages to more avidly ingest microbes, and produce more reactive oxygen species (ROS) and nitric oxide. If these activities fail to completely eliminate the inciting pathogens, ongoing recruitment and activation of macrophages leads to the formation of a *granuloma*, a collection of activated macrophages that effectively form a vigilant ring of inflammation to wall off the invader (Figure II.2.3.16). A potentially negative outcome of persistent granulomatous inflammation is tissue injury due to compression atrophy by the expanding sphere of activated macrophages, and by the surrounding fibrosis these activated cells will engender.

PATHOLOGY ASSOCIATED WITH IMMUNE RESPONSES

The chapter began with the assertion that innate and adaptive immune systems exist primarily to defend the human body against infection. Unfortunately, immune activation leads not only to the activation of host defenses and production of protective Ig and T cells, but also occasionally to the development of responses that may potentially damage host tissues.

Both innate and adaptive immune responses are implicated in causing disease states. As described above, in the setting of prolonged activation, macrophages will ultimately mediate tissue fibrosis and scarring. Indeed, the response to foreign materials – causing much of the local pathology associated with implants – is attributable to such persistent macrophage activation. Moreover, certain bacterial toxins (e.g., LPS) non-specifically stimulate macrophages (as well as other cell types), and

(A)

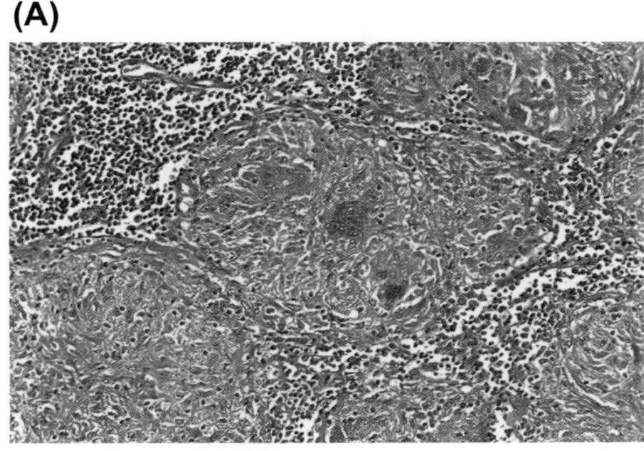

(B)

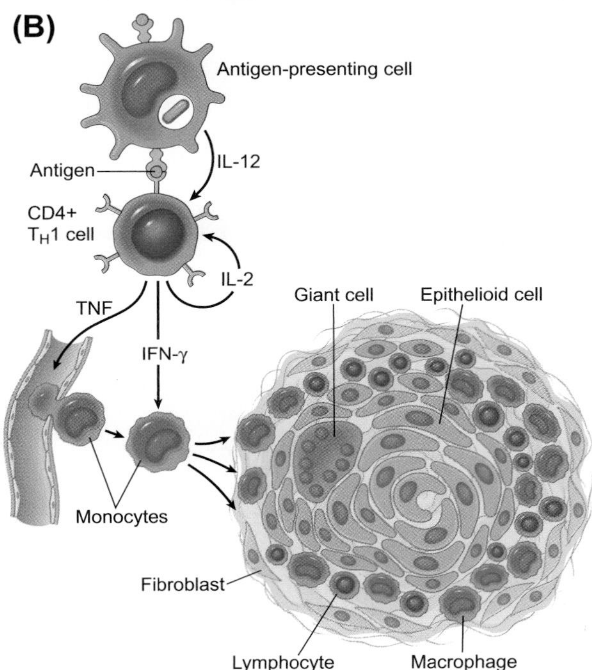

FIGURE II.2.3.16 Granulomatous inflammation. Schematic illustration of the events that lead to granuloma formation in response to large or persistent antigens. Antigen-presenting cells (APC) of the innate immune system process antigen and subsequently present it to CD4+ helper Th1 cells; the APC also provide interleukin-12 (IL-12) and other cytokines to drive T cell activation. Activated T cells, in turn, elaborate cytokines such as tumor necrosis factor (TNF) that will recruit inflammatory cells, and interferon-γ (IFN-γ) that will induce the activation of the recruited cells, in particular macrophages (large, activated macrophages are also called "epithelioid cells"). These cytokines can also induce macrophage fusion to generate multinucleated "giant cells." If the antigen is not effectively eliminated, the constant cycle of T cell and macrophage activation leads to the accumulation of an aggregate of activated cells. Activated macrophages will also elaborate mediators that result in tissue injury, as well as cytokines resulting in fibroblast activation and tissue fibrosis (see also Figure II.2.3.6). The end-result is loss of tissue function and scar formation. In the case of "inert" foreign bodies, absorption of host proteins onto the foreign body surface with subsequent denaturation and modification can lead to direct macrophage activation via the receptors involved in innate immunity. (Figure reprinted with permission from Kumar et al. (2005), *Robbins and Cotran Pathologic Basis of Disease,* 7th ed.)

result in systemic pathology from excessive cytokine elaboration.

By having increased specificity, adaptive immunity might be expected to lead overall to less secondary damage. Normally, an exquisite system of checks and balances optimizes the antigen-specific eradication of infecting organisms with only trivial innocent bystander injury. However, certain types of infection (e.g., virus) may require destroying host tissues to eliminate the disease (see Figure II.2.3.15B). Still other types of infections (e.g., tuberculosis) may only be controlled by a granulomatous response that walls off the offending agent with activated macrophages and scar, often at the expense of adjacent normal parenchyma (similar to foreign body responses).

Even when the host response to an infectious agent is a specific antibody, the antibody may occasionally cross-react with self-antigens (e.g., anti-cardiac antibodies following certain streptococcal infections, causing rheumatic heart disease). Immune complexes composed of specific antibody and circulating antigens can also precipitate at inappropriate sites (see below) and cause injury by activation of the complement cascade or by facilitating binding of neutrophils and macrophages (e.g., in a disease such as post-streptococcal glomerulonephritis). If the antibody made in response to a particular antigen is IgE, any subsequent response to that antigen will be immediate hypersensitivity (*allergy*), potentially culminating in *anaphylaxis*. Finally, not all antigens that attract the attention of lymphocytes are exogenous and infectious. The immune system occasionally (but fortunately, rarely), loses tolerance for endogenous self-antigens, which results in *autoimmune disease*.

All of these forms of immune-mediated injury are collectively denoted as **hypersensitivity reactions.** As discussed below and in Chapter II.2.4, they are traditionally subdivided into four types; three are variations on antibody (Ig)-mediated injury, while the fourth is T cell-mediated:

- IgE-mediated "immediate-type hypersensitivity" (allergy and anaphylaxis)
- antibody against circulating or tissue antigens
- antigen–antibody (immune complex)-mediated
- T cell-mediated (CTL and "delayed-type hypersensitivity" (DTH)).

PATHOGENESIS OF ANTIBODY-MEDIATED DISEASE

Antibodies involved in immune-mediated diseases can bind to antigenic determinants that are *intrinsic* to a particular tissue or cell, or which are *exogenous* and have been passively adsorbed (e.g., certain antibiotics or foreign proteins). Regardless of what they recognize or how they got there, antibodies bound to the surfaces of cells or to extracellular matrix cause injury by certain basic mechanisms.

IgE-Mediated (Immediate) Hypersensitivity (Kay, 2001a,b)

Mast cells and *basophils* express surface *Fc-receptors* that can bind the Fc constant region of immunoglobulin E (IgE). When circulating IgE binds to the Fc-receptors on these cells and is then cross-linked by specific *allergen* (typically a polyvalent antigen), this induces mast cell or basophil *degranulation* with release of preformed mediators, as well as synthesis of other potent effectors (Figure II.2.3.17):

- **pre-formed mediators:** amines such as histamine and serotonin (cause vasodilation and increased vascular permeability)
- **arachidonate, lipid, and cytokine mediators synthesized *de novo*:**
  - prostaglandins (e.g., PGD_2) that can affect vessel and airway contraction, and vascular permeability
  - leukotrienes (e.g., LTC_4, LTD_4, and LTE_4) that are exceptional vasoconstrictors and bronchoconstrictors
  - platelet activating factor (PAF), a rapidly catabolized phospholipid derivative that increases vascular permeability and diminishes vascular smooth muscle tone; it also causes bronchoconstriction
  - cytokines, in particular TNF (recruits sequential waves of neutrophils and monocytes), and IL-4 (interleukin 4, induces local epithelial and macrophage expression of chemokines like eotaxin, and also increases endothelial adhesion molecule expression: the combined effect will be to recruit *eosinophils*).

In most vascular beds, the overall result is vasodilation and increased vascular permeability, with a variable infiltrate classically predominated by *eosinophils*. Eosinophils are an inflammatory cell type classically associated with parasitic infections (especially worms), as well as with allergies; they contain specific granules with potent cytotoxic activity for a variety of cell types. In the respiratory tree, the net result of an allergic stimulus is increased mucus secretion and bronchoconstriction.

The nature of the symptoms in any particular instance will depend on the portal of antigen entry, e.g., cutaneous (hives and rash, although these can also occur with inhaled or ingested allergen), inhaled (wheezing, airway congestion), ingested (diarrhea, cramping) or systemic (hypotension). The associated diseases range from the merely annoying (seasonal rhinitis or "hay fever") to debilitating (asthma) to life-threatening (*anaphylaxis*). In the last case, systemic vasodilation leads to hypotension (low blood pressure), increased vascular permeability leads to increased tissue edema – most problematic around the larynx – and bronchoconstriction narrows the airways and impedes airflow; the combination can culminate in fatal anaphylactic shock.

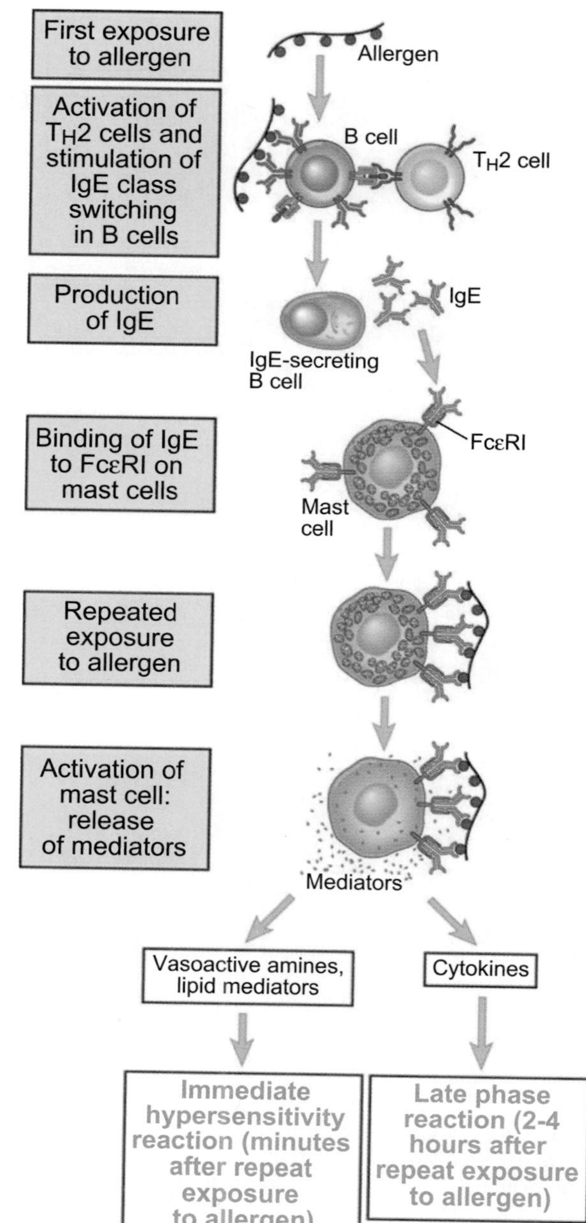

FIGURE II.2.3.17 Events in immediate-type hypersensitivity (allergy). Immediate hypersensitivity is initiated following contact with a specific *allergen* (typically a polyvalent antigen). For unclear reasons, allergens induce in a susceptible host a predominant Th2 response that ultimately promotes an IgE antibody response. IgE binds to mast cells in tissues (and *basophils* in the circulation, not shown) via specific IgE Fc receptors. Subsequent encounter with the relevant allergen results in IgE-Fc receptor cross-linking that activates the mast cells and basophils. Once activated, the cells release granules containing pre-formed mediators (e.g., histamine) causing the characteristic immediate response (vasodilation and increased vascular permeability, with bronchoconstriction). Over the next few hours (up to 24 hours), these activated cells will also synthesize and release additional mediators (prostaglandins, leukotrienes, platelet activating factor, and cytokines; see text). (Figure reprinted with permission from Abbas et al. (2007), *Cellular and Molecular Immunology,* 6th ed.)

Antibody Bound to Cell Surfaces or Fixed Tissue Antigens

Antibodies bound to either intrinsic or extrinsic tissue antigens can induce tissue injury by promoting complement activation, inducing opsonization, or by interacting with important cell surface molecules (Figure II.2.3.18):

- **Complement:** Recall that complement may induce injury either by direct cytolysis via the C5b–9 membrane-attack complex (MAC) punching holes in a cell's plasma membrane, or by opsonization (via the C3b fragment), enhancing phagocytosis by macrophages and neutrophils. In addition to direct cell killing, local activation of the complement cascade will result in the generation of complement fragments such as C3a and C5a that alter vascular tone and permeability (Figure II.2.3.4 and Chapter II.2.4).
- **Opsonization:** Antibody on the surface of cells will promote phagocytosis by cells bearing Fc receptors.
  - On *circulating blood cells*, bound complement may directly mediate cell lysis; in addition, bound antibody and opsonizing complement fragments induce efficient uptake and destruction by cells of the splenic and hepatic mononuclear phagocyte system.
  - Antibody binding in conjunction with C3b opsonization can also lead *indirectly* to tissue injury. Large, non-phagocytosable cells or tissue may promote "frustrated phagocytosis" by neutrophils or macrophages; the attempted intracellular lysis results instead in the extracellular release of proteases and toxic oxygen metabolites (Figures II.2.3.18B and II.2.3.19A).
- **Antibody binding to cell receptors:** Binding of antibodies to certain receptors can induce pathology even without causing tissue injury. For example, in the case of *Grave's disease*, antibodies bind to the thyroid stimulating hormone (TSH) receptor on thyroid epithelial cells and mimic authentic TSH ligand interaction; the result is autonomous stimulation of

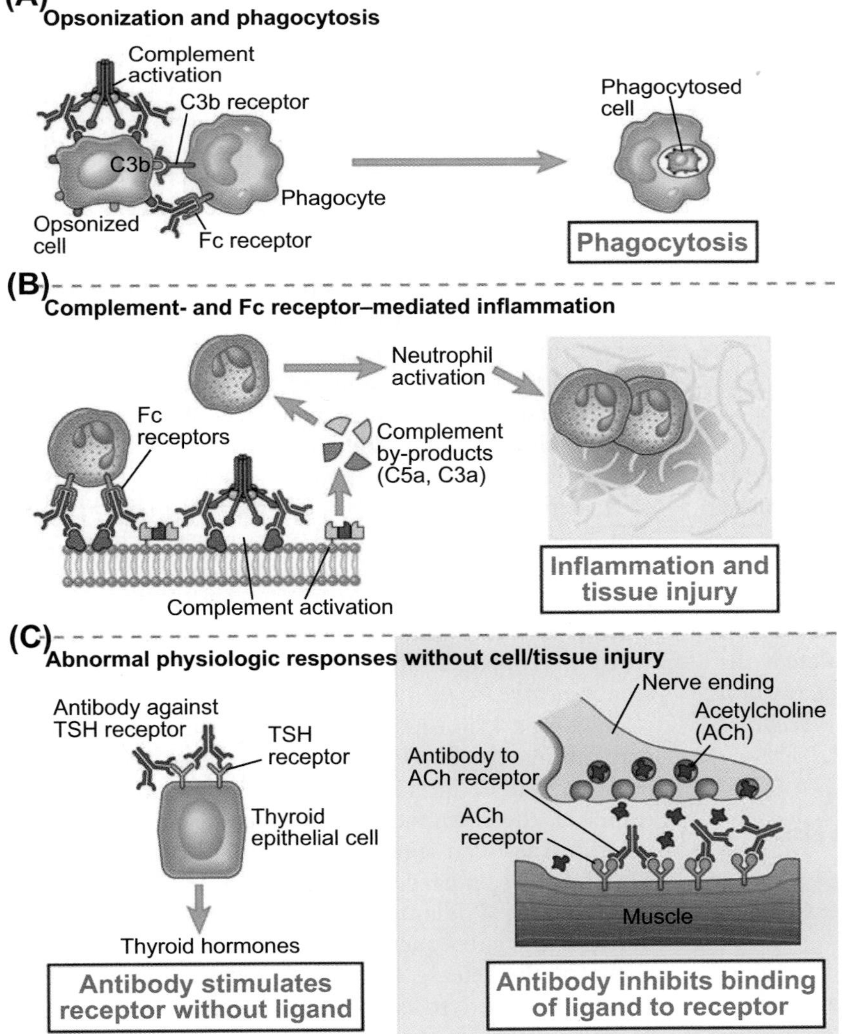

FIGURE II.2.3.18 Effector mechanisms in antibody-mediated disease. (A) Antibodies, with or without complement activation, will opsonize cells leading to phagocytosis and destruction. (B) Antibodies, and secondarily-generated complement fragments bound to large non-phagocytosable cells or tissues will recruit inflammatory cells such as neutrophils and macrophages. If these inflammatory cells cannot completely ingest the target, *frustrated phagocytosis* will result in the release of lysosomal contents and reactive oxygen intermediates into the tissues with subsequent cellular and matrix damage. (C) Antibodies can also elicit pathology without causing tissue damage. In the panel on the left, antibodies to the thyroid stimulating hormone (TSH) receptor can mimic authentic TSH to cause hyperstimulation of the thyroid (*Grave's disease*). In the panel on the right, antibodies to the acetylcholine (ACh) receptor at the neuromuscular junction can block normal ACh stimulation of muscle contraction leading to weakness (*myasthenia gravis*). (Figure reprinted with permission from Abbas et al. (2007), *Cellular and Molecular Immunology,* 6th ed.)

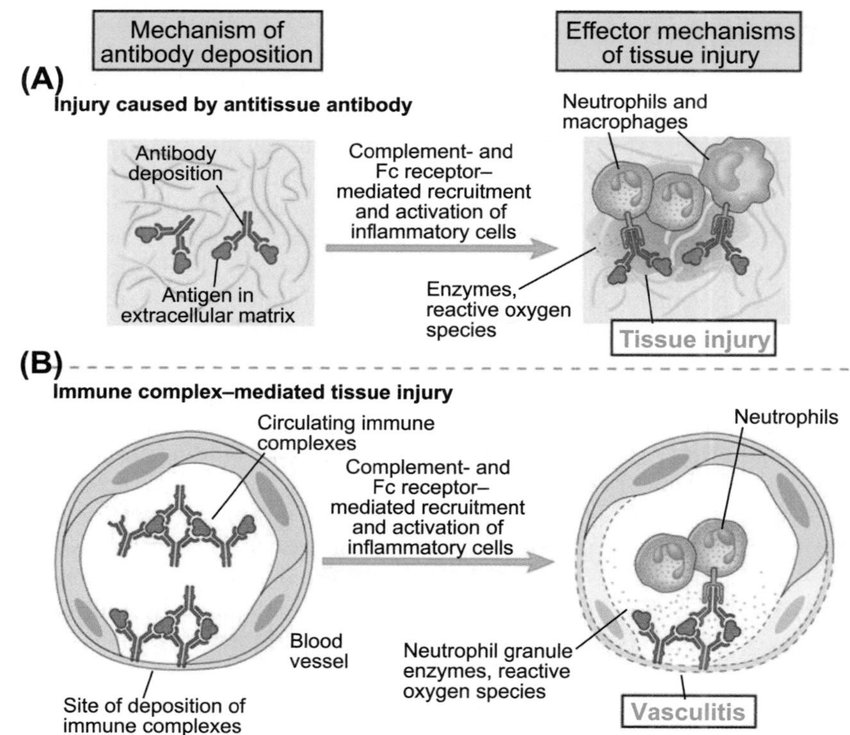

FIGURE II.2.3.19 Antibody-mediated pathology. (A) Direct binding of antibodies to tissue antigens will cause tissue injury by recruiting inflammatory cells and activating complement. (B) Circulating antigen–antibody complexes (also called *immune complexes*) can deposit in vessels and tissues also leading to inflammatory cell recruitment and complement activation. The classic vascular lesion in the setting of immune complex deposition is called *vasculitis*; such vessel inflammation will typically lead to thrombosis and tissue ischemia. (Figure reprinted with permission from Abbas et al. (2007), *Cellular and Molecular Immunology,* 6th ed.)

the gland with hyperthyroidism. Alternatively, antibodies that bind to the acetylcholine receptor at the nerve–muscle synapse can block binding of acetylcholine and result in the weakness seen in the disease *myasthenia gravis* (Figure II.2.3.18C).

Immune Complex (IC)-Mediated Injury

In many circumstances, circulating antigen and antibody combine to form insoluble aggregates called immune complexes (IC). These are usually efficiently cleared by macrophages in the spleen and liver, but can occasionally deposit in certain vascular beds. Once IC are deposited, the mechanism of injury is basically the same, regardless of where or for what reason IC have accumulated; the major sources of pathology are *complement activation* (see above) and *interactions with cells bearing Fc receptors* (e.g., neutrophils and macrophages) (Figure II.2.3.19B).

PATHOGENESIS OF T CELL-MEDIATED DISEASE

T cell-mediated responses are of two general types (Figure II.2.3.20):

T cell-mediated cytolysis: In CTL-mediated reactions, $CD8^+$ T lymphocytes recognize specific antigen in association with class I MHC and induce direct cytolysis. CTL-mediated cytolysis is *highly specific*, without significant "innocent bystander" injury.

Delayed-type hypersensitivity (DTH): In the case of cell-mediated immunity, $CD4^+$ helper T cells recognize specific antigen in the context of class II MHC, and respond by producing a variety of soluble *antigen-nonspecific* cytokines. These soluble mediators induce further T lymphocyte recruitment and proliferation, and attract and activate *antigen non-specific* macrophages and other inflammatory cells; at the site of a $CD4^+$ T cell-mediated response, the vast majority (greater than 90%) of newly recruited cells is not specific for the original inciting antigen. Although tightly regulated, the relatively non-specific effector components of cell-mediated immunity (cytokines and activated macrophages) are largely responsible for the injury seen in *delayed-type hypersensitivity* (DTH).

In comparison to CTL, cytokine-mediated immunity may ultimately develop an antigen non-specific component; that is, after the initial antigen-specific T cell response, the recruited antigen non-specific T cells and macrophages can cause significant bystander injury. Macrophages, in particular, are an important component of the recruited inflammatory cells in DTH, and mediate much of the subsequent immune effector responses. By virtue of the release of reactive oxygen intermediates, prostaglandins, lysosomal enzymes, and cytokines such as TNF (which, in turn, have potent effects, e.g., on the

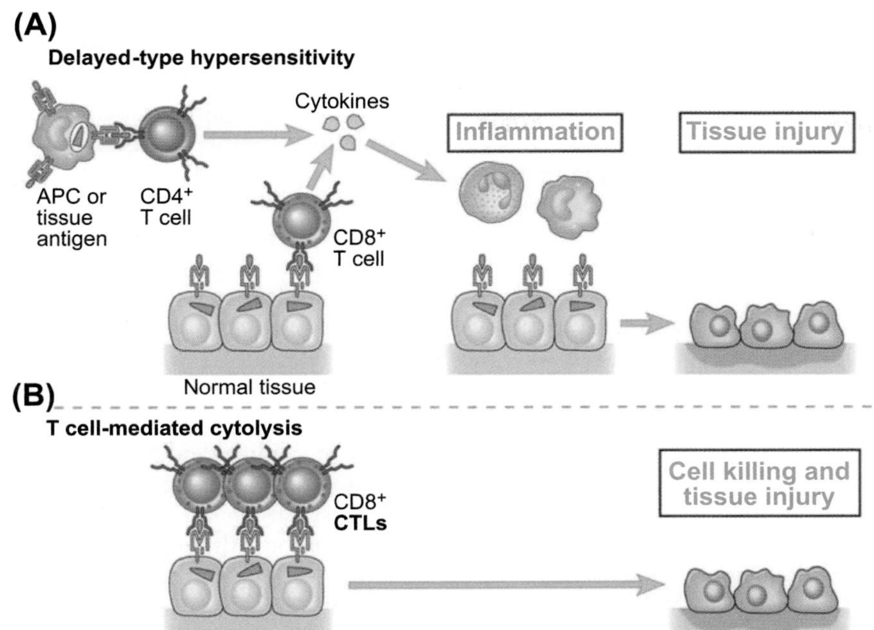

FIGURE II.2.3.20 Mechanisms of T cell-mediated disease. (A) In *delayed-type hypersensitivity* responses, T cells (typically $CD4^+$ helper T cells) respond to tissue or cellular antigens by secreting cytokines that stimulate inflammation, and ultimately promote tissue injury (APC, antigen-presenting cell). (B) In some diseases, $CD8^+$ cytotoxic T cells directly kill tissue cells. (Figure reprinted with permission from Abbas et al. (2007), *Cellular and Molecular Immunology,* 6th ed.)

synthetic function of fibroblasts, lymphocytes, and endothelium), activated macrophages can potentially wreak significant havoc.

Recall also that *granulomas* (nodules of granulomatous inflammation) are the characteristic response of the immune system to large, persistent, and/or non-degradable antigens. Besides antigenic stimulation, direct macrophage activation can also occur by binding to denatured or modified host proteins that have adsorbed on the surfaces of the foreign materials via the receptors used for innate immunity (Tang and Eaton, 1999; Anderson et al., 2008). Thus, foreign objects (such as implanted devices) can frequently elicit a granulomatous response that becomes a significant impediment to local tissue integrity and normal function. With persistent stimulation, for example, chronic macrophage activation leads to cytokines that culminate in a surrounding fibrosis. Injury associated with granulomas is also due to displacement, compression, and necrosis of adjacent healthy tissue. Granulomas associated with a variety of "autoimmune disorders" such as *temporal arteritis*, *Crohn's disease*, and *Wegener's granulomatosis* presumably reflect diseases with persistent antigen stimulation or a heightened DTH response to specific self-antigens.

IMMUNE RESPONSES TO TRANSPLANTED TISSUES, BIOMATERIALS, AND SYNTHETIC SUBSTANCES

When foreign cells or organs are transplanted into a new host, the histocompatibility proteins on the cell surfaces of the graft are recognized by the components of adaptive immunity as being non-self. Note that, except for minor genetic polymorphisms, most of the structural proteins and other molecular components in a graft are nearly identical to those that the host will also express (e.g., the contractile proteins in heart muscle, the collagenous extracellular matrix, the usual housekeeping proteins, etc.). The MHC molecules, however, are distinctly different between most humans (except identical twins) and will elicit helper and cytotoxic T cell activation, as well as B cell antibody production. Clearly, once these pathways have been activated, the usual physiologic effector mechanisms (direct cell killing, complement activation, phagocytosis, cytokine elaboration, etc.) will be brought to bear on the graft, and can effect its destruction. Besides direct killing of parenchymal cells (e.g., myocytes in a heart, hepatocytes in the liver, tubular epithelial cells in the kidney), the adaptive immune system and its various effectors also attack the graft vasculature, and can cause graft failure by vessel thrombosis and ischemia. Even in the absence of infarction or direct killing, cytokines and other inflammatory mediators such as prostaglandins and reactive oxygen species can cause significant graft dysfunction (i.e., a loss of myocardial contractility) that may be as clinically deleterious as frank injury.

Again, although components of innate immunity are recruited and activated in the process of graft damage, the initial recognition step and the driving force for *transplant rejection* is via the cells of adaptive immunity (you are also referred to the basic immunology texts for excellent overviews of the rejection phenomenon: Coico et al., 2003; Abbas et al., 2007; Murphy et al., 2007). To prevent or reverse such rejection requires a substantial armamentarium of immunosuppressive agents (e.g., cytotoxic drugs or agents such as *cyclosporine*), which put the recipient at risk of infections and malignancies.

The point is emphasized here because in the *immunologic* sense, synthetic devices are *not* rejected; moreover, if *non-cellular* biomaterials are derived from the same species (or sufficiently-related species so that there are no major antigenic differences in, e.g., collagen proteins), such engineered devices are also *not rejected*. Such devices do not elicit specific (adaptive) immune responses, and therefore will not have antibodies or lymphocytes that recognize the materials and cannot therefore drive the overall response. As a corollary statement, simply finding inflammatory cells (and even T cells and antibodies) does not in any way prove that the response is "rejection;" such elements will accrue at any site of injury in a non-specific way. For example, recall that some 90% of T cells in a DTH response are not antigen-specific, but are non-specifically recruited to the site of injury.

This is much more than a semantic point, in that synthetic or biomaterial device functions or longevity are not likely to benefit from specific immunosuppression. Of course, if a device incorporates viable cells in its manufacture (e.g., endothelial cells lining a vascular conduit), those *cells* will express MHC proteins, and will elicit adaptive immune responses that materially contribute to device failure. In that instance, it will either be necessary to engineer such devices using cells derived from the individual who will eventually receive the implant (making them an *autograft*), or to rely on long-term immunosuppression as is routinely required for organ transplants.

It should also be emphasized that although synthetic and biomaterials are *not* rejected in the immunologic sense, components of the immune system (particularly innate immunity) *can* contribute to device dysfunction and failure. In particular, and as described above, non-specific activation of macrophages and complement will lead to local tissue damage via proteolysis, accumulation of other inflammatory cells, and/or cytokine elaboration; in most cases, with an ongoing, persistent innate response to a device that cannot be eliminated, fibrous scar tissue will also occur in and around the implant (Anderson et al., 2008).

Thus, synthetic materials and biomaterials can have failure modes that are attributable to activation of the immune system (particularly innate immunity). An "inert" silastic-clad breast prosthesis, for example, can accumulate dense scar tissue around it (secondary to persistent macrophage activation) that is not aesthetically ideal. Similarly a titanium hip prosthesis can induce ongoing macrophage activation that in the bone will lead to cytokine production that ultimately drives bone resorption and prosthesis loosening. In some individuals, idiosyncratic IL-4 or IL-10 cytokine responses to stent grafts may underlie local aneurysm formation and "stent creep" associated with intravascular stenting (Levisay et al., 2008). Heavy metal hypersensitivities (e.g., to nickel) occur because the metal complexes with host proteins, creating a modified version of "self;" when these modified proteins are then processed and presented as part of normal immune surveillance, Th-mediated DTH responses can occur with associated inflammation and swelling. Notably, heavy metal hypersensitivities are localized to the immediate vicinity of the implant. While administration of steroids in any of these settings may have some beneficial effect by limiting macrophage activation, they may also induce complications since steroids (among other side-effects) also inhibit healing and increase susceptibility to infections.

Finally, there is a body of controversial work suggesting that denatured host proteins non-specifically adhering on device surfaces or complexing with fragmented and circulating implant materials can potentially break host self-tolerance for these molecules. In this hypothetical model, novel previously occult antigenic determinants become exposed (see Figure II.2.3.10). These can then elicit antibodies that can bind to damaged (or even normal) versions of the same or cross-reactive proteins, and cause local tissue damage or subsequent immune complex formation and deposition. Although the experiments that purport to support this are problematic, and the evidence for such a pathologic pathway is weak, such models are frequently invoked to explain new onset autoimmune disease following device implantation (or following infection, vaccine, or other external trigger) (Molina and Shoenfeld, 2005; Wolfram et al., 2008).

In deciding whether a particular response to biomaterials is driven by immune mechanisms, it may be well to consider applying an immunologic variant of Koch's Postulates. In the late 19th century, a microbiologist named Robert Koch proposed a set of objective criteria to be used in demonstrating that a particular organism was responsible for a particular disease (actually used the first time in proving the pathogenic basis of anthrax!). Koch's Postulates basically state that the presumed microbe should be routinely recovered from the pathologic lesions of the human host; should be culturable in a pure form and cause the same disease when injected into a new host; and should be recoverable again in a pure form from the secondary host.

Thus, Koch's *Immunologic* Postulates would provide a more rigorous basis for implicating host immunity in biomaterial failures or subsequent host pathology:

- antigen-specific elements (antibodies or cells) should be directly associated with the pathology of interest
- antibodies or cells from affected hosts should cause the same pathology by transfer into an appropriate secondary host
- adoptively transferred antibodies or cells should be recoverable (or at least demonstrable) in the pathology of the secondary host.

BIBLIOGRAPHY

Abbas, A., Lichtman, A., & Pober, J. (2000). *Cellular and Molecular Immunology* (4th ed.). Philadelphia, PA: WB Saunders.

Abbas, A., Lichtman, A., & Pillai, S. (2007). *Cellular and Molecular Immunology* (6th ed.). Philadelphia, PA: Saunders Elsevier.

Akira, S., Uematsu, S., & Takeuchi, O. (2006). Pathogen recognition and innate immunity. *Cell*, **124**, 783–801.
Anderson, J., Rodriguez, A., & Chang, D. (2008). Foreign body reaction to biomaterials. *Semin. Immunol.*, **20**, 86–100.
Carroll, M. (2004). The complement system in regulation of adaptive immunity. *Nat. Immunol.*, **5**, 981–986.
Coico, R., Sunshine, G., & Benjamini, E. (2003). *Immunology: A Short Course* (5th ed.). Wilmington, DE: Wiley-Liss.
Henry, G., & Garner, W. (2003). Inflammatory mediators in wound healing. *Surg. Clin. North Am.*, **83**, 483–507.
Hotchkiss, R., & Karl, I. (2003). The pathophysiology and treatment of sepsis. *N. Eng. J. Med.*, **348**, 138–150.
Kay, A. (2001a). Allergy and allergic diseases. *N. Eng. J. Med.*, **344**, 30–37.
Kay, A. (2001b). Allergy and allergic diseases. *N. Eng. J. Med.*, **344**, 109–113.
Klein, J., & Sato, A. (2000a). The HLA system. *N. Eng. J. Med.*, **343**, 782–786.
Klein, J., & Sato, A. (2000b). The HLA system. *N. Eng. J. Med.*, **343**, 702–709.
Kono, H., & Rock, K. (2008). How dying cells alert the immune system to danger. *Nat. Rev. Immunol.*, **8**, 279–289.
Korn, T., Oukka, M., Kuchroo, V., & Bettelli, E. (2007). Th17 cells: Effector T cells with inflammatory properties. *Semin. Immunol.*, **19**, 362–371.
Korn, T., Bettelli, E., Oukka, M., & Kuchroo, V. (2009). IL-17 and Th17 Cells. *Ann. Rev. Immunol.*, **27**, 485–517.
Kuhns, M., Davis, M., & Garcia, K. (2006). Deconstructing the form and function of the TCR/CD3 complex. *Immunity*, **24**, 133–139.
Kumar, V., Abbas, A., & Fausto, N. (2005). *Robbins and Cotran Pathologic Basis of Disease* (7th ed.). Philadelphia, PA: Saunders Elsevier.
Kumar, V., Abbas, A., Fausto, N., & Mitchell, R. (2007). *Robbins Basic Pathology* (8th ed.). Philadelphia, PA: Saunders Elsevier.
Levisay, J., Roth, R., & Schatz, R. (2008). Coronary artery aneurysm formation after drug-eluting stent implantation. *Cardiovasc. Revasc. Med.*, **9**, 284–287.
Litman, G., Cannon, J., & Dishaw, L. (2005). Deconstructing immune phylogeny: New perspectives. *Nat. Rev. Immunol.*, **5**, 866–879.
Martin, P., & Leibovich, S. (2005). Inflammatory cells during wound repair: The good, the bad and the ugly. *Trends Cell Biol.*, **15**, 599–607.
Medzhitov, R. (2007). Recognition of microorganisms and activation of the immune response. *Nature*, **449**, 819–826.
Medzhitov, R., & Janeway, C. (2000). Innate immunity. *N. Eng. J. Med.*, **343**, 338–344.
Meylan, E., Tschopp, J., & Karin, M. (2006). Intracellular pattern recognition receptors in the host response. *Nature*, **442**, 39–44.
Molina, V., & Shoenfeld, Y. (2005). Infection, vaccines and other environmental triggers of autoimmunity. *Autoimmunity*, **38**, 235–245.
Moss, R., Moll, T., El-Kalay, M., Kohne, C., Hoo, W. S., et al. (2004). Th1/Th2 cells in inflammatory disease states: Therapeutic implications. *Expert Opin. Biol. Ther.*, **4**, 1887–1996.
Murphy, K., Travers, P., & Walport, M. (2007). *Janeway's Immunobiology* (7th ed.). London, UK: Garland Science.
Palm, N., & Medzhitov, R. (2009). Pattern recognition receptors and control of adaptive immunity. *Immunol. Rev.*, **227**, 221–233.
Peter, H., & Warnatz, K. (2005). Molecules involved in T-B co-stimulation and B cell homeostasis: Possible targets for an immunological intervention in autoimmunity. *Expert Opin. Biol. Ther.*, **5**, S61–S71.
Pulendran, B. (2004). Modulating TH1/TH2 responses with microbes, dendritic cells, and pathogen recognition receptors. *Immunol. Res.*, **29**, 187–196.
Roozendaal, R., & Carroll, M. (2006). Emerging patterns in complement-mediated pathogen recognition. *Cell*, **125**, 29–32.
Segal, A. (2005). How neutrophils kill microbes. *Ann. Rev. Immunol.*, **23**, 197–223.
Tang, L., & Eaton, J. (1999). Natural responses to unnatural materials: A molecular mechanism for foreign body reactions. *Mol. Med.*, **5**, 351–358.
Walport, M. (2001a). Complement. First of two parts. *N. Eng. J. Med.*, **344**, 1058–1066.
Walport, M. (2001b). Complement. Second of two parts. *N. Eng. J. Med.*, **344**, 1140–1144.
Wolfram, D., Oberreiter, B., Mayerl, C., Soelder, E., Ulmer, H., et al. (2008). Altered systemic serologic parameters in patients with silicone mammary implants. *Immunol. Lett.*, **118**, 96–100.

CHAPTER II.2.4 THE COMPLEMENT SYSTEM

Richard J. Johnson
Baxter Healthcare Corporation, Round Lake, IL, USA

As discussed in the previous chapter, the immune system acts to protect each of us from the constant exposure to pathogenic agents such as bacteria, fungi, viruses, and cancerous cells that pose a threat to our lives. The sheer multitude of structures that the immune system must recognize, differentiate from "self," and mount an effective response against, has driven the evolution of this system into a complex network of proteins, cells, and distinct organs. An immune response to any foreign element involves all of these components, acting in concert, to defend the host from intrusion. Historically, the immune system has been viewed from two perspectives: cellular or humoral. This is a somewhat subjective distinction, since most humoral components (such as antibodies, complement components, and cytokines) are made by the various cells of the immune system and, in turn, often function to regulate the activity of the multiple types of immune cells. The focus of this chapter will be on the basic biochemistry and pathobiology of the complement system, a critical element of the innate immune response, and its relevance to biomaterials research and development.

INTRODUCTION

Complement is a term devised by Paul Ehrlich to refer to plasma components that were known to be necessary for antibody-mediated bactericidal activity (Ross, 1986). We now know that the complement system is composed of more than 30 distinct plasma- and membrane-bound proteins involving three separate pathways: classical; alternative; and the lectin pathway. The complement system directly and indirectly contributes to both innate inflammatory reactions, as well as cellular (i.e., adaptive) immune responses. This array of effector functions is due

TABLE II.2.4.1 Complement Activities

Activity	Complement Protein
Identification/opsonization of pathogens	C3, C4
Recruitment/activation of inflammatory cells	C3a, C5a
Lysis of pathogens/ cytotoxicity	C5b-9 (MAC)
Clearing immune complexes and apoptotic cells	C1q, C3b, C4b
Augment cellular immune responses (T and B cells)	C3, C4, C3a, C5a, DAF(CD55)

TABLE II.2.4.2 Proteins of the Classical Pathway of Complement

Protein	Molecular Weight	Subunits	Plasma Concentration (ug/ml)
C1q	410,000	6A, 6B, 6C	70
C1r	85,000	1	35
C1s	85,000	1	35
C2	102,000	1	25
C4	200,000	α, β, γ	600
C1-Inh	104,000	1	200
C4bp	570,000	8	230

to the activity of a number of complement components and their receptors on various cells. These activities are summarized in Table II.2.4.1, along with the responsible complement protein(s). One of the principal functions of complement is to serve as a primative self–non-self discriminatory defense system. This is accomplished by coating a foreign material with complement fragments and recruiting phagocytic cells that attempt to destroy and digest the "intruder." Although the system evolved to protect the host from the invasion of adventitious pathogens, the nonspecific and spontaneous nature of the alternative pathway permits activation by various biomaterial surfaces. Because complement activation can follow three distinct but interacting pathways, the various ways of activating the cascade will be outlined separately below.

One of the principal functions of complement is to serve as a primative self–non-self discriminatory defense system. This is accomplished by coating a foreign material with complement fragments and recruiting phagocytic cells that attempt to destroy and digest the "intruder." Although the system evolved to protect the host from the invasion of adventitious pathogens, the nonspecific and spontaneous nature of the alternative pathway permits activation by various biomaterial surfaces.

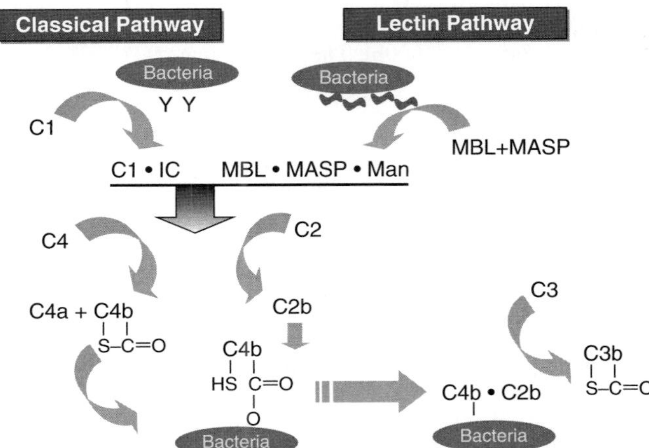

FIGURE II.2.4.1 Complement activation by the classical pathway (CP). Upon binding to the Fc region of an immune complex, C1 is activated and cleaves C4, exposing its thioester, which permits covalent attachment of C4b to the activating surface. C2 is cleaved, producing C2b, which binds to C4b to form the CP C3 convertase. C2b is a serine protease that specifically acts on C3 to generate C3b and C3a. The lectin pathway is also illustrated. MBL recognizes certain sugar residues (mannose, N-Acetylglucosamine) on the surface of an activator (bacteria). MASP-1 appears to activate MASP-2, which then cleaves both C4 and C2 of the CP, generating the CP C3 convertase.

CLASSICAL PATHWAY

The classical pathway (CP) is activated primarily by immune complexes (ICs) composed of antigen and specific antibody, although other proteins, such as C-Reactive Protein, Serum Amyloid Protein, and amyloid fibrils, as well as apoptotic bodies can also activate the CP (Bohlson et al., 2007; Cooper, 1985). The proteins of this pathway are C1, C2, C4, C1 inhibitor (C1-Inh), and C4 binding protein (C4bp). Some of their basic characteristics are summarized in Table II.2.4.2.

Complement activation by the CP is illustrated in Figure II.2.4.1. This system is an example of an enzyme cascade in which each step in the series, from initiation to the final product, involves enzymatic reactions (in this case, proteolytic cleavage reactions) that result in some degree of amplification. Research with knock-out mice (mice deficient in C1q, C2, C4 or IgG) has shown that the CP is in a state of continuous low level activation, essentially primed to react vigorously in the presence of an IC (Manderson et al., 2001). When an IC forms, the cascade is initiated when C1 binds to the Fc portion of an antigen–antibody complex. The C1 protein is composed of three different types of subunits called C1q, C1r, and C1s (Figure II.2.4.2). One end of C1q binds to an IC formed between an antigen and one molecule of (pentameric) immunoglobulin (Ig) M or several closely spaced IgG molecules. This interaction is believed to produce a conformational change in the C1q that results in activation (i.e., autocatalytic proteolysis) of the two C1r and then the two C1s subunits bound to the other end of the C1q protein. Both C1r and C1s are zymogen serine proteases that are bound to the C1q in a calcium-dependent manner that is inhibited by calcium chelators such as citrate or EDTA. The proteolysis of C1s completes the

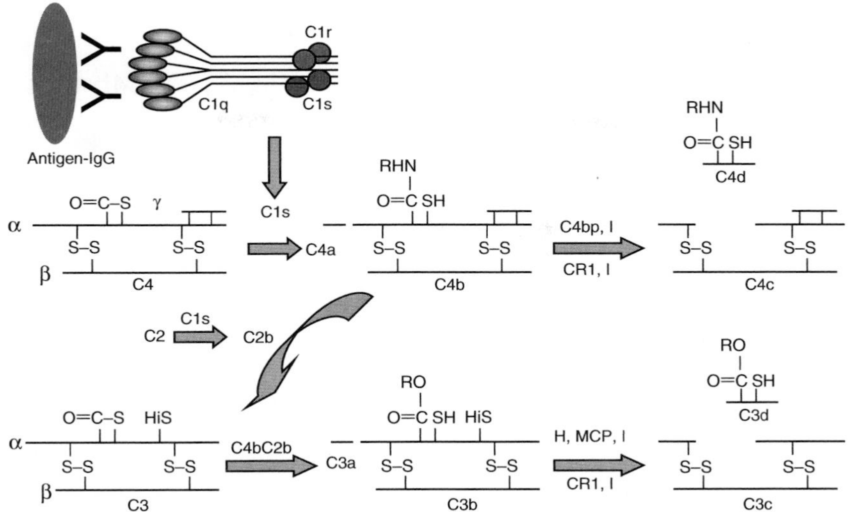

FIGURE II.2.4.2 Schematic illustration of C4 and C3 protein structures. O=C–S represents the reactive thioester bond that permits covalent attachment to surface nucleophiles (hydroxyl or amino groups). The pattern of proteolytic degradation and the resulting fragments are also shown. Although factor I is the relevant *in vivo* protease, some of these same fragments can be generated with trypsin, plasmin, and thrombin.

activation of C1, which then proceeds to act on the next proteins in the cascade, C4 and C2.

C4 is composed of three separate chains, α, β, and γ (Figure II.2.4.2), bound together by disulfide bonds. Activated C1s cleaves C4 near the amino terminus of the αchain, yielding a 77-amino acid polypeptide called C4a and a much larger (190,000 Da) C4b fragment. The C4 protein contains a unique structural element called a thioester (Figure II.2.4.2), that has only been detected in two other plasma proteins, α 2-macroglobulin and C3. Upon cleavage of C4, the buried thioester becomes exposed and available to react with a surface containing amino or hydroxyl moieties. About 5% of the C4b molecules produced react through the thioester, and become covalently attached to the surface. This represents the first amplification step in the pathway, since each molecule of C1 produces a number of surface-bound C4b sites.

The C4b protein, attached to the surface, acts as a receptor for C2. After binding to C4b, C2 becomes a substrate for C1s. Cleavage of C2 yields two fragments; a smaller C2a portion diffuses into the plasma, while the larger C2b remains bound to the C4b. The C2b protein is another serine protease that, in association with C4b, represents the classical pathway C3/C5 convertase.

As the name implies, the function of the C4b•C2b complex is to bind and cleave C3. The C3 protein sits at the juncture of the classical and alternative pathways, and represents one of the critical control points. Cleavage of C3 by C2b yields a 9000 Da C3a fragment and a 175,000 Dalton C3b fragment that is very similar to C4b in both structure and function. Cleavage of C3 produces a conformational change in the C3b protein that results in exposure of its thioester group (Figure II.2.4.2). Condensation with surface nucleophiles (such as carbohydrates) results in covalent attachment of 10–15% of the C3b to the surface of the activator. This is the second amplification step in the sequence, since as many as 200 molecules of C3b can become attached to the surface surrounding every C4b•C2b complex. Eventually one of the C3b molecules reacts with a site on the C4b protein, creating a C3b–C4b•C2b complex that acts as a C5 convertase (Figure II.2.4.3). Recent investigations have demonstrated that the C4b•C2b complex can also cleave C5 at very low rates, but that the addition of C3b to the complex increases the affinity for (and thus the turnover of) C5 by approximately 1000-fold (Rawal et al., 2008).

In contrast to C3, which can be cleaved in the fluid phase (see later discussion), proteolytic activation of C5 occurs only after it is bound to the C3b portion of the C5 convertase on the surface of an activator (e.g., the immune complex). Like C3, C5 is also cleaved by C2b to produce fragments designated C5a (16,000 Da) and C5b (170,000 Da). The C5b molecule combines with the proteins of the terminal components to form the membrane attack complex described later. C5a is a potent inflammatory mediator, and is responsible for many of the adverse reactions normally attributed to complement activation in various clinical settings.

LECTIN PATHWAY

In the 1990s, investigators working with a protein called mannose binding lectin (MBL) discovered a third pathway that leads to complement activation (Matsushita, 1996). This scheme is called the lectin pathway, and is composed of lectins like MBL and three MBL-associated serine proteases or MASPs (Table II.2.4.3). MBL is an acute phase protein, so its concentration in plasma increases substantially during an infection. MBL binds to terminal mannose, N-acetyl-glucosamine, and N-acetyl-mannosamine residues in complex carbohydrate structures. MBL has long been recognized as an opsonin, i.e., a protein that

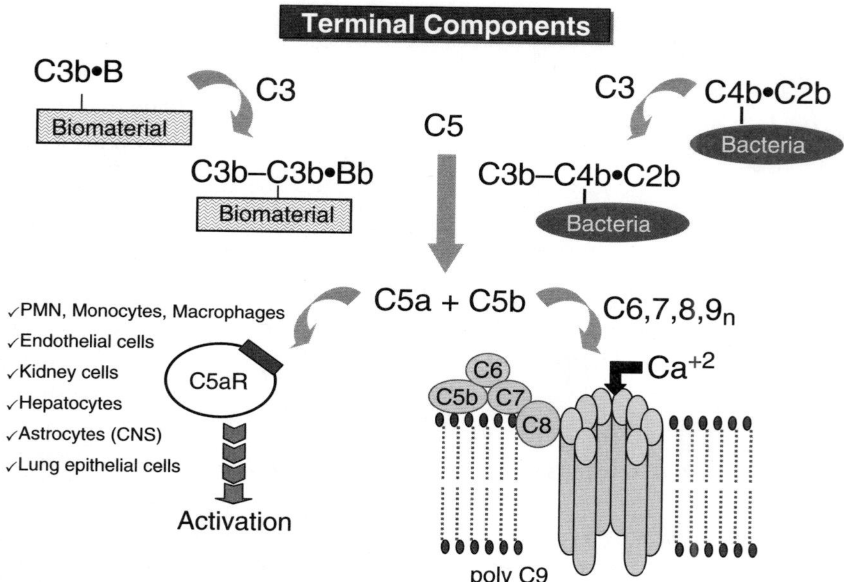

FIGURE II.2.4.3 Conversion of C5 produces C5a and C5b, the latter leads to formation of the membrane attack complex (MAC). C5a binds to receptors on a variety of cells and results in numerous activities. C5b, formed by the CP, lectin or the AP, binds C6 and C7 to form a complex that associates with the plasma membrane. This C5b67 multimer then binds C8, which results in the formation of a small hole in the lipid bilayer that allows small molecules to pass through. Association of multiple C9 proteins enlarges the pore, leading to loss of membrane integrity and cell death.

TABLE II.2.4.3 Proteins of the Lectin Pathway of Complement

Protein	Molecular Weight	Subunits	Plasma Concentration (ug/ml)
MBL	270–650,000	18	1–3
MASP-1	93,000	2 (H,L)	6
MASP-2	76,000	2 (H,L)	0.4
MASP-3	110,000	2 (H,L)	

facilitates phagocytosis of bacteria. Low concentrations of MBL in children are associated with recurrent bacterial infections. MBL is similar in structure to C1q, having an amino terminal domain with a collagen-like structure that binds the MASP proteins, followed by a globular carbohydrate recognition domain (CDR) that binds to sugar residues. There are three MASP proteins, called MASP-1, MASP-2, and MASP-3, which are very similar in structure to C1r and C1s (Wong et al., 1999). MASP-2 can cleave both C4 and C2, forming a CP C3 convertase (Figure II.2.4.1), while MASP-1 can target C2 and MASP-2 to help accelerate convertase formation (Takahashi et al., 2008).

ALTERNATIVE PATHWAY

The alternative pathway (AP) was originally discovered in the early 1950s by Pillemer et al. (1954). Pillemer's group studied the ability of a yeast cell wall preparation, called zymosan, to consume C3 without affecting the amount of C1, C2 or C4. A new protein, called properdin, was isolated and implicated in initiating C3 activation independent of the CP. This new scheme was called the properdin pathway. However, this work fell into disrepute when it was realized that plasma contains natural antibodies against zymosan, which implied CP involvement in Pillemer's experiments. The pathway was rediscovered in the late 1960s with the study of complement activation by bacterial lipopolysaccharide, and with the discovery of a C4-deficient guinea pig. The 1970s witnessed the isolation and characterization of each of the proteins of this pathway until it was possible to completely reconstruct the entire AP by recombining each of the purified proteins (Schreiber et al., 1978). Most biomaterials activate complement through the alternative pathway (AP), although there is evidence that the classical pathway (CP) can also contribute (presumably subsequent to IgG binding (Andersson et al., 2005)).

Most biomaterials activate complement through the alternative pathway (AP), although there is evidence that the classical pathway (CP) can also contribute (presumably subsequent to IgG binding (Andersson et al., 2005)).

The proteins of this pathway are described in Table II.2.4.4. Their actions can be conceptually divided into three phases: initiation; amplification; and regulation (Figures II.2.4.4 and II.2.4.5). Initiation is a spontaneous process that is responsible for the nonselective nature of complement. During this stage, a small portion of the C3 molecules in plasma undergo a conformational change that results in hydrolysis of the thioester group,

TABLE II.2.4.4 Proteins of the Alternative Pathway of Complement

Protein	Molecular Weight	Subunits	Plasma Concentration (ug/ml)
C3	185,000	α, β	1300
B	93,000	1	210
D	24,000	1	1
H	150,000	1	500
I	88,000	α, β	34
P	106–212,000	2–4	20

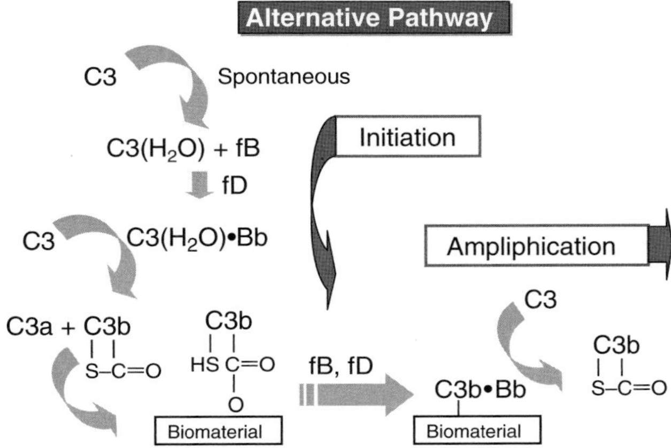

FIGURE II.2.4.4 Complement activation by the alternative pathway (AP). The spontaneous conversion of C3 to C3(H_2O) permits the continuous production of C3b from C3, a process called C3 tick-over. In the presence of an activating surface, the C3b is covalently bound and becomes the focal point for subsequent interactions. The bound C3b is recognized by factor B, which is then cleaved by factor D to produce a surface-bound C3 convertase (C3b•Bb). This results in amplification of the original signal to produce more convertase.

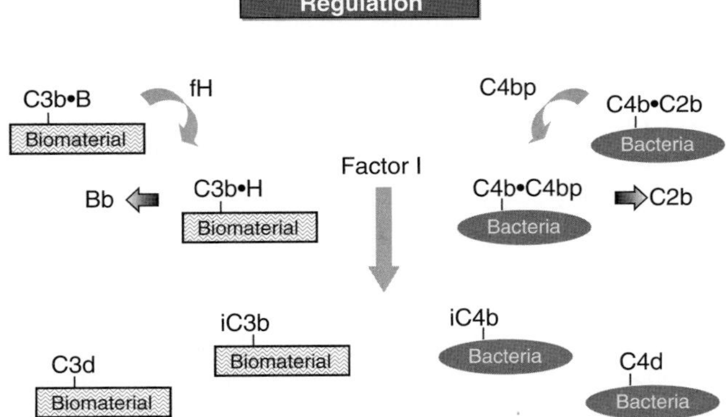

FIGURE II.2.4.5 Control of complement activation by factors H, I, and C4 binding protein. The extent to which complement activation occurs on different surfaces is dependent on the ability of fH or C4BP to recognize C3b or C4b on the surface. Degradation by factor I results in irreversible inactivation and the production of C3 and C4 fragments recognized by various complement receptors on white blood cell (WBC).

producing an activated form of C3 called C3(H_2O) ("C3-water"), that will bind to factor B. The C3(H_2O)•B complex is a substrate for factor D (another serine protease), which cleaves the B protein to form a solution phase alternative pathway C3 convertase: C3(H_2O)•Bb. Analogous to C2b in the CP, Bb is a serine protease that (in association with C3(H_2O)) will cleave more C3 to form C3b. Under normal physiological conditions, most of the C3b produced is hydrolyzed and inactivated, a process that has been termed "C3 tickover." C3 tickover is a continuous process that ensures a constant supply of reactive C3b molecules to deposit on foreign surfaces, such as cellulosic- or nylon-based biomaterials. Recognition of the C3b by factor B, cleavage by factor D, and generation of more C3 convertase leads to the amplification phase (Figure II.2.4.4). During this stage, many more C3b molecules are produced, bind to the surface, and in turn lead to additional C3b•Bb sites. Eventually, a C3b molecule attaches to one of the C3 convertase sites by direct attachment to the C3b protein component of the enzyme. This C3b–C3b•Bb complex is the alternative pathway C5 convertase and, in a manner reminiscent of the CP C5 convertase, converts C5 to C5b and C5a (Figure II.2.4.3).

Recent work with purified proteins and techniques to measure direct interactions with polymer surfaces has revealed an additional potential mechanism for alternative pathway activation (Andersson et al., 2002). Both C3b and C3 will adsorb to (not react with) polystyrene. A portion (about 10%) of the bound C3 or C3b binds factor B. This complex is recognized by factor D, which then catalyzes the formation of an AP C3 convertase. This process is facilitated by properdin, which increases the amount of convertase formed under these conditions. Interestingly, while the C3b•Bb convertase is controlled by factors H and I (see below), the surface bound C3•Bb convertase is not. The adsorption of C3 does not occur if the polystyrene surface is pre-coated with fibrinogen, so the extent to which this occurs in whole blood, where many other proteins can compete with C3 for binding to a biomaterial surface, has not been demonstrated.

MEMBRANE ATTACK COMPLEX

All three pathways lead to a common point: cleavage of C5 to produce C5b and C5a. C5a is a potent inflammatory mediator and is discussed later in the context of receptor-mediated white blood cell activation. The production of C5b initiates the formation of a macromolecular complex of proteins called the membrane attack complex (MAC) that disrupts the cellular lipid bilayer, leading to cell death (Table II.2.4.5). The sequence of events in MAC formation is outlined in Figure II.2.4.3.

Following cleavage of C5 by C5 convertase, the C5b remains weakly bound to C3b in an activated state in which it can bind C6 to form a stable complex called C5b6. This complex binds to C7 to form C5b67, which

TABLE II.2.4.5 Proteins of the Membrane Attack Complex

Protein	Molecular Weight	Subunits	Plasma Concentration (ug/ml)
C5	190,000	α, β	70
C6	120,000	1	60
C7	105,000	1	60
C8	150,000	α, β	55
C9	75,000	1	55
S-Protein	80,000	1	500

has ampiphilic properties that allow it to bind to, and partially insert into, lipid bilayers. The C5b67 complex then binds C8 and inserts itself into the lipid bilayer. The C5b678 complex disrupts the plasma membrane and produces small pores (r ~ 1 nm) that permit leakage of small molecules. The final step occurs when multiple copies of C9 bind to the C5b678 complex and insert into the membrane. This enlarges the pore to about 10 nm, and can lead to lysis and cell death. Even at sublytic levels, formation of MAC on host cells results in a number of activation responses (elevated Ca^{+2}, arachadonic acid metabolism, cytokine production).

In addition to the usual means of generating C5b (i.e., through C5 convertase activity), several groups (Vogt, 1996; Vogt et al., 1992) have shown that C5 can be modified by a variety of oxidizing agents (H_2O_2, superoxide anion, and others) to convert C5 into a C5b-like structure that will bind C6. The oxidized C5•C6 complex can bind C7, C8, and C9 to form lytic MAC. This mechanism of MAC formation may be relevant at sites where neutrophils and macrophages attempt to phagocytize a biomaterial, producing a variety of reactive oxygen species, or in hypoxia/reperfusion settings (angioplasty, CPB, etc.).

CONTROL MECHANISMS

Various types of control mechanisms have evolved to regulate the activity of the complement system at numerous points in the cascade (Liszewski et al., 1996). These mechanisms are shown in Figure II.2.4.6 and include: (1) decay (dissociation) of converatase complexes; (2) proteolytic degradation of active components that is facilitated by several cofactors; (3) protease inhibitors; and (4) association of control proteins with terminal components that interfere with MAC formation. Without these important control elements, unregulated activation of the cascade results in overt inflammatory damage to various tissues, and has been demonstrated to contribute to the pathology of many diseases (discussed below).

Starting at the top of the cascade, control of the classical and lectin pathway activation is mediated by a protein called C1 esterase inhibitor (C1-Inh). C1-Inh acts by binding to activated C1r and C1s subunits in C1, as

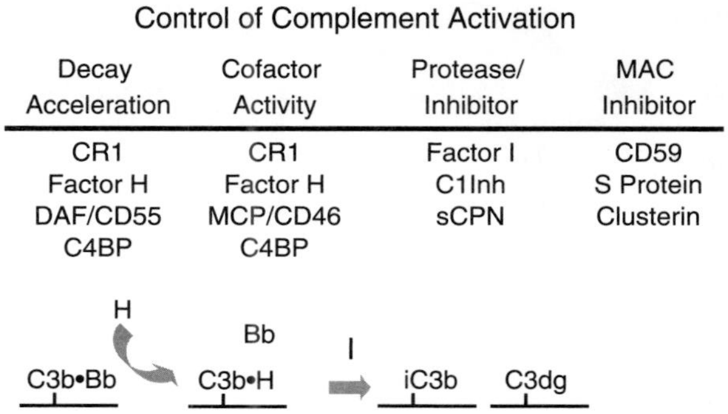

Control of Complement Activation

Decay Acceleration	Cofactor Activity	Protease/ Inhibitor	MAC Inhibitor
CR1	CR1	Factor I	CD59
Factor H	Factor H	C1Inh	S Protein
DAF/CD55	MCP/CD46	sCPN	Clusterin
C4BP	C4BP		

FIGURE II.2.4.6 Control of complement activation occurs by various mechanisms, and is facilitated by a number of different proteins in the plasma (fH, fI, C1 Inh, C4BP, sCPN, S protein, and clusterin) or on cell surfaces (CR1, DAF, MCP, and CD59). Decay acceleration refers to the increased rate of displacement of either C2b or fBb from CP or AP convertases. Cofactor activity refers to the increase rate of factor I-mediated proteolysis facilitated by some proteins.

well as MASP proteases bound to MBL. C1-Inh actually forms a covalent bond with these proteases, thus irreversibly inactivating these proteins. The effectiveness of this interaction is illustrated by the short half-life of C1s under physiological conditions (13 sec). The classical/lectin pathway C3/C5 convertase (C4b•C2b complex) spontaneously "decays" by dissociation of the C2b catalytic subunit. The rate of dissociation is increased by C4 binding protein (C4bp), which competes with C2 for a binding site on C4b. C4bp also acts as a cofactor for another control protein called factor I, which destroys the C4b by proteolytic degradation (Figures II.2.4.2 and II.2.4.6).

The alternative pathway is also highly regulated by mechanisms that are very similar to the CP. The intrinsic instability of the C3b thioester bond (half-life = 60 microseconds) ensures that most of the C3b (80–90%) is inactivated in the fluid phase. Once formed, the C3 convertase (C3b•Bb complex) also spontaneously dissociates, and the rate of "decay" is increased by factor H. Aside from accelerating the decay of C3 convertase activity, factor H also promotes the proteolytic degradation of C3b by factor I (Figures II.2.4.2 and II.2.4.6). Factors H and I also combine to limit the amount of active $C3(H_2O)$ produced in the fluid phase.

In addition to factor H, there are several cell membrane-bound proteins that have similar activities and structures (Figure II.2.4.7). These proteins act to limit complement-mediated damage to autologous, bystander cells. Decay-accelerating factor, or DAF, displaces Bb from the C3 convertase and thus destroys the enzyme activity. DAF is found on all cells in the blood (bound to the plasma membrane through a unique lipid group), but is absent in a disease called proximal nocturnal hemoglobinuria (PNH), which manifests a high spontaneous rate of red blood cell lysis. In addition to DAF, there are two other cell-bound control proteins: membrane cofactor protein (MCP); and complement receptor 1 (CR1; see later discussion). MCP is found on all blood cells except erythrocytes, while CR1 is expressed on most blood cells as well as cells in tissues such as the kidney. Both of these proteins display cofactor activity for the factor I-mediated cleavage of C3b. CR1 also acts like factor H and DAF to displace Bb from the C3 convertase. A soluble recombinant form of CR1 (sCR1) was originally described by Weisman et al. (1990), and later produced commercially (Avant Immunotherapeutics). A number of investigations have used sCR1 to limit complement activation in various disease models (Couser et al., 1995; Larsson et al., 1997).

In contrast to the inhibitory proteins discussed above, properdin, the protein originally discovered by Pillemer, functions by binding to surface-bound C3b and stabilizing the C3 and C5 convertase enzymes. Recent work from several labs has shown that properdin binds to a number of AP activators (Rabbit Erythrocytes, Zymosan, LPS, apoptotic and necrotic cells) independent of C3b. This has lead to an alternative model for AP activation on some surfaces, where properdin plays the initial role in surface recognition, binds C3b, facilitates Factor B binding, and helps drive AP activation (Spitzer et al., 2007; Kimura et al., 2008). A genetic deficiency of this protein in humans has been associated with an increased susceptibility to meningococcal infections, an observation corroborated in properdin knock-out mice.

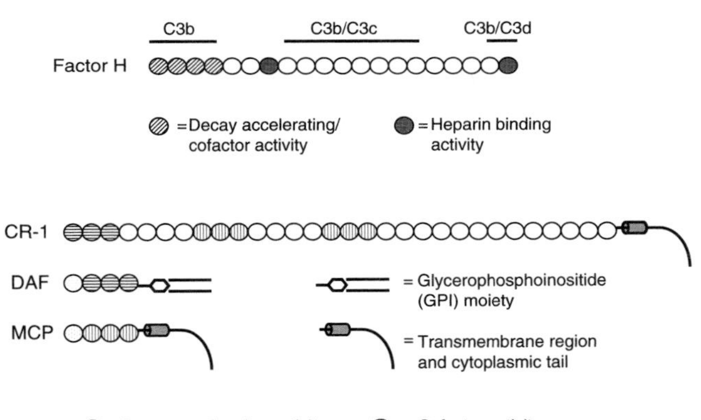

FIGURE II.2.4.7 Structure–activity relationships for complement control proteins. Each circle represents one Short Consensus Repeat (SCR) domain, made up of about 60 amino acids. These SCR domains are strung together to create the different structures shown.

As with the other stages of the cascade, there are several control mechanisms that operate to limit MAC formation and the potential for random lysis of "bystander" cells. The short half-life of the activated C5b (2 min), and the propensity of the C5b67 complex to self-aggregate into a nonlytic form, helps limit MAC formation. In addition, a MAC inhibitor, originally called S protein and subsequently shown to be identical to vitronectin, binds to C5b67 (also C5b678 and C5b6789) and prevents cell lysis. Another way to control MAC is through homologous restriction factors (HRF). HRFs control assembly of the MAC on autologous cells (i.e., human MAC on human cells), but do not stop heterologous interactions (e.g., guinea pig MAC on sheep red blood cells). One well-characterized member of this group is called CD59. It is widely distributed, found on erythrocytes, white blood cells, endothelial cells, epithelial cells, and hepatocytes. CD59 functions by interacting with C8 and C9, preventing functional expression of MAC complexes on autologous cells. Finally, cells are also able to form outward vesicles or exosomes, releasing a portion of their plasma membrane. These microvesicles have been shown to contain MAC, and MAC has been shown to stimulate vesicle formation, leading to the suggestion that this may be a mechanism used by cells to physically remove MAC from the plasma membrane.

COMPLEMENT RECEPTORS

Except for the cytotoxic action of the MAC, most of the biological responses elicited by complement proteins result from ligand–receptor-mediated cellular activation (Sengelov, 1995). These ligands are listed in Table II.2.4.6 and are discussed briefly here.

TABLE II.2.4.6 Receptors for Complement Proteins

Receptor Name/Ligand	Structure	Cellular Distribution/Response
CR1/C3b, C4b	200,000 Da single chain	RBC, PMN, monocytes, B, and T cells/clearance of immune complexes, phagocytosis, facilitates cleavage of C3b to C3dg by Factor I
CR2/C3dg	140,000 Da single chain	B cells/regulate B cell proliferation
CR4	150,000 Da α chain 95,000 Da β chain	PMN, platelets, B cells/leukocyte-endothelial cell interaction
CR3/ iC3b, ICAM-1, βglucan zfibrinogen, factor Xa	185,000 Da α chain 95,000 Da β chain	PMN, monocyte/phagocytosis of microorganisms, respiratory burst activity
$C5a/C5a_{des\ Arg}$	40,000 Da	PMN, monocytes, Mast cells, T cells, epithelial and endothelial cells, hepatocytes, CNS, fibroblasts/chemotaxis, degranulation, hyperadherence, respiratory burst, cytokine production (IL-6,IL-8)
$C3a/C3a_{des\ Arg}$	65,000 Da	Mast cells, eosinophils (various tissues)/histamine release, IL-6 production
C1q/C1q	70,000 Da	PMN, monocytes, B cells/respiratory burst activity
H/H	50,000 Da (3 chains)	B cells, monocytes/secretion of factor I, respiratory burst activity

The ability of complement to function in the opsonization of foreign elements is accomplished in large part by a set of receptors that recognize various C3 and C4 fragments bound to these foreign surfaces. These receptors are called Complement Receptor 1, 2, 3 or 4 (CR1, CR2, etc). CR1 is found on a variety of cells including RBCs, neutrophils, monocytes, B cells, and some T cells, and recognizes a site within the C3c region of C3b (Figure II.2.4.2). On neutrophils and monocytes, activated CR1 will facilitate the phagocytosis of C3b- and C4b-coated particles. On RBCs, CR1 acts to transport C3b-immune complexes to the liver for metabolism. As discussed above, CR1 is also a complement regulatory protein. CR2 is structurally similar to CR1 (with 16 SCR domains; see Figure II.2.4.7), but recognizes the C3d fragment of C3b that is bound to antigen. CR2 is expressed on antigen-presenting cells, such as follicular dendritic cells and B cells, where it facilitates the process of antigen-immune complex-driven B cell proliferation, providing a link between innate and adaptive immunity. CR3 represents another complement receptor that binds to iC3b and β-glucan structures found on zymosan (yeast cell wall). Also, on activated monocytes, CR3 has been shown to bind fibrinogen and factor Xa (of the coagulation cascade). CR3 is a member of the β2-integrin family of cell adhesion proteins that includes leukocyte functional antigen-1 (LFA-1) and CR4. LFA-1, CR3, and CR4 are routinely referred to as CD11a, CD11b, and CD11c respectively. Each of these proteins associates with a molecule of CD18 to form an α-β heterodimer that is then transported and expressed on the cell surface. These proteins help mediate the cell–cell interactions necessary for activities such as chemotaxis and cytotoxic killing. A genetic deficiency in CR3/LFA proteins leads to recurrent life-threatening infections. CR4 is found on neutrophils and platelets, and binds C3d and iC3b. CR4 may facilitate the accumulation of neutrophils and platelets at sites of immune complex deposition.

In contrast to the ligands discussed earlier, which remain attached to activating surfaces, C3a, C4a, and C5a are small cationic polypeptides that diffuse into the surrounding medium to activate specific cells. These peptides are called anaphylatoxins because they stimulate histamine release from mast cells and cause smooth muscle contraction, which can produce increased vascular permeability and lead to a fatal form of shock called anaphylactic shock. These activities are lost when the peptides are converted to their des Arg analogs (i.e., with the loss of their carboxyl terminal arginine residue, referred to as $C3a_{des\ Arg}$, $C5a_{des\ Arg}$, etc.). This occurs rapidly *in vivo*, and is catalyzed by serum carboxypeptidase N (Mueller-Ortiz et al., 2009).

In addition to its anaphylatoxic properties, C5a and $C5a_{des\ Arg}$ bind to specific receptors originally found on neutrophils and monocytes. The receptors for both C5a and C3a have been cloned and sequenced. C5aR (CD88) has been shown to be expressed on endothelial cells (EC), hepatocytes, epithelial cells (lung and kidney tubules), T cells, and cells in the CNS, as well as on the myeloid cell lines (Monk et al., 2007). In addition, expression levels of C5aRs are increased on EC and hepatocytes by exposure to LPS and IL-6. In myeloid cells (neutrophils and monocytes), the C5a-receptor interaction leads to a variety of responses, including chemotaxis of these cells into an inflammatory locus; activation of the cells to release the contents of several types of secretory vesicles and produce reactive oxygen species that mediate cell killing; increased expression of CR1, CR3, and LFA-1, resulting in cellular hyperadherence; and the production of other mediators such as various arachidonic acid metabolites and cytokines, e.g., IL-1, -6, and -8. Many of the adverse reactions seen during extracorporeal therapies, such as hemodialysis, are directly attributable to C5a production. C3aRs are expressed on a variety of cell types including eosinophils, neutrophils, monocytes, mast cells, and astrocytes (in the CNS), as well as γ-IFN-activated T cells. In eosinophils, C3a elicits responses similar to C5a, including intracellular calcium elevation, increases endothelial cell adhesion, and the generation of reactive oxygen intermediates.

CLINICAL CORRELATES

The normal function of complement is to mediate a localized inflammatory response to a foreign material. The complement system can become clinically relevant in situations where it either fails to function or where it is activated inappropriately; some of these settings are shown in Table II.2.4.7 (Markiewski and Lambris, 2007; Unsworth, 2008). In the first instance, a lack of activity due to a genetic deficiency in one or more complement proteins has been associated with increased incidence of recurrent infections (MBL deficiency in children), autoimmune disease (over 90% of C1 deficient patients develop SLE), and other pathologies (for example, a deficiency of C1 inhibitor is known to result in hereditary angioedema, where various soft tissues

TABLE II.2.4.7 Clinical Settings Involving Complement
Hemodialysis and cardiopulmonary bypass
Kidney disease (aHUS, glomerulonephritis)
Ischemia/reperfusion injury (e.g., trauma, surgical settings)
Sepsis and adult respiratory distress syndrome
Recurrent infections
Transplantation
Rheumatoid arthritis
Systemic lupus erythematosus
Asthma
Alzheimer's disease
Hereditary angioedema
Adult macular degeneration

become extremely swollen due to over-production of various vasoactive mediators). A recent genomic analysis has linked a mutation in Factor H (where a tyrosine at position 402 is replaced with a histidine residue) to various diseases. The strongest association has been made with the development of an eye disease called adult macular degeneration (AMD; Klein et al., 2005). The mutation has been shown to decrease the interaction of Factor H with cell surface sugars, linking this loss in cell surface control of complement to the development of AMD and other diseases (Schmidt et al., 2008). This is also an example of inappropriate activation, which can occur in a variety of circumstances. It is now recognized that endothelial cells exposed to hypoxic conditions (ischemia due to angioplasty or a blocked artery due to atherosclerosis) activate complement following reperfusion of the blocked vessel. This results in further damage to the vessel wall, and eventually to the surrounding tissue. Activation of the classical pathway by immune complexes occurs in various autoimmune diseases, such as systemic lupus erythematosus and rheumatoid arthritis. Glomerular deposition of immune complexes results in local inflammation that can contribute to a type of kidney damage called glomerulonephritis (GN). Quite a bit of experimental and clinical data has accumulated demonstrating that complement directly contributes to the initiation and/or progression of GN (Table II.2.4.8), resulting in the development of end stage renal disease, and the necessity of hemodialysis therapy.

One of the major settings where complement has been implicated in adverse clinical reactions is during extracorporeal therapies such as hemodialysis, cardiopulmonary bypass, and apheresis therapy. The same nonspecific mechanism, which permits the alternative pathway to recognize microbes, results in complement activation by the various biomaterials found in different medical devices. The following discussion summarizes the clinical experience with hemodialysis and cardiopulmonary bypass, but many of the observations concerning complement activation and WBC activation are relevant to other medical biomaterial applications.

TABLE II.2.4.8	Types of Studies Demonstrating a Role for Complement in Kidney Disease
Deficiency of loss of complement regulatory activity results in tissue damage	
Mechanistic and knock-out studies implicate complement and C5 in particular	
Ongoing glomerular disease is associated with various indices of complement activation	
Inhibition of complement activation attenuates tissue damage in model systems	

One of the major settings where complement has been implicated in adverse clinical reactions is during extracorporeal therapies such as hemodialysis, cardiopulmonary bypass, and apheresis therapy. The same nonspecific mechanism, which permits the alternative pathway to recognize microbes, results in complement activation by the various biomaterials found in different medical devices.

One of the most investigated materials (from the perspective of complement activation) is the cellulosic Cuprophan membrane used extensively for hemodialysis. Some of the adverse reactions that occur during clinical use of a Cuprophan dialyzer are listed in Table II.2.4.9. In 1977, Craddock et al. showed that some of these same manifestations (neutropenia, leukosequestration, and pulmonary hypertension) could be reproduced in rabbits and sheep when the animals were infused with autologous plasma that had been incubated *in vitro* with either Cuprophan or zymosan. This effect could be abrogated by treatment of the plasma to inhibit complement activation (heating to 56°C or addition of EDTA), thus linking these effects with complement. The development and use of specific radioimmunoassays (RIAs) to measure C3a and C5a by Dennis Chenoweth (1984) led to the identification of these complement components in the plasma of patients during dialysis therapy. A typical patient response to a Cuprophan membrane is shown schematically in Figure II.2.4.8. The C3a (and C5a) levels rise during the first 5–15 minutes, peaking between 10–20 minutes. For a Cuprophan membrane, typical peak C3a levels range from 4000 to 6000 ng/ml. During this period the white blood cells become hyperadherent and are trapped in the lung, resulting in a peripheral loss of these cells (neutropenia). As complement activation is controlled (e.g., by factor H), the C3a and C5a levels decrease to baseline levels, and the WBC returns to the peripheral circulation, now in a more activated (primed) state. This is a very consistent response, and many authors have noted a direct correlation between the extent of complement activation and the degree of

TABLE II.2.4.9	Clinical Symptoms Associated with Cuprophan-Induced Biocompatibility Reactions
Cardiopulmonary:	Pulmonary hypertension Hypoxemia Respiratory distress (dyspnea) Neutropenia (pulmonary leukosequestration) Tachycardia Angina pectoris Cardiac arrest
Other:	Nausea, vomiting, diarrhea Fever, chills, malaise Urticaria, pruritus Headache

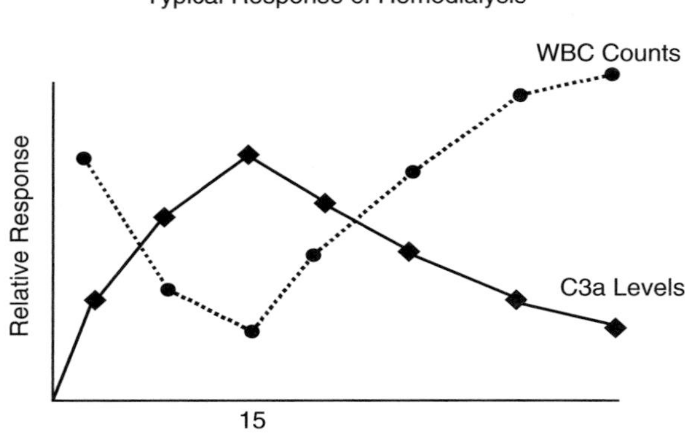

FIGURE II.2.4.8 A typical response pattern to dialysis with a complement-activating hemodialysis membrane. Many investigators have noted that the extent of C3a production is directly proportional to the degree of neutropenia at the same time points.

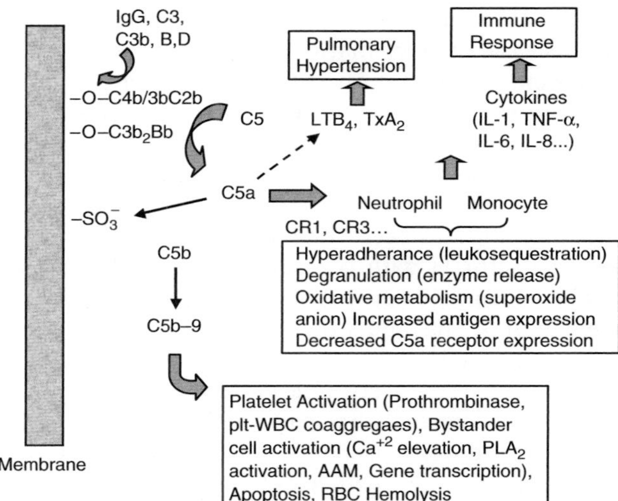

FIGURE II.2.4.9 The biochemical basis for complement-mediated adverse reactions during extracorporeal therapy. Production of C5a leads to receptor-dependent white blood cell activation. This results in profound neutropenia, increased concentrations of degradative enzymes, and reactive oxygen species that ultimately may lead to tissue damage and dysfunction of these important immune cells. Generation of secondary mediators, such as arachidonic acid metabolites (TxA2, LTB4) and cytokines can have profound consequences on whole organ systems. Finally, formation of the MAC (C5b-9) has been linked with increased hemolysis during cardiopulmonary bypass, formation of microparticles, and has been shown to increase platelet prothrombinase activity *in vitro*. This last observation suggests that surfaces that activate complement aggressively may be more thrombogenic.

neutropenia (as well as other responses such as CR3 expression) seen with various dialysis membranes.

Based on our understanding of the biochemistry of complement and its biological actions, the following scenario can be drawn (Figure II.2.4.9). Contact by blood with the membrane results in initial protein deposition, including IgG, C3, and especially C3b, eventually leading to the formation of C3 and C5 convertase enzymes. Conversion of C5 results in C5a production, which leads to receptor-mediated neutrophil and monocyte activation. Production of C5b leads to MAC formation, which binds to bystander cells and results in subsequent activation of these cells through calcium-dependent mechanisms. Recognition of biomaterial-bound C3 and C4 fragments by WBC results in cell adherence and further activation of these cells. These various responses account for much of the pathophysiology seen clinically. The critical role of C5a in mediating many of these adverse reactions was confirmed in experiments employing purified sheep C5a. Infusion of this isolated peptide into sheep, in a manner that would simulate exposure to this molecule during hemodialysis, produced dose-dependent responses identical to those seen when the sheep are subjected to dialysis (Johnson et al., 1996). In addition, numerous *in vivo* and *in vitro* studies have documented the relationship between complement activation by biomaterials, the extent of WBC activation, and the resulting inflammatory response illustrated in Figure II.2.4.9 (Lewis and Van Epps, 1987; Gemmell et al., 1996; Larsson et al., 1997; Tang et al., 1998).

In the same time frame that clinicians were linking complement with leukopenia in the hemodialysis setting, a number of cardiovascular scientists were demonstrating complement activation by the materials used to make cardiopulmonary bypass circuits. Typical levels of C3a produced in these procedures ranged from 300 to 2400 ng/ml. These investigations soon associated C3a and C5a production with a group of symptoms known as "post-perfusion" or "post-pump" syndrome (Table II.2.4.10). Further analysis showed that complement was activated by the materials in the circuit (such as the polypropylene membranes and the nylon filters), but was also activated during neutralization of the heparin anticoagulant with the protamine sulfate that was given to each patient at the end of the operation. This was further exacerbated by complement activation that occurred in the ischemic vascular bed upon reperfusion of the tissue that also occurred at the end of the procedure. The importance of complement activation, and C5 conversion in particular, to the clinical outcome of Cardiopulmonary Bypass (CPB) patients was clearly demonstrated in a study by Fitch et al. (1999). These investigators showed, using a single chain anti-C5 antibody fragment that inhibited C5a and MAC generation during the procedure, that this antibody fragment lowered WBC activation, blood loss, cognitive deficits, and

TABLE II.2.4.10 Post-Perfusion or Post-Pump Syndrome
Increased capillary permeability with accumulation of interstitial fluid
Blood loss requiring transfusions
Fever
Leukocytosis (increased WBC counts)
Organ dysfunction: Heart, liver, kidney, brain and GI tract

myocardial injury. Subsequent studies demonstrated that this complement inhibitor significantly reduced mortality in higher risk cardiac surgical patients (Haverich, 2006). These results are consistent with other studies (Velthuis et al., 1996; Hsu, 2001) using heparin-coated CPB circuits that demonstrate lower inflammatory indices (complement, cytokine, and elastase levels) that are associated with improved clinical outcomes (blood loss, shorter ICU stays, and morbidity).

The CPB experience with heparin-coated devices demonstrates that modification of device materials (or the blood-contacting surfaces of those materials) can dramatically limit complement activation and the subsequent inflammatory response. Based in part on this and similar observations, hemodialyzer/membrane manufacturers began developing new membranes to produce more biocompatible (i.e., less complement-activating) devices. These new membranes tend to fall into two groups: moderately activating modified cellulosics (such as cellulose acetate (CA), hemophane, and cellulose triacetate (CT)); and low activating synthetics (such as polyacrylonitrile (AN69), polymethylmethacrylate (PMMA), and polysulfone (PS)). Moderately activating modified cellulosics produce C3a levels and neutropenic responses that are about 50% of Cuprophan levels, while the synthetic materials display 0–20% activation compared to Cuprophan. Based on the known properties of complement, and the structures of these membranes, the reasons for the improved biocompatibility can be rationalized as follows. Most of these materials have a diminished level of surface nucleophiles. In theory, this should result in lower deposition of C3b, and in fact this has been verified experimentally. The diminished capacity to bind C3b results in lower levels of C3 and C5 convertase activity, and consequently a reduced production of C3a and C5a. Patient exposure to C5a is reduced even further by materials that allow for transport through the membrane to the dialysate (for example, high flux membranes such as polysulfone will do this) or by absorbing the peptide back onto the surface (the negatively charged AN69 has been shown to have a high capacity for binding cationic C5a). Thus, limiting C3b deposition and C5a exposure are two proven mechanisms of avoiding the clinical consequences of complement activation.

The same result can also be accomplished by facilitating the normal control of C3 convertase by factor H. Kazatchkine et al. (1979) have shown that heparin, coupled to either zymosan or Sepharose, limits the normal complement activation that occurs on these surfaces by augmenting C3b inactivation through factors H and I. Presumably, this accounts for the improved biocompatibility of heparin-coated circuits used in CPB described above. Mauzac et al. (1985) have prepared heparin-like dextran derivatives that are extensively modified with carboxymethyl and benzylamine sulfonate groups. These researchers have shown that these modifications diminish complement activation by the dextran substrate. A simple modification of cellulose membranes (Cuprophan) with maleic anhydride has been shown to limit the complement-activating potential of these materials by over 90% (Johnson et al., 1990). Again, increased binding of factor H to surface-bound C3b appears to account for the improved biocompatibility of maleated cellulose. Thus, materials that limit complement activation through normal regulatory mechanisms are on hand, and may prove to be the next generation of complement-compatible materials. In addition, pharmaceutical control of complement is now possible with agents that are FDA approved (Alexion's Solaris®) or now in clinical development.

Thus, materials that limit complement activation through normal regulatory mechanisms are on hand, and may prove to be the next generation of complement-compatible materials. In addition, pharmaceutical control of complement is now possible with agents that are FDA approved (Alexion's Solaris®) or now in clinical development.

SUMMARY AND FUTURE DIRECTIONS

The immune response to a biomaterial involves both humoral and cellular components. Activation of the complement cascade by either classical, lectin or alternative pathways leads to the deposition of C4b and C3b proteins. Recognition of these molecules by receptors on granulocytes can cause activation of these cells, leading to the production of degradative enzymes and destructive oxygen metabolites. Recognition of C4b or C3b by other proteins in the cascade leads to enzyme formation (C3 and C5 convertases), which amplifies the response and can lead to the production of a potent inflammatory mediator, C5a. C5a binds to specific receptors found on PMNs and monocytes. The interaction of C5a with these cells elicits a variety of responses, including hyperadherence, degranulation, superoxide production, chemotaxis, and cytokine production. Systemic exposure to C5a during extracorporeal therapies has been associated with neutropenia and cardiopulmonary manifestations (Tables II.2.4.9 and II.2.4.10) that can have pathologic consequences. The other portion of the C5 protein, C5b, leads to formation of a membrane attack complex that activates cells at sublytic levels and has cytotoxic potential if produced in large amounts. The control of these processes is understood well enough to begin designing materials that are more biocompatible. Limiting C3b deposition (nucleophilicity), adsorbing C5a to negatively charged surface groups, and facilitating the role of factors H and I are three approaches that have been shown to be effective. Translating the last mechanism into commercial materials is one of the major challenges facing the development of truly complement-compatible membranes.

BIBLIOGRAPHY

Andersson, J., Ekdahl, K. N., Lambris, J. D., & Nilsson, B. (2005). Binding of C3 frangments on top of adsorbed plasma proteins during complement activation on a model biomaterial surface. *Biomaterials*, **26**, 1477–1485.

Andersson, J., Ekdahl, K. N., Larson, R., Nilsson, U. R., & Nilsson, B. (2002). C3 absorbed to a polymer surface can form initiating alternative pathway convertase. *J. Immunol.*, **168**, 5786–5791.

Bohlson, S. S., Fraser, D. A., & Tenner, A. J. (2007). Complement proteins C1q and MBL are pattern recognition molecules that signal immediate and long-term protective immune functions. *Mol. Immunol.*, **44**, 33–43.

Chenoweth, D. E. (1984). Complement activation during hemodialysis: clinical observations, proposed mechanisms and theoretical implications. *Artificial Org.*, **8**, 231–287.

Couser, W. G., Johnson, R. J., Young, B. A., Yeh, C. G., Toth, C. A., & Rudolph, A. R. (1995). The effects of soluble complement receptor 1 on complement-dependent glomerulonephritis. *J. Amer. Soc. Nephrol.*, **5**, 1888–1894.

Craddock, P. R., Fehr, J., Brigham, K. L., Kronenberg, R. S., & Jacob, H. S. (1977). Complement and leukocyte-mediated pulmonary dysfunction in hemodialysis. *New Eng. J. Med.*, **296**, 769–774.

Fitch, J. C. K., Rollins, S., Matis, L., Alford, B., Aranki, S., Collard, C. D., Dewar, M., Elefteriades, J., Hines, R., Kopf, G., Kraker, P., Li, L., O'Hara, R., Rinder, C., Rinder, H., Shaw, R., Smith, B., Stahl, G., & Shernan, S. (1999). Pharmacology and biological efficacy of a recombinant humanized, single-chain antibody C5 complement inhibitor in patients undergoing coronary artery bypass graft surgery with cardiopulmonary bypass. *Circulation*, **100**, 2499–2506.

Gemmell, C. H., Black, J. P., Yeo, E. L., & Sefton, M. V. (1996). Material-induced up-regulation of leukocyte CD11b during whole blood contact: material differences and a role for complement. *J. Biomed. Mater. Res.*, **32**, 29–35.

Haverich, A., Shernan, S., Levy, J. H., Chen, J. C., Carrier, M., Taylor, K. M., Van de Werf, F., Newman, M. F., Adams, P. X., Todaro, T. G., van der Laan, M., & Verrier, E. D. (2006). Pexelizumab reduces death and myocardial infarction in higher risk cardiac surgical patients. *Ann. Thorac. Surg.*, **82**, 486–493.

Hsu, L. -C. (2001). Heparin-coated cardiopulmonary bypass circuits: current status. *Perfusion*, **16**, 417–428.

Johnson, R. J., Lelah, M. D., Sutliff, T. M., & Boggs, D. R. (1990). A modification of cellulose that facilitates the control of complement activation. *Blood Purif.*, **8**, 318–328.

Johnson, R. J., Burhop, K. E., & Van Epps, D. E. (1996). Infusion of ovine C5a into sheep mimics the inflammatory response of hemodialysis. *J. Lab. Clin. Med.*, **127**, 456–469.

Kazatchkine, M., Fearon, D. T., Silbert, J. E., & Austen, K. F. (1979). Surface-associated heparin inhibits zymosan included activation of the human alternative complement pathway by augmenting the regulatory action of control proteins. *J. Exp. Med.*, **150**, 1202–1215.

Kimura, Y., Miwa, T., Zhou, L., & Song, W. -C. (2008). Activator-specific requirement of properdin in the intiation and amplification of the alternative pathway complement. *Blood*, **111**, 732–740.

Klein, R. L., Zeiss, C., Chew, E. Y., Tsai, J. -Y., Sackler, R. S., Haynes, C., Henning, A. K., SanGiovanni, J. P., Mane, S. M., Mayne, S. T., Bracken, M. B., Ferris, F. L., Ott, J., Barnstable, C., & Hoh, J. (2005). Complement factor H polymorphism in age-related macular degneneration. *Science*, **308**, 385–389.

Larsson, R., Elgue, G., Larsson, A., Nilsson Ekdahl, K., Nilsson, U. R., & Nilsson, B. (1997). Inhibition of complement activation by soluble recombinant CR1 under conditions resembling those in a cardiopulmonary circuit: upregulation of CD11b and complete abrogation of binding of PMN to the biomaterial surface. *Immunopharmacology*, **38**, 119–127.

Lewis, S. L., & Van Epps, D. E. (1987). Neutrophil and monocyte alterations in chronic dialysis patients. *Am. J. Kidney Dis.*, **9**, 381–395.

Liszewski, M. K., Farries, T. C., Lubin, D. M., Rooney, I. A., & Atkinson, J. P. (1996). Control of the complement system. *Adv. In Immunol.*, **61**, 201–282.

Manderson, A. P., Pickering, M. C., Botto, M., Walport, M. J., & Parish, C. R. (2001). Continual low-level activation of the classical complement pathway. *J. Exp. Med.*, **194**, 747–756.

Markiewski, M. M., & Lambris, J. D. (2007). The role of complement in inflammatory diseases from behind the scenes into the spotlight. *Am. J. Pathol.*, **171**, 715–727.

Matsushita, M. (1996). The lectin pathway of the complement system. *Microbiol. Immunol.*, **40**, 887–893.

Mauzac, M., Maillet, F., Jozefonvicz, J., & Kazatchkine, M. (1985). Anticomplementary activity of dextran derivatives. *Biomaterials*, **6**, 61–63.

Monk, P. N., Scola, A. -M., Madala, P., & Fairlie, D. P. (2007). Function, structure and therapeutic potential of complement C5a receptors. *British J. Pharmacology*, **152**, 429–448.

Mueller-Ortiz, S. L., Wang, D., Morales, J. E., Li, L., Chang, J. -Y., & Wetsel, R. (2009). Targeted disruption of the gene encoding the murine small subunit of carboxypeptidase N (CPN1) causes susceptibility to C5a anaphylatoxin-mediated shock. *J. Immunol.*, **182**, 6533–6539.

Pillemer, L., Blum, L., Lepow, I. H., Ross, O. A., Todd, E. W., & Wardlaw, A. C. (1954). The properdin system and immunity. I. Demonstration and isolation of a new serum protein, properdin, and its role in immune phenomena. *Science*, **120**, 279–285.

Rawal, N., Rajagopalan, R., & Salvi, V. P. (2008). Activation of Complement Component C5; Comparison of the C5 convertases of the lectin and the classical pathway of complement. *J. Biol. Chem.*, **283**, 7853–7863.

Ross, G. D. (1986). *Immunobiology of the Complement System*. New York: Academic Press.

Schmidt, C. Q., Herbert, A. P., Hocking, H. G., Uhrin, D., & Barlow, P. N. (2008). Translational mini-review series on complement factor H: structural and functional correlations for factor H. *Clin. Exp. Immunol.*, **151**, 14–24.

Schreiber, R. D., Pangburn, M. K., Lesaure, P. H., & Muller-Eberhard, H. J. (1978). Initiation of the alternative pathway of complement: recognition of activators by bound C3b and assembly of the entire pathway from six isolated proteins. *Proc. Natl. Acad. Sci. U.S.A.*, **75**, 3948–3952.

Sengelov, H. (1995). Complement receptors in neutrophils. *Critical Review in Immunol.*, **15**, 107–131.

Spitzer, D., Mitchell, L. M., Atkinson, J. P., & Hourcade, D. E. (2007). Properdin can initiate complement activation by binding specific target surfaces and providing a platform for de novo convertase assembly. *J. Immunol.*, **179**, 2600–2608.

Takahashi, M., Iwaki, D., Kanno, K., Ishida, Y., Xiong, J., Matsushita, M., Endo, Y., Miura, S., Ishii, N., Sugamura, K., & Fujita, T. (2008). *J. Immunol.*, **180**, 6132–6138.

Tang, L., Liu, L., & Elwing, H. B. (1998). Complement activation and inflammation trigered by model biomaterial surfaces. *J. Biomed. Mater. Res.*, **41**, 333–340.

Unsworth, D. J. (2008). Complement deficiencies and disease. *J. Clin. Pathol.*, **61**, 1013–1017.

Velthuis, H., Jansen, P. G. M., Hack, E., Eijsman, L., & Wildevuur, C. R. H. (1996). Specific complement inhibition with heparin-coated extracorporeal circuits. *Ann. Thorac. Surg.*, **61**, 1153–1157.

Vogt, W., Zimmerman, B., Hesse, D., & Nolte, R. (1992). Activation of the fifth compenent of human complement, C5, without cleavage, by methionine oxidizing agents. *Mol. Immunol.*, **29**, 251–256.

Vogt, W. (1996). Complement activation by myeloperoxidase products released from stimulate human polymorphonuclear leukocytes. *Immunobiology*, **195**, 334–346.

Weisman, H. F., Bartow, T., Leppo, M. K., Marsch, H. C., Jr., Carson, G. R., Concino, M. F., Boyle, M. P., Roux, K. H., Weisfeldt, M. L., & Fearon, D. T. (1990). Soluble human complement receptor type 1: in vivo inhibitor of complement suppressing post-ischemic myocardial inflammation and necrosis. *Science*, **249**, 146–151.

Wong, N. K. H., Kojima, M., Dobo, J., Ambrus, G., & Sim, R. B. (1999). Activities of the MBL-associated serine proteases (MASPS) and their regulation by natural inhibitors. *Mol. Immunol.*, **36**, 853–861.

CHAPTER II.2.5 SYSTEMIC TOXICITY AND HYPERSENSITIVITY

Arne Hensten[1] *and Nils Jacobsen*[2]

[1]Department of Clinical Densitry, University of Tromsø, Tromsø, Norway

[2]Nordic Institute of Dental Materials Sognsveien, Ullevål Stadion, Oslo, Norway

Artificial implant devices comprise a variety of metallic alloys, polymers, ceramics, hydrogels or composites for a large number of purposes and with widely different properties. With the exception of drug delivery systems, sutures, and other degradable biomaterial systems (Chapter II.2.5), the implant devices are intended to resist chemical and biochemical degradation, and to have minimal leaching of structural components or additives. However, synthetic devices are influenced by chemical, and in some cases enzymatic, processes resulting in the release of biomaterials-associated components. Since there are no natural repair mechanisms parallel to natural tissues, degradation (biodegradation) is a "one-way" process that brings about microscopic and macroscopic surface and bulk changes of the devices, sometimes enhanced by the biomechanical and bioelectrical conditions that the devices are intended to resist. With the exception of pathologic calcification of certain polymer implants, the surface changes may not be significant for the mechanical strength of the implant, whereas in contrast the released substances very often have biological effects on the surrounding tissues or, possibly, at other remote locations. Inflammatory, foreign body, or other local host reactions, and tumorigenesis are discussed in Chapters II.4.1, II.4.2, and II.4.6. The following discussion is concerned with the possibility of systemic toxic reactions and/or hypersensitive reactions caused by biomaterials-derived xenobiotics.

KINETICS AND THE NATURE OF BIOMATERIALS COMPONENTS

Xenobiotic components derived from *in vivo* medical devices have parenteral contact with connective tissue or other specialized tissues such as bone, dentin, vascular or ocular tissue, whereas leachables from skin and mucosa contacting devices have to pass the epithelial lining of the oral mucosa, the skin, the gastrointestinal tract, or – for volatiles – the lung alveoli to get "inside" the body. In any of these cases, further distribution of foreign substances to other tissues and organs is dependent on membrane diffusion into blood capillaries and lymph vessels. The transport may be facilitated by reversible binding to plasma proteins, globulins (metal, metal compounds), and chylomicrons (lipophilic substances). Storage – and later release – may take place for certain components in tissues such as fat and bone.

In addition to particulate matter, the released components consist of chemical substances of different atomic and molecular size, solubility, and other chemical characteristics, depending on the mother material. Examples are metal ions from orthopedic implants or prosthodontic materials, or residual monomers, chemical initiators, inhibitors, plasticizers, antioxidants, etc., from polymer implants and dental materials. Other degradation products from inorganic, organic, and composite devices also "rub off" to the surrounding tissues. The kinetic mechanisms for biomaterials components are in part the same as those of xenobiotics introduced by food or environmental exposure, i.e., the released components are subject to oxidation, reduction, and hydrolysis, followed by conjugation mechanisms. All metabolic changes are, by their nature, intended to eliminate them by way of the urine, bile, lungs, and to a certain degree in salivary-, sweat-, and mammary glands, and hair (deBruin, 1981).

A key question is, do the released components or their metabolites have any systemic toxic effect on the host, and/or could they induce unwanted immunological reactions?

TOXICO-DYNAMIC CONSIDERATIONS

Systemic toxicity depends on toxic substances hitting a target organ with high sensitivity to a specific toxicant. Target organs are the central nervous system, the circulatory system, the hematopoietic system, and visceral organs such as liver, kidney, and lungs, in that order. The toxicity is based on interference with key cell functions, and depends on the dose, and the duration, of the exposure. Serious effects may be incompatible with continued life, but most effects cause local and reversible cell damage. However, some sublethal effects may include somatic cell mutation expressed as carcinogenesis, or germinal cell mutation, resulting in reproductive toxicity.

The key word in the evaluation of general toxicity is the *dose*, defined as the amount of substance an organism is exposed to, usually expressed as mg per kg bodyweight. Adverse effects of foreign substances are often the result of repeated, chronic exposure to small doses that, over a prolonged period of time, may have deleterious effects similar to one large, short-term exposure, provided that the repeated doses exceed a certain threshold level. This level is determined by the capacity for metabolism and elimination. Another important factor is the possibility of synergistic potentiating effects when several toxicants are present simultanously. Whatever mechanism is involved, the principle of systemic toxicity presupposes a dose-dependent reaction that may be measured and described, and that may be explained by specific reactions at distinct molecular sites (Eaton and Gilbert, 2008).

The components derived from biomaterials represent a large series of widely different foreign substances with few characteristics in common, and with a largely unknown concentration. Most of them have to be characterized as toxic *per se*, with large variations as regards their place on a ranking list of potential toxicity. This statement is relevant for metal ions and salts, as well as for polymeric components derived from biomaterials devices. Local inflammatory and other host reactions towards biomaterial implants are well known (Chapter II.4.2), whereas the significance of individual implant components is difficult to assess.

METALLIC MATERIALS

However, many studies have provided information on the presence of wear and corrosion debris from metallic orthopedic implants, often initiated by the aseptic loosening of total hip replacements. Oxides or hydroxides of cobalt, titanium, aluminium, iron, nickel, and chromium are found in the periprosthetic tissues, depending on the alloys used. Post-mortem data also indicate accumulation in regional lymph nodes and liver. Further vascular distribution of ions and extremely small particles has made it possible to estimate concentration of some of these components in whole blood. Such findings open up the possibility of evaluating the potential effect on different organs. Information in this area is growing by sophisticated cytotoxicity research, by experimental animal studies, and by extrapolation of occupational and environmental data. So far, the risk of systemic toxicity caused by metallic implant components is considered to be negligible, although it is recommended that the orthopedic research community pay attention to the basic functions of kidneys, reproductive organs, and the central nervous system among arthroplastic patients (Keegan et al., 2007).

Metallic dental material is another source of exposure. *In vitro* experiments have shown that chromium and nickel are released from base metal orthodontic appliances, although the amounts are not comparable with the amounts calculated in food intake (Park and Shearer, 1983). As witnessed by hypersensitive reactions, uptake of metal components by mucosa from fillings and prosthodontic devices does take place, but on a scale too low to be of interest in the systemic toxicity context.

An exception is the release and pulmonary uptake of volatile metallic mercury from dental amalgam fillings. Many studies have been able to quantify the concentration of mercury in plasma and urine depending on the burden of dental amalgam (Mackert and Berglund, 1997), and occupational studies have demonstrated that mercury accumulates in tissues belonging to the central nervous system (Nylander et al., 1989). Reproductive toxicity has been of specific concern for dental personnel.

However, similar to other metals such as chromium and nickel, mercury exposure also takes place in food and respiratory air. Careful scrutiny by national and international scientific committees of the large amount of (partly controversial) data has not resulted in a consensus conclusion that the application of mercury dental amalgam should be discontinued, although the environmental concern of mercury is recognized. The European Commission (2008), by its Scientific Committee of Emerging and Newly Identified Health Risks (SCENIHR), has recently investigated possible adverse effects of mercury dental amalgam on urinary, neurological, immunological, psychological, reproductive, and other systems, and concludes that there is little epidemiological evidence to show that mercury released by dental amalgam fillings contributes to the etiology of systemic diseases.

NON-METALLIC BIOMATERIALS

Non-metallic dental materials comprise numerous composite polymeric materials based on a variety of methacrylate monomers polymerized by specific chemical additives. Numerous elution experiments have demonstrated the release of monomers and additives, particularly during the first hours after polymerization (Michelsen et al., 2007). Degradation and erosion of the dental filling and prosthodontic materials over time is another source of released components. Most released substances have proven to be cytotoxic *in vitro* (Geurtsen et al., 1998). Some components have been of specific interest for their toxicity to the reproductive system. Examples are the UV-absorber oxybenzone, Bisphenol A, and phtalates, all released from different polymeric materials. Oxybenzone has an estrogenic potential, Bisphenol A binds estrogen receptors *in vitro* with the potential of impairing the reproductive system, and phtalates have antiandrogenic characteristics as judged by environmental research. Since Bisphenol A is released from fissure sealants, mainly applied in children, this issue has been of specific interest in the toxicological discussion of dental materials. However, current literature does not support any risk associated with these materials (Azarpazhooh and Main, 2008).

Minute quantities of phtalates and degradation products of chemical additives derived from freshly made (poly)methylmethacrylate dental prostheses have been demonstrated in saliva (Lygre et al., 1993). Larger amounts of plastiziser phtalates are released from temporary biomaterials such as PVC tubing; therefore the risk of deleterious effects on specific patient groups such as dialysis patients or premature neonatates has been discussed (Calafat et al., 2004). Another point in this context is the observation that (poly)methylmethacrylate monomer from setting bone cement may cause transient cardiovascular reactions. The dose applied in hip arthroplasty is large enough to cause hypotension, whereas the smaller dose associated with percutaneous vertebroplasty is not (Kaufmann et al., 2002).

Other clinically relevant data on the systemic toxicity of degradation products from dental polymeric materials and non-metallic implants are scarce. On this background a fair conclusion would be that there is no data indicating long-standing systemic toxicity caused by biomaterials-derived xenobiotics. However, current cytotoxic and animal research may reveal toxic mechanisms on the intracellular level that may prove relevant for released biomaterials components. In addition, the biomaterials field is characterized by the increasing number of synthetic materials on the market. Despite the premarketing testing programs, it is difficult to predict single or synergistic toxic effects of leachable components and degradation products in the future.

ADVERSE EFFECTS OF DEFENSE MECHANISMS

The low probability of direct systemic adverse effects on target organs caused by biomaterials products does not rule out deleterious effects by other mechanisms requiring only minute amounts of the foreign material. All substances not recognized as natural components of the tissues are subject to possible clearance by several mechanisms, e.g., phagocytic cells such as polymorphonuclear leucocytes, macrophages, and monocytes attempting to degrade and export the components. Larger foreign components are subject to more aggressive reactions by giant cells causing an inflammatory foreign-body reaction. Enzymes and other bioactive molecules associated with phagocytosis and foreign-body reaction may cause severe local tissue damage. In addition, phagocytic cell contact, and contact with the circulatory system of lymph and blood, opens up another way of neutralizing foreign substances by way of the immune system, introducing a biologic memory of previously encountered foreign substances, and an enhancement system for their neutralization.

HYPERSENSITIVITY AND IMMUNOTOXICITY

The immune system is an indispensable biological mechanism to fight potentially adverse invaders, most commonly of microbial origin. However, the immune system occasionally strikes invading molecules – adverse or not – with an intensity that stands in contrast to their modest concentration, and with the ability to cause host tissue damage. This phenomenon is called hypersensitivity. The resulting injury is part of a group of adverse reactions classified as immunotoxic.

In principle, immunologic hypersensitivity comprises two different mechanisms: allergy and intolerance. Allergy is an acquired condition resulting in an overreaction upon contact with a foreign substance, provided there is a genetic disposition and previous exposure to the substance. Allergic reactions may include asthma, rhinitis, urticaria, intraoral and systemic symptoms, and eczema. Intolerance is an inherited reaction that resembles allergy and has common mediators and potentiating factors, such as complement activation, histamine release, etc., but is not dependent on a previous sensitization process. Intolerance reactions have been associated with drugs such as acetyl-salicylic acid, whereas intolerance to leachable biomaterial components such as benzoic acid is conceivable but not known.

ALLERGY AND BIOMATERIALS

A foreign substance able to induce an allergic reaction is called an allergen. There is no acceptable way of predicting whether a substance or a compound is potentially allergenic on the basis of its chemical composition and/or structure alone. However, experimental evidence, and years of empirical results after testing substances causing allergic reactions, has given some leads, e.g., large foreign molecules such as proteins and nucleoproteins are strong allergens, whereas lipids are not. However, the strongest chemical allergens associated with biomaterials are often chemically active substances of low molecular weight, often less than 500 Da, such as lipid-soluble organic substances derived from polymer materials or metal ions and metal salts. These are called haptens, i.e., they become full allergens only after reaction or combination with proteins that may be present in macrophages and Langerhans cells of the host.

TYPES OF ALLERGIES

Allergies are most often categorized into four main groups (Type I–IV) according to the reaction mechanisms. Types I to III are associated with humoral antibodies initiated by B-lymphocytes that develop into immunoglobulin-producing plasma cells. The immunoglobulins are classified into five different classes, IgE, A, D, G, and M, according to their basic structure and size. A variable portion of the immmunoglobulin is specific for the antigen that induced its production. The Type IV reaction is a cell-mediated reaction caused by T lymphocytes.

Types II and III allergies comprise antigen–antibody encounters including complement activation, cell lysis, release of vasoactive substances, inflammatory reaction, and tissue damage. Necrosis of peri-implant tissue with histologic appearance and serum complement analyses consistent with Type III hypersensitivity has been observed in cases of atypical loosening of total hip prostheses (Hensten-Pettersen, 1993). However, according to the appropriate ISO and FDA documents for immunotoxicological testing of medical devices (ISO 10993-20, 2006, and FDA Immunotoxicity Testing Guidance, 1999) Type II and III reactions are omitted because they are relatively rare, and are less likely to occur with medical devices/materials, leaving Types I and IV relevant in the present context.

TYPE I HYPERSENSITIVITY

The Type I reaction is based on an interaction between an intruding allergen and IgE immunoglobulins located in mast cells, basophils, eosinophils, and platelets, resulting in release of active mediators such as histamine and other vasoactive substances. The results are local or systemic reactions seen within a short time (minutes). The symptoms depend on the tissue or organ subject to sensitization, e.g., inhaled allergens such as pollen, residual proteins associated with surgical latex gloves or other natural latex products may result in asthmatic seizures, swelling of the mucosa of the throat or worse: decreased blood pressure and anaphylactic shock. Food allergies may also give systemic symptoms. This type of host reaction is usually associated with full antigens. Since the potential allergens associated with biomaterials are small molecular haptens, the probablility of IgE-based allergic reactions is low, although IgE antibodies to chromium and nickel have been reported (Hensten-Pettersen, 1993). Reports on adverse reactions to orthopedic devices describe patients with urticarial reactions. Contact urticaria is a wheal and flare response to compounds applied on intact skin. The role of immunological contact urticaria in relation to medical devices is not clear.

TYPE IV HYPERSENSITIVITY

Cell-mediated hypersensitivity is referred to as "delayed," because it takes more than 12 hours to develop, often 24–72 hours. The T lymphocytes producing the response have been sensitized by a previous encounter with an allergen, and act in concert with other lymphocytes and mononuclear phagocytes to create three histologically different types, characterized by skin-related tissue reactions. The reactions are elicited by interaction of cells and mediators which comprise: (1) induration (the granulomatous type); (2) swelling and induration, and possibly fever (the tuberculin type); and (3) eczema (the contact type) (Britton, 2001).

Prolonged challenges of macrophage-resistant allergens, usually of microbial origin, may result in persistant immunological granuloma formation. Such reactions have been associated with occupational beryllium and zirconium exposure, but the main form of delayed hypersensitivity in relation to biomaterials is the contact form.

Allergic contact dermatitis is acquired through previous sensitization with a foreign substance. The hapten is absorbed by the skin or mucosa, and binds to certain proteins associated with the Langerhans cells, forming a complete antigen. The antigen stimulates the formation of activated, specialized T cells. Upon new exposure, the allergen–T cell encounter releases inflammatory mediators resulting in further production and attraction of T cells, causing tissue damage. The reactions are not necessarily limited to the exposure site.

The presence of allergic contact dermatitis is evaluated by epidermal or intradermal skin tests. A vast amount of information on the allergenic characteristics of biomaterials-related substances has been obtained in this way, especially as regards dental materials (Kanerva et al., 1995). However, many biomaterials employed in dentistry, such as metal alloys and resin-based materials, have medical counterparts which are encountered in everyday life. The sensitization process, therefore, often has taken place before the biomaterials contact.

ATOPY

Atopic individuals have a constitutional predisposition for IgE-based hypersensitive reactions caused by environmental and food allergens. The reaction includes histamine-mediated hay fever, asthma, gastrointestinal symptoms or skin rashes, and is more pronounced at an early age. Atopics have an increased risk of acquiring irritant contact dermatitis to external biomaterial devices, such as orthodontic appliances. The relation to allergic contact dermatitis is unclear (Lindsten and Kurol, 1997), so also is the relationship between atopy and allergens or haptens from biomaterials exposed parenterally.

IMMUNOLOGIC TOXICITY OF BIOMATERIALS

Biomaterial immunology includes reactions associated with oral rehabilitation, as well as medical devices. Of these, dental materials are by far the most common, and represent a large variety of metallic and polymer products. With the exception of some extraoral orthodontic equipment and latex utensils, the exposure site of these materials is oral tissues. The local immunological reactions include gingivitis or gingival lichenoids, mucosal inflammation and blistering or perioral dermal reactions, all of the delayed contact reaction type. In addition, dermal reactions may occur at remote locations. Occasionally IgE-based immediate reactions, such as angioedema, contact urticaria or anaphylaxis are seen, mostly related to temporary biomaterials, such as natural rubber latex.

Considering the vast number of dental patients, the incidence of immunotoxicity reactions is extremely low.

In contrast to medical implants, such reactions are easily observed in the dental clinic, but determination of the eliciting allergen is not easy. In Table II.2.5.1 some allergenic components associated with dental biomaterials are listed on the basis of empirical findings by epidermal patch tests. Some of these items have their counterparts in orthopedic implants or other medical devices.

Information on immunological adverse effects in dental medicine has been obtained by case reports, questionnaire studies, and reports from adverse reaction registry units and dermatological units. In some cases the cause–effect relationship has been assessed by clinical experience only. For these and other reasons, there are no reliable epidemiological statistics on adverse effects in terms of incidence data. The prevalence in published reports varies from 1:100 to 1:10,000 (Schedle et al., 2007), and comprises both local and remote hypersensitivity reactions. However, all reports indicate salts of nickel, gold, cobalt, palladium, mercury, and chromium as the main metallic allergens. Of the polymer alternatives, the monomers of HEMA, EGDMA, TEGDMA, and MMA, together with polymerization chemicals, such as benzoyl peroxide and dimethylparatoluidin, are prevalent. It is argued that the sensitization may have taken place by exposure to parallell allergens in everyday life, and that the oral encounter elicits the reaction. It is also accepted that, for anatomical reasons, oral tissues develop hypersensitive reactions less easily than dermal tissues.

TABLE II.2.5.1 Some Allergens in Dental Materials Based on Epicutaneous Testing

Metals	Monomers and Cross-Linking Agents
Nickel*	Methylmethacrylate (MMA)*
Cobalt*	Triethyleneglycol dimethacrylate (TEGDMA)
Chromium*	Urethane dimethacrylate (UEDMA)
Gold	Ethyleneglycoldimethacrylate (EGDMA)
Mercury	Bisphenol A diglycidyl dimethacrylate (BIS-GMA)
Copper*	1,4-Butanediol methacrylate (BUDMA)
Palladium	Bisphenol A methacrylate (BIS-MA)
Titanium*	2-hydroxyethylmethacrylate (HEMA)*
Aluminum	
Tin	
Beryllium	
Silver	
Platinum	
Zirconium	
Polymerization Chemicals and Additives	
Methylhydroquinone (stabilizer)*	
Camphoroquinone (initiator)	
Dibenzoylperoxide (initiator)*	
N,N-Dimethyl-4-paratoluidine (activator)*	
N,N-Dimethylaminoethyl methacrylate (activator)*	
4-Tolyldiethanolamine (activator)	
Tinuvin P (UV stabilizer)	
2-Hydroxy-4-methoxy benzophenone (UV stabilizer)	

*Sensitizers in orthopedic surgery. Other orthopedic metals are manganese, vanadium, molybdenum, aluminum, and zirconium, all with undecided allergen characteristics (Thomas et al., 2008; Geier et al., 2008). In addition, protein in natural rubber latex is a Type I allergen in dental and general medicine.

IMPLANT ALLERGY

The exact progress of hypersensitivity reactions associated with implants is not clear. It might include both the sensitization and the elicitation process, or just the latter. Reactions of this type are difficult to recognize unless they have dermal or systemic expressions, such as eczemas or fistulas. In addition, such reactions may be part of local atoxic and/or mechanically induced inflammation using similar mediators for tissue response. Because of a lack of more distinct descriptions, such reactions have been referred to as "deep tissue" reactions of Type IV hypersensitivity. Immunological toxicity to surface medical devices and external communicating devices (dialyzers, laparoscopes, etc.) may represent mechanisms of sensitization and hypersensitivity reactions similar to those of orally exposed biomaterials.

A vast battery of *in vitro* and *in vivo* experimental studies have been performed to study potential adverse effects of biomaterial devices, such as artificial joints, heart valves, and breast prostheses (Rodgers et al., 1997). Aseptic loosening of metallic hip prostheses has been associated with "biologic" causes, in addition to biomechanical factors and wear debris. However, it is unclear whether metal sensitivity is a contributing factor to implant failure (Hallab et al., 2001). In fact, it is argued that the loosening process enhances immunological sensitization, indicating that the cause–effect relation may be the reverse (Milavec-Puretic et al., 1998). German orthopedists and immunologists, discussing the complicated relationship between potential allergens and the presence of implant failures, fistulas, and dermal reactions, underline the value of an allergological anamnesis (Thomas et al., 2008). However, similar to dental medicine, there is no strict contraindication for using a biomaterial containing a known allergen for a particular patient. Such decisions have to be taken on the basis of individual evaluation. What is clear, however, is that local and general eczematous reactions have been observed following the insertion of metallic implants in patients subsequently shown to be allergic to cobalt, chromium, and nickel.

Many case reports also describe the immediate healing of dermal reactions after the removal of metal implants (Al-Saffar and Revell, 1999). In addition, metal allergy has been discussed as a contributing factor in the development of in-stent coronary restenosis, although there is little evidence for this effect (Hillen et al., 2002). However, established metal allergy in a patient does not as a rule seem to be accompanied by clinical reactions to implant alloys containing the metal. If this statement is true, it is in line with clinical observations made in surveys on the use of metallic alloys in prosthodontics and orthodontics. Inhomogeneiety or a mixture of alloys appears to determine the efflux of potentially hypersensitive metal ions, and hence increases the possibility of eliciting a hypersensitive reaction (Grimsdottir et al., 1992).

Methyl-methacrylate bone cement is another potential allergenic factor in orthopedic surgery, as is their counterpart in dentistry and cosmetic dentistry (Kaplan et al., 2002). As indicated in Table II.2.5.1, prime elicitors are monomers of MMA and HEMA, and additives such as benzoyl peroxide, N,N-dimethyl-p-toluidine, and hydroquinone (Thomas et al., 2006).

An extensive literature reflects clinical surveys and research activities related to natural latex used as a barrier material by the health professions. It is accepted that residual latex proteins and chemicals associated with the production process may cause immediate and delayed reactions in patients and health personnel (Turjanmaa et al., 1996).

OTHER INTERACTIONS

The FDA testing guidance referred to above (FDA, 1999) also lists other interactions of medical devices, extracts of medical devices or adjuvants with the immune system, such as impairment of the normal immunologic protective mechanisms (immunosuppression), and long-term immmunological activity (immunostimulation) that may lead to harmful autoimmune responses. The autoimmune reaction is explained by the biomaterial-associated agent acting as an adjuvant that is stimulating to antibody–complement-based tissue damage by cross-reactions with human protein. Chronic inflammatory, immune-related granuloma may take part in the development of autoimmune reactions.

CONCLUDING REMARKS

Biocompatibility issues related to medical devices form a multidimensional crossroads of technology and biology. One component is the various classes of biomaterials, such as plastics and other polymers, metals, ceramics, glasses, etc., depending on expert design to obtain maximal mechanical properties and minimal chemical dissolution. Another is the mode of application, ranging from skin and mucosal contact to totally submerged implants, with external communicating devices in-between. A third dimension is the duration of contact, ranging from minutes to the expected lifetime, and the fourth, and decisive component, is the biological reactions that can be expected. These circumstances prevent general statements on biomaterials. The present overview is aimed at students, and limited to a focus on collective mechanisms determining systemic toxicity and discussion of hypersensitivity reactions documented by clinical reports.

BIBLIOGRAPHY

Al-Saffar, N., & Revell, P. A. (1999). Pathology of the bone-implant interfaces. *J. Long-Term Effects Med. Implants*, **9**, 319–347.

Azarpazhooh, A., & Main, P. A. (2008). Is there a risk of harm or toxicity in the placement of pit and fissure sealant materials? A systematic review. *J. Can. Dent. Assoc.*, **74**, 179–183.

Britton, W. (2001). Hypersensitivity – Type IV. In I. Roitt, J. Brostoff, & D. Male (Eds.), *Immunology* (pp. 371–382). Edinburgh, London, New York, Philadelphia, St. Louis, Sydney, Toronto: Mosby. Chapter 24.

Calafat, A. M., Needham, L. L., Silva, M. J., & Lambert, G. (2004). Exposure to di-(2-ethylhexyl)phtalates among premature neonates in a neonatal intensive care unit. *Pediatrics*, **113**, 429–434.

deBruin, A. (1981). The metabolic fate of foreign compounds. In D. F. Williams (Ed.), *Fundamental Aspects of Biocompatibility* (Vol. 2) (pp. 3–43). Boca Raton, Fl: II. CRC Press.

Eaton, D. L., & Gilbert, S. G. (2008). Principles of toxicology. In C. D. Klaassen (Ed.), *Casaretts and Doull's Toxicology. The Basic Science of Poisons* (pp. 11–45). New York, NY: McGraw Hill Medical. Chapter 2.

FDA (Food and Drug Administration). (1999). *Immunotoxicity Testing Guidance.* Document issued May 6, US Department of Health and Human Services. (pp. 1–15).

Geier, J., Lessmann, H., Becker, D., & Thomas, P. (2008). Allergologische diagnostik bei verdacht auf implantatunverträglichkeit: Hinweise für die praxis. *Hautarzt*, **59**, 594–597.

Geurtsen, W., Lehmann, F., Spahl, W., & Leyhausen, G. (1998). Cytotoxicity of 35 dental resin composite monomers/additives in permanent 3T3 and three human primary fibroblast cultures. *J. Biomed. Mater. Res.*, **41**, 474–480.

Grimsdottir, M. R., Gjerdet, N. R., & Hensten-Pettersen, A. (1992). Composition and *in vitro* corrosion of orthodontic appliances. *Am. J. Dentofac. Orthop.*, **101**, 23–30.

Hallab, N., Merrit, K., & Jacobs, J. J. (2001). Metal sensitivity in patients with orthopedic implants. *J. Bone. Joint. Surg.*, **83**, 428–436.

Hensten-Pettersen, A. (1993). Allergy and hypersensitivity. In B. F. Morrey (Ed.), *Biological, Material, and Mechanical Considerations of Joint Replacement,* (pp. 353–361). New York, NY: Raven Press.

Hillen, U., Haude, M., Erbel, R., & Goos, M. (2002). Evaluation of metal allergies in patients with coronary stents. *Contact Dermatitis*, **47**, 353–356.

ISO (2006). ISO 10993-20. Biological evaluation of medical devices – Part 20: Principles and methods for immunotoxicology testing of medical devices. *International Organization for Standardization*, p. 17.

Kanerva, L., Estlander, T. & Jolanki, R. (1995). Dental problems. In J. D. Guin (Ed.), *Practical Contact Dermatitis,* (pp. 397–432). McGraw-Hill Health Profession Division.

Kaplan, K., Della Valle, C. J., Haines, H., & Zuckerman, J. D. (2002). Preoperative identification of a bone-cement allergy in a patient undergoing total knee arthroplasty. *J. Arthroplasty*, **17**, 788–791.

Keegan, G. M., Learmonth, I. D., & Case, C. P. (2007). Orthopaedic metals and their potential toxicity in the arthoplastic patient. A review of current knowledge and future strategies. *J. Bone Joint Surg., (Br)*, **89**, 567–573.

Kaufmann, T. J., Jensen, M. E., Ford, G., Gill, L. L., Marx, W. F., et al. (2002). Cardiovascular effects of polymethylmethacrylate use in percutaneous vertebroplasty. *Am. J. Neuroradiol.*, **23**, 601–604.

Lindsten, R., & Kurol, J. (1997). Orthodontic appliances in relation to nickel hypersensitivity. *J. Orofac. Orthop/Fortschr. Kieferorthop.*, **58**, 100–108.

Lygre, H., Klepp, K. N., Solheim, E., & Gjerdet, N. R. (1993). Leaching of additives and degradation products from cold-cured orthodontic resins. *Acta. Odontol. Scand.*, **52**, 150–156.

Mackert, J. R., & Berglund, A. (1997). Mercury exposure from dental amalgam fillings: Absorbed dose and the potential for adverse health effects. In I. A. Mjør, & G. N. Pakhomov (Eds.), *Dental Amalgam and Alternative Direct Restorative Materials* (pp. 47–60). Geneva: Oral Health Division of Noncommunicable Diseases, WHO.

Michelsen, B., Moe, G., Skalevik, R., Jensen, E., & Lygre, H. (2007). Quantification of organic eluates from polymerised resin-bases materials by use of GC/MS. *J. Chromatogr. B. Analyt. Technol. Biomed. Life Sci.*, **850**(1–2), 83–91.

Milavec-Puretic, V., Orlic, D., & Marusic, A. (1998). Sensitivity to metals in 40 patients with failed hip endoprosthesis. *Archs. Trauma. Surg.*, **117**, 383–386.

Nylander, M., Friberg, L., Eggleston, R., & Björkmann, L. (1989). Mercury accumulation in tissues from dental staff and controls in relation to exposure. *Swed. Dent. J.*, **13**, 225–245.

Park, H. Y., & Shearer, P. D. (1983). *In vitro* release of nickel and chromium from simulated orthodontic appliances. *Am. J. Orthod.*, **84**, 156–159.

Rodgers, K., Klykken., P., Jacobs, J., Frondoza, C., Tomazic, V., & Zelioff, J. (1997). Immunotoxicity of medical devices. Symposium overview. *Fundam. Appl. Toxical.*, **36**, 1–14.

Schedle, A., Örtengren, U., Eidler, N., Gabauer, M., & Hensten, A. (2007). Do adverse effects of dental materials exists? What are the consequences, and how can they be diagnosed and treated? *Clin. Oral Impl. Res.*, **18**(Suppl. 3), 232–256.

The European Commission. (2008). Opinion on the safety of dental amalgam and alternative dental restoration materials for patients and users. Adopted on 6 May 2008 by Scientific Committee on Emerging and Newly Identified Health Risks (SCENIHR), European Commission, Health and Consumer Protection Directorate-General (DG SANCO), Unit C7-Risk Assessment: http://www.ec.europa.eu/health/ph_risk/committees/04_scenihr/docs/scenihr_o_016.pdf.

Thomas, P., Shuh, A., Summer, B., Mazzochian, F., & Thomsen, M. (2006). Allergy towards bone cement. *Orthopäde*, **35**, 956–960.

Thomas, P., Schuh, A., Ring, J., & Thomsen, M. (2008). Orthopädisch-chirurgische implantate und allergien. *Hautarzt*, **59**, 220–229.

Turjanmaa, K., Alenius, H., Makinen-Kiljunen, S., Reunala, T., & Palosuo, T. (1996). Natural rubber latex allergy (Review). *Allergy*, **51**, 593–602.

CHAPTER II.2.6 BLOOD COAGULATION AND BLOOD–MATERIALS INTERACTIONS

Stephen R. Hanson and Erik I. Tucker
Division of Biomedical Engineering, School of Medicine,
Oregon Health & Science University, Portland, Oregon, USA

The hemostatic mechanism is designed to arrest bleeding from injured blood vessels. The same process may produce adverse consequences when artificial surfaces are placed in contact with blood. These events involve a complex set of interdependent reactions between: (1) the surface; (2) platelets; and (3) coagulation proteins, resulting in the formation of a clot or thrombus which may subsequently undergo removal by; (4) fibrinolysis. The process is localized at the surface by opposing activation and inhibition systems, which ensure that the fluidity of blood in the circulation is maintained. In this chapter, a brief overview of the hemostatic mechanism is presented. Although a great deal is known about blood responses to injured arteries and blood-contacting devices, important relationships remain to be defined in many instances. More detailed discussions of hemostasis and thrombosis have been provided elsewhere (Forbes and Courtney, 1987; Stamatoyannopoulos et al., 1994; Gresle et al., 2002; Esmon 2003; Colman et al., 2005).

PLATELETS

Platelets ("little plates") are non-nucleated, disk-shaped cells having a diameter of 3–4 μm, and an average volume of 10×10^{-9} mm^3. Platelets are produced in the bone marrow, circulate at an average concentration of about 250,000 cells per microliter of whole blood, and occupy approximately 0.3% of the total blood volume. In contrast, red cells typically circulate at 5×10^6 cells per microliter, and may comprise 40–50% of the total blood volume. As discussed below, platelet functions are designed to: (1) initially arrest bleeding through formation of platelet plugs; and (2) stabilize the initial platelet plugs by catalyzing coagulation reactions leading to the formation of fibrin.

Platelet structure provides a basis for understanding platelet function. In the normal (nonstimulated) state, the platelet discoid shape is maintained by a circumferential bundle (cytoskeleton) of microtubules (Figure II.2.6.1). The external surface coat of the platelet contains membrane-bound receptors (e.g., glycoproteins Ib, and IIb/IIIa) that mediate the contact reactions of adhesion (platelet–surface interactions) and aggregation (platelet–platelet interactions). The membrane also provides a phospholipid surface which accelerates important coagulation reactions (see below), and forms a spongy, canal-like (canalicular) open network which represents an expanded reactive surface to which plasma factors are

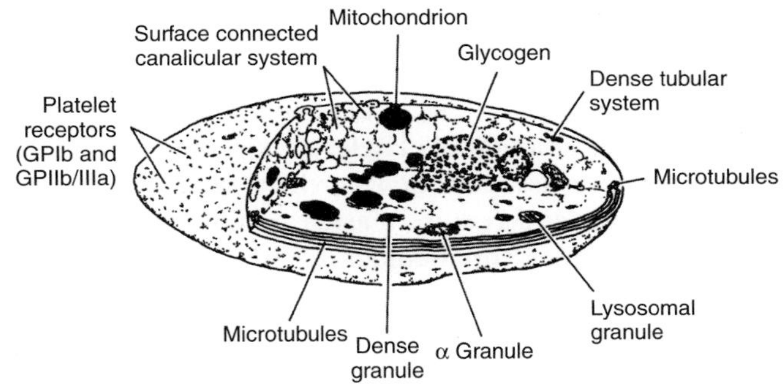

FIGURE II.2.6.1 Platelet structure.

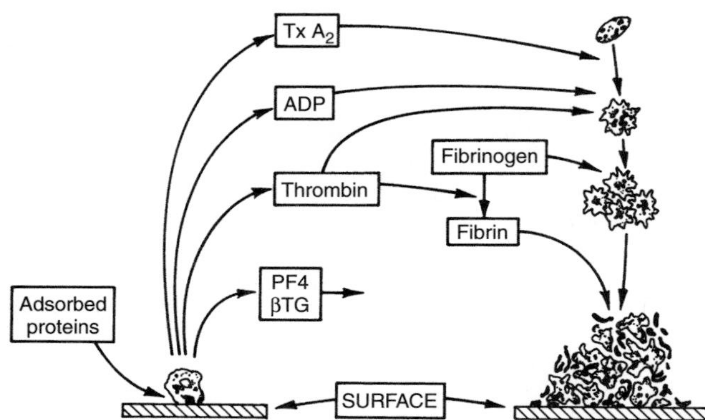

FIGURE II.2.6.2 Platelet reactions to artificial surfaces. Following protein adsorption to surfaces, platelets adhere and release α-granule contents, including platelet factor 4 (PF4) and β-thromboglobulin (β-TG), and dense granule contents, including ADP. Thrombin is generated locally through coagulation reactions catalyzed by procoagulant platelet surface phospholipids. Thromboxane A_2 (TxA_2) is synthesized. ADP, TxA_2, and thrombin recruit additional circulating platelets into an enlarging platelet aggregate. Thrombin-generated fibrin stabilizes the platelet mass.

selectively adsorbed. Platelets contain substantial quantities of muscle protein (e.g., actin, myosin) which allow for internal contraction when platelets are activated. Platelets also contain three types of cytoplasmic storage granules: (1) α-granules, which are numerous and contain the platelet-specific proteins platelet factor 4 (PF-4) and β-thromboglobulin (β-TG), and proteins found in plasma (including fibrinogen, albumin, fibronectin, coagulation factors V and VIII); (2) dense granules which contain adenosine diphosphate (ADP), calcium ions (Ca^{2+}), and serotonin; and (3) lysosomal granules containing enzymes (acid hydrolases).

Platelets are extremely sensitive cells that may respond to minimal stimulation. Activation causes platelets to become sticky and change in shape to irregular spheres with spiny pseudopods. Activation is accompanied by internal contraction and extrusion of the storage granule contents into the extracellular environment. Secreted platelet products such as ADP stimulate other platelets, leading to irreversible platelet aggregation and the formation of a fused platelet thrombus (Figure II.2.6.2).

Platelet Adhesion

Platelets adhere to artificial surfaces and injured blood vessels. At sites of vessel injury, the adhesion process involves the interaction of platelet glycoprotein Ib (GP Ib) and connective tissue elements, which become exposed (e.g., collagen) and require plasma von Willebrand factor (vWF) as an essential cofactor. GP Ib (about 25,000 molecules per platelet) acts as the surface receptor for vWF (Colman et al., 2005). The hereditary absence of GP Ib or vWF results in defective platelet adhesion and serious abnormal bleeding.

Platelet adhesion to artificial surfaces may also be mediated through platelet glycoprotein IIb/IIIa (integrin $\alpha_{IIb}\beta_3$), as well as through the GP Ib–vWF interaction. GP IIb/IIIa (about 80,000 copies per resting platelet) is the platelet receptor for adhesive plasma proteins which support cell attachment, including fibrinogen, vWF, fibronectin, and vitronectin (Gresle et al., 2002). Resting platelets do not bind these adhesive glycoproteins, an event which normally occurs only after platelet activation causes a conformational change in GP IIb/IIIa. Platelets which have become activated near artificial surfaces (for example, by exposure to factors released from already adherent cells) could adhere directly to surfaces through this mechanism (e.g., via GP IIb/IIIa binding to surface-adsorbed fibrinogen). Also, normally unactivated GP IIb/IIIa receptors could react with surface proteins which have undergone conformational changes as a result of the adsorption process (Chapter II.3.2). The enhanced adhesiveness of platelets toward surfaces preadsorbed with fibrinogen supports this view. Following adhesion, activation, and release reactions, the expression of functionally competent GP IIb/IIIa receptors may also support tight binding and platelet spreading through multiple focal contacts with fibrinogen and other surface-adsorbed adhesive proteins.

Platelet Aggregation

Following platelet adhesion, a complex series of reactions is initiated involving: (1) the release of dense granule ADP; (2) the formation of small amounts of thrombin (see below); and (3) the activation of platelet biochemical processes leading to the generation of thromboxane A_2. The release of ADP, thrombin formation, and generation of thromboxanes all act in concert to recruit platelets into the growing platelet aggregate (Figure II.2.6.2). Platelet stimulation by these agonists causes the expression on the platelet surface of activated GP IIb/IIIa, which then binds plasma proteins that support platelet aggregation. In normal blood, fibrinogen, owing to its relatively high concentration (Table II.2.6.1), is the most important protein supporting platelet aggregation. The platelet–platelet interaction involves Ca^{2+}-dependent bridging of adjacent platelets by fibrinogen molecules (platelets will not aggregate in the absence of fibrinogen, GP IIb/IIIa or Ca^{2+}). Thrombin binds directly to platelet thrombin receptors, and plays a key role in platelet aggregate formation by: (1) activating platelets which then catalyze the production of more thrombin; (2) stimulating ADP release and thromboxane A_2 formation; and (3) stimulating the formation of fibrin, which stabilizes the platelet thrombus.

Platelet Release Reaction

The release reaction is the secretory process by which substances stored in platelet granules are extruded from

TABLE II.2.6.1 Properties of Human Clotting Factors

Clotting Factor	Apparent Molecular Weight (Number of Chains)	Approximate Normal Plasma Concentration (μG/Ml)	Active Form
INTRINSIC CLOTTING SYSTEM			
Prekallikrein	86,000 (1)	50	Serine protease
High molecular weight kininogen	120,000 (1)	70–90	Cofactor
Factor XII	80,000 (1)	30–40	Serine protease
Factor XI	160,000 (2)	3–6	Serine protease
Factor IX	55,000 (1)	3–5	Serine protease
Factor VIII[a]	330,000 (1)	0.1–0.2	Cofactor
Von Willebrand factor[a]	250,000 (1)	10	Cofactor for platelet adhesion
EXTRINSIC CLOTTING SYSTEM			
Tissue factor	47,000 (1)	0[b]	Cofactor
Factor VII	50,000 (1)	0.5	Serine protease
COMMON PATHWAY			
Factor X	59,000 (2)	8–10	Serine protease
Factor V	330,000 (1)	5–12	Cofactor
Prothrombin	72,000 (1)	100–150	Serine protease
Fibrinogen	340,000 (6)	1500–4500	Fibrin polymer
Factor XIII	320,000 (4)	10	Transglutaminase

[a]In plasma, factor VIII is complexed with von Willebrand factor, which circulates as a series of multimers ranging in molecular weight from about 600,000 to 2×10^7.

[b]The tissue factor concentration in cell-free plasma is absent or minimal since tissue factor is an integral cell membrane-associated protein expressed by vascular and inflammatory cells.

the platelet. ADP, collagen, epinephrine, and thrombin are physiologically important release-inducing agents, and interact with the platelet through specific receptors on the platelet surface. Alpha-granule contents (PF-4, β-TG, and other proteins) are readily released by relatively weak agonists such as ADP. Release of the dense granule contents (ADP, Ca^{2+}, and serotonin) requires platelet stimulation by a stronger agonist, such as thrombin. Agonist binding to platelets also initiates the formation of intermediates that cause activation of the contractile–secretory apparatus, production of thromboxane A_2, and mobilization of calcium from intracellular storage sites. Elevated cytoplasmic calcium is probably the final mediator of platelet aggregation and release. As noted, substances which are released (ADP), synthesized (TxA_2), and generated (thrombin), as a result of platelet stimulation and release, affect other platelets and actively promote their incorporation into growing platelet aggregates. *In vivo*, measurements of plasma levels of platelet-specific proteins (PF-4, β-TG) have been widely used as indirect measures of platelet activation and release.

Platelet Coagulant Activity

When platelets aggregate, platelet coagulant activity is initiated, including expression of negatively charged membrane phospholipids (phosphatidyl serine) which accelerate two critical steps of the blood coagulation sequence: factor X activation; and the conversion of prothrombin to thrombin (see below). Platelets may also promote the proteolytic activation of factors XII and XI. The surface of the aggregated platelet mass thus serves as a site where thrombin can form rapidly in excess of the neutralizing capacity of blood anticoagulant mechanisms. Thrombin also activates platelets directly, and generates polymerizing fibrin which adheres to the surface of the platelet thrombus.

Platelet Consumption

In humans, platelets labeled with radioisotopes are cleared from circulating blood in an approximately linear fashion over time, with an apparent lifespan of approximately 10 days. Platelet lifespan in experimental animals may be somewhat shorter. With ongoing or chronic thrombosis that may be produced by cardiovascular devices, platelets may be removed from circulating blood at a more rapid rate. Thus steady-state elevations in the rate of platelet destruction, as reflected in a shortening of platelet lifespan, have been used as a measure of the thrombogenicity of artificial surfaces and prosthetic devices (Hanson et al., 1980, 1990).

COAGULATION

In the test tube, at least 12 plasma proteins interact in a series of reactions leading to blood clotting. Their designation as Roman numerals was made in order of discovery, often before their role in the clotting scheme was fully appreciated. Their biochemical properties are summarized in Table II.2.6.1. Initiation of clotting occurs either intrinsically by surface-mediated reactions or extrinsically through factors derived from

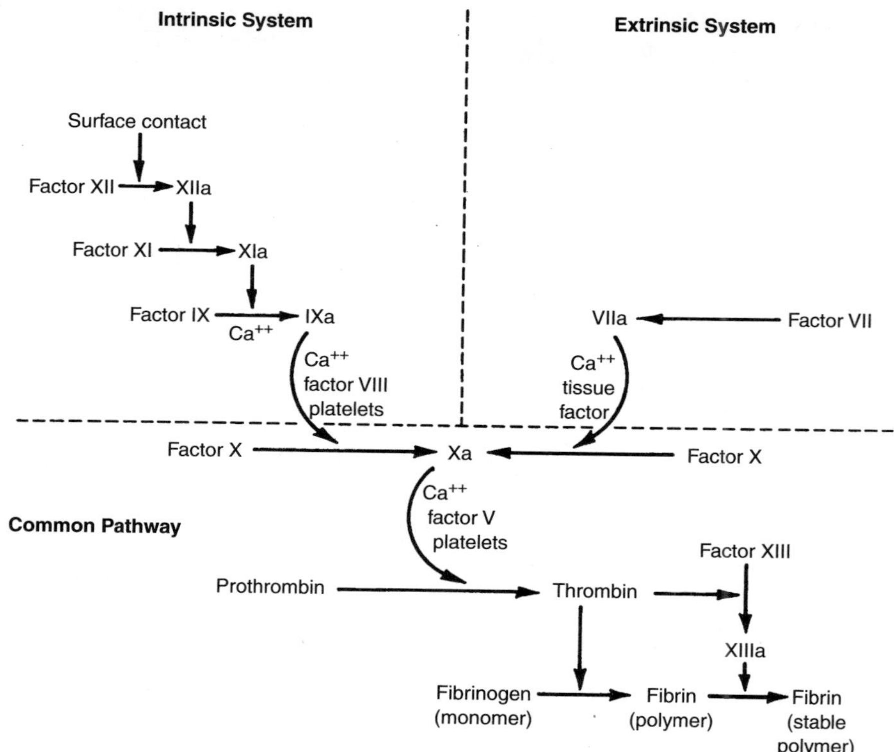

FIGURE II.2.6.3 Mechanisms of clotting factor interactions. Clotting factors (proenzymes), identified by Roman numerals, interact in a sequential series of enzymatic activation reactions (coagulation cascade) leading to the amplified production of the enzyme thrombin, which in turn cleaves fibrinogen to form a fibrin polymer that stabilizes the clot or thrombus. Clotting is initiated by either an intrinsic or extrinsic pathway with subsequent factor interactions which converge upon a final, common path. The underlined factors are all activatable by the enzyme thrombin. The dotted line highlights the importance of feedback activation of FXI by thrombin, which results in increased thrombin production.

tissues. The two systems converge upon a final common pathway which leads to the formation of thrombin and an insoluble fibrin gel when thrombin acts on fibrinogen.

Coagulation proceeds through a "cascade" of reactions by which normally inactive factors (e.g., factor XII) become enzymatically active following surface contact, or after proteolytic cleavage by other enzymes (e.g., surface contact activates factor XII to factor XIIa). The newly activated enzymes in turn activate other normally inactive precursor molecules (e.g., factor XIIa converts factor XI to factor XIa). Because this sequence involves a series of steps, and because one enzyme molecule can activate many substrate molecules, the reactions are quickly amplified so that significant amounts of thrombin are produced, resulting in platelet activation, fibrin formation, and the arrest of bleeding. The process is localized (i.e., widespread clotting does not occur) owing to dilution of activated factors by blood flow, the actions of inhibitors which are present or are generated in clotting blood, and because several reaction steps proceed at an effective rate only when catalyzed on the surface of activated platelets or at sites of tissue injury.

Figure II.2.6.3 presents a scheme of the clotting factor interactions involved in both the intrinsic and extrinsic systems, and their common path. Except for the contact phase, calcium is required for most reactions and is the reason why chelators of calcium (e.g., citrate) are effective anticoagulants. It is also clear that the *in vitro* interactions of clotting factors, i.e., clotting, is not identical with coagulation *in vivo*, which may be triggered by artificial surfaces and by exposure of the cell-associated protein, tissue factor. There are also interrelationships between the intrinsic and extrinsic systems, such that under some conditions "crossover" or reciprocal activation reactions may be important (Bennett et al., 1987; Colman et al., 2005).

MECHANISMS OF COAGULATION

In the intrinsic clotting system, contact activation refers to reactions following adsorption of contact factors onto a negatively charged surface. Involved are factors XII, XI, prekallikrein, and high molecular weight kininogen (HMWK) (Figure II.2.6.4). All contact reactions take place in the absence of calcium. Kallikrein also participates in fibrinolytic system reactions and inflammation (Bennett et al., 1987). Although these reactions are well-understood *in vitro*, their pathologic significance remains uncertain. For example, in hereditary disorders, factor XII deficiency is not associated with an increased bleeding tendency, and only a marked deficiency of factor XI produces abnormal bleeding.

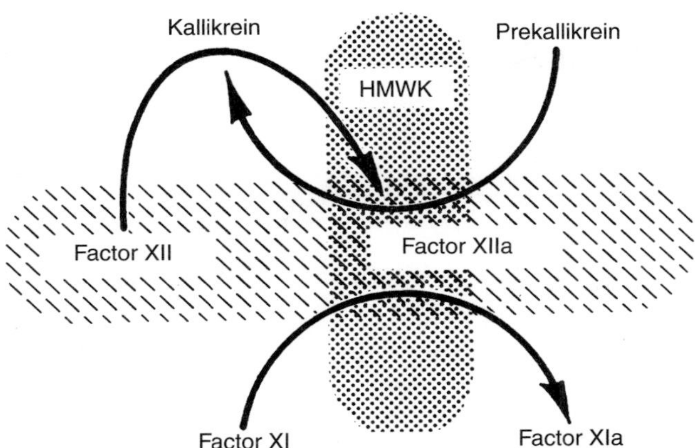

FIGURE II.2.6.4 Contact activation. The initial event *in vitro* is the adsorption of factor XII to a negatively charged surface (hatched, horizontal ovoid) where it is activated to form factor XIIa. Factor XIIa converts prekallikrein to kallikrein. Additional factor XIIa and kallikrein are then generated by reciprocal activation. Factor XIIa also activates factor XIa. Both prekallikrein and factor XI bind to a cofactor, high molecular weight kininogen (HMWK), which anchors them to the charged surface. Kallikrein is also responsible for the liberation of the vasoactive peptide bradykinin for HMWK, linking coagulation and inflammation.

A middle phase of intrinsic clotting begins with the first calcium-dependent step, the activation of factor IX by factor XIa. Factor IXa subsequently activates factor X. Factor VIII is an essential cofactor in the intrinsic activation of factor X, and factor VIII first requires modification by an enzyme, such as thrombin, to exert its cofactor activity. In the presence of calcium, factors IXa and VIIIa form a complex (the "tenase" complex) on phospholipid surfaces (expressed on the surface of activated platelets) to activate factor X. This reaction proceeds slowly in the absence of an appropriate phospholipid surface, and serves to localize the clotting reactions to the surface (versus bulk fluid) phase. The extrinsic system is initiated by the activation of factor VII. When factor VII interacts with tissue factor, a cell membrane protein which may also circulate in a soluble form, factor VIIa becomes an active enzyme which is the extrinsic factor X activator. Tissue factor is present in many body tissues; is expressed by stimulated white cells and endothelial cells, and becomes available when underlying vascular structures are exposed to flowing blood upon vessel injury.

The common path begins when factor X is activated by either factor VIIa-tissue factor by or the factor IXa–VIIIa complex. After formation of factor Xa, the next step involves factor V, a cofactor which (like factor VIII) has activity after modification by another enzyme, such as thrombin. Factor Xa–Va, in the presence of calcium and platelet phospholipids, forms a complex ("prothrombinase" complex) which converts prothrombin (factor II) to thrombin. Like the conversion of factor X, prothrombin activation is effectively surface catalyzed. The higher plasma concentration of prothrombin (Table II.2.6.1), as well as the biological amplification of the clotting system, allows a few molecules of activated initiator to generate a large burst of thrombin activity. Thrombin, in addition to its ability to modify factors V and VIII and activate platelets, acts on two substrates: fibrinogen and factor XIII. The action of thrombin on fibrinogen releases small peptides from fibrinogen (e.g., fibrinopeptide A) which can be assayed in plasma as evidence of thrombin activity. The fibrin monomers so formed polymerize to become a gel. Factor XIII is either trapped within the clot or provided by platelets, and is activated directly by thrombin. A tough, insoluble fibrin polymer is formed by interaction of the fibrin polymer with factor XIIIa.

CONTROL MECHANISMS

Obviously, the blood and vasculature must have mechanisms for avoiding massive thrombus formation once coagulation is initiated. At least four types of mechanisms may be considered. First, blood flow may reduce the localized concentration of precursors and remove activated materials by dilution into a larger volume, with subsequent removal from the circulation following passage through the liver. Second, the rate of several clotting reactions is fast only when the reaction is catalyzed by a surface. These reactions include the contact reactions, the activation of factor X by factor VII tissue factor at sites of tissue injury, and reactions which are accelerated by locally deposited platelet masses (activation of factor X and prothrombin). Third, there are naturally occurring inhibitors of coagulation enzymes, such as antithrombin III, which are potent inhibitors of thrombin and other coagulation enzymes (plasma levels of thrombin–antithrombin III complex can also be assayed as a measure of thrombin production *in vivo*). Another example of a naturally occuring inhibitor is tissue factor pathway inhibitor (TFPI), a protein that, in association with factor Xa, inhibits the tissue factor–factor VII complex. Fourth, during the process of coagulation, enzymes are generated which not only activate coagulation factors, but also degrade cofactors. For example, the fibrinolytic enzyme plasmin (see below) degrades fibrinogen and fibrin monomers, and can inactivate cofactors V and VIII. Thrombin is also removed when it binds to thrombomodulin, a protein found on the surface of blood vessel endothelial cells. The thrombin–thrombomodulin complex then converts another plasma protein, protein C, to an active form which can also degrade factors V and VIII. *In vivo*, the protein C pathway is a key physiologic anticoagulant mechanism (Esmon, 2003; Colman et al., 2005).

In summary, the platelet, coagulation, and endothelial systems interact in a number of ways which promote localized hemostasis while preventing generalized thrombosis. Figure II.2.6.5 depicts some of the relationships and inhibitory pathways which apply to blood

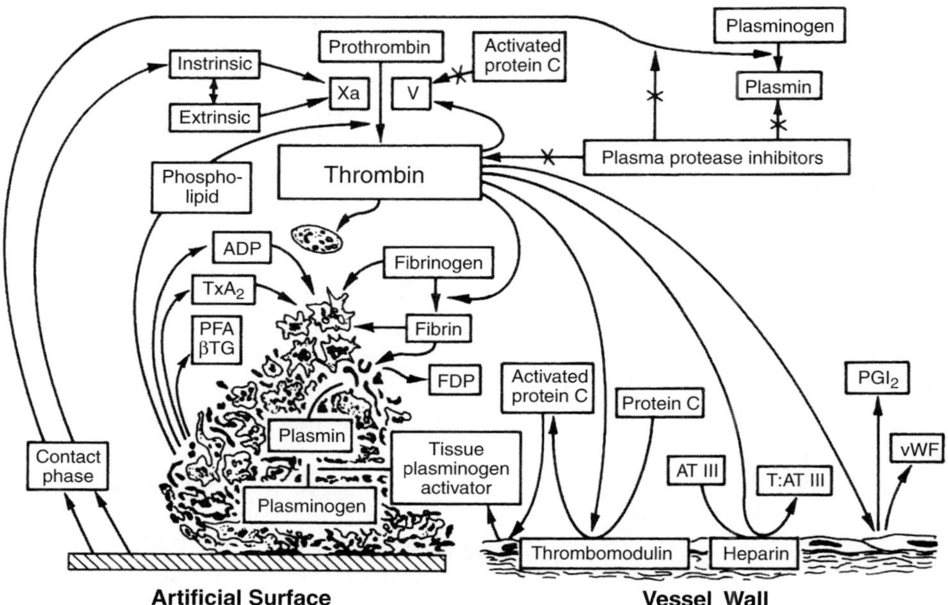

FIGURE II.2.6.5 Integrated hemostatic reactions between a foreign surface and platelets, coagulation factors, the vessel endothelium, and the fibrinolytic system.

reactions following contact with both natural and artificial surfaces.

Fibrinolysis

The fibrinolytic system removes unwanted fibrin deposits to improve blood flow following thrombus formation, and to facilitate the healing process after injury and inflammation. It is a multicomponent system composed of precursors, activators, cofactors, and inhibitors, and has been studied extensively (Forbes and Courtney, 1987; Colman et al., 2005). The fibrinolytic system also interacts with the coagulation system at the level of contact activation (Bennett et al., 1987). A simplified scheme of the fibrinolytic pathway is shown in Figure II.2.6.6.

The most well-studied fibrinolytic enzyme is plasmin, which circulates in an inactive form as the protein plasminogen. Plasminogen adheres to a fibrin clot, being incorporated into the mesh during polymerization. Plasminogen is activated to plasmin by the actions of plasminogen activators which may be present in blood or released from tissues, or which may be administered therapeutically. Important plasminogen activators occurring naturally in humans include tissue plasminogen activator (tPA) and urokinase. Following activation, plasmin digests the fibrin clot, releasing soluble fibrin-fibrinogen digestion products (FDP) into circulating blood, which may be assayed as markers of *in vivo* fibrinolysis (e.g., the fibrin D-D dimer fragment). Fibrinolysis is inhibited by plasminogen activator inhibitors (PAIs), and by a thrombin-activated fibrinolysis inhibitor (TAFI), which promotes the stabilization of fibrin and fibrin clots (Colman et al., 2005).

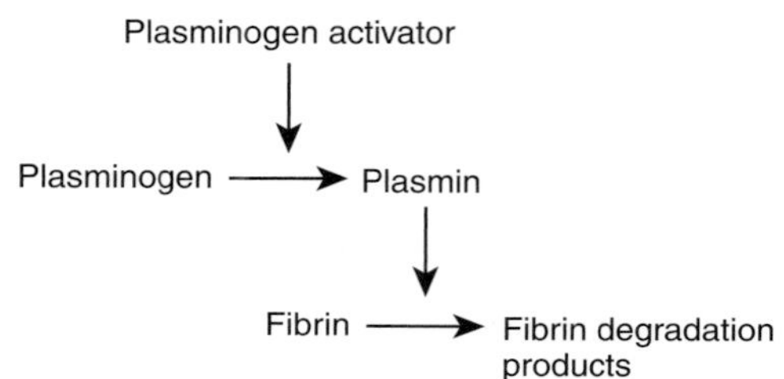

FIGURE II.2.6.6 Fibrinolytic sequence. Plasminogen activators, such as tissue plasminogen activator (tPA) or urokinase, activate plasminogen to form plasmin. Plasmin enzymatically cleaves insoluble fibrin polymers into soluble degradation products (FDP), thereby effecting the removal of unnecessary fibrin clot.

Complement

The complement system is primarily designed to effect a biological response to antigen–antibody reactions. Like the coagulation and fibrinolytic systems, complement proteins are activated enzymatically through a complex series of reaction steps (Bennett et al., 1987). Several proteins in the complement cascade function as inflammatory mediators. The end result of these activation steps is the generation of an enzymatic complex which causes irreversible damage (by lytic mechanisms) to the membrane of the antigen-carrying cell (e.g., bacteria).

Since there are a number of interactions between the complement, coagulation, and fibrinolytic systems, there has been considerable interest in the problem of complement activation by artificial surfaces, prompted in part by observations that devices having large surface areas (e.g., hemodialyzers) may cause: (1) reciprocal activation reactions between complement enzymes and white cells; and (2) complement activation which may mediate both white cell and platelet adhesion to artificial surfaces.

Further observations regarding the complement activation pathways involved in blood–materials interactions are likely to be of interest.

Red Cells

Red cells are usually considered as passive participants in processes of hemostasis and thrombosis, although under some conditions (low shear or venous flows) red cells may comprise a large proportion of the total thrombus mass. The concentration and motion of red cells have important mechanical effects on the diffusive transport of blood elements. For example, in flowing blood, red cell motions may increase the effective diffusivity of platelets by several orders of magnitude. Under some conditions, red cells may also contribute chemical factors that influence platelet reactivity (Turitto and Weiss, 1980). The process of direct attachment of red cells to artificial surfaces has been considered to be of minor importance, and has therefore received little attention in studies of blood–materials interactions.

White Cells

The various classes of white cells perform many functions in inflammation, infection, wound healing, and the blood response to foreign materials. White cell interactions with artificial surfaces may proceed through as yet poorly-defined mechanisms related to activation of the complement, coagulation, fibrinolytic, and other enzyme systems, resulting in the expression by white cells of procoagulant, fibrinolytic, and inflammatory activities. For example, stimulated monocytes express tissue factor, which can initiate extrinsic coagulation. Neutrophils may contribute to clot dissolution by releasing potent fibrinolytic enzymes (e.g., neutrophil elastase). White cell interactions with devices having large surface areas may be extensive (especially with surfaces that activate complement), resulting in their marked depletion from circulating blood. Activated white cells, through their enzymatic and other activities, may produce organ dysfunction in other parts of the body. In general, the role of white cell mechanisms of thrombosis and thrombolysis, in relation to other pathways, remains an area of considerable interest.

CONCLUSIONS

Interrelated blood systems respond to tissue injury in order to rapidly minimize blood loss, and later to remove unneeded deposits after healing has occurred. When artificial surfaces are exposed, an imbalance between the processes of activation and inhibition of these systems can lead to excessive thrombus formation, and an exaggerated inflammatory response. While many of the key blood cells, proteins, and reaction steps have been identified, their reactions in association with artificial surfaces have not been well-defined in many instances. Therefore, blood reactions which might cause thrombosis continue to limit the potential usefulness of many cardiovascular devices for applications in humans. Consequently, these devices commonly require the use of systemic anticoagulants, which present an inherent bleeding risk.

ACKNOWLEDGMENT

This work was supported by research grant HL-31469 and from the National Institutes of Health, U.S. Public Health Service, and by the ERC Program of the National Science Foundation under Award EEC-9731643.

BIBLIOGRAPHY

Bennett, B., Booth, N. A., & Ogston, D. (1987). Potential interactions between complement, coagulation, fibrinolysis, kinin-forming, and other enzyme systems. In A. L. Bloom, & D. P. Thomas (Eds.), *Haemostasis and Thrombosis* (2nd ed.). (pp. 267–282). New York, NY: Churchill Livingstone.

Colman, R. W., Marder, V. J., Clowes, A. W., George, J. N., & Goldhaber, S. Z. (Eds.) (2005). *Hemostasis and Thrombosis* (5th ed.). New York, NY: Lippincott.

Esmon, C. T. (2003). The protein C pathway. *Chest*, **124**(3 Suppl.), 26S–32S.

Forbes, C. D., & Courtney, J. M. (1987). Thrombosis and artificial surfaces. In A. L. Bloom, & D. P. Thomas (Eds.), *Haemostasis and Thrombosis* (2nd ed.). (pp. 902–921). New York, NY: Churchill Livingstone.

Gresle, P., Page, C. P., Fuster, F., & Vermylen, J. (2002). *Platelets in Thrombotic and Non-thrombotic Disorders* (1st ed.). Cambridge: Cambridge University Press.

Hanson, S. R., Harker, L. A., Ratner, B. D., & Hoffman, A. S. (1980). *In vivo* evaluation of artificial surfaces using a nonhuman primate model of arterial thrombosis. *J. Lab. Clin. Med.*, **95**, 289–304.

Hanson, S. R., Kotze, H. F., Pieters, H., Heyns, A., & du, P. (1990). Analysis of 111- Indium platelet kinetics and imaging in patients with aortic aneurysms and abdominal aortic grafts. *Arteriosclerosis*, **10**, 1037–1044.

Stamatoyannopoulos, G., Nienhuis, A. W., Majerus, P. W., & Varmus, H. (1994). *The Molecular Basis of Blood Diseases* (2nd ed.). Philadelphia, PA: W.B. Saunders.

Turitto, V. T., & Weiss, H. J. (1980). Red cells: Their dual role in thrombus formation. *Science*, **207**, 541–544.

CHAPTER II.2.7 TUMORS ASSOCIATED WITH BIOMATERIALS AND IMPLANTS

Frederick J. Schoen
Professor of Pathology and Health Sciences and Technology (HST), Hardvard Medical School, Executive Chairman, Department of Pathology, Brigham and Women's Hospital, Boston, MA, USA

Biomaterials have been shown to be potentially tumorigenic in laboratory animals, and malignancies have occurred rarely in association with metallic and other implants in pets and humans. Thus, the possibility that implant materials could cause tumors or promote tumor growth has long been a concern of clinicians and biomaterials researchers. This chapter describes general concepts in neoplasia (i.e., the process of tumor formation), the association of tumors with implants in humans and animals, the pathobiology of tumor formation adjacent to (and potentially stimulated by) biomaterials, and other pathophysiologic interactions of biomaterials with tissues that may simulate tumor formation.

GENERAL CONCEPTS

Neoplasia, which literally means "new growth," is the process of excessive and uncontrolled cell proliferation (Kumar et al., 2010). The new growth is called a *neoplasm* or *tumor* (i.e., a swelling, since most neoplasms are expansile, solid masses of abnormal tissue). Tumors are either *benign* (when their pathologic characteristics and clinical behavior are relatively innocent) or *malignant* (meaning harmful, often deadly). Malignant tumors are collectively referred to as *cancers* (derived from the Latin word for crab, to emphasize their obstinate ability to adhere to adjacent structures and spread in many directions simultaneously). The characteristics of benign and malignant tumors are summarized in Table. II.2.7.1. Types of malignant tumors are shown in Figure. II.2.7.1. Benign tumors do not penetrate (invade) adjacent tissues, nor do they spread to distant sites. They remain localized and, thus, surgical excision can be curative in many cases. In contrast, malignant tumors have a propensity to invade contiguous tissues. Moreover, owing to their ability to gain entrance into blood and lymph vessels, cells from a malignant neoplasm can be transported to distant sites, where subpopulations of malignant cells take up residence, grow, and again invade as satellite tumors (called *metastases*).

The primary descriptor of any tumor is its cell or tissue of origin. Benign tumors are identified by the suffix "oma," which is preceded by a reference to the cell or tissue of origin (e.g., adenoma – from an endocrine gland; chondroma – from cartilage). The malignant counterparts of benign tumors carry similar names, except that the suffix *carcinoma* is applied to cancers derived from epithelium (e.g., squamous- or adeno-carcinoma, from protective and glandular epithelia, respectively), and *sarcoma* (e.g., osteo- or chondro-sarcoma, producing bone and cartilage, respectively) is applied to those of mesenchymal (i.e., connective tissue or vascular) origin. Malignant neoplasms of the hematopoietic (i.e., blood cell producing) system, in which the cancerous cells circulate in blood, are called *leukemias*; solid tumors of lymphoid tissue are called *lymphomas*.

Cancer cells express varying degrees of resemblance to the normal precursor cells from which they derive. Thus, neoplastic growth entails both abnormal cellular proliferation, and modification of the structural and functional characteristics of the cell types involved. Malignant cells are generally less differentiated than their original normal counterparts. The structural similarity of cancer cells to those of the tissue of origin enables specific *diagnosis* (of both source organ and cell type); moreover, the degree of resemblance to the corresponding normal tissue usually predicts *prognosis* of the patient (i.e., expected outcome based on biologic behavior of the cancer surmised from its pathologic characteristics). Therefore, poorly differentiated tumors are generally more aggressive (i.e., display more malignant behavior) than those that are better differentiated. The degree to which a tumor mimics a normal cell or tissue type is called its *grade* of differentiation. The extent of spread and other effects on the patient determine its *stage*.

Neoplastic growth is unregulated. Neoplastic cell proliferation is therefore unrelated to the physiological requirements of the tissue, and is unaffected by removal of the stimulus which initially caused it. Thus, its growth is autonomous. These characteristics differentiate neoplasms from: (1) normal proliferations of cells during fetal development or postnatal growth; (2) normal

TABLE II.2.7.1 Characteristics of Benign and Malignant Tumors

Characteristics	Benign	Malignant
Differentiation	Well defined; structure may be typical of tissue of origin	Less differentiated, with bizarre (anaplastic) cells; often atypical structure
Rate of growth	Usually progressive and slow; may come to a standstill or regress; cells in mitosis are rare	Erratic, and may be slow to rapid; mitoses may be absent to numerous and abnormal
Local invasion	Usually cohesive, expansile, well-demarcated masses that neither invade nor infiltrate the surrounding normal tissues	Locally invasive, infiltrating adjacent normal tissues
Metastasis	Absent	Often present at distant sites; larger and more undifferentiated primary tumors more likely to metastasize

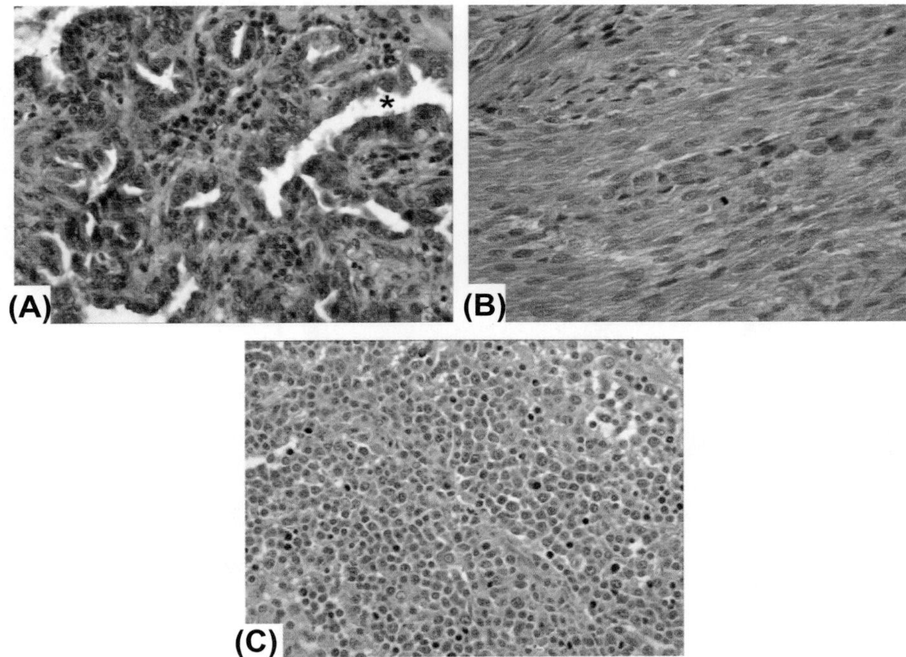

FIGURE II.2.7.1 Types of malignant tumors. (A) Carcinoma, exemplified by an adenocarcinoma (gland formation noted by asterisk). (B) Sarcoma (composed of spindle cells). (C) Lymphoma (composed of an aggregation of malignant lymphocytes). All stained with hematoxylin and eosin. (A, B and C courtesy of Robert F. Padera, M.D., Ph.D., Department of Pathology, Brigham and Women's Hospital, Boston, USA.)

wound healing following an injury; and (3) hyperplastic growth which adapts to a physiological need, but which ceases when the stimulus is removed.

Analogous to normal tissues (see Chapter II.1.5), benign and malignant tumors have two basic components: (1) proliferating neoplastic cells that constitute their *parenchyma*; and (2) supportive *stroma* made up of connective tissue and blood vessels. The parenchyma of neoplasms is characteristic of the specific cells of origin. Tumor growth and evolution are critically dependent on the non-specific stroma, which is usually composed of blood vessels, connective tissue, and inflammatory cells.

Neoplasms occurring at the site of implanted medical devices are rare, despite the large numbers of implants used clinically over an extended period of time. Nevertheless, both human and veterinary implant-associated tumors have been reported, most as case reports or small series.

ASSOCIATION OF IMPLANTS WITH HUMAN AND ANIMAL TUMORS

Neoplasms occurring at the site of implanted medical devices are rare, despite the large numbers of implants used clinically over an extended period of time. Nevertheless, both human and veterinary implant-associated tumors have been reported, most as case reports or small series (Pedley et al., 1981; Schoen, 1987; Black, 1988; Jennings et al., 1988; Keel et al., 2001; Visuri et al., 2006a,b; Balzer and Weiss, 2009). In all, approximately 100 cases of tumors associated with orthopedic and other surgical implants have been reported, and tumors related to other foreign materials (e.g., bullets, shrapnel, other metal fragments, sutures, bone wax, and surgical sponge) are also known. One review cites 46 cases of reported malignant tumors at the site of total hip arthroplasty from 1974 to 2002 (Visuri et al., 2006a). Tumors are also well-known to occur with environmental exposure and consequent pulmonary ingestion of foreign materials, especially asbestos, which has a cause-and-effect relationship with a special type of tumor called *mesothelioma* (Robinson et al., 2005).

The vast majority of malignant neoplasms associated with clinical fracture fixation devices, total joint replacements, vascular grafts, breast implants, and experimental foreign-bodies in both animals and humans are sarcomas (i.e., mesenchymal tumors). They comprise various histologic subtypes, including fibrosarcoma, osteosarcoma (osteogenic sarcoma), chondrosarcoma, malignant fibrous histiocytoma, angiosarcoma, etc., and are characterized by rapid and locally infiltrative growth. Therapeutic implant-related tumors have been reported both short- and long-term following implantation. In a more recent case series and review confined to orthopedic implant sarcomas (Keel et al., 2001), most presented less than 15 years following surgery; implant associated tumors presented as early as 0.5 years and as late as 30 years. An osteosarcoma has been reported forming adjacent to a clinical titanium total hip replacement for osteoarthritis three years following insertion in a 68-year-old man (illustrated in Figure. II.2.7.2) (Keel et al., 2001). Diagnosis was delayed, since the initial

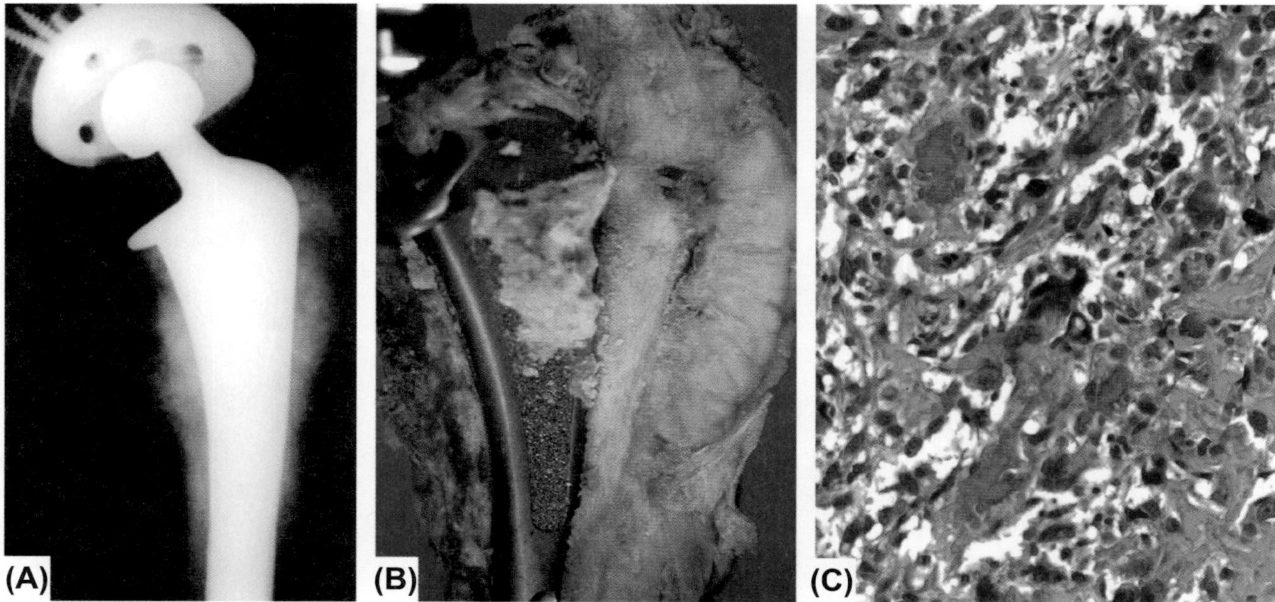

FIGURE II.2.7.2 Orthopedic implant-related osteosarcoma in a 68-year-old man three years following total hip replacement. (A) Radiograph demonstrating the osteoblastic (i.e., bone-making) tumor surrounding the implant. (B) Photo of the tumor, showing its relationship to the implant and partially destroyed proximal femur. (C) Photomicrograph of tumor, demonstrating high-grade osteosarcoma. The cytologically malignant cells surround coarse lace-like neoplastic bone. (Courtesy of Andrew E. Rosenberg, M.D., Department of Pathology, University of Miami, Miami, FL, USA. (A) and (B) reproduced by permission from Keel, S. B. et al. (2001). Orthopaedic implant-related sarcoma: A study of twelve cases. *Mod. Pathol.* **14**, 969–977.)

symptoms of the tumor were attributed to the original orthopedic problem. Carcinomas, reported in association with foreign-bodies far less frequently, have usually been restricted to situations where an implant has been placed in the lumen of an epithelium-lined organ. Illustrative reported cases are noted in Table. II.2.7.2; descriptions of others are available (Jennings, et al., 1988; Goodfellow, 1992; Jacobs, et al., 1992; Keel et al., 2001; Balzer and Weiss 2009, Mallick et al. 2009). Lymphomas have also been reported in association with the capsules surrounding breast implants (Keech and Creech, 1997; Gaudet et al., 2002; Sahoo et al., 2003; Visuri et al., 2006a; Mallick et al., 2009). A non-implant-related primary tumor (gastric cancer) with a metastasis to a total knee replacement also has been reported (Kolstad and Hgstorp, 1990).

> Whether there is a casual role for implanted medical devices in local or distant malignancy in general or in specific cases remains controversial. In an individual case, caution is necessary in implicating the implant in the formation of a neoplasm; clearly, demonstration of a tumor occurring adjacent to an implant does not necessarily prove that the implant caused the tumor.

Whether there is a causal role for implanted medical devices in local or distant malignancy in general or in specific cases remains controversial. In an individual case, caution is necessary in implicating the implant in the formation of a neoplasm; clearly, demonstration of a tumor occurring adjacent to an implant does not necessarily prove that the implant caused the tumor (Morgan and Elcock, 1995). Large-scale epidemiological studies and reviews of available data have concluded that there is no evidence in humans for tumorgenicity of non-metallic and metallic surgical implants (McGregor et al., 2000). Indeed, the risk in populations must be low, as exemplified by the recent cohorts of patients with both total hip replacement and breast implants that show no detectable increases in tumors at the implant site (Deapen and Brody, 1991; Berkel et al., 1992; Mathiesen et al., 1995; Brinton and Brown, 1997; Visuri, 2006b). Although cancers at most other (systemic) sites seem not to be increased, some studies suggest that there may be a slightly increased risk of the hematopoietic cancers (i.e., leukemias and lymphomas). Such cancers occur in organs where exogenous elements and particles may accumulate (e.g., lymph nodes); in this regard, one study suggests enhanced surveillance in total hip recipients with metal-on-metal articulations, in which the number of metallic particles is estimated to be much higher than with metal-on-plastic bearings (Visuri, 2006b), and the size of the particles smaller, with a correspondingly high reactive surface area. One clinical and experimental study suggested that the incidence of breast carcinoma may actually be decreased in women with breast implants (Su et al., 1995). Importantly, the presence of a breast implant does not impair the diagnosis of breast cancer (Brinton and Brown, 1997).

Moreover, neoplasms are common in both humans and animals, and can occur naturally at sites where biomaterials are implanted. Spontaneous human sarcomas in musculoskeletal sites are not unusual. Most clinical veterinary cases have been observed in dogs, a species

TABLE. II.2.7.2 Tumors Associated with Implant Sites in Humans: Representative Reports

Device (Adjacent Material)[a]	Tumor[b]	References	Postimplantation (Years)
Fracture Fixation			
Intramedullary rod (V)	L	MacDonald (1980)	17
Smith-Petersen (V)	OS	Ward et al. (1987)	9
Total Hip			
Charnley-Mueller (UHMWPE, PMMA)	MFH	Bago-Granell et al. (1984)	2
Mittlemeier (Al_2O_3)	STS	Ryu et al. (1987)	1+
Charnley-Mueller (UHMWPE)	OS	Martin et al. (1988)	10
Charnley-Mueller (SS, PMMA)	SS	Lamovec et al. (1988)	12
Unknown (porous Ti-cobalt alloy)	OS	Adams et al. (2003)	3
Total Knee			
Unknown (V)	ES	Weber (1986)	4
Vascular Graft			
Abdominal aortic graft (D)	MFH	Weinberg and Maini (1980)	1+
Abdominal aortic graft (D)	AS	Fehrenbacker et al. (1981)	12
Heart Valve Prosthesis			
St. Jude Medical (carbon, sizone-coated Dacron sewing cuff)	RS	Grubitzsch et al. (2001)	<1

[a]Materials: D: Dacron; PMMA: poly(methacrylate) bone cement; SS: stainless steel; Ti: titanium; UHMWPE: ultra-high molecular weight polyethylene; V: Vitallium.
[b]Tumor types: AS: angiosarcoma; ES: epithelioid sarcoma; L: lymphoma; MFH: malignant fibrous histiocytoma; OS: osteosarcoma; RS: rhapdomyosarcoma; SS: synovial sarcoma; STS: soft tissue sarcoma.

with a relatively high natural frequency of osteosarcoma, and other tumors at sites where orthopedic devices are implanted. However, since sarcomas arising in the aorta and other large arteries are particularly rare, the association of primary vascular malignancies with clinical polymeric grafts may be stronger than that with orthopedic devices (Weinberg and Maini, 1980).

A possible association between anaplastic large cell lymphoma (ALCL) and breast implants, which has received recent attention from the FDA (Center for Devices and Radiological Health, FDA, 2011), exemplifies the difficulties in assessing cause-effect relationships with implant-associated tumors. Firstly, it is not possible to confirm with statistical certainty that breast implants cause ALCL. ALCL is extremely rare, and women with breast implants may have a very small but increased risk of developing this disease in the scar capsule adjacent to the implant (and not in other sites within the breast tissue). No association of risk with type of implant (silicone versus saline) or a reason for implant (reconstruction versus augmentation) has been established. ALCL has been identified most frequently in patients undergoing implant revision operations for late onset, persistent peri-implant fluid collection (seroma), with implant durations reported to vary from 3 to 23 years. Some researchers have suggested that breast implant-associated ALCL may represent a new clinical entity with less-aggressive (indolent) behavior than ALCL not associated with implants (Miranda et al., 2009).

Benign but exuberant foreign-body reactions may simulate neoplasms. A fibrous lesion caused by a foreign-body reaction which masquerades as neoplasm is often called *pseudotumor*. For example, fibrohistiocytic lesions resembling malignant tumors may occur as a reaction to silica, previously injected as a soft tissue sclerosing agent (Weiss et al., 1978), and exuberant fibrohistiocytic lesions have been reported in associated with the capsules forming around breast implants (Balzer and Weiss, 2009) and a polytetraethylene sleeve used in association with a lung cancer resection (Fernandez et al., 2008). Moreover, regional lymphadenopathy (i.e., enlargement of lymph nodes) may result from an exuberant foreign-body reaction to material which has migrated from a prosthesis through lymphatic vessels. This has been documented in cases of silicone emanating from both finger joints (Christie et al., 1977) and breast prostheses (Hausner et al., 1978), as well as in association with polymeric replacements of the temporomandibular joint, and with conventional metallic, ceramic, and polymeric total replacements of large joints (Jacobs et al., 1995).

Exuberant fibro-inflammatory lesions have been reported recently with breast implants, metal-on-metal hip replacements, and other implants. For example, such lesions have been reported in associated with the capsules forming around breast implants (Blazer, 2009) and a polytetraethylene sleeve used in association with a lung cancer resection (Fernandez, 2008). Moreover, locally-destructive pseudotumors have been reported recently in association with failed metal-on-metal hip resurfacing arthroplasty and may occur in as many as 1% of patients who have had metal-on-metal resurfacings (Kwon et al, 2010; Ebramzadeh et al, 2011; Guyer et al, 2011). The cause is presently unknown, but a toxic and/or hypersensitivity reaction to a high rate of particulate wear debris is considered to be a central feature.

PATHOBIOLOGY OF FOREIGN-BODY TUMORIGENESIS

Considerable progress has been made over the last several decades toward understanding the molecular basis of cancer (Kumar et al., 2010). Four principles are fundamental and well-accepted: (1) neoplasia is associated with, and often results from, non-lethal genetic damage (or mutation), either inherited or acquired by the action of environmental agents such as physical effects (e.g., radiation, fibers or foreign bodies; Fry, 1989), chemicals or viruses; (2) the principal targets of the genetic damage are cellular regulatory genes (normally present and necessary for physiologic cell function, inducing cellular replication, growth, and repair of damaged DNA); (3) the tumor mass evolves from the clonal expansion of a single progenitor cell which has incurred the genetic damage; (4) tumorigenesis is a multistep process, generally owing to accumulation of successive genetic lesions. After a tumor has been initiated, the most important factors in its growth are the kinetics (i.e., balance of replication or loss) of cell number change and its blood supply. The formation of new vessels (called *angiogenesis*) is essential for enlargement of tumors, and for their access to the vasculature, and hence, metastasis.

> The pathogenesis of implant-induced tumors is not well-understood, yet most experimental data indicated that physical effects rather than the chemical characteristics of the foreign-body are the principal determinants of tumorigenicity.

The pathogenesis of implant-induced tumors is not well-understood, yet most experimental data indicate that physical effects rather than the chemical characteristics of the foreign-body are the principal determinants of tumorigenicity (Brand et al., 1975). Tumors are induced experimentally by a wide array of materials of diverse composition, including those that could be considered essentially nonreactive, such as certain glasses, gold or platinum, and other relatively pure metals and polymers. Indeed, one surgeon performed a much-maligned experiment in which coins inserted in rats yielded a rate of 60% sarcomas in 16 months (prompting the suggestion that probably all metallic coins were carcinogenic and should be discontinued!) (Moore and Palmer, 1997). Solid materials implanted in a configuration with high surface area are most tumorigenic. Materials lose their tumorigenicity when implanted in pulverized, finely shredded or woven form, or when surface continuity is interrupted by multiple perforations. This trend is often called the *Oppenheimer effect*. Thus, foreign-body neoplasia is generally considered to be a transformation process mediated by the physical state of implants; it is largely independent of the composition of the materials, unless specific carcinogens are present. However, recent evidence suggests that tumor formation induced by nanoparticles is dependent on both the physical and chemical properties in a particle-dependent preneoplasia-neoplasia model (Hansen et al., 2006). Clearly, more detailed studies regarding the physical and chemical particle–tissue interactions are of vital interest for our understanding of the use of nanoparticles in medicine.

Solid-state tumorigenesis depends on the development of a fibrous capsule around the implant, a process that involves myofibroblasts (Hinz et al., 2007), mesenchymal cells which could be involved in tumor formation. Tumorigenicity corresponds directly to the extent and maturity of tissue encapsulation of a foreign-body, and inversely with the degree of active cellular inflammation. Thus, an active, persistent inflammatory response inhibits tumor formation in experimental systems. Host (especially genetic) factors also affect the propensity to form tumors as a response to foreign-bodies. Humans are less susceptible to foreign-body tumorigenesis than rodents, the usual experimental model. In rodent systems, tumor frequency and latency depend on species, strain, sex, and age. Concern has been raised over the possibility that foreign-body neoplasia can be induced by the release of wear debris or needle-like elements from composites in a mechanism that is analogous to that of asbestos-related mesothelioma (Brown et al., 1990; Jurand, 1991). However, animal experiments suggest only particles with very high length-to-diameter ratios (>100) produce this effect. Particles with this high aspect ratio are highly unlikely to arise as wear debris from orthopedic implants.

Nevertheless, cancer at foreign-body sites may be mechanistically related to that which occurs in diseases in which tissue fibrosis is a prominent characteristic, including asbestosis (i.e., lung damage caused by chronic inhalation of asbestos), lung or liver scarring, or chronic bone infections (Brand, 1982; Robinson et al., 2005). However, in contrast to the mesenchymal origin of most implant-related tumors, other cancers associated with scarring are generally derived from adjacent epithelial structures (e.g., mesothelioma with asbestos).

Chemical induction effects are also possible. With orthopedic implants, the stimulus for tumorigenesis could be metal particulates released by wear of the implant (Harris, 1994). Indeed, implants of chromium, nickel, cobalt, and some of their compounds, either as foils or debris, are carcinogenic in rodents (Swierenza et al., 1987), and the clearly demonstrated widespread dissemination of metal debris from implants (to lymph nodes, bone marrow, liver and spleen, particularly in subjects with loose, worn joint prostheses) could not only cause damage to distant organs, but also could be associated with the induction of neoplasia (Case et al., 1994). Although unequivocal cases of metal particles or elemental metals provoking the formation of malignant tumors are not available, continued vigilance and further study of the problem in animal models is warranted (Lewis and Sunderman, 1996).

"Nonbiodegradable" and "inert" implants have been shown to contain and/or release trace amounts of substances such as remnant monomers, catalysts, plasticizers, and antioxidants. Nevertheless, such substances injected in experimental animals at appropriate test sites (without implants) in quantities comparable to those found adjacent to implants, are generally not tumorigenic. Moreover, chemical carcinogens, such as nitrosamines or those contained in tobacco smoke, may potentiate scar-associated cancers.

A chemical effect has been considered in the potential carcinogenicity of polyurethane biomaterials (Pinchuk, 1994). Under certain conditions (i.e., high temperatures in the presence of strong bases), diamines called 2,4-toluene diamine (TDA) and 4,4′-methylene dianiline (MDA) can be produced from the aromatic isocyanates used in the synthesis of polyurethanes. TDA and MDA are carcinogenic in rodents. However, it is uncertain whether: (1) TDA and MDA are formed *in vivo*; and (2) these compounds are indeed carcinogenic in humans, especially in the low dose rate provided by medical devices. Although attention has been focused on polyurethane foam-coated silicone gel-filled breast implants, one type of which contained the precursor to TDA, the risk is considered zero to negligible (Expert Panel, 1991).

Foreign-body tumorigenesis is characterized by a long latent period, during which the presence of the implant is required for tumor formation. Available data suggest the following sequence of essential developmental stages in foreign-body tumorigenesis: (1) cellular proliferation in conjunction with tissue inflammation associated with the foreign-body reaction (specific susceptible preneoplastic cells may be present at this stage); (2) progressive formation of a well-demarcated fibrotic tissue capsule surrounding the implant; (3) quiescence of the tissue reaction (i.e., dormancy and phagocytic inactivity of macrophages attached to the foreign body), but direct contact of clonal preneoplastic cells with the foreign-body surface; (4) final maturation of preneoplastic cells; and (5) sarcomatous proliferation. Support for this multistep hypothesis for foreign-body tumorigenesis comes from an experimental study by Kirkpatrick et al. (2000), in which premalignant lesions were frequently found in implant capsules. A spectrum of lesions was observed, from proliferative lesions without atypical calls, to atypical proliferation, to incipient sarcoma.

The essential hypothesis is that initial acquisition of neoplastic potential and the determination of specific tumor characteristics do not depend on direct physical or chemical interaction between susceptible cells and the foreign-body and, thus, the foreign-body *per se* probably does not initiate the tumor. However, although the critical initial event occurs early during the foreign-body reaction, the final step to neoplastic autonomy (presumably a genetic event) is accomplished only when preneoplastic cells attach themselves to the foreign-body surface. Subsequently, there is proliferation of abnormal mesenchymal cells in this relatively quiescent microenvironment, a situation not permitted with the prolonged active inflammation associated with less inert implants. Interestingly, a study, using a model of biomaterial-induced sarcoma formation in which preneoplastic changes in both tumors and the foreign-body-induced fibrous capsule can be studied by contemporary histologic and molecular techniques (Kirkpatrick et al., 2000), suggested that biomaterials-associated tumorigenesis does not occur by a specific genetic insult, and thereby is different from spontaneous tumorigenesis (Weber et al., 2009). The significance of this observation is not yet apparent.

Thus, the critical factors in sarcomas induced by foreign-bodies include implant configuration, fibrous capsule development and remodeling, and a period of latency long enough to allow progression to neoplasia in a susceptible host. The major role of the foreign-body itself seems to be that of stimulating the formation of a fibrous capsule conducive to neoplastic cell maturation and proliferation, by yet poorly-understood mechanisms. The rarity of human foreign-body-associated tumors suggests that cancer-prone cells are infrequent in the foreign-body reactions to implanted medical devices in humans.

A new set of considerations related to tumorigenesis has arisen with the use of stem cells in tissue engineering and regenerative therapeutics.

A new set of considerations related to tumorigenesis has arisen with the use of stem cells in tissue engineering and regenerative therapeutics. It is well known that similar mechanisms underly developmemtal processes, would healing and tumor formation (Friedl and Gilmour, 2009). The possibility of tumor stimulation by autologous stem cells is probably minimal or absent (Brittberg et al., 1994), but embryonic stem cells (ESC) isolated from the mouse and human are pleuripotent, and have robust tumorigenicity as each readily forms teratomas when transplanted *in vivo* (Blum and Benvenisty, 2008). Human-induced pleuripotent stem cells (IPSC) are predicted to have a tumorigenic potential at least as great as ESCs (Knoeppler, 2009). This issue is certain to generate considerable research and controversy over the next several years.

CONCLUSIONS

Neoplasms associated with therapeutic clinical implants in humans are rare, and causality is difficult to demonstrate in any individual case. Experimental implant-related tumors are induced by a large spectrum of materials and biomaterials, dependent primarily on the physical, and not the chemical, configuration of the implant. The mechanism of experimental tumor formation remains incompletely understood. The potential for tumors associated with therapeutic applications of stem cells must be considered.

BIBLIOGRAPHY

Adams, J. E., Jaffe, K. A., Lemons, J. E., & Siegal, G. P. (2003). Prosthetic implant associated sarcomas: A case report emphasizing surface evaluation and spectroscopic trace metal analysis. *Ann. Diagn. Pathol.*, **7**, 35–46.

Aladily, T. N., Mederious, L. J., Alayed, K., & Miranda, R. N. (2012). Breast implant-associated anaplastic large cell lymphoma: a newly recongnized entity that needs further refinement of its definition. *Leuk Lymphoma*, **53**, 749–750.

Bago-Granell, J., Aguirre-Canyadell, M., Nardi, J., & Tallada, N. (1984). Malignant fibrous histiocytoma at the site of a total hip arthroplasty. *J. Bone Joint Surg.*, **66B**, 38–40.

Balzer, B. L., & Weiss, S. W. (2009). Do biomaterials cause implant-associated mesenchymal tumors of the breast? Analysis of 8 new cases and review of the literature. *Human Path.*, **40**, 1564–1570.

Berkel, H., Birdsell, D. C., & Jenkins, H. (1992). Breast augmentation: A risk factor for breast cancer? *N. Engl. J. Med.*, **326**, 1649–1653.

Black, J. (1988). *Orthopedic Biomaterials in Research and Practice*. (pp. 1–394). New York, NY: Churchill-Livingstone.

Blum, B., & Benvenisty, N. (2008). The tumorigenicity of human embryonic stem cells. *Adv. Cancer Res.*, **100**, 133–158.

Brand, K. G. (1982). Cancer associated with asbestosis, schistosomiasis, foreign bodies and scars. In F. F. Becker (Ed.), *Cancer: A Comprehensive Treatise* (Vol. I, 2nd ed., pp. 661–692). New York, NY: Plenum Pub.

Brand, K. G., Buoen, L. C., Johnson, K. H., & Brand, I. (1975). Etiological factors, stages and the role of the foreign body in foreign body tumorigenesis: A review. *Cancer Res.*, **35**, 279–286.

Brinton, L. A., & Brown, S. L. (1997). Breast implants and cancer. *J. Natl. Cancer Inst.*, **89**, 1341–1349.

Brittberg, M., Lindahl, A., Nilsson, A., Ohlsson, C., Isaksson, O., et al. (1994). Treatment of deep cartilage defects in the knee with autologous chondrocyte transplantation. *N. Engl. J. Med.*, **331**, 889–895.

Brown, R. C., Hoskins, J. A., Miller, K., & Mossman, B. T. (1990). Pathogenetic mechanisms of asbestos and other mineral fibres. *Mol. Aspects. Med.*, **11**, 325–349.

Center for Devices and Radiological Health, & U.S. Food and Drug Administration. (2011). *Anaplastic Large Cell Lymphoma (ALCL) In Women with Breast Implants: Preliminary FDA Findings and Analyses.* http://www.fda.gov/MedicalDevices/ProductsandMedicalProcedures/ImplantsandProsthetics/BreastImplants/ucm239996.htm.

Case, C. P., Langkamer, V. G., James, C., Palmer, M. R., Kemp, A. J., et al. (1994). Widespread dissemination of metal debris from implants. *J. Bone Joint Surg [Br]*, **76-B**, 701–712.

Christie, A. J., Weinberger, K. A., & Dietrich, M. (1977). Silicone lymphadenopathy and synovitis. Complications of silicone elastomer finger joint prostheses. *JAMA*, **237**, 1463–1464.

Deapen, D. M., & Brody, G. S. (1991). Augmentation mammaplasty and breast cancer: A 5-year update of the Los Angeles study. *Mammaplast Breast Cancer*, **89**, 660–665.

Expert Panel on the Safety of Polyurethane-covered Breast Implants (1991). Safety of polyurethane-covered breast implants. *Can. Med. Assoc. J.*, **145**, 1125–1132.

Ebramzadeh, E., Campbell, P. A., Takamura, K. M., Lu, Z., Sangiorio, S. N., Kalma, J. J., De Smet, K. A., & Amustutz, H. C. (2011). Failure modes of 433 metal-on-metal hip implants: How, why, and wear. *Orthop. Clin. N. Am.*, **42**, 241–250.

Fehrenbacker, J. W., Bowers, W., Strate, R., & Pittman, J. (1981). Angiosarcoma of the aorta associated with a Dacron graft. *Ann. Thorac. Surg.*, **32**, 297–301.

Fernandez, S., de Castro, P. L., Tapia, G., & Astudillo, J. (2008). Pseudotumor associated with polytetrafluoroethylene sleeves. *Eur. J. Cardio. Surg.*, **33**, 937–938.

Friedl, P., & Gilmour, D. (2009). Collective cell migration in morphogenesis, regeneration and cancer. *Nat. Rev. Mol. Cell Biol.*, **10**, 445–457.

Fry, R. J. M. (1989). Principles of carcinogenesis: Physical. In V. DeVita (Ed.), *Cancer. Principles and Practice of Oncology* (3rd ed., pp. 136–148). Hagerstown, MD: Lippincott-Raven.

Gaudet, G., Friedberg, J. W., Weng, A., Pinkus, G. S., & Freedman, A. S. (2002). Breast lymphoma associated with breast implants: Two case-reports and a review of the literature. *Leuk. Lymphoma*, **43**, 115–119.

Goodfellow, J. (1992). Malignancy and joint replacement. *J. Bone Joint Surg.*, **74B**, 645.

Grubitzsch, H., Wollert, H. G., & Eckel, L. (2001). Sarcoma associated with silver coated mechanical heart valve prosthesis. *Ann. Thorac. Surg.*, *72*, 1730–1740.

Guyer, R. D., Shellock, J., MacLennan, B., Hanscom, D., Knight, R. Q., McCombe, P., Jacobs, J. J., Urban, R. M., Bradford, D., & Ohnmeiss, D. D. (2011). Early failure of metal-on metal artificial disc prostheses associated with lymphocytic reaction. *Spine*, *7*, E492–E497.

Hansen, T., Clermont, G., Alves, A., Eloy, R., Brochhausen, C., et al. (2006). Biological tolerance of different materials in bulk and nanoparticulate form in a rat model: Sarcoma development by nanoparticles. *J. R. Soc. Interface*, **3**, 767–775.

Harris, W. H. (1994). Osteolysis and particle disease in hip replacement. *Acta. Orth. Scand.*, **65**, 113–123.

Hausner, R. J., Schoen, F. J., & Pierson, K. K. (1978). Foreign body reaction to silicone in axillary lymph nodes after prosthetic augmentation mammoplasty. *Plast. Reconst. Surg.*, **62**, 381–384.

Hinz, B., Phan, S. H., Thannickal, V. J., Galli, A., Bochaton-Piallat, M. L., et al. (2007). The myofibroblast: One function, multiple origins. *Am. J. Pathol.*, **170**, 1807–1816.

Jacobs, J. J., Rosenbaum, D. H., Hay, R. M., Gitelis, S., & Black, J. (1992). Early sarcomatous degeneration near a cementless hip replacement. *J. Bone Joint Surg. Br.*, **74B**, 740–744.

Jacobs, J. J., Urban, R. M., Wall, J., Black, J., Reid, J. D., et al. (1995). Unusual foreign-body reaction to a failed total knee replacement: Simulation of a sarcoma clinically and a sarcoid histologically. *J. Bone Joint Surg.*, **77**, 444–451.

Jennings, T. A., Peterson, L., Axiotis, C. A., Freidlander, G. E., Cooke, R. A., et al. (1988). Angiosarcoma associated with foreign body material. A report of three cases. *Cancer*, **62**, 2436–2444.

Jurand, M. C. (1991). Observations on the carcinogenicity of asbestos fibers. *Ann. NY Acad. Sci.*, **643**, 258–270.

Keech, J. A., Jr., & Creech, B. J. (1997). Anaplastic T-cell lymphoma in proximity to a saline-filled breast implant. *Plast. Reconstr. Surg.*, **100**, 554–555.

Keel, S. B., Jaffe, K. A., Nielsen, G. P., & Rosenberg, A. E. (2001). Orthopaedic implant-related sarcoma. A study of twelve cases. *Mod. Pathol.*, **14**, 969–977.

Kirkpatrick, C. J., Alves, A., Kohler, H., Kriegsmann, J., Bittinger, F., et al. (2000). Biomaterial-induced sarcoma. A novel model to study preneoplastic change. *Am. J. Path.*, **156**, 1455–1467.

Kolstad, K., & Hgstorp, H. (1990). Gastric carcinoma metastasis to a knee with a newly inserted prosthesis. *Acta. Orth. Scand.*, **61**, 369–370.

Knoeppler, P. S. (2009). Deconstructing stem cell tumorigenicity: A roadmap to safe regenerative medicine. *Stem Cells*, **27**, 1050–1056.

Kumar, V., Fausto, N., Aster, J. C., & Abbas, A. (Eds.), (2010). *Cotran and Robbins Pathologic Basis of Disease* (8th ed.). Philadelphia, PA: W. B. Saunders.

Kwon, Y. M., Thomas, P., Summer, B., Pandit, H., Taylor, A., Beard, D., Murray, D. W., & Gill, H. S. (2010). Lymphocyte proliferation responses in patients with pseudotumors following metal-on-metal hip resurfacing arthoplasty. *J Orthop Res*, *28*, 444–450.

Lamovec, J., Zidar, A., & Cucek-Plenicar, M. (1988). Synovial sarcoma associated with total hip replacement. *J. Bone Joint Surg.*, **70A**, 1558–1560.

Lewis, C. G., & Sunderman, F. W., Jr. (1996). Metal carcinogenesis in total joint arthroplasty. *Clin. Orthop. Related Res.*, **329S**, S264–S268.

Mallick, A., Jain, S., Proctor, A., & Pandey, R. (2009). Angiosarcoma around a revision total hip arthroplasty and review of literature. *J. Arthroplasty*, **24**, 323.

Martin, A., Bauer, T. W., Manley, M. T., & Marks, K. H. (1988). Osteosarcoma at the site of total hip replacement. *J. Bone Joint Surg.*, **70A**, 1561–1567.

Mathiesen, E. B., Ahlbom, A., Bermann, G., & Lindsgren, J. U. (1995). Total hip replacement and cancer. A cohort study. *J. Bone Joint Surg. Br.*, **77B**, 345–350.

McDonald, W. (1980). Malignant lymphoma associated with internal fixation of a fractured tibia. *Cancer*, **48**, 1009–1011.

McGregor, D. B., Baan, R. A., Partensky, C., Rice, J. M., & Wilbourn, J. D. (2000). Evaluation of the carcinogenic risks to humans associated with surgical implants and other foreign bodies – a report of an IARC Monographs Programme Meeting. *Eur. J. Cancer*, **36**, 307–313.

Moore, G. E., & Palmer, Q. N. (1977). Money causes cancer. Ban it. *JAMA*, **238**, 397.

Morgan, R. W., & Elcock, M. (1995). Artificial implants and soft tissue sarcomas. *J. Clin. Epidemiol.*, **48**, 545–549.

Pinchuk, L. (1994). A review of the biostability and carcinogenicity of polyurethanes in medicine and the new generation of "biostable" polyurethanes. *J. Biomater. Sci. Polymer. Edn.*, **6**, 225–267.

Pedley, R. B., Meachim, G., & Williams, D. F. (1981). Tumor induction by implant materials. In D. F. Williams (Ed.), *Fundamental Aspects of Biocompatibility* (Vol. II, pp. 175–202). Boca Raton, FL: CRC Press.

Robinson, B. W., Musk, A. W., & Lake, R. A. (2005). Malignant mesothelioma. *Lancet*, **366**, 397–408.

Ryu, R. K.N., Bovill, E. G., Jr., Skinner, H. B., & Murray, W. R. (1987). Soft tissue sarcoma associated with aluminum oxide ceramic total hip arthroplasty. A case report. *Clin. Orth. Rel. Res.*, **216**, 207–212.

Sahoo, S., Rosen, P. P., Feddersen, R. M., Viswanatha, D. S., & Clark, D. A. (2003). Anaplastic large cell lymphoma arising in a silicone breast implant capsule: Case report and review of the literature. *Arch. Pathol. Lab. Med.*, **127**, e115–e118.

Schoen, F. J. (1987). Biomaterials-associated infection, tumorigenesis and calcification. *Trans. Am. Soc. Artif. Int. Organs*, **33**, 8–18.

Su, C. W., Dreyfuss, D. A., Krizek, T. J., & Leoni, K. J. (1995). Silicone implants and the inhibition of cancer. *Plast Reconstr. Surg.*, **96**, 513–520.

Swierenza, S. H. H., Gilman, J. P. W., & McLean, J. R. (1987). Cancer risk from inorganics. *Cancer Metas. Rev.*, **6**, 113–154.

Visuri, T., Pulkkinen, P., & Paavolainen, P. (2006a). Malignant tumors at the site of total hip prosthesis. Analytic review of 46 cases. *J. Arthroplasty*, **21**, 311–323.

Visuri, T. I., Pukkala, E., Pukkinen, P., & Paavolainen, P. (2006b). Cancer incidence and causes of death among total hip replacement patients: A review based on Nordic cohorts with a special emphasis on metal-on-metal bearings. *Proc. Imech. E.*, **220**, 399–407.

Ward, J. J., Dunham, W. K., Thornbury, D. D., & Lemons, J. E. (1987). Metal-induced sarcoma. *Trans. Soc. Biomater.*, **10**, 106.

Weber, P. C. (1986). Epithelioid sarcoma in association with total knee replacement. A case report. *J. Bone Joint Surg.*, **68B**, 824–826.

Weber, A., Strehl, A., Springer, E., Hansen, T., Schad, A., et al. (2009). Biomaterial-induced sarcomagenesis is not associated with microsatellite instability. *Virschow's Arch.*, **454**, 195–201.

Weinberg, D. S., & Maini, B. S. (1980). Primary sarcoma of the aorta associated with a vascular prosthesis. A case report. *Cancer*, **46**, 398–402.

Weiss, S. W., Enzinger, F. M., & Johnson, F. B. (1978). Silica reaction simulating fibrous histocytoma. *Cancer*, **42**, 2738–2743.

Wolkenhauer, O., Fell, D., De Meyts, P., Blűthgen, N., Herzel, H., et al. (2009). SysBioMed report: Advancing systems biology for medical applications. *IET Sys. Biol.*, **3**, 131–136.

CHAPTER II.2.8 BIOFILMS, BIOMATERIALS, AND DEVICE-RELATED INFECTIONS

Paul Stoodley[1], Luanne Hall-Stoodley[2,3], Bill Costerton[4,5], Patrick DeMeo[5], Mark Shirtliff[6], Ellen Gawalt[7], and Sandeep Kathju[8]

[1]National Centre for Advanced Tribology, University of Southampton, Southampton, UK

[2]Welcome Trust Clinical Research Facility, University Hospital, Southampton Foundation Trust, Southampton, UK

[3]NIHR Respiratory BRU, Faculty of Medicine, University of Southampton, UK

[4]Center for Genomic Sciences, Allegheny-Singer Research Institute, Pittsburgh, PA, USA

[5]Department of Orthopaedic Surgery, Allegheny General Hospital, Pittsburgh, PA, USA

[6]Department of Microbial Pathogenesis, University of Maryland, Baltimore, MD, USA

[7]Department of Chemistry and Biochemistry, Duquesne University, Pittsburgh, PA, USA

[8]Division of Plastic Surgery, University of Pittsburgh, School of Medicine, Pittsburgh, PA, USA

INTRODUCTION

The chapter "History of Biomaterials" at the start of this book documents the development and design of indwelling materials for medical and dental purposes. The initial design criteria in the choice of materials were pragmatic, and based on the necessary mechanical properties required to fashion a functional device. Orthopedic implants require strong materials for weight-bearing, and articulating surfaces such as joints require durability and resistance to wear. Stents and shunts require flexibility and patency, and sutures require a high tensile strength yet also must be flexible enough for intricate manipulation. As the devices became more sophisticated and developments in materials science provided more options for manufacture, implants are being used more frequently and with longer anticipated lifetimes. Concurrently, the design process increasingly incorporated biocompatibility and comfort into the design criteria. However, with longer lifetimes, the more frequent use of invasive surgical procedures involving indwelling devices and biologically-friendly materials, there has been a rise in the number of incidences of device-related infection. Urinary catheters have been estimated to account for 30% of all nosocomial infections (Gould et al., 2010). Between 66 and 88% of these occur after urinary catheterization (Wong, 1983). It is also reported that almost 100% of catheterized patients develop an infection in an openly draining indwelling catheter which has been in place for four days or more (Wong, 1983). For some procedures, such as orthopedic joint arthroplasties, the diagnosed surgical site infection rates are relatively low (between 1% and 2%; Hsieh et al., 2009); however, the increasing number of patients undergoing joint-replacement surgery translates to large numbers of patients afflicted with the consequences of complicating infections per year. Furthermore, infection of artificial joints can be devastating, since oral or IV antibiotic therapy frequently fails to resolve the infection, with the only remaining course of action being surgical debridement or partial or total revision. These two examples, the first with very high numbers of patients but of lesser severity in terms of impact to the individual, and the second, low numbers but severe patient impact, reflect the incentive to pursue a third design criteria – that of infection resistance – into materials and devices (Maki and Tambyah, 2001). In the following sections we will discuss the role of bacterial biofilms in infection, and the growing literature highlighting biofilms as an important cause of device-related infection.

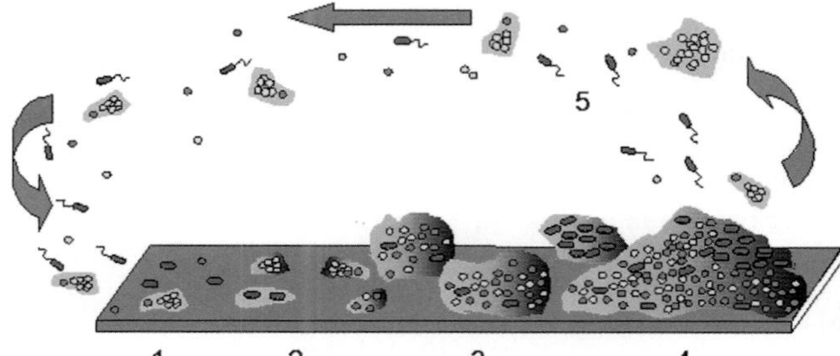

FIGURE II.2.8.1 Key processes in biofilm development based on the conceptual model described by Stoodley et al. (2002). For illustrative purposes this schematic shows a mixed biofilm consisting of rods capable of swimming motility (i.e., *P. aeruginosa* shown in red) and two types of non-motile cocci (i.e., *S. aureus* and *S. epidermidis* shown in green and yellow). The EPS slime matrix surrounding the cells is shown in green. (1) Initial attachment of single cells and clumps of detached biofilm bacteria. (2) Production of EPS to more firmly adhere cells to the surface. (3) Early development of biofilm clusters by clonal expansion (can be mixed or single species colonies). (4) Mature biofilm when the biofilm is in (pseudo) steady-state. (5) Dispersion of single cells by motility-driven swarming dispersion (*P. aeruginosa*) and the sloughing of biofilms clusters containing cells and EPS.

BACTERIAL BIOFILMS

What Are Biofilms, Where Are They Found, and Why Are They Problematic?

Bacterial biofilms are communities of bacteria which attach and subsequently grow on surfaces of abiotic materials, as well as host tissues (Stoodley et al., 2002; Hall-Stoodley and Stoodley, 2009) (Figure II.2.8.1). The bacteria embed themselves in a highly hydrated protective matrix termed “extracellular polymeric substance” (or sometimes slime) (EPS) (Donlan and Costerton, 2002). Biofilm development is an ancient adaptation of prokaryotes which is believed to have facilitated survival in hostile environments, and allowing colonization of new niches by active dispersal mechanisms (Hall-Stoodley and Stoodley, 2005; Purevdorj-Gage et al., 2005; Barraud et al., 2009; Davies and Marques, 2009; Kaplan, 2010).

Bacteria in biofilms also demonstrate coordinated behavior that culminates in the development of complex three-dimensional structures comprised of functionally heterogeneous bacterial communities on virtually all surfaces. Phenotypic heterogeneity and localized specialization in biofilms can be seen by differential expression of pili, fimbriae, flagella (Barken et al., 2008), carbohydrate (Vuong et al., 2004), adhesins (Hendrickx et al., 2009; Vergara-Irigaray et al., 2009), and genes associated with dormancy (Stewart and Franklin, 2008; Perez-Osorio et al., 2010) and antibiotic resistance (Bagge et al., 2004). Biofilm bacteria coordinate their behavior by cell–cell communication using secreted chemical signals which allow the bacteria to sense and respond to their environment by assessing cell density or environmental cues, resulting in modification of gene expression. This ability to adapt to and modify microniches at a surface interface allows bacteria in biofilms to facilitate survival at a population level.

Adaptation of Biofilm Structure for Survival in Changing Environments

Biofilms growing in highly varied environments from hot springs to urinary catheters appear to utilize similar startegies to attach and grow on surfaces, and can also show remarkable structural similarity suggesting a selective advantage that surface association offers. This suggests that structural specialization in different environments reflects an important survival strategy. For example, biofilm growth under high fluid shear often exhibits filamentous streamers, while biofilms grown in low shear environments form towers or mound-like structures which vary according to different nutrient conditions or mass transfer-determined localized growth patterns. These observations suggest that bacteria in biofilms can rapidly adapt to their local environment, to an extent not possible with multicellular eukaryotic organisms (Donlan and Costerton, 2002; Hall-Stoodley et al., 2004).

Resistance of Bacteria in Biofilms

Numerous studies document that bacteria in biofilms are more resistant to environmental stresses, such as dehydration, metals toxicity, and UV light exposure, than free floating or planktonic, bacteria. These strategies which have evolved for survival in the natural environment appear to be readily adapted to facilitate survival against human attempts to eradicate them with modern materials, antibiotics, and antimicrobials. The exact nature of the resistance is not fully understood, but the most commonly considered mechanisms are: (1) bacteria in the interior of the biofilm enter a dormant-like state (Klapper et al., 2007), possibly induced by nutrient depletion caused by consumption by bacteria on the periphery of the biofilm cell clusters (Fux et al., 2004); (2) reaction of the antimicrobial agent with the extracellular polymeric slime (EPS) matrix (binding and/or degradation) (Anderl et al., 2000); and (3) the development of resistant populations, such as slow growing small colony variants or "persister" cells which occur at greater frequency and in higher numbers in biofilm populations than in planktonic populations (Lewis, 2005).

The first and second mechanisms have been referred to as "tolerance" rather than "resistance," the latter of which is usually associated with a genetic basis (Harrison et al., 2007). In addition to providing a nidus for bacterial biofilm formation, foreign materials also suppress the efficacy of phagocytic clearance. The EPS matrix of carbohydrates, proteins, and extracellular bacterial DNA (eDNA) provides structural integrity, as well as protecting the bacterial cells within from phagocytic attack (Leid et al., 2002; Jesaitis et al., 2003; Bjarnsholt et al., 2005); although *in vitro* experiments suggest that the degree of protection also depends on the species and age of the biofilm (Guenther et al., 2009). The EPS also reduces the ability of IgG human antibodies to penetrate the biofilm (deBeer et al., 1997; Bryers and Drummond, 1998). Therefore, biofilm formation also appears to be a mechanism which helps bacteria to evade host immunity. Indeed, many diseases associated with biofilms (infection of the cystic fibrosis (CF) lung, otitis media, gingivitis) are also associated with chronic inflammation. Mounting evidence suggests that the "inappropriate" inflammatory response to biofilms often fails to eradicate the biofilm, and may in fact contribute to the pathology pathogenesis by degrading host tissue, which in turn provides nutrients to the bacteria.

BIOFILM MICROBIOLOGY AND INFECTIOUS DISEASE

Many of the concepts and techniques that have served microbiologists well in the understanding, diagnosis, and treatment of many acute epidemic bacterial diseases, are being re-evaluated with respect to device-related and other biofilm-associated infections. This section on biofilm microbiology will focus on how biofilm bacteria differ from their planktonic counterparts, and the importance of using appropriate biofilm *in vitro* assays to assess the potential bacterial colonization of biomaterials *in vivo*.

Bacterial Adhesion to Surfaces

Derjaguin, Landau, Verwey, and Overbeek (DLVO) theory is often applied to the study of bacterial adhesion to surfaces (van Loosdrecht et al., 1990). This classic concept of colloid behavior characterizes a planktonic bacterial cell as a smooth particle that interacts with a surface in a manner based on the charges on both surfaces, which overcome the basic repulsion of individual particles. However, direct observations of the surfaces of planktonic bacteria using electron microscopy have demonstrated that these surfaces are not smooth (Fux et al., 2005). In addition to protein appendages (i.e., flagella and pili) projecting as much as 2–6 μm from the cell, the entire surface of bacterial planktonic cells is covered by a structured matrix of hydrophobic exopolysaccharide (EPS) fibers and protein. The external glycocalyx of planktonic cells is anchored to the polysaccharide O antigen fibers that project from the lipopolysaccharide (LPS) of the outer membrane of gram-negative cells, and to the polysaccharide teichoic acid fibers that project from the cell wall of gram-positive cells. Elegant freeze-substitution microscopy preparations have shown that the actual surface of planktonic bacterial cells capable of interacting with a surface to be colonized consists of a 0.2–0.4 μm thick layer of protein and polysaccharide fibers. Thus, the planktonic bacterial cell is not a smooth-surfaced colloid particle, and the actual interaction of these cells with surfaces is based on the bridging of bacterial fibers with fibers adsorbed to the surface being colonized. Therefore, DLVO theory has a limited application in the study of bacterial adhesion.

Moreover, the reliance on bacteria isolated from the system of interest, but subcultured repeatedly in rich media, also does not serve us well in biofilm studies relevant to medical devices. This method, which dates from the 19th century, may select for strains of bacteria that exhibit rapid planktonic growth on rich media, since slow-growing strains might not be cultured or overgrown, and strongly adherent bacteria may not be readily transferred. Such strains may lack many surface structures that would be necessary for their survival in the real world on a surface where they are challenged by antibacterial agents, nutrient limitation, competing bacteria or host defenses. However, these laboratory-adapted strains used in studies of bacterial adhesion to biomaterials may be physically closer to the smooth-surfaced colloidal particles suggested in DLVO theory, potentially confounding data designed to further the understanding of device-related infections (Fux et al., 2005). This may help to explain why many novel biomaterials perform less successfully *in vitro* when tested with clinical isolates, and further highlights the necessity of using clinical strains in biomaterials testing. However, there are also challenges with using clinical isolates. Techniques such as rapid genomic sequencing (Hiller et al., 2007) and multi-locus strain typing (MLST) (Jefferies et al., 2004) have revealed an enormous amount of genomic and phenotypic diversity between strains of the same species, and indeed biofilms appear to accelerate the generation of variants (Webb et al., 2004; Starkey et al., 2009). Thus, data generated from any one clinical isolate might, like reference strains, also not be generalizable. Thus, ideally, anti-biofilm assays should be conducted with reference strains to allow benchmark comparisons between studies in different laboratories, as well as at least two or three low-passage clinical isolates.

Many researchers in the biomaterials field have intuitively speculated that a few key surface characteristics can inhibit (or favor) bacterial adhesion, and surfaces engineered with various combinations of these characteristics have been tried in the search for colonization-resistant biomaterials. Bacteria adhere equally well to very hydrophobic (e.g., Teflon™) and to very hydrophilic (e.g., PVC) surfaces, can colonize smooth surfaces as well as rough surfaces (Marrie et al., 1982; Sottile et al., 1986), and can colonize surfaces in relatively high physiologically-relevant shear flows (Leid et al., 2002). Indeed, Bos et al. have reported that hydrophobicity had little correlation with initial attachment of various bacterial strains to surfaces, but was correlated with attachment strength (Bos et al., 1999). Thus, currently, while there is no perfect biocompatible biomaterial surface which also inherently resists bacterial colonization, more sophisticated non-fouling surfaces which incorporate designer antimicrobial peptides to target and kill bacteria (Loose et al., 2006) or zwitterionic surfaces which are generally resistant to protein adhesion (Cheng et al., 2007) show potential for this application.

Processes of Biofilm Formation

When planktonic cells adhere to a surface, they exhibit behaviors that have been divided into "reversible" and "irreversible" patterns (Marshall et al., 1971). Motile organisms (e.g., *P. aeruginosa*) may use their flagella for initial attachment, followed by type IV pili for twitching motility that allow them to form elaborate structures (Barken et al., 2008). However, surface appendages and motility are not required for biofilm formation, staphylococci and streptococci are both capable of forming biofilms *in vitro* and *in vivo*, although in these cases it is assumed that biofilm structures develop from clonal growth. Movies showing the attachment of various biofilm bacteria are available on the Center for Biofilm Engineering (CBE) web site (www.erc.montana.edu). Such movies show that many cells that adhere to surfaces may also quickly detach, before they make the phenotypic switch to initiate irreversible attachment and subsequent biofilm formation.

When a bacterial cell colonizes a surface, the pattern of gene expression is profoundly different from the previous planktonic phenotype, resulting in a distinct biofilm phenotype that may differ by as much as 70% in the proteins expressed (Sauer and Camper, 2001). Among the first genes that are upregulated in adherent cells are those involved in the production of molecules associated with the EPS that forms the biofilm matrix and anchors the cell irreversibly to the surface. In mucoid *P. aeruginosa*, the upregulation of *algC*, which is a part of the alginate synthesis pathway, occurs within 18 minutes of initial cell adhesion (Davies and Geesey, 1995), and there is a secretion of bacterial biofilm matrix material by these cells within 30 minutes. Once attached, cells which have triggered the conversion to the biofilm phenotype and the formation of a multicellular community on the colonized surface begin to accrete larger numbers of cells through growth. As they increase in numbers and produce more EPS matrix material, the attached cells form microcolonies which constitute approximately 10% of the volume, with the EPS matrix occupying approximately 90% of the biofilm. Recent data suggest that the structure of the EPS is much more sophisticated than previously thought, with confocal and SEM (scanning electron microscope) images showing features such as "honeycombs" and 3D networks in *P. aeruginosa* and staphylococcal biofilms grown *in vitro* (Schaudinn et al., 2007). Borlee et al. (2010) have found evidence for secreted proteins which specifically cross-link extracellular polysaccharides, suggesting a possible mechanism for extracellular remodeling of the EPS.

The microcolonies can assume tower-like, frond, and mushroom-like structures in many natural and cultured biofilms, but other morphologies may be dictated by individual species characteristics, and by nutrient availability. As the biofilm matures and undergoes more phenotypic changes, the processes of growth and recruitment are balanced by detachment of planktonic cells from the sessile community and sloughing of larger pieces of biofilm. *In vitro* studies suggest that biofilms reach a mature thickness and a stable community structure within a week or two of their initiation of colonization (Stoodley et al., 2002). Numerous *in vitro* studies show that both gram-negative and gram-positive pathogens are clearly capable of forming extensive biofilms on commonly used medical materials, including devices themselves, such as surgical sutures and meshes (Engelsman et al., 2007), and orthopedic implants and catheters (Jacobsen et al., 2008; Machado et al., 2009).

DEVICE-RELATED INFECTION

Although biofilm formation may be an ancient adaptation of prokaryotic life, the full impact of biofilms on human health is only now being realized, through one of the fastest technologically advancing fields – the development of medical devices, which are all susceptible to colonization with microorganisms growing in biofilms.

Medical and surgical practices rely on an increasing array of biomedical implants, and tens of millions are implanted each year. Intravenous catheters, prosthetic heart valves, joint prostheses, peritoneal dialysis catheters, cardiac pacemakers, cerebrospinal fluid shunts, and endotracheal tubes certainly save millions of lives, but all carry the risk of surface-associated infections. Device-related infections were among the first clinical problems to be identified as having a bacterial biofilm etiology. Biofilms associated with medical devices were first noted in the literature in the early 1980s, although Bayston and Penny published a report in 1972 that correlated *Staphylococcus mucoidy* as a possible factor in the colonization of Holter shunts (Bayston and Penny, 1972). Electron microscopy revealed bacteria deposited on the surface of indwelling devices such as intravenous catheters and a cardiac pacemaker lead (Marrie et al., 1982). However, the relationship between medical devices and infection is arguably best established for catheters (O'Grady et al., 2002). Many types of catheters and fluid management devices, such as central venous catheters, dialysis catheters, ventilators, neurosurgical shunts, and drains are also commonly associated with infection (Rolighed, 2010).

Implant infections are common occurrences, with rates estimated between 4–5% for orthopedic implants and 7% for cardiovascular implants (Campoccia et al., 2006; Hetrick and Schoenfisch, 2006). Due to the increasing use of these interventions, infections affect millions of patients per year, making device related infections a serious clinical problem that is being addressed from many perspectives.

There is a clear correlation between implanted devices and risk of infection, which may not be surprising since any time the protective skin barrier is breached, the possibility for introduction of bacteria to normally sterile sites exists. A 1979 study by Fitzgerald (Fitzgerald, 1979), and discussed in Subbiahdoss et al. (2009), reported that during surgery a wound was exposed to approximately 270 bacterial-carrying particles per cm^2 per hour. No doubt with improved operating room (OR) facilities, such as laminar flow air handling and the implementation of more rigorous practices aimed at reducing the rates of nosocomial infections, this rate will have decreased. Nevertheless, the study illustrates that possible airborne contamination can occur within the "sterile field." Possibly of greater concern is the introduction of skin pathogens during the procedure. Higher rates of infection associated with more heavily microbially-colonized anatomical sites (such as the groin or armpit), suggest that transmission from the skin, despite disinfection of the skin, is an important route of transmission. Even devices which are not transdermal or surgically implanted, such as urinary catheters, are associated with a high risk for infection, in part due to the fact that these devices are placed in patients who are already immunocompromized, and it is likely that many catheters are being placed in an already infected site.

In the controlled condition of a hospital the device be implanted is sterile immediately out of the packaging. However, once it is exposed there is potential for the implant to be contaminated with bacteria that might also have inadvertently enter the wound during surgery. The probability of infection has been related to the relative time taken to colonize the artificial surface by host cells or infectious microorganisms. This was termed "The race to the surface" by Gristina et al. (1988), who hypothesized that if a surface was colonized with host cells before a biofilm could become established, the probability of subsequent infection was low, even if bacteria were introduced during the procedure. Subbiahdoss et al. (2009) experimentally demonstrated the importance of the "race to the surface" by showing that the successful colonization of a glass substratum by osteosarcoma cells was negatively influenced by preseeded *S. epidermidis* cells on the surface. Whether this observation is generalizable to the *in vivo* situation remains to be verified.

Although device-related infections are initially localized to the site where the device is inserted, these infections are often chronic in nature and can be a source of periodic acute sepsis. The direct link between the presence of a foreign body and bacterial sepsis in hemodialysis catheters is illustrated in a study by Feely (Feely et al., 2007). In a group of hemodialysis patients with a history of recurrent bacteremia, the rate of catheter-related bacteremia went down from 9.13 per 1000 catheter days to 1.04 after catheter-lock solutions (CLS) were instigated. When CLS was discontinued, the rate rose back up to 7.94 per 1000. Catheter lock is a technique whereby a controlled volume of antibiotic solution is introduced into a catheter so as to fill the catheter, but is not allowed to enter the bloodstream. Thus, much higher concentrations of antibiotics can be applied directly to the catheter than would otherwise be tolerated systemically. This study clearly implicates the foreign body as harboring the source of infection.

The organisms most frequently observed with medical devices are the staphylococci, both the coagulase negative staphylococci (particularly *Staphylococcus epidermidis*) and *S. aureus* followed by *P. aeruginosa* and other opportunistic bacteria that survive in the environment. Biofilm formation on medical implants, most notably by *S. epidermidis*, has led to the characterization of a new infectious disease: "chronic polymer-associated infection" (von Eiff et al., 1999; Gotz, 2002). *S. epidermidis*, a common colonizer of the skin, is therefore frequently found in wounds and implants. It was not considered even an opportunistic pathogen until device-related infections became problematic (Otto, 2009). Biofilm formation therefore can be considered a virulence factor in the context of device-related infections – a bacterial strategy that contributes to its ability to cause and maintain an infection.

Biofilm Formation by Staphylococci

Staphylococci are particularly problematic in biofilm-associated infections with surgical site infection of deep tissue implants. This is partly due to the propensity of these organisms to form biofilms, and the increasing spread of multi drug-resistant methicillin-resistant *S. aureus* (MRSA) and *S. epidermidis* (MRSE). Biofilm formation is characterized by two stages with staphylococci: (1) adhesion to a solid surface; and (2) growth-dependent cell accumulation resulting in multiple cell layers in microcolonies. In *S. epidermidis*, the cell–cell layer aggregation is specifically linked with a β-1,6-linked glycosaminoglycan polysaccharide complex know as polysaccharide intercellular adhesin (PIA) (Mack et al., 1996). Proteins involved in the synthesis of this matrix polysaccharide are regulated by the *ica* gene locus in *S. epidermidis* (Heilmann et al., 1997; Chang et al., 2010) which is conserved among phyologenetically related staphylococci. However, PIA and *ica* are not absolutely required for biofilm formation, since mutants lacking the *ica* locus have been isolated which can still form biofilms, suggesting complex, redundant mechanisms for biofilm development. For example, an *S. epidermidis* transposon mutant, unable to form biofilms on polystyrene, formed biofilms on glass, suggesting that bacterial adhesion mechanisms vary according to the substratum (Heilmann et al., 1997).

A further level of complexity comes from the numerous environmental factors that can affect biofilm formation. For example, biofilm formation in *S. epidermidis* is strongly influenced by sugar substrate, oleic acid, antibiotic treatment, oxygen levels, iron, osmolarity, temperature, and ethanol stress (Gotz, 2002). Also, the initial adhesion stage alone is multifactorial, depending on the physio-chemical properties of the biomedical polymer material, and the nature of the bacterial cell surface. Bacterial surface proteins contribute significantly to adhesion, and several key proteins have been identified that play a role in staphylococcal biofilm formation. In *S. aureus*, teichoic acid lacking D-alanine (dltA) resulted in a change in surface charge that affected several characteristics, including biofilm formation (Gross et al., 2001). Interestingly, dltA mutants were defective in the ability to attach to polystyrene and glass. While consistent with charge equilibrium of surface interactions, the biofilm defect was re-established by the addition of magnesium ions. Other researchers found that *S. epidermidis* biofilm formation was enhanced by Mg^{2+} and inhibited by EDTA (Dunne and Burd, 1992). Another surface protein in *S. aureus*, the biofilm-associated protein or bap, was shown to contribute to biofilm formation on microtiter plates, but more importantly showed increased pathogenesis in a mouse foreign-body infection model (Cucarella et al., 2002).

Detecting Device-Related Infections

Biofilm-associated infections are often difficult to detect using conventional culturing methods (Parsek and Singh, 2003), and many of the clinical studies which conclude that biofilms were present on a medical device rely on indirect evidence such as the recovery of bacteria removed using sonication (Tunney et al., 1998; Trampuz and Widmer, 2006; Bjerkan et al., 2009) or the culture of strains positive for slime production or biofilm formation (Arciola et al., 2006). Arguably, at the present time microscopic examination of the devices and associated tissue themselves, are still the only way to definitively demonstrate the presence of biofilm (Hall-Stoodley and Stoodley, 2009; Marrie et al., 1982). Confocal microscopy has an advantage over electron microscopy, in that it can be performed on fully hydrated specimens using water immersion objectives, which, in itself, demonstrates firm attachment of the bacteria to the surface, and is also compatible with a wide range of specific fluorescent staining techniques, including antibody labeling and fluorescent *in situ* hybridization (FISH), to identify bacterial species, or membrane integrity-based nucleic acid viability staining methods to detect live and dead bacteria. Figure II.2.8.2 shows an example of predominantly staphylococcal biofilm growing on a surgical screw.

Bacteria can also be identified through the detection of their DNA. First, a specimen or swab is digested to release any bacterial DNA. Next the polymerase chain reaction (PCR) is used to amplify specifically targeted fragments of DNA. In some cases the targets are specifically designed for a certain species, genus or strain. In other cases sections of universal genes, such as 16s ribosomal DNA (rDNA), are amplified using primers to highly conserved regions while the sequencing of the variable region between the primers can identify the bacteria.

While PCR-based 16S rDNA detection methods appear to be much more accurate than conventional clinical culture, which has increasingly been shown to produce high levels of false negatives in chronic biofilm-associated infections, PCR for targeted genes does not differentiate between biofilm and planktonic infections. PCR has also been criticized for being too sensitive, and producing false positives from contaminants which are not clinically relevant. However, the visual demonstration of clusters of bacteria firmly attached to devices demonstrates a "growth-in-place" process, which is unlikely to have resulted from contamination during the relatively short period of the surgery. Therefore, confocal microscopy, PCR, and reverse transcriptase PCR (RT-PCR) have been used in combination to provide strong evidence for the presence of active bacterial biofilms on various implant materials (Nucci et al., 2005; Stoodley et al., 2008, 2010, 2011) as well as host tissue (Hall-Stoodley et al., 2006; James et al., 2008; Schlafer et al., 2008; Bjarnsholt et al., 2009). However, a drawback of confocal microscopy-based diagnosis of biofilm is that it requires highly specialized equipment and training, along with laborious searching for clusters of bacteria with dimensions as low as 10s to 100s of microns on implants that might have large areas with convoluted surfaces.

While confocal microscopy might be useful to validate more rapid methods which are conducive to standardization in the setting of a clinical laboratory, it will require refinement of the existing technology before it can be used as a routine diagnostic tool. Ideally, the diagnosis of a biofilm infection would be made prior to surgery. One approach is to devise enzyme-linked immunosorbent

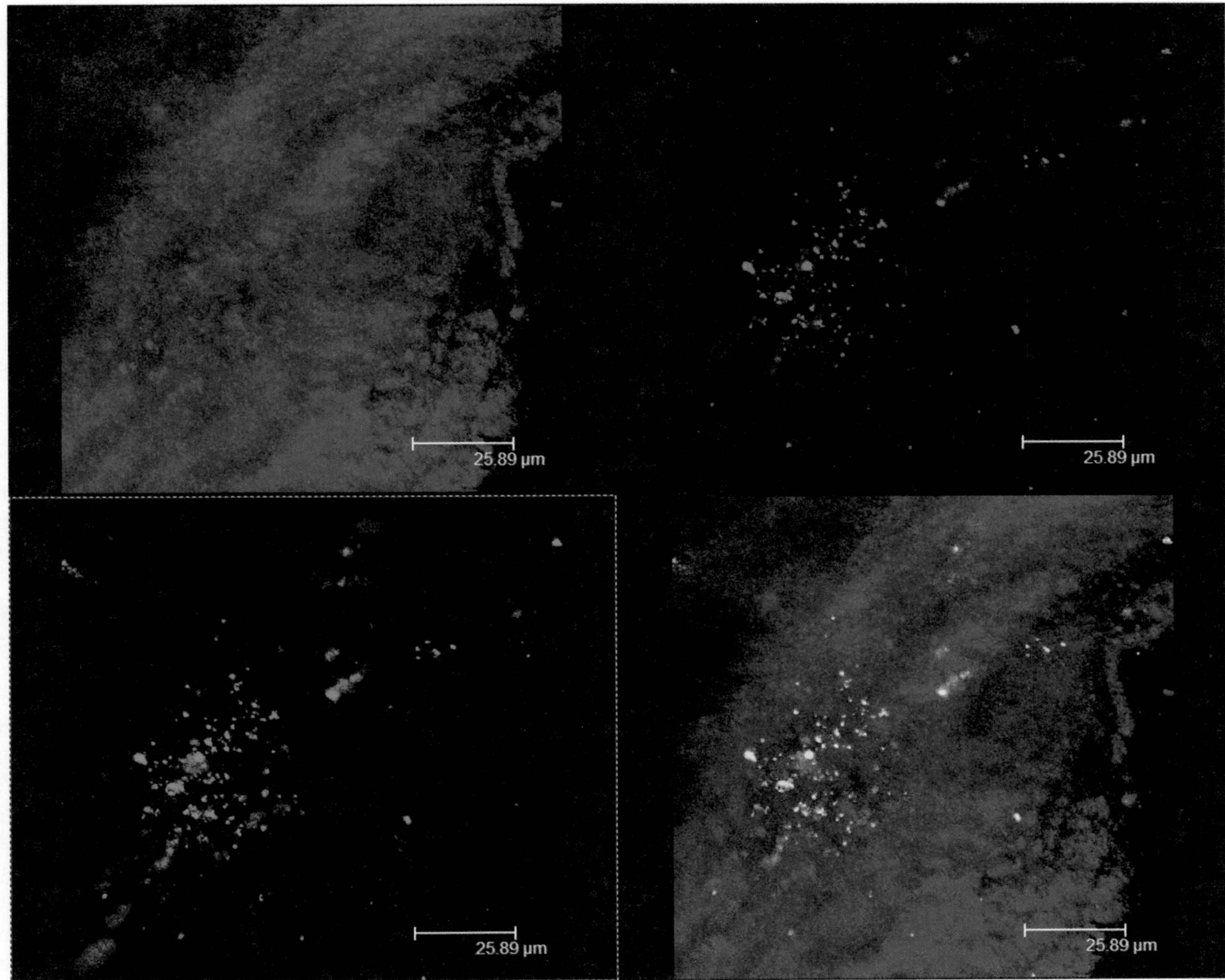

FIGURE II.2.8.2 Biofilm attached to orthopedic screws from a non-union fracture case. The surface of the screw and associated invested tissue are blue and imaged by reflected confocal microscopy. Staphylococci were stained red by fluorescence *in situ* hybridization (FISH) using the Sau probe. General bacteria were stained green with the Eub338 probe. The overlay (bottom left) shows that the biofilm consisted primarily of staphylococci with occasional other types of cocci. Scale bar = 25 µm.

array (ELISA) type assays to detect biofilm specific proteins through a simple blood test (Brady et al., 2007). Another approach is to develop medical imaging techniques, such as X-ray computed tomography (CT) scan or magnetic resonance imaging (MRI) scans, so that biofilms can be visualized in place. However, the small size of the biofilm, and the lack of contrast between biofilm and host tissue, makes imaging a challenge. In diagnostic imaging for cancer detection contrast agents are used which help distinguish the malignant cells from the healthy tissue. The importance of more effectively treating cancer through early detection has led to much research and development in the imaging of ever smaller tumors, and it is possible that techniques developed in oncology could be utilized for detection of biofilms. Imaging is not only useful for diagnostics, but also has the added advantage in that it can be invaluable in monitoring and planning surgical intervention.

Nucleic Acid-Based Detection Methods

PCR-based detection techniques generally fall into one of two categories. One approach is targeted PCR in which genus, species or strain-specific primers or combinations of primers are used to detect pathogens of interest. However, this approach does not provide any information regarding the presence or absence of non-targeted bacteria. A second approach is a "finger-printing method," by which primers are used to amplify universal DNA sequences (16S rDNA is commonly used). The amplicon is separated out as bands on a gel by techniques such as denaturing gel gradient electrophoresis (DGGE) (Li et al., 2005) or, after digestion using restriction enzymes, by restriction fragment length polymorphism (RFLP) (Dempsey et al., 2007). The separation patterns can be used to identify specific species. However, in polymicrobial infections the many overlapping and complex patterns of bands can be difficult to resolve.

A more specific approach is to sequence the various 16S amplimers from individual bacterial colonies in a clone library or directly by high-throughput sequencing techniques, such as 454 pyrosequencing (Hamady et al., 2008; McKenna et al., 2008). Individual species can be identified by comparing the cloned sequences with database sequences, using tools such as the Basic Local Alignment Search Tool (BLAST). The drawback of these approaches is that they are relatively time consuming, either in the preparation of the library or the data analysis, and are not yet conducive for utility where a rapid diagnosis is needed. An alternative approach is the combination of PCR with ion spray mass spectroscopy to precisely weigh each complementary strand of the amplimer. Bacterial species can be identified by mass due to differences in the nucleotide ratios. Multiple sets of primers are used to amplify sequences from genes which are universal to bacteria (i.e., 16S and 23S rDNA), and genes or alleles specific to certain species or strain (Ecker et al., 2008). More targeted primer sets can be designed to identify site-specific pathogens or the presence of the bacterial virulence genes, such as *mecA* which is responsible for methicillin resistance in *S. aureus* (MRSA) (Costerton et al., 2010; Stoodley et al., 2011). Primer sets could be designed to detect genes associated with biofilm formation, such as the *ica* operon which regulates EPS production in staphylococci.

CLINICAL EXAMPLES OF BIOFILM INFECTIONS

Infection Related to Surgical Repair Materials

Surgical repair materials, such as staples, sutures, and meshes, have also been identified as possible foci of infection in such diverse anatomic sites and procedures as abdominal wall repair and ocular and dental procedures (Nucci et al., 2005; Otten et al., 2005; Edmiston et al., 2006; Engelsman et al., 2007). The implication of the foreign body as a causative factor of infection is supported in a study by Wissing et al. (1987), who reported that the late post-operative incidence of suture sinus development associated with fascia repair after midline laparotomy was significantly greater with nonabsorbable sutures (nylon) than with absorbable sutures (polyglactin or polydioxanone). There was a direct relationship between material absorption time and incidence of sinus formation, which was 3–4% for polyglactin, which is fully absorbed after 70 days, 11% for polydioxanone, which is fully absorbed after 180 days, and 23% for nylon, which is permanent. This study nicely illustrates the compromise that has to be made between functionality of the material and the risk of infection, since there was an inverse relationship between the incident rate of incisional herniation and suture absorption time caused by the loss of tensile strength during the absorption process.

Bacterial Biofilms on Sutures

Bacterial adherence to, and colonization of, suture material has been of interest to the surgical community for decades. Thirty years ago Osterberg and Blomstedt observed that multifilament "capillary" suture material recovered from *S. aureus*-inoculated tissues in a rat model harbored substantially greater quantities of bacteria than "non-capillary" suture (Osterberg and Blomstedt, 1979). Their conclusion was that: "in the case of the capillary suture material, the bacteria would tend to be protected through their enclosure in the interstices of the material." In a subsequent study, Osterberg further commented that "bacteria which are enclosed in the interstices of multifilament suture material, and protected from the phagocytic activity of leukocytes, can sustain and prolong an infection" (Osterberg, 1983). These prescient early observations concur well with our present understanding of bacterial biofilm formation and persistence, but suture material as a host for pathogenic biofilm in patients has thus far been only occasionally examined (principally in the ophthalmologic literature). We have reported several cases where biofilm was documented on clinical specimens of explanted permanent suture material, with resolution of the infection on removal of the offending suture substratum (Kathju et al., 2009, 2010) (Figure II.2.8.3). Multiple studies have noted that "suture sinuses" are associated with nonabsorbable suture materials, presumably due to resident biofilm, although direct evidence was lacking (Hodgson et al., 2000; van't Riet et al., 2002).

Prevention of infectious complications by modifying the suture material itself has recently become an area of intense interest. A number of studies have focused on suture materials coated with the antimicrobial agent triclosan. *In vitro* and *in vivo* animal experiments have shown that coated polyglactin 910 suture (braided) can significantly decrease the number of viable bacteria recovered from inoculated suture material versus a non-coated control (Edmiston et al., 2006; Storch et al., 2004). Similar antimicrobial activity has also been reported with the addition of triclosan to poliglecaprone 25 (Ming et al., 2007) and polidioxanone (Ming et al., 2008) suture. Early clinical experience has also been promising: use of triclosan-coated polyglactin 910 suture versus uncoated polidioxanone suture in 2088 patients undergoing midline laparotomy resulted in a significantly reduced reported incidence of wound infection (Justinger et al., 2009).

Bacterial Biofilms on Surgical Mesh

Surgical mesh implants are routinely used to address the clinical problem of hernia. These implants typically restore structural domain to the abdominal/pelvic wall and prevent extrusion of visceral contents. Multiple complications have been associated with the use of surgical mesh, including infection, extrusion, and erosion into subjacent structures (i.e., development of

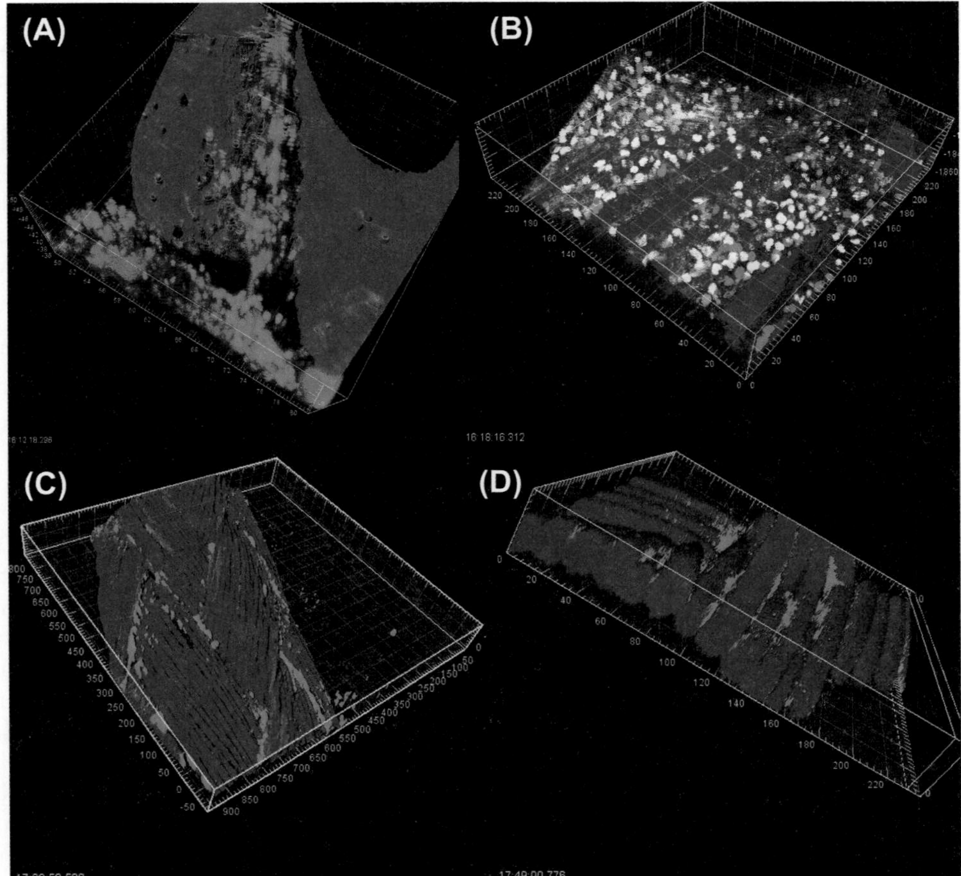

FIGURE II.2.8.3 Biofilm infected sutures. (A) Ti-cron braided polyester suture used for hernia repair following open gastric bypass surgery removed from a chronic draining wound was found to have a biofilm formed from live cocci (stained green with the BacLight™ LIVE/DEAD® kit). Diffuse green staining (arrow) is suggestive of eDNA in the biofilm slime matrix. The individual filaments of the suture are shown by autofluorescence (blue). Biofilm was particularly noted in the crevices formed from the braiding. Scale bar = 4 μm. (B) The same suture as in "A" but at a location a few mm away. In this location there were many inflammatory cells indicated by nuclear morphology (nuclei stained yellow and red). Scale bar = 20 μm. (C) *In vitro* biofilm grown on braided sutures from a clinical MRSA (stained green with the LIVE/DEAD® kit) isolated from a patient with an infected surgical repair mesh. Similar to the *ex vivo* image the biofilm preferentially grew in the protected crevices between individual filaments in the braid. Scale bar = 50 μm. (D) Higher magnification confocal image of the suture at a different location. Scale bar = 20 μm.

enterocutaneous fistula). Not infrequently, surgical mesh implants must be removed, often due to infectious complications, and recently an entirely new "current procedural terminology" code (11008) describing the explantation of abdominal wall mesh due to chronic and necrotizing infection has been recognized.

Multiple studies have examined the morphological and surface characteristics of various available meshes, and have documented the ability of both gram-negative and gram-positive organisms to form biofilms on these materials (Engelsman et al., 2008; Aydinuraz et al., 2009). The hydrophobicity of the mesh surface, as well as the presence of numerous niches (or pockets) in multifilament mesh configurations, appear to correlate with the extent of biofilm formation *in vitro* (Engelsman et al., 2008). We have been able to demonstrate biofilms present on clinical specimens of explanted mesh from patients with a variety of clinical presentations (Stoodley et al., manuscript in preparation).

A similar approach to modifying mesh materials (as has been done with suture) may prove beneficial to reducing the infectious complications noted with surgical mesh, but these studies are in their infancy. Monofilament polypropylene mesh modified with a coating to release nitric oxide in bactericidal quantities was found to significantly inhibit biofilm from multiple organisms *in vitro*, but had little effect in an *in vivo* animal model (Engelsman et al., 2009). Clinically, acellular dermal grafts composed entirely of organic materials and therefore presumably less susceptible to the establishment of persistent biofilms, have recently gained popularity as alternatives to mesh reconstruction in abdominal wall surgery.

Bacterial Biofilms in Orthopedic Prosthetic Joint Infection

Orthopedic implant infections differ clinically from the previous infections described, for several reasons. First, although they are chronic and persistent like all biofilm infections, they often develop months or years after the implant. Second, these infections are often difficult to detect in orthopedic implants via routine examination or culturing, thereby delaying diagnosis and treatment. Third, unlike the catheter lock solution treatment, there is no easy way to deliver high doses of antibiotics to the site. Therefore, orthopedic infections often require repeated antibiotic applications to resolve the infection, which increases the risk of the development of antibiotic resistance. Finally, treatment of chronic infections that fail to resolve with conventional antibiotic treatment may require sustained high dose antibiotic therapy, removal of the implant, and replacement of the orthopedic implant, resulting in pain and discomfort, immobility, and excessive costs to the patient and healthcare facility. The failure of antibiotics to effectively and fully treat these infections suggests that other strategies are necessary; approaches to resolve these problems range from enhancing hospital sterility protocols to the investigation of active release (Campoccia et al., 2006), non-adhesive or bactericidal coatings (Senaratne et al., 2005; Khoo et al., 2009; Zhao et al., 2009). Despite researchers' best efforts, often the only treatment for intractable implant infections is revision surgery, requiring removal of the infected implant and placement of a new one. However, secondary (revisional) total hip replacements have a higher infection rate (3.2%), and secondary total knee replacements exhibit a remarkable 5.6% re-infection rate. To compound the problem, in revision surgery a larger implant must be used, since bone is destroyed during the original implantation and subsequent explantation. This process may require splintering the surrounding bone, along with substantial soft tissue resection and debridement. The patient is then confined to absolute bed rest and continuous intravenous antibiotic therapy for several weeks. If the revisional surgery fails, limb amputation may occur. The number of total hip and knee replacements within the United States surpassed 1,000,000 in 2009 (http://www.cdc.gov/nchs/fastats/insurg.htm), and is expected to increase dramatically over the next several decades due to specific risk factors within the US population that include obesity, sports injuries, and longer lifespan. In addition to the physical and emotional costs, the financial cost of revisions is high, with explant surgery, post-procedural medical treatment, and re-implantational surgery exceeding $500,000 per patient. Conservative estimates put the current costs of revisional therapy in excess of two billion dollars annually.

Addressing this objective, the 2004 Association of Bone and Joint Surgeons Musculoskeletal Infection Workshop (Hanssen et al., 2005) suggested that orthopedic materials designed to prevent bacterial biofilm formation would be a relevant clinical strategy for reducing orthopedic device infections. Therefore, the investigation of antimicrobial surfaces that inhibit or retard biofilm development represents a significant new direction in the antimicrobial arsenal against bacterial biofilm formation and implant infections. These antimicrobial surfaces fall into two broad categories: (1) those surfaces that are made of materials that resist bacterial attachment; and (2) those materials that are impregnated or coated with antimicrobial compounds, which upon diffusion from the surface inhibit or abrogate bacterial growth at the implant site. This direction of research is at an early stage, and represents an important part of the unmet need of the pharmaceutical therapeutic market, but most saliently, the proven effectiveness of orthopedic materials specifically designed to resist bacterial colonization and biofilm formation would result in a significant advance in patient care.

Total joint replacement (TJR) is one of the most common and successful surgical procedures in orthopedics. The majority of patients undergoing TJR benefit from marked improvements in their pain, functional status, and overall quality of life (NIH statement 2003). Infection following primary TJR is uncommon, occurring with a prevalence of ~1–3%, but it can be a debilitating complication of TJR surgery. Infection following revision arthroplasty is significantly higher, approaching 10% (Cui et al., 2007). In the US, revisions are usually in two stages with a period of time between removal of the old hardware and replacement of the new which allows resolution of the infection. However, the fact that infection rates in revised cases are higher than for primary surgeries suggests that, although the symptoms might have resolved, the bacterial source of the underlying infection might not have been fully eradicated. The treatment of an infected joint prosthesis is both complicated and expensive. Diagnosis can itself be problematic, and the necessary intervention may vary depending on the clinical picture; e.g., acute postoperative infections may be initially managed with operative debridement, intravenous antibiotics, and retention of the implant, whereas late chronic infections will likely ultimately require more extensive surgery, typically in multiple stages. Even with aggressive measures there is a ~10% rate of failure to eradicate the infection, and the patient may proceed to resection of the arthroplasty, arthrodesis, or even amputation, in each case with significantly greater morbidity (Cui et al., 2007).

Although the concept of biofilm as a relevant physiological factor in prosthetic joint infections has gained some currency, there are actually few examples of experimentally imaged bacteria in biofilm configuration from any orthopedic implanted material. Gristina and Costerton first found bacteria in biofilms on orthopedic explants by scanning electron micrography (Gristina and Costerton, 1985). Confocal microscopy has visualized biofilm on the surface of an explanted hip prosthesis, and the microscopic identification of clusters of bacteria consistent with biofilms were present in sonicate from infected explanted prostheses, where the sonication was used to dislodge any extant attached biofilm (Neut et al., 2003). We have documented bacterial biofilm in the aspirate, on tissue, and on bone cement originally impregnated with gentamicin in a patient with a chronic infection after total elbow arthroplasty (Stoodley et al., 2008). The majority of reports imputing biofilms on orthopedic prostheses and surfaces, however, have been in *in vitro* systems, and a major argument for biofilm involvement in chronic orthopedic prosthetic joint infections remains the highly suggestive clinical features demonstrated by such infections, supported by our general understanding of the bacterial propensity to occupy available abiotic surfaces.

Nethertheless, surgeons and investigators have become increasingly interested in adjuvant techniques and technologies that minimize the risk of prosthetic joint infection, possibly by preventing adherence to and colonization of the implant in the first place. Orthopedic surgeons already routinely use antibiotic-loaded bone cement in surgery, and the use of other antibiotic-supplemented materials is under active investigation. Direct covalent attachment of an antibiotic moiety (vancomycin) to a titanium surface has been shown to inhibit both *S. aureus* and *S. epidermidis* colonization of the modified implant material *in vitro*, despite repeated challenges and over prolonged incubation times (Antoci et al., 2007a). The use of vancomycin-modified titanium implants in a rat model similarly showed evidence of reduced bacterial adherence (Antoci et al., 2007b). An alternative approach to direct modification that would still provide high local concentrations of antibiosis to an implant surface is the deployment of antibiotic-loaded conforming films or sheaths for orthopedic prostheses (Aviv et al., 2007).

PREVENTION AND TREATMENT

Biofilm-Resistant Biomaterials

In previous sections we have discussed biofilm-related complications associated with considerable morbidity and mortality. Here we will discuss some of the new agents and strategies that may give more effective control over the colonization of biomaterials, and the incidence of device-related infections. Finally, we will discuss new methods for the release/delivery of these agents at the surfaces of biomaterials. But first we discuss the evaluation of novel surfaces for resistance to bacterial adhesion, and the formation of bacterial biofilms.

Testing for Antibacterial and Antibiofilm Properties of Biomaterials

There are serious concerns with the utility and information content of some of the methods used to assess the biofilm resistance of biomaterials; namely, if a biomaterial gives a positive zone of inhibition test, what does this mean? Fundamentally, this widely-used test indicates that the biomaterial contained an antibacterial agent, which was released in the moist environment of the surface of an agar plate, was still active, and the agent could prevent the planktonic bacteria deposited on the agar from growing. The major parameter operative in the test is the diffusion of the antibacterial agent through the agar or in the fluid on the agar surface, as well as the effectiveness of the agent. A very potent agent may produce a small or no zone of inhibition, if it is not released or moves slowly through agar, and a less potent agent may produce a large zone if it diffuse rapidly through agar. Shake flask test such as the ASTM E2149-10 tests, in which candidate biomaterials are suspended in a growth medium that is simultaneously inoculated with planktonic bacteria, are equally naïve. If the biomaterials release enough of an antibacterial agent in the first few minutes of the test, all of the planktonic cells in the inoculum will be killed, resulting in a situation where there are no organisms left to colonize the biomaterial. So, an antibiotic-releasing biomaterial that releases all of its antibacterial agents in a few minutes may emerge from this test as a favorable candidate, although this may not reflect a physiologically relevant timescale over which the biomaterial is expected to perform. Generally, tests are conducted over a few hours in rich medium, in an essentially "closed" system, while *in vitro* the biomaterials might be expected to perform over days, or even years, in open systems where leached active agents can diffuse away. Therefore, such conventional tests are largely inappropriate, and may lead to the development of biomaterials that fail in biofilm resistance in animal and clinical trials.

More appropriate tests are ones that mimic the conditions in the systems in which the biomaterial is targeted for use. If the biomaterial will be subjected to flow or fluid exchange, the test should include these parameters. If the biomaterial will be used in a body fluid, such as blood or urine, an accurate simulation of that fluid should be used in the test, and the bacteria supplying the challenge should be adapted to the fluid. One of the pitfalls of laboratory testing is to base a culture medium on the relevant physiological chemistry, but then to add rich components, such as yeast or amino acid extracts, to stimulate growth to allow for rapid testing, typically over a 24-hour period or less. The concept is valid, that is, to accelerate natural processes; however, as discussed earlier, a rich medium and much higher concentrations of bacteria than might be physiologically relevant might also unintentionally interfere with surface chemistry and product efficacy.

Flow cells are useful for quantifying the rates of bacterial attachment and biofilm formation under a defined fluid shear stress, because they allow direct observation of the surface which can be monitored continuously. Confocal microscopy is particularly suited for biofilm assays since it allows the examination of living hydrated preparations, microbial cells that adhere to the biomaterial surface can be easily visualized (Figure II.2.8.4).

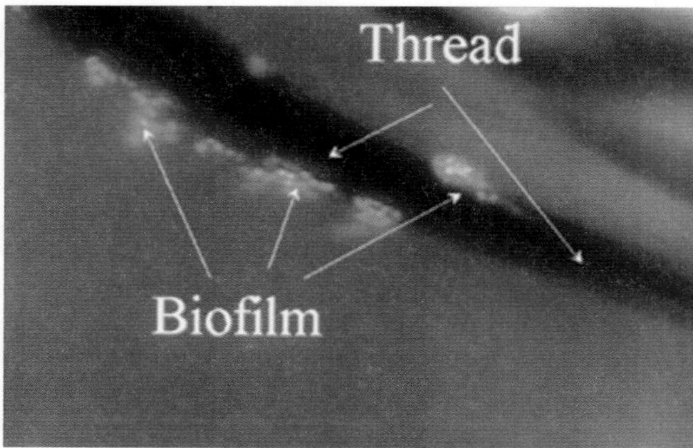

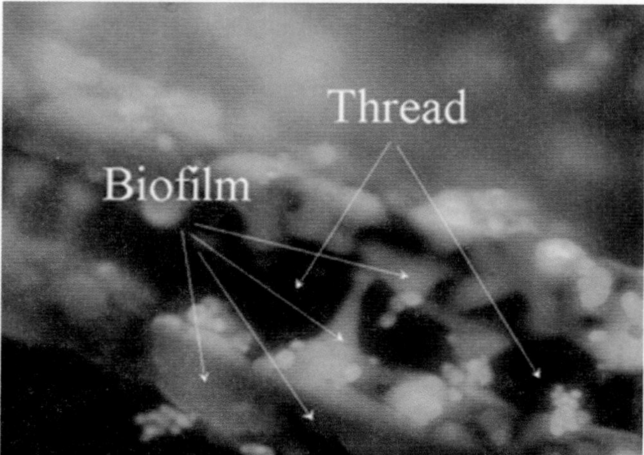

FIGURE II.2.8.4 Confocal microscope images of unfixed biofilms formed on individual fibers of the cloth-like material used to form the sewing cuffs of mechanical heart valves. The biofilms formed when fibers of a silver-coated medical device were exposed to planktonic cells of *Staphylococcus epidermidis*, in a flow cell (Cook et al., 2000) after 24 (top panel) and 48 hours (bottom panel). Staining with the LIVE/DEAD® BacLight™ probe showed that many cells were live (green) in marked contrast to dead cells (red). After 48 hours distinct biofilm cell clusters had formed on the fibers. Diffuse staining between and around the individual cells with the green nucleic acid stain Syto9 is associated with staining of the eDNA in the EPS slime matrix.

If the adherent cells survive, they will initiate biofilm formation, and the adherent cells will gradually form matrix-enclosed structured biofilm cell clusters (Figure II.2.8.4, bottom panel), within which the cells will be separated by EPS. The formation of biofilms requires that the adherent cells must be alive, so the observation of live biofilm on surfaces that make antibacterial and antibiofilm claims could have unfortunate clinical consequences (Cook et al., 2000).

Optimally the preference is for biomaterials which inherently resist attachment or kill colonizing bacteria, and do not retain these dead cells on the surface, i.e., prevention rather than treatment. For this reason, one of several available live/dead probes to ascertain the viability of adherent bacterial cells on biomaterials can be used. The Molecular Probes® BacLight™ LIVE/DEAD® kit has become a standard research method for distinguishing live cells on the basis of membrane integrity; live cells stain green while dead cells stain red. In practical terms, biomaterials set up in flow cells can be exposed to realistic fluid shear containing potentially pathogenic bacteria, and the colonization of its surface can be monitored by confocal microscopy.

Although direct observation is still the "gold standard" to test the resistance of biomaterials to bacterial colonization, confocal microscopes are complex and relatively expensive. Alternatively, sessile biofilm bacteria can be removed from colonized surfaces and enumerated by standard microbiological techniques; commonly referred to as "scraping and plating." The theory behind the technique is that it first removes all biofilm from a surface, then completely breaks up the clumps of bacteria into single unassociated cells, and finally spreads various dilutions of the cells on the surface of an agar plate so that each cell gives rise to one colony when the plate is incubated. Although this technique can drastically underestimate the actual surface cell concentration (usually expressed as colony forming units/cm^2, CFU/cm^2), microscopy can be used to validate the removal and homogenization steps which can be major sources of error. First, some cells may be left on the surface by the removal method, whether it is physical scraping, sonication or another method. Second, the removed biofilm must be completely homogenized to break up clumps of bacteria to ensure that each living bacterium gives rise to one colony on the agar plate. This step is usually achieved by vortexing, sonicating or use of a homogenizer. Sonication may kill some cells, so it is therefore important to calibrate the sonication time for each type of biofilm until microscopy shows that the resultant suspension is mostly single cells, and most of these cells are alive.

When scraping and plating are used without such validation, this method can yield data that are up to 4 logs lower than the bacterial numbers seen by direct microscopy. However, the scrape and plate method can yield accurate and consistent *in vitro* data when it is properly calibrated by microscopy, and the first biofilm method using this enumeration technique was ASTM Method E 2196-02. Darla Goeres and Marty Hamilton at the Center for Biofilm Engineering were instrumental in developing this method, and have been actively involved in developing other statistically robust experimental methods for biofilm evaluation which undoubtedly will form the basis for standard testing and claims substantiation in the anti-infective biomaterials industry. More rapid approaches include the development of conventional multi-well plates or biofilm-specific devices, used in combination with a plate reader to estimate biofilm biomass by general stains such as crystal violet staining or live/dead ratios.

Potential Agents for the Control of Microbial Colonization of Biomaterials

The strategy most commonly used in current antibacterial biomaterials is the incorporation of conventional antibiotics into the material, with the objective of killing incoming planktonic cells before they can adhere and initiate biofilm formation at the implant site (Campoccia et al., 2006). The release strategy has been used for years to infuse bone cements with gentamicin, and is actively pursued with researchers showing that release of NO, antibiotics, antibodies, and silver ion (Ag^+) from polymers and hydroxapatite materials reduce bacterial growth on plates in culture (Nablo et al., 2005; Hetrick et al., 2009; Zhao et al., 2009). This approach shows promise, and is being actively pursued. However, one problem is that these coatings are not synthetically flexible or covalently adhered to the surface, leading to bursts of antimicrobial activity and localized toxicity stemming from the introduction of non-resorbable coatings into the body (Zhao et al., 2009).

The quandary of balancing antimicrobial efficacy against the danger of promoting bacterial resistance does not affect the large cohort of bacterial manipulation molecules that is currently moving briskly toward the biomaterials market. Some of these biofilm control molecules are specifically targeted to interfere with the bacteria's own developmental signaling pathways to affect both biofilm formation and detachment. The theory is that by influencing natural signaling processes, bacteria can be manipulated not to form biofilms or to detach by triggering a dispersal event, thus reducing the likelihood of inducing defense responses or the generation of resistant mutants. Natural antibiofilm strategies developed by marine seaweeds, which use furanones to interfere with natural biofilm signaling pathways show promise since these defense strategies have been presumably developed over millions of years of coevolution, yet still remain effective (de Nys et al., 2006). The pivotal concept is that bacteria in contact with a biomaterial would be prevented from forming a biofilm on its surface, would be "locked" in the planktonic phenotype, and would be subsequently killed by host defenses (antibodies and activated leukocytes) and any antibiotics that might be present. The best described signal pathways are the acetyl homoserine lactone (AHL) systems of some gram-negative species (Davies et al., 1998), and the autoinducer two (AI2) signaling peptides in both gram-negative and gram-positive species (Camilli and Bassler, 2006). Of relevance to the design of biomaterials which utilize signal blocking is the finding that the effect of AI2 on biofilm is concentration dependent (Rickard et al., 2006), and while very low, μM concentrations added exogenously inhibited biofilm formation, higher concentrations showed no effect or actually stimulated biofilm formation.

A third class of signaling molecule which controls biofilm formation in staphylococci is RNAIII-inhibiting peptide (RIP) which targets the RNAIII-activating protein (RAP), and has been shown to prevent infection on implant materials in rats (Kiran et al., 2008). When specific antibiotics were administered to these test animals, while the challenging bacteria were locked in the planktonic phenotype, no live cells could be recovered from the biomaterial surfaces of the surrounding tissues. More recently, attention has been focused on the secondary messenger cyclic-di-GMP, which is not secreted into the surrounding medium but is intracellular (Hengge, 2009). Cyclic di-GMP is influenced by nitric oxide, but as with AI2, concentration is important. At higher concentrations, NO is toxic to the cells, but at nM levels appears to act as a dispersal signal (Barraud et al., 2009), thus highlighting the importance of precisely controlling the concentration to achieve the desired effect, perhaps a difficult challenge in an *in vivo* setting. Another potential biofilm dispersal signal is the fatty acid signal *cis*-2-decenoic acid, which is produced by *P. aeruginosa* (and possibly other species) and has been shown to be active *in vitro* against a broad range of organisms, over a wide range of concentrations (Davies and Marques, 2009). Such a signal makes a good candidate for antibiofilm biomaterials. However, while existence of a "universal" biofilm prevention or dispersal signal is attractive, the reality might be elusive. From an evolutionary perspective it seems unlikely that a diverse multitude of bacteria would so easily give up control of their biofilm development to a single molecule produced by competing species (Salta et al., 2010).

Delivery of Biofilm Control Agents at Biomaterial Surfaces

Killing the planktonic bacteria before they have time to initiate biofilm formation is the objective of many research programs, and this can be accomplished by three general strategies:

1. Systemic antibiotic therapy that produces bactericidal concentrations in the body fluids in the operative field;
2. Release of antibiotics and other bacterial manipulation molecules from the biomaterials to produce sustained effective concentrations of the agent in the immediate vicinity of the device;
3. Irrigation and other techniques that deliver antibiotics to the biomaterial surface after the device is installed, but before the operative wound is closed.

Many surgeons use systemic antibiotic therapy in the perioperative timeframe, and most also use this strategy in subsequent operations (including dental procedures) if a device has recently been installed and might not be fully epithelialized. The simplest manifestation of the antibiotic-releasing biomaterial strategy is a class of materials that can be "loaded," such as a sponge, by soaking them in a solution of the antibiotic in question immediately before the device is installed. In the case of filler or adhesive materials, such as poly-methylmethacrylate (PMMA), an active agent can be directly mixed prior to polymerization in the OR. Whenever a molecule is loaded into or onto a polymer there will be an immediate initial burst of diffusion in the early timeframe, followed by a tapering off over time as the concentration in this reservoir is depleted. These biomaterials are useful, but the low-level release of the agent for months or years after this effective timeframe cannot be controlled. This produces a prolonged period in which the agent is present at a sublethal concentration, near the device and sometimes in the whole body, and raises the specter of the development of acquired antibiotic resistance in many potentially dangerous bacterial species. Recently, several papers have even shown that sub-inhibitory concentrations of antibiotic may actually induce biofilm formation (Hoffman et al., 2005).

There is a similar problem with antibiotic impregnated coatings, since usually these coatings are not synthetically flexible or covalently adhered to the surface, leading to bursts of antimicrobial activity, and possibly localized toxicity stemming from the introduction of non-resorbable coatings into the body. The controlled release of active agents is one of the most problematic aspects of devices which rely on this technology for infection protection. One solution is to engineer surfaces to release active agents "on demand." Biomaterials can be coated with a molecular "skin," a self-assembled surface layer, that completely contains molecules loaded into an underlying plastic and can be temporarily deranged (by ultrasonic energy) to yield a controlled release (Kwok et al., 2001). Such a coating has been used to deliver insulin, as well as ciprofloxacin in controlled pulses in the laboratory (Norris et al., 2005). High concentrations of these agents could be released at the surfaces of medical devices perioperatively or at any preliminary signs of device-related infection.

The Bioelectric Effect as an Adjunct to Antibiotics

Coatings which release antibiotics at the site of potential infection have the advantage over systemically administered antibiotics, in that they can achieve much higher local concentrations and avoid toxicity effects, since the overall amount of antibiotic can be relatively small. However, due to difficulties with the controlled release and surfaces which are challenging to modify, it is desirable to be able to sensitize bacteria to conventional antibiotics. As discussed previously, one approach is to use signals in an attempt to induce a planktonic phenotype through cell signaling. Another promising area is to achieve planktonic-like antibiotic sensitivity by applying low voltage electrical fields (Khoury et al., 1992). The so called "bioelectric effect" has recently shown promise at treating *S. epidermidis* chronic foreign-body osteomyelitis in a rabbit model (Del Pozo et al., 2009a); however, the same group also reports that the efficacy is dependant on the pathogen and the class of antibiotic (Del Pozo et al., 2009b).

Biomaterials that Resist Bacterial Attachment and Biofilm Formation

An alternative approach to killing posited by the Association of Bone and Joint Surgeons Musculoskeletal Infection Workshop is to design coatings for implant surfaces that resist bacterial attachment. Most of the work in the literature has focused on the model systems of self-assembled monolayers (SAMs) on gold or polymeric coatings (Vaudaux et al., 1994; Sampedro and Patel, 2007). SAMs have many advantages as a flexible platform over macromolecules (Figure II.2.8.5). One major advantage of SAMs is their ease in engineering interfacial structures at the molecular level, due to the fact that they form defined structures on the surface and therefore present functional tail groups in a consistent, active manner (Laibinis et al., 1989; Hederos et al., 2005; Jiang et al., 2005). They are also synthetically flexible, which allows for the modification of the tail group for prevention of non-specific adsorption of cells. However, the SAM technology developed in model systems is not necessarily directly transferable to metal oxide systems used in biomaterials, and coating the surface with gold is not always practical. Alkanethiols form SAMs on gold and other noble metals, such as silver or copper, but not on oxide surfaces, such as those used in biomaterials (Ostuni et al., 2001). Functional groups can be immobilized through basic organic chemistry at the

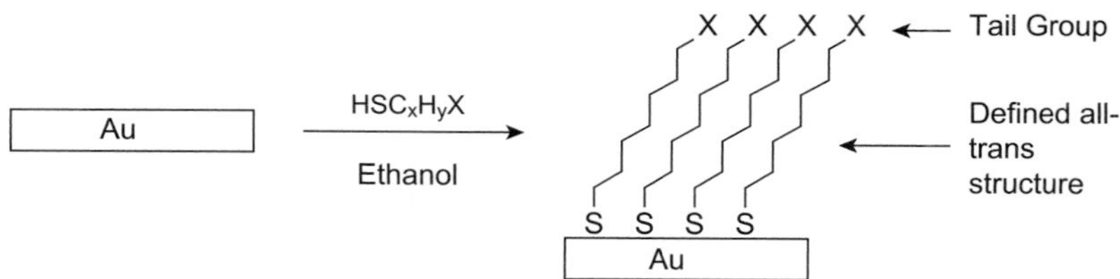

FIGURE II.2.8.5 Self-assembled monolayers form spontaneously on a surface placed in a solution. They consist of organic monomers with reactive head groups and exposed tail groups. Tail groups can be designed to accomplish a task, such as cell resistance, or to participate in chemical reaction for further functionalization of the surface. SAMs on gold form defined structures in which all of the carbon–carbon bonds are oriented "trans" to one another (at an angle of 109.5°). These monolayers are considered ordered.

tail groups of the monolayer or by synthesizing molecules with functional tail groups and forming a mixed SAM that includes multiple functionalities. Delivering multiple, active groups in a single polymeric coating is synthetically challenging (Kang et al., 2009). Polymeric coatings that are peptidomimetic have been successful in the laboratory at resisting bacteria adhesion, and could also be used as platforms for further functionalization (Jiang et al., 2005; Statz et al., 2005; Khoo et al., 2009). As a result, new chemistry for non-specific bacterial resistance on metal oxide surfaces must be developed that is appropriate for orthopedic implant materials.

The underlying mechanism by which SAMs render surfaces inert is under study. Many tail groups have been employed to render gold surfaces inert to protein and non-specific cellular adhesion. While oligo-ethylene glycol has been the standard by which inert surfaces are measured, and is the only SAM system used to mitigate bacterial adhesion, other groups such as mannitol, maltose, taurin (Espeland and Wetzel, 2001), and tertiary amine oxides have rendered gold surfaces inert to protein and cellular adhesion. A survey of literature suggests that formation of an inert surface utilizing self-assembled monolayers is a combination of many parameters including substrate wettability, tail functionality, lateral packing density, conformational flexibility of the molecule, structure of water on the surface, and kosmotropicity (the degree to which solutes stabilize water-to-water interactions, and hence affect molecular interactions) (Espeland and Wetzel, 2001). The underlying reason for their inert effect on surfaces may be due to their ability to order surrounding water molecules, excluding them from the surface and rendering the surface inert to protein adsorption. However, the molecular basis for the resistance of some surfaces continues to be debated, in part due to conflicting data in the literature, with some favoring the hydrophobic effect as the driving force (Dibdin et al., 1996), while others disagree. Due to the large interdependent, complex interactions between cells and organically modified surfaces, the understanding and harnessing of the phenomenon of controlling adhesion on oxide surfaces has been difficult.

Biomimicry or bioinspiration is another approach used by engineers in the design of inherently non-fouling surfaces (Salta et al., 2010). Nano-textured patterns found on the skins of sharks have been found to resist initial bacterial attachment in laboratory experiments. It is thought that this approach, originally designed as a control for marine biofouling (Schumacher et al., 2007), might be utilized for medical devices (Chung et al., 2007), illustrating the importance of interdisciplinary collaboration and thinking in the area of biomaterials design. A similar approach was used in the development of the natural and synthetic furanone cell signal blockers, which were first discovered to play a role in the resistance of red algae to marine biofouling, and are now being developed for infection protection of medical devices (de Nys et al., 2006).

CONCLUSIONS

Undoubtedly, medical devices will be increasingly used in patient management, whether temporarily or intermittently for fluids exchange, as in the case of catheters, or as permanent implants to repair loss of function, as is the case for orthopedic joints and cardiovascular stents. What is also ensured is that as more devices are used bacteria will take advantage of an access site into the body and with the convenience of an abiotic surface to colonize the spread of antibiotic resistant strains may likely increase. For permanent implants, overall rates of infection may drop as data show that there is an inverse correlation with surgical-site infection and numbers of procedures performed per institution and surgeon. However, the overall number of patients with surgical-site infections will inevitably rise, as more and more people undergo such surgeries, whether out of absolute necessity or to improve quality of life. To a certain extent, infection associated with permanent implants can be minimized through the controlled environment of the OR and the ever-increasing experience of the surgeon. Infections associated with catheters might present a greater challenge, since they require frequent manipulations and are usually managed in less controlled environments.

A challenge to biomaterials engineers and surface scientists is to design and develop materials for the manufacture of devices which are functional, biocompatible, resistant to infection, and do not add to the development of resistant strains. While 100% biofilm-proof surfaces may be a long way off, we at least now know our enemy much more intimately than even less than a decade ago. Observational studies of biofilms have begun to reveal the true complexities and coordination that pathogenic bacteria are capable of, and molecular techniques are revealing the mechanisms of biofilm development and dispersal, as well as revealing the full extent of diversity of biofilm populations, including the genetic and phenotypic divergence that develops in biofilms formed from a single strain in only a matter of hours to days. The challenge is formidable, but the rewards are great, with the potential to improve the quality of life of millions of people. To quote from *The Art of War* written by Sun Tzu in the 6th century: "So it is said that if you know your enemies and know yourself, you can win a hundred battles without a single loss. If you only know yourself, but not your opponent, you may win or may lose. If you know neither yourself nor your enemy, you will always endanger yourself." With our current understanding of biofilms we are only just beginning to know the enemy with respect to device-related biofilm infections. Even if the "biofilm-proof" surface proves elusive, the challenge is to effectively utilize what we have learned about biofilm biology to engineer playing fields which tip the balance for the race to the surface in favor of ourselves and away from the bacteria.

BIBLIOGRAPHY

Anderl, J. N., Franklin, M. J., & Stewart, P. S. (2000). Role of antibiotic penetration limitation in *Klebsiella pneumoniae* biofilm resistance to ampicillin and ciprofloxacin. *Antimicrob. Agents Chemother.*, **44**(7), 1818–1824.

Antoci, V., Jr., King, S. B., Jose, B., Parvizi, J., Zeiger, A. R., et al. (2007a). Vancomycin covalently bonded to titanium alloy prevents bacterial colonization. *J. Orthop. Res.*, **25**(7), 858–866.

Antoci, V., Jr., Adams, C. S., Hickok, N. J., Shapiro, I. M., & Parvizi, J. (2007b). Vancomycin bound to Ti rods reduces periprosthetic infection: Preliminary study. *Clin. Orthop. Relat. Res.*, **461**, 88–95.

Arciola, C. R., Campoccia, D., Baldassarri, L., Donati, M. E., Pirini, V., et al. (2006). Detection of biofilm formation in *Staphylococcus epidermidis* from implant infections. Comparison of a PCR-method that recognizes the presence of *ica* genes with two classic phenotypic methods. *J. Biomed. Mater. Res. A.*, **76**(2), 425–430.

Aviv, M., Berdicevsky, I., & Zilberman, M. (2007). Gentamicin-loaded bioresorbable films for prevention of bacterial infections associated with orthopedic implants. *J. Biomed. Mater. Res. A.*, **83**(1), 10–19.

Aydinuraz, K., Agalar, C., Agalar, F., Ceken, S., Duruyurek, N., et al. (2009). *In vitro S. epidermidis* and *S. aureus* adherence to composite and lightweight polypropylene grafts. *J. Surg. Res.*, **157**(1), e79–e86.

Bagge, N., Schuster, M., Hentzer, M., Ciofu, O., Givskov, M., et al. (2004). *Pseudomonas aeruginosa* biofilms exposed to imipenem exhibit changes in global gene expression and beta-lactamase and alginate production. *Antimicrob. Agents Chemother.*, **48**(4), 1175–1187.

Barken, K. B., Pamp, S. J., Yang, L., Gjermansen, M., Bertrand, J. J., et al. (2008). Roles of type IV pili, flagellum-mediated motility and extracellular DNA in the formation of mature multicellular structures in *Pseudomonas aeruginosa* biofilms. *Environ. Microbiol.*, **10**(9), 2331–2343.

Barraud, N., Schleheck, D., Klebensberger, J., Webb, J. S., Hassett, D. J., et al. (2009). Nitric oxide signaling in *Pseudomonas aeruginosa* biofilms mediates phosphodiesterase activity, decreased cyclic di-GMP levels, and enhanced dispersal. *J. Bacteriol.*, **191**(23), 7333–7342.

Bayston, R., & Penny, S. R. (1972). Excessive production of mucoid substance in staphylococcus SIIA: A possible factor in colonisation of Holter shunts. *Dev. Med. Child. Neurol. Suppl.*, **27**, 25–28.

Bjarnsholt, T. P. O., Jensen, M., Burmolle, M., Hentzer, J. A., Haagensen, H. P., et al. (2005). *Pseudomonas aeruginosa* tolerance to tobramycin, hydrogen peroxide and polymorphonuclear leukocytes is quorum-sensing dependent. *Microbiology*, **151**(Pt 2), 373–383.

Bjarnsholt, T., Tolker-Nielsen, T., Givskov, M., Janssen, M., & Christensen, L. H. (2009). Detection of bacteria by fluorescence *in situ* hybridization in culture-negative soft tissue filler lesions. *Dermatol. Surg.*, **35**(Suppl. 2), 1620–1624.

Bjerkan, G., Witso, E., & Bergh, K. (2009). Sonication is superior to scraping for retrieval of bacteria in biofilm on titanium and steel surfaces *in vitro*. *Acta. Orthop.*, **80**(2), 245–250.

Borlee, B. R., Goldman, A. D., Murakami, K., Samudrala, R., Wozniak, D. J., et al. (2010). *Pseudomonas aeruginosa* uses a cyclic-di-GMP-regulated adhesin to reinforce the biofilm extracellular matrix. *Mol. Microbiol.*, **75**(4), 827–842.

Bos, R., van der Mei, H. C., & Busscher, H. J. (1999). Physico-chemistry of initial microbial adhesive interactions: Its mechanisms and methods for study. *FEMS Microbiol. Rev.*, **23**(2), 179–230.

Brady, R. A., Leid, J. G., Kofonow, J., Costerton, J. W., & Shirtliff, M. E. (2007). Immunoglobulins to surface-associated biofilm immunogens provide a novel means of visualization of methicillin-resistant *Staphylococcus aureus* biofilms. *Appl. Environ. Microbiol.*, **73**(20), 6612–6619.

Bryers, J. D., & Drummond, F. (1998). Local macromolecule diffusion coefficients in structurally non-uniform bacterial biofilms using fluorescence recovery after photobleaching (FRAP). *Biotechnol. Bioeng.*, **60**(4), 462–473.

Camilli, A., & Bassler, B. L. (2006). Bacterial small-molecule signaling pathways. *Science*, **311**(5764), 1113–1116.

Campoccia, D., Montanaro, L., & Arciola, C. R. (2006). The significance of infection related to orthopedic devices and issues of antibiotic resistance. *Biomaterials*, **27**(11), 2331–2339.

Chang, Y. M., Jeng, W. Y., Ko, T. P., Yeh, Y. J., Chen, C. K., et al. (2010). Structural study of TcaR and its complexes with multiple antibiotics from *Staphylococcus epidermidis*. *Proc. Natl. Acad. Sci. USA*, **107**(19), 8617–8622.

Cheng, G., Zhang, Z., Chen, S., Bryers, J. D., & Jiang, S. (2007). Inhibition of bacterial adhesion and biofilm formation on zwitterionic surfaces. *Biomaterials*, **28**(29), 4192–4199.

Chung, K. K., Schumacher, J. F., Sampson, E. M., Burne, R. A., Antonelli, P. J., et al. (2007). Impact of engineered surface microtopography on biofilm formation of *Staphylococcus aureus*. *Biointerphases*, **2**(2), 89–94.

Cook, G., Costerton, J. W., & Darouiche, R. O. (2000). Direct confocal microscopy studies of the bacterial colonization *in vitro* of a silver-coated heart valve sewing cuff. *Int. J. Antimicrob. Agents*, **13**(3), 169–173.

Costerton, J. W., Post, J. C., Ehrlich, G. D., Hu, F. Z., Kreft, R., Nistico, L., Kathju, S., Stoodley, P., Hall-Stoodley, L., Maale, G. E., James, G. A., Shirtliff, M. E., Sotereanos, N. G., & DeMeo, P. J. (2011). New methods for the detection of orthopaedic and other biofilm infections. FEMS Immuno. *Microbiol.*, **61**(2), 133–140.

Cucarella, C., Tormo, M. A., Knecht, E., Amorena, B., Lasa, I., et al. (2002). Expression of the biofilm-associated protein interferes with host protein receptors of *Staphylococcus aureus* and alters the infective process. *Infect. Immun.*, **70**(6), 3180–3186.

Cui, Q., Mihalko, W. M., Shields, J. S., Ries, M., & Saleh, K. J. (2007). Antibiotic-impregnated cement spacers for the treatment of infection associated with total hip or knee arthroplasty. *J. Bone Joint Surg. Am.*, **89**(4), 871–882.

Davies, D. G., & Geesey, G. G. (1995). Regulation of the alginate biosynthesis gene algC in *Pseudomonas aeruginosa* during biofilm development in continuous culture. *Appl. Environ. Microbiol.*, **61**(3), 860–867.

Davies, D. G., & Marques, C. N. (2009). A fatty acid messenger is responsible for inducing dispersion in microbial biofilms. *J. Bacteriol.*, **191**(5), 1393–1403.

Davies, D. G., Parsek, M. R., Pearson, J. P., Iglewski, B. H., Costerton, J. W., et al. (1998). The involvement of cell-to-cell signals in the development of a bacterial biofilm. *Science*, **280**(5361), 295–298.

de Nys, R., Givskov, M., Kumar, N., Kjelleberg, S., & Steinberg, P. D. (2006). Furanones. *Prog. Mol. Subcell. Biol.*, **42**, 55–86.

deBeer, D., Stoodley, P., & Lewandowski, Z. (1997). Measurement of local diffusion coefficients in biofilms by micro-injection and confocal microscopy. *Biotechnol. Bioeng.*, **53**(2), 151–158.

Del Pozo, J. L., Rouse, M. S., Euba, G., Kang, C. I., Mandrekar, J. N., et al. (2009a). The electricidal effect is active in an experimental model of *Staphylococcus* epidermidis chronic foreign body osteomyelitis. *Antimicrob. Agents Chemother.*, **53**(10), 4064–4068.

Del Pozo, J. L., Rouse, M. S., Mandrekar, J. N., Sampedro, M. F., Steckelberg, J. M., et al. (2009b). Effect of electrical current on the activities of antimicrobial agents against *Pseudomonas aeruginosa, Staphylococcus aureus*, and *Staphylococcus epidermidis* biofilms. *Antimicrob. Agents Chemother.*, **53**(1), 35–40.

Dempsey, K. E., Riggio, M. P., Lennon, A., Hannah, V. E., Ramage, G., et al. (2007). Identification of bacteria on the surface of clinically infected and non-infected prosthetic hip joints removed during revision arthroplasties by 16S rRNA gene sequencing and by microbiological culture. *Arthritis. Res. Ther.*, **9**(3), R46.

Dibdin, G. H., Assinder, S. J., Nichols, W. W., & Lambert, P. A. (1996). Mathematical model of beta-lactam penetration into a biofilm of *Pseudomonas aeruginosa* while undergoing simultaneous inactivation by released beta-lactamases. *J. Antimicrob. Chemother.*, **38**(5), 757–769.

Donlan, R. M., & Costerton, J. W. (2002). Biofilms: Survival mechanisms of clinically relevant microorganisms. *Clin. Microbiol. Rev.*, **15**(2), 167–193.

Dunne, W. M., Jr., & Burd, E. M. (1992). The effects of magnesium, calcium, EDTA, and pH on the *in vitro* adhesion of *Staphylococcus epidermidis* to plastic. *Microbiol. Immunol.*, **36**(10), 1019–1027.

Ecker, D. J., Sampath, R., Massire, C., Blyn, L. B., Hall, T. A., et al. (2008). Ibis T5000: A universal biosensor approach for microbiology. *Nat. Rev. Microbiol.*, **6**(7), 553–558.

Edmiston, C. E., Seabrook, G. R., Goheen, M. P., Krepel, C. J., Johnson, C. P., et al. (2006). Bacterial adherence to surgical sutures: Can antibacterial-coated sutures reduce the risk of microbial contamination? *J. Am. Coll. Surg.*, **203**(4), 481–489.

Engelsman, A. F., van der Mei, H. C., Ploeg, R. J., & Busscher, H. J. (2007). The phenomenon of infection with abdominal wall reconstruction. *Biomaterials*, **28**(14), 2314–2327.

Engelsman, A. F., van der Mei, H. C., Busscher, H. J., & Ploeg, R. J. (2008). Morphological aspects of surgical meshes as a risk factor for bacterial colonization. *Br. J. Surg.*, **95**(8), 1051–1059.

Engelsman, A. F., Krom, B. P., Busscher, H. J., van Dam, G. M., Ploeg, R. J., et al. (2009). Antimicrobial effects of an NO-releasing poly(ethylene vinylacetate) coating on soft-tissue implants *in vitro* and in a murine model. *Acta. Biomater.*, **5**(6), 1905–1910.

Espeland, E. M., & Wetzel, R. G. (2001). Complexation, stabilization, and UV photolysis of extracellular and surface-bound glucosidase and alkaline phosphatase: Implications for biofilm microbiota. *Microb. Ecol.*, **42**(4), 572–585.

Feely, T., Copley, A., & Bleyer, A. J. (2007). Catheter lock solutions to prevent bloodstream infections in high-risk hemodialysis patients. *Am. J. Nephrol.*, **27**(1), 24–29.

Fitzgerald, R. H., Jr. (1979). Microbiologic environment of the conventional operating room. *Arch. Surg.*, **114**(7), 772–775.

Fux, C. A., Wilson, S., & Stoodley, P. (2004). Detachment characteristics and oxacillin resistance of *Staphyloccocus aureus* biofilm emboli in an *in vitro* catheter infection model. *J. Bacteriol.*, **186**(14), 4486–4491.

Fux, C. A., Shirtliff, M., Stoodley, P., & Costerton, J. W. (2005). Can laboratory reference strains mirror "real-world" pathogenesis? *Trends Microbiol.*, **13**(2), 58–63.

Gotz, F. (2002). *Staphylococcus* and biofilms. *Mol. Microbiol.*, **43**(6), 1367–1378.

Gould, C. V., Umscheid, C. A., Agarwal, R. K., Kuntz, G., & Pegues, D. A. (2010). Guideline for prevention of catheter-associated urinary tract infections 2009. *Infect. Control Hosp. Epidemiol.*, **31**(4), 319–326.

Gristina, A. G., & Costerton, J. W. (1985). Bacterial adherence to biomaterials and tissue. The significance of its role in clinical sepsis. *J. Bone Joint Surg. Am.*, **67**(2), 264–273.

Gristina, A. G., Naylor, P., & Myrvik, Q. (1988). Infections from biomaterials and implants: A race for the surface. *Med. Prog. Technol.*, **14**(3–4), 205–224.

Gross, M., Cramton, S. E., Gotz, F., & Peschel, A. (2001). Key role of teichoic acid net charge in *Staphylococcus aureus* colonization of artificial surfaces. *Infect. Immun.*, **69**(5), 3423–3426.

Guenther, F., Stroh, P., Wagner, C., Obst, U., & Hansch, G. M. (2009). Phagocytosis of staphylococci biofilms by polymorphonuclear neutrophils: *S. aureus* and *S. epidermidis* differ with regard to their susceptibility towards the host defense. *Int. J. Artif. Organs*, **32**(9), 565–573.

Hall-Stoodley, L., & Stoodley, P. (2005). Biofilm formation and dispersal and the transmission of human pathogens. *Trends Microbiol.*, **13**(1), 7–10.

Hall-Stoodley, L., & Stoodley, P. (2009). Evolving concepts in biofilm infections. *Cell Microbiol.*, **11**(7), 1034–1043.

Hall-Stoodley, L., Costerton, J. W., & Stoodley, P. (2004). Bacterial biofilms: From the natural environment to infectious diseases. *Nat. Rev. Microbiol.*, **2**(2), 95–108.

Hall-Stoodley, L., Hu, F. Z., Gieseke, A., Nistico, L., Nguyen, D., et al. (2006). Direct detection of bacterial biofilms on the middle-ear mucosa of children with chronic otitis media. *JAMA*, **296**(2), 202–211.

Hamady, M., Walker, J. J., Harris, J. K., Gold, N. J., & Knight, R. (2008). Error-correcting barcoded primers for pyrosequencing hundreds of samples in multiplex. *Nat. Methods*, **5**(3), 235–237.

Hanssen, A. D., Patel, R., & Osmon, D. R. (2005). Editorial comment. *Clin. Orthop. Relat. Res.*, **437**, 2.

Harrison, J. J., Ceri, H., & Turner, R. J. (2007). Multimetal resistance and tolerance in microbial biofilms. *Nat. Rev. Microbiol.*, **5**(12), 928–938.

Hederos, M., Konradsson, P., & Liedberg, B. (2005). Synthesis and self-assembly of galactose-terminated alkanethiols and their ability to resist proteins. *Langmuir*, **21**(7), 2971–2980.

Heilmann, C., Hussain, M., Peters, G., & Gotz, F. (1997). Evidence for autolysin-mediated primary attachment of *Staphylococcus epidermidis* to a polystyrene surface. *Mol. Microbiol.*, **24**(5), 1013–1024.

Hendrickx, A. P., van Luit-Asbroek, M., Schapendonk, C. M., van Wamel, W. J., Braat, J. C., et al. (2009). SgrA, a nidogen-binding LPXTG surface adhesin implicated in biofilm formation, and EcbA, a collagen binding MSCRAMM are two novel adhesins of hospital-acquired *Enterococcus faecium*. *Infect. Immun.*, **77**(11), 5097–5106.

Hengge, R. (2009). Principles of c-di-GMP signalling in bacteria. *Nat. Rev. Microbiol.*, **7**(4), 263–273.

Hetrick, E. M., & Schoenfisch, M. H. (2006). Reducing implant-related infections: Active release strategies. *Chem. Soc. Rev.*, **35**(9), 780–789.

Hetrick, E. M., Shin, J. H., Paul, H. S., & Schoenfisch, M. H. (2009). Anti-biofilm efficacy of nitric oxide-releasing silica nanoparticles. *Biomaterials*, **30**(14), 2782–2789.

Hiller, N. L., Janto, B., Hogg, J. S., Boissy, R., Yu, S., et al. (2007). Comparative genomic analyses of seventeen *Streptococcus pneumoniae* strains: Insights into the pneumococcal supragenome. *J. Bacteriol.*, **189**(22), 8186–8195.

Hodgson, N. C., Malthaner, R. A., & Ostbye, T. (2000). The search for an ideal method of abdominal fascial closure: A meta-analysis. *Ann. Surg.*, **231**(3), 436–442.

Hoffman, L. R., D'Argenio, D. A., MacCoss, M. J., Zhang, Z., Jones, R. A., et al. (2005). Aminoglycoside antibiotics induce bacterial biofilm formation. *Nature*, **436**(7054), 1171–1175.

Hsieh, P. H., Lee, M. S., Hsu, K. Y., Chang, Y. H., Shih, H. N., et al. (2009). Gram-negative prosthetic joint infections: Risk factors and outcome of treatment. *Clin. Infect. Dis.*, **49**(7), 1036–1043.

Jacobsen, S. M., Stickler, D. J., Mobley, H. L., & Shirtliff, M. E. (2008). Complicated catheter-associated urinary tract infections due to *Escherichia coli* and *Proteus mirabilis*. *Clin. Microbiol. Rev.*, **21**(1), 26–59.

James, G. A., Swogger, E., Wolcott, R., Pulcini, E., Secor, P., et al. (2008). Biofilms in chronic wounds. *Wound Repair Regeno*, **16**(1), 37–44.

Jefferies, J. M., Smith, A., Clarke, S. C., Dowson, C., & Mitchell, T. J. (2004). Genetic analysis of diverse disease-causing pneumococci indicates high levels of diversity within serotypes and capsule switching. *J. Clin. Microbiol.*, **42**(12), 5681–5688.

Jesaitis, A. J., Franklin, M. J., Berglund, D., Sasaki, M., Lord, C. I., et al. (2003). Compromised host defense on *Pseudomonas aeruginosa* biofilms: Characterization of neutrophil and biofilm interactions. *J. Immunol.*, **171**(8), 4329–4339.

Jiang, W., Zhitenev, N., Bao, Z., Meng, H., Abusch-Magder, D., et al. (2005). Structure and bonding issues at the interface between gold and self-assembled conjugated dithiol monolayers. *Langmuir*, **21**(19), 8751–8757.

Justinger, C., Moussavian, M. R., Schlueter, C., Kopp, B., Kollmar, O., et al. (2009). Antibacterial [corrected] coating of abdominal closure sutures and wound infection. *Surgery*, **145**(3), 330–334.

Kang, S. M., Rho, J., Choi, I. S., Messersmith, P. B., & Lee, H. (2009). Norepinephrine: Material-independent, multifunctional surface modification reagent. *J. Am. Chem. Soc.*, **131**(37), 13224–13225.

Kaplan, J. B. (2010). Biofilm dispersal: Mechanisms, clinical implications, and potential therapeutic uses. *J. Dent. Res.*, **89**(3), 205–218.

Kathju, S., Nistico, L., Hall-Stoodley, L., Post, J. C., Ehrlich, G. D., et al. (2009). Chronic surgical site infection due to suture-associated polymicrobial biofilm. *Surg. Infect. (Larchmt)*, **10**(5), 457–461.

Kathju, S., Lasko, L. A., Nistico, L., Colella, J. J., & Stoodley, P. (2010). Cutaneous fistula from the gastric remnant resulting from a chronic suture-associated biofilm infection. *Obes. Surg.*, **20**(2), 251–256.

Khoo, X., Hamilton, P., O'Toole, G. A., Snyder, B. D., Kenan, D. J., et al. (2009). Directed assembly of PEGylated-peptide coatings for infection-resistant titanium metal. *J. Am. Chem. Soc.*, **131**(31), 10992–10997.

Khoury, A. E., Lam, K., Ellis, B., & Costerton, J. W. (1992). Prevention and control of bacterial infections associated with medical devices. *ASAIO J.*, **38**(3), M174–M178.

Kiran, M. D., Giacometti, A., Cirioni, O., & Balaban, N. (2008). Suppression of biofilm related, device-associated infections by staphylococcal quorum sensing inhibitors. *Int. J. Artif. Organs*, **31**(9), 761–770.

Klapper, I., Gilbert, P., Ayati, B. P., Dockery, J., & Stewart, P. S. (2007). Senescence can explain microbial persistence. *Microbiology*, **153**(Pt 11), 3623–3630.

Kwok, C. S., Mourad, P. D., Crum, L. A., & Ratner, B. D. (2001). Self-assembled molecular structures as ultrasonically-responsive barrier membranes for pulsatile drug delivery. *J. Biomed. Mater. Res.*, **57**(2), 151–164.

Laibinis, P. E., Hickman, J. J., Wrighton, M. S., & Whitesides, G. M. (1989). Orthogonal self-assembled monolayers: Alkanethiols on gold and alkane carboxylic acids on alumina. *Science*, **245**(4920), 845–847.

Leid, J. G., Shirtliff, M. E., Costerton, J. W., & Stoodley, P. (2002). Human leukocytes adhere to, penetrate, and respond to *Staphylococcus aureus* biofilms. *Infect. Immun.*, **70**(11), 6339–6345.

Lewis, K. (2005). Persister cells and the riddle of biofilm survival. *Biochemistry (Mosc)*, **70**(2), 267–274.

Li, Y., Ku, C. Y., Xu, J., Saxena, D., & Caufield, P. W. (2005). Survey of oral microbial diversity using PCR-based denaturing gradient gel electrophoresis. *J. Dent. Res.*, **84**(6), 559–564.

Loose, C., Jensen, K., Rigoutsos, I., & Stephanopoulos, G. (2006). A linguistic model for the rational design of antimicrobial peptides. *Nature*, **443**(7113), 867–869.

Machado, J. D., Suen, V. M., Figueiredo, J. F., & Marchini, J. S. (2009). Biofilms, infection, and parenteral nutrition therapy. *J. Parenter. Enteral. Nutr.*, **33**(4), 397–403.

Mack, D., Fischer, W., Krokotsch, A., Leopold, K., Hartmann, R., et al. (1996). The intercellular adhesin involved in biofilm accumulation of *Staphylococcus epidermidis* is a linear beta-1,6-linked glucosaminoglycan: Purification and structural analysis. *J. Bacteriol.*, **178**(1), 175–183.

Maki, D. G., & Tambyah, P. A. (2001). Engineering out the risk for infection with urinary catheters. *Emerg. Infect. Dis.*, **7**(2), 342–347.

Marrie, T. J., Nelligan, J., & Costerton, J. W. (1982). A scanning and transmission electron microscopic study of an infected endocardial pacemaker lead. *Circulation*, **66**(6), 1339–1341.

Marshall, K. C., Stout, R., & Mitchell, R. (1971). Selective sorption of bacteria from seawater. *Can. J. Microbiol.*, **17**(11), 1413–1416.

McKenna, P., Hoffmann, C., Minkah, N., Aye, P. P., Lackner, A., et al. (2008). The macaque gut microbiome in health, lentiviral infection, and chronic enterocolitis. *PLoS Pathog.*, **4**(2), e20.

Ming, X., Nichols, M., & Rothenburger, S. (2007). *In vivo* antibacterial efficacy of MONOCRYL plus antibacterial suture (Poliglecaprone 25 with triclosan). *Surg. Infect. (Larchmt)*, **8**(2), 209–214.

Ming, X., Rothenburger, S., & Nichols, M. M. (2008). *In vivo* and *in vitro* antibacterial efficacy of PDS plus (polidioxanone with triclosan) suture. *Surg. Infect. (Larchmt)*, **9**(4), 451–457.

Nablo, B. J., Rothrock, A. R., & Schoenfisch, M. H. (2005). Nitric oxide-releasing sol-gels as antibacterial coatings for orthopedic implants. *Biomaterials*, **26**(8), 917–924.

Neut, D., van Horn, J. R., van Kooten, T. G., van der Mei, H. C., & Busscher, H. J. (2003). Detection of biomaterial-associated infections in orthopaedic joint implants. *Clin. Orthop. Relat. Res.*, **413**, 261–268.

NIH (2003). NIH Consensus statement on total knee replacement NIH. *NIH Consens State. Sci. Statements*, **20**(1), 1–34.

Norris, P., Noble, M., Francolini, I., Vinogradov, A. M., Stewart, P. S., et al. (2005). Ultrasonically controlled release of ciprofloxacin from self-assembled coatings on poly(2-hydroxyethyl methacrylate) hydrogels for *Pseudomonas aeruginosa* biofilm prevention. *Antimicrob. Agents Chemother.*, **49**(10), 4272–4279.

Nucci, C., Artini, M., Pasmore, M., Missiroli, F., Costerton, J. W., et al. (2005). A microbiological and confocal microscopy study documenting a slime-producing *Staphylococcus epidermidis* isolated from a nylon corneal suture of a patient with antibiotic-resistant endophthalmitis. *Graefes. Arch. Clin. Exp. Ophthalmol.*, **243**(9), 951–954.

O'Grady, N. P., Alexander, M., Dellinger, E. P., Gerberding, J. L., Heard, S. O., et al. (2002). Guidelines for the prevention of intravascular catheter-related infections. Centers for Disease Control and Prevention. *MMWR Recomm. Rep.*, **51**(RR-10), 1–29.

Osterberg, B. (1983). Influence of capillary multifilament sutures on the antibacterial action of inflammatory cells in infected wounds. *Acta. Chir. Scand.*, **149**(8), 751–757.

Osterberg, B., & Blomstedt, B. (1979). Effect of suture materials on bacterial survival in infected wounds. An experimental study. *Acta. Chir. Scand.*, **145**(7), 431–434.

Ostuni, E., Chapman, R. G., Liang, M. N., Meluleni, G., Pier, G., et al. (2001). Self-assembled monolayers that resist the adsorption of proteins and the adhesion of bacterial and mammalian cells. *Langmuir*, **17**(20), 6336–6343.

Otten, J. E., Wiedmann-Al-Ahmad, M., Jahnke, H., & Pelz, K. (2005). Bacterial colonization on different suture materials: A potential risk for intraoral dentoalveolar surgery. *J. Biomed. Mater. Res. B. Appl. Biomater.*, **74**(1), 627–635.

Otto, M. (2009). *Staphylococcus epidermidis*: The "accidental" pathogen. *Nat. Rev. Microbiol.*, **7**(8), 555–567.

Parsek, M. R., & Singh, P. K. (2003). Bacterial biofilms: An emerging link to disease pathogenesis. *Annu. Rev. Microbiol.*, **57**, 677–701.

Perez-Osorio, A. C., Williamson, K. S., & Franklin, M. J. (2010). Heterogeneous rpoS and rhlR mRNA levels and 16S rRNA/rDNA ratios within *Pseudomonas aeruginosa* biofilms, sampled by laser capture microdissection. *J. Bacteriol.*, **192**(12), 2991–3000.

Purevdorj-Gage, B., Costerton, W. J., & Stoodley, P. (2005). Phenotypic differentiation and seeding dispersal in non-mucoid and mucoid *Pseudomonas aeruginosa* biofilms. *Microbiology*, **151**(Pt 5), 1569–1576.

Rickard, A. H., Palmer, R. J., Jr., Blehert, D. S., Campagna, S. R., Semmelhack, M. F., et al. (2006). Autoinducer 2: A concentration-dependent signal for mutualistic bacterial biofilm growth. *Mol. Microbiol.*, **60**(6), 1446–1456.

Rolighed, T., Moser, C., Hall-Stoodley, L., & Stoodley, P. (2010). The Role of Bacterial Biofilms in Infections of Catheters and Shunts. In T. Bjarnsholt, P. O. Jensen, C. Moser, & N. Hoeby (Eds.), *Biofilm Infections*. New York, NY: Springer.

Salta, M., Wharton, J. A., Stoodley, P., Dennington, S. P., Goodes, L. R., et al. (2010). Designing biomimetic antifouling surfaces. *Phil. Trans. A., Royal. Society*, **368**(1929), 4729–4754.

Sampedro, M. F., & Patel, R. (2007). Infections associated with long-term prosthetic devices. *Infect. Dis. Clin. North Am.*, **21**(3), 785–819.

Sauer, K., & Camper, A. K. (2001). Characterization of phenotypic changes in *Pseudomonas putida* in response to surface-associated growth. *J. Bacteriol.*, **183**(22), 6579–6589.

Schaudinn, C., Stoodley, P., Kainović, A., O'Keeffe, T., Costerton, J. W., et al. (2007). Bacterial biofilms, other structures seen as mainstream concepts. *Microbe.*, **2**(5), 231–237.

Schlafer, S., Nordhoff, M., Wyss, C., Strub, S., Hubner, J., et al. (2008). Involvement of *Guggenheimella bovis* in digital dermatitis lesions of dairy cows. *Vet. Microbiol.*, **128**(1–2), 118–125.

Schumacher, J. F., Carman, M. L., Estes, T. G., Feinberg, A. W., Wilson, L. H., et al. (2007). Engineered antifouling microtopographies: Effect of feature size, geometry, and roughness on settlement of zoospores of the green alga Ulva. *Biofouling*, **23**(1–2), 55–62.

Senaratne, W., Andruzzi, L., & Ober, C. K. (2005). Self-assembled monolayers and polymer brushes in biotechnology: Current applications and future perspectives. *Biomacromolecules*, **6**(5), 2427–2448.

Sottile, F. D., Marrie, T. J., Prough, D. S., Hobgood, C. D., Gower, D. J., et al. (1986). Nosocomial pulmonary infection: Possible etiologic significance of bacterial adhesion to endotracheal tubes. *Crit. Care Med.*, **14**(4), 265–270.

Starkey, M., Hickman, J. H., Ma, L., Zhang, N., De Long, S., et al. (2009). *Pseudomonas aeruginosa* rugose small-colony variants have adaptations that likely promote persistence in the cystic fibrosis lung. *J. Bacteriol.*, **191**(11), 3492–3503.

Statz, A. R., Meagher, R. J., Barron, A. E., & Messersmith, P. B. (2005). New peptidomimetic polymers for antifouling surfaces. *J. Am. Chem. Soc.*, **127**(22), 7972–7973.

Stewart, P. S., & Franklin, M. J. (2008). Physiological heterogeneity in biofilms. *Nat. Rev. Microbiol.*, **6**(3), 199–210.

Stoodley, P., Sauer, K., Davies, D. G., & Costerton, J. W. (2002). Biofilms as complex differentiated communities. *Annu. Rev. Microbiol.*, **56**, 187–209.

Stoodley, P., Nistico, L., Johnson, S., Lasko, L. A., Baratz, M., et al. (2008). Direct demonstration of viable *Staphylococcus aureus* biofilms in an infected total joint arthroplasty. A case report. *J. Bone Joint Surg. Am.*, **90**(8), 1751–1758.

Stoodley, P., Braxton, E. E., Nistico, L., Hall-Stoodley, L., Johnson, S., et al. (2010). Direct demonstration of a *Staphylococcus* biofilm in an external ventricular drain in a patient with a history of recurrent ventriculoperitoneal shunt failure. *Pediatric. Neurosurgery*, **46**(2), 127–132.

Stoodley, P., Conti, S., DeMeo, P. J., Nistico, L., Melton-Kreft, R., Johnson, S., Darabi, Ehrlich, G. D., Costerton, J. W., & Kathju, S. (2011). Characterization of a mixed MRSA / MRSE biofilm in an explanted total ankle arthroplasty. *FEMS IMM*, **62**(1), 66–74.

Storch, M. L., Rothenburger, S. J., & Jacinto, G. (2004). Experimental efficacy study of coated VICRYL plus antibacterial suture in guinea pigs challenged with *Staphylococcus aureus*. *Surg Infect. (Larchmt)*, **5**(3), 281–288.

Subbiahdoss, G., Kuijer, R., Grijpma, D. W., van der Mei, H. C., & Busscher, H. J. (2009). Microbial biofilm growth vs. tissue integration: "The race for the surface" experimentally studied. *Acta Biomater.*, **5**(5), 1399–1404.

Trampuz, A., & Widmer, A. F. (2006). Infections associated with orthopedic implants. *Curr. Opin. Infect. Dis.*, **19**(4), 349–356.

Tunney, M. M., Patrick, S., Gorman, S. P., Nixon, J. R., Anderson, N., et al. (1998). Improved detection of infection in hip replacements. A currently underestimated problem. *J. Bone Joint Surg. Br.*, **80**(4), 568–572.

van Loosdrecht, M. C., Norde, W., & Zehnder, A. J. (1990). Physical chemical description of bacterial adhesion. *J. Biomater. Appl.*, **5**(2), 91–106.

van't Riet, M., Steyerberg, E. W., Nellensteyn, J., Bonjer, H. J., & Jeekel, J. (2002). Meta-analysis of techniques for closure of midline abdominal incisions. *Br. J. Surg.*, **89**(11), 1350–1356.

Vaudaux, P. E., Lew, D. P., & Waldvogel, F. A. (1994). Host Factors Predisposing to and Influencing Therapy of Foreign Body Infections. In A. L. Bisno, & F. A. Waldvogel (Eds.), *Infections Associated With Indwelling Medical Devices* (pp. 1–29). Washington, DC: American Society for Microbiology.

Vergara-Irigaray, M., Valle, J., Merino, N., Latasa, C., Garcia, B., et al. (2009). Relevant role of fibronectin-binding proteins in *Staphylococcus aureus* biofilm-associated foreign-body infections. *Infect. Immun.*, **77**(9), 3978–3991.

von Eiff, C., Heilmann, C., Herrmann, M., & Peters, G. (1999). Basic aspects of the pathogenesis of staphylococcal polymer-associated infections. *Infection*, **27**(Suppl. 1), S7–S10.

Vuong, C., Kocianova, S., Voyich, J. M., Yao, Y., Fischer, E. R., et al. (2004). A crucial role for exopolysaccharide modification in bacterial biofilm formation, immune evasion, and virulence. *J. Biol. Chem.*, **279**(52), 54881–54886.

Webb, J. S., Lau, M., & Kjelleberg, S. (2004). Bacteriophage and phenotypic variation in *Pseudomonas aeruginosa* biofilm development. *J. Bacteriol.*, **186**(23), 8066–8073.

Wissing, J., van Vroonhoven, T. J., Schattenkerk, M. E., Veen, H. F., Ponsen, R. J., et al. (1987). Fascia closure after midline laparotomy: Results of a randomized trial. *Br. J. Surg.*, **74**(8), 738–741.

Wong, E. S. (1983). Guideline for prevention of catheter-associated urinary tract infections. *Am. J. Infect. Control*, **11**(1), 28–36.

Zhao, L., Chu, P. K., Zhang, Y., & Wu, Z. (2009). Antibacterial coatings on titanium implants. *J. Biomed. Mater. Res. B. Appl. Biomater.*, **91**(1), 470–480.

SECTION II.3

Biological Testing of Biomaterials

SECTION II.3

Biological Testing of Biomaterials

CHAPTER II.3.1 HOW WELL WILL IT WORK? INTRODUCTION TO TESTING BIOMATERIALS

Buddy D. Ratner
Professor, Bioengineering and Chemical Engineering, Director of University of Washington Engineered Biomaterials (UWEB), Seattle, WA, USA

When an engineer designs a suspension bridge over a river, he or she will start with a precise engineering drawing and then, based on well-established mathematical models, assess the strength of materials and corrosion resistance needed to ensure the structure will stand for some anticipated lifetime, e.g., 100 years. The strength of materials and corrosion resistance are compiled in widely accepted handbooks containing data obtained through rigorous testing. The engineer can confidently design the bridge from computer models and handbook data. The bridge will stand and function appropriately for 100 years. Moreover, such a bridge manufactured according to a given set of design criteria and specifications should perform approximately the same whether it crosses a river in Seattle, Birmingham or Boston.

Now, consider an engineer who sets out to design a heart assist device or a hip prosthesis. This engineer will find the basic mechanical property data relevant to the materials used in handbooks, but little on specific biointeractions, biodurability, and the implications of these biological reactions for the ultimate performance of the complex device. Because of the limited amounts of validated and qualified data on how a given material will perform in the challenging *in vivo* environment, biomaterials scientists devise tests to observe the performance of materials (and devices), and assess potential biointeractions. These tests are often (but not always) focused on the mechanics and biology relevant to the specific application (orthopedic, cardiovascular, urological, etc.), and they do not consider the potential for different circumstances and possibly different biomaterials–tissue interactions in different patients. This section of the textbook focuses on useful tools and ideas for testing biomaterials.

After this introduction, the section leads off with a chapter titled "The Concept and Assessment of Biocompatibility (Chapter II.3.2)." Biocompatibility is possibly the most central concept in understanding biomaterials; and you cannot test for biocompatibility until you understand what it is. Chapter II.3.3 (*In Vitro* Assessment of Cell and Tissue Compatibility,) and Chapter II.3.4 (*In Vivo* Assessment of Tissue Compatibility) provide fine overviews of thinking *circa* early 21st century on testing for biocompatibility, primarily focused on toxicology (*in vitro*) and the foreign-body reaction (*in vivo*), and parallel the thinking by most national regulatory agencies (for example, the US Food and Drugs Administration, FDA). So, why do we need a chapter on the concept of biocompatibility? There are new developments that will probably change the definition of biocompatibility. For example, Chapter II.3.2 discusses a material (with no toxic leachables) fabricated as a solid film that heals with the classic foreign-body reaction, and the same material fabricated as a precise, porous structure that heals in a vascularized, afibrotic manner. Today's definitions and regulatory climate are not structured to appreciate the significance of these differences. Furthermore, there are other strategies to shift the healing of today's "biocompatible" biomaterials to a more tissue-integrated, vascularized reconstruction. To prepare this third edition of the *Biomaterials Science* textbook for changes that will occur in the next few years, we introduce the issues and concerns stimulating new thinking on biocompatibility in Chapter II.3.2. Chapter II.2.2 (Inflammation, Wound Healing, and the Foreign-Body Response) provides essential background to understand the material in Chapters II.3.2, II.3.3, and II.3.4.

Chapter II.3.5, Evaluation of Blood–Materials Interactions, looks at another face of the biological reaction to synthetic materials. Millions of medical devices are interfaced with the bloodstream each year. The reaction of blood with biomaterials is strikingly different from the reactions seen in soft tissue and bone in both kinetics (rapid) and outcomes (clot). Although blood compatibility assessment goes back to the 1930s, to this day we have no widely accepted list of biomaterials that are blood-compatible. Indeed, there is little agreement about the structural/chemical/molecular characteristics of a material that engender blood compatibility. Why this lack of clarity and consensus? Chapter II.3.5 introduces the issues in blood compatibility assessment (the nature of blood, the impact of flow, and the nature of the biomaterial), and illustrates how simple assessment schemes can lead to erroneous conclusions. Accepted blood compatibility tests are introduced but, more importantly, the chapter provides guidelines for drawing conclusions from the results of these tests. Some thoughts on which materials might be blood compatible are offered in this chapter.

Chapter II.3.6 and Chapter II.3.7 address the use of animals in biomaterials, and in medical device research. Animal implants (*in vivo* assessment) are usually required for preclinical testing of medical devices. Without animal testing, it is less likely that the research we perform will transition from laboratory bench to patient. The use of animals in testing is fraught with ethical issues – respect for the animal lives to be sacrificed and minimization of pain to the animal (see Chapter III.2.7). Also, a key issue is how significant are the results obtained in animals for understanding the performance of the medical device in humans? Humans are different genetically, biochemically, and biomechanically from animals. Also, humans are different genetically, biochemically, and biomechanically from each other, whereas animal models are genetically homogeneous. Can predictive models be developed? The answer is, in most cases, "yes," but not with some forethought and planning. Without careful consideration and understanding of specific animal models, animal lives will be sacrificed and research funds wasted, since the results obtained may not be useful for the human device. The biomechanics issue is particularly challenging. Almost any orthopedic device (spine, joints) will be subjected to different mechanical forces in the skeletal system of an animal walking on four legs versus an upright, bipedal human. Other considerations are: (1) should we go directly from *in vitro* testing to testing in human subjects (see Chapter III.2.7); and (2) what about developing devices for veterinary use (for example, about 5000 intraocular lenses are implanted in dogs each year)? Biomaterials researchers are largely dependent on animal models, and understanding them is essential – Chapters II.3.6 and II.3.7 provide a start in appreciating the complex issues surrounding animal models.

The section on testing concludes with a chapter on the basics of microscopy (Chapter II.3.8). Microscopic examination of tissues and implants is a key element in most biomaterials testing, *in vitro* and *in vivo*. To visually observe the cellular and tissue reactions at the interface between a biomaterial and a living system provides clues as to the functionality and prospects for success of a biomaterial or medical device. The range of microscopic techniques, and the resolution and information content of these methods, is now in a period of evolutionary development, so this chapter primarily addresses the basics for observing biomaterials and bioreactions at the micron scale and the nanoscale.

CHAPTER II.3.2 THE CONCEPT AND ASSESSMENT OF BIOCOMPATIBILITY

Buddy D. Ratner[1] and Frederick J. Schoen[2]
[1]Professor, Bioengineering and Chemical Engineering, Director of University of Washington Engineered Biomaterials (UWEB), Seattle, WA, USA
[2]Professor of Pathology and Health Sciences and Technology (HST), Harvard Medical School, Executive Vice Chairman, Department of Pathology, Brigham and Women's Hospital, Boston, MA, USA

What do we mean when we say a biomaterial is biocompatible?

- Is biocompatibility "yes" or "no," or is there a continuum of biocompatibilities ranging from "good" to "bad?"
- How can we measure biocompatibility?
- How do we improve or enhance the biocompatibility of a biomaterial?

The *idea* of biocompatibility is central to what makes a material a biomaterial. Also, new developments and concepts in cell biology are shifting thinking about biocompatibility, and this chapter offers an opportunity of a glimpse into the future of biocompatible biomaterials.

> Biocompatibility is so central to what makes a material a biomaterial, that a focused discussion of the subject is justified.

This chapter clarifies some of the issues in biocompatibility, and also raises questions that will likely impact the field in the coming years. In contrast to empirical approaches and practical considerations focused solely about the safety of implanted devices (for example, toxicology, the state of the art today), modern cell and molecular biology ideas may give us a useful "theory of biocompatibility" with quantifiable parameters, testable hypotheses, and validated engineering rules.

BIOCOMPATIBILITY TODAY

We start with an overview of the state of the art in biocompatibility today, i.e., in the first decade in the 21st century.

A definition for biocompatibility, widely used in the biomaterials/medical device community, has been presented previously in the textbook and is repeated here:

> *"the ability of a material to perform with an appropriate host response in a specific application"* (Williams, 1987).

This definition, though accurate and quite useful in the design, development, and application of biomaterials in medicine, nevertheless offers no insights into the mechanisms of biocompatibility, how to test the biocompatibility of a material or how to optimize or enhance the biocompatibility of a material. This section of the textbook will expand the definition and explore the philosophical and scientific ideas surrounding biocompatibility.

Specific concepts impacting biocompatibility are elaborated upon in Chapters II.2.2 (Inflammation, Wound Healing, and the Foreign-Body Response), II.3.3 (*In Vitro* Assessment of Cell and Tissue Compatibility), II.3.4 (*In Vivo* Assessment of Tissue Compatibility), and III.2.3 (Voluntary Consensus Standards) (and other sections throughout this textbook). Biocompatibility can be assessed using *in vitro* and *in vivo* assays. Although a variety of direct chemical and physical interactions may also be important, measurement of the consequences of leachable or secreted substances from biomaterials to cells in culture is the primary goal of *in vitro* biocompatibility assays. For example, cell proliferation inhibition or cell death are negative outcomes in such assays, and would be characteristic of materials that are not biocompatible. The *in vivo* (implantation) response to a "biocompatible" biomaterial is generally described as a no more than mild inflammatory reaction, which, after some 2–3 weeks, resolves itself into a thin fibrous capsule (largely scar as a result of the tissue trauma of implantation). This is sometimes referred to as a low level, long-term, minimal inflammatory response (i.e., macrophages are present at the implant surface even years after implantation, but the reaction site is relatively quiescent, and there is otherwise no indication of adverse local or systemic response). The presence of foreign-body giant cells at the interface may suggest lesser "biocompatibility." The composite reaction is termed the "foreign-body reaction," or FBR.

Four factors impact that which we refer to as "biocompatibility." These are:

1. Toxicology (the measurement and study of the effects of material leaching from biomaterials) (II.3.3 *In Vitro* Assessment of Tissue Compatibility and III.2.2 Standards).
2. Reactions related to products from extrinsic microbiologic organisms colonizing the biomaterial (for example, endotoxin contamination; see Chapters II.2.8 and III.1.2).
3. Mechanical effects such as rubbing, irritation, compression, and modulus mismatch. Closely related are size-related effects, for example, if the implant is much larger than a macrophage, comparable in size to a macrophage, and capable of being phagocytosed, or much smaller than a macrophage, i.e., nano-size.
4. A broad range of interactions with surrounding proteins, and cells, inducing cell–biomaterials interactions (and tissue–biomaterials interactions) that might direct longer-term *in vivo* bioreaction.

Four factors impact that which we refer to as "biocompatibility." These are:

- toxicology
- extrinsic organisms
- mechanical effects
- cell–biomaterial interactions.

Points 1, 2, and 3, above, are well-understood, and often applied in the design of biocompatible biomaterials – we understand the principles, have the ability to measure their impact, and we can design devices using clearly-defined principles to achieve good outcomes. Point 4 is less well-developed and concerns are expanded upon below.

TOXICOLOGY

Polymeric materials often contain extractable components, such as unreacted monomer, oligomers, initiator fragments, stabilizers, and other processing additives. Metals, glasses, and ceramics can release ions and other processing components. The type of reaction that will be considered from a toxicology standpoint is if these substances are released and negatively impact cells (*in vitro*) or adjacent tissues (*in vivo*), or if they affect an organism systemically. Biomaterials science, standards organizations, and allied fields have developed reliable methods to measure and identify leachates, and also sensitive and standardized methods to look at the reaction of tissues and cells to these leached substances. These methods are summarized in Chapter II.3.3. The ISO 10993 standards provide many specific, defined tests for toxicity associated with leachables.

THE PRODUCTS OF EXTRINSIC ORGANISMS COLONIZING THE BIOMATERIAL

Bacteria and their cell wall components are intense inflammatory activators. Fungi such as *Candida* are also inflammatory activators (Kojic and Darouiche, 2004). In the case of implants that are contaminated with fungi, bacteria or bacterial cell-wall endotoxin (lipopolysaccharide), an intense and usually long-term biological reaction is seen, characterized by large numbers of white cells (mostly neutrophils and macrophages, collectively called "pus") in the vicinity of the implant. In humans, this reaction would be described by the patient as producing pain, redness, and heat. This response to the contaminated implant can lead to exceptionally thick foreign-body capsules. High concentrations of extravascular white cells and thick, dense foreign-body capsules are characteristics of poor biocompatibility. Such extreme reactions to devices with contaminating organisms have been documented with breast implants (Pajkos et al., 2003), and with other implant devices. This is distinct from *infection* with live microbiologic organisms, such as bacteria or fungi, in which the organisms may proliferate and cause ongoing local tissue destruction and potentially systemic effects. Although infection with live organisms shares some morphologic features with, and can occasionally look like, "poor biocompatibility," we generally refer to this not as a biocompatibility issue, but rather a sterilization issue (see Chapter III.1.2).

MECHANICAL EFFECTS

If an implant is rubbing, abrading or moving in contact with tissue, or has sharp corners, undesirable reactions that are superficially classified as "non-biocompatible" will be observed. Mechanical mismatch between a hard biomaterial and a soft tissue can lead to damage or irritation to the soft tissue. Cell responses to mechanical forces are well-known and usually quite significant (Stamenović and Ingber, 2009). An excellent example of the mechanical effect on *in vivo* bioreaction was seen in 1976, where scientists implanted in rat muscle medical grade "biocompatible" materials having circular, triangular, and pentagonal shapes (Matlaga et al., 1976). The degree of reaction increased in the order: circle (lowest reaction), pentagon, triangle. The effect was attributed to micromotion associated with the acute angle of the triangle leading to the greatest tissue irritation. In general, it is the role of the implant designer to ensure that the device does not excessively rub or irritate tissue (rounded edges are better than sharp edges, for example). It is the role of the surgeon to appropriately place and anchor the device in the implant site to minimize such rubbing and irritation.

CELL–BIOMATERIALS INTERACTIONS

This topic dominates the literature in biomaterials science (see Chapter II.1.3, Cells and Surfaces *In Vitro*, for a detailed elaboration). It has been clear for 100 years or more that living cells interact and attach to different materials in different ways, and the nature of that interaction may have immense influence on cell fate, including attachment, spreading, proliferation, differentiation, activation, secretion, and detachment. It is also well-established that the adsorbed protein film always preceding cell interaction with surfaces directs and modulates the cell response (see Chapter II.1.2). Since inflammatory cells, such as neutrophils and macrophages, "interrogate" implanted materials shortly after implantation, and since different surfaces interact in different ways with proteins and cells, we would expect the biomaterial to impact the cell-driven *in vivo* reaction. However, let's examine this conclusion. *In vitro*, profound differences are seen in cell interactions between different materials. For example, a poly(2-hydroyxethyl methacrylate)(polyHEMA) hydrogel will not permit macrophages to adhere in cell culture, while a tissue culture polystyrene (TCPS) surface readily allows those same cells to adhere. Yet, if the polyHEMA and TCPS are implanted *in vivo*, both will heal similarly with an avascular, collagenous foreign-body capsule. In fact, all "biocompatible" materials, be they hydrophilic, hydrophobic, metallic, polymeric or ceramic, will heal similarly with a classic (and largely quiescent) foreign-body reaction (FBR) if there are: (1) no leachables; (2) no products from extrinsic organisms; and (3) no mechanical irritation. The explanation for this striking difference between *in vitro* bioreaction and *in vivo* bioreaction has yet to be identified, but does highlight the multi-cell complexity of the *in vivo* environment, in comparison to the relatively simple environment with one cell type *in vitro*.

The phenomena of: (1) frustrated phagocytosis; and (2) cytokine release are important to discuss here. Macrophages are programmed to engulf and digest foreign material. Phagocytosis occurs after the foreign-body has bound to molecules called "receptors" that are on the surface of the macrophage. The macrophage then stretches itself around the foreign-body and engulfs it. Once inside this phagocyte, the foreign particle is trapped in a compartment called a lysosome which has a battery of chemicals that attempt to degrade the foreign material. Frustrated phagocytosis is the term used to express the concept that the macrophage is incapable of "eating" a piece of biomaterial approximately larger than its size or certainly a macroscopic implant. In this situation the lysosomes (which may be in contact with the foreign material but open to the surroundings) may release their contents into the adjacent tissues. This may cause local tissue damage and inflammation. Moreover, to address the large surface, the macrophage might fuse with other macrophages and form multinucleated foreign-body giant cells, often used as a marker of a more severe FBR. Macrophages also release cytokines (diffusible signaling proteins) in response to biomaterials (Bonfield et al., 1992). Cytokines can be considered as proinflammatory (e.g., IL-1, TNF-a) or anti-inflammatory (e.g. IL-4, IL-10). A measurement of the cytokine shower from an implanted biomaterial may offer insights to biocompatibility.

SUMMARY OF IDEAS TO THIS POINT

Important points about biocompatibility as we understand it today are:

1. The biological reaction we call biocompatibility is negatively impacted by leachables, products of extrinsic organism surface contamination, and micromotion.
2. As long as leachables, extrinsic organism surface contamination, and micromotion are not impacting the reaction, all materials will give an approximately similar bioreaction *in vivo*, referred to as the normal FBR, composed of a thin fibrous capsule and minimal ongoing inflammation.
3. When the foreign body capsule is thin and the reaction site, after approximately one month, is relatively

quiescent, this is an acceptable FBR and we call the implant "biocompatible (or inert)."

4. Inert biomaterials lead to the reaction described in point 3 (above), and thus are called "biocompatible."
5. The favorable long-term interface between a biomaterial and the surrounding tissues in most medical devices is characterized by a thin, dense, collagenous capsule that isolates the biomaterial implant from the body.

NEW DEVELOPMENTS ARE CHANGING THE PARADIGM OF BIOCOMPATIBILITY

Although we implant millions of devices made of biocompatible biomaterials in humans every year, largely with much success, there are concerns with the way implants heal (the FBR), and new generations of materials are designed for applications in which something different than a quiescent FBR is desired.

For example, a dense fibrous capsule can inhibit diffusion of analytes to implanted sensors, interfere with release of drugs from implanted controlled drug release devices, and enhance the resistance of an electrical path, thereby blocking communication with tissues for implanted electrodes. Capsular contraction is a problem for some devices, such as breast implants, where the contraction of the fibrous capsule (i.e., all scars contract) distorts the soft implant, tending to make it spherical. Moreover, the lack of vascularity near the implant–tissue interface can slow the body's response to, and treatment of, bacterial invasion and related biofilm formation. Certainly, the capsule associated with the FBR can create surgical problems for device removal and revision. In many cases, a vascularized, integrated tissue reconstruction (more resembling normal tissue reconstruction) would be preferable to the avascular, dense capsule.

The potential for vascularized, non-fibrotic healing is now being realized. Such reconstructive healing can be achieved with extracellular matrix (ECM) components or with inert biomaterials with engineered porosity.

Decellularized ECM derived from a number of tissues, for example small intestinal submucosa (SIS), has been found to heal into the body with little or no fibrosis, excellent vascularity, and general tissue reconstruction (Badylak, 2007). SIS and other decellularized ECMs have been used in more than 1,000,000 human surgeries, largely with good results. Chapter II.6.16 elaborates on these decellularized tissues as biomaterials. If the SIS is cross-linked, it will heal in a proinflammatory manner, with a capsule and a classical FBR. The excellent healing of SIS is attributed to the ability of macrophages to degrade the ECM to bioactive peptides that actively promote healing. Importantly, the ECM structure is heavily infused with macrophages in the early stage of healing, and those macrophages have been shown to be in a phenotype (also referred to as a polarization) conducive to healing (M2), in contrast to the proinflammatory, fibrotic M1 phenotype (Mantovani, 2006; Badylak et al., 2008). There are also suggestions that soluble ECMs derived from neonatal cell culture might be used as coating for implants to aid in healing and integration (Naughton and Kellar, 2008).

Certain porous synthetic biomaterials will also heal in a minimally fibrotic, angiogenic fashion. Observations on the special characteristics of the healing of porous structures in the body go back to the early 1970s (Karp et al., 1973; Klawitter et al., 1976). Many studies discussed this porosity effect, and an extensive study in 1992 clearly demonstrated the importance of pore size to healing (Brauker et al., 1995). The concern with all these studies was that the materials used had a broad distribution of pore sizes, making it difficult to ascertain the effect on healing of a specific pore size. A method was developed to make materials with a single, consistent pore size, based on using solvent-soluble microspheres as templates to create uniform, interconnected pores. When such materials were implanted subcutaneously, it was noted that when pores were in the size range 30–40 microns, vascularized healing and reconstruction with little fibrosis was observed (Marshall et al., 2004; Madden et al., 2010). Also, these materials were heavily infused with macrophages during healing, and more of the macrophages were in the M2 phenotype (Madden et al., 2010). These same materials with 30–40 micron pores healed well in skin percutaneous sites (with dermal and epidermal reconstruction) (Fukano et al., 2010), heart muscle, and other tissues. A further discussion of porous materials can be found in Chapter I.2.15.

With relevance to biocompatibility, consider differences engendered solely as a result of physical factors. A synthetic hydrogel is fabricated as a solid slab or as a porous structure with 30–40 micron interconnected pores. The chemical compositions of both are identical. Also, they have similar mechanical properties, no leachables, and no endotoxin or bacteria. Yet one heals in a capsule with the classic FBR, while the other heals in a vascularized, reconstructed manner with little fibrosis. It seems challenging to use the word "biocompatible" for both, given the sharp differences in *in vivo* biological reaction, despite identical chemistries.

> A "biocompatible" material is fabricated as a porous structure or a solid slab. The porous structure heals in an avascular, pro-angiogenic manner. The solid slab is encapsulated with the classic FBR. How can we use the word "biocompatible" for two such different bioreactions?

CONCLUSIONS

Since the introduction of "modern" biomaterials to medical practice in the late 1940s and early 1950s, we have seen an evolution in sophistication and functionality (Williams, 2008). *First generation* biomaterials were simply inert (e.g., Titanium, Teflon™, Polyethylene). *Second generation* biomaterials (1970s, 1980s) interacted with and changed the biological environment in a controlled manner. Such materials might be engineered for biodegradability, poly(lactic acid) for example, or engineered to integrate into bone and induce bone healing, such as bioactive glasses, or to induce a controlled thrombotic deposit, such as a textured polymer surface on the internal surface of a left ventricular assist device. *Third generation* biomaterials (1990+) biospecifically orchestrate biological processes, and can direct regeneration and restore functionality, and/or respond to the environment in a proactive manner to favorably influence a tissue reaction. Examples include engineered extracellular matrices (Lutolf and Hubbell, 2005), sphere templated biomaterials (Madden et al., 2010), SIS (Badylak, 2007), and "smart" biomaterials (Reyes et al., 2007). Also, consider the published debate on the effectiveness of biomaterials guided by specific peptide signals (Williams, 2011). As biomaterials evolve, so too must the definition of biocompatibility.

Thus, the way we define biocompatibility may change over the next few years. Examples that challenge the present paradigm of biocompatibility are published, and materials that show unique healing are being applied in clinical medicine. Recent clarification on the diversity of macrophage phenotypes and understanding of resident tissue stem cell pools in the body may permit engineered "biocompatibility," with control of the extent, rate, and speed of integration. These new developments strengthen the science of biocompatibility, and bring biomaterials and tissue engineering closer together. Precision control of biocompatibility can lead to new biomaterials-based therapies with profound advantages for the patient.

For now, the medical device regulatory agencies subscribe almost exclusively to the definition of biocompatibility that we have been using since the first functionally successful implants in the 1950s. In learning the subject of biomaterials, it is important that one appreciates the accepted standards for biomaterials; standards that are used by the $300 billion plus medical device industry worldwide (estimated to reach $415 billion in 2016) and standards that impact millions of patients (see Chapters III.2.2 and III.2.3). Still, since we are on the cusp of a shift in thinking about this subject, it is useful for these newer ideas to be addressed and placed in the context of the word "biocompatibility."

> Understanding biocompatibility is central to understanding biomaterials.

BIBLIOGRAPHY

Badylak, S. F. (2007). The extracellular matrix as a biologic scaffold material. *Biomaterials*, **28**(25), 3587–3593.

Badylak, S., Valentin, J., Ravindra, A., Mccabe, G., & Stewart-Akers, A. (2008). Macrophage phenotype as a determinant of biologic scaffold remodeling. *Tissue Engineering Part A*, **14**(11), 1835–1842.

Bonfield, T. L., Colton, E., Marchant, R. E., & Anderson, J. M. (1992). Cytokine and growth factor production by monocyte/macrophages on protein preadsorbed polymers. *J. Biomed. Mater. Res.*, **26**, 837–850.

Brauker, J. H., Carr-Brendel, V. E., Martinson, L. A., Crudele, J., Johnston, W. D., et al. (1995). Neovascularization of synthetic membranes directed by membrane microarchitecture. *J. Biomed. Mater. Res.*, **29**, 1517–1524.

Fukano, Y., Usui, M. L., Underwood, R. A., Isenhath, S., Marshall, A. J., et al. (2010). Epidermal and dermal integration into sphere-templated porous poly(2-hydroxyethyl methacrylate) implants in mice. *Journal of Biomedical Materials Research Part A*, **94**(4), 1172–1186.

Karp, R. D., Johnson, K. H., Buoen, L. C., Ghobrial, H. K.G., Brand, I., et al. (1973). Tumorigenesis by millipore filters in mice: Histology and ultrastructure of tissue reactions as related to pore size. *Journal of the National Cancer Institute*, **51**(4), 1275–1279.

Klawitter, J. J., Bagwell, J. G., Weinstein, A. M., Sauer, B. W., & Pruitt, J. R. (1976). An evaluation of bone growth into porous high density polyethylene. *J. Biomed. Mater. Res.*, **10**(2), 311–323.

Kojic, E. M., & Darouiche, R. O. (2004). *Candida* infections of medical devices. *Clin. Microbiol. Rev.*, **17**(2), 255–267.

Lutolf, M. P., & Hubbell, J. A. (2005). Synthetic biomaterials as instructive extracellular microenvironments for morphogenesis in tissue engineering. *Nature Biotechnology*, **23**(1), 47–55.

Madden, L. R., Mortisen, D. J., Sussman, E. M., Dupras, S. K., Fugate, J. A., et al. (2010). Proangiogenic scaffolds as functional templates for cardiac tissue engineering. *Proc. Natl. Acad. Sci. USA*, **107**(34), 15211–15216.

Mantovani, A. (2006). Macrophage diversity and polarization: *In vivo* veritas. *Blood*, **108**(2), 408–409.

Marshall, A. J., Irvin, C. A., Barker, T., Sage, E. H., Hauch, K. D., et al. (2004). Biomaterials with tightly controlled pore size that promote vascular in-growth. *ACS Polymer Preprints*, **45**(2), 100–101.

Matlaga, B. F., Yasenchak, L. P., & Salthouse, T. N. (1976). Tissue response to implanted polymers: The significance of sample shape. *J. Biomed. Mater. Res.*, **10**(3), 391–397.

Naughton, G., & Kellar, R. (2008). Human ECM for devices and therapeutics. *MD&DI*, 102–109. May.

Pajkos, A., Deva, A. K., Vickery, K., Cope, C., Chang, L., et al. (2003). Detection of subclinical infection in significant breast implant capsules. *Plastic and Reconstructive Surgery*, **111**(5), 1605–1611.

Reyes, C. D., Petrie, T. A., Burns, K. L., Schwartz, Z., & García, A. J. (2007). Biomolecular surface coating to enhance orthopaedic tissue healing and integration. *Biomaterials*, **28**(21), 3228–3235.

Stamenović, D., & Ingber, D. E. (2009). Tensegrity-guided self assembly: From molecules to living cells. *Soft Matter*, **5**(6), 1137–1145.

Williams, D. F. (1987). Definitions in Biomaterials. In *Progress in Biomedical Engineering, 4*. Amsterdam: Elsevier. p 72.

Williams, D. F. (2008). On the mechanisms of biocompatibility. *Biomaterials*, **29**(20), 2941–2953.

Williams, D. F. (2011). The role of short synthetic adhesion peptides in regenerative medicine: The debate. *Biomaterials*, **32**(18), 4195–4197.

CHAPTER II.3.3 *IN VITRO* ASSESSMENT OF CELL AND TISSUE COMPATIBILITY

Michael F. Wolf[1], *Kelly P. Coleman*[2], *and Gregory M. Lewerenz*[2]

[1]Medtronic Inc., Corporate Technology and Innovation, Minneapolis, MN, USA

[2]Medtronic Inc., Physiological Research Laboratory, Minneapolis, MN, USA

INTRODUCTION

This chapter explains how *in vitro* methods may be used to assess the biocompatibility of medical device materials. While such methods unavoidably present a simplified view of the complex *in vivo* milieu, they are valuable because they provide insight to potential *in vivo* tissue and cellular responses. Many *in vitro* methods have been standardized through a global harmonization process. These standardized tests serve a key role in the biological safety evaluation of medical devices and materials. In the following pages, the scientific basis for each test is explored. Also presented are examples of *in vitro* techniques developed in response to a growing shift towards devices that combine drugs or biological agents to improve therapy. It should be noted that results of *in vitro* tests described in this chapter are often supported with additional *in vivo* testing (see Chapters II.5.3–II.5.5).

BACKGROUND CONCEPTS

At the cellular and molecular level, the presence of a biomaterial has potential to impart a perturbation in the local surroundings that may impact any number of proteins, genes, pathways, molecular modules and networks, and cells. The compatibility of such perturbations, that is the particular response of a cell or tissue to the material and *vice versa*, is then defined by the extent to which the response is deleterious to the microenvironment. Adding to this complexity is the wide range of tissue interfaces and potential materials used in today's medical device applications (see Table II.3.3.1).

Foreign materials or chemicals found inside the body are referred to as *xenobiotics*. Any chemical leaching or degrading from a device/material – or, in extreme circumstances, pieces of the material itself – falls under this definition. Thus, assessing the cellular and tissue compatibility of a device/material often involves examination of the potential xenobiotic chemicals that it may release. Whether these leachable chemicals trigger an unacceptable response in the host's biological microenvironment is key to determining the potential biocompatibility of the device/material.

Foreign materials or chemicals found inside the body are referred to as xenobiotics. Any chemical leaching or degrading from a medical device or material – or, in extreme circumstances, pieces of the material itself – falls under this definition. Thus, assessing the cellular and tissue compatibility of a device/material often involves examination of the potential xenobiotic chemicals that it may release.

Chemicals that possess potentially harmful or adverse properties are referred to as *toxicants*.[1] The response to a toxicant from a device/material is greatly influenced not only by the material itself, but also by the type of exposure. Exposure characteristics include the *route(s)* and *site(s)* of exposure (e.g., brain tissue versus subcutaneous tissue), *duration of exposure* (e.g., several hours versus permanent implantation), and the potential *frequency of exposure* over the treatment period or lifetime of the host. (Contrast, for example, the exposure a patient with degenerative heart disease receives from a single prosthetic valve implant to the exposure a diabetic individual receives from insulin pump infusion lines replaced every 72 hours.) *Chemical reactivity* of a toxicant is an additional key factor that will impact the degree and type of response in the host (Casarett et al., 2003). A particular reaction may occur because of an agent's chemical reactivity, or it may be due to a more specific recognition or reactivity towards a molecular structure in the biological environment. The compatibility of a chemically-reactive agent depends on whether it elicits undesirable cell responses or damages cellular structures.

The *degree of response* in the host in relation to the *amount of material or toxicant* present defines its *dose–response relationship*. This relationship is extremely important in the characterization of toxicants, as it can often depict a threshold dose below which the probability of an adverse response occurring in the host is either very low (the lowest-observable adverse effect level, or LOAEL) or not measureable (the no-observable adverse effect level, or NOAEL). This threshold dose is of key interest when studying the dose–response effect of a device material and attempting to assess its compatibility with cells and tissues. A classic example of the latter is demonstrated in the case of porcine bioprosthetic heart valves treated with glutaraldehyde to increase biostability and durability. Here, low levels of residual glutaraldehyde may produce negative responses in *in vitro* tests. Nonetheless, the device itself has proven to be safe and enormously beneficial in patients with cardiac valve disease.

Assessing cell and tissue biocompatibility has historically been evaluated with the *in vitro* tests described

[1]These are substances that are produced by, or are a by-product of, anthropogenic (man-made) activities, as opposed to *toxins* which are made by biological systems (Casarett et al., 2003).

TABLE II.3.3.1 International Organization for Standardization ISO 10993 *Biological Evaluation of Medical Devices* Standard (10) by Parts Relevant to In vitro Tests for Cell and Tissue Compatibility

Part*	Title (publication year: status)	Focus	In vitro Tests for Cell and Tissue Compatibility
3	Tests for genotoxicity (G), carcinogenicity (C), and reproductive toxicity (RT) (2003: CD)	In vitro and in vivo tests	• Bacterial Reverse Mutation Assay (G) • In vitro Mouse Lymphoma tk Assay (G) • Chromosomal Aberration in Mammalian Cells (G) • Syrian Hamster Embryo (SHE) Cell Transformation Assays (C) • Balb/c 3T3 Focus Transformation Assay (C) • Embryonic Stem Cell Test (RT)
4	Selection of tests for interactions with blood (2006: to be revised)	In vitro and in vivo tests	• Hemolysis • General suggestions regarding testing for thrombosis, coagulation, platelet function, hematology, and immunology.
5	Tests for in vitro cytotoxicity (1999: under revision)	In vitro tests	• Tests on Extracts - L929 Elution Test - Neutral Red Uptake Test (in Annex) - Colony Formation Test (in Annex) - MTT and related Tests (in Annex) • Direct Contact Test • Indirect Contact Tests - Agar Diffusion Test - Filter Diffusion
13, 14, 15	Identification and quantification of degradation products from polymeric medical devices (1998: under revision); from ceramics (2001); and from metals and alloys (2000), respectively	In vitro analytical tests	Information for Material Characterization and Toxicological Risk Assessment
17	Establishment of allowable limits for leachable substances (2002)		
18	Chemical characterization of materials (2005)		
19	Physicochemical, morphological, and topographical characterization of materials (2006)		

*Parts 1, 2, 7, 9, and 12 are general guidance documents on use of the standard, animal welfare, sterilization residuals, degradation products, and sample preparation, respectively; Parts 6, 10, 11, 16, and 20 are general guidance documents on in vivo testing relevant to local effects after implantation, irritation and delayed-type hypersensitivity, toxicokinetics study of degradation products and leachables, and principles of immunotoxicity testing, respectively. Part 8 was withdrawn and does not exist. CD refers to committee draft stage of the ISO standard development process.

in this chapter. Here, the focus has been on assays for DNA damage, cytotoxicity, cell proliferation, and quantification of specific proteins recognized to be an influential component or end product of a critical pathway or event.

USE OF MEDICAL DEVICE CHEMICAL COMPOSITION AND MATERIAL EXTRACTS FOR TOXICOLOGICAL RISK ASSESSMENT AND *IN VITRO* TESTING

Medical devices are primarily composed of polymers, metals, ceramics, and to a lesser extent, xenograft and allograft biological tissues and molecules. Assessing the *in vitro* cell and tissue compatibility of these solid multidimensional and multicompositional devices presents significant challenges. Not surprisingly, these challenges have lead to the development of an international standard on test material sample preparation (ISO 10993-12, 2007).

Challenges encountered in sample preparation include: (1) how to properly bring diverse material and device test samples into a standardized and appropriate level of contact with test systems; (2) how to treat composite materials and devices; (3) how to account for potential material interactions, given the frequency of composite devices; and (4) other confounding factors. Examples of the latter include device size, supplier variability, tissue contact differences, implant duration diversity, potential residues and contamination, and varying biological responses due to a patient's individual genetics, age, size, sex, and disease state. For *in vitro* testing, a simplifying premise is that compatibility is most readily verified by focusing on the impact of chemical constituents that may potentially be released from the device/material *in vivo*. For example, while most polymers used in medical devices are made from clinically proven, relatively inert, high molecular weight materials, it is their leachable components (i.e., residual monomers, oligomers, catalysts, processing aids,

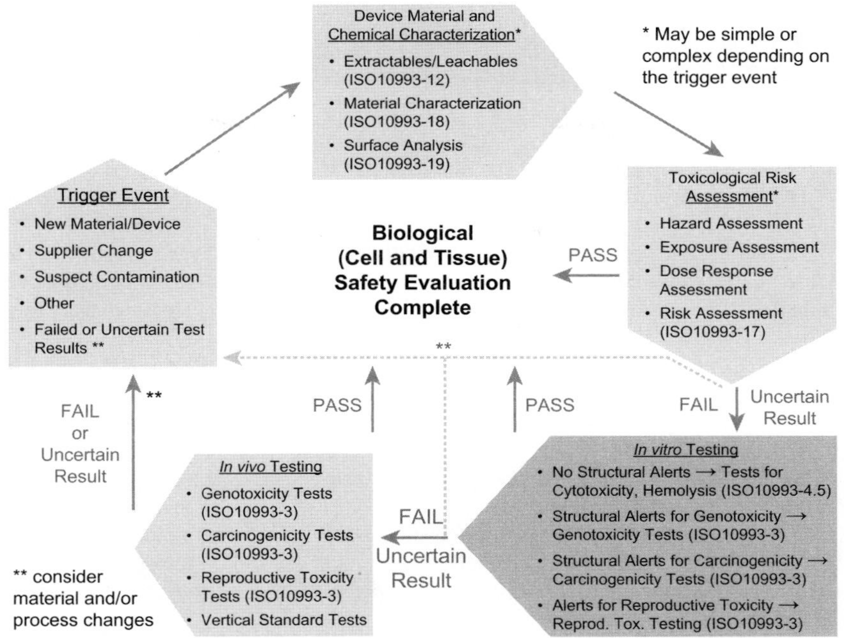

FIGURE II.3.3.1 Schematic depicting the process that occurs to establish the preclinical biological safety of a medical device material. Material characterization and toxicological risk assessment precede *in vitro* testing for cell and tissue compatibility. *In vivo* testing may follow, but only if warranted based upon *in vitro* findings. Note: *structural alerts* refers to chemical structures of varying concern that may be linked to known toxicological mechanisms.

additives, and contaminants) that often present a concern. The presence and extent of such leachables depends on a host of factors, such as the source of the material, differing manufacturing processes, and the intended function of the additive(s). In addition, new chemical species may be formed during secondary manufacturing processes, such as heat treatments, irradiation, welding, and sterilization.

In devices composed of inorganic materials such as metals, alloys, and ceramics, the release of metal ions may also be a cause for concern. Once free, metal ions have the potential to form complexes with proteins and/or DNA; these complexes may, in turn, disrupt a number of biological processes (ISO 10993-3, 2009).

The potential toxicity of a material is determined by a *toxicological risk assessment*. Such assessments involve: (1) identifying hazardous chemicals present in medical devices; (2) assessing the extent of a patient's potential exposure to these chemicals; (3) determining the chemical's dose–response relationship; and (4) combining this information to characterize the risk. Risk assessment calculations are based upon conservative worst-case exposure scenarios. Such scenarios often assume that all of a material's chemical constituents will be released into the patient during the first 24 hours. (In clinical practice this is unlikely to occur, but it is nevertheless a common conservative assumption designed to protect patients.) In these calculations, tolerable intakes (TI) for chemicals of concern are determined using safe exposure level values (NOAEL or LOAEL), and uncertainty factors (UF) to account for intraspecies variability among humans (UF_1), interspecies extrapolation from animals to humans (UF_2), and quality and relevance of the experimental data (UF_3). A margin of safety is then calculated (TI divided by the worst-case exposure dose) to verify that exposure to the chemical of concern does not present a health risk (ISO 10993-17, 2008). Supporting toxicity information is frequently found on free government websites or in proprietary commercial databases.[2] ISO 10993-1, 2009 recommends that toxicological risk assessments be conducted as a precursor to biological evaluation testing (see Figure II.3.3.1). This is because well-thought-out and thorough risk assessments may negate the need for expensive and time-consuming *in vitro* and *in vivo* testing.

The potential toxicity of a material is determined by a *toxicological risk assessment*. Such assessments involve: (1) identifying hazardous chemicals present in medical devices; (2) assessing the extent of a patient's potential exposure to these chemicals; (3) determining the chemical's dose–response relationship; and (4) combining this information to characterize the risk. A well-thought out and thorough risk assessment may negate the need for expensive and time-consuming *in vitro* and *in vivo* testing.

Detailed guidance for preparing *in vitro* testing samples is provided in ISO 10993 *Biological evaluation of*

[2]Free online sites include ToxNet, EPA-IRIS, ATSDR, IPCS-InChem & SIDS, IARC, ToxFile (Medline). Commercial databases include BIOSIS®Toxicology, ChemTox, IPA Toxicology, RTECS, Tox Center, and the Dictionary of Substances and Effects.

TABLE II.3.3.2 In vitro tests under ISO 10993 Part 3: Tests for Genotoxicity, Carcinogenicity, and Reproductive Toxicity

In vitro Test Category	Test*	Assay Basis	Information Gained
Genotoxicity	Bacterial Reverse Mutation Assay	*Reverse mutation assay;* Small changes in prokaryotic DNA induced by a test material or extract thereof, result in a quantifiable change in phenotypic expression of a target gene	Determines if a device material** has potential to react with DNA and cause altered DNA expression. Caveat: has potential for false positives
Genotoxicity	In vitro Mouse Lymphoma tk Assay	*Forward mutation assay;* Same as above but assay utilizes a eukaryotic cell line	Same as above. Note: this test may be more predictive than tests using prokaryotic cell lines.
Genotoxicity	Chromosomal Aberration	*Chromosome morphology analysis;* quantification and categorization of mammalian cell chromosome aberrations	Determines if a device material** shows potential to interfere with normal DNA replication and/or division processes leading to changes in normal chromosome morphology.
Carcinogenicity	Syrian Hamster Embryo (SHE) Cell Transformation Assays*	*Morphological transformation* of normal-karotype primary cells; assay designed to mimic 2-stage carcinogenesis model (initiation and promotion)	Determines if a device material** has potential to transform normal cells to cells with morphologies and behavior consistent with tumorigenic cells.
Carcinogenicity	Balb/c 3T3 Focus Transformation Assay*	*Macroscopic and microscopic morphological categorization* of foci; assay uses immortalized aneuploid fibroblastic cell line; assay can mimic 2-stage carcinogenesis model	Determines if a device material** has potential to transform cells into cells with morphologies consistent with tumorigenic cells; Caveat: time consuming and mixed reproducibility.
Reproductive Toxicity	Embryonic Stem Cell (ESC)Test*	*Biostatistically-derived, prediction-model equations* utilizing values for: 50% inhibition of ESC differentiation, 50% inhibition of cell growth of ESC cells, and 50% inhibition of cell growth of BALB/c 3T3 cells	Determines if a device material** has potential to be embryotoxic (teratogenic); presumes (with its limitations) that demonstration of toxicity at this stage may be indicative of concern for impact at other stages of reproduction.

*See Website for assay details
**Device material extract/leachable chemical(s), or a metabolic derivative(s) thereof.

medical devices, Parts 1, 3, and 12 (ISO 10993-1, 2009; ISO 10993-12, 2007; ISO 10993-3, 2009 Annex D). Sample preparation for genotoxicity tests requires careful attention due to each test's volume and extractable percentage requirements. Extraction procedures should: (1) simulate or exaggerate the clinical conditions of use; (2) take into account the chemical characteristics of the test sample; (3) not cause significant changes in the physicochemical properties of the sample; (4) test the device or material in its intended clinical use state (e.g., identically sterilized); (5) consider use of both a polar and a non-polar extraction solvent (see Table II.3.3.2); (6) be performed in chemically inert sterile closed containers; (7) be conducted for appropriate duration; and (8) be carried out at a suitable temperature, particularly when using cell culture media as an extraction vehicle.

IN VITRO ASSAYS TO ASSESS CELL AND TISSUE COMPATIBILITY IN BIOMATERIAL/ MEDICAL DEVICE EVALUATION FOR REGULATORY PURPOSES

The primary purpose of a preclinical biological safety evaluation of a material or device is to protect humans from potential biological risks arising from the use of the material or device (ISO 10993-1, 2009). Standardizing this process safeguards consumers, but also provides industry and governments with a common scientific basis for evaluating new medical technologies. The most widely recognized guidance documents for this purpose are the ISO 10993 series of standards[3] for the biological evaluation of medical devices developed by the International Organization for Standardization (ISO/TC 194, 2004; ISO, 2009 – see www.iso.org/iso/home.htm2009 for further information).

Three parts of the ISO 10993 standard describe certain considerations and particular methodologies for *in vitro* assessment of material/device compatibility with cells and tissues. These methods are discussed in further detail below. Table II.3.3.1 shows the various parts of the standard relevant to the specific *in vitro* assays prescribed as of 2009. As *in vitro* analytical tests focusing on the material characterization elements of the toxicological risk assessment (described earlier) are themselves

[3]Parts of the 10993 standard range from how to use the standard and how to prepare test samples (ISO 10993-1 and ISO 10993-12); understanding material stability (ISO 10993-9, -13, -14, and -15) (ISO/TC 194, 2004); establishment of allowable limits for leachable substances (ISO 10993-17, and -7) (ISO/TC 194, 2004); and, chemical characterization to support device material characterization and toxicological risk assessments (ISO 10993-18 and -19) (ISO/TC 194, 2004).

indirect tests for cell and tissue compatibility, these parts of the standard are included for reference. The *in vivo* approaches included in this standard, and the many device-specific *vertical standards*, are extremely important to proper cell and tissue compatibility testing. For further information on these topics the reader is referred to Chapters II.5.3, II.5.4, and II.5.5 of this book.

In Vitro Tests for Genotoxicity, Carcinogenicity, and Reproductive Toxicity: ISO 10993-3

Much of our understanding of cellular processes stems from the study of genetics and our ability to manipulate and characterize DNA and RNA (and the proteins they regulate). The ability to portray the impact of these manipulations has been crucial as well. Such information has increased our understanding of how genetic anomalies lead to human disease and defects, and has provided insight into the mechanisms of carcinogenesis. This awareness has also provided material providers, device manufacturers, and regulatory bodies with increasingly sophisticated tools and standards for screening materials and chemicals for potential health concerns.

In the context of biomaterial science, genotoxicity refers to the potential of the solid device material, and/or its constituent chemicals, biodegradation products, metabolites, leachables or contaminants to induce an undesirable effect on DNA itself, or on the genetic processes of a host cell or tissue (ISO 10993-3, 2009). To thoroughly screen the genotoxic *potential* of a new device or material, the standard relies on well-established *in vitro* tests from the Organization for Economic Cooperation and Development (OECD) guidelines for testing chemicals (OECD, 2009). These guidelines describe primary tests that look at gene mutations in prokaryote bacteria (OECD 471, 2007), cytogenetic evaluation of chromosomal damage in eukaryotic mammalian cells (i.e., chromosome aberration test; OECD 473, 1997; see Figure II.3.3.2), and gene mutation in mouse lymphoma cells (OECD 476, 1997). Taken together, when each is negative, these tests indicate a low potential for genotoxicity. Where one or more of these tests gives a positive result, the material or chemical is presumed to have a genotoxic potential. In the latter case, one or more *in vivo* assays are often considered to further evaluate the genotoxicity of the compound, material or device (OECD 474, 1997; OECD 475, 1997; ISO 10993-3, 2009). See website for details on the bacterial reverse mutation and chromosome aberration assays.

> In the context of biomaterials science, genotoxicity refers to the potential of the solid device material, and/or its constituent chemicals, biodegradation products, metabolites, leachables or contaminants to induce an undesirable effect on DNA itself, or on the genetic processes of a host cell or tissue (ISO 10993-3, 2009).

Along with tests for genotoxicity, carcinogenicity testing is also addressed in ISO 10993-3. Although materials indicating a genotoxic potential may be presumed to have a carcinogenic potential, a toxicological risk assessment focused on carcinogenic effects should occur before proceeding to carcinogenicity testing. This is based on the predictive potential of the *in vitro* assays, the need to abide by stipulations in ISO 10993-1, ISO 10993-2, and ISO 10993-6 (ISO/TC 194, 2004) to eliminate unnecessary use of animals, and on the known mechanisms of carcinogenesis. These mechanisms have clearly shown that not all genotoxins can also be classified as carcinogens. That is, some chemical carcinogens act through pathways that disrupt cellular metabolic processes (rather than DNA integrity) to induce uncontrolled cell division leading to tumor formation. Therefore, careful consideration must be given to a material's history of safe clinical use and chemical characterization *before* proceeding to carcinogenicity studies.

Given the complex processes involved in carcinogenesis, standard carcinogenicity testing (when called for) requires time-consuming and costly *in vivo* models. As a result, several non-OECD *in vitro* alternatives designed to mimic stages of the complex, multistep carcinogenic process have been developed for prescreening and are mentioned in the standard (Official Journal of the European Communities, 1988; LeBoeuf et al., 1999). The particular *in vitro* tests that have received the greatest attention are referred to as "cell transformation assays" (e.g., the Syrian Hamster Embryo (SHE) Cell Transformation Assay and the BALB/c 3T3 Focus Transformation Assay – see website for specific details). In these assays the transformed cells (resulting from exposure to test xenobiotics/test extracts) acquiring the characteristics of malignant cells have also demonstrated the ability to induce tumors in susceptible animals. Other cellular features indicative of transformation to a tumorigenic phenotype include alteration of individual cell morphology, development of anchorage-independent replication, and formation of disordered patterns of colony growth.

As in testing for carcinogenicity, safe history of clinical use and material/device chemical characterization requires careful consideration before proceeding to the final topic covered in this section, reproductive toxicity. Here, however, attention is directed at the potential exposure of the material/device, or any of its constituents, to the host reproductive tissues, an embryo/fetus or a potential nursing child. With sufficient and supportive toxicity data from absorption, distribution, metabolism, and excretion studies, there is often no need for reproductive toxicity testing (ISO 10993-3, 2009). This type of testing is furthermore not required where a toxicological risk assessment has established steps to appropriately mitigate situations of reproductive and developmental toxicity concern.

Current testing for reproductive and developmental toxicology relies heavily upon established *in vivo* models, given the complexity of the phenomena being investigated

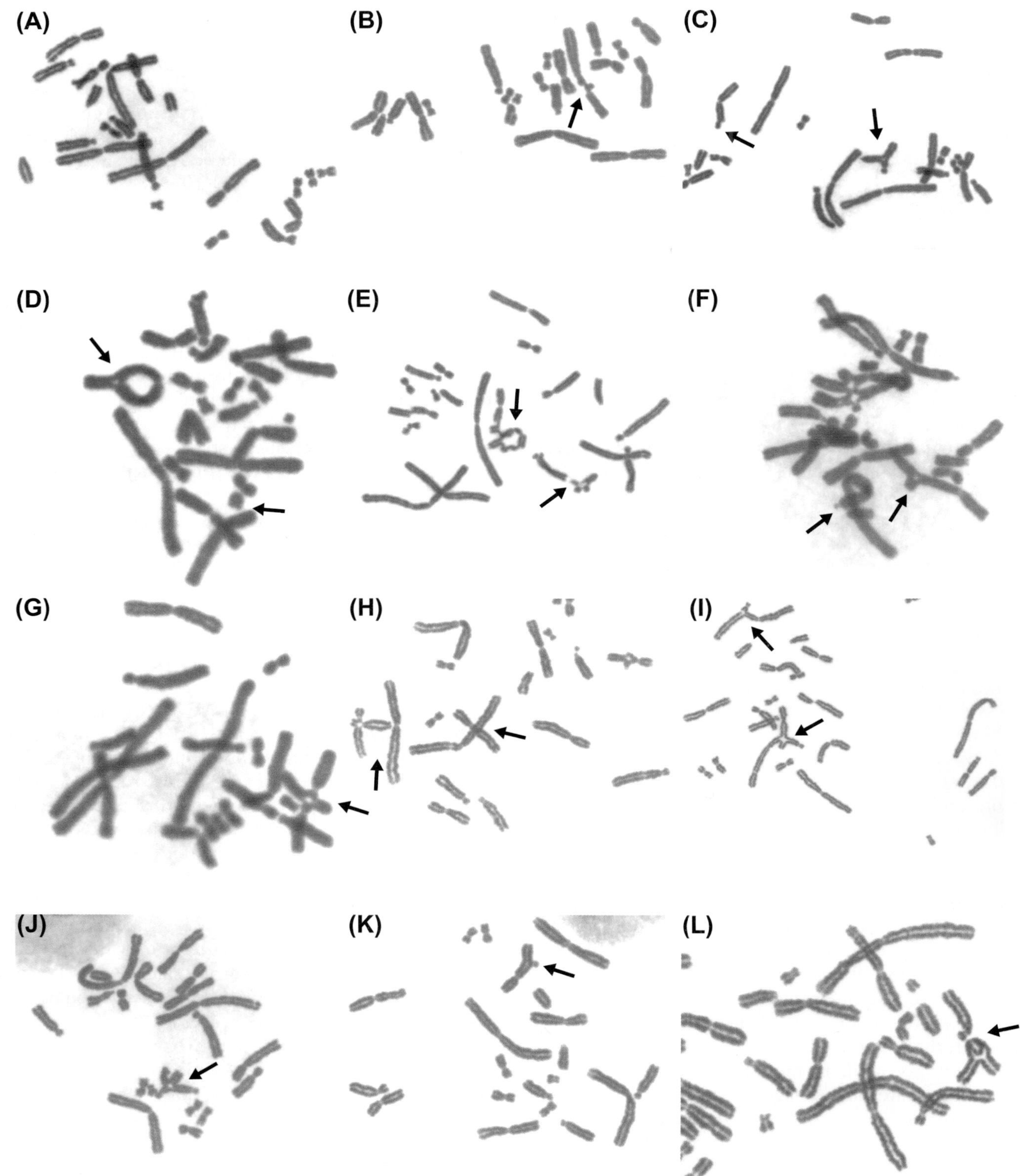

FIGURE II.3.3.2 Some common chromosome aberrations seen after exposure of CHO cells to extracts from test materials and/or positive controls. (A) Normal spread; (B) chromosome break; (C) chromatid break and triradial; (D) double minute and ring; (E) complex rearrangements and many breaks; (F) complex rearrangements and triradial; (G) quadraradial; (H) quadraradials; (I) triradial; (J) quadraradial; (K) interstitial deletion; and (L) chromosomal interchange. Note: in this type of testing chromatid gaps, isochromatid gaps, and uncoiled chromosomes are identified but are not computed in the analysis. Arrows point to each type of aberration. See website for further details.

(OECD 415, 1983; OECD 421, 1995; OECD 414, 2001; OECD 416, 2001). However, a variety of alternatives to whole animal testing are becoming increasingly available. The purely *in vitro* Embryonic Stem Cell Test (EST) has received particular attention following its correlation to primary embryonic cells and whole embryos. Here, a validation study funded by the European Centre for the Validation of Alternative Methods (ECVAM) showed strong correlation of three *in vitro* embryotoxicity tests across a diverse group of chemicals presenting different embryotoxic potentials (Brown, 2002; Genschow et al., 2002, 2004; ECVAM, 2008, 2009). Details of the EST test can be found on the website.

In the area of medical devices and materials, relatively few applications require reproductive toxicity consideration. However, noteworthy applications are intrauterine devices (IUDs) and energy-depositing radioactive seeds used in oncology that may come into close or direct contact with reproductive tissues, an embryo or fetus.

As discussed earlier, with medical devices typically composed of solid materials, testing often necessitates the use of extracts, rather than actual solid materials or their raw components. This is done to mimic the extraction that may occur in the body. The resultant extracts are often complex chemical mixtures, which is a notable difference from the single chemical assessment upon which the tests were based. These complex mixtures then require a number of refined analytical chemistry methods to identify specific agents of concern. In addition, analyses are further complicated by the fact that metabolic activation often plays a central role in the toxic potential of a material or chemical. To account for this, testing may require inclusion of a parallel set of test extracts exposed to an *in vitro* metabolic activation system.[4] This is done to capture chemicals called *promutagens* (i.e., compounds that are not mutagenic by themselves, but may be activated into mutagens by metabolic processes). In the simple Bacterial Reverse Mutation Assay (OECD 471), for instance, the prokaryotic cell lines clearly lack the metabolic capacity of normal intact mammalian tissue. Likewise, in testing involving eukaryotic cells, cells may present a metabolic deficiency and/or lack normal metabolic enzymes. Tests for genotoxicity, carcinogenicity, and reproductive toxicity (and cyotoxicity–see below) must therefore include a metabolic activator in order to detect the presence of metabolically-activated promutagens and toxicants in the device extracts.

Table II.3.3.2 briefly summarizes the *in vitro* tests for genotoxicity, carcinogenicity, and reproductive toxicity as mentioned in ISO 10993-3 (see website for further details on such methods). Recall that the *in vitro* methods described in Table II.3.3.2 are considered applicable *only* when the need to evaluate a medical device for potential genotoxicity, carcinogenicity, or reproductive toxicity has been established. This is determined through a toxicological risk assessment process and material characterization evaluation (see Figure II.3.3.1, and ISO 10993-1, -17, and -18 (ISO/TC 194, 2004)). When the risk assessment warrants *in vitro* testing, the limitations are that no single test is capable of detecting all relevant toxic agents, and that not all test methods are equally well-developed or validated for the testing of medical device materials. To effectively evaluate this category of material/device biological safety, the most objective assessment is often achieved by performing a battery of tests.

In Vitro Tests for Interactions with Blood: ISO 10993-4

A significant proportion of medical devices used today present a blood-contacting surface. These devices span a variety of limited, prolonged, and permanent contact applications, and the nature of the blood contact varies widely. Important factors at the blood interface include the material(s), blood flow, surface area(s), state of blood anticoagulation and dilution, and the presence of blood trauma and other tissues. Given the importance of this interface and the range of blood-contacting device applications, this area of testing is treated as a separate section within the ISO 10993 standard (ISO 10993-4). It is also given similar treatment in this book (see Chapters II.2.6 and II.3.5). Table II.3.3.3 identifies some of the most commonly applied *in vitro* tests by the key categories of concern in material/device interaction with blood. These assays focus on assessing the device or device material for potential to cause thrombosis, interact with platelets, bring about red blood cell damage and lysis, and stimulate alternative pathway immune reactions (i.e., complement pathway activation).

Just as there are a host of unique blood-contacting parameters associated with each device application, *in vitro* testing for blood interaction requires unique models and carefully designed and controlled experiments. Models must be designed to mimic the conditions expected in clinical use (e.g., precisely controlled temperature, blood flow, test material surface area:blood ratio (cm^2/mL blood), anticoagulant conditions, blood source, and contact duration). Examples of some application-specific *in vitro* models can be found in Chapter II.5.4 of this textbook. See also Münch et al., 2000.

Despite availability of various *in vitro* assays and models for characterization of materials with blood, lack of validation and questions on relevance to *in vivo* responses have made such testing more common in special characterization and feasibility studies, rather than as regulatory requirements. An exception here is the

[4]An *in vitro* metabolic activation system is commonly derived from a mammalian tissue homogenate (usually liver). The most commonly employed metabolic activation system is referred to as S9. This is a standard rat liver metabolic-activating system that refers to the supernatant fraction of a homogenate obtained after a 9000g centrifugation (Malling and Frantz, 1974).

TABLE II.3.3.3 Commonly applied in vitro tests by key category of material/device compatibility with the cellular and protein elements in blood

In vitro Test Area	Test	Assay Basis*	Information Gained
Thrombosis	Thrombus assessment	Assessment of extent and sites of thrombus formation using gross imaging and scanning electron microscopy (SEM)	Thrombus presence may indicate material or flow-related thrombogenesis. Caveat: may or may not be clinically relevant
Coagulation	Assays for coagulation proteins (examples include TAT and FPA)	*Enzyme-linked immunosorbant assay* for key coagulation factors. TAT is proportional to thrombin and FPA is proportional to fibrin.	Estimate of coagulation cascade activation. High levels may indicate extensive device or material-related thrombin activity and fibrin formation. Caveat: also reflects surgical trauma and/or injury, and levels may be subject dependent
Platelets and Platelet Function	Platelet counting, platelet degranulation assays, platelet stability assays	*Electronic particle counting and Enzyme-linked immunosorbant assay* for granule-release proteins e.g., βTG or PF4 *^{111}In-labelled platelet studies*	Estimates platelets being consumed or activated in response to a material or device. Caveat: platelet aggregation, activation and release are highly impacted by thrombin, e.g., from surgical trauma and/or injury
Hematology	Hemolysis Differential cell blood count (CBC) analysis	*Colorimetric assay* for plasma free hemoglobin (pfHb); *Electronic particle counting* for CBC	pfHb indicates red blood cell damage in response to a material or device, either from physical forces or extracts/leachants from the device material(s); CBC provides basic monitor of cell numbers
Immunology	Complement pathway activation C3a, C5a, and SC5b9 proteins	*Enzyme-linked immunosorbant assay* for key complement protein fragments	May indicate device material has potential to activate the alternative complement pathway; Caveat: surface area and surface chemistry dependent; may or may not have clinical significance

TAT = thrombin-antithrombinIII complex; FPA = fibrinopeptide A; βTG = β-thromboglobulin; PF4 = platelet factor 4; C3a, C5a and SC5b9 are components of the complement system. Note: TAT, FPA, βTG, PF4, and C5a assays are more commonly used in characterization studies rather than as standard required testing.
*Samples from proper in vitro or in vivo simulated use.

in vitro use of radiolabeled platelets for platelet stability studies on storage containers for transfusions (Holme et al., 1993; BEST Collaborative, 2006). Most often, a device in its final design is tested *in vivo*, with a subsection of the study dedicated to characterizing blood material/device interactions.

For basic material characterization for interaction with blood, the most common *in vitro* screening tests continue to be the direct and indirect hemolysis assays (ISO 10993-4, 2002; ASTM, 2009). These particular assays may be viewed as special cytotoxicity tests where human or animal red blood cells are used rather than the mouse L929 fibroblastic cells applied in standard cytotoxicity testing. The other common *in vitro* hematological assay applied today is for complement pathway activation. Here, ELISA assays (enzyme-linked immunosorbent assays) for C3a and SC5b9 protein fragments are used to estimate a material's potential for activation of the alternative pathway complement system (see Chapter II.4.4 of this textbook). Controversy exists over the clinical meaningfulness of such testing on small surface area devices and the actual adverse events due to complement activation.

In Vitro Tests for Cytotoxicity: ISO 10993-5

A general sequence of events is assumed to take place upon exposure of a toxicant to a host. First, the element can remain locally or be transported systemically by a combination of active transport processes, for example via entry into the vasculature and diffusion through tissues. The toxicant can then either: (1) alter the local biological environment causing a general molecular, organelle, cellular or tissue/organ dysfunction; or (2) interact with specificity to a particular endogenous target molecule such as a protein, membrane component or DNA. Both scenarios can lead to perturbation(s) in normal cell function and repair processes, leading to cytotoxicity.

The exposure of a single environmental toxicant to a host is a slightly different situation compared to exposure to a minute, single or mixture of potential toxicants leaching from a medical device/material. Nevertheless, similar complex biological perturbations are believed to take place when toxicant exposure comes from a medical device material. Classic examples include chemicals that alter local pH; solvents and detergents that disturb cell membranes and alter transmembrane gradients; elements that physically deter normal inter- or extracellular transport phenomena; chemicals that facilitate aberrant phosphorylation of proteins; xenobiotics that dysregulate electrically excitable cells; and agents that disrupt mitochondrial function (Chapter II.3.3 of this textbook) (Gregus and Klaassen, 2003). Mediators of tissue inflammation and fibrosis, as well as agents causing necrosis and apoptosis, are other well-known examples. The phenomenon of fibrotic tissue encapsulation that is

TABLE II.3.3.4 In vitro tests under ISO 10993 Part 5: Tests for Cytotoxicity

Assessment	Test[1]	Assay Basis	Information Gained[2]
Qualitative	L929 Elution Test	*Light microscopy and cell morphology scoring using mouse L929 fibroblast cells;* cells examined at 100× to evaluate cellular characteristics and the percent of apparent cell lysis according to a defined scoring system.	Determines if a device material *extract* has potential to cause cell morphology changes and/or lysis.
	Direct Contact Tests	*Light microscopy and cell morphology scoring following layering sample(s) onto mouse L929 fibroblast cell monolayers;* cells examined to evaluate cellular characteristics and the percent of apparent cell lysis according to a defined scoring system.	Determines if a device material *itself* has potential to be cytotoxic. Caveat: care must be taken to avoid confounding impact of uncontrolled physical interactions
	Indirect Contact Tests *Agar Diffusion Test* *Filter Diffusion Test*	Same as Direct Contact test but with test material physically separated from cell monolayer by a nominal 2 mm agarose layer or 0.45 μm pore cellulose acetate filter paper, respectively. Same scoring system as Direct contact.	Determines if a device material *itself* has potential to release agents that may be cytotoxic.
Quantitative[3]	Neutral Red Uptake (NRU) Test	*Colorimetric assay of NR uptake in BALB/c 3T3 cells.* NR acts via a non-cytotoxic penetration into live cells and accumulation in intracellular lysosomes. Cells exposed to a cytotoxic xenobiotic(s) show loss of integrity and function and decreased NR uptake and binding.	Determines if a device material extract has potential to be cytotoxic. Numerical assessment makes NRU assay less subjective than L929 test.
	V79 Colony Formation Assay	*Colony enumeration and calculation of plating (colony formation) efficiency using Chinese Hamster Lung cell (V79) culture system.* Plating efficiency of a test material is measured against the negative control group.	Determines if a device material extract has potential to be cytotoxic.
	MTT and Related Tests	*Microculture tetrazolium assay (MTA);* colorimetric assay using water-soluble MTT compound. Upon exposure to L929 cells, MTT is reduced enzymatically or through direct reaction with NADH or NADPH. A bight blue formazan product forms with intensity proportional to number of viable cells present.	Determines if a device material extract has potential to be cytotoxic. Note: other MTA tests that form similar formazan products are also used.

[1]See Website for assay details.
[2]Positive materials may require additional testing and justification to establish safe use.
[3]Common scoring criteria amongst the quantitative assessment methods determines non-cytotoxicity for plating or viability values >70% of negative control. Potential cytotoxic values fall below the 70% threshold.

invariably associated with implanted medical devices is yet another feature that distinguishes, and potentially limits, toxicant exposure associated with device materials (Chapters II.3.3 and II.4.2 of this textbook).

> Chemicals that possess potentially harmful or adverse properties are referred to as *toxicants*. When released from a medical device or material, such elements can remain locally or be transported systemically by a combination of active transport processes, such as entry into the vasculature and diffusion through tissues. A toxicant can then either: (1) alter the local biological environment causing a general molecular, organelle, cellular or tissue/organ dysfunction; or (2) interact with specificity to a particular endogenous target molecule, such as a protein, membrane component or DNA. Both scenarios can lead to perturbation(s) in normal cell function and repair processes leading to cytotoxicity.

Table II.3.3.4 presents the most commonly used *in vitro* tests to assess medical devices or their component materials for cytotoxicity. It should be noted that these particular tests assess general toxicity pathways and mechanisms. Thus, the nature of the tests is not inclusive of all toxicity pathways. These particular tests have, however, demonstrated appropriate usefulness in human risk assessment, the main objective of the ISO 10993 testing. As with any test, each has its limitations and criticisms. Typically, exposure is only of short duration, only a few toxicity pathways are tested, and false positives can occur. In addition, some tests utilize a subjective visual morphological scoring procedure (Figure II.3.3.3), whereas others use rapid analysis with colorimetric methods and numerical assessment. Finally, one should be aware that the 2009 version of ISO 10993-5 recommends the quantitative cytotoxicity assessment over the qualitative methods, and that such methods

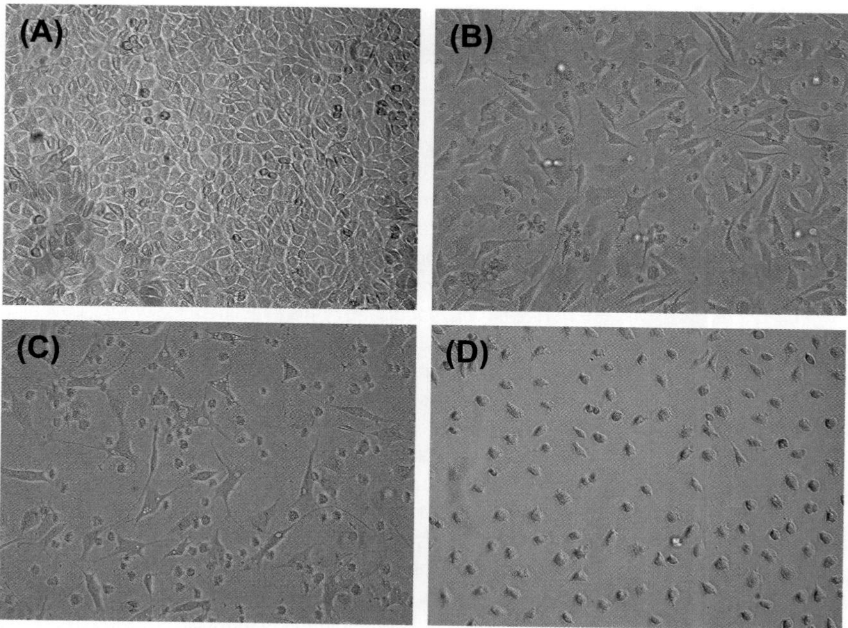

FIGURE II.3.3.3 Representative light microscopic images of L929 cells in the L929 Elution Test. (A) and (D) represent the negative and positive controls showing no cell cytoxicity and approximately 100% cytotoxicity, respectively. Images (B) and (C) represent approximately 50% and 80% cytotoxicity, respectively.

have a more stringent pass/fail criteria. See Table II.3.3.4 and website for further details of each test.

APPLICATION-SPECIFIC *IN VITRO* ASSAYS CONSIDERED IN PROOF-OF-CONCEPT TESTING

Classically, assessing the cell or tissue compatibility of a material or device has centered on applying *in vitro* and *in vivo* methodologies to establish a low risk for *adverse* impact on the host. However, biomaterials science is rapidly evolving in the area of proactive materials, that is those that actively present or impart *desirable* effects on certain cells and tissues. Concepts of this nature typically target one or more specific protein, cellular, or tissue reaction recognized to play a role in how the device or material interacts with the host in a specific application. The following three case examples are intended to illustrate this point, and to show how contemporary *in vitro* testing technology has played a role in demonstrating the capacity of special composite medical materials to improve certain cellular and tissue responses.

> Classically, assessing the cell or tissue compatibility of a material or device has centered on applying *in vitro* and *in vivo* methodologies to establish a low risk for *adverse* impact on the host. Biomaterials science, however, is rapidly evolving in the area of *proactive materials* (i.e., those that actively present or impart *desirable* effects on certain cells and tissues). Concepts of this nature typically target one or more specific protein, cellular or tissue reaction recognized to play a role in how the device or material interacts with the host in a specific application.

FUTURE CHALLENGES IN *IN VITRO* ASSESSMENT OF CELL AND TISSUE COMPATIBILITY

The task of regulatory agencies and device manufacturers is to select the appropriate combination of supporting information and testing that ensures product safety, device performance, and patient outcome. This task is becoming more complicated because of the expanding repertoire of device applications, specialty materials, and associated biological interfaces.

In light of this complexity, standard tests must be well-designed, scientifically sound, and broadly applicable and predictive. As a result, preclinical safety testing faces challenges which include: (1) questions about predictive power; (2) concerns with costs and completion times; (3) issues with the selection and design of appropriate animal models; (4) validation and acceptance of novel *in vitro* tests; and (5) lengthy development and approval timelines for new methods.

Thankfully, progress is being made in addressing these challenges. First, ISO 10993-1, which provides overall guidance for the biological evaluation of medical devices, was recently revised (2009). This version states that biological evaluation of medical devices or materials should follow a three-step process: (1) material characterization; (2) toxicological risk assessment; and (3) biological testing. Given the extensive information available today on medical device materials, including history of clinical use, ISO believes that focusing on characterization and risk assessment should reduce the amount of biological testing required.

In cases where new testing is indicated, ISO 10993-1 states that validated and reliable *in vitro* test methods: "shall

Case 1 Recombinant Human Bone Morphogenetic Protein (rhBMP-2) in Absorbable Collagen Sponge (ACS) Bone Grafts

When bone becomes damaged through traumatic or degenerative processes, one form of therapy involves replacement of the compromised area with bone-graft material. Autologous tissue is a common choice, with iliac crest bone the preferred material given its availability and degree of cellularity which favors healing. Unfortunately, harvesting this bone graft often results in post-operative pain and increased morbidity. To reduce these complications, smaller amounts of harvested bone have been supplemented with either synthetic calcium phosphate ceramics or demineralized allograft bone. Still, harvesting of autologous bone requires a second separate surgical procedure, and regardless of quantity of the harvest, surgeons and patients preferred an alternative method. That alternative arrived with the discovery of osteoinductive factors isolated from demineralized bone matrix (Urist and Strates, 1971), now known as bone morphogenetic proteins (BMPs). When the human BMP-2 protein was recombinantly manufactured (rhBMP-2) and delivered on an absorbable collagen sponge (ACS) scaffold, it was found to induce new bone formation. This was first demonstrated using simple *in vitro* tests to gauge the bioactivity of the material. In demonstrating chemotactic properties, *in vitro* studies showed that rhBMP-2 stimulated specific chemotactic migration of bone-forming cells (Lind et al., 1996; Fiedler et al., 2002). *In vitro* studies also showed that rhBMP-2 could increase the proliferation of several pluripotent cell lines, which are capable of differentiating into osteoblasts (Yamaguchi et al., 1991; Mayer et al., 1996; Puleo, 1997; Wilke et al., 2001; Akino et al., 2003). In addition, *in vitro* studies of rhBMP-2 demonstrated that rhBMP-2 binds to specific receptors on the surface of certain stem cells (e.g., mesenchymal stem cells or MSCs), and causes them to differentiate into bone-forming cells (Schmitt et al., 1999; Wilke et al., 2001). Together these studies showed the crucial role that BMP could have on new bone growth. Today, given the capacity of rhBMP-2 to induce alkaline phosphatase (AP) activity (Cheng et al., 2003; Luu et al., 2007), an *in vitro* AP activity test has been developed for quality control assessment of commercial product. This test uses controlled exposures of rhBMP-2 to the W20-17 bone marrow stromal cell line in combination with the synthetic AP substrate p-nitrophenol phosphate to assess inductive capacity. Clinically, rhBMP-2 combined with ACS is now used as an effective substitute to autologous bone in certain bone graft procedures (see Figure II.3.3.4).

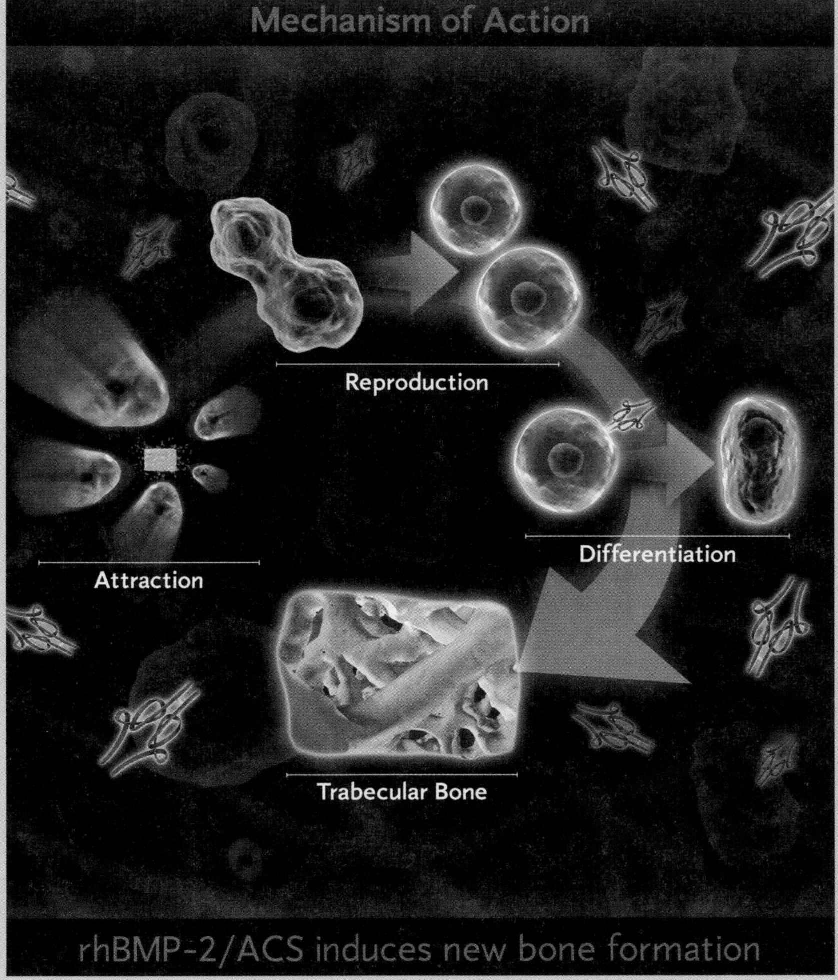

FIGURE II.3.3.4 Schematic showing the primary cell influences of BMP protein (Attraction, Reproduction, and Differentiation) that together lead to trabecular bone generation.

Case 2 Heparin Coatings in Blood-Contacting Medical Device Applications

Primarily from recognizing that processes associated with coagulation play a major factor in many of the adverse events observed with blood-contacting devices, material scientists have focused on ways materials can be made to better mimic the hemostatic properties of native endothelium. One early observation was that endothelial cell surfaces have high negative charges, and this characteristic may impart a portion of their hemostatic properties (i.e., by repelling the negatively-charged platelets (Coleman, 2000)). In addition, it was observed that *heparan* sulfate, a proteoglycan that is structurally and functionally similar to the anticoagulant heparin, is naturally present across the cell surface of vascular endothelium (Olsson et al., 2000). These molecules are chemically linked to endothelial cells in a manner that exposes the polysaccharide chains to protein molecules in the blood. One particular plasma protein, the serine protease inhibitor called antithrombin III (ATIII), upon binding to a specific chemical sequence on the heparan sulfate molecule, develops a high specificity for deactivating key proteins in the coagulation process, namely thrombin and factor Xa (Marcum et al., 1984; Coleman, 2000). This is precisely the mechanism of action of heparin itself, a polysaccharide that is not naturally found in the circulatory system, but rather is ubiquitously present in connective tissue. Here, heparin binds to ATIII at a unique binding site; the ATIII molecular configuration is altered, increasing the rate of enzyme-inhibitor complex formation by a factor of 1000 or more (Rosenberg et al., 2001); then, enzyme-inhibitor complex detaches from heparin leaving it available to interact with other AT molecules (Olsson et al., 2000). Heparin is not consumed in this process, but rather it acts as a catalyst (see Figure II.3.3.5).

Pioneering work by Larm et al. (1983) first described a method of attaching heparin to solid materials through reactive aldehyde groups

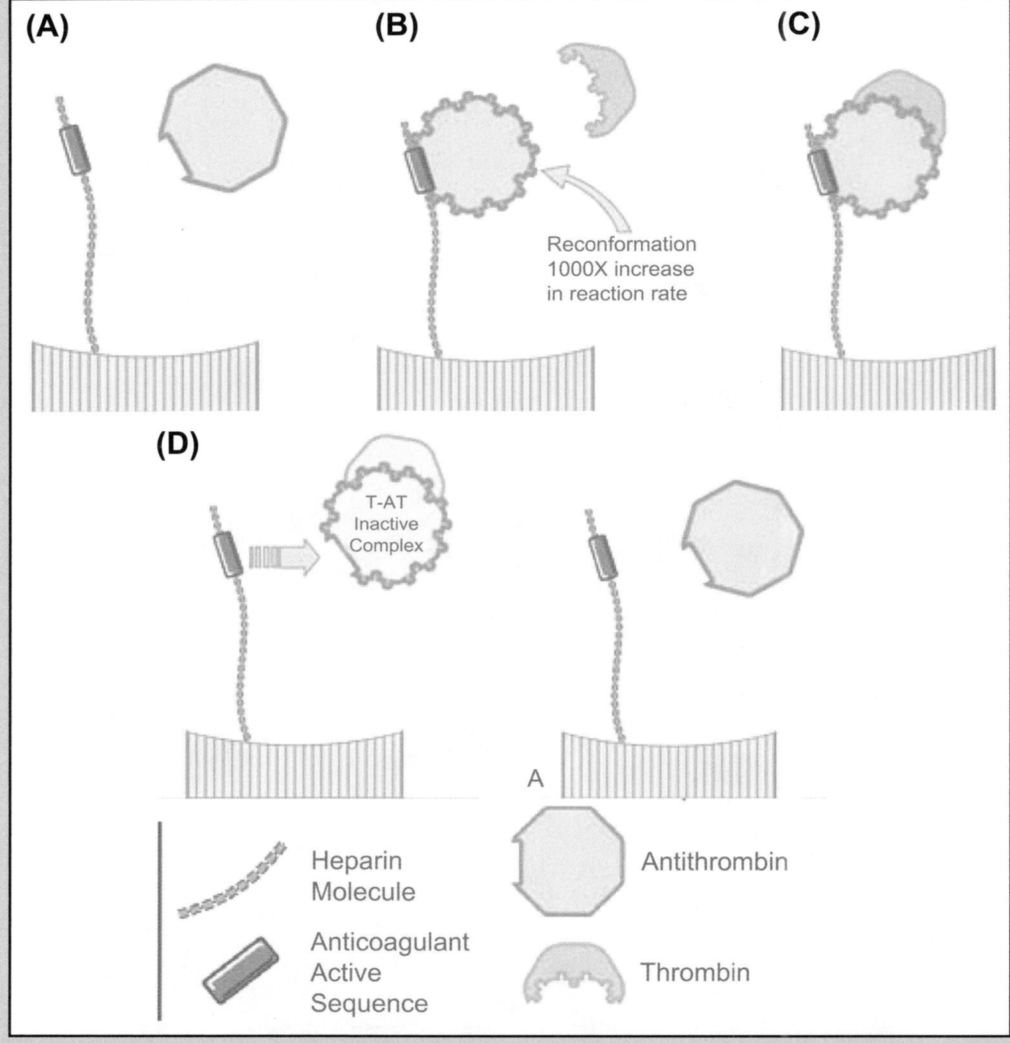

FIGURE II.3.3.5 A. Heparin attached to a solid material is oriented so that its active sequence can interact with the blood elements. B. When heparin binds to AT, its tertiary structure changes resulting in a heparin-AT complex that has a much greater affinity for coagulation factors than AT alone. The heparin-AT complex has a 1000-fold increased affinity for interaction with thrombin compared to AT alone. C. For example, an active coagulation factor, Factor IIs (thrombin) in the blood flowing past the material binds to the heparin-AT complex and becomes inactivated. D. The thrombin-AT complex detaches from the heparin molecule and is eventually metabolized by the body. E. Similar to naturally occurring heparin sulfate on the vascular endothelium, immobilized heparin is not consumed by this cycle and remains available to attach to other AT molecules.

Case 2 Heparin Coatings in Blood-Contacting Medical Device Applications—*cont'd*

present at the ends of heparin molecules. Using this approach, the immobilized heparin mimics the orientation of the heparin sulfate molecule on the vascular endothelium, with heparin's active sequence exposed and available to interact with blood elements. Importantly, the active ATIII binding region does not become involved in the surface attachment chemistry (see Figure II.3.3.6).

Numerous *in vitro* approaches have been applied to establish the hematologic compatibility of such heparin-modified surfaces. The starting approach, however, is a demonstration in an *in vitro* system that the immobilized heparin applied to the surface retains its intended biological activity. Here, heparinized materials or devices are first exposed to an excess of ATIII. This saturates the ATIII binding sites on the immobilized heparin. The uptake of this biologically active inhibitor is then quantified by subsequent exposure to a known quantity of active thrombin. Assessment of residual thrombin activity using a synthetic chromogenic substrate (e.g., S2238 or rhodamine 110, bis-[p-tosyl-L-glycyl-L-prolyl-L-arginine amide]) and analysis using a kinetic microplate reader completes the test. Immobilized heparin bioactivity is reported in units of international units of thrombin (T) deactivated per square centimeter of test material (IUT_{deac}/cm^2). Using this particular *in vitro* analysis, specific heparin activity can be measured on the various material components in cardiopulmonary bypass equipment that are coated with immobilized heparin.

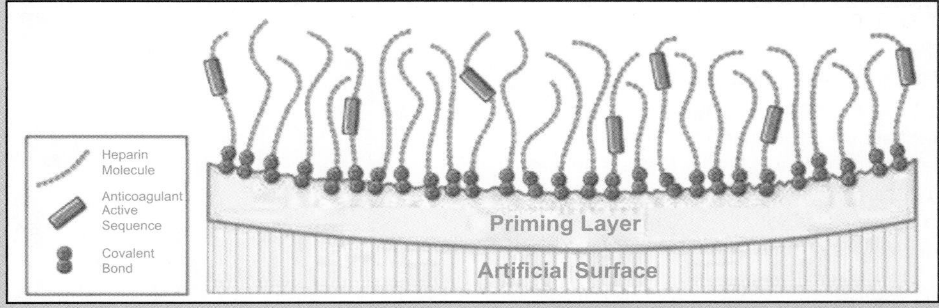

FIGURE II.3.3.6 Attachment of heparin to material surfaces uses a stable covalent bond to the end of the molecule and orients the heparin so that its anticoagulant active sequence is free to interaction with the blood.

be considered for use in preference to *in vivo* tests," and that: "whenever possible, *in vitro* screening shall be carried out before *in vivo* tests are commenced." This policy is clearly intended to minimize the use of *in vivo* testing.

A second area of progress involves the development of *in vitro* assays that are designed to replace *in vivo* methods. Over the past few years, news articles and published reports have pointed to a paradigm shift in toxicology testing involving *in vitro* cell-based assays (NRC, 2007; Schmidt, 2009; US EPA, 2009). In the United States, the TOX21 partnership between the Environmental Protection Agency, National Toxicology Program, National Institutes of Health, and the US Food and Drug Administration (FDA) is an example of this shift. TOX21 is focused on applying new scientific tools in computational, informational, and molecular science to identify and characterize chemicals on the basis of how they activate toxicity pathways inside cells (preferably of human origin). This move away from traditional whole animal toxicity models to cell-based assays seeks to identify new mechanisms of chemical activity in cells, and to develop better predictive models of human response to toxicants. Results of these models will help regulators make more informed decisions, based on actual human toxicity data instead of animal data, which is often suspect and controversial.

> The detailed information available today on medical device materials, often including extensive histories of clinical use, calls for evaluation of medical devices or materials to follow a three-step process: (1) material characterization; (2) toxicological risk assessment; and (3) biological testing. In the latter area, *in vivo* testing remains critical. However, technology is driving a move away from traditional whole animal toxicity models towards cell-based assays that identify actual mechanisms of chemical activity in cells. This, in turn, will lead to development of better predictive models of human response to toxicants. Results of these models will help regulators make more informed decisions, based on actual human toxicity data instead of animal data, which is often suspect and controversial.

The European Centre for the Validation of Alternative Methods is one of the leaders of the move towards *in vitro* methods. Over the past 10 years, ECVAM has validated or endorsed a number of alternative methods. These include the Embryonic Stem Cell Test (EST) for embryotoxicity, the Reduced Local Lymph Node Assay (rLLNA) for skin sensitization, along with the EpiSkin™ and EpiDerm™ *in vitro* tests for skin-irritation. To evaluate the skin-irritation assays, for example, ECVAM conducted a formal international validation study, with the goal of determining whether such human skin cell-based assays could accurately and reliably predict acute *in vivo* skin irritancy (Spielmann et al., 2007; Grindon et al., 2008a,b). Their multi-center blinded study and

Case 3 Steroid-Eluting Electrodes in Pacemaker Applications

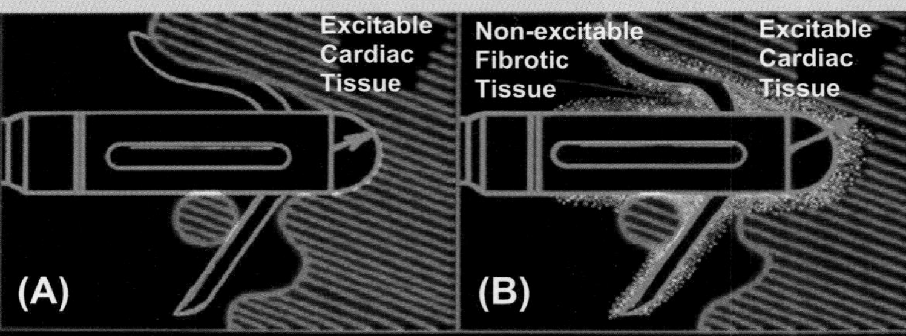

FIGURE II.3.3.7 Cross section of typical tissue response at the interface of a cardiac pacing lead electrode and endocardial tissue. A: immediately at implantation, and B: after eight or more weeks after implantation. Non-electrically-excitable fibrotic tissue is seen around the electrode.

One problem of early lead electrodes used in cardiac pacing was that the pacing threshold often increased over time. *Pacing threshold* is the minimum voltage that the electrode needs to deliver to cardiac tissue in order to consistently depolarize/excite the muscle into contraction. An increase in this threshold can cause pacing failure and/or escalate energy consumption on the battery – a factor that can ultimately reduce the longevity of the implanted device. Such an increase in threshold was found to be associated with the inflammatory reaction at the electrode tip insertion site (Timmis et al., 1983; Benditt et al., 1989). In particular, an inflammatory response-induced tissue capsule formed around the electrode, resulting in an increase in impedance at the electrode–tissue interface (see Figure II.3.3.7). To mitigate this problem a process for delivering anti-inflammatory steroid drugs from the electrode tip was developed (see Figure II.3.3.8). Through gradual elution, the steroid mitigates the foreign-body response at the electrode tip, and results in minimal to zero threshold rise as a function of implant time (Benditt et al., 1989). In this case, a drug-imparted control of the cell and tissue response at the minute tip of the pacing device was seen to control a deleterious response that leads to shorter device life and less accurate sensing.

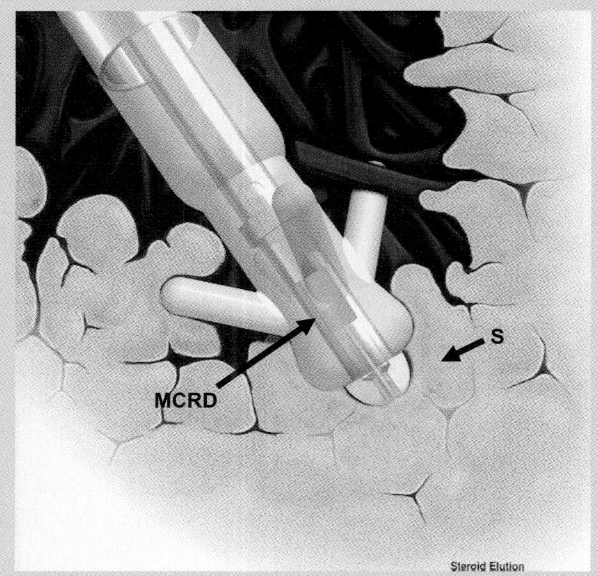

FIGURE II.3.3.8 Depiction of a tyned steroid-eluting pacing lead electrode positioned against cardiac tissue. The yellow zone represents steroid (S) eluting into the local tissue from a monolithic controlled release depot (MCRD) consisting of drug in a silicone matrix.

A critical development in this technology was demonstrating an appropriate and non-cytotoxic level of steroid delivered to the cardiac cells and tissues. To examine the delivery rate and prove this point, *in vitro* elution studies were conducted on test electrodes immersed in fresh pooled human plasma agitated at 37°C. Here the elution rate was seen to follow an exponential function that, when analyzed graphically, displayed a bimodal relationship. This is explained by considering the origin and the path the steroid must follow to reach the tissue. Initially, steroid within the electrode pores and on the surface of the silicone-steroid monolithic controlled-release device (MCRD "plug" located in the lead tip) readily dissolves and diffuses across a concentration gradient to the exterior. Then, upon acclimation to exposure to the blood plasma and moisture penetration into the MCRD, additional dissolved steroid from within the MCRD faces the extra barrier of passing through the silicone matrix. Importantly, throughout the elution time course, the amount of steroid released remains lower than the threshold of toxicological concern for the targeted patient population.

follow-up work found that both methods exhibited high sensitivity and specificity. These findings led ECVAM to endorse both methods as stand-alone replacements for the rabbit skin-irritation test. This new *in vitro* toxicity testing approach will likely replace the Draize rabbit skin-irritation test globally.

Other governmental organizations dedicated to the validation of alternative methods include the Interagency Coordinating Committee on the Validation of Alternative Methods (ICCVAM) in the United States, the Japanese Center for the Validation of Alternative Methods (JaCVAM), and the Korean Center for the Validation of Alternative Methods (KoCVAM). These groups are all part of a global consortium known as the International Cooperation on Alternative Test Methods (ICATM) which seeks to expand and strengthen cooperation, collaboration, and communications among national validation organizations on the scientific validation and evaluation of new alternative testing methods proposed for regulatory health and safety assessments (ICATM, 2009).

Efforts by these governmental organizations, along with work in academia and industry, will lead to the development and adoption of more sensitive and predictive human cell-based *in vitro* tests designed to replace *in vivo* methods (Basketter, 2008; Natsch and Emter, 2008).

Finally, as the ISO 10993 standards have been revised, they have recognized new bioanalytical techniques. One example is the inclusion of the MTT (methylthiazoltetrazolium assay in ISO 10993-3 as a supplement or replacement for traditional categorical cytotoxicity analysis. Also, while not fully incorporated into ISO 10993-4 yet, ELISA data for complement testing has become a routine part of many preclinical evaluations involving blood-contacting devices. Other *in vitro* tests measuring factors indicative of material-related coagulation and thrombosis may follow.

SUMMARY REMARKS

This chapter has focused on current *in vitro* methods used to evaluate the biological compatibility of medical devices and materials. The methods were presented from two perspectives: (1) *in vitro* assays that are accepted internationally for preclinical biological safety evaluations and risk assessments; and (2) specialized material- and application-specific *in vitro* assays designed for certain cell and tissue responses. As the latter area is complex and diverse, important methods were presented in contemporary case examples. These cases demonstrated that specifically-designed *in vitro* assays have proven to be critical in moving new medical device therapies into clinical practice.

Today, materials in new device concepts do not often go through an early evaluation using the ISO 10993 process. Instead, engineers and material scientists more often use other types of tests in proof-of-concept studies, in which the majority of prototypes use well-characterized biocompatible materials with long histories of safe use. This approach is driven by the desire to bring new therapies to the market quickly, with the influence of new technologies that allow subtle changes or additions to existing materials that improve device performance via promoting improved cell and tissue interaction properties.

As the range of medical device materials and applications expands, the challenges faced by product engineers, materials scientists, regulatory agencies, and standards developers will certainly grow. Nevertheless, progress will no doubt continue to be made in the development and application of *in vitro* assays to support new therapies, improved medical materials, and advances in patient care.

BIBLIOGRAPHY

Akino, K., Mineta, T., Fukui, M., Fujii, T., & Akita, S. (2003). Bone morphogenetic protein-2 regulates proliferation of human mesenchymal stem cells. *Wound Repair Regen.*, **11**, 354–360.

ASTM. (2009). Standard Practice for Assessment of Hemolytic Properties of Materials. Medical and Surgical Materials and Devices; Anesthetic and Respiratory Equipment; Manufacture of Pharmaceutical Products: American Society for Testing and Materials. F756–00.

Basketter, D. A. (2008). Nonanimal alternatives for skin sensitization: A step forward? *Toxicol. Sci.*, **102**, 1–2.

Benditt, D. G., Kriett, J. M., Ryberg, C., Gornick, C. C., Stokes, K. L., et al. (1989). Cellular electrophysiologic effects of dexamethasone sodium phosphate: Implications for cardiac stimulation with steroid-eluting electrodes. *Int. J. Cardiol.*, **22**, 67–73.

BEST Collaborative. (2006). Platelet radiolabelling procedure: The Biomedical Excellence for Safer Transfusion (BEST) Collaborative. *Transfusion*, **46**(Suppl.), 59–66.

Brown, N. A. (2002). Selection of test chemicals for the ECVAM international validation study on *in vitro* embryotoxicity tests. European Centre for the Validation of Alternative Methods. *Altern. Lab. Anim.*, **30**, 177–198.

Casarett, L. J., Klaassen, C. D., & Watkins, J. B. (2003). *Casarett and Doull's Essentials of Toxicology*. New York, NY: McGraw-Hill/Medical Pub. Div.

Cheng, H., Jiang, W., Phillips, F. M., Haydon, R. C., Peng, Y., et al. (2003). Osteogenic activity of the fourteen types of human bone morphogenetic proteins (BMPs). *J. Bone Joint Surg. Am.*, **85-A**, 1544–1552.

Coleman, R. (2000). Overview of hemostasis. In R. Coleman, A. Clowes, J. George, J. Hirsh, & V. Marder (Eds.), *Hemostasis and Thrombosis: Basic Principles and Clinical Practice* (4th ed.). Philadelphia, PA: Lippincott Williams & Wilkins.

ECVAM. (2009). European Commission Joint Research Centre. In *Institute for Health and Consumer Protection (IHCP) of the Joint Research Centre (JRC)*. Ispra, Italy: European Centre for the Validation of Alternative Methods.

Fiedler, J., Roderer, G., Gunther, K. P., & Brenner, R. E. (2002). BMP-2, BMP-4, and PDGF-bb stimulate chemotactic migration of primary human mesenchymal progenitor cells. *J. Cell Biochem.*, **87**, 305–312.

Genschow, E., Spielmann, H., Scholz, G., Pohl, I., Seiler, A., et al. (2004). Validation of the embryonic stem cell test in the international ECVAM validation study on three *in vitro* embryotoxicity tests. *Altern. Lab. Anim.*, **32**, 209–244.

Genschow, E., Speilmann, H., et al. (2002). The ECVAM international validation study on in vitro embryotoxicity tests: results of the definitive phase and evaluation of prediction models. European Centre for the Validation of Alternative Methods. *Alternatives to Laboratory Animals*, **30**(ATLA), 151–176.

Gregus, Z., & Klaassen, C. D. (2003). Mechanisms of toxicity. In C. D. Klaassen, & J. B. Watkins (Eds.), *Essentials of Toxicity*. New York, NY: McGraw-Hill.

Grindon, C., Combes, R., Cronin, M. T., Roberts, D. W., & Garrod, J. F. (2008a). Integrated decision-tree testing strategies for skin corrosion and irritation with respect to the requirements of the EU REACH legislation. *Altern. Lab. Anim.*, **36**(Suppl. 1), 65–74.

Grindon, C., Combes, R., Cronin, M. T., Roberts, D. W., & Garrod, J. F. (2008b). Integrated decision-tree testing strategy for skin sensitisation with respect to the requirements of the EU REACH legislation. *Altern. Lab. Anim.*, **36**(Suppl. 1), 75–89.

Holme, S., Heaton, A., & Roodt, J. (1993). Concurrent label method with ^{111}In and ^{51}Cr allows accurate evaluation of platelet viability of stored concentrates. *Br. J. Haematology*, **84**, 717–723.

ICATM. (2009). *Memorandum of Cooperation on International Cooperation on Alternative Test Methods (ICATM)*. http://iccvam.niehs.nih.gov/docs/about_docs/ICATM-MOC.pdf.

ISO 10993-1. (2009). Biological evaluation of medical devices – Part 1: Evaluation and testing. *Association for the Advancement of Medical Instrumentation.*

ISO 10993-3. (2009). Biological evaluation of medical devices – Part 3: Tests for genotoxicity, carcinogenicity and reproductive toxicity. *Association for the Advancement of Medical Instrumentation.*

ISO 10993-4. (2002). Biological evaluation of medical devices – Part 4: Selection of tests for interactions with blood. *Association for the Advancement of Medical Instrumentation.*

ISO 10993-5. (2009). Biological evaluation of medical devices – Part 5: Tests for *in vitro* cytotoxicity. *Association for the Advancement of Medical Instrumentation.*

ISO 10993-12. (2007). Biological evaluation of medical devices – Part 12: Sample preparation and reference materials. *Association for the Advancement of Medical Instrumentation.*

ISO 10993-17. (2008). Biological evaluation of medical devices – Part 17: Establishment of allowable limits for leachable substances. *Association for the Advancement of Medical Instrumentation.*

ISO. (2009). *International Standards for Business, Government and Society*. International Organization for Standardization. http://www.iso.org/iso/home.htm2009.

ISO/TC 194. (2004). *TC 194: Biological Evaluation of Medical Devices – Business Plan*. Geneva, Switzerland: International Organization for Standardization.

Larm, O., Larsson, R., & Olsson, P. (1983). A new non-thrombogenic surface prepared by selective covalent binding of heparin via a modified reducing terminal residue. *Biomater. Med. Devices Artif. Organs*, **11**, 161–173.

LeBoeuf, R. A., Kerckaert, K. A., Aardema, M. J., & Isfart, R. J. (1999). Use of Syrian hamster embryo and BALB/c3T3 cell transformation for assessing the carcinogenic potential of chemicals. *IARC Sci Publ*, **146**, 409–425.

Lind, M., Eriksen, E. F., & Bunger, C. (1996). Bone morphogenetic protein-2 but not bone morphogenetic protein-4 and -6 stimulates chemotactic migration of human osteoblasts, human marrow osteoblasts, and U2-OS cells. *Bone*, **18**, 53–57.

Luu, H. H., Song, W. X., Luo, X., Manning, D., Luo, J., et al. (2007). Distinct roles of bone morphogenetic proteins in osteogenic differentiation of mesenchymal stem cells. *J. Orthop. Res.*, **25**, 665–677.

Malling, H. V., & Frantz, C. N. (1974). Metabolic activation of dimethylnitrosamine and diethyl-nitrosamine to mutagens. *Mutat. Res.*, **25**, 179–186.

Marcum, J. A., McKenney, J. B., & Rosenberg, R. D. (1984). Acceleration of thrombin–antithrombin complex formation in rat hindquarters via heparin like molecules bound to the endothelium. *J. Clin. Invest.*, **74**, 341–350.

Mayer, H., Scutt, A. M., & Ankenbauer, T. (1996). Subtle differences in the mitogenic effects of recombinant human bone morphogenetic proteins -2 to -7 on DNA synthesis on primary bone-forming cells and identification of BMP-2/4 receptor. *Calcif. Tissue Int.*, **58**, 249–255.

Münch, K., Wolf, M. F., Fogt, E. J., Schroeder, P., Bergan, M., et al. (2000). Use of simple and complex *in vitro* models for multiparameter characterization of human blood-material/device interactions. *J. Biomaterials Sci. Polymer. Edn.*, **11**, 1147–1163.

Natsch, A., & Emter, R. (2008). Skin sensitizers induce antioxidant response element dependent genes: Application to the *in vitro* testing of the sensitization potential of chemicals. *Toxicol. Sci.*, **102**, 110–119.

NRC. (2007). Toxicity Testing in the 21st Century: A Vision and a Strategy. Committee on Toxicity Testing and Assessment of Environmental Agents, National Research Council & U.S. Environmental Protection Agency The National Academies Press.

OECD 414. (2001). *Prenatal Development Toxicity Study. OECD Guidelines for the Testing of Chemicals.* OECD Publishing.

OECD 415. (1983). *One-Generation Reproduction Toxicity Study. OECD Guidelines for the Testing of Chemicals.* OECD Publishing.

OECD 416. (2001). *Two-Generation Reproduction Toxicity Study. OECD Guidelines for the Testing of Chemicals.* OECD Publishing.

OECD 421. (1995). *Reproduction/Developmental Toxicity Screening Test.* OECD Guidelines for the Testing of Chemicals. OECD Publishing.

OECD 471. (2007). *Bacterial Reverse Mutation. OECD Guidelines for the Testing of Chemicals.* OECD Publishing.

OECD 473. (1997). *In Vitro Mammalian Chromosome Aberration Test. OECD Guidelines for the Testing of Chemicals.* OECD Publishing.

OECD 474. (1997). *Mammalian Erythrocyte Micronucleus Test. OECD Guidelines for the Testing of Chemicals.* OECD Publishing.

OECD 475. (1997). *Mammalian Bone Marrow Chromosome Aberration Test. OECD Guidelines for the Testing of Chemicals.* OECD Publishing.

OECD 476. (1997). *In Vitro Mammalian Cell Gene Mutation Test. OECD Guidelines for the Testing of Chemicals. OECD Publishing.*

OECD. (2009). *For a Stronger, Cleaner, Fairer World Economy.* Organization for Economic Cooperation and Development.

Official Journal of the European Communities. (1988). *In vitro cell transformation test.* Official Journal of the European Communities, L 133/73.

Olsson, P., Sanchez, J., Mollnes, T. E., & Riesenfeld, J. (2000). On the blood compatibility of end-point immobilized heparin. *J. Biomater. Sci. Polym. Ed.*, **11**, 1261–1273.

Preston, R. J., & Hoffmann, G. R. (2003). Genetic Toxicology. In C. D. Klassen, & J. B. Watkins (Eds.), *Casarett and Doull's Essentials of Toxicology*. New York, NY: McGraw-Hill.

Puleo, D. A. (1997). Dependence of mesenchymal cell responses on duration of exposure to bone morphogenetic protein-2 *in vitro*. *J. Cell. Physiol.*, **173**, 93–101.

Rosenberg, R., Edelberg, J., & Zhang, L. (2001). The heparin/antithrombin system: A natural anticoagulant mechanism. In R. Coleman, A. Clowes, J. George, J. Hirsh, & V. Marder (Eds.), *Hemostasis and Thrombosis: Basic Principles and Clinical Practice* (4th ed). Philadelphia, PA: Lippincott Williams & Wilkins.

Schmidt, C. W. (2009). TOX 21: New dimensions of toxicity testing. *Environ. Health Perspect.*, **117**, A348–353.

Schmitt, J. M., Hwang, K., Winn, S. R., & Hollinger, J. O. (1999). Bone morphogenetic proteins: An update on basic biology and clinical relevance. *J. Orthop. Res.*, **17**, 269–278.

Spielmann, H., Hoffmann, S., Liebsch, M., Botham, P., Fentem, J. H., et al. (2007). The ECVAM international validation study on *in vitro* tests for acute skin irritation: Report on the validity of the EPISKIN and EpiDerm assays and on the Skin Integrity Function Test. *Altern. Lab. Anim.*, **35**, 559–601.

Timmis, G., Gordon, S., Westveer, D., Stewart, J., Stokes, K., et al. (1983). A new steroid-eluting low-threshold pacemaker lead. In K. Steinbach (Ed.), *7th World Symposium on Cardiac Pacing*. Darmstadt: Steinkopff Verlag.

Urist, M. R., & Strates, B. S. (1971). Bone morphogenetic protein. *J. Dent. Res.*, **50**, 1392–1406.

US EPA (2009). *EPA 100/K-09/001: The U.S. Environmental Protection Agency's Strategic Plan for Evaluating the Toxicity of Chemicals.* Office of the Science Advisor Science Policy Council.

Wilke, A., Traub, F., Kienapfel, H., & Griss, P. (2001). Cell differentiation under the influence of rh-BMP-2. *Biochem. Biophys. Res. Commun.*, **284**, 1093–1097.

Yamaguchi, A., Katagiri, T., Ikeda, T., Wozney, J. M., Rosen, V., et al. (1991). Recombinant human bone morphogenetic protein-2 stimulates osteoblastic maturation and inhibits myogenic differentiation *in vitro*. *J. Cell Biol.*, **113**, 681–687.

CHAPTER II.3.4 *IN VIVO* ASSESSMENT OF TISSUE COMPATIBILITY

James M. Anderson[1] and Frederick J. Schoen[2]
[1]Department of Pathology, University of Cleveland, OH, USA
[2]Professor of Pathology and Health Sciences and Technology (HST), Harvard Medical School, Executive Vice Chairman, Department of Pathology, Brigham and Women's Hospital, Boston, MA, USA

INTRODUCTION

The goal of *in vivo* assessment of tissue compatibility of a biomaterial, prosthesis or medical device is to determine the biocompatibility (as a key component of safety) of the biomaterial, prosthesis or medical device in a biological environment. Indeed, biocompatibility is a critical element in the performance of an implanted medical device, and many complications of clinical devices derive from inadequate biocompatibility. Moreover, the concept of biocompatibility has evolved in recent years as biomaterials technology has become more sophisticated, particularly in permitting the biomaterial to effect complex, and potentially beneficial, interactions with the surrounding tissues (Hench and Pollak, 2002; Williams, 2008; and see Chapter II.3.2). Thus, the contemporary definition of biocompatibility focuses on the ability of a medical device to perform with an appropriate host response in a specific application, and biocompatibility assessment serves as a measurement of the magnitude and duration of the pathophysiologic mechanisms that determine the host response. From a practical perspective, the *in vivo* assessment of tissue compatibility of medical devices is carried out to determine that the device performs as intended, and presents no significant harm to the patient or user simulating clinical use. In this chapter, the term "medical device" will be used to describe biomaterials, prostheses, artificial organs, and other medical devices, and the terms "tissue compatibility assessment," "biocompatibility assessment," and "safety assessment" will be considered to be synonymous.

> The *in vitro* assessment of tissue compatibility is necessary to establish the biocompatibility safety and function (efficacy) of a medical device and its components under conditions of intended use. Intended use conditions are also utilized to select *in vitro* tests and animal models that will provide essential information to determine biocompatibility.

Recently, extensive efforts have been made by government regulatory agencies, e.g. the US Food and Drug Administration (FDA), and standards organizations, e.g. ASTM International, the International Organization for Standardization (ISO), and US Pharmacopeia (USP), to provide procedures, protocols, guidelines, and standards that may be used in the *in vivo* assessment of the tissue compatibility of medical devices. This chapter draws heavily on the ISO 10993 standard, Biological Evaluation of Medical Devices, in presenting a systematic approach to the *in vivo* assessment of tissue compatibility of medical devices.

The nature and extent of risk due to inadequate biocompatibility may be conceptualized as shown in Figure II.3.4.1. The first consideration in the selection of biomaterials to be used in device design should be the chemical, toxicological, physical, electrical, morphological, and mechanical properties of the biomaterial(s) to fulfill the intended use. Relevant to the overall *in vivo* assessment of tissue compatibility of a biomaterial or

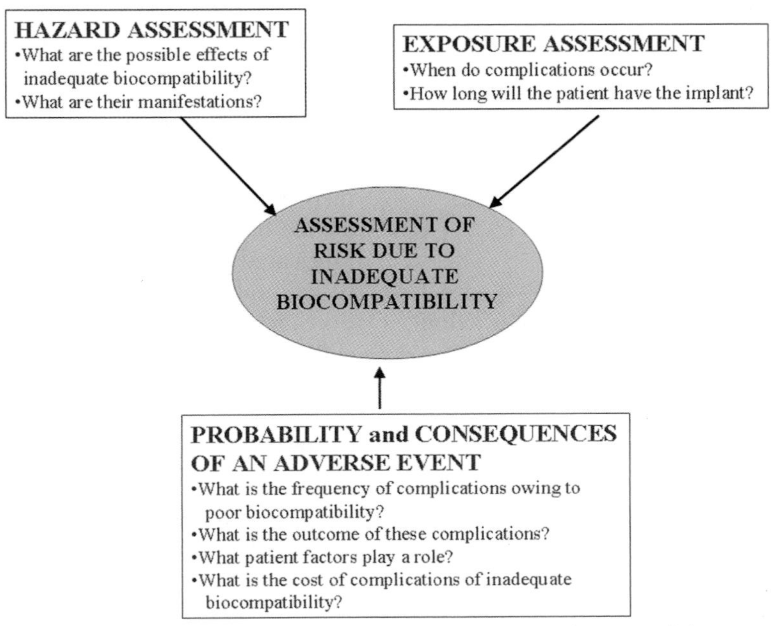

FIGURE II.3.4.1 The risks associated with inadequate biocompatibility are a composite function of the exposure (i.e., nature and duration), the hazard (i.e., effects on the recipient), and the probability and consequences of complications.

TABLE II.3.4.1 Biomaterials and Components Relevant to *In Vivo* Assessment of Tissue Compatibility
The material(s) of manufacture
Intended additives, process contaminants, and residues
Leachable substances
Degradation products
Other components and their interactions in the final product
The properties and characteristics of the final product

TABLE II.3.4.2 Medical Device Categorization by Tissue Contact and Contact Duration	
Tissue Contact	
Surface devices	Skin Mucosal membranes Breached or compromised surfaces
External communicating devices	Blood path, indirect Tissue/bone/dentin communicating Circulating blood
Implant devices	Tissue/bone Blood
Contact Duration	
	Limited: ≤24 hours Prolonged: >24 hours and <30 days Permanent: >30 days

device is knowledge of the chemical composition of the materials, and the nature, degree, frequency, and duration of exposure of the device and its constituents to the tissues in which it will be utilized. Table II.3.4.1 presents a list of biomaterial components and characteristics that may affect the overall biological responses of the medical device. The range of potential biological hazards is broad, and may include short-term effects, long-term effects or specific toxic effects, which should be considered for every material and medical device. However, this does not imply that testing for all potential hazards will be necessary or practical.

SELECTION OF *IN VIVO* TESTS ACCORDING TO INTENDED USE

As it is recognized that biomaterials–tissue interaction may vary with anatomic site and duration of exposure, and application-specific conditions, *in vivo* tests for assessment of tissue compatibility are chosen to simulate end-use applications. To facilitate the selection of appropriate tests, medical devices with their component biomaterials can be categorized by the nature of tissue contact of the medical device, and by its duration of contact. Table II.3.4.2 presents medical device categorization by nature of tissue contact and contact duration. The tissue contact categories and subcategories, as well as the contact duration categories, have been derived from standards, protocols, and guidelines utilized in the past for safety evaluation of medical devices. Certain devices may fall into more than one category, in which case testing appropriate to each category should be considered.

BIOMATERIAL AND DEVICE PERSPECTIVES IN *IN VIVO* TESTING

Two perspectives may be considered in the *in vivo* assessment of tissue compatibility of biomaterials and medical devices. The first perspective involves the utilization of *in vivo* tests to determine the general biocompatibility of newly developed biomaterials for which some knowledge of the tissue compatibility is necessary for further research and development. In this type of situation, manufacturing and other processes necessary to the development of a final product, i.e., the medical device, have not been carried out. However, the *in vivo* assessment of tissue compatibility at this early stage of development can provide additional information relating to the proposed design criteria in the production of a medical device. While it is generally recommended that the identification and quantification of extractable chemical entities of a medical device should precede biological evaluation, it is quite common to carry out preliminary *in vivo* assessments to determine if there may be unknown or as yet identified chemical entities that produce adverse biological reactions. Utilized in this fashion, early *in vivo* assessment of the tissue compatibility of a biomaterial may provide insight into the biocompatibility, and thereby may permit further development of this biomaterial into a medical device. Obviously, problems observed at this stage of development would require further efforts to improve the biocompatibility of the biomaterial, and to identify the agents and mechanisms responsible for the adverse reactions. As the *in vivo* assessment of tissue compatibility of a biomaterial or medical device is focused on the end-use application, it must be appreciated that a biomaterial considered compatible for one application may not be compatible for another.

The second perspective regarding the *in vivo* assessment of tissue compatibility of medical devices focuses on the biocompatibility of the final product, that is, the fabricated medical device in the condition in which it is to be implanted. Thus, issues related to desired fabrications, interactions between biomaterials, and to mechanical conductors, etc., may come into play. Although medical devices in their final form and condition are commonly implanted in carefully-selected animal models to determine function, as well as biocompatibility, it may be inappropriate to carry out all of the recommended tests necessary for regulatory approval on the final device. In these situations, some tests may initially be carried out on biomaterial components of devices that have been prepared under manufacturing and sterilization conditions, and other processes utilized in the final product (see Table II.3.4.3).

TABLE II.3.4.3 *In Vivo* Tests for Tissue Compatibility
Sensitization
Irritation
Intracutaneous reactivity
Systemic toxicity (acute toxicity)
Subchronic toxicity (subacute toxicity)
Genotoxicity
Implantation
Hemocompatibility
Chronic toxicity
Carcinogenicity
Reproductive and developmental toxicity
Biodegradation
Immune responses

SPECIFIC BIOLOGICAL PROPERTIES ASSESSED BY *IN VIVO* TESTS

In this section, brief perspectives on the general types of *in vivo* tests are presented. Details regarding these tests are found in the lists of standards provided at the end of this chapter. ISO 10993 standards advise that the biological evaluation of all medical device materials include testing for cytotoxicity, sensitization, and irritation. (Cytotoxicity tests are *in vitro*.) Beyond these fundamentals, the selection of further tests for *in vivo* biocompatibility assessment is based on the characteristics and end-use application of the device or biomaterial under consideration.

Sensitization, Irritation, and Intracutaneous (Intradermal) Reactivity

Exposure to, or contact with, even minute amounts of potential leachables from medical devices or biomaterials can result in allergic or sensitization reactions. Sensitization tests estimate the potential for contact sensitization to medical devices, materials, and/or their extracts. Symptoms of sensitization are often seen in skin, and tests are often carried out topically in guinea pigs. Test design should reflect the intended route (skin, eye, mucosa) and nature, degree, frequency, duration, and conditions of exposure of the biomaterial in its intended clinical use. While sensitization reactions are immune system responses to contact with chemical substances, ISO guidelines suggest irritation to be a local tissue inflammation response to chemicals, without a systemic immunological component. The most severely irritating chemical leachables may be discovered prior to *in vivo* studies by careful material characterization and *in vitro* cytotoxicity tests. Irritant tests emphasize utilization of extracts of the biomaterials to determine the irritant effects of potential leachables. Intracutaneous (intradermal) reactivity tests determine the localized reaction of tissue to intracutaneous injection of extracts of medical devices, biomaterials or prostheses in the final product form. Intracutaneous tests may be applicable where determination of irritation by dermal or mucosal tests is not appropriate. Albino rabbits are most commonly used.

Since these tests focus on determining the biological response of leachable constituents of biomaterials, their extracts in various solvents are utilized to prepare the injection solutions. Critical to the conduct of these tests is that the preparation of the test material and/or extract solution should be chosen to include testing for both water-soluble and fat-soluble leachables.

Systemic Toxicity: Acute, Subacute, and Subchronic Toxicity

Systemic toxicity tests estimate the potential harmful effects *in vivo* on target tissues and organs away from the point of contact (i.e., site of implantation) with either single or multiple exposures to medical devices, biomaterials, and/or their extracts. These tests evaluate the systemic toxicity potential of medical devices that release constituents into the body. These tests also include pyrogenicity testing, which assesses the induction of a systemic inflammatory response, often measured as fever.

In tests using extracts, the form and area of the material, the thickness, and the surface area to extraction vehicle volume are critical considerations in the testing protocol. Appropriate extraction vehicles, i.e., solvents, should be chosen to yield a maximum extraction of leachable materials for use in the testing. Mice, rats or rabbits are the usual animals of choice for the conduct of these tests, and oral, dermal, inhalation, intravenous, intraperitoneal or subcutaneous application of the test substance may be used, depending on the intended application of the biomaterial. Acute toxicity is considered to be the adverse effects that occur after administration of a single dose or multiple doses of a test sample given within 24 hours. Subacute toxicity (repeat dose toxicity) focuses on adverse effects occurring after administration of a single dose or multiple doses of a test sample per day during a period of from 14 to 28 days. Subchronic toxicity is considered to be the adverse effects occurring after administration of a single dose or multiple doses of a test sample per day given during a part of the lifespan, usually 90 days but not exceeding 10% of the lifespan of the animal.

Pyrogenicity tests are also included in the system toxicity category to detect material-mediated fever-causing reactions to extracts of medical devices or material. Although the rabbit pyrogen test has been the standard, the Limulus amebocyte lysate (LAL) reagent test has been used increasingly in recent years. It is noteworthy that no single test can differentiate pyrogenic reactions that are material mediated *per se* from those due to endotoxin contamination.

Genotoxicity

In vivo genotoxicity tests are carried out if indicated by the chemistry and/or composition of the biomaterial (see Table II.3.4.1) or if *in vitro* test results indicate potential genotoxicity (changes in deoxyribonucleic acid (DNA) that could lead to changes in cellular proliferation, differentiation, and/or function). Initially, at least three *in vitro* assays should be used, and two of these assays should utilize mammalian cells. The initial *in vitro* assays should cover the three levels of genotoxic effects: DNA destruction; gene mutations; and chromosomal aberrations (as assessed by cytogenetic analysis). *In vivo* genotoxicity tests include the micronucleus test, the *in vivo* mammalian bone marrow cytogenetic tests – chromosomal analysis, the rodent dominant lethal tests, the mammalian germ cell cytogenetic assay, the mouse spot test, and the mouse heritable translocation assay. Not all of the *in vivo* genotoxicity tests need be performed; the most common test is the rodent micronucleus test. Genotoxicity tests are performed with appropriate extracts or dissolved materials using appropriate media, as suggested by the known composition of the biomaterial.

Implantation

Implantation tests assess the local pathological effects on the structure and function of living tissue induced by a sample of a material or final product at the site where it is surgically implanted or placed into an implant site or tissue appropriate to the intended application of the biomaterial or medical device. In some cases, the anatomic site of implantation used for biocompatibility evaluation is not the same as the site of ultimate use, but has representative mechanisms and consequences of tissue–biomaterials interaction (e.g., subcutaneous implantation in rodents of bioprosthetic heart valve materials to study calcification that occurs as a major clinical limitation in humans; see Chapter II.4.5). The most basic evaluation of the local pathological effects is carried out at both the gross level and the microscopic level. Histological (microscopic) evaluation is used to characterize various biological response parameters (Table II.3.4.4). To address specific questions, more sophisticated studies may need to be done. Examples include immunohistochemical staining of histological sections to determine the types of cells present, and studies of collagen formation and destruction. For short-term implantation evaluation out to 12 weeks, mice, rats, guinea pigs or rabbits are the most common animals utilized in these studies. For longer-term testing in subcutaneous tissue, muscle or bone, animals such as rats, guinea pigs, rabbits, dogs, sheep, goats, pigs, and other animals with relatively long life expectancy are suitable. If a complete medical device is to be evaluated, larger species may be utilized so that human-sized devices may be used in the site of intended application. For example, substitute heart valves are usually tested as heart valve replacements in sheep, whereas calves are usually the animal of choice for ventricular assist devices and total artificial hearts.

TABLE II.3.4.4 Biological Response Parameters as Determined by Histological Assessment of Implants
Number and distribution of inflammatory cells as a function of distance from the material/tissue interface
Thickness and vascularity of fibrous capsule
Quality and quantity of tissue ingrowth (for porous materials)
Degeneration as determined by changes in tissue morphology
Presence of necrosis
Other parameters such as material debris, fatty infiltration, granuloma, dystrophic calcification, apoptosis, proliferation rate, biodegradation, thrombus formation, endothelialization, migration of biomaterials or degradation products

In all aspects of biocompatibility testing, it is important to recognize that the effects of the material on the surrounding tissues are generally superimposed on the events occurring during physiological wound repair induced by the surgery of implantation. This is particularly important in shorter-term experiments.

Hemocompatibility

Hemocompatibility tests evaluate effects on blood and/or blood components by blood-contacting medical devices or materials. *In vivo* hemocompatibility tests are usually designed to simulate the geometry, contact conditions, and flow dynamics of the device or material in its clinical application. From the ISO standards perspective, five test categories are indicated for hemocompatibility evaluation: thrombosis; coagulation; platelets; hematology; and immunology (complement and leukocytes). Two levels of evaluation are indicated: Level 1 (required); and Level 2 (optional). Regardless of blood contact duration, hemocompatibility testing is indicated for external communicating devices – blood path indirect; external communicating devices – circulating blood; and blood-contacting implant devices. Chapter II.5.4 gives further details on the testing of blood–material interactions.

Several issues are important in the selection of tests for hemocompatibility of medical devices or biomaterials. In particular, the hemocompatibility depends not only on the materials characteristics, but also on the fluid mechanics of the device (i.e., stasis promotes thrombus formation), and the coagulatability of the blood. Thus, *in vivo* testing in animals may be convenient, but anatomic differences among species and species-related differences in blood reactivity must be considered, and these may limit the predictability of any given test in the human clinical situation. While blood values and reactivity between humans and nonhuman primates are very similar, European Community law prohibits the use of nonhuman primates for blood compatibility and medical device testing. Hemocompatibility evaluation in animals is complicated by the lack of appropriate and adequate test materials, for example, appropriate antibodies for immunoassays. Use of human blood in hemocompatibility evaluation implies *in vitro* testing, which usually requires the use of anticoagulants that are not generally present with the device in the clinical situation, except for perhaps the earliest implantation period. Although species differences may complicate hemocompatibility evaluation, the utilization of animals in short- and long-term testing is considered to be appropriate for evaluating thrombosis and tissue interaction.

Chronic Toxicity

Chronic toxicity tests determine the effects of either single or multiple exposures to medical devices, materials, and/or their extracts during a period of at least 10% of the lifespan of the test animal, e.g., over 90 days in rats. Chronic toxicity tests may be considered an extension of subchronic (subacute) toxicity testing, and both may be evaluated in an appropriate experimental protocol or study.

Carcinogenicity

Carcinogenicity tests determine the tumorigenic potential of medical devices, materials, and/or their extracts from either single or multiple exposures or contacts over a period of the major portion of the lifespan of the test animal. Both carcinogenicity (tumorigenicity) and chronic toxicity may be studied in a single experimental study. With biomaterials, these studies focus on the potential for solid-state carcinogenicity, i.e., the Oppenheimer effect (see Chapter II.2.7). Thus, in carcinogenicity testing, controls of a comparable form and shape should be included; polyethyelene implants are a commonly used control material. The use of appropriate controls is imperative as animals may spontaneously develop tumors, and statistical comparison between the test biomaterial/device and the controls is necessary. To facilitate and reduce the time period for carcinogenicity testing of biomaterials, the FDA is exploring the use of transgenic mice carrying the human prototype c-Ha-ras gene as a bioassay model for rapid carcinogenicity testing.

Since tumors associated with clinical medical devices have been rare (see Chapter II.4.5) carcinogenicity tests should be conducted only if data from other sources suggest a tendency for tumor induction. However, considerations of carcinogenicity may become important in some future applications in which pleuripotential stem cells produced by any methodology are used.

Reproductive and Developmental Toxicity

These tests evaluate the potential effects of medical devices, materials, and/or their extracts on reproductive function, embryonic development (teratogenicity), and prenatal and early postnatal development. The application site of the device must be considered, and tests and/or bioassays should only be conducted when the device has a potential impact on the reproductive potential of the subject.

Biodegradation

Biodegradation tests determine the effects of a biodegradable material and its biodegradation products on the tissue response. They focus on the amount of degradation during a given period of time (the kinetics of biodegradation), the nature of the degradation products, the origin of the degradation products (e.g., impurities, additives, corrosion products, bulk polymer), and the qualitative and quantitative assessment of degradation products and leachables in adjacent tissues and in distant organs. The biodegradation of biomaterials may occur through a wide variety of mechanisms, which in part are biomaterial dependent, and all pertinent mechanisms related to the device and the end-use application of the device must be considered. Test materials comparable to degradation products may be prepared and studied to determine the biological response of degradation products anticipated in long-term implants. An example of this approach is the study of metallic and polymeric wear particles that may be present with long-term orthopedic joint prostheses. Moreover, for intentionally biodegradable scaffolds used in tissue engineering (see Chapter II.6.3), biodegradation and the attendant tissue response are important to follow histologically. Further insights on biodegradation are available in Chapters II.6.2 and II.6.3.

Immune Responses

Immune response evaluation is not a component of the standards currently available for *in vivo* tissue compatibility assessment. However, ASTM, ISO, and the FDA currently have working groups developing guidance documents for immune response evaluation where pertinent. Synthetic materials are not generally immunotoxic (see Chapter II.2.3). However, immune response evaluation is necessary with modified natural tissue implants such as collagen, which has been utilized in a number of different types of implants and may elicit immunological responses. The Center for Devices and Radiological Health of the FDA has released a draft immunotoxicity testing guidance document (Langone, 1998) whose purpose is to provide a systematic approach for evaluating potential adverse immunological effects of medical devices and constituent materials. Immunotoxicity is any adverse effect on the function or structure of the immune system or other systems as a result of an immune system dysfunction. Adverse or immunotoxic effects occur when humoral or cellular immunity needed by the host to defend itself against infections or neoplastic disease (immunosuppression) or unnecessary tissue damage (chronic inflammation, hypersensitivity or autoimmunity) is compromised. Potential immunological effects and responses that may be associated with one or more of these effects are presented in Table II.3.4.5.

Representative tests for the evaluation of immune responses are given in Table II.3.4.6. Table II.3.4.6 is not all-inclusive, and other tests that specifically consider possible immunotoxic effects potentially generated

TABLE II.3.4.5 Potential Immunological Effects and Responses

Effects
Hypersensitivity
Type I – anaphylactic
Type II – cytotoxic
Type III – immune complex
Type IV – cell-mediated (delayed)
Chronic Inflammation
Immunosuppression
Immunostimulation
Autoimmunity
Responses
Histopathological Changes
Humoral Responses
Host Resistance
Clinical Symptoms
Cellular Responses
T cells
Natural killer cells
Macrophages
Granulocytes

TABLE II.3.4.6 Representative Tests for the Evaluation of Immune Responses

Functional Assays	Phenotyping	Soluble Mediators	Signs of Illness
Skin testing	Cell surface markers	Antibodies	Allergy
Immunoassays (e.g., ELISA)	MHC markers	Complement	Skin rash
Lymphocyte proliferation	Chemokines	Immune complexes	Urticaria
Plaque-forming cells	Basoactive amines	Cytokine patterns (T-cell subsets)	Edema
Local lymph node assay	–	Cytokines (IL-1, IL-1ra, TNFα, IL-6, TGF-β, IL-4, IL-13)	Lymphadenopathy
Mixed lymphocyte reaction	–	–	–
Tumor cytotoxicity	–	–	–
Antigen presentation	–	–	–
Phagocytosis	–	–	–

ELISA: Enzyme-linked immunosorbent assay; IL: Interleukin; TNF: Tumor necrosis factor; TGF: Transforming growth factor; MHC: Major histocompatibility complex.

by a given device or its components may be applicable. Examples presented in Table II.3.4.6 are only representative of the large number of tests that are currently available. However, direct and indirect markers of immune responses may be validated and their predictive value documented, thus providing new tests for immunotoxicity in the future. Direct measures of immune system activity by functional assays are the most important types of tests for immunotoxicity. Functional assays are generally more important than tests for soluble mediators, which are more important than phenotyping. Signs of illness may be important in *in vivo* experiments, but symptoms may also have a significant role in studies of immune function in clinical trials and postmarket studies.

Combination devices where drugs (or cells) are utilized within medical devices should also be considered for immune response evaluation. Hypersensitivity reactions have been reported with drug-eluting coronary stents (DES). With DES, concomitantly prescribed medications such as clopidogrel (platelet inhibitor) have been considered the causative agent for hypersensitivity, as well as the DES itself (Nebeker et al., 2006).

SELECTION OF ANIMAL MODELS FOR *IN VIVO* TESTS

Animal models are used to predict the clinical behavior, safety, and biocompatibility of medical devices in humans (Table II.3.4.7). The selection of animal models for the *in vivo* assessment of tissue compatibility must consider

TABLE II.3.4.7 Animal Models for the *In Vivo* Assessment of Medical Devices

Device Classification	Animal
Cardiovascular	
Heart valves	Sheep
Vascular grafts	Dog, pig
Stents	Pig, rabbit, dog
Ventricular assist devices	Calf
Artificial hearts	Calf
Ex vivo shunts	Baboon, dog
Orthopedic/Bone	
Bone regeneration/ substitutes	Rabbit, dog, pig, mouse, rat Dog, goat, nonhuman primate
Total joints – hips, knees	Sheep, goat, baboon
Vertebral implants	Rabbit, pig, dog, nonhuman primate
Craniofacial implants	Rabbit, dog
Cartilage	Dog, sheep
Tendon and ligament substitutes	Goat
Neurological	
Peripheral nerve regeneration	Rat, cat, nonhuman primate
Electrical stimulation	Rat, cat, nonhuman primate
Ophthalmological	
Contact lens	Rabbit
Intraocular lens	Rabbit, monkey

the advantages and disadvantages of the animal model for human clinical application. Several examples follow, which exemplify the advantages and disadvantages of animal models in predicting clinical behavior in humans.

Preclinical testing in animal models is an important part of the regulatory process, used to determine the safety and efficacy of devices prior to human clinical trials. The choice of the animal model and the selection of *in vitro* tests should be made according to the intended use of the respective medical device, prosthesis or biomaterial.

A single test animal may not assess all pertinent clinically-important complications. For example, as described earlier, sheep are commonly used for the evaluation of heart valves (see Chapters II.3.7 and II.4.5). This is based on size considerations, and also the propensity to calcify tissue components of bioprosthetic heart valves and thereby be a sensitive model for this complication. Thus, the choice of this animal model for bioprosthetic heart valve evaluation is made on the basis of accelerated calcification, the major clinical problem, assessed in rapidly growing animals which has its clinical correlation in young and adolescent humans. Nevertheless, normal sheep may not provide a sensitive assessment of the propensity of a valve to thrombosis, which may be potentiated by the reduced flow seen in abnormal subjects, but diminished by the specific coagulation profile of sheep.

The *in vivo* assessment of tissue responses to vascular graft materials is an example in which animal models present a particularly misleading picture of what generally occurs in humans. Virtually all animal models, including nonhuman primates, heal rapidly and completely with an endothelial blood-contacting surface. Humans, on the other hand, do not show extensive endothelialization of vascular graft materials, and the resultant pseudointima from the healing response in humans has potential thrombogenicity. Consequently, despite favorable results in animals, small-diameter vascular grafts (less than 4 mm in internal diameter) usually yield early thrombosis in humans, the major mechanism of failure which is secondary to the lack of endothelialization in the luminal surface healing response.

Originally, the porcine coronary artery model was considered the model of choice for the evaluation of arterial stents. More recently, the rabbit iliac artery model for the evaluation of drug-eluting stents has been considered to be more realistic, as endothelialization is slower in the rabbit model than in the porcine model, and inflammation is not as extensive in the rabbit (Nakazawa et al., 2008). Thus, endothelialization, healing, and inflammation in the rabbit iliac artery model may be closer to these responses in humans than the porcine coronary artery model.

The use of appropriate animal models is an important consideration in the safety evaluation of medical devices that may contain potential immunoreactive materials. The *in vivo* evaluation of recombinant human growth hormone in poly(lactic-*co*-glycolic acid)(PLGA) microspheres demonstrates the appropriate use of various animal models to evaluate biological responses and the potential for immunotoxicity. Utilizing biodegradable PLGA microspheres containing recombinant human growth hormone (rhGH), Cleland et al. (1997) used Rhesus monkeys, transgenic mice expressing hGH, and normal control (Balb/C) mice in their *in vivo* evaluation studies. Rhesus monkeys were utilized for serum assays in the pharmacokinetic studies of rhGH release, as well as tissue responses to the injected microcapsule formulation. Placebo injection sites were also utilized, and a comparison of the injection sites from rhGH PLGA microspheres and placebo PLGA microspheres demonstrated a normal inflammatory and wound-healing response with a normal focal foreign-body reaction. To further examine the tissue response, transgenic mice were utilized to assess the immunogenicity of the rhGH PLGA formulation. Transgenic mice expressing a heterologous protein have been previously used for assessing the immunogenicity of structural mutant proteins. With the transgenic animals, no detectable antibody response to rhGH was found. In contrast, the Balb/C control mice had a rapid onset of high-titer antibody response to the rhGH PLGA formulation. This study points out the appropriate utilization of animal models to not only evaluate biological responses, but also one type of immunotoxicity (immunogenicity).

FUTURE PERSPECTIVES ON *IN VIVO* MEDICAL DEVICE TESTING

As presented earlier in this chapter, the *in vivo* assessment of tissue compatibility of biomaterials and medical devices is dependent on the end-use application of the biomaterial or medical device. In this sense, the development and utilization of new biomaterials and medical devices will dictate the development of new test protocols and procedures for evaluating these new products. Furthermore, it must be understood that the *in vivo* assessment of tissue compatibility of biomaterials and medical devices is open-ended, and new end-use applications will require new tests.

The future development of medical devices is anticipated to provide more complexity to the composition and construction of these devices and, thus, to the array of potential biomaterials–tissue interactions. Thus, further studies will require a more sophisticated approach to test protocols and methodologies that must clearly identify biocompatibility and function. In this regard, new tests, methods, and animal models may have to be developed to adequately and appropriately characterize the biocompatibility and function of these new devices. Therefore, the development of guidelines and standards is dynamic and constantly evolving, driven by the complexity of new devices developed for application in tissue engineering, regenerative medicine, and nanomedicine.

Over the past half-century, medical devices and biomaterials have generally been "passive" in their tissue interactions. That is, a mechanistic approach to biomaterials–tissue interactions has rarely been used in the development of biomaterials or medical devices. Heparinized biomaterials are an exception to this, but considering the five subcategories of hemocompatibility, these approaches have minimal impact on the development of blood-compatible materials.

In the past decade, increased emphasis has been placed on bioactivity and tissue engineering in the development of biomaterials and medical devices for potential clinical application. Rather than a "passive" approach to tissue interactions, bioactive and tissue-engineered devices have focused on an "active" approach in which biological or tissue components, i.e., growth factors, cytokines, drugs, enzymes, proteins, extracellular matrix components, and cells that may or may not be genetically modified, are used in combinations with synthetic, i.e., passive, materials to produce devices that control or modulate a desired tissue response. Obviously, *in vivo* assessment of the targeted biological response of a tissue-engineered device will play a significant role in the research and development of that device, as well as in its safety assessment. It is clear that scientists working on the development of tissue-engineered devices will contribute significantly to the development of *in vivo* tests for biocompatibility assessment, as these tests will also be utilized to study the targeted biological responses in the research phase of the device development.

Regarding tissue-engineered devices, it must be appreciated that biological components may induce varied effects on tissue in the *in vivo* setting. For example, a simplistic view of the potentially complex problems that might result from a device releasing a growth factor to enhance cell proliferation is presented. The presence of a growth factor may result in markedly different cell proliferation, differentiation, protein synthesis, attachment, migration, shape change, etc., which would be cell-type dependent. Thus, different cell-type dependent responses in an implant site, reacting to the presence of a single exogenous growth factor, may result in inappropriate, inadequate or adverse tissue responses. These perspectives must be integrated into the planned program for *in vivo* assessment of tissue compatibility of tissue-engineered devices. Moreover, a major challenge to the *in vivo* assessment of tissue compatibility of tissue-engineered devices is the use of animal tissue components in the early phase of device development, whereas the ultimate goal is the utilization of human tissue components in the final device for end-use application. Novel and innovative approaches to the *in vivo* tissue compatibility of tissue-engineered devices must be developed to address these significant issues. Finally, the development of clinically useful tissue-engineered devices will require enhanced understanding of the influence of the patient and biomechanical factors on the structure and function of healed and remodeled tissues. It will also require new methodology for assessment of biocompatibility, and the dynamic progression of remodeling *in vivo* (Mendelson and Schoen, 2006).

Careful studies of retrieved implants to establish biomarkers and mechanisms of structural evolution will be critical (see Chapter II.1.5).

BIBLIOGRAPHY

An, Y. H., & Friedman, R. J. (1999). *Animal Models in Orthopaedic Research*. Boca Raton, FL: CRC Press.

Association for the Advancement of Medical Instrumentation (1998). *AAMI Standards and Recommended Practices*, Vol. 4. *Biological Evaluation of Medical Devices, 1997*, Vol. 4S (Supplement).

Chapekar, M. S. (1996). Regulatory concerns in the development of biologic–biomaterial combinations. *J. Biomed. Mater. Res. Appl. Biomat.*, **33**, 199–203.

Cleland, J. L., Duenas, E., Daugherty, A., Marian, M., Yang, J., et al. (1997). Recombinant human growth hormone poly(lactic-*co*-glycolic acid) (PLGA) microspheres provide a long lasting effect. *J. Control Release*, **49**, 193–205.

FDA (US Food and Drug Administration) (1995). *Blue Book Memorandum G95-1: FDA-modified version of ISO 10993-Part 1, Biological Evaluation of Medical Devices – Part 1. Evaluation and Testing.*

Hench, L. L., & Pollak, J. M. (2002). Third-generation biomedical materials. *Biomaterials*, **295**, 1014–1017.

Langone, J. J. (1998). *Immunotoxicity Testing Guidance. Draft Document, Office of Science and Technology.* Center for Devices and Radiological Health, Food and Drug Administration.

Nakazawa, G., Finn, A. V., Ladich, E., Ribichini, F., Coleman, L., et al. (2008). Drug-eluting stent safety: Findings from preclinical studies. *Expert Rev. Cardiovasc. Ther.*, **6**(10), 1379–1391.

Nebeker, J. R., Virmani, R., Bennett, C. L., Hoffman, J. M., Samore, M. H., et al. (2006). Hypersensitivity cases associated with drug-eluting coronary stents. *J. Am. Coll. Cardiol.*, **47**, 175–181.

Mendelson, K., & Schoen, F. J. (2006). Heart valve tissue engineering: Concepts, approaches, progress, and challenges. *Ann. Biomed. Eng.*, **34**, 1799–1819.

Williams, D. F. (2008). On the mechanisms of biocompatibility. *Biomaterials*, **29**, 2941–2953.

Yamamoto, S., Urano, K., Koizumi, H., Wakana, S., Hioki, K., et al. (1998). Validation of transgenic mice carrying the human prototype c-Ha-ras gene as a bioassay model for rapid carcinogenicity testing. *Environ Health Perspect*, **106**(Suppl. 1), 57–69.

ISO STANDARDS

ISO 10993, Biological Evaluation of Medical Devices, International Standards Organization, Geneva, Switzerland.

ISO 10993-1 Evaluation and testing within a risk management system.

ISO 10993-2 Animal welfare requirements.

ISO 10993-3 Tests for genotoxicity, carcinogenicity, and reproductive toxicity.

ISO 10993-4 Selecton of tests for interactions with blood.

ISO 10993-5 Tests for *in vitro* cytotoxicity.

ISO 10993-6 Tests for local effects after implantation.

ISO 10993-7 Ethylene oxide sterilization residuals.

ISO 10993-9 Framework for the identification and quantification of potential degradation products.

ISO 10993-10 Tests for irritation and delayed-type hypersensitivity.

ISO 10993-11 Tests for systemic toxicity.
ISO 10993-12 Sample preparation and reference materials.
ISO 10993-13 Identification and quantification of degradation products from polymeric medical devices.
ISO 10993-14 Identification and quantification of degradation products from ceramics.
ISO 10993-15 Identification and quantification of degradation products from metals and alloys.
ISO 10993-16 Toxicokinetic study design for degradation products and leachables.
ISO 10993-17 Method for the establishment of allowable limits for leachable substances.
ISO 10993-18 Chemical characterization of materials.
ISO 10993-19 Physico-chemical morphological and topographical characterization of materials.
ISO-10993-20 Principles and methods for immunotoxicology testing of medical devices.
ISO 14971 Medical devices – application of risk management to medical devices.

ASTM INTERNATIONAL, 2008 ANNUAL BOOK OF ASTM STANDARDS, VOLUME 13.01

F 895 Agar Diffusion Cell Culture Screening for Cytotoxicity.
F 2382 Assessment of Intravascular Medical Device Materials on Partial Thromboplastin Time (PTT).
E 1397 *In Vitro* Rat Hepatocyte DNA Repair Assay.
E 1398 *In Vivo* Rat Hepatocyte DNA Repair Assay.
F 1983 Assessment of Compatibility of Absorbable/Resorbable Biomaterials for Implant Applications.
F 981 Assessment of Compatibility of Biomaterials for Surgical Implants with Respect to Effect of Materials on Muscle and Bone.
F 756 Assessment of Hemolytic Properties of Materials.
F 1027 Assessment of Tissue and Cell Compatibility of Orofacial Prosthetic Materials and Devices.
F 1877 Characterization of Particles.
F 813 Direct Contact Cell Culture Evaluation of Materials for Medical Devices.
F 749 Evaluating Material Extracts by Intracutaneous Injection in the Rabbit.
F 750 Evaluating Material Extracts by Systemic Injection in the Rabbit.
F 2148 Evaluation of Delayed Contact Hypersensitivity Using the Murine Local Lymph Node Assay (LLNA).
F 1906 Evaluation of Immune Responses in Biocompatibility Testing Using ELISA Tests, Lymphocyte Proliferation, and Cell Migration.
F 619 Extraction of Medical Plastics.
F 2147 Guinea Pig: Split Adjuvant and Closed Patch Testing for Contact Allergens.
F 748 Selecting Generic Biological Test Methods for Materials and Devices.
F 1905 Selecting Tests for Determining the Propensity of Materials to Cause Immunotoxicity.
F 763 Short-Term Screening of Implant Materials.
F 1408 Subcutaneous Screening Test for Implant Materials.
F 719 Testing Biomaterials in Rabbits for Primary Skin Irritation.
F 1903 Testing for Biological Responses to Particles *In Vitro*.
F 720 Testing Guinea Pigs for Contact Allergens: Guinea Pig Maximization Test.
F 2065 Testing for Alternative Pathway Complement Activation in Serum by Solid Materials.
F 2567 Testing for Classical Pathway Complement Activation in Serum by Solid Materials.
F 1984 Testing for Whole Complement Activation in Serum by Solid Materials.
F 1904 Testing the Biological Responses to Particles *In Vivo*.
E 1263 Conduct of Micronucleus Assays in Mammalian Bone Marrow Erythrocytes.
E 1202 Development of Micronucleus Assay Standards.
E 1262 Performance of Chinese Hamster Ovary Cell/Hypoxanthine Guanine Phosphoribosyl Transferase Gene Mutation Assay.
F 1439 Performance of Lifetime Bioassay for the Tumorigenic Potential of Implant Materials.
E 1280 Performing the Mouse Lymphoma Assay for Mammalian Cell Mutagenicity.

CHAPTER II.3.5 EVALUATION OF BLOOD–MATERIALS INTERACTIONS

Buddy D. Ratner[1] and Thomas A. Horbett[2]
[1]Professor, Bioengineering and Chemical Engineering, Director of University of Washington Engineered Biomaterials (UWEB), Seattle, WA, USA
[2]Bioengineering and Chemical Engineering University of Washington, Seattle, WA, USA

INTRODUCTION

Thousands of devices fabricated from synthetic materials or processed natural materials are interfaced with blood for a wide range of medical procedures (see Chapters II.5.3A–C, II.5.4, and II.5.5). Blood-contacting materials have a unique set of requirements, and most are quite different from implants in soft tissue or bone. The most obvious complication is thrombus or clot, but many other undesirable reactions can impact performance when a synthetic material is interfaced with blood. How can the biomaterials engineer know which materials might be best used in the fabrication of a blood-contacting device? It would be ideal if this chapter could end with a definitive list of "blood-compatible" biomaterials that might be used in blood-contacting medical devices (note that generalized suggestions are offered near the end of this chapter). The considerations driving the interpretation of blood interactions tests are sufficiently complex that such a list of materials cannot be presented. The reader should appreciate this point after working through Chapter II.3.5.

This chapter outlines methods and concerns in evaluating the blood-compatibility of biomaterials, and the blood-compatibility of medical devices. It does not automatically follow that if the materials comprising a device are blood-compatible, a device fabricated from those materials will also be blood-compatible. This important point should be clear upon completion of this chapter. Before considering the evaluation of materials and devices, the reader should be familiar with the protein and cellular reactions of blood coagulation, platelet responses, and fibrinolysis, as discussed in Chapter II.2.6. The history of methods to assess blood-compatibility is addressed in Ratner, 2000.

BACKGROUND AND PRINCIPLES OF BLOOD–MATERIALS INTERACTIONS ASSESSMENT

What is Blood-Compatibility?

Blood-compatibility assessment would be straightforward if, following the introductory paragraph, there were a list of standard tests that might be performed to evaluate blood-compatibility. By simply performing the tests outlined in such a list, a material could be rated "blood-compatible" or "not blood-compatible." Unfortunately, no widely-recognized, standard list of blood-compatibility tests exists. Due to the complexity of blood–materials interactions (BMI), there is a basic body of ideas that must be mastered in order to appreciate what blood interaction tests actually measure. This section introduces the rationale for BMI testing, and addresses a few important measurement schemes.

"Blood-compatibility" can be defined as the property of a material or device that permits it to function in contact with blood without inducing adverse reactions. But this simple definition offers little insight into what a blood-compatible material is. More useful definitions become increasingly complex. This is because there are many mechanisms for the body to respond to material intrusions into the blood. A material that will not trigger one response mechanism may be highly active in triggering an alternative pathway. The mechanisms by which blood responds to materials have been discussed in Chapter II.2.6. A more recent definition of blood-compatibility integrates a multiparameter assessment of BMI, where some of the parameters are defined quantitatively (Sefton et al., 2000). This chapter will integrate many of these ideas, and provide a framework for BMI (and blood compatibility) assessment.

> Blood compatibility is impacted by the biochemistry of coagulation, the mechanisms of blood–materials interactions (BMI), and the design and function of a device in the bloodstream. This chapter will integrate many of these ideas, and provide a framework for BMI and blood compatibility assessment.

Blood-compatibility can also be viewed from a different perspective, i.e., by considering a material that is *not* blood-compatible (a thrombogenic material). Such a material would produce specific adverse reactions when placed in contact with blood: formation of clot or thrombus composed of various blood elements; shedding or nucleation of emboli (detached thrombus); the destruction of circulating blood components; activation of the complement system and other immunologic pathways (Salzman and Merrill, 1987). Most often, in designing blood-contacting materials and devices, our aim is to minimize these generally undesirable blood reactions. Are these reactions always undesirable? Consider the case of a hemostatic device rapidly inducing clotting.

Why Measure Blood-Compatibility?

Many devices and materials are presently used in humans to treat, or to facilitate treatment of, health problems. Such devices include the extracorporeal pump-oxygenator (heart-lung machine) used in many surgical procedures, hollow fiber hemodialyzers for treatment of kidney failure, catheters for blood access and blood vessel manipulation (e.g., angioplasty), heart assist devices, stents for luminally supporting blood vessels, and devices for the permanent replacement of diseased heart valves (prosthetic heart valves) and arteries (vascular grafts). Since these and other blood-contacting devices have been successfully used in patients for 30 or more years, and are judged to be therapeutically beneficial, it is reasonable to ask: (1) is there a continued need for assessing BMI; and (2) are there important problems that remain to be addressed? The answer in both cases is an emphatic "yes."

To address point (1) above, the ongoing need for BMI measurement, consider that existing devices are frequently modified by incorporation of new design features or materials primarily intended to improve durability, physical and mechanical characteristics, i.e., devices may be modified to improve characteristics other than BMI. However, since these changes may also affect blood responses, and since BMI are not entirely predictable based on knowledge of device composition and configuration, blood-compatibility testing is nearly always required to document safety.

There are substantial problems with devices used today in the bloodstream (point 2, above). The performance of many existing devices is frequently less than optimal (McIntire et al., 1985; Salzman and Merrill, 1987; Williams, 1987; Ratner, 1993, 2000, 2007; Wang and Tsai, 2010). For example, prolonged cardiopulmonary bypass and membrane oxygenation can produce a severe bleeding tendency. Mechanical heart valves occasionally shed emboli to the brain, leading to stroke. Angiographic catheters can also generate strokes. Synthetic vascular grafts perform less well than grafts derived from natural arteries or veins; graft failure due to thrombosis can lead to ischemia (lack of oxygen) and death of downstream tissue beds; small diameter vascular grafts (<4 mm i.d.) cannot be made. Thus, while performance characteristics have been judged to be acceptable in many instances (i.e., the benefit/risk ratio is high), certain existing devices could be improved to extend their period of safe operation (e.g., oxygenators), and to reduce adverse BMI long-term (e.g., heart valves). Further, many devices are only "safe" when anticoagulating drugs are used (e.g., oxygenators, mechanical heart valves, hemodialyzers). Device improvements that would reduce adverse BMI, and thereby minimize the need for anticoagulant therapy, would have important implications both for health (fewer bleeding complications due to drug effects) and cost (complications can be expensive to treat). The reusability of devices that can undergo repeated blood exposure in individual patients (e.g., dialyzers) is also an important economic consideration.

For certain applications there are no devices presently available that perform adequately (due to adverse BMI), even when antithrombotic drugs are used. Thus, there are needs for devices that could provide long-term oxygenation for respiratory failure, cardiac support (total artificial heart), and intravascular measurement of physiologic parameters and metabolites (O_2, CO_2, pH, glucose). Also, there are demands for long-term, small diameter vascular grafts (<5 mm internal diameter) and other conduits (e.g., stents) for reconstruction of diseased arteries and veins. Overall, there is a compelling need for continued and improved methods for evaluating BMI.

What is Thrombogenicity?

A thrombogenic device may cause a localized accumulation of protein and cellular blood elements. Cardiovascular devices may also induce regions of disturbed flow or stasis that lead to the formation of blood clots. These accumulations or clots may compromise device functions, such as flow of blood through artificial blood vessels, the mechanical motions of heart valves, gas exchange through oxygenators, the removal of metabolic waste products through dialysis membranes, etc. Local blood reactions may also impact other parts of the host organism, i.e., systemic effects might be noted. Thus, thrombi may detach from a surface (embolize) and be carried downstream, eventually occluding a blood vessel of comparable size and impairing blood flow distal to the site of occlusion. Chronic devices may produce steady-state destruction or "consumption" of circulating blood elements, thereby lowering their concentration in blood (e.g., mechanical destruction of red cells by heart prostheses producing anemia or removal of platelets due to ongoing thrombus formation or platelet activation), with a concomitant rise in plasma levels of factors released from those blood elements (e.g., plasma hemoglobin, platelet factor 4). Mediators of inflammatory responses and vessel tone may also be produced or released from cells (e.g., platelets, white cells, the complement pathway) following blood–surface interactions that can affect hemodynamics and organ functions at other sites. Thus, "thrombogenicity" may be broadly defined as the extent to which a device, when employed in its intended use configuration, induces the adverse responses outlined above. While all artificial surfaces interact with blood, an acceptably nonthrombogenic device can be defined as one which would produce neither local nor systemic effects with significant health consequence to the host organism.

Thrombogenicity may be broadly defined as the extent to which a device, when employed in its intended use configuration, induces adverse responses.

With "thrombogenicity" now defined as adverse outcome events associated with device usage, the obvious goal is to design and improve devices using materials that are blood-compatible (nonthrombogenic) for specific applications. Ideally, the biomaterials engineer would like to consult a handbook for a list of materials useful in the fabrication of a blood-contacting device. Unfortunately, there is little consensus as to the materials that are blood-compatible (Ratner, 1984). Because of this lack of consensus, there is no "official" list of blood-compatible materials. As a result, in the construction of a new blood-contacting device the biomaterials engineer generally consults published studies or directly performs blood-compatibility assessment studies on candidate materials.

Understanding the blood-compatibility of specific materials used in blood-contacting devices is complex because:

(1) The types of blood-contacting devices used are numerous (see Chapter II.5.3), and the device design will impact the apparent thrombogenicity of materials used in those devices.
(2) Typically, blood-contacting devices are commercially manufactured, and manufacturers are, for competitive reasons, reluctant to discuss specific chemical compositions or changes made in response to the availability of materials.
(3) The possible blood responses are numerous, complex, dynamic, and often not fully-understood.
(4) It is difficult and expensive to measure device thrombogenicity (clinically significant local thrombosis or systemic effects) in a systematic way, in either experimental animals or humans (Williams, 1987, and this chapter).

Most tests purported to measure blood-compatibility in fact evaluate certain blood–material interactions (BMI), which are the events that occur (and observations that are made) when blood contacts a material. Figure II.3.5.1 illustrates how alternate interpretations can be applied to data from "blood-compatibility" tests. This concept is further expanded upon in Figure II.3.5.2. These alternate interpretations often invalidate or modify conclusions drawn from such tests. For accuracy, the term "BMI assessment" will be used for the remainder of this chapter, instead of "blood-compatibility test." Based upon the characteristics of the evaluation method (i.e., what is really being measured) the biomaterials scientist

Blood reactions during exposure
Appearance when evaluated
Test Device
expose to blood
(A)
(B)
(C)
(D)

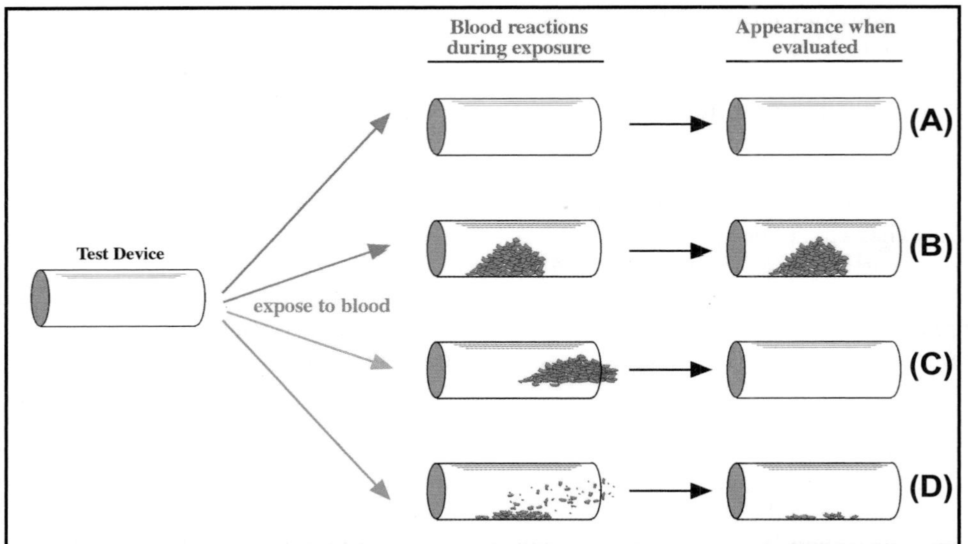

FIGURE II.3.5.1 Possible scenarios for blood–materials interactions and alternate interpretations of testing outcome. This example points out limitations of evaluating only local thrombus formation at fixed time points. (A) The device remains free of thrombus after exposure to blood; (B) a large thrombus forms, adheres tightly as is observed as an endpoint; (C) a large thrombus forms but detaches (embolizes); (D) the surface is highly reactive towards blood but deposited material is quickly removed through microembolism and/or lysis. Inspection of devices (C) and (D) could lead to the incorrect conclusion that these surfaces are blood compatible.

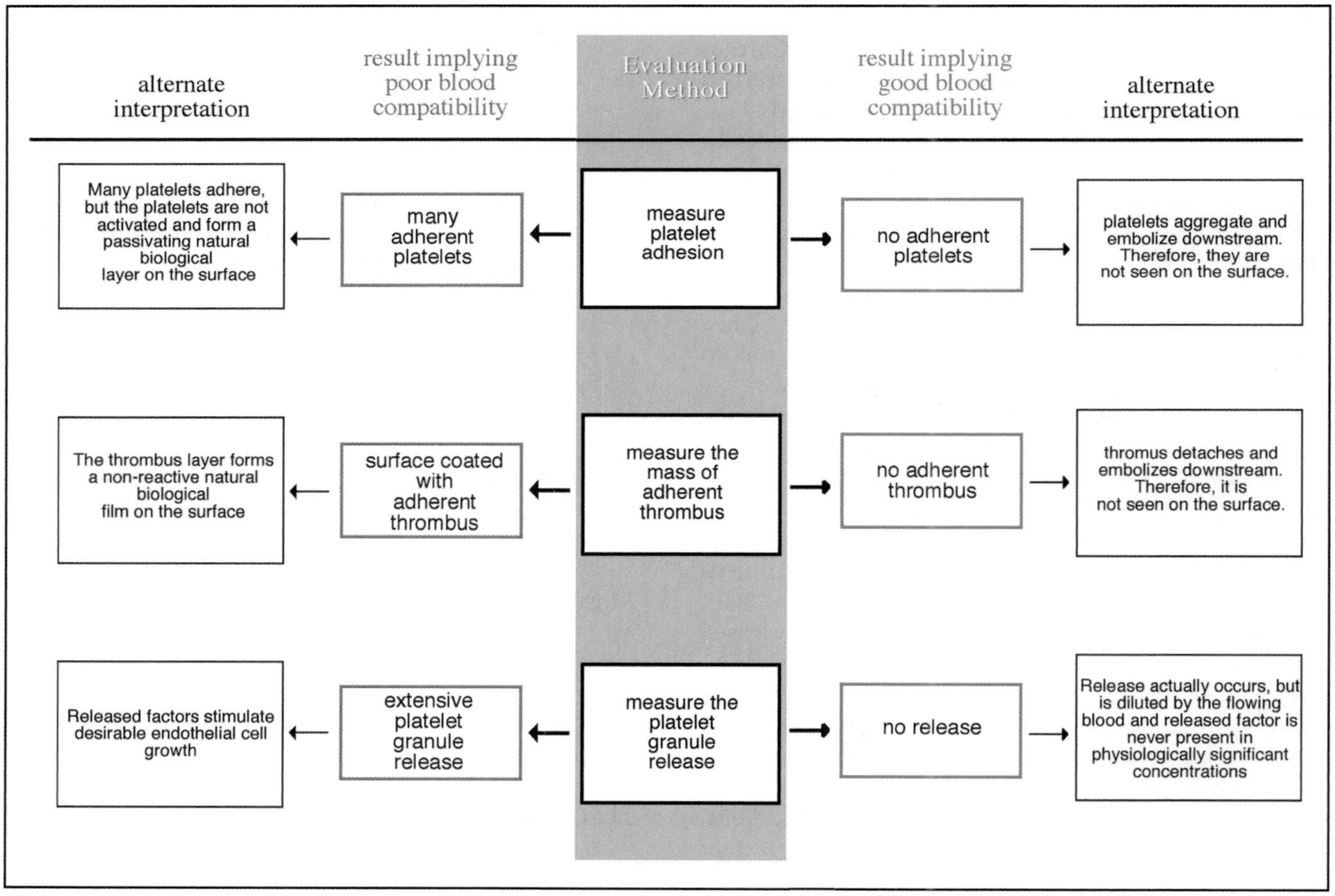

alternate interpretation
result implying poor blood compatibility
Evaluation Method
result implying good blood compatibility
alternate interpretation
Many platelets adhere, but the platelets are not activated and form a passivating natural biological layer on the surface
many adherent platelets
measure platelet adhesion
no adherent platelets
platelets aggregate and embolize downstream. Therefore, they are not seen on the surface.
The thrombus layer forms a non-reactive natural biological film on the surface
surface coated with adherent thrombus
measure the mass of adherent thrombus
no adherent thrombus
thromus detaches and embolizes downstream. Therefore, it is not seen on the surface.
Released factors stimulate desirable endothelial cell growth
extensive platelet granule release
measure the platelet granule release
no release
Release actually occurs, but is diluted by the flowing blood and released factor is never present in physiologically significant concentrations

FIGURE II.3.5.2 Alternate scenarios that can be applied for interpreting results of blood–material interaction assays.

must relate the significance of the events being observed (the BMI) to the blood-compatibility of the material or device. A solid understanding of the physical and biological mechanisms of blood–materials interactions is required to make this connection in a rational way.

Most tests purported to measure blood-compatibility in fact evaluate blood–material interactions (BMI), which are the events that occur when blood contacts a material. For example, is a material that adheres platelets not blood-compatible? Figures II.3.5.1 and II.3.5.2 demonstrate alternate interpretations of such data.

In more specific terms, BMI are the interactions (reversible and irreversible) between surfaces and blood solutes, proteins, and cells (e.g., adsorption, absorption, adhesion, denaturation, activation, spreading) that occur under defined conditions of exposure time, blood composition, and blood flow. Since each of these variables influence BMI, we generally cannot: (1) extrapolate results obtained under one set of test conditions to another set of conditions; (2) use short-term testing to predict long-term results; and (3) predict *in vivo* device performance based on BMI testing of materials *per se* in idealized flow geometries. Nonetheless, such tests have provided important insights into the mechanisms of thrombus formation in general, and the relationships between BMI and blood-compatibility. These studies also permit some general guidelines for device construction and, to a limited extent, may allow prediction of device performance in humans. These points are addressed in subsequent sections of this chapter.

The above considerations suggest that no material may be simply "blood-compatible" or "nonthrombogenic," since this assessment will depend strongly on details of the test system or usage configuration. In fact, under conditions of sluggish (low shear) blood flow or stasis, most if not all, polymeric materials may become associated with localized blood clotting and thus be considered "thrombogenic." This is because synthetic materials, unlike the vascular endothelium (the perfect "blood-compatible material" that lines all blood vessels), cannot actively inhibit thrombosis and clotting by directly producing and releasing inhibitors or by inactivating procoagulant substances. The possibility that there may be no "biomaterials solution" for certain situations or that device performance could be improved by emulating strategies found in nature, has led some investigators to consider coating devices with endothelial cells, antithrombotic drugs or anticoagulating enzymes. These approaches appear promising, and are being widely explored. As for conventional synthetic materials and devices, establishing the usefulness of biologic surfaces and drug delivery devices requires appropriate methods for evaluating their blood interactions. An important point is that we use millions of devices in contact with blood each year in patients with generally good outcomes. It is not that the problem of blood-compatibility is impossible to solve, but outcomes can often be improved by optimizing BMI.

We use millions of devices in contact with blood each year in patients with generally good outcomes. The problem of blood-compatibility is complex, but outcomes can often be improved by optimizing blood–materials interactions.

Key Considerations for BMI Assessment

In 1856 Rudolph Virchow implied that three factors contribute to the coagulation of the blood: blood chemistry, the blood-contacting surface, and the flow regime (commonly referred to as Virchow's triad – Figure II.3.5.3).

This assessment is still valid, and provides a framework for more formally introducing the variables important in any system intended to evaluate BMI. Also, the interaction time of blood with materials (ranging from seconds to years) has an impact on BMI. As described below, these variables may each profoundly influence the results and interpretation of BMI testing. It is assumed that the reader has reviewed the mechanisms of blood response to artificial surfaces, as outlined in Chapter II.2.6.

Blood: A Fragile Fluid that is Readily Compromised. The source of the blood and methods for its handling can strongly influence BMI. Human blood and blood from various animal species has been employed in BMI assessment *in vitro* and *in vivo*, both in the presence and absence of anticoagulants. Blood reactivity is also influenced by manipulations *in vitro*, the surface-to-volume ratio of blood in extracorporeal circuits, and the use of pumps for blood recirculation. These aspects are discussed below, and summarized in Table II.3.5.1.

Although there are general similarities in the blood chemistry of all mammals, there are also differences characteristic of each species. In particular, blood may vary with respect to blood proteins (concentrations and functionality), and cells that participate in coagulation, thrombosis, and fibrinolysis (Chapter II.2.6). The size of blood-formed elements may also differ. A comparison of blood chemistry between man and the more commonly used animal species has been published (McIntire et al., 1985). While human blood is obviously preferable for BMI, it is often impossible to use human blood in certain experiments. In addition, there are significant health concerns in experimenting with human blood, and animals are commonly used for both *in vitro* and *in vivo* studies. Unfortunately, most investigations have employed a

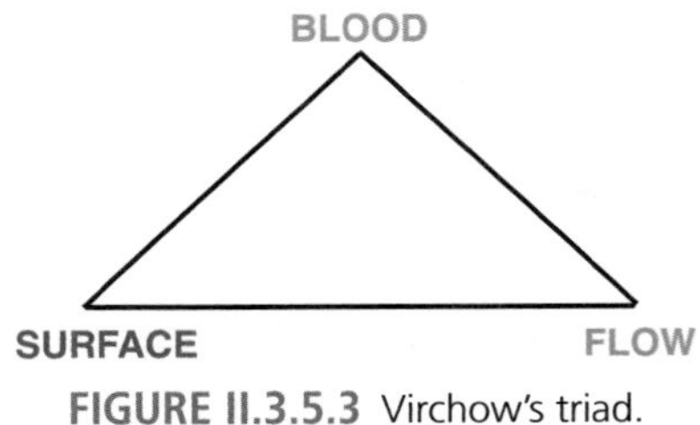

FIGURE II.3.5.3 Virchow's triad.

TABLE II.3.5.1	Factors Important in the Acquisition and Handling of Blood for BMI Experiments
• Species of the blood donor	
• Health, gender, and age of the blood donor	
• Blood reactivity of the donor (individual physiological differences)	
• Time interval between blood draw and BMI experiment	
• Care with which the puncture for the blood draw was made	
• Temperature (for blood storage and testing)	
• Anticoagulation	
• Drugs and anesthetics present in the blood	
• Blood damage due to centrifugation and separation operations	
• Blood damage due to contact with foreign surfaces prior to the BMI experiment (syringe, needle, blood bag, bottles, tubing, etc.)	
• Blood damage due to the air–blood interface	
• Blood damage due to pumping and recirculation	

single animal species or blood source. There have been few comparisons between human and animal blood responses for evaluating BMI in particular test situations (also see Chapter II.3.7). In many instances differences between human and animal blood responses are large, and such differences must be factored into interpreting experimental results. For example, the initial adhesiveness of blood platelets for artificial surfaces appears to be low in man and some primates, and high in the dog, rat, and rabbit (Grabowski et al., 1976). Following the implantation of chronic blood-contacting devices (e.g., vascular grafts) there may be large differences between man and other animal species in terms of device healing, which will be reflected as differences in the *time course* of BMI. Further, while laboratory animals serving as blood donors for BMI experiments may represent a relatively homogeneous population in terms of age, health status, blood responses, etc., the human donors of blood for research (and recipients of blood-contacting devices) may vary considerably in terms of these parameters. Consider, for example, human diet, drug use (e.g., aspirin, alcohol), overall health, and even donor sex (males and females were found to have different platelet responses in experiments in our laboratory). Thus, the results obtained with animal species must be viewed with caution if conclusions are to be drawn as to the significance of the results for humans. Even the blood drawn from one human can give misleading results when generalizations to all humans are attempted.

Despite these limitations, animal testing has been helpful in defining *mechanisms* of BMI and thrombus formation, and the interdependence of blood biochemical pathways, the nature of the surface, and the blood flow regime. In addition, while results of animal testing may not quantitatively predict results in man, in many cases results can be qualitatively similar. These aspects are discussed further below. In general, studies in lower animal species, such as the rabbit, rat, and guinea pig, may be useful to screen for profound differences between materials, for example, by incorporation of an antithrombotic drug delivery system into an otherwise thrombogenic device. Short-term screening to identify markedly reactive materials, and longer-term studies to evaluate healing and its impact on BMI can also be performed in species such as dog and pig. When differences in BMI are likely to be modest (for example, as a consequence of subtle changes in surface chemistry or device configuration) the ranking of materials based on tests with lower animal species may be unrelated to results that would be obtained in man; studies with primates, which are hematologically similar to man, are more likely to provide results which are clinically relevant. However, even the relationships between human and primate BMI are also not well-established, and should therefore be interpreted with caution. And experimentation with primates is expensive, and raises ethical issues.

In vitro testing generally requires anticoagulation of the blood that can have a profound effect on BMI. *In vivo* testing and the use of extracorporeal circuits are also commonly performed with anticoagulants. Two anticoagulants are frequently used: sodium citrate, a calcium ion chelator required for certain reactions of platelets and coagulation proteins; and heparin, a natural polysaccharide used to block the action of the coagulation protease thrombin (Chapter II.2.6). Both can markedly affect BMI. In particular, the removal of calcium ions may profoundly depress platelet-surface reactions, and the capacity of platelets to form aggregates and thrombi. Thus, the relationship between BMI in the presence of citrate anticoagulant, and "thrombogenicity" in the absence of anticoagulant, is questionable. Similar concerns apply to heparin anticoagulation. Although this agent is less likely to interfere with the earliest platelet reactions, platelet thrombus formation may be impaired by inhibition of thrombin activity. The use of heparin is appropriate for evaluation of devices where heparinization is used clinically (e.g., oxygenators, dialyzers). In general, results with anticoagulated blood cannot be used to predict performance in the absence of anticoagulants. A discussion of anticoagulants in the context of BMI has been presented (McIntire et al., 1985).

Blood is a fragile tissue that begins to change from the moment it is removed from the body. It may become more active (activated) or less active (refractory). Thus, BMI evaluations with blood externalized from the normal circulatory system for more than a few hours are questionable. If purified blood components or cells are used (e.g., platelets, fibrinogen), studies must be performed to ensure that they remain functionally normal. In most cases the volume of blood used, relative to test surface area, should be large. Similarly, the area of non-test surfaces, including exposure to air interfaces, should be minimized. Changes in blood temperature, test surface temperature or exposure to bright light sources (Haycox et al., 1991) can also produce artifactual results. When blood is pumped, the recirculation rate (fraction of total blood volume pumped per unit time) should be minimized, since blood pumping alone can induce platelet and red cell damage, platelet release reactions, and platelet refractoriness (Haycox and Ratner, 1993).

Blood is a complex, fragile fluid, and outcomes of blood–materials interactions studies will be strongly impacted by the animal species (or human donor), blood handling protocols, and storage time.

Flow: Blood Interactions Dictated by Shear and Mass Transport. Blood flow controls the rate of transport (by diffusion and convection) of cells and proteins in the vicinity of artificial surfaces and thrombi. This subject has been reviewed (Leonard, 1987; Turitto and Baumgartner, 1987; and Chapter II.1.6). While physiological blood shear forces probably do not damage or activate platelets directly, such forces can dislodge platelet aggregates and thrombi (embolization) to distal circulatory beds. Platelet diffusion in flowing blood, and early platelet attachment to surfaces, may be increased 50–100-fold by the presence of red blood cells that greatly enhance the movement of platelets across parallel streamlines. At higher shear forces, red cells may also contribute chemical factors that enhance platelet reactivity (Turitto and Baumgartner, 1987).

A number of studies using well-characterized flow geometries have suggested that the initial attachment of platelets to artificial surfaces increases with increasing blood flow or, more specifically, with increasing wall shear rate (the slope of the velocity profile at the surface). Under conditions of low wall shear rate flow (less than ~1000 sec^{-1}) early platelet adhesion (over the first minutes of exposure) may depend more upon the platelet arrival rate (i.e., platelet availability) than on substrate surface properties (Friedman et al., 1970). Under these conditions the platelet–surface reaction rate is said to be *diffusion controlled*. At higher shear rates, platelet adhesion may depend less upon platelet transport to the surface, and more on the reactions triggered by surface properties (*reaction controlled*, Schaub et al., 2000); thus, studies designed to assess the role of surface properties are best performed under flow conditions where platelet transport is not limiting. Following initial platelet adhesion, subsequent processes of platelet aggregation and *in vivo* thrombus formation (over minutes to hours) are predominantly *reaction controlled*. For example, platelet accumulation on highly thrombogenic artificial surfaces (e.g., fabric vascular grafts) or biologic surfaces (e.g., collagen) may be rapid, and dependent on both the substrate reactivity and factors influencing platelet availability (shear rate, hematocrit, and the platelet content of blood) (Harker et al., 1991). Under other circumstances, the rate of platelet-surface interactions may be almost entirely reaction controlled. For example, with smooth-walled artificial surfaces which

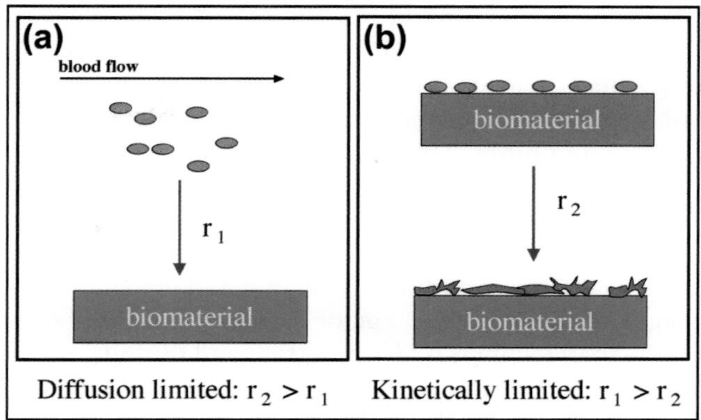

FIGURE II.3.5.4 The effects of flow and material surface properties on platelet transport to, and reaction with, surfaces. One can define two rate processes: r_1, the rate of platelet transport from blood to the surface (higher in rapidly flowing blood); and r_2, the rate of reaction of a platelet with the surface. (a) In low wall shear rate blood flow, platelets can be slow to reach the surface and r_1 dominates the kinetics of the reaction – for reactive surfaces, the surface can be "starved" of platelets for reaction. (b) In high wall shear rate blood flow, platelets are transported to the surface much more rapidly than they can react with the surface, and the intrinsic reactivity of the surface, r_2, can be observed.

repeatedly embolize small platelet aggregates continuously over days, the overall rate of platelet destruction depends strongly on material properties, but not on blood flow rate or circulating platelet numbers over wide ranges of these variables (Hanson et al., 1980). These concepts of diffusion and reaction control are further explained in Figure II.3.5.4.

It has been observed that under arterial flow conditions (high wall shear rate), thrombus that forms *in vivo* may be largely composed of platelets ("white thrombus"), while thrombus that forms under venous flow conditions (low shear rate) may contain mostly red cells entrapped in a fibrin mesh ("red thrombus"). The process of platelet thrombus formation may not be affected by administration of heparin (i.e., arterial thrombosis may be heparin-resistant), while venous thrombosis is effectively treated with heparin. This insensitivity of platelet reaction to heparin in high shear blood flow is somewhat surprising, since the procoagulant enzyme thrombin, one of the most potent activators of platelets, is strongly inhibited by heparin. These observations have been incorrectly interpreted to mean that arterial thrombosis and venous thrombosis are separable processes, with the former depending only on platelet reactions, and the latter depending only on surface protein-triggered coagulation events. However, while platelet-dependent (arterial) thrombosis may be little affected by heparin, it is blocked quite effectively by other inhibitors of thrombin (Hanson and Harker, 1988; Wagner and Hubbell, 1990), indicating that heparin is limited in its capacity to block the thrombin enzyme when thrombin is produced locally in high concentrations through reactions which may be catalyzed on the platelet surface (Chapter II.2.6). The formation of fibrin, due to the action of thrombin on fibrinogen, is also important for thrombus formation and stabilization since: (1) fibrinolytic enzymes can reduce platelet thrombus formation; and (2) arterial thrombi are often composed of alternating layers of platelets and fibrin. Thus, in most circumstances, thrombin is a key promotor of *local* platelet and fibrin accumulation (on surfaces) under both high shear and low shear conditions. Thrombi may differ in appearance, because under high flow conditions, thrombin and precursor procoagulant enzymes (e.g., Factor Xa) may be diluted sufficiently to prevent *bulk phase* clotting and trapping of red cells. These ideas have been addressed quantitatively (Basmadjian et al., 1997). Also, flow effects on surface-induced thrombosis have been reviewed, and the importance of pharmacological intervention on modulating this process has been described (Hanson and Sakaraissen, 1998). Thus, the shear dependency of surface thrombosis depends on the surface chemical composition. For example, anticoagulation effects may be more pronounced on a tissue factor-rich surface than on a collagen surface, particularly at low wall shear rates. Further, some platelet inhibitors such as aspirin are shear-dependent, while others seem not to be.

In summary, thrombus formation requires the transport by flow of platelets and coagulation proteins to surfaces. Fibrin polymerization, as well as local platelet activation and recruitment into growing thrombi, require conversion of prothrombin to thrombin, the end product of a sequential series of coagulation reactions that are also catalyzed by platelets, and may be amplified or inhibited by various feedback mechanisms (Chapter II.2.6). Blood flow regulates each reaction step, such that under low (venous) flow conditions fibrin formation is abundant; thrombi may resemble coagulated whole blood with many entrapped red cells. Under high (arterial) flows, platelets stabilized by much smaller amounts of fibrin may comprise the greater proportion of total thrombus mass.

> Thrombus formation requires the transport by flow of platelets and coagulation proteins to (and from) surfaces. Absence of flow, low shear flow, and high shear flow can each give dramatically different BMI results.

Surfaces: The Most Studied, and Least Well-Defined, of the BMI Variables. Many different artificial surfaces, in various device applications, are used in contact with blood. It is well-documented that surface physicochemical properties of materials can influence early events, for example, on protein adsorption (Chapter I.1.2) and platelet adhesion – yet how these effects relate to subsequent thrombus formation remains uncertain.

When placed in contact with blood most, if not all, artificial surfaces first acquire a layer of adsorbed blood proteins whose composition and mass may vary with time in a complex manner depending on substrate surface type (Chapter I.1.2). This layer mediates the subsequent attachment of platelets and other blood cells that can lead to the development of platelet aggregates and thrombi. The relationship between material properties, the protein layer, and the propensity of a material or device to accumulate thrombus is not well-understood because: (1) protein–surface reactions involve complex, dynamic processes of competitive adsorption, denaturation, and activation; (2) cell–surface interactions may modify the protein layer, i.e., cells may deposit lipid and protein "footprints" derived from the cell membrane; (3) the importance of specific adsorbed proteins for subsequent cell interactions, especially *in vivo*, is not well-defined; (4) there have been few *relevant* tests in which both protein adsorption and later thrombus formation have been assessed. Under conditions of low blood shear, the capacity of negatively charged surfaces (such as glass) to activate intrinsic coagulation (via Factor XII) can lead to thrombin production, with subsequent platelet deposition and fibrin clot formation. Under other circumstances the availability on surfaces of adhesive plasma proteins, such as fibrinogen, will be important for regulating cell attachment and subsequent thrombus formation (Horbett et al., 1986; Sivaraman and Latour, 2010).

With anticoagulated blood, initial platelet attachment to a variety of surfaces may be comparable and limited to a partial platelet monolayer, suggesting that surface properties may be "inconsequential" for early platelet adhesion, especially where platelet transport to the surface may be rate-limiting (Friedman et al., 1970). In the absence of anticoagulants, initial platelet attachment may vary, but no general relationship to substrate surface properties has been demonstrated. In attempts to establish such relationships, thrombus formation has been studied using devices implanted in animals and composed of various materials including polymers, metals, carbons, charged surfaces, and hydrogels. Correlations have been sought between the blood response and surface properties, such as charge (anionic-cationic), hydrophilicity, hydrophobicity, polarity, contact angle, wettability, and critical surface tension (Kaelble and Moacanin, 1977; McIntire et al., 1985; Salzman and Merrill, 1987; Williams, 1987). These parameters have not proven satisfactory for predicting device performance even in idealized test situations, reflecting the complexity of the phenomena being investigated, the limitations of animal experiments and, in some cases, inadequate characterization of material surface properties (see Chapters I.1.5 and III.1.4, and Ratner, 1993).

In many cases, material properties are constrained by the specific mechanical and morphological needs of the intended blood-contacting device application. For example, vascular grafts and the sewing ring of prosthetic heart valves are composed of fabric or porous materials to permit healing and tissue anchoring. Other materials must be permeable to blood solutes and gases (dialysis and oxygenator membranes) or distensible (pump ventricles, balloon catheters). These design constraints often necessitate complex flow geometries. In general, devices with flow geometries that cause regions of flow recirculation and stasis tend to produce localized clotting in the absence of heparin anticoagulation. On a microscopic scale, surface imperfections, cracks, and trapped air bubbles may serve as foci to initiate thrombus formation. While surface smoothness is usually desirable, many devices having a fabric or microporous surface (e.g., vascular grafts) function well if the layer of thrombus that forms is not thick enough to interfere with device function (Salzman and Merrill, 1987).

The lack of consensus on the blood-compatibility of materials was pointed out by Ratner (1984). This is a consequence of the complex, multifactorial nature of blood–materials interactions. Assigning a label of "blood-compatible" or ranking materials as to their suitability for use in blood is still fraught with uncertainty. This is illustrated by three studies. One study, using a number of different screening assays, questioned the effects of surface properties on blood interactions, since trends from each of the assays could not be correlated (Sefton et al., 2001). Another study, using fluorescent fiber optic microscopy to observe platelet deposition ranked materials as to the magnitude of their interaction with blood (Schaub et al., 2000). The third study used a cone and plate viscometer for assessment, but conclusions were difficult to draw due to poor statistics, contradictory data, and sample quality issues (Węgrzyn et al., 2010). These three studies were well thought out efforts and offered insights into testing blood interactions and measuring BMI. Yet, taken as a set, they do not allow us to conclude which surface properties or biomaterials will yield the most blood-compatible material.

Blood Interaction Times with Materials and Devices. Different events may occur at short and long BMI times. A test performed where blood contacts a device for seconds or minutes may yield a result that will have little meaning for devices used for hours or days, or which may be implanted chronically. Thus, measurements of protein adsorption (an early event) can, in some cases, predict levels of platelet adhesion. But platelet adhesion alone is not an adequate measure of thrombogenicity, and it does not predict local or systemic thrombogenic effects that could be harmful to the host organism. Still, several studies indicate that an early maximum in platelet thrombus accumulation may be seen within hours of device exposure, and this can be sufficient to produce device failure (e.g., small diameter vascular graft occlusion) (Harker et al., 1991). Therefore, short-term testing (over hours) may be appropriate for predicting the clinical usefulness of devices that can produce an acute, severe thrombotic response. In general, the nature and extent of BMI may change continuously over the entire period of device exposure. An exception to this rule may be chronic implants that do not undergo tissue coverage (e.g., heart valve struts, arteriovenous shunts), and which may interact with blood elements at a constant rate as shown, for example, by steady-state increases in rates of platelet consumption (Hanson et al., 1980).

EVALUATION OF BMI

A summary of some (or historically important) *in vitro* and *in vivo* testing procedures commonly used to evaluate BMI is presented in this section. A few relevant case studies with commentary are included. We emphasize that a thorough characterization of surface properties is critical for the interpretation of these tests (Chapters I.1.5 and I.2.12, and Ratner, 1993), since the surface composition of materials is often significantly different from "as received and labeled." A few specific tests that have been historically influential and tests that are used today are summarized in Table II.3.5.2.

BMI can be evaluated *in vitro* and *in vivo*. Also, we can look at the BMI of either the biomaterials in isolation in a test configuration or the blood interactions of real devices. There are typically unique apparatus and geometries for *in vitro*, *in vivo*, and device-based assays. However, there are commonalities in the measurements of blood parameters. For example, blood can be circulated through a tube *in vitro*, circulated through an implanted biomaterial shunt *in vivo* or circulated through a hemodiayzer (*in vivo* or *in vitro*). In all cases, the blood emerging from the test system might be assessed by a partial thromboplastin time (PPT) test, a laser scattering assay of emboli produced or a flow cytometric analysis of activated platelets. Thus, the methods for contacting blood with biomaterials, and the methods for assessing blood change, can often be considered independently.

TABLE II.3.5.2 A Partial List of BMI Tests
• Qualitative assessment (look at the thrombus) vena cava ring test (historical) thrombus adherent to a catheter in the bloodstream
• Count adherent blood elements various flow cells
• Observe and categorize platelet morphology
• Measure platelet release molecules bead column Chandler loop
• Clotting times and related coagulation assays Lee-White test (historical) activated partial thromboplastin time (APPT) catheter occlusion
• Measure soluble products of thrombosis
• Platelet consumption
• Fluorescence activated cell sorter (FACS) measures of activation
• Microparticles (FACS measurement)
• Embolus measurement renal embolus ring test (historical) light scattering Doppler ultrasound

In vitro Tests

In vitro BMI tests involve placing blood or plasma in a container composed of a test material, in a container containing the test material or recirculating blood through a flow system in which test materials contact blood under well-defined flow regimes that simulate physiologic flow conditions. Many flow geometries have been studied including tubes, parallel plates, packed beds, annular flows, rotating probes, and spinning disks. The historical (but still relevant) literature on these test methods has been reviewed (McIntyre et al., 1985; Turitto and Baumgartner, 1987), and has yielded considerable insight into how proteins and platelets are transported to, and react with, artificial surfaces. Such studies provided a wealth of morphologic information at the cellular level regarding details of platelet–surface and platelet–platelet interactions (Sakariassen et al., 1989). However, as discussed above, these tests are usually of short duration, and are strongly influenced by the blood source, handling methods, and the use of anticoagulants. Thus, *in vitro* test results generally cannot be used to predict longer-term BMI and *in vivo* outcomes, and can provide only the most general guidelines for the selection of materials for particular devices. However, *in vitro* tests may be useful for screening materials that are highly reactive towards blood.

Tests of the whole blood clotting time and variations thereof involve placing non-anticoagulated whole blood (or blood anticoagulated with sodium citrate which is then recalcified) into containers of test material (or adding the test material to the container), and measuring the time for a visible clot to form. Materials that quickly activate intrinsic coagulation and cause blood to clot within a few minutes (like glass surfaces) are probably unsuitable for use in devices with low shear blood flow or in the absence of anticoagulants. These tests are known by names such as whole blood clotting time (WBCT), activated partial thromboplastin time (APTT), and whole blood recalcification time (WBRCT). An older variant of these tests, frequently mentioned in the early biomaterials literature, is the Lee-White clotting time test.

Possibly the most commonly performed assay for BMI involves counting platelets adherent to surfaces and visualizing platelet reaction to surfaces. The limitations of these tests have been discussed, and are reiterated in this section. Still, these assessments are so commonly performed that some discussion is appropriate. The limitations of these platelet attachment tests are important to consider up front: (1) are the platelets mass transport-limited in arriving at the surface; (2) are dynamic events such as deadhesion, release, and/or emboli formation being missed due to limited time points for observation; (3) are the platelets active or refractory (unresponsive); (4) species differences in platelet adhesiveness can have large effects on the observed numbers of platelets; (5) platelets can spread into extremely thin surface films that are difficult to observe or count (Haycox and Ratner, 1993); (6) platelet adhesion studies are typically short-term, and thus generally have limited meaning for devices that reside for longer periods of time in blood. A promising approach to address the significance of surface adherent platelets for blood-compatibility is to consider not only platelet number, but also platelet morphology (Waples et al., 1996). Platelets on surfaces can appear discoid without dendrites (unactivated), and then might show various degrees of activation (spread, mildly dendritic, highly dendritic, etc.). Such platelet morphologies can be assessed using light microscopy, fluorescent methods, scanning electron microscopy or transmission electron microscopy.

Recirculation of heparinized blood or citrated blood through tubular devices and materials may lead to platelet activation and/or platelet deposition onto highly thrombogenic materials, with the appearance in plasma of proteins released from platelets (Kottke-Marchant et al., 1985; Haycox and Ratner, 1993). Thus, these and similar methods may identify materials which might cause rapid platelet accumulation *in vivo* over short time periods, and therefore be unsuitable for certain applications such as small diameter vascular grafts or blood conduits. Both recirculation tests and *in vitro* clotting assays can be considered for preliminary screening and identification of materials that could be highly thrombogenic. Most artificial surfaces in common use in blood-contacting devices would probably "pass" these tests. Since small differences in test results are of little significance for predicting material performance in actual use applications, these tests are not appropriate for optimizing or refining material properties. *In vivo* testing is more appropriate for discerning finer distinctions in blood reactivity, and such tests are discussed below.

Contemporary methods for evaluating damage to blood in *in vitro* test systems may expand the significance of these methods. For example, platelet-derived microparticles in blood have been used as a marker of blood damage induced by biomaterial surfaces (Gemmell, 2000; Blit et al., 2011).

In vivo Tests of BMI

Many studies have been performed in which test materials, in the form of rings, tubes, and patches, are inserted for short or long time periods into the arteries or veins of experimental animals (McIntire et al., 1985; Salzman and Merrill, 1987; Williams, 1987). Here are a few concerns and considerations for such tests.

(1) The timing and type of measurements may be such that important blood responses are unrecognized. In particular, the measurement of gross thrombus formation at a single point in time may lead to incorrect conclusions about local thrombus formation (e.g.,

Figure II.3.5.1), and does not provide assessment of systemic effects of thrombosis such as embolization and blood element consumption.
(2) With more commonly used animal species (e.g., dogs), blood responses may differ from humans both quantitatively and qualitatively.
(3) The blood flow conditions of the model may not be controlled or measured, and in fact, may not even be relevant for actual implant device geometries.
(4) There may be variable blood vessel trauma and tissue injury that can cause local thrombus formation through the extrinsic pathway of blood coagulation (Chapter II.2.6).

Evaluations of BMI may be performed in animals having arterio-venous (A-V) or arterio-arterial (A-A) shunts, i.e., tubular blood conduits placed between an artery and vein or between an artery and artery. A-V shunts have been studied in a variety of animals including baboons, dogs, pigs, and rabbits (McIntire et al., 1985). Qualitatively similar results have been obtained with shunts in dogs and baboons (Hanson et al., 1980; Sefton et al., 2000). An A-V shunt system is illustrated in Figure II.3.5.5. Once surgically established, shunts may remain patent (not occluded) for long periods of time (months) without the use of anticoagulants. Test materials or devices are simply inserted as extension segments or between inlet and outlet portions of the chronic shunt. These systems have many advantages: (1) blood flow is easily controlled and measured; (2) native or anticoagulated blood can be employed; (3) the animal's physiology removes damaged blood elements and makes new blood with each circulation through the body; and (4) both short-term and long-term BMI, including both local and systemic effects, can be evaluated. The downsides of these tests are demanding surgery, high expense, and ethical issues associated with chronic shunting of larger animals.

As an example, consider the A-V shunt model in baboons – baboons are used because they are hematologically similar to man. The blood responses to tubular biomaterials and vascular grafts have been quantitatively compared with respect to: (1) localized thrombus accumulation; (2) consumption of circulating platelets and fibrinogen; (3) plasma levels of factors released by platelets and coagulation proteins during thrombosis; and (4) embolization of microthrombi to downstream circulatory beds (Harker et al., 1991). These studies in primates are consistent with observations in man that certain commonly used polymers (e.g., polytetrafluoroethylene, polyethylene, plasticized poly(vinyl chloride), silicone rubbers) and some vascular grafts (e.g., polytetrafluoroethylene) are relatively nonthrombogenic in extracorporeal circuits and arteries. Thus, results with shunt models, particularly in higher animal species, may be predictive of BMI in humans when employed under comparable flow conditions (laminar unidirectional flow with arterial shear rates). Since extracorporeal shunts exclude modulating effects of blood vessel cells and tissue injury, results with these models may be less relevant to the behavior of devices which are placed surgically or whose responses may be mediated by interactions with the vessel wall as well as the blood (e.g., heart valves, grafts, indwelling catheters, and sensors).

In vivo Evaluation of Devices

Since the blood response to devices is complex and not well predicted by testing of materials *per se* in idealized configurations, animal testing, and ultimately clinical testing, of functioning devices is required to establish safety and efficacy. Broad guidelines, based on the type of device

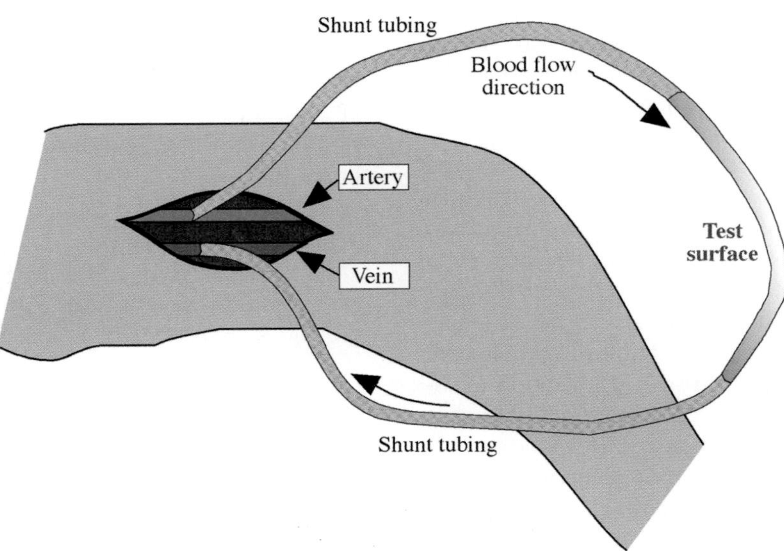

FIGURE II.3.5.5 Illustration of an arterio-venous (A-V) shunt placed between the femoral artery and vein (in the leg) of an experimental animal. Materials to be tested (in this case a tubular device) are interposed between inlet and outlet segments of the shunt.

TABLE II.3.5.3 Some Blood–Materials Responses and Their Evaluation

Blood Components	Blood Response	Assessment[a]
Thrombus	Clot	Direct visual and histologic evaluation; noninvasive imaging (angiography, ultrasound, radioisotope, magnetic resonance); evidence of device dysfunction.
	Thromboembolism	Emboli detection (ultrasound, laser); evidence of organ/limb ischemia, stroke.
Platelets	Consumption	Increased removal of radioisotopically labeled cells; reduced blood platelet count.
	Dysfunction[b]	Reduced platelet aggregation *in vitro*; prolonged bleeding time.
	Activation	Increased plasma levels of platelet factor 4 and B-thromboglobulin; platelet membrane alterations (e.g., by flow cytometry).
Red cells[b]	Destruction	Decreased red cell count; increased plasma hemoglobin.
White cells[b]	Consumption/activation	Decreased counts of white cell populations; increased white cell plasma enzymes (e.g., neutrophil elastase).
Coagulation factors	Consumption[b]	Reduced plasma fibrinogen, factor V, factor VIII.
	Thrombin generation	Increased plasma levels of prothrombin fragment 1.2 and thrombin:antithrombin III complex.
	Fibrin formation	Increased plasma level of fibrinopeptide A.
	Dysfunction[b]	Prolonged plasma clotting times.
Fibrinolytic proteins	Consumption[b]	Reduced plasma plasminogen level.
	Plasmin generation	Increased plasma level of plasmin:antiplasmin complex.
	Fibrinolysis	Increased plasma level of fibrin D-dimer fragment.
Complement proteins[b]	Activation	Increased plasma levels of complement proteins C3a, C3b, C5a, C5b-9.

[a]Radioimmunoassays (RIA) and enzyme-linked immunoassays (ELISA) may not be available for detection of non-human proteins.
[b]Tests which may be particularly important with long-term and/or large surface area devices.

being considered, are given below and apply to both animal and clinical testing. A summary of *in vivo* blood responses to devices, and of commonly used methods which have proven useful for evaluating those responses, is given in Table II.3.5.3. Table II.3.5.4 lists the wide range of screens that have been used in one group over a decade to understand blood interaction with devices, specifically ventricular assist devices (Wagner et al., 2000).

Devices that have relatively small surface areas and are exposed for short periods of time (hours to days) include catheters, guidewires, sensors, and some components of extracorporeal circuits. With these devices the primary concern is the formation of significant thrombus that could interfere with device function (e.g., the thrombus blocks the diffusion of an analyte to a sensor), occlude the vessel, embolize spontaneously or be stripped from the device surface when it is removed from the body (e.g., during catheter withdrawal through a vessel insertion site) producing occlusion of distal vessels and tissue ischemia. Devices exposed for short periods which have large surface areas (dialyzers) and complex circuitries (pump-oxygenators) may, in addition, produce: (1) a marked depletion of circulating blood cells and proteins (e.g., platelets and coagulation factors); (2) an immune/inflammatory response through activation of complement proteins and white cells; and (3) organ dysfunction mediated by hemodynamic, hematologic, and inflammatory reactions. Mechanical devices that are used for long periods of time (heart assist devices, extracorporeal membrane oxygenators) may produce profound systemic effects and organ dysfunction such that their use in man remains problematic.

With both long-term and short-term device applications, thrombus formation can be assessed directly and indirectly. Important indirect assessments include depletion from circulating blood of cells and proteins consumed in the process of thrombus formation, and the appearance in plasma of proteins generated in the process of thrombus formation (e.g., fibrinopeptide A, platelet factor 4). Direct assessment of blood flow rate,

TABLE II.3.5.4 Parameters Measured in the Assessment of Blood Interactions with Ventricular Assist Devices (VADs)

Flow Assessment
Laser Doppler anemometry, fluorescent image tracking velocimetry
Coagulation
Prothrombin fragment F1.2, thrombin–antithrombin (TAT)
Emboli
Transcranial Doppler ultrasound, flow cytometric assays for the quantification of circulating platelet-containing microaggregates
Fibrinolysis
D-dimer
Platelet Activation and Deposition
Platelet factor 4 (PF4), beta thromboglobulin (BTG), flow cytometric detection of p-selectin expression, platelet deposition
Complement
C3a, C5b-9
Leukocyte Activation
Flow cytometric detection of monocyte tissue factor expression, monocyte–platelet microaggregates, granulocyte–platelet microaggregates

(Based on Wagner et al., 2000).

flow geometry, and extent of flow channel occlusion can in many cases be achieved using sophisticated methods including angiography, ultrasound imaging, and magnetic resonance imaging. Devices that are removed from the circulation should be visually inspected to assess whether thrombus has formed at particular sites or on certain materials. Emboli in flowing blood may be detected using ultrasound and laser-based techniques (for example, Reynolds and Simon, 1980), although these methods are not used widely at present because of their complexity and expense (emboli detection will be elaborated upon below). Thrombus formation and rates of platelet destruction by both acutely placed and chronically implanted devices can be determined quantitatively by measurements of platelet lifespan and scintillation camera imaging of radioisotopically labeled blood elements (McIntire et al., 1985; Hanson et al., 1990).

In one study, a circulating loop system to assess materials under well-controlled conditions was directly compared to results obtained on an actual device in flowing blood (Münch et al., 2000). A Chandler loop was used as the well-defined test system. A cardiopulmonary bypass (CPB) device was used to observe blood reaction in a device. Both the recirculating loop system and the CPB device were surface treated with two types of heparin coatings. The parameters measured after blood flow were thrombin-antithrombin III (TAT) complex, platelet count, red blood cell count, white blood cell count, polymorphonuclear leucocyte elastase, and complement C3 activation. There were many observations reported, including some tests that showed no significant difference between the two surfaces in the loop model and statistically significant differences in the CPB model, and *vice versa*. The authors concluded: "In the more complex and realistic simulated CPB model, experimental design and cost factors prevented easy/optimum manipulation of critical variables such as blood donor (use of paired samples) and heparin level. Testing in the simpler loop model, on the other hand, readily offered manipulation of these variables, and produced endings which overlapped with observations from the more complex CPB model. Thus, the models described here complimented."

Finally, it is important to emphasize that thrombosis occurs dynamically, such that thrombi continuously undergo processes of both formation and dissolution. Device failure represents the imbalance of these processes. Older thrombi may also be reorganized considerably by the enzymatic and lytic mechanisms of white cells. While the initial consequences of surgical device placement include tissue injury, thrombosis due to tissue injury, and foreign-body reactions, the flow surface of long-term implants may become covered with a stable lining of cells (e.g., ingrowth of vascular wall endothelial and smooth muscle cells onto and into vascular grafts) or blood-derived materials (e.g., compacted fibrin). Certain reactions of blood elements (e.g., platelets, thrombin) may also stimulate the healing response. Ultimately, long-term devices, such as the small caliber graft, may fail due to excessive tissue ingrowth that could be largely unrelated to biomaterials properties or may be stimulated by the biomaterials.

We summarize this section on device testing recognizing that many device applications described above, as well as laboratory and clinical methods for evaluating their biologic responses, will be unfamiliar to the bioengineer. However, it is important to appreciate that: (1) each device may elicit a unique set of blood responses, both short-term and long-term; (2) methods are available to assess systemic changes in the blood and host organism which indirectly reflect thrombus formation; and (3) localized thrombus formation can usually be measured directly and quantitatively. Whenever possible, serial and dynamic studies should be performed to establish the time course of ongoing thrombus formation and dissolution. These measurements will ultimately predict device performance, and allow for the rational selection of biomaterials that will minimize adverse blood–device interactions.

Contemporary Concepts in BMI Evaluation

A few trends in the literature describe powerful methods to assess BMI. Two of these will be highlighted here: flow cytometry and emboli detection.

The use of flow cytometry to analyze activation of blood elements can be an effective method of measuring blood cell reaction and isolating specific pools of cells that have or have not undergone reaction. The basic technology behind flow cytometry (sometimes called fluorescence activated cell sorting, FACS) involves the laser separation of fluorescently labeled cells from a narrow flow stream of cells. The fluorescently labeled cells can be diverted from the flow stream to another flow channel where they are counted and harvested. One of the earliest studies using flow cytometry for blood-compatibility studies showed that in contact with blood-activating synthetic materials, thrombotic membrane fragments called microparticles are released (Gemmell et al., 1995). Since that important study, flow cytometry has been used to look at the upregulation of platelet P-selectin (CD62), a consequence of α-granule release, monocyte and neutrophil CD11b (Mac-1 receptor) upregulation (Gemmell, 2000), platelet activation by annexin V binding to the negative phospholipid found on activated platelets (Wagner et al., 2000), and a number of other biospecific factors relating to platelet activation, white cell activation, and platelet aggregate formation. The power of the flow cytometry method resides in its ability to pinpoint an event in a large pool of cells that can be identified by a fluorescent tag. The method is applicable to blood that has been contacted with synthetic materials *in vitro* or *in vivo*.

The ability to go beyond measurements of adhesive phenomena in blood (accumulating thrombus or platelets), and to expand studies to non-adhesive encounters leading to blood activation and damage, represents an important growth area for BMI evaluation. Cytometric studies allow us to do systemic measurements rather than just local assessment. Methods to study emboli production in real time permit a key, clinically relevant parameter, to be measured and quantified. Doppler ultrasound seems to have potential to do this, although the instrumentation is expensive. Other alternatives to study microemboli in blood include use of the Coulter counter on sampled blood, and filtration pressure methods. However, a method that seems to have excellent potential to advance our understanding of how surfaces induce emboli formation is laser light scattering. Initial developments making this method suitable for emboli analysis in whole blood were published over 20 years ago (Reynolds and Simon, 1980) and then used for the assessment of BMI in conjunction with a baboon shunt model (Garfinkle et al., 1984). New developments in light scattering instrumentation and data analysis should make laser microemboli detection a key tool for studying biomaterial reactions with blood (Solen et al., 2003).

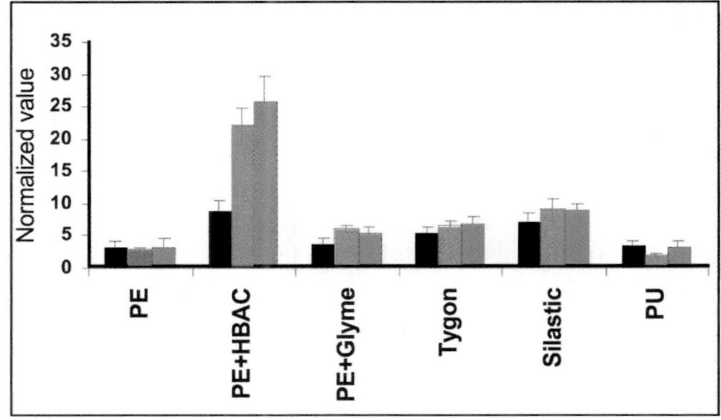

FIGURE II.3.5.6 Ratio of percentage of platelets expressing activation-dependent antibodies after exposure to biomaterials relative to percentage before any exposure to the biomaterial. ACD blood was used. Black bar: CD62p+; red: PAC-1+; green, CD62p+/PAC-1+. Results are shown for polyethylene (PE), heparinized polyethylene (PE + HBAC), an RF-plasma deposited PEG-like film (see Chapter I.2.12) (PE + Glyme), Tygon, Silastic, and polyurethane (PU). (Modified from Cao, 2006.)

Examples of BMI Evaluation

Five examples will now be presented that place many of the concepts discussed to this point in the context of real-world BMI testing.

Flow Cytometric Evaluation of Platelet Activation (Cao, 2006). Flow cytometry was used to measure platelet activation by polymer tubing filled with citrate anticoagulated blood. Tubes were agitated using a platform shaker at 100 rpm or 200 rpm for 1 hour at 37°C. After the incubation, blood samples were reacted with a dye labeled antibody for 20 minutes, fixed with paraformaldehyde, and analyzed on a FACS instrument. To restrict analysis to platelets only, only cells with a platelet-specific marker were counted. This was done with an antibody to CD61 antigen (GPIIIa), which is specific to platelets, but is not affected by platelet activation.

Antibodies specific to CD62 (antiCD62) and GPIIb/IIIa (PAC-1) were used to monitor exposure of these platelet activation-dependent markers. The PAC-1 antibody recognizes an epitope on the GPIIb/IIIa complex of activated platelets at or near the platelet fibrinogen receptor. It binds only to activated platelets, and is specific for this recognition site within GPIIb/IIIa. CD62 is a 140 kDa protein, also known as P-Selectin, GMP-140, and platelet activation-dependent granule membrane protein (PADGEM). CD62/P-selectin is stored in the α-granules of platelets, and is rapidly transported to the plasma membrane upon platelet activation.

Figure II.3.5.6 shows the increase in platelets expressing increased binding of PAC-1 and antiCD62 after exposure to various biomaterials (green bar). To correct for variations in the degree of platelet activation among donors, the activation for the biomaterials was normalized to that for the platelets in the starting blood before any biomaterials exposure. Thus, the ratio of percentage of platelets expressing activation-dependent antibodies after exposure to biomaterials at 200 rpm on a platform shaker relative to the percentage before any exposure to the biomaterial is plotted. As can be seen, there is considerable variation among the materials in the degree of activation they caused. The plot also shows the increase in platelets expressing only increased PAC-1 binding (red bars) or only increased binding of antiCD62 (black bars).

Platelet Adhesion Measured *In vitro* (Cao et al., 2006). Platelet adhesion measured under *in vitro* conditions is often used as an indicator of surface thrombogenicity, because adherent platelets can aggregate and promote clot formation.

Results of platelet adhesion to five different biomaterials are shown in Figure II.3.5.7. As shown, both the type of suspension used for the platelets and the biomaterial affect adhesion. This test indicated glyme treated tubes (an RF-plasma deposited PEG-like film, see Chapter I.2.12) were the least platelet adhesive, which would indicate they were likely to be more resistant to clot formation *in vivo*. The low reactivity of the glyme surfaces seen in *in vitro* adhesion assays was also seen *in vivo*, as noted below.

Recalcified Plasma Clotting Times (Cao et al., 2007). In these studies, citrate-anticoagulated human platelet-poor plasma is brought to 20 mM $CaCl_2$ by addition of calcium from a 1 M stock solution, and then added to the wells of a plate containing the surfaces to be tested and incubated at 37°C with shaking. The plasma clotting time is measured as the time it takes for the plasma to undergo gelation, detected by loss of movement of the plasma in response to the rotation and shaking.

As seen in Figure II.3.5.8, the recalcified plasma times are short for glass (ca. 3 minutes), a thrombogenic material used as a positive control in studies of this type. A

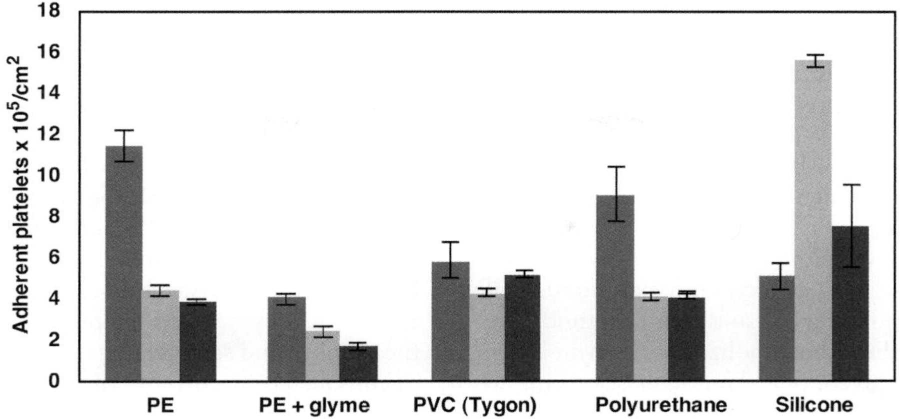

FIGURE II.3.5.7 Platelet adhesion under three conditions onto the surfaces of different tubing segments (6 cm). Error bar represents SD. N = 5. Platelet adhesion was measured using washed platelets (purple bars), platelet-rich plasma (orange bars) or whole blood (blue bars), all under static conditions. The platelet suspension (washed platelets, platelet-rich plasma or blood) was drawn into sample tubes and incubated under static conditions at 37°C for 1.5 hours. Samples incubated with washed platelets had been preadsorbed with 1% plasma to allow fibrinogen adsorption, and then incubated with albumin to block residual sites. The number of adherent platelets was determined using an assay to measure lactate dehydrogenase (LDH) released from the lysed platelets. (Modified from Cao et al., 2006.)

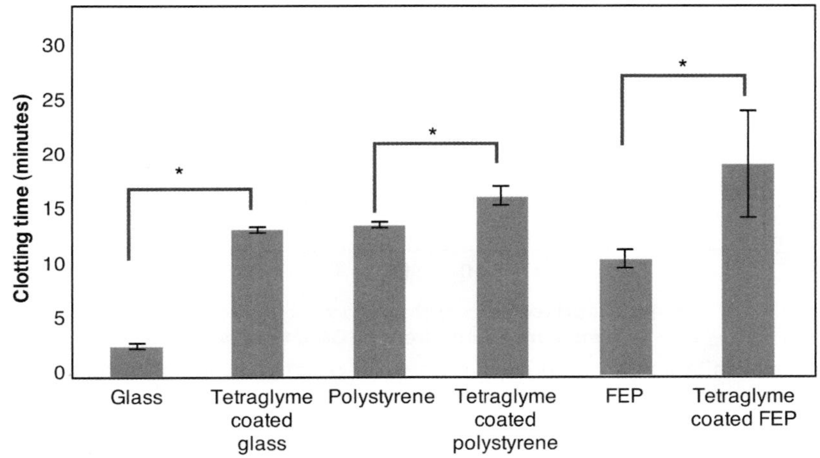

FIGURE II.3.5.8 Clotting time of recalcified platelet-poor plasma in the presence of different materials. All samples were put in the wells of a 24-well polystyrene plate. "Polystyrene" indicates the well only. (N = 6, *, $p < 0.01$, student's *t* test.) (Modified from Cao et al., 2007.)

second type of control must also be run, namely the clotting time for the container used to do the clotting time studies. In this case, the container was a multiwell polystyrene plate, and the data show that the clotting time was much longer (ca. 13 minutes) than for glass. Using a container with a long clotting time is important to make these assays useful. Thus, for example, clotting times for a polystyrene sample in a glass container would not be extended, i.e., the clotting time would still be short due to the glass, and the less thrombogenic nature of polystyrene could not be shown. Finally, as shown in Figure II.3.5.8, coating of polystyrene or FEP with the RF-plasma deposited glyme further extends the clotting time beyond that for the polystyrene dish control. Since tetraglyme coated materials strongly resist protein adsorption, and clotting by surfaces requires protein interactions with the surface, the results are as expected for a non-fouling surface.

Blood Interaction Studied Using an *Ex Vivo* Shunt. An effective and relevant way to evaluate blood interactions of a series of polymers is to insert them in the form of tubes into a permanent *ex vivo* shunt in an animal, as illustrated in Figure II.3.5.5. The *ex vivo* method can be used for both acute (short-term) and chronic (or steady-state) evaluations of blood–materials reactivity.

Acute phase studies typically last a few hours, but occasionally up to one day. At the end of each experiment the test segments are removed and analyzed to determine the deposited material on the test segment, often with SEM (Cao et al., 2006). Another short-term assay using the *ex vivo* shunt employs dynamic scintillation camera (gamma camera) imaging to continuously measure the deposition of radiolabeled platelets on surfaces inserted in the shunt or to look for shed emboli captured in downstream (distal) microcirculatory beds (Schneider et al., 1989).

To characterize chronic interactions, the platelet consumption method has been used. In this method, the excess rate of disappearance of radiolabeled platelets from circulation is measured by taking blood samples at periods up to one week after placement of the test sample. As shown in Figure II.3.5.9, the consumption of platelets is strongly dependent on surface chemistry for a series of polyurethanes.

Thromboembolism. Clots released from biomaterial surfaces are of course not detected with examination of the surface of the biomaterial, so special methods are needed to evaluate thromboembolization. Examples of two methods will be given, one using light scattering to detect flowing emboli, and the second using measurements of platelet accumulation in downstream organs of an animal.

As illustrated in Figure II.3.5.10 (Solen et al., 2003), light from a fiber optic illuminating flowing blood through tubing results in scattering of some of the light that can be detected with optical fibers placed at right angles to the incident beam. Red blood cells flowing through the tube generate a threshold background scattering level. Larger emboli increase the scattering intensity. The scattering intensity can be analyzed with a computer to count the number and size (related to size of the intensity peak) of emboli. The occurrence of thromboembolization leads

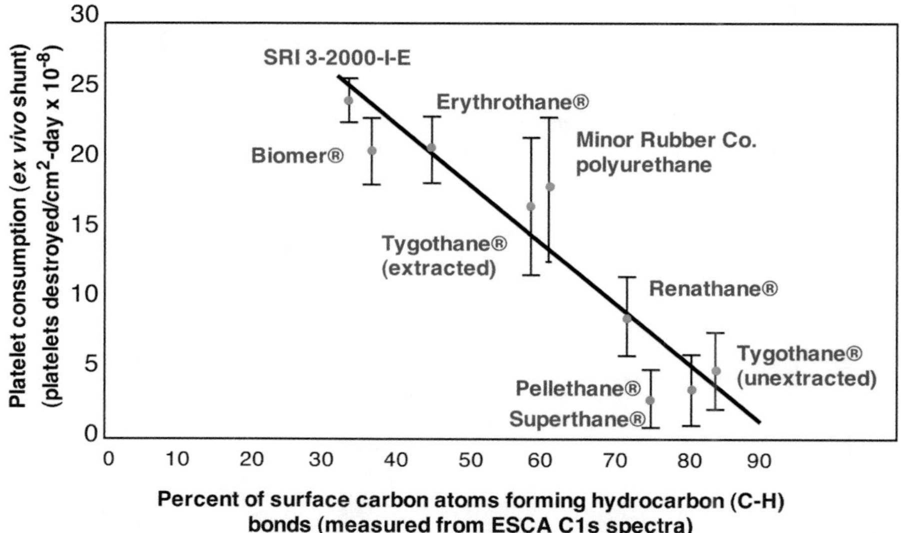

FIGURE II.3.5.9 Platelet consumption by a series of polyurethanes varying in surface hydrocarbon content. (Modified from Hanson et al., 1980.)

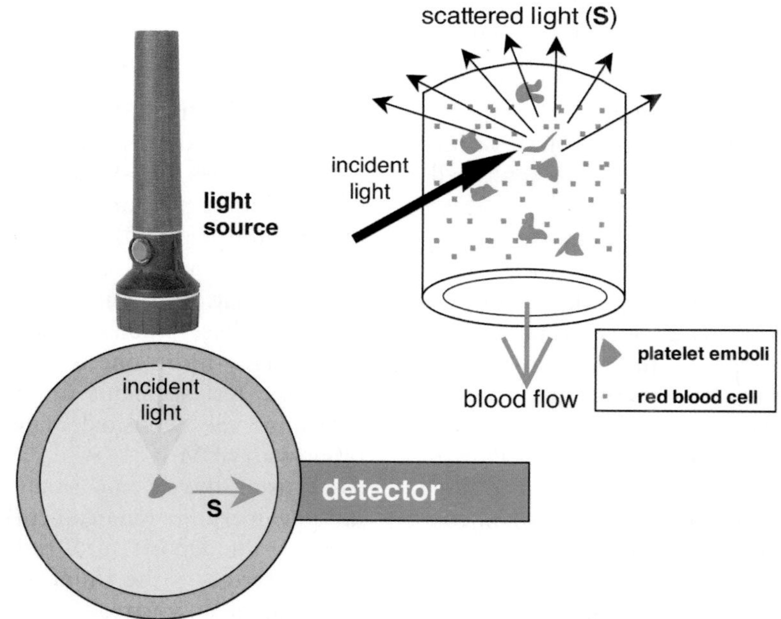

FIGURE II.3.5.10 A light scattering microparticle detector. (Based upon Solen et al., 2003.)

to accumulation of the emboli in downstream organs, which can be detected *in situ* using radiolabeled platelets and a gamma camera (Schneider et al., 1989).

What Materials Are Blood-Compatible?

The biomaterials community still awaits clear guidelines as to which materials are blood-compatible. But, from the perspective of one author of this article (Ratner), here are a few guidelines that you might use. First, the important distinction must be made that materials will perform differently in low wall shear rate blood flow (venous flow or static blood) and high wall shear rate blood flow (arterial flow). For low wall shear rate materials, heparinized surfaces are often good (although their period of performance may be limited), and hydrogels seem good. Excellent non-fouling surfaces are also promising (see Chapter I.2.10). For arterial blood flows, highly hydrophobic materials often show promise (silicones, fluoropolymers, and materials with surface-localized, long *n*-alkyl chains). Also, the exceptional non-fouling materials seem promising – these include good quality poly(ethylene glycol) surfaces, phosphatidyl choline-based surfaces, sulfobetaines, and carboxybetaines (see Chapter I.2.10). In all cases, the choice of materials will be dictated by the nature of the flow(s) in the actual device in the bloodstream, the mechanical properties required, the length of time in contact with the blood, the condition of the blood, and the consequences of embolization at the specific anatomical site. Trial and error optimization of suitable candidate materials may be needed to achieve performance appropriate to the specific application.

CONCLUSIONS

The most blood-compatible material known is the natural, healthy, living lining of our blood vessels. This "material" functions well by a combination of appropriate surface chemistries, good blood flow characteristics, and active biochemical processes involving removal of prothrombotic substances and secretion of natural anticoagulants. It seems unlikely that we will ever match this performance in a synthetic material or device, although attempts to imitate aspects of the natural system represent a promising strategy for developing a new generation of blood-compatible devices (for example see Chapters I.2.7 and I.2.8). At present, however, synthetic materials that perform less well than the vessel wall, but still satisfactorily, are needed. This chapter provides only a brief outline of the issues involved in evaluating materials and devices to find those that are minimally damaging or activating towards blood. The subject of blood-compatibility testing is complex, and advanced study of the subject is required before considering experiments intended to elucidate basic mechanisms or improve human health. Further discussion elaborating upon the complexity of the issues involved in BMI testing can be found in Ratner, 1984, 1993, and 2000. Detailed discussion on the characterization of materials for biomaterials application and on BMI testing can also be found in a publication coordinated by the Device and Technology Branch of the National Heart, Lung and Blood Institute, NIH (Harker et al., 1993). Also, see ISO standard 10993-4 (Biological evaluation of medical devices – Part 4: Selection of tests for interactions with blood). The importance of ISO standards is elaborated upon in Chapter II.2.2.

BIBLIOGRAPHY

Basmadjian, D., Sefton, M. V., & Baldwin, S. A. (1997). Coagulation on biomaterials in flowing blood: Some theoretical considerations. *Biomaterials*, **18**(23), 1511–1522.

Blit, P. H., McClung, W. G., Brash, J. L., Woodhouse, K. A., & Santerre, J. P. (2011). Platelet inhibition and endothelial cell adhesion on elastin-like polypeptide surface modified materials *Biomaterials*, **32**(25), 5790–5800.

Cao, L. (2006). *Tetraglyme glow discharge plasma modified biomaterials with ultralow protein adsorption and improved hemocompatibility*. PhD Thesis, University of Washington.

Cao, L., Ratner, B. D., & Horbett, T. A. (2006). Glow discharge plasma treatment of polyethylene tubing with tetraglyme results in ultralow fibrinogen adsorption and greatly reduced platelet adhesion. *J. Biomed. Mater. Res.*, **79A**, 788–803.

Cao, L., Chang, M., Lee, C. -Y., Castner, D., Sukavaneshvar, S., et al. (2007). Plasma deposited tetraglyme surfaces greatly reduce total blood protein adsorption, contact activation, platelet adhesion, platelet procoagulant activity, and *in vitro* thrombus deposition. *J. Biomed. Mater. Res.*, **81A**, 827–837.

Friedman, L. I., Liem, H., Grabowski, E. F., Leonard, E. F., & McCord, C. W. (1970). Inconsequentiality of surface properties for initial platelet adhesion. *Trans. Am. Soc. Artif. Intern. Organs*, **16**, 63–70.

Garfinkle, A. M., Hoffman, A. S., Ratner, B. D., Reynolds, L. O., & Hanson, S. R. (1984). Effects of a tetrafluoroethylene glow discharge on patency of small diameter Dacron vascular grafts. *Trans. Am. Soc. Artif. Int. Organs*, **30**, 432–439.

Gemmell, C. H. (2000). Flow cytometric evaluation of material-induced platelet and complement activation. *J. Biomater. Sci. Polymer Edn.*, **11**(11), 1197–1210.

Gemmell, C. H., Ramirez, S. M., Yeo, E. L., & Sefton, M. V. (1995). Platelet activation in whole blood by artificial surfaces: Identification of platelet-derived microparticles and activated platelet binding to leukocytes as material induced activation events. *J. Lab. Clin. Med.*, **125**, 276.

Gemmell, C. H., Yeo, E. L., & Sefton, M. V. (1997). Flow cytometric analysis of material-induced platelet activation in a canine model: Elevated microparticle levels and reduced platelet lifespan. *J. Biomed. Mater. Res.*, **37**, 176–181.

Grabowski, E. F., Herther, K. K., & Didisheim, P. (1976). Human versus dog platelet adhesion to cuprophane under controlled conditions of whole blood flow. *J. Lab. Clin. Med.*, **88**, 368–373.

Hanson, S. R., & Harker, L. A. (1988). Interruption of acute platelet-dependent thrombosis by the synthetic antithrombin D-phenylalanyl-L-prolyl-L-arginyl chloromethylketone. *Proc. Natl. Acad. Sci. USA*, **85**, 3184–3188.

Hanson, S. R., & Sakaraissen, K. S. (1998). Blood flow and antithrombotic drug effects. *Am. Heart. J.*, **135**, S132–S145.

Hanson, S. R., Harker, L. A., Ratner, B. D., & Hoffman, A. S. (1980). *In vivo* evaluation of artificial surfaces using a nonhuman primate model of arterial thrombosis. *J. Lab. Clin. Med.*, **95**, 289–304.

Hanson, S. R., Kotze, H. F., Pieters, H., & Heyns, A duP (1990). Analysis of 111-Indium platelet kinetics and imaging in patients with aortic aneurysms and abdominal aortic grafts. *Arteriosclerosis*, **10**, 1037–1044.

Harker, L. A., Kelly, A. B., & Hanson, S. R. (1991). Experimental arterial thrombosis in non-human primates. *Circulation*, **83**(6). IV-41–IV-55.

Harker, L. A., Ratner, B. D., & Didisheim, P. (1993). Cardiovascular biomaterials and biocompatibility: A guide to the study of blood–material interaction. *Cardiovasc. Pathol.*, **2**(Suppl. 3).

Haycox, C. L., & Ratner, B. D. (1993). *In vitro* platelet interactions in whole human blood exposed to biomaterial surfaces: Insights on blood compatibility. *J. Biomed. Mater. Res.*, **27**, 1181–1193.

Haycox, C. L., Ratner, B. D., & Horbett, T. A. (1991). Photoenhancement of platelet adhesion to biomaterial surfaces observed with epi-fluorescent video microscopy (EVM). *J. Biomed. Mater. Res.*, **25**, 1317–1320.

Horbett, T. A., Cheng, C. M., Ratner, B. D., Hoffman, A. S., & Hanson, S. R. (1986). The kinetics of baboon fibrinogen absorption to polymers: *In vitro* and *in vivo* studies. *J. Biomed. Mater. Res.*, **20**, 739–772.

Kaelble, D. H., & Moacanin, J. (1977). A surface energy analysis of bioadhesion. *Polymer*, **18**, 475–482.

Kottke-Marchant, K., Anderson, J. M., Rabinowitch, A., Huskey, R. A., & Herzig, R. (1985). The effect of heparin vs. citrate on the interaction of platelets with vascular graft materials. *Thromb. Haemost.*, **54**, 842–849.

Leonard, E. F. (1987). Rheology of Thrombosis. In R. W. Colman, J. Hirsh, V. J. Marder, & E. W. Salzman (Eds.), *Hemostasis and Thrombosis* (2nd ed., pp. 1111–1122). Philadelphia, PA: JB Lippincott.

McIntire, L. V., Addonizio, V. P., Coleman, D. L., Eskin, S. G., Harker, L. A., et al. (1985). *Guidelines for Blood–Material Interactions – Devices and Technology Branch, Division of Heart and Vascular Diseases, National Heart, Lung and Blood Institute*. NIH Publication No. 85–2185, revised July, 1985, U.S, Washington, DC: Department of Health and Human Services.

Münch, K., Wolf, M. F., Gruffaz, P., Ottenwaelter, C., Bergan, M., et al. (2000). Use of simple and complex *in vitro* models for multiparameter characterization of human blood–material/device interactions. *J. Biomater. Sci. Polymer Edn.*, **11**(11), 1147–1163.

Ratner, B. D. (1984). Evaluation of the Blood Compatibility of Synthetic Polymers: Consensus and Significance. In J. W. Boretos, & M. Eden (Eds.), *Contemporary Biomaterials: Materials and Host Response, Clinical Applications, New Technology and Legal Aspects* (pp. 193–204). Park Ridge, NJ: Noyes Publications.

Ratner, B. D. (1993a). The blood compatibility catastrophe. *J. Biomed. Mater. Res.*, **27**, 283–287.

Ratner, B. D. (1993b). Characterization of biomaterial surfaces. *Cardiovasc. Pathol.*, **2**(Suppl. 3), 87S–100S.

Ratner, B. D. (2000). Blood compatibility – A perspective. *J. Biomater. Sci. Polymer Edn.*, **11**(11), 1107–1119.

Ratner, B. D. (2007). The catastrophe revisited: Blood compatibility in the 21st century. *Biomaterials*, **28**(34), 5144–5147.

Reynolds, L. O., & Simon, T. L. (1980). Size distribution measurements of microaggregates in stored blood. *Transfusion*, **20**(6), 669–678.

Salzman, E. W., & Merrill, E. D. (1987). Interaction of Blood with Artificial Surfaces. In R. W. Colman, J. Hirsh, V. J. Marder, & E. W. Salzman (Eds.), *Hemostasis and Thrombosis* (2nd ed., pp. 1335–1347). Philadelphia, PA: JB Lippincott.

Sakariassen, K. S., Muggli, R., & Baumgartner, H. R. (1989). Measurements of platelet interaction with components of the vessel wall in flowing blood. *Methods Enzymol.*, **169**, 37–70. 1989.

Schaub, R. D., Kameneva, M. V., Borovetz, H. S., & Wagner, W. R. (2000). Assessing acute platelet adhesion on opaque metallic and polymeric biomaterials with fiber optic microscopy. *J. Biomed. Mater. Res.*, **49**, 460–468.

Schneider, P. A., Kotze, H. F., Heyns, A. P., & Hanson, S. R. (1989). Thromboembolic potential of synthetic vascular grafts in baboons. *Journal of Vascular Surgery*, **10**(1), 75–82.

Sefton, M. V., Gemmell, C. H., & Gorbett, M. B. (2000). What really is blood compatibility? *J. Biomater. Sci. Polymer Edn.*, **11**(11), 1165–1182.

Sefton, M. V., Sawyer, A., Gorbet, M., Black, J. P., Cheng, E., et al. (2001). Does surface chemistry affect thrombogenicity of surface modified polymers? *J. Biomed. Mater. Res.*, **55**(4), 447–459.

Sivaraman, B., & Latour, R. A. (2010). The relationship between platelet adhesion on surfaces and the structure versus the amount of adsorbed fibrinogen. *Biomaterials*, **31**(5), 832–839.

Solen, K., Sukavaneshvar, S., Zheng, Y., Hanrahan, B., Hall, M., et al. (2003). Light-scattering instrument to detect thromboemboli in blood. *J. Biomed. Optics*, **8**(1), 70–79.

Turitto, V. T., & Baumgartner, H. R. (1987). Platelet–Surface Interactions. In R. W. Colman, J. Hirsh, V. J. Marder, & E. W. Salzman (Eds.), *Hemostasis and Thrombosis* (2nd ed., pp. 555–571). Philadelphia, PA: JB Lippincott.

Wagner, W. R., & Hubbell, J. A. (1990). Local thrombin synthesis and fibrin formation in an *in vitro* thrombosis model result in platelet recruitment and thrombus stabilization on collagen in heparinized blood. *J. Lab. Clin. Med.*, **116**, 636–650.

Wagner, W. R., Schaub, R. D., Sorensen, E. N., Snyder, T. A., Wilhelm, C. R., et al. (2000). Blood biocompatibility analysis in the setting of ventricular assist devices. *J. Biomater. Sci. Polymer Edn.*, **11**(11), 1239–1259.

Wang, M., & Tsai, W. (2010). *Biomaterials in Blood Contacting Devices: Complications and Solutions*. Hauppauge, NY: Nova Scientific Publishers.

Waples, L. M., Olorundare, O. E., Goodman, S. L., Lai, Q. J., & Albrecht, R. M. (1996). Platelet–polymer interactions: Morphologic and intracellular free calcium studies of individual human platelets. *J. Biomed. Mater. Res.*, **32**, 65–76.

Węgrzyn, W., Jakieła, B., & Sanak, M. (2010). Assessment of hemocompatibility of materials with arterial blood flow by platelet functional tests. *Bull. Pol. Acad. Sci: Tech. Sci.*, **58**(2), 317–322.

Williams, D. (1987). *Blood Compatibility*. Boca Raton, FL: CRC Press.

CHAPTER II.3.6 ANIMAL SURGERY AND CARE OF ANIMALS

David Lee-Parritz
Department of Comparative Medicine, Genzyme Corporation, Framingham, MA, USA

INTRODUCTION

The appropriate use of animal models in biomaterials research allows prospective, controlled evaluation of disease processes and candidate therapeutics in a manner that is impossible in human patients or volunteers. Investigators must consider scientific, practical, and humane issues when developing studies that use animals. High-quality research requires close collaboration between veterinary and research professionals to guide model selection and development, minimize animal pain and distress, and to advance the scientific goals of the project. This chapter briefly reviews ethical and regulatory issues, describes available information resources and discusses the design of surgical facilities, animal selection, anesthesia, and analgesia.

ETHICAL AND REGULATORY OVERVIEW

Investigators and research institutions have an ethical and legal responsibility to consider animal welfare concerns in research using animals. Russell and Burch expressed the most widely understood ethical principles governing humane design of experiments using animals in 1959, summarized in the concept of the "three Rs:" *replacement*, *reduction*, and *refinement* (Russell and Burch, 1959). The first principle, *replacement*, states that non-animal models should be used instead of animals to the maximum extent possible. Although definitive safety and efficacy evaluation requires animal models, *in vitro* and *ex vivo* biocompatibility and efficacy screening methods are increasingly common in biomaterials research (Fujibayashi et al., 2003; Kirkpatrick et al., 2007; Pariente et al., 2000). These techniques reduce the total number of animals used, and may allow significant savings in time and resources compared to experimental surgical models. The principle of *reduction* states that investigators should use the minimum number of animals to allow statistically valid inferences to be drawn from the data. In this context, it is as important to avoid using too few animals as too many, because a study that is too small may require repetition. Important considerations in determining the statistical power of an animal experiment include inherent variability of the model, and expected efficacy of the test therapeutic. Consultation with a biostatistician may be useful to estimate required animal numbers. The most important of the "three Rs" is the principle of *refinement*, which states that investigators should use the least invasive and most modern techniques possible to minimize animal pain and distress (Orlans et al., 1998). Continuous refinement and improvement in animal husbandry and the diagnosis and control of infectious disease has greatly reduced nonexperimental morbidity and mortality in modern research facilities. Surgical models in particular have benefitted from advances in veterinary anesthesia, instrumentation, and monitoring, which have allowed further reduction and refinement of research animal use.

Widespread acceptance of the "three Rs" and continued public scrutiny of biomedical research oblige investigators and institutions to comply with strict regulatory standards governing all aspects of research animal use. In the United States, the United States Department of Agriculture (USDA), Public Health Service (PHS), and the Food and Drug Administration (FDA) are the primary agencies which regulate animal research. Federal animal welfare regulations embody the *US Government Principles for the Utilization and Care of Vertebrate Animals Used in Testing, Research, and Training* (National Research Council, 2010). *The Guide for the Care and Use of Laboratory Animals* (Guide) (National Research Council, 2010) elaborates upon the US Government Principles, and provides important reference materials to assist implementation. Federal regulations and the Guide assign primary responsibility for an animal research program to the Institutional Animal Care and Use Committee (IACUC). The institution's chief executive officer appoints the IACUC, which must minimally include a veterinarian, a scientist, a nonscientist, and an unaffiliated member. One member may fulfill more than one of these requirements. The veterinarian must have appropriate training and experience for the species and models in use. The IACUC has multiple responsibilities. It reviews and approves in advance all animal research protocols to ensure compliance with the US Government Principles. In addition, the IACUC regularly inspects the animal research facility, reviews the animal care and use program, and investigates allegations of animal mistreatment.

The specific focus of animal welfare regulations varies by agency. Well-publicized incidents (Wayman, 1966) of pet theft and their subsequent mistreatment in research laboratories prompted the United States Congress to enact the Animal Welfare Act (AWA) in 1966. The AWA required the Federal government to develop a mechanism to protect animals used in research, for exhibition, and sold as pets in interstate commerce. The AWA does not regulate the use of rats, mice or birds bred for research. In many institutions, the term "regulated species" refers to animals subject to the AWA. The USDA enforces the AWA through the Animal and Plant Health Inspection Service (APHIS). Registered facilities file annual reports of animal usage with USDA, and are subject to unannounced inspections of facilities and records by APHIS inspectors.

The Good Laboratory Practices (GLP) Act (21CFR58.90) requires the United States Food and Drug Administration to regulate preclinical research conducted in direct support of new drugs or medical devices.

The primary goal of the GLP regulations is to ensure reproducibility and integrity of data used to support new drug or device applications. The animal care provisions of the GLP are general. Animals must be "free of any disease or condition that might interfere … with the study." Standard operating practices (SOPs) must govern housing, feeding, and care of animals, and the food, water, and environment must be free of known contaminants.

The United States Public Health Service (PHS), through the National Institutes of Health (NIH), is the largest single sponsor of biomedical research using animals, and has adopted regulations to ensure humane and scientifically valid use of animals. The NIH Office of Laboratory Animal Welfare (OLAW) develops and implements standards for animal care and use at recipient institutions, as formulated in the Public Health Service Policy on Humane Care and Use of Laboratory Animals. Institutions receiving PHS support must file an "Animal Welfare Assurance" with OLAW indicating sufficient institutional resources to provide proper husbandry and veterinary care. Institutions must also indicate how they will approve and monitor research to ensure scientific integrity and prevent inhumane treatment of animals. Assurances are approved for five years, after which a new application must be submitted. The PHS Policy applies to all vertebrate animals, even those not regulated by the AWA (Office for the Protection from Research Risks, 1996).

The Association for the Assessment and Accreditation of Laboratory Animal Care, International (AAALAC) is a nongovernmental independent organization that assists and accredits research institutions seeking to maintain the highest standards of laboratory animal care and use. Approximately 770 research facilities in the United States and in 29 other countries hold AAALAC accreditation. AAALAC policies are determined by the Board of Trustees, representing more than 50 scientific, educational, and professional organizations involved with biomedical research use of animals. AAALAC uses the Guide and applicable governmental guidelines to evaluate the quality of an institutional animal care and use program. The AAALAC Board appoints the Council on Accreditation which reviews, grants or suspends accreditation based on a triennial program review and site visit by at least one Council member and an *ad hoc* consultant. OLAW accepts AAALAC accreditation as strong evidence that an institution's animal care and use program is in substantial compliance with the PHS Policy. AAALAC guidelines apply to all live vertebrate animals used in research.

Research using endangered species is generally allowed only for the direct benefit of animals with spontaneous disease. Although rare in biomedical research laboratories, the use of endangered species is regulated by the Endangered Species Act. Use of endangered species is governed by the Convention on International Trade in Endangered Species (CITES), and regulated by the Fish and Wildlife Service of the US Department of the Interior. The provisions of CITES apply to all animals, living or dead. Other agencies which regulate the use of research animals include the Centers for Disease Control (importation of non-human primates and other animals which may harbor zoonotic agents) and the International Air Transport Association (transportation of living animals by air) (National Research Council, 2010). The use of narcotics and barbiturates in experimental anesthesia and analgesia is regulated by the US Drug Enforcement Administration (DEA), which requires users to obtain proper permits, ensure appropriate drug security, and maintain proper records.

INFORMATION RESOURCES

Research animals require specialized care that recognizes anatomic and physiologic similarities and differences between species. These differences determine routine husbandry and behavioral needs, and may influence model selection and interpretation of data. Animals vary greatly in their response to anesthetics and other drugs, and anatomical differences often pose significant challenges to endotracheal intubation, surgical approach, and intravenous access.

Veterinarians offer important expertise to investigators using animals. Laboratory animal medicine is a subspecialty of veterinary medicine which involves additional training and acquisition of skills in the diagnosis and treatment of laboratory animal diseases, development of new experimental techniques, and provision of specialized preoperative, intraoperative, and postoperative care. Most large research institutions have at least one staff veterinarian trained and experienced in laboratory animal medicine. Smaller institutions may use consultant veterinarians. Large research programs may also include veterinarians as part of the research team. Other veterinary specialties of potential value in a biomaterials research program include surgery, anesthesia, pathology, and internal medicine. The Academy of Surgical Research (ASR) and the American Association of Laboratory Animal Science (AALAS) sponsor technician training and certification programs in routine and specialized animal care and use. Certified technicians and technologists can provide valuable support and management roles in the experimental surgery laboratory.

Many information resources are available to biomaterials investigators using animals. Several excellent textbooks discuss general laboratory animal medicine (Fox et al., 2002) and basic biology and methodology for rodents (Sharp and LaRegina, 1998), rabbits (Manning et al., 1994; Harkness and Wagner, 1995; Suckow and Douglas, 1997), swine (Bollen et al., 2000; Swindle, 2007) and small ruminants (Borkowski and Allen, 1999). Specialized texts provide detailed reviews of anesthetic techniques (Tranquilli et al., 2007; Fish et al., 2008), infectious disease (National Research Council,

1991; Straw et al., 2006), and drug dosages (Hawk et al., 2005; Plumb, 2005) for experimental animals. Finally, the Internet has excellent resources for laboratory animal users, including reference material, bibliographical databases, and discussion groups on a variety of technical and regulatory topics. The NetVet portal is an excellent entry point to web-based laboratory animal resources (http://netvet.wustl.edu/vet.htm). Other useful websites include OLAW (http://grants.nih.gov/grants/olaw) and the Animal Welfare Information Center (http://awic.nal.usda.gov/).

SURGICAL FACILITY DESIGN

Quality surgical research requires the use of appropriately designed, equipped, and managed surgical facilities. The Guide requires the use of aseptic technique for all survival surgical procedures on laboratory animals, including rodents. Essential components of aseptic techniques include patient preparation (clipping and disinfection of the surgical site), surgeon preparation (surgical attire, surgical hand scrub, sterile gloves), and the use of sterile instruments. Effective training and staffing levels are necessary to reduce the chance of contamination of instruments and the sterile field (draping, traffic control) (National Research Council, 2010).

Surgical suites represent significant institutional investments that must remain functional for many years. Design and configuration of operating room facilities requires careful planning and consultation with users, veterinarians, laboratory planners, and engineers to meet current scientific and regulatory requirements with appropriate flexibility for possible future changes.

Minimum functional components of the survival surgery suite include dedicated areas for surgery, animal preparation, surgeon's scrub, postoperative recovery, and surgical support. In the United States, Federal regulations require the use of dedicated surgical suites for major survival surgery on regulated species other than rats, mice, and birds. Surgical suites must allow aseptic surgery, animal preparation, and surgeon scrub to occur in separate dedicated rooms, arranged to facilitate entry and exit of animals and staff while minimizing unnecessary traffic. Animal recovery may often occupy the same room as anesthesia induction and preparation. The Guide permits rodent survival surgery in any procedure space provided necessary conditions for asepsis are present (National Research Council, 2010).

Many research institutions maintain centralized experimental surgery suites to maximize economies of scale and efficient use of skilled support staff. Support spaces in large surgical suites are generally configured to allow two or more operating rooms to run simultaneously from shared animal and surgeon preparation rooms. Smaller specialized facilities may be appropriate to accommodate unique experimental requirements. Principles of research animal operating room design have been described (Ruys, 1991; Hessler and Lehner, 2008; National Research Council, 2010).

Other functional areas commonly required in an experimental surgery suite are: pharmacy, including secure storage of controlled substances; instrument cleaning and sterilization; and record keeping. The use of inhalant anesthesia requires appropriate active scavenging capability. Secure data collection and storage require network connections, and possibly other information technology capabilities. Radiography and other imaging modalities are often necessary in experimental surgical studies, and may be conveniently located within or adjacent to the surgery suite.

PREOPERATIVE PREPARATION AND MONITORING

Careful preoperative preparation is necessary before invasive surgical procedures. Preoperative preparation includes fasting, premedication, establishment of vascular access, and surgical site preparation. Removal of food and contact bedding is necessary for dogs, swine, and ruminants to prevent regurgitation and aspiration pneumonia. Gastric dilatation may also occur in improperly fasted swine and ruminants. Preoperative fasting is not routine in rodents and rabbits. These species have a high metabolic rate, and are prone to hypoglycemia from prolonged withdrawal of food. Furthermore, they do not vomit and therefore are unlikely to develop aspiration pneumonia. Fasted or not, many species hoard food between the upper molars and the cheek. For this reason the anesthetist should carefully examine the oral cavity during anesthetic induction and remove materials which can be aspirated or pushed into the trachea during endotracheal tube placement. Premedication with analgesic and sedative drugs reduces animal anxiety, facilitates anesthetic induction, and prevents postoperative hyperalgesia and allodynia.

Hypothermia impairs wound healing, and predisposes animals to infection and other postoperative complications. Anesthesia itself is a major risk factor for hypothermia because anesthetic-induced peripheral vasodilation increases conductive heat loss. Many anesthetic and analgesic agents interfere with hypothalamic thermoregulatory systems and the shivering reflex. Cold dry anesthetic gases promote evaporative heat loss from the respiratory tract. Prolonged exposure of the abdomen or thorax during surgery allows heat loss through radiation. Small animals such as rats and mice develop hypothermia more rapidly than larger animals, because they have a large surface area in proportion to body weight. A comprehensive approach to maintenance of normothermia should include minimization of surgical time and exposure of body cavities, and the use of warmed fluids and humidified gases. Care and frequent body temperature monitoring are necessary to prevent hyperthermia, which can rapidly be fatal.

Supplementary heat sources including heated surgical tables and circulating warm air or water blankets are extremely valuable. A recent report evaluated rectal temperature in isoflurane-anesthetized Sprague-Dawley rats and CF-1 mice placed on a circulating warm water blanket or a reusable microwavable warming pad (SnuggleSafe®). Control animals were placed on a stainless steel surgical table with a surface temperature of 21°C–22°C. Surgery was not performed. All control animals became severely hypothermic by the end of the 60 minute test period, and mice lost significantly more heat than rats (−9.9°C versus −4.42°C from baseline). Animals which received supplemental heat support from either source remained essentially normothermic throughout the study (Taylor, 2007).

Preoperative placement of a venous cannula and intraoperative intravenous fluid administration maintains tissue perfusion and electrolyte balance, and allows delivery of emergency drugs when necessary. Appropriate fluids and administration rates for intraoperative maintenance include lactated Ringers solution or normal saline, 5–10 ml/kg/hr. Intraoperative hydration allays dehydration if delayed recovery or physiologic compromise delays spontaneous re-feeding after surgery. Intraperitoneal or subcutaneous fluid administration is a practical alternative to the intravenous route in low risk rats and mice, because intravenous cannulation is difficult in these species.

Minimum monitoring should include core temperature, heart rate, respiratory rate, oxygen saturation, and end-tidal carbon dioxide concentration. Cost-effective devices readily allow determination of these parameters for rabbits and larger animals. Specialized devices are available for reliable assessment of rats and mice. Monitoring of additional parameters such as electrocardiogram, expired anesthetic agent concentration, and arterial blood gases or hematologic or serum biochemical parameters may be indicated for specific studies and will require specialized instrumentation and training.

ANESTHESIA

Overview

Anesthesia provides reversible elimination of pain and is essential to all surgical procedures. The ideal anesthetic preserves cardiac output and other physiologic parameters, provides intraoperative analgesia that extends to the postoperative period, and is reversible and safe for the patient, surgical personnel, and environment. Experimental animal surgery almost always occurs under general anesthesia, which additionally renders the animal unconscious and immobile. Local or regional anesthetic techniques are often useful to reduce general anesthetic and postoperative analgesic requirements in major procedures.

Anesthetic agents for experimental surgery may be administered by injection or inhalation. Several considerations should determine the selection of anesthetic agents for experimental surgery. Balanced anesthetic techniques combine two or more sedative, hypnotic, and analgesic agents to reduce adverse effects and increase physiologic stability and analgesia compared to single-agent methods. The Guide requires institutional approval and veterinary participation in the selection of anesthetic agents for experimental surgery (National Research Council, 2010).

Selection of anesthetic technique will vary with the experiment. The anesthetist must maintain oxygenation, perfusion, metabolic balance, and normal core body temperature. Severe or prolonged alterations of these parameters may impair wound healing, predispose to infection or introduce significant scientific variation. Most surgical studies use young adult healthy animals which present minimal anesthetic risk for brief procedures. Unexpected disease may occur in any animal, however, and for this reason all animals should receive appropriate physical examination before anesthetic induction. Extensive procedures or studies with diseased animals pose significant anesthetic challenges and generally require an anesthetist with specialized training and experience.

The American Society of Anesthesiologists (ASA) anesthetic risk assessment categories classify anesthetic risks based on the animal's age, underlying health status, and other risk factors. Anesthetic, intraoperative and postoperative care and resource requirements can then be determined appropriately. A six-month-old normal Sprague-Dawley rat undergoing implantation of a permanent venous catheter, for example, would be considered ASA category I and require standard anesthetic protocols. By contrast, an 18-month-old Sprague-Dawley rat with advanced experimental myocardial insufficiency following coronary artery ligation would be considered ASA category III or IV, and require correspondingly more intense anesthetic support and monitoring (Tranquilli et al., 2007).

Anesthetics are potent medications which may confound experimental results, either directly or through alteration of normal physiology. Investigators should select agents carefully, and maintain anesthetic and analgesic protocols constant across groups and between experiments to minimize these effects. Pilot studies may be necessary to identify the best agent for specific models, as the following recent report illustrates. Weanling swine were used to evaluate colorimetry, transepidermal water loss, and laser Doppler perfusion imaging to determine the progress of wound healing and efficacy of candidate treatments. Animals were anesthetized with each of three different protocols: ketamine/xylazine (K/X); tiletamine/zolazepam/xylazine (T/X); or tiletamine/zolazepam/xylazine/isoflurane (T/X/I) three times at 24–48 hour intervals, separated by a seven day washout period. The T/X and T/X/I regimens depressed transepidermal water loss less than did K/X. The mechanism was unknown, but possibly associated with differential effects of anesthetic drugs on cutaneous blood flow. The study also identified a progressive reduction in cutaneous blood flow readings during the three day experiment, presumably the result of anesthetic accumulation following multiple doses (Graham et al., 2004).

Injectable anesthesia is popular in experimental surgery. The principal benefits of injectable anesthetic techniques are ease of administration and operator safety. Injectable techniques usually use a combination of agents given by the intramuscular or, for rodents, intraperitoneal routes. Intravenous anesthesia, by continuous or intermittent bolus infusion, may be appropriate in larger animals. Most injectable agents or combinations provide 15–30 minutes of anesthesia. Significant individual, species, and strain variation in the response to injected anesthetics is common. Injectable anesthetics are often used to induce anesthesia before endotracheal intubation and maintenance inhalation anesthesia. Prolonged recovery and physiologic imbalance may follow repeated administration of injectable agents to maintain anesthesia during long procedures. As reversal of injectable agents is often impossible, the operator must avoid administration of an accidental overdose. In addition, the operator must be careful not to breach aseptic technique when administering supplemental anesthetic. Many injectable anesthetic agents have a moderate-to-high abuse potential; therefore, investigators who use these agents in research must register with the United States Drug Enforcement Administration, and maintain proper records and storage facilities.

Inhalation anesthesia, usually with isoflurane or sevoflurane, allows the operator a high degree of control over anesthetic depth, and is the technique of choice for prolonged or invasive surgery. Inhalation anesthesia for major surgery requires a precision vaporizer and endotracheal intubation to allow proper control of anesthetic depth and the airway. Controlled ventilation further facilitates maintenance of normal tissue oxygenation and carbon dioxide balance, and allows rapid increase or decrease of anesthetic depth when necessary. Controlled ventilation is mandatory for intrathoracic procedures, because many laboratory animals lack a complete mediastinum and develop bilateral pneumothorax following thoracotomy. Endotracheal intubation requires special laryngoscopic equipment and training to prevent laryngeal trauma or esophageal placement.

Anesthetic administration by facemask may be appropriate for brief or noninvasive procedures. Prolonged inhalation by mask is undesirable because it does not protect the airway from aspiration in the event of vomiting or allow positive pressure ventilation in an emergency. A tight fitting mask and active scavenging are necessary to maintain anesthetic depth and reduce operator exposure to waste anesthetic agents.

Anesthetic Agents

Ketamine and Tiletamine. Ketamine and tiletamine are dissociative anesthetics commonly used in experimental surgery because they maintain cardiac output and provide good analgesia with a high margin of safety. Dissociative anesthetics are NMDA (N-Methyl-D-aspartate) antagonists which produce unconsciousness and analgesia through selective disruption of ascending impulses to conscious brain centers, rather than through generalized depression of the central nervous system. Dissociative anesthetics are almost always used in combination with other agents. When used as a sole agent, surgical anesthesia is rarely achievable at safe dosage levels. Other significant disadvantages include unacceptable muscle rigidity, salivation, and seizures, especially in dogs (Lin, 2007; Tranquilli et al., 2007; Meyer and Fish, 2008).

Ketamine solutions are acidic, and perivascular infiltration and large volume intramuscular injection will produce pain and tissue irritation. Telazol® is a commercial mixture of the dissociative agent tiletamine and the benzodiazepine zolazepam. The product is supplied in a sterile vial, which is reconstituted with 5 ml of sterile water. Reconstituted vials contain 50 mg/ml of each agent and may be kept for 48 hours at room temperature or for 14 days in a refrigerator. Telazol® has similar anesthetic efficacy as ketamine/xylazine or ketamine/diazepam. Telazol® may also be combined with xylazine or medetomidine for greater anesthetic depth.

Sedatives. Xylazine (Rompun®), medetomidine (Domitor®) and dexmedetomidine (Dexdomitor®) are common adjuncts to ketamine anesthesia. These agents are alpha-2 adrenergic agonists with potent sedative and analgesic activity. They lack adequate potency to be sole agents for general anesthesia, but may provide adequate sedation for dressing changes, suture removal, and other minor procedures. Coadministration with ketamine generally provides up to 30 minutes of anesthesia with excellent muscle relaxation, smooth recovery, and a moderate degree of postoperative analgesia in many species. These combinations are also useful for anesthetic induction prior to endotracheal intubation and subsequent inhalation anesthesia. Intramuscular administration is preferable, because transient hypertension and cardiac arrythmia can occur after rapid intravenous boluses. Xylazine is the most commonly used alpha-2 agonist in rodents. Medetomidine has increased specificity for the alpha-2 receptor, and increased analgesic potency compared to xylazine. Medetomidine is preferred over xylazine in dogs (Lemke, 2007). Medetomidine is a racemic mixture of the active R and inactive L isomers. Dexmedetomidine is a pure preparation of the R-isomer and produces physiologic and analgesic effects similar to medetomidine at approximately half the dose.

Several side-effects are associated with alpha-2 agonists and are more significant following IV administration. Xylazine commonly induces vomiting in dogs. All alpha-2 agonists significantly depress cardiac output at standard anesthetic doses and may cause severe bradycardia. Cardiac depression is frequently subclinical in young healthy animals. Decompensation may occur in aged animals or those with clinical or experimentally induced illness. The mechanism of alpha-2-agonist associated bradycardia is complex and may require combined anticholinergic (atropine or glycopyrollate) and specific antagonist (yohimbine or atipamezole) therapy.

A distinct advantage of the alpha-2 agonist sedatives is the availability of specific antagonist drugs. Yohimbine effectively reverses xylazine, and atipamezole is the recommended reversal agent for medetomidine (Lemke, 2007). These drugs effectively speed recovery following minor procedures, and may also be used in an anesthetic emergency.

Other adjuncts may also be used to allay animal anxiety and provide muscle relaxation. Acepromazine provides excellent sedation in many species, and protects the myocardium from catecholamine-induced arrhythmias. Disadvantages of acepromazine include hypotension secondary to adrenergic blockade and prolonged recovery. Benzodiazepines such as diazepam, midazolam, and zolazepam provide excellent short-term sedation with little effect on blood pressure (Lemke, 2007).

Propofol. Propofol is a non-narcotic, injectable sedative and hypnotic agent commonly used for anesthetic induction and for brief procedures in dogs, swine, and small ruminants. Propofol is insoluble in water and is available as an oil-based emulsion in single-dose ampules and multiple-dose vials for intravenous administration. Bacterial growth will occur rapidly in contaminated solutions, and vials should be used or discarded the day they are opened. Anesthetic induction and recovery are smooth. The duration of anesthesia following a single bolus intravenous injection in dogs is approximately 20 minutes. Prolonged anesthesia is possible using intermittent bolus injections or continuous rate infusions. Propofol has little analgesic activity, and should be used in combination with narcotics or alpha-2 agonists for painful procedures. Side-effects of propofol include apnea, bradycardia, and hypotension. The incidence of apnea is related to the total dose and rate of injection, and resolves spontaneously as the drug distributes. Manual ventilation may be necessary if apnea is prolonged. Human patients commonly report transient pain at the injection site. Unlike ketamine and the barbiturates, accidental perivascular propofol injection does not produce tissue damage (Branson, 2007).

Barbiturates. The barbiturates are among the oldest anesthetic agents and remain useful for some experimental applications. Barbiturates act through general depression of the central nervous system. There is dose-dependent depression of respiration and cardiac output. Hepatic metabolism terminates the activity of long-acting oxybarbiturates such as sodium pentobarbital. Rapid redistribution into adipose tissue followed by hepatic detoxification characterizes the metabolism of the short-acting thiobarbiturates thiamylal and thiopental.

The principal advantage of sodium pentobarbital for general anesthesia is the ability to rapidly induce deep anesthesia with a single agent. Several disadvantages limit sodium pentobarbital's utility in prolonged or invasive procedures. Progressive cardiovascular depression occurs with prolonged anesthesia. Despite sleeping times of 5–15 hours, surgical anesthesia is often present only for 30–60 minutes in most species because this agent provides very little analgesia. Stormy recovery with vocalization and an unstable gait is common. Rapid-acting intravenous thiobarbiturates provide about 10 minutes of general anesthesia, and are useful for anesthetic induction. Barbiturate solutions are strongly alkaline and require intravenous or intraperitoneal administration to avoid pain and tissue necrosis (Branson, 2007). The therapeutic index of most barbiturates is very low: most laboratories use these agents primarily for euthanasia.

Nonstandard Injectable Anesthetics. Specific experimental situations may require the use of nonstandard anesthetics. These agents include chloral hydrate, alpha-chloralose or urethane. Chloral hydrate and alpha-chloralose preserve motor and sensory nerve function, but are poor anesthetics with minimal analgesic effect except at very high dosages. Alpha-chloralose and urethane result in stable cardiovascular performance during prolonged anesthesia. Urethane is mutagenic and carcinogenic in experimental animals, and may be used only for nonsurvival procedures. Urethane may also be hazardous to research staff after prolonged contact. Because of these constraints, most IACUCs require investigators to justify the use of nonstandard agents on scientific grounds (Meyer and Fish, 2008).

Inhalation Anesthetics. Isoflurane and sevoflurane are the agents of choice for clinical and experimental inhalation anesthesia. These agents are poorly soluble in blood and undergo minimal metabolism. As a result, recovery after anesthetic discontinuation is rapid, even in animals with significant hepatic or renal impairment. Dose-dependent cardiac depression may occur during prolonged procedures. A balanced anesthetic technique combining narcotic analgesics and muscle relaxants to reduce the required concentration of isoflurane is suggested for long procedures on animals with cardiac disease. Unlike the obsolete agents halothane and methoxyflurane, isoflurane and sevoflurane do not sensitize the myocardium to catecholamine-induced arrhythmias (Steffey and Mama, 2007; Brunson, 2008). As sole agents, isoflurane and sevoflurane are relatively poor analgesics and should be used in combination with sedatives and narcotics or nonsteroidal anti-inflammatory drugs (NSAIDs) for optimal intraoperative and postoperative pain control. These anesthetics are halogenated drugs which must be scavenged with activated charcoal canisters or by connection to an active exhaust circuit (Steffey and Mama, 2007; Brunson, 2008).

Older inhalation anesthetics occasionally used in experimental surgery include diethyl ether, chloroform, and nitrous oxide. Although ether is inexpensive and readily available from chemical supply houses without prescription, significant disadvantages include flammability and the risk of explosion if peroxides form when ether evaporates to dryness. Mask induction with ether is unpleasant for the patient. High lipid solubility prolongs recovery and analgesia is poor. Repeated exposure to chloroform or methoxyflurane is toxic to the operator. Nitrous oxide provides slight analgesia when administered at high concentrations, but is not an effective anesthetic in animals. The maximum allowable concentration of nitrous oxide is 80% in oxygen, beyond which hypoxemia is likely. As activated charcoal does not absorb nitrous oxide, active exhaust scavengers are required when using nitrous oxide for anesthesia. Nitrous oxide diffuses out of the bloodstream to gas-filled body cavities such as the gastrointestinal tract. This property may cause significant distension of the rumen in sheep and goats. Most institutions strongly discourage the use of these agents without specific scientific justification (Brunson, 2008).

Bell jar administration of volatile anesthetics is no longer acceptable in most institutions. Chloroform, ether, and methoxyflurane have a low vapor pressure and achieve anesthetic concentrations when allowed to evaporate in a closed container, but these agents pose unacceptable risks to animals and operators. Isoflurane and similar modern agents are highly volatile and rapidly attain toxic levels in a nose cone or closed container. Investigators familiar with bell jar administration of obsolete agents commonly experience unacceptable animal mortality when attempting to use isoflurane in the same manner.

ANALGESIA

Prompt recognition and adequate treatment of postoperative pain are key responsibilities of experimental surgeons. Experimental surgical procedures rarely, if ever, directly benefit the research subject and impose a substantial moral requirement on investigators to prevent or minimize any discomfort related to the experiment. The scientific community has accepted that comparable mechanisms govern the production and response to pain in animals and humans, as the widespread use of animal models in pain research confirms. Animal welfare regulations and accreditation guidelines require the use of sedatives, analgesics or anesthetics for procedures which “may cause more than momentary or slight pain or distress” (Office for the Protection from Research Risks, 1996). Additionally, untreated pain increases catecholamine secretion and causes stress, which introduces experimental variation by impairing wound healing and immune function (Lee-Parritz, 2007). Animal welfare regulations require investigators to justify withholding of analgesics to the IACUC on scientific grounds.

Recognition of postoperative pain is difficult in most laboratory animal species. Anorexia, lethargy, piloerection, and wound hypersensitivity indicate the presence of moderate to severe pain, but absence of these signs does not necessarily prove that the animal is pain-free. Behavioral and physiological factors complicate the diagnosis of less severe pain, which may still require treatment to assure animal well-being. Rodents, for example, are nocturnal animals and are normally less active during the day than at night. Rodents typically nibble small amounts of food at frequent intervals. As a result, pain induced inhibition of activity or feeding is difficult for research staff to recognize. In addition, rodents issue distress calls at ultrasonic frequencies that are inaudible to human beings (Lee-Parritz, 2007). Most animals effectively conceal subtle signs of pain and distress from research staff. Animal husbandry staff may notice subtle signs of pain or distress in research animals that are not apparent to investigators. For these reasons, animal welfare regulations counsel investigators that surgical procedures that would induce pain in a human would produce pain in an experimental animal, unless convincing evidence to the contrary exists (National Research Council, 2010).

Narcotics remain the analgesics of choice for severe pain in most species. Adverse effects of narcotics include dose-dependent sedation, hypoventilation, anorexia, and constipation. These effects are more pronounced with pure agonists such as morphine than with partial agonists, such as butorphanol and buprenorphine. Buprenorphine is widely used in veterinary medicine because it provides excellent analgesia at a convenient dose interval (8–12 hours), and has minimal depressant effects on respiration and cardiac output (Lee-Parritz, 2007; Lamont and Mathews, 2007). Dosage and dose intervals of the narcotic analgesics vary widely between species because of the higher metabolic rate of the smaller species, as well as species-specific responses to the agents. In comparison to larger animals, rodents require very high doses of most narcotics in proportion to their body weight for effective analgesia (Lamont and Mathews, 2007; Lee-Parritz, 2007).

Fentanyl is a highly effective narcotic analgesic. Parenteral fentanyl has a very short half-life, and can be rapidly titrated to effect as a continuous rate infusion in a balanced anesthetic protocol. The fentanyl transdermal patch provides effective analgesia for up to 72 hours as a sole agent or in multimodal analgesic regimens. Several factors may affect transdermal fentanyl absorption, including temperature, skin thickness, and blood supply. Care and consistency in application is necessary to prevent toxicosis or inadequate analgesia. The patch should not be under an occlusive dressing, on a patient’s recumbent side or where it will be in contact with supplementary heat devices used during anesthesia, as these practices may cause significant increase in drug release from the patch. Patches should be placed at least 12 hours preoperatively to ensure adequate blood levels at the time of incision. Poorly applied patches may dislodge and be eaten, causing fatal fentanyl overdose. Facilities which use fentanyl patches should stock naloxone (0.1 mg/kg IV, repeated as necessary) for emergency use in case of accidental patch ingestion (Lamont and Mathews, 2007; Lee-Parritz, 2007).

Nonsteroidal anti-inflammatory drugs (NSAIDs) are most effective against moderate pain of musculoskeletal origin, including incisional pain from ventral abdominal incisions in swine and other large animals. NSAIDs are of particular value in orthopedic procedures. The principal side-effects of many NSAIDs include increased bleeding from reduced platelet activity or gastric irritation secondary to prostaglandin inhibition. Although clinically significant impairment of blood clotting ability is rare, gastric toxicity remains a concern, particularly in dogs (Lamont and Mathews, 2007; Lee-Parritz, 2007).

It is advisable to use multimodal analgesia consisting of local anesthetics, NSAIDs, and narcotics whenever possible. The use of agents from two or more analgesic classes will act at several complementary steps in the pain cascade. This strategy will generally have additive or synergistic activity with reduced potential for drug toxicity. Preoperative analgesic and local anesthetic administration will significantly reduce intraoperative anesthetic requirements, and likely reduce acute and chronic postoperative pain. Local anesthetics are most often administered to the incision or regionally at the time of surgery, but may be of significant value in the postoperative period as well.

The possible effect of analgesic treatment on wound healing is controversial. Most NSAIDs block COX-1 and COX-2 to a variable degree, and COX-2 plays an important role in wound healing. A study of the effect of continuous diclofenac administration for 10 days to rats with experimental dorsal skin incision demonstrated reduced fibroblast numbers in treated versus control rats, but no difference in epidermal thickness or clinical healing (Krischak et al., 2007). In another model, rats undergoing experimental medial collateral ligament transaction demonstrated improved healing at day 14 following 6 days' treatment with the NSAID piroxicam compared to controls. Animals which received naproxen, rofecoxib, acetaminophen or butorphanol at the same dose schedule demonstrated healing similar to controls (Hanson et al., 2005). A third study evaluated the effect of continuous celecoxib or indomethacin treatment versus controls on bone repair in a rat femur fracture model. Recipients of either NSAID had increased fibrous tissue compared to control at 4 and 8 weeks, but all groups demonstrated equivalent morphological and functional healing by 12 weeks (Brown et al., 2004). A rabbit spinal fusion model demonstrated equal fusion rate and strength at 8 weeks when animals received either ketoprofen or tramadol for 8 days after surgery (Urrutia et al., 2007). These results suggest that NSAIDs may have a small, agent-specific effect on experimental wound healing. The applicability of studies based on extended treatment protocols to the relatively brief treatment periods required for acute pain control requires further study. Narcotic analgesics have not been demonstrated to impair wound healing, and may actually speed recovery in specific settings (Peyman et al., 1994). Consistent administration of standard analgesic drugs across experimental groups, guided by pilot studies when necessary, should allow effective analgesic therapy without compromising the scientific validity of the study.

SPECIES-SPECIFIC RECOMMENDATIONS

The following sections provide general recommendations for the selection, care, and experimental use of laboratory animal species commonly used in biomaterials research. When appropriate, separate recommendations will be made for brief or noninvasive procedures. For detailed discussion of anesthesia for specific procedures or disease conditions, the reader is advised to consult the references or a veterinary specialist.

Rodent

Animal Selection and Preoperative Preparation. Specific pathogen-free rodents should be used for all surgical procedures to reduce morbidity and mortality from chronic respiratory disease. Infectious agents commonly implicated in chronic respiratory disease of rodents include *Mycoplasma pulmonis*, Sendai virus, and cilia-associated-respiratory (CAR) bacillus (National Research Council, 1991; Fox et al., 2002). Other infectious agents may alter immune function or impair detoxification of anesthetic or experimental drugs. Several commercial vendors supply common strains of laboratory rodents free from infection with these and other infectious agents. These animals are known as specific pathogen-free (SPF) animals, and should always be used for survival surgical procedures to minimize experimental variability and reduce surgical mortality. The specific panel of excluded agents may vary according to the vendor. The institution should adopt standard operating procedures to prevent introduction of rodent infectious agents, and should regularly survey all holding areas for evidence of infection. Introduction of SPF animals to rooms with enzootic viral infection may result in rapid onset of severe clinical disease. Cedar or pine shavings sometimes used for contact bedding contain aromatic compounds that induce hepatic microsomes and alter hepatic detoxification of anesthetics (Gaertner et al., 2008). To ensure uniform response to anesthetics and experimental drugs, rodents should receive heat-treated wood chip, corncob or cellulose bedding. A conditioning period of at least three days after purchase will ensure that animals have recovered from dehydration and stress associated with shipping. Rodents do not vomit, and preoperative fasting is not recommended. If otherwise indicated, fasting in rodents should be kept to a minimum (2–3 hours) to avoid hypoglycemia and shock (Flecknell et al., 2007).

General Anesthesia. Most surgical procedures in rodents are brief. Small body size and limited vascular access complicate anesthesia and intraoperative support of rodents. Nevertheless, skilled operators with appropriate instrumentation can accomplish delicate vascular surgery and other procedures in rats and mice with minimal postoperative mortality. Endotracheal intubation and positive pressure ventilation is also possible and requires the use of customized equipment (Gaertner et al., 2008).

Several anesthetic combinations are appropriate for brief, noninvasive procedures in rodents. Tribromoethanol (Avertin®) provides light anesthesia for 10–20 minutes in mice. Prepared solutions must be stored in the dark at 4°C to avoid production of gastric irritant decomposition compounds. The standard anesthetic dose is 0.2 ml/10 gm of a 1.2% solution. This anesthetic is most appropriate for brief, minimally painful procedures in mice, such as retro-orbital blood sampling, embryo transfer, vasectomy, and tail biopsy (Gaertner et al., 2008). Tribromoethanol is contraindicated in rats because peritoneal fibrosis and peritonitis is common following intraperitoneal injection in this species (Reid et al., 1999). Although propofol is intended only for intravenous administration, a recent report described effective surgical anesthesia of 20–30 minutes' duration and smooth recovery following intraperitoneal administration of propofol (75 mg/kg, mouse; 100 mg/kg rat), medetomidine (1–2 mg/kg, mouse; 0.1 mg/kg rat) and fentanyl (0.15–0.20 mg/kg, mouse; 0.1 mg/kg rat) to CD-1 mice and Wistar rats (Alves et al., 2009, 2010).

Intraperitoneal injection of ketamine (40–100 mg/kg) and xylazine (3–10 mg/kg) provides 20–60 minutes of surgical anesthesia in most rodent species. Muscle relaxation and analgesia are good. Anesthetic duration varies in a dose-dependent manner. If necessary, supplemental administration of ketamine will prolong anesthesia. Ketamine volumes commonly required for rats and mice require division of an IM (intramuscular) dose between two or more sites if the IP (intraperitoneal) route is contraindicated. Yohimbine (1–2 mg/kg IP) will reverse xylazine-associated sedation and speed recovery from anesthesia, but will also likely reverse residual xylazine-induced analgesia. Side-effects of general anesthesia in rodents may include hypercarbia, hypoxemia, and hypotension although these effects are less evident with ketamine and xylazine in comparison with sodium pentobarbital. Medetomidine and dexmedetomidine are safe but short acting in rodent species, and have not replaced xylazine in most models (Gaertner et al., 2008).

Barbiturates such as sodium pentobarbital (30–70 mg/kg IP) are occasionally used in rodent experimental surgery. Disadvantages of these agents include brief periods of effective anesthesia, prolonged recovery, and poor analgesia. In addition to environmental factors discussed earlier, rodents also display marked individual and strain variability in the response to sodium pentobarbital. Pretreatment with buprenorphine prior to incision prolongs the period of effective anesthesia, and lowers the required dose of sodium pentobarbital (Roughan et al., 1999). One report described the use of methohexitone (44 mg/kg of a 6.46 mg/ml solution IP) to achieve 2 minutes of chemical restraint for oral examination in C3H/Neu mice with recovery in 10–15 minutes. The major disadvantage of this technique was a very narrow therapeutic window: 40 mg/kg produced no immobility, whereas 50 mg/kg produced 40% mortality (Dorr and Weber-Frisch, 1999).

Isoflurane provides excellent anesthesia in rats and mice. This agent may be used both for brief restraint and for procedures of up to several hours' duration. Isoflurane requires the use of a precision vaporizer to maintain consistent anesthetic concentration. Rapid induction of anesthesia occurs following placement of animals into an induction chamber containing 3–4% isoflurane in oxygen. Most animals awake 1–2 minutes after removal from the chamber, which is sufficient time for retro-orbital blood sampling or tail biopsy. To maintain anesthesia for longer periods, the animal's head and nose may be placed into a customized nose cone connected to a non-rebreathing anesthetic circuit and scavenger. Concentrations of 2–3% isoflurane are commonly used for maintenance. The absence of the pedal withdrawal reflex confirms adequate anesthetic depth. Rodent nose cones are easily fashioned from funnels or disposable syringe barrels. For procedures requiring positive pressure ventilation, endotracheal intubation is easily accomplished. Techniques for endotracheal intubation for rodents have been described (Gaertner et al., 2008).

Analgesia. Buprenorphine is the most widely used analgesic in rodents. Buprenorphine is a mixed mu agonist/antagonist and a kappa receptor agonist with approximately 25–40 times the analgesic potency of morphine. The relative activity at each receptor may be dose-dependent. This agent provides safe and effective analgesia for 6–12 hours after dosing. The current recommended dose for most rodent indications is 0.03–0.05 mg/kg IP or SQ (subcutaneous) (Lee-Parritz, 2007; Curtin et al., 2009).

For procedures under isoflurane or sodium pentobarbital, buprenorphine should be given at least one hour before incision. The anesthetist should be aware of the anesthesia-sparing effects of buprenorphine, and be prepared to reduce the isoflurane vaporizer setting 30–50% as necessary (Brunson, 2008). For procedures conducted under ketamine/xylazine anesthesia, buprenorphine should be given only when the animal has regained sternal recumbency during recovery. Preoperative administration of buprenorphine or other narcotics may cause unpredictable increases in anesthetic depth when using ketamine/xylazine combinations for anesthesia. Local infusion of lidocaine (up to 1 ml/kg of 1.0% lidocaine or 0.25% bupivicaine) before surgical incision will provide additional intraoperative and postoperative analgesia (Lee-Parritz, 2007).

Clinically significant side-effects of buprenorphine include consumption of bedding (pica), and excessive licking or biting of the limbs and cage which may also be directed to the surgical incision. These effects are more common at high doses (0.1–0.3 mg/kg) and can often be managed through temporarily placing animals on a wire grid during the immediate postoperative period or substitution of a synthetic for a natural bedding substrate. A generalized increase in activity and a reduction of ventral grooming may also occur. These effects most likely represent a direct effect of the drug, and do not necessarily indicate the presence or absence of pain (Lee-Parritz, 2007).

The NSAIDs ketoprofen, carprofen, and meloxicam are the next most widely used analgesics. These drugs block the formation of inflammatory mediators associated with surgical injury, and also act centrally to inhibit secondary allodynia. The current recommended dose for meloxicam is 1 mg/kg PO (*per os* or orally) or SQ in rats, and up to 10 mg/kg in mice. The dose for ketoprofen and carprofen in rats is 5 mg/kg SQ. The dose interval for these agents has not been critically evaluated in rodents, but is 12–24 hours in other species. The NSAIDs are highly effective pre-emptive analgesics when administered one hour or more before surgery, but isoflurane vaporizer settings typically do not require adjustment (Lee-Parritz, 2007).

There are few reported adverse side-effects associated with NSAIDs in rodents. Diffuse intestinal ulceration may occur at high doses, especially in animals with concurrent disease. Most NSAIDs inactivate platelet function to some extent through cyclooxygenase-mediated inactivation of thromboxane. This may slightly prolong the bleeding time, but rarely causes adverse clinical effects in humans even when administered preoperatively (Lee-Parritz, 2007).

Acetaminophen (paracetamol) is a nonsteroidal analgesic which lacks significant anti-inflammatory properties. The drug is readily available in a palatable over-the-counter pediatric syrup formulation (Children's Tylenol®). Several studies suggest a possible role for this agent in rat analgesia, although specific evaluation for postoperative pain control is lacking. Reported effective dose ranges are 100–300 mg/kg twice a day. These studies suggest that acetaminophen may be effective for mild-to-moderate postoperative pain in rats. Although it is convenient to administer analgesics in drinking water, anorexia in the immediate postoperative period associated with anesthetic recovery and abdominal discomfort may reduce drug intake precisely when pain is maximal. Analgesic activity has not been assessed in mice, for which the LD50 (lethal dose to 50% of the subjects) is close to the doses evaluated for analgesia in rats (Lee-Parritz, 2007).

Rabbit

Animal Selection and Preoperative Preparation. Infection with the respiratory pathogen *Pasteurella multocida* is extremely common in conventional rabbits. Colonization of the upper respiratory tract may be clinically silent, but may spread to the lungs, middle ear, and brain in crowded or stressed conditions and produce characteristic disease syndromes (Harkness and Wagner, 1995; Lipman et al., 2008) and compromise long-term surgical studies. *Pasteurella*-negative rabbit colonies are available and should be used for all surgical protocols. Rabbits do not vomit, and aspiration pneumonia is therefore not a concern. Adult rabbits will tolerate an overnight fast without difficulty, and the resulting reduction in stomach contents may help maintain adequate oxygenation during spontaneous respiration under anesthesia. Rabbits weighing less than 2 kg should not be fasted, as hypoglycemia and metabolic acidosis may be significant (Lipman et al., 2008). Prolonged fasting may cause dehydration and subsequent intestinal motility disorders (Lipman et al., 2008). Traumatic lower-back fracture is common when inexperienced staff handle rabbits. The best way to handle rabbits is to grasp the scruff of the neck with one hand while supporting the rump and hind legs with the other. Sudden onset of flaccid paraplegia in rabbits is almost always the result of lower-back fracture, and warrants immediate euthanasia (Harkness and Wagner, 1995). Rabbits should never be lifted, moved or restrained by the ears.

Brief Procedures. Combined administration of ketamine (35–50 mg/kg IM) and xylazine (5–10 mg/kg IM) provides excellent anesthesia for a variety of applications in the rabbit. General anesthesia lasts for 30–60 minutes, and provides adequate analgesia and restraint for procedures of moderate intensity. Supplemental use of a narcotic analgesic such as buprenorphine (0.05 mg/kg IM) prolongs anesthesia and improves analgesia. Propofol administered by intravenous bolus or constant rate infusion produces light anesthesia suitable only for intubation or nonpainful procedures such as imaging. Telazol® should not be used for survival procedures in rabbits, because renal tubular damage is a common complication even at standard anesthetic dosages (Lipman et al., 2008).

General Anesthesia. The intramuscular ketamine/xylazine combination suggested for brief procedures is also an excellent induction agent prior to endotracheal intubation for inhalation anesthesia. Intravenous administration of ketamine and xylazine through the ear vein will achieve rapid induction, but care is required to minimize skin irritation from perivascular infiltration. Mask induction is rarely indicated because apnea, breath holding, bradycardia, and struggling are common when unsedated rabbits are exposed to isoflurane, and because operator exposure to waste anesthetic is difficult to avoid in this setting (Flecknell et al., 1999).

Propofol (10–20 mg/kg IV [intravenous]) is an effective rabbit anesthetic induction agent which produces significantly faster postoperative recovery than ketamine/xylazine in animals maintained on sevoflurane. In a recent report, rabbits anesthetized with propofol and maintained on sevoflurane were extubated 2 ±1 minutes, and achieved sternal recumbency 8 ±0.3 minutes after discontinuation of sevoflurane. Recovery following isoflurane anesthesia should be similarly rapid. By contrast, anesthetic recovery following ketamine/xylazine or ketamine/medetomidine may require up to 120 minutes. Animals should receive preoperative analgesics and be preoxygenated by mask before propofol administration. The initial dose of 10 mg/kg should be administered by hand over 60 seconds (~0.17 mg/kg/s), with additional small increments as required to allow endotracheal intubation. Propofol is not suitable as a sole agent because respiratory arrest may occur, particularly following rapid infusion (>0.25 mg/kg/s) (Allweiler et al., 2010).

Endotracheal intubation of the rabbit may be difficult because of several distinctive anatomic features. The prominent incisor teeth, long oropharynx and limited mobility of the temporomandibular joint hinder direct visualization of the larynx from the front, and require the use of a pediatric laryngoscope with a size 0–1 Wisconsin or size 1 Miller blade. Lidocaine spray on the vocal cords is necessary to prevent further narrowing of the larynx through laryngospasm. Benzocaine spray (Cetacaine®) produces methemoglobinemia in rabbits and should be avoided. The laryngeal opening is often smaller than the diameter of the trachea, requiring the use of a small endotracheal tube (2.5–4 mm). The tongue is short, friable, and difficult to grasp. Supine, prone or lateral positions are all suitable for endotracheal intubation. Hyperextension of the neck will straighten the larynx and facilitate proper tube placement. Blind intubation of the trachea is easily accomplished with practice. The tube is placed in the supraglottic region and advanced toward the larynx in coordination with respiration. Identification of normal breath sounds, visualization of condensate on a dental mirror or capnometry may be used to confirm proper tube placement (Harkness and Wagner, 1995; Lipman et al., 2008). Anesthetic maintenance with isoflurane usually requires a vaporizer setting of 1–4%. Absence of the pinna withdrawal reflex is the best indicator of a surgical anesthetic plane in the rabbit. Many rabbits retain the corneal and pedal withdrawal reflexes even under very deep anesthesia (Lipman et al., 2008; Muir, 2007).

If an anticholinergic agent is required, investigators should be aware that approximately 50% of rabbits produce atropinesterase (AtrE) as a genetically determined trait. Glycopyrrolate resists AtrE, and is therefore recommended over atropine for rabbits. Glycopyrrolate prevents xylazine-associated bradycardia in rabbits (Lipman et al., 2008).

Analgesia. Behavioral signs of postoperative pain in rabbits are vague and inconsistent, and may include lethargy, tooth grinding, and increased activity directed toward the painful area. Anorexia is a common sign of postoperative pain in rabbits. Untreated anorexia can create serious secondary disease, including generalized gastrointestinal stasis, rapid weight loss, and fatal hepatic lipidosis (Lipman et al., 2008).

The best indicator of GI (gastrointestinal) health in the rabbit is the quantity and consistency of fecal pellets. Dehydration and anorexia result in a reduced number of firm, dry pellets. Treatment of anorexic postoperative rabbits should be directed to relief of underlying pain if present and towards restoration of normal gastrointestinal function. A high fiber supplement such as Oxbow Critical Care or blenderized rabbit chow is preferred over low fiber nutritional gels such as Nutrical® to restore normal gut motility. Aggressive analgesia, fluid therapy, and force-feeding may be necessary in some cases.

Narcotic analgesics are preferred for treatment of moderate to severe postoperative pain in rabbits. Buprenorphine (0.01–0.05 mg/kg SQ BID/TID [twice or three times a day]) is the narcotic of choice, because it provides effective analgesia at a convenient dosing interval and is generally well-tolerated. The fentanyl patch (25 μg/hr) is an alternative narcotic analgesic option that provides continuous analgesia up to 72 hours without the requirement for regular injections. Fentanyl patches achieve effective blood levels approximately 12 hours after application, and should be applied the night before surgery. Alternatively, the patch may be applied at the time of surgery, with a single dose of buprenorphine to provide 8–12 hours of bridging analgesia. Clip the fur well and use tissue adhesive (Nexaband® or equivalent) if necessary around the edge of the patch to ensure good skin adhesion. Inadequate fur removal will greatly reduce drug absorption. Use of depilatories such as Neet® for hair removal results in increased drug absorption, and may cause toxicity (Foley et al., 2001).

The NSAIDs are also highly effective in rabbits, particularly for musculoskeletal pain or co-administered with narcotics for multimodal analgesia. Flunixin (1.1 mg/kg SQ q24h), carprofen (4 mg/kg SQ q24h), and meloxicam (0.3 mg/kg PO q 24h) all demonstrate good clinical efficacy in rabbits (Lipman et al., 2008). Oral meloxicam oral suspension (Metacam®, 1.5 mg/ml) has recently undergone pharmacokinetic evaluation in rabbits. Oral administration of 0.3 mg/kg once a day produced blood levels comparable to those shown to be clinically effective in other species for up to 24 hours. Repeated administration of 1.5 mg/kg/day for five days did not result in drug accumulation or drug toxicity. Although rabbits demonstrate more rapid metabolism of meloxicam than other species, delayed gastrointestinal absorption of the drug allows maintenance of effective blood levels throughout the dose period (Turner et al., 2006).

Dog

Animal Selection and Preoperative Preparation. Appropriate vendor selection and conditioning procedures are necessary to identify pre-existing cardiovascular or renal diseases, malnutrition or parasitism that can significantly complicate anesthesia and surgery. Heartworm disease, a result of *Dirofilaria immitis* infestation, causes eosinophilia and right heart failure. Intestinal parasites, including roundworms (*Toxocara canis*), hookworms (*Ancylostoma caninum*), and whipworms (*Trichuris vulpis*) cause eosinophilia, diarrhea, and general debilitation. All of these parasites are susceptible to common anthelmintics. A calm behavioral profile, reinforced by habituation to the research staff and regular training, is important for animals on long-term studies that call for frequent handling. Purpose-bred dogs have a known pedigree and health history, and often present a more consistent physiological profile than random-source dogs. Random-source dogs are significantly less expensive than purpose-bred dogs, but often have health and behavioral disorders which render them unsuitable for long-term studies (Dysko et al., 2002).

Brief Procedures. Calm dogs may be trained to cooperate in noninvasive clinical procedures, thereby avoiding the need for sedatives or anesthetics. Short acting anesthetics may be required for more invasive procedures. The preferred technique for short-term injectable anesthesia in dogs is coadministration of a dissociative anesthetic with a sedative or tranquilizer and an anticholinergic drug. All of these drugs are suitable for intravenous administration to achieve rapid induction and recovery, and provide adequate anesthetic depth to allow endotracheal intubation before prolonged procedures. Intravenous ketamine (5–10 mg/kg) and diazepam or midazolam (0.2–0.5 mg/kg) provides 5–15 minutes of light anesthesia suitable for dressing changes, radiography or other minor procedures. Medetomidine (10–40 μg/kg IM) or dexmedetomidine (5–20 μg/kg IM) are also appropriate for this purpose, and may be rapidly reversed with atipamezole (50–200 μg/kg) (Lemke, 2007). Supplementation of medetomidine or dexmedetomidine with ketamine (3 mg/kg) and an opioid such as butorphanol (0.2–0.4 mg/kg) is advisable before major procedures (Bednarski, 2007; Armitage-Chan, 2008).

For longer procedures, intramuscular administration of ketamine (10 mg/kg) and xylazine (0.7–1.0 mg/kg) produces 20–30 minutes of anesthesia. Telazol® (6–8 mg/kg), xylazine (0.7–1 mg/kg) and butorphanol (0.2 mg/kg) produce up to one hour of anesthesia. Anticholinergics (atropine 0.04 mg/kg or glycopyrrolate 5–10 μg/kg) are often useful when using ketamine or xylazine to counteract excessive salivation or bradycardia. Supplemental administration of one-third to half of the original dose will usefully prolong the effective anesthetic period by 30–50%. Ketamine is not acceptable as a sole anesthetic in dogs, because of excessive muscle tone, salivation, and the frequent occurrence of seizures (Bednarski, 2007; Armitage-Chan, 2008).

Slow intravenous bolus injection of propofol (6–8 mg/kg) provides safe and effective anesthesia for approximately 10 minutes followed by complete recovery within approximately 30 minutes. Intermittent boluses of approximately 0.5–2.0 mg/kg or a continuous rate infusion (0.15–0.40 mg/kg/min) are suitable for prolonged procedures. Apnea is a common consequence of rapid IV injection and requires assisted ventilation until the drug wears off. Premedication with acepromazine, medetomidine or morphine will reduce the induction and maintenance doses of propofol by 30–40%. Short-acting barbiturates such as sodium thiopental (8–12 mg/kg IV) or methohexital (4–8 mg/kg IV) provide about 15 minutes of light anesthesia, but have a lower margin of safety than dissociative anesthetics or propofol (Bednarski, 2007; Armitage-Chan, 2008).

General Anesthesia. Inhalant anesthetics such as isoflurane are most appropriate for prolonged or invasive procedures in the dog. Dogs are prone to vomiting under anesthesia, and proper preoperative fasting and endotracheal intubation are vital to prevent aspiration pneumonia. Premedication with the sedative and analgesic drugs previously discussed will allay anxiety, reduce postoperative pain, and reduce the required concentration of isoflurane for maintenance anesthesia. Rapid anesthetic induction may then be accomplished through slow intravenous administration of thiopental (8–12 mg/kg IV), methohexital (4–8 mg/kg IV) or propofol (2–4 mg/kg IV) to effect. Mask induction is rarely indicated in dogs, and should only be used in premedicated animals to reduce struggling (Bednarski, 2007; Armitage-Chan, 2008).

Endotracheal intubation of dogs is easily accomplished. A variety of endotracheal tube diameters should be available to accommodate individual and breed differences in tracheal diameter. Direct visualization and intubation of the larynx of most dogs is possible without a laryngoscope. A wire stylet and a laryngoscope with an appropriate-sized Miller or Bizarri-Guiffrida blade should be available to accommodate unexpected difficulties. The animal is placed in sternal recumbency, and an assistant holds the mouth open and extends the tongue to expose the larynx. Judicious application of lidocaine spray to the vocal cords will prevent laryngospasm and facilitate intubation in small dogs. After verifying accurate placement, the cuff is inflated and the tube is secured with gauze tied over the maxilla or behind the head (Hartsfield, 2007).

Inhalant anesthesia with isoflurane may be maintained for several hours. The anesthetic concentration should be varied according to the animal's clinical status. In general, a vaporizer setting of 1–3% will produce adequate anesthesia in dogs. Signs of adequate anesthesia include absent flexor withdrawal or corneal blink reflexes; stable heart rate and respiratory rates in response to surgical stimuli; and moderate to relaxed jaw tone. The eye will generally be central in very light or deep anesthesia and rotate ventromedially at medium anesthetic depth, but this sign is unreliable in animals which receive ketamine for induction (Muir, 2007). Paralytic agents are rarely indicated for experimental surgery in dogs. When these agents are used, the heart rate and blood pressure should be monitored and additional anesthetic administered as necessary to maintain homeostasis. Ketamine and xylazine used for induction will wear off after 30–60 minutes and require an increase in isoflurane concentration or use of supplemental analgesics to prevent sensation. The use of an "MLK" (morphine/lidocaine/ketamine) continuous rate infusion is an appropriate method to achieve balanced anesthesia at very low isoflurane vaporizer settings and maintain adequate perfusion without unnecessary prolongation of anesthetic recovery (Armitage-Chan, 2008). Isoflurane will accumulate in tissues during long procedures and multimodal analgesics often allow significant reduction of the isoflurane vaporizer settings. The vaporizer setting should generally be reduced toward the end of prolonged procedures, to avoid unnecessarily long recovery periods.

Analgesia. Diagnosis of pain in dogs is more reliable than in non-domestic species. Consistent signs of pain include vocalization, drooping of the head and neck ("hang-dog" expression), and site-specific behaviors (sensitivity to palpation, limping, biting or licking the incision). The Glasgow Composite Pain Scale (GCPS) is easily taught and demonstrates good interobserver reliability. A trained observer can evaluate a dog using this scale in approximately 5 minutes. Dogs receive a score of 0 (no pain) to 24 (severe pain), and should receive supplemental analgesia for a score greater than 6 (Hellyer et al., 2007).

Opioid analgesics are preferred for moderate to severe pain in dogs. The recommended dose of buprenorphine in dogs is 0.02 mg/kg IM every 8–12 hours. Fentanyl patches (2–5 μg/kg/h) are useful for postoperative pain longer than 24 hours' duration, but must be placed at least 12 hours before surgery to achieve effective blood levels. Epidural analgesia provides excellent pain relief with fewer systemic side-effects than parenteral injection. Epidural administration of analgesics can be readily accomplished by trained staff, and provides up to 12–24 hours of analgesia following a single injection. Morphine and oxymorphone are the most commonly used epidural analgesics. The alpha-2 agonists xylazine and medetomidine are also effective by this route, but may produce significant bradycardia and hypotension (Foley et al., 2001).

The NSAIDs are highly effective analgesics in dogs, and may provide effective analgesia up to 24 hours after a single dose. Adverse effects of the NSAIDs include gastritis, which may proceed to ulceration. For this reason, NSAIDs are most appropriate for acute postoperative pain control, and not for treatment of chronic pain conditions in dogs. Recommended NSAIDs for dogs include ketoprofen, meloxicam, and carprofen.

A recent report evaluated pain following routine ovariohysterectomy in dogs receiving buprenorphine (0.02 mg/kg single preoperative dose IM), carprofen (4.0 mg/kg single preoperative dose IM) or combined preoperative buprenorphine/carprofen. All dogs were premedicated with acepromazine, induced with propofol, and maintained on isoflurane. All treatments produced significant pain relief, but 10–20% of the dogs required rescue analgesia for breakthrough pain up to 6 hours after surgery. The carprofen group demonstrated the lowest mean GCPS score at all time points. There was no significant additive analgesic effect following combined buprenorphine/carprofen treatment. Dogs which received buprenorphine or combination analgesia required a lower induction dose of propofol, and experienced a quicker anesthetic recovery than dogs which received only carprofen. All animals were essentially pain-free by 24 hours after surgery (Shih et al., 2008). The duration of analgesic treatment required following orthopedic or major abdominal or thoracic procedures commonly used in biomaterials research is likely to be longer than the 24 hours required following ovariohysterectomy, and generally requires multimodal therapy for optimal control.

Pig

Animal Selection and Preoperative Preparation. Swine are commonly used in experimental surgery because of anatomic and physiologic similarity to humans in many models. Advantages to the use of pigs in the laboratory include low cost, ready availability, and ease of acclimation to the laboratory. Disadvantages include the uncooperative nature of pigs with respect to most clinical procedures, limited number of intravenous access sites, and relative difficulty of endotracheal intubation.

Because animals larger than 100 kg are difficult to handle in the laboratory, selective breeding has produced several types of miniature swine for research. The Gottingen minipigs and the Yucatan micropigs achieve a maximum bodyweight of 35–55 kg at 2 years of age. The Hanford and Yucatan minipigs are somewhat larger, and weigh 70–90 kg at 2 years of age. By contrast, adult crossbred farm pigs weigh 90–110 kg at 6 months of age, and 200–300 kg at 2 years of age (Bollen et al., 2000; Swindle, 2007). Juvenile crossbred farm pigs are less expensive than minipigs or micropigs, and are often used for short-term surgical studies. Juvenile farm pigs gain 2–4 kg/week and require larger pens as they grow, to maintain facility compliance with the Guide (National Research Council, 2010). The choice of research subject should include practical considerations such as length of study and maximum allowable body size, as well as the physiological characteristics of the different breeds (Bollen et al., 2000).

Careful evaluation of the vendor health program is required to reduce experimental morbidity from unrelated clinical conditions. Chronic respiratory disease is common in commercial swine operations. Although subclinical infections are common, transportation, anesthesia, and surgery may activate latent infections and result in excess morbidity and mortality. Causative agents in affected pigs include *Mycoplasma hyopneumoniae*, *Haeomophilus pleuropneumoniae*, *Bordetella bronchiseptica*, and *Actinobacillus pleuorpneumoniae* (Straw et al., 2006). Many commercial breeders of laboratory minipigs and micropigs maintain specific pathogen-free (SPF) herds that are free from infection with these agents. Rapid infection with respiratory pathogens occurs when SPF pigs are cohoused with conventional swine.

Malignant hyperthermia (MH) is an autosomal dominant trait that causes affected animals to develop marked hyperthermia (rectal temperature >41°C), and extensor muscle rigidity and necrosis after exposure to environmental extremes or halothane or isoflurane anesthesia. The disease has been described only in farm animals bred for rapid growth, and is becoming increasingly rare as commercial breeders identify and cull carrier pigs. Affected pigs are still encountered sporadically in farm pigs used for experimental surgery. The condition has not been described in minipigs or micropigs. MH has not been reported in other commonly used laboratory animal species. Known MH carriers should not be used for experimental surgery. Treatment of animals which develop MH during an experimental procedure requires immediate termination of anesthesia, whole body cooling, and administration of corcticosteroids, sodium bicarbonate, and dantrolene sodium (3–5 mg/kg IV) (Swindle, 2007; Smith et al., 2008).

Food and contact bedding must be removed 6–8 hours prior to general anesthesia (Swindle, 2007). Water may be offered until 2 hours before anesthesia. The presence of food in the stomach frequently results in gastric distension, hypoventilation, and tachycardia during prolonged general anesthesia, requiring prompt decompression through orogastric intubation (Thurmon and Smith, 2007).

Brief Restraint. Pigs may be easily trained to accept handling for physical examination, blood collection through a vascular access port, and dressing changes. The acclimation period following arrival to the facility before study initiation is an appropriate training time. Uncooperative pigs may require chemical restraint for minor clinical procedures. Slings and hammocks are available which will allow restraint of calm pigs up to 50 kg for up to several hours (Swindle, 2007). The Panepinto sling was specifically designed for the veterinary practice and laboratory environments, and is easy to operate and sanitize. Sedation may be necessary to facilitate initial placement of animals into the sling. Azaperone (4 mg/kg IM) is a useful agent for this purpose. For more invasive procedures, a mixture of Telazol®(4.4–6.6 mg/kg IM), xylazine (2.2 mg/kg IM), and atropine (0.05 mg/kg) provides approximately 30 minutes of anesthesia suitable for minor surgery, followed by smooth recovery. Endotracheal intubation for subsequent maintenance on isoflurane is also possible after Telazol® and xylazine induction (Bollen et al., 2000; Swindle, 2007). Alternative agents such as ketamine, ketamine/xylazine or Telazol® produce light anesthesia insufficient for surgery or intubation, and characterized by rough recovery (Ko et al., 1993, 1995).

Delayed recovery complicated by hindleg weakness may occur following Telazol®/xylazine administration in mature pigs, and is mediated by the zolazepam component of the proprietary Telazol® mixture. For this reason, many clinicians prefer to use a "cocktail" made by adding 2.5 ml of ketamine (100 mg/ml) and 2.5 ml of xylazine (100 mg/ml) to 1 vial of unreconstituted Telazol® powder. The resulting solution contains equivalent dissociative anesthetic potency to standard mixtures, with significantly less zolazepam, and is administered at a dose of 1 ml/35–75 kg IM (Thurmon and Smith, 2007).

General Anesthesia. Prolonged or invasive procedures are best conducted under isoflurane anesthesia. Endotracheal intubation is warranted for general anesthesia in swine to protect the airway and allow for controlled ventilation when required. Required equipment for endotracheal intubation of swine includes a laryngoscope with a 20–25 cm straight blade and a selection of cuffed tubes of appropriate size (4.5–8 mm). Endotracheal intubation is possible with the pig in dorsal, ventral or lateral recumbency. Two anatomic characteristics can complicate the procedure. First, the soft palate is long and must be displaced dorsally for visualization of the larynx. Second, the laryngeal diverticulum distal to the larynx may "trap" the tip of the endotracheal tube unless the tube is gently twisted as it passes over the epiglottis. After intubation, maintenance anesthesia usually requires an isoflurane vaporizer setting of 1.5%–2.5% in oxygen. The actual concentration of isoflurane will vary according to the anesthetic induction regimen, type of procedure, and concurrent use of other narcotic and sedative agents. Administration of Telazol® and xylazine for anesthetic induction has a substantial isoflurane-sparing effect that may last for the first 30–60 minutes of anesthesia. Adequate surgical anesthesia is indicated by absence of the pedal withdrawal reflex, minimal jaw tone, and stable heart rate and blood pressure (Thurmon and Smith, 2007; Smith et al., 2008).

Analgesia. Diagnosis of pain in swine may be challenging. Pain scoring systems reported by some investigators have not been validated in large-scale studies. Although an indirect measure, depressed appetite is the most consistent indicator of pain in postoperative pigs. In addition, pigs in pain generally exhibit impaired activity and depressed attitude compared to normal pigs.

Narcotics are commonly used postoperative analgesics in swine. Buprenorphine is the most widely used narcotic, because it provides excellent analgesia at a convenient dose interval with few side-effects. The fentanyl patch (5 μg/kg/hr) is effective in swine, and reduces the need to handle animals during the immediate postoperative period, when struggling may increase pain at the incision line or predispose to dehiscence. The 50 μg/hr patch produced therapeutic blood levels and clinically effective pain control for up to 48 hours in Yorkshire-cross white female pigs (26.2 ±2.1 kg) following experimental thoracotomy. Control pigs receiving buprenorphine (0.10 mg/kg) also experienced adequate pain control, but exhibited higher pain scores and required treatment at 4–8 hour intervals. Fentanyl patches should be applied in the dorsal mid-scapular region to prevent accidental dislodgement and ensure consistent absorption (Harvey et al., 2000). Group-housed pigs may remove and consume patches applied to cage mates and develop acute narcotic overdose.

The NSAIDs are highly effective analgesics in swine. Carprofen (2 mg/kg SQ or PO q24h), meloxicam (0.4 mg/kg SQ q24h), and flunixin (1–4 mg/kg SQ or IM q24h) have been used as sole agents or in combination with opioid local anesthetics. Gastric irritation has not been reported following short-term use, but may be a concern in animals which require chronic therapy (Thurmon and Smith, 2007; Smith et al., 2008).

Sheep and Goats

Animal Selection and Preoperative Preparation. Sheep and goats are desirable research animals, because the adult body weight is comparable to humans and they are hardy, inexpensive, and have a calm disposition. Sheep are particularly useful for reproductive research because investigators may easily obtain cohorts of pregnant animals with known gestational age and hysterotomy, and fetal manipulation is possible with a low postoperative abortion rate. Cardiovascular research also uses sheep and goats because the ratio of the size and weight of the heart and other thoracic organs to body weight is similar to humans (Riebold, 2007).

The quality of sheep and goats used in research often depends on the source. Surgical and transportation stress will reduce resistance to disease. Vendors should avoid mixing multiple sources of animals in one flock to ensure a consistent health profile. Similarly, research institutions should use animals from a single vendor to reduce the possibility of introducing unfamiliar infectious agents to stressed animals. Preventive health measures, including immunizations and anthelmintic treatments, should be provided at least one month before shipment. Recommended immunizations for sheep and goats include *Clostridia spp.*, *Pasteurella multocida*, *P. hemolytica*, contagious ecthyma, and parainfluenza III. A conditioning and quarantine period after arrival allows recovery from shipping stress, acclimation to the facility, and reduces transmission of infectious diseases to resident animals (Delano et al., 2002).

Facilities should adopt measures to prevent transmission of zoonoses from sheep and goats. Silent infection with *Coxiella burnetti*, the causative agent of human Q fever, is common in sheep. Fetal membranes and amniotic fluid from infected animals carry large numbers of hardy organisms. Human infection occurs from direct contact with infected materials or from fomites. Signs of Q fever infection in humans range from subclinical disease to severe flu-like symptoms, pneumonia, endocarditis, and death. Serologic evaluation of sheep is difficult because animals may shed large numbers of organisms in the absence of detectable antibody. Contagious ecthyma ("orf") is a poxvirus-induced papular disease of sheep and goats. Infected animals frequently have lesions on the mucocutaneous junctions of the head, which can spread to humans by direct contact. Affected animals and humans develop long-lasting immunity and usually recover in 10–14 days. Effective vaccines for sheep and goats are available (National Research Council, 1997).

Careful preoperative preparation of the rumen is required for safe anesthesia of sheep and goats. Adult animals require withdrawal of food and water for at least 12–24 hours before surgery to reduce rumen size and digestive activity. Rumen distension and hypoventilation are common when dorsal or lateral positioning of ruminants is required for anesthesia. Preoperative passage of a 1–2 cm diameter thick walled stomach tube into the rumen allows intraoperative aspiration of gas and fluid if required. Placement of a cuffed endotracheal tube should immediately follow anesthetic induction to prevent aspiration pneumonia. Aspiration pneumonia can be fatal after regurgitation because rumen fluid contains numerous anaerobic bacteria. Infant ruminants (<30 days of age) lack a functional rumen and do not require preoperative fasting (Riebold, 2007).

Brief Restraint. Sheep and goats are docile and rarely require sedation for blood sampling, dressing changes or other routine procedures. For more invasive procedures, animals may be sedated for 10–20 minutes with a low dose of xylazine (0.02–0.15 mg/kg IV or 0.05–0.3 mg/kg IM). Yohimbine (1 mg/kg IV) or atipamezole (0.02–0.06 mg/kg IV) (Muir, 2007) reverse xylazine-induced sedation when necessary. Research staff should be mindful of the much lower xylazine dose required for ruminants compared to other species. By contrast, the dose of medetomidine is similar in most laboratory animal species (30 μg/kg). Animals in late pregnancy should not receive xylazine to avoid fetal hypo-oxygenation secondary to depressed maternal cardiac output. The benzodiazepines diazepam (0.25 mg/kg IV) and midazolam (1.3 mg/kg IV) produce muscle relaxation and sedation in sheep and goats, with less depression of cardiac output than is seen with xylazine. Flumazenil (1 mg IV) reverses benzodiazepine-induced sedation. Ataxia and excitement are common following benzodiazepine administration (Riebold, 2007).

General Anesthesia. Isoflurane is the anesthetic of choice for general anesthesia of sheep and goats. Intravenous administration of ketamine (2.75 mg/kg) and diazepam (0.2 mg/kg) or xylazine (0.1 mg/kg) rapidly induces light anesthesia of 10–20 minutes duration suitable for endotracheal intubation. Propofol (4.0–6.0 mg/kg IV to effect) is a suitable alternative for brief procedures or to facilitate intubation and maintenance on isoflurane for longer procedures (Riebold, 2007). In a study comparing propofol to ketamine/xylazine/halothane for subcutaneous biomaterial implantation, propofol recipients exhibited better cardiovascular and respiratory parameters than ketamine/xylazine/halothane recipients. These animals also demonstrated a more rapid and smoother anesthetic recovery (Lin et al., 1997). Thiopental (25 mg/kg IV) is an alternative induction agent; however, significant disadvantages include regurgitation, profuse salivation, and irritation from perivascular infiltration (Dorr and Weber-Frisch, 1999). Profuse salivation follows the administration of most anesthetic drugs in ruminants. Administration of anticholinergic drugs is *not* recommended in ruminants, because these agents may increase salivary viscosity and impair gastrointestinal motility. Anesthetized ruminants should be positioned with the head down to encourage drainage of saliva away from the airway (Riebold, 2007; Thurmon and Smith, 2007).

Endotracheal intubation requires a laryngoscope with a 20–30 cm blade. Vinyl or silicone cuffed tubes 10–16 mm in diameter are suitable for most sheep and goats. An assistant positions the animal in sternal recumbency, extends the neck, and holds the mouth open. The anesthetist visualizes the epiglottis with the laryngoscope and intubates the trachea during inspiration. Prior application of 2% lidocaine spray to the vocal cords prevents laryngospasm. The use of a stylet may help deflect the tip of the tube into the larynx. A mouth gag is often necessary to prevent damage to the tube by sharp molar teeth. After inflation of the cuff, the tube is secured with gauze to the mandible. Isoflurane vaporizer settings of 0.75–1.0% are often sufficient to maintain general anesthesia in sheep and goats. Absence of chewing motions in response to stimulation indicates an adequate surgical plane. The presence of a centrally positioned eyeball with a dilated pupil and absent palpebral reflex indicates very deep anesthesia (Riebold, 2007).

Analgesia. The NSAIDs are highly effective in small ruminants, and are recommended for first-line treatment unless scientifically contraindicated. Flunixin (1.1–2.2 mg/kg), ketoprofen (3.3 mg/kg), and carprofen (0.7 mg/kg) provide effective analgesia for up to 24 hours after dosing. When used at a dose of 4.0 mg/kg, carprofen provides therapeutic blood levels for at least 72 hours (Thurmon and Smith, 2007). All NSAIDs should be administered parenterally in small ruminants. The intravenous route is preferred for animals with severe acute pain, but the subcutaneous and intramuscular routes have been widely reported.

Opioid analgesics may be indicated as a supplement or as primary analgesics in some cases. The epidural space is easily accessed in small ruminants. Epidural morphine (0.1 mg/kg diluted in sterile saline) administered in the lumbosacral or sacrocaudal spaces provides up to 12 hours of analgesia (Riebold, 2007). The transdermal fentanyl patch (50 μg/hr) produces effective blood levels in sheep and goats for up to 72 hours. The dose rate is lower than otherwise expected for an animal this size, because of significant recirculation from the rumen to the bloodstream, bypassing hepatic detoxification pathways, and animals should be observed for excessive sedation or excitement (Valverde and Doherty, 2008).

A recent report demonstrated the superiority of transdermal fentanyl over parenteral buprenorphine in a multimodal analgesic regimen following experimental tibial osteotomy and locking compression plate placement in mature Polypay-cross ewes. The fentanyl group received transdermal fentanyl patches (2 μg/kg/hr) on the lateral antebrachium 12 hours before surgery. Control animals received buprenorphine (0.01 mg/kg every 6 hours beginning at anesthetic induction). All animals received the NSAID phenylbutazone (2.2 mg/kg IV every 24 hours) for 3 days beginning at anesthetic induction. Blinded observers evaluated the animals for pain every 12 hours for 72 hours, including a final observation 12 hours after patch removal or the last dose of buprenorphine. Fentanyl recipients required a lower dose of diazepam during anesthetic induction, and exhibited lower pain scores than buprenorphine recipients throughout the observation period, although no animals required rescue analgesia. The buprenorphine group exhibited an increased pain score 12 hours after the last dose of analgesic, suggesting possible rebound hyperalgesia (Ahern et al., 2009).

Ruminants are more sensitive to systemic effects of local anesthetics than other species. The mean fatal dose of lidocaine in sheep is 30.8 mg/kg, compared to 80 mg/kg in dogs, and seizures may occur following administration of 10 mg/kg. The maximum suggested dose of lidocaine for sheep and goats is 0.5 ml/kg of a 2% solution. For wide infiltration, 2% lidocaine may be diluted to 1%. Signs of systemic local anesthetic overdose include seizures, respiratory depression, bradycardia, hypotension, and collapse.

SUMMARY

Animal models are an invaluable resource in biomaterials research. Surgery is often required in biomaterials models. Animal welfare regulations in most countries require scientists to use animal models only when necessary, and use the minimum number of animals consistent with good experimental design. All necessary steps must be used to prevent and treat pain or distress associated with experimental manipulation.

BIBLIOGRAPHY

Ahern, B. J., Soma, L. R., Boston, R. C., & Schaer, T. P. (2009). Comparison of the analgesic properties of transdermally administered fentanyl and intramuscularly administered buprenorphine during and following experimental orthopedic surgery in sheep. *American Journal of Veterinary Research*, **70**, 418–422.

Allweiler, S., Leach, M. C., & Flecknell, P. A. (2010). The use of propofol and sevoflurane for surgical anesthesia in New Zealand White rabbits. *Laboratory Animals*, **44**, 113–117.

Alves, H. C., Valentim, A. M., Olsson, I. A.S., & Antunes, L. M. (2009). Intraperitoneal anesthesia with propofol, medetomidine and fentanyl in mice. *Laboratory Animals*, **43**, 27–33.

Alves, H. C., da Silva, A. L.M., Olsson, I. A.S., Orden, J. M.G., & Antunes, L. M. (2010). Anesthesia with intraperitoneal propofol, medetomidine and fentanyl in rats. *JAALAS*, **49**, 454–459.

Armitage-Chan, E. (2008). Anesthesia and analgesia in dogs and cats. In R. E. Fish, M. J. Brown, P. J. Danneman, & A. Z. Karas (Eds.), *Anesthesia and Analgesia in Laboratory Animals* (2nd ed., pp. 365–384). London, UK: Academic Press.

Bednarski, R. M. (2007). Dogs and cats. In W. J. Tranquilli, J. C. Thurmon, & K. A. Grimm (Eds.), *Lumb and Jones' Veterinary Anesthesia* (4th ed., pp. 705–716). Ames, IA: Blackwell Publishing.

Bollen, J. A., Hansen, A. K., & Rasmussen, H. J. (2000). *The Laboratory Swine*. Boca Raton, FL: CRC Press.

Borkowski, G. L., & Allen, M. (1999). *The Laboratory Small Ruminant*. Boca Raton, FL: CRC Press.

Branson, K. R. (2007). Injectable and alternative anesthetic techniques. In W. J. Tranquilli, J. C. Thurmon, & K. A. Grimm (Eds.), *Lumb and Jones' Veterinary Anesthesia* (4th ed., pp. 273–300). Ames, IA: Blackwell Publishing.

Brown, K. M., Saunders, M. M., Kirsch, T., Donahue, H. J., & Reid, S. J. (2004). Effect of COX-2-specific inhibition on fracture-healing in the rat femur. *Journal of Bone and Joint Surgery: American Volume*, **86**, 116–123.

Brunson, D. (2008). Pharmacology of inhalation anesthetics. In R. E. Fish, M. J. Brown, P. J. Danneman, & A. Z. Karas (Eds.), *Anesthesia and Analgesia in Laboratory Animals* (2nd ed., pp. 83–96). London, UK: Academic Press.

Curtin, L. I., Grakowsky, J. A., Suarez, M., Thompson, A. C., DiPirro, J. M., et al. (2009). Evaluation of buprenorphine in a postoperative pain model in rats. *Comparative Medicine*, **59**, 60–71.

Delano, M. L., Mischler, S. A., & Underwood, W. J. (2002). Biology and diseases of ruminants: Sheep, goats and cattle. In J. G. Fox, L. C. Anderson, F. M. Loew, & F. W. Quimby (Eds.), *Laboratory Animal Medicine* (2nd ed., pp. 519–614). San Diego, CA: Academic Press.

Dorr, W., & Weber-Frisch, M. (1999). Short-term immobilization of mice by methohexitone. *Laboratory Animals*, **33**, 35–40.

Dysko, R. C., Nemzek, J. A., Levin, S. I., DeMarco, G. J., & Moalli, M. R. (2002). Biology and diseases of dogs. In J. G. Fox, L. C. Anderson, F. M. Loew, & F. W. Quimby (Eds.), *Laboratory Animal Medicine* (2nd ed., pp. 395–458). San Diego, CA: Academic Press. London, UK.

Fish, R. E., Brown, M. J., Danneman, P. J., & Karas, A. Z. (Eds.), (2008). *Anesthesia and Analgesia in Laboratory Animals* (2nd ed.). London, UK: Academic Press.

Flecknell, P. A., Roughan, J. V., & Hedenqvist, P. (1999). Induction of anaesthesia with sevoflurane and isoflurane in the rabbit. *Laboratory Animals*, **33**, 41–46.

Flecknell, P. A., Richardson, C. A., & Popovic, A. (2007). Laboratory Animals. In W. J. Tranquilli, J. C. Thurmon, & K. A. Grimm (Eds.), *Lumb and Jones' Veterinary Anesthesia* (4th ed., pp. 765–784). Ames, IA: Blackwell Publishing.

Foley, P. L., Henderson, A. L., Bissonette, E. A., Wimer, G. R., & Feldman, S. H. (2001). Evaluation of fentanyl patches in rabbits: Blood concentrations and physiologic response. *Comparative Medicine*, **51**, 239–244.

Fox, J. G., Anderson, L. C., Loew, F. M., & Quimby, F. W. (Eds.), (2002). *Laboratory Animal Medicine* (2nd ed.). San Diego, CA: Academic Press.

Fujibayashi, S., Neo, M., Kim, H., Kokubo, T., & Nakamura, T. (2003). A comparative study between *in vivo* bone ingrowth and *in vitro* apatite formation on Na_2O–CaO–SiO_2 glasses. *Biomaterials*, **24**, 1349–1356.

Gaertner, D. J., Hallman, T. M., Hankenson, F. C., & Batchelder, M. A. (2008). Anesthesia and analgesia for laboratory rodents. In R. E. Fish, M. J. Brown, P. J. Danneman, & A. Z. Karas (Eds.), *Anesthesia and Analgesia in Laboratory Animals* (2nd ed., pp. 239–298). London, UK: Academic Press.

Graham, J. S., Reid, F. M., Niemuth, N. A., Shumaker, S. M., & Waugh, J. D. (2004). Effects of three anesthetic regimens on bioengineering methods conducted on ventral abdominal skin of weanling swine. *Journal of Toxicology*, **23**, 105–118.

Hanson, C. A., Weinhold, P. S., Afshari, H. M., & Dahners, L. E. (2005). The effect of analgesic agents on the healing rat medial collateral ligament. *American Journal of Sports Medicine*, **33**, 674–679.

Harkness, J. E., & Wagner, J. E. (Eds.), (1995). *The Biology and Medicine of Rabbits and Rodents*. Baltimore, MD: Williams and Wilkins.

Hartsfield, S. M. (2007). Airway management and ventilation. In W. J. Tranquilli, J. C. Thurmon, & K. A. Grimm (Eds.), *Lumb and Jones' Veterinary Anesthesia* (4th ed., pp. 495–532). Ames, IA: Blackwell Publishing.

Harvey, C. J., Gilespie, K., & Riggs, K. W. (2000). Transdermal fentanyl compared with parenteral buprenorphine in post-surgical pain in swine: A case study. *Laboratory Animals*, **34**, 386–398.

Hawk, C. T., Leary, S. L., & Morris, T. H. (Eds.), (2005). *Formulary for Laboratory Animals* (3rd ed.). Ames, IA: Blackwell Publishing.

Hellyer, P. W., Robertson, S. A., & Fails, A. D. (2007). Pain and its management. In W. J. Tranquilli, J. C. Thurmon, & K. A. Grimm (Eds.), *Lumb and Jones' Veterinary Anesthesia* (4th ed., pp. 31–60). Ames, IA: Blackwell Publishing.

Hessler, J., & Lehner, N. D.M. (Eds.), (2008). *Planning and Designing Research Animal Facilities*. London, UK: Academic Press.

Kirkpatrick, C. J., Fuchs, S., Hermanns, M. I., Peters, K., & Unger, R. E. (2007). Cell culture models of higher complexity in tissue engineering and regenerative medicine. *Biomaterials*, **28**, 5193–5198.

Ko, J. C. H., Williams, B. L., Smith, V. L., McGrath, C. J., & Jacobson, J. D. (1993). Comparison of telazol, telazol-ketamine, telazol-xylazine, and telazol-ketamine-xylazine as chemical restraint and anesthetic induction combination in swine. *Laboratory Animal Science*, **43**, 476–480.

Ko, J. C. H., Williams, B. L., Rogers, E. R., Pablo, L. S., McCaine, W. C., et al. (1995). Increasing xylazine dose-enhanced anesthetic properties of telazol-xylazine combination in swine. *Laboratory Animal Science*, **45**, 290–294.

Krischak, G. D., Augat, P., Sorg, T., Blakytny, R., Kinzl, L., et al. (2007). Effects of diclofenac on periosteal callus maturation in osteotomy healing in an animal model. *Archives of Orthopaedic and Trauma Surgery*, **127**, 3–9.

Lamont, L. A., & Mathews, K. A. (2007). Opioids, nonsteroidal anti-inflammatories, and analgesic adjuvants. In W. J. Tranquilli, J. C. Thurmon, & K. A. Grimm (Eds.), *Lumb and Jones' Veterinary Anesthesia* (4th ed., pp. 241–272). Ames, IA: Blackwell Publishing.

Lee-Parritz, D. E. (2007). Analgesia for rodent experimental surgery. *Israel Journal of Veterinary Medicine*, **62**, 74–78.

Lemke, K. A. (2007). Anticholinergics and sedatives. In W. J. Tranquilli, J. C. Thurmon, & K. A. Grimm (Eds.), *Lumb and Jones' Veterinary Anesthesia* (4th ed., pp. 273–300). Ames, IA: Blackwell Publishing.

Lin, H. C. (2007). Dissociative anesthetics. In W. J. Tranquilli, J. C. Thurmon, & K. A. Grimm (Eds.), *Lumb and Jones' Veterinary Anesthesia* (4th ed., pp. 301–354). Ames, IA: Blackwell Publishing.

Lin, H. C., Purohit, R. C., & Powe, T. A. (1997). Anesthesia in sheep with propofol or with xylazine-ketamine followed by halothane. *Veterinary Surgery*, **26**, 247–252.

Lipman, N. S., Marini, R. P., & Flecknell, P. A. (2008). Anesthesia and analgesia of rabbits. In R. E. Fish, M. J. Brown, P. J. Danneman, & A. Z. Karas (Eds.), *Anesthesia and Analgesia in Laboratory Animals* (2nd ed., pp. 299–334). London, UK: Academic Press.

Manning, P. J., Ringler, D. H., & Newcomer, C. E. (Eds.), (1994). *The Biology of the Laboratory Rabbit* (2nd ed.). San Diego, CA: Academic Press.

Meyer, R. E., & Fish, R. E. (2008). Pharmacology of injectable anesthetics, sedatives and tranquilizers. In R. E. Fish, M. J. Brown, P. J. Danneman, & A. Z. Karas (Eds.), *Anesthesia and Analgesia in Laboratory Animals* (2nd ed., pp. 27–82). London, UK: Academic Press.

Muir, W. W. (2007). Considerations for general anesthesia. In W. J. Tranquilli, J. C. Thurmon, & K. A. Grimm (Eds.), *Lumb and Jones' Veterinary Anesthesia* (4th ed., pp. 7–30). Ames, IA: Blackwell Publishing.

National Research Council. (1991). *Infectious Diseases of Mice and Rats*. Washington, DC: National Academy Press.

National Research Council. (1997). *Occupational Health and Safety in the Care and Use of Research Animals*. Washington, DC: National Academy Press.

National Research Council. (2010). *Guide for the Care and Use of Laboratory Animals* (8th ed.). Washington, DC: National Academy Press.

Office for the Protection from Research Risks. (1996). *Public Health Service Policy on Humane Care and Use of Laboratory Animals*. Washington, DC: National Institutes of Health.

Orlans, F. B., Beauchamp, T. L., Dresser, R., Morton, D. B., & Gluck, J. P. (1998). *The Human Use of Animals: Case Studies in Ethical Choice*. New York, NY: Oxford University Press.

Pariente, J. L., Bordenave, L., Bareille, R., Baquey, Ch., & Le Guillou, M. (2000). Cultured differentiated human urothelial cells in the biomaterials field. *Biomaterials*, **21**, 835–839.

Peyman, G. A., Rahimy, M. H., & Fernandes, M. L. (1994). Effects of morphine on corneal sensitivity and epithelial wound healing: Implications for topical ophthalmic analgesia. *British Journal of Ophthalmology*, **78**, 138–141.

Plumb, D. C. (2005). *Veterinary Drug Handbook* (5th ed.). Ames, IA: Blackwell Publishing.

Reid, W. C., Carmichael, K. P., Srinivas, S., & Bryant, J. L. (1999). Pathologic changes associated with use of tribromoethanol (Avertin) in the Sprague-Dawley rat. *Laboratory Animal Science*, **49**, 665–667.

Riebold, T. W. (2007). Ruminants. In W. J. Tranquilli, J. C. Thurmon, & K. A. Grimm (Eds.), *Lumb and Jones' Veterinary Anesthesia* (4th ed., pp. 731–746). Ames, IA: Blackwell Publishing.

Roughan, J. V., Ojeda, O. B., & Flecknell, P. A. (1999). The influence of pre-anesthetic administration of buprenorphine on the anaesthetic effects of ketamine/medetomidine and pentobarbitone in rats and the consequences of repeated anaesthesia. *Laboratory Animals*, **33**, 234–242.

Russell, W. M.S., & Burch, R. L. (1959). *The Principles of Humane Experimental Technique*. Springfield, IL: Charles C. Thomas.

Ruys, T. (1991). *Handbook of Facilities Planning*. New York, NY: Van Nostrand Rhinehold, Inc.

Sharp, P., & LaRegina, M. (1998). *The Laboratory Rat*. Boca Raton, FL: CRC Press.

Shih, A. C., Robertson, S., Isaza, N., Pablo, L., & Davies, W. (2008). Comparison between analgesic effects of buprenorphine, carprofen, and buprenorphine with carprofen for canine ovariohysterectomy. *Veterinary Anesthesia and Analgesia*, **35**, 69–79.

Smith, A. C., Ehler, W. J., & Swindle, M. M. (2008). Anesthesia and analgesia in swine. In R. E. Fish, M. J. Brown, P. J. Danneman, & A. Z. Karas (Eds.), *Anesthesia and Analgesia in Laboratory Animals* (2nd ed., pp. 413–440). London, UK: Academic Press.

Steffey, E. P., & Mama, K. R. (2007). Inhalation anesthetics. In W. J. Tranquilli, J. C. Thurmon, & K. A. Grimm (Eds.), *Lumb and Jones' Veterinary Anesthesia* (4th ed., pp. 355–394). Ames, IA: Blackwell Publishing.

Straw, B. E., Zimmerman, J. J., D'Allaire, S., & Taylor, D. J. (2006). *Diseases of Swine* (9th ed.). Ames, IA: Blackwell Publishing.

Suckow, M. A., & Douglas, F. (1997). *The Laboratory Rabbit*. Boca Raton, FL: CRC Press.

Swindle, M. M. (2007). *Swine in the Laboratory: Surgery, Anesthesia, Imaging, and Experimental Techniques*. Boca Raton, FL: CRC Press.

Taylor, D. K. (2007). Study of two devices used to maintain normothermia in rats and mice during general anesthesia. *JAALAS*, **46**, 37–41.

Thurmon, J. C., & Smith, G. W. (2007). Swine. In W. J. Tranquilli, J. C. Thurmon, & K. A. Grimm (Eds.), *Lumb and Jones' Veterinary Anesthesia* (4th ed., pp. 747–764). Ames, IA: Blackwell Publishing.

Tranquilli, W. J., Thurmon, J. C., & Grimm, K. A. (2007). *Lumb and Jones' Veterinary Anesthesia* (4th ed.). Ames, IA: Blackwell Publishing.

Turner, P. V., Chen, H. C., & Taylor, W. M. (2006). Pharmacokinetics of meloxicam in rabbits after single and repeat oral dosing. *Comparative Medicine*, **56**, 63–67.

Urrutia, J., Mardones, R., & Quezada, F. (2007). The effect of ketoprophen on lumbar spinal fusion healing in a rabbit model. *Journal of Neurosurgery: Spine*, **7**, 631–636.

Valverde, A., & Doherty, T. J. (2008). Anesthesia and analgesia of ruminants. In R. E. Fish, M. J. Brown, P. J. Danneman, & A. Z. Karas (Eds.), *Anesthesia and Analgesia in Laboratory Animals* (2nd ed., pp. 365–384). London, UK: Academic Press. pp. 385–412.

Wayman, S. (1966). Concentration camps for dogs. *Life*, **60**, 22–29.

CHAPTER II.3.7 LARGE ANIMAL MODELS IN CARDIAC AND VASCULAR BIOMATERIALS RESEARCH AND ASSESSMENT

Richard W. Bianco, Karen R. Wasiluk, Jessica M. Voight, Matthew T. Lahti, Andrew L. Rivard, and Robert P. Gallegos
Department of Surgery, University of Minnesota, Minneapolis, MN, USA

INTRODUCTION

The purpose of using animal models in preclinical evaluation of biomaterials is to provide a preliminary assessment of human safety and efficacy relative to the manufacturer claims and comparable technology that can be used as an appropriate control standard. Although absolute unequivocal results cannot be obtained regarding human safety and efficacy, use of animal models have long been employed and have positively effected changes that have resulted in improved health and increased longevity for humankind. The importance of preclinical investigation has been realized by many great investigators including Charles Darwin, who poignantly describes this link in a letter to a Swedish professor of physiology in 1881 (Darwin, 1959): "I know that physiology cannot possibly progress except by means of experiments on living animals, and I feel the deepest conviction that he who retards the progress of physiology commits a crime against mankind."

The development of new technologies, including *in vitro* methods and computer modeling, that has occurred in the last few decades has enhanced the approach to the study of physiology. However, the use of *in vivo* models still remains a necessity in investigations into certain physiologic phenomena, especially those related to the pathophysiologic mechanisms of disease, as *in vitro* studies are limited to the assessment of durability of the test material and hemodynamic performance. Traditional small animal models (e.g., rats, mice, guinea pigs, hamsters, and rabbits) have provided the scientific community access to physiological *in vivo* models to study mechanisms relating to human disease and basic biological processes. For the evaluation of cardiac and vascular biomaterials, however, investigations using larger domestic animals have proven to be more appropriate as these models incorporate anatomical and physiological characteristics more closely resembling those observed in humans. These similarities allow for the development of models where devices are implanted in a site intended for clinical use (i.e., site-specific), and allow for a more accurate prediction of the safety and clinical efficacy of either a bioprosthetic device or a biomaterial.

In this chapter, we will focus on three commonly used animal models for *in vivo* biomaterials research and testing in cardiac and vascular surgery: pigs; cows; and sheep. The chapter will begin with current recommendations for *in vivo* preclinical evaluation of cardiac and vascular devices, in both traditional and minimally-invasive settings, will continue with a short discussion on responsible use of animals in the assessment of biomaterials and biomedical devices, and will be followed by specific considerations and examples of existing animal models for each species. Finally, we will discuss conventionally-placed devices and percutaneously-placed devices, as well as potential future directions. Figure II.3.7.1 shows representative cardiac valves that are widely used clinically.

RECOMMENDATIONS FOR PRECLINICAL ASSESSMENT

The purpose of preclinical *in vivo* assessment is to provide an estimation of device safety (primary) and efficacy (secondary) in humans. This assessment is a mandatory requirement prior to the initiation of the next phase of device evaluation, namely clinical trials. The International Standards Organization (ISO), in conjunction with the American National Standards Institute, Inc. (ANSI) and the Association for the Advancement of Medical Instrumentation (AAMI), has defined regulatory requirements governing preclinical studies which are specific to cardiovascular valve prostheses (ANSI/AAMI/ISO 5840:2005 (AAMI, 2005)), and to cardiovascular implants and tubular vascular prostheses (ANSI/AAMI/ISO 7198:1998/2001/(R)2004 (AAMI, 2001)). These documents govern all aspects of evaluation, from the initial *in vitro* assessment to the final packaging requirements. These requirements must be reviewed and, if necessary, revised every five years to reflect the advancements in the rapidly evolving fields of biomaterial and bioprosthetic device development. In addition, these requirements also strive to meet the challenge of ensuring that compliant preclinical assessments provide an accurate correlation of *in vitro* and *in vivo* performance, as well as a basis for comparison with future clinical results.

The preclinical assessment of biomaterials and bioprosthetic devices is the final investigational step prior to human implantation, thus evaluation must ensure that the devices assessed will perform at least as well in humans as those bioprostheses currently in use. We will limit our discussion to the established standards for *in vivo* preclinical evaluation of cardiac valvular and tubular vascular prostheses, both those placed conventionally and those placed percutaneously. The standards that apply to the experimental methodology used in the investigation and the specific study criteria that must be met, as well as required results documentation, will be delineated. Following this, a short discussion on the future directions of these standards and the predicted impact these changes may have on the study of these types of devices will be presented.

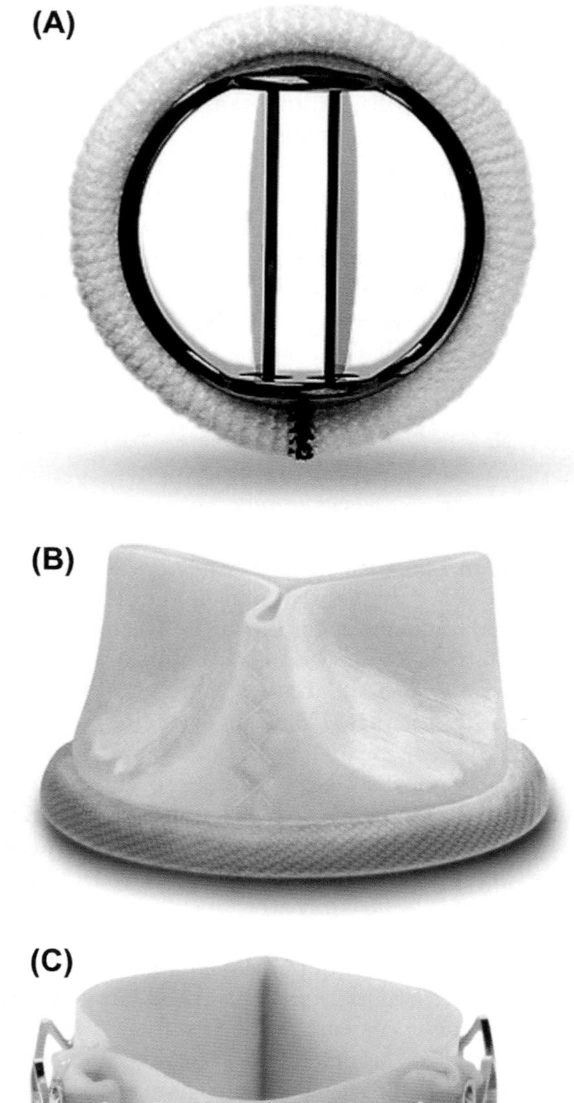

FIGURE II.3.7.1 Cardiac valves used clinically. (A) St. Jude Medical® Regent™ bileaflet mechanical aortic valve. (B) Sorin Mitroflow® bioprosthetic aortic valve. (C) Edwards Sapien bioprosthetic valve for percutaneous implantation. (Used with permission.)

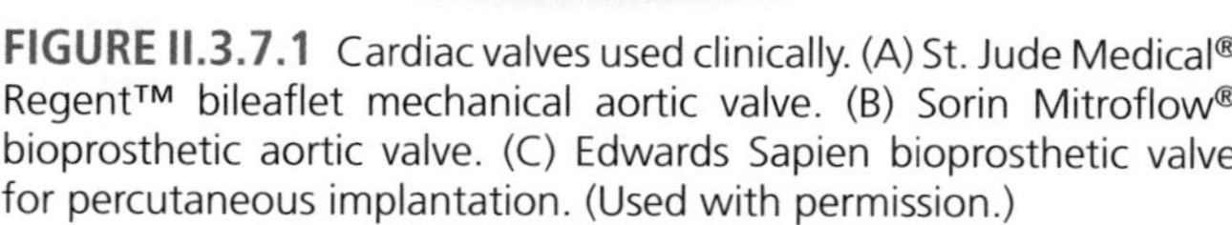

CURRENT RECOMMENDATIONS

Cardiac Devices

Rationale. Safety assessment is the primary outcome of preclinical studies, with an emphasis on risk analysis in humans. Prior to clinical trials, all new or modified heart valve designs must be evaluated in a standardized, reproducible animal model. The evaluation must use adequate numbers of the experimental devices and concurrent controls implanted in a site-specific fashion, and must involve the use of risk analysis. Neither endpoints nor a defined animal model of preclinical assessment are specified in regulatory documents. Inherent in the preclinical assessment is an evaluation of the performance characteristics of cardiac devices, such as surgical handling characteristics, hemodynamic performance, and the development of valve-related pathology. Although it is not feasible to assess the long-term durability of a cardiac device using an animal model, valuable hemodynamic performance and biological compatibility data can be obtained. Additionally, *in vivo* testing may expose unanticipated side-effects of the device. ISO 5840 guidelines were developed to guide designers and investigators, and to ensure that new or modified cardiac valvular prostheses are assessed in a uniform manner.

Major changes between ISO 5840:2005 (AAMI, 2005) and its previous iteration, ISO 5840:1996 (AAMI, 1996), include:

1. Incorporating risk-based analysis and placing the responsibility on the manufacturer to continually evaluate both known and theoretical risks of the device;
2. Detailing best practice methods for verification testing appropriate to heart valve substitute evaluation;
3. Requiring a collaborative environment between the device developer and the regulatory body regarding safety and device performance; and
4. Outlining a system to assist a surgeon in selecting a device of appropriate size for placement in a patient.

Methods. Current regulatory requirements defined in ISO 5840 have been revised to specifically address risk analysis in devices representing major design or material changes, recognizing that such changes may result in dramatic (sometimes catastrophic) changes in performance and safety. The changes to guidelines call for expansion of study design to include a minimum of 10–15 experimental devices of clinical quality, with 2–4 concurrent control devices implanted in each clinically-intended surgical site, followed by subsequent comprehensive observation for at least six months. ISO 5840 requirements state that control devices must be clinically-approved devices of a design similar to that of the experimental device, and must be constructed from the same materials, which can be problematic in the case of a newer polymeric material. Minor modifications to devices (i.e., a sewing ring alteration) may involve a less rigorous assessment, as determined by risk analysis, although it is important to accurately distinguish between a major and a minor modification (see Box 1).

In all preclinical assessments, performance criteria must take into consideration animal–device interactions, individual biological variation, and the small numbers of devices implanted for evaluation. In addition, a complete pathological examination of all animals evaluated

BOX 1

In an effort to render a mechanical heart valve more resistant to endocarditis, the sewing ring was redesigned to include impregnation with antibiotics. During *in vivo* preclinical assessment, unexpected thromboses were observed with each valve tested (see Figure II.3.7.2), resulting in the failure of this modified valve to proceed to clinical trials. Risk analysis is a critical component of preclinical assessment, and no assumptions can be made as to the importance of any modification to a device. Although there is biological variability, especially given the relatively low numbers of animals, it is important to separate outcomes that are device-related from those that are the result of the animal model chosen. However, a catastrophic failure, such as this example, must result in a complete review of the *in vivo* testing results, and reassessment of the suitability of the animal model chosen.

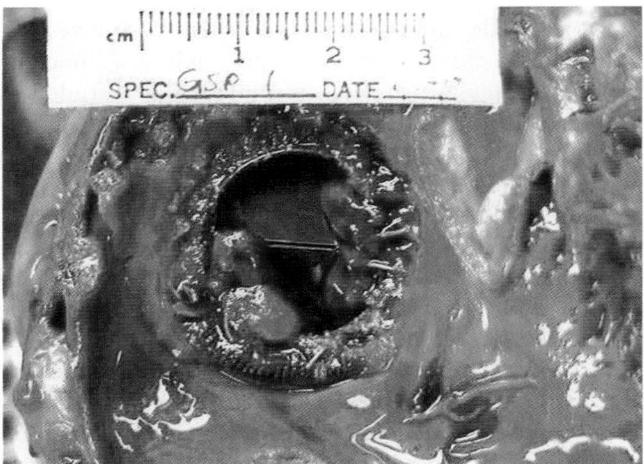

FIGURE II.3.7.2 Example of a thrombosed mechanical valve in which a minor modification to the sewing ring resulted in an unexpected outcome of the preclinical assessment and failure of the valve.

must be performed by a veterinary pathologist to distinguish model versus device complications. It is sometimes difficult to completely separate the outcomes due to the device from model-related outcomes. The exception is when the assessment of a device results in catastrophic failure.

Further specifications require each animal in which a cardiac device has been implanted to undergo a postmortem examination. This ensures that data will be obtained from all animals; whether or not they survive for the suggested minimum time of six months. An evaluation of hemodynamic performance of the device during or after the implantation period must be performed. This assessment should include measurement of the pressure gradient across the implanted device at a cardiac index of approximately 3 L/min/m^2 to mimic human conditions. Additionally, an assessment of resting cardiac output and prosthesis regurgitation should be performed. A pathological evaluation of the study animals should also be performed, and should include an assessment of the major organs for valve-related pathology; an evaluation for any device-related hematological consequences; and an assessment of any structural changes of the heart valve substitute.

ISO 5840 criteria for the evaluation in preclinical *in vivo* assessment are specific to acute or chronic settings, both of which provide valuable information. Acute settings focus on hemodynamic performance, ease of surgical handling, and acoustic characteristics of the implanted device. Chronic settings include the previously mentioned issues as well as hemolysis, potential thromboembolic complications, calcification, pannus formation/tissue ingrowth, structural deterioration and non-structural dysfunction, erosion caused by cavitation, and assessment of valve and non-valve related pathology (ANSI/AAMI/ISO 5840:2005, Annex G (AAMI, 2005)).

Previous ISO standards (ANSI/AAMI/ISO 5840:1996) relating to preclinical assessment of a device outlined requirements stating that at least six animals should be implanted with the experimental device or test article, and an additional two animals should be implanted with control devices for each assessment. A follow-up of at least six months was required for animals implanted with the test article which needed to include hemodynamic measurements, serial blood samples, and extensive pathological evaluation. No animal model was specified in the standards. These minimum requirements are rarely sufficient for adequate assessment of a bioprosthetic device, but may retain some value in the assessment of devices in which only minor modifications have been made, as long as concurrent controls are included and historical data is considered. The design of a preclinical assessment according to the updated regulatory requirements now must consider risk analysis, which is based on the device as well as the manufacturer claims attributed to it, and whether the device is novel or just an incremental modification of an existing device. Inherent in the risk-based approach outlined in the current ISO standards is encouragement for the manufacturer to strive for continued improvements in device design, as well as to ensure safety and efficacy of a device with less reliance on years of clinical assessment for verification of effectiveness.

Reporting. Documentation of results from preclinical evaluations is integral to this process and should be in the form of an official written report describing findings of the investigation in detail. This report should include the name of the institution(s) and investigator(s) involved in the study. A detailed description of the animal model used in the investigation, and the rationale for its use, should also be included. A pretest health assessment, including study animal age at the time of implantation, should also be included for each animal used in the study. A description of the operative procedure, including the suture technique, the orientation of the heart valve substitute, and any operative complications, must be detailed. Descriptions of any adjuvant procedures the animals underwent during the study period (e.g., phlebotomy, angiography) should also be provided. The results of all blood tests, including a statement of the time elapsed between implantation and procurement of the sample should be provided. Blood tests should include an evaluation of the hematologic profile (with emphasis on hemolysis), and blood chemistries. Additionally, the report should include a subjective assessment of specific surgical handling characteristics of the device and any accessories. This assessment should include a discussion of unusual and/or unique attributes the device might possess. The report must contain a complete list of medications administered to the animals during the study period. Finally, the report should contain the results obtained from the hemodynamic studies to assist in the assessment of hemodynamic performance.

Pathologic documentation should include a gross and microscopic report on each animal in which a heart valve substitute was implanted, including any animal that did not survive the minimum post-implantation period. Documentation should include visual records of the heart valve substitute *in situ*, and evidence of any thromboembolic events occurring in the major organs. The cause of death of any animal that was not euthanized at the study endpoint must be investigated and reported. Detailed examination of the explant must be performed, with specific attention to any structural changes in the device. If appropriate, further *in vitro* functional studies of the heart valve substitute should be undertaken, such as hydrodynamic testing.

Pathologic assessment is crucial in the evaluation of the safety profile of the test device, and input from both a veterinary pathologist and a physician pathologist are ideal and necessary for the complete assessment of device safety. It is the role of the veterinary pathologist to decipher whether any pathologic findings are due to device malfunction or the inherent physiology of the animal. It is then the role of the physician pathologist to assess the *in situ* findings, and correlate them with a potential clinical performance in humans.

Vascular Devices

Rationale. As with cardiac devices, the purpose of *in vivo* preclinical testing of vascular prostheses is to evaluate the characteristics of a prosthetic that are difficult, if not impossible, to obtain using *in vitro* testing models. The capacity of the prosthesis to maintain physiologic function when used in the circulatory system, the biologic compatibility of the prosthetic, and the surgical handling characteristics of the prosthesis are all important features that can be evaluated using an *in vivo* model and cannot be obtained with *in vitro* testing methods. This testing is not intended to demonstrate the long-term performance of the prosthesis, but rather the short-term (less than 20 weeks) response and patency of the prosthesis being investigated.

Methods. In designing an experimental protocol that will ensure clinical relevance, researchers should consider the intended application of the device being studied, as well as the necessary diameter and length of the prosthesis. Additionally, investigators should reflect on any specific biological characteristics of the chosen animal model, and the impact of those characteristics on their conclusions. Current ISO requirements state that a rationale for the use of a particular species, the site of implantation, the selection of a control prosthesis, the method and interval of patency observations, and the number of animals used in each group being studied be provided. Each type of prosthesis shall have been tested by implantation at the intended, or at an analogous, vascular site in not fewer than six animals for not less than 20 weeks in each animal, unless a justification for a shorter-term study can be provided. A prosthesis must not be tested in the species from which it was derived, unless appropriate justification can be provided. Angiography or Doppler should be used to monitor the duration of patency for each prosthesis, and the results recorded. Each investigation should include a control group in which a clinically-approved vascular prosthesis is implanted and studied in the same fashion as the experimental prosthesis.

Preoperative data should include the sex and weight of the animal, documentation of satisfactory preoperative health status, and any preoperative medications. Operative data must include the name of the implanting surgeon and a detailed description of the surgical procedure, to include the type and technique used in performing the proximal and distal anastomoses. The *in situ* length and diameter of the prosthesis, in addition to any adverse intraoperative events (e.g., transmural blood leakage), must also be documented. Postoperative medications, patency assessments (method and interval from implantation), and any adverse event or deviation from protocol must be noted. Loss of patency before the intended study duration does not necessarily exclude the animal from the study population used to assess prosthetic function and host tissue response. Reports from all animals implanted with either test or control prostheses, including those excluded from the final analyses, shall be recorded. Termination data should include an assessment of prosthesis patency (documenting the method used), and an assessment of prosthesis explant pathology.

Reporting. At the conclusion of the study, a detailed report should be compiled in which the study protocol is delineated. The report should include: the rationale for selection of animal species; implantation site; control prosthesis; method of patency assessment; and the intervals of observation. Additionally, documentation of sample size and pertinent operative, perioperative, and termination data must be presented. The results should include the patency rates and any adverse event that was encountered during the study interval, and should be compared between the study and control groups. All animals entered into the study must be accounted for in the report. If data are excluded a rationale must be provided. A summary of the gross and microscopic pathology, including photographs of the prosthesis *in situ* and micrographs, should also be provided. In addition to these objective results, a subjective statement of the investigator's opinion of the device must be included. Again, for this reason, it is important to include the input of a veterinary pathologist to decipher device malfunction from animal physiology, as well as a physician pathologist to correlate pathologic findings with potential clinical performance. Finally, conclusions drawn from the study and a summary of the data auditing procedures should be documented.

RESPONSIBLE USE OF ANIMALS

Introduction

As discussed in greater detail in Chapter II.3.6 of this book (Animal Surgery and Care of Animals), there are two basic principles governing the use of animals in research, education, and testing: (1) scientific reliance on the use of live animals should be minimized; (2) pain, distress, and other harm to laboratory animals should be reduced to the minimum necessary to obtain valid scientific data. Strict adherence to these basic ethical principles will not only enhance the quality of each *in vivo* preclinical evaluation, but will also ensure that current and future generations of scientists will be able to employ all valid tools available to predict clinical safety of either new or modified medical devices and other technology designed to improve human health.

Investigator and Institutional Responsibilities

When considering appropriate animal models for biomaterials research, the investigator and sponsoring institution should first assess their responsibilities with respect to federal, state, and local laws and regulations. Additionally, given the growing public and political sensitivity to the use of research animals, all laboratories using animals must recognize and comply with current definitions of humane animal care and use. The importance of this issue is reflected in a resolution adopted by the American Association for the Advancement of Science, February 19, 1990. This section briefly presents legal obligations and responsibilities of both the institutional administration's veterinary care program and the investigator, and provides some practical guidelines for responsible use of laboratory animals.

The Animal Welfare Act (Public Law 89-544, as amended) provides federal regulations governing laboratory animal care and use (Title 9 of the Code of Federal Regulations (CFR), Chapter 1, subchapter A: Animal Welfare, Parts 1, 2, 3) which are enforced by the Animal and Plant Health Inspection Service (APHIS), US Department of Agriculture. The Animal Welfare Act covers specific species including dogs, cats, non-human primates, guinea pigs, hamsters, rabbits, and "any other warm-blooded animal, which is used or is intended for use in research, teaching, testing, experimentation ..." (9 CFR subchapter A, Part 1, Section 1.1). Species specifically exempt from the Animal Welfare Act include birds, rats, and mice bred for use in research, as well as horses and livestock species used in agricultural research. Institutions using species covered by the Animal Welfare Act must be registered by APHIS. Continued registration is dependent on the submission of annual reports to APHIS by the institution, as well as maintaining a satisfactory rating upon inspection of the institution's animal facility during unannounced site visits by APHIS inspectors. The Animal Welfare Act provides specifications for animal procurement (i.e., from licensed suppliers), husbandry, and veterinary care that are used to determine compliance with the Animal Welfare Act. The annual report supplied to APHIS should contain a list of all species and the numbers of animals used by the institution in the previous year. The animals listed in the annual report must be categorized by the level of discomfort or pain they were thought to experience in the course of the research, education or testing process they were used for.

In 1985 amendments were made to the Animal Welfare Act (7 U.S.C. 2131, *et seq.*) and ultimately implemented on October 30, 1989, August 15, 1990, and March 18, 1991. These amendments extended the Act to require institutions to perform administrative review and provide a mechanism to control animal research programs (Federal Register, Vol. 54, No. 168, pp. 36112–36163; Vol. 55, No. 36, pp. 28879–28881; Vol. 56, No. 32, pp. 6426–6505). Specifically, the new regulation requires that all animal research protocols be reviewed and approved by an Institutional Animal Care and Use Committee (IACUC) before the initiation of the study. Furthermore, the submitted protocols must state in writing that less harmful alternatives were investigated, with justification as to why those methods are not suitable, and that the proposed research is not unnecessarily duplicative. Additional requirements increase the scope of husbandry requirements for laboratory dogs and non-human primates.

The other relevant federal body is the National Institutes of Health (NIH). The Health Research Extension Act of 1985 (Public Law 99-158) required the director of NIH to establish guidelines for the proper care of laboratory animals, and gave IACUC oversight of that care. Broad policy is described in the Public Health Service Policy on Humane Care and Use of Laboratory Animals (NIH, 2002), which identifies the Guide for the Care and Use of Laboratory Animals (NIH, 1985a) as the reference document for compliance. This policy applies to all activities involving animals either conducted or supported by the US Public Health Service (PHS). An institution cannot receive funding from the NIH or from any other PHS agency for research involving the use of animals, without a statement of assurance on file with the PHS Office for Protection from

Research Risks (OPRR) ensuring compliance with these guidelines.

As required by the Animal Welfare Act, PHS policy mandates that each institution provide an annual update on animal research use and IACUC review of animal protocols and facilities to be submitted to the Office of Laboratory Animal Welfare (OLAW) within the NIH. Unlike the Animal Welfare Act, however, the PHS policy covers all vertebrate species, does not include an enforcement arm for routine inspections (but does provide for inspection in cases of alleged misconduct), and penalizes noncompliant institutions by withdrawing funding support.

State and local laws and regulations may also affect animal research programs. This legislation imposes restrictions on the acquisition of cats and dogs from municipal shelters. Within the past decade several states have also enacted registration and inspection statutes similar to the Animal Welfare Act. In several states, court rulings have required IACUC reports and deliberations at state-supported institutions to be conducted in public (e.g., Florida, Massachusetts, North Carolina, and Washington).

Housing and Handling

The physical and psychological well-being of laboratory animals is determined in large part by their environment. Careful attention to housing and handling of any investigational animal is crucial to avoid improper handling or care which may result in exposure of the animal to undue stress. The Institute of Laboratory Animal Resources of the National Academy of Sciences publishes a detailed guide with specific recommendations for the housing of many laboratory animal species (*Guide for the Care and Use of Laboratory Animals*, Commission on Life Sciences, 2011). Adequate assessment of available facilities and their compliance with these regulations must be considered an essential component of any animal investigation. This assessment should be done not only for regulatory affairs compliance, but also for scientific accuracy, as data obtained from subjects that are not cared for properly may be inaccurate.

Euthanasia

In addition to housing and handling, investigators must become familiar with the methods appropriate to the practice of humane euthanasia techniques. Indications for the application of euthanasia in a study include such inviolable situations such as: (1) study protocol completion; or (2) undue suffering of an individual study subject that may be displaying cachexia, anorexia or a moribund state. In this phase of the investigation, as in all other phases of an animal investigation, great care should be taken to avoid subjecting the animal to undue stress. Humane euthanasia techniques used should result in rapid unconsciousness followed by cardiac and respiratory arrest. The 1993 Report of the American Veterinary Medical Association (AVMA, 1993) provides descriptions of acceptable techniques for euthanasia, and is an invaluable resource for investigators in the process of designing and performing a study; the most recent revision of these guidelines was in 2007, with a minor update in 2009 (AVMA, 2007; Nolen, 2009).

ANIMAL MODELS AND SPECIES CONSIDERATION

Choice of an Appropriate Animal Model

In determining which model is most appropriate for an *in vivo* assessment to best evaluate the likely human response to the device, it is first necessary and most important to outline the clinical goals of the bioprosthetic test device and choose the animal model that has been shown to permit control of the greatest number of variables. The closer the *in vivo* assessment parallels a clinical situation, the more clinically applicable are the results of the assessment. That being said, the sheep is often the superior model in cardiovascular research, as it is relatively easy to control many variables. In addition, the physiological parameters of sheep often approximate the human condition, especially with respect to thrombogenicity. In contrast, swine tend to be exquisitely sensitive to exogenous anticoagulation therapy in an unpredictable fashion. In our experience, when anticoagulation therapy is utilized in swine, the results generated are a series of anecdotes that relate more to each individual animal's differing level of thrombogenicity, rather than to a herd response to a device. If anticoagulation therapy is necessary, it is most important to consider the entire herd of test and control subjects, and dose them with equal amounts of anticoagulation therapy so that any effect this may have on the study outcome will be demonstrated in each of the animals. However, an ideal study protocol should incorporate no anticoagulation therapy for any animal, because it naturally crafts a more aggressive assessment of the test device and there is no need to address the variable effect of anticoagulation therapy in each animal. In general, although representative drugs and dosages may be stated in this chapter, the investigator with direct responsibility for animal studies is well advised to consult an experienced veterinarian for selections and dosages of drugs, as practice may change over time and unforeseen complications may arise.

Canine

The history of the use of canines in experimental surgery is extensive. Early experimental surgery required an animal model that was inexpensive, readily available, and easily managed. The solution to this problem came in the form of mongrel dogs. Dogs could be obtained from pounds inexpensively and were often familiar with humans.

Cardiopulmonary bypass, coronary artery bypass, and early valve studies were all developed primarily with the canine model. The evaluation of artificial heart valves began in the 1950s, using a canine model and valves created from materials such as polyurethane (Akutsu et al., 1974; Doumanian and Ellis, 1961). The canine model is suitable for testing both synthetic and biosynthetic vascular grafts used for coronary artery bypass. The canine model represents the gold standard for *in vivo* models of coronary artery bypass, and offers several advantages including: (1) a large historical database for comparison; (2) ease of cardiopulmonary bypass; and (3) the availability of autologous saphenous vein for grafting allows researchers to easily perform positive controls.

Interestingly, few studies involving PCI (percutaneous coronary intervention) for stent placement have been performed using this gold standard model (Schatz et al., 1987; Roubin et al., 1990). Because of the recent shift away from the canine as a model for chronic device evaluation, the current standard uses a swine model. Overall, the canine model has provided a long history of significant contributions to our understanding of cardiovascular physiology, and to the development and testing of cardiovascular biomaterials and devices. In more recent times, there has been a shift away from the canine model for many studies, because of public concerns and rising animal costs. Today, the emotional import of using canines for research has prohibitively increased their cost, and society no longer deems dogs appropriate for chronic studies. As a result, models using other large domestic animals have become widely used for ongoing biomaterials research.

Swine

Introduction. It is believed that the majority of the breeds of swine we now know are descended from the Eurasian wild boar (*Sus scrofa*). Archaeological evidence from the Middle East indicates domestication of the pig occurred as early as 9000 years ago, with some evidence for domestication even earlier in China. From here, the domestic pig spread across Asia, Europe, Africa, and ultimately North America (Towne and Wentworth, 1950; Mellen, 1952; Clutton-Brock, 1999a). Today, swine serve many purposes in society, ranging from food source and pets to biomedical research subjects. In this latter role, swine have proven useful during the past four decades in studying a variety of human ailments, including cardiovascular (Ramo et al., 1970; Cevallos et al., 1979), gastrointestinal (Kerzner et al., 1977; Leary and Lecce, 1978; Pinches et al., 1993), and hepatic diseases (Mersmann et al., 1972; Soini et al., 1992; Sielaff et al., 1997). Studies in swine have also provided extensive information in the areas of organ transplantation (Calne et al., 1974; Marino and De Luca, 1985; Grant et al., 1988; Al-Dossari et al., 1994; Granger et al., 1994).

Multiple authors have reported advantages of using swine in large animal models for biomedical research including, but not limited to, anatomic and physiologic similarities to humans, low cost, availability, and the ability of swine to produce large litters of hearty newborns (Stanton and Mersmann, 1986; Swindle, 1992; Tumbleson and Schook, 1996; Swindle, 2007). In addition, naturally occurring models of human disease (e.g., atherosclerosis) can be found in various strains of swine, making them ideal candidates for the study of these disease processes. Also helpful to investigators using swine is a large body of agricultural literature in the areas of swine nutrition, reproduction, and behavior that may have applications in biomedical research. For all these reasons, the use of swine has become increasingly popular in the biomedical research laboratory.

Comparative Anatomy and Physiology

Swine are the single readily-available species in which cardiovascular anatomy and physiology most closely resembles those in humans, likely because of a similar phylogenetic development that led to an omnivorous species with a cardiovascular system that has accommodated to a relative lack of exercise (McKenzie, 1996). The heart of an adult swine weighs 250–500 grams and comprises approximately 0.25% of the body weight (Ghoshal and Nanda, 1975), similar to that of humans. The coronary vasculature of a pig heart is nearly identical to that of humans in anatomic distribution, reactivity, and paucity of collateral flow (Hughes, 1986). The right and left coronary arteries are similar in size to each other. The left coronary artery supplies the greater part of the wall of the left ventricle and auricle, including the interventricular septum, by means of its circumflex and paraconal interventricular branches. The right coronary artery supplies the wall of the right ventricle and the auricle via its circumflex and subsinuosal branches (Ghoshal and Nanda, 1975). These similarities make swine ideal for studies of both ischemia/reperfusion and coronary stent technology.

There are certain cardiovascular anatomic differences between swine and humans that must be considered when using swine as investigational subjects. First, the left azygous vein drains directly into the right atrium in swine, compared to drainage into the brachiocephalic vein in humans (Ghoshal and Nanda, 1975). Second, there are only two arch vessels coming off the aorta in swine. The most proximal of these is the brachiocephalic trunk, which gives rise to the right subclavian and a common carotid trunk that bifurcates into the right and left carotid arteries. The second arch vessel is the left subclavian artery (Ghoshal and Nanda, 1975).

Electrophysiological parameters of pigs more closely resemble those of humans than any other non-primate animal. These parameters include similar P- and R-wave amplitudes, P–R intervals of 70–113 msec (longer in older animals), mean QRS duration of 39 msec, a Q–T interval of 148–257 (again longer in older animals), as well as ventriculoatrial (V–A) conduction when the right ventricular endocardium is stimulated faster than normal sinus rate (Hughes, 1986). These similarities make the pig a good model for studying pacemakers. Some investigators have found swine to be highly susceptible to ventricular arrhythmias, often requiring prophylactic pharmacologic protection (i.e., lidocaine, flecainide or bretylium). Alternatively, other investigators have exploited this ventricular instability to study "induced" tachyarrhythmias, and the pharmacologic agents and/or electrical interventions used in their management (Hughes, 1986).

Physiologically, swine have hemodynamic and metabolic values that are similar to those of humans (Tables II.3.7.1 and II.3.7.2). They also share similar

TABLE II.3.7.1 Hemodynamic Values of Species Used to Evaluate Cardiac and Vascular Bioprostheses

Hemodynamic Values	Bovine	Swine	Ovine	Human
Heart rate (beats/min)	119 ± 27	105 ± 10	95 ± 24.2	70 ± 14
Mean arterial pressure (mm Hg)	117 ± 18	102 ± 9	70 ± 24	95 ± 11
Systolic pressure (mm Hg)	137 ± 9	127 ± 8	85 ± 3	126 ± 14
Diastolic pressure (mm Hg)	108 ± 15	86 ± 7	58 ± 15	79 ± 10
Stroke volume (ml/beat/kg)	91.5 ± 23.3	1.34 ± 0.26	1.03 ± 0.78	1.14 ± 0.31
Cardiac index (ml/min/kg)	113 ± 42	99 ± 19.9	115 ± 30.8	93 ± 20
Systemic vascular resistance (dyn sec cm^{-5})	1168.5 ± 176.1	2759 ± 70	1463 ± 183	1200 ± 600
Pulmonary vascular resistance (dyn s cm^{-5})	73.3 ± 51.2	441 ± 62	180 ± 53	120 ± 30
Resting total O_2 consumption (O_2/min/kg)	Unknown	6.6 ± 1.27	6.7 ± 3.0	3.9 ± 1.9

Values represent mean ± standard deviation.
Table created from: McKenzie, J. E. (1996). Swine as a model in cardiovascular research. In M. E. Tumbleson and L. B. Schook (Eds.), *Advances in Swine in Biomedical Research*. Plenum Press; New York, NY, pp. 7–17; Gross, D. R. (1994). *Animal Models in Cardiovascular Research*, 2nd edn. Kluwer Academic Publishers: Dordrecht, The Netherlands; Kaneko, J. J., Harvey, J. W. & Bruss, M. L. (1997). *Clinical Biochemistry of Domestic Animals*, 5th edn. Academic Press: San Diego, CA; Fauci, A. S., Braunwald, E., Isselbacher, K. J., Wilson, J. D. & Martin, J. B. et al. (1998). *Harrison's Principles of Internal Medicine*. McGraw-Hill: New York, NY.

TABLE II.3.7.2 Metabolic and Hematologic Values of Species Used to Evaluate Cardiac and Vascular Biomaterials

Metabolic/Hematologic Values	Bovine	Swine	Ovine	Human
Arterial pH	7.45 ± 0.15	7.48 ± 0.03	7.45 ± 0.06	7.41 ± 0.03
Arterial pCO_2 (mm Hg)	42 ± 4	40 ± 2.3	38 ± 8.5	40 ± 5
Arterial pO_2 (mm Hg)	92.5 ± 7.5	82 ± 4.2	88 ± 7.5	90 ± 10.0
Arterial plasma HCO_3^- (mEq/L)	27 ± 3	29 ± 2.2	26.6 ± 5.5	25.5 ± 4.5
Sodium (mEq/L)	142 ± 10	138 ± 3.5	152 ± 12.0	140.5 ± 4.5
Potassium (mEq/L)	4.9 ± 0.9	4.4 ± 0.4	5.5 ± 1.3	4.3 ± 0.8
Chloride (mEq/L)	104 ± 7.0	106 ± 7.8	117 ± 4.0	102 ± 4.0
Bicarbonate (mEq/L)	27 ± 3	29 ± 2.2	27 ± 6.0	26 ± 5.0
Urea (mM/L)	15 ± 5 mg/dl	3.2 ± 1.2	5.0 ± 2.1	5.4 ± 1.8
Creatinine (mM/L)	1.5 ± 0.5 mg/dl	89 ± 19.5	137.0 ± 31.0	<133
Calcium (mg/dl)	10.8 ± 1.4	4.80 ± 0.29	12.15 ± 0.65	5.05 ± 0.55
Magnesium (mg/dl)	2.8 ± 0.3	1.4 ± 0.2	2.1 ± 0.3	2.0 ± 0.6
Phosphate (mg/dl)	Unknown	4.0 ± 0.6	6.2 ± 1.2	3.8 ± 0.8
Albumin (g/dl)	2.7 ± 0.3	7.6 ± 0.8	3.6 ± 0.9	4.5 ± 1.0
Globulin (mEq/L)	Unknown	6.1 ± 0.9	7.4 ± 1.7	6.8 ± 1.3
Glucose (mM/L)	45 ± 10 mg/dl	4.6 ± 0.7	3.6 ± 0.8	5.3 ± 1.1
Lactate (mM/L)	Unknown	1.0 ± 0.3	1.2 ± 0.2	1.2 ± 0.6
WBC (cells × 103/ml)	8 ± 4	14.8	8 ± 4	7.4 ± 3.4
Hemoglobin (g/dl)	11 ± 1.5	12	12 ± 3	15 ± 3
Hematocrit (%)	35 ± 11	41	36 ± 9	44.5 ± 7.5
Platelets (Platelets × 109/L)	450 ± 350	350 ± 150	550 ± 250	265 ± 135

Values represent mean ± standard deviation.
Table created from: Hannon, J. P., Bossone, C. A. & Wade, C. E. (1990). Normal physiologic values for conscious pigs used in biomedical research. *Laboratory Animal Science*, **40**(3), 293–298; McKenzie, J. E. (1996). Swine as a model in cardiovascular research. In: *Advances in Swine in Biomedical Research*. Plenum Press: New York, NY, pp. 7–17; Fauci, A. S., Braunwald, E., Isselbacher, K. J., Wilson, J. D. & Martin, J. B. et al. (1998). *Harrison's Principles of Internal Medicine*. McGraw-Hill: New York, NY; Kaneko, J. J., Harvey, J. W. & Bruss, M. L. (1997). *Clinical Biochemistry of Domestic Animals*, 5th edn. Academic Press: San Diego, CA; Gross, D. R. (1994). *Animal Models in Cardiovascular Research*, 2nd edn. Kluwer Academic Publishers: Dordrecht, The Netherlands.

pulmonary function parameters, such as respiratory rate and tidal volumes (Willette et al., 1996). Although they have lower levels of hemoglobin, circulating red cell volume, arterial oxygen saturation, and markedly lower venous oxygen saturations, they have similar platelet function and lipoprotein patterns. Other physiologic characteristics that may play a role in experimentation include a higher core body temperature, a contractile spleen that sequesters 20–25% of the total red cell mass, very fragile arteries and veins, a higher plasma pH, and a higher plasma bicarbonate level. Swine also have 50% more functional extracellular space than humans, and a slightly increased volume of total body water (Swindle et al., 1986; Hannon et al., 1990). However, individual swine test subjects tend to be more sensitive to anticoagulation therapy. Thus, if anticoagulation is utilized during the procedure, each animal will likely require a different dosage of anticoagulation, resulting in a series of individual anecdotes that are not easily summarized or generalized.

Perioperative Care

Anesthesia. Swine should be fasted at least 12 hours prior to the induction of anesthesia, longer if one desires the large bowel to be empty prior to surgery. Unless gastric surgery is to be performed, the animals may be given water without restriction. It is common to administer an intramuscular (IM) pre-anesthetic agent followed by an intravenous (IV) anesthetic agent after IV access has been obtained. IM injections can safely be administered in the neck just behind the ear, in the triceps muscle of the forelimb or in the semimembranosus–semitendinosus muscles of the hindlimb. IV access is most commonly obtained in the lateral auricular veins (Riebold and Thurmon, 1985; Swindle and Smith, 1994; Swindle, 1994; Smith et al., 2008). The pre-anesthetic/anesthetic protocol used in this laboratory involves the administration of an IM injection of a combination drug (Telazole) consisting of a dissociative anesthetic and a tranquilizer (tiletamine and zolazepam) at a dose of 2–4 mg/kg for sedation. Following this, IV thiopental sodium (6–25 mg/kg) is administered if necessary. Endotracheal (ET) intubation follows and anesthesia is maintained via inhaled isoflurane (1–2 volume %). Commonly used anesthetics, doses, and routes of delivery are given in Table II.3.7.3.

Analgesia. Postoperative analgesia is a very important component of any animal experiment. Appropriate use of postoperative analgesia involves rapid recognition of discomfort and appropriate therapy. When in pain or distress swine will demonstrate changes in social behavior, gait, posture, and a lack of bed-making. Most swine will become very reluctant to move when in pain and, if forced to move, will vocalize with even greater enthusiasm than they generally display (Gross, 1994). Buprenorphine is currently the analgesic of choice used in swine, effective when administered at 0.05–0.1 mg/kg every 8–12 hours (Swindle and Smith, 1994; Swindle, 1994). Standard narcotics such as morphine and fentanyl are effective pain relievers, but have a very short half-life in swine. Additional analgesic agents are listed in Table II.3.7.4.

EXISTING MODELS

Cardiac Devices

Orthotopic valve replacement in large animals is an important component of preclinical prosthetic valve assessment for the initial evaluation of surgical handling characteristics, hemodynamic performance, and valve-related pathology. Early investigators using swine for cardiovascular research described difficulties with venous access (Swan and Piermattei, 1971), anesthesia (Piermattei and Swan, 1970), cardiopulmonary bypass (Swan and Meagher, 1971), and several anatomic peculiarities (Swan and Piermattei, 1971). Subsequently, cardiovascular surgical procedures using swine have become technically feasible, and several investigators have conducted acute and chronic studies for the assessment of hemodynamic profiles of cardiac prostheses (Hasenkam et al., 1988a,b, 1989; Hazekamp et al., 1993). However, only a few long-term swine studies have been performed to examine the potential of prosthetic heart valve implants for thrombogenicity (Gross et al., 1997; Henneman et al., 1998; Grehan et al., 2000).

It has been reported that heart valve replacement in swine requires certain measures not needed with other species. These include the use of a crystalloid prime without plasma volume expanders (especially starch-based), prophylactic administration of pharmacological protection against ventricular arrhythmias, insurance of adequate hypothermic cardioprotection during the time of cross-clamp, administration of "shock" doses of corticosteroid just prior to reperfusion, and the use of inotropic support during bypass weaning (Gross et al., 1997). In our experience, (see later discussion) however, we have found some of these precautions unnecessary for the successful use of the pig in the evaluation of cardiac prostheses.

In our investigation, 22 swine underwent mitral valve replacement with no operative deaths. All animals were weaned from cardiopulmonary bypass without inotropic assistance. No cardioplegic protection was used, and only mild total body hypothermia was instituted. Pathologic analysis of valves explanted from animals surviving more than 30 days demonstrated extensive fibrous sheath formation leading to valve orifice obstruction and restriction of leaflet motion in a significant number of animals. The extensive fibrous sheath formation (and subsequent valvular dysfunction) represents a chronic tissue response observed in many species following mitral valve replacement. This chronic tissue response, however, was noted to have developed sooner in swine

TABLE II.3.7.3 Anesthetics and Sedatives Acceptable for Use in Species Used to Evaluate Cardiac and Vascular Biomaterials

	Dose/kg			Route of Delivery				
Drug Name	**Bovine**	**Swine**	**Ovine**	**PO**	**SQ**	**IM**	**IP**	**IV**
Preanesthetic anticholinergics								
Atropine	0.05 mg	0.05–0.5 mg	0.05 mg		X	X		X
Glycopyrrolate		0.011 mg				X		
Phenothiazine/buterophenone sedatives								
Acepromazine or chlorpromazine	0.05–0.2 mg	0.05–0.2 mg	0.05–0.2 mg		X	X		X
		0.5–2 mg		X				
Azaperone (Stresnil)		2.2 mg				X		
Benzodiazapene sedatives/anxiolytics								
Diazepam (Valium) – Class IV Controlled	0.25–0.5 mg	0.25–0.5 mg	0.25–0.5 mg					X
	2.5–5 mg	2.5–5 mg	2.5–5 mg			X	X	
Midazolam (Versed) – Class IV Controlled	0.5 mg	0.1–1 mg	0.5 mg			X		X
Zolazepam (*see* Telazol)								
Thiazine sedatives								
Xylazine	0.05–0.2 mg	2–4 mg	0.05–0.2 mg		X	X		X
Medetomidine		0.25 mg				X		
Narcotics/opiates								
Innovar-Vet (Fentanyl/droperidol) – Class II Controlled		0.1 ml				X		
Butorphanol (Torbugesic)	0.01–0.05 mg	0.05–0.5 mg	0.01–0.05 mg		X	X		X
		0.05–0.5 mg					X	
Morphine – Class II Controlled		0.5–1 mg			X	X		
Oxymorphone – Class II Controlled		0.15 mg			X	X		X
Barbituate anesthetics								
Phentobarbitol – Class II Controlled	15–30 mg	25–30 mg	15–30 mg					X
Thiamylal/thiopental – Class III Controlled	2–5 mg	6–10 mg	2–5 mg					X
Dissociative anesthetics/cataleptics								
Ketamine	10–30 mg	10–30 mg	10–30 mg			X		X
Telazol (Tiletamine/Zolazepam) – Class III Controlled	2.2 mg	5–10 mg	2.2 mg			X		X
Other anesthetics/hypnotics								
Propofol	7.5–15 mg							X
Alpha chloralose – Class IV Controlled	50–100 mg	55–86 mg	35–62 mg					X

PO, oral; SQ, subcutaneous; IM, intramuscular; IP, intraperitoneal; IV, intraveneous.
Table adapted from the *University of Minnesota Animal Care and Use Manual* 1997 reprint.

TABLE II.3.7.4 Analgesics Commonly Used when Testing Biomaterials in Large Animal Models

	Dose				Route of Delivery[b]			
Drug Name	**Bovine**	**Swine**	**Ovine**	**Frequency[a]**	**PO**	**SQ**	**IM**	**IV**
Nonsteroidal anti-inflammatory drugs								
Acetaminophen	–	–	–	TID	X			
Aspirin		25–50 mg/kg		BID	X			
Flunixin meglumine (Banamine)	1.1 mg/kg	1.1 mg/kg	1.1 mg/kg	QD			X	X
Ketoprofen	1–2 mg/kg	1–2 mg/kg	1–2 mg/kg	TID	X			X
Ketorolac (Toradol)	0.3–0.7 mg/kg	0.3–0.7 mg/kg	0.3–0.7 mg/kg	TID	X	X	X	X
Narcotic agonist/antagonists								
Buprenorphine (Buprenex)	0.005–0.02 mg/kg	0.1 mg/kg	0.005–0.02 mg/kg	BID–TID			X	
Butorphanol (Torbugesic)	0.2–0.4 mg/kg	0.2–0.4 mg/kg	0.2–0.4 mg/kg	q4h			X	
Nalbuphine (Nubain)	0.75–3 mg/kg	0.75–3 mg/kg	0.75–3 mg/kg	q4h			X	
Naloxone	0.04 mg/kg	0.04 mg/kg	0.04 mg/kg	q4h			X	
Narcotic agonist								
Morphine (15 mg/ml)	0.25–2 mg/kg	0.25–2 mg/kg	0.25–2 mg/kg	QID			X	X
Topical/regional anesthetics								
Lidocaine (xylocaine)	These are applied topically to painful wounds			TID				

[a]QD, once daily; BID, twice daily; TID, three times daily; QID, four times daily; q4h, every 4 hours.
[b]PO, oral; SQ, subcutaneous; IM, intramuscular; IV, intravenous.
Table adapted from the *University of Minnesota Animal Care and Use Manual* 1997 reprint.

than in our previous studies involving prosthetic valve implantation into other species (i.e., sheep). Additionally, we did not find this model to be useful in predicting device-related thrombogenicity, as there were no clinical or pathologic differences observed among the valve designs studied (Grehan et al., 2000). The utility of swine in predicting the thrombogenic potential of cardiac valvular prostheses is therefore limited.

Improved biotechnology has brought left ventricular assist devices (LVAD) into the clinical arena as a bridge to transplantation therapy for patients with end-stage cardiac disease or those suffering from acute myocardial infarction (Frazier et al., 1992; Oz et al., 1997; Park et al., 2000). Clinical observations of patients who have received LVAD have revealed that these patients have a propensity for developing right ventricular dysfunction. To study this phenomenon, a model of iatrogenic congestive heart failure (CHF) using seven days of rapid pacing to induce cardiomyopathy has been developed in swine (Chow and Farrar, 1992). This model has been used to investigate the hemodynamics and septal positioning of the assist devices that may contribute to right ventricular dysfunction (Chow and Farrar, 1992; Hendry et al., 1994).

Vascular Devices

Initially, the pig was developed as a model to study solid organ transplantation. However, more recently its use has made an impact on the study of vascular disease. The pig is currently the species of choice for *in vivo* evaluation of vascular stents, restenosis biology, and balloon injury. The advantages of using swine include: (1) easy access to coronary arteries with present catheterization and angioplasty techniques; (2) coronary arteries of sufficient size for catheters used in adult humans; (3) balanced coronary circulation that is anatomically similar to humans; (4) spontaneous development of atherosclerosis; and (5) comparable coagulation and fibrinolytic systems and lipid metabolism to humans (Fritz et al., 1980; White et al., 1989; Karas et al., 1992; Willette et al., 1996).

Numerous *in vitro* studies indicate that the coagulation and fibrinolytic systems in swine closely resemble those of humans (Wilbourn et al., 1993; Karges et al., 1994; Reverdiau-Moalic et al., 1996; Gross, 1997), which would suggest that this may be the ideal species for the preclinical evaluation of the thrombogenic potential of a prosthetic device. *In vivo* studies specifically looking at the thrombogenicity of implantable devices, both cardiac and noncardiac, have already been performed in swine (Rodgers et al., 1990; Walpoth et al., 1993; Scott et al., 1995). Previous *in vivo* studies of artificial valves identified a relationship between postoperative thrombus formation and bacteremia. Other significant factors impacting survival include a combination of preoperative antibiotics, postoperative anticoagulation, short bypass times, and aseptic blood sampling techniques (Bianco et al., 1986). Table II.3.7.5 outlines various antibiotics used in cardiovascular *in vivo* studies and their suggested dosing regimens.

Similar investigations have shown that stent thrombosis may occur early in swine *in vivo* models (within 6 hours of implantation), suggesting that if a pig survives the first 12 hours (coronary stent remains patent), it is highly likely that it will not suffer a stent occlusion later (Schwartz and Holmes, 1994). Observations of such rapid restenosis led Schwartz and Holmes to suggest that the pig should be used as an accelerated thrombotic model capable of predicting stent thrombosis within hours of implant (Schwartz and Holmes, 1994). Stenting or balloon inflation in the swine model leads to intimal smooth muscle cell proliferation that closely resembles the cell size, density, and histopathological appearance of that seen in human restenosis (Karas et al., 1992; Bonan et al., 1996; Willette et al., 1996). Additionally, as in humans, there is a predictable relationship between vascular injury and restenosis (i.e., more neointima is formed with deeper lesions) (Willette et al., 1996; Jordan et al., 1998).

Although investigation of stents and balloon injury has mainly been done using coronary arteries (Karas et al., 1992; Bonan et al., 1996; Willette et al., 1996), studies using other arteries, such as the iliacs, have also been performed (White et al., 1989). Swine have also been used to study the feasibility of percutaneous repair of abdominal aortic aneurysms (AAA). Jordan et al. (1998) developed an AAA model using rectus abdominus fascia, and demonstrated prolonged survival in stented AAA in comparison to unstented controls. Unfortunately, this AAA model did not accurately mimic aneurysms seen in humans, as the AAA created was saccular and did not reproduce the geometric configuration or collateral circulation and back bleeding observed in human disease (Jordan et al., 1998). Whitbread et al. (1996) created an AAA model in swine by interposing fusiform segments of glutaraldehyde-treated bovine internal jugular vein into the infrarenal aorta. This resulted in a pulsatile, nonthrombogenic AAA that approximates human dimensions (20 mm) and geometry (fusiform) (Whitbread et al., 1996). Both AAA models have been used to develop stents and stent placement technology with the goal of reducing the pulsatility and diameter of the aneurysm while maintaining location of the stent without occluding nearby vessels.

Swine models have become useful and popular to study biomaterials and the devices manufactured from these materials. Anatomic and physiologic similarities to humans have made swine the species of choice to study many cardiac and vascular prostheses, especially in acute studies looking at hemodynamic profiles of different devices. However, difficulties with husbandry and species temperament, and a rapid rate of somatic growth, have made swine less desirable as models for the chronic evaluation of devices in our experience.

TABLE II.3.7.5 Antibiotics Acceptable for Use in Large Animal Models

	Dose/kg				Route of Delivery[b]			
Drug Name	**Bovine**	**Swine**	**Ovine**	**Frequency[a]**	**PO**	**SQ**	**IM**	**IV**
Penicillins								
Amoxicillin	11–22 mg	11–22 mg	11–22 mg	BID-TID	X		X	
Amoxicillin/clavulanate	14 mg	14 mg	14 mg	BID-TID	X			
Ampicillin	11–22 mg	11–22 mg	11–22 mg	BID-TID	X		X	X
Penicillin benzathine	44,000 IU	44,000 IU	44,000 IU	q48h		X	X	
Penicillin procaine	44,000 IU	44,000 IU	44,000 IU	QD-BID		X	X	
Ticarcillin	50–300 mg	50–300 mg	50–300 mg	TID-QID				X
Cephalosporins								
Cephadroxil	22 mg	22 mg	22 mg	BID	X			
Ceftiofur	3–5 mg	3–5 mg	3–5 mg	QD		X	X	X
Cephalexin	10–30 mg	10–30 mg	10–30 mg	BID	X			
Cephalothin	20–35 mg	20–35 mg	20–35 mg	TID			X	X
Cephazolin	10–25 mg	10–25 mg	10–25 mg	TID			X	X
Aminoglycosides								
Amikacin	5 mg	5 mg	5 mg	TID			X	X
Gentamicin	2–3 mg	2–3 mg	2–3 mg	BID-TID			X	X
Neomycin	7–15 mg	7–15 mg	7–15 mg	QD	X			
Tetracyclines								
Chlortetracycline	10–20 mg	10–20 mg	10–20 mg	TID	X			
Oxytetracycline	20 mg	20 mg	20 mg	q48h			X	
Tetracycline	10–20 mg	10–20 mg	10–20 mg	TID	X			X
Chloramphenicol	45–80 mg	45–80 mg	45–80 mg	TID	X		X	X
Macrolides								
Erythromycin	10–15 mg	10–15 mg	10–15 mg	TID	X			
	2–8 mg	2–8 mg	2–8 mg	QD			X	
Tylosin	7–12 mg	7–12 mg	7–12 mg	QD-TID	X		X	
Lincosamides								
Clindamycin	5–40 mg	5–40 mg	5–40 mg	BID	X		X	X
Lincomycin	10–25 mg	10–25 mg	10–25 mg	BID	X		X	
Sulfonamides								
Ormetoprin-sulfadimethoxine	15–30 mg	15–30 mg	15–30 mg	QD	X			
Sulfadimethoxine	25–50 mg	25–50 mg	25–50 mg	QD	X			
Trimethoprim-sulfamethoxazole	15–30 mg	15–30 mg	15–30 mg	BID	X	X		X
Fluoroquinolones								
Ciprofloxacin	2.5–7.5 mg	2.5–7.5 mg	2.5–7.5 mg	BID	X			
Enrofloxacin	2.5–5 mg	2.5–5 mg	2.5–5 mg	BID	X		X	X
Metronidazole	50–60 mg	50–60 mg	50–60 mg	QD-BID	X			X

[a]QD, once daily; BID, twice daily; TID, three times daily; QID, four times daily; q48h, every 48 hours.
[b]PO, oral; SQ, subcutaneous; IM, intramuscular; IV, intravenous.
Table adapted from the *University of Minnesota Animal Care and Use Manual* 1997 reprint.

Bovine

Introduction. There are many species of cattle which are generally bred as livestock for meat and dairy production, as well as draft animals. Cattle have long been domesticated since Neolithic times as a distinct species as the genus *Bos*, and are often hybridized between closely related species. Cattle are raised in herds and graze in large areas of grassland or in feedlots. Cattle are of moderate intelligence and vocalize occasionally with a characteristic lowing sound, bawling or bellowing. The animals are usually mild mannered, slow moving, usually cooperative, and can be controlled behaviorally by education or with barriers.

Aside from antibody production, the cow is not often used in biomedical research, because of their size and husbandry needs. The cow's ability to produce milk is a potential source for genetically engineered proteins; that, however, remains ethically controversial. The calf has been used in cardiovascular research for both ventricular assist device testing and valve implantation. Ventricular assist devices are generally too large to be tested with smaller animals such as the dog, and pigs are not cooperative enough.

Comparative Anatomy and Physiology

Interestingly, the bovine genome has been recently mapped with about 80% of the genome similar to that of humans, despite obvious phenotypic differences. Cattle are ungulates and ruminants with a four-compartment stomach for digestion. Microbes in the rumen decompose cellulose into carbohydrates and fatty acids, and synthesize amino acids from urea and ammonia. The gestation period for a cow is nine months and a newborn calf weighs about 25–45 kg.

The bovine heart comprises 0.4%–0.5% of the body weight, which in the adult weighs about 2.5 kg. Aside from the normal mammalian structure of the heart, the bovine heart differs in that the great cardiac vein drains directly into the azygos. A semilunar valve (Thebesian valve) at the margin of the azygos and right atrium allows for unidirectional coronary venous flow. As the calf grows older, two bones called the *ossa cordis* develop in the fibrous aortic ring at the attachments of the right and left semilunar cusps. The bovine aortic arch consists of a common trunk that symmetrically branches into the right and left brachiocephalic arteries, supplying the head and upper extremities.

The bovine ventricles lend a conical appearance to the apex, with both an anterior and posterior interventricular groove, as well as an intermediate lateral groove for vessels. The left auricle is larger than the right atrial appendage. The left coronary artery is also larger than the right. The left coronary artery divides into the paraconal interventricular and circumflex branches. The circumflex artery courses in the coronary groove under the cardiac vein, and gives off a prominent subsinusoal interventicular branch which descends to the apex. The interventricular artery descends anteriorly to anastomose with a branch of the subsinuosal. As mentioned above, the right coronary artery is smaller than the left and lies in the right coronary groove covered by the right atrial appendage and terminates in small branches that anastomose with the subsinuosal interventricular artery.

Rheologic conditions are altered with different velocities of blood flow and blood stasis in animals following implantation of cardiac devices. As one of the fundamental homeostatic mechanisms of mammalian biology, the blood coagulation system establishes a delicate balance between the procoagulant and anticoagulant functions of blood and the vessel wall, thereby guarding against excesses in either direction and normally preventing unwanted hemorrhage or thrombosis. Despite all efforts to achieve effective anticoagulation, the risk of thromboembolic events is still approximately 30% in humans. The main reason for clot formation is the contact between blood components and foreign surfaces. System components induce a cascade of interacting proteases, accelerating steps that provide amplification, a variety of balanced feedback controls, and, most importantly, tightly regulated activation. Unfortunately, the range of biocompatibility assays for the evaluation of blood contacting materials is limited using the bovine model. Recent reports to quantify bovine circulating activated platelets have been developed using the monoclonal antibodies to BAQ125, GC5, and annexin V (Baker et al., 1998; Snyder et al., 2002). Additional assays to quantifying bovine platelets which have human analogs are now possible using platelet-expressing CD62P and CD63. CD62P (p-selectin, platelet activation dependent granule external membrane protein (PADGEM), granule membrane protein-140 (GMP-140)) and CD63 (lysosomal-membrane-associated glycoprotein 3 (LAMP-3), granulophysin, LIMP, PLTGP40, gp55) are both expressed on the surface of platelets following activation and degranulation (Snyder et al., 2007).

Perioperative Care

Anesthesia. Calves should be fasted for 12–18 hours preoperatively to decrease the likelihood of regurgitation during the induction of anesthesia and ET intubation. Typically, a sedative is administered IM to allow easy and safe placement of IV catheters for controlled administration of medication. Several sedatives are useful in the pre- and postoperative periods to aid in animal handling and for performing minor procedures (Table II.3.7.3). A common protocol uses ketamine, 10 mg/kg IM as a preanesthetic sedative. Anticholinergics or parasympatholytics, especially atropine, have been used on induction of anesthesia to decrease salivary and respiratory tract secretions. Table II.3.7.3 lists the recommended dosing of such agents. Once the sedative has taken effect, IV catheters can be easily introduced into the external jugular, cephalic or saphenous veins. Following placement of a secure IV catheter and the administration of short-acting relaxing agents (Table II.3.7.3), ET intubation can be safely performed using standard cuffed ET tubes 10–16 mm in internal diameter. Placement of the orogastric tube typically follows ET intubation. Orogastric intubation is especially important if preoperative fasting does not occur. The endotracheal tube can then be connected to a conventional anesthetic ventilator to maintain a deep plane of anesthesia. Preferred inhalation anesthetics for bovine are halothane (1.5–2.5%) or isoflurane (1.5–3%).

Analgesia. Calves are stoic and it is uncommon for them to vocalize when they are in pain. A reluctance to move, favoring the procedural site, excessive licking or scratching of the site, kicking, lack of appetite, and/or a depressed attitude are behaviors that may indicate a calf is experiencing pain. Scheduled medication administration and periodic evaluation of behavior can help to ensure adequate pain relief. Commonly used analgesics include narcotics administered via a subcutaneous, IV or epidural route. Nonsteroidal anti-inflammatory agents can be used in divided doses for additional analgesia. Table II.3.7.4 lists recommended dosing of additional analgesics.

EXISTING MODELS

Cardiac Devices

The calf model was first used extensively in cardiovascular research in the 1980s for the purpose of calcification studies (Schoen et al., 1985; Dewanjee et al., 1986a,b). A review of the current literature reflects diminished utilization of this model, accounting for only 6.5% reported animal studies, primarily because it shares many of the limitations noted with the pig model Figure II.3.7.3. A major limitation for chronic survival studies is the rapid somatic growth of the calf, contributing to both anatomic and hemodynamic alterations that undermine the validity of long-term follow-up studies. The rapid somatic growth in the calf is associated with dramatic elevation in cardiac output, well above the maximum observed in humans. In fact, if the bovine model were used for a standard six-month trial, the cardiac output (CO) would be expected to be nearly three times greater than that of the adult human at the end of six months.

Anatomically, the calf valve annulus is larger than that of humans (Tables II.3.7.1 and II.3.7.6) (Gallegos

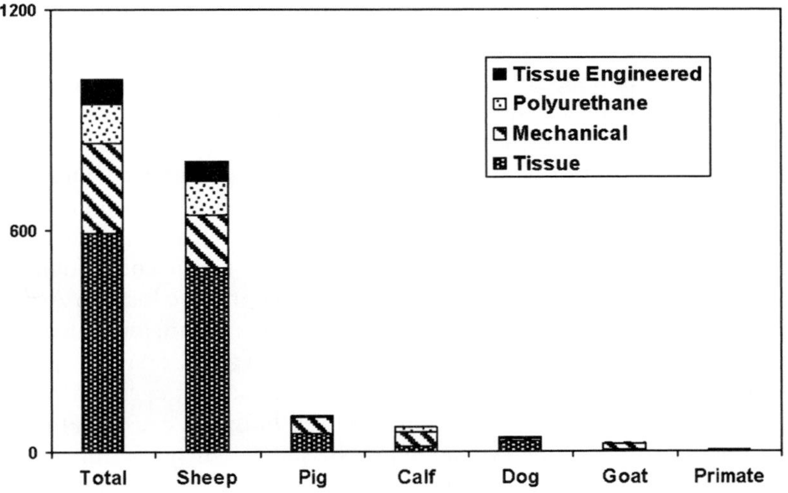

Valve type	Total	Total number of subjects in study					
		Sheep	Pig	Calf	Dog	Goat	Primate
Tissue (stented and stentless)	591 (58.4)	495 (83.7)	47 (7.9)	12 (2.0)	29 (4.9)	5 (0.8)	3 (0.5)
Mechanical	248 (24.5)	145 (58.4)	45 (18.1)	39 (15.7)	4 (1.6)	15 (6.0)	-
Polyurethane	106 (10.4)	92 (86.7)	-	14 (13.2)	-	-	-
Tissue-engineered	66 (6.5)	56 (84.8)	4 (6.0)	-	6 (9.0)	-	-
Total	1,011	788 (78.0)	96 (9.5)	65 (6.5)	39 (3.8)	20 (2.0)	3 (0.2)

FIGURE II.3.7.3 Reported preclinical *in vivo* assessment by species and valve type. Values in parentheses are percentage of total studies. (Adapted from Gallegos et al., 2005. Used with permission.)

TABLE II.3.7.6 Important Cardiac Parameters for Cardiac Valve Implantation Studies

Parameter	Bovine	Swine	Ovine	Human
Aortic annulus size (mm)	33.7 ± 2.7	26.6 ± 1.8	25.8 ± 1.3	26.4 ± 3.1
Cardiac output (L/min)	12 ± 6	2.36 ± 0.9	5.2 ± 1.4	6.0 ± 2
Left ventricular end fraction (%)	44 ± 7	59 ± 1	52 ± 8	69 ± 8
End diastolic volume (mL)	682 ± 137	54 ± 4	45 ± 7	115 ± 54

Values represent mean ± standard deviation.
Table adapted from Gallegos et al. 2005. Used with permission.

et al., 2005). Therefore, implantation of a prosthetic valve of a size that would be typically utilized in an adult human would result in patient prosthesis mismatch if placed in the calf aortic position. The result of patient prosthesis mismatch produces outflow obstruction, with valves sized for clinical relevance, and any additional growth may contribute to paravalvular leak development, as was observed in the pig model (Linden et al., 2003). It is well-known that a valve's effective orifice area (EOA) significantly impacts on the incidence of heart failure symptoms, adverse cardiac events, preoperative mortality, LV impairment, and is associated with short- and long-term mortality (Pibarot et al., 1998; Pibarot and Dumesnil, 2000, 2001; Blais et al., 2003). Indeed, this was actually reported in a bovine valve study whereby a majority of calves were actually in heart failure, given that the mean CO was only 12 ± 6 l/min with the study valve in place, when the expected normal value would be in the range of 18–20 l/min, based on animal weight and study duration (see Table II.3.7.6). Unfortunately, the calf model is not an ideal animal for testing prosthetic valves in a preclinical setting, because of the growth that causes an inherent patient prosthesis mismatch.

On the other hand, the bovine model is the most commonly used model for ventricular assist devices (Litwak et al., 2008). Calves are used in preclinical studies to provide information about the function, biocompatibility, and efficacy of ventricular assist devices (VAD). The VAD is a blood pump system designed to palliate severe congestive heart failure, and is a system composed of inflow and outflow cannnulae, connectors, and controllers. The blood contacting surfaces are procoagulant and activate blood clotting, so every attempt is made in the preclinical setting to minimize adverse events. Numerous VAD systems have been developed and implanted in calves with a variety of adverse events, such as thrombus, obstructions or kinks, mechanical, electrical or software failures that can occur within any component of the LVAD.

Clearly, valve replacement in the calf model is representative of the patient prosthesis mismatch condition, rather than an ideal *in vivo* model for replacement human valves. Choice of the animal model to study the replacement valve's hemodynamic characteristics should ideally reduce or eliminate confounding factors, such as patient prosthesis mismatch. Therefore, the calf model is not an ideal animal for testing prosthetic valves in a preclinical setting, because of the growth that causes an inherent patient prosthesis mismatch, as mismatch is strongly predictive of congestive heart failure following aortic valve replacement as well as re-operation (Ruel et al., 2004a,b). Thus, bovine models are best suited for acute studies in which somatic growth would not bias results or specifically to study the effect of patient–prosthesis mismatch itself.

Vascular Devices

There are no widely-used bovine *in vivo* models of vascular devices, so we will not discuss them further.

Ovine

Introduction. As a gentle, docile animal, sheep were among the first species to be domesticated, approximately 10,000 years ago as evidenced by pictures and statuettes depicting sheep in the Middle Eastern region. They remain popular to this day as a source of both food and goods. The prevailing theory for the origin of the domestic sheep is that they originated from the wild sheep known as mouflon (*Ovis musimon*). Currently, mouflon can be found in two separate geographic locales: Asiatic mouflon in Asia Minor and southern Iran; and European mouflon on the islands of Sardinia and Corsica (Clutton-Brock, 1999b). The rapid expansion of agriculture and human consumption has been both the cause and effect of routine crossbreeding of sheep. Currently more than 200 different breeds exist worldwide. Common breeds used in surgical research include smaller breeds commercially used for meat, such as the Hampshire and Dorset varieties, or the larger sheep, often used for wool commercially, such as the Merino and Rambouillet breeds.

Comparative Anatomy and Physiology

Specific organ differences between humans and sheep are notable (e.g., ruminant gut anatomy). However, in many instances the similarity in organ size of adult sheep to their human counterparts allows adequate approximation for the study of mechanical and bioprosthetic biomaterials and devices prior to human implantation. The ovine thorax has an exaggerated conical shape with a narrow spax (cranial aperture or thoracic inlet) and a caudal thoracic aperture that is six times wider. Cardiopulmonary anatomy is generally similar to that of humans, making sheep ideal models for cardiac research; however, a few anatomic differences require consideration (Hecker, 1974), for example, the single brachiocephalic artery arising from the aortic arch. In addition, tracheal anatomy varies slightly from humans, with the cranial lobe of the right lung arising directly from the trachea rather than the right mainstem bronchus. This variation can pose a difficulty during endotracheal intubation on induction of anesthesia, and can result in non-ventilation of a significant pulmonary segment (Carroll and Hartsfield, 1996).

Physiologic parameters (heart rate, blood pressure, cardiac index, and intracardiac pressures) have been established through numerous studies in both anesthetized and conscious sheep, including the percentage of blood flow to each major organ system (Matalon et al., 1982; Nesarajah et al., 1983; Newman et al., 1983). In most instances, the hemodynamic and metabolic values are similar to those of other large mammals, including humans (Tables II.3.7.1 and II.3.7.2).

In cardiovascular bioprosthesis studies, similarities between human and animal hematologic systems are important. Although there are differing blood groups, typing is unnecessary, as transfusion has never been necessary in our experience of over 30 years of working with sheep. There have been occurrences of a reaction, most often a dermatological response, which is easily treated with diphenhydramine. A comparison of the coagulation parameters and hematologic profiles between sheep and humans reveal some differences, which become important in assessing the thrombogenicity of implanted cardiovascular prostheses. Notably, sheep possess decreased fibrinolytic activity and increased platelet number and adhesiveness (Gajewski and Povar, 1971; Tillman et al., 1981; Karges et al., 1994).

Perioperative Care

Anesthesia. Sheep should be fasted preoperatively to decrease the likelihood of regurgitation on induction of anesthesia and ET intubation. The reticulum, the most proximal rumen, has no sphincter mechanism at its oral end and often needs intubation and suction decompression to prevent regurgitation and aspiration (Holmberg and Olsen, 1987; Carroll and Hartsfield, 1996). Orogastric intubation is especially important if preoperative fasting does not occur. Typically, a sedative is administered IM to allow easy and safe placement of IV catheters for controlled administration of medication. Several sedatives are useful in the pre- and postoperative periods to aid in animal handling, and for performing minor procedures as shown in Table II.3.7.3. Large named veins such as the internal or external jugular or femoral veins can be used to secure IV access for fluid administration and frequent blood draws. Specialized catheters such as pulmonary artery catheters or long-term tunneled venous catheters (Hickman) may also be placed in these easily accessible, large veins (Tobin and Hunt, 1996). Following placement of a secure IV catheter and the administration of short-acting relaxing agents, ET intubation can be safely accomplished using standard cuffed ET tubes 5–10 mm in internal diameter. Anticholinergics or parasympatholytics, especially atropine, have been used on induction of anesthesia to decrease salivary and respiratory tract secretions. Use of these agents, however, is controversial in ruminants such as sheep, goats, and cattle because they inhibit gastrointestinal smooth muscle activity, which can cause rumen stasis in sheep (Carroll and Hartsfield, 1996). Table II.3.7.3 lists the recommended dosing of anesthetic agents.

Analgesia. Scheduled medication administration and periodic evaluation of behavior can help to ensure adequate pain relief. Sheep should be assessed for changes from their preoperative temperament. Sheep tend to react stoically to pain, and their refusal to move can often indicate even minor discomfort (Carroll and Hartsfield, 1996). Other common clinical signs of postprocedural pain include tachycardia, tachypnea, grunting, grinding teeth, decreased appetite, vocalization on movement, and guarding of the operative site. Commonly used analgesics include narcotics administered via a subcutaneous, IV or epidural route. Nonsteroidal anti-inflammatory agents (NSAIDs) can be used in divided doses for additional analgesia. Table II.3.7.4 contains recommended dosing of analgesics.

EXISTING MODELS

Cardiac Devices

Because of the similar anatomic and physiologic characteristics of sheep and humans discussed earlier, sheep are often used in cardiovascular biomaterial research, and several models have been developed for studying compatibility and function of biomaterials. A total artificial heart model was developed in sheep at the University of Utah's Artificial Heart Laboratory in the 1980s (Holmberg and Olsen, 1987). Other studies examined the potential use of synthetic or xenograft pericardial substitutes to reduce adhesion formation following coronary artery bypass or valve replacement (Gabbay et al., 1989; Bunton et al., 1990).

Many investigators have employed sheep in the preclinical testing of mechanical valves in both the aortic and mitral positions (Barak et al., 1989; Vallana et al., 1992; Irwin et al., 1993; Cremer et al., 1995; Bhuvaneshwar et al., 1996; Okazaki et al., 1996). Our experience discovered that the use of juvenile sheep (younger than 20 weeks) results in an excellent long-term model for mechanical prosthetic valve evaluation. Careful standardization to prevent infectious complications, short cardiopulmonary bypass time, whole sheep blood transfusion to minimize anemia, and rumen decompression with an orogastric tube to prevent vena caval compression were found to be useful in the successful implementation of this model (Irwin et al., 1993).

Acute and chronic models of stented bioprosthetic valves are well established through several studies in both the mitral and aortic position in sheep (Gott et al., 1992; Liao et al., 1993; Bianco et al., 1996; Vyavahare et al., 1997; Northrup et al., 1998; Ouyang et al., 1998; Salerno et al., 1998; Spyt et al., 1998). Sheep grow at a rate that is comparable to human growth, which allows for comparative analysis. In contrast, larger ruminants such as cattle and large varieties of swine experience rapid and prolonged growth until reaching a large adult size (Bianco et al., 1996; Grehan et al., 2000; Gallegos et al., 2005). Such rapid growth can cause difficulties in size matching between prosthetic valves and the native valve annulus, resulting in paravalvular leaks and functional valve stenosis (Braunwald and Bonchek, 1967; Gallo and Frater, 1983; Spyt et al., 1998). Premature leaflet calcification can occur in stented bioprosthetic valves because of abnormal mechanical loading characteristics. This is most consistently reproduced in juvenile sheep (younger than 6 months). We and others currently employ a model in sheep younger than 20 weeks to study both the dynamic environment related to growth of the sheep and the chronic effect of calcification in stented porcine bioprosthetic valves in the aortic position (Gott et al., 1992; Liao et al., 1993; Bianco et al., 1996; Ouyang et al., 1998).

Sheep also provide an important model for the study of low-profile stentless aortic valves (David et al., 1988; Brown et al., 1991; Hazekamp et al., 1993; Schoen et al., 1994; Salerno et al., 1998). In this regard we found that, with careful surgical technique including a two-thirds transverse aortotomy for superior exposure and meticulous avoidance of coronary ostial obstruction by the scalloped valve edges, sheep provide an excellent *in vivo* preclinical model to test these aortic valves (Salerno et al., 1998).

In addition to the subcoronary implantation of these stentless devices, this laboratory has also developed a juvenile sheep model of aortic root replacement to allow for the preclinical evaluation of new or modified stentless devices using implantation techniques that parallel those seen in clinical practice. Again, careful attention to surgical technique proved to be very important in the success of this model, including creation of very generous-sized coronary buttons, inclusion of a posterior "rim" of native aorta in the proximal anastamosis, and the creation of a tension-free distal anastamosis by removing little if any native aorta (Grehan et al., 2001).

Vascular Devices

In addition to cardiac devices, researchers have studied the interface between blood and several biomaterial surfaces in the evaluation of biosynthetic vascular grafts placed in sheep. Several authors have noted the reduction in thrombogenicity provided by endothelial seeding of small-diameter synthetic vascular grafts constructed of expanded PTFE or woven Dacron® prior to placement in sheep (James et al., 1992; Taylor et al., 1995; Dunn et al., 1996; Poole-Warren et al., 1996; Jensen et al., 1997). Placement of vascular grafts composed of several engineered surface textures into sheep has also been performed in order to assess characteristics of the pseudointima layer (Fujisawa et al., 1999). Additionally, the important problem of intimal hyperplasia in the venous anastomosis of high-flow arteriovenous fistulas has recently been evaluated in a sheep model (Kohler and Kirkman, 1999).

Recent progress in the development of large-vessel endovascular stents has occurred in part due to research with specific vascular disease models in adult sheep, and has been increased by continued development of minimally invasive techniques. As in graft studies, there is significant interest in both the short- and long-term patency of endovascular stents. Stent incorporation into native tissue and the role of foreign-body reaction in normal arterial systems with respect to thrombogenicity and intimal hyperplasia has been evaluated in sheep aortic and iliac arteries (Rousseau et al., 1987; Neville et al., 1994; Schurmann et al., 1998; White et al., 1998). Current efforts have been directed at determining the role of endovascular stenting in the prevention of rupture of abdominal aortic aneurysms, and the prevention of complications such as spinal cord ischemia through the use of these stents (Boudghene et al., 1998; Beygui et al., 1999).

Current cardiovascular research has embraced the humane use of sheep for the testing of a variety of cardiac and vascular devices composed of many different biomaterials. In all of these investigations, sheep have proved a useful and reliable model, as evidenced by this laboratory's continued success using both juvenile and adult sheep for a variety of investigations. Figure II.3.7.3 depicts the relative use of various large animals for preclinical assessment of valves, with a reported 78% of studies using an *in vivo* sheep model. Current research indicates a trend towards an even broader application of emerging endovascular technologies. For many reasons, including their anatomic and physiologic similarities to humans, and a mild temperament that facilitates handling, the benefits of using sheep in preclinical studies continue to outweigh the use of other large animals in future investigations into biomaterials testing and development.

TESTING HIERARCHIES

Conventionally-Placed Devices

Conventional coronary artery bypass grafting (CABG) continues to be one of the mainstays of cardiovascular surgery (Wilson and Ferguson, 1995). Coronary artery disease remains the leading cause of death of people in developed countries, and an average of 800,000 CABG operations take place annually (Caparrelli et al., 2009). This requires the harvesting of one or more peripheral veins to reroute coronary vasculature and circumvent around an occlusion within the lumen of an artery while the patient is on cardiopulmonary bypass (Wilson and Ferguson, 1995). There are known risks for patients undergoing this surgical procedure, but the procedure itself is somewhat standardized and a common practice for surgeons. Similarly, the development and surgical implant of prosthetic valves dates back to the mid-20th century, and has proven to be another one of the greatest successes in surgery. Surgical valve replacement is well-established, and approximately 200,000 aortic valve replacements are performed annually worldwide (Fann et al., 2008). Advances in cardiopulmonary bypass, improved hemodynamic monitoring and perioperative care, and structural and hemodynamic improvements in prosthetic valves have aided in this success. Valve replacement surgery is now considered to have low operative mortality and constantly improving long-term survival rates (Fann et al., 2008; Christofferson et al., 2009). Its only limiting factor is the risk of conventional surgery on a patient with multiple comorbidities, a risk that obviously directly applies to patients undergoing CABG as well.

Percutaneously-Placed Devices

In cardiovascular surgery there is a current trend toward minimally-invasive placement of devices, rather than traditional surgical placement of devices (Mack et al., 2008). In addition, the responsibility of device placement is shifting from the cardiothoracic surgeon to the interventional cardiologist (Mack et al., 2008). The driving forces of this shift in medical practice are that the patient population as a whole is aging, and it is becoming increasingly difficult to care for their multiple comorbidities. Previously, many patients were deemed non-surgical candidates, due to the risks imposed by conventional cardiovascular surgical practices and the consequences of surgery in a patient with multiple comorbidities (Fann et al., 2008; Caparrelli et al., 2009) A percutaneous, or minimally-invasive, intervention is the process of inserting a needle and catheter into a vessel, and then utilizing a wire to guide the catheter through the lumen of the vessel to the desired site of intervention. The cardiac or vascular device to be placed in the patient can first be packaged inside of the catheter and released *in vivo* at the appropriate anatomical site. This technique facilitates the placement of devices such as prosthetic valves or vascular stents. Given the nature of this less invasive option, interventional cardiologists are optimistic about its possible use in patients who were previously not candidates for the conventional surgical approach due to its associated risks. Minimally-invasive placement of devices is generally thought to be less risky than conventional surgery, putting less overall stress on the patient, and dramatically shortening the recovery period (Fann et al., 2008; Caparrelli et al., 2009). However, little is known at this point regarding the long-term safety and efficacy of devices placed percutaneously, as well as the delivery method itself.

The shift from conventional coronary artery bypass grafting to the percutaneous placement of stents in the treatment of coronary artery disease was first recognized approximately 20 years ago (Wilson and Ferguson, 1995). There are now two options for the placement of stents in diseased arteries including drug-eluting stents, as well as bare metal stents.

The development of percutaneous valve replacement stems from the successes of other catheter-based vascular procedures, such as carotid and coronary stenting. The percutaneous placement of cardiac valves is currently under investigation in clinical trial; however, clinical trials were initiated prior to the development of an animal model for *in vivo* assessment (Fann et al., 2008). Thus, chronic performance remains unknown, as there is no chronic animal model representing *in vivo* valvular stenosis. Very little preclinical data is currently available on the effects of crimping the valve within the catheter during the actual procedure, the risk of migration of the device from its intended location, and the complications that arise from expanding and retaining the native diseased valve (Fann et al., 2008). In addition, there is always a chance that unexpected consequences may have been revealed by *in vivo* assessment, precisely the reason it is conducted. These clinical trials may have been initiated prematurely, as they must now function as an assessment of both the safety and efficacy of percutaneous valve placement (Gallegos et al., 2005). There are still several issues that need to be considered regarding the minimally-invasive placement of prosthetic valves, including device loading, vascular access, adequate device positioning and anchoring, ideal leaflet material, stent or frame characteristics for the valve, and optimal method of delivery (Huber et al., 2005; Flecher et al., 2007; Fann et al., 2008).

There are currently two classes of cardiac devices available for valve replacement, including bioprosthetic valves made of porcine aortic valve cusps or bovine pericardium, as well as mechanical heart valves (Gallegos et al., 2005). Each type of valve has known advantages and disadvantages based on outcomes in the conventional placement of these devices. For example, mechanical valves rarely require future replacements, regardless of the duration of the patient's life. However, patients with mechanical valves historically have been required to take lifelong prophylactic anticoagulation therapy in an effort to avoid a thrombotic event (Gallegos et al., 2005). Conversely, bioprosthetic (tissue) valves necessitate the use of anticoagulation far less frequently than mechanical valves, but these same tissue valves have a known risk of calcification and/or tearing, often necessitating replacement within

10–15 years (Gallegos et al., 2005). The known rate of thromboembolic complications in bioprosthetic valves is between 0.9% and 2.2% per patient year without anticoagulant therapy (Nowell et al., 2007), based on multiple retroactive studies, which is less frequent than that observed with mechanical valve replacement with subsequent long-term anticoagulation therapy (Gallegos et al., 2005). This makes bioprosthetic heart valves a desirable alternative to mechanical valves for patients who may not be candidates for lifelong anticoagulation therapy. However, if a patient outlives the 10–15 year lifespan of the bioprosthetic valve they will require continued surgical replacement of additional valves.

There are known specific developmental priorities for the improvement of heart valve prostheses. One is to design a mechanical valve with center flow and a trileaflet configuration that is superior to other valves in hemodynamics. Another is to modify tissue valves in such a way as to minimize calcification and prevent disintegration, which may require investigating new preservation chemistry. The last future developmental priority is to design a completely new synthetic valve made of new materials. New advancements are continuously being made in the method of delivery and placement of valves, and this should continue to drive the design of new valves (Gallegos et al., 2005).

Two key issues must be independently resolved with respect to new vascular stents and prosthetic valves placed using percutaneous minimally-invasive catheter techniques. These include: (1) delivery of the device to the intended site; and (2) chronic performance of the device within a physiologic human condition or preferably a model in lieu of this condition (Christofferson et al., 2009). It is important to separate these issues during the preclinical assessment, and it is not recommended that both be assessed in a single animal model. Arguably, chronic performance is the more important issue of the two, as it directly predicts patient outcomes and disease treatment. The only way to develop an understanding of the chronic performance of a device is to assess the placement in a large number of animals for a long period of time. It is immensely important to match the animal to the device, rather than to match the device to the selected animal in preclinical assessment. Generally, the most successful *in vivo* assessments of bioprosthetic (tissue) valves have been conducted using juvenile sheep, while the most successful assessment of vascular stents have used an *in vivo* swine model (Gallegos et al., 2005). It is also a challenge to correctly identify the control group in a preclinical assessment of the minimally-invasive placement of devices. Since percutaneous valve placement was intended only for patients deemed non-surgical candidates, should the control group be subjects undergoing the conventional placement or solely medical management of valvular disease? Again it is the role of the study director to determine which control group best aligns with the goals of the preclinical device assessment.

CURRENT RECOMMENDATIONS AND FUTURE DIRECTIONS

The use of animals to predict clinical safety is an essential (and required) phase in the development of new or modified medical device technology. It is extremely unlikely, however, that preclinical use of animal models to assess new technology will ever evolve to the point that absolute statements can be offered regarding future clinical performance. Therefore, the enhancement of the predictive qualities of human safety and efficacy based on preclinical *in vivo* evaluations should be our priority.

Enhanced prediction of human safety and efficacy can be achieved by developing increasingly more relevant models of human disease, and developing models in which new or modified devices are implanted orthotopically in each intended clinical location (site specific testing). Additionally, each preclinical study design should include an appropriate control "arm." This "arm" of the study should include animals implanted with a device that is currently approved and being used clinically. The control device should be of similar design (or preservation technique) as the device undergoing investigation, and should be implanted using the same techniques. The use of concurrent control implants offers several advantages. First, it provides a direct clinical bridge to assess performance of a new or modified device. Second, it provides an assessment of the model utilized in the investigation relative to a device with a known clinical history. Finally, it can reduce the number of implants required in each investigation by providing a correlation to historical animal and clinical data.

The use of animals to predict human safety will, and should, remain a crucial part of the regulatory requirements governing medical device development and modification. In order to enhance the predictive quality of preclinical *in vivo* evaluations, current models need to be refined and new models need to be developed that more closely mimic both human disease and the intended clinical use of the device under investigation. As these models are developed and perfected they should become part of the current standards to be met by investigators working in prostheses development. In addition, each preclinical evaluation should include the concurrent implantation of appropriate control devices, so that comparisons of the investigational data obtained can be made to that of concurrent and historic animal and clinical control data. This will allow more accurate assessments of both the investigational device and the model to be made.

ACKNOWLEDGMENTS

The authors thank the staff and student employees of the University of Minnesota, Department of Surgery, Division of Experimental Surgical Services for their commitment to achieving scientific accuracy and providing humane care to the investigational animals.

BIBLIOGRAPHY

AAMI (2005). ANSI/AAMI/ISO 5840:2005: *Cardiovascular implants: Cardiac valve prostheses*. Arlington, VA: AAMI.

AAMI (2001). ANSI/AAMI/ISO 7198:1998/2001/(R) 2004: *Cardiovascular implants: Tubular vascular prostheses*. Arlington, VA: AAMI.

Akutsu, T., Dreyer, B., & Kolff, W. J. (1974). Polyurethane artificial heart valves in animals. In ArchibaldJ., & J. V. LaCroix (Eds.), *Canine Surgery*. Santa Barbara, CA: American Veterinary Publications, Inc.

Al-Dossari, G. A., Kshettry, V. R., Jessurun, J., & Bolman, R. M., III (1994). Experimental large-animal model of obliterative bronchiolitis after lung transplantation. *Ann. Thorac. Surg.*, **58**, 34–39.

American National Standrards Institute, Inc./Association for the Advancement of Medical Instrumentation/International Standards Organization 5840 (1996). *Cardiovascular implants: Cardiac valve prostheses*.

AVMA (American Veterinary Medical Association) (1993). 1993 Report of the AVMA Panel on Euthanasia. *JAVMA*, **202**, 229–249.

AVMA (American Veterinary Medical Association). (2007). AVMA Guidelines on Euthanasia. <http://www.avma.org/issues/animal_welfare/euthanasia.pdf>.

Baker, L. C., Davis, W. C., Autieri, J., Watach, M. J., Yamazaki, K., et al. (1998). Flow cytometric assays to detect platelet activation and aggregation in device-implanted calves. *J. Biomed. Mater. Res.*, **41**, 312–321.

Barak, J., Einav, S., Tadmor, A., Vidne, B., & Austen, W. G. (1989). The effect of colloid osmotic pressure on the survival of sheep following cardiac surgery. *Int. J. Artif. Organs*, **12**, 47–50.

Beygui, R. E., Kinney, E. V., Pelc, L. R., Krievins, D., Whittmore, J., et al. (1999). Prevention of spinal cord ischemia in an ovine model of abdominal aortic aneurysm treated with a self-expanding stent-graft. *J. Endovasc. Surg.*, **6**, 287–284.

Bhuvaneshwar, G. S., Muraleedharan, C. V., Vijayan, G. A., Kumar, R. S., & Valiathan, M. S. (1996). Development of the Chitra tilting heart valve prosthesis. *J. Heart Valve Dis.*, **5**, 448–458.

Bianco, R. W., St. Cyr, J. A., Schneider, J. R., Rasmassen, T. M., Clack, R. M., et al. (1986). Canine model for long-term evaluation of prosthetic mitral valves. *J. Surg. Res.*, **41**, 134–140.

Bianco, R. W., Phillips, R., Mrachek, J., & Witson, J. (1996). Feasibility evaluation of a new pericardial bioprosthesis with dye mediated photo-oxidized bovine pericardial tissue. *J. Heart Valve. Dis.*, **5**, 317–322.

Blais, C., Dumesnil, J. G., Baillot, R., Simard, S., Doyle, D., et al. (2003). Impact of valve prosthesis–patient mismatch on short-term mortality after aortic valve replacement. *Circulation*, **108**, 983–988.

Bonan, R., Paiement, P., & Leung, T. K. (1996). Swine model of coronary restenosis: Effect of a second injury. *Cathet. Cardiovasc. Diagn.*, **38**, 44–49.

Boudghene, F. P., Sapoval, M. P., Bonneau, M., LeBlanche, A. F., Lavaste, F. C., et al. (1998). Abdominal aortic aneurysms in sheep: Prevention of rupture with endoluminal stent-grafts. *Radiology*, **206**, 447–454.

Braunwald, N. S., & Bonchek, L. I. (1967). Prevention of thrombus formation on rigid prosthetic heart valves by the ingrowth of autogenous tissue. *J. Thorac. Cardiovasc. Surg.*, **54**, 630–638.

Brown, W. M., 3rd, Jay, J. L., Gott, J. P., Pan-Chih, Dorsey, L. M., et al. (1991). Placement of aortic valve bioprostheses in sheep via a left thoracotomy: Implantation of stentless porcine heterografts. *Trans. Am. Soc. Artif. Intern. Organs.*, **37**, M445–M446.

Bunton, R. W., Xabregas, A. A., & Miller, A. P. (1990). Pericardial closure after cardiac operations. An animal study to assess currently available materials with particular reference to their suitability for use after coronary artery bypass grafting. *J. Thorac. Cardiovasc. Surg.*, **100**, 99–107.

Calne, R. Y., Bitter-Suermann, H., Davis, D. R., Dunn, D. C., Herbertson, B. M., et al. (1974). Orthotopic heart transplantation in the pig. *Nature*, **247**, 140–142.

Caparrelli, D. J., Ghazoul, M., & Diethrich, E. B. (2009). Indications for coronary artery bypass grafting in 2009: What is left to surgery. *J. Cardiovasc. Surg.*, **50**, 19–26.

Carroll, G. L., & Hartsfield, S. M. (1996). General anesthetic techniques in ruminants. *Vet. Clin. North Am. Food Anim. Pract.*, **12**(3), 627–661.

Cevallos, W. H., Holmes, W. L., Nyers, R. N., & Smink, R. D. (1979). Swine in atherosclerosis research: Development of an experimental animal model and study of the effect of dietary fats on cholesterol metabolism. *Atherosclerosis*, **34**, 303–317.

Chow, E., & Farrar, D. J. (1992). Right heart function during prosthetic left ventricular assistance in a porcine model of congestive heart failure. *J. Thorac. Cardiovasc. Surg.*, **104**, 569–578.

Christofferson, R. D., Kapadia, S. R., Rajagopal, V., & Ruzcu, E. M. (2009). Emerging transcatheter therapies for aortic and mitral disease. *Heart*, **95**, 148–155.

Clutton-Brock, J. (1999a). Pigs. In *A Natural History of Domesticated Mammals* (2nd ed., pp. 91–99). Cambridge UK: Cambridge University Press.

Clutton-Brock, J. (1999b). Sheep and goats. In *A Natural History of Domesticated Mammals* (2nd ed., pp. 69–80). Cambridge, UK: Cambridge University Press.

Commission on Life Sciences. (1996). Animal environment, housing and management. In *Guide for the Care and Use of Laboratory Animals* (pp. 21–55). Washington, DC: National Academy Press.

Cremer, J., Boetel, C., Fredow, G., Gebureck, P., & Haverich, A. (1995). Radiographic assessment of structural defects in Bjork-Shiley convexo-concave prostheses. *Eur. J. Cardiothorac. Surg.*, **9**, 373–377.

Darwin, C. (1959). The Life and Letters of Charles Darwin. In F. Darwin (Ed.), (pp. 382–383). New York, NY: Basic Books.

David, T. E., Ropchan, G. C., & Butany, J. W. (1988). Aortic valve replacement with stentless porcine bioprostheses. *J. Card. Surg.*, **3**, 501–505.

Dewanjee, M. K., Singh, S. K., Wooley, P. H., Mackey, S. T., Solis, E., et al. (1986a). Identification of new collagen formation with ^{125}I-labeled antibody in bovine pericardial tissue valves implanted in calves. *Int. J. Rad. Appl. Instrum. B*, **13**, 413–422.

Dewanjee, M. K., Solis, E., Mackey, S. T., Lenker, J., Edwards, W. D., et al. (1986b). Quantification of regional platelet and calcium deposition on pericardial tissue valve prostheses in calves and effect of hydroxethylene diphosphonate. *J. Thorac. Cardiovasc. Surg.*, **92**, 337–348.

Doumanian, A. V., & Ellis, F. H. (1961). Prolonged survival after total replacement of the mitral valve in dogs. *J. Thorac. Cardiovasc. Surg.*, **42**, 683–695.

Dunn, P. F., Newman, K. D., Jones, M., Yamada, I., Shayani, V., et al. (1996). Seeding of vascular grafts with genetically modified endothelial cells. Secretion of recombinant TPA results in decreased seeded cell retention *in vitro* and *in vivo*. *Circulation*, **93**, 1439–1446.

Fann, J. I., Chronos, N., Rowe, S. J., Michiels, R., Lyons, B. E., et al. (2008). Evolving strategies for the treatment of valvular heart disease: Preclinical and clinical pathways for percutaneous aortic valve replacement. *Catheter. Cardiovasc. Interv.*, **71**, 434–440.

Fauci, A. S., Braunwald, E., Isselbacher, K. J., Wilson, J. D., Martin, J. B., et al. (1998). *Harrison's Principles of Internal Medicine*. New York, NY: McGraw-Hill.

Flecher, E. M., Curry, J. W., Joudinaud, T. M., Kegel, C. L., Weber, P. A., et al. (2007). Coronary flow obstruction in percutaneous aortic valve replacement. An *in vitro* study. *Eur. J. Cardiothorac. Surg.*, **32**, 291–295.

Frazier, O. H., Rose, E. A., Macmanus, Q., Burton, N. A., Lefrak, E. A., et al. (1992). Multicenter clinical evaluation of the HeartMate 1000 IP left ventricular assist device. *Ann. Thorac. Surg.*, **53**, 1080–1090.

Fritz, K. E., Daoud, A. S., Augustyn, J. M., & Jarmolych, J. (1980). Morphological and biochemical differences among grossly-defined types of swine aortic atherosclerosis induced by a combination of injury and atherogenic diet. *Exp. Mol. Pathol.*, **32**, 61–72.

Fujisawa, N., Poole-Warren, L. A., Woodard, J. C., Bertram, C. D., & Schindhelm, K. (1999). A novel textured surface for blood-contact. *Biomaterials*, **20**, 955–962.

Gabbay, S., Guindy, A. M., Andrews, J. F., Amato, J. J., Seaver, P., et al. (1989). New outlook on pericardial substitution after open heart operations. *Ann. Thorac. Surg.*, **48**, 803–812.

Gajewski, J., & Povar, M. L. (1971). Blood coagulation values in sheep. *Am. J. Vet. Res.*, **32**, 405–409.

Gallegos, R. P., Nockel, P. J., Rivard, A. L., & Bianco, R. W. (2005). The current state of *in vivo* pre-clinical animal models for heart valve evaluation. *J. Heart Valve Dis.*, **14**, 423–432.

Gallo, I., & Frater, R. W. M. (1983). Experimental atrioventricular bioprosthetic valve insertion: A simple and successful technique. *Thorac. Cardiovasc. Surg.*, **31**, 288–290.

Ghoshal, N. G., & Nanda, B. S. (1975). Porcine heart and arteries. In R. Getty (Ed.), *Sisson and Grossman's The Anatomy of the Domestic Animals* (5th ed.). Philadelphia, PA: W. B. Saunders.

Gott, J. P., Pan-Chih, Dorsey, L. M., Jay, J. L., Jett, G. K., et al. (1992). Calcification of porcine valves: A successful new method of anti-mineralization. *Ann. Thorac. Surg.*, **53**, 207–216.

Granger, D. K., Matas, A. J., Jenkins, M. K., Moss, A. A., Chen, S. C., et al. (1994). Prolonged survival without post-transplant immunosuppression in a large animal model. *Surgery*, **116**, 236–241.

Grant, D., Duff, J., Zhong, R., Garcia, B., Lipohar, C., et al. (1988). Successful intestinal transplantation in pigs treated with cyclosporine. *Transplantation*, **45**(2), 279–284.

Gregoric, I. D., Eya, K., Tamez, D., Cervera, R., Byler, D., et al. (2004). Preclinical hemodynamic assessment of a new trileaflet mechanical valve in the aortic position in a bovine model. *J. Heart Valve Dis.*, **13**, 254–259.

Grehan, J. F., Hilbert, S. L., Ferrans, V. J., Droel, J. S., Salerno, C. T., et al. (2000). Development and evaluation of a swine model to assess the preclinical safety of mechanical heart valves. *J. Heart Valve Dis.*, **9**, 710–720.

Grehan, J. F., Casagrande, I., Oliveira, E. L., Santos, P. C., Pessa, C. J., et al. (2001). A juvenile sheep model for the long-term evaluation of stentless bioprostheses implanted as aortic root replacements. *J. Heart Valve Dis.*, **10**, 505–512.

Gross, D. R. (1994). Recognition of pain and the use of analgesics. In *Animal Models in Cardiovascular Research* (2nd revised ed.). Dordrecht, The Netherlands: Kluwer Academic Publishers.

Gross, D. R. (1997). Thromboembolic phenomena and the use of the pig as an appropriate animal model for research on cardiovascular devices. *Int. J. Artif. Organs*, **20**, 195–203.

Gross, D. R., Dewanjee, M. K., Zhai, P., Lanzo, S., & Wu, S. M. (1997). Successful prosthetic mitral valve implantation in pigs. *ASAIO. J.*, **43**, M382–M386.

Hannon, J. P., Bossone, C. A., & Wade, C. E. (1990). Normal physiologic values for conscious pigs used in biomedical research. *Lab. Anim. Sci.*, **40**(3), 293–298.

Hasenkam, J. M., Ostergaard, J. H., Pederson, E. M., Paulsen, P. K., Nygaard, H., et al. (1988a). A model for acute haemodynamic studies in the ascending aorta in pigs. *Cardiovasc. Res.*, **22**, 464–471.

Hasenkam, J. M., Pedersen, E. M., Ostergaard, J. H., Nygaard, H., Paulsen, P. K., et al. (1988b). Velocity fields and turbulent stress downstream of a biological and mechanical aortic valve prostheses implanted in pigs. *Cardiovasc. Res.*, **22**, 472–483.

Hasenkam, J. M., Nygaard, H., Pedersen, E. M., Ostergaard, J. H., Paulsen, P. K., et al. (1989). Turbulent stresses downstream of porcine and pericardial aortic valves implanted in pigs. *J. Cardiac. Surg.*, **4**, 74–78.

Hazekamp, M. G., Goffin, Y. A., & Huysmans, H. A. (1993). The value of the stentless biovalve prothesis: An experimental study. *Eur. J. Cardio-Thoracic. Surg.*, **7**, 514–519.

Hecker, J. F. (1974). *Experimental Surgery on Small Ruminants*. London, UK: Butterworths.

Hendry, P. J., Ascah, K. J., Rajagopalan, K., & Calvin, J. E. (1994). Does septal position affect right ventricular function during left ventricular assist in an experimental porcine model? *Circulation* (Suppl. 90), II353–II358.

Henneman, O. D., Van Rijk-Zwikker, Bruggemans, E. F., Rosendal, F. R., Delemarre, B. J., et al. (1998). The pig as an *in vivo* model for the evaluation of the thrombogenicity of mechanical heart valves. In *Workshop on Prosthetic Heart Valves: Future Directions* (p. 9). February 18–22, Hilton Head, SC.

Holmberg, D. L., & Olsen, D. B. (1987). Anesthesia and cardiopulmonary bypass technique in calves and sheep. *Vet. Surg.*, **16**, 463–465.

Huber, C. H., Cohn, L. H., & von Segesser, L. K. (2005). Direct-access valve replacement. A novel approach for off-pump valve implantation using valved stents. *J. Am. Coll. Cardiol.*, **46**, 366–370.

Hughes, H. C. (1986). Swine in cardiovascular research. *Lab. Anim. Sci.*, **36**, 348–350.

ISO (International Organization for Standardization). (1994). ISO 7198-1 and 2:1994: *Cardiovascular implants: Tubular vascular prostheses*.

Irwin, E., Lang, G., Clack, R., St. Cyr, J., Runge, W., et al. (1993). Long-term evaluation of prosthetic mitral valves in sheep. *J. Invest. Surg.*, **6**, 133–141.

James, N. L., Schindhelm, K., Slowiaczek, B. K., Milthorpe, B., Graham, A. R., et al. (1992). *In vivo* patency of endothelial cell-lined expanded polytetrafluroethylene prostheses in an ovine model. *Artif. Organs.*, **16**, 346–353.

Jensen, N., Lindblad, B., Ljungberg, J., Leide, S., & Bergqvist, D. (1997). Early attachment of leukocytes, platelets and fibrinogen in endothelial cell-seeded Dacron venous conduits. *Br. J. Surg.*, **84**, 52–57.

Jordan, W. D., Sampson, L. K., Iyer, S., Anderson, P. G., Lyle, K., et al. (1998). Abdominal aortic aneurysm repair via percutaneous endovascular stenting in the swine model. *Am. Surg.*, **64**, 1070–1073.

Kaneko, J. J., Harvey, J. W., & Bruss, M. L. (1997). *Clinical Biochemistry of Domestic Animals* (5th ed.). San Diego, CA: Academic Press.

Karas, S. P., Gravanis, M. B., Santoian, E. C., Robinson, K. A., Anderberg, K. A., et al. (1992). Coronary intimal proliferation after balloon-injury and stenting in swine: An animal model of restenosis. *J. Am .Coll. Cardiol.*, **20**, 467–474.

Karges, H. E., Funk, K. A., & Ronneberger, H. (1994). Activity of coagulation and fibrinolyis parameters in animals. *Arzneim-Forsch/Drug Res.*, **44**, 793–797.

Kerzner, B., Kelly, M. H., Gall, D. G., Butler, D. G., & Hamilton, J. R. (1977). Transmissible gastroenteritis: Sodium transport and the intestinal epithelium during the course of viral enteritis. *Gastroenterology*, **72**, 457–461.

Kohler, T. R., & Kirkman, T. R. (1999). Dialysis access failure: A sheep model of rapid stenosis. *J. Vasc. Surg.*, **30**, 744–751.

Leary, H. L., Jr., & Lecce, J. G. (1978). Effect of feeding on the cessation of transport of macromolecules by enterocytes of neonatal piglet intestine. *Biol. Neonate.*, **34**, 174–176.

Liao, K., Gong, G., & Hoffman, D. (1993). Spontaneous host endothelial growth on bioprosthetic valves and its relation to calcification. *Eur. J. Cardiothorac. Surg.*, **7**, 591–596.

Linden, B. C., Schumacher, C. W., MacIver, R. H., Mrachek, J. P., & Bianco, R. W. (2003). Paravalvular leaks around prosthetic valves implanted in the mitral position: Technical refinements in an ovine model. *J. Heart Valve Dis.*, **12**, 400–405.

Litwak, K. N., Unger, L. S., Fukamachi, K., & Saeed, D. (2008). Retrospective analysis of adverse events in preclinical ventricular assist device experiments. *ASAIO J.*, **54**, 347–350.

Mack, M. J., Prince, S. L., Herbert, M., Brown, P. P., Katz, M., et al. (2008). Current clinical outcomes of percutaneous coronary intervention and coronary artery bypass grafting. *Ann. Thorac. Surg.*, **86**, 496–508.

Marino, I. R., & De Luca, G. (1985). Orthotopic liver transplantation in pigs. An evaluation of different methods of avoiding the revascularization syndrome. *Transplantation*, **40**, 494–498.

Matalon, S., Nesarajah, M. S., Krasney, J. A., & Farhi, L. E. (1982). Pulmonary and circulatory changes in conscious sheep exposed to 100% O_2 at 1 ATM. *J. Appl. Physiol.*, **53**, 110–116.

McKenzie, J. E. (1996). Swine as a model in cardiovascular research. In M. E. Tumbleson, & L. B. Schook (Eds.), *Advances in Swine in Biomedical Research* (pp. 7–17). New York, NY: Plenum Press.

Mellen, I. M. (1952). *The Natural History of the Pig*. New York, NY: Exposition Press.

Mersmann, H. J., Goodman, J., Houk, J. M., & Anderson, S. (1972). Studies on the biochemistry of mitochondria and cell morphology in the neonatal swine hepatocyte. *J. Cell. Biol.*, **53**, 335–347.

Nesarajah, M. S., Matalon, S., Krasney, J. A., & Fahri, L. E. (1983). Cardiac output and regional oxygen transport in the acutely hypoxic conscious sheep. *Respir. Physiol.*, **53**, 161–172.

Neville, R. F., Jr., Bartorelli, A. L., Sidawy, A. N., & Leon, M. B. (1994). Vascular stent deployment in vein bypass grafts: Observations in an animal model. *Surgery*, **116**, 55–61.

Newman, J. H., Loyd, J. E., English, D. K., Ogletree, M. L., Fulkerson, W. J., et al. (1983). Effects of 100% oxygen on lung vascular function in awake sheep. *J. Appl. Physiol.*, **54**, 1379–1386.

Nolen, R. S. (2009). Revision process begins for AVMA euthanasia guidelines. *J. Am. Vet. Med. Assoc.*, **235**, 246–247.

Northrup, W. F., III, Mrachek, J. P., McClay, C., & Feeny, D. A. (1998). A novel annuloplasty system with rigid and flexible elements: Initial experimental results in sheep and the case for untreated autologous pericardium. *J. Heart Valve Dis.*, **7**, 62–71.

Nowell, J., Wilton, E., Markus, H., & Jahangiri, M. (2007). Antithrombotic therapy following bioprosthetic aortic valve replacement. *Eur. J. Cardiothorac. Surg.*, **31**, 578–585.

Okazaki, Y., Wika, K. E., Matasuyoshi, T., Fukamachi, K., Kunitoma, R., et al. (1996). Platelets are deposited early postoperatively on the leaflet of a mechanical heart valve in sheep without post-operative antiplatelet agents. *Am. Soc. Artif. Organs. J.*, **42**, M750–M754.

Ouyang, D. W., Salerno, C. T., Pederson, T. S., Bolman, R. M., III, & Bianco, R. W. (1998). Long term evaluation of orthotopically implanted stentless bioprosthetic aortic valves in juvenile sheep. *J. Invest. Surg.*, **11**, 175–183.

Oz, M. C., Argenziano, M., Catanese, K. A., Gardocki, M. T., Goldstein, D. J., et al. (1997). Bridge experience with long-term implantable left ventricular assist devices. Are they an alternative to transplantation? *Circulation*, **95**, 1844–1852.

Park, S. J., Nguyen, D. Q., Bank, A. J., Ormaza, S., & Bolman, R. M., III (2000). Left ventricular assist device bridge therapy for acute myocardial infarction. *Ann. Thorac. Surg.*, **69**, 1146–1151.

Pibarot, P., & Dumesnil, J. G. (2000). Hemodynamic and clinical impact of prosthesis–patient mismatch in the aortic valve position and its prevention. *J. Am. Coll. Cardiol.*, **36**, 1131–1141.

Pibarot, P., & Dumesnil, J. G. (2001). Patient–prosthesis mismatch can be predicted at the time of operation. *Ann. Thorac. Surg.*, **71**, S265–S268.

Pibarot, P., Dumesnil, J. G., Lemieux, M., Cartier, P., Metras, J., et al. (1998). Impact of prosthesis–patient mismatchon hemodynamic and symptomatic status, morbidity after aortic valve replacement with a bioprosthetic heart valve. *J. Heart Valve Dis.*, **7**, 211–218.

Piermattei, D. L., & Swan, H. (1970). Techniques for general anesthesia in miniature pigs. *J. Surg. Res.*, **10**, 587–592.

Pinches, S. A., et al. (1993). Preparation and characterization of basolateral membrane vesicles from pig and human colonocytes: the mechanism of glucose transport. *Biochem J. Sep 1*, **294**(Pt 2), 529–534.

Poole-Warren, L. A., Schindhelm, K., Graham, A. R., Slowiaczek, P. R., & Noble, K. R. (1996). Performance of small diameter synthetic vascular prostheses with confluent autologous endothelial cell linings. *J. Biomed. Mater. Res.*, **30**, 221–229.

Ramo, B. W., Peter, R. H., Ratliff, N., Kong, Y., McIntosh, H. D., et al. (1970). The natural history of right coronary arterial occlusion in the pig. Comparison with left anterior descending arterial occlusion. *Am. J. Cardiol.*, **26**, 156–161.

Reverdiau-Moalic, P., Watier, H., Vallee, I., Lebranchu, Y., Bardos, P., et al. (1996). Comparative study of porcine and human blood coagulation systems: Possible relevance in xenotransplantation. *Transplant. Proc.*, **28**, 643–644.

Riebold, T. W., & Thurmon, J. C. (1985). Anesthesia in swine. In M. E. Tumbleson (Ed.), *Swine in Biomedical Research* (pp. 243–254). New York, NY: Plenum Press.

Rodgers, G. P., Minor, S. T., Robinson, K., Cromeens, D., Wollbert, S. C., et al. (1990). Adjuvant therapy for intracoronary stents. Investigations in atherosclerotic swine. *Circulation*, **82**, 560–569.

Roubin, G. S., King, S. B., III, Douglas, J. S., Jr., Lembo, N. J., & Robinson, K. A. (1990). Intracoronary stenting during percutaneous transluminal coronary angioplasty. *Circulation*, **81**, IV92–100.

Rousseau, H., Puel, J., Joffre, F., Sigwart, U., Duboucher, C., et al. (1987). Self-expanding endovascular prosthesis: An experimental study. *Radiology*, **164**, 709–714.

Ruel, M., Kulik, A., Rubens, F. D., Bedard, P., Masters, R. G., et al. (2004a). Late incidence and determinants of reoperation in patients with prosthetic heart valves. *Eur. J. Cardiothorac. Surg.*, **25**, 364–370.

Ruel, M., Rubens, F. D., Masters, R. G., Pipe, A. L., Bedard, P., et al. (2004b). Late incidence and predictors of persistent or recurrent heart failure in patients with aortic prosthetic valves. *J. Thorac. Cardiovasc. Surg.*, **127**, 149–159.

Salerno, C. T., Droel, J., & Bianco, R. W. (1998). Current state of *in vivo* preclinical heart valve evaluation. *J. Heart Valve Dis.*, **7**, 158–162.

Schatz, R. A., Palmaz, J. C., Tio, F. O., Garcia, F., Garcia, O., et al. (1987). Balloon-expandable intracoronary stents in the adult dog. *Circulation*, **76**, 450–457.

Schoen, F. J., Levy, R. J., Nelson, A. C., Bernhard, W. F., Nashef, A., et al. (1985). Onset and progression of experimental bioprosthetic heart valve calcification. *Lab. Invest.*, **52**, 523–532.

Schoen, F. J., Hirsch, D., Bianco, R. W., & Levy, R. J. (1994). Onset and progression of calcification in porcine aortic bioprosthetic valves implanted as orthotopic mitral valve replacements in juvenile sheep. *J. Thorac. Cardiovasc. Surg.*, **108**, 880–887.

Schurmann, K., Vorwerk, D., Bucker, A., Grosskortenhaus, S., & Gunther, R. W. (1998). Single and tandem stents in sheep iliac arteries: Is there a difference in patency? *Cardiovasc. Intervent. Radiol.*, **21**, 411–418.

Schwartz, R. S., & Holmes, D. R. (1994). Pigs, dogs, baboons and man: Lessons for stenting from animal studies. *J. Interven. Cardiol.*, **7**, 355–368.

Scott, N. A., Robinson, K. A., Nunes, G. L., Thomas, C. N., Viel, K., et al. (1995). Comparison of the thrombogenicity of stainless steel and tantalum coronary stents. *Am. Heart J.*, **129**, 866–872.

Sielaff, T. D., Nyberg, S. L., Rollins, M. D., Hu, M. Y., Amiot, B., et al. (1997). Characterization of the three-compartment gel-entrapment porcine hepatocyte bioartificial liver. *Cell Biol. Toxicol.*, **13**, 357–364.

Smith, A. C., Ehler, W. J., & Swindle, M. M. (2008). Anesthesia and analgesia in swine. In R. E. Fish, M. J. Brown, P. J. Danneman, & A. Z. Karas (Eds.), *Anesthesia and Analgesia in Laboratory Animals* (2nd ed., pp. 413–440). London, UK: Academic Press.

Snyder, T. A., Litwak, K. N., Watach, M. J., & Wagner, W. R. (2002). Platelet activation, aggregation and life span in calves implanted with axial flow ventricular assist devices. *Ann. Thorac. Surg.*, **73**, 1933–1938.

Snyder, T. A., Tsukui, H., Kihara, S., Akimoto, T., Litwak, K. N., et al. (2007). Preclinical biocompatibility assessment of the EVAHEART ventricular assist device: Coating comparison and platelet activation. *J. Biomed. Mater. Res. A.*, **81**, 85–92.

Soini, H. O., Takala, J., Nordin, A. J., Makisalo, H. J., & Hockerstedt, K. A. (1992). Peripheral and liver tissue oxygen tensions in hemorrhagic shock. *Crit. Care. Med.*, **20**, 1330–1334.

Spyt, T. J., Fisher, J., Reid, J., Anderson, J. D., & Wheatley, D. J. (1998). Animal evaluation of a new pericardial bioprosthetic heart valve. *Artif. Organs*, **12**, 328–336.

Stanton, H. C., & Mersmann, H. J. (1986). Preface. In H. C. Stanton, & H. J. Mersmann (Eds.), *Swine in Cardiovascular Research* (Vol. I). Boca Raton, FL: CRC Press.

Swan, H., & Meagher, M. (1971). Total body bypass in miniature pigs. *J. Thorac. Cardiovasc. Surg.*, **61**, 956–967.

Swan, H., & Piermattei, D. L. (1971). Technical aspects of cardiac transplantation in the pig. *J. Thorac. Cardiovasc. Surg.*, **61**, 710–723.

Swindle, M. M. (1992). Preface. In M. M. Swindle (Ed.), *Swine as Models in Biomedical Research* (pp. ix–x). Ames, IA: Iowa State University Press.

Swindle, M. M. (1994). Anesthetic and perioperative techniques in swine: An update. In *Charles River Laboratories Technical Bulletin* (pp. 1–3). Wilmington, DE: Charles River Laboratories.

Swindle, M. M., & Smith, A. C. (1994). Swine: anesthesia and analgesia in Research Animal Anesthesia and Surgery. In A. C. Smith, & M. M. Swindle (Eds.), (pp. 107–110). Greenbelt, MD: Scientists Center for Animal Welfare.

Swindle, M. M. (2007). *Swine in the Laboratory: Surgery, Anesthesia, Imaging and Experimental Techniques*. Boca Raton, FL: CRC Press.

Swindle, M. M., Horneffer, P. J., Gardner, T. J., Gott, V. L., Hall, T. S., et al. (1986). Anatomic and anesthetic considerations in experimental cardiopulmonary surgery in swine. *Lab. Anim. Sci.*, **36**, 357–361.

Taylor, A., Ao, P., & Fletcher, J. (1995). Inhibition of intimal hyperplasia and occlusion in Dacron graft with heparin and low molecular weight heparin. *Int. Angio.*, **14**, 375–380.

Tillman, P., Carson, S. N., & Talken, L. (1981). Platelet function and coagulation parameters in sheep during experimental vascular surgery. *Lab. Anim. Sci.*, **31**, 262–267.

Tobin, E., & Hunt, E. (1996). Supplies and technical considerations for ruminant and swine anesthesia. *Vet. Clin. North Am. Food. Anim. Pract.*, **12**, 531–547.

Towne, C. W., & Wentworth, E. H. (1950). *Pigs from Cave to Corn Belt*. Norman, OK: University of Oklahoma Press.

Tumbleson, M. E., & Schook, L. B. (1996). Advances in swine in biomedical research. In M. E. Tumbleson, & L. B. Schook (Eds.), *Advances in Swine in Biomedical Research* (Vol. I, pp. 1–4). New York, NY: Plenum Press.

Ueda, Y., Ditakaze, M., Imakita, M., Ishibashi-Ueda, H., Minamino, T., et al. (1999). Glycoprotein IIb/IIIa antagonist FK633 could not prevent neointimal thickening in stent implantation model of canine coronary artery. *Arterioscler Thromb. Vasc. Biol.*, **19**(2), 343–347.

University of Minnesota Animal Care and Use Manual 1997 reprint. Web site: http://www.ahc.umn.edu/rar/housing.html

Vallana, F., Rinaldi, S., Galletti, P. M., Nguyen, A., & Piwnica, A. (1992). Pivot design in bileaflet valves. *ASAIO J.*, **38**, M600–M606.

Vyavahare, N., Hirsch, D., Lerner, E., Baskin, J. Z., Schoen, F. J., et al. (1997). Prevention of bioprosthetic heart valve calcification by ethanol preincubation: Efficacy and mechanisms. *Circulation*, **95**, 479–458.

Walpoth, B. H., Ammon, A., Galdikas, S., Ris, H. B., Schaffner, T., et al. (1993). Experimental assessment of thrombogenicity in vascular prostheses before and during prostaglandin E treatment. *Eur. J. Vasc. Surg.*, **7**(5), 493–499.

Whitbread, T., Birch, P., Rogers, S., Majeed, A., Rochester, J., et al. (1996). A new animal model for abdominal aortic aneurysms: Initial results using a multiple-wire stent. *Eur. J. Vasc. Endovasc. Surg.*, **11**, 90–97.

White, C. J., Ramee, S. R., Banks, A. K., Wiktor, D., & Price, H. L. (1989). The Yucatan miniature swine: An atherogenic model to assess the early patency rates of an endovascular stent. In M. M. Swindle (Ed.), *Swine as Models in Biomedical Research* (pp. 185–196). Ames, IA: Iowa State University Press.

White, J. G., Mulligan, N. J., Gorin, D. R., D'Agostino, R., Yucel, E. K., et al. (1998). Response of normal aorta to endovascular grafting: A serial histopathological study. *Arch. Surg.*, **133**, 246–249.

Wilbourn, B., Harrison, P., Lawvie, A., Savariau, E., Savidge, G., et al. (1993). Porcine platelets contain an increased quantity of ultra-high molecular weight von Willebrand factor and numerous alpha-granular tubular structures. *Br. J. Haematol.*, **83**, 608–615.

Willette, R. N., Zhang, H., Louden, C., & Jackson, R. K. (1996). Comparing porcine models of coronary restenosis. In M. E. Tumbleson, & L. B. Schook (Eds.), *Swine in Biomedical Research* (pp. 595–606). New York, NY: Plenum Press.

Wilson, J. M., & Ferguson, J. J. (1995). Revascularization therapy for coronary artery disease. Coronary artery bypass grafting versus percutaneous transluminal coronary angioplasty. *Tex. Heart Inst. J.*, **22**, 145–161.

CHAPTER II.3.8 MICROSCOPY FOR BIOMATERIALS SCIENCE

Kip D. Hauch[1] *and Buddy D. Ratner*[2]
[1]University of Washington, Seattle, WA, USA
[2]Professor, Bioengineering and Chemical Engineering, Director of University of Washington Engineered Biomaterials (UWEB), Seattle, WA, USA

The microscope is so widely used in all branches of science and technology that it is almost iconic of research or science. The scientist relies heavily on vision; thus the advent of the light microscope is among the earliest, and arguably most powerful, tools in the history of science. Today, microscopy in its various forms still remains as a pre-eminent tool at the forefront of scientific exploration. Thus, the biomaterials scientist makes extensive use of both light and electron microscopes to help fabricate and characterize new materials, coatings, and devices, and to study the behavior of cells and tissues at the biomaterials interface. This chapter will familiarize the reader with the more common microscopy tools used in current biomaterials research. The key concepts of magnification, resolution, and contrast are first introduced and then their meaning explored in the context of light microscopy. Digital imaging is very briefly addressed. Attention then moves to the electron microscopies, specifically scanning electron microscopy (SEM). Finally, new developments in optical microscopy are addressed that are shattering limitations of microscopy that have inhibited progress for hundreds of years. Fortunately, the ubiquitous importance of microscopy in the practice of science means that there exist numerous excellent resources to consult on these topics. A brief guide to some useful resources is found at the end of the chapter.

MAGNIFICATION, RESOLUTION, AND CONTRAST

The art of scientific instrumentation is to devise tools to observe natural phenomena and present them appropriately to our senses for interpretation (e.g., to our eyes as an image). Three concepts apply to this process: magnification; resolution; and contrast. The three concepts are interdependent, and frequently limitations in one aspect or another leads to trade-offs.

Magnification is the appropriate scaling of the phenomena to our detectors (ultimately, for our purposes here, the human visual system). In a standard light microscope this is accomplished through the millennia old traditions of refracting light through curved glass lenses (e.g., an objective lens or an ocular). In the electron microscope, as well as the other scanning microscopies (e.g., laser scanning confocal, multiphoton, scanning tunneling or atomic force microscopies), photons or electrons are used to probe the sample discretely at a single point. The location of this sampling point is then rastered across the sample. Magnification is achieved by displaying the combined results from this "probing" on a more appropriate scale.

Resolution is dependent on the information present in the collected data sufficient to allow one to achieve perception. In microscopy we are usually interested in spatial resolution – the ability to resolve one location from another in an image. Resolution is often limited by the physical nature of the probe and/or the measurement of the signal thereof.

Contrast is the ability to detect specific differences in the signal. More than just signal/noise, contrast implies some aspect of specificity – the ability to tell one particular part of an image or signal from another. Resolution and contrast are often related – one is often gained at the expense of the other. With respect to light microscopes, the fundamental limits of magnification and resolution have generally been reached long ago. New and varied methods for the generation of contrast are what lead to the many different modalities of light microscopy and, as will be shown here, continue to be a source of technological development.

CONFIGURATIONS

A specimen can be imaged by passing illuminating photons or electrons through the sample (transmission microscopy, diascopic illumination) or by reflecting the illumination off the sample (reflected microscopy, episcopic illumination) to a detector.

In transmission microscopy, the sample must be thin enough to allow the probe to pass through the sample and reach the detector. For example, in transmitted light microscopy (probe = photons), samples may be up to 100 μm thick (although for typical high-resolution histology work, sections of tissue are prepared that are nominally only 5 μm thick.) In transmission electron microscopy (TEM, probe = electrons), samples are prepared that are only 50 nm thick.

Reflected techniques are used when the sample is opaque. The sample must reflect the illumination or otherwise generate a signal that is directed from the surface back to a detector. Reflected light is how we are used to viewing the everyday world around us. Reflected light techniques include simple magnifying hand lenses, stereoscopes and dissecting/operating room microscopes, and compound microscopes commonly used in the material sciences or manufacturing sector. As will be discussed, fluorescence-based techniques are almost always performed with episcopic illumination. Scanning electron microscopy (SEM) can be considered in this category as well, although the signal generated is somewhat more complex than simply the detection of reflected electrons.

LIGHT MICROSCOPY

The biomaterials scientist relies daily on a variety of light microscopes for tasks, ranging from routine materials inspection to detailed complex biologic analyses. The interests of the biomaterials engineer are often focused on the biologic responses at the cell–biomaterial interface. These are studied *in vitro*, e.g., using cell culturing approaches or studied *in vivo* through the examination of a tissue–implant interface. In this regard, the microscopy needs of the biomaterials scientist overlap significantly with those of the molecular cell biologist or pathologist. The presence of a biomaterial in the sample is often just another (sometimes complicating) factor to be dealt with.

Platforms

Light microscopes can be configured on several different platforms – for different purposes and with varying levels of customization. Hand lenses, loupes, stereoscopes, and dissecting microscopes are essential for fabrication and inspection of small biomedical devices and coatings. Likewise, operating microscopes are often helpful for implant work in small animal models. An upright microscope is one in which the sample is observed from above by an objective lens. Upright microscopes are used most often for histology work, and are equipped with high-resolution but short working-distance optics. A typical sample might be a slice of tissue mounted between a microscope slide and a coverslip. An inverted microscope is the mainstay of the tissue culture laboratory, where specimens are viewed through objective lenses positioned underneath the specimen stage. This allows for the observation of live cells in culture, e.g., through the transparent bottoms of tissue culture flasks or multiwell plates. The large open stage of the inverted instrument allows for more experimental versatility, e.g., the use of complicated flow cells, micromanipulators or environmental enclosures. An inverted instrument is often equipped with slightly lower resolution, but longer working-distance optics. Either platform may be augmented for the specialized techniques to be discussed later, namely phase contrast, differential interference contrast, and fluorescence microscopy. With respect to research-grade instruments, there are more similarities than differences in the capabilities of the upright and inverted platforms. However, it is crucial that the choice of instrument platform and the experimental design receive mutual consideration early in the experimental planning stages.

Magnification in Light Microscopy

Magnification refers to the ratio of a feature dimension in the presented image to the corresponding feature dimension in the original specimen. In common practice, the range of magnification available in light microscopy reaches from 5× to about 1000× (a 100× objective lens and a 10× ocular). These values describe the magnification of the sample at the plane of the primary detector – our eyes at the ocular. Herein lies the danger for the modern microscopist, since invariably it is not our eyes, but rather a digital camera that acquires an image for dissemination to our colleagues. The camera does not use the 10× ocular; rather it uses a combination of extension tubes, transfer optics, intermediate magnifiers or reducers, camera coupling lenses, etc. Further, the acquired digital data are then displayed on computer monitors directly as images or embedded in mixed content files (.pdf) at different "zoom levels;" printed on paper for reports, reduced in journal pages or printed onto large poster boards for conferences; perhaps even displayed using data projectors onto huge screens in an auditorium. What then does magnification mean?

The days of conveying the level of magnification in an image by reporting the magnification of the objective lens alone or using vague terms such as "low power" or "high power," are long gone. In the digital age it is far preferable to include an accurate size scale bar as part of every published image. This is conveniently done by separately acquiring an image containing objects of known size, most commonly a stage micrometer available through any microscopy supply house. With an image containing an object of known dimension, it is then straightforward to establish the relationship between distance (μm) in the specimen and pixels, the individual element of the acquired digital image. A scale bar can then be embedded in the images used for presentation, and measurements of feature size reported directly in appropriate units. Further; no matter how the image is then later displayed, the viewer will have a direct feel for the size of the features in the image. Modern software makes these operations relatively painless, and the side-by-side presentation of images taken at macroscopic, microscopic, and nanoscopic scales makes this practice indispensable to the scientific audience.

Resolution in Light Microscopy

The Rayleigh criterion is often used to describe the resolving power in light microscopy:

$$d = 1.22\lambda / (NA_{condenser} + NA_{obj}) \text{ for brightfield}$$

and:

$$d = 0.61\lambda / NA_{obj} \text{ for fluorescence}$$

where λ is the wavelength of light. NA is the numerical aperture of the condenser or objective lens and is defined as:

$$NA = n \sin\Theta$$

where n is the index of refraction of the medium between the lens and the sample (1.0 for air, 1.515 for oil), and Θ is the half-angle of the cone of light collected by the lens (Figure II.3.8.1). Similarly, the depth of focus, which is the thickness in the sample within which features will all appear in focus, is given by:

$$Z = n\lambda / NA^2_{obj}$$

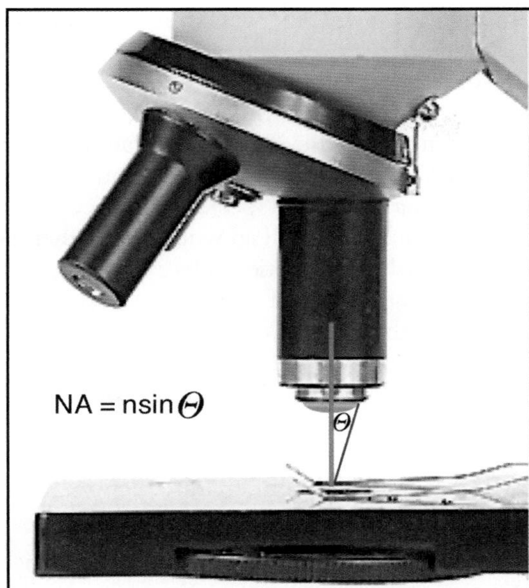

FIGURE II.3.8.1 The concept of numerical aperture (NA). The value, Θ, is half the acceptance angle of the lens, n is the refractive index of the medium between the sample and the objective, n = 1.00 for air, n = 1.515 for oil.

These equations are idealized and are based on the first overlap of the diffraction patterns of two infinitely small point sources in the specimen. For monochromatic light at 546 nm and a high-quality oil immersion lens with an NA of 1.3, this equates to a resolving power of about 0.26 μm, (often called the diffraction limit), easily adequate to resolve many subcellular organelles. Most modern research-grade microscopes are quite capable of realizing this resolution limit. However, the use of polychromatic white light may increase this value twofold. Further, the large values of NA_{cond}, which lead to optimal resolution, actually reduce the contrast available in the image. Thus, we may achieve superb resolution, but not be able to distinguish the features that we seek to resolve! Here we find one of the classic trade-offs: between resolution and contrast.

It is important to understand the resolution available in an image when considering the tools for digital image acquisition. Ideally, the physical size of the individual sensing elements (e.g., the pixel on a digital imager) will be less than half the smallest resolvable feature size, after magnification by the objective and any intermediate magnifiers between the microscope and the imaging plane. The appropriate matching of physical pixel size in the detector to the resolving power of the optics is crucial, and if overlooked can result in degraded system performance. One finds that our eyes, and high-quality film, are well suited to the task. Lower power but relatively higher NA objectives (e.g., 10×/0.4, 20×/0.75) are the most demanding, i.e., they require the smallest pixels, and hence largest number of pixels in the detector.

Note: New microscopy techniques are revolutionizing ideas about the resolution achievable in the visible range. They will be introduced toward the end of this chapter to highlight trends that are radically altering the future of microscopy.

Contrast and the Various Modes of Light Microscopy

The interaction of light with matter is what generates the contrast that we detect in a microscope image. Light can be characterized by its amplitude, wavelength, phase, and polarization. Unfortunately, our human eyes, and most other electronic detectors, are sensitive only to amplitude and wavelength! Thus, a light microscope must convert these many interactions of light and matter into differences in amplitude or wavelength that can be detected in the image. The various ways in which contrast is generated lead to the various light microscopy techniques or modalities. These are described here in quite general terms – the reader is encouraged to explore the many excellent introductory texts on optical microscopy for more thorough coverage.

Transmitted Light Microscopy

The simplest way to view a thin specimen is to shine light through the specimen and view the magnified image directly. In this method, brightfield, we rely on the sample to provide its own contrast. Different features in the specimen may absorb some of the light and/or do so unequally at various wavelengths, thus leading to amplitude or color contrast. In many instances this may work – but consider a layer of unadulterated cultured cells on a glass slide. These cells, if well spread, may only be a few microns thick; made mostly of water, they absorb very little light and provide almost no contrast. Indeed it may be very difficult to see the unstained cells at all!

Another technique is darkfield microscopy. In darkfield the sample is illuminated obliquely, at such glancing angles that the illuminating light, if undisturbed by the specimen, does not even enter the objective lens. Thus, the empty areas (the background areas or "surround") in the image are completely black. Only where elements in the sample scatter light into the objective lens is light collected. Thus, a darkfield image appears as pinpoints of light on a black background. Darkfield is particularly effective if the specimen contains small particulates.

Adding Contrast Agents: Chromophores and Fluorophores

How then can one generate more contrast in a specimen? The simplest answer is to cheat! Exogenous contrast agents can be added to the specimen. These are strongly absorbing or fluorescing large molecules or nanoparticles that are placed into the sample and ideally reside only in certain structures or features the investigator has chosen.

This, of course, is what is done in the staining of cells or tissue sections for histology. Many of these chromophores or stains date to ancient times when they were used to dye biologic materials, such as plant and animal tissues, skins or textile fibers. Perhaps the most common histologic stains are hematoxylin and eosin (H and E). Hematoxylin (which appears blue) is termed a basophilic dye and binds to basic components such as those found in the cell nucleus. Eosin (which appears pink and is also quite fluorescent) is acidophilic and generally stains cytoplasmic proteins. A plethora of histologic stains is available to stain a myriad of cells or structures within cells and tissues. Other common examples include Masson's trichrome, which is used to stain connective tissue or fibrosis around implants (collagen is blue or green); alizarin red, for calcific deposits; ver Hoeff's for elastin; Geimsa and Wright's stain for blood cells; and the Gram stain for bacteria.

It is easy to see the beauty in chromophores (both literally and figuratively), since not only do they generate contrast in the image, but they do so with specificity (i.e., they provide useful information). The investigator can see not only the cell, but can now also see which part is a nucleus, can determine the specific location of structural proteins, etc. This power of specificity can be enhanced by directly (covalently) linking chromophores or fluorophores to specific proteins or other molecules of interest, and then identifying and following the location of these reporter-linked conjugates on surfaces, inside cells, and in tissues.

Further, chromogenic substrates have been synthesized for certain key enzymes. These substrates are converted from colorless to colored (or otherwise undergo a color change) upon cleavage by the enzyme. Examples are the many substrates available for horseradish peroxidase, alkaline phosphatase, and β-galactosidase. These substrates can be used to detect enzymatic activity directly in certain cells and tissues.

Reaching yet another level of specificity, these enzymes can be conjugated to antibodies to impart molecular-level specificity in their binding. Thus, the location of specific molecular antigens can be elucidated by allowing these antibody conjugates to bind, and then revealing the location of the antibodies by using the chromogenic substrates. This process (immunohistochemistry) is directly analogous to forms of the enzyme-linked immunosorbent assays (ELISA) used on the benchtop. Alternatively, the antibodies may be labeled directly with a fluorophore (direct immunofluorescence) or their locations revealed by the binding of a second labeled antibody that recognizes the first (secondary or indirect immunofluorescence). Finally, similarly labeled nucleic acid strands can be used to probe for and reveal the presence of certain sequences of genomic material in cells (*in situ* hybridization) or on surfaces (so-called microarray analysis), and thus give a glimpse into the genomic profile of cells and tissues.

The immense power and versatility afforded by the use of chromophores and fluorophores is now made clear. Beyond simply generating contrast, they provide molecular-level specificity. This enables the investigator not only to simply visualize cells and structures, but also to begin to ask specific cell-physiology and bioengineering questions: is a particular species (a molecule, protein, receptor, enzyme) present? If so where is it, when does it appear, and in what quantities? Here in this generation of contrast lies the power of modern microscopy and its contribution to the cell biology and pathology fields so integral to the study of the biomaterials–biology interface. Because fluorophores play such a prominent role in these techniques, and since their use requires specialized technical approaches, fluorescence microscopy will be addressed in more detail in a separate section.

Optical Tricks: More Contrast – But No Cheating

So far, to generate contrast we have relied on the inherent contrast in the specimen (often inadequate) or the addition of contrast agents. Many of these contrast agents are toxic to cells. Histologic stains are typically used on tissue sections where cells have been fixed, embedded in a support matrix, and sliced open. Probes that bind to intracellular components have to cross the cell membrane; thus cells are permeabilized to allow access. Although there are "vital dyes" that can be used on living cells, many of the approaches described above interfere with the normal physiologic functioning of the cell. Thus, they are best suited for endpoint analysis. How then can one more safely generate contrast in living cells? How can one generate contrast without the addition of these exogenous agents? How can one generate contrast without "cheating?"

Fortunately, the answer lies in several types of optical "tricks." Cells have limited ability to cause absorbance or color changes. However, there is plenty of diffraction (bending of light around organelles), refraction (bending of light at interfaces of structures with different indices of refraction, e.g., membranes), and other optical phenomena that can be used to our advantage. Unfortunately, our eyes are insensitive to all but changes in light intensity and color. The tricks, therefore, will be used to convert these other phenomena into differences in intensity or color, which we then can detect. These techniques all involve optical elements placed in the light path that modulate the incoming light, as well as elements that modulate the light collected by the objective. As such, the sample usually resides in the middle of these two elements, and thus these can be thought of as the "sandwich techniques" (Foster, 1997).

Polarization

Molecules and materials that exhibit a strong regular repeat in their electronic structures are likely to induce orientationally specific changes in polarization during their interactions with light. Examples of such polarizing (birefringent) materials include metals, inorganic and organic crystals, collagen, certain organelles, muscle sarcomeres, and many others. Recall that light polarized by a polarizing filter will not be transmitted through a second polarizing filter oriented at 90° to the first (extinction). If the first filter is placed in the illumination path before reaching the sample, and the second filter is placed in the imaging path after sample, extinction will take place. No light will get through, provided that the sample does not impart any additional polarization to the transmitted light. However, if the sample *does* impart additional polarization, the conditions for extinction will be unsatisfied, and the location of the polarizing materials will be immediately revealed. The degree of this additional polarization induced by the sample can even be measured with yet a third polarizing element (a compensator). The directionality of the polarization can be determined by rotating the specimen stage.

Synthetic polymers, of course, can also contain strong regular repeats in their electronic structure. This can be very useful in locating and identifying remnants of implanted polymeric materials in explanted tissue. The polarization induced by polymers can also be a disadvantage, however, as the plastics used in common tissue culture ware make polarization microscopy nearly impossible using these substrates.

Phase Contrast and Hoffman Modulation Contrast

While there may be little amplitude change when light traverses a cell, there is plenty of diffraction of light by organelles, structures, and membranes. These rays that have interacted with the specimen experience a (unnoticeable) change in phase, having experienced an optical path-length difference $\lambda = (n_2 - n_1)$ (thickness). Typically, the light is retarded in phase by an amount approximately $-\lambda/4$. In the 1930s, Dutch physicist Fritz Zernicke invented a method to convert these phase changes into changes in intensity. The result was phase contrast microscopy, and the method helped foster the major advances in cell biology developed through the study of living cells in culture. For his work, Zernicke received the Nobel Prize in 1953.

The key to phase contrast is to somehow separate those rays of light that pass through the specimen undisturbed (undiffracted) and thus contribute to the background (called the surround, S), from those rays that have been diffracted by interaction with the specimen (D). This can be accomplished spatially at a special plane in the optical path, the diffraction plane. At this plane an image is not formed; rather one can envision that rays of light are separated spatially by their degree of diffraction. Here, each can be manipulated or treated selectively. In phase contrast, an annulus is placed in the illumination light path in a plane conjugate to this diffraction plane (Figure II.3.8.2). This results in a ring or cone of illumination approaching the specimen. Next, a special plate is placed at the diffraction plane, beyond the sample (actually located within the objective lens). This plate is constructed such that its thickness is less in a ring shaped region that matches exactly that of the ring in the annulus plate. Thus, rays of light that are not diffracted (S) will pass through this "phase ring," while diffracted rays (D) will pass through the remaining portions of the phase plate. This phase ring results in a relative $+\lambda/4$ advancement in the phase of light that is undiffracted. Further, this ring is made semitransparent, so that the light transmitted (i.e., the light that will contribute to the background in the final image) is reduced in intensity, and thus appears gray. The diffracted rays have already experienced a retardation in their phase, caused by their interactions with the specimen ($-\lambda/4$). When the surround (S) and the diffracted rays (D) are recombined in the real image plane, the resulting phase difference is $\lambda/2$ and destructive interference takes place. Now, the changes in phase are converted to a reduction in intensity – which is made manifest in the recombined image as areas of dark (black).

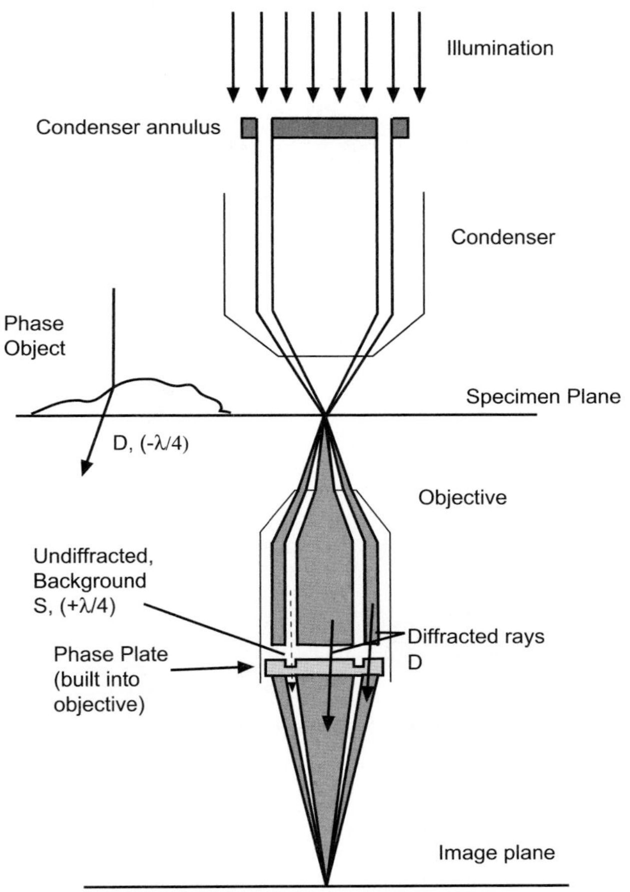

FIGURE II.3.8.2 Phase contrast microscopy in an inverted microscope.

We see that phase contrast is again a "sandwich technique," utilizing a phase ring in the illumination path and a matching ring in the image-forming path. The sample is in the middle. This latter ring is actually physically built into the objective lens, and thus phase contrast microscopy requires a special phase contrast objective and the use of a matching and carefully aligned phase annulus for each objective. Fortunately, phase contrast objectives can also be used for routine brightfield observation and most fluorescence applications, so for many laboratories they are good value.

Phase contrast is ideal for thinly spread cells, such as those in cell culture. They appear in high contrast, their organelles and pseudopodia revealed in areas of bright and dark on a gray background. Rounded cells, however, are surrounded by a strong halo of bright light. Simple phase contrast systems (often with fixed and prealigned annuli) are commonplace on inexpensive inverted microscopes found in tissue culture suites.

In Hoffman modulation contrast, use is made again of the spatial separation of diffracted and undiffracted rays at the diffraction plane (Figure II.3.8.3). This time, instead of a cone of illumination, an asymmetric slit is placed in the illumination path. Again, a modulation plate is built into the objective placed at the diffraction plane. Also again, a semitransparent region is constructed to catch the rays of light emanating from the illumination slit and transversing the specimen undisturbed, thus resulting in a gray background. To one side of this surround region in the modulation plate, the plate is made fully transparent, while to the other side the plate is made opaque. The result then is that rays that are diffracted slightly in one direction appear quite bright, while rays diffracted in the opposite direction appear dark. In the recombined real image these variations in intensity appear as shades from light to dark across structural boundaries in the specimen. Our eyes conveniently interpret these gradations as shadows. The specimen takes on a distinct and crisp three-dimensional appearance. Hoffman modulation contrast also requires the use of special objectives and illumination systems. Its use is compatible with tissue culture plasticware, unlike the other, more familiar shadow-like technique, DIC, which is discussed next.

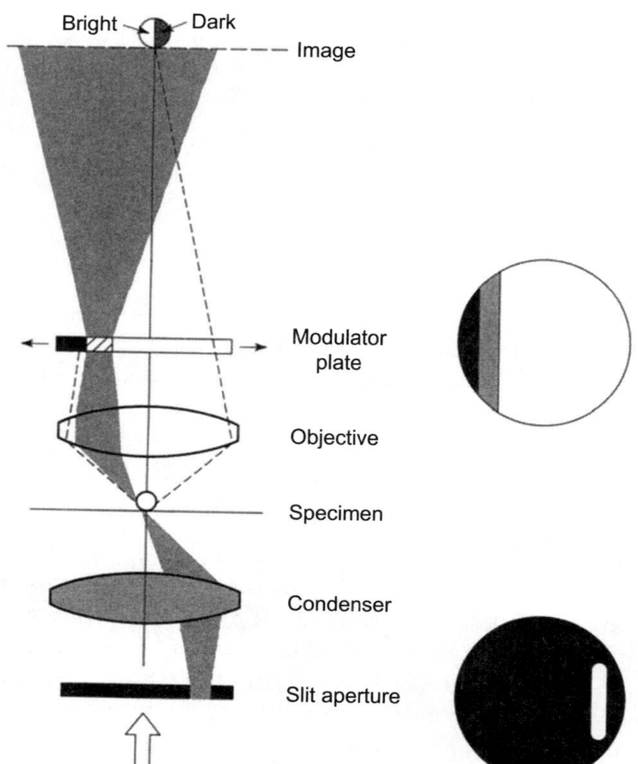

FIGURE II.3.8.3 Hoffman modulation contrast. Oblique illumination is provided by an off-axis slit in the condenser aperture. A modulator plate with matching complementary slit is placed in the objective back aperture and differentially blocks one sideband of diffracted light. (From Murphy, 2001.)

Differential Interference Contrast Microscopy

The techniques discussed so far have converted differences in the optical path length through the specimen into variations in image intensity by utilizing oblique illumination and tricks at the diffraction plane. In differential interference contrast (DIC) microscopy, also called Nomarski, gradients in optical path length are sensed in a slightly different way – by splitting the incoming light into two parallel rays and allowing each ray to sample nearby areas of the sample, before being recombined to form an image. DIC is akin to an imaging version of a dual-beam interferometer.

DIC starts with linearly polarized light; thus a polarizer is again the first extra element in the illumination path (Figure II.3.8.4). Next, mounted near the condenser is a modified Wollaston prism that separates the incoming light into two parallel rays. These two rays of light are physically separated by a small fraction of a micron – indeed, a distance less than the resolved distance of the microscope. These two rays traverse the specimen, and should they experience the same optical path through the specimen (i.e., no optical path difference), they are then recombined by a second prism located beyond the objective, again at the diffraction plane. This recombined light is again linearly polarized, and can be selectively rejected by the use of a second polarizer (termed an analyzer) oriented at 90° to the first. In the presence of an optical path difference (e.g., due to a gradient in the index of refraction in the sample), the two rays are recombined and result in an elliptically polarized beam, which is now only partially attenuated by the analyzer. Thus, gradients in refractive index present in the sample are converted into gradients in intensity in the image. Our eyes perceive these as shadows, and the sample takes on the striking appearance of three-dimensional relief. In practice, the objective prism is typically offset slightly so that the surround is not purely attenuated (black), but

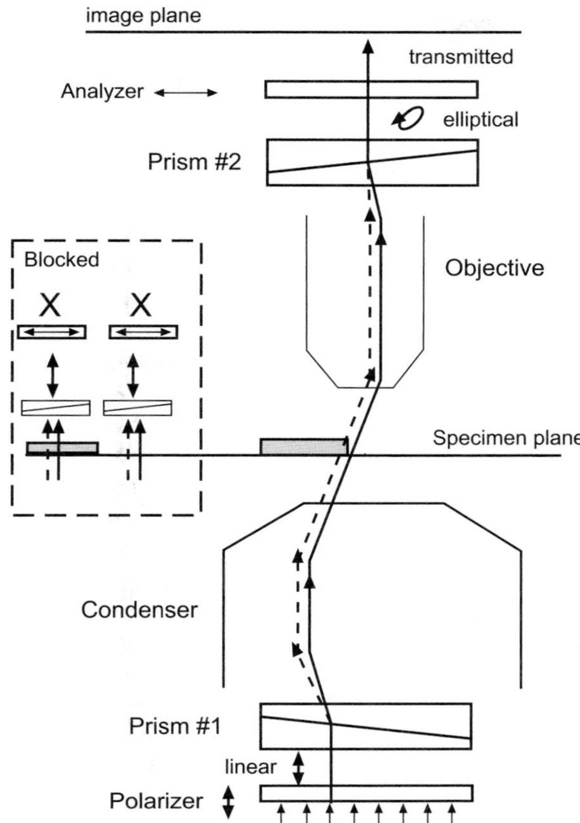

FIGURE II.3.8.4 Differential interference contrast on an upright microscope. Light is first linearly polarized, then split by the condenser prism (#1) into two spatially separate beams (shown as solid and dashed). If the two beams do not experience an optical path difference (inset), then they are recombined by the objective prism (#2) into a single ray that is again linearly polarized. This ray is then rejected by the second polarizer (the analyzer). If the two beams do experience an optical path difference in the sample, then the recombined ray is elliptically polarized and is partially transmitted by the analyzer. In this way spatial difference or gradients of refractive index in the sample are converted to differences in intensity. (After Murphy, 2001.)

rather appears in shades of gray along the diagonal of the image.

Since DIC utilizes polarization in generating the contrast, it is not suitable for use with tissue culture plasticware, which will induce extraneous polarization. DIC produces crisp, clear images in unstained specimens, with extraordinarily shallow depth of focus. This makes it ideal for optical sectioning tasks. DIC components are expensive, and must be made from strain-free glasses. DIC is particularly amenable to combination with fluorescence techniques – where the fluorescence labeling provides the spatial localization of specific intracellular features (e.g., a nucleus or cytoskeleton), and DIC provides clarity to the body of the cell and other organelles.

FLUORESCENCE MICROSCOPY

Like chromophores, fluorophores are (usually) exogenous large molecules whose unique spectral properties make them readily identifiable under the microscope, and thus generate contrast. Fluorescence techniques have had a wide-ranging and tremendous impact in the fields of molecular and cell biology, genomics, biophysics, bioengineering, and many others.

In the process of fluorescence, the fluorophore absorbs energy of a particular wavelength. While a small amount of this absorbed energy is lost to heat, the remainder is very quickly given up by the fluorophore as it re-emits the energy as light of a slightly longer (less energetic) wavelength. The difference between these excitation and emission wavelengths is called the Stokes shift, and provides the basis upon which the incoming and outgoing (emitted) light can be spectrally separated.

Fluorescein (known, together with its covalent linkage group isothiocyanate, as FITC) is perhaps the best known fluorophore. Today, there exist hundreds of new fluorophores, with various chemistries and spectral properties ranging from those excited by light in the UV range to the far red. The synthesis of new fluorophores with improved properties (e.g., increased quantum efficiency, resistance to photobleaching, narrower spectral characteristics) is an active enterprise. Excellent resources are available from vendors to help guide the choice of fluorophores for a myriad of applications (e.g., Molecular Probes, Inc.).

The applications of fluorescence are indeed numerous. As discussed earlier, fluorophores can be conjugated to a variety of biologically important molecules (lipids, membrane components, proteins, enzymes, antibodies, nucleotides, specific organelles) to serve a variety of reporting functions. Just a few common examples are the fluorophore DAPI to label cell nuclei and observe nuclear fragmentation during apoptosis; fluorescent conjugates of the mushroom toxin phalloidin which bind to f-actin in the cytoskeleton; and DiI for long-term cell tracking. The enzymatic activity of cells can be assessed by fluorescence, e.g., in the Live/Dead® cell assay (Molecular Probes, Inc.), where a masked nonfluorescent form of the green fluorophore calcein is cleaved by intracellular esterases in living cells, while dead cells fail to exclude the nuclear binding dye ethidium homodimer from their nuclei, which are then stained red. In this and many other systems, simultaneous labeling and imaging of multiple fluorophores is the norm.

Fluorophores can also be used to sense microenvironmental parameters, e.g., pH or the concentration of various ions such as Na^+, K^+, Cl^-, Ca^{2+}. The technique of FRAP, fluorescence recovery after photobleaching, can be used to study the intracellular dynamics or mobility of labeled macromolecules. In the technique named FRET, Förster resonance energy transfer, the proximity of two labeled molecules (or even two labeled domains on the same macromolecule) can be detected, as a donor fluorophore directly excites an acceptor fluorophore only when the two species are within a molecular distance of each other. This is truly an intermolecular ruler. Finally, the gene for the photoprotein from the jellyfish

Aqueorea victoria, green fluorescent protein (GFP), has been cloned. This has been used to create numerous fluorescent chimeric proteins that can be directly expressed in living cells. A number of mutations have yielded blue, cyan, and yellow variants. Another class of photoproteins, dsRed, from anemone is also in use. These advances give biologists unprecedented power to observe the processes of protein expression in cell populations.

Instrumentation for Fluorescence

Because of the small number of fluorophores often present and the potentially low quantum efficiency, fluorescence techniques require special, strong illumination sources. These are typically mercury or xenon arc lamps or lasers. Fluorescence benefits from the use of very efficient collection optics, i.e., objective lenses with the highest possible NA. Detection is typically achieved with sensitive imagers, e.g., cooled CCD (charge-coupled device) cameras or photomultiplier tubes (PMTs).

Fluorescence microscopy can be performed on upright or inverted microscopes equipped with an arc lamp, and appropriate filters and mirrors. This is known as wide-field fluorescence microscopy. (An alternative technique, laser scanning confocal and multiphoton instruments will be discussed in the next section.) In wide-field epi-fluorescence microscopy, illumination light from an Hg arc lamp is first passed through a band-pass filter that rejects all light except those wavelengths required to excite the fluorophore (the excitation band pass filter) (Figure II.3.8.5). Next, this excitation light hits a special mirror, the dichroic mirror, located directly behind the objective lens. The purpose of the dichroic mirror is to reflect shorter wavelength excitation light, but allow longer wavelength emission light to pass. The shorter wavelength excitation light is reflected by the dichroic mirror into the objective and delivered to the specimen. There the light is absorbed by the fluorophore and re-emitted at a longer wavelength. This light is emitted in all directions; however, a portion of the emitted light is collected again by the objective lens to form the image. This light passes back through the objective, and again hits the dichroic mirror. Now the longer wavelength emitted light is passed by the dichroic mirror (not reflected), and the beam travels on to one last filter. This last filter, the emission filter or barrier filter, rejects all but the emission wavelengths, thus leaving a clean signal that is composed only of light from the excited fluorophore. This illumination is apiscopic, a reflected light technique. The substrate can be opaque as long as the fluorophore is accessible at the surface.

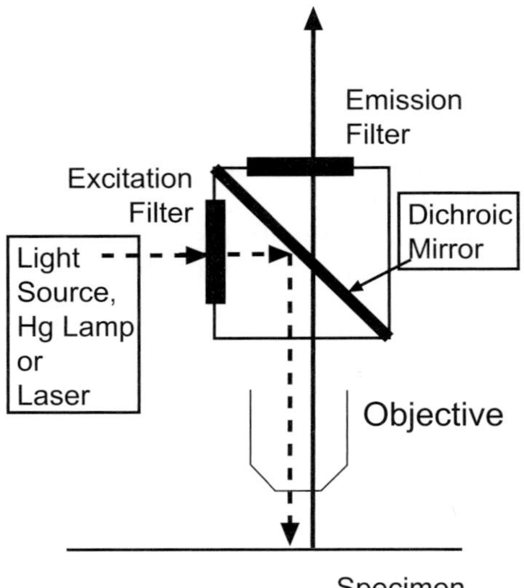

FIGURE II.3.8.5 Filter set. Dashed line represents shorter wavelength excitation light; solid line is longer wavelength emission light.

The three filter elements, the excitation filter, the dichroic mirror, and the emission filter, together form a filter set. Each filter set is specific to the spectral characteristics of one particular fluorophore. More complex filter sets that allow for the simultaneous use of two or more fluorophores are available, but it is usually more convenient to capture separate images, each comprising the signal from an individual fluorophore, and then overlay these images later.

Confocal Microscopy: A Special Tool for Fluorescence Microscopy

Excited fluorophores in the sample send their emitted light in all directions. In wide-field fluorescence microscopy it is inevitable that, in addition to the fluorescence emanating from the fluorophores residing in the in-focus plane, some fluorescence from slightly out-of-focus planes (above and below the desired focal plane) will also be collected and contribute to the image. The result can be a muddled or hazy image, especially in thick or highly scattering specimens. A solution is the confocal microscope.

In a laser scanning confocal microscope (LSCM), a physical barrier (a pinhole) is placed at a confocal location in the imaging path between the objective and the detector (Figure II.3.8.6). This pinhole allows only those rays of light that emanate from precisely the focal plane of the objective lens to pass through to the detector. The result is a very clean, if very weak, signal. Further, instead of flooding the field of view with illumination from an arc lamp, a laser is used to excite the fluorophore. The laser beam forms a small diffraction-limited spot in the plane of the specimen. This spot is then rastered across the specimen using scanning mirrors and optics. Instead of imaging the entire field of view at once with an imager, a simpler more sensitive PMT can be used in synchrony with the laser to simply count the photons arriving as a result of the laser illumination at each location in the specimen. The completed image is then composed by digitally presenting the measured intensities at each location

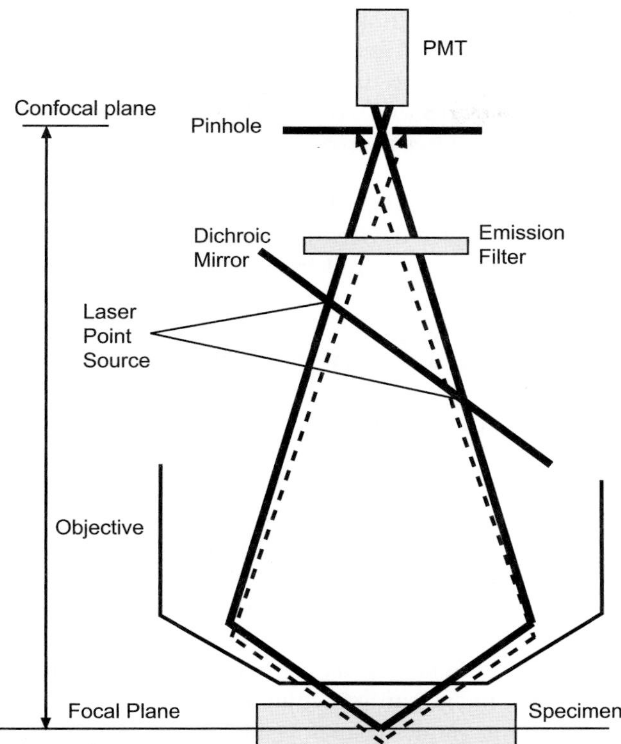

FIGURE II.3.8.6 Laser scanning confocal microscopy. The sample is illuminated by a laser that is reflected into the objective via a dichroic mirror. This beam forms a diffraction-limited spot in the specimen at the focal plane. This beam is rastered across the specimen by scanning mirrors (not shown). Fluorescence from the focal plane is captured by the objective, passes through the dichroic mirror and emission filters, and arrives precisely at the opening of the pinhole aperture, which is found directly in front of the detector (a photomultiplier tube). Fluorescence from an out-of-focus plane (a below-focus plane is shown, dotted) arrives to either side of the pinhole and is rejected. (After Murphy, 2001.)

(pixel) in the specimen. The diffraction-limited illumination spot provides for superb lateral resolution, and the confocal pinhole rejects all out-of-focus fluorescence contributions. The result is a crisp, clean image that represents a thin optical plane virtually sliced through the specimen. By repeating the process at different focal planes (with the help of nanometer-resolution stepper motors attached to the focus drive) one can collect images from hundreds of planes in thick specimens (up to 100–200 μm deep) and composite these into three-dimensional fluorescent representations of the sample.

Laser scanning confocal microscopes were once considered to require a dedicated operator to care for complex lasers, complicated electronics, and fragile detectors. Presently, the advent of solid-state lasers and simpler, more robust electronics, scanners, and detectors has led to new generations of confocal instruments that promise improved affordability and ease of use. Like wide-field fluorescence, laser scanning confocal microscopy can be implemented on either upright or inverted platforms.

Two new advances in confocal microscopy are worthy of note. The first is the use of multispectral array detectors and spectral unmixing to provide for the simultaneous use of up to eight fluorophores. This technology is featured in an instrument produced by Zeiss. The second is the advent of the multiphoton microscope.

A fluorophore can be excited equally well by one properly tuned excitation photon or by two excitation photons, each half the nominal excitation energy (twice the wavelength), or by three excitation photons, each a third the nominal excitation energy, etc. These multiple excitation photons must all reach the fluorophore molecule at precisely the same time, however; at least within the fluorescence decay lifetime of the fluorophore, which is measured in nanoseconds. This requires a femtosecond pulsed laser excitation source at long wavelengths (almost infrared). There are several advantages to this mode of illumination. First, longer wavelengths can penetrate deeper into cells and tissue, and do so with less biologic damage. Second, the pulsed nature of the excitation laser means that lower energy doses are delivered to the specimen. The effective excitation volume, the volume at the focal point of the objective where the spatial and temporal density of excitation photons is actually great enough to trigger a fluorescence emission, is exceedingly small – in the order of femtoliters. Thus, the emission signal produced and detected is always of very high spatial resolution, in both lateral as well as axial dimensions.

DIGITAL IMAGING

It is not enough that we build instruments to produce magnified and resolved images of the microscopic world. We must be able to detect these images, to visualize the information, record and archive images, and make useful quantitative measurements of our observations. This is the science of imaging.

Detectors: Eyes, Film, Digital Imagers

Our human eyes are by far the most commonly used and relied upon detectors in light microscopy. Together with the ocular lenses, the lens of the eye serves as part of the microscope system and the final image is focused on the retina, the detector. The human eye is a remarkable, yet limited, living detection instrument. Obviously, our eyes are acutely tuned to the visible portion of the light spectrum, with the greatest sensitivity found in the mid-green. The eye can be very sensitive (as few as 100 photons in the dark-adapted eye), and can adapt to a wide range of signal intensity from very dim to very bright. However, this adaptation takes time; and the intra-scene dynamic range, the range of dim to bright within one field of view, is only about 1000-fold. The individual sensing elements in the eye, the rods and cones, have an effective size of about 2 μm, which can be compared to the pixel size in other detectors. The signal response in the eye is nonlinear, and the signals from our eyes receive considerable postprocessing by the visual systems in the brain. This is particularly true with regard to edges, shadows, and objects in motion. We are all familiar with many "optical illusions" that prey upon some of the foibles of the human visual system. As discussed earlier, we make use of some of these illusions in light microscopy, e.g., the gradations of intensity in DIC and Hoffman are interpreted as shadows, when really there is little topographic information in the image.

Unfortunately, scientists with photographic memory are rare; thus the storage and archival capability of the human visual system is nil. The quantitative abilities of the eye are also limited. Although we can distinguish thousands of colors, we can barely discern between roughly 64–100 different gradations of gray (intensity)! Thus, data that come from human observations are qualitative or semiquantitative at best. The scientist must rely on careful subjective measurements, ideally with standardized scoring schemes, copious controls, blinded observers, and comparisons between multiple independent observers.

For more than a century the standard media for imaging was photographic film. Film can be made to detect broader spectral ranges than the eye, can be very sensitive, and has a dynamic range of about 2.5 log units. The grain size (individual sensing element or pixel) in film is typically 30 μm. The archival capability of film is excellent, and when properly used, the signal response can be made to be linear, and the data analyzed quantitatively. However, the cost, time, and chemical reagents required for film development are significant. The replacement of film with digital imagers in modern scientific imaging is essentially complete.

Digital Imaging Devices

Solid-state imaging devices are now a common part of our everyday world. Digital imaging devices are found in photocopiers, scanners, professional and consumer-grade video cameras and still cameras, surveillance cameras, toys, personal computers, and cell phones. The medical and scientific imaging fields have benefited greatly from this explosion of technology. The gaps in performance and sensitivity between consumer-grade and research-grade imagers are quickly closing.

The most common digital imaging device is the charge-coupled device (CCD). A CCD is a silicon chip that comprises a physical array of photodetectors, which correspond to the picture elements (pixels) in the image. As photons arrive at each photodiode they are converted to electrons, and the charges are accumulated and held locally on the chip in a potential well. Later, these groups of charges are transferred off the chip in a sequential manner, and the signal from each well is amplified and converted to a voltage. At this point the signal is still analog. An analog-to-digital converter is used to convert the steady stream of voltages into an array of numeric values that represent the relative intensity of photons at each photodiode in the detector array. It is this array of numbers that is the digital image.

Charge-coupled devices are available in several different architectures. In a full-frame camera, every spot on the chip is exposed to light. A physical shutter is used to control the exposure. After the exposure, the shutter is closed, and the charges are transferred and read. The operation is much like that found in a standard film camera. In a frame transfer camera, the CCD array is divided into two separate adjacent fields. The first is used for sensing light, while the second area on the chip is protected from light by an opaque mask and is used only to transfer and hold charges. The frame transfer camera has the advantage of speed. Thus, in a frame transfer camera, one image is acquired in the sensing region; then the charges are transferred quickly to the masked area. While the charges in the masked area are being read out, a new image can be acquired in the sensing region. Finally, many modern cameras take this concept one step further by simply masking every other line of photodiodes on a chip, a so-called interline camera. This allows for imaging at video frame rates; built-in microlenses can be used to funnel photons away from the masked pixels and to the sensing pixels.

The performance of a scientific imaging device depends on several important parameters. The resolution is determined by the number of sensing elements (photodiodes, i.e., pixels), as well as the physical size of each photodiode. The size of each element also determines the capacity of each well, i.e., the number of charges that can be held before reaching saturation. This "full-well capacity" determines the dynamic range, which is the range between the brightest and the darkest intensities possible in the image. The sensitivity is determined by the signal:noise ratio. The signal is affected by the quantum yield, which is the ratio of charges created to photons hitting the detector. The quantum yield can vary significantly with wavelength. The noise in the signal can arise from several factors. At very low light levels, the stochastic nature of the incoming photons themselves contributes some variability, the "shot noise." Thermal events also generate electrons in the chip. Thus, very sensitive detectors are often cooled with thermoelectric cooling devices or even liquid nitrogen to reduce this "thermal noise." Finally, noise can be generated in the process of transferring, reading, and amplifying the signals, and this "read noise" often scales with the speed of the reading process. Thus, very different and specialized imaging systems are used for various imaging tasks: e.g., very low-light fluorescence imaging; video-rate imaging or very high-speed imaging for following millisecond or microsecond events.

Although there is some variation in the quantum efficiency with wavelength, a CCD camera is basically a monochrome imaging detector. Often this is adequate. For example, in fluorescence imaging the signal is ideally composed of only single-emission-wavelength photons – fluorescence microscopy is actually a monochrome technique! But, as discussed earlier in this section, we often use multiple colored stains to create complex contrast in an image, and this requires a color imaging device.

A color CCD camera achieves color sensitivity by using filters to separate the image into its red (R), green (G), and blue (B) components. This can be accomplished in several different ways. In the first example, a separate tunable color filter is placed in front of a monochrome camera. The filter is sequentially set for red, green or blue, and three separate images are acquired. The images can then be compiled computationally to represent the color digital image. An advantage is that the full resolution of the imaging camera is maintained; however, the time required for the acquisition of three images means that the sample must hold still for a while (no sample motion). In the second approach, a prism is used to split the incoming image into three separate images, representing red, green, and blue. Three separate imaging chips are placed after the prism to collect the three separate images simultaneously. However, the third and most common approach is to use microprisms or colored filters mounted as a mask directly on the chip itself. In this arrangement, some photodiodes will be assigned to sense red, some green, and some blue. A computer algorithm keeps track of which pixels are which, and then combines the signals from adjacent R, G, and B sensing pixels and reports the result as one "virtual" pixel with R, G, and B values. Since the values from three (or more typically four) photodiodes are used to create one pixel, some resolution is sacrificed. Like all other areas of computing and electronics, imaging technology is advancing at a very rapid pace. Thus the reader is urged to consult current literature and vendor specifications for the latest advancements.

Digital Images, Image Processing, and Image Output

A full treatment of image processing and image analysis is beyond the scope of this chapter. However, a few thoughts are worthwhile for any scientist dealing with image data.

A digital image is ultimately an array of numbers where the value at each location in the array represents the intensity of one pixel in the image. A color image is represented by three values at each pixel. This array of numbers may be stored electronically in any of a myriad of proprietary file formats or in cross-platform formats such as bitmap (.bmp) or TIFF (.tif). In any event, the data should always be safely stored and archived in their raw, unadulterated form, before any image-processing steps or image-compression methods are applied. Subsequent processing steps should always be applied to a copy of the original data. Most vendors of imaging devices also offer software that controls image acquisition and offers some image-processing capabilities. Common image-processing packages include the NIH IMAGE software available from the US National Institutes of Health, Research Services Branch (http://rsb.info.nih.gov/nih-image), as well as programs primarily meant for the professional graphic arts world, such as Adobe® Photoshop®. There is often uncertainty about the scientific propriety of digital image manipulation. In general, the least amount of processing necessary is preferable, and no processing should be applied if quantitative data are being extracted from the images. Certain modest operations such as histogram stretching (which is scaling the intensity values in an image so that they span a wider range of output values) or gamma correction (a nonlinear display of data to accentuate dim features while maintaining bright features) are commonly applied for aesthetic images where quantitative analysis is not employed. Scientific guidelines are beginning to be formulated, and a good reference is the Microscopy Society of America's policy statement (www.microscopy.org). As with any treatment of scientific data, the methods of image manipulation must be documented and communicated such that others are capable of reproducing the findings from the raw data set or through independent experimental work.

Of course, the road to imaging productivity is not complete until the image is communicated to others in the scientific community. Image output must be considered as part of the scientific process. Digital images are embedded in other media files and viewed using a variety of means. Typical CRT computer monitors or LCD displays unfortunately have fairly limited resolution, displaying only about a million pixels at roughly 70 dots per inch (dpi). Scientific journals typically print at about 300 dpi. Standard data projectors also only display about a million pixels; the fact that the image is hugely magnified on a screen does not "create" added resolution. Likewise, images overly enlarged for printing on scientific posters can suffer if they do not contain sufficient resolution (pixels). If color information is important, then special care must be taken to ensure color fidelity throughout the image acquisition, display, and output processes. The responsibility for clear unbiased image presentation ultimately lies with the scientist. The image must be presented to the audience with adequate magnification, resolution, and contrast, such that an independent observer can identify the features that are important and independently reach the conclusions the author wishes to convey.

ELECTRON MICROSCOPY

The resolution of any probing technique is ultimately dependent upon the size of the probe and the nature of the probe–specimen interaction. The Rayleigh criterion above describes how resolving power is related to wavelength in light microscopy. How then can we achieve much greater resolution? Although we can certainly use shorter wavelengths, the visible light spectrum is actually quite limited. To make really large strides in the achievement of resolution, the answer is to reduce the size of the probe drastically, and use electrons. Electrons indeed provide extraordinary resolution. However, in addition to interacting with the samples, electrons are also absorbed and scattered by even the small molecules in the atmosphere. Thus, electron microscopy techniques must be conducted under high-vacuum conditions such as those described in Chapter II.1.4.

Electron microscopy can be performed in either transmission or reflected modes. In transmission electron microscopy (TEM), images are formed by diffraction or phase contrast mechanisms, much as has been described earlier for transmitted light microscopy. The detectors are phosphor screens coupled with digital imagers. Samples must be very, very thin (50 nm). This is achieved by embedding samples in hard plastic resins and using very sharp diamond knife ultramicrotomes. TEM provides extraordinary resolution of organelle and membrane structures in single cells. For biomaterials, there is perhaps no other instrument that provides a higher resolution image of the intimate cell membrane–biomaterial interface. Unfortunately, the challenging nature and limitations of TEM sample preparation mean that this technique is under-utilized in biomaterials. By far the most common electron microscopy tool is scanning electron microscopy (SEM).

Scanning Electron Microscopy

The scanning electron microscope was developed in the 1960s and 1970s, and has had a tremendous impact on the materials sciences and engineering fields, as well as the biologic community. SEM provides three-dimensional-like topographical information of bulk specimens at nanometer resolutions. It is remarkable that, while complex to generate, SEM images are easily accessible for even a child to interpret. The SEM has been truly instrumental in opening doors to the micro- and nanoworld for the general public. In biomaterials, the SEM is on an equal footing with Light Microscopy, as it is used to study the micro- and nanoscale structure of materials and devices, and cell–biomaterial interactions.

The scanning electron microscope (Figure II.3.8.7) is somewhat analogous to the scanning confocal microscope discussed earlier. Instead of a diffraction-limited laser beam, a beam of electrons is focused and swept across the surface of the specimen. The interaction of the electrons with the sample is localized and results in the generation of a signal, in this case other electrons escaping the specimen. These electrons are collected and counted by a detector, just as the photons are collected and counted by a photomultiplier tube (PMT)

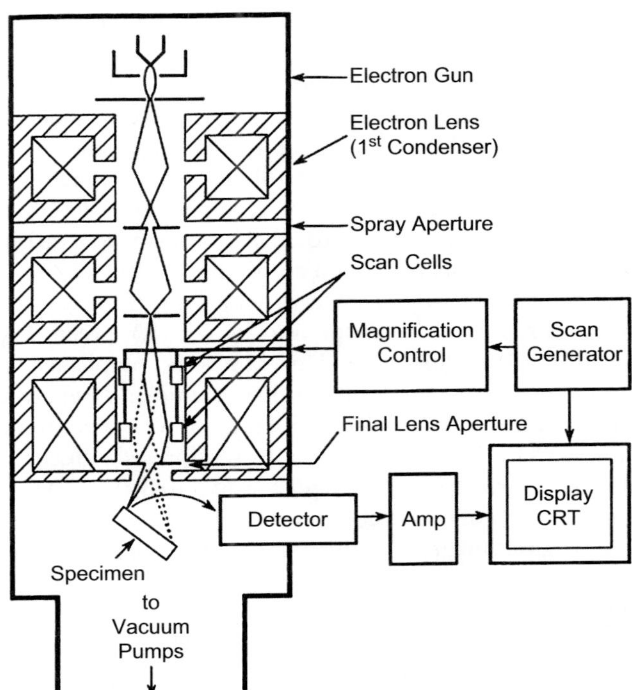

FIGURE II.3.8.7 Schematic drawing of the electron column in an SEM showing the electron gun, lenses, the scanning system, and the electron detector. (From Goldstein et al., 2003.)

in LSCM. The counts collected at each raster point are then compiled into a digital image. The LSCM has very shallow depth of focus, and three-dimensional information has to be reconstructed by taking many images at various focus planes. A key advantage of the SEM is its extraordinarily large depth of focus, allowing for the direct imaging of complex, even somewhat tortuous, microscale structures.

To start with, in SEM, we require a plentiful (bright) and narrow beam of electrons for illumination. Conventional SEMs generate electrons by heating a hairpin-shaped filament of tungsten. The emitted electrons are quickly focused to a point by the Wehnelt cap, and then accelerated toward an anode at high voltage. Alternatively, a crystal of LaB_6 can be used as a source – this provides 5–10× greater brightness (more electrons), but requires a higher vacuum (10^{-4} Pa) and is much more expensive. Many modern instruments now utilize an alternative source, the field emission (FE) gun. The element is again tungsten, but the filament is now shaped into a very sharp wire tip (<100 nm radius). Electrons are pulled from the tip by a very high electric field, either without (cold field emission) or with (Schottky or thermal FE) the aid of heating. An FE source can provide a very narrow beam that is 1000× brighter than the conventional thermal source. However, the vacuum required at the tip is much higher, in the order of 10^{-9} Pa.

Focusing the illuminating beam of electrons onto the sample is accomplished by lenses; however, these lenses are electromagnetic, not glass. Typically one or two condenser lenses begin to narrow the beam down, and a final "objective" lens focuses the beam onto the sample. Scanning coils are used to deflect the beam slightly from side-to-side, to accomplish the rastering of the beam across the specimen. Unlike glass lenses, the focal length of magnetic lenses can be changed simply by changing the magnetic field. A beam-limiting aperture is placed in the path, and is used to limit the diameter of the probe and ensure a narrow beam angle. A standard conically shaped objective lens can accommodate almost any size sample capable of fitting into the chamber. The lens may reside 1–20 mm away from the sample (the working distance). If the sample is small enough, it may actually be placed inside the objective lens field (an immersion lens instrument) with the advantages of lower aberrations, smaller probe size, and greater resolution.

Operation of the instrument requires balancing several operational parameters and making trade-offs. The diameter of the probe, the number of electrons in the probe (probe current), and the aperture angle are all crucial and quite interrelated parameters. Resolution is dependent upon small probe size, but there must be sufficient electrons to generate contrast (probe current): the familiar resolution/contrast trade-off again. Aperture angle also affects these, as well as determining the depth of focus. Thus, operational parameters such as source current, condenser strength, working distance, aperture size, and acceleration voltage are all adjusted depending upon the particular needs of the situation: e.g., high-resolution imaging, large depth of focus, a need for high current or surface sensitivity (low voltage).

The probe diameter of the incident electron beam may be as small as a few nanometers as it arrives at the surface. Unfortunately, the signal that is measured arises from interactions of this beam from within a much larger volume in the sample, the interaction volume. The incoming electrons may result in several measurable signals upon hitting the specimen (Figure II.3.8.8). First, the very energetic electrons may bounce off, or far more likely, bounce through the upper layers of the specimen, before getting turned around and again escaping the sample and returning to the chamber. These elastically scattered electrons are detected as backscattered electrons (BSEs). Depending on their energies, and the nature of the specimen (e.g., the atomic number), these backscattered electrons might travel as deep as 5–10 μm into the sample, and re-emerge as far as a micrometer or two away from their point of entry. Next, as these BSEs are bouncing through the specimen, much weaker secondary electrons (SEs) are generated. If these secondary electrons are generated near enough to the sample surface (say within five mean free path lengths), then they too will escape and can be detected as a secondary electron signal. Not only can secondary electrons be generated and escape near the point of entry of the incident beam, but secondary electrons can also be generated near the point of exit of a BSE, and these also escape and count toward the measured SE signal. Finally, escaped BSEs that hit the walls of the chamber can also generate SEs, and these too are counted. Most commonly, the signal used to form an image is the combination of counts from BSEs and all the SEs mentioned. Understanding the size and depth of the interaction volume in the specimen is crucial to interpreting the resolution and surface sensitivity in SEM imaging.

The detector is most commonly an Everhart–Thornly (E-T) detector located somewhere on the wall of the chamber. The E-T detector works by attracting escaped electrons (BSEs and all the SEs) with a positively biased grid placed directly in front of a positively charged scintillator. Electrons hitting the scintillator generate photons of light that are carried out of the chamber through a light guide to a PMT, where they are counted. A negative bias on the grid will result in only the BSEs being detected.

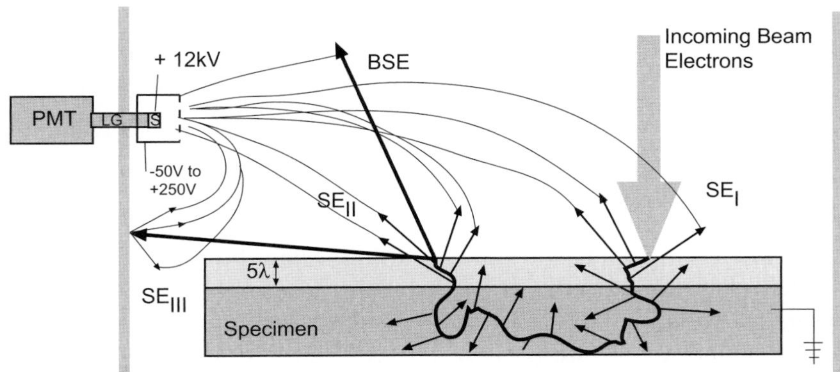

FIGURE II.3.8.8 Signal generation in SEM. Incident electrons enter the sample, collide with sample atoms, and may eventually exit the sample as backscattered electrons (BSE). Secondary electrons (SE_I, SE_{II}) are generated and can escape the sample if they are generated close enough to the surface (typically within five mean free path lengths). BSEs can also collide with the chamber walls or other components to generate even more secondary electrons (SE_{III}). The electrons are drawn to the E-T (Everhart–Thornly) detector by the positively charged Faraday cage and the positively charged coating on the scintillator (S). The electrons are converted to photons and carried by the light guide (LG) out of the chamber, where the photons are counted by a photomultiplier tube (PMT). (After Goldstein et al., 2003.)

All the incident electrons used for illumination have to be drained somewhere. Thus, samples need to be conductive (or at least semiconductive) and grounded to the instrument. When current is unable to be discharged adequately it builds up on the surface, and is manifest as "charging" in the image that degrades system performance. For this reason, many nonconducting samples (including biologic samples and polymers) are often coated with a very thin layer of metal, typically Au/Pd. The energetic incident electrons can also be damaging to some beam-sensitive samples, including polymers and delicate biologic structures.

Elemental Analysis in the SEM

In addition to the BSE and SE discussed above, incident electrons may also cause the emission of X-rays from the sample. Measurement of the energy and intensity of these X-rays can provide for elemental composition mapping across the specimen. This technique, energy dispersive X-ray (EDX) analysis, is conceptually the reverse of ESCA or XPS, discussed in the chapter on surface analysis. This technique is electrons in, X-rays out.

Low-Voltage Imaging

Another key advantage of instruments designed around FE sources is that they can be operated at very low accelerating voltages, as low as 100 eV as compared to the typical 5–30 KeV in conventional instruments. This results in a much smaller interaction volume and thus more sensitivity to surface topography. The samples receive less beam damage, and in fact many samples may be imaged without a metalized coating.

Variable Pressure and ESEM

A major drawback to SEM is the requirement of operating under high vacuum, 10^{-3}–10^{-6} Pa or even lower pressures. Biologic samples must be fixed and dehydrated in the hope that the structure will be preserved somehow under vacuum. Polymers too can undergo significant structural rearrangement when under high-vacuum conditions (e.g., hydrogels). Porous structures such as bone can be difficult to pump down to such vacuum levels.

Solutions are now appearing in the form of the Environmental SEM (ESEM) and numerous variable pressure (VP) SEM instruments. The approach in both instruments is the same: basically, to make a "leaky" SEM. These instruments are differentially pumped, meaning that the portions of the instrument housing the electron gun and optics are kept under high vacuum, while the sample chamber is allowed to reach some larger (but not really atmospheric) pressure. The opening between the optics column and the specimen chamber (the leak) is kept as small as possible, and the sample is kept as close to this opening as possible, so that the incident electrons have only a short distance to travel through the scattering gas atmosphere. The ESEM, which has several fairly elaborate differential pumping systems, can reach chamber pressures as high as 2700 Pa (20 torr) and utilizes a special SE detector capable of operating in the presence of such gas pressures. This pressure is the saturated vapor pressure of water at room temperature, and thus fully hydrated specimens can be viewed directly. All SEM vendors now offer some form of these variable-pressure systems with chamber pressures reaching 250 Pa (2 torr) and utilizing a variety of detectors. This pressure is not adequate to keep samples fully hydrated, but samples can be observed "partially hydrated."

The advantages are obvious. Cells and tissues have been imaged under fully hydrated conditions. Colloids, emulsions, and hydrated polymers are no longer excluded from SEM analysis. The resolution in this operating mode, however, does not equal that under high vacuum, and scattering of the incident beam from column to sample is the culprit. For this reason, standard thermal electron guns and high accelerating voltages are typically used. Fortunately, the presence of the gas near the sample means that charging is much reduced. The flexibility offered by this mode of operation is tremendously attractive, and "VP-capable" SEMs have become big sellers.

Focused Ion Beam Instruments

Instead of an incident beam of electrons, a focused beam of positively charged ions can be delivered to the specimen, e.g., gallium ions from a liquid metal ion gun. This beam can also be used to generate images of very high resolution, but the technique is destructive. However, this can be an advantage, in that the focused ion beam (FIB) can be used as an extremely precise machining tool at the microscopic and, indeed, nanoscopic scale. Several new instruments now combine a FIB column with a SEM column, to create the ultimate fabrication and imaging instrument for nanoscale engineering.

THE REVOLUTION IN OPTICAL MICROSCOPY

Since the time of Anton van Leeuwenhoek (1632–1723) (he is often referred to as the father of microscopy), microscopists striving to resolve smaller and smaller objects have been challenged by the so-called diffraction limit. Experimentally observed by Ernst Abbe in 1873, and then refined a few years later by Lord Rayleigh, for light microscopy the limit for x,y resolution is approximately 200–250 nm. In the last 20 years, new imaging modes have evolved that go well beyond the diffraction limit, with resolutions approaching 20 nm. Of course, we can always use energetic shorter wavelength radiation with the SEM and TEM, but we lose the advantage of seeing living cells in action, and the ability to use color (dyes, fluorophores) to enhance the information content of the image. So, this section will briefly address new optical methods permitting imaging of wet and living systems. Methods of this type are often categorized as super-resolution microscopy or nanoscopy.

The revolution in optical microscopy methods started with the laser scanning confocal microscope (LSCM), discussed earlier. The use of the scanned pinhole reduces the effect of diffraction on image formation, thereby increasing the possible x,y resolution to approximately 180 nm. The 4Pi laser scanning fluorescence microscope (first demonstrated in 1991) improved the image resolution to 80–150 nm by using two objective lenses focused to the same spot. These methods used clever technologies to improve resolution, but did not side-step the physics of the diffraction limit. Stimulated emission depletion microscopy (STED), first proposed in 1994, did indeed demonstrate that by using nonlinear de-excitation of fluorescent dyes, the diffraction limit could be significantly exceeded, and resolution to 16 nm could be observed. In 2006, photoactivated localization microscopy (PALM) was reported using photo-switchable fluorescent probes to reconstitute images from thousands of localized molecules. A related method, stochastic optical reconstruction microscopy (STORM), developed about the same time, also provides super-resolution images. Figure II.3.8.9 shows a STORM image of cell microtubules compared to an image recorded by conventional immunofluorescence methods. PALM and STORM images approach 20 nm x,y resolution. The physics permitting PALM, STORM, and other related methods to far exceed the diffraction limit is beyond the scope of this chapter, so a bibliography of review articles specifically focused on super-resolution microscopy is presented at the end of this chapter.

CONCLUSIONS

Although many of the techniques discussed here can trace their roots back for centuries, advances in modern microscopic imaging continue to abound. This section has not addressed scanning tunneling microscopy (STM) and atomic force imaging (AFM), important newer molecular imaging tools (see Chapter I.1.5). Surface analytical tools, such as ESCA and SIMS, are now being adapted to provide imaging capability at the microscopic scale. Conventional medical imaging tools, such as X-ray computed tomography (CT) and magnetic resonance imaging (MRI), are now being practiced at nearly microscopic resolutions. Electron tomography applied in high-voltage TEM instruments is revealing extraordinary three-dimensional representations of intracellular organelles and other structures. Modern computing power is reinvigorating the serial sectioning and reconstruction approaches of the past. The detection of intrinsic fluorescence via second harmonic generation in the multiphoton microscope is opening new doors to the imaging of biologic structures deeper in tissues. Optical coherence tomography promises important advances in tissue imaging for both clinical and research use. New nanoparticle fluorophores (quantum dots) provide extraordinarily bright emissions with narrow spectral signatures, enabling even further advancements in multilabel fluorescence imaging, biophotonics, and biosensors. Finally, new optical microscopy strategies for "breaking" the diffraction limit permit almost molecular-level views into living cells.

Microscopy is often assumed to be only about magnifying small objects. This is not totally inaccurate, but gives a poor description of how microscopy is used in modern science. The wealth of microscopy tools available to the modern scientist allows one to move beyond

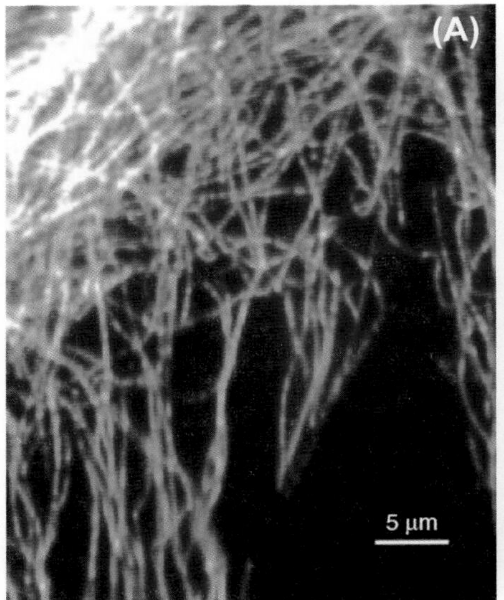

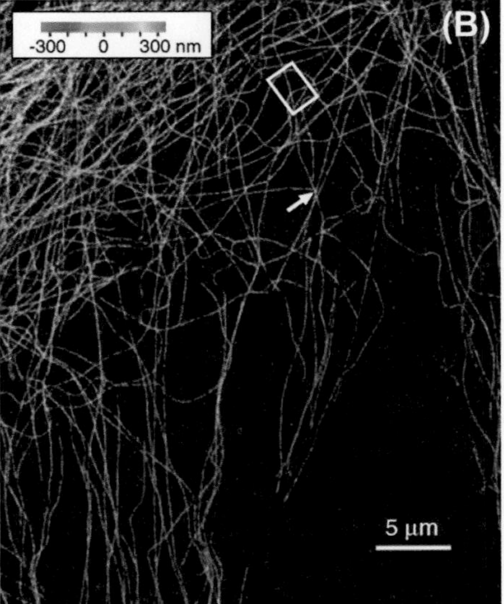

FIGURE II.3.8.9 Cell microtubules imaged by conventional fluorescence microscopy and STORM. (A) Cell microtubles by conventional indirect immunofluorescence imaging; (B) 3D STORM image of the same area, with the z-position information indicated by the color scale bar. (From Huang et al., 2008. *Science* **319,** 810–813.)

simple visual exploration, and to begin to ask tailored mechanistic questions, pose hypotheses, and prove them with solid quantitative data. This is essential for scientific publication, as well as for the documentation required by regulatory agencies. The detailed description of structure; the localization of species in two and three dimensions with molecular specificity; the ability to follow and visualize molecular biological process such as gene expression over time – these are all powerful tools in the biomaterial scientist's arsenal. To be successful, the scientist must include microscopy and imaging considerations from the very onset of any experimental planning.

BIBLIOGRAPHY

Web Resources

http://www.histochemicalsociety.org (The Histochemical Society)
http://www.microscopy.fsu.edu/primer/index.html
http://www.microscopyu.com/index.html
http://swehsc.pharmacy.arizona.edu/exppath/micro (extensive microscopy and imaging resources)
http://www.microscopy.org
http://www.invitrogen.com/site/us/en/home/brands/Molecular-Probes.html (Note: Molecular Probes, Inc., a division of Life Technologies, is a leading vendor of fluorophores and home of the *Molecular Probes Handbook*, available online)
http://www.nsh.org (The National Society for Histotechnology USA)
http://www.rms.org.uk (The Royal Microscopy Society UK)

Books

Chandler, D., & Robertson, R. W. (2008). *Bioimaging: Current Techniques in Light and Electron Microscopy*. Burlington, MA: Jones & Bartlett Publishers.

Foster, B. (1997). *Optimizing Light Microscopy for Biological and Clinical Laboratories*. Dubuque, IA: Kendall/Hunt Publishing Company.

Goldstein, J., Newbury, D., Joy, D., Lyman, C., Echlin, P., et al. (Eds.), (2003). *Scanning Electron Microscopy and X-Ray Microanalysis*. New York, NY: Kluwer Academic/Plenum Publishers.

Kim, M. K. (2010). *Digital Holographic Microscopy: Principles, Techniques, and Applications*. New York, NY: Springer Publishing.

Inouá, S., & Spring, K. R. (1997). *Video Microscopy: The Fundamentals* (2nd ed.). New York, NY: Plenum Press.

Mason, W. T. (Ed.), (1993). *Fluorescence and Luminescence Probes for Biological Activity*. New York, NY: Academic Press.

Murphy, D. B. (2001). *Fundamentals of Light Microscopy and Electronic Imaging*. New York, NY: Wiley-Liss.

Pawley, J. B. (Ed.), (1995). *Handbook of Biological Confocal Microscopy* (2nd ed.). New York, NY: Plenum Press.

Russ, J. C. (Ed.), (2002). *The Image Processing Handbook* (4th ed). Boca Raton, FL: CRC Press.

Williams, D. B., & Carter, B. C. (2009). In *Transmission Electron Microscopy: A Textbook for Materials Science* (4 Volume set). New York, NY: Springer Publishing.

Super-Resolution Microscopy

Hänninen, P., Hell, S., Salo, J., Soini, E., & Cremer, C. (1995). Two-photon excitation 4Pi confocal microscope: Enhanced axial resolution microscope for biological research. *Applied Physics Letters*, **66**, 1698–1700.

Hell, S. W. (2007). Far-field optical nanoscopy. *Science* (New York, NY), **316**(5828), 1153–1158.

Huang, B., Wang, W., Bates, M., & Zhuang, X. (2008). Three-dimensional super-resolution imaging by stochastic optical reconstruction microscopy. *Science*, **319**, 810–813.

Huang, B. (2010). Super-resolution optical microscopy: multiple choices. *Current Opinion in Chemical Biology*, **14**(1), 10–14.

Huang, B., Babcock, H., & Zhuang, X. (2010). Breaking the diffraction barrier: Super-resolution imaging of cells. *Cell*, **143**(7), 1047–1058.

Sigrist, S. J., & Sabatini, B. L. (2012). Optical super-resolution microscopy in neurobiology. *Current Opinion in Neurobiology*, **22**(1), 86–93.

Schermelleh, L., Heintzmann, R., & Leonhardt, H. (2010). A guide to super-resolution fluorescence microscopy. *The Journal of Cell Biology*, **190**(2), 165–175.

Vogelsang, J., Cordes, T., Forthmann, C., Steinhauer, C., & Tinnefeld, P. (2010). Intrinsically resolution enhancing probes for confocal microscopy. *Nano Letters*, **10**(2), 672–679.

SECTION II.4

Degradation of Materials in the Biological Environment

SECTION

II.4

Degradation of Materials in the Biological Environment

CHAPTER II.4.1 INTRODUCTION: THE BODY FIGHTS BACK – DEGRADATION OF MATERIALS IN THE BIOLOGICAL ENVIRONMENT

Buddy D. Ratner
Professor, Bioengineering and Chemical Engineering, Director of University of Washington Engineered Biomaterials (UWEB), Seattle, WA, USA

The biological environment, seemingly a mild, aqueous salt solution at 37°C, is, in fact, surprisingly aggressive and can lead to rapid or gradual breakdown of many materials. Some mechanisms of biodegradation have evolved over millennia specifically to rid the living organism of invading foreign substances – these same mechanisms now attack our contemporary biomaterials. Other breakdown mechanisms have their basis in well-understood chemical and physical principles, and will occur in a living organism or in a beaker on a laboratory bench. After this introduction, four chapters (II.4.2, II.4.3, II.4.4, and II.4.5) directly address degradation. The first three of these consider breakdown in the biological environment. Chapter II.4.5 describes another type of degradation, calcification, which can lead to device failure and can exacerbate other degradation mechanisms. In addition, many of the textbook chapters address degradation in other contexts. Chapter I.2.6 reviews the chemistry of polymers designed to be biodegradable. Chapter III.1.4 addresses device failure, sometimes related to unintentional degradation. Most of the device-specific chapters consider degradation issues.

The biomaterials of medical devices are usually exposed to varying degrees of cyclic or periodic stress (humans ambulate and the cardiovascular system pumps). Abrasion and flexure may also take place. Such mechanical challenges occur in an aqueous, ionic environment that can be electrochemically active to metals, and plasticizing (softening) to polymers. It is well-known that a material under mechanical stress will degrade more rapidly than the same material that is not under load.

Specific biological mechanisms are also invoked. Proteins adsorb to the material and can enhance the corrosion rate of metals. Cells (especially macrophages) adhere to materials via those interfacial proteins, and can be activated to secrete powerful oxidizing agents and enzymes intended to digest or dissolve the material. The secreted, potent degradative agents are concentrated in the space between the adherent cell and the biomaterial upon which they act, undiluted by the surrounding aqueous medium. Also, bacteria, bacterial biofilms (Chapter II.2.8) and yeast can enhance degradation and corrosion rates.

To understand the biological degradation of implant materials, synergistic pathways must be considered. For example, cracks associated with stress crazing open up fresh surface area to reaction. Swelling and water uptake can similarly increase the number of sites for reaction, and provide an access route for degradative agents into the "core" of the biomaterial. Amorphous material at metal (and polymer) grain boundaries can degrade more rapidly, leading to increases in surface area and localized stresses. Degradation products can alter the local pH, catalyzing further reaction. Hydrolysis of hydrophobic polymers can generate hydrophilic species, leading to polymer swelling and providing an entry mechanism for degrading species to transport into the bulk of the polymer. Cracks might also serve as sites for the initiation of calcification.

Biodegradation is a term that is used in many contexts. It can be used for reactions that occur over minutes or over years. It can be engineered to happen at a specific time after implantation or it can be an unexpected long-term consequence of the severity of the biological environment. Implant materials can solubilize, crumble, become rubbery or become rigid with time. The products of degradation may be toxic or irritating to the body or they may be designed to perform a pharmacologic function.

Calcification, a process we strive for in bone healing, is undesirable in most soft tissue contexts. Calcific mineral can interfere with the mechanical function of devices, induce cracking in polymers and embolize, leading to complications downstream. Implants based on natural tissue are particularly subject to calcification, but calcification is reasonably common in synthetic polymer devices.

Here are a few interesting biomaterial degradation issues that might stimulate further thinking on this subject in conjunction with the tutorial chapters in this section.

- Consider strategies used to create materials that degrade at controlled rates, versus strategies for synthesizing biostable materials intended for long-term performance in the body.
- Consider the degradation of materials commonly used in medicine that do not have well-defined breakdown mechanisms. Some examples include poly(ethylene glycol), hydroxyapatite, and some polysaccharides. How does the body deal with these common materials?
- A new class of biomaterials is now under development that degrades on cue. The cue might be thermal, photonic or enzymatic. Ingenious chemical design principles are being applied to create such materials, but how might the body react to the products generated by a sudden breakdown of the structure?
- Learn about new strategies to stabilize materials against degradation, for example, vitamin E loading of orthopedic polymers, and incorporation of polyisobutylene segments into elastomers.
- Endovascular stents are among the most widely used of all medical devices (Chapter II.5.3.B). A new generation of biodegradable stents is expected to have huge impact on cardiovascular therapies. Consider how biodegradable poly(lactic acid) or magnesium or iron will perform in the complex intra-vascular environment.
- For a medical device intended for years of service, especially a device where failure can lead to death, how can we test and qualify the device for the expected period of service? Are there useful *in vitro* tests? Are there relevant and justified animal models?
- Henry Petroski and other authors have discussed the important role of failure in advancing engineering design. Consider medical device failure, past and present, associated with degradation, and how these unintended complications will lead to better medical devices. A few examples include the degradation of polyurethane pacemaker leads, the breakdown of a protective sheath on the tailstring of the Dalkon Shield IUD, and the wear debris associated with the oxidation of ultra-high molecular weight polyethylene in hip prostheses.

Degradation in biological environments is seen with metals, polymers, ceramics, and composites. It is observed to some degree in most long-term implants, and even in some medium-term and short-term implants. Often, its initiation, mechanism, and consequences are incompletely defined. Biodegradation as a subject is broad in scope, and critical to device performance. It rightfully should command considerable attention for the biomaterials scientist. This section introduces biodegradation issues for a number of classes of materials, and provides a basis for further study on this complex but critical subject.

CHAPTER II.4.2 CHEMICAL AND BIOCHEMICAL DEGRADATION OF POLYMERS INTENDED TO BE BIOSTABLE

Arthur J. Coury
Coury Consulting Services, Boston, MA, USA

Biodegradation is the chemical breakdown of materials by the action of living organisms that leads to changes in physical properties. It is a concept of vast scope, ranging from decomposition of environmental waste involving microorganisms to host-induced deterioration of biomaterials in implanted medical devices. Yet it is a precise term, implying that specific biological processes are required to effect such changes (Williams, 1989). This chapter, while grounded in biodegradation, addresses other processes that contribute to the often complex mechanisms of polymer degradation. Its focus is the unintended chemical breakdown in the body of synthetic solid-phase polymers. (See Chapters I.2.6 and II.4.3 for a description of systems engineered to break down in the body.) The factors impacting the undesired biodegradation of polymeric implants are largely well-defined, although some recent progress (to be noted) has been made in mitigating such degradation.

POLYMER DEGRADATION PROCESSES

Pre-Implant Degradation

Polymeric components of implantable devices are generally reliable for their intended lifetimes. Careful selection and extensive preclinical testing of the compositions, fabricated components, and devices usually establish functionality and durability. However, with chronic, indwelling devices, it is not feasible during qualification (typically short-term testing) to match all implant conditions in real time for years or decades of use. The accelerated aging, animal implants, and statistical projections employed cannot expose all of the variables that may cause premature deterioration of performance. The ultimate measure of the acceptability of a material for a medical device is its functionality for the device's intended lifetime as ascertained in human post-implant surveillance (Coury, 1999).

No polymer is totally impervious to the chemical processes and mechanical action of the body. Generally, polymeric biomaterials degrade because body constituents attack the biomaterials directly or through other device components, sometimes with the intervention of external factors.

Numerous operations are performed on a polymer from the time of its synthesis to its use in the body (see

TABLE II.4.2.1 Typical Operations on an Injection-Moldable Polymer Biomaterial
Polymer: Synthesis, extrusion, pelletizing
Pellets: Packaging, storage, transfer, drying
Components: Injection-molding, post-mold finishing, cleaning, inspecting, packaging storage
Device: Fabrication storage (presterilization) cleaning, inspecting, packaging, storage (packaged), sterilization, storage (sterile), shipment, storage (pre-implant), implantation, operation in body

Table II.4.2.1). Table II.4.2.1 lists mechanisms of physical and chemical deterioration, which may occur alone or in concert at various stages of a polymer's history. Moreover, a material's treatment prior to implantation may predispose it to stable or unstable end-use behavior (Brauman et al., 1981; Greisser et al., 1994; Ling et al., 1998). A prominent example of biomaterial degradation caused by pre-implant processing is the gamma irradiation sterilization of ultra-high molecular weight polyethylene used in total joint prostheses. The process generates free radicals within the material that react with oxygen to produce undesirable oxidation products. Chain oxidation, scission, and cross-linking can occur for periods of months to years, causing loss of strength and embrittlement with limited shelf-life (Furman and Li, 1995; McKellop et al., 1995; Weaver et al., 1995; Daly and Yin, 1998; Blanchet and Burroughs, 2001; Kurtz et al., 2005; Rimnac and Pruitt, 2008). Recent attempts to mitigate degradation effects by gamma sterilization in inert atmospheres instead of air and the use of anti-oxidants such as Vitamin E have provided some improvement in the long-term *in vivo* performance of polyethylene joint components (Booth, 1995; Medel et al., 2009; Crowninshield and Muratoglu, 2008). Polypropylene and polytetrafluoroethylene are also notable as polymers that are predominately, but not entirely (Liebert et al., 1976), chemically stable in the body, but can be severely degraded (during processing) by sterilization with ionizing radiation (Williams, 1982; Portnoy, 1997). Gamma irradiation may also cause optical changes such as darkening of poly(methyl methacrylate) intraocular lenses (Hoffman, 1999). It is crucially important, therefore, that appropriate and rigorous processing and characterization protocols be followed for all operations (Coury et al., 1988; Shen et al., 1999).

Post-Implant Degradation Forces

After a device has been implanted, adsorption and absorption processes occur. Polymeric surfaces in contact with body fluids immediately adsorb proteinaceous components, and the bulk begins to absorb soluble components such as water, ions, proteins, and lipids. Cellular elements subsequently attach to the surfaces and initiate chemical processes. With biostable components, this complex interplay of factors is of little functional consequence. At equilibrium fluid absorption, there may be some polymer plasticization causing dimensional and mechanical property changes (Coury et al., 1988). On the surface, a powerful acute attack by cells and many chemical agents, including oxidants and enzymes, will have been substantially withstood, although inconsequential, superficial surface changes may occur. With the resolution of this acute inflammatory phase, a fibrous capsule will likely have formed over the device, and the rate of release of powerful chemicals from activated cells will have markedly decreased.

For those polymers subjected to chemical degradation *in vivo*, few if any reports have comprehensively described the sequential processes and interactions that produce the deleterious effects. Rather, explant analysis and occasionally metabolite evaluation are used to infer reaction pathways. Analysis of chemically degraded polymers has almost always implicated either hydrolysis or oxidation as an essential component of the process.

HYDROLYTIC BIODEGRADATION

Structures of Hydrolyzable Polymers

Hydrolysis is the scission of susceptible molecular functional groups by reaction with water. Hydrolysis may be catalyzed by acids, bases, salts or enzymes. It is a single-step process in which the rate of chain scission is directly proportional to the rate of initiation of the reaction (Schnabel, 1981). A polymer's susceptibility to hydrolysis is the result of its chemical structure, its morphology, its mass and dimensions, and the body's environment.

In a commonly used category of hydrolyzable polymeric biomaterials, functional groups consist of carbonyls bonded to heterochain elements (O, N, S). Examples include esters, amides, urethanes, carbonates, and anhydrides (Figure II.4.2.1). Other polymers containing groups such as ether, acetal, nitrile, phosphonate, sulfonate, sulfonamide or active methylenes hydrolyze under certain conditions (Figure II.4.2.1). Hydrolytically susceptible groups exhibit differing rates of degradation which are dependent on the intrinsic properties of the functional group, and on other molecular and morphological characteristics. Among carbonyl polymers with oxygen hetero-atoms attached, anhydrides display the highest hydrolysis rates followed, in order, by esters and carbonates. Polymers containing such groups, in fact, comprise many of the intentionally-resorbable devices (Chapter II.4.3). Other carbonyl-containing groups such as urethane, imide, amide, and urea can demonstrate long-term stability *in vivo* if contained in a hydrophobic backbone or highly crystalline morphologic structure. Groups that are normally very stable to hydrolysis are indicated in Figure II.4.2.2.

The rate of hydrolysis tends to increase with one or more of the following: a high proportion of hydrolyzable

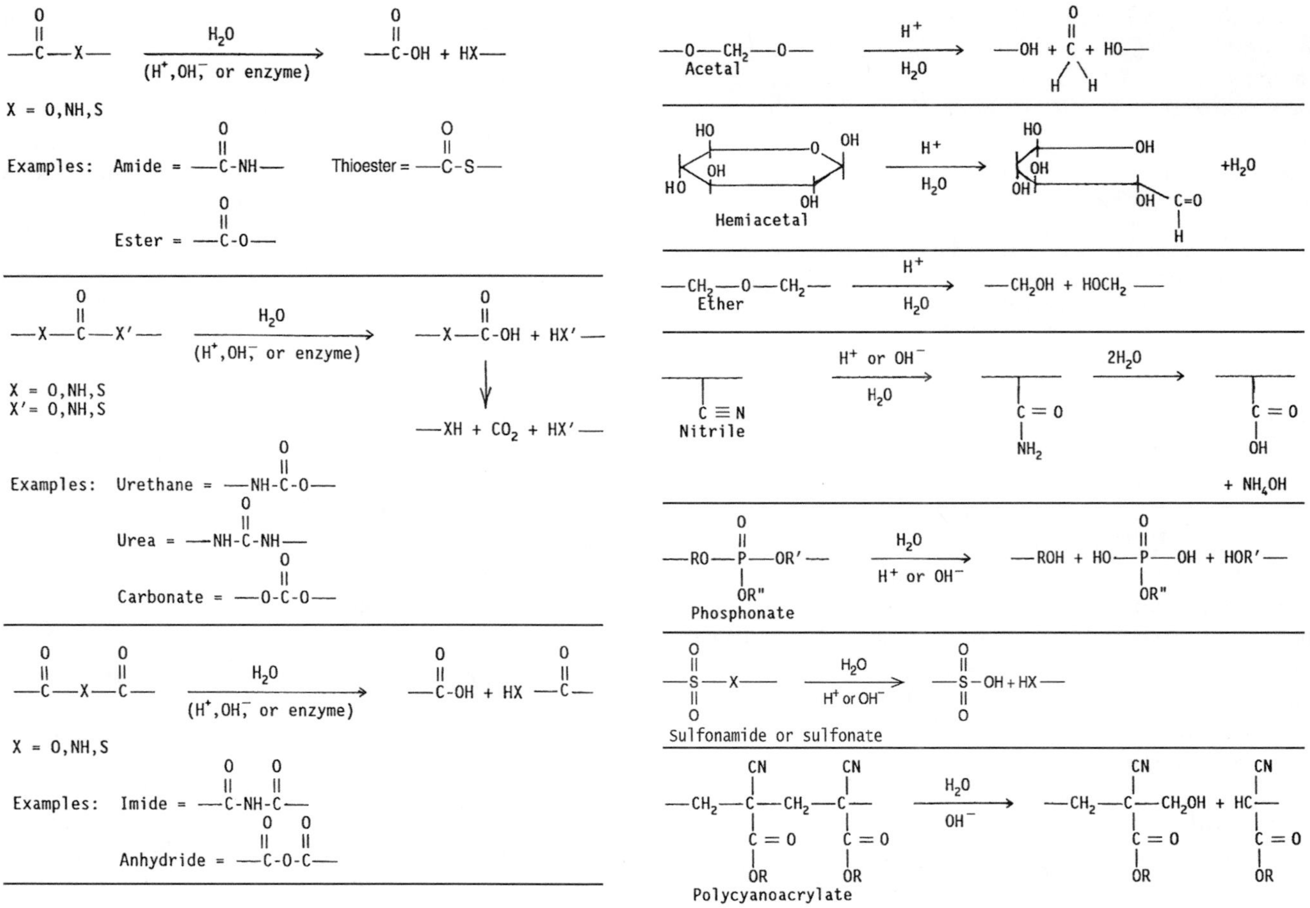

—C(=O)—X— $\xrightarrow[(H^+, OH^-, \text{ or enzyme})]{H_2O}$ —C(=O)-OH + HX—

X = O,NH,S

Examples: Amide = —C(=O)-NH— Thioester = —C(=O)-S—

Ester = —C(=O)-O—

—X—C(=O)—X'— $\xrightarrow[(H^+, OH^-, \text{ or enzyme})]{H_2O}$ —X—C(=O)-OH + HX'— → —XH + CO_2 + HX'—

X = O,NH,S
X'= O,NH,S

Examples: Urethane = —NH-C(=O)-O—

Urea = —NH-C(=O)-NH—

Carbonate = —O-C(=O)-O—

—C(=O)—X—C(=O)— $\xrightarrow[(H^+, OH^-, \text{ or enzyme})]{H_2O}$ —C(=O)-OH + HX —C(=O)—

X = O,NH,S

Examples: Imide = —C(=O)-NH-C(=O)—

Anhydride = —C(=O)-O-C(=O)—

—O—CH_2—O— (Acetal) $\xrightarrow[H_2O]{H^+}$ —OH + C(=O)(H)(H) + HO—

Hemiacetal $\xrightarrow[H_2O]{H^+}$ (ring-opened form, C=O, H) $+H_2O$

—CH_2—O—CH_2— (Ether) $\xrightarrow[H_2O]{H^+}$ —CH_2OH + $HOCH_2$—

C≡N (Nitrile) $\xrightarrow[H_2O]{H^+ \text{ or } OH^-}$ C=O, NH_2 $\xrightarrow{2H_2O}$ C=O, OH + NH_4OH

—RO—P(=O)(OR")—OR'— (Phosphonate) $\xrightarrow[H^+ \text{ or } OH^-]{H_2O}$ —ROH + HO—P(=O)(OR")—OH + HOR'—

—S(=O)(=O)—X— (Sulfonamide or sulfonate) $\xrightarrow[H^+ \text{ or } OH^-]{H_2O}$ —S(=O)(=O)-OH + HX—

—CH_2—C(CN)(C=O, OR)—CH_2—C(CN)(C=O, OR)— (Polycyanoacrylate) $\xrightarrow[OH^-]{H_2O}$ —CH_2—C(CN)(C=O, OR)—CH_2OH + HC(CN)(C=O, OR)—

FIGURE II.4.2.1 Hydrolyzable groups in polymer biomaterials.

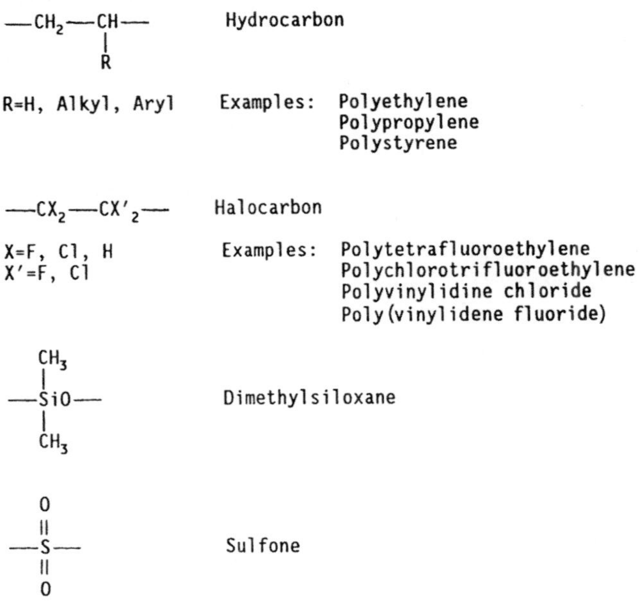

—CH_2—CH(R)— Hydrocarbon

R=H, Alkyl, Aryl Examples: Polyethylene, Polypropylene, Polystyrene

—CX_2—CX'_2— Halocarbon

X=F, Cl, H
X'=F, Cl

Examples: Polytetrafluoroethylene, Polychlorotrifluoroethylene, Polyvinylidine chloride, Poly(vinylidene fluoride)

—Si(CH_3)(CH_3)O— Dimethylsiloxane

—S(=O)(=O)— Sulfone

FIGURE II.4.2.2 Groups highly stable to hydrolysis.

groups in the main or side chain; other polar groups which enhance hydrophilicity; low crystallinity; low or negligible cross-link density; a high ratio of exposed surface area to volume; and, in some cases, mechanical stress. Porous hydrolyzable structures undergo rapid property loss relative to monolithic structures, because of their large surface area. Factors that tend to suppress hydrolysis rate include hydrophobic molecular moieties (e.g., hydrocarbon or fluorocarbon), cross-linking, high crystallinity due to chain order, thermal annealing or orientation, low stress, and compact shape (low surface-to-volume ratio). While the molecular weight of linear polymers *per se* may not have a great effect on the hydrolysis induced degradation rate, physical property losses may be retarded for a given number of chain cleavage events with relatively high molecular weight polymers. Property loss caused by chain cleavage is more pronounced in polymers with weak intermolecular bonding forces.

Host-Induced Hydrolytic Processes

The body is normally a highly controlled reaction medium. Through homeostasis, the normal environment of most implants is maintained at isothermal (37°C), neutral (pH 7.4), aseptic, and photo-protected aqueous steady-state. By *in vitro* standards, these conditions may appear mild. However, complex interactions of humoral and cellular components of body fluids involving activators, receptors, inhibitors, etc., produce aggressive responses to any foreign-bodies through the processes of cell adhesion, cell activation, chemical reaction, and particulate transport.

Several scenarios leading to hydrolysis in the host can be considered. For all scenarios, hydrolysis can only occur at a site other than the surface of a polymer mass after water permeates to the site. The first scenario considers water at pH 7. Water is capable of hydrolyzing certain polymers (e.g., polyglycolic acid) at a significant rate (Chapters I.2.6 and II.4.3, and Zaikov, 1985). However, this simple hydrolysis mechanism is unlikely to be significant in polymer compositions selected for long-term *in vivo* biostability.

Next, ion-catalyzed hydrolysis offers a likely scenario in body fluids. Extracellular fluids contain ions such as: H^+, OH^-, Na^+, Cl^-, HCO_3^-, PO_4^{3-}, K^+, Mg^{2+}, Ca^{2+}, and SO_4^{2-}. Organic acids, proteins, lipids, lipoproteins, etc., also circulate as soluble or colloidal components. It has been shown that certain ions (e.g., PO_4^{3-}) are effective hydrolysis catalysts, enhancing, for example, reaction rates of polyesters by several orders of magnitude (Zaikov, 1985). Ion catalysis may be a surface effect or a combined surface–bulk effect, depending on the hydrophilicity of the polymer. Very hydrophobic polymers (e.g., those containing <2% water of saturation) absorb negligible concentrations of ions. Hydrogels, on the other hand, which can absorb large amounts of water (>15% by weight) are essentially "sieves," allowing significant levels of ions to permeate and be absorbed with consequent bulk hydrolysis via acid, base or salt catalysis.

Localized pH changes in the vicinity of the implanted device, which usually occur during acute inflammation or infection, can cause catalytic rate enhancement of hydrolysis (Zaikov, 1985). Organic components, such as lipoproteins, circulating in the bloodstream or in extracellular fluid, appear to be capable of transporting catalytic inorganic ions into the polymer bulk by poorly defined mechanisms.

Enzymes generally serve a classic catalytic function, altering reaction rate (via ion or charge transfer) without being consumed by modifying activation energy, but not thermodynamic equilibrium. While enzymes function in extracellular fluids, they are most effectively transferred onto target substrates by direct cell contact (e.g., during phagocytosis). Hydrolytic enzymes or hydrolases (e.g., proteases, esterases, lipases, glycosidases) are named for the molecular structures they affect. They are cell-derived proteins which act as highly specific catalysts for the scission of water-labile functional groups.

Enzymes contain molecular chain structures and develop conformations that allow "recognition" of chain sequences (receptors) predominately found on biopolymers. Complexes form between chain segments of the enzyme and the biopolymer substrate which result in enhanced bond cleavage rates. Lacking the recognition sequences of susceptible natural polymers, most synthetic polymers are more resistant to enzymatic degradation than natural polymers. Nevertheless, comparative studies have shown some enhancement of hydrolysis rates by enzymes, particularly with synthetic polyesters and polyamides (Kopecek and Ulbrich, 1983; Zaikov, 1985; Smith et al., 1987). Apparently the enzymes can recognize and interact with structural segments of the polymers, or more accurately, of the polymers coated with serum proteins, to initiate their catalytic action *in vivo* (Pitt, 1992).

Enzymes with demonstrated effects on hydrolysis rates can be quite selective in the presence of several hydrolyzable functional groups. For example, poly(ether urethane ureas) and poly(ester urethane ureas) exposed to hydrolytic enzymes (an esterase, cholesterol esterase, and a protease, elastase) were observed for rate of hydrolysis and hydrolytic site. Enzyme catalysis was clearly observed for the ester groups, while the hydrolytically susceptible urea, urethane, and ether groups did not show significant hydrolysis, as indicated by release of radio-labeled degradation products (Santerre et al., 1994; Labow et al., 1995, 2002a).

Many enzymes exert predominantly a surface effect because of their molecular size, which prevents absorption. Even hydrogels (e.g., poly(acrylamide)), which are capable of absorbing certain proteins, have molecular weight cut-offs for absorption, and can exclude large enzymes. However, as the degrading surface becomes roughened or fragmented, enzymatic action may be enhanced as a result of increased surface area if the substrates remain accessible to phagocytic cells that express the active enzymes. Implanted devices that are in continuous motion relative to neighboring tissue can provoke inflammation, stimulating enzyme release.

Hydrolysis: Preclinical and Clinical Experience

A discussion of *in vivo* responses of several prominent polymer compositions known to be susceptible to hydrolysis follows. The structures of these polymers are described in Chapter I.2.2.

Polyesters. Poly(ethylene terephthalate) (PET), in woven, velour or knitted fiber configurations, remains a primary choice of cardiovascular surgeons for large-diameter vascular prostheses, arterial patches, valve sewing rings, etc. It is a strong, flexible polymer, stabilized by high crystallinity as a result of chain rigidity and orientation, and is often considered to be biostable. Yet, over several decades, there have been numerous reports of long-term degeneration of devices *in vivo*, owing to breakage of fibers and device dilation. Proposed causes have been structural defects, processing techniques, handling procedures, and hydrolytic degradation (Cardia and Regina, 1989).

Systematic studies of PET implants in healthy dogs have shown slow degradation rates, which were estimated to be equivalent to those in humans. For woven patches implanted subcutaneously, a mean total absorption time by the body of 30 ± 7 years, with 50% deterioration of fiber strength in 10 ± 2 years was projected. In infections in dogs, however, where localized pH dropped to as low as 4.8, degradation was enhanced exponentially, with complete loss of properties within a few months (Zaikov, 1985). Human implant retrieval studies have shown significant evidence of graft infection (Gumargalieva et al., 1982; Vinard et al., 1991). Besides the obvious pathological consequences of infection, the enhanced risk of polymer degradation due to local pH and oxidative agents is a cause for concern.

Aliphatic polyesters are most often intended for use as biodegradable polymers, with polycaprolactone, for example, undergoing a significant decrease in molecular weight, as indicated by a drop of 80–90% in relative viscosity within 120 weeks of implant (Kopecek and Ulbrich, 1983).

Poly(ester urethanes). The earliest reported implants of polyurethanes, dating back to the 1950s, were cross-linked, aromatic poly(ester urethane) foam compositions (Bloch and Hastings, 1972; Blais, 1990). Their use in plastic and orthopedic reconstructive surgery initially yielded promising results. Acute inflammation was low. Tissue ingrowth promoted thin fibrous capsules. However, within months they were degraded and fragmented, producing unsatisfactory chronic effects (Bloch and Hastings, 1972). Foci of initial degradation of those polymers are generally considered to be the polyadipate ester soft segments that undergo hydrolysis (Figure II.4.2.3). By comparison, corresponding poly(ether urethanes) are very resistant to hydrolysis, although more susceptible to oxidation (see the section, below, on oxidative biodegradation). Whether such hydrolytically degraded poly(ester urethanes) subsequently produce meaningful levels of aromatic amines (suspected carcinogens) by hydrolysis of urethane functions *in vivo* is an unresolved subject (Blais, 1990; Szycher and Siciliano, 1991).

It is noteworthy that poly(ester urethane) foam-coated silicone mammary implants have survived as commercial products for decades (Blais, 1990), despite their known propensity to degrade. Apparently, the type of fibrous capsules formed, initially by ingrowth into the degradable foam, was favored by some clinicians over those caused by smooth-walled silicone implants. In large devices, unstabilized by tissue ingrowth, the frictional effects of sliding may cause increased capsule thickness and contraction (Snow et al., 1981), along with extensive chronic inflammation.

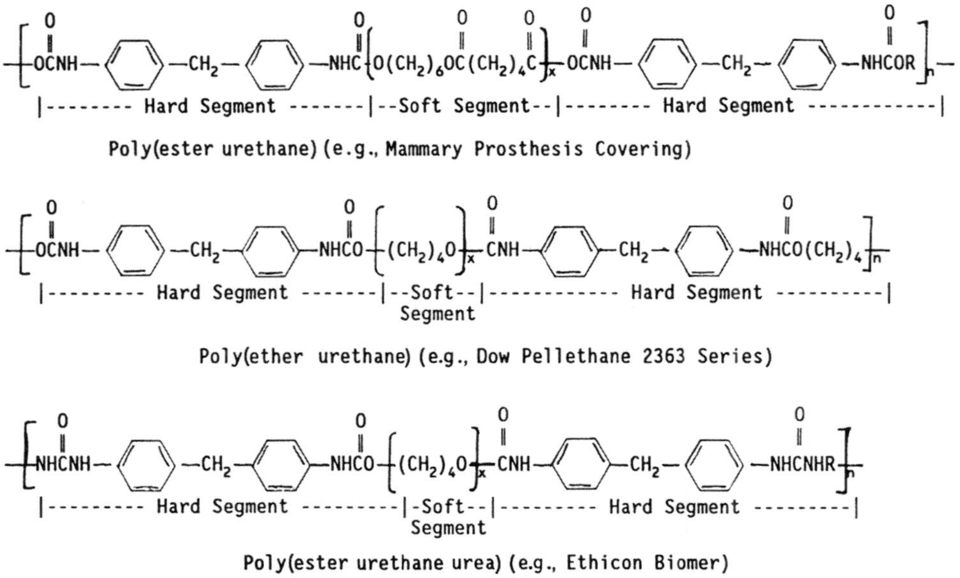

|-------- Hard Segment -------|--Soft Segment--|-------- Hard Segment ------------|

Poly(ester urethane) (e.g., Mammary Prosthesis Covering)

|--------- Hard Segment -------|--Soft--|------------ Hard Segment ----------|
Segment

Poly(ether urethane) (e.g., Dow Pellethane 2363 Series)

|--------- Hard Segment ---------|-Soft--|--------- Hard Segment ---------|
Segment

Poly(ester urethane urea) (e.g., Ethicon Biomer)

FIGURE II.4.2.3 Structure of implantable poly(ester urethane), poly(ether urethane), and poly(ester urethane urea).

Polyamides. Nylon 6 (polycaproamide) and nylon 6,6 [poly(hexamethylene adipamide)] contain a hydrolyzable amide connecting group, as do proteins. These synthetic polymers can absorb 9–11% water, by weight, at saturation. It is predictable, then, that they degrade by ion-catalyzed surface and bulk hydrolysis (Figure II.4.2.1). In addition, hydrolysis due to enzymatic catalysis leads to surface erosion (Zaikov, 1985). Quantitatively, nylon 6,6 lost 25% of its tensile strength after 89 days, and 83% after 726 days in dogs (Kopecek and Ulbrich, 1983). An example of polyamide degradation of particular consequence involved the *in vivo* fragmentation of the nylon 6 tail string of an intrauterine contraceptive device. This string consisted of a nylon 6-sheath around nylon 6 multifilaments. The combination of fluid absorption (>10%) and hydrolysis was claimed to produce environmental stress cracking. The cracked coating allegedly provided a pathway for bacteria to travel from the vagina into the uterus, resulting in significant pelvic inflammatory disease (Hudson and Crugnola, 1987).

Degradation of a polyarylamide intended for orthopedic use (the fiber-reinforced polyamide from *m*-xylylene diamine and adipic acid) was also shown in a rabbit implant study. Although the material provoked a foreign-body reaction comparable to a polyethylene control, surface pitting associated with resolving macrophages was noted at 4 weeks, and became more pronounced by 12 weeks. This result was not predicted, since polyarylamides are very resistant to solvents and heat (Finck et al., 1994).

Polyamides with long aliphatic hydrocarbon chain segments (e.g., polydodecanamide) are more hydrolytically stable than shorter chain nylons, and correspondingly degrade slower *in vivo*.

Poly(alkyl cyanoacrylates). This class of polymers used as tissue adhesives is noteworthy as a rare case in which carbon–carbon bonds are cleaved by hydrolysis (Figure II.4.2.1). This occurs because the methylene ($-CH_2-$) hydrogen in the polymer is highly activated inductively by electron-withdrawing neighboring groups. Formation of the polymer adhesive from monomers is base catalyzed, with adsorbed water on the adherend being basic enough to initiate the reaction.

Catalysts for equilibrium reactions affect the reverse, as well as the forward reaction. Therefore, water associated with tissue can induce polycyanoacrylate hydrolysis by a "reverse Knoevenagel" reaction (Figure II.4.2.1). More basic conditions and (as suggested by *in vitro* cell culture or implant studies) enzymatic processes are much more effective at inducing hydrolysis. In chick embryo liver culture (a rich source of a variety of enzymes), poly(methyl cyanoacrylate) degraded much faster than in cell culture medium alone. In animal implants, poly(methyl cyanoacrylate) was extensively degraded within 4–6 months (Kopecek and Ulbrich, 1983). Higher alkyl (e.g., butyl) homologs degraded slower than the methyl homolog, and were less cytotoxic (Hegyeli, 1973; Vauthier et al., 2003). Octyl cyanoacrylate polymer, introduced to the device field as a dermal tissue adhesive (Singer et al., 2008), promises to be the most stable cyanoacrylate device composition to date, based on increased hydrophobicity. Preclinical and clinical studies have yielded promising results with use of octyl cyanoacrylate as a surgical sealant (Barbarini Ferraz et al., 2009; Carr, 2011), and recently, a composition containing the monomers octyl cyanoacrylate and butyl lactoyl cyanoacrylate was FDA (US Food and Drug Administration) approved as a blood vessel anastomotic sealant (US FDA, 2010).

Polymers Containing Hydrolyzable Pendant Groups

Certain polymers intended for long-term implantation consist of biostable main chain sequences and hydrolyzable pendant groups. Poly(methyl methacrylate) (PMMA) used in bone cements and intraocular lenses is an example of a hydrophobic polymer with a stable hydrocarbon main chain and hydrolyzable ester side groups. It has been proven, over decades of use, to provide reliable, stable service, with hydrolysis rates being so slow as to be inconsequential.

Another polymer system with a hydrocarbon backbone, poly(methyl acrylate-*co*-2-hydroxyethyl acrylate) also contains hydrolyzable ester side groups. This polymer, which forms hydrogels in an aqueous environment, has been used as a "scleral buckling" device for retinal detachment surgery. Basically, the dry polymer, shaped as a band or ring, placed as a "belt" around the sclera, expands through hydration to create an indentation in the zone of the retinal detachment to re-establish retinal contact. The device is left in place as a permanent implant (or "exoplant" as it is sometimes called because it is external to the sclera) (Braunstein and Winnick, 2002). This hydrogel device, introduced into clinical practice in the 1980s (Refojo and Leong, 1981; Colthurst et al., 2000), apparently performed satisfactorily for years as an approved product. However, in the 1990s, reports of long-term complications of these hydrogel scleral buckles began to surface (Hwang and Lim, 1997; Roldan-Pallares et al., 1999). The hydrogel structures resumed swelling, sometimes with fragmentation, after maintaining stable dimensions for years. One report described a difficult explantation 13 years after implantation (Braunstein and Winnick, 2002). An article described three unique complications from hydrogel sclera buckle use: orbital cellulitis mimicry; fornical shortening with orbital prosthetic intolerance; and orbital pseudo-tumor (Bernardino et al., 2006). Those buckles remained in place for 7 to 15 years, with a mean time of 10.7 years. In a study of 15 patients with 17 scleral buckles, all reported complications within 4 to 14 years (Figures II.4.2.4, II.4.2.5, II.4.2.6). Removal of the buckles was technically difficult and post-operative complications were significant, although immediate palliative relief was experienced after surgery. Pressures applied to the eye by the swelling have led to blindness and loss of the eyeball. Hydrogel scleral buckles are no longer used in retinal surgery (Watt, 2001).

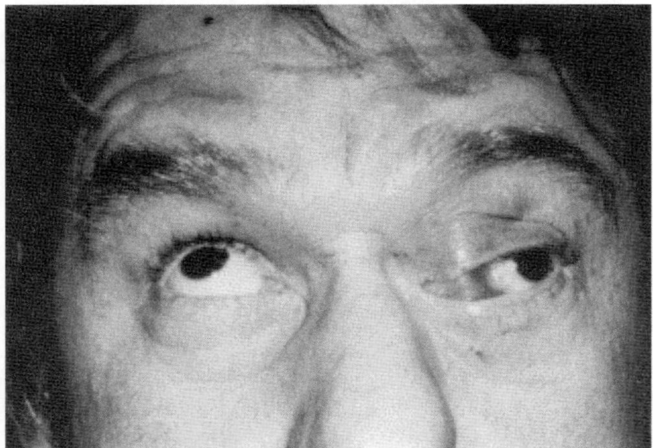

FIGURE II.4.2.4 A subconjunctival and subpalpebral hydrogel explant interferes with ocular motility, especially on attempted gaze up and to the right. (From J. J. Kearney et al. (2004). *Am. J. Ophthal.,* 137.)

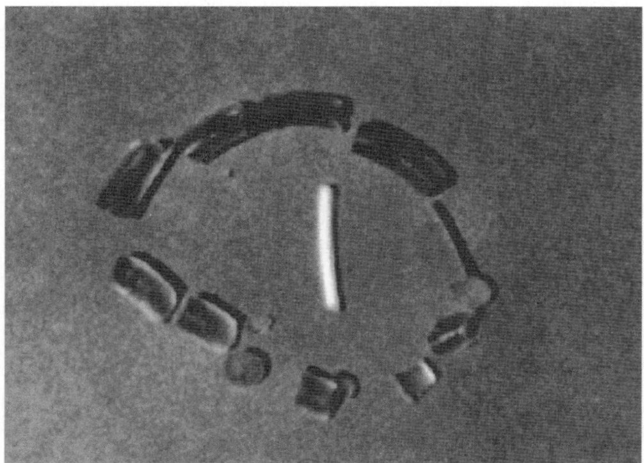

FIGURE II.4.2.5 A washed 4 mm hydrogel explant showing multiple fragments of a swollen explant and a 3 mm silicone sponge explant for comparison. The sponge was removed at the same surgical sitting as the hydrogel explant. (From J. J. Kearney et al. (2004). *Am. J. Ophthal.,* 137.)

Very little speculation has been provided in published articles about the mechanism of failure of acrylate scleral buckling devices other than that chemical degradation has occurred (Roldan-Pallares et al., 1999). I suggest that a likely mechanism involves hydrolysis of the ester side groups enhanced by the hydrophilic nature of the polymer (as contrasted to hydrophobic polymers such as PMMA). Hydrolysis of either of the two acrylate esters in the polymer chain provides an acrylic acid moiety. Linear poly(acrylic acid) is fully water soluble, and each hydrolytic event renders the polymer more hydrophilic and subject to enhanced swelling. This process is slow but inexorable in the case of the scleral buckling device. The valuable lesson is that devices with intrinsically susceptible groups can eventually degrade by predictable mechanisms. This may take longer than is required for pivotal preclinical qualification studies (typically two-year animal implants). In fact, Bernardino et al. (2006), in discussing long-term hydrogel implants, warn: "Patients with newer uses of hydrogel, such as orbital expanders, should also be observed for long-term complications." If late degradation is suspected, therefore, accelerated aging studies should be performed *in vitro* with correlations made to *in vivo* studies. Such efforts give valuable, if not completely trustworthy, information (see the section above "Polymer Degradation Processes").

OXIDATIVE BIODEGRADATION

Oxidation Reaction Mechanisms and Polymer Structures

While much is known about the structures and reaction products of polymers susceptible to oxidative biodegradation, confirmation of the individual reaction steps has not yet been demonstrated analytically. Still, mechanistic inferences are possible from extensive knowledge of physiological oxidation processes and polymer oxidation *in vitro*.

The polymer oxidation processes to be discussed may be consistent with a homolytic chain reaction or a heterolytic mechanism. Species such as carbonyl, hydroxyl, and chain scission products are detectable in these types of processes. Classic initiation, propagation, and

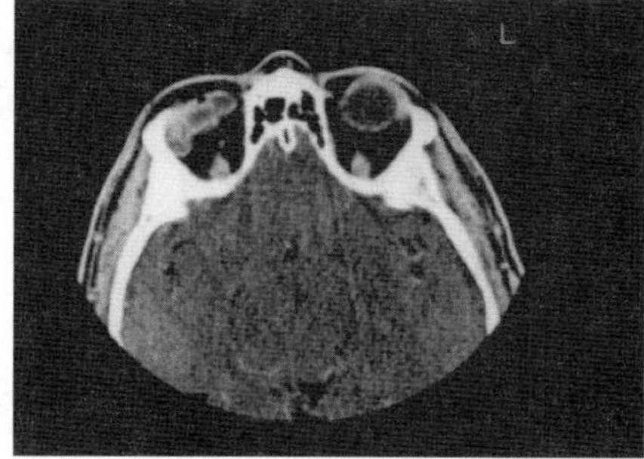

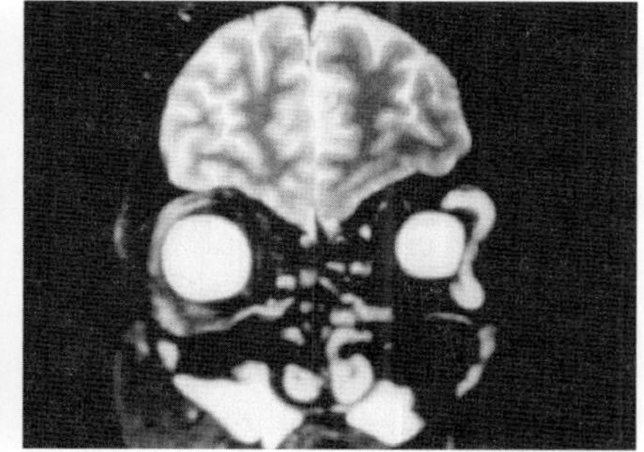

FIGURE II.4.2.6 Left panel: Computed tomography scan showing a hydrogel implant in the anterior orbit of the right eye. Right panel: Magnetic resonance image of the left orbit shows an expanded hydrogel explant. (From J. J. Kearney et al. (2004). *Am. J. Ophthal.,* 137.)

termination events for homolysis and ionic heterolytic processes are detailed in Figure II.4.2.7.

The principles of polymer degradation resistance stated in the section on hydrolyzable polymers (e.g., group frequency, crystallinity, hydrophobicity) are valid for predicting relative oxidation resistance of polymers, except where particularly oxidation-susceptible groups are present. Sites favored for initial oxidative attack, consistent with a homolytic or heterolytic pathway, are those that allow abstraction of an atom or ion, and provide resonance stabilization of the resultant radical or ion. Figure II.4.2.8 provides a selection of readily oxidized groups and the atom at which initial attack occurs. In Figure II.4.2.9, examples of radical and ion stabilization by resonance in ether and branched hydrocarbon structures are provided. Peroxy, carbonyl, and other radical intermediates are stabilized by similar resonance delocalization of electrons from the elements C, O, H or N.

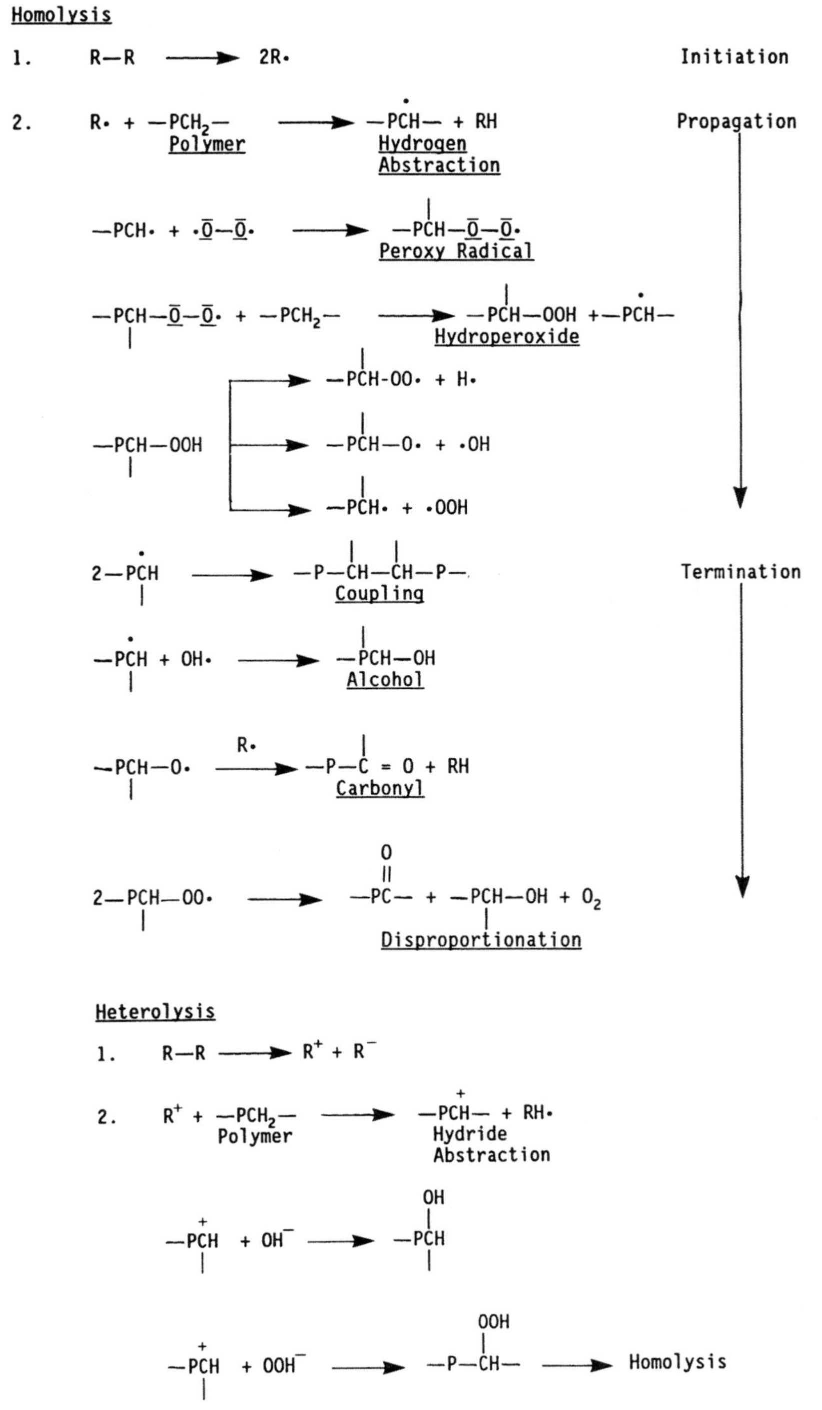

FIGURE II.4.2.7 Proposed homolytic chain reaction and heterolytic oxidation mechanisms.

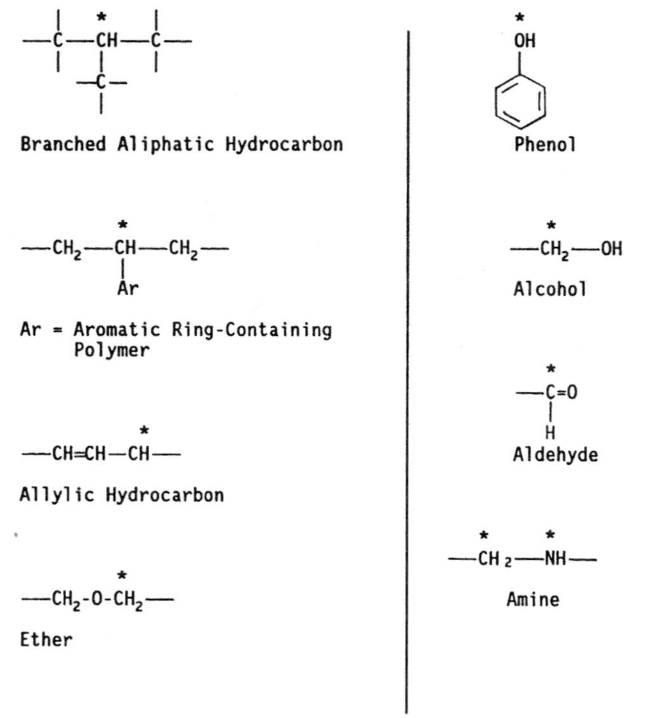

Branched Aliphatic Hydrocarbon

Ar = Aromatic Ring-Containing Polymer

Allylic Hydrocarbon

Ether

Phenol

Alcohol

Aldehyde

Amine

FIGURE II.4.2.8 Readily oxidizable functional groups (* is site of homolysis or heterolysis).

Two general categories of oxidative biodegradation, based on the source of initiation of the process, are direct oxidation by the host and external environment-mediated oxidation.

Direct Oxidation by Host

In these circumstances, host-generated molecular species effect or potentiate oxidative processes directly on the polymer. Current thinking, based on solid analytical evidence, is that such reactive molecules are derived from activated phagocytic cells responding to the injury and the properties of the foreign-body at the implant site (Zhao et al., 1991). The two major types of these cells are the neutrophils (polymorphonuclear leukocytes, PMNs) and the monocytes. The latter, which are found in circulation, can differentiate upon attachment to tissue and divide into macrophage and foreign-body giant cell (FBGC) phenotypes (see Chapter II.2.2) (Ziats, 1988).

Much work is underway to elucidate the sequence of events leading to phagocytic oxidation of biomaterials. Certain important processes of wound healing in the presence of biologically derived foreign-bodies such as bacteria and parasites are showing some relevance to biomaterial implants (Northup, 1987).

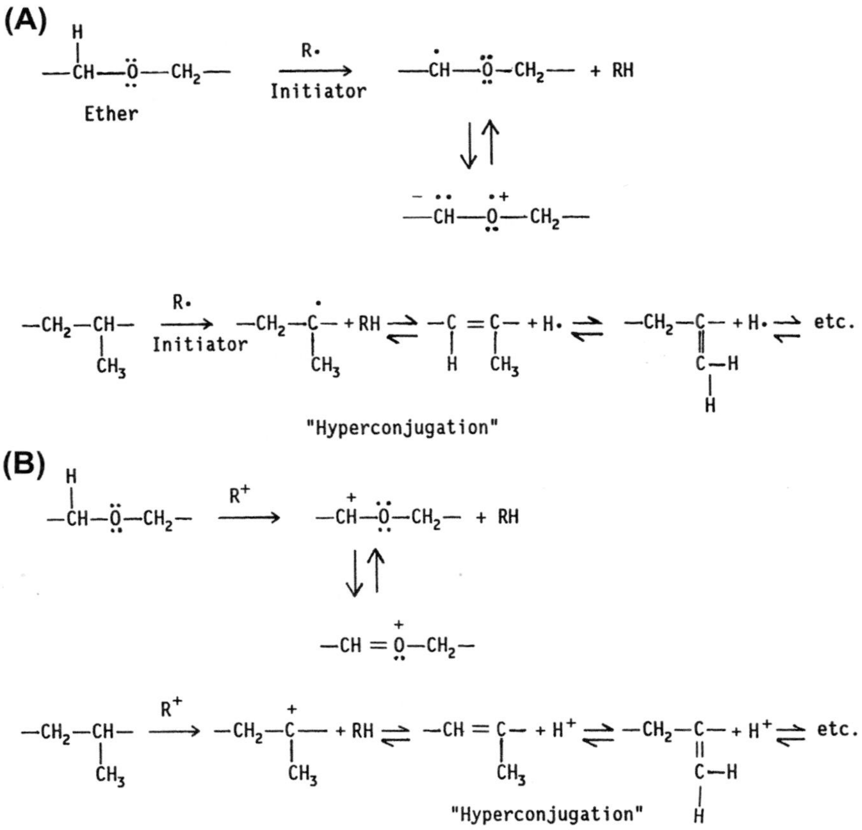

(A)

Ether

Initiator

"Hyperconjugation"

(B)

"Hyperconjugation"

FIGURE II.4.2.9 (A) Resonance stabilization of ether and hydrocarbon radicals; (B) Resonance stabilization of ether and hydrocarbon cations.

$$\text{Neutrophil or Macrophage} + O_2 + e^- \xrightarrow{\text{Activating Factors}} O_2^{\cdot -} \text{ (Superoxide Anion)}$$

$$2O_2^{\cdot -} + 2H^+ \xrightarrow{\text{SOD}} O_2 + H_2O_2 \text{ (Hydrogen Peroxide)}$$

$$O_2^{\cdot -} + Fe^{3+} \text{ (Ferric Ion)} \longrightarrow O_2 + Fe^{2+}$$

$$O_2^{\cdot -} + H^+ \longrightarrow HO_2\cdot$$

$$H_2O_2 + Fe^{2+} \longrightarrow Fe^{3+} + OH\cdot \text{ (Hydroxyl Radical)} + OH^-$$

$$H_2O_2 + Cl^- + H^+ \xrightarrow{\text{MPO}} HOCl \text{ (Hypochlorous Acid)} + H_2O$$

$$HOCl + R_2NH \rightleftharpoons R_2N\text{-}Cl \text{ (Chloramine)} + H_2O$$

$$NO\cdot \text{ (Nitric Oxide)} + O_2^{\cdot -} \longrightarrow ONOO^- \text{ (Peroxynitrite Anion)}$$

FIGURE II.4.2.10 Generation of potential oxidants by phagocytic processes.

Neutrophils, responding to chemical mediators at the wound site, mount a powerful but transient chemical attack within the first few days of injury (Test and Weiss, 1986; Northup, 1987). Chemically susceptible biomaterials may be affected if they are in close apposition to the wound site (Sutherland et al., 1993). Activated macrophages subsequently multiply and subside within days at a benign wound site or in weeks if stimulants such as toxins or particulates are released at the site. Their fusion products (foreign-body giant cells) FBGCs, can survive for months to years on the implant surface. Macrophages also remain resident in the collagenous foreign-body capsule for extended periods.

While it is recognized that the mechanism of cellular attack and oxidation of biomaterials is as yet unconfirmed, the following discussion attempts to provide logical biological pathways to powerful oxidants capable of producing known degradation products.

Both PMNs and macrophages metabolize oxygen to form a superoxide anion (O_2^-). This intermediate can undergo transformation to more powerful oxidants, and conceivably can initiate homolytic reactions on the polymer. Superoxide dismutase (SOD), a ubiquitous peroxidase enzyme, can catalyze the conversion of superoxide to hydrogen peroxide which, in the presence of myeloperoxidase (MPO) derived from PMNs, is converted to hypochlorous acid (HOCl). A potent biomaterial oxidant in its own right (Coury et al., 1987), hypochlorite (ClO^-) can oxidize free amine functionality (e.g., in proteins) to chloramines that can perform as long-lived sources of chlorine oxidant (Test and Weiss, 1986; Figures II.4.2.10, II.4.2.11). Hypochlorite can oxidize other substituted nitrogen functional groups (amides, ureas, urethanes, etc.) with potential chain cleavage of these groups.

The following paragraphs describe potential cooperative reactions involving acquired peroxidase and free ferrous ions. Macrophages contain essentially no MPO, so their hydrogen peroxide is not normally converted to HOCl. However, PMN-derived MPO can bind securely to foreign-body surfaces (Locksley et al., 1982), and serve as a catalyst reservoir for macrophage- or FBGC-derived HOCl production. If free ferrous ion, which is normally present in negligible quantities in the host, is released to the implant site by hemolysis or other injury, it can catalyze the formation of the powerfully oxidizing hydroxyl radical via the Haber–Weiss cycle (Klebanoff, 1982; Figure II.4.2.10).

Figure II.4.2.11 shows radical and ionic intermediates of HOCl that may initiate biomaterial oxidation. Figure II.4.2.12 is a diagram showing a leukocyte phagocytic process that employs endogenous MPO catalysis of HOCl formation. In a more general sense, the MPO may come from within or outside of the cell.

The foregoing discussion of sources of direct oxidation focused primarily on acute implant periods in which bursts of PMN activity followed by macrophage activity normally resolve within weeks. However, since the

Equilibrium Products

$$HOCl + Na^+ \underset{\longleftarrow}{\overset{pH\ 7\text{-}8}{\longrightarrow}} NaOCl + H^+ \longrightarrow Na^+ + OCl^-$$

~50% ~50%

Radical Intermediates

$$HOCl \longrightarrow HO\cdot + Cl\cdot$$

$$HOCl \xrightarrow{RR'NH} RR'N\text{-}Cl + H_2O \longrightarrow RR'N\cdot + Cl\cdot$$

Chloramine

$$HOCl \xrightarrow{HOCl} Cl_2O + H_2O \longrightarrow ClO\cdot + Cl\cdot$$

Ionic Intermediates

$$HOCl + Cl^- + H^+ \rightleftharpoons Cl_2 + H_2O \rightleftharpoons Cl^+Cl^-$$

$$HOCl \rightleftharpoons H^+ + OCl^-$$

$$HOCl \rightleftharpoons HO^- + Cl^+$$

FIGURE II.4.2.11 Hypochlorous acid: Formation and potential reaction intermediates.

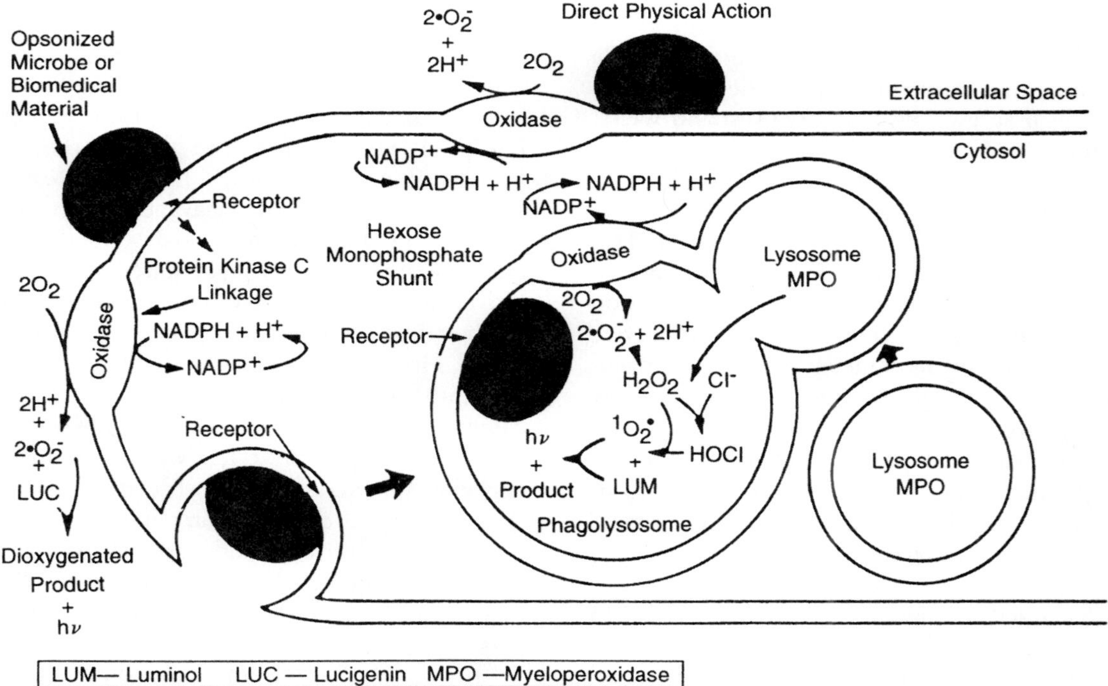

FIGURE II.4.2.12 Activation of phagocyte redox metabolism: Chemiluminigenic probing with luminol and lucigenin. (From R. C. Allen, personal communication, 1991.)

foreign-body subsequently remains implanted a sustained if futile attempt to phagocytose an implanted device provides a prolonged release of chemicals onto the biomaterial. This phenomenon, called exocytosis, occurs over months to possibly years (Zhao et al., 1990), and results primarily from the macrophage-FBGC line. It can contribute to long-term chemical degradation of the polymer.

The oxidation processes induced by phagocytes are the result of oxidants produced by general foreign-body responses, not direct receptor–ligand catalysis by oxidase enzymes. Attempts to degrade oxidatively susceptible polymers by direct contact with oxidase enzymes have produced short-range or limited effects (Sutherland et al., 1993; Santerre et al., 1994).

Macrophages mediate other processes, such as fibrous capsule formation, around the device. Their release of cellular regulatory factors stimulates fibroblasts to populate the implant site and produce the collagenous sheath. Any knowledge of the effects of factors such as fibroblasts or fibrous capsules on rates and mechanisms of polymer degradation is, at this time, rudimentary.

Stress Cracking

An important category of host-induced biodegradation with an oxidative component is stress cracking as manifested in poly(ether urethane) elastomers implants. It differs from classic environmental stress cracking (ESC), which involves a susceptible material at a critical level of stress in a medium that may permeate and swell the polymer but does not dissolve the polymer. Classic ESC is not accompanied by significant chemical degradation (Stokes, 1988). In contrast, stress cracking of polyurethanes is characterized by surface attack of the polymer, and by chemical changes induced by relatively specific *in vivo* or *in vitro* oxidizing conditions. Conditions relevant to stress cracking of certain poly(ether urethane) compositions are stated in Table II.4.2.3.

Experimental information on the stress cracking of poly(ether urethanes) and poly(ether urethane ureas) (e.g., the polymers described in Figure II.4.2.3) has provided insights which may be valid for these and other compositions that can be oxidized, for example, polypropylene (Liebert et al, 1976; Altman et al., 1986) or polyethylenes (Wasserbauer et al., 1990; Zhao et al., 1995).

Poly(ether urethanes), which are resistant to hydrolysis *in vivo*, are used as connectors, insulators, tines, and adhesives for cardiac pacemakers and neurological stimulators (Figure II.4.2.13). They have performed with high reliability in chronic clinical applications since 1975. Certain poly(ether urethane) pacing leads have displayed surface cracks in their insulation after residence times *in vivo* of months to years. These cracks are directly related in frequency and depth to the amount of residual stress (Figures II.4.2.14, II.4.2.15) and the ether (soft segment) content of the polyurethane (Coury et al., 1987; Martin et al., 2001).

Morphologically, the rough-walled cracks display regular patterns predominately normal to the force vectors, occasionally with "tie fibers" bridging the gaps, indicative of ductile rather than brittle fracture (Figures II.4.2.16, II.4.2.17). Infrared (IR) analysis indicates that oxidation does not take place detectably in the bulk, but only on the surface where extensive loss of ether functionality (as seen in the ether IR stretch, 1110 cm^{-1}) and enhanced absorption in the hydroxyl and carbonyl

TABLE II.4.2.2 Mechanisms Leading to Degradation of Polymer Properties[a]

Physical	Chemical
Sorption	Thermolysis
Swelling	Radical scission
Softening	Depolymerization
Dissolution	Oxidation
Mineralization	Chemical
Extraction	Thermooxidative
Crystallization	Solvolysis
Decrystallization	Hydrolysis
Stress cracking	Alcoholysis
Fatigue fracture	Aminolysis, etc.
Impact fracture	Photolysis
	Visible
	Ultraviolet
	Radiolysis
	Gamma rays
	X-rays
	Electron beam
	Fracture-induced radical reactions

[a]Some degradation processes may involve combinations of two or more individual mechanisms.

TABLE II.4.2.3 Characteristics of Poly(ether urethanes) that Cracked *In Vivo*

Components contained residual processing and/or applied mechanical stresses/strains
Components were exposed to a medium of viable cellular and extracellular body constituents
Polymers had oxidatively susceptible (aliphatic ether) groups
Analysis of polymers showed surface oxidation products

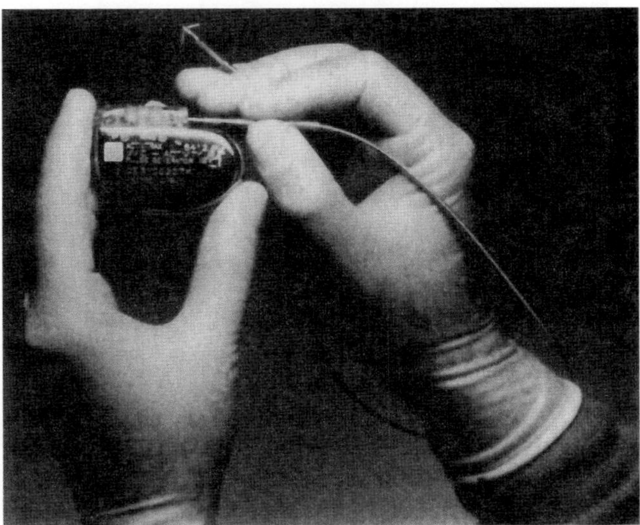

FIGURE II.4.2.13 Cardiac pacemaker with polyurethane lead, tine, and connector. (Courtesy of Medtronic, Inc.)

FIGURE II.4.2.14 Pellethane 2463–80A pacemaker lead tubing with high applied radial stress showing total breach.

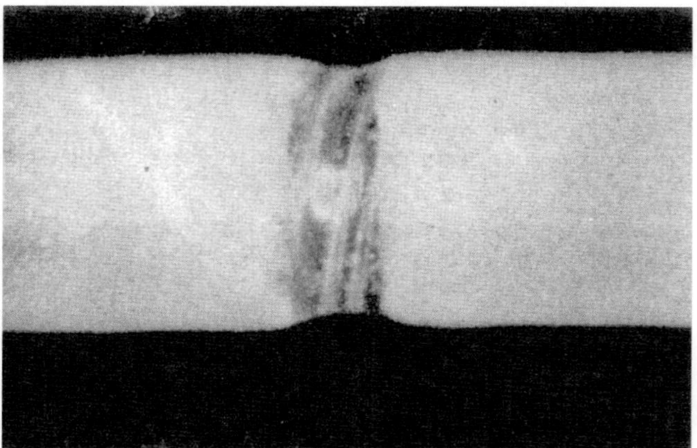

FIGURE II.4.2.15 Pellethane 2363–80A pacemaker lead tubing showing "frosting" due to stress from tight ligature.

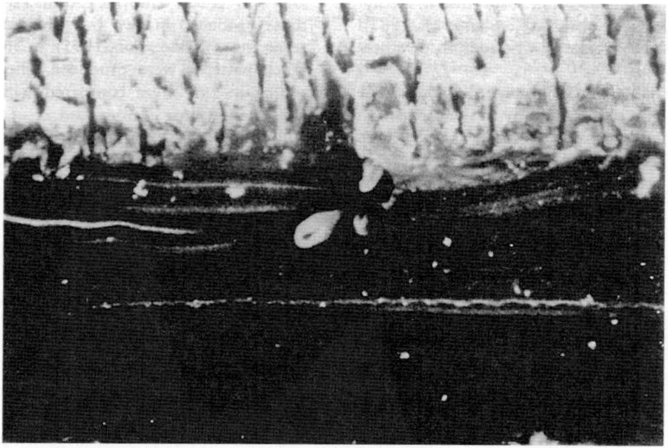

FIGURE II.4.2.16 Stress crack pattern (frosting) near tight ligature (×14).

regions are observed (Stokes et al., 1987). Possible mechanisms for the oxidative degradation of ethers are presented in Figure II.4.2.18. The participation of molecular oxygen in the degradation mechanism is supported by studies which showed that poly(ether urethane urea)

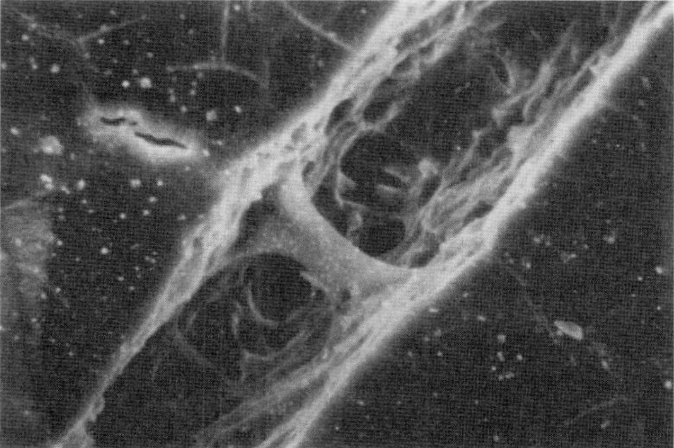

FIGURE II.4.2.17 Single stress crack in pacemaker lead tubing with rough walls and "tie fibers" indicative of ductile fracture (×700).

degradation *in vitro* correlated with oxygen diffusion into the polymer bulk after surface oxidation was initiated by hydrogen peroxide/cobalt chloride (Schubert et al., 1997a,b).

In a seminal study, Zhao et al. (1990) placed polyurethane tubing under strain in cages permeable to fluids and cells (therefore under high initial stress, which was subject to subsequent stress relaxation), and implanted them in rats. In certain cases, anti-inflammatory steroids or cytotoxic polymers were co-implanted in the cages. Implants of up to 15 weeks were retrieved. The only prestressed samples to crack were those that did not reside in the cages with the cytotoxic co-implants. The authors concluded that adherent cells caused the stress cracking, and cell necrosis or deactivation inhibited crack induction.

Subsequently, viable phagocytic cells were implicated as a cause of crack initiation *in vivo* (Zhao et al., 1991). By removing adherent foreign-body giant cells after a 10 week implantation of a curved poly(ether urethane urea) film in a wire cage, exposed foreign-body cell "footprints" showed localized surface cracking in the order of several microns deep and wide. Adjacent areas of polymer which were devoid of attached cells were not cracked. Owing to relatively low stresses in the implanted film, deep crack propagation was not observed.

In vitro studies of strained (Stokes, 1988) and unstrained poly(ether urethane) films (Phua et al., 1987; Ratner et al., 1988; Bouvier et al., 1991; Wiggins et al., 2003) using oxidants, enzymes, etc., have sought to duplicate *in vivo* stress cracking. Although some surface chemical degradation with products similar to those seen *in vivo* was demonstrated, stress crack morphology was most closely matched *in vitro* in two studies. A test which involves immersing stressed poly(ether urethane) tubing in a medium of glass wool, hydrogen peroxide, and cobalt chloride produces cracks which duplicate those produced *in vivo*, but with rate acceleration of up to seven times (Zhao et al., 1995). These investigators also showed that human plasma proteins,

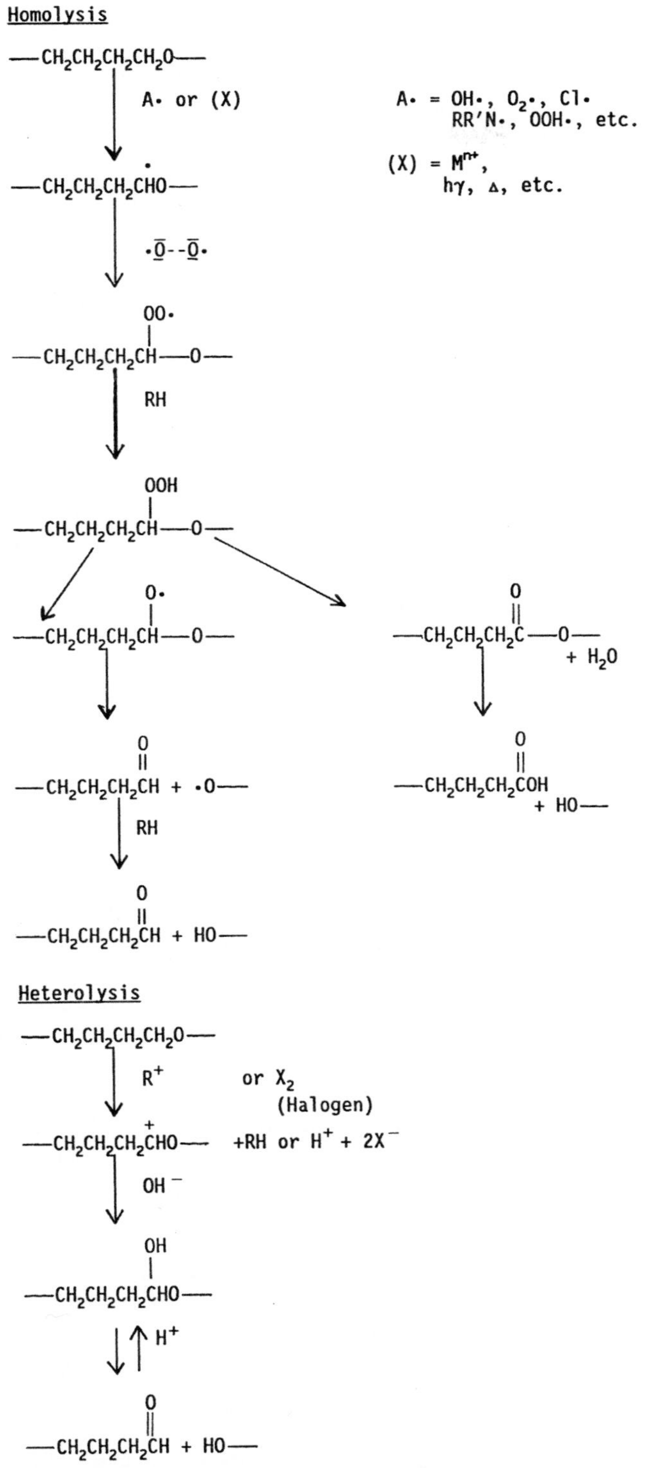

Homolysis

$—CH_2CH_2CH_2CH_2O—$

↓ A• or (X)

A• = OH•, O_2•, Cl•, RR′N•, OOH•, etc.

(X) = M^{n+}, hγ, Δ, etc.

$—CH_2CH_2CH_2\dot{C}HO—$

↓ •Ō–Ō•

$—CH_2CH_2CH_2CH(OO•)—O—$

↓ RH

$—CH_2CH_2CH_2CH(OOH)—O—$

↙ ↘

$—CH_2CH_2CH_2CH(O•)—O—$ | $—CH_2CH_2CH_2C(=O)—O—$ + H_2O

↓ ↓

$—CH_2CH_2CH_2CH(=O)$ + •O— | $—CH_2CH_2CH_2C(=O)OH$ + HO—

↓ RH

$—CH_2CH_2CH_2CH(=O)$ + HO—

Heterolysis

$—CH_2CH_2CH_2CH_2O—$

↓ R^+ or X_2 (Halogen)

$—CH_2CH_2CH_2\overset{+}{C}HO—$ +RH or $H^+ + 2X^-$

↓ OH^-

$—CH_2CH_2CH_2CH(OH)O—$

↓↑ H^+

$—CH_2CH_2CH_2CH(=O)$ + HO—

FIGURE II.4.2.18 Pathways for oxidative fragmentation of polyethers.

particularly alpha-2-macroglobulin and ceruloplasmin, enhance *in vitro* stress cracking by oxidants in patterns morphologically similar to those observed *in vivo* (Zhao et al., 1993). The potential of macrophages to contribute to stress cracking of poly(ether urethanes) was verified in an *in vitro* study which succeeded in potentiating macrophage oxidative effects with ferrous chloride, and inhibiting them with the anti-inflammatory steroid dexamethasone (Casas et al., 1999). In another study, comparable crack patterns were produced when specimens of stressed tubing in rats were compared with those incubated with PMNs in culture (Sutherland et al., 1993). Moreover, this study revealed a difference in chemical degradation products with time of implants which correlated with products from oxidants generated primarily by PMNs (HOCl) and macrophages ($ONOO^-$). Early implant times, activated PMNs, and HOCl caused preferential decrease in the urethane oxygen stretch peak, while longer implant times and $ONOO^-$ caused selective loss of the aliphatic ether stretch peak (by infrared spectroscopy).

Taken together, the foregoing observations are consistent with a hypothesized two-step mechanism for stress cracking *in vivo*. In the first step, surface oxidation induces very shallow, brittle micro-cracks. The second step involves propagation of the cracks in which specific body fluid components act on the formed cracks to enhance their depth and width, without inducing major detectable bulk chemical reactions. Should this hypothesis prove correct, the term "oxidation-initiated stress cracking" would be reasonably descriptive.

The above description of stress cracking has generally considered static stress, such as that entrained in polymers during the cooling of molten parts or the assembly of components. Dynamic stresses and strains such as those occurring during the operation of diaphragm or bladder heart pumps or artificial joints can cause related cracking in areas of high flex. The cracking has been purported to increase with time of device operation, but to display only minor surface chemical changes (Tomita et al., 1999; Wu et al., 1999).

The stress cracking related in entrained stress in the polymer has been controlled by reducing residual stress, isolating the polymer from cell contact (Tang et al., 1994), protecting the polymer from stress-cracking media or using stress crack-resistant polymers (e.g., in the case of polyurethanes, ether-free compositions) (Coury et al., 1990; Takahara et al., 1994; Tanzi et al., 1997), and use of antioxidants such as hindered phenols (e.g., vitamin E, Monsanto Santowhite powder) (Schubert et al., 1997a,b). Stress cracking is next compared with another type of degradation, metal ion-induced oxidation.

Device- or Environment-Mediated Oxidation

Metal Ion–Induced Oxidation. A process of oxidative degradation that has, thus far, only been reported clinically for poly(ether urethane) pacemaker leads, requires, as does stress cracking, a very specific set of conditions. The enabling variables and fracture morphology are quite different from stress cracking, although oxidative degradation products are similar. Biodegradation of implanted devices through stress cracking always occurs on polymer surfaces exposed to cells, and provides characteristic rough-walled fissures (indicative of ductile fracture) oriented perpendicular to the stress vector (Figures II.4.2.14–II.4.2.17). Metal ion-induced oxidation initiates on the enclosed inner surfaces of pacing lead insulation near corroded metallic components and their entrapped corrosion products. Smooth crack walls and microscopically random crack orientation is indicative of brittle fracture (Figures II.4.2.19, II.4.2.20). Macroscopically, crack patterns that track metal component configurations may be present (Figure II.4.2.21). Degradation products that may be found deeper in the bulk than with stress cracking are again indicative of brittle fracture.

This phenomenon called metal ion-induced oxidation has been confirmed by *in vitro* studies, in which poly(ether urethanes) were aged in metal ion solutions of different standard oxidation potentials. Above an oxidation potential of about +0.77, chemical degradation was severe. Below that oxidation potential, changes in the polymer that are characteristic of simple plasticization were seen (Coury et al., 1987; Table II.4.2.4). This technique also showed that metal ion-induced oxidation was proportional to the ether content of the polyurethane (Coury et al., 1987; Table II.4.2.5).

FIGURE II.4.2.19 Random crack pattern of Pellethane 2363–80A lead insulation caused by metal ion-induced oxidation (×480).

The effect of various metals on oxidation *in vitro* and *in vivo* has also been studied. Different metallic components of pacing lead conductors were sealed in poly(ether urethane) (Dow Pellethane 2363–80A) tubing and immersed in 3% hydrogen peroxide at 37°C for up to six months (Stokes et al., 1987) or implanted in rabbits for up to two years (Stokes et al., 1990). Both techniques resulted in corroded metals and degraded tubing lumen surfaces under certain conditions within 30 days. In particular, the *in vivo* interaction of body fluids with cobalt and its alloys resulted in oxidative cracking of the polymer.

The metal ion-induced oxidation process clearly involves corrosion of metallic elements to their ions, and subsequent oxidation of the polymer. In operating devices, the metal ion may be formed by solvation, galvanic or electrolytic corrosion, or chemical or biochemical oxidation (Figure II.4.2.22). In turn, these metal ions develop oxidation potentials that may well be enhanced in body fluids over their standard half-cell potentials. As strong oxidants, they produce intermediates or attack the polymer to initiate the chain reaction (Figure II.4.2.23). Metal ion-induced oxidation is therefore the result of a highly complex interaction of the device, the polymer, and the body.

Should metal ion-induced oxidation be a possibility in an implanted device, several approaches are available to control this problem. They are not universally applicable, however, and should be incorporated only if functionality and biocompatibility are retained. Potentially useful techniques include using corrosion-resistant metals, "flushing" corrosive ions away from the susceptible polymer, isolating the metals and polymer from electrolyte solutions, incorporating appropriate antioxidants, and using oxidation-resistant polymers.

To address oxidation susceptibility, polyurethane elastomers with enhanced oxidation stability have been developed. They should be ether- and ester-free, and generally contain unconventional soft segments, including,

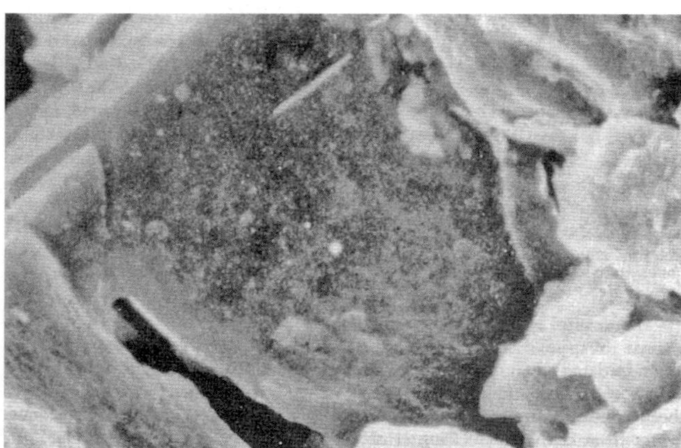

FIGURE II.4.2.20 Smooth crack wall indicative of brittle fracture caused by metal ion-induced oxidation (×830).

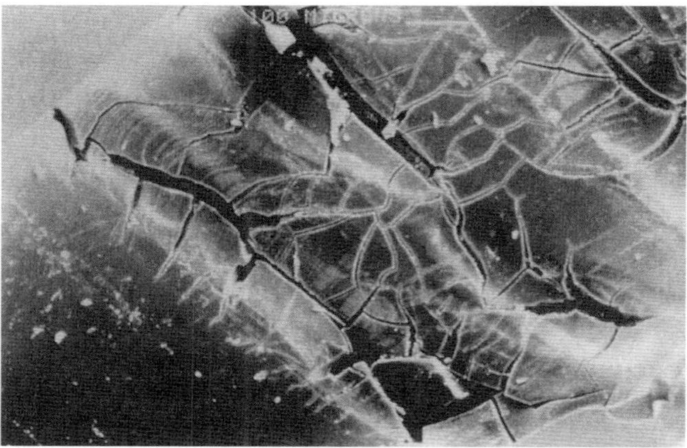

FIGURE II.4.2.21 Crack pattern on inner lumen of polyether urethane lead insulation tracking coil indicative of metal ion-induced oxidation (×100).

TABLE II.4.2.4 Effect of Metal Ion Oxidation Potential on Properties of Poly(ether urethane) (Pellethane 2363-80A)[a]

Aqueous Solution	Standard Oxidation Potential	Change in Tensile Strength (%)	Change in Elongation (%)
$PtCl_2$	Ca+1.2	−87	−77
$AgNO_3$	+0.799	−54	−42
$FeCl_3$	+0.771	−79	−10
Cu_2Cl_2	+0.521	−6	+11
$Cu_2(OAc)_2$	+0.153	−11	+22
$Ni(OAc)_2$	−0.250	−5	+13
$Co(OAc)_2$	−0.277	+1	+13

[a]Conditions: 0.1 *M* solutions/90°C/35 days vs controls aged in deionized water; ASTM (D-1708) microtensile specimens; specimens were tested wet.

TABLE II.4.2.5 Effect of Ether Content of Poly(ether urethane) on Susceptibility to Metal Ion–Induced Oxidation[a]

Poly(ether urethane)	Polyether Content	Change in Tensile Strength (%)	Change in Elongation (%)
Pellethane 2363-80A	High	−54	−42
Pellethane 2363-55D	Low	−23	−10
Model segmented polyurethane	None	+9	+3

[a]Conditions: 0.1 *M* $AgNO_3$/90°C/35 days vs controls aged in deionized water; ASTM (D-1708) microtensile specimens.

$$M^{\circ} \xrightarrow{\text{Electrolysis}} M^{+n} + ne^{-}$$

$$2M^{\circ} + 2H^{+} \longrightarrow 2M^{+} + H_2$$

$$4M^{\circ} + O_2 + 2H_2O \longrightarrow 4M^{+} + 4OH^{-}$$

$$M^{\circ} + HOOH \longrightarrow M^{+} + HO^{-} + HO\cdot$$

$$M^{\circ} + HOCl \longrightarrow M^{+} + HO\cdot + Cl^{-}$$

etc.

FIGURE II.4.2.22 Formation of metal ion from metal.

$$M^{+n} + H_2O \longrightarrow M^{+(n-1)} + HO\cdot + H^{+}$$

$$M^{+n} + \text{—PH—} \longrightarrow M^{+(n-1)} + \text{—}\underset{|}{P}\cdot + H^{+}$$

(PH = Polymer)

$$M^{+n} + O_2 \longrightarrow M^{+(n+1)} + O_2^{\cdot-}$$

$$M^{+n} + HOCl \longrightarrow M^{+(n+1)} + HO\cdot + Cl^{-}$$

$$M^{+n} + \text{—PH—} \longrightarrow M^{+(n-1)}H + \text{—}P^{+}\text{—}$$

$$M^{+n} + H_2O_2 \longrightarrow M^{+(n+1)} + OH^{-} + HO\cdot$$

$$M^{+n} + H_2O_2 \longrightarrow M^{+(n-1)} + HO_2\cdot + H^{+}$$

FIGURE II.4.2.23 Initiation of oxidation pathways by metal ions.

for example, hydrogenated polybutadiene, polyisobutylene, polydimethylsiloxane, polycarbonate, and dimerized fatty acid derivatives (Coury et al., 1990; Pinchuk et al., 1991; Takahara et al., 1991, 1994; Kato et al., 1995; Mathur et al., 1997; Hernandez et al., 2007; Kang et al., 2010; Desai et al, 2011). In implant tests, they have shown reduced tendency to stress crack, and some of them have shown high resistance to metal ion oxidants *in vitro*. Early attempts to stabilize polyurethanes by laminating more biostable polymers (such as silicone rubbers) to tissue-facing surfaces have met with limited success in dynamic applications, due to the delamination tendencies (Pinchuk, 1992). In 2006, the first new polyurethane composition in decades was introduced as pacemaker/defibrillator lead insulation It is an aromatic polyurethane with soft segments of mixed polysiloxane/polyether composition (Simmons et al., 2004). Indications are that it demonstrates physical properties superior to silicone elastomers (Parmar, 2012). Some studies show improved, but not total degradation resistance of this material when compared to poly(ether urethanes) (St Jude Medical, Inc., 2012), but, since cardiac leads can be implanted for decades, verification of long-term durability in humans must await longer term implantation.

More recent approaches to stabilizing polyurethanes to oxidative attack *in situ* have involved the use of surface modifying macromolecules (SMMs) (Santerre et al., 2000), surface modifying end groups (SMEs) (Ward et al., 1995, 1998), and, more recently, self-assembling monolayer end groups (SAMEs) (Ward, 2008). SMMs, typically fluorocarbon-based polymers, are blended with the polyurethanes during processing, and migrate to the surface prior to implantation. SMEs are moieties (typically polysiloxane) bonded to polyurethane as end groups. SAMEs are formed from molecules appended to the backbone of polymers during synthesis. Such molecules consist of three components: a chemically reactive group for conjugation to the polymer as it is being made; typically, a hydrophobic spacer chain, to provide self-assembly at the surface of the formed article; and a head group from a variety of surface chemistries to provide specific bioactivity. One potential function is to stabilize the polymer against biodegradation (Ward, 2008).

The covalently modified polyurethanes may be used in bulk or as additives to conventional polyurethanes. Both approaches have provided enhanced *in vivo* stability for polyurethane implants; however, the long-term effects of these treatments are not, as yet, known. SMMs have been covalently modified with bioactive agents, such as antioxidants, to provide further degradation resistance (Ernsting et al., 2002).

All of the "barrier" strategies to protecting polyurethanes described above appear to have validity, at least for protecting polyurethanes in the short-term. The long-term (multiyear) benefits of these approaches remain to be seen in light of issues such as surface dynamics, interfacial interactions, and coating durability.

All of these polyurethane modifications, while potentially providing enhanced resistance to biodegradation, still allow susceptibility to attack by biological components, often at slow rates. With poly(carbonate urethanes), for example, superior oxidation resistance has been observed in several studies (Mathur et al., 1997; Tanzi et al., 1997). In other studies, these compositions were shown to be susceptible to oxidation (Christenson et al., 2004). Additionally, in aqueous media *in vitro* and *in vivo*, slow degradation attributable to simple hydrolysis was also detected (Zhang et al., 1997). The body fluid environment provides a relatively stable long-term hydrolytic medium, generally less subject to "respiratory bursts" that strongly enhance oxidative processes. Although phagocytic processes may also produce hydrolytic enzymes, their effects on synthetic polymers are specific and limited (Labow et al., 2002a). Hydrolysis, therefore, may be expected to take place continuously with poly(carbonate urethane) integrity susceptible to a combination of mechanical stress and vigorous oxidizing conditions (Faré et al., 1999; Labow et al., 2002b). Studies of up to three years implantation have indicated detectable degradation (Seifalian et al., 2003). Only long-term implant studies (e.g., five years or greater) would predict the acceptability of poly(carbonate urethanes) or, for that matter, other new polymers having potentially susceptible groups.

Oxidative Degradation Induced by External Environment

Under very limited circumstances, the body can transmit electromagnetic radiation that may affect the integrity of implanted polymers. For example, the cornea and vitreous humor of the eye, as well as superficial skin layers, allow the passage of long-wave (320–400 nanometer) "ultraviolet A" radiation. Absorption of ultraviolet radiation causes electron excitation that can lead to photo-oxidative degradation. This process has been suggested in the breakdown of polypropylene components of intraocular lenses (Altman et al., 1986; Jongebloed and Worst, 1986).

(A)

$-O-C(=O)-NH-C_6H_4-CH_2-C_6H_4-NH-C(=O)-O-$

↓ [O]

$-O-C(=O)-NH-C_6H_4-CH=C_6H_4=N-C(=O)-O-$ (Quinone-imide)

↓ [O]

$-O-C(=O)-N=C_6H_4=C=C_6H_4=N-C(=O)-O-$ (Diquinone-imide)

(B)

$-NHC(=O)OCH_2- \xrightarrow{h\nu} -NHC(=O)\cdot + \cdot O-$ → Oxidation Products

$-NHC(=O)OCH_2- \xrightarrow{h\nu} -NH\cdot + \cdot C(=O)OCH_2-$

↓

$CO_2 + \cdot CH_2-$

↓

Oxidation Products

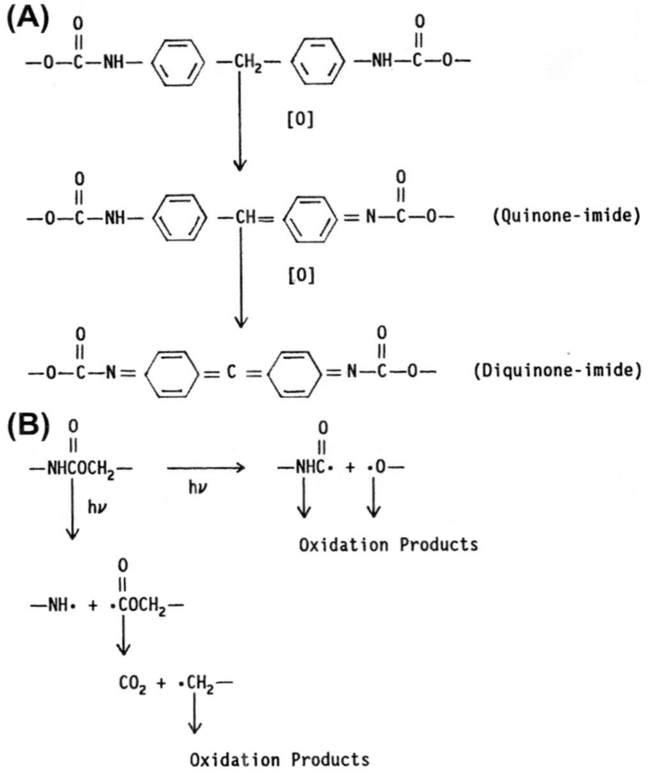

FIGURE II.4.2.24 Photo-oxidative reactions of aromatic polyurethanes. (A) Formation of quinone-imide from aromatic polyurethane. (From A. J. Coury et al. (1988). *J. Biomater. Appl.,* **3.**) (B) Photolytic cleavage of urethane link. (From S. K. Brauman et al. (1981). *Ann. Biomed. Eng.,* **9.**)

In maxillofacial exo- and very likely endoprostheses, elastomers may undergo undesirable changes in color and physical properties as a consequence of exposure to natural sunlight-frequency radiation (Craig et al., 1980). Photo-oxidation mechanisms involving the aromatic units and the urethane of aromatic poly(ether urethanes) or poly(ester urethanes) are shown in Figure II.4.2.24. Antioxidants and ultraviolet absorbers provide limited protection for these materials.

CONCLUSION

Polymers that are carefully chosen for use in implanted devices generally serve effectively for their intended lifetimes if they are properly processed and device–material–host interactions are adequately addressed. In certain limited circumstances, excessive hydrolytic or oxidative biodegradation occurs. This may be induced by direct attack by the host or by non-biological factors in the environment surrounding the implant. With susceptible polymers, protective measures can be taken to ensure extended efficacy, although new, biodegradation resistant polymers that are on the horizon may require less protection. Knowledge of biodegradation mechanisms and the employment of appropriate countermeasures such as proper material selection or modification, optimal component design, protection from environmental attack, and careful handling and implantation procedures will promote continued progress in the development of polymers as long-term implantable biomaterials.

ACKNOWLEDGMENTS

The author is very grateful to Dr. R. C. Allen for providing the drawing on activated phagocyte redox metabolism. For their technical advice and contributions, I sincerely thank James Anderson, Ken Stokes, Jonathan Sears, John Eaton, Allan Hoffman, John Mahoney, Maurice Kreevoy, Grace Picciolo, Buddy Ratner, and Bob Ward. For the preparation of the original manuscript, I am deeply indebted to my "computer wizard," Mrs. Jayne McCaughey. For help in updating literature sources, I thank Ms. Mari Ferentinos.

BIBLIOGRAPHY

Allen, R. C. (1991). Activation of phagocyte redox metabolism: Chemiluminigenic probing with luminol and lucigenin. *Drawing provided.*

Altman, J. J., Gorn, R. A., Craft, J., & Albert, D. M. (1986). The break-down of polypropylene in the human eye: Is it clinically significant? *Ann. Ophthalmol.*, **18**, 182–185.

ASTM D1708 – 10 (2010) Standard Test Method for Tensile Properties of Plastics by Use of Microtensile Specimens: http://www.astm.org/Standards/D1708.htm

Barbarini Ferraz, L. C., Schellini, S. A., Wludarski, S. L., Padovani, C. R., Selva, D., et al. (2009). Extraocular muscle fixation to porous polyethylene orbital implants using 2-octyl cyanoacrylate. *Eur. J. Ophthalmol. (Jul–Aug)*, **19**(4), 527–529.

Bernardino, C. R., Mihora, L. D., Fav, A. M., & Rubin, P. A. (2006). Orbital complications of hydrogel scleral buckles. *Ophthal. Plast. Reconstr. Surg.*, **22**(3), 206–208.

Blais, P. (1990). Letter to the editor. *J. Appl. Biomater.*, **1**, 197.

Blanchet, T. A., & Burroughs, B. R. (2001). Numerical oxidation model for gamma radiation-sterilized UHMWPE: Consideration of dose-depth profile. *J. Biomed. Mater. Res.*, **58**(6), 684–693.

Bloch, B., & Hastings, G. (1972). *Plastics Materials in Surgery* (2nd ed.). Springfield, IL: Charles C. Thomas. 97–98.

Booth, A. E. (1995). Industrial sterilization technologies: New and old trends shape manufacturer choices. *Med. Dev. Diagn. Ind.*, February, 64–72.

Bouvier, M., Chawla, A. S., & Hinberg, L. (1991). *In vitro* degradation of a poly (ether urethane) by trypsin. *J. Biomed. Mater. Res.*, **25**, 773–789.

Brauman, S. K., Mayorga, G. D., & Heller, J. (1981). Light stability and discoloration of segmented polyether urethanes. *Ann. Biomed. Eng.*, **9**, 45–58.

Braunstein, R. A., & Winnick, M. (2002). Complications of Miragel: Pseudotumor. *Arch. Ophthalmol.*, **120**, 228–229.

Cardia, G., & Regina, G. (1989). Degenerative Dacron graft changes: Is there a biological component in this textile defect? A case report. *Vasc. Surg.*, **23**(3), 245–247.

Carr, J. (2011). The intracorporeal use of 2-octyl cyanoacrylate resin to control air leaks after lung resection. *Eur. J. Cardiothorac. Surg.*, **39**(4), 579–583.

Casas, J., Donovan, M., Schroeder, P., Stokes, K., & Untereker, D. (1999). *In vitro* modulation of macrophage phenotype and inhibition of polymer degradation by dexamethasone in a human macrophage/Fe/stress system. *J. Biomed. Mater. Res.*, **46**, 475–484.

Christenson, E. M., Anderson, J. M., & Hiltner, A. (2004). Oxidative mechanisms of poly(carbonate urethane) and poly(ether urethane) biodegradation: *In vivo* and *in vitro* correlations. *J. Biomed. Mater Res. A.*, **70**, 245–255.

Colthurst, M. J., Williams, R. L., Hiscott, P. S., & Grierson, I. (2000). Biomaterials used in the posterior segment of the eye. *Biomaterials*, **21**, 649–665.

Coury, A. J. (1999). Biostable polymers as durable scaffolds for tissue engineered vascular prostheses. In P. Zilla, & H. Greisler (Eds.), *Tissue Engineering of Vascular Prosthetic Grafts* (Vol. 43, pp. 469–480). Austin, TX: R. G. Landes Company.

Coury, A. J., Slaikeu, P. C., Cahalan, P. T., & Stokes, K. B. (1987). Medical applications of implantable polyurethanes: Current issues. *Prog. Rubber Plastics Tech.*, **3**(4), 24–37.

Coury, A. J., Slaikeu, P. C., Cahalan, P. T., Stokes, K. B., & Hobot, C. M. (1988). Factors and interactions affecting the performance of polyrethane elastomers in medical devices. *J. Biomater. Appl.*, **3**, 130–179.

Coury, A. J., Hobot, C. M., Slaikeu, P. C., Stokes, K. B., & Cahalan, P. T. (1990). A new family of implantable biostable polyurethanes. *Trans. 16th Annual Meeting Soc. for Biomater.*, **20–23**, 158. May.

Craig, R. G., Koran, A., & Yus, R. (1980). Elastomers for maxillofacial applications. *Biomaterials*, **1**(Apr.), 112–117.

Crowninshield, R., & Muratoglu, O. (2008). How have new sterilization techniques and new forms of polyethylene influenced wear in total joint replacement? *J. Am. Acad. Orthop. Surg.*, **Vol. 16**(no. Suppl 1), S80–S85. June.

Daly, B. M., & Yin, J. (1998). Subsurface oxidation of polyethylene. *J. Biomed. Mater. Res.*, **42**, 523–529.

Desai, S., Boden, M., DeRoche, S., Foster, A., & Reddy, S. (2011). US Patent Application 20110054580. *Polyisobutylene Urethane Urea and Urethane/Urea Copolymers and Medical Leads Containing the Same.* March 3.

Ernsting, M. J., Santerre, J. P., & Labow, R. S. (2002). Surface modification of a polycarbonate-urethane using a Vitamin E derivatized fluoroalkyl surface modifier. *Trans. 28th Annual Meeting Soc. Biomater.*, 16. April 24–27.

Faré, S., Petrini, P., Motta, A., Cigada, A., & Tanzi, M. C. (1999). Synergistic efforts of oxidative environments and mechanical stress on *in vitro* stability of polyetherurethanes and polycarbona-teurethanes. *J. Biomed. Mater. Res.*, **45**, 62–74.

Finck, K. M., Grosse-Siestrup, C., Bisson, S., Rinck, M., & Gross, U. (1994). Experimental *in vivo* degradation of polyarylamide. *Trans. 20th Annual Meeting Soc. for Biomater.*, 210. April 5–9.

Furman, B., & Li, S. (1995). The effect of long-term shelf life aging of ultra high molecular weight polyethylene. *Trans. 21st Annual Meeting Soc. for Biomater.*, 114. March 18–22.

Greisser, H. J., Gengenbach, T. R., & Chatelier, R. C. (1994). Longterm changes in the surface composition of polymers intended for biomedical applications. *Trans. 20th Annual Meeting Soc. for Biomater.*, 19. April 5–9.

Gumargalieva, K. Z., Moiseev, Y. V., Daurova, T. T., & Voronkova, O. S. (1982). Effect of infections on the degradation of polyethylene terephthalate implants. *Biomaterials*, 3(3), 177–180. July.

Hegyeli, A. (1973). Use of organ cultures to evaluate biodegradation of polymer implant materials. *J. Biomed. Mater. Res.*, 7, 205–214.

Hernandez, R., Weksler, J., Padsalgikar, A., & Runt, J. (2007). *In vitro* oxidation of high polydimethylsiloxane content biomedical polyurethanes: Correlation with the microstructure. *J. Biomed. Mater. Res.*, 87(2), 546–556.

Hoffman, A. (1999). *Personal Communication.*

Hudson, J., & Crugnola, A. (1987). The *in vivo* biodegradation of nylon 6 utilized in a particular IUD. *J. Biomater. Appl.*, **1**, 487–501.

Hwang, K. I., & Lim, J. I. (1997). Hydrogel explant fragmentation 10 years after scleral buckling surgery. *Arch. Ophthalmol.*, **115**, 1205–1206.

Jongebloed, W. L., & Worst, J. F.G. (1986). Degradation of polypropylene in the human eye: A sem-study. *Doc. Ophthalmol.*, **64**, 143–152.

Kang, J., Erdodi, G., Brendel, C. M., Ely, D., & Kennedy, J. P. (2010). Polyisobutylene-based polyurethanes. V. Oxidative-hydrolytic stability and biocompatibility. *J. Polym. Sci. A: Polymer Chemistry*, **48**(10), 2194–2203.

Kato, Y. P., Dereume, J. P., Kontges, H., Frid, N., Martin, J. B., et al. (1995). Preliminary mechanical evaluation of a novel endoluminal graft. *Trans. 21st Annual Meeting Soc. for Biomater*, 81. March 18–22.

Kearney, J. J., Lahey, J. M., Borirakchanyavat, S., Schwartz, D. M., Wilson, D., et al. (2004). Complications of hydrogel explants used in scleral buckling surgery. *Am. J. Ophthal.*, **137**(1), 96–100.

Klebanoff, S. (1982). Iodination catalyzed by the xanthine oxidase system: Role of hydroxyl radicals. *Biochemistry*, **21**, 4110–4116.

Kopecek, J., & Ulbrich, K. (1983). Biodegradation of biomedical polymers. *Prog. Polym. Sci.*, **9**, 1–58.

Kurtz, S. M., Rimnac, C. M., Hozack, W. J., Turner, J., Marcolongo, M., et al. (2005). *In vivo* degradation of polyethylene liners after gamma sterilization in air. *J. Bone Joint Surg. (Am.)*, **87**, 815–823.

Labow, R. S., Erfle, D. J., & Santerre, J. P. (1995). Neutrophil-mediated degradation of segmented polyurethanes. *Biomaterials*, **16**, 51–59.

Labow, R. S., Tang, Y., McCloskey, C. B., & Santerre, J. P. (2002a). The effect of oxidation on the enzyme-catalyzed hydrolytic degradation of polyurethanes. *Can. J. Biomater. Sci., Polymer Ed.*, **13**(6), 651–665.

Labow, R. S., Meek, E., Matherson, L. A., & Santerre, J. P. (2002b). Human macrophage-mediated biodegradation of polyurethanes: assessment of candidate enzyme activities. *Biomaterials*, **23**(19), 3969–3975.

Liebert, T. C., Chartoff, R. P., Cosgrove, S. L., & McCuskey, R. S. (1976). Subcutaneous implants of polypropylene filaments. *J. Biomed. Mater. Res.*, **10**(6), 939–951.

Ling, M. T.K., Westphal, S. P., Qin, S., Ding, S., & Woo, L. (1998). Medical plastics failures from heterogeneous contamination. *Med. Plast. Biomater.*, **5**(2), 45–49.

Locksley, R., Wilson, C., & Klebanoff, S. (1982). Role of endogenous and acquired peroxidase in the toxoplasmacidal activity of murine and human mononuclear phagocytes. *J. Clin. Invest.*, **69**(May), 1099–1111.

Martin, D. J., Poole Warren, L. A., Gunatillake, P. A., McCarthy, S. J., Meijs, G. F., et al. (2001). New methods for the assessment of *in vitro* and *in vivo* stress cracking in biomedical polyurethanes. *Biomaterials*, **22**(9), 973–978.

Mathur, A. B., Collier, T. O., Kao, W. J., Wiggins, M., Schubert, M. A., et al. (1997). *In vivo* biocompatibility and biostability of modified polyurethanes. *J. Biomed. Mater. Res.*, **36**, 246–257.

McKellop, H., Yeom, B., Campbell, P., & Salovey, R. (1995). Radiation induced oxidation of machined or molded UHMWPE after seventeen years. *Trans. 21st Annual Meeting Soc. Biomater.*, 54. March 18–22.

Medel, F. J., Kurtz, S. M., Hozack, W. J., Parvizi, J., Purtill, J. J., et al. (2009). Gamma inert sterilization: A solution to polyethylene oxidation? *J. Bone Joint Surg. (Am.)*, **91**, 839–849.

Northup, S. (1987). Strategies for biological testing of biomaterials. *J. Biomater. Appl.*, **2**, 132–147.

Parmar, A. (2012). Who makes the secret sauce in St. Jude's Optim technology? *MedCity News*, January 24.

Phua, S. K., Castillo, E., Anderson, J. M., & Hiltner, A. (1987). Biodegradation of a polyurethane *in vitro*. *J. Biomed. Mater. Res.*, **21**, 231–246.

Pinchuk, L. (1992). Adhesiveless bonding of silicone rubber to polyurethanes and the use of bonded materials. *US Patent*, **5**. 147, 725, September 15.

Pinchuk, L., Esquivel, M. C., Martin, J. B., & Wilson, G. J. (1991). Corethane: A new replacement for polyether urethanes for longterm implant applications. *Trans. 17th Annual Meeting Soc. Biomater.*, 98. May 1–5.

Pitt, C. G. (1992). Non-microbial degradation of polyesters: Mechanisms and modifications. In M. Vert, J. Feijin, A. Albertson, G. Scott, & E. Chiellini (Eds.), *Biodegradable Polymers and Plastics* (pp. 1–19). Cambridge, UK: R. Soc. Chem.

Portnoy, R. (1997). Clear, radiation-tolerant autoclavable polypropylene. *Med. Plast. Biomater.*, **4**(1), 40–48.

Ratner, B. D., Gladhill, K. W., & Horbett, T. A. (1988). Analysis of *in vitro* enzymatic and oxidative degradation of polyurethanes. *J. Biomed. Mater. Res.*, **22**, 509–527.

Refojo, M. F., & Leong, F. L. (1981). Poly(methylacrylate-*co*-hydroxyethyl acrylate) hydrogel implant material of strength and softness. *J. Biomed. Mater. Res.*, **15**, 497–509.

Rimnac, C., & Pruitt, L. (2008). How do material properties influence wear and fracture mechanisms? *Am. Acad. Orthop. Surg.*, **16**(Suppl. 1), S94–S100.

Roldan-Pallares, M., del Castillo, J. L., Awad-El Susi, S., & Refojo, M. F. (1999). Long- term complications of silicone and hydrogel explants in retinal reattachment surgery. *Arch. Ophthalmol.*, **177**, 197–201.

Santerre, J. P., Labow, R. S., Duguay, D. G., Erfle, D., & Adams, G. A. (1994). Biodegradation evaluation of polyether- and polyesterurethanes with oxidative and hydrolytic enzymes. *J. Biomed. Mater. Res.*, **28**, 1187–1199.

Santerre, J. P., Meek, E., Tang, Y. W., & Labow, R. S. (2000). Use of fluorinated surface modifying macromolecules to inhibit the degradation of polycarbonate-urethanes by human macrophages. *Trans. 6th World Biomaterials Congress*, **77**.

Schnabel, W. (1981). *Polymer Degradation Principles and Practical Applications*. New York, NY: Macmillan. pp. 15–17, 179–185.

Schubert, M. A., Wiggins, M. J., Anderson, J. M., & Hiltner, A. (1997a). Comparison of two antioxidants for poly(etherurethane urea) in an accelerated *in vitro* biodegradation system. *J. Biomed. Mater. Res.*, **34**, 493–505.

Schubert, M. A., Wiggins, M. J., Anderson, J. M., & Hiltner, A. (1997b). Role of oxygen in biodegradation of poly(etherurethane urea) elastomers. *J. Biomed. Mater. Res.*, **34**, 519–530.

Seifalian, A. M., Salacinski, H. J., Tiwari, A., Edwards, E., Bowald, S., et al. (2003). *In vivo* biostability of a poly(carbonateurea) urethane graft. *Biomaterials*, **24**(14), 2549–2557.

Shen, F. W., Yu, Y. J., & McKellop, H. (1999). Potential errors in FTIR measurement of oxidation in ultra-high molecular weight polyethylene implants. *J. Biomed. Mater. Res. (App. Biometer.)*, **48**, 203–210.

Simmons, A., Hyvarinen, J., O'Dell, R., Martin, D., et al. (2004). Long-term in vivo biostability of poly(dimethylsiloxane)/poly(hexamethylene oxide) mixed macrodiol-based polyurethane elastomers. *Biomaterials*, **25**, 4887–4900.

Singer, A. J., Quinn, J. V., & Hollander, J. E. (2008). The cyanoacrylate topical skin adhesives. *Am. J. Emerg. Med.*, **26**(4). May 490–496.

Smith, R., Oliver, C., & Williams, D. F. (1987). The enzymatic degradation of polymers *in vitro*. *J. Biomed. Mater. Res.*, **21**, 991–1003.

Snow, J., Harasaki, H., Kasick, J., Whalen, R., Kiraly, R., et al. (1981). Promising results with a new textured surface intrathoracic variable volume device for LVAS. *Trans. Am. Soc. Artif. Intern. Organs XXVII*, 485–489.

St. Jude Medical, Inc.. (2012). *Optim® Insulation, A New Material for a New Generation of Cardiac Leads.* Product Information: http://www.sjmprofessional.com/Products/US/CRT-Systems/Optim-Insulation.aspx.

Stokes, K. (1988). Polyether polyurethanes: Biostable or not? *J. Biomater. Appl.*, 3(Oct.), 228–259.

Stokes, K., Coury, A., & Urbanski, P. (1987). Autooxidative degradation of implanted polyether polyurethane devices. *J. Biomater. Appl.*, **1**(Apr.), 412–448.

Stokes, K., Urbanski, P., & Upton, J. (1990). The *in vivo* autooxidation of polyether polyurethane by metal ions. *J. Biomater. Sci. Polymer Edn.*, **1**(3), 207–230.

Sutherland, K., Mahoney, J. R., II, Coury, A. J., & Eaton, J. W. (1993). Degradation of biomaterials by phagocyte-derived oxidants. *J. Clin. Invest.*, **92**, 2360–2367.

Szycher, M., & Siciliano, A. (1991). An assessment of 2,4-TDA formation from Surgitek polyurethane foam under stimulated physiological conditions. *J. Biomater. Appl.*, **5**, 323–336.

Takahara, A., Coury, A. J., Hergenrother, R. W., & Cooper, S. L. (1991). Effect of soft segment chemistry on the biostability of segmented polyurethanes. I. *in vitro* oxidation. *J. Biomed. Mater. Res.*, **25**, 341–356.

Takahara, A., Coury, A. J., & Cooper, S. L. (1994). Molecular design of biologically stable polyurethanes. *Trans. 20th Annual Meeting Soc. Biomater.*, 44. April 5–9.

Tang, W. W., Santerre, J. P., Labow, R. S., Waghray, G., & Taylor, D. (1994). The use of surface modifying macromolecules to inhibit biodegradation of segmented polyurethanes. *Trans. 20th Annual Meeting Soc. Biomater.*, 62. April 5–9.

Tanzi, M. C., Mantovani, D., Petrini, P., Guidoin, R., & Laroche, G. (1997). Chemical stability of polyether urethanes versus polycarbonate urethanes. *J. Biomed. Mater. Res.*, **36**, 550–559.

Test, S., & Weiss, S. (1986). The generation of utilization of chlorinated oxidants by human neutrophils. *Adv. Free Radical Biol. Med.*, **2**, 91–116.

Tomita, N., Kitakura, T., Onmori, N., Ikada, Y., & Aoyama, E. (1999). Prevention of fatigue cracks in ultra-high molecular weight polyethylene joint components by the addition of vitamin E. *J. Biomed. Mater. Res. (App. Biomater.)*, **48**, 474–478.

US Food and Drug Administration Document: Medical Devices. (2010). *Device Approvals*: http://www.fda.gov/medicaldevices/productsandmedicalprocedures/deviceapprovalsandclearances/recently-approveddevices/ucm215106.htm2010.

Vauthier, C., Dubernet, C., Fattal, E., Pinto-Alphandary, H., & Couvreur, P. (2003). Poly (alkylcyanoacrylates) as biodegradable materials for biomedical applications. *Adv. Drug Deliv. Rev.*, **55**(4), 519–548.

Vinard, E., Eloy, R., Descotes, J., Brudon, J. R., Giudicelli, H., et al. (1991). Human vascular graft failure and frequency of infection. *J. Biomed. Mater. Res.*, **25**, 499–513.

Ward, R. S. (2008). New horizons for biomedical polymers. *Med. Des. Tech.*, **19**(5), 26–28. 30–31.

Ward, R. S., White, K. A., Gill, R. S., & Wolcott, C. A. (1995). Development of biostable thermoplastic polyurethanes with oligomeric polydimethylsiloxane end groups. *Trans. 21st Annual Meeting Soc. Biomater.*, 268. March 18–22.

Ward, R. S., Tian, Y., & White, K. A. (1998). Improved polymer biostability via oligomeric end groups incorporated during synthesis. *Polymeric Mater. Sci. Eng.*, **79**, 526–527.

Wasserbauer, R., Beranova, M., Vancurova, D., & Dolezel, B. (1990). Biodegradation of polyethylene foils by bacterial and liver homogenates. *Biomaterials*, **11**(Jan.), 36–40.

Watt, D. R. (2001). *Miragel Sponge Complications*: www.retinadoc.com/scripts/retina.pl?function=viewquestions&;forum=retina, 11/07/012001.

Weaver, K. D., Sauer, W. L., & Beals, N. B. (1995). Sterilization induced effects on UHMWPE oxidation and fatigue strength. *Trans. 21st Annual Meeting Soc. Biomater.*, 114. March 18–22.

Wiggins, M. J., Anderson, J. M., & Hiltner, A. (2003). Biodegradation of polyurethane under fatigue loading. *J. Biomed. Mater. Res. Part A.*, **65A**(4), 524–535.

Williams, D. F. (1982). Review: Biodegradation of sugical polymers. *J. Mater. Sci.*, **17**, 1233–1237.

Williams, D. F. (1989). *Definitions in Biomaterials.* Amsterdam: Elsevier.

Wu, L., Weisberg, D. M., Runt, J., Felder, G., III, Snyder, A. J., et al. (1999). An investigation of the *in vivo* stability of poly(ether urethaneurea) blood sacs. *J. Biomed. Mater. Res.*, **44**, 371–380.

Zaikov, G. E. (1985). Quantitative aspects of polymer degradation in the living body. *JMS-Rev. Macromol. Chem. Phys.*, **C25**(4), 551–597.

Zhang, Z., Marois, Y., Guidoin, R., Bull, P., Marois, M., et al. (1997). Vascugraft® polyurethane arterial prosthesis as femoro-popliteal and femoro-peroneal bypass in humans: Pathological, structural and chemical analyses of four excised grafts. *Biomaterials*, **18**, 113–124.

Zhao, Q., Agger, M., Fitzpatrick, M., Anderson, J., Hiltner, A., et al. (1990). Cellular interactions with biomaterials: *In vivo* cracking of pre-stressed pellethane 2363–80A. *J. Biomed. Mater. Res.*, **24**, 621–637.

Zhao, Q., Topham, N., Anderson, J. M., Hiltner, A., Lodoen, G., et al. (1991). Foreign-body giant cells and polyurethane biostability: *In vivo* correlation of cell adhesion and surface cracking. *J. Biomed. Mater. Res.*, **25**, 177–183.

Zhao, Q. H., McNally, A. K., Rubin, K. R., Renier, M., Wu, Z., et al. (1993). Human plasma α2-macroglobulin promotes *in vitro* oxidative stress cracking of Pellethane 2363-80A. *Biomed. Mater. Res.*, **27**, 379–389.

Zhao, Q., Casas-Bejar, C., Urbanski, P., & Stokes, K. (1995). Glass wool-$H_2O_2/COCl_2$ for *in vitro* evaluation of biodegradative stress cracking in polyurethane elastomers. *J. Biomed. Mater. Res.*, **29**, 467–475.

Ziats, N., Miller, K., & Anderson, J. (1988). *In vitro* and *in vivo* interactions of cells with Biomaterials. *Biomaterials*, **9**(Jan.), 5–13.

CHAPTER II.4.3 THE BIODEGRADATION OF BIODEGRADABLE POLYMERIC BIOMATERIALS

Chien-Chi Lin[1] *and Kristi S. Anseth*[2]

[1]Department of Biomedical Engineering, Indiana University-Purdue, University at Indianapolis, Indianapolis, IN, USA

[2]Department of Chemical and Biological Engineering, Howard Hughes Medical Institute and University of Colorado, Boulder, CO, USA

INTRODUCTION

The application of biodegradable polymers in medicine can be dated back to the early 1960s when polymers (e.g., poly(glycolic acid) or PGA), that were initially found unstable due to the rapid degradation, were used as suture materials (Morgan, 1969; Dardik et al., 1971; Vert, 1989). Since then, numerous biodegradable biomaterials, both of synthetic and natural origin, have been investigated and continue to play a critical role in clinical applications, as well as laboratory research. While almost all polymers will degrade eventually, only those with therapeutically-relevant degradation rates are considered biodegradable, and are the focus of this section. Among all, synthetic biodegradable polymers are particularly intriguing. In contrast to natural biopolymers, synthetic polymers can be synthesized in a highly controllable and reproducible manner, with defined material properties and user-designed degradability (Middleton and Tipton, 2000). The lack of immunogenicity of synthetic polymers also enhances their potential in clinical applications. The past few decades have seen great advances in the design and applications of synthetic polymeric biodegradable materials (Vert et al., 1992; Anderson and Shive, 1997; Langer and Tirrell, 2004). Through theoretical and experimental investigations, these biodegradable polymers have found numerous applications, particularly in the fields of drug delivery (Siepmann and Gopferich, 2001) and tissue engineering (Freed et al., 1994; Langer and Tirrell, 2004). This chapter discusses the fundamental aspects of biodegradation of synthetic polymeric biomaterials, including: (1) the types of biodegradable chemical bonds that are commonly designed into biomaterials; (2) the degradation modes, mechanisms, and kinetics of biodegradation; and (3) the effect of biodegradation on material properties relevant to their biomedical applications, including drug delivery and tissue engineering.

CHARACTERISTICS OF BIODEGRADABLE POLYMERS

In general, the process of degradation involves a series of cleavage events, leading to polymer chain scission (Gopferich, 1996). Consequently, the molecular weight of the polymer decreases to form lower molecular weight oligomers and monomers, and ultimately results in the loss of mechanical stability of the material. While polymers can be degraded through a variety of mechanisms such as thermal, mechanical, and photodegradation, the most common modes of degradation in a therapeutically-relevant *in vivo* environment are either hydrolytic or enzymatic degradation. The term "biodegradation" is commonly used when degradation involves biological processes, such as body fluids, cellular activities, and enzymatic reactions. The conditions of the biological environment, therefore, determine the rate of polymer degradation to a great extent. For example, the degradation of an implanted polymeric biomaterial may proceed faster in a highly inflammatory environment, due to excessive biological and biochemical activity. The pH and temperature of the local biological environment, enzyme concentrations, and cell infiltration/recruitment to the implant sites all affect the rate of polymer degradation.

In addition to the biological environment, the chemical, physical, and morphological properties of degradable polymers also govern the mode and rate of biodegradation. For example, hydrophobic polymers limit water accessibility, and typically have decreased hydrolytic degradation rates compared to their more hydrophilic counterparts (Gopferich, 1996). Further, amorphous polymers usually degrade faster than semicrystalline polymers. Similarly, polymers with high glass transition temperature (T_g) degrade at slower rates compared to polymers with lower T_g (Pitt et al., 1981a,b; Gopferich, 1996). One must note that biodegradation is a dynamic process, as the polymer properties change with degradation. For instance, the crystallinity of poly(L-lactic acid) or L-PLA increases during degradation, due to a faster degradation rate in the amorphous regions than the crystalline regions of the polymer. This results in a decreased degradation rate over time, due to enhanced polymer crystallinity. On the other hand, the generation of acidic monomers (e.g., lactic acid) can decrease the local pH and result in auto-accelerated degradation of poly(α-hydroxy) esters (Cha and Pitt, 1989; Gopferich, 1996).

Hydrolytically Biodegradable Polymer Bonds

The type of chemical bond in the backbone of a biodegradable polymer plays a major role in dictating the rate of bond cleavage. A brief list of biodegradable polymer bonds is shown in Figure II.4.3.1. In general, anhydride and ortho-ester bonds are among the faster degrading polymer backbone chemical bonds, followed by ester and thioether bonds. The degradation rates of these chemical bonds, however, are largely affected by several factors, such as the presence of hydrophobic functional groups, the altered local pH due to the degradation products, molecular weights of the polymer, and the composition of the copolymer.

Polyanhydrides. The anhydride linkage is among the least hydrolytically stable, so polyanhydrides are typically synthesized from hydrophobic diacids (Rosen et al.,

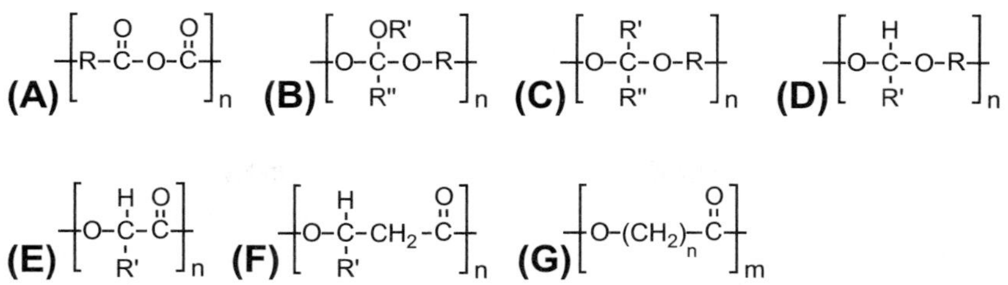

FIGURE II.4.3.1 Typical repeating units in common biodegradable polymers: (A) polyanhydrides; (B) poly(ortho esters); (C) polyketals; (D) polyacetals; (E) poly(α-hydroxyl esters); (F) poly(β-hydroxyl esters); (G) poly(ε-caprolactone) (an example of specific polymer with degradable ester linkages).

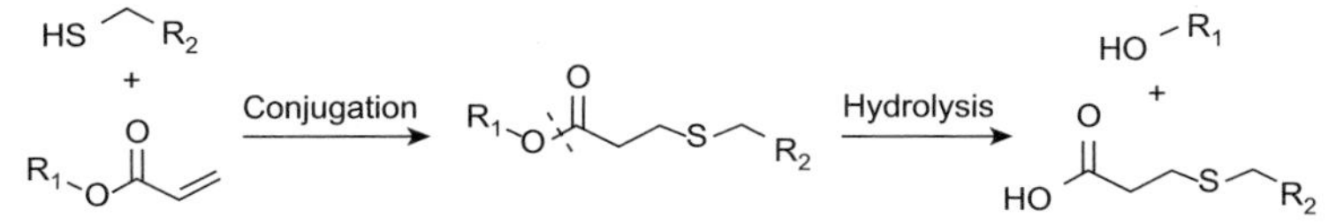

FIGURE II.4.3.2 Conjugation and hydrolysis of a thioether-ester bond. (Adapted from DuBose et al., 2005.)

1983). The hydrophobicity limits water penetration and the fast hydrolysis typically leads to surface erosion, producing di-acid monomers which can be metabolized by the body. Due to the fast degrading anhydride bonds (days to weeks) and the non-toxic acid degradation products, polymers made from polyanhydrides have been used for short-term drug release for more than two decades (Rosen et al., 1983; Leong et al., 1986) (see Chapter II.5.16.C). These unique properties of polyanhydrides result in near zero-order drug release profiles (Tamada and Langer, 1992; Tabata et al., 1993) (surface erosion and bulk degradation are discussed in more detail below).

Poly(ortho esters). Compared to polyanhydrides, poly (ortho esters) are more hydrophobic. They also undergo surface erosion when degrading, albeit with slower degradation rates compared to polyanhydrides (Heller et al., 2002). The degradation of poly(ortho esters) produces esters as initial degradation products, followed by further degradation into acids and alcohols. The use of poly(ortho esters) as controlled release devices can be dated back to the early 1970s (Heller and Barr, 2004). Since the polymer is surface-eroding, the drug release rates are controlled by the rate of polymer erosion, which can be tuned by the polymer composition.

Poly(α-hydroxy esters). Polymers in this category include poly(lactic acid) or PLA, poly(glycolic acid) or PGA, and their copolymers poly(lactic-co-glycolic acid) or PLGA. Poly(α-hydroxy esters) were among the first synthetic polymers with clinical applications (Gilding and Reed, 1979; Reed and Gilding, 1981). For example, PGA has long been used as biodegradable suture material due to its rapid degradation rate. The hydrolysis of poly(α-hydroxy esters) occurs at the ester bonds in the polymer backbone, and is governed by the accessibility of water to those bonds. L-PLA, for example, degrades more slowly than PGA, because of the presence of the hydrophobic methyl side group that hinders water attack to the ester bonds, as well as increasing crystallinity during degradation (Vert et al., 1994). The hydrolysis/degradation rates of amorphous PLGA are faster than either semicrystalline homopolymers of PLA or PGA, and they can be tuned by adjusting the compositions of the monomers (Gilding and Reed, 1979; Reed and Gilding, 1981; Zhu et al., 1991). It should be noted, however, that PLGA with a 50:50 monomer ratio exhibits the fastest degradation because it has the lowest crystallinity. Further, the acidic degradation products (i.e., lactic acid and glycolic acid) can lower the local pH and accelerate the hydrolytic degradation rate (Cha and Pitt, 1989).

Thioether-ester. An ester bond next to a thioether group is much more susceptible to hydrolytic degradation compared to an unmodified ester bond (Figure II.4.3.2) (DuBose et al., 2005; van de Wetering et al., 2005). Thioether-ester bonds can be formed from addition reactions between unsaturated acrylates and thiols (Elbert et al., 2001; van de Wetering et al., 2005). This so-called "Michael addition" reaction has been used to form degradable hydrogels for controlled release (Elbert et al., 2001) and tissue engineering applications (Elbert and Hubbell, 2001). The degradation rate of the thioether-ester bonds is tuned by altering the neighboring functional groups, such as charge (Jo et al., 2009) and the length of the sulfide containing linker (Schoenmakers et al., 2004).

Modes of Biodegradation

The mode of biodegradation often determines the therapeutic performance of polymeric devices. In general, two types of degradation have been described, including surface erosion and bulk degradation (Figure II.4.3.3) (Gopferich, 1996; Siepmann and Gopferich, 2001). Surface erosion mainly occurs in hydrophobic polymers, such as polyanhydrides (Tamada and Langer, 1992; Tabata et al., 1993) and poly(ortho esters) (Heller et al., 2002; Heller and Barr, 2004), in which the rate of bond cleavage is faster than the rate of water diffusion into the polymer device. Since the degradation process is primarily limited to the surface of the device, the bulk polymer

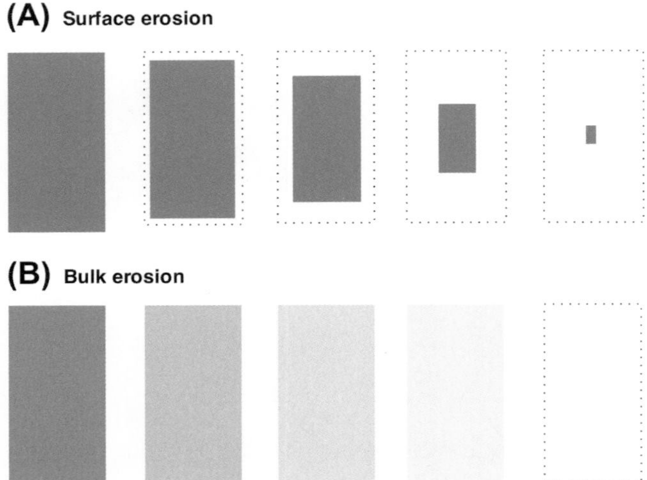

FIGURE II.4.3.3 Schematics of: (A) surface erosion; and (B) bulk degradation.

structure remains relatively unchanged until the polymer degrades completely (Figure II.4.3.3A). Surface erosion is useful in drug delivery applications, as the release rate of drugs from eroding polymers is independent of diffusion, and is easily altered by changes in the device geometry. Typically, surface eroding systems produce a facile means to achieve zero-order release (Tamada and Langer, 1992; Tabata et al., 1993).

In contrast to surface erosion, bulk degradation occurs when bonds are cleaved throughout the polymeric device (Figure II.4.3.3B). The direct consequence of bulk degradation is the change in molecular weight of the polymer, followed by increasing water content, the libration of soluble monomers and chain fragments, and dramatic changes in mechanical properties. The degradation of poly(α-hydroxy esters), such as PLA, PGA, and PLGA typically belong to this category (Pitt et al., 1981a,b). Since degradation takes place throughout the device, the mechanical properties of the device can decrease dramatically, even at early stages of the degradation process. For controlled release applications, the release rate of the loaded drug is governed by both diffusion and mass loss of the polymer. The prediction of drug release, therefore, requires sophisticated theoretical models (Siepmann and Gopferich, 2001).

In addition to hydrolytic degradation, polymers can also be designed to degrade enzymatically. Since bond cleavage in such polymers is mediated by enzymatic reactions, the mode of degradation is determined by the location and concentration of the enzyme, as well as the kinetics of the enzymatic reaction. Enzymatic degradation, therefore, can be either surface erosion or bulk degradation, and is mainly determined by the accessibility of the enzyme to the interior of the polymer. For example, the degradation of poly(ε-caprolactone) or PCL is greatly accelerated by the presence of lipases, and polymers containing PCL components have been reported to degrade via surface erosion or bulk degradation, depending on the formulation of the polymer. PCL has been fabricated as polymer blends (Tsuji and Ikada, 1998; Tsuji et al., 1998; Broz et al., 2003), block co-polymers (Allen et al., 2000; Qian et al., 2000; Ahmed and Discher, 2004), and hydrogels (Rice et al., 2006; Rice and Anseth, 2007) for drug delivery and tissue engineering applications.

Polymers containing natural enzyme substrates, typically oligopeptides, can be degraded via cell-secreted enzymes. Such hybrid synthetic–biologic macromolecules are becoming increasingly important in the field of tissue engineering. For example, matrix metalloproteinases (MMPs) are secreted by a variety of cells that trigger the degradation of extracellular matrix (ECM) proteins (Stamenkovic, 2003; Mott and Werb, 2004). This type of enzymatic degradation is extremely important in cell migration, proliferation, angiogenesis, and differentiation in native tissues. Inspired by this natural biological phenomenon, researchers have developed numerous synthetic polymers containing MMP-degradable peptides that are degraded by cell-secreted MMPs (West and Hubbell, 1999; Lutolf et al., 2003; Seliktar et al., 2004; Fairbanks et al., 2009). This mode of degradation is largely determined by the location and concentration of enzymes (which is related to number and proximity of cells), the enzymatic kinetics, and the chemistry and concentration of the substrate. Another approach to synthetic polymers that can be enzymatically degraded involves the incorporation of disulfide (S-S) bonds in the backbone chain. These polymers have been developed for gene delivery and intracellular drug delivery where, upon cellular uptake, S-S bonds are degraded in the intracellular reducing environment by glutathione and thioredoxine reductases.

Role of Water in Biodegradation: Hydrophobic Polymers versus Hydrogels

Regardless of the mode of hydrolytic biodegradation (i.e., surface erosion or bulk degradation), the presence of water is a prerequisite for degradation to occur. Thus, the hydrophilicity of the polymer chains and the transport of water into the device play important roles in determining the rate of biodegradation. The rate of degradation is often governed by the accessibility of the chemical bonds to water, rather than their inherent degradability (Gopferich, 1996). Thus, degradable bonds with hydrophobic side groups will degrade much more slowly than the same chemical bonds with hydrophilic side groups (e.g., hydrophobic PLA versus more hydrophilic PGA). Hydrophobic degradable polymers include polyesters, polyanhydrides, polyphosphazenes, and poly(ortho esters), while hydrophilic degradable polymers often contain low molecular weight segments of poly(ethylene glycol) (PEG), poly(vinyl alcohol) (PVA), poly(acrylic acid) (PAA), poly(hydroxyethyl methacrylate) (pHEMA), and/or poly(N-isopropyl acrylamide) (pNIPAM), although these polymers are not, by themselves, biodegradable. Note that pNIPAM is a temperature-responsive polymer that is hydrophilic below its lower critical solution temperature or LCST which becomes hydrophobic above its LCST (see Chapter I.2.11).

As has been mentioned, the mode of biodegradation of hydrophobic polymers can be either surface erosion or bulk-degradation, depending on the polymer backbone (Gopferich, 1996; Siepmann and Gopferich, 2001). For example, polyanhydrides usually degrade via a surface erosion mechanism, as the penetration of water into the interior of the polymer is inhibited by the highly hydrophobic polymer. On the other hand, the degradation of polyesters, such as PLA, PGA, and co-polymer PLGA, proceed through bulk-degradation due to the presence of water molecules throughout the polymer. In contrast to hydrophobic polymers where the molarity of water within the polymer is usually much lower than 55.36 M (molarity of pure water at room temperature), hydrophilic polymers usually contain water at much higher concentrations. For example, the concentration of water in hydrogels is usually at a molarity very close to that of pure water, due to the small weight percentage of the polymer in the entire device. As a result, in computational models of hydrophilic polymer hydrolytic degradation, the water concentration is often viewed as constant and combined with the rate constant of polymer bond cleavage.

KINETICS OF BIODEGRADATION

As briefly discussed above, the type of chemical bond determines, to a great extent, the rate of biodegradation. Generally, anhydride and orthoester bonds have higher hydrolytic bond cleavage rates than ester bonds. Other considerations include, but are not limited to: hydrophobicity of the monomer; steric hindrance of the side group; co-polymer compositions; glass transition temperature (T_g); crystallinity of the polymer; and pH of the local environment (Gopferich, 1996).

Hydrolytic Degradation

Surface Erosion. The kinetics of surface-eroding polymers has been well-studied. The hydrolysis of surface-eroding polymers starts at the surface and proceeds gradually to the interior (Figure II.4.3.3A). If a molecule is entrapped in the polymer, the release rates are often correlated directly to the mass loss. To facilitate experimental observation, surface-eroding polymers, such as polyanhydrides, are often fabricated as disks with high aspect ratios (e.g., large diameter compared to the height of the polymer), such that the surface area of the polymer remains roughly constant throughout degradation. Thus, the rate of polymer erosion can be expressed as:

$$\frac{dM}{dt} = -k \tag{1}$$

where M is the polymer mass at any given time t, and k is the kinetic rate constant of the degradation. Assuming one-dimensional erosion and constant density, Eq. (1) can be rearranged into:

$$\frac{dl}{dt} = -k' \tag{2}$$

where l is the dimension of the polymer in the direction of degradation front at any given time t, and k' is the adjusted rate constant.

Equation (2) can be easily solved to obtain:

$$l = l_0 - k't \tag{3}$$

Equation (3) illustrates a linear decrease in polymer thickness as the polymer degrades. This linear relationship is very useful in controlled release applications, as drug release rates can be stringently controlled by the rate of polymer mass loss and approach zero-order release. Equations (1) to (3) dictate the kinetics of surface erosion when the device is fabricated as thin slabs. For other geometries, such as cylinders and spheres, similar equations can be derived accordingly, and once k' is known, multidimensional degradation profiles can be predicted as well.

Bulk Degradation. For hydrolytic bulk degradation, a majority of studies focus on the mechanism of polyester degradation, particularly PLA and its co-polymers with PGA. As an example, the mechanism of polyester biodegradation is discussed in this section. Hydrolysis of polyester is usually described as a random polymer chain scission throughout the polymer, and is auto-accelerated by increasing carboxylic acid end groups. The hydrolysis of ester bonds to produce carboxylic acids can be expressed as:

$$-\frac{d[Ester]}{dt} = k[Ester][H_2O] \tag{4}$$

Here, $[Ester]$ is the concentration of ester bonds in the polymer; t is the degradation time; k is the kinetic rate constant; and $[H_2O]$ is the concentration of water. As a simplifying assumption for bulk degrading polyesters, the water concentration is often considered constant. Then, Eq. (4) simplifies to:

$$-\frac{d[Ester]}{dt} = k'[Ester] \tag{5}$$

where k' is a pseudo-first order rate constant. Integrating Eq. (5) yields the following equation:

$$\frac{[Ester]}{[Ester]_0} = e^{-k't} \tag{6}$$

where $[Ester]_0$ is the initial ester bond concentration.

Since $[Ester]$ is related to the molecular weight (M_n) of the polymer at any time during degradation, Eq. (6) can be expressed as:

$$\frac{M_n}{M_{n0}} = e^{-k't} \tag{7}$$

It can be seen from Eq. (7) that the molecular weight of the degrading polymer decreases exponentially. It should be noted that Eq. (7) only holds until significant degradation causes the loss of oligomers and carboxylic acids. It has been observed that this relationship is approximately true until the molecular weight of the polymer decreases to below 5000 Da (Pitt et al., 1981b). Substitution of the Mark–Houwink equation ($[\eta] = KMn^{\alpha}$) into Eq. (7) yields:

$$[\eta] = [\eta]_0 \cdot e^{-\alpha k' t} \tag{8}$$

Here, η is the intrinsic viscosity of the polymer, and α is the Mark–Houwink exponent, which depends on the solvent in which the polymer degrades. Equation (8) facilitates the observation of polymer degradation via measuring the change in its intrinsic viscosity.

Equations (4) to (8) only illustrate the basic kinetics of ester bond hydrolysis in bulk-degrading polymers. The prediction of mass loss kinetics in the bulk-degrading polymers, however, often requires more sophisticated mathematical models, as simple cleavage of bonds does not accurately reflect loss of mass from the polymer device. For example, not only monomers, but also oligomers with different molecular weights can be soluble and liberated, leading to difficulties in predicting mass loss from bulk-degrading polymers.

Bulk Degradation in Hydrogels. The degradation of block co-polymer hydrogel networks differs significantly from the degradation of many of the common linear polyesters (Metters et al., 2000a,b, 2001; Martens et al., 2001). For example, the degradation kinetics of networks formed from PLA-*b*-PEG-*b*-PLA precursors (polymerized through acrylate end groups) differs substantially from that of pure PLA degradation. The highly swollen hydrogel network, together with the use of buffered solution, facilitates the removal of relatively low concentrations of acidic degradation products. As a result, the kinetic equation for ester bond cleavage is identical to Eq. (5). For pure polyesters, a significant change in polymer properties occurs at relatively low levels of ester hydrolysis, as a small number of cleavage events leads to a dramatic change in polymer molecular weight, even though the mass loss can be quite small, as significant liberation of monomers and oligomers often occurs at much later stages of hydrolysis. Likewise, for PLA-*b*-PEG-*b*-PLA hydrogels, the cleavage of a single ester bond within a PLA block of the co-polymer is sufficient to cause microscopic gel structural change. The [*Ester*] in Eq. (5) can be replaced by the [*PLA*], the concentration of PLA blocks in the co-polymer, because hydrolyzing one ester bond per PLA block is sufficient to cause the detachment of the degradable block from the cross-linked network. Equation (5) becomes:

$$-\frac{d[PLA]}{dt} = k'[PLA] \tag{9}$$

Integrating and solving for the [*PLA*] in Eq. (9) gives an exponentially decaying PLA concentration as:

$$[PLA] = [PLA]_0 \cdot e^{-k't} \tag{10}$$

Combining Eq. (10) and the structural information about the PEG-*b*-PLA hydrogel, the cleavage of PLA units can be related to mass loss from the hydrogel. Metters, Bowman, and Anseth have devised statistical kinetic models to predict the degradation behavior of such hydrogels. The following equation describes the probability (P) that any random PLA unit has been hydrolyzed (Metters et al., 2000a,b, 2001):

$$P = 1 - f_{PLA} = 1 - e^{-k't} \tag{11}$$

where f_{PLA} is the fraction of total hydrolyzed PLA units, which is equal to $[PLA]/[PLA]_0$.

As illustrated in Figure II.4.3.4, each PEG chain is attached to the cross-linked network by two PLA units. At any time during gel degradation, each PEG is in one of the three states: (1) completely attached; (2) only one PLA unit hydrolyzed; and (3) both PLA units hydrolyzed. Given the probability of PLA units being hydrolyzed at any time during degradation (P), the fraction (y) of each of the three states can be described as (Metters et al., 2000a,b, 2001):

(1) Fraction with both PLA units intact:

$$y_1 = (1 - P)^2 \tag{12}$$

(2) Fraction with only one PLA unit hydrolyzed:

$$y_2 = 2P(1 - P) \tag{13}$$

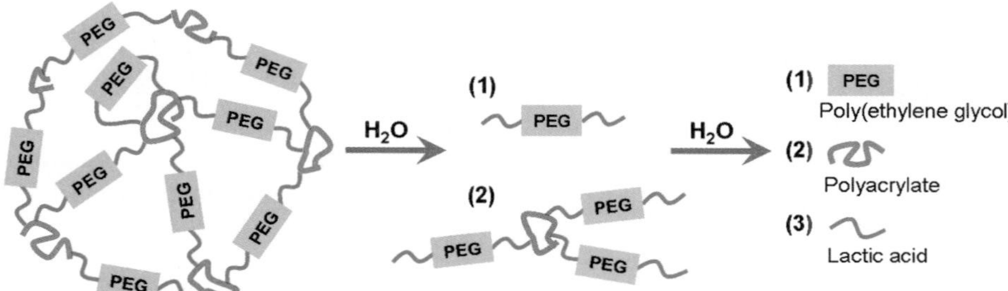

FIGURE II.4.3.4 Schematic of a PLA-*b*-PEG-*b*-PLA hydrogel and its hydrolytic degradation process and degradation products.

(3) Fraction with both PLA units hydrolyzed:

$$y_3 = P^2 \tag{14}$$

The mass loss of PEG-*b*-PLA hydrogels involves the loss of PLA-PEG-PLA block co-polymers and/or polyacrylate kinetic chains. Since multiple PLA units are attached to a single polyacrylate kinetic chain, at least one PLA unit along each cross-link must be hydrolyzed in order for a polyacrylate chain to be released from the degradation gel. The fraction of polyacrylate (F_{PA}) chains to be released from the degradation gel can be described as:

$$F_{PA} = (y_2 + y_3)^N \tag{15}$$

where N is the number of cross-links initially attached to each polyacrylate kinetic chain. Substituting Eqs. (13) and (14) into Eq. (15) yields:

$$F_{PA} = [1 - (1 - P)^2]^N \tag{16}$$

To liberate PLA-PEG-PLA blocks, one of the following two situations has to occur: (1) two PLA units, one at each end of the block, need to be hydrolyzed; or (2) one PLA unit is hydrolyzed while the other is attached to an unhydrolyzed PLA unit, which is part of a releaseable polyacrylate chain. Given the probability of PLA unit hydrolysis (P) and the fraction of releaseable polyacrylate chain from the degrading network (F_{PA}), the fraction of PEG units being released from the network is:

$$F_{PEG} = P^2 + F_{PA}\frac{y_2}{2} \tag{17}$$

Equations (16) and (17) give the necessary information about the mass loss from PEG-*b*-PLA hydrogels:

$$\% \text{ mass loss} = (W_{PA}F_{PA} + W_{PEG}F_{PEG}) \tag{18}$$

Here, W_{PA} and W_{PEG} are the mass percent of the cross-linked network contained in the polyacrylate chains and in the PLA-PEG-PLA segments, respectively.

It should be noted that while the network degradation can be modeled using Eqs (12) to (18), the mass loss behavior will deviate from the prediction once sufficient cross-links are hydrolyzed, causing the entire gel network to be completely water soluble (Metters et al., 2000a,b). To account for this, a "reverse gelation" concept can be implemented.

$$X(t) = [1 - P(t)]^2 \tag{19}$$

At any given time during degradation, $X(t)$ is the fraction of cross-links that remain unhydrolyzed, and $P(t)$ is the fraction of hydrolyzed PLA units. Reverse gelation occurs when $X(t)$ is equal to or less than the gel point conversion (X_c), which is defined as:

$$X_c = \frac{2}{f_{avg}} \tag{20}$$

where f_{avg} is the average functionality of the monomer. For reverse gelation of a cross-linked hydrogel system that degrades via a bulk-degradation mechanism, f_{avg} equals the number of cross-links (N) attached to each kinetic backbone chain. Combining Eqs. (19) and (20) and the information of f_{avg} (= N) yields the critical degradation conversion at which reverse gelation occurs (P_c):

$$P_c = 1 - \left[\frac{2}{N}\right]^{1/2} \tag{21}$$

From the above equations, time (t_c) at which reverse gelation occurs is:

$$t_c = \frac{\ln(1 - P_c)}{-k'} = \frac{\ln\left(1 - \left[1 - \frac{2}{N}\right]^{1/2}\right)}{-k'} \tag{22}$$

where N is the weight average number of cross-links per kinetic chain, and k' is the pseudo first-order hydrolysis rate constant. Figure II.4.3.5 shows a theoretical prediction of mass loss profiles of PLA-PEG-PLA hydrogels without (curve A), and with (curve B), reverse gelation (Metters et al., 2000b).

Enzymatic Degradation

Enzymatic Degradation of Synthetic Polymers. Several synthetic polymers are susceptible to enzymatic degradation. For example, amorphous poly(L-lactide) can be degraded by proteinase K, while both amorphous and crystalline poly(ε-caprolactone) can be degraded by lipases of various origins (Liu et al., 2000). Enzymatic degradation of synthetic polymers can be either surface erosion or bulk degradation, depending on the location and stability of the acting enzyme. However, due to the limited water accessibility of most hydrophobic polymers, surface erosion is the dominant degradation mechanism and the kinetics are similar to the hydrolytic surface erosion described in an earlier section of this

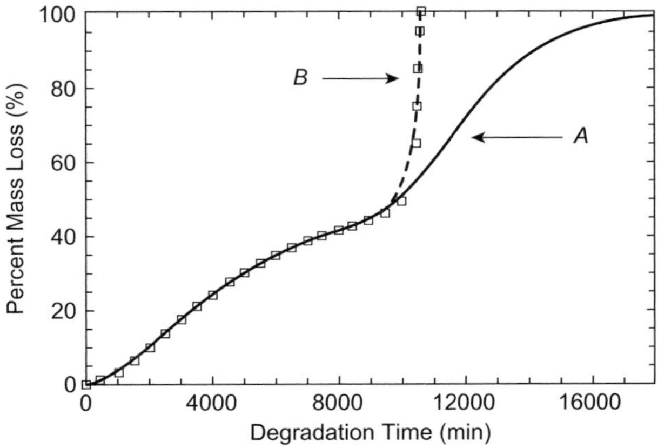

FIGURE II.4.3.5 Prediction of mass loss of a PLA-*b*-PEG-*b*-PLA hydrogel versus degradation time. Curve A: without reverse gelation; Curve B: with reverse gelation. Model parameters: W_{PA} = W_{PEG} = 50 wt %; N = 1000 cross-links per backbone chain; and k' = 0.0003 min^{-1}. (From Metters et al. (2000b). *J. Phys. Chem. B.*)

chapter. For hydrophilic polymers permitting the diffusion of the enzyme to the interior of the polymer, the degradation is also mediated by surface erosion mechanism if the rate of enzymatic polymer bond cleavage is faster than the rate of enzyme diffusion.

Bulk degradation of synthetic polymers induced by enzymatic activity may occur under two conditions: (1) the enzyme is able to infiltrate and distribute uniformly throughout the bulk of the polymer; and (2) the rate of enzymatic bond cleavage is slower than the diffusion of the enzyme. Typically, hydrogels containing enzymatic cleavable units in their polymer backbone fall into this category of degradation. Under these assumptions, Michaelis–Menten enzymatic kinetics are commonly used to predict the enzymatic degradation rates of synthetic polymers via bulk degradation mechanism:

$$v_0 = v_{max}\frac{[S]}{K_M + [S]} \tag{23}$$

Here, v_0 and v_{max} are the initial and maximum reaction rate of degradation, respectively. $[S]$ is the degradable polymeric substrate concentration, and K_M is the Michaelis–Menten constant.

In Eq. (23), v_{max} can also be expressed as:

$$v_{max} = k_{cat}[E] \tag{24}$$

where k_{cat} is the catalytic constant describing the rate of enzymatic degradation of polymer bonds and $[E]$ is the concentration of enzyme catalytic sites.

As in other enzymatic reactions, both k_{cat} and K_M are important parameters in characterizing the enzymatic degradation of polymers. While k_{cat} represents the sensitivity of an enzyme for a specific polymeric substrate, K_M is the substrate concentration needed to achieve a half-maximum enzyme velocity. Factors affecting these parameters will also determine the rate of polymer degradation.

While simple Michaelis–Menten kinetics can be used to describe enzymatic polymer bond cleavage, they do not provide information regarding the mass loss behavior of the polymers. Sophisticated mathematical models are often required in order to correlate microscopic enzymatic bond cleavage to macroscopic polymer mass loss. For example, to describe the mass loss of cross-linked hydrogels containing enzymatic cleavable substrate (e.g., polycaprolactone), a statistical model (similar to the one described earlier in the hydrolytic degradation of hydrogels) integrating the structural information of the hydrogels was developed (Rice et al., 2006). In this particular example, Eqs. (11) to (14) can be used again except that f_{PLA} is now replaced with f_{CAP}, where CAP represents caprolactone subunits. In addition to the network information, the Michaelis–Menten reaction equations and the first-order decay in active enzyme (e.g., lipase) concentration can be described by the following equations:

$$Lipase + CAP \rightleftharpoons E*S \xrightarrow{k_2} Lipase + D \tag{25}$$

$$Lipase + CAP \xrightarrow{k_d} Inactive\ Lipase \tag{26}$$

$$\frac{[Lipase]}{[Lipase]_0} = e^{-k_d t} \tag{27}$$

Here, $E*S$ is the enzyme–substrate complex, D is the degradation product, CAP is the caprolactone block, k_2 is CAP degradation late constant and k_d is the first order rate constant for deactivation of the lipase.

Time-derivatives of lipase and CAP concentrations can be expressed as (Rice et al., 2006):

$$\frac{d[Lipase]}{dt} = -k_1[CAP][Lipase] + k_{-1}[E*S] + k_2[E*S] - k_d[Lipase] \tag{28}$$

$$\frac{d[CAP]}{dt} = -k_1[CAP][Lipase] + k_{-1}[E*S] \tag{29}$$

where k_1 and k_{-1} are the forward and backward enzymatic reaction rate constants for lipase and CAP units, respectively. Solving equations (27) to (29) yields an equation related to the fraction of CAP unit degradation (Rice et al., 2006):

$$\frac{N_{CAP}}{N_{CAP_0}} = exp\left[\frac{k*[Lipase]_0}{k_d}(e^{-k_d t} - 1)\right] \tag{30}$$

Figure II.4.3.6 illustrates the mass loss profiles as a function of time by solving Eq. (30). It can be seen that the rates of PEG-PCL hydrogel mass loss are functions of several factors, including kinetic parameters (k^*) and the half-lives of the acting enzyme. One must note that the assumption of bulk degradation will not hold at high enzyme concentrations (i.e., high reaction rates)

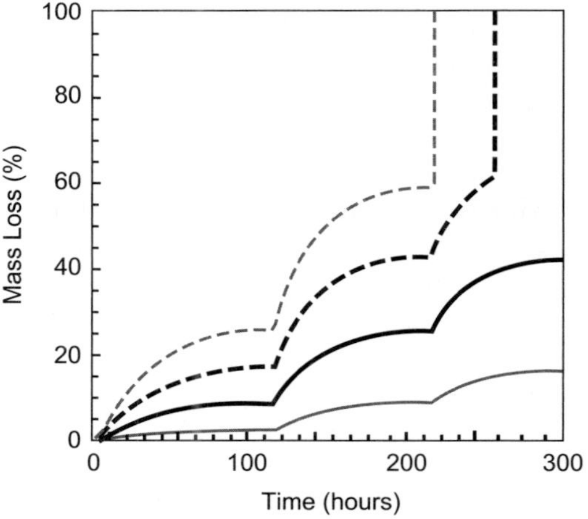

FIGURE II.4.3.6 Calculated mass loss profiles for varying values of k^* or lipase concentration. Baseline values are shown in solid black line, with $k^* = 31.4\ L{\cdot}mol^{-1}{\cdot}min^{-1}$, lipase concentration = 0.2 mg/mL, and active lipase half-life = 20.4 h (corresponding to $k_d = 5.66 \times 10^{-4}\ min^{-1}$). Baseline values for each variable were multiplied by 0.5 (solid gray line), 1.5 (dashed black line), and 2.0 (dashed gray line). $n_0 = 40$ in all calculated profiles. Enzyme was refreshed at 116 hr and 216 hr. (From Rice et al. (2006). *Biomacromolecules*.)

or in thick gels where the diffusion timescale is large. Under these conditions significant surface erosion will occur, which complicates the interpretation of modeling results.

Enzymatic Degradation of Peptide Substrates in Hydrogels. Recently, hydrogels composed of synthetic polymer backbones and enzymatically degradable peptide cross-linkers have been synthesized and studied extensively. These degradable hydrogels are of great importance in controlled release, tissue engineering, and regenerative medicine applications. For example, Hubbell and colleagues pioneered the development of enzymatic degradable hydrogels formed by Michael-type conjugation reactions (West and Hubbell, 1999; Elbert and Hubbell, 2001; Elbert et al., 2001; Lutolf et al., 2003) (also see Chapter I.2.17). The step-growth polymerization mechanism yields cross-linked gel with a regular structure (Figure II.4.3.7), compared to hydrogels formed by chain-growth polymerizations as shown in Figure II.4.3.4. These Michael-type hydrogels are commonly synthesized from the reaction of multi-arm macromers (e.g., PEG) with di-functional peptide linkers and the final networks can be degraded by enzymes (e.g., matrix metalloproteinases). An important application of this type of gel is to encapsulate cells that secrete MMPs to degrade the synthetic matrices through enzymatic action, a degradation process that proceeds only in the presence of encapsulated cells. The cleavage of peptide cross-linkers leads to a decrease in the local gel cross-linking density and the erosion of the polymer chains only in the vicinity of the cells.

While the degradation kinetics of these gels can also be described by the Michaelis–Menten reaction equations, it should be noted that the kinetic parameters (k_{cat} and K_M) in cross-linked hydrogel systems may differ substantially from those in systems with soluble peptide substrates. Firstly, the presence of PEG chains on both sides of the peptide substrate may create steric hindrance for the enzyme, hence increasing K_M and decreasing the rate of degradation (Lutolf et al., 2003). Further, it was found that k_{cat} is higher when peptide substrates are cross-linked within hydrogels, compared to their soluble counterparts (Lutolf et al., 2003).

INFLUENCE OF BIODEGRADATION ON PROPERTIES OF BIODEGRADABLE POLYMERS AND THEIR APPLICATION

Drug Delivery: Influence of Biodegradation in Diffusion

The benefits of biodegradable polymers, including the versatility of degradation mechanisms and tunable degradation rates, have led to the development of numerous drug delivery and controlled release formulations (Amsden, 1998; Peppas et al., 2000; Anseth et al., 2002; Lin and Metters, 2006; Lin and Anseth, 2009) (Chapter II.5.16). Similar to the mechanisms of biodegradation described earlier, the mode of drug release from biodegradable polymers, in general, can also be categorized into the following mechanisms: surface erosion mediated release; bulk-degradation mediated release; and pendent chain cleavage mediated release.

Drug Release Mediated by Surface Erosion. The release of therapeutics from biodegradable polymers is mediated by surface erosion when: (1) the polymers are degraded via a surface erosion mechanism; and (2) diffusion of the therapeutics is not significant in the release process until degradation occurs. As described earlier, hydrophobic polymers (e.g., polyanhydrides, PCL) with limited water infiltration are typically used as base formulations for release mechanism mediated by surface erosion (Gopferich, 1996). An important characteristic of surface erosion mediated release is that the release profiles correlate well with the mass loss profiles of the degrading polymers. This characteristic is particularly attractive in clinical applications, as the delivery of many drugs requires a controllable and sustained manner. Since the release is mediated by surface erosion, the prediction of the drug release rates is relatively easy, and almost identical to that used to predict polymer erosion. The following empirical equation derived by Hopfenberg and colleagues can be used, and is valid for surface-eroding polymers in slabs, cylindrical, and spherical geometries (Hopfenberg, 1976; Siepmann and Gopferich, 2001):

$$\frac{M_t}{M_\infty} = 1 - \left(1 - \frac{k_0 \cdot t}{c_0 \cdot a}\right)^n \qquad (31)$$

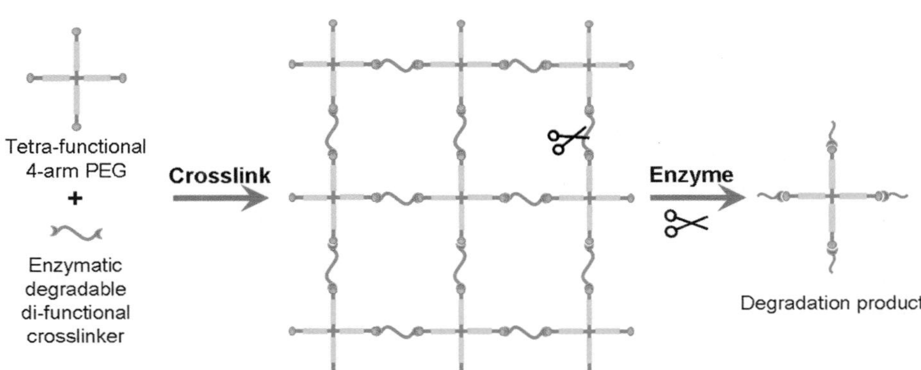

FIGURE II.4.3.7 Schematic of cross-linking and enzymatic degradation of hydrogels formed via conjugation reactions.

Here, M_t and M_∞ are the cumulative amounts of drug release at any time and infinite time, respectively; k_0 is the rate constant of polymer erosion; c_0 is the initial drug loading concentration; and a is the half-thickness of a slab or the radius of a cylinder or a sphere. Finally, n is defined as a "shape factor" and is 1, 2 or 3, for a slab, cylindrical or spherical geometry, respectively. It is worth noting that this empirical model assumes that all mass transfer events involved in drug release are considered together in a zero-order process confined to the surface of the device, and that it neglects all edge and end effects. Clearly, when the polymers are in a slab geometry ($n = 1$), the resulting drug release rate is independent of time. On the other hand, the release rates of drugs in polymers with cylindrical or spherical geometry decrease with time, as the surface area of the cylindrical or spherical device decreases with increasing degradation time.

Drug Release Mediated by Bulk Degradation. Compared to surface erosion mediated release, the release of therapeutics from bulk-degrading polymers is more complicated as, in addition to polymer degradation, diffusion of the therapeutics plays an important role and cannot be neglected. Siepmann and Gopferich review the mathematical models, both non-Monte Carlo and Monte Carlo models, that are often used to predict drug release from degradable polymers (Siepmann and Gopferich, 2001). For example, in a non-Monte Carlo approach, the diffusion of any species involved in a poly(ortho ester)-based delivery system that contains an acid-generating species to accelerate hydrolysis can be expressed as (Thombre and Himmelstein, 1985; Joshi and Himmelstein, 1991):

$$\frac{\partial C_i}{\partial t} = \frac{\partial}{\partial x}\left(D_i(x,t) \cdot \frac{\partial C_i}{\partial x}\right) + \upsilon_i \tag{32}$$

where i represents any of the multiple components involved in the degradation process, such as water, acidic degradation product, and drug. C_i and D_i are the concentration and diffusion coefficient of the diffusing component, respectively, x is the space variable, and v_i is the sum of the production and degradation rate of species i. During the degradation process, the permeability of the system typically increases, leading to diffusion coefficient (D_i) that varies with the polymer concentration, C_D:

$$D_i = D_i^0 \cdot exp\left[\frac{\mu(C_D^0 - C_D)}{C_D^0}\right] \tag{33}$$

Here, D_i^0 is the initial diffusion coefficient of species i prior to polymer degradation, C_D^0 is the initial concentration of polymer, and μ is a constant. These partial differential equations can be solved numerically.

Degradable hydrogels are another important area of research where significant efforts have been dedicated to both theoretical modeling and experimental verification. The understanding of network structural evolution during degradation is critical in developing mathematical models for predicting drug release profiles from hydrogel networks. For example, given an experimentally determined volumetric swelling ratio (Q) of a degrading hydrogel, one can easily calculate the average molecular weight between cross-link ($\overline{M}_c$) and the volume fraction of the polymer ($\upsilon_{2,s}$). Other parameters important for drug release from these degrading hydrogels can also be derived. An important one is the mesh size of a hydrogel network (ξ) that has been described by Canal and Peppas (Canal and Peppas, 1989; Peppas et al., 2000):

$$\xi = \upsilon_{2,s}^{-1/3}\left(\overline{r_0^{\,2}}\right)^{1/2} = Q^{1/3}\left(\overline{r_0^{\,2}}\right)^{1/2} \tag{34}$$

where $\left(\overline{r_0^{\,2}}\right)^{1/2}$ is the root mean squared end-to-end distance of the polymer chains in the unperturbed state.

Under the assumption that the degradation of the polymer bond follows pseudo-first order kinetics, both Q and $\overline{M}_c$ increase exponentially during hydrogel degradation. Consequently, the degrading hydrogel network mesh size will also increase exponentially (Mason et al., 2001):

$$\xi = Q^{1/3}\left(\overline{r_0^{\,2}}\right)^{1/2} \sim \overline{M}_c^{7/10} = \left[\overline{M}_c\,|_{t=0}e^{2jk_E' t}\right]^{7/10} = \eta e^{(7/5)jk_E' t} \tag{35}$$

where η is a constant. For a non-degradable hydrogel, solute diffusivity can be estimated with a model developed by Lustig and Peppas (Lustig and Peppas, 1988):

$$\frac{D_g}{D_0} = \left(1 - \frac{r_s}{\xi}\right) exp\left(-Y\left(\frac{\upsilon_{2,s}}{1-\upsilon_{2,s}}\right)\right) \tag{36}$$

where D_g is the solute diffusivity in the swollen hydrogel, D_0 is the solute diffusivity in water, r_s is the size of the solute, and Y is the ratio of the critical volume required for a successful translational movement of the solute and the average free volume per molecule of solvent. A good approximation for Y for PEG hydrogels is unity.

Clearly, D_g in the above equation is a constant in a non-degrading network. However, for a degradable hydrogel the increasing mesh size (ξ) and decreasing polymer volume fraction ($\upsilon_{2,s}$) with network degradation alter the diffusion coefficient (D_g) of a solute in the swollen and degrading hydrogel network. Due to the highly swollen network structure where $\upsilon_{2,s}$ is relatively small (<0.1), Eq. (36) can be simplified into (Mason et al., 2001):

$$\frac{D_g}{D_0} = \left(1 - \frac{r_s}{\xi}\right) exp\left(1 - \frac{r_s}{\eta e^{(7/5)jk_E' t}}\right) \tag{37}$$

From the above equation, one can see that as the network degrades, solute diffusivity increases and approaches D_0.

Drug Release Mediated by Pendent Chain Cleavage. The release of a drug conjugated to a polymer network through a cleavable linker (i.e., pro-drug approach) can be modeled by accounting for both the kinetics of linker cleavage and drug diffusion. In general,

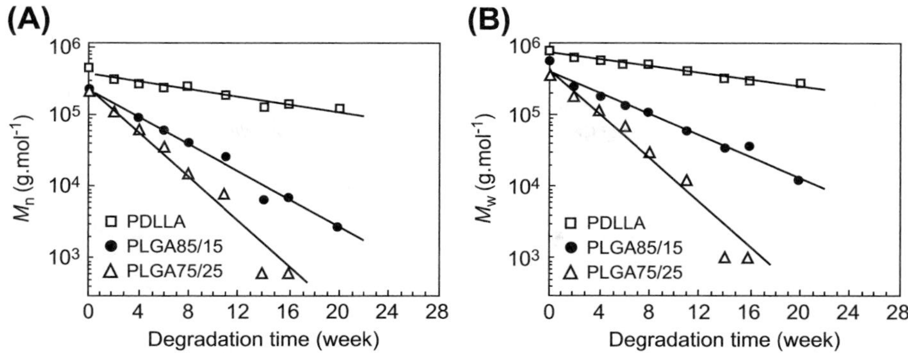

FIGURE II.4.3.8 Changes in the: (A) number (M_n) and (B) average molecular weight (M_w) of polymers composing porous scaffolds with different initial copolymer compositions degraded in PBS solution at 37°C. (From Wu & Ding, 2004.)

the following Fickian diffusion equation can be used to describe one-dimensional diffusion of a drug from a hydrogel network:

$$\frac{\partial [P](z,t)}{\partial t} = D_P \frac{\partial^2 [P](z,t)}{\partial z^2} \tag{38}$$

where $[P]$ is the concentration of the cleaved drug and D_P is its diffusivity, t is the degradation time, and z is the space variable. The rate of product formation $\partial P/\partial t$ can be assumed to be identical to that of the bond cleavage. If the linker is hydrolytically degradable, pseudo-first order degradation kinetics can be incorporated into the above governing equation for highly swollen gels. On the other hand, if the linker is enzymatically degradable, Michaelis–Menton kinetics can be employed.

Tissue Engineering: Influence of Biodegradation on Scaffold Mechanics

Tissue engineering (see Section II.6), as stated by Langer and Vacanti, is: "an interdisciplinary field that applies the principles of engineering and the life sciences toward the development of biological substitutes that restore, maintain or improve tissue functions" (Langer and Vacanti, 1993). Since the conception of tissue engineering, many synthetic biodegradable polymers have lent themselves in the development of "biological substitutes" or "scaffolds" for many biomedical applications, including cartilage and bone regeneration, angiogenesis, nerve regeneration, cardiovascular reconstruction, etc. While in many cases the use of non-degradable scaffolds is favorable (e.g., encapsulation and transplantation of pancreatic islet), degradable polymers are required in numerous applications where the scaffold only serves as a temporary structural support, and degrades gradually as the new tissue is formed. From this perspective, the mechanisms and kinetics of polymer degradation influence greatly the process of tissue regeneration, including cell proliferation and migration.

The most direct effect of polymeric scaffold degradation is the decrease in both the polymer molecular weight and the scaffold mechanical properties, such as storage and loss moduli. For example, the number (M_n) and average molecular weight (M_w) of bulk-degrading polymers (e.g., PLGA) decrease exponentially with degradation time, and can be expressed as (Wu and Ding, 2004):

$$\log M = \log M_0 - kt \tag{39}$$

Here, M_0 and M are the molecular weights of the degrading polymer prior to and during degradation, respectively; k is the apparent degradation rate constant that can be determined experimentally, and t is the degradation time. Figure II.4.3.8 illustrates the relationship between polymer molecular weight and degradation time (Wu and Ding, 2004).

While these average molecular weights of the degrading polymers are found to decrease monotonically with degradation time, the changes in the moduli of the bulk-degrading polymers are more profound. As observed experimentally, the polymer modulus will increase at early stages of degradation, followed by a plateau phase and finally a dramatic decrease (Wu and Ding, 2004). This behavior is believed to be a result of polymer shrinkage that occurs during the beginning of degradation, which leads to the initial increase, but the changes in properties that are observed during the late stages of degradation are a direct result of the significant decrease in the molecular weight of the polymer molecular chains. In many tissue engineering strategies, biodegradable polymers are processed into highly porous structures to allow cell infiltration. Theoretically, the moduli of degrading polymers can be correlated to their overall porosity (Hou et al., 2003; Zhang et al., 2005; Baker et al., 2009):

$$E_t = E_0(1-P)^m \tag{40}$$

$$P = \left(1 - \frac{\rho}{\rho_0}\right) \cdot 100 \tag{41}$$

where E_t and E_0 are the moduli of the polymer at given time and prior to degradation, respectively. P is the porosity of the degrading polymer, ϱ is the density of the polymer, and m is a characteristic constant for the specific polymer ($1<m<3$, e.g., $m = 2.37$ for PLGA (Zhang et al., 2005) and 2.59 for PCL (Hou et al., 2003)). Since the porosity (void volume) of the degrading polymer can be easily measured *in vitro*, the change in the scaffold moduli can also be estimated using these predictions.

In other tissue engineering approaches, biodegradable hydrogels are often used to encapsulate cells and direct tissue regeneration (e.g., encapsulating chondrocytes in hydrogels for cartilage regeneration). For highly swollen hydrogels, the relationships between scaffold degradation and mechanical properties differ significantly from that of hydrophobic porous polymers. For example, the mechanical properties of PEG hydrogels are typically described by rubber elasticity theory and, under certain circumstances (e.g., extremely low temperature), the theory of viscoelasticity (Flory, 1953; Treloar, 1975; Aklonis and MacKnight, 1983; Anseth et al., 1996). In particular, hydrogels in their swollen state usually satisfy the theory of rubber elasticity (Anseth et al., 1996):

$$G_s = \frac{\rho RT}{\overline{M}_c}\left(1 - \frac{2\overline{M}_c}{\overline{M}_n}\right)(v_{2,s})^{1/3} = \frac{\rho RT}{\overline{M}_c}\left(1 - \frac{2\overline{M}_c}{\overline{M}_n}\right)\frac{1}{Q^{1/3}} \tag{42}$$

Here, G_s is the shear modulus of a swollen gel; ϱ is polymer density; R is the gas constant (8.314 $\mathrm{Jmol^{-1}K^{-1}}$); M_n is the number-average molecular weight; and T is temperature (K). This relationship correlates the shear modulus of a swollen gel to its cross-linking density ($\overline{M}_c$) and degree of swelling (Q). Clearly, a hydrogel's modulus decreases with an increasing degree of swelling and decreasing gel cross-linking density (i.e., $\overline{M}_c$ is decreased). To simplify the above equation, the Flory–Rehner equation can be substituted into Eq. (42), and assuming a high swelling ratio ($Q>10$) (Metters et al., 2000a,b; Metters and Lin, 2007):

$$G_s = \frac{\rho RT}{\overline{M}_c}\frac{1}{\left[\beta\left(\overline{M}_c^{3/5}\right)\right]^{1/3}} = \frac{\gamma}{\overline{M}_c^{6/5}} \tag{43}$$

where β and γ are constants.

For hydrolytically degradable hydrogels (e.g., PLA-*b*-PEG-*b*-PLA hydrogels), $\overline{M}_c$ increases exponentially with time (Anseth et al., 2002). Therefore, Eq. (43) dictates a time-dependent exponentially decreasing modulus for a bulk-degrading hydrogel. Figure II.4.3.9 illustrates the *in vitro* degradation behavior, as well as theoretical verification, of a PLA-*b*-PEG-*b*-PLA hydrogel where the swelling increases and the compressive modulus decreases exponentially with degradation (Anseth et al., 2002). By tailoring the degradation kinetics, these degradable hydrogels have found successful applications in both drug delivery and the regeneration of cartilage (Anseth et al., 2002).

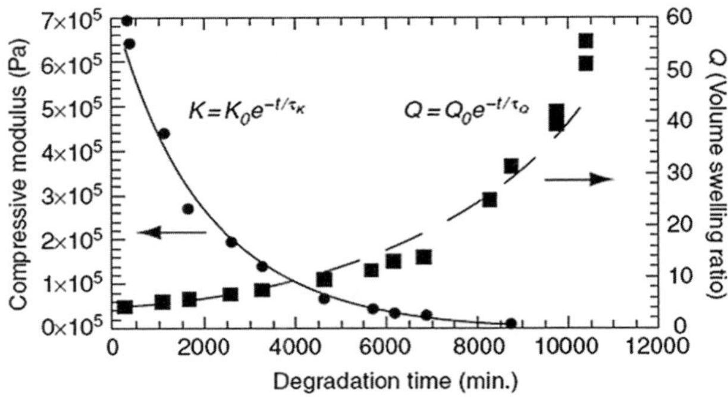

FIGURE II.4.3.9 Changes in material properties during the *in vitro* degradation of a PLA-*b*-PEG-*b*-PLA hydrogel. The solid and dashed lines are exponential curves fit to compressive modulus (•) and volumetric swelling ratio (■) with time constants of τ_K = 2000 min and τ_Q = 4200 min, respectively. (From Anseth et al. 2002.)

SUMMARY

The focus of this chapter has been on developing both a fundamental and quantitative understanding of biodegradation of polymeric biomaterials. Understanding the mechanisms and kinetics of biodegradation is critical in designing biodegradable polymers for many biomedical applications, including matrices for drug delivery and scaffolds for tissue engineering. Designing biodegradable polymeric biomaterials can be challenging, as the design space can be large, with a wide variety of polymer chemistries, degradable moieties, and degradation mechanisms. Further, biodegradation influences numerous material properties, and hence the performance of the final biomedical devices. One important factor that this chapter does not take into account is the complication of the *in vivo* environment. The biodegradation of materials described in this chapter can deviate substantially from *in vitro* observations when placed in a complex and dynamic *in vivo* environment (Chapter II.4.2 addresses key issues in *in vivo* degradation). Nonetheless, understanding the basic principles of biodegradation *in vitro* certainly provides valuable lessons on which biomaterial design is based.

BIBLIOGRAPHY

Ahmed, F., & Discher, D. E. (2004). Self-porating polymersomes of PEG-PLA and PEG-PCL: Hydrolysis-triggered controlled release vesicles. *J. Control. Release*, **96**, 37–53.

Aklonis, J. J., & MacKnight, W. J. (1983). *Introduction to Polymer Viscoelasticity*. New York, NY: Wiley-Interscience.

Allen, C., Han, J. N., Yu, Y. S., Maysinger, D., & Eisenberg, A. (2000). Polycaprolactone-b-poly(ethylene oxide) copolymer micelles as a delivery vehicle for dihydrotestosterone. *J. Control. Release*, **63**, 275–286.

Amsden, B. (1998). Solute diffusion within hydrogels. Mechanisms and models. *Macromolecules*, **31**, 8382–8395.

Anderson, J. M., & Shive, M. S. (1997). Biodegradation and biocompatibility of PLA and PLGA microspheres. *Adv. Drug Deliv. Rev.*, **28**, 5–24.

Anseth, K. S., Bowman, C. N., & BrannonPeppas, L. (1996). Mechanical properties of hydrogels and their experimental determination. *Biomaterials*, **17**, 1647–1657.

Anseth, K. S., Metters, A. T., Bryant, S. J., Martens, P. J., Elisseeff, J. H., et al. (2002). *In situ* forming degradable networks and their application in tissue engineering and drug delivery. *J. Control. Release*, **78**, 199–209.

Baker, S. C., Rohman, G., Southgate, J., & Cameron, N. R. (2009). The relationship between the mechanical properties and cell behaviour on PLGA and PCL scaffolds for bladder tissue engineering. *Biomaterials*, **30**, 1321–1328.

Broz, M. E., VanderHart, D. L., & Washburn, N. R. (2003). Structure and mechanical properties of poly(D, L-lactic acid)/poly(ε-caprolactone) blends. *Biomaterials*, **24**, 4181–4190.

Canal, T., & Peppas, N. A. (1989). Correlation between mesh size and equilibrium degree of swelling of polymeric networks. *J. Biomed. Mater. Res.*, **23**, 1183–1193.

Cha, Y., & Pitt, C. G. (1989). The acceleration of degradation-controlled drug delivery from polyester microspheres. *J. Control. Release*, **8**, 259–265.

Dardik, H., Dardik, I., & Laufman, H. (1971). Clinical use of polyglycolic acid polymer as a new absorbable synthetic suture. *Am. J. Surg.*, **121**, 656–660.

DuBose, J. W., Cutshall, C., & Metters, A. T. (2005). Controlled release of tethered molecules via engineered hydrogel degradation: Model development and validation. *J. Biomed. Mater. Res. A*, **74**, 104–116.

Elbert, D. L., & Hubbell, J. A. (2001). Conjugate addition reactions combined with free-radical cross-linking for the design of materials for tissue engineering. *Biomacromolecules*, **2**, 430–441.

Elbert, D. L., Pratt, A. B., Lutolf, M. P., Halstenberg, S., & Hubbell, J. A. (2001). Protein delivery from materials formed by self-selective conjugate addition reactions. *J. Control. Release*, **76**, 11–25.

Fairbanks, B. D., Schwartz, M. P., Halevi, A. E., Nuttelman, C. R., Bowman, C. N., et al. (2009). A versatile synthetic extracellular matrix mimic through thiol-ene photopolymerization. *Adv. Mater.*, **21**, 5005–5010.

Flory, P. J. (1953). *Principles of Polymer Chemistry*. Ithaca, NY: Cornell Univeristy Press.

Freed, L. E., Vunjaknovakovic, G., Biron, R. J., Eagles, D. B., Lesnoy, D. C., et al. (1994). Biodegradable polymer scaffolds for tissue engineering. *Bio-Technology*, **12**, 689–693.

Gilding, D. K., & Reed, A. M. (1979). Biodegradable polymers for use in surgery: Polyglycolic-poly(actic acid) homopolymers and copolymers. 1. *Polymer*, **20**, 1459–1464.

Gopferich, A. (1996). Mechanisms of polymer degradation and erosion. *Biomaterials*, **17**, 103–114.

Heller, J., & Barr, J. (2004). Poly(ortho esters): From concept to reality. *Biomacromolecules*, **5**, 1625–1632.

Heller, J., Barr, J., Ng, S. Y., Abdellauoi, K. S., & Gurny, R. (2002). Poly(ortho esters): Synthesis, characterization, properties and uses. *Adv. Drug Deliv. Rev.*, **54**, 1015–1039.

Hopfenberg, H. B. (1976). Controlled release from erodible slabs, cylinders, and spheres. *ACS Sym. Ser.*, 26–32.

Hou, Q. P., Grijpma, D. W., & Feijen, J. (2003). Porous polymeric structures for tissue engineering prepared by a coagulation, compression moulding and salt leaching technique. *Biomaterials*, **24**, 1937–1947.

Jo, Y. S., Gantz, J., Hubbell, J. A., & Lutolf, M. P. (2009). Tailoring hydrogel degradation and drug release via neighboring amino acid controlled ester hydrolysis. *Soft Matter*, **5**, 440–446.

Joshi, A., & Himmelstein, K. J. (1991). Dynamics of controlled release from bioerodible matrices. *J. Control. Release*, **15**, 95–104.

Langer, R., & Tirrell, D. A. (2004). Designing materials for biology and medicine. *Nature*, **428**, 487–492.

Langer, R., & Vacanti, J. P. (1993). Tissue engineering. *Science*, **260**, 920–926.

Leong, K. W., Kost, J., Mathiowitz, E., & Langer, R. (1986). Polyanhydrides for controlled release of bioactive agents. *Biomaterials*, 7, 364–371.

Lin, C. C., & Anseth, K. S. (2009). PEG hydrogels for the controlled release of biomolecules in regenerative medicine. *Pharm. Res.*, **26**, 631–643.

Lin, C. C., & Metters, A. T. (2006). Hydrogels in controlled release formulations: Network design and mathematical modeling. *Adv. Drug Deliv. Rev.*, **58**, 1379–1408.

Liu, L. J., Li, S. M., Garreau, H., & Vert, M. (2000). Selective enzymatic degradations of poly(L-lactide) and poly(ε-caprolactone) blend films. *Biomacromolecules*, **1**, 350–359.

Lustig, S. R., & Peppas, N. A. (1988). Solute diffusion in swollen membranes. 9. Scaling laws for solute diffusion in gels. *J. Appl. Polym. Sci.*, **36**, 735–747.

Lutolf, M. P., Lauer-Fields, J. L., Schmoekel, H. G., Metters, A. T., Weber, F. E., et al. (2003). Synthetic matrix metalloproteinase-sensitive hydrogels for the conduction of tissue regeneration: Engineering cell-invasion characteristics. *Proc. Natl. Acad. Sci. USA*, **100**, 5413–5418.

Martens, P., Metters, A. T., Anseth, K. S., & Bowman, C. N. (2001). A generalized bulk-degradation model for hydrogel networks formed from multivinyl cross-linking molecules. *J. Phys. Chem. B.*, **105**, 5131–5138.

Mason, M. N., Metters, A. T., Bowman, C. N., & Anseth, K. S. (2001). Predicting controlled-release behavior of degradable PLA-b-PEG-b-PLA hydrogels. *Macromolecules*, **34**, 4630–4635.

Metters, A. T., & Lin, C. C. (2007). Biodegradable hydrogels: Tailoring properties and function through chemistry and structure. In J. Y. Wong, & J. D. Bronzino (Eds.), *Biomaterials*. Boca Ratyon, FL: CRC Press.

Metters, A. T., Anseth, K. S., & Bowman, C. N. (2000a). Fundamental studies of a novel, biodegradable PEG-b-PLA hydrogel. *Polymer*, **41**, 3993–4004.

Metters, A. T., Bowman, C. N., & Anseth, K. S. (2000b). A statistical kinetic model for the bulk degradation of PLA-b-PEG-b-PLA hydrogel networks. *J. Phys. Chem. B.*, **104**, 7043–7049.

Metters, A. T., Anseth, K. S., & Bowman, C. N. (2001). A statistical kinetic model for the bulk degradation of PLA-b-PEG-b-PLA hydrogel networks: Incorporating network non-idealities. *J. Phys. Chem. B.*, **105**, 8069–8076.

Middleton, J. C., & Tipton, A. J. (2000). Synthetic biodegradable polymers as orthopedic devices. *Biomaterials*, **21**, 2335–2346.

Morgan, M. N. (1969). New synthetic absorbable suture material. *Br. Med. J.*, **2**, 308.

Mott, J. D., & Werb, Z. (2004). Regulation of matrix biology by matrix metalloproteinases. *Curr. Opin. Cell Biol.*, **16**, 558–564.

Peppas, N. A., Bures, P., Leobandung, W., & Ichikawa, H. (2000). Hydrogels in pharmaceutical formulations. *Eur. J. Pharm. Biopharm.*, **50**, 27–46.

Pitt, C. G., Chasalow, F. I., Hibionada, Y. M., Klimas, D. M., & Schindler, A. (1981a). Aliphatic polyesters. 1. The degradation of poly(ε-caprolactone) *in vivo*. *J. Appl. Polym. Sci.*, **26**, 3779–3787.

Pitt, C. G., Gratzl, M. M., Kimmel, G. L., Surles, J., & Schindler, A. (1981b). Aliphatic polyesters. 2. The degradation of poly(dl-lactide), poly(ε-caprolactone), and their copolymers *in vivo*. *Biomaterials*, **2**, 215–220.

Qian, H. T., Bei, J. Z., & Wang, S. G. (2000). Synthesis, characterization and degradation of ABA block copolymer of L-lactide and ε-caprolactone. *Polym. Degrad. Stab.*, **68**, 423–429.

Reed, A. M., & Gilding, D. K. (1981). Biodegradable polymers for use in surgery: Poly(glycolic)-poly(lactic acid) homo and co-polymers. 2. *In vitro* degradation. *Polymer*, **22**, 494–498.

Rice, M. A., & Anseth, K. S. (2007). Controlling cartilaginous matrix evolution in hydrogels with degradation triggered by exogenous addition of an enzyme. *Tissue Eng.*, **13**, 683–691.

Rice, M. A., Sanchez-Adams, J., & Anseth, K. S. (2006). Exogenously triggered, enzymatic degradation of photopolymerized hydrogels with polycaprolactone subunits: Experimental observation and modeling of mass loss behavior. *Biomacromolecules*, 7, 1968–1975.

Rosen, H. B., Chang, J., Wnek, G. E., Linhardt, R. J., & Langer, R. (1983). Bioerodible polyanhydrides for controlled drug delivery. *Biomaterials*, **4**, 131–133.

Schoenmakers, R. G., van de Wetering, P., Elbert, D. L., & Hubbell, J. A. (2004). The effect of the linker on the hydrolysis rate of drug-linked ester bonds. *J. Control. Release*, **95**, 291–300.

Seliktar, D., Zisch, A. H., Lutolf, M. P., Wrana, J. L., & Hubbell, J. A. (2004). MMP-2 sensitive, VEGF-bearing bioactive hydrogels for promotion of vascular healing. *J. Biomed. Mater. Res. A.*, **68A**, 704–716.

Siepmann, J., & Gopferich, A. (2001). Mathematical modeling of bioerodible, polymeric drug delivery systems. *Adv. Drug Deliv. Rev.*, **48**, 229–247.

Stamenkovic, I. (2003). Extracellular matrix remodelling: The role of matrix metalloproteinases. *J. Pathol.*, **200**, 448–464.

Tabata, Y., Gutta, S., & Langer, R. (1993). Controlled delivery systems for proteins using polyanhydride microspheres. *Pharm. Res.*, **10**, 487–496.

Tamada, J., & Langer, R. (1992). The development of polyanhydrides for drug delivery applications. *J. Biomater. Sci. Polym. Ed.*, **3**, 315–353.

Thombre, A. G., & Himmelstein, K. J. (1985). A simultaneous transport-reaction model for controlled drug delivery from catalyzed bioerodible polymer matrices. *AIChE J.*, **31**, 759–766.

Treloar, L. R. G. (1975). *Physics of Rubber Elasticity*. Oxford, UK: Clarendon Press.

Tsuji, H., & Ikada, Y. (1998). Blends of aliphatic polyesters. II. Hydrolysis of solution-cast blends from poly(L-lactide) and poly(ε-caprolactone) in phosphate-buffered solution. *J. Appl. Polym. Sci.*, **67**, 405–415.

Tsuji, H., Mizuno, A., & Ikada, Y. (1998). Blends of aliphatic polyesters. III. Biodegradation of solution-cast blends from poly(L-lactide) and poly(ε-caprolactone). *J. Appl. Polym. Sci.*, **70**, 2259–2268.

van de Wetering, P., Metters, A. T., Schoenmakers, R. G., & Hubbell, J. A. (2005). Poly(ethylene glycol) hydrogels formed by conjugate addition with controllable swelling, degradation, and release of pharmaceutically active proteins. *J. Control. Release*, **102**, 619–627.

Vert, M. (1989). Bioresorbable polymers for temporary therapeutic applications. *Angew. Makromol. Chem.*, **166**, 155–168.

Vert, M., Li, S. M., Spenlehauer, G., & Guerin, P. (1992). Bioresorbability and biocompatibility of aliphatic polyesters. *J. Mater. Sci. Mater. Med.*, **3**, 432–446.

Vert, M., Mauduit, J., & Li, S. M. (1994). Biodegradation of PLA/GA polymers: Increasing complexity. *Biomaterials*, **15**, 1209–1213.

West, J. L., & Hubbell, J. A. (1999). Polymeric biomaterials with degradation sites for proteases involved in cell migration. *Macromolecules*, **32**, 241–244.

Wu, L. B., & Ding, J. D. (2004). *In vitro* degradation of three-dimensional porous poly(D, L-lactide-co-glycolide) scaffolds for tissue engineering. *Biomaterials*, **25**, 5821–5830.

Zhang, J. C., Wu, L. B., Jing, D. Y., & Ding, J. D. (2005). A comparative study of porous scaffolds with cubic and spherical macropores. *Polymer*, **46**, 4979–4985.

Zhu, J. H., Shen, Z. R., Wu, L. T., & Yang, S. L. (1991). In vitro degradation of polylactide and poly(lactide-co-glycolide) microspheres. *J. Appl. Polym. Sci.*, **43**, 2099–2106.

CHAPTER II.4.4 DEGRADATIVE EFFECTS OF THE BIOLOGICAL ENVIRONMENT ON METALS AND CERAMICS

David F. Williams[1] *and Rachel L. Williams*[2]

[1]Wake Forest Institute of Regeneration Medicine Winston-Salem, North Carolina, USA

[2]Clinical Engineering, Institute of Ageing and Chronic Disease, University of Liverpool, Liverpool, UK

Metallic and ceramic biomaterials continue to be used in many medical device applications. Several of these materials now have more than 40 years' clinical experience, and the factors that control their performance and the mechanisms by which they interact with the human body are well-known and understood. However, since the biological environment to which they are exposed is both very aggressive and very susceptible to material-provoked adverse responses (Williams, 2008), it is important that the potential for degradative effects on these materials is kept clearly in mind as new materials and new applications are introduced.

The environment to which biomaterials are exposed during prolonged use (i.e., the internal milieu of the body) can be described as an aqueous medium containing various anions, cations, organic substances, and dissolved oxygen. The anions are mainly chloride, phosphate, and bicarbonate ions. The principal cations are Na^+, K^+, Ca^{2+}, and Mg^{2+}, but with smaller amounts of many others. Table II.4.4.1 presents the range of values for the anion and cation concentrations in blood plasma and extracellular fluid (Bundy, 1994). This represents an environment with a chloride concentration of approximately a third of that of sea water (Hanawa, 2002). The concentration of dissolved oxygen also influences the aggressive nature of the environment, and in venous blood is approximately a quarter of that in air. The organic substances include low molecular weight species, as well as relatively high molecular weight proteins and lipids. Table II.4.4.2 gives examples of the concentration of various organic components of blood plasma. The protein content of the environment is known to have a significant influence on the corrosive nature of body fluids (Williams, 1985; Khan et al., 1999a; Mueller et al., 2009). The pH in this well-buffered system is around 7.4, although because of inflammation it may change for short periods following surgery to as low as 4 or 5 (Bundy, 1994). The temperature remains constant at around 37°C.

On the basis of existing knowledge of the stability of materials in various environments, we should predict that metals, as a generic group, should be relatively susceptible to corrosion in this biological environment, whereas ceramics should display a varying susceptibility, depending on solubility. This correlates fairly well with experimental

TABLE II.4.4.1 Ionic Concentrations in Blood Plasma and Extracellular Fluid

Anion, Cation	Blood Plasma (m*M*)	Extracellular Fluid (m*M*)
Cl^-	96–111	112–120
HCO_3^-	16–31	25.3–29.7
HPO_4^{2-}	1–1.5	193–102
SO_4^{2-}	0.35–1	0.4
$H_2PO_4^-$	2	—
Na^+	131–155	141–15
Mg^{2+}	0.7–1.9	1.3
Ca^{2+}	1.9–3	1.4–1.55
K^+	35–5.6	3.5–4

(Bundy, 1994)

TABLE II.4.4.2 Major Proteins and Other Organic Constituents of Blood Plasma

Major Proteins and Organic Molecules in Blood Plasma (gl^{-1} Unless Stated Otherwise)	
Albumin	30–55
α-Globulins	5–10
β-Globulins	6–12
γ-Globulins	6.6–15
α-Lipoproteins	3.5–4.5
Fibrinogen	1.7–4.3
Total cholesterol	1.2–2.5
Fatty acids	1.9–4.5
Glucose	0.65–1.1
Lactate	0.5–2.2 m*M*
Urea	3–7 m*M*

(Bundy, 1994)

observations and clinical experience, since it is well-known that all but the most corrosion-resistant metals will suffer significant and destructive attack upon prolonged implantation. Also, even the most noble of metals, and those that are most strongly passivated (i.e., naturally protected by their own oxide layer), will still show some degree of interaction. The important passivating implant alloys and their compositions are presented in Table II.4.4.3.

There are some ceramics that have a combination of very strong partially ionic, partially covalent bonds that are sufficiently stable to resist breakdown within this environment, such as the pure simple oxide ceramics, and others in which certain of the bonds are readily destroyed in an aqueous medium so that the material essentially dissolves, for example certain calcium phosphates. Typical implant ceramics and their compositions are presented in Table II.4.4.4.

With these general statements in mind, we have to consider the following questions in relation to the corrosion and degradation of metals and ceramics:

1. Within these groups, how does the susceptibility to corrosion and degradation vary, by what precise mechanisms do the interfacial reactions take place, and how is material selection (and treatment) governed by this knowledge?
2. Are there variables within this biological environment other than those described above that can influence these processes?
3. What are the consequences of such corrosion and degradation phenomena?

TABLE II.4.4.3 Chemical Composition of Implant Alloys

Implant Alloys	Composition
316L stainless steel	Cr, Mo, Ni, Mn, C, S, Si, P, Fe
Co–Cr based alloys	Cast Co–Cr–Mo
	Wrought Co–Cr–Mo
	Wrought Co–Cr–W–Ni
	Wrought Co–Cr–Mo–Ni
	Wrought Co–Cr–Mo–Ni–Fe
	Wrought Co–Cr–Mo–Ni–W–Fe
Cp titanium	Ti + traces of O, N, C, H, Fe
Titanium alloys	Ti-6Al-4V
	Ti-6Al-7Nb
	Ti-15Mo
	Ti-12Mo-6Zr-2Fe
	Ti-13Nb-13Zr
	Ti-15Mo-2.8Nb-0.2Si-0.26O
	Ti-16Nb-10Hf
	Ti-15Mo-5Zr-3Al
	Ti–Ni

TABLE II.4.4.4 Chemical Composition of Common Implant Ceramics

Implant Ceramics	Composition
Alumina	$-Al_2O_3$ + <0.3 wt% MgO
Zirconia	Yttria stabilized tetragonal zirconia ZrO_2 + 2–3 mol% Y_2O_3
Calcium phosphates	$Ca_3(PO_4)_2$–α- or β-tricalcium phosphate $Ca_{10}(PO_4)_6(OH)_2$—hydroxyapatite

We review each of these questions in this chapter. It is particularly important to bear in mind some general points as these questions are discussed.

1. Material selection cannot be governed solely by considerations of stability, and mechanical and physical properties especially may be of considerable importance. Since corrosion is a surface phenomenon, however, it may be possible to optimize corrosion resistance by attention to, or treatment of, the surface rather than by manipulation of the bulk chemistry (Trepanier et al., 1999, Singh and Dahotre, 2007). This offers the possibility of developing sufficient corrosion resistance in materials of excellent bulk mechanical and physical properties. Thus, noble metals such as gold and platinum are rarely used for structural applications (apart from dental restoration) because of their inferior mechanical properties,

even though they have excellent corrosion resistance; instead, base metal alloys with passivated or protected surfaces offer better all-around properties.

2. Medical devices are not necessarily used in mechanically stress-free conditions, and indeed the vast majority of those using metals or ceramics are structurally loaded. It is well-known that mechanical stress plays a very important role in the corrosion of metals (Gilbert et al., 1993; Jacobs et al., 1998) and the degradation process in ceramics (Piconi and Maccauro, 1999), both potentiating existing effects and initiating others. This has to be taken into consideration.
3. We cannot expect the biological environment to be constant. Within the overall characteristics described earlier, there are variations (with time, location, activity, health status, etc.) in, for example, oxygen levels, availability of free radicals, and cellular activity, all of which may cause variations in the corrosive nature of the environment (Tengvall et al., 1989; Fonseca and Barbosa, 2001). Most important, corrosion is not necessarily a progressive homogeneous reaction with zero-order kinetics. Corrosion processes can be quiescent, but then become activated, or they can be active, but then become passivated and localized, with transient fluctuations in conditions playing a part in these variations.
4. The effects of corrosion or degradation may be twofold. First, and in the conventional metallurgical sense the most obvious, the problem can lead to loss of structural integrity of the material and function. This may be undesirable, as in the case of many long-term prostheses, or desirable, as in devices intended for short-term function (e.g., ceramics for drug delivery systems) or where the material is replaced by tissue during the degradation process, as with ceramic bone substitution. In addition to this, however, and usually of much greater significance with biomaterials, when released into the tissue, the corrosion or degradation products can have a significant and controlling effect on that tissue (Bravo et al., 1990; Jacobs et al., 1998). Indeed, it is likely that the corrosion process is the most important mediator of the tissue response to metallic materials. It is therefore important that we know both the nature of the reaction products and their rate of generation. In this respect it is important to recognize that a very small release of certain metallic ions that cause adverse biological reactions may be more significant than a larger amount of a less stimulating byproduct of corrosion or degradation. The implications of these effects on biocompatibility and the future uses of biomaterials have recently been discussed by Williams (Williams, 2008, 2009).

METALLIC CORROSION

Basic Principles

The most pertinent form of corrosion related to metallic biomaterials is aqueous corrosion. This occurs when electrochemical reactions take place on a metallic surface in an aqueous electrolyte. There are always two reactions that occur: the anodic reaction, which yields metallic ions, for example, involving the oxidation of the metal to its salt:

$$M \rightarrow M(n+) + n(\text{electrons}) \quad (1)$$

and the cathodic reaction, in which the electrons so generated are consumed. The precise cathodic reaction will depend on the nature of the electrolyte, but two of the most important in aqueous environments are the reduction of hydrogen:

$$2H^+ + 2e^- \rightarrow H_2 \quad (2)$$

and the reduction of dissolved oxygen:

$$O_2 + 4H^+ + 4e^- \rightarrow 2H_2O \quad (3)$$

in acidic solutions:

$$O_2 + 2H_2O + 4e^- \rightarrow 4OH^- \quad (4)$$

or in neutral or basic solutions.

In all corrosion processes, the rate of the anodic or oxidation reaction must equal the rate of the cathodic or reduction reaction. This is a basic principle of electrochemically-based metallic corrosion. It also explains how variations in the local environment can affect the overall rate of corrosion by influencing either the anodic or cathodic reactions. The whole corrosion process can be arrested by preventing either of these reactions.

From a thermodynamic point of view, first consider the anodic dissolution of a pure metal isolated in a solution of its salt. The metal consists of positive ions closely surrounded by free electrons. When the metal is placed in a solution, there will be a net dissolution of metal ions since the Gibbs free energy (ΔG) for the dissolution reaction is less than for the reverse reaction. This leaves the metal with a net negative charge, thus making it harder for the positive ions to leave the surface, and increasing the ΔG for the dissolution reaction. There will come a point when the ΔG for the dissolution reaction will equal the ΔG for the reverse reaction. At this point, a dynamic equilibrium is reached, and a potential difference will be set up across the charged double layer surrounding the metal. The potential difference will be characteristic of the metal and can be measured against a standard reference electrode. When this is done against a standard hydrogen electrode in a 1 N solution of its salt at 25°C, it is defined as the standard electrode potential for that metal (Table II.4.4.5). The position of a metal in the electrochemical series primarily indicates the order with which metals displace each other from compounds, but it also gives a general guide to reactivity in aqueous solutions. Those at the top are the noble, relatively unreactive metals, whereas those at the bottom are the more reactive. This is the first guide to corrosion resistance, but, as we shall see, there are major difficulties related to the use and interpretation of reactions from this simple analysis.

TABLE II.4.4.5 Electrochemical Series

Metal	Potential (V)
Gold	1.43
Platinum	1.20
Mercury	0.80
Silver	0.79
Copper	0.34
Hydrogen	0
Lead	−0.13
Tin	−0.14
Molybdenum	−0.20
Nickel	−0.25
Cobalt	−0.28
Cadmium	−0.40
Iron	−0.44
Chromium	−0.73
Zinc	−0.76
Aluminum	−1.33
Titanium	−1.63
Magnesium	−2.03
Sodium	−2.71
Lithium	−3.05

Now consider a system in which the metal is in an aqueous solution that does not contain its ions. In this situation, the electrode potential at equilibrium (i.e., when the rate of the anodic reaction equals the rate of the cathodic reaction) will be shifted from the standard electrode potential and can be defined by the Nernst equation:

$$E = E_0 + (RT/nF \ln(a_{anod}/a_{cath})$$

where E_0 is the standard electrode potential, RT/F is a constant, n is the number of electrons transferred, and a is the activity of the anodic and cathodic reactants. At low concentrations, the activity can be approximated to the concentration. In this situation, there is a net dissolution of the metal and a current will flow. At equilibrium, the rate of the metal dissolution is equal to the rate of the cathodic reaction, and the rate of the reaction is directly proportional to the current density by Faraday's law; therefore:

$$i_{anodic} = i_{cathodic} = i_{corrosion}$$

and the Nernst equation can be rewritten:

$$E = E_0 = \pm\beta \ln(i_{corr}/i_0)$$

where β is a constant and i_0 is the exchange current density, which is defined as the anodic (or cathodic) current density at the standard electrode potential. Current density is the current, measured in amperes, normalized to the surface area of the metal.

These conditions represent convenient models for the basic mechanisms of corrosion, but they are hardly realistic. Indeed, in this situation of a homogeneous pure metal existing within an unchanging environment, an equilibrium is reached in which no further net movement of ions takes place. In other words, the corrosion process takes place only transiently, but is effectively stopped once this equilibrium is reached.

In reality, we usually have neither entirely homogeneous surfaces nor solutions, nor complete isolation of the metal from other parts of the environment, and this equilibrium is easily upset. If the conditions are such that the equilibrium is displaced, the metal is said to be polarized. There are several ways in which this can happen. Two main factors control the behavior of metals in this respect, and determine the extent of corrosion in practice. The first concerns the driving force for continued corrosion (i.e., the reasons why the equilibrium is upset and the nature of the polarization), and the second concerns the ability of the metal to respond to this driving force.

It is self-evident that if either the accumulating positive metal ions in the surrounding media or the accumulating electrons in the metal are removed, the net balance between the dissolution and the replacement of the ions will be disturbed. This will occur in the biological environment surrounding implanted alloys, due to the interaction of the proteins with the metal ions. Metal ions can form complexes with proteins (Steinemann, 1996; Jacobs et al., 1998; Büdinger and Hertl, 2000; Munoz and Mischler, 2007) and these complexes can be transported away from the immediate vicinity. This removes the metal ions from the charged double layer at the interface allowing further release of metal ions to re-establish the equilibrium. Similarly, relative movement between the implant and the tissue, for example, at a bearing surface or on a cyclically loaded implant, will cause mixing at the interface and will modify the composition of the electrolyte, and may modify the surface of the alloy (Khan et al., 1999b). The equilibrium is established precisely because of the imbalance of charge, so that if the charge balance is disturbed, further corrosion will occur to attempt to re-establish the balance. The result will be continued dissolution as the system attempts to achieve this equilibrium, in other words, sustained corrosion. An environment that allows the removal of electrons in contact with the metal or stirring of the electrolyte will achieve this.

The process of galvanic corrosion may be used to demonstrate this effect. Consider a single homogeneous pure metal, A, existing within an electrolyte (Figure II.4.4.1). The metal will develop its own potential, V_A, with respect to the electrolyte. If a different metal electrode, B, is placed into the same electrolyte, but without contacting A, it will develop its own potential V_B. If V_A is not equal to V_B, there will be a difference in the numbers of excess free electrons in each. This is of no consequence if A and B are isolated from each other, but should they be placed in electrical contact, electrons will flow from that metal with the greater potential in an attempt to make the two electrodes equipotential. This upsets the equilibrium and causes continued and accelerated corrosion of the more

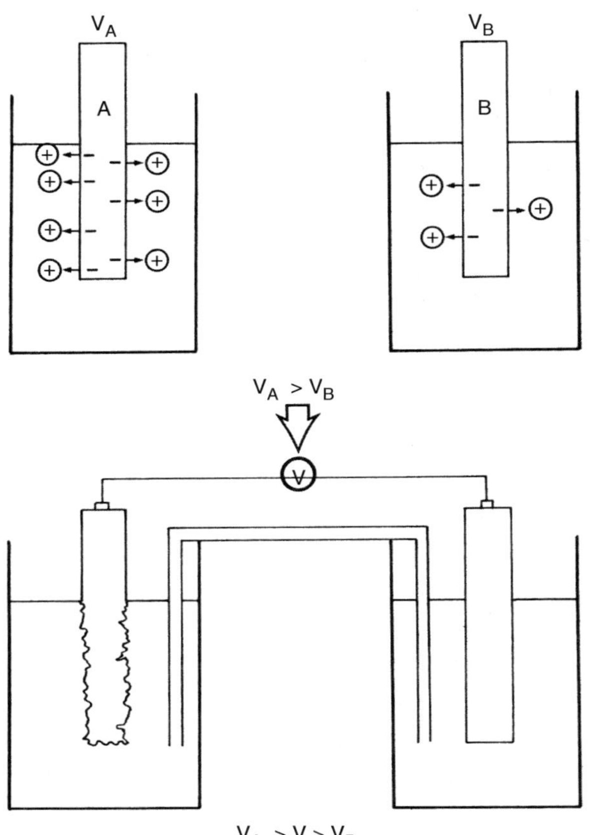

FIGURE II.4.4.1 When electrical contact is made between electrodes A and B, electrode B acts as an electron sink, thus upsetting the equilibrium and causing continued dissolution of A.

active metal (anodic dissolution), and protects the less active metal (cathodic protection).

Galvanic corrosion may be seen whenever two different metals are placed in contact in an electrolyte. It has been frequently observed with complex, multicomponent surgical implants such as modular total joint designs consisting of titanium alloy femoral stems and cobalt alloy femoral heads (Jacobs et al., 1998; Virtanen et al., 2008). It is not necessary for the components to be macroscopic, monolithic electrodes for this to happen, and the same effect can be seen when there are different microstructural features within one alloy, such as the multiphase microstructure evident in implants of sensitized stainless steel where the grain boundaries become depleted in chromium and corrode preferentially to the remaining surface (Disegi and Eschbach, 2000). In practice, it is the regional variations in electrode potential over an alloy surface that are responsible for much of the generalized surface corrosion that takes place in metallic components.

Many of the commonly used surgical alloys contain highly reactive metals (i.e., with high negative electrode potentials), such as titanium, aluminum, and chromium. Because of this high reactivity, they will react with oxygen upon initial exposure to the atmosphere. This initial oxidation leaves an impervious oxide layer firmly adherent to the metal surface; thus all other forms of corrosion may be significantly reduced because the oxide layer acts as a protective barrier, passivating the metal. The manufacturing process for implant alloys may include a passivating step to enhance the oxide layer prior to implantation, for example nitric acid treatment of 316L stainless steel (Fraker and Griffith, 1985; Yoo et al., 2008).

In summary, the basic principles of corrosion determine that:

1. In theory, corrosion resistance can be predicted from standard electrode potentials. This explains the nobility of some metals and the considerable reactivity of others, but is not useful for predicting the occurrence of corrosion of most alloy systems in practice.
2. Irrespective of standard electrode potentials, the corrosion resistance of many materials is determined by their ability to become passivated by an oxide layer that protects the underlying metal.
3. Corrosion processes in practice are influenced by variations in surface microstructural features, and in the environment that disrupt the charge transfer equilibrium.

INFLUENCE OF THE BIOLOGICAL ENVIRONMENT

It is reasonable to assume that the presence of biological macromolecules will not cause a completely new corrosion mechanism. However, they can influence the rate of corrosion by interfering in some way with the anodic or cathodic reactions discussed earlier. Four ways in which this could occur are discussed next.

1. The biological molecules could upset the equilibrium of the corrosion reactions by consuming one or other of the products of the anodic or cathodic reaction. For example, proteins can bind to metal ions and transport them away from the implant surface. This will upset the equilibrium across the charged double layer and allow further dissolution of the metal; in other words, it will decrease ΔG for the dissolution reaction.
2. The stability of the oxide layer depends on the electrode potential and the pH of the solution. Proteins and cells can be electrically active and interact with the charges formed at the interface, and thus affect the electrode potential (Bundy, 1994). Bacteria (Laurent et al., 2001) and inflammatory cells (Fonseca and Barbosa, 2001; Hanawa, 2002; Messer et al., 2009) can alter the pH of the local environment through the generation of acidic metabolic products that can shift the equilibrium.
3. The stability of the oxide layer is also dependent on the availability of oxygen. The adsorption of proteins and cells onto the surface of materials could limit the

diffusion of oxygen to certain regions of the surface. This could cause preferential corrosion of the oxygen-deficient regions and lead to the breakdown of the passive layer. Alternatively, the biomolecule adsorption layer could act as a capacitor, preventing the diffusion of molecules from the surface (Hiromoto et al., 2002).

4. The cathodic reaction often results in the formation of hydrogen, as shown earlier. In a confined locality, the build-up of hydrogen tends to inhibit the cathodic reaction, and thus restricts the corrosion process. If the hydrogen can be eliminated, then the active corrosion can proceed. It is possible that bacteria in the vicinity of an implant could utilize the hydrogen, and thus play a crucial role in the corrosion process.

There is sufficient evidence to support the premise that the presence of proteins and cells can influence the rate of corrosion of some metals (Williams, 1985; Khan et al., 1999a,b; Hanawa, 1999, 2002). Studies have examined these interactions electrochemically, and have found very few differences in many of the parameters measured (e.g., electrode potential, polarization behavior, and current density at a fixed potential). However, analysis of the amount of corrosion through weight loss or chemical analysis of the electrolyte has shown significant effects from the presence of relatively low concentrations of proteins. These effects have varied from several-fold increases for some metals under certain conditions, to slight decreases under other conditions.

It has been shown that proteins adsorb onto metal surfaces and that the amount adsorbed appears to be different in a range of metals (Williams and Williams, 1988; Wälivaara et al., 1992). Similarly, proteins have been shown to bind to metal ions, and it is suggested that they are transported away from the local site as a protein–metal complex and distributed systemically in the body (Jacobs et al., 1998). It is therefore likely that proteins will influence the corrosion reactions that occur when a metal is implanted, although there is no direct evidence to explain the mechanism of the interaction at this time.

CORROSION AND CORROSION CONTROL IN THE BIOLOGICAL ENVIRONMENT

The need to ensure minimal corrosion has been the major determining factor in the selection of metals and alloys for use in the body. Two broad approaches have been adopted. The first has involved the use of noble metals, that is, those metals and their alloys for which the electrochemical series indicates excellent corrosion resistance. Examples are gold, silver, and the platinum group of metals. Because of cost and relatively poor mechanical properties, these are not used for major structural applications, although it should be noted that gold and its alloys are extensively used in dentistry; silver is sometimes used for its antibacterial activity; and platinum-group metals (Pt, Pd, Ir, Rh) are used in electrodes.

The second approach involves the use of the passivated metals. Of the three elements that are strongly passivated (i.e., aluminum, chromium, and titanium), aluminum cannot be used on its own for biomedical purposes because of toxicity problems; however, it has an important role in several titanium alloys. Chromium is very effectively protected, but cannot be used in bulk. It is, however, widely used in alloys, especially in stainless steels and in the cobalt–chromium-based alloys, where it is normally considered that a level of above 12% gives good corrosion resistance, and about 18% provides excellent resistance. Titanium is the best in this respect, and is used as a pure metal or as the major constituent of alloys (Long and Rack, 1998; Geetha et al., 2009). In alloys the passivating layer promoting the corrosion resistance is predominantly composed of one of these metal oxides. For example, chromium oxide passivates 316L stainless steel and Co–Cr-based alloys, and Ti oxide passivates Ti alloys. The other alloying elements may be present in the surface oxide and this can influence the passivity of the layer (Sittig et al., 1999). Careful pretreatment of the alloys can be used to control the passivity of these alloys (Trepanier et al., 1998; Shih et al., 2000). In particular, production procedures need to be controlled because of their influence on the surface oxides, for example, cleaning (Aronsson et al., 1997) and sterilization (Thierry et al., 2000) procedures.

Although these metals and alloys have been selected for their corrosion resistance, corrosion will still take place when they are implanted in the body. Two important points have to be remembered. First, whether noble or passivated, all metals will suffer a slow removal of ions from the surface, largely because of local and temporal variations in microstructure and environment. This need not necessarily be continuous, and the rate may either increase or decrease with time, but metal ions will be released into that environment. This is particularly important with biomaterials, since it is the effect of these potentially toxic or irritant ions that is the most important consequence of their use. Even with a strongly passivated metal, there will be a finite rate of diffusion of ions through the oxide layer, and possibly dissolution of the layer itself. It is well-known that titanium is steadily released into the tissue from titanium implants (Jacobs et al., 1998; Hanawa et al., 1998). Second, some specific mechanisms of corrosion may be superimposed on this general behavior; some examples are given in the next section.

Pitting Corrosion

The stainless steels used in implantable devices are passivated by the chromium oxide that forms on their surface. It has been shown, however, that in a physiological saline environment, the driving force for repassivation

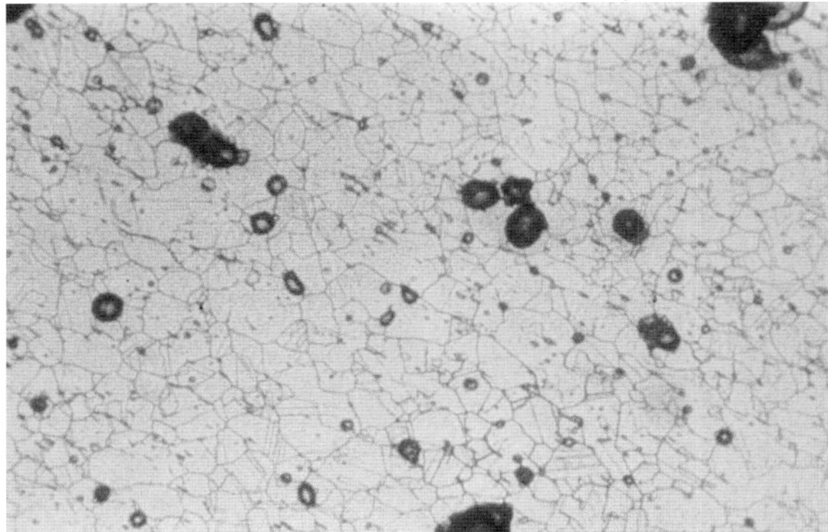

FIGURE II.4.4.2 This etched metallographic micrograph demonstrates the pitting corrosion of stainless steel.

of the surface is not high (Seah et al., 1998). Thus, if the passive layer is broken down, it may not repassivate and active corrosion can occur.

Localized corrosion can occur as a result of imperfections in the oxide layer, producing small areas where the protective surface is removed (Rondelli and Vicentini, 1999). These localized spots will actively corrode, and pits will form in the surface of the material. This can result in a large degree of localized damage, because the small areas of active corrosion become the anode and the entire remaining surface becomes the cathode. Since the rate of the anodic and cathodic reactions must be equal, it follows that a relatively large amount of metal dissolution will be initiated by a small area of the surface, and large pits may form (Figure II.4.4.2).

Fretting Corrosion

The passive layer may be removed by a mechanical process (Khan et al., 1999b; Okazaki, 2002). This can be a scratch that does not repassivate, resulting in the formation of a pit, or a continuous cyclic process in which any reformed passive layer is removed. This is known as fretting corrosion, and it is suggested that this can contribute to the corrosion observed between a fracture fixation plate and the bone screws attaching the plate to the bone. There are four reasons why fretting can affect the corrosion rate. The first is due to the removal of the oxide film as just discussed. The second is due to plastic deformation of the contact area; this can subject the area to high strain fatigue and may cause fatigue corrosion. The third is due to stirring of the electrolyte, which can increase the limited current density of the cathodic reaction. The fourth, especially seen with titanium, is the influence of fretting on hydrogen absorption, and the consequential hydrogen embrittlement (Rodrigues et al., 2009).

Crevice Corrosion

The area between the head of the bone screw and countersink on the fracture fixation plate can also be influenced by the crevice conditions that the geometry creates (Figure II.4.4.3) (Cook et al., 1987). Porous coated implants may also demonstrate crevice corrosion (Seah et al., 1998). Accelerated corrosion can be initiated in a crevice by restricted diffusion of oxygen into the crevice. Initially, the anodic and cathodic reactions occur uniformly over the surface, including within the crevice. As the crevice becomes depleted of oxygen, the reaction is limited to metal oxidation, balanced by the cathodic reaction on the remainder of the surface. In an aqueous sodium chloride solution, the build-up of metal ions within the crevice causes the influx of chloride ions to balance the charge by forming the metal chloride. In the presence of water, the chloride will dissociate to its insoluble hydroxide and acid. This is a rapidly accelerating process, since the decrease in pH causes further metal oxidation.

Intergranular Corrosion

As mentioned earlier, stainless steels rely on the formation of chromium oxides to passivate the surface. If some areas of the alloy become depleted in chromium, as can happen if carbides are formed at the grain boundaries, the regions adjacent to the grain boundaries become depleted in chromium. The passivity of the surface in these regions is therefore affected, and preferential corrosion can occur (Figure II.4.4.4). Although this problem can easily be overcome by heat treating the alloys (Disegi and Eschbach, 2000), it has been observed on retrieved implants (Walczak et al., 1998) and can cause severe problems, since once initiated it will proceed rapidly and may well cause fracture of the

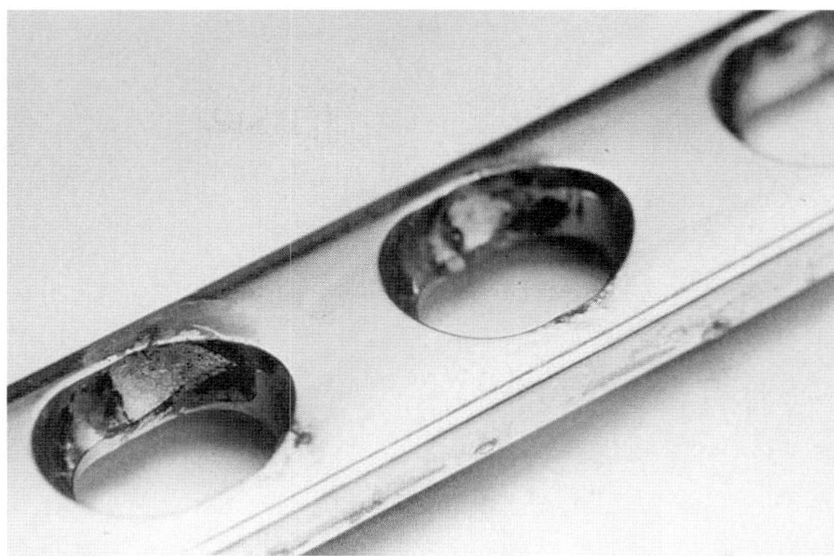

FIGURE II.4.4.3 Crevice corrosion is evident in the screw hole in this fracture fixation plate.

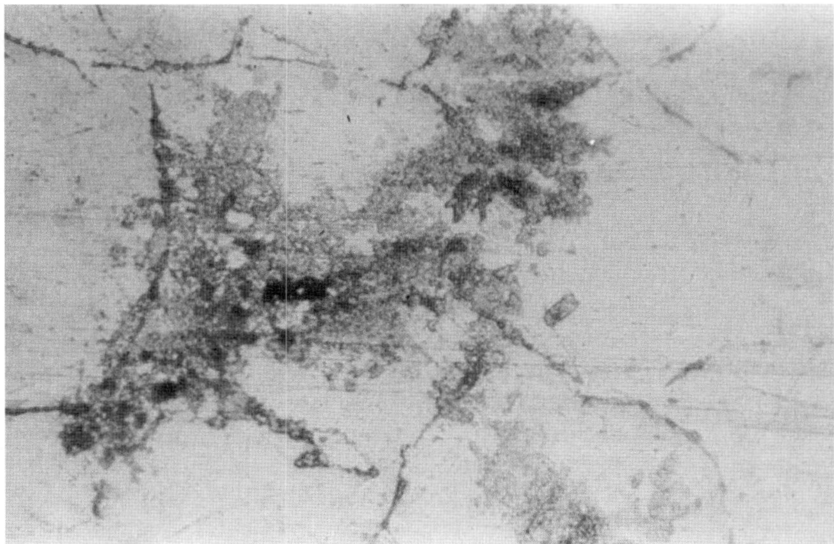

FIGURE II.4.4.4 Intergranular corrosion is demonstrated on this etched stainless steel specimen.

implant and the release of large quantities of corrosion products into the tissue.

Stress Corrosion Cracking

Stress corrosion cracking is an insidious form of corrosion, since an applied stress and a corrosive environment can work together and cause complete failure of a component, when neither the stress nor the environment would be a problem on their own. The stress level may be very low, possibly only residual, and the corrosion may be initiated at a microscopic crack tip that does not repassivate rapidly. Incremental crack growth may then occur, resulting in fracture of the implant. Industrial uses of stainless steels in saline environments have shown susceptibility to stress corrosion cracking, and therefore it is a potential source of failure for implanted devices, although to date there has been little evidence that it occurs extensively in clinical practice.

Galvanic Corrosion

If two metals are independently placed within the same solution, each will establish its own electrode potential with respect to the solution. If these two metals are placed in electrical contact, then a potential difference will be established between them, electrons passing from the more anodic to the more cathodic metal. Thus, equilibrium is upset and a continuous process of dissolution from the more anodic metal will take place. This accelerated corrosion process is galvanic corrosion. It is important if two different alloys are used in

FIGURE II.4.4.5 Extensive corrosion on the titanium stem of a modular hip prosthesis.

an implantable device when the more reactive may corrode freely.

Whenever stainless steel is joined to another implant alloy, it will suffer from galvanic corrosion. If both alloys remain within their passive region when coupled in this way, the additional corrosion may be minimal. Some modular orthopedic systems are made of titanium alloys and cobalt-based alloys on the basis that both should remain passive, but evidence of corrosion has been reported (Gilbert et al., 1993). Certainly, as shown in Figure II.4.4.5, titanium stems of modular prostheses can exhibit extensive corrosion. Galvanic corrosion may also take place on a microscopic scale in multiphase alloys where phases are of considerably different electronegativity. In dentistry, some amalgams may show extensive corrosion because of this mechanism.

Intentional Corrosion and Resorption of Implantable Alloys

The vast majority of applications of metallic systems in medical devices are predicated on the assumption that the device will remain in the patient's body for their full life or until it is intentionally removed surgically. As discussed elsewhere in this book, there are some situations where permanence is not required, and indeed could be disadvantageous. Two situations may be mentioned here: the use of temporary supports during tissue healing (as with bone fracture plates), and the use of intravascular stents that support blood vessels after angioplasty procedures. In both of these situations, an intentionally degradable and resorbable material may be considered as an alternative. With bone fracture plates, a range of biodegradable polymers have been used clinically, and there is some good clinical success. With stents, clinical experience is only just beginning, and again attention is being concentrated on polymers. However, there is increasing interest in the possibility of using magnesium alloys that are known to degrade in the body, and are potentially resorbable.

As reviewed by Staiger et al. (Staiger et al., 2006) and Witte (Witte, 2010), magnesium itself corrodes readily in the body, and magnesium samples can disappear in relatively short times. The rapid loss of strength, and the doubtful biocompatibility of such a rapidly degrading structure, has to be viewed with caution, even though it would appear that the magnesium ions are relatively well-tolerated, magnesium being an essential element in humans. Developments with this element have concentrated on alloying additions that reduce the corrosion rate, without inducing significant cytotoxicity or radically altering the mechanical properties (Xu et al., 2007; Li et al., 2008; Kannan and Raman, 2008). Alloys of Mg–Mn–Zn and Mg–Ca, wherein all alloying additions are also essential elements, figure prominently in these developments. It is a little early to predict the likely clinical success with these alloys.

CERAMIC DEGRADATION

The rate of degradation of ceramics within the body can differ considerably from that of metals, in that they can be either highly corrosion resistant or highly soluble. As a general rule, we should expect to see a very significant resistance to degradation with ceramics and glasses. Since the corrosion process in metals is one of a conversion of a metal to a ceramic structure (i.e., metal to a metal oxide, hydroxide, chloride, etc.) we must intuitively conclude that the ceramic structure represents a lower energy state, in which there would be less driving force for further structural degradation. The interatomic bonds in a ceramic, being largely ionic but partly covalent, are strong directional bonds and large amounts of energy are required for their disruption. As extraction metallurgists know, it takes a great deal of energy to extract aluminum metal from the ore aluminum oxide, but as we have seen, the reverse process takes place readily by surface oxidation. Thus, we should expect ceramics such as Al_2O_3, ZrO_2, TiO_2, SiO_2, and TiN to be stable under normal conditions (Dalgleish and Rawlings, 1981). This is what is observed in clinical practice. There is limited evidence to show that some of these ceramics (e.g., polycrystalline Al_2O_3 and ZrO_2) do show "aging" phenomena (Piconi and Maccauro, 1999; Marti, 2000), with reductions in some mechanical properties, but the significance of this is unclear. With some variations in structure, there is clear evidence that problems may arise if incorrect structural forms are used. Zirconia is often used in the transformation toughened state; as with most oxide ceramics, the material is inherently brittle but this can be ameliorated by the use of doping elements which cause the formation and retention of a metastable state. This is achieved by the addition of small amounts of the rare earth oxide yttria to the zirconia. The toughening arises from the transformation of crystallographic structure at a crack tip under strain, which causes blunting of the crack and prevents its catastrophic propagation. However, this metastable state can transform under other conditions. Many batches of improperly heat-treated zirconia hip replacement components underwent such changes over a short time after implantation, causing significant structural failure under load (Deville et al., 2005; Chevalier, 2006).

Alternatively, there will be many ceramic structures that, although stable in the air, will dissolve in aqueous environments. Consideration of the classic fully ionic ceramic structure NaCl and its dissolution in water demonstrates this point. It is possible, therefore, on the basis of the chemical structure, to identify ceramics that will dissolve or degrade in the body, and the opportunity exists for the production of structural materials with controlled degradation.

Since any material that degrades in the body will release its constituents into the tissue, it is necessary to select anions and cations that are readily and harmlessly incorporated into metabolic processes, and utilized or eliminated. For this reason, it is compounds of sodium, and especially calcium, including calcium phosphates and calcium carbonates, that are primarily used.

The degradation of such compounds will depend on chemical composition and microstructure (Bohner, 2000). For example, tricalcium phosphate [$Ca_3(PO_4)_2$] is degraded fairly rapidly, while calcium hydroxyapatite [$Ca_{10}(PO_4)_6(OH)_2$] is relatively stable. Within this general behavior, however, porosity will influence the rates so that a fully dense material will degrade slowly, while a microporous material will be susceptible to more rapid degradation.

In general, dissolution rates of these ceramics *in vivo* can be predicted from behavior in simple aqueous solution. However, there will be some differences in detail within the body, especially with variations in degradation rate seen at different implantation sites. It is possible that cellular activity, either by phagocytosis or the release of free radicals, could be responsible for such variations.

In between the extremes of stability and intentional degradability lie a small group of materials in which there may be limited activity. This is particularly seen with a number of glasses and glass ceramics, based on Ca, Si, Na, P, and O, in which there is selective dissolution on the surface involving the release of Ca and P, but in which the reaction then ceases because of the stable SiO_2-rich layer that remains on the surface. This is of considerable interest because of the ability of such surfaces to bond to bone, and this subject is dealt with elsewhere in this book.

On the basis of this behavior, bioceramics are normally classified under three headings:

- Inert or "nearly inert" ceramics
- Resorbable ceramics
- Ceramics of controlled surface reactivity.

This area is discussed in detail in other chapters within this book.

SUMMARY

This chapter has attempted to demonstrate that metals are inherently susceptible to corrosion, and that the greatest care is needed in using them within the human body. In general, ceramics have much less tendency to degrade, but care still has to be taken over aging phenomena. The human body is very aggressive toward all of these materials.

BIBLIOGRAPHY

Aronsson, B.-O., Lausmaa, J., & Kasemo, B. (1997). Glow discharge plasma treatment for surface cleaning and modification of metallic biomaterials. *J. Biomed. Mater. Res.*, **35**, 49–73.

Bohner, M. (2000). Calcium orthophosphates in medicine: From ceramics to calcium phosphate cements. *Injury*, **31**(S-4), 37–47.

Bravo, I., Carvalho, G., Barbosa, M., & de Sousa, M. (1990). Differential effects of eight metal ions on lymphocyte differentiation antigens *in vitro*. *J. Biomed. Mater. Res.*, **24**, 1059–1068.

Büdinger, L., & Hertl, M. (2000). Immunological mechanisms in hypersensitivity reactions to metal ions: An overview. *Allergy*, **55**, 108–115.

Bundy, K. J. (1994). Corrosion and other electrochemical aspects of biomaterials. *Crit. Rev. Biomed. Eng.*, **22**(3/4), 139–251.

Chevalier, J. (2006). What future for zirconia as a biomaterial? *Biomaterials*, **27**, 535–543.

Cook, S. D., Tomas, K. A., Harding, A. F., Collins, C. L., Haddad, R. J., et al. (1987). The *in vivo* performance of 250 internal fixation devices: A follow up study. *Biomaterials*, **8**, 177–184.

Dalgleish, B. J., & Rawlings, R. D. (1981). A comparison of the mechanical behaviour of aluminas in air and simulated body environments. *J. Biomed. Mater. Res.*, **15**, 527–542.

Deville, S., Gremillard, L., Chevalier, J., & Fantozzi, G. (2005). A critical comparison of methods for the determination of the ageing sensitivity in biomedical grade yttria-stabilised zirconia. *J. Biomed. Mater. Res. (Part B, Appl. Biomater.)*, **72**, 239–245.

Disegi, J. A., & Eschbach, L. (2000). Stainless steel in bone surgery. *Injury*, **31**(suppl 4), 2–6.

Fonseca, C., & Barbosa, M. A. (2001). Corrosion behaviour of titanium in biofluids containing H_2O_2 studied by electrochemical impedance spectroscopy. *Corr. Sci.*, **43**, 547–559.

Fraker, A. C., & Griffith, C. D. (Eds.), (1985). *Corrosion and Degradation of Implant Materials. ASTM S.T.P. No. 859*. Philadelphia, PA: American Society for Testing and Materials.

Geetha, M., Singh, A. K., Asokamari, R., & Gogia, A. K. (2009). Ti based biomaterials, the ultimate choice for orthopaedic implants: A review. *Prog. Mater. Sci.*, **54**, 397–425.

Gilbert, J. L., Buckley, C. A., & Jacobs, J. J. (1993). *In vivo* corrosion of modular hip prosthesis components in mixed and similar metal combinations. The effect of crevice, stress, motion and alloy coupling. *J. Biomed. Mater. Res.*, **27**, 1533–1544.

Hanawa, T. (1999). *In vivo* metallic biomaterials and surface modification. *Mater. Sci. Eng.*, **A267**, 260–266.

Hanawa, T. (2002). Evaluation techniques of metallic biomaterials *in vitro*. *Sci. Technol. Adv. Mater.*, **3**, 289–295.

Hanawa, T., Asami, K., & Asaoka, K. (1998). Repassvation of titanium and surface oxide film regeneration in simulated bioliquid. *J. Biomed. Mater. Res.*, **40**, 530–538.

Hiromoto, S., Noda, K., & Hanawa, T. (2002). Development of electrolytic cell with cell-culture for metallic biomaterials. *Corr. Sci.*, **44**, 955–965.

Jacobs, J. J., Gilbert, J. L., & Urban, R. M. (1998). Current concepts review corrosion of metal orthopaedic implants. *J. Bone Joint Surg.*, **80A**, 268–282.

Kannan, M. B., & Raman, R. K.S. (2008). *In vitro* degradation and mechanical integration of calcium containing magnesium alloys in modified simulated body fluids. *Biomaterials*, **29**, 2306–2314.

Khan, M. A., Williams, R. L., & Williams, D. F. (1999a). The corrosion behaviour of Ti-6Al-4V, Ti-6Al-7Nb and Ti-13Nb-13Zr in protein solutions. *Biomaterials*, **20**, 631–637.

Khan, M. A., Williams, R. L., & Williams, D. F. (1999b). Conjoint corrosion and wear in titanium alloys. *Biomaterials*, **20**, 765–772.

Laurent, F., Grosgogeat, B., Reclaru, L., Dalard, F., & Lissac, M. (2001). Comparison of corrosion behaviour in presence of oral bacteria. *Biomaterials*, **22**, 2273–2282.

Li, Z., Gu, X., Lou, S., & Zhang, Y. (2008). The development of binary Mg–Ca alloys for use as biodegradable metals within bone. *Biomaterials*, **29**, 1329–1344.

Long, M., & Rack, H. J. (1998). Titanium alloys in total joint replacement: A materials science perspective. *Biomaterials*, **19**, 1621–1639.

Marti, A. (2000). Inert bioceramics (Al_2O_3, ZrO_2) for medical applications. *Injury*, **31**(S-4), 33–36.

Messer, R. L.W., Tackas, J. B., Mickalonis, J., Brown, Y., Lewis, J. B., et al. (2009). Corrosion of machined titanium dental implants under inflammatory conditions. *J. Biomed. Mater. Res., (Part B, Appl. Biomater.)*, **88**, 474–481.

Mueller, W. D., De Mele, L., Nascimento, M. L., & Zeddies, M. (2009). Degradation of magnesium and its alloys: Dependence on the composition of the synthetic biological media. *J. Biomed. Mater. Res., Part A*, **90**, 487–495.

Munoz, A. I., & Mischler, S. (2007). Interactive effects of albumin on phosphate ions on the corrosion of CoCrMo implant alloy. *J. Electrochem. Soc.*, **154**, C562–570.

Okazaki, Y. (2002). Effect of friction on anodic polarization properties of metallic biomaterials. *Biomaterials*, **23**, 2071–2077.

Piconi, C., & Maccauro, G. (1999). Zirconia as a ceramic biomaterial. *Biomaterials*, **20**, 1–25.

Rodrigues, D. C., Viban, R. M., Jacobs, J. J., & Gilbert, J. L. (2009). *In vivo* severe corrosion and hydrogen embrittlement of retrieved modular body titanium alloy hip implant. *J. Biomed. Mater. Res. (Part B, Appl. Biomater.)*, **88**, 206–219.

Rondelli, G., & Vicentini, B. (1999). Localized corrosion behaviour in simulated human body fluids of commercial Ni–Ti orthodontic wires. *Biomaterials*, **20**, 785–792.

Seah, K. H. W., Thampuran, R., & Teoh, S. H. (1998). The influence of pore morphology on corrosion. *Corr. Sci.*, **40**, 547–556.

Shih, C.-C., Lin, S.-J., Chung, K.-H., Chen, Y.-L., & Su, Y.-Y. (2000). Increased corrosion resistance of stent materials by converting current surface film of polycrystalline oxide into amorphous oxide. *J. Biomed. Mater. Res.*, **52**, 323–332.

Singh, R., & Dahotre, N. B. (2007). Corrosion degradation and prevention by surface modification of biometallic materials. *J. Mater. Sci.: Mater. Med.*, **18**, 725–751.

Sittig, C., Textor, M., Spencer, N. D., Wieland, M., & Vallotton, P.-H. (1999). Surface characterization of implant materials c.p.Ti, Ti-6Al-7Nb and Ti-6Al-4V with different pretreatments. *J. Mater. Sci.: Mater. Med.*, **10**, 35–46.

Staiger, M. P., Pietak, A. M., Huadma, J., & Dias, G. (2006). Magnesium and its alloys as orthopaedic biomaterials: A review. *Biomaterials*, **27**, 1728–1734.

Steinemann, S. G. (1996). Metal implants and surface reactions. *Injury*, **27**(S-3), 16–22.

Tengvall, P., Lundström, I., Sjökvist, L., Elwing, H., & Bjursten, L. (1989). Titanium–hydrogen peroxide interactions: Model studies of the influence of the inflammatory response on titanium implants. *Biomaterials*, **10**, 166–175.

Thierry, B., Tabrizian, M., Savadogo, O., & Yahia, L'H. (2000). Effects of sterilization processes on NiTi alloy: Surface characterization. *J. Biomed. Mater. Res.*, **49**, 88–98.

Trepanier, C., Tabrizian, M., Yahia, L'H., Bilodeau, L., & Piron, D. (1998). Effect of modification of oxide layer on NiTi stent corrosion resistance. *J. Biomed. Mater. Res.*, **43**, 433–440.

Trepanier, C., Leung, T. K., Tabrizian, M., Yahia, L'H., Bienvenu, J.-G., Tanguay, J.-F., Piron, D. L., & Bildeau, L. (1999). Preliminary investigation of the effects of surface treatments on biological response to shape memory NiTi stents. *J. Biomed. Mater. Res. (Part B, Appl. Biomater.)*, **48**, 165–171.

Virtanen, S., Milosev, I., Gomez-Barrena, E., Trebse, R., Salo, J., et al. (2008). Special modes of corrosion under physiological and simulated physiological conditions. *Acta. Biomater.*, **4**, 468–476.

Walczak, J., Shahgaldi, F., & Heatley, F. (1998). *In vivo* corrosion of 316L stainless-steel hip implants: Morphology and elemental compositions of corrosion products. *Biomaterials*, **19**, 229–237.

Wälivaara, B., Askendal, A., Elwing, H., Lundström, I., & Tengvall, P. (1992). Antisera binding onto metals immersed in human plasma *in vitro*. *J. Biomed. Mater. Res.*, **26**, 1205–1216.

Williams, D. F. (1985). Physiological and microbiological corrosion. *Crit. Rev. Biocompat.*, **1**(1), 1–24.

Williams, D. F. (2008). On the mechanisms of biocompatibility. *Biomaterials*, **29**, 2941–2953.

Williams, D. F. (2009). On the nature of biomaterials. *Biomaterials*, **30**, 5897–5909.

Williams, R. L., & Williams, D. F. (1988). The characteristics of albumin adsorption on metal surfaces. *Biomaterials*, **9**(3), 206–212.

Witte, F. (2010). The history of biodegradable magnesium alloys. *Acta. Biomater.* **6**(8), 1680–1692.

Xu, L., Yu, G., Zhang, E., Pan, F., & Yang, K. (2007). *In vivo* corrosion behavior of Mg–Mn–Zn alloy for bone implant applications. *J. Biomed. Mater. Res., Part A*, **83**, 703–711.

Yoo, Y. R., Jang, S. G., Oh, K. T., Kim, J. G., & Kim, Y. S. (2008). Influences of passivating elements on the corrosion and biocompatibility of super stainless steels. *J. Biomed. Mater. Res., (Part B, Appl. Biomater.)*, **86**, 710–720.

CHAPTER II.4.5 PATHOLOGICAL CALCIFICATION OF BIOMATERIALS

Frederick J. Schoen[1] *and Robert J. Levy*[2]
[1]Professor of Pathology and Health Sciences and Technology (HST), Harvard Medical School, Executive Vice Chairman, Department of Pathology, Brigham and Women's Hospital, Boston, MA, USA
[2]Abramson Pediatric Research Center, The Children's Hospital of Philadelphia, University of Pennsylvania, Philadelphia, PA, USA

INTRODUCTION

Formation of nodular deposits of calcium phosphate or other calcium-containing compounds may occur on biomaterials and prosthetic devices used in the circulatory system and at other sites. This process, known as *calcification* or *mineralization*, has been encountered in association with both synthetic and biologically derived biomaterials in diverse clinical and experimental settings, including chemically-treated tissue (*bioprosthetic*) and homograft cardiac valve substitutes (Mitchell et al., 1998; Schoen and Levy, 1999, 2005), blood pumps used as cardiac assist devices (Schoen and Edwards, 2001), breast implants (Peters et al., 1998; Legrand et al., 2005), intrauterine contraceptive devices (Patai et al., 1998), urological stents (Vanderbrink et al., 2008), intraocular lenses (Neuhann et al., 2008; Nakanome et al., 2008; Rimmer et al., 2010), and scleral buckling materials (Brockhurst et al., 1993; Yu et al., 2005). Vascular access grafts for hemodialysis and synthetic vascular replacements composed of Dacron® or expanded polytetrafluoroethylene (e-PTFE) also calcify in some patients (Tomizawa et al., 1998; Schlieper et al., 2008). Calcification can lead to important clinical complications, such as stiffening and/or tearing of tissue heart valve substitutes, encrustation with blockage of a urinary stent or clouding of intraocular lenses (Table II.4.5.1).

Deposition of mineral salts of calcium (as calcium phosphates, especially calcium hydroxyapatite) occurs normally in bones and teeth, and is a critical determinant of their strength (called *physiologic* mineralization). Mineralization of skeletal tissues is both controlled and restricted to specific anatomic sites. The mature mineral phase of bone is a poorly crystalline calcium phosphate known as calcium hydroxyapatite, which has the chemical formula $Ca_{10}(PO_4)_6(HO)_2$. Mineralization of some implant biomaterials is desirable for proper function, e.g., osteoinductive materials used for orthopedic or dental applications (Habibovic and de Groot, 2007), and materials used to engineer skeletal and dental tissues (Kretlow and Mikos, 2007; Bueno and Glowacki, 2009). However, severe consequences can occur if mineralization occurs in regions that do not normally calcify (*pathologic* or *ectopic* mineralization) (Kirsch, 2008).

Since the biomaterials used in medical devices outside of the musculoskeletal and dental systems are not intended to calcify, calcification of these biomaterials is pathologic. The mature mineral phase of biomaterial-related and other forms of pathologic calcifications is a poorly crystalline calcium phosphate that closely resembles the calcium hydroxyapatite present in bones and teeth. Indeed, as we will see later, biomaterials-related calcification shares many features with other conditions of pathologic calcification and physiologic mineralization. Pathologic calcification of natural structures is also a common feature of important disease processes; for example, in native arteries and heart valves, calcification occurs as an important feature of the serious diseases atherosclerosis and calcific aortic stenosis, respectively (Mitchell and Schoen, 2010; Schoen and Mitchell, 2010).

Pathologic calcification is further classified as either *dystrophic* or *metastatic*, depending on its setting (Kumar et al., 2010). In dystrophic calcification the deposition of calcium salts (usually calcium phosphates) occurs in damaged or diseased tissues or related to biomaterials; moreover, dystrophic calcification usually occurs in the setting of normal systemic calcium metabolism (generally defined by a normal product of the serum levels of calcium and phosphorus). In contrast, metastatic calcification comprises the deposition of calcium salts in otherwise normal tissues in individuals who have deranged mineral metabolism, usually with markedly elevated blood calcium or phosphorus levels. Conditions favoring dystrophic and metastatic calcification can act synergistically. Thus, the rate and extent of dystrophic mineralization within abnormal tissues are accelerated when calcium and/or phosphorus serum levels are high, for example, in kidney failure or calcium supplementation (Umana et al., 2003; Peacock, 2010), and in osteoporosis (Hofbauer et al., 2007; Hjortnaes et al., 2010). Moreover, the ability to form physiologic mineral (e.g.,

TABLE II.4.5.1 Representative Prostheses and Devices with Clinical Consequences Due to Biomaterials Calcification

Configuration	Biomaterial	Clinical Consequence
Cardiac valve prostheses	Glutaraldehyde-pretreated porcine aortic valve or bovine pericardium, and allograft aortic/pulmonary valves	Valve obstruction or incompetency
Cardiac ventricular assist bladders	Polyurethane	Dysfunction by stiffening or cracking
Vascular grafts	Dacron® grafts and aortic allografts	Graft obstruction or stiffening
Soft contact lens	Hydrogels	Opacification
Intrauterine contraceptive devices	Silicone rubber, polyurethane or copper	Birth control failure by dysfunction or expulsion
Urinary prostheses	Silicone rubber or polyurethane	Incontinence and/or infection

in bone) is regulated through adjustment of enhancing and inhibiting substances, many of which circulate in the blood (Weissen-Planz et al., 2008). In young individuals (especially into early adulthood) the balance appropriately favors bone formation; moreover, the biochemical environment that favors physiologic bone formation in children also promotes calcification of biomaterials (Bass and Chan, 2006; Peacock, 2010).

The cells and extracellular matrix of dead tissues are the principal sites of pathologic calcification. Calcification of an implant biomaterial can occur within the tissue (*intrinsic* calcification) or at the surface, usually associated with attached cells and proteins (*extrinsic* calcification) or in the implant fibrous capsule. An important instance of extrinsic calcification is that associated with prosthetic valve infection (prosthetic valve endocarditis) or thrombi.

THE SPECTRUM OF PATHOLOGIC BIOMATERIALS AND MEDICAL DEVICE CALCIFICATION

Tissue Heart Valves

Calcific degeneration of glutaraldehyde-pretreated porcine bioprosthetic and bovine pericardial heart valves (Figure II.4.5.1) is a clinically significant and well-characterized dysfunction of a medical device due to biomaterials calcification (Schoen and Levy, 1999, 2005). The predominant pathologic process is intrinsic calcification of the valve cusps, largely initiated in the deep cells and the tissue from which the valve was fabricated and often involving collagen. Calcification leads to valve failure most commonly by causing cuspal tears, less frequently by cuspal stiffening, and rarely by inducing distant emboli. Overall, approximately half of porcine bioprostheses fail within 12–15 years. Calcification is more rapid and aggressive in the young; to exemplify, the rate of failure of bioprostheses is approximately 10% in 10 years in elderly recipients, but is nearly uniform in less than four years in most adolescent and preadolescent children (Chan et al., 2006).

In some young individuals with congenital cardiac defects or acquired aortic valve disease, human *allograft/homograft* aortic (or pulmonary) valves surrounded by a sleeve of aorta (or pulmonary artery) are used. Allograft valves are valves removed from a person who has died and transplanted to another individual; the tissue is usually cryopreserved, but not chemically cross-linked. Allograft vascular segments (without a valve) can be used to replace a large blood vessel. Allograft vascular tissue, whether containing an aortic valve or nonvalved, can undergo severe calcification, particularly in the wall; calcification can lead to allograft valve dysfunction or deterioration (Mitchell et al., 1998).

Calcific degeneration of glutaraldehyde-pretreated porcine bioprosthetic and bovine pericardial heart valves (Figure II.4.5.1) is a clinically significant and well-characterized dysfunction of a medical device due to biomaterials calcification (Schoen and Levy, 1999, 2005). The predominant pathologic process is intrinsic calcification of the valve cusps, largely initiated in the deep cells and the tissue from which the valve was fabricated and often involving collagen. Calcification leads to valve failure most commonly by causing cuspal tears, less frequently by cuspal stiffening, and rarely by inducing distant emboli. Overall, approximately half of porcine bioprostheses fail within 12–15 years.

Polymeric Heart Valves and Bladders in Blood Pumps

Calcification has also complicated the use of heart valves and artificial mitral valve *chordae tendineae* composed of polymers (e.g., polyurethane) (Fukunaga et al., 2010; Hilbert et al., 1987; Schoen et al., 1992a; Schoen and Levy, 1999; Wang et al., 2010) and the flexing bladder surfaces of blood pumps used as ventricular assist systems or total artificial hearts (Schoen and Edwards, 2001) (Figure II.4.5.2). Massive deposition of mineral leading to failure has been noted in experimental animals, but a lesser degree of calcification has been encountered following extended human implantation. Mineral deposits can result in deterioration of pump or valve

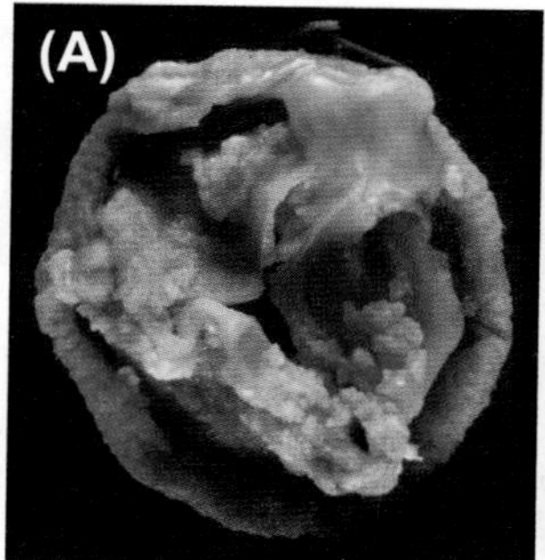

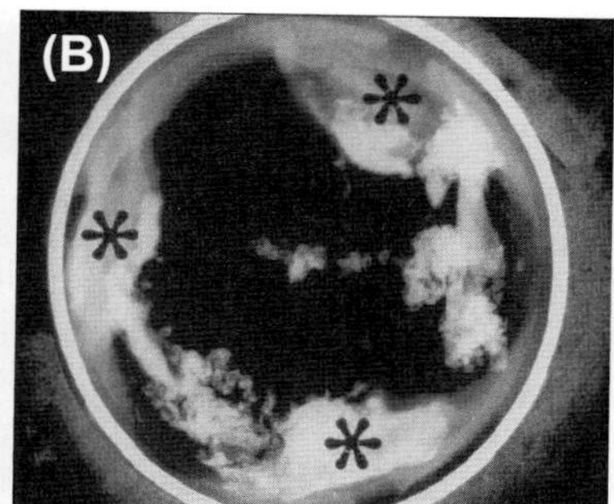

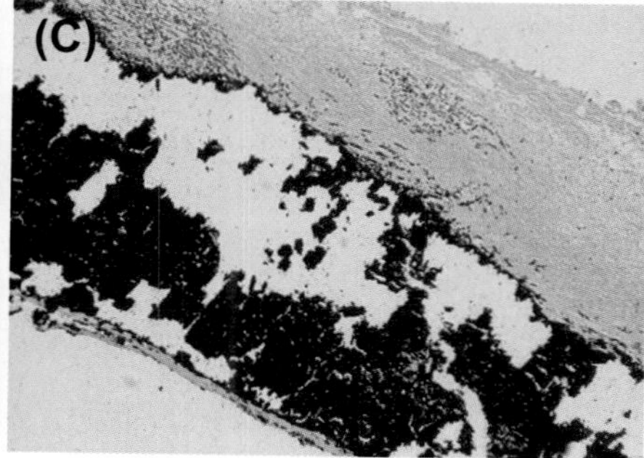

FIGURE II.4.5.1 Calcification of a pericardial bioprosthetic heart valve, implanted in a person for many years. (A) Gross photograph, demonstrating marked thickening of cusps by nodular calcific deposits; (B) radiograph of another long-term valve, demonstrating predominant deposits at commissures; (C) photomicrograph demonstrating calcific nodule deeply embedded within cuspal tissue; von Kossa stain (calcium phosphates black) 100×.

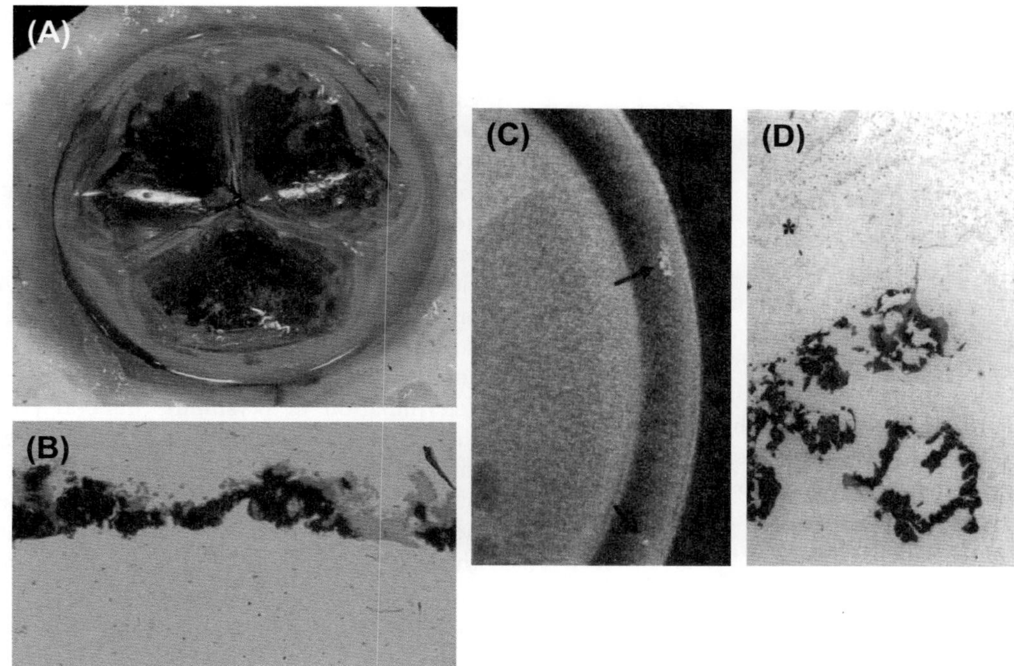

FIGURE II.4.5.2 Calcification of an experimental polymeric (polyurethane) heart valve. (A) Gross photograph of valve; (B) photomicrograph calcified material at surface of polymer (polymer at bottom of photo); and calcification of the flexing bladder of a ventricular assist pump removed from a person after 257 days. (C) Gross photograph. Calcific masses are noted by arrows. (D) photomicrograph (B) and (D) von Kossa stain (calcium phosphates black) 100×.

performance through loss of bladder pliability or the initiation of tears. Blood pump calcification, regardless of the type of polyurethane used, generally predominates along the flexing margins of the diaphragm, emphasizing the important potentiating role of mechanical factors in this system (Coleman et al., 1981; Harasaki et al., 1987; Kantrowitz et al., 1995).

Calcific deposits associated with polymeric heart valve or blood pump components can occur either within the adherent layer of deposited thrombus, proteins, and cells (*pseudointima*) on the blood-contacting surface (extrinsic mineralization) or is below the surface (intrinsic calcification) (Joshi et al., 1996). In some cases, calcific deposits are associated with microscopic surface defects, originating either during bladder fabrication or resulting from cracking during function.

Breast Implants

Calcification of silicone-gel breast implant capsules occurs as discrete calcified plaques at the interface of the inner fibrous capsule with the implant surface (Peters and Smith, 1998; Gumus, 2009). Capsular calcification has also been encountered with breast implants in patients with silicone envelopes filled with saline. Calcification could interfere with effective tumor detection and diagnosis, which could potentially delay treatment, particularly in patients who have breast implants following reconstructive surgery for breast cancer. In a study of breast implants removed predominantly for capsular contraction, 16% overall demonstrated calcific deposits, including 26% of implants inserted for 11–20 years, and all those inserted for more than 23 years (Peters and Smith, 1995). Another study demonstrated calcification associated with virtually all implants examined after more than 20 years (Legrand et al., 2005).

Ivalon® (polyvinyl alcohol) sponge prostheses, used quite extensively during the 1950s, were also frequently associated with calcification. In Japan, where augmentation mammoplasty was frequently performed using injection of foreign material (liquid paraffin from approximately 1950 until 1964, and primarily liquid silicone injections thereafter), the incidence of calcification has been much higher. One study showed calcification in 45% of breast augmentations which were done by injection (Koide and Katayama, 1979).

Intrauterine Contraceptive Devices

Intrauterine contraceptive devices (IUDs) are composed of plastic or metal, and placed in a woman's uterus chronically to prevent implantation of a fertilized egg. Device dysfunction due to calcific deposits can be manifested as contraceptive failure or device expulsion. For example, accumulation of calcific plaque could prevent the release of the active contraception-preventing agent – either ionic copper from copper-containing IUDs or an active agent from hormone-releasing IUD systems. Studies of explanted IUDs using transmission and electron microscopy coupled with X-ray microprobe analysis have shown that surface calcium deposition is virtually ubiquitous, but highly variable among patients (Khan and Wikinson, 1985; Patai et al., 1998).

Urinary Stents and Prostheses

Mineral crusts form on the surfaces of urinary stents and nephrostomy tubes, which are used extensively in urology to alleviate urinary obstruction or incontinence (Goldfarb et al., 1989; Vanderbrink et al., 2008). Observed in both male and female urethral implants and artificial ureters, this calcification can lead to obstruction and device failure. The mineral crust typically consists of either calcium oxalate or calcium phosphate mineral such as hydroxyapatite or struvite, an ammonium- and magnesium-containing phosphate mineral derived from urine. There is some evidence that encrustation may both result from and predispose to bacterial infection.

Intraocular and Soft Contact Lenses and Scleral Buckles

Calcium phosphate deposits can opacify intraocular and soft contact lenses, typically composed of poly-2-hydroxyethyl-methacrylate (HEMA) (Bucher et al., 1995; Nakanome et al., 2008; Neuhann et al., 2008; Rimmer et al., 2010). Calcium from tear fluid is considered to be the source of the deposits found on HEMA contact lens, and calcification may be potentiated in patients with systemic and ocular conditions associated with elevated tear calcium levels (Klintworth et al., 1977). Encircling scleral bands, used in surgery for retinal detachment, and composed of silicone or hydrogel materials, also calcify (Lane et al., 2001).

ASSESSMENT OF BIOMATERIALS CALCIFICATION

Calcific deposits are investigated using morphologic and chemical techniques (Table II.4.5.2). Morphologic techniques facilitate detection and characterization of the microscopic and ultrastructural sites and distribution of calcific deposits, and their relationship or tissue or biomaterials structural details. Such analyses directly yield important qualitative (but not quantitative) information. In contrast, chemical techniques, which require destruction of the tissue specimen, permit identification and quantitation of both bulk elemental composition and determination of crystalline mineral phases. However, such techniques generally cannot relate the location of the mineral to the details of the underlying tissue structure. The most comprehensive studies use several analytical modalities to simultaneously characterize both morphologic and chemical aspects of calcification. Moreover, newer techniques are available for non-destructive and potentially non-invasive characterizing of calcification, both in specimens (microcomputer tomography (micro CT)) (Ford-Hutchinson, 2003; Neues and Epple, 2008), and *in vivo* using molecular imaging (Aikawa et al., 2007). Micro CT has been used extensively in studies of bone regenerative biomaterials (Jiang et al., 2009). Molecular imaging, which probes *biomarkers* of particular targets or pathways of the cellular and molecular mechanisms of calcification, is particularly exciting in this context to enable the visualization of the ongoing and dynamic process of calcification, potentially quantitatively and repetitively in living organisms (New and Aikawa, 2011).

Morphologic Evaluation

Morphologic assessment of calcification is done by means of several readily available and well-established

TABLE II.4.5.2 Methods for Assessing Calcification

Technique	Sample Preparation	Analytical Results
Morphologic Procedures		
Gross examination	Gross specimen	Overall morphology
Radiographs	Gross specimen	Calcific deposit distribution
Micro-computed tomography	Gross specimen	Three-dimensional reconstruction of calcific deposit morphology, localization, and quantification
Light microscopy – von Kossa or alizarin red stains	Formalin or glutaraldehyde fixed	Microscopic phosphate or calcium distribution, respectively
Transmission electron microscopy	Glutaraldehyde fixed	Mineral ultrastructure
Scanning electron microscopy with electron microprobe	Glutaraldehyde fixed	Elemental localization and quantitation
Electron energy loss spectroscopy	Glutaraldehyde fixed or rapidly frozen	Elemental localization and quantitation (high sensitivity)
Chemical Procedures		
Atomic absorption	Ash or acid hydrolyzate	Bulk calcium
Colorimetric phosphate analysis	Ash or acid hydrolyzate	Bulk phosphorus
X-ray diffraction	Powder	Nature of crystal phase
Infrared spectroscopy	Powder	Carbonate mineral phase

techniques that range from macroscopic (gross) examination and radiographs (X-rays) of explanted prostheses to sophisticated electron energy loss spectroscopy. Each technique has advantages and limitations; several techniques are often used in combination to obtain an understanding of the structure, composition, and mechanism of each type of calcification.

Careful visual examination of the specimen, often under a dissecting (low power) microscope, and radiography assess distribution of mineral. Specimen radiography typically involves placing the explanted prosthesis on an X-ray film plate and exposing to an X-ray beam in a special device used for small samples (e.g., we use the Faxitron, Hewlett-Packard, McMinnville, CA with an energy level of 35 keV for 1 minute for valves). Deposits of mineral appear as bright densities which have locally blocked the beam from exposing the film (see Figure II.4.5.1B).

Light microscopy of calcified tissues is widely used. Identification of mineral is facilitated through the use of either calcium- or phosphorus-specific stains, such as alizarin red (which stains calcium) or von Kossa (which stains phosphates brown-black) (Figure II.4.5.2B and Figure II.4.5.3). These histologic stains are readily available, can be easily applied to tissue sections embedded in either paraffin or plastic, and are most useful for confirming and characterizing suspected calcified areas which have been noted by routine hematoxylin and eosin staining techniques. Sectioning of calcified tissue which has been embedded in paraffin often leads to considerable artifacts due to fragmentation; embedding of tissue with calcific deposits in a harder medium such as glycolmethacrylate polymer can yield superior section quality.

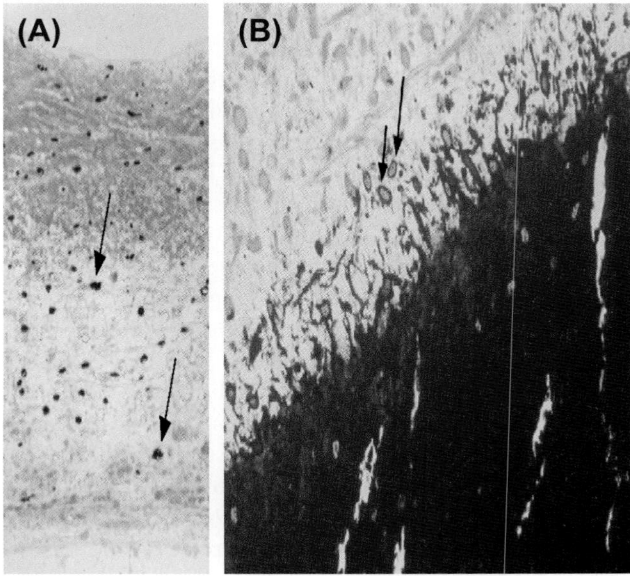

FIGURE II.4.5.3 Light microscopic appearance of calcification of experimental porcine aortic heart valve tissue. Note cell-based orientation of initial deposits (arrows). (A) Implanted subcutaneously in three-week-old rats for 72 hours; (B) implanted in growing sheep for five months, demonstrating predominant site of growing edge of calcification in cells of the residual porcine valve matrix (arrows). ((B) Reproduced by permission from Schoen, F. J. (2001). Pathology of heart valve substitution with mechanical and tissue prostheses, In: *Cardiovascular Pathology,* 3rd edn., Silver, M. D., Gotlieb, A. I. & Schoen, F. J. (Eds.). W. B. Saunders: Philadelphia, PA, pp. 629–677.)

Electron microscopic techniques, which involve the bombardment of the specimen with a highly focused electron beam in a vacuum, have much to offer in the determination of early sites of calcific deposits. In transmission electron microscopy (TEM), the beam traverses an ultra-thin section (0.05 μm) (Figure II.4.5.4); observation of the ultrastructure (submicron tissue features) of calcification by TEM facilitates the understanding of the mechanisms by which calcific crystals form. Scanning electron microscopy (SEM) images the specimen

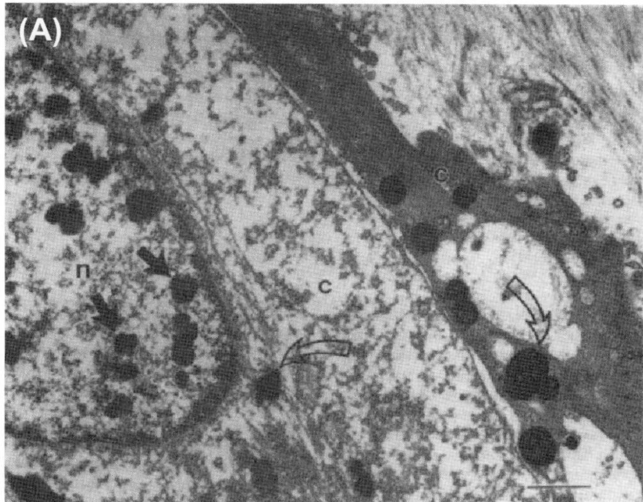

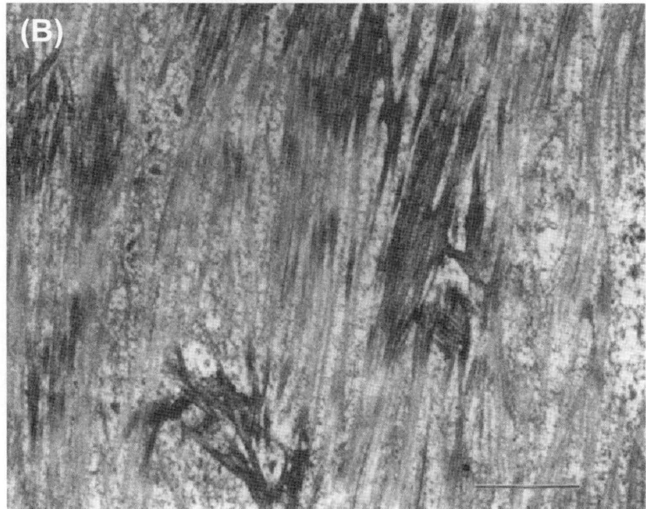

FIGURE II.4.5.4 Transmission electron microscopy of calcification of experimental porcine aortic heart valve implanted subcutaneously in three-week-old rats. (A) 48-hour implant demonstrating focal calcific deposits in nucleus of one cell (closed arrows) and cytoplasm of two cells (open arrows), n, nucleus; c, cytoplasm. (B) 21-day implant demonstrating collagen calcification. Bar = 2 μm. Ultrathin sections stained with uranyl acetate and lead citrate. (Figure (A) reproduced with permission from Schoen, F. J. et al. (1985), p.521.)

surface, and can be coupled with elemental localization by energy-dispersive X-ray analysis (EDXA), allowing a semi-quantitative evaluation of the local progression of calcium and phosphate deposition in a site-specific manner. Electron energy loss spectroscopy (EELS) couples transmission electron microscopy with highly sensitive elemental analyses to provide a most powerful localization of incipient nucleation sites and early mineralization (Webb et al., 1991). In general, the more highly sensitive and sophisticated morphologic techniques require more demanding and expensive preparation of specimens to avoid unwanted artifacts. Forethought about, and careful planning of, specimen handling optimizes the yield provided by the array of available techniques, and allows multiple approaches to be used on specimens from a particular experiment.

Chemical Assessment

Quantitation of calcium and phosphorus in biomaterial calcifications permits characterization of the progression of deposition, comparison of severity of deposition among specimens, and determination of the effectiveness of preventive measures (Levy et al., 1983a, 1985a; Schoen et al., 1985, 1986, 1987; Schoen and Levy, 1999). However, such techniques destroy the specimen (and hence obliterate any spatial information) during preparation. Calcium has been quantitated by atomic absorption spectroscopy of acid-hydrolyzed or ashed samples. Phosphorus is usually quantitated as phosphate, using a molybdate complexation technique with spectrophotometric detection. The specific crystalline form of the mineral phase can be determined by X-ray diffraction. Carbonate-containing mineral phases may also be analyzed by infrared spectroscopy.

MECHANISMS OF BIOMATERIALS CALCIFICATION

Regulation of Pathologic Calcification

The determinants of biomaterial mineralization include factors related to: (1) host metabolism; (2) implant structure and chemistry; and (3) mechanical factors. Natural cofactors and inhibitors may also play a role (see below). The most important host metabolic factor relates to young age, with more rapid calcification taking place in immature patients or experimental animals (Levy et al., 1983a). Although the relationship is well-established, the mechanisms accounting for this effect are uncertain. An important implant factor for bioprosthetic tissue is pretreatment with glutaraldehyde, done to preserve the tissue (Golomb et al., 1987; Grabenwoger et al., 1996). It has been hypothesized that the cross-linking agent glutaraldehyde stabilizes and perhaps modifies phosphorous-rich calcifiable structures in the bioprosthetic tissue. These sites seem to be capable of mineralization upon implantation when exposed to the comparatively high calcium levels of extracellular fluid. Calcification of the two principal types of tissue used in bioprostheses – glutaraldehyde-pretreated porcine aortic valve or glutaraldehyde-pretreated bovine pericardium – is similar in extent, morphology, and mechanisms (Schoen and Levy, 2009).

> The determinants of biomaterial mineralization include factors related to: (1) host metabolism; (2) implant structure and chemistry; and (3) mechanical factors. Natural cofactors and inhibitors may also play a role.

Calcification has typically been considered a passive, unregulated, and degenerative process. However, the observations of matrix vesicles, hydroxyapatite mineral, and bone-related morphogenetic and non-collagenous proteins in situations of pathological calcification has suggested that the mechanisms responsible for pathologic calcification may be regulated, similarly to physiologic mineralization of bone and other hard tissues, and that some of the mechanisms regulating these paradoxically different processes may be shared (Johnson et al., 2006; Demer and Tintut, 2008; Persy and D'Haese, 2009). In normal blood vessels and valves, inhibitory mechanisms outweigh procalcification inductive mechanisms; in contrast, in bone and pathologic tissues, inductive mechanisms dominate. Naturally-occurring inhibitors to crystal nucleation and growth (of which at least 11 have been identified), may also play a role in biomaterial and other cardiovascular calcification (Weissen-Plenz et al., 2008). Specific inhibitors in this context include osteopontin (Steitz et al., 2002) and phosphocitrate (Tew et al., 1980). Naturally-occurring mineralization cofactors, such as inorganic phosphate (Jono et al., 2000), bone morphogenic protein (a member of the transforming growth factor (TGF) beta family) (Bostrom et al., 1993), pro-inflammatory lipids (Demer, 2002), and other substances (e.g., cytokines) may also play a role in pathologic calcification. The non-collagenous proteins osteopontin, TGF-beta1, and tenascin-C involved in bone matrix formation and tissue remodeling have been demonstrated in clinical calcified bioprosthetic heart valves, natural valves, and atherosclerosis, suggesting that they play a regulatory role in these forms of pathologic calcification in humans (Srivasta et al., 1997; Bini et al., 1999; Jian et al., 2001, 2003), but a direct relationship has not yet been demonstrated.

Mechanical factors also regulate calcification. Both intrinsic and extrinsic mineralization of a biomaterial is generally enhanced at the sites of intense mechanical deformations generated by motion, such as the points of flexion in heart valves (Thubrikar et al., 1983). The mechanisms underlying mechanical potentiation of calcification associated with biomaterials are incompletely understood, but the effect mimics the well-known Wolf's

TABLE II.4.5.3 Experimental Models of Calcification

Type	System	Typical Duration
Calcification of bioprosthetic or other tissue heart valve	*In vitro* incubation of tissue fragment or flexing valves	Days to weeks
	Rat subdermal implant of tissue fragment	3 weeks
	Calf or sheep orthotopic valve replacement	3–5 months
	Rat or sheep descending aorta	1–5 months
Calcification of polyurethane	Rat subdermal implant of material sample	1–2 months
	Calf or sheep artificial heart implant	5 months
	Trileaflet polymeric valve implant in calf or sheep	5 months
Calcification of hydrogel	Rat subdermal implant of material sample	3 weeks
Calcification of collagen	Rat subdermal implant of material sample	3 weeks
Urinary encrustation	*In vitro* incubation	Hours to days
	In vivo bladder implants (rats and rabbits)	10 weeks

law in bone, in which formation and adaptation occurs in response to the mechanical forces that it experiences (Chen et al., 2010). Moreover, the enhancement is seen in systems where static but not dynamic mechanical strain is applied (Levy, 1983a), and in systems where live stem cells are subjected to a spectrum of mechanical environments (Yip et al., 2009).

Experimental Models for Biomaterials Calcification

Animal models have been developed for the investigation of the calcification of bioprosthetic heart valves, aortic homografts, cardiac assist devices, and trileaflet polymeric valves (Table II.4.5.3). The most widely used experimental models used to investigate the pathophysiology of bioprosthetic tissue calcification, and as a preclinical screen of new or modified materials and design configurations, include tricuspid or mitral replacements or conduit-mounted valves in sheep or calves, and isolated tissue (i.e., not in a valve) samples implanted subcutaneously in mice, rabbits or rats (Levy et al., 1983a; Schoen et al., 1985, 1986). In both circulatory and non-circulatory models, bioprosthetic tissue calcifies progressively with a morphology similar to that observed in clinical specimens, but with markedly accelerated kinetics. *In vitro* models of biomaterials calcification have been investigated, but have not been useful in studying mechanisms or preventive strategies (Schoen et al., 1992a; Mako and Vesely, 1997).

Compared with the several years normally required for calcification of clinical bioprostheses, valve replacements in sheep or calves calcify extensively in three to six months (Schoen et al., 1985, 1986, 1994). However, expense, technical complexity, and stringent housing and management procedures pose important limitations to all the circulatory models using large animals. In addition, implantation in the heart requires the use of complex surgical procedures utilizing cardiopulmonary bypass, as well as a high level of surgical expertise and postoperative care. These limitations stimulated the development of subdermal (synonym *subcutaneous* – under the skin) implant models. Subdermal bioprosthetic implants in rats, rabbits, and mice provide the following useful features: (1) calcification occurs at a markedly accelerated rate in a morphology comparable to that seen in circulatory explants; (2) the model is economical so that many specimens can be studied with a given set of experimental conditions, thereby allowing quantitative characterization and statistical comparisons; and (3) specimens are easily and quickly retrieved from the experimental animals, facilitating the careful manipulation and rapid processing required for detailed analyses (Levy et al., 1983; Schoen et al., 1985, 1986).

Thus, the subcutaneous model is a technically convenient and economically advantageous vehicle for investigating host and implant determinants and mechanisms of mineralization, as well as for screening potential strategies for its inhibition (*anticalcification*). Promising approaches may be investigated further in a large animal valve implant model. Large animal implants as valve replacements are also used: (1) to elucidate further the processes accounting for clinical failures; (2) to evaluate the performance of design and biomaterials modifications in valve development studies; (3) to assess the importance of blood–surface interactions; and (4) to provide data required for approval by regulatory agencies (Schoen, 1992b and 1994). Polyurethane calcification has also been studied with subdermal implants in rats (Joshi et al., 1996).

Pathophysiology of Bioprosthetic Heart Valve Calcification

Data from valve explants from patients and subdermal and circulatory experiments in animal models using bioprosthetic heart valve tissue have elucidated the pathophysiology of this important clinical problem, and enhanced our understanding of pathologic calcification in general (Figure II.4.5.4). The similarities of calcification in the different experimental models and clinical bioprostheses suggest a common pathophysiology, independent of implant site. Calcification appears to depend on exposure of a susceptible substrate to extracellular fluid; mechanical factors and local implant-related or

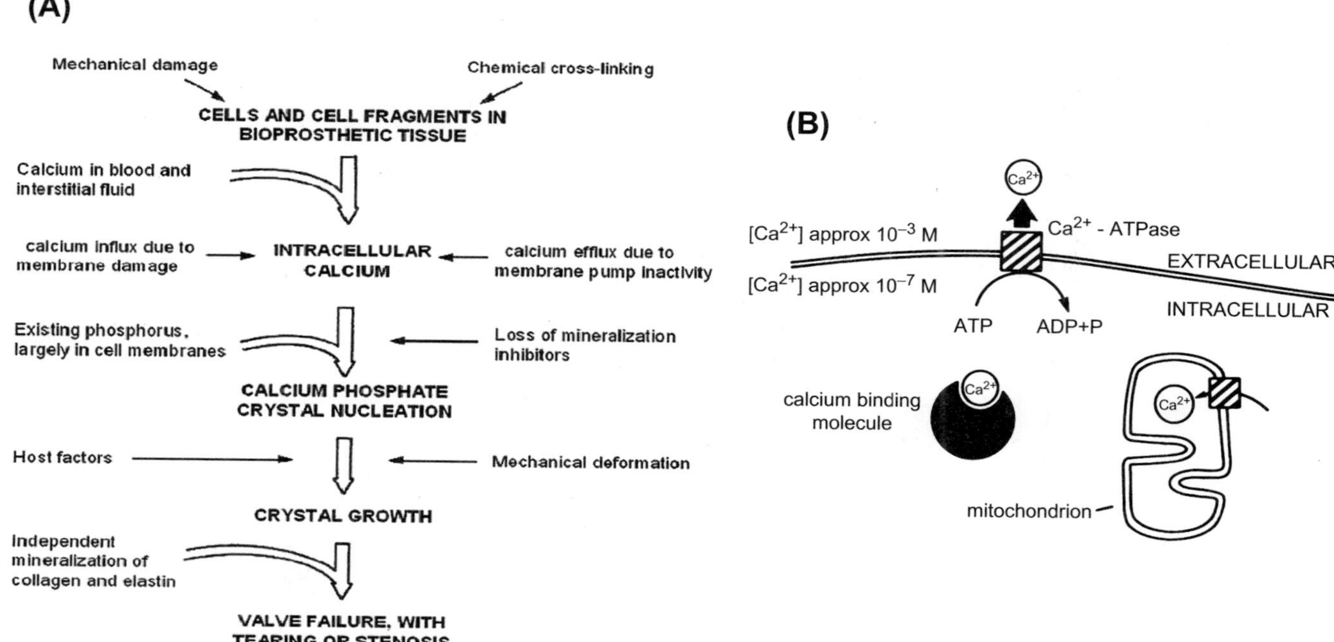

FIGURE II.4.5.5 Extended hypothetical model for the calcification of bioprosthetic tissue. (A) Overall model, which considers host factors, implant factors, and mechanical damage, and relates initial sites of mineral nucleation to increased intracellular calcium in residual cells and cell fragments in bioprosthetic tissue. The ultimate result of calcification is valve failure, with tearing or stenosis. The key contributory role of existing phosphorus in membrane phospholipids and nucleic acids in determining the initial sites of crystal nucleation is emphasized, and a possible role for the independent mineralization of collagen is acknowledged. Mechanical deformation probably contributes to both nucleation and growth of calcific crystals. (B) Events at the cell membrane and other calcium-binding structures. There is a substantial physiologic (normal) gradient of free calcium across the cell membrane (10^{-3} M outside, 10^{-7} M inside) which is maintained as an energy-dependent process. With cell death or membrane dysfunction, calcium phosphate formation can be initiated at the membranous cellular structures. ((A) Reproduced by permission from Schoen, F. J. & Levy, R. J. (2005). Calcification of tissue heart valve substitutes: Progress toward understanding and prevention. *Ann. Thorac. Surg.*, **79**, 1072–1080. (B) Reproduced by permission from Schoen, F. J. et al. (1988). Biomaterials-associated calcification: Pathology, mechanisms, and strategies for prevention. *J. Appl. Biomater.*, **22**, 11–36.)

circulating substances may play regulatory roles. However, since the morphology and extent of calcification in subcutaneous implants is analogous to that observed in clinical and experimental circulatory implants, despite the lack of dynamic mechanical activity characteristic of the circulatory environment, it is clear that dynamic stress promotes, but is not prerequisite for, calcification of bioprosthetic tissue. Interestingly, in the subcutaneous model, calcification is enhanced in areas of tissue folds, bends, and areas of shear, suggesting that static mechanical deformation also potentiates mineralization (Levy et al., 1983a). Although these data suggest that local tissue disruption mediates the mechanical effect, the precise mechanisms by which mechanical factors influence calcification are uncertain.

Although a potential role for inflammatory and immune processes has been postulated by some investigators (Love, 1993; Human and Zilla, 2001a,b), no definite role has been demonstrated for circulating cells, and many lines of evidence suggest that neither nonspecific inflammation nor specific immunologic responses appear to favor bioprosthetic tissue calcification. Proponents of an immunological mechanism for failure cite that: (1) experimental animals can be sensitized to both fresh and cross-linked bioprosthetic valve tissues; (2) antibodies to valve components can be detected in some patients following valve dysfunction; and (3) some failed tissue valves have brisk mononuclear inflammation, no causal immunologic basis has been demonstrated for bioprosthetic valve calcification. However, calcification in either circulatory or subcutaneous locations is not usually associated with inflammation. Moreover, in experiments in which valve cusps were enclosed in filter chambers that prevent host cell contact with tissue, but allow free diffusion of extracellular fluid and implantation of valve tissue in congenitally athymic ("nude") mice who have essentially no T-cell function, calcification morphology and extent are unchanged (Levy et al., 1983a,b). Clinical and experimental data detecting antibodies to valve tissue after failure may reflect a secondary response to valve damage, rather than a cause of failure.

Glutaraldehyde pretreatment of tissues or severe anoxia/hypoxia/ischemia leading to cell death in natural tissues or tissue-based biomaterials engenders cell injury. Consistent with a dystrophic mechanism, the initial calcification sites in bioprosthetic tissue are predominantly dead cells and cell membrane fragments (Levy et al., 1983a; Schoen et al., 1985, 1986; Schoen and Levy, 1999) (see Figure II.4.5.4). In non-functional cells or cells which have been rendered non-viable by glutaraldehyde fixation, the normal cellular handling of calcium ions is disrupted. Normally, plasma calcium concentration is 1 mg/ml (approximately 10^{-3} M); since the membranes of healthy cells pump calcium out, the concentration of calcium in the cytoplasm is 1000–10,000 times lower (approximately 10^{-7} M). The cell membranes and other intercellular structures are high in phosphorus (as phospholipids, especially phosphatidyl serine, which can bind calcium); they can serve as nucleators. Mitochondria are also enriched in calcium. Other initiators under various circumstances include collagen and elastic fibers of the extracellular matrix, denatured proteins, phosphoproteins, fatty acids, blood platelets and, in the case of infection, bacteria. We have previously hypothesized that cells calcify after glutaraldehyde pretreatment because this cross-linking agent stabilizes all the phosphorous stores, but the normal mechanisms for elimination of calcium from the cells are not available in glutaraldehyde-pretreated tissue (Schoen et al., 1986). Initial calcification deposits eventually enlarge and coalesce, resulting in grossly mineralized nodules that cause prostheses to malfunction.

Calcification of the adjacent aortic wall portion of glutaraldehyde-pretreated porcine aortic valves, and valvular allografts and vascular segments, is also observed clinically and experimentally. Mineral deposition occurs throughout the vascular cross-section, but is accentuated in the dense bands at the inner and outer media, and cells and elastin (itself not a prominent site of mineralization in cusps) are the major sites. In non-stented porcine aortic valves which have greater portions of aortic wall exposed to blood than in currently used stented valves, calcification of the aortic wall could stiffen the root, altering hemodynamic efficiency, causing nodular calcific obstruction, potentiating wall rupture or providing a nidus for emboli. Some anticalcification agents (see later), including 2-amino-oleic acid (AOA) and ethanol, prevent experimental cuspal but not aortic wall calcification (Chen et al., 1994a,b).

Calcification of Collagen and Elastin

Calcification of the extracellular matrix structural proteins collagen and elastin has been observed in clinical and experimental implants of bioprosthetic and homograft valvular and vascular tissue, and has been studied using a rat subdermal model. Collagen-containing implants are widely used in various surgical applications, such as tendon prostheses and surgical absorptive sponges, but their usefulness is compromised owing to calcium phosphate deposits and the resultant stiffening. Cross-linking by either glutaraldehyde or formaldehyde promotes the calcification of collagen sponge implants made of purified collagen, but the extent of calcification does not correlate with the degree of cross-linking (Levy et al., 1986). Elastin calcification has also been studied (Vyavahare et al., 1999; Lee et al., 2006).

PREVENTION OF CALCIFICATION

Three strategies have been investigated for preventing calcification of biomaterial implants: (1) systemic therapy with anticalcification agents; (2) local therapy with implantable drug delivery devices; and (3) biomaterial modifications, whether by removal of a calcifiable component, addition of an exogenous agent or chemical alteration. The subcutaneous model has been widely used to screen potential strategies for calcification inhibition (*anticalcification* or *antimineralization*). Promising approaches have been investigated further in a large animal valve implant model. However, strategies that appeared efficacious in subcutaneous implants have not always proven favorable when used on valves implanted into the circulation (see below).

Analogous to any new or modified drug or device, a potential antimineralization treatment must be effective and safe. The treatment should not impede valve performance such as hemodynamics. Investigations of an anticalcification strategy must demonstrate not only the effectiveness of the therapy, but also the absence of adverse effects (Schoen et al., 1992b). Adverse effects in this setting could include systemic or local toxicity, tendency toward thrombosis on infection, and induction of immunological effects or structural degradation, with either immediate loss of mechanical properties or premature deterioration and failure. Indeed, there are several examples whereby an antimineralization treatment contributed to unacceptable degradation of the tissue (Jones et al., 1989; Gott et al., 1992; Schoen, 1998). The treatment should not impede normal valve performance, such as hemodynamics and durability. As summarized in more detail in Table II.4.5.4, a rational approach for preventing bioprosthetic calcification must integrate safety and efficacy considerations with the scientific basis for

TABLE II.4.5.4 Criteria for Efficacy and Safety of Antimineralization Treatments

Efficacy

- Effective and sustained calcification inhibition

Safety

- Adequate valve performance (i.e., unimpaired hemodynamics and durability)
- Does not cause adverse blood–surface interactions (e.g., hemolysis, platelet adhesion, coagulation protein activation, complement activation, inflammatory cell activation, binding of vital serum factors)
- Does not enhance local or systemic inflammation (e.g., foreign body reaction, immunologic reactivity, hypersensitivity)
- Does not cause local or systemic toxicity
- Does not potentiate inflection

(Modified from Schoen F.J. et al. (1992). Antimineralization treatments for bioprosthetic heart valves. Assessment of efficacy and safety. *J. Thorac. Cardiovasc. Surg.*, **104**,1285–1288.)

TABLE II.4.5.5 Preclinical Efficacy and Safety Testing of Antimineralization Treatments

Type of Study	Information Derived
Subcutaneous implantation in rats	• Initial efficacy screen • Mechanisms • Dose–response • Toxicity
Biomechanical evaluation	• Hemodynamics • Accelerated wear
Morphologic studies of unimplanted valves	• Structural degradation assessed by light and transmission electron microscopy • Scanning electron microscopy
Circulatory implants in large animals	• Device configuration, surgical technique, *in vivo* hemodynamics, explant valve pathology • Durability, thrombi, thromboembolism, hemolysis, cardiac and systemic pathology

inhibition of calcium phosphate crystal formation. This will of necessity involve the steps summarized in Table II.4.5.5, before appropriate clinical trials can be undertaken (Schoen et al., 1992b; Vyavahare et al., 1997a).

Experimental studies using bioprosthetic tissue implanted subcutaneously in rats have clearly demonstrated that adequate doses of systemic agents used to treat clinical metabolic bone disease can prevent its calcification (Levy et al., 1987). However, systemic therapy with anticalcification agents may be efficacious, but is unlikely to be safe. Sufficient doses of systemic agents used to treat clinical metabolic bone disease, including calcium chelators (e.g., diphosphonates such as ethane hydroxybisphosphonate (EHBP)), can prevent the calcification of bioprosthetic tissue implanted subcutaneously in rats. However, because they also interfere with physiologic calcification (i.e., bone growth), systemic drug administration is associated with many side-effects in calcium metabolism, and animals receiving doses sufficient to prevent bioprosthetic tissue calcification suffer growth retardation. Thus, the principal disadvantage of the systemic use of anticalcification agents for preventing pathologic calcification is the consequent inhibition of bone formation and body growth. To avoid this difficulty, approaches for preventing the calcification of bioprosthetic heart valves have been based on co-implants of a drug delivery system adjacent to the prosthesis, in which the effective drug concentration would be confined to the site where it is needed (i.e., the implant), and systemic side-effects would be prevented (Levy et al., 1985b). A localized anticalcification effect would be particularly attractive in young people. Studies incorporating EHBP in nondegradable polymers, such as ethylene-vinyl acetate (EVA), polydimethylsiloxane (silicone), silastic, and polyurethanes have shown the effectiveness of this strategy in animal models. This approach, however, has been difficult to implement clinically.

TABLE II.4.5.6 Prototypical Agents for Mechanism-Based Prevention of Calcification in Bioprosthetic Heart Valves

Mechanisms	Strategy/Agent
Inhibition of hydroxyapatite formation	Ethane hydroxybisphosphonate (EHBP)
Inhibition of calcium uptake	2-Alpha-amino-oleic acid (AOA™)
Inhibition of Ca-P crystal growth; inhibition of alkaline phosphate; chemical modification of elastin	Ferric/aluminum chloride exposure
Phospholipid extraction	Sodium dodecyl sulfate (SDS)
Phospholipid extraction and collagen conformation modification	Ethanol exposure
Eliminate glutaraldehyde potentiation of calcification: • amino acid neutralization of glutaraldehyde residues • polyepoxide (polyglycidal ether), acyl azide, carbodiimide, cyanimide and glycerol cross-linking • dye-mediated photooxidation	Modification of (alternatives to) glutaraldehyde fixation

AOA™ (2-α-amino-oleic acid) is a trademark of Biomedical Designs, Inc., of Atlanta, GA, USA.

The approach which is most likely to yield an improved clinical valve in the short-term involves modification of the substrate, either by removing or altering a calcifiable component or binding an inhibitor. Forefront strategies should also consider: (1) a possible synergism provided by multiple anticalcification agents and approaches used simultaneously; (2) new materials; and (3) the possibility of tissue-engineered heart valve replacements. The agents most widely studied, for efficacy, mechanisms, lack of adverse effects, and potential clinical utility are summarized below and in Table II.4.5.6. Combination therapies using multiple agents may provide synergy of beneficial effects (Levy et al., 2003), potentially permitting simultaneous prevention of calcification in both cusps and aortic wall, particularly beneficial in stentless aortic valves.

Inhibitors of Hydroxyapatite Formation

Bisphosphonates. Ethane hydroxybisphosphonate (EHBP) has been approved by the FDA for human use to inhibit pathologic calcification, and to treat hypercalcemia of malignancy. Compounds of this type probably inhibit calcification by poisoning the growth of calcific crystals. Either cuspal pretreatment or systemic or local therapy of the host with diphosphonate compounds inhibits experimental bioprosthetic valve calcification (Levy et al., 1985, 1987; Johnston et al., 1993). Recently, controlled clinical trials which have orally administered bisphosphonates have demonstrated the ability to stabilize osteoporosis. These agents, such as Alendronate (Phosphomax®, Merck, Inc.) are hypothesized to act by stabilizing bone mineral.

Trivalent Metal Ions. Pretreatment of bioprosthetic tissue with iron and aluminum (e.g., $FeCl_3$ and $AlCl_3$) inhibits calcification of subdermal implants with glutaraldehyde-pretreated porcine cusps or pericardium (Webb et al., 1991; Carpentier et al., 1997). Such compounds are hypothesized to act through complexation of the cation (Fe or Al) with phosphate, thereby preventing calcium phosphate formation. Both ferric ion and the trivalent aluminum ion inhibit alkaline phosphatase, an important enzyme used in bone formation. This may be related to their mechanism for preventing initiation of calcification. Furthermore, research from our laboratories has demonstrated that aluminum chloride prevents elastin calcification through a permanent structural alteration of the elastin molecule. These compounds are also active when released from polymeric controlled-release implants.

Calcium Diffusion Inhibitor

Amino-oleic Acid. 2-alpha-amino-oleic acid (AOA™, Biomedical Designs, Inc., Atlanta, GA) bonds covalently to bioprosthetic tissue through an amino linkage to residual aldehyde functions, and inhibits calcium flux through bioprosthetic cusps (Chen et al., 1991, 1994a). AOA™ is effective in mitigating cusp, but not aortic wall, calcification in rat subdermal and cardiovascular implants. This compound is used in FDA-approved porcine aortic valves (Fyfe and Schoen, 1999; Celiento et al., 2012; El-Hamamsy et al., 2011).

Removal/Modification of Calcifiable Material

Surfactants. Incubation of bioprosthetic tissue with sodium dodecyl sulfate (SDS) and other detergents extracts the majority of acidic phospholipids (Hirsch et al., 1993); this is associated with reduced mineralization, probably resulting from suppression of the initial cell-membrane oriented calcification (Figure II.4.5.6). This compound is used in an FDA-approved porcine valve (David et al., 1998; Bottio et al., 2003).

Ethanol. Ethanol preincubation of glutaraldehyde-cross-linked porcine aortic valve bioprostheses prevents calcification of the valve cusps in both rat subdermal implants, and sheep mitral valve replacements (Vyavahare et al., 1997b, 1998). Eighty percent ethanol pretreatment: (1) extracts almost all phospholipids and cholesterol from glutaraldehyde-cross-linked cusps; (2) causes a permanent alteration in collagen conformation as assessed by Attenuated Total Reflectance-Fourier Transform Infrared Spectroscopy (ATR-FTIR); (3) affects cuspal interactions with water and lipids; and (4) enhances cuspal resistance to collagenase. Ethanol is in clinical use as a porcine valve cuspal pretreatment in both Europe and the United States, and use in combination with aluminum treatment of the aortic wall of a stentless valve is under consideration.

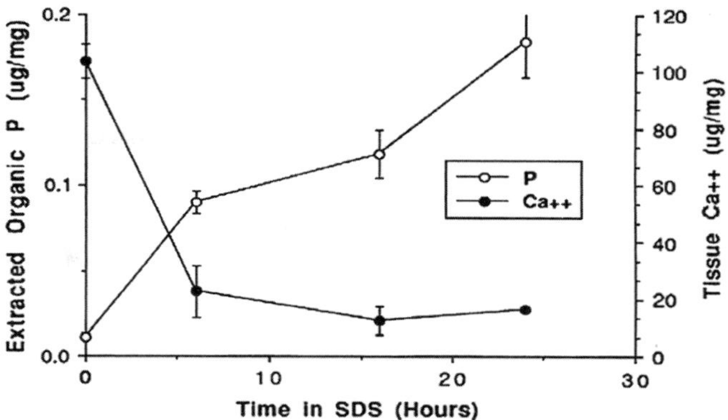

FIGURE II.4.5.6 Reduction of calcification of bioprosthetic tissue by preincubation in 1% SDS demonstrated in a rat subcutaneous model of glutaraldehyde cross-linked porcine aortic valve. These results support the concept that phospholipid extraction is an important but perhaps not the only mechanism of SDS efficacy. (Reproduced by permission from Schoen, F. J., Levy, R. J. & Piehler, H. R. (1992). Pathological considerations in replacement cardiac valves. *Cardiovasc. Pathol.*, **1**, 29–52.)

Decellularization. Since the initial mineralization sites are devitalized connective cells of bioprosthetic tissue, these cells may be removed from the tissue, with the intent of making the bioprosthetic matrix less prone to calcification (Courtman et al., 1994; Wilson et al., 1995).

Use of Tissue Fixatives Other than Glutaraldehyde, and Modification of Glutaraldehyde Fixation

Since previous studies have demonstrated that conventional glutaraldehyde fixation is conducive to calcification of bioprosthetic tissue, several studies have investigated modifications of, and alternatives to, conventional glutaraldehyde pretreatment. For example, and paradoxically, fixation of bioprosthetic tissue by extraordinarily high concentrations of glutaraldehyde (5–10× those normally used) appear to inhibit calcification (Zilla et al., 1997, 2000); moreover, residual glutaraldehyde residues in bioprosthetic tissue can be neutralized ("detoxified") by treatment with lysine or diamine, this inhibits calcification of subdermal implants (Grabenwoger et al., 1992; Zilla et al., 2000, 2005; Trantina-Yates et al., 2003).

Nevertheless, there is presently no mechanistic explanation for these glutaraldehyde-related observations. Non-glutaraldehyde cross-linking of bioprosthetic tissue with epoxides, carbodiimides, acylazides, and other compounds reduces their calcification in rat subdermal implant studies (Xi et al., 1992; Myers et al., 1995), and with triglycidylamine (TGA), an epoxy cross-linker

(Connolly et al., 2011) and photooxidative preservative in large animal implants in sheep (Moore and Phillips, 1997).

Alternative Materials

Polyurethane trileaflet valves have been fabricated and investigated as a possible alternative to bioprostheses or mechanical valve prostheses (Ghanbari et al., 2009). Despite versatile properties, such as superior abrasion resistance, hydrolytic stability, high flexural endurance, excellent physical strength, and acceptable blood compatibility, the use of polymers has been hampered by calcification, thrombosis, tearing, and biodegradation. Although the exact mechanism of polyurethane calcification is as yet unclear, it is believed that several physical, chemical, and biologic factors (directly or indirectly) play an important role in initiating this pathologic disease process (Schoen et al., 1992c; Joshi et al., 1996; Hyde et al., 1999).

CONCLUSIONS

Calcification of biomaterial implants is an important pathologic process affecting a variety of tissue-derived biomaterials, as well as synthetic polymers in various functional configurations. The pathophysiology has been partially characterized with a number of useful animal models; a key common feature is the involvement of devitalized cells and cellular debris. Clinically useful preventive approaches based on either modifying biomaterials or local drug administration appear to be promising in some contexts.

BIBLIOGRAPHY

Aikawa, E., Nahrendorf, M., Figueiredo, J. L., Swirski, F. K., Shtatland, T., et al. (2007). Osteogenesis associates with inflammation in early-stage atherosclerosis evaluated by molecular imaging *in vivo*. *Circulation*, **116**, 2841–2850.

Anderson, H. C. (1988). Mechanisms of pathologic calcification. *Rheum. Dis. Clin. N. Am.*, **14**, 303–319.

Anderson, H. C. (1989). Mechanism of mineral formation in bone. *Lab. Invest.*, **60**, 320–330.

Barrere, F., van Blitterswijk, C. A., & de Groot, K. (2006). Bone regeneration: Molecular and cellular interactions with calcium phosphate ceramics. *Int. J. Nanomed.*, **1**, 317–332.

Bass, J. K., & Chan, G. M. (2006). Calcium nutrition and metabolism during infancy. *Nutrition*, **22**, 1057–1066.

Bechtel, J. F., Muller-Steinhardt, M., Schmidtke, C., Bruswik, A., Stierle, U., et al. (2003). Evaluation of the decellularized pulmonary valve homograft (SynerGraft). *J. Heart Valve Dis.*, **12**, 734–739.

Bini, A., Mann, K. G., Kudryk, B. J., & Schoen, F. J. (1999). Noncollagenous bone proteins, calcification and thrombosis in carcinoid artery atherosclerosis. *Arterio Sci. Thromb. Vasc. Biol.*, **19**, 1852–1861.

Bonucci, E. (1987). Is there a calcification factor common to all calcifying matrices? *Scanning Microscopy*, **1**, 1089–1102.

Bostrom, K., Watson, K. E., Horn, S., Wortham, C., Herman, I. M., et al. (1993). Bone morphogenetic protein expression in human atherosclerotic lesions. *J. Clin. Invest.*, **91**, 1800–1809.

Bottio, T., Thiene, G., Pettenazzo, E., Ius, P., Bortolotti, U., et al. (2003). Hancock II bioprostheses: A glance at the microscope in mid-long-term explants. *J. Thorac. Cardiovasc. Surg.*, **126**, 99–105.

Brockhurst, R. J., Ward, P. C., Lou, P., Ormerod, D., & Albert, D. (1993). Dystrophic calcification of silicone scleral buckling implant materials. *Am. J. Ophthalmol.*, **115**, 524–529.

Bucher, P. J., Buchi, E. R., & Daicker, B. C. (1995). Dystrophic calcification of an implanted hydroxyethylmethacrylate intraocular lens. *Arch. Ophthalmol.*, **113**, 1431–1435.

Bueno, E. M., & Glowacki, J. (2009). Cell-free and cell-based approaches for bone regeneration. *Nat. Rev. Rheumatol.*, **5**, 685–697.

Carpentier, S. M., Carpentier, A. F., Chen, L., Shen, M., Quintero, L. J., et al. (1997). Calcium mitigation in bioprosthetic tissues by iron pretreatment: The challenge of iron leaching. *Ann. Thorac. Surg.*, **63**, 1514–1515.

Celiento, M., Ravenni, G., Milano, A. D., Pratali, S., Scioti, G., Nardi, C., et al. (2012). Aortic valve replacement with Medtronic Mosaic bioprosthesis: a 13-year follow-up. *Ann. Thorac. Surg*, **93**, 510–515.

Chan, V., Jamieson, W. R., Germann, E., Chan, F., Miyagishima, R. T., et al. (2006). Performance of bioprostheses and mechanical prostheses assessed by composites of valve-related complications to 15 years after aortic valve replacement. *J. Thorac. Cardiovasc. Surg.*, **131**, 1267–1273.

Chen, J. -H., Liu, C., You, L., & Simmons, C. A. (2009). Boning up on Wolff's Law: Mechanical regulation of the cells that make and maintain bone. *J. Biomech*, **43**, 108–118.

Chen, W., Kim, J. D., Schoen, F. J., & Levy, R. J. (1994a). Effect of 2-amino oleic acid exposure conditions on the inhibition of calcification of glutaraldehyde crosslinked porcine aortic valves. *J. Biomed. Mater. Res.*, **28**, 1485–1495.

Chen, W., Schoen, F. J., & Levy, R. J. (1994b). Mechanism of efficacy of 2-amino oleic acid for inhibition of calcification of glutaraldehyde-pretreated porcine bioprosthetic valves. *Circulation*, **90**, 323–329.

Cheng, P. T. (1988). Pathologic calcium phosphate deposition in model systems. *Rheum. Dis. Clin. N. Am.*, **14**, 341–351.

Coleman, D., Lim, D., Kessler, T., & Andrade, J. D. (1981). Calcification of nontextured implantable blood pumps. *Trans. Am. Soc. Artif. Intern. Organs*, **27**, 97–103.

Connolly, J. M., Bakay, M. A., Alferiev, I. S., Gorman, R. C., Gorman, J. H., Kruth, H. S., Ashworth, P. E., Kutty, J. K., Schoen, F. J., Bianco, R. W., & Levy, R. J. (2011). Triglycidylamine cross-linking combined with ethanol inhibits bioprosthetic heart valve calcification. *Ann. Thorac. Surg.*, **92**, 858–865.

Courtman, D. W., Pereira, C. A., Kashef, V., McComb, D., Lee, J. M., et al. (1994). Development of a pericardial acellular matrix biomaterial: Biochemical and mechanical effects of cell extraction. *J. Biomed. Mater. Res.*, **28**, 655–666.

Courtman, D. W., Pereira, C. A., Omar, S., Langdon, S. E., Lee, J. M., et al. (1995). Biomechanical and ultrastructural comparison of cryopreservation and a novel cellular extraction of porcine aortic valve leaflets. *J. Biomed. Mater. Res.*, **29**, 1507–1516.

David, T. E., Armstrong, S., & Sun, Z. (1998). The Hancock II bioprosthesis at 12 years. *Ann. Thorac. Surg.*, **66**, S95–S98.

Demer, L. L. (2002). Vascular calcification and osteoporosis: Inflammatory responses to oxidized lipids. *Intl. J. Epidemiol.*, **31**, 737–741.

Demer, L. L., & Tintut, Y. (2008). Vascular calvification. Pathobiology of a multifaceted disease. *Circulation*, **117**, 2938–2948.

Discher, D. E., Mooney, D. J., & Zandstra, P. W. (2009). Growth factors, matrices, and forces combine and control stem cells. *Science*, **324**, 1673–1677.

El-Hamamsy, I., Clark, L., Stevens, L. M., Sarang, Z., Melina, G., Takkenberg, J. J., et al. (2010). Late outcomes following freestyle versus homograft aortic root replacement: results from a prospective randomized trial. *J. Am. COIL Cardiol.*, **55**, 368–376.

Evaerts, F., Torrianni, M., van Luyn, M. J., Van Wachem, P. B., Feijen, J., et al. (2004). Reduced calcification of bioprostheses, cross-linked via an improved carbodiimide based method. *Biomaterials*, **25**, 5523–5530.

Flameng, W., Ozaki, S., Meuris, B., Herijgers, P., Yperman, J., et al. (2001). Antimineralization treatments in stentless porcine bioprostheses. An experimental study. *J. Heart Valve Dis.*, **10**, 489–494.

Fradet, G., Bleese, N., Busse, E., Jamieson, E., Raudkivi, P., et al. (2004). The Mosaic valve clinical performance at seven years: Results from a Multicenter Prospective Clinical Trial. *J. Heart Valve Dis.*, **13**, 239–247.

Ford-Hutchinson, A. F., Cooper, D. M., Hallgrimsson, B., & Jirik, F. R. (2003). Imaging skeletal pathology in mutant mice by microcomputed tomography. *J. Rheumatol.*, **30**, 2659–2665.

Fukunaga, S., Tomoeda, H., Ueda, T., Mori, R., Aovagi, S., & Kato, S. (2010). Recurrent mitral regurgitation due to calcified synthetic chordae. *Ann. Thorac. Surg.*, **89**, 955–957.

Fyfe, B., & Schoen, F. J. (1999). Pathologic analysis of removed non-stented Medtronic Freestyle™ aortic root bioprostheses treated with amino oleic acid (AOA). *Semin. Thorac. Cardiovasc. Surg.*, **11**(4), 151–156.

Ghanbari, H., Viatge, H., Kidane, A. G., Burriesci, G., Tavakoli, M., et al. (2009). Polymeric heart valves: New materials, emerging hopes. *Trends Biotechnol.*, **27**, 359–367.

Giachelli, C. M. (1999). Ectopic calcification: Gathering hard facts about soft tissue mineralization. *Am. J. Pathol.*, **154**, 671–675.

Goldfarb, R. A., Neerhut, G. J., & Lederer, E. (1989). Management of acute hydronephrosis of pregnancy by urethral stenting: Risk of stone formation. *J. Urol.*, **141**(4), 921–922.

Golomb, G., Schoen, F. J., Smith, M. S., Linden, J., Dixon, M., et al. (1987). The role of glutaraldehyde-induced crosslinks in calcification of bovine pericardium used in cardiac valve bioprostheses. *Am. J. Pathol.*, **127**, 122–130.

Gott, J. P., Chih, P., Dorsey, L. M., Jay, J. L., Jett, G. K., et al. (1992). Calcification of porcine valves: A successful new method of antimineralization. *Ann. Thorac. Surg.*, **53**, 207–216.

Grabenwoger, M., Grimm, M., Ebyl, E., Leukauf, C., Müller, M. M., et al. (1992). Decreased tissue reaction to bioprosthetic heart valve material after L-glutamic acid treatment. A morphological study. *J. Biomed. Mater. Res.*, **26**, 1231–1240.

Grabenwoger, M., Sider, J., Fitzal, F., Zelenka, C., Windberger, U., et al. (1996). Impact of glutaraldehyde on calcification of pericardial bioprosthetic heart valve material. *Ann. Thorac. Surg.*, 62,772–62,777.

Grunkemeier, G. L., Jamieson, W. R., Miller, D. C., & Starr, A. (1994). Actuarial versus actual risk of porcine structural valve deterioration. *J. Thorac. Cardiovasc. Surg.*, **108**, 709–718.

Gümüş, N. (2009). Capsular calcification may be an important factor for the failure of breast implant. *Plast. Reconstr. Aesthet. Surg.*, **62**, e606–e608.

Habibovic, P., & de Groot, K. (2007). Osteoinductive biomaterials – properties and relevance in bone repair. *J. Tissue Eng. Regen. Med.*, **1**, 25–32.

Harasaki, H., Moritz, A., Uchida, N., Chen, J. F., McMahon, J. T., et al. (1987). Initiation and growth of calcification in a polyurethane coated blood pump. *Trans. Am. Soc. Artif. Intern. Organs*, **33**, 643–649.

Hendriks, M., Eveaerts, F., & Verhoeven, M. (2001). Alternative fixation of bioprostheses. *J. Long-Term Eff. Med. Implants*, **11**, 163–183.

Hilbert, S. L., Ferrans, V. J., Tomita, Y., Eidbo, E. E., & Jones, M. (1987). Evaluation of explanted polyurethane trileaflet cardiac valve prostheses. *J. Thorac. Cardiovasc. Surg.*, **94**, 419–429.

Hirsch, D., Drader, J., Thomas, T. J., Schoen, F. J., Levy, J. T., et al. (1993). Inhibition of calcification of glutaraldehyde pretreated porcine aortic valve cusps with sodium dodecyl sulfate: Preincubation and controlled release studies. *J. Biomed. Mater. Res.*, **27**, 1477–1484.

Hjortnaes, J., Butcher, J., Figueiredo, J.-H., Riccio, M., Kohler, R. H., et al. (2010). Arterial and aortic valve calcification inversely correlates with osteoporotic bone remodeling: A role for inflammation. *Eur. Heart J.*, **31**(16), 1975–1984.

Hofbauer, L. C., Brueck, C. C., Shanahan, C. M., Schoppet, M., & Dobnig, H. (2007). Vascular calcification and osteoporosis – clinical observation towards molecular understanding. *Osteoporos. Int.*, **18**, 251–259.

Human, P., & Zilla, P. (2001a). Inflammatory and immune processes: The neglected villain of bioprosthetic degeneration? *J. Long-Term Eff. Med. Implants*, **11**, 199–220.

Human, P., & Zilla, P. (2001b). The possible role of immune responses in bioprosthetic heart valve failure. *J. Heart Valve Dis.*, **10**, 460–466.

Human, P., Bezuidenhout, D., Torrianni, M., Hendriks, M., & Zilla, P. (2002). Optimization of diamine bridges in glutaraldehyde treated bioprosthetic aortic wall tissue. *Biomaterials*, **23**, 2099–2103.

Hyde, J. A., Chinn, J. A., & Phillips, R. E., Jr. (1999). Polymer heart valves. *J. Heart Valve Dis.*, **8**, 331–339.

Jian, B., Jones, P. L., Li, Q., Mohler, E. R., 3rd, Schoen, F. J., et al. (2001). Matrix metalloproteinase-2 is associated with tenascin-C in calcific aortic stenosis. *Am. J. Pathol.*, **159**, 321–327.

Jian, B., Narula, N., Li, Q. Y., Mohler, E. R., 3rd, & Levy, R. J. (2003). Progression of aortic valve stenosis: TGF-beta 1 is present in calcified aortic valve cusps and promotes aortic valve interstitial cell calcification via apoptosis. *Ann. Thorac. Surg.*, **75**, 457–465.

Jiang, X., Zhao, J., Wang, S., Sun, X., Zhang, X., et al. (2009). Mandibular repair in rats with premineralized silk scaffolds and BMP-2 modified bMSCs. *Biomaterials*, **30**, 4522–4532.

Johnson, R. C., Leopold, J. A., & Loscalzo, J. (2006). Vascular calcification. Pathobiological mechanisms and clinical implications. *Circ. Res.*, **99**, 1044–1059.

Johnston, T. P., Webb, C. L., Schoen, F. J., & Levy, R. J. (1992). Assessment of the *in vitro* transport parameters for ethanehydroxy diphosphonate through a polyurethane membrane. A potential refillable reservoir drug delivery device. *ASAIO*, **38**, M611–M616.

Johnston, T. P., Webb, C. L., Schoen, F. J., & Levy, R. J. (1993). Site-specific delivery of ethanehydroxy diphosphonate from refillable polyurethane reservoirs to inhibit bioprosthetic tissue calcification. *J. Control Rel.*, **25**, 227–240.

Jones, M., Eidbo, E. E., Hilbert, S. L., Ferrans, V. J., & Clark, R. E. (1989). Anticalcification treatments of bioprosthetic heart valves: *In vivo* studies in sheep. *J. Cardiac. Surg.*, **4**, 69–73.

Jono, S., McKee, M. D., Murry, C. E., Shiroi, A., Nishizawa, Y., et al. (2000). Phosphate regulation of vascular smooth muscle cell calcification. *Circ. Res.*, **87**, e10–e17.

Joshi, R. R., Underwood, T., Frautschi, J. R., Phillips, R. E., Jr., Schoen, F. J., et al. (1996). Calcification of polyurethanes implanted subdermally in rats is enhanced by calciphylaxis. *J. Biomed. Mater. Res.*, 31,201–31,207.

Kantrowitz, A., Freed, P. S., Zhou, Y., Mandell, G., DeDecker, P., Riddle, J., et al. (1995). A mechanical auxiliary ventricle. Histologic responses to long-term, intermittent pumping in calves. *AS70 J*, **41**, M340–M345.

Khan, S. R., & Wilkinson, E. J. (1985). Scanning electron microscopy, X-ray diffraction, and electron microprobe analysis of calcific deposits on intrauterine contraceptive devices. *Hum. Pathol.*, **16**, 732–738.

Kirsch, T. (2008). Determinants of pathologic mineralization. *Crit. Rev. Eukaryot Gene. Expr.*, **18**, 1–9.

Klintworth, G. K., Reed, J. W., Hawkins, H. K., & Ingram, P. (1977). Calcification of soft contact lenses in patient with dry eye and elevated calcium concentration in tears. *Invest. Ophthalmol. Vis. Sci.*, **16**, 158–161.

Koide, T., & Katayama, H. (1979). Calcification in augmentation mammoplasty. *Radiology*, **130**, 337–338.

Kretlow, J. D., & Mikos, A. G. (2007). Mineralization of synthetic polymer scaffolds for bone tissue engineering. *Tissue Eng.*, **13**, 927–938.

Kumar, V., Fausto, N., Aster, J. C., & Abbas, A. (2010). *Robbins/Cotran Pathologic Basis of Disease* (8th ed.). Philadelphia, PA: W. B. Saunders.

Lane., J. I., Randall, J. G., Campeau, N. G., Overland, P. K., McCannell, C. A., Matsko, T. A., et al. (2001). Imaging of hydrogel episcleral buckle fragmentation as a late complication after retinal reattachment surgery. *AJNRAm. J. Neuroradiol.*, **22**, 1199–1202.

Langer, R., & Vacanti, J. P. (1993). Tissue engineering. *Science*, **260**, 920–926.

Lee, J. S., Basalyga, D. M., Simionescu, A., Isenburg, J. C., Simionescu, D. T., et al. (2006). Elastin calcification in the rat subdermal model is accompanied by up-regulation of degradative and osteogenic cellular responses. *Am. J. Pathol.*, **168**, 490–498.

Legrand, A. P., Marinov, G., Pavlov, S., Guidoin, M. F., Famery, R., et al. (2005). Degenerative mineralization in the fibrous capsule of silicone breast implants. *J. Mater. Sci. Mater. Med.*, **16**, 477–485.

Lentz, D. L., Pollock, E. M., Olsen, D. B., & Andrews, E. J. (1982). Prevention of intrinsic calcification in porcine and bovine xenograft materials. *Trans. Am. Soc. Artif. Intern. Organs*, **28**, 494–497.

Levy, R. J., Schoen, R. J., Levy, J. T., Nelson, A. C., Howard, S. L., et al. (1983a). Biologic determinants of dystrophic calcification and osteocalcin deposition in glutaraldehyde-reserved porcine aortic valve leaflets implanted subcutaneously in rats. *Am. J. Pathol.*, **113**, 142–155.

Levy, R. J., Schoen, F. J., & Howard, S. L. (1983b). Mechanism of calcification of porcine aortic valve cusps: Role of T-lymphocytes. *Am. J. Cardiol.*, **52**, 629–631.

Levy, R. J., Hawley, M. A., Schoen, F. J., Lund, S. A., & Liu, P. Y. (1985a). Inhibition by diphosphonate compounds of calcification of porcine bioprosthetic heart valve cusps implanted subcutaneously in rats. *Circulation*, **71**, 349–356.

Levy, R. J., Wolfrum, J., Schoen, F. J., Hawley, M. A., Lund, S. A., et al. (1985b). Inhibition of calcification of bioprosthetic heart valves by local controlled-released diphosphonate. *Science*, **229**, 190–192.

Levy, R. J., Schoen, F. J., Sherman, F. S., Nichols, J., Hawley, M. A., et al. (1986). Calcification of subcutaneously implanted type I collagen sponges: Effects of glutaraldehyde and formaldehyde pretreatments. *Am. J. Pathol.*, **122**, 71–82.

Levy, R. J., Schoen, F. J., Lund, S. A., & Smith, M. S. (1987). Prevention of leaflet calcification of bioprosthetic heart valves with diphosphonate injection therapy. Experimental studies of optimal dosages and therapeutic durations. *J. Thorac. Cardiovasc. Surg.*, **94**, 551–557.

Levy, R. J., Vyavahare, N., Ogle, M., Ashworth, P., Bianco, R., et al. (2003). Inhibition of cusp and aortic wall calcification in ethanol- and aluminum-treated bioprosthetic heart valves in sheep: Background, mechanisms, and synergism. *J. Heart Valve Dis.*, **12**, 209–216.

Love, J. W. (1993). *Autologous Tissue Heart Valves*. Austin, TX: R. G. Landes.

Luo, G., Ducy, P., McKee, M. D., Pinero, G. J., Loyer, E., et al. (1997). Spontaneous calcification of arteries and cartilage in mice lacking matrix GLA protein. *Nature*, **386**, 78–81.

Mako, W. J., & Vesely, I. (1997). *In vivo* and *in vitro* models of calcification in porcine aortic valve cusps. *J. Heart Valve Dis.*, **6**, 316–323.

Mayer, J. E., Jr., Shin'oka, T., & Shum-Tim, D. (1997). Tissue engineering of cardiovascular structures. *Curr. Opin. Cardiol.*, **12**, 528–532.

McGonagle-Wolff, K., & Schoen, F. J. (1992). Morphologic findings in explanted Mitroflow pericardial bioprosthetic valves. *Am. J. Cardiol.*, **70**, 263–264.

Meuris, B., Phillips, R., Moore, M. A., & Flameng, W. (2003). Porcine stentless bioprostheses: Prevention of aortic wall calcification by dye-mediated photooxidation. *Artif. Organs*, **27**, 537–543.

Mitchell, R. N., Jonas, R. A., & Schoen, F. J. (1998). Pathology of explanted cryopreserved allograft heart valves: Comparison with aortic valves from orthotopic heart transplants. *J. Thorac. Cardiovasc. Surg.*, **115**, 118–127.

Mitchell, R. N., & Schoen, F. J. (2010). Blood vessels. In V. Kumar, N. Fausto, J. C. Aster, & A. Abbas (Eds.), *Robbins/Cotran Pathologic Basis of Disease* (8th ed.). (pp. 487–528). Philadelphia: W.B. Saunders.

Moore, M. A., & Phillips, R. E. (1997). Biocompatibility and immunologic properties of pericardial tissue stabilized by dye-mediated photooxidation. *J. Heart Valve Dis.*, **6**, 307–315.

Myers, D. J., Nakaya, G., Girardot, G. M., & Christie, G. W. (1995). A comparison between glutaraldehyde and diepoxide-fixed stentless porcine aortic valves: Biochemical and mechanical characterization and resistance to mineralization. *J. Heart Valve Dis.*, **4**, S98–S101.

Nakanome, S., Watanabe, H., Tanaka, K., & Tochikubo, T. (2008). Calcification of Hydroview H60M intraocular lenses: Aqueous humor analysis and comparisons with other intraocular lens materials. *J. Cataract Refract. Surg.*, **34**, 80–86.

Neues, F., & Epple, M. (2008). X-ray microcomputer tomography for the study of biomineralized endo- and exoskeletons of animals. *Chem. Rev.*, **108**, 4734–4741.

Neuhann, M., Kleinmann, G., & Apple, D. J. (2008). A new classification of calcification of intraocular lenses. *Ophthalmology*, **115**, 73–79.

New, S. E., & Aikawa, E. (2011). Molecular imaging insights into early inflammatory stages of arterial and aortic valve calcification. *Circ. Res.*, **108**, 1381–1391.

Ogle, M. F., Kelly, S. J., Bianco, R. W., & Levy, R. J. (2003). Calcification resistance with aluminum-ethanol treated porcine aortic valve bioprostheses in juvenile sheep. *Ann. Thorac. Surg.*, **75**, 1267–1273.

Parhami, F., Basseri, B., Hwang, J., Tintut, Y., & Demer, L. L. (2002). High-density lipoprotein regulates calcification of vascular cells. *Circ. Res.*, **91**, 570–576.

Patai, K., Berényi, M., Sipos, M., & Noszál, B. (1998). Characterization of calcified deposits on contraceptive intrauterine devices. *Contraception*, **58**, 305–308.

Peacock, M. (2010). Calcium metabolism in health and disease. *Clin. J. Am. Soc. Nephrol.*, **5**, S23–S30.

Persy, V., & D'Haese, P. (2009). Vascular calcification and bone disease: The calcification paradox. *Trends Mol. Med.*, **15**, 405–416.

Peters, W., Pritzker, K., Smith, D., Fornasier, V., Holmyard, D., et al. (1998). Capsular calcification associated with silicone breast implants: Incidence, determinants, and characterization. *Ann., Plast. Surg.*, **41**, 348–360.

Peters, W., & Smith, D. (1995). Calcification of breast implant capsules: incidence, diagnosis, and contributing factors. *Ann. Plast. Surg.*, **34**, 8–11.

Rimmer, T., Hawkesworth, N., Kirkpatrick, N., Price, N., Manners, R., et al. (2010). Calcification of Hydroview lenses implanted in the United Kingdom during 2000 and 2001. *Eye (Lond.)*, **24**, 199–200.

Schlieper, G., Krűger, T., Djuric, Z., Damjanovic, T., Markovic, N., et al. (2008). Vascular access calcification predicts mortality in hemodialysis patients. *Kidney Int.*, **74**, 1582–1587.

Schoen, F. J. (1998). Pathologic findings in explanted clinical bioprosthetic valves fabricated from photooxidized bovine pericardium. *J. Heart Valve Dis.*, **7**, 174–179.

Schoen, F. J. (2001). Pathology of heart valve substitution with mechanical and tissue prothesis. In M. D. Silver, A. I. Gotlieb, & F. J. Schoen (Eds.), *Cardiovascular Pathology* (3rd ed., pp. 629–677). Philadelphia, PA: W. B. Saunders.

Schoen, F. J., & Edwards, W. D. (2001). Pathology of cardiovascular interventions, including endovascular therapies, revascularization, vascular replacement, cardiac assist/replacement, arrhythmia control and repaired congenital heart disease. In M. D. Silver, A. I. Gotlieb, & F. J. Schoen (Eds.), *Cardiovascular Pathology* (3rd ed., pp. 678–723). Philadelphia, PA: W. B. Saunders.

Schoen, F. J., & Levy, R. J. (1999). Tissue heart valves: Current challenges and future research perspectives. *J. Biomed. Mater. Res.*, **47**, 439–465.

Schoen, F. J., & Levy, R. J. (2005). Calcification of tissue heart valve substitutes: Progress toward understanding and prevention. *Ann. Thorac. Surg.*, **79**, 1072–1080.

Schoen, F. J., Levy, R. J., Nelson, A. C., Bernhard, W. F., Nashef, A., et al. (1985). Onset and progression of experimental bioprosthetic heart valve calcification. *Lab. Invest.*, **52**, 523–532.

Schoen, F. J., & Mitchell, R. N. (2010). The heart. In V. Kumar, N. Fausto, J. C. Aster, & A. Abbas (Eds.), *Robbins/Cotran Pathologic Basis of Disease* (8th ed.). (pp. 529–587). Philadelphia: W.B. Saunders.

Schoen, F. J., Tsao, J. W., & Levy, R. J. (1986). Calcification of bovine pericardium used in cardiac valve bioprostheses. Implications for mechanisms of bioprosthetic tissue mineralization. *Am. J. Pathol.*, **123**, 143–154.

Schoen, F. J., Kujovich, J. L., Webb, C. L., & Levy, R. J. (1987). Chemically determined mineral content of explanted porcine aortic valve bioprostheses: Correlation with radiographic assessment of calcification and clinical data. *Circulation*, **76**, 1061–1066.

Schoen, F. J., et al. (1988). Biomaterials-associated calcification: Pathology, mechanisms, and strategies for prevenion. *J. Appl. Biomater.*, **22**, 11–36.

Schoen, F. J., Golomb, G., & Levy, R. J. (1992a). Calcification of bioprosthetic heart valves: A perspective on models. *J. Heart Valve Dis.*, **1**, 110–114.

Schoen, F. J., Levy, R. J., Hilbert, S. L., & Bianco, R. W. (1992b). Antimineralization treatments for bioprosthetic heart valves. Assessment of efficacy and safety. *J. Thorac. Cardiovasc. Surg.*, **104**, 1285–1288.

Schoen, F. J., Levy, R. J., & Piehler, H. R. (1992c). Pathological considerations in replacement cardiac valves. *Cardiovasc. Pathol.*, **1**, 29–52.

Schoen, F. J., Hirsch, D., Bianco, R. W., & Levy, R. J. (1994). Onset and progression of calcification in porcine aortic bioprosthetic valves implanted as orthotopic mitral valve replacements in juvenile sheep. *J. Thorac. Cardiovasc. Surg.*, **108**, 880–887.

Shinoka, T., Ma, P. X., Shum-Tim, D., Breuer, C. K., Cusick, R. A., et al. (1996). Tissue-engineered heart valves. Autologous valve leaflet replacement study in a lamb model. *Circulation*. 94, II-164–II-168.

Simon, P., Kasimir, M. T., Seebacher, G., Weigel, G., Ullrich, R., et al. (2003). Early failure of the tissue engineered porcine heart valve SYNERGRAFT in pediatric patients. *Eur. J. Cardiothorac. Surg.*, **23**, 1002–1006.

Speer, M. Y., & Giachelli, C. M. (2004). Regulation of vascular calcification. *Cardiovasc. Pathol.*, **13**, 63–70.

Speer, M. Y., McKee, M. D., Guldberg, R. E., Liaw, L., Yang, H.-Y., et al. (2002). Inactivation of the osteopontin gene enhances vascular calcification of matrix Gla protein-deficient mice: Evidence for osteopontin as an inducible inhibitor of vascular calcification *in vivo*. *J. Exp. Med.*, **196**, 1047–1055.

Srivasta, S. S., Maercklein, P. B., Veinot, J., Edwards, W. D., Johnson, C. M., et al. (1997). Increased cellular expression of matrix proteins that regulate mineralization is associated with calcification of native human and porcine xenograft bioprosthetic heart valves. *J. Clin. Invest*, **5**, 996–1009.

Steitz, S. A., Speer, M. Y., McKee, M. D., Liaw, L., Almeida, M., et al. (2002). Osteopontin inhibits mineral deposition and promotes regression of ectopic calcification. *Am. J. Pathol.*, **161**, 2035–2046.

Stock, U. A., Nagashima, M., Khalil, P. N., Nollert, G. D., Herden, T., et al. (1999). Tissue engineered valved conduits in the pulmonary circulation. *J. Thorac. Cardiovasc. Surg.*, **119**, 732–740.

Tew, W. P., Mahle, C., Benavides, J., Howard, J. E., & Lehninger, A. L. (1980). Synthesis and characterization of phosphocitric acid, a potent inhibitor of hydroxylapatite crystal growth. *Biochemistry*, **19**, 1983–1988.

Thoma, R. J., & Phillips, R. E. (1995). The role of material surface chemistry in implant device calcification: A hypothesis. *J. Heart Valve Dis.*, **4**, 214–221.

Thubrikar, M. J., Deck, J. D., Aouad, J., Nolan, S. P. (1983). Role of mechanical stress in calcification of aortic bioprosthetic valves. *J. Thorac. Cardiovasc. Surg.*, **86**, 115–125.

Tomizawa, Y., Takanashi, Y., Noishiki, Y., Nishida, H., Endo, M., et al. (1998). Evaluation of small caliber vascular prostheses implanted in small children: Activated angiogenesis and accelerated calcification. *ASAIO J.*, **44**, M496–M500.

Trantina-Yates, A. E., Human, P., & Zilla, P. (2003). Detoxification on top of enhanced, diamine-extended glutaraldehyde fixation significantly reduces bioprosthetic root calcification in the sheep model. *J. Heart Valve Dis.*, **12**, 93–100.

Umana, E., Ahmed, W., & Alpert, M. A. (2003). Valvular and perivalvular abnormalities in end-stage renal disease. *Am. J. Med. Sci.*, **325**, 237–242.

Vanderbrink, B. A., Rastinehad, A. R., Ost, M. C., & Smith, A. D. (2008). Encrusted urinary stents: Evaluation and endourologist management. *J. Endocrinol.*, **22**, 905–912.

Van Wachem, P. B., Brouwer, L. A., Zeeman, R., Dijkstra, P. J., Feijen, J., et al. (2000). *In vivo* behavior of epoxy-crosslinked porcine heart valve cusps and walls. *J. Biomed. Mater. Res.*, **53**, 18–27.

Van Wachem, P. B., Brouwer, L. A., Zeeman, R., Dijkstra, P. J., Feijen, J., et al. (2001). Tissue reactions to epoxy-crosslinked porcine heart valves post-treated with detergents or a dicarboxylic acid. *J. Biomed. Mater. Res.*, **55**, 415–423.

Vyavahare, N. R., Chen, W., Joshi, R., Lee, C.-H., Hirsch, D., et al. (1997a). Current progress in anticalcification for bioprosthetic and polymeric heart valves. *Cardiovasc. Pathol.*, **6**, 219–229.

Vyavahare, N., Hirsch, D., Lerner, E., Baskin, J. Z., Schoen, F. J., et al. (1997b). Prevention of bioprosthetic heart valve calcification by ethanol preincubation. Efficacy and mechanism. *Circulation*, **95**, 479–488.

Vyavahare, N. R., Hirsch, D., Lerner, E., Baskin, J. Z., Zand, R., et al. (1998). Prevention of calcification of glutaraldehyde-crosslinked porcine aortic cusps by ethanol preincubation: Mechanistic studies of protein structure and water–biomaterial relationships. *J. Biomed. Mater. Res.*, **40**, 577–585.

Vyavahare, N., Ogle, M., Schoen, F. J., & Levy, R. J. (1999). Elastin calcification and its prevention with aluminum chloride pretreatment. *Am. J. Pathol.*, **155**, 973–982.

Wada, T., McKee, M. D., Steitz, S., & Giachelli, C. M. (1999). Calcification of vascular smooth muscle cell cultures. Inhibition by osteopontin. *Circ. Res.*, **84**, 166–178.

Wang, Q., McGoron, A. J., Bianco, R., Kato, Y., Pinchuk, L., & Schoephoerster, R. T. (2010). In-vivo assessment of a novel polymer (SIBS) trileaflet heart valve. *J. Heart Valve Dis.*, **19**, 499–505.

Weissen-Plenz, G., Nitschke, Y., & Rutsch, F. (2008). Mechanisms of arterial calcification: Spotlight on the inhibitors. *Adv. Clin. Chem.*, **46**, 263–293.

Webb, C. L., Benedict, J. J., Schoen, F. J., Linden, J. A., & Levy, R. J. (1988). Inhibition of bioprosthetic heart valve calcification with aminodiphosphonate covalently bound to residual aldehyde groups. *Ann. Thorac. Surg.*, **46**, 309–316.

Webb, C. L., Schoen, F. J., Flowers, W. E., Alfrey, A. C., Horton, C., et al. (1991). Inhibition of mineralization of glutaraldehyde-pretreated bovine pericardium by $AlCl_3$. Mechanisms and comparisons with $FeCl_3$ $LaCl_3$ and $Ga(NO_3)_3$ in rat subdermal model studies. *Am. J. Pathol.*, **138**, 971–981.

Wilson, G. J., Courtman, D. W., Klement, P., Lee, J. M., & Yeger, H. (1995). Acellular matrix: A biomaterials approach for coronary artery and heart valve replacement. *Ann. Thorac. Surg.*, **60**, S353–S358.

Xi, T., Ma, J., Tian, W., Lei, X., Long, S., et al. (1992). Prevention of tissue calcification on bioprosthetic heart valve by using epoxy compounds: a study of calcification tests *in vitro* and *in vivo*. *J. Biomed. Mater. Res.*, **26**, 1241–1251.

Yip, C. Y.Y., Chen, J. -H., Zhao, R., & Simmons, C. A. (2009). Calcification by valve interstitial cells is regulated by the stiffness of the extracellular matrix. *Arterioscler. Thromb. Vasc. Biol.*, **29**, 936–942.

Yu, S.-Y., Viola, F., Christoforidis, J. B., & D'Amico, D. J. (2005). Dystrophic calcification of the fibrous capsule around a hydrogel explant 13 years after scleral buckling surgery: Capsular calcification of a hydrogel explant. *Retina*, **25**, 1104–1107.

Zilla, P., Weissenstein, C., Bracher, M., Zhang, Y., Koen, W., et al. (1997). High glutaraldehyde concentrations reduce rather than increase the calcification of aortic wall tissue. *J. Heart Valve Dis.*, **6**, 490–491.

Zilla, P., Weissenstein, C., Human, P., Dower, T., & von Oppell, U. O. (2000). High glutaraldehyde concentrations mitigate bioprosthetic root calcification in the sheep model. *Ann. Thorac. Surg.*, **70**, 2091–2095.

Zilla, P., Weissenstein, C., Bracher, M., & Human, P. (2001). The anticalcific effect of glutaraldehyde detoxification on bioprosthetic aortic wall tissue in the sheep model. *J. Card. Surg.*, **16**, 467–472.

Zilla, P., Bezuidenhout, D., Torrianni, M., Hendriks, M., & Human, P. (2005). Diamine-extended glutaraldehyde- and carbodiimide crosslinks act synergistically in mitigating bioprosthetic aortic wall calcification. *J. Heart Valve Dis.*, **14**, 538–545.

SECTION II.5

Applications of Biomaterials

SECTION

II.5

Applications of Biomaterials

CHAPTER II.5.1 INTRODUCTION: APPLICATIONS OF BIOMATERIALS

Frederick J. Schoen[1] *and Jack E. Lemons*[2]

[1]Professor of Pathology and Health Sciences and Technology (HST), Harvard Medical School, Executive Vice Chairman, Department of Pathology, Brigham and Women's Hospital, Boston, MA, USA

[2]University Professor, Schools of Dentistry, Medicine and Engineering, University of Alabama at Birmingham, Birmingham, AL, USA

Most students of biomaterials have a strong interest in medical or dental applications. Biomaterials are used for the construction of components in an extensive array of devices across a wide range of medical disciplines. When considering the applications of biomaterials as a section focus, a primary consideration is on outcomes of treatments. Outcomes are evaluated in terms of the discipline and the specific biomaterial properties needed for a highly specific application to improve outcomes of patients with specific clinical problems. For example, a total joint replacement has very different considerations than a tooth root replacement, although both anchor in bone for function. Similarly, requirements for a heart valve are very different compared to a vessel replacement or an endovascular stent, although all have extensive blood contact and some biomaterial properties are in common.

The following chapters in Part II present a broad spectrum of biomaterials applications and the key properties needed for specific physiological environments. The Cardiovascular section starts with Nonthrombogenic Biomaterials (II.5.2) by Sefton and Cardiovascular Medical Devices (II.5.3). Subsections A–D describe Valves (II.5.3A), Endovascular Stents, Vascular grafts, and Stent Grafts (II.5.3B), Other Cardiovascular Devices (II.5.3C), all by Schoen and Padera, and Implantable Cardiac Assist Devices and IABPs (II.5.3D) by Simon, Borovetz and Wagner. The emphasis shifts to artificial cells (II.5.4) by Chang, and Extracorporeal Artificial Organs (II.5.5) by Ritchie. Two chapters with musculoskeletal emphasis are Orthopedic and Dental Applications (II.5.6 and II.5.7) by Hallab and Jacobs, and Lemons and Misch, respectively. Adhesives and Sealants (II.5.8) are presented by Watts. Ophthalmologic Applications (II.5.9) and subdivisions on contact lenses (II.5.9A), IOLs (II.5.9B), Corneal Inlays (II.5.9C), Glaucoma Drains (II.5.9D), and Retinal Prostheses (II.5.9E) are presented by Steinert and Jain; Jacob; Patel; Cunanan; Cunanan; and Humayun et al. respectively. Chapters follow on Bioelectrodes (II.5.10), Cochlear Prostheses (II.5.11) and Stimulating Electrodes (II.5.12) implants, Biosensors (II.5.13), Burn Dressings and Skin Substitutes (II.5.14), and Sutures (II.5.15) by Venugopalan and Ideker; Spelman; Peckhan, Ackermann, and Moss; LaFleur and Yager; Helm, Orgill B., Ogawa and Orgill D.; and Taylor and Shalaby, respectively. Chapters on applications are extended in the next series on Drug Delivery Systems (II.5.16), with subsections on Injected Nanocarriers PEGylation, Targeting, Polymer–Drug Conjugates, Liposomes, Polymeric Micelles, Dendrimers, Nucleic Acid Delivery, and Polymeric and Albuminated Drug Nanoparticles, collectively by Hoffman, and Gombotz; Pun; Stayton, Ghosn, and Wilson J. The drug delivery systems (DDS) chapters continue with Injected Degradable Depot DDS (II.5.16C), Implants and Inserts (II.5.16D), Smart DDS (II.5.16E), Transdermal DDS (II.5.16F), and Oral DDS (II.5.16G) by Gombotz; Wright and Kleiner; Hoffman; Cleary; and Wilson C., respectively. This section closes with chapters on Diagnostic Applications (II.5.17) by Domingo, Hawkins, Peck, and Weigl, and Silicones (II.5.18) by Curtis and Colas.

A central theme is the generation and use of design criteria based on desired functionality, potentially deleterious biomaterials–tissue interaction mechanisms, pathologies of the underlying conditions for which the implant is needed, and the basic properties of the various biomaterials available or needing to be developed. It should not be surprising that considerable research and development has led to clinically used devices with active mechanical, electrical, biologic or mass exchange functions.

Biomaterials applications and surgical implant technology assume a key role in current clinical practice. In 2000, it was estimated that approximately 20 million individuals had an implanted medical device. The number is growing, but is difficult to assess precisely. A recent article in *The Wall Street Journal* (July 18, 2011) summarized the five most frequently implanted medical devices in the

US: artificial eye lenses (>2.5 million per year); tympanostomy ear tubes (>0.75 million per year); coronary arterial stents (>0.5 million per year); prosthetic knee joints (>0.5 million per year); and metal screws, pins, plates, and rods (>450,000 per year). The economic impact is massive; in 2000 costs associated with prostheses and organ replacement therapies exceeded $300 billion US per year, and comprised >1% of the US gross domestic product (GDP) and nearly 8% of total healthcare spending worldwide (Lysaght and O'Loughlin, 2000). Thus, medical devices contribute to the expense associated with modern healthcare in the United States, which is presently in excess of 17% of GDP, continues to grow, and has become a significant public policy issue. For the widely used artificial eye lenses, at 2.5 million surgeries performed annually at a rate of about $3200 to $4500 per eye, the total expenditure is estimated at between $8 billion and $10 billion per year. The use of health technology assessment tools can assist those in leadership positions in making rational decisions as to which new technologies to adopt, based on evaluation of clinical effectiveness, cost-effectiveness, and risk to patients.

Most implants serve their recipients well for extended periods by alleviating the conditions for which they were implanted. Considerable effort is expended in understanding biomaterials–tissue interactions and eliminating patient–device complications (the clinically important manifestations of biomaterials–tissue interactions). Moreover, many patients receive substantial and extended benefit, despite complications. For example, heart valve disease is a serious medical problem affecting over 30,000 people per year in the United States. Patients with aortic stenosis (the most common form of heart valve disease) have a 50% chance of dying within approximately three years without surgery. Surgical replacement of a diseased valve leads to an expected survival of 70% at 10 years, a substantial improvement over the natural course. However, of these patients whose longevity and quality of life have clearly been enhanced, approximately 60% will suffer a serious valve-related complication within 10 years after the operation. Thus, long-term failure of biomaterials leading to a clinically significant event does not preclude clinical success, for a significant duration and overall.

The range of tolerable risk of adverse effects varies directly with the medical benefit obtained by the therapy. Benefit and risk go hand-in-hand, and clinical decisions are made to maximize the ratio of benefit to risk. The tolerable benefit–risk ratio may depend on the type of implant and the medical problem it is used to correct. Thus, more risk can be tolerated with a heart assist device (a life-sustaining implant) than with a prosthetic hip joint (an implant that relieves pain and disability and enhances function), and much more risk than with a breast implant (an implant with predominantly cosmetic benefit). As an example, total hip arthroplasties (THAs) with metal-on-metal (MoM) or more correctly cobalt alloy-on-cobalt alloy articulating surfaces, have been used clinically since the 1950s. Applications of recent generation THAs exceed hundreds of thousands. Very recently, metallic debris products in larger quantities have been associated with adverse foreign-body reactions, need for revisions, and a recall action of one design/product; for updated information on this topic, students are referred to the US Food and Drug Administration (FDA) web site (http://www.fda.gov/MedicalDevices/ProductsandMedicalProcedures/ImplantsandProsthetics/MetalonMetalHipImplants/default.htm), the American Academy of Orthopaedic Surgeons website (AAOS.org) and/or the American Society for Testing and Materials website (ASTM.org). An ASTM symposium Standard Technical Publication (STP) on this topic should be available in late 2012.

In summary, this section explores the most widely used applications of materials in medicine, biology, and artificial organs. The progress made in many of these areas has been substantial. In most cases, the individual chapters describe a device category from the perspective of the clinical need, the armamentarium of devices available to the practitioner, the results and complications, and the challenges to the field that limit success.

BIBLIOGRAPHY

US Food and Drug Administration (FDA). (2012). http://www.fda.gov/MedicalDevices/ProductsandMedicalProcedures/ImplantsandProsthetics/MetalonMetalHipImplants/default.htm.

Lysaght, M. J., & O'Loughlin, J. A. (2000). Demographic scope and economic magnitude of contemporary organ replacement therapies. *ASAIO J.*, **46**, 515–521.

The Wall Street Journal. (2011). http://247wallst.com/2011/07/18/the-eleven-most-implanted-medical-devices-in-america/. July 18.

CHAPTER II.5.2 NONTHROMBOGENIC MATERIALS AND STRATEGIES: CASE STUDY

Michael V. Sefton
Institute of Biomaterials and Biomedical Engineering, University of Toronto, Ontario, Canada

ISO Standard 10993-4, *Biological Evaluation of Medical Devices Part 4: The Effects on Blood* (ISO/AAMI 1995), which manufacturers of medical devices need to use as guidance to register their products, includes thrombosis and coagulation among the tests that need to be done. However, specific test methods are not detailed. With a view to clarifying this question, a series of plasma-modified tubes (along with an unmodified control and other commercially available tubing) were prepared, surface characterized, and exposed to heparinized whole blood (1 U/mL heparin) for one hour at 37°C (Sefton et al., 2001). The surface modifications included several different plasma vapors (H_2O, CF_4, and fluorine). The 1.5 mm ID (internal diameter) tubing was incubated with whole blood on a rocking platform to gently agitate the blood and keep the cells from overtly settling. This system does not probe the effect of shear on cell activation; rather the agitation and long incubation time are thought sufficient to create "well-mixed" conditions.

Some of the results from this study are shown in Figure II.5.2.1 and Table II.5.2.1. One of the conclusions from this study is that the materials with the lowest levels of platelet and leukocyte activation (microparticle formation, CD11b upregulation) were the unmodified materials (polyethylene, Pellethane™, a polyurethane), and that the surface modifications tested here had either no effect or only made things worse from the perspective of platelet and leukocyte activation. The scanning electron micrographs showing little cellular deposit on the polyethylene or Pellethane™ were consistent with the flow cytometry findings. Except for those materials that were worse, the other materials expressed similar levels of activation for platelets, leukocytes, and the other markers tested; i.e., the majority of the "inert" materials (in the absence of bioactive components like

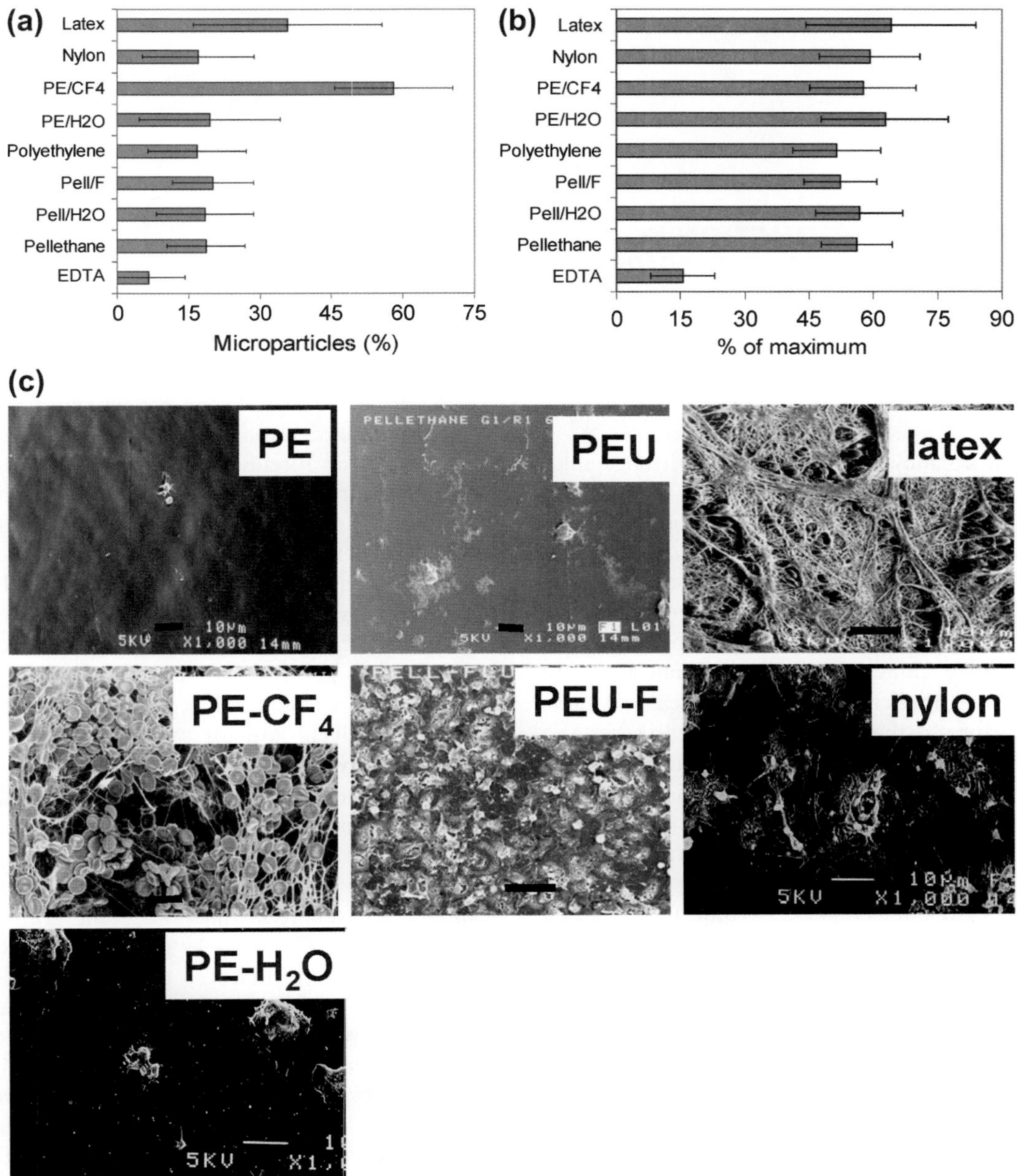

FIGURE II.5.2.1 Cell activation results after 60 min contact with whole blood (1 unit/mL heparin) at 37°C. Flow cytometry results are mean ±SD. (a) Platelet microparticle levels (percentage of platelet events); (b) Leukocyte CD11b up-regulation (expressed relative to the maximum upregulation obtained with a phorbol ester). (c) Scanning Electron Micrographs of biomaterial surfaces following exposure to blood. Scale bar is 10 μm. PE: Polyethylene; Pell: Pellethane.

TABLE II.5.2.1 Most Biomaterials are the Same

Parameter	EDTA* Control (or Equivalent)	Value for Most Materials
Platelet count loss (%)	0 (by definition)	25–35
P-selectin (% positive)	6.4	~8–9
Platelet–leukocyte aggregates (fluorescent intensity)	46	~200–250
CD11b upregulation (% of maximum)	15	~50–60
L-selectin shedding	11	~70–90

Most materials, despite very different non-specific chemical modifications, resulted in similar levels of platelet and leukocyte activation. These were higher than the corresponding negative control values. The similarity of CD11b upregulation values is seen in Figure II.5.2.1b, while the presence of exceptions that were more activating is seen in the microparticle results in Figure II.5.2.1a.

*Ethylenediaminetetraacetic acid.

heparin) appear to have a similar non-specific effect on blood.

The results from this study contrast sharply with other studies (many of them cited in this chapter) showing large differences in blood interactions between different biomaterials. This highlights the lack of consensus about blood–materials interactions. Reasons for this lack of consensus are discussed in Chapter II.3.5.

BIBLIOGRAPHY

ISO/AAMI. (1995). ISO Standard 10993-4, Biological Evaluation of Medical Devices. In: *AAMI Standards and Recommended Practice*, pp. 45–68, Vol. 4. Washington, DC: American Association for Medical Instrumentation.

Sefton, M. V., Sawyer, A., Gorbet, M., Black, J. P., Cheng, E., Gemmell, C., & Pottinger-Cooper, E. (2001). Does surface chemistry affect thrombogenicity of surface modified polymers? *J. Biomed. Mater. Res.*, 55, 447–459.

CHAPTER II.5.3 INTRODUCTION TO CARDIOVASCULAR MEDICAL DEVICES

Frederick J. Schoen

Professor of Pathology and Health Sciences and Technology (HST), Harvard Medical School, Executive Vice Chairman, Department of Pathology, Brigham and Women's Hospital, Boston, MA, USA

The past several decades have witnessed a virtual explosion in the number and scope of innovative surgical and interventional diagnostic and therapeutic procedures performed on patients with cardiovascular diseases. Data from the National Center for Health Statistics and the American Heart Association indicate that approximately seven million major cardiac and vascular operations are done annually in the United States. Concurrent with and integral to the broad application of these surgical and interventional procedures is the use of various prostheses and medical devices. Data from 2006 (reported in 2009) show 641,000 percutaneous coronary interventions (almost all using endovascular bare-metal and drug-eluting stents), 253,000 coronary artery bypass graft procedures, 104,000 cardiac valve procedures (using approximately 85,000 substitute heart valves, pacemakers (418,000), implanted cardioverter-defibrillators (114,000) and their leads, and many cardiac assist devices, vascular grafts, umbrellas, patches, and others (Lloyd-Jones et al., 2009).

Thus, cardiovascular prostheses and medical devices, and their constituent biomaterials, are of critical importance to interventional cardiologists, and cardiac and vascular surgeons. The number and complexity of devices permit choices among surgical or catheter-based interventional options that optimize short- and long-term patient management. The recognition and understanding of complications of these devices, many of them related to the biomaterials that comprise them, has led to iterative efforts to improve their performance and safety through biomaterials and device research and development that has led to improvements which have been translated into improved patient care. The nature, frequency, and pathologic anatomy of their complications, as well as the responsible blood–tissue–biomaterials interaction mechanisms have been published for widely used devices used for many years, but are less well-appreciated for recently introduced or modified devices, and those currently in development (Schoen, 2001; Schoen and Edwards, 2001).

This section, composed of 4 sub-chapters, summarizes key considerations in cardiovascular medical devices, including the underlying pathology of the conditions they are designed and used to treat, relevant biomaterials research, and the most important complications that need to be circumvented. The first chapter (II.5.3.A) summarizes cardiac valve prostheses, which have been used extensively and for approximately a half century, are clinically important; their outcomes and pathological descriptions of complications encountered with many different types of valve prostheses are well-known. The second chapter (II.5.3.B) discusses devices used for vascular repair and replacement (including vascular grafts and endovascular stents [and stent grafts]). The third chapter (II.5.3.C) discusses pacemakers and implantable cardioverter-defibrillators, cardiac assist and replacement devices and miscellaneous cardiovascular devices, including percutaneous catheter-based techniques to treat cardiovascular disease in a minimally invasive manner, such as septal defect closure devices, filters to prevent pulmonary embolism and left atrial occlusion devices, and devices to minimize the consequences of a dilated, failing heart. Finally, in the fourth chapter (II.5.3.D), specific biomaterials and engineering design issues related to implantable cardiac assist devices and artificial hearts, a complex and evolving set of technologies, are discussed.

BIBLIOGRAPHY

Lloyd-Jones, D., Adams, R., Carnethon, M., et al. (2009). Heart disease and stroke statistics 2009 update. *Circulation*, **119**, e21–180.

Schoen, F. J. (2001). Pathology of heart valve substitution with mechanical and tissue prostheses. In M. D. Silver, A. I. Gotlieb, & F. J. Schoen (Eds.), *Cardiovascular Pathology 3rd Ed.* (pp. 629–677): WB Saunders.

Schoen, F. J., & Edwards, W. D. (2001). Pathology of cardiovascular interventions, including endovascular therapies, revascularization, vascular replacement, cardiac assist/replacement, arrhythmia control and repaired congenital heart disease. In M. D. Silver, A. I. Gotlieb, & F. J. Schoen (Eds.), *Cardiovascular Pathology 3rd Ed.* (pp. 678–723): WB Saunders.

A. SUBSTITUTE HEART VALVES

Frederick J. Schoen and Robert F. Padera, Jr
Department of Pathology, Brigham and Women's Hospital and Harvard Medical School, Boston, MA, USA

INTRODUCTION

Used extensively for approximately half a century, cardiac valve prostheses are a clinically important achievement of biomaterials science and biomedical engineering. Indeed, the prestigious 2007 Lasker Award for Clinical Medical Research was granted to Drs. Albert Starr and Alain Carpentier to recognize the importance of cardiac valve replacement as a major clinical success (Chaikof, 2007; Lifton, 2007). Starr performed the first successful valve replacement in the heart by implanting a caged-ball mechanical valve prosthesis in the mitral position (Starr, 2007); the history of cardiac valve replacement is discussed in more detail in Chapter III.2.9 of this book. Carpentier fabricated a "bioprosthesis," combining chemically-treated biologic tissue and a mechanical structure to create a tissue-based (although non-living) heart valve replacement (Carpentier, 2007). Relevant outcome data and pathological descriptions of complications of many different types of valve prostheses are well-known (Vongpatanasin et al., 1996; Schoen, 2001; Bonow et al., 2006).

In this chapter we review the past history, present status, and future directions of cardiac valve replacement, focusing on biomaterials challenges and opportunities in substitute heart valves, a widely used medical device.

> Used extensively for approximately half a century, cardiac valve prostheses are a clinically important achievement of biomaterials science and biomedical engineering.

HEART VALVE FUNCTION AND DYSFUNCTION

The four valves play a critical role in ensuring unidirectional forward blood flow through the heart. The tricuspid valve allows one-way flow from the right atrium to the right ventricle, and correspondingly, the pulmonary valve from the right ventricle to the pulmonary artery, the mitral valve from the left atrium to the left ventricle, and the aortic valve from the left ventricle to the aorta. The heart valves open and close with each cardiac cycle, approximately once per second, which equates to approximately 40 million times per year and 3 billion times in a 75-year lifetime. Detailed insights about the biomechanical function of the natural heart valves have recently emerged (Sacks and Yoganathan, 2007; Sacks et al., 2009).

Disorders of heart valves can cause stenosis (i.e., obstruction to flow) or regurgitation (i.e., reverse flow across the valve) (Sharma et al., 2008; Schoen and Mitchell, 2010). Sometimes, both stenosis and regurgitation are present in the same valve. Some disease processes such as infective endocarditis (infection of a heart valve) can cause rapid (in days) destruction of the affected valve, and can lead to abrupt heart failure and death, while others such as calcific aortic stenosis can take many decades to develop clinical manifestations. Progress has been made in recent years toward elucidating a conceptual framework that integrates the dynamic functional structure of heart valves from macro- to micro- to ultra-structure, the biomechanical properties, and the pathobiological behavior of the cardiac valves (Schoen, 2008, 2012).

There are several major forms of valvular heart disease (Figure A.1); most involve the aortic and/or the mitral valve. The most common type of valve disease and most frequent indication for valve replacement overall is calcific aortic stenosis – obstruction at the aortic valve secondary to age-related calcification of the cusps of a valve that was previously anatomically normal (Figure A.1A) (Carabello and Paulus, 2009). Although the precise mechanisms are not yet completely understood, calcific aortic stenosis occurs owing to formation of calcific nodules in the valve cusps, which does not allow them to fully open. This causes pressure overload of the left ventricle that induces hypertrophy (enlargement of the mass) of the walls of this chamber (Rajamannan et al., 2011). This condition takes decades to develop, and typically produces symptoms in approximately the 7th–8th decades of life. Although the normal aortic valve has three cusps, 1–2% of all individuals are born with a bicuspid aortic valve (i.e., with only two cusps), a condition called congenital bicuspid aortic valve. A congenitally bicuspid valve generally works well initially and into adulthood, but persons who have this condition develop valve dysfunction

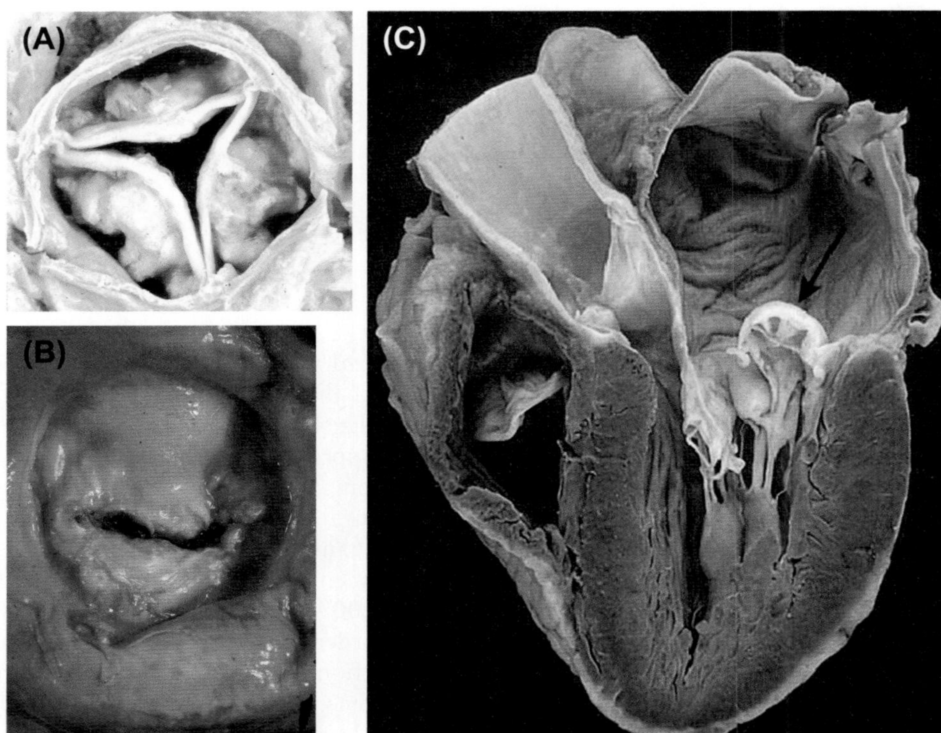

FIGURE A.1 Types of heart valve disease. (A) Severe degenerative calcification of a previously anatomically normal tricuspid aortic valve, the predominant cause of aortic stenosis, and the leading form of valvular heart disease. (B) Chronic rheumatic heart disease, manifest as mitral stenosis, viewed from the left atrium. (C) Myxomatous degeneration of the mitral valve, demonstrating hooding with prolapse of the posterior mitral leaflet into the left atrium (arrow). ((A), (B): Reproduced by permission from Schoen, F. J. & Edwards, W. D. (2001). Valvular heart disease: General priciples and stenosis. In: Cardiovascular Pathology, 3rd edn, Silver, M. D., Gotlieb, A. I. & Schoen, F. J. (eds.). Churchill Livingstone: New York. (C): Reproduced by permission from Schoen, F. J. & Mitchell, R. N. (2010). The heart. In: Robbins Pathologic Basis of Disease, 8th edn, Kumar, V. et al (eds.). W.B. Saunders: Philadelphia.)

and thereby symptoms at younger ages – approximately 10 years earlier than in a patient having a valve with three cusps. Aortic regurgitation (also known as insufficiency) is a less frequent (but nevertheless important) problem, most often caused by dilation of the aortic root. This prevents complete and effective closure of the cusps, allowing backflow across the valve and leading to volume overload of the left ventricle (Goldbarg, 2008). Mitral stenosis (Figure A.1B) has a single predominant cause – chronic rheumatic heart disease – which leads to scarring and stiffening of the mitral leaflets. This condition usually occurs many years following an episode of acute rheumatic fever secondary to streptococcal pharyngitis (a common form of childhood throat infection) (Chandrashekhar et al., 2009). However, mitral regurgitation results from many different conditions; the most frequent is myxomatous valve degeneration (also known as floppy mitral valve), in which the strength of the mitral valve tissue is deficient, thereby causing the valve leaflets to deform excessively (Carabello, 2008) (Figure A.1C). Conditions in which the left ventricle is abnormally dilated and/or scarred, and consequently the valve is not supported properly and infective endocarditis (i.e., infection of the valve) are among the other major causes. Diseases of the right-sided valves (tricuspid and pulmonic) are much less common than those of the left-sided valves. However, in children with congenital heart disease, there is a need for valves in the pulmonary position (in addition to left-sided valves); replacement of valves in congenital anomalies accounts for approximately 5% of valve replacements (Barnett and Ad, 2009). The major clinical complication of valvular heart disease is cardiac failure secondary to changes in the myocardium induced by pressure or volume overload of the chambers, either upstream or downstream of the diseased valve.

HEART VALVE REPLACEMENT AND REPAIR

Heart valve disease is common and serious, and individuals with its various forms have significant mortality and morbidity. For example, the mortality of non-surgically treated critical aortic stenosis, the most deleterious functional abnormality, is approximately 50% at 2–3 years following the onset of symptoms; this natural history is more severe than that of many cancers. Each year in the United States, 20,000 individuals die directly of valvular heart disase; aortic valve disease accounts for approximately 60% of these (Barnett and Ad, 2009). Valve replacement is a highly beneficial

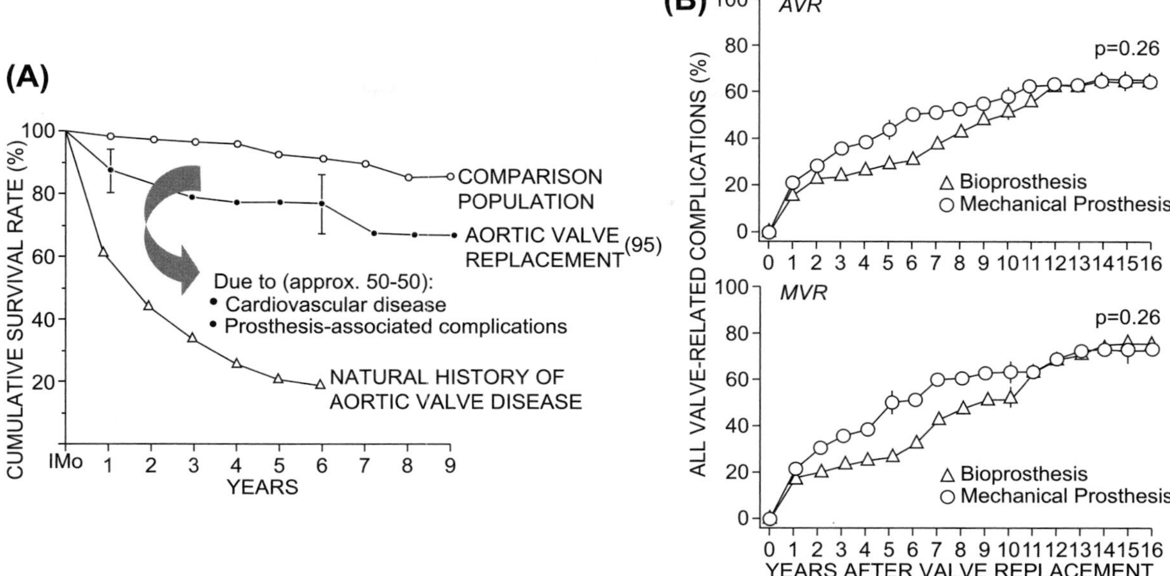

FIGURE A.2 Outcome following cardiac valve replacement. (A) Survival curves for patients with untreated aortic valve stenosis (natural history of valve disease) and aortic valve stenosis corrected by valve replacement, as compared with an age-matched control population without a history of aortic valve stenosis. The numbers presented in this figure for survival following valve replacement nearly four decades ago remain accurate today. This reflects the fact that improvements in valve substitutes and patient management have been balanced by a progressive trend toward operations on older and sicker patients with associated medical illnesses. (Reproduced by permission from Roberts, L. et al. (1976). Long-term survival following aortic valve replacement. Am. Heart J. 91: 311–317.) (B) Frequency of valve related complications for mechanical and tissue valves following mitral valve replacement (MVR) and aortic valve replacement (AVR). (Reproduced by permission from Hammermeister, K. et al. (2000). Outcomes 15 years after valve replacement with a mechanical versus a bioprosthetic valve: Final report of the Veterans Affairs Randomized Trial. J. Am. Coll. Cardiol. 36: 1152–1158.)

therapy for such patients; survival following valve replacement is 50–70%, and serious complication-free survival is approximately 30–50% at 10–15 years (Rahimtoola, 2003). Operative mortalities for aortic and mitral valve replacement are 3% and 6% respectively. While valve replacement thus provides a substantial improvement over the natural history of disease, patients with artificial valves still do not fare as well as similarly aged individuals without valve disease; complications related to the device are a major reason for this difference (Figure A.2).

The surgical treatments available for valvular heart disease include replacement of the valve with a prosthesis and repair of the existing abnormal valve tissue to make it functional (Fedak et al., 2008). Reconstructive/repair procedures to eliminate mitral insufficiency and to minimize the severity of rheumatic mitral stenosis are now highly effective and commonplace. A recent survey of practice in the US showed that 69% of mitral valve operations for mitral regurgitation, and 46% of mitral valve operations overall, compared with only 3% of aortic valve operations, currently use repair rather than replacement (Gammie et al., 2009). Whenever possible, repair of a valve is preferable over replacement; advantages of repair relate to the elimination of both the risk of prosthesis-related complications and the need for chronic anticoagulation (which will be discussed later) that is required in many patients with substitute valves, particularly mechanical valves. Surgical valve repair is often accompanied by stabilization of the annulus (valve ring) with or without implantation of a prosthetic annuloplasty ring. Regrettably, repair is usually not possible for most forms of aortic valve disease.

When repair is not possible, severe symptomatic valvular heart disease is treated by surgical valve replacement, which comprises excision of part or all of the diseased valve, and replacement by a functional substitute. From a design standpoint, the ideal replacement valve would be nonthrombogenic, non-hemolytic, infection resistant, chemically inert, durable, and easily inserted. It would open fully and close quickly and completely, heal appropriately in place, and not be noticed by the patient (for example, it would be noise-free) (Harken et al., 1962; Sapirstein, 2001). Additional criteria for valve design also emphasize the patient's point of view (Rahimtoola, 2010).

The evolution of prosthetic heart valves and related cardiovascular surgical technology was enabled during the first half of the 20th century, by multiple key developments including cardiac catheterization, innovative surgical techniques, the cardiopulmonary bypass, and the anticoagulant heparin (Chaikoff, 2007). In the late 1950s, stimulated by collaborations established between surgeons and biomedical engineers, innovative procedures and device technology which matured in the surgical research laboratory were translated to clinical practice. These developments fostered new opportunities to replace dysfunctional cardiovascular components with biologic or synthetic prostheses. A key step in modern valve replacement technology was the Hufnagel ball valve, designed to be implanted rapidly into the descending thoracic aorta with the use of proximal and distal fixation rings (Butany et al., 2002). However, with this valve regurgitant flow from the lower body was prevented, but cardiac work was only partially relieved and coronary flow was not improved. Subsequently, Albert Starr and his colleagues, along with a mechanical engineer, Lowell Edwards, fabricated a valve consisting of a stainless steel cage, a heat-cured Silastic ball, and a base surrounded by a Teflon fabric sewing cuff, the latter component permitting the surgeon to suture the valve in place orthotopically (i.e., in the anatomically appropriate location within the heart). The three generic components described above: 1) moving part (either synthetic or biologic); 2) superstructure to guide the motion of the moving occluder; and 3) sewing cuff (anchored at the anastomotic site); comprise the key parts of all previous and present heart valve prostheses.

The achievements of Starr and Carpentier provided the foundation on which the clinical success of heart valve replacement is built. Biomedical engineering technological developments have contributed to markedly improved materials and designs for mechanical and biological valves. Today, more than 80,000 valve replacement procedures are performed each year in the United States, and more than 275,000 per year are done worldwide. Moreover, devices and techniques for minimally invasive and percutaneous (catheter-based) valve replacement and repair and other interventional techniques are undergoing rapid innovation and development, and there has been exciting progress toward the creation of a living tissue-engineered heart valve replacement (Kidane, 2008).

... devices and techniques for minimally invasive and percutaneous (catheter-based) valve replacement and repair and other interventional techniques are undergoing rapid innovation and development, and there has been exciting progress toward the creation of a living tissue-engineered heart valve replacement.

MECHANICAL AND TISSUE VALVE REPLACEMENT DEVICES: TYPES AND COMPLICATIONS

Hundreds of designs of substitute heart valve replacement devices have been explored experimentally and in patients; most have been abandoned owing to design and material deficiencies that manifest in complications which became apparent only in clinical use (Dewall, 2000; Edmunds, 2001). Today's cardiac valvular substitutes are of two generic types, mechanical valves and biological tissue valves. The choice of which valve to use in a particular patient is often difficult (El Oakley, 2008; Rahimtoola, 2010).

Mechanical prosthetic heart valves (Figure A.3) are composed of non-physiologic biomaterials that employ rigid, mobile occluders in a metallic cage (cobalt-chrome or titanium alloy), as in the Bjork-Shiley, Medtronic-Hall, and OmniScience valves or two carbon hemidisks in a carbon housing, as in the St. Jude Medical, CarboMedics CPHV, and the Medical Carbon Research Institute or On-X prostheses. Today, all mechanical valve occluders are fabricated from pyrolytic carbon. As described in Chapter I.2.8, pyrolytic carbon has high strength, fatigue and wear resistance, and exceptional biocompatibility, including thromboresistence. The opening and closing of a prosthetic valve is purely passive, with the moving parts (occluder or disc(s)) responding to changes in pressure and blood flow within the chambers of the heart and great vessels. Patients receiving mechanical valves must be treated with lifelong anticoagulation to reduce the risk of thrombosis and thromboembolic events (see below).

Tissue valves (Figure A.4) resemble natural valves more than do mechanical prostheses. The term "bioprostheses" describes a special type of tissue valve composed of three cusps of tissue derived from animals – most frequently either a porcine (pig) aortic valve or bovine (cow) pericardium – each treated with glutaraldehyde. Glutaraldehyde fixation preserves the tissue and decreases its (already relatively low) immunological reactivity, and kills the cells within the valve tissue. No immunosuppression is generally used for these xenografts as is required for whole organ transplants (e.g., kidney, liver or heart). However, since these valves no longer contain viable cells, the cusps themselves cannot remodel or respond to injury as normal tissue does. Bioprosthetic valve cusps are mounted on a metal or plastic stent with three posts (or struts) to simulate the geometry of a native semilunar valve. As with mechanical valves, the base ring is covered by a Dacron or Teflon covered sewing cuff to facilitate surgical implantation and healing. Commonly used bioprosthetic valves are the Hancock porcine, Carpentier-Edwards porcine, and Carpentier-Edwards pericardial tissue valves.

Also used occasionally are tissue valves derived from human cadaveric aortic or pulmonary valves, with or without the associated vascular conduit (called allografts or homografts). These valves have good hemodynamic profiles, a low incidence of thromboembolic complications without chronic anticoagulation, and a low re-infection rate following an original valve replacement for endocarditis (O'Brien, 2001). Several decades ago, when the use of valve allografts began, they were sterilized and/or preserved with chemicals or irradiation; such valves suffered a high rate of leaflet calcification and rupture. Subsequent technical developments have led to the current practice, allografts that are cryopreserved rather than chemically preserved. Freezing is performed with protection from ice crystal formation using dimethyl-sulfoxide. The valves are subsequently stored until use at −196°C in liquid nitrogen. Contemporary allograft valves yield

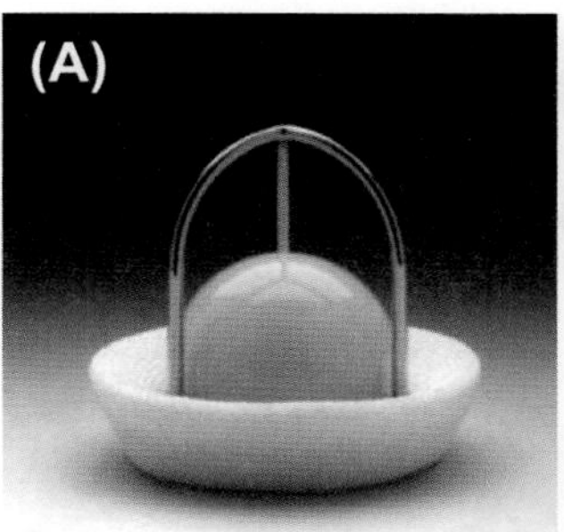

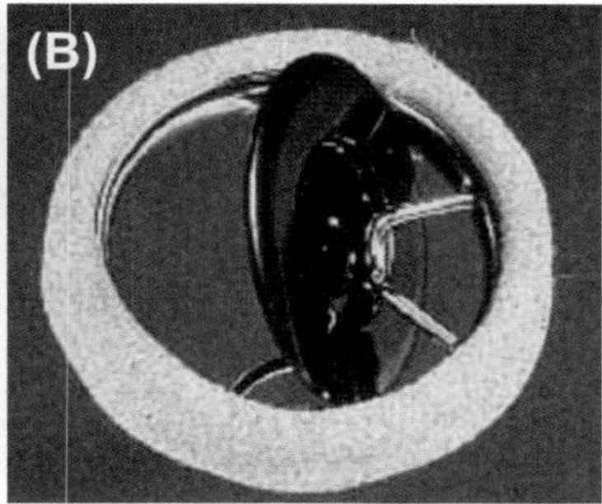

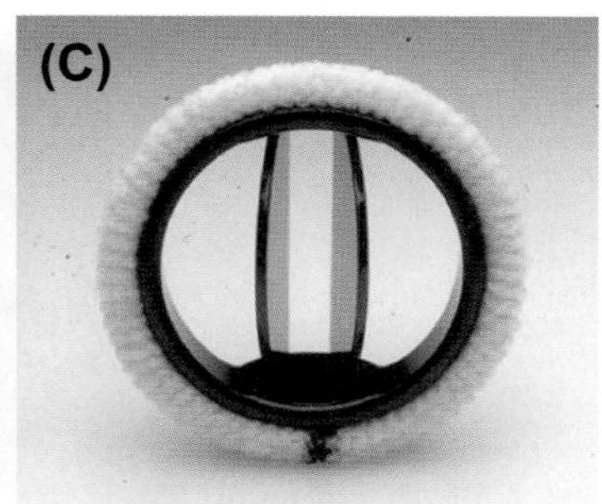

FIGURE A.3 Mechanical prosthetic heart valves. (A). Starr-Edwards caged-ball valve. (B) Bjork-Shiley tilting disk valve. (C) St. Jude Medical bileaflet tilting disk heart valve. (Reproduced by permission from Schoen, F. J. (2001). Pathology of heart valve substitution with mechanical and tissue prostheses. In: Cardiovascular Pathology, 3rd edn, Silver, M. D., Gotlieb, A. I. & Schoen, F. J. (eds.). Churchill Livingstone: New York.)

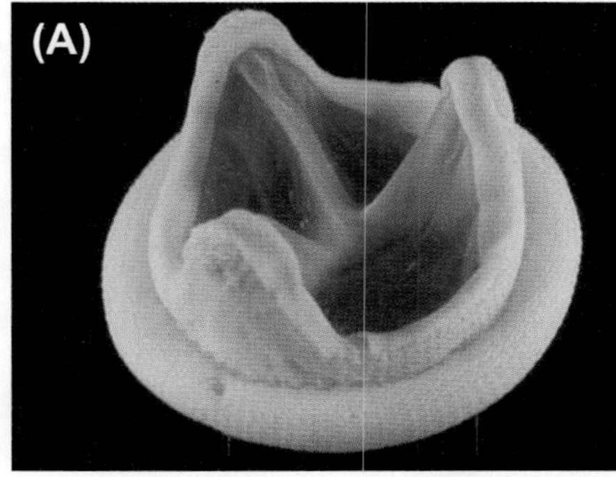

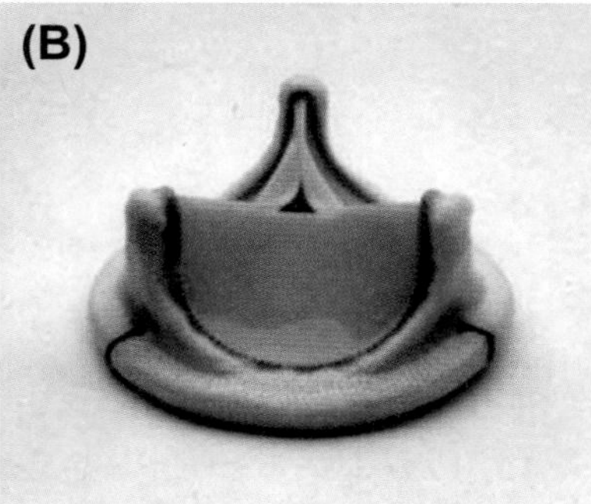

FIGURE A.4 Tissue heart valve replacement devices. (A) Hancock porcine valve. (B) Carpentier-Edwards bovine pericardial valve. (Reproduced by permission from Schoen, F. J. (2001). Pathology of heart valve substitution with mechanical and tissue prostheses. In: Cardiovascular Pathology, 3rd edn, Silver, M. D., Gotlieb, A. I. & Schoen, F. J. (eds.). Churchill Livingstone: New York.)

freedom from degeneration and tissue failure equal to or better than those of conventional porcine bioprosthetic valves, but their use is limited by availability, difficulty in obtaining the proper size, and a more complex surgical procedure for implantation.

The reliability of a valve prosthesis and its interactions with the host and local tissues play a major role in patient outcome. Four categories of valve-related complications (Figure A.5) are most important: thrombosis and thromboembolism; infection; structural dysfunction (i.e., failure or degeneration of the biomaterials comprising a prosthesis); and non-structural dysfunction (i.e., miscellaneous complications and modes of failure not encompassed in the previous groups). The major advantages of tissue valves compared to mechanical prostheses are their pseudo-anatomic central flow and relative nonthrombogenicity; consequently, patients with tissue valves usually do not require anticoagulant therapy unless they have atrial fibrillation or another specific propensity to thrombose the valve. As reflected in overall heart valve replacement industry data, innovations in tissue valve technologies and design have stimulated this segment of the market to grow disproportionately in the last decade by expanding indications for tissue valve use (Fann et al., 2001), and as will be noted later in this chapter, potentially enhanced durability. Thus, there has been a trend toward an increasing fraction of tissue valves implanted relative to mechanical valves (see later).

> The reliability of valve prosthesis and its interactions with host tissue play a major role in patient outcome. Thromboembolic complications are the major cause of mortality and morbidity after cardiac valve replacement with mechanical valves. In contrast, structural dysfunction has been a frequent cause of failure of the most widely used bioprostheses.

Thromboembolic complications are the major cause of mortality and morbidity after cardiac valve replacement with mechanical valves. As discussed in Chapter II.2.6, no synthetic or modified biological surface produced by man is as resistant to thrombosis (*thromboresistant*) as normal unperturbed endothelium. As in the cardiovascular system in general, Virchow's triad (surface thrombogenicity, hypercoagulability, and locally static blood flow) largely predicts the relative propensity of a device to thrombus formation and location of thrombotic deposits with cardiovascular prostheses (Bennett et al., 2009). Exposure of blood to an artificial surface can induce thrombosis, embolization, and consumption of platelets and plasma coagulation factors, as well as the systemic effects of activated coagulation, complement products, and platelets. Thus, patients who have received mechanical substitute heart valves require lifetime therapeutic anticoagulation with warfarin derivatives, which induces a risk of hemorrhage, potentially serious and in some cases fatal (Vahanian, 2008). Thrombotic deposits

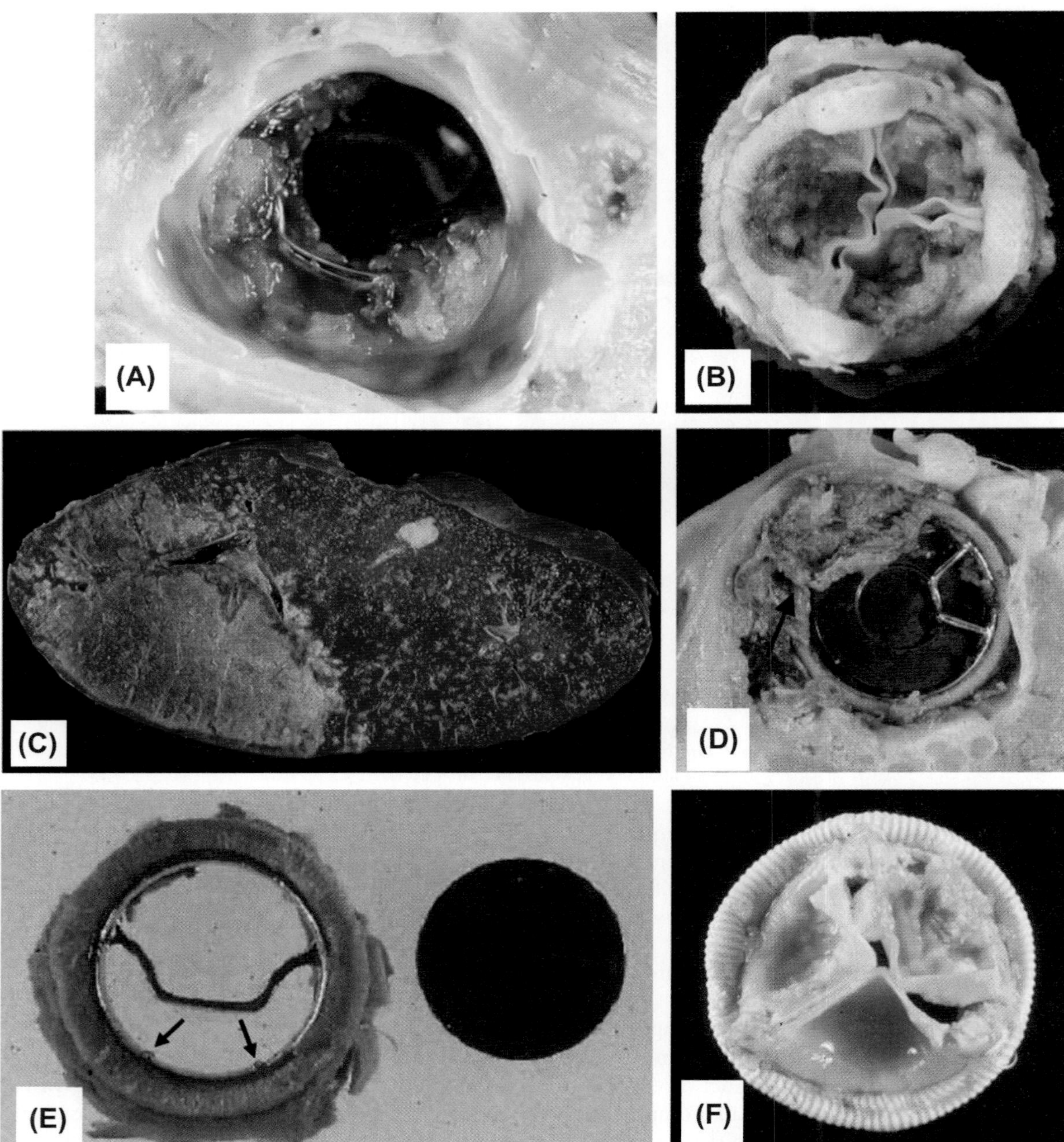

FIGURE A.5 Complications of prosthetic heart valves. (A) Thrombosis on a Bjork-Shiley tilting disk aortic valve prosthesis, localized to outflow strut near minor orifice, a point of flow stasis. (B) Thrombosis of Hancock porcine bioprosthetic valve. (C) Thromboembolic infarct of the spleen (light area at left) secondary to embolus from valve prosthesis. (D) Prosthetic valve endocarditis with large ring abscess (arrow), viewed from the ventricular aspect of an aortic Bjork-Shiley tilting disk aortic valve. (E) Strut fracture of Bjork-Shiley valve, showing valve housing with single remaining strut and adjacent disk. Sites of prior attachment of missing fractured strut designated by arrows. (F) Structural valve dysfunction (manifest as calcific degeneration with tear) of porcine valve. ((D): Reproduced by permission from Schoen, F. J. (1987). Cardiac valve prostheses: Pathological and bioengineering considerations. J. Card. Surg. 2: 65; (A) and (E): Reproduced by permission from Schoen, F. J., Levy, R. J., Piehler, H. R. (1992). Pathological considerations in replacement cardiac valves. Cardiovasc. Pathol. 1: 29.)

forming on valve prostheses can immobilize the occluder or shed emboli to downstream arterial beds (Figures A.5A–C).

Prosthetic valve infection (endocarditis) occurs in 3–6% of recipients of substitute valves (Figure A.5D). When endocarditis was the reason for the original valve replacement, the risk is markedly increased. Rates of infection of bioprostheses and mechanical valves are similar. However, since mechanical valve biomaterials cannot themselves become infected, endocarditis on mechanical valves is localized to the prosthesis–tissue junction at the sewing ring, with accompanying tissue destruction in this area (Piper, 2001). While bioprosthetic valve endocarditis can also be localized to the host tissue–prosthesis junction, biological tissue, despite being chemically fixed, can support growth of bacteria and other microorganisms, and thus the cusps are involved in some cases. The most frequent portals of entry include the mouth via dental procedures, urologic infections and interventions, and indwelling catheters; all comprise breaches of a natural mucosal membrane that may release organisms into the blood. Prosthetic valve endocarditis can occur either early (less than 60 days post-operatively) or late (can be years). The microbial etiology of early prosthetic valve endocarditis is dominated by the staphylococcal species *S. epidermidis* and *S. aureus*, even though prophylactic antibiotic regimens used routinely at the time of implantation are targeted against these microorganisms. The clinical course of early prosthetic valve endocarditis tends to be fulminant. The most common organisms in late prosthetic valve endocarditis are *S. epidermidis*, *S. aureus*, *Streptococcus viridans*, and *enterococci*. Prosthetic valve endocarditis usually necessitates surgical reintervention.

Prosthetic valve dysfunction owing to materials degradation can necessitate reoperation or cause prosthesis-associated death. Many valve models have been withdrawn from clinical use because of poor durability. Durability considerations vary widely for mechanical valves and bioprostheses, for specific types of each, for different models of a particular prosthesis utilizing different materials or having different design features, and even for the same model prosthesis placed in the aortic rather than the mitral site. Fractures of metallic or carbon components of mechanical valve prostheses are fortunately rare, but are usually catastrophic and life-threatening (Figure A.5E). Contemporary single-leaflet or bileaflet tilting disk valves with pyrolytic carbon occluders and either metallic struts or carbon housing have generally favorable durability. Fractures related to past design defects are noteworthy in two valve cohorts. In one instance, the Bjork-Shiley single leaflet tilting disk valve was redesigned with the intention of enhancing disk opening and relieving obstruction and thromboembolic complications that occurred with the original and widely used model. The resultant Bjork-Shiley 60° and 70° convexoconcave tilting disk valves suffered fractures of the welded metallic outlet strut and separation from the valve, leading to frequently fatal disk escape. Over 80,000 valves of this model were implanted, and at least 600 fractured in this manner. The underlying problem was due to the unanticipated consequence of disk closure at a higher velocity and force, causing over-rotation and an excessively hard contact with the metallic outlet strut. When the outlet strut stresses exceeded its endurance limit fatigue fracture occurred, most frequently in the region of the welds anchoring this strut to the housing. In another instance, fractures of carbon valve components (hemi-disk or housing) occurred in implanted Edwards (previously Hemex)-Duromedics bileaflet tilting disk valves. At least 37 valves of this type failed in this manner. Studies

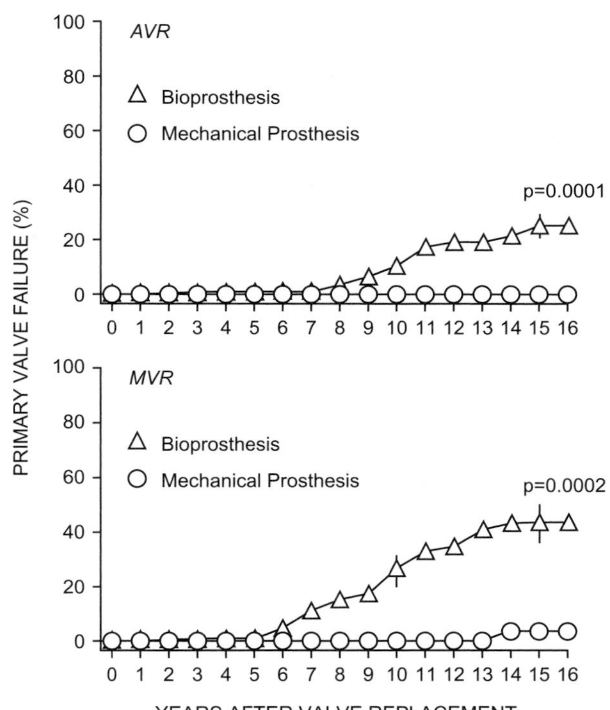

FIGURE A.6 Frequency of primary valve failure (nonthrombotic valve obstruction or central valvular regurgitation) for mechanical and tissue valves following mitral valve replacement (MVR) and aortic valve replacement (AVR). Cuspal mineralization is the major responsible pathologic mechanism with regurgitation through tears the most frequent failure mode. (Reproduced by permission from Hammermeister, K. et al. (2000). Outcomes 15 years after valve replacement with a mechanical versus a bioprosthetic valve: Final report of the Veterans Affairs Randomized Trial. J. Am. Coll. Cardiol. 36: 1152–1158.)

of these explants suggest that valve fracture with leaflet escape resulted from variable combinations of five factors: (1) microporosity in the pyrolytic carbon coating in the leaflets; (2) cavitation bubbles impacting on the carbon surfaces during function; (3) unusual combinations of dimensional tolerances; (4) poor shock-absorbing qualities of the annular tissues in some patients (perhaps due to calcification-induced rigidity); and (5) structural defects in the valve prosthesis induced by fabrication or surgical mishandling. Fractures of carbon components have been encountered only rarely with other carbon bileaflet tilting disk valves, such as the St. Jude Medical valve.

In contrast structural dysfunction is frequent, and is the major cause of failure of the most widely used bioprostheses (Figure A.5F). Bioprosthetic valve structural tissue failure usually results in progressive symptomatic deterioration, which requires reoperation (Schoen and Levy, 2005). Within 15 years following implantation, 30–50% of porcine aortic valves implanted as either mitral or aortic valve replacements require replacement because of primary tissue failure (Figure A.6). Cuspal mineralization is the major responsible pathologic mechanism, with regurgitation through tears the most frequent failure mode in porcine valves (see Chapter II.4.5). Bovine pericardial valves also suffer design-related tearing, and/or calcification. Calcification is markedly accelerated in younger patients, with children and adolescents having an especially accelerated course.

Within the group of complications causing non-structural failure are those that relate to healing of the valve in the site of implantation, either too little or too much. Inadequate healing can cause paravalvular leaks, which permit reverse flow usually through a relatively small hole at the junction of prosthesis and host tissue when the valve is closed. Paravalvular leaks may be clinically inconsequential, may cause hemolysis (see below), or can cause heart failure through regurgitation. In contrast, over-exuberant healing, called tissue overgrowth (or pannus), can block occluder motion or lead to secondary thrombus. Hemolysis (destruction of red blood cells) is unusual with contemporary valves, but paravalvular leaks or dysfunction owing to materials degeneration may induce clinically important hemolysis through mechanical destruction of red blood cell membranes by the high shear stresses that are engendered by blood being forced at high velocity through small spaces.

Various incremental improvements to valve prostheses are being investigated in preclinical studies and clinical research and implementation. For example, methods are being actively studied, and some are being used clinically to prevent calcification of bioprosthetic valves (see Chapter II.4.5). The confidence engendered by early data that these methods may have extended the durable lifetime of bioprosthetic valves has led to a dramatic resurgence of their use, particularly in the USA (Figure A.7). Indeed, in the AHRQ database for 2005, bioprostheses were used in 50% of aortic and 40% of mitral valve replacements, respectively (Barnett and Ad, 2009). Tissue valve use continues to expand; in the database of

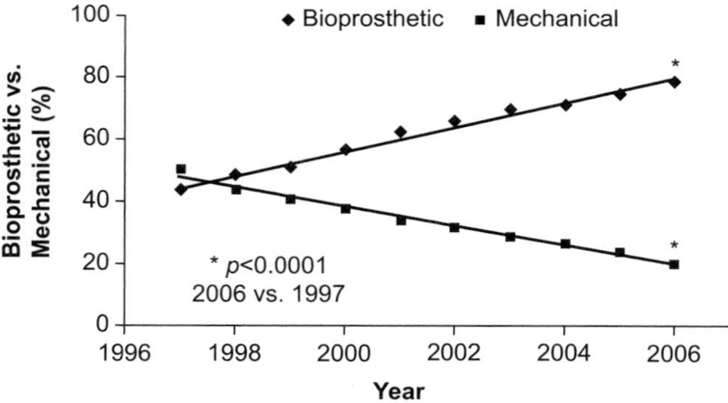

FIGURE A.7 Percentage use of bioprosthetic valves relative to mechanical valves from 1997 through 2006. Bioprosthetic valve use increased progressively during 10 years. Asterisk indicates $P < 0.000001$. (Reproduced by permission from Brown, J. M., O'Brien, S. M., Wu, C. et al. (2009). Isolated aortic valve replacement in North America comprising 108,687 patients in 10 years: Changes in risks, valve types, and outcomes in the Society of Thoracic Surgeons National Database. J. Thorac. Cardiovasc. Surg. 137: 82–90.)

the Society for Thoracic Surgeons, contemporary utilization of bioprosthetic tissue valves is estimated to be about 80% of all aortic and 69% of all mitral substitute heart valves used (Brown et al., 2009; Gammie, 2009). As bioprosthetic valve structural degeneration is accelerated in younger patients, the trend toward increasing use of bioprostheses (relative to mechanical valves) is especially high in older recipients.

Other approaches to provide improved valves include modifications of bioprosthetic valve stent design and tissue mounting techniques to reduce cuspal stresses, tissue treatment modifying or alternative to conventional glutaraldehyde pretreatment to enhance durability and post-implantation biocompatibility, non-stented porcine valves, minimally cross-linked autologous pericardial valves, and flexible trileaflet polymeric (polyurethane) prostheses (Kidane, 2009).

TRANSCATHETER VALVE REPLACEMENT

The surgically implanted bioprosthetic and mechanical valves discussed above are excellent therapies for valvular heart disease, and have a long and proven track record of success. However, some patients with valvular heart disease are currently unable to receive these valves due to other significant medical conditions (called co-morbidities) that create a prohibitively high surgical risk with unacceptable operative mortality. These patients would benefit from a less invasive approach, which would carry very low procedural mortality compared to the traditional surgical one, and might provide significant benefit in treating the valvular disease. Percutaneous balloon valvuloplasty (where a balloon threaded via a catheter to the inside of the aortic valve is expanded under pressure, thereby stretching the valve tissue and fracturing calcific deposits) alone provides insufficient benefit. New catheter-based interventional techniques for inserting foldable prosthetic valves within stenotic aortic and pulmonary valves, and for emulating surgical repair of regurgitant mitral valves, are in various stages of preclinical development, clinical testing, and clinical use (Vassiliades et al., 2005; Block, 2006) (Figure A.8). Percutaneous implantation of a heart valve that can be mounted on an expandable stent, delivered percutaneously through standard catheter-based techniques, and implanted within a diseased valve annulus, has been demonstrated to be feasible. Such valves are expanded in-vivo either by a balloon or are self-expanding (see Figure A.8). Percutaneous valve replacement is most likely to be used in patients with severe aortic stenosis deemed otherwise inoperable, and in congenital heart disease where percutaneous pulmonary valve replacement may find a distinct niche to obviate the morbidity of reoperation to replace malfunctioning pulmonary conduits. In patients with repaired congenital heart disease, complications of an obligatory right ventricle to pulmonary artery conduit are frequent, and the opportunity for palliation to

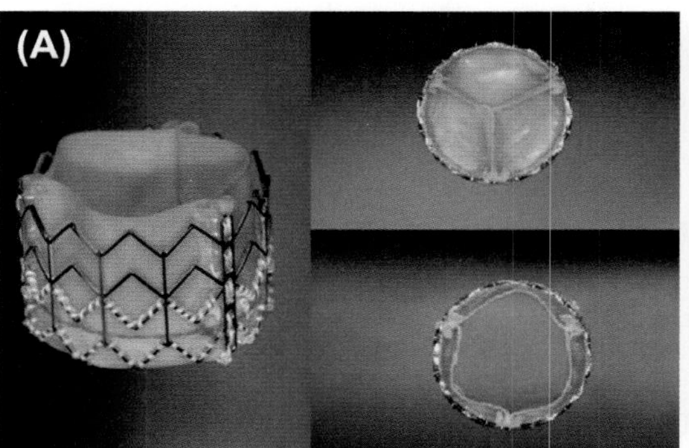

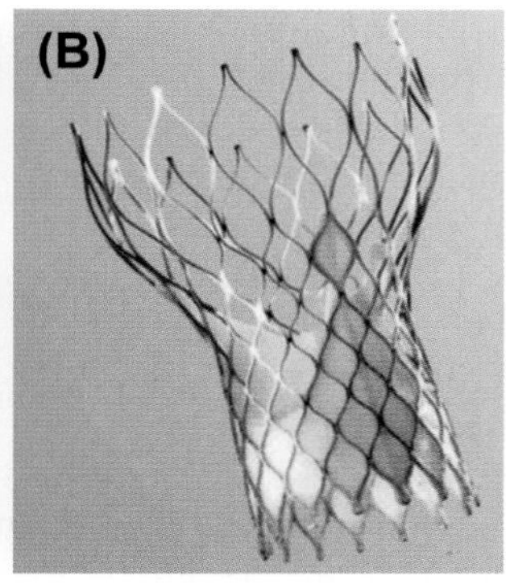

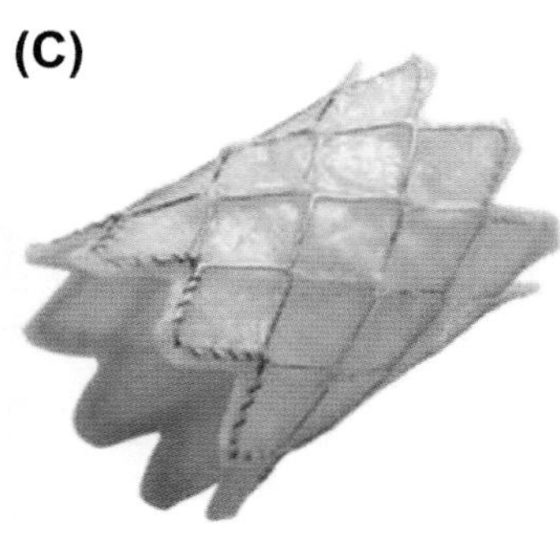

FIGURE A.8 Percutaneous valve replacement technology. (A) The Edwards-SAPIEN™ balloon-expandable aortic valve replacement designed for percutaneous implantation, constructed from bovine pericardium attached to a stainless steel stent. A fabric sealing cuff covers the ventricular aspect to prevent leaks between the prosthesis and surrounding tissues. The valve is mechanically crimped onto a valvuloplasty balloon catheter and expanded within the aortic annulus to displace and exclude the stenotic native from the circulation. (B) Corevalve™ aortic bioprosthesis, constructed of bovine pericardium attached to a self-expanding nickel-titanium alloy (nitinol) stent. The ventricular portion has a high radial force to compress the native valve. The midportion is tapered to avoid interference with the coronary arteries. The aortic portion is flared to provide additional fixation against the wall of the ascending aorta. Nitinol can be made soft at cold temperatures, allowing the stent to be tightly compressed within a delivery sheath. Once positioned within the native valve the sheath is withdrawn, allowing the stent to assume its predetermined shape. There is adequate radial force to compress the native valve. (C) The Melody™ pulmonary valve is constructed from a bovine jugular venous valve attached with sutures to a platinum-iridium alloy stent. The relatively delicate venous valve functions well in the pulmonary circulation, but is too fragile for use in the systemic. Although often referred to as a pulmonary valve, its maximum expanded diameter of 22 mm largely limits its use to surgically-constructed right ventricular to pulmonary artery conduits in the pediatric population. ((A) and (B) reproduced by permission from Schoen, F. J. & Webb, J. G. (2008). Prosthetics and the Heart. In: Atlas of Cardiovascular Pathology for the Clinician, McManus, B. M. & Braunwald, E. (eds.), pp. 241–256. Current Medicine: Philadelphia, PA).

delay or prevent surgery is attractive in many cases. For aortic valve disease, the excellent clinical success with surgical aortic valve replacement presently restricts the use of percutaneous valve replacement to non-surgical candidates or as a bridge to valve replacement in patients in whom surgery needs to be delayed.

> New catheter-based interventional techniques for inserting foldable prosthetic valves within stenotic aortic and pulmonary valves, and for emulating surgical repair of regurgitant mitral valves, are in various stages of preclinical development, clinical testing, and early clinical use.

The devices used as percutaneous valve replacements have an outer stent-like structure which contains leaflets. The stent holds open a valve annulus or segment of a prosthetic conduit, and resists the tendency of a vessel, valve annulus or diseased native leaflets to recoil following balloon dilation, supports the valve leaflets, and provides the means for seating the prosthesis in the annulus or vessel. The types of tissues used for the valve component include bovine, equine or porcine pericardium, and bovine jugular venous valves. The stents can be made from self-expandable or shape-memory materials such as nickel-titanium alloys (e.g., Nitinol) or from balloon-expandable materials such as stainless steel, platinum-iridium or other alloys. The delivery strategy involves collapsing the device and placing it within a catheter-based sheath; for balloon expandable devices, they must be collapsed over a balloon. The catheter containing the device can be inserted through a local incision into the femoral artery (or vein as appropriate to the target site) using essentially the same technique for deployment of coronary artery stents and other endovascular and cardiac devices. In the case of aortic stenosis, the catheter containing the valve device is passed from the femoral artery retrograde up the aorta to the aortic valve, and deployed between the cusps of the calcified aortic valve, pushing the diseased cusps as far as possible out of the way. Alternatively, the device can be deployed in an antegrade fashion using a minimally invasive surgical approach through the apex of the left ventricle. This transapical approach is favored in patients with significant atherosclerotic disease of the femoral artery or aorta, since manipulation of catheters through such sites can dislodge atherosclerotic debris, leading to emboli. For children with failed right ventricular to pulmonary artery devices used to correct certain types of congenital heart defects, the valve is threaded from the femoral vein to the inferior vena cava through the right side of the heart.

Several catheter-based devices are currently in various stages of development and clinical use in the aortic and pulmonary position (Chiam, 2009; Lurz, 2009; Ye, 2009; Zajarias, 2009). The two transcatheter aortic valves with the largest clinical experience are the Edwards SAPIEN device and the CoreValve ReValving system. The FDA approved SAPIEN device is composed of a balloon expandable stainless steel stent which houses a bovine pericardial trileaflet valve. The stent has a low profile and is designed to be placed in the subcoronary position (Figure A.8A). There is a polymer skirt circumferentially attached to the stent to reduce paravalvular leaks.

The CoreValve device, which at the time of this writing, has achieved the CE Mark in Europe but not FDA approval, is composed of a self-expandable Nitinol stent which houses a porcine pericardial trileaflet valve (Figure A.8B). These devices have been used in approximately 40000 patients worldwide since the first human experience was reported in, and further use is expected grow rapidly (Rodés-Cabau, 2012). The FDA approved Medtronic Melody transcatheter pulmonary valve is composed of a balloon expandable platinum-iridium alloy stent which houses a segment of bovine jugular vein containing its native venous valve (Figure A.8C) (Lurz et al., 2009). The Melody was designed to be used in patients with congenital heart disease, usually children or young adults, who have received surgically implanted right ventricular outflow tract conduits that are failing, either due to stenosis or regurgitation. Catheter-based valves may also play a role in the treatment of surgically implanted bioprosthetic valves that are failing due to stenosis or regurgitation, in a so-called "valve-in-valve" application in which a new prosthesis is inserted directly into a prior one.

ENGINEERED HEART VALVES

Recent scientific and technological progress has stimulated the goal of generating a "tissue-engineered" heart valve, a living valve replacement that would obviate the complications of conventional valve replacement, adapt to changing environmental conditions in the recipient, and potentially grow with a growing patient (Breuer et al., 2004; Vesely, 2005; Mendelson and Schoen, 2006; Sacks et al., 2009). Innovative work toward this objective is active in many laboratories, and may eventually lead to clinical application, though many challenges exist. The clinical need, opportunities, progress, and challenges of heart valve tissue engineering are described in detail in Chapter II.6.10.

BIBLIOGRAPHY

Barnett, S. C., & Ad, N. (2009). Surgery for aortic and mitral valve disease in the United States: A trend of change in surgical proactice between 1998 and 2005. *J. Thorac. Cardiovasc. Surg.*, **137**, 1422–1429.

Bennett, P. C., Silverman, S. H., Gill, P. S., & Lip, G. Y. (2009). Peripheral arterial disease and Virchow's triad. *Thromb. Haemost.*, **101**, 1032–1040.

Block, P. C. (2006). Percutaneous transcatheter repair for mitral regurgitation. *J. Interv. Cardiol.*, **19**, 547–551.

Bonow, R. O., Carabello, B. A., Kanu, C., et al. (2006). ACC/AHA 2006 guidelines for the management of patients with valvular heart disease: A report of the American College of Cardiology/American Heart Association Task Force on Practice Guidelines (writing committee to revise the 1998 Guidelines for the Management of Patients With Valvular Heart Disease): Developed in collaboration with the Society of Cardiovascular Anesthesiologists: Endorsed by the Society for Cardiovascular Angiography and Interventions and the Society of Thoracic Surgeons. *Circulation*, **114**, e84–e231.

Breuer, C. K., Mettler, B. A., Anthony, T., Sales, V. L., Schoen, F. J., & Mayer, J. E. (2004). Application of tissue-engineered principles toward the development of a semilunar heart valve substitute. *Tissue Eng.*, **10**, 1725.

Brown, J. M., O'Brien, S. M., Wu, C., et al. (2009). Isolated aortic valve replacement in North America comprising 108,687 patients in 10 years: Changes in risks, valve types, and outcomes in the Society of Thoracic Surgeons National Database. *J. Thorac. Cardiovasc. Surg.*, **137**, 82–90.

Butany, J., Ahluwalia, M. S., Payet, C., et al. (2002). Hufnagel valve: The first prosthetic mechanical valve. *Cardiovasc. Pathol.*, **11**, 351–353.

Carabello, B. A. (2008). The current therapy for mitral regurgitation. *J. Am. Col. Cardiol.*, **52**, 319–326.

Carabello, B. A., & Paulus, W. J. (2009). Aortic stenosis. *Lancet*, **373**, 956–966.

Carpentier, A. (2007). Lasker Clinical Research Award. The surprising rise of nonthrombogenic valvular surgery. *Nat. Med.*, **13**, 1165–1168.

Chaikoff, E. L. (2007). The development of prosthetic heart valves – lessons in form and function. *New Engl. J. Med.*, **357**, 1368–1371.

Chandrashekhar, Y., Westaby, S., & Narula, J. (2009). Mitral stenosis. *Lancet*, **374**, 1271–1283.

DeWall, R. A., Qasim, N., & Carr, L. (2000). Evolution of mechanical heart valves. *Ann. Thorac. Surg.*, **69**, 1612–1621.

Edmunds, L. H., Jr. (2001). Evolution of prosthetic heart valves. *Am. Heart J.*, **141**, 849–855.

El Oakley, R., Kleine, P., & Bach, D. S. (2008). Choice of prosthetic heart valve in today's practice. *Circulation*, **117**, 253–256.

Fann, J. I., & Burdon, T. A. (2001). Are the indications for tissue valves different in 2001 and how do we communicate these changes to our cardiology colleagues? *Curr. Opin. Cardiol.*, **16**, 126–135.

Fedak, P. W., McCarthy, P. M., & Bonow, R. O. (2008). Evolving concepts and technologies in mitral valve repair. *Circulation*, **117**, 963–974.

Gammie, J. S., Sheng, S., Griffith, B. P., et al. (2009). Trends in mitral valve surgery in the United States: Results from the Society of Thoracic Surgeons Adult Cardiac Surgery Database. *Ann. Thorac. Surg.*, **87**, 1431–1437.

Goldbarg, S. H., & Halperin, J. L. (2008). Aortic regurgitation: Disease progression and management. *Nat. Clin. Pract. Cardiovasc. Med.*, **5**, 269–279.

Harken, D. F., Taylor, W. J., LeFemine, A. A., et al. (1962). Aortic valve replacement with a caged ball valve. *Am. J. Cardiol.*, **9**, 292–299.

Hoerstrup, S. P., Sodian, R., Daebritz, S., et al. (2000). Functional living trileaflet heart valves grown in-vitro. *Circulation*, **102**. III-44.

Hoerstrup, S. P., Cummings, I., Lachat, M., et al. (2006). Functional growth in tissue engineered living vascular grafts: Follow up at 100 weeks in a large animal model. *Circulation*, **114**, 159.

Isom, O. W. (2002). Mitral commissurotomy and valve replacement for mitral stenosis: Observations on selection of surgical procedures. *Adv. Cardiol.*, **39**, 114–121.

Kidane, A. G., Burriesci, G., Cornejo, P., et al. (2009). Current developments and future prospects for heart valve replacement therapy. *J. Biomed. Mater. Res. B Appl. Biomater.*, **88**, 290–303.

Lifton, R. P. (2007). Lasker Award to heart valve pioneers. *Cell*, **130**, 971–974.

Lurz, P., Bonhoeffer, P., & Taylor, A. M. (2009). Percutaneous pulmonary valve implantation: An update. *Expert Rev. Cardiovasc. Ther.*, **7**, 823–833.

Matheny, R. G., Hutchison, M. L., Dryden, P. E., Hiles, M. D., & Shaar, C. J. (2000). Porcine small intestine submucosa as a pulmonary valve leaflet substitute. *J. Heart Valve Dis.*, **9**, 769.

Mendelson, K. A., & Schoen, F. J. (2006). Heart valve tissue engineering: Concepts, approaches, progress, and challenges. *Ann. Biomed. Engin.*, **34**, 1799–1819.

O'Brien, M. F., Harrocks, S., Stafford, E. G., et al. (2001). The homograft aortic valve: A 29-year, 99.3% follow up of 1,022 valve replacements. *J. Heart Valve Dis.*, **10**, 334–344.

Piper, C., Korfer, R., & Horstkotte, D. (2001). Prosthetic valve endocarditis. *Heart*, **85**, 590–593.

Rahimtulla, S. H. (2003). Choice of prosthetic heart valve for adult patients. *J. Am. Coll. Cardiol.*, **41**, 893–904.

Rahimtoola, S. H. (2010). Choice of prosthetic heart valve in adults: An update. *J. Am. Coll. Cardiol.*, **55**, 2413.

Rajamannan, N. M., Evans, F. J., Aikawa, E., Grande-Allen, K. J., Demer, L. L., Heistad, D. D., Simmons, C. A., Masters, K. S., Mathieu, P., O'Brien, K. D., Schoen, F. J., Towler, D. A., Yoganathan, A. P., & Otto, C. M. (2011). Calcific aortic valve disease: Not simply a degenerative process. A Review and Agenda for Research from the National Heart and Lung and Blood Institute Aortic Stenosis Working Group. *Circulation*, **124**, 1783–1791.

Rodés-Cabau, J. (2012). Transcatheter aortic valve implantation: Current and future approaches. *Nat. Rev. Cardiol.*, **9**, 15.

Sapirstein, J. S., & Smith, P. K. (2001). The "ideal" replacement heart valve. *Am. Heart J.*, **141**, 856–860.

Sacks, M. S., & Yoganathan, A. P. (2007). Heart valve function: A biomechanical perspective. *Philos. Trans. R Soc. Lond. B Biol. Sci.*, **362**, 1369–1391.

Sacks, M. S., Schoen, F. J., & Mayer, J. E., Jr. (2009). Bioengineering challenges for heart valve tissue engineering. *Annu. Rev. Biomed. Eng.*, **11**, 289–313.

Sacks, M. S., Merryman, W. D., & Schmidt, D. E. (2009). On the biomechanics of heart valve function. *J. Biomech.*, **42**, 1804–1824.

Schoen, F. J. (2001). Pathology of heart valve substitution with mechanical and tissue prostheses. In M. D. Silver, A. I. Gotlieb, & F. J. Schoen (Eds.), *Cardiovascular Pathology* (3rd ed., pp. 629–677): WB Saunders.

Schoen, F. J. (2008). Evolving concepts of heart valve dynamics. The continuum of development, functional structure, pathology and tissue engineering. *Circulation*, **118**, 1864–1880.

Schoen, F. J. (2012). Mechanisms of function and disease in natural and replacement heart valves. *Ann. Rev. Path. Mech. Dis.*, **7**, 161.

Schoen, F. J., & Levy, R. J. (2005). Calcification of tissue heart valve substitutes: Progress toward understanding and prevention. *Ann. Thorac. Surg.*, **79**, 1072–1080.

Schoen, F. J., & Mitchell, R. N. (2008). The heart. In V. Kumar, A. Abbas, N. Fausto, & R. N. Mitchell (Eds.), *Robbins Basic Pathology* (8th ed., pp. 379–419): WB Saunders.

Schoen, F. J., & Webb, J. G. (2008). Prosthetics and the Heart. In B. M. McManus, & E. Braunwald (Eds.), *Atlas of Cardiovascular Pathology for the Clinician* (pp. 241–256). Philadelphia: Current Medicine.

Sharma, S., Mehra, A., & Rahimtoola, S. (2008). Valvular heart disease: A century of progress. *Am. J. med.*, **121**, 664–673.

Simon, P., Kasimir, M. T., Seebacher, G., et al. (2003). Early failure of the tissue engineered porcine heart valve SYNERGRAFT in pediatric patients. *Eur. J. Cardiothorac. Surg.*, **23**, 1002–1006.

Starr, A. (2007). Lasker Clinical Medical Research Award. The artificial heart valve. *Nat. Med.*, **10**, 1160–1164.

Vahanian, A. (2008). Antithrombotic therapy for patients with valvular heart disease. *Herz*, **33**, 44–51.

Vassiliades, T. A., Block, P. C., Cohn, L. H., et al. (2005). The clinical development of percutaneous heart valve technology. A position statement of the Society of Thoracic Surgeons (STS), the American Association of Thoracic Surgery (AATS), and the Society for Cardiovascular Angiography and Interventions (SCAI). *J. Thorac. Cardiovasc. Surg.*, **129**, 970–976.

Vesely, I. (2005). Heart valve tissue engineering. *Circ. Res.*, **97**, 743.

Vongpatanasin, W., Hillis, L. D., & Lange, R. A. (1996). Prosthetic heart valves. *New Engl. J. Med.*, **335**, 407–416.

Zajarias, A., & Cribier, A. G. (2009). Outcomes and safety of percutaneous aortic valve replacement. *J. Am. Coll. Cardiol.*, **53**, 1829–1886.

B. ENDOVASCULAR STENTS, VASCULAR GRAFTS, AND STENT GRAFTS

Frederick J. Schoen and Robert F. Padera, Jr.
Department of Pathology, Brigham and Women's Hospital and Harvard Medical School, Boston, MA, USA

INTRODUCTION

Considerable progress on the impact of atherosclerosis-related disease has been made over the last several decades in the United States and elsewhere. Between 1963 (the peak year) and 2000 there was an approximately 50% decrease in the death rate from the consequences of coronary artery atherosclerosis and a 70% decrease in death from strokes. From 1995–2005 alone, death rates from cardiovascular disease declined more than 25%. Three factors have contributed to this impressive improvement: (1) prevention of atherosclerosis through changes in lifestyle, including reduced cigarette smoking, altered dietary habits with reduced consumption of cholesterol and saturated animal fats, and control of hypertension; (2) improved methods of treatment of myocardial infarction and other complications of atherosclerosis-related disease, especially coronary artery bypass graft surgery, and manipulation of atherosclerotic blood vessels through catheters placed into blood vessels in the groin and threaded up to the heart (a collection of procedures called percutaneous coronary intervention (PCI), which includes percutaneous transluminal coronary angioplasty and placement of stents (see below)); and (3) prevention of recurrence in patients who have previously suffered serious atherosclerosis-related clinical events.

> Vascular pathologies – and their downstream sequelae – are responsible for more morbidity and mortality than any other category of human disease (Mitchell and Schoen, 2010). Cardiovascular disease overall accounted for one-third of the nearly 2.5 million deaths in the United States in 2006; coronary heart disease alone caused about one of every five deaths (Lloyd-Jones, 2009).

This chapter summarizes the major types of treatments and related challenges and opportunities to manage the two principal mechanisms causing clinically-important vascular disease states:

- Narrowing (stenosis) or complete obstruction of vessel lumina;
- Weakening of vessel walls.

Most vascular disease occurs through the process of *atherosclerosis*, a slowly progressive disease in which asymmetric focal thickenings (also called *atheroma* or *plaques*) form in and obstruct small arteries or weaken larger arteries, beginning in childhood and typically causing symptoms in middle age or later life (Mitchell and Schoen, 2010). The characteristic lesion of atherosclerosis is the atheroma or plaque that forms through thickening of the inner layer of the vessel wall (called the *intima*). Plaque formation occurs by a mechanism involving smooth muscle cell proliferation, production of collagen, and accumulation of lipids in the arterial wall (Libby, 2005). Vascular pathology caused by atherosclerosis is shown in Figure B.1.

Atherosclerosis primarily affects the large elastic arteries, and large and medium-sized muscular arteries of the systemic circulation, particularly near branches and sharp curvatures. Although veins are usually spared from atherosclerosis, venous bypass grafts interposed within branches of the arterial system (such as saphenous vein grafts) frequently develop intimal thickening, and ultimately atherosclerotic obstructions. Moreover, some arteries used as arterial grafts, such as the internal mammary artery, are largely spared. The principal sites of involvement of atherosclerosis are the abdominal aorta, coronary arteries, popliteal arteries, descending thoracic aorta, and internal carotid arteries.

In small arteries, atherosclerotic plaques can grow sufficiently large to occlude lumens, and thereby compromise blood flow to downstream organs, leading to ischemia, a pathologic condition caused by inadequate blood flow, and the resulting tissue damage. Obstruction, which can occur either progressively (e.g., by atherosclerosis) or precipitously (e.g., by thrombosis or embolism), diminishes flow (*perfusion*) to the downstream tissue. Thus, if a coronary artery is blocked, the perfusion via coronary flow may be inadequate to meet the metabolic needs of the downstream heart muscle (myocardium), leading to a physiologic condition called *ischemia*. If the blockage is severe and prolonged, then

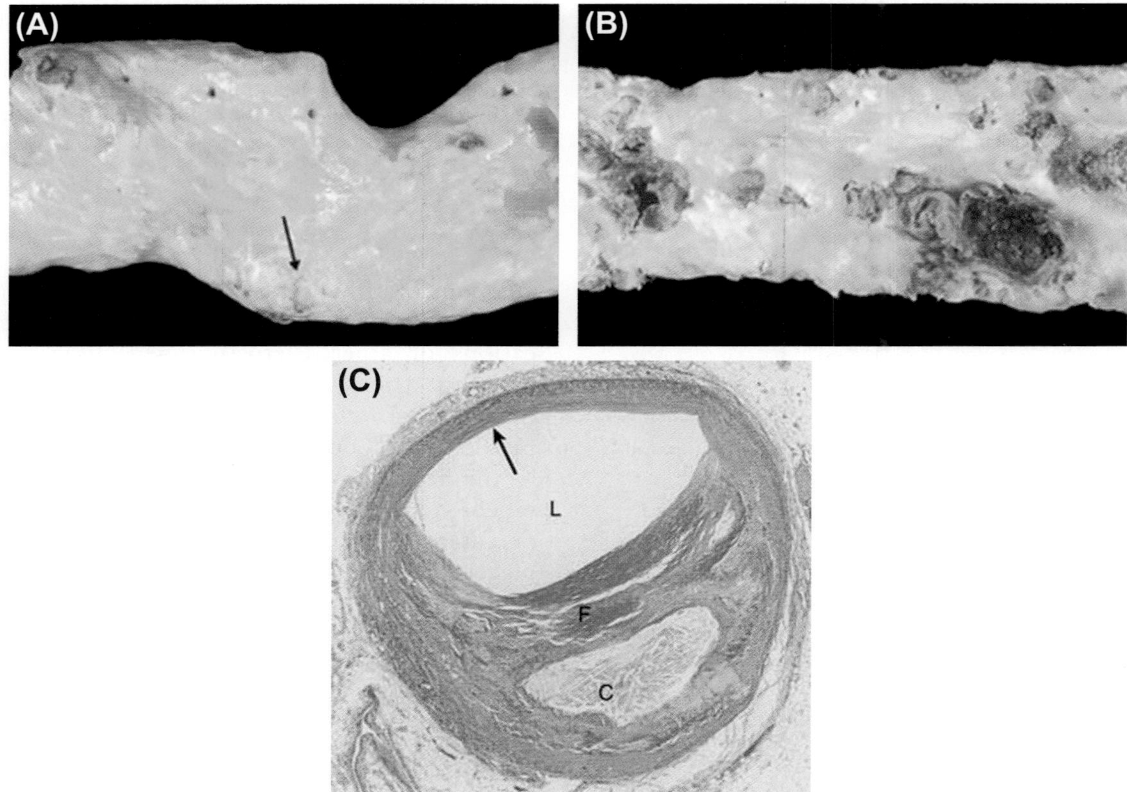

FIGURE B.1 Atherosclerosis. (A) and (B) Gross views of atherosclerosis in the aorta. (A) Mild atherosclerosis composed of fibrous plaques, one of which is denoted by the arrow. (B) Severe disease with diffuse and complicated lesions (with wall destruction and superimposed thrombosis). (C) Histologic features of atherosclerotic plaque in the coronary artery, demonstrating fibrous cap (F), and a central necrotic (largely lipid) core (C). The lumen (L) has been moderately compromised. An uninvolved portion of the arterial wall is designated by an arrow. In this section, collagen has been stained blue (Masson's trichrome stain). (Reproduced by permission from Mitchell, R. N. & Schoen, F. J. (2010). Blood vessels. In: Robbins/Cotran Pathologic Basis of Disease, 8th ed., Kumar, V., Fausto, N., Abbas, A. et al. (eds.), p. 487. W.B. Saunders: Philadelphia, PA.)

tissue dies and the result is a heart attack (*myocardial infarction*); if in a cerebral artery neurons die, and the result is a stroke (*cerebrovascular accident*).

In the aorta and other large vessels, where plaque does not typically cause obstruction, advanced atherosclerosis can weaken the arterial wall, resulting in an aneurysm or dissection, either of which can rupture. An *aneurysm* is a localized abnormal dilation of a blood vessel that forms as a result of destruction and weakening of the vascular wall beneath an atherosclerotic plaque. The most frequent type of aneurysm is the abdominal aortic aneurysm (AAA), a dilation of the distal aorta immediately before it bifurcates into the iliac arteries to provide the blood supply to the legs (Sakalihasan, 2005); aneurysms located in the chest, either as ascending or descending thoracic aortic aneurysms (TAA), can also occur. Another type of arterial disease is a *dissection*, which arises when blood enters the wall of the artery to form a hematoma dissecting between its layers. This is often due to conditions such as hypertension or intrinsic genetic deficiency of the quality of the collagen of the arterial wall, which weakens its mechanical properties (Prêtre, 1997). Approximately 45,000 people die annually in the United States from diseases of the aorta and its branches.

> Mechanical therapies to prevent and treat the complications of vascular disease in general and atherosclerotic obstructions in particular include endovascular manipulation such as balloon angioplasty and stents, vascular bypass, vascular replacement using grafts, and various combinations of these approaches.

ANGIOPLASTY AND STENTS

Percutaneous cardiovascular interventions (PCI), including balloon dilation of an artery blocked by atherosclerotic plaque and/or thrombotic deposits (called percutaneous transluminal coronary angioplasty (PTCA)) and endovascular stents are used to restore blood flow through a diseased portion of the coronary circulation, and are shown in Figure B.2 (Landau, 1994; Daemen et al, 2007, 2 parts). In PTCA, a procedure developed and implemented first

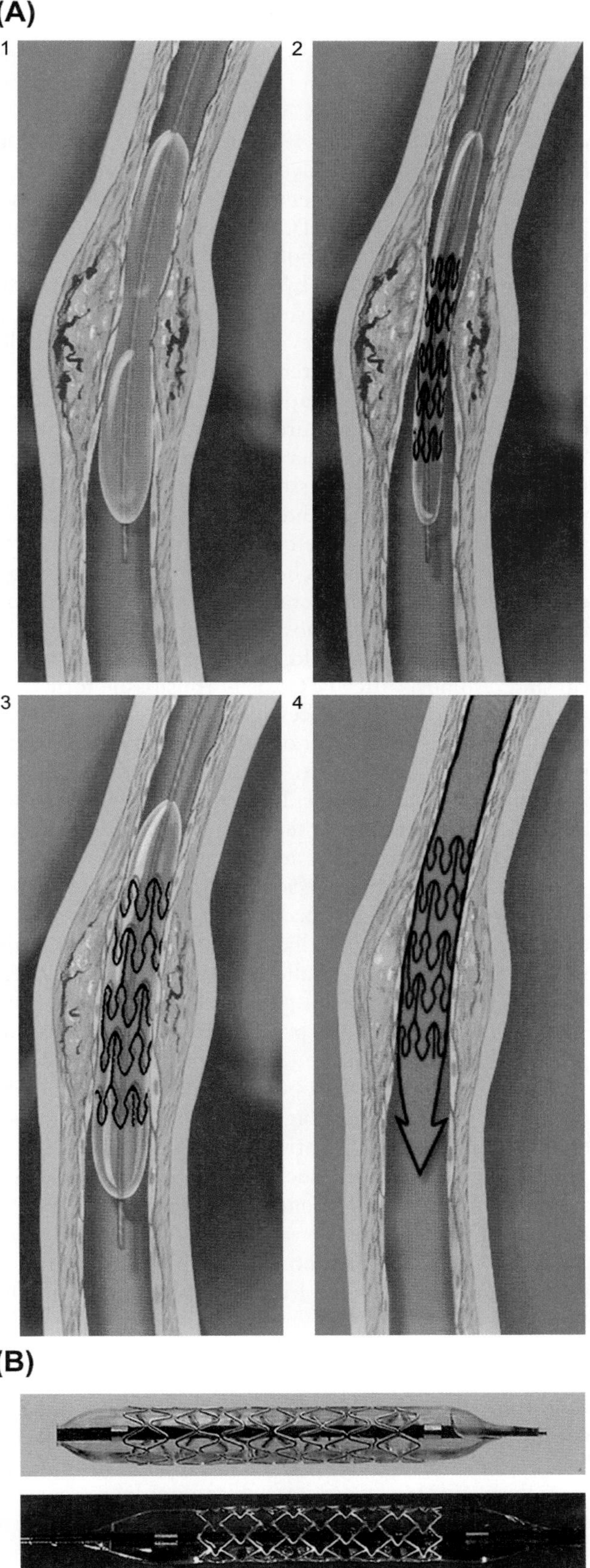

FIGURE B.2 Balloon angioplasty and endovascular stenting. (A) Catheter-based interventions for opening occluded coronary arteries: 1, Percutaneous transluminal coronary angioplasty (PTCA); and successively, 2, 3, and 4, balloon-expandable stent placement. A balloon-tipped catheter is positioned in the coronary artery narrowing and inflated (1). The stent is positioned at the site of the coronary narrowing (2). When the balloon is inflated, the stent expands and presses against the arterial wall (3). The balloon is deflated and removed. The stent remains permanently in place, helping to keep the artery open (4). Within a few weeks, new tissue will grow over the stent struts and cover them. (B) Metallic stents on expanded balloons. ((A) Reproduced by permission from Michaels, A. D. & Chatterjee, K. (2003). Circulation 106: e187; (B) Reproduced by permission from Al Suwaidi, J., Berger, P. B. & Holmes, D. R. (2000). Coronary artery stents. JAMA 284: 1828.)

by Andreas Gruntzig in the late 1970s, a long catheter is passed from the femoral artery up the aorta to the openings (called *ostia*) of the coronary arteries that arise from the aorta immediately distal to the aortic valve cusps. Using radio-opaque dye and fluoroscopy, areas of stenosis can be identified. In PTCA a deflated balloon is passed over a guidewire to a site of stenosis, where the balloon is inflated using progressive and substantial expansile force (~10 atm). Enlargement of the lumen and increased blood flow induced by PTCA occurs by plaque reduction via compression, embolization or redistribution of the plaque contents, and by overall mechanical expansion of the vessel wall (Virmani, 1994). Although early angioplasty balloons were made from polyvinyl chloride and polyethylene, current balloons are composed of nylon or polyethylene teraphthalate. Short-term failure of this procedure (i.e., closure of the treated vessel within hours to days) can occur via several mechanisms, including elastic recoil of the vessel wall, acute thrombosis at the site of angioplasty, and acute dissection or blood within the vessel wall itself in the area of angioplasty. Nevertheless, the "Achilles heel" of PTCA is the development of progressive, proliferative restenosis which occurs in 30–50% of patients, most frequently within the first 4–6 months (Haudenschild, 1993). The usual process causing restenosis after PTCA is fibrous tissue formation in the lumen owing to excessive medial smooth muscle proliferation as an exaggerated response to angioplasty-induced injury, similar to features of atherosclerosis itself, and to stent and vascular graft healing (see below). Systemic pharmacologic therapies, by themselves, have not effectively mitigated restenosis after PTCA.

Stents mitigate at least some of the negative consequences of PTCA. Developed in the late 1980s, stents are expandable tubes of metallic mesh that splint open the vessel wall at the site of balloon angioplasty or vascular injury. Today, virtually all patients undergoing PCI will also receive a stent. Stents preserve luminal patency, and provide a larger and more regular lumen by acting as a scaffold to support the disrupted vascular wall and thereby reduce the impact of postangioplasty restenosis (Daemen et al, 2007a, 2007b). Stenting may also be done in the peripheral vasculature, such as the femoro-popliteal system (Schillinger and Minar, 2009; Yustein et al, 2009). Bare metal stents are short tubular segments of metal mesh composed of balloon-expandable 316L stainless steel or nitinol that range from 2.5 to 4.0 mm in diameter. Other metals and polymers are being investigated.

Stents can have complications, particularly thrombosis, occurring early or late, and late restenosis (Figure B.3). Early thrombosis, occurring in 1 to 3% of patients within 7 to 10 days of the procedure (see Figure B.3A) has largely been overcome by aggressive multidrug treatment with antiplatelet agents such as clopidogrel, aspirin, and glycoprotein IIb/IIIa inhibitors. The major long-term complication of bare metal stenting with devices is in-stent restenosis (Figure B.3B), which occurs in 50% of patients within six months (Virmani, 1999). The causes of stent thrombosis and restenosis are complex, and are largely due to stent–tissue interactions, including inflammation, which may potentially interfere with healing and re-endothelialization (Nakazawa, 2009). Damage to the endothelial lining and stretching of the vessel wall stimulate adherence and accumulation of platelets, fibrin, and leukocytes (white blood cells). Stent wires may eventually become completely embedded in an endothelium-lined layer of intimal fibrosis composed of smooth muscle cells in a matrix of collagen, which is essentially scar. This tissue may thicken secondary to the release of growth factors, chemotactic factors, and inflammatory mediators from platelets and other inflammatory cells that result in increased migration and proliferation of smooth muscle cells, and increased production of extracellular matrix molecules, narrowing the lumen and resulting in restenosis.

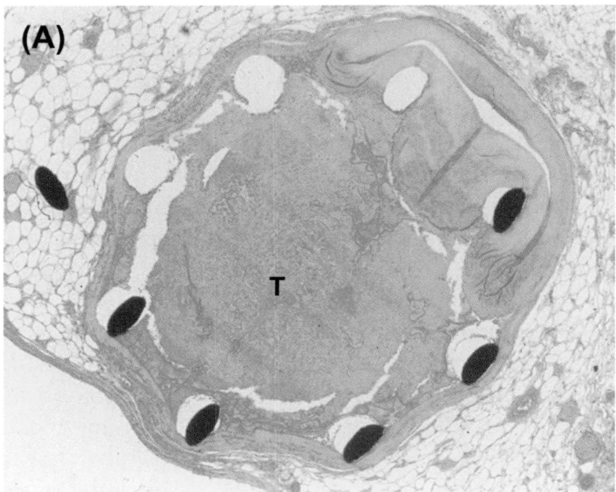

FIGURE B.3 Major complications of endovascular stenting. (A) Early thrombosis (T) associated with a metallic coronary artery stent. (B) Late proliferative restenosis with fibrous tissue (R). Stent wire cross sections are black in this histologic section. Filling the lumen inside of the stent. (Reproduced by permission from Schoen, F. J. & Edwards W. D. (2001). Pathology of cardiovascular interventions. In: Cardiovascular Pathology, 3rd edn, Silver, M. D., Gotlieb, A. I. & Schoen, F. J. (eds.). Churchill Livingstone: New York.)

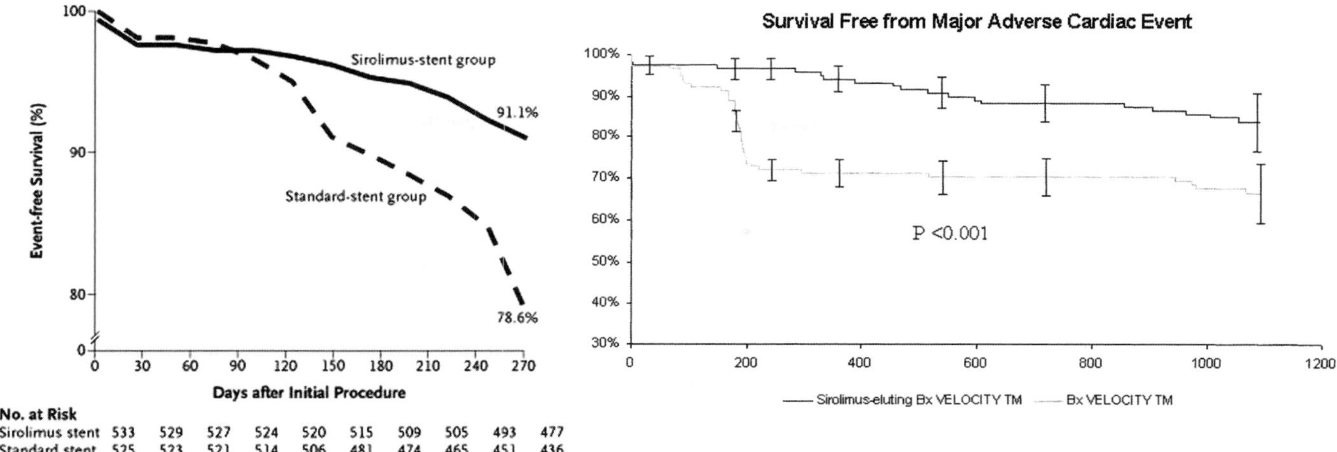

FIGURE B.4 Patient outcomes after standard bare metal stenting compared with those following drug eluting stent placement. (A) Actuarial rate of event-free survival. (B) Survival free from major adverse coronary events (sirolimus-eluting stent, solid line, bare metal stent dashed line). ((A) Reproduced by permission from Moses, J. W. et al. (2003). Sirolimus-eluting stents versus standard stents in patients with stenosis in a native coronary artery. N. Engl. J. Med. 349: 1315; (B) Reproduced by permission from Fajadet, J. et al. (2005). Maintenance of long-term clinical benefit with sirolimus-eluting coronary stents. Circulation 111: 1040.)

Stent technologies have undergone a relatively rapid evolution, with three distinct key stages of biomaterials development: (1) bare metal stents (BMS); (2) polymer-coated drug-eluting stents (DES); both used extensively in contemporary clinical interventional cardiology, and more recently; (3) completely resorbable/biodegradable stents (RBS), which are presently in clinical trials. Stent development has focused on mitigation of in-stent thrombosis and flexibility and ease of implantation, thereby allowing the treatment of a greater number and variety of diseased sites that occur within a particular patient and among patients. The choice of stent is based on several factors, including the characteristics of a given plaque, such as its diameter, length, and location within the coronary arterial anatomy, and the experience of the interventional cardiologist with a particular type of stent.

The most effective approach to reduce the incidence of in-stent restenosis occurring with BMS is the use of DES (Figure B.4) (Serruys, 2006; Daemen, 2007a, 2007b; Kukreja, 2008; Pedon, 2008; Moore, 2009; Moses, 2003). In DES, the drugs are embedded in a polymer matrix (such as a mixture of poly-n-butyl methacrylate and polyethylene-vinyl acetate (EVA) or a gelatin-chondroitin sulfate coacervate film) that is coated onto the stent wires. The drugs used most widely are rapamycin (sirolimus) (Sousa et al, 2003) and paclitaxel (Ong and Serruys, 2005) in the Cypher (Cordis) and Taxus (Boston Scientific) stents, respectively. Rapamycin, a drug used for immunosuppression in solid organ transplant recipients, inhibits proliferation, migration, and growth of smooth muscle cells and extracellular matrix synthesis. Paclitaxel, a drug used in chemotherapeutic regimens for several types of cancer, also has similar anti-smooth muscle cell activities. The drugs are released by diffusion and/or polymer degradation over varying periods of time that can be controlled by engineering the specifics of the polymer-drug system. At the time of this writing, it is estimated that more than six million people worldwide have received DES, including about three million in the United States. Indeed, stents are the most common device used to treat heart disease currently, and their insertion has become one of the most common medical procedures of any kind.

> The most effective approach to reduce the rate of in-stent restenosis encountered with bare metal stents is the use of polymer-coated drug-eluting stents.

Although DES inhibits in-stent restenosis (Moses, 2003), data from recent studies suggest a small but significant increased risk of late (>1 year post-implantation) stent thrombosis causing myocardial infarction and/or sudden death in patients who have DES. Although animal studies and some clinical data suggest that late stent thrombosis may be related to DES-induced inhibition of stent endothelialization, the significance and causes of late stent thrombosis are controversial (Holmes et al., 2007). The risk of late stent thrombosis does not seem to negate the overall advantages in patient outcomes of DES over BMS.

In contrast to both BMS and DES, which are permanent implants, RBS ultimately resorb to effect removal of foreign material that may potentiate a thrombotic event. RBS may permit more versatility in subsequent therapies, and do not interfere with the diagnostic evaluation by non-invasive imaging such as cardiac magnetic resonance and CT. Several RBS variants are in development or in clinical trials. The key challenges are to enhance biocompatibility, control the kinetics of degradation needed to maintain sufficient mechanical strength to limit recoil, and minimize inflammation (Ramcharitar, 2008). Recent studies have used several innovative features, including: (1) an everolimus eluting poly-L-lactide stent; (2) a novel locking mechanism to overcome recoil; (3) a combination of an antiproliferative drug with endothelial progenitor cell capturing antibodies to facilitate endothelialization; (4) dual drug eluting stents having an antiproliferative drug and polymeric salicyclic acid to limit inflammation; and (5) a metal RBS composed of magnesium (Ramcharitar et al, 2007; Sharkawi et al, 2007; Brown et al, 2009).

VASCULAR GRAFTS

Vascular grafts are used to bypass an obstructed vessel or to replace a segment of vessel that has formed an aneurysm or dissection. To ensure a high likelihood of clinical success and long-term patency, vascular grafts must be 1) resistant to thrombosis, smooth muscle cell-caused intimal thickening, fatigue, and aneurysm development, 2) have material compliance properties similar to the normal and diseased vessels to which they will be anastomosed, and 3) have a sufficient level of suturability. Although animal studies strongly suggest that this combination of specifications is achievable, translation to clinical practice remains a challenge, particularly in grafts <6–8 mm diameter.

The concept of using synthetic material as a conduit in the vascular system dates back to the early 1900s, when animal experiments were carried out using aluminum, silver, glass, and lucite tubes as vascular replacements. Fabrics such as Vinyon N, a nylon fabric used in parachutes, were employed in the mid-1950s as vascular conduits that could be fashioned from commercially available textiles. Current synthetic vascular grafts, such as those shown in Figure B.5, are typically fabricated from polyethylene terephthalate (Dacron®) or expanded polytetrafluoroethylene (ePTFE®), with the Dacron® grafts being used for larger vessel applications and the ePTFE® used to bypass smaller vessels (<8 mm diameter). These grafts can be made porous to enhance healing, but may be impregnated with connective tissue proteins (such as gelatin or albumin) to aid clotting, reduce the blood loss through the pores implantation and stimulate tissue ingrowth, and with antibiotics to reduce the risk of infection. Loosely woven or porous synthetic grafts that are not impregnated need to be preclotted with the patient's own blood before implantation, in order to minimize hemorrhaging through the graft interstices. Modern techniques for peripheral and coronary bypass differ little from the first leg vascular bypass (1948) or the first successful coronary artery bypass (1967). The expanded polytetrafluoroethylene (ePTFE) material used in Scribner's first chronic-use prosthetic blood vessel (1961) is largely identical to the graft used currently. Functional, lasting hemodialysis access for patients with kidney failure is a particularly challenging application for vascular grafts in which many "solutions" exist, but none is ideal (Scott, 2007). There is considerable ongoing research and development to make more sophisticated, safe, and effective vascular substitute materials (de Mel et al, 2008; Aper et al, 2009; Ravi et al, 2009).

> Synthetic fabric grafts perform well in large-diameter, high-flow, low-resistance locations such as the aorta and the iliac and proximal femoral arteries, with grafts used for aortofemoral bypass having 5–10 year patency rates of 90% (Clagett, 2002). In contrast, synthetic small-diameter vascular grafts (<6 to 8 mm in diameter) generally perform less well, with five-year patency less than 50%.

In general, the longer the interposition or bypass graft, and the smaller the diameter of the recipient vessel (with a corresponding increase in resistance to flow), the less favorable are both short- and long-term patency rates. For this reason, a major superficial vein in the leg, the saphenous vein, is typically removed surgically (from the patient needing the bypass) and moved to the site needed to bypass blockages in the coronary circulation (for coronary artery bypass grafting) or in the extremities (e.g., for femoropopliteal bypass grafting) when an adequate length of disease-free vein segment can be harvested for the necessary application. Other natural vessels such as the internal mammary artery may be used to bypass blockages in the coronary circulation.

When a synthetic graft is implanted, the luminal surface of the graft becomes coated with an adsorbed layer of plasma proteins, primarily fibrinogen. This layer

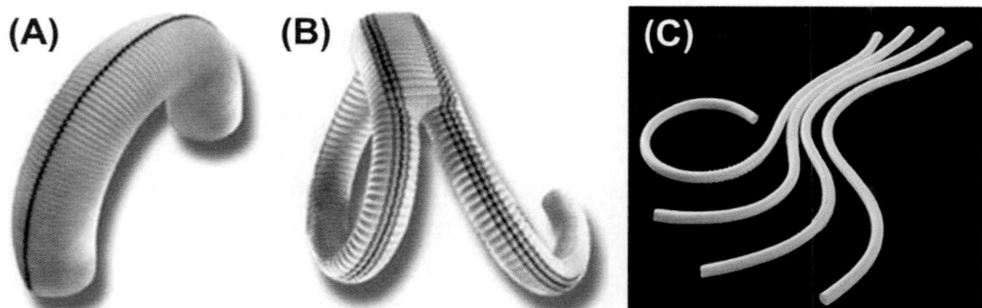

FIGURE B.5 Vascular grafts. (A) Dacron aortic interposition graft (Terumo). (B) Dacron aorto-iliac graft (Terumo). (C) ePTFE small vessel grafts (Gore).

develops over time into a platelet-fibrin aggregate (not covered by endothelium), termed a *pseudointima*. When endothelial cells cover this layer, simulating the inner layer of a native blood vessel and serving as a nonthrombogenic surface, the entire tissue thickness is termed a *neointima*. A neointima is generally considered more favorable than a pseudointima. Healing of a vascular graft by neointima is generally incomplete over its internal surface, despite long intervals following implantation (Figure B.6). Tissue lining the inner wall of a vascular graft or other biomaterial implanted in blood has three possible sources: (1) overgrowth from the host vessel across anastomotic sites; (2) tissue ingrowth through sufficiently large interstices; and (3) deposition of functional endothelial cells and/or multipotential stem cells from the circulating blood (see Figures B.6A and B.6B). Experience suggests that humans have a limited ability to endothelialize vascular grafts from the circulation, resulting

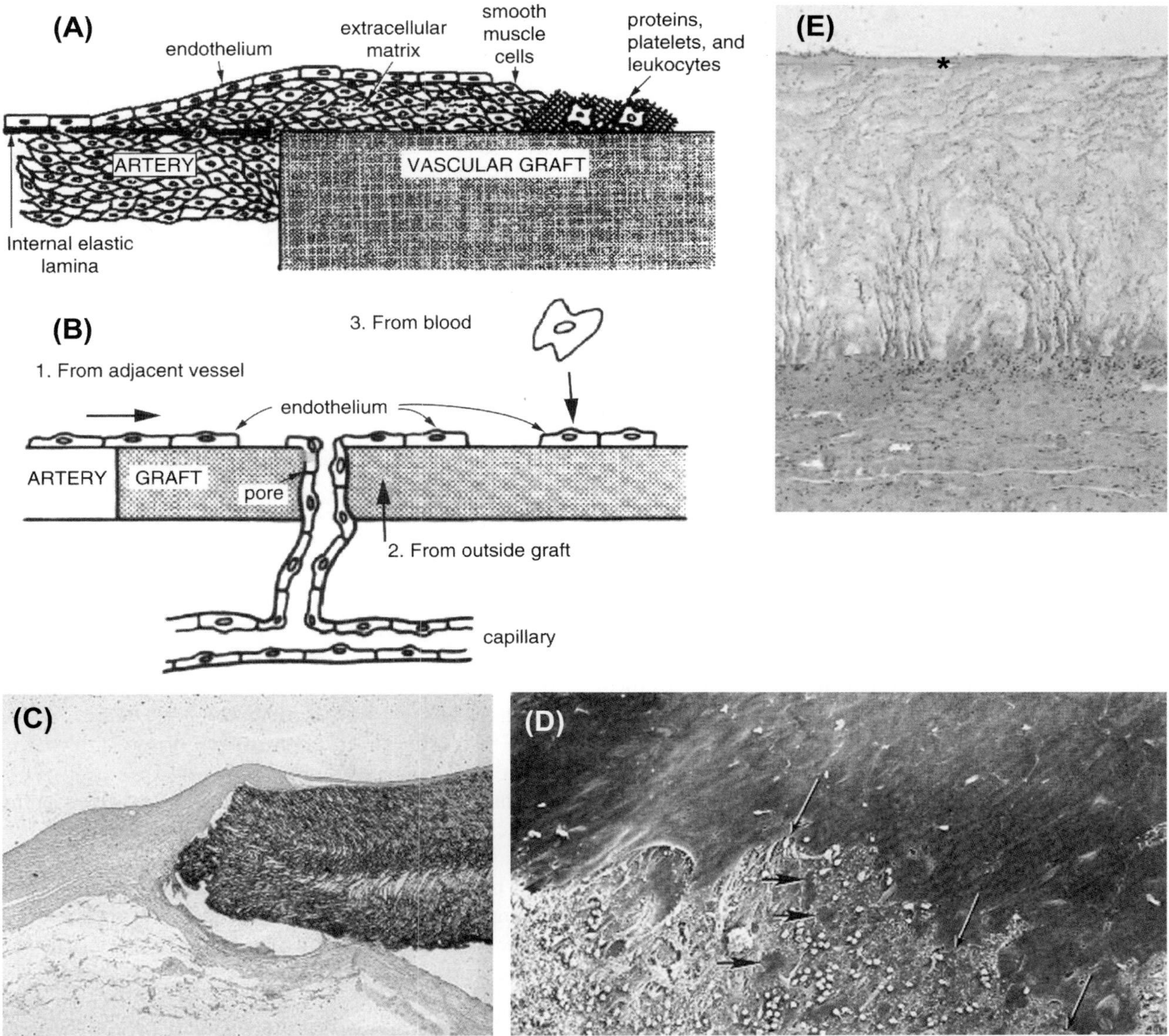

FIGURE B.6 Vascular graft healing at the anastomosis (site where the vascular graft is sutured to the native artery). (A) Schematic diagram of pannus formation, the major mode of graft healing with currently available vascular grafts. Smooth muscle cells migrate from the media to the intima of the adjacent artery, and extend over and proliferate on the graft surface; this smooth muscle cell layer is covered by a proliferating layer of endothelial cells, forming an endothelialized (neointimal) layer. (B) Possible sources of endothelium on the blood-contacting surface of a microporous knit or velour Dacron vascular graft. (C) Low-power photomicrograph of histologic section of ePTFE graft anastomosis, demonstrating thin overgrowth of tissue from adjacent normal artery (similar to (A)). (D) Scanning electron micrograph at the limit of endothelial coverage shown schematically in (A) (native artery toward upper right, graft center lower left). (E) Graft center at a distance from the anastomosis of ePTFE graft, showing a thin layer of non-endothelialized thrombus (a pseudointima) at asterisk. ((A) and (B) Reproduced by permission from Schoen, F. J. (1989). Interventional and Surgical Cardiovascular Pathology: Clinical Correlations and Basic Principles. WB Saunders: Philadelphia, PA. (C) and (E) Reproduced by permission from Schoen, F. J. & Edwards W. D. (2001). Pathology of cardiovascular interventions. In: Cardiovascular Pathology, 3rd edn, Silver, M. D., Gotlieb, A. I. & Schoen, F. J. (eds.). Churchill Livingstone: New York. (D) Reproduced by permission from in confluent endothelium covering only a 10–15 mm zone adjacent to the anastomosis (i.e., the sutured connection of the graft to the native artery). Thus, except adjacent to an anastomosis, a compacted platelet-fibrin pseudointima comprises the inner lining of clinical fabric grafts, even after long-term implantation (see Figure B.6C–E). Firm adherence of such linings to the underlying graft is difficult to achieve, and dislodgment of the lining leading to distal embolization or formation of a flap obstructing flow may occur. In a graft with interstices large enough to permit ingrowth of fibrovascular tissue, endothelial cells may originate from capillaries extending from outside to inside the graft and migrate to the luminal surface at a large distance from the anastomosis. However, since most porous knit clinical vascular grafts have insufficient porosity for tissue ingrowth, although they can bleed through and are therefore pre-clotted to avoid hemorrhage, existing grafts and other fabrics used as cardiovascular implants heal primarily by ingrowth of endothelium and smooth muscle cells from the cut edges of the adjacent artery or other tissue. The ability of grafts to heal over anastomoses depends on largely poorly-understood factors related to the graft material, local hemodynamics, potential other aspects of the host, and species; for example, sheep tend to heal grafts better than humans (Figure B.7). Also, circulating endothelial progenitor cells (EPC) derived from the bone marrow with the ability to differentiate into mature endothelial cells are thought to be recruited into the peripheral blood in situations of vascular repair or angiogenesis (Sata 2006; Tilki et al, 2009). EPC are believed to exert their function using two main strategies: activating the local endothelial cells and/or differentiating into mature endothelial cells themselves that integrate into the damaged vessel. To do this, EPC must home to "angiogenic active" sites, adhere to the activated/damaged endothelial cells or to the extracellular matrix, and participate in the endothelial activation/repair process. However, the role of EPC in typicl graft healing is lokely small. However, harnessing this mechanism represents a potential therapeutic opportunity.

An implanted graft also becomes encapsulated in the surrounding connective tissue and elicits a typical foreign-body reaction. The tissue covering the graft on its exterior surface, separating it from normal tissue, consists of a layer of inflammatory cells including giant cells adjacent to the material, covered by collagen, fibroblasts, blood vessels, and other cellular and extracellular connective tissue elements. This foreign-body capsule extends from the outside graft surface to the surrounding undisturbed body tissues. Graft interstices may be filled with fibrin or connective tissue elements, including cells and extracellular matrix and, where ingrowth has occurred, blood vessels.

The major complications of vascular grafts are thrombosis/thromboembolism, infection, a fluid collection outside the graft separating it from surrounding tissues

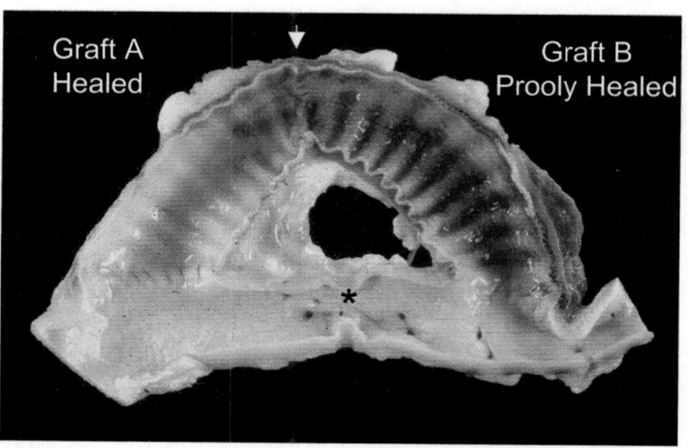

FIGURE B.7 Gross appearance of opened composite graft (2 pieces, each a different graft material, sewn together at the midpoint, denoted by arrow,) implanted in a sheep as a bypass conduit adjacent to a surgically constricted (by a ligating suture, now removed) descending thoracic aorta and removed after six months. One graft segment (left) has partial neointima formation extending over the graft; the other (right) has minimal neointima formation. The graft segment on the right has porosity sealed with albumin and the one on the left is gelatin sealed. The aortic constriction is indicated by an asterisk. (Reprinted by permission from Kadoba, K. et al. (1992). Experimental comparison of albumin-sealed and gelatin-sealed knitted Dacron conduits. J. Thorac. Cardiovasc. Surg. 103: 1059.)

(called a *seroma*), an extravascular hematoma resulting from a defect in the graft or a disrupted anastomosis and that communicates with the intravascular space (called a *pseudoaneurysm*), intimal hyperplasia (similar to the process of restenosis caused by SMC proliferation and scarring), and structural degeneration of the component biomaterials. Small diameter blood vessel grafts (<6-8 mm) have an especially high attrition rate (Figure B.8). In general, tissue grafts perform better than synthetic grafts (see Figure B.8A). Failure of synthetic small-diameter vascular prostheses is most frequently due to occlusion by thrombus formation (see Figure B.8B and B.8C) or generalized or anastomotic growth of fibrous tissue (called anastomotic fibrous hyperplasia, similar to intimal hyperplasia, above). As in any cardiovascular site "Virchow's triad" of surface thrombogenicity, hypercoagulability of the patient's blood, and locally static or low blood flow largely predict the propensity towards thrombus formation (see Figure B.8D and B.8E). Virchow's triad predicts thrombosis in small-diameter synthetic grafts where there can be low flow states due to poor flow out of the end of the graft (termed *run-off*), usually owing to atherosclerotic obstructions beyond the site of anastomosis. In larger vessels with higher blood flow such as the aorta, the surface thrombogenicity of the graft is usually overcome by brisk flow to minimize macroscopic thrombus formation. Contributing factors to anastomotic hyperplasia include: (1) surface thrombogenesis; (2) delayed or incomplete endothelialization of the fabric; (3) disturbed flow across the anastomosis; and (4) mechanical factors, especially compliance mismatch, at the junction of implant and host tissues. In vein grafts, intimal hyperplasia is often diffuse along the

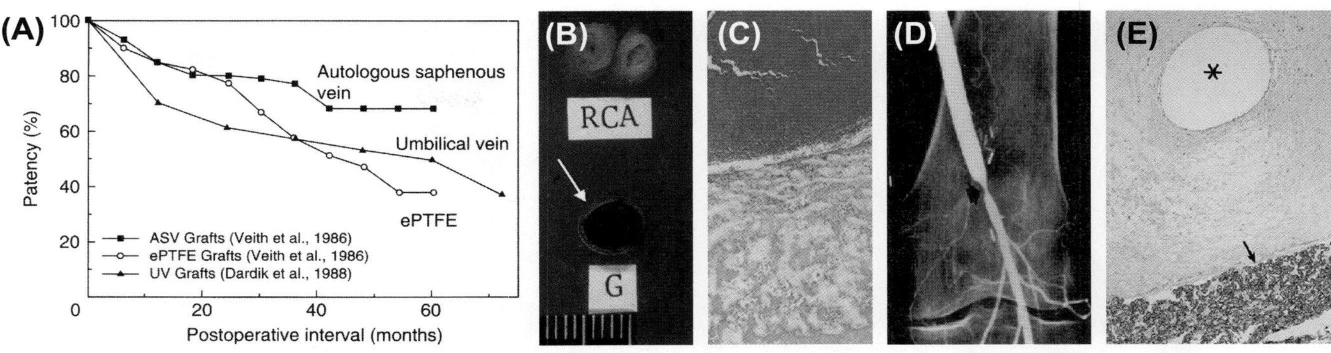

FIGURE B.8 Performance of small diameter vascular grafts. (A) Comparison of patency rates of autologous saphenous vein grafts, umbilical vein grafts and ePTFE small diameter vascular grafts. (B) and (C) Early (days) thrombosis of clinical ePTFE graft (approximately 4 mm diameter) used as aortocoronary bypass graft. (B) Gross photograph (RCA: native right coronary artery (RCA) with atherosclerotic obstruction; G: graft; RCA is native atherosclerotic right coronary artery). (C) Photomicrograph of thrombus (top) attached to graft (below). (D) and (E) Anastomotic hyperplasia of the distal anastomosis of a synthetic material femoropopliteal graft. (D) Angiogram demonstrating constriction (arrow). (E) Photomicrograph demonstrating ePTFE graft (arrow) with prominent intimal proliferation and very small residual lumen (*). ((A) and (B) Reproduced by permission from Schoen, F. J. (1989). Interventional and Surgical Cardiovascular Pathology: Clinical Correlations and Basic Principles. WB Saunders: Philadelphia, PA; (D) and (E) Reproduced by permission from Mitchell, R. N. & Schoen, F. J. (2010). Blood vessels. In: Robbins/Cotran Pathologic Basis of Disease, 8th ed., Kumar, V., Fausto, N., Abbas, A. et al. (eds.), p. 487. W.B. Saunders: Philadelphia, PA.)

graft length, leading to progressive luminal reduction of the entire graft. In contrast, synthetic vascular prostheses tend to develop intimal hyperplasia predominantly at or near anastomoses, particularly at the distal site.

Prophylactic systemic antibiotics at the time of surgery limit infection of implanted vascular prostheses. Early infections typically are related to the surgical procedure or to perioperative complications such as wound infection. Late infections usually occur secondary to seeding of the synthetic material by microorganisms from the blood in patients with low-grade bacteremia, often secondary to dental or gastrointestinal procedures. Since the anastomotic suture line is often involved in vascular graft infection, an infected vascular graft usually has a partially disrupted connection to the natural artery. Thus, rupture with hemorrhage at the graft site (i.e., a pseudoaneurysm, as described above) may bring the patient to clinical attention. Surgical removal of an infected graft is usually necessary to manage the infection. So far as we know from a clinical perspective, all graft materials or fabric textures are approximately equally susceptible to infection.

Progressive deterioration of a synthetic vascular graft can cause mechanical failure at the anastomotic site or in the body of the prosthesis, leading to aneurysm or pseudoaneurysm formation. The causes of delayed failure of a synthetic prosthesis include chemical, thermal or mechanical damage to polymeric yarn materials during manufacture, fabric defects induced during manufacture (e.g., dropped stitches), damage by vascular clamps during the insertion procedure and, rarely, postoperative degradation of graft material by mechanisms that are poorly-understood.

In view of the complications associated with vascular grafts, current research has focused on improvement of synthetic vascular grafts and on alternatives such as tissue-engineered blood vessels (see below). Attempts have been made to covalently modify the luminal surface of the grafts with various biologically active compounds in order to: (1) prevent coagulation; (2) prevent platelet adhesion/aggregation; (3) promote fibrinolysis; (4) inhibit smooth muscle cell adhesion/proliferation; and (5) promote endothelial cell adhesion and proliferation. Endothelialization of the entire graft would improve the thrombo-resistance of the graft, and may also help prevent bacterial attachment to the graft and subsequent infection.

STENT GRAFTS

Aortic aneurysms and dissections have been traditionally treated with open surgical procedures involving a large thoracic or abdominal incision and the replacement of the diseased portion of aorta by a synthetic graft. It is thus not surprising that, although such procedures are very effective, aortic aneurysm and dissection repair and aorto-bifemoral bypass grafting can have significant associated complications in certain patient populations, and the recovery time can be long. In 1999, aortic stent grafts for the treatment of abdominal aortic aneurysms that could be placed via catheter rather than open surgery became commercially available (Greenhalgh and Powell, 2008).

Stent grafts are composed of a synthetic fabric tube reinforced by a stent, whose struts facilitate rapid and stable expansion during insertion. This type of implant combines the features of stents and vascular grafts, and can be deployed endovascularly via a catheter (Figure B.9). The stent portion is manufactured from stainless steel, cobalt chromium alloy or nickel alloys (Nitinol) (Soor, 2008). The graft portion, usually polyester or ePTFE, can be located either on the luminal or abluminal

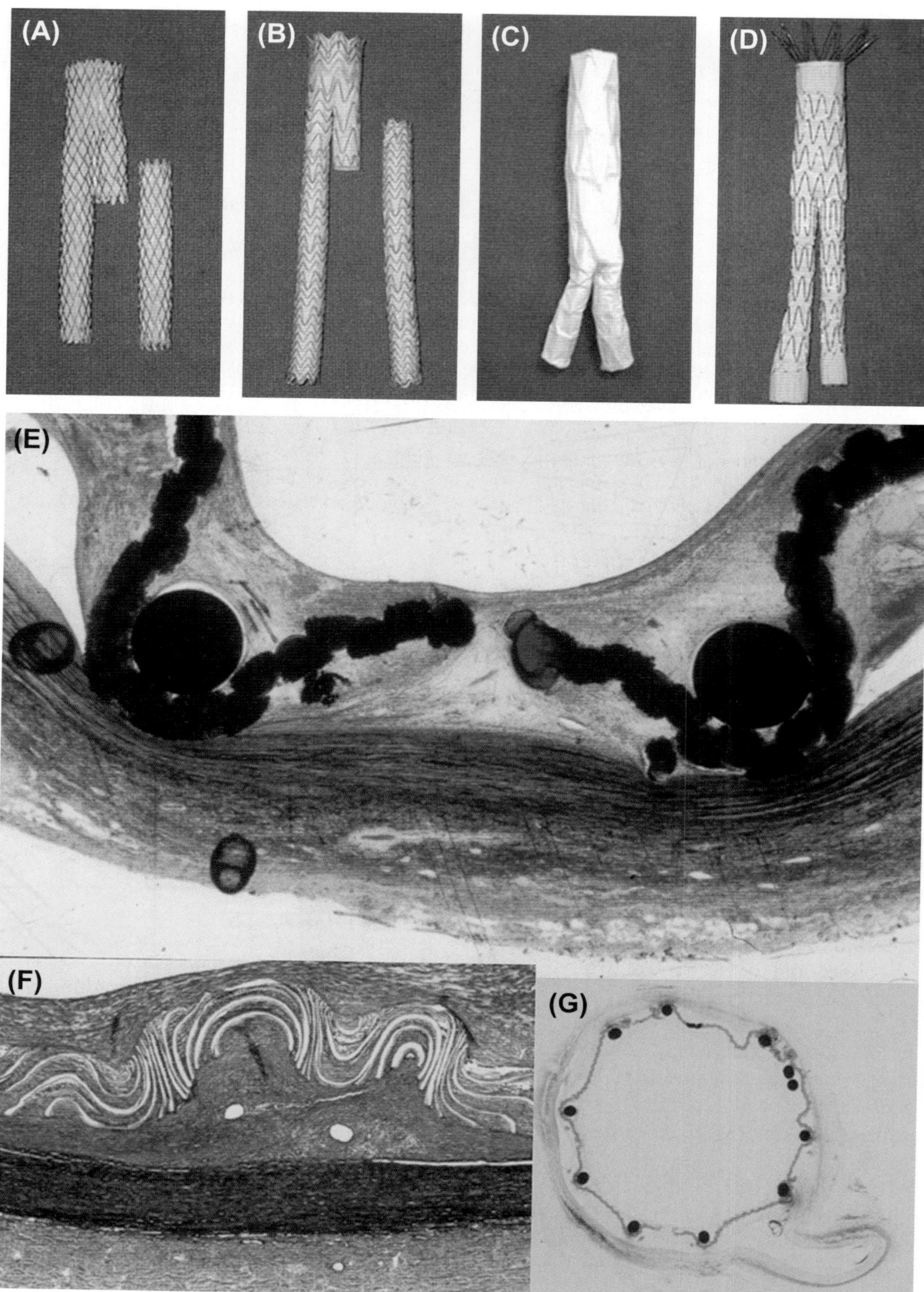

FIGURE B.9 Stent grafts. (A)–(D) Configuration of various commercially available stent graft devices showing composite metal and fabric portions (see original source for specific models and manufacturers). (A) and (B) are modular bifurcated stent graft composed of a nitinol exoskeleton and polyester or ePTFE linings. The stent-graft shown in (A) relies on radial force to fix the device into place. The stent-graft shown in (B) has proximal barbs to anchor the graft at its proximal end. The stent-graft shown in (C) is a unibody device made of polytetrafluoroethylene graft and a cobalt chromium alloy stent inside the graft. The stent graft shown in (D) is formed from several segments with a stainless steel stent on the outside of the graft. (E) High power photomicrograph of the cross-section of a well-healed experimental stent graft device explanted from a dog aorta. The lumen is widely patent, and the fabric and metal components are visible. (F) high and (G) low power photomicrographs of the cross sections of stent graft showing interaction with the vascular wall, and demonstrating mild intimal thickening. ((A)–(D) Reproduced by permission from Baril, D. T. et al. (2007). Surgery insight: Advances in endovascular repair of abdominal aortic aneurysms. Nature Clinical Practice Cardiovascular Medicine 4, 206; (E)–(G) Courtesy Jagdish Butany, MD, University of Toronto.)

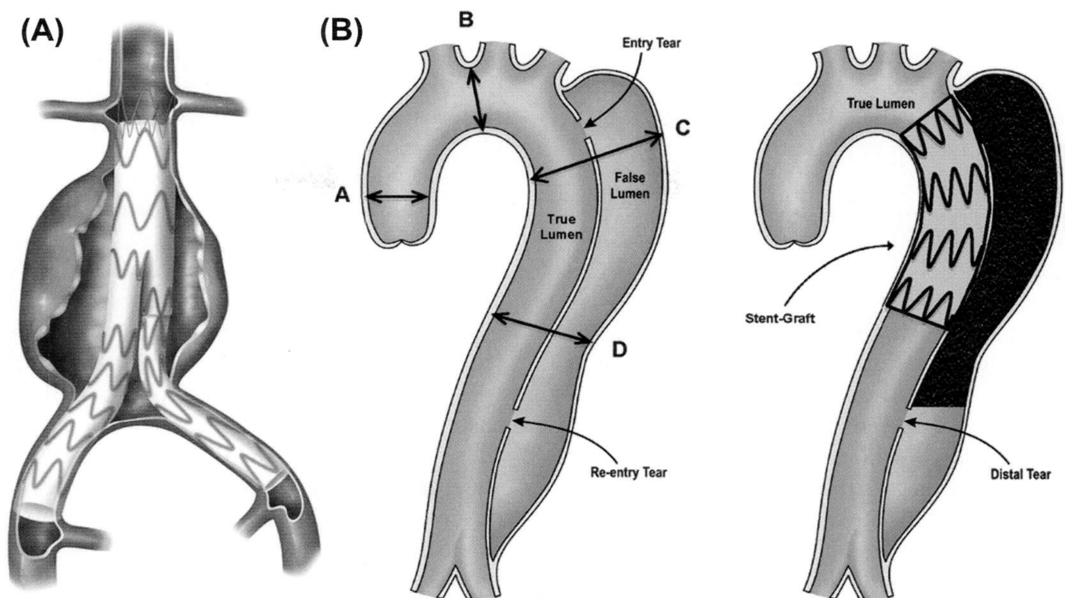

FIGURE B.10 Use of stent grafts in vascular disease. (A) Deployed endovascular stent graft in abdominal aortic aneurysm, with the proximal end immediately distal to the renal arteries and bifurcated into the iliac arteries bilaterally. (B) Endovascular stent graft in dissection. Left panel demonstrates the typical features of a thoracic dissection with flow in both the true and the expanded false lumen resulting from a major proximal entry tear. Right panel shows a stent graft placed to scaffold the dissected aorta and to seal the entry to the false lumen, resulting in reconstruction of the true lumen with subsequent false lumen thrombosis. The thrombosed segment of the original vessel eventually scars shut. ((B) Reproduced by permission from Nienader, C. A. et al. (2009). Randomized comparison of strategies for Type B aortic dissection. Circulation 120: 2519.)

(outside) aspect of the metallic stent, and is intended to provide a mechanical barrier to prevent intravascular pressure from being transmitted to the weakened wall of the aneurysm, thus excluding the aneurysm from the flow of blood. Once in place, a stent graft acts as an artificial lumen for blood to flow through, not into the surrounding aneurysm sac. The stent graft seals tightly upstream and downstream of the aneurysm or dissection. Fixation in some models is enhanced by hooks or barbs. Endovascular stent grafting can be used to treat some AAAs, as well as some thoracic aortic aneurysms and dissections. In the case of AAA or TAA, deployment of a stent graft and exclusion of blood from the aneurysm immediately takes the pressure off the aneurysm wall. In the case of a dissection, the stent portion holds the vessel open, and compresses and squeezes blood out of the false lumen in the wall of the artery. Stent grafts are deployed in a similar manner to stents in the coronary circulation, either as self-expanding Nitinol-based units or over an inflatable balloon. A stent graft used for a given application is selected by diameter, length, and geometry of the lesion, and location of side branches or branch points.

Endovascular aneurysm repair (EVAR) surgery is done via a catheter introduced in the peripheral vasculature, threaded to the diseased portion of aorta, and a stent graft deployed. EVAR has many advantages over conventional surgery, chief among which are lower operative complication and mortality rates. This is particularly important when determining treatment options for elderly and high-risk patients with concurrent medical conditions. Many models have been introduced, and the range of vascular problems amenable to this therapy has been expanding with generally favorable results (Matsumoto, 2008; Matsumura et al, 2008; Svensson et al, 2008; Espinosa et al, 2009). Figure B.10 shows the concept for repair of both an abdominal aortic aneurysm and a thoracic aortic dissection. Considerable development of new devices is underway (Desai et al, 2010).

Although open surgery is avoided with EVAR, endovascular repair is a complex interventional procedure accompanied by potential procedural and device-related complications. The major device-related problem is leakage of blood through or around the graft (called an *endoleak*), which occurs in approximately 20% of stent graft recipients. First-generation devices have also suffered from problems such as stroke at the time of insertion, ascending aortic dissection or aortic penetration from struts, vascular injury, graft collapse, graft material failure, continued aneurysm expansion or rupture, migration of the device from the location of original deployment, and kinking.

Stents and stent grafts have been especially successful in treating sub-totally occluded short (5–10 cm) segments of the iliac artery that can cause significant chronic lower extremity ischemia, and in the treatment of stenosis of the renal arteries, the smaller arteries of the lower extremity (femoral, popliteal and tibial arteries), the carotid arteries, the celiac artery, and the superior mesenteric artery. However, stents and stent grafts in these sites suffer mechanical failure at a greater rate than those in the coronary circulation. Failure modes related to the stent graft biomaterials include fracture of the stent struts, fabric erosion by external calcific deposits, and fatigue fracture of the stent wires (Jacobs et al, 2003; Chuter, 2009; Diehm et al, 2009).

ENGINEERED VASCULAR GRAFTS

Engineered blood vessels are described in detail in Chapter II.6.9 of this book. Therefore, only a brief overview is included here. Conventional vascular grafts lined *in vitro* with autologous endothelial cells have been used in patients needing bypass of obstructed leg arteries, but who had no suitable saphenous vein available. Patency of endothelialized grafts was good, and explants are purported to show an endothelium following years of implantation (Deutsch et al, 2009).

Several approaches to fully-engineered vascular grafts have been pursued. Bell and colleagues postulated that cell-seeded living grafts could be grown *in vitro*. Vessel "equivalents" composed of collagen and cultured bovine fibroblasts, smooth muscle cells, and endothelial cells have been investigated, but despite reinforcement with a Dacron mesh such grafts have been unable to withstand burst strengths for *in vivo* applications (Weinberg and Bell, 1986). An approcah used by several groups has used vascular cells applied onto tubular resorbable polymer scaffolds matured *in vitro* in a bioreactor prior to *in vivo* implantation (Gong and Niklason, 2006). Shin'oka and colleagues applied the cell-scaffold-bioreactor approach clinically with successful use of cell-seeded polymers to repair congenital defects in the low-pressure pulmonary outflow tracts of pediatric patients. However, none of these approaches have yielded sufficient mechanical strength to warrant application of an engineered graft to adult arterial bypass. Exposure to pulsatile physical forces during *in vitro* bioreactor maturation generally enhances graft properties; pulsed grafts are thicker, have greater suture retention, higher cell and collagen density than nonpulsed engineered grafts, and have a histological appearance similar to that of native arteries (Solan et al, 2009). Another approach to vascular graft engineering utilizes naturally derived matrices with or without cell repopulation prior to implantation (Kaushall et al, 2001). Vascular grafts fabricated from small intestine submucosa used experimentally in dogs were reported to be completely endothelialized and histologically similar to arteries (Lantz et al, 1993).

Another approach to vascular engineering extends the concept of Sparks' silicone mandril-grown graft used clinically in the 1970s, in which a collagenous tube was formed as the fibrous capsule reactive to an implanted foreign-body (silicone mandril) adjacent to a diseased vessel; the mandril was subsequently removed, yielding an autologous tissue tube (Sparks, 1973). However, owing to the variability of the quality of tissue generated in older patients in areas of circulatory insufficiency, such vascular replacements frequently developed aneurysms when used clinically. Grafts grown as the reactive tissue that forms around silicone tubing inserted into the peritoneal cavity of rats, rabbits, and dogs, everted (so that mesothelial cells became the blood-contacting surface), and grafted into the carotid artery of the same animal remained patent for up to four months (Hoenig et al, 2005).

Other investigators have fabricated mechanically-sound engineered tissue vascular grafts by constructing a cohesive cellular sheet of smooth muscle cells, rolling this sheet to form the vessel media, analogous to a "jelly roll," and wrapping a sheet of human fibroblasts around the media to serve as an adventitia, and seeding endothelial cells in the lumen (L'Heureux et al, 1998). Cell-sheet technology seems to yield high cell density, which ensures synthesis and deposition of the proper amount of extracellular matrix. The technique involves rolling a monolayer of living fibroblast cells into a tubular tissue construct. The matured construct formed by the fused cell layers was then decellularized. A second concentric layer was fabricated by rolling a second monolayer of living fibroblast cells around the outside of the first tube. Finally, the tissue-engineered vascular tube biofabricated from two concentric acellular and cellular layers is seeded with autologous endothelial cells. Cell-sheet technology has been used to generate clinical arteriovenous fistulas for hemodialysis access and for other tissues (Elloumi-Hannachi, 2010). Recently, McAllister and co-workers reported a six-month follow-up for the first 10 patients implanted with a fully biological tissue-engineered graft used as access for hemodialysis, developed with cell-sheet tissue-engineering technology (McAllister et al, 2009).

REFERENCES

Agostoni, P., Valgimigli, M., Biondi-Zoccai, G. G., Abbate, A., Garcia Garcia, H. M., Anselmi, M., Turri, M., McFadden, E. P., Vassanelli, C., Serruys, P. W., & Colombo, A. (2006). Clinical effectiveness of bare-metal stenting compared with balloon angioplasty in total coronary occlusions: insights from a systematic overview of randomized trials in light of the drug-eluting stent era. *Am. Heart J.* **151**, 682–689.

Aper, T., Haverich, A., Teebken, O. (2009). New developments in tissue engineering of vascular prosthetic grafts. *Vasa.* **38**, 99–122.

Brown, D. A., Lee, E. W., Loh, C. T., & Kee, S. T. (2009). A new wave in treatment of vascular occlusive disease: Biodegradable stents – clinical experience and scientific principles. *J. Vasc. Interv. Radiol.*, **20**, 315–325.

Chuter, T. A. (2009). Durability of endovascular infrarenal aneurysm repair: when does late failure occur and why? *Semin. Vasc. Surg.*, **22**, 102–110.

Clagett, G. P. (2002). What's new in vascular surgery? *J. Am. Coll. Surg.*, **194**, 165–201.

Daemen, J., & Serruys, P. W. (2007a). Drug-eluting stent update 2007: Part I. A survey of current and future generation drug-eluting stents: meaningful advances or more of the same? *Circulation*, **116**, 316–328.

Daemen, J., & Serruys, P. W. (2007b). Drug-eluting stent update 2007: part II: Unsettled issues. *Circulation*, **116**, 961–968.

de Mel, A., Jell, G., Stevens, M. M., & Seifalian, A. M. (2008). Biofunctionalization of biomaterials for accelerated in situ endothelialization: a review. *Biomacromolecules*, **9**, 2969–2979.

Desai, M., Eaton-Evans, J., Hillery, C., Bakhshi, R., You, Z., Lu, J., Hamilton, G., & Seifalian, A. M. (2010). AAA stent- grafts: Past problems and future prospects. *Ann. Biomed. Eng.*, **38**, 1259–1275.

Deutsch, M., Meinhart, J., Zilla, P., Howanietz, N., Gorlitzer, M., Froeschl, A., Stuempflen, A., Bezuidenhout, D., Grabenwoeger, M. (2009). Long-term experience in autologous in vitro endothelialization of infrainguinal ePTFE grafts. *J Vasc Surg*, **49**, 352–362.

Diehm, N., Dick, F., Katzen, B., Do, D. D., Baumgartner, I. (2009). Endovascular repair of abdominal aortic aneurysms: only a mechanical solution for a biological problem? *J. Endovasc. Ther.*, **1**(16 Suppl), 119–126.

Durand, E., Lemitre, M., Couty, L., Sharkawi, T., Brasselet, C., Vert, M., Lafont, A. (2012). Adjusting a polymer formulation for an optimal bioresorbable stent: a 6-month follow-up study. *Euro Intervention*, **8**: 242–249.

Elloumi-Hannachi, I., Yamato, M., Okano, T. (2010). Cell sheet engineering: a unique nanotechnology for scaffold-free tissue reconstruction with clinical applications in regenerative medicine. *J. Intern. Med.*, **267**: 54–70.

Espinosa, G., Ribeiro, A. M., Ferreira, C. M., Dzieciuchowicz, L., & Santos, S. R. (2009). A 10-year single-center prospective study of endovascular abdominal aortic aneurysm repaire with the latent stent-graft. *J. Endovasc. Ther.*, **16**, 125–135.

Fajadet, J., Morice, M. -C., Bode, C., et al. (2005). Maintenance of long-term clinical benefit with sirolimus-eluting coronary stents. *Circulation*, **111**, 1040–1044.

Farb, A., Sangiorgi, G., Carter, A. J., et al. (1999). Pathology of acute and chronic coronary stenting in humans. *Circulation*, **99**, 44–52.

Faries, P., Morrissey, N. J., Teodorescu, V., et al. (2002). Recent advances in peripheral angioplasty and stenting. *Angiology*, **53**, 617–626.

Gong, Z., & Niklason, L. E. (2006). Blood vessels engineered from human cells. *Trends Cardiovasc. Med.*, **16**, 153.

Greenhalgh, R. M., & Powell, J. T. (2008). Endovascular repair of abdominal aortic aneurysm. *New Engl. J. Med.*, **358**, 494–501.

Haudenschild, C. C. (1993). Pathobiology of restenosis after angioplasty. *Am. J. Med.*, **94**(Suppl), 40–44.

Hoenig, M. R., Campbell, G. R., Rolfe, B. E., & Campbell, J. H. (2005). Tissue-engineered blood vessels: alternative to autologous grafts? *Arterioscler. Thromb. Vasc. Biol.*, **25**, 1128.

Holmes, D. R. Jr., Kereiakes, D. J., Laskey, W. K., Colombo, A., Ellis, S. G., Henry, T. D., Popma, J. J., Serruys, P.W., Kiruma, T., Williams, D. O., Windecker, S., Krucoff, M. W. (2007). Thrombosis and drug-eluting stents: an objective appraisal *J. Am. Coll. Cardiol.*, **50**, 109–118.

Isenberg, B. C., Williams, C., & Tranquillo, R. T. (2006). Small-diameter artificial arteries engineered in vitro. *Circ. Res.*, **98**, 25.

Jacobs, T. S., Won, J., Gravereaux, E. C., et al. (2003). Mechanical failure of prosthetic human implants: a 10-year experience with aortic stent graft devices. *J. Vasc. Surg*, **37**, 16–26.

Kaushall, S., Amiel, G. E., Gulesarian, K. J., et al. (2001). Functional small diameter neovessels using endothelial progenitor cells expanded ex-vivo. *Nature Med.*, **7**, 1035.

Landau, C., Lange, R. A., & Hillis, L. D. (1994). Percutaneous transluminal coronary angioplasty. *New Engl. J. Med*, **330**, 981–993.

Lantz, G. C., Badylak, S. F., Hiles, M. C., et al. (1993). Small intestine submucosa as a vascular graft: a review. *J. Invest. Surg.*, **3**, 297.

Lenfant, C. (2003). Clinical research to clinical practice – lost in translation? *New Engl. J. Med*, **349**, 868.

L'Heureux, N., Paquet, S., Labbe, R., Germain, L., & Auger, F. A. (1998). A completely biological tissue-engineered human blood vessel. *FASEB J.*, **12**, 447.

Libby, P., & Theroux, P. (2005). Pathophysiology of coronary artery disease. *Circulation*, **111**, 3481.

Lloyd-Jones, D., Adams, R., Carnethon, M., et al. (2009). Heart Association Statistics Committee and Stroke Statistics Subcommittee Heart Disease and Stroke Statistics 2009 Update: A Report From the American Heart Association Statistics Committee and Stroke Statistics Subcommittee. *Circulation 119*, e21–e181.

Matsumura, J. S., Cambria, R. P., Dake, M. D., Moore, R. D., Svensson, L. G., & Snyder, S. (2008). International controlled clinical trial of thoracic endovascular eneurysm repair with the Zenith TX2 endovascular graft: 1-year results. *J. Vasc. Surg.*, **47**, 247–257.

Matsumoto, A. H. (2008). What randomized controlled trials tell us about endovascular repair of abdominal aortic aneurysms. *J. Vasc. Interv. Radiol.*, **19**, S18–S21.

McAllister, T. N., Maruszewski, M., Garrido, S. A., Wystrychowski, W., Dusserre, N., Marini, A., et al. (2009). Effectiveness of haemodialysis access with an autologous tissue-engineered vascular graft: a multicentre cohort study. *Lancet*, **373**, 1440–1446.

Meinhart, J. G., Deutsch, M., Fischlein, T., Howanietz, N., Froschl, A., & Zilla, P. (2001). Clinical autologous in vitro endothelialization of 153 infrainguinal ePTFE grafts. *Ann. Thorac. Surg.*, **71**(suppl 1), S327.

Mitchell, R. N., & Schoen, F. J. (2010). Blood vessels. In V. Kumar, N. Fausto, J. C. Aster, & A. Abbas (Eds.), *Robbins/Cotran Pathologic Basis of Disease* (8th Ed., pp. 487–528). Philadelphia: W.B. Saunders.

Moore, J. E., Jr. (2009). Biomechanical issues in endovascular device design.. *J. Endovasc. Ther*, **16**(Suppl 1), I1–11.

Moses, J. W., Leon, M. B., Popma, J. J., et al. (2003). Sirolimus-eluting stents versus standard stents in patients with stenosis in a native coronary artery. *New Engl. J. Med.*, **349**, 1315–1323.

Nakazawa, G., Vorpahl, M., Finn, A. V., Narula, J., Virmani, R. (2009). One step forward and two steps back with drug-eluting-stents: from preventing restenosis to causing late thrombosis and nouveau atherosclerosis. *JACC Cardiovasc Imaging*, **2**, 625–628.

Nerem, R. M., & Ensley, A. E. (2004). The tissue engineering of blood vessels and the heart. *Am. J. Transplant.*, **4**, 36.

Ong, A. T., & Serruys, P. W. (2005). An overview of research in drug-eluting stents. *Nat. Clin. Pract. Cardiovasc. Med*, **2**, 647–658.

Pedon, L., Zennaro, M., Calzolari, D., & Zanchetta, M. (2008). Strut fracture: a further concern with drug-eluting stents. *J. Cardiovasc. Med.*, **9**(9), 949–952.

Prêtre, R., & Von Segesser, L. K. (1997). Aortic dissection. *Lancet*, **349**, 1461–1464.

Poon, M., Badimon, J. J., & Fuster, V. (2002). Overcoming restenosis with sirolimus: from alphabet soup to clinical reality. *Lancet*, **359**, 619–622.

Rafii, S. (2010). Circulating endothelial precursors: Mystery, reality, and promise. *J. Clin. Invest.*, **105**, 17–19.

Ramaswami, G., & Marin, M. L. (1999). Stent grafts in occlusive arterial disease. *Surg. Clin. North Am.*, **79**, 597–609.

Ramcharitar, S., & Serruys, P. W. (2008). Fully biodegradable coronary stents: progress to date. *Am. J. Cardiovasc. Drugs*, **8**, 305–314.

Ravi, S., Qu, Z., Chaikof, E. L. (2009). Polymeric materials for tissue engineering of arterial substitutes. *Vascular*, **17**, S45–S54.

Sakalihasan, N., Limet, R., & Defawe, O. O. (2005). Abdominal aortic aneurysm. *Lancet*, **365**, 1577–1589.

Sata, M. (2006). Role of circulating vascular progenitors in angiogenesis, vascular healing, and pulmonary hypertension. lessons from animal models. *Arterioscler. Thromb. Vasc. Biol.*, **26**, 1008.

Schillinger, M., & Minar, E. (2009). Past, present and future of femoropopliteal stenting. *J. Endovasc. Ther.*, **16**, I147–I152.

Scott, E. C., & Glickman, M. H. (2007). Conduits for hemodialysis access. *Sem. Vasc. Surg.*, **20**, 158–163.

Serruys, P. W., de Jaegere, P., Kiemeneij, F., et al. (1994). A comparison of balloon-expandable-stent implantation with balloon angioplasty in patients with coronary artery disease. *New Engl. J. Med.*, **331**, 489–495.

Sharkawi, T., Cornhill, F., Lafont, A., Sabaria, P., Vert, M. (2007). Intravascular bioresorbable polymeric stents: a potential alternative to current drug eluting metal stents. *J. Pharm. Sci.*, **96**, 2829–2837.

Shin'oka, T., Imai, Y., & Ikada, Y. (2001). Transplantation of a tissue-engineered pulmonary artery. *New Engl. J. Med.*, **344**, 532.

Solan, A., Dahl, S. L., & Niklason, L. E. (2009). Effects of mechanical stretch on collagen and cross-linking in engineered blood vessels. *Cell Transplant*, **18**, 915–921.

Soor, G. S., Chakrabarti, M. O., Abraham, J. R., Leong, S. W., Vukin, I., Lindsay, T., & Butany, J. (2008). Aortic stent grafts. *J. Clin. Pathol.*, **61**, 794–801.

Sousa, J. E., Serruys, P. W., Costa, M. A. (2003). New frontiers in cardiology: drug-eluting stents: Part I. *Circulation.* **107**, 2274–2279.

Sparks, C. H. (1973). Silicone mandril method for growing reinforced autogenous femoro-popliteal artery graft in situ. *Ann Surg*, **177**, 293.

Stone, G. W., Grines, C. L., Cox, D. A., et al. (2002). Comparison of angioplasty with stenting, with or without abciximab, in acute myocardial infarction. *New Engl. J. Med*, **346**, 957–966.

Svensson, L. G., Kouchoukos, N. T., & Miller, D. C. (2008). Expert consensus document on the treatment of descending thoracic aortic disease using endovascular stent-grafts. *Ann. Thorac. Surg.*, **85**, S1–S41.

Tilki, D., Hohn, H. P., Ergűn, B., Rafii, S., & Ergűn, S. (2009). Emerging biology of vascular wall progenitor cells in health disease. *Trends Mol. Med.*, **15**, 501–509.

Virmani, R., Farb, A., & Burke, A. P. (1994). Coronary angioplasty from the perspective of atherosclerotic plaque: Morphologic predictors of immediate success and restenosis. *Am. Heart J.*, **127**, 163–179.

Virmani, R., & Farb, A. (1999). Pathology of in-stent restenosis. *Curr. Opin. Lipid*, **10**, 499–506.

Weinberg, C. B., & Bell, E. (1986). A blood vessel model constructed from collagen and cultured vascular cells. *Science*, **231**, 397.

Welt, F. G., & Rogers, C. (2002). Inflammation and restenosis in the stent era. *Arterioscler. Thromb. Vasc. Biol.*, **22**, 1769–1776.

Yustein, A. S., Schultz, D., Neuland, C., Buckles, D. S., Nipper, J. C., Stephenson, R. A., & Gonzalez, G. U.S. (2008). Food and Drug Administration and off-label use of expandable metal biliary stents within the peripheral vasculature, **19**, 965–969.

C. OTHER CARDIOVASCULAR DEVICES

Robert F. Padera, Jr. and Frederick J. Schoen
Department of Pathology, Brigham and Women's Hospital and Harvard Medical School, Boston, MA, USA

INTRODUCTION

There are many life-saving and live-sustaining medical devices in clinical use and development to treat a wide variety of cardiovascular conditions; these devices are the subject of the current chapter. The treatment of arrhythmias by pacemakers and implantable cardioverter-defibrillators has been one of the major successes in modern medicine. Mechanical circulatory support devices, from cardiopulmonary bypass machines first introduced in the 1950s to the latest generation of ventricular assist devices, have both enabled development of advanced surgical procedures such as valve replacement and coronary artery bypass grafting (discussed in the previous sections), and have provided new tools in the armamentarium of heart failure therapies. Many devices take advantage of the development of percutaneous catheter-based techniques to treat cardiovascular disease in a minimally invasive manner, such as septal defect closure devices, filters to prevent pulmonary embolism and left atrial occlusion devices, and devices to minimize the consequences of a dilated, failing heart.

PACEMAKERS AND ICDS (FOR CARDIAC ARRHYTHMIAS)

The normal cardiac electrical cycle (Figure C.1A) begins with an impulse initiated by the sinoatrial (SA) node, the heart's natural pacemaker, which is located in the right atrium near the junction with the superior vena cava. The impulse spreads through the muscle of both left and right atrial walls, causing depolarization of the cardiac myocytes that result in atrial contraction. The impulse arrives at the atrioventricular (AV) node, which is in the posterior right atrium enclosed by the ostium of the coronary sinus, located in the septal leaflet of the tricuspid valve and the membranous portion of the interatrial septum (called the triangle of Koch). After a short delay within the AV node, the impulse passes to the bundle of His, and into the left and right bundle branches, located in the intraventricular septum. The impulse spreads through the right and left ventricular myocardium, causing a wave of myocyte depolarization and thereby coordinated ventricular contraction. The SA and AV nodes and the bundles of His and its right and left bundle branches are composed of cardiac muscle cells specialized for conduction.

Cardiac arrhythmias (Huikuri et al., 2001) reflect disturbances of either impulse initiation or impulse conduction. Foci of impulse-generating (automatic) cells outside the SA node, called ectopic foci, may initiate cardiac impulses that generate suboptimal ventricular

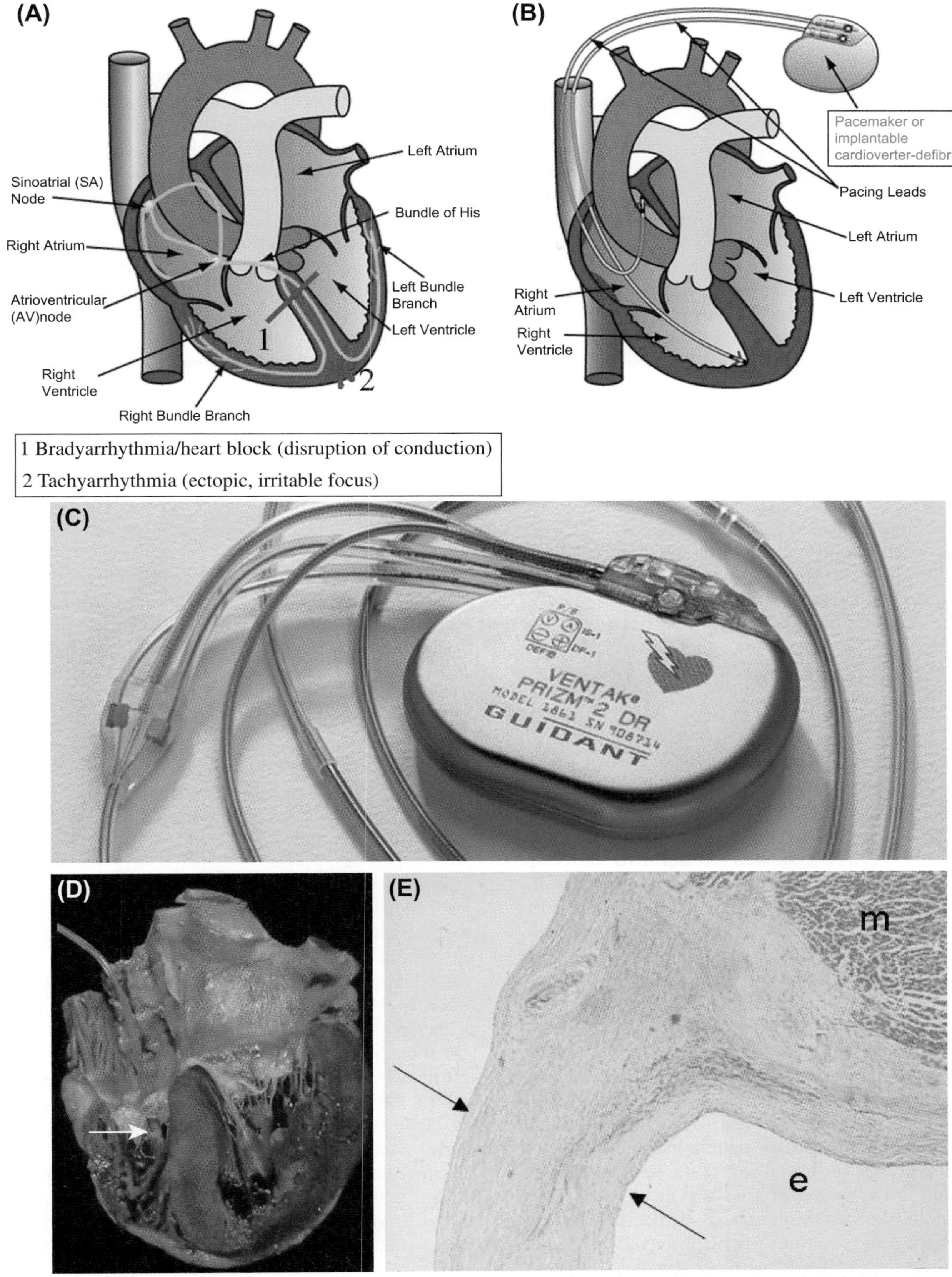

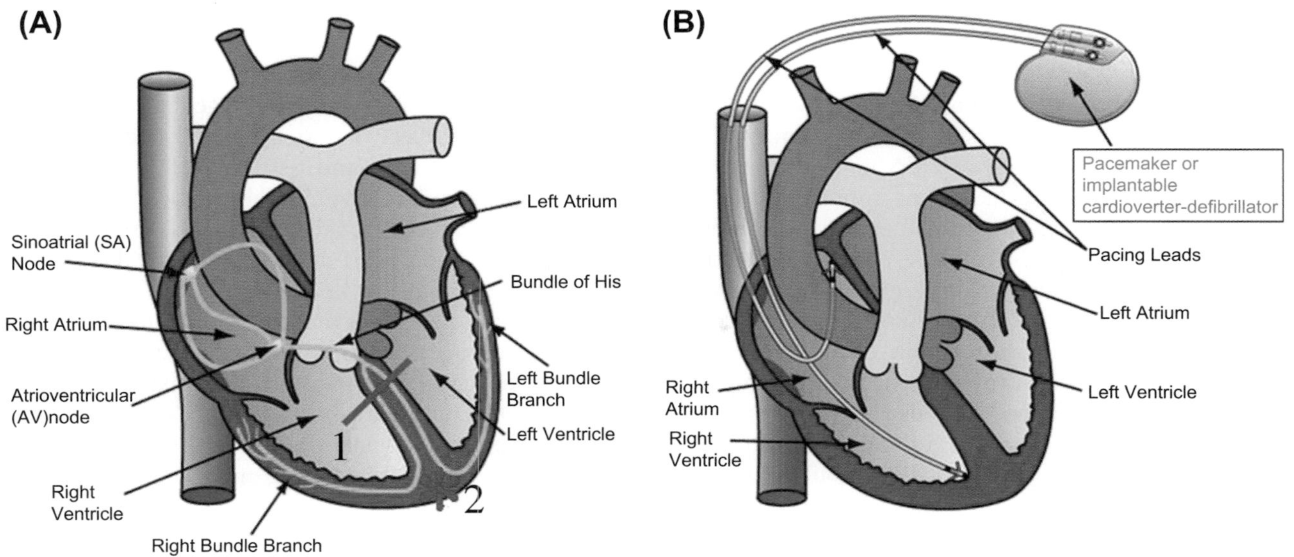

FIGURE C.1 (A) The normal cardiac electrical cycle showing schematically both conduction blocks and ectopic foci of impulse generation. (B) Schematic demonstrating ICD lead placement in right ventricle. (C) Guidant Prizm II DR ICD, introduced to the US market in 2000, and withdrawn in 2005. (D) Transvenous pacing lead placed in right ventricle demonstrating fibrosis of the distal portion of the lead (arrow). (E) Fibrous capsule surrounding pacemaker electrode in right ventricle (see FIGURE C.1D). Low power photomicrograph demonstrating space previously occupied by electrode (e), fibrous tissue separating electrode from blood in the right ventricular chamber (between arrows) and extending around enlectrode to separate it from myocardium (m), potentially creating a barrier to conduction of the pacing impulse. ((E) Reproduced with permission from Schoen, F. J. & Edwards, W. D. (2001). Pathology of cardiovascular interventions, including endovascular therapies, revascularization, vascular replacement, cardiac assist/replacement, arrhythmia control and repaired congenital heart disease. In: Cardiovascular Pathology, 3rd edn., Silver, M. D., Gotlieb, A. I. & Schoen, F. J. (eds.), p.678. WB Saunders: Philadelphia, PA.)

contractions. These arrhythmias are usually fast, i.e., tachyarrhythmias, that can result in ventricular fibrillation, which can be fatal. Intrinsic SA node dysfunction can also account for disturbances of impulse initiation. In contrast, disturbances of impulse conduction mainly consist of conduction blocks or re-entry. Conduction blocks constitute a failure of propagation of the usual impulse through the specialized muscle as a result of a disease process (such as ischemia or inflammation) or certain drugs. Blocks can be complete (no impulse propagation) or incomplete (impulse propagates more slowly than normal), and can be permanent or transient. Re-entry is said to occur when a cardiac impulse traverses a loop of cardiac fibers and re-excites previously excited tissue without a second impulse from the SA node. For patients in whom these cardiac arrhythmias cannot be controlled pharmacologically by antiarrhythmic drugs, two therapeutic options are available: (1) electrical therapy to control the cardiac rhythm, such as direct current cardioversion or implantable devices such as pacemakers and implantable cardioverter-defibrillators; and (2) interventional/surgical therapy to remove the affected tissue or interrupt the abnormal pathway, such as endocardial resection, cryoablation or radiofrequency ablation.

Cardiac Pacemakers

Cardiac pacemakers are medical devices that provide impulses to the conduction system to initiate contraction. The first cardiac pacemaker was implanted (Atlee and Bernstein, 2001) in 1958, and since then cardiac pacing has become a well-established therapeutic tool. The first pacemakers were large (40–200 cm^3) by today's standards (9–45 cm^3), and contained few of the features that are standard in current devices. Over a million patients in the United States currently have pacemakers (Figure C.1B and C.1C), and over 250,000 new permanent pacemakers are implanted each year; pacemaker placement, revision or removal is a commonly performed procedure. Most cardiac pacemakers are implanted in patients older than 60 years, but they are also used in children, including infants, when necessary. The most common indications for permanent cardiac pacing are various types of conduction block. Some conduction blocks lead to bradycardia (abnormally low heart rate), while others, predominantly in the left or right bundles, will result in ventricular dyssynchrony and inefficient ventricular contraction in the setting of a normal heart rate. These conduction blocks can result in decreased cardiac output and the signs and symptoms of congestive heart failure, but can be well treated by cardiac pacing.

Modern cardiac pacing (Kusumoto and Goldschlager, 2002), either temporary or permanent, is achieved by a system of interconnected components consisting of: (1) a pulse generator which includes a power source and circuitry to initiate the electric stimulus and to sense cardiac electrical activity; (2) one or more electrically insulated conductors leading from the pulse generator to the heart, with a bipolar electrode at the distal end of each; and (3) a tissue or blood and tissue interface between electrode and adjacent stimulatable myocardial cells. The pacemaker delivers a small current (2–4 mA) to the myocardium via the electrodes, resulting in depolarization and contraction of the heart.

Temporary pacing is most frequently used for patients with acute myocardial infarction that is complicated by cardiac conduction system disturbances that could progress to complete heart block. Leads for temporary cardiac pacing are generally directed transvenously into the apex of the right ventricle, and the pulse generator is located outside the body. In the context of cardiac surgery when the epicardium is already exposed, temporary pacing is achieved by placing insulated wires with bare ends to the epicardial surfaces of the atria or ventricles with the leads emerging transthoracically from the anterior chest to permit easy withdrawal. Ultimately, the temporary pacemaker is either replaced by a permanent device or discontinued.

Permanent cardiac pacing involves long-term implantation of both pulse generator and electrode leads. The generator is placed in a tissue pocket beneath the skin on the left anterior chest with the leads advanced transvenously through the left subclavian vein to terminate at the endocardial surface of the heart. The tips of the electrodes are typically placed within the right atrium and/or right ventricle, depending on the pacing modality.

A single chamber pacemaker delivers a stimulus based on a programmed timing interval. The pacemaker also senses intrinsic cardiac activity, and can be inhibited from providing unnecessary or inappropriate stimuli. This "demand" pacing is valuable in a patient whose problem is intermittent. A dual chamber pacemaker with electrodes in both the atrium and ventricle delivers the sequential atrial and ventricular signals to approximate the timing of the normal heartbeat. This device also senses intrinsic atrial and ventricular depolarizations, and delivers stimuli at the appropriate time to maintain proper synchrony of the chambers.

Patients with ventricular conduction delays, such as left bundle branch block, may suffer from heart failure due to dyssynchrony of ventricular contraction, where the right and left ventricles do not contract simultaneously. Cardiac resynchronization therapy (CRT) via biventricular pacing is a recently implemented intervention in which pacing electrodes are placed in the right atrium, right ventricle, and coronary sinus. The coronary sinus electrode stimulates the lateral wall of the left ventricle to allow for simultaneous excitation of the right and left ventricles, and for more uniform contraction of the entire left ventricle. CRT has been shown to significantly improve cardiac function in these patients (McAlister et al., 2007).

Permanent implantable pacemakers are powered by lithium-iodide batteries with a finite lifespan of 5–8 years, requiring removal and reimplantation of a new device when the battery is exhausted. In fact, that first patient to receive an implantable pacemaker in 1958 required 22 different pulse generators until his death in 2001 at the age of 86. Improving battery technology to allow for longer lifespan would minimize the number of reimplantations that a patient would require, and the complications that arise from these procedures.

The interface between the electrode and depolarizable myocardial tissue is of critical importance in the proper functioning of the pacemaker (Figure C.1D). Typically, a layer of nonexcitable fibrous tissue induced by the electrode forms around the tip of the electrode, which is undesirable as it increases the strength of the threshold pacing stimulus required to initiate myocyte depolarization (Figure C.1E). Strategies to reduce this fibrosis include improved lead designs, and the use of slow local release of corticosteroids to minimize the thickness of fibrous tissue formed after lead implantation (Mond and Grenz, 2004). The practical point is that, if the pulse generator output is not set sufficiently high in the early post-implantation phase, loss of pacing with potentially fatal consequences can result. By contrast, maintaining output at such high levels once thresholds have stabilized greatly shortens battery life. Thus, pacemakers with adjustable variations in output have been developed.

An ideal endocardial pacing lead should provide stable fixation immediately from the time of implantation, achieve and maintain a minimal threshold for stimulation, maximize sensing, and function reliably for many years. Electrode fixation to the endocardium may be active or passive. In active fixation, the electrode is designed to grasp the endocardial surface to achieve immediate fixation at implantation. A very effective aid to passive fixation is the addition of projecting "tines" or fins, in the region of the electrode tip. A different approach to improving fixation has been the development of electrodes with porous metal surfaces to foster tissue ingrowth. An endocardial pacemaker lead may require a special design if it is implanted at a particular site. One example is the J-shaped atrial lead, which is curved to facilitate placing the electrode tip in the right atrial appendage, inherently the most stable site for fixation.

Implantable Cardioverter-Defibrillators (ICDs)

The first implantable cardioverter-defibrillator was placed in 1980; currently more than 100,000 ICDs are implanted annually in the United States. The goal of ICDs is to prevent sudden death in patients with certain life-threatening arrhythmias by resetting the heart's electrical activity and stimulating a normal cardiac rhythm. ICDs have been shown to revert sustained ventricular tachycardia (abnormally high ventricular rate) and ventricular fibrillation (uncoordinated electrical/myocardial activity) in multiple prospective clinical trials. Benefit in overall mortality has been well-documented (The Antiarrhythmics versus Implantable Defibrillators (AVID) Investigators, 1997).

An ICD consists of similar components to a pacemaker, namely a pulse generator and leads for tachydysrhythmia detection and therapy. The pulse generator is a self-powered, self-contained computer with one or two 3.2 V lithium-silver vanadium oxide batteries used to power all components of the system, including aluminum electrolytic storage capacitors. The devices have a service life of 3–5 years, at which point they require removal and implantation of a new device. The lead is generally placed in the right ventricle through a transvenous approach. The ICD constantly monitors the ventricular rate, and when the rate exceeds a certain value, provides therapy. Current devices will initially provide a short burst of rapid ventricular pacing that terminates some types of ventricular tachyarrhythmias without providing a large shock. This approach can terminate up to 96% of episodes of ventricular tachycardia without the need for a shock. If this pacing fails to break the arrhythmia, the ICD delivers a shock of 10–30 Joules between the electrode in the right ventricle and the surface of the pulse generator to terminate the dysrhythmic episode. These devices also keep a running record of arrhythmias and treatment results. ICDs are indicated in patients at high risk for ventricular arrhythmias (primary prevention), and in patients who have already had an episode of aborted sudden cardiac death (secondary prevention).

Complications of Pacemakers and ICDs

These devices share many of the same complications, many of them requiring device removal and replacement. Like many cardiovascular devices, these are life sustaining technologies, and the implications of device failure can be fatal due to a lack of appropriate cardiac pacing (for pacemakers) or inability to sense or deliver appropriate therapy for a lethal arrhythmia (for ICDs). While normal device end-of-service from a depleted battery may not be technically considered a device malfunction, it certainly requires device replacement, and may happen prematurely due to increased fibrosis at the lead–tissue interface requiring a higher stimulus threshold. Failures of the hardware, including the battery/capacitor and charge circuit, connectors and leads are the most common device malfunctions, with software problems being less prevalent. Some mechanical failures include electrode dislodgment, lead fractures, electrode corrosion, and insulation failure. Complications related to leads may be related to the body of the lead, as distinct from the lead-device pack interface or the electrodes. Several devices and components have been recalled in recent years for these modes of failure (Amin and Ellenbogen, 2010). Lead improvements over the years have included helical coil and multifilament designs, to decrease electrical resistance and enhance flexibility and durability (Haqqani and Mond, 2009). In the past, many reports appeared on interference with pacemaker function by devices ranging from electric razors, toothbrushes, and microwave ovens at home, to electrosurgical and diathermy apparatus in hospitals. Fortunately, recent generations of cardiac pacemakers have been greatly improved with regard to their resistance to electromagnetic interference.

Many complications relate to the interaction of the device biomaterials with the host tissue. These include infection, thrombosis, and thromboembolism, myocardial penetration or perforation, pressure necrosis of the skin overlying the pulse generator, and migration or rotation of the pulse generator. Infection is a dreaded complication of implantable devices in general, and is this is certainly true for pacemakers and ICDs (Uslan, 2008). The infection may originate in the subcutaneous pocket and track along the lead, which acts as a contaminated foreign-body. Alternatively, it may occur by implantation of bacteria on traumatized endocardium or thrombus contiguous with the lead. The most common organisms responsible for these infections are coagulase-negative *Staphylococcus* species such as *S. epidermidis*. Septicemia may develop, and septic pulmonary emboli may occur. The fundamental therapeutic principle in device-related endocarditis is treatment of the infection with antibiotics, followed by removal of at least the lead and, when the pacemaker pocket is involved, the entire pacing system (Baddour et al., 2010).

ICDs contain more extensive hardware than pacemakers, and this may contribute an increased relative frequency of complications. Several additional considerations are specific to ICDs. The consequences of repeated defibrillations can cause the following effects: (1) direct effect of repeated discharges on the myocardium and vascular structures; and (2) possible thrombogenic potential of the indwelling intravascular electrodes. Another major complication of ICDs from the standpoint of the patient, other than the inability to sense or terminate an arrhythmia leading to sudden death, is an inappropriate shock. In addition to being startling and quite painful at the time of the shock, patients receiving multiple inappropriate shocks have been known to develop post-traumatic stress disorder symptoms.

As mentioned above, the leads are designed to optimize their interactions with the adjacent myocardium; this can be problematic when a complication arises in which the leads must be removed. Some leads can be removed by prolonged gentle traction, but many require additional tools and techniques to free them from the venous wall, through which the body of the lead travels, and from the myocardium, to which they are often tenaciously adherent (Smith and Love, 2008). Recourse to cardiotomy with cardiopulmonary bypass may be needed if the lead is densely incarcerated in fibrous tissue.

CARDIAC ASSIST AND REPLACEMENT DEVICES (FOR HEART FAILURE)

Heart Failure and Treatment Options

Congestive heart failure (Jessup and Brozena, 2003) is a deficiency of the pumping function of the heart, and is an extremely common condition, affecting approximately 5.8 million Americans. Each year in the US, congestive heart failure is the principal cause of death in 60,000 individuals, a contributing factor in over 280,000 deaths, and the primary discharge diagnosis in over 1.1 million hospitalizations, all increases over previous years. Cardiac transplantation is a potential solution for some of these patients (Mancini and Lietz, 2010). However, the increasing discrepancy between the number of acceptable donor hearts (only 2500 per year) and the large number of patients who might benefit from cardiac transplantation (estimated at greater than 100,000 per year) has prompted efforts towards the development of mechanical devices to augment or replace cardiac function (Baughman and Jarcho, 2007; Boilson et al., 2010; Krishnamani et al., 2010).

Congestive heart failure is the final common pathway of many cardiac conditions, including valvular heart disease, coronary artery atherosclerosis with resultant ischemic heart disease, and diseases that affect the cardiac muscle directly (termed cardiomyopathies). Heart failure can occur precipitously, as in myocardial infarction or viral myocarditis, or it can be a slow, progressive worsening of exercise tolerance and shortness of breath over many months or years owing to ongoing deterioration of the heart muscle. It can manifest itself in the post-operative period after both cardiac surgery (e.g., valve replacement, cardiac transplantation) and non-cardiac surgery (e.g., abdominal aortic aneurysm repair).

As one might expect, therefore, the natural history of heart failure depends on the cause and progression of the underlying disease process. For example, patients with heart failure after cardiac surgery (called postcardiotomy shock) often recover the vast majority of their cardiac function after a short period of time if they are otherwise sustained by mechanical circulatory support. In contrast, patients with dilated cardiomyopathy, one of the most common indications for cardiac transplantation, often need long-term mechanical support; studies have shown that at least 50% of such individuals would die in 3–5 years from their disease without it. One must take these clinical considerations into account when designing mechanical support systems, as different devices may best serve patients with different problems (DiGiorgi et al., 2003).

Cardiopulmonary Bypass

First used in 1953 by Dr. John H. Gibbon, cardiopulmonary bypass devices pump blood external to the body, and thereby permit complex cardiac surgical procedures to be done safely and effectively. Bypass machines are useful in extracorporeal membrane oxygenation (ECMO) to assist in the transport of oxygen and carbon dioxide for patients (especially neonates and infants) with pulmonary diseases such as the respiratory distress syndrome (Alpard and Zwischenberger, 2002).

The basic operating principles of the current heart-lung machines are quite straightforward, and have changed little in the past half century. Deoxygenated blood returning from the systemic circulation into the right atrium is withdrawn by gravity siphon into a cardiotomy reservoir, and is then pumped into an oxygenator. The most common type of oxygenator is a membrane oxygenator, where oxygen is passed through the tube side of a shell-and-tube type device, while the blood passes through the shell side. Oxygen and carbon dioxide are exchanged via diffusion through synthetic membranes (usually polypropylene or silicone) with high permeability to these respiratory gases. The oxygenated blood is then passed through a heat exchanger to adjust the temperature of the blood, and the blood is returned to the systemic circulation via the aorta. At the beginning of the procedure, the patient is anticoagulated with heparin to reduce the risk of thrombosis within the device; as the patient is weaned from bypass, the anticoagulation can be quickly reversed by the use of a drug called protamine. During an operation, the heat exchanger lowers the temperature of the blood and therefore the core body temperature, decreasing the metabolic requirements of the body and protecting the organs (including the heart) against ischemic damage. At the end of the operation, the blood can be warmed to normal physiologic temperature as the patient is weaned from the bypass machine. A specially trained perfusionist controls the operation of the heart-lung machine, allowing the surgeon and anesthesiologist to concentrate on their respective tasks. This device, therefore, provides the function of both the heart (maintaining systemic blood flow and pressure) and the lungs (oxygenating blood and removing carbon dioxide), allowing the heart to be effectively stopped for delicate surgical procedures that would be more difficult or impossible to perform on a beating, moving heart.

Many improvements to the original design of cardiopulmonary bypass machines have been made since their inception. One of the problems with the original heart-lung machines was the trauma that they would cause to the blood cells. Hemolysis of red blood cells would lead to functional anemia and loss of oxygen carrying capacity of the blood; damage to platelets would lead to thrombocytopenia (low numbers of or dysfunctional platelets), resulting in bleeding problems. The problem of blood cell damage has been largely overcome with advanced pump designs and the use of the membrane oxygenators. Roller pumps and centrifugal pumps are commonly used, because they cause a lower degree of hemolysis and shear forces; it is important in the design of these pumps to determine the optimum balance between pumping function and hemolysis/shear stress to the formed blood elements. Bubble oxygenators, which directly pass bubbles of oxygen gas through the blood, cause more hemolysis, protein denaturation, and platelet dysfunction than membrane oxygenators, and are currently less frequently used. In addition, newer devices allow blood that has escaped the circulation within the sterile operating field around the heart to be processed and returned to the patient, reducing the need for blood transfusion during the procedure.

Cardiopulmonary bypass can result in many pathophysiologic changes, including complement activation from the prolonged interaction of blood with synthetic surfaces, platelet and neutrophil activation and aggregation, changes in systemic vascular resistance, and expression of other proinflammatory mediators (Levy and Tanaka, 2003). When these changes are severe, the use of the heart-lung machine can result in complications including confusion, renal insufficiency, pulmonary dysfunction, low-grade hepatic dysfunction, and increased susceptibility to infection. Together, these manifestations are termed the postperfusion syndrome. The last decade has seen the development of mini-extracorporeal circuit (MECC) cardiopulmonary bypass systems (Curtis et al., 2010) with a goal of providing cardiopulmonary bypass with a reduction in this harmful systemic inflammatory response. The MECC has a greatly reduced tubing length, smaller priming volumes, reduction in the blood–air interface, and fewer components than the standard bypass systems, and it utilizes heparin-coating components and centrifugal blood pumps. These systems show a reduction in post-operative cytokine levels, organ damage, post-operative complications, and the need for blood transfusions, compared to standard circuits (Vohra et al., 2009).

Intra-Aortic Balloon Pumps

Since the original use of the intra-aortic balloon pump (IABP) in 1968 by Kantrowitz, the basic design and function of the current device has remained relatively similar during the ensuing decades, but the indications for IABP have increased; an estimated 42,000 patients received IABPs in the US in 2002 (Lee and Makkar, 2006). The typical patient is critically ill with acute cardiogenic shock, mechanical complications of myocardial infarction such as ventricular septal or papillary muscle rupture, unrelenting ventricular arrhythmias or advanced heart failure. Also, patients undergoing high-risk cardiac surgical procedures or percutaneous revascularization may benefit from the use of devices in the peri-procedure period. IABPs (Figure C.2) are catheter-based polyethylene or polyurethane balloons with volumes of 25–50 ml, although smaller devices are used in the pediatric population. Helium is most often used as the inflating gas;

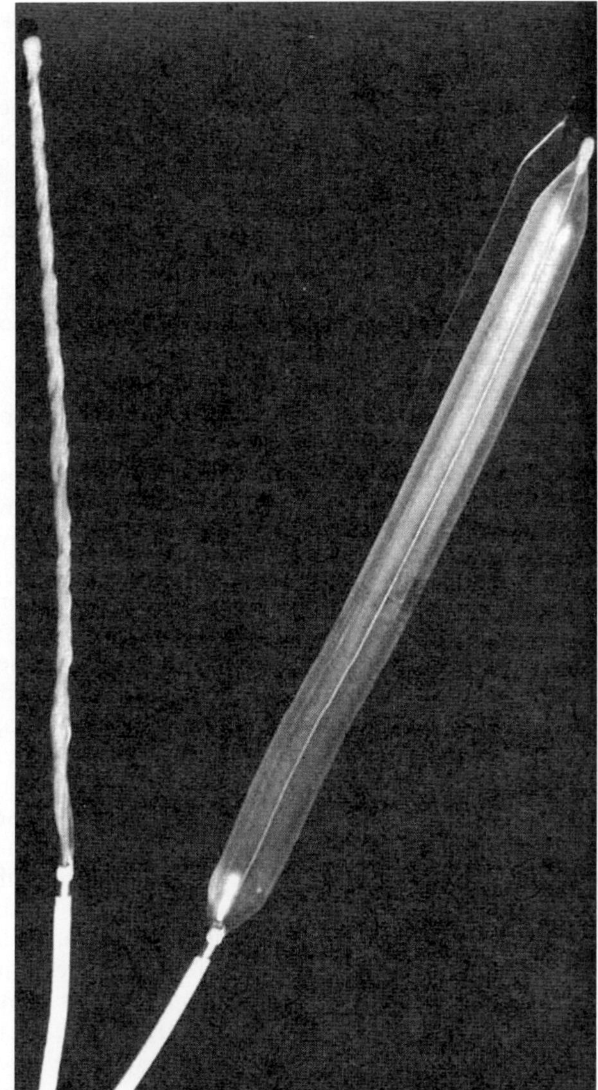

FIGURE C.2 Percutaneous intra-aortic balloon pump. Left: Balloon deflated for insertion. Right: Balloon inflated. (Courtesy S. Volvek, Datascope Corp., Oakland, NJ.)

its low viscosity allows for rapid inflation and deflation, and it is rapidly dissolved in the bloodstream in the event of inadvertent balloon rupture.

IABPs (Baskett et al., 2002; De Sousa et al., 2010) are generally positioned under fluoroscopic guidance in the descending thoracic aorta after percutaneous insertion via the femoral artery. They are timed to inflate during diastole (ventricular filling) and deflate during systole (ventricular contraction), using the patient's electrocardiogram or arterial pressure curve for synchronization. This is counter-pulsation (Trost and Hillis, 2006), which is out of phase with the patient's heartbeat, and causes volume displacement of blood proximally and distally within the aorta. Several beneficial effects serve to improve cardiac function. Coronary blood flow (the majority of which occurs in diastole) is increased by the rise in diastolic pressure, delivering more oxygenated blood to the myocardium. In addition, left ventricular afterload (the pressure the myocardium must attain to pump blood into the aorta) is decreased, reducing the workload and therefore the oxygen requirement of the myocardium. The combination of these two hemodynamic factors therefore improves the balance between myocardial oxygen supply and demand, and results in improved cardiac performance. The device also directly improves systemic circulation to a modest degree (approximately 10%).

IABP therapy permits the heart to rest and recover enough function to support adequate circulation after the device has been removed, usually after only a few days. The major contraindications for IABP use include severe peripheral vascular disease including aneurysms, aortic valve regurgitation, and aortic dissection (owing to the need to thread the balloon through peripheral arteries and the aorta). Complications, which occur in approximately 7% of patients with IABPs in a recent registry study, include limb ischemia from insertion site problems, bleeding, thrombosis with embolization, aortic dissection, balloon rupture, and sepsis.

Ventricular Assist Devices and Total Artificial Hearts

Ventricular assist devices are used primarily in three settings: (1) for potentially reversible heart failure, in which cardiac function is likely to recover with cardiac rest (e.g., postcardiotomy shock); (2) for end-stage cardiac failure not likely to recover, and where mechanical support will provide a bridge to transplantation; and (3) for long-term cardiac support for patients with end-stage congestive heart failure that are not transplant candidates (so-called "destination therapy") (Christiansen et al., 2008). LVADs are also being investigated as a "bridge-to-recovery" in patients with congestive heart failure to induce ventricular changes that might improve cardiac function and eventually allow device removal. Research in this area focuses on the mechanisms of cardiac recovery, identification of patients who could achieve recovery, and specifics such as the timing and duration of therapy (Maybaum et al., 2008; Birks, 2010).

Ventricular assist devices, first successfully employed by DeBakey in 1963, can replace ventricular function for extended periods, in contrast to the short-term duration of cardiopulmonary bypass or IABP use. The earliest devices used for bridge-to-transplant or destination therapy (Hunt and Frazier, 1998) were large, pulsatile systems, with the inflow cannula of the device generally connected to the left ventricular apex, and the outflow cannula connected to the ascending aorta. The pump itself would either be implanted in the peritoneal cavity with a driveline traversing the skin to provide power and controller functions or would remain extracorporeal with the inflow and outflow cannulae each traversing the skin. Over the past decade there has been a seismic

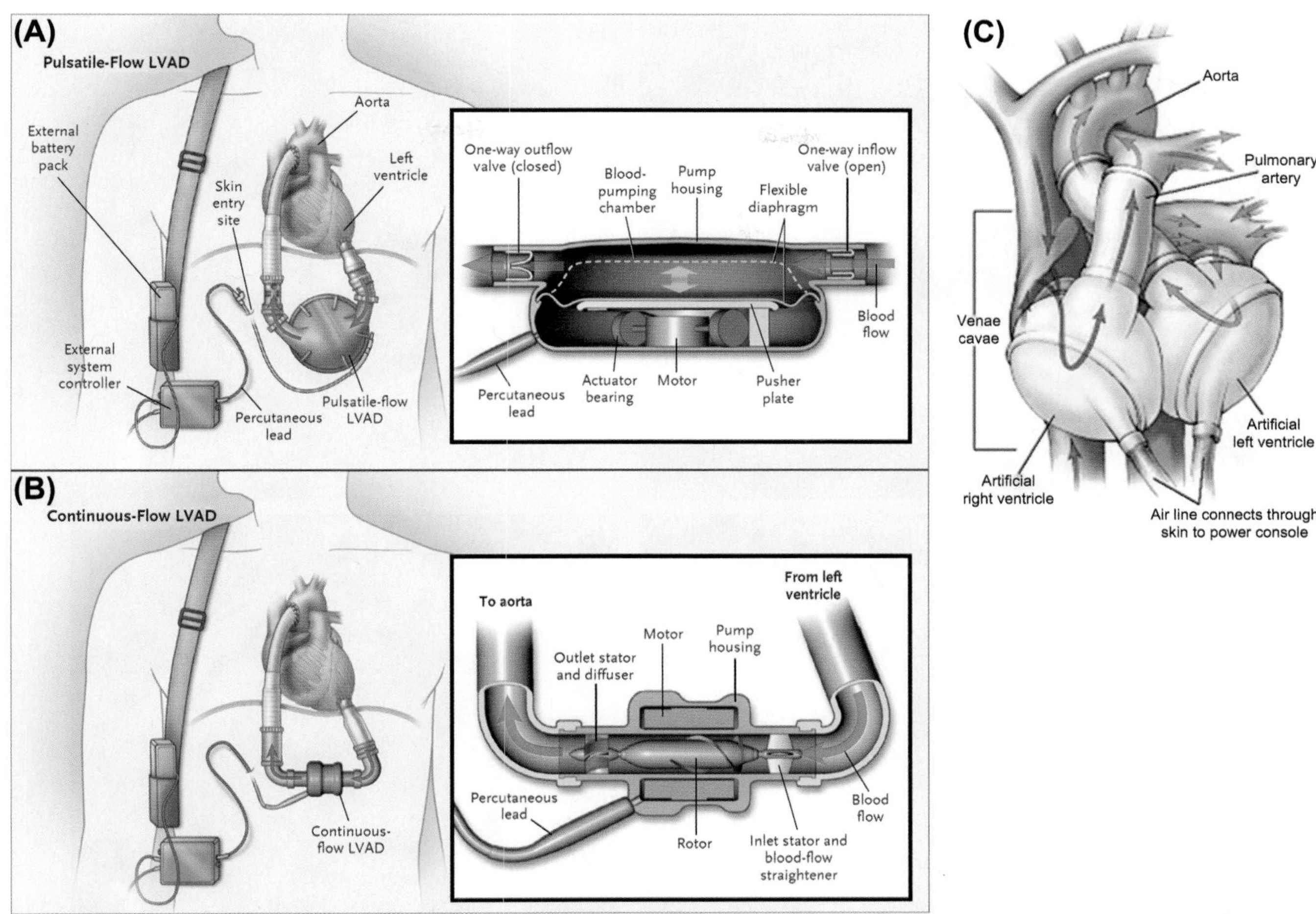

FIGURE C.3 Diagrams of pulsatile (A) and continuous flow (B) left ventricular assist devices, and total artificial heart (C). ((A), (B) Reproduced with permission from Slaughter, M. S. et al. (2009). Advanced heart failure treated with continuous-flow left ventricular assist device. N. Engl. J. Med. 361: 2241–2251; (C) Reproduced with permission from Copeland, J. G. et al. (2004). Cardiac replacement with a total artificial heart as a bridge to transplantation. N. Engl. J. Med. 351: 859–867.)

shift away from these pulsatile devices to so-called "continuous flow" devices, both for bridge-to-transplant and destination therapy (Figure C.3A and C.3B) (Slaughter et al., 2009). These devices, which are comprised of axial or centrifugal pumps along with similar inflow and outflow cannulae and drivelines to the pulsatile systems, are much smaller, more easily implanted, and provide continuous flow for augmentation of cardiac output. In contrast to VADs, where the native heart remains in place, a total artificial heart is composed of two pumping chambers which together replace the entire heart and provide both right and left ventricular function, analogous to heart transplantation (Figure C.3C). Percutaneous VADs, which can be quickly deployed in a manner similar to IABPs, have been developed for short-term applications such as acute cardiogenic shock, and support for high-risk interventional procedures (Sarkar and Kini, 2010).

The major complications of cardiac assist devices, as they have been since the inception of this therapy, are hemorrhage, thrombosis/thromboembolism, infection and device component failure, including the pump and electrical systems (Figure C.4). Hemorrhage continues to be a problem in device recipients, with many predisposing factors including: (1) anticoagulation therapy and its management along with coagulopathy secondary to liver dysfunction and poor nutritional status; (2) contact of the blood with the device, resulting in intrinsic platelet dysfunction and acquired von Willebrand disease; and (3) the extensive nature of the required surgery. In the absence of adequate anticoagulation, and despite the development of minimally thrombogenic blood-contacting surfaces and appropriate blood flow characteristics, thrombi can form in areas of disturbed blood flow such as connections of conduits and other components to each other and to the natural heart. These thrombi can detach and lead to catastrophic embolic events such as ischemic stroke. Infectious complications have been a major limiting factor in the prolonged use of cardiac assist devices (Padera, 2006). Infection can occur within the device, but may also be associated with the percutaneous pneumatic or electrical lines. These infections are a significant cause of morbidity and mortality, and are often resistant to antibiotic therapy and host defenses.

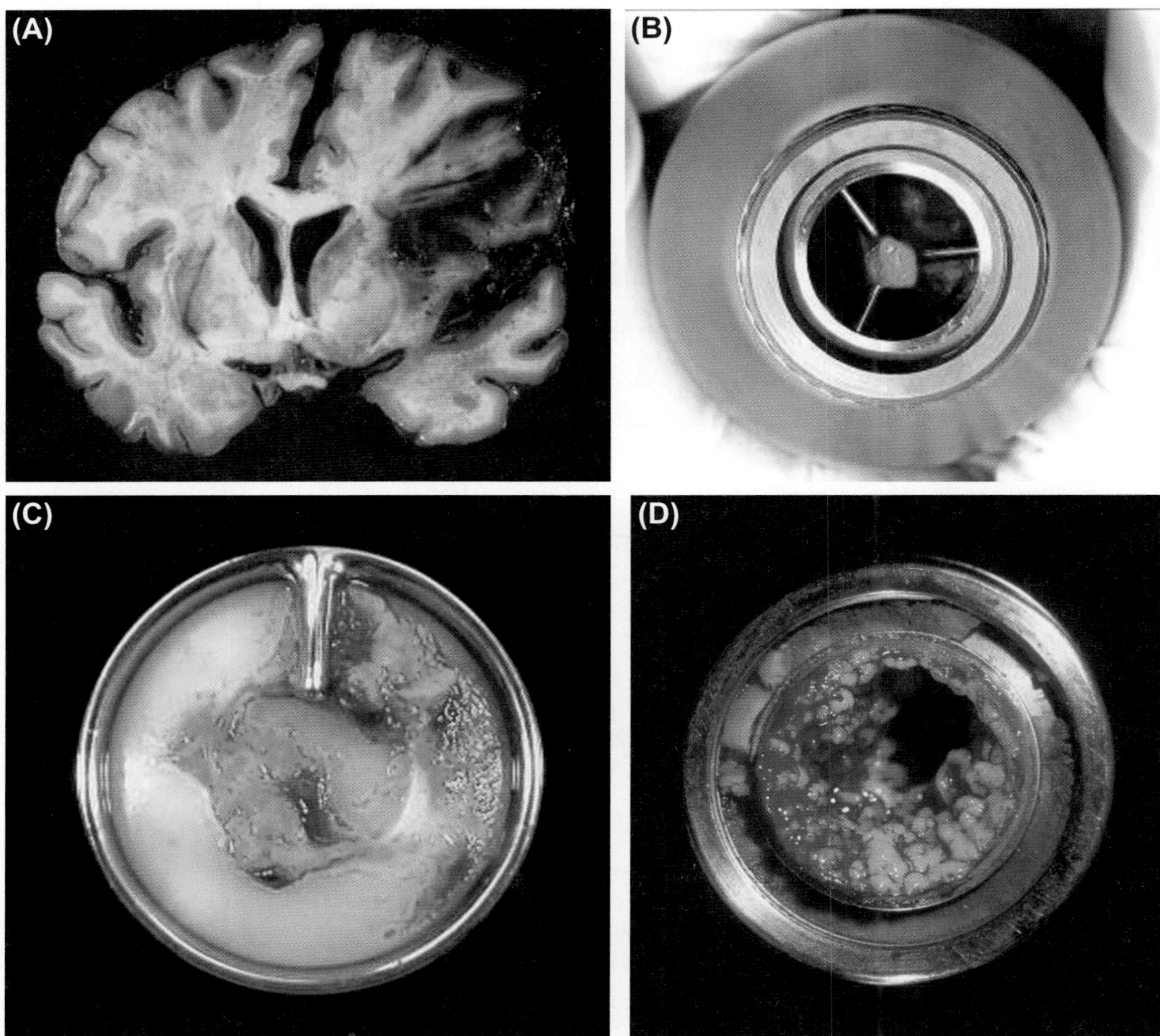

FIGURE C.4 Complications of cardiac assist devices. (A) Hemorrhage into the brain in a patient with an LVAD. (B) Thrombus on the inflow flow straightener and impeller of the HeartMate II LVAD. (C) Thrombus on LVAD valve. (D) Fungal infection in LVAD outflow graft. ((D) Reproduced by permission from Schoen, F. J. & Edwards, W. D. (2001). Pathology of cardiovascular interventions. In: Cardiovascular Pathology, 3rd edn., Silver, M. D., Gotlieb, A. I. & Schoen, F. J. (eds.), WB Saunders: Philadelphia, PA.)

MISCELLANEOUS CARDIOVASCULAR DEVICES

Closure Devices: Patent Ductus Arteriosus, Patent Foramen Ovale, Atrial, and Ventricular Septal Defects

In prenatal life, the circulation is different than it is in postnatal life (Schoen, 1999). The lungs of the fetus are not expanding, and oxygenation of fetal blood is provided via the placenta and maternal circulation. This requires two important shunts that need to close immediately after birth. The foramen ovale, a hole in the fetal intra-atrial septum, allows oxygenated blood returning to the right atrium from the placenta to preferentially pass into the left atrium. This blood passes through the mitral valve into the left ventricle, and is pumped out through the aorta into the systemic circulation. The ductus arteriosus, present between the pulmonary artery and aorta, allows deoxygenated blood pumped from the right ventricle to bypass the lungs and directly re-enter the systemic circulation, as the prenatal pulmonary circulation has a high vascular resistance (owing to the non-expanded lungs). After birth, these functional shunts should close to completely separate the right and left circulations; failure to do so results in a patent foramen ovale (PFO) or patent ductus arteriosus (PDA) that can allow inappropriate shunting of blood in the postnatal circulation. In addition, atrial septal defects (ASDs) or ventricular septal defects (VSDs) can also result from abnormal formation of the atrial septum or ventricular septum. While these defects can be closed via an open surgical procedure (sutures and/or fabric patches for PFO, ASD or VSD, ligation for PDA), efforts have been made to allow closure of these defects using a minimally invasive approach (Hornung et al., 2002). The decision to close a PFO, ASD, VSD or PDA depends on the size of the shunt and the symptoms of the patient.

The first catheter-based closure of a PDA was performed in 1967 by Porstmann using an Ivalon plug to occlude flow through the ductus arteriosus. Of the many PDA closure devices that have been developed over the years, most are metal-based devices that work by causing thrombosis of the PDA with subsequent organization and fibrosis, permanently preventing flow through the residual ductus arteriosus. Two of the more commonly used devices are the Gianturco coil (a stainless steel coil containing polyester fibers to promote thrombosis), and the Amplatzer duct occluder (a conical device consisting of Nitinol wires and a polyester fiber patch to promote thrombosis and tissue integration).

Mills and King reported the first transcatheter closure of an ASD in 1976 using a double umbrella device that covered the opening both from the right and left atrial sides. Their occlusion device consisted of a skeleton of ePTFE coated wire supporting an occluder of Dacron fabric delivered through a catheter. Improvements over the years include better device fixation methods and smaller caliber introducers. Several designs of PFO/ASD closure devices are currently in use (Kim et al., 2007). The Amplatzer device (AGA Medical Corp., Golden Valley, MN) consists of double Nitinol discs filled with polyester patches connected by a small waist; the waist sits within the defect to connect the discs, which sit on either side of and are selected to be larger than the defect. The StarFLEX (NMT Medical, Boston, MA) device is a double umbrella device with metallic arms covered in polyester fabric; the two umbrellas are connected centrally. The Helex (W. L. Gore and Associates, Flagstaff, AZ) is a single Nitinol wire in a helical configuration which is covered by an expanded polytetrafluoroethylene patch. These devices are shown in Figure C.5. Advantages of nonsurgical closure devices such as these include a shorter hospital stay, more rapid recovery, and no residual thoracotomy scar. With the experience gained in the transcatheter closure of PFOs and ASDs, this interventional technology is being extended to the closure of some VSDs, particularly in patients thought to be poor operative risks.

Several types of complications have been reported for closure devices. The most straightforward is the failure to fully close the defect, resulting in residual shunting. These devices work to close the defects at least in part via thrombosis; if the thrombosis extends beyond the defect on the device, thromboemboli may result. Inadequate fixation of the device within the defect or a device-defect size mismatch can result in device embolization. Fractures of various device components, air embolism at the time of device deployment, infection, device erosion through adjacent tissues with perforation, and development of new arrhythmias have also been reported. A number of devices are in development, with the trends being defect-specific design and minimization of the amount of foreign material left in the patient, including the use of biodegradable components (Majunke and Sievert, 2007).

Inferior Vena Cava Filters: Deep Venous Thrombosis and Pulmonary Embolus

Venous thromboembolic disease is a significant cause of morbidity and mortality, largely due to the complication of pulmonary embolism (PE). The most common scenario is for a thrombus to form in the deep venous system of the lower extremities (so-called deep venous thrombus or DVT), become detached from the wall of the vein, travel through the inferior vena cava to the right side of the heart, and lodge as an embolus in one of the large branches of the pulmonary artery. First-line therapy for patients with DVT or PE involves anticoagulation with warfarin and/or low molecular weight heparin. However, when anticoagulation is contraindicated due to active or threatened bleeding or when there is recurrent DVT/PE despite adequate anticoagulation, placement of an inferior vena cava (IVC) filter is often indicated (Crowther, 2007).

The concept of placing a barrier to catch DVTs destined for the pulmonary circulation dates back to Trousseau in the 1860s, but such filters did not become clinical reality until the late 1960s (Greenfield and Michna, 1988). Prior to the development of these devices, surgical ligation of the IVC during an open abdominal operation was performed. Interventional radiologists began to place filters using a minimally invasive percutaneous approach in the 1980s, and currently about 50,000 devices are implanted in the US annually. Five types of permanent IVC filters are commonly used in current clinical practice whose common design elements include metallic wires in a configuration to catch emboli in the bloodstream, a mechanism to anchor the device securely to the wall of the IVC, and an ability to be deployed through a catheter percutaneously. These devices (Streiff, 2000) include the two Greenfield-style filters (one titanium and one stainless steel), the Bird's nest filter (stainless steel), the Simon nitinol filter (nickel-titanium allow with thermal memory properties), and the Vena Tech filter (alloy of cobalt, chromium, iron, nickel, molybdenum, magnesium, carbon, and beryllium). Retrievable IVC filters (Figure C.6) have been developed which can be left in place indefinitely or removed after a period of time percutaneously. Contraindications for anticoagulation (and thus the indications for the device) are often temporary, so retrievable IVC filters will allow device removal after safe and adequate anticoagulation is

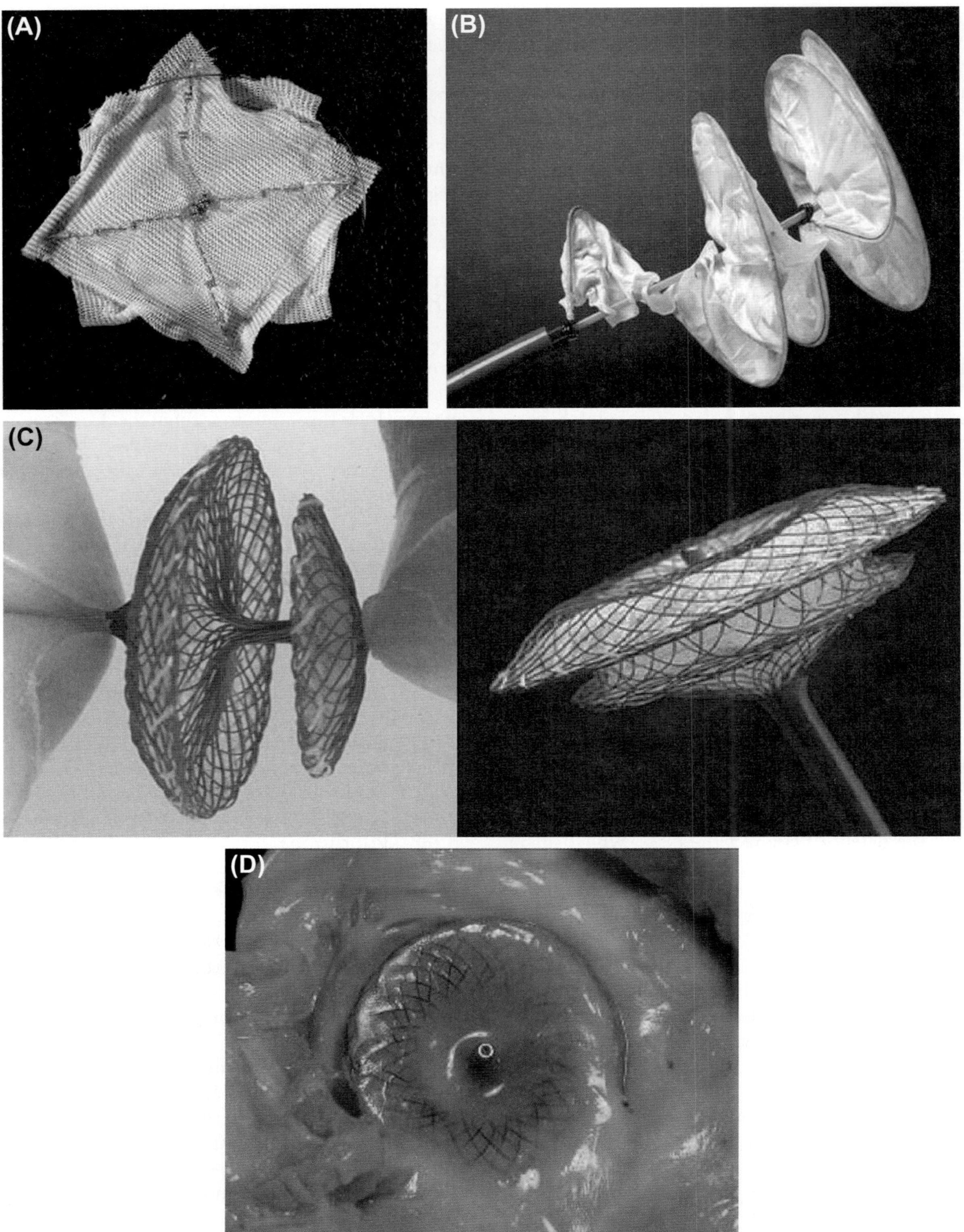

FIGURE C.5 (A) CardioSEAL septal repair device. Helex septal occluder device. (C) (B) Amplatzer ASO occluder device, constructed of woven nitinol with a fabric core. Implantation requires an adequate rim of tissue around the defect to secure the device and prevent migration or embolism of the device. Tissue ingrowth into the interstices of the fabric typically provides effective sealing within a few months following implantation. Amplatzer™ occluders specifically designed for closure of defects of the atrial or ventricular septum, patent ductus and patent foramen ovale are available. (D) Amplatzer ASO occuder device with fibrous tissue following long term implantation. ((A), (B) and (D) Reproduced with permission from Kim, M. S., Klein, A. J. & Carroll, J. D. (2007). Transcatheter closure of intracardiac defects in adults. J. Interven. Cardiol. 20: 524–545; (C) Schoen, F. J. & Webb, J. G. (2008). Prosthetics and the Heart. In: Atlas of Cardiovascular Pathology for the Clinician, McManus, B. M. & Braunwald, E. (eds.), pp. 241–256. Current Medicine: Philadelphia, PA.)

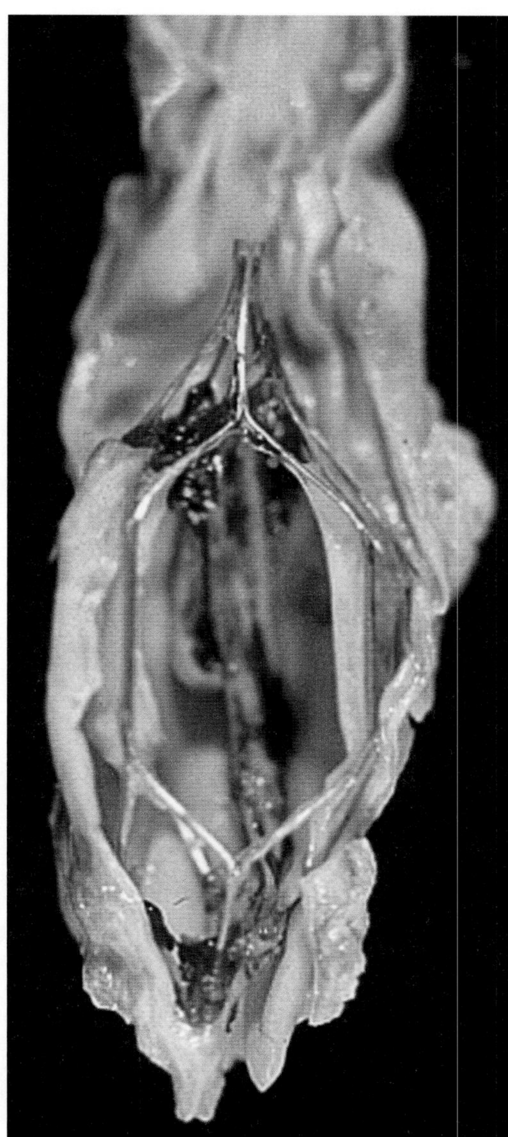

FIGURE C.6 Vena caval filter with clot.

possible for the patient, thereby reducing the possibility of developing device-related complications (Tschoe et al., 2009).

Complications of these devices include thrombosis at the insertion site, thrombi forming on the filter itself, thrombosis and obstruction of the IVC, migration or tilting of the device, filter fracture, and penetration of the IVC wall. Current research involves creating devices: (1) with a lower profile to make insertion easier and reduce insertion site thrombosis; (2) that could be converted into a stent when no longer needed; and (3) that could degrade and potentially release therapeutics. Research is also ongoing to evolve the existing devices with the goal of catching potentially lethal emboli while maintaining adequate vena caval blood flow. As these devices have become safer and more effective, the indications for their use have expanded and clinical trials are ongoing to determine their most appropriate use.

Left Atrial Appendage Occlusion Devices

Atrial fibrillation (AF) affects more than three million individuals in the US, making it the most common cardiac arrhythmia. Instead of orderly atrial contraction initiated by the SA node, rapid disorganized electrical activity in the atria causes these upper chambers of the heart to quiver or fibrillate, resulting in poor contractile function and irregular flow within the chamber. As would be predicted by Virchow's triad, thrombosis may occur due to these flow abnormalities within the atria, especially within the atrial appendage. Atrial appendage thrombi are thought to be the source of thromboemboli, explaining why patients with AF are at a five-fold greater risk for embolic stroke than people in sinus rhythm. Anticoagulation with warfarin is effective in reducing the risk of atrial thrombosis and stroke, but this therapy has many drawbacks including a narrow therapeutic window, variability in metabolism of the drug, interactions with other drugs and metabolites, need for frequent monitoring by blood drawing, poor patient compliance and, most importantly, the risk of life-threatening bleeding. An approach for reducing the risk of thromboembolic stroke in patients with AF is to remove or ligate the left atrial appendage, first proposed in the 1930s and first performed in 1949, via a surgical approach. Nonsurgical approaches to close the left atrial appendage have been developed, and include several devices that can be deployed percutaneously to occlude the opening to the appendage and isolate it from the blood in the left atrium (Maisel, 2009). The Watchman (Figure C.7) left atrial appendage system (Atritech Inc., Plymouth, MN) is a percutaneously deployed parachute-shaped device consisting of a Nitinol cage with a polytetrafluoroethylene membrane on its surface, and fixation barbs along the perimeter which allow it to anchor in the atrial appendage. This device has been shown to be non-inferior to standard warfarin therapy in a recent clinical trial (Holmes et al., 2009). Other devices that have been used or are in development include the Amplatzer (AGA Medical Corp., Golden Valley, MN) cardiac plug, the PLAATO (ev3 Endovascular, Plymouth, MN) system (which is no longer available), and the biodegradable Transcatheter Patch (Custom Medical Devices, Athens, Greece), along with several epicardial clip systems (Singh and Holmes, 2010).

Cardiac Support Devices

Patients with chronic congestive heart failure and a low ejection fraction show evidence of left ventricular remodeling, resulting in a dilated, spherical-shaped ventricle rather than the normal elliptical configuration. While this anatomic shape change has long been known to be

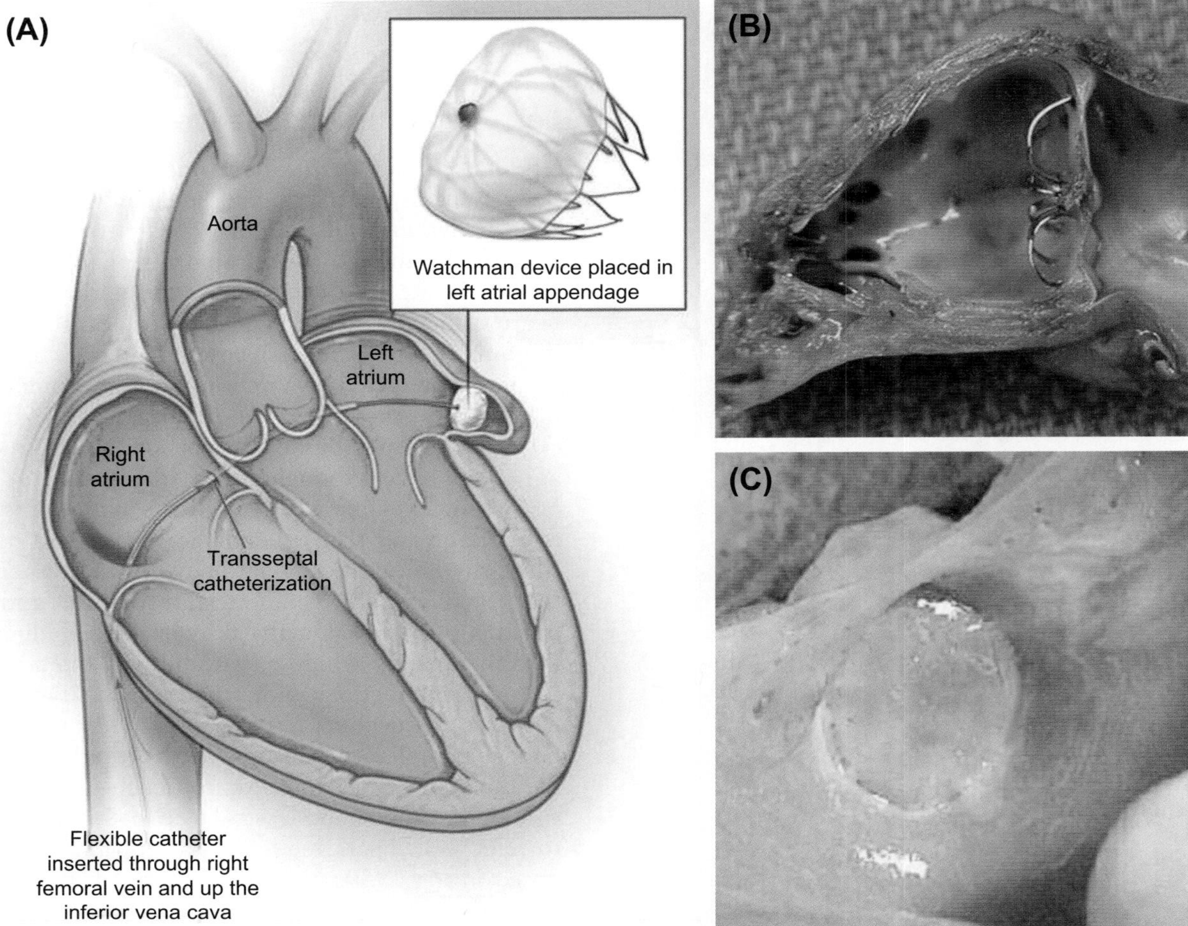

FIGURE C.7 The Watchman left atrial appendage occluder device. (A) Diagrammatic. (B) Left atrial appendage occlusion device in an animal model 45 days after implantation (Watchman™, Atritech Inc.) (Onalan, O. & Crystal, E. (2007). Left atrial appendage exclusion for stroke prevention in patients with nonrheumatic atrial fibrillation. Stroke 38: 624–630). The appendage is thought to be a major source of stasis and thrombus formation in the setting of atrial fibrillation. Surgical removal or exclusion of the appendage may reduce the risk of thromboembolic stroke. Percutaneous occlusion can be achieved by implantation of a catheter delivered device using a femoral vein and transseptal approach to the left atrium. ((A) Reproduced with permission from Maisel, W. H. (2009). Left atrial appendage occlusion – closure or just the beginning? N. Engl. J. Med. 360: 2601–2603; (B) Reproduced from Schoen, F. J. & Webb, J. G. Prosthetics and the Heart. In: Atlas of Cardiovascular Pathology for the Clinician, McManus, B. M. & Braunwald, E. (eds.), pp. 241–256. Current Medicine: Philadelphia, PA.)

a marker of heart failure, recent evidence suggests that it may actually contribute to disease progression, and that therapies aimed at reducing wall stress and myocyte stretch may halt or reverse the detrimental remodeling process, resulting in improved cardiac function (Sabbah, 2003). The CorCap Cardiac Support Device (Acorn Cardiovascular, St. Paul, MN) is a mesh-like device that is surgically implanted around the heart to provide diastolic support to the ventricles, thereby reducing wall stress (Figure C.8A) (Starling and Jessup, 2004). It is composed of a polyester knit with an open weave construction, such that it will stretch more in the apex-to-base direction than in the circumferential direction, prompting the failing heart to return to a more natural elliptical configuration. The device was shown to be safe in a recent clinical trial, and effective in promoting reverse remodeling characterized by reduced left ventricular volumes and a more elliptical left ventricular shape, along with improvements in patients functional status and quality of life (Starling et al., 2007). The Myosplint device (Figure C.8B) used another approach, with three transventricular tension members preventing further remodeling of the left ventricle.

CONCLUSION

Cardiovascular medical devices have been used for more than half a century for myriad applications that save, prolong, and enhance the quality of life for countless individuals. Still, complications from these devices can cause significant morbidity and mortality for individual patients, even years after implantation. Ongoing research aims to improve existing devices, reduce the frequency and severity of complications, and develop novel approaches to the treatment of cardiovascular disease in this exciting field truly at the interface of medicine and technology.

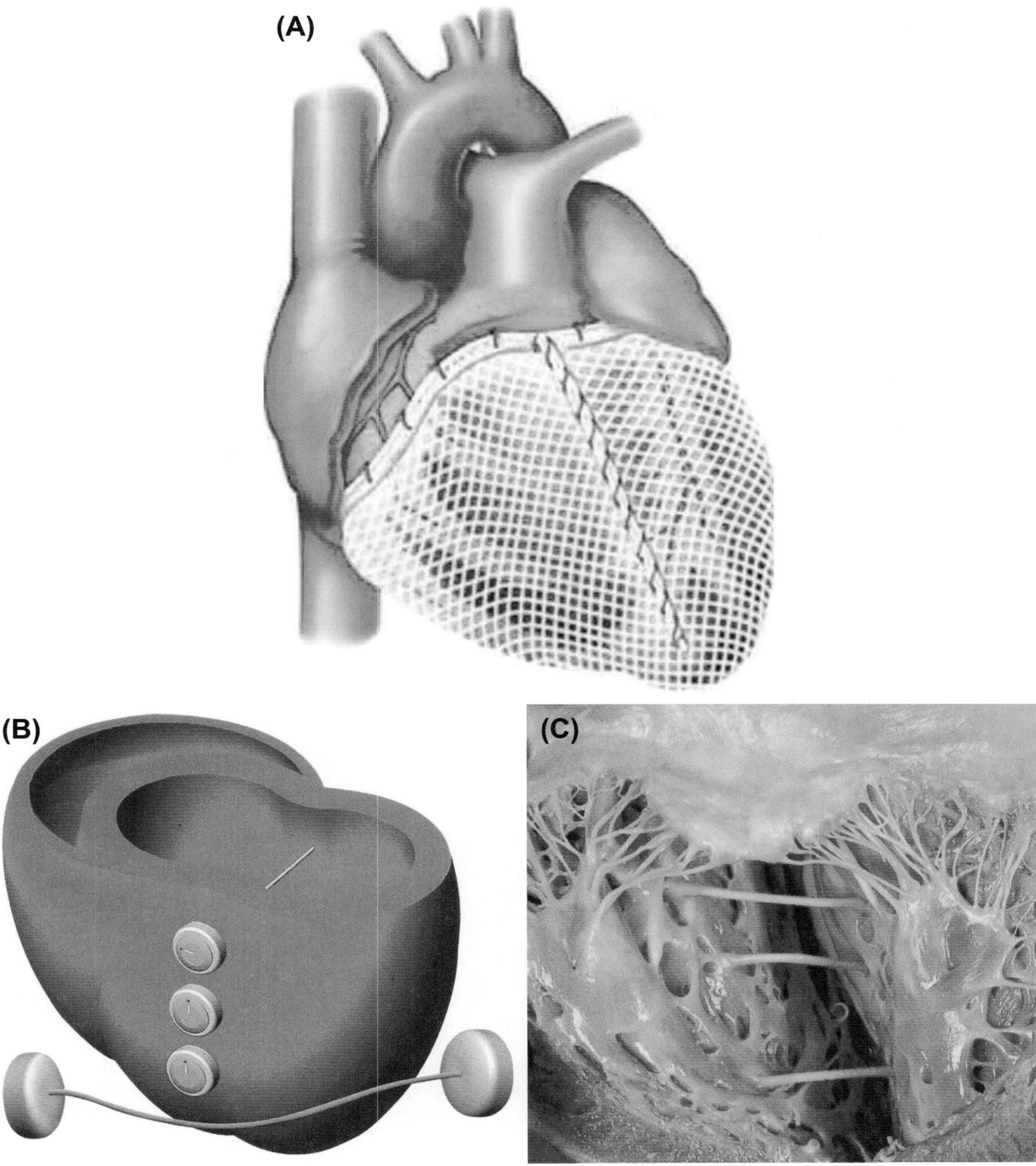

FIGURE C.8 Devices to mitigate the effects of left ventricular dialation in congestive heart failure. (A) The CorCap Cardiac Support Device. (B) and (C) The cardiac Myosplint, (Myocor®, Maple Grove, MN, USA), which consists of implantable transventricular tension members fixed to two epicardial pads that secure the tension member in place. (B) Schematic placement. (C) A human explant showing three tension members following implantation. ((A) Reproduced with permission from Sabbah, H. N. (2003). The Cardiac Support Device and the Myosplint: Treating heart failure by targeting left ventricular size and shape. Ann. Thorac. Surg. 75: S13–19.)

REFERENCES

Alpard, S. K., & Zwischenberger, J. B. (2002). Extracorporeal membrane oxygenation for severe respiratory failure. *Chest Surg. Clin. N. Am.*, **12**, 355–378.

Amin, M. S., & Ellenbogen, K. A. (2010). The effect of device advisories on implantable cardioverter-defibrillator therapy. *Curr. Cardiol. Rep.*, **12**, 361–366.

Atlee, J. L., & Bernstein, A. D. (2001). Cardiac rhythm management devices (part I): indications, device selection, and function. *Anesthesiology*, **95**, 1265–1280.

Baddour, L. M., Epstein, A. E., Erickson, C. C., Knight, B. P., Levison, M. E., Lockhart, P. B., Masoudi, F. A., Okum, E. J., Wilson, W. R., Beerman, L. B., Bolger, A. F., Estes, N. A., 3rd, Gewitz, M., Newburger, J. W., Schron, E. B., & Taubert, K. A. (2010). Update on cardiovascular implantable electronic device infections and their management: a scientific statement from the American Heart Association. *Circulation*, **121**, 458–477.

Baskett, R. J. F., Ghali, W. A., Maitland, A., & Hirsch, G. M. (2002). The intraaortic balloon pump in cardiac surgery. *Ann. Thorac. Surg.*, **74**, 1276–1287.

Baughman, K. L., & Jarcho, J. A. (2007). Bridge to life – cardiac mechanical support. *N. Engl. J. Med.*, **357**, 846–849.

Birks, E. J. (2010). Myocardial recovery in patients with chronic heart failure: is it real? *J. Card. Surg.*, **25**, 472–477.

Boilson, B. A., Raichlin, E., Park, S. J., & Kushwaha, S. S. (2010). Device therapy and cardiac transplantation for end-stage heart failure. *Curr. Prob. Cardiol.*, **35**, 8–64.

Christiansen, S., Kolcke, A., & Autschback, R. (2008). Past, present and future of long-term mechanical circulatory support in adults. *J. Card. Surg.*, **23**, 664–676.

Crowther, M. A. (2007). Inferior vena cava filters in the management of venous thromboembolism. *Am. J. Med.*, **120**, S13–S17.

Curtis, N., Vohra, H. A., & Ohri, S. K. (2010). Mini extracorporeal circuit cardiopulmonary bypass system: a review. *Perfusion*, **25**, 115–124.

De Sousa, C. F., Brito, F. D., de Lima, V. C., & Carvalho, A. C. (2010). Percutaneous mechanical assistance for the failing heart. *J. Interven. Cardiol.*, **23**, 195–202.

DiGiorgi, P. L., Rao, V., Naka, Y., & Oz, M. C. (2003). Which patient, which pump? *J. Heart Lung Transplant*, **22**, 221–235.

Greenfield, L. J., & Michna, B. A. (1988). Twelve-year clinical experience with the Greenfield vena cava filter. *Surgery*, **104**, 706–712.

Haqqani, H. M., & Mond, H. G. (2009). The implantable cardioverter-defibrillator lead: Principles, progress and promises. *Pacing Clin. Electrophysiol.*, **32**, 1336–1353.

Holmes, D. R., Reddy, V. Y., Turi, Z. G., Doshi, S. K., Sievert, H., Buchbinder, M., Mullin, C. M., & Sick, P. (2009). Percutaneous closure of the left atrial appendage versus warfarin therapy for prevention of stroke in patients with atrial fibrillation: a randomized non-inferiority trial. *Lancet*, **374**, 534–542.

Hornung, T. S., Benson, L. N., & McLaughlin, P. R. (2002). Catheter interventions in adult patients with congenital heart disease. *Curr. Cardiol. Rep.*, **4**, 54–62.

Huikuri, H. V., Castellanos, A., & Myerburg, R. J. (2001). Sudden death due to cardiac arrhythmias. *N. Engl. J. Med.*, **345**, 1473–1482.

Hunt, S. A., & Frazier, O. H. (1998). Mechanical circulatory support and cardiac transplantation. *Circulation*, **97**, 2079–2090.

Jessup, M., & Brozena, S. (2003). Heart failure. *N. Engl. J. Med.*, **348**, 2007–2018.

Kim, M. S., Klein, A. J., & Carroll, J. D. (2007). Transcatheter closure of intracardiac defects in adults. *J. Interven. Cardiol.*, **20**, 524–545.

Krishnamani, R., DeNofrio, D., & Konstam, M. A. (2010). Emerging ventricular assist devices for long-term cardiac support. *Nat. Rev. Cardiol.*, **7**, 71–76.

Kusumoto, F. M., & Goldschlager, N. (2002). Device therapy for cardiac arrhythmias. *JAMA*, **287**, 1848–1852.

Lee, M. S., & Makkar, R. R. (2006). Percutaneous left ventricular support devices. *Cardiol. Clin.*, **24**, 265–275.

Levy, J. H., & Tanaka, K. A. (2003). Inflammatory response to cardiopulmonary bypass. *Ann. Thorac. Surg.*, **75**, S715–S720.

Maisel, W. H. (2009). Left atrial appendage occlusion – closure or just the beginning? *N. Engl. J. Med.*, **360**, 2601–2603.

Majunke, N., & Sievert, H. (2007). ASD/PFO devices: What is in the pipeline? *J. Interven. Cardiol.*, **20**, 517–523.

Mancini, D., & Lietz, K. (2010). Selection of cardiac transplantation candidates in 2010. *Circulation*, **122**, 173–183.

Maybaum, S., Kamalakannan, G., & Murthy, S. (2008). Cardiac recovery during mechanical assist device support. *Semin. Thorac. Cardiovasc. Surg.*, **20**, 234–246.

McAlister, F. A., Ezekowitz, J., Hooton, N., Vandermeer, B., Spooner, C., Dryden, D. M., Page, R. L., Hlatky, M. A., & Rowe, B. H. (2007). Cardiac resynchronization therapy for patients with left ventricular systolic dysfunction: a systematic review. *JAMA*, **297**, 2502–2514.

Mond, H. G., & Grenz, D. (2004). Implantable transvenous pacing leads: the shape of things to come. *Pacing Clin. Electrophysiol.*, **27**, 887–893.

Padera, R. F. (2006). Infection in ventricular assist devices: the role of biofilm. *Cardiovasc. Pathol.*, **15**, 264–270.

Sabbah, H. N. (2003). The Cardiac Support Device and the Myosplint: treating heart failure by targeting left ventricular size and shape. *Ann. Thorac. Surg.*, **75**, S13–S19.

Sarkar, K., & Kini, A. S. (2010). Percutaneous left ventricular support devices. *Cardiol. Clin.*, **28**, 169–184.

Schoen, F. J. (1999). The heart. In R. S. Cotran, V. Kumar, & T. Collins (Eds.), *RobbinsPathologic Basis of Disease* (6th ed.). Philadelphia: W.B. Saunders.

Singh, I. M., & Holmes, D. R. (2010). Left atrial appendage closure. *Curr. Cardiol. Rep.*, **12**, 413–421.

Slaughter, M. S., Rogers, J. G., Milano, C. A., Russell, S. D., Conte, J. V., Feldman, D., Sun, B., Tatooles, A. J., Delgado, R. M., Long, J. W., Wozniak, T. C., Ghumman, W., Farrar, D. J., & Frazier, O. H. (2009). Advanced heart failure treated with continuous-flow left ventricular assist device. *N. Engl. J. Med.*, **361**, 2241–2251.

Smith, M. C., & Love, C. J. (2008). Extraction of transvenous pacing and ICD leads. *Pacing Clin. Electrophysiol.*, **31**, 736–752.

Starling, R. C., & Jessup, M. (2004). Worldwide clinical experience with the CorCap Cardiac Support Device. *J. Card Fail.*, **6**(Suppl), S225–S233.

Starling, R. C., Jessup, M., Oh, J. K., Sabbah, H. N., Acker, M. A., Mann, D. L., & Kubo, S. H. (2007). Sustained benefits of the CorCap Cardiac Support Device on left ventricular remodeling: three year follow-up results from the Acorn Clinical Trial. *Ann. Thorac. Surg.*, **84**, 1236–1242.

Streiff, M. B. (2000). Vena caval filters: a comprehensive review. *Blood*, **95**, 3669–3677.

The Antiarrhythmics versus Implantable Defibrillators (AVID) Inevstigators (1997). A comparison of antiarrhythmic-drug therapy with implantable defibrillators in patients resuscitated from near-fatal ventricular arrhythmias. *N. Engl. J. Med.*, **337**, 1576–1583.

Trost, J. C., & Hillis, L. D. (2006). Intra-aortic balloon counterpulsation. *Am. J. Cardiol.*, **97**, 1391–1398.

Tschoe, M., Kim, H. S., Brotman, D. J., & Streiff, M. B. (2009). Retrievable vena cava filters: a clinical review. *J. Hosp. Med.*, **4**, 441–448.

Uslan, D. Z. (2008). Infections of electrophysiologic cardiac devices. *Expert Rev. Med. Devices*, **5**, 183–195.

Vohra, H. A., Whistance, R., Modi, A., & Ohri, S. K. (2009). The inflammatory response to miniaturised extracorporeal circulation: a review of the literature. *Mediators Inflamm.*, 707042.

D. IMPLANTABLE CARDIAC ASSIST DEVICES AND IABPs

Marc A. Simon[1,4], *Harvey S. Borovetz*[2,3], *William R. Wagner*[2,3]

[1]University of Pittsburgh, Departments of Medicine, Pittsburgh, PA, USA

[2]University of Pittsburgh, Departments of Surgery, Pittsburgh, PA, USA

[3]University of Pittsburgh, Departments of Bioengineering, & Chemical Engineering, Pittsburgh, PA, USA

[4]McGowan Institute for Regenerative Medicine, Pittsburgh, PA, USA

CLINICAL NEED AND APPLICATIONS

Heart failure results in over 250,000 deaths per year in the United States. There are approximately 550,000 new cases diagnosed annually, and those diagnosed with this condition have a mortality rate of 70% to 80% at eight years (Thom et al., 2006). When the severity of heart failure is graded, those in the sickest class (New York Heart Association class IV) have an even poorer prognosis, with a survival rate of only about 25% at two years (Schocken et al., 1992). Despite survival improvements with medical therapies, survival with advanced disease is still reported to be 20% at five years (Ammar et al., 2007). For these end-stage patients, heart transplantation has become an effective treatment. Actuarial survival rates for heart transplant patients are 85% at one year, 80% at two years, and 75% at five years (Ammar et al., 2007).

The success of heart transplantation is limited, however, by an inadequate and stagnant donor supply. Although 35,000–64,000 patients could potentially benefit from heart transplantation annually (Funk, 1991), only about 2200–2300 cardiac transplants are performed in the US annually. There has not been a significant change in cardiac transplant volume over the ten years from 1998–2007 (US Department of Health and Human Services, 2008). While waiting for scarce donor organs to become available, approximately 10% of individuals listed for heart transplantation die annually (Ammar et al., 2007). To address the need to support the circulation in patients with end-stage heart failure, a wide variety of mechanical devices have been developed over the past several decades. Current devices that are in clinical use will be discussed in this chapter, as will circulatory support devices that may enter the market in the near future. In the vast majority of cases these devices do not physically replace the heart (as would a total artificial heart); rather they function as ventricular assist devices (VADs), providing flow assistance to one side of the heart.

Roles for Ventricular Assist Devices

As a bridge to cardiac transplantation, VADs have provided circulatory support to end-stage heart failure patients who otherwise would have a low likelihood of surviving until a donor organ became available. Nearly 70% of patients undergoing VAD support have survived until heart transplantation (Mehta et al., 1995), with more recent devices having a 79% survival to transplant (Pagani et al., 2009). Recent data have shown that six months after VAD implantation there was 17% mortality, with 51% receiving a heart transplant (Kirklin et al., 2008). Patients undergoing VAD "bridging" generally do not undergo further deterioration in their condition and, in fact, demonstrate markedly improved end-organ function, and an improvement in their health status due to improved perfusion (Frazier et al., 1994). Ironically, the success of VAD bridging has been implicated in intensifying the donor shortage by including recipients who would otherwise not have survived until transplantation (Massad et al., 1996). A second result of this trend is the need for increasingly extended periods of VAD support prior to organ availability.

With the increasingly broad and positive experience of supporting patients until transplantation with VADs, the concept of using VADs as a permanent source of circulatory support or as a "destination therapy" has been advanced. A number of factors led to the consideration and investigation of permanent support: the limited donor organ supply; a significant patient population that fails to meet age and medical eligibility requirements for transplant listing; and improved quality of life with recent portable VAD designs. This has now been approved for one device in the US after the Randomized Evaluation of Mechanical Assistance for the Treatment of Congestive Heart Failure (REMATCH) trial (Rose et al., 2001) demonstrated survival benefit for patients deemed not to be cardiac transplant candidates compared to medical therapy with a two year survival of 23% versus 8%, respectively. Major concerns regarding this option are primarily the complications that remain associated with VAD support. As one might expect, biomaterial performance, device reliability, and biocompatibility are central to the perceived limits of utilizing VADs as alternatives to transplant in end-stage heart failure patients. In the REMATCH trial such complications included: bleeding (42%); infection (28% at three months, with 25% dying of sepsis); stroke (24% with neurological event); and peripheral thromboembolism (Rose et al., 2001), although complications appear to be decreasing with experience (Long et al., 2005).

The improvement in end-organ function generally associated with extended circulatory support has lead investigators to evaluate whether the heart muscle itself may undergo recovery from the disease process of heart failure during the support period. A variety of recent studies in VAD-supported patients have shown a reduction in the inflammatory mediators associated with heart failure, as well as decreased myocyte necrosis and apoptosis, improved myocyte contractility, and indications of improved function in the left ventricle (Mann et al., 1998). These reports have lead to the investigation of

VAD support as a "bridge to recovery," where patients undergo device explantation without heart transplant due to sufficient recovery of myocardial function while mechanically supported. The attractiveness in bridging to recovery lies in freedom from transplantation and immunosuppressive therapy for the patient and an increased donor organ supply for the community. Early clinical experience with this procedure indicates that, although success can be achieved, patients must be carefully screened for this procedure and the majority of VAD patients are unlikely to meet selection criteria (Loebe et al., 1999). Small studies have reported an incidence of myocardial recovery sufficient for VAD removal in 5–10% of adult patients (Simon, 2005). A multicenter prospective study funded by the National Institutes of Health is currently underway to investigate myocardial recovery during VAD support.

VENTRICULAR ASSIST DEVICE DESIGN AND BLOOD-CONTACTING MATERIALS

Six currently available devices are approved by the FDA for the provision of circulatory support while patients await a donor heart. All but one of these works as a VAD (the other is a total artificial heart). The general flow path of a VAD can be seen in Figure D.1. The inflow conduit of the device is connected to the apex of the left ventricle. Blood from the patient's ventricle enters the VAD conduit and flows into a pumping sac. For pulsatile pumps, a unidirectional valve in the inflow conduit prevents flow reversal during pump activation. A valve

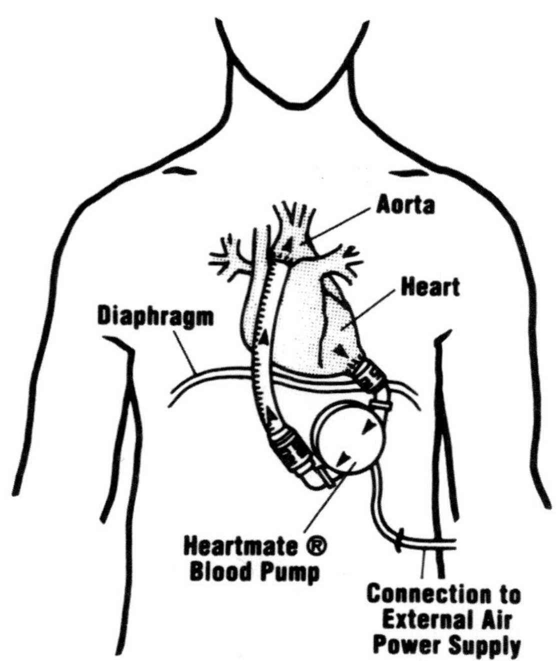

FIGURE D.1 ThermoCardiosystems HeartMate VAD demonstrating anatomical placement and blood flow. (Reprinted from Arabia, F. A., Smith, R. G., Rose, D. S., Arzouman, D. A., Sethi, G. K. and Copeland, J. G. (1996). *ASAIO J.* 42, M542–M546.)

is not necessary for the growing number of rotary, continuous flow pumps. The pumping sac in pulsatile pumps can be triggered to compress by several methods: sensing pump filling, compressing at a constant frequency or by triggering off the patient's electrocardiogram. Upon compression the inflow valve is closed and the outflow valve opens, permitting blood flow through the outflow conduit toward the patient's ascending aorta. The VAD thus "unloads" the left ventricle of the heart and provides the necessary work to periodically propel blood into the arterial tree.

This pulsatile or first generation design is giving way to continuous flow (second and third generation) pumps (Olsen, 1999). Rotary blood pumps have a number of advantages over pulsatile VADs. A continuous flow pump requires no bladder to create a stroke volume comparable to the native ventricle, rather flow is accomplished with an impeller, in either an axial flow (straight through) or centrifugal (center to tangential edge) configuration. The rotary pumps are significantly smaller (on the order of an AA cell battery in certain designs), have fewer moving parts, and are free from valves and cyclic actuators. The major sources of wear are the bearings which are blood- or plasma-immersed in second generation designs. Third generation designs have reduced mechanical wear further by suspending the impellers magnetically. This innovative design is felt to increase overall reliability and durability. Lower overall masses and volumes of implanted materials make these second and third generation systems applicable to patients generally excluded from current implantable pulsatile VADs, e.g., smaller women and children. Also, with the smaller pump size, less of a pocket needs to be created in the body for implantation, and this may reduce bleeding and infection rates. Without valves and the compression of a pumping sac, the rotary pumps are also much quieter than current pulsatile flow generating devices.

Although free from some of the shortcomings of VADs that generate pulsatile flow, rotary pumps introduce some unique biocompatibility concerns. The high revolutions per minute (RPM) required of rotors to generate ~5 liters per minute flow rates in these pumps (~5000–10,000 RPM) subjects blood elements to high peak shear stresses not encountered physiologically. Although these stresses are of very short duration, concerns arise regarding hemolysis and platelet activation due to this exposure. Biocompatibility testing with rotary blood pumps invariably involves assessment of hemolytic indices, evaluation of thrombotic deposition and, in animal studies, assessment of potential end-organ embolic damage. To minimize the occurrence of shear-related blood trauma associated with regions of flow turbulence or stagnation in these pumps, fluid flow over turbine blade designs is often visualized during pumping or computationally modeled (Kerrigan et al., 1996; Burgreen et al., 1999). A second hemodynamic concern regards the reduced pulsatility of blood flow that occurs when a continuous flow, rotary pump is connected to empty the left ventricle and pump blood to the aorta. The chronic effects of diminished pulsatility on long-term patient health are not known, although the routine use of continuous flow pumps for cardiopulmonary bypass, acute circulatory support, and extracorporeal membrane oxygenation (ECMO) have not lead to major physiologic complications, and large animal implants appear to tolerate this type of circulation for periods beyond six months. Data on this topic was reviewed by Saito (2004).

The mechanical requirements inherent to small rotary blood pumps have resulted in the selection of metals, primarily titanium alloys, as the material of choice for impellers, flow straighteners, and pump housings. In some current rotary systems inflow and outflow cannulae do not differ remarkably in design and materials selection from those used with pulsatile VADs, while in others the departure is significant, illustrated by designs wherein the pump is moved into the ventricle (Westaby et al., 1998; Wampler et al., 1999; Frazier et al., 2002). Design and control concerns arise in association with the inflow orifice of rotary pumps. When ventricular pressures are reduced, the ventricular septum has the potential to be drawn into or around the inflow orifice, obstructing flow and potentially damaging the ventricular wall (Amin et al., 1998). Such concerns do not arise with the passive filling, pulsatile flow generating VADs currently used clinically.

The design of inflow and outflow pump bearings (which support the spinning rotor and resist axial and radial thrust loads) represents a major challenge for rotary blood pump designers. The bearings used in this setting must effectively dissipate heat, exhibit acceptably low wear, support radial and axial loads, and be readily machined to very high tolerances. Bearing designs currently under investigation utilize magnetic suspension, blood lubrication or lubrication provided by a purge fluid. Since sliding contact may occur during transients in pressure and flow, starts and stops, and unexpected patient acceleration and deceleration, it is important that bearing materials be wear-resistant. With blood or plasma contact, it is also important that these bearings exhibit appropriate blood biocompatibility. Current reported materials that have been evaluated for rotary pump bearings include zirconium-niobium alloy (Zr-2.5Nb), titanium-zirconium-niobium alloy (Ti-13Zr-13Nb), as well as carbon coating techniques (Golding et al., 1998).

First Generation Blood Pumps

As described above, these pumps were some of the first designed and provide pulsatile flow via compression of a pumping sac. Such designs can either replace the patient's heart (as would a total artificial heart) or work as a ventricular assist device (VAD). Clinically, VADs are used much more frequently, although one total artificial heart (TAH) is approved by the FDA (CardioWest™ Temporary Total Artificial Heart, Syncardia Systems, Inc.). TAH design began in the early to mid-1980s with the first implants of the Jarvik-7 total artificial heart intended for permanent support (DeVries et al., 1984). Medical complications such as infection and thromboembolism, as well as manufacturing concerns, lead to the discontinuation of TAH use as permanent support. Also at this time the success of immunosuppressive therapies was being reported, raising heart transplantation to the role of preferred therapy for end-stage heart failure. The TAH was subsequently utilized in the bridge to transplant role (Guy, 1998) for which it is now approved (Arabia et al., 1999; Copeland et al., 2004).

In the 1990s several VADs received FDA approval for use as a bridge to cardiac transplantation, including the Thoratec PVAD™, HeartMate XVE (both Thoratec Corporation, Pleasanton, CA), and Novacor Left Ventricular Assist System (World Heart Inc, Salt Lake City, UT). The Novacor device has since been removed from the market by World Heart in favor of developing the third generation Levacor™ VAD. The Thoratec PVAD™ (Thoratec Corporation, Pleasanton, CA) is based upon the Pierce-Donachy VAD originally developed in the mid-1970s at Pennsylvania State University, and currently has been implanted in over 4000 patients. In Figure D.2 the interface of this VAD with the circulatory system is presented. Of particular note is the paracorporeal nature of this device. Both inflow and outflow conduits cross the skin, and the

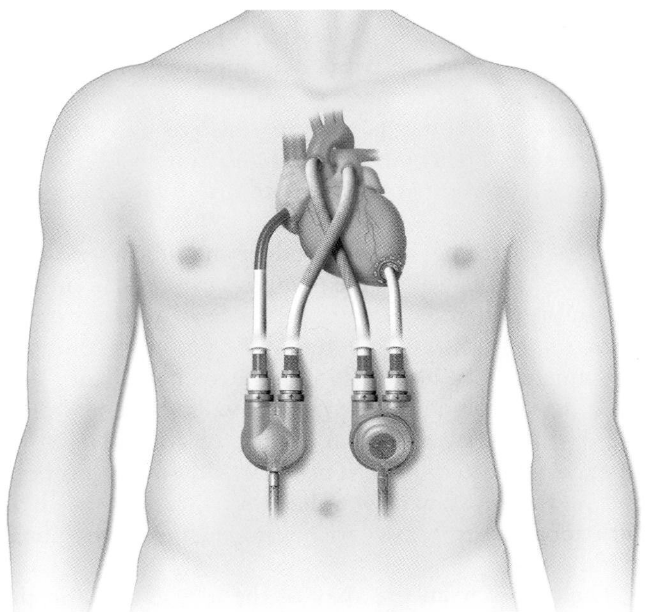

FIGURE D.2 Thoratec VADs utilized for bi-ventricular support. The right ventricular assist device (RVAD) draws blood from the right atrium, while the left ventricular assist device (LVAD) is connected to the apex of the left ventricle. Both pumping sacs rest outside of the body with inflow and outflow conduits crossing the skin. (Reprinted from Hunt, S. A. and Frazier, O. H. (1998). *Circulation* 97, 2079–2090.)

pumping sac rests on the lower abdomen. One advantage of this type of design is the potential to implant smaller patients. Also of note is the potential for this device to be used in a bi-ventricular support mode with two devices implanted in tandem, as shown in Figure D.2. Although it has been found that the majority of patients experiencing end-stage heart failure require only left ventricular assistance (Kormos et al., 1996), for those who require support of the right ventricle, the Thoratec pump can be placed in conjunction with another Thoratec pump or other VAD supporting the left ventricle.

The Thoratec VAD has the potential to draw blood from cannulae designed to interface with the ventricle or atrium, with the former being more commonly utilized for extended left ventricular support. The inflow cannula and the pumping sac are composed of Thoralon™, a proprietary polyurethane elastomer blended with a surface modifying agent. The surface modifying agent is designed to increase the potential for surface molecular rearrangement and reduce thrombogenicity (Farrar et al., 1988). The outflow cannula is composed proximally of Thoralon™ polyurethane that fuses distally with a Dacron graft, to allow suturing to the ascending aorta. The Dacron graft is preclotted with patient blood and thrombin or cryoprecipitate and thrombin prior to implantation to seal the graft pores. Where the cannulae cross the skin polyester velour is present on external surfaces of the Thoralon™ to encourage tissue integration. The opposing valves controlling flow into the pumping sac are tilting disc mechanical valves (Bjork-Shiley Monostrut). Compression of the full pumping sac produces a stroke volume of approximately 65 mL and is accomplished pneumatically, requiring connection to a controller providing positive and negative air pressure. A portable pneumatic system allows for reasonable mobility and the potential for discharge to home.

The HeartMate XVE, including the inflow and outflow conduits as well as the pumping sac, resides entirely within the patient, as seen in Figure D.1. An older pneumatic model utilized an external air pump to compress the pumping sac in a fashion akin to that of the Thoratec device, while the currently available vented electric design uses an integral electric motor powered by an external battery pack. The HeartMate XVE has been one of the most widely utilized VAD design to date, with more than 4500 patients having been implanted worldwide.

The inflow from the apex of the left ventricle in the Heartmate XVE design occurs through a titanium alloy (Ti-6Al-4V) cannula attached to a low-porosity Dacron graft. The blood- and tissue-contacting surfaces of the titanium cannula are surface coated with sintered titanium microspheres 50–75 μm in diameter, as seen in Figure D.3A. The Dacron portion of the inflow conduit houses a 25 mm porcine xenograft valve (Medtronic). The outflow conduit is also a valved Dacron conduit with an identical opposing valve at the exit of the pumping sac. The pumping sac is comprised on one side of a titanium alloy surface with sintered titanium microspheres. On the opposing side the flexible pusher plate diaphragm is made of Biomer polyurethane that has been textured on its blood-contacting surface with surface integral fibrils approximately 18 μm in diameter and 300 μm in length, as seen in Figure D.3B (Menconi et al., 1995). The maximum stroke volume of the pumping sac is 83 mL.

The implication of having textured surfaces contacting the blood in the HeartMate VAD is that these

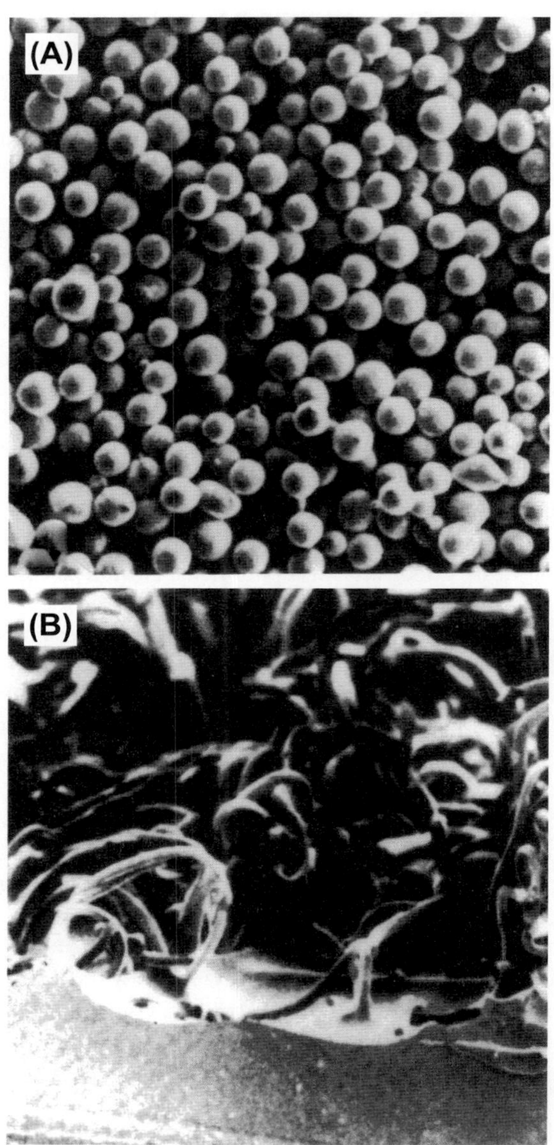

FIGURE D.3 (A) Blood-contacting surface of the rigid housing in the HeartMate VAD with sintered titanium microspheres (spheres are 50–75 μm in diameter). (B) Integrally textured polyurethane surface of the flexible diaphragm (fibrils are approximately 18 μm in diameter). (Reprinted from Menconi, M. J., Pockwinse, S., Owen, T. A., Dasse, K. A., Stein, G. S. and Lian, J. B. (1995). *J Cell Biochem* 57, 557–573.)

surfaces rapidly clot upon device placement. While this may seem like a negative scenario for the interior surfaces of a VAD, the concept behind the design is that while clots rapidly form on these surfaces, these thrombi are densely adherent and do not appear to embolize into the bloodstream in a clinically relevant manner. Over time additional blood cell interaction occurs, similar to an inflammatory reaction (see Chapter II.2.2). A heterogeneous surface containing platelets, monocytes, macrophages, foreign-body giant cells, lymphocytes, and multipotent circulating cells is deposited. It is postulated that multipotent circulating cells differentiate into fibroblasts, myofibroblasts, and in some cases endothelial cells, which are reported to populate the surface (Frazier et al. 1993; Rafii et al., 1995; Spanier et al., 1999). The fibroblastic cells may then secrete extracellular matrix components such as collagen, which is routinely found on textured VAD surfaces following extended implants. Unlike the inflammatory response associated with porous cardiovascular biomaterial surfaces, the cells populating the interior HeartMate VAD surface are likely derived entirely from the passing blood. Cellular migration from tissue onto these surfaces in the timeframe that the cellular deposits are observed is considered unlikely. The blood-contacting biological interface that develops from the initial surface coagulum after approximately one week is referred to as the pseudointimal layer (Figure D.4), and has been shown not to grow in thickness exceeding 150 μm over periods of implantation in the order of one year (Menconi, 1995).

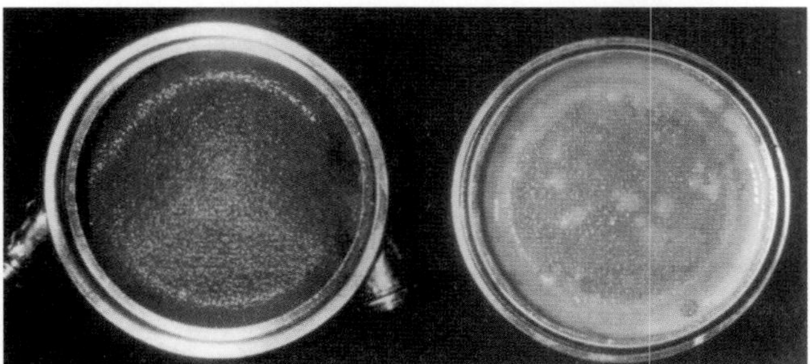

FIGURE D.4 Pseudointima developed on the interior textured surfaces of the HeartMate VAD after an implantation period of 243 days. On the left is the titanium surface with sintered microspheres and on the right is the textured polyurethane surface. (Reprinted from Poirier, V. L. (1999). *Thorac Cardiovasc Surg* 47 (suppl), 316–320.)

Second Generation Blood Pumps

Second generation blood pumps are generally rotary pumps with bearings immersed in blood or plasma. They deliver diminished pulsatile or continuous blood flow. These include those described by Butler et al. (1997) seen in Figure D.5, DeBakey (1997) illustrated in cross-section in Figure D.6 (Westaby et al., 1998; Wampler et al., 1999). While the HeartMate II® (Thoratec Corp.), Jarvik 2000® (Jarvik Heart, Inc.), HeartAssist 5™ (MicroMed Cardiovascular Inc.), INCOR® (Berlin Heart GmbH) and the Synergy® (CircuLite, Inc.) utilize a rotary pump design, the EVAHEART (EVAHEART MEDICAL USA) and the CorAide™ (Arrow International, Inc.) are centrifugal pumps. These pumps are not completely implantable, having a percutaneous line for power transmission, pump monitoring, and speed (RPM) control. Most, if not all of these pumps have now entered clinical trials including one, HeartMate II, approved by the FDA for bridge-to-cardiac transplantation based upon favorable trial completion (Miller et al., 2008). Clinical trials have focused on bridge-to-cardiac transplantation, although HeartMate II is completing a destination therapy trial. In addition to these long-term implantable pumps, continuous-flow technology has allowed for the development of shorter duration, less invasive assist devices utilized in emergency situations (such as organ perfusion during cardiogenic shock) or to provide support during high risk surgical procedures. Examples of these devices are the

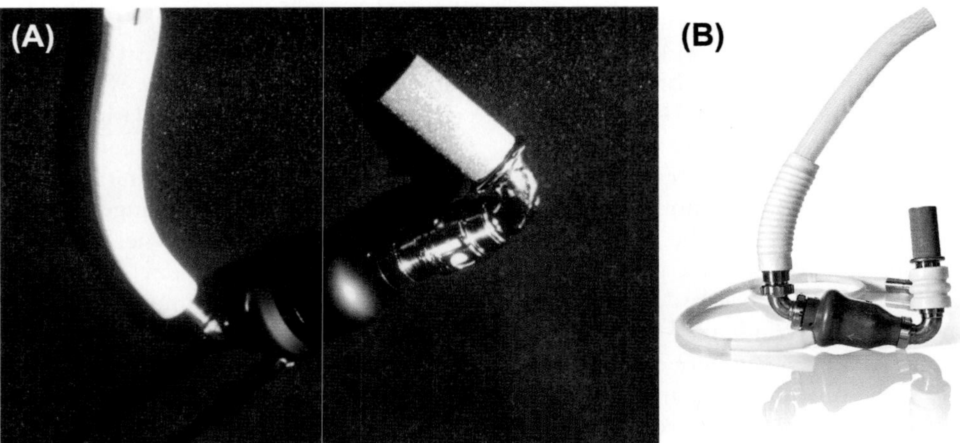

FIGURE D.5 (A) The Nimbus-Pittsburgh Innovative Ventricular Assist System axial flow blood pump with rotor assembly shown above. This device was developed into what is currently the HeartMate II Left Ventricular Assist System (B). ((A) Kindly provided by K.C. Butler, President, Nimbus, Inc.; (B) Kindly provided by Thoratec Corporation).

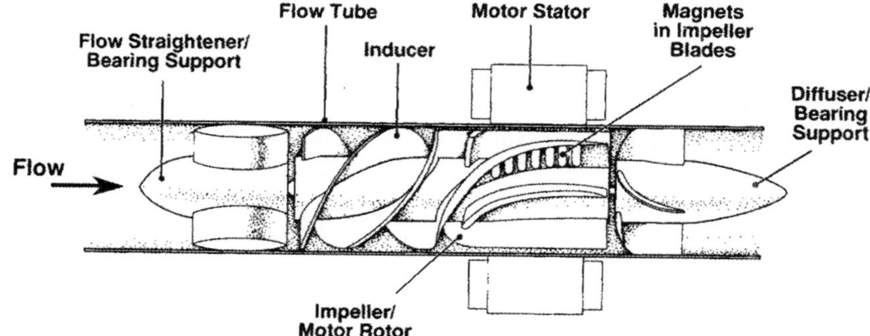

FIGURE D.6 Cross-section of the DeBakey VAD. (Reprinted from DeBakey, M. E. (1999). *Ann Thorac Surg* 68, 637–640.)

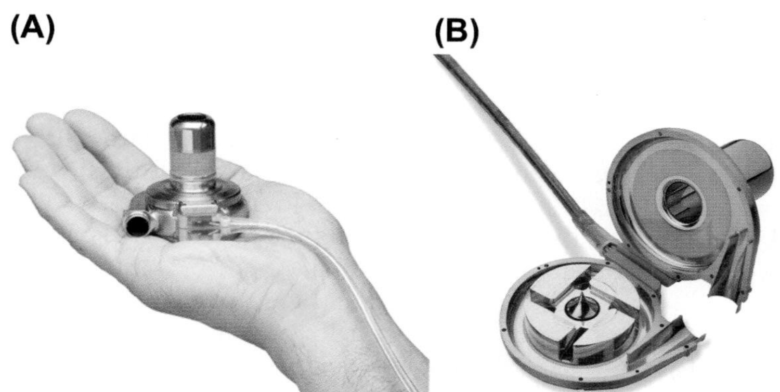

FIGURE D.7 (A) The third generation HeartWare HVAD is small enough to be contained in the pericardial cavity and has an integrated inflow conduit (the outflow graft is not included in the photograph). (B) The HeartWare HVAD is open to reveal the levitated impeller and blood-contacting surfaces. (Kindly provided by HeartWare International Inc.).

Impella® 2.5 & 5.0 (ABIOMED, Inc.) and the TandemHeart® (Cardiac Assist Inc.).

Third Generation Blood Pumps

Third generation pumps are characterized by magnetically or hydrodynamically suspended impellers. Similar to second generation pumps, they can deliver continuous or diminished pulsatile blood flow. Examples are the DuraHeart™ (Terumo Heart, Inc.), HeartWare® HVAD™ (HeartWare International, Inc.) [Figure D.7] and the short-term Thoratec® Centrimag® (Thoratec Corp).

Complications and VAD Biocompatibility Issues. Complications associated with the implantation and operation of VADs include those that may be related to biomaterial-centered processes discussed earlier in this text, including thromboembolism, infection, and bleeding. It is worth noting, however, that the mechanisms behind these complications vary among patients and are multifactorial. Many of the processes contributing to the complication may be independent of the biomaterial implant. Of particular relevance here is the health status of the patient at the time of device implantation. Patients in end-stage heart failure suffer from a variety of conditions secondary to the poor perfusion of their organs. They undergo a variety of invasive treatments and monitoring that precede VAD insertion, and have often spent extended periods sedentary in the hospital environment.

Independent of VAD implantation end-stage heart failure patients are at risk of infection due to poor perfusion, compromised immune function, extended exposure to nosocomial infections, and invasive monitoring devices. Thromboembolism can occur in this patient group as a result of ventricular arrhythmia and ventricular mural thrombi, poor distal hemodynamics, and systemic vascular disease. Bleeding risks are elevated as a result of compromised liver function secondary to poor organ perfusion, and invasive monitoring and therapy short of VAD implantation. With these considerations in mind, there is certainly a relationship between the extensive biomaterial surface implantation of a VAD and the above complications. For first generation pumps blood-contacting surface areas range from 400–500 cm^2; and while second and third generation pumps have reduced this contact area, these devices still represent a substantial biomaterial implantation for an extended period. The major complications will be discussed below.

Thromboembolism

Thromboembolic rates for VAD support vary widely among the specific devices studied and the reporting centers. Rates can vary simply due to the average implant period of the VAD, the health of the patients selected for support, and the patient medical management routine. Thromboembolism can be defined in a conservative manner to include only obvious neurologic events (strokes) lasting for more than 24 hours that are coupled with imaging evidence of a recent cerebral infarction, and which cannot be explained as originating from a source other than the VAD. Alternatively, the definition can be liberalized to include suspected neurologic events such as blurred vision, weakness on one side, compromised field of view, and other events that are transient in nature ("transient ischemic attacks;" with symptoms resolving within 24 hours), that may or may not be followed up with diagnostic imaging, and that may potentially be explained by some other risk factor. It is also possible to investigate peripheral embolization with imaging techniques that would not likely be employed in the majority of clinical centers (Schmid et al., 1998). Finally, although not generally included in reported thromboembolic rates, tissue infarcts attributable to VAD support can be quantified at the time of autopsy in patients who die while on the device. With these limitations to reported thromboembolic rates in mind, historically reported rates for first generation pumps have been 5–30% (Mehta et al., 1995; DeRose et al., 1997; El-Banayosy et al., 1999a). Interest in standardizing definitions, and therefore the reported incidence of adverse events, has resulted in the formation of the Interagency Registry for Mechanically Assisted Circulatory Support (INTERMACS), which is an NIH-sponsored registry of US FDA approved mechanical circulatory support devices. While this only includes FDA-approved devices, it has been mandated that all implanting centers participate in order to receive reimbursement from the Centers for Medicare and Medicaid Services. INTERMACS recently reported 199 neurologic events, 14 arterial non-neurologic events, and 33 venous thrombotic events in a total of 483 patients over an approximately two-year period (Holman et al., 2009). In this report, neurologic events were the leading cause of death (11%), which further emphasizes the how persistent the problem of thromboembolism is for blood pumps.

To explain the principle mechanisms of thromboembolism in VAD patients one need recall the mechanisms of thrombosis on artificial surfaces discussed in Chapter II.2.6 and consider the role of flow on this thrombotic process as discussed in Chapter II.2.6 (Hanson & Tucker). Low pump volumetric outputs have been directly associated with thrombotic deposition on the valves of first generation pumps (Wagner et al., 1993). This design was subsequently altered to reduce the potential for low blood velocities to develop within the pumping sac and valve system. With all VAD designs, excessively low volumetric flow is considered to be a risk factor for thromboembolism.

The hemostatic system is affected with the implantation of all types of VADs. With the HeartMate pulsatile device it has been shown that platelets, coagulation, and fibrinolysis are all activated early in the implant period, at two hours after the operation (Livingston et al., 1996). This activation is in excess of the marked elevations resulting from the extensive biomaterial contact involved in cardiopulmonary bypass for the implant surgery. The textured, prothrombotic surfaces of this VAD likely contribute to this effect. Longer-term elevations in hemostatic markers have also been reported for all VAD designs. Ongoing elevations in platelet activation, coagulation, and fibrinolysis suggest that the device surfaces remain active during the implant period in bridge to transplantation patients (Spanier et al., 1996; Dewald et al., 1997; Wilhelm et al., 1999).

Attempting to relate activation of coagulation, platelets or fibrinolysis with the risk for thromboembolism remains difficult for VAD patients. Although thromboembolic events are uncomfortably common, most centers do not have the numbers of patients to collect the number of observations necessary to statistically relate specific hemostatic alterations to thromboembolic events. As a result, markers of thromboembolism that do not result in an obvious clinical event (e.g., stroke) have been utilized to relate to measurable thrombotic markers in the blood (as discussed in Chapter II.2.6 – Hanson & Tucker). Microembolic signals detected in the cerebral arteries of VAD patients using ultrasound techniques have been explored as a potential indicator of stroke risk in VAD patients. Some reports have suggested that microembolic signals increase on days when patients experience a clinically obvious thromboembolic event (Schmid et al., 1998). Others have shown that thrombin generation increases on days when large numbers of microemboli are detected in the cerebral arteries (Wilhelm et al., 1999).

As with all cardiovascular devices, anticoagulation and antiplatelet therapy are options to control the reactions at the device–blood interface. A variety of specific anticoagulation and antiplatelet drug and monitoring regimens have been developed for VAD patients. Most commonly patients are managed acutely with heparin, and chronically with oral anticoagulants (i.e., warfarin) and antiplatelet agents (i.e., aspirin, clopidogrel). Anticoagulation is monitored with measurement of the activated partial thromboplastin time during heparin therapy and with the prothrombin time or international normalized ratio during oral anticoagulation. Antiplatelet therapy effectiveness can be monitored with thromboelastography. For the HeartMate, patients generally are not given chronic anticoagulation due to the unique lining of this device, and in only a fraction of the patient population are antiplatelet agents administered.

The fact that HeartMate VAD patients do not require chronic anticoagulation, yet still appear to have a markedly lower rate for thrombembolic complications, merits discussion and has important ramifications for our definition of material biocompatibility in this setting. In much of the clinical literature the lower rate for thromboembolic complications in HeartMate VAD patients is attributed to the textured surfaces discussed above which lead to the formation of a biological pseudointima. This concept of encouraging well-anchored clot formation as a means of ultimately encouraging a biological lining goes back to the mid-1960s and earlier when Sharp, Hall, and other investigators researched this concept (Sharp et al., 1964; Hall et al., 1967).

There are few reports to date regarding the mechanisms through which the HeartMate pseudointima might act to reduce thromboembolism. Investigators have shown the mature surface to be dominated by inflammatory and fibroblastic cell lineages expressing tissue factor, inflammatory cytokines, and laying down matrices that include type I collagen (Menconi et al., 1995; Spanier et al., 1999). Such a surface would not appear to have anti-thrombotic properties. Indeed, Spanier et al. (1999) suggested that such a surface may work by creating a sustained prothrombotic and potentially proinflammatory environment that triggers ongoing coagulation through the tissue factor pathway. The body's extended fibrinolytic response to the ongoing coagulation on this surface may serve as auto-anticoagulation to prevent large potential thromboemboli from developing. Alternative explanations for the HeartMate VADs impressive clinical results should also be considered. The development of small thrombi on the surface of the pump may occur with regularity throughout the implant period, but these thrombi may have a much lower likelihood of embolizing due to stronger adhesion to the extracellular matrix present on the surface. Deposition of the pseudointima may also act to "smooth" discontinuities in pump surfaces that otherwise would serve as a nidus for thrombi formation. The macroscopic flow patterns within the pumping chamber should also be considered. This VAD has been designed to have a "wandering vortex" which may act to minimize the size of thromboemboli generated from pumping chamber surfaces (Slater et al., 1996). Unfortunately, this surface design has not been able to be successfully translated to smaller volume, continuous flow pumps, and likely the smaller blood-contacting surface areas of these pumps offsets any potential increased risk. In fact, it has been reported that the HeartMate II has a stroke rate of 8.9% (plus an additional 2% with transient ischemic attacks), with the event rate dropping to 0.05 events per patient year (Pagani et al., 2009).

Infection

Infection is considered by many physicians and investigators to be the most serious complication facing VAD patients and a major obstacle to the implementation of current VAD designs as permanent implants to treat end-stage heart failure. As implant periods for bridge to transplant patients increase, the primary cause of death becomes infection (El-Banayosy et al., 1999a). Historically, the reported infection rates in VAD patients has varied widely between centers due to the criteria selected to classify this complication, as well as patient selection and medical management differences. Infection rates can be reported in a very general sense to include all patient infections, usually defined as clinical evidence of infection reflected by elevated leukocyte count, fever or both in the presence of a positive culture. With this general definition, infection rates generally fall near 50% for all three aforementioned types of VADs (El-Banayosy et al., 1999a; McBride et al. 1999; Sun et al., 1999). Alternatively, only device-related infections may be reported. These infections can be classified as positive cultures from percutaneous drivelines and cannulae, and from the VAD pocket or mediastinum (Figure D.8). With this definition the infection rates tend to fall in the 20–30% range. As noted above, interest in standardizing definitions and reported outcomes has resulted in the formation of the Interagency Registry for Mechanically Assisted Circulatory Support (INTERMACS). Reporting information on implantations of FDA-approved devices is mandatory in the US in order to receive reimbursement from the Centers for Medicare and Medicaid Services. INTERMACS has defined major infection as an event requiring antimicrobial treatment that occurs in the setting of pain, fever, drainage, and/or leukocytosis, preferably with positive culture results. Infection is subdivided into localized non-device infection (no systemic involvement), percutaneous site and/or pocket infection (positive culture from the skin and/or tissue surrounding the drive line or from the tissue surrounding the external housing of a pump implanted within the body), internal pump infection (involving the blood-contacting surfaces of the VAD documented by positive culture), and sepsis (positive blood cultures and/or hypotension). INTERMACS recently reported 479 episodes of infection in 193 out of 483 patients (Holman et al., 2009).

FIGURE D.8 *Candida albicans* vegetations partially obstructing the inflow valve of a Novacor VAD patient at the time of valve replacement. (Kindly provided by R. L. Kormos, M.D, Director, and S. Winowich of the University of Pittsburgh Medical Center Mechanical Circulatory Support Program.).

Infection is responsible for the death of 4.3% bridge to transplant VAD patients in recent INTERMACS data (Holman et al., 2009). The seriousness of an infection is largely related to its location. Infections of the blood-contacting interior surfaces of the VAD, confirmed at device explant, are associated with high rates of mortality (~50%). Infection may also contribute to embolic damage and death by the generation of septic emboli or through enhancement of the thromboembolic process. Patient infections not directly related to the device are not as serious, with the exception of positive blood cultures, which may indicate interior VAD surface vegetations. As discussed in Chapter II.2.8, there are distinct bacterial and fungal species which tend to infect biomaterial surfaces and often these infections include multiple organisms. In Figure D.9 the composition of infections in a multi-center study are presented in terms of the infected site. Common skin bacteria that are traditionally associated with biomaterial infections dominate the VAD-infecting strains.

Management of infection in patients can be accomplished to a reasonable degree with antibiotics and antifungal agents. It is fortunate that the patient population remains dominated by individuals awaiting device removal at the time of transplantation. Surprisingly, a history of infection during VAD support is not a counter-indication for heart transplantation and studies have shown that, despite the use of anti-rejection drugs in the post-transplant period, these patients do well (Argenziano et al., 1997). For those patients with suspected infection of the interior VAD surfaces, it has been suggested that device replacement may be an effective method of treatment if transplant is not a short-term option (Argenziano et al., 1997).

Prevention of infection is a primary goal during device implantation and later patient medical management. Prophylactic antibiotics and, in some instances, antifungal agents are utilized in the perioperative period. Careful cleaning and attention to the percutaneous driveline and cannulae sites are also of primary importance. From a biomaterials perspective two important design issues arise. First, it is critical for the exterior surfaces of the drivelines and cannulae to rapidly encourage tissue healing and to discourage bacterial colonization. Second, the mechanical properties of the percutaneous lines are of relevance, in that large mismatches in stiffness between the skin and lines places stress at the biomaterial–tissue interface. The resulting abrasion may prevent or retard adequate wound healing. While the elimination of percutaneous lines altogether through the introduction of transcutaneous power transmission and control systems has been extensively studied (Weiss et al., 1999), this technology has not been perfected to the point where it would be attractive for widespread adoption.

Bleeding

Bleeding remains a potentially lethal complication associated with VAD implantation, but one that generally occurs in the early postoperative period, when numerous wound sites exist. Reported bleeding rates range from 30–35% (El-Banayosy et al., 1999b; McBride et al., 1999) to about 40% of patients (537 events in 201 out of 483 patients, with the majority occurring in the first 30 days) in the recent INTERMACS report (Holman et al., 2009). Bleeding can be defined in terms of the administration of blood products, chest tube drainage after surgery or surgical interventions to treat hemorrhage. Subsequent trips to the operating room to address bleeding place patients at risk for bacterial and fungal colonization, linking this complication to infection risk. Also of concern in this patient population is the exposure and sensitization of transplant candidates to human antigens associated with blood products which can limit organ matching (Massad et al., 1997). The use of leukocyte-filtered products has been utilized to address this latter concern.

A number of mechanisms contribute to hemorrhage in VAD patients. As mentioned previously, VAD

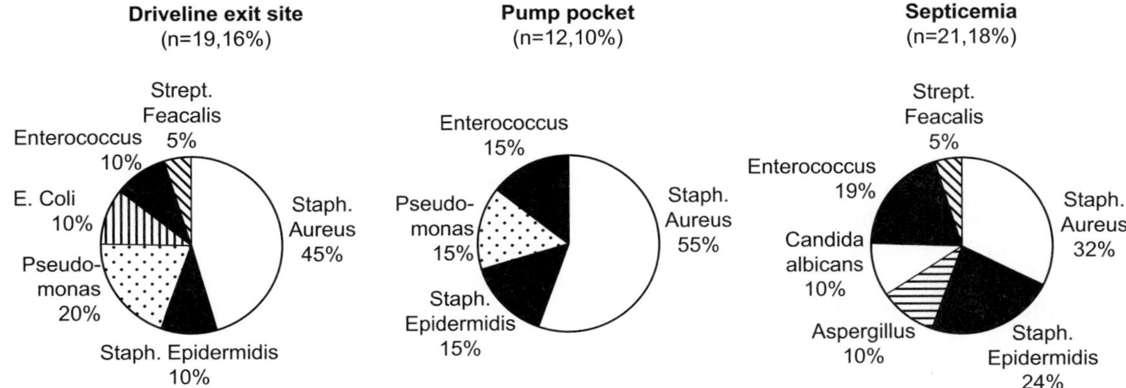

FIGURE D.9 Incidence of infectious organisms grouped by site. Cultures were performed of driveline exit site infection, pump pocket infection, and blood (septicemia) in a group of 118 Novacor VAD patients. The number of patients with infection for each site are presented above each pie chart as is the percentage of total patients experiencing infection at this site. (Reprinted from A. El-Banayosy, M. Deng, D. Y. Loisance, H. Vetter, E. Gronda, M. Loebe, and M. Vigano, *Eur J Cardiothorac Surg* **15**:835–841, 1999.).

implantation surgery is associated with consumption and activation of patients' platelets and coagulation system, as well as activation of the fibrinolytic pathway (Livingston et al., 1996). End-stage heart failure patients undergoing VAD implantation have deficiencies in coagulation factors, even in the absence of obvious liver dysfunction (Wang et al., 1995). After post-operative stabilization, patients are placed on chronic anticoagulation therapy which carries an ongoing risk for over-anticoagulation (and possible bleeding). Of particular concern in the extended support setting are cerebral bleeds leading to neurologic complications. Anticoagulation in these VAD patients is thus carefully monitored throughout the implant period.

Mechanical Failure

In the extensive clinical experience with VADs over the past several decades there have been very few reported incidents of mechanical failure related to material issues. This is likely due to the long development phases through which all three aforementioned FDA-approved devices have passed. The pulsatile VAD systems subject internal diaphragms, cams, bearings, and springs to cyclic loads that would likely lead to more wear-associated failure were these devices asked to perform beyond their design specifications. In bridge to transplant implantations bearing wear has been reported, but has not lead to catastrophic failure (Sun et al., 1999). Other complications reported involve tearing of Dacron inflow conduits in the HeartMate device (Scheld et al., 1997) (Figure D.10). This complication was likely due to wear from the metal support cage rubbing on the graft surface, and the design has since been altered to constrain cage motion. Other complications can involve the external drive units, electric power supply, batteries, controller or cables. Second and third generation pumps are expected to have significantly improved pump longevity, which is needed for patients being permanently supported. INTERMACS has recently reported a total of 81 device malfunctions in 62 out of 483 patients (Holman et al., 2009).

A variety of diagnostic techniques have been suggested to non-invasively assess the wear process in implanted VAD electrical and mechanical components short of pumping failure that would result in clinical symptoms. Acoustic and power consumption monitoring have been reported as innovative methods for evaluating wear (Dohmen et al., 1999). Durability and reliability testing *in vitro* will continue to be essential for discovering the typical time course of electromechanical failure modes, and to provide the opportunity to develop such diagnostic methods. This failure can be readily detected and may thus prevent unnecessary replacement of devices at fixed implant periods, while allowing detection of failure and remediation in those devices that may develop early signs of this wear (Lee et al., 1999). With knowledge of the failure mode engineers are provided a focal point to work on redesign of the device, leading to further improvements in long-term durability and reliability.

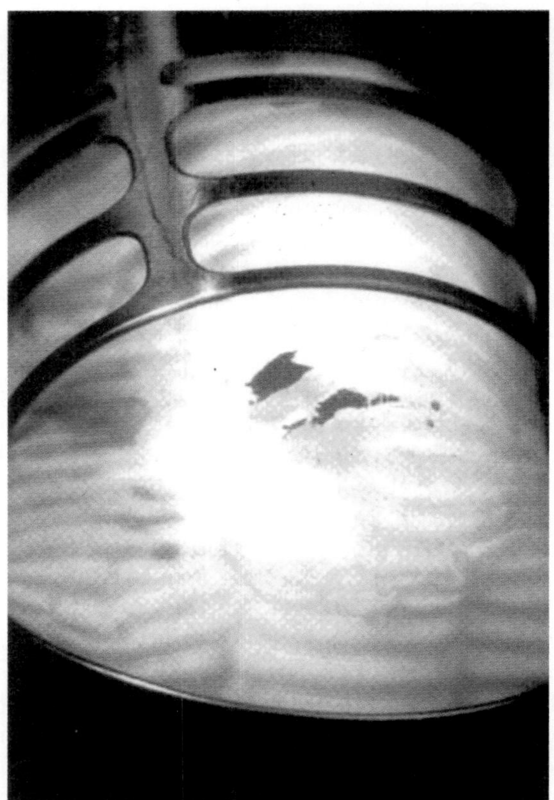

FIGURE D.10 Ruptured Dacron inflow conduit of a HeartMate VAD that lead to sudden blood loss after 22 days of pump implantation. (Reprinted from H. H. Scheld, R. Soeparwata, C. Schmid, M. Loick, M. Weyand, and D. Hammel, *J Thorac Cardiovasc Surg* **114**: 287–289, 1997.).

CONCLUSIONS

Major advances have occurred in mechanical circulatory support technology over the past several decades. Bridging patients to cardiac transplantation has now become routine, with seven devices currently approved by the FDA to fill this role in the United States. This success has set the stage for the application of VADs as alternatives to cardiac transplantation in the transplant-ineligible population, and one device is already approved by the FDA for this indication. As one would expect with an extensive blood-contacting biomaterial implantation accompanied by percutaneous drivelines, thromboembolism, infection, and bleeding remain complications of primary concern in VAD patients. Rotary blood pumps and perhaps completely implantable VADs may serve to reduce these complications, and improve device durability and reliability. In the next decade the challenges associated with supporting patients for years, as opposed to months, will be faced. Biomaterial and biocompatibility issues will likely be central to the problems and solutions that arise as mechanical circulatory support devices continue to evolve.

ACKNOWLEDGMENTS

The authors wish to thank Dr. Robert L. Kormos, Director of the University of Pittsburgh Medical Center Mechanical Circulatory Support Program, for his expert guidance and innumerable insights on circulatory support issues, as well as the engineers and staff of this program for their assistance on this manuscript. Special thanks must be extended to our VAD patients and their loved ones who have always encouraged us to better understand the challenges and opportunities of implantable circulatory assist devices.

BIBLIOGRAPHY

American Heart Association. (1998). *1999 Heart and Stroke Statistical Update*. Dallas, TX: American Heart Association.

Amin, D. V., Antaki, J. F., Litwak, P., Thomas, D., Wu, Z. J., & Watach, M. (1998). Induction of ventricular collapse by an axial flow blood pump. *ASAIO J.*, **44**, M685–M690.

Ammar, K. A., Jacobsen, S. J., Mahoney, D. W., Kors, J. A., Redfield, M. M., et al. (2007). Prevalence and prognostic significance of heart failure stages: Application of the American College of Cardiology/American Heart Association heart failure staging criteria in the community. *Circulation*, **115**, 1563–1570.

Arabia, F. A., Copeland, J. G., Smith, R. G., Banchy, M., Foy, B., Kormos, R., Tector, A., Long, J., Dembitsky, W., Carrier, M., Keon, W., Pavie, A., & Duveau, D. (1999). CardioWest total artificial heart: A retrospective controlled study. *Artif. Organs*, **23**, 204–207.

Argenziano, M., Catanese, K. A., Moazami, N., Gardocki, M. T., Weinberg, A. D., Clavenna, M. W., Rose, E. A., Scully, B. E., Levin, H. R., & Oz, M. C. (1997). The influence of infection on survival and successful transplantation in patients with left ventricular assist devices. *J. Heart Lung Transplant.*, **16**, 822–831.

Burgreen, G. W., Antaki, J. F., Wu, J., le Blanc, P., & Butler, K. C. (1999). A computational and experimental comparison of two outlet stators for the Nimbus LVAD. *ASAIO J.*, **45**, 328–333.

Butler, K., Thomas, D., Antaki, J., Borovetz, H., Griffith, B., Kameneva, M., Kormos, R., & Litwak, P. (1997). Development of the Nimbus/Pittsburgh axial flow left ventricular assist system. *Artif. Organs*, **21**, 602–610.

Copeland, J. G., Smith, R. G., Arabia, F. A., Nolan, P. E., Sethi, G. K., Tsau, P. H., McClellan, D., & Slepian, M. J. (2004). Cardiac replacement with a total artificial heart as a bridge to transplantation. *N. Engl. J. Med.*, **351**, 859–867.

DeBakey, M. E. (1999). A miniature implantable axial flow ventricular assist device. *Ann. Thorac. Surg.*, **68**, 637–640.

DeRose, J. J., Jr., Argenziano, M., Sun, B. C., Reemtsma, K., Oz, M. C., & Rose, E. A. (1997). Implantable left ventricular assist devices: An evolving long-term cardiac replacement therapy. *Ann. Surg.*, **226**, 461–470.

Dewald, O., Fischlein, T., Vetter, H. O., Schmitz, C., Godje, O., Gohring, P., & Reichart, B. (1997). Platelet morphology in patients with mechanical circulatory support. *Eur. J. Cardiothorac. Surg.*, **12**, 634–641.

DeVries, W. C., Anderson, J. L., Joyce, L. D., Anderson, F. L., Hammond, E. H., Jarvik, R. K., & Kolff, W. J. (1984). Clinical use of the total artificial heart. *N. Engl. J. Med.*, **310**, 273–278.

Dohmen, P. M., Laube, H., de Jonge, K., Baumann, G., & Konertz, W. (1999). Mechanical circulatory support for one thousand days or more with the Novacor N100 left ventricular assist device. *J. Thorac. Cardiovasc. Surg.*, **117**, 1029–1030.

El-Banayosy, A., Korfer, R., Arusoglu, L., Minami, K., Kizner, L., Fey, O., Schutt, U., & Morshuis, M. (1999b). Bridging to cardiac transplantation with the Thoratec ventricular assist device. *Thorac. Cardiovasc. Surg.*, **47**(suppl), 307–310.

Farrar, D. J., Litwak, P., Lawson, J. H., Ward, R. S., White, K. A., Robinson, A. J., Rodvien, R., & Hill, J. D. (1988). *In vivo* evaluations of a new thromboresistant polyurethane for artificial heart blood pumps. *J. Thorac. Cardiovasc. Surg.*, **95**, 191–200.

Frazier, O. H., Baldwin, R. T., Eskin, S. G., & Duncan, J. M. (1993). Immunochemical identification of human endothelial cells on the lining of a ventricular assist device. *Tex. Heart Inst. J.*, **20**, 78–82.

Frazier, O. H., Macris, M. P., Myers, T. J., Duncan, J. M., Radovancevic, B., Parnis, S. M., & Cooley, D. A. (1994). Improved survival after extended bridge to cardiac transplantation. *Ann. Thorac. Surg.*, **57**, 1416–1422.

Frazier, O. H., Myers, T. J., Gregoric, I. D., Khan, T., Delgado, R., Croitoru, M., Miller, K., Jarvik, R., & Westaby, S. (2002). Initial clinical experience with the Jarvik 2000 implantable axial-flow ventricular assist system. *Circulation*, **105**, 2855–2860.

Funk, M. (1991). Epidemiology of end-stage heart disease. In J. R. Hogness, & M. VanAntwerp (Eds.), *"The Artificial Heart: Prototypes, Policies, and Patients* (pp. 251–261). Washington, D.C: National Academy Press.

Golding, L., Medvedev, A., Massiello, A., Smith, W., Horvath, D., & Kasper, R. (1998). Cleveland Clinic continuous flow blood pump: Progress in development. *Artif. Organs*, **22**, 447–450.

Goldstein, D. J., Seldomridge, J. A., Chen, J. M., Catanese, K. A., DeRosa, C. M., Weinberg, A. D., Smith, C. R., Rose, E. A., Levin, H. R., & Oz, M. C. (1995). Use of aprotinin in LVAD recipients reduces blood loss, blood use, and perioperative mortality. *Ann. Thorac. Surg.*, **59**, 1063–1067.

Guy, T. S. (1998). Evolution and current status of the total artificial heart: The search continues. *ASAIO J.*, **44**, 28–33.

Hall, C. W., Liotta, D., Ghidoni, J. J., DeBakey, M. E., & Dressler, D. P. (1967). Velour fabrics applied to medicine. *J. Biomed. Mater. Res.*, **1**, 179–196.

Holman, W. L., Skinner, J. L., Waites, K. B., Benza, R. L., McGiffin, D. C., & Kirklin, J. K. (1999). Infection during circulatory support with ventricular assist devices. *Ann. Thorac. Surg.*, **68**, 711–716.

Holman, W. L., Pae, W. E., Teutenberg, J. J., Acker, M. A., Naftel, D. C., Sun, B. C., Milano, C. A., & Kirklin, J. K. (2009). INTERMACS: Interval analysis of registry data. *J. Am. Coll. Surg.*, **208**, 755–761.

Hosenpud, J. D., Bennett, L. E., Keck, B. M., Fiol, B., Boucek, M. M., & Novick, R. J. (1998). The registry of the International Society for Heart and Lung Transplantation: Fifteenth official report – 1998. *J. Heart Lung Transplant.*, **17**, 656–668.

Kerrigan, J. P., Yamazaki, K., Meyer, R. K., Mori, T., Otake, Y., Outa, E., Umezu, M., Borovetz, H. S., Kormos, R. L., Griffith, B. P., Koyanagi, H., & Antaki, J. F. (1996). High-resolution fluorescent particle-tracking flow visualization within an intraventricular axial flow left ventricular assist device. *Artif. Organs*, **20**, 534–540.

Kirklin, J. K., Naftel, D. C., Stevenson, L. W., Kormos, R. L., Pagani, F. D., Miller, M. A., Ulisney, K., & Young, J. B. (2008). INTERMACS database for durable devices for circulatory support: First annual report. *J. Heart Lung Transplant*, **27**, 1065–1072.

Kormos, R. L., Gasior, T. A., Kawai, A., Pham, S. M., Murali, S., Hattler, B. G., & Griffith, B. P. (1996). Transplant candidate's clinical status rather than right ventricular function defines need for univentricular versus biventricular support. *J. Thorac. Cardiovasc. Surg.*, **111**, 773–782.

Lee, J., Miller, P. J., Chen, H., Conley, M. G., Carpenter, J. L., Wihera, J. C., Jassawalla, J. S., & Portner, P. M. (1999). Reliability model from the *in vitro* durability tests of a left ventricular assist system. *ASAIO J.*, **45**, 595–601.

Livingston, E. R., Fisher, C. A., Bibidakis, E. J., Pathak, A. S., Todd, B. A., Furukawa, S., McClurken, J. B., Addonizio, V. P., & Jeevanandam, V. (1996). Increased activation of the coagulation and fibrinolytic systems leads to hemorrhagic complications during ventricular assist implantation. *Circulation*, **94**(suppl II). II-227-II-234.

Loebe, M., Muller, J., & Hetzer, R. (1999). Ventricular assistance for recovery of cardiac failure. *Curr. Opin. Cardiol.*, **14**, 234–248.

Long, J. W., Kfoury, A. G., Slaughter, M. S., Silver, M., Milano, C., Rogers, J., Delgado, R., & Frazier, O. H. (2005). Long-term destination therapy with the HeartMate XVE left ventricular assist device: Improved outcomes since the REMATCH study. *Congest Heart Fail*, **11**, 133–138.

Mann, D. L., & Willerson, J. T. (1998). Left ventricular assist devices and the failing heart: A bridge to recovery, a permanent assist device, or a bridge too far? *Circulation*, **98**, 2367–2369.

Massad, M. G., Cook, D. J., Schmitt, S. K., Smedira, N. G., McCarthy, J. F., Vargo, R. L., & McCarthy, P. M. (1997). Factors influencing HLA sensitization in implantable LVAD recipients. *Ann. Thorac. Surg.*, **64**, 1120–1125.

Massad, M. G., McCarthy, P. M., Smedira, N. G., Cook, D. J., Ratliff, N. B., Goormastic, M., Vargo, R. L., Navia, J., Young, J. B., & Stewart, R. W. (1996). Does successful bridging with the implantable left ventricular assist device affect cardiac transplantation outcome? *J. Thorac. Cardiovasc. Surg.*, **112**, 1275–1283.

McBride, L. R., Naunheim, K. S., Fiore, A. C., Moroney, D. A., & Swartz, M. T. (1999). Clinical experience with 111 Thoratec ventricular assist devices. *Ann. Thorac. Surg.*, **67**, 1233–1239.

Mehta, S. M., Aufiero, T. X., Pae, W. E., Jr., Miller, C. A., & Pierce, W. S. (1995). Combined registry for the clinical use of mechanical ventricular assist pumps and the total artificial heart in conjunction with heart transplantation: Sixth official report – 1994. *J. Heart Lung Transplant.*, **14**, 585–593.

Menconi, M. J., Pockwinse, S., Owen, T. A., Dasse, K. A., Stein, G. S., & Lian, J. B. (1995). Properties of blood-contacting surfaces of clinically implanted cardiac assist devices: Gene expression, matrix composition, and ultrastructural characterization of cellular linings. *J. Cell. Biochem.*, **57**, 557–573.

Miller, L. W., Pagani, F. D., Russell, S. D., John, R., Boyle, A. J., Aaronson, K. D., Conte, J. V., Naka, Y., Mancini, D., Delgado, R. M., MacGillivray, T. E., Farrar, D. J., & Frazier, O. H. (2007). Use of a continuous-flow device in patients awaiting heart transplantation. *N. Engl. J. Med.*, **357**, 885–896.

Olsen, D. B. (1999). Rotary blood pumps: A new horizon. *Artif Organs*, **23**, 695–696.

Pagani, F. D., Miller, L. W., Russell, S. D., Aaronson, K. D., John, R., Boyle, A. J., Conte, J. V., Bogaev, R. C., MacGillivray, T. E., Naka, Y., Mancini, D., Massey, H. T., Chen, L., Klodell, C. T., Aranda, J. M., Moazami, N., Ewald, G. A., Farrar, D. J., & Frazier, O. H. (2009). Extended mechanical circulatory support with a continuous-flow rotary left ventricular assist device. *J. Am. Coll. Cardiol.*, **54**, 312–321.

Rafii, S., Oz, M. C., Seldomridge, J. A., Ferris, B., Asch, A. S., Nachman, R. L., Shapiro, F., Rose, E. A., & Levin, H. R. (1995). Characterization of hematopoetic cells arising on the textured surface of left ventricular assist devices. *Ann. Thorac. Surg.*, **60**, 1627–1632.

Rose, E. A., Gelijns, A. C., Moskowitz, A. J., Heitjan, D. F., Stevenson, L. W., Dembitsky, W., Long, J. W., Ascheim, D. D., Tierney, A. R., Levitan, R. G., Watson, J. T., Meier, P., Ronan, N. S., Shapiro, P. A., Lazar, R. M., Miller, L. W., Gupta, L., Frazier, O. H., Desvigne-Nickens, P., Oz, M. C., & Poirier, V. L. (2001). Long-term mechanical left ventricular assistance for end-stage heart failure. *N. Engl. J. Med.*, **345**, 1435–1443.

Saito, S., Nishinaka, T., & Westaby, S. (2004). Hemodynamics of chronic nonpulsatile flow: Implications for LVAD development. *Surg. Clin. N. Am.*, **84**, 61–74.

Scheld, H. H., Soeparwata, R., Schmid, C., Loick, M., Weyand, M., & Hammel, D. (1997). Rupture of inflow conduits in the TCI-HeartMate system. *J. Thorac. Cardiovasc. Surg.*, **114**, 287–289.

Schmid, C., Weyand, M., Nabavi, D. G., Hammel, D., Deng, M. C., Ringelstein, E. B., & Scheld, H. H. (1998). Cerebral and systemic embolization during left ventricular support with the Novacor N100 device. *Ann. Thorac. Surg.*, **65**, 1703–1710.

Schocken, D. D., Arrieta, M. I., Leaverton, P. E., & Ross, E. A. (1992). Prevalence and mortality rate of congestive heart failure in the United States. *J. Am. Coll. Cardiol.*, **20**, 301–306.

Sharp, W. V., Finelli, A. F., Falor, W. H., & Ferraro, J. W., Jr. (1964). Latex vascular prosthesis: Patency rate and neo-intimization related to prosthesis lining and electrical conductivity. *Circulation*, **29**(suppl), 165–170.

Simon, M. A., Kormos, R. L., Murali, S., Nair, P., Heffernan, M., Gorcsan, J., Winowich, S., & McNamara, D. M. (2005). Myocardial recovery using ventricular assist devices: Prevalence, clinical characteristics, and outcomes. *Circulation*, **112**, I32–I36.

Slater, J. P., Rose, E. A., Levin, H. R., Frazier, O. H., Roberts, J. K., Weinberg, A. D., & Oz, M. C. (1996). Low thromboembolic risk without using advanced-design left ventricular assist devices. *Ann. Thorac. Surg.*, **62**, 1321–1328.

Spanier, T., Oz, M., Levin, H., Weinberg, A., Stamatis, K., Stern, D., Rose, E., & Schmidt, A. M. (1996). Activation of coagulation and fibrinolytic pathways in patients with left ventricular assist devices. *J. Thorac. Cardiovasc. Surg.*, **112**, 1090–1097.

Spanier, T. B., Chen, J. M., Oz, M. C., Stern, D. M., Rose, E. A., & Schmidt, A. M. (1999). Time-dependent cellular population of textured-surface left ventricular assist devices contributes to the development of a biphasic systemic procoagulant response. *J. Thorac. Cardiovasc. Surg.*, **118**, 404–413.

Sun, B. C., Catanese, K. A., Spanier, T. B., Flannery, M. R., Gardocki, M. T., Marcus, L. S., Levin, H. R., Rose, E. A., & Oz, M. C. (1999). 100 long-term implantable left ventricular assist devices: The Columbia Presbyterian interim experience. *Ann. Thorac. Surg.*, **68**, 688–694.

Thom, T., Haase, N., Rosamond, W., Howard, V. J., Rumsfeld, J., et al. (2006). Heart disease and stroke statistics – 2006 update: A report from the American Heart Association Statistics Committee and Stroke Statistics Subcommittee. *Circulation*, **113**, e85–151.

U.S. Department of Health and Human Services. (2008). *2008 Annual report of the U.S. Organ Procurement and Transplantation Network and the Scientific Registry of Transplant Recipients: Transplant data 1998-2007.*. Rockville, MD: Health resources and Services Administration, Healthcare Systems Bureau, Division of Transplantation.

United Network for Organ Sharing. (1999). Transplant Patient Data Source.

Wagner, W. R., Johnson, P. C., Kormos, R. L., & Griffith, B. P. (1993). Evaluation of bioprosthetic valve-associated thrombus in ventricular assist device patients. *Circulation*, **88**, 2023–2029.

Wampler, R., Lancisi, D., Indravudh, V., Gauthier, R., & Fine, R. (1999). A sealless centrifugal blood pump with passive magnetic and hydrodynamic bearings. *Artif Organs*, **23**, 780–784.

Wang, I. W., Kottke-Marchant, K., Vargo, R. L., & McCarthy, P. M. (1995). Hemostatic profiles of HeartMate ventricular assist device recipients. *ASAIO J.*, **41**, M782–M787.

Weiss, W. J., Rosenberg, G., Snyder, A. J., Pierce, W. S., Pae, W. E., Kuroda, H., Rawhouser, M. A., Felder, G., Reibson, J. D., Cleary, T. J., Ford, S. K., Marlotte, J. A., Nazarian, R. A., & Hicks, D. L. (1999). Steady state hemodynamic and energetic characterization of the Penn State/3M Health Care Total Artificial Heart. *ASAIO J.*, **45**, 189–193.

Westaby, S., Katsumata, T., Houel, R., Evans, R., Pigott, D., Frazier, O. H., & Jarvik, R. (1998). Jarvik 2000 Heart: Potential for bridge to myocyte recovery. *Circulation*, **98**, 1568–1574.

Wilhelm, C. R., Ristich, J., Knepper, L. E., Holubkov, R., Wisniewski, S. R., Kormos, R. L., & Wagner, W. R. (1999). Measurement of hemostatic indices in conjunction with transcranial doppler sonography in ventricular assist device patients. *Stroke*, **30**, 2554–2561.

CHAPTER II.5.4 ARTIFICIAL CELLS

Thomas Ming Swi Chang
Director, Artificial Cells and Organs Research Centre, Departments of Physiology, Medicine and Biomedical Engineering, Faculty of Medicine, McGill University, Montreal, QC, Canada

BASIC FEATURES OF ARTIFICIAL CELLS

The initial research on artificial cells (Chang, 1964) forms the basic principle that has been extended for use in many areas by many groups (Chang, 2005a, 2007, 2010; Liu and Chang, 2010). Very brief examples of some of these basic features include (Figure II.5.4.1):

1. Artificial cells can contain the same biological material as biological cells. In addition, they are more versatile since adsorbents, magnetic materials, drugs, cells, stem cells, enzymes, multienzyme systems, hemoglobin, microorganisms, vaccines, gene-for-gene therapy, genetically engineered cells, hormones, peptides, and many other materials can also be included separately or in combination (Figure II.5.4.1).
2. The membrane of an artificial cell separates its content from the outside, but at the same time the membrane can be prepared to selectively allow different types of molecules to cross (Figure II.5.4.1). For example, one can prepare artificial cell membranes that selectively allow the movement of molecules according to molecular size, lipid solubility, affinity to carrier mechanisms, etc. By selecting the proper membrane material, the permeability can range from a membrane that does not allow any molecules to cross, to those that allow even very large molecules such as proteins to cross. The membrane material includes polymer, biodegradable polymer, lipid, cross-linked protein, lipid–polymer complex, lipid–protein complex, and membrane with transport carriers.
3. Surface properties of artificial cell membrane can be varied by: (1) incorporation of negative or positive charge; (2) incorporation of albumin to increase blood compatibility; (3) incorporation of antigens to bind antibodies or antibodies to bind antigen; (4) incorporation of polysaccharide such as heparin or polyethylene glycol (PEG) to increase compatibility or retention time in circulation.
4. In addition to being of cellular dimensions in the micron range, they can also be in the macro range, in the nano range or in the nanobiotechnological range (Figure II.5.4.1).
5. The artificial cell membranes can be ultrathin and yet strong. There is a large surface area to volume relationship. For example, 10 ml of 20 μm diameter artificial cells has a total surface area of 2500 cm^2, which is the same as that in an artificial kidney machine. Since the artificial cell membrane is also 100 times thinner, permeant molecules can potentially move across 10 ml of 20 μm diameter artificial cells 100 times faster than across the artificial kidney machine (Chang, 1966). In addition, the microscopic size of artificial cells allows material to diffuse rapidly inside the artificial cells.

RESEARCH INTO THE APPLICATIONS OF ARTIFICIAL CELLS

This includes hemoperfusion, immunosorbents, drug delivery, blood substitutes, enzyme therapy, cell and stem cell therapy, biotechnology and nanobiotechnology, nanomedicine, regenerative medicine, agriculture, industry, aquatic culture, nanocomputers, nanorobotics, nanosensors, and other areas (Table II.5.4.1). The following is a brief overview of some examples of artificial cells.

ARTIFICIAL CELLS IN HEMOPERFUSION

This is the first large-scale US Food and Drug Administration (FDA) approved routine clinical application in patients based on artificial cells. As mentioned above, the microscopic dimensions of artificial cells result in a large surface-to-volume relationship. This, together with the ultrathin membranes, allows artificial cells containing bioadsorbents to be much more effective when compared to standard hemodialysis in removing toxins and drugs from the blood of patients (Chang, 1966, 1969, 1972a, 1975, 2007; Chang et al., 1973). In this method, thousands of 90 micron diameter artificial cells containing adsorbents are retained by screens at either end, inside a column the size of a tea cup (Figure II.5.4.2). Blood from the patients containing toxins or drug can perfuse through the screens to come into contact with the

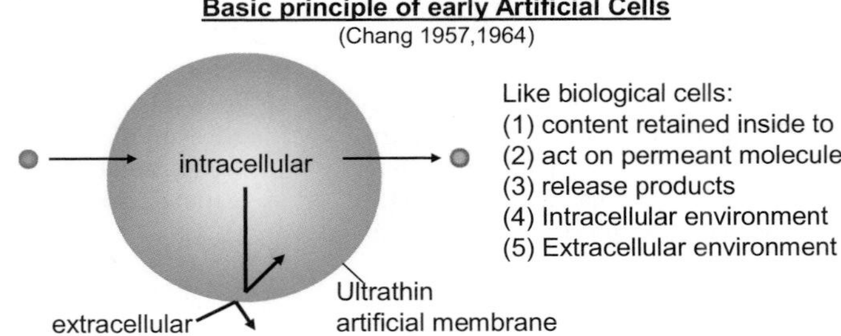

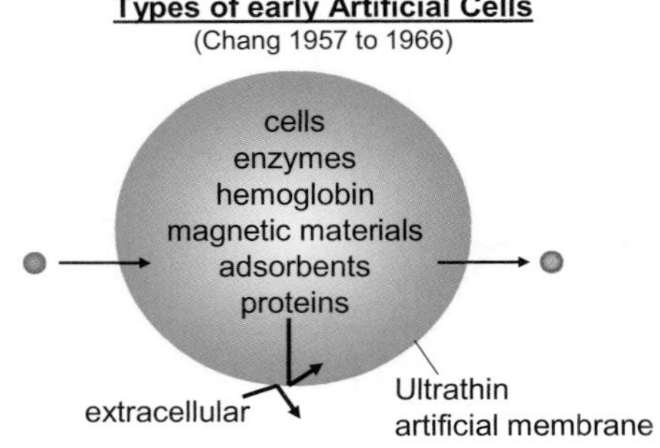

Present status of Artificial Cells
(Chang 2005 Nature Rev Drug Disc)

Contents
Cells
Stem cells
Enzymes
Multienzyme systems
hemoglobin
magnetic materials
Microorganism
vaccines
gene for gene therapy
genetically engineered cells
biotechnological products
adsorbents
drug release
hormones
peptides
proteins

Membrane:
Polymeric
Biodegradable
Lipid
Xlinked protein
Conjugated
Carriers
etc

dimensions
Macro
Micron
Nano

FIGURE II.5.4.1 Upper: Basic principle of early artificial cells. Center: Different types of early artificial cells based on this basic principle. Lower: Present status of artificial cells with wide variations in contents, membrane material, and dimensions. (Figure from Chang (2007), Monograph with permission from World Scientific Publisher.)

artificial cells. Toxins or drugs diffusing into the artificial cells are removed by the adsorbents inside artificial cells. This cleanses the blood that returns to the patient.

Routine Clinical Uses in Patients with Suicidal Drug Overdose

The most common routine use of this approach is the use of microscopic polymeric artificial cells encapsulating activated charcoal (Chang, 1969, 1975, 2007; Chang et al., 1973) (Figure II.5.4.2). This solves the major problems of release of embolizing particles and damage to blood cells when bioadsorbents are used without artificial cell membranes. This was first successfully used in suicidal overdose patients (Chang et al., 1973) (Figure II.5.4.3). Since then, this has been become a routine treatment worldwide for acute poisoning in adults and children, especially in suicidal overdose (Chang, 1975; Chang et al., 1973; Winchester, 1996; Kawasaki et al., 2000; Lopez Lago et al., 2002; Lin et al., 2002, 2004; Singh et al., 2004; Peng et al., 2004). This is particularly useful in places where dialysis machines are not

TABLE II.5.4.1 Artificial Cell: Applications

Artificial Cell: Applications
• Hemoperfusion
• Immunosorbents
• Drug delivery
• Blood substitutes
• Enzyme and gene therapy
• Cell and stem cell therapy
• Biotechnology and nanobiotechnology
• Nanomedicine
• Regenerative medicine
• Agriculture, industry, aquatic culture
• Nanocomputers and nanorobotics
• Nanosensors, etc.

Chang, 2005a
Chang, 2007
Chang, 2009
www.artcell.mcgill.ca

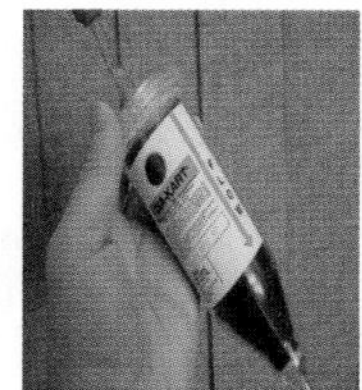

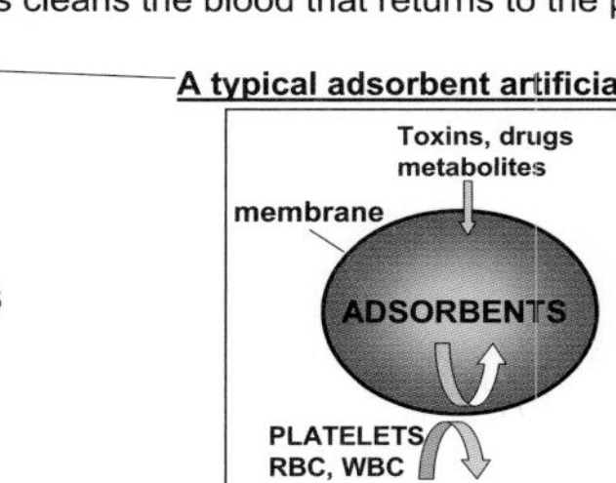

FIGURE II.5.4.2 Upper right: A hemoperfusion device held in the hand. Left: Schematic representation of the hemoperfusion device. Right: Schematic representation of an artificial cell containing adsorbent. (Figure from Chang (2007), with copyright permission from World Scientific Publisher.)

easily available, and the hemoperfusion devices are less costly, especially outside North America. Hemoperfusion has saved the lives of thousands of suicidal drug overdose patients in some countries outside North America.

Immunosorbents

The success in the clinical uses of artificial cells containing bioadsorbents for detoxification has led to increasing interest in research and development in many other

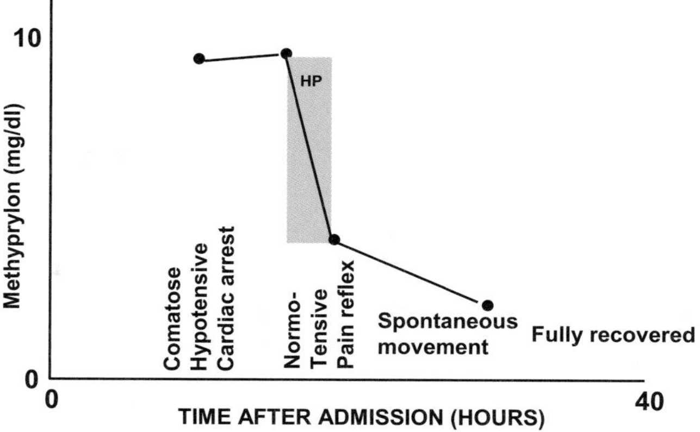

FIGURE II.5.4.3 Clinical and laboratory results of hemoperfusion in a patient with severe suicidal methyprylon overdose. (Figure from Chang (2007), with copyright permission from World Scientific Publisher.)

areas. One of these is in artificial cells containing immunoadsorbents (Chang, 1980).

Albumin can bind tightly to the ultrathin collodion membrane of adsorbent artificial cells (Chang, 1969). This was initially used to increase the blood compatibility of the adsorbent artificial cells for hemoperfusion (Chang, 1969). We also applied this albumin coating to synthetic immunosorbents resulting in blood compatible synthetic blood group immunoadsorbents (Chang, 1980). This albumin-coated synthetic adsorbent has been applied clinically for removing blood group antibodies from plasma for bone marrow transplantation (Bensinger et al., 1981). In addition, albumin-coated collodion activated charcoal (ACAC) was found to effectively remove antibodies to albumin in animal studies (Terman et al., 1977). This has become the basis of one line of research in which other types of antigens or antibodies are applied to the collodion coating of the activated charcoal to form immunoadsorbents. Other immunoadsorbents based on this principle have also been developed for the treatment of human systemic lupus erythrematosus, removal of antiHLA antibodies in transplant candidates, treatment of familial hypercholesterolemia with monoclonal antibodies to low-density lipoproteins, and other conditions (Terman et al., 1979; Hakim et al., 1990; Wingard et al., 1991; Yang et al., 2004).

NANOBIOTECHNOLOGY FOR PARTIAL ARTIFICIAL RED BLOOD CELLS

There is much recent interest in nanobiotechnology and nanomedicine. This is a large and complex area that embraces many diverse approaches. One of these is to make the original artificial cells smaller, using the same basic principle and methods. This includes biodegradable nanoparticles, nanosphere, and nanocapsules. Examples include nano-artificial red blood cells with lipid membrane or biodegradable polymeric membranes. A later section will summarize other examples used in drug delivery systems. Much smaller nanobiotechnological complexes can be prepared by the assembling of biological molecules. Examples include the assembling of hemoglobin molecules into soluble polyhemoglobin, and the assembling of hemoglobin molecules and enzymes into soluble polyhemoglobin–enzyme complexes.

Nanobiotechnology and Artificial Cells

Nanobiotechnology is the assembling of biological molecules into nanodimension complexes of different configurations. These include nanostructures with nano-range diameters, membranes with nanodimension thickness or nanotubules with nanodimension diameters.

The first nanobiotechnology approach reported is the cross-linking of hemoglobin into an ultrathin polyhemoglobin (PolyHb) membrane with nanodimension thickness (Chang, 1964, 1965) (Figure II.5.4.4). This is used to form the membrane of artificial red blood cells (Chang, 1964, 1965, 1972a). If the emulsion is made very small, then the whole submicron artificial cells can be cross-linked into PolyHb of nanodimension. Glutaraldehyde can cross-link hemoglobin into soluble PolyHb of nanodimension, each consisting of an assembly of 4–5 hemoglobin molecules (Chang 1972b; Keipert and Chang, 1985, 1987) (Figure II.5.4.4).

Two groups have independently developed this 1971 (Chang, 1971b) basic method of glutaraldehyde cross-linking for clinical use. One is glutaraldehyde human PolyHb (PolyHb) (Gould et al., 1998, 2002). Their phase III clinical trial shows that this can replace blood lost in trauma surgery by keeping the blood hemoglobin at an acceptable level. More recently they have carried out clinical trials in prehospital ambulance patients (Moore et al., 2009). The second PolyHb is glutaraldehyde-cross-linked bovine PolyHb which has been tested in phase III clinical trials (Pearce and Gawryl, 1998; Sprung et al., 2002; Pearce et al., 2006; Jahr et al., 2008). South Africa has approved this for routine clinical use in patients. Unlike red blood cells, there is no blood group in PolyHb, and thus PolyHb can be given on the spot, without waiting for typing and cross-matching in the hospital. PolyHb is also free from infective agents such as HIV, hepatitis C, bacteria, parasites, and so on. Furthermore, whereas donor blood has to be stored at 4°C and is only good for 42 days, PolyHb can be stored at room temperature for more than one year. Thus, PolyHb can have important uses in a number of clinical conditions, notably for surgery when there is no potential for ischemia-reperfusion, as will be discussed below (Chang, 2006a, 2009).

Nanobiotechnology and the Assembling of Hemoglobin with Antioxidant Enzyme

As PolyHb can be kept at room temperature and used immediately, it can have potential for treating severe bleeding (hemorrhagic shock). However, this has to be done as soon as possible, since if there is much delay PolyHb alone might result in the production of oxygen radicals that cause tissue injury (ischemia-reperfusion injuries). Antioxidant enzymes normally present in red blood cells are insufficient to prevent this problem. We use glutaraldehyde cross-linking to assemble a soluble nanobiotechnology complex by cross-linking hemoglobin, superoxide dismutase, and catalase into PolyHb-SOD-CAT (D'Agnillo and Chang, 1998; Chang 2006b, 2007, 2008) (Figure II.5.4.4). This way, one can increase the antioxidant enzymes to much higher levels than those in red blood cells.

Obstruction of arteries due to clots or other causes can result in stroke (cerebral ischemia) or heart attack (myocardial infarction). As it is a solution, PolyHb can more easily perfuse partially obstructed vessels. However, if there is prolonged lack of oxygen, reperfusion with PolyHb alone may result in damaging oxygen radicals resulting in ischemia-reperfusion injuries. Thus, in patients with coronary heart disease and myocardial ischemia, infusion with polyhemoglobin can result in ischemia-reperfusion injury that can be severe enough to result in heart attack (myocardial infarction). In severe trauma, one is often faced with hemorrhagic shock from severe bleeding, in addition to cerebral ischemia due to head injuries. In a hemorrhagic shock and stroke rat model, after 60 minutes of ischemia, reperfusion with PolyHb resulted in significant increase in the breakdown of the blood–brain barrier, and an increase in brain water (brain edema) (Powanda and Chang, 2002). On the other hand, PolyHb-SOD-CAT did not result in these adverse changes (Powanda and Chang, 2002). In sustained hemorrhagic shock due to severe loss of blood, ischemia-reperfusion can result in irreversible shock due to leakage of bacteria and endotoxins from the intestine into the circulating blood. Our study shows that in an ischemia-reperfusion rat inestinal model, polyhemoglobin resulted in significant increases in oxygen radicals. On the other hand, the use of PolyHb-SOD-CAT did not result in any significant increases in oxygen radicals (Razack et al., 1997; Chang, 2007).

Nanobiotechnology for the Assembling of Hemoglobin with Other Enzymes

The microcirculation structure in tumors is abnormal, and as a result there is a decrease in perfusion by oxygen carrying red blood cells. This results in a lower oxygen tension in the tissue of many types of tumors (Pearce and Gawryl, 1998). It is known that radiation therapy and some chemotherapy can work better when the tissue oxygen tension is higher. PolyHb can more easily perfuse the abnormal microcirculation of tumors to supply oxygen needed for chemotherapy or radiation therapy. With a circulation half-time of 24 hours, the effect is not long-term and can be adjusted to the duration of chemotherapy or radiation therapy. When used together with chemotherapy, PolyHb decreases tumor growth and increases the lifespan in a rat model of gliosarcoma brain tumor

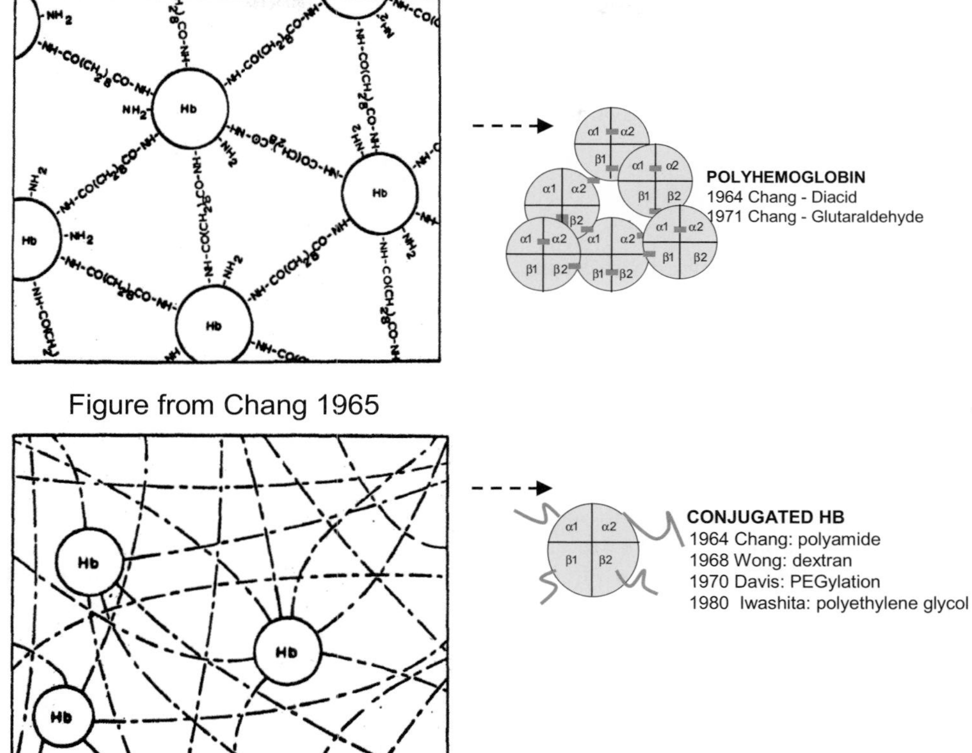

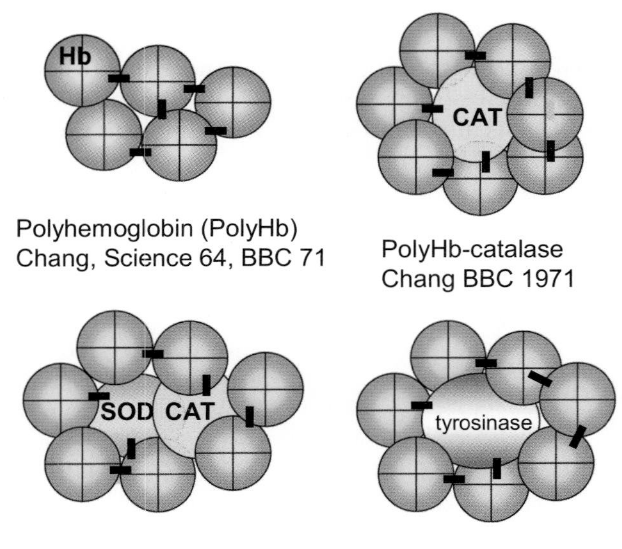

FIGURE II.5.4.4 Nanoartificial cells can be prepared in the nanodimensions as membrane enclosed nanoartificial cells or by the use of nanobiotechnology to assemble biological molecules together into nanodimension structures. Upper: An example of assembling of biological molecules to form PolyHb and conjugated Hb. Lower: Examples of different types of nanobiotechnology based PolyHb-enzymes. (Figure from Chang (2007), with copyright permissions from World Scientific Publisher.)

(Pearce and Gawryl, 1998). We have recently cross-linked tyrosinase with hemoglobin to form a soluble PolyHb-tyrosinase complex (Yu and Chang, 2004) (Figure II.5.4.4). This has the dual function of supplying the needed oxygen, and at the same time lowering the systemic levels of tyrosine needed for the growth of melanoma. Intravenous injections delayed the growth of the melanoma without causing adverse effects in the treated animals (Yu and Chang, 2004).

Polyhemoglobin-Fibrinogen

Blood is a multifunctional fluid. As a blood substitute, PolyHb is limited by its lack of platelets and/or coagulation properties. In situations of high blood volume loss, large volumes of PolyHb have to be infused to replace the lost blood. Replacing the loss of red blood cells without replacing the lost platelets and coagulation factors can result in the inability of the blood at the injured sites to clot, resulting in continued severe bleeding. As a result, platelets and coagulation factors have to be infused in these situations. Unfortunately, donor platelets are very hard to obtain. There has been development in platelet substitutes to combat thrombocytopenia, but with limited success. We therefore prepared a novel blood substitute that is an oxygen carrier with platelet-like activity. This is formed by cross-linking fibrinogen to hemoglobin to form polyhemoglobin-fibrinogen (PolyHb-Fg) (Wong and Chang, 2007). This was studied and compared to PolyHb for its effect on coagulation, both *in vitro* and *in vivo*. In the *in vitro* experiments PolyHb-Fg showed similar clotting times as whole blood, whereas PolyHb showed significantly higher clotting times. This result was confirmed in *in vivo* experiments using an exchange transfusion rat model. Using PolyHb, exchange transfusion of 80% or more increased the normal clotting time (1–2 mins) to >10 minutes. Partial clots formed with PolyHb did not adhere to the tubing wall. With PolyHb-Fg, a normal clotting time (1–2 mins) is maintained even with 98% exchange transfusion.

NANOBIOTECHNOLOGY FOR COMPLETE ARTIFICIAL RED BLOOD CELLS

Original Micron Dimension Complete Artificial Red Blood Cells

The first artificial red blood cells (RBCs) prepared in 1957 (Chang, 1957) have an oxygen dissociation curve similar to red blood cells. We continued to study a number of artificial cell membranes including cellulose, silicone rubber, 1,6-hexamethylenediamine, cross-linked protein, phospholipid–cholesterol complexes on cross-linked protein membrane or polymer (Chang, 1964, 1965, 1972a). However, these artificial RBCs, even with a diameter down to one micron, survived for a short time in the circulation after intravenous injection. Our study showed that the long circulation time of RBCs is due to the presence of neuraminic acid on the membrane (Chang, 1965, 1972a). This led us to study the effects of changing surface properties of the artificial RBCs (Chang, 1964, 1965, 1972a). This has resulted in significant increases in circulation time, but is still insufficient for clinical use.

Submicron Hemoglobin Lipid Vesicles as Artificial Red Blood Cells

We prepared larger artificial cells with lipid membrane by supporting the lipid in the form of lipid–protein membrane and lipid–polymer membrane (Chang, 1972a). Others later reported the preparation of submicron (0.2 micron) diameter artificial RBCs using lipid membrane vesicles to encapsulate Hb (Djordjevich and Miller, 1980). This increased the circulation time significantly, although it was still rather short. Many investigators have since carried out research to improve the preparation and circulation time. The most successful approach to improve the circulation time is to incorporate polyethylene glycol (PEG) into the lipid membrane of artificial RBCs, resulting in a circulation half-time of more than 30 hours (Philips et al., 1999). Another group in Japan has for many years carried out extensive research and development, commercial development, and preclinical animal studies. Their extensive studies are available in many publications and reviews (Kobayashi et al., 1997; Sakai et al., 2004; Tsuchida, 1998).

Biodegradable Polymeric Nanodimension Completely Artificial Red Blood Cells

We have used a biodegradable polymer, polylactic acid (PLA), for the microencapsulation of Hb, enzymes, and other biologically active material since 1976 (Chang, 1976). More recently, we started to prepare artificial RBCs of 100 nanometer mean diameter using PLA, PEG–PLA membrane, and other biodegradable polymers (Yu and Chang, 1994, 1996; Chang, 1997a, 2006c, 2007; Chang et al., 2003). A typical electron micrograph for the biodegradable polymer Hb nanocapsules prepared with d,l-PLA shows that they are spherical and homogeneous. Their diameter ranges from 40–120 nm, with a mean diameter of 80 nm. The membrane thickness is 5–15 nm. We have replaced most of the 6 g/dl of lipid membrane in Hb lipid vesicles with 1.6 g/dl of biodegradable polymeric membrane material (Figures II.5.4.5 and II.5.4.6). This marked decrease in the lipid component would lessen the effects on the reticuloendothelial system (RES), and lessen lipid peroxidation in ischemia-reperfusion.

We can increase the Hb content in the PLA nano-artificial cell suspension from 3 g/dl to 15 g/dl (the same as whole blood). This has normal P_{50}, Hill's coefficient, and Bohr coefficient. Nanocapsules can be prepared

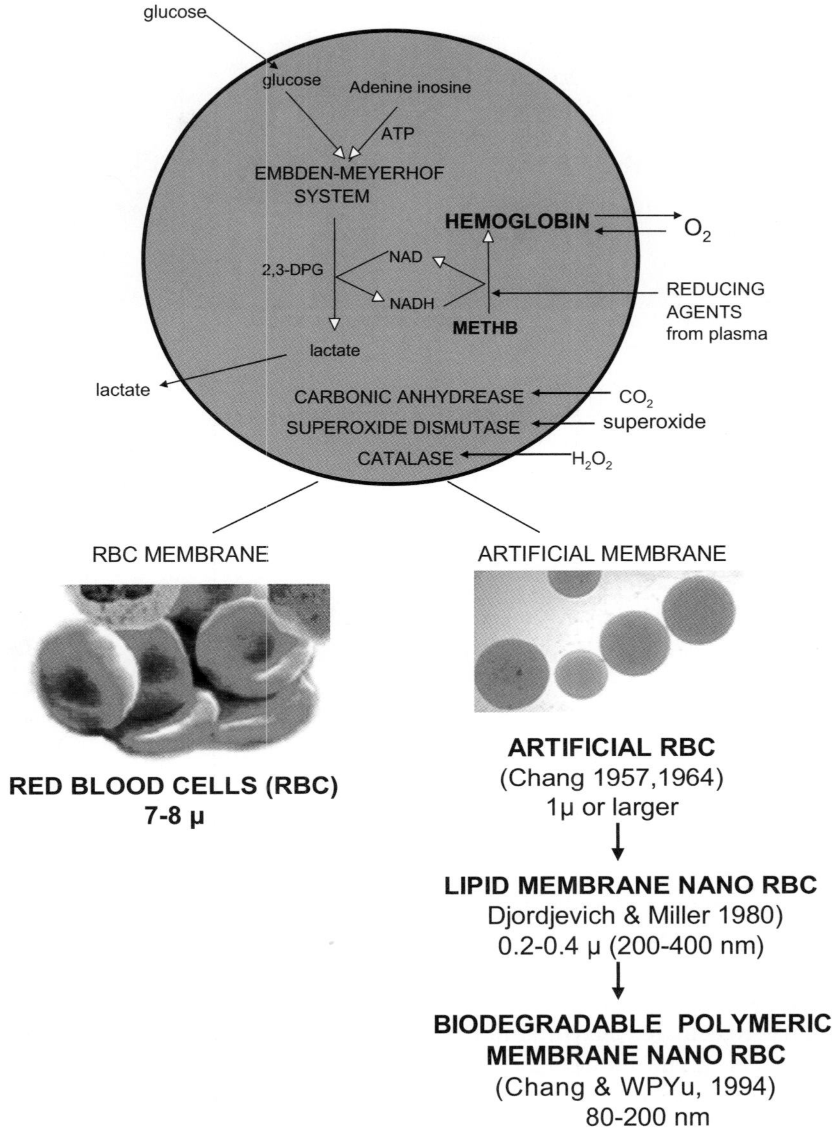

FIGURE II.5.4.5 Upper: Composition of red blood cell and artificial red blood cells. Lower Left: Red blood cells. Lower Right: First artificial RBC of 1 micron or larger diameter, first lipid membrane nanodimension artificial RBC, first nanodimension biodegradable polymeric membrane artificial RBC. (Figure from Chang (2007), with copyright permissions from World Scientific Publisher.)

with up to 15 g/dl Hb concentration. The preparation of PLA nano-artificial RBCs does not have adverse effects on the Hb molecules. This is shown by the following experimental results. There is no significant difference in the oxygen affinity (P_{50}) of Hb nano-artificial RBC and the original Hb used for the preparation, the Hill coefficient is 2.4 to 2.9 and the Bohr effect is −0.22 to −0.24 (Yu and Chang, 1994, 1996; Chang, 1997a, 2006c, 2007; Chang et al., 2003).

We have earlier carried out basic research on artificial cells containing multienzyme systems with cofactor recycling (Chang, 1987). A number of enzymes such as carbonic anhydrase, catalase, superoxide dismutase and the MetHb reductase system, normally present in RBC have been encapsulated within nano-artificial RBC and retain their activities (Figure II.5.4.5) (Chang et al., 2003; Chang, 2007). Unlike lipid membrane, biodegradable polymeric membrane is permeable to glucose. Thus, the inclusion of an RBC MetHb reductase enzyme system prevents MetHb formation even at 37°C, and we can also convert MetHb to Hb at 37°C. In addition, unlike lipid membrane, the nanocapsule membrane allows plasma factors such as ascorbic acid to enter the nanocapsules to prevent MetHb formation (Figure II.5.4.5).

In order to increase the circulation time, we synthesized a new PEG–PLA copolymer for the artificial RBC membrane (Chang et al., 2003). After a 30% blood volume toploading using PolyHb (10 g/dl), the best PolyHb can only attain a maximal Hb concentration of 3.35 g/dl. The best PEG–PLA nano-artificial RBC,

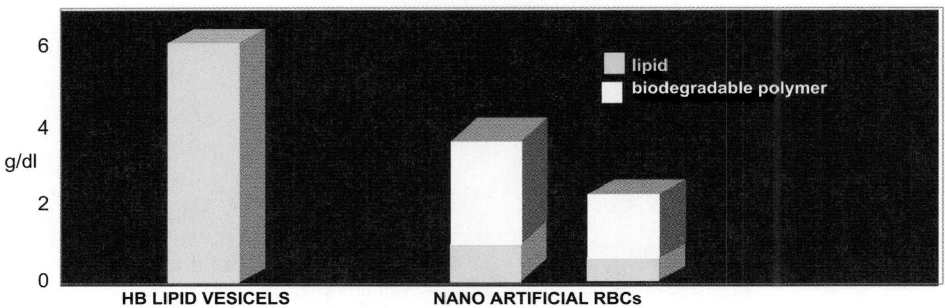

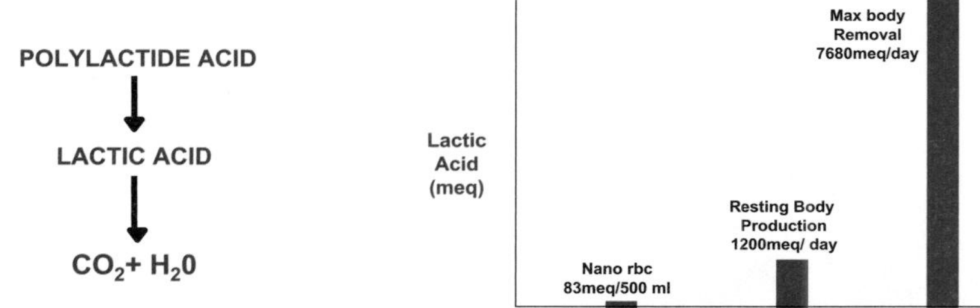

FIGURE II.5.4.6 Top: Amount of membrane material in hemoglobin lipid vesicles compared to polylactide membrane nano RBC. Bottom: Fate of polylactide membrane in PLA nano RBC compared to PLA metabolism in the body. (Figure from Chang (2007), with copyright permission from World Scientific Publisher.)

on the other hand, can reach a maximal Hb concentration of 3.60 g/dl. After extensive research, we now have a circulation time in rats that is double that of glutaraldehye cross-linked PolyHb. Since the RES in rat is much more efficient in removing particulate matter compared to humans, it is likely that the half-time would be longer in humans. Recent long-term studies in rats show no adverse effects in the kidney and the reticuloendothelial systems (liver and spleen) (Liu and Chang, 2008).

CELLS, ISLETS, STEM CELLS, GENETICALLY ENGINEERED CELLS, AND MICROORGANISMS

Introduction

The first artificial cells containing intact biological cells were reported in 1964 based on a drop method (Chang, 1964), and it was proposed that: "protected from immunological process, encapsulated endocrine cells might survive and maintain an effective supply of hormone" (Chang et al., 1966) (Figure II.5.4.1).

Artificial Cells Containing Islets, Hepatocytes, and Other Cells

With Chang's encouragement and initial consultation, Sun from Conaught Laboratory and his collaborator developed Chang's original drop method (Chang, 1964, 1965, 1972a; Chang et al., 1966) by changing to alginate-polylysine-alginate (APA) as the artificial cell membrane (Lim and Sun, 1980). They showed that after implantation, the insulin secreting islets inside artificial cells indeed remained viable and continued to secrete insulin to control the glucose levels of diabetic rats (Lim and Sun, 1980; Sun et al., 1996). Extensive research has been carried out by many laboratories around the world since that time and is available in reviews (Calafiore, 1999; De Vos et al., 2002; Orive et al., 2003, 2004; Hunkeler, 2003; Chang, 2005a,b, 2007). The major hurdle is the need for long-term function after implantation.

We have been studying the use of artificial cells containing liver cells (hepatocytes) for liver support. Implanting these increases the survival of acute liver failure rats (Wong and Chang, 1986); lowers the high bilirubin level in congenital Gunn rats (Bruni and Chang, 1989); and prevents xenograft rejection (Wong and Chang, 1988). We developed a two-step cell encapsulation method to improve the APA method, resulting in improved survival of implanted cells (Wong and Chang, 1991). Artificial cells containing hepatocytes effectively lower the systemic bilirubin in hyperbilirubinemia Gunn rats (Bruni and Chang, 1989, 1995). We used the two-step method plus the co-encapsulation of stem cells and hepatocytes into artificial cells (Liu and Chang, 2000). This results in further increases in the viability of encapsulated hepatocytes, both in culture and after implantation (Liu and Chang, 2002). One implantation of artificial cells containing both hepatocytes and stem cells into Gunn rats lowers the systemic bilirubin levels and maintains this low level for two months (Liu and Chang, 2003). Without stem cells, implanted hepatocytes in artificial cells can only maintain a low level for one month.

Genetically Engineered Cells and Microorganisms

Many groups have carried out extensive research on artificial cells containing genetically engineered cells. This is a very important area which includes potential applications in amyotrophic lateral sclerosis, Dwarfism, pain treatment, IgG_1 plasmacytosis, Hemophilia B, Parkinsonism and axotomized septal cholinergic neurons, tumor suppression, and other areas (Winn et al., 1994; Saitoh et al., 1995; Aebischer et al., 1996; Al-Hendy et al., 1996; Basic et al., 1996; Tan et al., 1996; Hagihara et al., 1997; Dalle et al., 1999; Bachoud-Lévi et al., 2000; Lorh et al., 2001; Xu et al., 2002; Cirone et al., 2002; Bloch et al., 2004). To avoid the need for implantation, we studied the oral use of artificial cells containing microorganisms for the removal of cholesterol (Garofalo and Chang, 1991) or conversion of phenol to tyrosine (Lloyd-George and Chang, 1995). We have also studied the use of genetically engineered nonpathogenic *E. coli* DH5 cells to lower systemic urea in renal failure rats (Prakash and Chang, 1993, 1996; Chang, 1997b, 2005a,b). Genetically engineered *E. coli* allows for proof of principle. For safe use, we have studied the use of a modified *Lactobacillus*, similar to those used in yoghurt, for enclosure in artificial cells to remove urea (Chow et al., 2003).

General

Cell bioencapsulation for cell therapy has been extensively developed by many groups, especially using artificial cells containing endocrine tissues, hepatocytes, genetically engineered cells, and stem cells (Chang, 1972a; Orive et al., 2003; Chang, 2005a,b, 2007). The major hurdle is the need to develop systems that can function on a long-term basis after implantation.

ARTIFICIAL CELLS CONTAINING STEM CELLS IN REGENERATIVE MEDICINE

The above examples of artificial cells containing islets, cells, genetically engineered cells, and other cells require long-term treatment. Developments are needed before artificial cells are ready for long-term implantation use. To allow for more immediate use, one area of research in this laboratory is to look at the use of artificial cells in regenerative medicine. For this we have been studying the use of artificial cells containing bone marrow stem cells for liver regeneration. Marked decrease in liver function due to fulminant hepatic failure or extensive liver resection for metastatic cancer or severe injury can result in the death of the patient. Since liver can regenerate if given the required time and conditions, we therefore studied whether artificial cells containing bone marrow stem cells can maintain the experimental animal alive long enough to allow the liver to regenerate and the animal to recover (Figure II.5.4.7).

Artificial Cells Containing Bone Marrow Stem Cells in 90% Hepatectomized Rats

In the present study we injected intraperitoneally artificial cells containing bone marrow stem cells into 90% hepatectomized rats (Liu and Chang, 2005, 2006) (Figure II.5.4.7). In the hepatectomized groups that received no treatment, only 20% of the animals survived by the second day. In the hepatectomized animals receiving free bone marrow stem cells, only 20% of the animals survived on the seventh day. There was no significant difference in survival rate between the normal control group (with no hepatectomy) and the group receiving artificial cells containing bone marrow stem cells. The survival for both groups was 100% when followed for 14 days. Thus it would appear that in rats, artificial cells containing bone marrow stem cells are effective in improving the survival rates of 90% hepatectomized rats. For those rats that survived at week 2 post hepatectomy, the remnant livers were removed and the weights were measured. In the 90% hepatectomized rats that received artificial cells containing bone marrow stem cells, the liver wet weights had recovered to the same size as that of the normal control group (Liu and Chang, 2006). On the other hand, those few animals in the other groups that survived had very low liver weights.

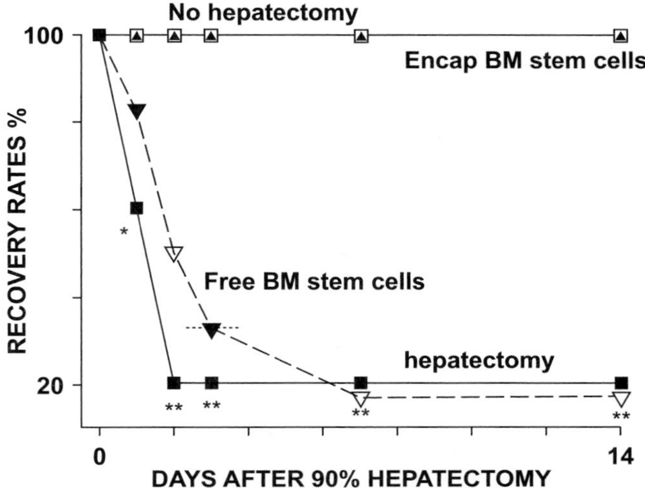

FIGURE II.5.4.7 Survival rates of rats with 90% of liver surgically removed, hepatectomy, compared to those without removal of liver tissues, no hepatectomy. One peritoneal injection of artificial cells containing bone marrow stem cells resulted in the survival of the 90% hepatectomized rats. On the other hand, free bone marrow stem cells did not significantly increase the survival rates (Liu and Chang, 2006). (Figure from Chang (2007), with permission from World Scienctific Publisher.)

Plasma Hepatic Growth Factor (HGF) Levels

In the hepatectomized groups that received artificial cells containing bone marrow stem cells, the blood levels of HGF peaked at day 2 and 3 post-surgery, and were significantly higher than the other groups (Liu and Chang, 2006). After this, the levels decreased and returned to pre-surgery levels on day 14. In the hepatectomized groups that received: (1) no treatment; or (2) free bone marrow stem cells, the blood HGF levels were much lower.

Laparotomy, Histology, and Immunocytochemistry

Two weeks after implantation, laparotomy shows that in the group that received artificial cells containing bone marrow stem cells, the artificial cells remained in the peritoneal cavity. They were found freely disseminated throughout the peritoneal cavity, aggregated behind the liver or under the spleen, or attached to the large omentum. Histological examination of bone marrow stem cells in the artificial cells showed that before transplantation most cells were polygonal, star-like, some with tail shaped cytoplasm. When they were retrieved two weeks after transplantation, most cell morphology transformed to round or oval (Liu and Chang, 2006).

Before implantation, the bone marrow stem cells recovered from artificial cells did not show positive immunochemistry stain. However, when bone marrow stem cells were recovered from artificial cells at week 2 after implantation, there were scattered cells positively stained with hepatocyte markers CK8 and CK18, and also albumin production. There were also cells that stained positively with AFP (Liu and Chang, 2006). In the bone marrow stem cells recovered from artificial cells retrieved 2 weeks post-transplantation, there were cells stained positive for PAS, indicating that there were also cells capable of glycogen production.

Possible Mechanisms Responsible for the Recovery of 90% Hepatectomized Rat Model

What are the mechanisms responsible for the survival and recovery of the 90% hepatectomized rat model in our present study? Most likely this is due to two combined mechanisms (Chang, 2007).

1. Transdifferentiation of bone marrow stem cells in the artificial cells into hepatocytes. This is supported by our immunochemistry stain studies which show that some of the bone marrow stem cells recovered from artificial cells transdifferentiated into hepatocyte-like cells that expressed ALB, CK8, CK18, and AFP, which are typical markers of hepatocytes. They also produced albumin and glycogen. However, this is a slow process that takes time, and is most likely only responsible for the later phases of the recovery of the animals.
2. Hepatic growth factor (HGF) is an important factor in liver regeneration (Rokstad et al., 2002) and also in stimulating the transdifferentiation of bone marrow cells into hepatocytes (Spangrude et al., 1988). There are two subgroups of HGF, one >100,000 Da the other <10,000 Da (Michalopoulos et al., 1984). Alginate-polylysine-alginate membrane artificial cells allow the passage of molecules of 64,000 Da or less, and retain those >64,000 Da (Ito and Chang, 1992). HGF of >100,000 mw secreted by hepatocytes are retained and accumulated in the artificial cells, thus helping to increase the regeneration of hepatocytes in the artificial cells (Kashani and Chang, 1988, 1991) and also the transdifferentiation of BMCs. The smaller molecule weight HGF of <10,000 can diffuse out of the artificial cells to stimulate the regeneration of the remaining 10% liver mass in the 90% hepatectomized rats. Thus, the 90% hepatectomized rats that received artificial cells containing bone marrow stem cells had significantly higher blood HGF levels than the other groups. Our result shows that artificial cells containing bone marrow stem cells stayed in the peritoneal cavity throughout the 14 days of the study. This way, HGF secreted can be drained into the portal circulation to reach the 10% remaining liver mass to stimulate its regeneration. On the other hand, intraperitoneal injection of free bone marrow stem cells did not increase survival rates. Most likely this is because the free bone marrow stem cells are rapidly removed from the peritoneal cavity through the lymphatic drainage, and any HGF released does not drain into the portal circulation to have the most effective stimulating effect on the remnant liver.

Artificial Cells Containing Stem Cells in Regenerative Medicine

It would appear from the above study that implantation of artificial cells containing bone marrow stem cells results in the regeneration of the 90% hepatectomized liver, and the survival of the animal. These observations could stimulate further investigation of the potential for the treatment of acute liver failure or extensive liver resection to allow the liver to regenerate. The use of artificial cells containing stem cells could also be investigated in other areas of regeneration medicine.

GENE AND ENZYME THERAPY

Artificial cells have been studied for use in gene and enzyme therapy (Chang, 1964, 1972a, 2005a, 2007). The enclosed enzyme would not leak out, but could act on external permeant substrates. This would avoid protein sensitization, anaphylactic reaction or antibody production with repeated injection (Figure II.5.4.1). Implanted urease artificial cells convert systemic urea into ammonia (Chang, 1964, 1965). Implanting artificial cells containing catalase replaces the defective enzyme in mice with a congenital defect in catalase – acatalasemia (Chang and Poznansky, 1968). The artificial cells protect the enclosed enzyme from immunological reactions (Poznansky and Chang, 1974). Artificial cells containing asparaginase implanted into mice delay the onset and growth of lymphosarcoma (Chang, 1971b).

Giving enzyme artificial cells by mouth avoids the need for repeated injection. For example, artificial cells containing urease and ammonia adsorbent can lower the systemic urea level (Chang, 1972a). In Lesch-Nyhan disease, enzyme defect results in the elevation of hypoxanthine to toxic levels. Given by mouth, artificial cells containing xanthine oxidase lower the toxic systemic hypoxanthine levels in an infant with this disease (Chang, 1989; Palmour et al., 1989). Phenylketonuria is a more common congenital enzyme defect. Artificial cells containing phenylalanine ammonia lyase given by mouth lower the systemic phenylalanine levels in phenylketonuria (PKU) rats (Bourget and Chang, 1985, 1986). This has led to investigation into recombinant sources of this enzyme (Safos and Chang, 1995; Sarkissian et al., 1999; Liu et al., 2002) that are now being tested in clinical trials.

Multienzyme Systems with Cofactor Recycling

Most enzymes in the body function as multienzyme systems with cofactor recycling. After basic research on artificial cells containing multienzyme systems (Campbell and Chang, 1976; Cousineau and Chang, 1977; Yu and Chang, 1982; Chang 1985; Ilan and Chang, 1986) we looked into their possible use. Thus, artificial cells containing three different enzymes can convert metabolic waste such as urea and ammonia into essential amino acids (Gu and Chang, 1988, 1990). The needed cofactor, NADH, can be recycled and retained inside the artificial cell by cross-linking to dextran or by the use of a lipid–polymer membrane. As discussed earlier, under artificial red blood cells, all the multienzyme system in red blood cells can be incuded inside nanodimensional artificial red blood cells (Chang, 2007; Chang et al., 2003) (Figure II.5.4.5).

DRUG DELIVERY

Gene and enzyme therapy is a form of drug delivery using artificial cells. However, the genes and enzymes are retained inside the artificial cells at all times, and act on substrates diffusing into the artificial cells. Artificial cells in different modified forms have also been used in drug delivery. However, these are used to separate the drug from the external environment, and to release the drug at a specific site and a specific rate when and where it is needed. Drug delivery is an extremely large and wide area, and many excellent reviews and books are available. In this chapter, only artificial cell-related systems will be very briefly summarized.

Polymeric Membrane Artificial Cells of Microscopic Dimensions

Luzzi (1970) used nylon membrane artificial cells, microcapsules, prepared as reported (Chang, 1964), to microencapsulate drugs for slow release for oral administration. Others have also extended this approach. However, the modern approaches in drug delivery systems are based on injectable biodegradable systems.

Biodegradable Polymeric Artificial Cells (Microparticles, Nanoparticles, Microcapsules, Nanocapsules)

Biodegradable polylactide membrane artificial cells have been prepared to contain enzymes, hormones, vaccines, and other biologicals (Chang, 1976). The polylactide polymer can degrade in the body into lactic acid, and finally into water and carbon dioxide. Variations in preparation can result in artificial cells that release insulin at different rates (Chang, 1976) (Figure II.5.4.8).

We have also used these for the slow release of prostaglandin E2 (Zhou and Chang, 1988) and ciprofloxacin (Yu et al., 1998, 2000). Biodegradable drug delivery systems are now used widely in different forms, ranging from microscopic to nanodimensions. They are also known as nanoparticles, nanocapsules, polymersomes, nanotubules, etc. Langer's group has written an excellent review on this topic (LaVan et al., 2002). Copolymers of polyethylene glycol (PEG) and polylactic acid (PLA) have been used to increase the circulation time of nanodimensional artificial cells. As described earlier in this chapter, this also forms the basis for preparing nanodimension PEG–PLA membrane artificial red blood cells (Yu and Chang, 1996; Chang et al., 2003; Chang, 2005a, 2006a).

Liposomes Evolved into Lipid Membrane Artificial Cells and Then Back into Polymeric Membrane Artificial Cells

Bangham (1965) reported the preparation of liposomes each consisting of microspheres of hundreds of concentric lipid bilayers – multi-lamellar. This was initially used as membrane models in basic research. Gregoriadis (1976) first reported the use of liposomes as drug delivery systems. However, the large amount of lipid in the multi-lamellar liposome limits the amount of water-soluble drugs that can be enclosed. Thus, the basic principle and method of preparing artificial cells using ether as the dispersing phase (Chang, 1957, 1964) was extended by researchers into what they call an "ether evaporation method" to form single bilayer (unilamellar) lipid membrane liposomes (Deamer and Bangham, 1976). These lipid membrane artificial cells have since

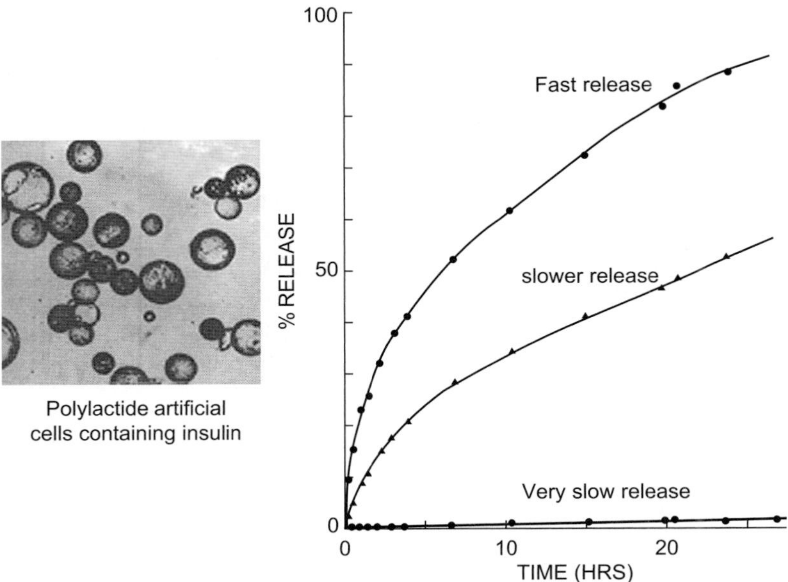

FIGURE II.5.4.8 Biodegradable membrane artificial cells have been prepared to contain enzymes, hormones, vaccines, and other biologicals (Chang, 1976). This figure summarizes the result of polylactide artificial cells prepared by double emulsion method. Variations in the molecular weight of polylactide and the thickness of the membrane can result in artificial cells that release insulin at different rates. Faster release comes from the encapsulation of insulin solution at high concentration. Very slow release comes from the encapsulation of insulin crystals (Chang, 1976). (Figure from Chang (2007), with copyright permission of World Scientific Publisher.)

been extensively studied for use as drug delivery systems (Torchilin, 2005). Surface charges have also been incorporated into liposomes for possible targeting of drug, and more recently the use of positively charged lipid to complex with DNA. Polyethylene glycol (PEG) has also been incorporated to the liposome surface to result in longer circulation time. Thus, lipid vesicles are becoming more like the lipid–polymer membrane artificial cells (Chang, 1972a), and are no longer pure lipid vesicles. Further development led to the incorporation of antibodies into the lipid membrane, to allow for targeting to cells with the corresponding antigens. The principle of loading magnetic particles into artificial cells (Chang, 1966) has also been used for loading into liposome, allowing for magnetic targeting. One major advantage of lipid vesicles is in their ability to fuse with cellular membrane or membranes of intracellular organelles. This allows for much versatility in their ability to deliver drugs to different sites of the cells. A number of drugs in PEG–lipid vesicles have already been approved for clinical use or are in clinical trial (Torchilin, 2005). Discher's group (Discher et al., 1999; Photos et al., 2003) tried to increase the strength of the PEG–lipid membrane artificial cells by using self-assembling of copolymer to form a membrane of PEG polymer. This significantly increased the circulation time and strength when compared to PEG–lipid membrane artificial cells Thus, multilamellar liposome has evolved into lipid membrane artificial cells, then polymer(PEG)–lipid membrane artificial cells, and finally back to the original polymeric membrane artificial cells (Chang, 1964) now called polymersomes.

OTHER AREAS OF ARTIFICIAL CELLS

Artificial Cells Containing Magnetic Materials

When magnetic material is included in artificial cells (ACs) containing bioactive materials, one can use an external magnetic field to direct the artificial cells (Chang, 1966). This way, the magnetic field can: (1) direct the movement of the AC; (2) remove the AC after reaction; (3) retain the AC at specific site of action; (4) stir or agitate the AC as in bioreactors. This principle is now being used very extensively in bioreactors, in removing specific materials from a mixture as in diagnositcs kits, in drug delivery systems, and in other areas of application.

Nanobiosensors

Nanobiosensors is an area that is of increasing interest, and different approaches are being investigated. One of the many approaches is to use a biosensor where a lipid bilayer is "tethered" on ultrathin polymeric support to form a lipid–polymer complex. In this form, different "channels" can be inserted into the membrane to allow for selective movement of specific solute for detection, somewhat similar to the principle reported earlier (Chang, 1969, 1972a; Rosenthal and Chang, 1971, 1980). Another approach that is also possible is to encapsulate enzymes inside artificial cells of microscopic or nanodimension. This way, the product of enzymatic reaction can be followed by fluorescence or other methods. The ability to prepare artificial cells with intracellular compartmentation (Chang, 1965, 1972a; Chang et al., 1966) would allow multi-step enzyme reaction to occur and be detected separately. Depending on the type of reaction being followed, one can use either polymeric membrane artificial cells, lipid membrane artificial cells or lipid–polymer membrane artificial cells.

Artificial Cells Containing Radioisotopes or Radio-Opaque Material

The general principle of artificial cells could be explored in many other areas. Thus, artificial cells containing radioactive isotopes or antimetabolites might be used for intra-arterial injection into tumors. In this case, some of the microcapsules might lodge at the tumor site, while others would be carried by lymphatic channels to metastases in regional lymph nodes. Artificial cells containing radio-opaque material would provide a contrast medium. Provided they can circulate readily in the bloodstream, they might be used as vehicles for contrast materials in angiography.

Nonmedical Uses of Artificial Cells

There are many developments and uses of the principle of artificial cells for agriculture, bioreactors, cosmetics, food production, aquatic culture, nanocomputers, and nanorobotics. However, these are not within the scope of this chapter.

THE FUTURE OF ARTIFICIAL CELLS

The 1972 monograph *Artificial Cells* (Chang, 1972a) predicted that: "Artificial Cell is not a specific physical entity. It is an idea involving the preparation of artificial structures of cellular dimensions for possible replacement or supplement of deficient cell functions. It is clear that different approaches can be used to demonstrate this idea." This prediction is already out of date since, in the last 50 years (Chang, 1957), artificial cells have progressed way beyond this 1972 prediction. Artificial cells can now be of macro-, micro-, nano-, and molecular dimensions. There are also unlimited possibilities in variations for artificial cell membranes and contents (Figure II.5.4.1). We have just touched the surface of the potential of artificial cells.

ACKNOWLEDGMENTS

The ongoing support of the Canadian Institutes of Health Research (whose original name was the Medical Research Council of Canada) over the last many years is gratefully acknowledged. Other support included the Quebec Hemovigillance and Transfusion Medicine Program under FRSQ.

BIBLIOGRAPHY

Many of the references and books from the authors are available free from the McGill University public service website at www.artcell.mcgill.ca.

Aebischer, P., Pochon, N. A., Heyd, B., Déglon, N., Joseph, J. M., et al. (1996). Gene therapy for amyotrophic lateral sclerosis (ALS) using a polymer encapsulated xenogenic cell line engineered to secrete hCNTF. *Hum. Gene. Ther.*, **1**, 851–860.

Al-Hendy, A., Hortelano, G., Tannenbaum, G. S., & Chang, P. L. (1996). Growth retardation: An unexpected outcome from growth hormone gene therapy in normal mice with microencapsulated myoblasts. *Hum. Gene. Ther.*, **7**, 61–70.

Bachoud-Lévi, A. C., Déglon, N., Nguyen, J. P., Bloch, J., Bourdet, C., et al. (2000). Neuroprotective gene therapy for Huntington's disease using a polymer encapsulated BHK cell line engineered to secrete human CNTF. *Hum. Gene. Ther.*, **11**, 1723–1729.

Bangham, A. D. (1965). Diffusion of univalent ions across the lamellae of swollen phospholipids. *J. Mol. Biol.*, **13**, 238–252.

Basic, D., Vacek, I., & Sun, A. M. (1996). Microencapsulation and transplantation of genetically engineered cells: A new approach to somatic gene therapy. *Artif. Cells Blood Substit. Immobil. Biotechnol.*, **24**, 219–255.

Bensinger, W., Baker, D. A., Buckner, C. D., Clift, R. A., & Thomas, E. D. (1981). Immunoadsorption for removal of A and B blood-group antibodies. *New Engl. J. Med.*, **314**, 160–162.

Bloch, J., Bachoud-Lévi, A. C., Déglon, N., Lefaucheur, J. P., Winkel, L., et al. (2004). Neuroprotective gene therapy for Huntington's disease, using polymer-encapsulated cells engineered to secrete human ciliary neurotrophic factor: Results of a phase I study. *Hum. Gene. Ther.*, **15**, 968–975.

Bourget, L., & Chang, T. M. S. (1985). Phenylalanine ammonia-lyase immobilized in semipermeable microcapsules for enzyme replacement in phenylketonuria. *FEBS Lett.*, **180**, 5–8.

Bourget, L., & Chang, T. M. S. (1986). Phenylalanine ammonia-lyase immobilized in microcapsules for the depleture of phenylalanine in plasma in phenylketonuric rat model. *Biochim. Biophys. Acta.*, **883**, 432–438.

Bruni, S., & Chang, T. M. S. (1989). Hepatocytes immobilized by microencapsulation in artificial cells: Effects on hyperbilirubinemia in Gunn rats. *J. Biomater. Artif. Cells Artif. Organs.*, **17**, 403–412.

Bruni, S., & Chang, T. M. S. (1995). Kinetics of UDP-glucuronosyltransferase in bilirubin conjugation by encapsulated hepatocytes for transplantation into Gunn rats. *J. Artif. Organs.*, **19**, 449–457.

Calafiore, R. (1999). Transplantation of minimal volume microcapsules in diabetic high mammalians. *Ann. NY. Acad. Sci.*, **875**, 219–232.

Campbell, J., & Chang, T. M. S. (1976). The recycling of NAD+ (free and immobilized) within semipermeable aqueous microcapsules containing a multi-enzyme system. *Biochem. Biophys. Res. Commun.*, **69**, 562–569.

Chang, E. J., Lee, T. H., Mun, K. C., Kim, H. C., Suh, S. I., et al. (2004a). Effects of polyhemoglobin-antioxidant enzyme complex on ischemia-reperfusion in kidney. *Transplant Proc.*, **36**, 1952–1954.

Chang, E. J., Lee, S. H., Mun, K. C., Suh, S. I., Bae, J. H., et al. (2004b). Effect of artificial cells on hepatic function after ischemia-reperfusion injury in liver. *Transplant Proc.*, **36**, 1959–1961.

Chang, T. M. S. (1957). Hemoglobin corpuscles. Report of a research project for Honours Physiology, Medical Library, McGill University. Also reprinted 1988 as part of "30th Anniversary in Artificial Red Blood Cells Research." *Biomater. Artif. Cells Artif. Organs.*, **16**, 1–9.

Chang, T. M. S. (1964). Semipermeable microcapsules. *Science*, **146**, 524–525.

Chang, T. M. S. (1965). *Semipermeable Aqueous Microcapsules.* Ph.D. thesis, McGill University.

Chang, T. M. S. (1966). Semipermeable aqueous microcapsules ("artificial cells"): With emphasis on experiments in an extracorporeal shunt system. *Trans. Am. Soc. Artif. Intern. Organs.*, **12**, 13–19.

Chang, T. M. S. (1969). Removal of endogenous and exogenous toxins by a microencapsulated absorbent. *Can. J. Physiol. Pharmacol.*, **47**, 1043–1045.

Chang, T. M. S. (1971a). The *in vivo* effects of semipermeable microcapsules containing L-asparaginase on 6C3HED lymphosarcoma. *Nature*, **229**, 117–118.

Chang, T. M. S. (1971b). Stabilisation of enzymes by microencapsulation with a concentrated protein solution or by microencapsulation followed by cross-linking with glutaraldehyde. *Biochem. Biophys. Res. Commun.*, **44**, 1531–1536.

Chang, T. M. S. (1972a). *Artificial Cells.* Springfield, IL: Charles C. Thomas (out of print but available for free online viewing at www.artcell.mcgill.ca).

Chang, T. M. S. (1972b). Haemoperfusions over microencapsulated adsorbent in a patient with hepatic coma. *Lancet*, **2**, 1371–1372.

Chang, T. M. S. (1974). Platelet–surface interaction: Effect of albumin coating or heparin complexing on thrombogenic surfaces. *Can. J. Physiol. Pharmacol.*, **52**, 275–285.

Chang, T. M. S. (1975). Microencapsulated adsorbent hemoperfusion for uremia, intoxication and hepatic failure. *Kidney Int.*, **7**, S387–S392.

Chang, T. M. S. (1976). Biodegradable semipermeable microcapsules containing enzymes, hormones, vaccines, and other biologicals. *J. Bioeng.*, **1**, 25–32.

Chang, T. M. S. (1980). Blood compatible coating of synthetic immunoadsorbents. *Trans. Am. Soc. Artif. Intern. Organs.*, **26**, 546–549.

Chang, T. M. S. (1985). Artificial cells with regenerating multienzyme systems. *Meth. Enzymol.*, **112**, 195–203.

Chang, T. M. S. (1987). Recycling of NAD(P) by multienzyme systems immobilised by microencapsulation in artificial cells. *Meth. Enzymol.*, **136**, 67–82.

Chang, T. M. S. (1989). Preparation and characterization of xanthine oxidase immobilized by microencapsulation in artificial cells for the removal of hypoxanthine. *Biomater. Artif. Cells Artif. Organs.*, **17**, 611–616.

Chang, T. M. S. (1997a). *Red Blood Cell Substitutes: Principles, Methods, Products and Clinical Trials.* (Vol. I (Monograph)). Basel, Switzerland: Karger/Landes Systems. (Available for free online viewing at www. artcell.mcgill.ca or www.artificialcell.info.)

Chang, T. M. S. (1997b). Artificial cells. In D. Renalo (Ed.), *Encyclopedia of Human Biology* (2nd ed., pp. 457–463). San Diego, CA: Academic Press, Inc.

Chang, T. M. S. (2001). Bioencapsulated hepatocytes for experimental liver support. *J. Hepatol.*, **34**, 148–149.

Chang, T. M. S. (2005a). Therapeutic applications of polymeric artificial cells. *Nat. Rev: Drug Discov.*, **4**, 221–235.

Chang, T. M. S. (2005b). Methods for microencapsulation of enzymes and cells. *Meth. Biotechnol.*, **17**, 289–306.

Chang, T. M. S. (2006a). Blood substitutes based on bionanotechnology. *Trends Biotechnol.*, **24**, 372–377.

Chang, T. M. S. (2006b). Polyhemoglobin-enzyme complexes. In R. Winslow (Ed.), *Blood Substitutes* (pp. 451–459). London, UK: Elsevier Academic Press.

Chang, T. M. S. (2006c). PEG-PLA biodegradable hemoglobin nanocapsules as RBC substitutes. In R. Winslow (Ed.), *Blood Substitutes* (pp. 523–531). London, UK: Elsevier Academic Press.

Chang, T. M. S. (2007). *Artificial Cells: Biotechnology, Nanotechnology, Blood Substitutes, Regenerative Medicine, Bioencapsulation, Cell/Stem Cell Therapy.* Singapore and London: World Scientific Publisher/Imperial College Press. 435 pages.

Chang, T. M. S. (2008). Nanobiotechnological modification of hemoglobin and enzymes from this laboratory. *Biochimica. et. Biophysica. Acta: Proteins and Proteomics*, **1784**, 1435–1144.

Chang, T. M. S. (2009). Nanobiotechnology for hemoglobin based blood substitutes. *Critical Care Clinics*, **25**, 373–382.

Chang, T. M. S. (2010). Blood replacement with nanobiotechnologically engineered hemoglobin and hemoglobin nanocapsules. *Wiley. Interdiscip. Rev. Nanomed. Nanobiotechnol.*, **2**, 418–430.

Chang, T. M. S., & Malave, N. (1970). The development and first clinical use of semipermeable microcapsules (artificial cells) as a compact artificial kidney. *Trans. Am. Soc. Artif. Intern. Organs.*, **16**, 141–148.

Chang, T. M. S., & Poznansky, M. J. (1968). Semipermeable microcapsules containing catalase for enzyme replacement in acatalsaemic mice. *Nature*, **218**, 242–245.

Chang, T. M. S., & Prakash, S. (1998). Microencapsulated genetically engineered cells: Comparison with other strategies and recent progress. *Mol. Med. Today.*, **4**, 221–227.

Chang, T. M. S., & Yu, W. P. (1992). *Biodegradable polymer membrane containing hemoglobin as potential blood substitutes.* British Provisional Patent No. 9219426.5 (issued Sept. 14, 1992).

Chang, T. M. S., Macintosh, F. C., & Mason, S. G. (1966). Semipermeable aqueous microcapsules: I. Preparation and properties. *Can. J. Physiol. Pharmacol.*, **44**, 115–128.

Chang, T. M. S., Coffey, J. F., Barre, P., Gonda, A., Dirks, J. H., et al. (1973). Microcapsule artificial kidney: Treatment of patients with acute drug intoxication. *Can. Med. Assoc. J.*, **108**, 429–433.

Chang, T. M. S., Langer, R., Sparks, R. E., & Reach, G. (1988). Drug delivery systems in biotechnology. *J. Artif. Organs.*, **12**, 248–251.

Chang, T. M. S., Bourget, L., & Lister, C. (1995). A new theory of enterorecirculation of amino acids and its use for depleting unwanted amino acids using oral enzyme artificial cells, as in removing phenylalanine in phenylketonuria. *Artif. Cells Blood Substit. Immobil. Biotechnol.*, **25**, 1–23.

Chang, T. M. S., D'Agnillo, F., & Razack, S. (1998). Cross-linked hemoglobin superoxide dismutase-catalase: A second generation hemoglobin based blood substitute with antioxidant activities. In T. M. S. Chang (Ed.), *Blood Substitutes: Principles, Methods, Products and Clinical Trials.* (Vol. 2, pp. 178–196). Basel, Switzerland: Karger.

Chang, T. M. S., D'Agnillo, F., Yu, W. P., & Razack, S. (2000). New generations of blood subsitutes based on polyhemoglobin-SOD-CAT and nanoencapsulation. *Adv. Drug Delivery Rev.*, **40**, 213–218.

Chang, T. M. S., Powanda, D., & Yu, W. P. (2003). Analysis of polyethyleneglycolpolylactide nano-dimension artificial red blood cells in maintaining systemic hemoglobin levels and prevention of methemoglobin formation. *Artif. Cells Blood Substit. Biotechnol.*, **31**, 231–248.

Chow, K. M., Liu, Z. C., Prakash, S., & Chang, T. M. S. (2003). Free and microencapsulated *Lactobacillus* and effects of metabolic induction on Urea Removal Artificial Cells. *Blood Substit. and Biotechnol.*, **4**, 425–434.

Cirone, P., Bourgeois, M., Austin, R. C., & Chang, P. L. (2002). A novel approach to tumor suppression with microencapsulated recombinant cells. *Hum. Gene. Ther.*, **13**, 1157–1166.

Cousineau, J., & Chang, T. M. S. (1977). Formation of amino acid from urea and ammonia by sequential enzyme reaction using a microencapsulated multienzyme system. *Biochem. Biophys. Res. Commun.*, **79**, 24–31.

D'Agnillo, F., & Chang, T. M. S. (1998). Polyhemoglobin-superoxide dismutasecatalase as a blood substitute with antioxidant properties. *Nature Biotechnol.*, **16**, 667–671.

Dalle, B., Payen, E., Regulier, E., Deglon, N., Rouyer-Fessard, P., et al. (1999). Improvement of the mouse ß-thalassemia upon erythropoietin delivery by encapsulated myoblasts. *Gene. Ther.*, **6**, 157–161.

De Vos, P., Hamel, A. F., & Tatarkiewicz, K. (2002). Considerations for successful transplantation of encapsulated pancreatic islets. *Diabetologia*, **45**, 159–173.

Deamer, D. W., & Bangham, A. D. (1976). Large-volume liposomes by an ether vaporization method. *Biochim. Biophys. Acta.*, **443**, 629–634.

Discher, B. M., Won, Y., & David, S. E. (1999). Polymersomes: Tough vesicles made from diblock copolymers. *Science*, **284**, 1143–1144.

Djordjevich, L., & Miller, I. F. (1980). Synthetic erythrocytes from lipid encapsulated hemoglobin. *Exp. Hematol.*, **8**, 584.

Garofalo, F. A., & Chang, T. M. (1991). Effects of mass transfer and reaction kinetics on serum cholesterol depletion rates of free and immobilized. *Pseudomonas pictorum. Appl. Biochem. Biotech.*, **27**, 75–91.

Gould, S. A., et al. (1998). The clinical development of human polymerized hemoglobin. In T. M. S. Chang (Ed.), *Blood Substitutes: Principles, Methods, Products and Clinical Trials* (Vol. 2, pp. 12–28). Basel, Switzerland: Karger.

Gould, S. A., Moore, E. E., Hoyt, D. B., Ness, P. M., Norris, E. J., et al. (2002). The life-sustaining capacity of human polymerized hemoglobin when red cells might be unavailable. *J. Am. Coll. Surg.*, **195**, 445–452.

Greenburg, A. G., & Kim, H. W. (1992). Evaluating new red cell substitutes: A critical analysis of toxicity models. *Biomater. Artif. Cells Immobil. Biotechnol.*, *20*, 575–581.

Gregoriadis, G. (1976). *Drug Carriers in Biology and Medicine*. New York, NY: Academic Press, Inc.

Gu, G., & Chang, T. M. S. (2009). Extraction of erythrocyte enzymes for the preparation of Polyhemoglobin-catalase-superoxide dismutase. *Artif. Cells, Blood Substit. and Biotechnol.*, **37**, 69–77.

Gu, K. F., & Chang, T. M. S. (1988). Conversion of alpha-ketoglutarate into L-glutamic acid with urea as ammonium source using multienzyme system and dextran-NAD+ immobilised by microencapsulation with artificial cells in a bioreactor. *J. Bioeng. Biotechnol.*, **32**, 363–368.

Gu, K. F., & Chang, T. M. S. (1990). Production of essential L-branched-chained amino acids, in bioreactors containing artificial cells immobilized multienzyme systems and dextran-NAD+. *Appl. Biochem. Biotechnol.*, **26**, 263–269.

Hagihara, Y., Saitoh, Y., Iwata, H., Taki, T., Hirano, S., et al. (1997). Transplantation of xenogeneic cells secreting beta-endorphin for pain treatment: Analysis of the ability of components of complement to penetrate through polymer capsules. *Cell Transplan.*, **6**, 527–530.

Hakim, R. M., Milford, E., Himmelfarb, J., Wingard, R., Lazarus, J. M., et al. (1990). Extracorporeal removal of antiHLA antibodies in transplant candidates. *Am. J. Kidney. Dis.*, **16**, 423.

Hunkeler, D. L. (2003). Bioartificial organ grafts: A view at the beginning of the third millennium. *Artif. Cells Blood Substit. Immobil. Biotechnol.*, **31**, 365–382.

Ilan, E., & Chang, T. M. S. (1986). Modification of lipid–polyamide microcapsules for immobilization of free cofactors and multienzyme system for the conversion of ammonia to glutamate. *Appl. Biochem. Biotechnol.*, **13**, 221–230.

Ito, Y., & Chang, T. M. S. (1992). *In vitro* study of multicellular hepatocytes spheroid formed in microcapsules. *J. Artif. Organs.*, **16**, 422–426.

Jahr, J. S., Mackenzie, C., Pearce, L. B., Pitman, A., & Greenburg, A. G. (2008). HBOC-201 as an alternative to blood transfusion: Efficacy and safety evaluation in a multicenter phase III trial in elective orthopaedic surgery. *J. Trauma.*, **64**, 1484–1497.

Kashani, S. A., & Chang, T. M. S. (1988). Release of hepatic stimulatory substance from cultures of free and microencapsulated hepatocytes: Preliminary report. *J. Biomater. Artif. Cells Artif. Organs.*, **16**, 741–746.

Kashani, S., & Chang, T. M. S. (1991). Effects of hepatic stimulatory factor released from free or microencapsulated hepatocytes on galactosamine induced fulminant hepatic failure animal model. *Biomater. Artif. Cells Immobil. Biotechnol.*, **19**, 579–598.

Kawasaki, C., Nishi, R., Uekihara, S., Hayano, S., & Otagiri, M. (2000). Charcoal hemoperfusion in the treatment of phenytoin overdose. *Am. J. Kidney Dis.*, **35**, 323–326.

Keipert, P. E., & Chang, T. M. S. (1985). Pyridoxylated polyhemoglobin as a blood substitute for resuscitation of lethal hemorrhagic shock in conscious rats. *Biomater. Med. Dev. Artif. Organs*, **13**, 1–15.

Keipert, P. E., & Chang, T. M. S. (1987). *In vivo* effects of total and partial isovolemic exchange transfusion in fully conscious rats using pyridoxylated polyhemoglobin solution as a colloidal oxygen-delivery blood substitute. *Vox Sang.*, **53**, 7–14.

Kobayashi, K., Izumi, Y., Yoshizu, A., Horinuchi, H., Park, S. I., et al. (1997). The oxygen carrying capability of hemoglobin vesicles evaluated in rat exchange transfusion models. *Artif. Cells Blood Substit. Immobil. Biotechnol.*, **25**, 357–366.

LaVan, D. A., Lynn, D. M., & Langer, R. (2002). Moving smaller in drug discovery and delivery. *Nat. Rev. Drug Discov.*, **1**, 77–84.

Lim, F., & Sun, A. M. (1980). Microencapsulated islets as bioartificial endocrine pancreas. *Science*, **210**, 908–909.

Lin, C. C., Chou, H. L., & Lin, J. L. (2002). Acute aconitine poisoned patients with ventricular arrhythmias successfully reversed by charcoal hemoperfusion. *Am. J. Emerg. Med.*, **20**, 66–67.

Lin, C. C., Chan, T. Y., & Deng, J. F. (2004). Clinical features and management of herb induced aconitine poisoning. *Ann. Emerg. Med.*, **43**, 574–579.

Liu, J., Jia, X., Zhang, J., Hu, W., & Zhou, Y. (2002). Study on a novel strategy to treatment of phenylketonuria. *Artif. Cells Blood Substit. Immobil. Biotechnol.*, **30**, 243–258.

Liu, Z. C., & Chang, T. M. S. (2000). Effects of bone marrow cells on hepatocytes: When cocultured or co-encapsulated together. *Artif. Cells Blood Substit. Immobil. Biotechnol.*, **28**, 365–374.

Liu, Z. C., & Chang, T. M. S. (2002). Increased viability of transplantation hepatocytes when coencapsulated with bone marrow stem cells using a novel method. *Artif. Cells Blood Substit. Immobil. Biotechnol.*, **30**, 99–11.

Liu, Z. C., & Chang, T. M. S. (2003). Coencapsulation of stem sells and hepatocytes: *In vitro* conversion of ammonia and *in vivo* studies on the lowering of bilirubin in Gunn rats after transplantation. *Int. J. Artif. Organs.*, **26**, 491–497.

Liu, Z. C., & Chang, T. M. S. (2005). Transplantation of bioencapsulated bone marrow stem cells improves hepatic regeneration and survival of 90% hepatectomized rats: A preliminary report. *Artif. Cells Blood Substit. Biotechnol.*, **33**, 405–410.

Liu, Z. C., & Chang, T. M. S. (2006). Transdifferentiation of bioencapsulated bone marrow cells into hepatocyte-like cells in the 90% hepatectomized rat model. *J. Liver Transplant.*, **12**, 566–572.

Liu, Z. C., & Chang, T. M. S. (2008). Long term effects on the histology and function of livers and spleens in rats after 33% toploading of PEG-PLA-nano-artificial red blood cells. *Artif. Cells, Blood Substit. and Biotechnol.*, **36**, 513–524.

Liu, Z. C., & Chang, T. M. S. (2009). Preliminary study on intrasplenic implantation of artificial cell bioencapsulated stem cells to increase the survival of 90% hepatectomized rats. *Artif. Cells, Blood Substit. and Biotechnol.*, **37**(1), 53–55.

Liu, Z. C., & Chang, T. M. S. (2010). Artificial cell microencapsulated stem cells in regenerative medicine, tissue engineering and cell therapy. *Adv. Exp. Med. Biol.*, **670**, 68–79.

Lloyd-George, I., & Chang, T. M. S. (1995). Characterization of free and alginate polylysine-alginate microencapsulated *Erwinia herbicola* for the conversion of ammonia, pyruvate and phenol into L-tyrosine and L-DOPA. *J. Bioeng. Biotechnol.*, **48**, 706–714.

Lopez Lago, A. M., Rivero Velasco, C., Galban Rodríguez, C., Mariño Rozados, A., Piñero Sande, N., et al. (2002). Paraquat poisoning and hemoperfusion with activated charcoal. *Ann. Intern. Med.*, **19**, 310–312.

Lorh, M., Hoffmeyer, A., & Kroger, J. C. (2001). Microencapsulated cell-mediated treatment of inoperable pancreatic carcinoma. *Lancet*, **357**, 1591–1592.

Luzzi, L. A. (1970). Preparation and evaluation of the prolonged release properties of nylon microcapsules. *J. Pharm. Sci.*, **59**, 338.

Michalopoulos, G., Houck, K. D., Dolan, M. L., & Luetteke, N. C. (1984). Control of hepatocyte replication by two serum factors. *Cancer Res.*, **44**, 4414–4419.

Moore, E. E., Moore, F. A., Fabian, T. C., Bernard, A. C., Fulda, G. J., et al. (2009). Human polymerized hemoglobin for the treatment of hemorrhagic shock when blood is unavailable: The USA Multicenter Trial. *J. Am. Coll. Surg.*, **208**, 1–13.

Orive, G., Hernández, R. M., Gascón, A. R., Calafiore, R., Chang, T. M., et al. (2003). Cell encapsulation: Promise and progress. *Nat. Med.*, **9**, 104–107.

Orive, G., Hernández, R. M., Gascón, A. R., Calafiore, R., Chang, T. M., et al. (2004). History, challenge and perspectives of cell microencapsulation. *Trends Biotechnol.*, **22**, 87–92.

Palmour, R. M., Goodyer, P., Reade, T., & Chang, T. M. S. (1989). Microencapsulated xanthine oxidase as experimental therapy in Lesch-Nyhan disease. *Lancet*, **2**, 687–688.

Pearce, L. B., & Gawryl, M. S. (1998). Overview of preclinical and clinical efficacy of Biopure's HBOCs. In T. M. S. Chang (Ed.), *Blood Substitutes: Principles, Methods, Products and Clinical Trials* (Vol. 2, pp. 82–98). Basel, Switzerland: Karger.

Pearce, L. B., Gawryl, M. S., Rentko, V. T., et al. (2006). HBOC-201 (Hb Glutamer-250 (Bovine), Hemopure): Clinical studies. In R. Winslow (Ed.), *Blood Substitutes* (pp. 437–450). San Diego, CA: Academic Press.

Peng, A., Meng, F. Q., Sun, L. F., Ji, Z. S., & Li, Y. H. (2004). Therapeutic efficacy of charcoal hemoperfusion in patients with acute severe dichlorvos poisoning. *Acta. Pharmacol. Sin.*, **25**, 15–21.

Philips, W. T., Klipper, R. W., Awasthi, V. D., Rudolph, A. S., Cliff, R., et al. (1999). Polyethylene glyco-modified liposome-encapsulated hemoglobin: A long circulating red cell substitute. *J. Pharm. Exp. Ther.*, **288**, 665–670.

Photos, P. J., Bacakova, L., Discher, B., & Bates, F. S. (2003). Polymer vesicles *in vivo*: Correlations with PEG molecular weight. *J. Control Release.*, **90**, 323–334.

Powanda, D., & Chang, T. M. S. (2002). Cross-linked polyhemoglobin-superoxide dismutase-catalase supplies oxygen without causing blood brain barrier disruption or brain edema in a rat model of transient global brain ischemia-reperfusion. *Artif. Cells Blood Substit. Immobil. Biotechnol.*, **30**, 25–42.

Poznansky, M. J., & Chang, T. M. S. (1974). Comparison of the enzyme kinetics and immunological properties of catalase immobilized by microencapsulation and catalase in free solution for enzyme replacement. *Biochim. Biophys. Acta.*, **334**, 103–115.

Prakash, S., & Chang, T. M. S. (1993). Genetically engineered *E. coli* cells containing *K. aerogenes* gene, microencapsulated in artificial cells for urea and ammonia removal. *Biomater Artif. Cells Immobil. Biotechnol.*, **21**, 629–636.

Prakash, S., & Chang, T. M. S. (1996). Microencapsulated genetically engineered live *E. coli* DH5 cells administered orally to maintain normal plasma urea level in uremic rats. *Nature Med.*, **2**, 883–887.

Razack, S., D'Agnillo, F., & Chang, T. M. S. (1997). Cross-linked hemoglobin superoxide dismutase-catalase scavenges free radicals in a rat model of intestinal ischemia-reperfusion injury. *Artif. Cells Blood Substit. Immobil. Biotechnol.*, **25**, 181–192.

Rokstad, A. M., Holtan, S., Strand, B., Steinkjer, B., Ryan, L., et al. (2002). Microencapsulation of cells producing therapeutic proteins: Optimizing cell growth and secretion. *Cell Transplant*, **11**, 313–324.

Rosenthal, A. M., & Chang, T. M. S. (1971). The effect of valinomycin on the movement of rubidium across lipid coated semipermeble microcapsules. *Proc. Canad. Fed. Biol. Soc.*, **14**, 44.

Rosenthal, A. M., & Chang, T. M. S. (1980). The incorporation of lipid and Na+-K+-ATPase into the membranes of semipermeable microcapsules. *J. Membrane Sciences*, **6**(3), 329–338.

Safos, S., & Chang, T. M. S. (1995). Enzyme replacement therapy in ENU2 phenylketonuric mice using oral microencapsulated phenylalanine ammonialyase: A preliminary report. *Artif. Cells Blood Substit. Immobil. Biotechnol.*, **23**, 681–692.

Saitoh, Y., Taki, T., Arita, N., Ohnishi, T., & Hayakawa, T. (1995). Cell therapy with encapsulated xenogeneic tumor cells secreting beta-endorphin for treatment of peripheral pain. *Cell Transplant*, **S1**, S13–S17.

Sakai, H., Masada, Y., Horinouchi, H., Yamamoto, M., Ikeda, E., et al. (2004). Hemoglobin-vesicles suspended in recombinant human serum albumin for resuscitation from hemorrhagic shock in anesthetized rats. *Crit. Care. Med.*, **32**, 539–545.

Sarkissian, C. N., Shao, Z., Blain, F., Peevers, R., Su, H., et al. (1999). A different approach to treatment of phenylketonuria: Phenylalanine degradation with recombinant phenylalanine ammonia lyase. *Proc. Natl. Acad. Sci.*, **96**, 2339–2344.

Singh, S. M., McCormick, B. B., Mustata, S., Thompson, M., & Prasad, G. V. (2004). Extracorporeal management of valproic acid overdose: A large regional experience. *J. Nephrol.*, **17**, 43–49.

Spangrude, G. J., Heimfeld, S., & Weissman, I. L. (1988). Purification and characterization of mouse hematopoietic stem cells. *Science*, **241**, 58–62.

Sprung, J., Kindscher, J. D., Wahr, J. A., Levy, J. H., Monk, T. G., et al. (2002). The use of bovine Hb glutamer-250 (Hemopure) in surgical patients: Results of a multicenter, randomized, singleblinded trial. *Anesth. Analg.*, **94**, 799–808.

Sun, Y. L., Ma, X. J., & Zhou, D. B. (1996). Normalization of diabetes in spontaneously diabetic cynomologus monkeys by xenografts of microencapsulated porcine islets without immunosuppression. *J. Clin. Invest.*, **98**, 1417–1422.

Tan, S. A., Déglon, N., Zurn, A. D., Baetge, E. E., Bamber, B., et al. (1996). Rescue of motoneurons from axotomy-induced cell death by polymer encapsulated cells genetically engineered to release CNTF. *Cell Transplant*, **5**, 577–587.

Terman, D. S., Tavel, T., Petty, D., Racic, M. R., & Buffaloe, G. (1977). Specific removal of antibody by extracorporeal circulation over antigen immobilized in colodion charcoal. *Clin. Exp. Immunol.*, **28**, 180–188.

Terman, D. S., Buffaloe, G., Mattioli, C., Cook, G., Tiilquist, R., et al. (1979). Extracorporeal immunoabsorption: Initial experience in human systemic lupus erythematosus. *Lancet*, **2**, 824.

Torchilin, V. P. (2005). Recent advances with liposomes as pharmaceutical carriers. *Nat. Rev. Drug Discov.*, **4**, 145–160.

Tsuchida, E. (Ed.), (1998). *Present and Future Perspectives. Blood Substitutes* (Vol. 1, pp. 267). Amsterdam: Elsevier.

Winchester, J. F. (Ed.), (1996). *Replacement of Renal Function by Dialysis* (4th ed.). Boston, MD: Kluwer Academic Publishers.

Wingard, R. L., Lee, W. O., & Hakim, R. O. (1991). Extracorporeal treatment of familial hypercholesterolemia with monoclonal antibodies to low-density lipoproteins. *Am. J. Kidney Dis.*, **18**, 559.

Winn, S. R., Hammang, J. P., Emerich, D. F., Lee, A., Palmiter, R. D., et al. (1994). Polymer-encapsulated cells genetically modified to secrete human nerve growth factor promote the survival of axotomized septal cholinergic neurons. *Proc. Natl. Acad. Sci. USA*, **91**, 2324–2328.

Wong, H., & Chang, T. M. S. (1986). Bioartificial liver: Implanted artificial cells microencapsulated living hepatocytes increases survival of liver failure rats. *Int. J. Artif. Organs.*, **9**, 335–336.

Wong, H., & Chang, T. M. S. (1988). The viability and regeneration of artificial cell microencapsulated rat hepatocyte xenograft transplants in mice. *J. Biomater Artif. Cells Artif. Organs.*, **16**, 731–740.

Wong, H., & Chang, T. M. S. (1991). A novel two-step procedure for immobilizing living cells in microcapsule for improving xenograft survival. *Biomater Artif. Cells Immobil. Biotechnol.*, **19**, 687–698.

Wong, N., & Chang, T. M. S. (2007). Polyhemoglobin-fibrinogen: A novel blood substitute with platelet-like activity for extreme hemodilution. *Artif. Cells, Blood Substit. and Biotechnol.*, **35**, 481–489.

Xu, W., Liu, L., & Charles, I. G. (2002). Microencapsulated iNOS-expressing cells cause tumor suppression in mice. *FASEB J.*, **16**, 213–215.

Yang, L., Cheng, Y., Yan, W. R., & Yu, Y. Y. (2004). Extracorporeal whole blood immunoadsorption of autoimmune myasthenia gravis by cellulose tryptophan adsorbent. *Artif. Cells, Blood Substit. Biotechnol.*, **32**, 519–518.

Yu, B. L., & Chang, T. M. S. (2004). *In vitro* and *in vivo* effects of poly hemoglobintyrosinase on murine B16F10 melanoma. *Melanoma. Res. J.*, **14**, 197–202.

Yu, W. P., & Chang, T. M. S. (1994). Submicron biodegradable polymer membrane hemoglobin nanocapsules as potential blood substitutes: A preliminary report. *J. Artif. Cells Blood Substit. Immobil. Biotechnol.*, **22**, 889–894.

Yu, W. P., & Chang, T. M. S. (1996). Submicron polymer membrane hemoglobin nanocapsules as potential blood substitutes: Preparation and characterization. *Artif. Cells Blood Substit. Immobil. Biotechnol.*, **24**, 169–184.

Yu, W. P., Wong, J., & Chang, T. M. S. (1998). Preparation and characterization of polylactic acid microcapsules containing ciprofloxacin for controlled release. *J. Microencapsul.*, **15**, 515–523.

Yu, W. P., Wong, J. P., & Chang, T. M. S. (2000). Sustained drug release characteristics of biodegradable composite poly(d, l) lactic acid poly(l)lactic acid microcapsules containing ciprofloxacin. *Artif. Cells Blood Substit. Immobil. Biotechnol.*, **28**, 39–56.

Yu, Y. T., & Chang, T. M. S. (1982). Immobilization of multienzymes and cofactors within lipid-polyamide membrane microcapsules for the multistep conversion of lipophilic and lipophobic substrates. *Enzyme. Microb. Technol.*, **4**, 327–331.

Zhou, M. X., & Chang, T. M. S. (1988). Control release of prostaglandin E2 from polylactic acid microcapsules, microparticles and modified microparticles. *J. Microencapsul.*, **5**, 27–36.

CHAPTER II.5.5 EXTRACORPOREAL ARTIFICIAL ORGANS

Alastair Campbell Ritchie
Department of Mechanical, Materials and Manufacturing Engineering, University of Nottingham, Nottingham, UK

INTRODUCTION

Extracorporeal circulation is, by definition, any procedure in which blood is taken from a patient, treated, and then returned. The vast majority of these procedures are performed intermittently, or as a temporary partial or total replacement. In this chapter, we will consider only those artificial organs which undertake mass transfer to and from the blood. This includes devices used for dialysis in renal replacement therapy, plasma separation (plasmapheresis), and extracorporeal oxygenation.

All these procedures involve the contact of blood with biomaterials, mechanical pumping of the blood, and mass transfer to and from the blood. The issues of biocompatibility are essential to the success or failure of a procedure; readers are advised to refer to the previous chapter on blood–biomaterials interaction for a full discussion of the effects of blood–biomaterial contact, as in this chapter only those aspects specific to extracorporeal therapy will be discussed.

Renal Replacement Therapy

Renal replacement therapy describes treatments such as peritoneal dialysis, hemodialysis, and hemofiltration, used to detoxify blood in cases of renal insufficiency. Renal insufficiency may be caused by acute renal failure, typically of rapid onset and due to external causes or by chronic renal failure, usually due to damage to the tissues of the kidneys. Chronic renal failure may be caused by infection, toxins or hypertension, and may also arise as a complication in diabetes mellitus. The number of prevalent (i.e., alive and receiving treatment) patients with end-stage renal disease in the USA between 1980 and 2007 is given in Figure II.5.5.1 (USRDS, 2009). As can be seen from the graph, the incidence of chronic renal failure has risen steadily since 1980, and diabetes has become the prominent cause of renal failure, accounting for some 37.4% of all cases in 2007, followed by glomerulonephritis, and renal vascular disease. Of the 527,283 surviving end-stage renal disease (ESRD) patients listed by the United States Renal Data System in that year,[1] 158,739 (30.1% of the total) had received a transplant that was still functioning. The other patients were dependent upon renal replacement therapy.

In 2007, there were 368,544 patients receiving renal replacement therapy in the United States, of whom 92.5% received hemodialysis treatment (USRDS, 2009). Blood may be treated by extracorporeal hemodialysis (familiarly known as dialysis), peritoneal dialysis, and extracorporeal hemofiltration (a form of plasmapheresis) (Coles and El Nahas, 2000). Figure II.5.5.2 shows the numbers of patients treated by blood purification therapies in the USA between 1980 and 2007 (USRDS, 2009). As can be seen, as the patient population has increased,

[1]Data from US Renal Data System, USRDS, 2009, *Annual Data Report: Atlas of End-Stage Renal Disease in the United States*, National Institutes of Health, National Institute of Diabetes and Digestive and Kidney Diseases, Bethesda, MD, 2009. The data reported here have been supplied by the United States Renal Data System (USRDS). The interpretation and reporting of these data are the responsibility of the author, and in no way should be seen as an official policy or interpretation of the US government.

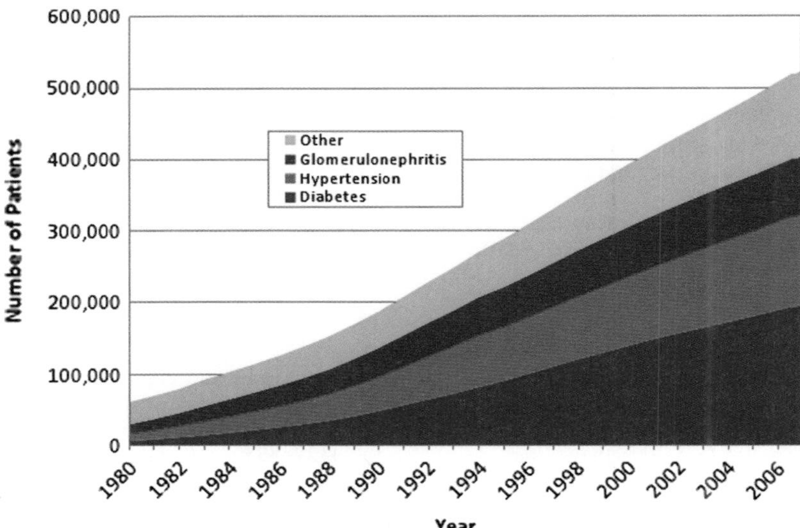

FIGURE II.5.5.1 Prevalence and causes of chronic renal failure in the USA, 1980–2007 (Source: United States Renal Data System (USRDS) Annual Data Report, 2009. Data is as at December 31st of stated year.)

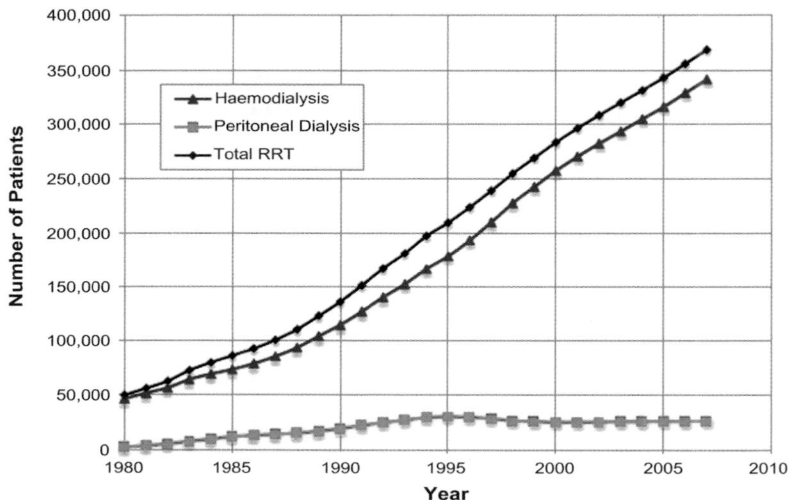

FIGURE II.5.5.2 Number of patients receiving regular dialysis (hemodialysis and peritoneal dialysis) in the USA, 1980–2007 (Source: United States Renal Data System (USRDS) Annual Data Report, 2009.)

the number and proportion of patients treated by hemodialysis has increased similarly. The number of patients treated by peritoneal dialysis has actually decreased, from 9580 in 1995 to 6734 in 2007. The proportion of patients treated by hemofiltration is very small, as this is predominantly used in intensive care medicine, particularly in the treatment of multiple organ failure (Coles and El Nahas, 2000).

Function of the Kidney. The kidney is responsible for maintaining the osmotic and water balance of the body; for detoxification and the removal of metabolic waste; and for endocrine functions. It operates by a combination of filtration, selective reabsorption, and active secretion. The functional unit of the kidney is the nephron, of which there are approximately 1 million in each kidney. Figure II.5.5.3 shows a schematic representation of a nephron, and its associated specialized blood vessels.

In the glomerulus, the whole blood is filtered. The filtrate contains water and solutes with a molecular weight below 60,000 (small proteins, amino acids, glucose, urea, and ionic solutes), of similar composition to blood plasma. Approximately 10% of the blood volume entering the glomerulus is filtered here. The remaining blood passes into the peritubular capillaries and vasa recta, while the filtrate enters the proximal convoluted tubule. Within the remaining structures of the kidneys, water, small proteins, and solutes are reabsorbed by active and passive processes, while certain metabolites are secreted directly into the tubular fluid. These processes result in a net reabsorption of approximately 98–99% of the water filtered, and total reabsorption of glucose, proteins,

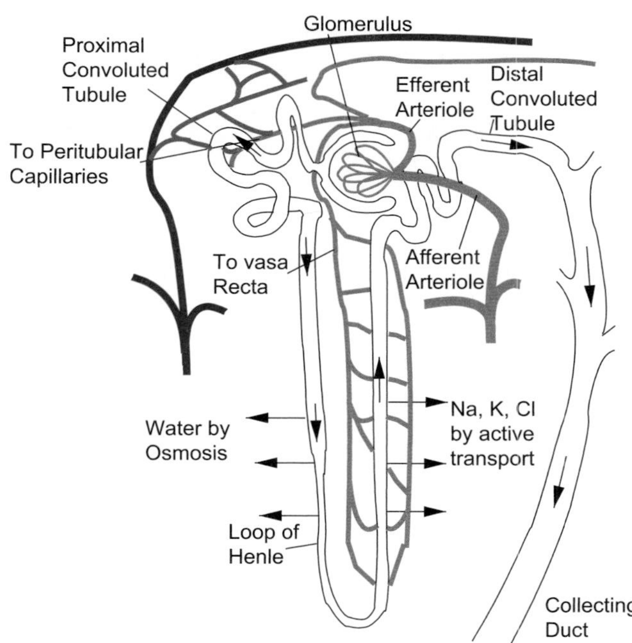

FIGURE II.5.5.3 Structure of a renal nephron.

and amino acids from the filtrate. A healthy individual will excrete 1–2 liters of concentrated urine per day, on average.

In addition to its excretory function, the kidney also has regulatory and endocrine functions. It controls the acid–base balance of the blood, and the water balance of the body, through selective secretion of H^+ into the urine to control pH, and the action of antidiuretic hormone, which controls the urine concentration. The kidney plays a role in the regulation of blood pressure through secretion of renin and tissue hormones (prostaglandins). The kidney also secretes erythropoietin, which promotes red blood cell formation in the bone marrow.

Treatment of Renal Failure. In its initial stages, chronic renal failure (end-stage renal disease or ESRD) is treated conservatively. As the disease progresses, blood purification is needed to remove toxins from the blood (Wing and Jones, 2000). The severity of chronic renal failure may be reduced by restricting a patient's protein, potassium (K^+), and water intake through carefully managed diets, and the administration of bicarbonate supplements to maintain a healthy blood pH. Erythropoietin supplements are given to maintain the patient's haematocrit (Coles and El Nahas, 2000). As the disease progresses, some form of renal replacement therapy becomes necessary.

Peritoneal Dialysis. Peritoneal dialysis (PD) uses the peritoneum, the membrane lining the walls of the abdominal cavity, and the abdominal organs for mass exchange. Dialysate is introduced into the peritoneal cavity through a catheter, as shown in Figure II.5.5.4. While in the peritoneal cavity, metabolic wastes such as urea and creatinine, toxins, and excess solutes (K^+, for example) are transferred from the blood into the dialysate by dialysis, while water is removed by osmosis (Sargent and Gotch, 1996). Used dialysate is then removed.

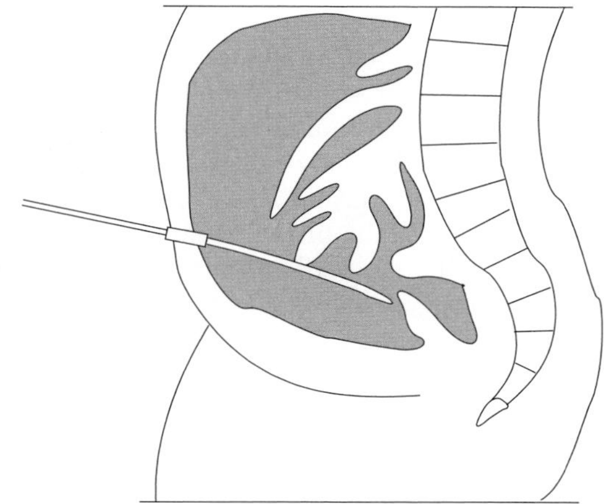

FIGURE II.5.5.4 Positioning of catheter in peritoneal cavity for peritoneal dialysis.

Peritoneal dialysis is dependent on a number of factors. The rate of solute removal is dependent on the blood flow in the capillaries of the peritoneum and its permeability, as well as on concentration gradients between the blood and the dialysate. In addition, the suitability of a patient for treatment by peritoneal dialysis is often dependent on the amount of residual renal function.

There are two principal forms of peritoneal dialysis therapy. *Continuous Ambulatory Peritoneal Dialysis* (CAPD) is the most commonly used form, particularly in younger patients who can manage their own therapy. In this treatment method, the dialysate is infused into the peritoneal cavity, allowed to dwell for a set length of time, then drained and replaced with fresh dialysate. Typically, a patient would perform between five and six exchanges per day (Malchesky, 2004). It has the advantage that it offers patients a certain amount of independence, as they are not tied to any equipment for the majority of the time.

In *Automated Peritoneal Dialysis* (APD), also known as continuous cycler-assisted peritoneal dialysis (CCPD), a machine (cycler) performs the exchanges, usually at night while the patient is sleeping. APD can increase the clearance of metabolites considerably, particularly in patients with rapid rates of exchange between the peritoneum and the dialysate (high or fast transporters). In APD treatment, the cycler will perform several complete exchanges in a night, with a complete cycle of infusion, dwell, and drainage being completed in 30 minutes or less, and treatment for 8–12 hours in total. APD is sometimes referred to as *intermittent peritoneal dialysis*, if the APD is only performed at night. Many patients use a combination of CAPD during the day and APD at night.

Toxin removal in peritoneal dialysis is the sum of the patient's residual renal function and the toxin removal by hemodialysis. The overall removal of toxins, metabolites, and water must be sufficient to keep the patient healthy.

As water removal is dependent on the osmotic pressure of the dialysate, dialysate is supplied in three different strengths: 3.86% glucose; 2.27% glucose; and 1.36% glucose. All three strengths are hypertonic, and will result in water transfer to the dialysate; the rate of transfer is dependent on the osmotic gradient. A more concentrated dialysate will increase the ultrafiltration rate, and hence the removal of solutes as well as water will be enhanced; however, the stronger dialysate concentrations are only used when necessary, and most transfers are performed using the standard (1.36%) dialysate strength.

Peritoneal dialysis has the advantage that the patient's blood does not come into contact with any artificial surfaces; hence there is no need for anticoagulation. It has the disadvantage that the dialysate has to be infused into a body cavity, with the attendant risk of infection (Malchesky, 2004). It is vital to ensure the sterility of the dialyzing solutions, and to prevent infections at the access site. The most common complication of peritoneal dialysis is peritonitis, and this is also the most common reason for the discontinuation of PD. In most cases where peritoneal dialysis has to be discontinued, patients will begin treatment by hemodialysis.

Hemodialysis. Dialysis is the process of solute transfer across a semi-permeable membrane, and is similar in principle to diffusion. It is a passive process, with solute molecules that are small enough to pass through the membrane traveling from an area of high solute concentration to an area of low solute concentration. Typical hemodialysers use membranes with a molecular weight cut-off between 5000 (low-flux membranes) and 20,000 (high-flux membranes) (Boure and Vanholder, 2004), and a surface area of between 0.5 and 2.1 m^2.

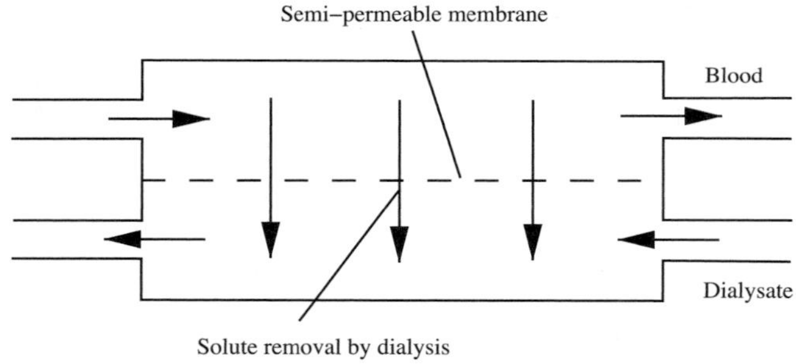

FIGURE II.5.5.5 Principle of dialysis.

In the hemodialyser, blood flows through membrane hollow fibers, with dialysate flowing in counter-current fashion, as shown in Figure II.5.5.5. Metabolic wastes such as urea and creatinine are transferred from the blood into the dialysate, due to the concentration gradient. The dialysate must be carefully formulated to ensure that normal concentrations of glucose and blood solutes are maintained. The blood and dialysate are osmotically balanced; hence any solute removal is by ultrafiltration, due to the differential pressure (P_B – P_D) between the blood and dialysate compartments.

Hemodialysis is an intermittent therapy, and is typically administered every 48 hours, with sessions lasting up to six hours at a time (Malchesky, 2004). The limiting factor for hemodialysis session duration is the rate of water removal, as water can only be drawn out of the blood plasma, and rates of fluid transfer from the intracellular fluid and interstitial fluid pools to the plasma (plasma refilling) must be sufficient to maintain normal blood pressure. A rare complication of hemodialysis, dialysis disequilibrium syndrome, should not be confused with the maintenance of blood pressure by plasma refilling.

Figure II.5.5.6(a) shows a typical hemodialysis circuit, incorporating a blood pump, dialyser, and arterial line filter. Blood is drawn from a dialysis access (Figure II.5.5.6b), usually located in the arm, and pumped through the dialyser (Figure II.5.5.7) at rates varying between 50 and 400 ml/min, depending on the condition of the patient, particularly the blood flow and blood pressure within the access vessel, and the required solute clearance and ultrafiltration rates. The dialysate is either

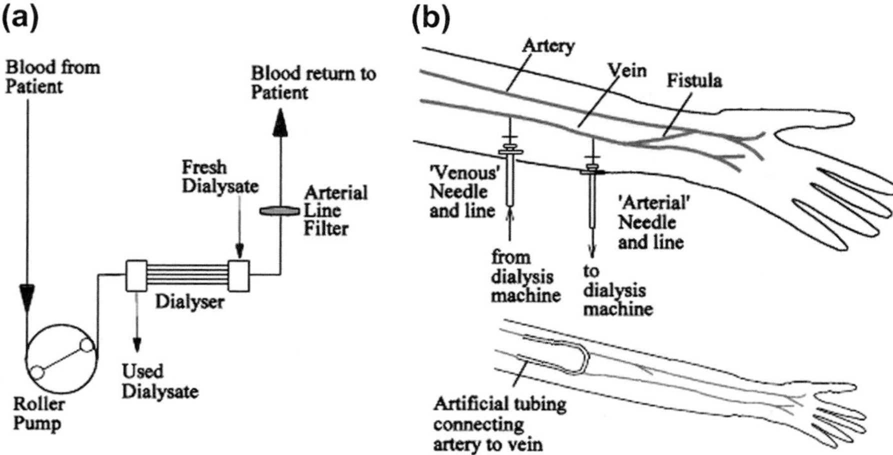

FIGURE II.5.5.6 (a) Circuit for hemodialysis. (b) Hemodialysis access from arteriovenous fistula (top) and from a graft (bottom). Arteriovenous fistula is most commonly used outside the USA (Wing and Jones, 2000).

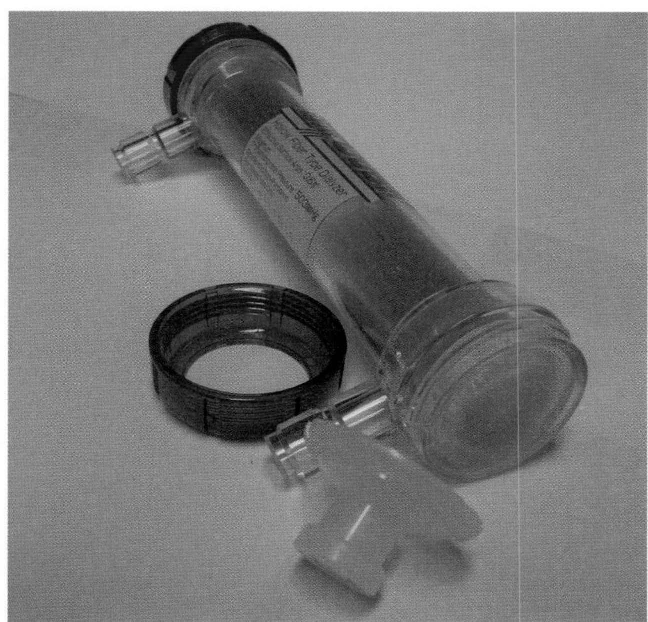

FIGURE II.5.5.7 Modern hollow fiber dialyser (Terumo Clirans C061) with end cap removed to show arrangement of hollow fibers.

prepared in batches or continuously, typically by mixing concentrate and processed mains water. Dialysate flow rates vary within the range of 300–800 ml/min, with most procedures performed at 500 ml/min. A higher dialysate flow rate will increase the clearance of solutes from the blood, but this increase in performance has to be balanced against the increased cost of the dialysate used.

The intermittent nature of treatment means that there is a build-up of metabolites and fluids between treatments, in comparison to the normal continuous function of the kidney. At the time of writing, the standard frequency of treatment is once every 48 hours for most patients, with sessions lasting up to six hours, although it has been shown that more frequent treatment gives better clinical outcomes. Frequency of hemodialysis treatments is therefore a balance between efficacy of treatment, cost, and convenience.

Membrane Materials. The first membranes to be used for hemodialysis were sausage skins made of regenerated cellulose (cellophane), as this was the only suitable membrane material available to Willem Kolff at the time (Broers, 2006). Regenerated cellulose thus became the standard hemodialysis membrane, and remains in clinical use. Although mass transport through these membranes is satisfactory, the molecular weight cut-off is only 5000 (low flux), and there are a number of biocompatibility issues, particularly complement activation (Boure and Vanholder, 2004; Kay and Raij, 1987). These membranes are normally referred to as unmodified cellulosic membranes. In order to improve the biocompatibility of cellulosic membranes, chemical processes are used to mask hydroxyl groups on the surface of the membrane – these membranes are known as modified cellulose.

TABLE II.5.5.1 Types of Hemodialysis Membranes (Boure and Vanholder, 2004)

Unmodified Cellulose	Modified Cellulose	Synthetic
Low-flux	Low-flux	Low-flux
Cuprophan®	Hemophan®	Polysulfone
Cellulose diacetate		Polycarbonate
Cuprammonium rayon		
	High-flux	High-flux
	Cellulose triacetate	Polysulfone
		Polyamide
		Polyethersulfone
		Polyacrylonitrile
		Polymethylmethacrylate

Synthetic membranes such as polysulfone and polyamide have also been developed. These membranes have better biocompatibility than unmodified cellulosic membranes, but poorer than modified cellulose. They can be produced in both low flux and high flux membrane types. Table II.5.5.1 gives some examples of the polymers used in hemodialysis membranes.

The principal advantage of synthetic membranes over modified cellulosic membranes is the range of pore sizes available. A further advantage of these synthetic membranes is their adsorption capability, particularly for the adsorption of bacterial by-products which may enter the dialysate circuit. However, due to the low membrane areas present, this adsorption capability is limited (Boure and Vanholder, 2004).

As with all blood-contacting biomaterials, the following compatibility issues are present: complement activation; platelet adhesion; and thrombus formation. In artificial organs, complement activation normally takes place through the alternate pathway, leading to systemic inflammation in the patient and a post-procedural drop in leukocyte count (leukopenia) (Kay and Raij, 1987; Tayama et al., 1999). Contact with artificial materials will initiate coagulation via the intrinsic pathway, through activation of factor XII and the release of platelet phospholipids (Mackay and Arrowsmith, 2004; Hanson, 2004). Thrombus formation is prevented by the use of heparin, which accelerates the action of antithrombin III in blocking the coagulation pathway, preventing the formation of a fibrin clot. It should be noted that heparin will not prevent platelet adhesion to the membrane and platelet activation. Platelet activation and aggregation can lead to the formation of thrombi, and can also lead indirectly to leukocyte activation. All blood returning to the patient is filtered to ensure that thrombi cannot enter the patient's bloodstream.

In addition to the membranes of the hemodialyser, the blood will also come into contact with the tubing and connectors used in the extracorporeal circuit. Commonly used materials for blood tubing include polyvinyl chloride

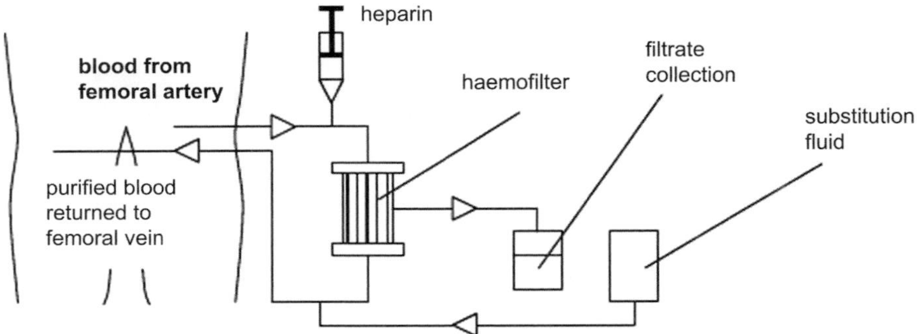

FIGURE II.5.5.8 Schematic showing access and extracorporeal circuit for hemofiltration.

(PVC), polyurethane, and silicone rubber (PDMS, polydimethyl siloxane). However, the area of these materials, and the exposure time, is much lower than for the functional membrane surface.

Extracorporeal Hemofiltration. Extracorporeal hemofiltration is mostly used in the treatment of acute renal failure, particularly where renal failure is accompanied by fluid overload. The toxins and excess water are removed by ultrafiltration, and fluid balance is maintained by addition of an isotonic substitution fluid "downstream" of the filter (see Figure II.5.5.8). This method is particularly effective in the removal of toxins, although as the substitution fluid is mixed directly with the blood, and hence enters the body directly, it is very expensive to prepare. Its principal advantage over dialysis is the enhanced removal of solutes from the blood, of particular importance in cases of poisoning and multiple organ failure.

The most modern hemofiltration techniques use pure hemofiltration (convective transfer) in combination with dialysis (diffusive transfer) to achieve the desired blood purification. This technique is usually referred to as hemodiafiltration or high flux dialysis; the set-up is similar to that for conventional dialysis, although the vascular access is still artero-venous (Malchesky, 2004).

Hemoperfusion

The term hemoperfusion applies to the direct perfusion of whole blood over a sorbent bed or reactor. The sorbent used may be activated charcoal, non-ionic or ionic resins, or immunosorbents, while reactors may contain enzymes, cells or tissues. The purpose of the treatment is either to remove specific toxins or metabilites, or to carry out specific biochemical reactions. Issues with biocompatibility, safety, and immunological isolation have limited the application of hemoperfusion to the use of charcoal and resins as sorbents (Kambic and Nose, 1993).

Sorption of the solute into the sorbent is based on chemical affinity, rather than molecular size. Due to the nature of the sorbent granules, the surface area for sorption is very high, but it also makes it possible for particulates (fines) to be released from the reactor into the blood. The nonspecific nature of charcoal sorption can also lead to the removal of beneficial compounds from the blood. For these reasons, hemoperfusion has largely been superceded by plasma treatment (plasma perfusion) and sorbent dialysis, which have a similar therapeutic effect with fewer potential hazards.

The very large surface area of sorbents also leads to extensive blood–biomaterial contact, resulting in complement activation and initiation of the coagulation cascade. These events will cause a reduction in leukocyte and platelet levels in the blood.

Therapeutic Apheresis

"Apheresis" describes any technique in which a fraction of the blood (platelets, plasma, red blood cells, leukocytes) is removed, and the remaining blood is returned to the donor. This can be divided into "donor apheresis," in which volunteers give blood fractions for the treatment of others, and "therapeutic apheresis," where blood fractions are selectively removed to achieve a therapeutic result (Zydney, 2006).

Therapeutic apheresis can be used to treat diseases where abnormal blood proteins or cells are present in the bloodstream, and these proteins or cells are implicated in the condition's progression. In plasmapheresis, the plasma is separated and either replaced (plasma exchange) or treated prior to recombination with the blood (plasma treatment). In cytapheresis, one or more of the cellular components of the blood is selectively removed. Cytapheresis is overwhelmingly used as a donation procedure, for collection of blood proteins, platelets, and immune cells, with only 0.8% of the 1562 patient cases (13 cases) studied in the 2007 International Apheresis Registry[2] treated with therapeutic cytapheresis.

Plasmapheresis. Plasmapheresis has been used with considerable success in the treatment of a number of conditions, notably myasthenia gravis, leukemia, multiple myeloma, non-Hodgkins lymphoma, and other disorders affecting the blood and blood-forming organs (Zydney, 2006; Kambic and Nose, 1993). Plasmapheresis is also

[2]The 2007 International Apheresis Registry is drawn from 20 participating centers offering apheresis treatment; these numbers should not be taken as estimates of the total number of patients worldwide.

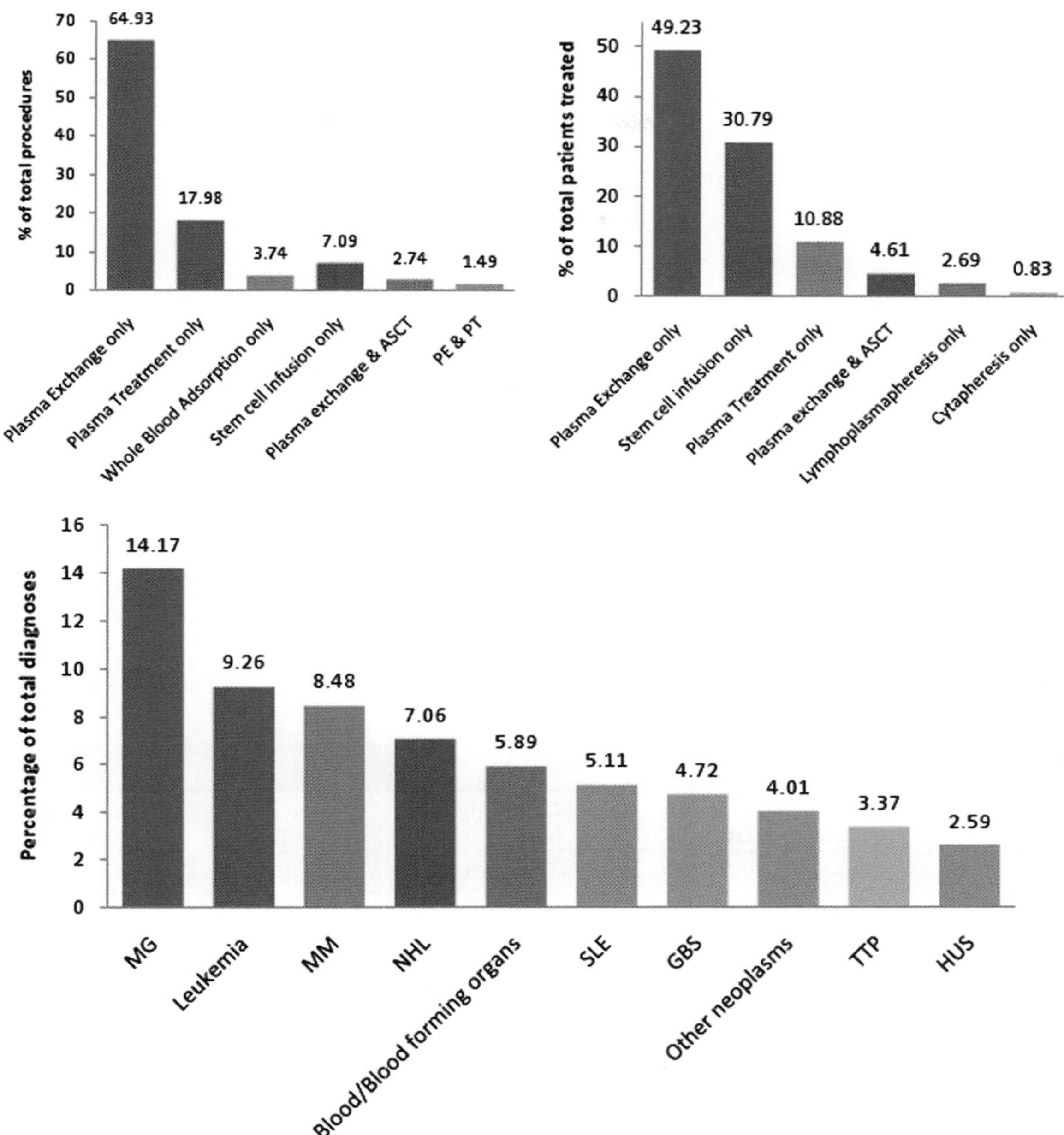

FIGURE II.5.5.9 Modes of treatment and conditions treated by apheresis, from the 2007 International Apheresis Registry (Malchesky et al., 2009). ASCT = Autologous Stem Cell Therapy; GBS = Guillan-Barre Syndrome; HUS = Hemolytic Uremic Syndrome; MG = Myasthenia Gravis; MM = Multiple Myeloma; NHL = Non-Hodgkins lymphoma; PE & PT = Plasma Exchange and Plasma Treatment; SLE = Systemic Lupus erythematosus; TTP = Thrombotic thrombocytopenic purpura.

used in the management of possible rejection of transplanted tissue due to ABO mismatch. Figure II.5.5.9 gives a breakdown of the number of treatments, number of patients treated by therapeutic apheresis, and the type of treatment used in the 2007 International Apheresis Registry (Malchesky et al., 2009). Note that stem cell infusion was recorded in the survey, although it is not an apheresis technique. Myasthenia gravis was the most common condition treated, both in terms of the number of patients and the number of treatments (it is a chronic condition, hence repeated treatments would be necessary), followed by leukemia (this figure does not include Hodgkins disease and non-Hodgkins lymphoma), multiple myeloma, and non-Hodgkins lymphoma.

Plasma Separation

Plasma separation can be achieved by centrifugal or membrane methods. Centrifugal methods can be batch or continuous. In a batch method, anticoagulated blood is removed from the patient, and then centrifuged to separate the plasma, red blood cells, and buffy coat (lymphocytes and platelets). In a continuous process, shown schematically in Figure II.5.5.10, blood is fed

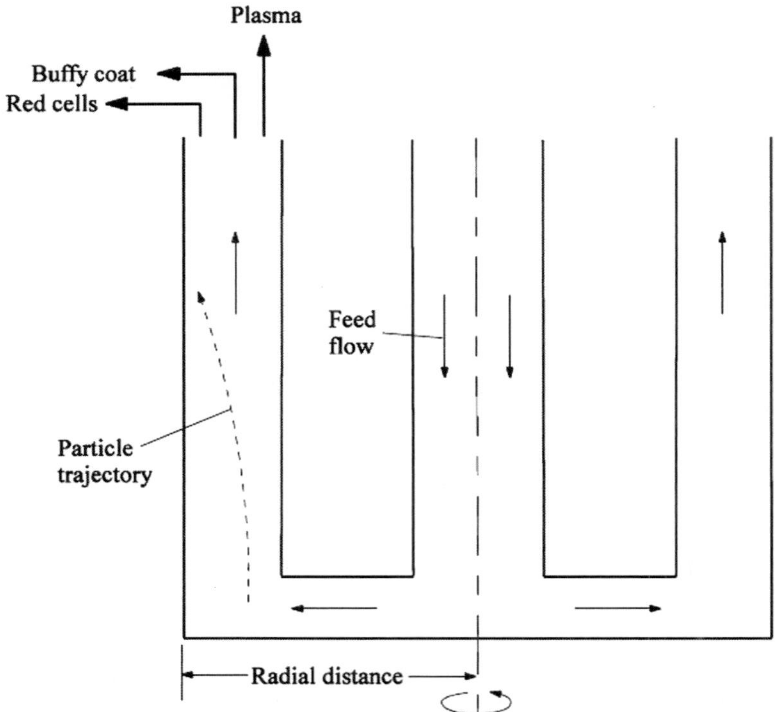

FIGURE II.5.5.10 Centrifugal separation of blood into plasma, buffy coat, and red cells (adapted from Malchesky, 2004).

TABLE II.5.5.2 Cell Properties in Plasmapheresis

	Normal Concentration in Blood (per Liter)	Cell Diameter (μm)	Mean Density (g/ml)	Sedimentation Coefficient (×10^7)
Red blood cell	$4.2–6.2 \times 10^{12}$	8	1.098	12.0
White blood cells	$4.0–11.0 \times 10^{9}$	–	–	1.2
Lymphocyte	$1.5–3.5 \times 10^{9}$	7–18	1.072	–
Granulocyte	$2.5–8.0 \times 10^{9}$	10–15	1.082	–
Monocyte	$0.2–0.8 \times 10^{9}$	12–20	1.062	–
Platelet	$150–400 \times 10^{9}$	2–4	1.058	0.032
Plasma	–	–	1.027	–

Membrane separation is dependent on cell diameter, while centrifugal separation is dependent on sedimentation coefficients.

into the center of a rotating bowl, and centrifugal forces (in the order of 100 G) cause separation into three distinct streams of red blood cells, buffy coat, and plasma (Malchesky, 2004; Zydney, 2006). Table II.5.5.2 gives the concentration, cell diameter, density, and sedimentation coefficient for the constituents of whole blood. The most dense element of the blood is the red blood cells, and hence these will be most affected by the centrifugal forces, followed by the white blood cells, and then the platelets. One stage separation can isolate the red blood cells, buffy coat, and the plasma, but two or more stages are required if concentration of specific white blood cell types is to be achieved. As the velocities used in the separation must not be too high to prevent damage to the cells, it is usual for the plasma fraction to contain some cells – usually platelets.

Membrane plasma separation is dependent on the pore size of the membrane, and it is usual to use pore sizes much smaller than the cells to prevent sieving. Figure II.5.5.11 shows the principle of membrane separation. If the pressure on the blood side is higher than the pressure on the plasma side, then there will be ultrafiltration of water and solutes with a molecular weight lower than the membrane cut-off. Due to the nature of the process, the separated plasma will not contain any cells, and the concentration of macromolecular solutes in the filtrate will be almost identical to the concentration in the plasma.

The production of particulate-free filtrate by membrane plasma separation is considered to be a key advantage of this process over centrifugal plasma separation. The absence of these particulates is particularly important if the plasma is to be treated and then returned to the patient, as the particulates may interfere with the treatment process.

Plasma Exchange. Plasma exchange has already been introduced in the section on hemofiltration. It is the most

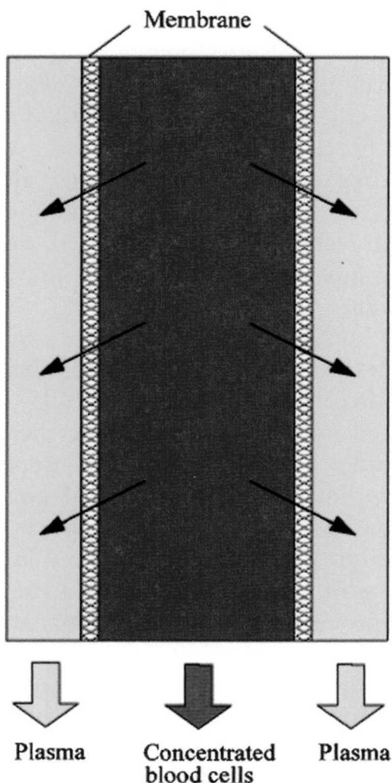

FIGURE II.5.5.11 Membrane plasma separation (adapted from Malchesky, 2004).

used form of plasmapheresis, and is very effective in the removal of toxins from the bloodstream. However, it has the disadvantage that it is indiscriminate – in addition to the toxins and metabolites removed, large quantities of beneficial or physiological proteins including albumins, immunoglobulins, and clotting factors will be removed as well. This limits the volume of plasma that can be removed in a single session, and the infusion of substitution fluids carries the risk of allergic reactions or viral infections. Plasma exchange would usually be used when the specific pathogenic factors in the plasma are unknown or where a specific technique to remove these factors is not available (Malchesky, 2004). It could also be used if a combination of factors is to be removed, and plasma exchange is the simplest and most effective method to achieve this. The substitution fluid typically contains ionic solutes and albumin, but in some cases donated whole plasma or reconstituted plasma containing plasma fractions such as immunoglobulins can be used.

Plasma Treatment. Plasma may be treated by dialysis, membrane filtration, sorption, and processing by enzyme or biological reactions. Dialysis of fractionated plasma is similar to the dialysis of whole blood, and achieves the same result – removal of small solutes with a molecular weight below the membrane's cut-off. In membrane filtration, the plasma is filtered through one or more membranes following separation from the blood. The membrane cut-off of these membranes would be larger than a dialysis membrane, but smaller than the cut-off of the membrane plasma separator, so that larger molecules in the plasma would be filtered out. It is also possible to perform membrane filtration at temperatures below physiological levels (cryofiltration) and above physiological levels (thermofiltration). Cyrofiltration is used to remove serum solutes which aggregate at lower temperatures, while thermofiltration is used to selectively remove low density lipoproteins (LDL) in the treatment of hyperlipidemia.

The principle of plasma treatment by sorption is similar to hemoperfusion over a sorbent column. The plasma is perfused through a sorbent column to remove the targeted solutes, then recombined with the concentrated cell fraction and returned to the patient. Sorption has been used to treat poisoning, drug overdose, uremia, liver insufficiency, autoimmune disorders, and familial hypercholesterolemia.

Cytapheresis. Although, as noted earlier, cytapheresis is usually a donation procedure, it has been used clinically to treat leukemia by leukocytapheresis (selective removal of the white blood cells). Leukocytapheresis is normally performed by centrifugation of the whole blood, rather than a membrane method, due to the similar cell diameters of white and red blood cells. Lymphocytapheresis has been investigated as a means to treat autoimmune diseases, including rheumatoid arthritis, and to prevent rejection in renal transplantation. Erthyrocytapheresis (red blood cell removal) has been used to treat sickle cell anaemia and malaria. In addition, thrombocytapheresis has been used to reduce very high levels of platelets in the blood (normally over 10^{12}/liter) (Malchesky, 2004).

While cytapheresis can be performed by centrifugation, it is also possible to use filters for therapeutic cytapheresis, particularly those with an affinity for certain types of blood cell – for example, lymphocytes will adhere preferentially to acrylic and polyester fibers, while granulocytes will adhere to cotton. In these cases, whole blood is perfused through the device and the cells are removed.

Photophoresis can be used to remove white blood cells in the treatment of T-cell lymphoma. The patient is given oral methoxsalen, which binds to the DNA in the nucleus of white blood cells. Blood is taken from the patient and centrifuged to separate the plasma and buffy coat, which are irradiated with UVA light. The UVA radiation activates the drug, which blocks replication of the white blood cells. After treatment, the blood is recombined and returned to the patient (Malchesky, 2004).

Sorbent Dialysis

The main area of application for sorbent dialysis is the treatment of acute liver failure, a condition for which the medical prognosis is usually poor (Bosch, 2005; Galletti and Jauregui, 2006). There has been considerable research into the development of a bioartificial liver (extra-corporeal liver assist device or ELAD); however, at the time of writing (January 2012), clinical trials of ELADs have not shown significant improvement in patient outcome over conventional intensive care medicine.

Improvement in patient condition has been shown as a result of treatment by albumin dialysis (molecular adsorbent recirculating system or MARS), and by fractionated plasma separation and adsorption (Prometheus), a form of plasmapheresis, discussed above.

The MARS system passes the patient's blood through a high flux dialyser, and albumin-bound toxins in the patient's blood diffuse through the membrane to bind with albumin recirculating in the primary dialysate compartment. The dialysate coming out of the dialyser is regenerated in a series of sorbent columns (typically activated charcoal and anion exchange resin) which will remove toxins from the albumin, and a low-flux dialyser which removes water-soluble toxins. Prospective randomized controlled trials of MARS treatment in acute-on-chronic hepatic failure have shown significantly improved patient survival compared to the control group (patients receiving conventional ICU care). A meta-analysis of MARS trials has shown that further studies are needed before a final judgment can be made regarding the superiority of the MARS treatment (Bosch, 2005).

Blood Oxygenation

Applications of Extracorporeal Oxygenation. Extracorporeal oxygenation is required if respiratory gas exchange in the lungs is insufficient, usually due to acute respiratory distress syndrome or ARDS. If used to treat ARDS over a prolonged period, this is termed extracorporeal life support or ECLS. It is also used to maintain vital function during isolation of the heart and lungs for surgery, termed cardiopulmonary bypass or CPB, with a duration of a few hours (Murphy and Bryan, 2004; Mulholland, 2008; Lim, 2006).

Extracorporeal life support is often subdivided into extracorporeal membrane oxygenation or ECMO, and extracorporeal CO_2 removal, $ECCO_2R$. In ECMO, the oxygenator supplies all of the patient's metabolic oxygen demand and removes CO_2. $ECCO_2R$ is a partial bypass technique in which the oxygenator is used to remove the metabolically generated CO_2 from the blood. Only 20–30% of the cardiac output is treated in the oxygenator, which is sufficient to remove all the CO_2 generated metabolically. The lungs are kept filled with oxygen by low-frequency positive-pressure ventilation, and uptake of oxygen from the lungs supplies the remainder of the patient's oxygen demand. In all cases of ECLS, whether treating neonates or adults, the aim of treatment is to promote the recovery of function in the patient's lungs.

In cardiopulmonary bypass, the heart and pulmonary circulation are isolated and vital function is maintained by a pump and oxygenator. Blood is taken from the venous return to the heart in the vena cava, and returned to the arterial tree in the aorta. Blood scavenged from the operation site is also oxygenated and returned to the patient (Murphy and Bryan, 2004; Mulholland, 2008; Lim, 2006; Galletti and Colton, 2006). Cardiopulmonary bypass procedures are typically used during coronary artery bypass grafting procedures, although they are also used for heart or lung transplantation.

As surgical techniques have advanced, the number of procedures requiring the maintenance of vital function by an oxygenator has reduced. In 1995, there were an estimated 573,000 coronary artery bypass graft (CABG) procedures in the USA, whereas in 2003 there were only 466,369 (source: OECD Indicators, 2005). In addition to the reduced number of procedures, there are also a growing number of CABG procedures done on a beating heart, without the extracorporeal circuit (usually termed off-pump CABG). The reduction in numbers of CABG procedures can also be attributed in part to better monitoring of patients at risk, and a rise in coronary angioplasty procedures, from an estimated 433,000 in 1995 to 1,237,000 in 2003 (OECD, 2005). Membrane oxygenators can also be used to maintain vital function during procedures to correct congenital defects in neonates. Numbers of ECLS procedures are much smaller, as this is an acute therapy used in the treatment of adult and neonatal respiratory failure.

Equipment Used in Extracorporeal Oxygenation. The circuit for extracorporeal oxygenation includes a pump, reservoir, heat exchanger, and oxygenator, as shown schematically in Figure II.5.5.12. Additional pumps used for scavenging of blood from the operation site and venous return are not shown. Most of the oxygenators available at the time of writing (January 2012) combine a reservoir, heat exchanger, and oxygenator compartment, as shown in Figure II.5.5.13. The heat exchanger is used to control the patient's temperature. While ECLS is usually carried out in normothermic conditions, deep core hypothermia is induced in CPB procedures to reduce the patient's metabolic oxygen demand. In addition to the advantages of reducing oxygen demand, reducing blood temperature will also increase the volume of oxygen that can be transported by the blood for a given O_2 partial pressure.

Within the oxygenator, the blood and gas are separated by a membrane which is permeable only to gas. The membrane may be either microporous or homogeneous. The microporous membranes rely on the surface tension of the blood and the hydrophobic nature of the membrane material to prevent liquid intrusion into the pores. Homogeneous membranes operate on the principle of solution-diffusion driven by concentration gradients, and although mass flux through these membranes is lower, they can be used for sustained periods. A third class of membranes, composite or hybrid membranes, incorporates a thin homogeneous layer on a microporous support.

The gas transfer processes in microporous and homogeneous membranes are shown schematically in

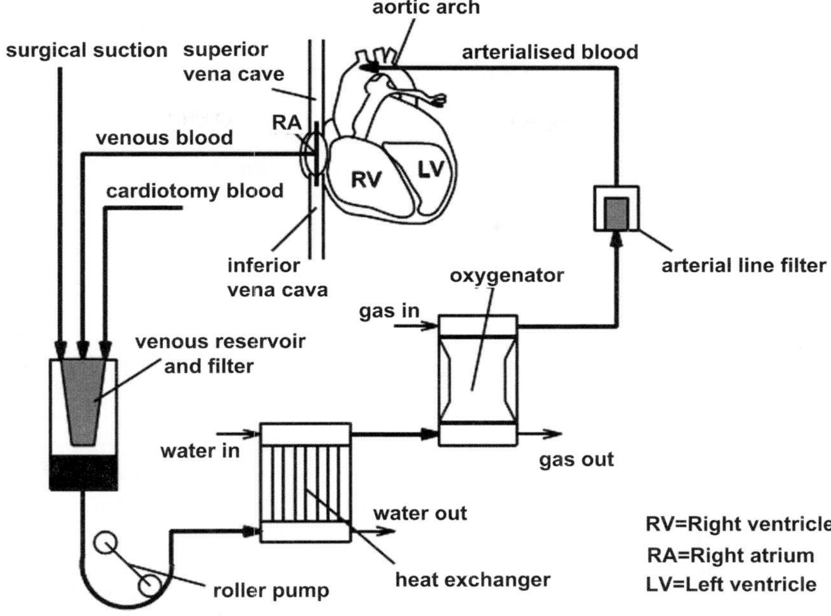

FIGURE II.5.5.12 Circuit for cardiopulmonary bypass.

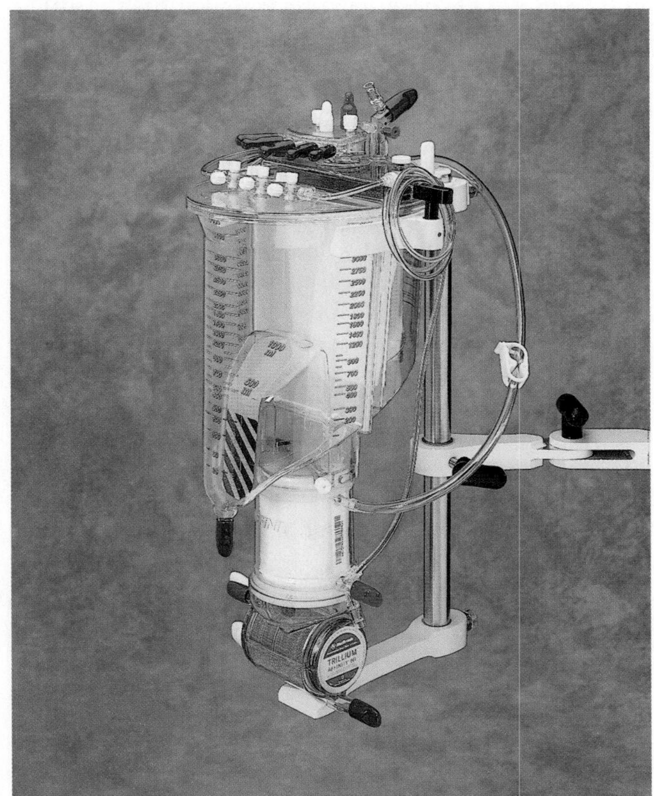

FIGURE II.5.5.13 The Medtronic Affinity® oxygenator with integral heat exchanger and attached hard shell blood reservoir. Note the mesh filter within the reservoir to prevent passage of macroemboli through the heat exchanger and oxygenator. (Image courtesy of Medtronic, Inc., Minneapolis, MN.)

Figure II.5.5.14. In the microporous membrane, gas transport through the pores is by diffusion, driven by concentration gradients. At the gas–blood interface, oxygen must first dissolve into the blood before diffusing

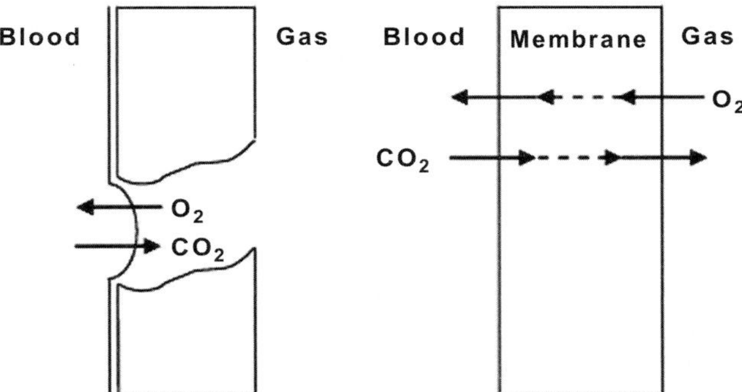

FIGURE II.5.5.14 Principle of gas exchange in microporous and homogeneous membranes (Campbell Ritchie et al., 1996).

through the blood plasma and binding to hemoglobin within the red blood cells. Similarly, CO_2 is released from the bound state (bound to plasma proteins and hemoglobin) into the plasma, and then from the plasma into the gas within the pore. The limiting factors on the gas transfer are the solubility and mobility of the gas molecules in blood plasma. CO_2 is both more soluble and more mobile, and hence the limiting factor in microporous membrane oxygenators is the oxygenation rate. The pore diameter and void fraction of the membrane are also factors that influence the performance of the oxygenator.

The porous nature of the membrane gives rise to the phenomenon of "pore wetting," where the blood plasma intrudes into the pores, eventually causing plasma to leak through the pore into the gas compartment. It has been shown that this phenomenon occurs more rapidly if there is a high concentration of polar phospholipids in the blood, and it has been ascribed to a loss of hydrophobicity of the membrane as phospholipids coat the membrane surface, and particularly the pores of the membrane. As microporous membranes are run in conditions of overpressure, where the blood is at a higher pressure than the gas, the blood pressure will drive liquid through the pores if the hydrophobic repulsion is overcome. Overpressure ensures that in the event of membrane rupture or leakage, blood will leak out rather than gas leaking in, with the attendant risk of gaseous emboli entering the systemic circulation.

For this reason, membrane oxygenators incorporating microporous membranes are only approved for short-term procedures such as cardiopulmonary bypass. For extracorporeal life support, homogeneous membranes are required. The most commonly used homogeneous membranes are made from polydimethylsiloxane or PDMS, usually reinforced by a fabric mesh support. In homogeneous membranes, gas molecules dissolve into the solid matrix before diffusing through the material. Composite membranes, with a very thin homogeneous layer on top of a microporous support, have also been developed (Karichev and Muler, 2001; Nogawa, 2002).

The permeability coefficients of a number of materials of importance in membrane oxygenators are given in Table II.5.5.3. As can be seen from the table, PDMS is the most permeable membrane material, and has completely displaced polyethylene and PTFE (polytetrafluoroethylene), which were used in early membrane oxygenators (Lim, 2006; Galletti and Colton, 2006). Microporous membranes rely on diffusion through the air in the pores, a much faster process than solution-diffusion through PDMS, (Campbell Ritchie et al., 1996) and it can be seen from Table II.5.5.3 that a microporous membrane with a typical porosity of 30–40% and thickness of 30 μm will have a much higher gas flux than a homogeneous membrane of similar thickness, and hence the membrane areas of homogeneous oxygenators are much larger (most adult oxygenators using microporous membranes have a membrane area of approximately 2.0 m^2, compared to 3.5 m^2 for homogeneous membranes). There is therefore a balance between the intrinsically more biocompatible nature of homogeneous membranes, where there is no direct air–blood contact, and the effect of exposure to a larger area of a non-biological material.

Biocompatibility. Extracoporeal oxygenation is the most challenging of the extracorporeal procedures in clinical use, due to the high blood flow rates required, extent of biomaterial exposure, and the tissue trauma inherent in the surgical procedure. During a cardiopulmonary bypass procedure, performed on an open heart, there will be extensive trauma to skin, bone, and cardiovascular tissues, resulting in the release of tissue factor and activation of the extrinsic pathway of coagulation. Exposure of the blood to the biomaterials of the extracorporeal circuit will result in platelet activation, activation of the intrinsic pathway of coagulation, and complement activation via the alternate pathway (Tayama et al., 1999; Hanson, 2004). In addition to these issues, there will also be mechanical damage to the formed elements of the blood due to the high flow rates required (Lever, 2005) and the action of the blood pump, which is discussed in more detail below.

As with hemodialysis, heparin is used to prevent thrombus formation (Murphy and Bryan, 2004; Mulholland, 2008). As heparin blocks the activation of fibrinogen at a late point in the coagulation cascade, there will be platelet reduction and depletion of blood factors. This can result in postoperative bleeding or coagulopathy. The action of heparin is reversed by protamine once the procedure is completed (Mulholland, 2008).

Complement and platelet activation will result in a systemic inflammatory response: cytokine release; activation of endothelial cells; and migration of activated macrophages into peripheral tissues. Activation of lymphocytes and lymphocyte destruction or sequestration in the extracorporeal circuit results in a post-procedural drop in circulating lymphocytes to below 50% of the preoperative level. This challenge to the immune system can make patients more vulnerable to postoperative infection.

In cardiopulmonary bypass, some thrombus formation is inevitable, and although the arterial line filter will

TABLE II.5.5.3 O_2 and CO_2 Permeability Coefficients for Materials Important to Blood Oxygenation

Material	Common Name	O_2 Permeability (ml mm min^{-1} m^{-2} atm^{-1})	$\frac{P_m(CO_2)}{P_m(O_2)}$
Air	–	1.27×10^6	0.8
Polydimethylsiloxane	Silicone rubber	27.94	5
Water	–	3.810	18
Polystyrene	–	1.397	5
Polyisopropene	Natural rubber	1.270	6
Polybutadiene	Butyl rubber	1.016	7
Polyethylene	–	0.3048	5
Polytetrafluoroethylene	Teflon (PTFE)	0.2032	3
Polyvinylidene chloride	–	2.540×10^{-4}	6

The permeability coefficient is defined as the product of the diffusion coefficient, *D*, and the Henry's law solubility coefficient, α. (Adapted from Galletti and Colton, 2006).

remove emboli >40 μm, other emboli can pass through and lodge in smaller blood vessels, resulting in widespread low level cell necrosis.

Blood Pumps in Extracorporeal Circulation

The common factor in all the extracorporeal circuits described in this chapter is the need for a pump to move the blood through the circuit. In apheresis and hemodialysis, roller pumps are used as blood flow rates are relatively small and the flexible tubing required for pumping can be easily incorporated into a single-use tubing kit. In extracorporeal oxygenation, much higher blood flow rates are needed, and a variety of pumps are used.

While a full discussion of the operating principles and efficiency of the pumps used in extracorporeal circulation lies beyond the scope of this chapter, mechanical pumping of the blood will have an effect on its condition, and may add to the complications of the procedure.

Roller Pumps. In a roller pump, a section of flexible tubing is mounted against a rigid outer support, as shown in Figure II.5.5.15. The rollers compress a section of tubing, preventing backflow, and pushing the blood forwards. The flow from a roller pump is pulsatile, which can have a beneficial effect on mass transfer, and if the rollers occlude the tubing completely the flow rate is dependent on the speed of rotation of the roller head, and relatively independent of the back-pressure due to the flow resistance of the extracorporeal circuit.

The roller pump is very versatile, and flow can be reversed if this is desired. It has the disadvantage that it can damage the blood, due to high shear rates induced in the vicinity of the rollers during occlusion, and particularly during the reopening of the tubing once the roller has passed. During reopening, there may be momentary backflows with very high velocities. High shear on the cell membranes can lead to premature aging of the red blood cells, which leads to clumping, and in extreme cases, to hemolysis of the red blood cells, releasing hemoglobin into the plasma.

Platelets and leukocytes are more susceptible to shear damage than red blood cells, and hence the mechanical trauma of pumping will add to the effects of blood–biomaterial contact, particularly platelet activation and leukopenia. The degree of trauma is considerably increased when high flow rates are needed – hence the roller pump is most often used in applications where flow rates are low, such as dialysis and apheresis.

Centrifugal Pumps. The earliest type of centrifugal pump used in cardiopulmonary bypass was the conventional vaned type, where vanes on the rotating impeller accelerate the blood, pushing it out into the volute, where it decelerates. The kinetic energy is converted to potential energy, providing the driving pressure (pressure head) to push the blood through the extracorporeal circuit. The effect of the whirling vanes on the flow patterns and static pressure of the blood can be seen from the velocity and pressure plots in Figure II.5.5.16.

The centrifugal pump is more efficient and less damaging than the rotary pump (Nishinaka et al., 1996), but has the disadvantage that it is designed to operate over a smaller range of operating conditions (usually termed the design point, a specified combination of pressure head and flow rate). Operating a turbine type pump under conditions away from the design point can lead to an increase in the shear imparted to the blood, and in extreme conditions to cavitation, the formation of vapor bubbles in areas of very low pressure. Cavitation will cause severe hemolysis. In addition, the high velocity of the rotor tips will add to the shear-induced damage to the blood.

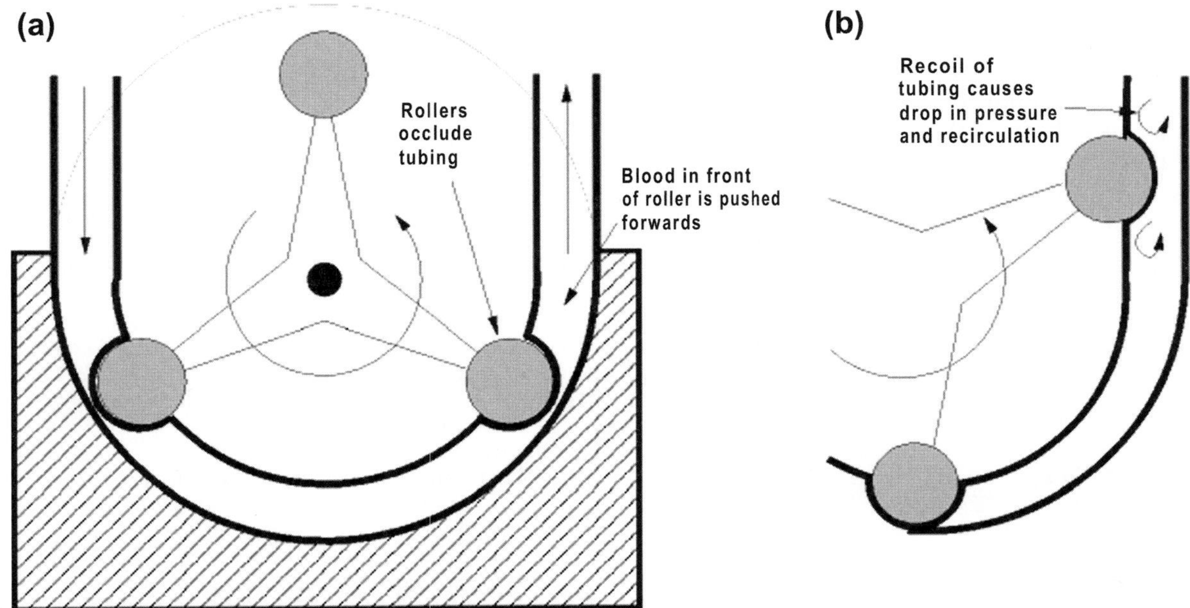

FIGURE II.5.5.15 Principle of operation of a roller pump.

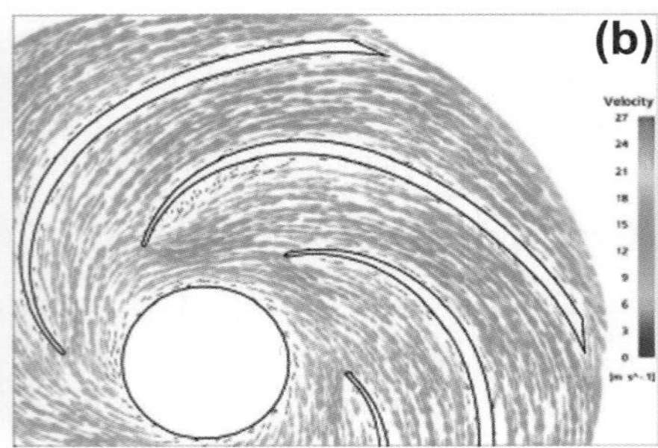

FIGURE II.5.5.16 (a) Pressure and (b) velocity for flow through a vaned centrifugal pump operating at the design point (Cheah, 2009). (Images courtesy of Cheah Kean Wee, with permission.)

FIGURE II.5.5.17 The Bio-Pump® Plus vaneless centrifugal pump. (Image courtesy of Medtronic, Inc., Minneapolis, MN.)

To address these shortcomings the vaneless centrifugal pump which uses a spinning cone, rather than a vaned impeller, was developed to accelerate the fluid. The Medtronic Bio-Pump®, shown in Figure II.5.5.17, is one example of a vaneless centrifugal pump. As the blood flows through the pump, the cone diameter increases, and the surface velocity increases accordingly. The viscosity of the blood is key to the operation of the Bio-Pump, as it relies on the friction between the spinning cone and the blood. The Bio-Pump is much less damaging to the blood than a conventional vaned pump, at the cost of reduced pumping efficiency.

Most centrifugal pumps use magnetic coupling between the drive motor and the rotating impeller, which allows the pumping head to be a disposable sealed unit, suitable for single use only.

BIBLIOGRAPHY

Bosch, T. (2005). Therapeutic apheresis – State of the art in the year 2005. *Therapeutic Apheresis and Dialysis*, **9**, 459.

Boure, T., & Vanholder, R. (2004). Which dialyser membrane to choose? (Editorial comment), *Nephrology Dialysis Transplantation*, **19**, 293.

Broers, H. (2006). *Inventor for Life – the Story of W. J. Kolff, Father of Artificial Organs*. Kampen, The Netherlands: Trans. K. Ashton, B&V Media Publishers.

Campbell Ritchie, A., Bowry, K., Fisher, A. C., & Gaylor, J. D.S. (1996). A novel automated method for the determination of membrane permeability in gas–liquid transfer applications. *Journal of Membrane Science*, **121**, 169.

Cheah, K. W. (2009). *Numerical Simulation of Centrifugal Pump*. PhD Thesis: National University of Singapore.

Coles, G. A., & El Nahas, A. M. (2000). Mechanisms and Clinical Management of Chronic Renal Failure. In A. M. El Nahas (Ed.), *Mechanisms and Clinical Management of Chronic Renal Failure* (2nd ed.). Oxford, UK: Oxford University Press.

Galletti, P. M., & Colton, C. K. (2006). Artificial Lungs and Blood-Gas Exchange Devices. In J. D. Bronzino (Ed.), *The Biomedical Engineering Handbook: Tissue Engineering and Artificial Organs* (3rd ed.). Boca Raton, FL: CRC Press.

Galletti, P. M., & Jauregui, H. O. (2006). Liver Support Systems. In J. D. Bronzino (Ed.), *The Biomedical Engineering Handbook: Tissue Engineering and Artificial Organs* (3rd ed.). Boca Raton, FL: CRC Press.

Gifford, D., & Gray, S. J. (2004). Cardiopulmonary Equipment. In J. H. Mackay, & J. E. Arrowsmith (Eds.), *Core Topics in Cardiac Anaesthesia* (pp. 253–258). Cambridge, UK: Greenwich Medical Media.

Hanson, S. R. (2004). Blood Coagulation and Blood-Materials Interactions. In B. D. Ratner, A. S. Hoffman, F. J. Schoen, & J. E. Lemons (Eds.), *Biomaterials Science: An Introduction to Materials in Medicine* (2nd ed.). San Diego, CA: Academic Press.

Kambic, H. E., & Nose, Y. (1993). Plasmapheresis: Historical perspective, therapeutic applications, and new frontiers. *Artificial Organs*, **17**, 850.

Karichev, Z. R., & Muler, A. L. (2001). Composite hollow-fiber membranes in blood oxygenation. *Theoretical Foundations of Chemical Engineering*, **35**, 383.

Kay, N. E., & Raij, L. (1987). Differential effect of hemodialysis membranes on human lymphocyte natural killer function. *Artificial Organs*, **11**, 165.

Lever, M. J. (2005). Artificial Exchange Systems. In L. L. Hench, & J. R. Jones (Eds.), *Biomaterials, Artificial Organs and Tissue Engineering*. Cambridge, UK: Woodhead.

Lewandowski, K. (2000). Extracorporeal membrane oxygenation for severe acute respiratory failure. *Critical Care*, **4**, 156.

Lim, M. W. (2006). Historical Note: The history of extracorporeal oxygenators. *Anaesthesia*, **61**, 984–995.

Mackay, J. H., & Arrowsmith, J. E. (2004). Coagulopathy during Cardiopulmonary Bypass. In J. H. Mackay, & J. E. Arrowsmith (Eds.), *Core Topics in Cardiac Anaesthesia*. Cambridge, UK: Greenwich Medical Media.

Malchesky, P. S. (2004). Extracorporeal Artificial Organs. In B. D. Ratner, A. S. Hoffman, F. J. Schoen, & J. E. Lemons (Eds.), *Biomaterials Science: An Introduction to Materials in Medicine* (2nd ed.). San Diego, CA: Academic Press.

Malchesky, P. S., Koo, A. P., Skibinski, C. I., Hadsell, A. T., & Rybicki, L. A. (2009). Apheresis technologies and clinical applications: The 2007 International Apheresis Registry. *Therapeutic Apheresis and Dialysis*, published online: Aug 11 2009.
Mulholland, J. W. (2008). Cardiopulmonary Bypass. *Surgery (Oxford)*, **26**(12), 486–488.
Murphy, G. J., & Bryan, A. J. (2004). Cardiopulmonary Bypass. *Surgery (Oxford)*, **22**(6), 126–128.
Nishinaka, T., Nishida, H., Endo, M., Miyagishima, M., Ohtsuka, G., et al. (1996). Less blood damage in the impeller centrifugal pump: A comparative study with the roller pump in open heart surgery. *Artificial Organs*, **20**, 707.
Nogawa, A. (2002). The future of artificial lungs: An industry perspective. *Journal of Artificial Organs*, **5**, 211.
Organisation for Economic Co-operation and Development. (2005). *Health at a Glance – OECD Indicators 2005*: OECD Publishing.
Sargent, J. A., & Gotch, F. A. (1996). Principles and Biophysics of Dialysis. In C. Jacobs, C. M. Kjellstrand, K. M. Koch, & J. F. Winchester (Eds.), *Replacement of Renal Function by Dialysis* (4th ed.). Netherlands: Springer.
Tayama, E., Hayashida, N., Oda, T., Tomoeda, H., Akasu, K., et al. (1999). Recovery from lymphocytopenia following extracoporeal circulation: Simple indicator to assess surgical stress. *Artificial Organs*, **23**, 736.
United States Renal Data Service (USRDS). (2009). *USRDS 2009 Annual Data Report: Atlas of End-Stage Renal Disease in the United States*. Bethesda, MD: National Institutes of Health, National Institute of Diabetes and Digestive and Kidney Diseases.
Wing, A. J., & Jones, E. (2000). Epidemiology of End-Stage Failure: A Global Perspective. In A. M. El Nahas (Ed.), *Mechanisms and Clinical Management of Chronic Renal Failure* (2nd ed.). Oxford, UK: Oxford University Press.
Zydney, A. L. (2006). Therapeutic Apheresis and Blood Fractionation. In J. D. Bronzino (Ed.), *The Biomedical Engineering Handbook: Tissue Engineering and Artificial Organs* (3rd ed.). Boca Raton, FL: CRC Press.

CHAPTER II.5.6 ORTHOPEDIC APPLICATIONS

Nadim James Hallab and Joshua James Jacobs
Department of Orthopedic Surgery, Rush University Medical Center, Chicago, IL, USA

INTRODUCTION

Orthopedic biomaterials are enormously successful in restoring mobility and quality of life to millions of individuals each year. Orthopedic implants include reconstructive implants, fracture management products, spinal products, rehabilitation products, arthroscopy products, electrical stimulation products, and casting products. These products are generally used for fracture fixation enhancement, joint replacement or dynamic stabilization. More specific orthopedic applications within these categories are listed here.

Fracture fixation devices:
- Spinal fixation devices
- Fracture plates
- Wires, pins, and screws
- Intramedullary devices
- Artificial ligaments

Joint replacement (Figure II.5.6.1):
- Hip arthroplasty
- Knee arthroplasty
- Spine arthoplasty
- Ankle arthroplasty
- Shoulder arthroplasty
- Elbow arthroplasty
- Wrist arthroplasty
- Finger arthroplasty

Dynamic stabilization devices (new):
- Spine stabilization devices.

Orthopedic Biomaterials Market

The overwhelming success of orthopedic biomaterials is exemplified by their worldwide market, dominating biomaterial sales at approximately $24 billion in 2007, with an expected growth rate of 7–9% annually. Global sales of trauma fracture management products only totaled approximately $3.7 billion in 2007, whereas $10 billion was spent on knee and hip joint replacements (Figure II.5.6.1). Global sales of knee implant products equaled approximately $5.8 billion in 2007, representing approximately 1.5 million knee replacement surgeries which include first-time joint replacement procedures and revision procedures for replacement, repair or enhancement of an implant product or component from a previous procedure. Revision procedures are growing at an accelerated rate of approximately 60% in the United States.

Orthopedic Biomaterials

Orthopedic biomaterials are generally limited to those materials that withstand cyclic loadbearing applications. While metals, polymers, and ceramics are used in orthopedics, it remains metals, which have over the years uniquely provided appropriate material properties such as high strength, ductility, fracture toughness, hardness, corrosion resistance, formability, and biocompatibility necessary for most loadbearing roles required in fracture fixation and total joint arthroplasty (TJA). The use of orthopedic biomaterials generally falls into one of three surgical specialty categories: upper extremity; spine; or lower extremity; and each specialty is typically divided into three general categories: pediatric; trauma; and reconstruction. Despite these numerous specialties and the hundreds of orthopedic applications, there are only a few orthopedic metals,

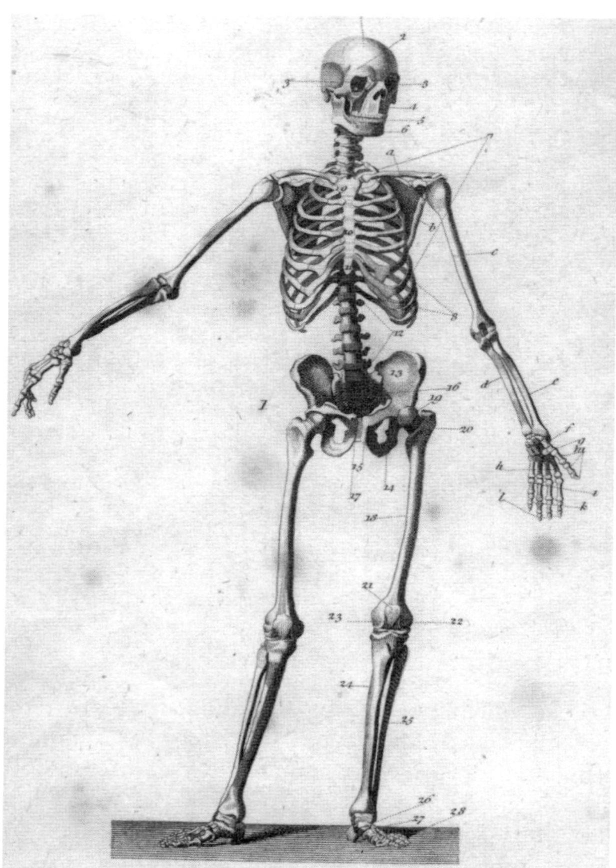

FIGURE II.5.6.1 Total joint arthroplasties (TJA) are currently used to replace hip, knee, shoulder, etc. (Courtesy of BioEngineering Solutions Inc.)

ceramics, and polymers which dominate all implants. Knowing the general properties, uses, and limitations of the "primary" orthopedic biomaterials is requisite to understanding what is required to improve the performance of current implant materials, and why only a few dominate the industry. A summary of seven more prevalent orthopedic biomaterials and their primary use(s) are listed in Table II.5.6.1.

Orthopedic Biomaterials Design

New biomaterials for orthopedic purposes face the same concerns present in current implants: (1) the material must not adversely affect its biological environment; (2) in return the material must not be adversely affected by the surrounding host tissues and fluids; and (3) new materials must exceed the performance of present materials. Thus, understanding the interrelationship between the structure and properties of the natural tissues that are being replaced is important. An appreciation of the "form–function" relationship in calcified tissues will help provide insight into critical factors determining implant design, as well as deciding which materials best meet a specific orthopedic need.

TABLE II.5.6.1 Most Common Orthopedic Biomaterials

Material	Primary Use(s)
Metals	
Ti alloy (Ti-6%Al-4%V)	Plates, screws, TJA components (non-bearing surface)
Co–Cr–Mo alloy	TJA components
Stainless Steel	TJA components, screws, plates, cabling
Polymers	
Polymethylmethacrylate (PMMA)	Bone cement
Ultra-high molecular weight polyethylene (UHMWPE)	Low friction inserts for bearing surfaces in TJA
Ceramics	
Alumina (Al_2O_3)	Bearing surface TJA components
Zirconia (ZrO_2)	Bearing surface TJA components

STRUCTURE AND PROPERTIES OF CALCIFIED TISSUES

There are several different calcified tissues in the human body, and several different ways of categorizing them. All calcified tissues have one thing in common: in addition to the principal protein component, collagen, and small amounts of other organic phases, they all have an inorganic component hydroxyapatite (abbreviated OHAp, HA or $Ca_{10}(PO4)_6(OH)_2$). In the case of long bones, such as the tibia or femur, an understanding of the organization of these two principal components is central to characterization. It has been convenient to treat the structure of compact cortical bone (e.g., the dense bone tissue found in the shafts of long bones) using four levels of organization.

The first, or molecular, level of organization is the collagen triple helical structure (tropocollagen) and OHAp crystallography. It forms a hexagonal unit cell with space group symmetry $P6_3/m$ and lattice constants $a = 9.880$ Å and $c = 6.418$ Å, containing two molecular units, $Ca_5(PO4)_3OH$, per unit cell. How cells produce this mineral phase, and whether it is the first calcium phosphate laid down, are subjects of considerable research at present. Because of its small crystallite size in bone (approximately 2 × 20 × 40 nm), the X-ray diffraction pattern of bone exhibits considerable line broadening, compounding the difficulty of identifying additional phases. A Ca-bearing inorganic compound in one of the components of calcified tissues has led to the development of a whole class of ceramic and glass–ceramic materials that are osteophilic within the body (i.e., they present surfaces that bone chemically attaches to). As yet, we do not know fully how the two components, collagen and OHAp, are arranged and held together at this molecular level. Whatever the arrangement, when it is interfered with (as is apparently the case in certain bone pathologies in which the collagen structure is altered during formation), the result is a bone that is formed or remodeled with seriously compromised physical properties.

The second or ultrastructural level may be loosely defined as the structural level observed with transmission electron microscopy (TEM) or high magnification scanning electron microscopy (SEM). Here, too, we have not yet achieved a full understanding of the collagen–OHAp organization. It appears that the OHAp can be found both inter- and intrafibrillarly within the collagen. At this level, we can model the elastic properties of this essentially two-component system by resorting to some sort of linear superposition of the elastic moduli of each component, weighted by the percent volume concentration of each.

The third or microstructural level of organization is where these fibrillar composites form larger structures, fibers, and fiber bundles, which then pack into lamellar-type units that can be observed with both SEM and optical microscopy. The straight lamellar units forming the plexiform (lamellar) bone are found generally in young quadruped animals, the size of cats and larger. This is the structural level that is described when the term "bone tissue" is used or when histology is generally being discussed. At this level, composite analysis can also be used to model the elastic properties of the tissue, thus providing an understanding of the macroscopic properties of bone. Unfortunately, this modeling is very complex and a complete description lies beyond the scope of this chapter (and the authors). Interested readers are referred to some of the original sources (Katz, 1980a,b).

The fourth level is that of each of the macroscopic levels of each bone sample or large section of bone. Since a significant portion of bone is composed of collagen, it is not surprising to find that, in addition to being anisotropic and inhomogeneous, bone is also viscoelastic like all other biological tissues. Duplicating such properties with long-lasting synthetic biomaterials remains an unrealized goal of orthopedic biomaterials, where the history of implant development has been characterized by the elimination of available candidate materials based on their poor performance, rather than production of biocompatible synthetic bone-mimetic materials.

Perhaps the best example of how orthopedic biomaterials have undergone implant design improvements over the past 100 years resulting in widespread success is total hip replacement (THR) or total hip arthroplasty (THA). Newer types of total joint arthroplasties, such as those currently used for disc replacements (total disc arthroplasty, TDA), have benefitted from the arduous history of the total joint arthroplasty. Many, if not all, of the biomaterial-related issues (both mechanical and biological) that impact the performance of the THR's are applicable to other orthopedic implants. Therefore, this chapter will detail where we have come from, using the total hip arthroplasty, and where we are now, using the example of current total disc arthroplasty designs, general current orthopedic materials technology, and future developments as a proxy of clinical concerns of orthopedic biomaterial development and current technology. The history of total hip arthroplasty is particularly pertinent to biomaterials science, because it is one of the best illustrations of how an implant first used over a century ago has evolved to the highly successful status it has, primarily because of advances in biomaterials.

BIOMATERIALS DEVELOPMENT: A HISTORY OF TOTAL HIP ARTHROPLASTY

The earliest attempts to restore mobility to painful and deformed hip joints took place in the 1820s (White in 1822 and Barton in 1827), and centered on simply removing the affected femoral and acetabular bone involved. This evolved in the 1830–1880s into ghastly attempts to restore mobility using interpositional membranes between the femoral head and acetabulum, where such materials as wooden blocks and animal (e.g., pig) soft tissue were tried. The first prosthetic hip replacement is dated to 1890, when Gluck published a description of a carved ivory femoral head replacement using bone cement-like materials such as pumice and plaster of paris to secure the implants in place (Walker, 1978; Stillwell, 1987).

The interpositional membrane strategy continued from the 19th into the 20th century, where the use of new implant materials in the early 1900s (1900–1920) included organic materials (e.g., pig bladders and peri-implant soft tissues), and inorganic materials such as gold foil. The use of the individual's own soft tissues was the most popular method of interpositional membrane hip surgery. The limited success of this procedure prevented widespread use, and thus the treatment (rather than surgery) of painful, disfigured, and "frozen" (ankylosed) hip joints remained commonplace into the 1920s.

Mold Arthroplasty

It was not until 1923 when Marius Smith-Peterson was credited with ushering in the modern era of total joint replacement with his development of the "mold" arthroplasty (Figure II.5.6.2), made of glass, inspired by a shard of glass found in a patient's back with a benign synovial-like membrane around it. This mold or cup arthroplasty was designed as a cup that fitted in between the femoral head and the acetabular cup, and articulated on both surfaces prompting a "tissue-engineered" synovial/cartilage-like layer. This was the first widespread attempt to develop a better interpositional membrane, a technique that had been in practice for the previous 100 years. The efforts of Smith-Peterson and his colleagues over the years from 1923–1938 were spent improving the fracture resistance of the glass mold arthroplasty cup design, using materials such as early polymers (e.g., celluloid or phenol-formeldehyde Bakelite® or Formica®) and improved glass, e.g., Pyrex®. But it was not until

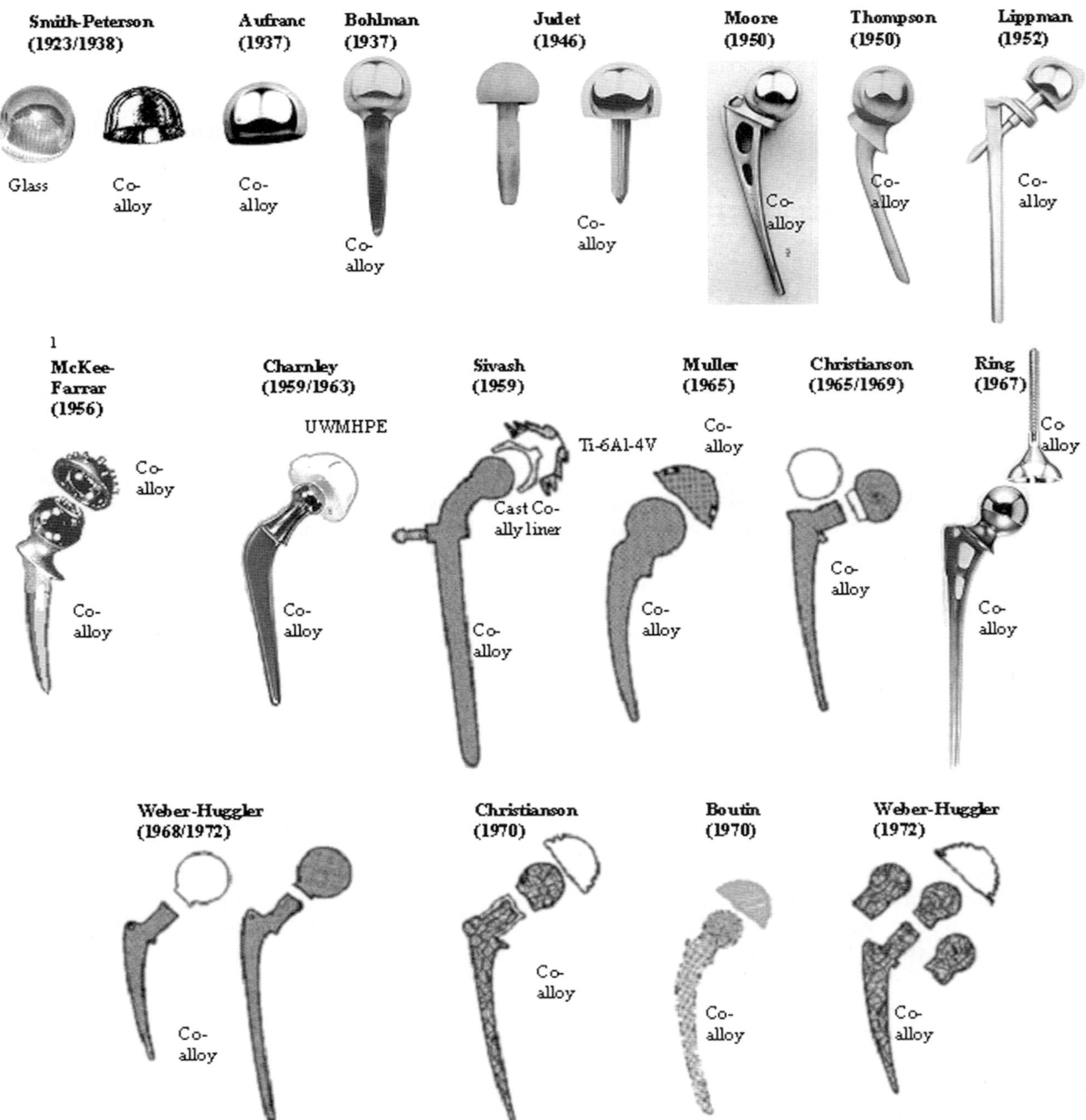

FIGURE II.5.6.2 The history of total hip arthroplasty is particularly pertinent to biomaterials science because it is one of the best illustrations of how an implant first used over a century ago has evolved to the highly successful status it has, primarily because of advances in biomaterials.

1939 when the first metal implant, a cobalt alloy termed Vitallium®, became available and, as used by Venable, Stuck, and Beach, showed that the corrosion resistance of the hip arthroplasty provided sufficient biocompatibility and performance to be incorporated into future popular implants such as the Judet, Moore and Thompson hip arthroplasties.

In 1937 Venable, Stuck, and Beach published a landmark article that was the first to analyze in a systematic fashion the electrolytic effects of various metals and alloys on bone and tissues (e.g., aluminum, copper, iron, nickel, lead, gold, magnesium, silver, stainless steel, and other alloys). They arrived at the conclusion that Vitallium® (a cobalt–chromium alloy) was superior to the other metals in corrosion resistance and in the mechanical properties required for implants (Venable et al., 1937; Charnley, 1979). By observing the effects of corrosion, and proposing guidelines for performance Venable, Stuck, and Beach set the standard by which future metallic alloys were selected for use in hip and other types of implants.

The superior material properties of the Vitallium® alloy facilitated further design modifications of the mold arthroplasty by Otto E. Aufranc (Figure II.5.6.2), where the rim of the Smith-Peterson mold was removed (which was often the cause of adhesions and cup "freezing," with subsequent pain and immobility), and matching curves on the inner and outer surface were machined to meet at the rounded outer edge. Despite the high short-term success rates (<4 years) reported by Aufranc (>82%), the overall failure rate remained high (>50%). Another design modification of the mold arthroplasty in the 1940–1950s was the fixation of the mold to the acetabulum rim with screws by such physicians as Albee-Pearson and Gaenslen. Although used in only four cases, Gaeslen reported using a cobalt alloy mold fixed to the acetabulum and another fixed to the femoral head, creating a metal-on-metal total hip replacement. The popularity of mold arthroplasties endured into the 1970s, when they remained touted as the treatment of choice for traumatic arthritis of the hip by leading orthopedic surgeons (Harris, 1969). However, back in the 1930s the natural progress in THA development was the progression from mold arthroplasty to short-stem prosthesis.

Femoral Head Prostheses/Short-Stem Prostheses

Femoral head prosthetics were first made of such materials as ivory (Gluck in 1890) and rubber (Delbet in 1919), and were cemented (using a plaster-like cement) for stability (Walker, 1978; Stillwell, 1987). At about the same time these replacement heads were first fitted with a short-stem by Earnest Hey Groves, who used an ivory nail to replace the articular surface of the femur. These types of implants were rare and remained unpopular compared to mold arthroplasties until 1937, when Harold Bohlman, using the work of Venable and Stuck, designed a corrosion-resistant cobalt–chrome alloy femoral head replacement with a short-stem. This design was popularized by the Judet brothers in Paris in 1946; they used polymethylmethacrylate (PMMA), which was presumed biologically inert *in vivo*, to manufacture short-stemmed prostheses (Figure II.5.6.2). Initial good results were soon replaced with problems of implant fracture and excessive wear debris, and by the early 1950s these implants were losing favor and being removed by surgeons. Vitallium® (cobalt–chrome alloy) eventually replaced acrylic in several other short-stem designs. However, there were sound short-stem designs as early as 1938, when Wiles introduced the cobalt alloy femoral shell attached to the femur with a central nail. This design was later popularized by Peterson in 1950, where he used a similar Vitallium® shell design with a central nail and a plate attached to the nail for added stability. Others adopted and adapted the Judet brothers design using Vitallium®, such as J. Thompson (1951) and Rossignal (1950). Rossignal designed large threads onto the stem to aid in fixation. These short-stem designs were subject to what was deemed high shear stress, and resulted in early loosening and failure in some patients. Short-stem designs were gradually replaced by longer-stem designs that provided less stress concentration.

Long-Stem Prostheses

Long-stem prostheses continued the trend established by short-stemmed prostheses, that is, more and more weight-bearing forces were transferred to the femur though an intramedullary stem. The pattern for a long-stem prosthesis was established in 1940 by Bohlman in collaboration with Austin T. Moore, in which they implanted a 12 inch Vitallium® prosthesis that replaced the femoral head and had long supports that were screwed into the outside of the femoral shaft (Moore, 1943). And while there were innovations in long-stem design in the 1940s, such as the door knob design of Earl McBride where a threaded stem was screwed into the intramedullary canal of the femur for fixation and load transferal, these designs were not popular. It was not until 1950 with the designs of Frederick R. Thompson and Austin T. Moore that long-stemmed prostheses became popular (Figure II.5.6.2). These designs were cast in Vitallium® (cobalt–chrome alloy), and required the removal of the femoral head but only part of the neck. The design of Moore differed from that of Thompson in that it had fenestrations through the implant to allow bone growth, and it had a rear vane to enhance rotational stability. Initially, these implants were used without bone cement. Evidence for the successful designs of the Thompson and Moore prostheses is proved by their continued use today, with only slight variations from the original. Despite the excellent design of these early long-stemmed prostheses they were primarily successful when used in place of diseased femoral heads, and did not work well when acetabular reaming was required. Therefore, this inadequacy prompted the development of the total hip replacement arthroplasty.

Total Hip Replacement Arthroplasty

Philip Wiles is credited with first total hip arthroplasty in 1938, when he used a stainless steel ball secured to the femur with a bolt and a stainless acetabular liner secured with screws (Wiles, 1953). The results of this design were disappointing, because of the poor corrosion resistance of early stainless steel *in vivo*, and the high stress concentrations of short-stemmed prostheses. An adaptation of this design that proved successful was developed by G. K. McKee and J. Watson-Farrar in 1951. They used a stain-less steel cup and long-stemmed prosthesis (Thompson stem) which failed rapidly due to the poor corrosion resistance of the stainless steel, and was then replaced by cobalt–chrome alloy with greater success. The McKee–Farrar prosthesis evolved quickly to incorporate a true spherical femoral head that was undercut at the neck to reduce the impingement of the head on the rim of the acetabular prosthesis to provide a greater range of mobility (Figure II.5.6.2) (McKee and Watson-Farrar, 1943).

The next milestone in the evolution of modern total hip arthroplasty was the advent/popularization of acrylic dental bone cement, first used by Sven Kiar in 1950 to attach a plastic prosthesis to bone (Charnley, 1964). Later that year the Hospital for Joint Diseases in New York used polymethylmethacrylate (acrylic) bone cement as a means of fixation in total hip arthroplasties (Wilson and Scales, 1970). The development of acrylic bone cement dramatically reduced the rates of loosening associated with metal–metal total hip arthroplasty. The Stanmore metal–metal design which used a horseshoe-shaped cup was popular, but it led to excessive wear and was replaced by a complete cup. McKee and Watson-Farrar adapted their design to facilitate bone cement with a land-mine-like studded acetabular cup intended to maximize mechanical fixation.

The 1950s marked the introduction and popularization of the total hip arthroplasty where it became simple and reliable enough to be practiced on a wide scale by the average orthopedic surgeon. However, the squeaking reported to occur in Judet and some later metal prostheses was identified by Charnley to be a result of the relatively high frictional forces in the joint. These high torque and frictional forces resulted in the generation of significant metallic debris, which purportedly resulted in early loosening. In 1960 Charnley developed a "low friction arthroplasty" device using shells of polytetrafluoroethylene, PTFE (commonly called Teflon® in related publications) on the femoral and acetabular sides, which resulted in early/immediate failures because of massive debonding and wear debris. This was quickly followed by a thick-walled PTFE acetabular component articulating on a small head designed to reduce the shearing forces and torque. However, this design also generated excessive wear debris, which produced immediate and severe inflammation and failure of the prosthesis. Charnley then replaced the PTFE with high density polyethylene which was not as friction-free as Teflon®, but was 1000-times more wear resistant. This prototype of total hip arthroplasty developed in 1962 was the basis for future designs which remain the most popular form of total hip arthroplasty performed today (Figure II.5.6.2).

The basic design of Charnley was modified by Muller with variable neck sizes and larger heads. At the same period metal-on-metal designs by Smith, Ring, and others (Ring, 1968) were unsuccessful challengers to the basic Charnley metal-on-polymer design. Other currently adopted design modifications were developed by Ling, Aufranc, Turner, Amstutz, Harris, and Galante, which include such innovations as femoral prosthesis geometrical modification for increases in stability and mobility, modular components for increased customization, porous coatings, surface texturing/coating to increase fixation and bone ingrowth, etc. Charnley is often deified in orthopedic literature as the metaphorical spark that lit the flames of innovation in prosthetic design. This is a typical surgeon-centered over-glorification. For one thing, other implant designs which predate Charnley, such as the all metal McKee-Farrar THA implant, have enjoyed similar success rates to those reported by Charnley. More importantly, total hip arthroplasty is, perhaps, the best example of how orthopedic biomaterials and implant success have evolved over the last century through the innovation and hard work of many scientists and physicians, and advances in areas of materials technology, biomechanics, biochemistry, immunology, infectious diseases, thrombosis, and pharmacology, to name a few.

NEW DEVELOPMENTS: TOTAL DISC ARTHROPLASTY

In contrast to THA, total disc arthroplasty (TDA) is a relatively recent development which has yet to become a mainstream option for treating disc degeneration when compared with fusion. Spinal fixation device usage is steadily increasing, where the number of cervical and lumber fusions increased 111% from the years 1993 to 2003, to roughly 105 fusions per 100,000 people in the US which is about 305,000 fusions per year (Cowan et al., 2006). The ultimate goal of intervertebral disc replacement technology is to replace spine fusion, eliminate pain, and restore structure and mobility. The first disc arthroplasties were cobalt alloy spheres, implanted as early as 1957 in between vertebrae without any method of fixation (Harmon, 1963). A decade later, stainless steel metal spheres termed Fernström balls were used in 103 patients (Figure II.5.6.3) starting in 1969 (Fernstrom, 1966; McKenzie, 1995). Other clinicians around the same time tried using polymer balls made of polymethylmethacrylate, but the results were disastrous (Hamby Wallace, 1959). The era of modern disc arthroplasty began in 1982 with the first functional artificial intervertebral disc, the SB Charité™, at the Charité Hospital in Berlin (Figure II.5.6.4) (Büttner-Janz, 1992). This design used the low-friction arthroplasty principle of John Charnley, which by that time had been successfully used in total hip replacement for over 20 years. The Charité TDA consisted of a UHMWPE sliding core, which articulated between two metal endplates with multiple teeth-like projections for fixation to the vertebral endplates (Figure II.5.6.4). The following sections detail the different kinds of material selection used in different modern total disc replacement (TDR) designs in the bourgeoning world of both lumbar and cervical total disc replacements.

In addition to what is termed total disc arthroplasty, there are several disc sparing and motion preservation devices that, while not technically TDAs, are a new type of implant called "dynamic stabilization" implants. However, none of these dynamic stabilization implants are approved for general use. The following

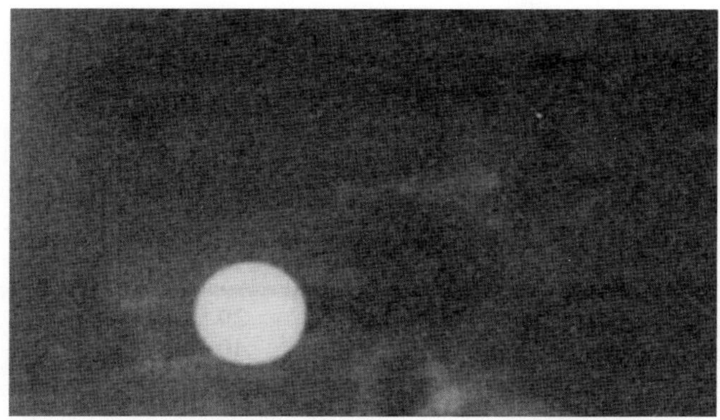

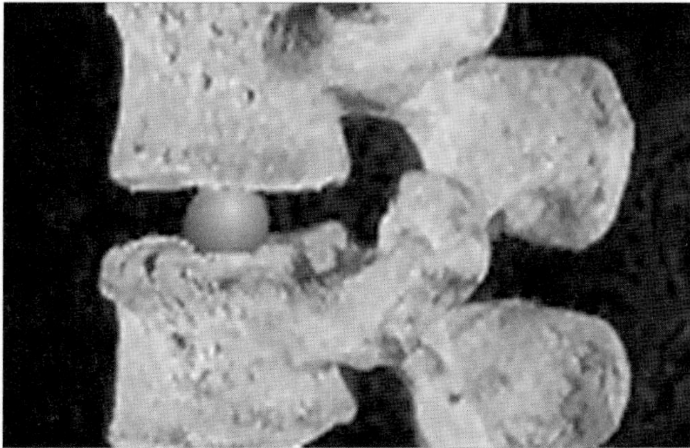

FIGURE II.5.6.3 Fernström Ball implants (stainless steel metal spheres) were the first disc arthroplasties. (Courtesy of BioEngineering Solutions Inc.)

list summarizes three central types of motion preserving spine implants, their articulation couple, and primary material constituents:

1. Lumbar Posterior Motion Sparing Technology (Figure II.5.6.5)
   Stabilimax NZ®, Applied Spine Technologies, Inc. (metal-on-metal: articulation: cobalt alloy-on-cobalt alloy, titanium)
   TOPS®, Impliant, Inc. (metal-polymer-metal with elastic core: cobalt alloy-polyurethane-cobalt alloy)
   Dynesis®, Zimmer, Inc. (metal-polymer-metal with elastic core-like structures: cobalt alloy-polyurethane-cobalt alloy)
   DIAM™, Medtronic (all polymer, polyester, silicone)
2. Cervical Disc Arthroplasty (Figure II.5.6.6)
   Bryan® Cervical Disc, Medtronic (metal-polymer-metal: elastic core: cobalt alloy-polyurethane-cobalt alloy)
   PCM®, Cervitech (metal-on-polymer: articulation: cobalt alloy-on-UWMWPE)
   PRESTIGE® Cervical Disc, Medtronic (metal-on-metal: articulation: stainless steel-on-stainless steel)

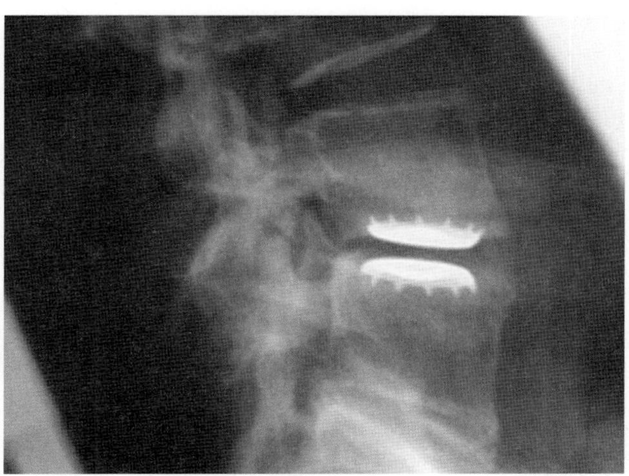

FIGURE II.5.6.4 In 1982, Schellnack and Büttner-Janz developed the SB Charité™ artificial disc, which consisted of a UHMWPE sliding core articulating unconstrained between two highly polished stainless steel metal endplates. (Courtesy of Depuy Spine, Inc.)

   PRODISC-C®, Synthes, Inc. (metal-on-polymer: articulation: cobalt alloy-on-UWMWPE)
   Secure®-C, Globus Medical (metal-on-polymer: articulation: cobalt alloy-on-UWMWPE)
3. Lumbar Disc Arthroplasty (Figure II.5.6.7)
   NUBAC™, Pioneeer Surgical, Inc. (polymer-on-polymer: articulation: PEEK on PEEK)
   SB Charité®, DePuy Spine, Inc. (metal-on-polymer: articulation: cobalt alloy-on-UWMWPE)
   Prodisc™ II, Synthes, Inc. (metal-on-polymer: articulation: cobalt alloy-on-UWMWPE)
   Maverick®, Medtronic (metal-on-metal: articulation: cobalt alloy-on-cobalt alloy)
   eDisc™, Theken (metal-polymer-metal: elastic core: cobalt alloy-polyurethane-cobalt alloy)
   Freedom™ Lumbar Disc, Axiomed, (metal-polymer-metal: elastic core: cobalt alloy-polyurethane-cobalt alloy)
   Activ L, Asculap (metal-on-polymer: articulation: cobalt alloy-on-UWMWPE)

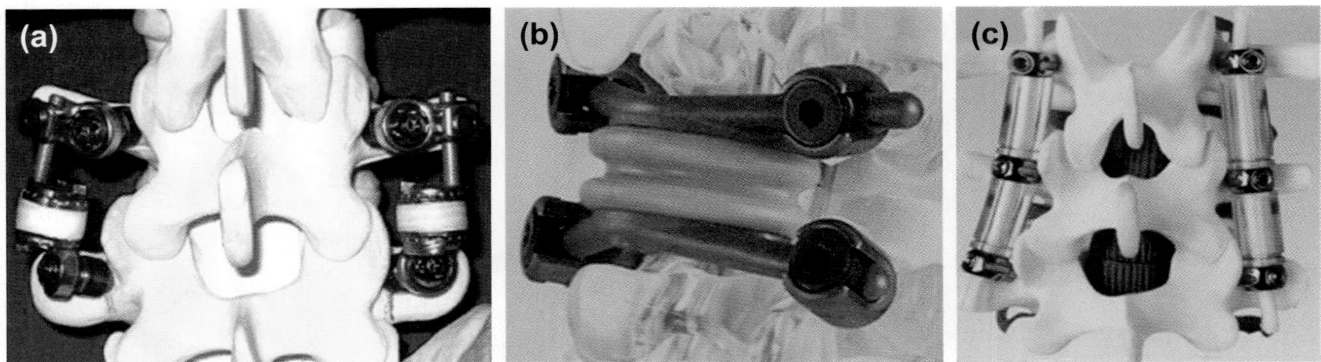

FIGURE II.5.6.5 Dynamic Spine Stabilization Implants: A new class of motion preservation implants are under investigational use in the lumbar spine: (a) Stabilimax NZ® (Applied Spine Technologies, Inc.); (b) TOPS® (Impliant, Inc.); (c) Dynesis® (Zimmer, Inc.); all of these use a combination of articulation and spring or elastomeric interior components to provide both articulation and resistance force back a neutral position. (Courtesy of BioEngineering Solutions, Inc.)

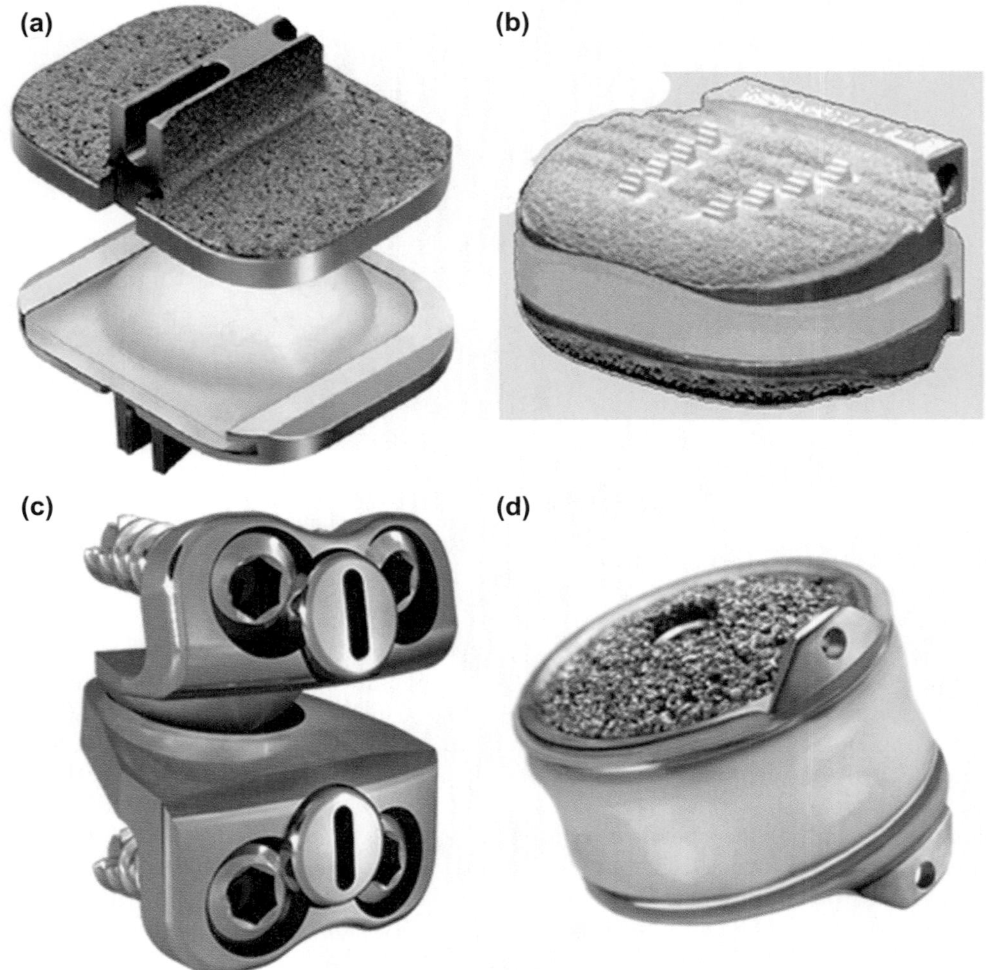

FIGURE II.5.6.6 Cervical IDRs (Intervertebral Disc Replacement): There are a number of cervical total disc replacements in use which employ the three primary types of disc joint articulation. 1. Metal-on-Polymer Articulation: (a) The PRODISC-C® (Synthes, Inc.); and (b) PCM® (Cervitech); use Co-alloy endplates that articulate on a polymeric (UHMWPE) core that is mechanically fixed to one of the endplates and articulates in a ball-and-socket type manner. 2. Metal-on-Metal Articulation: (c) The PRESTIGE® Disc (Medtronic) has metal-on-metal articulation where the end plates are constructed of stainless steel. 3. Elastic Core Articulation: (d) The Bryan® Cervical Disc System (Medtronic) is axially symmetric and incorporates cobalt-chrome alloy clamshell-shaped endplates, which flex upon a loadbearing polymeric (polyurethane-based) nucleus core. A novel feature in the design of this component is the polyurethane flexible membrane that surrounds the entire articulation and forms a sealed space containing a saline lubricant to reduce friction and prevent migration of any wear and corrosion debris. (Pictures provided courtesy of DePuy Spine, Inc., Medtronic, Spine Solutions, Inc., and BioEngineering Solutions, Inc.)

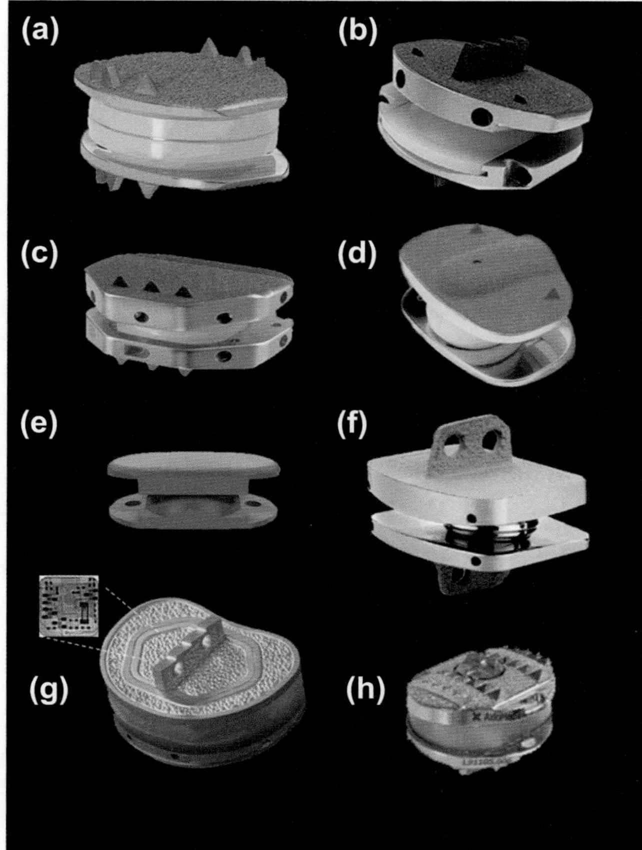

FIGURE II.5.6.7 Lumbar total disc replacements include metal-on-polymer, metal-on-metal, polymer-on-polymer, and flexible core technologies. (a) Metal-on-Polymer Articulation: *The LINK® SB Charité III* (DePuy Spine, Inc.) cobalt–chrome alloy endplates articulate on a mobile bearing ultra-high molecular weight polyethylene core. The endplates are covered with an osteoconductive surface of titanium/calcium-phosphate double coating under the trade name "TiCaP®." (b) Metal-on-Polymer Articulation: The Prodisc (Synthes, Inc.) lumbar TDR is composed of cobalt–chrome–molybdenum alloy and covered with a porous titanium alloy and articulates on a central core of UHMWPE. (c) Metal-on-Polymer Articulation: ActivL (Aesculap) uses a polymeric center core intended to allow both translation and rotation and to more closely approximate physiological motion. (d) Metal-on-Polymer Articulation: Dynardi (Zimmer, Inc.) is a disc replacement implant with two opposing cobalt alloy (Co–Cr–Mo) endplates coated with porous pure titanium for bone ingrowth, that articulate on a semi-constrained UHMWPE core. (e) Polymer-on-Polymer Articulation: NUBAC™ (Pioneer Surgical, Inc.) is a polymer-on-polymer disc arthroplasty device and the first polyetheretherketone (PEEK)-on-PEEK articulated disc arthroplasty device. (f) Metal-on-Metal articulation: The Maverick Disc (Medtronic) uses metal-on-metal articulation where the end plates are constructed of Co-alloy. (g) Elastic Core Articulation: The Theken eDisc™ (Theken Disc) represents another step in the evolution orthopedic implant devices in that as well as containing an elastic polyurethane based core it provides measured *in vivo* load information to the surgeon and patient via electronic sensors and transmitters. (h) Elastic Core Articulation: The Freedom™ Lumbar Disc (Axiomed) uses a viscoelastic polymer (like polyurethane) to replicate the native function of a natural disc. The elastic core in combination with the implant design provides a three-dimensional motion that functions within the natural biomechanics of the spine. (Pictures provided courtesy of DePuy Spine, Inc., Medtronic, Spine Solutions, Inc., and courtesy of BioEngineering Solutions, Inc.)

In general, current TDA designs have a THA-like primary articulation (or motion) using polymer-on-metal, polymer-on-polymer, metal-on-metal or some form of an all elastic core technology. However, because the emerging area of motion preservation in spine orthopedics facilitates the use of a wide array of implant designs and materials, the following sections will detail some of these designs and the materials they are comprised of.

Metal-on-Polymer Articulation. Examples of metal-on-polymer articulating TDA implants are as follows:

SB Charité III Artificial Disc (Depuy Spine, Inc.). The Charité III includes thick cobalt chrome endplates (Figure II.5.6.7). The metal endplates are coated with an osteoconductive surface of porous titanium beads and calcium-phosphate double coating under the trade name "TiCaP®." This coating is also used in non-cemented total joint arthroplasty such as femoral stems, acetabular cups, ankle joint prostheses, and dental implants (Liefeith et al., 2003).

Prodisc (Synthes, Inc.). The lumbar TDA Prodisc reflects how hip and knee designs have shaped TDA designs (Figure II.5.6.7). The two cobalt-chrome-molybdenum alloy endplates are covered with a porous titanium alloy, and articulate on a UHMWPE core. The locking polyethylene core provides ball and socket movement which is different from other "mobile" bearing designs, e.g., the Charité. The cobalt-chrome alloy endplates are also coated with pure titanium (Plasmapore), in similar fashion to the Charité.

Metal-on-Metal Articulation. Examples of metal-on-metal articulating TDA are currently used in both lumbar and cervical disc replacement applications, and dynamic stabilization devices of the lumbar spine.

Dynamic Stabilization: ***Stabilimax NZ®*** (Applied Spine Technologies, Inc.). The Stabilimax NZ® dynamic spine system is an implant that provides stabilization of the lumbar spine for patients with spine surgery that results in less superstructure for stability (e.g., bone removal for spinal stenosis; Figure II.5.6.5). It is composed of two ball-and-socket joints with cobalt alloy on cobalt alloy (ASTM F-75) articulation, two titanium alloy (Ti-6Al-4V) screws, and an interposed spring of cobalt alloy in between the two ball-and-socket joints (Figure II.5.6.5a). This type of device attempts to offer a less invasive surgical procedure than fusion or disc replacement.

Total Disc Arthroplasty: ***Prestige Disc.*** The Prestige Disc (Medtronic) uses metal-on-metal articulation where the end plates are currently constructed of stainless steel (and in the near future cobalt alloy) (Figure II.5.6.6). The articulation mimics that of a ball-and-socket construct with a more constrained center of rotation, i.e., not the joint kinematics of mobile bearing. Because the wear resistance of stainless steel exceeds that of polyethylene, the amount of total particulate debris generated by metal-on-metal couples is an order of magnitude less than that produced by metal-on-polymer couples. The location of this implant next to vital tissues raises concerns associated with the release of metallic debris. However, the importance of metal debris will be resolved primarily through careful clinical follow-up and peer reviewed analysis.

Polymer-on-Polymer Articulation. The lower loads of the spine when compared to hip and knee replacements has enabled the use of polymer-on-polymer articulation designs of total disc arthroplasty devices, which if successful will be a new phenomena in the world of total joint arthroplasty.

NUBAC™ (Pioneer Surgical, Inc.) is a polymer-on-polymer disc arthroplasty device made of polyetheretherketone (PEEK). This implant aims to maintain or restore the disc height and mechanical function by replacing only the nucleus of a spine disc, theoretically using a less invasive procedure than other total disc arthroplasty implants, i.e., it is inserted into a partially resected disc, and has partial structural support of part of the remaining disc (the annulus around the outside of the implant). It does not restrict any physiological rotational motions, leaving constraint and stability to the retained surrounding annulus of the disc and ligaments. The risk of implant extrusion is low, because biomechanics of motion tend to push it in the opposite direction (Figure II.5.6.7).

Elastic Intervertebral Replacement Devices. The advent of elastic core total disc replacement implants represents another pioneering effort in the world of total joint implants. Elastic core total joint replacement (TJR) technology is feasible because there are relatively small motions associated with disc arthroplasty implants when compared to a hip or knee. There are generally two different kinds of total disc arthroplasty designs with elastic "core "components: (1) elastic cores interposed between two endplates; and (2) all elastic devices for nucleus replacement.

Total Disc Arthroplasty: Bryan® Cervical Disc Prosthesis (Medtronic). The Bryan® Cervical Disc System was approved by the United States FDA for distribution on May 12, 2009, and incorporates cobalt-chrome alloy endplates, which sandwich and flex on a load-bearing polymeric (polyurethane-based) nucleus core. The endplates have a porous coating of 250 micron titanium beads sintered to the cobalt alloy endplates. There is also a polyurethane flexible membrane that surrounds the entire articulation, which forms a sealed space to prevent migration of any wear and corrosion debris. This sheath also aims to prevent the intrusion of connective tissue or the creation of a pseudo-capsule over time. The inclusion of a sheath component to address the clinical concern of wear and corrosion debris (discussed later) is a significant development in the design of orthopedic joint arthroplasty components.

All these spinal implants seek to replace fusion as the operative solution for disc degeneration or injury. Many of these designs are new to orthopedics in terms of materials and/or design, and have not withstood the test of time; it is likely that only a few of these designs will dominate and become more popular over time. However, these disc replacement implants represent how imaginative new designs and materials are being used to address age-old orthopedic problems that remain unresolved in this new 21st century.

CURRENT BIOMATERIALS IN TOTAL ARTHROPLASTY

Today, the archetype of the total hip implant remains much as it was in the 1970s, albeit with a wider variety of implant materials and geometries (Figure II.5.6.8). Current THA is typically constructed of a titanium or cobalt–chromium alloy femoral stem (cemented with polymethylmethacrylate, PMMA, or press fit into place; Figure II.5.6.9), connected to a "modular" cobalt–chrome alloy or ceramic head that articulates on a ultra-high molecular weight polyethylene (UHMWPE) or ceramic acetabular cup fitted into a titanium or cobalt–chromium cup liner which is cemented, screwed or press fitted into place (Figure II.5.6.9). Despite this simple archetype of the total hip replacement, there are hundreds of variations on this theme offered to today's orthopedic surgeons, with little in terms of absolute guidelines as to which type of implant is (or which of the over 10 major manufacturers has) the best for well-defined orthopedic disease states. However, there are some general guidelines. Typically, implants in older individuals (>80 years of age) are cemented into place with PMMA bone cement, because the chance for revision is minimal when compared to younger individuals (<60 years), and removing bone cement is both technically challenging and may compromise the availability of bone stock. Generally, there are choices of surface roughness, coatings, geometry, material composition, etc., and each manufacturer claims that its product is superior to the rest. This, in combination with little or no publicly available information tracking the performance of each type of implant in patients, precludes accurate scientific analysis of which implant materials and designs perform best. Additionally, competition between manufacturers, and the requisite attention to marketing required to compete in the marketplace, has resulted in a dizzying array of new implants released each year claiming to be an improvement over last year's model. These claims are suspect, because the typical total hip replacement enjoys a success rate of over 90% at seven years, therefore in most cases a minimum of seven to ten years must pass

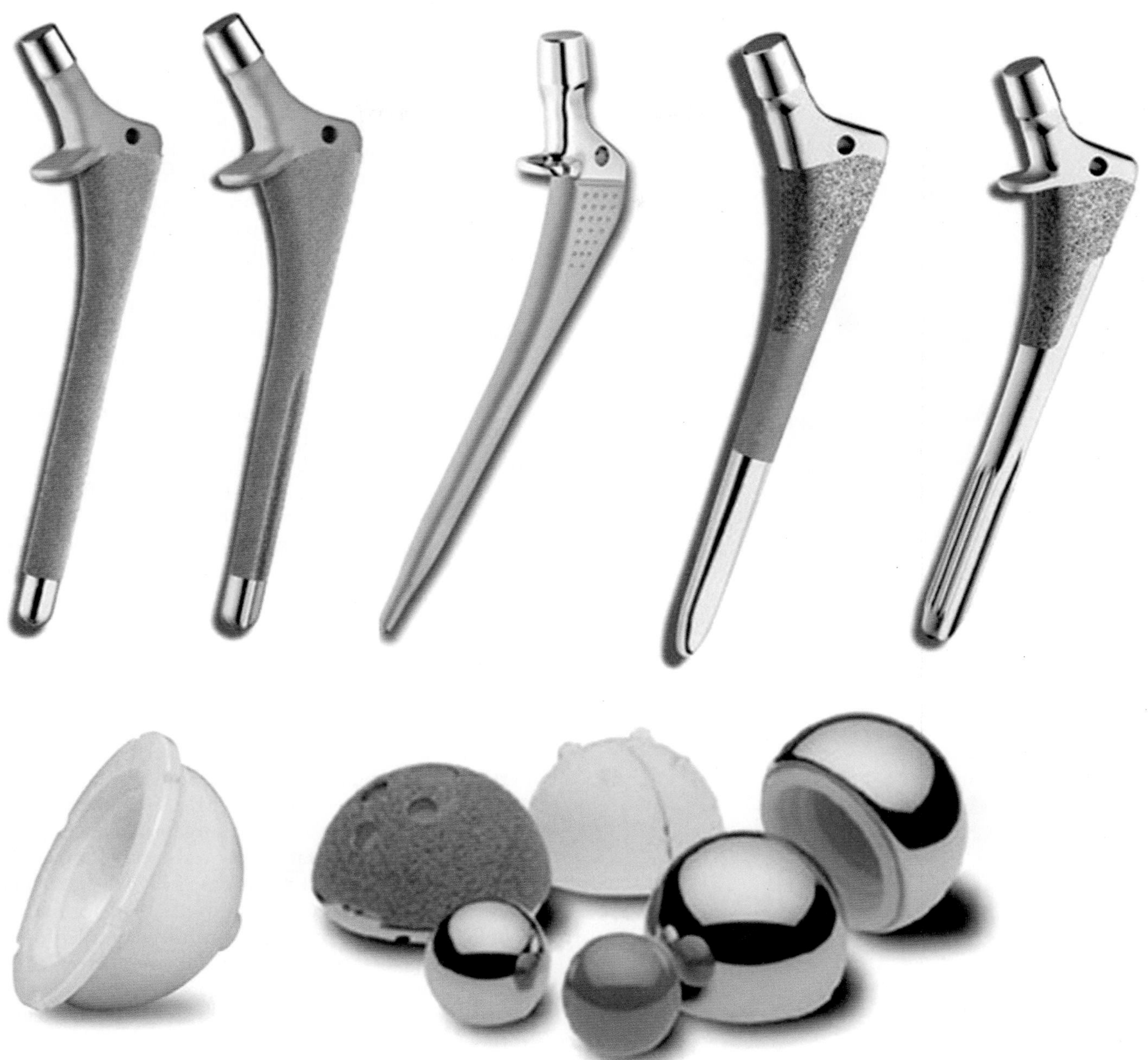

FIGURE II.5.6.8 Examples of typical current total hip arthroplasty (THA) components available from a single manufacturer: titanium alloy stem, with a cobalt-base alloy (ASTM F-75) modular head bearing on an ultra-high molecular weight polyethylene (UHMWPE) liner within a titanium alloy cup. Also shown are a ceramic head and three acetabular sockets with various surfaces for both cemented and cementless fixation. From left to right the stems are all components of the VerSys® Hip System (Zimmer, Inc, Warsaw, IN, USA), and from left to right are designated Beaded Fullcoat, Beaded Fullcoat with distal flutes, Cemented, Fiber Metal Taper, and Fiber Metal Midcoat. (Photographs courtesy of Zimmer, Inc.)

before such claims can be substantiated, and even then proof of superior performance is compromised by a myriad of external factors, such as the surgeon, region of the country, average activity of patient populations, etc. This conflict between science, marketing, and market share may (in the opinion of this author) represent the single biggest obstacle to the scientific determination of superior implant design and progress. The unenviable responsibility rests with the FDA to prevent the zeal of economic pressure from undermining implant design in a regressive fashion. Today, optimal implant selection is primarily based on which material couple may best suit the individual (Figure II.5.6.10).

Polymers

Polymers are most commonly used in orthopedics as articulating bearing surfaces of joint replacements

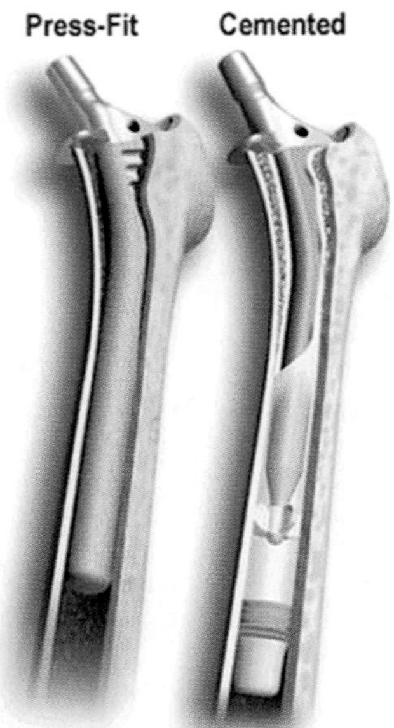

FIGURE II.5.6.9 Examples of cemented and non-cemented stems showing the plug and centralizer used with a cemented stem. (Picture courtesy of Zimmer, Inc.)

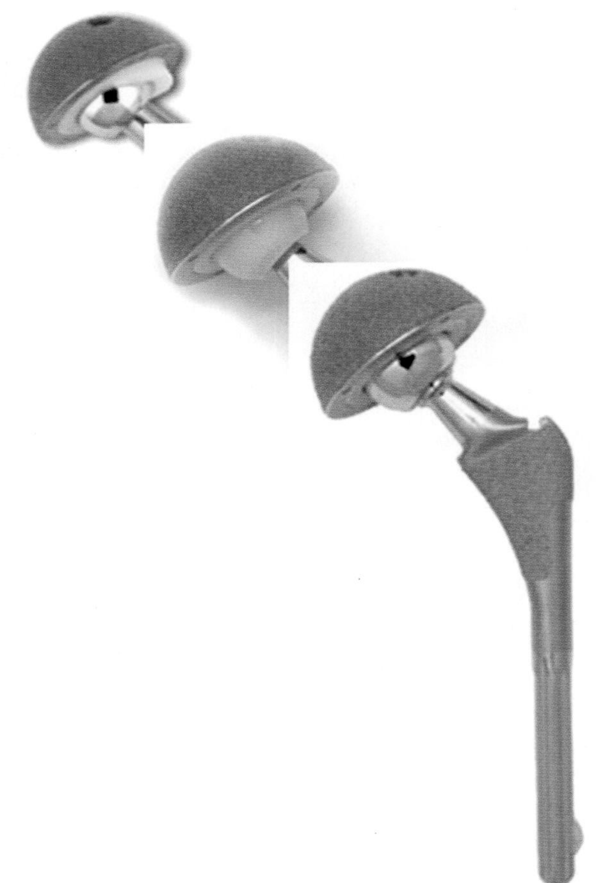

FIGURE II.5.6.10 Examples of the three types of bearing couples used in modern TJA. From top to bottom: metal-on-polymer, ceramic-on-ceramic, and metal-on-metal (Lineage™ line from Wright Medical Technology, Inc. Arlington, TN, USA).

(Figure II.5.6.8), and as an interpositional cementing material between the implant surface and bone (Figure II.5.6.8). Polymers used as articulating surfaces must have a low coefficient of friction, and low wear rates when in articulating contact with the opposing surface, which is usually made of metal. Initially, John Charnley used polytetrafluoroethylene (PTFE) for the acetabular component of his total hip arthroplasty (Figure II.5.6.2). However, its accelerated creep and poor stress degradation (for the material he used) caused it to fail *in vivo*, requiring replacement with his ultimate choice, ultra-high molecular weight polyethylene (UHMWPE).

Polymers used for fixation as a structural interface between the implant component and bone tissue require the appropriate mechanical properties of a polymer, which can be molded into shape and cured *in vivo*. The first type to be used, polymethylmethacrylate (PMMA), was again popularized by Charnley, who borrowed it from the field of dentistry. He adapted dental PMMA as a "grouting" material to fix both the stem of the femoral component and the acetabular component in place, and thus distributed the loads more uniformly from the implant to the bone. Since high interfacial stresses result from the accommodation of a high modulus prosthesis within the much lower modulus bone, the use of a lower modulus interpositional material has been a goal of those seeking to improve upon PMMA fixation. Thus, polymers such as polysulfone have been tried as porous coatings on the implant's metallic core to permit mechanical interlocking through bone and/or soft tissue ingrowth into the pores. However, to date PMMA remains the substance of choice for orthopedic surgeons. This requires polymers that have surfaces that resist creep under the stresses found in clinical situations, and that have high enough yield strengths to minimize plastic deformation. As indicated earlier, the important mechanical properties of orthopedic polymers are yield stress, creep resistance, and wear rate. These factors are controlled by such parameters as molecular chain structure, molecular weight, and degree of branching or (conversely) of chain linearity.

One of the more prevalent polymerics used in orthopedics today is a highly cross-linked ultra-high molecular weight polyethylene (UHMWPE), which is typically used in total joint arthroplasty as a loadbearing articulating surface, designed to provide low friction loadbearing articulation. Polyethylene is available commercially in three different grades: low density; high density; and UHMWPE. The better packing of linear chains within UHMWPE results in increased crystallinity, and provides improved mechanical properties required for orthopedic use even though there is a decrease in both ductility and fracture toughness. In total hip arthroplasty applications, an acetabular cup of UHMWPE typically articulates against a femoral ball of cobalt–chromium alloy. The predominant problem presented by these metal–polymer articulating surfaces is the production of wear particles, i.e., polymer debris. The resultant wear of the polyethylene bearing purportedly produces billions of sub-micron sized wear particles annually, in the <1–10 micron range. Producing greater cross-linking of polyethylene, using chemical and radiation techniques, has only recently improved its wear resistance in orthopedic applications. Wear tests have shown that the wear resistance of UHMWPE is improved by cross-linking with gamma irradiation at 2.5–5.0 Mrad and below, as evidenced by simulator studies; however, this can negatively affect such physical properties as tensile strength (McKellop et al., 2000). Therefore, care must be taken to minimize any negative oxidative effects, while preserving high wear resistance characteristics. Although newer more highly cross-linked polyethylene has generally been accepted as superior to previous implant UHMWPE, there remains incomplete data regarding its ultimate long-term performance. In order to maximize the performance characteristics of polyethylene, it is cross-linked prior to fabrication into its final form, e.g., an acetabular cup. Typically, an extruded bar of polyethylene is cross-linked using conventional gamma irradiation, and then heat treated to reduce residual free radicals.

Ceramics

In recent years, ceramics and glass ceramics have played an increasingly important role in implants. Although used in Europe for over a quarter century, the FDA has only recently (3 February, 2003) approved the first ceramic-on-ceramic bearing hip implant to be used in total hip replacement procedures (Figure II.5.6.10). The primary reason for the introduction of this alternative bearing surface is the superior wear resistance of ceramics when compared to metal–metal or metal–polymer bearing surfaces. This, and other improved properties such as resistance to further oxidation (implying inertness within the body), high stiffness, and low friction require the use of full-density, controlled, small, uniform grain size (usually less than 5 μm) ceramic materials. The small grain size and full density are important, since these are the two principal bulk parameters controlling the ceramic's mechanical properties. Any voids within the ceramic's body will increase stress, degrading the mechanical properties. Grain size controls the magnitude of the internal stresses produced by thermal contractions during cooling. In ceramics, such thermal contraction stresses are critical, because they cannot be dissipated as they can in ductile materials via plastic deformation.

Alumina (Al_2O_3) and zirconia (ZrO_2) ceramics have been used in orthopedic THA for the past 30 years. The first ceramic couple (alumina–alumina) was implanted in 1970 by Pierre Boutin. Since the outset, the theoretical advantage of hard-on-hard articulating surfaces was low wear. Ceramics, because of their ionic bonds and chemical stability, are also relatively biocompatible. Initial concerns about fracture toughness and wear have been addressed by reducing grain size, increasing purity, lowering porosity, and improving manufacturing techniques (e.g., hot isostatic pressing, HIP). Early failures of these couples were plagued with both material-related and surgical errors. The very low wear rates combined with steadily decreasing rates of fracture (now estimated to occur 1 in 2000 over 10 years) have resulted in the growing popularity of all ceramic bearings.

Zirconia was introduced in 1985 as a material alternative to Al_2O_3 for ceramic femoral heads, and was gaining market share because of its demonstrably enhanced mechanical properties in the laboratory when compared to alumina. Femoral heads of zirconia can typically withstand 250 kN (or 25 tons), a value generally exceeding that possible with alumina or metal femoral heads. However, mechanical integrity of all ceramic components are extremely dependent on manufacturing quality controls, as evidenced in the recall of thousands of zirconia ceramic femoral heads by their manufacturer, St. Gobain Desmarquest, in 2001. This was because of *in vivo* fracture of some components due to a slight unintended variation in the manufacturing sintering process, caused when the company bought a newer high-throughput assembly line type oven. In general, ceramic particulate debris is chemically stable and biocompatible, and causes no untoward biologic responses at high concentrations.

There have been recent attempts to take advantage of the osteophilic surface of certain ceramics and glass ceramics. These materials provide an interface of such biological compatibility with osteoblasts (bone-forming cells) that these cells lay down bone in direct apposition to the material in some form of direct chemicophysical bond. Special compositions of glass ceramics, termed bioglasses, have been used for implant applications in orthopedics. The model proposed for the "chemical" bond formed between glass and bone is that the former undergoes a controlled surface degradation, producing an SiO-rich layer, and a Ca, P-rich layer at the interface. Originally amorphous, the Ca, P-rich layer eventually crystallizes as a mixed hydroxycarbonate apatite structurally integrated with collagen, which permits subsequent bonding by newly-formed mineralized tissues. There is still an entirely different series of inorganic compounds that have also been shown to be osteophilic. These include OHAp, which is the form of the naturally occurring inorganic component of calcified tissues, and calcite, CaCO, and its Mg analog, dolomite, among others being studied. The most extensive applications in both orthopedics and dentistry have involved OHAp. This has been used as a cladding for metal prostheses for the former, and in dense, particulate form for the latter. The elastic properties (modulus) of OHAp and related compounds are compared with those of bone, dentin, and enamel in Table II.5.6.2. The use of both OHAp and glass ceramics as cladding on the metallic stems of hip prostheses is still another method of providing fixation instead of using PMMA.

TABLE II.5.6.2 Mechanical Properties of Dominant Orthopedic Biomaterials

Orthopedic Biomaterial	ASTM Designation	Trade Name and Company	Elastic Modulus (Young's Modulus)	Yield Strength (Elastic limit)	Ultimate Strength	Fatigue Strength (Endurance Limit)	Hardness	Elongation at Fracture
		(Examples)	*(GPa)*	*(MPa)*	*(MPa)*	*(MPa)*	*HVN*	*(%)*
Cortical Bone[$]								
Low strain			15.2	114t	150c/90t	30–45	–	–
High strain			40.8	–	400c–270t	–	–	–
Polymers								
UHMWPE			0.5–1.3	20–30	30–40t	13–20	60–90 (Mpa)	130–500
PMMA			1.8–3.3	35–70	38–80t	19–39	100–200 (Mpa)	2.5–6
Ceramics								
Al_2O_3			366	–	3790c/310t	–	20–30 (Gpa)	–
ZrO_2			201	–	7500c/420t	–	12 (Gpa)	–
Metals								
Stainless steels	ASTM F138	Protusul S30, Sulzer	190	792	930t	241–820	130–180	43–45
Co–Cr Alloys								
	ASTM F75	Alivium, Biomet CoCrMo, Biomet Endocast SIL, Krupp Francobal, Benoist Girard Orthochrome, DePuy Protosul 2, Sultzer Vinertia, Deloro Vitallium C, Howmedica VitalliumFHS, Howmedica Zimaloy, Zimmer Zimalloy, Micrograin	210–253	448–841	655–1277t	207–950	300–400	4–14
	ASTM F90	Vitallium W, Howdmedica	210	448–1606	1896t	586–1220	300–400	10–22
	ASTM F562	HS25I, Haynes Stellite MP35N, Std Pressed Steel Corp.	200–230	300–2000	800–2068t	340–520	8–50 (RC)	10–40
	ASTM 1537	TJA 1537, Allvac Metasul, Sulzer	200–300	960	1300t	200–300	41 (RC)	20
Ti Alloys								
CPTi	ASTM F67	CSTi, Sulzer	110	485	760t	300	120–200	14–18
Ti-6Al-4V	ASTM 136	Isotan, Aesculap Werke Protosul 64WF, Sulzer Tilastan, Waldemar Link Tivaloy 12, Biomet Tivanium, Zimmer	116	897–1034	965–1103t	620–689	310	8

ASTM: American Society for Testing and Materials (ASTM International).
$: Cortical bone is both anisotropic and viscoelastic thus properties listed are generalized.
c: Compression.
t: Tension.
RC: Rockwell Hardness Scale.

In these cases, the fixation is via the direct bonding of bone to the cladding surface.

Metals

Since the principal function of the long bones of the lower body is to act as loadbearing members, it was reasonable that the initial materials introduced to replace joints, such as artificial hips, were metals. Both stainless steel, such as 316L, and cobalt–chromium alloys became the early materials of choice, because of their relatively good corrosion resistance and reasonable fatigue life within the human body. Of course, their stiffness, rigidity, and strength exceeded those of bone considerably. However, in certain applications, owing to size restrictions and design limitations (e.g., in rods used to straighten the spine in scoliosis), fatigue failures did occur. Metals remain the central material component of state-of-the-art total hip arthroplasties. Metals provide appropriate material properties, such as high strength, ductility, fracture toughness, hardness, corrosion resistance, formability, and biocompatibility necessary for use in loadbearing roles required in fracture fixation and total joint arthroplasty (TJA). Implant alloys were originally developed for maritime and aviation uses, where mechanical properties such as high strength and corrosion resistance are paramount. There are three principal metal alloys used in orthopedics and particularly in total joint replacement: (1) titanium-based alloys; (2) cobalt-based alloys; and (3) iron-(stainless steel) based alloys. The alloy's specific differences in strength, ductility, and hardness generally determine which of these three alloys is used for a particular application or implant component. However, it is the high corrosion resistance of all three alloys, more than anything, which has led to their widespread use as loadbearing implant materials. These material properties of metals (Table II.5.6.2) are due to the miraculous nature of the metallic bond, atomic microstructure, and elemental composition of alloys.

Stainless Steel Alloys

Stainless steels were the first metals to be used in orthopedics in 1926. However, it was not until 1943, when ASTM 304 was recommended as a standard implant alloy material, that steels were reliable as an implant alloy. All steels are comprised of iron and carbon, and may typically contain chromium, nickel, and molybdenum. Trace elements such as manganese, phosphorous, sulfur, and silicon are also present. Carbon and the other alloy elements affect the mechanical properties of steel through alteration of its microstructure.

The form of stainless steel most commonly used in orthopedic practice is designated 316LV (American Society for Testing and Materials F138, ASTM F138; others include F139, F899, F1586, F621, etc). "316" classifies the material as austenitic, the "L" denotes the low carbon content, and "V" defines the vacuum under which it is formed. The carbon content must be kept at a low level to prevent carbide (chromium–carbon) accumulation at the grain boundaries.

Although the mechanical properties of stainless steels are generally less desirable than those of the other implant alloys (lower strength and corrosion resistance), stainless steels do possess greater ductility, indicated quantitatively by a three-fold greater "percentage of elongation at fracture" when compared to other implant metals (Table II.5.6.2). This aspect of stainless steel has allowed it to remain popular as a material for cable fixation components in total knee arthroplasty, and a low cost alternative to titanium and cobalt alloys.

Cobalt–Chromium Alloys

Of the many Co–Cr alloys available, there are currently only two predominantly used as implant alloys (Table II.5.6.3). These two are: (1) cobalt–chromium–molybdenum (CoCrMo), which is designated ASTM F75 and F76; and (2) cobalt–nickel–chromium–molybdenum (CoNiCrMo) designated as ASTM F562. Other cobalt alloys approved for implant use include one which incorporates tungsten (W) (CoCrNiW, ASTM F90), and another with iron (CoNiCrMoWFe, ASTM F563). Co–Ni–Cr–Mo alloys which contain large percentages of nickel (25–37%) promise increased corrosion resistance, yet raise concerns of possible toxicity and/or immunogenic reactivity (discussed later) from released nickel. The biologic reactivity of released nickel from Co–Ni–Cr alloys is a cause for concern under static conditions, and due to their poor frictional (wear) properties Co–Ni–Cr alloys are also inappropriate for use in articulating components. Therefore the dominant implant alloy used for total joint components is CoCrMo (ASTM F75).

Cobalt alloys are generally cast into their final shape, because they are susceptible to work-hardening at room temperatures. The improvements in strength and hardness gained by cold-working are not worth the loss in fracture toughness. Thus, Co–Cr–Mo alloy hip implant components are predominantly manufactured using lost wax (investment) casting methods.

Although Co–Cr–Mo alloys are the strongest, hardest, and most fatigue resistant of the alloys used for joint replacement components, care must be taken to maintain these properties, because the use of finishing treatments can also reduce these same properties (Table II.5.6.2). For example, sintering of porous coatings onto femoral or tibal TJA Co–Cr–Mo stems can decrease the fatigue strength of the alloy from 200–250 MPa to 150 MPa after heating (annealing) the implant at 1225°C.

Titanium Alloys

Titanium alloys were developed in the mid-1940s for the aviation industry, and were first used in orthopedics

TABLE II.5.6.3 Approximate Weight Percent of Different Metals Within Popular Orthopaedic Alloys

Alloy	Ni	N	Co	Cr	Ti	Mo	Al	Fe	Mn	Cu	W	C	Si	V
Stainless steel (ASTM F138)	10–15.5	<0.5	*	17–19	*	2–4	*	61–68	*	<0.5	<2.0	<0.06	<1.0	*
Co–Cr–Mo alloys (ASTM F75)	<2.0	*	61–66	27–30	*	4.5–7.0	*	<1.5	<1.0	*	*	<0.35	<1.0	*
(ASTM F90)	9–11	*	46–51	19–20	*	*	*	<3.0	<2.5	*	14–16	<0.15	<1.0	*
(ASTM F562)	33–37	*	35	19–21	<1	9.0–11	*	<1	<0.15	*	*	*	<0.15	*
Ti alloys														
CPTi (ASTM F67)	*	*	*	*	99	*	*	0.2–0.5	*	*	*	<0.1	*	*
Ti-6Al-4V (ASTM F136)	*	*	*	*	89–91	*	5.5–6.5	*	*	*	*	<0.08	*	3.5–4.5
45TiNi	55	*	*	*	45	*	*	*	*	*	*	*	*	*
Zr Alloy (95% Zr, 5% Nb)	*	*	*	*	*	*	*	*	*	*	*	*	*	*

* Indicates less than 0.05%.
Note: Alloy compositions are standardized by the American Society for Testing and Materials (ASTM vol. 13.01).

TABLE II.5.6.4 Electrochemical Properties of Implant Metals (Corrosion Resistance) in 0.1 M NaCl at pH=7

Alloy	ASTM Designation	Density	Corrosion Potential (vs Calomel)	Passive Current Density	Breakdown Potential
		(g/cm³)	*(mVolts)*	*(mAmps/cm²)*	*(mVolts)*
Stainless steel	ASTM F138	8.0	−400	0.56	200–770
Co–Cr–Mo Alloys					
	ASTM F75	8.3	−390	1.36	420
Ti Alloys					
CPTi	ASTM F67	4.5	−90 to −630	0.72–9.0	>2000
Ti-6Al-4V	ASTM 136	4.43	−180 to −510	0.9–2.0	>1500
Ti5Al2.5Fe	**	4.45	−530	0.68	>1500
Ni45Ti	**	6.4–6.5	−430	0.44	890

** No current ASTM standard.
The Corrosion Potential represents the open circuit potential (OCP) between the metal and a calomel electrode. The more negative the OCP, the more chemically reactive and thus the less corrosion resistance. Generally low current density indicates greater corrosion resistance. The higher the breakdown potential the better, (i.e. the more elevated the breakdown potential, the more stable the protective layer).

around the same time. Two post-World War II alloys, commercially pure titanium (CPTi) and Ti-6Al-4V, remain the two dominant titanium alloys used in implants. Commercially pure titanium (CPTi, ASTM F67) is 98–99.6% pure titanium. While CPTi is most commonly used in dental applications, the stability of the oxide layer formed on CPTi, and consequently its high corrosion resistance (Table II.5.6.4), and its relatively higher ductility (i.e., the ability to be cold-worked) compared to Ti-6Al-4V, has led to the use of CPTi in porous coatings (e.g., fiber metal) of TJA components. Generally, joint replacement components (i.e., TJA stems) are made of Ti-6Al-4V (ASTM F136) Nb rather than CPTi, because of its superior mechanical properties (Table II.5.6.2).

Titanium alloys are particularly good for THA components because of their high corrosion resistance compared with stainless steel and Co–Cr–Mo alloys. A passive oxide film (primarily of TiO_2) protects both Ti-6Al-4V and CPTi alloys. Generally, Ti-6Al-4V has mechanical properties that exceed those of stainless steel, with a flexural rigidity less than stainless steel and Co–Cr–Mo alloys. The torsional and axial stiffness (moduli) of Ti alloys are closer to those of bone, and theoretically provide less stress shielding than cobalt alloys and stainless steel. However, titanium alloys are particularly sensitive to geometrical factors, in particular notch sensitivity. This reduces the effective strength of a component by increasing the material's susceptibility to crack initiation and propagation through the component. Therefore, care is taken both in the design geometry and in the fabrication of titanium alloy components. Perhaps the greatest drawback of titanium alloys is their relative softness compared with Co–Cr–Mo alloys (Table II.5.6.2), and their relatively poor wear and frictional properties. Ti-6Al-4V is >15% softer than Co–Cr–Mo alloy, and also results in significantly more wear than Co–Cr–Mo when used in applications requiring articulation, e.g., total knee arthroplasty (TKA) or THA femoral heads. Thus, titanium alloys are seldom used as materials where hardness or resistance to wear is the primary concern.

NEW ALLOYS AND SURFACE COATINGS

The quest for new THA metal alloys with improved biocompatibility and mechanical properties remains an ongoing one. The use of Ti alloys, Co–Cr–Mo alloys or stainless steels in a specific application generally involves trade-offs of one desirable property for another. Some examples of this are the sacrifice of chemical inertness for hardness (wear resistance), as is the case with titanium alloy for Co–Cr–Mo in TJA bearing surfaces, and the compromise of strength for ductility when using stainless steel instead of titanium and Co–Cr–Mo alloys for bone fixation cables. Although new alloys claim to be just that, "new," they are often merely variations of the three categories of implant metals previously described (which are already approved for use). These improved alloys usually contain only the minor addition of new elements to protect assertions of substantial equivalence to existing ASTM and FDA approved alloys, therefore easing the burden of regulatory approval. These new alloys generally fall under one of four categories, i.e.: (1) titanium alloys; (2) cobalt alloys; (3) stainless steels; and less approved (4) refractory group metals and alloys.

New Zirconium and Tantalum Alloys

Zirconium (Zr) and tantalum (Ta) are characterized as refractory metals (others include molybdenum and tungsten) because of their relative chemical stability (passive oxide layer) and high melting points. Because of its high strength, chemical stability, and resistance to wear, alloys such as zirconium (e.g., Oxinium™) are likely to gain popularity as orthopedic biomaterials. Because of their surface oxide layer's stability, zirconium and tantalum are highly corrosion resistant. Corrosion resistance generally correlates with biocompatibility (although not always), because more stable metal oxides tend to be less chemically active and/or biologically available and are thus less participatory in biologic processes. This enhanced biocompatibility is produced by the relatively thick surface oxide layer (approximately 5 micrometers), and the ability to extend ceramic-like material properties (i.e., hardness) into the material through techniques such as oxygen enrichment. This has resulted in the production of new implant components using these alloys (see Figure II.5.6.11) (e.g., oxidized zirconium TKA femoral components, Smith and Nephew, Inc.). Although, new zirconium alloys such as Oxinium™ generally possess high levels of hardness (12 GPa) and wear resistance (approximately 10-fold that of cobalt and titanium alloys, using abrasion testing), which makes them well-suited for bearing surface applications, they are costly

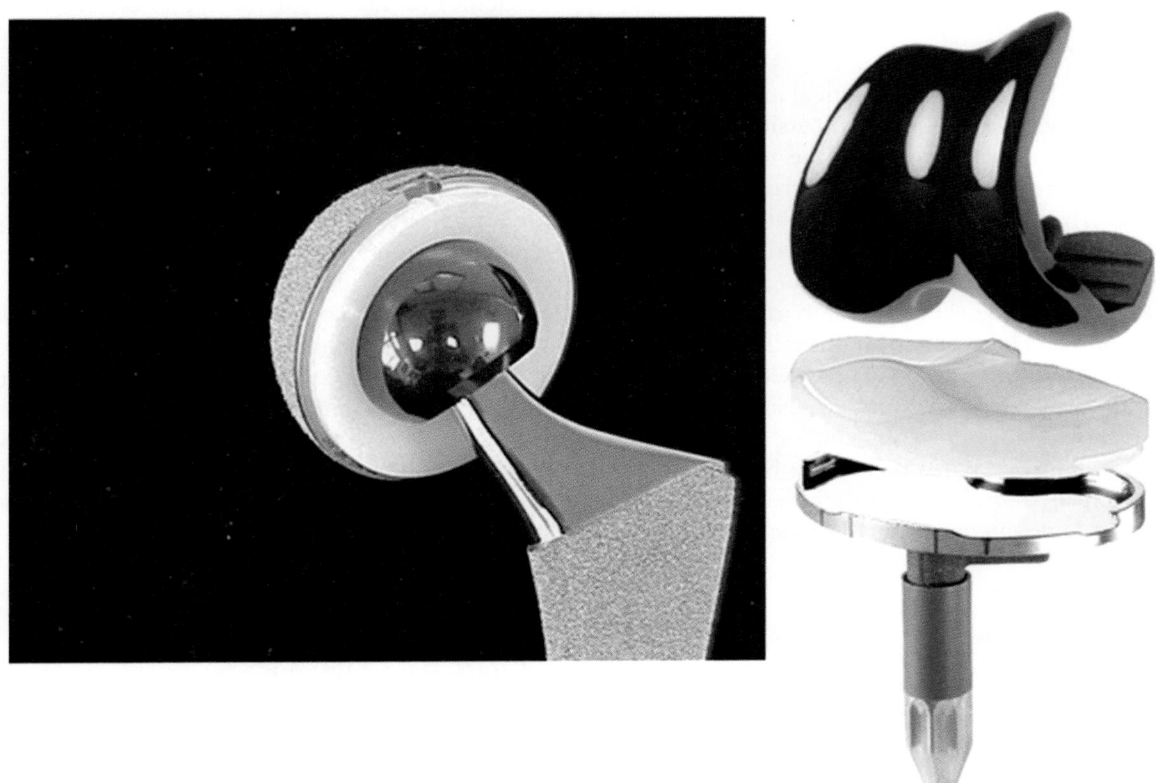

FIGURE II.5.6.11 Examples of new THA and TKA oxidized zirconium components currently gaining popularity because of enhanced mechanical and biocompatibility properties (Oxinium™, Smith and Nephew, Inc., Memphis, TN, USA). (Photographs courtesy of Smith and Nephew, Inc.)

to manufacture and currently are sought after in special circumstances where issues such as a metal allergy (or more accurately metal hypersensitivity) require particular attention to biocompatibility. As difficulties associated with the cost of forming and machining these metals are overcome, the use of these materials is expected to grow (Black, 1992).

New Titanium Alloys

One new group of titanium alloys put forward for orthopedic component uses molybdenum at concentrations greater than 10%. The addition of molybdenum acts to stabilize the BCC (Body Centered Cubic) (beta) phase at room temperature; thus these alloys are referred to as beta titanium alloys. These beta titanium alloys promise 20% lower moduli, which are closer to bone and thus provide better formability with maintenance of other mechanical properties typical of Ti-6-4.

Other attempts at improving traditional Ti-6Al-4V alloys seek to improve biocompatibility and mechanical properties by the substitution of vanadium (a relatively toxic metal) with other less toxic metals. Two such titanium alloys include Ti5Al2.5Fe and Ti6Al7Nb, which substitute iron and niobium for vanadium, respectively. These alloys have similar properties to traditional Ti-6-4, yet they claim higher fatigue strength and a lower modulus, thus enhancing bone to implant load transfer.

New Cobalt Alloys

Some "newer" cobalt alloys are identical in composition to traditional alloys, but use novel processing techniques to manipulate the microstructure of the implant materials to improve their mechanical properties. One such example recently patented, TJA-1537, although compositionally identical to ASTM F75, claims enhanced wear resistance and fatigue strength through elimination of carbide, nitride, and second phase particles (Allegheny Technologies). These particles normally form at the grain boundaries within a standard F75 CoCrMo alloy, and act to decrease wear and fatigue resistance. Other new cobalt alloys under development for use in orthopedics seek to improve biocompatibility by eliminating nickel, and improve mechanical properties by reducing the carbon content, thus avoiding carbide precipitation at grain boundaries.

New Stainless Steels

Because of the desirable cost, machinability, and ductility of stainless steels, there are still efforts to improve its mechanical properties to compete with cobalt and titanium alloys. The relatively poor corrosion resistance and biocompatibility of stainless steels when compared to titanium and Co–Cr–Mo alloys provide incentives for development of improved stainless steels. New alloys such as BioDur® 108 (Carpenter Technology Corp.) attempt to solve the problem of corrosion with an essentially nickel-free austenitic stainless alloy. This steel contains a high nitrogen content to maintain its austenitic structure, and boasts improved levels of tensile yield strength, fatigue strength, and improved resistance to pitting corrosion and crevice corrosion, as compared to nickel-containing alloys such as Type 316L (ASTM F138).

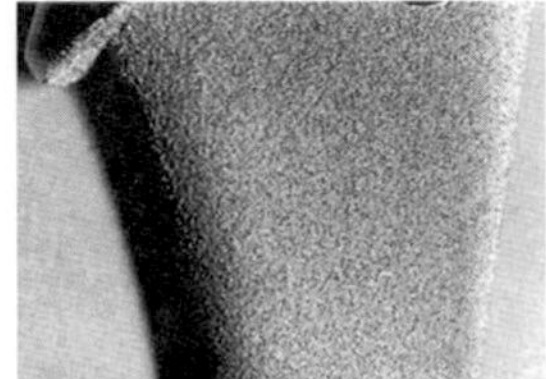

Co-Cr alloy THA femoral stem with a Co-Cr beaded surface

Pure-titanium fiber metal coating on a Ti-alloy THA stem

Plasma sprayed titanium surface on a Ti-alloy stem

Hydroxyapatite coating on a roughened titanium alloy THA stem

FIGURE II.5.6.12 Examples of currently used surface coatings on stems of THA to enhance both short- and long-term fixation.

Surfaces and Coatings

A variety of surface coatings are currently used to enhance the short- and long-term performance of implants by encouraging bone ingrowth and providing enhanced fixation. These different surfaces include roughened titanium, porous coatings made of cobalt chromium or titanium beads, titanium wire mesh (fiber mesh), plasma-sprayed titanium, and bioactive non-metallic materials such as hydroxyapatite or other calcium phosphate compositions (Figure II.5.6.12). Currently osteoconductive and osteoinductive growth factors such as transforming growth factor beta (TGF beta) are being developed for use as osteogenic surface coating treatments to enhance orthopedic implant fixation.

ORTHOPEDIC BIOMATERIALS: CLINICAL CONCERNS

Implant biocompatibility/performance is dependent on the type and amount of degradation produced by wear and electrochemical corrosion. Host response to orthopedic implant debris is central to clinical performance (Willert and Semlitsch, 1977). Implant loosening due to aseptic osteolysis accounts for over 75% of TJA implant failure, and is the predominant factor limiting the longevity of current total joint arthroplasties; other reasons include infection (7%), recurrent dislocation (6%), periprosthetic fracture (5%), and surgical error (3%) (Holt et al., 2007). Properly positioned implants tend to wear at predictable rates. However, there are variable amounts of debris-induced bone loss around implants in patients with similar rates of implant wear (i.e., debris generation). It is commonly noted that some individuals with severely worn components can demonstrate little periprosthetic bone loss, while others with modest amounts of wear can demonstrate extensive osteolysis and implant loosening (Willert et al., 1990; Huo et al., 1992; Jacobs et al., 1992; Jasty et al., 1994; Huk et al., 1994; Schmalzried et al., 1994; Harris, 1995; Thompson and Puleo, 1995; Yao et al., 1995; Goodman et al., 1998; Granchi et al., 1998; Jones et al., 1999; von Knoch et al., 2000). Debris-induced immune reactivity, aseptic inflammation, and subsequent early failure can be as high as 4–5% at 6–7 years post-operatively in current generation metal-on-metal total hip arthroplasties (Jacobs and Hallab, 2006; Korovessis et al., 2006; Milosev et al., 2006). The benefits provided to patients by orthopedic implants in terms of pain, mobility, and quality of life, are immeasurable. Therefore, the following sections which focus on the problems associated with implants, seek to provide the student of biomaterials with a foundation for understanding the relevant issues for orthopedic research and are not intended to serve as an indictment of orthopedic materials.

Despite their overwhelming success over the long term (>7 years), orthopedic biomaterials have been associated with adverse local and remote tissue responses. It is generally the degradation products of orthopedic biomaterials (generated by wear and electrochemical corrosion) which mediate these adverse effects. This debris may be present as particulate wear, colloidal nanometer size complexes (specifically or non-specifically bound by protein), free metallic ions, inorganic metal salts/oxides or in an organic storage form such as hemosiderin. Clinical aspects of biocompatibility regarding polymer and metal release from orthopedic prosthetic devices have taken on an increasing sense of urgency, due to the escalating rates of people receiving implants and the recognition of extensive implant debris within local and remote tissues. Particulate debris has enormous specific surface areas available for interaction with the surroundings and chronic elevations in serum metal content. Clinical issues associated with biomaterial degradation can be broken down into four basic questions: (l) how much material is released from the implant; (2) where is the material transported to and in what quantity; (3) what is the chemical form of the released degradation products (e.g., inorganic precipitate versus soluble organometallic complex); and (4) what are the pathophysiological interactions and consequences of such degradation? The answers to these questions, over the long term, remain largely unknown. There is a growing body of literature addressing the issues associated with the first two questions. However, little is currently known with regard to the latter two questions. The remainder of this chapter will focus on that which is known (and of orthopedic clinical concern) regarding biomaterial degradation (through wear and electrochemical corrosion), dissemination of debris, and consequent local/systemic effects.

Orthopedic Biomaterial Wear

The generation of wear debris, and the subsequent tissue reaction to such debris, is central to the longevity of total joint replacements. In fact, particulate debris is currently extolled as the primary factor affecting the long-term performance of joint replacement prostheses, and the primary source of orthopedic biomaterial degradation (based on overall implant mass or volume lost). Particulate debris generated by wear, fretting or fragmentation induces the formation of an inflammatory reaction, which at a certain point promotes a foreign–body granulation tissue response that has the ability to invade the bone–implant interface. This commonly results in progressive local bone loss that threatens the fixation of both cemented and cementless devices (Jacobs et al., 1994b; Jacobs, 1995; Jacobs et al., 2001).

Mechanisms of Wear Debris Generation

Wear involves the loss of material in particulate form as a consequence of relative motion between two surfaces. Two materials placed together under load will only contact over a small area of the higher peaks or asperities. Electro-repulsive and atomic binding interactions occur at the individual contacts and, when the two surfaces slide relative to one another, these interactions are disrupted; this results in the release of material in the form of particles (wear debris). The particles may be lost from the system, transferred to the counterface or remain between the sliding surfaces. There are primarily three processes which can cause wear: (1) abrasion – by which a harder surface "plows" grooves in the softer material; (2) adhesion – by which a softer material is smeared onto a harder counter surface forming a transfer film; and (3) fatigue – by which alternating episodes of loading and unloading result in the formation of subsurface cracks which propagate to form particles that are shed from the surface.

Wear Rates. During an initial "wearing in" period, the relative motion of surfaces causes a large number of asperities to break, resulting in a high wear rate. After this initial period, the actual contact area increases and the two surfaces can be said to have adapted to one other. Over time, the wear rates decrease and eventually become linearly dependent on the contact force and sliding distance represented by the steady-state wear equation:

$$V = KFx$$

where V is volumetric wear (mm^3/year), K is a material constant of the material couple, F is the contact force (N) and x is the distance of relative travel (mm).

Different types of orthopedic materials and couples produce different amounts and kinds of wear debris. Hard-on-hard material couples such as metal-on-metal articulations generally produce less wear (weight loss) than metal-on-polymer (see Figure II.5.6.13). There is a great deal of variability associated with *in vivo* wear rates of orthopedic biomaterials, which are generally measured by radiographic follow-up studies. Radiographic wear measurements are expressed as linear wear rates, whereas *in vitro* studies generally report volumetric wear. Volumetric wear can be directly related to the number of wear particles released into periprosthetic fluids (typically in the order of 1×10^9 of particles per year). The most common wear couple for hip and knee arthroplasty currently in use in the US is a cobalt-based alloy head (most commonly a Co–Cr–Mo alloy ASTM F75) bearing on an ultra-high molecular weight polyethylene (UHMWPE) cup or liner. The linear wear rates of this couple are generally in the order of 0.1 mm/year, with particulate generation as high as 1×10^6 particles per step or per cycle. Clinically, implant wear rates have been found to increase with the following: (1) physical activity; (2) weight of the patient; (3) size of the femoral head (32 versus 28 millimeters); (4) roughness of the metallic counterface; and (5) oxidation of the polyethylene (Jacobs et al., 1994a; Jacobs, 1995; Jacobs et al., 2001).

Although well established in the hip and knee, newer spinal implants are now undergoing the same types of analyses to try to assess which type of implant works best. The following two basic types of articulating bearing designs are available in spine, knee, and hip arthroplasties.

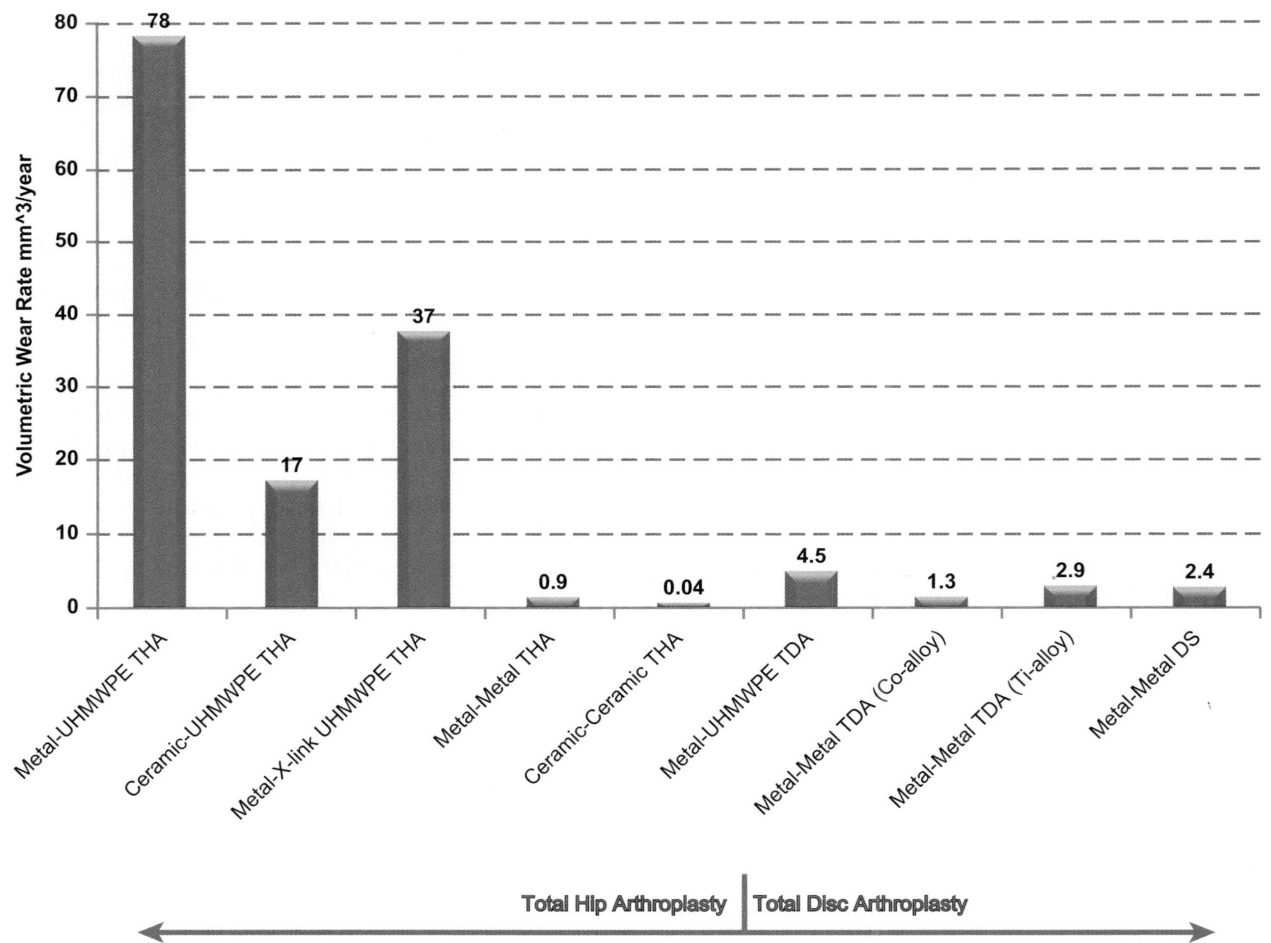

FIGURE II.5.6.13 A comparison of the amount of wear debris generated from different types of total joint arthroplasties. There is relatively less (10×) polymeric debris generated by a total disc arthroplasty with a metal-on-polymer articulation. Note: Figure References: Metal-Poly (Holt et al., 2007); Ceramic-Poly (Jacobs et al., 1994c); Metal-X-linked Poly (Martinon et al., 2006); Metal-X-linked Poly (Jacobs et al., 1994c); Metal-X-linked Poly (Hallab, 2009); Metal-X-linked Poly (Hallab, 2009); Metal-Metal (Huk et al., 1994); Ceramic-Ceramic (Hallab, 2009); Metal-UHMWPE TDA (Hallab, 2009); Metal-Metal TDA (Hallab, 2009).

Metal-on-polymer spine arthroplasty wear. In vitro analysis of wear has demonstrated wear rates of metal-on-polymer bearing lumbar total disc arthroplasty devices range from 2 to 20.8 mm^3 per million cycles (Pare et al., 2007; Popoola et al., 2007), where the size of the wear debris generally ranges from 0.1 μm to 100 microns in diameter UHMWPE (Anderson et al., 2003; Popoola et al., 2007; van Ooij et al., 2007; Hallab et al., 2008). This amount of debris is 10-fold less wear than THAs and TKAs that are composed of metal on highly cross-linked polyethylene (x-UHMWPE) bearing surfaces (Greenwald and Garino, 2001; Kurtz et al., 2005).

Metal-on-metal spinal arthroplasty wear. In general the wear of metal-on-metal TJA is well below that of metal-on-polymer (Figure II.5.6.13) (Wroblewski et al., 1996; Saikko et al., 2002; Tipper et al., 2002; Catelas et al., 2004; Heisel et al., 2004; Kurtz et al., 2005; Minoda et al., 2005; Callaghan et al., 2007). The few published reports on the volumetric wear rates of metal-on-metal disc arthroplasty prostheses indicate a wear rate of 0.93–1.26 mm^3 per million cycles (cobalt alloy) (Firkins et al., 2001; Pare et al., 2007). Another study of an all titanium-6%Al-4%V alloy disc arthroplasty found wear rates to be as high as 3 mm^3 per million cycles (Hellier et al., 1992). These values are similar to those reported for metal-on-metal hip replacements, which have been shown to range from approximately 0.05 to 6 mm^3 per million cycles (cobalt alloy) (McKellop et al., 1996; Clarke et al., 2000; Catelas et al., 2004). Long-term follow-up of patients undergoing total disc arthroplasty is required to assess how intimately wear will correlate with inflammation and poor implant performance.

Orthopedic Biomaterial Corrosion

Electrochemical corrosion occurs to some extent on all metallic surfaces including implants. This is undesirable for two primary reasons: (1) the degradative process may reduce structural integrity of the implant; and (2) the release of products of degradation is potentially toxic to the host. Metallic biomaterial degradation may result from electrochemical dissolution phenomena or wear, but most commonly occurs through a synergistic combination of the two. Electrochemical processes include generalized corrosion uniformly affecting an entire surface, and localized corrosion affecting either areas of a device relatively shielded from the environment (crevice corrosion) or seemingly random sites on the surface (pitting corrosion). Additionally, these electrochemical and other mechanical processes interact, potentially causing premature structural failure and/or accelerated metal release (e.g., stress corrosion cracking, corrosion fatigue, and fretting corrosion) (Brown and Merritt, 1981; Cook et al., 1983; Bundy et al., 1991; Brown et al., 1992; Collier et al., 1992b; Gilbert and Jacobs, 1997).

Corrosion Mechanisms

Corrosion of orthopedic biomaterials is a multifactorial phenomenon, and is dependent on five primary factors: (1) geometric variables (e.g., taper crevices in modular component hip prostheses); (2) metallurgical variables (e.g., surface microstructure, oxide structure and composition); (3) mechanical variables (e.g., stress and/or relative motion); (4) solution variables (e.g., pH, solution proteins, enzymes); and (5) the mechanical loading environment (e.g., degree of movement, contact forces, etc.). Current investigational efforts to minimize the corrosion of orthopedic biomaterials deal directly with the complex interactions of these factors.

There are two essential features associated with how and why a metal corrodes. The first has to do with thermodynamic driving forces, which cause corrosion (oxidation/reduction) reactions. In general, whether or not corrosion will take place under the conditions of interest depends on the chemical driving force (ΔG), and the charge separation. This separation contributes to what is known as the electrical double layer (Figure II.5.6.14) which creates an electrical potential across the metal–solution interface (much like a capacitor):

$$\Delta G = -nF\Delta E$$

where n is the valence of the ion, F is the Faraday constant (95,000 coulombs/mole electrons), and E is the voltage across the metal solution interface. This potential is a measure of the reactivity of the metals or the driving force for metal oxidation. Therefore, the more negative the potential of a metal in solution, the more reactive it will tend to be (i.e., the greater is ΔG for reduction).

The second factor governing the corrosion process of metallic biomaterials is the kinetic barrier to corrosion (e.g., surface oxide layer). Kinetic barriers prevent corrosion not by energetic considerations, but by physically limiting the rate at which oxidation or reduction processes can take place. The well-known process of passivation or the formation of a metal–oxide passive film on a metal surface is one example of a kinetic limitation to corrosion. In general, kinetic barriers to corrosion prevent either the migration of metallic ions from the metal to the solution, the migration of anions from solution to metal, or the migration of electrons across the metal–solution interface. Passive oxide films are the most well-known forms of kinetic barriers in corrosion, but other kinetic barriers exist, including polymeric coatings (Gilbert and Jacobs, 1997; Jacobs et al., 1998a).

Passivating Oxide Films

Most alloys used in orthopedic appliances rely on the formation of passive films to prevent significant oxidation from taking place. These films consist of metal oxides, which form spontaneously on the surface of the

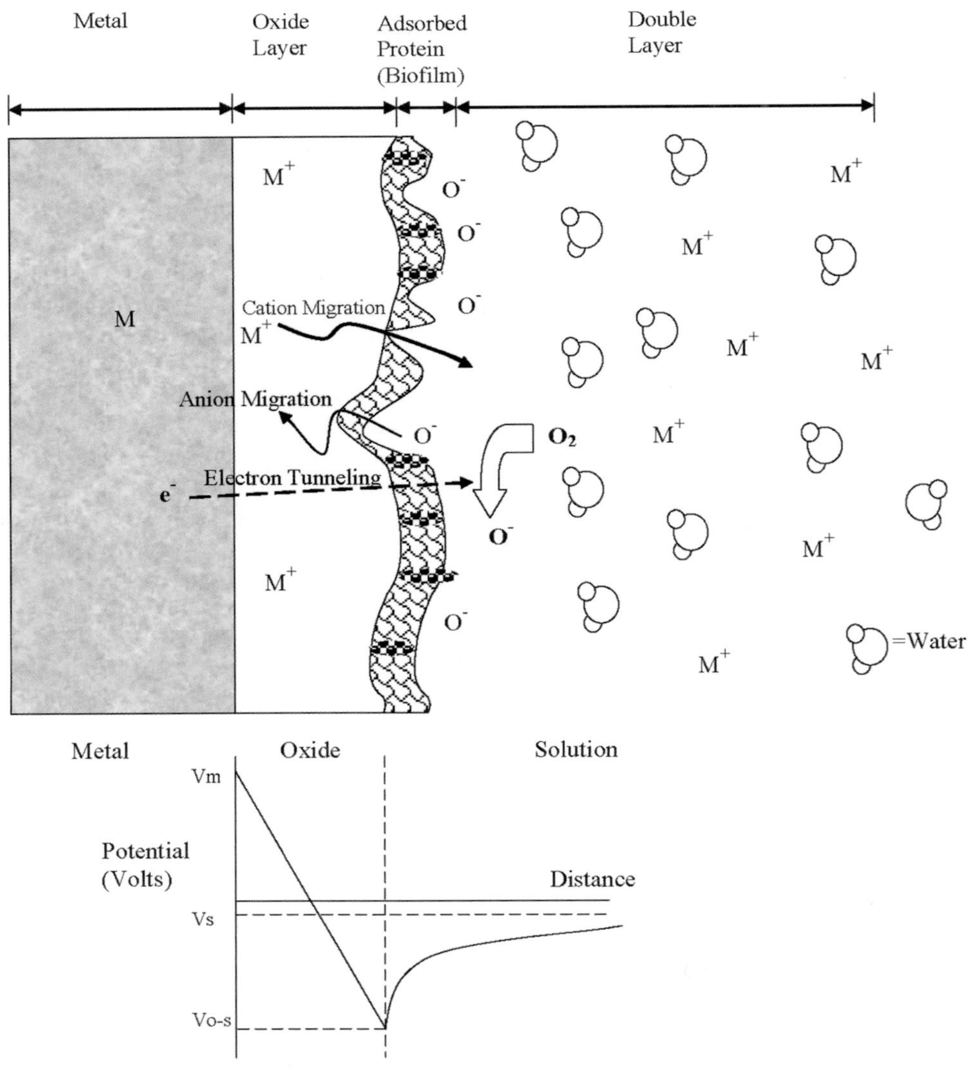

FIGURE II.5.6.14 Schematic of the interface of a passivating alloy surface in contact with a biological environment.

metal in such a way that they prevent further transport of metallic ions and/or electrons across the film. Passive films must have certain characteristics to be able to limit further oxidation. The films must be compact and fully cover the metal surface, they must have an atomic structure which limits the migration of ions and/or electrons across the metal–oxide–solution interface, and they must be able to remain on the surface of these alloys even with the mechanical stress and abrasion which can be expected with orthopedic devices.

Passivating oxide films, which spontaneously grow on the surface of many metals and alloys, have five primary structural and physical characteristics, which are particularly relevant to implant degradation processes:

1. First, these oxide films are very thin, typically in the order of 5–70 Å which depends on the potential across the interface as well as solution variables (e.g., pH). Furthermore, the oxide structure may be amorphous or crystalline. Since the potentials across the metal solution interface for these reactive metals are typically 1–2 volts, the electric field across the oxide is very high, in the order of 10^6–10^7 V/cm. One of the more widely accepted models is based on the theory of Mott and Cabrera, which states that oxide film growth depends on the electric field across the oxide. If the potential across the metal–oxide–solution interface is decreased (i.e., made closer to the electrochemical series potential), then the film thickness will decrease by reductive dissolution processes at the oxide. If the interfacial potential is made sufficiently negative or the pH of the solution is made low enough, then these oxide films will no longer be thermodynamically stable, and will undergo reductive dissolution without which corrosion will increase (Gilbert and Jacobs, 1997; Jacobs et al., 1998a).

2. Second, oxide films have the characteristics of semiconductors with an atomic defect structure, which determines the ability for ionic and electronic transport across films. Metal cations and oxygen anions require the presence of cationic or anionic vacancies (respectively) in the oxide for transport of these species across the film. If there is a deficit of metal ions in the oxide film (i.e., there are cationic vacancies), for example, then metal ion transport is possible and these oxides are known as p-type semiconductors. Chromium oxide (Cr_2O_3) is such a metal-deficit oxide. On the other hand, if there is an excess of metal ions in the oxide (or a deficit of anions) then cation transport is limited, but anion transport can occur. These oxides will also have excess electrons and are known as n-type semiconductors. TiO_2 spontaneously formed on titanium alloy implant (Ti-6Al-4V) surfaces is one such n-type semiconductor. The greater the number of defects (vacancies or other valence species), the less able is the oxide film to prevent migration of ionic species, and the lower is the kinetic barrier to corrosion. TiO_2 is very close to being stoichiometric (chemically homogeneous) and hence does not have many ionic defects, resulting in an increased resistance to ionic transport. Other defects may be present in these passive oxide films which may alter their ability to limit corrosion. For instance, the addition of other metal ions with valence states which are different from the native metal ions can alter both the electronic and ionic transport of charge across the interface. These additions may enhance or degrade the ability of the oxide to prevent corrosion, depending on the nature of the oxide. One example of improved corrosion resistance from mixed oxides comes from what is known as a spinel. Spinels are typically mixed oxides of the form $(A_2O_3)BO$, where A and B are +3 and +2 valence metal ions. In Co–Cr alloys, for instance, a spinel of $(Cr_2O_3)CoO$ can form on the surface. Spinels are typically known to have higher strengths and better resistance to diffusion of ions compared to single metal ion oxides. Therefore, a high concentration of spinels in the oxide layer will act to resist dissolution of a metal implant (Gilbert and Jacobs, 1997; Jacobs et al., 1998a).
3. Third, the ratio of the "oxide specific volume" to metal alloy specific volume (i.e., Pilling Bedworth ratio) will determine if the oxide will adhere to the metal or not. If there is too great a mismatch between the metal and oxide lattice parameters, then consequential stresses will be generated between the two. The magnitude of the internal stress will vary with the thickness of the oxide. Too great an oxide thickness will thus result in spontaneous fracture or spalling of the oxide, lowering the kinetic barrier effect of the oxide to corrosion (Gilbert and Jacobs, 1997; Jacobs et al., 1998a).
4. Fourth, the morphology of these oxide films is not one of a smooth, flat, continuous sheet of adherent oxide covering the metal. Transmission electron microscopy (TEM) and atomic force microscopy (AFM) techniques have shown that oxides of titanium, for instance, consist of needle or dome shapes. The size and shape of these oxide domes change with applied potential when immersed in oxalic and other acids (Gilbert and Jacobs, 1997; Jacobs et al., 1998a).
5. Finally, mechanical factors such as fretting, micromotion or applied stresses may abrade or fracture oxide films. When an oxide film is ruptured from the metal substrate, fresh unoxidized metal is exposed to solution. When these films reform or repassivate, the magnitude of the repassivation currents which are subsequently generated may be large. This is because large driving forces exist for oxidation, and when the kinetic barrier is removed these large driving forces can operate to cause oxidation. However, the extent and duration of the oxidation currents will depend on the repassivation kinetics for oxide film formation. Hence, the mechanical stability of the oxide films, as well as the nature of their repassivation process, are central to the performance of oxide films in orthopedic applications (Gilbert and Jacobs, 1997; Jacobs et al., 1998a).

Corrosion at Modular Interfaces of Joint Replacements

One issue associated with orthopedic alloys is the corrosion observed in the taper connections of retrieved modular joint replacement components (Figure II.5.6.15). With the growing number of total joint designs which use metal-on-metal conical tapers as modular connectors between components, the effects of crevices, stress, and motion take on increasing importance. Severe corrosion attack can take place in the crevices formed by these tapers *in vivo*. Gilbert et al. (1993) have shown that, of 148 retrieved implants, approximately 35% showed signs of moderate to severe corrosion attack in the head–neck taper connections of total hip prostheses. This attack was observed in components which consisted of Ti-6Al-4V alloy stems and Co–Cr heads, as well as Co–Cr stems on Co–Cr heads. This corrosion process is the result of a combination of stress and motion at the taper connection, and the crevice geometry of the taper; the stresses resulting from use cause fracturing and abrasion of the oxide film covering these passive metal surfaces. This, in turn, causes significant changes in the metal surface potential (makes it more negative) and in the crevice solution chemistry as the oxides continuously fracture and repassivate. These changes may result in deaeration (loss of O_2) of the crevice solution, and a lowering of the pH in the crevice as is expected in crevice corrosion attack. The ultimate result of this process is a loss of the oxide film and its kinetic barrier effect, and an increase in

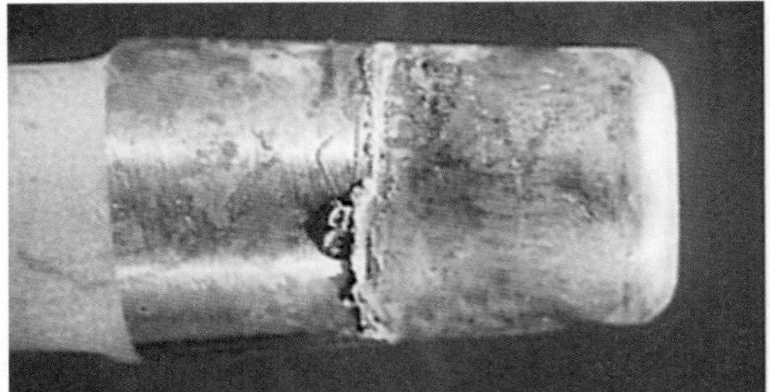

FIGURE II.5.6.15 Modular junction taper connection of a total hip arthroplasty showing corrosion of the taper connections. Macrograph of deposits of $CrPO_4$ corrosion particles on the rim of a modular cobalt–chrome femoral head.

the rate of corrosive attack in the taper region. The corrosion processes in the Co–Cr alloys have been observed to consist of intergranular corrosion, etching, selective dissolution of cobalt, and the formation of Cr-rich particles. In isolated cases, this occurs to such an extent that intergranular corrosion caused fatigue failure in the neck of a Co–Cr stem. Corrosion attack of titanium alloy stems has also been observed in some cases.

Very little is known about the mechanical stability of passive oxide films and the electrochemical reactions (e.g., ion and particle release) which occur when the oxide film is fractured. What is known is that when the oxide films of these orthopedic alloys are abraded or removed from the surface by rubbing, the open circuit potential can decrease to as low as −500 mV (versus Standard Calomel Electrode). These voltage potentials may be significant and prolonged enough to cause changes in the oxide structure and stability by bringing the interface potential into the active corrosion range of the alloy, thereby dramatically accelerating the corrosion rate and decreasing implant performance (Brown et al., 1992; Collier et al., 1992a; Gilbert et al., 1993; Bobyn et al., 1994; Brown et al., 1995).

Implant Debris Types: Particles and Ions

The degradation products of all orthopedic implants are one of two basic types: particles or soluble debris (metal ions). While there is a large distinction between the two (ions versus particles), the difference between them blurs as the size of particles decreases into the nanometer range and becomes essentially "in solution." Typically, particulate wear debris (metal, ceramic or polymer) exists from the submicron size up to thousands of microns (mm), while so-called "soluble debris" is limited to metal ions (or nanoparticles that are too small to be distinguished from ions) that are bound to plasma proteins.

Particulate Debris. Different types of joint arthroplasty designs not only produce different amounts and kinds of wear debris, but also different sizes and shapes of debris that are specific to the type of implant materials used for the bearing interfaces. For instance, hard-on-hard material couples such as metal-on-metal articulations generally produce smaller-sized (submicron), fairly round debris, whereas traditional metal-on-polymer bearings produce larger (micron-sized) debris that is more elongated in shape (see Figure II.5.6.16).

As is evident in Figure II.5.6.16, hard-on-hard material couples (e.g., metal-on-metal) produce smaller debris than do hard-on-soft material couples (e.g., metal on polyethylene). The particles produced from articulating bearing in any metal-on-polymer bearing implants are dominated by polymer particles, with little metallic debris unless there are other sources of metal release, such as corrosion at metal–metal connections. Polymeric particles produced from implants generally fall into the range from 0.23 to 1 μm (Figure II.5.6.16). Past investigations, primarily of UHMWPE wear debris in peri-implant tissues, have shown that 70–90% of recovered particulates were submicron, with the mean size being approximately 0.5 μm (Maloney et al., 1993; Jacobs et al., 1994a; Campbell et al., 1995). Metal and ceramic particles have generally been characterized as an order of magnitude smaller than polymer particles (at approximately 0.05 μm in diameter).

Histological Identification of Particles *In Vivo*

The tissues surrounding modern implants may include areas of osseointegration, fibrous encapsulation, and a variable presence of the foreign-body response to polyethylene and cement debris in joint replacement devices. Absent is any specific histologic evidence of the slow release of metallic species that is known to occur with all metallic implants. However, accelerated corrosion and a tissue response that can be directly related to identifiable corrosion products can be demonstrated in the tissues surrounding multi-part devices (Urban et al., 2000).

1. **Stainless steels**: Histological sections of the tissues surrounding stainless steel internal fixation devices generally show encapsulation by a fibrous membrane, with little or no inflammation over most of the device. At screw plate junctions, however, the membranes often contain macrophages, foreign-body giant cells, and a variable number of lymphocytes in association

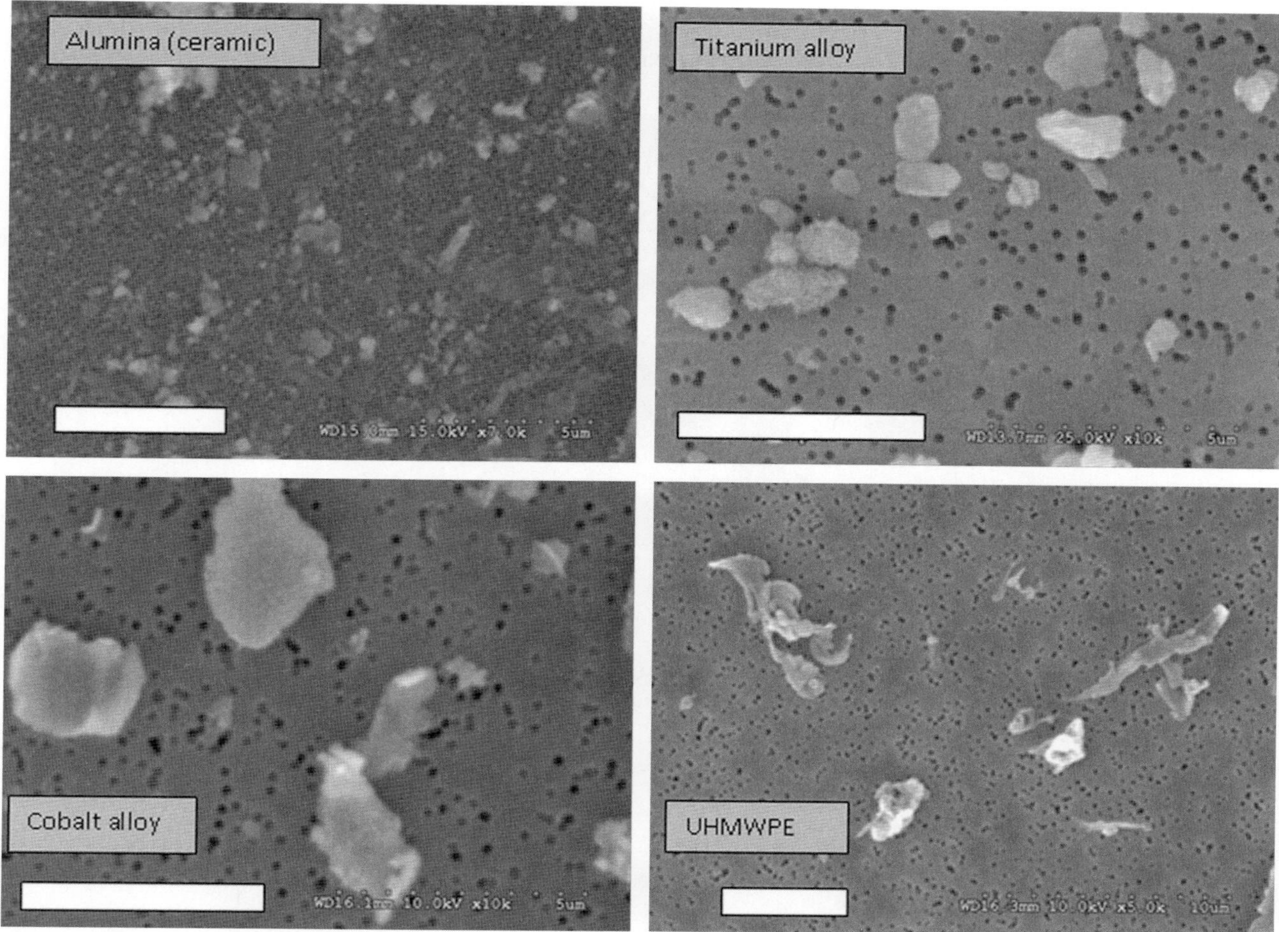

FIGURE II.5.6.16 Implant debris from four types of materials are shown where the metal (cobalt alloy and titanium) and ceramic (alumina) debris are more rounded versus the polymeric (UHMWPE) debris which is more elongated in shape. Note: Bar = 5 μm.

with two types of corrosion products. The first consists of iron-containing granules. The second, termed microplates, consists of relatively larger particles of a chromium compound. Microplates are found within the tissues as closely packed, plate-like particle aggregates ranging in size from 0.5 mm to 5.0 mm.

Hemosiderin–like granules often surround the collections of microplates, but the granules are also found alone. The granules are yellow–brown, mainly spherical, and 0.1–3 or more micrometers in diameter. They are predominantly intracellular, most often in macrophages, but may also be found in other periprosthetic cells (e.g., fibroblasts). X–ray diffraction has indicated that the granules consist of a mixture of two or more of the iron oxides, αFe_2O_3 and σFe_2O_3, and the hydrated iron oxides, $\alpha Fe_2O_3{\cdot}H_2O$ and $\sigma Fe_2O_3{\cdot}H_2O$.

2. **Cobalt–based alloys:** The nature of corrosion at modular connections are similar, whether modular heads are mated with cobalt–chromium alloy or Ti–6Al–4V alloy femoral stems. The principal corrosion product identified by electron microprobe energy-dispersive X–ray analysis and Fourier transform infrared microprobe spectroscopy is a chromium-phosphate ($Cr(PO_4)4H_2O$) hydrate-rich material termed "orthophosphate." This corrosion product can be found at the modular head–neck junction and as particles within the joint capsules, at the bone–implant interfacial membranes, and at sites of femoral osteolytic lesions. Particles of the orthophosphate material have been found at the bearing surface of the UHMWPE acetabular liners, suggesting their participation in three-body wear and an increased production of polyethylene debris. Particles of the chromium orthophosphate hydrate-rich corrosion product found in the tissues ranged in size from submicron to aggregates of particles up to 500 micrometers.
3. **Titanium-based alloys:** The degradation products observed in histologic sections of tissues adjacent to titanium-based alloys are of a different nature than the precipitates associated with stainless steel and cobalt-based alloys. Despite the remarkable corrosion resistance of titanium-based alloys, there have been persistent reports of tissue discoloration due to metallic debris in the periprosthetic tissues. These particulates observed in local tissues surrounding titanium alloy implants have the same elemental composition

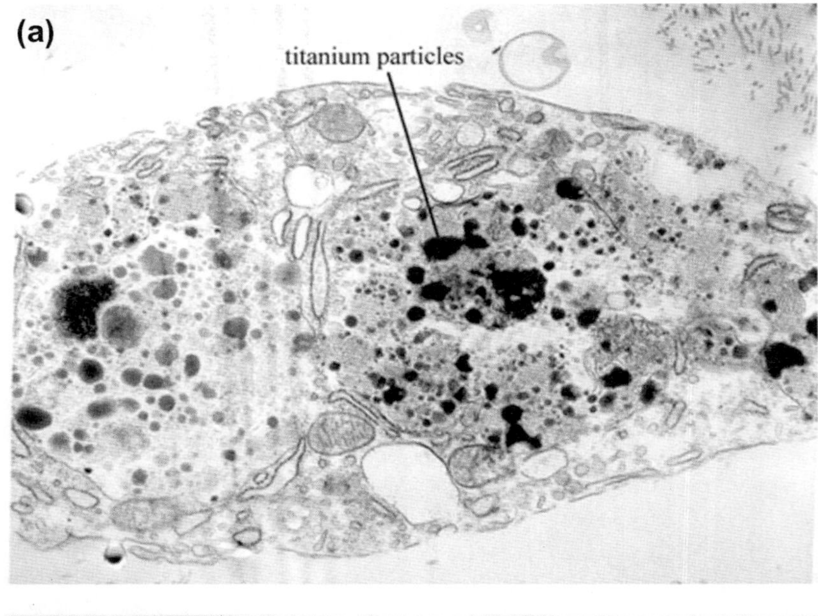

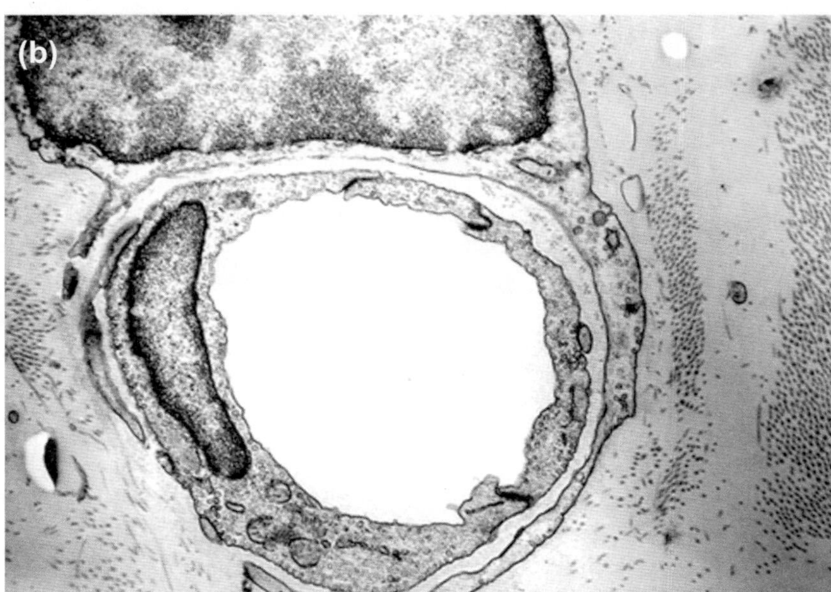

FIGURE II.5.6.17 Transmission Electron Photomicrographs: (a) Macrophage containing phagocytized titanium particles. (b) Endothelial cell lining with embedded titanium debris. These specimens were obtained from a tissue sample overlying the posterolateral fusion mass (sixteen-week autograft + titanium) (TEM magnification = 20,000 ×).

as the parent alloy, as opposed to precipitated corrosion products which occur with stainless steel and cobalt–chromium alloys (Figure II.5.6.17). However, wear debris presents an enormous surface area for electrochemical dissolution, which, in all likelihood, is a major factor contributing to observed systemic elevations in titanium of patients with titanium implants (Urban et al., 1996a, 1997).

Particle Characterization

Traditionally particle characterization uses methods such as Scanning Electron Microscopy (SEM) or Transmission Electron Microscopy (TEM), both of which are number-based counting methods. These methods have indicated that the majority of the wear (mass loss) from an implant is comprised of particles in the nanometer-to-submicron range. This understanding stems from the relatively low numbers of particles, (e.g., 100s–1000s) that are counted using image-based analysis techniques such as SEM. Newer analytical techniques, such as low angle laser diffraction (LALLS) have the capability of sampling millions to billions of particles, counted as they pass in front of and scatter a laser light beam proportionally to their size. Thus, as millions of particles flow by, LALLS analysis can detect the one-in-a-million large particle that comprises a significant portion of the total mass loss (i.e., total debris). This brings up the confusing concept that particle size of

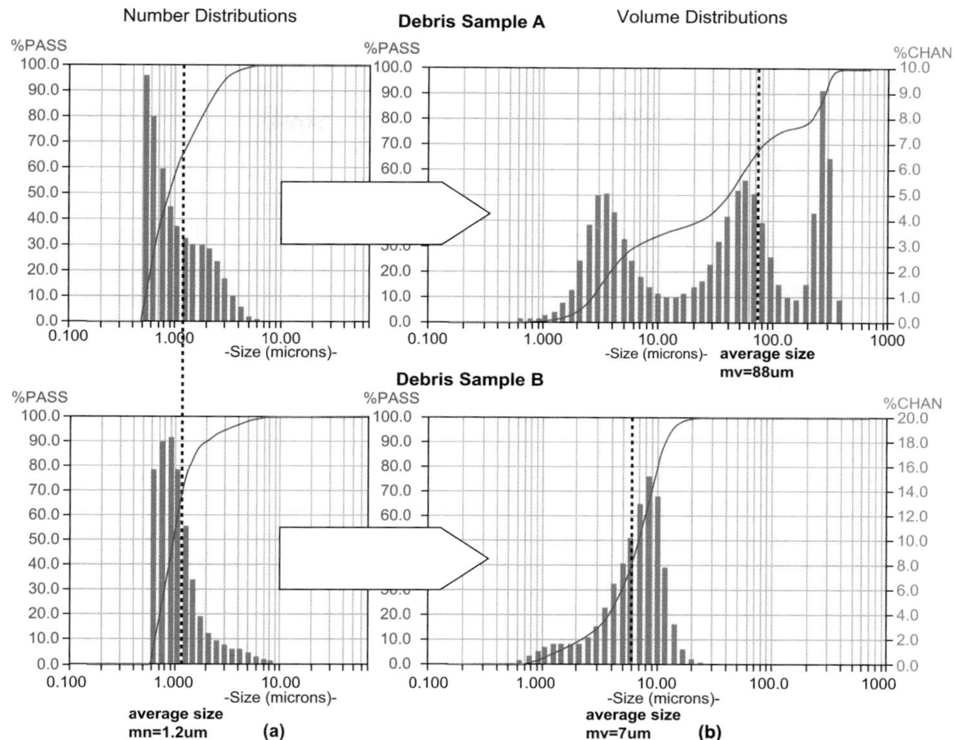

FIGURE II.5.6.18 Analyses of: (a) volume; and (b) number distributions of two debris samples demonstrate that similar number distributions can result from very different actual size distributions. Note: The *x*-axis is particle diameter and the *y*-axis is (i) percentage of total number of particles in each size range; and (ii) the percentage of total mass in each size range. (Courtesy of BioEngineering Solutions, Inc.)

any given distribution depends on the method of evaluation. There is no one particle size. For example, the average size of 500 marbles and 5 basketballs is approximately the size of the marbles on a number basis, and approximately the size of the basketball on a volume basis. Thus, the question "what is the average size of particles that comprises approximately 50% of the total volume of particles?" is another way of asking what is the mean size (or average size) on a volume basis. Asking "what is the average size of the particles that comprises 50% of the total number of particles?" is the average size based on number basis. The stark differences between a volume and number-based analysis of implant debris are shown in Figure II.5.6.18. When comparing the volume and number distributions in this figure, the dominant contribution of larger particles to the total debris mass (volume) is evident, yet it is insignificant compared to the total number of particles (Figure II.5.6.18). Thus, as can be seen in both the simple example of marbles and basketballs and in the LALLS analysis of implant debris particles in Figure II.5.6.18, volume-based analysis tends to represent the largest (most massive) particles, and number analysis tends to depict the size characteristics of the most numerous.

The ability to comprehensively characterize implant debris is important to the new designs and bearing surfaces used in new spinal implants. This multi-analysis approach is necessary because a given amount of wear debris (weight loss from the implant) after a year of use could be attributed to the loss of a relatively few large particles or hundreds of millions of small particles (e.g., approx 0.2 mm^3 volume loss after a million cycles of use could be from approx 400 particles of 100 microns diameter or 400 million particles only 1 micron in diameter). The bias of techniques limited to number analysis is that very similar number-based distributions can look very different from a volume perspective. This phenomenon is illustrated in Figure II.5.6.18 when comparing samples A and B, where different samples of particles look like very different volume-based distributions, however they look like very similar number-based distributions. This shows how important it is to have both number- and volume-based distributions of the same particles to fully characterize the types of particles in the mix, if there is enough particulate mass (>0.05 mg of particles) for a LALLS-type analysis. Unfortunately, in implant debris analysis there is usually less than the 0.05 mg of debris required for obtaining an accurate volume distribution, and thus historically SEM and TEM analyses have been used to characterize debris.

Particulate Debris Reactivity Characterization

Macrophages are immune cells that are involved in the phagocytosis of implant debris and the resulting inflammatory responses. Once debris is ingested by macrophages a host of biologic reactions can occur, such as activation of T-cells through antigen presentation (Hallab et al., 2001a), release of pro-inflammatory mediators (Glant and Jacobs, 1994; Shanbhag et al., 1994; Ingham et al., 2000; Matthews et al., 2000; Cunningham et al., 2002), cytotoxicity (Hallab et al., 2005), DNA damage (Nagaya et al., 1989; Savarino et al., 2000; Hallab et al., 2005), and oxidative stress (Soloviev et al., 2005). Macrophage reactions to debris are responsible for mediating debris-induced inflammation that is the leading cause of implant loosening over time.

Despite the large number of studies on metallic, polymeric, and ceramic particles' effects on peri-implant cells (e.g., macrophages, fibroblasts, osteoblasts, and osteoclasts), there are surprisingly few guidelines on what type of debris is most deleterious or bioreactive. However, there are a few general particle characteristics on which local inflammation has been shown to depend: (1) particle load (which depends on both particle size and total volume); (2) aspect ratio (the shape of the particles); and (3) chemical reactivity (the chemical composition). Thus, theoretically, a bioreactivity index of particulate would take the mathematical form:

$$\text{Particle Pro} - \text{inflammatory Index} \approx K_{\text{Load}}(\text{particle load}) \times K_{\text{Shape}}(\text{aspect ratio}) \times K_{\text{Material}}(\text{material type}) \quad (1)$$

where K's are constants that depend on the testing environment, and particle load is a function of both particle size and total debris volume.

1. **Greater particle load: (size and volume) increases inflammation.** An inflammatory response *in vivo* is proportional to the particle load (the concentration of phagocytosable particles per tissue volume, which is characterized by both the size and total volume) (González et al., 1996; Matthews et al., 2000). While this seems obvious, the ramifications and the conditions under which this remains true are important and not so obvious. If a given amount of debris (mass loss from an implant) is comprised of small diameter particles, there will be far greater numbers than if that same mass of debris was composed of larger diameter particles (Matthews et al., 2000). The degree to which equal numbers (doses) of large particles (e.g., 10 micron diameter) and smaller particles (1 micron diameter) induce a proinflammatory response has not been thoroughly investigated, and remains unknown.
2. **Aspect ratio: elongated (fibers) particles are more proinflammatory than round particles** (Laquerriere et al., 2003; Sieving et al., 2003). That fibers are more reactive than round debris was well-established over 30 years ago, with studies of asbestos fibers (Bruch, 1974). However, it remains unknown at what aspect ratio (aspect ratio: length/width) in the transition from round particles to fibers that elevated inflammation is generally initiated, and thus to date there is no "guideline" aspect ratio for implant debris particles to remain below.
3. **More chemically reactive particles are more proinflammatory.** There is a growing consensus that metal particles are more proinflammatory when compared to other materials such as polymers (Ramachandran et al., 2006). However, this is not a unanimous opinion; others have concluded polymers are more proinflammatory than metals (von Knoch et al., 2000), and other reports have shown no differences between similar metal and polymer particles (Baumann et al., 2004, 2005, 2006). Despite these reports to the contrary, there is a growing consensus that metallic particles are more proinflammatory because they are capable of corroding and releasing ions that have been associated with hypersensitivity responses, cytotoxicity, and DNA damage (Hallab et al., 2001a, 2005; Caicedo et al., 2007).

Controversial Particle Characteristics

Does particle size matter? Absolutely. *In vitro* inflammatory responses generally require particles <10 microns in diameter to be phagocytosed. Thus, to produce an *in vitro* inflammatory response, particles need to be less than 10 μm, that is, within a phagocytosable range. Purportedly, particles with a mean size of 0.2–10 μm are the most proinflammatory. Within this range there is no consensus as to which specific size(s) and/or dose of particles (particles/cell or particles/tissue volume) are maximally inflammatory (Shanbhag et al., 1995; Rader et al., 1999; Matthews et al., 2000; Ingram et al., 2002).

The effects of bacteria products (or endotoxin) on implant debris particles is presumably important because it has been found in periprosthetic tissue of failed implants, even in the absence of clinical signs of infection (Nalepka et al., 2006). Furthermore, the bacteria that are present at below clinically detectable levels have been shown to affect implant performance, because antibiotic-eluting bone cement and systemically administered antibiotics reportedly reduce the frequency of long-term failure, i.e., aseptic loosening (Espehaug et al., 1997).

Metal Ions (Soluble Debris)

There is continuing concern regarding the release of chemically active metal ions which bind to proteins and remain in solution from which they can then disseminate into the surrounding tissues, bloodstream, and remote organs. Particulate metallic wear debris presents an enormous surface area for electrochemical dissolution, which, in all likelihood, is a major factor contributing to observed systemic elevations in metals of patients with titanium implants (Urban et al., 1996b, 1997, 1998; Jacobs et al., 1998b; Urban et al., 2000). Normal human serum levels of prominent implant metals are approximately 1–10 ng/ml Al, 0.15 ng/ml Cr, <0.01 ng/ml V, 0.1–0.2 ng/ml Co, and <4.1 ng/ml Ti. Following total joint arthroplasty, levels of circulating metal have been

TABLE II.5.6.5 Approximate Average Concentrations of Metal in Human Body Fluids With and Without Total Joint Replacements (Michel et al., 1984; Stulberg et al., 1994; Jacobs et al., 1998a,b, 1999,)

		(ng/ml or ppb)						
Fluid		**Ti**	**Al**	**V**	**Co**	**Cr**	**Mo**	**Ni**
Serum	**Normal**	**2.7**	**2.2**	**<0.8**	**0.18**	**0.05–0.15**	*	**0.4–3.6**
	THA	4.4	2.4	1.7	0.2-0.6	0.3	*	<9.1
	THA-F	8.1	2.2	1.3	*	0.2	*	*
	THA-F	8.1	2.2	1.3	*	0.2	*	*
	TKA	3.2	1.9	<0.8	*	*	*	*
	TKA-F	135.6	3.7	0.9	*	*	*	*
Urine	**Normal**	**<1.9**	**6.4**	**0.5**	*	**0.06**	*	*
	TJA	3.55	6.53	<0.4	*	0.45	*	*
Synovial fluid	**Normal**	**13**	**109**	**5**	**5**	**3**	**21**	**5**
	TJA	556	654	62	588	385	58	32
Joint capsule	**Normal**	**723**	**951**	**122**	**25**	**133**	**17**	**3996**
	TJA	1540	2053	288	1203	651	109	2317
	TJA-F	19173	1277	1514	821	3329	447	5789
Whole blood	**Normal**	**17**	**13**	**6**	**0.1–0.1.2**	**2.0–4.0**	**0.5–1.8**	**2.9–7.0**
	TJA	67	218	23	20	110	10	29

Normal: Subjects without any metallic prosthesis (not including dental).
THA: Subjects with well-functioning total hip arthroplasty.
THA-F: Subjects with a poorly-functioning total hip arthroplasty (needing surgical revision).
TKA: Subjects with well-functioning total knee arthroplasty.
TKA-F: Subjects with a poorly-functioning total knee arthroplasty (needing surgical revision).
TJA: Subjects with well-functioning total joint arthroplasty.
TJA-F: Subjects with a poorly-functioning total joint arthroplasty (needing surgical revision).
*Not tested.

shown to increase (Table II.5.6.5). The values in this table show that following successful primary total joint replacement there are measurable elevations in serum and urine cobalt, chromium, and titanium. The clinical ramifications of metal ion release are discussed in the next few sections.

Local Tissue Effects of Wear and Corrosion

Implant debris limits the long-term performance of total joint replacement by causing a local inflammatory response that leads to bone erosion and implant loosening. Normal bone maintenance relies in the balance of bone formation and bone resorption, which mainly involves the coordinated function of osteoblasts and osteoclasts. Thus, either a decrease in osteoblastic bone formation or an increase in osteoclastic bone resorption can result in net bone loss and osteolysis. Bone loss (i.e., osteolysis) around an implant is the primary concern associated with the local effects of orthopedic implant degradation. This osteolysis causing implant debris occurs through both wear and corrosion mechanisms. Osteolysis is observed either as diffuse cortical thinning or as a focal cyst-like lesion. It was initially thought that reaction to particulate polymethylmethacrylate (PMMA) bone cement produced osteolytic lesions based on histological studies demonstrating cement debris associated with macrophages, giant cells, and a vascular granulation tissue. Recently, however, osteolysis has been recognized in association with loose and well-fixed uncemented implants, demonstrating that the absence of acrylic cement does not preclude the occurrence of osteolysis (Jacobs et al., 2001; Vermes et al., 2001a,b).

Implant debris causes low grade inflammation that ultimately leads to implant failure. Exactly how this happens remains unclear. Over the past 40 years implant debris-induced inflammation has been characterized *ad nauseam*, where debris-induced localized inflammation is caused in large part by macrophages which upregulate NFκβ and secrete inflammatory cytokines like IL-1β, TNFα, IL-6, and IL-8 (Jacobs et al., 2001) (Figure II.5.6.19). Other anti-inflammatory cytokines such as IL-10 modulate the inflammatory process. Other factors involved with bone resorption include the enzymes responsible for catabolism of the organic component of bone. These include matrix metalloproteinases collagenase and stromelysin. Prostaglandins, in particular PGE_2, are also known to be important intercellular messengers in the osteolytic cascade produced by implant debris. More recently, several mediators known to be involved in stimulation or inhibition of osteoclast differentiation and maturation, such as RANKL (also referred to as osteoclast differentiation factor) and osteoprotegerin, respectively, have been suggested as key factors in the development and progression of bone loss (osteolytic lesions) produced from implant debris. Over the past

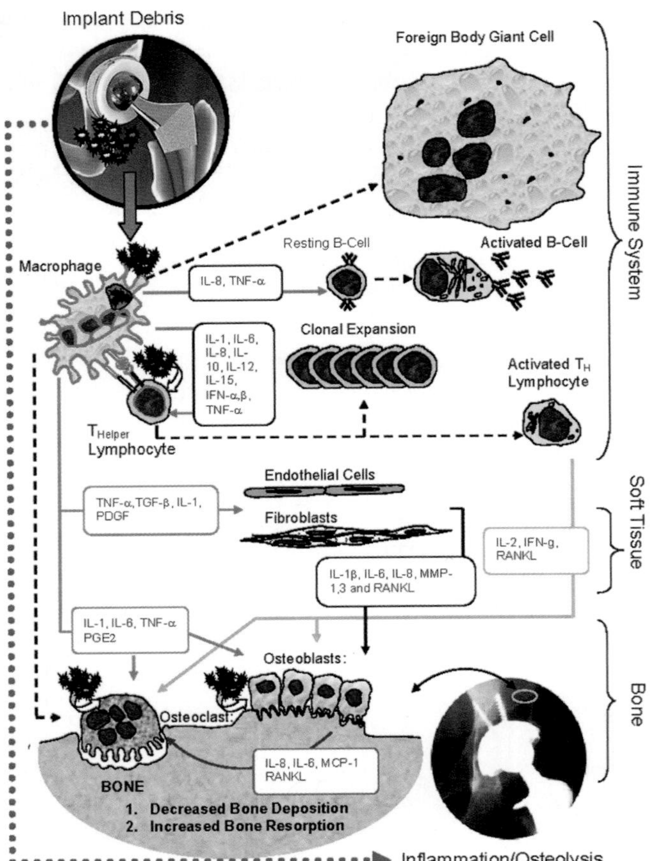

FIGURE II.5.6.19 This schematic shows the numerous pro-inflammatory mediators produced by peri-implant tissue and immune cells reacting to implant debris, which can negatively affect bone turnover. The proinflammatory cytokines IL-1, IL-6, and TNF-α are thought to be some of the most potent cytokines in this cascade of signaling. These cytokines produced by cells react to implant debris acting through a variety of pathways to negatively affect bone turnover.

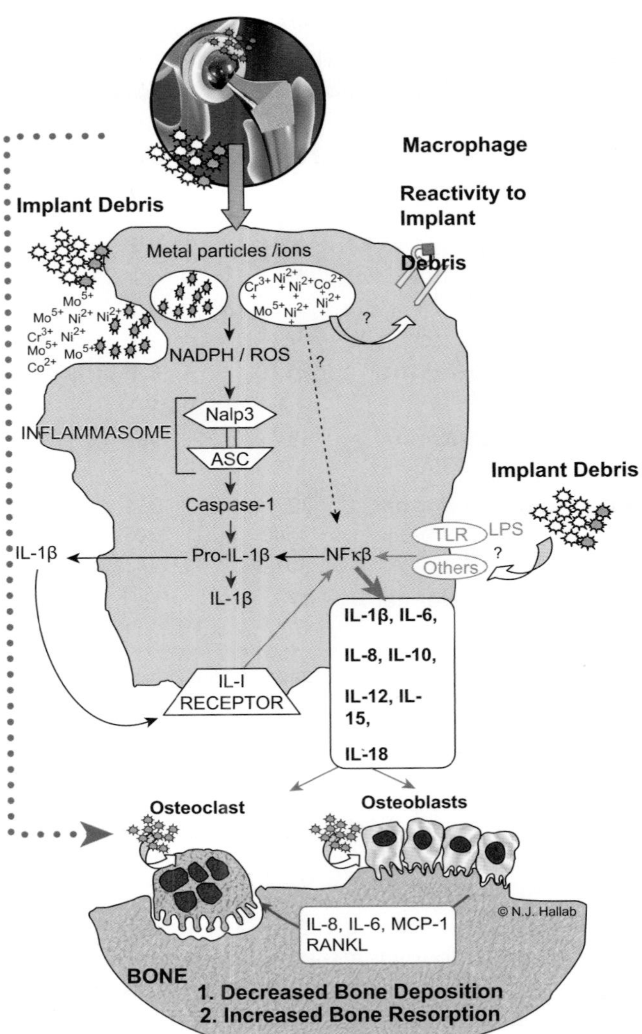

FIGURE II.5.6.20 The inflammasome pathway within cells such as macrophages has recently been reported to be central to implant debris-mediated proinflammatory reactivity. (Picture courtesy of BioEngineering Solutions, Inc.)

30 years we have understood that these mediators act to promote inflammation that decreases bone remodeling and is associated with the pathogenesis of osteolysis. However, we are only beginning to understand how implant debris could actually induce this immune system response at the cellular level.

Implant debris is typically sterile, relatively inert, and does not "look" like a pathogen in any molecularly recognizable way. How then can implant debris provoke an inflammatory response? That is, how do intracellular mechanisms sense and respond to sterile nonbiological challenge agents such as implant debris? This question has remained unanswered for the past 40 years, but recently progress has implicated the "inflammasome," a danger signaling pathway (Figure II.5.6.20) (Caicedo et al., 2009).

In 1996 the discovery of specific pattern recognition receptors (PRRs) in the membrane and cytosol of human immune cells, such as macrophages, identified toll-like receptors (Taguchi et al., 1996) and their role in recognizing specific bacterial glycoproteins, now called "pathogen-associated molecular patterns" or PAMPs. We now understand that these receptors to highly conserved pathogen-associated molecular patterns (Mariathasan and Monack, 2007) include toll-like receptors (TLRs), mannose receptors (MR), and NOD-like receptors (NLRs) (Mariathasan and Monack, 2007). Upon pathogen/cell contact these PRRs initiate a downstream cascade of events that activate the cell and induce the secretion of proinflammatory cytokines, leading to a broader inflammatory response.

In 2005 danger signal pathways were discovered where nonpathogenic-derived stimuli were found to activate immune cells, similarly to PAMPs. Key components in this pathway were named the "inflammasome," and the activating stimuli were termed "danger-associated molecular patterns," or DAMPs (Martinon et al., 2006). The paradigm for immune system activation now includes reliance on specific receptors that recognize both pathogen-associated molecular patterns (PAMPs) and danger-associated molecular patterns (DAMPs) (Medzhitov, 2008; Ting et al., 2008). The inflammasome complex of proteins were the first pathway to explain how cells transduce sterile, nonpathogen-derived stimuli (e.g., cell stress and cell necrosis), into an inflammatory response (Mariathasan et al., 2004; Mariathasan and Monack, 2007). Nonpathogen-derived danger includes such nonbiological stimuli as UV light, particulate adjuvants present in modern vaccines (Dostert et al., 2008; Hornung et al., 2008), and, as it turns out, implant debris (Caicedo et al., 2010).

When the inflammasome pathway is activated it causes the release of IL-1β, IL-18, IL-33 and other cytokines. How this happens is as follows:

Debris → Phagocytosis → Lysosome damage → ROS(reactive oxygen species) → Inflammasome(NALP3/ASC) → Caspase1 → IL-1ß (and other IL-1-family) cytokines (Figure II.5.6.20).

Once ingested by immune cells, DAMPs, such as asbestos and implant debris, etc., (i.e., macrophages) induce some degree of lysosomal destabilization. This causes an increase in NADPH (nicotinamide adenine dinucleotide phosphate oxidase), and an increase in reactive oxygen species (ROS). This is not surprising, given the protease and acid rich extreme environment inside lysosomes used to digest and breakdown ingested particles/bacteria, etc. The release of these intracellular contents are sensed by the intracellular multi-protein inflammasome complex which is composed of NALP3 protein (NACHT-, LRR- and pyrin domain-containing protein 3), in association with ASC (apoptosis-associated speck-like protein containing a CARD domain) (Mariathasan and Monack, 2007; Petrilli et al., 2007). Activation of the inflammasome (NALPs-ASC complex) leads to the cleavage of pro-caspase-1 into active caspase-1. Active caspase-1 is required for the processing and subsequent release of active proinflammatory cytokines such as IL-1β and IL-18 (and others) by cleaving intracellular pro-IL-1β, pro-IL-18, etc., into their mature forms, IL-1β and IL-18, etc.

How different implant debris can cause different immune responses through specific mechanisms, such as the inflammasome, remains unknown and is currently under study. These new understandings facilitate both direct targeting for drugs and can enhance diagnostic measurement for improving, measuring, and predicting when implant debris will result in loose total joint arthroplasties.

Goldring et al. (1983) were among the first to describe the synovial-like character of the bone implant interface in patients with loose total hip replacements, and determine that the cells within the membrane have the capacity to produce large amounts of bone resorbing factors PGE_2 and collagenase. However, since studies typically can only document the end stage of the loosening process, rather than the initiating processes, pharmacologic interventions have been limited.

Osteolysis associated with total knee arthroplasty has been reported less frequently than that associated with total hip arthroplasty. It is unclear why. However, in addition to obvious factors such as implant/bone mechanical loading environments, other more subtle differential mechanisms of hip and knee wear, and differences in interfacial barriers to migration of debris have been postulated to account for this apparent disparity.

Although polyethylene particles are generally recognized as the most prevalent particles in the periprosthetic milieu, metallic and ceramic particulate species are also present in variable amounts, and may have important repercussions. The bulk of this debris originates from the articular surface and has easy access to local bone. When present in sufficient amounts, particulates generated by wear, corrosion or a combination of these processes can induce the formation of an inflammatory, foreign-body granulation tissue with the ability to invade the bone–implant interface (Figure II.5.6.21). Localized osteolytic lesions in these areas are common, but their clinical significance is limited unless large granulomatous lesions develop.

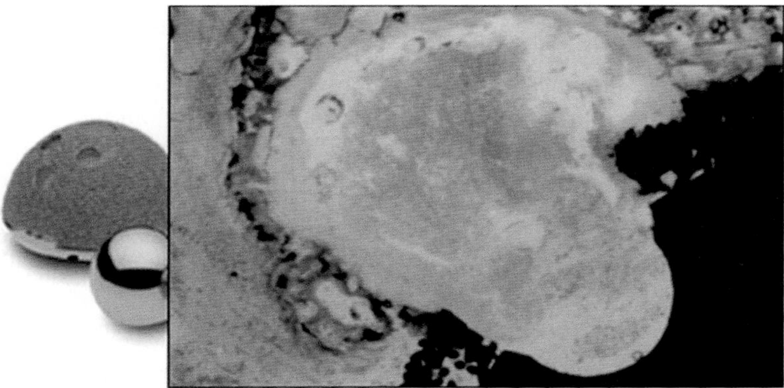

FIGURE II.5.6.21 Photomicrograph (5 ×) of a section through an acetabular section of a femoral stem retrieved at autopsy, 89 months after implantation. Note the periprosthetic cavity surrounded development of a granuloma emanating from an unfilled screw hole.

The common observance of particle-induced osteolysis remote from the articulation surfaces has shown there is substantial particle migration between the joint space and the distal regions of the THA implant space. Autopsy specimens of retrieved implants have demonstrated the presence of connective tissue macrophages (histiocytes) in cavities surrounding regions of the femoral component. While the overall incidence of femoral osteolysis associated with THA tends to be proximal in the initial stages, over time it tends to progress distally. The volume of debris generated from THA polyethylene is related to a number of variables, including the smoothness of the concave metallic surface of the acetabular component, the tolerance between polyethylene and metal shell, and the relative stability of the insert (LaBerge et al., 1998; Shanbhag et al., 1998; Wimmer et al., 1998).

Remote and Systemic Effects of Wear and Corrosion

Implant surfaces and wear debris generated from the implant may release chemically active metal ions into the surrounding tissues. While these ions may stay bound to local tissues, there is an increasing recognition that released metal products bind to specific protein moieties, and are transported in the bloodstream and/or lymphatics to remote organs. The concern about the release and distribution of metallic degradation products is justified by the known potential toxicities of the elements used in modern orthopedic implant alloys: titanium; aluminum; vanadium; cobalt; chromium; and nickel. In general terms, metal toxicity may occur through: (1) metabolic alterations; (2) alterations in host/parasite interactions; (3) immunologic interactions of metal moieties by virtue of their ability to act as haptens (specific immunological activation) or anti-chemotactic agents (non-specific immunological suppression); and (4) by chemical carcinogenesis (Luckey and Venugopal, 1979; Beyersmann, 1994; Goering and Klaasen, 1995; Britton, 1996; Hartwig, 1998).

Cobalt, chromium, and possibly nickel and vanadium, are essential trace metals in that they are required for certain enzymatic reactions. In excessive amounts, however, these elements may be toxic. Excessive cobalt may lead to polycythemia, hypothyroidism, cardiomyopathy, and carcinogenesis. Excessive chromium can lead to nephropathy, hypersensitivity, and carcinogenesis. Nickel can lead to eczematous dermatitis, hypersensitivity, and carcinogenesis. Vanadium can lead to cardiac and renal dysfunction, and has been associated with hypertension and depressive psychosis.

Biologically non-essential metallic elements also possess specific toxicities. Titanium, although generally regarded as inert, has been associated with pulmonary disease in patients with occupational exposure, and with platelet suppression in animal models. Aluminum toxicity is well-documented in renal failure, and has been associated with anemia, osteomalacia, and neurological dysfunction, possibly including Alzheimer's disease. However, when considering the variety of documented toxicities of these elements, it is important to keep in mind that the toxicities generally apply to soluble forms of these elements, and may not apply to the chemical species that result from prosthetic implant degradation.

At this time, the association of metal release from orthopedic implants with any metabolic, bacteriologic, immunologic or carcinogenic toxicity is conjectural, since cause and effect have not been well-established in human subjects. However, this is due in large part to the difficulty of observation, in that most symptoms attributable to systemic and remote toxicity can be expected to occur in any population of orthopedic patients (Jacobs et al., 1999b).

Metal Ion Release

In the long clinical experience of permanent and temporary metallic implants there has always been concern with local tissue reactions. There is a considerable literature concerning serum and urine chromium (Cr), cobalt (Co), and nickel (Ni) levels following total joint replacement, but relatively fewer studies examining titanium (Ti), aluminum (Al), and vanadium (V) levels. Many investigations have been hampered by technical limitations of the analytical instrumentation. Normal human serum levels of prominent implant metals are approximately: 1–10 ng/ml Al; 0.15 ng/ml Cr; <0.01 ng/ml V; 0.1–0.2 ng/ml Co; and <4.1 ng/ml Ti. Following total joint arthroplasty, levels of circulating metal (Al, Cr, Co, Ni, Ti, and V) have been shown to increase (Table II.5.6.5).

Multiple studies have demonstrated chronic elevations in serum and urine cobalt and chromium following successful primary total joint replacement. In addition, transient elevations of urine and serum nickel have been noted immediately following surgery. This hypernickelemia/hypernickeluria may be unrelated to the implant itself, since there is such a small percentage of nickel used in these implant alloys. Rather, this may be related to the use of stainless steel surgical instruments or the metabolic changes associated with the surgery itself.

Chronic elevations in serum titanium and chromium concentrations are found in subjects with well-functioning titanium and/or chromium containing THR components without measurable differences in urine and serum aluminum concentrations. Vanadium concentrations have not been found greatly elevated in patients with TJA (Table II.5.6.5) (Michel et al., 1984; Dorr et al., 1990; Jacobs et al., 1994c; Stulberg et al., 1994; Jacobs et al., 1998b).

Metal ion levels within serum and urine of TJA patients can be affected by a variety of factors. For example, patients with total knee replacement components containing titanium-based alloy and carbon fiber reinforced polyethylene wear couples demonstrated

TABLE II.5.6.6 Concentrations of Metal in Body Tissue of Humans With and Without Total Joint Replacements (μg/g)

		Cr	Co	Ti	Al	V
Skeletal Muscle	**Normal**	**<12**	**<12**	*	*	*
	TJA	570	160	*	*	*
Liver	**Normal**	**<14**	**120**	**100**	**890**	**14**
	TJA	1130	15200	560	680	22
Lung	**Normal**	*	*	**710**	**9830**	**26**
	TJA	*	*	980	8740	23
Spleen	**Normal**	**10**	**30**	**70**	**800**	**<9**
	TJA	180	1600	1280	1070	12
Psuedocapsule	**Normal**	**150**	**50**	**<65**	**120**	**<9**
	TJA	3820	5490	39400	460	121
Kidney	**Normal**	**<40**	**30**	*	*	*
	TJA	<40	60	*	*	*
Lymphatic tissue	**Normal**	**690**	**10**	*	*	*
	TJA	**690**	**390**	*	*	*
Heart	**Normal**	**30**	**30**	*	*	*
	TJA	90	280	*	*	*

TJA: Subjects with a well-functioning total joint arthroplasty.
* Not tested.

a 10-fold elevation in serum titanium concentrations at an average of four years after implantation. Up to a hundred times higher than normal control values of serum titanium elevations have also been reported in patients with failed metal-backed patellar components where unintended metal/metal articulation was possible. However, even among these TJA patients there was no elevation in serum or urine aluminum, serum or urine vanadium levels, or urine titanium levels. Mechanically assisted crevice corrosion in patients with modular femoral stems from total hip arthroplasty has been associated with elevations in serum cobalt and urine chromium. It has been previously assumed that extensively porous coated cementless stems would give rise to higher serum and urine chromium concentrations, due to the larger surface area available for passive dissolution. Recent studies suggest that disseminated chromium can predominantly come from fretting corrosion of the modular head–neck junction. However, wear of the articulating surface remains the purported predominant source of metallic implant debris (Jacobs et al., 1998a,b, 1999b).

Homogenates of remote organs and tissue obtained postmortem from subjects with cobalt-based alloy total joint replacement components have indicated that significant increases in cobalt and chromium concentrations occur in the heart, liver, kidney, spleen, and lymphatic tissue (Table II.5.6.6). Similarly, patients with titanium-based alloy implants demonstrated elevated titanium, aluminum, and vanadium levels in joint pseudocapsules (with up to 200 ppm of titanium six orders of magnitude greater than that of controls, 880 ppb of aluminum, and 250 ppb of vanadium). Spleen aluminum levels and liver titanium concentrations can also be markedly elevated in patients with failed titanium-alloy implants (Jacobs et al., 1994c).

Systemic Particle Distribution

Variables influencing accumulation of wear debris in remote organs are not clearly identified. When the magnitude of particulate debris generated by a prosthetic device is increased, it seems likely that a corresponding elevation in both the local and systemic burden of particles may be expected. Thus, component loosening, duration of implantation, and the modular designs of contemporary hip and knee replacement prostheses provide the potential for increased generation of metallic and polymeric debris (Figure II.5.6.22). Wear particles found disseminated beyond the periprosthetic tissue are primarily in the submicron size range. Numerous case reports document the presence of metallic, ceramic, or polymeric wear debris from hip and knee prostheses in regional and pelvic lymph nodes (Figure II.5.6.23), along with the findings of lymphadenopathy, gross pigmentation due to metallic debris, fibrosis (build-up of fibrous tissue), lymph node necrosis, and histiocytosis (abnormal function of tissue macrophages), including complete effacement of nodal architecture. The inflammatory response to metallic and polymeric debris in lymph nodes has been demonstrated to include immune activation of macrophages and associated production of cytokines. Metallic wear particles have been detected in the para-aortic lymph nodes in up to 70% of patients with total joint replacement components.

Lymphatic transport is thought to be a major route for dissemination of wear debris. Wear particles may migrate via perivascular lymph channels as free or phagocytosed particles within macrophages. Within the abdominal para-aortic lymph nodes, the majority of disseminated particles are submicron in size; however, metallic particles as large as 50 micrometers, and polyethylene particles as large as 30 micrometers, have also been identified. These particles may further disseminate to the liver or spleen where they are found within macrophages or, in some cases, as epithelioid granulomas throughout the organs. Within liver and spleen, the maximum size of metallic wear particles are nearly an order of magnitude less than that in lymph nodes, indicating there may be additional stages of filtration preceding the lymphatic system or alternate routes of particle migration. In the liver and spleen, as in the lymph nodes, cells of the mononuclear–phagocyte system may accumulate small amounts of a variety of foreign materials without apparent clinical significance. However, accumulation of exogenous particles can induce granulomas or granulomatoid lesions in the liver and spleen (Figure II.5.6.23). It is likely that the inflammatory reaction to particles in the liver, spleen, and lymph nodes is modulated, as it is in other tissues by: (1) material composition; (2) the number of particles; (3) their rate of accumulation; (4) the duration that they are present; and (5) the biologic reactivity of cells to these particles. Metallic particles in the liver or spleen have been more prevalent in subjects with previously failed arthroplasties when compared with cases of well-functioning primary joint replacements. Metal particles, unlike polyethylene debris, can be characterized using an electron microprobe, which allows identification of individual, submicron metallic wear particles against a background of particulates from environmental or sources other than the prosthetic components. Overall, the smallest identifiable disseminated particles using the microprobe are approximately 0.1 micrometers in diameter. However, metallic wear debris may extend into the nanometer range, suggesting that additional methods of specimen preparation and analytic instrumentation may be required to more fully define the high burden of metallic wear particles in remote tissues (Urban et al., 2000).

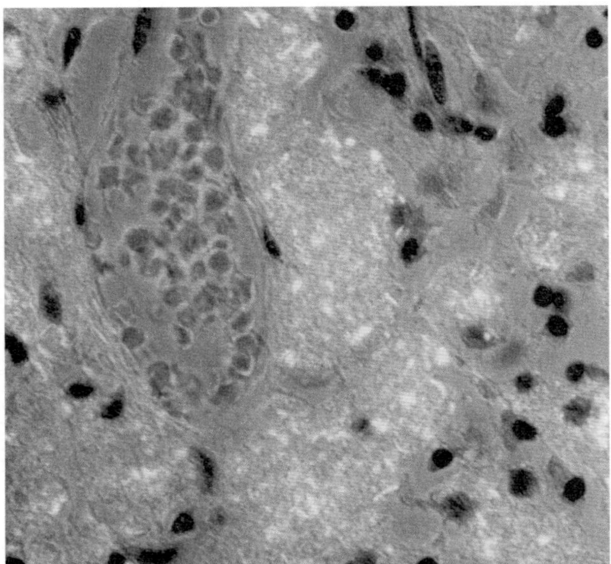

FIGURE II.5.6.22 Polarized light micrograph (190 ×) of para-aortic lymph node demonstrates the abundance and morphology of birefringent particles within macrophages. The larger filamentous particles were identified by infrared spectroscopy to be polyethylene.

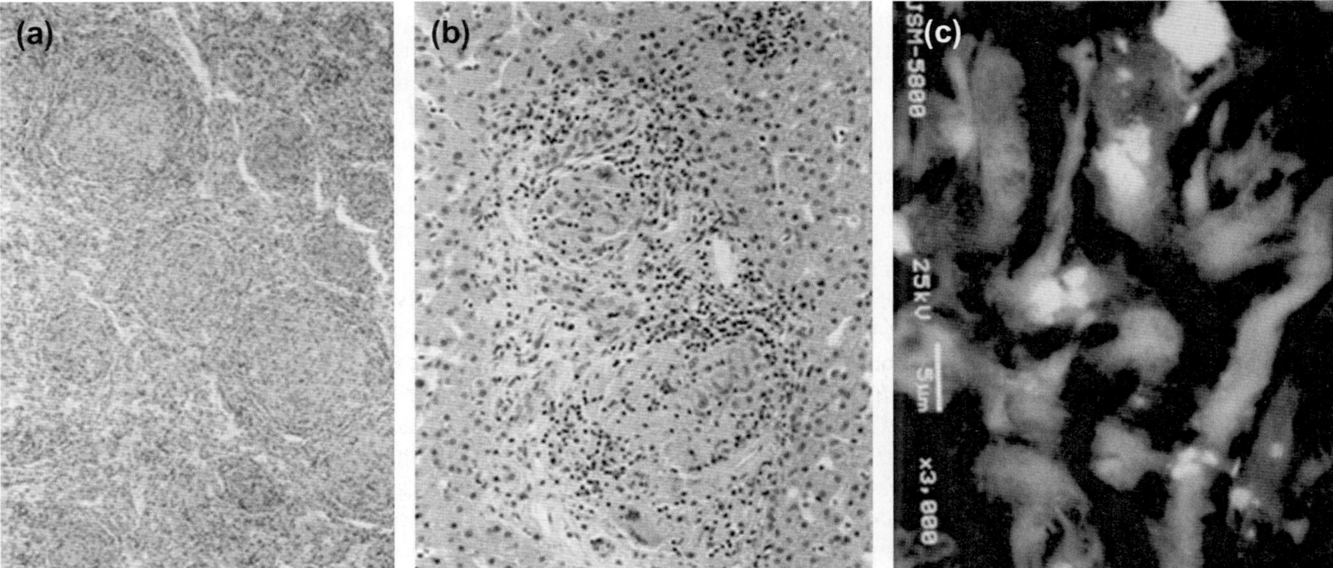

FIGURE II.5.6.23 Epithelioid granulomas: (a) within the portal tract of the liver (40 ×); and (b) within the splenic parenchyma (15 ×) in a patient with a failed titanium-alloy total hip replacement and symptomatic hepatitis. (c) Backscattered SEM of a granuloma in the spleen (3000 ×) demonstrating titanium alloy particles.

Polyethylene particles comprise a substantial fraction of the disseminated wear particles both in subjects with revision and primary TJAs. While the presence of these polyethylene particles in lymph nodes can be confirmed by Fourier Transform Infrared Spectroscopy microanalyses, polyethylene particulates in liver and spleen have so far precluded unequivocal identification. In these sites, the size of wear particles may be much smaller than 0.1 micrometers, making differentiation impossible by polarized light microscopy or infrared spectroscopy.

Diseases which cause obstruction of lymph flow through lymph nodes, such as metastatic tumor, or which cause generalized disturbances of circulation, such as chronic heart disease or diabetes, may be expected to decrease particle migration to remote organs. Other diseases, such as acute or chronic-active inflammation in the periprosthetic tissues may increase particle migration (Urban et al., 1995; Jacobs et al., 1999b, 2001).

Hypersensitivity

Some adverse responses to orthopedic biomaterials are subtle, and continue to foster debate and investigation. One of these responses is "metal allergy" or hypersensitivity to metallic biomaterials. Released ions, while not sensitizers on their own, can activate the immune system by forming complexes with native proteins. These metal–protein complexes are considered to be candidate antigens (or allergens) in human clinical applications. Polymeric wear debris is not easily chemically degraded *in vivo* and has not been implicated as sources of allergic type immune responses. This is presumably due to the relatively large degradation products associated with the mechanical wear of polymers *in vivo*, which may be large enough to prevent the formation of polymer–protein haptenic complexes with human antibodies (Hallab et al., 2000a,b, 2001a,b).

Metal hypersensitivity is a well-established phenomenon. Moreover, dermal hypersensitivity to metal is common, affecting about 10–15% of the population. Dermal contact and ingestion of metals have been reported to cause immune reactions which most typically manifest as skin hives, eczema, redness, and itching. Although little is known about the short- and long-term pharmacodynamics and bioavailability of circulating metal degradation products *in vivo*, there have been many reports of immunologic type responses temporally associated with implantation of metal components. Individual case reports link hypersensitivity immune reactions with adverse performance of metallic clinical cardiovascular, orthopedic, plastic surgical, and dental implants.

Metals accepted as sensitizers (haptenic moieties in antigens) include beryllium, nickel, cobalt, and chromium, while occasional responses have been reported to tantalum, titanium, and vanadium. Nickel is the most common metal sensitizer in humans, followed by cobalt and chromium. Cross-sensitivity reactions between metals are common. Nickel and cobalt are, reportedly, the most frequently cross-reactive.

Type IV Delayed Type Hypersensitivity (DTH) is a cell mediated type of response with which orthopedic implant associated hypersensitivity reactions (metal sensitivity or metal allergy) are generally associated. Metal-antigen sensitized T-DTH lymphocytes release various cytokines which result in the accumulation and activation of macrophages. The majority of DTH participating cells are macrophages. Only 5% of the participating cells are antigen specific T lymphocytes (T-DTH cells), with a fully developed DTH response. The effector phase of a DTH response is initiated by contact of sensitized T cells with antigen. In this phase T cells, which are antigen-activated, are termed T-DTH cells and secrete a variety of cytokines that recruit and activate macrophages, monocytes, neutrophils, and other inflammatory cells. These released cytokines include IL-3 and GM-CF, which promote hematopoesis of granulocytes; monocyte chemotactic activating factor (MCAF) which promotes chemotaxis of monocytes toward areas of DTH activation; INF-γ and TNF-β which produce a number of effects on local endothelial cells facilitating infiltration; and migration inhibitory factor (MIF), which inhibits the migration of macrophages away from the site of a DTH reaction. Activation, infiltration, and eventual migration inhibition of macrophages is the final phase of the DTH response. Activated macrophages, because of their increased ability to present class II MHC (Major Histocompatibility Complex) and IL-2, can trigger the activation of more T-DTH cells, which in turn activates more macrophages, which activates more T-DTH cells, and so on. This DTH self-perpetuating response can create extensive tissue damage.

The first apparent correlation of eczematous dermatitis to metallic orthopedic implants was reported in 1966 by Foussereau and Laugier (1966), where nickel was associated with hypersensitivity responses. Over the past 20 years, growing numbers of case reports link immunogenic reactions with adverse performance of metallic cardiovascular orthopedic, plastic surgical, and dental implants. In some instances clinical immunological symptoms have led directly to device removal. In these cases reactions of severe dermatitis (inflammation of the skin), urticaria (intensely sensitive and itching red round wheels on the skin), and/or vasculitis (patch inflammation of the walls of small blood vessels) have been linked with the relatively more general phenomena of metallosis (metallic staining of the surrounding tissue), excessive periprosthetic fibrosis, and muscular necrosis. The temporal and physical evidence leaves little doubt that the phenomenon of hypersensitivity to metal released from orthopedic implants does occur in some patients. These cases of severe metal sensitivity raise the greatest concern.

Incidence of Hypersensitivity Responses Among Patients With Metal Implants

The incidence of metal sensitivity among patients with both well- and poorly-functioning implants is roughly twice as high as that of the general population, approximately 25% (Figure II.5.6.24). Furthermore, the average incidence of metal sensitivity among patients with a "failed" implant (in need of revision surgery) is approximately 50–60% (Figure II.5.6.24). This is greater than five times the incidence of metal sensitivity found in the general population, and two to three times that of patients with metal implants. This increased prevalence of metal sensitivity among patients with a loose prosthesis has prompted the speculation that immunological processes may be a factor in implant loosening.

Specific types of implants with greater propensity to release metal *in vivo* may be more prone to induce metal sensitivity. Failures of total hip prostheses with metal-on-metal bearing surfaces were associated with greater incidence of metal allergy than similar designs with metal-on-ultra-high molecular weight polyethylene bearing surfaces. Alternatively, several published reports have indicated that after total joint replacement with metallic components some patients show an induction of metal tolerance; that is, previously diagnosed metal sensitivity abated after implantation of a metallic prosthetic.

Additionally confounding to any clear connection between metal sensitivity and implant failure is the lack of any reported correlation between incidence of metal sensitivity and implant residence time, infection, reason for removal or pain. This lack of causal evidence implicating cell-mediated immune responses has prompted some to conclude that implantation of cemented metal-to-plastic joint prosthesis is safe, even in the case of a pre-existing metal allergy. However, this is not a consensus opinion. At this time, however, it is unclear whether metal sensitivity causes implant loosening or whether implant loosening results in the development of metal sensitivity.

The majority of investigations conclude that metal sensitivity can be a contributing factor to implant failure. Such cases include instances in which clinical immunological symptoms lead directly to the need for device removal. In these or similar cases there have been reported reactions of severe dermatitis, urticaria, and/or vasculitis, all presumably linked to what has been reported as metallosis, excessive periprosthetic fibrosis, and muscular necrosis. The clinical observation of apparent immune sensitivity to metallic implants is not limited to orthopedic surgery. Some case reports suggest metal sensitivity to pacemakers, heart valves, reconstructive, dental, and general surgical devices. The temporal and physical evidence associated with such cases leaves little doubt that the phenomenon of metal-induced hypersensitivity does occur in some cases, currently accepted within the orthopedic community to be <1% of patients. However, it is currently unclear whether metal sensitivity exists only as an unusual complication in a few susceptible patients or

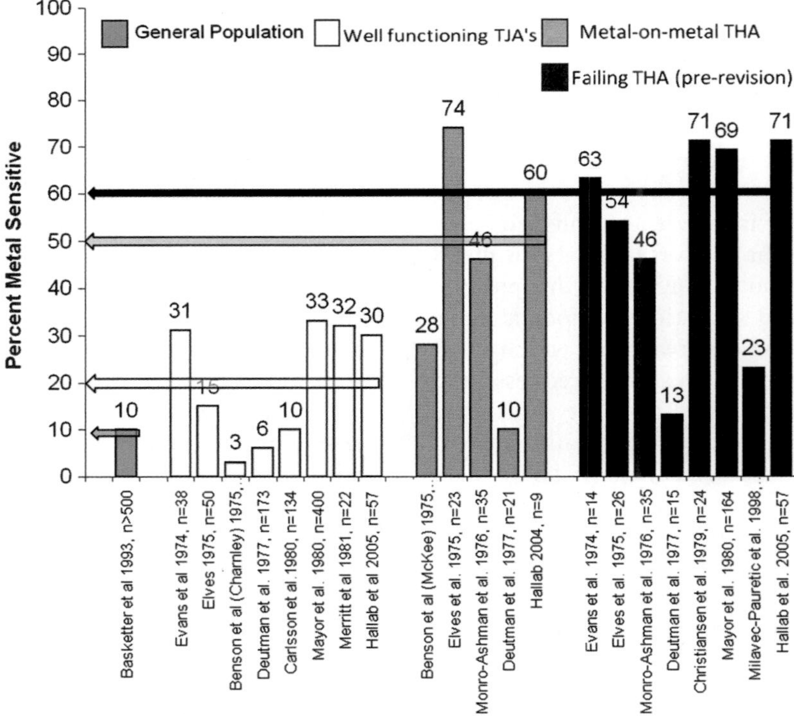

FIGURE II.5.6.24 A compilation of investigations show the averaged percentages of metal sensitivity among the general population for nickel, cobalt, and chromium, among patients after receiving a metal-containing implant, and among patient populations with failed implants. All subjects were tested by means of a patch or metal-LTT (lymphocyte transformation test).

is a more subtle and common phenomenon, which over time plays a significant role in implant failure. It is likely that cases involving implant-related metal sensitivity have been under-reported, because alternate causes were attributed to the failure of the implant. Mechanisms by which *in vivo* metal sensitivity occurs have not been well-characterized. Thus, the degree to which a precondition of metal hypersensitivity may elicit an over-aggressive immune response in a patient receiving an implant remains unpredictable. Continuing improvements in immunologic testing methods will likely enhance future assessment of patients susceptible to hypersensitivity responses (Hallab et al., 2000a,b, 2001a,b).

Carcinogenesis

The carcinogenic potential of the metallic elements used in TJA remains an area of concern. Animal studies have documented the carcinogenic potential of orthopedic implant materials. Small increases in rat sarcomas were noted to correlate with high serum cobalt, chromium or nickel content from metal implants. Furthermore, lymphomas with bone involvement were also more common in rats with metallic implants. Implant site tumors in dogs and cats – primarily osteosarcoma and fibrosarcoma – have been associated with stainless steel internal fixation devices.

Initially, epidemiological studies implicated cancer incidence in the first and second decades following total hip replacement. However, larger more recent studies have found no significant increase in leukemia or lymphoma; although these studies did not include as large a proportion of subjects with a metal-on-metal prosthesis. There are constituitive differences in the populations with and without implants that are independent of the implant itself, which confound the interpretation of epidemiological investigations.

The association of metal release from orthopedic implants with carcinogenesis remains conjectural, since causality has not been definitely established in human subjects. The identification of such an association depends both on the availability of comparative epidemiology, and on the ability to perform tests on the patient before and after device removal. The actual number of cases of tumors associated with orthopedic implants is likely under-reported. However, with respect to the number of devices implanted on a yearly basis, the incidence of cancer at the site of implantation is relatively rare. Continued surveillance and longer-term epidemiological studies are required to fully address these issues (Gillespie et al., 1988; Visuri and Koskenvuo, 1991; Matheisen et al., 1995; Nyren et al., 1995).

Preventive Strategies and Future Directions

Current strategies designed to address the problem of biomaterial-related implant failure are primarily aimed at decreasing the amount of periprosthetic particulate burden and any subsequent effects. Recently there has been a great deal of innovation regarding stronger, more wear-resistant polyethylene. These more highly cross-linked UHMWPE polymers are currently in various phases of clinical trials. However, initial results show demonstrable decrease in polyethylene wear, with potential for less particulate-induced bioreactivity/osteolysis, and therefore greater implant performance. In the same vein, femoral heads with diameters of 32 mm have been associated with increased volumetric polyethylene wear; to combat this smaller 28 mm heads are currently extolled as more biocompatible. Manufacturing flaws, such as fusion defects and foreign-body inclusions, have also been suggested to contribute to adverse polyethylene wear properties. The elimination of polyethylene is another approach being investigated clinically in various centers. With the realization that early problems may have been related to the design and not the articulation, there has been a renewed interest in the application of metal–metal and ceramic–ceramic bearings. Future designs which attempt to reduce wear include: improved tolerances between polyethylene inserts and their metal backing; improved surface finish on the metallic concave surfaces; secure locking mechanisms; and the avoidance of holes on the convex portion of the acetabular prosthesis.

Metallic wear is also being addressed through techniques such as nitriding and nitrogen ion implantation to decrease the potential for abrasive wear and fretting in titanium alloy and cobalt alloy stems. Fabrication of metallic bearing surfaces with extremely low roughness can be expected to decrease articular wear rates. A polished metal head can be made as smooth as a ceramic head. Polishing of the stem will remove surface asperities and decrease particle generation from stem/bone fretting. In addition, polishing will minimize silicate contamination.

New metallic biomaterials are being proposed which attempt to improve load transfer to the bone and reduce the incidence of loosening and thigh pain. Currently used alloys (Co–Cr–Mo alloy, E = 227 GPa and Ti-6Al-4V alloy, E = 115 GPa) have relatively high elastic moduli which limit smooth transfer of load to the surrounding bone in THA. Designs to improve load transfer can use a reduced cross-sectional area to increase flexibility, but at the expense of adequate stability of the implant within the bone. Additionally, the stresses may exceed the relatively low fatigue strength of Co–Cr–Mo implant alloy. Lower modulus, more corrosion-resistant implant alloys are being developed. A Ti-13Nb-13Zr (E = 79 GPa) alloy is one such alloy which contains fewer elements of questionable cell response (i.e., Co, Cr, Mo, Ni, Fe, Al, V), and which possesses comparable strength and toughness to existing Ti-6Al-4V implant alloy. The niobium and zirconiun constituents seek to improve bone biocompatibility and corrosion resistance.

Additionally, novel surface treatments on implant alloys (such as the diffusion hardening (DH) treatment proposed for the Ti-13Nb-13Zr alloy), can produce a hardened surface with wear resistance superior to that of Co–Cr–Mo alloy, currently the industry leader. These enhanced surface properties may lead to an improvement in the resistance to micro-fretting occurring within femoral head–neck taper regions and modular interfaces of current implant designs.

Electrochemical corrosion of orthopedic implants remains a significant clinical concern. Although the freely corroding implants used in the past have been replaced with modern corrosion resistant "superalloys," deleterious corrosion processes have been observed in certain clinical settings. Attention to metallurgical processing variables, tolerances of modular connections, surface processing modalities, and appropriate material selection can diminish corrosion rates and minimize the potential for adverse clinical outcomes. For example, nitriding can reportedly significantly reduce the magnitude of fretting corrosion of Ti-6Al-4V devices. A need to further investigate the mechanical–electrochemical interactions of metal oxide surfaces in implants persists. Characterization of the stresses and motion needed to fracture passivating oxide films, as well as the effects of repeated oxide abrasion on the electrochemical behavior of the interface and ultimately the implant, remain avenues of active investigation.

The clinical significance of elevated metal content in body fluids and remote organs of patients with metallic implants needs to be further elucidated. Considerably more work will be required to discern the specific chemical forms and distribution of metal degradation products associated with the various forms of implant degradation. Additionally uncharacterized is how these degradation products interact with proteins *in vivo* in terms of: (1) metal-ion/protein complexes; (2) nanometer-particle/protein complexes (ion-like particles); and (3) particle/protein-biofilm complexes. Although much has been revealed regarding the deleterious end effects of particulate debris (e.g., osteolysis) there remain very few ways of testing people to determine which type of implant and materials are right for them. Therefore, both an understanding of the constituents of orthopedic implant degradation and their biological effects is necessary to ultimately determine threshold levels of debris and circulating metal ions, and measures of biologic reactivity (e.g., metal-LTT testing) that can be used to clinically determine when intervention is required to fix a downward spiral before too much bone loss and inflammation occurs. The importance of this evaluation of orthopedic biomaterial performance is growing as the use of orthopedic biomaterials is increasing, as new orthopedic implants are being developed (Figure II.5.6.7), and as expectations of implant durability and performance increase (Black, 1996; Jacobs et al., 1996).

BIBLIOGRAPHY

Anderson, P. A., Rouleau, J. P., Bryan, V. E., & Carlson, C. S. (2003). Wear analysis of the Bryan Cervical Disc prosthesis. *Spine*, **28**(20), S186–S194.

Basketter, D. A., Briatico-Vangosa, G., Kaestner, W., Lally, C., & Bontinck, W. J. (1993). Nickel, cobalt and chromium in consumer products: a role in allergic contact dermatitis? *Contact Dermatitis*, **28**, 15–25.

Baumann, B., Rader, C. P., Seufert, J., Noth, U., Rolf, O., et al. (2004). Effects of polyethylene and TiAlV wear particles on expression of RANK, RANKL and OPG mRNA. *Acta. Orthop. Scand.*, **75**(3), 295–302.

Baumann, B., Seufert, J., Jakob, F., Noth, U., Rolf, O., et al. (2005). Activation of NF-kappaB signalling and TNFalpha-expression in THP-1 macrophages by TiAlV- and polyethylene-wear particles. *J. Orthop. Res.*, **23**(6), 1241–1248.

Baumann, B., Rolf, O., Jakob, F., Goebel, S., Sterner, T., et al. (2006). Synergistic effects of mixed TiAlV and polyethylene wear particles on TNFalpha response in THP-1 macrophages. *Biomed. Tech. (Ber.)*, **51**(5–6), 360–366.

Benson, M. K., Goodwin, P. G., & Brostoff, J. (1975). Metal sensitivity in patients with joint replacement arthroplasties. *British Medical Journal*, **4**, 374–375.

Beyersmann, D. (1994). Interactions in metal carcinogenicity. *Toxicol. Lett.*, **72**(1–3), 333–338.

Black, J. (1992). *Biomaterials* (2nd ed.). New York, NY: Marcel Dekker, Inc.

Black, J. (1996). *Prosthetic Materials*. New York, NY: VCH Publishers, Inc.

Bobyn, J. D., Tanzer, M., Krygier, J. J., Dujovne, A. R., & Brooks, C. E. (1994). Concerns with modularity in total hip arthroplasty. *Clin. Orthop.*, **298**, 27–36.

Britton, R. S. (1996). Metal-induced hepatoxicity. *Seminars in Liver Disease*, **16**(1), 3–12.

Brown, S. A., & Merritt, K. (1981). Fretting corrosion in saline and serum. *J. Biomed. Mater. Res.*, **15**(4), 479–488.

Brown, S. A., Flemming, C. A. C., Kawalc, J. S., Vassaux, C. J., Payer, J. H., et al. (1992). Fretting acclerated crevice corrosion of modular hips. *Trans. Soc. Biomater. Implant Retrieval Symposium*, **15**, 59.

Brown, S. A., Flemming, C. A., Kawalec, J. S., Placko, H. E., Vassaux, C., et al. (1995). Fretting corrosion accelerates crevice corrosion of modular hip tapers. *J. Appl. Biomater.*, **6**(1), 19–26.

Bruch, J. (1974). Response of cell cultures to asbestos fibers. *Environ. Health Perspect.*, **9**, 253–254.

Bundy, K. J., Williams, C. J., & Luedemann, R. E. (1991). Stress-enhanced ion release: The effect of static loading. *Biomaterials*, **12**(7), 627–639.

Büttner-Janz, K. (1992). *The development of the artificial disc SB Charité*. Hundley & Associates.

Caicedo, M., Reddy, A., Samee, I., Jacobs, J. J., & Hallab, N. (2007). Cobalt ions and Co-Cr-Mo alloy particles induce human monocyte co-stimulatory molecules CD-86, ICAM-1 and the cytokine IL-8: Implications for innate activation of adaptive immune responses. *6th Combined Meeting of the Orthopedic Research Societies*, **6**, 535.

Caicedo, M. S., Desai, R., McAllister, K., Reddy, A., Jacobs, J. J., et al. (2009). Soluble and particulate Co-Cr-Mo alloy implant metals activate the inflammasome danger signaling pathway in human macrophages: A novel mechanism for implant debris reactivity. *J. Orthop. Res.*, **27**, 847–854.

Caicedo, M. S., Pennekamp, P. H., McAllister, K., Jacobs, J. J., & Hallab, N. J. (2010). Soluble ions more than particulate cobalt-alloy implant debris induce monocyte costimulatory molecule expression and release of proinflammatory cytokines critical to metal-induced lymphocyte reactivity. *J. Biomed. Mater. Res. A.*, **93**, 1312–1321.

Callaghan, J. J., Rosenberg, A., & Rubash, H. (2007). *The Adult Hip*. New York, NY: Lippincott Williams & Wilkins.

Campbell, P., Ma, S., Yeom, B., McKellop, H., Schmalzried, T. P., et al. (1995). Isolation of predominantly submicron-sized UHMWPE wear particles from periprosthetic tissues. *J. Biomed. Mater. Res.*, **29**(1), 127–131.

Carlsson, A. S., Macnusson, B., & Moller, H. (1980). Metal sensitivity in patients with metal-to-plastic total hip arthroplasties. *Acta. Orthop. Scand.*, **51**, 57–62.

Catelas, I., Medley, J. B., Campbell, P. A., Huk, O. L., & Bobyn, J. D. (2004). Comparison of *in vitro* with *in vivo* characteristics of wear particles from metal-metal hip implants. *J. Biomed. Mater. Res. B. Appl. Biomater.*, **70**(2), 167–178.

Charnley, J. (1960). Anchorage of the femoral head prosthesis to the shaft of the femur. *J. Bone Joint Surg. (Br.)*, **42**, 28.

Charnley, J. (1964). The bonding of prosthesis to bone by cement. *J. Bone Joint Surg. (Br.)*, **46**, 518.

Charnley, J. (1979). *Low Friction Arthroplasty of the Hip, Theory and Practice*. Berlin: Springer-Verlag.

Christiansen, K., Holmes, K., & Zilko, P. J. (1980). Metal sensitivity causing loosened joint protheses. *Ann. Rheum. Dis.*, **39**(5), 476–480.

Clarke, I. C., Good, V., Williams, P., Schroeder, D., Anissian, L., et al. (2000). Ultra-low wear rates for rigid-on-rigid bearings in total hip replacements. *Proc. Inst. Mech. Eng. (H)*, **214**(4), 331–347.

Collier, J. P., Mayor, M. B., Jensen, R. E., Surprenant, V. A., Surprenant, H. P., et al. (1992a). Mechanisms of failure of modular prostheses. *Clin. Orthop.*, **285**, 129–139.

Collier, J. P., Surprenant, V. A., Jensen, R. E., Mayor, M. B., & Surprenant, H. P. (1992b). Corrosion between the components of modular femoral hip prostheses. *J. Bone Joint Surg. (Am.)*, **74-B**, 511–517.

Cook, S. D., Gianoli, G. J., Clemow, A. J., & Haddad, R. J. J. (1983). Fretting corrosion in orthopedic alloys. *Biomater. Med. Devices Artif. Organs*, **11**(4), 281–292.

Cowan, J. A., Jr., Dimick, J. B., Wainess, R., Upchurch, G. R., Jr., Chandler, W. F., et al. (2006). Changes in the utilization of spinal fusion in the United States. *Neurosurgery*, **59**(1), 15–20.

Cunningham, B. W., Orbegoso, C. M., Dmitriev, A. E., Hallab, N. J., Sefter, J. C., et al. (2002). The effect of titanium particulate on development and maintenance of a posterolateral spinal arthrodesis: An *in vivo* rabbit model. *Spine*, **27**(18), 1971–1981.

Deutman, R., Mulder, T. H., Brian, R., & Nater, J. P. (1977). Metal sensitivity before and after total hip arthroplasty. *J. Bone Joint Surg[Am]*, **59-A**, 862–865.

Dorr, L. D., Bloebaum, R., Emmanual, J., & Meldrum, R. (1990). Histologic, biochemical and ion analysis of tissue and fluids retrieved during total hip arthroplasty. *Clin. Orthop. Relat. Res.*, **261**, 82–95.

Dostert, C., Petrilli, V., Van, B. R., Steele, C., Mossman, B. T., et al. (2008). Innate immune activation through Nalp3 inflammasome sensing of asbestos and silica. *Science*, **320**(5876), 674–677.

Elves, M. W., Wilson, J. N., Scales, J. T., & Kemp, H. B. (1975). Incidence of metal sensitivity in patients with total joint replacements. *British Medical Journal*, **4**, 376–378.

Espehaug, B., Engesaeter, L. B., Vollset, S. E., Havelin, L. I., & Langeland, N. (1997). Antibiotic prophylaxis in total hip arthroplasty. Review of 10,905 primary cemented total hip replacements reported to the Norwegian arthroplasty register, 1987 to 1995. *J. Bone Joint Surg. (Br.)*, **79**(4), 590–595.

Evans, E. M., Freeman, M. A., Miller, A. J., & Vernon-Roberts, B. (1974). Metal sensitivity as a cause of bone necrosis and loosening of the prosthesis in total joint replacement. *The Journal of bone and Joint Surgery*, **56-B**, 626–642.

Fernstrom, U. (1966). Arthroplasty with intercorporal endoprothesis in herniated disc and in painful disc. *Acta. Chir. Scand. Suppl.*, **357**, 154–159.

Firkins, P. J., Tipper, J. L., Saadatzadeh, M. R., Ingham, E., Stone, M. H., et al. (2001). Quantitative analysis of wear and wear debris from metal-on-metal hip prostheses tested in a physiological hip joint simulator. *Biomed. Mater. Eng.*, **11**(2), 143–157.

Foussereau, J., & Laugier, P. (1966). Allergic eczemas from metallic foriegn bodies. *Trans. St John's Hosp. Derm. Soc.*, **52**, 220–225.

Gilbert, J. L., & Jacobs, J. (1997). The mechanical and electrochemical processes associated with taper fretting crevice corrosion: A review. *ASTM STP 1301 Modularity of Orthopedic Implants*, (pp. 45–59) Philadelphia, PA: ASTM.

Gilbert, J. L., Buckley, C. A., & Jacobs, J. J. (1993). *In vivo* corrosion of modular hip prosthesis components in mixed and similar metal combinations. The effect of crevice, stress, motion, and alloy coupling. *J. Biomed. Mater. Res.*, **27**(12), 1533–1544.

Gillespie, W. J., Frampton, C. M., Henderson, R. J., & Ryan, P. M. (1988). The incidence of cancer following total hip replacement. *J. Bone Joint Surg. (Br.)*, **70**(4), 539–542.

Glant, T. T., & Jacobs, J. J. (1994). Response of three murine macrophage populations to particulate debris: Bone resorption in organ cultures. *J. Orthop. Res.*, **12**, 720–732.

Goering, P. L., & Klaasen, C. D. (1995). *Hepatoxicity of Metals*. New York, NY: Academic Press.

Goldring, S. R., Schiller, A. L., Roelke, M., Rourke, C. M., O'Neill, D. A., et al. (1983). The synovial-like membrane at the bone–cement interface in loose total hip replacements and its proposed role in bone lysis. *J. Bone Joint Surg.*, **65A**, 575–584.

González, O., Smith, R. L., & Goodman, S. B. (1996). Efffect of size, concentration, surface area, and volume of polymethylmethacrylate paticles on human macrophages *in vitro*. *J. Biomed. Mater. Res.*, **30**, 463–473.

Goodman, S. B., Lind, M., Song, Y., & Smith, R. L. (1998). *In vitro, in vivo*, and tissue retrieval studies on particulate debris. *Clin. Orthop.*, **352**, 25–34.

Granchi, D., Verri, E., Ciapetti, G., Stea, S., Savarino, L., et al. (1998). Bone-resorbing cytokines in serum of patients with aseptic loosening of hip prostheses. *J. Bone Joint Surg. (Br.)*, **80**(5), 912–917.

Greenwald, A. S., & Garino, J. P. (2001). Alternative bearing surfaces: The good, the bad, and the ugly. *J. Bone Joint Surg. (Am.)*, **83-A**(Suppl. 2 Pt 2), 68–72.

Hallab, N. J., Mikecz, K., & Jacobs, J. J. (2000a). A triple assay technique for the evaluation of metal-induced, delayed-type hypersensitivity responses in patients with or receiving total joint arthroplasty. *J. Biomed. Mater. Res.*, **53**(5), 480–489.

Hallab, N. J., Jacobs, J. J., Skipor, A., Black, J., Mikecz, K., et al. (2000b). Systemic metal-protein binding associated with total joint replacement arthroplasty. *J. Biomed. Mater. Res.*, **49**(3), 353–361.

Hallab, N., Merritt, K., & Jacobs, J. J. (2001a). Metal sensitivity in patients with orthopedic implants. *J. Bone Joint Surg. (Am.)*, **83-A**(3), 428–436.

Hallab, N. J., Mikecz, K., Vermes, C., Skipor, A., & Jacobs, J. J. (2001b). Differential lymphocyte reactivity to serum-derived metal-protein complexes produced from cobalt-based and titanium-based implant alloy degradation. *J. Biomed. Mater. Res.*, **56**(3), 427–436.

Hallab, N. J., Anderson, S., Stafford, T., Skipor, A., Campbell, P., & Jocabs, J. J. (2004). Correlation between lymphocyte reactivity and metal ion levels in patients with metal-on-metal hip arthroplasty. *Trans 50th Orthopaedic Research Society*, **49**.

Hallab, N. J., Anderson, S., Caicedo, M., Brasher, A., Mikecz, K., et al. (2005). Effects of soluble metals on human peri-implant cells. *J. Biomed. Mater. Res. A.*, **74**(1), 124–140.

Hallab, N. J., Khandha, A., Malcolmson, G., & Timm, J. P. (2008). *In vitro* assessment of serum-saline ratios for fluid simulator testing of highly modular spinal implants with articulating surfaces. *SAS Journal*, **2**(4), 171–183.

Hallab, N. J. (2009). A review of the biologic effects of spine implant debris: Fact from Fiction. *SAS journal*, **3**, 143–160.

Hamby Wallace, B. (1959). Replacement of spinal intervertebral discs with locally polymerizing methyl methacrylate. *J. Neurosurg.*, **16**, 311–313.

Harmon, P. H. (1963). Anterior excision and vertebral body fusion operation for intervertebral disc syndromes of the lower lumbar spine. *Clin. Orthop.*, **26**, 107–111.

Harris, W. H. (1969). Traumatic arthritis of the hip after dislocation and acetabular fractures: Treatment by mold arthroplasty. An end-result study using a new method of result evaluation. *J. Bone Joint Surg. (Am.)*, **51**(4), 737–755.

Harris, W. H. (1995). The problem is osteolysis. *Clin. Orthop.*, **311**, 46–53.

Hartwig, A. (1998). Carcinogenicity of metal compounds: Possible role of DNA repair inhibition. *Toxicol. Lett.*, *102–103*, 235–239.

Heisel, C., Silva, M., la Rosa, M. A., & Schmalzried, T. P. (2004). Short-term *in vivo* wear of cross-linked polyethylene. *J. Bone Joint Surg. (Am.)*, **86-A**(4), 748–751.

Hellier, W. G., Hedman, T. P., & Kostuik, J. P. (1992). Wear studies for development of an intervertebral disc prosthesis. *Spine*, **17**(Suppl. 6), S86–S96.

Holt, G., Murnaghan, C., Reilly, J., & Meek, R. M. (2007). The biology of aseptic osteolysis. *Clin. Orthop. Relat. Res.*, **460**, 240–252.

Hornung, V., Bauernfeind, F., Halle, A., Samstad, E. O., Kono, H., et al. (2008). Silica crystals and aluminum salts activate the NALP3 inflammasome through phagosomal destabilization. *Nat. Immunol.*, **9**(8), 847–856.

Huk, O. L., Bansal, M., Betts, F., Rimnac, C. M., Lieberman, J. R., et al. (1994). Polyethylene and metal debris generated by non-articulating surfaces of modular acetabular components. *J. Bone Joint Surg. (Br.)*, **76**(4), 568–574.

Huo, M. H., Salvati, E. A., Lieberman, J. R., Betts, F., & Bansal, M. (1992). Metallic debris in femoral endosteolysis in failed cemented total hip arthroplasties. *Clin. Orthop.*, **276**, 157–168.

Ingham, E., Green, T. R., Stone, M. H., Kowalski, R., Watkins, N., et al. (2000). Production of TNF-alpha and bone resorbing activity by macrophages in response to different types of bone cement particles. *Biomater.*, **21**(10), 1005–1013.

Ingram, J., Matthews, J. B., Tipper, J., Stone, M., Fisher, J., et al. (2002). Comparison of the biological activity of grade GUR 1120 and GUR 415HP UHMWPE wear debris. *Biomed. Mater. Eng.*, **12**(2), 177–188.

Jacobs, J. J. (1995). Particulate wear. *JAMA*, **273**, 1950–1956.

Jacobs, J. J., & Hallab, N. J. (2006). Loosening and osteolysis associated with metal-on-metal bearings: A local effect of metal hypersensitivity? *J. Bone Joint Surg. (Am.)*, **88**(6), 1171–1172.

Jacobs, J. J., Urban, R. M., Schajowicz, F., Gavrilovic, J., & Galante, J. O. (1992). Particulate-associated endosteal osteolysis in titanium-base alloy cementless total hip replacement. In: *Particulate Debris from Medical Implants*. Philadelphia, PA: American Society for Testing and Materials.

Jacobs, J. J., Gilbert, J. L., & Urban, R. M. (1994a). Corrosion of metallic implants. In R. N. Stauffer (Ed.), *Advances in Orthopedic Surgery* (Vol. 2, pp. 279–319). St. Louis, IL: Mosby.

Jacobs, J. J., Shanbhag, A., Glant, T. T., Black, J., & Galante, J. O. (1994b). Wear debris in total joints. *J. Amer. Acad. Orthop. Surg.*, **2**, 212–220.

Jacobs, J. J., Skipor, A. K., Urban, R. M., Black, J., Manion, L. M., et al. (1994c). Systemic distribution of metal degradation products from titanium alloy total hip replacements: An autopsy study. *Trans. Orthop. Res. Soc.*, **838**.

Jacobs, J. J., Skipor, A. K., Doorn, P. F., Campbell, P., Schmalzried, T. P., et al. (1996). Cobalt and chromium concentrations in patients with metal on metal total hip replacements. *Clin. Orthop.*, **329**(Suppl.), S256–S263.

Jacobs, J. J., Gilbert, J. L., & Urban, R. M. (1998a). Corrosion of metal orthopedic implants. *J. Bone Joint Surg. (Am.)*, **80**(2), 268–282.

Jacobs, J. J., Skipor, A. K., Patterson, L. M., Hallab, N. J., Paprosky, W. G., et al. (1998b). Metal release in patients who have had a primary total hip arthroplasty. A prospective, controlled, longitudinal study. *J. Bone Joint Surg. (Am.)*, **80**(10), 1447–1458.

Jacobs, J., Goodman, S., Sumner, D. R., & Hallab, N. (1999a). Biologic response to orthopedic implants. In *Orthopedic Basic Science* (pp. 402–426). Chicago, IL: American Academy of Orthopedic Surgeons.

Jacobs, J. J., Silverton, C., Hallab, N. J., Skipor, A. K., Patterson, L., et al. (1999b). Metal release and excretion from cementless titanium alloy total knee replacements. *Clin. Orthop.*, **358**, 173–180.

Jacobs, J. J., Roebuck, K. A., Archibeck, M., Hallab, N. J., & Glant, T. T. (2001). Osteolysis: Basic science. *Clin. Orthop.*, **393**, 71–77.

Jasty, M., Bragdon, C., Jiranek, W., Chandler, H., Maloney, W., et al. (1994). Etiology of osteolysis around porous-coated cementless total hip arthroplasties. *Clin. Orthop.*, **308**, 111–126.

Jones, L. C., Frondoza, C., & Hungerford, D. S. (1999). Immunohistochemical evaluation of interface membranes from failed cemented and uncemented acetabular components. *J. Biomed. Mater. Res.*, **48**(6), 889–898.

Katz, J. L. (1980a). Anisotropy of Young's modulus of bone. *Nature*, **283**(5742), 106–107.

Katz, J. L. (1980b). The structure and biomechanics of bone. *Symp. Soc. Exp. Biol.*, **34**, 137–168.

Korovessis, P., Petsinis, G., Repanti, M., & Repantis, T. (2006). Metallosis after contemporary metal-on-metal total hip arthroplasty. Five- to nine-year follow-up. *J. Bone Joint Surg. (Am.)*, **88**(6), 1183–1191.

Kurtz, S. M., Hozack, W., Turner, J., Purtill, J., MacDonald, D., et al. (2005). Mechanical properties of retrieved highly cross-linked crossfire liners after short-term implantation. *J. Arthroplasty*, **20**(7), 840–849.

LaBerge, M. (1998). Wear. In J. Black, & M. C. Hastings (Eds.), *Biomaterial Properties* (pp. 364–405). London, UK: Chapman & Hall.

Laquerriere, P., Grandjean-Laquerriere, A., Jallot, E., Balossier, G., Frayssinet, P., et al. (2003). Importance of hydroxyapatite particles characteristics on cytokines production by human monocytes *in vitro*. *Biomaterials*, **24**(16), 2739–2747.

Liefeith, K., Hildebrand, G. & Schade, R. (2003). *In vitro* and *in vivo* evaluation of a fully resorbable calcium phosphate coatin deposited on TPS coated implants. Third International Essen Symposium on the Working Group on Biomaterials and Tissue Compatibility, Essen, 3.

Luckey, T. D., & Venugopal, B. (1979). *Metal Toxicity in Mammals*. New York, NY: Plenum.

Maloney, W. J., Smith, R. L., Castro, F., & Schurman, D. J. (1993). Fibroblast response to metallic debris *in vitro*. Enzyme induction cell proliferation and toxicity. *J. Bone Joint Surg. (Am.)*, **75**(6), 835–844.

Mariathasan, S., & Monack, D. M. (2007). Inflammasome adaptors and sensors: Intracellular regulators of infection and inflammation. *Nat. Rev. Immunol.*, **7**(1), 31–40.

Mariathasan, S., Newton, K., Monack, D. M., Vucic, D., French, D. M., et al. (2004). Differential activation of the inflammasome by caspase-1 adaptors ASC and Ipaf. *Nature*, **430**(6996), 213–218.

Martinon, F., Petrilli, V., Mayor, A., Tardivel, A., & Tschopp, J. (2006). Gout-associated uric acid crystals activate the NALP3 inflammasome. *Nature*, **440**(7081), 237–241.

Matheisen, E. B., Ahlbom, A., Bermann, G., & Lindgren, J. U. (1995). Total hip replacement and cancer. *J. Bone Joint Surg. (Br.)*, **77-B**(3), 345–350.

Matthews, J. B., Besong, A. A., Green, T. R., Stone, M. H., Wroblewski, B. M., et al. (2000). Evaluation of the response of primary human peripheral blood mononuclear phagocytes to challenge with *in vitro* generated clinically relevant UHMWPE particles of known size and dose. *J. Biomed. Mater. Res.*, **52**(2), 296–307.

Mayor, M. B., Merritt, K., & Brown, S. A. (1980). Metal allergy and the surgical patient. *The American Journal of Dermatogolgy*, **139**, 477–479.

McKee, G. K., & Watson-Farrar, J. (1943). Replacement of the arthritic hips to the McKee-Farrar replacement. *J. Bone Joint Surg. (Br.)*, **48**, 245.

McKellop, H., Park, S. H., Chiesa, R., Doorn, P., Lu, B., et al. (1996). *In vivo* wear of three types of metal on metal hip prostheses during two decades of use. *Clin. Orthop.*, **329**(Suppl.), S128–S140.

McKellop, H., Shen, F. W., Lu, B., Campbell, P., & Salovey, R. (2000). Effect of sterilization method and other modifications on the wear resistance of acetabular cups made of ultra-high molecular weight polyethylene. A hip-simulator study. *J. Bone Joint Surg. (Am.)*, **82-A**(12), 1708–1725.

McKenzie, A. H. (1995). Fernström intervertebral disc arthroplasty: A long-term evaluation. *Orthopedics International Edition*, **3B**, 313–324.

Medzhitov, R. (2008). Origin and physiological roles of inflammation. *Nature*, **454**(7203), 428–435.

Merritt, K., & Brown, S. (1981). Metal sensitivity reactions to orthopedic implants. *International Journal of Dermatology*, **20**, 89–94.

Michel, R., Hoffman, J., Loer, F., & Zilkens, J. (1984). Trace element burdening of human tissue due to corrosion of hip-joint prostheses made of cobalt-chromium alloys. *Arch. Orthop. Trama. Surg.*, **103**, 85–95.

Milavec-Puretic, V., Orlic, D., & Marusic, A. (1998). Sensitivity to metals in 40 patients with failed hip endoprosthesis. *Arch Orthop Trauma Surg*, **117**(6–7), 383–386.

Milosev, I., Trebse, R., Kovac, S., Cor, A., & Pisot, V. (2006). Survivorship and retrieval analysis of Sikomet metal-on-metal total hip replacements at a mean of seven years. *J. Bone Joint Surg. (Am.)*, **88**(6), 1173–1182.

Minoda, Y., Kobayashi, A., Iwaki, H., Miyaguchi, M., Kadoya, Y., et al. (2005). Polyethylene wear particle generation *in vivo* in an alumina medial pivot total knee prosthesis. *Biomater*, **26**(30), 6034–6040.

Moore, A. T. (1943). Metal hip joint: A case report. *J. Bone Joint Surg. (Am.)*, **25**, 688.

Munor-Ashman, D., & Miller, A. J. (1976). Rejection of metal to metal prosthesis and skin sensitivity to cobalt. *Contact Dermatitis*, **2**, 65.

Nagaya, T., Ishikawa, N., & Hata, H. (1989). Sister chromatid exchange analysis in lymphocytes of workers exposed to hexavalent chromium. *Br. J. Ind. Med.*, **46**(1), 48–51.

Nalepka, J. L., Lee, M. J., Kraay, M. J., Marcus, R. E., Goldberg, V. M., et al. (2006). Lipopolysaccharide found in aseptic loosening of patients with inflammatory arthritis. *Clin. Orthop. Relat. Res.*, **451**, 229–235.

Nyren, O., Mclaughlin, J. K., Anders-Ekbom, G. G., Johnell, O., & Fraumeni, A. H. (1995). Cancer risk after hip replacement with metal implants: A population-based cohort study in Sweden. *Journal of the National Cancer Institute*, **87**, 28–33.

Pare, P. E., Chan, F., Powell, M. L., & Mathews, H. H. (2007). *Wear characterization of the MAVERICK total disc replacment*. Trans 7th Annual Meeting Spine Arthroplasty Society (Berlin), **59**.

Petrilli, V., Dostert, C., Muruve, D. A., & Tschopp, J. (2007). The inflammasome: A danger sensing complex triggering innate immunity. *Curr. Opin. Immunol.*, **19**(6), 615–622.

Popoola, O. O., Shen, M., Heller, M., & Seebeck, J. (2007). *In vitro wear of UHMWPE inlays in Dynardi and Prodisc spine disc replacment implants*. Trans 7th Annual Meeting Spine Arthroplasty Society (Berlin), **49**.

Rader, C. P., Sterner, T., Jakob, F., Schutze, N., & Eulert, J. (1999). Cytokine response of human macrophage-like cells after contact with polyethylene and pure titanium particles. *J. Arthroplasty*, **14**(7), 840–848.

Ramachandran, R., Goodman, S. B., & Smith, R. L. (2006). The effects of titanium and polymethylmethacrylate particles on osteoblast phenotypic stability. *J. Biomed. Mater. Res. A.*, **77**(3), 512–517.

Ring, P. A. (1968). Complete replacement arthroplasty of the hip by the Ring prosthesis. *J. Bone Joint Surg. (Br.)*, **50**, 720.

Saikko, V., Calonius, O., & Keranen, J. (2002). Wear of conventional and cross-linked ultra-high-molecular-weight polyethylene acetabular cups against polished and roughened CoCr femoral heads in a biaxial hip simulator. *J. Biomed. Mater. Res.*, **63**(6), 848–853.

Savarino, L., Stea, S., Granchi, D., Visentin, M., Ciapetti, G., et al. (2000). Sister chromatid exchanges and ion release in patients wearing fracture fixation devices. *J. Biomed. Mater. Res.*, **50**(1), 21–26.

Schmalzried, T. P., Wessinger, S. J., Hill, G. E., & Harris, W. H. (1994). The Harris-Galante porous acetabular component press-fit without screw fixation. Five-year radiographic analysis of primary cases. *J. Arthroplasty*, **9**(3), 235–242.

Shanbhag, A. S., Jacobs, J. J., Black, J., Galante, J. O., & Glant, T. T. (1994). Macrophage/particle interactions. Effect of size, composition and surface area. *J. Biomed. Mater. Res.*, **28**, 81–90.

Shanbhag, A. S., Jacobs, J. J., Black, J., Galante, J. O., & Glant, T. T. (1995). Human monocyte response to particulate biomaterials generated *in vivo* and *in vitro*. *J. Orthop. Res.*, **13**, 792–801.

Shanbhag, A. S., Hasselman, C. T., Jacobs, J. J., & Rubash, H. E. (1998). Biological Response to Wear Debris. In J. J. Callaghan, A. G. Rosenberg, & H. Rubash (Eds.), *The Adult Hip* (pp. 279–288). Philadelphia, PA: Lippincott-Raven Publishers.

Sieving, A., Wu, B., Mayton, L., Nasser, S., & Wooley, P. H. (2003). Morphological characteristics of total joint arthroplasty-derived ultra-high molecular weight polyethylene (UHMWPE) wear debris that provoke inflammation in a murine model of inflammation. *J. Biomed. Mater. Res.*, **64A**(3), 457–464.

Soloviev, A., Schwarz, E. M., Darowish, M., & O'Keefe, R. J. (2005). Sphingomyelinase mediates macrophage activation by titanium particles independent of phagocytosis: A role for free radicals, NFkappaB, and TNFalpha. *J. Orthop. Res.*, **23**(6), 1258–1265.

Stillwell, W. T. (1987). *The Art of Total Hip Arthroplasty*. Orlando, FL: Grune & Stratton, Inc.

Stulberg, B. N., Merritt, K., & Bauer, T. (1994). Metallic wear debris in metal-backed patellar failure. *J. Biomed. Mat. Res. Applied Biomaterials*, **5**, 9–16.

Taguchi, T., Mitcham, J. L., Dower, S. K., Sims, J. E., & Testa, J. R. (1996). Chromosomal localization of TIL, a gene encoding a protein related to the *Drosophila* transmembrane receptor Toll, to human chromosome 4p14. *Genomics*, **32**(3), 486–488.

Thompson, G. J., & Puleo, D. A. (1995). Effects of sublethal metal ion concentrations on osteogenic cells derived from bone marrow stromal cells. *J. Appl. Biomater.*, **6**(4), 249–258.

Ting, J. P., Willingham, S. B., & Bergstralh, D. T. (2008). NLRs at the intersection of cell death and immunity. *Nat. Rev. Immunol.*, **8**(5), 372–379.

Tipper, J. L., Hatton, A., Nevelos, J. E., Ingham, E., Doyle, C., et al. (2002). Alumina-alumina artificial hip joints. Part II: Characterisation of the wear debris from *in vitro* hip joint simulations. *Biomaterials*, **23**(16), 3441–3448.

Urban, R. M., Hall, D. J., Sapienza, C. I., Jacobs, J. J., Sumner, D. R., et al. (1998). A comparative study of interface tissues in cemented vs. cementless total knee replacement tibial components retrieved at autopsy. *Trans. SFB*, **21**.

Urban, R. M., Jacobs, J. J., Tomlinson, M. J., Gavrilovic, J., & Andersen, M. (1995). *Migration of Corrosion Products from the Modular Head Junction to the Polyethylene Bearing Surface and Interface Membranes of Hip Prostheses*. New York, NY: Raven Press.

Urban, R. M., Jacobs, J. J., Sumner, D. R., Peters, C. L., Voss, F. R., et al. (1996a). The bone–implant interface of femoral stems with non-circumferential porous coating: A study of specimens retrieved at autopsy. *J. Bone Joint Surg. (Am.)*, **78-A**(7), 1068–1081.

Urban, R. M., Jacobs, J. J., Tomlinson, M. J., Black, J., Turner, T. M., et al. (1996b). Particles of metal alloys and their corrosion products in the liver, spleen and para-aortic lymph nodes of patients with total hip replacement prosthesis. *Orthop. Trans.*, **19**, 1107–1108.

Urban, R. M., Jacobs, J., Gilbert, J. L., Rice, S. B., Jasty, M., et al. (1997). Characterization of solid products of corrosion generated by modular-head femoral stems of different designs and materials. In D. E. Marlowe, J. E. Parr, & M. B. Mayor (Eds.), *STP 1301 Modularity of Orthopedic Implants* (pp. 33–44). Philadelphia, PA: ASTM.

Urban, R. M., Jacobs, J. J., Tomlinson, M. J., Gavrilovic, J., Black, J., et al. (2000). Dissemination of wear particles to the liver, spleen, and abdominal lymph nodes of patients with hip or knee replacement. *J. Bone Joint Surg. (Am.)*, **82**(4), 457–476.

van Ooij, A., Kurtz, S. M., Stessels, F., Noten, H., & van Rhijn, L. (2007). Polyethylene wear debris and long-term clinical failure of the Charite disc prosthesis: A study of 4 patients. *Spine*, **32**(2), 223–229.

Venable, C. S., Stuck, W. G., & Beach, A. (1937). The effects on bone of the presence of metals; based upon electrolysis. An experimental study. *Annals of Surgery*, **105**, 917.

Vermes, C., Chandrasekaran, R., Jacobs, J. J., Galante, J. O., Roebuck, K. A., et al. (2001a). The effects of particulate wear debris, cytokines, and growth factors on the functions of MG-63 osteoblasts. *J. Bone Joint Surg. (Am.)*, **83**(2), 201–211.

Vermes, C., Glant, T. T., Hallab, N. J., Fritz, E. A., Roebuck, K. A., et al. (2001b). The potential role of the osteoblast in the development of periprosthetic osteolysis: Review of *in vitro* osteoblast responses to wear debris, corrosion products, and cytokines and growth factors. *J. Arthroplasty*, **16**(8 Suppl. 1), 95–100.

Visuri, T., & Koskenvuo, M. (1991). Cancer risk after Mckee-Farrar total hip replacement. *Orthopedics*, **14**, 137–142.

von Knoch, M., Engh, C. A.S., Sychterz, C. J., Engh, C. A.J., & Willert, H. G. (2000). Migration of polyethylene wear debris in one type of uncemented femoral component with circumferential porous coating: An autopsy study of 5 femurs. *J Arthroplasty*, **15**(1), 72–78.

Walker, P. S. (1978). *Human Joints and Their Artificial Replacements*. Springfield, IL: Charles C. Thomas.

Wiles, P. (1953). The surgery of the osteoarthritic hip. *Brit. J. Surg.*, **45**, 488.

Willert, H. G., & Semlitsch, M. (1977). Reactions of the articular capsule to wear products of artificial joint prostheses. *J. Biomed. Mater. Res.*, **11**, 157–164.

Willert, H. G., Bertram, H., & Buchhorn, G. H. (1990). Osteolysis in alloarthroplasty of the hip. The role of ultra-high molecular weight polyethylene wear particles. *Clin. Orthop.*, **258**, 95–107.

Wilson, J. N., & Scales, J. T. (1970). Loosening of total hip replacements with cement fixation. Clinical findings and laboratory studies. *Clin. Orthop.*, **72**, 145–160.

Wimmer, M., Berzins, A., Kuhn, H., Bluhm, A., Nassutt, R., et al. (1998). Presence of multiple wear directions in autopsy retrieved acetabular components. *Trans. ORS*, **23**.

Wroblewski, B. M., Siney, P. D., Dowson, D., & Collins, S. N. (1996). Prospective clinical and joint simulator studies of a new total hip arthroplasty using alumina ceramic heads and cross-linked polyethylene cups. *J. Bone Joint Surg. (Br.)*, **78**(2), 280–285.

Yao, J., Glant, T. T., Lark, M. W., Mikecz, K., Jacobs, J. J., et al. (1995). The potential role of fibroblasts in periprosthetic osteolysis: Fibroblast response to titanium particles. *J. Bone Miner. Res.*, **10**(9), 1417–1427.

CHAPTER II.5.7 DENTAL IMPLANTATION

Jack E. Lemons[1] *and Carl E. Misch*[2]

[1]University Professor, Schools of Dentistry, Medicine and Engineering, University of Alabama at Birmingham, Birmingham, AL, USA

[2]DDS, MDS, Misch International Institute, Beverly Hills, MI, USA

PATIENT PROFILES, DENTAL NEEDS, AND SURGICAL IMPLANTS: 1950S–2010S

Functional, aesthetic, and general health compromises have been correlated with the loss of oral dentition. The dental profession has developed a wide range of treatments to deal with dentition losses and oral diseases; however, a significant percentage of the world population continues to lose teeth progressively with dental diseases and aging. In recent decades, since the 1950s, the modern era of treatments based on surgical implants has evolved. Significant advances in quality and quantity have occurred during each decade. In the USA, a larger population of completely edentulous individuals existed in the 1950s, and many implant treatments were initially designed to support full arch implant supported removable dentures (Misch, 1999). A prominent design was the subperiosteal type (Figure II.5.7.1A,B,C), where a cast cobalt alloy metallic framework was fabricated and implanted under the periosteum and fitted to surface features of the bone anatomy (Rizzo, 1988). Posts extended through the gingival and mucosal soft tissues and the implant denture was directly supported on bone without significant soft tissue contact. The surgical and implant fabrication procedures were technically demanding for the various subperiosteal implant designs. A group of dentists and supporting staff emerged as the experts in this subdiscipline.

Early subperiosteal systems were shown to function through implant-to-soft tissue interfaces and many

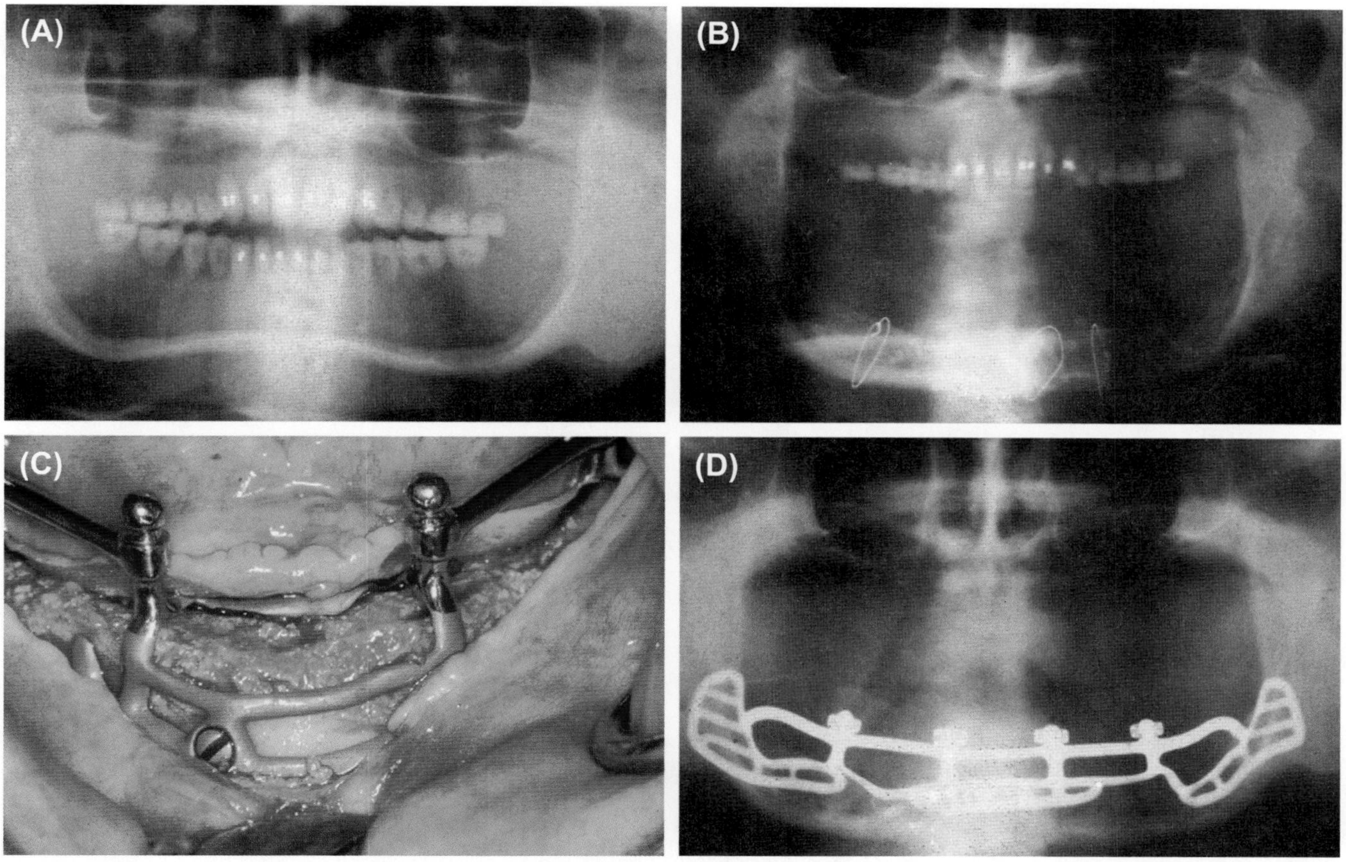

FIGURE II.5.7.1 A, A panoramic radiograph of a severely resorbed mandible, measuring less than 3 mm in height. B, A postoperative panoramic radiograph of an iliac crest bone graft wired around the atrophic mandible (1983). C, Reentry after 6 months into the iliac crest bone graft and a subperiosteal implant inserted (1983). D, Postoperative panoramic radiograph of the iliac crest graft after maturity and the subperiosteal implant (1987).

supported the relative benefits of fibrous tissue integration (James, 1983). A subdiscipline network of education evolved with the subperiosteal designs; however, many dental professionals started to support the merits of other types of implant designs. In all situations, the dental implant "experts" were required to coordinate a wide range of technologies, including implant design, fabrication, finishing, and placement, plus the various aspects of intra-oral prosthetic restorations and long-term maintenance. Some supported the concept of plate (blade) form implant designs (Figure II.5.7.2), where the body of the implant was placed into a surgically prepared slot in the bone (Weiss, 1986). In part due to implant design, and in part due to "immediate restoration" and oral function, many of the plate form and related designs of dental implant systems were shown to function through fibrous tissue and/or fibrous–osseous type tissue interfaces (James and Keller, 1974). Some called the interfacial tissue zone a "pseudo-ligament," and described properties somewhat similar to the natural tooth periodontal ligament. One aspect of the plate form implant was the intention to have a condition like the tooth-to-bone interfacial zone. With systems evolution, treatments extended from fully endentulous to partially edentulous regions of the mandibular and maxillary arches.

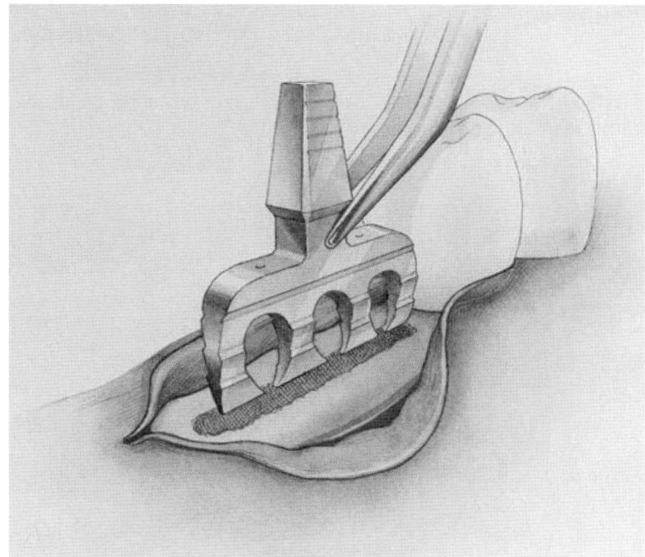
FIGURE II.5.7.2 Plate (blade) form implant design.

The dental community followed multiple avenues for dental implant designs and intraoral restorations, depending on their backgrounds, the patient population being treated, and recognized needs for improved treatments. One system was called the Ramus Frame (Figure II.5.7.3A,B,C), another the Transosseous and/or the Staple design (Figure II.5.7.4). Each had relative merits which have been described in articles and books associated with dental implants.

As another extension of pre-1950s studies, some in the dental implant profession supported root-form type designs (Figure II.5.7.5 and Figure II.5.7.6), often made in the shape of helical formed wires, pins, rods, screws or plateaus for the body sections of the implants that

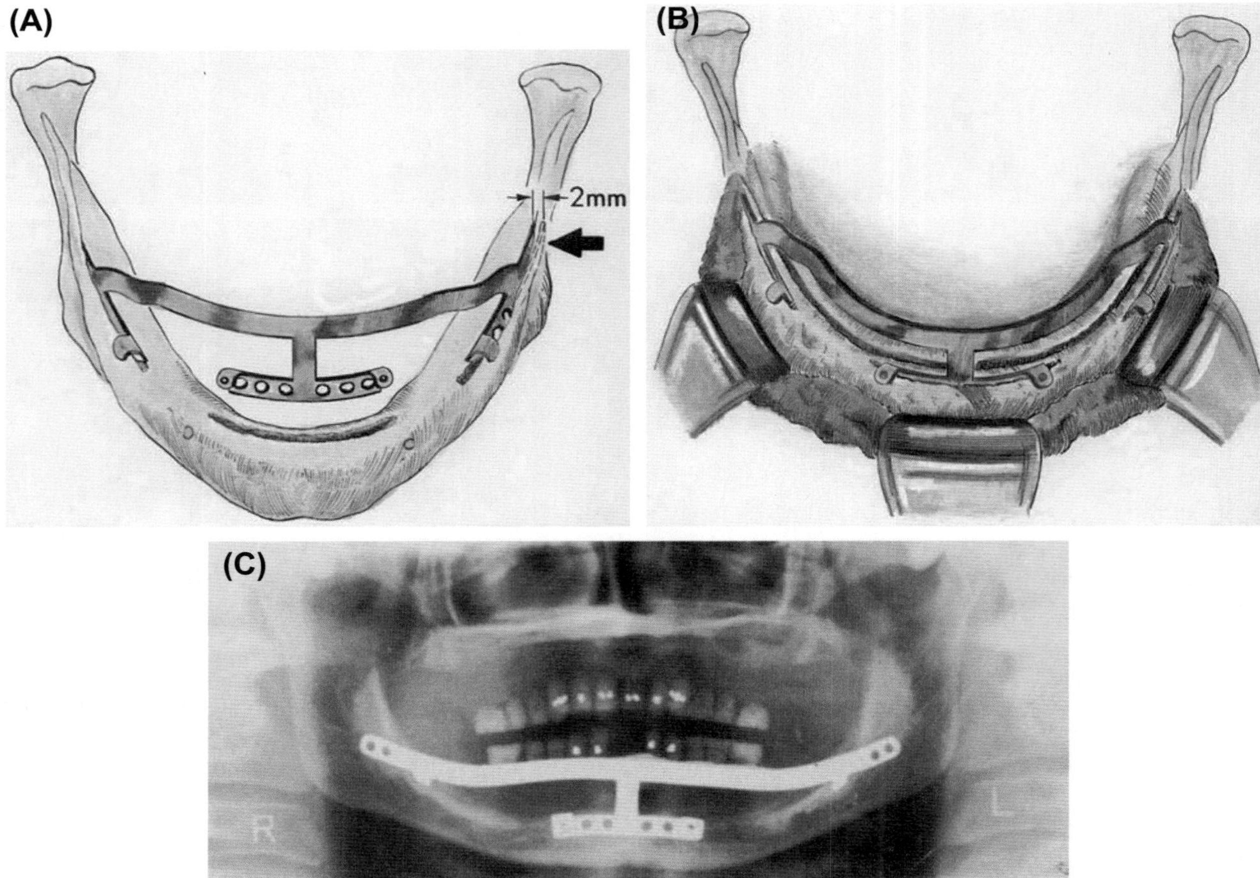

FIGURE II.5.7.3 (A) Schematic of insertion into Mandibular Bone; (B) Schematic of Ramus Frame Seated into Bone; (C) Radiograph of Ramus Frame Mandibular Inplant.

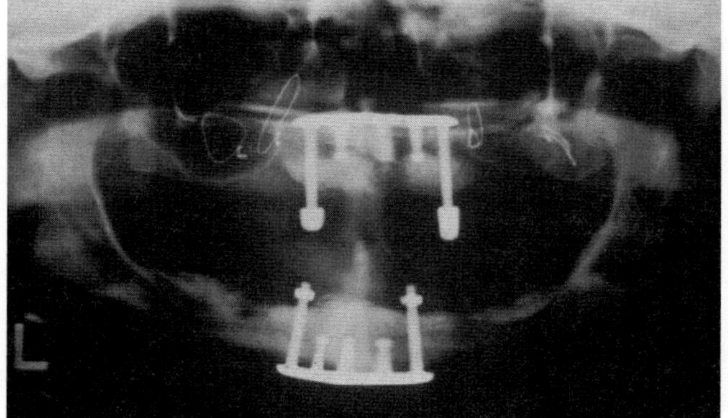

FIGURE II.5.7.4 Transosseous and/or the Staple design.

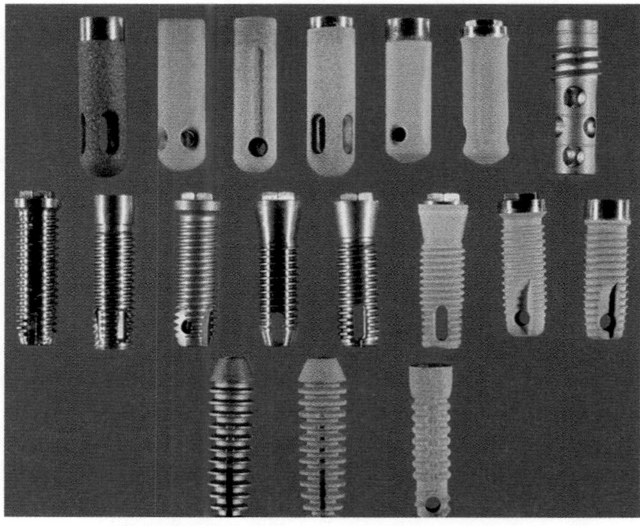

FIGURE II.5.7.5 Root-form type designs.

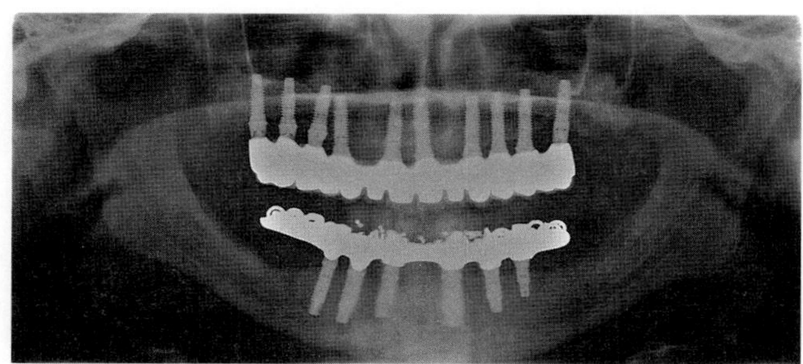

FIGURE II.5.7.6 Radiograph of Root-Form Implants Supporting Intraoral Bridge Reconstructions.

were placed into the bone (called endosseous or endosteal devices) (Cranin, 1970; Branemark et al., 1977; Small, 1980; Schnitman, 1987; Misch, 1999). Early in development, dental implants were primarily uniblock (single piece) systems combined with existing (known) intraoral restorative procedures for removable, fixed, and fixed-removable intraoral prosthetics.

The range of materials and biomaterials used for dental implants included available surgical implant grade metallics, ceramics, polymerics, mechanical mixtures, and composites. During this period, groups championed different concepts where the implants were made from metallics, and one of the more popular root forms supported primarily by dental specialists was the two-stage endosteal screw type implant. The abutment and implant components were assembled after a controlled period of post-surgical healing of bone and soft tissues. One of these early designs was fabricated from vitreous carbon, and initial studies supported the merits of this biomaterial for this application (Rizzo, 1988; Meffert et al., 1992). These approaches were popularized internationally, and the use of dental implants continued to increase with experience and time. During the 1980s and 1990s multiple clinical trials were completed showing the relative merits of implant-based treatments. One aspect of this evolution was the impact from the biomaterial and biomechanical disciplines which strongly influenced the overall enhancements of implant bulk and surface properties (Lemons, 1999).

As an example of interdisciplinary coordinations, the original subperiosteal and plate form dental implants have been redesigned in shape, surface biomaterial, and the surgical-restorative techniques to provide an osseous integrated dental implant system (now called Custom Osseous Integrated Implant or COII for these systems) (Baker et al., 2010). This same philosophy is being extended into on-going and planned (future) investigations in implant dentistry.

ANATOMICAL AND IMAGING CONSIDERATIONS

The overall maxillary and mandibular bone anatomies and regions for possible dental implant treatments are described in prior books. In general, the subperiosteal implants were placed onto the surface anatomy of the fully edentulous and axillary or mandibular bone. In part because of bone structure, subperiosteal implants in mandibular regions, when compared to maxillary implants, were more stable over time. The plate-form implants were initiated primarily for use in the edentulous posterior mandible, while the anterior, maxillary, and mandibular regions were selected for initial studies of root-form implant placement. This was in part because of available dimensions of bone, the magnitudes of functional forces, and avoidance of neurovascular regions.

Most of the different implant designs were eventually utilized wherever edentulous regions existed. Overall, the intent was to provide a stable, functional, and aesthetic dentition. Access to the oral regions and the number of teeth per individual (up to 32) resulted in larger numbers of dental implants compared to other types (hips, knees, etc.) of implant-based treatments for musculoskeletal disorders.

To define the bone anatomies, dentists in the earlier periods utilized radiographic imaging procedures that were normally available for the evaluation of teeth and supporting bone. Periapical (zone) radiographs were extended to full arch panographic images to better describe the overall bone and tooth relationships. With the advance of the science and technology of imaging, dental radiographic imaging evolved to routine use of computed tomography, with cone bean and other technologies to minimize radiation dose to the head and neck regions. This imaging has now been extended to computer-based software and hardware systems for design and manufacture (CAD-CAM). This approach also now extends from treatment planning, surgical guides, and abutment connections, to multiple types of intraoral prosthetics, and procedures for longer-term maintenance (Misch, 1999).

BIOMATERIALS

The various biomaterials used for the construction of dental implant systems are summarized in Tables II.5.7.1–II.5.7.4. This summary includes the nominal bulk compositions, mechanical properties, and surface conditions as described in national and international standards. A wide range of surface modifications have been, and continue to be, utilized for the body sections of dental implants. Many have been based on calcium phosphate compounds which are summarized in Table II.5.7.5. A review of the biomaterials and surface alterations of dental implant systems shows the wide range of surface conditions that continue to be preferred by the profession. To simplify these considerations, the surface modifications have been categorized as those that subtract or add mass to the substrate. More details on bulk and surface characteristics will be included in the next section. Multiple articles and books, again, exist about the relative merits of different biomaterials and surface modifications.

TABLE II.5.7.1 Engineering Properties of Metals and Alloys Used for Surgical Implants*

Material	Nominal Analysis (w/o)	Modulus of Elasticity GN/m² (psi μ 10⁶)	Ultimate Tensile Strength MN/m² (ksi)	Elongation to Fracture (%)	Surface
Titanium	99+Ti	97 (14)	240–550 (25–70)	>15	Ti oxide
Titanium–aluminum–vanadium	90Ti–6Al–4V	117 (17)	869–896 (125–130)	>12	Ti oxide
Cobalt–chromium–molybdenum (casting)	66Co–27Cr–7Mo	235 (34)	655 (95)	>8	Cr oxide
Stainless steel (316L)	70Fe–18Cr–12Ni	193 (28)	480–1000 (70–145)	>30	Cr oxide
Zirconium	99⁺Zr	97 (14)	552 (80)	20	Zr oxide
Tantalum	99⁺Ta	—	690 (100)	11	Ta oxide
Gold	99⁺Au	97 (14)	207–310 (30–45)	>30	Au
Platinum	99⁺Pt	166 (24)	131 (19)	40	Pt

GN/m², Giganewton per meter squared; *ksi*, thousand pounds per inch squared; *MN/m²*, meganewton per meter squared; *psi*, pounds per inch squared; *w/o*, weight percent.
*Minimum values from the American Society for Testing and Materials Committee F4 documents are provided. Selected products provide a range of properties.

TABLE II.5.7.2 Engineering Properties of Some Inert Ceramics Used as Biomaterials*

Material	Modulus of Elasticity GN/m² (psi μ 10⁶)	Ultimate Bending Strength MPa (ksi)	Surface
Aluminum oxide			
Polycrystalline	372 (54)	300–550 (43–80)	Al_2O_3
Single crystal (sapphire)	392 (56)	640 (93)	Al_2O_3
Zirconium oxide zirconia (PSZ)	195–210 (28–30)	500–650 (72–94)	ZrO_2
Titanium oxide (titania)	280 (41)	69–103 (10–15)	TiO_2

GN/m², Giganewton per meter squared; *psi*, pounds per inch squared; *MPa*, megapascal; *ksi*, thousand pounds per inch squared.
*These high ceramics have 0% permanent elongation at fracture.

TABLE II.5.7.3 Engineering Properties of Bioactive and Biodegradable Ceramics*

Material	Modulus of Elasticity GPa (psi μ 10⁶)	Ultimate Bending Strength MPa (ksi)	Surface
Hydroxyapatite	40–120 (6–17)	40–300 (6–43)	$Ca_{10}(PO_4)_6(OH)_2$
Tricalcium phosphate	30–120 (4–17)	15–120 (2–17)	$Ca_3(PO_4)_2$
Bioglass or Ceravital	40–140 (6–20)	20–350 (3–51)	$CaPO_4$
AW ceramic	124 (18)	213 (31)	$CaPO_4 + F$
Carbon	25–40 (4–6)	150–250 (22–36)	C
Carbon–silicon (LTI)	25–40 (4–6)	200–700 (29–101)	CSi

GPa, Gigapascal; *psi*, pounds per inch squared; *MPa*, megapascal; *ksi*, thousand pounds per inch squared; *LTI*, low-temperature isotropic.
*These ceramics and carbons have 0% permanent elongation at fracture.

TABLE II.5.7.4 Engineering Properties of Polymers (Some Medical Grades)*

Material	Modulus of Elasticity GPa (psi μ 10⁵)	Ultimate Tensile Strength MPa (ksi)	Elongation to Fracture (%)
PTFE	0.5–3 (0.07–4.3)	17–28 (2.5–4)	200–600
PET	3 (4.3)	55 (8)	50–300
PMMA	3 (4.3)	69 (10)	2–15
PE	8 (1.2)	48 (7)	400–500
PP	9 (1.3)	35 (5)	500–700
PSF	3.5 (5)	69 (10)	20–100
SR	0.1 (0.014)	5 (1.1)	300–900
POM	3 (4.3)	70 (10.1)	10–75

GPa, Gigapascal; *psi*, pounds per inch squared; *MPa*, megapascal; *ksi*, thousand pounds per inch squared; *PTFE*, polytetrafluoroethylene; *PET*, polyethylene terephthalate; *PMMA*, polymethylmethacrylate; *PE*, polyethylene; *PP*, polypropylene; *PSF*, polysulfone; *SR*, silicone rubber; *POM*, polyoxymethylene (IME, intra mobile element).
*Polymer properties exhibit a wide range depending on processing and structure. These values have been taken from general tables.

TABLE II.5.7.5 Names, Formulae, and Atomic Ratios for Some Calcium Phosphate Materials

Mineral or General Name	Formula	Ca:P Ratio	Applications
Monetite (DVP)	$CaHPO_4$	1	Nonceramic bone substitute particulate
Brushite (DCPD)	$CaHPO_4\ 2H_2O$	1	Phase of some $CaPO_4$ biomaterials
Octacalcium phosphate (OCP)	$Ca_8(HPO_4)_2(PO_4)\ 5H_2O$	1.33	Phase of some $CaPO_4$ biomaterials
Whitlockite (WH)	$Ca_{10}(HPO_4)(PO_4)_6$	1.43	Phase of some $CaPO_4$ biomaterials
Beta-tricalcium phosphate (b-TCP)	$Ca_3(PO_4)_2$	1.48	Biodegradable $CaPO_4$ ceramic for bone substitute and coatings; also a phase of some $CaPO_4$ biomaterials
Defective hydroxyapatite (DOHA) biomaterials	$Ca_9(HPO_4)(PO_4)_5(OH)$	1.5	Component of some $CaPO_4$ biomaterials
Hydroxyapatite (HA)	$Ca_{10}(PO_4)_6(OH)_2$	1.67	Major mineral phase of bone; when fired as a ceramic, named *HA*

TISSUE INTEGRATION: BIOMATERIAL AND BIOMECHANICAL ASPECTS

The science, technology, and clinical development of surgical implant-to-bone integration has for decades been a central area of emphasis (Misch, 1999; Rizzo, 1988). Practical aspects of the inability to physically separate bone-to-implant interfaces for ceramic and metallic oxide implant interfaces led to considerable interest in why this occurred, and thereby to many basic and applied studies. One part of these studies was based on element and force transfers for conditions of stable *in vivo* function. Dental implant treatments, over four decades have, in part, contributed to a major change in the philosophy of dental implant-based treatments for replacement of tooth loss. From a historical perspective, early reports of tissue integration for dental implants, in general, were complicated by the factors of clinical placement and restoration. Most implants were placed under variable conditions, and were immediately (hours to weeks) subjected to biomechanical function. Immediate intraoral restoration resulted in force transfers (stresses and strains) between the implant body and the supporting tissues during the period of initial tissue healing. Therefore, this discussion of tissue integration of dental implants will include both material and mechanical aspects of device function.

Experience prior to 1960 for a wide range of materials showed implant-to-tissue regions, where significant higher magnitude loads were transferred, that were fibrous soft tissue (scar-like) interfaces (James and Keller, 1974; James, 1983; Weiss, 1986). These zones were relatively dense collagenous structures that contained limited numbers of cells and blood vessels. Dental implants fabricated from steel, gold, platinum, and cobalt alloys, upon review of tissue interfaces after months to years of clinical function, showed soft tissue scar-like zones of contact (Lemons, 1999). In contrast, early experience with some dental implants fabricated from higher purity aluminum oxides (alumina and sapphire forms), carbons, calcium phosphate compounds, and reactive group metals and alloys (primarily titanium in the 1960s) demonstrated a mixture of direct bone-to-implant contact and soft tissues when evaluated histologically by optical microscopy methods. These interfacial conditions were described as osseo- or osteo-integrated and special conditions to routinely achieve these conditions were developed (Rizzo, 1988). A central focus developed for analyses of titanium, with oxidized surfaces contiguous with endosteal regions of oral bones. Multiple journal articles, books, conference proceedings, etc., described these and other conditions of tissue integration (Misch, 1999).

Several central conditions were defined, i.e., bulk and surface conditions for synthetic biomaterials that did not elicit a foreign-body response; surgical procedures to minimally traumatize the surgical site; and functional stresses and strains at the implant-to-bone interface where biomechanical microstrains were within the physiological limits of the bone during healing and longer-term function.

Considering these conditions, multiple requirements were developed for all levels of implant devices including diagnosis, surgery, restoration, and maintenance of dental implants. Analysis of the interface zone at nano-, micro-, and macrolevels of resolution demonstrated the importance of careful control of implant bulk and surface properties, and cleanliness, surgical methods and site dimensions, times and conditions of intraoral abutment placement, intraoral restoration, and maintenance (cleaning) of the percutaneous transition zone. As an example, of biomaterial properties, if unit area of implant, bone, and force were modeled, the interfacial strain magnitude would depend directly on any applied forces (the force vector direction and magnitude), the smoothness/roughness of the contact zone, attachment (bonding), and the elastic moduli of the implant biomaterial and the bone. Limited attachment (bonding) has been shown for some bioactive surfaces (e.g., calcium phosphates, glasses, etc.); however, most ceramics such as alumina, zirconia, and titania, and most surface oxidized metallics have demonstrated contact, but minimal or no chemical-type bonding. Most have reported that chromium oxides found on iron and cobalt alloys related to passivation were separated from the bone by a zone of non-osseous (fibrous) tissue. Surface roughness (irregularities) were shown to provide conditions for interdigitation under conditions of microscopic contact with mineralized bone, and microscopic regions of force transfer, plus potential advances for early healing (depositions of blood clot, fibrin, integrins, etc.) Microtopography was proposed to be critical for conditions of loading as a resistance to microscopic shear at the interface.

Returning to the role of implant and tissue moduli, conditions of unit area and force led to a method for evaluating relative strain magnitudes at the interfacial zone. If the interface is non-interdigitated and non-chemically bonded, strain magnitudes of the contact zone for each part would be proportional to the elastic modulus of each part. On a comparative basis of the biomaterial relative to compact bone, ceramics of alumina and zirconia have moduli of more than 10×, alloys 5–10×, with unalloyed titanium the lower magnitude, carbon and calcium phosphates about 0.5–2×, and polymerics lower by 50–1000× difference. The polymerics in general have moduli that are more similar to soft tissues. Clearly, the overall macroscopic shape, size, and geometry of the implant, the implant biomaterial and surface, the macro- and microanatomy of the bone, and the clinical conditions of function have been shown to directly influence the specific properties and conditions at the device interface. In this regard, dental implants over the years have included a very wide range of the circumstances listed above. However, in general, longer-term bone integration has been shown for implants fabricated from ceramic and metallic biomaterials which have included a range of physical features (macro- and microgeometries, microtopographies over a range of micro-dimensions and shapes often classified as pins, rods, screws, finns, plates, etc.) with and without surface modifications.

Implant survival statistics support high percentages (above 90%) remaining functional for decades. In the 2010s dental implants are expanding as a treatment of choice; where a tooth is lost, an implant is placed. Many propose this as the best and most cost-efficient way to maintain dental health over the longer-term. Therefore, the demand for dental implant treatments in society at large has increased, and the professional society and dental school programs are now offering basic education and training at the dental medicine (DMD) undergraduate level. These treatments have evolved to a "standard of care" status.

Many different implant biomaterials and designs have been studied; however, the larger numbers of applications have now become root-form designs placed, restored, and maintained for function through osteointegrated interfaces with bone. The biomaterials are now mostly titanium and alloys with a wide range of surface modifications for influencing interfacial tissue regions. Surface modifications include compositional (primarily oxides, calcium phosphates, and fluoride) and microtopographical features (degrees of roughness). These concepts also extend to the gingival and mucosal soft tissue to implant interfacial systems.

Interest has also been ever increasing in the use of synthetic-active biologic combinations for enhancing tissue quantity, quality, and healing characteristics. Many dentists now restore the dental implants to limited functional use immediately after surgical placement. Some call this "back to the future." Research and development at many levels supports possibilities for hard and soft tissue and tooth regeneration through tissue-engineered regenerative medicine approaches. If accomplished, the need for current dental implants based on synthetic biomaterials may be decreased accordingly. Most say that considerable science needs to be completed to accomplish the regeneration of anatomical replicates of functional and healthy teeth. Early research is most promising, and rapid advancements in regenerative approaches are anticipated.

BIBLIOGRAPHY

Baker, M., Eberhardt, A., Martin, D. M., McGuin, G., & Lemons, J. (2010). Bone Properties Surrounding Hydroxyapatite-Coated Custom Osseous Integrated Dental Implants. *J. Biomed. Mater. Res.*, **95B**, 218–224.

Branemark, P. I., Hanssen, B. O., Adell, R., Brien, U., Lindstrom, J., et al. (1977). Osseointegrated implants in the treatment of the edentulous jaw. *Scand. J. Plast. Reconstr. Surg.* (Suppl. 16).

Cranin, A. N. (1970). Some philosophic comments on the endosteal implant. *Dent. Clin. N. Am.*, **14**, 173–175.

James, R. A. (1983). Subperiosteal implant designs based on peri-implant tissue behavior. *NY J. Dent.*, **53**, 407.

James, R. A., & Keller, E. E. (1974). A histopathological report on the nature of the epithelium and underlying connective tissue which surrounds oral implants. *J. Biomed. Mater. Res.*, **8**, 373–383.

Lemons, J. (1999). Biomaterials for Dental Implants, In: Misch, C.E. (Ed.). *Contemporary Implant Dentistry*, **3e**, Mosby Elsevier, St. Louis, MA, Ch24.

Meffert, R. M., Langer, B., & Fritz, M. E. (1992). Dental implants: A review. *J. Periodental.*, **63**, 859–870.

Misch, C. E. (1999). *Contemporary Implant Dentistry* (2nd ed.). St. Louis, IL: Mosby. 94–106.

Rizzo, A. A., (1988). Proceedings of the 1988 Consensus Development Conference on Dental Implants. *J. Dent. Educ.*, **52**, 678–827.

Roberts, H. D., & Roberts, R. A. (1970). The ramus endosseous implant. *J. S. Calif. Dent. Assoc.*, **38**, 571.

Schnitman, P. A. (1987). Diagnosis, treatment planning, and the sequencing of treatment for implant reconstructive procedures. *Alpha Omegan*, **80**, 32.

Small, I. A. (1980). Benefit and risk of mandibular staple bone-plates. In *Dental Implants: Benefit and Risk* (pp. 139–152). Washington, DC: US Public Health Service. PHS Pub. 81–1531.

Strock, A. E. (1939). Experimental work on direct implantation in the alveolus. *Am. J. Orthol. Oral. Surg.*, **25**, 5.

Weiss, C. M. (1986). Tissue integration of dental endosseous implants: Description and comparative analysis of fibro-osseous integration and osseous integration systems. *J. Oral. Implantol.*, **12**, 169.

CHAPTER II.5.8 ADHESIVES AND SEALANTS

David Christopher Watts
Professor of Biomaterials Science, The University of Manchester, School of Dentistry & Photon Science Institute, Manchester, UK

INTRODUCTION

Description and Definition of Adhesives, and Related Terminology

In the context of biomaterials, adhesion science and technology assume high importance. It is useful to commence with the slightly wider concept of *bonding*. It is evident that many biomaterial devices are structured from multiple components, and these need to retain their mutual integrity during clinical service. In this context, the bonds in question may include metallic welds, although these would not be considered within the domain of adhesion, any more than screw attachments that have a role in some implants.

The most challenging issue is the creation of a durable interfacial bond between a biomaterial and its host tissue. In addition to the function of bonding, *per se*, an adhesive biomaterial may be required to fulfil a *space-filling* role – replacing some or all of any lost natural tissue. In the former case, the function may be termed *grouting*. This is, for example, the primary function of bone cements in orthopedics. A further ideal function of an adhesive is that of *sealing*; that is, the prevention of ingress of moisture, air, biological fluids, bacteria or other species through the adhesively bonded zone.

Adhesive is a general term, and in specific contexts may be replaced by designations such as cement, glue, paste, fixative, and bonding agent. Some adhesives may be designed to exhibit further functions, such as antibacterial action, delivery of drugs or beneficial ions, such as the antibacterial ion Ag^+ or fluoride (F^-). Fluoride is a component of dental preventative treatments which initiates the partial replacement of hydroxyapatite (HA) – the tooth enamel's normal crystalline composition – with fluorapatite (a related crystal which incorporates F^-). Fluorapatite is more resistant to decay than HA.

Sometimes the converse of adhesion, namely *abhesion*, is necessary in clinical treatment. This "non-stick" feature is required with blood-compatible biomaterials, and is exhibited by non-fouling surfaces covered with polyethylene glycol (PEG).

During the past half-century there has been a strong convergence of the ancient technologies of adhesives and the modern science of adhesion – based especially on advances in surface science and the understanding of molecular adhesive mechanisms. This progress has been stimulated by widespread industrial uses and biomedical applications of adhesives. The former have been driven by advances in polymer science, and the adoption of lightweight alloy and non-metallic composite materials in the aerospace and automobile industries. Nevertheless, the stringent biomedical requirements for adhesives have led to some specialized developments.

The diverse biomedical *contexts* (host environments) of adhesives each require appropriate materials and techniques. Biological hosts differ principally as to whether they consist of *hard* or *soft* tissues. Connected with this are the *timescales* required for adhesive/bonding durability. These may vary from days, in the case of wound-closure adhesives, to decades of years, in the cases of bone cements and dental restorations. The aim of this chapter is to give a sound introduction to the physico-chemical and materials science aspects of adhesives and sealants, together with a description of current systems utilized in a range of surgical practice. It does not attempt a detailed study of the molecular biological interactions of adhesives.

More details on the background to adhesion and adhesives can be found in a number of texts (Kinloch, 1987; Comyn, 1997; Pocius, 1997; Mittal and Pizzi, 1999; Eliades et al., 2005; Matinlinna and Mittal, 2009). General information on wound closure and surgical adhesives is given in Sierra and Salz (1996) and Chu et al. (1997). The proceedings of a conference on adhesion in dentistry provide an important set of contemporary reviews (Armstrong et al., 2010; Braga et al., 2010; Mair and Padipatvuthikul, 2010; Marshall et al., 2010; Perdigão, 2010; Scherrer et al., 2010; Söderholm, 2010; Tagami et al., 2010; Van Meerbeek et al., 2010).

THE LOGIC OF ADHESION PROCEDURES

Inspection of a wide range of industrial, domestic, and biomedical adhesives shows the generally common presentation as some form of fluid agent that incorporates a means of conversion into a solid form. The vast majority of such adhesives are designed for application to *adherends* that are fairly rigid solid surfaces. However, there is often an educational gap in the understanding of why *fluid* adhesives are necessary for application to solid adherends, and why they must undergo *solidification* reactions. Hence, we shall outline the straightforward logic of such an adhesive design strategy.

Our starting point – in chemistry – is the existence of *electromagnetic forces* between, as well as within, molecules. Additional to the *primary* bond types (covalent, ionic, and metallic), these *secondary* intermolecular forces include hydrogen bonds and van der Waals bonds, both dipolar and dispersion types. There is a wide variation in the relative strength of these bonds, covalent bonds being the strongest; even weak bonds can have a massive cumulative effect if they have a high number density, as with the H bonds between water molecules.

However, it is characteristic of intermolecular forces and interaction energies that they diminish rapidly with separation, often depending on the inverse 6th power of the separation distance (r). For example, the net interaction energy (U) between two permanent dipole moments

(μ_A, μ_B) within neighboring molecules or surfaces is determined by the overall average of the product:

$$\left[U(r) \times e^{-U(r)/kT}\right]$$

where k is Boltzmann's constant and T is temperature. When:

$$\left(\mu_A \cdot \mu_B \big/ \varepsilon' \cdot r^3 \right) \ll KT$$

then:

$$U_{average} = -\frac{2}{3} \frac{\mu_A^2 \cdot \mu_B^2}{r^6} \times \frac{1}{KT}$$

Where ε is the permittivity of the medium.

Hence, if two solid surfaces are brought into sufficiently close proximity, at moderate temperatures, there is the possibility of interaction energy between them.

Nevertheless, there is a further topological factor; namely that even macroscopically smooth surfaces are microscopically rough. Hence, two solids in close proximity at the nanoscale may be akin to a superimposed pair of mountain ranges. The actual contact regions between the two solids are restricted to the "mountain peaks," and a major proportion of each total solid area is not in contact and is held apart *beyond* the distance zone over which the interaction energies are significant. The *effective intimate contact area* between the two surfaces is then too small to *directly* realize the potential of the intermolecular forces. It follows that rigid contact between the two solids fails to deliver an appreciable net interaction force. However, there are some exceptions to this outcome when one or more of the contacting solids exhibits plastic flow under mechanical stress. Thus, pieces of gold foil may become self-adherent under compressive stress. Similarly, waxes can mutually bond when temperature is raised slightly.

To achieve adhesive bonding in the general case of two rigid solids, such as a tooth enamel surface and an orthodontic bracket, it is necessary to apply a fluid adhesive between them. Moreover, the fluid must be of appropriate chemical formulation to initially wet both surfaces, exhibiting a low *contact angle*. One or both surfaces may have been subjected to some form of pre-treatment or *conditioning* with an *etchant* or *primer* that, *inter alia*, may have modified surface porosity. In this case, the adhesive fluid may be drawn into the solid surface layers by capillary action.

The presence of a suitable fluid between two solids greatly enhances the potential for intermolecular force interactions at each solid–fluid boundary. Further, if for example, the solid is a calcified tissue and the fluid incorporates carboxylic acid groups, then *primary* bonds may be created at the interface. Such primary and/or secondary bonds can immediately generate a measureable resistance to *tensile* loading of the solid interface. This may be augmented by the Laplace pressure difference arising from the fluid surface tension. Nevertheless, such solid pairs weakly bonded by a fluid can be readily disrupted by *shear* forces.

In the general case, therefore, to achieve an adhesive bond zone stable against shear forces, it is imperative to convert the adhesive agent from a fluid to a solid. In this process, it is desirable for the adhesive agent to exhibit high dimensional stability in relation to the interface with one or more adherends. There are several physicochemical mechanisms available for achieving adhesive solidification, or *setting*, including:

- Phase transformation on cooling;
- Solvent evaporation;
- Polymerization (and cross-linking) of fluid monomers or oligomers;
- Acid–base reaction.

Phase transformation has not been applied with biomedical adhesives, as the requisite temperatures (normally above 100°C) are beyond the biologically-tolerable range. Polymerization of monomer systems is the most widely used approach. However this is susceptible to *polymerization shrinkage phenomena* and thus stress-development (see the section on Stress-Development in Adhesive Joints due to Polymerization Shrinkage). Acid–base reactions are present in the setting of polycarboxylate and glass–ionomer cements.

The resultant interfacial bonds, generated by the above adhesion strategy, can be analyzed into different micro- or nanostructural contributions. These may include *micromechanical bonds* created by the interlocking of solidified adhesive "tags" (extensions) into surface porosity or roughness of the adherend. Where the microadaptation of the solid adhesive and the adherend remains molecularly intimate, the intermolecular forces can contribute; and where specific primary bonds are generated, especially with metallic adherends, their contribution is significant.

Given the foregoing general scheme of how adhesive bonds can be generated, many of the factors that can weaken or compromise these bonds will be obvious. These include: air voids; contaminants; and weak boundary layers. This leads to consideration of *surface pretreatments* designed to avoid these problems.

ADHEREND SURFACE PRETREATMENTS TO ENHANCE BOND STRENGTH AND DURABILITY

Biomedical adhesive systems for use as, or with, biomaterials are usually accompanied by detailed instructions for use. Frequently this printed information and advice is supplemented by audiovisual material relating to the mechanisms of action and to detailed applicatory steps. The steps involve meticulous preparation of the adherend surfaces, especially in the case of hard tissues. In practice, there can be a tension between the complexity and effectiveness of the pretreatment *versus* the desire for clinical simplicity and overall speed of application.

Preparation may involve cleansing the tissues from contaminants such as blood or saliva, for if these substances remain *in situ* they can have a negative, abhesive effect. However, the goal is often not merely to remove a contaminant, but also to enhance surface free energy and thereby promote improved wetting, spreading, and penetration of the adhesive agents. In the case of hard dental tissues, a well-established treatment is application of an aqueous phosphoric acid solution to etch and/or demineralize the outer tissue layers (as considered in the sections on Hard Tissue Adhesives: Bone and Tooth Cements; and Acid-Etch Bonding to Enamel). A converse experimental approach in former times was to hyper-mineralize calcified tissues. This was successful *in vitro*, but required clinically unfeasible pretreatment times.

When bonding to synthetic biomaterials, such as ceramics, roughening of a "fitting" surface – to be bonded – increases the effective surface area. This is often combined with application of chemical primers, such as silane coupling agents. In the case of many implants and oral prostheses, the bonding surface may undergo plasma spraying to deposit a well-attached ceramic layer, perhaps of hydroxyapatite (HA). This radically changes the chemical nature of the surface to which adhesive bonding must then be achieved.

HARD TISSUE ADHESIVES: BONE AND TOOTH CEMENTS

Auto-Polymerizing PMMA Bone Cement

Historical Background. In 1936 it was noted that mixing ground polymethylmethacrylate (PMMA) powder with the monomer (MMA) produced a dough which could be manipulated and molded; hence it became one of the first biomaterials (Weber and Chapman, 1984; Donkerwolcke et al., 1998; Kühn, 2000; Webb and Spencer, 2007). Early applications were in dentistry. This was initially employed by orthopedic surgeons as a cement or grout to improve implant fixation in 1953. However, the major breakthrough in the use of PMMA/MMA bone cement in total hip replacement (THR) was proposed by Smith and Charnley, who used it to secure fixation of the acetabular and femoral components and to transfer loads to bone (Charnley, 1960, 1970) (Figure II.5.8.1). More recent applications include percutaneous vertebroplasty applied to the osteoporotic spine, which can provide significant and prolonged relief of pain (Lewis, 2006; Webb and Spencer, 2007).

Mechanism of Setting of PMMA/MMA Dough. Methyl methacrylate (MMA) liquid (Figure II.5.8.2) is rapidly absorbed into PMMA powder, creating a tacky material that progresses to a non-tacky dough stage. These can be considered merely *physical* changes. However, benzoyl peroxide (BPO) initiator, mixed in with the powder, can be chemically activated to form free radicals, either by heating to above 60°C or by the incorporation of a suitable amine activator into the monomer, to enable room temperature auto-polymerization or "cold-cure." As activation generates free radicals from BPO, these start to react with the MMA molecules. The free radical polymerization reaction thus continues through the *propagation* stage – where chain growth occurs from multiple radical initiation sites – until this reaches the *termination* stage, where polymerization is either fully complete or the free radicals are consumed. MMA polymerization is a highly exothermic process, arising from the conversion of C=C bonds. Hence, in the constrained interfemoral environment there is normally a significant temperature rise, which auto-accelerates the setting process, peaking above 56°C for 2–3 minutes *in vivo*. Sometimes higher temperatures are reached, up to 90°C, which can volatilize unreacted monomer (boiling point of MMA is 101°C). This creates voids that can later lead to mechanical failure.

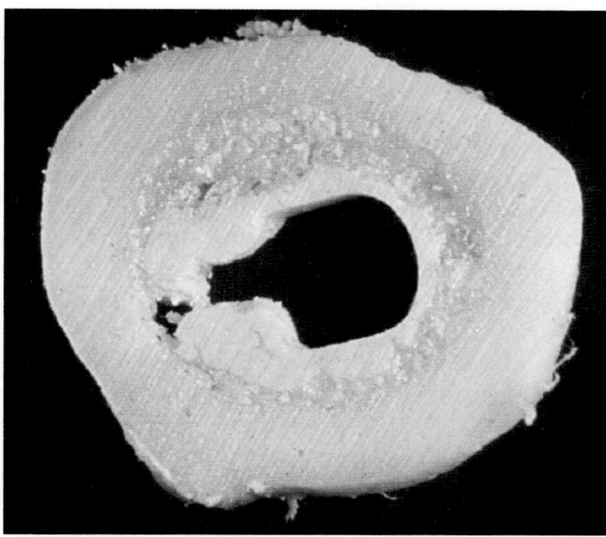

FIGURE II.5.8.1 Cross-section of femur and interlocking bone cement showing good adaptation of the cement to the prosthesis (now removed), and a zone of interaction with the inner surface of the compact bone. (with courtesy of J. Charnley, personal communication)

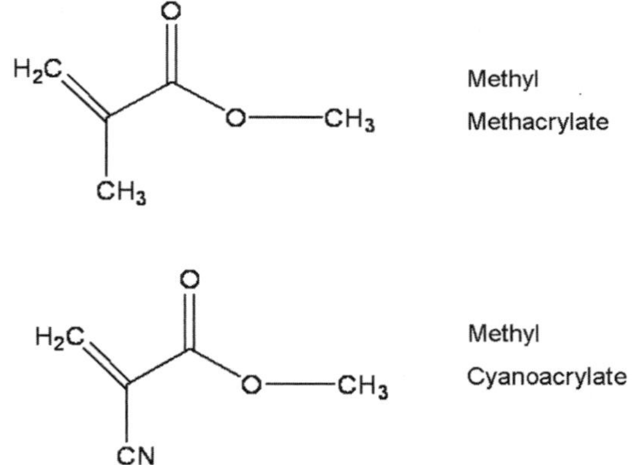

FIGURE II.5.8.2 Methyl methacrylate and methyl cyanoacrylate.

Rheological Factors in Cement Delivery. While the physical absorption and chemical polymerization processes take place, the material at the early dough stage is forcibly injected into the femoral space around the metallic femoral prosthesis. The rheology of the injection process is complex. Bone cements can be formulated to yield different viscosity ranges: low; medium; and high. High viscosity cements have a short waiting/sticky phase, followed by a long working phase that is more suitable for many surgeons.

Cement in its liquid phase of curing behaves as a non-Newtonian (pseudoplastic) fluid, with viscosity decreasing as shear rate is increased. However, the viscosity of all cements increases appreciably during polymerization as the polymer chains lengthen until the material is solidified. Setting times range from 5 to 13 minutes from start of mixing (Kühn, 2000).

Mechanism of "Bonding" or Grouting. As the cement flows under pressure to penetrate the interstices of cancellous bone and adapts to the surface of the femoral stem, it can achieve micro-interlock with the bone when solidified. Nevertheless, PMMA bone cement undergoes a degree of shrinkage on polymerization. Hence another benefit of pressurized injection is partial compensation for this effect. Shrinkage is a potential major source of porosity (Gilbert et al., 2000).

Influence of Pore Distribution in Bone Cements on the Cement/Implant Interface

Implant fixation has been achieved with acrylic bone cement for many years. Although this is a successful application overall, the bond between the implant and bone cement has been shown to be the most fragile link in the construct in femoral components, with failure likely only a short time after implantation. Interfacial porosity and the formation of microcracks have been identified as factors detrimental to cement/implant performance, with pores serving as nucleation sites for these microcracks.

To decrease interfacial porosity and improve the longevity of the cement/implant bond, several authors have suggested preheating implants. Although cements react differently to the curing environments, the most prevalent trend was increased mechanical properties when cured at 50°C versus room temperature. Pores were shown to gather near the surface of cooler molds, and near the center in warmer molds for all cement brands. Pore size was also influenced. Small pores were more often present in cements cured at cooler temperatures, with higher-temperature molds producing more large pores. The mechanical properties of all cements were above the minimum regulatory standards. This supports the practice of heating cemented implants to influence interfacial porosity.

Pelletier, M. H., Lau, A. C. B., Smitham, P. J., Nielsen, G., & Walsh, W. R. (2010). Pore distribution and material properties of bone cement cured at different temperatures. *Acta Biomaterialia*, **6**, 886–891. Elsevier. DOI: 10.1016/j.actbio.2009.09.016

Alternative Bone-Cements: Calcium Phosphate. In orthopedic surgery there are some alternatives to PMMA bone cement, and there is also the option of cementless fixation. Inorganic cements, notably from the Ca–P system, include HA [$Ca_{10}(PO_4)_6OH_2$] and TCP [$Ca_3(PO_4)_2$]. The biocompatibility of HA and its similarities to bone mineral have led to the study of dense HA for the augmentation of osseous defects.

Classical and Modern Dental-Bonding Cements: Conventional Acid–Base Cements. Dental cements are, traditionally, fast-setting pastes obtained by mixing solid and liquid components. Most of these materials set by an acid–base reaction, and subsequently developed resin cements harden by polymerization (Smith, 1971, 1991, 1998). The composition of these materials is shown in Table II.5.8.1. The classes and typical mechanical properties are given in Table II.5.8.2. The

TABLE II.5.8.1 Classification and Composition of Tissue Adhesives

Type	Components	Setting Mechanism
Hard Tissue Adhesives		
Bone		
Acrylic bone cement	Methyl methacrylate and polymethylmethacrylate	Peroxide – amine initiated polymerization
Teeth		
Dental cements: Zinc phosphate	Zinc oxide powder, phosphoric acid liquid	Acid–base reactions; Zn complexation
Zinc polycarboxylate	Zinc oxide powder, aqueous poly(acrylic acid)	Acid–base reactions; Zn complexation
Glass ionomer (polyalkenoate)	Ca, Sr, Al silicate glass powder aqueous poly(acrylic acid-itaconic acid)	Acid–base reactions; Metal ion complexation
Resin modified glass ionomer	Dimethacrylate monomers. Aqueous poly(acrylic acid-methacrylate) co-monomers. Silicate or other glass fillers	Peroxide-amine or photo-initiated polymerization
Resin-based	Aromatic or urethane dimethacrylates, HEMA	Photoinitiated addition polymerization
Dentin adhesive	Etchant: Phosphoric acid (aq.) Primer: HEMA in ethanol or acetone Bond resin: Dimethacrylate monomers	Photoinitiated addition polymerization
Soft Tissue Adhesives		
Cyanoacrylate	Butyl or isobutyl cyanoacrylate	Addition polymerization
Fibrin sealants	A. Fibrinogen, Factor XIII B. Thrombin, CaC12	Fibrin clot formation
GRF glue	Gelatin, resorcinol, formaldehyde	Condensation polymerization
Hydrogel	Block copolymers of PEG, poly(lactic acid) and acrylate esters	Photoinitiated addition polymerization

TABLE II.5.8.2 Properties of Dental Cements and Sealants

Dental Cement/Sealant	Strength: Compressive (MPa)	Strength: Tensile (MPa)	Elastic Modulus (GPa)	Fracture Toughness (MN$^{-1.5}$)
Zinc phosphate	80–100	5–7	13	~0.2
Zinc polycarboxylate	55–85	8–12	5–6	0.4–0.5
Glass ionomer	70–200	6–7	7–8	0.3–0.4
Dimethacrylate sealant unfilled	90–100	20–25	2	0.3–0.4
Dimethacrylate sealant filled	150	30	5	–
Dimethacrylate cement	100–200	30–40	4–6	–
Dimethacrylate composite	350–400	45–70	15–20	1.6

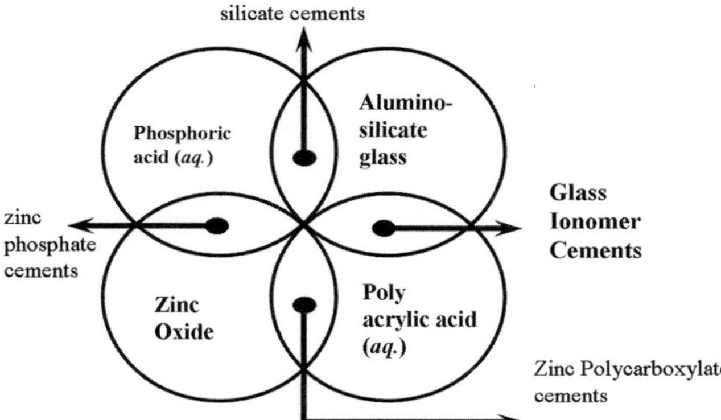

FIGURE II.5.8.3 Classification of major types of acid–base dental cement based upon the combinations of basic powders of either zinc oxide or alumino-silicate glass with two aqueous acidic liquids.

FIGURE II.5.8.4 Structure of the polyelectrolyte: poly(acrylic acid), as present in aqueous solution, in polycarboxylate and glass ionomer cements. The carboxyl anions COO^- are bridged by divalent or trivalent metal ions during the setting process, forming a salt matrix.

FIGURE II.5.8.5 Neutralization of poly(acrylic acid) by zinc oxide to form zinc polycarboxylate cement, in which Zn^{2+} functions as a bridging ion between pairs of carboxylate groups.

relationship between principal categories is illustrated in Figure II.5.8.3.

Zinc phosphate cement is the traditional standard. This material is composed primarily of zinc oxide powder and a 50% phosphoric acid solution containing aluminum and zinc. The mixed material sets to a hard, rigid cement (Table II.5.8.1) by formation of an amorphous zinc phosphate binder. Although the cement is gradually soluble in oral fluids and can cause pulpal irritation, it is clinically effective over 10 to 20 year periods. The bonding arises entirely from penetration into mechanically produced irregularities on the surface of the prepared tooth and the fabricated restorative material. Some interfacial leakage occurs because of cement porosity and imperfect adaptation, but this is usually acceptable since the film thickness is generally below 100 μm.

Poly-Electrolyte Cements: Zinc Polycarboxylates and Glass Ionomers. Poly(carboxylic acid) cements were developed in 1967 (Smith, 1967, 1998) to provide materials with properties comparable to those of phosphate cements, but with adhesive properties to calcified tissues. Zinc polyacrylate (polycarboxylate) cements are formed from zinc oxide and aqueous poly(acrylic acid) solution (Figure II.5.8.4). The metal ion cross-links the polymer structure via carboxyl groups, and other carboxyl groups complex to Ca ions in the surface of the tissue (Figure II.5.8.5). The zinc polycarboxylate cements have adequate physical properties, excellent biocompatibility in the tooth, and proven adhesion to enamel and dentin (Smith, 1991), but are opaque. The need for a translucent material led to the development of the glass-ionomer cements (GIC).

GICs are also based on poly(acrylic acid) or its copolymers with itaconic or maleic acids, but utilize a calcium aluminosilicate glass powder instead of zinc oxide. GICs set by cross-linking of the polyacid with calcium and aluminum ions from the glass, together with formations of a silicate gel structure. The set structure and the residual glass particles form a stronger, more rigid cement (Table II.5.8.1), but with similar adhesive properties to the zinc polyacrylate cements. Both cements are widely used clinically.

Neutralization of Bacterial Acids. Acid–base cement formulations are non-stoichiometric, having an excess of the basic powder. This is beneficial in intra-oral applications as the net basicity of the set cement retains the

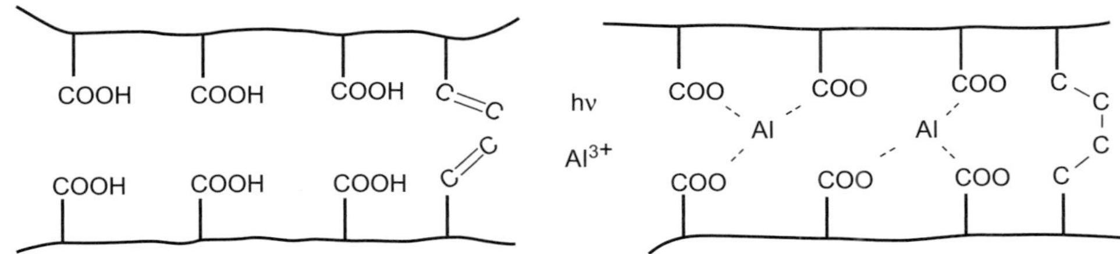

FIGURE II.5.8.6 Formation of the matrix of a "resin modified" glass ionomer cement (RMGIC), by means of C=C bond polymerization, concurrently with ionic salt bridges with metal ions derived from an ion-leachable glass.

capability of neutralizing *in situ* erosive acids produced by bacterial metabolism of dietary sugars.

Resin-Modified Glass-Ionomer Cements. In these cements, the polyacid molecule contains both ionic carboxylate and polymerizable methacrylate groups (Figure II.5.8.6). It is induced to set by both an acid–base reaction and visible light polymerization. These dual-cure cements are widely used clinically. Adhesive bonding but not complete sealing is obtained, because of imperfect adaptation to the bonded surfaces under practical conditions.

Dual-Setting Resin-Based Cements. Resin cements are fluid or paste-like monomer systems based on aromatic or urethane dimethacrylates. Silanated ceramic fillers are usually present to yield a composite composition. They are normally two-component materials that are mixed to induce setting. They may also be light-cured. These set materials are strong, hard, rigid, insoluble, cross-linked polymers (Table II.5.8.1). Bonding is achieved by mechanical interlocking to surface roughness.

Self-Etching Resin-Based Cements. In recent materials, reactive adhesive monomers may also be present to avoid the necessity of using a separate dentine-bonding system, as described in the section on Bonding to Dentin via the Hybrid Inter-Phase).

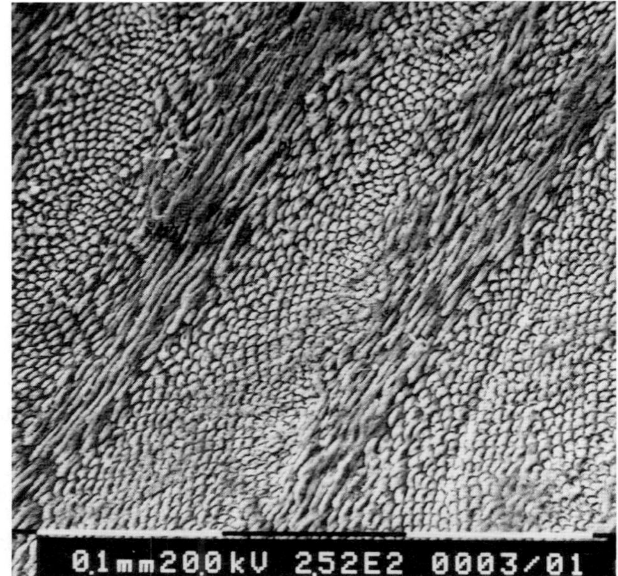

FIGURE II.5.8.7 Surface of acid-etched dental enamel. In this image, there are alternating regions where the bundles of prism structures are viewed: (i) transversely (end-on); or (ii) semi-longitudinally. In the case of the transverse prisms, the prism peripheries have been preferentially etched, leaving the prism cores visble.

Acid-Etch Bonding to Enamel

Dental enamel is a hard, stiff substrate of generally uniform chemical composition. Bonding to enamel is required either in conjunction with bonding to dentin, for cavity restoration or for bonding veneers or orthodontic brackets directly to enamel of teeth in the upper or lower dental arch. In both cases, the modern approach was pioneered by Buonocore (Roulet and Degrange, 2000; Eliades et al., 2005). This involves conditioning with an aqueous phosphoric acid etchant for about 60 seconds. Subsequent water-rinsing and air-drying leaves a visibly matt surface on the enamel – now differentially etched, with either the enamel-prism cores or the prism peripheries exposed (Figure II.5.8.7). Application of an unfilled resin (monomer mixture) or a moderate-viscosity resin-composite allows this to flow across the enamel, penetrating surface porosity. Subsequent hardening of the resin results in its retention via a multitude of microscopic tag-like resin extensions into the enamel surface.

Bonding to Dentin via the Hybrid Inter-Phase

Intense Research Effort and Advances in this Area. Bonding restorative biomaterials to dentin is much more challenging than to enamel, due to the more complex, hydrated substrate. Not only may this involve caries-weakened dentin, but the cut dentin surface has a weakly attached "smear" layer. This is a layer of mixed composition, including denatured collagen, generated by the thermo-mechanical process of cutting dentin by high-speed burs. This material is mechanically weak and is smeared all over the cut surface, covering the sound dentine underneath. This must be either removed or significantly modified.

Nevertheless, recent decades of intensive worldwide research have made remarkable advances both in understanding the ultra-structure of the resultant bonds, and then devising improved formulations (Nakabayashi and Pashley, 1998; Roulet and Degrange, 2000; Eliades et al., 2005; Moszner and Salz, 2007; Matinlinna and

Mittal, 2009). At present, the outstanding challenges are to enhance the long-term durability of adhesive bonds against both intra-oral stresses and biochemical degradation (Breschi et al., 2008; Watts et al., 2009). There is also a secondary goal of formulating simpler, and thus time-saving, application protocols.

Early Unsuccessful Approaches

It was recognized early that simply applying hydrophobic resins to dentin gave little benefit. Moreover, for several decades it was deemed clinical malpractice to apply an acid etchant to dentin. Development of adhesives proceeded slowly until the critical insights concerning dentin hybridization and the hybrid-layer were achieved, as discussed in the section titled Hybrid-Layer Creation via Three-Stage Approach: Etch; Prime; Bond.

Chemistry of Etchants, Primers, and Bonding Agents

Modern effective dentine bonding systems can be considered as incorporating a three-fold set of agents, although two or even three of these functions may be combined in a simplified system. The *etchant* or *conditioner* function is provided by either an aqueous phosphoric acid gel or by acidic monomers. The *primer* function is classically achieved by a solution of a hydrophilic monomer, such as HEMA, in a solvent such as acetone or ethanol. Subsequently, a "*bonding agent*" consisting of unfilled resin monomers is applied to the dentine.

Hybrid-Layer Creation via Three-Stage Approach: Etch; Prime; Bond

The optimal approach is to use a separate acid etchant, and to apply this to the dentin for about 10 seconds, which removes the smear layer (Figure II.5.8.8). This demineralizes the outer dentin layers, leaving residual type I collagen fibrils and proteoglycans. Washing the dentine removes the acid and salt residues. However, drying results in collapse or compaction of the collagen. The second agent, the primer, has the role of re-expanding the collagen, thereby restoring its porosity (Figure II.5.8.9). This permits the third agent – unfilled hydrophobic "bond" monomers – to permeate and seal the collagen. When these are photo-cured they form an entangled interpenetrating network (IPN) with the collagen fibers. The resultant interfacial IPN region, typically 4 μm in thickness, is known as the hybrid zone. Effectively, the polymerized resin is taking the place of the original mineral phase in this region. The strength of this bonding zone depends on the continuing integrity of its component elements, principally the collagen fibrils and the polymerized monomers.

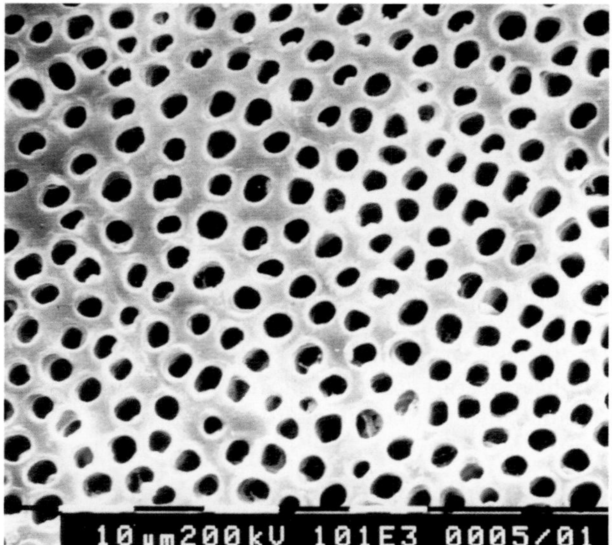

FIGURE II.5.8.8 Dentin surface following etching with phosphoric acid solution. The dentine tubules are clearly visible.

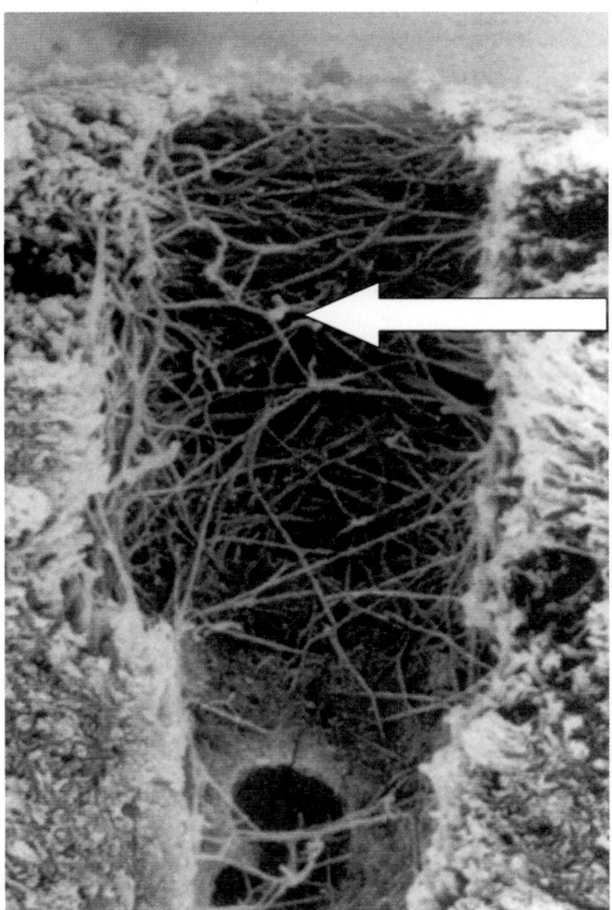

FIGURE II.5.8.9 Cross-section of acid-etched and demineralized dentin. The collagen fibrils are visible (arrowed) within a tubule.

The Quest for Simplified Clinical Procedures

Clinical success depends on understanding the rationale of each step and meticulous observance of manufacturer's instructions. The market has led to availability of simplified systems, where two or more steps may be combined into one, including self-etching systems (Figure II.5.8.10). Earlier "one step" systems of this type gave reduced performance, but newer products incorporating clever delivery devices are more successful. Nevertheless, there are many systems available where a simplification of clinical application procedures is done to the detriment of bonding efficacy.

Aging and Stability of the Bonded Interface

When properly applied, good adaptation and bonding of the adhesive and restorative material to dentin can be achieved in the short- and medium-term. Also, a good initial seal of the interfacial zone can be achieved. However, a number of potential weaknesses have been identified (Breschi et al., 2008).

1. *Insufficient resin impregnation of dentin:* There are circumstances where the demineralized collagen is not fully supported by infiltrated resin, leaving a mechanically weak zone.
2. *High permeability of the bonded interface:* The hybrid zone is susceptible to "nano-leakage" by small molecules and ions, which may provide a pathway for degradation reactions over time.
3. *Sub-optimal polymerization:* The efficiency of polymerization of the infiltrated monomer may be compromised by the presence of water and dissolved oxygen, thereby weakening the IPN structure.
4. *Phase separation*: Monomer mixtures *in situ* of varying hydrophilicity may spontaneously phase separate upon polymerization.
5. *Activation of endogenous collagenolytic enzymes:* Recent studies have revealed that dentin matrices may be slowly degraded over time by dentin-derived proteolytic enzymes, particularly matrix metalloproteinases (MMPs). Collagen integrity within the hybrid layers may be promoted by incorporation of chlorhexidine, an anti-bacterial agent with MMP-inhibiting properties.

Inhibitors for the Preservation of the Hybrid Interfacial Zone Between Adhesives and Human Dentin

The outstanding resistance of the collagenous dentin matrix against thermal and proteolytic disruption has been attributed to the high degree of intermolecular cross-linking and tight mechanical weave of this specialized connective tissue. However, great attention to the potential proteolytic activity of dentin has been raised since complexed and active forms of matrix metalloproteinases (MMPs) were identified in either non-mineralized or mineralized compartments of human dentin. MMPs belong to a group of zinc- and calcium-dependent enzymes that have been shown to be able to cleave native collagenous tissues at neutral pH in the metabolism of all connective tissues. Galardin is a synthetic MMP-inhibitor with potent activity against MMP-1, -2, -3, -8, and -9. It also partially preserved the mechanical integrity of the hybrid layer created by a two-step etch-and-rinse adhesive after artificial aging.

Breschi, L., Martin, P., Mazzoni, A., Nato, F., & Carrilho, M. et al. (2010). Use of a specific MMP-inhibitor (galardin) for preservation of hybrid layer. *Dent. Mater.*, **26**, 571–578. Elsevier: DOI: 10.1016/j.dental.2010.02.007

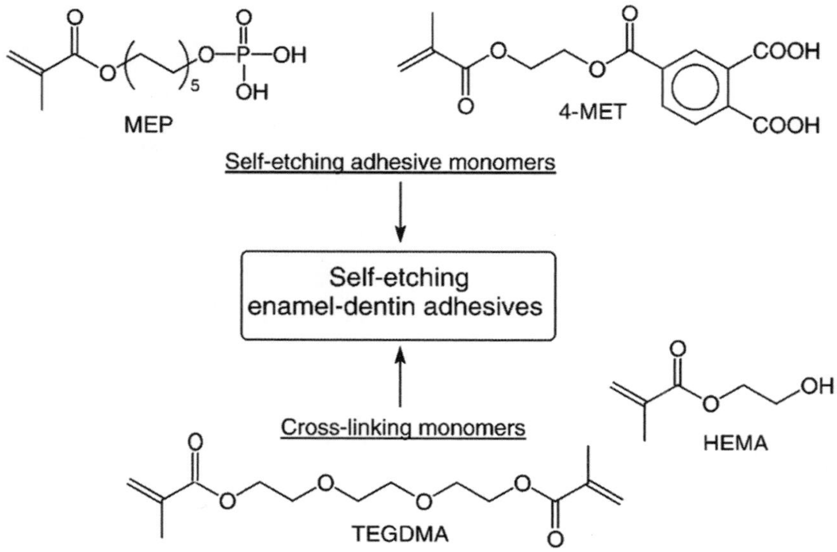

FIGURE II.5.8.10 Components of self-etching enamel-dentin adhesives. (After Moszner and Salz, 2007.) The adhesive systems usually incorporate monofunctional co-monomers and additives, including photo-initiators, solvents, stabilizers, and fillers. (Recent Developments of New Components for Dental Adhesives and Composites. Macromolecular Materials and Engineering. **292**, 245–271. 2007. Wiley. DOI: 10.1002/mame.200600414)

Incorporation of Anti-Bacterial Functionality

Efforts have been made for some time to incorporate anti-bacterial agents in either composites or dentin-adhesives, with the latter showing greater promise (Imazato, et al. 1998, 1999; Imazato, 2003).

Stress-Development in Adhesive Joints due to Polymerization Shrinkage

Inevitability When Using Methacrylate Matrix Composite Biomaterials. As already noted, adhesive layers, when solidified by polymerization, are accompanied by shrinkage phenomena. In restorative dentistry, tooth cavities – caused by carious decay – are now principally repaired by tooth-coloured polymer–ceramic composites, instead of silver amalgam fillings. These bulk materials also exhibit setting shrinkage which exacerbates the interfacial stresses on the adhesive joints. This is an inescapable feature of the currently dominant dimethacrylate "resin" composites. Despite numerous attractive properties of these biomaterials, many adverse clinical problems can arise if this shrinkage problem is not adequately managed (Watts et al., 2009). It is a major potential cause of adhesive failure.

Molecular Origins of Shrinkage and Polymerization-Kinetics

When methacrylate monomers undergo polymerization, this involves the conversion of C=C double bonds to C–C single bonds. The original monomer molecules are no longer separated by a mean intermolecular (van der Waals) distance, but become connected via the C–C bonds. This implies a closer packing of the molecular units – or in other terms, molecular densification. For example, with methyl methacrylate (MMA), the volume change per mole of methacrylate groups is:

$$\Delta V_{C=C} = 22.5\ cm^3\ mole^{-1}$$

This localized molecular shrinkage normally occurs throughout the bulk of the material and so produces a macroscopic shrinkage.

For multi-methacrylate monomers of higher C=C functionality (f), and for a *degree of conversion* (DC) that may be less than 100%, the volumetric shrinkage is given by:

$$\varepsilon_{vol} = 22.5 \times DC \times f \times \frac{V}{V_m} \times 100$$

where: V is volume and V_m is molar volume. This can easily be generalized for a mixture of monomers (Silikas et al., 2005).

Both methacrylate and acrylate polymerization occur by a free radical mechanism. This may be activated by various means, although *photo-activation* using visible blue light is the most widespread option in dentistry, with *chemical activation* being essential for bone cements (see Mechanism of Setting of PMMA/MMA Dough).

In dimethacrylate dental monomers, there are two C=C polymerizable groups at either end of the molecule. Statistically, both C=C groups react in a high proportion of cases, generating a tightly cross-linked matrix network. However, some monomers only react via one C=C group, leaving the other C=C group unreacted and pendant. This is because following a rapid auto-acceleration phase, the material vitrifies and the chain-end free radicals become locally trapped within the glassy network. The *rate* of polymerization then rapidly diminishes (auto-deceleration) towards zero (Watts, 2005). Thus, at 37°C, the *degree of conversion* of monomer is only about 65%. Nevertheless, physical properties are otherwise satisfactory.

Factors that Affect the Magnitude and Vector Direction of Shrinkage Stress

When an adhesively-bonded resin-composite is placed between two or more opposing cavity walls, then the material is not able to shrink freely upon polymerization. Therefore, providing adhesive bonding is achieved to the walls, a state of "shrinkage-stress" will be generated in the restored tooth system. However, material at or in proximity to a free, un-bonded surface will be able to shrink freely, with minimal local-stress generation.

Different tooth cavity sizes and shapes involve differing cavity configuration factors (C_f) of bonded to non-bonded interfacial surface areas.

Hence:

$$C_f = Area_{\text{bonded}} / Area_{\text{non-bonded}}$$

other factors being equal, cavities with a high C-factor will be more highly prone to shrinkage stress (Figure II.5.8.11).

Nevertheless, it has been recognized recently that other factors are seldom equal. In particular, the mass (or volume) of setting material commonly varies between different cavity shapes or designs. It is the composite mass that drives the shrinkage stress magnitude, and so if this is reduced appreciably, the shrinkage stress will also be reduced (Watts et al., 2009).

In complex cavity shapes, Finite Element Analysis (FEA) has been fruitfully applied to determine the distribution and vector directions of shrinkage stress. This requires realistic tooth and cavity shapes, and allowance being made for the visco-elastic character of both the setting material and dentin tooth tissue. In the interior of a cavity, the repair biomaterial is subject to a state of generalized plane strain, whereas close to

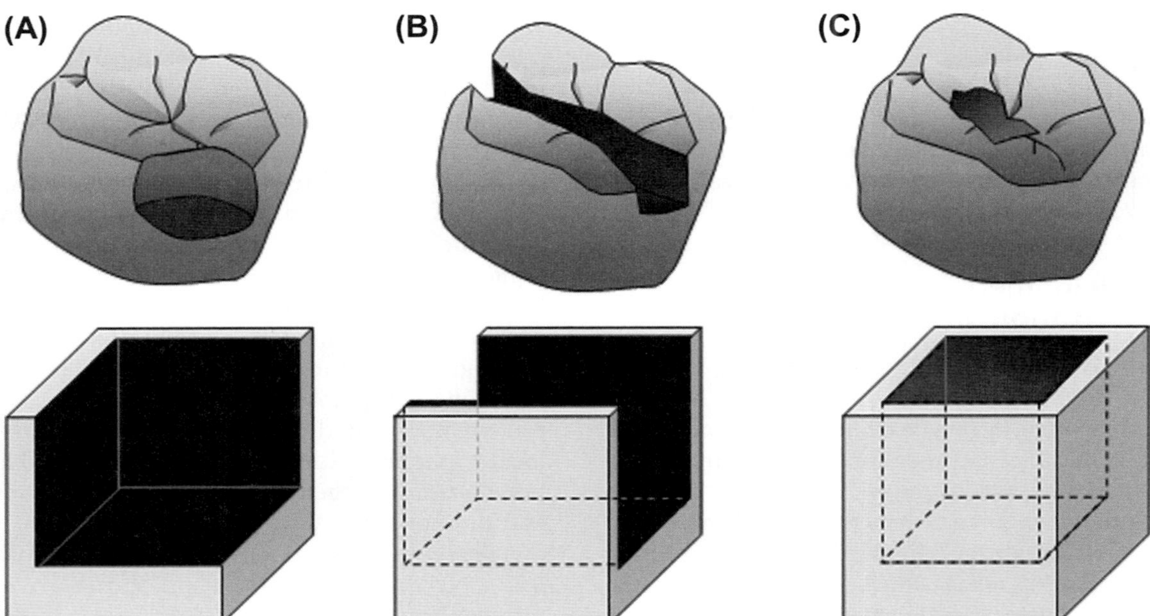

FIGURE II.5.8.11 Schematic cavities in molar teeth. Cavity (C) shows a greater ratio of bonded to non-bonded area than is the case with cavities (A) and (B). This has consequences for the magnitude of shrinkage stress established at the biomaterial–tissue interfaces within each cavity, for a given mass of a specific material.

the cavity opening, the material is subject to a state of plane stress.

Strategies to Minimize Shrinkage Stress in Bonded Cavities of Low Compliance

The *compliance* of the cavity walls is a further significant factor. When the cavity walls are relatively thick, then the compliance will be low and strategies must be sought to minimize shrinkage-stress. One common clinical practice is to apply the material in a succession of layers, either parallel, or at an angle, to the base of the cavity. These layers are adhesively bonded to the walls and photo-cured in sequence. This can reduce the net effect of shrinkage stress. The effect of shrinkage stress within restored molar teeth can be directly measured by detection of inward displacement of cavity walls and tooth cusps (Watts et al., 2009).

Development and Chemistry of New Low Shrinkage Monomers, Including Siloranes

The perceived significance of the above shrinkage problem has led to intense efforts to produce improved formulations, including novel monomer chemistry. An innovative approach is the development of four branched silorane monomers (Figure II.5.8.12) that are photo-cured via a cationic polymerization mechanism (Weinmann et al., 2005). These materials exhibit less than 1% shrinkage on setting, and are strongly hydrophobic. They require a specialized system adhesive to bond to enamel and dentin.

Soft Tissue Adhesives and Sealants

Performance Requirements. Most soft tissue adhesives are intended to be temporary. That is, they are removed or degrade when wound healing is sufficiently advanced for the tissue to maintain its integrity. Effective adhesion can be obtained on dry skin or wound surfaces by using wound dressing strips with acrylate-based adhesives. However, on wound surfaces that are wet with tissue fluid or blood, the adhesive must be able to be spread easily on such a surface, provide adequate working time, develop and maintain adhesion, desirably provide hemostasis, facilitate wound healing, and maintain biocompatibility. Positive antimicrobial action would be an additional advantage (Ikada, 1997).

Historical Overview. Few, if any, systems comply with all these requirements. Currently, there are two principal systems in widespread clinical use – cyanoacrylate esters and fibrin tissue adhesives. Another system, based on a gelatin-resorcinol-formaldehyde (GRF) combination, still receives limited use (Table II.5.8.3). For many years there have been studies undertaken using bioadhesives, especially polypeptides from marine organisms (mussel adhesive) (Sierra and Salz, 1996; Chu et al., 1997). An important new development by Messersmith and co-workers (Brubaker et al., 2010) has led to synthesis of branched polyethylene glycol (PEG) cores with endgroups derivatized with catechol, a functional group abundant in mussel adhesive proteins.

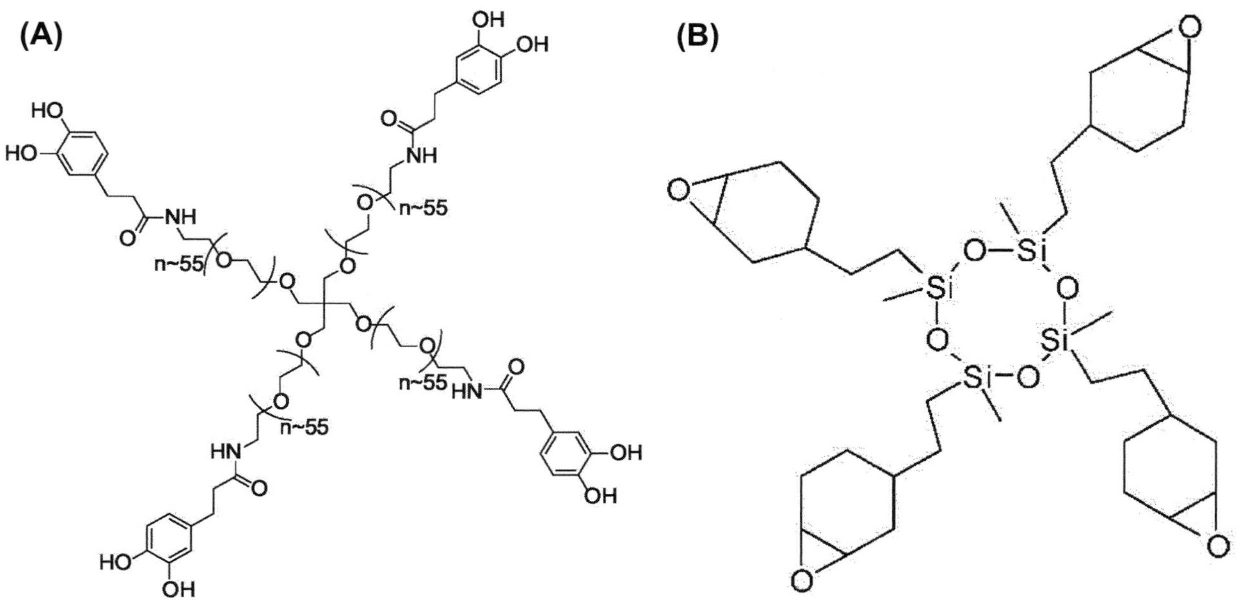

FIGURE II.5.8.12 (A) Structure of a 4–branched silorane molecule. The 3–member functional epoxide groups react via a cationic mechanism to give a ring–opening polymerization. This by itself creates molecular expansion upon setting, and can be used to formulate low–shrinkage materials. (B) A four arm catechol–terminated PEG (cPEG) adhesive precursor inspired by the protein glues of marine mussels. This rapidly forms adhesive hydrogels under oxidizing solution conditions.

The Relationship Between Soft Tissue Adhesion and Drug Delivery

Soft tissue adhesives are not only required for surgical wound healing, but also for the controlled delivery of drugs to specific sites, of which the mucosal membranes are particularly suitable. The process of mucoadhesion involving a polymeric drug delivery platform is a complex one that includes wetting, adsorption, and interpenetration of polymer chains among various other processes. The success and degree of mucoadhesion bonding is influenced by various polymer-based properties, such as the degree of cross-linking, chain length, and the presence of various functional groupings. The attractiveness of mucosal-targeted controlled drug delivery of active pharmaceutical ingredients has led formulation scientists to engineer numerous polymeric systems for such tasks. Formulation scientists have at their disposal a range of *in vitro* and *in vivo* mucoadhesion testing set-ups in order to select candidate adhesive drug delivery platforms. As such, mucoadhesive systems have found wide use throughout many mucosal covered organelles for delivery for local or systemic effect. Evolution of such mucoadhesive formulations has developed from first-generation charged hydrophilic polymer networks to more specific second-generation systems based on lectin, thiol, and various other adhesive functional groups.

Andrews, G. P., Laverty, T. P., & Jones, D. S. (2009). Mucoadhesive polymeric platforms for controlled drug delivery. *Eur. J. Pharm. Biopharm.*, **71**, 505–518.

TABLE II.5.8.3 Characteristics of Soft Tissue Adhesives

Property	Cyanoacrylate	Fibrin Glue	GRF
Ease of application	Poor	Excellent	Poor
Set time	Short	Medium	Medium
Tissue bonding	Good	Poor	Excellent
Pliability	Poor	Excellent	Poor
Toxicity	Medium	Low	High
Resorbability	Poor	Good	Poor
Cell infiltration	Poor	Excellent	Poor

(After Ikada, 1997.)

Cyanoacrylate Esters

Chemistry. These esters are fluid, water-white monomers that polymerize rapidly by an anionic mechanism in the presence of weak bases such as water or NH_2 groups. Initially, methyl cyanoacrylate (Figure II.5.8.2) was used, but in the past two decades isobutyl and n-butyl cyanoacrylate have been found more acceptable. The higher cyanoacrylates spread more rapidly on wound surfaces, and polymerize more rapidly in the presence of blood. Furthermore, they degrade more slowly over several weeks, in contrast to the methyl ester, which hydrolyzes rapidly yielding formaldehyde that results in an acute inflammatory response.

Performance. These materials achieve rapid hemostasis, as well as a strong bond to tissue. However, the polymer film is somewhat brittle and can be dislodged on mobile tissue, and the materials can be difficult to apply to large wounds. Because of adverse tissue response and production of tumors in laboratory animals, cyanoacrylates are not approved for routine clinical use in the United States, although a commercial material based on n-butyl cyanoacrylate is approved by several other countries.

The current uses are as a surface wound dressing in dental surgery, especially in periodontics, and in life-threatening applications such as brain arteriovenous malformations. Reports of sarcomas in laboratory animals (Reiter, 1987), evidence of *in vitro* cytotoxicity (Ciapetti et al., 1994), and lack of regulatory approval have restricted their further use, in spite of work on synthesis of new types of cyanoacrylate.

Fibrin Sealants

Formulation, Presentation, and Setting Processes. Fibrin sealants involve the production of a synthetic fibrin clot as an adhesive and wound-covering agent. The concept of using fibrin dates back to 1909, but was placed on a specific basis by Matras in 1972 (Matras, 1972). The commercial materials first available consisted of two solutions that are mixed immediately before application to provide a controlled fibrin deposition. Later a "ready-to-use" formulation (Tisseel Duo) was introduced (Schlag and Redl, 1987). The essential components of these solutions are as follows:

Solution A:	**Solution B:**
Fibrinogen	Thrombin
Factor XIII	$CaCl_2$

Fibrinogen is at a much higher concentration (~70 mg/ml) than that in human plasma. On mixing the two solutions, using a device such as a twin syringe with a mixing nozzle, a reaction similar to that of the final stages of blood clotting occurs. Polymerization of the fibrinogen to fibrin monomers and a white fibrin clot is initiated under the action of thrombin and $CaCl_2$. Aprotinin, an inhibitor of fibrinolysis, may also be included in solution A. The composition may be adjusted to promote hemostasis, for example, or to minimize persistence of the clot to avoid fibrosis.

Advantages and Applications. There are now several commercial products available in Western Europe, Japan, and Canada; but they were not approved for use in the USA until 1998 due to the FDA concerns about viral contamination. Fibrinogen for these commercial materials is manufactured from the pooled plasma of selected donors, using processes such as cryoprecipitation. The material is subjected to in process virus activation, and routinely screened for hepatitis and HIV. To minimize these risks recent processes produce the fibrinogen in a "closed" (single donor) blood bank or utilize the patient's own blood. This autologous fibrin glue is preferred, but its quality is partly determined by the fibrinogen level in the donor plasma (Ikada, 1997). Other complications include formation of antibodies and thrombin inhibitors, as well as potential risks of BSE (bovine spongiform encephalopathy) if bovine thrombin is used. More recently, human-derived thrombin has been employed.

Originally developed during World War II to stop bleeding from battle injuries, fibrin sealants are presently used during surgery for several different purposes:

- to control bleeding in the area where the surgeon is operating;
- to speed wound healing;
- to seal off hollow body organs or cover holes made by standard sutures;
- to provide slow release delivery of medication to tissues exposed during surgery.

Fibrin sealants have several advantages over older methods of hemostasis. They speed up the formation of a stable clot; they can be applied to very small blood vessels and to areas that are difficult to reach with conventional sutures; they reduce the amount of blood lost during surgery; they lower the risk of postoperative inflammation or infection; and they are biodegradable by the body during the healing process. They are particularly useful for minimally invasive procedures, and for treating patients with blood clotting disorders.

The adhesive strength is not as high as that of cyanoacrylates, but it is adequate for many clinical situations. Thorough mixing of the ingredients and application techniques or devices that allow uniform spreading, are essential to success.

The material has been used in a wide variety of surgical techniques for hemostasis and sealing involving thoracic–cardiovascular, neurologic, plastic, and ophthalmic surgery, and as a biodegradable adhesive scaffold for meshed skin grafts in burn patients (Sierra and Salz, 1996; Ikada, 1997; Feldman et al., 1999). A symposium has reviewed clinical uses (Spotnitz, 1998). An auxiliary aspect of fibrin sealants is their use as a delivery vehicle at local sites for antibiotics and growth factors (Sierra and Salz, 1996).

Attempts to modify fibrin sealant have been directed toward: (1) improvements in ease of application and control of setting by use of a one-component light-activated product (Scardino et al., 1999); (2) improvements in strength and performance by addition of fibrillar collagen (Sierra and Salz, 1996); and (3) development of a formulation containing gelatin (Ikada, 1997).

Gelatin-Resorcinol-Aldehyde Glues

Formulations. Gelatin is an animal protein with adhesive properties and chemical similarity to connective tissue. This glue was developed in the 1960s by Falb and co-workers (Falb and Cooper, 1966; Cooper et al., 1972) as a less toxic material than methyl cyanoacrylate. The material is fabricated by warming a 3:1 mixture of gelatin and resorcina, and adding an 18% formaldehyde solution. (This glue is called "GRF.") Cross-linking of the gelatin and resorcinol by the formaldehyde takes place in about 30 seconds.

Limited Applications. The material was used in a variety of soft tissue applications, but technical problems and toxicity have limited its application in recent years to aortic dissection (Ikada, 1997). In attempts to overcome the toxicity and potential mutagenecity/carcinogenicity of the formaldehyde component, modified formulations have been developed in which other aldehydes such as glutaraldehyde and glyoxal (Ennker et al., 1994a, b) are substituted for the formaldehyde. Favorable results with this material (GR-DIAL) have been reported (Ennker et al., 1994a, b). Concerns over toxicity remain, however, and this material has not received FDA approval for commercial use. Less toxic gelatin cross-linking agents have been investigated *in vitro* without substantial improvement.

The three aforementioned types of adhesives have undergone extensive clinical trials in Europe and Japan. Fibrin sealant is the most widely used material currently. However, all of these systems have significant deficiencies. This is illustrated by their relative characteristics listed in Table II.5.8.3 (Ikada, 1997).

Bioadhesives

Bioadhesives are involved in cell-to-cell adhesion, adhesion between living and nonliving parts of an organism, and adhesion between an organism and foreign surfaces. Adhesives produced by marine organisms, such as the barnacle and the mussel, have been extensively investigated over the past 40 years because of their apparent stable adhesion to a variety of surfaces under adverse aqueous conditions. These studies have shown that these organisms secrete a liquid acidic protein adhesive that is cross-linked by a simultaneously secreted enzyme system. The bonding probably involves hydrogen bonding and ionic bonding from the acidic groups (Waite, 1989).

The adhesive from the mussel has been identified as a polyphenolic protein, molecular mass about 130,000 Daltons, which is cross-linked by a catechol oxidase system in about 3 minutes. A limiting factor in the practical use of this material is the difficulty of extraction from the natural source. The basic unit of the polyphenolic protein has been identified as a specific decapeptide (Waite, 1989; Green, 1996). Recombinant DNA technology and peptide synthesis have been used in attempts to produce an affordable adhesive with superior properties. Little information has been reported on the performance (including biocompatibility) of these materials.

Hydrogel Sealants

A new approach is the development of synthetic sealants based on poly(ethylene glycol) (PEG) hydrogels that are derived from the original work of Sawhney, Pathak, and Hubbell (1993). A family of fully synthetic, implantable, resorbable hydrogels intended for use as surgical sealants has been developed (Moody et al., 1996; Ranger et al., 1997; Alleyne et al., 1998; Tanaka et al., 1999), for barrier coatings (West et al., 1996), and for drug delivery matrices (Lovich et al., 1998). The hydrogels are formed by *in situ* deposition of aqueous formulations based on specialized dendritic macromolecules, followed by photopolymerization to highly cross-linked structures.

The macromers are reactive block copolymers consisting of a water-soluble core, such as polyoxyethylene, flanking biodegradable oligomers such as poly(lactic acid) or poly(trimethylene carbonate), and polymerizable end caps such as acrylate esters (Sawhney et al., 1993). Control of the physical properties and degradation rates is achieved by specifying the molecular structures and concentration of the reactants in the formulation (Dai et al., 2011). Photopolymerization can be effected by ultraviolet or visible light using appropriate photoinitiators. Typically, the initiator system eosin Y spirit-soluble (EYss)/triethanolamine (TEA) is employed with visible illumination in the 450–550 nm range.

In most applications for these hydrogels, strong bonding to tissue is required. This is achieved by use of a two-part sealant system consisting of primer and topcoat. Strong, durable bonding to a wide variety of internal tissues has been demonstrated (Coury et al., 1999).

New Research Directions: Biomimetic Approaches

As a result of the experience of recent decades, the problems involved in developing an adhesive system for both soft and hard tissues have been addressed and identified. A practical limitation in many systems remains ease of manipulation and application. For example, the effectiveness of the fibrin sealant is critically dependent on proper mixing of the ingredients and uniform application. It has proved difficult to reconcile short- and long-term biocompatibility needs with chemical adhesion mechanisms that use a reactive monomer system.

Where relatively temporary (less than 30 days) adhesion is required, as in wound healing, systems based on natural models that allow biodegradation of the adhesive and interface, and subsequent normal tissue remodeling, appear to merit further development.

For longer-term (years) durability in both soft and hard tissues, hydrophilic monomers and polymers of low toxicity that can both diffuse into the tissue surface, and form ionic bonds across the interface, seem to be a promising approach. Evidence has been obtained of the need for hydrophobic–hydrophilic balance in adhesive monomer systems (Nakabayashi and Pashley, 1998), and the use of hydrophilic monomers such as hydroxyethyl methacrylate in commercial materials has facilitated surface penetration.

Recent development trends in soft tissue adhesives appear to reflect these approaches. Newer materials (Chu, 1997) comprise cross-linked collagen materials, light-cured polymerizable and biodegradable polyethylene glycols, photopolymerized derivatized collagen (DeVore, 1999), a bioresorbable hemostatic collagen-derived matrix with thrombin, and serum albumin cross-linked with a derivatized poly(ethylene glycol) sealant. The development of synthetic peptides and materials based on human recombinant components are being investigated. However, the most promising development for hydrogels inspired by mussel adhesives appears to be the approach of the Messersmith group (Brubaker et al., 2010). Rather than using natural polypeptides, the synthesis of catechol-terminated PEG solutions allows a hydogel to be formed in about 30 seconds. Upon implantation, the cPEG adhesive elicited minimal acute or chronic inflammatory response in C57BL6 mice, and maintained an intact interface with supporting tissue for up to one year. *In situ* cPEG adhesive formation was shown to efficiently immobilize transplanted islets at the epididymal fat pad and external liver surfaces, permitting normoglycemic recovery and graft revascularization. These findings establish the use of synthetic, biologically-inspired adhesives for islet transplantation at extrahepatic sites.

Another significant deployment of PEG block copolymers and hyaluronic structures to make photopolymerizable formulations is exemplified by the Tirelli group (Tirelli et al., 2002; Park et al., 2003).

On calcified surfaces, the use of hydrophilic electrolytes, such as the polycarboxylates, has demonstrated that proven ionic bonding *in vitro* can also be achieved *in vivo* (Smith, 1998). An advantage of such systems is that surface molecular reorientations can improve bonding with time (Peters et al., 1974). Encouraging preliminary results have been obtained with new glass ionomer hybrid systems, and there is considerable scope for the future developments of modified polyelectrolyte cements.

The development of more efficient adhesives and sealants that, in addition to enhancing the durability of current applications, would permit new applications such as osteogenic bone space fillers, percutaneous and permucosal seals, and functional attachment of prostheses is still a challenging problem for the future.

Finally, a major biomimetic initiative is the creation of self-healing polymer composites (Trask et al., 2007). Most materials in nature are themselves self-healing composite materials. The concept of an autonomic self-healing material, where initiation of repair is integral to the material, is now being considered for engineering applications. This bio-inspired concept offers the designer an ability to incorporate secondary functional materials capable of counteracting service degradation, whilst still achieving the primary, usually structural, requirement. So far this research has been mainly directed towards structural materials (Brown et al., 2005a,b), but it is a short step to apply these concepts and designs to the structural adhesives being designed for hard tissue bonding.

Case Study — Nano-Controlled Molecular Interaction at Adhesive Interfaces for Hard Tissue Reconstruction[1]

Since 2008 the use of dental amalgam to restore teeth has been forbidden in Norway, and since June 2009 in Sweden also, mainly because of environmental issues and potential health risks related to its mercury content. Other countries may soon follow, so that composites are definitely the materials of choice to directly restore teeth in the least invasive way. Although decayed/fractured teeth can be reconstructed minimally invasively and nearly invisibly using adhesive technology, the clinical longevity of composite restorations is still too short. Bonding is also indispensable in the treatment of root caries lesions, the current worldwide prevalence of which increases dramatically with age. However, tooth bonding in the relatively aggressive oral environment is far from perfect. Within a time frame of 3–5 years adhesive restorations lose their marginal seal, leading first to unesthetic discoloration, and eventually caries recurrence. This forces dentists to replace restorations too often, leading with each new intervention to further weakening of the patient's tooth and, naturally, also to higher public-health costs.

Also in orthopedic medicine, one of the main causes of aseptic loosening of cemented hip replacements is the lack of a compound stable against hydrolysis between the hydrophobic bone cement and the hydrophilic acetabular bone stock. Current hydrophobic bone cements cannot chemically bond to the hydrophilic osseous surface, and thus undergo long-term hydrolytic degradation processes, causing the bone–bone cement interface to de-bond. Bone bonding cements containing functional monomers that can directly adhere to bone, and thus can complement the current purely mechanical stabilization of orthopedic implants, would therefore be of great benefit in orthopedic surgery.

In an attempt to answer questions of direct importance to the mechanisms of bond degradation/stability, the complex biochemical interplay at the adhesive–tooth interface was studied by analyzing the chemical interaction of the functional monomers phenyl-P and 4-MET with HAp, using correlatively solid-state nuclear magnetic resonance (NMR) and X-ray diffraction (XRD). Clinical data confirmed the overall better performance of a 10-MDP-based adhesive over its precursor, a phenyl-P-based adhesive, while a 4-MET-based adhesive performed somewhat in between the other two. Direct comparison of the interfacial molecular interactions of the functional monomers phenyl-P and 4-MET with similar data obtained before for the best performing 10-MDP was expected to reveal how functional monomers interact at the adhesive interface. The hypothesis was tested that no difference in chemical interaction with HAp was found among the three functional monomers.

A time-dependent molecular interaction at the interface with stable ionic bond formation of the monomer to hydroxyapatite was found, competing in time with the deposition of less stable calcium phosphate salts. The advanced tooth–biomaterial interaction model gave not only an insight into the mechanisms of bond degradation, but also provided a basis to develop functional monomers for more durable tooth reconstruction.

[1]Yoshihara et al. (2010). *Acta Biomaterialia*, doi:10.1016/j.actbio.2010.03.024

GLOSSARY OF TERMS

BPO: Benzoyl peroxide
C_f: Cavity Configuration Factor
FEA: Finite element analysis
GIC: Glass ionomer cement
GRF: Gelatin-resorcinol-formaldehyde
HA: hydroxyapatite
H-bonds: Hydrogen bonds
HEMA: Hydroxy ethyl methacrylate
IPN: Inter-penetrating network
MEP: 10-methacryloyloxydecyl dihydrogen phosphate
4-MET: 4 methacrloyloxyethyl trimellitic acid
MMA: Methyl methacrylate
MMP: Matrix metalloproteinases
PEG: Polyethylene glycol
PMMA: Polymethylmethacrylate
RMGIC: Resin Modified Glass Ionomer Cement
TCP: Tricalcium phosphate
TEGDMA: Triethylene glycol dimethacrylate
THR: Total hip replacement.

BIBLIOGRAPHY

Alleyne, C. H., Jr., Cawley, C. M., Barrow, D. L., Poff, B. C., Powell, M. D., et al. (1998). Efficacy and biocompatibility of a photopolymerized synthetic, absorbable hydrogel as a dural sealant in a canine craniotomy model. *J. Neurosurg.*, **88**, 308–313.

Andrews, G. P., Laverty, T. P., & Jones, D. S. (2009). Mucoadhesive polymeric platforms for controlled drug delivery. *Eur. J. Pharm. Biopharm.*, **71**, 505–518.

Armstrong, S., Geraldeli, S., Maia, R., Raposo, L. H. A., Soares, C. J., et al. (2010). Adhesion to tooth structure: A critical review of micro bond strength test methods. *Dent. Mater.*, **26**, e50–e62.

Braga, R. R., Meira, J. B. C., Boaro, L. C. C., & Xavier, T. A. (2010). Adhesion to tooth structure: A critical review of macro test methods. *Dent. Mater.*, **26**, e38–e49.

Breschi, L., Mazzoni, A., Ruggeri, A., Cadenaro, M., Di Lenarda, R., et al. (2008). Dental adhesion review: Aging and stability of the bonded interface. *Dent. Mater.*, **24**, 90–101.

Breschi, L., Martin, P., Mazzoni, A., Nato, F., Carrilho, M., et al. (2010). Use of a specific MMP-inhibitor (galardin) for preservation of hybrid layer. *Dent. Mater.*, **26**, 571–578.

Brown, E. N., White, S. R., & Sottos, N. R. (2005a). Retardation and repair of fatigue cracks in a microcapsule toughened epoxy composite. Part I: Manual infiltration. *Composites Science and Technology*, **65**, 2466–2473.

Brown, E. N., White, S. R., & Sottos, N. R. (2005b). Retardation and repair of fatigue cracks in a microcapsule toughened epoxy composite. Part II: *In situ* self-healing. *Composites Science and Technology*, **65**, 2474–2480.

Brubaker, C. E., Kissler, H., Wang, L. -J., Kaufman, D. B., & Messersmith, P. B. (2010). Biological performance of mussel-inspired adhesive in extrahepatic islet transplantation. *Biomaterials*, **31**, 420–427.

Charnley, J. (1960). Anchorage of the femoral head prosthesis to the shaft of the femur. *J. Bone Joint Surg. (B)*, **42**, 28–30.

Charnley, J. (1970). *Acrylic Cement in Orthopedic Surgery*. Edinburgh, UK: E. S. Livingstone.

Chu, C. C. (1997). New emerging materials for wound closure. In C. C. Chu, J. A. von Fraunhofer, & H. P. Greisler (Eds.), *Wound Closure Biomaterials and Devices* (pp. 347–384). Boca Raton, FL: CRC Press.

Chu, C. C., von Fraunhofer, J. A., & Greisler, H. P. (1997). *Wound Closure Biomaterials and Devices*. Boca Raton, FL: CRC Press.

Ciapetti, G., Stea, S., Cenni, E., Sudanese, A., Marraro, D., et al. (1994). Toxicity of cyanoacrylates *in vitro* using extract dilution assay on cell cultures. *Biomaterials*, **15**, 92–96.

Comyn, J. (1997). *Adhesion Science*. London, UK: Royal Society of Chemistry.

Cooper, C. W., Grode, G. A., & Falb, R. D. (1972). The chemistry of cyanoacrylate adhesives. In T. Matsumoto (Ed.), *Tissue Adhesives in Surgery*. New York, NY: Medical Examination Publishing Company.

Coury, A., Hebida, P., Mao, J., Medalie, D., Barman, S., et al. (1999). *In vivo* bonding efficacy of PEG-based hydrogels to a variety of internal tissues. *Trans. Soc. Biomater.*, **25**, 45.

Dai, X., Chen, X., Yang, L., Foster, S., Coury, A. J., & Jozefiak, T. H. (2011). Free radical polymerization of poly(ethylene glycol) diacrylate macromers: Impact of macromer hydrophobicity and initiator chemistry on polymerization efficiency. *Acta Biomaterialia*, 7(5), 1965–1972.

DeVore, D. P. (1999). Photopolymerized collagen-based adhesives and surgical sealants. *Trans. Soc. Biomater.*, **25**, 161.

Donkerwolcke, M., Burny, F., & Muster, D. (1998). Tissues and bone adhesives: Historical aspects. *Biomaterials*, **19**, 1461–1466.

Eliades, G., Watts, D. C., & Eliades, T. (2005). *Dental Hard Tissues and Bonding: Interfacial Phenomena and Related Properties*. Berlin: Springer-Verlag.

Ennker, J., Ennker, I. C., Schoon, D., Schoon, H. A., Dorge, S., et al. (1994a). The impact of gelatine-resocinal-formaldehyde glue on aortic tissue: A histomorphologic examination. *J. Vasc. Surg.*, **20**, 34–43.

Ennker, I. C., Ennker, J., Schoon, D., Schoon, H. A., Rimpler, M., et al. (1994b). Formaldehyde free collagen glue in experimental lung gluing. *Ann. Thorac. Surg.*, **57**, 1622–1627.

Falb, R. D., & Cooper, C. W. (1966). Adhesives in surgery. *New Scientist*, 308–309.

Feldman, D. S., Barker, T. H., Blum, B. E., Kilpadi, D. V., & Reddon, R. A. (1999). Fibrin as a tissue adhesive and scaffold for meshed skin grafts in burn patients. *Trans. Soc. Biomater.*, **25**, 42.

Gilbert, J. L., Hasenwinkel, J. M., Wixson, R. L., & Lautenschlager, E. P. (2000). A theoretical and experimental analysis of polymerization shrinkage of bone cement: A potential major source of porosity. *J. Biomed Mater. Res.*, **52**, 210–218.

Green, K. (1996). Mussel adhesive protein. In D. H. Sierra, & R. Salz (Eds.), *Surgical Adhesives and Sealants* (pp. 19–28). Lancaster, PA: Technomic Publishing.

Ikada, Y. (1997). Tissue adhesives. In C. C. Chu, J. A. von Fraunhofer, & H. P. Greisler (Eds.), *Wound Closure Biomaterials and Devices* (pp. 317–346). Boca Raton, FL: CRC Press.

Imazato, S. (2003). Antibacterial properties of resin composites and dentin bonding systems. *Dent. Mater.*, **19**, 449–457.

Imazato, S., Ehara, A., Torii, M., & Ebisu, S. (1998). Antibacterial activity of dentine primer containing MDPB after curing. *J. Dent.*, **26**, 267–271.

Imazato, S., Ebi, N., Tarumi, H., Russell, R. R.B., Kaneko, T., et al. (1999). Bactericidal activity and cytotoxicity of antibacterial monomer MDPB. *Biomaterials*, **20**, 899–903.

Kinloch, A. J. (1987). *Adhesion and Adhesives*. London, UK: Chapman and Hall.

Kühn, K.-D. (2000). *Bone Cements*. Berlin: Springer-Verlag.

Lewis, G. (2006). Injectable bone cements for use in vertebroplasty and kyphoplasty: State-of-the-art review. *J. Biomed. Mater. Res. Part B: Appl. Biomater.*, **76B**, 456–468.

Lovich, M. A., Philbrook, M., Sawyer, S., Weselcouch, E., & Edelman, E. R. (1998). Aterial heparin deposition: Role of diffusion, convection and extravascular space. *Am. J. Physiol., 275 Heart Circ. Physiol.*, **44**, H2236–H2242.

Mair, L., & Padipatvuthikul, P. (2010). Variables related to materials and preparing for bond strength testing irrespective of the test protocol. *Dent. Mater.*, **26**, e17–e23.

Marshall, S. J., Bayne, S. C., Baier, R., Tomsia, A. P., & Marshall, G. W. (2010). A review of adhesion science. *Dent. Mater.*, **26**, e11–e16.

Matinlinna, J., & Mittal, K. L. (2009). *Adhesion Aspects in Dentistry*. Leiden: Brill Academic Publishers.

Matras, H. (1972). Suture-free interfascicular nerve transplantation in animal experiments. *Wiener Medizinische Wochenschrift*, **122**, 517–523.

Mittal, K. L., & Pizzi, A. (Eds.), (1999). *Adhesion Promotion Techniques*. New York, NY: Marcel Dekker.

Moody, E. W., Levine, M. A., Rodowsky, R., & Sawhney, A. (1996). A synthetic photopolymerized biodegradable hydrogel for sealing arterial leaks. *5th World Biomater Congress*, **587**.

Moszner, N., & Salz, U. (2007). Recent developments of new components for dental adhesives and composites. *Macromol. Mater. Eng.*, **292**, 245–271.

Nakabayashi, N., & Pashley, D. H. (1998). *Hybridization of Dental Hard Tissues*. Tokyo: Quintessence Publishing Co, Ltd.

Park, Y. D., Tirelli, N., & Hubbell, J. A. (2003). Photopolymerized hyaluronic acid-based hydrogels and interpenetrating networks. *Biomaterials*, **24**, 893–900.

Pelletier, M. H., Lau, A. C. B., Smitham, P. J., Nielson, G., & Walsh, W. R. (2010). Pore distribution and material properties of bone cement cured at different temperatures. *Acta Biomaterialia*, **6**, 886–891.

Perdigão, J. (2010). Dentin bonding: Variables related to the clinical situation and the substrate treatment. *Dent. Mater.*, **26**, e24–e37.

Peters, W. J., Jackson, R. W., & Smith, D. C. (1974). Studies of the stability and toxicity of zinc polyacrylate (polycarboxylate) cements (PAZ). *J. Biomed. Mater Res.*, **8**, 53.

Pocius, A. V. (1997). *Adhesion and Adhesives Technology: An Introduction*. Munich: Hanser Publishers.

Ranger, W. R., Halpin, D., Sawhney, A. S., Lyman, M., & Locicero, J. (1997). Pneumostastis of experimental air leaks with a new photopolymerized synthetic tissue sealant. *Am. Surg.*, **63**, 788–795.

Reiter, A. (1987). Induction of sarcomas by the tissue-binding substance Histoacryl-blau in the rat. *Z. fiir Exper. Chirg. Transpl. Kunsliche Organe*, **20**, 55–60.

Roulet, J. -F., & Degrange, M. (Eds.), (2000). *Adhesion: The Silent Revolution in Dentistry*. Chicago, IL: Quintessence Publishing Company, Inc.

Sawhney, A. S., Pathak, C. P., & Hubbell, J. A. (1993). Bioerodible hydrogels based on photopolymerized polyethylene glycol)-copoly(x-hydroxy acid) diacrylate macromers. *Macromolecules*, **26**, 581–587.

Scardino, M. S., Swain, S. F., Morse, G. S., Sartin, E. A., Wright, J. C., et al. (1999). Evaluation of fibrin sealants in cutaneous wound closure. *J. Biomed. Mater. Res. (Appl. Biomater.)*, **48**, 315–321.

Scherrer, S. S., Cesar, P. F., & Swain, M. V. (2010). Direct comparison of the bond strength results of the different test methods: A critical literature review. *Dent. Mater.*, **26**, e78–e93.

Schlag, G., & Redl, H. (1987). Fibrin Sealant in Operative Medicine. *Plastic Surgery, Maxillo Facial and Dental Surgery*. (Vol. 4). Berlin: SpringerVerlag.

Sierra, D. H., & Salz, R. (Eds.), (1996). *Surgical Adhesives and Sealants*. Lancaster, PA: Technomic Publishing.

Silikas, N., Al-Kheraif, A., & Watts, D. C. (2005). Influence of P/L ratio and peroxide/amine concentrations on shrinkage-strain kinetics during setting of PMMA/MMA biomaterial formulations. *Biomaterials*, **26**(2), 197–204.

Smith, D. C. (1967). A new dental cement. *Brit. Dent. J.*, **123**, 540–541.

Smith, D. C. (1971). Medical and dental applications of cements. *J. Biomed. Mater. Res. Symp.*, **1**, 189–205.

Smith, D. C. (1991). Dental cements. *Curr. Opin. Dent.*, **1**, 228–234.

Smith, D. C. (1998). Development of glass-ionomer cement systems. *Biomaterials*, **19**, 467–478.

Söderholm, K.-J. (2010). Review of the fracture toughness approach. *Dent. Mater.*, **26**, e63–e77.

Spotnitz, W. D. (Ed.), (1998). *Topical issue on clinical use of fibrin sealants and other tissue adhesives. J. Long Term Effects Med. Implants,* **8**, 81–174.

Tagami, J., Nikaido, T., Nakajima, M., & Shimada, Y. (2010). Relationship between bond strength tests and other *in vitro* phenomena. *Dent. Mater.*, **26**, e94–e99.

Tanaka, K., Tadamoto, S., Ohtsuda, T., & Kotsuda, Y. (1999). Advaseal for acute aortic dissection: Experimental study. *Eur. J. Cardio-Thoracic. Surg.*, **15**, 114–115.

Tirelli, N., Lutolf, M. P., Napoli, A., & Hubbell, J. A. (2002). Poly(ethylene glycol) block copolymers. *Reviews Mol. Biotech.*, **90**, 3–15.

Trask, R. S., Williams, H. R., & Bond, I. P. (2007). Self-healing polymer composites: Mimicking nature to enhance performance. *Bioinspiration and Biomimetics*, **2**(1), 1–9.

Van Meerbeek, B., Peumans, M., Poitevin, A., Mine, A., Van Ende, A., et al. (2010). Relationship between bond-strength tests and clinical outcomes. *Dent. Mater.*, **26**, e100–e121.

Waite, J. H. (1989). The glue protein of ribbed mussels (*Genkenska denissa*): A natural adhesive with some features of collagen. *J. Comp. Physiol., MI* **159**(5), 517–525.

Watts, D. C. (2005). Reaction kinetics and mechanics in photopolymerized networks. *Dent. Mater.*, **21**(1), 27–35.

Watts, D. C., Schneider, L. F. J., & Marghalani, H. Y. (2009). Bond-disruptive stresses generated by composite polymerization in dental cavities. *J. Adhesion Sci. Technol.*, **23**, 1023–1042.

Webb, J. C.J., & Spencer, R. F. (2007). The role of polymethylmethacrylate bone cement in modern orthopedic surgery. *J. Bone Joint Surg.*, **89-B**(7), 851–857.

Weber, S. C., & Chapman, M. W. (1984). Adhesives in orthopedic surgery. *Clin. Orthop. Relat. Res.*, **191**, 249–261.

Weinmann, W., Thalacker, C., & Guggenberger, R. (2005). Siloranes in dental composites. *Dent. Mater.*, **21**(1), 68–74.

West, J. L., Chowdhury, S. M., Sawhney, A. S., Pathak, C. P., Dunn, R. C., et al. (1996). Efficacy of adhesion barriers. *J. Reproduct. Med.*, **1**, 149–154.

Yoshihara, K., Yoshida, Y., Nagaoka, N., Fukegawa, D., Hayakawa, S., Mine, A., Nakamura, M., Minagi, S., Osaka, A., Suzuki, K., & Van Meerbeek, B. (2010). Nano-controlled molecular interaction at adhesive interfaces for hard tissue reconstruction. *Acta Biomaterialia*, **6**(9), 3573–3582. DOI: 10.1016/j.actbio.2010.03.024.

CHAPTER II.5.9 OPHTHALMOLOGIC APPLICATIONS: INTRODUCTION

Roger Steinert and Rakhi Jain
Implant R&D, Abbott Medical Optics, Inc., Santa Ana, CA, USA

OVERVIEW OF EYE ANATOMY

The function of the eye is to provide sight. Briefly, images are seen when reflected or emitted light rays from an object are focused by the cornea and lens, and converted by photochemical receptors in the retina to electrical nerve impulses, which are transmitted to the visual cortex of the brain where complex, multi-layer neuronal processing creates the perception of an image. The nerve impulses from the retina are not pixel representation, but rather a complex set of data including shape, movement, and on-off encoded information that is assembled in the visual cortex.

The eye is approximately a sphere, commonly ranging in diameter from approximately 23 to 26 mm. The axial length of the globe is the length from the apex of the cornea to the retina at the macula. The eye is contained in the orbit or eye socket, within the skull. In the orbit, the eyeball is attached to six muscles for movement and surrounded by a cushion of fat. Tenon's capsule is a layer of connective tissue that encapsulates the globe and the muscles, extending to the back of the orbit. The optic nerve carries several million nerve fibers from their cell bodies in the retina to the first connection point in the brain, known as the lateral geniculate body. The outer protective tissue of the eye globe is the sclera, the "white," opaque collagen wall of the eye. The thickness of the sclera varies with the location of the globe and with age, typically ranging from 0.4 mm to 1.3 mm. The sclera is a connective tissue that protects the globe, while maintaining its shape. A schematic diagram of the cross-section of the globe is shown in Figure II.5.9.1, and ocular dimensions are summarized in Table II.5.9.1.

Light enters the eye through the dome shaped cornea that functions both as the window of the eye and the dominant lens. The cornea is the only transparent tissue in nature. The corneal diameter typically ranges from 10.5 to 12.5 mm. The anterior surface of the cornea provides the majority of the refractive component of the eye. The refractive index of the cornea is approximately 1.34. The anterior surface of cornea contributes approximately 48 diopters (D) of positive power towards the convergence of a given image on the retina. (A diopter is a unit of measurement that describes the refractive power of a lens, defined as the reciprocal of the focal length (f) of the lens in meters ($D = 1/f$)).

The cornea is comprised of five structurally distinct layers. From the anterior surface of the cornea to the posterior surface, the layers are epithelium, Bowman's layer, stroma, Descemet's membrane, and endothelium. The cornea is typically 500–600 microns thick, and is non-vascularized. The outermost layer of the cornea, the epithelium, is 5–6 cell layers of about 50 microns thickness. Bowman's layer is 10–15 microns of acellular collagen, separating the corneal epithelium from the cellular stroma. The stroma comprises approximately 90% of the corneal thickness. It is composed of extracellular collagen fibers, keratocyte cells that produce and repair the collagen, and glycosaminoglycans, complex chemicals interspersed between the collagen fibers. Descemet's membrane is the basement membrane of the corneal endothelium that lines the inner surface of the cornea. The corneal endothelium is a single layer of cells. In humans, the corneal endothelium does not regenerate. The distance from the posterior surface of the cornea to the anterior surface of the natural crystalline lens is referred to as the anterior chamber depth.

The endothelium is a physiologically leaky layer that allows water, ions, and glucose to pass from the aqueous humor of the anterior chamber into the stroma. The endothelium has membrane pumps that then eject the water molecules back into the anterior chamber, leaving

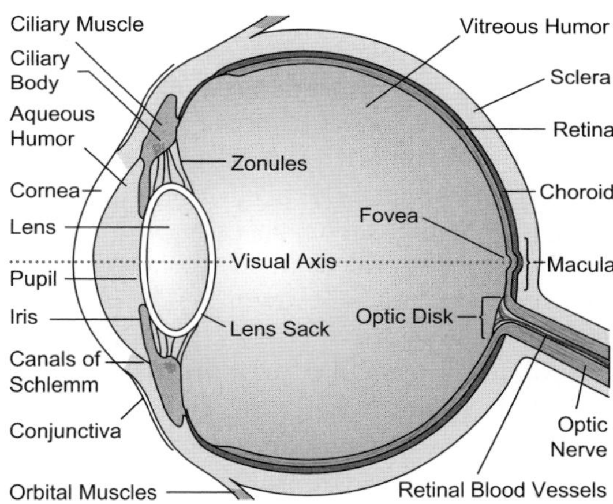

FIGURE II.5.9.1 Schematic representation of the cross-section of the eye (www.99main.com/~charlief/Blindness.htm).

TABLE II.5.9.1 Approximate Dimensions of Globe and Ocular Structures

Measurement	Length, Thickness or Diameter (mm)
Axial length of globe	23–26
Corneal diameter (white-to-white)	10.5–12.5
Equatorial diameter	23–26
Anterior chamber depth (ACD)	3–4
Anterior chamber diameter	12.0–12.5
Sulcus diameter	11.0–11.7
Natural crystalline lens diameter	9.0–10.2
Natural crystalline lens thickness	3.5–4.5
Capsular bag diameter after lens extraction	9.2–10.5
Capsular thickness	0.004–0.014
Retinal thickness	0.2–0.25

behind the nutrients necessary for the keratocytes to maintain corneal health. This mechanism is necessary to maintain the physiology of the cornea in the absence of the blood supply that nourishes all other tissues in the body. The transparency of the cornea is due to: (1) the highly organized and tightly packed collagen structure of the stroma; (2) the relative dehydration of the cornea due to the endothelial pump, preventing osmotic swelling of the stroma leading to light scattering; (3) the presence of the glycosaminoglycans, reducing light scattering by occupying spaces between collagen fibers; and (4) the lack of blood vessels.

The iris is a pigmented, vascularized tissue. The iris is responsible for controlling the amount of light that reaches the retina by varying the pupillary opening. The iris separates the anterior chamber from the posterior chamber. The ciliary body is also a pigmented, vascularized tissue. The ciliary body consists of the ciliary muscle and the ciliary processes. The ciliary body has two primary functions: (1) production of aqueous humor; and (2) accommodation, which will be discussed later. The aqueous humor is a clear fluid that fills the anterior chamber. The aqueous humor is produced by the ciliary processes in the posterior chamber of the eye and provides nutrients to the crystalline lens and cornea, as both structures do not have direct blood supplies. The composition of aqueous humor is mostly water and similar to blood plasma. The refractive index is approximately 1.34. The aqueous humor travels around the lens, through the pupil, bathes the back of the cornea, and drains into Schlemm's canal through the trabecular meshwork. The flow rate of aqueous humor is approximately 2.4 μL/min. In a simplistic view, the intraocular pressure (IOP) of the eye is maintained by the balance of the rate of aqueous humor production by the ciliary processes, and the rate of drainage through the trabecular meshwork. Normal IOPs are usually 10–20 mmHg.

The natural crystalline lens is in the posterior chamber of the eye, and resides behind the iris and pupil. The crystalline lens is connected to the ciliary body by fine suspending ligaments known as zonules. The natural crystalline lens, ciliary body, and zonules are critical structures in accommodation. Accommodation is the ability to change focus from distant to near objects. As a person's age increases, the ability of the natural crystalline lens to accommodate decreases. Hermann V. Helmholtz described the accommodative mechanism in 1855, and his theory is generally accepted today, although other theories of accommodation persist. Briefly, when the ciliary muscle contracts, the zonular fibers relax their natural state of tension on the periphery of the lens and this allows accommodation. As is well-known from geometry, the shape with the maximal volume for a given external surface area is a sphere. When zonular tension is reduced, the ovoid shape of the lens naturally moves toward a sphere, creating a greater curvature of the anterior and posterior surfaces of the lens. The lens power is therefore increased, allowing for the focus of objects closer to the eye. This will be discussed in greater detail in Chapter II.5.9B.

The natural crystalline lens is encompassed by a thin collagen membrane referred to as the lens capsule (Figure II.5.9.2). The capsule is a thin, elastic basement membrane. The thickness of the capsule varies by location. Beneath the anterior capsule and at the equator of the lens, there is a single layer of lens epithelial cells. In the natural non-cataractous lens, no cells are located on the posterior surface of the lens. The center of the lens is comprised of the nucleus, which is surrounded by lens fiber cells. Lens fiber cells synthesize proteins, crystallins, which are responsible for the refractive index of the lens. The refractive index of lens is approximately 1.42.

The retina covers approximately two-thirds of the inner wall of the posterior surface of the eye (Figure II.5.9.3). The innermost layer of the retina is in contact with the vitreous body. The retinal vasculature provides oxygen and nutrients to the inner layer of the retina. The choroid, the outermost vascular layer behind the retina, supplies oxygen and nutrients to the retinal pigment epithelium and the photoreceptors. The layers of the retina and choroid are shown in Figure II.5.9.3. There are two types of photoreceptors next to the retinal pigment epithelium: rods and cones. Cones are concentrated in the macula, the central region of the retina, and rods dominate in the peripheral region of the retina. Cones are responsible for central vision and daytime (photopic) vision, whereas rods are responsible for peripheral and dim light (scotopic) vision.

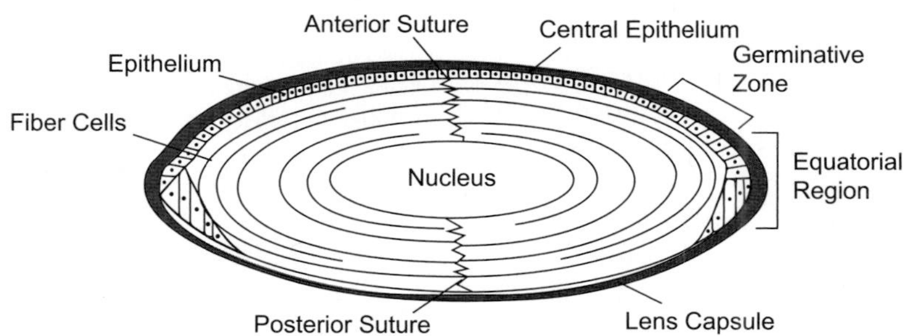

FIGURE II.5.9.2 Schematic representation of the natural crystalline lens. (Bhat et al., 2001.)

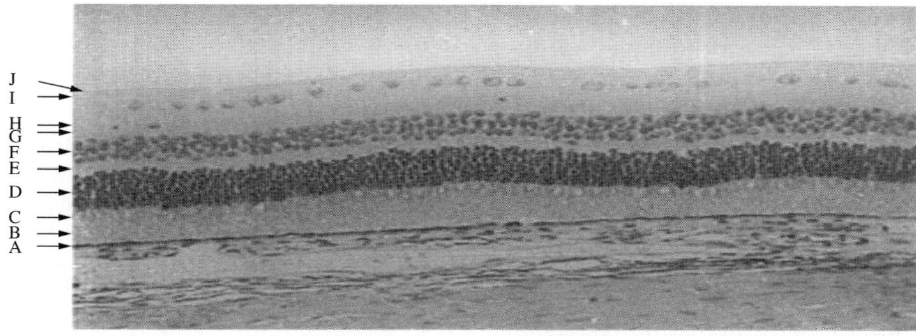

FIGURE II.5.9.3 Histological section of retina and choroid (Colthurst et al., 2000) demonstrating the 10 layers from outermost layer to inner layer: (A) retinal pigment epithelium; (B) outer and (C) inner segments of photoreceptors; (D) outer nuclear layer; (E) outer plexiform layer; (F) inner nuclear layer; (G) inner plexiform layer; (H) ganglion cell layer; (I) nerve fiber layer; and (J) internal limiting membrane.

EYE-RELATED CONDITIONS AND STATISTICS

In the United States, "legal" blindness is defined as the best-corrected visual acuity of 6/60 or 20/200 or worse in the better-seeing eye, and vision impairment (low vision) is defined as the best-corrected visual acuity less than 6/12 or <20/40 in the better-seeing eye (The Eye Diseases Prevalence Research Group, 2004).

In the US, Prevent Blindness America and the National Eye Institute estimate that approximately 2.4 million Americans over the age of 40 are visually impaired, and approximately 1 million are blind (Table II.5.9.2). The World Health Organization (WHO) estimates that approximately 269 million people live with low vision, and 45 million people live with blindness worldwide.

In the US, the leading causes of vision impairment and blindness are primarily age-related diseases, such as cataract, glaucoma, age-related macular degeneration, and diabetic retinopathy. In the US, the annual cost of the ocular problems in adults age 40 or older is approximately $51 billion (Prevent Blindness America, 2002). The most common ocular disorders are refractive errors: myopia; hyperopia; astigmatism; and presbyopia. In a normal eye (emmetrope), light is focused directly on the retina to produce an image. If the light is focused in front or behind the retina, then either an object at distance or a near object will be blurry or out-of-focus. An individual with 20/20 vision can see near and distance objects clearly. Most refractive errors can be corrected with spectacles or contact lenses. Contact lenses will be discussed in detail in Chapter II.5.9.A.

Myopia, commonly referred to as nearsightedness, is an ocular disorder in which one can focus well on near objects, but cannot focus on objects in the distance. An object in the distance will appear out-of-focus or blurry. In myopic eyes, the image is focused in front of the retina, as opposed to directly on the retina. The axial length of a myopic eye is generally longer than that of an emmetrope. Hyperopia is referred to as farsightedness, as one cannot focus on near objects, but can focus on objects in the distance. That is, in hyperopic eyes, the image is formed behind the retina. A schematic of image formation in myopic and hyperopic eyes is shown in Figure II.5.9.4. Astigmatism is an ocular disorder that describes a change in the curvature of the cornea, such that the cornea is no longer spherical. Astigmatism affects approximately 30% of Americans. Presbyopia occurs as a result of a loss of accommodation. Presybyopic individuals with good distance vision, whether naturally or with glasses, contact lenses or refractive surgery, have difficulty in focusing on near objects. Presbyopia typically becomes symptomatic at age 40–45.

TABLE II.5.9.2 US Statistics on Ocular Conditions

Ocular Disease	Number of Americans Affected
Vision impairment/low vision	2.4 million (over age 40)
Blindness	1 million (over age 40)
Refractive error:	
Myopia (> 1D)	30.5 million (over age 40)
Hyperopia (> 3 D)	12.0 million (over age 40)
Astigmatism	Affects 30% of population
Presbyopia	Affects almost all over age 50
Cataracts	21 million (over age 40)
Primary open-angle glaucoma	2.2 million (over age 40)
Age-related macular degeneration (AMD)	1.6 million (over age 50) (late AMD)
Diabetic retinopathy	5.3 milllion (over age 18)

Source: Prevent Blindness America, 2002.

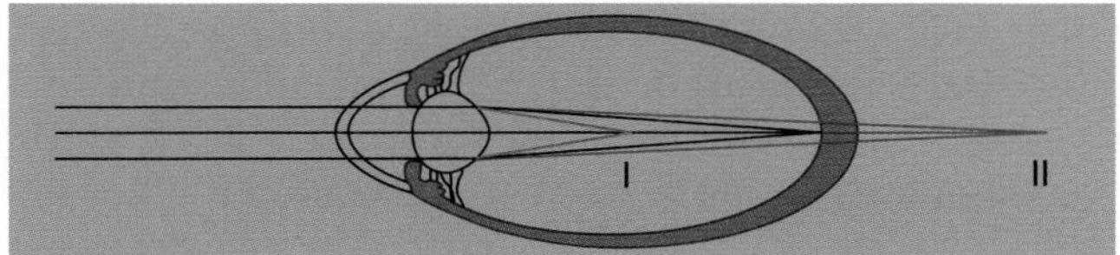

FIGURE II.5.9.4 Schematic of formation of images in myopes and hyperopes as compared to emmotropes. (I) The image is formed in front of the retina in myopia. (II) The image in formed behind the retina in hyperopia. (Kohnen et al., 2008.)

Cataract is the progressive clouding of the natural lens with age. Cataracts can also be due to trauma or injury, congenital defect, or can result from other underlying pathologies or medication, such as long-term steroid use. Cataracts are the leading cause of low vision in the US, and are highly treatable. Cataracts affect approximately 20.5 million people in the US over the age of 40 years. In the US, approximately 2 million cataract surgeries are performed each year. Cataract surgery is one of the most successful surgical procedures, as the success rate is above 90%. In cataract surgery, the natural crystalline lens of the eye is replaced by an artificial intraocular lens (IOL), referred to as a "pseudophakic IOL."

In cataract surgery, the natural crystalline lens of the eye is usually extracted with a procedure known as phacoemulsification. Ultrasound energy is delivered to the cataractous lens to break it up into small pieces that can be aspirated out. There are many types of IOLs: monofocal, multifocal; toric; and accommodating. Certain IOLs can also be placed in the eye without removal of the lens; these are referred to as phakic IOLs. Phakic IOLs correct refractive errors for myopes or hyperopes, while leaving the natural lens intact in individuals with a clear lens. Intraocular lenses are discussed in greater detail in Chapter II.5.9B.

Primary open angle glaucoma is the most common form of glaucoma. Other forms of glaucoma include angle-closure, secondary, and congenital. Glaucoma is typically characterized by elevated intraocular pressure (IOP), changes in the optic nerve head (increased cupping), and/or visual field loss. Vision loss is due to the death of nerve cells. The vision lost from glaucoma cannot be restored. The clinical strategy to treat glaucoma usually begins with a topical pharmaceutical treatment to lower IOP. If IOP is not managed with medication, either laser treatment or surgery to implant a glaucoma shunt or stent is performed. Two types of laser treatment are argon laser trabeculoplasty (ALT) and selective laser trabeculoplasty (SLT). Both laser treatments use a laser to create space in the trabecular meshwork to increase aqueous humor outflow. A glaucoma or aqueous shunt is comprised of a plate and a tube with or without a valve. The tube is placed in the anterior chamber, and the plate resides in the sub-conjunctival space, where the aqueous humor flows out through a sieve into the tissues where it is absorbed. The shunt also aids in aqueous humor outflow by creating another outflow path. Glaucoma treatments are discussed in Chapter II.5.9.D.

There are two types of age-related macular degeneration (AMD): dry AMD and wet AMD. Dry AMD accounts for approximately 90% of all cases and is the milder form of AMD. It develops gradually over time and usually causes mild to moderate loss of vision. Often small yellow deposits, called drusen, accumulate in the retinal pigmented epithelium (RPE) cells adjacent to the macula. The dry form may eventually develop into the wet form of AMD.

While wet age-related macular degeneration accounts for approximately 10% of all cases, it is responsible for approximately 90% of cases of severe vision loss associated with AMD. Wet AMD involves the growth of abnormal blood vessels, known as choroidal neovascularization (CNV). These abnormal blood vessels invade from the choroid under the macula, hemorrhage, and lead to major cellular damage. This causes very sudden loss of central vision. Bleeding is followed by the formation of scar tissue that leads to vision loss.

Diabetic retinopathy (DR) also affects the retina and is associated with both diabetes type 1 (juvenile onset diabetes) or type 2 (adult onset diabetes). Diabetic retinopathy affects approximately 5.3 million Americans. There are two types of diabetic retinopathy: non-proliferative or proliferative. Non-proliferative DR is less severe. Blurred vision is a result of blood vessels leaking edema fluid into the retina. Proliferative DR is usually more severe. Neovascularization (formation of new blood vessels) occurs on the surface of the retina, and sometimes reaches up into the vitreous. The new vessels are delicate and ultimately hemorrhage, resulting in profound vision loss due to blurring as a result of the blood cells. The blood in the vitreous often stimulates a fibrotic scar reaction in the vitreous which pulls on the retina, ultimately leading to traction retinal detachment. The current approved treatments for AMD and DR are pharmaceutical injections into the vitreous, and laser treatments of the hypoxic retina and/or the abnormal vessels. New implants and drug delivery systems are in development, as discussed in Chapter II.5.9.E.

Some Special Consideration for Ophthalmic Biomaterials and Commonly Used Materials

Many considerations for ophthalmic biomaterials are similar to the considerations in other disciplines, such as biocompatibility, mechanical properties and stability, material degradation, manufacturability, and implant design optimization. Some common ophthalmic biomaterials are listed in Table II.5.9.3.

Unique to ophthalmology, the functionality of ocular biomaterials is often based on the optical properties of the material. The optical properties are critical for contact lenses, artificial corneas, corneal inlays or onlays, as well as intraocular lenses. The materials for these devices must be transparent and have a refractive index equal to or greater than that of the tissue it is replacing. These materials often incorporate an ultraviolet (UV) blocker to protect the retina from UV light. Additionally, the stability of the material in the eye, lack of toxicity, and long-term functionality of the product or device over time is critical. Optical devices need to be mechanically stable, because any tilt or decentration of these optical medical devices will lead to unfavorable visual performance. As the ophthalmic industry is moving to smaller surgical

TABLE II.5.9.3 Biomaterials Commonly Used in Ophthalmology

Ophthalmic Implants	Materials Commonly Used
Contact lenses	Poly(methyl methacrylate) (PMMA), 2-hydroxyethyl methacrylate (HEMA) copolymers, silicone hydrogels
Inlays or onlays	Hydrogels, collagen, permeable membranes
Intraocular lenses	Optic: PMMA, hydrophobic acrylic, silicone, hydrophilic acrylic Haptic: polypropylene, PMMA, polyimide, polyvinylidene fluoride (PVDF)
Ophthalmic viscosurgical device (OVD)	Chrondroitin sulfate, sodium hyaluronate, hyaluronic acid, hydroxypropyl methylcellulose (HPMS), polyacrylamide, collagen, or combinations of these materials
Glaucoma shunts	Plates: silicone (impregnated with barium), polypropylene Tubing: silicone
Vitreous replacements	Silicone oil, gases

incisions, the materials for ophthalmic implants must be able to recover their shape after folding and unfolding, with no affect on the optical properties. This is clearly demonstrated with intraocular lenses.

Any material used for a corneal application, such as contact lenses or corneal inlays, must have acceptable oxygen and nutrient permeability. As the cornea is avascular tissue, the metabolic needs of the cornea depend on permeability to oxygen. For contact lenses, the surface wettability is important for a tear film for patient comfort. As contact lenses are implants exposed to proteins and lipid deposits in the tear film, the material must allow removal and/or be resistant to these deposits.

When designing a new ocular material or modifying an existing material, special attention should be given to the relevant standards. For example, for intraocular lenses, the biocompatibility of the material must meet the guidelines in ISO 10993-1: 2003, Biological Evaluation of Medical Devices. Part 1: Evaluation and testing, and ISO 11979-5(E): 2006, Ophthalmic Implants: Intraocular Lenses. Part 5: Biocompatibility.

BIBLIOGRAPHY

Bhat, S. P. (2001). The ocular lens epithelium. *Bioscience Reports*, **21**(4), 537–563.

Colthurst, M. J., Williams, R. L., Hiscott, P. S., & Grierson, I. (2000). Biomaterials used in the posterior segment of the eye. *Biomaterials*, **21**, 649–665.

ISO 10993-1. (2003). *Biological Evaluation of Medical Devices. Part 1: Evaluation and testing*.

ISO 11979-5(E). (2006). *Ophthalmic Implants: Intraocular lenses. Part 5: Biocompatibility*.

Kohnen, T., Strenger, A., & Klaproth, O. K. (2008). Basic knowledge of refractive surgery: Correction of refractive errors using modern surgical procedures. *Dtsch Arztebl Int.*, **105**(9), 163–172.

Prevent Blindness America and National Eye Institute. (2002). *Vision Problems in the U.S.: Prevalence of Adult Vision Impairment and Age-Related Eye Disease in America*.

Prevent Blindness America. (2007). *The Economic Impact of Vision Problems*.

The Eye Diseases Prevalence Research Group (2004). Causes and prevalence of visual impairment among adults in the United States. *Arch. Ophthalmol.*, **122**, 477–485.

World Health Organization. (May 2009). *Visual Impairment and Blindnesss (Fact Sheet N 282)*. (http://www.who.int/mediacentre/factsheets/fs282/en/).

A. BIOMATERIALS: CONTACT LENSES

Jean Jacob

LSU Eye Center, New Orleans, LA, USA

INTRODUCTION

Variations in eye length can disrupt the ability of the refractive elements of the eye, cornea, and interior lens to focus the light image correctly on the retina, resulting in blurred vision. In order to correct such abnormalities, lenses such as spectacles and contact lenses are added to the front of the cornea to adjust the light refraction. Contact lenses which are placed directly upon the cornea can be made of various biomaterials. Additionally, since contact lenses are in intimate contact with living tissues, they are subject to the same federal regulations that govern the use of implant materials. Currently over 128 million people use contact lenses worldwide, including between 35 and 38 million in the US (Morgan et al., 2010). People choose to wear contact lenses for many reasons, often due to the practicality of the lenses or their personal appearance. When compared to spectacles, contact lenses are less affected by wet weather, do no steam up, and provide a wider field of vision. They are also more suitable for a number of sporting activities. Additionally, some corneal conditions such as keratoconus and aniseikonia can be more accurately corrected with contact lenses than with spectacles. The materials used in contact lenses will be reviewed herein. The materials used in spectacle lenses are outside the scope of this chapter.

GENERAL PROPERTIES AND CORNEAL REQUIREMENTS

Materials used for contact lenses must meet a variety of specific physical requirements in terms of optical, mechanical, permeation, and surface properties. Since contact lenses are optical devices, 100% transmission of visible light is a key factor. In addition to transmitting light, lens materials must be chemically stable and wettable, with high oxygen permeability and low or competitive manufacturing cost.

Since the contact lens seats in the tear film while residing against the cornea, the lens material needs have a wettable surface which allows the tear film to maintain its normal structure, both above and below the lens. Additionally, the lens material surface structure should be such that adsorption of the proteins and lipids from the tear film, biofouling, is minimized and allows any adsorbed biological factors to be removed with standard cleaning solutions. Biofouling is the process of uncontrolled accumulation of biological material, such as proteins and lipids, onto the surface of a material in contact with a biological media (Kingshott et al., 2000) (see Chapters I.2.10 and II.2.8). A common reaction to materials placed in biological environments, biofouling is especially important with regard to contact lenses, as protein and lipid deposition decreases optical quality, wear time, comfort, and even permeability. Hydrophilic surfaces, in general, are more resistant to biofouling than hydrophobic surfaces.

Conventional hydrogel materials have inherent wettability when fully hydrated. However, on the eye, two other polymer/thermodynamic factors influence material surface performance, anterior surface dehydration, and polymer chain rearrangement. First, the anterior surface of the lens can progressively lose water, especially under adverse environmental conditions such as low humidity, high wind, and high temperatures. Second, hydrated polymer chains are capable of rotating in response to changes at their surface interface, depending on their cross-link density and side-group size. When in contact with aqueous fluid such as the tear film, hydrophilic groups rotate to the surface. Conversely, when the environment becomes more hydrophobic, such as with air or tear lipid exposure during eyelid blink, the hydrophilic groups can be thermodynamically driven into the material bulk, and a more hydrophobic material surface is exposed. Both of these factors also contribute to the degree of biofouling and final material-on-eye performance.

The cornea requires oxygen to maintain its functional metabolism and transparency. Since the cornea is avascular, it receives oxygen from the ambient air during open eye conditions and the palpebral conjunctival vessels during closed eye conditions. Oxygen from both of these sources diffuses through the tear layer to the cornea. When worn, contact lenses become barriers to the transmission of oxygen to the cornea. Two factors are critical to decreasing the barrier effects of the lens: (1) transmission of oxygen through the lens; and (2) tear exchange under the lens.

The oxygen permeability of a lens material is represented by the term Dk. D is the diffusivity in cm^2/sec and k is the Henry's law solubility coefficient in cm^3 (O_2STP)/(cm^3 (polymer) mmHg). The lens material oxygen transmissibility (Dk/L) is a better representation of the ability of oxygen to cross the lens, shown by the oxygen permeability coefficient of the material divided by the average thickness of the lens (L) (Holden et al., 1990). In general, though, the literature and contact lens companies report lens oxygen permeability in terms of Dk, the units of which are called barrers (Stern, 1968). Oxygen transmissibility, Dk/t, is reported in barrers/mm.

The cornea is exposed maximally to the mean partial pressure O_2 in ambient air (about 160 mmHg at sea level) and palpebral conjunctiva vessels (approximately 55 mmHg) when the eye is open and closed, respectively. The ideal contact lens would allow complete passage of the available oxygen to the cornea. Lack of oxygen to the cornea, or hypoxia, is characterized by corneal swelling (edema), and can result in a number of other physiological conditions such as limbal hyperemia, neovacularization, corneal acidosis, epithelial keratitis, and endothelial phylemgethism. Holden and Mertz determined the critical lens oxygen transmissibility to avoid corneal edema for daily lens wear to be $24.1 \pm 2.7 \times 10^{-9}$ (cm × ml O_2)/(sec × ml × mmHg) and $87.0 \pm 3.3 \times 10^{-9}$ (cm × ml O_2)/(sec × ml × mmHg) for overnight wear (Holden and Mertz, 1984). More recent experiments and computations, taking into account additional factors, have suggested the critical oxygen transmissibility needed for corneal health might be higher still (Harvitt and Bonanno, 1999).

CONTACT LENS MATERIALS

In general, all current contact lens materials consist of carbon–carbon backbone polymeric chains with pendant groups which impart function to the material. Imagine the carbon–carbon chain as the "spine" of the polymer material, and the pendant chemical groups as the appendages; arms and legs. The final properties of the material are dependent on the number and type of pendant groups. Hydrophilic pendant groups (–OH, –COOH, –pyrrolidone, etc.) primarily attract and bind water, forming hydrogels. Pendant groups of silicon–oxygen increase oxygen permeability, as do fluorine groups. Fastening the spines or backbones together through arms or cross-links gives the polymer greater physical stability, and produces a three-dimensional network. These arms or cross-links, such as ethylene glycol dimethacrylate (EGDMA) or tetraethylene glycol dimethacrylate (TEGMA), in and of themselves may also contain chemical pendant groups to improve function. In addition to the monomers and cross-linkers, a lens formulation may also contain other monomers, such as n-vinyl pyrrolidone (VP), as a physical property modifier and solvent to facilitate processing (Refojo, 1973; Peppas, 1986,1987,1987a; Refojo, 1996).

There are essentially two main categories of contact lenses that are grouped by their basic mechanical properties: hard and soft. Hard contact lenses are rigid materials with a high modulus. Current rigid lenses are non-hydrogel materials, generally containing silicone or fluorine derivatives for high oxygen permeability, and commonly referred to as rigid gas permeable (RGP) lenses.

Soft contact lenses are generally hydrogel materials with a low modulus. Within the soft hydrogel lens category there are two basic types of lens material groups: (1) homogeneous hydrogels (see Chapter I.2.5); and (2) heterogeneous hydrogels which generally consist of a water-poor hydrophobic phase of siloxane and/or fluorocarbon moieties dispersed within a water-rich hydrophilic phase similar to the standard hydrogels. These heterogeneous hydrogels have highly enhanced oxygen transmissibility compared with the standard hydrogels of similar hydration.

Hard Contact Lenses

Hard contact lenses ride loosely on the cornea, and move with the eyelid blink more or less freely over the tear film that separates the lens from the corneal surface. The mechanical properties of hard contact lenses must be such that any flex on the lens induced by the blink recovers instantaneously at the end of the blink. Until approximately 1938, all contact lenses were made from glass, despite glass's impermeability to oxygen (see Box 1). These contact lenses were made wearable by fenestrations (small holes) in the peripheral areas which allowed limited oxygen and tear fluid exchange. With the development of clear plastics, after 1938, commercial lenses were of plexiglass, polymethylmethacrylate or PMMA.

Although relatively impermeable to oxygen, PMMA had the advantage of being lightweight and more durable to the lathing and molding techniques needed for manufacture, in addition to its excellent light transmission and biofouling resistance (Refojo, 1973). Because of the material durability, PMMA lenses were also able to be made smaller, so that they could move more freely on the cornea during blink, and allow increased exchange of the tear fluid and oxygen behind the lens. While oxygen transmission was now less of an issue for hard PMMA lenses, due to the high degree of tear exchange behind the lens with every blink, lens materials manufacturers moved on to develop more biocompatible materials that are gas permeable and more wettable.

BOX 1

In the late 1880s, Adolf Eugen Fick, Eugene Kalt and August Müller all produced glass scleral lenses. In 1889 Müller, myopic and frustrated with his spectacles, felt that it would be possible to place thin spectacle-like lenses directly on the eye (Pearson and Efron, 1989). He ground and fitted himself with large, thick, glass contact lenses to prove his point. Unfortunately, he was only able to wear the lenses for around a half hour before pain drove him to remove them. Müller hypothesized, correctly it turns out, that the material properties of the glass interfered with required corneal functions. We now know the glass lenses which covered the entire cornea and did not move with eyelid blinks blocked the exchange of oxygen and tear fluid behind the lens, essentially starving the cornea. A frustrated Müller ultimately abandoned his work in this area. However, the quest to develop contact lenses which could be worn comfortably and continually for hours, day, weeks, and even months had begun.

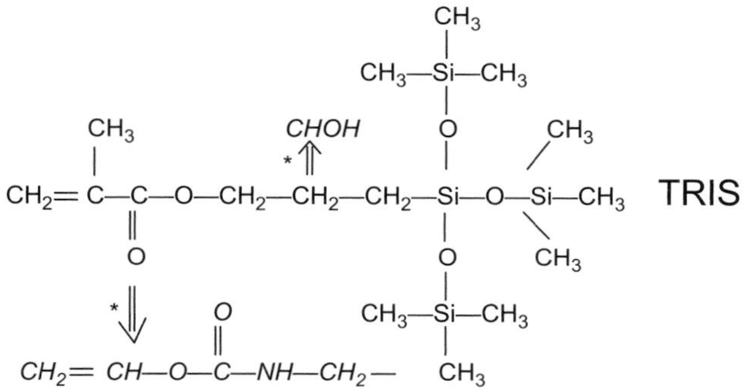

FIGURE A.1 Chemical structure of TRIS (methacryloxypropyl tris(trimethylsiloxy)silane); * hydrophilic modifications of TRIS developed for formation of silicone hydrogels. (Kunzler and Ozark, 1994; McCabe et al., 2004.)

Rigid Gas Permeable Contact Lenses

The development of copolymers of PMMA with functionalized silicone and fluorine containing macromers produced the current RGP lenses. The introduction of siloxanylalkyl methacrylates, most often methacryloxypropyl tris(trimethylsiloxy)silane (TRIS; Figure A.1), into the PMMA lens significantly increased the oxygen permeability while also increasing the hydrophobic nature of the lens. Hydrophilic co-monomers such as methacrylic acid (MAA) are commonly used in these lenses to counteract the increased hydrophobicity. For these RGP siloxanylalkyl methacrylate contact lens materials, the Dk is inversely proportional to the density of the lens (Refojo, 1996). Subtle changes in the chemical make-up of the lens polymer can strongly affect its clinical performance. Silcone is lipophilic and can have some effect on the extent of biofouling. Fluoro-derivates also improve oxygen permeability, but are more resistant to biofouling. Most currently marketed RGP lenses consist of perfluoroalkyl-siloxanylalkyl-methyl methacrylate type materials which have a high Dk and good surface properties (Table A.1; see Box 2). However, their Dk is not directly related to their density as with the original RGP lenses. Some newer RGP contact lenses made with styrene derivates have a high enough Dk to be used for overnight/extended wear (Benjamin and Cappelli, 2002). RGP lenses make up about 26% of all dispensed contact lenses (Morgan and Efron, 2008). Since hard contact lenses ride loosely on the cornea, they are good for patients with significant corneal astigmatism or nonspherical corneal surfaces. However, the rigid nature of these lenses is, at least initially, somewhat uncomfortable and requires some patient adaptation.

Soft Hydrogel Contact Lenses

Due to the low modulus of hydrogels, soft contact lenses drape over the cornea, conforming to its surface. Besides

TABLE A.1 Selected Hard and Rigid Gas Permeable Lens Materials and Their Properties

USAN/ Trade Name#	Oxygen Permeability (Barrers)	Density	Refractive Index	Polymer/Principal Monomer Types
PMMA[t]	0.5	1.20	1.49	Polymethyl methacrylate
Siflufocon A/ Silsoft[1]	126	1.13	1.44	Silicone
Itafocon A/ Boston II[1]	12	1.19	1.43	TRIS with methylmethacrylate (MMA), dimethyl itaconate, methyl acrylic acid (MAA), tetraethylene glycol dimethacrylate (TEGDMA)
Paflufocon C/ Fluoroperm 30[2]	30	1.18	1.43	TRIS with 2,2,2,-trifluoroethyl methacrylate, MAA, MMA, 1-vinyl-2-pyrrolidone (VP), EGDMA
Optifocon A/ Boston Equalens II[1]	125	1.24	1.42	Co-Siloxy-fluoromethacrylate
Paflufocon D/ Fluoroperm 151[2]	151	1.10	1.44	TRIS with 2,2,2,-trifluoroethyl methacrylate, MAA, MMA, siloxy-based polyether macromer, EGDMA
Melafocon A/ Menicon SFP[3]	159	1.12	N/A	TRIS with 2,2, 2,-trifluoroethyl methacrylate, VP, MAA, EDGMA
Tisilfocon A/ Menicon Z[3]	163-250	N/A	N/A	Co-Fluoro-siloxanylstyrene

[t]No USAN name.
#Manufacturer: [1]Bausch and Lomb; [2]Paragon Vision Science; [3]Menicon Co.

BOX 2

Lens materials have three names: (1) the scientific chemical name; (2) the USAN name; and (3) the trade name. Contact lens materials are rarely called by their scientific chemical name, but rather by their USAN name in the scientific literature. The United States Adopted Names Council (USANs), which provides the generic or nonproprietary names of all drugs and therapies marketed in the United States, also names all contact lens materials. The generic name is officially called the United States Adopted Name (USAN), and serves as an important and unique designation for the active ingredient or material. It appears with the company's trade name on labels, advertisements, and other information, and is also requested on FDA applications and in drug substance chemical descriptions. Because most journals request that authors use the generic name in clinical trial reports, it is used in publishing scientific data about investigational and established therapies/materials. The USAN is also much shorter, easier to remember, and more pronounceable than the scientific chemical name. For example, Etafilcon A is the USAN name for poly2-hydroxyethyl methacrylate-co-methacrylic acid marketed under the trade name of Acuvue by Vistakon (a division of Johnson & Johnson, Inc.).

the comfort this affords, this behavior limits the number of lens curvatures needed to fit the general population.

Standard Hydrogel Contact Lenses

The first commercialized hydrogel contact lens material, poly 2-hydroxyethylmethacrylate (PHEMA) with an equilibrium water content (EWC, percentage of water by weight) of approximately 40%, was based on work by Wichterle and Lim in 1959. The advantage of this material over other hydrogels, such as polyacrylamide, was its stability to pH, temperature, and osmolarity found under physiological conditions. Additionally, Wichterle and Lim developed a spin-casting technique for hydrogel lens production in 1961 that revolutionized the contact lens manufacturing industry (Wichterle, 1978).

A wide range of hydrophilic monomers, besides HEMA, have been investigated and developed as soft contact lens material components. N-vinyl pyrrolidone (NVP), n,n-dimethyl acrylamide (DMA), vinyl acetate (VA) (post-hydrolyzed to vinyl alcohol units), and methacrylic acid (MAA) have proved to be particularly successful (Tighe, 2007). Monomers that mimic biological molecules, such as a methacrylate of phosphorylcholine, have also been commercially successful. Table A.2 lists the some of the currently marketed contact lenses and their composition and properties.

The Dk of standard hydrogel materials increases exponentially with the water content (Refojo, 1979). Therefore, these hydrogel materials have a hypothetical limit in oxygen permeability that approaches the oxygen permeability of pure water (Dk/t of water = 125 barrer). However, high water content hydrogels (>50%), while better for increased O_2 permeability, have some disadvantages in terms of increased biofouling, dehydration, and fragility due to very low modulus.

One approach to increase the effective O_2 transmission of these materials is to decrease the lens thickness. This ultrathin modality takes advantage of the law of diffusion, which states that the gas flux through a material is inversely proportional to the thickness. When applied to contact lenses, this guarantees that the O_2 flux through the lens will double when the thickness is halved for any lens type under the same wear conditions. In general, the Dk of conventional hydrogel lenses ranges between 8 and 44 barrer.

Due to their moderate oxygen permeability, conventional hydrogel lenses are limited in terms of overnight and extended wear. Sleeping in these lenses increases the risk for adverse events more than four-fold (Sankaridurg et al., 1999; Stapleton et al., 2006). However, there is a large market demand for lenses that can be worn continuously.

TABLE A.2 Some Soft Hydrogel Contact Lens Materials and Their Properties

USAN/ Trade Name#	Equilibrium Water Content (%)	Oxygen Permeability (Barrers)	FDA Group	Principal Monomer Types
Polymacon/ Soflens 38[1]	38	9	I. Low Water (<50%) Non-Ionic Polymer	2-Hydroxyethyl methacrylate (HEMA), ethylene glycol dimethacrylate (EGDMA)
Hilafilcon B/ Softlens 59[1]	59	22	II. High Water >50%) Non-Ionic Polymer	HEMA, 1-vinyl-2-pyrrolidone (VP), EGDMA, 2-vinylethyl methacrylate (VEMA)
Bufilcon A/ Hydrocurve II 45[2]	45	12	III. Low Water <50%) Ionic Polymer	HEMA, N-(1,1-dimethyl-3-oxobutyl)acrylamide, 2-ethyl-2(hydroxymethyl)-1,3-propanediol trimethacrylate
Etafilcon A/ Acuvue[3]	58	28	IV. High Water (>50%) Ionic Polymer	HEMA, methacrylic acid (MAA), EGDMA
Ocufilcon D/ Biomedics 55[4]	55	19.7	IV. High Water (>50%) Ionic Polymer	HEMA, sodium methacrylate (Na-MA), 2-ethyl-2-(hydroxymethyl)-1,3-propanediol trimethacrylate
Vilifilcon A/ Focus Softcolors[2]	55	16	IV. High Water (>50%) Ionic Polymer	MMA, HEMA, VP, EGDMA

#Manufacturer: [1]Bausch and Lomb; [2]CIBA Vision; [3]Vistakon; [4]CooperVision.

TABLE A.3 Silicone Hydrogel Lens Materials and Their Properties

USAN/ Trade Name#	Equilibrium Water content (%)	Oxygen Permeability (Barrers)	Tensile modulus (psi)	Initial modulus (MPa)	Surface Modification	Principal Monomer Types
Balafilcon A/ PureVision[1]	36	110	148	1.1	Plasma Oxidation	VP, TPVC, NCVE, PBVC
Lotrafilcon A/ Focus Night & Day[2]	24	140	238	1.4	Plasma Coating	DMA, TRIS, siloxane monomer
Lotrafilcon B/ O_2 (Air) Optix[2]	33	110	190	1.2	Plasma Coating	DMA, TRIS, siloxane monomer
Galyfilcon A/ Acuvue Advance[3]	47	60	65	0.4	Internal Wetting Agent (PVP)	MPDMS, DMA, HEMA, EGDMA, siloxane macromer, PVP
Senofilcon A/ Acuvue Oasys[3]	38	103	92	0.6	Internal Wetting Agent (PVP)	MPDMS, DMA, HEMA, TEGDMA, siloxane monomer, PVP
Narafilcon A/ TruEye[3]	54	100	96	0.7	Internal Wetting Agent (PVP)	MPDMS, DMA, HEMA, TEGDMA, siloxane monomer, PVP
Comifilcon A/ Bioinfinity[4]	48	128	105	0.8	Undisclosed	VP, MVA, IBM, TAIC, M3U, FM0411M, HOB
Enfilcon A/ Avaira[4]	46	100	80	0.5	Undisclosed	VP, MVA, IBM, TAIC, M3U, FM0411M, HOB
Asmofilcon A/ PremiO[5]	40	129	121	0.9	Plasma Treatment	SIMA, SIA, DMA, pyrrolidone derivative

Abbreviations: VP: N-vinylpyrrolidone; TPVC: tris-(trimethyl siloxysilyl) propylvinyl carbamate; NCVE: *N*-carboxyvinyl ester; PBVC: poly(dimethysiloxy) di(silylbutanol) bis(vinyl carbamate); DMA: *N,N*-dimethylacrylamide; TRIS: methacryloxypropyl tris(trimethylsiloxy)silane; MPDMS: monofunctional polydimethylsiloxane; HEMA: hydroxyethyl methacrylate; EGDMA: ethyleneglycol dimethacrylate; PVP: polyvinyl pyrrolidone; TEGDMA: tetraethyleneglycol dimethacrylate; MVA: N-methyl-N-vinyl acetamide; IBM: isobornyl methacrylate; TAIC: 1,3,5-triallyl-1,3,5-triazine-2,4,6(1*H*,3*H*,5*H*)-trione; M3U: bis (methacryloyloxyethyl iminocarboxy ethyloxypropyl)-poly(dimethylsiloxane)-poly(trifluororpropylmethylsiloxane)-poly(methoxy-poly[ethyleneglycol] propylmethylsiloxane); FM0411M: methacryloyloxyethyl iminocarboxyethyloxypropyl-poly(dimethylsiloxy)-butyldimethylsilane; HOB: 2-hydroxybutyl methacrylate; SIMA: siloxanyl methacrylate; SIA: siloxanyl acrylate.

#Manufacturer: [1]Bausch and Lomb; [2]CIBA Vision; [3]Vistakon; [4]CooperVision; [5]Menicon.

Silicone Hydrogel Contact Lenses

In an effort to enhance the O_2 transmissibility, while maintaining the comfort of the standard hydrogel lens, the contact lens industry developed siloxane-hydrogel contact lenses with sufficiently high Dk and water permeability to allow continuous wear for over two weeks (Table A.3). These lenses are made essentially by copolymerizing hydrophilic methacrylate or vinyl macromers with methacrylate or vinyl end-capped polydimethylsiloxanes (Figure A.2), novel co-monomers, and/or hydrophilically modified tris(trimethylsiloxy)silanes (Figure A.1). By functionalizing the silicone macromers and the

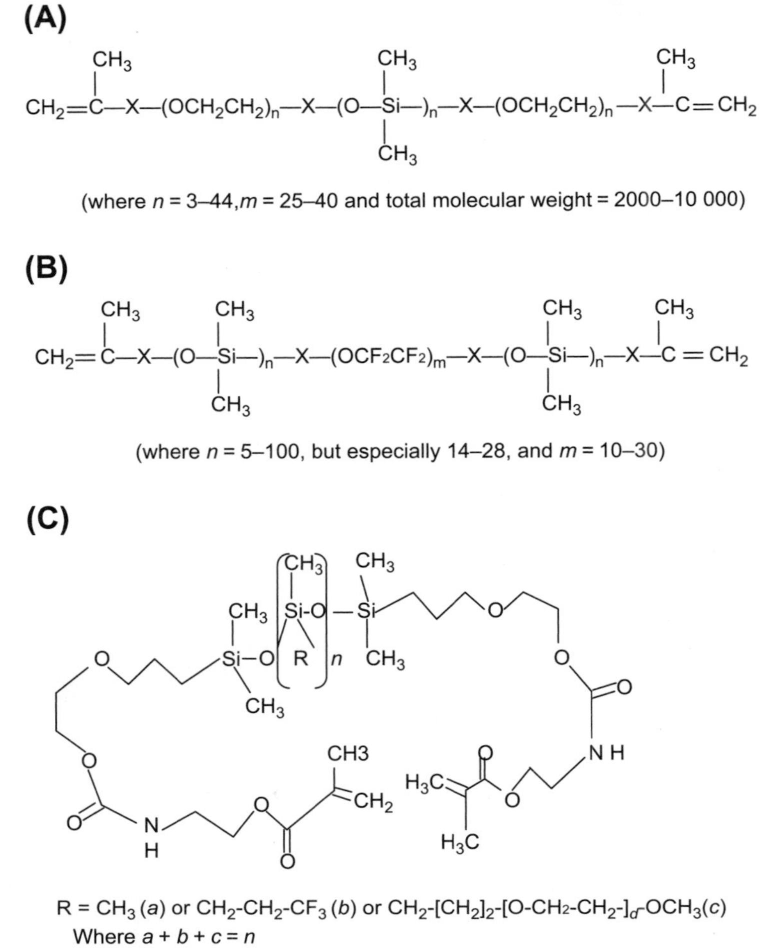

(A)

$$CH_2{=}C(CH_3){-}X{-}(OCH_2CH_2)_n{-}X{-}(O{-}Si(CH_3)_2{-})_n{-}X{-}(OCH_2CH_2)_n{-}X{-}C(CH_3){=}CH_2$$

(where $n = 3$–$44, m = 25$–40 and total molecular weight = 2000–10 000)

(B)

$$CH_2{=}C(CH_3){-}X{-}(O{-}Si(CH_3)_2{-})_n{-}X{-}(OCF_2CF_2)_m{-}X{-}(O{-}Si(CH_3)_2{-})_n{-}X{-}C(CH_3){=}CH_2$$

(where $n = 5$–100, but especially 14–28, and $m = 10$–30)

(C)

R = CH_3 (*a*) or CH_2-CH_2-CF_3 (*b*) or CH_2-$[CH_2]_2$-$[O$-CH_2-CH_2-$]_d$-OCH_3(*c*)
Where *a* + *b* + *c* = *n*

FIGURE A.2 Additional building blocks of silicone hydrogel materials: (A) and (B) silicone and fluorine macromers (Nicholson et al., 1996); and (C) novel co-monomers, Comfilcon A.

TRIS-like monomers with hydrophilic groups, the siloxane rich monomers become more compatible with the hydrophilic components of the hydrogel. The resulting lens materials have oxygen permeability values ranging from 60 to 163 barrers (Efron et al., 2010). Unlike conventional hydrogel lens materials, the relationship between Dk and the EWC is inversely proportional for silicone hydrogels (Figure A.3). In silicone hydrogels, the oxygen permeation occurs mainly through the siloxane-rich areas of the material, due to the bulkiness and chain mobility of the siloxane group ($-Si(CH_3)_2-O-$), while the water regions of the hydrogel primarily allow water and ion permeability for the lens-on-eye movement (Nicholson and Vogt, 2001; see Box 3).

Despite the increase in oxygen permeability, the first silicone hydrogels to market, balifilcon A and lotrafilcon A, and most silicone hydrogels as a group, still possess some inherent surface incompatibility. In general, silicone hydrogels have a low surface-free energy, hydrophobic and lipophilic properties, and poor wettability. Additionally, the siloxane moieties in the lens material can migrate to the lens surface, causing these moieties to accumulate on the lens portion that interact the most

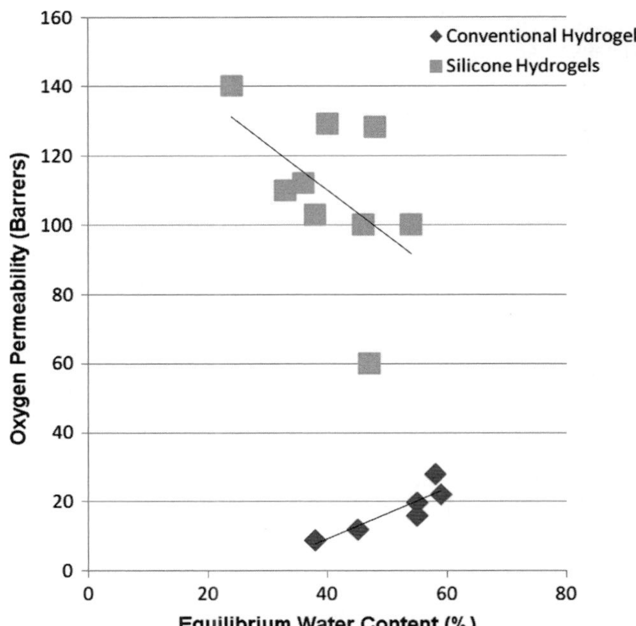

FIGURE A.3 Relationship between oxygen permeability (Dk) versus equilibrium water content (EWC) of the silicone and conventional hydrogel lenses in Tables A.1 and A.2.

BOX 3

The development of a silicone hydrogel lens was intensely pursued by the contact lens industry for over 20 years before a successful material was achieved. The early versions of lenses stuck to the cornea surface within the first hour of wear, despite having wettable surfaces. This "suction cup" type of behavior was attributed to the lack of ion permeability and water transport through, as well as around, the lens. Unlike oxygen, ions, such as Na^{++}, require a water molecule "shell" for transport. In high water content materials the difference between water and ion permeability is negligible. However, in low water situations, as in early silicone hydrogel materials, the ionic permeability becomes critical, since the tear film on both sides of the lens behaves as a dilute salt solution. The critical minimum ion permeability for lens movement was determined by Nicholson and co-workers to be 0.2×10^{-6} cm^2s^{-1}, which is very similar to the reported value for PHEMA, 0.18×10^{-6} cm^2s^{-1}. They attributed the wearability of their Lotrafilcon A material, which has a water content significantly below that of PHEMA, to a unique physicochemical biphasic silicone–water structure where the block copolymer forms micro, channel-like co-continuous areas of pure tris-silicone and pure hydrogel, which allow the preferential exchange of oxygen and ions/water through the respective channels of the lens (Nicholson et al., 1996). Silicone hydrogel materials with water contents similar to PHEMA do not need this unique biphasic structure for on eye lens movement.

with the surrounding environment. These variant surface properties cause various biocompatibility issues, such as increased biofouling on the lens, and require surface modification to increase the hydrophilicity of the surface.

SURFACE MODIFICATIONS

Conventional hydrogel materials were not thought to need any surface modification, due to their inherent hydrophilicity. However, it was found that specific bulk lens material side chain moieties could impart enhanced surface properties (see Box 4). This finding encouraged the industry to develop materials with surface active side groups, surface modifying endcaps or interpenetrating networks, which not only increase surface hydrophilicity, but also attempt to direct the tear fluid/lens surface interactions for increased biocompatibility and decreased biofouling (Portoles et al., 1993). Omafilcon A (Proclear) lenses contain a hydrophilic methacrylate monomer with a zwitterion side group modeled after the phosphatidyl

BOX 4 Effect of contact lens surface structure on biocompatibility

In the mid-1980s, Vistakon developed Etafilcon A (a PHEMA-based lens polymer with 3% methacrylic acid) as a daily wear lens. Initial *in vivo* wear showed a high percentage of protein adsorbing to the lens surface; however, the lenses also had a significantly decreased rate of infections compared to other lenses on the market. Investigation into the situation revealed that the pendant carboxyl groups of the methacrylic acid monomers imparted a negative charge to the lens surface. This charged surface preferentially adsorbed lysozyme, a small, positively-charged protein in the tear film with antibacterial properties (Boles et al., 1992). The coating of adsorbed lysozyme induced antibacterial properties to the surface and ultimately increased the lens biocompatibility and usefulness.

choline head-group (Figure A.4). This charged side group preferentially holds water to the lens surface, and was the first material to be approved by the FDA for use in patients with dry eyes.

The first silicone hydrogel materials to market in 1998 were surface treated with high energy gases or plasma treatments with gas mixtures (see Chapter I.2.12) to modify the surfaces without affecting the bulk material properties. Balafilcon A (Pure Vision), for example, is plasma oxidized, which causes glassy, silicate islands to form on the surface. These islands create bridges crossing over the hydrophobic balafilcon regions. The balafilcon regions that are left exposed may also be modified and have lower hydrophobicity. Furthermore, these islands are separated from each other to a degree that allows the lens material to maintain its flexibility. Lotrifilcon A & B (Night 'n Day & O_2 Optix), on the other hand, use a plasma coating (deposition) to combat the hydrophobicity. This treatment uniformly coats the lens surface with a 25 nm thick hydrophilic polymer produced by using glass plasma (Sweeney, 2004). Both resultant balafilcon and lotrafilcon lens surfaces have low molecular mobility, which minimizes the migration of hydrophobic silcone moieties to the surface. However, the lack of chain mobility also increases the lens modulus, creating "stiffer" materials.

Vistakon skirted the plasma treatment issues by developing silicone hydrogel materials with internal wetting agents. Galyfilcon A (Acuvue Advance) and senofilcon A (Acuvue Oasys) both use poly(N-vinyl pyrrolidone) (PVP) as an internal wetting agent which is preferentially

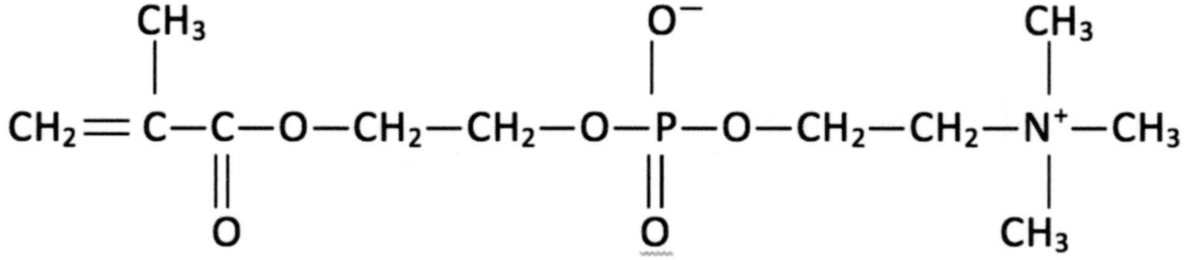

$$CH_2{=}C(CH_3){-}C({=}O){-}O{-}CH_2{-}CH_2{-}O{-}P(O^-)({=}O){-}O{-}CH_2{-}CH_2{-}N^+(CH_3)_2{-}CH_3$$

FIGURE A.4 Hydrogel monomer with surface active side group based on the head-group of phosphatidyl choline.

associated with the material surface. Galyfilcon A was released as the precursor to an extended wear version, senofilcon A. Both materials are based on copolymerization of a hydrophilic variant of TRIS, HEMA, DMA, and a siloxy macromer. The incorporation of PVP into the lens produces a wettable, lubricious surface without subsequent surface treatment. Galyfilcon A lenses also have PVP in their packing solution. For senofilcon A, the PVP is incorporated as polymeric end-caps that preferentially migrate to the lens surface (McCabe et al., 2004). The results of these formulations are materials with increased surface hydrophilicity and markedly lower modulus (Table A.1).

Comfilcon A, Bioinfinity, is one of the highest water content silicone hydrogel materials (EWC = 48%), and has no surface treatment or internal wetting agent. The first marketed material produced without the use of the TRIS monomer or its variants, comfilcon is produced by polymerization of two siloxy macromers of different sizes, one of which has only one vinyl polymerization site, as well as vinyl amides such as n-methyl-n-vinyl acetamide. Comfilcon A's unexpectedly high Dk, considering its water content, is attributed to its longer siloxane chains.

SPECIALTY LENSES

There are a number of specialty applications of contact lenses, both prescriptive and cosmetic in nature. There are only three types of contact lens materials FDA approved for dry eye patients, omafilcon A, senofilcon A, and comfilcon A. The surfaces of these materials have a stabilizing effect on the tear film. Toric and astigmatic lenses are used to treat patients with irregularly shaped corneas, and are generally made out of hard or silicone hydrogel materials. Multifocal lenses were developed for the aging contact lens wearer who began to need reading glasses (presbyopia). These lenses are generally made of silicone hydrogels. Through the use of designed material geometries, either concentric rings of varying refractive index (RI) or a change in refractive index progressively across the lens surface, these lenses have varying visual correction across different focal planes. Orthokeratatomy lenses are RGP lenses used at night to reshape the cornea surface so the patient does not need visual correction during the day. Additionally, plano lenses are sold in varying colors and with multiple designs imprinted on them for cosmetic purposes.

CONTACT LENS SOLUTIONS

Effective disinfection and cleaning of contact lenses can normally be achieved using approved chemical or hydrogen peroxide systems. Multipurpose cleaning solutions are used to remove lens deposits that are adsorbed from the tear film, disinfect the lenses, prevent future biofouling, and allow lens storage. The basic steps seen with these products are cleaning, disinfecting, and rinsing. Product efficiency varies depending on the composition, manufacturer, and length of time for surfactant cleaning and storage. Most systems use hydrophilic polymers to increase cleaning efficiency and lens surface hydration. Some of the hydrophilic polymers used, such as Poloxamer 407, actually act on the bacteria surface and not the lens to inhibit surface adhesion (Portoles et al., 1994). Hydrogel lenses are categorized by the United States Food and Drug Administration into four groupings for the purpose of evaluating effects of accessory products, i.e., cleaning solutions and rewetting drops, upon the lens material. Lenses with less than 50% water content are considered to be "low water," and those with >50% are "high water." Surfaces without fixed charges are called "non-ionic" while more reactive materials that are charged at physiological pH are termed "ionic."

BIBLIOGRAPHY

Benjamin, W. J., & Cappelli, Q. A. (2002). Oxygen permeability (Dk) of thirty-seven rigid contact lens materials. *Optometry & Vision Science*, 79(2), 103–111.

Boles, S. F., Refojo, M. F., & Leong, F. L. (1992). Attachment of *Pseudomonas* to human-worn, disposable etafilcon A contact lenses. *Cornea*, **11**(1), 47–52.

Efron, N., Morgan, P. B., Malodonado-Codina, C., & Brennan, N. A. (2010). Contact lenses: The search for superior oxygen permeability. In T. Chirilia (Ed.), *Biomaterials and Regenerative Medicine in Ophthalmology* (pp. 280–303). CRC Press LLC.

Harvitt, D. M., & Bonanno, J. A. (1999). Re-evaluation of the oxygen diffusion model for predicting minimum contact lens Dk/t values needed to avoid corneal anoxia. *Optometry and Vision Science*, 76(10), 712–719.

Holden, B. A., & Mertz, G. W. (1984). Critical oxygen levels to avoid corneal edema for daily and extended wear contact lenses. *Investigative Ophthalmology & Visual Science*, **25**(10), 1161–1167.

Holden, B. A., Newton-Howes, J., Winterton, L., Fatt, I., Hamano, H., La Hood, D., Brennan, N. A., & Efron, N. (1990). The Dk project: An interlaboratory comparison of Dk/L measurements. *Optometry & Vision Science*, 67(6), 476–481.

Kingshott, P., St John, H. A., Chatelier, R. C., & Griesser, H. J. (2000). Matrix-assisted laser desorption ionization mass spectrometry detection of proteins adsorbed *in vivo* onto contact lenses. *Journal of Biomedical Materials Research*, **49**(1), 36–42.

Kunzler, J., & Ozark, R. (1994). Fluorosilicone hydrogels. *US Patent, 5321108*.

McCabe, K., Molock, F., Azaam, A., Steffen, R. B., Vanderlaan, D. G., & Young, K. A. (2004). To Johnson & Johnson Vision Care Inc. Biomedical devices containing internal wetting agents. *US Patent 6822016*.

Nicholson, P., Baron, R., Chabrecek, P., et al. (1996). To CIBA Vision. Extended wear ophthalmic lens, *WO 96/31792*.

Nicolson, P. C., & Vogt, J. (2001). Soft contact lens polymers: An evolution. *Biomaterials*, **22**(24), 3273–3283.

Morgan, P. B., & Efron, N. (2008). The evolution of rigid contact lens prescribing. *Contact Lens & Anterior Eye*, **31**(4), 213–214.

Morgan, P. B., Efron, N., Helland, M., Itoi, M., Jones, D., Nichols, J. J., van der Worp, E., & Woods, C. A. (2010). Demographics of international contact lens prescribing. *Contact Lens & Anterior Eye*, 33(1), 27–29.

Pearson, R. M., & Efron, N. (1989). Hundredth anniversary of August Müller's inaugural dissertation on contact lenses. *Surv Ophthalmol*, **34**(2), 133–141.

Peppas, N. A. (Ed.), (1986). *Hydrogels in Medicine and Pharmacy. Fundamentals*. (Vol. 1). CRC Press.

Peppas, N. A. (Ed.), (1987). *Hydrogels in Medicine and Pharmacy. Polymers*. (Vol. 2). CRC Press.

Peppas, N. A. (Ed.), (1987a). *Hydrogels in Medicine and Pharmacy. Properties and Applications*. (Vol. 3). CRC Press.

Portoles, M., Refojo, M. F., & Leong, F. L. (1993). Reduced bacterial adhesion to heparin-surface-modified intraocular lenses. *Journal of Cataract & Refractive Surgery*, **19**(6), 755–759.

Portoles, M., Refojo, M. F., & Leong, F. L. (1994). Poloxamer 407 as a bacterial adhesive for hydrogel contact lenses. *Journal of Biomedical Materials Research*, **28**(3), 303–309.

Refojo, M. F. (1973). Contact lens materials. *International Ophthalmology Clinics*, **13**(1), 263–277.

Refojo, M. F. (1979). Mechanism of gas transport through contact lenses. *Journal of the American Optometric Association*, **50**(3), 285–287.

Refojo, M. F. (1996). Polymers, Dk, and contact lenses: Now and in the future. *CLAO Journal*, **22**(1), 38–40.

Sankaridurg, P. R., Sweeney, D. F., Sharma, S., Gora, R., Naduvilath, T., Ramachandran, L., Holden, B. A., & Rao, G. N. (1999). Adverse events with extended wear of disposable hydrogels: Results for the first 13 months of lens wear. *Ophthalmology*, **106**(9), 1671–1680.

Stapleton, F., Stretton, S., Papas, E., Skotnitsky, C., & Sweeney, D. F. (2006). Silicone hydrogel contact lenses and the ocular surface. *The Ocular Surface*, **4**(1), 24–43.

Stern, S. A. (1968). The barrer permeability unit. *Journal of Polymer Science Part A-2: Polymer Physics*, **6**(11), 1933–1934.

Sweeney, D. (Ed.), (2004). *Silicone Hydrogels* (2nd ed.). Butterworth-Heinemann.

Tanaka, K., Takahashi, K., Kanada, M., & Toshikawa, T. (1979). *Methyl di(trimethysiloxy)silylpropyl glycerol methacrylate*. Japan: To Toyo Contact Lens Co. Ltd. US Patent, 4 139 548.

Tighe, B. J. (2007). Contact Lens Materials. In A. J. Philips, & L. Speedwell (Eds.), *Contact Lenses* (5th ed., pp. 59–78). Butterworths.

Tighe, B. J. (2010). Extended wear contact lenses. In T. Chirilia (Ed.), *Biomaterials and Regenerative Medicine in Ophthalmology* (pp. 304–336). CRC Press LLC.

Wang, J., Fonn, D., Simpson, T. L., Sorbara, L., Kort, R., & Jones, L. (2003). Topographical thickness of the epithelium and total cornea after overnight wear of reverse-geometry rigid contact lenses for myopia reduction. *Investigative Ophthalmology & Visual Science*, **44**(11), 4742–4746.

Wichterle, O. (1978). The beginning of the soft lens. In M. Ruben (Ed.), *Soft Contact Lenses, Clinical and Applied Technology* (pp. 3–5). Bailliére Tindall.

Wichterle, O., & Lim, D. (1960). Hydrophilic gels for biological use. *Nature*, **185**(4706), 117–118.

B. INTRAOCULAR LENS IMPLANTS: A SCIENTIFIC PERSPECTIVE

Anil S. Patel
Retired Vice President of Research for Surgical Products for Alcon, Seattle, WA, USA

INTRODUCTION TO INTRAOCULAR LENS IMPLANTS, THE OPTICS OF THE EYE AND CATARACTS

Intraocular lenses (IOLs) are polymeric devices implanted in the globe of the eye and intended to replace the cloudy, cataractous natural lens. There are now at least 34 companies manufacturing these lenses for humans and animals, and more than 10,000,000 lenses are implanted each year. Thus, IOLs are one of the most widely used medical implants and an understanding of their biomaterials, design and application are important for the biomaterials scientist.

Vision is considered to be our dominant sense. We see in the brain through optic nerves, which carry information of the optical image formed in the eye. The physiological optics of the eye for formation of the image, the photochemical transduction in the neurosensory retina at the optical image plane, and the subsequent neural processing in the retina and the brain are quite complex and beyond the scope of this chapter (LeGrand and El Hage, 1980). As an introduction, a brief description of optical imaging of the eye is necessary.

The eye is schematically illustrated in Figure B.1. The average healthy adult eye is about 24.2 mm long axially from the anterior surface of the cornea to the retina. It can be considered as a camera, with the cornea and the natural crystalline lens as its two lenses, while the iris forming the pupil is an aperture between them, and the retina is the imaging plane. The image of an outside object of interest is formed on the retinal photoreceptors by the fixed refractive power of the cornea in combination with the variable refractive power of the natural lens, permitting imaging of any object at various far and near distances from the eye. The fixed refractive power of the cornea is about 43 diopters, which is about 70% of the total refractive power of the eye. The refractive power of the natural lens is about 20 diopters for seeing far objects. For seeing near objects, it increases as required to an upper limit, which gradually diminishes with age. When young, the refractive power of the natural lens can increase by up to 14 diopters, but by about the age of 42, this capability is less than the 3 diopters required for near reading distance. The physiology of variable refractive power of the natural lens is as follows.

As shown in Figure B.1, the natural lens has a capsular bag containing a nucleus surrounded by the cortex. It is attached by the suspensory ligaments called zonules to the ciliary muscle. For an average healthy eye, the refractive power of the cornea and its natural lens are, in combination, adequate to form a focused image on the retina for any object at 20 feet or greater distance when the ciliary muscle is relaxed. In this physiologically relaxed condition of the eye, the zonules are in tension, and the natural lens is also in tension, providing about 20 diopters of required refractive power for distant vision. For seeing an object

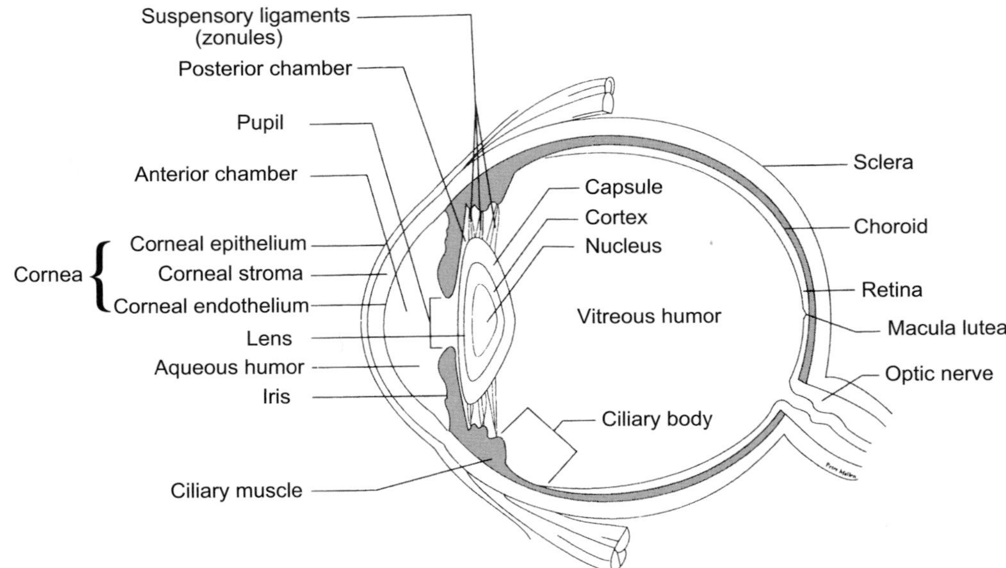

FIGURE B.1 Schematic representation of the eye.

at any nearer distance, the ciliary muscle appropriately contracts radially and moves anteriorly, thereby relaxing the zonules, which in turn relaxes the natural lens. This results in the needed change in the shape of the lens, and thus increases its refractive power as required for near vision. This complex mechanism, which results in the ability to increase the refractive power of the natural lens for near vision, is called accommodation (Kaufman, 1992). This ability gradually diminishes with age as the lens grows, and eventually becomes less than the 3 diopters required for near reading distance, and the eye is considered presbyopic. This presbyopia of the eye is treated with eyeglasses, contact lenses or evolving refractive surgery, in order to restore the ability to see near objects.

Further aging of the lens results in loss of its clarity as opacities begin to appear in its structures, and optical image formation deteriorates. Eventually, such a compromised natural lens is unacceptable and is called cataractous, since it is transformed to an unacceptable structure for the primary function – to allow formation of a clear image on the retina (US Department of Health and Human Services, 1983). This normal aging process related to cataract formation eventually leads to blindness, and is the leading cause of blindness in the world. In the developed world, depending upon the need for vision, cataract surgery is carried out much more promptly before the total blindness stage when the degradation in the vision by the cataractous lens reaches an unacceptable level (AAO, 1989). During a relatively safe and short-duration surgery, not only is the cataractous lens removed, but in its place a biomaterial medical device called an intraocular lens is implanted to provide the clear optical imaging function of the normal lens (Alpar and Fechner, 1986). While currently worldwide an estimated 10 million IOLs are implanted, primarily in the developed world and emerging countries, the lack of resources and rapid increase in life expectancy accounts for an estimated 20 million cataract-related blind people in the underdeveloped world. Cataract blindness is preventable by a simple surgical procedure using the IOL implant, and the number of such procedures is expected to grow. Even then, cataracts are expected to remain the leading cause of blindness, with an estimated 40 million cases by the year 2020 (Brian and Taylor, 2001).

WHY ARE IOLS SUCCESSFUL?

Within a span of only 35 years, the number of cataract surgeries and IOL implantations has rapidly expanded for good reasons, both in terms of improvements in surgical procedure and visual outcome from the patient's perspective, as summarized below.

Prior to IOL implantation, thick spectacles of about 10 diopters refractive power were needed to attempt to restore the function of the natural lens after cataract surgery. Imaging by such external spectacle lenses and the cornea creates a 25% magnification of the resultant image on the retina, compared with imaging by the natural lens and the cornea. Thus, when a cataract develops in one eye of the patient, its functional restoration by cataract surgery and a spectacle lens is unacceptable to the brain since, for fusion of images of the two eyes, the brain requires them to be no more than 6% different in size. Thus, restoration of vision by spectacles after cataract removal required waiting for maturation of cataracts in both eyes before surgeries were carried out to achieve equal image sizes for both eyes. This resulted in significantly poor vision in the unoperated cataractous eye for a prolonged period before cataract surgery was undertaken. Additionally, cataract surgery 35 years ago was associated with other risks and poor visual outcome (as compared in Table B.1), and was thus relatively

TABLE B.1 Why Are IOLs Successful? Comparison of Cataract Surgery in the Developed World for Restoration of Vision 35 Years Ago with Spectacles and Today with Foldable IOLs

35 years ago	Foldable IOL today	Improvements (2008+)
• Ten days of hospitalization	• Same-day outpatient surgery in a few hours	• Convenience and cost
• Use of general anesthesia or retrobulbar injection in the back of the eye with long needle	• Use of topical drops and intracameral anesthetic or peribulbar injection if needed	• Lowered risk and complications
• Large >12 mm incision requiring significant stitching and resultant astigmatism	• Small 3 mm or less incision requiring no stitching	• No stitching-related surgically-induced astigmatism
• Use of alpha-chymotrypsin enzyme to weaken zonules and cryoextraction of the entire cataractous lens through the large incision	• Use of ultrasonic phacoemulsifier and irrigating/aspirating instruments to break up and remove the nucleus and cortex, leaving behind the capsular bag with an anterior opening for insertion of IOL in the bag	• Significantly reduced complications of retinal detachment, excessive inflammation, and infection
• No protection of cornea and intraocular tissues during surgery	• Use of a viscoelastic biomaterial to protect the corneal endothelium and other intraocular tissues	• Significantly reduced risk of losing corneal clarity and inflammation of the uveal tissues
• Thick spectacles to restore function of the natural lens resulting in a 25% image magnification, which is incompatible with monocular cataract surgery; also resultant reduced field of vision	• No restriction on monocular cataract surgery	• No delay in cataract surgery in the first eye since IOL provides natural unchanged size of image; also unchanged field of vision
• Need for spectacles for correction of astigmatism and to have vision for all distances	• Toric and presbyopic correcting multifocal and accommodative IOLs can eliminate need for spectacles	• Truly spectacle-free vision is now possible
• Overall unsatisfying vision achieved several months after surgery; anxiety and fear of cataract surgery	• Satisfying vision next day after surgery; minimal anxiety and fear of cataract surgery and IOL implantation	• Minimal duration of compromised vision and interruption from occupation; prompt restoration of excellent vision

discouraged until too late an age, when the compromised vision was totally unacceptable in a relatively advanced state of dysfunction, compared with today's practice in the developed world. Binocular vision with spectacles also had a significantly reduced field of vision and severe astigmatism, on account of the required large size of the incision for cataract surgery.

The size of the optic of the IOL and its intraocular location eliminated image size magnification and reduced visual field problems, which were present with spectacles. Foldable IOLs encouraged the development of small-incision, safer, cataract surgery techniques without surgically induced astigmatism, and thus achieved quicker and better vision after cataract surgery (Apple et al., 2000a). Table B.1 captures all the reasons for the overwhelming success of IOLs, where needed resources are available.

EMERGING FUNCTIONAL VARIATIONS OF IOLS

As a replacement of the cataractous natural lens, the majority of currently used IOLs is monofocal IOLs of fixed refractive spherical diopteric power, which cannot correct existing astigmatism as is needed for some eyes or restore accommodation as was provided by the pre-presbyopic natural lens. Aspheric IOLs, toric IOLs, multifocal IOLs, and emerging accommodative IOLs are relatively recent developments, and they will be discussed later as IOLs with emerging variations of the optical function (Maxwell and Nordan, 1990; Sanders et al., 1992; Samalonis, 2002; Mester et al., 2003).

The eye with its natural lens is called the phakic eye. The eye after removal of the natural lens, either clear or cataractous, is called an aphakic eye. The eye with replacement of the natural lens with an IOL is called a pseudophakic eye. An ophthalmic implant placed in a phakic eye without removal of its natural lens, for example, an additional IOL in order to correct refractive error, is called a phakic IOL (Obstbaum, 2003). Surgical implantation of a phakic IOL is a form of refractive surgery that provides needed refractive correction in a myopic or hyperopic eye as an option to eyeglasses, contact lenses or corneal refractive surgery, such as LASIK. The current status of phakic IOLs will also be discussed later as a variation of optical function.

Thus, the IOL as an ophthalmic intraocular implant is now evolving to provide a variety of optical functions. The biomaterials used for such a variety of IOLs have been many, and new materials are emerging in order to meet the requirements of their various optical functions and designs.

BIOMATERIALS FOR IOLS

Biomaterial selection for IOLs is based on specialized requirements to maintain a stable, clear path for optical

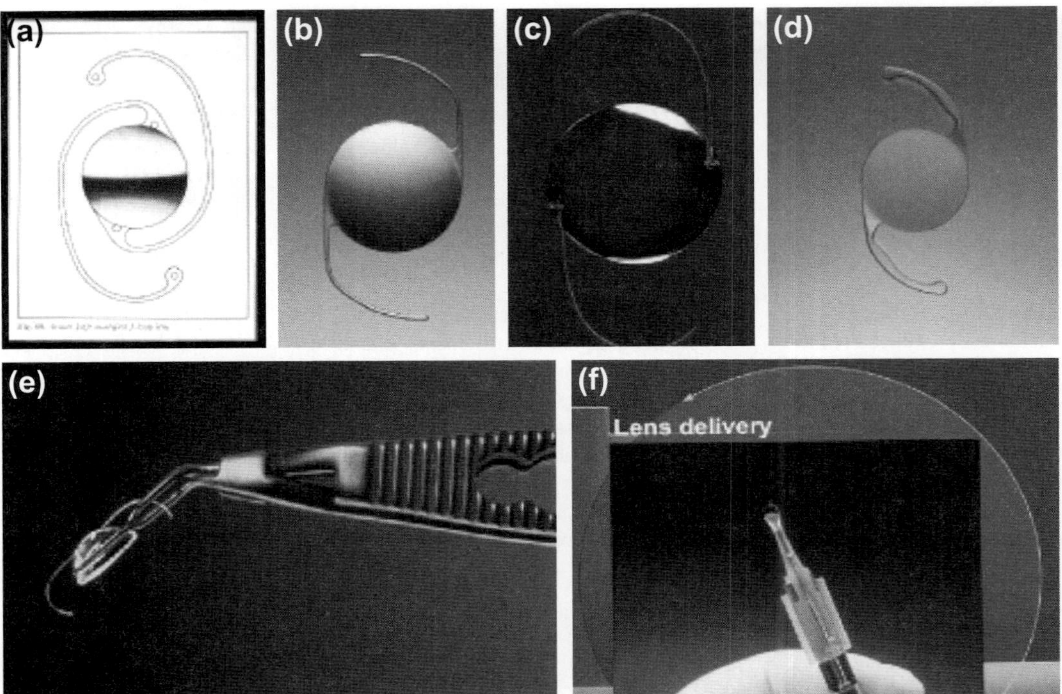

FIGURE B.2 IOL transition occurring in developed world is best illustrated here: (a) Past: three-piece PMMA IOL; (b) Current: single-piece PMMA IOL; (c) Past: three-piece foldable IOL; (d) Current: single-piece foldable IOL; (e) Past: insertion forceps held folded IOL; (f) Current: folded IOL emerging from an injector tip cartridge.

imaging, as well as long-term safe acceptance as a permanent implant by the unique biology of the eye (Ratner, 1998; Mamalis, 2002). Additionally, the selected biomaterial is required to meet IOL design objectives, such as targeted incision size for insertion and needed mechanical fixation, typically via anchoring components called loops or haptics. The haptics permit the IOL to achieve stable location within the eye, and also permit the targeted optics design to achieve the functional goal for the IOL. Figure B.2 A-D illustrates various lens haptics.

Historical Beginning of Poly(Methyl Methacrylate) (PMMA) as an IOL Material

Harold Ridley of London, in 1949, was the first to implant an IOL made from a PMMA sheet originally formulated for canopies of British Royal Air Force airplanes (Ridley, 1951). His selection of the material was based on an accidental implantation of pieces of the canopy in pilot's eyes, which were observed by him to be quietly tolerated without noticeable inflammatory reaction or other unacceptable biological consequence. Also, since it remained optically transparent, he chose this Imperial Chemical Industry's PMMA formulation, known as Perspex CQ (Clinical Quality), to create a biconvex 10.5 mm diameter disk-shaped IOL, a historical first. This Ridley IOL was too big and bulky, weighing about 110 mg in air, compared with current IOLs, weighing about 20 mg on average (Patel et al., 1999). It was difficult to mechanically fix such a bulky IOL in the eye, and serious surgical and postoperative complications led to the discontinuation of its use. However, this concept of the IOL implantation is recognized as a pioneering breakthrough. Also, PMMA remained the dominant IOL material for the next 40 years until the 1990s, when foldable IOLs of other materials emerged.

Evolution of Monofocal IOL Designs and Biomaterials

Although the Ridley IOL was unacceptable, it stimulated many other surgeons who created several generations of monofocal IOL designs with a much smaller optic made of PMMA, and a variety of fixation means and materials (Apple et al., 1989).

The anatomical confines of the eye's anterior chamber (AC) were selected for fixation of IOLs. Several AC-IOL designs emerged between 1952 and 1962, with a relatively heavy footplate extension of the optic which rested in the angle formed by the peripheral iris structure and the collagen fibers of the scleral tissue at the junction with the cornea (Choyce, 1958; Strampelli, 1961). While fixation of the IOL was ensured by these AC-IOL designs, the unique biocompatibility needs of the nearby cornea and iris uveal tissue were revealed. Contact of the IOL with the cornea, for example by rubbing, damaged the corneal endothelium and resulted in loss of corneal clarity. The uveal tissue responded with chronic inflammation. Since exact anatomical sizing was difficult, missized AC-IOLs aggravated the foregoing complications by protrusive erosion or loose rotation within the anterior chamber (Ellingson, 1978). These AC-IOL designs were abandoned, and iris-fixated or -supported designs emerged between 1959 and 1973 (Binkhorst, 1959). These iris-supported IOLs introduced fixation using several biomaterials attached to the PMMA optic. Fixation structures were shaped as round loops of various materials including nylon 6, nylon 66, and polypropylene suture materials. Even platinum–iridium loops and titanium clips were used. Nylon suture degenerated in the eye (Kronenthal, 1977). Metal loops or clips were heavy, causing injury to the iris and hemorrhage within the anterior chamber (Shephard, 1977). Only iris-supported IOL designs with polypropylene loops created by Cornelius Binkhorst seemed viable. His two-loop iridocapsular IOL was implanted with the loops supported by the iris, while the optic was captured by the capsular opening (Alpar and Fechner, 1986). The nucleus and the cortex of the cataractous lens were removed earlier during surgery through this opening. This method of cataract surgery is called extracapsular cataract extraction (ECCE). Prior to this surgical technique, removal of the entire cataractous lens using an enzyme to weaken the zonules and a cryoextractor was more prevalent. This method of cataract surgery is called intracapsular cataract surgery (ICCE). Iridocapsular supported Binkhorst two-loop IOL implantation after ECCE gave encouraging results, but with the passage of 6 to 10 years late decompensation of the cornea occurred, often requiring corneal transplantation. Also, erosion of the iris, IOL dislocation with dilation of the pupil, and difficulty of implantation discouraged its use (Nicholson, 1982; Obstbaum, 1984). The Binkhorst two-loop iridocapsular IOL with ECCE is considered the forerunner of current capsular-fixated posterior chamber IOLs (PC-IOL) with ECCE using phacoemulsification.

Failure of iris-supported IOLs led to revisiting AC-IOLs in the 1970s and 1980s, with a thinner footplate and the evolution of haptics, much thinner and somewhat flexible extensions of the optic, for fixation. These haptics were an integral part of the IOL made from the same PMMA material, or the haptic could be formed using other biomaterials. Polypropylene suture material, extruded monofilament from other PMMA formulations, and polyamide materials were used for haptics. Even these newer AC-IOL designs gave unsatisfactory long-term outcomes, with undesirable corneal decompensation, chronic uveal inflammation, and eventual glaucoma (Apple et al., 1987). Many AC-IOLs with finely looped haptics were associated with micromovement and rubbing with sensitive tissues of the anterior chamber, resulting in loss of corneal endothelial cells, iris chafing, and adhesions around the haptics. While the latest AC-IOL design with a one-piece PMMA optic with integral open haptics, which were wider and yet somewhat flexible, improved the results by avoiding the complications just discussed, the search intensified for fixating the IOL in the posterior chamber where the natural lens is located.

Finally, the search for a PC-IOL led to an IOL by Stevens Shearing in 1979 with a PMMA optic and two J-shaped polypropylene haptics designed for placement behind the iris in the anatomical space called the ciliary sulcus, in front of the capsular bag (Shearing, 1979; similar to Figure B.2C). Soon, a variety of shapes of haptics emerged, with a different degree of flexibility and contact area for ease of implantation and stable fixation. Also, PMMA monofilament as a haptic material replaced polypropylene, after some concern for its degradation was raised (Apple et al., 1984). The PC-IOL designs achieved such success that manufacturing processes even created one-piece PMMA IOLs with integral flexible haptics (Figure B.2A,B).

PC-IOL implantation required the ECCE cataract surgery method and, while large numbers of IOLs were implanted with this method from 1977 to 1992, several refinements emerged to address complications of the sulcus-fixated PC-IOL. Inflammation on account of iris chafing, decentration of the IOL when one haptic was fixated in the sulcus and the other unintentionally went into the bag, and posterior capsule opacification emerged as unacceptable complications of the sulcus-fixated IOL. Placement of the entire IOL (both haptics and the optic) within the capsular bag avoided iris chafing, reduced inflammation, and improved central fixation. C-shaped haptics and the surgical technique of opening the capsular bag by manually controlled tearing called anterior curvilinear capsulorhexis provided reliable implantation of the PC-IOL in the capsular bag with good centration (Gimbel and Neuhann, 1990). As a result, a large number of surgeons with varying skill levels began to implant PC-IOLs.

By the early 1980s, Perspex CQ PMMA was replaced by most IOL manufacturers with differing PMMA formulations with UV chromophore additives, in order to protect the retina from damage by UV radiation.

Viscoelastic materials, such as sodium hyaluronate, were introduced and used to protect the corneal endothelium and other intraocular tissues (Miller and Stegmann, 1982). Even coatings for IOLs and surface modifications for IOLs were investigated in order to minimize damage to corneal endothelium due to accidental touch during insertion into the eye, as well as to minimize foreign-body reaction on the surface of the IOL. Heparin surface-modified PMMA IOLs were introduced, but since they did not solve the posterior capsular opacification complication, and also required a longer incision than needed for foldable IOLs, their use was minimal and short-lived (Philipson et al., 1990; Winther-Nielsen et al., 1998).

The ECCE method of cataract removal, which is required for PC-IOL implantation, retains the entire capsular bag except the capsulorhexis-excised anterior capsule face. There is a region near the equator of the lens capsule where the residual lens epithelial cells (LEC) with potency for mitosis and migration remain. These LECs proliferate and migrate and spread onto the anterior capsule, as well as onto the posterior capsule, and even potentially onto the IOL surface. These LECs undergo morphological changes, reducing their transparency, resulting in anterior capsule opacification (ACO) and posterior capsular opacification (PCO), as well as contraction of the capsular bag (Amon, 2001; Kruger et al., 2001). PCO is the complication that reduces clarity of the optical path and causes degradation of the quality of image formation, to the extent that it is also called "secondary cataract" formation. While making an adequate opening in the cloudy posterior capsule using an Nd-YAG laser beam without surgically opening the eye restored the clarity of the optical path and the image quality on the retina, there are several disadvantages of this surgical procedure. They include the cost of the procedure, the degraded vision prior to it, the possibility of damaging the lens, and the potential for inducing retinal detachment because of the shockwave created by the focused laser. These disadvantages stimulated approaches to prevent PCO after PC-IOL implantation in the capsular bag. The shape of the optic was changed from convexo-plano to biconvex, and haptics were angulated to achieve complete contact of the posterior surface of the PC-IOL to the posterior capsule. Ridges were also added on the posterior surface to stop migration of LECs, without much success. Thus, Nd-YAG laser-created posterior capsulotomy remained a necessary procedure for "secondary cataracts" due to PCO. The rate of need for this posterior capsulotomy depends on the type of IOL material and design, success in surgical removal of much of the cortical contents of the lens, age-related vigor of remaining LEC, and other inflammatory stimulating factors. With some foldable IOL materials, as discussed later, and using the latest surgical techniques for ECCE with phacoemulsification, the PCO formation rate and the posterior capsulotomy as a necessary second surgical procedure have been significantly reduced.

Rationale and Emergence of Soft Biomaterials for Foldable IOLs

While PC-IOLs of PMMA material were implanted in very large numbers in the early 1980s, another ECCE surgical method drew attention. A probe with a titanium needle that is axially oscillated by an ultrasonic instrument achieved removal of the cataractous nucleus by fragmenting and aspirating it. Cortical material was aspirated out by another probe as a component of the instrument system known as an ultrasonic phacoemulsifier. It also provided necessary simultaneous irrigation into the eye to prevent collapse of the anterior chamber and damage to the cornea. Advances in this technology, with further refinement in the surgical techniques, resulted in eyes with less inflammation postoperatively. Also, the incision length required to enter and remove the cataract by this technology is only 3 mm or less, compared with an approximately 9 mm incision required for the manual ECCE extraction method for removal of the nucleus of the cataract. This reduction in incision length resulted in significant reduction in surgically-induced astigmatism. Additionally, ECCE using the phacoemulsifier spared mechanical trauma to intraocular tissue encountered during the manual ECCE extraction method, thus minimizing tissue irritation and inflammation.

Postoperative large astigmatism induced by the relatively large 9 mm incision required for manual ECCE was degrading the optical image formation and resultant vision (Luntz and Livingston, 1977). While it was managed by surgical and suturing techniques, and by appropriate eyeglass correction, it was recognized that the surgically-induced astigmatism by a 3 mm incision to enter the eye for ECCE by phacoemulsification was negligible, but subsequent enlargement to insert the PMMA IOL with the preferred 6 mm diameter optic size increased surgically-induced astigmatism (Kohnen et al., 1995; Olson and Crandall, 1998). A soft biomaterial for designing foldable IOLs for insertion through a 3 mm incision was needed to eliminate this surgically-induced astigmatism. Additionally, foldable IOL insertion through a 3 mm incision after ECCE with phacoemulsification promised elimination of the need for stitching, and retention of the minimal postsurgical inflammation provided by this method of cataract surgery (Gills and Sanders, 1991).

In the early 1980s, a transparent polysiloxane, a formulation originally created for other industrial applications, was used as a biomaterial for foldable IOLs by Starr, Inc. (Mazzocco et al., 1986). In the subsequent 15 years, several formulations of polysiloxane with different refractive indexes, mechanical properties, and UV transmission characteristics emerged from other companies. PolyHEMA, a hydrogel material for contact lenses, was also used as a biomaterial for foldable IOLs, but fixation was difficult and it produced unacceptable postoperative complications after Nd-YAG posterior capsulotomy.

In 1994, the first new soft biomaterial for IOLs was introduced by Alcon Laboratories Inc. as AcrySof IOL. This biomaterial was specifically created to meet the desired optical, mechanical, and biological properties for foldable IOLs (Anderson et al., 1993). This copolymer of phenylethyl acrylate and phenylethyl methacrylate with a cross-linking agent and a bonded UV absorbing chromophore provided a higher refractive index of 1.55. This allowed the design of a thinner foldable IOL from this hydrophobic soft acrylate material. Its tailored mechanical properties resulted in slow unfolding of the IOL, unlike the rapid, uncontrolled, spring-like opening of early silicone IOLs with plate haptics. Early three-piece AcrySof IOLs with PMMA monofilament C-shaped haptics were folded and inserted using forceps, similar to three-piece silicone IOL designs (Figure B.2E). Subsequently IOL injectors were developed, and are currently used for most foldable IOLs (Figure B.2F). Also, AcrySof IOL is now mostly used as a single-piece foldable IOL with integral haptics of the same biomaterial as optic.

The AcrySof IOL soon became the most widely used foldable IOL, and millions of these IOLs had been implanted globally with overall satisfactory outcomes. Besides being a small-incision IOL, it unfolds gently in the eye, remains well centered, and postoperatively results in "quieter" eyes without significant inflammation. Some minimal glistening, which is judged cosmetic, and edge-related optical photic phenomena were reported, and now have been minimized through improvements. The newer generation of silicone foldable IOLs and other hydrophobic, as well as hydrophilic soft acrylate foldable IOLs, also emerged (see Table B.2) (Apple et al., 2000a; Werner et al 2008). This success for soft, foldable IOLs, and in particular, AcrySof, is attributed not only

TABLE B.2 Types of Biomaterials for Representative IOLs in the Market or Clinical Investigation by Key Leading and Newly Emerging Manufacturers

Manufacturer	Type of IOL	Biomaterials
Abbott Medical Optics(AMO)	Monofocal IOLs	PMMA with UV absorber
	Foldable IOLs: Monofocal Aspheric Tecnis® Multifocal ReZoom® Phakic Verisyse®	Proprietary polysiloxane and also hydrophobic copolymer of acrylates with UV absorber for optic and single piece haptics PMMA blue core monofilament haptics for multipiece IOLs
Alcon	Monofocal IOLs	PMMA with UV absorber PMMA monofilament haptics for three-piece IOLs
	Foldable IOLs: Monofocal Aspheric AcrySof® IQ®, Multifocal ReSTOR® Toric Phakic	Proprietary hydrophobic copolymer of acrylates with UV absorber only or with additional proprietary bonded yellow dye for blue-light filtering AcrySof® Natural® IOL material
Bausch and Lomb	Monofocal IOLs Foldable IOLs Monofocal Aspheric SoftPort® AO Aspheric Akreos™ AO Akreos™ MI-60 Foldable accommodative IOL movement-based Crytalens® AT-50 Phakic AC IOL	PMMA with UV absorber Proprietary polysiloxane with UV and violet light absorber, Polyamide haptics for Crystalens only Proprietary hydrophilic copolymer of acrylates with UV absorber
Calhoun Vision	Foldable monofocal Light Adjustable power IOL	Proprietary photopolymerizable polysiloxane macromers dispersed within polymerized polysiloxane matrix
Carl Zeiss Meditec	Foldable IOLs Monofocal XL Stabi®, Hydromax®, Aspheric Acri. Smart® Multifocal Acri.Lisa® Toric Phakic	Proprietary hydrophilic and hydrophobic acrylates and polysiloxane materials with UV and optional violet filtering absorbers Polyvinylidene fluoride (PVDF) haptic material
Hoya Surgical Optics	Foldable IOLs Monofocal AF-1® Aspheric AF-1 FY-AD Multifocal	Proprietary hydrophobic copolymer of acrylates with UV and optional blue filtering bonded absorbers. PMMA haptic bonded to foldable optic
Human Optics, AG	Foldable accommodative IOL movement-based	Proprietary copolymer of acrylates with UV absorber (hydrophilic)
Medenium	Foldable monofocal IOL Matrix® IOL	Proprietary hydrophobic copolymer of acrylates with UV absorber Polyvinylidene fluoride (PVDF) haptic material
Ophtec	Monofocal IOL Phakic AC-IOL Foldable Monofocal IOL Toric phakic IOL	PMMA with UV absorber Proprietary polysiloxane with bonded UV and orange absorber as optic, brown PMMA haptics Iris-fixated Artisan® IOL
Rayner IOL Ltd., (Manufacturer of Ridley IOL)	Monofocal IOL Foldable IOLs Monofocal Toric Multifocal	PMMA with UV absorber Proprietary hydrophilic copolymer of acrylates with UV absorber
Visiogen (now acquired by Abbott Medical Optics)	Foldable dual optic single piece, movement based Accommodative IOL Synchrony	Proprietary polysiloxane for two optics and a connecting spring haptic

to the above-discussed good short-term postoperative results, but also, most importantly, to long-term reduction in PCO formation and the necessary Nd-YAG posterior capsulotomy (Hollick et al., 1999). The Nd-YAG capsulotomy rate for AcrySof IOL is <5%, the lowest thus far reported for all IOLs (Apple et al., 2001). Scientific investigations have explained this, based on both the properties of its biomaterial and its design. The surface property of the material and the square-edged design of the optic are attributed as factors for reduction of PCO (Linnola, 1997; Linnola et al., 2000). These investigations led to several more IOLs from other companies (AMO, Hoya, Zeiss-Meditec, etc.) beside AcrySof IOL, which are also made of hydrophobic acrylate materials with square-edged designs. Long-term clinical evaluations are ongoing to compare these IOLs with AcrySof IOL, which is regarded as a reference IOL because of its long-term successful track record.

In addition to several proprietary hydrophobic soft acrylates and polysiloxanes, several hydrophilic soft acrylates and newer hydrogels have also been introduced as foldable IOLs. But their disadvantages, such as greater tendency for fibrotic membrane formation over the anterior surface of the IOL, ACO, and postoperative surface or deep calcification, higher PCO rates, and lack of proven advantage over hydrophobic soft acrylates or polysiloxane IOLs, had reduced their acceptance (Koch et al., 1999; Apple et al., 2000b; Werner et al., 2001; Izak et al., 2003). But recently Rayner, Bausch and Lomb, Zeiss-Meditec, and others have introduced newer hydrophilic materials with around 26% water content, and have designed IOLs with a variety of square-edged designs. These IOLs are also being evaluated for long-term performance related to calcification possibility and PCO rates (Neuhann et al., 2006; Nishi et al., 2007).

Currently, hydrophobic acrylates and polysiloxane foldable IOLs are predominantly used in developed countries. Special insertion tools have been designed and used to implant them through a 3 mm or smaller incision, the same as required for ECCE cataract extraction by phacoemulsifier. Recent improvements have further reduced the incision size for phacoemusification to 2.2 mm, with the original coaxial phacoemulsification hand piece design, and down to 1.4 mm for a bimanual technique requiring two separate incisions and the use of a sleeveless phacoemulsification tip for breaking up cataract through one very small incision, while providing irrigation separately through another small incision. This incision size is subsequently slightly enlarged depending upon the biomaterial, design, and the power in diopters of the IOL injected through them. With this improvement, foldable IOLs and their injectors are available for implantation through 1.8 to 2.4 mm for the average eye's requirement.

Though the AcrySof IOL is widely used, many surgeons find silicone IOLs also as acceptable for most cataract patients without other ocular pathology, especially when the cost of the IOL is a consideration. Silicone IOLs are made by several manufacturers. Concern has been expressed about optical complications associated with silicone oil adhesion on silicone IOLs if the use of silicone oil is required in the future for vitreo-retinal surgery (Khawly et al., 1998; Kaushik et al., 2001). IOLs fabricated from other soft hydrophobic acrylates are also now in the market (Werner et al., 2008).

Monofocal IOLs with fixed spherical refractive power, discussed earlier, remained the dominant IOL implant for replacement of the cataractous lens until recently, when aspheric foldable IOLs have gained significant momentum in the developed world. This optical function enhancement is described in the next section. Transition from hard PMMA IOLs to soft foldable IOLs for small-incision cataract surgery has mostly occurred in the developed world. This was primarily driven by improved postoperative results for patients. Therefore, this transition will be ongoing in the developing world when the necessary resources and skills are available.

IOLS WITH VARIATIONS OF OPTICAL FUNCTION

Monofocal Aspheric IOLs

Since the image formation on the retina in a pseudophakic eye is determined by the two optical lenses consisting of cornea and IOL, it can be optimized in terms of reducing overall spherical aberration by designing appropriate aspheric optics for the IOL to compensate for the corneal spherical aberration. Average corneal spherical aberration compensating AMO's Tecnis IOL (created in 2003 by Pharmacia Company), was the first such FDA-approved IOL. Night-time driving simulations, and other critical visual performance, showed some improvement leading to a "New Technology" designation and a $50 higher reimbursement by Medicare in the USA. Soon Alcon, with its AcrySof IQ, and Bausch and Lomb, with its AO Port IOL, followed by their own aspheric IOL designs for reducing spherical aberrations, also achieved the "New Technology" designation. Besides different biomaterials, they also have slightly different asphericity. Rigorous comparative evaluations of the three designs are ongoing to delineate the patient's corneal shape-dependent advantage of one design over other (Montes-Mico et al., 2009).

Monofocal Toric IOLs

Since monofocal IOLs with spherical refractive power do not correct pre-existing corneal astigmatism, it is corrected either surgically by certain techniques during the cataract surgery or postoperatively using spherocylindrical power spectacles. With the same biomaterials as those used for monofocal IOLs, toric IOLs with toric optics have recently emerged as another alternative for the more precise correction of significant pre-existing corneal astigmatism. The axis of the cylinder for the spherocylindrical optic of the monofocal toric IOL is marked on the IOL for necessary proper alignment during cataract surgery. Postoperative rotation of a toric silicone IOL with plate haptics (from Staar Surgical) led to its redesign, since such rotation significantly compromised the optical objective of the toric IOL. Another toric single piece IOL design with somewhat flexible haptics (from Alcon Inc.) that can provide improved stable fixation has been clinically found in a comparative study to give much better results without significant rotation (Chang, 2008). AcrySof Toric IOL is now approved by various regulatory agencies, including the FDA in USA and CE marking in Europe. Its use is rapidly increasing, especially since it is considered as more profitable "premium IOL" for reimbursement purposes in the US, as will be discussed in the "Overall Summary and Future of IOLs."

Multifocal IOLs

Since monofocal IOLs optically focus on an object at only one chosen distance, with a relatively narrow depth of focus, postoperatively patients are required to wear spectacles for other distances. Even though almost all elderly cataract patients are accustomed to wearing spectacles because of their presbyopia prior to cataract surgery, the investigation of the opportunity to eliminate spectacles using multifocal IOLs began in the 1980s. With the same biomaterials as for monofocal IOLs, several multifocal optical designs using refractive optical zones or diffractive optical structures have been investigated, implementing simultaneous optical imaging for objects located at a distance, as well as near objects (Maxwell and Nordan, 1990). This optical design principle is called simultaneous vision, and has the potential to eliminate halo and other unwanted optical consequences, depending on the optical design. Multifocal IOLs with refractive optical zones are sensitive to pupil size in terms of their optical performance. The FDA-approved Array Multifocal IOL by AMO, Inc. has a radial array of alternating variable optical power to provide both distant and near optics for any pupil size (Steinert et al., 1999). Although for pupils of 3 mm diameter or larger such a design does provide multifocal vision, the near vision is reported to be suboptimal for a smaller pupil size. Also, halos appear with night-time driving when the pupil is larger, and these remain a potentially unacceptable complication for some patients (Haring et al., 2001). A redesigned Array Multifocal IOL with improvements on the earlier design is currently on the market as ReZoom IOL.

Earlier, the 3M Corporation, and, more recently, AMO (who acquired Pharmacia Corporation's multifocal IOL and now has been acquired by Abbott Medical Optics, still retaining the familiar abbreviation AMO) introduced a Tecnis diffractive optic design that provides pupil size-independent distant and near optics, but still retains potentially unacceptable halos with night-time driving. With better understanding of the cause of such halos, a unique optical design that combines a limited central apodized diffractive optical zone with refractive optics beyond it was introduced by Alcon, Inc., as the AcrySof ReSTOR® multifocal IOL to minimize night-time halo and retain pupil-independent optics for day-time (Dublineau, 2003). Carl Zeiss-Meditec also has a variety of diffractive/refractive combination optics for Acri.Lisa® multifocal IOLs. Thus, there are now many designs of multifocal IOLs (Maxwell et al., 2009).

The simultaneous vision principle based multifocal IOLs demand a more accurate selection of the necessary IOL power in order to achieve spectacle-free vision post-operatively. Also, since the vision achieved is not an actual accommodation, it is called pseudoaccommodation. Since they also correct presbyopia, they are also called "presbyopia-correcting." When used for replacement of cataract in Medicare patients in the US, they too are considered as "premium IOLs" (Maxwell et al., 2008).

Applications of multifocal IOLs for treatment of eyes without cataract, but with presbyopia, hyperopia or myopia, are now clinically carried out in certain suitable patients. For treatment of any one of these refractive errors of the eye, its noncataractous, clear natural lens is surgically removed and replaced by a multifocal IOL (Packer et al., 2002). This lensectomy is carried out by the same surgical procedure as that required for cataract surgery. The unpredictability of achieving necessary targeted refraction, cost, and acceptance of unavoidable optical effects of the simultaneous vision optics of multifocal IOLs remain key hurdles for its widespread use for treatment of refractive errors. Since the prominent use of computers in daily life has developed, satisfactory intermediate distance has become a requirement, as well as near and distance vision provided by the multifocal IOLs, and thus significant improvement has been achieved in designs by adjusting near add power, and incorporating aspheric and toric optical enhancements discussed above for monofocal IOLs.

Phakic IOLs

Currently, for correction of refractive errors of myopia or hyperopia, treatment alternatives include spectacles, contact lenses, and a variety of corneal refractive surgery procedures including LASIK. As another modality of treatment, the phakic IOL, which is an IOL to correct the refractive error of the eye, is placed in the phakic eye in addition to the natural crystalline lens. Phakic IOLs have been clinically investigated for many years with continuous improvements to achieve long-term safety and, it is hoped, widespread use. Unique designs with newer biomaterials are currently being investigated for achieving this goal.

For correction of myopia, the angle-supported phakic AC-IOL designs of PMMA, similar to earlier AC-IOLs for cataract surgery, resulted in complications of damaged corneal endothelial cells, cataract formation, and pupil ovalization (Chen et al., 2008). More flexible, soft, and foldable phakic AC-IOL designs of hydrophobic soft acrylate material (Alcon's AcrySof Cachet Phakic AC-IOL), when properly sized, are reported to have improved acceptable results in long-term clinical investigations, and are approved in international markets except the US where FDA-required additional studies are ongoing (Knorz, 2003). Iris-fixated phakic AC-IOLs of silicone material (AMO's Verisyse®) are approved by the US FDA, and when properly implanted have also been reported to have acceptable results (Moshirfar et al., 2007). The anterior chamber of the hyperopic eye is anatomically relatively small, and thus less suitable for a phakic AC-IOL.

Phakic posterior chamber IOL designs using silicone and hydrogel materials are placed between the iris and the natural lens to correct myopic and hyperopic refractive errors. They have been clinically investigated over the past few years. Their anatomical placement has potential for iris chafing and the related complication of pigmentary glaucoma. Also, cataractogenesis of the natural lens by either metabolic disturbances or accidental touch of a surgical instrument has been reported, especially with silicone phakic PC-IOLs (Sanchez-Galeana et al., 2003). A hydrogel phakic PC-IOL (Starr's Visian ICL), after several design iterations, in its final design is reported to have fewer complications and US FDA has approved it (Sanders et al., 2004).

Accommodative IOLs

In order to provide true accommodation after cataract surgery, as well as to treat refractive errors of myopia, hyperopia, and presbyopia, a safe and effective accommodative IOL with three or more diopters of accommodation ability to replace the cataract or the natural lens has been long sought.

Designs implementing the concept of forward movement of the optic of the IOL to achieve accommodation were pursued by C&C Vision, Inc., of the US, and Human Optics of Germany (Samalonis, 2002). The Crystalens AT-45 and AT-50 of C&C Vision, which has now been acquired by Bausch and Lomb, is a silicone IOL with polyamide haptics and two hinges that permit axial forward movement of the optic for accommodation and subsequent backward movement for disaccommodation (Cumming et al., 2001). Although only 1 to 2 diopters of accommodation is reported, the results remain variable, questionable, and without scientific understanding of such amplitude from movement of a single optic (Findl and Leydolt, 2007). Since there is no other significant safety issue, even this limited success is encouraging, and the lens is approved by the FDA, and is also a "premium IOL" for reimbursement by Medicare in the US.

In order to achieve a greater magnitude of accommodation, a mechanically coupled pair of optic-based accommodative IOLs were investigated in an animal model by Hara in Japan (Hara et al., 1992), and then by Sarfarazi (Sarfarazi, 2003) in the US. A new design based on this concept called "Synchrony" has been created in silicone material by Visiogen Company (now acquired by Abbott Medical Optics (AMO)). It is currently available in Europe and under clinical investigation in the US for FDA approval. This dual optic design has an anterior +32 diopter power optic which moves and is coupled by a spring haptic to a minus power stationary optic to create much higher scientifically understandable overall +2 to +2.5 diopters of accommodation (McLeod et al., 2007).

Another concept under investigation in animal models requires removal of the contents of the natural cataractous or clear lens through the smallest possible anterior capsulotomy, and subsequently filling by an appropriate silicone polymer to achieve the needed geometrical shape and optical power (Haefliger et al., 1987). Other gel biomaterials have also been investigated, but a host of surgical, biological, and optical issues remain as hurdles to this approach to restore accommodation using the natural capsular bag, zonules, and ciliary muscle for a physiology-based truly accommodative IOL. In another variation of this approach, Nishi of Japan has reported a new design for an accommodative IOL by refilling of the capsular bag with a liquid silicone polymer (Nishi et al., 2008).

Two new concepts for accommodating IOLs based on shape changes to achieve accommodation have emerged recently. NuLens Ltd., of Israel, has a concept for NuLens which consists of a shape change using a diaphragm and a silicone gel with a claim of 8 to 10 diopters of accommodation. Power Vision Inc., of the US, has its accommodative IOL concept based on movement of a biomaterial fluid from a reservoir. Thus, a search for truly accommodative IOL may challenge interdisciplinary resources, including newer biomaterials for some of the above concepts.

Adjustable Power IOLs

For achieving the targeted postoperative refraction for any pseudophakic eye, careful measurement of the axial length of the eye and refractive power of its cornea are required before IOL implantation surgery. Subsequently, using one of many available IOL power selection mathematical formulas, the necessary IOL power is selected by the surgeon for implantation. Errors in this procedure, and deviation of the actual axial position of the IOL from that anticipated by the formula, result in postoperative residual refractive error. Most of the time, postoperative spectacle correction to address this error is acceptable, but occasionally excessive error requires explantation of the IOL to address an unacceptable imbalance of refraction between the two eyes. Also, to achieve the goal of spectacle-free vision with multifocal IOLs or phakic IOLs, achievement of targeted refraction is more critical and necessary. Thus, for many years, concepts for a postoperative adjustable power IOL have been investigated. Several concepts requiring surgical re-entry into the eye were investigated, but did not result in products because of concern for safety and cost. A new, unique biomaterial-based light adjustable IOL by Calhoun Vision, Inc. is being pursued clinically in the human eye (Schwartz, 2003; Olson et al., 2006). This approach is attractive, since it does not require surgical re-entering of the eye for adjustment of the refractive power of the already implanted IOL. This new biomaterial technology consists of an IOL which is implanted as a polymerized silicone matrix containing nonpolymerized silicone macromers, with an attached photoinitiator dispersed within it for subsequent staged photopolymerization (Maloney, 2003). The IOL is implanted as per current practice for IOL power selection for the targeted refraction. Postoperatively, the achieved refraction is measured. Depending upon sign and magnitude of the refractive error, for necessary adjustment of power of the IOL, the IOL is judiciously exposed to long UV radiation from a slit-lamp based LAL® light delivery instrument, leading to partial photopolymerization. Spatial intensity distribution is used for either exposing the central optical zone, to increase the power of the IOL or exposing the peripheral optical zone, to decrease the power of the IOL. After the required duration for redistribution of the remaining unpolymerized macromers in the IOL, the patient is re-examined for the consequent change in the shape and, hence, power of the IOL toward the targeted correction of the refractive error. If the IOL power adjustment is not achieved, the procedure of partial polymerization is repeated with needed adjustment until targeted refraction is achieved. At this stage, as a final step, the entire optic is exposed to UV radiation for complete "lock-in" of the shape of the IOL. After the safety of this IOL was established in rabbit eyes, clinical investigation required to establish safety and efficacy in terms of accuracy, precision, and reproducibility were done in Mexico, and US FDA-required clinical investigations are being pursued (Olson et al., 2006). It is also now available and used in Germany and certain other European countries. Additional applications of this technology for custom aspheric monofocal and aspheric multifocal IOLs are being identified and pursued.

Visible Light Filtering IOLs

Earlier, the Menicon and Hoya corporations of Japan, for their PMMA IOLs, and in 2003 Alcon, Inc., for its AcrySof Natural IOL and Hoya for its AF-1 IOL and several other manufacturers, have each incorporated a yellow dye into their foldable IOL materials in order to filter visible blue light, in addition to complete blocking of UV radiation (Brockmann et al., 2008) for protection of the retina. Bausch and Lomb Company have introduced a violet-light filtering IOL as a more conservative approach to retinal protection. The most widely used AcrySof Natural IOL mimics the UV-visible transmission of a young adult human natural lens (Cionni, 2003). In 2002 a significant scientific discovery found previously unknown intrinsically photosensitive retinal ganglion cells (ipRGC) which play a significant role in photoentrainment of the circadian rhythm. A concern was raised for possible filtering of light required by these cells when yellow tinted blue light-filtering IOLs are implanted. This was carefully investigated (Patel and Dacey, 2009), with the conclusion that since the action spectrum for photoentrainment of the circadian rhythm is sufficiently wide, and peaks at much higher wavelengths, there is no such concern. The scientific rationale for such a yellow-tinted IOL is to prevent unnatural excessive blue color perception, and to protect the more susceptible aging retina after cataract surgery (Davison and Patel, 2005). Blue light toxicity to the retina is mediated by age-related accumulation of lipofuscin (Carson et al., 2008), and is considered one of the potential risk factors of age-related macular degeneration (AMD), a serious sight-threatening disease. A recent European Eye Study has found an association between blue light and AMD (Fletcher et al., 2008). While a prospective randomized trial is needed to confirm this relationship, there is an increasing concern for light as a risk factor for AMD in eyes with IOLs, since they remain for a much longer duration after their implantation in pediatric cataract patients, as well as in younger adults for presbyopia correction. Also, an increasing number of elderly patients live much longer after cataract surgery as it is done today worldwide. A recent paper (Davison et al., 2011) have reviewed all recent studies related to the consequences on photoreception, as well as evidence for potential photoprotection provided by the blue light-filtering IOLs. This review provides a contemporary clinical perspective for their use as a potential simple measure against the dreaded AMD, for which we have currently only expensive treatment with limited effectiveness. Thus, a relatively simple incorporation of a safe bonded yellow dye into a biomaterial for IOLs provides an illustration of how IOL biomaterials can be tailored to achieve a potentially significant impact on the prevention of a sight-threatening disease as an additional function, along with the primary function of the IOL for optical imaging.

OVERALL SUMMARY AND FUTURE OF IOLS

In a relatively brief span of a few decades, IOLs have emerged as one of the most successful biomaterial-based implants. They are considered a safe and effective means to prevent the leading cause of age-related degradation of vision due to cataracts, which eventually leads to functional blindness. Currently, an estimated 10 million IOLs are implanted annually for cataract replacement, mostly in the developed world and in some regions of the developing world where resources are available. If the prevailing rate of cataract surgery in the US is projected for the approximately six billion population of the world, then it is estimated that 45 million IOLs are needed annually! Few ophthalmologists and little money have led to delayed (or no) cataract treatment in the developing world, resulting in a tragic backlog of 20 million cataract-related blind people. Thus, with expected continuous global development and simultaneous increases in population, as well as life expectancy, and with increasing demand for good vision, the future of IOLs for cataract surgery is ensured.

The successful developments of IOLs have been the outcome of interdisciplinary team efforts by creative surgeons, scientists, and engineers in both academia and industry, and also economics of reimbursement by healthcare systems. In the year 2005, the Center for Medicare and Medicaid Services (CMS) in the US made a momentous ruling which shifted their policy (Mamalis, 2005) related to reimbursement of IOLs. This ruling allowed cost sharing between Medicare (Government) and the patient for significantly improved IOLs which was not allowed prior to this ruling, and Medicare paid as if it was a standard monofocal IOL. CMS now lists several significantly improved IOLs, such as presbyopia-correcting (multifocal and accommodative) IOLs and astigmatism-correcting IOLs. They are called "premium IOLs" at a free market price set by the manufacturer, and the patient will pay any cost difference between it and the reimbursement by CMS which will continue, but is only limited to that for a conventional standard monofocal IOL (Maxwell et al., 2008). This new CMS policy thus encourages IOL manufacturer's to invest in R&D for competitive new innovations, since they can get their return with a free market price of IOLs without the limit of affordability set by the US government.

Starting from a biomaterial selection based on the accidental implantation of pieces of PMMA from the canopies of fighter airplanes, a variety of newer biomaterials have been specifically created to achieve the various designs and functional goals of IOLs, as outlined in Table B.2. Creative interactions and an understanding of the opportunities and issues of biomaterials, optical and mechanical designs, and evolving surgical techniques, along with improved understanding of the unique biology of the eye, have spearheaded the sophistication we see in modern IOLs. These efforts are expected to accelerate with the above described new CMS ruling in the US as an incentive for new innovative IOLs. Specifically, efforts will continue to achieve the development of truly accommodative IOLs with the capability of mimicking the physiological function of the natural lens. Multidisciplinary efforts will also continue to achieve improved phakic IOLs to correct the refractive errors of presbyopia, myopia, and hyperopia. When long-term safety and efficacy are demonstrated, phakic IOLs or truly accommodative IOLs have the potential to replace the current alternatives of spectacles, contact lenses, and a variety of corneal refractive surgery procedures, including LASIK. Thus, globally, the number of IOLs required for all functional goals could be many times greater than the given projection for IOLs needed for cataract replacement.

Further opportunities are expected for improved optical designs of IOLs for correction of corneal aberrations beyond the first generation of current aspheric IOLs. With further understanding of the optical needs for achieving optimum vision, eventually a custom IOL may evolve for each individual eye for all of the projected functions of IOLs for aphakic and phakic eyes (Kohnen, 2003). Even incorporating the additional non-optical function of postoperatively needed drug delivery is being explored as a potential opportunity. Achievement of these goals will require advances in biomaterials for IOLs, and biomaterials scientists working closely with physicists, chemists, engineers, surgeons, pharmacologists, regulatory affairs experts, and business people. The success of IOLs in improving vision worldwide genuinely showcases the strength of the biomaterials endeavor, and the highly multidisciplinary nature of this field.

BIBLIOGRAPHY

Alpar, J. J., & Fechner, P. U. (1986). *Fechner's Intraocular Lenses*. Thieme Inc., New York.

American Academy of Ophthalmology. (1989). *Cataract in the Otherwise Healthy Adult Eye*. San Francisco, CA: AAO.

Amon, M. (2001). Biocompatibility of intraocular lenses. *J. Cataract Refract Surg.*, **27**, 178–179.

Anderson, C., Koch, D. D., Green, G., Patel, A., & Vannoy, S. (1993). Alcon AcrySof Acrylic Intraocular Lens. In R. G. Martin, J. P. Gills, & D. R. Sanders (Eds.), *Foldable Intraocular Lenses* (pp. 161–177). Thorofare, NJ: Slack.

Apple, D. J., Mamalis, N., Brady, S. E., Loftfield, K., Kavka-Van Norman, D., & Olson, R. J. (1984). Biocompatibility of implant materials: A review and scanning electron microscopic study. *J. Am. Intraocular Implant Soc., I*, 53–66.

Apple, D. J., Brems, R. N., Park, R. B., Norman, D. K., Hansen, S. O., Tetz, M. R., Richards, S. C., & Letchinger, S. D. (1987). Anterior chamber lenses. Part I: Complications and pathology and a review of design. *J. Cataract Refractive Surgery*, **13**, 157–174.

Apple, D. J., Kincaid, I., Mamalis, N., & Olson, R. J. (1989). *Intraocular Lenses: Evolution, Designs, Complications and Pathology*. Baltimore, MD: Williams and Wilkins.

Apple, D. J., Auffarth, G. U., Peng, Q., & Visessook, N. (2000a). *Foldable Intraocular Lenses: Evolution, Clinicopathologic, Correlations and Complications*. Thorofare, NJ: Slack.

Apple, D. J., Werner, L., Escobar-Gomez, M., & Pandey, S. K. (2000b). Deposits on the optical surfaces of Hydroview intraocular lenses (letter). *J. Cataract Refract Surg.*, **26**, 1773–1777.

Apple, D. J., Peng, Q., Visessook, N., Werner, N., Pandey, S. K., et al. (2001). Eradication of posterior capsule opacification: Documentation of a marked decrease in Nd-YAG laser posterior capsulotomy rates noted in analysis of 5,416 pseudophakic human eyes obtained post-mortem. *Ophthalmology*, **108**, 505–518.

Binkhorst, C. D. (1959). Iris-supported artificial pseudophakia: A new development in intraocular artificial lens surgery (iris clip lens). *Trans. Ophthalmol. Soc. UK*, **79**, 859–584.

Brian, G., & Taylor, H. (2001). Cataract blindness: Challenges for the 21st century. *Bull World Health Org.*, 249–256.

Brockmann, C., Schultz, M., & Laube, T. (2008). Transmittance characteristics of ultraviolet and blue-light-filtering intraocular lenses. *Cataract Refract Surg.*, **34**, 1161–1166.

Carson, D., Margrain, T. H., & Patel, A. (2008). New approach to evaluate retinal protection by intraocular lenses against age-related lipofuscin accumulation-mediated retinal phototoxicity. *J. Cataract Refract Surg.*, **34**, 1785–1792.

Chen, L., Chang, Y., Kuo, J. C., Rajagopal, R., & Azar, D. T. (2008). Meta-anlysis of cataract development after phakic intraocular lens surgery. *J. Cataract Refract Surg.*, **34**, 1181–1200.

Choyce, D. P. (1958). Correction of uni-ocular aphakia by means of anterior chamber acrylic implants. *Trans. Ophthalmol. Soc. UK*, **78**, 459–470.

Cionni, R. J. (2003). Clinical study results of the AcrySof Natural IOL. In *Abstracts, Symposium on Cataract, IOL & Refractive Surgery* (pp. 11). Fairfax, VA: ASCRS.

Cumming, J. S., Slade, S. G., & Chayet, A. (2001). Clinical evaluation of the model AT-45 silicone accommodative intraocular lens; results of feasibility and the initial phase of a Food and Drug Administration clinical trial: The AT-45 Study Group. *Ophthalmology*, **108**, 2005–2009; discussion by Werblin, T. P, p. 2010.

Davison, J. A., & Patel, A. S. (2005). Light normalizing intraocular lenses. *International Ophthalmology clinics (Ultraviolet Radiation and the Eye)*, **45**, 55–106.

Davison, J. A., Patel, A. S., Cunha, J. P., Schwiegerling, J., & Muftuoglu, O. (2011). Recent studies provide a contemporary clinical perspective of blue light-filtering IOL. *Graefes. Arch. Clin. Exp. Ophthalmol.*, **249**, 957–968.

Dublineau, P. (2003). Experience with the AcrySof RESTOR IOL. In *Abstracts, Symposium on Cataract, IOL & Refractive Surgery* (pp. 57). Fairfax, VA: ASCRS.

Ellingson, F. T. (1978). The uveitis–glaucoma–hyphema syndrome associated with the Mark VIII Anterior Chamber Lens Implant. *Am. Intraocular Implant Soc. J.*, **4**, 50–53.

Findl, O., & Leydolt, C. (2007). Meta-analysis of accommodating intraocular lenses. *J. Cataract Refract Surg.*, **33**, 522–527.

Fletcher, A. E., Bentham, G. C., Agnew, M., Young, I. S., Augood, C., et al. (2008). Sunlight exposure, antioxidants and age-related macular degeneration. *Arch. Ophthalmol.*, **120**, 1396–1403.

Gills, J. P., & Sanders, D. R. (1991). Use of small incision to control induced astigmatism and inflammation following cataract surgery. *J. Cataract Refract Surg.*, **17**(Suppl), 740–744.

Gimbel, H. V., & Neuhann, T. (1990). Development, advantages and methods of the continuous circular capsulorhexis technique. *J. Cataract Refract Surg.*, **16**, 31–37.

Haefliger, E., Parel, J. M., Fantes, F., Norton, E. W., Anderson, D. R., et al. (1987). Accommodation of an endocapsular silicone lens (phaco-ersatz) in the non-human primate. *Ophthalmology*, **94**, 471–477.

Hara, T., Hara, T., Yasuda, A., Mizumoto, Y., & Yamada, Y. (1992). Accommodative intraocular lens with spring action: Part 2. Fixation in the living rabbit. *Ophthalm. Surg.*, **23**, 632–635.

Haring, G., Dick, H. B., Krummenaur, F., Weissmantel, U., & Uroncke, W. (2001). Subjective photic phenomena with refractive multifocal and monofocal intraocular lenses; results of a multicenter questionnaire. *J. Cataract Refract Surg.*, **27**, 245–249.

Hollick, E. J., Spalton, D. J., & Ursell, P. G. (1999). The effect of polymethyl methacrylate, silicone and polyacrylic intraocular lenses on posterior capsule opacification three years after cataract surgery. *Ophthalmology*, **106**, 49–55.

Izak, A. M., Werner, L., Pardey, S. K., & Apple, D. J. (2003). Calcification of modern foldable hydrogel intraocular lens designs. *Eye*, **17**, 393–406.

Kaufman, P. L. (1992). Accommodation and presbyopia: Neuromuscular and biophysical aspects. In W. M. HartJr (Ed.), *Adler's Physiology of the Eye: Clinical Application* (9th ed, pp. 391–411). St. Louis, MO.

Kaushik, R. J., Brar, G. S., & Gupta, A. (2001). Neodymium-YAG capsulotomy rates following phacoemulsification with implantation of PMMA, silicone and acrylic intraocular lenses. *Ophthalmic Surg. Lasers*, **32**, 375–382.

Khawly, J. A., Lambert, R. J., & Jaffe, J. G. (1998). Intraocular lens changes after short- and long-term exposure to intraocular silicone oil: An *in vivo* study. *Ophthalmology*, **105**, 1227–1233.

Knorz, M. (2003). Three-year phase I clinical results of the AcrySof phakic ACL. In *Abstracts, Symposium on Cataract, IOL and Refractive Surgery* (pp. 25). Fairfax, VA: ASCRS.

Koch, M. U., Kalicharan, D., & Vanderwant, J. J.L. (1999). Lens epithelial cell formation related to hydrogel foldable intraocular lenses. *J. Cataract Refract Surg.*, **25**, 1637–1640.

Kohnen, T., Dick, B., & Jacobi, K. W. (1995). Comparison of the induced astigmatism after temporal clear corneal tunnel incision of different size. *J. Cataract Refract Surg.*, **21**, 417–424.

Kohnen, T. (2003). Aberration-correcting intraocular lenses. *J. Cataract Refract Surg.*, **29**, 627–628.

Kronenthal, R. L. (1977). Intraocular degradation of nonabsorbable suture. *Am. Intraocular Implant Soc. J.*, **3**, 222–228.

Kruger, A. J., Amon, M., Schauersberger, J., Abela-Formanek, C., Schild, G., et al. (2001). Anterior capsule opacification and lens epithelial outgrowth on the intraocular lens surface after curettage. *J. Cataract Refract Surg.*, **127**, 1987–1991.

LeGrand, Y., & El Hage, S. G. (1980). Physiological Optics. *Springer-Verlag Series in Optical Sciences*. (Vol. 13). New York, NY: Springer-Verlag.

Linnola, R. J. (1997). The sandwich theory: A bioactivity-based explanation for posterior capsule opacification after cataract surgery. *J. Cataract Refract Surg.*, **23**, 1539–1542.

Linnola, R. J., Werner, L., Pandey, S. K., Escobar-Gomez, M., Znoiko, S. L., et al. (2000). Adhesions of fibronectin, vitronectin, laminin and collagen-type IV to intraocular lens materials in pseudophakic human autopsy eyes. Part II: Explanted IOLS. *J. Cataract Refract Surg.*, **26**, 1807–1818.

Luntz, M. H., & Livingston, D. G. (1977). Astigmatism in cataract surgery. *Br. J. Ophthalmol.*, **61**, 360–365.

Maloney, R. (2003). The changing shape of customized IOLs. *Rev. Ophthalmol.*, 10, 01.

Mamalis, N. (2002). IOL biocompatibility. (editorial), *J. Cataract Refract Surg.*, **28**, 1–2.

Mamalis, N. (2005). Additional payments for Presbyopia-correcting intraocular lenses. (editorial), *J. Cataract Refract Surg.*, **31**, 1467–1468.

Maxwell, W. A., & Nordan, L. T. (Eds.), (1990). *Current Concepts of Multifocal Intraocular Lenses*. Thorofare, NJ: Slack.

Maxwell, W. A., Waycaster, C. R., D'Souza, A. O., Meissner, B. L., & Hileman, K. (2008). A United States cost-benefit comparison of an apodized, diffractive, Presbyopia-correcting, multifocal intraocular lens and a conventional monofocal lens. *J. Cataract Refract Surg.*, **34**, 1855–1861.

Maxwell, W. A., Lane, S. S., & Zhou, F. (2009). Performance of presbyopia-correcting intraocular lenses in distance optical bench tests. *J. Cataract Refract Surg.*, **35**, 166–171.

Mazzocco, T. R., Rajacich, G. M., & Epstein, E. C. (1986). *Soft Implant Lenses in Cataract Surgery*. Thorofare, NJ: Slack.

McLeod, S. D., Vargas, L. G., Portney, V., & Ting, A. (2007). Synchrony dual-optic accommodating intraocular lens. Part 1: Optical and biomechanical principles and design considerations. *J. Cataract Refract Surg.*, **33**, 37–46.

Mester, V., Dillinger, P., & Anterist, N. (2003). Impact of a modified optic design on visual function: Clinical comparative study. *J. Cataract Refract Surg.*, **29**, 652–660.

Miller, D., & Stegmann, R. (1982). The use of Healon in intraocular lens implantation. *Int. Ophthalmol. Clin.*, **22**, 177–187.

Montes-Mico, R., Ferrer-Blasco, T., & Cervinom, A. (2009). Analysis of the possible benefits of aspheric intraocular lenses: Review of the literature. *J. Cataract Refract Surg.*, **35**, 172–181.

Moshirfar, M., Hulk, H. A., & Davis, D. K. (2007). Two-year follow-up of Artisan/Verisyse iris-supported phakic intraocular lens for the correction of high myopia. *J. Cataract Refract Surg.*, **33**, 1392–1397.

Neuhann, I. M., Stodulka, P., Werner, L., Mamlis, N., Pandey, S. K., et al. (2006). Two opacification patterns of the same hydrophilic acrylic polymer: Case reports and clinicopathological correlation. *J. Cataract Refract Surg.*, **32**, 879–886.

Nicholson, D. H. (1982). Occult iris erosion: A treatable case of recurrent hyphema in iris-supported intraocular lenses. *Ophthalmology*, **84**, 113–120.

Nishi, O., Nishi, K., Nishi, Y., & Chang, S. (2008). Capsular bag refilling using a new accommodating intraocular lens. *J. Cataract Refract Surg.*, **34**, 302–309.

Nishi, Y., Rabsilber, T. M., Limberger, I., Reuland, A. J., & Auffarth, G. U. (2007). Influence of 360-degree enhanced optic edge design of a hydrophilic acrylic intraocular lens on posterior capsule opacification. *J. Cataract Refract Surg.*, **33**, 227–231.

Obstbaum, S. A. (1984). Complications of intraocular lenses. Membranes, discolorations, inflammation and management of the posterior capsule. In J. M. Engelstein (Ed.), *Cataract Surgery: Current Options and Problems* (pp. 509–533). Orlando, FL: Grune and Stratton.

Obstbaum, S. A. (2003). Emergence of the role of cataract and IOL surgery in the correction of refractive errors. (editorial), *J. Cataract Refract Surg.*, **29**, 857.

Olson, R. J., & Crandall, A. S. (1998). Prospective randomized comparison of phacoemulsification cataract surgery with a 3.2-mm versus a 5.5-mm sutureless incision. *Am. J. Ophthalmol.*, **125**, 612–620.

Olson, R., Mamalis, N., & Haugen, B. (2006). A light adjustable lens with injectable optics. *Ophthalmology Clinics of North America*, **19**, 135–142.

Packer, M., Fine, I. H., & Hoffman, R. S. (2002). Refractive lens exchange with the Array multifocal intraocular lens. *J. Cataract Refract Surg.*, **26**, 421–422.

Patel, A. S., Carson, D. R., & Patel, P. H. (1999). Evaluation of an unused 1952 Ridley intraocular lens. *J. Cataract Refract Surg.*, **25**, 1535–1539.

Patel, A. S., & Dacey, D. M. (2009). Relative effectiveness of a blue light-filtering intraocular lens for photoentrainment of the circadian rhythm. *J. Cataract Refract Surg.*, **35**, 529–539.

Philipson, B., Fagerholm, P., Calel, B., Grunge, A., Hallnäs, K., et al. (1990). Heparin surface modified intraocular lenses: A one-year followup of a safety study. *Aceta Ophthalmol. (Copenhagen)*, **68**, 601–603.

Ratner, B. D. (1998). Ophthalmologic biocompatibility: Anachronism or oxymoron? (guest editorial), *J. Cataract Refract Surg.*, **24**, 288–290.

Ridley, H. (1951). Intra-ocular acrylic lenses. *Trans. Ophthalmol. Soc. UK*, **71**, 617–621.

Sarfarazi, F. M. (2003). *Optical and mechanical design for human implantation of the Sarfarazi elliptical accommodating IOL.* Abstracts, Symposium on Cataract, IOL & Refractive Surgery, pp. 189. Fairfax, VA: ASCRS.

Samalonis, L. B., (Ed.). (2002). The 21st century IOL. *Eye World News*, 7, 30. ASCRS: Fairfax, VA.

Samalonis, L. B., (Ed.). (2002). Accommodative IOLs coming a long way. *Eyeworld*, March. American Society of Cataract and Refractive Surgery: Fairfax, VA.

Sanchez-Galeama, C. A., Smith, R. J., Sanders, D. R., Rodríguez, F. X., Litwak, S., et al. (2003). Lens opacities after posterior chamber phakic intraocular lens implantation. *Ophthalmology*, **110**, 781–785.

Sanders, D. R., Grabow, H. B., & Shepard, J. (1992). The toric IOL. In J. P. Gills, R. G. Martin, & D. R. Sanders (Eds.), *Sutureless Cataract Surgery: An Evolution Towards Minimally Invasive Technique* (pp. 183–197). Thorofare, NJ: Slack.

Sanders, D. R., Doney, K., & Poco, M. (2004). United State Food and Drug Administration clinical trial of implantable Collamer lens (ICL) for moderate to high myopia: Three-year follow-up. *Ophthalmology*, **111**, 1683–1692.

Schwartz, D. M. (2003). Light-adjustable lens. *Trans. Am. Ophthalmol. Soc.*, **101**, 417–436.

Shearing, S. P. (1979). Mechanism of fixation of the Shearings posterior chamber intra-ocular lens. *Contact Intraocular Lens Med. J.*, **5**, 74–77.

Shepard, D. D. (1977). The dangers of metal-looped intraocular lenses. *Am. Intraocular Implant Soc. J.*, **3**, 42.

Steinert, R. F., Aker, B. L., Trentacost, D. L., Smith, P. J., & Tarantino, N. (1999). A prospective comparative study of the AMO ARRAY zonal-progressive multifocal silicone intraocular lens and a monofocal intraocular lens. *Ophthalmology*, **106**, 1243–1255.

Strampelli, B. (1961). Anterior chamber lenses: Present technique. *Arch. Ophthalmol.*, **66**, 12–17.

U.S. Department of Health and Human Services. (1983). *Cataract in Adults: Management of Functional Impairment, Clinical Practice.* Washington, DC: Guideline No. 4.

Werner, L., Apple, D. J., Kaskaloglu, M., & Pardey, S. K. (2001). Dense opacification of the optical component of a hydrophilic acrylic intraocular lens: A clinicopathologic analysis of 9 explanted lenses. *J. Cataract Refract Surg.*, **27**, 1485–1492.

Werner, L., Muller, M., & Tetz, M. (2008). Evaluating and defining the sharpness of intraocular lenses. Microedge structure of commercially available square-edged hydrophobic lenses. *J. Cataract Refract Surg.*, **34**, 310–317.

Winther-Nielsen, A., Johansen, J., Pederson, G. K., & Corydon, L. (1998). Posterior capsular opacification and neodymium-YAG capsulotomy with heparin surface modified intraocular lenses. *J. Cataract Refract Surg.*, **24**, 940–944.

C. CORNEAL INLAYS AND ONLAYS

Crystal Cunanan
Tissue Engineering at Boston Scientific, Los Gatos, CA, USA

A successful corneal inlay or onlay must satisfy a number of different requirements, and these can be categorized as optical, mechanical, chemical, and biological. In addition, it must make business sense; that is, it must be capable of volume manufacture, have a suitable profit margin, and meet an unmet or under-served market need. To date, no device is approved in the United States which meets or exceeds all the requirements in each of these categories. This chapter will describe what we have learned from previous experiences about the requirements of an ideal corneal inlay or onlay, as well as outline future directions. Hopefully, this chapter will inspire the student to take on the challenge of developing a successful corneal inlay or onlay, as such a device would provide unique advantages over other types of refractive therapies.

HISTORY OF CORNEAL INLAYS AND ONLAYS

Since 1949, ophthalmic surgeons have tried to harness the power of the cornea to change the refraction of a patient. Jose Barraquer, considered by many to be the Father of Modern Refractive Surgery, invented the microkeratome and the cryolathe, in order to perform more accurate corneal refractive procedures. Using these tools, he introduced the techniques of keratophakia, keratomileusis, and epikeratophakia, all surgical procedures performed on corneal tissue to change the curvature of the cornea, and thus alter the refractive state of the eye (Barraquer, 1949). In the case of additive procedures, such as epikeratophakia, donor corneal tissue was shaped into the desired form and then added to the patient's cornea to alter refraction. In subtractive procedures, small parts of the patient's own corneal tissue were removed to change corneal curvature.

While most of the techniques developed by Barraquer are performed by the surgeon, Drs. Herbert Kaufman and Marguerite McDonald encouraged one company to develop and commercialize a standardized epikeratophakia lenticule made from human donor corneal tissue that was unsuitable for transplant. The physicians were impressed with the wide range of refractive errors that theoretically could be treated by this approach, and they argued successfully that a standardized product manufactured by a licensed medical device company would improve clinical outcomes for patients. Finally, a device that was removable and reversible represented a unique advantage that other refractive device products and procedures could not provide. Products were developed for adult aphakia (McDonald, 1987), pediatric aphakia (Morgan, 1987), myopia (McDonald, 1987b), and keratoconus (Dietze, 1988). Despite the rather good clinical success of the product line, particularly for pediatric aphakia and keratoconus, and the repeated recommendation for approval by the Ophthalmic Device Advisory Panel of the Food and Drug Administration (ISRK, 1989), the product application was withdrawn before it obtained final FDA approval.

Epikeratophakia is still practiced today by a handful of clinicians skilled in cryolathing human donor corneal tissue, and remains an alternative for children born with cataracts who are contact lens intolerant (Halliday, 1990). The lenticule becomes repopulated by the cells from the host cornea, making it a "living contact lens" or what could also be described as a tissue-engineered corneal scaffold (Yoon, 1998; Cahill, 1999). While refractive surgical procedures in general are associated with a loss of corneal sensitivity in the region associated with the surgery (Kohlhaas, 1998), some epithelial axon terminals have been reported in primates implanted with epikeratophakia lenticules, and some measure of corneal sensitivity is restored in patients in a time-dependent manner (Koenig, 1983; Kaminski, 2002). The lenticule is removable, if necessary, and clinicians have been able to perform other refractive procedures in eyes which had undergone previous epikeratophakia (Kaminski, 2003). For children, the lenticule grows with them as they grow, avoiding the need for continued adjustment of device power with growth, as would be required for an intraocular lens implant.

Today, laser-assisted *in situ* keratomileusis or LASIK, and PhotoRefractive Keratectomy or PRK, are the most common corneal refractive procedures performed, due to their refractive predictability and rapid recovery. In the LASIK procedure, the initial corneal flap is created with either a blade system (very similar to the original Barraquer design) or with a femtosecond laser, such as the IntraLase™ system. In either case, after the corneal flap is lifted away from the corneal surface, an excimer laser is then used to reshape the corneal surface, and the corneal flap is returned over the reshaped bed. Approximately 1.2 million LASIK procedures are performed every year in the United States, with a high degree of satisfaction for most patients (95–98%), although the penetration of LASIK into the available patient population is still relatively low, at about 10% (Kalorama, 2009). Clearly there are still hurdles which remain to be addressed before refractive surgery can fulfill its promise to the 500 million people worldwide who require refractive correction.

Terminology

Don't let the language of ophthalmology scare you! Just as new techniques are developed, so are new words which describe them. Once you know the meaning of the parts, you can usually figure out the meaning of any word, no matter how long it is.

Kerato- (KEHR-uh-toh). Prefix: pertaining to the cornea

A- (AY). Prefix: without.

Epi- (EP-ee). Prefix: on top of

Photo- (FOE-toe). Prefix: With light.

Pseudo- (SU-doh). Prefix: False or imitation.

-Conus (KOH-nus). Description: Cone-shaped.

-Phakia (FAY-key-a). Description: Refers to a lens.

-Plasty (PLAS-tee). Surgical procedure. General term.

-Tectomy (TEK-tuh-mee): Surgical procedure. Removal of tissue.

-Totomy (TAH-tuh-mee): Surgical procedure. Incision into tissue.

Now put it together:

Epikeratophakia: Lens placed on top of the cornea.

Keratoconus: A cone-shaped cornea. Pathological condition where the cornea thins centrally and bulges progressively outwards.

Photorefractive keratectomy: Surgical procedure using light (laser) to remove a portion of the cornea to alter the refractive state of the eye. Also called PRK.

(Adapted from Cassin, B. (2006). *Dictionary of Eye Terminology*, 5th edn, M. L. Rubin, MD, ed. Triad Communications, Inc.)

SYNTHETIC BIOMATERIALS IN THE CORNEA

Although the human donor corneal lenticules used in the epikeratophakia procedure can be considered a biomaterial, this is not a commercially viable biomaterial. In the 1990s, the United States passed laws which restricted the fees which could be charged for devices fabricated from human donor tissues to coverage of a "processing fee," and prohibited any profit to be made from donated tissues. There are also other non-financial disadvantages of using human donor tissues as a source material, including the lack of reproducible starting material, the potential for transmission of disease through the implant, the inability to control the material during its history (for example, time to harvest the cornea after death, the method of collection and storage, and the patient's own ocular history), and the limited supply of donor human corneas. Taken together, these disadvantages outweigh the advantages of matched physical/chemical/biological properties, removable, replaceable, and excellent healing response, although a synthetic version of an epikeratophakia lenticule will need to provide those advantages as well, if it is to be successful.

OPTICAL REQUIREMENTS

The first requirement of a corneal inlay or onlay is that it must provide optical correction of the patient's refractive error. Myopia or nearsightedness, and hyperopia or farsightedness, are two of the most common refractive errors. Astigmatism and presbyopia are also common refractive errors. Refraction is the deflection of a ray of light as a result of passing from one medium to another of different optical density (refractive index). Maximum refraction occurs when the difference between the optical densities of the two media are greatest. The cornea is the most powerful refracting surface of the human eye, because of the large difference in the refractive index between the air and the cornea. The cornea is responsible for about 70% of the refractive power of the eye. The crystalline lens, responsible for accommodation and fine focusing of an image, is responsible for about 30% of the refractive power of the eye. Table C.1 contains the refractive index values for a variety of ocular tissues and media.

Corneal inlays and onlays can provide refraction by: (1) changing the power of the cornea, using a material with a higher refractive index; and/or (2) changing the shape of the anterior surface of the cornea. Both approaches have been tried in the past 50 years, with varying success. In Barraquer's original studies, he implanted synthetic materials in the cornea, in addition to corneal reshaping with surgical methods. The only materials available to him at that time, sintered glass and polymethyl methacrylate (PMMA), were hard, impermeable materials with a high refractive index. When he placed lenticules made from these materials into the cornea, however, he found that the corneal tissue above the implant became necrotic, and eventually the entire cornea over the implant disappeared, and the lens was extruded from the cornea.

TABLE C.1 Refractive Index Values of Various Ocular Tissues and Fluids Compared to Air and Water (Hecht, 1987)

Medium	Refractive Index (n_d)
Air	1.000
Water	1.333
Cornea	1.376
Aqueous humor	1.336
Crystalline lens*	1.386–1.406
Vitreous humor	1.337

*Note the natural crystalline lens has a progressive gradient refractive index of 1.386 in its outer, less dense, layers and 1.406 in its denser nucleus.

BIOLOGICAL REQUIREMENTS

The reason for Barraquer's failure was not understood until 1961, when Knowles demonstrated the importance of corneal permeability on corneal physiology (Knowles, 1961). Knowles placed impermeable materials within the cornea, and observed the corneal tissue directly above the implant to begin to break down, become necrotic, and eventually dissolve over the implant. Soon it became appreciated that the cornea receives its hydration and nutrients from the aqueous humor, and not from the tear film, as had been previously believed. Glucose, in particular, was identified as an important metabolite which was provided solely from the aqueous humor. Thus, an important material property for a synthetic lens is the permeability of the material to glucose, which can be measured experimentally using a diffusion chamber.

A hydrogel is water permeable, and thus is a more appropriate plastic material for intrastromal implantation. Early studies by Dohlman and Mester demonstrated long-term implants of both high (88%) and low (38%) water content hydrogels in rabbit and cat corneas (Dohlman, 1967; Mester, 1972). Surprisingly, Dohlman experienced some corneal necrosis with the 88% water content materials, while Mester reported no cases of necrosis with the 38% polyHEMA materials he used.

The reasons for this may relate to the actual dimensions of the implants themselves. Dohlman reported using implants ranging in thickness from 0.19 mm to 0.57 mm, while Mester used implants which were 0.2 mm thick. Thus, while Dohlman was working with a higher water content material, he still saw nutritional necrosis due to the thickness of some of his lenses. While Mester used a material with a lower water content, his lenses were much thinner, and thus he saw no nutritional deprivation.

Another potential difference between the two studies included the influence of the shape of the device on the corneal tissues. Dohlman cast his material as a flat sheet, and punched disks of material from this sheet for implantation. Mester, however, cast his material in molds which were curved, to better match the curvature of the cornea.

While no controlled studies have been reported to look at the influence of lens mechanical mismatch on corneal tissue, similar observations have been reported in other types of implants in other implant locations (Matlaga, 1976).

Updated known requirements

Optical:
- Provide correction of spherical aberration with high degree of accuracy
- Prevent increase in Higher Order Aberrations (HOA)
- Ideally, enable correction of eye's natural HOA
- Adequate contrast sensitivity, color
- Avoid glare, halos, night problems
- Ideally, correct for presbyopia
- Predictable, stable vision

Surgical:
- Highly accurate procedure in any surgeon's hands
- Rapid recovery of vision
- Little pain/discomfort
- Ideally, outpatient procedure with topical anesthesia
- Ideally, removable and replaceable

Biological:
- Excellent corneal response to material
- No encapsulation, reactivity, fibrosis or haze
- No long-term hydrolysis of material
- No immunogenic or antigenic response

Chemical:
- Long-term stability, including UV stability

Mechanical:
- No dimensional change with conditions (ambient temperature, lid closure/open, eyedrop use)
- Design does not elicit undesirable biological response (deposits, melting, etc.)

Other:
- Does not interfere with diagnostic procedures
- Does not limit clinician's ability to use alternative or supplemental therapies.

COMMERCIAL ATTEMPTS AT SYNTHETIC CORNEAL INLAYS AND ONLAYS

Given the early successes of Dohlman and Mester, and armed with the new understanding of corneal physiology, the ophthalmic industry embraced the notion of synthetic corneal inlays and onlays with a passion. At least four major ophthalmic companies and a number of start-up ventures began developing synthetic implants for refractive keratoplasty. These programs took one of two approaches: (1) mimic the cornea, and change refraction by changing the anterior corneal curvature or (2) change the power of the cornea with a high refractive index material.

Bernard McCarey, at Emory University, became a nearly universal resource in helping industry study their implants, and understand their materials and the requirements of the cornea. Over the 15 years that this research was conducted, Dr. McCarey worked with many world-class researchers and clinicians as he utilized the platform of intracorneal lenses (abbreviated to ICLs) to continue to advance our understanding of the cornea. Simultaneously, the interest in corneal physiology and its importance in developing a commercially successful ICL stimulated independent research in leading centers around the globe. Much of this work forms the cornerstone of our understanding today of corneal physiology and its reaction to foreign materials.

PERMEABLE INTRACORNEAL LENSES

As may be expected from Barraquer's failure with impermeable materials, most companies settled on permeable materials and utilized their understanding of hydrogels for contact lenses to develop their programs. These materials can all be described as ionically neutral, since ionically-charged hydrogels can change physical dimensions with changing pH (McCarey, 1982); high water content to match the water content of the cornea, since materials with water contents higher than the cornea are not dimensionally stable in the cornea (Beekhuis, 1985); and hopefully, adequate glucose permeability, since implants with insufficient permeability result in anterior stromal necrosis.

While no hydrogel intracorneal inlay is available for widespread correction of refractive error in the United States, there were significant learnings which occurred during those early studies. For example, the surgical technique that is utilized can strongly influence the response of the cornea to the implanted lens, and a microkeratome technique is more predictable compared to a free-handed dissection (Beekhuis, 1986; McCarey, 1986). The animal model can have an influence on the reactivity of the cornea to the implant, with rabbit eyes being the most reactive (McCarey, 1981, 1982b) followed by cat eyes (Climenhaga, 1988). The nonhuman primate model is preferred for intracorneal implant studies for several reasons, including the presence of a true Bowman's membrane, which is lacking in the rabbit, and the ability to assess refractive predictability (Watsky, 1985; McCarey, 1989, 1990). The primate model also allows long-term follow-up, and studies have been reported of long-term hydrogel intracorneal lens implants at five and eight years (Parks, 1991; Werblin, 1992; McDonald, 1993). Because of the difficulties in using cats and nonhuman primates in animal research, most studies today utilize the pig model when a long-term implant is needed. The pig cornea is similar in size and curvature, making it possible to use clinical instruments and methods for flap formation.

Animal studies have enabled histological evaluation to be conducted at various time points and correlated to the clinical presentation of the eye. For example, scattered zones of abnormal fibroblast activity at the posterior interface of the lens with the stroma are fairly typical in the rabbit model (McCarey, 1982b). Keratocytes may also line the anterior stromal face, but an acceptable reaction would not contain irregular areas or masses which would scatter light and interfere with vision, and the ultrastructure of these cells remains normal. By contrast, development of intrastromal corneal crystals is not a normal occurrence (Parks, 1993). These crystals are composed of lipid, and can occur in eyes that have experienced surgical trauma or chronic metabolic stress (Rodrigues, 1990). Animal studies are also good at detecting material problems, such as inadequate biocompatibility of the material (Beekhuis, 1987). A biocompatibility problem would occur shortly after implantation, 3–12 days in the primate model, and would present as severe edema and inflammation throughout the cornea. Fibrin clots can form in the anterior chamber with marked iritis and conjunctival hyperemia. Significant stromal edema can be followed by the development of lipid crystals in the cornea. Gradual stromal thinning is a concern that, if it does not stabilize, will lead to ulceration. Stromal thinning may lead to epithelial thinning by two weeks postoperatively, with ulceration occurring by the sixth week. Causes of biocompatibility problems can include chemical or biological contamination, which may trace back to the source polymer itself. Since many companies used contact lens formulations, there may be contaminants in the contact lens material which are tolerated when the lens is on the tear film, but are irritating to the eye when implanted. Care must be taken to recognize the higher standards of quality and purity that an implanted lens requires, compared to a contact lens. Another common source of contaminants is from the manufacturing process, either residual solvent or more likely, residual cleaning or polishing agents. Such lessons are well-known to those accustomed to making intraocular lenses, but not necessarily at a contact lens company or a start-up venture with inexperienced personnel.

Surgical complications can also arise, and difficulty with the microkeratome incision in animals is more frequent than one would care to admit. Today, the possibility of using a femtosecond laser to create the flap may lead to a more precise placement of the lens, and less surgical trauma. These early studies used large diameter, thick lenses, and sutures were used to secure the corneal flaps. Sutures can create irritation, and occasionally result in additional complications, particularly if the suture penetrates the anterior chamber. Today, using modern LASIK procedures should improve the quality of ICL implantation, without the need for sutures. Finally, there is the potential to include foreign-bodies in the lamellar bed, such as fragments or burrs from the microkeratome blade, powder from gloves, and residues from surgical instruments. While these contaminants can also develop in a standard LASIK procedure, that situation is different because there is living tissue surrounding the foreign-body which, with time, will ensure its removal (either through phagocytosis or extrusion). When a lens is present, however, the foreign-body is trapped against an inert surface, preventing its removal and causing increased inflammation in the underlying stroma. Haze, edema, and inflammation posterior to the lens (but not anterior to it) could be signs of trapped debris at the interface. While the refractive surgery suite is not an operating room *per se*, in the case of ICL implantation, each case should be treated as an intraocular procedure.

Other complications which can occur include mechanical complications, resulting in implant fibrosis, formation of an iron line in the epithelial layer, and decentration of the implant with or without extrusion. Optical complications can also be studied in the primate model, and one can detect overcorrection or undercorrection, induced corneal astigmatism, and development of a refractive nomogram prior to human cases, which should increase the success rate of early clinical trials. Two high water content hydrogels were studied in clinical trials after extensive animal studies, and both studies demonstrated the biocompatibility of the two materials in the human cornea for as long as six years (Steinert, 1996; Barraquer, 1997). ICL technology did not provide stable vision for myopic patients, although the aphakic patients in both series were well-corrected. Both studies encountered difficulties with the surgical procedure, as these studies were conducted before LASIK became popular. Figure C.1 shows the appearance of an eye with an implant two years postoperatively. Note the crystal-clear cornea and the rather deep implantation of the lens. Because of the concurrent development and success of the excimer laser, both companies involved in these clinical trials decided to close out the studies and focus their efforts on laser vision correction.

Intracorneal implants have given us additional tools to study the processes of nutrient transport in the cornea (McCarey, 1990b), and to build upon the work of Klyce (1979) and Baum (1984) in understanding the roles of each of the cell types within the cornea in corneal hydration and normal physiology. Recently, Fischbarg

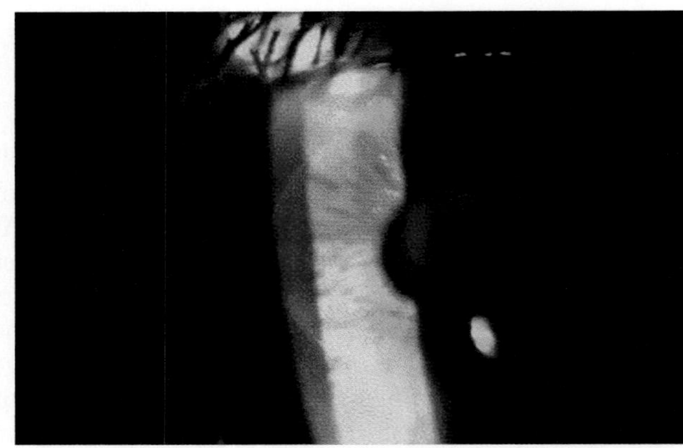

FIGURE C.1 Slit-lamp photomicrograph of an eye after two years implantation of a high water content hydrogel material, implanted for hyperopia correction. Note the clear lens in the slit beam with only a trace amount of scattered light around the edges of the implant. This example highlights the excellent biocompatibility that can be obtained with synthetic materials in the human cornea. (From Ismail (2006), reprinted with permission.)

(2004) and Li (2006) have expanded upon these earlier studies to examine the roles of receptors in corneal physiology, and to model the transport of a number of ions across the cornea in normal maintenance and in response to physiological stresses. Larrea et al. have modeled simultaneously the oxygen and glucose fluxes through the cornea to determine the optimum depth of an intracorneal lens implant (Larrea et al., 2007). Continued work in this area may someday help us to better understand the development of lipid and iron deposits in the cornea, as well as to develop early predictors of corneal stress in cases which may be in danger of stromal ulceration or excessive haze development.

IMPERMEABLE INTRACORNEAL LENSES

Despite the evidence associated with stromal necrosis and impermeability, several companies continued to pursue impermeable, high refractive index materials, most notably poly methyl methacrylate (PMMA) and polysulfone. Given the refractive indices of 1.489 and 1.633, respectively, this is an appreciable difference compared to high water hydrogels with refractive indices of around 1.385, and would enable a greater degree of optical correction compared to a hydrogel ICL. Polysulfone intraocular lenses were being implanted in Europe with suitable biocompatibility (Lane, 1986), and as intracorneal lenses they demonstrated a predictable and consistent change in refraction (McCarey, 1987).

To determine if an impermeable lens could be tolerated in the cornea at all, a systematic study of the influence of lens diameter and implant depth was conducted in cats (Climenhaga, 1988). Lenses of 4, 5, 6, and 7 mm diameter were implanted at depths ranging from 45% to 95% of corneal thickness. These lenses averaged 180 microns thick at the center and had an edge thickness of less than 100 microns. The lenses were implanted into a freehand dissected pocket and followed for up to 20 months. With time, central corneal opacities developed in 100% of the eyes, following a diameter-dependent pattern where 7 mm lenses developed stromal opacities within one month, and 4 mm lenses required 12 months to develop opacities in the majority of lenses. Ultimately, however, anterior stromal deposits developed in nearly all lenses by 13 months, and an average of 28% of eyes developed stromal melting. The stromal deposits which developed appeared first as a fine, stippled haze that became increasingly confluent as granular deposits coalesced into a continuous white opacity in the anterior stroma centered over the lens. In some eyes the opacities developed fine, elongated, spicule-shaped crystals over the center of the lens, and sometimes behind the lens. These deposits developed from months 1 to 14, depending on lens diameter, and continued to grow with time. In contrast, peripheral deposits developed at a different rate, with 40% of all lenses exhibiting a hazy gray ring around the lens periphery by 2 months. By 9 months, 100% of lenses exhibited edge deposits. Both central and peripheral deposits were consistent with lipid deposition in the cornea. The authors concluded that because opacities developed in all lens diameters and implantation depths, the metabolic demands of the cornea were not met, and thus no safe limit existed at which these ICLs could be used predictably in cat eyes.

The clinical study in Europe on polysulfone ICLs eventually reached the same conclusion after 12 years' implantation (Horgan, 1996). Research has shown that fenestrated polysulfone lenses can survive in the cornea of cat eyes at one year, however, with 100% anterior stromal clarity over the fenestrated portion of the lens (Lane, 1991). The fenestrations ultimately filled in with normal stromal tissue and keratocytes. Due to the difference in refractive index between the polysulfone lens and the surrounding tissue, however, these fenestrated lenses were not clinically useful.

SYNTHETIC MATERIALS FOR CORNEAL ONLAYS

Despite the extensive development of tissue lenses for epikeratophakia, and their rather good clinical outcomes, progress on a synthetic corneal onlay or total synthetic cornea has met with far less success. This is not for lack of trying, and the industry's efforts to create a successful synthetic epikeratophakia lens have been substantial, but this investment was insufficient to overcome the technical hurdles that were encountered.

Many of the technical requirements of a synthetic onlay mirror that of a synthetic inlay, such as permeability, optical clarity, dimensional stability, and manufacturability. The distinct difference, however, is in the added requirement of supporting a robust and healthy epithelium across the lens, and this is the hurdle which has proved insurmountable, at least up to this point.

When the synthetic lenticule is placed on the cornea, it is placed directly onto Bowman's membrane, after the corneal epithelium has been removed. This creates a breech in the sterile barrier function of the eye, and it is paramount that the lens supports rapid and stable re-epithelialization in order to restore the integrity of the sterile barrier of the eye. This creates a real conundrum, as the hydrogel materials that have been used for intracorneal implants do not support cell adhesion and growth (Cunanan, 1991). While surface modification of these materials was tried extensively (Cunanan, 1996), these efforts failed to provide a robust stable surface treatment that corneal epithelial cells would recognize and adhere to long-term. Although many treatments looked satisfactory in an organ culture system designed to test the outgrowth potential of different materials (Pettit, 1990), these treatments failed to support cell migration and adhesion in an animal implant model. The ideal conditions in the cell culture system did not adequately challenge the materials being tested, and it is believed that the biomolecules used to encourage cell migration and attachment were degraded by proteases in the tear film, which are an integral part of the eye's natural defense mechanisms. Protease inhibitor eye drops were developed with some success at protecting the biological signaling proteins tethered to the lens surface; however, as soon as the eye drops were discontinued, the epithelium sloughed off the lens. Such an approach is a band-aid effort to stop the proteolytic degradation occurring on the lens surface, and it was insufficient to address the fundamental problem of cell migration and adhesion.

CORNEAL INLAYS AND ONLAYS TODAY

Despite the long history of failures, and the many challenges that have been encountered, development of a successful corneal inlay or onlay still represents a very attractive market opportunity. Because the implant is additive, rather than subtractive, such as with PRK or LASIK, the implant can be removed and the patient is usually left in their original condition. This is highly desirable to help overcome the fears of many patients who have not undergone LASIK surgery today because of their reluctance to undergo an irreversible procedure (Kalorama, 2009). These additive technologies do not interfere with the clinicians' use of other refractive therapies, as an adjunct or replacement technology, as the patient's needs change with aging. This is an important factor in the adoption of any new technology.

Today, there are four basic approaches being taken in developing implants for refractive correction. The first device concept, thin intracorneal rings or ring segments fashioned from PMMA, is used in the peripheral anterior stroma to alter the corneal curvature (Figure C.2). These implants do not experience permeability failure because they are only a narrow ring of material, and leave the central cornea untouched. While these devices had been originally developed as a treatment for myopia, the advent of the excimer laser and PRK/LASIK have limited the use of these implants to patients who are contraindicated for LASIK surgery, such as keratoconus patients and in patients with thin corneas that may develop keratoconus after LASIK surgery, also called keratectasia (Ertan, 2007; Pinero, 2009). These ring segments provide a mechanical reshaping of the cornea, and thus can only correct low degrees of myopia, typically up to 2 diopters. Sometimes they can induce chronic pain, presumably by pressing against a corneal nerve (Randleman, 2006). Much like the other corneal implants, however, the intracorneal rings and segments can be removed, and the patient usually returns to their original refraction. Today, there is a program with the IntraLase™ femtosecond laser to create the circular tracks for the ring segments (Pinero, 2009b). Long-term results in this unique patient population reveal that the intrastromal ring segments are well-tolerated and provide stable vision, without late complications or loss of effect (Kymionis, 2006).

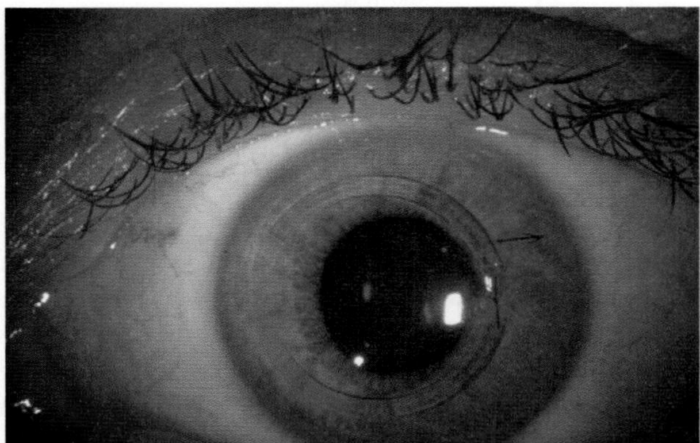

FIGURE C.2 Photomicrograph of an eye after Intacs® implantation. Note the clear corneal tissue around the implant, indicating excellent biocompatibility. This patient experienced pain after implantation due to the physical compression of an ocular nerve by the rigid PMMA ring. (From Randleman et al., (2006), reprinted with permission.)

The AlphaCor™ artificial cornea is another niche product that has met with some success in a limited patient population for patients with extensive corneal haze or pain who are at high risk of graft failure with a donor cornea (Hicks, 2003). The AlphaCor™ device is a one-piece button of polyHEMA, with a clear 4.5 mm optical portion in the center surrounded by a 1.25 mm opaque spongy rim. The outer rim allows cellular ingrowth from the host tissue, thereby locking the device into place, but the surgical procedure is complicated, and the device must be implanted in a complex two-stage process completed over a 12-week timeframe. The first procedure involves placing the device in a midstromal pocket which is hand-dissected, with a 3 mm trephination through Descemet's membrane to create a clear zone in the posterior cornea. The second procedure is completed several months later, after host ingrowth has been accomplished, and involves a simple anterior stromal 3 mm trephination to create a clear optical zone in the anterior cornea. Implantation of the device can also be combined with cataract surgery, as many traumatic cases of corneal injury may also require lens replacement (Eguchi, 2004).

Complications associated with the device include problems with its poor mechanical strength and flexibility, discoloration, and formation of deposits (Hicks, 2004), infection and melting of the hydrogel (Chow, 2007), development of a retroprosthetic membrane (Chalam, 2007), development of intractable glaucoma (Ngakeng, 2008), focal calcification, and spoilage of the hydrogel material by various ophthalmic formulations that might be used for patient care (Morrison, 2006). As this device continues to be studied, there are reports of an immunologic aspect of the device failure, associated with cellular response to the porous skirt

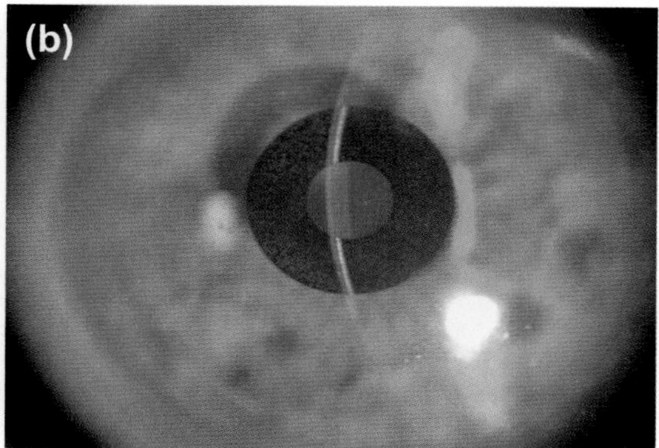

FIGURE C.3 (a) The AcuFocus intracorneal inlay shown next to a contact lens. Note the overall small size and opaque nature of the ring. The stippling seen on the inlay is really the laser-drilled holes which prevent nutrient deprivation of the anterior cornea. (b) The inlay in a patient's eye. (From Yilmaz et al., (2008), reprinted with permission.)

(Coassin, 2007). At least some cases of retroprosthetic membrane formation can be attributed to concomitant diabetes, indicating a systemic influence on the device (Hicks, 2005).

A new type of intracorneal implant under development is the AcuFocus KAMRA™ Corneal Inlay for the treatment of presbyopia. The KAMRA™ inlay has an overall diameter of 3.8 mm, which is smaller than a contact lens. It has a 1.6 mm center aperture, which is a small opening in the center that creates a pinhole effect. The device is 5 microns thick, made of polyvinylidene fluoride and carbon, and it has 8400 random holes for oxygen and nutrition flow. The AcuFocus corneal inlay is designed to improve depth of focus, allowing the eye to see near and intermediate objects more clearly. Figure C.3 compares the KAMRA™ lens to a standard contact lens and shows the inlay implanted in a human eye (Yilmaz, 2008). Like other corneal implants, the AcuFocus implant can be removed and the patient's refraction returns to baseline levels. This device received the CE mark in 2005, and is under active clinical investigation in the United States.

Two start-up companies continue to pursue the use of hydrogel materials for intracorneal lenses. ReVision Optics has developed a 2 mm lens made from a high water content material, relying on changing the corneal curvature to provide refractive correction. Presbia is developing a 3.2 mm lens made from a low water content material, placing the implant deep in the cornea to prevent nutritional deprivation of the overlying stroma. Both companies have focused their efforts on the treatment of presbyopia, which requires a smaller refractive correction compared to other types of refractive errors.

To control centration, ReVision has worked with IntraLase™ to develop a custom pocket which holds the tiny implant, thus improving the surgical technique. Figure C.4 shows the microlens on the tip of a finger, while Figure C.5 shows the excellent biocompatibility of the microlens in a porcine cornea. Today, ReVision Optics has completed pivotal clinical trials in the US and awaits FDA approval, while it commercializes the lens in Europe. Patients receiving the ReVision lens demonstrate an impressive improvement in their near-vision acuity (90% at J1; 100% at J2) at one year. The Presbia lens has no visual correction in the central portion of the lens, surrounded by a positive-powered ring. The Presbia lens creates a bifocal effect with the high refractive index implant, and does not change the corneal curvature. The Presbia lens is CE mark approved and is undergoing US clinical trials. Table C.2 contains comparative information about the intracorneal inlays in clinical use and in development today.

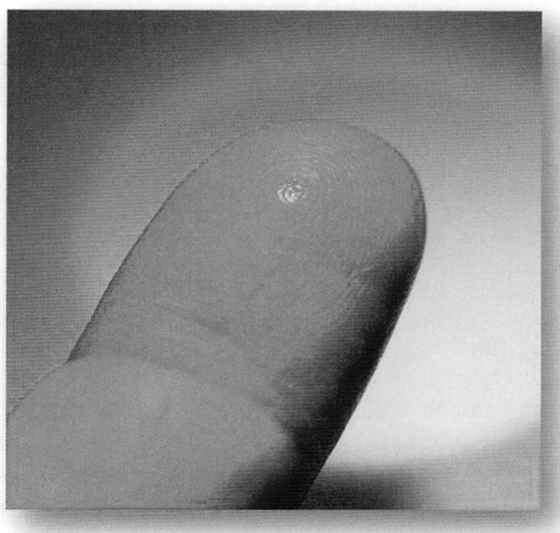

FIGURE C.4 Photo of the ReVision Optics micro-inlay on the tip of a finger. Note the small size of the inlay, which is optically clear and made from a high-water content hydrogel material. (Photo provided by ReVision Optics, printed with permission.)

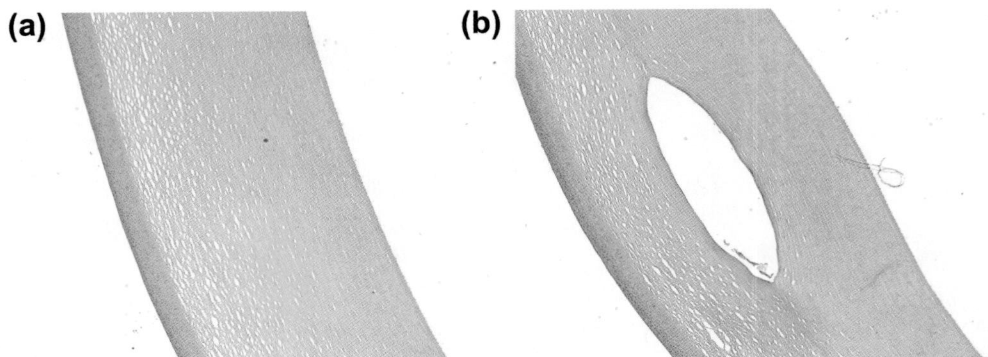

FIGURE C.5 (a) A histological cross-section of an unimplanted pig cornea (H&E stain, 10×). (b) A histological cross-section of the contralateral eye of the pig after 12 months implantation of the RVO corneal inlay (H&E stain, 10×). The polymer material was lost from the section during the histological processing, leaving an open space in the corneal section. The biocompatibility of the implant is evident in both the stromal areas surrounding the implant, as well as the overlying epithelium, which is neither hypertrophic nor thinned. (Photo provided by ReVision Optics, printed with permission.)

TABLE C.2 Comparison Between the Various Intracorneal Implants in Use and Under Development Today

Attribute	Intacts® Rings	AlphaCor™ Implant	AcuFocus KAMRA™	ReVision Optics	Presbia Microlens
Polymer type	PMMA	PolyHEMA	PVDF/carbon black	Acrylic hydrogel	Acrylic hydrogel
Water content	0%	38%	0%	high	low
Implant location	Anterior periphery	Central, partial to full thickness	Anterior central	Anterior central	Posterior central
Implant exerts refractive effect by:	Changes corneal curvature by bulking up the periphery	Higher refractive index than cornea	Pinhole optics	Changes corneal curvature with central implant	High refractive index
Regulatory status:	FDA approved; CE mark	FDA approved; CE mark	CE mark; in FDA trial	CE mark; FDA trial submitted for approval	CE mark; in FDA trial

THE FUTURE OF CORNEAL INLAYS AND ONLAYS

As long as people continue to require refractive correction, there will be an incentive to develop the optimum corneal inlay/onlay. As we continue to update the requirements of these products, we can see that the field as a whole is making progress, and that there is room for more than one product or technology. New enabling technologies, such as the use of a laser to provide final, minute lens adjustments (Peyman, 2005), using advanced imaging systems capable of capturing the unique optical abnormalities inherent in each of us (Warden, 2008) may be the refinement needed to provide discriminating patients with the highest quality of vision possible. Continued advances in surgical techniques and procedures (Zarbin, 2004) may provide the rapid, pain-free visual recoveries that patients expect. Given the plethora of biomaterials currently available and in development (Xie, 2006; Myung, 2008; Pinchuk, 2008; Deng, 2009; Liu, 2009), a successful corneal inlay or onlay may be closer than you think. Some approaches may be considered as tissue engineering (Liu, 2008) and the reader is referred to Section II.6, "Applications of Biomaterials in Functional Tissue Engineering," for the latest information on these advances.

BIBLIOGRAPHY

Barraquer, J. I. (1949). Queratoplastia Refractive. *Estudios e Informaciones Oftalmoloqiaces*, **2**, 10.

Barraquer, J. I., & Gomez, M. L. (1997). Permalens hydrogel intracorneal lenses for spherical ametropia. *J. Refract. Surg.*, **13**, 342–348.

Baum, J. P., Maurice, D. M., & McCarey, B. E. (1984). The active and passive transport of water across the corneal endothelium. *Exp. Eye Res.*, **39**, 335–342.

Beekhuis, W. H., & McCarey, B. E. (1985). Hydration stability of intracorneal hydrogel implants. *Invest. Ophthalmol. Vis. Sci.*, **26**, 1634–1636.

Beekhuis, W. H., McCarey, B. E., Waring, G. O., & van Rij, G. (1986). Hydrogel keratophakia: A microkeratome dissection in the monkey model. *Br. J. Ophthalmol.*, **70**, 192–198.

Beekhuis, W. H., McCarey, B. E., van Rij, G., & Waring, G. O., 3rd (1987). Complications of hydrogel intracorneal lenses in monkeys. *Arch. Ophthalmol.*, **105**, 116–122.

Cahill, M., Condon, P., & O'Keefe, M. (1999). Long-term outcome of epikeratophakia. *J. Cataract Refract. Surg.*, **25**, 500–507.

Cassin, B. (2006). Dictionary of Eye Terminology. In M. L. Rubin, & M.D. (Eds.), (5th ed.). Triad Communications, Inc..

Chalam, K. V., Chokshi, A., Agarwal, S., & Edward, D. P. (2007). Complications of AlphaCor keratoprosthesis: A clinicopathologic report. *Cornea*, **26**, 1258–1260.

Chow, C. C., Kulkarni, A. D., Albert, D. M., Darlington, J. K., & Hardten, D. R. (2007). Clinicopathologic correlation of explanted AlphaCor artificial cornea after exposure of implant. *Cornea*, **26**, 1004–1007.

Climenhaga, H., Macdonald, J. M., McCarey, B. E., & Waring, G. O., 3rd. (1988). Effect of diameter and depth on the response to solid polysulfone intracorneal lenses in cats. *Arch. Ophthalmol.*, **106**, 818–824.

Coassin, M., Zhang, C., Green, W. R., Aquavella, J. V., & Akpek, E. K. (2007). Histopathologic and immunologic aspects of AlphaCor artificial corneal failure. *Am. J. Ophthalmol.*, **144**, 699–704.

Cunanan, C. M., Tarbaux, N. M., & Knight, P. M. (1991). Surface properties of intraocular lens materials and their influence on *in vitro* cell adhesion. *J. Cataract Refract. Surg.*, **17**, 767–773.

Cunanan, C. M., Graham, R. S., Manesis, N. J., et al. (1996). Hydrolysis of a Neutral Hydrogel and Biomolecule Attachment to Increase Cell Adhesion and Migration. In B. D. Ratner, & D. G. Castner (Eds.), *Surface Modification of Polymeric Biomaterials* (pp. 129–134). Plenum Press.

Deng, C., Li, F., Hackett, J. M., Chaudhry, S. H., Toll, F. N., Toye, B., Hodge, W., & Griffith, M. (2009). Collagen and glycopolymer based hydrogel for potential corneal application. *Acta Biomater*, **6**(1), 187–194.

Dietze, T. R., & Durrie, D. S. (1988). Indications and treatment of keratoconus using epikeratophakia. *Ophthalmology*, **95**, 236–246.

Dohlman, C. H., Refojo, M. F., & Rose, J. (1967). Synthetic polymers in corneal surgery. I. Glyceryl methacrylate. *Arch. Ophthalmol.*, **77**, 252–257.

Eguchi, H., Hicks, C. R., Crawford, G. J., Tan, D. T., & Sutton, G. R. (2004). Cataract surgery with the AlphaCor artificial cornea. *J. Cataract Refract. Surg.*, **30**, 1486–1491.

Ertan, A., & Colin, J. (2007). Intracorneal rings for keratoconus and keratectasia. *J. Cataract Refract. Surg.*, **33**, 1303–1314.

Fischbarg, J., & Maurice, D. M. (2004). Review: An update on corneal hydration control. *Exp. Eye Res.*, **78**, 537–541.

Halliday, B. L. (1990). Epikeratophakia for aphakia, keratoconus, and myopia. *Br.J.Ophthalmol*, **74**, 67–72.

Hecht, E. (1987). *Optics* (2nd ed.). Addison Wesley, Boston, MA.

Hicks, C. R., Crawford, G. J., Lou, X., Tan, D. T., Snibson, G. R., Sutton, G., Downie, N., Werner, L., Chirila, T. V., & Constable, I. J. (2003). Corneal replacement using a synthetic hydrogel cornea, AlphaCore: Device, preliminary outcomes and complications. *Eye*, **17**, 385–392.

Hicks, C. R., Chirila, T. V., Werner, L., Crawford, G. J., Apple, D. J., & Constable, I. J. (2004). Deposits in artificial corneas: Risk factors and prevention. *Clin. Experiment. Ophthalmol.*, **32**, 185–191.

Hicks, C. R., & Hamilton, S. (2005). Retroprosthetic membranes in the AlphaCor patients: Risk factors and prevention. *Cornea*, **24**, 692–698.

Horgan, S. E., Fraser, S. G., Choyce, D. P., & Alexander, W. L. (1996). Twelve year follow-up of unfenestrated polysulfone intracorneal lenses in human sighted eyes. *J. Cataract Refract. Surg.*, **22**, 1045–1051.

Ismail, M. M. (2006). Correction of hyperopia by intracorneal lenses: Two-year follow-up. *J Cataract Refract. Surg.*, **32**, 1657–1660.

(ISRK) International Society of Refractive Keratoplasty, the Board of Directors. (1989). Statement of epikeratoplasty. *Refract. Corneal Surg.*, **5**, 33–38.

Kalorama Information. (2009). *Advances in Ophthalmology: Markets in the Treatment of Eye Disorders and Corrective Vision.* MarketResearch.com, Inc.

Kaminski, S. L., Biowski, R., Lukas, J. R., Koyuncu, D., & Grabner, G. (2002). Corneal sensitivity 10 years after epikeratoplasty. *J. Refract. Surg.*, **18**, 731–736.

Kaminski, S. L., Biowski, R., Koyuncu, D., Lukas, J. R., & Grabner, G. (2003). Ten-year follow-up of epikeratophakia for the correction of high myopia. *Ophthalmology*, **110**, 2147–2152.

Klyce, S. D., & Russell, S. R. (1979). Numerical solution of coupled transport equations applied to corneal hydration dynamics. *J. Physiol.*, **292**, 107–134.

Knowles, W. F. (1961). Effect of intralamellar plastic membranes on corneal physiology. *Am. J. Ophthalmol.*, **51**, 1146–1156.

Koenig, S. B., Berkowitz, R. A., Beuerman, R. W., & McDonald, M. B. (1983). Corneal sensitivity after epikeratophakia. *Ophthalmology*, **90**, 1213–1218.

Kohlhaas, M. (1998). Corneal sensation after cataract and refractive surgery. *J. Cataract Refract. Surg.*, **24**, 1399–1409.

Kymionis, G. D., Tsiklis, N. S., Pallikaris, A. I., Kounis, G., Diakonis, V. F., Astyrakakis, N., & Siganos, C. S. (2006). Long-term follow-up of intacs for post-LASIK corneal ectasia. *Ophthalmology*, **113**, 1909–1917.

Lane, S. L., Lindstrom, R. L., Cameron, J. D., Thomas, R. H., Mindrup, E. A., Waring, G. O., 3rd, McCarey, B. E., & Binder, P. S. (1986). Polysulfone corneal lenses. *J. Cataract Refract. Surg.*, **12**, 50–60.

Lane, S. S., & Lindstrom, R. L. (1991). Polysulfone intracorneal lenses. *Int. Ophthalmol. Clin.*, **31**, 37–46.

Larrea, X., DeCourten, C., Feingold, V., Burger, J., & Buchler, P. (2007). Oxygen and glucose distribution after intracorneal lens implantation. *Optom. Vis. Sci.*, **84**, 1074–1081.

Li, L. Y., & Tighe, B. (2006). Numerical simulation of corneal transport processes. *J. R. Soc. Interface*, **3**, 303–310.

Liu, W., Merrett, K., Griffith, M., Fagerholm, P., Dravida, S., Heyne, B., Scaiano, J. C., Watsky, M. A., Shinozaki, N., Lagali, N., Munger, R., & Li, F. (2008). Recombinant human collagen for tissue engineered corneal substitutes. *Biomaterials*, **29**, 1147–1158.

Liu, W., Deng, C., McLaughlin, C. R., Fagerholm, P., Lagali, N. S., Heyne, B., Scaiano, J. C., Watsky, M. A., et al. (2009). Collagen-phosphorylcholine interpenetrating network hydrogels as corneal substitutes. *Biomaterials*, **30**, 1551–1559.

Matlaga, B. F., Yasenchak, L. P., & Salthouse, T. N. (1976). Tissue response to implanted polymers: The significance of sample shape. *J. Biomed. Mater. Res.*, **10**, 391–397.

McCarey, B. E., & Andrews, D. M. (1981). Refractive keratoplasty with intrastromal hydrogel lenticular implants. *Invest. Ophthalmol. Vis. Sci.*, **21**, 107–115.

McCarey, B. E., & Wilson, L. A. (1982). pH, osmolarity and temperature effects on the water content of hydrogel contact lenses. *Contact Intraocul. Lens Med. J.*, **8**, 158–167.

McCarey, B. E., Andrews, D. M., Hatchell, D. L., & Pederson, H. (1982b). Hydrogel implants for refractive keratoplasty: Corneal morphology. *Curr. Eye Res.*, **2**, 29–38.

McCarey, B. E., van Rij, G., Beekhuis, W. H., & Waring, G. O., 3rd (1986). Hydrogel keratophakia: A freehand pocket dissection in the monkey model. *Br. J. Ophthalmol.*, **70**, 187–191.

McCarey, B. E., Waring, G. O., & Street, D. A. (1987). Refractive keratoplasty in monkeys using intracorneal lenses of various refractive indexes. *Arch. Ophthalmol.*, **105**, 123–126.

McCarey, B. E., McDonald, M. B., van Rij, G., Salmeron, B., Pettit, D. K., & Knight, P. M. (1989). Refractive results of hyperopic hydrogel intracorneal lenses in primate eyes. *Arch. Ophthalmol.*, **107**, 724–730.

McCarey, B. E., Storie, B. R., van Rij, G., & Knight, P. M. (1990). Refractive predictability of myopic hydrogel intracorneal lenses in nonhuman primate eyes. *Arch. Ophthalmol.*, **108**, 1310–1315.

McCarey, B. E., & Schmidt, F. H. (1990b). Modeling glucose distribution in the cornea. *Curr. Eye Res.*, **9**, 1025–1039.

McDonald, M. B., Kaufman, H. E., Aquavella, J. V., Durrie, D. S., Hiles, D. A., Hunkeler, J. D., Keates, R. H., Morgan, K. S., & Saunders, D. R. (1987). The nationwide study of epikeratophakia for aphakia in adults. *Am. J. Ophthalmol.*, **103**, 358–365.

McDonald, M. B., Kaufman, H. E., Aquavella, J. V., Durrie, D. S., Hiles, D. A., Hunkeler, J. D., Keates, R. H., Morgan, K. S., & Sanders, D. R. (1987b). The nationwide study of epikeratophakia for myopia. *Am. J. Ophthalmol.*, **103**, 375–383.

McDonald, M. B., McCarey, B. E., Storie, B., Beuerman, R. W., Salmeron, B., van Rij, G., & Knight, P. M. (1993). Assessment of the long-term corneal response to hydrogel intrastromal lenses implanted in monkey eyes for up to five years. *J. Cataract Refract. Surg.*, **19**, 213–222.

Mester, U., Roth, K., & Dardenne, M. U. (1972). Versuche mit 2-Hydroxyaethyl-methycrylatlinsen als Kerotophakiematerial. *Ber. Ophthalmol. Ges.*, **72**, 326–327.

Morgan, K. S., McDonald, M. B., Hiles, D. A., Aquavella, J. V., Durrie, D. S., Hunkeler, J. D., Kaufman, H. E., Keates, R. H., & Sanders, D. R. (1987). The nationwide study of epikeratophakia for aphakia in children. *Am. J. Ophthalmol.*, **103**, 366–374.

Morrison, D. A., Gridneva, Z., Chirila, T. V., & Hicks, C. R. (2006). Screening for drug-induced spoliation of the hydrogel optic of the AlphaCor artificial cornea. *Contact Lens & Anterior Eye*, **29**, 93–100.

Myung, D., Farooqui, N., Waters, D., Schaber, S., Koh, W., Carrasco, M., Noolandi, J., Frank, C. W., & Ta, C. N. (2008). Glucose-permeable interpenetrating polymer network hydrogels for corneal implant applications: A pilot study. *Current Eye Research*, **22**, 29–43.

Ngakeng, V., Hauck, M. J., Price, M. O., & Price, F. W., Jr. (2008). AlphaCor keratoprosthesis: A novel approach to minimize the risks of long-term postoperative complications. *Cornea*, **27**, 905–910.

Parks, R. A., & McCarey, B. E. (1991). Hydrogel keratophakia: Long-term morphology in the monkey model. *CLAO J.*, **17**, 216–222.

Parks, R. A., McCarey, B. E., Knight, P. M., & Storie, B. R. (1993). Intrastromal crystalline deposits following hydrogel keratophakia in monkeys. *Cornea*, **12**, 29–34.

Pettit, D. K., Horbett, T. A., Hoffman, A. S., & Chan, K. Y. (1990). Quantitation of rabbit corneal epithelial cell outgrowth on polymeric substrates *in vitro*. *Invest Ophthal Vis Sci*, **31**(11), 2269–2277.

Peyman, G. A., Beyer, C. F., Bezerra, Y., Vincent, J. M., Arosemena, A., Friedlander, M. H., Hoffmann, L., Kangeler, J., & Roussau, D. (2005). Photoablative inlay laser *in situ* keratomileusis (PAI-LASIK) in the rabbit model. *J. Cataract Refract. Surg.*, **31**, 389–397.

Pinchuk, L., Wilson, G. J., Barry, J. J., Schoephoerster, R. T., Parel, J. M., & Kennedy, J. P. (2008). Medical applications of poly(styrene-block-isobutylene-block-styrene) ("SIBS"). *Biomaterials*, **29**, 448–460.

Pinero, D. P., Alio, J. L., Uceda-Montanes, A., El Kady, B., & Pascual, I. (2009). Intracorneal ring segment implantation in corneas with post-laser *in situ* keratomileusis keratectasia. *Ophthalmology*, **116**, 1665–1674.

Pinero, D. P., Alio, J. L., El Kady, B., Coskunseven, E., Morbelli, H., Uceda-Montanes, A., Maldonado, M. J., Cuevas, D., & Pascual, I. (2009b). Refractive and aberrometric outcomes of intracorneal ring segments for keratoconus: Mechanical versus femtosecond-assisted procedures. *Ophthalmology*, **116**, 1675–1687.

Randleman, J. B., Dawson, D. G., Larson, P. M., Russell, B., & Edelhauser, H. F. (2006). Chronic pain after intacs implantation. *J. Cataract Refract. Surg.*, **32**, 875–878.

Rodrigues, M. M., McCarey, B. E., Waring, G. O., Hidayat, A. A., & Kruth, H. S. (1990). Lipid deposits posterior to impermeable intracorneal lenses in rhesus monkeys: Clinical, histochemical and ultrastructural studies. *Refract. Corneal Surg.*, **6**, 32–37.

Steinert, R. F., Storie, B., Smith, P., McDonald, M. B., van Rij, G., Bores, L. D., Colin, J. P., Durrie, D. S., Kelley, C., Price, F., Jr., Rostron, C., Waring, G. O., 3rd, & Nordan, L. T. (1996). Hydrogel intracorneal lenses in aphakic eyes. *Arch. Ophthalmol.*, **114**, 135–141.

Xie, R. Z., Evans, M. D., Bojarski, B., Hughes, T. C., Chan, G. Y., Nguyen, X., Wilkie, J. S., McLean, K. M., Vannas, A., & Sweenet, D. F. (2006). Two-year preclinical testing of perfluoropolyether polymer as a corneal inlay. *Invest Ophthalmol Vis Sci. Feb*, **47**(2), 574–581.

Warden, L., Liu, Y., Binder, P. S., Dreher, A. W., & Sverdrup, L. (2008). Performance of a new binolcular wavefront aberrometer based on a self-imaging diffractive sensor. *J. Refract. Surg.*, **24**, 188–196.

Watsky, M. A., McCarey, B. E., & Beekhuis, W. H. (1985). Predicting refractive alterations with hydrogel keratophakia. *Invest. Ophthalmol. Vis. Sci.*, **26**, 240–243.

Werblin, T. P., Peiffer, R. L., Binder, P. S., McCarey, B. E., & Patel, A. S. (1992). Eight years experience with Permalens intracorneal lenses in nonhuman primates. *Refract. Corneal Surg.*, **8**, 12–22.

Yilmaz, O. F., Bayraktar, S., Agca, A., Yilmaz, B., McDonald, M. B., & van de Pol, C. (2008). Intracorneal inlay for the surgical correction of presbyopia. *J. Cataract Refract. Surg.*, **34**, 1921–1927.

Yoon, Y. D., Waring, G. O., 3rd, Stulting, R. D., Edelhauser, H. F., & Grossniklaus, H. E. (1998). Keratocyte repopulation in epikeratoplasty specimens. *Cornea*, **17**, 180–184.

D. OPHTHALMOLOGIC APPLICATIONS: GLAUCOMA DRAINS AND IMPLANTS

Crystal Cunanan
Tissue Engineering at Boston Scientific, Los Gatos, CA, USA

Although biomaterials occupy a central role in any medical device, there is possibly no other area that has seen such an explosion of new materials and devices as in the treatment of glaucoma. Moving away from the larger devices made from silicone or polypropylene, the implants under development today are uniformly focused on minimally-invasive surgical methods for implantation, and include a wide variety of different materials, including metals, polymers, and biologics. Most of these devices still function primarily as drainage devices, although some new approaches have been proposed. Given the high cost of glaucoma to society, it is likely that new materials and devices will continue to be investigated to curb the onset and progression of this painless, yet debilitating, disease.

HISTORICAL PERSPECTIVE ON THE TREATMENT OF GLAUCOMA

Historically, the treatment of glaucoma has focused on the use of pharmaceutical agents, either alone or in combination, and today the pharmaceutical industry enjoys over $4 billion annual sales for glaucoma (Stuart, 2010). Given the high cost of these pharmaceuticals, which require regular oversight from a trained glaucoma specialist, the use of these agents is focused primarily in the developed countries. Sadly, people in India and Africa and other developing countries still go blind due to glaucoma without ever being treated. This is in marked contrast to patients in the United States and Europe, where the glaucoma specialist will have exhausted every pharmaceutical, surgical, and device approach available, and patients rarely lose their vision entirely. Nonetheless, the difficulties associated with the current treatments have generated a strong interest in improving the treatment of glaucoma in all markets, especially for patients of African descent, in whom glaucoma can be most severe (Sommer, 1996; Leske, 1997).

Glaucoma is called the "silent" disease because it typically attacks eyesight subtly over time and without noticeable symptoms. It is a controllable disease if detected early, although an estimated 40% of the 65–70 million people with glaucoma do not know they have the disease, and over seven million people go blind from glaucoma each year (Quigley, 1996). Glaucoma can occur in young children, primarily due to a malformation of the anatomical structures of the eye or acutely, due to injury or a temporary closure of the ocular drainage pathways. The most common form of glaucoma is called Primary Open Angle Glaucoma (POAG) and is associated with aging, race, and other co-morbidities such as ocular and systemic hypertension. In a landmark clinical trial to determine if early treatment of patients having elevated intraocular pressure (IOP) with pharmaceutical treatments would protect against developing glaucoma, patients who received early treatment had a 54% relative reduction in risk of developing POAG within five years, compared to patients who were monitored but received no early treatment (Kass, 2002).

The hallmark of the disease is the loss of ocular neurons, which cannot be replaced, leading to loss of vision. Some neuronal loss commonly occurs with aging, and the diagnostic challenge is to determine if the neuronal loss is occurring at a higher-than-average rate, and thus poses a threat to vision. Most patients with glaucoma have higher-than-average intraocular pressure (IOP), and thus most treatments are focused on lowering pressure. A disturbing 25% of patients with glaucoma experience loss of vision without demonstrating elevated IOP, and these normal pressure glaucoma (NPG) patients pose a significant diagnostic and therapeutic challenge for clinicians. Figure D.1 shows a schematic diagram of the eye and key ocular structures. As glaucoma affects the optic nerve and is frequently associated with higher pressures, the clinician can monitor the disease in part through a careful examination of the optic nerve head, visible through the patient's pupil. In a normal patient, the optic nerve head is flat, and is termed a "disk," while in patients with elevated pressure, the optic nerve head is pushed backwards, and forms a depression called a "cup." Clinicians will monitor the extent of cupping and the ratio of cup-to-disk during regular office visits.

The traditional treatment paradigm for a glaucoma patient is shown in Figure D.2, and involves taking topical drops daily to manage disease, increasing the dose and/or adding additional agents as the initial efficacy of the drops wears off. Within two years, 75% of all patients will be on at least two medications to try to control their IOP (Lichter, 2001). Prostaglandins are the most commonly used pharmacologic agent, which increase uveal-scleral outflow, but also turn blue eyes brown. Beta-blockers are the second agent of choice, which suppress aqueous production, but they can also have systemic effects, particularly on the cardiovascular system. Alpha-agonists and carbonic anhydrase inhibitors are older drugs that are used less often due to their side-effects, but a patient may still find themselves on several of these agents within a few years of being diagnosed with glaucoma. Even more complicating, the patient may have different drug regimens and dosing schedules for each eye, making glaucoma a patient-compliance nightmare, which only adds to the problem.

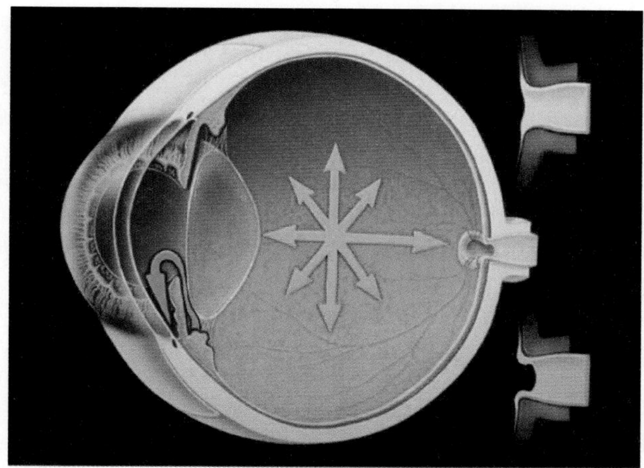

FIGURE D.1 Schematic diagram of the eye, showing the production of aqueous humor (AH) by the ciliary body, and the normal flow out through the trabecular meshwork. In a non-diseased eye, the optic nerve head is flat and is referred to as a "disk." In a glaucomatous eye, the optic nerve head is deformed due to the pressure, forming a depression, or "cup."

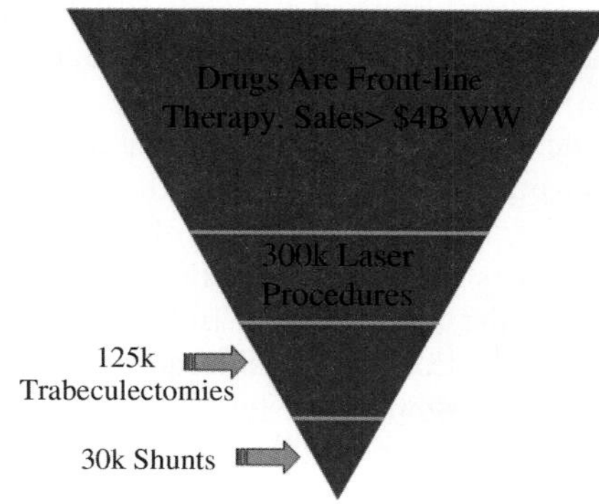

FIGURE D.2 Traditional treatment paradigm where device implants are used as a last resort. Recent studies (Gedde, 2012) and new devices are challenging this paradigm, and clinicians are re-examining the use of devices earlier in the disease process.

Of the nearly three million Americans with glaucoma, the traditional treatment paradigm shows only 10% of them will move on to require surgery. The primary surgical tools include Argon Laser Trabeculoplasty (ALT) or Selective Laser Trabeculoplasty (SLT), and the average patient will require two or three different laser treatment sessions to manage their disease. Typically patients can reduce the number of medications after LT, but within a few years, they may need to add back additional pharmaceutical treatments to try and maintain IOP control. By five years, over 80% of patients treated with laser trabeculoplasty will have failed to have adequately controlled IOP (Juzych, 2004). Only about 5% of glaucoma patients are advanced into the surgical treatment phase, and the majority of these surgical cases are trabeculectomy procedures. Traditionally considered the "gold standard" for IOP management, trabeculectomy is a very invasive surgical procedure which redirects outflow drainage through a hole punched directly through the sclera. The wound healing response of the body is inhibited with anti-neoplastic agents like mitomycin C (MMC) or 5-fluorouracil (5-FU), and a successful procedure results in the formation of a bubble-like structure under the conjunctiva called a filtering bleb. Because of the MMC/5-FU that is used, the wound usually does not heal, thus providing a new pathway for outflow of fluid from the eye, with a large surface area for subsequent reabsorption of the fluid into the systemic drainage system. Bleb management and infection control are difficult challenges with these patients who must visit the glaucoma specialist up to several times a week for care. Although trabeculectomy is considered the gold standard for glaucoma management, over 45% of patients fail to demonstrate adequate IOP control after three years (Yalvac, 2004).

Traditionally, less than 1% of all glaucoma patients in the US are treated with surgical implants. The existing approved devices are large and bulky, designed to imitate the bleb of a trabeculectomy patient, but replacing the patient's tissue with silicone or polypropylene, which cannot be degraded. Figure D.3 contains an example of a typical shunt used in these cases. Patients undergoing traditional shunt implantation have had decades of declining vision, escalating treatment regimens, and increasing complication rates and severity. It is no surprise that today's patients do not consider this acceptable and are looking for new solutions to manage their disease. A recent clinical trial has demonstrated that implantation of a glaucoma drainage implant (GDI) has equivalent IOP control compared to the surgical trabeculectomy procedure, with a higher success rate and fewer complications (Gedde, 2012). This study points to the utility of using an implanted device rather than a surgical procedure for more repeatable outcomes with fewer complications. It is the impetus for looking at a change in the treatment paradigm where patients are treated with implants much earlier in the disease progression, with the potential to radically change the rate of progression of vision loss for millions of glaucoma patients worldwide.

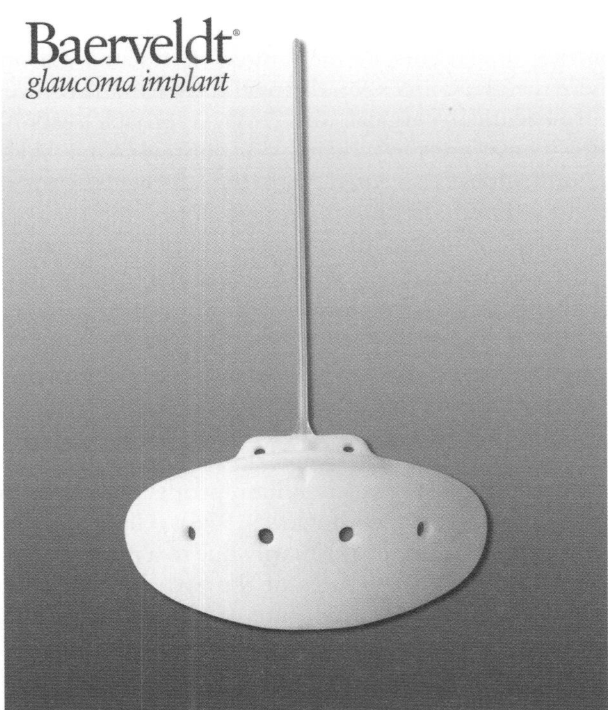

FIGURE D.3 Photograph of the Baerveldt® shunt, one of the most popular glaucoma device implants in use today. The tube is inserted into the anterior chamber, while the wide plate area sits on top of the sclera to provide an area for outflow. (Photo reprinted with permission from Abbott Medical Optics. All rights reserved.)

NEW DRAINAGE DEVICES AND MATERIALS UNDER DEVELOPMENT

Table D.1 contains a list of a number of new devices being developed to manage glaucoma patients, and compares them with the characteristics of the larger GDIs approved for use today. Before discussing their differences, it is notable to discuss their similarities. First, all new devices under development are small and are focused on using minimally-invasive procedures for implantation. The ocular trauma from the older technologies results in high complication rates, which can cloud the performance of a new device. It is unlikely that a company would succeed in obtaining regulatory approval today for a product that carried a 30–50% rate of complications (Gedde, 2012). Second, all new devices under development do not contain valves, relying instead on their small size to control outflow to an acceptable level and avoid hypotony (low IOP). The last similarity between the different new devices is that they all focus on improving outflow, although they differ widely in where and how they do this. Some devices have focused on enhancing natural outflow mechanisms through the trabecular meshwork to Schlemm's canal, while other devices are shunting fluid to the subconjunctival space, as is done with today's older and larger implants. Finally, there are other devices under development which strive to shunt fluid to new locations, such as in the supra-choroidal space.

TABLE D.1

Device/Company	Material	Bleb? MMC used?	Implant Location	Size (W x L x T; mm)	Mechanism of Action
Ahmed/New World Medical	Silicone with polypropylene valve casing	Yes; Yes	Subconjunctival	13 x 41 x 0.9 184 mm² plate area	Drains AH from the AC through tube inserted into the angle; plate keeps bleb open to allow fluid drainage.
Baerveldt®/Abbott Medical Optics	Silicone	Yes; Yes	Subconjunctival	10 x 64 x 0.84 250 mm² or 350 mm² plate area	Drains AH from the AC through tube inserted into the angle; plate keeps bleb open to allow fluid drainage.
ExPRESS®/Alcon	Stainless Steel	Yes; Yes	Trans-scleral	2.0 x 2.6 x 0.4	Drains AH from the AC to the subconjunctival bleb, as a part of a trabeculectomy procedure
MIDI Arrow/InnFocus	SIBS	Yes; Yes	Trans-scleral	1.2 x 9.5 x 0.35	Drains AH from the AC to the subconjunctival bleb. SIBS material does not trigger fibrotic healing, which can limit utility of bleb.
Aquesys®/Aquesys	ECM-based	Yes; No	Ab interno placement from angle to subconjunctival space	N.R. Fits within 25 Ga needle	Atraumatic placement with degradable biologic creates a new drainage channel from the AC to the subconjunctival space.
iStent®/Glaukos	Heparin-coated Titanium	No; No	Ab interno placement from angle to Schlemm's canal	0.5 x 1.0 x 0.25	Bypasses trabecular meshwork for direct drainage from AC to Schlemm's canal. Heparin prevents clotting on device to maintain patency.
Hydrus™ Aqueous Implant/Ivantis	Nitinol	No; No	Intra-cannicular stent in Schlemm's canal	N.R.	Stents open Schlemm's canal to 9xs normal area (approx. 150 micron diameter). Open structure prevents blockage of collector channels. Does not extend into the AC
GMS/SOLX	24-kt Gold	No; No	Suprachoroidal space	3.2 x 5.2 x 0.04	Drains AH directly from AC to the suprachoroidal space. Inert material does not trigger inflammatory response in highly vascularized suprachoroidal space.

ECM = Extra cellular matrix. N.R. = Not Reported

Aqueous Humor Production

Figure D.4 shows the ocular structures important in glaucoma. Aqueous humor (AH) is produced by the ciliary body at a rate of about 2 μL/minute (Brubaker, 1991). The AH is secreted into the anterior chamber, which has a volume of approximately 250 μL. Outflow occurs directly through the trabecular meshwork (TM), which is a layered meshwork of increasingly smaller pore size, where each trabecular beam is covered with phagocytic TM cells. It is believed that these phagocytic cells help to "clean" the AH before it passes through the TM, engulfing cells and large particles to avoid blockage of the smaller pores. After flowing through the TM, the AH flows out through Sclemm's canal, that runs circularly around the limbus of the eye. Once in the Sclemm's canal, the AH flows to collector channels, that drain into the venous system. Aqueous humor can also flow indirectly through the uveo-scleral pathway, although this accounts for only about 25% of the AH outflow in healthy eyes (Mishima, 1997).

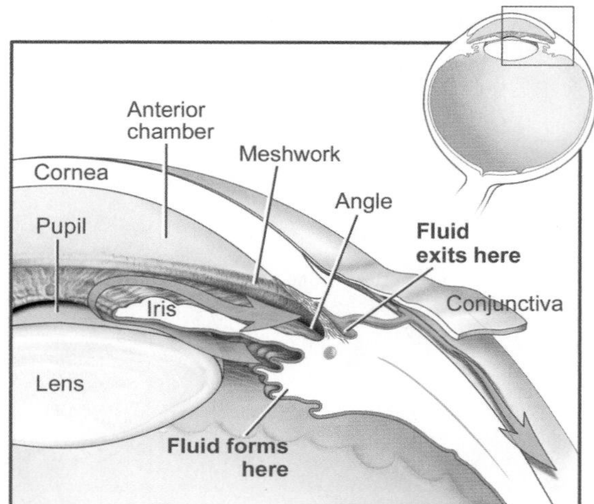

FIGURE D.4 Schematic diagram of the ocular structures important in aqueous fluid production and outflow. Aqueous humor is produced by the ciliary body and flows around the iris and out of the eye, passing through the trabecular meshwork and into Schlemm's canal and the collector channels. (Reprinted with permission, NEI-NIH.gov, ref NEA11.)

Micro-Invasive Glaucoma Surgery (MIGS) Devices

The Alcon ExPRESS® shunt. The Alcon ExPRESS® shunt is a first generation MIGS device that relies on a small bore to the drainage pathway to avoid hypotony. The device is made from stainless steel and is placed trans-sclerally as a part of the trabeculectomy procedure. The device is 2 mm wide and 2.6 mm long, with a 50 or 200 micron inner bore diameter. The device has a partial spearhead design to keep the implant from migrating outwards, with a flat plate against the sclera to prevent the implant from migrating inwards (see Figure D.5a). The ExPRESS device was originally developed by Optonol,

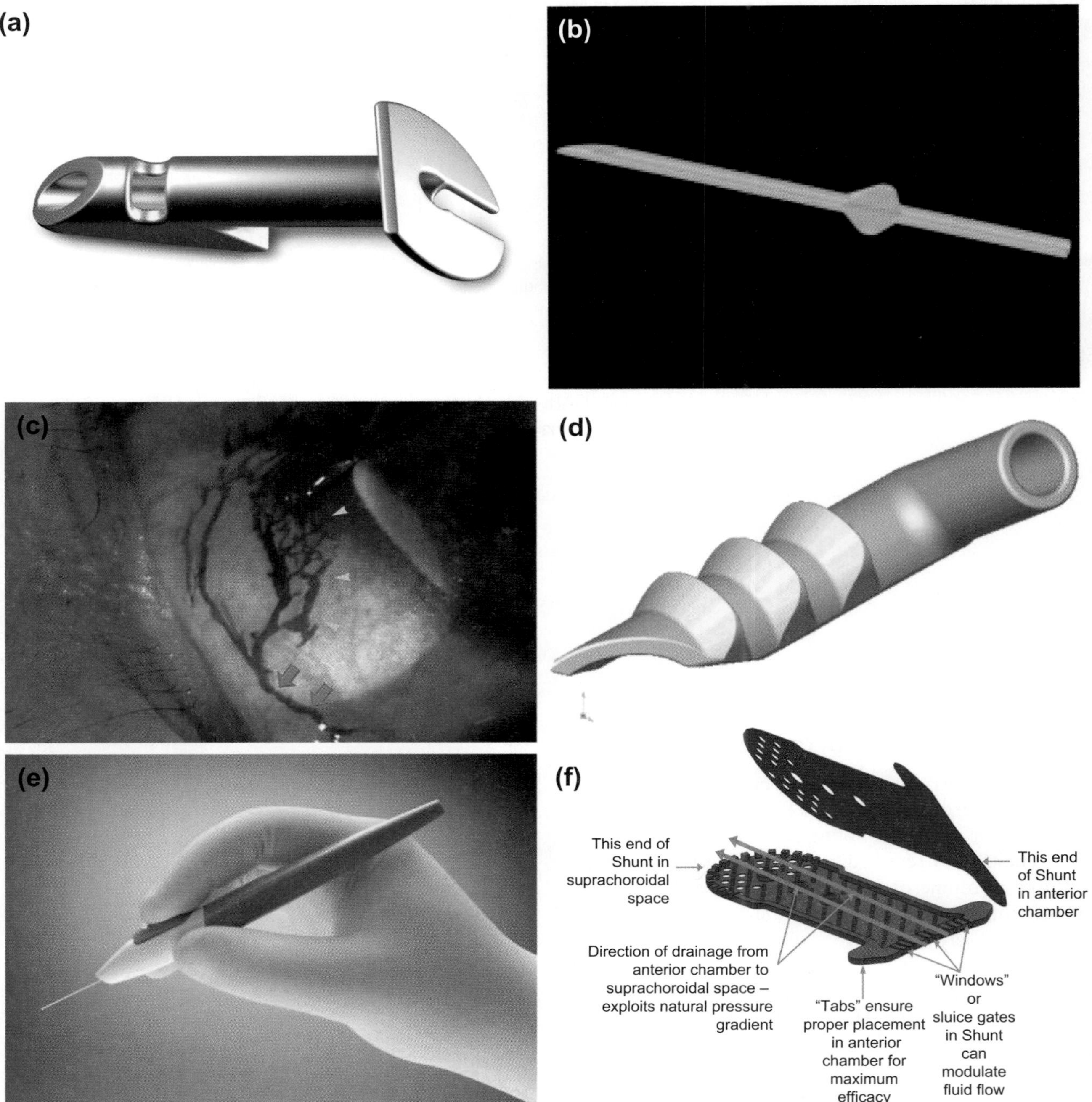

FIGURE D.5 An overview of the various devices under development today, and their reported advantages. (a) The ExPRESS® mini shunt from Alcon Laboratories (reprinted with permission); (b) The MIDI Arrow from InnVision (reprinted with permission); (c) Normal conjunctival lymphatics in a monkey, demonstrated by subconjunctival trypan blue injection. Initial lymphatics (yellow arrowheads) flow into larger structures forming collecting lymphatics (blue arrows). The distribution of conjunctival lymphatic vessels is different than conjunctival blood vessels, and is important in successful longterm drainage. The Aquesys® device has demonstrated long-term maintenance of both lymphatics and blood vessels in the conjunctiva of animals (figure reprinted with permission; Yu, 2009). (d) the iStent® trabecular meshwork bypass implant, from Glaukos Corporation (reprinted with permission); (e) the iStent inserter (reprinted with permission); (f) The SOLX GMS shunt fabricated from 24 carat gold (reprinted with permission).

and was acquired by Alcon in 2009. Today it is typically implanted with the use of mitomycin C (MMC) to inhibit proliferation and maintain a diffuse bleb. The device is approved worldwide for use in end-stage patients who have failed one or more surgical procedures.

The MIDI Arrow. The MIDI Arrow, developed by InnFocus LLC, is a microtube fabricated from the SIBS polymer (see Figure D.5b). The styrene-isobutylene-styrene block copolymer is a stable, biocompatible material with a long, successful history of use in the Boston Scientific TAXOL™ drug-eluting stent (Pinchuk, 2008). The design is simple, with a thin tube with side fins to prevent migration of the device. The MIDI device drains the AH from the Anterior Chamber (AC) to the subconjunctival space, which is first treated with MMC to prevent fibroproliferation and maintain an open bleb. The MIDI device has been successful in significantly reducing IOP in end-stage patients.

The Aquesys® device. The Aquesys® device is a resorbable material, such as gelatin, which is placed across the trabecular meshwork through to the subconjunctival space with an *ab interno* procedure (Yu, 2003). The biocompatible tube, combined with the minimally traumatic procedure, enables the formation of a stable drainage channel with a healthy conjunctival bleb containing normal blood vessels and lymphatic drainage vessels (see Figure D.5c). Formation of a drainage network has been associated with long-term stability of the procedure in animal models (Yu, 2009). The Aquesys device is currently undergoing clinical trials.

The Glaukos iStent®. The Glaukos iStent® is a heparin-coated titanium implant that shunts fluid from the AC directly into Schlemm's canal. The device is placed into the canal through an *ab interno* approach, typically in conjunction with cataract surgery, when the surgeon will already be working inside the eye. In that case, the implantation adds little to the length of the procedure. The device is implanted with a specialized tool, similar to an intraocular lens inserter used for lens replacement after cataract surgery. Figures D.5d and D.5e show the unique snorkel-shaped device with its inserter. The device is coated with heparin to minimize deposition of fibrin and cells, as sometimes a small amount of blood can enter the anterior chamber upon implantation. The device is marketed in Europe, and recently received FDA approval.

The Ivantis Hydrus™. The Ivantis Hydrus™ device is a nitinol stent that is placed within Schlemm's canal to keep the canal open and allow drainage. In that regard it is similar to the Glaukos iStent in concept, although how the two devices keep the canal open are entirely different. The Ivantis device is still under development, but recent reports demonstrate the utility of the device in reducing IOP in patients also undergoing cataract extraction (Samuelson, 2012). The nitinol stent is biocompatible when tested in the orthotopic position in New Zealand albino rabbits for six months. Non-human primates also demonstrated biocompatibility after 13 weeks implantation, although both implanted and sham-operated controls registered some response to the mechanical distortion which occurs during the implantation procedure (Grierson, 2012).

The SOLX GMS. The SOLX GMS device is fabricated from 24 carat gold, the rationale being to take advantage of the inert qualities of gold. As seen in Figure D.5f, the device is shaped like an elongated plate which contains many microchannels. The device is placed with the footplate in the anterior chamber to drain fluid to the suprachoroidal space. The suprachoroidal space is not the major outflow pathway for aqueous humor in the eye, but it is pressure-independent, operating constantly, albeit at lower levels. Because the device drains to the suprachoroidal space, it does not require a subconjunctival bleb nor the use of antimetabolites. The SOLX device is marketed in Europe and is under clinical investigation in the United States.

SUMMARY

The field of glaucoma therapies is undergoing a period of tremendous change and innovation, sparked by the high failure rates and complications of surgical procedures. The promising results observed with devices, as well as new surgical procedures, fewer complications, and a higher success rate are noted. As in most areas of medicine, the focus on today's devices is towards minimally-invasive procedures that elicit reduced trauma and inflammation compared to older treatments. The field remains divided as to the best course of action to take for filtration, with roughly equal developments occurring in the bleb-based subconjunctival drainage devices, Schlemm's canal stents, and the drainage plates that shunt fluid to the suprachoroidal spaces. The field also remains divided as to the best material(s) to use, and new devices employ metals such as stainless steel, gold, titanium, and nitinol; polymers such as SIBS and silicone; and biologics, either as a scaffold (gelatin), or a coating (heparin). The upcoming years promise to be exciting ones in the area of biomaterials for glaucoma devices, as the various materials and devices move into the clinic and their efficacy can be measured.

REFERENCES

Brubaker, R. F. (1991). Flow of aqueous humor in humans [The Friedenwald Lecture]. *Invest Ophthalmol. Vis. Sci.*, **32**(13), 3145–3166.

Gedde, S. J., Schiffman, J. C., Feuer, W. J., et al. (2012). Treatment Outcomes in the Tube Versus Trabeculectomy (TVT) Study After Five Years of Follow Up. *Am. J. Ophthalmol.*, **153**(5), 789–803.

Grierson, I., Johnstone, M., Toris, C., et al. (2012). *In-Vivo Biocompatibility Evaluation of a Novel Nickel-Titanium Schlemm's Canal Scaffold.* New York: Poster presentation at the 22nd annual meeting of The American Glaucoma Society. March 1, 2012.

Jukych, M. S., Chopra, V., Banitt, M. R., et al. (2004). Comparison of long-term outcomes of selective laser trabeculoplasty versus argon laser trabeculoplasty in open angle glaucoma. *Ophthalmology*, **111**, 1853–1859.

Kass, M. A., Heuer, D. K., Higginbotham, E. J., et al. (2002). The ocular hypertension treatment study: a randomized trial determines that topical ocular hypotensive medication delays or prevents the onset of primary open-angle glaucoma. *Arch. Ophthalmol.*, **120**, 701–713.
Leske, M. C., Connel, A. M., Wu, S. Y., et al. (1997). Distribution of intraocular pressure. The barbados eye study. *Arch. Ophthalmol.*, **115**, 1051–1057.
Lichter, P. R., Musch, D. C., Gillespie, B. W., et al. (2001). Interim clinical outcomes in the collaborative initial glaucoma treatment study comparing initial treatment randomized to medications or surgery. *Ophthalmology*, **108**, 1943–1953.
Mishima, H. K., Kiuchi, Y., Takamatsu, M., et al. (1997). Circadian intraocular pressure management with latanoprost: diurnal and nocturnal intraocular pressure reduction and increased uveoscleral outflow. *Surv. Ophthalmol.*, **41**, S139–S144.
Pinchuk, L., Wilson, G. J., Barry, J. J., et al. (2008). Medical applications of poly(styrene-block-isobutylene-block-styrene) ("SIBS"). *Biomaterials*, **29**, 448–460.
Quigley, H. A., Tielsch, J. M., Katz, J., & Sommer, A. (1996). Rate of progression in open-angle glaucoma estimated from cross-sectional prevalence of visual field damage. *Am. J. Ophthlamol.*, **122**, 355–363.
Samuelson, T. W., Lorenz, K., Pfeiffer, N. (2012). *Six-Month Results from a Prospective, Multicenter Study of a Nickel-Titanium Schlemm's Canal Scaffold for IOP Reduction After Cataract Surgery in Open Angle Glaucoma*. New York: Poster presentation at the 22nd annual meeting of the American Glaucoma Society. March 1st, 2012.
Sommer, A. (1996). Glaucoma risk factors observed in the Baltimore eye survey. *Curr. Opin. Ophthalmol.*, **7**, 93–98.
Stuart, M. (2010). In Glaucoma, Devices Go Eye-To-Eye with Drugs. *Start-Up Magazine*, September 1, 2010.
Yalvac, I. S., Sahin, M., Eksioglu, U., et al. (2004). Primary viscocanalostomy versus trabeculectomy for primary open angle glaucoma: three year prospective, randomized clinical trial. *J. Cataract. Refract. Surg.*, **30**, 2050–2057.
Yu, D.-Y., & Morgan, W. H. (2003). Biological microfistula tube and implantation method and apparatus. *US Patent 6,544,249*. Issued April 8, 2003.
Yu, D.-Y., Morgan, W. H., Su, X., et al. (2009). The critical role of the conjunctiva in glaucoma filtration surgery. *Progress in Retinal & Eye Research*, **28**, 303–328.

E. THE DEVELOPMENT OF A RETINAL PROSTHESIS: A SIGNIFICANT BIOMATERIALS CHALLENGE

Mark S. Humayun, Adrian P. Rowley, John J. Whalen III, James D. Weiland, and Armand R. Tanguay, Jr.
University of Southern California, Los Angeles, CA, USA

INTRODUCTION

Restoring functional vision using electrical stimulation has been a goal of ophthalmologists and vision scientists for more than two centuries (LeRoy, 1755). The difficulty confronted by scientists is that the visual system consists of roughly 10^{10} neurons, each of which may receive input from multiple, perhaps even thousands, of neurons. In spite of this, cutting across traditional interdisciplinary boundaries of medicine and engineering has enabled rapid progress in the field of prosthetic vision over the last decade, reconnecting some blind patients to the visual world. Millions of people worldwide lose their vision as a consequence of photoreceptor damage stemming from retinal degeneration (e.g., retinitis pigmentosa (RP) or age-related macular degeneration (AMD)) (Augustin 2009; Shintani 2009). The feasibility of an implantable retinal prosthesis or "bionic eye" that could partially restore vision by direct electrical stimulation of retinal neurons is supported by several studies. Under development are devices that bypass the degenerated photoreceptors and apply electrical stimulation at different locations along the visual pathway (Rizzo 2001; Asher 2007; Chader 2009). These microelectronic prosthetic devices rely on the principle of electrical activation of nerves, which is also the concept behind other successful neural stimulators, such as the cochlear implant.

The intent of this chapter on the development of a retinal prosthesis is to provide an introduction to the materials used in current devices, and the biomaterial challenges that limit the implementation of more sophisticated implants. The chapter outlines the various approaches that numerous research groups worldwide have taken to build such a device. Technical challenges remain with respect to mechanical design, hermetic packaging, polymers, metal electrodes, surgical techniques, attachment methods, stimulus parameters, and effectiveness (Bertschinger, 2008). Several research groups around the world have already implanted prototype retinal prosthetic devices in blind humans. At present, retinal prostheses are still experimental, although they are expected to become commercially available in the near future. This chapter introduces the anatomy and physiology of the visual system, especially the retina. It describes the architecture of a retinal prosthesis, and the biomaterials that scientists use to design this ground-breaking medical device.

OVERVIEW OF THE VISUAL SYSTEM

Visual perception is the ability to interpret information from the visible light spectrum that reaches the human eye. Incident photons cause phototransduction in the retina. The human eye can perceive wavelengths from 380 to 760 nm (Kaufman, 2003). The visual system is the part of the central nervous system (CNS) that enables one to assimilate information from the surrounding environment. The visual pathway includes the retina, optic nerve, optic chiasm, lateral geniculate nucleus, numerous visual nuclei, and the visual cortex, which is located in the occipital lobe. These anatomical structures accomplish a number of complex tasks, including filtering of undesirable wavelengths and reception of desirable ones, the formation of binocular vision from monocular inputs, the construction of images over time (temporally), the identification of objects, and the evaluation of distance. The neural control of eye movement allows for resolution of the visual field during movement, the ability to resolve an object's motion, and also enables the exploration of space (Kaufman, 2003).

The eye includes a variety of tissue types that work together to facilitate sight. The eyeball, or globe, is spherical in shape and about 2.5 cm in diameter (Figure E.1). The sclera is the tough, white, outer tunic that protects the internal structures and helps maintain the shape of the eyeball. At the front of the eye is an equally strong but clear structure called the cornea. The cornea is responsible for allowing light into the eye and, combined with the crystalline lens, allows focusing of images onto the retina. The vitreous humor is a clear gel that forms more than two-thirds of the ocular volume, and occupies the vitreous cavity between the crystalline lens and the retina (Bron, 2001).

Internal to the sclera in the posterior portion of the eye is the choroid, which is a vascular structure that carries the blood to nourish the internal structures of the eye, mainly the outer photoreceptor layer of the neural retina. Finally, the retina lines the posterior two-thirds of the inside of the eye. The retina contains neurons (photoreceptors) that are sensitive to photons, as well as more conventional neurons that process the photo-stimulated signals. The macula is an oval zone of yellow coloration within the central retina. This area of the retina has the highest number of cone photoreceptors, and is reponsible for both high resolution vision and the perception of color.

The Retina

The retina is a thin (approximately 200 μm), fragile, light sensitive, neuronal tissue lining the inner surface of the posterior segment of the eye. Because of its importance with regard to the retinal prostheses, it will be discussed in further detail. The refractive power of the cornea and crystalline lens of the eye focuses light on the retina, which serves a similar function as the sensor in a video camera.

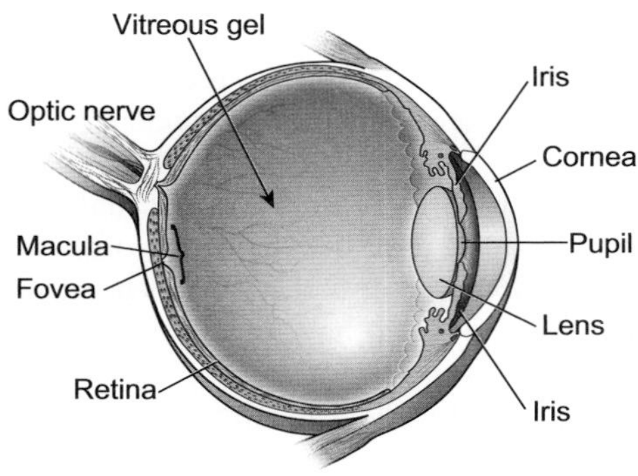

FIGURE E.1 Sketch of the eye showing the major anatomical structures. *(Courtesy* http://www.nei.nih.gov.*)*

Light striking the retina's photoreceptors initiates a cascade of chemical and electrical events. This physiological process ultimately triggers a hyperpolarization in the photoreceptors, followed by a depolarization in the bipolar and then ganglion cells. These signals are sent to various visual centers of the brain via the ganglion cell axons, which collectively form the optic nerve. The vertebrate retina has 10 distinct layers, in which there are three major layers of neurons interconnected by synapses (Figure E.2). The outermost layer of neurons (photoreceptors) is sensitive to light. There are two types of photoreceptors, rods and cones. Rods function in low light (scotopic vision) and provide relatively low resolution black-and-white vision, while cones function in bright light (photopic vision) and allow for sharp, color vision in the central field.

Anatomy of the Vertebrate Retina

The vertebrate retina has 10 distinct layers. From innermost to outermost, they include:

1. *Inner limiting membrane* – interfaces with the vitreous humor;
2. *Nerve fiber layer* – axons of the ganglion cells exit the eye as the optic nerve;
3. *Ganglion cell layer* – contains nuclei of ganglion cells and gives rise to optic nerve fibers;
4. *Inner plexiform layer* – synaptic layer between the bipolar cells and ganglion cells;
5. *Inner nuclear layer* – contains cell bodies of the bipolar cells;
6. *Outer plexiform layer* – synaptic layer between photoreceptors and bipolar cells;
7. *Outer nuclear layer* – contains cell bodies of the photoreceptors;
8. *External limiting membrane* – separates the inner segment portions of the photoreceptors from their cell nuclei;
9. *Photoreceptor layer* – rods and cones;
10. *Retinal pigment epithelium* – nourishes the retinal neurons.

Critical to the design of a retinal prosthesis is the location of the remaining viable neurons in the affected retina. In retinal degenerative diseases, the primary target for pathology is the retinal photoreceptor cells. The pathological processes cause the photoreceptors to dysfunction and die. The bipolar and retinal ganglion neurons, however, are relatively spared. Therefore, when an image falls on the retina, it is unable to be processed into an electrical signal. The rationale of a retinal prosthesis

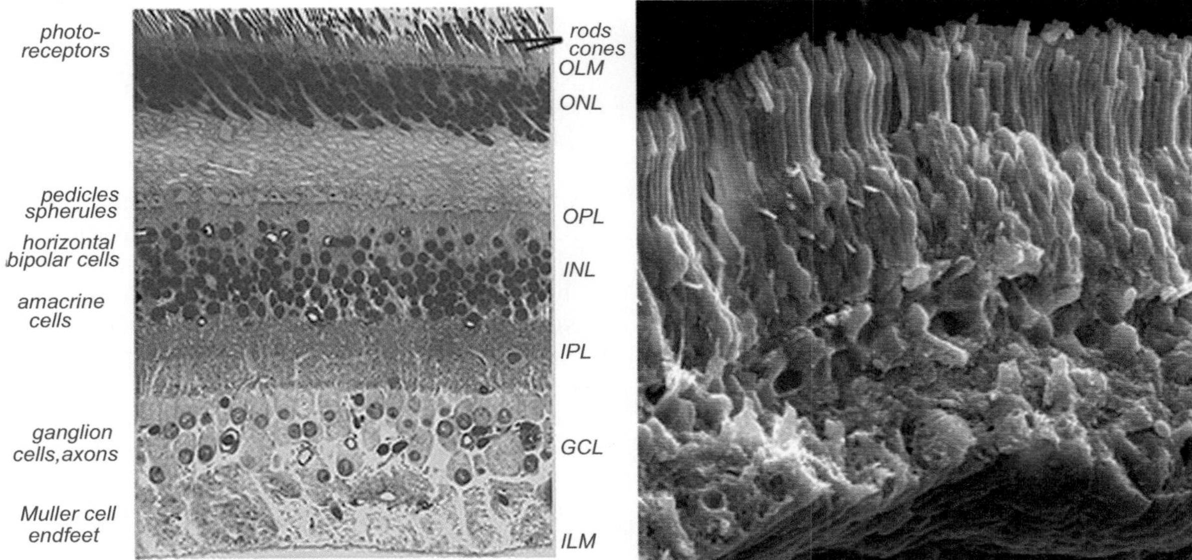

FIGURE E.2 A photomicrograph of a retinal cross-section shows the highly laminar structure. From top to bottom, rods and cones, outer limiting membrane (OLM), outer nuclear layer (ONL), outer plexiform layer (OPL), inner nuclear layer (INL), inner plexiform layer (IPL), ganglion cell layer (GCL), inner limiting membrane (ILM). *(Courtesy* http://webvision.med.utah.edu/.*) SEM image of a retinal cross-section of rabbit retina.*

then, is to bypass the damaged photoreceptor cells and directly stimulate the remnant, secondary neurons (bipolar cells and/or retinal ganglion cells).

Although in RP and AMD patients it is the photoreceptors that are destroyed, the remaining cells of the retina also undergo extensive and complex degenerative remodeling (Marc, 2003). Despite the fact that patients with RP and AMD suffer from the destruction of the photoreceptors, up to 80% of bipolar cells and up to 30% of ganglion cells remain viable (Humayun, 1999).

RETINAL PROSTHESES

The retina not only responds to photons, but also to electrical stimulation. Moreover, the central nervous system is able to interpret the electrical stimulation generated by electrical signals (such as from a retinal prosthesis), and convert it into an identifiable image. Recent studies on one device showed that even limited sight restoration involved a slow learning process that took months for improvement. In this particular study, light perception was eventually restored in all six patients (Chader, 2009). Thus, both basic and clinical work now has firmly established the scientific feasibility of sight restoration through electronic prosthetic devices. A number of companies around the world are working towards clinically useful devices (Chader, 2009).

Approaches to Stimulating the Retinal Tissue

Progress in the field of microelectronic prostheses has converged with advances in ophthalmic surgery to enable several groups to develop epiretinal prostheses (Caspi, 2009; Roessler, 2009). The term "epiretinal" specifies that the electrode array is intraocular and lies on the surface of the retina, in close proximity to the retinal ganglion cells. Other groups have been developing subretinal prostheses (Stett, 2007; Besch, 2008). The subretinal space is the gap created between the photoreceptors and the retinal pigment epithelium when the retina is purposefully detached. Arrays can be placed to stimulate the nearby bipolar cells in the inner nuclear layer. A wide variety of subretinal techniques exist, ranging from passive photosensitive diode arrays (Peyman, 1998) to neurotransmitter-based stimulation methods (Peterman, 2003), and to more active stimulation, using optical-to-electrical signal amplification (Palanker, 2005). The aim of subretinal stimulation is to maximize the neuronal processing of the remaining retinal circuitry by targeting the remaining functional bipolar cells.

Some researchers have directly stimulated the optic nerve, which is the cylindrical anatomical structure that exits the back of the eye. The optic nerve contains ganglion cell axons running from the retina to synapse with neurons in the lateral geniculate nucleus (LGN). The optic nerve can be reached surgically; however, the high density of the axons (1.2 million) may make it difficult to achieve a detailed pattern of stimulation (Shandurina, 1996; Veraart, 1998). A few researchers have used extraocular disc electrodes that have been sutured to the sclera (Chowdury, 2005), while others have performed transretinal electrical stimulation from the suprachoroidal space (Sakaguchi, 2004).

Select patients have been implanted with cortical and intracortical electrode arrays. There are several advantages to directly stimulating the visual cortex; however, it is difficult to achieve simple perception even when stimulating large numbers of neurons in the case of intracortical microstimulation. The convoluted surface of the brain cortex makes for more difficult implantation, and the resultant complications can be fatal (Dobelle, 1976; Bak, 1990).

The majority of efforts in developing visual prostheses for the blind have focused on electrical stimulation of the retina. It is well-understood that there is a substantial amount of computational processing within the LGN (Dan, 1996) and the visual cortex that control the visual signal (Hubel, 1962). Targeting stimulation early in the visual pathway (e.g., the surviving bipolar and retinal ganglion cells) better utilizes the remainder of the visual system.

The Architecture of a Retinal Prosthesis

Common features of retinal prostheses include hermetic casing of electronics, electrode array interfaces, and a cable that connects the two components. The following describes the architecture of one particular design of an epiretinal prosthesis, shown in Figure E.3 (Argus™ I, manufactured by Second Sight Medical Products, Inc.). This specific model consists of an implanted unit and an external unit. The external unit consists of a small camera, similar to a webcam, which is worn in a pair of glasses and connects to a belt-worn visual processing computer (not shown). Power and signal information are sent from this processor through a wire cable to an external transmitter coil. This coil couples electromagnetically (wirelessly) to a secondary coil that is implanted under the skin. Other groups have designed the secondary coil and electronics to be sutured around the eye (Roessler, 2009). The coils allow information to be sent inductively. The implanted unit consists of an extraocular (electronic case) and an intraocular component (electrode array). From the secondary coil, power and signal information are sent through a subcutaneous cable across the eye wall to the intraocular electrode array which, in this particular case, is attached to the epiretinal surface via a retinal tack. The electrode array consists of disc shaped platinum electrodes in a square layout (Margalit, 2002). Stimulation can be presented to the patient using two different protocols. The first is the camera mode, which uses real-time video captured from the video camera. The visual processing computer then encodes the brightness of the various regions of the camera's field of view into various amplitudes of stimulation current. The other method is the direct stimulation mode, which allows the researchers to send stimulation signals to each electrode independently. This is important as it enables neuroscientists to perform precisely controlled psychophysical tests (Horsager, 2009).

Role of Materials in Retinal Implants

The role of materials in retinal implants is to ensure that the electronic system can effectively deliver stimulating current to the retina, while not significantly damaging the surrounding tissue. Delicate retinal tissue can be affected by infection and inflammation, and thus any introduced device can cause acute and delayed (foreign-body) types of immunologic reactions. Also, patients can have hypersensitivity reactions to polymers and metals. Toxic substances from fabrication equipment, such as oxidants,

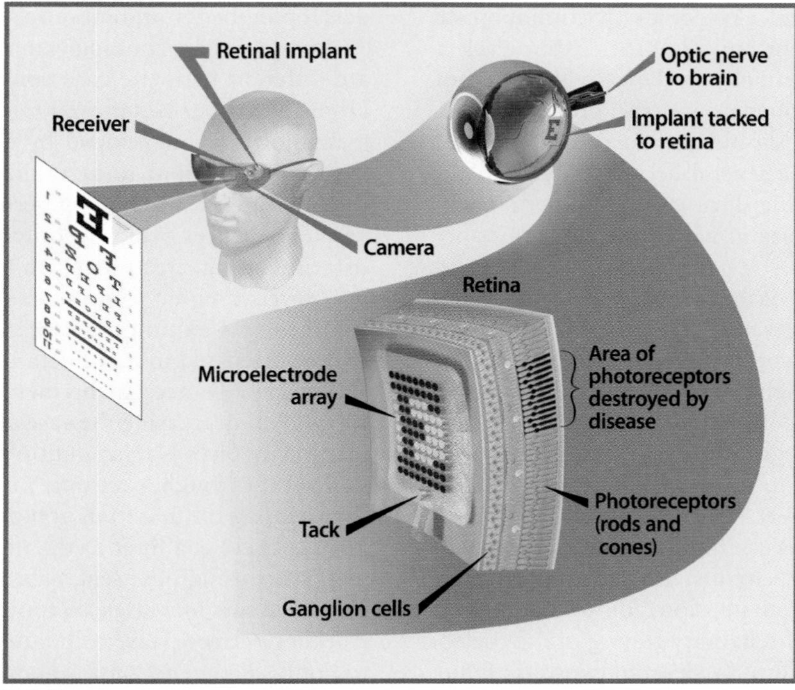

FIGURE E.3 Illustration of epiretinal prosthesis system. *(Courtesy Department of Energy 2008.)*

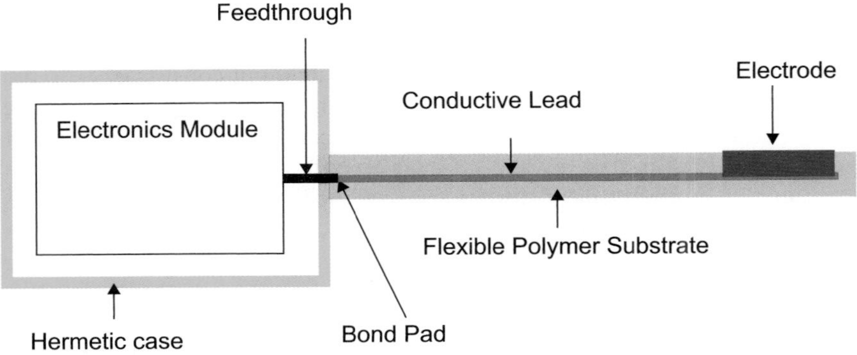

FIGURE E.4 The components of a flexible electrode array. *(Adapted from* Kelly, 2009.*)*

catalysts, and contaminants, may be left as residues on implants and can cause inflammation (Margalit, 2002). The tissue can have deleterious effects on the implant, including degradation of the polymers, corrosion of the metals (Jansen, 1991), mechanical dislocation of the implant (Yuen, 1990), and damage to the electronic components from the saline environment, should the hermetic case fail. The main implant components to be discussed are the hermetic case, the conducting feedthroughs, the bond pads, the conductive leads, the electrodes, and the flexible polymer substrate (Figure E.4). Other implant components are primarily electrical (e.g., microchips), and will not be discussed in this chapter.

Hermetic Case

The microelectronic case is one of the most critical elements involved in the development of a retinal prosthesis. It must provide two primary functions: (1) protect the microelectronics from the surrounding environment; and (2) allow the internal electronics to communicate with both the environment and the other external electronic components. All retinal prostheses will consist of various electronic components, and the protective casing for each component should last for the lifetime of the patient, which may be several decades after implantation. In most implantable devices, the hermetic package takes the form of a case in which the electronics are placed.

The first basic function of the microelectronic case is to form a barrier between the various semiconductor devices and the saline environment (Tummala, 1989; Nichols, 1994; Najafi, 2003). A high quality barrier that is virtually impermeable is considered hermetic. In practice, hermeticity is a matter of degree, and standards have been established, for example, to measure the rate at which a case will leak water vapor. For implantable microelectronics, the environment is essentially saline. Sodium will irreparably damage microcircuits in a matter of days, through processes such as corrosion, ion diffusion, and delamination. The field of reliability testing of microelectronics has established failure modes and exposure limits for water vapor (Nichols, 1994; Najafi, 2003). The casing material is typically formed from ceramic, metal or glass, since these materials are among the least permeable to water. Examples of materials commonly used for the case include titanium, which forms a stable oxide (TiO_2), alumina (Al_2O_3), and glass (e.g., Pyrex) (Tummala, 1989).

The hermetic case also has a second basic function; it must allow electrical signals to communicate between circuitry inside the case and the neural interface outside the case. This communication is facilitated by feedthroughs (also called vias or interconnects), which are electrical conductors. The most vulnerable points in the casing are at the multiple conductor feedthroughs and their interface with the insulating case material (Margalit, 2002). The materials that surround feedthroughs must be electrically insulating to prevent cross-talk among feedthroughs, so ceramic or glass is commonly used for this portion of the case. Metal can be used for the case, provided that it is not in contact with the feedthroughs. Although this may seem like a relatively simple part of an active implantable medical device, companies exist solely for the purpose of developing better and more reliable feedthroughs.

A variety of technologies has been developed to create different hermetic case constructs (Tummala, 1989; Honma, 2002; Najafi, 2003). Much of the original technology was developed by the cardiovascular device industry. Titanium and stainless steel cans have been the benchmark for creating hermetic containers. Cardiovascular devices typically only employ a small number of stimulating electrodes, and the devices themselves are several cubic centimeters in volume; therefore, feedthroughs exiting the case can be large, a topic that will be discussed in more detail later.

Recent advances in hermetic case design have been focused on decreasing the case and feedthrough size. One method involves placing multiple wires comprised of platinum (Pt) through a ceramic insulator (e.g., Al_2O_3), and then heating the materials at high temperatures (Backous, 2002). There is a limit to the number of conductors that can pass through the seal, based on area. As the number of conductors increases, so too does the chance of a leak. Another hermetic case technique uses a high electric field to bond silicon and glass (anodic bonding) (Ziaie, 1997). Solder bonding can be used to adhere aluminum/silicon to glass. This hermetic method provides more than 10 mega Pascals of bonding strength (Cheng, 2000). More recently, yttria-stabilized tetragonal zirconia polycrystals (Y-TZPs) and a titanium alloy Ti-6Al-4V have also been used as the ceramic and metal components of the seal. Both materials have strong mechanical properties and biocompatibility. A brazing method involving titanium nickel (TiNi)-clad braze filler material is used to create a tight hermetic seal (Jiang, 2005).

Metal Feedthroughs

Metallic feedthroughs are the conducting elements of a hermetic casing that allow the internal microelectronics to communicate with the outside environment or other electronic components. The microelectronics industry has investigated and developed a number of feedthrough fabrication approaches (Tummala, 1989; Honma, 2002; Najafi, 2003). However, these advances do not easily translate to biomedical applications like the retinal prosthesis, because only certain materials can be safely implanted and exposed to tissue for sustained periods of time. For example, electrochemical deposition has been investigated by the microelectronics industry as a method of producing high-aspect ratio, micron-scaled feedthroughs (Honma, 2002). This work has focused primarily on materials like copper, aluminum or silver, which are not considered biocompatible materials for electrically active components like feedthroughs. Acceptable materials, like platinum and iridium, are difficult to process due to their high melting temperatures (>1400°F) and chemical nobility.

A novel high-density feedthrough approach for medical implants has been proposed (Najafi, 2003). This design uses silicon micromachining to pattern high-density feedthroughs on a silicon wafer. A glass case is anodically bonded to the wafer to protect the microelectronics. This approach significantly increases the two-dimensional area of the case size above the chip size, because the feedthroughs are coplanar with the microchip. The state-of-the-art in microelectronics casing has not yet been able to provide a biocompatible case design that incorporates feedthroughs with ultra-high packing densities (~50 feedthroughs mm^{-2}).

Bond Pads

The term "bond pad" refers to a metal trace feature on any kind of circuit element that will be bonded or interconnected to another part, such as a metal trace feature that will be bonded with a gold wirebond or connected with solder to a component. In the retinal prosthesis electrode array, one end of the array is called the electrode end and the other end, the bond pad end. Physically, they can be similar, that is, openings in the insulation layer to open up an exposed metal area. In the context of the electronics case, the feedthroughs are exposed on the bottom. A thin film metallization layer can cover the feedthrough and route a trace some distance away to the bonding site or bond pad. Challenges relate to the development of thin bond pads on a flexible substrate that connects to a rigid case.

Conductive Leads

The conductive leads have a separate set of design constraints. Each conducting element must individually connect a single electrode to a single output on the processor chip. The leads must be electrically isolated from each other to prevent cross-talk. Lead insulation must prevent signal leakage, and also prevent fluid penetration that might lead to corrosion and/or electrical shorting. Lastly, the composite size of all the leads for all of the electrodes must be minimized to prevent weight problems; however, the leads cannot be so small that the electrical resistance of the leads creates too much load or impedance for the system. Recently, a high density lead construct has been fabricated by thin film deposition of platinum leads on flexible parylene substrates (Rodger, 2006).

Electrodes

The significant biomaterials' challenges associated with fabricating electrodes for a retinal prosthesis are: (1) maintaining biocompatibility and electrode integrity during electrical stimulation; and (2) selecting a material that can be processed into electrode arrays with the required geometries. A variety of electrode materials have been designed and their properties reported (Robblee, 1990; Stieglitz, 2004).

Biocompatibility during Electrical Stimulation. Electrical stimulation is an electrochemical process in which current carried through the metal electrodes is transferred to the surrounding electrolyte solution (i.e., the retinal tissue) (Merrill, 2005). Electrochemical processes at the electrode–electrolyte interface determine how much charge can be transferred safely from the electrode to the surrounding tissue. Reversible charge transfer is the only way that current can be passed safely from an electrode to tissue. Irreversible reactions can result in damage to the tissue or electrode. Materials that have reversible reactions in the absence of significant irreversible reactions are well-suited for neural stimulation. In practice, any material will undergo irreversible reactions with a high enough electrode surface potential. Biphasic current pulses were created to prevent the electrode surface potential from exceeding this safe charge transfer potential range (Brummer, 1975, 1983; Roblee, 1990; Stieglitz, 2004; Merrill, 2005). The current is reversed just prior to hydrogen or oxygen gas evolution, which are irreversible reactions. The potential range over which reactions are reversible is commonly known as the "water window," because it is the window inside which water does not hydrolyze into either hydrogen gas or oxygen gas. Charge capacity is the ability of a material to reversibly store charge via non-faradaic or reversible faradaic processes. Electrodes with higher charge capacity allow for larger current amplitudes or longer pulse widths to be used.

The goal is to identify electrodes that can transfer maximum charge, reversibly. This can be achieved through selecting a material and geometry that undergoes high double layer charging via high surface areas, through high reversible faradaic charging, and oxidizes into a conductive oxide that does not dissolve (leave the electrode surface) in solution. The noble metals were the first class of materials to be investigated for implantable electrode applications, because of their corrosion resistance properties (Brummer, 1975; Robblee, 1990). Platinum and iridium have the unique characteristic that their oxides are conductive. This means that current can be passed through these electrodes to much higher potentials. There is evidence that the precious metals (platinum, iridium, rhodium, gold, and palladium) corrode under certain conditions of electrical stimulation (McHardy, 1980), but it has also been reported that under typical stimulus regimens the dissolution levels would be non-toxic (Hibbert, 2000). Capacitive electrodes, such as tantalum oxide (Ta_2O_5), titanium dioxide (TiO_2), and barium titanate ($BaTiO_3$), are another area that has been researched, in which an oxide with a very high dielectric constant serves as a barrier to charge transfer at the electrode surface (Rose, 1985). However, because of the associated low capacitance, the charge injection limits of such capacitive electrodes are still not comparable to iridium oxide (Weiland, 2002; Wang, 2006). Iridium oxide (IrO_x) electrodes, termed "valence change oxides," have proven to be very resistant to corrosion, and can withstand more than 2 billion 10 mA current pulses without degradation (Ziaie, 1997). Using simple waveforms, a conservative charge density limit for chronic stimulation with Pt is 100 $\mu C/cm^2$ (McCreery, 1988). A titanium nitride (TiN) thin film electrode has demonstrated charge injection limits of 0.87 mC/cm^2, which is higher than both platinum and IrO_x. TiN electrodes appear to have better mechanical properties than those made from IrO_x, but they appear to reduce retinal cell survival (Guenther, 1999). Carbon nanotubes (CNTs) have interesting electrochemical and mechanical properties. They are about five times stronger than steel and recently vertically-aligned multiwalled CNT pillars were used as microelectrodes (Wang, 2006). A new material, platinum grey, has also been described that has better long-term survivability than iridium oxide with similar charge limits (Zhou, 2009; US Pat #6,974,533).

Processing Characteristics of Electrodes. In addition to the charge injection properties of the electrode material, the ability to machine, shape or otherwise process the electrode material into the desired form, geometry, and spacing is a critical aspect in materials selection. Currently, the leading candidates for stimulating electrodes are platinum and iridium oxide electrodes (Merrill, 2005). However, these materials have melting temperatures in the area of 1400°F, and hence are very difficult to deposit and process. The cost of these materials also requires that processing be highly efficient. A variety of techniques can be used. For example, metal casting has been used to make larger (>500 μm diameter) disc electrodes used in early retinal prostheses (Humayun, 2003). Thin film processing techniques (evaporation and sputter coating) have been used to fabricate microelectrodes of much smaller size (Kovacs, 1987; Rozman, 1991; Nisch, 1994; Owens, 1995; Maynard, 1997). Thin film techniques deposit material over the entire deposition chamber, and hence high vacuum techniques are not efficient. Electrochemical deposition of platinum is another area investigated for deposition of electrodes (Frazier, 1993; Whalen, 2006).

Significant research has been performed in the study of microelectrode materials for sustained implantable use. For a retinal prosthesis, the added requirements that the electrodes be fabricated in the order of <100 μm diameter with several hundred electrodes arranged in a two-dimensional, curved array with spacings in the order of 10 to 100 μm, create another level of complexity. Recent work has been reported on novel techniques for fabricating microelectrode arrays on flexible substrates meeting these criteria (Rodger, 2006). Microprocessed electrodes can be machined with these features, but their charge injection properties are limited to the mass of platinum that can be deposited. Electrochemical deposition to increase charge capacity is a real solution (Whalen, 2006; Zhou, 2007).

Polymers

An important consideration regarding the safety of electrical stimulation is the mechanical design of the intraocular electrode array. This array is in direct contact with the retinal tissue, and therefore has the potential to damage the tissue mechanically, chemically, and thermally (Piyathaisere, 2001). The retina is approximately 200 μm thick, and extremely delicate, with a Young's modulus of only 20 to 40 kPa (Chen, 2009). Therefore, any polymer material that will chronically contact such a fragile biological tissue must also have a low Young's modulus, and be extremely thin to avoid tearing, compressing or detaching the retina. In some cases, if the polymer material is too thin, it may retain sharp edges that could cut the retina. It has been shown that with optimized designs and thin polymer arrays, safe implantation can be achieved with certain materials (Ameri, 2009). The following polymers have been studied in retinal prosthesis research: PDMS; polyimide; and parylene.

- Polydimethylsiloxane (PDMS) is a group of polymeric organosilicon compounds, referred to as silicones. PDMS is well-known for its rheological properties, and it is the most common material currently used in most breast implants. At high temperatures, PDMS acts like a viscous liquid (viscoelastic). However, at low temperatures it acts like an elastic solid, similar to rubber. As the retina is extremely delicate, the softer the interfacing polymer, the less damage the array will inflict upon the retina. PDMS has unique flexibility characteristics, with a shear elastic modulus G ~250 kPa, and a very low transition temperature Tg ~125°C. These factors, plus the ease with which it can be used in a cleanroom setting, make PDMS extremely well-suited for incorporation into microelectronic electrode arrays, such as those required for the retinal prosthesis (Lotters, 1997).

- Thermosetting polyimides are commercially available in multiple forms, such as resins, sheets, and machine parts. When polyimides first became available in the 1960s, chip carriers were based on multilayered ceramic cases. However, use of such carriers caused signal delays. Polyimides are currently being used as dielectrics for microelectronic circuits, because of their higher signal transmission speed. Thermoplastic polyimides are very often called *pseudothermoplastic*, and they are generally either aromatic heterocycles or linear. Polyimides are able to achieve a low dielectric constant while maintaining mechanical strength, thermal stability, and chemical resistance (Ghosh, 1996).
- Parylene is the generic name used for a variety of chemical vapor deposited poly(p-xylylene) polymers. Using vapor deposition polymerization, parylene can be made ultrathin, uniform, conformal, and pinhole free. It has high tensile strength, low permeability to moisture, and dielectric barrier properties. Among the parylenes, Parylene C is inert, clear, hydrophobic, biocompatible, and is the most commonly used (Kazemi, 2004).

All three of the polymers above can be metallized, and are capable of being molded into array designs that conform to the curvature of the retina. However, both polyimide and parylene have the ability to cut the retina when produced as thin sheets. Due to this issue, some retinal prosthesis designs involve coating the parylene or polyimide with silicone, which is a much softer polymer. A silicone–polyimide hybrid microelectrode array has been shown to have moderate physical properties that are suitable for implantation (Kim, 2009). The viscoelastic nature of silicone is ideally suited to interface with delicate biological tissues such as the retina. A recent study on the biocompatability of materials for use in subretinal prostheses demonstrated that poly(ethylene glycol) PEG, parylene, and poly(vinyl pyrrolidone) PVP produced less histological disruption than polyimide or polyimide coated with either amorphous aluminum oxide or amorphous carbon (Montezuma, 2006).

OTHER MATERIALS CONCERNS

Attachment

One of the disadvantages of the epiretinal prosthesis is the resulting distance from their target cells. The thickness of the nerve fiber and ganglion cell layers is at least 20–200 and 20–40 μm, respectively (Greenberg, 1998). In order to reduce the charge density to within safe limits for long-term stimulation, researchers can use either larger electrodes and thus lose resolution, or get closer to the target cells.

Retinal tacks, bioadhesives, and magnets have been some of the methods examined for epiretinal attachment. In one study, metal alloy retinal tacks (composed of cobalt–nickel–chromium–molybdenum–tungsten–iron) remained firmly affixed to the retina for up to 2.5 years. All tacks were ultimately surrounded by connective and/or glial tissue induced by firm retinal adherence; however, the retinal architecture was normal within 1 mm of the scar (Ohira, 1998). One *in vivo* study examined nine commercially available compounds for their suitability as intraocular adhesives in rabbits. One type of adhesive (SS-PEG hydrogel, Shearwater Polymers, Inc.) proved to be strongly adherent and non-toxic to the retina (Margalit, 2000). Although tacks can be surgically removed with minimal further damage to the eye, bioadhesives have remained difficult to detach from the delicate retina. However, recent studies have shown that the reversible thermosensitive glue polymerized N-isopropyl acrylamide (pNIPAM) can be used in combination with polyimide, parylene C, and PDMS. It provides effective *in vitro* retinal adhesion between 32 and 38°C, and this adhesion is completely reversible by lowering the temperature of the physiologic medium (i.e., saline) to 22°C (Tunc, 2007). Another reversible method of attachment may be achievable by immobilizing "sticky" peptides onto the surface of polymers. Disintegrins are cysteine-rich peptides ranging from 45 to 84 amino acids in length, and many of them contain an Arg-Gly-Asp (RGD) sequence, a recognition site for cell membrane adhesion proteins (integrins). Due to their extremely high integrin binding affinity, these peptides, when attached to polymers, form a strong attachment to cells (Massia, 1990). The use of low concentrations of proteolytic enzymes cleaves the peptides, and may enable microelectrode arrays to interface with the retina reversibly.

Materials Issues Unique to Subretinal Implants

Several subretinal implant designs have used microphotodiode arrays in the subretinal space. Microphotodiode array technology is completely compatible with semiconductor device integration. Microphotodiode arrays develop current in the photodiodes in response to light, and deliver it to the overlying retina. Most are designed

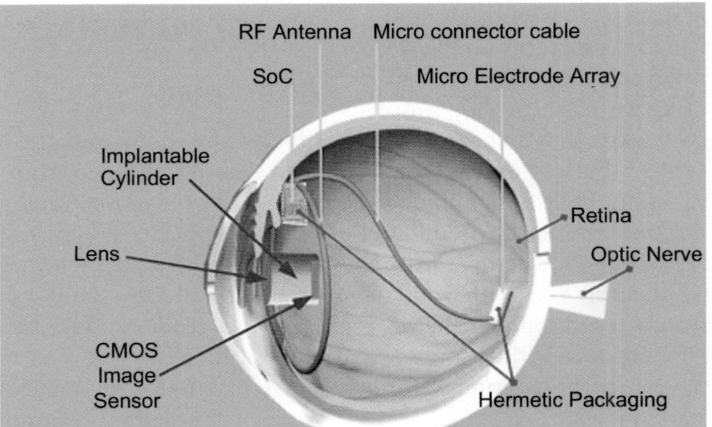

FIGURE E.5 Intraocular camera and associated retinal microstimulator array.

to be subretinal devices. One such example is the Artificial Silicon Retina (ASR), which does not have any external connections and is powered by incident light with wavelengths of 500 to 1100 nm (Peyman, 1998). It contains approximately 3500 microscopic light detectors called "microphotodiodes." Each has a stimulating electrode and is designed to convert light energy into electrical impulses. Early microphotodiode arrays used gold as the electrode material, which dissolved during the postoperative period. Semiconductor-based silicon microphotodiodes utilizing iridium, iridium oxide (IrOx), and platinum (Pt) electrodes have also been implanted. However, several biocompatibility issues to the subretinal approach have been encountered. These include corrosion of the device, limitation of the implant power generation by size requirements (Chow, 2002), and creation of a mechanical barrier between the retina and the choroid (Zrenner, 1999). The vascular choroidal layer provides nourishment to the outer layers of the retina. The viability of using infrared enhancement of the stimulation power has been proven (Schubert, 1999). Minimizing the stimulation electrode surface area is another partial solution for this same problem. More recently, researchers have studied the effects of four coatings (silicon oxide, iridium oxide, parylene, and titanium nitride) on the functioning of subretinal devices. Implants coated with iridium oxide and parylene were generally well-tolerated in the subretinal space, inducing only mild gliotic responses. The silicon oxide coatings produced the formation of a significant fibrotic seal around the implants, and dissolved within a period of 6 to 12 months. Titanium nitride stimulating electrodes were well preserved *in vivo* (Hämmerle, 2002; Butterwick, 2009).

DIRECTIONS FOR THE FUTURE

Intraocular Camera

Current intraocular retinal prostheses driven by an external head-mounted camera, as described above, require slow and unnatural head movements for mobility and navigation tasks. To allow for normal foveation, a more natural coupling of eye and head movements, and an expanded depth of field, one group is developing a novel intraocular camera (IOC) to work in conjunction with the epiretinal microstimulator array in a fully intraocular retinal prosthesis (Nasiatka, 2007).

The intraocular camera is designed for implantation in the crystalline lens sac, in a manner similar to that of an intraocular lens (IOL). This configuration necessitates an extremely compact, lightweight package (3.0 × 4.5 mm, <150 mg) with a focal length of approximately 2 mm (500 diopters) and an *f*/# close to unity. While these constraints appear highly aggressive, visual psychophysics experiments have provided optimal pixellation and image pre- and post-processing requirements that in turn have yielded significantly relaxed camera design constraints (Figure E.5).

The intraocular camera housing is designed to be both biocompatible and hermetic, and currently comprises a thin-walled titanium housing with a fused silica window. The CMOS image sensor array and VLSI control chips incorporated within the hermetic housing are silicon based, with oxide, nitride, and metallic overlayers. If required, feedthroughs will be provided by a ceramic (alumina) plate with platinum or palladium vias. Haptic elements to support the intraocular camera within the crystalline lens sac will be fabricated from polymeric materials (such as PMMA) similar to those used in intraocular lenses.

SUMMARY

Attempts at implanting electronic devices at various anatomical locations along the visual pathway have proven to be extremely challenging. As knowledge increases about how to stimulate neurons with microelectronics, and as both microelectronics and materials science continue to evolve, it is hopeful that one day the restoration of vision to the blind will become a scientific reality.

ACKNOWLEDGMENTS

This work was supported in part by Grants from the National Science Foundation, BMES ERC, NIH, US Department of Energy DE-PS02-09ER09-02, as well as a research fund from the Alcon Research Institute.

BIBLIOGRAPHY

Ameri, H., Ratanapakorn, T., Ufer, S., Eckhardt, H., Humayun, M. S., et al. (2009). Toward a wide-field retinal prosthesis. *J. Neural. Eng.*, 6(3). 035002.

Asher, A., Segal, W. A., Baccus, S. A., Yaroslavsky, L. P., & Palanker, D. V. (2007). Image processing for a high-resolution optoelectronic retinal prosthesis. *IEEE Trans. Biomed. Eng.*, **54**(6 Pt 1), 993–1004.

Augustin, A. J., & Kirchhof, J. (2009). Inflammation and the pathogenesis of age-related macular degeneration. *Expert Opin. Ther. Targets*, **13**(6), 641–651.

Backous, D., Dunford, R., Segel, P., Muhlocker, C., Carter, P., et al. (2002). Effects of hyperbaric exposure on the integrity of the internal components of commercially available cochlear implant systems. *Otology & Neurotology*, **23**(4), 463–467.

Bak, M., Girvin, J. P., Hambrecht, F. T., Kufta, C. V., Loeb, G. E., et al. (1990). Visual sensations produced by intracortical microstimulation of the human occipital cortex. *Med. Biol. Eng. Comput.*, **28**, 257–259.

Bertschinger, D. R., Beknazar, E., Simonutti, M., Safran, A. B., Sahel, J. A., et al. (2008). A review of *in vivo* animal studies in retinal prosthesis research. *Graefes Arch. Clin. Exp. Ophthalmol.*, **246**(11), 1505–1517.

Besch, D., Sachs, H., Szurman, P., Gülicher, D., Wilke, R., et al. (2008). Extraocular surgery for implantation of an active subretinal visual prosthesis with external connections: Feasibility and outcome in seven patients. *Br. J. Ophthalmol.*, **92**(10), 1361–1368. Epub 2008 Jul 28.

Bron, A. J., Tripathi, R. C., & Tripathi, B. J. (2001). *Wolff's Anatomy of the Eye and Orbit* (8th ed.). London, UK: Arnold. p. 225 and p. 455.

Brummer, S. B., & Turner, M. J. (1975). Electrical stimulation of the nervous system: The principle of safe charge injection with noble metal electrodes. *Bioelectrochemistry and Bioenergetics*, **2**, 13–25.

Brummer, S. B., Robblee, L. S., & Hambrecht, F. T. (1983). Criteria for selecting electrodes for electrical stimulation: Theoretical and practical considerations. *Annals of the Academy of Sciences*, **405**, 159–171.

Butterwick, A., Huie, P., Jones, B. W., Marc, R. E., Marmor, M., et al. (2009). Effect of shape and coating of a subretinal prosthesis on its integration with the retina. *Exp. Eye Res.*, **88**(1), 22–29.

Caspi, A., Dorn, J. D., McClure, K. H., Humayun, M. S., Greenberg, R. J., et al. (2009). Feasibility study of a retinal prosthesis: Spatial vision with a 16-electrode implant. *Arch. Ophthalmol.*, **127**(4), 398–401.

Chader, G. J., Weiland, J., & Humayun, M. S. (2009). Artificial vision: Needs, functioning, and testing of a retinal electronic prosthesis. *Prog. Brain Res.*, **175**, 317–332.

Chen, K., Rowley, A. P., & Weiland, J. D. (2010). Elastic properties of porcine ocular posterior soft tissues. *J. Biomed. Mater. Res. A.*, **93**, 634–645.

Cheng, Y., Lin, L., & Najafi, K. (2000). Fabrication and hermeticity testing of a glass-silicon package formed using localized aluminum/silicon-to-glass bonding. *IEEE 13th Ann. Internat. Conf. Microelectromechanical Sys.*, 757–762.

Chow, A. Y., & Chow, V. Y. (1997). Subretinal electrical stimulation of the rabbit retina. *Neurosci. Lett.*, **225**, 13–16.

Chow, A. Y., & Peachey, N. S. (1998). The subretinal microphotodiode array retinal prosthesis. *Ophthalmic Res.*, **30**, 195–198.

Chow, A. Y., Pardue, M. T., Perlman, J. I., Ball, S. L., Chow, V. Y., et al. (2002). Subretinal implantation of semiconductor-based photodiodes: Durability of novel implant designs. *J. Rehabil. Res. Dev.*, **39**(3), 313–321.

Chowdhury, V., Morley, J. W., & Coroneo, M. T. (2005). Evaluation of extraocular electrodes for a retinal prosthesis using evoked potentials in cat visual cortex. *Journal of Clinical Neuroscience*, **12**(5), 574–579.

Dan, Y., Atick, J., & Reid, R. (1996). Efficient coding of natural scenes in the lateral geniculate nucleus: Experimental test of a computational theory. *J. Neurosci.*, **16**, 3351–3362.

Dobelle, W. H., Mladejovsky, M. G., Evans, J. R., Roberts, T. S., & Girvin, J. P. (1976). "Braille" reading by a blind volunteer by visual cortex stimulation. *Nature*, **259**, 111–112.

Frazier, A. B., O'Brien, D. P., & Allen, M. G. (1993). Two-dimensional metallic microelectrode arrays for extracellular stimulation and recording of neurons. *IEEE Proceedings of the Micro Electro Mechanical Systems Conference*, 195–200.

Ghosh, M. K., & Mittal, K. L. (1996). *Polymides Fundamentals and Applications*. New York, NY: Marcel Dekker, Inc. p. 759.

Greenberg, R. J. (1998). Analysis of electrical stimulation of the vertebrate retina – work towards a retinal prosthesis. PhD Dissertation, Baltimore, MD: The Johns Hopkins University.

Guenther, E., Troger, B., Schlosshauer, B., & Zrenner, E. (1999). Long-term survival of retinal cell cultures on retinal implant materials. *Vision Res.*, **39**, 3988–3994.

Hämmerle, H., Kobuch, K., Kohler, K., Nisch, W., Sachs, H., et al. (2002). Biostability of micro-photodiode arrays for subretinal implantation. *Biomaterials*, **23**(3), 797–804.

Hibbert, D. B., Weitzner, K., Tabor, B., & Carter, P. (2000). Mass changes and dissolution of platinum during electrical stimulation in artificial perilymph solution. *Biomaterials*, **21**, 2177–2182.

Honma, H., & Watanabe, H. (2002). Advanced Plating Technology for Electronics Packaging. In J. W. Schultze, T. Osaka, & M. Datta (Eds.), *Electrochemical Microsystem Technologies* (pp. 224–244). New York, NY: Taylor & Frances, Inc.

Horsager, A., Greenwald, S. H., Weiland, J. D., Humayun, M. S., Greenberg, R. J., et al. (2009). Predicting visual sensitivity in retinal prosthesis patients. *Invest Ophthalmol. Vis. Sci.*, **50**(4), 1483–1491. Epub 2008 Dec 20.

Hubel, D. H., & Wiesel, T. N. (1962). Receptive fields, binocular interaction and functional architecture in the cat's visual cortex. *J. Physiol.*, **160**, 106–154.

Humayun, M. S., Prince, M., de Juan, E., Jr., Barron, Y., Moskowitz, M., et al. (1999). Morphometric analysis of the extramacular retina from postmortem eyes with retinitis pigmentosa. *Invest Ophthalmol. Vis. Sci.*, **40**, 143–148.

Humayun, M. S., Weiland, J. D., Fujii, G. Y., Greenberg, R., Williamson, R., et al. (2003). Visual perception in a blind subject with chronic microlectronic retinal prosthesis. *Vision Research*, **43**, 2573–2581.

Jansen, B., Schumacher-Perdreau, F., Peters, G., & Pulverer, G. (1991). Evidence for degradation of synthetic polyurethanes by *Staphylococcus epidermidis*. *Zentralbl Bakteriol*, **276**, 36–45.

Jiang, G., Mishler, D., Davis, R., Mobley, J. P., & Schulman, J. H. (2005). Zirconia to Ti-6Al-4V braze joint for implantable biomedical device. *J. Biomed. Mater. Res. B. Appl. Biomater*, **72**(2), 316–321.

Kaufman, P. L., & Alm, A. (2003). *Adler's Physiology of the Eye* (10th ed.). St Louis, IL: Mosby. p. 578 and p. 830.

Kazemi, M., Basham, E., Sivaprakasam, M., Wang, G., Rodger, D., et al. (2004). A test microchip for evaluation of hermetic packaging technology for biomedical prosthetic implants. *Conf. Proc. IEEE Eng. Med. Biol. Soc.*, **6**, 4093–4095.

Kelly, S. K., Shire, D. B., Chen, J., Doyle, P., Gingerich, M. D., et al. (2009). Realization of a 15-channel, hermetically-encased wireless subretinal prosthesis for the blind. *Proc. IEEE Engineering in Medicine and Biology Conference*, p. 200–203.

Kim, E. T., Kim, C., Lee, S. W., Seo, J. M., Chung, H., et al. (2009). Feasibility of micro electrode array (MEA) based on silicone-polyimide hybrid for retina prosthesis. *Invest Ophthalmol Vis Sci.*, **50**, 4337–4341.

Kovacs, G. T.A., Stephanides, M. C., Knapp, W. R., McVittie, J. P., & Rosen, J. M. (1987). Design of two-dimensional neural prosthesis microelectrode arrays. Boston, MA: Proceedings of the 9th Annual Conference, IEEE Engineering in Medicine and Biology Society. Nov 13–16, 1034–1035.

LeRoy, C. (1755). Ou l'on rend compte de quelques tentatives que l'on a faites pour guerir plusieurs maladies par l'electricite. *Hist Acad Roy Sciences (Paris)*, **60**, 87, 95.

Lötters, J. C., et al. (1997). The mechanical properties of the rubber elastic polymer polydimethylsiloxane for sensor applications. *J. Micromech. Microeng.*, **7**, 145–147.

Marc, R. E., Jones, B. W., Watt, C. B., & Strettoi, E. (2003). Neural remodeling in retinal degeneration. *Prog. Retin Eye Res.*, **22**, 607–655.

Margalit, E., & Sadda, S. R. (2003). Retina and optic nerve diseases. *Artif Organs*, **27**(11), 963–974.

Margalit, E., Fujii, G., Lai, J., Gupta, P., Chen, S. J., et al. (2000). Bioadhesives for intraocular use. *Retina*, **20**, 469–477.

Margalit, E., Maia, M., Weiland, J. D., Greenberg, R. J., Fujii, G. Y., et al. (2002). Retinal prosthesis for the blind. *Survey of Ophthalmology*, **47**(4), 335–356.

Massia, S. P., & Hubbell, J. A. (1990). Covalently attached GRGD on polymer surfaces promotes biospecific adhesion of mammalian cells. *Ann. NY Acad Sci.*, **589**, 261–270.

Maynard, E. M., Nordhausen, C. T., & Normann, R. A. (1987). The Utah intracortical electrode array: A recording structure structure of potential brain-computer interfaces. *Electroenceph Clinical Neurophysiology*, **102**, 228–239.

McCreery, D. B., Agnew, W. F., Yuen, T. G., & Bullara, L. A. (1988). Comparison of neural damage induced by electrical stimulation with faradaic and capacitor electrodes. *Ann. Biomed. Eng.*, **16**, 463–481.

McHardy, J., Robblee, L. S., Marston, J. M., & Brummer, S. B. (1980). Electrical stimulation with pt electrodes. IV. Factors influencing Pt dissolution in inorganic saline. *Biomaterials*, **1**, 129–134.

Merrill, D. R., Bikson, M., & Jeffreys, J. G. R. (2005). Electrical stimulation of excitable tissue: Design of efficacious and safe protocols. *Journal of Neuroscience Methods*, **141**, 171–198.

Montezuma, S. R., Loewenstein, J., Scholz, C., & Rizzo, J. F., 3rd (2006). Biocompatibility of materials implanted into the subretinal space of Yucatan pigs. *Invest Ophthalmol. Vis. Sci.*, **47**(8), 3514–3522.

Najafi, K. (2003). Micropackaging technologies for integrated microsystems: Applications to MEMS and MOEMS (Plenary Paper). *Proceedings SPIE*, **4979**, 1–19.

Nasiatka, P. J., Hauer, M. C., Stiles, N. R. B., Lue, L., Takahashi, S., et al. (2007). An intraocular camera for retinal prostheses. Irvine, CA: *2nd Frontiers in Biomedical Devices Conference*.

Nichols, M. F. (1994). The challenges for hermetic encapsulation of implanted devices: A review. *Critical Reviews in Biomedical Engineering*, **1**, 39–67.

Nisch, W., Bock, J., Egert, U., Hammerle, H., & Mohr, A. (1994). A thin film microelectrode array for monitoring extracellular neuronal activity *in vitro*. *Biosensors and Bioelectronics*, **9**, 737–741.

Ohira, A., de Juan, E., & Tsai, M. (1991). Long-term histologic and electrophysiologic evaluation of the alloy retinal tack. *Graefes Arch. Clin. Exp. Ophthalmol.*, **229**(1), 95–98.

Owens, A. L., Denison, T. J., Versnel, H., Rebbert, M., Peckerar, M., et al. (1995). Multi-electrode array for measuring evoked potentials from surface of ferret primary auditory cortex. *J. Neuroscience Methods*, **58**, 209–220.

Palanker, D., Vankov, A., Huie, P., & Baccus, S. (2005). Design of a high resolution optoelectronic retinal prosthesis. *Journal of Neural Engineering*, **2**, S105–S120.

Peterman, M. C., Mehenti, N. Z., Bilbao, K. V., Lee, C. J., Leng, T., et al. (2003). The artificial synapse chip: A flexible retinal interface based on directed retinal cell growth and neurotransmitter stimulation. *Artif Organs*, **27**, 975.

Peyman, G., Chow, A. Y., Liang, C., Chow, V. Y., Perlman, J. I., et al. (1998). Subretinal semiconductor microphotodiode array. *Ophthalmic Surg. Lasers*, **29**, 234–241.

Piyathaisere, D. V., Margalit, E., Chen, S. J., et al. (2001). Effects of short-term exposure to heat on the retina. *Invest Ophthalmol Vis. Sci.*, **42**(Suppl), S814.

Rizzo, J. F., 3rd, Wyatt, J., Humayun, M., de Juan, E., Liu, W., et al. (2001). Retinal prosthesis: An encouraging first decade with major challenges ahead. *Ophthalmology*, **108**(1), 13–14.

Robblee, L. S., & Rose, T. L. (1990). Electrochemical Guidelines for Selection of Protocols and Electrode Materials for Neural Stimualtion. In W. F. Agnew, & D. B. McCreery (Eds.), *Neural Prostheses: Fundamental Studies*. Englewood Cliffs, NJ: Prentice Hall, Inc..

Rodger, D. C., Li, W., Fong, A. J., Ameri, H., Meng, E., et al. (2006). Flexible microfabricted parylene multielectrode arrays for retinal stimulation and spinal cord field modulation. Okinawa, Japan: Proceedings of the 4th International IEEE-EMBS SpecialTopic Conference on Microtechnologies in Medicine and Biology (MMB '06). May 9–1231–34.

Rodger, D. C., Weiland, J. D., Humayun, M. S., & Tai, Y. C. (2006). Scalable high lead-count parylene package for retinal prostheses. *Sensors and Actuators B: Chemical*, **117**, 107–114.

Roessler, G., Laube, T., Brockmann, C., Kirschkamp, T., Mazinani, B., et al. (2009). Implantation and explantation of a wireless epiretinal retina implant device: Observations during the EPIRET3 prospective clinical trial. *Invest Ophthalmol Vis. Sci.*, **50**(6), 3003–3008. Epub 2009 May 6.

Rose, T. L., Kelliher, E. M., & Robblee, L. S. (1985). Assessment of capacitor electrodes for intracortical neural stimulation. *Journal of Neuroscience Methods*, **12**, 181–193.

Rozman, J. (1991). Microelectrode cuff for extraneural selective stimulation of nerve fibers. *Proceedings of the 13th Annual International Conference of the IEEE Engineering in Medicine and Biology Society*, **13**, 914–915.

Sakaguchi, H., Fujikado, T., Fang, X., Kanda, H., Osanai, M., et al. (2004). Transretinal electrical stimulation with a suprachoroidal multichannel electrode in rabbit eyes. *Jpn. J. Ophthalmol.*, **48**, 256–261.

Schubert, M., Hierzenberger, A., Lehner, H., & Werner, J. (1999). Optimizing photodiodes arrays for the use as retinal implants. *Sensors Actuators*, **74**, 193–197.

Shandurina, A. N., Panin, A. V., Sologubova, E. K., Kolotov, A. V., Goncharenko, O. I., et al. (1996). Results of the use of therapeutic periorbital electrostimulation in neurological patients with partial atrophy of the optic nerves. *Neurosci. Behav. Physiol.*, **26**, 137–142.

Shintani, K., Shechtman, D. L., & Gurwood, A. S. (2009). Review and update: Current treatment trends for patients with retinitis pigmentosa. *Optometry*, **80**(7), 384–401.

Stett, A., Mai, A., & Herrmann, T. (2007). Retinal charge sensitivity and spatial discrimination obtainable by subretinal implants: Key lessons learned from isolated chicken retina. *J. Neural. Eng.*, **4**(1), S7–16. Epub 2007 Feb 20.

Stieglitz, T. (2004). Electrode Materials for Recording and Stimulation. In: K. W. Horch, & G. S. Dhillon (Eds.), *Neuroprosthetics: Theory and Practice. Series on Bioengineering & Biomedical Engineering*. (Vol. 2). World Scientific Publishing Co.

Tummala, R. R., & Rymaszewski, E. J. (1989). *Microelectronics Packaging Handbook*. New York, NY: Van Nordstrand Reinhold. Inc. 14–32.

Tunc, M., Cheng, X., Ratner, B. D., Meng, E., & Humayun, M. (2007). Reversible thermosensitive glue for retinal implants. *Retina*, **27**(7), 938–942.

Veraart, C., Raftopoulos, C., Mortimer, J. T., Delbeke, J., Pins, D., et al. (1998). Visual sensations produced by optic nerve stimulation using an implanted self-sizing spiral cuff electrode. *Brain Res.*, **813**, 181–186.

Wang, K., Fishman, H. A., Dai, H., & Harris, J. S. (2006). Neural stimulation with a carbon nanotube microelectrode array. *Nano Lett.*, **6**(9), 2043–2048.

Weiland, J. D., Anderson, D. J., & Humayun, M. S. (2002). *In vitro* electrical properties of iridium oxide vs. titanium nitride stimulating electrodes. *IEEE Transactions on Biomedical Engineering*, **49**(12), 1574–1579.

Whalen, J. J., III, Young, J., Weiland, J. D., & Searson, P. C. (2006). Electrochemical characterization of charge injection at electrodeposited platinum electrodes in phosphate buffered saline. *Journal of the Electrochemical Society*, **153**(12), C834–C839.

Yuen, T., Agnew, W. F., Bullara, L., & McCreery, D. B. (1990). Biocompatibilty of electrodes and materials in the central nervous system. In: W. F. Agnew, & D. B. McCreery (Eds.), *Neural Prostheses, Fundamental Studies* (pp. 197–223). New Jersey: Prentice-Hall.

Zhou, D. M. (2007). Second Sight Medical Products, Inc. Platinum Electrode Surface Coating and Method for Manufacturing the Same. International Patent Pub No. WO/2007/050212.

Ziaie, B., Nardin, M. D., Coghlan, A. R., & Najafi, K. (1997). A single-channel implantable microstimulator for functional neuromuscular stimulation. *IEEE Trans. Biomed. Eng.*, **44**, 909–920.

Zrenner, E., Stett, A., Weiss, S., Aramant, R. B., Guenther, E., et al. (1999). Can subretinal microphotodiodes successfully replace degenerated photoreceptors? *Vision Res.*, **39**, 2555–2567.

CHAPTER II.5.10 BIOELECTRODES

Ramakrishna Venugopalan[1] *and Ray Ideker*[2]

[1]Auroru Burlington Clinic, Burlington, WI, USA

[2]The University of Alabama at Birmingham, Birmingham, AL, USA

INTRODUCTION

Bioelectrodes are sensors used to transmit information into or out of the body. This chapter: (1) introduces the reader to the basic reactions that occur at an electrode–electrolyte interface; (2) explains how these processes define the type of electrode; (3) explores some simple equivalent circuit models for the electrode–electrolyte interface; (4) reviews the factors influencing material selection for electrodes; and finally (5) looks at some applications of such electrodes. Although the fundamentals discussed apply to both recording and stimulating electrodes, this chapter focuses on stimulating electrodes in the material selection and application sections, because of their recent emergence as effective treatment options (see pacemakers and defibrillators in Chapter II.5.3.C.) Also, see Chapter II.5.12 that addresses this subject from a somewhat different perspective.

Surface or transcutaneous electrodes used to monitor or measure electrical events that occur in the body are considered as monitoring or recording electrodes. Typical applications for recording electrodes include electrocardiography, electroencephalography, and electromyography. Electrodes used to transmit information into the body influence specific processes that occur in the body, and are considered as stimulator electrodes. In all of these applications the electrodes are used to transmit voltage or current waveforms to specific target areas in the body for either direct control of a bioelectric event or for indirect influence on the target area through a stimulated chemical change. Such stimulator electrodes are used in cardiac pacemakers and defibrillators to maintain or restore sinus rhythm, in transcutaneous electronic nerve stimulators for pain suppression, in neural stimulation systems for applications ranging from epilepsy control to auditory augmentation, in polarizing devices for intranscutaneous drug delivery, and in stimulators for tissue healing/regeneration.

ELECTRODE–ELECTROLYTE INTERFACE

In 1800 Volta demonstrated that the electrode–electrolyte interface was the source of electrical potential, and initiated his research on direct current electricity. The nonlinearity of this interface with current led Georg Ohm (1826) to use a thermopile as a source of electrical potential, and resulted in the law that bears his name. However, the origin of the term "electrode" is attributed to one of the most prolific researchers in electrochemistry, Faraday (1834). In 1879 Helmholtz established the first model for the electrode–electrolyte interface (Helmholtz, 1879). Several more complex models of this interface were proposed by Gouy (1910) and Stern (1924). This section will explain the basic reactions at the electrode–electrolyte interface, how they define a specific electrode type, and finally the make-up of this interface.

Faradaic and Nonfaradaic (Capacitive) Processes

Bioelectrodes can interact with the body fluids (electrolytes) in two primary ways – faradaic and nonfaradaic (summarized in Figure II.5.10.1). An electrode may establish ohmic contact with the surrounding environment, thereby transferring electrons across the electrode–electrolyte interface via oxidation and/or reduction reactions. All such charge-transfer processes are governed by Faraday's law (i.e., the amount of chemical

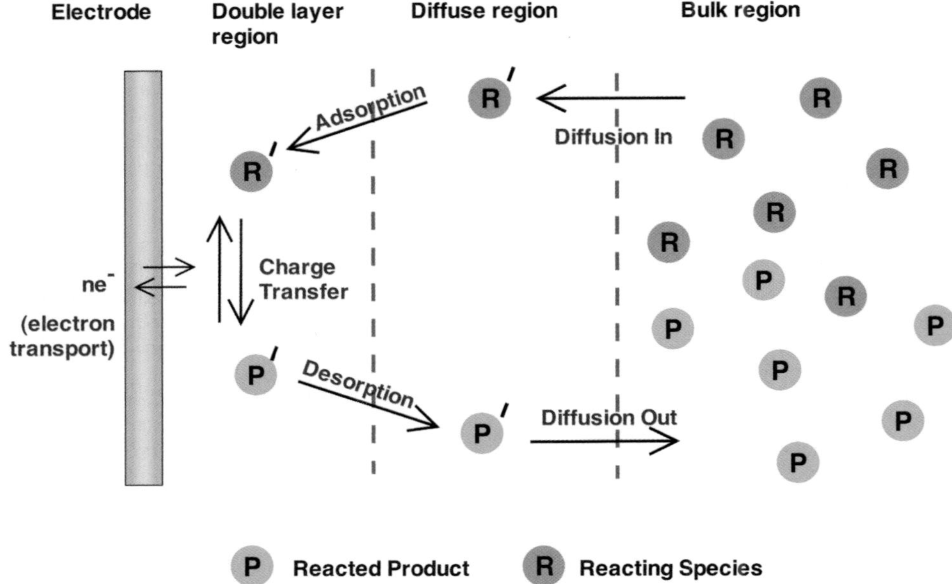

FIGURE II.5.10.1 Faradaic and nonfaradaic pathways of general electrode reactions. The charge transfer reaction at the surface is faradaic. The diffusion and the adsorption/desorption processes are nonfaradaic.

change occurring at an electrode–electrolyte interface is directly proportional to the current that flows through that interface), and hence are called faradaic processes. Electrodes at which faradaic processes occur are also known as charge-transfer electrodes. Under certain thermodynamically or kinetically unfavorable conditions an electrode–electrolyte interface may exhibit an absence of charge-transfer reactions. However, the electrode–electrolyte interface may change because of adsorption or desorption processes that occur at the interface, and transient external currents can flow due to changes in the environment. These processes at the electrode–electrolyte interface are known as nonfaradaic processes.

Polarizable and Nonpolarizable Electrodes

Electrodes in which charge transfer can occur in an unhindered manner represent truly faradaic, or perfectly nonpolarizable, electrodes; conversely, electrodes in which no faradaic charge transfer can occur are called perfectly polarizable electrodes. In the latter case, the half-cell potential of the electrode results in the formation of an electrical double layer of charge akin to two parallel plates of a capacitor, and charging or discharging of this capacitor monitors or influences a bioelectric event. Although an electrode can behave as either a polarizable or a nonpolarizable electrode under specific conditions at the interface, it will also exhibit a minimal amount of the other process (secondary) during clinical use.

Electrical Double Layer

At a given potential there will exist a charge on the metal electrode (q_M), and a charge in the solution (q_S).

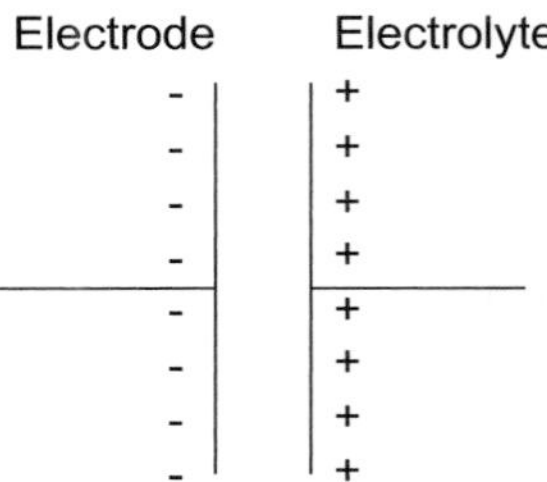

FIGURE II.5.10.2 Simplistic parallel-plate Helmholtz model for electrical double layer.

The potential across the electrode–electrolyte interface and the composition of the electrolyte will determine if the charge on the electrode is positive or negative. The sum total of the charge at the interface will, however, be equal to zero at all times, i.e., $q_M = -q_S$. The charge on the metal is a result of an excess or a deficiency of electrons, and the charge in the electrolyte is due to an excess of anions or cations in close proximity to the electrode. Helmholtz (1879) established a model for this aligned array of charges, referred to as the electrical double layer (EDL). This model is illustrated in Figure II.5.10.2.

Under realistic conditions this interface is more complex than the simplified parallel-plate model. The metal electrode still forms one charge array at the electrode–electrolyte interface. On the aqueous solution side of the interface polar solvent molecules, and specifically adsorbed ions or molecules, make up the first layer. This layer, referred to as the compact, Stern or Helmholtz layer extends to the locus of the electrical centers of the specifically adsorbed ions (inner Helmholtz plane). Charged ions in the electrolyte attract their own sheath of solvent molecules, and form what are called micelles or solvated ions. Because these ions are well-insulated

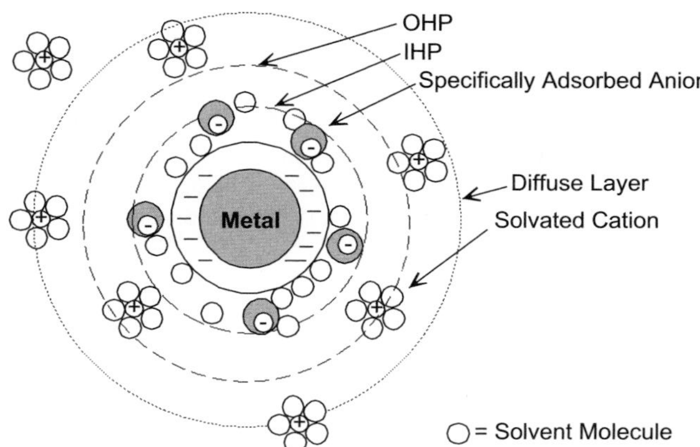

FIGURE II.5.10.3 Multilayer model for the electrode–electrolyte interface. IHP, inner Helmholtz plane; OHP, outer Helmholtz plane.

they interact with the conducting electrode surface, primarily through long-range electrostatic forces. The size of these ions and their nonspecific adsorption mechanism results in the formation of a three-dimensional diffuse outer layer. This outer layer extends from the locus of the electrical centers of the solvated ions (outer Helmholtz plane) into the bulk of the solution. The resulting interfacial structure is illustrated in Figure II.5.10.3.

EQUIVALENT CIRCUIT MODELS

Helmholtz (1879) not only described the nature of the electrode–electrolyte interface, but also proposed the first equivalent circuit model for it, i.e., an idealized parallel-plate capacitor. Warburg (1899), Fricke (1932), Randles (1947), and Sluyters-Rehbach (Sluyters-Rehbach and Sluyters, 1970) contributed significantly to continued development of equivalent circuit models for specific conditions at the interface. However, their models did not account for the passage of direct current through this interface. In 1968 Geddes and Baker applied the concept of a shunting (parallel) faradaic leakage resistance across the non-faradaic components that resulted in two equivalent circuit models for the bioelectrode–electrolyte interface. These models accounted for the passage of direct current through this interface. Subsequent experimental investigations by Onaral and Schwan (1982) very strongly corroborated this concept of a shunting faradaic leakage resistance at low frequencies. Detailed analysis of different types of recording and stimulation electrodes, and corresponding equivalent circuits, are discussed by Geddes and Baker (1989) in their seminal textbook on the principles of biomedical instrumentation. A review article by Geddes (1997) provides a less rigorous historical perspective on the evolution of such circuits, their advantages, and their limitations.

Basic Circuit Elements

A voltage is developed across the electrical double layer, and is referred to as the half-cell potential for

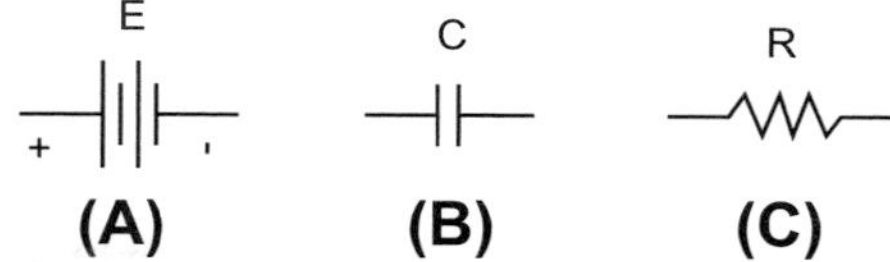

FIGURE II.5.10.4 Basic circuit elements: (A) the half-cell potential or battery circuit element; (B) simplistic double-layer or parallel-plate capacitor circuit element; (C) hindrance to faradaic transfer or the resistive circuit element.

that electrode under specific environmental factors. It is impossible to measure the potential developed at a single electrode. Hence, an arbitrary standard electrode is used to close the circuit, and the total voltage is measured as the difference of the half-cell potentials of the two electrodes. This can be represented in terms of a voltage source element or battery, represented in Figure II.5.10.4A. The electrical double layer in its most simplistic form is represented by two charged parallel plates with an insulated gap, i.e., a capacitor circuit element (Figure II.5.10.4B). The hindrance to the charge transfer that occurs at a faradaic electrode–electrolyte interface is represented by a resistive circuit element, illustrated in Figure II.5.10.4C.

Simple Equivalent Circuits

The electrolyte/tissue surrounding an electrode will result in a resistive element, referred to as the uncompensated or series resistance (R_u), that will be a contributing factor in all cases. An ideally polarizable electrode (nonfaradaic) will be represented by a series combination of the established double layer capacitance (C_{dl}) and R_u. This is illustrated in Figure II.5.10.5A. However, this model predicts infinite impedance for direct current (0 Hz). But it is known that a direct current can be passed between an electrode–electrolyte interface. This disparity can be explained by placing a charge-transfer resistance (R_{ct}) or a faradaic leakage resistance across the model illustrated in Figure II.5.10.5A, as shown in Figure II.5.10.5B. Most physiological electrolytes diffuse in and out of the tissue surrounding an electrode. A polarizable electrode with such mass transport limitations (partial diffusion control) will have an impedance term called the Warburg impedance (Z_w) associated with it, and will result in the model illustrated in Figure II.5.10.5C. Finally, a half-cell potential (E) is associated with every electrode placed in an electrolyte, and it is placed in series with the circuit in Figure II.5.10.5C, as shown in Figure II.5.10.5D.

Electrodes on a Subject

The equivalent circuit in Figure II.5.10.5D is a very good approximation of a single electrode–electrolyte interface. However, it is known that at least a two-electrode system is required either to make measurements of or to initiate a local biological event (E_b). The equivalent circuit models

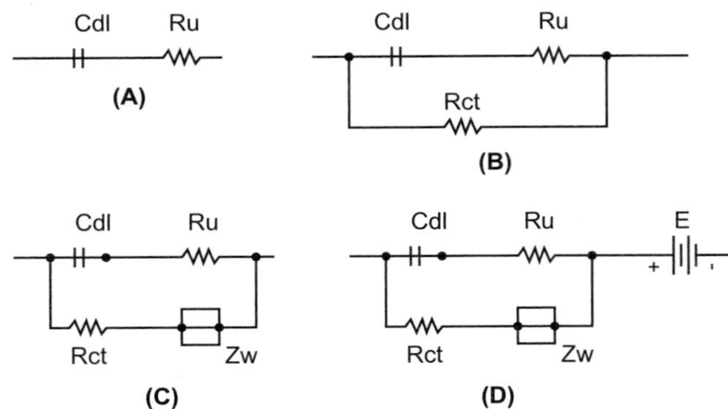

FIGURE II.5.10.5 Simple equivalent circuits: (A) equivalent circuit for an ideally polarizable electrode; (B) equivalent circuit element for a practical electrode; (C) a polarizable electrode subject to mass-transport limitations; (D) a polarizable circuit with mass-transport limitations and the half-cell potential of the electrode.

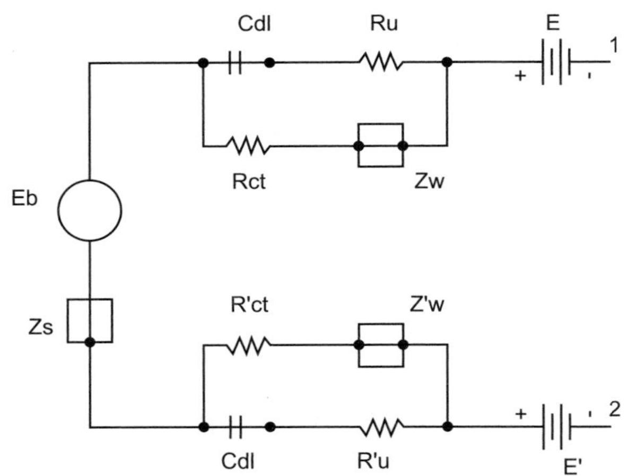

FIGURE II.5.10.6 The equivalent circuit model for a two-electrode scheme on a subject. E_b is the local biological event being monitored or initiated, and Z_S is the simplified black-box model used to represent the generic subject impedance.

to represent the subject are quite complex, because of the varying nature of tissue microstructure. Thus, a simplistic black box model (Z_s) is used to represent the generic subject impedance. The whole equivalent circuit comprising the subject impedance, the biological event, and the two electrodes is illustrated in Figure II.5.10.6.

FACTORS INFLUENCING MATERIAL SELECTION

While recording electrodes are not subject to active degradation, because of the inherent nature of the application, they are subject to degradation by the environment. Stimulation electrodes, on the other hand, are subject to degradation by the very excitation applications they are used for. All the factors listed in Table II.5.10.1 are of specific importance to material selection for electrodes (Bard and Faulkner, 1980; Fontana, 1986).

TABLE II.5.10.1 Factors Influencing Material Selection for Electrodes

Electrode	Surface area, geometry, and surface condition
Electrical	Potential, current, and quantity of charge
Environmental	Mass-transfer variables and solution variables
Engineering	Availability, cost, strength, and fabricability

Stimulating electrodes work by producing transient electric fields in their vicinity. The uniformity and precision of such electrical fields are dependent on the charge density distribution created by the stimulation electrode. Although ideally polarizable electrodes can create such fields with no chemical changes at the electrode–tissue interface, they are typically limited to less than 20 $\mu C/cm^2$ of true surface area for commonly used metallic electrodes (Robblee and Sweeney, 1996). Typical clinical applications require higher charge density distribution, and hence, most stimulation electrodes rely on predominantly faradaic charge transfer mechanisms for direct stimulation of a bioelectric event or a chemical reaction that would stimulate a bioelectric event. Some faradaic processes are reversible, while other processes are not (Brummer and Turner, 1977). Irreversible reactions initiate the electrolysis of water (oxygen or hydrogen evolution), oxidation of chloride ions, and corrosion or dissolution of an electrode that produces soluble metal–ion complexes (Geddes and Baker, 1989). This is undesirable, as it can result in multiple local and systemic biological responses that may change the properties of surrounding tissue or the type of surrounding tissue itself, i.e., fibrous tissue encapsulation to isolate a corroding, non-biocompatible object (Akers et al., 1997; Grill and Mortimer, 1994).

This change in property or type of surrounding tissue significantly influences the impedance at the electrode–tissue interface (Beard et al., 1992; Glikson et al., 1995; Danilovic and Ohm, 1998). This has necessitated the estimation of a property called the "reversible charge limit." The charge injection limit is the maximum quantity of charge that can be injected or transferred before irreversible chemical changes occur. The charge injection limit of any bioelectrode will depend on the reversible processes available during stimulation (environmental variables), the shape and surface morphology of the electrode (electrode variables), and finally, the geometry and duration of the stimulation waveform. An excellent review article by Robblee and Rose (1990) discusses the relative importance of the aforementioned factors in achieving reversible charge injection.

Even if careful consideration is given to the reversible charge injection limit, any implanted device (foreign body) will produce local changes in the tissue, as discussed before. Consequently, regulated-voltage stimulation (resultant current is dependent on tissue impedance) is susceptible to complex nonreproducible fluctuations, in comparison to regulated-current stimulation (direct application of current/charge) for implanted electrode systems. The stimulation waveforms in their simplest configuration may be pulsatile or symmetric sinusoidal. The charge injected (q) by a pulsatile waveform (say a rectangular pulse) is proportional to the pulse current (i) and the time duration (t) for which it is applied, i.e., $q = it$. The charge can thus be controlled very precisely, and is independent of the frequency with which it is delivered. In contrast, the charge delivered by a symmetric sinusoidal waveform is dependent on its frequency, i.e., current amplitudes that result in an acceptable charge density at high frequency may deliver an unacceptably high charge density at low frequency. But, symmetric sinusoidal waveforms possess no net DC component, and thus have reduced residual effect on the periprosthetic tissue. The basics of simulation waveform geometry, duration, and frequency, and how they relate to a specific application can be found in the chapter on stimulation electrodes by Geddes and Baker (1989) in their textbook on biomedical instrumentation. The design of stimulation waveforms is also one of the strongest emerging research areas in interventional clinical research (Hillsley et al., 1993; Walcott et al., 1995, 1998).

ELECTRODE MATERIALS

Noble Metals

Platinum is the most popular electrode material, because it is stable, inert, and corrosion resistant. However, pure platinum is a soft metal and is typically alloyed with iridium to improve its mechanical properties, thereby facilitating construction or manufacture of small electrodes. The amount of iridium in a platinum–iridium alloy can vary from ~2% to ~30% without limiting the charge injection ability of the alloy (Robblee and Sweeney, 1996). Small diameter intracortical stimulation needles that require flexural rigidity to achieve appropriate placement are examples of bioelectrodes that would use platinum–iridium alloys. Although platinum and platinum–iridium alloys exhibit extremely high corrosion resistance, even at high injection current densities (50–150 μC/cm^2) without gas evolution or electrolyte oxidation, metal dissolution is not avoidable, and occurs at all charge densities (McHardy et al., 1980; Robblee et al., 1980).

Iridium metal can be used instead of platinum in applications that require superior mechanical properties. Although the iridium metal electrodes possess no advantage over the platinum electrodes in terms of charge injection ability, iridium oxide films exhibit significantly reduced interface impedance compared to bare metal (Glarum and Marshall, 1980). This can facilitate greater efficiency of charge flow, thus contributing to prolonged battery life of an implanted stimulation device. Multilayer oxides can be formed on iridium electrodes by cycling the electrochemical iridium potential between predetermined values versus a standard electrode. These multilayer oxides can transfer or inject significant charge densities by reversible transitions between two stable valence levels or oxides (Robblee et al., 1983), but they are susceptible to limiting charge injection densities due to the shape or type of stimulation waveform. For example, when subjected to a predominantly cathodal pulse of 0.2 msec the electrode can inject only 1000 μC/cm^2. But the electrode can inject almost 3500 μC/cm^2 when the cathodal pulses are interspersed with an anodic bias voltage. The anodic bias voltage returns the electrode/oxide to its highest stable valence state between treatments, thereby increasing the charge injection ability of the electrode (Agnew et al., 1986; Kelliher and Rose, 1989). Also, transport and conductivity limitations during the short treatment timeframes restrict the utilization of the large charge injection capacity of such electrodes.

Non-Noble Metals

Intramuscular electrodes that require high mechanical and fatigue strength may be made from non-noble metals/alloys, such as 316LVM stainless steel or nickel–cobalt-based Elgiloy and MP35N. These materials rely on thin passive films for corrosion resistance, and inject charge by reduction or oxidation of their passive films. Although large anodic pulses that push these electrodes into their transpassive[1] behavior can lead to significant corrosion, it is important to note that corrosion can result at even relatively low cathodal pulses. These electrodes are capable of injecting ~40 μC/cm^2, and are susceptible to corrosion-related failures even at that level (White and Gross, 1974).

Titanium and tantalum can form insulation oxide films during anodic polarization, and can be used for manufacturing capacitive electrodes (electrodes that inject charge without faradaic reactions at the electrode–electrolyte interface). While the titanium oxide possesses a significantly higher dielectric strength compared to tantalum oxide it is susceptible to significant DC leakage, limiting its use in electrode applications. Capacitive electrodes based on tantalum–tantalum pentoxide have limited charge storage capacity (100–150 μC/cm^2), and can be used only in applications above a particular geometric size (0.05 cm^2) for neural prosthesis applications (Rose et al., 1985).

Also, the translation of micro- or nanofabrication technology (Wise and Najafi, 1991) from the semiconductor to the biomedical community has resulted in multielectrode arrays (Mastrototaro et al., 1992; Johnstone et al., 2010; Jones et al., 2010) to either control or monitor different functions of a single event or multiple events.

[1]If the potential is high enough, passive metals can enter a transpassive state with current/voltage characteristics different from the passive state.

APPLICATIONS

Cardiology

The pumping of the heart is controlled by regular electrical impulses generated by a group of special cells in the sinus node. The propagation of these impulses in a specific sequence causes the heart to contract, pump blood, and subsequently relax/expand once the signal passes. Any disorganization of this complex generation and propagation of electrical impulses (transmembrane potentials) or abnormal change of heart rate results in a condition called an arrhythmia (James, 1994).

When an arrhythmia results in slowing down of the heart – called *bradycardia* – due to a loss of pacemaker function or disorganized, ineffective propagation (a condition called atrial fibrillation) an electronic pacemaker is used to provide electrical impulses – essentially small shocks – to pace the heart to its normal rate (Ellenbogen et al., 1995) (see Chapter II.5.3.C). When an arrhythmia results in a very rapid heart rate with disorganization of initiation and propagation of the electrical wave, leading to asynchrony of contraction, a condition called ventricular *tachycardia* (i.e., rapid rate) or ventricular fibrillation (i.e., rapid, disorganized rate) occurs (Epstein and Ideker, 1995). Ventricular fibrillation is known to cause more than 325,000 deaths in the United States annually. This condition can be rectified using shock therapy, and the corrected rhythm often can be maintained using a combination of medication and/or an electronic device called the implantable cardioverter/defibrillator (ICD), a device that administers a large shock to the heart to "reset" the rhythm to normal (Glikson and Friedman, 2001; van Rees et al., 2011; and Chapter II.5.3.C).

For the first few years after the ICD became commercially available, the electrodes were placed directly on the external heart surface during open-chest surgery (Ideker et al., 1991). These electrodes were approximately 20–60 cm^2 and constructed out of titanium mesh. More recently, the ICD electrodes have been mounted on catheters that are inserted into the wall of right ventricular chamber of the heart by way of the venous system, so that major surgery to open the chest is no longer necessary (Singer et al., 1998). These electrodes, typically 5–8 cm long, are frequently made of titanium wire wound around the catheter to form a tight coil with no space between the turns. The sealed, titanium-coated can housing the circuits and battery is used as one of the defibrillation electrodes in ICDs (Figure II.5.10.7). Articles elaborating on contemporary usage of ICD electrodes and leads are available (Haqqani and Mond, 2009; Welsenes et al., 2010).

Ventricular tachycardia or fibrillation is sometimes caused by an abnormal focus in the heart that fires prematurely or by an abnormal path of conduction of electrical impulses within the heart muscle. Ablation of that abnormal focus or abnormal path may be obtained via a catheter-based procedure that therapeutically destroys ("ablates") a small area of tissue (Iskos et al., 1997; Tung et al., 2010). Ablation was originally performed by administering an intense DC shock through a catheter electrode placed next to the target tissue and a large plate on the back of the patient. This shock-damaged tissue by the direct effects of the large electric field and barotrauma caused by sudden gas bubble formation when arcing occurred between the shocking electrodes. Currently, a radio-frequency pulse applied through a platinum or platinum–iridium catheter electrode is used to kill the tissue by ablating or heating it during the application of the pulse. This technique is more localized, and causes less damage to tissue surrounding the area of anomaly.

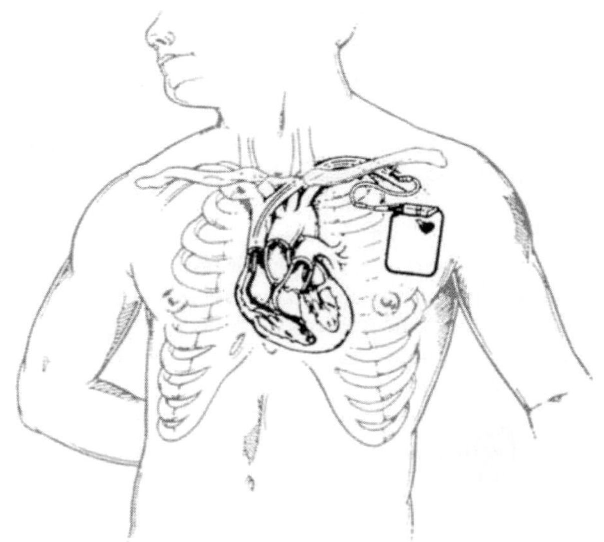

FIGURE II.5.10.7 Schematic illustrating the placement configuration of an implantable cardioverter/defibrillator device.

Neurology

Significant advances in imaging modalities, minimally invasive instrumentation, and robotic guidance/placement systems (Fenton et al., 1997; Hefti et al., 1998) are facilitating three-dimensional positioning of stimulation electrodes in the brain and other sites. Electrical stimulation of the brain, spine, peripheral nerves or muscles can be used for a variety of applications including pain management, treatment for Parkinson's disease, skeletal muscle rehabilitation after stroke, swallowing rehabilitation, cerebral palsy therapies, improving posture, ankle instability treatment, multiple sclerosis, muscle therapy for chronic obstructive pulmonary disease (COPD), preventing pressure ulcers, bladder and bowel management, building muscle strength, and this is just a partial list of applications. Also see Chapters II.5.11, II.5.12, and II.5.9.E.

Stimulation within the brain must be executed without any charge transfer across the electrode–tissue interface to prevent permanent tissue damage. Capacitor electrodes are ideal for this application. The tissue acts as the dielectric between the capacitive plates and, since the surfaces of these electrodes are oxide coated, negligible charge transfer occurs (Rose et al., 1985). Several clinical studies provide evidence that deep brain stimulation can be used to control medically refractory essential tremor (Lyons et al., 1998; Ondo et al., 1998) or untreatable Parkinsonian tremor (Ghika et al., 1998; Krack et al., 1998). Motor cortex stimulation has been shown to reduce post-stroke pain in patients when muscle contraction was inducible (Katayama et al., 1998).

The spinal cord can also be stimulated to control pain in various clinical conditions (Segal et al., 1998). Percutaneous spinal electrodes are placed in patients who are not suitable candidates for coronary bypass grafting for treatment of ischaemic pain condition such as angina pectoris (Andersen, 1997). They may also be used to control chronic back pain if surgical outcomes are not successful (Rainov et al., 1996). Transcutaneous electrical stimulation (TES) of peripheral nerves has been used to control chronic pain due to spinal cord injury or stroke damage (Taub et al., 1997). Localized stimulation of nerve endings using retinal and cochlear implants has also been attempted to enhance the loss of vision or hearing, respectively (Heetderks and Hambrecht, 1988) (see Chapters II.5.9.E and II.5.11).

Functional neuromuscular simulation (FNS) refers to the electrical stimulation of skeletal muscle for artificially controlled exercise or restoration of motor function lost due to spinal cord injury or disease (Sweeney, 1992). FNS systems are designed to provide upper-extremity control (Hoshimiya et al., 1989), lower-extremity control (Graupe, 1989) or both. FNS takes advantage of the fact that most motor neurons below the level of injury remain intact. Thus, electrical stimulation of their axons can be used to cause muscle contraction. For example, the phrenic nerve can be stimulated to control respiration or the sacral nerves can be stimulated to regulate the detrusor muscle in the bladder. Although FNS designers strive to mimic the body's control strategies, it is still extremely difficult to attain the fine gradation of force required to achieve adequate function. Review articles by Heetderks and Hambrecht (1988) and Maffiuletti (2010) are excellent sources of information on neuromuscular stimulation along with Chapter II.5.12.

Other

Catheter ablation electrodes can be used via the urethra or perineal cavity to damage/kill benign or malignant hyperplasia in the prostate (Dixon, 1995). Ablation has also been used to reduce the size of metastatic liver tumors and osteoid osteomas by damaging the irregular cells (Rosenthal et al., 1995; Solbiati et al., 1997). Cryosurgery, i.e., ablation by extreme cooling, has been used to treat refractory hepatic neuroendocrine metastases (Bilchik et al., 1997). Stimulating electrodes have also been explored to enhance tooth movement in orthodontics or the growth of teeth. In this case, platinum and noble metals are not useful because they irritate the gingival tissue. Composite electrodes are used with biocompatible gels to keep the electrode separated from the moist gingival tissue (Beard et al., 1992).

Future Trends

Healing and Regeneration. The US Food and Drug Administration has approved a number of electrical bone-growth stimulating devices for treating nonunion fractures, and for enhancing spinal fusion (Polk, 1995), although there is still controversy on the efficacy of these methods (Mollon et al., 2008). These devices generally use one of three approaches to apply low-intensity fields to the tissues: direct current applied to the wound site through surgically implanted wire electrodes; high-frequency (>20 kHz) "capacitatively coupled" sine-wave signals applied through skin electrodes; and low-frequency pulsed electromagnetic fields (PEMFs) applied by Helmholtz-type coils strapped to the limb. The PEMF devices do not conduct current directly to the tissues and, in the strictest sense, may not be considered as electrodes.

The implanted electrodes usually consist of stainless steel or titanium wire, but may include other metals such as silver or platinum. The wires, serving as cathodes, are connected to small batteries supplying a continuous current of 2–20 μA. It is uncertain whether the resulting stimulation of bone growth is due to electrochemical changes in the tissue surrounding the cathodes or to the electric current itself (Black, 1987). Spadaro (1997) used a rabbit femur model to compare the bone growth stimulation by different metallic cathodes at the same current density. The observed differences in the degree of bone formation would seem to argue for an electrochemical, rather than an electrical, basis for the effects. Yet the same results might also be attributed to the mechanical properties of the wires; as the animals moved about, the stiffer wires would transmit more mechanical force to the bone, and this micromotion might itself elicit an osteoinductive response.

The capacitatively coupled devices use 2–3 cm diameter electrodes adjoined to the skin through a conductive gel. These are powered by a small battery pack, and produce voltage gradients in bone estimated at 1–100 mV/cm at current densities in the $\mu A/cm^2$ range. As with the implanted electrodes, the precise basis for their effects are not completely understood, but appear to involve the distribution of cations, particularly calcium, at the cell surface (Zhuang et al., 1997).

The range of these applications should expand considerably when the molecular basis for electric field–tissue interactions is better understood. Electrical stimulation can be clinically useful for stimulating wound healing (Ennis et al., 2011) and nerve regeneration. For example, platinum cuff electrodes that surround the nerve or penetrating needle electrodes may be used to stimulate peripheral nerve regeneration (Heiduschka and Thanos, 1998). An electrode array with polymeric guidance tubules or a DC electrical field may also be applied to direct growth of new peripheral nerve axons. The nerves grow toward the cathode of the bioelectrode (Heiduschka and Thanos, 1998). It must be kept in mind, however, that this area of research remains controversial, because of frequent difficulties in reproducing experimental results and the absence of a generally accepted model.

Brain–Computer Interface (BCI). The ability to interface a computer directly to the brain via electrodes has the potential to restore functionality to those paralyzed or with vision or speech deficits. Such BCIs convert intent (i.e., the electrical signals generated in the brain associated with "I want to pick up that object") to electrical signals that can be interpreted by a computer and used to control a robotic arm to pick up the object. BCIs are classified as non-invasive (a helmet sits on the head to pick up signals) or invasive (implanted). The implanted electrode arrays can be placed on the surface of the brain (electrocorticography, ECoG) or implanted penetrating into brain tissue. The more electrodes there are in an array, the more precisely interrogation of brain electrical signs can be performed. Obviously there are material challenges associated with the mechanically soft nature of brain tissue (modulus mismatch), the geometry of the brain, the spacing of electrodes, and electrode biocompatibility (consider how biological encapsulation might impact the detection of signals that are generally less than 100 μV). Array electrodes can be fashioned from silicon, platinum, platinum–iridium, conducting polymers or from surface modified versions of these materials. An ECoG electrode was fabricated from carbon nanotubes and titanium nitride on a flexible polyimide support, demonstrating the application of many biomaterials to this complex problem (Sauter-Starace et al., 2009). Advanced BCI devices will require collaboration between biomaterials scientists, neurosurgeons, electrical engineers, computer specialists, and psychologists. Though only a few BCI devices have been implanted in humans, to date, important studies in primates have been performed demonstrating that a monkey, using only its thoughts, can control a robotic arm (Velliste et al., 2008). A few references on BCI devices and brain recording electrodes are provided here for further reading on this evolving, complex field that may, in the future, have tremendous impact for new biomaterials-based devices (Kipke et al., 2008; Kim et al., 2010; Konrad and Shanks, 2010; Lebedev et al., 2011; Schalk and Leuthardt, 2011).

SUMMARY

Although significant progress has been made in our understanding of the bioelectrode–tissue interface, it is still not possible to predict all of its properties with certainty. New materials, micro- or nanofabricating capability, computer processing power, and a better understanding of biological processes over the past two decades have significantly improved our ability to make bioelectrodes that more than adequately monitor or influence a biological event. However, it is becoming apparent at this time that even the most advanced synthetic prosthesis cannot totally restore normal function. Hence, the human impact of interventional and potential regenerative procedures made possible due to current biolectrodes will form the basis for further advances in this old, but still nascent, field.

BIBLIOGRAPHY

Agnew, W. F., Yuen, T. G.H., McCreedy, D. B., & Bullara, L. A. (1986). Histopathologic evaluation of prolonged intracortical electrical stimulation. *Exper. Neurol.*, **92**, 162–185.

Akers, J. M., Peckman, P. H., Keith, M. W., & Merritt, K. (1997). Tissue response to chronically stimulated implanted epimysial and intramuscualr electrodes. *IEEE Trans. Rehab. Eng.*, **5**(2), 207–220.

Andersen, C. (1997). Complications in spinal cord stimulation for treatment of angina pectoris. Differences in unipolar and multipolar percutaneous inserted electrodes. *Acta Cardiol.*, **52**(4), 325–333.

Atie, J., Maciel, W., Pierobon, M. A., & Andrea, E. (1996). Radiofrequency ablation in patients with Wolff–Parkinson–White syndrome and other accessory pathways. *Arquivos Brasileiros de Cardiologia*, **66**(Suppl 1), 29–37.

Bard, A. J., & Faulkner, L. R. (1980). Introduction and overview of electrode processes. In *Electrochemical Methods* (pp. 1–43). New York, NY: John Wiley and Sons.

Baynham, T. C., & Knisley, S. B. (1998). Combating heart disease with FEA. *Mech. Eng.*, 70–72. Oct.

Beard, R. B., Hung, B. N., & Schmukler, R. (1992). Biocompatibility considerations at stimulation electrode interfaces. *Ann. Biomed. Eng.*, **20**, 395–410.

Bilchik, A. J., Sarantou, T., Foshag, L. J., Giuliano, A. E., & Ramming, K. P. (1997). Cryosurgical palliation of metastatic neuroendocrine tumors resistant to conventional therapy. *Surgery*, **122**(6), 1040–1047.

Black, J. (1987). *Electrical Stimulation. Its Role in Growth, Repair, and Remodeling of the Musculoskeletal System.* New York, NY: Praeger. 92–96.

Brummer, S. B., & Turner, M. J. (1977). Electrochemical considerations for safe electrical stimulation of nervous system with platinum electrodes. *IEEE Trans. Biomed. Eng., BME-24*, 59–63.

Danilovic, D., & Ohm, O. J. (1998). Pacing impedance variability in tined steriod eluting leads. *Pacing Clin. Electrophysiol.*, **21**(7), 1356–1363.

Dixon, C. M. (1995). Transurethral needle ablation for the treatment of benign prostatic hyperplasia. *Urol. Clin. N. Amer.*, **22**(2), 441–444.

Ellenbogen, K. A., Kay, G. N., & Wilkoff, B. L. (1995). *Clinical Cardiac Pacing*. Philadelphia, PA: W.B. Saunders.

Ennis, W. J., Lee, C., Plummer, M., & Meneses, P. (2011). Current status of the use of modalities in wound care: Electrical stimulation and ultrasound therapy. *Plast. Reconstr. Surg.*, **127**, 93S–102S.

Epstein, A. E., & Ideker, R. E. (1995). Ventricular fibrillation. In D. P. Zipes, & J. Jalife (Eds.), *Cardiac Electrophysiology: From Cell to Bedside* (pp. 927–934). Philadelphia, PA: W. B. Saunders.

Faraday, M. (1834). Experimental researches in electricity, 7th series. *Phil. Trans. R. Soc. Lond.*, **124**, 77–122.

Fenton, D. S., Geremia, G. K., Dowd, A. M., Papathanasiou, M. A., Greenlee, W. M., et al. (1997). Precise placement of sphenoidal electrodes via fluoroscopic guidance. *Am. J. Neuroradiol.*, **18**(4), 776–778.

Fontana, M. G. (1986). Corrosion principles. In *Corrosion Engineering* (3rd ed., pp. 12–38). New York, NY: McGraw-Hill.

Fotuhi, P. C., Epstein, A. E., & Ideker, R. E. (1999). Energy levels for defibrillation: What is of real clinical importance? *Am. J. Cardiol.*, **83**, 240–330.

Fricke, H. (1932). The theory of electrolytic polarization. *Phil. Mag.*, **14**, 310–318.

Geddes, L. A. (1997). Historical evolution of circuit models for the electrode–electrolyte interface. *Ann. Biomed. Eng.*, **25**, 1–14.

Geddes, L. A., & Baker, L. E. (Eds.) (1989). *Principles of Applied Biomedical Instrumentation* (3rd ed, pp. 315–452). New York, NY: Wiley-Interscience.

Ghika, J., Villemure, J. G., Fankhauser, H., Favre, J., Assal, G., et al. (1998). Efficiency and safety of bilateral contemporaneous pallidal stimulation (deep brain stimulation) in levodopa-responsive patients with Parkinson's disease with severe motor fluctuations: A 2-year follow-up review. *J. Neurosurg.*, **89**(5), 713–718.

Glarum, S. H., & Marshall, J. H. (1980). The A-C response of iridium oxide films. *J. Electrochem. Soc.*, **127**, 1467–1474.

Glikson, M., & Friedman, P. A. (2001). The implantable cardioverter defibrillator. *The Lancet*, **357**(9262), 1107–1117.

Glikson, M., von Feldt, L. K., Suman, V. J., & Hayes, D. L. (1995). Clinical surveillance of an active fixation, bipolar, polyurethane insulated pacing lead. Part II: The ventricular lead. *Pacing Clin. Electrophysiol.*, **18**(2), 374–375.

Gouy, M. (1910). Sur la constitution de la charge electricque a la surface d'un electrolyte. *J. Phys. (Paris)*, **9**, 457–468.

Graupe, D. (1989). EMG pattern analysis for patient-responsive control of FES in paraplegics for walker-supported walking. *IEEE Trans. Biomed. Eng.*, **36**(7), 711–719.

Grill, W. M., & Mortimer, J. T. (1994). Electrical properties of implant encapsulation tissue. *Ann. Biomed. Eng.*, **22**(1), 23–33.

Haissaguerre, M., Cauchemez, B., Marcus, F., Le Metayer, P., Lauribe, P., et al. (1995). Characteristics of the ventricular insertion sites of accessory pathways with anterograde decremental conduction properties. *Circulation*, **91**(4), 1077–1085.

Haqqani, H. M., & Mond, H. G. (2009). The implantable cardioverter-defibrillator lead: Principles, progress, and promises. *Pacing Clin. Electrophysiol.*, **32**(10), 1336–1353.

Heetderks, W. J., & Hambrecht, F. T. (1988). Applied neural control in the 90s. *Proc. IEEE*, **76**, 1115–1121.

Hefti, J. L., Epitaux, M., Glauser, D., & Frankhauser, H. (1998). Robotic three-dimensional positioning of electrode in the brain. *Comput. Aided Surg.*, **3**(1), 1–10.

Heiduschka, P., & Thanos, S. (1998). Implantable bioelectric interfaces for lost nerve functions. *Prog. Neurobiol.*, **55**(5), 433–461.

Helmholtz, H. (1879). Studien über electrische Grenzchichten. *Ann. Phys. Chem.*, **7**, 337–382.

Hillsley, R. E., Walker, R. G., Swanson, D. K., Rollins, D. L., Wolf, P. D., et al. (1993). Is the second phase of a biphasic defibrillation waveform the defibrillating phase? *Pacing Clin. Electrophysiol.*, **16**(7 Pt 1), 1401–1411.

Hoshimiya, N., Naito, A., Yajima, M., & Handa, Y. (1989). A multichannel FES system for the restoration of motor functions in high spinal cord injury patients: A respiration-controlled system for multijoint upper extremity. *IEEE Trans. Biomed. Eng.*, **36**(7), 754–760.

Ideker, R. E., Chen, P. S., & Zhou, X. H. (1990). Basic mechanisms of defibrillation. *J. Electrocardiol.*, **23**(Suppl), 36–38.

Ideker, R. E., Wolf, P. D., Alferness, C., Krassowska, W., & Smith, W. M. (1991). Current concepts for selecting the location, size and shape of defibrillation electrodes. *Pacing Clin. Electrophysiol.*, **14**(2 Pt 1), 227–240.

Iskos, D., Fahy, G. J., Lurie, K. G., Sakaguchi, S., Adkisson, W. O., et al. (1997). Nonpharmacologic treatment of atrial fibrillation: Current and evolving strategies. *Chest*, **112**(4), 1079–1090.

Jais, P., Haissaguerre, M., Shah, D. C., Takahashi, A., Hocini, M., et al. (1998). Successful irrigated-tip catheter ablation of atrial flutter resistant to conventional radiofrequency ablation. *Circulation*, **98**(9), 835–838.

James, M. J. (1994). *Mechanisms of Arrhythmias*. New York, NY: Futura Publishing Company Inc.

Johnstone, A. F.M., Gross, G. W., Weiss, D. G., Schroeder, O. H.U., Gramowski, A., et al. (2010). Microelectrode arrays: A physiologically based neurotoxicity testing platform for the 21st century. *Neurotoxicology*, **31**(4), 331–350.

Jones, I. L., Livi, P., Lewandowska, M. K., Fiscella, M., Roscic, B., et al. (2010). The potential of microelectrode arrays and microelectronics for biomedical research and diagnostics. *Anal. Bioanal. Chem.*, **399**(7), 2313–2329.

Katayama, Y., Fukaya, C., & Yamamoto, T. (1998). Poststroke pain control by chronic motor cortex stimulation: Neurological characteristics predicting a favorable response. *J. Neurosurg.*, **89**(4), 585–591.

Kelliher, E. M., & Rose, T. L. (1989). Evaluation of charge injection properties of thin film redox materials for use as neural stimulation electrodes. *MRS Symp. Proc.*, **110**, 23–27.

Kim, D. -H., Wiler, J. A., Anderson, D. J., Kipke, D. R., & Martin, D. C. (2010). Conducting polymers on hydrogel-coated neural electrode provide sensitive neural recordings in auditory cortex. *Acta Biomater.*, **6**(1), 57–62.

Kipke, D. R., Shain, W., Buzsaki, G., Fetz, E., Henderson, J. M., et al. (2008). Advanced neurotechnologies for chronic neural interfaces: New horizons and clinical opportunities. *J. Neurosci.*, **28**(46), 11830–11838.

Konrad, P., & Shanks, T. (2010). Implantable brain computer interface: Challenges to neurotechnology translation. *Neurobiol. Dis.*, **38**(3), 369–375.

Krack, P., Pollak, P., Limousin, P., Hoffmann, D., Benazzouz, A., et al. (1998). Inhibition of levodopa effects by internal pallidal stimulation. *Movement Disord.*, **13**(4), 648–652.

Kunze, K. P., Hayen, B., & Geiger, M. (1998). Ambulatory catheter ablation. Indications, results and risks. *Herz*, **23**(2), 135–140.

Lebedev, M. A., Tate, A. J., Hanson, T. L., Li, Z., O'Doherty, J. E., et al. (2011). Future developments in brain–machine interface research. *Clinics (São Paulo, Brazil)*, **66**(Suppl 1), 25–32.

Lyons, K. E., Pahwa, R., Busenbark, K. L., Troster, A. I., Wilkinson, S., et al. (1998). Improvements in daily functioning after deep brain stimulation of the thalamus. *Movement Disord.*, **13**(4), 690–692.

Maffiuletti, N. A. (2010). Physiological and methodological considerations for the use of neuromuscular electrical stimulation. *Eur. J. Appl. Physiol.*, **110**(2), 223–234.

Mastrototaro, J. J., Massoud, H. Z., Pilkington, T. C., & Ideker, R. E. (1992). Rigid and flexible thin-film multielectrode arrays for transmural cardiac recording. *IEEE Trans. Biomed. Eng.*, **39**(3), 271–279.

McHardy, J., Robblee, L. S., Marston, J. M., & Brummer, S. B. (1980). Electrical stimulation with Pt. electrodes. IV. Factors influencing Pt dissolution in inorganic saline. *Biomaterials*, **1**, 129–134.

Michaiel, J. J. (1993). *Mechanisms of Arrhythmias*. Armonk, NY: Futura Publishing Company.

Mollon, B., da Silva, V., Busse, J. W., Einhorn, T. A., & Bhandari, M. (2008). Electrical stimulation for long-bone fracture-healing: A meta-analysis of randomized controlled trials. *J. Bone. Joint Surg.*, **90**(11), 2322–2330.

Ohm, G. S. (1826). Bestimmung des Gesetzes nach welchem Metalle die Contaktelectricität leiten. *Schweiggers J. Chem. Phys.*, **46**, 137–166.

Onaral, B., & Schwan, H. P. (1982). Linear and nonlinear properties of platinum electrode polarization. *Med. Biol. Eng. Comput.*, **20**, 299–300.

Ondo, W., Jankovic, J., Schwartz, K., Almaguer, M., & Simpson, R. K. (1998). Unilateral thalamic deep brain stimulation for refractory essential tremor and Parkinson's disease tremor. *Neurology*, **51**(4), 1063–1069.

Polk, C. (1995). Therapeutic applications of low-frequency sinusoidal and pulsed electric and magnetic fields. In J. D. Bronzino (Ed.), *Handbook of Biomedical Engineering* (pp. 1404–1416). Boca Raton, FL: CRC Press.

Rainov, N. G., Heidecke, V., & Burkert, W. (1996). Short test-period spinal cord stimulation for failed back surgery syndrome. *Minim. Invas. Neurosurg.*, **39**(2), 41–44.

Randles, E. B. (1947). Rapid electrode reactions. *Discuss. Faraday Soc.*, **1**, 11–19.

Robblee, L. S., & Rose, T. L. (1990). Electrochemical guidelines for selection of protocols and electrode materials for neural stimulation. In W. F. Agnew, & D. B. McCreedy (Eds.), *Neural Prostheses: Fundamental Studies* (pp. 25–66). Upper Saddle River, NJ: Prentice Hall.

Robblee, L. S., & Sweeney, J. D. (1996). Bioelectrodes. In B. D. Ratner, A. S. Hoffman, F. J. Schoen, & J. E. Lemons (Eds.), *Biomaterials Science* (1st edn, pp. 371–375,). San Diego, CA: Academic Press.

Robblee, L. S., McHardy, J., Marston, J. M., & Brummer, S. B. (1980). Electrical stimulation with Pt electrodes. V. The effects of protein on Pt dissolution. *Biomaterials*, **1**, 135–139.

Robblee, L. S., Lefko, J. L., & Brummer, S. B. (1983). An electrode suitable for reverse charge injection in saline. *J. Electrochem. Soc.*, **130**, 731–733.

Rose, T. L., Kelliher, E. M., & Robblee, L. S. (1985). Assessment of capacitor electrodes for intracortical neural stimulation. *J. Electrochem. Soc.*, **130**, 731–733.

Rosenthal, D. I., Springfield, D. S., Gebhardt, M. C., Rosenberg, A. E., & Mankin, H. J. (1995). Osteoid osteoma: Percutaneous radio-frequency ablation. *Radiology*, **197**(2), 451–454.

Sauter-Starace, F., Bibari, O., Berger, F., Caillat, P., & Benabid, A. L. (2009). ECoG recordings of a non-human primate using carbon nanotubes electrodes on a flexible polyimide implant. In *Proceedings: 4th International IEEE/EMBS Conference on Neural Engineering, NER '09*, April 29 2009–May 2 2009 (pp. 112–115).

Schalk, G., & Leuthardt, E. C. (2011). Brain–computer interfaces using electrocorticographic signals. *IEEE Reviews in Biomedical Engineering*, **4**, 140–154.

Sebag, C., Lavergne, T., Motte, G., & Guize, L. (1997). Radiofrequency ablation of accessory atrioventricular pathways. *Archives des Maladies du Coeur et des Vaisseaux*, **90**(Spec No 1), 11–17.

Segal, R., Stacey, B. R., Rudy, T. E., Baser, S., & Markham, J. (1998). Spinal cord stimulation revisited. *Neurol. Res.*, **20**(5), 391–396.

Singer, I., Barold, S. S., & Camm, A. J. (1998). *Nonpharmacological Therapy of Arrhythmias for 21st Century: The State of the Art. Futura Publishing*. NY: Armonk.

Sluyters-Rehbach, M., & Sluyters, J. H. (1970). Sine wave methods in the study of electrode processes. *Electroanal. Chem.*, **4**, 1–121.

Solbiati, L., Goldberg, S. N., Ierace, T., Livraghi, T., Meloni, F., et al. (1997). Hepatic metastases: Percutaneous radio-frequency ablation with cooled-tip electrodes. *Radiology*, **205**(2), 367–373.

Spadaro, J. A. (1997). Mechanical and electrical interactions in bone remodeling. *Bioelectromagnetics*, **18**, 193–202.

Stancak, B., Pella, J., Resetar, J., Palinsky, M., & Bodnar, J. (1996). Ablation of supraventricular tachydysrhythmias with direct and radiofrequency current. *Vnitrni Lekarstvi*, **42**(11), 779–783.

Stern, O. (1924). Zur theory der electrolytischen Doppelschicht. *Z. Elektonchem.*, **30**, 508–516.

Sweeney, J. D. (1992). Skeletal muscle responses to electrical stimulation. In J. P. Reilly (Ed.), *Electrical Stimulation and Electropathology* (pp. 285–327). Cambridge, UK: Cambridge University Press.

Taub, E., Munz, M., & Tasker, R. R. (1997). Chronic electrical stimulation of the gasserian ganglion for the relief of pain in a series of 3 and 4 patients. *J. Neurosurg.*, **86**(2), 197–202.

Tung, R., Boyle, N. G., & Shivkumar, K. (2010). Catheter ablation of ventricular tachycardia. *Circulation*, **122**(3), e389–e391.

van Rees, J. B., de Bie, M. K., Thijssen, J., Borleffs, C. J.W., Schalij, M. J., et al. (2011). Implantation-related complications of implantable cardioverter-defibrillators and cardiac resynchronization therapy devices. *J. Amer. Coll. Cardiol.*, **58**(10), 995–1000.

Velliste, M., Perel, S., Spalding, M. C., Whitford, A. S., & Schwartz, A. B. (2008). Cortical control of a prosthetic arm for self-feeding. *Nature*, **453**(7198), 1098–1101.

Volta, A. (1800). On the electricity excited by the mere contact of conducting substances of different kinds. A letter from Alexander Volta, F.R.S. (Professor of Natural Philosophy, University of Pavia) to Rt. Hon. Sir Joseph Banks, K.B.P.R.S.. *Phil. Trans. R. Soc. Lond.*, **90**, 744–746.

Walcott, G. P., Walker, R. G., Cates, A. W., Krassowska, W., Smith, W. M., et al. (1995). Choosing the optimal monophasic and biphasic waveforms for ventricular defibrillation. *J. Cardiovasc. Electrophysiol.*, **6**(9), 737–750.

Walcott, G. P., Knisley, S. B., Zhou, X., Newton, J. C., & Ideker, R. E. (1997). On the mechanism of ventricular defibrillation. *Pacing Clin. Electrophysiol.*, **20**(2 Pt 2), 422–431.

Walcott, G. P., Melnick, S. B., Chapman, F. W., Jones, J. L., Smith, W. M., et al. (1998). Relative efficacy of monophasic and biphasic waveforms for transthoracic defibrillation after short and long durations of ventricular fibrillation. *Circulation*, **98**(20), 2210–2215.

Warburg, E. (1899). Über das verhalten sogenanter unpolarsbarer Elektroden gegen Wechselstrons. *Ann. Phys. Chim.*, **67**, 493–499.

Warburg, E. (1901). Über die Polarizationscapacitat des Platins. *Ann. Phys.*, **6**, 125–135.

Welsenes, G. H., Borleffs, C. J.W., Rees, J. B., Atary, J. Z., Thijssen, J., et al. (2010). Improvements in 25 years of implantable cardioverter defibrillator therapy. *Nether. Heart J.*, **19**(1), 24–30.

White, R. L., & Gross, T. J. (1974). An evaluation of the resistance to electrolysis of metals for use in biostimulation probes. *IEEE Trans. Biomed. Eng., BME-21*, 487–490.

Wise, K. D., & Najafi, K. (1991). Microfabrication techniques for integrated sensors and microsystems. *Science*, **254**, 1335–1342.

Zhuang, H., Wang, W., Seldes, R., Tahernia, A., Fan, H., et al. (1997). Electrical stimulation induces the level of TGF-beta 1 mRNA in osteoblastic cells by a mechanism involving calcium/calmodulin pathway. *Biochim. Biophys. Res. Commun.*, **237**, 225–229.

CHAPTER II.5.11 COCHLEAR PROSTHESES

Francis A. Spelman
Department of Bioengineering, University of Washington, Seattle, WA, USA

INTRODUCTION

Cochlear prostheses help the deaf to contact the auditory environment (NIH, 1995). The problem of sensorineural deafness is immense. The probable number of users has been estimated by some as 900,000 in the US alone (Levitt and Nye, 1980). Other estimates range as high as 2,000,000 sensorineural deaf in the US. The treatment for sensorineural deaf patients is the cochlear implant (NIH, 1995). Sensorineural deaf subjects cannot perceive sound without extraordinary aid. This is because the basic transduction system is lost as a result of damage or destruction of either the cochlea or the auditory nerve (Levitt and Nye, 1980).

This chapter introduces cochlear prostheses and some of the materials issues that must be faced by their designers. The concentration of the chapter is on cochlear implants, rather than on implants in the brainstem or the midbrain. In the US approximately 120,000 people use cochlear implants, while less than 1000 have brainstem implants (Brackmann et al., 1993; NIDCD, 2003; McCreery, 2008; Wilson and Dorman, 2008). At present, cochlear implants are in clinical use, while brainstem implants are largely experimental, with limited clinical trials (McCreery, 2008). There is no discussion of tactile prostheses for the deaf (Martin, 1985). This chapter introduces the physiology of the auditory system, describes cochlear prostheses, and some of the issues related to biomaterials that bioengineers face in the design of such apparatuses.

OVERVIEW OF THE AUDITORY SYSTEM

The auditory system can be divided into its peripheral organs and the centers in the central nervous system that process the signals produced by the peripheral organs. This is a greatly abbreviated presentation. More details are found in texts like those of Geisler (Geisler, 1998) and Dallos and Popper (Dallos et al., 1996). Geisler refers to several websites that demonstrate the behavior of the periphery, e.g., http://www.neurophys.wisc.edu/animations. The sketch in Figure II.5.11.1 shows the auditory periphery with a cochlear prosthesis implanted in it (courtesy of Cochlear Pty Ltd, Sydney, Australia).

The Periphery

The auditory periphery consists of the outer ear, middle ear, and inner ear. The external ear, the *pinna*, collects changes in pressure (condensations and rarefactions) produced by the auditory signals. Those signals are generated by sound sources in the environment of the listener. Acoustic signals are guided to the middle ear along the ear canal, an entry into the head of the subject that is lined by soft tissue. The ear canal is open at its peripheral end, and bounded by the eardrum, *the tympanic membrane*, at its inner end. The length of the ear canal is about 3 cm in the human (Geisler, 1998).

The middle ear is bounded distally by the tympanic membrane and proximally by the *cochlea*, where the foot plate of the *stapes* contacts the oval window of the cochlea.[1] In the middle ear are three tiny bones that comprise the *ossicular chain*: the *malleus* (hammer), *incus* (anvil), and *stapes* (stirrup). The bones have flexible connections. They provide a mechanical advantage so that the eardrum can be driven by air and, in turn, can drive the dense fluids that are found in the cochlea. The motion of the foot plate of the stapes is about 75% of that of the tympanic membrane in the human (Geisler, 1998). Further mechanical advantage is provided by the relative areas of the tympanic membrane and the foot plate of the stapes: the tympanic membrane has about 20 times the area of the stapes (Geisler, 1998). Small pressure changes in air are transformed into larger pressures in the fluids of the cochlea. Conversely, large displacements of the eardrum produce small displacements of the footplate of the stapes.

The *cochlea* is a snail-shaped organ that is located bilaterally in the temporal bones of the head. The cochlea is oriented such that its wide base faces in a medial and posterior direction, while the axis of its spiral points laterally and anteriorly (Figure II.5.11.1).

The cochlea contains three spiraling chambers or *scalae*: the *scala tympani*, *scala vestibuli*, and *scala media*. The *organ of Corti* lies within the *scala media* on the *basilar membrane*. The *hair cells* are located on the basilar membrane (Dallos et al., 1996; Geisler, 1998). The basilar membrane is an elegant structure that acts as a mechanical Fourier analyzer; specific regions of the membrane vibrate maximally in response to the frequency of the sound waves that are imposed on the stapes. The membrane displacements produce maxima for high frequencies at the basal end and for low frequencies at the apical end of the cochlea. The hair cells residing on the membrane have *cilia* that are bent when the membrane vibrates. The hair cells synapse with the peripheral processes of the auditory nerve, the *hearing nerve* in Figure II.5.11.1. There are 25–30,000 afferent neurons that synapse with the hair cells (Geisler, 1998). The organization of the auditory nerve is by the frequencies of the acoustic signals that it receives. Indeed, the entire auditory system is tonotopically organized, that is, by frequency (Popper and Fay, 1991; Geisler, 1998).

[1]The terms *proximal* and *distal* refer to the brain in this case. Proximal is nearer to the brain and distal is further from the brain.

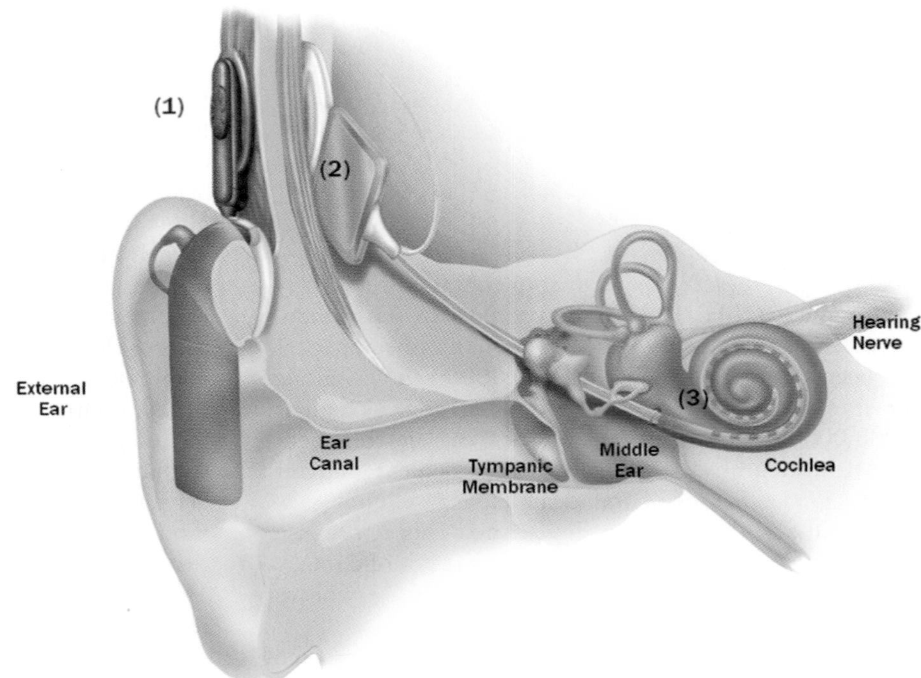

FIGURE II.5.11.1 Sketch of a cochlear implant in the ear of a human patient (Figure reproduced by courtesy of and with permission from Cochlear Ltd. PTY, Sydney, Australia). The numbered circles show components of the implanted device. (1) Is the external receiver and signal processor; (2) is the internal processor hat transmits signals via a multi-conductor cable to (3) the electrode array that resides in the scala tympani of the patient's cochlea.

Tonotopic Organization of the Auditory System

The entire auditory system parses frequency from the signals that it receives. The peripheral system begins the process (Geisler, 1998), and that anatomical organization proceeds at least to the level of the *inferior colliculus*, in the midbrain (Snyder et al., 2004). As has been illustrated dramatically by Bierer, the organization can be detected in the auditory cortex (Bierer and Middlebrooks, 2004). This organization permits investigators to test the efficacy of simulation designs experimentally in animals (Bierer and Middlebrooks, 2002, 2004; Snyder et al., 2004). Their work has enabled others to design and test midbrain implants in human subjects who suffer from tumors of the auditory nerve (Lim et al., 2008). While the tonotopic organization of the auditory system is an overarching part of its design, many subtleties, e.g., temporal analysis of fine structure, accompany the organization by frequency. Those subtleties appear to underlie the ability of the system to extract signals from noise and to decode music (Rubinstein et al., 1999; Litvak and Narayan, 2007; Wilson and Dorman, 2008).

Critical to the design of the cochlear prosthesis is the location of the peripheral processes of the auditory neurons, the cells of the auditory or hearing nerve.[2] The neurons are bipolar cells, and their peripheral processes are found in and under the *bony spiral lamina* (Popper and Fay, 1991; Spelman and Voie, 1996; Geisler, 1998), an osseous or bony structure that extends from the *modiolar* (medial) wall of the scala tympani in the cochlea. The cell bodies of the VIII nerve are located in Rosenthal's canal, a hollow structure in the modiolar bone.

The anatomy of the scala tympani has led to the design of the cochlear implant. The prosthesis is designed to stimulate the auditory neurons electrically. Placing the sites of electrodes near the neurons without violating the bony wall of the modiolus requires the electrode arrays of the implants to be located in the scala tympani (*vide infra*).

Highlights of the Central Auditory System

The tonotopic structure of the auditory system is found throughout the system. The frequency-dependent structure of the auditory signal forms the responses of single neurons in the auditory nerve (Sachs and Young, 1979). As signals are produced binaurally, those signals travel from each cochlea to the *cochlear nucleus*, where the neurons send data to the *contralateral trapezoid body*, the *ipsilateral olivary complexes*, the *inferior colliculus*, the *medial geniculate nucleus*, and the *auditory cortex* (Rubel and Dobie, 1989). The properties of the auditory nerve are well-understood, and have been for more than a decade. The properties of the cochlear nucleus are under active investigation, as are those of the higher centers of the auditory system.

This chapter focuses on the auditory periphery. It is in the periphery where cochlear prostheses are most effective, although auditory prostheses have been used in the cochlear nucleus, notably for patients who suffer from damage to the VIII nerve (Shannon et al., 1993; McCreery et al., 1997; Lim et al., 2008; McCreery, 2008).

[2]The auditory nerve is also known as the VIII Cranial Nerve.

Damage to the Periphery

Sensorineural deafness caused peripherally can result from serious damage to the hair cells or to the auditory nerve. Clearly, if the neurons of the hearing nerve are damaged, their peripheral processes cannot be driven, and stimulation from sites in the cochlea will not work. In those cases, central prostheses have been used experimentally (Shannon et al., 1993; McCreery et al., 1997; Lim et al., 2008; McCreery, 2008).

Damage to the hair cells can result from a number of causes. Pyman et al. cite eleven root causes in people over six years of age, and seven causes in people less than six years of age (Pyman et al., 1990). Their population was 65 people in the former case, and 29 in the latter. Large numbers of subjects had unknown causes of deafness, but there were cases of meningitis, otosclerosis, and trauma that caused the problems (Pyman et al., 1990). In another study, Hinojosa and Marion analyzed 65 ears and found six causes of congenital deafness in 19 subjects, and nine causes of acquired deafness in 46 subjects. In the latter population, otosclerosis caused the greatest damage, followed closely by bacterial infections (Hinojosa and Marion, 1983).

Damage to the hair cells from loud sounds requires special mention, since the popularity of painful audio systems in automobiles, and as portable sources of entertainment, is increasing. Hair cells can be damaged by intense sounds (Popper and Fay, 1991); chronic exposure to loud sounds should be avoided, despite the relatively small numbers cited by Hinojosa and Pyman (Hinojosa and Marion, 1983; Pyman et al., 1990).

Neural Plasticity

The auditory system responds to stimulation, both anatomically and physiologically. The central nervous system can reorganize itself in response to auditory signals (Brugge, 1991). The system is more plastic in children than in adults. That understanding has changed the application of cochlear prostheses from a focus only on adults to a large distribution of instruments for children as well (Clark, 1996). Deaf children are treated at two years of age, and in some cases at 12 months (Osberger, 1997; Skinner, 2001).

COCHLEAR PROSTHESES

Cochlear prostheses present one of the remarkable success stories of biomedical engineering. The idea of electrical stimulation of the peripheral auditory system is credited to Volta (1806), and cited in Simmons (Simmons, 1966). More modern approaches to solve the problem of deafness with a neural prosthesis were offered in the latter part of the 20th century (Djourno and Eyries, 1957; Simmons, 1966; House and Berliner, 1991).

Architecture of a Cochlear Prosthesis

General Architecture. The architecture of a cochlear prosthesis is shown in Figure II.5.11.2. The architecture shown here follows the textual architecture described on the NIDCD website. Here the receiver/stimulator described on that site is divided into a signal processor and controlled current sources (NIDCD, 2003). A microphone is the transducer that converts the auditory signal into an electrical signal. The microphone's signal is sent to an external signal processor. That signal processor decomposes the electrical information into amplitude and frequency data; the signal is filtered and analyzed to produce data about the envelope of the information within a particular frequency band (Loizou, 1999). Today's clinical processors have a small external processor behind the ear and an internal processor that is housed in a titanium case that is surgically implanted into the temporal bone of the patient (Figure II.5.11.1).

Processing techniques have been introduced commercially to provide information about the temporal fine structure of auditory signals (*vide infra*). Several bands are analyzed simultaneously to develop a vocoder model of the audio signal (Gold and Rader, 1967). The data are transferred across the skin as digital signals. An internal processor takes those data and converts them to current drive signals for the electrodes of the multichannel electrode array. In some implementations, single current sources are used and switched between contacts, while in others multiple sources can be driven simultaneously (Loizou, 1999). The currents that are sent to the electrodes can be either analog or pulsatile signals.

Cochlear prostheses are multi-channel systems. The first implants employed single electrodes, while present devices use up to 24 contacts inside the scala tympani (House and Berliner, 1991; Loizou, 1999; Spelman, 1999). In most processing strategies, one electrode is driven at a time, although analog drives present currents to all electrodes simultaneously (Osberger and Fisher, 1999).

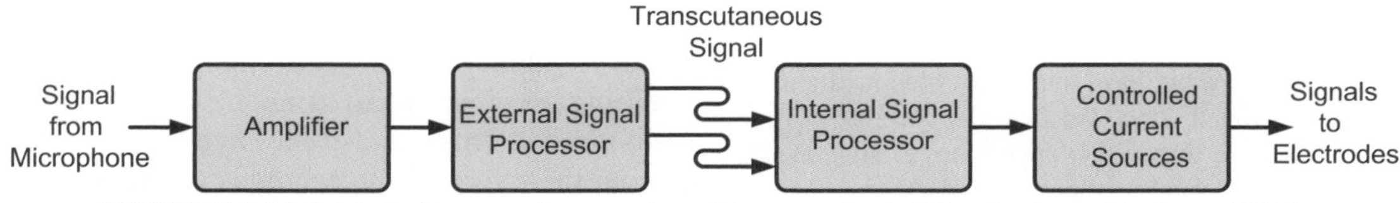

FIGURE II.5.11.2 Block diagram of a generic cochlear prosthesis (After Spelman (Spelman 1999)).

Recent Advances in Signal Processing

Combined Electrical and Acoustic Implants. Gantz and Turner have undertaken a series of studies of patients who have 6 mm or 10 mm cochlear electrode arrays inserted in the basal turns of the cochlea, leaving the apical turns available for acoustic stimulation (Gantz and Turner, 2004). In nine subjects they found that hearing scores doubled over those obtained with acoustic stimulation alone. The study demonstrates that, when short electrode arrays are used, acoustic information can be used to deliver low-frequency information, while high-frequency information is obtained via electrical stimulation. This approach may reduce the low-frequency limits of cochlear implants that are imposed by the anatomical necessity of placing their electrodes nearer to neurons that are sensitive to high-frequency stimuli.

Bilateral Implants. Several patients have received bilateral cochlear implants. The technique provides some improvement of the ability of patients to acquire sound in a noisy environment; that improvement is attributed mainly to the head shadow effect that is present with binaural hearing (Brown and Balkany, 2007). The major disadvantages of the approach are that it requires a second surgery, and that it doubles the cost of the hardware that is implanted.

Cochlear Nerve Implants

The family of cochlear nerve implants proposed early on by the group at Stanford introduced wire bundle arrays into the VIII cranial nerve via the modiolus of the cochlea (Simmons et al., 1979; White et al., 1984). Unfortunately, when microfabrication techniques were employed, delamination of the arrays occurred. Recent advances in microfabrication and the longevity of electrode arrays that are produced with the techniques used by integrated circuit manufacturers have led to tests of arrays that are designed to penetrate the VIII nerve, placing electrode contacts into close proximity with the target neurons (Hillman et al., 2003). The approach may permit stimulation with less current and greater specificity of excitation. The use of coatings of neural growth factors may provide specificity for both penetrating arrays in the auditory nerve and scala tympani arrays (Volckaerts et al., 2007).

Active Electrode Arrays

Groups at the University of Michigan and the Cochlear Implants Centre in Belgium have proposed building flexible integrated circuit arrays with large numbers (ca. 128) of electrodes and distributed signal processing, to reduce the number of wires that must be carried by the arrays. Up until this time, those arrays have been fabricated and tested in the laboratory, but have not been tested in chronic applications (Wise and Najafi, 2004; Volckaerts et al., 2007). The approach is more promising than earlier approaches to integrated electrode arrays that often delaminated or shattered.

Brainstem and Central Nervous System (CNS) Implants

Auditory brainstem implants were proposed for patients with acoustic neuromas in the early 1990s (Brackmann et al., 1993; Shannon et al., 1993). Those implants have been demonstrated to be safe, but not nearly as effective as cochlear implants (McCreery et al., 1997; Shannon et al., 1997; Schwartz et al., 2003; McCreery, 2008; Schwartz et al., 2008). The surgeries are more difficult, and the results not as spectacular as those produced by cochlear implants (McCreery, 2008). About 500 patients have received brainstem implants (McCreery, 2008). Surgeons have implanted electrode arrays in the cochlear nucleus within the central nervous system (CNS). Results are similar to those obtained with brainstem implants, demonstrating that the surgeries are possible and the electrode arrays are safe, but the hearing comprehension of both of the groups of subjects is similar to that achieved with single-channel cochlear implants (Lim et al., 2008).

Commercially Available Systems

The following sections are brief descriptions of the strategies that are used by the three manufacturers of cochlear prostheses that are presently available to deaf patients. All of the manufacturers have websites that provide additional information.

The Nucleus® Implant. Cochlear Pty Ltd is the largest producer of cochlear prostheses in the world. For further information on the Nucleus® prosthesis, visit www.cochlear.com. The Nucleus® has an electrode array that uses 22 or 24 electrodes. Twenty-two electrodes are placed in the scala tympani. The 24-electrode array uses two external electrodes that can be used as distant return electrodes. They permit true monopolar excitation of the internal electrodes. The Nucleus® C124R electrode array is curved in order to approximate the modiolar wall of the scala tympani, bringing the electrodes of the array into apposition with the cells of the auditory nerve. The array employs Advance Off Stylet (AOS). AOS has the curved array inserted with a straight stylet for the insertion into the first turn of the cochlea. The array is moved from the stylet into the second and possibly the third turn, where it hugs the modiolar wall (Patrick et al., 2006). The Freedom Processor is a behind-the-ear processor which can be programmed with SPEAK, ACE or Continuous Interleaved Stimulation (CIS) (see CIS described below). SPEAK and ACE are processors that extract frequency information with a high-speed Digital Signal Processing (DSP) that can sample at 180 million samples per second (Patrick et al., 2006). The auditory signal to determine the frequency bands containing the maximum energy during an interval can range from about 2 msec to a few hundred microseconds. The envelopes of the signals taken from the filters whose energy is greatest are sampled. The data are distributed to electrodes within the array, driving between five and ten electrodes within a given sequence. One electrode or a single electrode pair is driven at each sample. The electrodes are driven with biphasic rectangular pulses that repeat rapidly (Patrick et al., 2006). As of the time of the cited 2006 article by Patrick et al., no conclusions could be drawn about the benefits of the processing used in the Freedom.

The silicone rubber electrode array, when straightened, approximates a truncated cone whose diameter varies from 0.4 mm at the apical end to 0.6 mm at the basal end. The array is composed of half-segments of platinum rings, designed to face the peripheral processes of the auditory nerve. Each ring is approximately 0.35 mm in width, and the edge-to-edge separation is about 0.4 mm. The insulated platinum/iridium wires that connect to each electrode are led back through the silicone rubber carrier of the electrodes (Patrick et al., 1990). During insertion, the array is straightened with an internal stylet (Patrick et al., 2006). The array is placed into the scala tympani via a small hole that is drilled through the temporal bone. After an initial insertion of 8.5 mm, the stiffener is withdrawn while the array is gently inserted by the surgeon. In the past, the implanted electrode array resided near the lateral wall of the scala tympani when it was implanted in the ear of a human subject (Skinner et al., 1994).

The Clarion® and the Harmony™. The Clarion® cochlear implant has been supplanted by the Harmony™ implant, with the HiResolution® signal processor of the Advanced Bionics Corporation (Sylmar, CA). For the most recent information about the Clarion® implant, visit www.bionicear.com.

The device offers several processing strategies to deliver signals to patients (Kessler, 1999). Its processor can operate with several different processing strategies. The strategies include Simultaneous Analog Stimulation (SAS), the Paired Pulsatile Sampler (PPS), and the Continuous Interleaved Sampler (CIS), as well as the MP3000 strategy (Litvak and Narayan, 2007). The microphone and amplifier drive an analog/digital converter that employs signal-compressing software, most often logarithmic compression. The compressed signal is decomposed into a set of digitally filtered signals. The digitized and filtered outputs are transmitted across the skin with an RF link to an internal processor dedicated to the acoustic signal (Koch et al., 2004). The data are demodulated and demultiplexed internally, and delivered to current sources that drive electrodes directly with the compressed and filtered analog signals (Koch et al., 2004). The sampling rate is 70,000 samples per second per channel, with an aggregate sampling rate of 90,000 samples per second. The SAS system operates similarly to the Compressed Analog (CA) systems that have been used earlier (Kessler, 1999; Loizou, 1999; Spelman, 1999). The SAS system drives several electrodes simultaneously. It still suffers from the interference that occurs as a result of field interactions among electrodes.

The Harmony™ processor provides other strategies as well. Continuous Interleaved Stimulation (CIS) was developed by Wilson and his colleagues to overcome the field interactions between channels in the Ineraid implant (Wilson et al., 1991). Biphasic pulses are delivered one-pair-at-a-time to the electrodes of the implanted array in the cochlea. The pulses are interleaved (multiplexed) in time to eliminate interactions among the electric fields in the cochlea. The rates of excitation can be varied to optimize the signals delivered to specific patients. The acoustic signal is amplified, compressed, and band-pass filtered. The outputs of the filtered signals are rectified and low-pass filtered to obtain envelope information for each of the band-pass filters. The amplitudes of the pulse pairs are varied in proportion to the magnitudes of the envelopes of the signals in specific frequency bands at the time of analysis, driving the electrodes within whose bandwidths the greatest signals are found (Litvak and Narayan, 2007). The system is similar to the SPEAK strategy in that regard, although it selects a single frequency in one band rather than a range of frequencies. The widths of the pulses delivered are constant. The Harmony™ processor updates the signals as much as 2800 times per second. Eight or sixteen electrodes are usually driven (Koch et al., 2004).

The processor developed by Advanced Bionics allows the use of paired pulsatile stimulation (PPS) (Kessler, 1999). Paired Pulsatile Simulation (PPS) drives two electrodes simultaneously with paired biphasic pulses, while maintaining the maximum physical separation between the driven electrodes. Physical separation limits field interactions between electrodes (Kessler, 1999; Zimmerman-Phillips and Murad, 1999). Multiple Pulsatile Stimulation (MPS) can drive more than two electrodes simultaneously (Koch, 2000).

Koch et al. reported on 51 postlingually deafened adults who used both HiResolution™ and conventional sound processing in the new HiRes120 system of Advanced Bionics (Koch et al., 2004). As has also been found with other systems, HiRes120 users who had relatively poor performance showed greater improvement than users who had done very well. However, 96% of the patients in the study preferred the HiRes system (Koch et al., 2004).

Advanced Bionics introduced a method to appose the electrodes of their array to the cells of the VIII Cranial Nerve in Rosenthal's Canal, which lies behind the modiolar wall of the scala tympani. They produced a pre-shaped electrode that was used jointly with a polymeric Electrode Positioning System (EPS). The EPS applied pressure to the lateral wall of the cochlea, positioning the electrode array against the modiolus. The system was used for three years and then withdrawn from production in the summer of 2002.

The Med-El Implant. The SonataTI100™ implant is offered by Med-El Corporation (Innsbruck, Austria, www.medel.com). The Med-El electrode array is slim, having 24 contacts. Med-El provides data that indicate that their array can be placed within the entire length of the human scala tympani, 31 mm.

Med-El offers Continuous Interleaved Stimulation, and at this time they offer a high multiplexing rate of 50,704 pulses per second, extracting data with a Hilbert transform approach (Anonymous, 2003) or using zero crossings of the high-frequency data (Anonymous, 2008). Med-El has drawn on its experience using EAS (Electro-Acoustical Stimulation), and applies time and place code information for the fundamental frequency of sound to the two-to-three lower frequency channels (the apical channels). The company calls the approach Channel-Specific Sampling Sequences (CSSS), and uses it in its Opus1™ and Opus2™ processors in combination with its Fine Structure Processing (FSP) that employs Hilbert transforms to gain both envelope and fine structure from filtered signals (Anonymous, 2003).

Optical Stimulation of Auditory Neurons

Richter and his co-workers have demonstrated the feasibility of stimulating auditory neurons with optical pulses of minimum durations of 5 μsec using wavelengths of about 2 μm (Izzo et al., 2007, 2008; Richter et al., 2008). They demonstrate that the excitation of neurons is limited in spatial extent, combining that excitation with acoustic masking stimulation (Izzo et al., 2007). The approach is promising, but raises questions about signal distribution, since the investigators have used optical fibers whose diameters are 200 μm. The large diameters probably contribute to the specificity of the excitation, since the beam width of a radiator is inversely proportional to the ratio of its diameter to the wavelength of the signal transmitted. That is, the aperture of a radiating structure must be large with respect to the wavelength of the signal transmitted, if the signal must have a narrow beam (Terman, 1955). However, it is possible to distribute optical signals of 2 μm wavelengths, using optical waveguides the widths of which span 0.5 μm (Baehr-Jones et al., 2008). Since the waveguides could be either silicon-based or polymer-based, small distribution systems of large numbers of waveguides might be placed in the cochlea. However, the small dimensions of the waveguides may necessitate antenna analogs at the locations of the excitable cells to limit the spread of the optical signals, because of the small apertures of the waveguides. Another issue is that of power requirements. See the Problems section later in this chapter to address the question of the possible power consumption of optically-based cochlear stimulation.

MATERIALS AND ELECTRODE ARRAYS

Materials issues are salient in the design and construction of electrode arrays. The polymers that are used to insulate the arrays must be compatible with the tissues of the scala tympani; the electrode sites must use metals that can deliver appropriate currents to excite neurons; the mechanical characteristics of the arrays must allow safe and easy insertion into a space that is both small and complex in shape. Rebscher and co-workers describe techniques and tests designed to ensure safe insertion (Rebscher et al., 1999). Later, Wright and co-workers demonstrated substantial insertion depth, about 17 mm, with the "Thin Lateral" electrode array, and about 16 mm with the "Helix II" electrode array of Advanced Bionics. No damage was reported to the temporal bone specimens that were used (Wright et al., 2005).

While other parts of the prosthesis are of concern, the concerns are small compared to those that surround the electrode arrays. This chapter addresses issues of electrode design and choice of materials, the electrical and mechanical properties of insulating materials, and some of the issues that are related to the tissue sheaths that surround electrode arrays in the scala tympani of the cochlea.

Electrode Arrays

Electrode arrays change electrical currents into the ionic currents that can stimulate neurons. The number of electrode sites and the strategy used to drive the contacts determines the electric fields that are generated in the scala tympani and, ultimately, the number of independent channels that can be driven simultaneously. Several people have suggested ways by which electrical currents could be combined to focus the neural stimuli. Suesserman and co-workers suggested an approach that was supported by a lumped element, electrical model of the inner ear (Suesserman and Spelman, 1993). Later, Jolly and co-workers performed experiments to demonstrate that the predicted fields could be produced in the inner ear of the guinea pig (Jolly et al., 1996), and that it might be possible to stimulate independent groups of neurons (Jolly et al., 1997). Later, Middlebrooks and Bierer, and Bierer and Middlebrooks, showed clearly that focused fields produced focused excitation of neurons in the auditory system (Bierer and Middlebrooks, 2002; Middlebrooks and Bierer, 2002).

Contacts and Focusing. Simple models of point sources of current can be used to illustrate some of the effects of driving tissue with point sources of electric current. Consider a conductive medium that is semi-infinite in extent, and bounded by a perfectly insulating surface. The current source lies on the insulating boundary at location x_0, y_0, z_0. The dimensions are given in meters. The method of images (Kong, 1986) can be used to show that the potential field produced by the source is:

$$V(x,y,z) = \frac{\rho I}{2\pi} \frac{1}{[(x-x_0)^2 + (y-y_0)^2 + (z-z_0)^2]^{0.5}}$$

where:

ρ is the resistivity of the medium, given in Ohm-m;

I is the magnitude of current, given in Amperes;

$V(x,y,z)$ is the electrical potential at location x, y, z, given in Volts.

The equation above ignores the properties of the electrode, e.g., its polarization impedance, since they are overlooked entirely by the point source approximation (Macdonald, 1987).

Point sources can't be produced. However, the equation above describes a hemispherical source of radius, r, placed on the insulating surface with the source's center at the location of the point source, x_0, y_0, z_0. As long as the potential is computed for radial distances that are greater than r, the equation above is valid, and approximates the electrode in the absence of electrochemical effects.

A special case of the potential is illustrative. Let the location of the source be at the origin of a Cartesian coordinate system. Compute the field produced along the z-axis. Let the radius of the hemisphere be 50 μm. Consider a material, like perilymph, the fluid that fills the scala tympani, whose resistivity is 0.63 Ohm-m. Then:

$$V(z) = 0.1\frac{I}{z} \quad Volts, \qquad z > 50\,\mu m$$

Let the current be 100 μA, a reasonable threshold for a 200 μsec pulse in an experimental animal. The potential decreases monotonically from 200 mV at the surface to 50 mV at a distance of 200 μm.

Problem 3 (at the end of this chapter) illustrates additional issues that can be addressed with this simple model. With a straightforward model of this sort, the engineer can gain a quick understanding of the benefits and limitations of using multiple sources to focus the electric fields in the cochlea. The solution to the problem shows that combinations of positive and negative currents can change the widths of fields, but that narrowing the fields will reduce peak potentials. While simple, analytical solutions give results that can be understood in ideal geometries, the more complex geometries of the cochlea require more complicated solutions (Finley et al., 1989; Suesserman and Spelman, 1993; Frijns et al., 1995).

Thinking about dipole fields leads to that conclusion. A theoretical dipole consists of a pair of positive and negative point sources of current, placed at an infinitesmal distance from each other. The potential field decreases rapidly for distances that are much larger than the separation between the electrodes. The point of the experiment is to suggest that the smaller the distance between sources of opposite polarities, the smaller will be their peak potentials at a distance. Hence, sources and sinks of electric current must be used judiciously to focus potential fields.

The point source model of an electrode is suitable for spherical electrodes if their fields are modeled for radial distances greater than the radial dimension of the sphere. The same model works for a hemispherical electrode that is attached to a planar insulator. The electrodes that are used in cochlear prostheses are often more like finite, planar surfaces, and their solutions are not straightforward except in the simplest cases (Rubinstein et al., 1987; Pearson, 1990). The potential field that is generated by a circular electrode has been understood since the time of Weber (Rubinstein et al., 1987), and was described by Wiley and Webster more than 20 years ago (Wiley and Webster, 1982). The critical finding for the latter work is that the current density of such an electrode is non-uniform over its surface, and becomes singular at the boundaries of the electrode. Current density is singular at the edges of many kinds of planar electrodes, as was shown in an elegant demonstration by Rubinstein (Rubinstein, 1988). Recessing the electrode can eliminate the singularities of current density at its metallic surface, while singularities are found at the aperture of the recessed electrode (Rubinstein et al., 1987). However, shaping the aperture can eliminate those singularities (Suesserman et al., 1991).

Consideration of surface-mounted and recessed electrodes of finite size leads to an understanding of the potential for corrosion of such electrodes. Corrosion will be greater at the edges where the current and charge densities are high. Further, thinking about the boundary condition at the surface of a metallic electrode that the potential is constant on the surface, suggests that the field produced by an electrode of finite size will differ from that produced by a point source. Finite electrodes produce a second effect, not seen for point sources. As the observer moves closer to the source, the field approaches a constant value over the surface of the source. That is a result of the boundary condition requiring superconductors to have uniform potentials on their surfaces. Of course, the potential decreases as the observer moves laterally away from the source. Another difference between point sources and finite sources is that unused point sources have no effect on the potential fields that are near them, while finite sources must sustain constant potentials at their surfaces.

Contact Materials

What materials are suitable to make the electrode arrays that are used in cochlear implants? Noble metals, that is, gold, platinum, iridium, and some of their alloys have been used extensively (Robblee and Rose, 1990). The arrays that are available clinically employ alloys of platinum and iridium, usually 90% Pt and 10% Ir. Those alloys have been studied and used for decades (Brummer and Turner, 1977a,b; Robblee and Rose, 1990; McCreery et al., 1992). The platinum provides a material that can carry charge efficiently, while the iridium provides structural strength.

More recently, the oxides of iridium have been investigated, and have demonstrated characteristics that are markedly superior to those of platinum (Robblee and Rose, 1990; Meyer et al., 2001; Weiland et al., 2003). Electrode materials are studied with standard electrochemical techniques, notably cyclic voltammetry (Robblee and Rose, 1990) and electrochemical impedance spectroscopy (EIS) (Macdonald, 1987). The former technique provides direct evidence that shows the amount of charge density that a given material can support. EIS gives indirect evidence of the charge-carrying capacity, but provides direct information about the voltage that is required to drive a specific current.

Charge density is a critical variable to consider in the design of a cochlear electrode array. The charge density calculation is usually based on a "geometric" surface area, which assumes that the current density is uniform across the metallic surface. While that assumption is not valid in most cases, there are few calculations of geometries that result in mixed boundary problems. For the purpose of this chapter, we will assume that geometric areas suffice, but warn the reader to be wary when calculating surface areas. That said, the measurement of the charge capacity of a particular electrode is certainly valid when it is done appropriately (Meyer et al., 2001). The charge density calculation may be in error unless great care is taken to ensure that the electrode's surface is smooth, and that the current density profile is accounted for (Robblee and Rose, 1990; Meyer et al., 2001; Weiland et al., 2003).

Concern for the charge carrying capacity of electrode sites is critical for the design of cochlear electrode arrays, since it bears directly on their safety and longevity. Platinum-iridium arrays have followed the work of Brummer and Turner (Brummer and Turner, 1977a,b), usually applying a safety factor of at least two. The charge densities are held below 150 μCoul/cm^2. Increasing the surface area of a platinum electrode with platinum black can increase the charge density by a factor of 30 (Jolly et al., 1996), but the approach has not been used outside of the laboratory.

Oxides of iridium, both activated and plated, offer real promise for cochlear electrode arrays. The charge densities measured cover a range from 4 mCoul/cm^2 to 27 mCoul/cm^2 (Meyer et al., 2001; Weiland et al., 2003; Cogan, 2008). Those charge densities are not used consistently, but a safety factor is applied for long-term operation. The charge density can be reduced by as much as a factor of 20 for long-term tests (Meyer et al., 2001). Even so, the promise of IrOx is great: the potential increase in charge density, and thus in maximum stimulus current, is nearly a factor of 10. If Pt-Ir can tolerate stimuli whose charges lie below 150 μCoul/cm^2 and IrOx can withstand a long-term stimulus of 1.2 mCoul/cm^2, it is easy to compute maximum currents if we assume that we can deal with geometric surface areas. If two electrodes whose areas are 10^{-4} cm^2 are used, and they are driven with square pulses whose durations are 200 μsec, the maximum safe current is 75 μA for the Pt-Ir electrode, and 600 μA for the IrOx electrode. The IrOx electrode can excite the auditory nerve over its full dynamic range, while the Pt-Ir electrode can barely meet a threshold value (Vollmer et al., 2001).

Another concern for cochlear implant designers is the geometry of their electrode arrays. The developers of thin-film arrays must be concerned with the traces that are used on the thin films. As the arrays get smaller, the traces must also decrease in size. While implant designers may think that the currents that they use are small (they are!), the traces are small as well. Temperature increases in the traces may well be a problem as the traces decrease in size (Brooks, 1998). UltraCAD Design, Inc. offers freeware to solve printed circuit board (PCB) trace problems with a regression relation (www.UltraCAD.com). For example, a copper trace that is 7.5 cm long, 12.5 μm wide, and 1 μm thick will have a temperature increase of nearly 1°C when it carries 800 μA. If several traces carry similar currents, temperature in the cochlea could increase artificially, possibly causing thermal damage to the tissue.

Material Properties

The choice of materials for cochlear prostheses extends beyond the selection of materials that will be used on the surfaces of electrode contacts. At issue are the compatibility of arrays with the tissues of the cochlea, and compatibility of the materials that cover the internal processors that are placed in the temporal bones of the recipients of implants. This chapter does not consider the latter, since the ceramic and titanium cases and their silicone cases have not created measurable problems at the time of this writing, at least since some early problems with infection were taken care of in the late 1970s and early 1980s. The materials used as carriers for the electrode arrays have presented issues that will be discussed below.

Advanced Bionics recalled its HiFocus™ electrode array; some attributed that recall to potential concerns with the electrode positioning system (www.leifcabraser.com/cochlear.htm), a polymeric system that placed the silicone array in contact with the modiolar wall.

Placement issues had been raised before in relation to other implants (Skinner et al., 1994; Rebscher et al., 2001). Spiral tomographs of the implanted ear showed that the electrode arrays entered the scala vestibuli by way of the basilar membrane. The composition of the fluids in the scala tympani and the scala vestibuli are dramatically different. The perilymph in the scala tympani is rich in sodium ions, while the endolymph in the scala vestibuli is rich in potassium ions (Dallos et al., 1996). Mixing the two damages, and can kill, hair cells. While that is not critical in the case of the sensorineural deaf patient, the electrode arrays are distant from the peripheral processes of the auditory neurons, and the leakage of potassium-rich endolymph will damage those neurons as well.

The silicone compounds used as substrates for the electrode arrays are well-tolerated by both humans and animals. Early investigations demonstrated that an inflammatory reaction occurs in the implanted scala tympani of the inner ear; the result is a thin cellular sheath that surrounds the electrode array in both animals and humans (Clark, 2003).

In a study of the electrode arrays produced by Cochlear Corporation and Advanced Bionics Corporation, four surgeons implanted both new and old designs of arrays into cadaveric human temporal bones. The investigators assessed the amount of trauma observed, the insertion depth, and the proximity to the modiolar wall. The arrays showed no significant difference in trauma produced by the arrays of the two corporations; the greatest difference was found among the surgeons who did the implants (Rebscher et al., 2001). The Spiral™ could be inserted further than any of the other designs, and of the new designs, the Contour™ had the furthest insertion depth. Both the Contour™ and the HiFocus™ arrays apposed the modiolar wall (Rebscher et al., 2001).

The paragraph above raises the issue of insertion depth. The advantage of a cochlear electrode array that reaches the more apical turns of the cochlea is that it can reach the tonotopic locations that decode frequencies below 1 kHz. For example, insertion of 70% of the cochlear spiral reaches the 300 Hz region of the basilar membrane (Geisler, 1998). Since the range of frequencies necessary to decode speech is 300–3000 Hz (Levitt and Nye, 1980), it is desirable to reach the low-frequency regions of the cochlea with cochlear implants.

That problem, and the need to place electrode contacts near to the modiolar wall of the cochlea, has led designers to build arrays with mechanical characteristics that make the implants conform to the anatomy of the cochlea. Rebscher and his colleagues suggested a design that elegantly made use of the mechanical properties of the wires that interconnected the electrode sites and the drive electronics (Rebscher et al., 1999). They built an array that arranged the wires to form a central, vertical beam. The beam bent easily to conform to the spiral shape of the cochlea, and the pitch of the spiraling array could be controlled by the geometry of the beam in a basal–apical direction. Advanced Bionics has employed the design in two of their arrays. *In vitro* measures of the stiffness of Cochlear and Advanced Bionics arrays showed a clear anisotropy for the latter, but not for the former (Rebscher et al., 2001). Another approach suggested the use of shape-memory wire, e.g., Nitinol™, within the array to match the cochlear spiral. The wire was shaped to fit the modiolar wall (Spelman et al., 1998), straightened for insertion into the cochlear electrode array, and electrically heated beyond its transition temperature after the array was inserted into the scala tympani. The system has not been commercialized.

The stiffness of a cochlear electrode array is important for ease of insertion, placement near the target neurons and, possibly, for insertion trauma. The arrays are mechanical beams, and their mechanical properties can be investigated with classical methods in some cases, at least to obtain information about the relative stiffness of arrays that employ different materials (Boyd, 1935; Enderle et al., 2000). For a cantilevered beam of uniform cross-section and homogeneous material, the maximum deflection that occurs when the beam is loaded at its free end is:

$$y_{max} = \frac{Pl^3}{3EI}$$

where P is the load in N; l is the length in m; E is the modulus of elasticity in Pa (N/m^2); and I is the cross-sectional area moment of inertia, m^4.

The above equation is a gross oversimplification of a complex problem. The assumptions that underlie the relationship are that the deflection produces small angles, that the cross-section is uniform, and that the material is linear and homogeneous. Cochlear electrode arrays are inhomogeneous: they contain both metallic conductors and polymeric substrates; they may be laminated. Cochlear electrode arrays are tapered. The number of conductors inside them varies with length. The angles of bending can be large, particularly in the apical turns of the cochlea. The arrays are likely to be anisotropic. Thus, for a complete analysis, more sophisticated mathematics is necessary, and may not be amenable to closed-form mathematical solutions. Numerical techniques, e.g., finite element analysis, are likely to be required for better understanding.

Measurements

Cochlear prostheses are sophisticated instruments that require thorough measurement and analysis. Since they are Class III devices from the point of view of the FDA (US Food and Drug Administration), they must be carefully tested, and the tests documented before they are used.

Each component of a cochlear implant is tested before the assembly can be applied to a human subject. For example, the processing section of the prosthesis must be tested to ensure that it can parse the auditory signal and reassemble the decomposed aggregate into a new auditory signal that can be understood readily by hearing listeners. While that is not a definitive test for its ultimate users, the deaf, it gives confidence that the processor is not introducing anomalous information. Such tests also indicate the data rates that must be transmitted across the skin to an internal processor.

The data transfer link must be tested to ensure that it can sustain the necessary data rates across an appropriate thickness of skin (about 1 cm in the adult human). Animal tests can provide confidence in this case.

The internal processor must be tested to learn whether it can select electrode sites unambiguously and reliably. The current drivers are tested to discover their linearity, repeatability, and voltage range over the full life of the battery. As batteries sag during use, does the transfer function of the current driver remain constant? Is the voltage range sufficient to drive the full dynamic range of every electrode?

The electrode array is tested carefully throughout the design to the manufacturing process. The tests include long-term soaking, electrochemical tests (Parker et al., 1999), field tests, and neurophysiological tests (Jolly et al., 1996, 1997; Bierer et al., 2003; Snyder et al., 2004). As an electrode design is introduced, it must be placed in physiological saline solutions for several weeks to discover whether the insulating carriers promote leakage between the conductors of the arrays.

Electrochemical tests include tests of open-circuit potential, whose stability may indicate shorting across traces (Parker et al., 1999), as well as electrochemical impedance spectroscopy, which can also indicate open-circuited and short-circuited electrode sites, as well as poorly plated electrode sites (Macdonald 1987; Parker et al., 1999). (Note that Parker et al. can be downloaded from http://www.eng.monash.edu.au/ieee/ieeebio1999/p41.htm.) If more complete information about the characteristics of the electrode sites is needed, cyclic voltammetry is in order (Goodisman, 1987; Cogan, 2002).

Measurement of the electric fields produced by electrodes is useful during the initial design phase. It shows the characteristics of the potential fields that are produced by the excitation of single or multiple electrodes (Suesserman et al., 1991; Jolly et al., 1996). It also can show which electrodes have parasitic conductive paths that lie between them (Prochazka and Spelman, unpublished results).

When bench testing and *in vitro* testing are complete, and a cochlear prosthesis is ready for clinical trials, behavioral testing begins. That testing ranges from psychophysical tests (Shannon, 1992; Pfingst et al., 1997; Vandali et al., 2000; Won et al., 2007), to tests of monosyllabic words and simple sentences (Osberger and Fisher 1999; Skinner, 2001), to investigations of the understanding of contextual information (Vandali et al., 2000). Behavioral testing is the true "gold standard," since it determines the success or failure of the cochlear prosthesis.

Focusing Fields and the Interaction with Tissue

The cochlear prostheses that are in use today drive electrodes singly, with the exception of the Advanced Bionics system. That is a result of the interference between the fields that are produced by the electrodes of the devices, fields that are defined by the size of the electrodes used, their distances from the target cells, and the magnitudes and phases of the signals that are applied to each contact (Spelman et al., 1995). If fields are focused, it is likely that multiple groups of neurons can be driven simultaneously and independently (Jolly et al., 1997). It is not clear whether such independence would be preferred by patients, and whether it will provide improved speech perception (Pfingst et al., 1997). It seems logical from the operation of the normal auditory system (Sachs and Young, 1979), but Pfingst's work argues against it (Pfingst et al., 1997).

Both modeling and experimental studies suggest clear benefits that can result from perimodiolar placement of electrode arrays. The modeling studies show that the fields are clearly more limited in extent when the electrode contacts are near to the target cells (Spelman et al., 1995; Frijns et al., 1996, 2001). Use of the simple model given in the earlier section on Contacts and Focusing can demonstrate the narrowing of the field that takes place when a point source is placed near its target.

Finite-sized electrodes do not produce infinitesimally wide fields as they are approached; the fields approach the width of the electrode. However, a point can be made for either point sources or finite sources; the further from the source, the wider the field. That statement is clear for a single electrode, driven as a monopole; less clear for a tripolar configuration, driven as a quadrupole (Spelman et al., 1995). In the latter case, the width of the potential field changes less with distance than for the former case. In both cases, the peak potential in the field is larger the closer the contact or contacts are to the target. Thus, intervening tissue between the electrodes and their target neurons must be minimized or avoided entirely.

If scar tissue surrounds the implanted electrode array, then several problems can arise as a result of the foreign-body response. First, the distance between the electrodes and their target cells will increase. Second, the sheaths that surround implanted electrodes have higher resistivities than the solutions of the scala tympani, and they are anisotropic (Spelman, 2004). Third, tissue sheaths can cause an increase in the impedances of the electrodes, and a concomitant increase in the voltage required to achieve excitation, with accompanying increases in power consumption.

The effect of increasing the distance becomes greater the closer is the electrode array to the neurons of the auditory nerve. Considering the simple case of a point source driven as a monopole, the electric field is proportional to the inverse of the distance between the source and the target. For a distance of 100 μm, an increase of 50 μm represents a 50% change in the distance between the cells and the target. If there is a distance of 500 μm initially, the change is 10%. In experimental animals, sheaths of 50 μm thickness were found after a few months of implantation (Shepherd et al., 1983; Leake et al., 1990).

The introduction of connective tissue sheaths introduces multiple layers into the analysis of fields in the inner ear. Even a simple three-layered problem produces a complicated result that is not a closed-form solution, and will not be addressed here (Spelman, 1989). It is enough to say that the signal produced by the stimulating current will be attenuated and, in the case where the sheath is anisotropic, will likely direct the excitation to undesirable locations (Spelman, 2004). Other effects that could affect the excitation of tissue include motion of the tissue sheath; motion that might induce an intervening layer of perilymph between the electrode and the sheath, increasing the attenuation. However, motion within the cochlea should be small, because the organ is not subject to vasomotion and the electrode array is fixed in place with respect to the surrounding bony tissues.

If cochlear electrode arrays could have their surfaces treated appropriately, they might resist the adhesion of cells and the growth of surrounding tissue sheaths (Dalsin et al., 2003). That approach to combating the growth of cells that are produced by inflammatory responses may prove to be successful for future developments of electrode arrays. The approach can protect the surface of the substrate of the array, but not the metallic surfaces of the electrodes themselves.

Tissue on the surfaces of the electrodes must increase their impedances. The membranous tissues behave like leaky capacitors, introducing materials with specific capacitances of 33.8 μF/cm^2 (Junge, 1977). The introduction of such capacitances can affect the impedance of the electrodes of the array, and their abilities to carry currents and stimulate cells. Thus, the materials to prevent cell adhesion to cochlear electrode arrays must be designed to maintain current-carrying capacity and low impedance, while they prevent cell adhesion and the growth of undesirable tissue sheaths.

Measures of electrode impedances have been variable with time, often increasing initially and decreasing after about four weeks. The causes are not known definitively. It may be that while sheaths of cells will increase impedances, positional changes within the cochlea cause those impedances to decrease, since the effective resistivity of the surrounding tissue decreases when the electrode array is more distant from the bony walls of the cochlea. Experiments have not been done to establish sheathing and separation effects definitively.

Costs and Benefits

That more than 120,000 people use cochlear prostheses is testimony to their effectiveness. The prostheses have improved monotonically over the past several decades, with substantial improvements in comprehension that occurred when multi-channel prostheses were developed, and more gradual improvements otherwise (Rubinstein and Miller, 1999). At the same time, cochlear prostheses have been found to be cost effective. In careful cost analyses, Cheng and his co-workers investigated cochlear implants, and found that they were competitive with other surgical interventions (Cheng and Niparko, 1999). The study accounted for the costs of surgery and rehabilitation, and considered change in the quality of life as an intangible.

FUTURE DIRECTIONS

Clearly, multi-channel cochlear prostheses will be produced and implanted in large numbers, and will be used extensively worldwide. The three major producers of implants are likely to move to more cosmetic devices. All of them presently make behind-the-ear prostheses, and are pursuing totally implantable devices. At this time, the greatest two problems faced by the latter are battery and microphone technologies. Implantable microphones will require great attention to the issue of tissue encapsulation around the diaphragm of the microphone.

Another likely improvement will be the number of electrodes that are used in the array. Investigators have pursued a 72-contact array (Spelman et al., 1998), a device that demands a quantum change in processor technology (Clopton et al., 2002). High-density electrode arrays will require great care and attention to be paid to the issues of tissue growth and compatibility. If a high-density array and its processor are available, they could provide precise phase information that is unavailable now and that will likely make possible successful binaural cochlear prostheses (Clopton and Lineaweaver, 2001).

Optical stimulation may reduce interactions among stimulus sites. If so, it is likely to provide greater numbers of excitation sites.

Bilateral implants are likely to increase in numbers, as are combined electrical and acoustic implants. The results of tests with both modalities are quite promising (Wilson and Dorman, 2008).

SUMMARY

Cochlear implants are used by more than 120,000 people worldwide. They represent a remarkable achievement by bioengineers. This chapter has provided a brief introduction to the design of cochlear prostheses. First, the chapter reviews the operation of the normal auditory system. That review highlights the approach taken by bioengineers to the initial design of cochlear prostheses. Understanding the tonotopic organization of the auditory system has led to the design of very successful prostheses.

The chapter reviews the architecture of the hardware of the devices, and outlines some of the issues that are being addressed currently, including signal processing techniques, which attempt to stress the need to deliver information about acoustic fine structure to users, in order that they may decode sound in a noisy background and potentially hear music.

Commercial systems are reviewed, with brief descriptions of the offerings of the world's major manufacturers. The chapter addresses the material issues inherent to cochlear electrode arrays and their contacts to the auditory nerve. Finally there is a brief mention of the costs and benefits of the auditory implant and a mention of directions for the future.

ACKNOWLEDGMENTS

This work was supported in part by Grants DC005531, NS37944, and DC04614 from the National Institutes of Health.

GLOSSARY OF TERMS

Pinna: the external ear.

Tympanic membrane: the inner boundary of the external ear canal. The membrane vibrates in response to acoustic stimulation.

The bones of the inner ear: the stirrup (stapes), hammer (malleus), and anvil (incus) provide a mechanical advantage so that low-pressure, large-displacement vibrations of the tympanic membrane can be converted into high-pressure, small-displacement signals at the oval window of the cochlea.

Cochlea: the organ that converts auditory signals into the signals that drive the central nervous system.

Scalae: the chambers of the cochlea. The hair cells reside on top of the basilar membrane within the scala media. A cochlear electrode array is placed within the scala tympani, where it drives the peripheral processes of the auditory nerve cells. The cell bodies of the auditory nerve reside in Rosenthal's Canal.

Modiolus: the axis of the cochlear spiral. The auditory nerve courses through the modiolus into the internal auditory canal, and thence to the brain.

Nuclei: several centers of the brain process signals from the inner ear. Signals pass into the cochlear nucleus, the trapezoid body, the inferior colliculus, the medial geniculate nucleus, and the auditory cortex.

BIBLIOGRAPHY

Anonymous. (2003). Med-El Web Page. Innsbruck, Austria: Med-Elwww.medel.com2003.

Anonymous. (2008). Fine Hearing Technology. Innsbruck, Austria: Med-El.

Baehr-Jones, T., Hochberg, M., et al. (2008). All-optical modulation in a silicon waveguide based on a single-photon process. *IEEE J. Sel. Top. Quant. Electron.*, **14**(5), 1335–1342.

Bierer, J. A., & Middlebrooks, J. C. (2002). Auditory cortical images of cochlear-implant stimuli: Dependence on electrode configuration. *Jour. Neurophysiol.*, **87**(1), 478–492.

Bierer, J. A., & Middlebrooks, J. C. (2004). Cortical responses to cochlear implant stimulation: Channel interactions. *J. Assoc. Res. Otolaryngol.*, **5**(1), 32–48.

Bierer, J. A., Litvak, L., et al. (2003). Effects of electrode configuration on psychophysical measures of channel interaction in cochlear implant subjects. *Society for Neuroscience*.

Boyd, J. E. (1935). *Strength of Materials*. New York, NY: McGraw-Hill.

Brackmann, D. E., Hitselberger, W. E., Nelson, R. A., Moore, J., Waring, M. D., et al. (1993). Auditory brainstem implant: I. Issues in surgical implantation. *Otolaryng. Head Neck Surg.*, **108**(6), 624–633.

Brooks, D. (1998). Temperature rise in PCB traces. *Proceedings of the PCB West Design Conference, March 1998*, Miller Freeman, Inc.

Brown, K. D., & Balkany, T. J. (2007). Benefits of bilateral cochlear implantation: A review. *Otol. Neurotol.*, **15**(5), 315–318.

Brugge, J. F. (1991). An overview of central auditory processing. In A. N. Popper, & R. R. Fay (Eds.), *The Mammalian Auditory Pathway: Neurophysiology* New York, NY: Springer-Verlag. (2, pp. 1–34).

Brummer, S. B., & Turner, M. J. (1977a). Electrical stimulation with Pt electrodes I: A method for determination of real electrode areas. *IEEE Trans. Biomed. Eng.*, **24**(5), 436–440.

Brummer, S. B., & Turner, M. J. (1977b). Electrical stimulation with Pt electrodes II: Estimation of maximum surface redox (theoretical non-gassing) limits. *IEEE Trans. Biomed. Eng.*, **24**(5), 440–444.

Cheng, A. K., & Niparko, J. K. (1999). Cost-utility of the cochlear implant in adults. *Arch. Otolaryngol. Head Neck Surg.*, **125**(11), 1214–1218.

Clark, G. M. (1996). Electrical stimulation of the auditory nerve: The coding of frequency, the perception of pitch and the development of speech processing strategies for profoundly deaf people. *Clin. Exp. Pharmacol. Physiol.*, **23**, 766–776.

Clark, G. M. (2003). *Cochlear Implants: Fundamentals and Applications*. New York, NY: Springer-Verlag.

Clopton, B. & Lineaweaver, S. (2001). The importance of phase information for the encoding of speech.

Clopton, B. M., Lineaweaver, S. K.R., et al. (2002). *Method of processing auditory data*. USA: U.S. Patent Office. Advanced Cochlear Systems.

Cogan, S. F. (2002). *Stability of electro-active materials and coatings for charge-injection electrodes*. Second Joint EMBS-BMES Conference 2002, Houston, TX: IEEE Press.

Cogan, S. F. (2008). Neural stimulation and recording electrodes. *Annu. Rev. Biomed. Eng.*, **10**, 275–309.

Dallos, P., Popper, A. N., & Fay, R. R. (Eds.), (1996). *The Cochlea. Springer Handbook of Auditory Research*. New York, NY: Springer.

Dalsin, J. L., Hu, B. H., Lee, B. P., & Messersmith, P. B. (2003). Mussel adhesive protein mimetic polymers for the preperation of nonfouling surfaces. *J. Am. Chem. Soc.*, **125**(14), 4253–4258.

Djourno, A., & Eyries, C. (1957). Prothèse auditive par excitation électrique à distance du nerf sensoriel à l'aide d'un bobinage inclus à demeure. *Presse. Med.*, **35**, 14–17.

Enderle, J., Blanchard, S., et al. (2000). *Introduction to Biomedical Engineering*. San Diego, CA: Academic Press.

Finley, C. C., Wilson, B. S., et al. (1989). Models of neural responsiveness to electrical stimulation. In F. A. Spelman (Ed.), *Cochlear Implants: Models of the Electrically Stimulated Ear* (1, pp. 55–96). New York, NY: Springer-Verlag.

Frijns, J. H. M., de Snoo, S. L., & Schoonhoven, R. (1995). Potential distirbutions and neural excitation patterns in a rotationally symmetric model of the electrically stimulated cochlea. *Hear Res.*, **87**, 170–186.

Frijns, J. H. M., de Snoo, S. L., & ten Kate, J. H. (1996). Spatial selectivity in a rotationally symmetric model of the electrically stimulated cochlea. *Hear Res.*, **95**, 33–48.

Frijns, J. H. M., Briaire, J. J., & Grote, J. J. (2001). The importance of human cochlear anatomy for the results of modiolus-hugging multichannel cochlear implants. *Otol. & Neurol.*, **22**(3), 340–349.

Gantz, B. J., & Turner, C. (2004). Combining acoustic and electrical speech processing: Iowa/Nucleus hybrid implant. *Acta. Otolaryngol.*, **124**(4), 344–347.

Geisler, C. D. (1998). *From Sound to Synapse: Physiology of the Mammalian Ear*. New York, NY: Oxford University Press.

Gold, B., & Rader, C. M. (1967). The channel vocoder. *IEEE Trans. Audio Electroacoust.*, 148–161. AU-15(Dec. 1967).

Goodisman, J. (1987). *Electrochemistry: Theoretical Foundations*. New York, NY: Wiley Interscience.

Hillman, T., Badi, A. N., Normann, R. A., Kertesz, T., & Shelton, C. (2003). Cochlear nerve stimulation with a 3-dimensional penetrating electrode array. *Otol. Neurotol.*, **24**(5), 764–768.

Hinojosa, R., & Marion, M. (1983). Histopathology of profound sensorineural deafness. *Ann. NY Acad. Sci.*, **405**, 458–484.

House, W. F., & Berliner, K. I. (1991). Cochlear Implants: From Idea to Clinical Practice. In H. Cooper (Ed.), *Cochlear Implants: A Practical Guide* (1, pp. 9–33). San Diego, CA: Singular Publishing Group, Inc.

Izzo, A. D., Suh, E., Pathria, J., Walsh, J. T., Whitlon, D. S., et al. (2007). Selectivity of neural stimulation in the auditory system: A comparison of optic and electric stimuli. *J. Biomed. Opt.*, **12**(2). 021008.

Izzo, A. D., Walsh, J. T., Jr., Ralph, H., Webb, J., Bendett, M., et al. (2008). Laser stimulation of auditory neurons: Effect of shorter pulse duration and penetration depth. *Biophys. J.*, **94**(8), 3159–3166.

Jolly, C. N., Spelman, F. A., & Clopton, B. M. (1996). Quadrupolar stimulation for cochlear prostheses: Modeling and experimental data. *IEEE Trans. Biomed. Eng.*, **43**(8), 857–865.

Jolly, C. N., Clopton, B. M., Spelman, F. A., & Lineaweaver, S. K. (1997). Guinea pig auditory nerve response triggered by a high density electrode array. *Med. Prog. Technol.*, **21**(Suppl.), 13–23.

Junge, D. (1977). *Nerve and Muscle Excitation*. Sunderland, MA: Sinauer and Associates, Inc.

Kessler, D. K. (1999). The CLARION multi-strategy cochlear implant. *Ann. Otol. Rhinol. Laryngol. Suppl.*, **177**(Apr.), 8–16.

Koch, D. B. (2000). Cochlear implants: An overview. www.audiologyonline.com/articles/pf_article_detail.asp?article_id=222.

Koch, D. B., Osberger, M. J., Segel, P., & Kessler, D. (2004). HiResolution and conventional sound processing in the HiResolution bionic ear: Using appropriate outcome measures to assess speech recognition ability. *Audiol Neurootol.*, **9**(4), 214–223.

Kong, J. A. (1986). *Electromagnetic Wave Theory*. New York, NY: John Wiley & Sons.

Leake, P. A., Kessler, D. K., et al. (1990). Application and Safety of Cochlear Prostheses. In W. F. Agnew, & D. B. McCreery (Eds.), *Neural Prostheses: Fundamental Studies* (1, pp. 253–296). Englewood Cliffs, NJ: Prentice-Hall.

Levitt, H., & Nye, P. W. (1980). Sensory Training Aids for the Hearing Impaired (Prologue). In H. Levitt, J. M. Pickett, & R. A. Houde (Eds.), *Sensory Aids for the Hearing Impaired* (1, pp. 3–28). New York, NY: IEEE Press.

Lim, H. H., Lenarz, T., Anderson, D. J., & Lenarz, M. (2008). The auditory midbrain implant: Effects of electrode location. *Hear Res.*, **242**(1–2), 74–85.

Litvak, L., & Narayan, L. (2007). *Encoding fine time structure in presence of substantial interaction across an electrode array*. United States: Advanced Bionics Corporation.

Loizou, P. C. (1999). Signal-processing techniques for cochlear implants. *IEEE Eng. Med. Biol. Mag.*, **18**(3), 34–46.

Macdonald, J. R. (1987). *Impedance Spectroscopy*. New York, NY: Wiley.

Martin, M. C. (1985). Alternatives to cochlear implants. In R. A. Schindler, & M. M. Merzenich (Eds.), *Cochlear Implants*. New York, NY: Raven Press.

McCreery, D. B. (2008). Cochlear nucleus auditory prostheses. *Hear Res.*, **242**, 64–73.

McCreery, D. B., Yuen, T. G., Agnew, W. F., & Bullara, L. A. (1992). Stimulation with chronically implanted microelectrodes in the cochlear nucleus of the cat: Histologic and physiologic effects. *Hear Res.*, **62**(1), 42–56.

McCreery, D. B., Shannon, R. V., et al. (1997). *The Feasibility of a Cochlear Nucleus Auditory Prosthesis Based on Microstimulation*. Pasadena, CA: Huntington Medical Research Institutes and House Ear Institute.

Meyer, R. D., Cogan, S. F., Nguyen, T. H., & Rauh, R. D. (2001). Electrodeposited iridium oxide for neural stimulation and recording electrodes. *IEEE Trans. Neural Sys. & Rehab. Eng.*, **9**(1), 2–11.

Middlebrooks, J. C., & Bierer, J. A. (2002). Auditory cortical images of cochlear-implant stimuli: Coding of stimulus channel and current level. *J. Neurophysiol.*, **87**(1), 493–507.

NIDCD (2003). *Health Information: Cochlear Implants*. www.nidcd.nih.gov/health/hearing/coch.asp. NIH (US Government).

NIH (1995). *NIH Consensus Statement: Cochlear Implants in Adults and Children*. Bethesda, MD: National Institutes of Health.

Osberger, M. J. (1997). Current issues in cochlear implants in children. *The Hearing Review*, **4**(October, 1997), 28–31.

Osberger, M. J., & Fisher, L. (1999). SAS-CIS preference study in postlingually deafened adults implanted with the Clarion cochlear implant. *Ann. Otol. Rhinol. Laryngol.*, **108**(4). 74–70.

Parker, J. R., Duan, Y. Y., et al. (1999). *Testing of thin-film electrode arrays for cochlear implants of the future*. Victoria, NSW, Australia: Inaugural Conference of the Victorian Chapter of the IEEE EMBS.

Patrick, J. F., Seligman, P. M., et al. (1990). Engineering Cochlear Prostheses. In G. M. Clark, Y. C. Tong, & J. F. Patrick (Eds.), Cochlear Prosthesis (pp. 99–125). Edinburgh, UK: Churchill Livingstone. 1.

Patrick, J. F., Busby, P. A., et al. (2006). The development of the Nucleus(R) Freedom™ cochlear implant system. *Trends in Amplif.*, **10**(4), 175–200.

Pearson, C. E. (1990). *Handbook of Applied Mathematics*. New York, NY: Van Nostrand Reinhold.

Pfingst, B. E., Zwolan, T. A., & Holloway, L. A. (1997). Effects of stimulus configuration on psychophysical operating levels and on speech recognition with cochlear implants. *Hear. Res.*, **112**(October, 1997), 247–260.

Popper, A. N., & Fay, R. R. (1991). *The Mammalian Auditory Pathway: Neurophysiology*. New York, NY: Springer-Verlag.

Pyman, B. C., Brown, A. M., et al. (1990). Preoperative evaluation and selection of adults. In G. M. Clark, Y. C. Tong, & J. F. Patrick (Eds.), *Cochlear Prostheses* (pp. 125–135). Melbourne, Australia: Churchill Livingstone.

Rebscher, S. J., Heilmann, M., Bruszewski, W., Talbot, N. H., Synder, R. L., et al. (1999). Strategies to improve electrode positioning and safety in cochlear implants. *IEEE Trans. Biomed. Eng.*, **46**(3), 340–352.

Rebscher, S. J., Wardrop, P. J., et al. (2001). *Insertion trauma, mechanical performance and optimum dimensions: refining a second generation of cochlear implant electrodes*. Pacific Grove, CA: 2001 Conference on Implantable Auditory Prostheses.

Richter, C.-P., Bayon, R., Izzo, A. D., Otting, M., Suh, E., et al. (2008). Optical stimulation of auditory neurons: Effects of acute and chronic deafening. *Hearing Res.*, **242**(1–2), 42–51.

Robblee, L. S., & Rose, T. L. (1990). Electrochemical guidelines for selection of protocols and electrode materials for neural stimulation. In W. F. Agnew, & D. B. McCreery (Eds.), *Neural Prostheses: Fundamental Studies* (1, pp. 25–66). Englewood Cliffs, NJ: Prentice Hall.

Rubel, E. W., & Dobie, R. A. (1989). The Auditory System: Central Auditory Pathways. In H. D. Patton, A. F. Fuchs, B. Hille, A. M. Scher, & R. Steiner (Eds.), *Textbook of Physiology: Excitable Cells and Neurophysiology* (pp. 386–411). Philadelphia, PA: W. B. Saunders.

Rubinstein, J. T. (1988). *Quasi-Static Analytical Models for Electrical Stimulation of the Auditory Nervous System*. Seattle, WA: Department of Bioengineering. University of Washington. 96.

Rubinstein, J. T., & Miller, C. A. (1999). How do cochlear prostheses work? *Curr. Opin. Neurophysiol.*, **9**, 399–404.

Rubinstein, J. T., Spelman, F. A., Soma, M., & Suesserman, M. F. (1987). Current density profiles of surface mounted and recessed electrodes for neural prostheses. *IEEE Trans. Biomed. Eng.*, *BME*-**34**(11), 864–875.

Rubinstein, J. T., Wilson, B. S., Finley, C. C., & Abbas, P. J. (1999). Pseudospontaneous activity: Stochastic independence of auditory nerve fibers with electrical stimulation. *Hearing Res.*, **127**(1–2), 108–118.

Sachs, M. B., & Young, E. D. (1979). Encoding of steady-state vowels in the auditory nerve: Representation in terms of discharge rate. *J. Acoustic. Soc. Am.*, **667**(2), 470–479.

Schwartz, M. S., Otto, S. R., Brackmann, D. E., Hitselberger, W. E., & Shannon, R. V. (2003). Use of a multichannel auditory brainstem implant for neurofibromatosis type 2. *Stereotact. Funct. Neurosurg.*, **81**(1–4), 110–114.

Schwartz, M. S., Otto, S. R., Shannon, R. V., Hitselberger, W. E., & Brackmann, D. E. (2008). Auditory brainstem implants. *Neurotherapeutics, Device Therapy*, **5**(1), 128–136.

Shannon, R. V. (1992). Temporal modulation transfer function in patients with cochlear implants. *J. Acoust. Soc. Am.*, **91**, 2156–2164.

Shannon, R. V., Fayad, J., Moore, J., Lo, W. W., & Otto, S. R. (1993). Auditory brainstem implant: II. Postsurgical issues and performance. *Otolaryng. Head Neck Surg.*, **108**(6), 634–642.

Shannon, R. V., Moore, J., et al. (1997). *The Feasibility of a Cochlear Nucleus Auditory Prosthesis Based on Microstimulation*. Los Angeles, CA: House Ear Institute and Huntington Medical Research Institutes.

Shepherd, R. K., Clark, G. M., Black, R. C., & Patrick, J. F. (1983). The histopathological effects of chronic electrical stimulation of the cat cochlea. *J. Laryngol. Otol.*, **97**(4), 333–341.

Simmons, F. B. (1966). Electrical stimulation of the auditory nerve in man. *Arch. Otolaryng.*, **84**(July, 1966), 24–76.

Simmons, F. B., Mathews, R. G., Walker, M. G., & White, R. L. (1979). A functioning multichannel auditory nerve stimulator. A preliminary report on two human volunteers. *Acta. Otolaryngol.*, **87**(3–4), 170–175.

Skinner, M. W. (2001). *Cochlear implants in children: What direction should future research take?* Pacific Grove, CA: 2001 Conference on Implantable Auditory Prostheses.

Skinner, M. W., Ketten, D. R., Vannier, M. W., Gates, G. A., Yoffie, R. L., et al. (1994). Determination of the position of Nucleus cochlear implant electrodes in the inner ear. *Am. J. Otol.*, **15**(5), 644–651.

Snyder, R. L., Bierer, J. A., & Middlebrooks, J. C. (2004). Topographic spread of inferior colliculus activation in response to acoustic and intracochlear electric stimulation. *J. Assoc. Res. Otolaryngol.*, **5**(3), 305–322.

Spelman, F. A. (1989). Determination of Tissue Impedances of the Inner Ear: Models and Measurements. In J. M. Miller, & F. A. Spelman (Eds.), *Cochlear Implants: Models of the Electrically Stimulated Ear* (1, pp. 422). New York, NY: Springer-Verlag.

Spelman, F. A. (1999). The past, present and future of cochlear prostheses. *IEEE Engineering in Medicine and Biology Magazine*, **18**(3), 27–33.

Spelman, F. A. (2004). Cochlear Implants. In P.-Å Öberg, T. Togawa, & F. A. Spelman (Eds.), *Biomedical Instrumentation* (pp. 309–335). Berlin: Wiley.

Spelman, F. A., & Voie, A. H. (1996). *Fascicles of the auditory nerve in the human cochlea: Measurements in the region between the spiral ganglion and the osseous spiral lamina.* St. Petersburg, FL: Nineteenth Annual Midwinter Meeting of the Association for Research in Otolaryngology.

Spelman, F. A., Pfingst, B. E., Clopton, B. M., Jolly, C. N., & Rodenhiser, K. L. (1995). The effects of electrode configuration on potential fields in the electrically-stimulated cochlea: Models and measurements. *Ann. Otol. Rhinol. & Laryngol.*, **104**(Suppl. 166), 131–136.

Spelman, F. A., Clopton, B. M., et al. (1998). *Cochlear Implant with Shape Memory Material and Method for Implanting the Same*. USA: MicroHelix, Inc.

Suesserman, M. F., & Spelman, F. A. (1993). Lumped-parameter model for *in vivo* cochlear stimulation. *IEEE Trans. Biomed. Eng.*, **40**(3), 234–235.

Suesserman, M. F., Spelman, F. A., & Rubenstein, J. T. (1991). *In vitro* measurement and characterization of current density profiles produced by nonrecessed simple recessed and radially varying recessed stimulating electrodes. *IEEE Trans. Biomed. Eng.*, **38**(5), 401–408.

Terman, F. E. (1955). *Electronic and Radio Engineering*. New York, NY: McGraw-Hill Book Company, Inc.

Vandali, A. E., Whitford, L. A., Plant, K. L., & Clark, G. M. (2000). Speech perception as a function of electrical stimulation rate: Using the Nucleus 24 cochlear implant system. *Ear & Hear*, **21**(6), 608–624.

Volckaerts, B. B., Corless, A. R., Mercanzini, A., Silmon, A. M., Bertsch, A., et al. (2007). Technology developments to initiate a next generation of cochlear implants. *Conf. Proc. IEEE Eng. Med. Biol. Soc.*, 515–518.

Vollmer, M., Beitel, R. E., & Synder, R. L. (2001). Auditory detection and discrimination in deaf cats: Psychophysical and neural thresholds for intracochlear electrical signals. *J. Neurophysiol.*, **86**(5), 2330–2343.

Weiland, J. D., Anderson, D. J., & Humayun, M. S. (2003). *In vitro* electrical properties for iridium oxide versus titanium nitride stimulating electrodes. *IEEE Trans. Biomed. Eng.*, **49**(12), 1574–1579.

White, M. W., Merzenich, M. M., & Gardi, J. N. (1984). Multichannel cochlear implants: Channel interactions and processor design. *Arch. Otolaryngol.*, **110**(August, 1984), 493–501.

Wiley, J. D., & Webster, J. G. (1982). Analysis and control of the current distribution under circular dispersive electrodes. *IEEE Trans. Biomed. Eng.*, *BME*-**29**(5), 381–385.

Wilson, B. S., & Dorman, M. F. (2008). Cochlear implants: A remarkable past and a brilliant future. *Hearing Research*, **242**(1–2), 3–21.

Wilson, B. S., Finley, C. C., Lawson, D. T., Wolford, R. D., Eddington, D. K., et al. (1991). Better speech recognition with cochlear implants. *Nature*, **352**, 236–238. (July 18, 1991).

Wise, K. D., & Najafi, K. (2004). *Annual Report of the Engineering Research Center for Wireless Integrated Microsystems (WIMS)*. Ann Arbor, MI: University of Michigan.

Won, J. H., Drennan, W. R., & Rubenstein, J. T. (2007). Spectral-ripple resolution correlates with speech reception in noise in cochlear implant users. *J. Association for Research in Otolaryngology*, **8**, 384–392.

Wright, C. G., Roland, P. S., & Kuzma, J. (2005). Advanced bionics thin lateral and helix II electrodes: A temporal bone study. *The Laryngoscope*, **115**(11), 2041–2045.

Zimmerman-Phillips, S., & Murad, C. (1999). Programming features of the CLARION multi-strategy cochlear implant. *Ann. Otol. Rhinol. Laryngol., Suppl.*, **177**, 17–21. (April).

CHAPTER II.5.12 THE ROLE OF BIOMATERIALS IN STIMULATING BIOELECTRODES

P. Hunter Peckham[1], *D. Michael Ackermann, Jr.*[2], *and Christa W. Moss*[1]

[1]Biomedical Engineering, Case Western Reserve University, Cleveland, Ohio, OH, USA

[2]Biodesign, Stanford University, Stanford, CA, USA

INTRODUCTION

Bioelectrodes are devices used to produce or measure electrical activity in the body for electrophysiological stimulation or monitoring. This chapter will focus on electrodes for use in stimulation. Stimulation of electrically excitable tissue is achieved by establishing electrical fields in the tissue, using electrodes. These electrodes transduce electrons (the charge carrier for the electrical circuitry used in therapeutic stimulation devices) to ions (the charge carrier in the body), allowing electric charge to be injected into tissue. The materials used for these electrodes are integral to ensuring that this charge injection occurs safely and effectively. See also Chapter II.5.10, which addresses some aspects of bioelectrodes using a different and complementary approach.

NEUROSTIMULATION

Neurostimulation therapies work by activating or inhibiting excitable tissue, such as the nervous system or heart, through the delivery of electrical currents through electrodes. Current and emerging neurostimulation therapies include devices for treating cardiac arrhythmias (Ellenbogen et al., 2000), reducing pain (Cameron, 2004), alleviating symptoms of movement disorders (Starr et al., 1998), restoring motor control to the paralyzed (Peckham, 2005), restoring sensory function to the profoundly deaf (Clark, 2003) and blind (Humayun et al, 2003; Chow et al., 2004), treating epilepsy (Binnie, 2000; Theodore, 2004), restoring lost bladder and bowel function (Jarrett, 2004; Brazzelli, 2006; Rijkhoff, 1998), and treating psychiatric disorders (Kopell, 2004). These devices have restored lost function and have extended and improved the life of hundreds of thousands of individuals worldwide.

As shown in Figure II.5.12.1, most neurostimulators conform to a similar design and consist of three primary components: an *implantable pulse generator* (*IPG*); one or more *leads*; and one or more *electrodes* (Peckham and Ackermann, 2009). The implantable pulse generator is hermetically sealed and houses the stimulation circuitry, a battery, and a telemetry system for communicating with an external device (e.g., for receiving programming or command signals). The lead is a conductor that carries electrical currents to the electrode, which interfaces directly with the excitable tissue.

Electrical stimulation of neurons induces *action potentials* (and therefore functional outcomes) by depolarizing the neuronal cell membranes (Ranck, 1975; Rattay, 1989), and thus activating voltage-sensitive ion channels responsible for action potential generation (Hodgkin and Huxley, 1952). The electric field generated in the tissue by the injected electrode current is directly responsible for this membrane depolarization. The type of material, shape of the electrode, and stimulation parameters all affect the size and strength of the electric field. Therefore, proper electrode design and material selection is imperative to delivering an effective therapy.

FUNDAMENTAL REQUIREMENTS OF A BIOELECTRODE

In general, an electrode has three primary components (see Figure II.5.12.1, inset), each with distinct material properties: the charge delivery surface; the insulator; and a conductor which connects to the charge delivery surface (often the same conductor from the lead). This chapter will focus on the *charge delivery surface* and the *insulator*. The charge delivery surface is the electrical interface between the implanted stimulator and the tissue. The insulator ensures the lead has the desired mechanical properties and

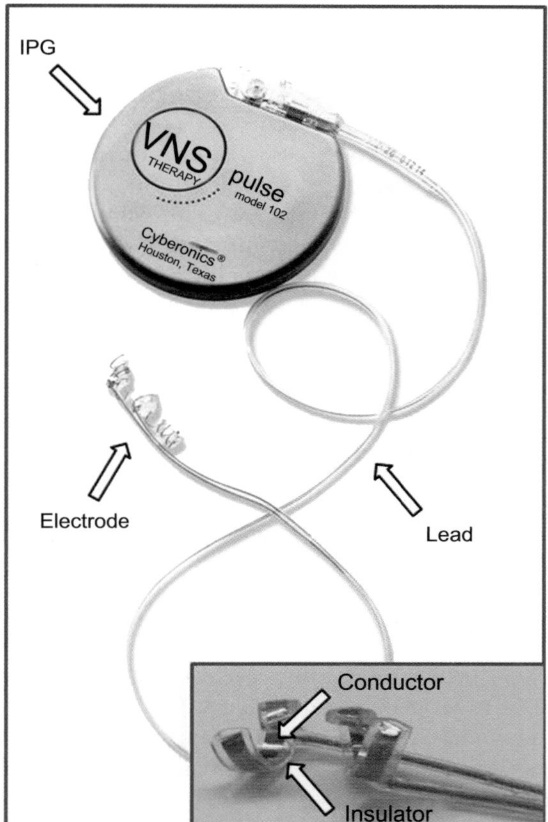

FIGURE II.5.12.1 Implantable neurostimulation system for vagus nerve stimulation. This figure shows the three principal components of most neurostimulators: the implantable pulse generator (IPG); the lead; and the electrode. (Photograph courtesy of Cyberonics, Inc. Inset shows a magnified image of the Huntington-type electrode with markers showing the conductor and insulator.)

isolates current flow to the desired tissue. Generally speaking, for an electrode to be a safe and effective component of a neurostimulation therapy, it must meet three fundamental requirements: the electrode must be efficacious; the electrode must not damage the tissue; and the electrode must remain intact for the life of the implant. Proper material selection is integral to fulfilling each of these. Each of these requirements is considered in more detail.

1. *Efficacy*: The electrode must deliver sufficient current to activate the target tissue. For some applications the electrode must selectively activate the target tissue without activating other excitable structures which may generate undesired side-effects (e.g., pain (Falowski et al., 2008) or undesired movements or sensation (Tamma, 2002)). Additionally, the electrode must have adequate mechanical properties: it must be of a size and shape sufficient for implantation (e.g., cochlear electrodes for the hearing-impaired must conform to the very small, spiral-shaped cochlea), and must be rigid enough for insertion into the tissue.
2. *Tissue Safety*: The electrode must not cause damage to the tissue in which it has been implanted. It must also be implanted and potentially explanted without damage to the neural tissue. Tissue damage can occur due to *active chemical processes* that occur when stimulating current is being delivered through an electrode, and *passive chemical processes* that occur when stimulating current is not being delivered through an electrode (passive biocompatibility).
3. *Electrode Integrity*: Electrode failure can occur due to mechanical or chemical stresses. Some neuroprosthetic therapies involve implantation of electrodes and leads in regions of the body where motion subjects them to large degrees of mechanical cycling (Kilgore et al., 2003; Kondziolka, 2002), making them prone to failure. As will be discussed in more detail in the section on electrochemical reversal below, some chemical reactions involved in charge injection can result in dissolution/corrosion of the electrode. Proper material selection, electrode size, and stimulation paradigm can help ensure that the electrode will not undergo irreversible corroding reactions.

PRINCIPLES OF CHARGE INJECTION

In order for a stimulating electrode to activate neural tissue, it must generate an electric field within that tissue. This electric field is generated when the electrode delivers charge to the neural tissue in the vicinity of the electrode. It is important to note that, because of the principles of conservation of charge and current (as elucidated by James Maxwell and Gustav Kirchoff), at least two electrodes are always present in a neurostimulation system: one to deliver cathodic current; and one to receive an equal and opposite anodic current. Generally speaking, cathodic currents activate neural tissue and anodic currents inhibit neural tissue (Ranck, 1975). Therefore, one of these two electrodes is often located far from the neural tissue being stimulated (e.g., the titanium case of the IPG located in a subcutaneous pocket in the chest or abdomen). This is called monopolar stimulation.

In the metal-conducting surface of the electrode, charge is carried by electrons. In the tissue, ions such as Na^+, K^+, Cl^-, and others are the charge carrier. As shown in Figure II.5.12.2, when a stimulating current is applied, the surface of the electrode acts as a transducer, converting current flow in the form of electrons in the electrode into ionic current flow in the tissue (or *vice versa*). This transduction of current occurs via two mechanisms of charge transfer at the electrode–electrolyte interface: capacitive charge transfer and Faradaic charge transfer. This chapter will present a brief review of these processes and their implications for neurostimulation. A more in depth review has been written by Merrill et al. (2005), and is recommended for the interested reader.

- *Capacitive Charge Transfer*: A capacitor consists of two conductors which are separated by a non-conductive material. When an electrode is placed into an aqueous environment (e.g., tissue), a so-called "capacitive double-layer" is formed at the electrode–electrolyte interface. As shown in Figure II.5.12.2, the two conductive layers of the capacitor are the electrode material, and the diffuse ion-containing aqueous solution. Charge is physically separated at the interface of these two components of the double layer, primarily because halide ions adsorb to the surface of the electrode material, acting to separate charge, and because water molecules are polar and adsorb to the surface, orienting themselves based on charge (Grahame, 1947). Capacitive charge transfer occurs when a stimulating current is delivered to an electrode, changing its net charge. Charged species in the aqueous environment react to this net change in charge by moving towards or away from the electrode. For example, at the cathode, a negative current delivers electrons to the electrode surface. These electrons result in a net negative change in charge on the electrode. This net negative charge attracts positively-charged species in the electrolyte, and repels negatively-charged species (as shown in Figure II.5.12.2). This flow of electrons and charged species within the electrolyte constitutes the capacitive electrode current. For sufficiently small stimulation currents, all delivered charge can occur via the capacitive mechanism. Some materials, such as tantalum, allow for very large capacitive current delivery (Rose et al., 1985).
- *Faradaic Charge Transfer*: Faradaic charge transfer results from reduction and oxidation reactions at the electrode–electrolyte interface. During these reactions, electrons are transferred from the electrode material to species in the electrolyte for cathodic

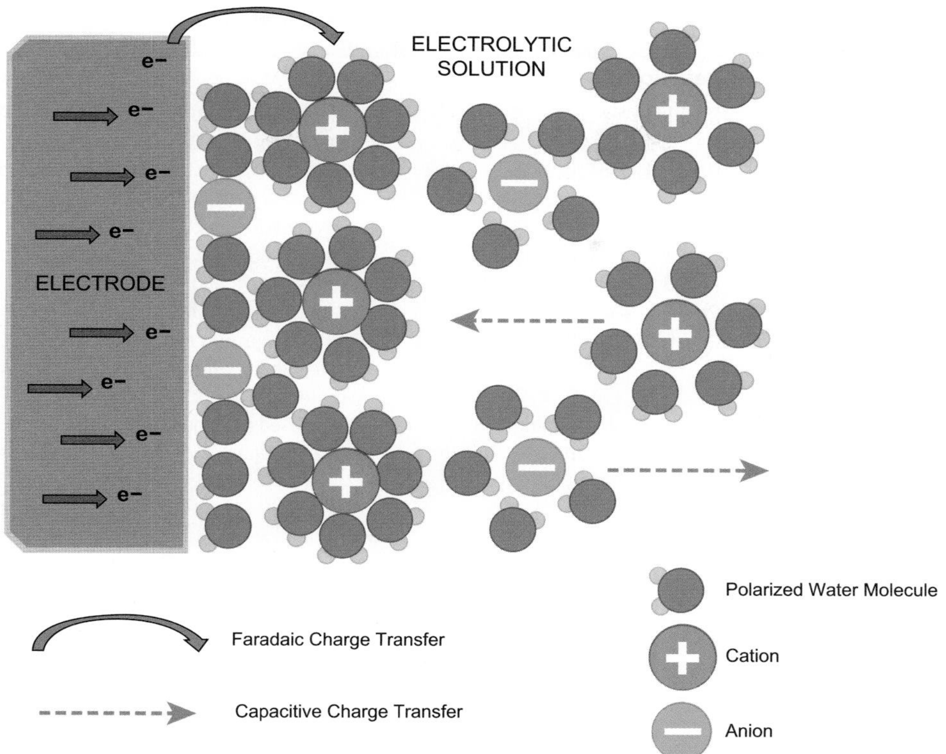

FIGURE II.5.12.2 Representation of the electrode–electrolyte interface during cathodic current delivery. Both Faradaic and capacitive charge transfer are depicted.

reactions (see Figure II.5.12.2) or from species in the electrolyte to the electrode material for anodic reactions. Unlike capacitive charge transfer, Faradaic exchange of electrons results in a change in the chemical species. For example, a silver electrode is oxidized to generate an electron and an aqueous silver ion in the following reaction:

$$Ag \rightleftarrows Ag^{+} + e^{-}$$

These reduction–oxidation reactions occur when an applied electrode potential drives the energy of electrons on one side of the electrode–electrolyte interface sufficiently high or low such that it becomes equal to that of the reduction–oxidation pair species on the other side of the interface. During *cathodic stimulation* (where the electrode is driven to a negative potential), the energy level of the free electrons in the metal electrode is raised, allowing them to be transferred to unoccupied orbitals of a reactant species. During *anodic stimulation* (where the electrode is driven to a positive potential), the opposite is true. The energy level of the free electrons in the metal electrode is lowered, allowing electrons from aqueous species to be transferred to the metal interface. While noble metals are often the preferred choice of Faradaic electrode material, because of their resistance to corrosion, non-noble metals such as 316 stainless steel are also used (Memberg et al., 1989). These materials are sometimes combined with other metals to create alloys with enhanced mechanical properties, such as platinum-iridium. Surface coatings such as iridium oxide are also sometimes used to enhance Faradaic charge delivery characteristics (Beebe and Rose, 1988). Table II.5.12.1 shows a list of electrode materials and the Faradaic reactions associated with each of them.

ACTIVE CHEMICAL PROCESSES AND ELECTROCHEMICAL REVERSAL

Faradaic electrode currents are generally conceptualized as reversible or irreversible (for the conditions under which stimulation occurs). A reversible reaction is one where the Faradaic reaction products can be recovered with a reversal in electrode current (i.e., the reverse reaction occurs in the same magnitude as the forward reaction). *Electrochemical reversal* is an important concept in electrical stimulation for two reasons: (1) Reactions that result in dissolution of the electrode material into the electrolyte can ultimately lead to the destruction of the electrode if the dissolved species are not recovered with a reversal reaction (Cogan, 2008); (2) these *active chemical processes* (referring to chemical reactions that occur due to charge injection) can produce species that may be deleterious to tissue health if not reversed. For example, free radicals produced in some reactions can damage axonal myelin sheaths (Chan et al., 1982) and DNA (Buettner, 1993). While it has been shown that some degree of irreversibility is not only tolerated by the body, but is likely the norm for chronic stimulation (Merrill et al., 2005), a primary goal in safely stimulating

TABLE II.5.12.1 Common Electrode Materials and Associated Faradaic Reactions

Material	Process	Reaction	Comments	Use
Tantalum/ Tantalum Oxide	Capacitive	Double Layer Charging	• Electrode surface area has to be greater than $0.5 cm^2$	BION® injectable micro-stimulator
Titanium	Capacitive	Double Layer Charging	• Material is mechanically and chemically stable in the body • Low charge injection capacity	Cardiac pacing applications that require lower charge injection
Platinum	Capacitive Reversible Faradaic: • Oxidation/ Reduction • H Atom Plating	Double Layer Charging $Pt + H_2O \leftrightarrow PtO + 2H^+ + 2e^-$ $Pt + H_2O + e^- \leftrightarrow Pt\text{-}H + OH$ $Pt + H^+ + e^- \leftrightarrow Pt\text{-}H$	• Pseudocapacitive (combination of capacitive and faradaic) processes allow for larger charge injection at safe levels than purely capactive materials	Cochlear implants Nerve cuff electrodes
Iridium	Capacitive Reversible Faradaic: • Oxidation/ Reduction	Double Layer Charging $Ir + 2H_2O \leftrightarrow Ir(OH)_2 + 2H^+ + 2e^-$	• Increased mechanical strength over platinum	Pt/Ir electodes: Spinal cord stimulation, neuromuscular stimulation
Iridium Oxide	Reversible Faradaic: • Valence Change Oxide	$Ir(OH)_n \leftrightarrow IrO_x(OH)_{n-x} + xH^+ + xe^-$	• Multilayer film formed on electrode at tissue interface • Increases charge injection capacity of an Ir or Pl electrode by an order of magnitude	Microelectrodes

neural tissue is to recover as much reversible charge as possible by using a biphasic stimulation pulse, as shown in Figure II.5.12.3. For most neurostimulation systems, the first phase (usually cathodic for the stimulating electrode) activates the neural tissue, and the second phase reverses the capacitive and/or Faradaic charge transfer processes. Electrode material selection is also quite important in ensuring that electrochemical reversal is possible. Several electrode material properties contribute to the ability to achieve electrochemical reversibility:

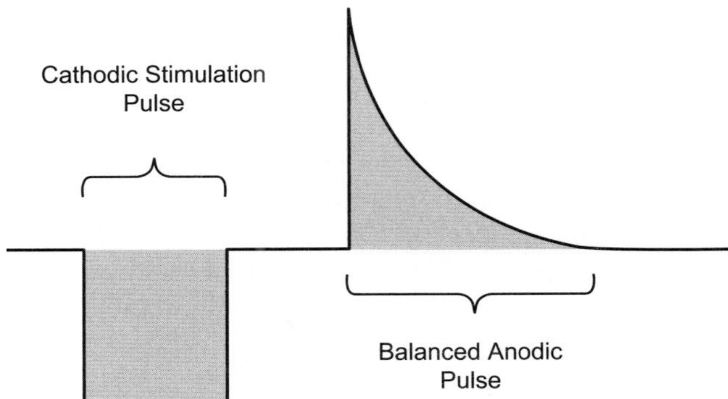

FIGURE II.5.12.3 An example of a balanced biphasic stimulation waveform. The waveform is representative of a typical voltage or current controlled stimulation pulse. The first phase of the pulse is usually cathodic, followed by an interphase wait interval, followed by a passive anodic recharge phase. The recharge phase ensures electrochemical reversal.

- *Electrode Capacitance*: A large electrode capacitance translates into a larger fraction of charge transfer occurring via capacitive mechanisms. Capacitively transferred charge is completely reversible, since no electron exchanging chemical reaction occurs. Electrode capacitance is governed by the inherent properties of the electrode material and the area of the electrode. Materials such as titanium nitride (TiN) (Weiland et al., 2002) and tantalum/tantalum pentoxide (McCreery, 1988) can be used to produce electrodes with an interface capacitance that is sufficient for purely capacitive stimulation (without any Faradaic reaction). For some applications, these materials may be practically limited by surface area, because large surface areas are required for delivery of large amounts of charge. However, these materials have been found to be quite useful in the BION® injectible microstimulator. This stimulator is shown in Figure II.5.12.4. It is very small, and combines both the stimulation circuitry and electrode into a single unit that can be implanted with a large needle. This device minimizes the number of electrical components used to reduce its size. As a safety feature, implantable stimulators require an output coupling capacitor to ensure that there is no direct current path from the implanted circuitry to the tissue that could allow tissue damage to occur in the event of an internal failure (Peckham and Ackermann, 2009). By using tantalum for the stimulating surface material the electrode itself is a capacitor, allowing for omission of a discrete capacitor (Loeb and Richmond, 2001).
- *Surface Area*: A large electrode surface area can help ensure electrochemical reversal, because both the electrode capacitance and the amount of electrode material available for reversible Faradaic reactions

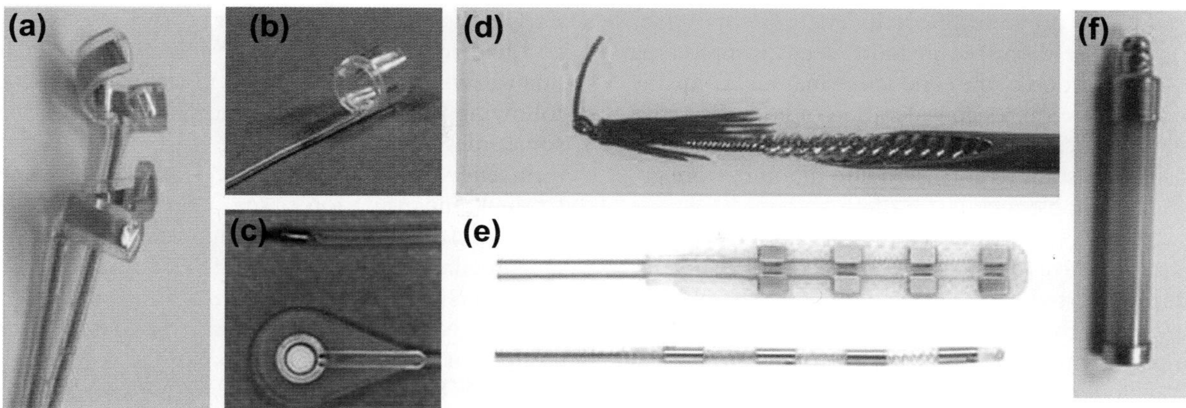

FIGURE II.5.12.4 Neurostimulation electrodes. (a) Huntington-type spiral electrode; (b) Case Western spiral nerve cuff electrode; (c) Intramuscular and epimysial electrodes; (d) Peterson type electrode (photograph courtesy of J. Thomas Mortimer, Case Western Reserve University, OH, USA); (e) Medtronic Specify paddle-type and Pisces™ Quad percutaneous-type spinal cord stimulation electrodes. (Photo courtesy of Medtronic, Inc.); (f) BION® microstimulator.

are both proportional to the electrode's surface area. Unfortunately, there is an inherent trade-off regarding electrode surface area: while large surface areas allow for the safe delivery of large currents, they hinder the ability to selectively activate particular portions of neural tissue, because of the physical size of the electrode. One method to circumvent this issue is to modify the surface area of the electrode to give it a microscopically three-dimensional texture. This three-dimensional texture increases the capacitance of the electrode surface without increasing the geometric size of the electrode surface. For example, platinum gray (platinum with an area-increasing surface modification) has a surface area five times greater than standard platinum foil (Zhou, 2005).

- *Electrode-Pseudocapacity*: Electrode materials that are capable of undergoing Faradaic reactions whose products remain bound to the electrode or diffuse away from the electrode–electrolyte interface at a negligibly slow rate are said to have the property of *pseudo-capacity*. The phenomenon is referred to as such because the Faradaic reaction behaves much like a capactive transfer: the slow diffusion of reaction products ensures that most, if not all, of the Faradaic charge transfer is reversible. Platinum is perhaps the most commonly used of all electrode materials, primarily because of its pseudo-capacitive property, owed to the following reaction:

$$Pt + H^+ + e^- \rightleftarrows PtH$$

PASSIVE CHEMICAL PROCESSES AND MECHANICAL INTERACTION OF ELECTRODE AND TISSUE

Like other medical implants, an implanted bioelectrode will elicit some foreign-body response when implanted. This foreign-body response occurs due to passive chemical and mechanical interactions between the implanted device and the tissue (passive chemical interactions refer to those which occur even when stimulating currents are not delivered). The response, which involves macrophage migration to the implant site followed by fibrous encapsulation of the foreign-body, is reviewed in more detail in previous chapters of this book. There are two primary factors which contribute to the foreign-body response of an implanted bioelectrode: chemical biocompatibility and mechanical biocompatibility. Each of these will be considered in more detail.

Biocompatibility: Passive Chemical Processes

The chemical biocompatibility of an electrode is dependent on both the ability of the tissue to withstand the passive chemical interaction with the implant, and the ability of the implant to withstand the environment it is subjected to within the body. Traditionally, successful materials used in implanted devices are inert, and chemically interact as little as possible with their surroundings. A bioelectrode has additional constraints placed on the specific materials used in order for it to effectively stimulate neural tissue. As shown in the Figure II.5.12.1 insert, a bioelectrode is composed of conductive materials that deliver electrical current to target tissue, and insulating materials that limit the region of stimulation to the intended tissues. The insulating material also often serves the additional functional role of maintaining the overall shape and mechanical stability of the electrode. For example, the electrode needs to conform to the anatomy which it is intended to stimulate, and needs to have sufficient rigidity to allow for surgical implantation. As a result, the insulating material often can represent a large fraction of total electrode volume. To minimize the foreign-body response, biocompatible materials that also serve these functional requirements are often used. Commonly used materials include silicone elastomer (Silastic), pellethanes, and polytetrafluorethylene homopolymer (Teflon®); they are discussed in more detail in preceding chapters. The selection of the conducting material is equally as important, and requires consideration of both the active and passive chemical biocompatibility. Generally, the electrode must be resistant to corrosion and minimize foreign-body response. Noble metals are often preferred, because they resist surface corrosion. Platinum and its alloys with iridium are the most widely used and studied materials. The reader is referred to (Robblee and Rose, 1990) for a further review of these metals and their associated behaviors in response to various stimulation paradigms.

Biocompatibility: Mechanical Interaction of Electrode and Tissue

The three functional requirements of a bioelectrode (efficacy, maintenance of electrode integrity, and maintenance of tissue safety) are well-represented in the mechanical structure of a bioelectrode. To be mechanically effective, the shape of a stimulating electrode must be appropriate for the anatomy of the tissue being stimulated, must maintain the orientation of the stimulating electrode(s) to the neural tissue, must have sufficient rigidity to allow for implantation, must not migrate once implanted, must not damage tissue, and must maintain a stable electrical impedance.

As discussed in the "Neurostimulation Applications" section below, the shape of the electrode must sufficiently conform to the anatomy of the tissue. This shape will depend on the type of tissue being stimulated. For example, nerve cuff electrodes (Figure II.5.12.4) generally wrap around or encircle a nerve, and have a cylindrical shape. Paddle-type electrodes (Figure II.5.12.4) are implanted in the epidural space above the spinal cord electrode for pain control, and have dimensions which allow them to fit into this thin space. The geometric orientation of the stimulating contacts of an electrode is designed to maximize the functional outcomes of a therapy. For example, the inter-contact spacing and orientation of contacts on a spinal cord stimulation electrode is sometimes optimized to minimize total current delivery or to minimize the activation of undesired structures, such as dorsal spinal roots which can cause pain when activated (Holsheimer and Struijk, 1997).

Choosing a material with an appropriate stiffness is an important aspect of electrode design. The electrode must be sufficiently stiff to allow for manipulation by a surgeon, and to maintain its anatomy-conforming shape. However, if an electrode is too stiff it can induce an inflammatory response or damage tissue (Naples et al., 1990). For example, a nerve cuff (Figure II.5.12.4a) which surrounds a peripheral or cranial nerve needs to maintain the conducting electrode contacts in close contact with a nerve to be most effective. However, the cuff needs to be loosely fitting enough to prevent occlusion of the nerve's vascular supply, which has been shown to cause axonal degeneration (Cuoco Jr. and Durand, 2000). Early designs of fixed diameter cuffs were often very stiff and had to be larger than the nerve to prevent damage due to swelling that occurred in the period immediately following implantation (Ducker and Hayes, 1968). This sometimes resulted in damage to the nerve before long-term encapsulation took place, and likely contributed to thick encapsulation tissue between the electrode and nerve. Thick encapsulation can result in increased electrical impedance or changes in the electrode–nerve orientation which can change the effectiveness of a stimulation therapy. Newer nerve cuff designs implement structural forms that allow the cuff to expand in response to nerve swelling. These include the Case Western Reserve University self-sizing nerve cuff electrode (Figure II.5.12.4b), and the Huntington helical nerve cuff electrode (Figure II.5.12.4a) (Naples et al., 1988; Agnew et al., 1989).

Mechanical impedance matching between electrode and tissue is particularly important when interfacing with small populations of neurons, as occurs with cortical interfacing. In these applications, even thin layers of tissue encapsulation can interfere with the ability to record or stimulate the cells of interest. As a result, materials research in this field has focused on the development of polymers which will not generate a substantial encapsulation (Seymour and Kipke, 2007). Other electrodes, including percutaneous spinal cord stimulation electrodes (Figure II.5.12.4d), achieve a mechanical impedance differential by mating with a stiff metal needle for implantation (North et al., 2002). The stylus is inserted into a lumen within the electrode, allowing for implantation and steering of an electrode into the proper anatomical location by an implanting physician. Once the electrode is in place, the stylus is removed and the mechanically compliant electrode remains.

NEUROSTIMULATION APPLICATIONS

The use of electrical stimulation for medical purposes is not a new idea. It dates back to the recommended use of the electrically active torpedo fish to treat pain from headache or gout by a roman physician in 46 AD. By the 16th century, additional indications for the torpedo fish included melancholy (depression) and epilepsy. Luigi Galvani was the first to demonstrate the use of electric current to elicit a movement, by connecting a bimetallic rod to a muscle in a frog's leg (McNeal, 1977). With the advent of electric generators in the 18th century, the use of electric shocks to treat a variety of ailments surfaced.

Today, electrical stimulation of the spinal cord, peripheral nerves, and cranial nerves is used to treat a variety of diseases and disorders. This section reviews applications for electrical stimulation in clinical practice, and the electrodes used for these applications. The following applications are reviewed: peripheral stimulation for neuromuscular applications, cranial nerve stimulation, spinal cord stimulation for pain, peripheral nerve stimulation and block for pain, and genitourinary stimulation.

Neuromuscular Stimulation

The use of electrical stimulation to elicit functional movement of skeletal muscle is often called Functional Electrical Stimulation (FES). This type of stimulation has primarily been used to restore motor function in individuals suffering from paralysis resulting from injury or disease (primarily spinal cord injury or stroke). Motor FES systems target the peripheral nervous system by electrically stimulating motor neurons that innervate specific muscles of interest (Grill and Kirsch, 2000). By stimulating these motor neurons, the muscles they innervate can be activated in a coordinated fashion to re-establish functional movements, such as grasping with a paralyzed hand (Kilgore et al., 2008; Peckham et al., 2001), standing with the aid of a walker (Fisher et al., 2008; Agarwal et al., 2003), stepping (Marsolais and Kobetic, 1987), and support of a paralyzed torso through activation of the trunk musculature (Triolo et al., 2009).

While surface and percutaneous FES systems have been developed, an implantable FES system will be the focus of this section. Like other implantable neural stimulators, this type of system is composed of an IPG, leads, and electrodes (with the exception of FES systems utilizing the BION®, which will be discussed below). Uniquely, these implants may have as many as 16 leads emerging from the IPG, since this type of system must sometimes activate many muscles in locations around the body (Smith et al., 1998). Figure II.5.12.5 shows a 16-channel FES system developed at Case Western Reserve University. Several types of electrodes have been developed for use in motor FES: epimysial electrodes, intramuscular electrodes, nerve cuff electrodes, and the BION® electrode/microstimulator. The physical form, material composition, and therapeutic indications of each of these electrodes will be considered.

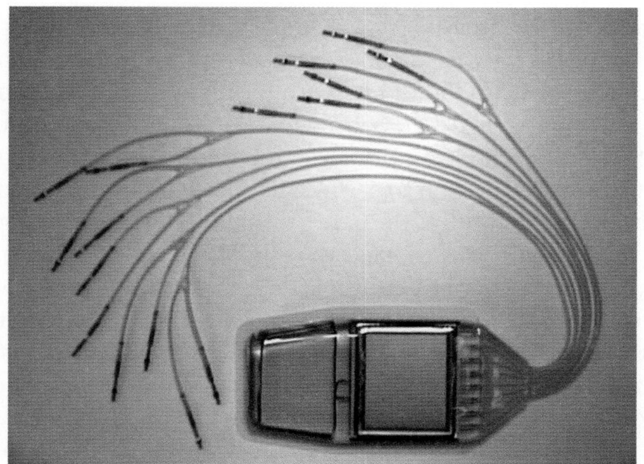

FIGURE II.5.12.5 16-channel implantable motor functional electrical stimulation system developed at Case Western Reserve University OH, USA.

Epimysial Electrodes

Epimysial electrodes are sutured to the surface of a muscle near the motor point, the location where a nerve inserts into the muscle. Epimysial electrodes were first considered in the 1980s and 1990s for use with an implantable FES system to restore hand grasp to individuals with spinal cord injury at Case Western Reserve University (Peckham, 2005; Keith et al., 1989). These electrodes were demonstrated to have a simple implantation procedure, and showed good mechanical integrity during muscle contraction and movement (Smith et al., 1988). An 8-channel hand grasp system marketed under the name Freehand was briefly available and implanted in more than 250 people worldwide (Peckham, 2005). This system is no longer commercially available today; however, advanced systems built upon that technology continue to use epimysial electrodes for stimulation of muscles in the upper and lower extremities.

These electrodes are constructed using molded silicone elastomer (a Dacron®-reinforced Silastic® sheeting molded in Silastic MDX4-4210 elastomer, Dow Corning, Midland, MI, USA) with a 10 mm^2 diameter stimulating disc made from 90% Platinum (Pt) and 10% Iridium (Ir) (Uhlir et al., 2004). A photograph of an epimysial electrode can be seen in Figure II.5.12.4c. The electrode leads are made in a variety of lengths, to accommodate muscles located at different distances from an implanted stimulator.

Intramuscular Electrodes

Intramuscular-type electrodes incorporate barbed structures, and are inserted into the belly of a muscle near the motor point. The small profile and suture-free implantation facilitates a less invasive percutaneous surgical approach. This type of electrode has been implemented for several applications, including the restoration of respiratory and motor function to the paralyzed (DiMarco et al., 2005; Peterson et al., 1989).

Early intramuscular electrodes were developed for restoring respiratory function in spinal cord injury. These types of systems can reduce the amount of time spent on a ventilator for someone with diaphragm paralysis. The Peterson pacing electrode is designed for insertion into the diaphragm muscle, and is composed of two helically wound, Teflon®-insulated, multi-stranded 316 stainless steel wires. A photograph is shown in Figure II.5.12.4d. Each wire is partially deinsulated to form a stimulating tip. Polymer barbs at the distal end of the electrode surround the circumference of the tip, and are designed to anchor the electrode in place, reducing electrode migration that could result from motion within the muscle (Peterson, 1989). An elastic polymer core is threaded through the center of the helix, making the electrode more stable for insertion. This can be removed after implantation to promote tissue ingrowth into the lead's open helix, which improves the mechanical stability of the lead (Peterson, 1989).

A similar intramuscular electrode was created for use with the implantable motor FES systems. For these applications, intramuscular electrodes enable implanting surgeons to reach small or deep muscles that are challenging to reach with epimysial electrodes (Memberg et al., 1989). The primary difference between this electrode and the Peterson electrode is that Silastic® tubing is placed around the outside of the coil, rather than using a polymer core. A photograph is shown in Figure II.5.12.4e. The tubing prevents fibrotic ingrowth that may increase stiffness. Maintaining flexibility in an electrode lead is important in applications such as a hand grasp restoration, for which leads must cross the elbow joint. The electrode tip is created by wrapping 316 stainless steel, which is continuous with the lead, around the outside of the Silastic® tubing. The lead contains two redundant conductors which are insulated with Teflon® and wound together in a double helix. A polypropylene anchor is attached to the end of a tip to secure the electrode in the muscle (Memberg et al., 1989).

Nerve Cuff Electrodes

A nerve can also be stimulated directly via a cuff electrode that wraps circumferentially around the nerve. This method has been implemented with implantable clinical devices since the late 1960s, when it was used in a phrenic nerve pacing system for the restoration of respiratory function in spinal cord injury (Glenn et al., 1970). The phrenic nerve innervates the diaphragm, and like diaphragm pacing, cyclic phrenic nerve stimulation generates paced contractions for breathing.

Many nerve cuff electrodes have been designed and implemented for various neuromuscular applications, including phrenic nerve pacing for respiration (Glenn et al., 1970), peroneal nerve stimulation for the treatment of post-stroke hemiplegia (paralysis on one side of the body) (Waters et al., 1975), and restoration of motor function to individuals with spinal cord injury (Fisher et al., 2008; Polasek et al., 2007). Nerve cuff electrodes have several advantages over epimysial or intramuscular electrodes, including the ability to simultaneously activate multiple muscles using a single channel of stimulation, a reduced number of required surgical incisions, and lower power requirements to achieve muscle contraction (Naples et al., 1988; Veraart et al., 1993). However, these electrodes have not yet replaced epimysial or intramuscular electrodes, because it is difficult to selectively activate the muscles of interest in a nerve which innervates multiple muscles. New cuff electrodes, not yet implanted chronically in humans, are designed to reshape the nerve, and may help improve selectivity (Schiefer et al., 2008; Tyler and Durand, 2002).

As discussed in the section on mechanical biocompatibility above, nerve cuff electrodes interface directly with a nerve, and therefore careful mechanical design is imperative. Designs that allow for post-surgical swelling of a nerve without occlusion of vasculature have been the most successful (Naples et al., 1988; Agnew et al., 1989). These electrodes tend to be made of molded Silastic®, with one or more platinum or platinum-iridium electrode contacts exposed within the cuff. Pictures of the self-sizing nerve cuff electrode designed and built at Case Western Reserve University and the Huntington spiral nerve cuff electrode are shown in Figure II.5.12.4.

BION® Microstimulator

The BION® microstimulator is a miniature, self-contained, implantable neurostimulator. First designed at the University of Southern California and Illinois Institute of Technology as a platform technology, the BION® has been used in a variety of clinical research studies. One of several versions of the BION® is displayed in Figure II.5.12.4f. Motor applications for the microstimulator include stimulation of the peroneal nerve to correct footdrop in incomplete SCI, and stimulation of the upper extremity muscles to improve grasp in stroke (Wilson et al., 1988; Schulman et al., 2004). The design incorporates both the IPG and the electrode into a small capsule measuring 28 mm in length and 3.2 mm in diameter that can be injected into a muscle using a large needle (Carbunaru et al., 2004). The first generation BION® utilized an external RF power and communication coil placed on the skin over the microstimulator. A BION® that incorporates an integrated battery is currently manufactured by Bioness, Inc. (Valencia, CA, USA). The miniature device contains a stimulating microchip, in addition to the stimulating electrodes. An external controller allows the user to control the device during typical use.

Spinal Cord Stimulation

Stimulation of the dorsal column fibers in the spinal cord to treat pain was first introduced following the publication of a seminal "gate control" theory of pain by Melzack and Wall (Melzack, 1965). With this new theory in mind, Shealy et al. implanted the first dorsal column stimulator in an individual with terminal metastatic cancer at Western Reserve University, USA (Shealy et al., 1967). The initiating principle for the therapy was that selective stimulation of large afferent fibers in the dorsal column would "shut the gate" to small afferent fibers signaling pain. It is now clear that the mechanism of action for spinal cord stimulation, SCS, is likely more complex than the gate theory predicts, and is still not fully-understood (Falowski et al., 2008). Current research suggests that SCS affects processes at several levels of the nervous and vascular system that may play a role in pain relief. The reader is referred to the following publication for a further review of the mechanisms thought to be involved in SCS (Oakley and Prager, 2002). The most successful indications for SCS include intractable angina

(90% effective), complex regional pain syndromes I and II (84% effective), ischemic limb pain (77% effective), and low back and leg pain (62% effective) (Cameron, 2004; DeJongste, 2000).

Spinal cord stimulation therapies are currently available from three companies: Medtronic, Inc.; Boston Scientific, Inc.; and Advanced Neuromodulation Systems (ANS) – a St. Jude Medical subsidiary. Like other neurostimulation systems, each system consists of an IPG, one or more leads, and an electrode or electrode array. Two types of electrodes are used for spinal cord stimulation: percutaneous-type electrodes and paddle-type electrodes. Each of these electrodes will be discussed in further detail below.

Percutaneous Electrodes

Percutaneous electrodes were initially developed to screen potential candidates for an implantable spinal cord stimulator with paddle electrodes (North et al., 2002). The procedure was so successful that these electrodes were then adopted for long-term use. These electrodes are placed using a fluoroscopically-guided needle procedure, similar to that performed for epidural drug injection. These electrodes tend to have either four or eight cylindrical contacts, as shown in Figure II.5.12.4e. The contacts are circumferential, resulting in a stimulation current that is distributed around the entire circumference of the electrode. These circumferential contacts are required, since the orientation of the electrode cannot be easily determined during the implantation procedure, but cause current flow to non-neural tissue and a resulting higher power cost than a paddle electrode (Falowski et al., 2008). Two, or in rare cases, three electrodes can be placed in parallel to one another in order to use current steering methods to shape the electric field. Figure II.5.12.6 shows a representation of percutaneous-type electrodes using a heterogeneous field to stimulate

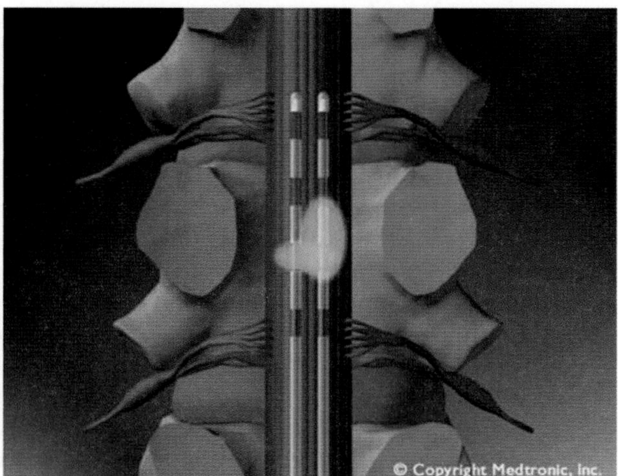

FIGURE II.5.12.6 Representation of two percutaneous-type electrodes exhibiting a heterogeneous field to stimulate a particular volume of the spinal cord. (Photograph courtesy of Medtronic, Inc.)

a particular volume of the spinal cord. Medtronic has three different versions of the Pisces™ Quad and Octad electrode available (four/eight contacts). These differ with respect to the spacing between electrode contacts. Each of these contacts is platinum-iridium, with polyurethane insulation between contacts. The lead is insulated in silicone rubber. Boston Scientific offers a similar lead with eight contacts, the Precision Plus™ lead. These contacts are closely spaced (1 mm separation distance).

While percutaneous electrodes have enjoyed much clinical success, their primary mode of failure is electrode migration after implantation (Rosenow, 2006). An anchor is used to tether the electrode to the surrounding ligamentous tissue to help prevent this migration. Despite electrode migration and the superior average control of stimulation with a paddle electrode, percutaneous electrodes continue to be the more popular therapeutic option, due to the relatively simple and noninvasive implantation procedure.

Paddle-Type Electrodes

Paddle electrodes were the first type of electrode used for SCS. Paddle electrodes are multi-contact arrays of platinum or platinum-iridium with a silicone rubber base. Many paddle-type electrodes have been implemented, but they generally consist of one or more columns of contacts that are oriented along the rostral–caudal dimension of the epidural space of the spinal column. The molded silicone of the electrode guarantees a fixed orientation of the multiple electrode contacts in a paddle electrode. This is ideal for utilizing so-called "current steering" techniques, where currents of varied amplitude are applied to multiple contacts to shape an electric field that will maximize therapeutic benefit. Since each of the electrodes is fixed in the same structure, they do not migrate with respect to one another, an advantage of a paddle-type electrode over percutaneous-type electrodes (which are generally used in pairs). This reportedly contributes to the better therapeutic results seen with this type of electrode (North et al., 2002). Additionally, lower current amplitudes, relative to those used with a percutaneous electrode, can sometimes be used to achieve a given therapeutic effect. This is possible because the conducting contacts of a paddle electrode are oriented towards the spinal cord, which acts to direct the current to the tissue of interest (Falowski et al., 2008). The contacts in a percutaneous electrode are omnidirectional, and thus tend to require higher stimulation currents. In addition, in a controlled randomized study of paddle verses cylindrical percutaneous electrodes for SCS, subjects with paddle electrodes reported significantly better results in terms of pain coverage by parasthesia (North et al., 2002).

Medtronic currently offers the Specify™ lead with 8 electrodes in two parallel line configurations or the Specify™ 5-6-5 lead with 16 electrodes spaced into three columns. Boston Scientific has a similar paddle lead available known as the Artisan™, with 16 contacts aligned in two parallel columns of 8 contacts each. The first transverse tri-polar paddle electrode was made available from ANS, and consists of three columns of electrode contacts that run rostral–caudally. Tri-polar stimulation is thought to better confine stimulation to the dorsal column fibers, without activating spinal root fibers that can cause discomfort (Holsheimer and Struijk, 1997).

The primary disadvantage of a paddle electrode, and the reason that percutaneous electrodes are more frequently used, is the required implantation procedure. A paddle electrode requires a laminectomy (removal of portions of vertebrae) to be placed in the epidural space. This procedure can be performed under local anesthetic, but requires a skilled surgeon. Alternatively, percutaneous electrodes are frequently implanted by a trained anesthesiologist/pain specialist, and do not require a laminectomy.

Peripheral Nerve Stimulation and Block for Pain

Peripheral nerve stimulation (PNS) for the relief of pain has been in use since the 1960s, when Sweet and Wall first stimulated their own infraorbital nerves with needle electrodes to show decreased pain sensation (Wall and Sweet, 1967). There are two broad types of strategies for treating pain by delivery of electrical currents to a peripheral nerve: the first is to *induce* action potential activity in sensory fibers that *do not* transmit pain; and the second is to *block* action potential activity in sensory fibers that *do* transmit pain.

Activating nerve stimulation (commonly referred to as peripheral nerve stimulation, PNS) therapies involve the delivery of current pulses to an electrode near, around or within a peripheral nerve to activate large sensory afferents that do not produce a painful sensation when stimulated. These strategies produce a tingling parasthesia sensation that is associated with decreased pain in the stimulated area. The mechanisms of action are a current topic of research, but are thought to involve suppression of pain-related neural activity at the level of the spinal cord (Wall and Sweet, 1967), and/or excitation of central pain processing pathways (Bartsch and Goadsby, 2002; Le Doare et al., 2006). Activating PNS has been employed to treat numerous neuropathic pain conditions (those arising due to injury or disease of the nervous system), such as postherapeutic neuralgia (Dunteman, 2002), occipital neuralgia (Weiner and Reed, 1999), transformed migraine (Charles and Popeney, 2003), cluster headache (Burns et al., 2007), inguinal neuralgia (Stinson, 2001), fibromyalgia (Thimineur, 2007), post-surgical pain (Waisbrod, 1985), complex regional pain syndromes (Hassenbusch, 1996), hemicrania continua (Bartsch et al., 2009), and coccygodynia (Kothari, 2007). PNS has not seen the same widespread clinical deployment as spinal cord stimulation (which is an increasingly common therapy for intractable pain conditions), but there has been a recent surge in both the clinical prescription and research of PNS, as it continues to show therapeutic efficacy (Slavin, 2008).

Three types of electrodes are used for PNS: percutaneous cylindrical electrodes; paddle-type electrodes; and nerve cuff electrodes (Waisbrod, 1985) (which are less commonly used). The cylindrical and paddle-type electrodes are the very same type of electrodes which are used for spinal cord stimulation. Percutaneous electrodes are the most commonly used, due to their suitability for minimally invasive implantation with a fluoroscopically guided needle procedure. Paddle-type electrodes provide advantages over percutaneous electrodes, because they generally result in stable stimulation thresholds (Jones, 2003) and minimal electrode migration. Despite these functional advantages, this type of electrode requires a more invasive surgical procedure for implantation, and is therefore less commonly used.

Percutaneous and paddle-type electrodes are generally placed transverse to a nerve of interest to maximize the likelihood of having at least one contact in proximity to the nerve of interest (Popeney and Alo, 2003). Peripheral nerve cuff electrodes (see the previous "Neuromuscular Stimulation" section for more detail on nerve cuff electrodes) can provide significant selectivity for the nerve of interest (Tarler and Mortimer, 2004), and generally require lower currents to activate nerves than cylindrical or paddle-type electrodes (Naples et al., 1990). However, nerve cuff electrodes are not commonly used for PNS, due to the complexity of surgical implantation and because of a lack of commercially available electrodes which can be used with available IPGs.

The second strategy for treating pain (or other disorders associated with pathological or undesired peripheral nerve activity) by delivery of electrical currents to peripheral nerves is to block peripherally arising pain signals from reaching the central nervous system with an electrical nerve block. Electrical nerve block techniques are not yet used in the clinical environment, but are an area of current research with increasing interest. These techniques generally operate by electrically inactivating the sodium channels responsible for action potential conduction. Methods for achieving this include using direct (Bhadra, 2004) or alternating high frequency currents (Kilgore, 2004), and a technique referred to as "collision block," which involves proximally generated unidirectional action potentials to annihilate those traveling from the periphery (Van Den Honert and Mortimer 1981).

These techniques employ significantly different electrode designs than other PNS therapies. While other electrode designs have been used successfully (Joseph et al., 2007; Ackermann et al., 2010), most peripheral nerve block studies utilize nerve cuff electrodes. Electrodes for nerve block have a fundamentally different requirement than electrodes for activation: they must not only block peripherally generated action potentials, but they must not generate any action potentials that will travel proximally to the central nervous system and relay a pain signal. This is generally accomplished by careful design of the blocking current waveform (Van Den Honert and Mortimer, 1981), and using a nerve cuff design that will contain the blocking currents within the nerve cuff, limiting "leakage currents" that can cause activation of fibers (Van Den Honert and Mortimer, 1981). The configuration of the conductive contacts within the nerve cuff influences the field seen by the nerve fibers, and can therefore be optimized to block the nerve with lower amplitude currents. Lower amplitudes are generally associated with less power consumption and a safer therapy, and are therefore preferred (Merrill et al., 2005).

Cranial Nerve Stimulation

Cranial nerves differ from other nerves in the body, in that they emerge directly from the brainstem, rather than from the spinal cord. These nerves stem from or innervate cells in nuclei of the brainstem, hence electrical stimulation of these nerves may have a direct impact on neural pathways in the central nervous system. Three specific cranial nerves have been identified as therapeutic targets for electrical stimulation: the cochlear nerve to restore hearing in individuals with deafness; the vagus nerve to decrease epileptic seizures and to treat depression; and the trigeminal nerve, to decrease epileptic seizures and to treat facial pain. Cochlear implants are discussed in Chapter II.5.11. Applications related to vagus nerve stimulation and trigeminal nerve stimulation are considered here.

Vagus Nerve Stimulation (VNS). Vagus nerve stimulation for the treatment of epilepsy was first proposed by Dr. Jacob Zabara in 1985 (Terry et al., 1991). The mechanism is not well-understood; however, studies suggest that stimulation of the afferent neurons in the vagus nerve alter brain function to help suppress seizures (Groves and Brown, 2005). PET scans during stimulation have shown an increase in blood flow in certain areas of the brainstem, hypothalamus, and insular cortices, and a subsequent decrease in blood flow in other deep brain structures (Mapstone, 2008). The blood flow changes to these deeper areas of the brain persist in chronic VNS, and thus may be related to the underlying mechanism of the therapy.

Approved by the FDA (US Food and Drug Administration) in 1997 after two large clinical trials, VNS therapy is used to treat pharmacologically intractable epilepsy in individuals over the age of 12. For approximately 40% of the population with epilepsy, drug therapies are unsuccessful at diminishing seizure frequency (Mapstone, 2008). This VNS therapy is targeted to this population. An implantable VNS device is available from Cyberonics, Inc. The IPG is implanted in the chest, and a lead is tunneled under the skin to the electrode that is implanted on the vagus nerve. Electrical stimulation is applied to the nerve using a Huntington helical cuff electrode. Typically, 30 seconds of stimulation at 30 Hz is applied at five minute intervals (Guberman, 2004). Studies indicate that after chronic stimulation of the recommended duty cycle, 35–45% of the patients see a decrease frequency of seizures exceeding 50%; however, only approximately 2% become seizure free (Mapstone, 2008).

VNS therapy using the Cyberonics device has also been FDA-approved for treatment-resistant depression (Critchley et al., 2007). Again, the mechanism is unclear, but the structures that exhibit blood flow changes in VNS such as the thalami are also involved in mood regulation. VNS has been used in research studies in an attempt to treat a variety of indications for which it does not yet have FDA approval. These include treatment for anxiety, cognitive deficits, and migraines (Groves and Brown, 2005).

The electrode currently used by Cyberonics for VNS stimulation is a modification of the Huntington helix discussed above (see Figure II.5.12.4a) (Agnew et al., 1989). The Huntington electrode, like the self-sizing cuff, is an open helix made of silicone rubber elastomer that can change in size in response to nerve swelling. The conducting surfaces of the electrodes are platinum-iridium ribbons that are helically wound around the nerve. The initial design used seven turns with two contacts. The current version uses two contacts with only a few turns of a helix per contact, and has a gap between the two electrode contacts (Figure II.5.12.4c). Fewer turns per helix makes it more likely for the electrode to slip off the nerve during movement, but minimizes mechanical injury to the nerve by decreasing the length of the individual helix. A third helix located proximal to the electrode contacts tethers the electrode to the nerve, and provides additional mechanical stability. The electrode is anchored in place with sutures. Two helix diameters are available: 2.0 mm and 3.0 mm.

Trigeminal Nerve Stimulation (TNS). Trigeminal nerve stimulation, TNS, has been used as a therapy to treat both intractable neuropathic pain conditions and epilepsy. Epilepsy studies have not utilized implanted electrodes, and therefore will not be considered in further detail. The technique was first used to treat pain in the 1960s (Wall and Sweet, 1967). In 1965, Melzack and Wall published the "gate control" theory of pain, which suggested that stimulation of large diameter cutaneous afferent fibers may reduce pain (Melzack, 1965). Wall adopted this idea, and in 1967 published the first study that used implanted electrodes in the infraorbital region to stimulate a cutaneous branch of the trigeminal nerve to treat trigeminal neuralgia (Wall and Sweet, 1967). Stimulation induced parethesia in particular regions of the face, eliminating pain during stimulation, and in some cases for a period after stimulation had ended as well.

Four or eight contact percutaneous-type electrodes (the same type used for spinal cord stimulation) are generally used for the TNS pain therapy. Like in spinal cord stimulation, the electrodes are first placed for a trial period to test the effectiveness of the therapy in treating the individual's pain. The electrodes are passed under the skin but above the muscles, and pass perpendicular to the path of the nerve (Slavin et al., 2006). These electrodes are then connected to an external stimulator device. If there is at least a 50% reduction in pain, a permanent IPG system is implanted. Commonly used electrodes include the Medtronic Quad (4-channel) or the Octad (8-channel) (Medtronic, Minneapolis, MN, USA).

Genitourinary Stimulation

Electrical stimulation was first considered as a method for treating urinary retention in 1878, when Saxtroph attempted to stimulate the muscles in the bladder wall using an electrode mounted on a catheter inserted into the bladder (reviewed in Madersbacher (1990)). It was not until the middle of the 20th century that electrical stimulation for the treatment of urine voiding dysfunction became the first area of significant research in the neuromodulation field.

The lower urinary tract has two major functions: to store urine in the bladder (continence) and to void urine (micturition). The muscular layer in the bladder is innervated by parasympathetic neurons stemming from the sacral spinal segments S2–S4. Sufficient bladder emptying requires not only contraction of the bladder, but also relaxation of the urethral sphincter. Various injuries, disorders, and diseases can interrupt these pathways and cause voiding dysfunction. Research studies have explored numerous targets for electrical stimulation to improve bladder management and eliminate reliance on indwelling catheters. This section will review a variety of electrodes that have been used in the past and present in attempt to control bladder function.

Disc Electrodes

The component of the micturition circuit initially targeted for electrical stimulation with an implanted, RF powered device was the bladder wall muscle. Susset and Boctör reported success with an implant that incorporated 8 disc electrodes around the dome of the bladder of an individual with lower motorneuron (LMN) injury (Susset and Boctor, 1967). However, in individuals with a neurogenic bladder (spastic with contractions triggered at low filling volumes) from spinal cord injury, this treatment was not successful, because bladder stimulation also activated the urethral sphincter, preventing voiding in this population (Gaunt and Prochazka, 2009).

Book Electrodes

The first commercially available stimulator for bladder dysfunction, the Finetech-Brindley Bladder System, uses so-called book electrodes implanted intradurally on the S2–S4 sacral roots of the spinal cord. These sacral roots are stimulated in a cyclical fashion to generate voiding from the bladder (Brindley, 1977). It's slotted "book" shape is designed to confine current to target spinal root without activating neighboring roots (Brindley et al., 1982). The Finetech system uses two book electrodes: an upper book and a lower book. The lower book has a single slot with three electrode contacts for tripolar root stimulation. The S4, and in some cases, S5 sacral roots were placed in the slot of the lower book. The upper book contains three slots, each with three contacts for tripolar nerve stimulation. The S3 roots are placed in the center slot. Bilateral S2 roots were placed in the lateral slots. These lateral slots are electrically connected to apply identical stimulation parameters to bilateral S2 roots. The electrode contacts are made of platimum foil and the books are formed from silicone rubber. The lead cable conductor is a platinum-iridium alloy insulated with polyimide and then coated with silicone (Brindley et al., 1986).

An issue with the initial intradural electrodes was CSF leakage out of the subdural space (Brindley et al., 1986). This complication was prevented in subsequent implants with the use of a grommet composed of fine woven polyester on the outside and silicone rubber on the inside that was then glued around the cables (Brindley et al., 1986). Alternate extradural electrodes have also been used successfully with the Brindley system, and are currently used with the version of the device marketed in the US.

Cuff Electrodes

Cuff-type electrodes have also been implemented for the Finetech-Brindley bladder voiding system. These electrodes can be implemented using a less invasive extradural electrode placement which was required for FDA approval of the device in the US. The first Finetech device with extradural electrodes was implanted in 1986 (Sauerwein et al., 1990). For this type of implant, the leads are connected to the implanted receiver and connect to three tripolar helical electrodes. The electrodes are wrapped around the roots using a strip of Dacron®-reinforced silicone rubber that is sutured closed. The leads are encapsulated in silicone rubber and the electrode contacts are platinum-iridium, like the lead cables for the book electrodes (Egon et al., 1998).

This procedure, unfortunately, requires a dorsal rhizotomy (cutting of the afferent fibers returning to the spinal cord) to limit reflex contractions of the bladder. While the rhizotomy reduces reflex bladder contractions, it also has several negative side-effects that can include the loss of reflex erection and defecation. These side-effects have limited the widespread deployment of this system (Gaunt and Prochazka, 2009), although it has been implemented in more than 2500 people worldwide, and is still commercially available in Europe.

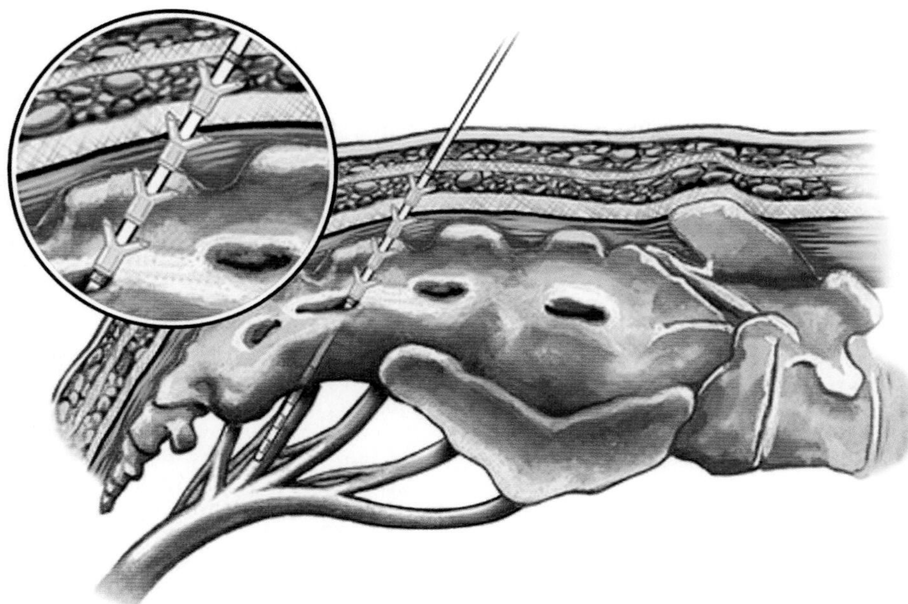

FIGURE II.5.12.7 Illustration of implanted tined lead for Medtronic Interstim® sacral nerve stimulation therapy. Illustration shows sagittal view of sacrum with intraforaminal electrode placement. (Courtesy of Medtronic, Inc.)

Percutaneous Electrodes

In the 1980s, Tanagho et al. first showed that continence could be controlled by low-frequency, low-amplitude stimulation of the sacral roots to maintain sphincter contraction without initiating bladder contractions (Tanagho and Schmidt, 1988). Sacral nerve stimulation can successfully treat non-neurogenic bladder dysfunction including incontinence, urgency-frequency, and urinary retention. Women with Fowler's syndrome, characterized by relaxation of the urethral sphincter and frequent urine leakage, seem to respond especially well to this type of stimulation. Continuous stimulation or a cycling mode of electrical pulses is used to inhibit neural reflexes that may trigger bladder contraction (Kessler and Fowler, 2008). Results from 10 studies, reviewed in Kessler and Fowler, suggest success rates between 55 and 100%.

The sacral nerve can be reached through a relatively non-invasive procedure through the sacral foramina. Medtronic has commercialized the Interstim® therapy, which utilizes a percutaneous electrode technology similar to those developed for spinal cord stimulation. For this therapy, an IPG is typically placed in a subcutaneous pocket in either the lower abdomen or posterior pelvis. The IPG is connected to quadripolar in-line leads that are implanted next to the sacral nerve on one side of the body. Bilateral implementation showed some evidence of improved outcomes (Hohenfellner et al., 1998), but unilateral systems remain the trend to minimize the complexity of the procedure (Kessler and Fowler, 2008). As shown in Figure II.5.12.7, integrated tines fixed to the lead proximal to the cylindrical stimulating electrodes help to prevent migration. The electrode contacts are made of a platinum-iridium alloy, and the tines are made of polyurethane. The insulation for the conductor wires is fluoropolymer.

FUTURE DIRECTIONS

Stimulating bioelectrodes have evolved into highly functional, safe, and reliable implantable medical devices. The use and development of new biocompatible materials holds promise to advance neural interfacing in the spinal cord and peripheral nervous system. Specifically, developments in mechanical impedance matching, nerve tissue selectivity, and improvements in charge density capacity of electrodes are likely to result in improved neural interfaces in the near future. For example, novel polymer materials are being developed with dynamic material stiffness that are stiff when implanted, and become compliant due to changes in temperature or the presence of water once implanted (Capadona et al., 2008). These new materials could allow for ease of electrode implantation and manipulation, while maintaining a minimal foreign-body response. Progress is also being made to improve spatial stimulation selectivity within a nerve. New electrode technologies, such as the Flat Interface Nerve Cuff Electrode (FINE) (Tyler and Durand, 2002) and the University of Utah Array (McDonnall et al., 2004), may be able to selectively recruit individual fascicles within a nerve, and improve motor control in a variety of applications and reduce surgical complexity. Novel materials that allow for increased safe charge injection capacity of small electrodes will also contribute to smaller, and therefore more selective, nerve interfaces.

BIBLIOGRAPHY

Ackermann, D., Foldes, E. L., Bhadra, N., & Kilgore, K. L. (2010). Conduction block of peripheral nerve using high frequency alternating currents delivered through an intrafascicular electrode. *Muscle and Nerve*, **41**(1), 117–119.

Agarwal, S., Triolo, R. J., Kobetic, R., Miller, M., Bieri, C., Kukke, S., Rohde, L., & Davis, J. R., Jr. (2003). Long-term user perceptions of an implanted neuroprosthesis for exercise, standing, and transfers after spinal cord injury. *J. Rehabil. Res. Dev.*, **40**(3), 241–252.

Agnew, W. F., McCreery, D. B., Yuan, T. G., & Bullara, L. A. (1989). Histologic and physiologic evaluation of electrically stimulated peripheral nerve: Considerations for the selection of parameters. *Ann. Biomed. Eng.*, **17**(1), 39–60.

Bartsch, T., & Goadsby, P. J. (2002). Stimulation of the greater occipital nerve induces increased central excitability of dural afferent input. *Brain*, **125**(Pt 7), 1496–1509.

Bartsch, T., Paemeleire, K., & Goadsby, P. J. (2009). Neurostimulation approaches to primary headache disorders. *Curr. Opin. Neurol.*, **22**(3), 262–268.

Beebe, X., & Rose, T. L. (1988). Charge injection limits of activated iridium oxide electrodes with 0.2 ms pulses in bicarbonate buffered saline (neurological stimulation application). *IEEE Trans. Biomed. Eng.*, **35**(6), 494–495.

Bhadra, N., (2004). Direct current electrical conduction block of peripheral nerve. *IEEE Trans. Neural Syst. Rehabil. Eng.*, **12**(3), 313.

Binnie C. D., (2000). Vagus nerve stimulation for epilepsy: A review. *Seizure*, **9**(3), 161.

Brazzelli, M., (2006). Efficacy and safety of sacral nerve stimulation for urinary urge incontinence: A systematic review. *J. Urol.*, **175**(3), 835.

Brindley, G. S. (1977). An implant to empty the bladder or close the urethra. *J. Neurol. Neurosurg. Psychiatry*, **40**(4), 358–369.

Brindley, G. S., Polkey, C. E., & Rushton, D. N. (1982). Sacral anterior root stimulators for bladder control in paraplegia. *Paraplegia*, **20**(6), 365–381.

Brindley, G. S., Polkey, C. E., Rushton, D. N., & Cardozo, L. (1986). Sacral anterior root stimulators for bladder control in paraplegia: the first 50 cases. *J. Neurol. Neurosurg. Psychiatry*, **49**(10), 1104–1114.

Buettner, G. R. (1993). The pecking order of free radicals and antioxidants: Lipid peroxidation, alpha-tocopherol, and ascorbate. *Arch. Biochem. Biophys.*, **300**(2), 535–543.

Burns, B., Watkins, L., & Goadsby, P. J. (2007). Treatment of medically intractable cluster headache by occipital nerve stimulation: Long-term follow-up of eight patients. *The Lancet*, **369**(9567), 1099–1106.

Cameron, T. (2004). Safety and efficacy of spinal cord stimulation for the treatment of chronic pain: A 20-year literature review. *J. Neurosurg.*, **100**(3 Suppl Spine), 254–267.

Capadona, J. R., Shanmuganathan, K., Tyler, D. J., Rowan, S. J., & Weder, C. (2008). Stimuli-responsive polymer nanocomposites inspired by the sea cucumber dermis. *Science*, **319**(5868), 1370–1374.

Carbunaru, R., Whitehurst, T., Jaax, K., Koff, J., & Makous, J. (2004). Rechargeable battery-powered BION microstimulators for neuromudulation. *Conf. Proc. IEEE Eng. Med. Biol. Soc.*, **6**, 4193–4196.

Chan, P. H., Yurko, M., & Fishman, R. A. (1982). Phospholipid degradation and cellular edema induced by free radicals in brain cortical slices. *J. Neurochem.*, **38**(2), 525–531.

Charles, A., & Popeney, K. M. A. (2003). Peripheral neurostimulation for the treatment of chronic, disabling transformed migraine. *Headache: The Journal of Head and Face Pain*, **43**(4), 369–375.

Chow, A., Chow, V. Y., Packo, K. H., Pollack, J. S., Peyman, G. A., & Schuchard, R. (2004). The artificial silicon retina microchip for the treatment of vision loss from retinitis pigmentosa. *Arch. Opthalmol.*, **122**(4), 460–469.

Clark, G. (2003). *Cochlear Implants, Fundamentals and Applications*. New York, NY: Springer-Verlag.

Cogan, S. F. (2008). Neural stimulation and recording electrodes. *Annu. Rev. Biomed. Eng.*, **10**, 275–309.

Critchley, H. D., Lewis, P. A., Orth, M., Josephs, O., Deichmann, R., Trimble, M. R., & Dolan, R. J. (2007). Vagus nerve stimulation for treatment-resistant depression: Behavioral and neural effects on encoding negative material. *Psychosom. Med.*, **69**(1), 17–22.

Cuoco, F. A., Jr., & Durand, D. M. (2000). Measurement of external pressures generated by nerve cuff electrodes. *IEEE Trans. Rehabil. Eng.*, **8**(1), 35–41.

DeJongste, M. J. (2000). Spinal cord stimulation for ischemic heart disease. *Neurol. Res.*, **22**(3), 293–298.

DiMarco, A. F., Onders, R. P., Ignagni, A., Kowalski, K. E., & Mortimer, J. T. (2005). Phrenic nerve pacing via intramuscular diaphragm electrodes in tetraplegic subjects. *Chest*, **127**(2), 671–678.

Ducker, T. B., & Hayes, G. J. (1968). Experimental improvements in the use of Silastic cuff for peripheral nerve repair. *J. Neurosurg.*, **28**(6), 582–587.

Dunteman, E. A., (2002). Peripheral nerve stimulation for unremitting ophthalmic postherpetic neuralgia. *Neuromodulation*, **5**(1), 32.

Egon, G., Barat, M., Colombel, P., Visentin, C., Isambert, J. L., & Guerin, J. (1998). Implantation of anterior sacral root stimulators combined with posterior sacral rhizotomy in spinal injury patients. *World J. Urol.*, **16**(5), 342–349.

Ellenbogen, K., Kay, G., & Wilkoff, B. (2000). *Clinical Cardiac Pacing and Defibrillation*, 2nd ed. Philadelphia, PA: W.B. Saunders Company.

Falowski, S., Celii, A., & Sharan, A. (2008). Spinal cord stimulation: An update. *Neurotherapeutics*, **5**(1), 86–99.

Fisher, L. E., Miller, M. E., Bailey, S. N., Davis, J. A., Jr., Anderson, J. S., Rhode, L., Tyler, D. J., & Triolo, R. J. (2008). Standing after spinal cord injury with four-contact nerve-cuff electrodes for quadriceps stimulation. *IEEE Trans. Neural Syst. Rehabil. Eng.*, **16**(5), 473–478.

Gaunt, R., & Prochazka, A. (2009). Transcutaneously coupled, high-frequency electrical stimulation of the pudendal nerve blocks external urethral sphincter contractions. *Neurorehab. Neural Repair*, **23**, 615.

Glenn, W. W., Holcomb, W. G., Gee, J. B., & Rath, R. (1970). Central hypoventilation; long-term ventilatory assistance by radiofrequency electrophrenic respiration. *Ann. Surg.*, **172**(4), 755–773.

Grahame, D. C., (1947). The electrical double layer and the theory of electrocapillarity. *Chem. Rev.*, **41**(3), 441.

Grill, W. M., & Kirsch, R. F. (2000). Neuroprosthetic applications of electrical stimulation. *Assist. Technol.*, **12**(1), 6–20.

Groves, D. A., & Brown, V. J. (2005). Vagal nerve stimulation: A review of its applications and potential mechanisms that mediate its clinical effects. *Neurosci. Biobehav. Rev.*, **29**(3), 493–500.

Guberman, A. (2004). Vagus nerve stimulation in the treatment of epilepsy. *CMAJ*, **171**(10), 1165–1166.

Hassenbusch, S. J. (1996). Long-term results of peripheral nerve stimulation for reflex sympathetic dystrophy. *J. Neurosurg.*, **84**(3), 415.

Hodgkin, A. L., & Huxley, A. F. (1952). A quantitative description of membrane current and its application to conduction and excitation in nerve. *J. Physiol.*, **117**(4), 500.

Hohenfellner, M., Schultz-Lampel, D., Dahms, S., Matzel, K., & Thüroff, J. W. (1998). Bilateral chronic sacral neuromodulation for treatment of lower urinary tract dysfunction. *J. Urol.*, **160**(3 Pt 1), 821–824.

Holsheimer, J., & Struijk, J. (1997). *US Patent 5,643,330*. Multichannel apparatus for epidural spinal cord stimulation, USPTO, Editor.

Humayun, M., et al. (2003). Chronically implanted intraocular retinal prosthesis in two blind subjects. *Invest. Ophth. Vis. Sci.*, **44**, 4206.

Jarrett, M. E., (2004). Systematic review of sacral nerve stimulation for faecal incontinence and constipation. *Brit. J. Surg.*, **91**(12), 1559.

Jones, R. L. (2003). Occipital nerve stimulation using a medtronic resume II electrode array. *Pain Physician*, **6**(4), 507.

Joseph, L., Haeffele, B. D., & Butera, R. J. (2007). Conduction block induced by high frequency AC stimulation in unmyelinated nerves. In *Engineering in Medicine and Biology Society, 2007. EMBS 2007. 29th Annual International Conference of the IEEE*, IEEE.

Keith, M. W., Peckham, P. H., Thrope, G. B., Stroh, K. C., Smith, B., Buckett, J. R., Kilgore, K. L., & Jatich, J. W. (1989). Implantable functional neuromuscular stimulation in the tetraplegic hand. *J. Hand Surg. (Am.)*, **14**(3), 524–530.

Kessler, T. M., & Fowler, C. J. (2008). Sacral neuromodulation for urinary retention. *Nat. Clin. Pract. Urol.*, **5**(12), 657–666.

Kilgore, K., (2004). Nerve conduction block utilising high-frequency alternating current. *Medical*, **42**(3), 394.

Kilgore, K., Peckham, P. H., Keith, M. W., Montague, F. W., Hart, R. L., Gazdik, M. M., Bryden, A. M., Snyder, S. A., & Stage, T. G. (2003). Durability of implanted electrodes and leads in an upper-limb neuroprosthesis. *J. Rehab. Res. Dev.*, **40**(6), 457–468.

Kilgore, K. L., Hoyen, H. A., Bryden, A. M., Hart, R. L., Keith, M. W., & Peckham, P. H. (2008). An implanted upper-extremity neuroprosthesis using myoelectric control. *J. Hand Surg. (Am.)*, **33**(4), 539–550.

Kondziolka, D. (2002). Hardware-related complications after placement of thalamic deep brain stimulator systems. *Stereot. Func. Neuros.*, **79**(3-4), 228.

Kopell, B. H. (2004). Deep brain stimulation for psychiatric disorders. *J. Clin. Neurophysiol.*, **21**(1), 51.

Kothari, S. (2007). Neuromodulatory approaches to chronic pelvic pain and coccygodynia. In *Operative Neuromodulation* (pp. 365–371).

Le Doare, K., Akerman, S., Holland, P. R., Lasalandra, M. P., Bergerot, A., Classey, J. D., Knight, Y. E., & Goadsby, P. J. (2006). Occipital afferent activation of second order neurons in the trigeminocervical complex in rat. *Neurosci. Lett.*, **403**(1-2), 73–77.

Loeb, G., & Richmond, F. (2001). BION implants for therapeutic and functional electrical stimulation, Chapter 3. In *Neural Prostheses for Restoration of Sensory and Motor Function*. CRC Press.

Madersbacher, H. (1990). Intravesical electrical stimulation for the rehabilitation of the neuropathic bladder. *Paraplegia*, **28**(6), 349–352.

Mapstone, T. B. (2008). Vagus nerve stimulation: Current concepts. *Neurosurg. Focus*, **25**(3), E9.

Marsolais, E. B., & Kobetic, R. (1987). Functional electrical stimulation for walking in paraplegia. *J. Bone Joint Surg. Am.*, **69**(5), 728–733.

McCreery, D. B. (1988). Comparison of neural damage induced by electrical stimulation with faradaic and capacitor electrodes. *Ann. of Biomed. Eng.*, **16**(5), 463.

McDonnall, D., Clark, G. A., & Normann, R. A. (2004). Selective motor unit recruitment via intrafascicular multielectrode stimulation. *Can. J. Physiol. Pharmacol.*, **82**(8-9), 599–609.

McNeal, D. R. (1977). 2000 years of electrical stimulation. In F. T. Hambrecht, & J. B. Reswick (Eds.), *Functional Electrical Stimulation*. New York: Marcel Dekker, Inc..

Melzack, R. (1965). Pain mechanisms: A new theory. *Science*, **150**(3699), 971.

Memberg, W. D., et al. (1989) A surgically-implanted intramuscular electrode for an implantable neuromuscular stimulation system. *Engineering in Medicine and Biology Society, 1989. Images of the Twenty-First Century. Proceedings of the Annual International Conference of the IEEE Engineering in.*

Merrill, D. R., Bikson, M., & Jefferys, J. G.R. (2005). Electrical stimulation of excitable tissue: Design of efficacious and safe protocols. *J. Neurosci. Meth.*, **141**(2), 171–198.

Naples, G. G., Mortimer, J. T., Scheiner, A., & Sweeney, J. D. (1988). A spiral nerve cuff electrode for peripheral nerve stimulation. *IEEE Trans. Biomed. Eng.*, **35**(11), 905–916.

Naples, G. G., Mortimer, J. T., & Yuen, T. G. (1990). Overview of peripheral nerve electrode design and implantation. In W. F. Agnew, & D. B. McCreery (Eds.), *Neural Prostheses Fundamental Studies*. Englewood Cliffs, NJ: Prentice Hall.

North, R. B., Kidd, D. H., Olin, J. C., & Sieracki, J. M. (2002). Spinal cord stimulation electrode design: Prospective, randomized, controlled trial comparing percutaneous and laminectomy electrodes - part I: Technical outcomes. *Neurosurgery*, **51**(2), 381–389. Discussion 389-90.

Oakley, J. C., & Prager, J. P. (2002). Spinal cord stimulation: Mechanisms of action. *Spine*, **27**(22), 2574–2583.

Peckham, P. (2005). Functional electrical stimulation for neuromuscular applications. *Ann. Rev. Biomed. Eng.*, **7**(1), 327.

Peckham, P., & Ackermann, D. (2009). Implantable neural stimulators. In E. Krames, P. Peckham, & A. Rezai (Eds.), *Neuromodulation*: Elsevier.

Peckham, P. H., Keith, M. W., Kilgore, K. L., Grill, J. H., Wuolle, K. S., Thrope, G. B., et al. (2001). Efficacy of an implanted neuroprosthesis for restoring hand grasp in tetraplegia: A multicenter study. *Arch. Phys. Med. Rehabil.*, **82**(10), 1380–1388.

Peterson, D. K. (1989). Chronic intramuscular electrical activation of the phrenic nerve. Department of Biomedical Engineering, Cleveland, OH: Case Western Reserve University.

Peterson, D. K., Stellato, T., Nochomovitz, M. L., DiMarco, A. F., Abelson, T., & Mortimer, J. T. (1989). Electrical activation of respiratory muscles by methods other than phrenic nerve cuff electrodes. *Pacing Clin. Electrophysiol.*, **12**(5), 854–860.

Polasek, K. H., Hoyen, H. A., Keith, M. W., & Tyler, D. J. (2007). Human nerve stimulation thresholds and selectivity using a multi-contact nerve cuff electrode. *IEEE Trans. Neural Syst. Rehabil. Eng.*, **15**(1), 76–82.

Popeney, C. A., & Alo, K. M. (2003). Peripheral neurostimulation for the treatment of chronic, disabling transformed migraine. *Headache*, **43**(4), 369–375.

Ranck, J. B. (1975). Which elements are excited in electrical stimulation of mammalian central nervous system: A review. *Brain Res.*, **98**(3), 417.

Rattay, F. (1989). Analysis of models for extracellular fiber stimulation. *IEEE Trans. Biomed. Eng.*, **36**(7), 676–682.

Rijkhoff, N. J. M. (1998). Urinary bladder control by electrical stimulation: Review of electrical stimulation techniques in spinal cord injury. *J. Urol.*, **160**(3P1), 961.

Robblee, L. S., & Rose, T. L. (1990). Electrochemical guidelines for selection of protocols and electrode materials for neural stimulation. In W. F. Agnew, & D. B. McCreery (Eds.), *Neural Prostheses: Fundamental Studies*. Englewood Cliffs, NJ: Prentice Hall.

Rose, T. L., Kelliher, E. M., & Robblee, L. S. (1985). Assessment of capacitor electrodes for intracortical neural stimulation. *J. Neurosci. Meth.*, **12**(3), 181–193.

Rosenow, J. M. (2006). Failure modes of spinal cord stimulation hardware. *J. Neurosurg. Spine*, **5**(3), 183.

Sauerwein, D., Ingunza, W., Fischer, J., Madersbacher, H., Polkey, C. E., Brindley, G. S., Colombel, P., & Teddy, P. (1990). Extradural implantation of sacral anterior root stimulators. *J. Neurol. Neurosurg. Psychiatry*, **53**(8), 681–684.

Schiefer, M. A., Triolo, R. J., & Tyler, D. J. (2008). A model of selective activation of the femoral nerve with a flat interface nerve electrode for a lower extremity neuroprosthesis. *IEEE Trans. Neural Syst. Rehabil. Eng.*, **16**(2), 195–204.

Schulman, J., Mobley, J. P., Wolfe, J., Regev, E., Perron, C. Y., Anath, R., Matei, E., Glukhovsky, A., & Davis, R. (2004). Battery powered BION FES network. *Proc. 26th Ann. Intl. Conf. of IEEE EMBS*. San Francisco, CA.

Seymour, J. P., & Kipke, D. R. (2007). Neural probe design for reduced tissue encapsulation in CNS. *Biomaterials*, **28**(25), 3594–3607.

Shealy, C. N., Mortimer, J. T., & Reswick, J. B. (1967). Electrical inhibition of pain by stimulation of the dorsal columns: Preliminary clinical report. *Anesth. Analg.*, **46**(4), 489–491.

Slavin, K. V. (2008). Peripheral nerve stimulation for neuropathic pain. *Neurotherapeutics*, **5**(1), 100–106.

Slavin, K. V., Colpan, M. E., Munawar, N., Wess, C., & Nersesyan, H. (2006). Trigeminal and occipital peripheral nerve stimulation for craniofacial pain: A single-institution experience and review of the literature. *Neurosurg. Focus*, **21**(6), E5.

Smith, B., Tang, Z., Johnson, M. W., Pourmehdi, S., Gazdik, M. M., Buckett, J. R., & Peckham, P. H. (1988). An externally powered, multichannel, implantable stimulator-telemeter for control of paralyzed muscle. *IEEE Trans. Biomed. Eng.*, **45**(4), 463–475.

Smith, B., Tang, Z., Johnson, M. W., Pourmehdi, S., Gazdik, M. M., Buckett, J. R., & Peckham, P. H. (1998). An externally powered, multichannel, implantable stimulator-telemeter for control of paralyzed muscle. *IEEE Trans. Biomed. Eng.*, **45**(4), 463–475.

Starr, P. A., Vitek, J. L., & Bakay, R. A. (1998). Deep brain stimulation for movement disorders. *Neurosurg. Clin. of N. Am.*, **9**(2), 381.

Stinson, L. W. (2001). Peripheral subcutaneous electrostimulation for control of intractable post-operative inguinal pain: A case report series. *Neuromodulation*, **4**(3), 99.

Susset, J. G., & Boctor, Z. N. (1967). Implantable electrical vesical stimulator: Clinical experience. *J. Urol.*, **98**(6), 673–678.

Tamma, F., (2002). Anatomo-clinical correlation of intraoperative stimulation-induced side-effects during HF-DBS of the subthalamic nucleus. *Neurol. Sci.*, **23**(0), 109.

Tanagho, E. A., & Schmidt, R. A. (1988). Electrical stimulation in the clinical management of the neurogenic bladder. *J. Urol.*, **140**(6), 1331–1339.

Tarler, M. D., & Mortimer, J. T. (2004). Selective and independent activation of four motor fascicles using a four contact nerve-cuff electrode. *IEEE Trans. Neural Syst. Rehabil. Eng.*, **12**(2), 251–257.

Terry, R. S., Tarver, W. B., & Zabara, J. (1991). The implantable neurocybernetic prosthesis system. *Pacing Clin. Electrophysiol.*, **14**(1), 86–93.

Theodore. (2004). Brain stimulation for epilepsy. *The Lancet Neurology*, **3**(2), 111.

Thimineur, M., (2007). C2 area neurostimulation: A surgical treatment for fibromyalgia. *Pain Med.*, **8**(8), 639.

Triolo, R. J., Boggs, L., Miller, M. E., Nemunaitis, G., Nagy, J., & Bailey, S. N. (2009). Implanted electrical stimulation of the trunk for seated postural stability and function after cervical spinal cord injury: A single case study. *Arch. Phys. Med. Rehabil.*, **90**(2), 340–347.

Tyler, D. J., & Durand, D. M. (2002). Functionally selective peripheral nerve stimulation with a flat interface nerve electrode. *IEEE Trans. Neural Syst. Rehabil. Eng.*, **10**(4), 294–303.

Uhlir, J. P., Triolo, R. J., Davis, J. A., Jr., & Bieri, C. (2004). Performance of epimysial stimulating electrodes in the lower extremities of individuals with spinal cord injury. *IEEE Trans. Neural Syst. Rehabil. Eng.*, **12**(2), 279–287.

Van Den Honert, C., & Mortimer, J. T. (1981). A technique for collision block of peripheral nerve: Single stimulus analysis. *IEEE Trans. Biomed. Eng.*, **BME-28**(5), 373–378.

Veraart, C., Grill, W. M., & Mortimer, J. T. (1993). Selective control of muscle activation with a multipolar nerve cuff electrode. *IEEE Trans. Biomed. Eng.*, **40**(7), 640–653.

Waisbrod, H. (1985). Direct nerve stimulation for painful peripheral neuropathies. *J. Bone Joint Surg. Br.*, **67**(3), 470.

Wall, P. D., & Sweet, W. H. (1967). Temporary abolition of pain in man. *Science*, **155**(3758), 108–109.

Waters, R. L., McNeal, D., & Perry, J. (1975). Experimental correction of footdrop by electrical stimulation of the peroneal nerve. *J. Bone Joint Surg. Am.*, **57**(8), 1047–1054.

Weiland, J. D., Anderson, D. J., & Humayun, M. S. (2002). *In vitro* electrical properties for iridium oxide versus titanium nitride stimulating electrodes. *IEEE Trans. Biomed. Eng.*, **49**(12), 1574–1579.

Weiner, R. L., & Reed, K. L. (1999). Peripheral neurostimulation for control of intractable occipital neuralgia. *Neuromodulation*, **2**(3), 217–221.

Wilson, B. S., Finley, C. C., Farmer, J. C., Jr., Lawson, D. T., Weber, B. A., Wolford, R. D., et al. (1988). Comparative studies of speech processing strategies for cochlear implants. *The Laryngoscope*, **98**(10), 1069.

Zhou, D. (2005). *Platinum Electrode and Method for Manufacturing the Same*. United States Patent.

CHAPTER II.5.13 MEDICAL BIOSENSORS

Lisa LaFleur and Paul Yager

Department of Bioengineering, University of Washington, Seattle, WA, USA

INTRODUCTION

The term biosensor is sometimes broadly defined as a sensor used in the detection or monitoring of any medically relevant parameter, but for this chapter we will define it more narrowly to mean *a sensor that uses biological molecules, tissues, organisms or principles to measure chemical or biochemical concentrations*. Biosensors can be used in many medical and non-medical applications. Biomedical sensors are sensors that detect medically relevant parameters; these could range from simple physical parameters like blood pressure or temperature, to analytes for which biosensors are appropriate (e.g., blood glucose). This is summarized in Figure II.5.13.1. We will focus on the overlap between these two fields that we are calling medical biosensors: sensors that fulfill the criteria for both classes of sensors.

Biosensor development has been a significant activity in the last 50 years. Between 1984 and 1990 there were roughly 3000 scientific publications and 200 patents on biosensors. The number of patents doubled through 1997. The introduction of nanotechnology into the broader field of sensor development led to over 6000 articles and 1100 patents to be issued through 2004 (Luong et al., 2008).

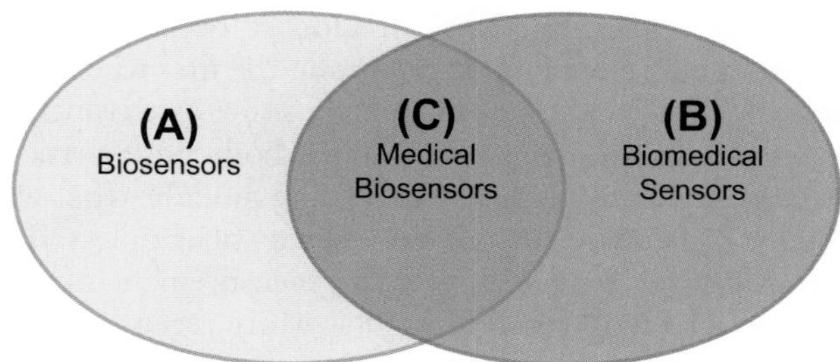

FIGURE II.5.13.1 The Venn diagram of the relationship between biosensors (A) and biomedical sensors (B). In (A) are those sensors that use biological molecules, tissues, organisms or principles in performing its designed function. Examples might include nucleic acid-based sensors for bacteria found in a water supply or enzyme-based sensors for the sugar levels in food. In (B) are sensors used to monitor living organisms, like those for blood pressure, blood pH, blood gases, temperature, and even optical methods for monitoring tissue glucose by observing the near-infrared absorption of glucose. In (C) are the medical biosensors, which meet both criteria, such as sensors for blood glucose that utilize biomolecules like the enzyme glucose oxidase.

Nanotechnology Applied to Biosensors

In the last decade some old, and some new, phenomena have been collectively labeled "nanotechnology." Some nanostructures and nanodevices have introduced new and useful capabilities to the biosensing world. Nanomaterials fall in the realm between individual atoms and bulk materials, leading to different physical and electrical properties than their bulk relatives. Some nanomaterials are uniquely suited to biosensing, including nanotubes and nanoparticles. Nanotubes are one-dimensional structures whose high surface-to-volume ratio causes their electrical conductivity to be exquisitely sensitive to surface adsorption. As such, nanotubes have been widely used in biosensors (Liu, 2008). One common application is carbon nanotube field effect transistors; antibodies or DNA strands are immobilized on the surface of the nanotube, then any corresponding antigen or complementary DNA strand-binding event can be detected by a change in the electrical conductance of the nanotube.

Another nanotechnology in common use for biosensing is nanoparticles functionalized with biomolecules. Nanoparticles are 1–100 nm in diameter, usually monodisperse, and most commonly made of Au modified by the adsorption of proteins such as antibodies. Au nanoparticles are commonly used in immunoassays. Their unique optical scattering and absorption properties can be exploited in a variety of imaging methods (Aslan et al., 2005). They have also been used to participate in enzyme reactions for which the enzyme is immobilized on the surface of the nanoparticle, and the Au reduces the enzyme as part of a desired reaction (Pandey et al., 2008). Nanotechnology promises to continue to be at the forefront of biosensor development.

BASICS OF BIOSENSING

Both biosensors and biomedical sensors fall into two general categories: physical and chemical. Physical parameters of biomedical importance include pressure,

TABLE II.5.13.1 Chemical Indicators of Health

Small, Simple	pH (acidity)
↓	Electrolytes (ions)
↓	Blood gases (O_2, CO_2, etc.), including general anesthetics
↓	Drugs and neurotransmitters
↓	Hormones
↓	Proteins (antibodies and enzymes)
↓	Viruses
↓	Bacteria
↓	Parasites
Large, Complex	Tumors

volume, flow, electrical potential, and temperature, of which pressure, temperature, and flow are generally the most clinically significant, and lend themselves to the use of small *in vivo* sensors. Chemical sensing generally involves the determination of the concentration of a chemical species in a volume of gas, liquid or tissue. The species can vary in size from the hydronium ion to a live pathogen (see Table II.5.13.1); when the analyte is complex, an interaction with another biological entity may be required to recognize it. In general, it is necessary to distinguish this chemical species from a number of similar interferents, which can be technologically challenging, but this is an area in which biosensors excel.

Glucose Sensors: First and Foremost

Early biosensor technology was almost exclusively geared towards monitoring blood glucose in diabetics, largely because of the number of patients with the disease, and the potential for reducing morbidity and mortality by adjusting the dosage of insulin to better control the analyte. According to the World Health Organization, there were approximately 171 million people suffering from diabetes worldwide in 2000; that number is expected to at least double by 2030 (Wild et al., 2004). In the late 1950s, several paper-based biosensors were developed for semi-quantitative glucose detection by monitoring a color change when the paper was dipped in urine (Comer, 1956; Free et al., 1957). The paper was embedded with glucose oxidase and other enzymes, so that the reaction between glucose, glucose oxidase, and the other enzymes produced a color. However, urine is not an ideal sample, as there is a substantial and uncontrolled lag between the concentration of glucose in urine and the critical blood glucose level. Later, using this same chemistry, optical absorbance was utilized for more quantitative measurements (Grady and Lamar, 1959; Kingsley and Getchell, 1960). Some even went so far as to demonstrate how a device could continuously perform *in vivo* measurements (Weller et al., 1960).

The first electrochemical glucose assay was described in 1961 (Malmstadt and Pardue, 1961). It was a potentiometric sensor, monitoring the electrical potential difference between two electrodes. Several amperometric sensors – sensors that monitor changes in electrical current in the system – were also developed in the late 1960s (Updike and Hicks, 1967). The first commercialized self-contained glucose analyzer was introduced by Yellow Springs Instrument Company in 1975; it was an amperometric sensor that monitored the production of hydrogen peroxide resulting from the breakdown of glucose by the enzyme glucose oxidase (Heller and Feldman, 2008). This demonstration that biomolecules could be part of a practical biomedical sensor spurred basic and applied research efforts for next 30 years.

Major advances have been made since the first commercial biosensors, leading to many simple self-tests that give patients the ability to control their blood sugar. The vast majority of these tests utilize blood collected from finger-sticks, which can be painful and usually must be done many times per day. Many less-painful testing techniques have been promoted over the years, but few have proven clinically or commercially viable. The optimal scenario, however, is continuous monitoring linked to therapeutic systems and centralized medical databases. Medtronic, Inc. has produced a device called the MiniMed Paradigm® that does operate continuously and connects to an automatic insulin pump. The glucose sensor is percutaneous and lasts for up to three days.

The most common mechanism for the operation of glucose sensors is the enzymatic oxidation of glucose by the enzyme glucose oxidase, as shown in Figure II.5.13.2. Potentiometric sensors measure the difference in electrical potential between two electrodes. An early potentiometric glucose sensor by Updike et al., shown schematically in Figure II.5.13.3, provides a simple example of the design and operation of such sensors (Updike and Hicks, 1967). In the case of the Updike sensor, the electrodes detected a decrease in the concentration of oxygen due to reaction with glucose.

The working platinum electrode was surrounded by a semi-permeable dialysis membrane to limit the molecules that could reach the electrode. Technological advances in materials science have led to a wide variety of very specific semipermeable membranes, including low molecular weight filters, ion-selective membranes, and others.

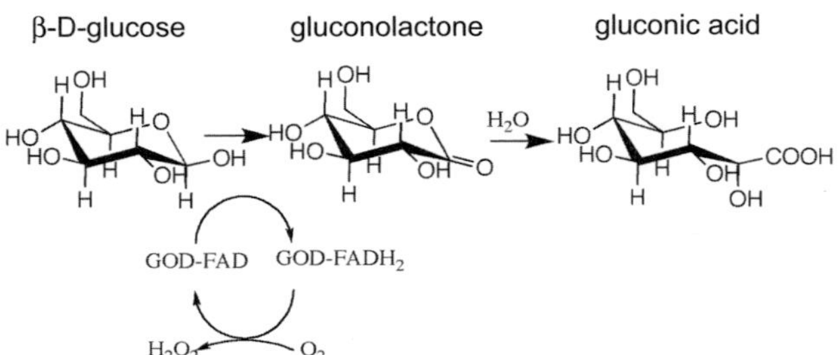

FIGURE II.5.13.2 The reaction where glucose is converted to gluconic acid in the presence of oxygen with hydrogen peroxide as a byproduct.

Electrochemical methods of glucose detection using trapped glucose oxidase were among the first to be tried clinically, but were thought to have some disadvantages for *in vivo* use: electrical wires in the body pose a risk of electrical shock. An alternative sensing modality that was thought to be safer for *in vivo* use was fiber optics. The first successful *in vivo* fiber optic measurement of pO_2 was reported by Peterson in 1984. More recently, many variations on fiber optic biosensors have been applied to *in vivo* glucose monitoring, including near-infrared non-invasive imaging (Maruo et al., 2003), fluorescence resonance energy transfer (Liao et al., 2008), and a dual sensor system by Pasic et al. that uses a reference fiber to compensate for local variations in oxygen concentration that affect the sensor's signal output (Pasic et al., 2007). This sensor system, shown in Figure II.5.13.4, has a diffusion barrier covering an immobilized enzyme, and also

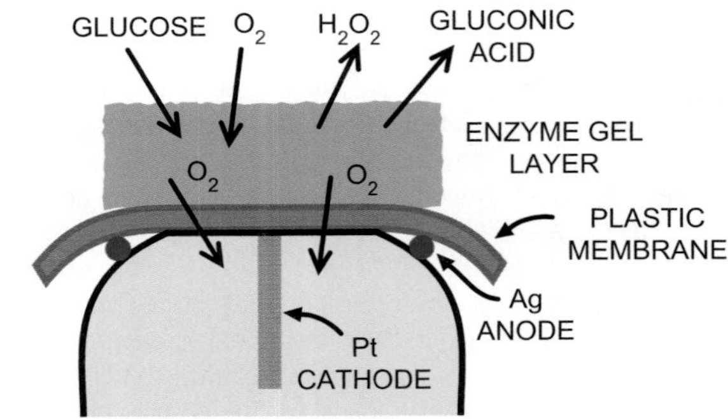

FIGURE II.5.13.3 A schematic representation of the potentiometric glucose sensor developed by Updike and Hicks in 1967. Glucose oxidase is immobilized in the enzyme gel layer outside a semipermeable membrane that allows oxygen gas to pass through. The sensor detects a decrease in oxygen resulting from the reaction of glucose and oxygen to form gluconic acid and hydrogen peroxide.

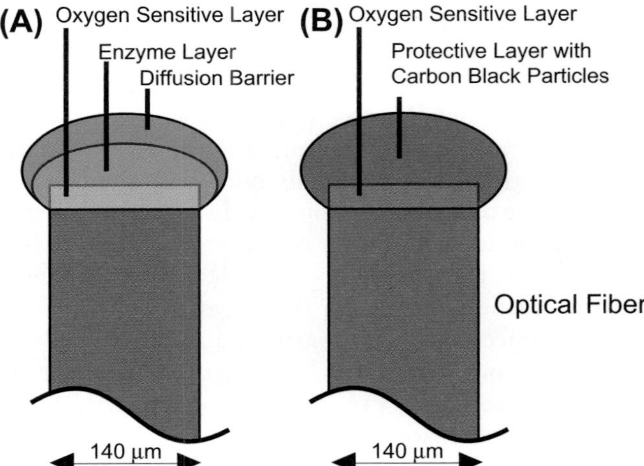

FIGURE II.5.13.4 Optical fibers forming a glucose sensor system. (A) Fiber uses immobilized glucose oxidase covered by a diffusion barrier to monitor oxygen depletion. (B) Fiber measures local oxygen concentration. Signal reported is the difference between the two, accounting for variations in local oxygen concentration. (Redrawn from Pasic et al. (2007). *Sensors and Actuators B.*)

utilizes glucose oxidase. This sensor was designed to be implanted subcutaneously.

Due to the limited space in this chapter we can't present examples of every type of biosensor. However, Table II.5.13.2 provides an overview of the different sensing modalities, and some of the potential biosensing applications. The biosensors described throughout the rest of this chapter provide more in-depth examples of how these sensing modalities can be applied. For a more thorough review, we recommend recent reviews of the field.

Interaction of Sensor with Environment

One helpful way to classify sensors is to consider the relationship between the sensor and the analyte, as shown schematically in Figure II.5.13.5. The more intimate the contact between the sensor and the analyte, the more complete is the information about the nature of the chemical species being measured. However, obtaining greater chemical information may involve some hazard to the physical condition of the system being studied. This is not a trivial problem when dealing with human subjects.

Noncontacting Sensors. Noncontacting sensors produce only a minimal perturbation of the sample to be monitored. In general, such measurements are limited to the use of electromagnetic radiation such as light or sampling the gas or liquid phase near a sample. It may even be necessary to add a probe molecule to the sample to make the determination.

Contacting Sensors may be either Noninvasive or Invasive. Direct physical contact with a sample allows a rich exchange of chemical information; much effort has been put into developing practical contacting sensors for biomedical purposes. A temperature probe can be in either category, but with the exception of removing or adding small quantities of heat, it does not change the environment of the sample. Few chemical sensors approach the non-perturbing nature of physical sensors. All invasive sensors damage the biological system to a certain extent, and physical damage invariably leads to at least localized chemical change. Tissue response can, in turn, lead to spurious sampling. Furthermore, interfacial phenomena and mass transport govern the function of sensors that require movement of chemical species into and out of the sensor. Restrictions on the size of the invasive sensor allowable in the biological system can limit the types of measurements that can be made; even a 1 mm diameter pH electrode is of no use in measuring intracapillary pH values.

Sample Removal Sensors. Most contacting sensors are derived from chemical assays first developed as sample removal sensors. While it is certainly invasive and traumatic to remove blood or tissue from a live animal, removal of some fluids, such as urine and saliva, can be achieved

TABLE II.5.13.2 Applications of Different Sensing Modalities to Biosensors

Sensing Modalities	Description of Operation	Potential Applications	Advantages/Disadvantages
Acoustic	Binding of mass to oscillating surfaces allows propagation of acoustic waves along or under the surface or resonant condition of device	Detection of mass accumulation on surface in real-time; made selective by coating surface with polymers or binding molecules	Excellent mass sensitivity but no inherent sensitivity to chemicals; vacuum and air are best media due to inherent drag in liquids
Cantilever motion or deflection	Static deflection of cantilever caused by binding to one side or alteration in resonant frequency of cantilever	Binding of molecules to a prepared surface or presence of particles in fluids	Excellent mass sensitivity. Static deflection is slow, alteration of resonant frequency with wet samples requires encapsulation of fluid path
Amperometric	Current generated by a reaction at an electrode	Monitor redox reactions, enzyme-catalyzed reactions or generation of any charged species	Good sensitivity, wide range of detectable species; currents between electrodes can interfere with biological activity
Thermal	Coupling of reactions that produce or consume heat to detect presence of analytes or when coupled to binding reactions, as a method to detect binding	Usually in laboratory-based reactors where "thermal noise" can be minimized	Good generic sensing modality, but generally insensitive
Optical	Change in light frequency, phase intensity or lifetime in simple evanescent fields	Many configurations including placing sensitive chemistry at terminus of optical fiber material, monitoring fluorescent reaction products, and detection of nanoparticles	Excellent broad applicability, many direct and probe-based methods available. Can generate unwanted heat in a sample; fluorescent species susceptible to photobleaching; potential safety concerns over use of nanoparticles
Potentiometric	Potential changes at a surface; may be in a conventional two-electrode circuit or at the surface of a field-effect transistor	Monitor change in concentration of charged species, typically small molecules or pH changes	Relatively insensitive

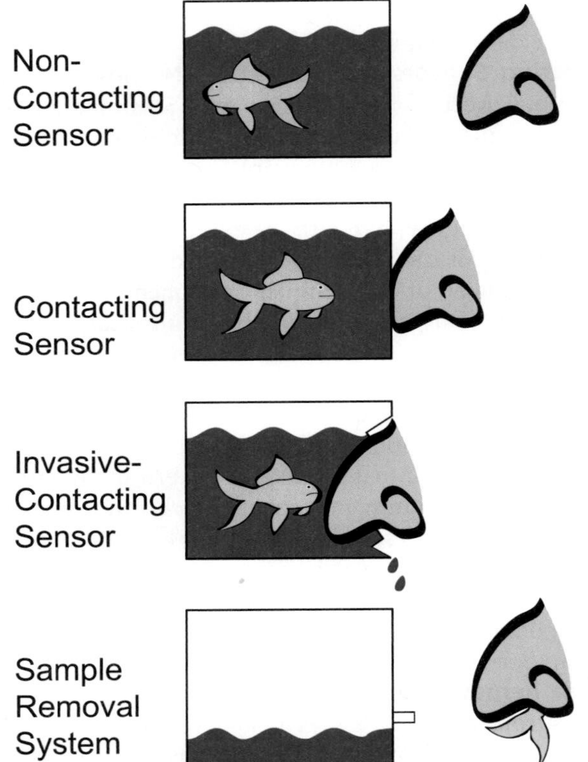

FIGURE II.5.13.5 Comparisons of the interaction between the sensor (nose) and analyte (fish) for different types of detection.

without trauma. Once removed, a fluid can be pretreated to make it less likely to adversely affect the functioning of a sensor. For example, heparin can easily be added to blood to prevent clotting in an optical measurement cell. Cells that might interfere with optical assays can also be removed prior to measurement. Toxic reagents and probe molecules can be added at will, and samples can be fractionated to remove interfering species. The sensor and associated equipment can be of any size, be at any temperature, and use as much time as necessary to make an accurate measurement. Further, a sensor outside the body is much easier to calibrate. This approach to chemical measurement allows the greatest flexibility in sensor design, and avoids many biocompatibility problems.

Biosensor Design Goals and Constraints

All biosensors are governed by specific chemical and biological reactions that need to be identified and closely controlled for accurate biosensing. Numerical output from a biosensor only provides information indicative of the rate-limiting step of any reaction, so the most effective biosensors must be designed to ensure that there is only one potential rate-limiting step. For example, in the potentiometric and fiber-optic glucose sensors described earlier in this chapter, the sensors were intended to measure changes in glucose, so oxygen must be in excess and the glucose oxidase concentration constant for these sensors to produce a reliable output. Table II.5.13.3 summarizes the primary design considerations that must be considered when developing a biosensor.

TABLE II.5.13.3 Overview of Biosensor Design Considerations
What is a biochemical species that responds predictably to changes in the chemical state (i.e., glucose concentration) you want to study?
Is there a convenient method of chemical-to-electrical transduction?
Does binding of the biochemical species perturb the system?
Does measurement of the biochemical species consume it?
What other species intefere with measurement?
How accurate and precise are the measurements?
What is the response time of the sensor?
Is sensor response stable and can it be calibrated?
What are acceptable operating conditions?
Can the sensor be made at low cost?
What impact does the presence or operation of the sensor have on the patient?

Selectivity Inherent to Biosensing

An advantage biosensors have over other types of sensors is that they utilize the billions of years of evolution that have produced extreme selectivity and sensitivity in the binding between biomolecules. Antigen–antibody complexes, complementary DNA strands or DNA/RNA complexes, and other specific biomolecule pairs like biotin-streptavidin are commonly used in biosensing, because of their high selectivity and sensitivity. The human body is capable of producing 10^8 distinct antibody species (Nelson and Cox, 2005). These antibodies can bind to antigens ranging from small molecules to proteins to viral capsids to bacterial cell walls. All biomolecular interactions are characterized by rates of association (binding) and dissociation (unbinding); however, the rates of dissociation for useful complexes are far less than the rates of association. Careful choice of highly selective biomolecules is central to good biosensor design.

CHALLENGES IN BIOSENSING

First, Do No Harm

The introduction to the body of a sensor is a traumatic event, although the degree of trauma depends on the site of placement. The gastrointestinal tract can clearly be less traumatically accessed than the pulmonary artery. Critically ill patients often must have catheters placed into their circulatory systems for monitoring of blood pressure and administration of drugs, fluids, and food, so that no *additional* trauma is caused by including a small flexible sensor in that catheter.

When short-term implantation in tissue is possible, there is initial trauma at the site of insertion. Longer-term implantation increases the risk of infection along the surface of the implant. When the site of implantation of the probe is the circulatory system, the thrombogenicity of the probe is of paramount importance. Surface chemistry, shape, and placement within the vessel have all been shown to be of great importance in reducing the risk of embolism. Also, sensors based on chemical reactions often contain or produce toxic substances during the course of their operation, so great care must be taken to ensure that these are either not released or are realeased at low enough levels to avoid significant risk to the patient.

Two major questions in the design of any sensor are how often and for how long it is expected to be used. There are several factors to be considered.

Length of Time for which Monitoring is Required. Determination of blood glucose levels must be made for the entire lifetime of diagnosed diabetics, while intra-arterial blood pressure monitoring may only be needed during a few hours of surgery.

Frequency of Measurements. Blood cholesterol only needs to be tested every few years, but for a diabetic blood glucose measurements must be done several times a day, and are most critical after meals and after administration of insulin.

Reusability of the Sensor. Some chemical sensors contain reagents that are consumed in a single measurement. Such sensors are usually called "dip-stick sensors," such as are now found in pregnancy testing kits and other rapid immunoassays. High affinity antibodies, for example, generally bind their antigens so tightly that they cannot be reused. On the other hand, most physical and chemical sensors are capable of measuring the concentration of their analyte on a continuous basis, and are therefore inherently reusable.

Lifetime of the Sensor. Chemical sensors all have limited lifetimes, because of such unavoidable processes as oxidation, and while these may be extended through low-temperature storage, *in vivo* conditions are a threat to the activity of the most stable biochemical. Most sensors degrade with time, and the requirement for accuracy and precision usually limit their practical lifetime, particularly when recalibration is not possible.

Appropriateness of Repeated Use. The need for sterility is the most important reason to avoid reuse of an otherwise reusable sensor. If it is not logistically possible or economically feasible to completely sterilize a used sensor, it will only be used on a single individual, and probably only once.

Biocompatibility. If the performance of a sensor is degraded by continuous contact with biological tissue or if the risk to the health of the patient increases with the time in which a sensor is in place, the lifetime of the sensor may be much shorter *in vivo* than *in vitro*.

As a consequence of all of these factors in the design of an integrated sensing system, the probe – that part of the sensor that must be in contact with the tissue or blood – is often made disposable. Probes must therefore be as simple and as inexpensive to manufacture as possible, although it is often true that it is the sale of consumables such as probes that can be more profitable than the sale of the device itself.

All Sensors are Affected by Biofouling

After careful choice of target species and designing your sensor to keep this species the rate-limiting factor, the next most important consideration for biosensor design is preventing sensing inaccuracy due to sampling a micro-environment created by biofilm formation. It is difficult to prevent biofilm formation on surfaces exposed to active media, such as bacterial or eukaryotic cell cultures. The growth of a biofilm on a sensor degrades its performance; in the most extreme case, the sensor can become sensitive only to conditions in the biofilm. Biofilms often consist of bacteria embedded in a secreted matrix of complex polysaccharides. While the microenvironment of these films may be beneficial to the bacteria, it is detrimental to the function of the sensor for several reasons. First, the chemistry of the film may damage structural or active components of the sensor. Second, if the film completely encloses the sensor, only the film's microenvironment is sensed, rather than the solution that surrounds it. If the living components of the film metabolize the analyte to be sensed, it may never reach the sensor. Even a "dead" film may exclude certain analytes by charge or size, and thereby lower the concentration available for sensing at the surface below the film. If a sensor used *in vitro* is fouled, it can often be removed, cleaned, and restored to its original activity. For example, carbon electrodes containing immobilized enzymes can be restored by simply polishing away the fouled surface. When sensors are used *in vitro* for monitoring the chemistry of body fluids, preprocessing can be used to reduce the accretion of biofilms that might impede the function of the sensor (Gifford et al., 2006; Li et al., 2007).

In vivo sensing has all the problems associated with *in vitro* sensing, plus the additional effects of the sensor on the body. *In vivo* sensors do not have the luxury of being able to be easily removed, cleaned, resterilized or recalibrated, then returned to the sensing site. The combination of microenvironment formation and damage to the body often limit the utility of *in vivo* sensors to short use. In some cases, it has been found that the problems are almost immediate, producing spurious results from the outset. Subcutaneous glucose needle electrodes have been found to give accurate results *in vitro* before and after producing erroneous values *in vivo*. The biological environment may simply make it impossible to perform accurate chemical measurements with certain types of sensors.

Sample removal systems are not immune to all the problems of biofouling that plague *in vivo* and *in vitro* biosensors. Whole blood samples still contain encapsulating eukaryotic cells, but even the removal of these still poses a biofouling problem for all biosensors. Human biological samples contain a wide variety of proteins, fats, mucins, carbohydrates, and other highly fouling substances. The biological materials most responsible for biofouling, and some of their adhesion and encapsulation mechanisms, are described thoroughly throughout this text.

BIOFOULING PREVENTION METHODS

Methods to minimize the effects of biofouling in biosensors are similar to those used in other biomaterials applications. However, a critical requirement in many biosensors is that the "working" surface still allows components of the sample to reach the (reactive) surface below. Therefore, inclusion of holes through the non-fouling surface can be vital.

Blocking for Biofouling Prevention

Blocking is a nearly ubiquitous method for coating the non-functionalized regions of a biosensor to prevent non-specific binding and delay biofilm formation. There are many ways to implement this process, but all include the same basic principle: an unreactive molecule is applied to portions of the biosensing surface or non-sensing regions of the biosensor, to prevent other molecules from binding to the sensing region. Blockers can be adsorbed to the surface, covalently attached, joined using a self-assembled monolayer or by any other means of keeping the surface coated. A common theme among blockers is that they present hydrophilic groups to the area around the biosensor. Some of the most common blocking materials include polyethylene glycol (PEG), non-reactive globular proteins and surfactants. Application method and uniformity of the blocking materials can have a large impact on antifouling effectiveness.

PEG. There are many variations of PEG used as blockers in biosensing: a range of molecular weights have been shown to be effective (500–2000), and a variety of structures including straight chains and branched "brushes." PEG is a hydrophilic polymer and is believed to be an effective blocker, because its structure prevents the release of bound surface water, a key component in adsorption at surfaces. Other chapters in this book provide a much more comprehensive investigation of PEG as a non-fouling material. It should be noted that PEG has been shown to *adsorb* preferentially, when dry; this is an important consideration for biosensor design (Foley et al., 2005).

Proteins as Blockers

Another common method of preventing biofouling is blocking with non-reactive proteins. Bovine serum album (BSA) and casein, a protein found in milk, are often used for this application, because of their historical effectiveness and low cost. Commercial blocking solutions using proprietary protein formulations are also available. After functionalizing the sensing surface, the blocker can be applied by soaking the biosensor or sensing surface in a protein solution or spraying the protein solution on and then rinsing off unbound protein. Different proteins will have different success at preventing biofouling for each specific sensing surface and sensor design, so it is common to test multiple blocking solutions to find the best fit for each particular application. More about proteins for biofouling prevention can be found throughout this text.

Other Coating Materials

There are many other types of coatings that can be used to prevent undesired binding to a biosensor surface, many of which are elaborated on throughout this text. A few common examples include surfactants and hydrogels. However, for some biosensors this restricts the designer's ability to functionalize the surface in a way that is necessary to make sensing measurements. This remains an area of active research.

EFFECTS OF BIOFOULING ON SAMPLE REMOVAL SYSTEMS

For many biosensing systems, accurate and effective sensing is accomplished by a combination of preprocessing of the sample to remove problematic compounds, and thorough blocking to prevent additional non-specific binding. Non-specific binding can block target binding sites, decreasing potential signal and creating false readings, and interfere directly with the target analyte in the sample, preventing proper sensing. The rest of this section describes a variety of sensing modalities used in sample removal systems, and illustrates how biofouling prevention has been attempted in each.

Colorimetric Antigen Detection

The Yager group recently adapted a colorimetric immunoassay for the detection of antigens produced by malarial parasites: *Plasmodium falciparum* histidine rich protein II (PfHRPII), produced only by *Plasmodium falciparum*, the most deadly pathogen, and an aldolase antigen, which is produced by all four major forms of *Plasmodium* (Lafleur et al., 2008). This assay was implemented in microfluidic cards; the antigen detection was performed on a porous nitrocellulose membrane substrate. An overview of the system and biomolecular interaction is illustrated in Figure II.5.13.6.

One element of novelty in this assay was the porous substrate, which had a high surface area with the potential for high capture molecule-binding and high signal. The capture molecules anti-PfHRPII IgM and anti-aldolase IgG were spotted onto the nitrocellulose membranes, and then dried. The membranes were then soaked in a blocking solution of casein intended to bind to the rest of the membrane surface and prevent non-specific binding of the subsequent reagents. The next fluid introduced was the sample. In the preliminary work, this was antigen spiked into pooled human plasma. Plasma is a realistic matrix for this sort of assay, and contains a wide variety of other proteins that are potential interferents. The last reagent flowed through the membrane was a suspension of Au nanoparticles conjugated to anti-PfHRPII secondary antibodies (Stevens et al., 2008).

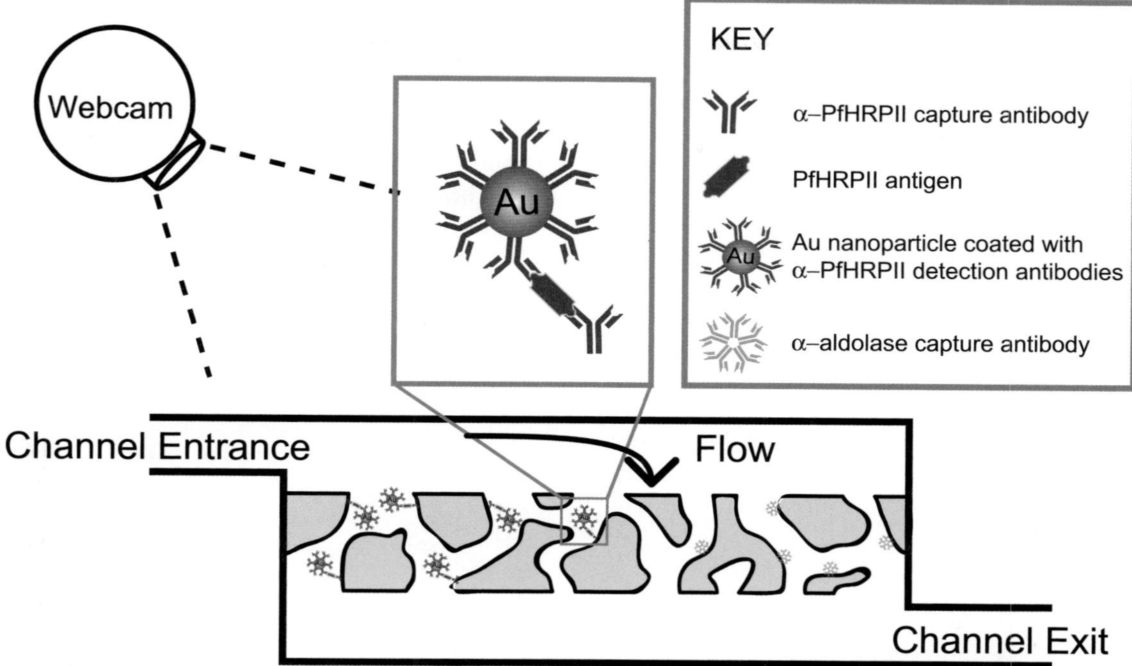

FIGURE II.5.13.6 A high surface area porous membrane is used as the substrate for a multiplexed malaria detection assay. Capture antibodies are pre-spotted onto the membrane. Sample, washes and Au-labeled detection antibodies are flowed sequentially through the membrane forming a "stack" that localizes the Au-labeled detection antibodies. Detection is performed with an inexpensive webcam.

An effective biosensor has to be both sensitive and selective, and that selectivity, in this case, largely constitutes rejection of binding of Au-labeled secondary antibody non-specifically. Nitrocellulose was chosen as the material for the substrate partly because, in our system, it bound *less* protein than some other materials of similar porosity. This decreased binding capacity affected how much capture antibody could be spotted, but also decreased how susceptible the membrane substrate was to non-specific binding of antigens, interfering proteins in the plasma or secondary antibodies, all of which affect the resulting assay signal. This sort of trade-off is very indicative of the types of choices one has to make for effective biosensor design.

Small Molecule Detection Using SPR

Surface plasmon resonance (SPR) is an optical technique that measures changes in index of refraction caused by changes in mass near specific metallic surfaces, and changes in the bulk refractive index of the fluid above the surface. The most common substrate for SPR is thin films of gold, whose loose valence electrons allow surface plasmons to be excited by appropriate wavelengths of light coupled at appropriate angles. SPR has the advantage of not requiring the target molecules to need labeling of any kind (Brockman et al., 2000).

Yager et al. developed a portable SPR system with disposable microfluidic cards for the rapid detection of phenytoin, and anti-epileptic drug, from saliva samples, shown in Figure II.5.13.7 (Yager et al., 2006; Fu et al., 2007). Saliva is a complex matrix with many interferents,

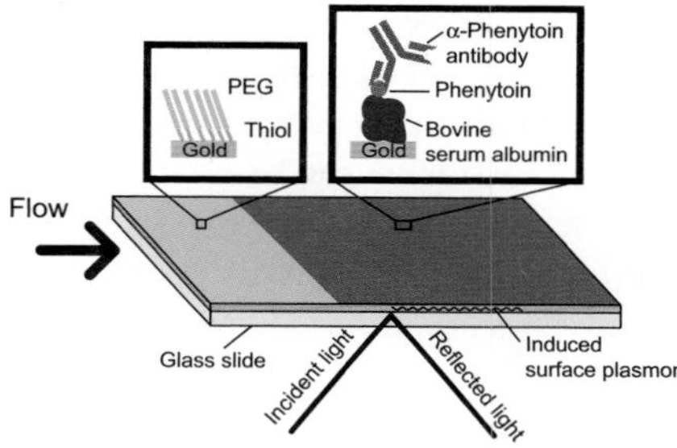

FIGURE II.5.13.7 The surface plasmon resonance (SPR) system employed by Yager et al. to detect phenytoin in saliva samples (Fu et al., 2007). The gold-coated glass substrate was covered with PEG blocker in upstream regions and capture molecules in the detection region. The signal component from the bulk material was the measurement in the PEG-coated region; it was subtracted from the total signal measured in the detection region, yielding an accurate response to the binding of anti-phenytoin antibodies.

but many small molecule concentrations track closely with blood concentrations making it a non-invasive alternative. Phenytoin was detected using an indirect immunoassay; the SPR signal did not measure binding of phenytoin itself, but the reduced rate of binding of an anti-phenytoin antibody. The saliva samples were premixed with a known concentration of anti-phenytoin antibody and then flowed over a gold substrate coated with immobilized phenytoin. Only free anti-phenytoin antibodies bound to the surface, allowing a computer to back-calculate the concentration of phenytoin in the original sample.

A combination of methods was used to combat non-specific binding in this system. Initially, the saliva sample was processed upstream of the detection surface to remove mucins and other interferents (Helton et al., 2008). The capture surface contained an upstream region coated only with PEG-thiol blocker (the thiol allows the PEG to be immobilized on the gold substrate), followed by the detection region coated with bovine serum albumin (BSA)-phenytoin. As the sample flowed over the upstream PEG-coated region, it produced a bulk refractive index change signal that is not indicative of non-specific binding, but would contribute to a "background" signal level. This signal was subtracted from the signal detected in the BSA-phenytoin region. Many biosensors only utilize blocking to prevent erroneous signal, but this SPR detection system took advantage of sample preconditioning, blocking and subtracting the signal generated by the bulk refractive index in the PEG blocked upstream region to accurately quantify the concentration of phenytoin in saliva samples.

IgG Detection Using Resonating Cantilevers

The last example is a cantilever-based sensor. Many static cantilever approaches have been described in the last decade (Waggoner and Craighead, 2007); they are based on deflection of a cantilever due to binding to only one of its surfaces. Most such sensors have been relatively impractical. The sensor in question is based on a hollow resonating cantilever that is surrounded by air or vacuum; the signal is measured as a shift in the resonant frequency due to changes in the mass of the cantilever (Burg and Manalis, 2003). Manalis et al. have developed a resonating cantilever biosensor that allows small volumes of sample to flow through the resonating cantilever and detect biomolecular-binding events, reporting the signal as frequency shift proportional to mass of biomolecules that have been bound. By monitoring the changes in frequency with time, real-time binding can be detected. The authors demonstrated a system to detect goat IgG, and showed that human IgG and BSA would not non-specifically bind in their device (Burg et al., 2007).

The cantilevers were fabricated using traditional silicon microfabrication, with silicon dioxide walls. The analyte in this system was goat IgG, so the authors needed to immobilize an anti-goat IgG on the surface. The immobilization was achieved through a multi-step process which can be seen in Figure II.5.13.8. As each set of molecules was added, the mass changes resulted in a frequency shift. Firstly, poly-L-lysine conjugated to PEG conjugated to biotin was electrostatically adsorbed in the desired orientation (with the biotin pointing up). This layer was rinsed with buffer, and the next link in this immobilizing chain was neutravidin, a variant of the better-known streptavidin. This was again rinsed with buffer, then the biotinylated anti-goat IgG was added and rinsed.

Once the cantilever was functionalized with immobilized anti-goat IgG, samples of goat IgG ranging from 7 nM to 7 μM were tested. The sample volumes were ~10 pL. The authors reported mass detection resolution of 10^{-19} g/μm^2, which is equivalent to 1 protein per square micron. The authors reported that they corrected for changes in the frequency due to the density of the sample by subtracting out the shift due to just the buffer. They also tested how susceptible the sensor was to non-specific binding by testing human IgG and BSA as samples. Neither produced a significant frequency shift, indicating that neither molecule was binding to the anti-goat IgG or any other surface of the sensor.

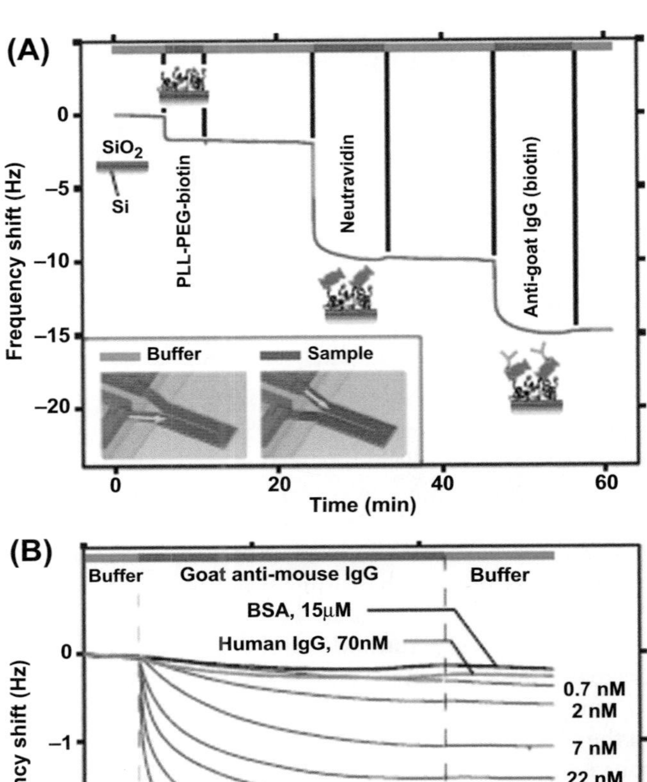

FIGURE II.5.13.8 Graphical representation of the cantilever-based biosensor developed by Manalis et al. and data collected using this type of biosensor. (A) The multi-step layer addition required to immobilize anti-goat IgG on the sensor inner surfaces. The red segments show the different molecules being added, and the accompanying frequency shifts. The blue segments show buffer flowing, which doesn't correspond to a significant frequency shift. (B) The goat IgG samples ranging from 7 nM to 7 μM being flowed through the sensor and their corresponding frequency shifts. Human IgG and BSA produce no noticeable frequency shift, showing the sensor's resistance to non-specific binding (Reprinted by permission from Macmillan Publishers Ltd. *Weighing of biomolecules, single cells and single nanoparticles in fluid* (Burg et al., 2007)).

POINT-OF-CARE MEASUREMENTS ENABLING DISTRIBUTED DIAGNOSIS AND HOME HEALTHCARE

Continued developments in the field of biosensing are leading to more small, rapid, portable sensing options for the hospital bedside, home use, and keeping on one's person. These devices are often described as point-of-care, because detection is brought to the patient, instead of the patient being brought to a detector. More frequent measurements taken by the patient are an inexpensive way to establish a more personalized baseline and improve overall care. Well-established information networks allow easy communication between these devices and central medical databases. New smart cell phones are providing a perfect platform for data input and delivery. Together, these advances are creating a new phase of personalized medicine, sometimes called distributed diagnosis and home healthcare (D_2H_2), illustrated in Figure II.5.13.9. Continuous glucose monitors that communicate with insulin pumps are only a few steps away from a realized D_2H_2 system.

A major contributor to the ability of biosensors to become sufficiently inexpensive and small for personal use has been the development of micro-electromechanical systems built on the design and fabrication techniques of the semiconductor industry. Many of the microchips (including microprocessors, RF chips, digital signal processors, etc.) are already being mass-produced very cheaply; more unique components can be built using the same technology much more cheaply than previous generations of biosensors. Lab-on-a-chip (LOC) design is now contributing greatly to point-of-care biosensing. As a field it now has its own very popular journal. The basic principle of LOC diagnosis is to take a bench-top process usually performed in a laboratory (i.e., determine an HIV patient's CD4 count) and reduce all the steps to an automated process performed in a disposable, self-contained device. The devices are designed to be as cheap as possible for one-time use; and a more expensive and complicated system is used to control the simple disposable. LOC has the advantage of not requiring a trained technician to operate, making it ideal for POC use and a potentially large contributor to the future of home healthcare.

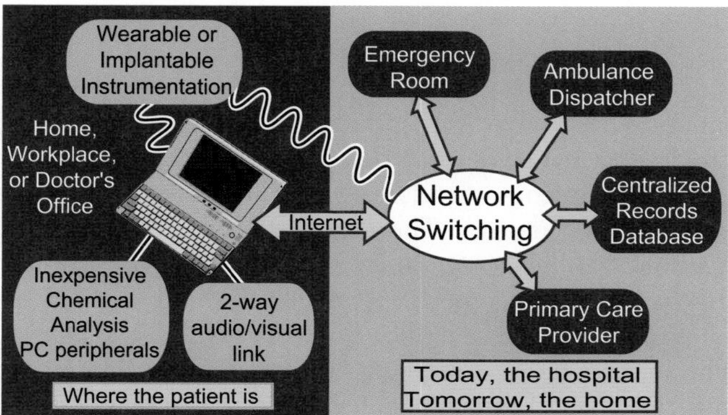

FIGURE II.5.13.9 An evolving doctor–patient interface as part of a vision of distributed diagnosis and home healthcare (D_2H_2). By utilizing the existing infrastructure for wired and wireless data communication, as well as existing capabilities for storing large amounts of patient data, it is becoming possible to allow patients to maintain an up-to-date record of many different health parameters without frequent visits to hospitals and physicians.

SUMMARY

Medical biosensors play a large and growing role in medical diagnostics and patient monitoring; this chapter is merely a brief overview of biosensors. The use of biological molecules, tissues, and organisms has made biosensors uniquely suited to many diagnostic and real-time detection challenges, and their high potential for sensitivity and specificity promises to continue this trend. Biosensors are still plagued by the ever-present problem of biocompatibility, particularly for *in vivo* use which is more complicated for sensing use because of the need for *permeable* non-fouling surfaces. Researchers are rapidly advancing the state-of-the-art in biofouling prevention, making dramatic strides in short-term implantable sensors like those described in this chapter for percutaneous glucose sensing. As biocompatibility is better-understood and controlled, longer-term, more accurate *in vivo* sensors will become commonplace.

ACKNOWLEDGMENTS

The authors would like to thank the students of Yager lab, past and present, and all the other University of Washington students who have helped us clarify our vision of this field. We would specifically like to thank the researchers, collaborators, and funding sources for the two Yager lab projects described in this chapter. The collaborators on the Malaria detection project include: PATH, Micronics, Inc., Nanogen, and the Pat Stayton lab at the University of Washington. The project was funded by The Bill and Melinda Gates Foundation. The SPR phenytoin detection project was funded by *National Institute of Dental and Craniofacial Research, Washington Research Fund and the Technology Gap Innovation Fund*. The students and post-doc contributors were: Gayathri Balasubramanian, Daryl Campbell, Thayne Edwards, Jennifer Foley, Elain Fu, Beatrice Gandara, Richard Garzon, Neil Geisler, Melissa Hasenbank, Kenneth Hawkins, Kristen Helton, Afshin Mashadi-Hossein, Turgut Kosar, Maris Lemba, Michael Look, Katherine McKenzie, Kjell Nelson, Jennifer Osborn, Paolo Spical-Mihalic, Dean Stevens, and Karman Tandon.

BIBLIOGRAPHY

Aslan, K., Lakowicz, J. R., & Geddes, C. D. (2005). Plasmon light scattering in biology and medicine: New sensing approaches, visions and perspectives. *Current Opinion In Chemical Biology*, **9**, 538–544.

Brockman, J. M., Nelson, B. P., & Corn, R. M. (2000). Surface plasmon resonance imaging measurements of ultrathin organic films. *Annual Review of Physical Chemistry*, **51**, 41–63.

Burg, T. P., & Manalis, S. R. (2003). Suspended microchannel resonators for biomolecular detection. *Applied Physics Letters*, **83**, 2698.

Burg, T. P., Godin, M., Knudsen, S. M., Shen, W., Carlson, G., Foster, J. S., Babcock, K., & Manalis, S. R. (2007). Weighing of biomolecules, single cells and single nanoparticles in fluid. *Nature*, **446**, 1066–1069.

Comer, J. P. (1956). Semiquantitative specific test paper for glucose in urine. *Analytical Chemistry*, **28**, 1748–1750.

Foley, J., Schmid, H., Stutz, R., & Delamarche, E. (2005). Microcontact printing of proteins inside microstructures. *Langmuir*, **21**, 11296–11303.

Free, A. H., Adams, E. C., Kercher, M. L., Free, H. M., & Cook, M. H. (1957). Simple specific test for urine glucose. *Clinical Chemistry*, **3**, 163–168.

Fu, E., Chinowsky, T., Nelson, K., Johnston, K., Edwards, T., Helton, K., Grow, M., Miller, J. W., & Yager, P. (2007). SPR imaging-based salivary diagnostics system for the detection of small molecule analytes. *Annals of the New York Academy of Sciences*, **1098**, 335–344.

Gifford, R., Kehoe, J. J., Barnes, S. L., Kornilayev, B. A., Alterman, M. A., & Wilson, G. S. (2006). Protein interactions with subcutaneously implanted biosensors. *Biomaterials*, **27**, 2587–2598.

Grady, H. J., & Lamar, M. A. (1959). Glucose determination by automatic chemical analysis. *Clinical Chemistry*, **5**, 542–550.

Heller, A., & Feldman, B. (2008). Electrochemical glucose sensors and their applications in diabetes management. *Chemical Reviews*, **108**, 2482–2505.

Helton, K. L., Nelson, K. E., Fu, E., & Yager, P. (2008). Conditioning saliva for use in a microfluidic biosensor. *Lab on a Chip*, **8**, 1847–1851.

Kingsley, G. R., & Getchell, G. (1960). Direct ultramicro glucose oxidase method for determination of glucose in biologic fluids. *Clinical Chemistry*, **6**, 466–475.

Lafleur, L., Lutz, B., Stevens, D., Spicar-Mihalic, P., Osborn, J., Mckenzie, K., & Yager, P. (2008). Air-driven point-of-care malaria immunoassays for the developing world. *Bioengineering Applications to Address Global Health Conference*: Duke University.

Li, C. M., Dong, H., Cao, X., Luong, J. H.T., & Zhang, X. (2007). Implantable electrochemical sensors for biomedical and clinical applications: Progress, problems, and future possibilities. *Current Medicinal Chemistry*, **14**, 937–951.

Liao, K. C., Hogen-Esch, T., Richmond, F. J., Marcu, L., Clifton, W., & Loeb, G. E. (2008). Percutaneous fiber-optic sensor for chronic glucose monitoring *in vivo*. *Biosensors and Bioelectronics*, **23**, 1458–1465.

Liu, A. H. (2008). Towards development of chemosensors and biosensors with metal-oxide-based nanowires or nanotubes. *Biosensors and Bioelectronics*, **24**, 167–177.

Luong, J. H. T., Male, K. B., & Glennon, J. D. (2008). Biosensor technology: Technology push versus market pull. *Biotechnology Advances*, **26**, 492–500.

Malmstadt, H., & Pardue, H. L. (1961). Quantitative analysis by an automatic potentiometric reaction rate method – specific enzymatic determination of glucose. *Analytical Chemistry*, **33**. 1040.

Maruo, K., Tsurugi, M., Chin, J., Ota, T., Arimoto, H., Yamada, Y., Tamura, M., Ishii, M., & Ozaki, Y. (2003). Noninvasive blood glucose assay using a newly developed near-infrared system. *IEEE Journal of Selected Topics in Quantum Electronics*, **9**, 322–330.

Nelson, D. L., & Cox, M. M. (2005). *Lehniger Principals of Biochemistry*. New York: W. H. Freeman And Company.

Pandey, P., Datta, M., & Malhotra, B. D. (2008). Prospects of nanomaterials in biosensors. *Analytical Letters*, **41**, 159–209.

Pasic, A., Koehler, H., Klimant, I., & Schaupp, L. (2007). Miniaturized fiber-optic hybrid sensor for continuous glucose monitoring in subcutaneous tissue. *Sensors and Actuators B – Chemical*, **122**, 60–68.

Peterson, J. I., Fitzgerald, R. V., & Buckhold, D. K. (1984). Fiber-optic probe for in vivo measurement of oxygen partial pressure. *Analytical Chemistry*, **56**(1), 62–67.

Stevens, D., Petri, C. R., Osborn, J., Spicar-Mihalic, P., Mckenzie, K., & Yager, P. (2008). Enabling a microfluidic immunoassay for the developing world by integration of on-card dry-reagent storage. *Lab on a Chip*, **8**(12), 2038–2045.

Updike, S. J., & Hicks, G. P. (1967). Enzyme electrode. *Nature*, **214**. 986.

Waggoner, P. S., & Craighead, H. G. (2007). Micro- and nanomechanical sensors for environmental, chemical, and biological detection. *Lab on a Chip*, **7**, 1238–1255.

Weller, C., Linder, M., Macaulay, A., Ferrari, A., & Kessler, G. (1960). Continuous *in vivo* determination of blood glucose in human subjects. *Annals of the New York Academy of Sciences*, **87**, 658–668.

Wild, S., Roglic, G., Green, A., Sicree, R., & King, H. (2004). Global prevalence of diabetes – estimates for the year 2000 and projections for 2030. *Diabetes Care*, **27**, 1047–1053.

Yager, P., Edwards, T., Fu, E., Helton, K., Nelson, K., Tam, M. R., & Weigl, B. H. (2006). Microfluidic diagnostic technologies for global public health. *Nature*, **442**, 412–418.

CHAPTER II.5.14 BURN DRESSINGS AND SKIN SUBSTITUTES

Douglas L. Helm[1], Britlyn D. Orgill[1], Rei Ogawa[2], and Dennis P. Orgill[1]

[1]Division of Plastic and Reconstructive Surgery, Brigham and Women's Hospital, Boston, MA, USA

[2]Division of Plastic, Reconstructive and Aesthetic Surgery, Nippon Medical School, Tokyo, Japan

SKIN: THE LARGEST ORGAN

Skin protects the body from microorganisms and external forces, integrates complex sensory nervous and immune systems, controls fluid loss, and serves important aesthetic functions. The two layers, the dermis and the epidermis, are linked by epidermal derivatives or "appendages," such as sebaceous glands and hair follicles that invaginate into the dermis. The dermis hosts a rich vascular network that helps regulate temperature. For most minor injuries, such as a superficial paper cut, the skin is able to self-repair without scarring (Dunkin et al., 2007). Deeper skin injuries due to deep cuts, burns or degloving injuries can cause significant physiological derangement, expose the body to a risk of systemic infection, and become a life-threatening problem. Large skin deficits have motivated the development of improved technologies to replace and restore skin. Although many of these technologies were originally developed to treat burns, these technologies have been extended to other areas, including chronic wounds and acute surgical wounds. In this chapter we will explore the factors that need to be considered when developing wound dressings and skin substitutes, and also review the technologies currently available to treat patients.

> Over 500,000 patients require treatment for burns each year in the United States. Of these 40,000 will be hospitalized and 4000 will succumb to burn injuries. Query of the National Center for Health Statistics database revealed over three million traumatic wound admissions each year. It is estimated that 10% of these will require advanced wound care, representing 300,000 wounds (American Burn Association, 2007).

The Use of Skin Substitutes and Burn Dressings Depends on the Depth of the Wound

Epidermal appendages are lined with keratinocytes that migrate across the wound and divide to form a multilayered epidermis with a cornified barrier layer. This healing response is commonly seen in split-thickness donor sites with depths of 0.008–0.020 inches. When these injuries are kept clean and moist, healing occurs within 7–21 days. Similarly, superficial and mid-dermal burns heal over similar time periods, if they do not become infected and are dressed properly. Injuries and burns that extend into the deep dermis or through the entire dermis will heal only after a prolonged time or may become chronic wounds. To treat skin loss, autologous skin grafts are often useful, and can be provided as full-thickness or split-thickness skin grafts. Full-thickness grafts are applied to small wounds, and consist of the entire epidermis and dermis; the donor site is closed with sutures. Split-thickness grafts are applied to larger wounds and can be expanded through a meshing process that allows them to cover a larger area. Microscopically, blood vessels from the wound revascularize the graft. Donor sites cause scars and are painful, and for large skin losses there may not be enough donor skin to cover all of the excised burn. Therefore, special burn dressings and skin substitutes have been designed to help treat such cases. The ideal characteristics of a skin substitute are listed in Table II.5.14.1. Fluid loss and bacterial infection are two major concerns when treating a patient with massive skin loss, and these two factors have influenced the technology of burn dressings and skin substitutes. Several practical issues, including ease of application, storage time, and cost, also factor into the choice made by the clinician.

TABLE II.5.14.1 Qualities of Skin Substitutes

The ideal skin substitute is:
• inexpensive
• long lasting
• a bacterial barrier
• semipermeable to water
• elastic
• easy to apply
• painless to the patient
• non-antigenic and non-toxic
• has a durable shelf-life

Conventional Treatment: Partial-Thickness Burns

Conventional therapy for treating burns focuses on preventing wound infection through the use of topical antimicrobial agents. Silver-based dressings, such as Silver Sulfadiazine (Silvadine), have been used since their introduction in 1968 because of their infection prevention qualities. These dressings are relatively inexpensive, but tend to adhere to the wound surface and require daily dressing changes and impede epithelial growth. They can temporarily depress the white blood cell count. Silver nitrate solution is effective at treating many bacteria, but requires dressings to be continuously wet, and they tend to stain the skin and sheets black. Silver nitrate can lead to electrolyte disturbances. Mafenide acetate has excellent penetration of burn eschar, but can be painful. It has been classically used on areas where eschar is left on for a prolonged period, such as ears. These classical burn dressings do not prevent fluid loss through evaporation or exudative loss.

> Depending on the thickness, we can classify a burn as being one of three degrees. First degree burns are superficial wounds affecting only the epidermis. Second degree burns are deeper than first degree burns, and extend into the dermal layers of the skin. Second degree burns vary in seriousness, and there is debate over treatment of these wounds. Third degree burns are the deepest, and extend past the dermis into the fat.

Advanced Dressing: Partial-Thickness Burns

Acticoat™ (Smith & Nephew) is a silver antimicrobial dressing containing nano-crystalline silver that offers a slow release of silver ions over time. This product treats a broad spectrum of bacteria and is changed every 2–4 days. Indicated in both full- and partial-thickness wounds, the product is kept moist with sterile water that activates release of the silver. Studies have shown that silver-impregnated dressings such as Acticoat reduced wound-healing time and pain level in comparison to traditional silver sulfadiazine treatment (Wasiak et al., 2008).

Aquacel® Ag (ConvaTec) is a silver hydrofiber dressing. It has been useful in treating second degree burns, chronic wounds, and donor sites. In addition to the advantages of preventing wound infection, the dressing also maintains a moist wound environment by controlling wound exudate. Since it does not have to be applied as often as silver sulfadiazine (only once every other day versus twice a day), Aquacel has received many positive reviews because the product reduces pain to the patient (Connor-Ballard, 2009).

Biobrane® (UDL Laboratories, Inc., Rockford, IL) is a temporary biosynthetic burn dressing made from porcine collagen-coated nylon mesh embedded in silicone. Designed to adhere to a clean and debrided burn wound, studies have shown this dressing to be at least as effective as topical silver sulfadiazine in the areas of pain and wound healing time. Also, the total cost of the dressing tends to be less with Biobrane than with traditional silver sulfadiazine dressing (Wasiak et al., 2008). However, other studies have shown Biobrane not to be as effective as OrCel™ or allogeneic cultured keratinocytes (Pham et al., 2007). No complications have been reported (Pham et al., 2007).

TransCyte® (Advanced Biohealing) is a transparent covering created by culturing neonatal fibroblasts on a porcine collagen coated nylon mesh, which is bonded to a semipermeable silicone membrane. TransCyte simulates the epidermis and the dermal matrix. It has FDA (US Food and Drug Administration) approval for treatment of full- and partial-thickness burns. Studies have shown that TransCyte requires fewer dressing changes when compared with silver sulfadiazine, and has overall better results in wound healing. No deaths or complications associated with the use of TransCyte have been reported (Wasiak et al., 2008).

Apligraf® (Organogenesis, Canton, MA, and Novartis, East Hanover, NJ) is created from neonatal foreskin fibroblasts and keratinocytes, and has both dermal and epidermal components. First, the dermal component is made with cultured fibroblasts with bovine type I collagen. Then, cultured keratinocytes are added, and over time are allowed to epithelialize and form a stratum corneum. Apligraf is currently FDA-approved for use in treating diabetic foot and venous ulcers, although it is also used in burn treatment. Apligraf has a cost of $51/cm^2$ and a 10-day shelf-life. It is used as a covering over meshed, expanded autografts, and is designed to be applied weekly. Studies have shown that this product has a better wound healing result than with an autograft alone. No complications or deaths have been cited due to Apligraf (Pham et al., 2007).

OrCel™ (OrCel International, New York, NY), similar to Apligraf, is an allogeneic, bilayered substitute comprised of cultured human fibroblasts and keratinocytes. However, the fibroblasts in OrCel are seeded onto a preformed matrix instead of being co-cultured with collagen in solution. Also, OrCel lacks Apligraf's stratum corneum. OrCel is FDA-approved for use on split-thickness skin graft donor sites in burn victims, and in reconstructive hand surgery of patients with epidermolysis bullosa. A study performed by Still et al. found OrCel to be more effective than Biobrane with regards to wound closure time and scar appearance (Still et al., 2003). It is not currently available for clinical use.

SKIN SUBSTITUTES

Although burn dressings are successful in effectively treating some burn injuries, some wounds require more support than a temporary dressing or they are not candidates for an autograft. Therefore, a variety of skin substitutes are used to help treat full-thickness burns. Skin substitutes literally substitute either part or the entirety of the skin, and therefore have dermal and/or epidermal components. Originally, these constructs were designed for life-saving effects on very large burns. Subsequently it has been shown that both functional and aesthetic reconstructions can occur using these products. However, even the best skin substitutes available today still lack epidermal appendages, intact vasculature, immune cells, and melanocytes.

Epidermal Replacement

CellSpray® (C3) is an aerosol spray solution consisting of cells cultured from a small biopsy of the patient's own skin. The solution of cells is sprayed onto the burn injury to create an epidermal layer. It is sold as a kit that can be used by the surgeon in the office or operating room to provide cell suspensions from biopsies. It is currently available in Europe and Australia.

Epicel® (Genzyme Biosurgery, Cambridge, MA) is the best known cultured epidermal autograft on the market. The patient's own keratinocytes are expanded *ex vivo* in co-culture with murine 3T3 fibroblast feeder cells. A neoepidermis consisiting of keratinocyte layers of variable thickness is arranged into sheets that are shipped back to the patient in need. Epicel is expensive and has a 24-hour shelf-life at room temperature.

Dermal Replacement

Allograft (cadaver skin) is harvested from recently deceased individuals and used fresh or cryopreserved. It is tested for infectious agents such as hepatitis, and needs to be carefully tracked to the source. It has good take, but will undergo rejection at about 10–14 days. The rejection is primarily the epidermis; dermal elements can remain and be used as a base for thin skin grafts of cultured epidermal grafts (Cuono et al., 1987).

Integra™ (Integra Lifesciences Corp., Plainsboro, NJ) is an acellular, purely biosynthetic dermal substitute. The Integra Dermal Regeneration Template consists of two layers: the dermal layer made from (bovine) type I collagen and chondroitin-6-sulfate; and the epidermal layer made of silicone. The collagen–chondroitin dermal matrix allows in-growth of cells from the wound bed, and the artificial silicone epidermis regulates heat and fluid loss, and provides some antimicrobial protection. After integration with the wound bed, the silicone backing is removed and the neodermis acts as a recipient bed for a split-thickness autograft. Integra was the first skin substitute to be FDA-approved, and is approved for use in full- and partial-thickness burn wounds. Integra should not be used in patients who are allergic to bovine collagen or who have infected wounds. Reports on the efficacy of Integra have been mixed. Studies have shown Integra to be better than autograft, allograft, and xenograft in the parameter of wound healing time. However, Integra was shown to be inferior to autograft, allograft, xenograft, and Biobrane with regards to wound infection and graft take.

Matriderm® (Suwelack, Germany) is a collagen–elastin matrix used for dermal replacement. In a comparative rat study it had similar wound performance to Integra (Schneider et al., 2009). It does not have a silicone layer over it, and several clinicians have used it as a one-stage replacement using a thin skin graft over the matrix acutely. It is available for use in Europe.

AlloDerm® (Lifecell, Branchberg, NJ) is a dermal matrix derived from cadaveric skin. The cadaveric tissue is aseptically processed to remove the epidermis and cellular components of the dermis that are known to lead to graft failure. A scaffold composed of collagen, elastin, fibronectin, proteoglycans, and the three-dimensional framework of the microvasculature (without the actual vessels) remains. Alloderm is used in a range of applications, including breast reconstruction, urogynecological, orthopedic, and trauma.

Cultured Epithelial Autograft was the first cultured skin substitute, and was pioneered in the 1970s by Rheinwald and Green (Rheinwald and Green, 1975). The Cultured Epithelial Autograft (CEA) enabled generation of an autograft from a very small skin sample, which would prove useful in treating patients with large surface area burns. However, the use of CEAs has been limited due to expense, friability, and lack of a dermal layer. Although still used (e.g., Epicel), other available biological dressings for patients with extensive burns have made the use of CEAs impractical.

COMPOSITE AUTOLOGOUS TISSUE AND SKIN TRANSFER

In certain cases of severe burns, the underlying muscle and bone can be heavily damaged and suffer necrosis. Although this may happen with thermal burns, electrical burns can cause deep tissue destruction beyond the level of the skin, as the electricity conducts throughout the body on its way to a grounded surface. When muscle and bone is needed, local and free tissue transfer of muscle, fat, and bone with the overlying skin may be utilized to cover an injury site. Countless free tissue grafts have been surgically described, but the most commonly used is radial arm muscle with overlying skin to cover dorsal or palmar hand surface burns. In addition, commonly used gracilis, rectus, and latissimus muscles can also be reliably transferred, with the aid of microvascular anastamoses to a robust blood supply, to injury sites around the body.

COMBINED THERAPY WITH VACUUM ASSISTED CLOSURE (VAC)

In many cases, more than one clinical therapy must be combined to give the optimal treatment to a patient. A common situation in the area of wound healing is to combine a dressing or skin substitute with Vacuum Assisted Closure (VAC) therapy. The VAC consists of a polyurethane sponge, which is placed into the wound bed and then sealed off, so that suction can be applied through a tube. This device has had optimal results with wound healing since the suction allows excess fluid from edema to leave the wound, which prevents bacteria overgrowth, while at the same time a moist wound environment is maintained. The apparatus creates macrodeformations of pulling the sides of the wound together, and the pores in the sponge create microdeformations, stretching the cells in the wound bed and inducing proliferation and angiogenesis. The VAC device has proven to be helpful in burn care when it can be paired with a skin substitute to help hold the substitute in place and stimulate the healing response of the wound bed (Stiefel et al., 2009). Through this the VAC helps the wound site adhere to the skin substitute and have better take.

THE MARKETPLACE FOR BURN DRESSINGS AND SKIN SUBSTITUTES

Cost does play a role in the development of burn dressings and skin substitutes. Physicians must determine the optimal treatment they can provide to a patient considering the economic resources available. Also, although the development of dressings and skin substitutes for the treatment of burn injury is rewarding, our current marketplace has not proven to be supportive of this industry. One of the problems is that burn injury correlates to a very small market, and in order for a dressing or skin substitute to be truly successful it must also be able to be applied to a wider market such as that for general wound care, or the company must sell other products. For example, the company Integra, which began with one product, the Integra Dermal Regeneration Template, now also sells technologies for orthopedics and neurosciences, as well as medical instruments. Burn dressings are also very costly to produce, and it is difficult to quantify the success of a burn dressing or skin substitute because randomized clinical trials are difficult to conduct in a surgical field. Therefore, it is difficult to show quantified results to potential investors to acquire funding. Logistical factors such as time lapses in receiving FDA approval can also hold a company back from developing products (Garfein, 2009). Therefore, in the future of developing technology for burn injuries, it is imperative that cost be considered so that the dressings can continue to be produced and patients can receive the benefit of the treatment.

BIBLIOGRAPHY

American Burn Association. (2007). *Burn Incidence and Treatment in the US: 2007 Factsheet*: http://www.ameriburn.org/resources_factsheet.php. Date 25.06.08 Accessed.

Connor-Ballard, P. (2009). Understanding and managing burn pain: Part 2. *Am. J. Nurs.*, **109**, 54–62.

Cuono, C. B., Langdon, R., Birchall, N., Barttelbort, S., & McGuire, J. (1987). Composite autologous-allogeneic skin replacement: Development and clinical application. *Plast. Reconstr. Surg.*, **80**, 626–637

Dunkin, C. S., Pleat, J. M., Gillespie, P. H., Tyler, M. P., Roberts, A. H., & McGrouther, D. A. (2007 May). Scarring occurs at a critical depth of skin injury: Precise measurement in a graduated dermal scratch in human volunteers. *Plast Reconstr Surg*, **119**(6), 1722–1732.

Garfein, E. (2009). Skin replacement products and markets. In D. P. Orgill, & C. Blanco (Eds.), *Biomaterials for Treating Skin Loss*. Cambridge, UK: Woodhead Publishing Limited.

Pham, C., Greenwood, J., Cleland, H., Woodruff, P., & Maddern, G. (2007). Bioengineered skin substitutes for the management of burns: A systematic review. *Burns*, **33**, 946–957.

Rheinwald, J. G., & Green, H. (1975). Serial cultivation of strains of human epidermal keratinocytes: The formation of keratinizing colonies from single cells. *Cell*, **6**, 331–343.

Schneider, J., Biedermann, T., Widmer, D., Montano, I., Meuli, M., et al. (2009). Matriderm versus Integra: A comparative experimental study. *Burns*, **35**, 51–57.

Stiefel, D., Schiestl, C. M., & Meuli, M. (2009). The positive effect of negative pressure: Vacuum-assisted fixation of Integra artificial skin for reconstructive surgery. *J. Pediatr. Surg.*, **44**, 575–580.

Still, J., Glat, P., Silverstein, P., Griswold, J., & Mozingo, D. (2003). The use of a collagen sponge/living cell composite material to treat donor sites in burn patients. *Burns*, **29**, 837–841.

Wasiak, J., Cleland, H., & Campbell, F. (2008). Dressings for superficial and partial thickness burns. *Cochrane. Database Syst. Rev.*, CD002106.

CHAPTER II.5.15 SUTURES

*M. Scott Taylor[1] and Shalaby W. Shalaby**

[1]Poly-Med, Inc., Anderson, SC, USA

GENESIS AND COMMON USES

A suture, by definition, is any strand of material that is used to ligate (tie) blood vessels or approximate tissue. Ligatures are used to achieve hemostasis or to close a structure to prevent leakage. Suture literally means *to sew*, which is derived from the Latin *sutura*. The earliest examples of this technology trace back thousands of years, when Egyptians, Indians, and other early suture developers used readily available materials, such as silk, linen, cotton, horsehair, and animal tendons and intestines. Documented descriptions of suture materials and techniques trace to *Susruta*, written approximately 500 BCE in India (Mackenzie, 1973).

Suture needles share a parallel history, with initial versions being prepared from available materials such as sharpened pieces of bone. There have been many innovations in needle materials and design, leading to the current offering of needles that are tailored for use in various tissues and surgical sites.

The late 1800s, with the advent of synthetic "plastics," filaments, and improvements in surgical techniques, ushered in a new wave of suture materials development. Aseptic technique was introduced in 1885, and rubber gloves in 1890. Many of the most common synthetic polymers were developed in the last 100 years, and synthetic absorbable sutures were introduced in the early 1970s. Currently, sutures comprise the largest portion of wound closure technologies, with the yearly market for sutures approaching a billion dollars in the United States alone. Ethicon is the largest manufacturer of sutures; however, many small and large companies have emerged to compete with similar and novel suture offerings.

The suture device is comprised of: (1) the suture strand; (2) the surgical needle; and (3) the packaging material used to protect the suture and needle during storage. The suture strand is the only part of the device that remains in place after the surgical procedure is accomplished, and is solely responsible for securing the tissue. The needle provides a means to insert the suture strand, and the package holds the suture and needle, while protecting them from damage and providing a sterile barrier.

A common misconception is that further suture development is unnecessary due to the wide range of products currently available. In fact, there is always an opportunity for development of improved sutures and other wound closure devices. The design of surgical sutures combines knowledge of polymer chemistry, metallurgy, current surgical practice, and the proposed indication for the suture. Current advances in these fields allow for continued advances and improvements in suture technologies.

DESCRIPTION OF SURGICAL SUTURE

The suture is comprised of three distinct components. Table II.5.15.1 further describes each component, along with example permutations. The first component that identifies a suture, and the most descriptive for the performance of the device, is the strand.

The major distinguishing factor in suture strands is the absorbability of the suture. Absorbable (occasionally

*Dr. Shalaby Shalaby (1938–2010) was Founder, President, and Director of R&D at Poly-Med, Inc. In his career he served as the Director of the R&D team at Ethicon, and managed the Johnson and Johnson Polymer Technology Center. He also served as a professor of Bioengineering at Clemson University, and received the Technology Innovation and Development Award from the Society of Biomaterials for his long history of development of absorbable biomaterials for drug delivery systems, tissue adhesives, wound closure devices, and other biomedical innovations. The editors and the biomaterials community will miss Shalaby very much.

TABLE II.5.15.1 Components and Permutations of the Suture Device

Component	Description and Permutations
Suture strand	Component of device that provides securement, only part of device that remains in tissue after implantation: • Absorbable or non-absorbable • Sized according to USP[1] or EU[2] guidelines • Braided or monofilament • Coated or uncoated • Natural or dyed
Needle	Component of device that pulls suture through tissue: • Size is based on suture diameter • Taper point, cutting, spatula, specialty • Straight, ¼ circle, ⅜ circle, ½ circle, ⅝ circle, compound curved • Coated or uncoated • Swaged or rolled
Packaging	Component that provides a sterile barrier for storage and maintains storage conditions: • Tyvek® or foil • Single suture or multipack • Dry or with plasticizing fluid • Plastic or paper folder

[1]United States Pharmacopeia.
[2]European Union.

denoted resorbable, biodegradable or erodible) materials are those which undergo a chemical degradation. In suture materials, this is most commonly the result of hydrolysis. With surgical gut sutures, however, the absorption is a result of enzymatic degradation. In the past, absorbable sutures have been defined as losing a significant portion of strength within a two-month period (Swanson and Tromovitch, 1982), but current advances in materials have yielded absorbable sutures maintaining measurable strength for six months or longer. Absorbable sutures lose strength long before losing mass; it may take months or years after the suture loses strength to fully absorb.

Suture is manufactured in many sizes, and is selected by the surgeon based on a number of factors, including: surgical site; patient age and weight; immune history and response; presence of disease or infection; and personal history with suture materials (Ethicon, 2001). Size is based on diameter, as defined in the *United States Pharmacopeia* (USP) (Table II.5.15.2), and often materials will exhibit slightly different performance characteristics based on size. Monofilament sutures exhibit lower tissue drag compared to braided multifilament sutures; however, many materials having high modulus and stiffness are not compliant enough to be useful as monofilament. Also, any defect caused during handling can create a weak region in the suture strand. For these reasons, multifilament braided sutures were created. As a combination of fibers with much smaller diameter, relatively stiff materials in a composite structure have a perceived compliance much lower than material properties would suggest.

TABLE II.5.15.2 Suture Diameter Requirements for Commonly Used Sutures[1]

USP Suture Size	Collagen Suture Maximum Diameter (mm)	Synthetic Suture Maximum Diameter (mm)
6–0	0.10	0.07
5–0	0.15	0.10
4–0	0.20	0.15
3–0	0.30	0.20
2–0	0.35	0.30
0	0.40	0.35
1	0.50	0.40
2	0.60	0.50

[1]*United States Pharmacopeia*, 2008.

Although the multifilament allows creation of suture from relatively stiff materials, there are possible drawbacks to this design. Void space in the braid creates capillarity, and potential areas to harbor bacteria. Increased surface roughness associated with a braided construction can cause damage to the implant site due to increased tissue drag. To counteract this and increase lubricity, almost all braided and a few monofilament sutures are coated. Additionally, sutures are supplied as either natural or dyed to increase visual contrast. Table II.5.15.3 identifies many of the most common sutures and examples of manufacturers.

Needles are used to place the suture in tissues, and are primarily supplied pre-attached to the suture strand. They are supplied in a variety of shapes, each of which is suited to certain applications. Needle and suture are connected at the end of the needle, which can be attached mechanically by swaging (in which the drilled end of the needle is crimped around the suture strand), rolled in a rolled-end needle or attached using an adhesive (e.g., epoxy resin or cyanoacrylate), but they can also be supplied packaged individually as an eyed needle. The basic needle description includes shape and point geometry (Table II.5.15.4 and Figure II.5.15.1). Needle shapes include straight, curved (described as a fraction of a circle), and compound curved. Figure II.5.15.2 provides an illustration of a curved needle, along with pertinent dimensional descriptions. The ideal needle is inert, as slim as possible without compromising strength, stable when held in the needle holder, sharp enough to penetrate tissue with minimum material resistance, rigid enough to not allow bending, ductile enough to not break during use, sterile, and corrosion resistant (Ethicon, 2001).

The primary purpose of the suture package is to maintain sterility during storage. For absorbable sutures, the package serves a second function as a barrier to protect the suture from moisture, which would initiate the degradation process. In the case of gut sutures, a plasticizing fluid is used to maintain compliance. Inside the pouch, the suture is contained in either a paper folder or a formed plastic tray.

TABLE II.5.15.3 Common Sutures and Materials Used

Material	Origin/ Composition	Form[1]	Ultimate Strength, Straight (MPa)	Ultimate Strength, Knot (MPa)	Ultimate Elongation (%)	Young's Modulus (GPa)	Functionality	Clinical Applications	Example Devices
Non-absorbable									
Silk	p-Gomore silk worm	B	370–570	240–290	9–31	8.4–12.9	1 ± years	General suturing, ligation	Ethicon Permahand™
Cotton	Plant-based	Tw	280–390	160–320	3–6	5.6–10.9	2 ± years	Not commonly used	Cotton
PET	Synthetic polyester	B	510–1060	300–390	8–42	1.2–6.5	Permanent	Heart valves, vascular prostheses, general	Ethicon Ethibond™ Teleflex Tevdek®
PP	Synthetic polypropylene	M	410–760	300–570	30–60	2.1–4.1	Permanent	General, vascular	Ethicon Prolene™ Covidien Surgipro™
Nylon	Synthetic Nylon-6,6 or Nylon-6	M	460–710	300–330	17–65	1.8–4.5	Permanent	Skin	Ethicon Covidien Monosof™
Stainless steel	300-series stainless steel	M, B	540–780	420–710	29–65	200	Permanent	Sternal closure, tendon repair, orthopedic	Ethicon Ethisteel™ Covidien Flexon™
UHMWPE	Synthetic polyethylene	B	1130–1590	680–908	<10	1.4–4.8	Permanent	Orthopedic	Teleflex Force Fiber® Arthrex FiberWire®
Absorbable									
PGA	Glycolide	B	760–920	310–590	18–25	7–14	2–4 weeks	Subcutaneous, peritoneal	Teleflex Bondek®
PGLA	90/10 Glycolide/ Lactide	B	570–910	300–400	18–25	7–14	3–4 weeks	General, subcutaneous	Ethicon Vicryl™
PGLA (continued)	90/10 Glycolide/ Lactide	B	570–910	300–400	18–25	7–14	2 weeks	General, subcutaneous, skin	Ethicon Vicryl™ Rapide
PGLA (continued)	88/12 Lactide/ Glycolide	B	410–760	200–400	30–50	1.4–4.1	4–6 months	Orthopedic	Biomet Osteoprene®
PDO	p-dioxanone	M	450–560	240–340	30–38	1.2–1.7	4–6 weeks	General, cardiovascular, opthalmic	Ethicon PDS™ II Angiotech PDO
PG-Cl	75/25 Glycolide/ Caprolactone	M	628	315	39	0.8	2–3 weeks	General	Ethicon Monocryl™ Angiotech Monosorb
PG-TMC	67/33 Glycolide/ TMC	M	540–610	280–480	26–38	3.0–3.4	6 weeks	General, cardiovascular	Covidien Maxon™
Natural gut	Bovine serosa or submucosa of sheep or goat intestine	Tw M	310–380	110–210	15–35	2.4	1–2 weeks	General, subcutaneous	Ethicon Plain Gut
Chromic gut	Bovine serosa or submucosa of sheep or goat intestine treated in chromium solution	Tw M	310–380	110–210	15–35	2.4	2–3 weeks	General, subcutaneous	Ethicon Chromic Gut

(Burg and Shalaby, 1999; Roby and Kennedy, 2004; Teleflex Medical OEM, 2009.)
[1]B = Braid, M = Monofilament, Tw = Twisted fiber.

The Ideal Suture is...

- Biocompatible
- Sterile
- Compliant
- Adequate knot/straight strength
- Secure and stable knot
- Strength and mass loss profiles adequate for proposed usage
- Low friction
- Adequate needle attachment strength
- Atraumatic needle design
- Non-electrolytic
- Non-capillary
- Non-allergenic
- Non-carcinogenic
- Minimally reactive
- Uniform and predictable performance

TABLE II.5.15.4 Common Suture Needles

Type and Description
Taper point: Circular cross-section gradually increasing in diameter from the point to the needle body
Conventional cutting: Triangular cross-section at the needle tip to provide cutting edges, allowing easy penetration into the tissue, which gradually transitions to a square needle body. One cutting edge is on the inside curvature of the needle
Reverse cutting: Triangular cross-section at the needle tip to provide cutting edges. Flat surface on the inside curvature of the needle to minimize risk of needle tearing through the tissue

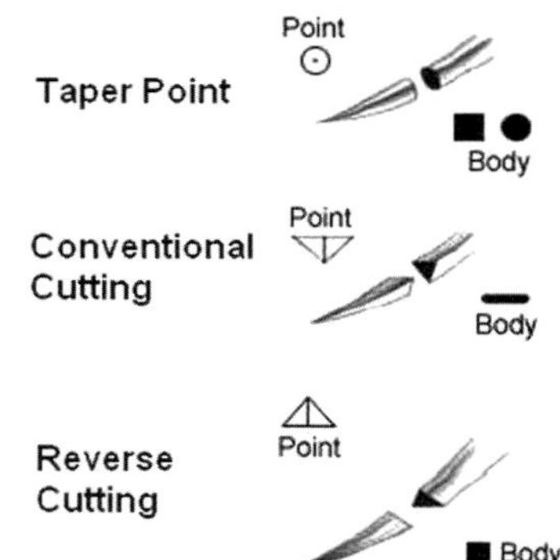

FIGURE II.5.15.1 Cross-section and point shape of common suture needles.

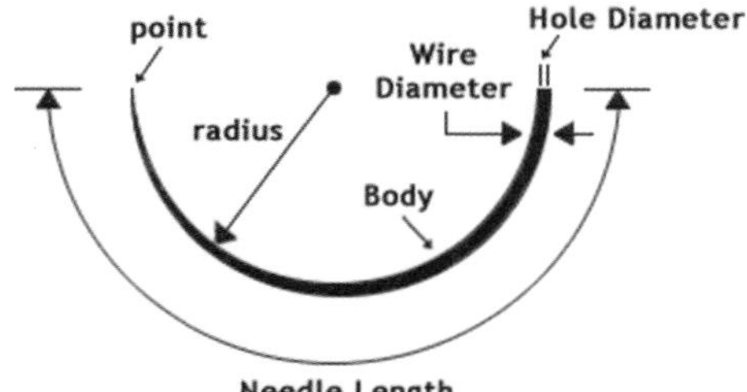

FIGURE II.5.15.2 Anatomy of a curved suture needle.

MANUFACTURING PROCESS AND INTENDED USE

Manufacture of the suture strand varies greatly, depending on the type of material. Natural materials are harvested in various ways and formed into fibers, but do not require extrusion, while synthetic materials require extrusion to form fibers. Most often melt extrusion is used, but some materials require gel spinning for fiber formation. Monofilament (single fiber) or multifilament extrusion is followed by orientation, a process in which the extrudate is heated and oriented, by drawing to reduce fiber size and align the molecular chains in the amorphous region and the crystallites therein to increase tensile strength along the fiber axis. Multifilament fibers are combined by braiding to form a suture strand, which is further stretched to tighten the construction. Monofilament and braided strands are heat treated, which could include annealing and/or relaxation, to increase dimensional stability and relieve internal stresses associated with previous processing. The suture strand, in some cases, is then surface coated to increase lubricity, improve tie down characteristics (during knot formation) or impart other desired attributes, such as reduction of capillarity and the inclusion of antibacterial agents.

Needles are typically attached to the suture strand by swaging (or crimping) the end of the needle to the strand. For small size sutures, adhesive attachment is preferred. The needle/strand combination is placed in a folder, and then in a pouch. The pouch is sealed and sterilized by high-energy irradiation or ethylene oxide gas (denoted EO or EtO). For absorbable sutures sterilized by ethylene oxide, the pouch is dried to a very low moisture level (<500 ppm), and sealed a second time to create an occlusive moisture barrier, providing protection from moisture.

Surgical Gut Sutures

Gut sutures are primarily derived from bovine intestinal serosa, but also from the submucosa of sheep or goat intestine. The base material is primarily collagen, and requires purification and strengthening before it is suitable as a suture. After harvesting, the material is cleaned and cut into ribbons. The cleaned ribbons are further processed with a dilute formaldehyde solution, which increases strength and slows the degradation rate by modifying the activity of the end groups of the collagen (Chu, 1997a). The ribbons are plied (twisted together), ground, and polished to create the final suture diameter. Natural (or plain) gut suture is now ready for packaging. Chromic gut sutures are further processed by tanning in the presence of chromium salts, which slows the rate of degradation (Chu, 1997a).

The suture is sized by centerless grinding, in which two grinding wheels remove material around the diameter of the suture. Due to the grinding process, fibrils at the strand surface can be damaged, which can lead to weak points and inconsistent strength. Also, the fiber is easily frayed during tying (Benicewicz and Hopper, 1990). Since gut sutures are a naturally derived material, there are slight inconsistencies in performance, including initial strength and strength retention profile. The grinding and polishing process can, ultimately, result in fraying and inconsistent diameter, leading to weak spots along the suture strand. Coating gut sutures to lower their friction coefficient and propensity to fray was described by Totakura and Shalaby (1997a,b). Recent studies have discouraged the use of gut sutures, and created very stringent manufacturing requirements in Europe and Japan due to concerns over bovine spongiform encephalopathy (mad cow disease) (European Commission, 1998). These materials, however, are inexpensive compared to their synthetic counterparts, and have a long history of use, which accounts for their continued popularity. Gut sutures can be sterilized by irradiation or ethylene oxide gas permeation.

Silk Sutures

Silk suture is a protein fiber typically derived from the *Bombyx mori* silkworm (Chu, 1997a). The silk is stripped of its natural waxes and gums, and braided or twisted to construct the suture thread. As with all braided sutures, it is coated. Typically, this is in the form of a beeswax-based coating or, in some cases, a siloxane-based coating is used. Successful efforts to substitute the highly tissue-reactive beeswax-based suture with practically non-reactive synthetic polyesters led to the development of a low-reactivity silk suture which can be made antimicrobial (Nagatomi et al., 2008). Although silk is not considered an absorbable suture, the strand loses strength and mass after implantation (Van Winkle and Hastings, 1972; Stashak and Yturraspe, 1978; Benicewicz and Hopper, 1990). This material is broken down by proteolysis, and is typically undetectable after two years (Lai et al., 2009). Silk sutures can be sterilized by irradiation and ethylene oxide gas permeation.

Cotton Sutures

Cotton has only been widely used as a suture material since World War II, due to the scarcity of silk (Harms, 1948). Cotton is a cellulose material and is prepared from long-staple yarn. A wax coating is applied to improve lubricity and help reduce fraying. An interesting note is that cotton gains strength when wet (Herrmann, 1971; Stashak and Yturraspe, 1978). Cotton elicits a mild tissue response, similar to silk, and exhibits a gradual loss in tensile strength over about a two-year period (Van Winkle and Hastings, 1972). Cotton sutures are sterilized by ethylene oxide. Although not as strong as silk, cotton sutures are still available today, and are supplied as a twisted multifilament yarn.

Polyester Sutures

Non-absorbable polyester sutures are made from polymers such as poly(ethylene terephthalate) (PET) and poly(butylene terephthalate) (PBT), which are produced by condensation reaction. This involves condensation of a glycol with terephthalic acid or an ester of terephthalic acid. The repeat unit of PET is shown in Table II.5.15.5. PBT is relatively compliant compared to PET, although both are extruded into multifilament fibers and braided into suture. Both of these braided sutures are coated to improve their lubricity. PET, introduced in the 1950s, is the most commonly used polyester suture, and has well-documented properties. Polyester sutures are sterilized mostly by gamma irradiation.

Nylon Sutures

The polyamides Nylon-6 and Nylon-6,6 are both used as suture materials. Structures for both repeat units are shown in Table II.5.15.5. Nylon materials were discovered in the 1930s at DuPont, who first manufactured Nylon-6,6 through a condensation reaction using hexamethylene diamine and adipic acid. Other companies developed the similar Nylon-6, which is formed by the ring-opening polymerization of caprolactam. Nylon-6,6 is slightly limper (softer hand) than its counterpart, although mechanical properties are similar for both materials.

Nylon sutures are made into braided or monofilament sutures, although the monofilament sutures may not tie as well, and require additional knots to secure them (Stashak and Yturraspe, 1978; Bellenger, 1982; Guttman and Guttman, 1994). To counteract this, monofilaments can be packaged with a plasticizing fluid to soften the hand (Capperauld and Bucknall, 1984). Although nylon is a non-absorbable suture, the amide bond is susceptible to hydrolysis and can lose strength over time (Chu, 1997a). Nylon sutures are sterilized by gamma irradiation.

TABLE II.5.15.5 Repeat Units for Common Synthetic Non-Absorbable Sutures

Monomer Names	Repeat Unit Structure	Homo-polymer
Ethylene	$[CH_2\text{-}CH_2]_n$	Polyethylene (PE)
Propylene	$[CH(CH_3)\text{-}CH_2]_n$	Polypropylene (PP)
Hexamethylene diamine and adipic acid	$[CO\text{-}(CH_2)_4\text{-}CO\text{-}NH\text{-}(CH_2)_6\text{-}NH]_n$	Nylon-6,6
Caprolactam	$[CO\text{-}(CH_2)_5\text{-}NH]_n$	Nylon-6
Ethylene terephthalate	$[CO\text{-}C_6H_4\text{-}CO\text{-}O\text{-}CH_2\text{-}CH_2\text{-}O]_n$	Polyethylene terephthalate (PET)

Polypropylene Sutures

Polypropylene can be made with an atactic, isotactic or syndiotactic chain conformation. The material tacticity determines how the material will crystallize, and the mechanical properties of the final device. Material used for suture applications is practically isotactic, with an intermediate density. Polymerization is accomplished by polymerization from propylene monomer using a Zeigler-Natta catalyst; the repeat unit can be seen in Table II.5.15.5. Polypropylene sutures elicit very low tissue response, and are not susceptible to hydrolysis. These sutures can also be difficult to tie securely, due to the smoothness and lubricity of the suture strand (Stashak and Yturraspe, 1978; Bellenger, 1982; Guttman and Guttman, 1994). Polypropylene is easily melt-extruded into monofilament or multifilament yarn, and is well-suited for the production of monofilament suture. Polypropylene sutures are usually sterilized by ethylene oxide, and occasionally by high energy radiation.

Ultra-High Molecular Weight Polyethylene (UHMWPE) Sutures

A relatively new material that has gained considerable popularity with orthopedic surgeons is ultra-high molecular weight polyethylene (UHMWPE) suture, which means they are a very high molecular weight polymer of ethylene (structure seen in Table II.5.15.5). The molecular weight can be difficult to determine directly, but is typically in the millions of Daltons. Introduced in the early 2000s, these ultra-strong sutures are 10–100 times stronger by weight than steel. As opposed to other synthetic suture materials, which exhibit about 40% crystallinity, UHMWPE can obtain crystallinity of greater than 85%, and can be highly oriented. The high molecular weight prevents the use of melt extrusion. Therefore, gel spinning techniques are employed to create the multifilament fibers. Gamma irradiation weakens this suture; for this reason they are sterilized by ethylene oxide (Deng et al., 1996).

Ultra-high strength sutures are much stronger than standard ones, and a smaller size will have a similar breaking load to a larger size standard suture. This allows the surgeon to use a smaller diameter suture for the same application, leaving less foreign material in the implant site. The inherent lubricity of the material results in increased tendency to slip compared to polyester sutures (Barber et al., 2009). Molded UHMWPE will also creep under load, although sutures are not under the same loading conditions and will therefore not exhibit the same level of creep as other devices made of the same materials, as in replacement joint bearing surfaces.

In addition to sutures made of only UHMWPE, there are several composite sutures that are prepared by co-braiding UHMWPE with PET, nylon or other materials. Ultra-high polyethylene suture is difficult to dye during extrusion, and its high crystallinity makes post-extrusion solvent dying difficult, so manufacturers exploited other materials that can be easily dyed as a component in the suture braid, as a means to create contrast with surrounding tissue (Teleflex Medical OEM, 2009). Also, textures imparted by combining different materials allow for more secure tying, which can be a problem with inherently lubricious UHMWPE sutures (Burkart, 2004).

Stainless Steel Sutures

For surgical procedures requiring high strength materials, such as sternal fixation and tendon repair, stainless steel is a common selection. These sutures are most often prepared from 300-series stainless steels (316L is a common variety and is used in Covidien STEEL sutures, for example). Stainless steel sutures are inert, do not exhibit capillarity, and are very strong. The high modulus of stainless steel, however, can cause irritation at the free end of the suture. These sutures are also prone to kinks, and surgeons must use extra caution during handling (Bellenger, 1982; Capperauld and Bucknall, 1984; Guttman and Guttman, 1994). Stainless steel sutures are available as a braid or monofilament (for small sizes).

Synthetic Absorbable Sutures

Monomers and Preparation of Polymers. Natural and synthetic non-absorbable materials are prepared from common materials that are also used in their fiber forms commonly used in many other industries. Synthetic absorbable sutures, which account for the largest segment of suture sales, were designed specifically for suture applications, and are currently the area with the most development possibilities. These materials are polymerized from a select group of constituent monomers, which can be combined in a vast array of configurations exhibiting unique properties. These monomers include glycolide, lactide (*d*-, *l*-, and *dl*-), *p*-dioxanone, trimethylene carbonate (TMC), ε-caprolactone, and polyethylene glycol (PEG), which become repeat units in the final polymer. Table II.5.15.6 includes the repeat unit structure of commonly used absorbable polymers, all of which degrade by hydrolysis of ester bonds. Of recent interest are sutures containing PEG, as in polyether-esters, which impart a certain degree of "swellability" to the suture. Polymers used to prepare absorbable sutures are crystalline, tough, and able to be oriented along the long axis of the fiber. Most of these are in the form of random copolymers, although some are made as block or segmented copolymers.

Polyglycolide and predominantly glycolide-based copolymers are highly crystalline stiff materials which are very susceptible to hydrolysis, leading to a faster absorption profile. Lactide-based polymers can be crystalline, but are slightly softer than glycolide-based ones, and are not as susceptible to hydrolysis. *p*-Dioxanone polymers are crystalline, but are softer than most of the glycolide- or lactide-based polymers, and are between glycolide and lactide in sensitivity to hydrolysis. These three provide the major constituents for all synthetic absorbable sutures. TMC, ε-caprolactone, and PEG represent minor constituents in a number of sutures, and serve to alter the material properties to improve some aspect of suture performance. For example, many monofilaments are prepared from copolymers made primarily with glycolide as a hard segment and TMC or caprolactone as a softer segment, making the copolymer more flexible, while retaining strength.

Ethylene oxide is the standard method for sterilization due to the tendency of the copolymer chains to undergo scission when irradiated. Synthetic absorbable polymers degrade by hydrolysis, and are therefore sensitive to moisture. For this reason, the packaging is not only important for providing a sterile barrier, but also for protecting the suture from moisture.

TABLE II.5.15.6 Repeat Units for Common Synthetic Absorbable Sutures

Monomer Names	Repeat Unit Structure	Homopolymer
p-Dioxanone	$[O\text{-}CH_2\text{-}CH_2\text{-}O\text{-}CH_2\text{-}CO]_n$	Polydioxanone (PDO)
Glycolide	$[CH_2\text{-}COO\text{-}CH_2\text{-}COO]_n$	Polyglycolide (PGA)
l-Lactide	$[CH(CH_3)\text{-}COO\text{-}CH(CH_3)\text{-}COO]_n$	Polylactide (PLA)
ε-Caprolactone	$[O\text{-}(CH_2)_5\text{-}CO]_n$	Polycaprolactone (PCL)
Trimethylene carbonate	$[(CH_2)_3\text{-}O\text{-}CO\text{-}O]_n$	Polytrimethylene carbonate (PTMC)
Ethylene oxide	$[CH_2\text{-}CH_2\text{-}O]_n$	Polyethylene glycol (PEG, used as polyether-ester constituent in copolymers)

Absorbable sutures are described primarily by their strength retention profile, then by absorption profile. The strength retention profile indicates the useful life of the device, and is the primary factor for selection in a particular application. The strength retention and absorption profiles for several common sutures are shown in Figures II.5.15.3 and II.5.15.4, respectively.

PGA

Dexon™, the first synthetic absorbable suture, is prepared from a homopolymer made of glycolic acid, and was released in 1971 by Davis & Geck, which is now Covidien (Benicewicz and Hopper, 1990). PGA suture is available in braided form, and is coated to improve the handling properties. Current PGA sutures exhibit about 30–50% crystallinity and retain about 65% of the initial strength after two weeks implantation, and 35% after three weeks. Absorption is essentially complete after 60–90 days (Covidien, 2005).

PDO Sutures

First introduced by Ethicon in 1983 as PDS™, monofilament sutures comprised of *p*-dioxanone (PDO) combine strength, flexibility, and a longer absorption profile than PGA sutures. After the initial introduction of PDS, Ethicon released PDS™ II. Reportedly, slight changes in the heat treating process resulted in greater flexibility and better handling characteristics (Chu, 1997b). PDS™ II suture retains 25% of its initial strength after 6 weeks, making it the longest-lasting commonly used

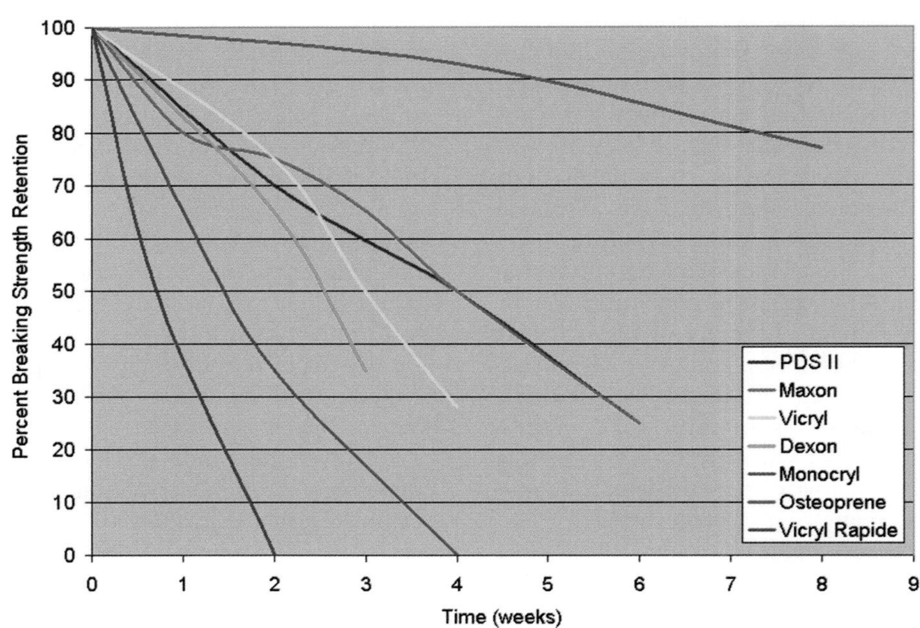

FIGURE II.5.15.3 Breaking strength retention profiles of several common synthetic absorbable sutures (Ethicon, 2001, 2008; Covidien, 2009a,b; Taylor, 2010).

suture currently available, allowing it to be used in slow healing tissue that previously required non-absorbable sutures (Ethicon, 2001; Capperauld and Bucknall, 1984). Also of interest are sutures prepared from PDO-based polymers, as recently discussed by Carpenter and Shalaby (2002).

High Glycolide Copolymeric Sutures

The largest segment of absorbable sutures is prepared from copolymers containing primarily glycolide in the main polymer chain. The modulus and absorption profile varies widely, based on the copolymer composition, and is quite sensitive to changes in the molecular formula.

The most well-known synthetic absorbable suture is braided Vicryl™, which is a random copolymer containing 90% glycolide and 10% lactide (Guttman and Guttman, 1994). Braided Vicryl™ retains up to 25% of its initial strength for 4 weeks, and absorption is essentially complete in 56–70 days (Ethicon, 2001). An alternate version of Vicryl™, Vicryl™ Rapide, was introduced as a faster-absorbing version of the same material. This was accomplished by reducing the molecular weight of the fiber through a radiation sterilization process. Vicryl™ Rapide was designed as an alternative to Plain Gut suture, and loses strength within 2 weeks of implantation (Ethicon, 2008).

Most customization in this class of synthetic absorbable sutures has been in the area of monofilament sutures. Maxon™ suture, introduced in 1985 by Davis & Geck (now Covidien) is a tri-block copolymer prepared from 67.5% glycolide and 32.5% TMC (Benicewicz and Hopper, 1991). The central block is composed of a random copolymeric initiator made of 85% TMC and 15% glycolide, while the two end grafts are primarily glycolide (Casey and Roby, 1984). This composition and structure is a result of a study comparing various block structures and ratios of the constituent monomers, results of which can be seen in Table II.5.15.7. This work highlights the sensitivity of the copolymers to structure, as well as to monomer composition. Increased TMC in the central chain reduces the tensile modulus, extends the strength retention profile, and increases the time required for absorption. The small amount of glycolide in the middle block is necessary

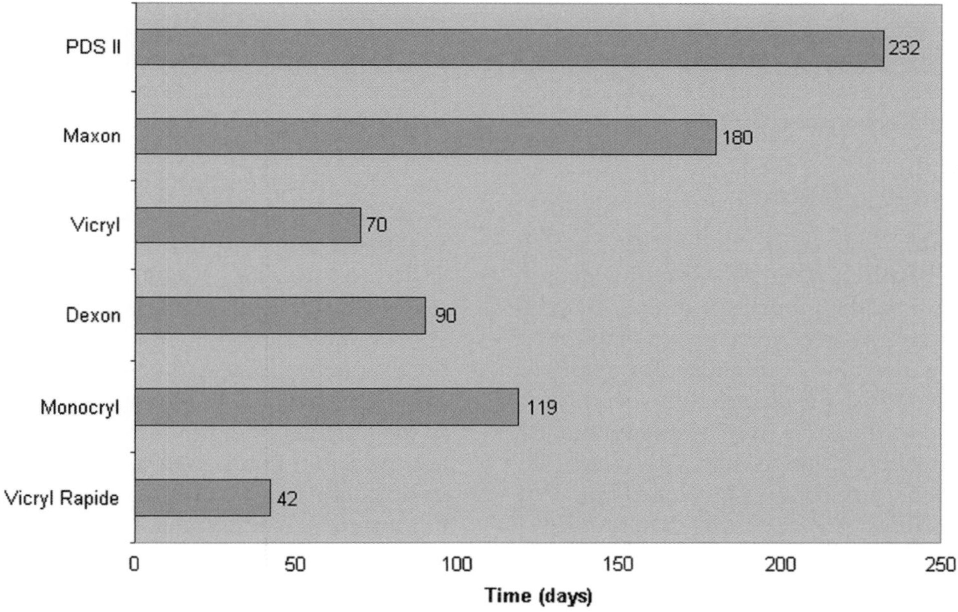

FIGURE II.5.15.4 Maximum time for sutures to be essentially absorbed (Ethicon, 2001, 2008; Covidien, 2009a,b).

TABLE II.5.15.7 Effect of Block Structure on PGA-co-TMC Copolymer Properties[1]

Block Type	Glycolide Content (wt%)	Melting Peak (°C)	Tensile Strength (kpsi)	Tensile Modulus (kpsi)	*In Vivo* BSR[2] at 3 Weeks (%)	*In Vivo* Absorption (Days)
Homopolymer	100	227	122	2100	20	<90
Random	88	209	65	970	3	<90
Block	65	216	42	610	53	<180
Modified block	65	213	61	260	69	<180

[1]Roby et al. (1997).
[2]Breaking strength retention.

because TMC absorbs very slowly; the glycolide allows the material to degrade into short segments that can be metabolized. Maxon™ suture retains 25% of its initial strength up to 6 weeks post-implantation, and essentially absorbs by 6 months. Other unique polymer compositions that are suitable for suture applications include Monoprene®, TGS, and Glycoprene® II, which are of particular interest due to the branched structure of the polymer compared to the linear structure of all other suture polymers, providing a softer hand compared to linear polymers with similar composition and strength retention.

To fill the need for a shorter-term monofilament, Ethicon developed Monocryl™ suture. This material was developed as an alternative to gut suture (Bezwada et al., 1995), and is a tri-block copolymer containing 75% glycolide and 25% caprolactone. The initiator is reported to be a random copolymer containing 55% glycolide and 45% caprolactone, while the end blocks are primarily made of glycolide (Bezwada et al., 1995). Monocryl™ suture retains 30–40% of its initial strength after 2 weeks of implantation, and is essentially absorbed between 91 and 119 days.

This is not inclusive of all sutures currently available, but is an example of a few technologies. Major suture manufacturers are continuing development of other high glycolide copolymers, and focus remains on bringing novel, proprietary sutures to market with unique strength retention and absorption profiles.

Lactide Copolymeric Sutures

Copolymers prepared primarily of lactide exhibit extended strength retention profiles, sometimes retaining measurable strength at 6 months or longer. For this reason, lactide copolymers have been used in orthopedic applications to prepare screws and plates, for example. For slow-healing wounds, an absorbable suture having this type of extended profile may be advantageous. This was the aim when Ethicon released Panacryl™ suture, a material composed of 95% lactide and 5% glycolide. The absorption profile of this material, however, can extend to 2½ years. As of 2006, Panacryl™ has not been marketed as a suture device, although it may have use as part of other Ethicon products. An alternative to Panacryl™ is Osteoprene®, currently the only other high-lactide suture, which is used as part of a suture anchor kit, although there are other lactide-based copolymers in development for suture applications (Carpenter et al., 2002; Anneaux et al., 2003; Arthrotek, 2006).

Dyes

Sutures are provided in *natural* or *dyed* colors. The natural color of suture materials is most often white to light tan in color, with the exception of chromic gut which is a darker brown color. When sutures are placed in tissue, the contrast between the suture and host tissue is low. This effect is magnified when in the presence of blood, which can make differentiation between the suture and tissue difficult. For this reason, dyed sutures are often preferred. The use of colorants is restricted by the FDA (US Food and Drug Administration), and there are a select few dyes that are approved for use in specific sutures. New dyes can be used with approval through a Color Additive Petition, which requires extensive biocompatibility and toxicology data to receive approval. For this reason, sutures are available in a select color palette, as seen in Table II.5.15.8. Synthetic sutures are dyed during the extrusion process by adding a small amount (typically less than 0.2% by weight) of powdered colorant to the polymer just prior to extrusion. Natural materials are dyed in a color solution. Processing conditions can also prohibit the use of some dyes in the suture strand. For instance, D&C Violet #2 is not used in sutures sterilized by gamma irradiation, because the dye is not stable, degrades, and loses color during exposure.

TABLE II.5.15.8 FDA-Approved Dyes for Use in Sutures[1]

FDA-Approved Dye	Color	Used in Sutures Containing
Pyrogallol/Ferric ammonium citrate	Blue	Gut
Logwood extract	Black	Silk, nylon
Chromium-cobalt-aluminum oxide	Blue-green	Polyethylene
D&C Blue No. 9	Blue	Cotton, Silk
D&C Green No. 5	Green	Nylon
[Phthalocyaninato(2-)]Copper	Blue	Polybutester, polybutylene terephthalate, poly(vinylidene fluoride), poly(vinylidene fluoride-co-hexafluoropropylene)
FD&C Blue No. 2	Blue	Nylon
D&C Blue No. 6	Blue	Various
D&C Green No. 6	Green	Various
D&C Violet No. 2	Violet	PDO, PGA, glycolide copolymers

[1]FDA (2007): 21 CFR Parts 73 and 74.

Coatings

Braided and monofilament sutures have been coated with a wide variety of materials for the primary purpose of reducing tissue drag. Coatings also fill void space, thus reducing capillarity and tendency to wick, and can act as carriers for antimicrobial agents such as triclosan. Early coatings comprised of beeswax and other natural waxes. Other coating materials for non-absorbable sutures include silicone- and PTFE-based systems, as used on Deknatel's Tevdek® and Polydek® polyester sutures. Coatings for absorbable sutures are typically based on absorbable copolymers, and can include polyurethanes, PEG, caprolactone copolyers, and lactide/glycolide copolymers (Mattei, 1980; Bichon et al., 1984; Ethicon, 2001). Calcium stearate and other lubricants may also be included to improve friction properties.

Coatings are generally applied through solution dip coating, and are applied at a level of 2–10 wt% of the suture strand for braids, and less than 1 wt% for monofilaments. In the case of antimicrobial sutures, a small amount of triclosan is dissolved in the coating solution, and sutures are coated in the typical fashion. If the suture is coated with too little coating, the suture will display high tissue drag and the knot will not tie smoothly, possibly fraying the suture. With a high level of coating, knot security may be compromised and the strand can loosen, thereby not securing the wound site.

Needles and Attachment

Suture needles are made from high-quality stainless steel. Typically, needles are prepared from 400-series stainless steel, though some specialties prefer 300-series needles (Hu-Friedy Mfg. Co., Inc., 2007). The manufacturing process starts with wire, which is formed to shape, and precision ground and polished to provide the final geometry. The needle end on the suture side is most typically drilled to accept the suture strand (known as a *drilled end* needle). Needles are cleaned to remove any process residue, and some are coated with silicone to reduce friction when passing through tissue.

Needles are permanently attached by a *swaging* process. Monofilament sutures are placed directly into the drilled end for swaging, but braided sutures require *tipping* to harden the end of the strand so the swaging process holds firmly onto the end of the suture. Tipping is performed with a cyanoacrylate-based adhesive on the last ½ inch of the suture strand. Also, for small size sutures, epoxy and cyanoacrylate are commonly used for needle attachment.

Packaging

Sutures are sold sterile in individual or multi-packs. Packaging for non-absorbable sutures can be in Tyvek® packaging; however, synthetic absorbable sutures are always packaged in hermetically sealed foil packaging. The sealed foil maintains a sterile barrier, as well as protecting the suture from oxygen and moisture which would initiate the degradation of the suture. This is known as a *hermetically* sealed package. Seals are designed to provide a secure seal, but also for easy opening in the surgical suite. This is often obtained by using a chevron-style peel pouch.

Inside the package is a carrier, which houses and protects the suture and needle. The carrier is made of paper or plastic, and is designed so that suture can be removed easily without tangling or otherwise damaging the suture strand. Carrier materials must be made of materials that do not adversely affect the suture. For absorbable materials, this includes drying paper carriers before insertion so that the paper does not introduce moisture in the package, leading to decreased shelf-life stability of the suture.

Common Uses of Suture

Besides wound closure and ligation, sutures are used in a variety of other medical procedures. In particular, many implants require sutures to maintain placement. Grafts require sutures for attachment at *anastomoses* (junctions between two structures, such as between an artery and a graft), and different graft types have a higher chance of surgical success with certain suture types. PET (e.g., Dacron®) grafts are typically woven and provide a tough, resilient, non-absorbable implant. Due to this construction, the suture type is not very critical. For other types of grafts, such as Gore-Tex® or naturally derived materials, the selection of monofilament sutures is indicated to reduce the chance of suture tearing (W. L. Gore & Associates, 2007). For joint repair, sutures are used in conjunction with suture anchors to reposition bone, which requires sutures with high strength, and fatigue and abrasion resistance.

> Surgeons use many types of knots, depending on the tissue type and location, suture type, and placement technique. Knots can be classified as: (1) non-sliding; (2) sliding; or (3) sliding locking. The knot should be as small as possible to minimize tissue reaction, but should be strong and stable. Examples of common surgical knots include: Surgeon's Knot; Square Knot; Revo Knot; SMC Knot; Duncan Loop; and Tennessee Slider. Often, two to four additional throws are added on top of these knots to enhance knot security.
>
> Forceps are used to assist in suture placement and knot tying. The free ends of suture are trimmed with scissors close to the knot to minimize material left at the implant site. For internal placement, it is necessary to use additional surgical tools to securely tie knots. This includes a *knot pusher*, which is available in many sizes for uses in laparoscopic, arthroscopic, and other types of procedures.

PERFORMANCE EVALUATION

There are many factors defining the performance of a suture, covering every component of the suture device. The suture must have suitable mechanical properties, which can vary widely depending on the intended application. The suture material must be sterile and biocompatible throughout the life of the device, and must be stable so that material can be stored before use. There is a large variety of analysis to prove the safety and effectiveness of the suture device. Some of the analyses are standard test protocols that are applied to every biomedical device, and some are designed to analyze specific functionalities of the suture.

Physical Properties

The function of the suture strand is to provide a holding force while maintaining a stable position in the implant location. There are, however, many aspects to tissue securement. The *United States Pharmacopeia* (USP) is a list of standards and requirements for the testing and production of medical devices, including several guidance documents pertaining to sutures.

Suture is defined by USP size, which is defined numerically on a scale from size 7 (largest) to 11–0 (smallest); however, the majority used is between 6–0 and 2. For size 0 and smaller, the number of 0s in the suture size increases as the diameter decreases, e.g., 00 is written as 2–0 and 0000 is written as 4–0. Smaller diameter means the suture will exhibit lower breaking load, and all sutures must meet requirements set out in the USP. The accepted practice is to use the smallest diameter suture that will adequately support the wound. This minimizes trauma caused by the placement, and ensures the minimum mass of foreign material is left in the body.

There are two basic tensile strength tests associated with sutures. USP requirements are based on the knotted tensile strength, which is always lower than the unknotted tensile strength. This ensures that the suture is strong enough to be tied during implantation without breaking. Typically, knot tensile strength is 60–80% of the unknotted (straight) tensile strength. Unknotted, or straight, tensile strength provides a good measure of the comparative mechanical properties. Beyond ultimate elongation, the tensile modulus, for monofilaments, provides a good measure of the material stiffness.

Another mechanical requirement set forth in the USP is the needle attachment strength. Suture strands are placed by pulling a needle through the tissue. The needle is fixed to the suture strand, and this junction must meet a minimum strength to ensure that the needle will not separate from the strand under typical implantation conditions.

In Vitro and *In Vivo* Performance

For absorbable sutures, strength retention profile is the key to material functionality. There are two methods for determining the strength loss profile. *In vivo* strength is typically analyzed by implanting strands of the suture subcutaneously in Sprague-Dawley rats. *In vitro* analysis involves placing suture in a physiologic buffered solution, e.g., 7.4 pH phosphate buffered solution (PBS) at 37°C. At predetermined time points, samples are collected and tested for ultimate tensile strength. Information is reported as a percentage of the initial load. *In vitro* analysis must always be followed up with at least limited *in vivo* analysis, due to the difficulty of directly correlating performance between the two degradation conditions.

The suture strand must remain securely in place after implantation. To accomplish this, the knot needs to be stable. A single surgeon's knot is typically not sufficient to hold the suture securely, particularly in coated sutures. Additional knots are tied to increase the holding power of the knot. Contributing factors to the knot stability are friction properties associated with the suture strand. For this reason, friction is another important performance parameter.

The ideal suture provides not only tissue securement, but is also sterile and biocompatible. Medical devices must meet stringent requirements for safety, with guidance by the International Standards Organization (ISO) standard 10993-1 (ISO, 2009). According to this document, the suture strand must be tested for cytotoxicity, genotoxicity, sensitization, irritation, systemic toxicity, and sub-chronic toxicity. This represents an extensive set of *in vitro* and *in vivo* analyses. For absorbable sutures, the hydrolysis by-products, along with the polymeric material, are tested.

Suture devices are labeled *sterile*, and manufacturers must continually prove this by monitoring the effectiveness of every sterilization batch. Sterility, while not a factor in the functionality of the device, is paramount to safety. Manufacturers use spore strips, a biological indicator containing a certain number of spores, to perform this test. For the standard sterilization processes used in suture manufacturing: (1) ethylene oxide gas permeation; and (2) radiation, *bacillus atrophaeus* and *bacillus pumilus* spore strips are used, respectively, to verify sterility.

Also, for absorbable sutures, the amount of time for the tissue to absorb is a contributing factor to performance. To analyze this property, samples can be placed *in vitro* in 7.4 pH PBS at 37°C, with samples being removed at intervals and dried to constant weight. Mass loss profile is reported as a percentage of the original weight. *In vitro* test methods are only appropriate for those materials that degrade by hydrolysis, but not for sutures that are only affected by enzymatic degradation, such as gut sutures. A less quantitative method to determine absorption profile involves implantation of the suture material intramuscularly. Histological evaluation over time reveals the absorption profile. Absorption is defined as the time the suture material takes to *essentially absorb*, or lose greater than approximately 90% of its initial mass or cross-sectional area. Above this level, mass loss is difficult to analyze. Another outcome of this histological evaluation is the measure of tissue reaction to the suture, typically sampled at one week post-implantation.

In vivo analysis of sutures involves implantation in an animal model. To determine absorption and strength retention properties, a Sprague-Dawley rat model is often used. Other animal models are often employed to study the effect of surgical technique, the application of a particular suture or the suture as part of a device. This includes porcine models to compare suture technique on vascular anastomoses (Tozzia et al., 2001), rats to analyze the effectiveness of Achilles tendon repair (Kraemer et al., 2009), and rabbits to determine the effect of different suture materials in bladder surgery (Hanke et al., 1994).

Additional Performance Considerations

There are many considerations dealing with the handling of sutures that are not covered in the above testing. These include the suture *hand*, or the perceived handling characteristics. Suture hand includes the suture *drape*, which is the suture limpness (think raw versus cooked spaghetti), and the ease of knot tying. It is difficult to quantify these properties, and results are subjective.

For a medical device to be useful, it must have a proven shelf-life. Shelf-life studies track functional suture properties over standard storage conditions to determine the expiration date. Accelerated conditions are also analyzed to provide early indications of shelf-life, as specified in ASTM F1980, (ASTM, 2011). Common parameters analyzed include diameter, knot strength, and strength retention for absorbable sutures.

Thermal transitions are a contributing factor for processing conditions. Melting temperature and rheological analysis provide guidance for the melt extrusion. Glass transition temperature and information about the crystallinity endotherm, best measured with dynamic mechanical analysis (DMA) and differential scanning calorimetry (DSC) respectively, provide insight into the process temperatures to use during orientation and heat treatment. To measure the effectiveness of heat treatment and stabilization, strands of fiber are heated and measured for dimensional stability. Dimensional stability is critical, particularly for sutures sterilized by ethylene oxide gas permeation, so that fiber properties do not change due to exposure to sterilization or storage temperature.

REGULATORY CONSIDERATIONS

In order to bring a new suture to market in the United States, it must be cleared for use by the US Food and Drug Administration (FDA). Sutures are primarily considered a Class II device by the FDA, meaning new devices can be submitted using a 510(k) approach instead of a premarket approval (PMA), which significantly reduces the time-to-market (FDA, 2003). Documentation for a 510(k) submission provides data proving the new device is *substantially equivalent* to a device that is already legally marketed for the same use (known as a *predicate device*). Novel suture materials, those that do not have a substantially equivalent predecessor, may be considered a Class III device, in which case a PMA would be necessary. The purpose of a PMA is to provide adequate safety and efficacy information for a new material or a new indication for an existing material, which requires extensive pre-clinical and clinical testing, resulting in longer development times and elevated cost.

The FDA provides guidance documents to assist with the assembly of the submission packet. Accordingly, it is expected that the submission will include: (1) the design for package labeling, instructions for use (IFU), description of the device, indications, contraindications, warnings, and precautions; (2) identification of the risks associated with the use of the device and mitigation measures; (3) description of test methods and standards used during design and for release testing; (4) proof of biocompatibility following the recommendation of ISO 10993-1 standard for the biological evaluation of medical devices; and (5) a validated sterilization method (FDA, 2003). Each part of the submission can be very involved, and evidence must be provided that each section is substantially equivalent to the predicate device. For absorbable sutures, part of the above is providing evidence for the strength retention and mass loss profile. Also of interest are the various validations and studies that must be performed to show that the manufacturing process produces suture with consistent properties, and the packaged material is stable after final packaging (known as *shelf-life* or *package stability*).

The above illustrates the extent of testing necessary to bring a suture product to market, even after the product has left the initial phases of research and development. Commonly, new suture materials will take several years from conception to reach market using the 510(k) approach. When designing novel suture devices, a major consideration in material selection is the ability to show substantial equivalence for materials and indications that an extended approval process using a PMA will not be necessary.

NEWER TRENDS AND FUTURE DEVELOPMENT

In the early 2000s, Ethicon released the first of their *Plus* line of sutures containing a small amount of the antibiotic/antifungal agent Irgacare®, a brand name for triclosan. Since this time, other agents have been incorporated into sutures and explored for antimicrobial effectiveness, including chlorhexadine and octenidine (Matl et al., 2009). Also, various coating polymers have been applied to improve or extend the length of triclosan from the suture (Nagatomi et al., 2007). Therapeutic release applications have not been limited to antimicrobials, and drug release from the suture coating could be applied to a wide variety of soluble drugs and other active agents (Zhukovsky, 2003). Modified sutures have also been used as carriers in non-traditional applications. An instance of this includes the use of suture strands to position radioactive brachytherapy seeds to treat the cancerous margins after lung resection (Fernando et al., 2005).

Medical devices are now being developed with the specific goal to repair the injured area to pre-wound functionality, and a great deal of research is developing the idea of *active* wound healing. Devices can accomplish this by initially supporting the wound through the initial healing phases, then transferring strain to the tissue to exercise and thereby improve localized healing. In other technologies, active wound healing technologies are becoming more common; an example is the recently released TIGR™ hernia mesh marketed by Novus Scientific. In the case of sutures, several patents have been issued with suture designs exhibiting biphasic functionality (Shalaby et al., 2006a). This novel class of sutures, currently in development, would provide stiffness through the initial phase of wound healing, after which the suture loosens to allow strain transfer to the wound, promoting improved wound healing.

New extrusion techniques are allowing for design of non-traditional surgical threads. Multicomponent co-extrusion is allowing for filaments with *island in the sea* and other geometries that could provide enhanced properties. In one instance, a PDO polymer was co-extruded with a softer copolymer containing *p*-dioxanone, trimethylene carbonate, and caprolactone, the fiber of which exhibited a much higher knot security compared to a standard PDO suture (Im et al., 2007). Also of interest is the creation of ultra-high strength absorbable sutures. Currently, high-lactide copolymers are being investigated for this use, with proposed applications in orthopedic areas.

A novel class of synthetic absorbable sutures includes those that swell in contact with water (Lindsey et al., 2007; Ingram et al., 2009). These materials, based on polyether-esters (such as PEG), will fill the void created by the suture needle. This technology could prove particularly useful in cardiovascular surgery, where minimized leakage at anastomosis sites is desired. Also in this class of sutures, currently under development, is USLG, a suture exhibiting an extended strength retention profile (25% strength retention after 18 weeks in 37°C PBS) (Taylor, 2010).

While traditional sutures depend on knots to secure placement of the suture strand, a novel technique involves the use of a suture with an integral securement mechanism, or barb, as seen in Figure II.5.15.5. Quill™ SRS barbed suture, marketed by Angiotech, was approved for use in 2004, and is gaining popularity in many soft tissue procedures. This alternative approach to wound closure provides a faster method of approximation, and avoids extraneous foreign material in the wound due to the bulk of a surgical knot (Angiotech Pharmaceuticals, 2009).

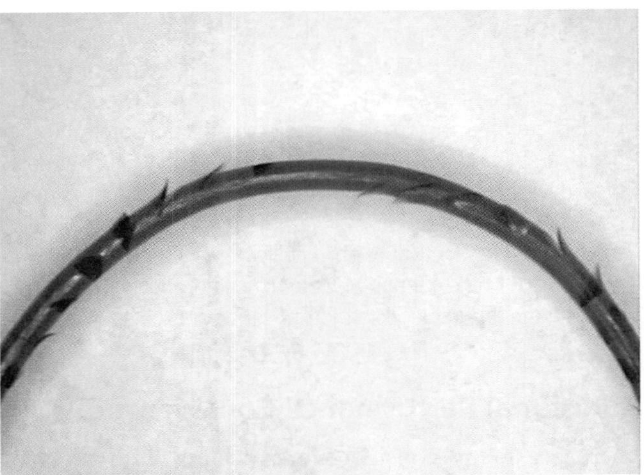

FIGURE II.5.15.5 Image of spiral cut barbs in Quill™ SRS suture, marketed by Angiotech Pharmaceuticals.

Beyond suture applications, these monofilaments and braids are being incorporated into a host of medical devices. For example, suture-type materials are in use as components in hemostasis devices (FemoSeal®, marketed by St. Jude), in marker devices (Hologic SecurMark® Biopsy Site Marker), in hernia mesh, and tissue engineering scaffolding.

While sutures are most commonly sterilized by gamma irradiation or ethylene oxide, there are new sterilization methods under development that may not degrade sensitive suture materials, as can be the case with both of the standard methods. Of particular interest is a new method developed and used experimentally by Shalaby and co-workers for the sterilization of radiation sensitive sutures such as polypropylene and absorbables (Shalaby, 2003, 2010). The new method is denoted radiochemical sterilization, and is a combination of low-dose gamma radiation (5–10 kGy) and radiolytically-generated formaldehyde from an oligomeric or polymeric precursor.

Alternatives to Suture

It is important to note that sutures are not the only device used for ligation and approximation. Alternative methods include staples, tapes, clips, tissue sealants, and tissue adhesives. Staples and clips can be applied quickly, and are made of non-absorbable, as well as absorbable materials. These devices can provide securement faster than sutures, but are limited in the areas of use, whereas sutures can be used for a much wider range of applications.

The most interesting of these emergent technologies is tissue adhesive. Topical tissue adhesives based on cyanoacrylates, such as Histoacryl®, Indermil®, Dermabond®, and TissueMend™ II, quickly form a strong bond to tissues. The film of adhesive, in addition to securing the wound, can act as a bacterial barrier, and affect hemostasis. There is currently no tissue adhesive approved for internal use, but development of new absorbable tissue adhesives, such as Tissuemend™ II and HTS, may be used for just such an application. Additionally, the above absorbable tissue adhesives may be useful in areas that are traditionally difficult to treat with sutures, such as in the lung and liver (Shalaby et al., 2006b).

BIBLIOGRAPHY

Angiotech Pharmaceuticals. (2009). *Quill™ SRS Manual.*

Anneaux, B. L., Carpenter, K. A., Greene, D. D., Taylor, M. S., Shalaby, M., et al. (2003). Effect of composition on physical properties of segmented copolylactides and BSR of suture braids therefrom. *Trans. Soc. Biomater.*, **26**, 321.

Arthrotek. (2006). *Charlotte™ Shoulder System.* http://www.biomet.com/sportsMedicine/getFile.cfm?id=1837&rt=inline. Date accessed 5/5/2010.

ASTM. (2011). ASTM F1980-07 (2011): Standard Guide for Accelerated Aging of Sterile Barrier Systems for Medical Devices. ASTM International.

Barber, F. A., Herbert, M. A., & Beavis, R. C. (2009). Cyclic load and failure behavior of arthroscopic knots and high strength sutures. *Arthroscopy*, **25**, 192–199.

Bellenger, C. R. (1982). Sutures. Part I: The purpose of sutures and available suture materials. *Comp. Ed. Pract. Vet.*, **4**, 507–515.

Benicewicz, B. C., & Hopper, P. K. (1990). Polymers for absorbable surgical sutures: Part I. *J. Bioactive Compat. Polymers*, **5**, 453–472.

Benicewicz, B. C., & Hopper, P. K. (1991). Polymers for absorbable surgical sutures: Part II. *J. Bioactive Compat. Polymers*, **6**, 64–94.

Bezwada, R. S., Jamiolkowski, D. D., Lee, I. Y., Agarwal, V., Persivale, J., et al. (1995). Monocryl suture, a new ultrapliable absorbable monofilament suture. *Biomat.*, **16**(15), 1141–1148.

Bichon, D., Borloz, W., & Cassano-Zoppi, A. L. (1984). *In vivo* evaluation of a new polyurethane coated catgut suture. *Biomat.*, **5**, 255–263.

Burg, K. J. L., & Shalaby, S. W. (1999). Absorbable materials and pertinent devices. In A. F. Von Recum (Ed.), *Handbook of Biomaterials Evaluation: Scientific, Technical, and Clinical Testing of Implant Materials* (2nd ed., pp. 99–110). Philadelphia, PA: Taylor & Francis. Chapter 6.

Burkart, S. S. (2004). Arthroscopic knots: The optimal balance of loop security and knot security. *Arthroscopy*, **20**, 489–502.

Capperauld, I., & Bucknall, T. E. (1984). Sutures and dressings. In T. E. Bucknall, & H. Ellis (Eds.), *Wound Healing for Surgeons* (pp. 75–93). Eastbourne, UK: Baillie're Tindall.

Carpenter, K. A., & Shalaby, S. W. (2002). Dioxanone- and dioxepanone-based absorbable polymers. In A. Atala, & R. Lanza (Eds.), *Methods for Tissue Engineering* (pp. 591). New York, NY: Academic Press.

Carpenter, K. A., Anneaux, B. L., Shalaby, W. S. W., Pilgrim, J. A., Linden, D. E., et al. (2002). Segmented lactide copolymers as monofilament sutures: A preliminary report. *Trans. Soc. Biomater.*, **25**, 660.

Casey, D. J. & Roby, M. S. (1984). Synthetic Copolymer Surgical Articles and Method of Manufacturing the Same. U.S. Patent No. 4,429,080 (to American Cyanamid Co.).

Chu, C. C. (1997a). Chemical structure and manufacturing processes. In C. C. Chu, J. Von Fraunhofer, & H. P. Greisler (Eds.), *Wound Closure Biomaterials and Devices* (pp. 65–106). Boca Raton, FL: CRC Press. Chapter 5.

Chu, C. C. (1997b). New emerging materials for wound closure. In C. C. Chu, J. Von Fraunhofer, & H. P. Greisler (Eds.), *Wound Closure Biomaterials and Devices* (pp. 347–384). Boca Raton, FL: CRC Press. Chapter 2.

Covidien. (2005). *IFU for Dexon™ II Coated, Braided Synthetic Absorbable Sutures.*

Covidien (2009a). *Maxons™ Suture.* http://www.covidien.com/imageServer.aspx?contentID=14354&contenttype=application/pdf. Date accessed 5/5/2010.

Covidien (2009b). *Dexon™ S Suture.* http://www.covidien.com/imageServer.aspx?contentID=14356&contenttype=application/pdf. Date accessed 5/5/2010.

Deng, M., Tian, Y., Latour, R. A., Jr., & Shalaby, S. W. (1996). Effects of gamma irradiation on ultrahigh molecular weight polyethylene fibers. *Proc. Fifth World Biomater. Cong.*, **2**, 352.

Ethicon. (2001). *Ethicon Wound Closure Manual.*

Ethicon. (2008). *IFU for Vicryl Rapide™ Sterile Synthetic Absorbable Suture.*

European Commision Scientific Committee on Medicinal Products and Medical Devices. (1998). Opinion and report on the equivalency of alternative products to intestines of animal origin for use as surgical sutures.

Fernando, C., Santos, R. S., Benfield, J. R., Grannis, F. W., Keenan, R. J., et al. (2005). Lobar and sublobar resection with and without brachytherapy for small stage 1A and non-small cell lung cancer. *J. Thorac. Cardiovasc. Surg.*, **129**, 261–267.

FDA (US Food and Drug Administration) (2003). Guidance for Industry and FDA Staff – Class II Special Controls Guidance Document: Surgical Sutures.

FDA (US Food and Drug Administration) (2007). Summary of Color Additives for Use in United States in Foods, Drugs, Cosmetics, and Medical Devices. 21 CFR Parts 73 and 74.

Guttman, B., & Guttman, H. (1994). Sutures: Properties, uses and clinical investigation. In S. Dumitriu (Ed.), *Polymeric Biomaterials* (pp. 325–346). New York, NY: Marcel Dekker.

Hanke, P. R., Timm, P., Falk, G., & Kramer, W. (1994). Behavior of different suture materials in the urinary bladder of the rabbit with special reference to wound healing, epithelization and crystallization. *Urol. Int.*, **52**, 26–33.

Harms, M. T. (1948). Preparation and use of cotton sutures. *Am. J. Nursing*, **48**, 651–652.

Herrmann, J. B. (1971). Tensile strength and knot security of surgical suture materials. *Am. Surg.*, **37**, 209–217.

Hu-Friedy Mfg. Co., Inc (2007). *Perma Sharp® Sutures.* http://www.hu-friedy.com/resource/viewResource.aspx?ResourceID=277. Date accessed 5/5/2010.

Im, J. N., Kim, J. K., Kim, H. K., Lee, K. Y., & Park, W. H. (2007). Characteristics of novel monofilament sutures prepared by conjugate spinning. *J. Biomed. Mater. Res. B: App. Biomat.*, **83B**, 499–504.

Ingram, D. R., Taylor, M. S., Corbett, J. T., Shalaby, W. S. W., & Shalaby, S. W. (2009). Hydroswellable absorbable braided sutures preliminary report. *Trans. Soc. Biomater.*, **31**, 533.

ISO (International Standards Organization). (2009). ISO 10993-1:2009. Biological Evaluation of Medical Devices–Part 1.

Kraemer, R., Lorenzen, J., Rotter, R., Vogt, P. M., & Knoblock, K. (2009). Achilles tendon suture deteriorates tendon capillary blood flow with sustained tissue oxygen saturation: An animal study. *J. Orth. Surg. Res.*, **12**(4), 32.

Lai, S., Becker, D., & Edlich, R. (2009). *Sutures and Needles.* http://emedicine.medscape.com/article/884838-overview. Date accessed 12/2/2009.

Lindsey, J. M., III, Ingram, D. R., Taylor, M. S., Linden, D. E., & Shalaby, S. W. (2007). Absorbable swellable monofilament fibers: A preliminary report. *Trans. Soc. Biomater.*, **30**, 344.

Mackenzie, D. (1973). The history of sutures. *Med. Hist.*, **17**, 158–168.

Matl, F. D., Zlotnyk, J., Obermeier, A., Friess, W., Vogt, S., et al. (2009). New anti-infective coatings of surgical sutures based on a combination of antiseptics and fatty acids. *J. Biomat. Sci.*, **20**, 1439–1449.

Mattei, F. V. (1980). *Absorbable coating composition for sutures.* U.S. Patent No. 4,201,216 (to Ethicon, Inc.).

Nagatomi, S. D., Tate, P. L., Linden, D. E., Hucks, M. A., & Shalaby, S. W. (2007). Antimicrobial absorbable multifilament braided sutures: A preliminary report. *Trans. Soc. Biomater.*, **30**, 182.

Nagatomi, S., Linden, D., Vaughn, M., Corbett, J. & Shalaby, M. et al. (2008). New composite silk braids (ligaprene silk) as low-reactivity and antimicrobial sutures. *8th World Biomaterials Congress*.

Roby, M. S., Bennett, S., Kokish, M. and Jiang, Y. (1997). A new synthethic monofilament absorbable suture from block copolymers of trimethylene carbonate, dioxanone and glycolide. Part I: Synthesis and Processing. Trans 23rd Annual Meeting Soc. for Biomater., April 30-May 4,1997. p.338.

Roby, M. S., & Kennedy, J. (2004). Sutures. In B. D. Ratner, A. S. Hoffman, F. J. Schoen, & J. E. Lemons (Eds.), *Biomaterials Science, an Introduction to Materials in Medicine* (2nd ed., pp. 615–627). San Diego, CA: Elsevier Academic Press. Chapter 7.13.

Shalaby, S. W. (2003). Radiochemical sterilization and its use for sutures. *Nuclear Instr. Methods*, **B 208**, 110.

Shalaby, S. W. (2010). Package Component for Radiochemical Sterilization. U.S. Patent Application 11/228,719 and 12/157,516 (to Poly-Med, Inc.).

Shalaby, S. W., Peniston, S. J. & Carpenter, K. A. (2006a). Selectively Absorbable/Biodegradable, Fibrous Composite Constructs and Applications Thereof. U.S. Patent Application 60/860,033.

Shalaby, S. W., Nickelson, D., Vaughn, M. A., Kennedy, T., & Tate, P. L. (2006b). Use of radiochemically sterilized, absorbable tissue adhesive for cat and dog lung repair. *Trans. Soc. Biomater.*, **29**, 427.

Stashak, E. S., & Yturraspe, D. J. (1978). Considerations for selection of suture materials. *J. Vet. Surg.*, **7**, 48–55.

Swanson, N. A., & Tromovitch, T. A. (1982). Suture materials, 1980s: Properties, uses and abuses. *Int. J. Dermatol.*, **21**, 373–378.

Taylor, M. S. (2010). USLG and Osteoprene *In Vitro* and *In Vivo* Studies. *Poly-Med, Inc. Internal Memorandum.*

Teleflex Medical, OEM (2009). *Force Fiber® Surgical Suture.* http://www.teleflexmedicaloem.com/brands/deknatel/product-Areas/sutures/productGroups/nonabsorbableSutures/products/forceFiber/documents/FORCE%20FIBER%20SUTURE.pdf. Date accessed 5/5/2010.

Totakura, N. & Shalaby, S. W. (1997a). Polymeric Compositions. U.S. Patent No. 5,607,866 (to U.S. Surgical).

Totakura, N. & Shalaby, S. W. (1997b). Suture Coating and Tubing Fluid. U.S. Patent No. 5,584,857 (to U.S. Surgical).

Tozzia, P., Hayozb, D., Ruchata, P., Cornoa, A., Oedmana, C., et al. (2001). Animal model to compare the effects of suture technique on cross-sectional compliance on end-to-side anastomoses. *Eur. J. Cardiothorac. Surg.*, **19**, 477–481.

United States Pharmacopeia (2008). United States Phamacopeia and National Formulary (USP 31-NF 26). United States Pharmacopeia Convention; 3 volumes, **3**, 788 pages.

Van Winkle, W., Jr., & Hastings, J. C. (1972). Considerations in the choice of suture material for various tissues. *Surg. Gyn. Obstet.*, **135**, 113–126.

W.L Gore, & Associates (2007). *Gore-Tex® Vascular Grafts IFU.*

Zhukovsky, V. (2003). Bioactive surgical sutures. *AUTEX Res. J.*, **3**, 41–45.

CHAPTER II.5.16 DRUG DELIVERY SYSTEMS

Allan S. Hoffman, Editor

A. INTRODUCTION

Allan S. Hoffman
Professor of Bioengineering and Chemical Engineering, UWEB Investigator, University of Washington, Seattle, WA, USA

(Includes contributions from the Second Edition by Jorge Heller (deceased).)

INTRODUCTION: PRINCIPLES, ORIGINS, AND EVOLUTION OF CONTROLLED DRUG DELIVERY SYSTEMS (CDDS)

A wide variety of polymeric biomaterials are components of the various formulations and devices that are routinely used for delivering drugs to the body. This section reviews the current state-of-the-art of drug delivery systems (DDS) that release drugs in a controlled manner. The major advantage of developing systems that release drugs in a controlled manner can be appreciated by examining Figure A.1, which is a plot of drug concentration in blood (plasma) versus time after delivery of the drug. This curve is known as a pharmacokinetic or PK curve. It can be seen that the blood plasma level following a single dose administration of a therapeutic agent rapidly rises, and then gradually decays as the drug is metabolized and eliminated from the body. The separate stages of this rise and fall have been identified by pharmaceutical scientists as the *LADME* sequence, or *Liberation* of the drug from the formulation, *Absorption* of the drug into the blood, *Distribution* of the drug throughout the body, including action of the drug at various sites – especially at and within cells, *Metabolism* of the drug, usually in the liver, and finally, *Elimination* of the drug from the body, usually by excretion through the kidneys in the urine. The figure also shows drug concentrations above which the drug produces undesirable (e.g., toxic) side-effects, and below which it is not therapeutically effective. The ratio of these two levels is known as the *Therapeutic Index*, or TI, which is usually based on the dose–response of the median 50% of a large population.

Using only a single dose administration, the time during which the concentration of the drug is above the minimum effective level can be extended by increasing the size of the dose. However, when this is done, blood plasma concentrations may rise and extend into the toxic response region, an undesirable situation. One can also administer safe doses at periodic intervals by cycling the dose in order to maintain a desired drug concentration level; however, this is inconvenient, and patient compliance is often poor. For these reasons, there has been great interest in developing controlled-release formulations and devices that can maintain a desired blood plasma level within the TI for long periods of time. These DDS

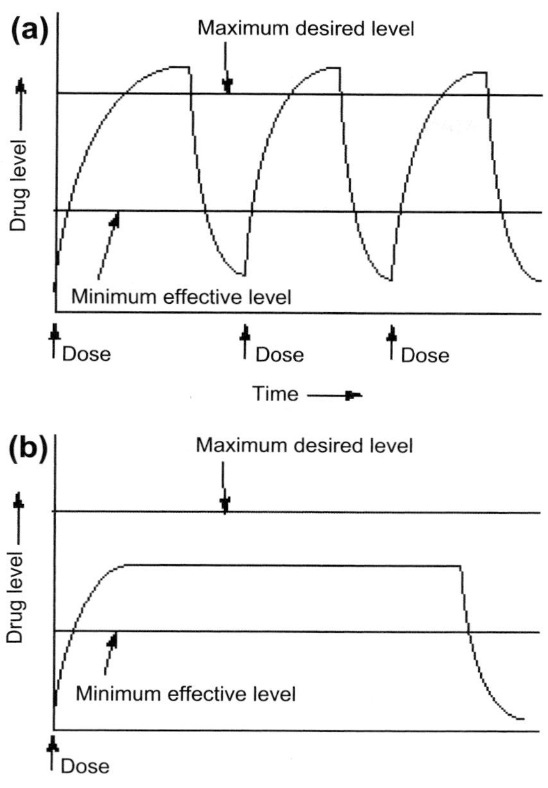

FIGURE A.1 Pharmacokinetic (PK) curves of plasma concentration of a drug versus time for two types of drug delivery systems: (a) Typical bolus PK for multiple dosing with oral tablets or injections; (b) "Zero Order" PK for one dose of controlled drug delivery from a specific formulation or device.

are called "zero order" systems, since they release drug at a constant rate.

Aside from the "assumed" therapeutic advantage of zero order controlled-release products (Note: zero order PK may not be ideal for all drugs and medical conditions, but historically that was the goal in the early days when controlled DDS were being developed), there are also compelling business reasons for the development of such products. Because of increasingly stringent US Food and Drug Administration (FDA) regulations, the cost for introducing new drug formulations has escalated to more than $200 million US dollars for each drug. It is not uncommon for this development to require more than 15–20 years of research and development work, including preclinical animal studies and human clinical trials. Thus, it is reasonable for pharmaceutical companies to attempt to maximize their financial return for each drug by research aimed at extending the patent lifetime of the drug, thereby postponing the inevitable introduction of the generic drug. One means of doing this is to patent a new controlled-release formulation with the same drug. However, the commercial feasibility of such a strategy is predicated on a demonstration that the controlled-release formulation is indeed superior in safety and efficacy to the single bolus dose formulation, and most importantly that the cost of the controlled-release formulation is low enough to ensure a reasonable market penetration.

Successful efforts to produce pharmaceutical formulations that would prolong the action of therapeutic agents for any particular dose, thus reducing the need for frequent dosing, go back to the early 1950s with the introduction of the commercial product known as Spansules®. This product was designed to increase the duration of action of orally administered drugs, and consisted of small drug spheres coated with a soluble polymer coating. By using coatings with varying thickness, capsule dissolution times could be varied, thus prolonging the action of one dose of the therapeutic agent. Such formulations are now known as "sustained release" or "prolonged release" products. However, the pharmacokinetics of such products depended greatly on the local *in vivo* patient environment and as such, varied from patient to patient. These reasons were among the most important driving forces that led to the birth of the field of "controlled drug delivery" (CDD) in the mid- to late-1960s. There was great interest in the development of products that were capable of releasing drugs by reproducible and predictable kinetics, especially at a constant rate, yielding constant plasma drug concentrations over many hours that were safely within the TI. Ideally, such products should be designed to be unaffected by the local *in vivo* environment, greatly reducing patient-to-patient variability. Allan Hoffman has published an interesting history of the origins and early days of the controlled drug delivery field (Hoffman, 2008). The early days of this interesting history are briefly described below.

In the mid- to late-1960s, Alejandro Zaffaroni, a pharmaceutical chemist and also an entrepreneur, saw a need and a market for products that would release drugs with reproducible and predictable kinetics, independent of the patient. At the same time, and independent of each other, Dr. Judah Folkman, MD at Harvard University, published an article (Folkman and Long, 1964) in which his research suggested that a capsule made of silicone rubber (SR) could be filled with a drug and implanted in the body, where it would release drug. This concept led to the first "zero order" reservoir DDS. Zaffaroni was just in the process of founding Alza Corp. in 1968 in Palo Alto, CA, the first company dedicated to the concept of controlled drug delivery. Folkman was enlisted by Zaffaroni to head Alza's Scientific Advisory Board (see http://en.wikipedia.org/wiki/Alza).

The first CDD products developed by Alza were designs with reservoirs of constant drug concentration enclosed or covered by rate-controlling membranes. The membranes were made of one of two polymers, SR or poly(ethylene-co-vinyl acetate) (EVA). They exhibited "zero order" PK profiles (Figure A.1). A key advantage of CDDS can be appreciated by examining this figure, which illustrates changes in blood plasma levels following a single dose administration of a therapeutic agent. As shown, the blood plasma level rapidly rises, and then exponentially decays as the drug is metabolized and/or eliminated from the body. The figure also shows drug concentrations above which the drug produces undesirable (e.g., toxic) side-effects, and below which it is not therapeutically effective. The difference between these two levels is known as the *Therapeutic Index* (TI), which is a recommended range of minimum and maximum drug concentrations in blood plasma, based on the average (or median) response of 50% of a large population to a range of specific drug doses.

Using only a single dose administration, the time during which the concentration of the drug is above the minimum effective level can be extended by increasing the size of the dose. However, when this done, blood plasma concentrations may extend into the toxic response region, an undesirable situation. Furthermore, there is still a significant fraction of the dose below the minimum effective level, which is wasted. One can also administer safe doses cyclically, e.g., by ingesting pills or administering injections at periodic intervals, in order to maintain a desired drug concentration level; however, this is inconvenient, and patient compliance is often poor. For these reasons, there has been great interest in developing controlled-release formulations and devices that can maintain a desired blood plasma level in the middle of the TI for long periods of time without reaching a toxic level or dropping below the minimum effective level. These DDS are called "zero order" systems since they release drug at a constant rate and it appears in the blood at a fixed concentration over the duration of drug delivery.

One can categorize the evolution of CDDS from the 1970s in terms of the size scale of the DDS being developed, and this is shown in Table A.1. All of the first drug delivery systems (DDS) were macro-scale devices that could be held in the hand, and each was designed to exhibit constant or zero order drug delivery rates, leading to constant plasma drug concentrations over long time durations of drug delivery. Alza applied the zero order concept to a variety of useful designs and devices and patented many of them in the 1970s, receiving FDA approval for them in the 1980s. They included drug-loaded IUDs, ocular inserts, and skin patches.

The CDD field became very popular in the 1980s, and grew to include drug-loaded, biodegradable polymer microparticles and phase-separated masses, both as degradable "depot" DDS. They were based on glycolic and lactic acid polyesters, known today as PLA (poly(lactic acid)) and PLGA (poly(lactic-*co*-glycolic acid)), and were not developed at Alza, which was pursuing another degradable polymer DDS based on polyorthoesters. These degradable DDS did not exhibit zero order kinetics, but yielded sustained release kinetics with a gradually decreasing delivery rate over time. The DDS size scale had decreased to "microscopic" with the degradable microparticles. This scale also applied to and

TABLE A.1 Categorizing the Evolution of Controlled DDS by Size Scale

Macroscale DDS ("Zero Order" Constant Delivery Rate DDS)
• Implants (e.g., subcutaneous or intramuscular)
• Inserts (e.g., vaginal, ophthalmic)
• Ingested DDS (e.g., osmotic pumps, hydrogels)
• Topical DDS (e.g., skin patches)
Macroscale and Microscale DDS (Site-Specific, Sustained Delivery Rate DDS)
• Surface-coated DDS (e.g., oral tablets, catheters, drug-eluting stents)
• Injected DD depots (e.g., degradable microparticles and phase-separated masses)
Nanoscale DDS (Targeted DDS)
• Injected nanocarrier DDS (e.g., PEGylated drugs, polymer-drug conjugates, PEGylated liposomes, PEGylated polymeric micelles, and drug nanoparticles, sometimes targeted by monoclonal antibodies or cell membrane receptor ligands)

included the microscopically thin, heparinized surface coatings which were developed in the 1960s, and drug-loaded polymer coatings on stents, which came much later.

In parallel with this, interest and activity was rapidly growing in nanoscopic or nanoscale injectable nanocarriers, including PEGylated drugs and PEGylated, drug-loaded polymeric micelles and liposomes. Many drug delivery scientists were also attaching antibodies to the nanocarriers, to target them to specific cells, but active targeting evolved much more slowly into the clinic. The enhanced permeability and retention effect (EPR) also provided strong rationale for the use of nanocarriers to target solid tumors.

In summary, the CDD field has grown to a multi-billion dollar, worldwide business over the past 55 years, and the scale of CDD devices and formulations themselves has shrunk from macroscopic devices to microscopically thin coatings and microparticles, and down to nanoscopic, molecular CDDS. In parallel with this, the materials and designs progressed from off-the-shelf polymeric materials and devices of simple designs, to degradable, drug-loaded, polymeric microparticles, and down to the complex world of controlled molecular weight block graft copolymers and dendrimers, intracellular druggable targets, such as nuclear DNA, cytosolic mRNA, ribosomes, and apoptotic pathways, with monoclonal antibodies and peptide ligands being used to target the drugs to specific cells.

Aside from the clear therapeutic advantage of controlled-release products, there are also compelling business reasons for the development of such devices. Because of increasingly stringent FDA regulations, the cost for introducing new drug entities has escalated to hundreds of millions of US dollars for each drug, and it is not uncommon for this development to require more than 10 years of research and development work and clinical trials. Thus, it is reasonable for pharmaceutical companies to attempt to maximize their financial return for each drug by research aimed at extending the patented lifetime of the drug, and one means of doing this is to patent a new controlled-release formulation with the same drug. However, the commercial feasibility of such a strategy requires that the controlled release formulation is demonstrated to be superior in safety and efficacy to the single dose formulation, and most importantly, that the cost of the controlled-release formulation is low enough to ensure a reasonable market penetration.

BIBLIOGRAPHY

Hoffman, A.S. (2008). The Origins and evolution of "controlled" drug delivery systems, J. Contr. Release, **132**, 153–163.

Folkman, J. and Long, D. M. (1964). The use of Silicone Rubber as a Carrier for Prologned Drug Therapy, J. Surgical Research, **4**, 139.

B. INJECTED NANOCARRIERS

Allan S. Hoffman, Wayne R. Gombotz, and Suzie H. Pun, Editors

B.1. INTRODUCTION

Allan S. Hoffman
Professor of Bioengineering and Chemical Engineering, UWEB Investigator, University of Washington, Seattle, WA, USA

Drugs may interact with the body in three sites: (1) within the circulation or interstitial space; (2) at membrane receptors on cell surfaces or (3) at various sites within the cell. Nanoscale drug delivery systems are colloidal solutions or dispersions of nanoparticles containing drugs. The particles usually range from several tens to a few hundreds of nm in diameter. Drug-loaded nanocarriers are usually endocytosed by cells. The endosome has a proton pump in its membrane, causing the pH to drop from 7.4 to as low as 5.0. Within one to a few hours after endocytosis the endosome is "trafficked" to the lysosome, which is also at a lowered pH. The lysosome contains enzymes that are active at the low pH of that vesicle. If the drug is enzyme-susceptible, its efficacy will be determined by how efficiently it can escape the endosome before it ends up in the lysosome, where it may be enzymatically degraded. Enzymatically susceptible drugs, such as DNA, RNA, antisense oligonucleotides, peptides or proteins must escape the endosome to be effective at targets within the cell.

Nanocarriers include PEGylated drug systems where the drug may be a protein, peptide, small organic drug or a nucleic acid drug. Other nanocarrier drug systems include polymeric micelles, liposomes, polymer–drug conjugates, polymer–drug polyelectrolyte complexes (known as polyplexes or lipoplexes), dendrimers, and drug nanoparticles. These Drug Delivery Systems (DDS) are often PEGylated to enhance circulation times. Most nanosize DDS have been designed to target tumors, although there are other specific cellular targets in the body that are not in tumors. Nanosize drug carriers are injected intravenously and distributed throughout the circulation, where they may be taken up into the tumor tissue via the Enhanced Permeability and Retention (EPR) effect. This has been called "passive" targeting. Sometimes the drug action may be enhanced by targeting of specific cancer cells with antibodies or cell receptor ligands, which is designed to stimulate endocytosis into specific target cells. (This is like the "magic bullet" concept of Ehrlich, e.g., see Ehrlich, 1900, and Winau et al., 2004.) This is called "active" targeting.

Depending on the indication, an ideal nanocarrier delivery system should have some of the following properties to facilitate targeted delivery of drugs (Petrak, 1993):

- Prolonged duration of delivery due to longer circulation times compared to free drug
- Reduced drug toxicity in non-targeted tissues
- Reduced adsorption of proteins, which reduces opsonization and recognition by the immune system
- Minimal loss of drug from the nanocarrier during circulation
- Increased drug concentration at the desired site of action
- Enhanced uptake by the target tumor or by the target cells
- Release of the drug at the target site at an appropriate rate.

BIBLIOGRAPHY

Ehrlich, P. (1900). On immunity with special reference to cell life. *Proc. R. Soc. London*, **66**, 424.

Petrak, K. (1993). Design and properties of particulate carriers for intravascular administration. In A. Rolland (Ed.), *Pharmaceutical Particulate Carriers* (pp. 275–297). Marcel Dekker.

Winau, F., Westphal, O., & Winau, R. (2004). Paul Ehrlich – in search of the magic bullet. *Microbes Infect.*, **6**, 786–789.

B.2. PEGYLATION OF DRUGS AND NANOCARRIERS

Allan S. Hoffman and Suzie H. Pun
Department of Bioengineering, University of Washington, Seattle, WA, USA

Poly(ethylene glycol) (PEG) is an unusual molecule, with good solubility in both aqueous and organic media. It has been shown that it binds approximately two water molecules per oxygen ether group (Antonsen and Hoffman, 1992). Above molecular weights (MWs) of 500–1000 it may coil randomly and transiently entrap such "bound" water molecules within the coil, which is sometimes called the "excluded volume" of PEG. MWs of 3–40 (2 × 20) kDa are preferred for protein PEGylation. The excluded volume of a PEG coil in water may prevent close approach of and recognition by the immune system and, more simply, it may prevent opsonization by protein binding and subsequent removal by the reticuloendothelial system (RES) in the liver. (The concept of PEGylation of drugs originated in the late 1960s by Frank Davis (see Davis, 2002) when he was looking for a way to reduce recognition of the new recombinant protein drugs by the immune system.) Kopecek and co-workers showed that another hydrophilic polymer, hydroxypropyl methacrylamide (HPMA), is a non-immunogenic drug carrier similar to PEG (Minko et al., 2000).

Another special property of PEG is claimed to be its inability to be a proton donor in a hydrogen bond. Both Merrill (Merrill and Salzman, 1983) and Whitesides (Ostuni et al., 2001) and their collaborators have cited this property as important to its "repulsion" of proteins. All of these effects lead to reduced renal clearance, immunogenicity and enzymolysis, and enhanced solubility, stability, and circulation times, especially for biomolecular drugs such as proteins.

Most nanocarrier drug delivery systems are comprised of a macromolecular carrier conjugated to a drug or a prodrug through a linker. These macromolecular delivery systems offer several potential advantages compared with the free drug. For protein drugs, conjugation to hydrophilic polymers such as poly(ethylene glycol) (PEG) or HPMA results in improved pharmacokinetic profiles and reduced immunogenicity. For small molecule drugs, water-soluble polymer carriers can improve drug solubility, and may also be used to target the drug passively to tumors (via the enhanced permeability and retention (EPR) effect) or actively to specific cells (via conjugation of cell specific ligands to the hydrophilic polymer). The primary application for small molecule delivery with molecular carriers has therefore been cancer treatment. Chemotherapeutic drugs may be effluxed from multidrug resistant (MDR) cells by the P-glycoprotein (P-gp) mechanism, and they are also free to extravasate and diffuse throughout the body after administration, resulting in unwanted side-effects and low therapeutic indices. By increasing the effective molecular weight of the drug through conjugation to a macromolecular carrier, P-gp efflux from cancer cells and renal clearance rates are both reduced; thus, improved tumor treatment can be achieved by a combination of enhanced circulation times plus the EPR effect. Sometimes drug efficacy can be further enhanced by active targeting of the nanocarrier to specific cancer cells. Because most chemotherapeutic drugs are hydrophobic, polymer–drug conjugates tend to form micelle-like structures in aqueous solutions and have dimensions of ~5–20 nm. In this section, several polymer nanocarriers will be discussed.

BIBLIOGRAPHY

Antonsen, K. P., & Hoffman, A. S. (1992). Water structure of PEG solutions by DSC measurements. In J. M. Harris (Ed.), *Polyethylene Glycol Chemistry* (pp. 15). NY: Plenum Press.

Davis, F. F. (2002). The origin of pegnology. *Adv. Drug Del. Revs.*, **54**, 457–458.

Merrill, E. W., & Salzman, E. W. (1983). Polyethylene oxide as a biomaterial. *ASAIO J.*, **6**, 60.

Minko, T., Kopecková, P., & Kopecek, J. (2000). Efficacy of the chemotherapeutic action of HPMA copolymer-bound doxorubicin in a solid tumor model of ovarian carcinoma. *Int. J. Cancer*, **86**, 108–117.

Ostuni, E., Chapman, R. G., Homlin, R. E., Takayama, S., & Whitesides, G. M. (2001). A survey of structure–property relationships of surfaces that resist the adsorption of protein. *Langmuir*, **17**, 5605–5620.

B.3. TARGETING

Patrick S. Stayton, Bilal Ghosn, and John T. Wilson
Department of Bioengineering, University of Washington, Seattle, WA, USA

INTRODUCTION

Despite great potential, many promising drugs remain limited by low stability, toxicity, inefficient administration, and the need for multiple doses. For example, while chemotherapy is effective for the treatment of malignancies, it is widely known for harmful side-effects on healthy tissues. Moreover, such factors contribute not only to poor efficacy, but also reduce patient compliance. Targeted drug delivery systems offer great promise for improving the specificity of new therapeutics and revitalizing conventional therapies limited by adverse side-effects. In this section, we discuss methods of carrier targeting and their applications in drug delivery. While conjugation chemistries are not thoroughly reviewed, more detailed information is available in other literature (Hermanson, 2008).

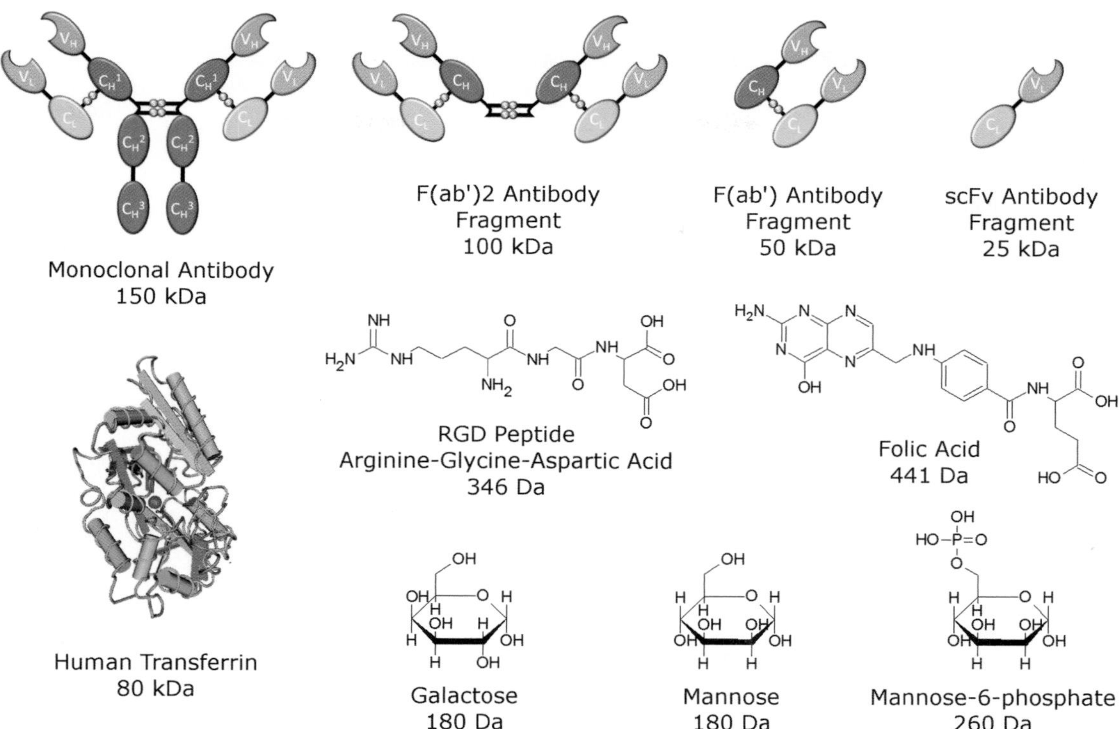

FIGURE B.3.1 Targeting ligands used in drug delivery and their molecular weights.

There are two general classes of targeting, passive and active. Passive targeting is dependent on the physicochemical properties of the drug/carrier system. Targeting is achieved by tailoring the properties of the delivery system to exploit physiological properties of target tissues, while simultaneously minimizing uptake into undesired tissues. This is often achieved by altering surface charge, hydrophobicity, and size and shape of the carrier system (Cai et al., 2007). Examples of passive targeting include cationic microparticles for enhanced delivery to phagocytic cells (Kanke et al., 1983), surface modification with polyethylene glycol (PEG) to provide "stealth" behavior from immune recognition (Abuchowski et al., 1977), and use of larger macromolecules or nanoparticles to exploit the enhanced permeability and retention (EPR) effect (Matsumura and Maeda, 1986).

While passive targeting can enhance delivery to tumors and/or phagocytic cells, it does not provide the high specificity found in active targeting. Active targeting is based on molecular recognition of specific receptors on cellular or extracellular targets where the effect of the drug is desired. Ligand–receptor and antigen–antibody interactions are two targeting modalities commonly employed for drug delivery. Active targeting moieties can range widely in size from small molecules such as folic acid, peptides, and carbohydrates to larger macromolecules such as antibodies and glycoproteins. Here, we will discuss the use of such moieties for targeted delivery ranging from organ-specific to cell-specific levels. Structures and molecular weights of the ligands discussed are shown in Figure B.3.1. Examples of various carriers with each ligand are summarized in Table B.3.1.

IMMUNOTARGETING

Monoclonal Antibodies

The proposed use of antibodies for drug targeting, which has been described as a magic bullet due to their high specificity, has existed for over a century (Ehrlich, 1900). Key characteristics which make antibodies a promising targeting agent include their high specificity and affinity for target, and their capability to induce specific immune responses, which include antibody-dependent cell-mediated cytotoxicity and cytolysis. The development of monoclonal antibody (mAb) technology in 1975 was a key breakthrough that enabled the use of antibodies as drugs and targeting agents (Kohler and Milstein, 1975). The first US Food and Drug Administration (FDA)-approved use of monoclonal antibodies occurred in 1986 with Muromonab CD3 (Thistlethwaite et al., 1984). While most early clinically-approved mAb therapies were purely antibody-based therapies, a few antibody–conjugate therapies have received approval since 2000, including Mylotarg and Cimzia (Schreiber et al., 2005).

Monoclonal Antibody-Targeted Liposomes

The development of antibodies for targeting nanoparticle-based drug carriers has grown greatly over the past two decades, in part due to the ability of nanoparticles to

TABLE B.3.1 Targeted Carriers Used in Drug Delivery

Ligand Type	Carrier Type	Targeting Moiety	Drug	Target Cell/ Tissue	Reference
Monoclonal Antibody (mAb)	Liposome	Anti-H-2k IgG2A Anti-Thy1.1 IgG2A	Methotrexate-y-aspartate	L929 cells	Heath et al., 1983
	HPMA (Polymer)	Anti-θ	γ-globulin	Lymphocytes	Rihova and Kopocek, 1985
	Pluronic 85 (Micelle)	Anti-ADG Anti-GFA	Haloperidol	Brain	Kabanov et al., 1989
Antibody Fragments (Fab, $(Fab)_2$, scFv)	Liposome	Anti-HEr2 (Fab, scFV)	Doxorubicin	BT-474, MDA-MB-453, MCF-7, MCF-7/ HER2	Park et al., 2002
	HPMA (Polymer)	Anti-OV-TL16 (F(ab′))	Adriamycin (ADR), Mce_6	OVCAR-3	Omelyanenko et al., 1996
Galactose	Liposome	Lactose	Inulin	Hepatocytes	Spanjer and Scherphof, 1983
	Poly-L-lysine (Polymer)	Asialoorosomucoid (ASOR)	pDNA	Hep G2	Wu and Wu, 1987
	Poly(γ-glutamic acid)-poly(lactide) (Micelle)	Galactose	Paclitaxel	Hepatoma Tumor	Liang et al., 2006
Mannose	Liposome	Oligomannose	Ova	E.G7-OVA tumors (*in vivo*)	Kojima et al., 2008
	Chitosan (Polymer)	Mannose	pDNA	Colon adenocarcinoma	Kim et al., 2006
	PEG-PLGA (Micelle)	Mannose	Amphoterecin B (AmB)	Macrophages (J774.A1)	Nahar and Jain, 2009
Folate	Liposome	PEG-Folate	Anti-EGFR antisense ODN	KB cells	Wang et al., 1995
	pDMAEMA (Polymer)	PEG-Folate	pDNA	OVCAR-3	van Steenis et al., 2003
	PEG-PLGA (Micelle)	Folate	Doxorubicin	KB cells	Yoo and Park, 2004
Transferrin	Liposome	Transferrin-PEG	Doxorubicin	C6 glioma cells	Eavarone et al., 2000
	PEI (Polymer)	Transferrin-PEG	pDNA	Human primary melanoma cells	Wightman et al., 1999
	PEG-g-PEI (micelle)	Transferrin	Antisense ODN	KBv cells	Vinogradov et al., 1999
Peptide	Liposome	RGD-PEG	Doxorubicin	Angiogenic epithelial cells	Schiffelers et al., 2002
	PEI (Polymer)	RGD	pDNA	HeLa and MRC-5 cells	Erbacher et al., 1999
	PCL-b-PEG (Micelle)	Cyclic-RGD	Doxorubicin	Kaposi's sarcoma cells	Nasongkla et al., 2004

HPMA: hydroxypropyl methacrylate
PCL: polycaprolactone
pDMAEMA: polydimethylaminomethyl methalrylate
PEG: polyethylene glycol
PEI: polyethylenimine
PLGA: poly(lactic-*co*-glycolic acid)
PGD: arginine-glycine-aspartic acid

carry high drug payloads relative to molecular conjugates. As early as the 1980s, introduction of antibodies to liposomes demonstrated cell-specific uptake of liposomes coated with an anti-H-2k antibody (Huang et al., 1980). Chemical conjugation of monoclonal antibodies to liposomes for targeted delivery of encapsulated small molecules demonstrated up to 10-fold increases in tumor growth inhibition (Heath et al., 1983). Further progress with antibody-targeted liposomes, often referred to as immunoliposomes, has continued for nearly a quarter of a century. More recently, immunoliposomes have employed a polyethylene glycol (PEG)-spacer for improved circulation time and specificity. For example, DOPE liposomes utilizing PEG-linked anti-EGFR antibodies and encapsulating Gemcitabine have demonstrated significant anti-tumor effects relative to untargeted and free drug (Kim et al., 2009).

Monoclonal Antibody-Targeted Polymeric Carriers

The use of antibodies conjugated to polymeric delivery systems has also been investigated, in part due to the tailorability afforded by polymers. This allows for numerous conjugation schemes in combination with other delivery-enhancing properties. An early study of hydroxypropyl methacrylate (HPMA) polymers conjugated to an anti-θ mAb demonstrated a 70-fold increase in the activity of gamma globulin against targeted lymphocytes (Rihova and Kopocek, 1985). Several examples of targeting with the polymer HPMA have been previously reviewed (Kopecek and Kopeckova, 2010; Minko, 2010; Ulbrich and Subr, 2010). Conjugation of anti-thrombomodulin mAbs to the cationic polypeptide poly-L-lysine increased uptake in lung cells both *in vitro* and *in vivo*. Antibody targeting of polyethylenimine (PEI) nanoparticles also enhanced transfection efficiency and specificity for numerous cancer cell types (Chiu et al., 2004; Germershaus et al., 2006). Targeting with monoclonal and polyclonal antibodies has also been studied for polymeric micellar carriers. Initial studies in 1989 demonstrated that targeted micelles prepared with Pluronic 85 increased haloperidol efficacy by nearly 500-fold relative to free drug (Kabanov et al., 1989). Recently, conjugation of human monoclonal antibody against hypoxia inducible factor-1α (HIF-1α) to Pluronic P123 micelles demonstrated selective cytotoxicity to tumor cells when delivering Paclitaxel (Song et al., 2010).

Antibody fragments. While monoclonal antibodies have shown promise as targeting moieties, their large size (~150 kDa) may reduce penetration into certain tissues. Furthermore, mAbs may impart immunogenic side-effects, including cytokine release, thus leading to cytotoxicity. To circumvent these limitations, targeting with antibody fragments has been evaluated in a variety of forms. The two most common types of antibody fragments explored today are Fab and Fv (or scFv) fragments. Fab fragments represent the antigen binding fragment of an intact antibody containing both the variable and constant regions of both heavy and light chains. Fv, or the variable fragments, represent the portion of the antibody containing variable regions of both chains, which are shown in Figure B.3.1.

Antibody fragments have been thoroughly investigated for their potential to enhance the specificity of numerous carrier systems. Initial development of a conjugation scheme for Fab fragments to liposomes was published in 1980 using rabbit anti-human erythrocyte $F(ab')_2$ fragment, resulting in a 200-fold increase in erythrocyte binding (Heath et al., 1980). Later studies focused on the use of anti-HER2 targeted immunoliposomes with both F(ab′) and scFv fragments for enhancing delivery of the cancer drug doxorubicin (Park et al., 2002). More recently, it has been shown that the circulation time of immunoliposomes is enchanced for both anti-CD19 Fab and scFv fragments relative to whole mAb, an effect responsible in part for the increased mean survival time relative to all other groups in a murine B-cell lymphoma model (Cheng and Allen, 2008).

Antibody fragment-targeted polymeric carriers have also been investigated. In initial studies, the conjugation of F(ab′) antibody fragments to HPMA demonstrated a nearly 10-fold increase in conjugate circulation times, although no *in vivo* tumor targeting was reported (Seymour et al., 1991). Later, however, conjugation of anti-OV-TL16 F(ab′) to HPMA resulted in more homogeneous affinity to target cells relative to whole antibody–polymer conjugates, resulting in enhanced plasmid DNA (pDNA) expression within targeted lung epithelium and submucosal glands (Omelyanenko et al., 1996). Recent conjugation of anti-OV-TL16 antibody fragments to PEI-*graft*-PEG polymers for gene delivery (Merdan et al., 2003) improved expression levels by up to 80-fold compared to untargeted carriers in OA3-expressing OVCAR-3 cells. Fab (anti-envelope glycoprotein, gp160) and scFv (anti-ErbB2) antibody fragments conjugated to protamine, a cationic polymer, demonstrated enhanced and specific uptake into target cells expressing the appropriate receptors (Song et al., 2005). Thus, antibody fragments conjugated to polymeric carriers provide a highly specific and effective method for targeting potent macromolecular drugs.

CARBOHYDRATES

Carbohydrates have also been evaluated as potential targeting moieties for cell- and tissue-specific drug delivery. Interactions of carbohydrates with highly-specific cell membrane receptors known as lectins provide a powerful mechanism for targeting. It has been four decades since the discovery of the first mammalian lectin (Van Den Hamer et al., 1970). In this landmark study, removal of the terminal sialic acid residues from ceruloplasmin exposed galactosyl residues and resulted in rapid clearance of the intravenously administered dose. The removal of the terminal sialic acid residues results in glycoproteins known as asialoglyocoproteins. It was later determined that the majority of accumulation for these asialoglycoproteins was in the liver (Pricer and Ashwell, 1971). The receptor was later termed the asialoglycoprotein receptor (ASGP-R), and noted to internalize the β-linked galactose-bearing ligands (Ashwell and Morell, 1974).

Several lectins demonstrating specific affinity for carbohydrates such as galactose, mannose, fucose, and lactose have been identified to date, and may contain monovalent or multivalent binding sites. ASGP receptors are classified as part of the C-class lectins, which is predominantly due to the requirement of Ca^{2+} for binding to carbohydrates. The various classifications of lectin hold unique characteristics and have previously been reviewed (Wadhwa and Rice, 1995).

Galactose-Targeted Liposomes

Targeting moieties for liposomes have most commonly utilized monosaccharides or oligosaccharides of galactose or mannose to provide specific targeting to the liver and immune cells, respectively. One of the first evaluations of galactose-based targeting was shown by Gregoriadis and Neerunjun in 1975 (Gregoriadis and Neerunjun, 1975). In these studies, incorporation of fetuin, a desialylated glycoprotein, to the surface of liposomes increased uptake of carriers in the liver following intravenous administration. From these early studies emerged the development of galactose-targeted liposomes as tissue-specific drug delivery systems for a variety of molecules such as siRNA (Sato et al., 2007).

Galactose-Targeted Polymeric Carriers

Targeting of lectins such as ASGP-R has also been researched in a variety of polymeric systems for delivery of drugs. One of the earliest investigations again involved HPMA functionalized with D-galactosamine, which demonstrated nearly 70% uptake into the liver within 1 hour following administration relative to 10% of control conjugates functionalized with either D-mannosamine or D-glucosamine (Duncan et al., 1983). Conjugation of asialoorosomucoid (ASOR), a desialylated glycoprotein, to poly-l-lysine effectively delivered pDNA into ASGP-R positive Hep G2, but not into ASGP-R negative SK-Hep 1 cells (Wu and Wu, 1987). Recently, galactose-targeting with the polymer PBAVE demonstrated effective localization of siRNA/polymer conjugates into hepatocytes following intravenous administration (Rozema et al., 2007). The evaluation of galactose-targeting for polymeric micelles has also shown great promise for improved delivery to the liver. Introduction of galactosamine to the surface of poly(γ-glutamic acid)-poly(lactide) nanoparticles encapsulating Paclitaxel effectively enhanced uptake of nanoparticles in hepatoma tumors (Liang et al., 2006), demonstrating the potential of galactose-targeting for polymeric micellar carriers.

Mannose-Based Targeting. Mannose and its oligomeric form mannan have been examined as ligands for specific delivery to macrophages and dendritic cells. Much like the ASGP receptor, the mannose-based receptors come in many forms, and are predominantly found in cells involved in inflammation and immunity such as macrophages, dendritic cells, and endothelial cells (Irache et al., 2008). A thorough review of the receptor biology is available (Irache et al., 2008). The specificity of mannose-binding lectins on macrophages was demonstrated by evaluating the uptake of a variety of glycoproteins by rat alveolar macrophages (Stahl et al., 1980). It was noted that glycoproteins with an exposed mannose, glucose or n-acetyl glucosamine were bound by the alveolar macrophages, while glycoproteins with a terminal galactose were not.

Mannose-Targeted Liposomes

One of the earliest studies with mannose-targeted delivery demonstrated that incorporation of 6-aminomannose directed liposomes to leukocytes following subcutaneous injection, and to the lungs following intravenous injection (Mauk et al., 1980). Later studies determined that the mannosylation of liposomes resulted in improved binding and an adjuvant or immunoenhancement effect from the carriers when used to deliver tetanus toxoid, thus providing great potential for use in vaccine delivery (Garcon et al., 1988). More recently, mannosylated liposomes encapsulating ciproflaxin (CPFX) were developed for pulmonary treatment of intracellular parasitic infections of the respiratory system (Chono et al., 2008). Dual targeting of liposomes with both mannose and transferrin has also been evaluated for improved delivery across the blood–brain barrier (Ying et al., 2010).

Mannose-Targeted Polymeric Carriers

Mannose-based targeting of polymeric carriers has also been evaluated. Introduction of mannosamine to HPMA was shown to increase uptake by liver macrophages following intravenous administration in mice (Seymour et al., 1987). Later, mannosylated poly-l-lysine conjugates were shown to effectively increase specific uptake of the anti-leishmanial drug allopurinol riboside, improving drug efficacy 50-fold for the treatment infection with *L. donovani* (Negre et al., 1992). Kim and colleagues recently reported enhanced treatment of a colon adenocarcinoma model by delivery of pDNA encoding for the cytokine IL-12 via a mannosylated chitosan carrier (Kim et al., 2006). Targeted delivery with mannose analogs has also been evaluated with polymeric micellar carriers. Mannose conjugation via a PEG-spacer to PLGA (poly(lactic-*co*-glycolic acid) micelles encapsulating Amphoterecin B was shown to enhance drug uptake in J774.A1 macrophages, and *in vivo* uptake in the liver, spleen, and lymph nodes (Nahar and Jain, 2009).

NUTRIENT-BASED TARGETING

Folate/Folic Acid

Folate, or vitamin B9, is a small molecule that binds to folate-binding proteins (FBP) with a high affinity. Exploitation of folate as a targeting ligand for delivery of macromolecules was initially demonstrated in 1991 to increase macromolecule uptake in KB cells, a human carcinoma line (Leamon and Low, 1991). Interest in folate as a targeting ligand increased as it was demonstrated that folate receptors were overexpressed in ovarian cancer cells (Campbell et al., 1991), and subsequently many other carcinomas (Weitman et al., 1992). Given these characteristics and the relatively small size of folate, the promise for this vitamin as a targeting ligand for cancer therapeutics and diagnostics has spawned a plethora of research in the area of material science.

Folate-Targeted Liposomes and Polymeric Carriers

While initially studied as a direct conjugate to macromolecules, folate has been evaluated for targeting in a variety of drug delivery systems. Initial studies with folate conjugated to liposomes demonstrated a 37-fold increase in cellular uptake by KB cells relative to untargeted liposomes (Lee and Low, 1994). Folate-targeted liposomes have also been shown to increase uptake and therapeutic efficacy when used to deliver doxorubicin (Lee and Low, 1995) and antisense oligonucleotides (Wang et al., 1995). More recently, uses of folate-targeted liposomes have also been shown to serve as potential carriers for molecular probes, such as quantum dots for cancer diagnosis (Yang et al., 2009).

Conjugation of folate to polymeric carrier systems also enhances targeted uptake. Early studies of folate-targeted polymeric carriers for pDNA include folate-poly-l-lysine (Mislick et al., 1995) and folate/PEI (Guo and Lee, 1999). Conjugation of folate via PEG spacers to poly-dimethylaminomethyl methacrylate (pDMAEMA)/pDNA polyplexes resulted in significant two-fold increases in gene expression levels in OVCAR-3 relative to untargeted polyplexes (van Steenis et al., 2003). Similar conjugation of folic acid to micelles has also been studied with a variety of polymers. Introduction of folate as a targeting moiety to PEG-PLGA diblock micelles also enhanced doxorubicin-mediated cytotoxicity to KB cells, reducing the IC_{50} for the drug by nearly 30% relative to untargeted micelles, and reduced tumor cells *in vivo* by nearly 30% as compared to free DOX/untargeted micelles (Yoo and Park, 2004).

Transferrin

Transferrin (Tf), an 80 kDa glycoprotein found in serum, binds and transports iron into cells via receptor-mediated endocytosis. Unlike many other receptor-mediated pathways, following endocytosis and release of its iron payload, both the receptor and transferrin are rapidly exocytosed from the cell and released to repeat the process (Ciechanover et al., 1983). Similar to a number of receptors, overexpression of the Tf recepter, (TfR) has been noted in numerous malignant cell lines, and has thus been proposed as a potential targeting ligand for cancer therapy (Trowbridge, 1988; Thorstensen and Romslo, 1993). As early as 1985, transferrin conjugation to the diphtheria toxin was shown to significantly enhance its efficacy (Okeefe and Draper, 1985). A nearly eight-fold increase in the circulatory half-life of the chemotherapeutic drug neocarzinostatin was also demonstrated following its conjugation to transferrin (Kohgo et al., 1988). These early studies demonstrated the potential for transferrin-targeted delivery.

Transferrin-Targeted Drug Delivery Systems

One of the earliest studies of transferrin-targeted carriers demonstrated the use of transferrin conjugated to the surface of both pH-sensitive and insensitive liposomes, where results demonstrated that enhancement of efficiency was dominated by the presence of transferrin (Brown and Silvius, 1990). Transferrin introduced to the surface of phosphatidylethanolamine directed effective delivery to leukemia HL60 cells (Sarti et al., 1996), while transferrin-polylysine conjugated liposomes were effective at enhancing the delivery of α-interferon (α-INF) (Liao et al., 1998). PEGylation of liposomes can further enhance efficacy when using the transferrin-targeting agent (Eavarone et al., 2000; Soni et al., 2008).

Transferrin-targeting has also been applied to numerous polymeric carriers for drug delivery. Early evaluation of transferrin conjugated to HPMA resulted in a nine-fold increase in conjugate uptake compared to untargeted HPMA carriers (Flanagan et al., 1989). Transferrin conjugated to PEI demonstrated enhanced pDNA transfection up to two orders of magnitude greater than untargeted carrier in primary melanoma cell lines (Wightman et al., 1999). Transferrin is perhaps most promising as a targeting ligand for carrier delivery across the blood–brain barrier, such as with transferrin-targeted PEG-polyamidoamine dendrimer for the delivery of pDNA to the brain (Huang et al., 2007). The conjugation of transferrin to PEG-g-PEI polymeric micelles has also been demonstrated to increase polyion complex accumulation by 80-fold compared to free oligonucleotide (Vinogradov et al., 1999), thus further demonstrating the potential of transferrin as a targeting ligand.

PEPTIDE-BASED TARGETING

Peptides represent another potential avenue for targeted drug delivery which has received a great deal of attention over the past 20 years. The discovery of the arginine-glycine-aspartate (RGD) peptide as a cell adhesion sequence in 1994 has led to a flurry of research not only on its function, but also on the potential of RGD for enhancing drug delivery (Pierschbacher and Ruoslahti, 1984). In 1997 it was later discovered that RGD was an essential sequence for numerous extracellular matrix (ECM) proteins and membrane proteins, thus making it a major player for cellular targeting (Pfaff, 1997).

Peptide-Targeted Drug Delivery Systems

RGD-containing peptides have been used in a variety of drug carrier systems over the past two decades. Incorporation of a GRGDSPC peptide sequence to liposomes provided a nine-fold increase in platelet cells relative to untargeted liposomes (Nishiya and Sloan, 1996). Later studies evaluated the potential of RGD-targeted PEG-coated-liposomes targeted specifically for the integrin $\alpha_v\beta_3$ which is overexpressed in angiogenic epithelial cells, demonstrating enhanced circulation time and uptake of doxorubicin into angiogenic epithelial cells, increasing binding efficiencies by seven-fold (Schiffelers et al., 2002). Recent studies have demonstrated effective delivery of a variety of agents using RGD-containing peptide targeted liposomes to a diversity of cell types, such as colon cancer cells (Garg et al., 2009).

The introduction of an RGD peptide to a small poly-l-lysine polymer for targeted DNA delivery represented one of the earliest RGD-targeted polymeric carriers demonstrating selective uptake based on the integrin-binding capacity of the carrier (Hart et al., 1995). Similarly, conjugation of RGD to PEI has demonstrated increased *in vitro* gene expression of 10- to 100-fold relative to both plain PEI and RGE-targeted (non-specific) PEI carriers (Erbacher et al., 1999). PEGylated PEI/DNA complexes modified with either RGDC or HIV-1 TAT peptides demonstrated up to a 14-fold increase in gene expression, while also mediating enhanced vector endosomal escape in RGD-targeted conjugates (Suk et al., 2006). RGD-targeting can also enhance the delivery efficiencies of polymeric micelles, such as the surface functionalization of PCL-b-PEG micelles with a cyclic RGD peptide which resulted in a 30-fold increase of doxorubicin internalization by Kaposi's sarcoma cells relative to unmodified micelles (Nasongkla et al., 2004).

SUMMARY

The goal of targeted drug delivery continues to be increasing target specificity, an evident need in the world of medicine and biotechnology. A wide range of biomolecular derivatives have been investigated with each possessing unique properties and advantages. These can include tissue or disease specificity, immunogenicity, size, ease of synthesis or isolation, and stability. Despite a century of major advances since Erlich's "magic bullet" theory, further understanding and development of active targeting remains necessary and crucial to improving drug delivery, and is vital to successful clinical translation of liposomal, polymeric, and micellar drug delivery technology.

BIBLIOGRAPHY

Abuchowski, A., van Es, T., Palczuk, N. C., & Davis, F. F. (1977). Alteration of immunological properties of bovine serum albumin by covalent attachment of polyethylene glycol. *J. Biol. Chem.*, **252**(11), 3578–3581.

Ashwell, G., & Morell, A. G. (1974). The role of surface carbohydrates in the hepatic recognition and transport of circulating glycoproteins. *Adv. Enzymol. Relat. Areas Mol. Biol.*, **41**(0), 99–128.

Brown, P. M., & Silvius, J. R. (1990). Mechanisms of delivery of liposome-encapsulated cytosine arabinoside to CV-1 cells *in vitro*. Fluorescence-microscopic and cytotoxicity studies. *Biochim. Biophys. Acta*, **1023**(3), 341–351.

Cai, S., Vijayan, K., Cheng, D., Lima, E. M., & Discher, D. E. (2007). Micelles of different morphologies – advantages of worm-like filomicelles of PEO-PCL in paclitaxel delivery. *Pharm. Res.*, **24**(11), 2099–2109.

Campbell, I. G., Jones, T. A., Foulkes, W. D., & Trowsdale, J. (1991). Folate-binding protein is a marker for ovarian cancer. *Cancer Res.*, **51**(19), 5329–5338.

Cheng, W. W., & Allen, T. M. (2008). Targeted delivery of anti-CD19 liposomal doxorubicin in B-cell lymphoma: A comparison of whole monoclonal antibody, Fab′ fragments and single chain Fv. *J. Control. Release*, **126**(1), 50–58.

Chiu, S. J., Ueno, N. T., & Lee, R. J. (2004). Tumor-targeted gene delivery via anti-HER2 antibody (trastuzumab, Herceptin) conjugated polyethylenimine. *J. Control. Release*, **97**(2), 357–369.

Chono, S., Tanino, T., Seki, T., & Morimoto, K. (2008). Efficient drug targeting to rat alveolar macrophages by pulmonary administration of ciprofloxacin incorporated into mannosylated liposomes for treatment of respiratory intracellular parasitic infections. *J. Control. Release*, **127**(1), 50–58.

Ciechanover, A., Schwartz, A. L., Dautry-Varsat, A., & Lodish, H. F. (1983). Kinetics of internalization and recycling of transferrin and the transferrin receptor in a human hepatoma-cell line – effect of lysosomotropic agents. *J. Biol. Chem.*, **258**(16), 9681–9689.

Duncan, R., Kopecek, J., Rejmanova, P., & Lloyd, J. B. (1983). Targeting of N-(2-hydroxypropyl)methacrylamide copolymers to liver by incorporation of galactose residues. *Biochim. Biophys. Acta*, **755**(3), 518–521.

Eavarone, D. A., Yu, X., & Bellamkonda, R. V. (2000). Targeted drug delivery to C6 glioma by transferrin-coupled liposomes. *J. Biomed. Mater. Res.*, **51**(1), 10–14.

Ehrlich, P. (1900). On immunity with special reference to cell life. *Proc. R. Soc. London*, **66**, 424.

Erbacher, P., Remy, J. S., & Behr, J. P. (1999). Gene transfer with synthetic virus-like particles via the integrin-mediated endocytosis pathway. *Gene Ther.*, **6**(1), 138–145.

Flanagan, P. A., Kopeckova, P., Kopecek, J., & Duncant, R. (1989). Evaluation of protein-N-(2-hydroxypropyl)methacrylamide copolymer conjugates as targetable drug carriers. 1. Binding, pinocytic uptake and intracellular distribution of transferrin and anti-transferrin receptor antibody conjugates. *Biochim. Biophys. Acta*, **993**(1), 83–91.

Garcon, N., Gregoriadis, G., Taylor, M., & Summerfield, J. (1988). Mannose-mediated targeted immunoadjuvant action of liposomes. *Immunology*, **64**(4), 743–745.

Garg, A., Tisdale, A. W., Haidari, E., & Kokkoli, E. (2009). Targeting colon cancer cells using PEGylated liposomes modified with a fibronectin-mimetic peptide. *Int. J. Pharm.*, **366**(1-2), 201–210.

Germershaus, O., Merdan, T., Bakowsky, U., Behe, M., & Kissel, T. (2006). Trastuzumab-polyethylenimine-polyethylene glycol conjugates for targeting Her2-expressing tumors. *Bioconjug. Chem.*, **17**(5), 1190–1199.

Gregoriadis, G., & Neerunjun, E. D. (1975). Homing of liposomes to target cells. *Biochem. Biophys. Res. Commun.*, **65**(2), 537–544.

Guo, W., & Lee, R. L. (1999). Receptor-targeted gene delivery via folate-conjugated polyethylenimine. *AAPS Pharm. Sci.*, **1**(4), E19.

Hart, S. L., Harbottle, R. P., Cooper, R., Miller, A., Williamson, R., & Coutelle, C. (1995). Gene delivery and expression mediated by an integrin-binding peptide. *Gene Ther.*, **2**(8), 552–554.

Heath, T. D., Fraley, R. T., & Papahadjopoulos, D. (1980). Antibody targeting of liposomes: Cell specificity obtained by conjugation of F(ab′)2 to vesicle surface. *Science*, **210**(4469), 539–541.

Heath, T. D., Montgomery, J. A., Piper, J. R., & Papahadjopoulos, D. (1983). Antibody-targeted liposomes: Increase in specific toxicity of methotrexate-gamma-aspartate. *Proc. Natl. Acad. Sci. USA*, **80**(5), 1377–1381.

Hermanson, G. T. (2008). *Bioconjugate Techniques*. San Diego: Academic Press.

Huang, A., Huang, L., & Kennel, S. J. (1980). Monoclonal antibody covalently coupled with fatty acid. A reagent for *in vitro* liposome targeting. *J. Biol. Chem.*, **255**(17), 8015–8018.

Huang, R. Q., Qu, Y. H., Ke, W. L., Zhu, J. H., Pei, Y. Y., & Jiang, C. (2007). Efficient gene delivery targeted to the brain using a transferrin-conjugated polyethyleneglycol-modified polyamidoamine dendrimer. *Faseb J.*, **21**(4), 1117–1125.

Irache, J. M., Salman, H. H., Gamazo, C., & Espuelas, S. (2008). Mannose-targeted systems for the delivery of therapeutics. *Expert Opin. Drug Deliv.*, **5**(6), 703–724.

Kabanov, A. V., Chekhonin, V. P., et al. (1989). The neuroleptic activity of haloperidol increases after its solubilization in surfactant micelles. Micelles as microcontainers for drug targeting. *FEBS Lett.*, **258**(2), 343–345.

Kanke, M., Sniecinski, I., & DeLuca, P. P. (1983). Interaction of microspheres with blood constituents: I. Uptake of polystyrene spheres by monocytes and granulocytes and effect on immune responsiveness of lymphocytes. *J. Parenter. Sci. Technol.*, **37**(6), 210–217.

Kim, T. H., Jin, H., Kim, H. W., Cho, M. H., & Cho, C. S. (2006). Mannosylated chitosan nanoparticle-based cytokine gene therapy suppressed cancer growth in BALB/c mice bearing CT-26 carcinoma cells. *Mol. Cancer Ther.*, **5**(7), 1723–1732.

Kim, I. Y., Kang, Y. S., Lee, D. S., Park, H. J., Choi, E. K., Oh, Y. K., Son, H. J., & Kim, J. S. (2009). Antitumor activity of EGFR targeted pH-sensitive immunoliposomes encapsulating gemcitabine in A549 xenograft nude mice. *J. Control. Release*, **140**(1), 55–60.

Kohgo, Y., Kato, J., et al. (1988). Targeting chemotherapy with transferrin-neocarzinostatin conjugate. *Gan To Kagaku Ryoho*, **15**(4 Pt 2-1), 1072–1076.

Kohler, G., & Milstein, C. (1975). Continuous cultures of fused cells secreting antibody of predefined specificity. *Nature*, **256**(5517), 495–497.

Kojima, N., Biao, L., Nakayama, T., Ishii, M., Ikehara, Y., & Tsujimura, K. (2008). Oligomannose-coated liposomes as a therapeutic antigen-delivery and an adjuvant vehicle for induction of *in vivo* tumor immunity. *J. Control. Release*, **129**(1), 26–32.

Kopecek, J., & Kopeckova, P. (2010). HPMA copolymers: Origins, early developments, present, and future. *Adv. Drug Deliv. Rev.*, **62**(2), 122–149.

Leamon, C. P., & Low, P. S. (1991). Delivery of macromolecules into living cells: A method that exploits folate receptor endocytosis. *Proc. Natl. Acad. Sci. USA*, **88**(13), 5572–5576.

Lee, R. J., & Low, P. S. (1994). Delivery of liposomes into cultured KB cells via folate receptor-mediated endocytosis. *J. Biol. Chem.*, **269**(5), 3198–3204.

Lee, R. J., & Low, P. S. (1995). Folate-mediated tumor cell targeting of liposome-entrapped doxorubicin *in vitro*. *Biochim. Biophys. Acta*, **1233**(2), 134–144.

Liang, H. F., Chen, C. T., Chen, S. C., Kulkarni, A. R., Chiu, Y. L., Chen, M. C., & Sung, H. W. (2006). Paclitaxel-loaded poly (gamma-glutamic acid)-poly(lactide) nanoparticles as a targeted drug delivery system for the treatment of liver cancer. *Biomaterials*, **27**(9), 2051–2059.

Liao, W. P., Dehaven, J., Shao, J., Chen, J. X., Rojanasakul, Y., Lamm, D. L., & Ma, J. K. (1998). Liposomal delivery of alpha-Interferon to murine bladder tumor cells via transferrin receptor-mediated endocytosis. *Drug Deliv.*, **5**(2), 111–118.

Matsumura, Y., & Maeda, H. (1986). A new concept for macromolecular therapeutics in cancer chemotherapy: Mechanism of tumoritropic accumulation of proteins and the antitumor agent smancs. *Cancer Res.*, **46**(12 Pt 1), 6387–6392.

Mauk, M. R., Gamble, R. C., & Baldeschwieler, J. D. (1980). Targeting of lipid vesicles: Specificity of carbohydrate receptor analogues for leukocytes in mice. *Proc. Natl. Acad. Sci. USA*, **77**(8), 4430–4434.

Merdan, T., Callahan, J., Petersen, H., Kunath, K., Bakowsky, U., Kopeckova, P., Kissel, T., & Kopecek, J. (2003). Pegylated polyethylenimine-Fab′ antibody fragment conjugates for targeted gene delivery to human ovarian carcinoma cells. *Bioconjug. Chem.*, **14**(5), 989–996.

Minko, T. (2010). HPMA copolymers for modulating cellular signaling and overcoming multidrug resistance. *Adv. Drug Deliv. Rev.*, **62**(2), 192–202.

Mislick, K. A., Baldeschwieler, J. D., Kayyem, J. F., & Meade, T. J. (1995). Transfection of folate-polylysine DNA complexes: Evidence for lysosomal delivery. *Bioconjug. Chem.*, **6**(5), 512–515.

Nahar, M., & Jain, N. K. (2009). Preparation, characterization and evaluation of targeting potential of amphotericin B-loaded engineered PLGA nanoparticles. *Pharm. Res.*, **26**(12), 2588–2598.

Nasongkla, N., Shuai, X., Ai, H., Weinberg, B., Pink, J., Boothman, D., & Gao, J. (2004). cRGD-functionalized polymer micelles for targeted doxorubicin delivery. *Angew. Chem. Int. Ed.*, **43**, 6323–6327.

Negre, E., Chance, M. L., Hanboula, S. Y., Monsigny, M., Roche, A. C., Mayer, R. M., & Hommel, M. (1992). Antileishmanial drug targeting through glycosylated polymers specifically internalized by macrophage membrane lectins. *Antimicrob. Agents Chemother.*, **36**(10), 2228–2232.

Nishiya, T., & Sloan, S. (1996). Interaction of RGD liposomes with platelets. *Biochem. Biophys. Res. Commun.*, **224**(1), 242–245.

Okeefe, D. O., & Draper, R. K. (1985). Characterization of a transferrin-diphtheria toxin conjugate. *J. Biol. Chem.*, **260**(2), 932–937.

Omelyanenko, V., Kopeckova, P., Gentry, C., Shiah, J. G., & Kopecek, J. (1996). HPMA copolymer-anticancer drug-OV-TL16 antibody conjugates. 1. Influence of the method of synthesis on the binding affinity to OVCAR-3 ovarian carcinoma cells *in vitro*. *J. Drug Target.*, **3**(5), 357–373.

Park, J. W., Hong, K., et al. (2002). Anti-HER2 immunoliposomes: Enhanced efficacy attributable to targeted delivery. *Clin. Cancer Res.*, **8**(4), 1172–1181.

Pfaff, M. (1997). Recognition sites of RGD-dependent integrins. In J. A. Eble (Ed.), *Integrin-Ligand Interaction* (pp. 101–121). Heidelberg: Springer-Verlag.

Pierschbacher, M. D., & Ruoslahti, E. (1984). Cell attachment activity of fibronectin can be duplicated by small synthetic fragments of the molecule. *Nature*, **309**(5963), 30–33.

Pricer, W. E., Jr., & Ashwell, G. (1971). The binding of desialylated glycoproteins by plasma membranes of rat liver. *J. Biol. Chem.*, **246**(15), 4825–4833.

Rihova, B., & Kopocek, J. (1985). Biological properties of targetable poly[N-(2-hydroxypropyl)-methacrylamide]-antibody conjugates. *J. Control. Release*, **2**, 289–310.

Rozema, D. B., Lewis, D. L., et al. (2007). Dynamic PolyConjugates for targeted *in vivo* delivery of siRNA to hepatocytes. *Proc. Natl. Acad. Sci. USA*, **104**(32), 12982–12987.

Sarti, P., Ginobbi, P., D'Agostino, I., Arancia, G., Lendaro, E., Molinari, A., Ippoliti, R., & Citro, G. (1996). Liposomal targeting of leukaemia HL60 cells induced by transferrin-receptor endocytosis. *Biotechnol. Appl. Biochem.*, **24**(Pt 3), 269–276.

Sato, A., Takagi, M., Shimamoto, A., Kawakami, S., & Hashida, M. (2007). Small interfering RNA delivery to the liver by intravenous administration of galactosylated cationic liposomes in mice. *Biomaterials*, **28**(7), 1434–1442.

Schiffelers, R. M., Molema, G., ten Hagen, T. L., Janssen, A. P., Schraa, A. J., Kok, R. J., Koning, G. A., & Storm, G. (2002). Ligand-targeted liposomes directed against pathological vasculature. *J. Liposome Res.*, **12**(1-2), 129–135.

Schreiber, S., Rutgeerts, P., Fedorak, R. N., Khaliq-Kareemi, M., Kamm, M. A., Boivin, M., Bernstein, C. N., Staun, M., Thomsen, O. Ø., & Innes, A. (2005). A randomized, placebo-controlled trial of certolizumab pegol (CDP870) for treatment of Crohn's disease. *Gastroenterology*, **129**(3), 807–818.

Seymour, L. W., Duncan, R., et al. (1987). Potential of sugar residues attached to n-(2- hydroxypropyl)methacryl amide copolymers as targeting groups for the selective delivery of drugs. *J. Bioact. Compat. Pol.*, **2**(2), 97–119.

Seymour, L. W., Flanagan, P. A., al-Shamkhani, A., Subr, V., Ulbrich, K., Cassidy, J., & Duncan, R. (1991). Synthetic polymers conjugated to monoclonal antibodies: vehicles for tumour-targeted drug delivery. *Sel. Cancer Ther.*, **7**(2), 59–73.

Song, E., Zhu, P., Lee, S. K., Chowdhury, D., Kussman, S., Dykxhoorn, D. M., Feng, Y., Palliser, D., Weiner, D. B., Shankar, P., Marasco, W. A., & Lieberman, J. (2005). Antibody mediated *in vivo* delivery of small interfering RNAs via cell-surface receptors. *Nat. Biotechnol.*, **23**(6), 709–717.

Song, H., He, R., Wang, K., Ruan, J., Bao, C., Lin, N., Ji, J., & Cui, D. (2010). Anti-HIF-1alpha antibody-conjugated pluronic triblock copolymers encapsulated with Paclitaxel for tumor targeting therapy. *Biomaterials*, **31**(8), 2302–2312.

Soni, V., Kohli, D. V., & Jain, S. K. (2008). Transferrin-conjugated liposomal system for improved delivery of 5-fluorouracil to brain. *J. Drug Target.*, **16**(1), 73–78.

Spanjer, H. H., & Scherphof, G. L. (1983). Targeting of lactosylceramide-containing liposomes to hepatocytes *in vivo*. *Biochim. Biophys. Acta*, **734**(1), 40–47.

Stahl, P., Schlesinger, P. H., Sigardson, E., Rodman, J. S., & Lee, Y. C. (1980). Receptor-mediated pinocytosis of mannose glycoconjugates by macrophages: Characterization and evidence for receptor recycling. *Cell*, **19**(1), 207–215.

Suk, J. S., Suh, J., Choy, K., Lai, S. K., Fu, J., & Hanes, J. (2006). Gene delivery to differentiated neurotypic cells with RGD and HIV Tat peptide functionalized polymeric nanoparticles. *Biomaterials*, **27**(29), 5143–5150.

Thistlethwaite, J. R., Jr., Cosimi, A. B., Delmonico, F. L., Rubin, R. H., Talkoff-Rubin, N., Nelson, P. W., Fang, L., & Russell, P. S. (1984). Evolving use of OKT3 monoclonal antibody for treatment of renal allograft rejection. *Transplantation*, **38**(6), 695–701.

Thorstensen, K., & Romslo, I. (1993). The transferrin receptor – its diagnostic-value and its potential as therapeutic target. *Scand. J. Clin. Lab. Inv.*, **53**, 113–120.

Trowbridge, I. S. (1988). Transferrin receptor as a potential therapeutic target. *Prog. Allergy*, **45**, 121–146.

Ulbrich, K., & Subr, V. (2010). Structural and chemical aspects of HPMA copolymers as drug carriers. *Adv. Drug Deliv. Rev.*, **62**(2), 150–166.

Van Den Hamer, C. J., Morell, A. G., Scheinberg, I. H., Hickman, J., & Ashwell, G. (1970). Physical and chemical studies on ceruloplasmin. IX. The role of galactosyl residues in the clearance of ceruloplasmin from the circulation. *J. Biol. Chem.*, **245**(17), 4397–4402.

van Steenis, J. H., van Maarseveen, E. M., Verbaarn, F. J., Verrijk, R., Crommelin, D. J., Storm, G., & Hennink, W. E. (2003). Preparation and characterization of folate-targeted pEG-coated pDMAEMA-based polyplexes. *J. Control. Release*, **87**(1-3), 167–176.

Vinogradov, S., Batrakova, E., Li, S., & Kabanov, A. (1999). Polyion complex micelles with protein-modified corona for receptor-mediated delivery of oligonucleotides into cells. *Bioconjug. Chem.*, **10**(5), 851–860.

Wadhwa, M. S., & Rice, K. G. (1995). Receptor mediated glycotargeting. *J. Drug Target.*, **3**(2), 111–127.

Wang, S., Lee, R. J., Cauchon, G., Gorenstein, D. G., & Low, P. S. (1995). Delivery of antisense oligodeoxyribonucleotides against the human epidermal growth factor receptor into cultured KB cells with liposomes conjugated to folate via polyethylene glycol. *Proc. Natl. Acad. Sci. USA*, **92**(8), 3318–3322.

Weitman, S. D., Lark, R. H., et al. (1992). Distribution of the folate receptor GP38 in normal and malignant cell lines and tissues. *Cancer Res.*, **52**(12), 3396–3401.

Wightman, L., Patzelt, E., et al. (1999). Development of transferrin-polycation/DNA based vectors for gene delivery to melanoma cells. *J. Drug Target.*, **7**(4), 293–303.

Wu, G., & Wu, C. (1987). Receptor-mediated *in vitro* gene transformation by a soluble DNA carrier system. *J. Biol. Chem.*, **262**, 4429–4432.

Yang, C., Ding, N., et al. (2009). Folate receptor-targeted quantum dot liposomes as fluorescence probes. *J. Drug Target.*, **17**(7), 502–511.

Ying, X., Wen, H., et al. (2010). Dual-targeting daunorubicin liposomes improve the therapeutic efficacy of brain glioma in animals. *J. Control. Release*, **141**(2), 183–192.

Yoo, H., & Park, T. (2004). Folate receptor targeted biodegradable polymeric doxorubicin micelles. *J. Control. Rel.*, **96**, 273–283.

B.4. POLYMER–DRUG CONJUGATES

Suzie H. Pun and Allan S. Hoffman
Department of Bioengineering, University of Washington, Seattle, WA, USA

INTRODUCTION

The concept of synthetic polymer–drug conjugates was developed in the 1970s. In the late 1960s and early 1970s, Frank Davis at Rutgers University NJ, USA conceived of the idea of reducing the potential immunogenicity of the new genetically-engineered recombinant proteins by PEGylating them (Davis, 2002). Helmut Ringsdorf published a prescient article on polymer–drug conjugates in 1975 that included concepts of conjugation of cell ligands and "trafficking" molecules (e.g., lipids and polyethylene glycol (PEG)) to the polymer backbone (Ringsdorf, 1975). In the mid-1970s Jindrich Kopecek and colleagues in Prague prepared poly(HPMA)–drug conjugates with a pendant, degradable tetrapeptide spacer to the drug (Kopecek et al., 1977). Later, in a collaboration with Ruth Duncan, they designed and synthesized a hydroxypropyl methacrylamide (HPMA) copolymer with doxorubicin conjugated to the HPMA backbone by a tetrapeptide spacer that was a substrate for the lysosomal enzyme, cathepsin B (Duncan, 2009). They also later conjugated a targeting ligand (galactose) to the same spacer, producing one of the first examples of a targeted, polymer–drug conjugate. This section reviews many of the different polymer–drug conjugates other than PEG, which was discussed earlier in this section.

POLY(HPMA) AS A DRUG CARRIER

The first synthetic polymer drug conjugate to be tested clinically was an HPMA copolymer (N-2-hydroxypropyl methacrylamide) conjugated to doxorubicin (Figure B.4.1A). HPMA polymers were developed by Kopecek and Bazilova (Kopecek and Bazilova, 1973), and used by Duncan and co-workers for drug delivery applications due to its biocompatibility and low blood protein-binding properties (Duncan, 2009). Functionalized copolymers were synthesized by copolymerization of HPMA with

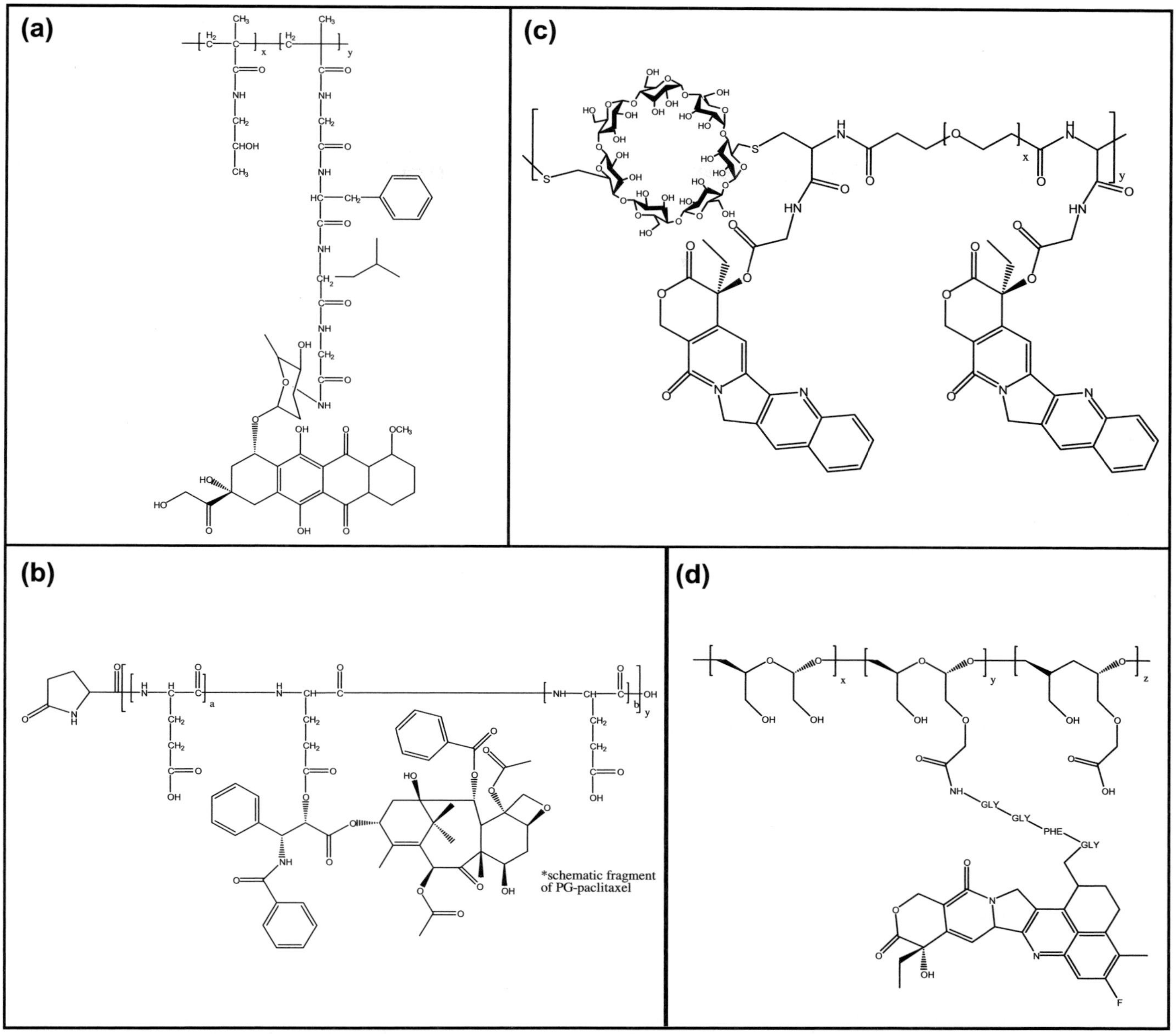

FIGURE B.4.1 Four polymer-drug conjugates: (A) Poly(hydroxypropyl methacrylamide) (PHPMA) conjugated to doxorubicin; (B) Poly(glutamic acid) (PG) conjugated to paclitaxel; (C) Cyclodextrin backbone polymer conjugated to camptothecin; (D) Polyacetal conjugated to camptothecin.

a methacryloylated-peptidyl-nitrophenylester; this co-monomer was used for drug conjugation through an aminolysis reaction. Many clinically-tested conjugates use the Gly-Phe-Leu-Gly peptidyl linker introduced through the second co-monomer. This peptide sequence is a substrate for lysosomal thiol-dependent proteases (e.g., cathepsin B), so that drug release is designed to occur after endocytosis of the drug conjugate and trafficking to the lysosomal compartment. Because the HPMA backbone is not biodegradable, the molecular weight of the polymers is typically limited to <40 k, so that materials can be cleared renally.

Several other chemotherapeutic drugs, such as doxorubicin, paclitaxel, camptothecin, and platinates, have been conjugated to HPMA and evaluated in clinical trials with varying success. The paclitaxel and camptothecin conjugates developed by Pharmacia were halted after Phase I trials due to toxicity issues. Doxorubicin conjugates licensed to Pharmacia entered Phase II trials, and a galactose-modified copolymer was also evaluated for targeted delivery to hepatocarcinoma. Prolonged plasma circulation was achieved with this conjugate compared with free doxorubicin, and tumor accumulation was observed in metastatic breast cancer patients (Duncan, 2009). An HPMA–platinate conjugate has been evaluated in Phase II clinical trials by Access Pharmaceuticals. They reported drug efficacy with favorable safety profiles in patients (Nowotnik and Cvitkovic, 2009).

POLY(GLUTAMIC ACID) (PG) AS A DRUG CARRIER

Poly(L-glutamic acid) (PG) is a biodegradable polymer that is degraded by the lysosomal enzyme cathepsin B (Figure B.4.1B). PG has been conjugated to various anti-cancer agents, including doxorubicin, paclitaxel, and camptothecin (Li, 2002). Several drug–polymer linkages were also tested, including amides, hydrolyzable ester, hydrazone bonds, and enzymatically degradable peptide spacers. Drug activity was optimized by varying the release kinetics from the degrading polymer and evaluating drug efficacy.

Two PG–drug conjugates, PG–paclitaxel and PG–camptothecin, were evaluated clinically by Cell Therapeutics, Inc. The first to enter the clinic was a PG conjugated to paclitaxel via a glycine linker. The polymers had an average molecular weight of 38 kD with 37% by weight paclitaxel conjugated in the 2′ position of the glutamic acid by an ester bond. The drug conjugate was shown to be relatively stable in the blood circulation, and to release active drug primarily through lysosomal cathepsin B degradation of the polymer backbone after cellular uptake (Paz-Ares et al., 2008). The PG–paclitaxel conjugate delivered similar amounts of active paclitaxel compared with equivalent doses of standard paclitaxel (the "area under the curve" or AUC values were similar), but the prolonged distribution of the polymer–drug conjugate resulted in decreased maximum plasma concentration (C_{max}) compared with the standard paclitaxel dose, and also reduced myelosuppression and alopecia (Bonomi, 2007). Interestingly, Phase III studies showed that the overall survival of patients treated with PG–paclitaxel did not differ significantly from patients treated with free paclitaxel. However, a subset of patients, specifically premenopausal women, responded more favorably to the drug conjugates, perhaps due to higher estrogen levels that correlate with cathepsin B activity. An application was submitted for marketing to the European Medicines Agency by Cell Therapeutics in 2008 for this formulation; PG–paclitaxel may therefore be the first polymer–anticancer drug conjugate to be marketed. Cell Therapeutics also investigated clinically a PG-camptothecin conjugate. However, development of this material was halted after Phase I/II studies.

CYCLODEXTRIN POLYMERS AS DRUG CARRIERS

Cyclodextrins (Figure B.4.1C) are cyclic oligomers of glucose, and have been used as drug solubilizers in US Food and Drug Administration (FDA)-approved formulations. Linear polymers of cyclodextrin synthesized by condensation polymerization of bifunctionalized cyclodextrin monomers with a second comonomer have been investigated as drug carriers in clinical trials. The first material evaluated clinically by Insert Therapeutics is a cyclodextrin polymer–camptothecin conjugate (IT-101; Figure B.4.1C) (Davis, 2009a,b). Camptothecin was conjugated to the carboxylate groups of the cyclodextrin polymer through a glycine linker, resulting in a hydrolyzable ester bond (Cheng, 2003). IT-101 has molecular weight ~70 k, and is renally excreted. In aqueous solutions, IT-101 self-assembles into particles with sizes ~30–40 nm in diameter. Unlike the micellar structure depicted in Figure B.4.1, the self-assembly of IT-101 particles is likely driven by inclusion complex formation between the cyclodextrin and camptothecin molecules. Phase I clinical trials of IT-101 for treatment of advanced solid tumors showed long half-life of IT-101 in humans (~40 hours) and disease stabilization in several patients.

POLYACETALS AS DRUG CARRIERS

A biodegradable polyacetal poly(1-hydroxymethylethylene hydroxymethylformal) (Figure B.4.1D) conjugated to camptothecin (XMT-1001) has also been investigated in clinical trials. This molecular carrier, under development by Mersana Therapeutics, delivers camptothecin via a two-phase drug release mechanism (Yurkovetskiy et al., 2004). A camptothecin derivative is first released from the polymeric backbone; this lipophilic derivative is active but not as potent as camptothecin. This derivative is then hydrolyzed into camptothecin. It is hypothesized that the dual-release approach may result in improved tumor delivery. XMT-1001 entered Phase I clinical trials in 2007.

BIBLIOGRAPHY

Bonomi, P. (2007). Paclitaxel poliglumex (PPx, CT-2103): Macromolecular medicine for advanced non-small-cell lung cancer. *Exp. Rev. Anticancer Ther.*, 7, 415–422.

Cheng, J., Khin, K. T., Jensen, G. S., Liu, A., and Davis, M. E. (2003). Synthesis of linear, β-cyclodextrin-based polymers and their camptothecin conjugates. *Bioconj. Chem.* **14**, 1007–1017.

Davis, M. E. (2009a). The first targeted delivery of siRNA in humans via a self-assembling, cyclodextrin polymer-based nanoparticle: From concept to clinic. *Mol. Pharm.*, **6**(3), 659–668.

Davis, M. E. (2009b). Design and development of IT-101, a cyclodextrin-containing polymer conjugate of camptothecin. *Adv. Drug Del. Rev.*, **61**, 1189–1192.

Davis, F. F. (2002). The origin of pegnology. *Adv. Drug Del. Rev.*, **54**, 457–458.

Duncan, R. (2009). Development of HPMA copolymer-anticancer conjugates: Clinical experience and lessons learnt. *Adv. Drug Del. Rev.*, **61**, 1131–1148.

Kopecek, J., & Bazilova, H., (1973). Poly[HPMA]. 1. Radical polymerization and copolymerization. *Eur. Polym. J.*, **9**, 7–14.

Kopecek, J., Ulbrich, K., Vacik, J., Strohalm, J., Chytry, V., Drobnik, J., & Kalal, J. (1977). Copolymers based on n-substituted acrylamides and methacrylamides, and n,n disubstituted acrylamides, and the method of their manufacturing. *US Patent 4,062,831*. Dec 13, 1977.

Li, C. (2002). Poly(L-glutamic acid)-anticancer drug conjugates. *Adv. Drug Del. Rev.*, **54**, 695–713.

Nowotnik, D. P., & Cvitkovic, E. (2009). ProLindac (AP5346): A review of the development of an HPMA DACH platinum polymer therapeutic. *Adv. Drug Del. Rev.*, **61**, 1214–1219.

Paz-Ares, L., Ross, H., O'Brien, H., Riviere, A., Gatzemeier, U., von Pawel, J., et al. (2008). Phase III trial comparing paclitaxel poliglumex vs docetaxel in the second-line treatment of non-small-cell lung cancer. *Br. J. Cancer*, **98**, 1608–1613.

Ringsdorf, H. (1975). Structure and properties of pharmacologically active polymers. *J. Polym. Sci. Polym. Symp.*, **51**, 135–153.

Yurkovetskiy, A. V., Hiller, A., Syed, S., Yin, M., Lu, X. M., Fischman, A. J., Papisov, M. I., et al. (2004). Synthesis of a macromolecular camptothecin conjugate with dual phase drug release. *Mol. Pharm.*, **5**, 375–382.

B.5. LIPOSOMES

Wayne R. Gombotz
Immune Design Corp., Seattle, WA, USA

The first observation that ordered structures are obtained when water-insoluble lipids, such as phospholipids, are mixed with an excess of water was made by Bangham and co-workers in the 1960s (see historical review by Bangham, 1983). These ordered structures eventually self-assembled as concentric, closed spherical membranes known as liposomes or vesicles. Such liposomes can consist of one or a multiplicity of bilayer membranes. Liposomes with a single lipid bilayer are typically 25–100 nm in size, and are called small unilamellar vesicles (SUV) as shown in Figure B.5.1. Multilamellar vesicles (MLV) have several concentric bilayers and are larger in size, with diameters up to several microns. A good review has recently been published (Gregoriades, 2003).

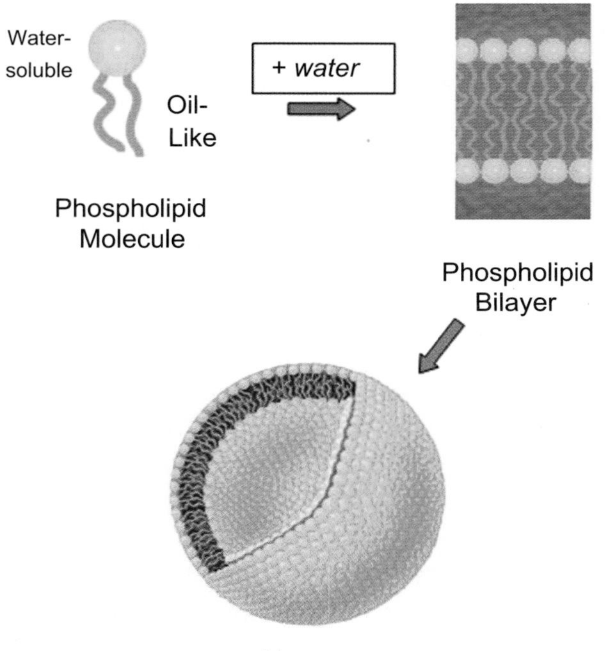

FIGURE B.5.1 Formation of a single unilamellar vesicle (SUV) liposome. (NanolifeNutra® website.) Note that the addition of "water" could be an aqueous solution of the drug, which will end up in the core of the liposome if it is very hydrophilic or in the lipid bilayer if it is very hydrophobic, or it may partition between those two phases depending on its relative hydrophilic/hydrophobic character.

Stable liposomes from phospholipids are formed only at temperatures above the gel-to-liquid crystalline phase transition temperature (Tc) of the phospholipid, which represents the melting point of the acyl chains. The Tc of phospholipids is also dependent upon the nature of the polar head group, and on the length and degree of unsaturation of the acryl chains (Gregoriadis and Senior, 1980). A variety of phospholipids have been used to prepare liposomes, the most common ones being distearoyl (with a saturated hydrocarbon chain) and dioleoyl phosphatidylcholine (with an unsaturated hydrocarbon chain) (Scherphof et al., 1979).

Liposomes are employed as drug delivery vehicles for their ability to protect encapsulated drugs from degradation, and to improve the biodistribution of drugs, thereby reducing systemic toxicity. Liposomes are able to entrap water-soluble solutes in the aqueous inner core. Lipophilic drugs may be entrapped in the lipid bilayers by combining such drugs with the phospholipids used during preparation of the vesicles. The encapsulation within or association of drugs with liposomes alters the pharmacokinetic properties of the drug, and this may be exploited to achieve targeted therapies with improved efficacy.

Vesicles larger than about 200 nm are rapidly cleared from the blood, and end up in the macrophages of the reticuloendothelial system (RES). This can be exploited to deliver antiparasitic and antimicrobial drugs to the mononuclear phagocytic system (Alving et al., 1978). To reduce phagocytic uptake of liposomes and increase their circulation times, liposomes can be coated with polyethylene glycol (PEG) or sialic acid surfactants. In the former case, PEG is conjugated to a phospholipid such as phosphatidylethanolamine, which inserts into the liposome's lipid bilayer, "PEGylating" the liposome. Sialic acid surfactants may be similarly used. These polymers form a water-retaining coating around the liposome, and prevent recognition by the reticuloendothelial system (RES) (Gregoriades, 1995a). Such liposomes have been termed "stealth liposomes." To impart cellular specificity to liposomes, antibodies may be conjugated directly to them (Torchilin et al., 1979). PEGylated or stealth liposomes with target recognition properties have also been synthesized (Gregoriades, 1995b), and they are the most commonly used today. Such "active targeting" requires the identification of suitable receptors on the surface of the target cells and appropriate ligands on the surface of the liposome that can recognize this target. When the liposome is PEGylated, the targeting ligand is conjugated to the outer end of the PEG molecules.

Four major factors control the *in vivo* behavior and biodistribution of liposomes. First, liposomes will leak drug if cholesterol is not included in the vesicle membrane. Cholesterol increases the packing density of phospholipids in the lipid bilayer and stabilizes the liposome structures (Damen, 2005). Second, large liposomes are cleared more rapidly than small liposomes, due to rapid elimination by the RES system (Senior, 1987). Third, charged liposomes are cleared more rapidly than uncharged liposomes, again typically by RES removal. Finally, the half-life of a liposome increases as the liposome dose is increased (Ostro and Cullis, 1989). Since the RES is the primary mode of clearance, saturation of the phagocytic cells with high liposome doses can increase circulation time of remaining liposomes.

Many anticancer drugs exhibit toxic side-effects, which make them ideal candidates for targeted delivery by liposomes. Some of these drugs are also subject to P-glycoprotein (P-gp) efflux from cancer cells, which are multi-drug resistant (MDR) cells. The use of nanocarriers such as liposomes avoids this effect, since the carriers are entrapped within the tumor tissue by the enhanced permeability and retention (EPR) effect, and subsequently may be taken up into cancer cells by endocytosis, avoiding the P-gp efflux mechanism. An early publication on PEGylated liposomes was by Klibanov and coworkers (Klibanov et al., 1990). The first liposomal product was approved by the US Food and Drug Administration (FDA) in 1995. It was a PEGylated liposome-doxorubicin product called Doxil® for the treatment of ovarian cancer (Muggia et al., 1997; Lyass et al., 2000). Doxorubicin-loaded "stealth" liposomes circulate for prolonged periods of time and accumulate within tumors (Papahadjopoulos et al., 1991). Non-stealth liposomes made from high Tc lipids such as distearoyl phosphatidylcholine also accumulate within tumors, and have been marketed as the product called DaunoXome®, a liposomal daunorubicin formulation, for the treatment of Kaposi's sarcoma (Forssen, 1997).

Liposomes have been developed successfully as vaccine adjuvants, potentiating both cell-mediated and humoral immunity (Gregoriades et al., 1996). Liposomal adjuvants act by slowly releasing antigen upon intramuscular injection, and by passively accumulating within regional lymph nodes. They can be prepared by encapsulating soluble antigens, microbes, cytokines or DNA. There are two liposome-based vaccines approved in Europe, Epaxal® for the prevention of hepatitis A and Inflexa® V for immunization against influenza. Both of these vaccines contain two influenza viral antigens, hemagglutinin and neuraminidase. These antigens are exposed on the liposome surface and also anchored in the liposome membrane, and this design mimics the outer coat of the influenza virus. In addition, Epaxal® contains the whole cell hepatitis A virus RG-SB strain adsorbed to the surface of the liposome, while Inflexal® V has HIN1, H3N2, and B2 influenza strains on its surface.

Table B.5.1 summarizes many of the current commercially available liposome products. Of the four therapeutic products listed, all of the drugs are encapsulated within the liposome except AmBisome®, which consists of a unilamellar bilayer liposome with the hydrophobic drug, amphotericin B, intercalated within the lipophilic membrane. Amphoterecin B, a polyene antibiotic, is associated with extensive renal toxicity in the treatment of systemic fungal infections. By passively targeting the liver and spleen, AmBisome® reduces renal and general toxicity of the drug (Gray and Morgan, 1991). Another product, Abelcet®, although technically not a true liposome, consists of amphotericin B complexed with two phospholipid molecules in a 1:1 drug-to-lipid molar ratio. The two phospholipids, L-α-dimyristoylphosphatidylcholine and L-α-dimyristoylphosphatidylglycerol, are present in a 7:3 molar ratio. Abelcet® is used to treat a variety of serious fungal infections, and acts in a similar manner to AmBisome®; it also exhibits reduced kidney toxicity. It is often used in patients who cannot tolerate or who do not respond to regular amphotericin treatment.

Cationic liposomes have also been used for the delivery of DNA, a topic that is discussed in more detail in the section in this chapter on gene therapy, and polyplexes and lipoplexes for nucleic acid drug delivery.

TABLE B.5.1 Examples of Commercially Approved Liposome Products

Product	Company	Active Agent	Indication	Year Approved
Epaxal®	Crucell	Inactivated hepatitis A virions	Hepatitis A	1993
Inflexal® V	Crucell	H1N1, H3N2, and B strains	Influenza	1997
AmBisome®	Gilead Sciences/Astellas/ Fujisawa Healthcare	Amphoterecin B	Systemic fungal infections	1997
DaunoXome®	Gilead Sciences	Daunorubicin	Kaposi's sarcoma	1996
Doxil®	Johnson & Johnson/Alza	Doxorubicin	Ovarian cancer	1995
Myocet®	Elan	Doxorubicin	Metastatic breast cancer	2000

BIBLIOGRAPHY

Alving, C. R., Steck, E. A., Chapman, W. L., Jr., Waits, V. B., Hendricks, L. D., Swartz, G. M., Jr., & Hanson, W. L. (1978). Therapy of leishmaniasis: Superior efficacies of liposome-encapsulated drugs. *PNAS*, **75**, 2959–2963.

Bangham, A. D. (1983). Liposomes: A historical perspective. In M. J. Ostro (Ed.), *Liposomes*. New York: Marcel Dekker.

Damen. J. (2005). BBA, 665, 558.

Forssen, E. (1997). The design and development of DaunoXome for solid tumor targeting *in vivo*. *Adv. Drug Del. Rev.*, **24**, 133–150.

Gray, A., & Morgan, J. (1991). Liposomes in hematology. *Blood Rev.*, **5**, 258–271.

Gregoriades, G. (1995a). Fate of liposomes *in vivo*: A historical perspective. In D. Lasic, & F. Martin (Eds.), *Stealth Liposomes* (pp. 7–12). Boca Raton, FL: CRC Press.

Gregoriades, G. (1995b). Engineering of targeted liposomes: Progress and problems. *Trends Biotechnol.*, **13**, 527–537.

Gregoriades, G. (2003). Liposomes in drug and vaccine delivery. *Drug Deliv. Syst. Sci.*, **21**, 91–97.

Gregoriades, G., & Senior, J. (1980). The phospholipids component of small unilamellar liposomes controls the rate of clearance of entrapped solutes from the circulation. *FEBS Lett.*, **119**, 43–46.

Gregoriades, G., Gursel, I., Gursel, M., & McCormack, B. (1996). Liposomes as immunological adjuvants and vaccine carriers. *J. Control. Rel.*, **41**, 49–56.

Klibanov, A. L., Maruyama, K., Torchilin, V. P., & Huang, L. (1990). Amphipathic polyethyleneglycols effectively prolong the circulation time of liposomes. *FEBS 08706*, **268**, 235–237.

Lyass, O., Uziely, B., Ben-Yosef, R., Tzemach, D., Heshing, N., Lotem, M., Brufman, G., & Gabizon, A. (2000). Correlation of toxicity with pharmacokinetics of pegylated liposomal doxorubicin (Doxil) in metastatic breast carcinomas. *Cancer*, **89**, 1037–1047.

Muggia, F., Hainsworth, J., Jeffers, S., Miller, P., Groshen, S., Tan, M., Roman, L., Uziely, B., Muderspach, L., Garcia, A., Burnett, A., Greco, F., Morrow, C., Paradiso, L., & Liang, L. (1997). Phase II study of liposomal doxorubicin in refractory ovarian cancer: Antitumor activity and toxicity modification by liposomal encapsulation. *J. Clin. Oncol.*, **15**, 987–993.

Ostro, M., & Cullis, P. (1989). Use of liposomes as injectable-drug delivery systems. *Am. J. Health Sys. Pharmacy*, **46**, 1576–1587.

Papahadjopoulos, D., Allen, T., Gabizon, A., Mayhew, E., Matthay, K., & Huang, S. (1991). Sterically stabilized liposomes: Improvements in pharmacokinetics and antitumor therapeutic efficacy. *PNAS USA*, **88**, 11460–11464.

Scherphof, G., Morselt, H., Regts, J., & Wilschut, J. (1979). The involvement of the lipid phase transition in the plasma-induced dissolution of multilamellar phosphatidyl-choline vesicles. *Biochem. Biophys. Acta*, **556**, 196–207.

Senior, J. H. (1987). *Crit. Rev. Ther. Drug Carr. Sys.*, **3**, 123.

Torchilin, V. P., Khaw, B. A., Smirnov, V. N., & Haber, E. (1979). Preservation of antimyosin antibody activity after covalent coupling to liposomes. *Biochem. Biophys. Res. Commun.*, **89**, 1114–1119.

B.6. POLYMERIC MICELLES

Wayne R. Gombotz[1] and Allan S. Hoffman[2]

[1]Immune Design Corp., Seattle, WA, USA

[2]Professor of Bioengineering and Chemical Engineering, UWEB Investigator, University of Washington, Seattle, WA, USA

INTRODUCTION

Amphiphilic (or amphipathic) block copolymers can self-assemble into supramolecular structures in aqueous media, adopting a variety of morphologies that include spheres, rods, vesicles, and lamellae. This section focuses on spherical polymeric micelles because of the extensive investigation of these materials as drug delivery vehicles. For details on other structures formed from amphiphilic block copolymers, the reader is directed to excellent published reviews (Discher et al., 2007; Letchford and Burt, 2007).

Most micelles are assembled from amphiphilic, A–B, block copolymers, i.e., block copolymers composed of a hydrophilic and a hydrophobic segment, although micelles can also be constructed from graft copolymers with a soluble hydrophilic main chain and insoluble hydrophobic side chains. The earliest papers on A–B block copolymer micelles designed for drug delivery were jointly developed by the groups of Kataoka and Okano (e.g., Yokoyama et al., 1990a,b). Polymeric micelles have a fairly narrow size distribution, and have a core shell structure where the core contains the hydrophobic segments surrounded by an outer shell of the hydrophilic segments; the outer shell is often referred to as the "corona." Micelle formation is driven by the decrease in system free energy resulting from the sequestering of hydrophobic segments from aqueous environments as they release their hydrophobically bound water, with an associated large entropy gain (for a detailed review see Allen et al., 1999). Above the critical micelle concentration (CMC), micelles exist in equilibrium in solution with single polymer chains (unimers). Below the CMC, micelles dissociate into unimers, and it has often been speculated that when polymeric micelles containing drugs are injected into the blood they are rapidly diluted below the CMC and dissociate. This is a concern with their clinical use. Some micelles are kinetically stable below the CMC due to their slow dissociation.

Polymeric micelles have excellent potential as drug delivery systems (DDS), because they offer the ability to solubilize poorly water-soluble drugs and thus increase drug bioavailability (Sutton et al., 2007). Micelles are typically 10–100 nm in size, and therefore they can passively target tumors due to the enhanced permeability and retention (EPR) effect. Like liposomes, micelles can also be actively targeted by attachment of specific ligands at the ends of the hydrophilic block segment (Torchilin, 2004; Sutton et al., 2007). Many types of targeting ligands have been conjugated to micelles including small organic molecules, peptides, carbohydrates, antibodies, and aptamers. Table B.6.1 provides a summary of the types of ligand-targeted micelles that have been reported in the literature for cancer treatment. Several excellent reviews on polymeric micelles in drug delivery have

TABLE B.6.1 Ligand-Targeted, PEGylated Micelle Formulations for Cancer Treatment

Ligand Type	Ligand	Polymer	Drug	Reference
Small organic molecule	Folic acid	PEG-PLGA	DOX	Yoo and Park, 2004a
	Folic acid	PEG-Dox	DOX	Yoo and Park, 2004b
	Folic acid	PEG-PCL	Paclitaxel	Park et al., 2005
Peptide	cRGD	PEG-PCL	DOX	Nasongkla et al., 2004
	cRGD	PLA-PEO	Paclitaxel	Hu et al., 2008
Carbohydrate	Galactose	Poly(L-benzyl l-glutamate)-PEG	Paclitaxel	Jeong et al., 2005
	Lactose, galactose, mannose, glucose	PEG-PLA		Yasugi et al., 1999
Antibody	Anti-GFA Ab	Pluronic®	Haloperidol	Kabanov et al., 1989
	mAb 2C5 and mAb 2G4	PEG-PE	Paclitaxel	Torchilin et al., 2003
RNA Aptamer	Anti-PSMA aptamer	PEG-PLA	Docetaxel	Farokhzad et al., 2006

PCL: polycaprolactone
PE: polyethylene
PEG: polyethylene glycol
PEO: polyethylene oxide
PLA: poly(lactic acid)
PLGA: poly(lactic-co-glycolic acid)

been published (Kataoka et al., 2001; Torchilin, 2001). Encapsulation of a drug within a stable polymeric micelle will significantly alter drug pharmacokinetics, and can lead to an increase in drug targeting to tumor tissues. Once a micelle reaches the target site by either passive targeting (e.g., EPR) or active targeting via ligands, it becomes critically important that the drug is released from the micelle carrier and elicits its desired cytotoxic effect. Several strategies have been developed in which the micelles are engineered to trigger drug release by certain stimuli at the tumor site. These include drug release facilitated by a pH change, temperature change, ultrasound, acid-labile bonds in the linkage between the drug and the polymer forming the micelle, an ionizable component within the micelle structure, and chemical reactivity (Sutton et al., 2007).

Polymeric micelles designed for delivery of drugs and imaging agents can be divided into four classes that share a similar molecular architecture (Koo et al., 2005a). These include: (1) phospholipid micelles; (2) Pluronic® micelles (which are A–B–A triblock copolymers, in contrast to the others which are A–B structures); (3) poly (L-amino acid) micelles; and (4) polyester micelles.

In contrast to typical phospholipids used to form liposomes, such as phosphatidylcholine, polyethylene glycol (PEG)-conjugated phospholipids self-assemble into nanosize micelles. These micelles are simple to prepare and have relatively low toxicity; they also have prolonged circulation times (Sethi et al., 2003). A number of therapeutic agents have been incorporated and stabilized in these micelles, including paclitaxel (Krishnadas et al., 2003), diazepam (Ashok et al., 2004), camptothecin (Koo et al., 2005b), and vasoactive intestinal peptide (Onyuksel et al., 1999).

Although other hydrophilic polymers can be used to solubilize and stabilize hydrophobic drugs (Torchilin and Papisov, 1994), poly(ethylene glycol) or PEG remains the polymer of choice. However, a variety of polymers have been used to form the water-insoluble block in the micelle core. This block may be a water-insoluble hydrophobic polymer or a water-insoluble ionic complex, such as between a nucleic acid drug and a polycation block. The insoluble core blocks thus may be comprised of poly(propylene oxide) (Kabanov et al., 1989), poly(L-lysine)-nucleic acid complexes (Katayose and Kataoka, 1998), hydrophobic esters of poly(aspartic acid) (Yokoyama et al., 1990a,b), γ-benzoyl-L-aspartate (Kwon et al., 1997), γ-benzyl-L-glutamate (Jeong et al., 1998), polycaprolactone (Kim et al., 1998), poly(D,L-lactic acid) (Ramaswamy et al., 1997), and poly(ortho esters) (Toncheva et al., 2003). Polyion complex cores were described by Harada and Kataoka (1995), and metal ion complex cores were published in 2003 (Nishiyama et al., 2003).

A recent generation of targeted micelles is based on systems with multiple functionalities. Bae, Kataoka, and collaborators conjugated doxorubicin to the aspartic acid residues of a PEG-polyaspartic acid (PEG-pAsp) copolymer via a hydrazone linkage to improve drug loading and create a pH-triggered drug release system (Bae et al., 2003). Folic acid was then added as a targeting ligand to increase uptake and ultimate cytotoxicity of the formulation to folate receptor presenting cells (Bae et al., 2005). Torchilin designed a pH-controlled and targeted micelle system with two targeting ligands on the PEG-phospholipid micelles. The first is an antibody attached to the end of a longer 3.4 kD PEG chain, the second is a targeting ligand such as TAT peptide attached to shorter PEG 2 kDa chains (Sawant et al., 2006). Filamentous micelles with nanometer diameters, but lengths up to several microns, are worth mentioning because of their improved *in vivo* biodistribution compared with spherical micelles. Discher and co-workers have shown that these "filomicelles" have a wormlike structure that can remain in the circulation for at least a day after intravenous (IV) injection (Christian et al., 2009). In their studies, filomicelles formed from PEG-polycaprolactone (PCL) diblock copolymers were loaded with paclitaxel. Intravenous injection of the paclitaxel filomicelles nearly doubled the maximum tolerated dose (MTD) of the drug in normal mice compared to paclitaxel-loaded spherical micelles. They were also shown to effectively deliver both paclitaxel and a near-infrared imaging agent to solid tumors, where they produced sustained tumor shrinkage and tumor cell apoptosis.

Polymeric micelles have principally been used as drug carriers in tumor-targeting applications. Clinical data have been reported on three polymer micelle systems, SP1049C (Danson et al., 2004), NK911 (Matsumura et al., 2004), and Genexol-PM (Kim et al., 2004). All three micelle formulations have stabilizing PEG coronas to minimize opsonization and uptake by the reticuloendothelial system (RES) of the liver, enhancing blood circulation times.

Pluronics® or Poloxamer® polyols are thermally-responsive tri-block copolymers of polyethylene glycol (PEG or PEO) and polypropylene glycol (PPO) with the formula PEO–PPO–PEO. They have been used as polymeric micelle drug carriers (Wang and Johnston, 1991). While many of these block copolymers have transition temperatures (called "cloud points") well above body temperature, when aqueous solution concentrations rise above around 16% some copolymer compositions will exhibit thermally-induced transitions around body temperature. Pluronic polyols L61 and F127 have recently been used as polymeric micelles in Phase I clinical trials for delivering anticancer drugs, where they may avoid the membrane-bound P-glycoprotein (P-gp) efflux pump, and thereby overcome the multi-drug resistance (MDR) of certain cancer cells (Kabanov et al., 1989). P-gp efflux is a phenomenon whereby free anti-neoplastic drugs are pumped out of a cell, resulting in failures of chemotherapy regimes (Batrakova et al., 1996). Nanocarriers in general can avoid this phenomenon; however, one major disadvantage of Pluronic® polyol polymeric micelles in particular is that they are not degradable.

SP1049C is a Pluronic®-micelle encapsulating doxorubicin. In a Phase I study it exhibited a slower clearance of doxorubicin compared to free doxorubicin (Danson et al., 2004). A subsequent Phase II study showed that among 19 eligible patients with adenocarcinoma of the esophagus, nine partial responses (47%) and eight stable diseases (42%) were achieved (Armstrong et al., 2006). The investigators concluded that the micellar doxorubicin was active in this group of patients, and recommended studies combining doxorubicin with other active agents. In an earlier Phase II study this system was shown to avoid the P-gp drug efflux mechanism.

> Cells that exhibit the P-glycoprotein (P-gp) efflux phenomenon with free drugs are called multi-drug-resistant or MDR cells. It has generally been concluded that drugs delivered by nanocarriers are sequestered in tumor tissues due to the EPR effect, and the carriers can then be endocytosed into cancer cells, avoiding the P-gp efflux effect.

NK911 is comprised of doxorubicin encapsulated in micelles based on a block copolymer of poly(ethylene glycol)-poly(aspartic acid), with a fraction of doxorubicin covalently bound and a fraction ionically complexed to the aspartic acid units (Nakanishi et al., 2001). The drug was given intravenously to patients with solid tumors every three weeks in a Phase I clinical trial. A total of 23 patients were enrolled, and a maximum tolerated dose (MTD) of 67 mg/m^2 and a dose-limiting toxicity of neutropaenia were observed.

Genexol-PM (where PM = polymeric micelle), a formulation comprised of paclitaxel encapsulated in a polymeric PEG-(PLA) (poly D,L lactic acid) micelle, was approved in 2008 in South Korea for the treatment of breast and lung cancer (Kim et al., 2004; Lee et al., 2006). In a current Phase IIa pancreatic cancer trial in the US, Genexol-PM showed similar pharmacokinetics to conventional paclitaxel formulated in Cremophor® EL. In contrast, marked improvement in patient morbidity was observed. Moreover, a lower degree of myelosupression was observed with Genexol-PM, which allowed for a considerable increase in the MTD, with an MTD of 390 mg/m^2 compared to 230 mg/m^2 for Cremophor® EL.

BIBLIOGRAPHY

Allen, C., Maysinger, D., & Eisenberg, Adi (1999). Nano-engineering block copolymer aggregates for drug delivery. *Colloid. Surface. B*, **16**, 3–27.

Armstrong, A., Brewer, J., Newman, C., Alakhov, V., Pietrzynski, G., Campbell, S., Corriem, P., Ranson, M., & Valle, J. (2006). SP1049C as first-line therapy in advanced (inoperable and metastatic) adenocarcinoma of the oesophagus: a phase II window study. *J. Clin. Oncol. (Meeting Abstracts)*, **24**, 4080.

Ashok, B., Arleth, L., Hjelm, R., Rubenstein, I., & Onyuksel, H. (2004). *In vitro* characterization of PEGylated phospholipids micelles for improved drug solubilization: Effects of PEG chain length and PC incorporation. *J. Phar. Sci.*, **93**, 2476–2487.

Bae, Y., Fukushima, S., Harada, A., & Kataoka, K. (2003). Design of environment-sensitive supramolecular assemblies for intracellular drug delivery: Polymeric micelles that are responsive to intracellular pH change. *Angew. Chem. Int. Ed. Engl.*, **42**, 4640–4643.

Bae, Y., Jang, W., Nishiyame, N., Fukushima, S., & Kataoka, K. (2005). Multifunctional polymeric micelles with folate mediated cancer cell targeting and pH-triggered drug releasing properties for active intracellular drug delivery. *Mol. Biosyst.*, **1**, 242–250.

Batrakova, E. V., Dorodynch, T. Y., Klinskii, E. Y., Kliuschnenkova, E. N., Shemchukova, O. B., Goncharova, O. N., Arjakov, S. A., Alakhov, V. Y., & Kabanov, A. V. (1996). Anthracycline antibiotics non-covalently incorporated into block copolymer micelles: *In vivo* evaluation of anticancer activity. *Br. J. Cancer*, **74**, 1545–1552.

Christian, D., Cai, S., Garbuzenko, O., Harada, T., Zajac, A., Minko, T., & Discher, D. (2009). Flexible filaments for *in vivo* imaging and delivery: Persistent circulation of filomicelles opens the dosage window for sustained tumor shrinkage. *Mol. Pharm.*, **6**, 1343–1352.

Danson, S., Ferry, D., Alakhov, V., Marigison, J., Kerr, D., Jowle, D., Brampton, M., Halbert, G., & Ranson, M. (2004). Phase I dose escalation and pharmacokinetic study of pluronic polymer-bound doxorubicin (SP1049C) in patients with advanced cancer. *Br. J. Cancer*, **90**, 2085–2091.

Discher, D., Ortiz, V., Srinivas, G., Klein, M., Kim, Y., Christian, D., Cai, S., Photos, P., & Ahmed, F. (2007). Emerging applications of polymersomes in delivery: From molecular dynamics to shrinkage of tumors. *Prog. Pol. Sci.*, **32**, 838–857.

Farokhzad, O., Cheng, J., Teply, B., Sherifi, I., Jon, S., Kantoff, P., Richie, J., & Langer, R. (2006). Targeted nanoparticle-aptimer bioconjugates for cancer chemotherapy *in vivo*. *Proc. Natl. Acad. Sci. USA*, **103**, 6315–6320.

Harada, A., & Kataoka, K. (1995). Formation of polyion complex micelles in aqueous media from a pair of oppositely-charged block copolymers with PEG segments. *Macromolecules*, **28**, 5294–5299.

Hu, Z., Luo, F., Pan, Y., Hou, C., Ren, L., Chen, J., Wang, J., & Zhang, Y. (2008). Arg-Gly-Asp (RGD) peptide conjugated poly(lactic acid)-poly (ethylene oxide) micelle for targeted drug delivery. *J. Biomed. Mater. Res.*, **85A**, 797–807.

Jeong, Y. I., Cheon, J. B., Kim, S. H., Nah, J. W., Lee, Y. M., Sung, Y. K., Akaike, T., & Cho, C. S. (1998). Clonazepam release from core-shell type nanoparticles *in vitro*. *J. Control. Release*, **51**, 169–178.

Jeong, Y., Seo, S., Park, I., Lee, H., Kang, I., Akaike, T., & Cho, C. (2005). Cellular recognition of paclitaxel loaded polymeric nanoparticles composed of poly(gamma-benzyl L-glutamate) and poly(ethylene glycol) diblock copolymer endcapped with galactose moiety. *Int. J. Pharm.*, **296**, 151–161.

Kabanov, A. V., Chekhonin, V. P., Alakhov, V., Yu, Batrakova, E. V., Lebedev, A. S., Melik-Nubarov, N. S., Arzhakov, S. A., Levashov, A. V., Morozov, G. V., Severin, E. S., & Kabanov, V. A. (1989). The neuroleptic activity of haloperidol increases after its solubilization in surfactant micelles. *FEBS Lett.*, **258**, 343–345.

Kataoka, K., Harada, A., & Nagasaki, Y. (2001). Block copolymer micelles for drug delivery: Design, characterization and biological significance. *Adv. Drug Del. Rev.*, **47**, 113–131.

Katayose, A., & Kataoka, K. (1998). Remarkable increase in nuclease resistance of plasmid DNA through supramolecular assembly with poly(ethylene glycol)–poly(L-lysine) block copolymer. *J. Pharm. Sc.*, **87**, 160–163.

Kim, S. Y., Shin, I. G., Lee, Y. M., Cho, C. G., & Sung, Y. K. (1998). Methoxy poly(ethylene glycol) and ε-caprolactone amphiphilic block copolymers micelle containing indomethacin. II. Micelle formation and drug release behavior. *J. Control. Release*, **51**, 13–22.

Kim, T., Kim, D., Chung, J., Shin, S., Kim, S., Heo, D., Kim, N., & Bang, Y. (2004). Phase I and pharmacokinetic study of Genexol-PM, a cremophor-free, polymeric micelle-formulated paclitaxel, in patients with advanced malignancies. *Clin. Cancer Res.*, **10**, 3708–3716.

Koo, O., Rubenstein, I., & Onyuksel, H. (2005a). Role of nanotechnology in targeted drug delivery and imaging: A concise review. *Nanomed. Nanotechnol. Biol. Med.*, **1**, 193–212.

Koo, O., Rubenstein, I., & Onyuksel, H. (2005b). Camptothecin in sterically stabilized phospholipids micelles; a novel nanomedicine. *Nanomedicine*, **1**, 77–84.

Krishnadas, A., Rubenstein, I., & Onyuksel, H. (2003). Sterically stabilized phospholipids mixed micelles: *In vitro* evaluation as a novel carrier for water-insoluble drugs. *Pharm. Res.*, **20**, 297–302.

Kwon, G. S., Naito, M., Yokoyama, M., Okano, T., Sakurai, Y., & Kataoka, K. (1997). Block copolymer micelles for drug delivery: Loading and release of doxorubicin. *J. Control. Release*, **48**, 195–201.

Lee, K., Chung, H., Im, S., Park, Y., Kim, S., Kim, C., & Ro, J. (2006). Multicenter phase II study of a cremophor-free polymeric micelle-formulated paclitaxel in patients with metastatic breast cancer. *J. Clin. Oncol.*, **24**(Suppl. 18), 10520.

Letchford, K., & Burt, H. (2007). A review of the formation and classification of amphiphilic block copolymer nanaoparticulate structures: Micelles, nanospheres, nanocapsules and polymersomes. *Eur. J. Pharm. Biopharm.*, **65**, 259–269.

Matsumura, Y., Hamaguchi, T., Ura, T., Muro, K., Yamada, Y., Shimada, Y., Shirao, K., Okusaka, T., Ueno, H., Ikeda, M., & Wanatebe, N. (2004). Phase I clinical trial and pharmacokinetic evaluation of NK911, a micelle-encapsulated doxorubicin. *Br. J. Cancer*, **91**, 1775–1781.

Nakanishi, T., Okamoto, K., Suzuki, M., Matsumura, Y., Yokoyama, M., Okano, T., Sakurai, Y., & Kataoka, K. (2001). Development of the polymer micelle carrier system for doxorubicin. *J. Control. Release*, **74**, 295–302.

Nasongkla, N., & Shuai, X., et al. (2004). cRGD-functionalized polymer michelles for targeted doxorubicin delivery. *Angewandte Chemie-International Edition*, **43**(46), 6323–6327.

Nishiyama, C., Okazaki, S., Cabral, H., Miyamoto, M., Kato, Y., Sugiyama, Y., Nishio, K., Matsumura, Y., & Kataoka, K. (2003). Novel cisplatin-incorporated polymeric micelles can eradicate solid tumors in mice. *Cancer Res.*, **63**, 8977–8983.

Onyuksel, H., Ikezaki, H., Patel, M., Gao, X., & Rubenstein, I. (1999). A novel formulation of VIP in sterically stabilized micelles amplifies vasodilation *in vivo*. *Pharm. Res.*, **16**, 155–160.

Park, E., Kim, S., Lee, S., & Lee, Y. (2005). Folate-conjugated methoxy poly(ethylene glycol)/poly(epsilon-caprolactone) amphiphilic block copolymeric micelles for tumor-targeted drug delivery. *J. Control. Release*, **109**, 158–168.

Ramaswamy, M., Zhang, X., Burt, H. M., & Wasan, K. M. (1997). Human plasma distribution of free paclitaxel and paclitaxel associated with diblock copolymers. *J. Pharm. Sci.*, **86**, 460–464.

Sawant, R., Hurley, J., Salmaso, S., Kale, A., Tolcheva, E., Levchenko, T., & Torchilin, V. (2006). "SMART" drug delivery systems: double-targeted pH-responsive pharmaceutical nanocarriers. *Bioconjugate Chem.*, **17**, 943–949.

Sethi, V., Onyuksel, H., & Rubenstein, I. (2003). Enhanced circulation half-life and reduced clearance of vasoactive intestinal peptide (VIP) loaded in sterically stabilized micelles (SSM) in mice with collagen-induced arthritis (CIA). *AAPS Pharm. Sci.*, **5**, M1045.

Sutton, D., Nasongkla, N., Blanco, E., & Gao, J. (2007). Functionalized micellar systems for cancer targeted drug delivery. *Pharm. Res.*, **24**, 1029–1046.

Toncheva, V., Schacht, E., Ng, S. Y., Barr, J., & Heller, J. (2003). Use of block copolymers of poly(ortho esters) and poly(ethylene glycol) as micellar drug carriers for tumor targeting applications. *J. Drug Targeting*, **11**, 345–353.

Torchilin, V. P. (2001). Structure and design of polymeric surfactant-based drug delivery systems. *J. Control. Release*, **73**, 137–172.

Torchilin, V. P. (2004). Targeted polymeric micelles for delivery of poorly soluble drugs. *Cell. Mol. Life Sci.*, **61**, 1–11.

Torchilin, V. P., & Papisov, M. (1994). Why do polyethylene glycol-coated liposomes circulate so long? *J. Liposome Res.*, **4**, 725–739.

Torchilin, V. P., Lukyanov, A., Gao, Z., & Papahadjopoulos-Sternberg, B. (2003). Immunomicelles: Targeted pharmaceutical carriers for poorly soluble drugs. *Proc. Natl. Acad. Sci. USA*, **100**, 6039–6044.

Wang, P., & Johnston, T. P. (1991). Kinetics of sol-to-gel transition for Poloxamer polyols. *J. Appl. Polymer Sci.*, **43**, 283–292.

Yasugi, K., Nakamura, T., Nagasaki, Y., Kato, M., & Kataoka, K. (1999). Sugar-installed polymer micelles: Synthesis and micellization of poly(ethylene glycol)-poly(D, L-lactide) block copolymers having sugar groups at the PEG chain end. *Macromolecules*, **32**, 8024–8032.

Yokoyama, M., Miyauchi, M., Yamada, N., Okano, T., Sakurai, Y., Kataoka, K., & Inoue, S. (1990a). Polymer micelles as novel drug carriers. *J. Control. Release*, **11**, 269–278.

Yokoyama, M., Miyauchi, M., Yamada, N., Okano, T., Sakurai, K., Kataoka, K., & Inoue, S. (1990b). Characterization and anti-cancer activity of the micelle-forming polymeric anticancer drug adriamycin-conjugated poly(ethylene glycol)–poly(aspartic acid) block copolymers. *Cancer Res.*, 1693–1700.

Yoo, H. S., & Part, T. G. (2004a). Folate receptor targeted biodegradable polymeric doxorubicin micelles. *J. Control. Release*, **96**(2), 273–283.

Yoo, H. S., & Part, T. G. (2004b). Folate receptor targeted delivery of polymeric doxorubicin nanoaggregates stabilized by doxorubicin-PEG-folate conjugate. *J. Control. Rel.*, **100**, 247–256.

B.7. DENDRIMERS

Wayne R. Gombotz
Immune Design Corp., Seattle, WA, USA

Dendrimers are a unique class of macromolecules comprised of a highly branched three-dimensional architecture around an inner core with low polydispersity and a high degree of functionality. The word "dendrimer" comes from the Greek "dendros", meaning trees, and "meros", meaning part. Dendrimers can range in diameter from 1 to 10 nm. On a molecular level the dendritic branching results in semi-globular structures with a high density of surface functionalities and a small molecular volume. Higher generation dendrimers occupy a smaller hydrodynamic volume compared to corresponding linear polymers of the same molecular weight as a result of their globular structure (Boas and Heegaard, 2004). The dendritic structure is characterized by layers between each focal point (or cascade) that are called generations (Figure B.7.1).

Dendrimers were first introduced in the late 1970s by Vögtle and co-workers (Buhleier et al., 1978). A "cascade" synthesis was described, in which an exhaustive Michael-type addition of acrylonitrile to an amine, followed by the reduction of the nitrile groups to primary amines, could in theory be repeated *ad infinitum* to produce highly branched macromolecules. Current dendrimer designs are based on a wide variety of linkages, including polyamines (PPI dendrimers) (Buhleier et al., 1978), polyamides and amine mixtures (polyamidoamine or PAMAM dendrimers) (Tomalia et al., 1985), poly(aryl ether) subunits (Hawker and Fréchet, 1990), carbohydrate core structures (Turnbull and Stoddart, 2002) or elements such as silicon or phosphorous (Majoral and Caminade, 1999) (Figure B.7.2).

The physical characteristics of dendrimers, including their water solubility, monodispersity, encapsulation ability, and large number of functionalizable peripheral groups, makes them ideal candidates for use as drug delivery systems (Morgan et al., 2006). Considerable interest has been shown applying dendrimers for targeted delivery of drugs, peptides, oligonucleotides, carbohydrates, and vaccines, and also as permeability enhancers able to promote oral and transdermal drug delivery (Boas and Heegard, 2004). Drugs can be physically encapsulated in the void spaces of the dendrimer's interior or chemically or physically bound onto the dendrimer surface along with targeting ligands. Several reviews have been published that discuss the potential for using dendrimers as drug delivery systems (Liu and Fréchet, 1999; Esfand and Tomalia, 2001; Boas and Heegard, 2004; Florence, 2005). Since these systems are still relatively early in their development there are no commercially available injectable products using dendrimer technology. One barrier to commercialization may be the difficulty of carrying out a large-scale synthesis and purification process that reproducibly yields a uniform composition and molecular weight of the dendrimer product.

In an *in vitro* study, a polyester dendrimer composed of glycerol and succinic acid was used to encapsulate several camptothecin anticancer drugs (Morgan et al., 2006). The cytotoxicity of the dendrimer–drug complex toward four different human cancer cell lines was evaluated, and low nmol/L IC_{50} values were measured. Significant increases in both cellular uptake and drug retention within the cells were also reported. PAMAM dendrimers conjugated to *cis*-platin were shown to act as a macromolecular carrier for platinum. The dendrimer–platinate provided a slower release of the platin, while exhibiting higher accumulation in solid tumors and lower toxicity compared to *cis*-platin alone (Malik et al., 1999). In another study, dendrimers based on a 1,4,7,10-tetra,azacyclododecane core having primary amines at its surface have been partially modified with 1-bromoacetyl-5-fluorouracil to form a labile imide linkage (Zhuo et al., 1999). Under physiological conditions, hydrolysis of the imide released the 5-fluorouracil *in vitro*.

VivaGel® is the first dendrimer drug to be tested in the clinic. The product manufactured by, Australia-based (Melbourne) Starpharma is a gel-based dendrimer–drug formulation under development as a topical vaginal microbicide to prevent the transmission of sexually transmitted infections, including HIV and genital herpes. To date, the limited clinical experience using dendrimers makes it impossible to designate any particular chemistry intrinsically nontoxic. Key factors in determining the wide adoption of future dendrimer technologies for drug delivery applications will center on reducing the costs and complexity of their manufacture, increasing reproducibility of the product, and confirming their biocompatibility and safety in man (Terrillion, 2008).

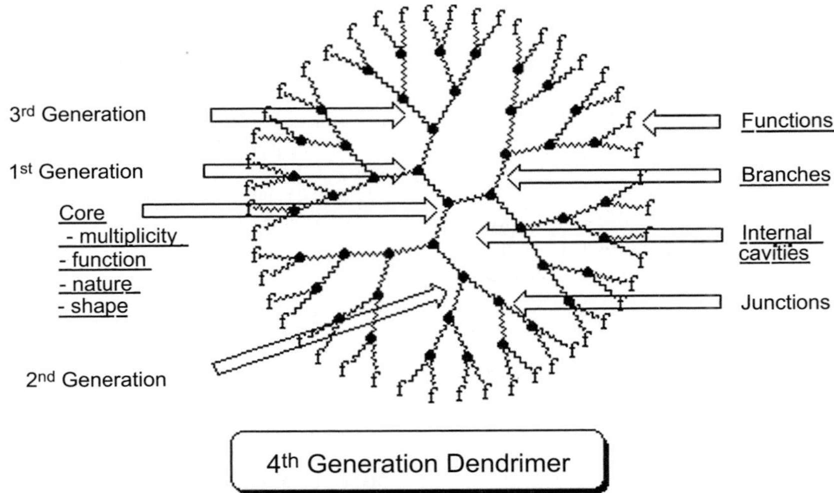

FIGURE B.7.1 Schematic of a fourth generation dendrimer.

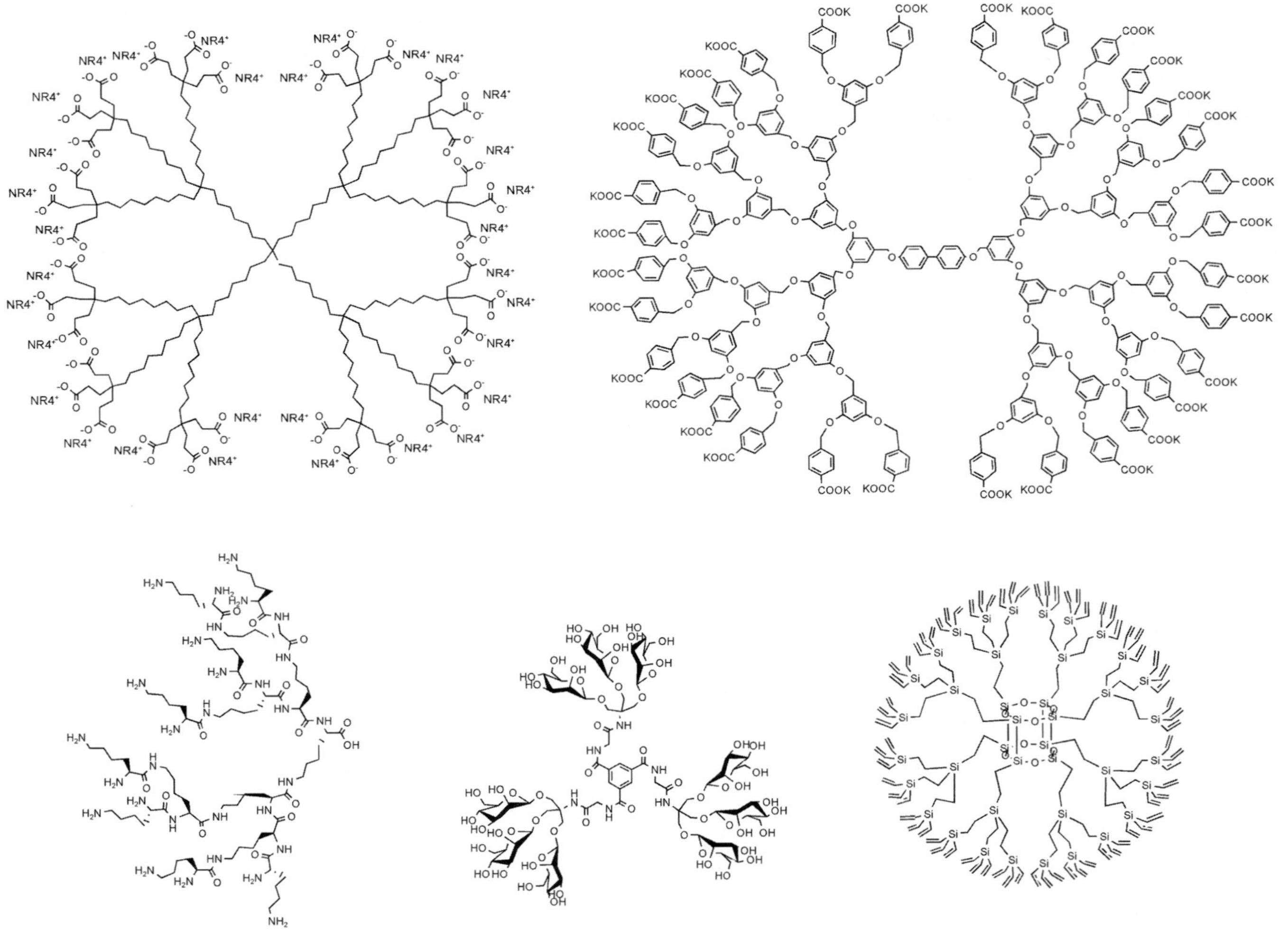

FIGURE B.7.2 Schematic representation of different types of dendrimers. Top left: Unimolecular micelle. Top right: Poly(aryl ether). Bottom left: Polylysine. Bottom middle: Carbohydrate. Bottom right: Silicon based. (Boas and Heegaard, 2004.)

BIBLIOGRAPHY

Boas, U., & Heegaard, M. (2004). Dendrimers in drug research. *Chem. Soc. Rev.*, **33**, 43–63.

Buhleier, E., Wehner, W., & Vögtle, F. (1978). Cascade and nonskid-chain-like synthesis of molecular cavity topologies. *Synthesis*, **2**, 155–158.

Esfand, T., & Tomalia, D. (2001). Poly(amidoamine)(PAMAM) dendrimers; from biomimicry to drug delivery and biomedical applications. *Drug Discov. Today*, **6**, 427–436.

Florence, A. (2005). Dendrimers: A versatile targeting platform. *Adv. Drug Del. Rev.*, **57**, 2104–2286.

Hawker, C., & Fréchet, J. (1990). A new convergent approach to monodisperse dendritic molecules. *J. Chem. Soc. Chem. Comm.*, **15**, 1010–1012.

Liu, M., & Fréchet, J. (1999). Designing dendrimers for drug delivery. *Pharm. Sci. Technol. Today*, **2**, 393–401.

Majoral, J., & Caminade, A. (1999). Dendrimers containing heteroatoms (Si, P, B, Ge, or Bi). *Chem. Rev.*, **99**, 845–880.

Malik, N., Evagorou, E., & Duncan, R. (1999). Dendrimer-platinate: A novel approach to cancer chemotherapy. *Anti-cancer Drugs*, **10**, 767–776.

Morgan, M., Nakanishi, Y., Kroll, D., Griset, A., Carnahan, M., Wathier, M., Oberlies, N., Manikumar, G., Wani, M., & Grinstaff, M. (2006). Dendrimer-encapsulated camptothecins; increased solubility, cellular uptake, and cellular retention affords enhanced anticancer activity *in vitro*. *Cancer Res.*, **66**, 11913–11921.

Terrillion, M. (2008). Poised to branch out. *Nature Biotech.*, **26**, 729–732.

Tomalia, D., Baker, H., Dewald, J., Hall, M., Kallos, G., Martin, S., Roeck, J., Ryder, J., & Smith, P. (1985). A new class of polymers: Starburst-dendritic macromolecules. *Polym. J.*, **17**, 117–132.

Turnbull, W., & Stoddart, J. (2002). Design and synthesis of glyco-dendrimers. *Rev. Mol. Biotechnol.*, **90**, 231–255.

Zhuo, R., Du, B., & Lu, R. (1999). *In vitro* release of 5-fluorouracil with cyclic core dendritic polymer. *J. Cont. Rel.*, **65**, 441–473.

B.8. NUCLEIC ACID DELIVERY

Suzie H. Pun and Allan S. Hoffman

Department of Bioengineering, University of Washington, Seattle, WA, USA

INTRODUCTION

Nucleic acid therapy offers the potential to treat diseases with higher specificity and efficacy than traditional small molecule drugs. Nucleic acid-based drugs modulate the protein expression profiles of cells by interfering either with gene transcription or translation. This class of drugs offers high specificity of action and the potential to address the genetic causes of disease.

GENE EXPRESSION

In diseases such as hemophilia or ADA-SCID (adenosine deaminase-linked SCID), patients lack a functional protein due to a gene mutation. Expression of the missing protein can be restored by introducing the gene or messenger RNA transcript back into the desired cells. Viruses can be engineered to deliver replacement genes to cells. In non-viral delivery systems, protein replacement is typically achieved by delivering a plasmid coding for the desired gene under the control of an appropriate promoter.

GENE KNOCKDOWN

Poorly-controlled expression of proteins or expression of mutated proteins can also lead to disease. In addition, infections involve undesirable pathogen proteins in host cells. There are several potential approaches for selective gene silencing in cells.

- In traditional antisense therapy, a synthetic antisense oligonucleotide (AS-ODN) (either DNA or RNA) complementary to an mRNA transcript is introduced to cells. This drug binds to and inactivates its mRNA target. The first US Food and Drug Administration (FDA)-approved nucleic acid drug, *fomivirsen*, is an antisense therapy for cytomegalovirus retinitis. More recently, antisense oligonucleotides complementary to microRNA (miRNA) have been developed for disease treatment. miRNAs are short, noncoding RNA that regulate gene expression through partial recognition of mRNA targets.
- Catalytic DNA- or RNA-based enzymes, called DNAzymes or ribozymes, respectively, recognize target RNA through complementary base pairing, cleave the RNA, and then dissociate and become available to seek new targets. While ribozymes are naturally occurring, DNAzymes were first developed through *in vitro* selection, and offer advantages of stability and ease of synthesis over ribozymes (Tafech et al., 2006).
- Short, interfering RNA (siRNA) are ~21 base pair RNA duplexes that trigger the RNA interference (RNAi) mechanism used by microRNA to silence genes. After cytosolic delivery, siRNA are loaded into an RNA-induced silencing complex (RISC), a protein machine that recognizes and cleaves mRNA complementary to the incorporated guide strand of siRNA. Synthetic siRNA can be delivered as the nucleic acid therapeutic, or be generated in the cell by delivery of DNA that is transcribed to short hairpin RNA. Gene knockdown by the RNAi mechanism has been shown to be more potent than antisense oligonucleotides, DNAzymes or ribozymes (Grunweller et al., 2003).

Several of the oligonucleotide drugs currently in clinical trials are chemically modified in order to

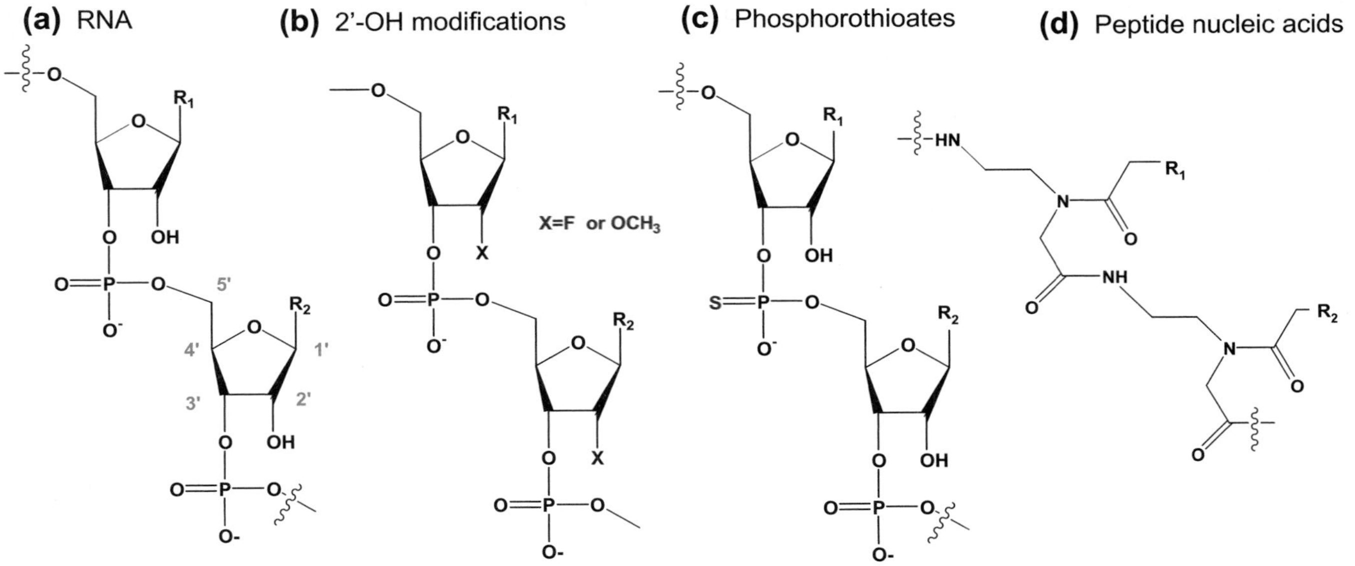

(a) RNA (b) 2'-OH modifications (c) Phosphorothioates (d) Peptide nucleic acids

X=F or OCH_3

FIGURE B.8.1 Structure of RNA (a) and commonly used chemical modifications: methylation or fluorination of 2′-OH (b) phosphorothioate backbone (c). The synthetic peptide nucleic acid polymers (d) are also used as RNA mimics.

improve their bioavailability (Watts et al., 2008). Some common modifications are: replacement of the phosphodiester backbone to phosphorothioates, phosphorodiamidates or polyamides to reduce nuclease susceptibility; modification of the 2′-OH in RNA by methylation or fluorination to increase serum stability; and conjugation of ligands such as cholesterol or steroids to assist in cell permeation (Figure B.8.1). However, there are often trade-offs associated with these chemical modifications. For example, phosphorothioates tend to exhibit higher toxicity and immunogenicity *in vivo*, and 2′-O-methylation has been shown to reduce siRNA activity in some cases. Recent work has also shown that multimeric siRNA prepared by cross-linking siRNA with reducible disulfide bonds results in enhanced gene silencing without significant immunogenicity (Mok et al., 2010).

LIGAND BINDING

DNA or RNA aptamers are single-stranded oligonucleotides that fold into three-dimensional structures and are able to bind to targets such as proteins. Pegaptanib is an FDA-approved polyethylene glygol-modified aptamer that binds to $VEGF_{165}$ for treatment of age-related macular degeneration.

Rapid translation of siRNA technology from worms to humans: in 1998, RNA interference was first demonstrated in the nematode *C. elegans* by Andrew Z. Fire and Craig C. Mello. For this discovery, they were awarded the Nobel Prize in Physiology or Medicine in 2006. McCaffrey and Kay were the first to report in 2002 that siRNAs delivered by hydrodynamic injection could be used to suppress gene expression *in vivo* in adult mice. In 2004, the first siRNA-based drug entered clinical trials as a treatment for age-related macular degeneration. Within five years, several more siRNA-based drugs were in clinical trials for diseases including hepatitis B, respiratory syncytial virus infection, and cancer (Castanotto and Rossi, 2009). The first targeted siRNA delivery formulation entered clinical trials in 2008 (Davis, 2009). There are also many diseases for which siRNA are being tested in preclinical trials, including hypercholesterolaemia and Parkinson's disease. Efficient, non-toxic, and specific delivery systems for siRNA drugs remain an important need for broad translation of this promising technology.

BIOMATERIALS AND NUCLEIC ACID DELIVERY

Viral and non-viral delivery vectors have been used successfully to deliver nucleic acids both in cell cultures and in several animal models. The basic premise is that these delivery vectors help to condense and also to protect the nucleic acid from degradation. When formulated as nanoparticles with diameters typically <200 nm, the small size of the vehicles allow for systemic administration. The delivery systems can also facilitate cell targeting and intracellular trafficking. Ideally, the biomaterials used in delivery systems should be both non-toxic and non-immunogenic. The major classes of delivery vectors are discussed in the following sections.

VIRAL DELIVERY

Several families of viruses, including retrovirus, adenovirus, adeno-associated virus, and poxvirus, have been evaluated as gene delivery vectors in clinical trials. Viruses are natural pathogens that efficiently deliver

DNA to specific cells. Viral vectors contain a modified genome within the virion structure. The modified genome contains only essential viral sequences and the desired transcription unit of the exogenous gene, thereby rendering the virus replication-incompetent (Smith, 1995). Engineered viruses are generally very efficient in overcoming the intracellular barriers to delivery, but tend to be more immunogenic than synthetic carriers. In addition, the application of some viral types may be limited due to their natural tropism. Ongoing research in the field includes efforts to reduce immunogenicity through capsid engineering or regulated expression cassettes (Nayak and Herzog, 2010). Viruses have also been retargeted to new cell types by vector engineering approaches such as pseudotyping (swapping viral attachment proteins within a given family of virus) or genetically engineering to introduce peptides for cell-targeting (Schaffer et al., 2008). Synthetic polymers have been used to help address both of the above concerns (Cattaneo et al., 2008). Hydrophilic polymers such as poly(ethylene glycol) (PEG) and poly(N-(2-hydroxypropyl)methacrylamide) (pHPMA) have been conjugated to the surface of viruses to reduce recognition by neutralizing antibodies. Bifunctional polymers have been used to both shield the surface and present new targeting ligands for changing cell tropism.

NON-VIRAL DELIVERY: INTRODUCTION

Gene therapy (pDNA), antisense oligonucleotide (asODN), and silencing RNA (siRNA) delivery systems have become very actively researched in the past several years. In particular, since DNA, RNA, and asODNs are all highly negatively charged molecules, cationic polymers and cationic liposomes have been used to ionically complex and condense in size these nucleic acid drugs for delivery into cells. Plasmid DNA complexes with cationic polymers are called polyplexes, and complexes with cationic liposomes are called lipoplexes. Together, these two methods are called non-viral delivery systems. They are much less efficient than viral carriers, but the latter may cause undesirable toxic and immunogenic responses in the body (although there is also some toxicity associated with polycations in the body).

It is not clear how and when the polycation actually releases the DNA. It is probably not an efficient process, since the polycation and the DNA must each be "reunited" with their counterions in that process. This may be one reason why non-viral systems are so inefficient. Some polyplexes and lipoplexes also tend to be unstable in serum, possibly due to disruption by proteins and lipoproteins in the circulation, and also potentially due to degradation of the DNA by nucleases. Another challenge to the stability of the polyplexes or lipoplexes may be heparan sulfate groups on cell surfaces that can compete with the nucleic acid drug for the cationic groups on the polymer or liposome, leading to some dissociation of the delivery complex.

Lipoplexes (Cationic Lipid–Nucleic Acid Complexes)

Cationic lipids are amphiphilic molecules that have a cationic head group and a hydrophobic tail group connected by either stable or degradable linkages (Hirko et al., 2003). Felgner and co-workers first demonstrated the use of cationic lipids for DNA delivery in 1987 (Felgner et al., 1987). Many cationic lipids since then have been synthesized and evaluated for nucleic acid delivery. Common head groups used in nucleic acid delivery agents include quaternary ammonium salt lipids and lipoamines. Tail groups are usually 12–18 carbon long alkyl chains or cholesteryl groups. Typical nucleic delivery formulations include a cationic lipid, a neutral "helper" lipid, and the nucleic acid. Figure 1h.2 shows the structure of DOTAP (1,2-diolyoxy-3-(trimethylammonio)propane), a lipid that was used in clinical trials, and DOSPA (N,N-dimethyl-N-([2-sperminecarboxamido]ethyl)-2,3-bis(dioleyloxy)-1-propaniminium pentahydrochloride), a major component of the commercially available Lipofectamine™. When mixed with anionic nucleic acids, the resulting self-assembled complex is called a "lipoplex." Some lipoplexes undergo a phase transition in the endosome that assists with membrane destabilization and translocation into the cytoplasm. Although cationic lipids are some of the most efficient synthetic materials used for *in vitro* delivery, *in vivo* applications have been limited due to toxicity and susceptibility to protein-induced aggregation in the blood. Lipoplexes have primarily been used in clinical trials for treatment of cancer or cystic fibrosis (Caplen et al., 1995). PEGylation of lipoplexes by including a PEG–lipid conjugate during formulation can lead to increased plasma circulation times, but often at the expense of cellular uptake efficiency. PEG conjugated to lipids via a labile bond can be incorporated in the lipoplexes; the kinetics of PEG release in these formulations depend on the length of the lipid that serves as the lipoplex anchor (Guo and Szoka, 2001). Recently, cationic lipids and lipid-like materials that facilitate siRNA delivery *in vivo* at low doses were identified through both rational design and combinatorial synthesis and screening, respectively (Love et al., 2010; Semple et al., 2010).

Polyplexes (Cationic Polymer–Nucleic Acid Complexes)

Several classes of cationic polymers have been used successfully for nucleic acid transfer *in vitro*; two commonly-used polymers, poly(L)-lysine (PLL) and polyethylenimine (PEI) are shown in Figure B.8.2 (Wu and Wu, 1987; Boussif et al., 1995; Pack et al., 2005). Other polycations include chitosan, poly(vinyl imidazole), poly(amidoamines), and acrylic copolymers of *N,N*-dimethyl aminoethyl methacrylate (DMAEMA). The polyvalency of cationic polymers allows for high-affinity interactions with polyanionic nucleic acids. In addition, the functional groups present

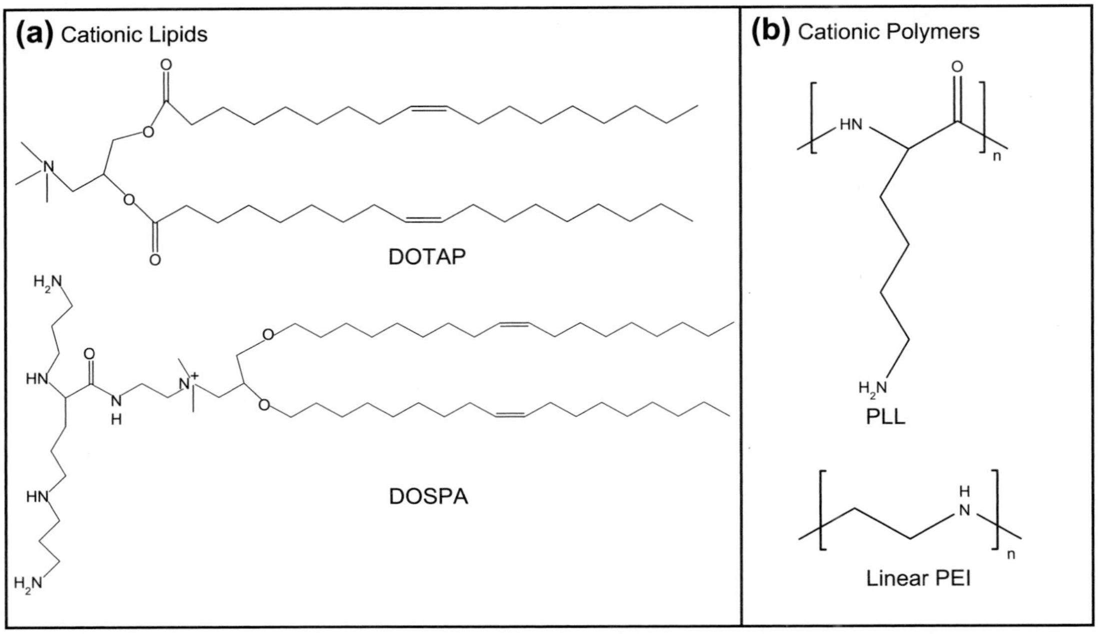

(a) Cationic Lipids

DOTAP

DOSPA

(b) Cationic Polymers

PLL

Linear PEI

FIGURE B.8.2 Structures of some biomaterials commonly used in nucleic acid delivery vehicles. DOTAP: 1,2-Dioleoyl-3-trimethylammonium propane; DOSPA: 2,3-dioleyloxy-N-[2(spermine carboxamido)ethyl]-N,N-dimethyl-1-propanaminium trifluoroacetate; PLL: poly(L)lysine; PEI: poly(ethylenimine).

in polymers can be used to further modify the polymers with moieties to assist in vehicle trafficking. When mixed with nucleic acids in low salt conditions, cationic polymers will self-assemble with nucleic acids by electrostatic interactions into complexes called "polyplexes" that are typically ~50–200 nm in size. These complexes impart nuclease resistance to the packaged nucleic acids. The resultant polyplexes are typically slightly positively charged nanoparticles of ca. 50–200 nm in diameter.

Polyplexes can be internalized by cells through electrostatic interactions with the cell surface or, when functionalized with a targeting ligand, through specific ligand–receptor interactions. Once they are taken up into a cell by endocytosis, some of the amino groups on the polymers that have the appropriate pK have been postulated to buffer the acidic pH within the endosome by binding protons, and thereby causing sodium ions to enter to maintain electrical neutrality. This leads to osmotic swelling and bursting of the endosome, releasing the polyplex (or lipoplex) to the cytosol. This is called the "proton sponge" effect (Boussif, 1995), and it may be one step in the mechanism by which polyplexes (or lipoplexes) can transfect cells. (See "Escaping from endosome" for further discussion of research based on this effect.)

For *in vivo* applications, cationic polymers are typically modified, either through stable or labile bonds, with hydrophilic polymers such as PEG and pHPMA that decrease non-specific protein adsorption and impart salt stability to the colloids (Murthy et al., 2003; Walker et al., 2005). In addition, cationic polymers are often functionalized with ligands for cell-specific targeting. *In vivo* toxicity and low transfection efficiencies remain major challenges for this class of materials. To address the former issue, several degradable cationic polymers have been synthesized to reduce *in vivo* toxicity. Polyplexes have been used in conjunction with cationic lipids in formulations termed lipopolyplexes (Gao and Huang, 1996).

Inorganic Particles

More recently, several classes of inorganic nanoparticles have been applied for nucleic acid delivery, including colloidal gold (Rosi et al., 2006), silica nanoparticles (Bharali et al., 2005), and quantum dots (Derfus et al., 2007). Nucleic acids have been either directly conjugated to the nanoparticle surface or nanoparticles have been surface modified to be cationic for electrostatic complexation with nucleic acids. In addition, protonatable groups have been conjugated to inorganic nanoparticle surfaces to mediate increased endosomal release (Yezhelyev et al., 2008). One advantage of these particles is their small size (often <40 nm) which facilitates more efficient delivery. However, the slow rate or lack of clearance of these materials *in vivo* remains a concern.

LOCAL GENE DELIVERY

Nucleic acids or nucleic acid delivery vehicles can be integrated with biomaterials for localized delivery. Examples of potential applications for this approach include controlled release of nucleic acids from stents for anti-restenosis treatment or incorporation of nucleic acids in tissue engineering scaffolds to guide new tissue formation. Nucleic acids have been successfully integrated with substrates by embedding, direct surface immobilization or integration in layer-by-layer assemblies. Both unencapsulated nucleic acids and nucleic acid delivery vehicles have been embedded by including these components during fabrication of biomaterials such as poly(lactide-co-glycoide), collagen or fibrin matrices (Kong and Mooney, 2007). Direct surface immobilization of nucleic acids has been accomplished by adsorption of vectors to biomaterial surfaces, covalent conjugation of vectors to surfaces, binding of biotinylated vectors to avidin-modified surfaces or recognition of vector epitopes by antibody-conjugated surfaces (Bengali and Shea, 2005). Finally, layer-by-layer assemblies of oppositely-charged films of polymers allow for programmed release of nucleic acids from surfaces during film erosion (Jewell and Lynn, 2008).

EXTRACELLULAR REQUIREMENTS FOR EFFICIENT NUCLEIC ACID DELIVERY

After *in vivo* administration, nucleic acid drugs need to remain stable and active until reaching their site of action. Nucleic acids need to avoid rapid elimination, either by nuclease degradation or by renal excretion. Nuclease degradation for both DNA and RNA has been shown to occur in the order of minutes *in vitro* (Abdelhady et al., 2003; Bartlett and Davis, 2006). In addition, oligonucleotides such as siRNA also undergo rapid plasma clearance, because their size is significantly less than the renal filtration cut-off (~30 kDa). For example, unmodified siRNA has a plasma half-life of only six minutes in mice (Soutschek et al., 2004).

To address these issues, nucleic acid drugs are typically either chemically modified as discussed previously or packaged in a delivery vehicle. Delivery vehicles should be designed to avoid eliciting host immune responses, including opsonization and phagocytosis, activation of the complement cascade, and antibody recognition of vehicle motifs. At the first level, recognition of the delivery vehicles results in rapid clearance by the reticuloendothelial system. More seriously, a systemic inflammatory response can lead to organ failure. Because opsonization and activation of the complement cascade both involve adsorption of proteins to vehicle surfaces, vehicles that minimize non-specific protein binding are desirable. One approach that has been used to reduce protein interactions *in vivo* is PEGylation of the delivery vehicle. Vehicles that do not contain foreign proteins are also more likely to be biocompatible.

Most delivery vehicles are nanoparticulate to allow injection into the bloodstream. These colloidal particles, therefore, should also have high critical flocculation concentration to avoid salt-induced aggregation in the bloodstream and remain stable during circulation (Rozema et al., 2007; Meyer et al., 2009). For targets outside of the vasculature after systemic administration, the formulations have to extravasate in order to reach their targets. Vascular fenetrations in the liver, kidney, and in some tumor sites allow for extravasation and tissue delivery. Within solid tumors such extravasation leads to "enhanced permeability and retention," or the enhanced permeability and retention (EPR) effect (Maeda et al., 2000). Another approach is to develop formulations that can undergo transcytosis in endothelial cells. For intracellular delivery, nucleic acids or their delivery vehicles need to penetrate the tissue and be internalized by their target cells. Cell binding can occur by non-specific interaction with cell surfaces, such as by electrostatic binding or by specific binding of a ligand-modified vehicle with a cellular receptor.

INTRACELLULAR REQUIREMENTS FOR EFFICIENT NUCLEIC ACID DELIVERY

With a possible exception of aptamers that may be used as cell-targeting agents, the site of action for nucleic acid-based drugs is intracellular. After internalization into the cell, a series of barriers must be overcome before the drugs reach their subcellular target (Figure B.8.3). Nucleic acids that require transcription act in the cell nucleus, while nucleic acids that require or interfere with translation are active in the cytosol. Nucleic acid drugs or their carriers are typically internalized by an endocytic pathway into vesicles. The specific uptake mechanism depends on the cell type and properties of the carrier, including size, composition, and the presence of targeting ligands. The route of internalization is important, because it affects the intracellular trafficking pathway (Khalil et al., 2006).

ESCAPING FROM THE ENDOSOME

Internalized nucleic acid drugs must be released from endosomes in order for them to be active. A common strategy used for endosome release takes advantage of the acidic environment in early and late endosomes. pH-sensitive proteins or peptides can be incorporated in delivery vehicles for selective lipid membrane disruption at low pH (Bergen and Pun, 2005).

pH-sensitive synthetic polymers that become hydrophobic in acidic conditions have also been shown to be effective in mediating endosome release, and this has led to a delivery system that mimics viruses. Viruses have evolved an efficient system for delivering DNA or RNA to the cytosol that utilizes the lowered pH within the endosome. Certain peptide sequences on the surface of the virus become hydrophobic when their acidic amino acids are protonated, causing those sequences to fuse with or cause pore formation in the endosomal membrane, allowing the virus to deliver its genomic contents to the cytosol. The Stayton/Hoffman research group has mimicked this natural action of the virus by designing and synthesizing pH-sensitive polymers, such as poly(propylacrylic acid), PPAA, that would similarly become hydrophobic at the endosomal pHs and disrupt the endosomal

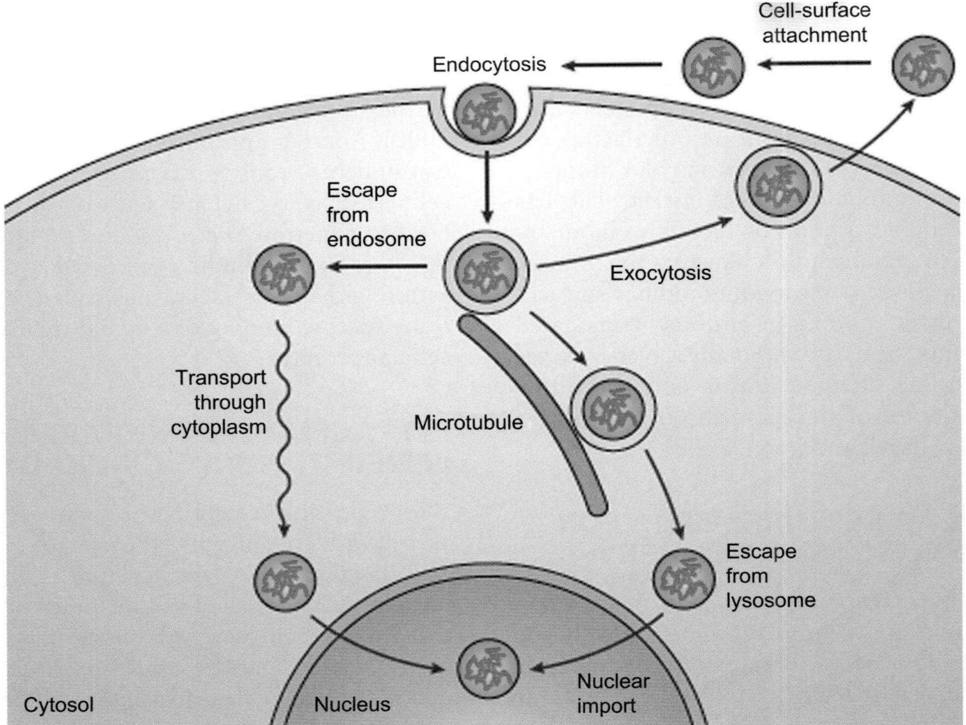

FIGURE B.8.3 Schematic of intracellular barriers to nucleic acid delivery. A nanoparticle carrier is used here for illustration. After cell binding and internalization the drugs must exit endocytic vesicles, be transported through the cytoplasm and be available for performing its intracellular action either in the cytosol or nucleus. Figure from Pack et al. (2005). Nat. Reviews Drug Discovery, **4**, 581–593.

membrane (Murthy et al., 1999, 2001, 2003; Stayton et al., 2000; Cheung et al., 2001; Hoffman and Stayton, 2002; Kyriakides et al., 2002; Henry et al., 2006; Stayton and Hoffman, 2008; Convertine et al., 2009). A pendant glutathione-responsive disulfide group has been added to this acid-responsive polymer, such that when the polymer–drug complex or conjugate escapes the endosome, it releases the drug after reduction by glutathione in the cytosol (Bulmus et al., 2003). Other approaches by these same researchers have been to design and synthesize membrane-disruptive polymers that are PEGylated via acid-degradable acetal bonds; these bonds degrade within the endosome, exposing the backbone which is membrane-disruptive, leading to the release of the drug to the cytosol (Murthy et al., 2003).

Finally, pH-sensitive systems that contain a high concentration of protonatable groups with pK_a between 5–7 can enhance endosomal release through the "proton-sponge effect," whereby the pH in endosomes is buffered, eventually resulting in vesicle swelling and rupture (Boussif et al., 1995). If packaged within a delivery vector, the nucleic acids need to be accessible to their target after endosomal release. Inefficient vector unpackaging has also been shown to be a barrier to gene delivery by synthetic biomaterials (Schaffer et al., 2000). For synthetic materials that assemble with nucleic acids through charge interactions, the mechanism of intracellular unpackaging is not clear, but may occur through competitive displacement by cytosolic peptides, proteins or nucleic polyanions such as cellular RNA (Bertschinger et al., 2006).

NUCLEAR ENTRY AND DELIVERY

For successful nuclear delivery, nucleic acids must be transported through the crowded cytoplasm and past the nuclear membrane. Several different mechanisms of nuclear delivery have been reported for various viruses, including transport of intact virus through the nuclear pore complex (NPC), and docking at the NPC followed by release of DNA. However, transfection of non-dividing cells is a major challenge for most non-viral systems.

There have been several approaches tested for facilitating nuclear delivery of plasmids. The maximum cargo size reported to date that can be successfully actively transported through NPCs is ~40 nm (Pante and Kann, 2002). First, delivery vehicles with dimensions <30 nm have been formulated in an attempt to prepare materials that can traverse nuclear pores. Second, several researchers have demonstrated varying levels of success in facilitating nuclear delivery by conjugating nuclear localization sequences either to their non-viral carrier or directly to the DNA cargo (Nagasaki et al., 2003). Finally, DNA sequences that bind to proteins with nuclear localization signal (NLS) sequences, such as transcription factors, can be included in the plasmid construct.

BIBLIOGRAPHY

Abdelhady, H. G., Allen, S., Davies, M. C., Roberts, C. J., Tendler, S. J., & Williams, P. M. (2003). Direct real-time molecular scale visualisation of the degradation of condensed DNA complexes exposed to DNase I. *Nucleic Acids Research*, **31**(14), 4001–4005.

Bartlett, D. W., & Davis, M. E. (2006). Insights into the kinetics of siRNA-mediated gene silencing from live-cell and live-animal bioluminescent imaging. *Nucleic Acids Research*, **34**(1), 322–333.

Bengali, Z., & Shea, L. D. (2005). Gene delivery by immobilization to cell-adhesive substrates. *MRS Bulletin*, **30**(9), 659–662.

Bergen, J. M., & Pun, S. H. (2005). Peptide-enhanced nucleic acid delivery. *MRS Bulletin*, **30**(9), 663–667.

Bertschinger, M., Backliwal, G., Schertenleib, A., Jordan, M., Hacker, D. L., & Wurm, F. M. (2006). Disassembly of polyethylenimine-DNA particles *in vitro*: Implications for polyethylenimine-mediated DNA delivery. *Journal of Controlled Release*, **116**(1), 96–104.

Bharali, D. J., Klejbor, I., Stachowiak, E. K., Dutta, P., Roy, I., Kaur, N., Bergey, E. J., Prasad, P. N., & Stachowiak, M. K. (2005). Organically modified silica nanoparticles: a nonviral vector for *in vivo* gene delivery and expression in the brain. *Proceedings of the National Academy of Sciences of the United States of America*, **102**(32), 11539–11544.

Boussif, O., Lezoualcl'h, F., Zanta, M. A., Mergny, M. D., Scherman, D., Demeneix, B., & Behr, J. P. (1995). A versatile vector for gene and oligonucleotide transfer into cells in culture and *in vivo*: polyethylenimine. *Proceedings of the National Academy of Sciences of the United States of America*, **92**, 7297–7301.

Bulmus, V., Woodward, M., Lin, L., Stayton, P. S., & Hoffman, A. S. (2003). A new pH-responsive and glutathione-reactive, membrane-disruptive polymeric carrier for intracellular delivery of biomolecular drugs. *J. Contr. Rel.*, **93**, 105–120.

Caplen, N. J., Alton, E., Middleton, P. G., Dorin, J. R., Stevenson, B. J., Gao, X., Durham, S. R., Jeffery, P. K., Hodson, M. E., Coutelle, C., et al. (1995). Liposome-mediated CFTR gene-transfer to the nasal epithelium of patients with cystic-fibrosis. *Nature Medicine*, **1**(1), 39–46.

Castanotto, D., & Rossi, J. J. (2009). The promises and pitfalls of RNA-interference-based therapeutics. *Nature*, **457**(7228), 426–433.

Cattaneo, R., Miest, T., Shashkova, E. V., & Barry, M. A. (2008). Reprogrammed viruses as cancer therapeutics: targeted, armed and shielded. *Nature Reviews Microbiology*, **6**(7), 529–540.

Cheung, C. Y., Murthy, N., Stayton, P. S., & Hoffman, A. S. (2001). A pH-sensitive polymer that enhances cationic lipid gene transfer. *Bioconjugate Chemistry*, **12**, 906–910.

Convertine, A. J., Benoit, D. S.W., Duvall, C. L., Hoffman, A. S., & Stayton, P. S. (2009). Development of a novel endosomolytic diblock copolymer for siRNA delivery. *Journal of Controlled Release*, **133**, 221–229.

Davis, M. E. (2009). The first targeted delivery of siRNA in humans via a self-assembling, cyclodextrin polymer-based nanoparticle: from concept to clinic. *Molecular Pharmaceutics*, **6**(3), 659–668.

Derfus, A. M., Chen, A. A., Min, D. H., Ruoslahti, E., & Bhatia, S. N. (2007). Targeted quantum dot conjugates for siRNA delivery. *Bioconjugate Chemistry*, **18**(5), 1391–1396.

Felgner, P. L., Gadek, T. R., Holm, R., Roman, R., Chan, H. W., Wenz, M., Northrop, J. P., Ringold, G. M., & Danielsen, M. (1987). Lipofection – A highly efficient, lipid-mediated DNA-transfection procedure. *Proceedings of the National Academy of Sciences of the United States of America*, **84**(21), 7413–7417.

Gao, X., & Huang, L. (1996). Potentiation of cationic liposome-mediated gene delivery by polycations. *Biochemistry*, **35**, 1027–1036.

Grunweller, A., Wyszko, E., Bieber, B., Jahnel, R., Erdmann, V. A., & Kurreck, J. (2003). Comparison of different antisense strategies in mammalian cells using locked nucleic acids, 2′-O-methyl RNA, phosphorothioates and small interfering RNA. *Nucleic Acids Research*, **31**(12), 3185–3193.

Guo, X., & Szoka, F. C. (2001). Steric stabilization of fusogenic liposomes by a low-pH sensitive PEG-diortho ester-lipid conjugate. *Bioconjugate Chemistry*, **12**(2), 291–300.

Henry, S. M., El-Sayed, M. E. H., Pirie, C. M., Stayton, P. S., & Hoffman, A. S. (2006). pH-responsive poly(styrene-alt-maleic anhydride) copolymers for intracellular drug delivery. *Biomacromolecules*, **7**, 2407–2414.

Hirko, A., Tang, F. X., & Hughes, J. A. (2003). Cationic lipid vectors for plasmid DNA delivery. *Current Medicinal Chemistry*, **10**(14), 1185–1193.

Hoffman, A. S., Stayton, P. S., Press, O., Murthy, N., Lackey, C. A., Cheung, C., Black, F., Campbell, J., Fausto, N., Kyriakides, T. R., & Bornstein, P. (2002). Design of "smart" polymers that can direct intracellular drug delivery. *Polymers for Advanced Technologies*, **13**, 992–999.

Jewell, C. M., & Lynn, D. M. (2008). Multilayered polyelectrolyte assemblies as platforms for the delivery of DNA and other nucleic acid-based therapeutics. *Advanced Drug Delivery Reviews*, **60**(9), 979–999.

Khalil, I. A., Kogure, K., Akita, H., & Harashima, H. (2006). Uptake pathways and subsequent intracellular trafficking in nonviral gene delivery. *Pharmacological Reviews*, **58**(1), 32–45.

Kong, H. J., & Mooney, D. J. (2007). Microenvironmental regulation of biomacromolecular therapies. *Nature Reviews Drug Discovery*, **6**(6), 455–463.

Kyriakides, T. R., Cheung, C. Y., Murthy, N., Bornstein, P., Stayton, P. S., & Hoffman, A. S. (2002). pH-sensitive polymers that enhance intracellular drug delivery in vivo. *Journal of Controlled Release*, **78**, 295–303.

Love, K. T., Mahon, K. P., et al. (2010). Lipid-like materials for low-dose, *in vivo* gene silencing. *Proceedings of the National Academy of Sciences of the United States of America*, **107**(5), 1864–1869.

Maeda, H., Wu, J., Sawa, T., Matsumura, Y., & Hori, K. (2000). Tumor vascular permeability and the EPR effect in macromolecular therapeutics: A review. *Journal of Controlled Release*, **65**(1-2), 271–284.

Meyer, M., Dohmen, C., Phillipp, A., Kiener, D., Maiwald, G., Scheu, C., Ogris, M., & Wagner, E. (2009). Synthesis and biological evaluation of a bioresponsive and endosomolytic siRNA-polymer conjugate. *Molecular Pharmaceutics*, **6**(3), 752–762.

Mok, H., Lee, S. H., Park, J. W., & Park, T. G. (2010). Multimeric small interfering ribonucleic acid for highly efficient sequence-specific gene silencing. *Nature Materials*, **9**(3), 272–278.

Murthy, N., Campbell, J., Fausto, N., Hoffman, A. S., & Stayton, P. S. (2003). Bioinspired pH-responsive polymers for the intracellular delivery of biomolecular drugs. *Bioconjugate Chemistry*, **14**(2), 412–419.

Murthy, N., Campbell, J., Fausto, N., Hoffman, A. S., & Stayton, P. S. (2003). Design and synthesis of pH-responsive polymeric carriers that target uptake and enhance the intracellular delivery of oligonucleotides. *Journal of Controlled Release*, **89**, 365–374.

Murthy, N., Chang, I., Stayton, P. S., & Hoffman, A. S. (2001). pH-sensitive hemolysis by random copolymers of alkyl acrylates and acrylic acid. *Macromol Symposia*, **172**, 49–55.

Murthy, N., Robichaud, J. R., Tirrell, D. A., Stayton, P. S., & Hoffman, A. S. (1999). The design and synthesis of polymers for eukaryotic membrane disruption. *Journal of Controlled Release*, **61**, 137–143.

Nagasaki, T., Myohoji, T., Tachibana, T., Futaki, S., & Tamagaki, S. (2003). Can nuclear localization signals enhance nuclear localization of plasmid DNA? *Bioconjugate Chemistry*, **14**(2), 282–286.

Nayak, S., & Herzog, R. (2010). Progress and prospects: Immune responses to viral vectors. *Gene Therapy*, **17**(3), 295–304.

Pack, D. W., Hoffman, A. S., Pun, S., & Stayton, P. S. (2005). Design and development of polymers for gene delivery. *Nature Reviews Drug Discovery*, **4**(7), 581–593.

Pante, N., & Kann, M. (2002). Nuclear pore complex is able to transport macromolecules with diameters of similar to 39 nm. *Molecular Biology of the Cell*, **13**(2), 425–434.

Rosi, N. L., Giljohann, D. A., Thaxton, C. S., Lytton-Jean, A. K., Han, M. S., & Mirkin, C. A. (2006). Oligonucleotide-modified gold nanoparticles for intracellular gene regulation. *Science*, **312**(5776), 1027–1030.

Rozema, D. B., Lewis, D. L., et al. (2007). Dynamic PolyConjugates for targeted *in vivo* delivery of siRNA to hepatocytes. *Proceedings of the National Academy of Sciences of the United States of America*, **104**(32), 12982–12987.

Schaffer, D., Fidelman, N., Dan, N., & Lauffenburger, D. A. (2000). Vector unpacking as a potential barrier for receptor-mediated polyplex gene delivery. *Biotechnology and Bioengineering*, **67**, 598–606.

Schaffer, D., Koerber, J. T., & Lim, K. L. (2008). Molecular engineering of viral gene delivery vehicles. *Annual Review of Biomedical Engineering*, **10**, 169–194.

Semple, S. C., Akinc, A., Chen, J., Sandhu, A. P., Mui, B. L., Cho, C. K., Sah, D. W., Stebbing, D., et al. (2010). Rational design of cationic lipids for siRNA delivery. *Nature Biotechnology*, **28**(2), 172–U118.

Smith, A. E. (1995). Viral vectors in gene therapy. *Annual Review of Microbiology*, **49**, 807–838.

Soutschek, J., Akinc, A., Bramlage, B., Charisse, K., Constien, R., Donoghue, M., Elbashir, S., Geick, A., et al. (2004). Therapeutic silencing of an endogenous gene by systemic administration of modified siRNAs. *Nature*, **432**(7014), 173–178.

Stayton, P. S., & Hoffman, A. S. (2008). Smart pH-Responsive Carriers for Intracellular Delivery of Biomolecular Drugs. In V. Torchilin (Ed.), *Multifunctional Pharmaceutical Nanocarriers*. Springer.

Stayton, P. S., Hoffman, A. S., Murthy, N., Lackey, C., Cheung, C., Tan, P., Klumb, L. A., Chilkoti, A., Wilbur, F. S., & Press, O. W. (2000). Molecular engineering of proteins and polymers for targeting and intracellular delivery of therapeutics. *Journal of Controlled Release*, **v65**, 203–220.

Tafech, A., Bassett, T., Sparanese, D., & Lee, C. H. (2006). Destroying RNA as a therapeutic approach. *Current Medicinal Chemistry*, **13**(8), 863–881.

Walker, G. F., Fella, C., Pelisek, J., Fahrmeir, J., Boeckle, S., Ogris, M., & Wagner, E. (2005). Toward synthetic viruses: endosomal pH-triggered deshielding of targeted polyplexes greatly enhances gene transfer *in vitro* and *in vivo*. *Molecular Therapy*, **11**(3), 418–425.

Watts, J. K., Deleavey, G. F., & Damha, M. J. (2008). Chemically modified siRNA: tools and applications. *Drug Discovery Today*, **13**(19-20), 842–855.

Wu, G., & Wu, C. (1987). Receptor-mediated *in vitro* gene transformation by a soluble DNA carrier system. *Journal of Biological Chemistry*, **262**, 4429–4432.

Yezhelyev, M. V., Qi, L. F., et al. (2008). Proton-sponge coated quantum dots for siRNA delivery and intracellular imaging. *Journal of the American Chemical Society*, **130**(28), 9006–9012.

B.9. POLYMERIC AND ALBUMINATED DRUG NANOPARTICLES

Wayne R. Gombotz
Immune Design Corp., Seattle, WA, USA

Nanoparticles may be defined as solid colloidal particles ranging in size from 10 to 100s of nms. The nanoparticles used in drug delivery systems (DDS) are comprised of the active drug as a simple nanoparticle or polymeric nanoparticles in which the drug is dissolved, entrapped or encapsulated; usually the nanoparticles are stabilized by a surfactant (e.g., Kreuter, 2004). Although the concept of nanoparticles for drug-targeting and diagnostic purposes has been studied for decades, and a significant amount of research has been devoted to preclinical studies with a variety of different systems, there are relatively few products that have been commercially approved. An excellent review by Kreuter describes the historical development of nanoparticles, and highlights some of the major scientific contributions to the field, especially by Spieser (see Kreuter, 2007). Some of the noteworthy systems include nanoparticles made from poly(lactic-co-glycolic) acid, poly(cyanoacrylates) (Couvreur et al., 1977, 1979), and poly(isohexyl cyanoacrylate) (Kattan et al., 1992).

The French company, BioAlliance Pharma has developed a nanoparticle product called Doxorubicin-Transdrug® based on poly(isohexyl cyanoacrylate) for the treatment of resistant hepatocellular carcinomas. The product progressed into Phase II/III clinical trials, and in 2006 was granted orphan drug status by the European Medicines Agency (EMA) in Europe and the US Food and Drug Administration (FDA) in the USA. In 2008 the trials were suspended. There was an observed clinical benefit, but also more frequent and more severe pulmonary damage was observed at the maximum tolerated dose (MTD) of 35 mg/m^2. The premature end to the Doxorubicin-Transdrug® clinical program illuminates some of the challenges still associated with the development of nanoparticles as delivery systems for therapeutic agents.

Albumin is a major protein component in serum, and albumin nanoparticles containing lipophilic drugs have been developed for cancer therapy. Abraxane® is the first FDA-approved albumin-based nanoparticulate drug carrier containing paclitaxel for the treatment of breast cancer. It was developed to retain the therapeutic benefits of paclitaxel while eliminating the toxicities associated with the emulsifier Cremophor® EL in the paclitaxel formulation (Taxol®) (Sparreboom et al., 2005). The product has a mean particle size of 130 nm, and is supplied as a sterile lyophilized powder. Following reconstitution in 0.9% sodium chloride, Abraxane® is administered by intravenous (IV) infusion. The maximum tolerated dose of Abraxane® was 70–80% higher than that reported for Taxol®.

BIBLIOGRAPHY

Baker, R. W., & Lonsdale, H. K. (1974). Controlled release: Mechanisms and rates. In A. C. Tanquary, & R. E. Lacey (Eds.), *Controlled Release of Biologically Active Agents* (pp. 15–71). New York: Plenum Press.

Cooper, E. R. (2010). Nanoparticles: a personal experience for formulating poorly water soluble drugs. *J. Control. Release*, **141**, 300–302.

Couvreur, P., Tulkens, P., Roland, M., Trouet, A., & Speiser, P. (1977). Nanocapsules: a new type of lysosomotropic carrier. *FEBS Lett.*, **84**, 323–326.

Couvreur, P., Kante, B., Roland, M., Guiot, P., Baudhin, P., & Speiser, P. (1979). Polycyanoacrylate nanocapsules as potential lysosomotropic carriers: preparation, morphological and sorptive properties. *J. Phar. Pharmacol.*, **31**, 331–332.

Drucker, D., Buse, J., Taylor, K., Kendall, D., Trautmann, M., Zhuang, D., & Porter, L. (2008). Exenatide once weekly versus twice daily for the treatment of type 2 diabetes: a randomized, open-label, non-inferiority study. *The Lancet*, **372**, 1240–1250.

Folkman, J., & Long, D. M. (1964). The use of silicone rubber as a carrier for prolonged drug therapy. *J. Surg. Res.*, **4**, 139.

Higuchi, T. (1961). Rates of release of medicaments from ointment bases containing drugs in suspension. *J. Pharm. Sci.*, **50**, 874–875.

Hoffman, A. S. (2008). The origins and evolution of "controlled" drug delivery systems. *J. Control. Release*, **132**, 153–163.

http://en.wikipedia.org/wiki/Alza March, 2011.

Kattan, J., Droz, J., Couvreur, P., Marino, J., Boutan-Laroze, A., Rougier, P., Brault, P., Vranckx, H., Grognet, J., Morge, X., & Sancho-Garnier, H. (1992). Phase I clinical trial and pharmacokinetic evaluation of doxorubicin carried by polyisohexylcyanoacrylate nanoparticles. *Invest. New Drugs*, **10**, 191–199.

Khanna, S. C., & Speiser, P. P. (1969). Epoxy resin beads as a pharmaceutical dosage form 1: Methods of preparation. *J. Pharm. Sci.*, **58**, 1114–1117.

Kreuter, J. (2004). Nanoparticles as drug delivery systems. In H. S. Nalwa (Ed.), *Encyclopedia of Nanoscience and Nanotechnology* (Vol. 7), (pp. 161–180). American Scientific Publishers.

Kreuter, J. (2007). Nanoparticles: A historical perspective. *Int. J. Pharm.*, **331**, 1–10.

Nakayama, M., & Okano, T. (2005). Polymer terminal group effects on properties of thermoresponsive polymeric micelles with controlled outer-shell chain lengths. *Biomacromolecules*, **6**, 2320–2327.

Sahay, G., Alakhova, D. Y., Alexander, V., & Kabanov, A. V. (2010). Review: endocytosis of nanomedicines. *J. Contr. Release*, **145**, 182–195.

Sparreboom, A., Scripture, C., Trieu, V., Williams, P., De, T., Yang, A., Beals, B., Figg, W., Hawkins, W., & Desai, N. (2005). Comparative preclinical and clinical pharmacokinetics of a Cremophor-free, nanaparticle albumin-bound paclitaxel (ABI-007) and paclitaxel formulated in Cremophor (Taxol). *Clin. Cancer Res.*, **11**, 4136–4143.

Subramanian, G., Fiscella, M., Lamouse-Smith, A., Zeuzem, S., & McHutchison, J. (2007). Albinterferon alfa-2b: a genetic fusion protein for the treatment of chronic hepatitis C. *Nature Biotechnol.*, **25**, 1411–1419.

Torchilin, V. P., Trubetskoy, V., Whiteman, K., Caliceti, P., Ferruti, P., & Veronese, F. (1995). New synthetic amphiphilic polymers for steric protection of liposomes *in vivo*. *J. Pharm. Sci.*, **84**, 1049–1053.

C. INJECTED DEPOT DDS

Wayne R. Gombotz[1] *and Allan S. Hoffman*[2]

[1]Immune Design Corp., Seattle, WA, USA

[2]Professor of Bioengineering and Chemical Engineering, UWEB Investigator, University of Washington, Seattle, WA, USA

INJECTED DEPOT SYSTEMS

Injected depot systems include degradable polymers as microparticles or phase separating polymer/drug solutions. The therapeutic agent is dispersed or dissolved in the polymer matrix. A mass of polymer and drug is then deposited in the tissue space, and the drug is subsequently delivered as the polymer degrades over time.

The release of drug from injected delivery systems can be governed by several mechanisms including diffusion of drug through the polymer matrix and out, surface erosion of the polymer and/or swelling and bulk erosion of the polymer by the aqueous medium. The erosion is usually due to hydrolysis of the polymer backbone; enzymolysis is probably a rare event. Often the release rate is determined by a combination of mechanisms. For example, in a degradable microsphere system, release of the drug is often initially controlled by desorption from the surface of the microsphere, followed by diffusion of the drug through the porous channels of the polymer matrix, which in turn is influenced by the swelling rate of the system (Gombotz and Pettit, 1995). Acidic byproducts of the degradation of polyesters like poly(lactic-co-glycolic acid) (PLGA) can accelerate the degradation rate, and the bulk sometimes degrades faster than the surface due to the buffering by the aqueous medium at the surface. The physical state of an injected drug delivery system can also change as it degrades, which can further complicate release kinetics. Park has reported that water hydration in microspheres allowed the polymer morphology to change from a glass to a rubbery state by lowering the glass transition temperature. This in turn led to a faster degradation rate (Park, 1994).

Injected Microparticle Depots

The most common type of polymers used to make injected microparticle depots or microspheres are poly(lactic acid) (PLA) and poly(lactic-co-glycolic acid) (PLGA) copolymers. These polymers currently occupy a dominant place among biodegradable drug delivery systems (Heller, 1984). The literature contains many reports describing the use of PLGA microspheres for drug delivery, and many techniques have been used to fabricate microspheres (Watts et al., 1990). Several common microsphere fabrication methods include:

- Single emulsion technique
- Double emulsion technique
- Hydrophobic ion pairing
- Prolease®-low temperature casting.

The method of choice will depend on the type of drug to be delivered, and its stability in various organic solvents. Hydrophobic drugs, for example, lend themselves well to a single emulsion technique where the drug and polymer are co-solubilized in the organic phase. Hydrophilic drugs such as proteins are not readily soluble in an organic phase, and the Prolease® method is more amenable to this type of drug (Gombotz et al., 1991). A wide variety of additives can also be included with the drugs in the microspheres. These can improve stability, modify release kinetics, increase loading efficiencies, and enhance "syringability." Terminal sterilization by gamma irradiation is preferable but for highly radiation-sensitive drug formulations, aseptic processing is necessary and can be technically challenging. The final step in microsphere fabrication involves some type of drying process such as lyophilization or vacuum drying. Prior to administration the microspheres are reconstituted and suspended in a viscous diluent such as sodium carboxymethyl cellulose (CMC). A surfactant is often incorporated in the diluent to prevent aggregation of the microspheres prior to injection through a narrow gauge needle.

Table C.1 lists the commercially approved microsphere products. LHRH analogs are the most common, and the success of these products is due to several factors that are related to both the drug and the polymers used in the systems. Chronic administration of LHRH and its analogs causes a reversible chemical shutdown of the pituitary gland, resulting in regression of hormone-responsive tumors, including prostate and breast carcinomas. The desired clinical effect is one of downregulation, and a well-defined delivery profile is not necessary as long as a sufficient quantity of drug is provided. Since the drugs were originally administered by once-daily injection, the development of a once monthly or once every three months system was desirable from a patient compliance viewpoint. An initial burst of drug was not a problem due to their low toxicity, and the LHRH analogs are stable during the microsphere fabrication process and *in vivo* after injection. Finally the PLGA polymers used in the microspheres required minimal toxicology testing.

In the late 1970s the first PLGA microspheres containing LHRH were developed and patented by Southern Research Institute in conjunction with Syntex (Sanders et al., 1985; Kent et al., 1987). This technology was licensed from Syntex by the Takeda-Abbott Pharmaceutical Co., and marketed as Lupron® Depot. The microspheres are manufactured using a water/oil/water double emulsion technique using a PLGA (75:25) copolymer with a molecular weight of approximately 14,000 Da (Ogawa et al., 1988). The initially approved product released drug for a period of one month (Okada et al., 1991) and since that time three- and four-month products have been introduced into the market.

During the 1980s, Southern Research Institute developed a [D-Trp-6 LHRH]/PLGA microsphere for DebioPharm, the first clinical PLGA–drug product. The product, called Decapeptyl®, was commercially launched in Europe in 1986 as a one-month sustained-release formulation. In October 2009, DebioPharm announced that a six-month formulation had successfully completed its European decentralized registration procedure. This product is still on the market.

In 1999 Alkermes Corp., in conjunction with Genentech, launched a PLGA-based microsphere product containing recombinant human growth hormone (rhGH) called Nutropin Depot® for the treatment of growth deficiency. The product used the Prolease® process, where micronized protein is dispersed in a lactide–glycolide solution in methylene chloride, and then atomized into liquid nitrogen that contains a frozen layer of ethanol at the bottom of the vessel. The small particles of polymer

TABLE C.1 Commercially Approved Degradable Microsphere Depot Drug Delivery Systems (DDS) Based on PLGA Copolymers

Product	Company	Active Agent	Indication	Year Approved
Decapeptyl®	DebioPharm	Triptorelin	Prostate cancer	1986
Lupron Depot®	Takeda Abbott Pharmaceuticals	Leuprolide Acetate	Prostate cancer/ Early puberty/ anemia	1989
Procrin®	Abbott	Leuprolide Acetate	Prostate cancer/ Early puberty/ anemia	1989
Enantone®	Takeda	Leuprolide Acetate	Prostate cancer/ Early puberty/ anemia	1989
Sandostatin- LAR®	Novartis	Octreotide	Acromegaly	1998
Nutropin Depot®*	Genentech/Alkermes	hGH	Growth deficiency	1999
Arestin™	OarPharma/J&J	Minocycline HCl	Periodontitis	1999
Trelstar™	Debiopharm	Triptorelin	Prostate cancer	2000
Risperdal Consta™	Eli Lilly/Alkermes	Risperdal	Schizophrenia	2003
Vivitrol™	Cephalon/Alkermes	Naltrexone	Alcohol dependence	2006
Somatulin®	Beaufour Ipsen	Lanreotide	Acromegaly	2007

*Nutropin Depot® has been withdrawn.

solution with dispersed protein are instantly frozen and settle on top of the frozen ethanol layer. The liquid nitrogen is then evaporated, the ethanol melts, and the particles settle into the cold ethanol, where they harden as the methylene chloride is extracted, and they are then harvested (Khan et al., 1992).

This cryogenic method was used to encapsulate rhGH using an 8 kDa, 50/50 copolymer of lactic and glycolic acid. The protein was stabilized during the encapsulation process and during *in vivo* release by complexing with zinc (Johnson et al., 1996). The product was eventually withdrawn from the market due to poor sales and the high cost of manufacturing. Another important factor was that it was painful to inject, and many patients were children.

Alkermes has also worked with several other companies to develop PLGA-based microsphere delivery systems. Together with Eli Lilly and Company they developed Risperdal® Consta, a once-weekly formulation of risperidone for the treatment of schizophrenia, and for the longer-term treatment of Bipolar I Disorder. The drug was approved in 2003. In April 2006 Alkermes and Cephalon, Inc. announced the approval of Vivitrol™, a once-monthly injectable naltrexone formulation for the treatment of alcohol dependence. The product is comprised of 337 mg of naltrexone per gram of 75:25 PLG microspheres. In May 2009, Alkermes, in collaboration with Amylin and Eli Lilly, filed an NDA for Exenatide LAR, a once-weekly injectable formulation of byetta, a synthetic peptide hormone used for the treatment of type 2 diabetes. The product exhibited significantly greater improvements in glycemic control than byetta given twice a day (Drucker et al., 2008).

PLGA microspheres have received considerable attention for the controlled release of injectable protein vaccines (Wise et al., 1987). These systems can potentially reduce the number of inoculations, reduce the total antigen dose required to achieve immune protection, and enhance the immune response (Cohen et al., 1994; McGee et al., 1994). It is important to control the diameter of the microspheres used in vaccine delivery to 5 μm or less, in order to allow them to be phagocytized by macrophages or dendritic cells.

There a several other types of biodegradable polymers used for depot drug delivery systems, including polyanhydrides, poly(orthoesters) and poly(phosphoesters). Many of these depot systems have been fabricated into surgically implanted discs, such as the Gliadel®, a polyanhydride wafer containing BCNU (1,3-bis(2-chloroethyl)-1-nitrosourea) approved for the treatment of glioblastoma multiforme. There are no approved injectable products based on these polymers (see Chapter II.5.16.D).

Alza, and later Jorge Heller, along with the company AP Pharma, developed a family of degradable polyorthoesters, one of which, (APF530), contained a drug for the prevention of chemotherapy-induced nausea and vomiting. This product was submitted to the US Food and Drug Administration (FDA), and the company subsequently received a Complete Response Letter on the APF530 NDA in March 2010. Subsequent to that date, they were in the process of preparing a resubmission to the FDA.

In the 1990s biodegradable polyester was developed by Jan Feijen at Twente University, in The Netherlands, based on a copolymer of poly(ethylene glycol terephthalate) and poly(butylene terephthalate)(Bezedmer et al., 2000). A microparticle depot formulation of this polymer, called "Polyactive®," contains α-interferon (Locteron®) and is currently in clinical trials for the treatment of hepatitis C (De Leede et al., 2008).

Injected Phase-Separating Depot Systems

Injectable, *in situ* phase separating drug delivery systems are made from biodegradable components that are injected with a syringe into the body and, once injected, solidify to form a semi-solid depot. These systems have several advantages over injected microparticulate systems, including ease of manufacture, no need for reconstitution and suspension in a diluent prior to injection, and relatively little migration from the site of injection. A comprehensive review of these systems has been published (Hatefi and Amsden, 2002). Table C.2 shows the different categories of injectable phase separated systems based on their mechanism of solidification. The following discussion will review some of these systems.

Thermoplastic Pastes

Thermoplastic pastes are polymer systems that are injected as a melt, and upon cooling to body temperature form a semi-solid depot. A variety of polymers could potentially be used to form thermoplastic pastes, but only a few have been evaluated *in vivo*. Disadvantages of these systems include: (1) the temperature needed to form the melt upon injection, which could cause some pain and tissue necrosis; and (2) the very slow rate of drug release. Zhang et al. developed a thermoplastic triblock PLA-PEG-PLA copolymer of blends of PLA and poly(caprolactone) (PCL) for the local delivery of Taxol™ to tumors in rats (Zhang et al., 1996). Drug was delivered over months from both these systems which adversely impacted the ability of Taxol™ to inhibit tumor growth. Other groups have increased the release rate of Taxol™ from PCL by adding water soluble additives including gelatin, albumin, methylcellulose, dextran (Dordanoo et al., 1997), and methoxyPEG (Winternitz et al., 1996).

A thermoplastic injectable system comprised of a PEG-PCL copolymer was developed for delivery of drugs to the eye (Davis and Cousins, 1995). When heated to 50°C the system could be injected through a 25 gauge needle.

TABLE C.2 Types of Injected, Phase-Separated Depot DDS

Category	Polymer	Reference
1. Thermoplastic pastes	PLA-PEG-PLA, PLA/poly(caprolactone)	Zhang et al., 1996
	Poly(caprolactone)/excipients	Winternitz et al., 1996; Dordanoo et al., 1997
	PCL-PEG	Davis and Cousins, 1995
2. *In situ* cross-linked systems		
Thermosets	PLA-PCL	Dunn et al., 1994
Photocrosslinked gels	PEG-oligoglycolyl-acrylate	Hubbell et.al., 1995
Ion-mediated gelation	Alginate	Gombotz and Wee, 1998
		Cohen et al., 1997
3. *In situ* polymer precipitation		
Solvent-removal	PLA/PGLA/PCL	Dunn et al., 1990, 1995; Duenas et al., 2001; Eliaz and Kost, 2001
	Poly(acrylic acid)	Haglund et al., 1996
	HPMC-carbopol and PMA-PEG	Ismail et al., 2000
	Sucrose acetate isobutyrate	Smith and Tipton, 1996; Tipton and Hall, 1998; Lu et al., 2008;
4.Thermally-induced gels	Pluronics® (PEO-PPO-PEO)	Johnston et al., 1992; Miyazaki et al., 1992; Paavola et al., 1995; Bhardwaj and Blanchard, 1996; Paavola et al., 1999; Desai et al., 1998; Veyreis et al., 1999
	(PCL-co-glycolide)-PEG-(PCL-co-glycolide)	Jiang et al., 2007, 2009
	Poly(*N*-isopropyl acrylamide)	Heskins and Guillet, 1968; Hirotsu et al., 1987
	PEO-PLGA-PEO	Jeong et al., 1999a,b, 2000
	Also PLGA-PEO-PLGA (Re-Gel® of MacroMed)	
	PEO, PEO copolymers/Cyclodextrin	Li et al., 2003a,b
	Chitosan	Chenite et al., 2000
5. *In situ* solidifying organogels	Glycerol esters of fatty acids	Yim et al., 1989; Ericsson et al., 1991; Engstrom et al., 1992; Gao et al., 1998

In Situ Cross-Linked Systems

These injectable systems include thermoset polymers which are "cured" or set into their final shape after heating by the formation of covalent cross-links between polymer chains. Few applications of these systems exist for the delivery of drugs, since the conditions for curing the system can contain potentially toxic monomers or solvents, and generate temperatures that can cause tissue necrosis. Dunn et al. used prepolymers of PLA and PCL that were terminated with acrylic ester groups (Dunn et al., 1994). Prior to injection, curing was initiated by the addition of either benzoyl peroxide or *N*,*N*-dimethyl-*p*-toluidine, depending which ingredient of the co-catalyst was missing. The system was able to deliver flurbiprofen *in vitro* over a week.

Photocross-linked gels represent another type of *in situ* cross-linked injectable delivery system. Prepolymers are introduced to the desired site and then photocured *in situ* with fiber optic cables. Polymerization rates occur rapidly under physiological conditions. Hubbell et al. developed a photopolymerizable system based on PEG-oligoglycolylacrylates that gelled upon exposure to an eosin dye, a photosensitive initiator, and a light source (Hubbell et al., 1995). The system was shown to be best suited for delivery of high molecular weight protein-based therapeutic agents. Another recently developed *in situ* forming drug delivery system is based on photocross-linked poly(ε-caprolactone fumarate) (PCLF) networks loaded with tamoxifen citrate (Sharifi et al., 2009). Networks were made from PCLF macromers, a photo-initiator and accelerator, and N-vinyl-2-pyrrolidone as a cross-linker. Cytotoxicity assays showed that while the photocross-linked PCLF network exhibited no significant cytotoxicity against MCF-7 and L929 cell lines, 40–60% of the MCF-7 cells were killed after incubation with the tamoxifen citrate loaded system.

Alginate, a linear polysaccharide that is extracted from red-brown kelp, is the most common ion-mediated gelling system that has been investigated for drug delivery, and many reports have been published on the formation of alginate microparticle formation *in vitro* (Gombotz and Wee, 1998). Since alginates form a gel upon contact with divalent cations such as calcium ions, this system can only gel upon injection where local tissue concentrations of calcium are high. The human eye is one site where the concentration of $CaCl_2$ is high enough (0.008%) to facilitate gelation. Cohen et al. demonstrated that an aqueous solution of sodium alginate could gel in the eye and subsequently be used to deliver pilocarpine in a sustained fashion (Cohen et al., 1997).

In Situ Polymer Precipitation

Another means of developing *in situ*-forming drug delivery systems is polymer precipitation in a poor solvent. Such an injectable delivery system is prepared from a water-insoluble biodegradable polymer dissolved in a water-miscible, biocompatible solvent. When this system is mixed with a drug and injected into the physiologic environment, the solvent diffuses out into the surrounding tissues while water diffuses in, and the polymer precipitates, forming a solid polymeric implant containing entrapped drug. This method has been used in humans (Dunn et al., 1990), and in veterinary applications (Dunn et al., 1995).

The polymer most commonly used is a poly(lactic-co-glycolic acid) copolymer dissolved in *N*-methylpyrrolidone (NMP). This system is known as Atrigel®. Four formulations of PLGA dissolved in *N*-methyl-pyrrolidone and containing leuprolide acetate have been used for the treatment of prostate cancer and are marketed by Sanofi Aventis under the name Eligard®. These formulations provide patients with the option of being dosed every 1, 3, 4 or 6 months.

There are a number of other systems, such as: (1) poly(lactic acid) dissolved in a mixture of benzyl benzoate and benzyl alcohol, known as PLAD (Duenas et al., 2001); (2) poly(lactic-co-glycolic acid) copolymer dissolved in glycofurol (Eliaz and Kost, 2001); (3) poly(lactic-co-glycolic acid) copolymer dissolved in benzyl benzoate known as Alzamer®; and (4) sucrose acetate isobutyrate (SAIB) dissolved in NMP or Miglyol, known as SABER (Smith and Tipton, 1996; Tipton and Hall, 1998). A recent study evaluated the *in vivo* release of risperidone from a SAIB polymer system dissolved in ethanol combined with PLA as a release regulator (Lu et al., 2008). The addition of 10% PLA mw 9000 was able to significantly reduce the burst effect, and led to a stable plasma concentration over 25 days.

Polyacrylic acid and its derivatives have also been used as *in situ* precipitating polymers. Poly(methacrylic acid) (PMAAc) and PEG were solubilized in 50–80% ethanol and used to deliver albumin and pheniramine (Haglund et al., 1996). Upon injection the ethanol diffuses out of the system and water diffuses in, resulting in precipitation of the dissolved polymer network. A similar approach used a hydroxypropylmethylcellulose-Carbopol system and a PMAAc-PEG system for the delivery of plasmid DNA (Ismail et al., 2000). Both systems release DNA intact, but with a 35–70% drug burst effect in the first two hours followed by little to no subsequent release, indicating a strong interaction between the DNA and the polymer matrix.

Thermally-Induced Gels

A well-known example of a polymer that undergoes a sharp, thermally-induced phase separation is poly(*N*-isopropyl acrylamide) (Heskins and Guillet, 1968; Hirotsu et al., 1987). Its transition at 32°C can be shifted to higher or lower temperatures by copolymerization with more hydrophilic or hydrophobic comonomers, respectively. Thus, this material could be of interest in drug delivery applications; however, it may not be suitable due to potential cell toxicity of the homopolymer and its copolymers, although this issue has not been thoroughly investigated.

Of more interest are triblock copolymers of poly(ethylene glycol) and poly(propylene glycol), PEO–PPO–PEO, known as Pluronics® or Poloxamer® polyols, some compositions of which have already been approved for use in the body (Wang and Johnston, 1991). While many of these block copolymers have transition temperatures well above body temperature, they do exhibit transitions at body temperature in solutions at concentrations above 16 wt%. Since Pluronic® F127 has been reported to be the least toxic of these block copolymers, it has been used the most extensively for drug delivery studies with a number of low molecular weight compounds including vancomycin (Veyreis et al., 1999), mitomycin C (Miyazaki et al., 1992), melanotan-I (Bhardwaj and Blanchard, 1996), lidocaine (Paavola et al., 1999), ibuprofen (Paavola et al., 1995), and pilocarpine (Desai and Blanchard, 1998). As a result of its nondenaturing effects on proteins, Pluronic F127 gels have also been used to deliver several proteins including interleukine-2 (Morikawa et al., 1987; Johnston et al., 1992), urease (Fults and Johnston, 1990), rat atrial natriuretic factor (Juhasz et al., 1989), and transforming growth factor β-1 (Puolakkainen et al., 1995). The copolymer consists by weight of approximately 70% ethylene oxide and 30% propylene oxide, with an average molecular weight of 11,500 Da. Although Pluronic® F127 is not metabolized by the body, the gels do slowly dissolve over time and the polymer is eventually cleared.

More recently, block copolymer injectable gelling systems have been developed with biodegradable polymeric components. Such materials can be constructed from ABA block copolymers based on poly(ethylene glycol) and poly(lactic-*co*-glycolic acid) copolymers. The triblock may be designed as PLGA-PEO-PLGA or as PEO-PLGA-PEO, where the former will have a more pronounced thermal gelation capability. PLA has also been used as the hydrophobic, degradable block segment. When the block lengths and relative amounts are correctly chosen, a material is obtained that is soluble in water at room temperature, and forms a firm gel at body temperature (Jeong et al., 1999a, 2000). Such materials can be injected as a solution, having a viscosity not much different from a saline solution. In addition, since only water is used with this material, there are no organic solvent–water interface problems, and proteins can be incorporated without loss of activity. These materials are available as dry powders that are reconstituted prior to use. While the exact mechanism of thermogelling is not known with absolute certainty, it has been postulated that the hydrophobic blocks (PLGA or PLA) of the amphiphilic block copolymers lose their hydrophobically-bound water and aggregate together to form micelles that are in equilibrium with monomeric polymer chains (Jeong et al., 1999b). Then, as the temperature is increased, the equilibrium shifts to micelle formation, and above the critical gelation temperature, the micelles pack together to occupy the entire volume, resulting in gel formation. Evidence for such a mechanism is provided by ^{13}C NMR studies, dye solubilization studies, and light-scattering studies, which all show an abrupt change in micellar diameter and aggregation at the critical gelation temperature (Jeong et al., 1999b). It has also been postulated that the water content of the gel will determine its degradation rate and release kinetics of the incorporated therapeutic agent (Jeong et al., 1999b).

A triblock copolymer based on poly(ethylene glycol) and poly(lactic-*co*-glycolic acid) copolymer is under development under the trade name ReGel®. A formulation of Paclitaxel in ReGel® is known as OncoGel®, and a Phase II dose escalation study for the treatment of patients with esophageal cancer has recently been completed with no dose-limiting toxicities, and a reported reduction in tumor burden and tumor size (DuVall et al., 2009). The company BTG is currently enrolling a Phase IIb clinical study for the same indication that should be completed in 2010.

Another recently reported biodegradable thermogelling triblock copolymer is comprised of poly(CL-co-glycolide)-PEG-poly(CL-co-glycolide) (Jiang et al., 2007, 2009). The copolymer aqueous solution (25 wt %) underwent sol–gel transition at 35°C, and as the temperature increased formed a stable gel at body temperature. After incubation in physiological buffer the gel completely degraded into a viscous liquid after 14 weeks. The gel formed an *in vitro* controlled release depot with delivery times of 12, 32, and 25 days for isoniazid, rifampicin, and bovine serum albumin respectively.

Chitosan is another naturally derived polymer that is obtained by deacetylation of chitin, a component of shrimp and crab shells. Chitosan remains dissolved in aqueous solution up to a pH of 6.2, and forms a hydrated gel-like precipitate upon neutralization to a higher pH. Chenite et al. added polyol salts bearing a single anionic head such as glycerol-, sorbitol-, fructose- or glucose-phosphate to chitosan aqueous solutions to create thermally sensitive pH-dependent gel-forming solutions (Chenite et al., 2000). This system was used for the *in vivo* delivery of biologically active growth factors, as well as encapsulation of chondrocytes for tissue engineering.

An injectable supramolecular hydrogel has been developed based on PEO or PEO block copolymers and α-cyclodextrin (CD) (Li et al., 2003a). The hydrogel formation is based on a physical cross-linking induced by supramolecular self-assembling with no chemical cross-linking reagents. An inclusion complex is formed between CD and the PEO molecules, resulting in necklace-like supramolecular structures (Li et al., 2003b). Rheologic studies of the hydrogels showed that these systems are thixotropic and display a decreased viscosity when exposed to the shear stress of an injection through a fine needle. Components of the hydrogels are biocompatible and non-toxic, and drugs can be directly incorporated into the hydrogels at room temperature without any contact with organic solvents.

In Situ Solidifying Organogels

Organogels are composed of water-insoluble amphiphilic lipids that swell in water and form lyotropic liquid crystals. The nature of the liquid crystalline phase formed depends on the structure of the lipid, the type of drug incorporated, and the amount of water in the system (Engstrom and Engstrom, 1992). The most commonly used amphiphilic lipids for drug delivery include glycerol esters of fatty acids such as glycerol monooleate, glycerol monopalmitostearate, and glycerol monolinoleate. The liquid crystalline structure formed upon injection into an aqueous medium is gel-like and highly viscous. These systems have been studied for delivery of water-soluble and -insoluble drugs, including lidocaine (Engstrom and Engstrom, 1992), somatostatin (Ericsson et al., 1991), interferon-α (Yim et al., 1989), levenorgestrel, and ethinyl estradiol (Gao et al., 1998). These systems can be formulated with a low concentration of water, and the viscosity can be reduced by mixing with vegetable oils. Reducing the viscosity in this way provides for better injectability and increases the duration of release, particularly for lipophilic drugs. Organogels are biodegradable, and degradation occurs through the action of lipases.

BIBLIOGRAPHY

Bezemer, J., Radersma, R., Grijpma, D., Dijkstra, P., Van Blitterswijkt, C., & Feijen, J. (2000). Microspheres for protein delivery prepared from amphiphilic multiblock copolymers: 1. Influence of preparation techniques on particle characteristics and protein delivery. *J. Control. Release*, **67**, 233–248.

Bhardwaj, R., & Blanchard, J. (1996). Controlled release delivery system for the α-MSH analog melanotan-I using Polyoxamer 407. *J. Pharm. Sci.*, **85**, 915–919.

Chenite, A., Chaput, C., Wang, D., Combes, C., Buschmann, M., Hoemann, C., Leroux, J., Atkinson, B., Binette, F., & Selmani, A. (2000). Novel injectable neutral solutions of chitosan form biodegradable gels in-situ. *Biomaterials*, **21**, 2155–2161.

Cohen, S., Alonso, M., & Langer, R. (1994). Novel approaches to controlled-release antigen delivery. *Int. J. Tech. Assoc. Health Care*, **10**, 121–130.

Cohen, S., Lobel, E., Trevgoda, A., & Peled, Y. (1997). A novel *in-situ* forming drug delivery system from alginates undergoing gelation in the eye. *J. Control. Release*, **44**, 201–208.

Davis, P., & Cousins, S. (1995). Biodegradable injectable drug delivery polymer. *US Patent 5,384,333*.

De Leede, L. G., Humphries, J. E., Bechet, A. C., Van Hoogdalem, E. J., Verrijk, R., & Spencer, D. G. (2008). Novel controlled release Lemma-derived IFN-α2b (Locteron): pharmacokinetics, pharmacodynamics, and tolerability in a phase I clinical trial. *J. Interf. Cytok. Res.*, **28**, 113–122.

Desai, S., & Blanchard, J. (1998). Evaluation of Pluronic F-127 based sustained release ocular delivery systems for pilocarpine using the albino rabbit eye model. *J. Pharm. Sci.*, **87**, 1190–1195.

Dordanoo, S., Oktaba, A., Hunter, W., Min, T., Cruz, T., & Burt, H. (1997). Release of taxol from poly caprolactone pastes: Effect of water soluble additives. *J. Control. Release*, **44**, 87–94.

Duenas, E., Okumu, F., Daugherty, A., & Cleland, J. (2001). Sustained delivery of rhVEGF from a novel injectable liquid. *Proc. Int. Symp. Control. Rel. Bioact. Mater.*, **28**, 1014–1015.

Dunn, R. L., English, J. P., Cowsar, D. R., & Vanderbelt, D. P. (1990). *US Patent No. 4,938,763*.

Dunn, R., English, J., Cowsar, D., & Vanderbelt, D. (1994). Biodegradable *in situ* forming implants and methods for producing the same. *US Patent 5,340,849*.

Dunn, R. L., Hardee, G., Polson, A., Bennett, Martin, S., Wardley, R., Moseley, W., Krinick, N., Foster, T., Frank, K., & Cox, S. (1995). *In situ* forming biodegradable implants for controlled release veterinary applications. *Proc. Int. Symp. Control. Rel. Bioact. Mater.*, **22**, 91–92.

DuVall, G., Tarabar, D., Seidel, R., Elstad, N., & Fowers, K. (2009). Phase 2: A dose-escalation study of OncoGel (ReGel/paclitaxel), a controlled-release formulation of paclitaxel, as adjunctive local therapy to external-beam radiation in patients with inoperable esophageal cancer. *Anti-Cancer Drugs*, **20**, 89–95.

Eliaz, R., & Kost, J. (2001). *US Patent No. 6,206,921 B1*.

Engstrom, S., & Engstrom, L. (1992). Phase behavior of the lidocaine-monoolein-water system. *Int. J. Pharm.*, **79**, 113–122.

Ericsson, B., Ericsson, P., Lofroth, J., & Engstrom, S. (1991). Cubic phases as drug delivery systems for peptide drugs. *ACS Symp. Ser.*, **469**, 251–265.

Fults, K., & Johnston, T. (1990). Sustained-release of urease from a polyoxamer gel matrix. *J. Parenter. Sci. Technol.*, **44**, 58–65.

Gao, Z., Crowley, W., Shukla, A., Johnson, J., & Reger, J. (1998). Controlled release of contraceptive steroids from biodegradable and injectable gel formulations: *In vivo* evaluation. *Pharm. Res.*, **12**, 864–868.

Gombotz, W. R., & Pettit, D. K. (1995). Biodegradable polymers for protein and peptide drug delivery. *Bioconjug. Chem.*, **6**, 332–351.

Gombotz, W., & Wee, S. (1998). Protein release from alginate matrices. *Adv. Drug Del. Rev.*, **31**, 267–285.

Gombotz, W., Healy, M., & Brown, L. (1991). Very low temperature casting of controlled release microspheres. *US Patent 5,019,400*.

Haglund, B., Rajashree, J., & Himmelstein, K. (1996). An *in-situ* gelling system for parenteral delivery. *J. Control. Release*, **41**, 229–235.

Hatefi, A., & Amsden, B. (2002). Biodegradable injectable *in situ* forming drug delivery systems. *J. Control. Release*, **80**, 9–28.

Heller, J. (1984). Biodegradable polymers in controlled drug delivery. *CRC Crit. Rev. Therap. Drug Carrier Syst.*, **1**, 39–90.

Heskins, H., & Guillet, J. E. (1968). *J. Macromol. Sci. Chem. A2*, **6**, 1209.

Hirotsu, S., Hirokawa, Y., & Tanaka, T. (1987). Volume-phase transitions of ionized *N*-isopropylacrylamide gels. *J. Chem. Phys.*, **87**, 1392–1395.

Hubbell, J., Pathak, C., Sawhney, A., Desai, N., & Hill, J. (1995). Photopolymerizable biodegradable hydrogels as tissue contacting materials and controlled release carriers. *US Patent 5,410,016*.

Ismail, F., Napaporn, J., Hughes, J., & Brazeau, G. (2000). *In situ* gel formulations for gene delivery: Release and myotoxicity studies. *Pharm. Dev. Technol.*, **5**, 391–397.

Jiang, Z., You, Y., Deng, X., & Hao, J. (2007). Injectable hydrogels of poly(ε-caprolactone-co-glycolide)-co-poly(ethylene glycol)-poly(ε-caprolactone-co-glycolide) triblock copolymer aqueous solutions. *Polymer*, **48**, 4786–4792.

Jiang, Z., Hao, J., You, Y., Gu, Q., Cao, W., & Deng, X. (2009). Biodegradable thermogelling hydrogel of p(CL-GL)-PEG-p(CL-GL) triblock copolymer: degradation and drug release behavior. *J. Pharm. Sci.*, **98**, 2603–2610.

Jeong, B., Choi, Y. K., Bae, Y. H., Zentner, G., & Kim, S. W. (1999a). New biodegradable polymers for injectable drug delivery systems. *J. Control. Release*, **62**, 109–114.

Jeong, B., Bae, Y. H., & Kim, S. W. (1999b). Biodegradable thermosensitive micelles of PEG-PLGA-PEG triblock copolymers. *Colloids Surf. B: Interfaces*, **16**, 185–193.

Jeong, B., Kibbey, M. R., Birnhaum, J. C., Won, Y. -Y., & Gutowska, A. (2000). Thermogelling biodegradable polymers with hydrophilic backbones: PEG-g-PLGA. *Macromolecules*, **33**, 8317–8322.

Johnson, O. F. L., Cleland, J. F., Lee, H. J., Charnis, M., Duenas, E., Jaworowicz, W., Shepard, D., Shazamani, A., Jones, A. J. S., & Putney, C. D. (1996). A month-long effect from a single injection of microencapsulated human growth hormone. *Nat. Med.*, **2**, 795–799.

Johnston, T., Punjabi, M., & Froelich, C. (1992). Sustained delivery of interleukin-2 from Polyoxamer 407 gel matrix following intraperitoneal injection in mice. *Pharm. Res.*, **9**, 425–434.

Juhasz, J., Lenaerts, V., Raymond, P., & Ong, H. (1989). Diffusion of rat atrial natriuretic factor in thermoreversible polyoxamer gels. *Biomaterials*, **10**, 265–268.

Kent, J., Lewis, D., Sanders, L., & Tice, T. (1987). Microencapsulation of water soluble active polypeptides. *US Patent 4,675,189*.

Khan, M. A., Healy, M. D. S., & Bernstein, H. (1992). Low temperature fabrication of protein loaded microspheres. *Proc. Int. Symp. Control. Rel. Bioact. Mater.*, **19**, 518–519.

Li, J., Ni, X., & Leong, K. (2003a). Injectable drug-delivery systems based on supramolecular hydrogels formed by poly(ethylene oxide)s and α-cyclodextrin. *J. Biomed. Mater. Res.*, **65A**, 196–202.

Li, J., Ni, X., & Leong, K. (2003b). Preparation and characterization of inclusion complexes of biodegradable amphiphilic poly(ethylene oxide)-poly[(*R*)-3-hydroxybutyrate]–poly(ethylene oxide) triblock copolymers with cyclodextrins. *Macromolecules*, **36**, 1209–1214.

Lu, Y., Tang, X., Cui, Y., Zhang, Y., Qin, F., & Lu, X. (2008). *In vivo* evaluation of risperidone-SAIB *in situ* system as a sustained release delivery system in rats. *Eur. J. Pharm. and Biopharm.*, **68**, 422–429.

McGee, J., Davis, S., & O'Hagan, D. (1994). The immunogenicity of a model protein entrapped in poly(lactide-co-glycolide) microparticles prepared by a novel phase separation technique. *J. Control. Release*, **31**, 55–60.

Miyazaki, S., Ohkawa, Y., Takeda, M., & Attwood, D. (1992). Antitumor effect of Pluronic F-127 gel containing Mitomycin C on sarcoma-180 ascites tumor. *Chem. Pharm. Bull.*, **40**, 2224–2226.

Morikawa, K., Okada, O., Hosokawa, M., & Kobayashi, H. (1987). Enhancement of therapeutic effects of recombinant interleukin-2 on a transplantable rat fibrosarcoma by the use of a sustained release vehicle, pluronic gel. *Cancer*, **47**, 37–41.

Ogawa, Y., Yamamoto, M., Okada, H., Yashiki, T., & Shimamoto, T. (1988). A new technique to efficiently trap leuprolide acetate into microcapsules of polylactic acid or copoly(lactic/glycolic) acid. *Chem. Pharm. Bull.*, **36**, 1095–1103.

Okada, H., Inoue, Y., Heya, T., Ueno, H., Ogawa, Y., & Toguchi, H. (1991). Pharmacokinetics of once-a-month injectable microspheres of leuprolide acetate. *Pharm. Res.*, **8**, 787–791.

Paavola, A., Yliruusi, J., Kajimoto, Y., Kalso, E., Wahlstrom, T., & Rosenberg, P. (1995). Controlled release of lidocaine from injectable gels and efficacy in rat sciatic nerve block. *Pharm. Res.*, **12**, 1997–2002.

Paavola, A., Tarkkila, P., Xu, M., Wahlstrom, T., Yliruusi, J., & Rosenberg, P. (1999). Controlled release gel of ibuprofen and lidocaine in epidural use – analgesia and systemic absorption in pigs. *Pharm. Res.*, **15**, 482–487.

Park, T. G. (1994). Degradation of poly(D, L-lactic acid) microspheres: effect of molecular weight. *J. Control. Release*, **30**, 161–173.

Puolakkainen, P., Twardzik, D., Ranchalis, J., Pankey, S., Reed, M., & Gombotz, W. (1995). The enhancement in wound healing by transforming growth factor ß1 (TGF-ß1) depends on the topical delivery system. *J. Surg. Res.*, **58**, 321–329.

Sanders, L., McRae, G., Vitale, K., & Kell, B. (1985). Controlled delivery of an LHRH analogue from biodegradable injectable microspheres. *J. Control. Release*, **2**, 187–195.

Sharifi, S., Mirzadeh, H., Imani, M., Rong, Z., Jamshidi, A., Shokrgozar, M., Atai, M., & Roohpour, A. (2009). Injectable *in situ* forming drug delivery system based on poly(ε-caprolactone fumarate) for tamoxifen citrate delivery: gelation characteristics, *in vitro* drug release and anti-cancer evaluation. *Acta Biomater.*, 5, 1966–1978.

Smith, D. A., & Tipton, A. (1996). A novel parenteral delivery system. *Pharm. Res.*, **13**, 300.

Tipton, A. J., & Hall, R. J. (1998). *US Patent No. 5,747,058.*

Veyreis, M., Couarraze, G., Gieger, S., Agnely, L., Massias, B., Kunzli, B., Faurisson, F., & Rouveiux, B. (1999). Controlled release of vancomycin from Polyoxamer 407 gels. *Int. J. Pharm.*, **192**, 183–193.

Watts, P. J., Davies, M. C., & Melia, C. D. (1990). Microencapsulation using emulsification/solvent evaporation: an overview of techniques and applications. *CRC Crit. Rev. Ther. Drug Carrier Sys.*, 7, 235–259.

Winternitz, C., Jackson, J., Oktaba, A., & Burt, H. (1996). Development of a polymeric surgical paste formulation for taxol. *Pharm. Res.*, **13**, 368–375.

Wise, D., Trantolo, D., Marino, R., & Kitchell, J. (1987). Opportunities and challenges in the design of implantable biodegradable polymeric systems for the delivery of antimicrobial agent and vaccines. *Adv. Drug Del. Rev.*, **1**, 19–39.

Yim, Z., Zupon, M., & Chaudry, I. (1989). Stable oleaginous gel. *US Patent 4,851,220.*

Zhang, X., Jackson, W., Wong, W., Min, W., Cruz, T., Hunter, W., & Burt, H. (1996). Development of biodegradable polymeric paste formulations for taxol: an *in vivo* and *in vitro* study. *Int. J. Pharm.*, **137**, 199–208.

D. IMPLANTS AND INSERTS

Lothar W. Kleiner[1] *and Jeremy C. Wright*[2]

[1]Drug Delivery and Medical Device Polymer Consultant, Los Altos, CA, USA

[2]DURECT Corporation Cupertino, CA, USA

INTRODUCTION

Implantable drug delivery systems (IDDS) originated in the 1960s when silicones were used to prolong the effectiveness of a therapy. From this beginning, the potential was recognized that this mode of delivery could overcome the problems associated with oral administration of specific therapies. Despite considerable effort since the beginning, initial progress has been slow to commercialize safe and effective implants. Some of the major hurdles that needed to be overcome were stability, reproducibility, toxicity, lack of biocompatibility, carcinogenicity, lack of compatibility between drug and carrier leading to burst release or shutdown, and physician and patient acceptability. However, the appeal, activity, and approval of IDDS accelerated when the silicone-based device Norplant® was approved by the US Food and Drug Administration (FDA) in 1990. Good evidence is the proliferation of academic and industrial research, the increase in the number of published articles, and the increased number of commercialized products.

Classification of IDDS is difficult, because there will be exceptions and hybrids that may fall under more than one category. When the term IDDS is used in this section it will also refer to "inserted" implants (e.g., intraocular); however, as much as possible this section will use the classifications that are historically used: drug implants and implantable pumps that contain and deliver drug. Drug implants can be further subdivided into non-degradable and degradable systems. Other DDS will also be included, such as inserts.

History

The concept and research of implantable drug delivery systems started with Deanesly and Parkes (1937) working with chickens. IDDS formulations with drug release rates controlled by a "rate-controlling membrane" (RCM) were pioneered by Folkman and Long (1964). They proposed an implanted Silastic® capsule containing a drug for prolonged, steady-state systemic drug administration. Silicone rubber capsules containing a variety of different drugs were prepared and implanted into the cardiac muscle of dogs. These rudimentary IDDS succeeded in the controlled release of many different classes of drugs, and were shown to be "biocompatible" over the period of implantation.

Since these early days, the research using IDDS has proliferated with the use of many different carriers, biostable and bioerodible, a wide variety of different classes of drugs, many different implantation techniques and implantation sites. A vast number of review articles can be found in the literature, originating with veterinary applications and quickly followed with human therapies (Hsieh and Langer, 1981; Jaffe et al., 1981; Lewis, 1981; McCormick et al., 1981; Wheeler and Friel, 1981; Danckwerts and Fassihi, 1991; Hoffmann, 2008; del Valle et al., 2009).

Rationale

To mitigate or avoid the problems associated with oral dosage form administration, other modes of delivery such as implants and inserts evolved. Numerous design approaches have been pursued, depending upon the specific drug therapy under development (Vernon and Wegner, 2004). The primary end-goal of any design approach is to improve safety and efficacy by carefully controlling dose, and dose rate at the desired site and over the desired duration of drug delivery.

Some of the issues with oral dosage forms that can be overcome with IDDS are described in Table D.1. Benefits of IDDS are described in Table D.2 and potential drawbacks are summarized in Table D.3.

TABLE D.1	Summary of Issues with Some Oral Dosage Forms
•	Drug bioavailability – Drug is not soluble enough to be absorbed by the gastro-intestinal system (GITS) or there is extensive first pass hepatic metabolism.
•	Drug stability – Drug is not stable in the gastrointestinal system.
•	Drug toxicity – Efficacious concentrations at the target site require systemic concentrations that are toxic or produce unacceptable side-effects.
•	Duration of release – In order to be effective, the drug must be able to be delivered in a controlled and sustained manner for more than 24 hours.
•	Drug half-life is too short – Drug loses its potency too quickly to be effective.

TABLE D.2	Summary of Benefits of IDDS
•	Patient compliance – Patient does not need to think about taking medication throughout the dosing interval of the IDDS. Less dosing required.
•	Fewer side-effects – Controlled release and usually much lower dose with improved control at site of action for extended periods of time; adverse effects away from site of action are minimized; peaks and valleys in plasma drug concentration from repeated immediate release dosing are avoided.
•	Lower dose – If action is site specific, drug has to overcome fewer biological barriers, such as first pass hepatic effects, before coming to the active receptor.
•	Drug stability improved – Protection of drug slows metabolization.
•	Suitability over intravenous– Hospital stay may not be required for chronic illnesses.
•	Easy removal – If allergic or other adverse reaction to drug is experienced, immediate removal is possible.

TABLE D.3	Summary of Potential Drawbacks of IDDS
•	Regulatory issues – IDDS are more complex than oral dosage forms – this makes the regulatory path for approval longer and more expensive.
•	Surgical needs – Some IDDS require minor surgery for implant and explant – this lowers patient acceptance and raises the desire for a less invasive alternative.
•	Cost/benefit ratio – The IDDS may not be cost effective enough for insurance coverage. Patient and doctor acceptance is lowered.
•	Training – Implantation usually requires specialized training.
•	Pain and discomfort – If an IDDS has this effect, patient acceptance is lowered.

Despite all the hurdles that need to be overcome, the examples of IDDS provided in the next sections indicate that they can be manufactured cost-effectively, and administered to a desired site to achieve the goal of reliability, safety, efficacy, and patient acceptance. The examples that are provided are a small representation and not comprehensive, since it is impossible to cover all the approved applications and the ever-increasing amount of research within universities and industry. More expansive reviews can be found (Lewis, 1981, 1990; Danckwerts and Fassihi, 1991; Dash and Cudworth, 1998; Pizzi et al., 2004; Hoffman, 2008).

IMPLANTS

In the next two sections for non-degradable and degradable IDDS, most or all the key components are produced from polymeric materials. In addition to drug delivery, the components may have additional functions, such as structural support and improvement of biocompatibility or stability. Lists and descriptions of the key polymers used in approved IDDS may be found in this and other chapters in this book covering degradable and non-degradable polymers used in biomedical devices and implants (see also Mani et al., 2007; Lyu and Untereker, 2009).

Non-Degradable Systems

Several types of non-degradable implants are presently commercially available, but by far the most common are reservoir and matrix DDS systems. Matrix systems are also called monolithic systems. The non-degradable polymers used in these systems include silicones, acrylates and their copolymers, ethylene vinyl acetate copolymers (PEVA), vinylidene fluoride copolymers, and urethanes.

Applications. In this section three applications will be highlighted. They include Norplant®, Implanon®, and drug-eluting stents (DES) with a non-degradable drug delivery (DD) coating. Space limitations preclude including additional applications, therefore the reader is referred to the review by Dash and Cudworth (1998) as a good starting point to obtain additional information.

Norplant®. The most common reservoir non-degradable system is Norplant®. This IDDS was developed and trademarked by the Population Council in 1980, introduced in certain countries worldwide in 1983, approved by the US FDA in December 1990, and marketed in the US starting February 1991. Norplant® is a five-year contraceptive system consisting of the hormone levonorgestrel (LNG) filled into six thin, flexible, silicone (Silastic®) tubes that are implanted subcutaneously on the inside of a woman's upper arm, as seen in Figure D.1 (Population

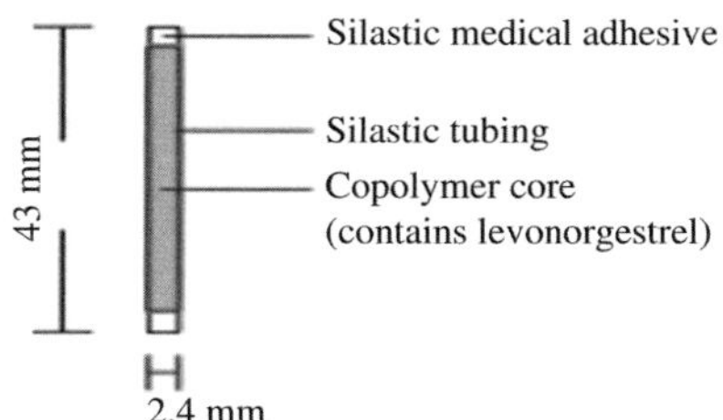

FIGURE D.1 Norplant® capsule. (Source: Population Council, 1990.)

Council, 1990). As of 2009, this device is approved in over 60 countries.

Later in the US, the six-rod Norplant® proved to be cumbersome and became associated with implantation and retrieval problems due to operator inexperience. Therefore, Norplant® was removed from the US market in 2002, but it is still available in other countries with successful use by over 60 million women.

Implanon®. A similar, more recent FDA-approved IDDS system is Implanon®, which entered the international market in 1998 and the US in 2006 (Alam et al., 2008). This single-rod IDDS is 4 cm long and 2 mm wide, and consists of a PEVA core (reservoir) containing 68 mg of etonogestrel. The rod is covered with a PEVA rate-controlling membrane, and is designed to release drug to prevent ovulation for a three-year period. Protection from pregnancy can continue longer if the implant is removed and immediately replaced with a new one. This newer, long-acting system is intended for easier subcutaneous insertion and removal than Norplant®, and therefore has gained improved acceptance by patients and their providers.

Drug-Eluting Stent (DES). Another important non-degradable IDDS application is the drug-eluting stent (DES) that has revolutionized the treatment of coronary artery disease (CAD) by reducing restenosis by 60–75% across all patients when compared to bare metal stents (Kukreja et al., 2009). DES is now the standard of care in the treatment of coronary artery disease (opening a blockage in a coronary artery). A DES consists of a scaffold (the stent) coated with a polymer plus drug mixture. The stent keeps the coronary artery patent, and the coating releases a drug that inhibits smooth muscle cell proliferation and the resultant neointimal growth. The mode of release from these coatings is diffusion-controlled.

Table D.4 highlights the four major globally-marketed DES systems. All are based on a polymer coating on a metallic stent that releases drug (Virmani et al., 2009). Current DES systems are examples of matrix or monolithic drug delivery systems.

Even though DES has become the standard of care for CAD, considerable research continues to be focused on the formulation and design of DES that create less inflammation. In addition, new systems are being developed to eliminate late stent thrombosis, the evaluation of which can only be done prospectively after a particular system has been in the clinic for many years. Therefore, the reader is cautioned to remember that materials currently in use may become out of favor years later, as a result of data from registries and follow-up studies. Even more caution should be exercised in considering materials for clinical use if only preclinical or early clinical studies are available. This gives an example of the benefits of using biomaterials that have a long history of clinical use.

Biodegradable Systems

There is a strong motivation to develop or replace a non-degradable therapeutic IDDS with a biodegradable one any time the drug delivery is required for only a limited period of time. The normal healing process can be aided when the tissue surrounding an implant can return to its native state. Moreover, the explantation or retrieval procedure is eliminated for those implants that require surgery or minor surgery (referred to as "minimally invasive procedures") for implantation. The major limitations of implanted biodegradable IDDS are complexity, development cost, regulatory requirements, and availability of polymers with the exact physical properties needed (including mechanical properties and degradation kinetics). The *in vivo* degradation kinetics, degradation byproducts, and mechanism must be understood. They can be highly variable from patient to patient, depending upon the age of the patient and the state and type of the medical problem.

Many polymers that are degradable have been explored as components of IDDS; however, only a few have been commercialized. The most widely investigated in regard to available clinical data are the aliphatic polyesters based upon lactic and glycolic acids. These materials have received considerable attention since the 1970s, first as sutures, then as excipients for drug delivery, and finally as IDDS. Their desirable features include acceptable degradation products, ability to tune degradation kinetics by LA/GA ratio, and ease of fabrication. Other

TABLE D.4 Summary of Four Major, Globally-Marketed, Polymer/Drug-Coated, Drug-Eluting Stents (DES)

Company	Trade Name	Drug	Polymer (Drug Release Coating)	Platform[1]
Johnson & Johnson	Cypher®	Sirolimus	PEVA blend with PBMA with PBMA topcoat and Parylene C primer	SS, 140 μm
Boston Scientific	Taxus®	Paclitaxel	SIBS	SS, 132 μm
Medtronic	Endeavor®	Zotarolimus	PC	CoCr, 91 μm
Abbott Vascular	Xience V®	Everolimus	PVDF·HFP with PBMA primer	CoCr, 81 μm

Abbreviations used: PEVA: Polyethylene-co-vinyl acetate; PBMA: Poly n-butyl methacrylate; SIBS: Poly(styrene-b-isobutylene-b-styrene); PC: a copolymer composed of 2-methacryloyloxyethyl phosphorylcholine, lauryl methacrylate, hydroxypropyl methacrylate, and trimethoxysilylpropyl methacrylate; PVDF·HFP: Poly(vinylidene fluoride-*co*-hexafluoropropylene); SS: stainless steel; CoCr: Cobalt/Chromium alloy.

[1]All micrometer (μm) dimensions represent stent strut thickness.

important biodegradable polymers include PEG-b-PLGA [poly(ethylene glycol)-block-poly(lactide-*co*-glycolide)] or PEG-PLLL [poly(ethylene glycol)-block-poly(L-lactide)] block copolymers, polyanhydrides, poly(ortho esters), and poly(phosphoesters) (see Chapters I.2.6 and II.5.16; Lewis, 1990; Cooper et al., 2004).

Applications. In this section only a few representative commercial biodegradable IDDS products and those that are currently in human clinical trials can be highlighted, due to space limitations. (Such implantable DDS include injectable microparticles and other depot DDS; these systems are covered in Chapter II.5.16.G). Only those systems that are fully absorbable are emphasized instead of non-degradable systems with biodegradable coatings. However, examples of biodegradable polymer coatings on a non-degradable substrate are included in the bibliography (Mani et al., 2007; Acharya and Park, 2006; Chan-Seng et al., 2009; Ormiston and Serruys, 2009; Virmani et al., 2009).

Although biodegradable IDDS have been used for the delivery of a wide range of therapeutics, the most common biodegradable IDDS include those for the treatment of cancer and cancer pain. A more recent development in 2009 is a fully bioresorbable drug-eluting stent, which is currently undergoing human clinical trials.

An example of a biodegradable IDDS for the treatment of brain tumors is the Gliadel® wafer (Perry and Schmidt, 2006). This device was approved by the FDA on June 14, 1996, and is now marketed by Eisai Co., Ltd. (Eisai Co., Ltd., 2009). Gliadel® wafers are dime-sized biodegradable polyanhydride disks, 1.45 cm in diameter and 1.0 mm thick, designed to deliver the chemotherapeutic drug, BCNU or carmustine, directly into the surgical cavity created after the tumor (high-grade malignant glioma) has been surgically excised. Up to eight wafers are implanted along the walls and floor of the cavity that once contained the tumor. Each wafer has been designed to deliver a precise dose of carmustine to the surrounding cells (Eisai Co., Ltd., 2009).

The trade name of the biodegradable polyanhydride copolymer is polifeprosan 20, it consists of poly[bis(p-carboxyphenoxy) propane: sebacic acid] in a 20:80 molar ratio and is used to control the local delivery of carmustine. Each Gliadel® wafer contains 192.3 mg of polifeprosan 20 and 7.7 mg of carmustine (1,3-bis (2-chloroethyl)-1-nitrosourea or BCNU), which is the oncolytic agent uniformly distributed in the copolymer matrix. The chemical structure of polifeprosan 20 is shown in Figure D.2.

Upon exposure to the aqueous environment in the surgical cavity, the anhydride bonds hydrolyze to release

$$\left[-O-\overset{O}{\overset{\|}{C}}-C_6H_4-O(CH_2)_3O-C_6H_4-\overset{O}{\overset{\|}{C}}- \right]_m \left[-O-\overset{O}{\overset{\|}{C}}-CH_2(CH_2)_6CH_2-\overset{O}{\overset{\|}{C}}- \right]_n$$

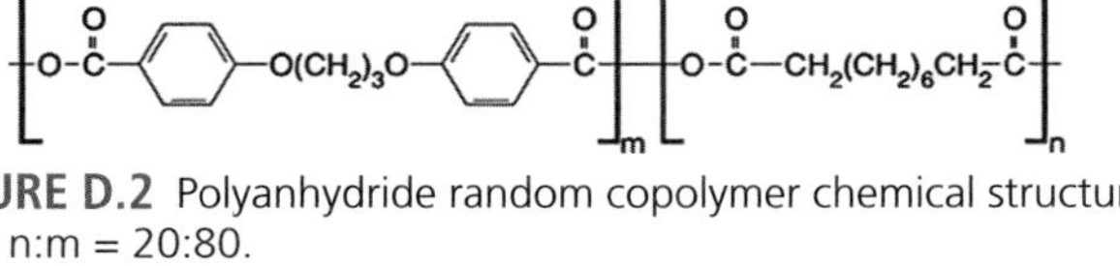

FIGURE D.2 Polyanhydride random copolymer chemical structure; ratio n:m = 20:80.

carmustine, carboxyphenoxypropane, and sebacic acid. The carmustine released from the Gliadel® wafer diffuses into the surrounding brain tissue. More than 70% of the copolymer biodegrades within three weeks back to constituent monomers.

Another example of a biodegradable IDDS is Zoladex®, which is indicated for the palliative treatment of advanced carcinoma of the prostate, and is manufactured by AstraZeneca. It is supplied as a sterile biodegradable product containing the drug goserelin acetate (equivalent to 3.6 mg of goserelin) dispersed in a matrix of poly(lactide-co-glycolide) (PLGA). Goserelin acetate is a potent synthetic peptide analog of luteinizing hormone-releasing hormone (LHRH). Zoladex® is designed for subcutaneous injection, with continuous release of drug over a period of 28 days. The encapsulated drug is in the form of a 1 mm diameter cylinder preloaded in a single use syringe with a 16 gauge needle, and packaged in a sealed, light, moisture-proof aluminum foil laminate pouch containing a desiccant capsule. The encapsulated drug is released by a combination of diffusion and erosion-controlled mechanisms. Studies have shown that the PLGA is completely biodegradable and exhibits no demonstrable antigenic potential (Zoladex® IFU, 2003; Kotwal et al., 2007).

An active development area for biodegradable IDDS are biodegradable DES, which most commonly come in two versions: a biodegradable coating on a permanent scaffold and a fully bioresorbable system. Table D.5 describes some of the many types of platforms currently under development (Isenbarger and Resar, 2005; Ormiston and Serruys, 2009; Virmani et al., 2009). The rationale for this extensive activity can be summarized under the heading of patient outcome or safety. Even though DES have revolutionized the treatment of coronary artery disease, there are still late adverse effects in a very minor subset of patients. A fully bioresorbable drug-eluting vascular scaffold (BVS) has many potential advantages over a non-degradable DES. A drug-eluting BVS is designed to perform all the functions of a DES when needed for scaffolding, and then is naturally resorbed and metabolized by the body when no longer needed. This potentially reduces the need for long-term antiplatelet therapy, and allows the coronary artery to return to its native state. Late stent thrombosis is thereby abolished or greatly reduced. In addition, the disappearance of the BVS may avoid a future need for reintervention. More comprehensive discussion of BVS therapeutic advantages are found in recent literature (Oberhauser et al., 2009; Ormiston and Serruys, 2009; Serruys et al., 2009; Virmani et al., 2009).

Clinical Trial on Bioresorbable Stents. The recently published two-year outcome of human clinical trials involving drug-eluting bioresorbable vascular scaffolds (BVS) (Oberhauser et al., 2009; Serruys et al., 2009) with 30 patients showed that the BIOSORB scaffolding of the vessel lasted long enough, and that it was also transient. The device was fully bioresorbed, late lumen

TABLE D.5 Summary of Some Biodegradable Drug Eluting Stents Under Development

Company	Trade Name	Drug	Polymer (drug release coating)	Platform
Biosensors International, Ltd.	BioMatrix	Biolimus A9	PDLLA: asymmetric abluminal coating	SS, 112 μm
Terumo Cardiovascular Systems	Nobori [1]	Biolimus A9	PDLLA: asymmetric abluminal coating	SS, 112 μm
Boston Scientific	JACTAX	Paclitaxel	PDLLA: microdots on abluminal side	SS, 96 μm
Johnson & Johnson	NEVO	Sirolimus	PLGA in small reservoirs, each acting as a depot	CoCr alloy, 89 μm
Abbott Vascular	BVS	Everolimus	PDLLA	PLLA, 152 μm
Biotronik GmbH	Lekton Motion	No drug[2]	No polymer coating[2]	Mg alloy – WE 43, 165 μm
REVA Medical[3]		No drug and Paclitaxel	No polymer coating	Tyrosine-derived polycarbonate, 200 μm
Bioabsorbable Therapeutics		Sirolimus salicylate	Salicylate linked by adipic acid	PAE salicylic acid, 200 μm

Abbreviations: BVS: Bioresorbable Vascular Scaffold; CoCr: Cobalt/Chromium alloy; PAE: poly(anhydride ester); PDLLA: Poly(D,L-lactide); PLGA: poly(lactide-co-glycolide); PLLA: Poly(L-lactide); SS: Stainless steel.
[1]DES technology licensed from Biosensors.
[2]Drug and polymer under investigation.
[3]Strategic alliance with Boston Scientific.

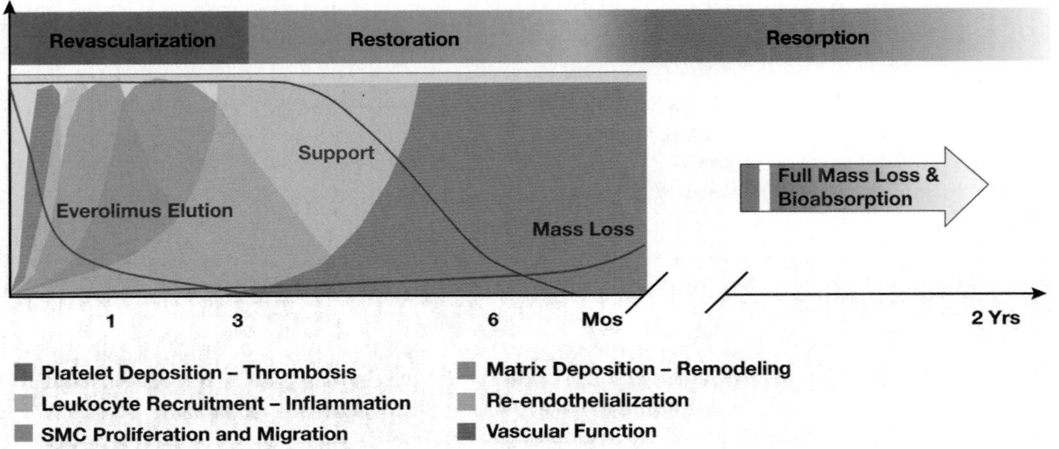

FIGURE D.3 Time progression of stent functionality and physiological responses after implantation of a biodegradable vascular stent (BVS) made of PLLA. (Oberhauser et al., 2009.)

enlargement was associated with reduced plaque burden and vasomotion was restored similar to the native state of a coronary artery. Absence of foreign body reaction and vasomotion restoration are indications of a healthy vessel and suggested that late stent thrombosis risk was eliminated. The above findings need to be confirmed in larger studies that are currently underway (Serruys et al., 2009). These results also suggest that such bioresorbable scaffolds might be used more generally as a device for tissue regeneration, sometimes called "tissue engineering" (see Figure D.3).

The unique progression of scaffold area post-implantation (also called PCI, for percutaneous coronary intervention) obtained by IVUS (intravascular ultrasound) is detailed in the literature (Serruys et al., 2009). When a non-degradable DES is compared to a BVS after two years, the BVS exhibits positive remodeling and full resorption of the scaffold, while the non-degradable implants prevent the lumen from one day exhibiting positive remodeling.

> Early results suggest that as biodegradable vascular stent (BVS) implants fully resorb, the vascular tissue may remodel to the original healthy native state (Nikolsky et al., 2009; Oberhauser et al., 2009; Serruys et al., 2009). It is likely that this will not occur with non-degradable drug-eluding stents (DES).

PUMP-BASED DDS

Introduction

Fully implanted pump DDS offer a number of advantages, both potential and realized. Obviously, the barrier of the GI tract and the first pass effect are avoided. Pump systems offer convenience over repeated syringe injections. Patient compliance is ensured. Release rates can be faster than diffusion-limited systems. If localized therapy is desired, the system can be implanted at the site of action; in some cases the pump can be implanted nearby the site in an easily accessible location, and an attached catheter can be directed to the site of action. Remote adjustment of the delivery rate can be achieved with electronic transmitters attached on the skin, or an implanted sensor can be electronically-coupled to the pump to create feedback-control of drug delivery.

Osmotic Pumps

Osmosis is the movement of water through a semi-permeable membrane in response to high solute (usually ionic) concentration on the opposing side of the membrane. This transport increases the volume of solution on the opposing side of the membrane and can be described by the equation:

$$\frac{dV}{dt} = \frac{A}{l} L_p [\sigma \Delta \pi - \Delta P]$$

where dV/dt is the volume flux of water across the membrane, $\Delta\pi$ and ΔP are, respectively, the osmotic and hydrostatic pressure differences across the semipermeable membrane, L_p is the membrane hydraulic permeability coefficient, σ is the reflection coefficient (usually ~1), and A and l are respectively the membrane area and thickness (Theeuwes and Yum, 1976). Osmotic volume expansion has been utilized as the basis for a number of designs of implanted and ingested DDS.

> In osmotic drug delivery systems, the drug delivery rate is controlled by the diffusion of water into the osmotic device, as opposed to most other DDS, where control of the drug delivery rate is controlled by diffusion of drug out of the device.

DUROS® Implant

A schematic of the DUROS® implantable osmotic system is shown in Figure D.4. It resembles a miniature syringe, and consists of an outer cylindrical reservoir, a water diffusion rate-controlling membrane at one end, the "osmotic engine" (tablets containing primarily NaCl combined with other pharmaceutical excipients), an elastomeric piston, the drug formulation in the drug reservoir, and an exit port at the other end (diffusion moderator/orifice in Figure D.4). The outer reservoir can be

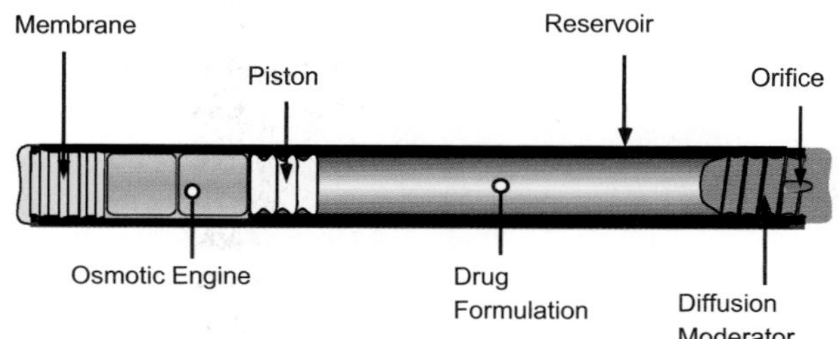

FIGURE D.4 Cross-sectional diagram of the DUROS® Implant. *(Reprinted from Wright, 2001.)*

constructed from titanium alloy or a biocompatible rigid polymer.

The DUROS® osmotic DDS can be implanted subcutaneously (e.g., the inside of the upper arm) or into other body sites. The rate-controlling membrane is composed of a polyurethane copolymer with PEG segments, designed to provide a specific water permeability and the ability not to foul during *in vivo* operation. The drug formulation consists of the drug dissolved or suspended in a pharmaceutically-acceptable solvent or solution (e.g., DMSO). The drug must remain stable in the formulation for the duration of delivery; thus, the formulation components are specific to the properties of the drug being delivered. The device is usually contained in a titanium-based metal cylinder. For site-directed therapy, a catheter can be attached to the exit port and directed to a nearby target site. When implanted subcutaneously, dimensions have ranged from 4 mm OD × 45 mm L to 10 mm OD × 45 mm L, although smaller or larger systems are possible, depending on the site of implantation, and the required drug loading and resultant duration of therapy.

The rate of drug delivery from the system, dM/dt, is given by:

$$\frac{dM}{dt} = \frac{dV}{dt} C$$

where C is the concentration of the drug in the formulation. The system delivers at a constant rate because the osmotic pressure of a saturated solution of sodium chloride (i.e., the solution present in the osmotic engine compartment after a small amount of water is imbibed) is much greater than the osmotic pressure of body tissue. The system can be designed for durations of one month or less, up to one year or even longer. The DUROS® osmotic pump system has been commercialized for the delivery of the GnRH analog leuprolide for the palliative treatment of prostate cancer (Viadur® system). The system is under development for the delivery of interferon (OMEGA DUROS® system) and sufentanil (Chronogesic® system), and site-targeted delivery of other opioids (Wright et al., 2001, 2003; Wright and Culwell, 2008; Yang et al., 2008).

INFUSION PUMPS

Another important example of an IDDS where the drug is delivered subcutaneously is a palm-sized infusion pump for the treatment of insulin-dependent diabetes. In this case, the cannula portion of the total delivery system is implanted, while the pump portion is worn outside the body. This first of many subsequent pumps was first introduced in 1983 by Pacesetter Systems Infusion Division, which in 1985 was officially spun-off as MiniMed, and acquired by Medtronic in 2001. Recent programmable versions integrate the pump with continuous glucose monitoring. The pump is capable of delivering subcutaneous infusions of very minute amounts of insulin with automatic changes in delivery rate or dose at specified intervals. Pumps can also record infusion data history. Some pumps offer a short-syringe option, where the syringe connector is recessed inside the pump for greater patient comfort. Insulin is delivered subcutaneously through a soft cannula that sits under the skin for up to three days after being inserted virtually painlessly. Continuous glucose monitoring is made possible through a tiny glucose sensor worn up to three days at a time. It is also inserted subcutaneously with an automated insertion device. Sensor data is sent to a transmitter attached to the sensor. The transmitter sends the data to the pump through wireless technology (MiniMed Infusion Pumps, 2009; MiniMed Infusion Pump History, 2009; MiniMed Product Information, 2009). The reader is encouraged to review the prolific scientific literature on infusion pumps (e.g., Panteleon et al., 2006).

Positive Displacement Pumps

The Infusaid® pump is a fully implantable, fixed rate pump that contains a chamber for a fluorocarbon propellant separated from the drug formulation chamber by a metal bellows in a disc-shaped titanium housing (Danckwerts and Fassihi, 1991). The system contains a flow restrictor, and the vapor pressure of the propellant creates a constant pressure source at fixed temperature. The pump is approximately 9 cm in diameter by 3 cm in height. This pump has been utilized for insulin delivery, anticoagulant therapy, and cancer chemotherapy.

Peristaltic Pumps

The Synchromed® pump developed by Medtronic, Inc., is an example of a fully implantable peristaltic pump that features external microelectronic control of the delivery rate. It has been utilized for the intrathecal delivery of opioids for pain management, baclofen for the treatment of severe spasticity, and other therapies. The pump consists of an outer titanium casing that houses the pumping mechanism, the controller, the drug solution reservoir, and the battery. There is a silicone rubber septum on the system so that it can be refilled with a needle and syringe. The pump is 8.8 cm in diameter by 2.5 cm in thickness, and is implanted abdominally. A silicone catheter is tunneled to the intrathecal space. Battery life is given as 6–7 years (Krames, 1996; Medtronic, Inc., 2007; Medtronic, Inc., 2009).

Micro-Fabricated Pumps

A number of implantable delivery systems based on micro-electromechanical systems (MEMS) and microelectronic fabrication technologies are being researched or are in development (Gardner, 2006). In one system under development by MicroCHIPS, tiny microliter-sized reservoirs are etched into a silicon wafer and filled with drug or multiple drugs. The reservoirs are covered by gold films (or other films) that are electrically connected on the chip surface. After implantation of the chip, a voltage can be applied to the gold foil of an individual reservoir, with resulting electrochemical dissolution of the film and release of the drug from the reservoir. Controlled, pulsatile or patterned release of single or multiple drugs are possible (Santini et al., 1999).

INSERTS

DDS can be placed or "inserted" into a number of body sites. Local therapy can be achieved (e.g., delivery of drugs into the eye via ocular inserts) or systemic therapy can be achieved via release of the drug and transport across a permeable biological surface into the bloodstream (e.g., sublingual tablets or mucoadhesive buccal tablets). A number of sites or routes that have been utilized for insert therapy, including ocular, transurethral, vaginal, and intrauterine. DDS for several of these routes are discussed below.

Intraocular Inserts

The Ocusert® is a reservoir and membrane system that was developed and commercialized in the 1970s by ALZA Corporation for the treatment of glaucoma. It is an elliptical-shaped planar system that is inserted into the cul-de-sac of the eye. The system consists of an inner drug reservoir containing pilocarpine, and a polyethylene vinyl acetate (PEVA) outer rate-controlling membrane. The system delivers drug at a constant, zero order rate, for one week (Heilmann, 1978).

The Vitrasert® implant was developed and commercialized for the treatment of cytomegalovirus (CMV) retinitis. The system releases ganciclovir. It consists of a compressed tablet of the drug, overcoated with polyvinyl alcohol (PVOH), then partially overcoated with PEVA, and then affixed to a PVOH suture stub. It is designed to release the drug over 5–8 months (Bausch and Lomb, Inc., 2005; Lee et al., 2008). An extensive array of other ocular implants and inserts has been investigated (Bourges et al., 2006).

Vaginal and Intrauterine Inserts

Site-specific delivery to the female reproductive tract has been pursued for treatment of diseases and disorders, and also for contraception. One of the first systems developed for contraception was the Progestasert® (ALZA Corporation). This T-shaped DD device was inserted into the uterus. The body of the system consisted of a rate-controlling cylindrical PEVA reservoir containing progesterone dispersed in silicone oil. The arms of the "T" were also constructed of PEVA. The drug was released by diffusion over a period of one year (Heilmann, 1978).

A more recent intrauterine system is the Mirena® intrauterine system, which delivers levonorgestrol. The system consists of a T-shaped polyethylene frame. The vertical part of the "T" has a reservoir of a mixture of the drug and polydimethylsiloxane. The reservoir is surrounded by a silicone membrane, and has delivery duration of ≥5 years (Bayer Healthcare Pharmaceuticals, Inc., 2009). Additionally, there are several intrauterine systems that release copper (Benagiano et al., 2008).

Vaginal rings are another insertable contraceptive delivery system. The vaginal mucosa is permeable to a number of substances. The NuvaRing® is constructed of PEVA and delivers two drugs: ethynyl estradiol and etonorgestrol. The outer diameter of the ring is 54 mm, and it has a cross-sectional diameter of 4 mm. The ring is inserted by the user after day 4 of the cycle and removed 21 days later (Benagiano et al., 2008).

Vaginal delivery systems are also being investigated for prevention of sexually-transmitted diseases (Loxley et al., 2010). pH-responsive DDS may be of special utility in this application (Gupta et al., 2007).

THE FUTURE

Some of the most pressing problems of therapeutic drug delivery still have not been addressed adequately. Two major, interrelated problems are patient compliance (which could be solved with appropriate DDS) versus the complexity and expense of developing DDS that control the rate, dose, delivery rate, and duration of delivery to a specific site (e.g., del Valle et al., 2009). Major pharmaceutical companies may not be prepared to absorb the costs of developing a new DDS unless reimbursement is negotiated beforehand. This sometimes results in smaller companies having to accept the risk and expenses incurred during the development and approval process of new DDS.

The key drivers for new DDS applications in the future will be for more cost-effective devices with improved efficacy and patient acceptance. Some DDS may include a diagnostic feature with remote feedback control. Future devices will be smaller, less invasive, more site-specific, and more efficient in drug dosing. Expert opinions on this subject can be found in numerous journal articles (Santini et al., 1999; Telleman, 2001; LaVan et al., 2003; Martin and Grove, 2001; Pizzi et al., 2004; Gardner, 2006; McCoy et al., 2007; Hoffman, 2008; del Valle et al., 2009).

Conflict of Interest Statement. One of the authors, Lothar Kleiner, was affiliated at Abbott Vascular until January 2012, where he was working on fully resorbable drug eluting devices that serve as temporary scaffolds for diseased coronary arteries. The content of the section on drug eluting stents was obtained from peer reviewed references, human clinical trials and conference proceedings. He currently consults in the area of drug delivery systems and medical devices. The other author, Jeremy Wright, currently works at Durect, which has rights to selected DUROS® applications.

BIBLIOGRAPHY

Acharya, G., & Park, K. (2006). Mechanisms of controlled drug release from drug-eluting stents. *Advanced Drug Delivery Reviews*, **58**(3), 387–401.

Alam, S., Baldwin, J., Jordan, B., Tombros Korman, A., Rinehart, B., Shields, W. C., & Swann, A. M. (2008). The single-rod contraceptive implant. *Clinical Proceedings from Association of Reproductive Health Professionals*. July, 7–9, 2008.

Bausch and Lomb, Inc. (2005). *Vitrasert Package Insert*. http://www.bausch.com/en_US/package_insert/surgical/vitrasert_pkg_insert.pdf. Date accessed 14/11/09.

Bayer HealthCare Pharmaceuticals, Inc. (2009). *Mirena Package Insert*. http://berlex.bayerhealthcare.com/html/products/pi/Mirena_PI.pdf. Date accessed 14/11/09.

Benagiano, G., Gabelnick, H., & Farris, M. (2008). Contraceptive devices: Intravaginal and intrauterine delivery systems. *Expert Review of Medical Devices*, **5**(5), 639–654.

Bourges, J. L., Bloquel, C., Thomas, A., Froussart, F., Bochot, A., Azan, F., Gurny, R., BenEzra, D., & Behar-Cohen, F. (2006). Intraocular implants for extended drug delivery: Therapeutic applications. *Advanced Drug Delivery Reviews*, **58**(11), 1182–1202.

Chan-Seng, D., Ranganathan, T., Zhang, X., Tang, Y., Lin, Q., Kleiner, L., & Emrick, T. (2009). Aliphatic polyester terpolymers for stent coating and drug elution: Effect of polymer composition on drug solubility and release. *Drug Delivery*, **16**(6), 304–311.

Cooper, S. L. et al. (2004). Classes of materials used in medicine. In Buddy D. Ratner, et al. (Ed.), *Biomaterials Science, An Introduction of Materials in Medicine*, pp. 79, Elsevier, Academic Press, Boston.

Danckwerts, M., & Fassihi, A. (1991). Implantable controlled release drug delivery systems: A review. *Drug Development and Industrial Pharmacy, 1*, **7**(11), 1465–1502.

Dash, A. K., & Cudworth, G. C. (1998). Therapeutic applications of implantable drug delivery systems. *Journal of Pharmacological and Toxicological Methods*, **40**(1), 1–12.

Deanesly, R., & Parkes, A. S. (1937). Biological properties of some new derivatives of testosterone. *Biochemical Journal*, **31**, 1161–1164.

del Valle, E. M.M., Galan, M. A., & Carbonell, R. G. (2009). Drug delivery technologies: The way forward in the new decade. *Industrial and Engineering Chemistry Research*, **48**(5), 2475–2486.

Eisai Co., Ltd. (2009). Gliadel® Wafer (polifeprosan 20 with carmustine implant) Rx only, http://www.eisai.com/pdf_files/gliadel_pi.pdf. Accessed Oct 22, 2009.

Folkman, J., & Long, D. M. (1964). The use of silicone rubber as a carrier for prolonged drug therapy. *Journal of Surgical Research*, March **4**(3), 139–142.

Gardner, P. (2006). Microfabricated nanochannel implantable drug delivery devices: Trends, limitations and possibilities. *Expert Opinion on Drug Delivery*, **3**(4), 479–487.

Gupta, K. M., Barnes, S. R., Tangaro, R. A., Roberts, M. C., Owen, D. H., Katz, D. F., & Kiser, P. F. (2007). Temperature and pH sensitive hydrogels: An approach towards smart semen-triggered vaginal microbicidal vehicles. *Journal of Pharmaceutical Sciences*, **96**(3), 670–681.

Heilmann, K. (1978). *Therapeutic Systems*. Goerg Thieme Publishers, Stuttgart.

Hoffman, A. S. (2008). The origins and evolution of controlled drug delivery systems. *Journal of Controlled Release*, **132**(3), 153–163.

http://www.gliadel.com/. Accessed October 21, 2009.

Hsieh, D. S. T., & Langer, R. (1981). Experimental approaches for achieving both zero-order and modulated controlled release from polymer matrix systems. *Controlled Release of Pesticides and Pharmaceuticals (Proceedings of International Symposium)*, 5–15. 7th Meeting Date 1980.

Implanon (etonogestrel implant) *Physician Insert.* (2006). Roseland, NJ: Organon USA, Inc.

Isenbarger, D. W., & Resar, J. R. (2005). Drug-eluting versus third-generation bare metal stents: The US strategy. *International Journal of Cardiovascular Interventions*, 7(4), 171–175.

Jaffe, H., Giang, P. A., Hayes, D. K., Miller, J. A., & Stroud, B. H. (1981). Implantable systems for delivery of insect growth regulators to livestock. II. *Controlled Release of Pesticides and Pharmaceuticals (Proceedings of International Symposium)*, 303–310. 7th Meeting Date 1980.

Kotwal, V. B., Saifee, M., Inamdar, N., & Bhise, K. (2007). Biodegradable polymers: Which, when and why? *Indian Journal of Pharmaceutical Sciences*, **69**(5), 616–625.

Krames, E. S. (1996). Intraspinal opiod therapy for chronic nonmalignant pain: Current practice and clinical guidelines. *Journal of Pain and Symptom Management*, **11**(6), 333–352.

Kukreja, N., Onuma, Y., & Serruys, P. W. (2009). Future directions of drug-eluting stents. *Journal of Interventional Cardiology*, **22**(Suppl. 1), S96–S105.

LaVan, D. A., McGuire, T., & Langer, R. (2003). Small-scale systems for *in vivo* drug delivery. *Nature Biotechnology*, **21**(10), 1184–1191.

Lee, S. S., Yuan, P., & Robinson, M. R. (2008). Ocular implants for drug delivery. In G. E. Wnek, & G. L. Bowlin (Eds.), *Encyclopedia of Biomaterials and Biomedical Engineering, Second Edition.* (Vol. 3). USA: Informa Healthcare. 1981–1995.

Lewis, D. H. (Ed.), (1981). *Controlled Release of Pesticides and Pharmaceuticals* (pp. 340). Plenum Press, New York.

Lewis, D. H. (1990). Controlled release of bioactive agents from lactide/glycolide polymers. In M. Chasin, & R. Langer (Eds.), *Biodegradable Polymers and Drug Delivery Systems, Drugs and the Pharmaceutical Sciences* (Vol. 45, pp. 1–41). Marcel Dekker.

Loxley, A., McConnell, J., Okoh, O., Morgan, M., Clark, M., Friend, D., & Mitchnick, M. (2010). *Controlled release of antiretroviral drugs from ethylene-vinylacetate intravaginal rings to protect women from HIV transmission.* Portland, OR: Poster presented at 37th Meeting of Controlled Release Society. July 2010. http://www.particlesciences.com/docs/Controlled%20Release%20of%20Antiretroviral%20Drugs%20from%20Ethylenevinylacetate%20Intravaginal%20Rings%20to%20Protect%20Women%20From%20HIV%20Transmission.pdf, accessed March 5, 2011.

Lyu, S. P., LaVan, D. A., McGuire, T., & Langer, R. (2003). Small-scale systems for *in vivo* drug delivery. *Nature Biotechnology*, **21**(10), 1184–1191.

Lyu, S. P., Untereker, D. (2009). Degradability of polymers for implantable biomedical devices. *International Journal of Molecular Sciences*, **10**(9), 4033–4065.

Mani, G., Feldman, M. D., Patel, D., & Agrawal, C. M. (2007). Coronary stents: A materials perspective. *Biomaterials*, **28**(9), 1689–1710.

Martin, F. J., & Grove, C. (2001). Microfabricated drug delivery systems: Concepts to improve clinical benefit. *Biomedical Microdevices*, **3**(2), 97–107.

McCormick, C. L., Anderson, K. W., Pelezo, J. A., & Lichatowich, D. K. (1981). Controlled release of metribuzin, 2,4–D, and model aromatic amines from polysaccharides and poly(vinyl alcohol). *Controlled Release of Pesticides and Pharmaceuticals (Proceedings of International Symposium)*. 7th Meeting Date 1980.

McCoy, C. P., Donnelly, L., Edwards, C. R., Gray, J. L., & Rooney, C. (2007). Triggered release of therapeutics from medical polymers. *Gummi, Fasern, Kunststoffe*, **60**(3), 158–163.

MedMarket Diligence, LLC. (2009). *Drug-Eluting Stents Dominate Coronary Market., For Now.* Sept 9, 1–2.

Medtronic, Inc. (2007). *SYNCHROMED® II Programmable Pumps (Health Professional Instructions).*

Medtronic, Inc. (2009). *SYNCHROMED® II Infusion System Patient Manual.*

MiniMed Infusion Pumps. (2009). http://www.msdonline.com/biomed/meh/MINIMED.HTM, accessed November 24, 2009.

MiniMed Infusion Pump History. (2009). http://www.minimed.com/about/history.html# accessed November 24, 2009

MiniMed Product Information. (2009). http://www.medtronic-diabetes-me.com/CGM-MiniMed-Paradigm-REAL-Time.html, accessed November 24, 2009.

Myschik, J., Mcburney, W. T., Hennessy, T., Phipps-Green, A., Rades, T., & Hook, S. (2008). Immunostimulatory biodegradable implants containing the adjuvant Quil-A – Part II: *In vivo* evaluation. *Journal of Drug Targeting*, **16**(3), 224–232.

Nikolsky, E., Lansky, A. J., Sudhir, K., Doostzadeh, J., Cutlip, D. E., Piana, R., Su, X., White, R., Simonton, C. A., & Stone, G. W. (2009). SPIRIT IV trial design: A large-scale randomized comparison of everolimus-eluting stents and paclitaxel-eluting stents in patients with coronary artery disease. *American Heart Journal*, **158**(4), 520–526. e2.

Oberhauser, J., Hossainy, S., & Rapoza, R. (2009). Design principles and performance of bioresorbable polymeric coronary scaffolds. *EuroIntervention Supplement*, **Vol. 5**(Supplement F). F-15–F22.

Ormiston, J. A., & Serruys, P. W. S. (2009). Bioabsorbable coronary stents. *Circulation: Cardiovascular Interventions*, **2**(3), 255–260.

Panteleon, A. E., Loutseiko, M., Steil, G. M., & Rebrin, K. (2006). Evaluation of the effect of gain on the meal response of an automated closed-loop insulin delivery system. *Diabetes*, **55**(7), 1995–2000.

Perry, A., & Schmidt, R. E. (2006). Cancer therapy-associated CNS neuropathology: An update and review of the literature. *Acta Neuropathologica*, **111**(3), 197–212.

Pizzi, M., De Martiis, O., & Grasso, V. (2004). Fabrication of self assembled micro reservoirs for controlled drug release. *Biomedical Microdevices*, **6**(2), 155–158.

Population Council. (1990, 1995). *Norplant® Implants.* http://www.reproline.jhu.edu/english/1fp/1methods/1ni/ni0a901.ppt.

Santini, J., Cima, M. J., & Langer, R. (1999). A controlled-release microchip. *Nature (London)*, **397**(6717), 335–338.

sNDA # 20–637 GLIADEL® WAFER. http://www.fda.gov/ohrms/dockets/ac/01/slides/3815s2_04_Shapiro.ppt: Dec 6, 2001 and web site accessed on Oct 23, 2009.

Serruys, P. W., Ormiston, J. A., Onuma, Y., Regar, E., Gonzalo, N., Garcia-Garcia, H. M., Nieman, K., Bruining, N., Dorange, C., Miquel-Hebert, K., Veldhof, S., Webster, M., Thuesen, L., & Dudek, D. (2009). A bioabsorbable everolimus-eluting coronary stent system (ABSORB): 2-year outcomes and results from multiple imaging methods. *Lancet*, **373**(9667), 897–910.

Telleman, P. (2001). Micro technology: The future for life sciences. *Probe Microscopy*, **2**(2), 203–211.

Theeuwes, F. (1975). Elementary osmotic pump. *Journal of Pharmaceutical Sciences*, **64**(12), 1987–1991.

Theeuwes, F., & Yum, S. I. (1976). Principles of the design and operation of generic osmotic pumps for the delivery of semisolid or liquid drug formulations. *Annals of Biomedical Engineering*, **4**(4), 343–353.

Vernon, B., & Wegner, M. (2004). Controlled release. In G. E., Wnek, and G. L., Bowlin (Eds.), *Encyclopedia of Biomaterials and Biomedical Engineering 1* (pp. 384–391), Marcel Dekker.

Virmani, R., Finn, A. V., & Kolodgie, F. D. (2009). A review of current devices and a look at new technology: Drug eluting stents. *Expert Review of Medical Devices*, **6**(1), 33–42.

Wheeler, R. G., & Friel, P. G. (1981). Release of drugs from IUDs using an ethylene vinyl acetate matrix. *Controlled Release of Pesticides and Pharmaceuticals (Proceedings of International Symposium)*, 111–124. 7th Meeting Date 1980.

Wright, J. C., & Culwell, J. (2008). Long-term controlled delivery of therapeutic agents by the osmotically driven DUROS® Implant. In M. J. Rathbone, J. Hadgraft, M. S. Roberts, & M. E. Lane (Eds.), *Modifed-Release Drug Delivery Technology* (2nd ed.). *Informa. Healthcare* (Vol. 2, pp. 143–149).

Wright, J. C., Leonard, S. T., Stevenson, C. L., Beck, J. C., Chen, G., Jao, R. M., et al. (2001). An *in vivo/in vitro* comparison with a leuprolide osmotic implant for the treatment of prostate cancer. *Journal of Controlled Release*, **75**, 1–10.

Wright, J. C., Johnson, R. M., & Yum, S. I. (2003). DUROS® osmotic pharmaceutical systems for parenteral and site-directed therapy. *Drug Delivery Technology*, **3**, 3–11.

Yang, B., Rohloff, C., Mercer, R., Horwege, K., Negulescu, C., Lautenbach, S., et al. (2008). Continuous delivery of stabilized proteins and peptides at consistent rates for at least 3 months from the DUROS® device. *American Association of Pharmaceutical Scientists Annual Meeting, T3150.*

Zoladex® IFU (Instructions for use). (2003). December.

E. SMART DDS

Allan S. Hoffman
Professor of Bioengineering and Chemical Engineering, UWEB Investigator, University of Washington, Seattle, WA, USA

ENVIRONMENTALLY-RESPONSIVE SYSTEMS

This section is focused only on "smart" drug delivery systems, and is a companion section to Chapter I.2.11, "Applications of "Smart Polymers" as Biomaterials."

Temperature-Responsive, Phase-Separating Polymer Systems

A well-known polymer that undergoes a sharp, thermally-induced phase separation is poly(*N*-isopropylacrylamide) (PNIPAAm) (Heskins and Guillet, 1968; Schild, 1992). Its transition at 32°C can be shifted to higher or lower temperatures by copolymerization with more hydrophilic or hydrophobic comonomers, respectively. Thus, this material could be of interest in drug delivery applications; however, since acrylamide monomer is toxic, many worry about the potential cell toxicity of NIPAAm monomer, and by extension, of the polymer itself, although this issue has not been thoroughly investigated.

Also of interest as thermally-responsive, nanocarrier drug delivery systems (DDS) are the ABA tri-block copolymers of PEO-PPO-PEO, where either PEG or PEO symbolizes poly(ethylene glycol), and either PPG or PPO symbolizes poly(propylene glycol). These polymers are known as Pluronic® or Poloxamer® polyols. They are discussed further in the sub-section on polymeric micelle drug nanocarriers in Section B above.

One of the more recent trends in thermogelling materials is the development of biodegradable systems suitable as injectable and gelling depot drug delivery systems. Early work on degradable PEO-polyester block copolymers was carried out by Gilding and Reed (1979), with polyethylene terephthalate (PET) as the polyester. Interesting degradable biomaterials can be constructed from ABA triblock copolymers based on poly(ethylene glycol) and poly(lactic-*co*-glycolic acid) copolymers. Sung Wan Kim and co-workers have pioneered with degradable triblock polymers as DDS (Jeong et al., 1997; Shim et al., 2007). The triblock may be designed as PLGA-PEO-PLGA or as PEO-PLGA-PEO, where the former will have a more pronounced thermal gelation capability. PLA and PCL have also been used as the hydrophobic, degradable block segment. When the block lengths and relative amounts are correctly chosen, a material is obtained that is soluble in water at room temperature and forms a firm gel at body temperature (Jeong et al., 1999a; Rohr et al., 2002). Drugs may be dissolved or dispersed, and the resulting mixtures can be injected as a solution or a dispersion having a viscosity not much different from a saline solution. In addition, since only water is used with this material, there are no organic solvent–water interface problems, and protein or nucleic acid drugs can be incorporated without loss of activity. Sometimes these materials are available as dry powders that may be reconstituted prior to use. While the exact mechanism of thermogelling is not known with absolute certainty, it has been postulated that the hydrophobic blocks (PLGA or PLA) of the amphiphilic block copolymers aggregate together to form micelles that are in equilibrium with monomeric polymer chains (Jeong et al., 1999b, 2000). Then, as the temperature is increased, the equilibrium shifts to micelle formation, and above the critical gelation temperature the micelles pack together, resulting in gel formation. Evidence for such a mechanism is provided by ^{13}C NMR studies, dye solubilization studies, and light-scattering studies, which all show an abrupt change in micellar diameter and aggregation at the critical gelation temperature (Jeong et al., 1999b). It has also been postulated that the water content of the gel will determine its degradation rate, and the related release kinetics of the incorporated therapeutic agent (Jeong et al., 1999b).

Kam Leong and Jun Li have developed some very interesting degradable triblock polymer DDS using degradable blocks of poly(hydroxybutyrate) (PHB) as the hydrophobic block in a PEO-PHB-PEO triblock that is further condensed with cyclodextrin (CD) to form a poly(pseudorotaxane) hydrogel DDS (Li et al., 2001, 2003, 2006).

Solvent-Responsive, Phase-Separating Polymer Depot DDS

Another means of developing *in situ* forming drug delivery systems is polymer precipitation in a poor solvent such as aqueous body fluids. Such an injectable delivery system is prepared from a water-insoluble biodegradable polymer dissolved in a water-miscible, biocompatible solvent. When this system is mixed with a drug and injected into the physiologic environment, water diffuses in to dilute the solvent, which also diffuses out into the surrounding tissues, and the polymer precipitates, forming a solid polymeric implant containing entrapped drug. This method has been used in human (Dunn et al., 1990) and in veterinary applications (Dunn et al., 1995). The polymer that has been most commonly used is a poly(lactic-*co*-glycolic acid) copolymer dissolved in *N*-methylpyrrolidone (NMP). This system is known as Atrigel®. Three formulations of PLGA dissolved in *N*-methyl-pyrrolidone and containing leuprolide acetate, have been used for the treatment of prostate cancer.

There are a number of other similar systems, such as: (1) poly(lactic acid) dissolved in a mixture of benzyl benzoate and benzyl alcohol, known as PLAD (Duenas et al., 2001); (2) poly(lactic-*co*-glycolic acid) copolymer dissolved in glycofurol (Eliaz and Kost, 2001); (3) poly(lactic-*co*-glycolic acid) copolymer dissolved in benzyl benzoate, known as Alzamer®; and (4) sucrose acetate isobutyrate dissolved in NMP or Miglyol, known as SABER™ (Tipton and Holl, 1998).

Temperature- and pH-Responsive Hydrogel DDS

In this smart DDS, drugs are imbibed into lightly cross-linked gels which release the drugs as the gel shrinks in response to changes in temperature or pH. Cross-linked poly(*N*-isopropylacrylamide) or PNIPAAm, is a typical temperature-responsive gel that releases a drug upon collapse at the LCST (lower critical solution temperature) (Hoffman et al., 1986; Dong and Hoffman, 1987; Hoffman, 1987; Okano et al., 1990). Such materials can deliver biological molecules when triggered by temperature changes, but this is not very practical due to the isothermal condition in the body. Furthermore, such systems are not very effective, since the drug may be gradually released by slow diffusion from the hydrogel before the stimulus causes it to release in a burst. However, one unique way to utilize such a thermally-responsive hydrogel in the body was developed by Dong and Hoffman (1991). They prepared a combined pH- and temperature-sensitive hydrogel matrix, based on a random copolymer of NIPAAm and acrylic acid (AAc), and it was shown to release indomethacin, a model NSAID (non-steroidal anti-inflammatory drug), linearly over a four hour period as the pH went from the low, acidic pH found in the gastric region to the physiologic pH in the enteric region of the GI tract. At 37°C body temperature the NIPAAm component was above its LCST, and was trying to maintain the gel in the collapsed state, while as the pH went from acidic (e.g., as it would be in the gastric region) to neutral conditions (e.g., as it would be in the enteric region), the AAc component was becoming ionized, forcing the gel to gradually swell and slowly release the drug. This is similar to so-called enteric coatings, but in this case the hydrophobic component was thermally-responsive, and the formulation matrix was composed entirely of the copolymer, rather than just being coated.

Temperature-responsive smart polymer hydrogels might also be useful for topical delivery to open wounds, skin, and mucosal surfaces such as the eye or nose. The temperatures of such surfaces may be a few degrees below 37°C, but they are still well above ambient temperature, and that difference could be utilized to deliver a drug from a thermally-sensitive polymer formulated with the drug and applied to the skin or mucosal surface.

REFERENCES

Dong, L. C., & Hoffman, A. S. (1987). Thermally reversible hydrogels: swelling characteristics and activities of copoly(NIPAAm-AAm) gels containing immobilized asparaginase. In P. Russo (Ed.), *Reversible Polymeric Gels and Related Systems ACS Symposium Series* (Vol 350, pp. 236–244). Washington, D.C: ACS.

Dong, L. C., & Hoffman, A. S. (1991). A novel approach for preparation of pH- and temperature-sensitive hydrogels for enteric drug delivery. *J. Contr. Rel.*, **15**, 141–152.

Duenas, E., Okumu, F., Daugherty, A., & Cleland, J. (2001). Sustained delivery of rhVEGF from a novel injectable liquid. *Proc. Int. Symp. Control. Rel. Bioact. Mater.*, **28**, 1014–1015.

Dunn, R. L., Hardee, G., Polson, A., Bennett, Martin, S., Wardley, R., Moseley, W., Krinick, N., Foster, T., Frank, K., & Cox, S. (1995). In-situ forming biodegradable implants for controlled release veterinary applications. *Proc. Int. Symp. Control. Rel. Bioct. Mater.*, **22**, 91–92.

Dunn, R. L., English, J. P., Cowsar, D. R., & Vanderbelt, D. P. (1990). *U.S. Patent No. 4,938,763*.

Eliaz, R., & Kost, J. (2001). *U.S. Patent No. 6,206,921 B1*.

Gilding, D. K., & Reed, A. M. (1979). Biodegradable polymers for use in surgery–poly(ethylene oxide)-poly(ethylene terephthalate) (PEO/PET) copolymers: I, 20, 1454–1458.

Heskins, H., & Guillet, J. E. (1968). Solution properties or poly(*N*-isopropyl acrylamide). *J. Macromol. Sci. Chem. A2*, **6**, 1209.

Hoffman, A. S. (1987). Applications of thermally reversible polymers and hydrogels in therapeutics and diagnostics. *J. Controlled Release*, **6**, 297–305.

Hoffman, A. S., Afrassiabi, A., & Dong, L. C. (1986). Thermally reversible hydrogels: II. Delivery and selective removal of substances in aqueous solutions. *J. Controlled Release*, **4**, 213–222.

Jeong, B., Bae, Y. H., Lee, D. S., & Kim, S. W. (1997). Biodegradable block copolymers as injectable drug-delivery systems. *Nature*, **388**, 860–862.

Jeong, B., Choi, Y. K., Bae, Y. H., Zentner, G., & Kim, S. W. (1999a). New biodegradable polymers for injectable drug delivery systems. *J. Controlled Release*, **62**, 109–114.

Jeong, B., Bae, Y. H., & Kim, S. W. (1999b). Biodegradable thermosensitive micelles of PEG-PLGA-PEG triblock copolymers. *Colloids Surf. B: Interfaces*, **16**, 185–193.

Jeong, B., Kibbey, M. R., Birnhaum, J. C., Won, Y. -Y., & Gutowska, A. (2000). Thermogelling biodegradable polymers with hydrophilic backbones: PEG-g-PLGA. *Macromolecules*, **33**, 8317–8322.

Li, J., Li, X., Zhou, Z., Ni, X., & Leong, K. W. (2001). Formation of supramolecular hydrogels induced by inclusion complexation between pluronics and α-cyclodextrin. *Macromolecules*, **34**, 7236–7237.

Li, J., Li, X., Zhou, Z., Ni, X., Wang, X., Li, H., & Leong, K. W. (2006). Self-assembled supramolecular hydrogels formed by biodegradable PEO–PHB–PEO triblock copolymers and α-cyclodextrin for controlled drug delivery. *Biomaterials*, **27**, 4132–4140.

Li, J., Ni, X., Zhou, Z., & Leong, K. W. (2003). Preparation and characterization of polypseudorotaxanes based on block-selected inclusion complexation between poly(propylene oxide)-poly(ethylene oxide)-poly(propylene oxide) triblock copolymers and α-cyclodextrin. *J. Am. Chem. Soc.*, **125**, 1788–1795.

Okano, T., Bae, Y. H., Jacobs, H., & Kim, S. W. (1990). Thermally on-off switching polymers for drug permeation and release. *J. Contr. Rel.*, **11**, 255–265.

Rohr, U. D., Markmann, S., Oberhoff, C., Janat, M. M., Schindler, A. E., McRea, J. C., & Zentner, G. (2002). Oncogel: a technological breakthrough in local treatment of paclitaxel sensitive cancers. *Proc. Int. Symp. Controlled Release Biaoct. Mater.*, **29**, 330–334.

Schild, H. G. (1992). Poly(N-isopropylacrylamide): experiment, theory, and application. *Prog. Polym. Sci.*, **17**, 163–249.

Shim, W. S., Kim, J. H., Kim, K., Kim, Y. S., Park, R. W., Kim, I. S., Kwon, I. C., & Lee, D. S. (2007). pH- and temperature-sensitive, injectable, biodegradable block copolymer hydrogels as carriers for paclitaxel. *Intl. J. Pharmaceutics*, **331**, 11–18.

Tipton, A. J., & Holl, R. J. (1998). *U.S. Patent No. 5,747,058*.

F. TRANSDERMAL DDS

Gary Cleary
Corium Inc. Menlo Park, CA, USA

INTRODUCTION

In this section we describe the technologies and materials that are being used in "Passive" and "Active" Transdermal Drug Delivery Systems (TDDS), as well as the biological barriers to drug delivery through the skin. Passive TDDS refers to delivery of drugs by diffusion across the stratum corneum (SC), while Active TDDS includes physical methods that disrupt or bypass the SC for such delivery. From 1980 to 2010 only 20 small drugs, ranging in molecular weight from 62 Da to 359 Da, were approved by the US Food and Drug Administration (FDA) for Passive TDDS.

1. It has been proposed that the upper limit of molecular weight is below 500 Da for reasonable drug diffusion rates through the skin.
2. Thus, Passive TDDS methods are mainly applicable for delivery of small, potent, lipophilic drugs.

> **Passive Transdermal Drug Delivery Systems** are systems that depend on molecular diffusion of drugs across the stratum corneum. These systems are most suitable for transdermal delivery of small, potent, mostly lipophilic drugs from skin patches.
>
> **Active Transdermal Drug Delivery Systems** are systems that utilize some form of mechanical or electromagnetic energy to create micropores or temporary pathways in the stratum corneum for delivery of large biomolecular drugs.

With the advent of molecular biology and biotechnology in the 1980s, it became apparent that the large biomolecular drugs such as proteins, peptides, and nucleic acids would not penetrate intact SC (which is a reflection of the protective behavior of the SC). From the 1990s to the present time, a number of enabling technologies have been developed that use various forms of energy to disrupt the SC and allow passage of larger drugs into the body. These new approaches are called "Active" TDDS, and they can deliver drug molecules greater than 500 Da across the skin (Barry, 2001; Prausnitz et al., 2004; Prausnitz and Langer, 2008).

Skin and the Stratum Corneum (SC)

The top layer of the epidermis (Figure F.1) is the stratum corneum (SC), which is the outermost layer of skin. The SC is packed with dying or dead cells called corneocytes. They have a "brick and mortar-like" structure, and the mortar in the intercellular space is made of lamellar lipid bilayers surrounding the aqueous cores of the dying corneocytes. There are four potential routes through the SC that drug molecules can take as they migrate through the skin and into the systemic circulation. These routes include: (1) molecular diffusion through a tortuous pathway of the intercellular lipophilic space between the corneocytes; (2) electrophoretic transport of charged drugs, which may be drawn through the hair follicles and sweat ducts; (3) various physical methods that can form transient pores in the SC, opening a path directly through the SC for larger biomolecular drugs; and (4) diffusion through holes in the SC, punctured by microneedles (Figure F.2). This section will describe these four methods in more detail below.

The intercellular bilayers in the SC provide alternating regions of interconnected lipid and aqueous layers forming tortuous pathways between the corneocytes. The lipids found in the lipid bilayer domain are free fatty acids, cholesterol, cholesterol sulfate, and ceramides. The corneocyte's outer cell layer of cross-linked protein and intracellular keratin fibers provide a chemical resistant layer and strength as a mechanical barrier to injury. Just below the stratum corneum is the viable epidermis, which is made up of hydrophilic cells called keratinocytes and specialty cells such as langerhans cells (LCs). The LCs play an important role in the immune system, and their location in the viable epidermis is important when vaccines are delivered by microneedles. The outer two layers of skin, the stratum corneum and the viable epidermis, do not contain the sensory nerve endings or the blood vessels that can cause pain and bleeding (Bodde et al., 1989; Barry, 2001;

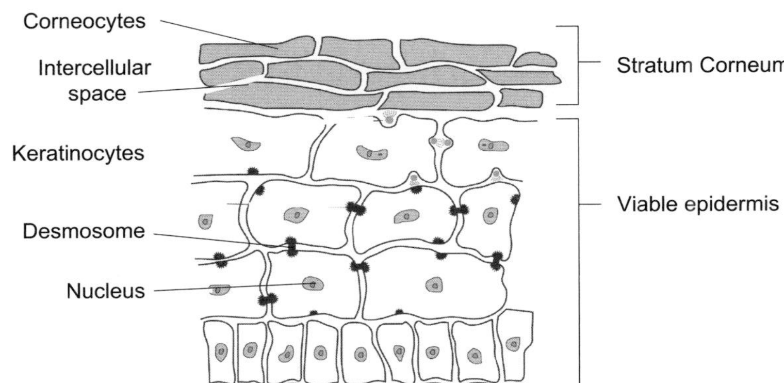

FIGURE F.1 A schematic of the cross section of the stratum corneum (SC, the main barrier in skin) and the viable epidermis beneath it, showing their major components. (Bodde, Verhoeven and van Driel, 1989.)

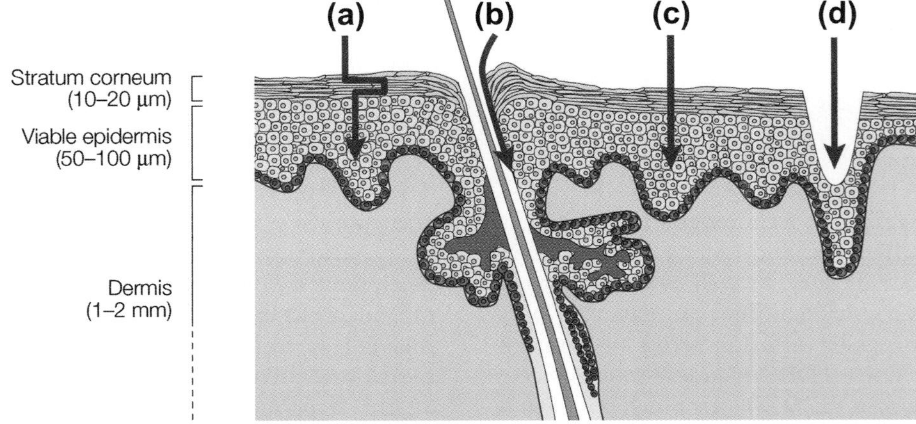

FIGURE F.2 The diagram illustrates four potential entry points through the main skin barrier, the stratum corneum: (a) tortuous lipophilic pathways between the corneocytes cells; (b) pathways through and around the hair follicles and sweat ducts; (c) transient pathways created by high electrical voltage pulses (electroporation); and (d) micro-pore pathways through the stratum corneum created by mechanical (microneedles) or electromagnetic (thermal ablation) methods. (Prausnitz, Mitragotri and Langer, 2004.)

Prausnitz and Langer, 2008). Below the viable epidermis layers is the dermis, where the capillary blood vessel beds, dendritic cells, and neural cells are located (Bodde et al., 1989; Cleary, 1993; Naik et al., 2000; Menon, 2002). The locations of these cells and the capillary bed are important to consider when delivering vaccines to the lymphatic system, and to minimize or eliminate bleeding and pain when using Active TDDS technologies.

The **Stratum Corneum** consists of the outer cell layers of skin with a composite of the corneocytes (approximately 10–15 μ thick), with intracellular pathways made of lipid bilayers. The stratum corneum layer is the main barrier to skin permeation.

PASSIVE TRANSDERMAL DELIVERY SYSTEMS (PASSIVE TDDS)

There are various types of topical ointments, lotions, creams, and sprays for applying and delivering drugs passively to the skin. These methods are usually limited to drug molecules with broad therapeutic windows (e.g., testosterone, estradiol). More potent drugs with narrow therapeutic windows are not suitable for such topical applications. With these drugs it is necessary to have a patch with a fixed surface area (footprint), and a controlled release rate.

The challenge with Passive TDDS is to formulate a patch that achieves a therapeutic blood level fairly rapidly, and that also can adhere to the skin without irritation for from one day to a week. Many different components go into a successful TDDS patch, including pressure sensitive adhesives, penetration and diffusion enhancers, solvents to improve drug solubility and partitioning into the SC, barrier membranes, and rate-controlling membranes (Menon, 2002). The patches that have reached the US marketplace are in the form of laminate film structures that consists of layers that adhere to one another within the delivery system, and ultimately adhere to the skin for a day to a week. The drug molecule and other components of this delivery system should not be irritating or cause the patch to fall off from the skin prematurely or impact on drug stability. Selection of biomaterials for Passive TDDS needs to consider the various conditions that the patch will face during manufacturing, storage, and wear by the user for up to several days (Cleary, 1984; Hadgraft and Lane, 2006). The components must meet several tests for biocompatibility, irritation, sensitization, drug and adhesive stability, human wear time, the amount and duration of drug delivery, and other quality testing required by the regulatory agencies.

Components and Biomaterials used in Passive TDDS Patches

Types and Structures of the Skin Patch. This section will focus on the more predominant structured laminate systems found in the marketed Passive TDDS patches. The typical patch is comprised of 2 or 5 laminated layers with a total thickness from around 100 to 500 μ (not including the release liner), and areas ranging typically from 2 to 40 cm^2. There are three basic laminate structure designs used in passive transdermal systems, shown in Figure F.3 (Cleary, 1984). Type (a) is a single adhesive plus drug reservoir/skin contact layer design. It has no rate-controlling membrane (RCM), and the SC is the rate-limiting barrier for the diffusing drug. It is often called a matrix or drug-in-adhesive patch. The type (b) design has a backing layer, a viscoelastic or gel-like reservoir, and a middle layer that can serve as a rate-controlling and/or structural membrane, and an adhesive layer. The middle layer is sometimes a non-woven film that serves as a structural support or a very porous structure that can prevent or slow down the adhesive cold flow. The third design, Figure F.3c, is similar to type (b), but it can have a liquid drug reservoir rather than a viscoelastic adhesive reservoir. The liquid reservoir patch is sometimes referred to as a liquid "fill and seal" patch, and may also be called a "liquid fill patch". Both types (b) and (c) are rate-controlled membrane "reservoir patches" (Cleary, 1984, 1991; Van Buskirk et al., 1997). The reservoir in type (c) is essentially a drug that is formulated as liquid, then sealed between two layers of film, one of which is a rate-controlling membrane. Types (b) and (c) can deliver drug at a rate that is constant (zero order). Each of these three designs utilizes polymeric films that are laminated together by either pressure sensitive adhesives (PSA) or by structural adhesives that make a permanent heat-seal bond by melting the adhesive between two non-adhesive polymer films (Cleary, 1984, 1991; Ventrakaman and Gale, 1998). The drug and excipients are typically formulated together in the drug-adhesive layers in these patches.

A **Rate-Controlling Membrane** is a membrane (either contiguous or porous) that provides a specific rate of drug diffusion from a device such as a skin patch. It is "rate-controlling" when it controls the rate at a lower rate than that across the stratum corneum, so that the control of drug delivery is built into the patch.

Backing Layer. This layer functions as the primary structural element of the device, it provides the device with much of its flexibility and drape (how the film lays on the curvature of the skin), and skin occlusivity (the degree of breathability of the backing film). Examples of materials useful for the backing layer can be occlusive films such as polyethylene terephthalate (PET), high

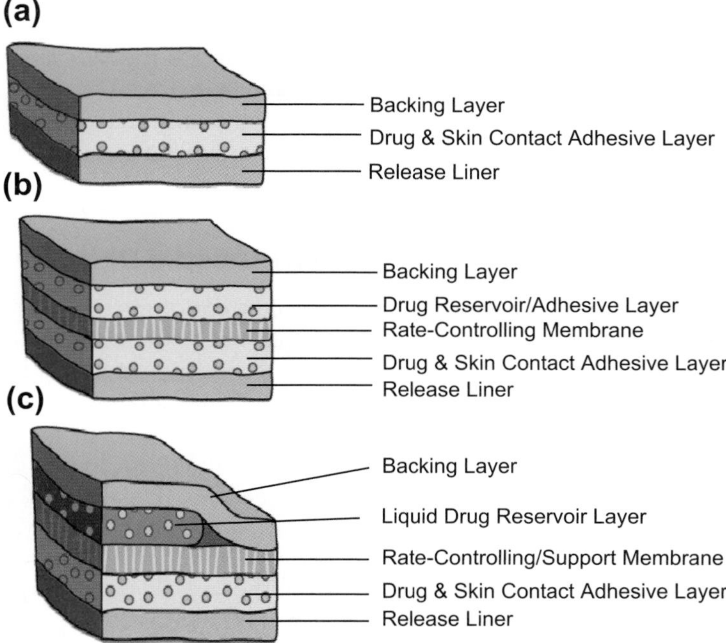

FIGURE F.3 Three common passive transdermal drug delivery systems' designs and structures include (a) a three layer patch: backing layer, single adhesive reservoir/skin contact layer, and release liner; (b) a five layer patch: backing layer, adhesive reservoir, a laminate structure as rate-controlling membrane, drug and skin contact adhesive layer, and release liner; (c) a five layer patch: backing layer, a sealed liquid reservoir, a rate-controlling membrane, drug and skin contact adhesive layer, and a release liner. (Van Buskirk et al., 1997; Cleary, 1984, 1991.)

density polyethylene (HDPE), and polypropylene (PP) or more breathable or better drape films, such as polyurethanes (PU) and poly(ether-amide)s (PEA).

Reservoir Layer. In the adhesive reservoir design the reservoir layer is comprised of a polymeric pressure-sensitive adhesive that contains the drug, solubilizers, and enhancers. The liquid reservoir system will have a liquid or hydrogel medium (such as a hydrogel or ethanolic/water gel using carboxymethyl cellulose or hydroxymethyl cellulose) contained in a sealed pouch-like area where one side is an impermeable backing layer, and sealed on the other side is a rate-controlling membrane. This layer may also contain adhesive in order to adhere the rate-controlling layer to the backing layer.

Penetration Enhancers. Penetration enhancers are key components that enable the drug to diffuse from the system and through the skin at higher flux rates. There have been literally thousands of skin permeation enhancers that have been tested and published in the literature and patents. However, there are only a small number of enhancers that have been approved by the regulatory agencies. These enhancers must be non-irritating and non-sensitizing to the skin, and should not interact with other components of the formulation (Bodde et al., 1989; Barry, 2004; Hadgraft and Lane, 2006).

Enhancers make it feasible for a drug molecule to permeate through the SC when the drug by itself does not have ideal physicochemical properties. Barry (2004) suggests that chemical penetration enhancers act through three mechanisms: (1) disruption of the tortuous intercellular lipid pathway between the corneocytes (examples of this type of enhancer are alcohols, glycols, fatty acids, and fatty acid esters); (2) opening up the dense keratin protein structure making it more permeable (examples of these enhancers are ionic surfactants such as decylmethyl sulfoxide and dimethyl sulfoxide); and (3) increasing the partitioning of the drug into the SC (an example of this type of enhancer is ethanol, which increases the SC penetration of nitroglycerin and estradiol) (Barry, 2004; Songkro, 2009).

Skin Contact Adhesive Layer. This layer is adhered to the release liner prior to use. Once the release liner is removed, it adheres to the skin site of application. The skin contact adhesive is responsible for maintaining the system at the same site for anywhere from one to seven days. Pressure-sensitive adhesives are sticky (tacky) to the slight touch of a finger, and have a range of cold-flow characteristics. Usually a very light touch of a finger is sufficient to adhere them to a surface such as skin, while they still may be removed easily from the skin without leaving any residue on its surface. Pressure-sensitive adhesives are based on three families of adhesives – polyacrylates (acrylates), polyisobutylene (PIB) or polydimethylsiloxane (silicone). Drugs usually have low diffusivity in the PIB and acrylate polymers, while silicone has a higher order of magnitude of diffusivity depending on the drug lipophilicity. These effects are related to the large "free volume" within the adhesive molecular structure of silicones (Cleary, 1984; Ventrakaman and Gale, 1998; Kandavilli et al., 2002).

Release Liner Layer. This layer is a disposable layer which serves only to protect the device prior to application. It is removed prior to application of the device to the skin. The liner should allow easy handling, be user-friendly, and endure the manufacturing processes. The release liner is impermeable to the drug, liquid vehicles or enhancers that could migrate into it while still in the package. Typical examples are films of PET, HDPE, PP, and polycarbonate (PC) ranging in thickness from 50 to 125 microns. Release liners are often coated with a very thin layer of 0.5 nm to several microns of silicone or perfluoroether, to facilitate ease of removal of the liner before applying the patch to the skin.

ACTIVE TRANSDERMAL DELIVERY SYSTEMS (ACTIVE TDDS)

In this section we describe two technologies in depth, microneedle devices and iontophoresis devices. They are examples of mechanical- and electromagnetic-based TDDS. In general, Active TDDS utilize some type of energy to form a transient or a physical micropore through the SC. (One exception is iontophoresis, which utilizes natural, existing skin pores such as hair follicles and sweat glands.) The sources of energy used may be: (1) mechanical, such as with microneedles, microabrasion, solid rods, micro-liquid jets (used at moderate speeds), and micro-solid jets (used at ballistic speeds); (2) electromagnetic, such as with iontophoresis, electroporation, and sonophoresis; and (3) thermal, which uses electric current, radio frequency current or coherent laser light to ablate the SC. Table F.1 describes these eleven different types of Active TDDS technologies. Iontophoresis is the only Active TDDS that has reached the marketplace, while the others are still in different stages of development.

Microneedle Delivery Technology

The concept of microneedle patches was patented nearly 40 years ago (Gerstel and Place, US Patent, 1976). However, technologies that were needed to fabricate microneedles economically were not available until the 1990s. There are a number of different strategies that are being explored today to deliver drugs through the skin using microneedles (Higaki et al., 2003; Cross and Roberts, 2004; Banga, 2009; Cleary, 2009). Figure F.4 illustrates the various compositions and designs of microneedles from the late-1990s to the present day. In the 1990s they were initially made of glass or metals such as stainless steel and titanium (Prausnitz, 2004; Cleary, 2009). In the early 2000s, newer methods of microfabrication, such as heat embossing or microinjection molding, have been used to make polymeric microneedles from polysulfone

TABLE F.1 Active Transdermal Drug Delivery Systems

Active Transdermal Delivery Systems		Microporation Mechanisms Through the Stratum Corneum	References
Mechanically-Based Energy	Microneedles	Micropores formed using mechanical force and velocity of an array of microstructures ranging from 100 to 1000 microns long and 40 to 200 microns wide.	Prausnitz, 2004; Gill et al., 2006; Banga, 2009; Cosman et al., 2010; Kalluri and Banga, 2010; Sullivan et al., 2010; Wendorf et al., 2010
	Microabrasion	Removes > several mm^2 of stratum corneum by abrasion, using sand paper or inert sharp-edged microparticles flowing in a gas at a moderate velocity, where skin is masked forming a 200 micron diameter micropore.	Glenn et al., 2007; Gowrishankar et al., 2009
	Solid formulated rod	A solid millimeter-sized rod, formulated with a drug, is inserted into the stratum corneum at a moderate force and velocity.	Bennett and Potter, 2006
	Micro-solid jets	A range of 20 to 100 micron-sized solid particles are inserted into and through the stratum corneum at supersonic speeds using compressed gas from a hand-held delivery device.	Burkoth et al., 1999; Higaki et al., 2003; Cross and Roberts, 2004; Dean and Chen, 2004
	Micro-liquid jets	A large number of pulsed liquid nl-sized droplets are pulsed through the stratum corneum at a velocity greater than 100 m/s.	Arora et al., 2007
Electromagnetically-Based Energy	Iontophoresis	Charged drug molecules migrate through hair follicles and sweat glands along an ion-based pathway using low level milliamp/mm^2 electric current density.	Panchagnula and Pillai, 2000; Higaki et al., 2003; Cross and Roberts, 2004; Kalia et al., 2004; Sathyan et al., 2005; Subramony et al., 2006; Power, 2007
	Electroporation	Short microsecond pulses of high voltage AC (>100 V) creates micropores through the stratum corneum.	Higaki et al., 2003; Cross and Roberts, 2004; Denet et al., 2004; Gowrishankar et al., 2009
	Sonophoresis	Ultrasonic energy with frequencies less than 100 kHz initiates cavitation that temporarily disrupts the lipid structure of the stratum corneum, providing increased permeability.	Higaki et al., 2003; Cross and Roberts, 2004; Mitragotri and Kost, 2004
	Electric current	A brief surge of electric current (<60 Hz) occurs with ohmic resistance followed by a rapid increase in heat in microelectrodes. An array of micropores is then created from ablation of the stratum corneum to form micropores.	Badkar et al., 2007; Banga, 2009
	Radio frequency current	Thermal ablation occurs using a high frequency alternating current (~100 kHz) that heats an array of of 100 microelectodes/cm^2 to create micropores.	Sintov et al., 2003; Banga, 2009
	Laser	An erbium:YAG laser emits light at 2.94 μm (H_2O absorption peak) that focuses only on water and not on surrounding tissue. A laser light scanner causes water to vaporize within the stratum corneum to form an array of micropores.	Higaki et al., 2003; Cross and Roberts, 2004; Kalia et al., 2008; Lee et al., 2008; Banga, 2009

(PS), poly(methyl methacrylate) (PMMA), and PC. Figure F.4d shows some newer materials that are biodegradable or dissolvable. Microneedles have been fabricated from degradable biomaterials such as poly(lactic-co-glycolic acid) (PLGA), polyglycolic acid (PGA), and polycaprolactone (PCL) (Figures F.4 and F.5) (Prausnitz, 2004; Cleary, 2009; Kalluri and Banga, 2010; Sullivan et al., 2010).

Solid microneedles can be used or applied to the skin in the following ways: (1) as a pretreatment modality where the microneedles are inserted and removed, with the subsequent application of a drug formulation that diffuses through the pores; (2) by first coating the microneedles with the drug, which dissolves off the microneedle tips as they are inserted (in this case, the patch is applied to the skin with a special applicator device, e.g., Zosano®, 3M); (3) by fabricating the needles from a biodegradable drug-containing polymer (in this case, the patch is inserted in the skin with a separate or integrated

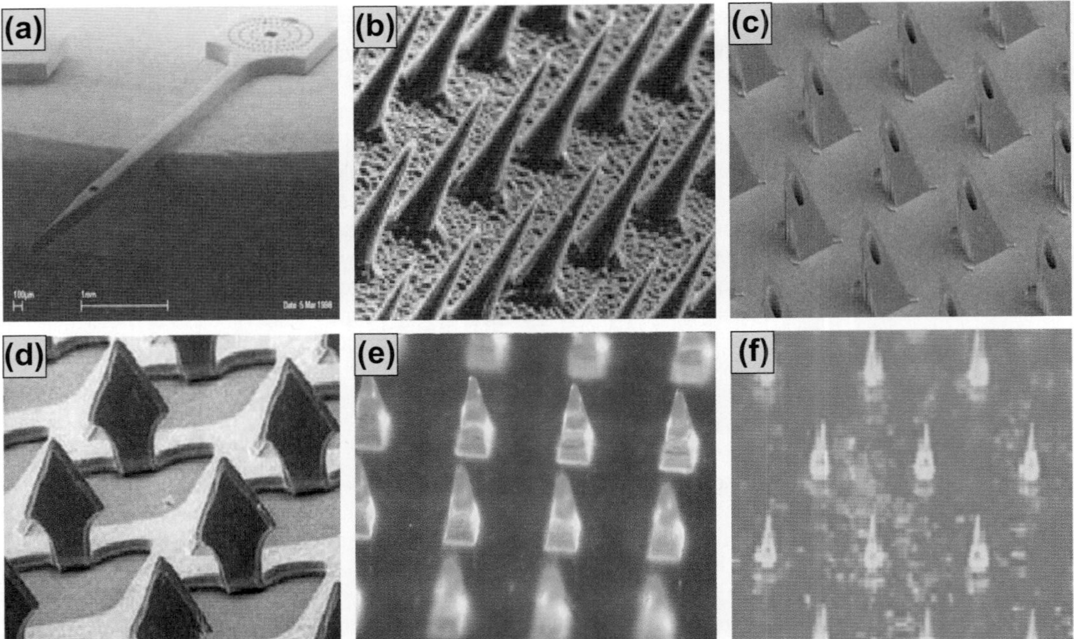

FIGURE F.4 Examples of biomaterials used to make microneedles that are in-plane designs (a) silicon with hollow structure (University of California, Berkeley); and solid out-of-plane designs made of (b) silicon, solid (Georgia Tech); (c) silicon, hollow (University of Twente); (d) titanium, metal, (Zosano); (e) dissolvable and non-degradable polymers, polycarbonate (3M); and (f) biodegradable polymers (Corium). Non-degradable structures (metal, silicon, polycarbonate) are typically coated at the tip of the structure (e) (red tip). A dissolvable microneedle formulation that includes the drug mixed into the dissolving/biodegradable structures (f) – the drug is compounded together with a polymer in the whole structure. (Cleary, 2009; Wendorf et al., 2010.)

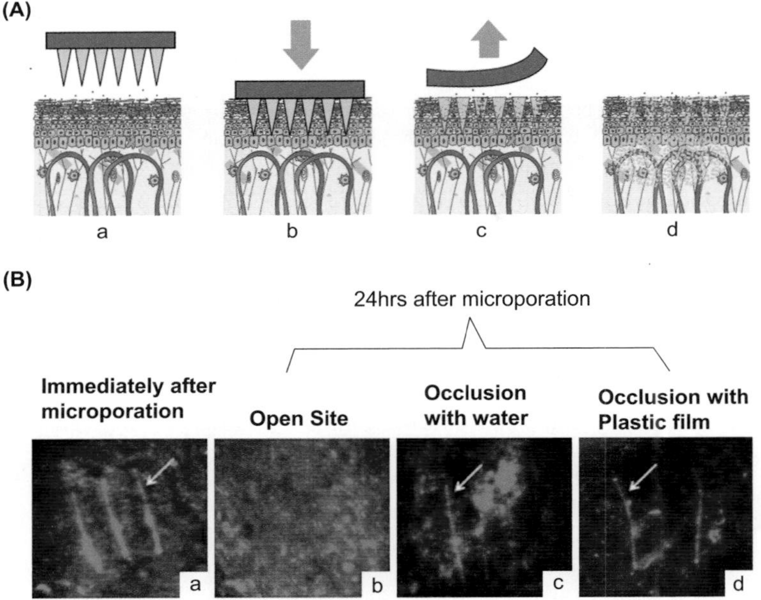

FIGURE F.5 (A) Schematic diagram of a degradable polymer microneedle array showing the process of applying the microstructures to the skin: (a) prior to biodegrading; (b) inserted into skin; (c) beginning to degrade, backing is removed; and (d) releasing the drug as they degrade. (B) A side view of the micropore in the skin where a dissolvable maltose microneedle is inserted in the skin using Casein imaging to show presence or absence of the micropore after 24 hours following insertion. Micropores are shown by the arrows pointing towards near vertical white lines: (a) immediately after insertion; (b), (c), and (d) 24 hours later under the following three different conditions: (b) site was kept open (unoccluded); (c) site occluded with water; and (d) site occluded with a plastic film. (Kalluri and Banga, 2010.)

applicator system, e.g., TheraJect®, Corium); or (4) by fabricating the patch with hollow microneedles, which are inserted into the skin, followed by forcing a liquid drug solution through the hollow central bores with a syringe-type applicator, for delivery through the pores (e.g., NanoPass®, 3M) (Prausnitz, 2004; Banga, 2009; Kalluri and Banga, 2010) (Figure F.4).

More recently, the microneedles have been fabricated from solids such as titanium, stainless steel, silicon, and glass, and coated with drug containing soluble polymers. Upon insertion, the coated polymers and their drugs are solubilized by the skin's interstitial fluid, releasing the active ingredients into the body. Figure F.5 provides a side view of the micropore after the microneedle is inserted into the skin. Note that the micropore comes to closure within a few hours when not occluded or kept moist, but persists when the micropores are covered with an impermeable polymer film (Kalluri and Banga, 2010; Sullivan et al., 2010; Wendorf et al., 2010).

Another recent development is that of microneedles entirely made of biodegradable polymers, loaded with the drug (Figure F.6). The microneedles may be designed to dissolve within 15 minutes, releasing soluble drug into the micropore at the same time. With the microneedle and the drug now gone, it becomes easier to dispose of the patch, which is simpler than disposing of a syringe and needle in hazardous waste disposal cans. It also eliminates the possibility of accidental finger sticks.

Microneedles have been used and validated in animal and human clinical studies by determination of therapeutic blood levels, pain sensation, and the time it takes for closure of the micropore (Gill et al., 2006; Kalluri and Banga, 2010). Human clinical studies with a microneedle patch delivering a polypeptide (parathyroid hormone) have shown pharmacokinetics similar to that of a subcutaneous injection using a syringe and needle. However, the C_{max} may be reached earlier with microneedles compared to a syringe and needle (Cosman et al., 2010). Other human studies have shown that microneedles ranging from 200 to 750 microns in length are painless and bloodless. The micropores created by the microneedles take from 4 to 8 hours to reach closure if they are left without any covering (non-occluded). When they are covered (occluded) it takes about 16 to 22 hours for complete resealing (Kalia et al., 2008; Cosman et al., 2010).

Iontophoresis Technology

Iontophoresis utilizes small electric fields (producing current densities around 0.5 mA/cm^2 or less) to increase the flux of an ionized drug across the intact skin. Iontophoretic delivery is primarily through the appendages: hair follicles, sebaceous glands, sweat glands, and any skin imperfections. These access pores reach into the dermis, and once the water-soluble drug reaches the aqueous viable epidermis and dermis, the drug diffusion continues until it is picked up by the capillary bed in the upper level of the dermis (Kalia et al., 2004; Power, 2007) (Figure F.7).

An electrophoresis device has a power source and two electrodes, one positive (the anode) and the other negative (the cathode), typically silver-based, $Ag^{(+)}$ and $AgCl^{(-)}$ respectively. The current flows from the anode to the cathode, and this draws positively-charged drug molecules (D^+) through the skin and towards the cathode,

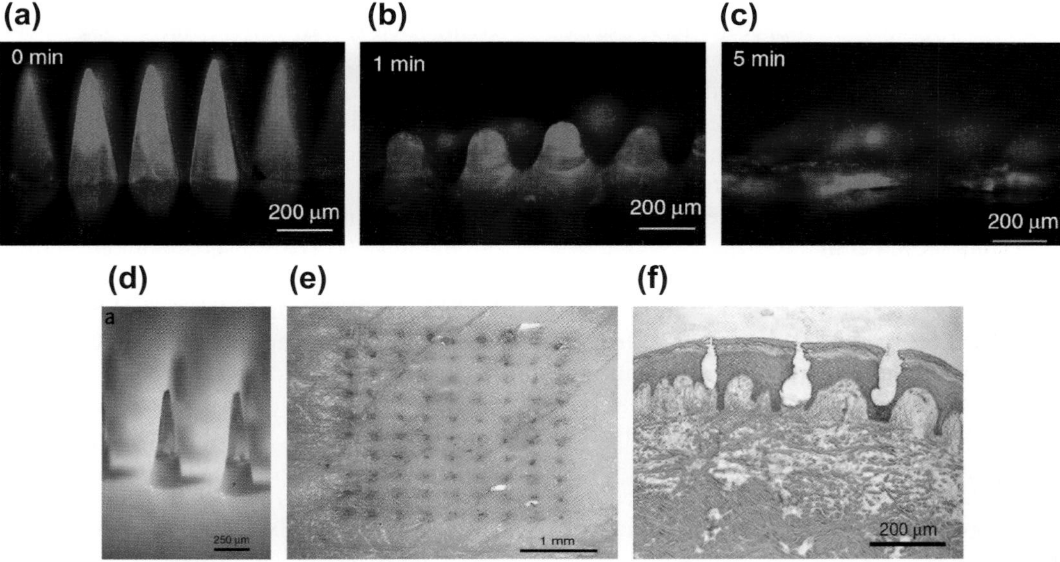

FIGURE F.6 Example of dissolvable microneedles prior to and after they were inserted into skin. These figures show the microneedles at various times: at (a) 0; (b) 1; and (c) 5 minutes after insertion into the skin. (d) View of the microneedle array prior to erosion. (e) Top view of porcine cadaver skin after the microneedles are inserted and dissolved leaving the red Sulforhodamine dye in each hole. (f) Side view of the micropore left by the dissolving microneedles that were inserted into the skin and dissolved in the viable epidermis (Brightfield micrography with H&E staining). (Sullivan et al., 2010.)

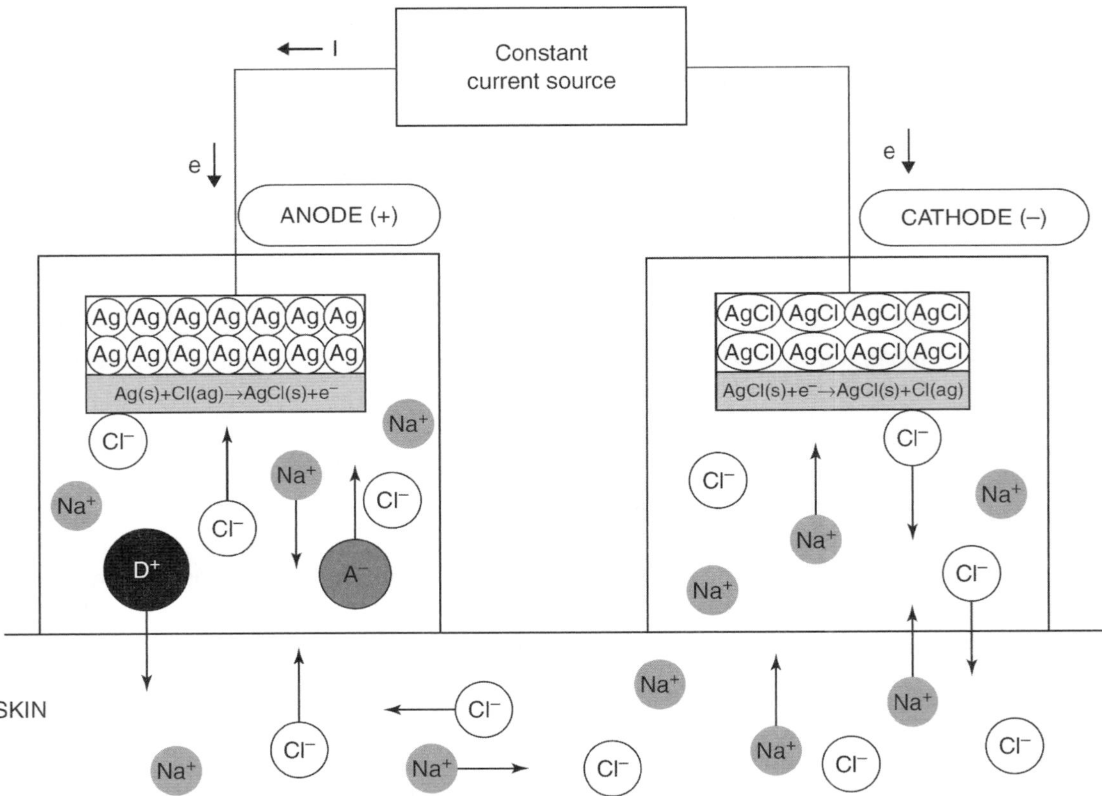

FIGURE F.7 This schematic diagram describes the electrode system of an iontophoretic drug delivery device that uses an Ag/AgCl electrode where the current is flowing to the anode (+). There are two electrode compartments: the anode (+) that contains the Drug, D^+A^-, (here the drug D^+ is placed in the reservoir of the same charge, e.g., D is lidocaine or fentanyl) and the cathode (–) or "indifferent" electrode (placed at some other site on the skin). (Kalia et al., 2004; Power, 2007.) See text for more detail on the delivery mechanism.

as the negatively-charged counterion (Cl^-) moves in the opposite direction, toward the anode. There are two reservoirs (or compartments) next to the two electrodes, where one of the reservoirs (the anode) contains the drug of the same charge (in this case, it is +) either as an aqueous liquid or a formulated hydrogel. The total observed flux of the drug from the iontophoretic device is made up of two types of electrokinetic phenomena: electromigration and electro-osmosis. Passive diffusion is considered negligible, particularly for higher molecular weight drug molecules (Kalia et al., 2004).

Iontophoretic devices that have reached the market place are sold either without the drug ("no drug") or come already prefilled with the drug ("with drug") where the physician or paramedic adds the drug solution to a reservoir before treatment. A major advance in the capabilities of iontophoretic systems has come with the availability of microprocessors. Microprocessors allow precise control of drug dosing and delivery profile, and provide increased system safety through monitoring of system functions such as current levels and battery voltage. The ability of microprocessor-controlled iontophoretic systems to offer minimally invasive and controlled administration of medications to patients has spurred an increased interest in the use of iontophoresis for systemic delivery of drugs, especially those with limited oral bioavailability or those that require a quick onset or cannot be delivered with passive transdermal technology (Kalia et al., 2004; Sathyan et al., 2005; Subramony et al., 2006; Power, 2007) (Figure F.8).

Investigators have used many different types of polymers as "gelling agents," which form hydrogel reservoir films in iontophoretic devices. Examples of such cross-linked gels are agar-agar, gelatin, polyvinyl alcohol (PVA), poly(vinylpyrollidone) (PVP), poly(dimethylaminopropyl acrylamide) (PDAA), cross-linked poly(hydroxyethyl methacrylate) (PHEMA), methylcellulose (MC), and hydroxypropylmethylcellulose (HPMC) (Panchagnula and Pillai, 2000; Peppas et al., 2000; Wang et al., 2005).

Strategies to increase the permeation of large and small drug molecules have combined iontophoresis with other technologies, such as electroporation, sonophoresis or microneedles, or with other materials, such as ion exchange materials that may also serve to improve drug stability and shelf-life. Also, chemical enhancers such as fatty acids, propylene glycol (PG), polyethylene glycol (PEG), oleic acid, terpenes, and surfactants may be added to the formulations for electrophoresis reservoirs. These combination techniques may provide improved delivery of some large drug molecules or poorly water-insoluble

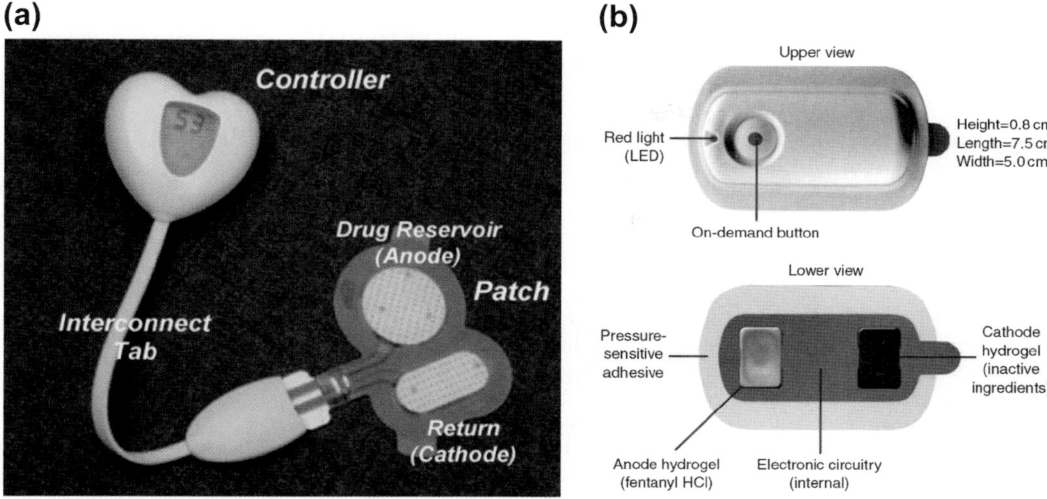

FIGURE F.8 Two examples of today's electronic iontophoretic devices with "drug on board". These devices include microprocessors, on-demand controls, etc. along with an anode hydrogel drug reservoir for use in local and systemic pain control. (a) LidoSite® (Vyteris), a device that delivers lidocaine locally for dermal pain; and (b) Ionsys® (J&J), a device that delivers fentanyl transdermally to reach the systemic circulation. (Power, 2007.)

drug molecules (Panchagnula and Pillai, 2000; Peppas et al., 2000; Wang et al., 2005).

GLOSSARY

EVA is poly(ethylene-co-vinylacetate), a polymer used in Passive TDDS as a non-porous rate-controlling membrane.

FDA is the abbreviation of the US Food and Drug Administration.

HDPE is high density polyethylene, a thermoplastic polymer that is used as a release liner in Passive TDDS.

HPMC is hydroxypropylmethylcellulose, and is used as a hydrogel for the drug reservoir and counterion reservoir in ionotophoresis.

PC is polycarbonate and is melt-extruded to make arrays of microneedle structures; also used as a release liner.

PCL is polycaprolactone, a degradable polymer used to make microneedle structures.

PEA is poly(ether-amide), a polymer with breathable and elastic properties. It has been considered as a backing or rate-controlling membrane in Passive TDDS.

PEG is polyethylene glycol, a biocompatible water soluble polymer that is used as a solubilizer of components in Passive TDDS.

PET is polyethylene terephthalate, a thermoplastic polymer that has been used as a backing and a release liner in Passive TDDS laminate structures.

PG is propylene glycol, which is used as a solubilizer and skin permeation enhancer for drugs.

PGA is polyglycolic acid, a bioerodible polymer used in making bioerodible microneedle structures in Active TDDS.

PHEMA is poly(hydroxylethyl methacrylate), used in hydrogel reservoirs for iontophoresis.

PIB is polyisobutyl rubber, a hydrocarbon-based pressure-sensitive adhesive that is used as an adhesive layer in Passive TDDS.

PLGA is poly(lactic-co-glycolic acid), a bioerodible polymer used in making degradable microneedle structures in Active TDDS.

PMMA is poly(methyl methacrylate), a thermoplastic polymer used to make microneedles.

PP is polypropylene, a semi-crystalline, thermoplastic polymer that has been used as a release liner. It may also be used as a microporous rate-controlling membrane.

PS is polysulfone, it has thermoplastic properties and is a polymer that is used to make microneedles using melt-compression methodologies.

PSA is the abbreviation for pressure sensitive adhesive; PSA polymers are adhesives that bond or adhere to surfaces when pressure is applied. These adhesives have tack (sticky properties when touched), and are used to adhere Active and Passive TDDS to skin and the laminated structures found in TDDS.

PU is polyurethanes, elastomeric polymers that as a film have breathability properties, flexibility, and drape when used as a backing layer.

PVA is polyvinyl alcohol, a water-soluble polymer that is used to make hydrogel reservoirs in iontophoresis drug delivery.

BIBLIOGRAPHY

Arora, A., Itzhak, H., Baxter, J., Rathnasingham, R., Srinivasan, R., Fletcher, D. A., & Mitragotri, S. (2007). Needle-free delivery of macromolecules across the skin by nanoliter-volume pulsed microjets. *Proc. Nat. Acad. Sci.*, **104**, 4255–4260.

Badkar, A. V., Smith, A. M., Eppstein, J. A., & Banga, A. K. (2007). Transdermal delivery of interferon Alpha-2B using microporation and iontophoresis in hairless rats. *Pharm. Res.*, **24**, 389–395.

Banga, A. K. (2009). Microporation applications for enhancing drug delivery. *Expert Opin. Drug Del.*, **6**, 343–354.

Barry, B. W. (2001). Novel mechanisms and devices to enable successful transdermal drug delivery. *Eur. J. Pharm. Sci.*, **14**, 101–114.

Barry, B. W. (2004). Breaching the skin barrrier to drugs. *Nat. Biotechnol.*, **22**, 165–167.

Bennett, S., & Potter, C. (2006). Pushing the boundaries of needle-free injection drug delivery. Report. *Autumn/Winter*, 24–28.

Bodde, H. E., Verhoeven, J., & van Driel, L. M. J. (1989). The skin compliance of transdermal drug delivery systems. *Crit. Rev. Therap. Drug Carrier Systems*, **6**, 87–115.

Bos, J. D., & Meinardi, M. H.M. (2000). The 500 dalton rule for the skin penetration of chemical compounds and drugs. *Exp. Dermatol.*, **9**, 165–169.

Burkoth, T. L., Bellhouse, B. J., et al. (1999). Transdermal and transmucosal powdered drug delivery. *Crit. Rev. Therap. Drug Carrier Systems*, **16**, 331–384.

Cleary, G. W. (1984). Transdermal Controlled Release Systems. In R. Langer, & D. L. Wise (Eds.), *Medical Applications of Controlled Release* (Vol. 1, pp. 203–251): CRC Press.

Cleary, G. W. (1991). Transdermal drug delivery. *Cosmet. Toiletries*, **5**, 97–109.

Cleary, G. W. (1993). Transdermal Delivery Systems. In V. P. Shah, & H. I. Maibach (Eds.), *A Medical Rational in Topical Drug Bioavailability, Bioequivalence and Penetration* (pp. 17–63): Plenum.

Cleary, G. W. (2009). Emergence of active transdermal drug delivery. Plenary Presentation at the 36th Annual Meeting and Exposition of the Controlled Release Society. Copenhagen: Denmark. July 18–22.

Cosman, F., Lane, N. E., Bolognese, M. E., Zanchetta, J. R., Garcia-Hernandez, P. A., Sees, K., Matriano, J. A., Gaumer, K., & Daddona, P. E. (2010). Effect of transdermal teriparatide administration on bone mineral density in postmenopausal women. *J. Clin. Endocrinol. Metab.*, **95**, 151–158.

Cross, S. E., & Roberts, M. S. (2004). Physical enhancement of transdermal drug application: Is delivery technology keeping up with pharmaceutical development? *Curr. Drug Del.*, **1**, 81–92.

Dean, H. J., & Chen, D. (2004). Epidermal powder immunization against influenza. *Vaccine*, **23**, 681–686.

Denet, A., Vanbever, R., & Preat, V. (2004). Skin electroporation for transdermal and topical delivery. *Adv. Drug Del. Rev.* **56**, 659–674.

Gerstel, M. S., & Place, V. A. (1976). Drug Delivery Device. *US. Patent No. 3,964,482*.

Gill, H. S., Denson, D. D., Burris, B. A., & Prausnitz, M. R. (2006). Effect of microneedle design on pain in human volunteers. *Clin. J. Pain*, **24**, 585–594.

Glenn, G. M., Flyer, D. C., Ellingsworth, L. R., S.A., Frerichs, D. M., Seid, R. C., & Yu, J. (2007). Transcutaneous immunization with heat-labile enterotoxin: Development of a needle-free vaccine patch. *Expert Rev. Vaccines*, **17**, 1354–1359.

Gowrishankar, T. R., Herndon, T. O., & Weaver, J. C. (2009). Transdermal drug delivery by localized intervention: Field-confined skin electroporation and dermal microscissioning. I *EEE Eng. Med. Biol.*, Jan/Feb.

Hadgraft, J., & Lane, M. E. (2006). Passive transdermal drug delivery systems. *Amer J. Drug Del.*, **4**, 153–160.

Higaki, K., Amnuaikit, C., & Kimura, T. (2003). Strategies for overcoming the stratum corneum: Chemical and physical approaches. *Am. J. Drug Del., Healthcare Technol. Rev.*, **1**, 187–214.

Kalia, Y. N., Naik, A., Garrison, J., & Guy, R. H. (2004). Iontophoretic drug delivery. *Adv. Drug Del. Rev.*, **56**, 619–658.

Kalia, Y. N., Bachhav, Y. G., et al. (2008). P.L.E.A.S.E. (Painless Laser Epidermal System): A new laser microporation technology. *Drug Del. Technol.*, **8**, 25–31.

Kalluri, H., & Banga, A. K. (2010). Formation and closure of microchannels in skin following microporation. *Pharm. Res.*, **28**(1), 82–94http://dx.doi.org/10.1007/s11095-010-0122-x. Published online: March 31, 2010.

Kandavilli, S., Nair, V., & Panchagnula, R. (2002). Polymers in transdermal drug delivery systems. *Pharma. Technol.* May, 62–80.

Lee, W., Pan, T., Wang, P. W., Zhuo, R. Z., Huang, C. M., & Fang, J. Y. (2008). Erbium: YAG laser enhances transdermal peptide delivery and skin vaccination. *J. Control. Release*, **128**, 200–208.

Menon, G. K. (2002). New insights into skin structure: Scratching the surface. *Adv. Drug Del. Rev.*, **54**, S3–S17.

Mitragotri, S., & Kost, J. (2004). Low-frequency sonophoresis – A review. *Adv. Drug Del. Rev.*, **56**, 589–601.

Naik, A., Kalia, Y. N., & Guy, R. H. (2000). Transdermal drug delivery: Overcoming the skin's barrier function. *Pharm. Sci. Technol. Today*, **3**, 318–326.

Panchagnula, R., & Pillai, O. (2000). Transdermal iontophoresis revisited. *Curr. Opin. Chem.*, **4**, 468–473.

Peppas, N. A., Bures, P., Leobandung, W., & Ichikawa, H. (2000). Hydrogels in pharmaceutical formulations. *Eur. J. Pharm. Biopharm.*, **50**, 27–46.

Power, I. (2007). Fentanyl HCl iontophoretic transdermal system (ITS): Clinical application of iontophoretic technology in the management of acute postoperative pain. *Brit. J. Anesth.*, **98**, 4–11.

Prausnitz, M. R. (2004). Microneedles for transdermal drug delivery. *Adv. Drug Del. Rev.*, **56**, 581–587.

Prausnitz, M. R., & Langer, R. (2008). Transdermal drug delivery. *Nat. Biotech.*, **26**, 1261–1268.

Prausnitz, M. R., Mitragotri, S., & Langer, R. (2004). Current status and future potential of transdermal drug delivery. *Nat. Rev. Drug Discov.*, **3**, 115–124.

Sathyan, G., Zomordi, K., Gidwani, S., & Gupta, S. (2005). The effect of dosing frequency on the pharmacokinetics of fentanyl HCl patient-controlled transdermal system (PCTS). *Clin. Pharmacokinet.*, **44**, 17–24.

Sintov, A. C., Krymberk, I., Daniel, D., Hannan, T., Sohn, Z., & Levin, G. (2003). Radio frequency-driven skin microchanneling as a new way for electrically assisted transdermal delivery of hydrophilic drugs. *J. Control. Release*, **89**, 311–320.

Songkro, S. (2009). An overview of skin penetration enhancers: Penetration enhancing activity, skin irritation potential and mechanism of action. *Songklanakarin J. Sci. Technol.*, **31**, 299–321.

Subramony, J. A., Sharma, A., & Phipps, J. B. (2006). Microprocesor controlled transdermal drug delivery. *Intern. J. Pharm.*, **317**, 1–6.

Sullivan, S. P., Koutonanos, D. G., Del Pilar Martin, M., Lee, J. W., Zarnitsyn, V., Choi, S. O., Murthy, N., Compans, R. W., Skountzou, I., & Prausnitz, M. R. (2010). Dissolving polymer microneeedle patches for influenza vaccine. *Nat. Med.*, **16**(8), 915–920. http://dx.doi.org/10.1038/nm.2182. Published online: July 18, 2010.

Van Buskirk, G. A., Gonzalez, M. A., et al. (1997). Scale-up of adhesive transdermal delivery systems. *Pharm. Res.*, **14**, 848–852.

Ventrakaman, S., & Gale, R. (1998). Skin adhesives and skin adhesion 1. Transdermal drug delivery Systems. *Biomaterials*, **19**, 1119–1136.

Wang, Y., Thakur, R., Fan, Q., & Michniak, B. (2005). Transdermal iontophoresis: Combination strategies to improve transdermal iontophoretic drug delivery. *Eur. J. Pharm. Biopharm.*, **60**, 179–191.

Wendorf, J. R., Ghartey-Tagoe, E. B., Williams, S. C., Enioutina, E., Singh, P., & Cleary, G. W. (2010). Transdermal delivery of macromolecules using solid-state biodegradable microstructures. *Pharm. Res.* http://dx.doi.org/1007/s11095-010-0174-y. Published online: June 10, 2010.

G. ORAL DRUG DELIVERY

Clive G. Wilson
Strathclyde Institute of Pharmacy and Biomedical Sciences, Glasgow, Scotland, UK

Oral drug delivery remains the most popular and most utilized method of patient therapies, although the newest candidate drugs are becoming increasingly difficult to formulate for effective oral delivery. This route of oral delivery represents an important area of innovation in pharmaceutical formulations, where there is a need to maintain the stability of the new biomolecular drugs, increase the solubility and absorption of hydrophobic drugs, and enhance the bioavailability of both types of drugs.

FEATURES OF THE GASTROINTESTINAL TRACT

The arrangement of the human gut – called the GI tract – is illustrated in Figure G.1, showing the esophagus, continuing down into the stomach, and from there into the small intestines through the duodenum, jejunum, and ileum, and then to the ascending, transverse, and descending colon, and rectum. The major area for drug absorption is denoted by shading. Characteristics of the absorption routes through the GI tract are summarised in Table G.1.

pH of the Gastrointestinal (GI) Tract

Drugs used for oral medication are often weak electrolytes and many are salts of bases; thus, the pH change between the stomach and the small intestine will exert important effects on drug absorption. At rest, the stomach pH varies: in one study, the median basal pH for females was 2.79 ± 0.18, and that for males was 2.18 ± 0.18 (Feldman and Barnett, 1991). In adults, the population of parietal cells decreases, leading to somewhat higher gastric pH in the elderly. Moreover, differences will emerge when considering absorption from the fasted state compared to the fed state. If a medication is taken with water, the pH will be elevated temporarily by dilution, returning to baseline around 20 minutes after imbibing the drug. The change in the pH environment of the upper gastrointestinal tract is very important in oral drug delivery (Table G.2).

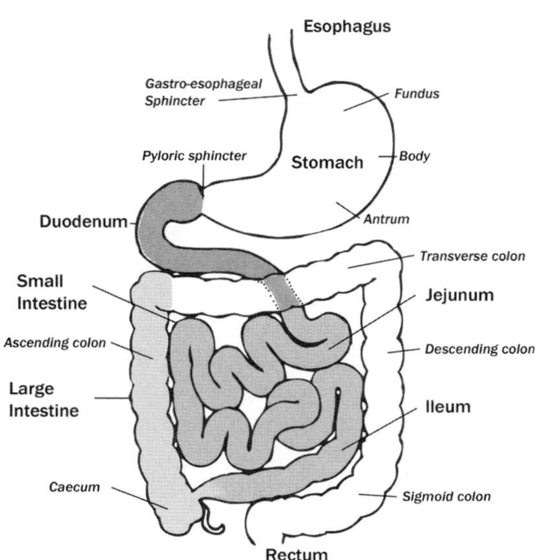

FIGURE G.1 Schematic diagram of the gastrointestinal (GI) tract.

Often it is desirable to avoid releasing the drug in the stomach. In this case, the dosage form must remain intact until it reaches the small intestines, so an enteric coating is applied, and it must be insoluble in the gastric fluids. Two types of copolymers are commonly used as enteric coatings: cellulose acetate phthalate (pH range 4.8–7) and methacrylic acid-based copolymers. Control of coating polymer composition and thickness allows the appropriate release location in the small intestine, and the pharmacokinetics to be achieved. If the formulation is designed for colonic delivery, there are two options: coat with a polymer that breaks down by bacterial enzyme attack or use a thick coating of a slowly dissolving polymer.

Gastric Emptying

The duodenum regulates the emptying of the stomach into the small intestine. Fat, high salinity, and highly acid solutions cause the duodenal wall pressure to increase and slow down the exit of the gastric contents. The diameter of the pyloric valve opening varies according to the nature of the gastric contents. When a tablet is taken with water, the time of gastric emptying of the tablet will be highly variable according to posture, volume of fluid taken, and the calorific value of food taken before or with the dosing. If the formulation disintegrates and partially dissolves in the stomach, pulses of material will appear regularly in the small intestine at a rate determined by the meal. As tablets travel down the gut the movement slows, and periods of stasis are common just before the tablet leaves the ileum and enters the large intestine. Eating food later on will cause a gastrocolic reflex, enabling the contents to move from the small intestine to the large bowel.

Pellets and disintegrated dosage forms empty from the stomach as either a series of pulses when fasted or distributed in the meal when fed. The emptying of pellets is much more predictable in the fasted state, but large, intact tablets will stay in the stomach for prolonged and variable periods of time following dosing in the morning. Eating a light meal reduces the variability in emptying. Smaller particulates are emptied in the mass of the food, and the presentation of the dose in a dispersed system is a function of calorific load and mass of the gastric contents (Reilly et al., 1987). If food is eaten throughout the day after dosing with a heavy breakfast, then in some individuals an

TABLE G.1 Absorption Characteristics in the GI Tract

GI Tract Segment	Epithelial Type	Mucus	Absorption Patterns
Oral Cavity	Stratified Squamous	Dilute	Passive absorption of smaller molecules mol.wt up to around 800 daltons, avoidance of first pass as venous drainage goes to heart directly. Proportion of dose swallowed.
Esophagus	Stratified Squamous	Mainly Salivary	Passive, low surface area. Insignificant contribution to absorption.
Stomach	Secretory Columnar	Thick, Adherent	Uptake of acids into tissue but little escapes into the circulation. Secretion of strong bases from plasma to gastric lumen (follows pH-partition).
Duodenum	Simple columnar	Thin, adherent	Active & Passive absorption over range of polarity up to around 2 kDalton: important route for absorption of drugs released from dose form in stomach.
Jejenum	Simple columnar	Thin, adherent	High Active transport, efflux & metabolism, & passive transport along concentration/ion gradients. Beginning of absorption of moderate lipophiles, lymphatic absorption starts.
Ileum	Simple columnar	Thin, adherent	As jejenum, completion of active absorption, absorption of particulates in Peyers' patches. Transit begins to slow.
Colon & Rectum	Columnar dominated	Thick, increasing	Generally reverts to passive absorption, driven by pH-partition. Absorption of non- by mucus-secreting towards rectum polar drugs in caecum, slowing in transverse ,ceasing as colon contents solidify. Rectal glands route generally DOES NOT avoid first pass (except at anal sphincter). Approximately 1/10TH rate of small intestine.

TABLE G.2 Characteristics of Segments of the Gastrointestinal Tract

GI Tract Segment	Approx. Surface Area	Approx. Segment Length	Approx. Residence Time	Approx. pH of Segment	Principal Catabolic Activities	Bacterial Count* (cfu/ml)
Oral cavity	100 cm^2	–	Seconds to minutes	6.5	Polysaccharidases	–
Esophagus	200 cm^2	23–25 cm	4–8 seconds**	–	–	–
Stomach	3.5 m^2 (variable)	0.25 m (variable)	90 minutes (variable)	1–2	Proteases, lipases	10^2
Duodenum	1.9 m^2	0.35 m	30–40 minutes	4–5.5	Polysaccharidases, oligosaccharidases, proteases, peptidases, lipases	10^2
Jejunum	184 m^2	2.8 m	1.5–2 hours	5.5–7.0	Oligosaccharidases, peptidases, lipases	10^5
Ileum	276 m^2	4.2 m	5–7 hours	7.0–7.5	Oligosaccharidases, peptidases, lipases	10^7
Colon and rectum	1.3 m^2	1.5 m	1–60 hours (35–38 hours on average)	7.0–7.5	Broad spectrum of bacterial enzymes	10^{12}

Marsny, R. J. (1997). In: *Controlled Drug Delivery*, Park, K. (Ed.). ACS, Washington, DC.
*Kendall, R. A. and Basit, A. W. (2006). In: *Polymers in Drug Delivery*, Uchegbu, I. F. and Schatzlein, A. G. (Eds.). Taylor and Francis, London.
**Perkins et al., 2001a,b.

enteric-coated tablet may remain in the stomach for longer than twelve hours (Wilson et al., 1989).

When drug is ejected from the stomach, it will pass into the duodenum, which is a highly permeable section of the intestine. Thus, the duodenum is a desirable region to deliver drug; however, transit through this region occurs rapidly. Therefore, it is probable that significant duodenal absorption occurs only when the dose is retained in the stomach for a period of time, and slowly exits into the duodenum. This may be accomplished with gastric retention polymers or devices (see below, and Park et al., 2006).

Immediate Release Dosage Forms. Administration of an active compound must be carried out as a unit dosage form; however, once in the body, the work of the formulator must be undone. The formulation usually contains fillers and binders, which aid physical processing during manufacture, including wet and dry granulation, filling into capsules, and formulation into advanced dosage forms. Typical fillers and manufacturing aids include lactose, mannitol, and di-calcium phosphate; binding materials include microcrystalline celluloses and carboxymethylcellulose, starch and gums; fillers include cellulosics such as hydroxypropyl methyl cellulose (HPMC);

and coating materials include enteric copolymers, as described above. To ensure disintegration, the active formulation may contain swelling or fast dissolving agents, and in addition lubricants are added to improve powder flow during manufacture. The usual objective is to produce rapid dissolution in the small intestine. Rapid dissolution is measured as >85% dissolution in 30 minutes in 900 ml media. A compound is classified as highly soluble if it dissolves in 250 ml media.

Lipid Dosage Forms. A solution to the poor aqueous solubility of many drugs is to utilize an alternative solvent, which forms emulsions or supersaturated solutions on dilution with the gut contents. Surfactants may also be incorporated to aid dispersion on release of the drug. The "container" of such formulations is usually soft or hard gelatin capsules, and some formulations may be chewable or enteric-coated, soft gelatin-coated products. Gelatin formulations are often preferred by patients as they are easier to swallow (although if they are buoyant, the reverse may be true, especially in the elderly) (Perkins et al., 1999). The use of gelatin in hard capsule shells is gradually being replaced by HPMC, but in soft capsules highly plasticized gelatin remains prominent.

Gastric Retention Devices. There are regions of the gastrointestinal tract where high absorption occurs, and efforts have been made to design formulations that deliver drugs in those regions. These formulations utilize copolymers that act by several mechanisms, including ligand association and bioadhesion to the local lumen (Park et al., 2006). A few important drugs show an apparent window of absorption, with best permeability in the duodenal segment. Since transit through this region of the gut is very rapid – typically less than 5 minutes – the formulator must attempt to keep the delivery device for a prolonged period of time in the stomach, such that the duodenum is continually bathed in drug. Methods to do this often employ gastric retention devices. A simple test was proposed to test the usefulness of such devices – simply sipping a formulation over a prolonged period to examine a change in the pharmacokinetic parameters (Burke and Wilson, 2006). Other gastro-retention methods have been studied, including use of mucoadhesion, flotation, and obstruction devices, such as superporous hydrogels (Chen et al., 2000; Kim et al., 2004; Park et al., 2006). Floating systems are either effervescent or non-effervescent systems in which gas is either generated or trapped during rapid swelling of a polymeric hydrogel to increase the buoyancy. The control of fluid penetration into the device, and therefore drug dissolution and release, is controlled by the formation of a cross-linked gel barrier on the exterior. A general strategy is to float or to expand due to the liberation of gas into a gelling structure, such as alginate or acrylamide hydrogels, and therefore is most successful if the patient is fed and upright. There are problems if the subject is recumbent and turns onto the left, as the floating layer will empty out of the stomach ahead of the rest of the gastric contents (Jenkins et al., 1983).

Intestinal Transit: How Constant is it?

The primary area of the gut for drug absorption is the small intestine, having an absorptive flux around 10 to 20 times that of the large bowel (Table G.1). The slow absorption associated with compounds of poor solubility or intrinsic slow dissolution still prompts drug formulation scientists to look for methods of increasing small intestinal transit time. Imaging techniques, such as gamma scintigraphy and magnetic moment imaging, show that movement gradually slows as the dosage form moves from the duodenum to the ileum, with periods of stasis and sluggish movement. Sluggish movement in the gastrointestinal tract is often associated with blockages – called "bezoars," due to ingestion of fruit (especially unripe persimmons, phytobezoars), hair (trichobezoars), and mixtures of tablets with an anticholinergic effect (pharmacobezoars).

The opposite extreme – fast transit – is also sometimes evident. In volunteer studies with the drug gefitinib, the pharmacokinetics was shown to be highly variable (Swaisland et al., 2005). Subsequently, it was appreciated that there was a subgroup in the population with consistently faster intestinal transit times (156 minutes for the subgroup versus 204 minutes for the average), resulting in lower drug absorption (Wilson et al., 2009).

> The mechanisms of drug delivery from controlled-release oral dosage forms can be classified into these five main areas: (1) diffusion control; (2) swelling and dissolution control; (3) osmotic pump control; (4) ion exchange control; and (5) bacterial enzyme control.

CONTROLLED RELEASE IN THE GI TRACT

In functional terms, progress in controlled release technology in the GI tract has branched into three main directions: deposition control (control of the site or region of delivery); temporal control (control of the time of delivery, also called chronotherapy or chronopharmaceutics); and barrier control (control of the rate of delivery). Deposition control refers to the coverage of tissue achieved by the device, and its dispersion (spread) and accumulation are the parameters measured. Temporal control refers to devices that delay release of the active component after ingestion, eventually releasing the formulation further down the gut. This will generally result in less dispersion when compared to devices that release in the stomach. Barrier control refers to the construction of devices in which excipients or membranes control the release of the drug. This is especially important for drugs exhibiting a short half-life, and hence requiring three or more doses per day.

The extension of dosing intervals to facilitate increased compliance has been a successful strategy in therapeutics. This relies on the ability to sustain drug delivery for an extended time, using one or more delivery retarding mechanisms of matrix, coating or pump. These mechanisms may be used in combination, and in a variety of geometries. A slowly dissolving drug will display a flat and variable plasma concentration–time profile, and here the objective may also be first to control the solubility of the drug, and second to modulate its rate of release. This principle has been used to target more distal parts of the gut by controlling the matrix polymer composition and/or the coating polymer composition and thickness. Bimodal distributions of the plasma concentration–time profile (PK) may occur in slowly eroding matrices of conventional tablets, which break up as they go through the ileo-caecal junction (Wilson et al., 1989; Gabacz et al., 2008).

The bacterial enzyme-controlled drug delivery mechanism in the colon depends on the action of the bacteria in the ascending colon to degrade polymer–drug conjugates or polymer matrices or coatings of a tablet as it moves through the colon.

Colon Drug Delivery

Targeted delivery of drugs to the colon, the terminal region of the GI tract, has been employed to achieve one of several major therapeutic objectives: (1) to deliver high local concentrations of a drug for treatment of colonic diseases; (2) to delay delivery of a drug until the colon is reached (where there is a long residence time) in order to treat acute phases of a disease at the appropriate time of day; (3) to utilize sustained delivery to reduce dosing frequency; and (4) to deliver fragile biotherapeutics such as protein drugs, in the hope that the colon is less hostile metabolically to protein-based therapeutics.

> Oral pulsed release systems are generally divided into five major types of systems:
>
> Drug delivery systems (DDS) with rupturable coating layers
> DDS with swellable/erodible layers
> DDS with increasingly permeable coating layers
> Capsule-shaped systems with controlled-release plugs
> Osmotic systems
>
> (Maroni et al., 2005).

Osmotic Pumps

The osmotic pump has become an important delivery vehicle for oral therapeutics, and is also widely used in pharmacological research. The principle was first demonstrated by Rose and Nelson as a long-term implantable pump that would be of much value in pharmacology; for example for long-term, site-specific delivery to the rat (Rose and Nelson, 1955). A wide variety of osmotic pumps was further famously developed by Alza Corporation scientists, and now there are many different forms, including push-pull, a pulse-release system that works by the generation of sufficient pressure to rupture a controlling membrane. A push-pull system can typically maintain a zero order, constant rate of release for 80% or more of its theoretical content, and was strikingly effective as a system for delivery of nifedipine (ProCardia XL®). The principle advantage of this type of technology is that the system loads up with water prior to entry into the relatively dry region of the large bowel and, providing transit is prolonged, the delivery of drug can be sustained. The major issue is the variation in normal bowel habit, and all delayed systems run into the problem of the limited amount of water in the colon – at best around 30 mL in the caecum – with a problem of dispersion if the release occurs in the transverse colon and beyond.

> A variety of oral osmotic pump designs and coatings were developed at Alza Corporation in the 1970s and 1980s for controlling the site and rate of release of a drug. Zero order controlled-release kinetics are achievable with these pumps, which are usually ingested as small tablets. Some osmotic pumps are also designed to release two drugs in sequence (Theeuwes and Yum, 1976; Theeuwes, 1983; Theeuwes et al., 1985).

Bacteria and Bacterial Enzymes in the Colon. Biopolymers used for colonic drug delivery are plant-based polysaccharides, which are digestible by the bacterial enzymes of the colon and are almost untouched by mammalian enzymes secreted by the gut wall. As a consequence such materials, particularly those that have pronounced swelling properties, have been investigated as potential excipients in the formulation of controlled-release colonic drug delivery systems. The colonic microflora secretes a number of enzymes that are capable of hydrolytic cleavage of glycosidic bonds. These include β-D-glucosidase, β-D-galactosidase, amylase, pectinase, xylanase, α-D-xylosidase, and dextranases. In addition there are scission reactions catalyzed by azo-reductases that are secreted by the anaerobes that can degrade azo-containing polymers and hydrogels (e.g., Miyata, 2010). Such azo-polymers are also used as prodrugs, e.g., 5-ASA adducts such as balsalazide and olsalazine.

Biodegradable polysaccharides can be incorporated into the formulation matrix or as a coat, alone and in combination with other polymers. Since many of these polysaccharides have limited release-control properties due to their high water solubility, they are used in combination with synthetic non-biodegradable polymers, especially acrylates, or are modified to reduce their solubility (Wilson et al., 2008). Typical polysaccharides include pectin, guar gum, amylose, chitosan, and inulin.

BIBLIOGRAPHY

Burke, M. D., & Wilson, C. G. (2006). Clinical protocol design: Gastroretentive dosage forms. *Drug Deliv.*, **6**(8), 26–31.

Chen, J., Blevins, W. E., Park, H., & Park, K. (2000). Gastric retention properties of superporous hydrogel composites. *J. Contr. Rel.*, **64**, 39–51.

Feldman, M., & Barnett, C. (1991). Fasting gastric pH and its relationship to true hypochlorhydria in humans. *Dig. Dis. Sci.*, **36**, 866–869.

Gabacz, G., Wedermeyer, R. S., Stefan, N., Giessmann, T., Monnikes, H., Wilson, C. G., Sigmund, W., & Weischies, W. (2008). Irregular absorption profiles observed from diclofenac extended release tablets can be predicted using a dissolution apparatus that mimics *in vivo* stresses. *Eur. J. Pharm. Biopharm.*, **70**, 421–428.

Jenkins, J. R.F., Hardy, J. G., & Wilson, C. G. (1983). Monitoring antacid preparations in the stomach using gamma scintigraphy. *Int. J. Pharmaceut.*, **14**, 143–148.

Kendall, R. A., & Basit, A. W. (2006). In I. F. Uchegbu, & A. G. Schatzlein (Eds.), *Polymers in Drug Delivery*. London: Taylor and Francis.

Kim, D. J., Seo, K., & Park, K. (2004). Polymer composition and acidification effects on the swelling and mechanical properties of poly(AAm-co-AAc) superporous hydrogels. *J. Biomater. Sci., Polym. Ed.*, **15**, 189–199.

Maroni, A., Zema, L., Cerea, M., & Sangalli, M. E. (2005). Oral pulsatile drug delivery systems. *Expert Opin. Drug Deliv.*, **2**(5), 855–871.

Marsny, R. J. (1997). In K. Park (Ed.), *Controlled Drug Delivery*. Washington DC: ACS.

Miyata, T. (2010). Biomolecule-responsive hydrogels. In R. M. Ottenbrite, K. Park, & T. Okano (Eds.), *Biomedical Applications of Hydrogels Handbook* (pp. 65–86). NY: Springer.

Park, H., Park, K., & Kim, D. (2006). Preparation and swelling behavior of chitosan-based superporous hydrogels for gastric retention applications. *J. Biomed. Mater. Res., Part A.*, **76A**, 144–150.

Perkins, A. C., Wilson, C. G., Vincent, R. M., Frier, M., Blackshaw, P. E., Dansereau, R. J., Juhlin, K. D., Bekker, P. J., & Spiller, R. C. (1999). Esophageal transit of risedronate cellulose-coated tablet and gelatin capsule formulations. *Int. J. Pharm.*, **186**, 169–175.

Perkins, A. C., Wilson, C. G., Frier, M., Blackshaw, P. E., Juan, D., Dansereau, R. J., Hathaway, S., S., Li, Z., Long, P., & Spiller, R. C. (2001a). Oesophageal transit, disintegration and gastric emptying of a film-coated risedronate placebo tablet in gastro-oesophageal reflux disease and normal control subjects. *Aliment. Pharm. Ther.*, **15**, 115–121.

Perkins, A. C., Wilson, C. G., Frier, M., Blackshaw, P. E., Dansereau, R. J., Vincent, R., R., Wenderoth, D., Hathaway, S., Li, Z., & Spiller, R. C. (2001b). The use of scintigraphy to demonstrate the rapid esophageal transit of the oval film-coated placebo riedronate tablet compared to a round uncoated placebo tablet when administered with minimal volumes of water. *Int.J. Pharm.*, **222**, 295–303.

Reilly, S., Wilson, C. G., & Hardy, J. G. (1987). The influence of food on gastric emptying of multiparticulate dosage forms. *Int. J. Pharmaceut.*, **34**, 213–216.

Rose, S., & Nelson, J. F. (1955). A continuous long-term injector. *Aust. J. Exp. Biol.*, **33**, 415.

Stevens, H. N.E., Wilson, C. G., Welling, P. G., Bakhsaheee, M., Binns, J. S., Perkins, A. C., Frier, M., Blackshaw, E. P., Frame, M. W., Nichols, D. J., Humphrey, M. J., & Wicks, S. R. (2002). 2002 Evaluation of Pulsincap to provide regional delivery of dofetilide to the human GI tract. *Int. J. Pharm.*, **236**, 27–34.

Swaisland, H. C., Smith, R. P., Laight, A., Kerr, D. J., Ranson, M., Wilder-Smith, C. H., & Duvauchelle, T. (2005). Single-dose clinical pharmacokinetic studies of gefitinib. *Clin. Pharmacokinet.*, **44**, 1165–1177.

Theeuwes, F. (1983). Evolution and design of "rate controlled" osmotic forms. *Curr. Med. Res. Opin.*, **8**, S2, 20–27. (see also http://informahealthcare.com/doi/abs/10.1185/03007998309109820).

Theeuwes, F., & Yum, S. I. (1976). Principles of the design and operation of generic osmotic pumps for delivery of liquid or semi-solid drug formulations. *Ann. Biomed. Eng.*, **4**, 343–353.

Theuwes, F., Swanson, D. R., Guitttard, G., Ayer, A., & Khanna, S. (1985). Osmotic delivery systems for the beta-adrenoreceptor antagonists metoprolol and oxprenolol: Design and evaluation of systems for once-daily administration. *Br. J.Clin. Pharmacol.*, **19**, 69–76S.

Wilson, C. G., Washington, N., Greaves, J. L., Kamali, F., Rees, J. A., Sempik, A. K., & Lampard, J. F. (1989). Bimodal release of drug in a sustained release Ibuprofen formulation: A scintigraphic and pharmacokinetic open study in healthy volunteers under different conditions of food intake. *Int. J. Pharmaceut.*, **50**, 155–161.

Wilson, C. G., Mukherji, G., & Shah, H. K. (2008). Biopolymers and colonic delivery. In M. Rathbone, J. Hadgraft, & M. Roberts (Eds.), *Modified Release Drug Delivery Technology* (2nd ed., pp. 315–329). New York: Roberts Drugs and the Pharmaceutical Sciences Series, J. Swarbrick (Ed.). Marcel Dekker, Inc.

Wilson, C. G., O'Mahony, B., Connolly, S. M., Cantarini, M. V., Farmer, M. R., Dickinson, P. A., Smith, R. P., & Swaisland, H. C. (2009). Do gastrointestinal transit parameters influence the pharmacokinetics of gefitinib? *Int. J. Pharm.*, **376**, 7–12.

CHAPTER II.5.17 DIAGNOSTIC APPLICATIONS OF BIOMATERIALS

Gonzalo Domingo, Kenneth R. Hawkins, Roger B. Peck, and Bernhard H. Weigl

All authors are affiliated with PATH (www.path.org)

OVERVIEW OF DIAGNOSTICS

Medical diagnostics are devices that aid in the diagnosis of a disease or condition. While there are many different classes of diagnostic devices, ranging from blood pressure cuffs to MRIs, biomaterials are most often used as components of *in vitro* diagnostics (IVDs).

IVDs generally measure the presence, count or concentration of an *analyte* – a chemical or biochemical constituent or bioparticle – in a *sample* such as blood, saliva, urine, other bodily fluids, stool, and sometimes tissues. Analysis is done *in vitro* (i.e., in a controlled environment outside the body) to minimize noise and potential interferents. This is an important difference from biosensors (see Chapter II.5.13 for a discussion of biosensors). Common analytes include biomarkers for disease presence and progression, components of pathogens themselves, and xenobiotic substances.

The field of medical diagnostics is vast, and this chapter cannot provide an overview of all major classes of diagnostics. Therefore, the authors focus on types of diagnostics that frequently incorporate biomaterials, and especially on devices that are used in lower-throughput, point-of-care (POC) settings.

Requirements of Diagnostics

Diagnostic tests, first and foremost, must provide value to the patient by assisting the healthcare provider in improving the accuracy of a diagnosis beyond what one could achieve based solely on observable symptoms, epidemiology, and statistics (*syndromic diagnosis*).

Diagnostics, like any other medical intervention, are used within an overall healthcare and economic context. Given infinite resources and patients' willingness to submit to frequent testing, an enormous number of chemical and biological parameters could be measured with current technology to provide information about a patient's current and likely future health. These range from full genomes and time profiles of thousands of known biomarkers to the presence of every foreign organism in or on a patient's body. However, resources are always finite – especially in developing countries that hold the majority of the world's population – and trade-offs always have to be made. The World Health Organization (WHO) has published a list of desirable attributes for diagnostics specifically for global health (*low-resource*) settings. According to the WHO ***A.S.S.U.R.E.D.*** model, (Mabey et al., 2004) a diagnostic should be:

A = Affordable by those at risk of disease.
S = Sensitive (able to detect the analyte at a clinically significant level).
S = Specific (not responsive to sample constituents other than the analyte).
U = User-friendly (simple to perform with minimal training).
R = Rapid/robust (to enable action at point-of-care).
E = Equipment free (or requiring as little equipment as possible).
D = Deliverable to those who need it (requiring temperature and mechanical stability in transport).

Different healthcare settings value these characteristics according to their priorities. In global health settings, cost, simplicity, and stability of the biomaterials are paramount. In POC settings in developed countries, simplicity is still critical, together with sensitivity and specificity. In laboratory settings, time to result is very important, because a diagnostic parameter may be needed to guide emergency or surgical treatments. In central laboratory settings, a parameter not on the WHO list – throughput – is critical, because a single device may be used to process hundreds of samples daily.

Biomaterials can affect the performance of diagnostics in many of these areas. Affordability is affected by material cost, as well as its stability. The performance of a sensing biomaterials layer will affect both sensitivity and specificity. The response time of a sensing layer (or the resuspension time of a dried reagent layer) affects time to result. A biomaterial that visibly changes in response to contact with a biochemical analyte allows the design of instrument-free diagnostics. The stability of biomaterials determines if a diagnostic can be stored at room temperature, used in hot and humid environments or stored for long enough periods to allow use in remote settings.

The Three Main Aspects of Diagnostics

In general, diagnostic processes all naturally divide into three main aspects: *pre-analytical*; *analytical*; and *interpretation* (Figure II.5.17.1). Each aspect has its own challenges and tools, and can be considered autonomously, although the needs of one will affect the others. Very few diagnostic processes can be performed with a single device from sample collection to output of result. Most often, diagnostic devices perform a pre-analytical, analytical or interpretation function, or some combination thereof. The role of biomaterials in these diagnostic devices varies widely to reflect the huge diversity of *in vitro* diagnostic processes that contain biomaterials.

Case Study 1 Biomaterials in a Lateral Flow Immunoassay Strip

The lateral flow immunoassay strip seems to be a simple diagnostic device; however, it requires at least *four distinct biomaterials*, serving seven different functions:

1. In the pre-analytical domain, an *anti-coagulant-coated glass blood collection tube* connected to a puncture device is often used for sample collection.
2. The sample is applied to the *surfactant-impregnated sample pad* of a lateral flow rapid immunoassay test strip where the sample is processed, in this case simply by retaining the majority of red cells, and thus allowing the remaining fraction of the sample to flow along the strip.
3. The sample also interacts with an *immunoreagent attached to a mobile, visible particle* in the sample pad. The target analyte molecules are "labeled" with this immunoreagent.
4. In the analytical part of the diagnostic device – the strip – the sample first encounters *strip paper treated with a blocking agent* to prevent nonspecific binding.
5. The sample next encounters a band of *paper-immobilized immunoreagent* that captures the labeled analyte molecules.
6. Shortly thereafter, the sample encounters a second band of *paper-immobilized immunoreagents* (*control line*) as it flows along the strip. The control line captures another generic component of the sample.
7. For interpretation, the bands on the strip paper can be visually observed or inserted into an optical, electrochemical or magnetic reader for machine-aided interpretation. At this phase, the optical properties of the strip paper become important.

With this case study in mind, a more in-depth consideration of the details of each of the three main aspects of diagnostics should implicitly suggest many more ways in which biomaterials can influence the final performance of an IVD device.

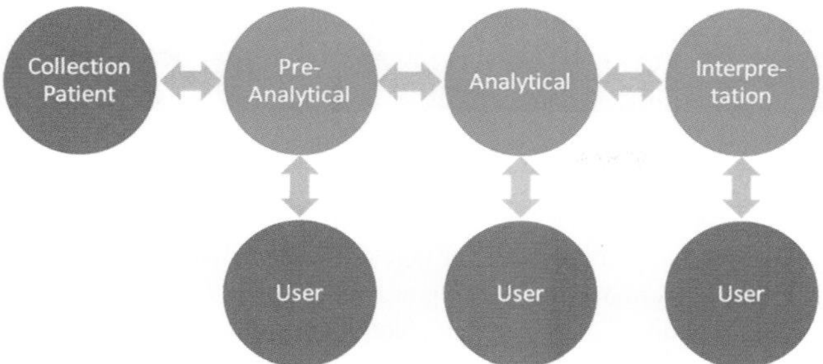

FIGURE II.5.17.1 A schematic representation of the three main aspects of any diagnostic process: pre-analytical; analytical; and interpretation. The three phases are linear in nature. Pre-analytical addresses sample collection and preparation; analytical comprises capture and/or detection of the target moiety; and interpretation takes into consideration calibration, comparison, and classification of the results. Each phase interacts with the aspect immediately before or after it. Additionally, each aspect has interactions with samples, patients or users. It is important to note that the users that interface with each aspect may represent very diverse backgrounds with varying levels of training, experience, and expertise.

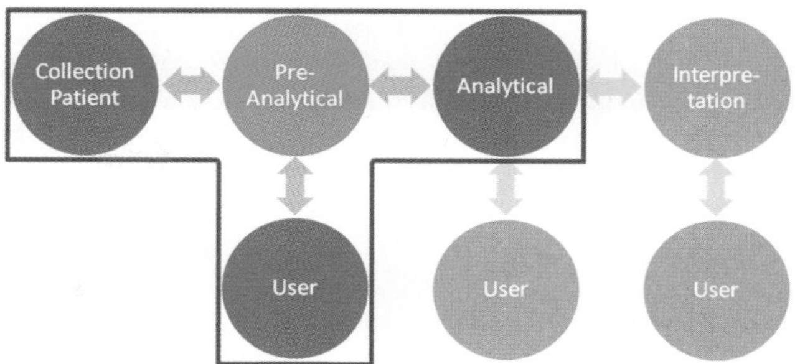

FIGURE II.5.17.2 The pre-analytical phase is the first phase of the diagnostic process. In this phase samples are collected from the patient and prepared for the analytical stage. Depending on the level of integration in the diagnostic device, users that may interact with this phase could include the patient, healthcare provider or laboratory technician. Biomaterials may be used to process the sample, but the pre-analytical phase always results in a biomaterial that is appropriate for use in the analytical phase.

THE PRE-ANALYTICAL PHASE

The pre-analytical phase of a diagnostic test refers to all of the steps required to deliver the analyte from the sampled environment or patient to the analytical assay. First, the test must sample the matrix in which the target analyte is present. Next, the matrix is processed to perform one or several of the following: (1) concentration of the analyte; (2) purification of the analyte; and (3) conditioning of the analyte for the downstream analytical testing. The performance of a diagnostic test, especially its limit of detection (LOD) and reproducibility, is highly influenced by the pre-analytical phase.

Technologies resolving the pre-analytical phase of the diagnostic test interface with the patient, from whom the specimen is obtained, the user, who utilizes the technology to collect the specimen and process it, and the analytical test, which requires the analyte to be appropriately conditioned. Each of these interfaces must be considered when selecting the appropriate method and biomaterial for the task (Figure II.5.17.2). Different contexts for diagnostics result in different combinations of interfaces (Table II.5.17.1).

Factors Determining Specimen, Specimen Volume, and Specimen Conditioning

In the context of medical diagnostics – and more specifically, infectious disease diagnostics – matrix sampling typically is referred to as *specimen collection*. The major factors contributing to specimen selection are:

- *Analyte abundance in the specimen.* Ideally one would use the specimen in which the highest concentration of the target analyte is found.
- *Clinical relevance of analyte in the specimen.* The presence of a specific analyte in one specimen may not indicate disease, while in another specimen it will.
- *Ease of specimen collection.* Noninvasive sources of specimens, such as urine, are preferable.
- *Complexity of the specimen.* A less complex specimen simplifies downstream processing.

Specimen collection entails direct interaction with the patient. For POC diagnostics, sample collection minimizes active roles for biomaterials in the host–device interaction, and thus minimizing biocompatibility constraints. Biomaterials are typically restrained to an active role in the specimen receptacle or container and the downstream applications.

The specimen volume or sample size required for a diagnostic test is determined by the analyte concentration in the specimen, the efficiency of analyte purification and/or concentration, the LOD of the analytical phase, and the target LOD for the final diagnostic test. These concepts are illustrated in Case Study 2 at the end of this section.

The pre-analytical phase encompasses purification, enrichment or concentration of the target analyte for downstream analysis. Purification is required to remove substances that inhibit, compete or interfere with the analytical phase; enrichment could be required to improve signal-to-noise ratios; and concentration of the target analyte from large specimen volumes to smaller volumes may be required to meet the appropriate clinical sensitivities. In all of these processes, biomaterials can play a key role.

The downstream analytical assay determines the level and nature of analyte preconditioning that must be performed. For example, for a diagnostic assay using polymerase chain reaction (PCR) as the analytical test, key functions for the pre-analytical phase are release of DNA or RNA from organelles (by sample cell lysis), removal of reverse-transcriptase and polymerase inhibitors and, often, concentration of DNA or RNA into a

TABLE II.5.17.1 Common Specimen Types and Sampling Methods and their Suitability in Various Contexts

Specimen	Sampling Method	Appropriate for Point of Care	Example Analyte and Comments
<50 μL blood	Finger-prick/heel stick	Yes	Complete blood count (CBC); blood glucose monitoring; bloodborne pathogens, antigens, and nucleic acid; host immune response such as IgM and IgG; xenobiotics
>50 μL blood	Venipuncture	No	Pathogen culture as well as CBC; bloodborne pathogens, antigens, and nucleic acid; host immune response such as IgM and IgG; xenobiotics
Saliva, oral fluid or oral mucosal transudate	Swab or salivette	Yes	Hormones, xenobiotics. Many analytes present only at very low concentration
Sputum	Expectorated	Yes	Lower respiratory tract infections such as tuberculosis
Stool	Self-collected, rectal swab or diaper swab	Yes	Diarrheal infections, fecal occult blood, pathogen culture, nucleic acid, and antigens
Central nervous system	Spinal tap	No	Encephalitis, pathogen culture, nucleic acid, antigen, immune response
Nasal swabs	Swab	No	Upper respiratory infections, pathogen culture, nucleic acid, antigen
Urine	Self-collected	Yes	Hormones, xenobiotics
Urogenital	Provider or self-collected	Yes	Reproductive tract infections

small volume. Purification of the specific target DNA or RNA is not critical, because the PCR reaction itself can be highly specific. In contrast, a diagnostic test based on the nonspecific staining of the entire class of molecules to which the analyte belongs requires a high degree of purification during the pre-analytical phase, because the assay itself has only limited specificity at the analytical phase.

Affinity Capture for Concentration and Purification

One of the most common tools for analyte purification and concentration is affinity capture, whereby physical, chemical or biochemical features of the target analyte are used to capture or exclude the target analyte. In these techniques it is common to immobilize the capture species on a solid support, and then introduce the sample analyte in a liquid or mobile phase.

Charge-based interaction between the phosphate backbone of these biopolymers and silica is used to purifiy DNA and RNA from complex materials (Figure II.5.17.3), whereas specific purification and enrichment of target nucleic acid molecules or subclasses of these molecules can be mediated through complementary strand hybridization. A common example is the use of a poly-T-oligonucleotide immobilized to a surface for the capture of eukaryotic messenger RNA, which is typically polyadenylated.

Case Study 2 Assay Specifications Drive Biomaterial Choices

A diagnostic test is under development that is specified to detect at least 50 copies/mL of HIV viral RNA in blood plasma. The analytical test is a highly sensitive real-time PCR system with a demonstrated LOD of five copies per reaction. Given the precision of the assay, specimens should be tested in duplicate. The designers would like to use a finger-stick to collect the specimen. Finger-sticks can be expected to yield a maximum of ~500 μL of whole blood, which after spinning down should yield ~275 μL of plasma.

There are two candidate affinity-based biomaterials suitable for the RNA extraction device. The first material is a traditional solid matrix which, due to its kinetic and thermodynamic properties, is capable of extracting 70% of the RNA in the plasma. The second material is a thermally responsive smart polymer system which, due to its ability to change phases in a controlled fashion, is capable of releasing almost all of the RNA it captures, and is therefore 95% efficient. The smart polymer system is far more expensive on a per test basis. The assay designer needs to determine if the extra cost is justified. Performing the unit analysis shows that:

$$\frac{copies}{test} \times \frac{test}{extraction} \times \frac{mL\ plasma}{copies} \times \frac{1}{extraction\ efficiency} \times \frac{\mu L}{mL} = \frac{\mu L\ plasma}{extraction}$$

$$\text{For biomaterial \#1}: \frac{5 \times 2 \times 1000}{50 \times 0.7} = \frac{286\ \mu L\ plasma}{extraction}$$

$$\text{For biomaterial \#2}: \frac{5 \times 2 \times 1000}{50 \times 0.95} = \frac{211\ \mu L\ plasma}{extraction}$$

The traditional solid matrix material will not be able to meet the LOD specifications with the volume of plasma expected to be available from a finger-stick. The more efficient smart polymer is required to meet specification; therefore, the additional cost is justified.

Step 1: Lyse sample

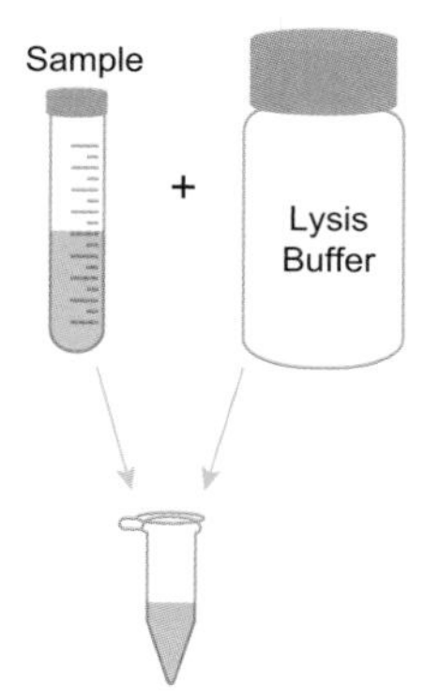

Step 2: Adsorb sample to capture matrix

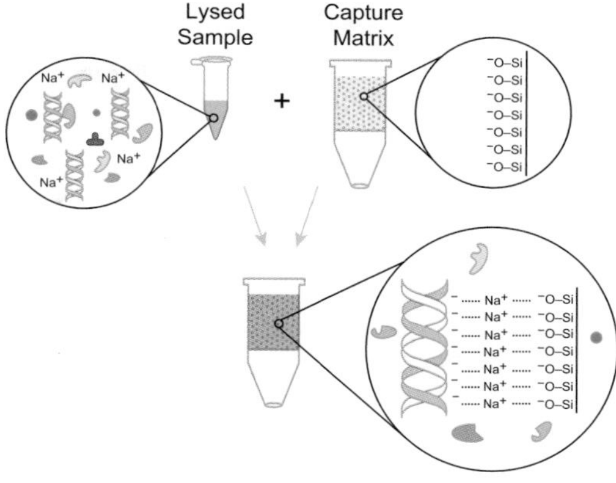

Step 3: Wash contaminants from capture matrix

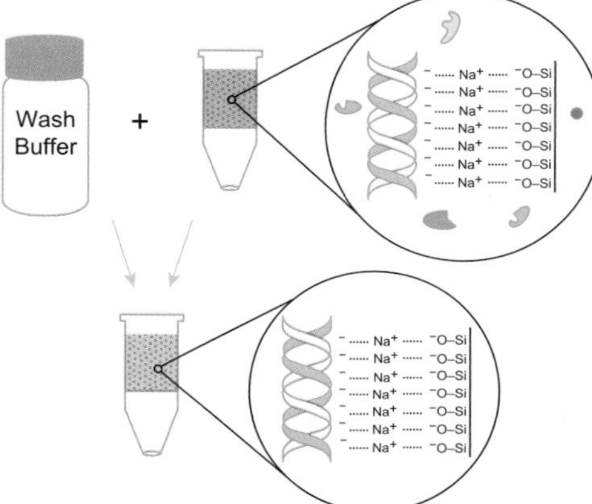

Step 4: Elute DNA from capture matrix

Elution Buffer

+

FIGURE II.5.17.3 Charge-based interaction system for concentration and purification of DNA.

Directed polyclonal or monoclonal antibodies are common tools used for the capture of a wide variety of analytes, ranging from small molecules to proteins, and even whole cells.

Biomaterials can play a key role in optimizing the binding kinetics for effective analyte capture and conditioning. The effective k_{on} rate for a binding event in a complex mixture includes the diffusion rate between the analyte and its cognate receptor. Presentation of the receptors in microfluidic platforms with large surface-to-volume ratios or on nanoparticles that simulate free solution conditions significantly accelerates the overall binding rate and efficiency. Paramagnetic nanoparticles that can be silicanized for nucleic acid capture or conjugated to antibodies for analyte capture offer the benefits of free suspension kinetics during the binding events, and the possibility of analyte concentration through exposure to a magnetic field. A similar concept applies to stimuli-responsive "smart" polymers, which behave as solubilized reagents under a certain pH or temperature, but act as large aggregates when the pH or temperature is modified.

THE ANALYTICAL PHASE

Over the years, many different types of IVD assays have been developed, and new types are being developed daily. To review them all exceeds the scope of this chapter. (A summary of a few of the most common assays and the biomaterials they employ can be found in Table II.5.17.2.) Nevertheless, diagnostic assay design has its lexicon, which concisely describes the general principles that underlie the design. These terms and concepts are briefly reviewed here, and can be used to differentiate the bewildering array of extant assay types. In the sections that follow, the term *sample* will be assumed to mean the output of the pre-analytical processes discussed above.

Technologies resolving the analytical phase of the diagnostic test interface with the pre-analytical phase where the sample is prepared or purified, when the user performs the steps required to conduct the assay, and at interpretation of the results. Each of these interfaces must be considered when selecting the appropriate method and biomaterial for the task (Figure II.5.17.4). Failure to

TABLE II.5.17.2 Several Common Assay Types, Biomaterials Used, and their Suitability in Various Contexts

Platform	Types of Tests	Biomaterials Utilized	Where Used
PCR	Nucleic acid detection Genetic testing High-sensitivity applications	Nuclease-free lab ware Oligonucleotides Fluorescent labels	Central or reference laboratory
Enzyme-linked immuno-sorbent assay (ELISA)	Protein detection Antibody detection	Coated lab ware Antigens Antibodies Blockers Enzymes/substrate	Central, reference or general laboratory
Flow cytometry	Cell detection and differentiation	Antibodies Fluorescent labels	Central, reference or specialized laboratory
Microscopy	Direct observation Cell differentiation Cell morphology	Slides (surface prep.) Dyes Antibodies	Central, reference or clinic laboratory
Lateral flow strip test	Antibody detection Protein detection Chemistry	Engineered "papers" (strips and pads) Antigens Antibodies Blockers Colormetric labels	POC or near patient (e.g., home use, bedside, doctor's office)
Arrays	Genetic testing Protein detection Complex algorithm	Functionalized glass or plastic substrates Antigens Antibodies Oligonucleotides Fluorescent labels	Central or reference laboratory

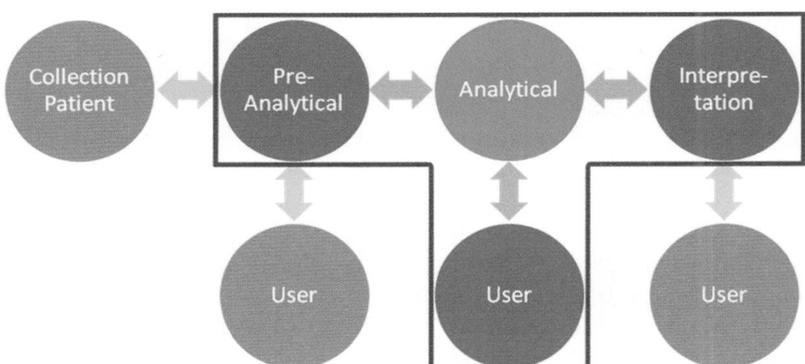

FIGURE II.5.17.4 The analytical phase of the diagnostic process receives biomaterial from the pre-analytical phase, and captures and/or detects the target moiety generating a signal or output for interpretation. The user may be a laboratory technician involved in detailed procedures of the test or if the analytical process is integrated into the test, the device user may be detached from the analytical phase. Biomaterials used in the analytical process may include antigens, antibodies, nucleic acid sequences, proteins, and enzymes.

address these interfaces will result in an analytical system that is inadequate.

Any analytical system can be described abstractly as a series of elements that interact to produce the system performance (Figure II.5.17.5). These modules include signal generation, signal modulation, signal transduction, noise filtering, amplification, response extraction, dose response modeling, and others (Christian, 2004; Strobel and Heineman, 1989). The designer manipulates some or all of these elements to achieve the specifications of the assay. There are many different ways to categorize assays. For example, they can be *homogeneous* (all of the steps needed to generate the response occur in the same compartment as the sample) or *heterogeneous* (separation steps isolate the response to a different compartment). They can be *qualitative* (returning a yes or no answer for analyte presence or identity) or *quantitative* (returning a cardinal value for analyte concentration or mass). All of these divisions are useful when discussing assays in particular contexts, but a general treatment should be careful to qualify the specific divisions used as an example.

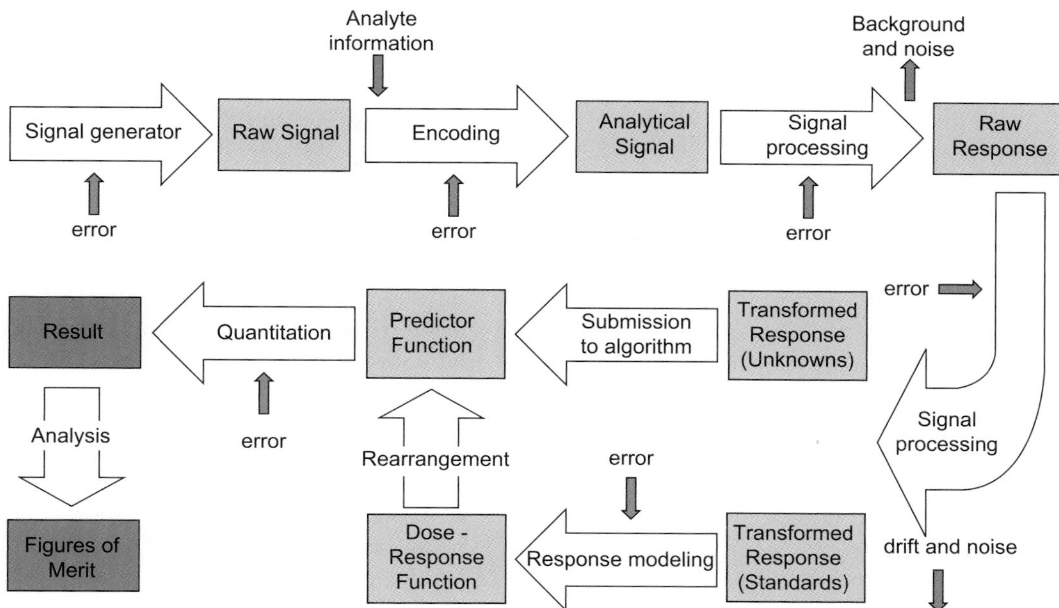

FIGURE II.5.17.5 A block diagram of the various modules or elements that comprise a generalized diagnostic assay design. The raw signal is encoded with the analyte information in the sample. The analyte response is extracted from the complex analytical signal. Various signal processing steps (amplification, noise filtering, data transformation, background correction, etc.) may be required to achieve a response-to-noise ratio favorable to facile modeling and quantization. Empirical response modeling (calibration) is enabled by the assay of standards. The dose–response function (standard curve) that results is used to determine the predictor function that is used to quantitate unknowns. Error is introduced at each step. Validation is demonstrated by the figures of merit. Predictor functions and validation are considered in the "Interpretation" section of this chapter. *(Reproduced with permission* (Hawkins, 2007).*)*

Capture and Detection

The most common types of IVDs are heterogeneous. At the heart of most heterogeneous diagnostic technologies is the ability to both *capture* and *detect* a specific analyte from a biological matrix. This is typically done via affinity-binding reactions (as discussed in Box 1 above). As described above, pre-analytical steps will improve analyte availability from the biological matrix. Technologies that eliminate or integrate pre-analytical steps to reduce assay complexity for the end user are increasing – especially in resource-limited locations where high demands are placed on the healthcare provider's time. These demands on the assay performance enhance the importance of analyte capture and detection capabilities, especially because it serves an enrichment and purification step as well.

Capture and detection are the processes that generate the analytical signal in the scheme shown in Figure II.5.17.5. Together they are sometimes referred to as the *assay principle.* (This term can be applied to other types of assays as well, sometimes with a different meaning.) These are the steps that are most visible to the clinician and medical technologist, while signal amplification and processing are frequently "hidden" in instrumentation or analysis software. While capture and detection must work in concert with the analyte of interest, we will first consider them as separate entities, and then examine how they combine to form a fully-functioning assay system.

BOX 1 The Kinetics and Thermodynamics of Affinity Capture

Key aspects to affinity capture and labeling of the analyte are the strength of the interactions between the analyte and the binding molecule defined by the dissociation equilibrium constant (K_d, in M) and the kinetics of binding defined by rate constants for association (k_{on}, in $M^{-1}min^{-1}$) and dissociation (k_{off}, in min^{-1}). For the interaction between a binding molecule B and the analyte A, at equilibrium, $A + B = AB$, $k_{on}[A][B] = k_{off}[AB]$, and $K_D = [A][B]/[AB] = k_{on}/k_{off}$. From these relationships one can infer that small dissociation constants denote strong affinity between A and B, and that the proportion of analyte bound to the capture molecule is related to the available concentrations of analyte and capture molecule. To capture a significant portion of the analyte in purification, the sample must be exposed to an excess of high-affinity binding sites. It is also generally true that a high k_{on} rate (for fast binding) and a low k_{off} rate (so that the complex remains associated) is optimal. If release of the analyte is required, then the capture biomaterial must be selected such that the affinity of the binding reaction can be reduced through a change in properties of the mobile-phase diluent (e.g., a pH or ionic strength change) which elutes the analyte. Consideration must then be given to harmonize the elution conditions to the requirements of the downstream analytical operations. For many of the highest affinity interactions, however, such elution is not possible without damaging the analyte. In these cases, the last step of the pre-analytical process (*concentration/purification*) must coincide with the first step of the analytical process (*analyte capture*). As such, the K_d of the binding molecule can represent a boundary to the LOD of an analytical assay, k_{on} can affect the throughput, and k_{off} can affect the stability of the response once it has been generated.

Most heterogeneous diagnostic platforms have a solid phase and a mobile phase, both of which can be engineered biomaterials. The solid phase is a functionalized material that will capture the analyte of interest. The mobile phase is a solution containing the analyte, other matrix components that may remain after the pre-analytical process, and frequently other biomolecules or biomaterials necessary for the assay principle.

Examples of solid phase materials include the polystyrene plates used in the ELISA, nitrocellulose or nylon membranes used in lateral flow strip tests, and glass surfaces such as slides for microscopy or chips for arrays. Capture of the target analyte can occur via specific binding or nonspecific binding. Nonspecific binding can occur by exploiting the natural properties of the solid phase, such as the ability of nitrocellulose to bind protein or by tethering a reagent that will nonspecifically bind the target analyte to the solid phase. If, for example, one is developing an assay to detect human antibodies directed toward a specific target, the solid phase may consist of a protein, such as protein A or an anti-human immunoglobulin G (IgG), used to nonspecifically capture all IgG in the sample. Specific capture almost exclusively uses a protein or nucleic acid sequence, such as an oligonucleotide or aptamer, to capture the analyte of interest with high-affinity while being relatively insensitive to other species in the sample. Examples include affinity-purified antibodies to the target or a protein that expresses the desired epitope to detect the antibody of interest. These capture biomolecules are tethered to the solid phase in a manner that allows their presentation to the sample in the mobile phase.

Similar to capture, the detector molecule can be either specific or nonspecific in its analyte recognition. Typically, the detector is in the mobile phase. Most detectors also include a signal transduction step (discussed in more detail later in this section). The fundamental assay principle (capture and detection through specific molecular-binding reactions) is transduced into a different type of signal (e.g., fluorescence) by a bifunctional detector molecule (e.g., an antibody labeled with a fluorophore). Assays that are interpreted visually or spectrophotometrically utilize a detector labeled with a dye, colored particle, fluorescent reporter or enzyme. This reporter is tethered to the detector. Linkage to the detector will depend on the chemistry of the reporter and detector. It may be able to occur via ionic bonding or through covalent linkages made possible with a number of chemical cross-linkers.

Assay Formats

A successful assay can combine several of the components mentioned above. As examples, we will describe some of the common heterogeneous immunoassay configurations and examples of the biomaterials used in each. The assay principles illustrated in these formats can be applied to specific recognition molecules other than antibodies. The extension of these formats by using oligonucleotides, aptamers, and molecularly-imprinted polymers is a very active area of research. Signal types may also vary, although colorimetric and fluorometric methods are still the most common.

1. *Antigen detection assays*: These assay formats focus on the detection of a specific antigen in a sample by using its specific cognate antibody. The analyte may be a protein, protein fragment, lipid, carbohydrate or nucleic acid. The analyte may be produced by the host or by an organism infecting the host or it may be indicative of a substance in the host. No matter what the source of the analyte, it must be validated as a marker predictive of the disease, condition or state being assessed. See Figure II.5.17.6.
2. *Serological detection*: Here, *serology* refers specifically to the detection of antibodies in a sample. Antibodies are produced in a host's blood in response to a foreign antigen or infection. Several classes of antibodies are produced, and each has its own half-life relative to the host's state, disease or condition. The most common antibody is IgG, but immunoglobulin A (IgA) and immunoglobulin M (IgM) are also produced, and can be important in diagnostic tests, especially when there is a need to differentiate new infections from previous or latent infections. Since the production of antibodies is a systemic response by the host to what can be a local infection, serology tests are often useful when it is impossible or difficult to test for or collect a sample containing the antigen or nucleic acid. Additionally, serology is useful for determining if a host has been exposed to an antigen or infection after the host has successfully cleared the source of the antigen or infection. See Figure II.5.17.7.

Assay Platforms

It is beyond the scope of this chapter to describe every platform utilized in diagnostic testing. New platforms continue to be developed using the assay formats described in the previous section. Examples of several heterogeneous assay formats include flow-through devices (Figure II.5.17.8), agglutination devices (Figure II.5.17.9), solid-phase devices (Figure II.5.17.10), and lateral flow strip devices (Figure II.5.17.11). Each of these formats have been successfully commercialized and are suitable for use in POC diagnostics. The application of biomaterials in these platforms is extensive and has already been discussed (Table II.5.17.2). Platforms for gold-standard testing in high-resource laboratories tend to use more homogeneous assay principles (although bead-based heterogeneous assays are also common), and are usually focused on very precise, high-throughput

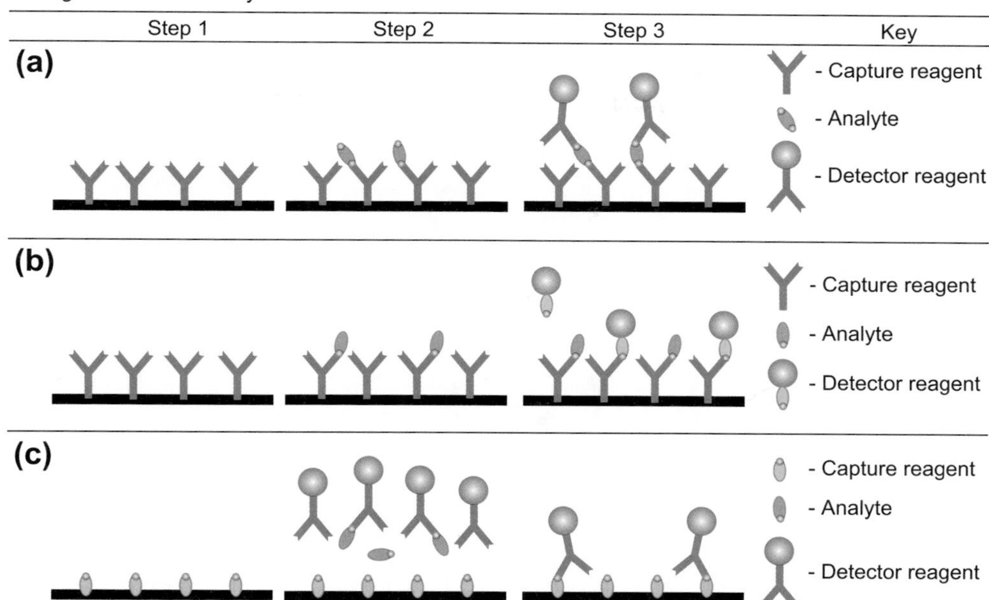

FIGURE II.5.17.6 Antigen detection assay formats: (a) Sandwich assay. Step (1) A capture reagent specific to the analyte is immobilized on the solid phase. Step (2) The sample is introduced through the mobile phase. If the analyte is present in the sample, it will bind to the capture reagent on the solid phase through its cognate binding site. Step (3) The detector reagent that is specific to the analyte is introduced through the mobile phase. If the analyte is present in the sample, the detector reagent will specifically bind to it, forming a sandwich and resulting in a signal. If no sample is present, the detector reagent has no place to bind and is washed away. This format works well when the analyte has at least two distinct epitopes or multiple repeats of the same epitope. (b) Competitive assay. Step (1) A capture reagent specific to the analyte is immobilized on the solid phase. Step (2) The sample is introduced through the mobile phase. If the analyte is present in the sample, it will bind to the capture reagent on the solid phase through its cognate binding site. Step (3) The detector reagent is introduced through the mobile phase. In this case, the detector reagent presents an epitope analogous to that of the analyte. If the analyte is present in the sample it will compete with the detector reagent and no signal (or a reduced signal) will be present. If no sample is present, the capture reagent is available to bind the detector reagent, resulting in a maximal signal. Interpretation of this format is somewhat counterintuitive, since the analytical signal is greatest when the analyte is not present in the sample. This format is well-suited for analytes with only one epitope, such as a small protein or xenobiotic. (c) Inhibition assay. Step (1) A capture reagent analogous to that of the analyte is immobilized on the solid phase. Step (2) The detector reagent is specific to the analyte. The sample is premixed with the detector reagent prior to introduction to the solid phase. If the analyte is present in the sample, it will bind to the capture reagent on the solid phase through its cognate binding site. Step (3) If the sample contains the analyte, it will bind to the detector reagent and inhibit its ability to be captured on the solid phase, resulting in no signal or a reduced signal. If no analyte is present in the sample, the detector reagent is available to bind to the capture reagent on the solid phase, resulting in a maximal signal. Similar to the competitive assay, interpretation of this format is somewhat counterintuitive, since the analytical signal is greatest when the analyte is not present in the sample. This format is well-suited for analytes with only one epitope, such as a small protein or xenobiotic.

testing achieved with robotic liquid handling of solutions and parallel processing.

Amplification

For many analytes assayed on conventional assay platforms, the required assay performance dictates that direct labeling and simple detection is inadequate. These analytes require amplification. Amplification can be of two distinct types:

1. *Target or marker amplification* is most frequently found in nucleic acid testing, where a specific nucleic acid sequence is replicated exponentially, thus increasing the amount available for detection and increasing both sensitivity and specificity of the assay. The most widely used nucleic acid amplification is PCR, the details of which can be found in any number of textbooks (McPherson and Müller, 2000; Meltzer, 1998). This method utilizes specific unique enzymes, primers, reagents, and temperature cycling to amplify the target. PCR is especially sensitive to contaminants and inhibitors; thus, pre-analytical methods are very important. Other isothermal nucleic acid amplification methods recently have been developed that utilize similar principles to specifically increase nucleic acid sequences, but frequently use enzymes that are less susceptible to contamination. Biomaterial designs intended to speed up these processes or localize their products must consider how reactant, heat transport, and enzyme kinetics are altered by reaction geometry and biomaterial chemistry.
2. *Signal amplification* is often done for protein, antibody, and xenobiotic tests.

Some signal amplification is obvious, e.g., electronic gain and optical magnification in a microscope.

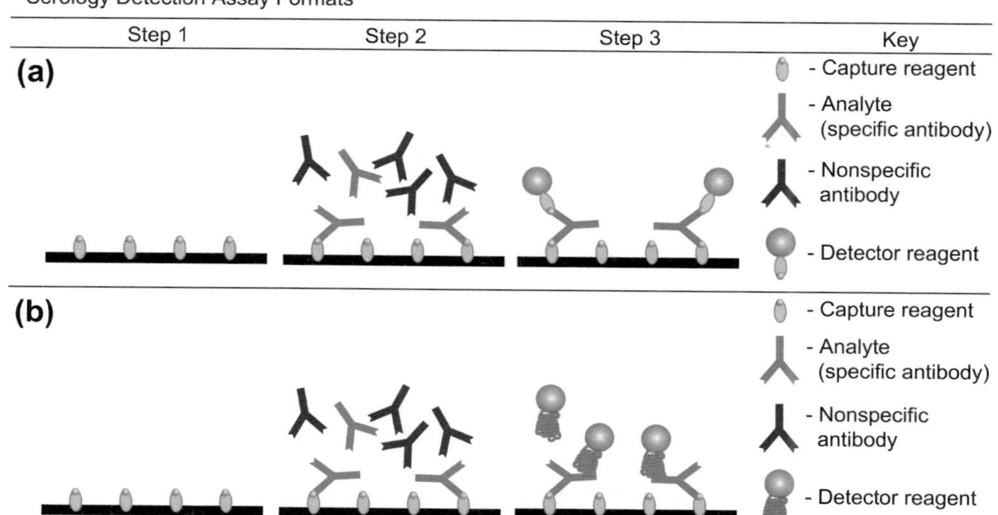

FIGURE II.5.17.7 Serology detection assay formats. (a) Antigen–antibody sandwich assay. Step (1) A capture reagent with the epitope specific to the antibody of interest is immobilized on the solid phase. Step (2) The sample is introduced through the mobile phase. If the specific antibody of interest is present in the sample, it will bind to the capture reagent on the solid phase through its cognate binding site. Nonspecific antibodies present in the sample will not bind to the solid phase and will be washed away. Step (3) Similar to the capture reagent, the detector reagent contains the epitope specific to the antibody of interest. The detector reagent will bind only to the specific antibody of interest, forming a sandwich and resulting in a signal. If the specific antibody is not present, the detector reagent has nothing to bind to and will be washed away, resulting in no signal. (b) Indirect assay. Step (1) A capture reagent with the epitope specific to the antibody of interest is immobilized on the solid phase. Step (2) The sample is introduced through the mobile phase. If the specific antibody of interest is present in the sample, it will bind to the capture reagent on the solid phase through its cognate binding site. Nonspecific antibodies present in the sample will not bind to the solid phase and will be washed away. Step (3) The detector reagent contains an antibody-binding domain that will bind all of the antibodies in the sample – those specific to the epitope of interest and those not directed to the epitope of interest. While the detector binds to all antibodies, due to the specificity of the capture reagent and the wash steps samples with the antibody of interest will generate a signal, while samples without the antibody of interest will have no signal. This assay format can also be reversed, with the specific epitope on the detector and the nonspecific antibody-binding domain immobilized on the solid phase acting as the capture reagent.

However, signal amplification can also be chemical, like the amplification that results when an enzyme is coupled to a detector molecule in a heterogeneous assay. The amount of product produced by the catalytic activity of the enzyme is proportional to the amount of enzyme immobilized by the analyte capture (in the presence of an excess of the enzymes substrate), but can produce a much larger signal.

Enzyme amplification is at the heart of several interesting homogeneous assay principles – the enzyme-multiplied immunoassay technique (EMIT) and competitive enzyme donor immunoassay (CEDIA) – both workhorses in xenobiotic testing. In these techniques, the activity of an enzyme conjugated to an analog of the analyte is changed when bound to a cognate antibody. Competition between analyte in the sample and the analog labeled with enzyme for antibody binding sites serves to modulate the analytical signal.

Regardless of amplification method, signal amplification often amplifies the noise and variation in the analytical signal, as well as the analyte response – sometimes preferentially. Assay contrast is only increased if amplification steps increase the response-to-noise ratio. In cases where the assay principle produces an analytical signal that is inherently difficult to amplify, signal transduction may be desirable.

Signal Transduction

Frequently, the method used to detect the analyte does not generate an analytical signal that is amenable to measurement or amplification, and transduction of the signal into another mode is desirable, especially when simplicity and cost reduction are an end goal. As discussed above, the analyte-specific molecular-binding events that are the core detection principal of an immunoassay or nucleic acid hybridization assay are not easily detected without some linked reporter. Thus, the analyte-modulated chemical-binding signal is converted into a more easily measurable signal with the desired signal-to-noise ratio. Reporters typically used in diagnostics fall under several general categories, including electromagnetic, optical, and physical/mechanical. Several of these modes may be used in series. Signal transduction techniques remain an active area of research, as investigators strive to improve the performance and reduce the complexity or cost of diagnostic devices and methods.

Electromagnetic transduction is attractive, due to the ubiquity of electronic devices and the resulting advanced understanding of electronics. By placing electrodes on either side of a flow cell for liquid reactants or immobilizing specific analyte-capture molecules on an electrode surface, using a biomaterial, potentiometry, amperometry,

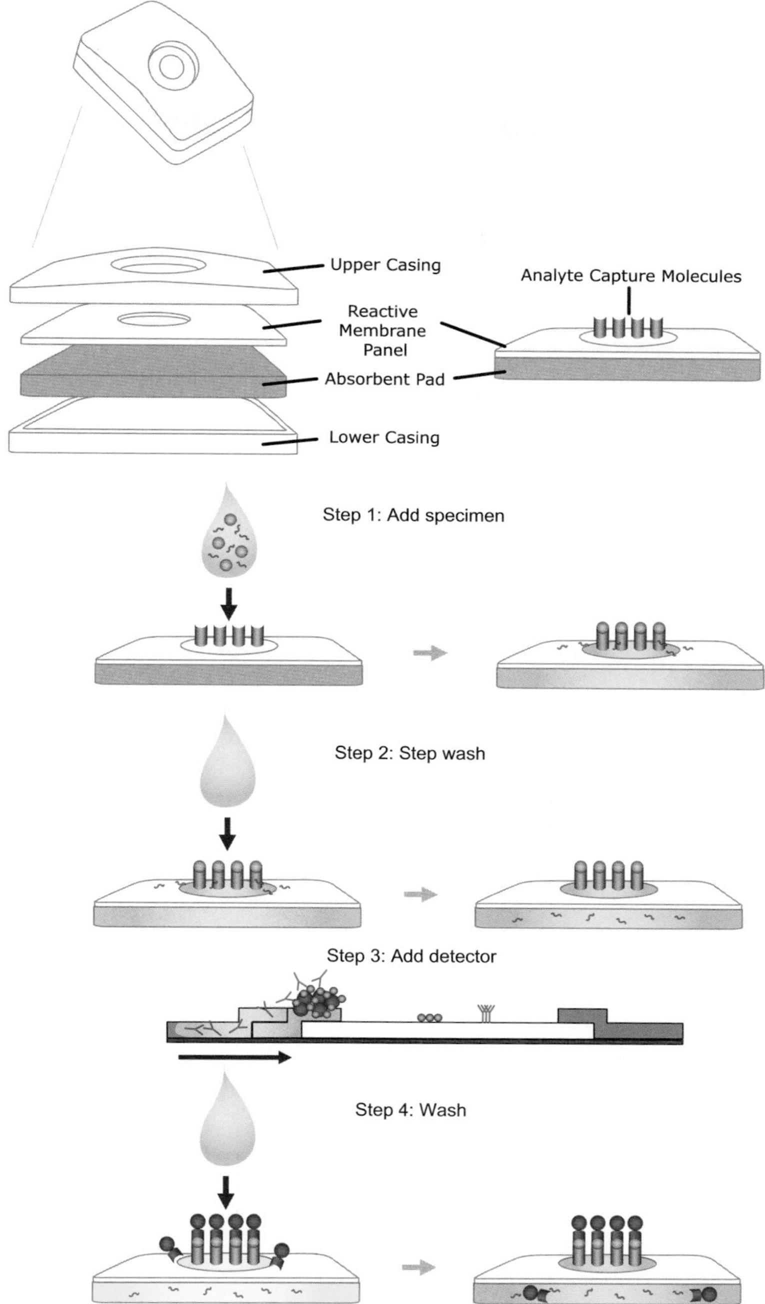

FIGURE II.5.17.8 Flow-through devices. Step (1) The specimen is diluted in buffer and dropped on the surface of the membrane. While the liquid flows through the membrane to the absorbent pad, the analyte is captured on the analyte capture molecule. Step (2) The membrane is washed to remove non-specific background binding. The wash buffer flows through the membrane, washing away material that is not bound by the analyte capture molecule. Step (3) A solution containing the detection molecule is dropped onto the membrane. The detection molecule consists of two components: one that binds the analyte specifically and another that facilitates visual detection. In most modern tests, visualization is due to the use of microscopic particles, such as latex beads or colloidal gold, which are visible when they aggregate. Some tests, however, utilize traditional enzyme immunoassay chemistry to provide a precipitate that is visible to the naked eye. As the liquid flows through the membrane into the absorbent pad, the detection molecule binds to the captured analyte. Step (4) The membrane is washed a second time. The wash buffer flows through the membrane, removing detector molecules that are not bound by the analyte capture molecule. A spot is visible on the membrane surface if analyte was present in the specimen.

and cyclic voltammetry can all be modulated by analyte concentration. Native electronic properties of the analyte, charge or spin labels coupled to specifically-binding reporter molecules, and changes in solution electromagnetic properties due to aggregation have all been exploited.

Optical methods have their roots in both microscopy and spectroscopy. Microscopic methods may be as straightforward as traditional examination of a fluorescently stained specimen by a trained pathologist or may involve image capture and analysis through complicated

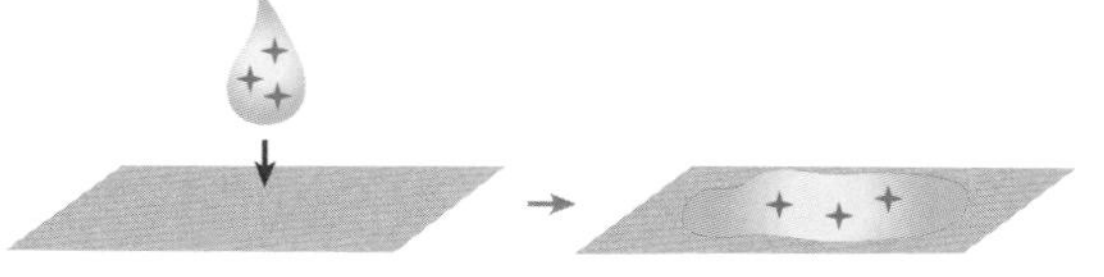

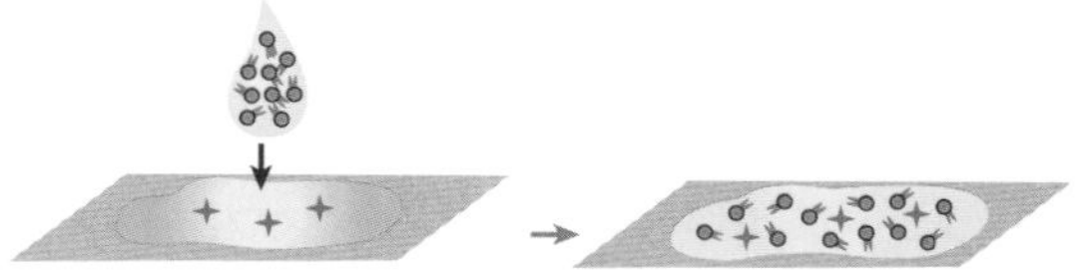

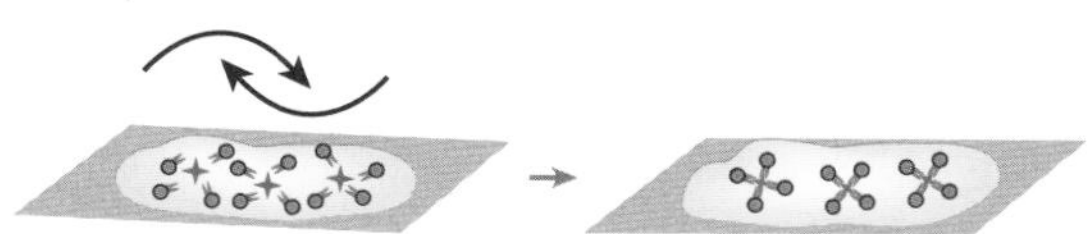

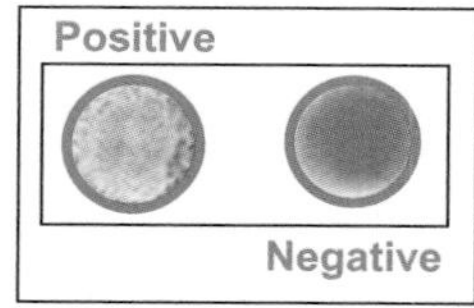

FIGURE II.5.17.9 Agglutination devices. Step (1) A drop of specimen is placed on a card. The analyte is floating freely in solution. Step (2) A solution that contains an analyte-specific detector molecule attached to a particle is added to the specimen on the card. Analyte and detector molecules initially float free of one another, causing the color and texture of the drop on the card to be uniform throughout. Step (3) The mixture is gently agitated, which causes the analyte and detector molecules to interact. Step (4) The interaction of the analyte and detector molecules forms visibly detectable aggregates. If analyte is not present in the specimen, aggregates do not form, and the color and texture of the mixture remains uniform throughout.

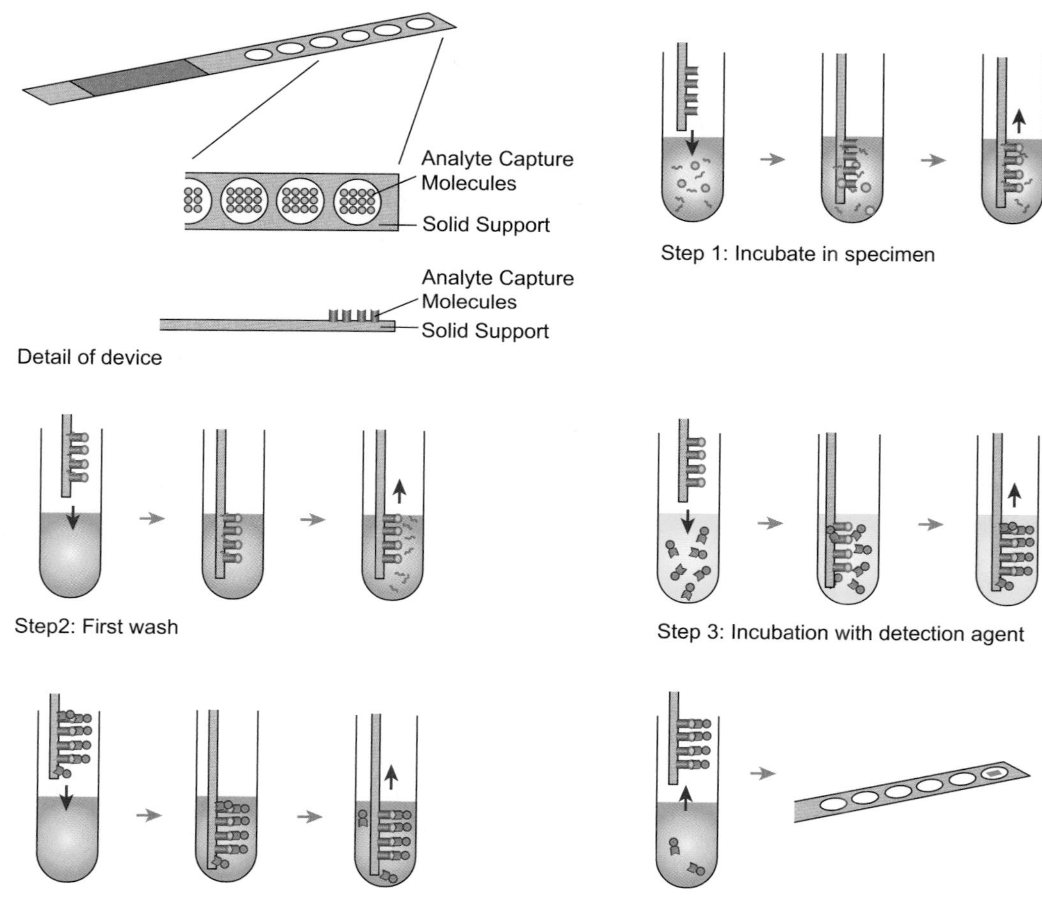

FIGURE II.5.17.10 Solid-phase devices. Step (1) The dipstick is placed in the specimen and incubated to allow analyte (if present) to bind to the analyte capture molecule. Nonspecific binding can occur with molecules other than analyte. Step (2) Nonspecific binding molecules are washed away by incubating the dipstick in wash buffer. Step (3) The dipstick is incubated in a solution containing the detection molecule. The detection molecule consists of two components: one that binds the analyte specifically; and another that facilitates visual detection. In most modern tests, visualization is due to the use of microscopic particles such as latex beads or colloidal gold, which are visible when they aggregate. Some tests, however, utilize traditional enzyme immunoassay chemistry to produce a precipitate that is visible to the naked eye. Some of the analyte detection molecule might stick nonspecifically to the solid substrate. Step (4) Nonspecific binding of detector molecule is washed away by a second incubation in wash buffer. Step (5) The dipstick is removed from the wash solution and read. A spot is visible if analyte is present in the specimen.

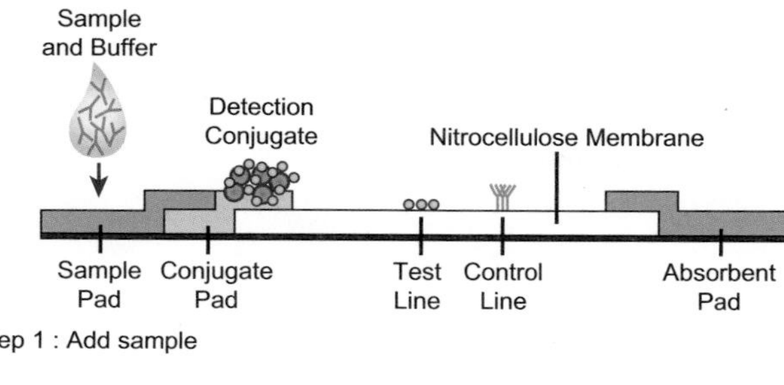

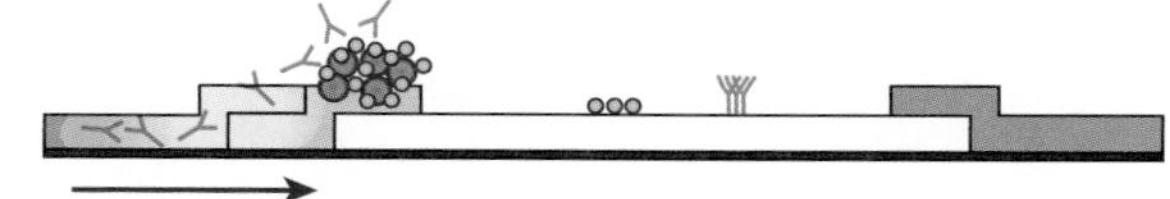

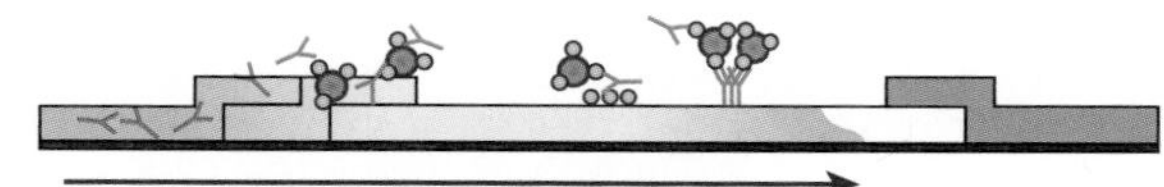

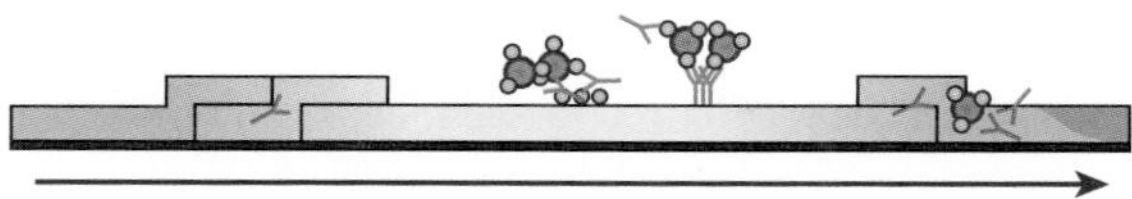

FIGURE II.5.17.11 Lateral flow strip devices. Step (1) The sample is placed on the sample pad at one end of the strip. The sample may be used alone, as is commonly done with urine- or serum-compatible tests or it may be mixed with a buffer specific to the test. This buffer may simply be a diluent or running buffer or it may be much more complex. It may have specific components or properties, such as a cell lyses buffer, that are required in order for the strip to perform properly. Step (2) With the addition of the sample, the detector molecules are solubilized. Once dissolved, the detector molecules mix with and bind to analyte present in the sample. Step (3) Capillary action draws the fluid mixture up the sample pad and into the membrane. The sample and detector molecule mix continues to move up the membrane until it reaches the analyte capture molecule. In these lines, a second and sometimes third antibody or antigen, immobilized as a thin stripe in the nitrocellulose, captures the complex if it is positive for the target analyte. The control line should always show as a visible line, otherwise the test is invalid and must be repeated. If the test is positive, a colored (typically pink or purple) line develops along with the control line. Step (4) When the colored line appears, the strip is fully developed. Excess buffer, along with reagents not captured at the test of control line, move into the absorbent pad. This description applies to a gold conjugate detector – a very common detection method. Details for other detectors may vary slightly.

expert system algorithms. A recent trend is to use image analysis for classification and counting of engineered analyte-specific reporters, such as quantum dots and fluorescent barcodes, frequently in a multiplexed assay. Spectroscopic interrogation is most frequent in the visual spectrum (*colorimetry*). Light absorption is modulated by the analyte through a linked chemical reaction or specific reporter (e.g., gold nanoparticles). Much greater analytical sensitivity is achievable through fluorescence spectroscopy (*fluorimetry*), primarily because of the superior signal-to-noise ratio inherent in properly configured fluorescence optics. However, when *low cost* and *disposable* are design objectives, much of this advantage can be lost. To minimize cost and complexity, simple visual readouts of colored or fluorescent reporters are frequently used, e.g., the indicating line on an immunochromatography strip (ICS). Here a chemical signal (*specific antibody-binding*) is transduced into a visual readout by the accumulation and aggregation of absorbing gold nanoparticles at the indicating line.

A number of other diverse transduction techniques are primarily physical or mechanical in nature. The simplest of these is the spatial segregation of the analyte from the rest of the sample through localization of a specific binding event. Spots on a microarray or the accumulation profile of a diffusion immunoassay are examples. More complicated mechanical transduction methods include mass sensing through changes in the resonant frequencies or non-oscillatory bending of piezoelectric structures or microcantilevers. Mechanical methods usually require another transduction step to enable an optical or electronic readout. This readout generally becomes more complex in proportion to the complexity of the mechanical transduction. For example, quartz crystal microbalance mass detection requires complex electronics to generate the oscillations and detect the change in quality factor associated with the specific binding, but it can detect exquisitely small masses of analyte.

Biomaterials play a key role in diagnostic device signal transduction. Polymers are used frequently as inexpensive and rugged diagnostic device materials; however, fluorescence detection methods can be confounded by the photobleaching of highly autofluorescent plastics. Surface immobilized patterned biomolecules can be used to spatially segregate analyte with great specificity. Anisotropic functionalization of a microcantilever enables entropy changes from specific binding to one side to provide the energy to distort the microcantilever. Equally important is the need to create non-fouling surfaces between the specific recognition sites, to ensure that nonspecific adsorption of the analyte to an attractive surface does not erode the desired specificity. (See Box 2 on blockers.) The choice of signal transduction method may hinge on the properties of the biomaterials used. New biomaterials create new methods for signal transduction that have the potential to overcome traditional limits on diagnostic devices and change the face of the field.

Response Algorithms and Noise Filtering

Once the final analytical signal has been generated, the response of the system to the analyte must be abstracted from the complex analytical signal (i.e., separated from noise and background). When this is done "by eye" – with operator discretion – the steps of the algorithm are

BOX 2 Blockers

The ideal diagnostic test is both highly sensitive and highly specific. Sensitivity and specificity are very closely related to the biomaterials selected for capture and detection of the analyte. Most biomaterials available for diagnostics have characteristics that innately allow nonspecific binding of materials other than the analyte of interest. This is true for the solid phase, capture reagent, and detector reagent. If appropriate cleaning procedures are not identified and followed, the performance of functional devices can be eroded by the deposit of biofilms. In order to minimize nonspecific binding of materials that are not of interest, diagnostic test developers go to great lengths to block these reactions.

Blocking can be grouped into three major categories: protein, surfactant, and chemical modification.

Protein blocking utilizes a protein, protein fragment or peptide to block nonspecific binding. The ultimate goal is to use the blocker protein to mimic the nonspecific binding. Utilizing a blocker that allows the analyte of interest to bind, and is not recognized by the detector, will maximize the analyte available for specific detection.

Surfactant blocking uses a chemical surfactant or detergent to prevent nonspecific binding from occurring. Ionic and nonionic detergents are commonly used. Effective concentrations of surfactant are generally below the critical micelle concentration for the surfactant. The harshness of the surfactant required for a system will generally be dictated by the robustness of the other biomaterials in the system. Surfactant blocking is generally most effective when implemented in the liquid phase.

Chemical modification involves changing the chemical structure of the solid phase, capture or detector molecule surface or surfaces available to the sample, to make them less likely to nonspecifically bind materials that are not of interest. The methods and procedures for chemical modification depend entirely on the chemical structure of the surface to be modified, and how it interacts with and reacts to its environment during the process of running the assay.

BOX 3 Immunoassay Calibration with the Log/Logit Model

A common dose–response model used in immunoassays is the *four-parameter log/logit* equation (Maciel, 1985).

$$R = \frac{R_0 - R_\infty}{1 + \left(\frac{[Ag]}{c}\right)^b} + R_\infty$$

Where R_0 is the response at vanishing analyte concentration, R_∞ is the response at infinite analyte concentration, *c* is the point on the antigen concentration *[Ag]* axis where the point of inflection is located, and *b* is the slope of the curve at the point of inflection. In order to empirically determine the standard curve, at least four calibration standards must be assayed: a reagent blank containing no analyte (R_0); a very high *[Ag]* standard to estimate R_∞; and two intermediate *[Ag]* standards to account for the degrees of freedom in *b* and *c*. For a full calibration of an immunoassay, six calibrators are typical. A properly chosen set of six calibrators will result in an interpolated function that adequately captures the inherently sigmoid character of the immunoassay dose response. This model can now be fitted easily with the nonlinear regression routines in Excel® or statistical software packages. Its adoption in the days before desktop computing partly arose from the fact that the native variables can be transformed to make the equation linear. By plotting R'_i versus *log* [*Ag*], where:

$$R'_i = \log \frac{R_0 - R_i}{R_i - R_\infty}$$

the dose response is expected to be linear. If appropriate log/log graph paper is used, a best-fit standard curve can be drawn by eye with a straight edge. The user can use this standard curve to quantify the *[Ag]* from any response in the dynamic range, simply by noting the point on the curve associated with that ordinate (R'_i), and projecting down to the axis to find the abscissa.

frequently cryptic, uncontrolled, and vary from operator to operator. Operator training can minimize this variation. Analytical instruments are more consistent, and sometimes more sensitive, but instruments must also be "trained" by the judicious application of response abstraction methods. Identifying the best algorithms to maximize the response-to-noise ratio can be one of the most challenging tasks of the diagnostic assay designer. At first, algorithms may not seem of interest to the biomaterial scientist; however, consideration of the nature and type of these algorithms suggests how biomaterials can impact robust calculations.

The response algorithm defines the dose response relationship, i.e., how the analytical signal responds to changes in the amount of analyte, both in terms of magnitude and variation. A good response algorithm yields a dose response that is predictable (low noise) and is modeled by a tractable monotonic mathematical function (*monotonicity* is discussed in the "Interpretation" section below). Molecular binding assays, such as immunoassays and nucleic acid hybridization assays, all have regions where the dose response is not constantly proportional, due to transport limitations and saturation of binding sites.

Sigmoidal functions, like the log/logit function, can serve as excellent models to describe the dose response of an immunoassay over its entire range. With an appropriate coordinate transformation, they can be fit as a simple line (see Box 3). The response may be fundamentally multivariate. While forcing a multivariate response into a univariate dose response model has advantages for simple interpretation and calculation, a large part of the usable modulation may be discarded.

Noise filtering is advisable at many steps of the response algorithm. The error in a diagnostic device may be systematic (background signal, signal anisotropy) or random (electronic noise, stochastic variation in molecular affinities). Systematic error is generally corrected for by an algorithm suggested by the nature of the error, e.g., in quantitative fluorescence microscopy, material autofluorescence can be removed by subtracting an image of the device before the assay materials are introduced. Random noise can be digitally filtered from the raw data sequence by electronic or digital filters. The right filter technique for the frequency of the noise can be found in any good text on signal processing.

Care must be taken to apply the noise filtering at the right point in the response algorithm; some transformations can preferentially amplify noise relative to the response.

Very low temporal frequency random noise manifests itself as test-to-test or run-to-run, variation. This type of error is best corrected by some form of analytical signal normalization. In order to normalize, a standard signal must be present. If part of the analytical signal is not affected by the response, it can be used as an internal standard, and all parts of the signal are then rescaled appropriately. Sometimes internal standards are introduced into the analytical signal to enable normalization, e.g., the internal standards used in analytical chromatography to rescale the chromatogram to correct for run-to-run differences in the column and detector response. External standards are also frequently used, both to calibrate and monitor the performance of a diagnostic assay. Calibration and standardization are discussed further in the "Interpretation" section below.

Review of this section suggests how the biomaterials used can be tuned to improve the performance of a diagnostic assay. The goal should be to identify biomaterials that increase the contrast in the assay – that is, steepen the dose–response curve (greater changes in response per unit change in analyte amount). An obvious way to increase contrast is to emphasize the difference between the blank response (no analyte) and the positive analyte case. This can be as simple as using the whitest paper possible for an immunochromatographic assay or finding a polymer for a microfluidic flow cell that has the lowest autofluorescence. Given the many possibilities for response abstraction and noise filtering, however, other methods for increasing contrast are suggested.

Any contribution to the analytical signal by the biomaterial should be easily correctable by being confined to a different domain than the analyte response. For example, if a polymer substrate used for immobilization of capture molecules had a grainy background signal, the spot size of the capture molecules should be made significantly larger than the grain size of the background. This allows a filter to be used to eliminate spatial frequency components of the analytical signal associated with the background, without significantly affecting the response to the analyte at the spots. Similarly, when intense illumination is used for quantitative fluorescence microscopy, some photobleaching of the flow cell is unavoidable. A flow cell material should be chosen that allows significant fluorescence changes to occur more slowly than the time during which the analytical signal is being gathered. In this way, an image subtraction (a simple high-pass filter) is still a valid correction. In the end, the ability to filter the noise will serve to increase the contrast as well, because the response will stand out of the background more prominently.

INTERPRETATION

The interpretation phase involves taking the response from the analytical phase, placing it in context, and assigning meaning to the response. Technologies that resolve the interpretation phase of the diagnostic test interface with the analytical phase, which generated the signal, and the user, who is seeking results of the diagnostic test (Figure II.5.17.12).

Calibration

A common approach in many diagnostics assays is to correct for run-to-run and day-to-day variation with routine recalibration, rather than attempting to eliminate that variation altogether. In a calibration routine, the dose–response curve is redefined using a set of standards (*calibrators*). Pre-existing knowledge about the expected dose–response model is applied to fit a curve to this calibrator set – a *standard curve* – for quantification of unknowns. Thus, the response of unknowns is normalized to the response of the standards. Depending on the form of the model, the magnitude of the variation being corrected for, and the assay dynamic range required, the calibration set may contain one standard or

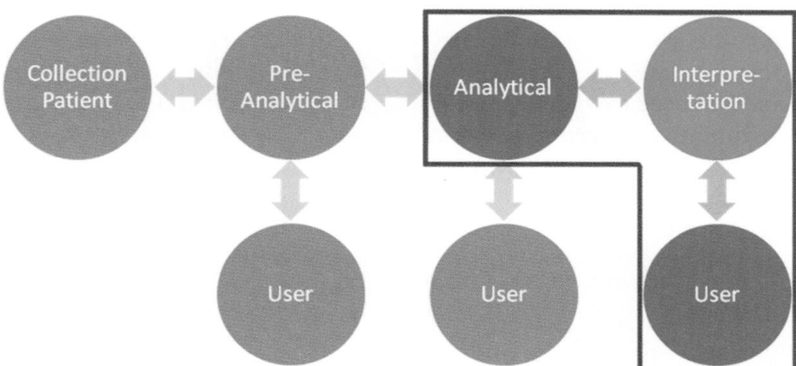

FIGURE II.5.17.12 Interpretation is the final phase in the diagnostic process. Input is received from the analytical phase. Assay calibration and controls are taken into consideration to classify the diagnostic test results in a manner that is meaningful to the user which may be a laboratory technician, healthcare provider or the patient.

many. Regression is done to fit the model, either in the native coordinate space or using transformed coordinates to simplify fitting of the curve. One of the most exciting recent developments in diagnostics is the engineering of a biomaterial to present calibration data contemporaneously with each unknown as it is assayed. This "instant recalibration" can result in greatly increased accuracy and precision.

In order to use the standard curve to quantify unknowns, the dose versus response standard curve must be rotated in "assay space" to become a response versus dose curve (i.e., the abscissa and ordinate must be reversed). This is usually referred to as a predictor function. The quantification algorithm uses the response of an unknown as an argument, and returns an analyte concentration. We can now see the importance of being able to apply a monotonic function for the dose–response model. Any function with a localized minimum or maximum on the dose versus response space when rotated into response versus dose space will have multiple possible analyte values for a single response – an obviously unacceptable situation. Any tuning of a responsive biomaterial must respect the absolute requirement for a monotonic response.

Readout and Classification

Most of the discussion in the last few sections focuses on quantitative diagnostic assays – assays in which a concentration, mass or count is the result. Many diagnostic assays, particularly those for infectious diseases, can be qualitative in nature. Although classification as positive or negative does not require routine calibration and construction of predictor functions by the end user, the assay designer and developer still need to consider all of the quantitative discussions above. No assay, qualitative or quantitative, can function without some dose–response relationship, and that relationship is usually continuous, not binary. But for qualitative classification, display, and readout purposes, the last step of a response algorithm may include a binary simplification. Even the development of a visible line on a lateral flow strip can be considered a binary simplification of the continuous relationship of analyte concentration and binding equilibrium – with the reader's vision as a threshholding instrument. End users, however, *will* want to be able to relate that classification to a known cut-off value, and still need to understand the limits of detection – both quantitative concepts. The best qualitative assays are good quantitative assays up to the last binary classification step, and it behooves a designer and developer to adopt that attitude.

Controls

Assay controls are an important aspect of diagnostics. Controls assist the user in ensuring that the steps of the assay were conducted, and that the results of the assay are valid and meaningful. A comprehensive discussion of good diagnostic quality control practices is beyond the scope of this chapter; however, this topic is so integral to good diagnostic practice that a summary is needed.

There are two types of controls: internal and external. Internal controls can take two forms:

1. A *procedural control* indicates that certain steps or events in the assay occurred, but does not determine if the assay was run correctly, e.g., a lateral flow strip test that develops a control signal when reagents hydrate with a sample diluent, but cannot differentiate between diluent with or without sample.
2. An *adequacy control* demonstrates that the procedures of the assay were properly conducted, e.g., a lateral flow strip test may have a control line that only develops a signal if the minimum amount of specimen is added to the device.

Of the two types of internal controls, an adequacy control is most meaningful for the device user. As suggested by the examples, engineering of the biomaterials in an assay can be a powerful way to include internal controls in the assay design.

For assays that have multiple channels and can run several samples in parallel, internal controls may not be possible (either due to the nature of the platform or the cost or logistics of running an internal control in each channel). These types of assays are typically best suited for external controls. External controls are generally used to determine the need for recalibration and to assess the proficiency of the system, operator or laboratory. Biomaterials science can contribute greatly to diagnostic technology by creating new ways to include controls and calibration in the assay of every sample.

SUMMARY

Biomolecules are *the* critical components of *in vitro* diagnostics. Ultimately, the interaction of one or more biomolecular species (antibodies, antigens, nucleic acids or enzymes) contained in the diagnostic device with biomolecular species in the sample (analyte) defines the capability, performance, and use conditions of the assay.

In POC assays, where reagents are frequently stored in a disposable container, diagnostic biomolecules are often contained in a biomaterial layer that interacts with a liquid sample. Most POC assays with biomaterial reagent layers that are used in clinical practice are either electrochemical or optical biosensors (e.g., glucometers), rapid strip assays (e.g., pregnancy strips) or slightly more complex cartridge-based assays that can handle assays for multiple targets.

Novel Techniques at Various Stages

POC diagnostics, and the biomaterials that comprise and enable them, are a very active area of research and development. To date, most POC diagnostics have been based on electrochemical or optical sensors, rapid strip tests or more complex cartridges.

An extremely successful example of a sensor-based test is the now ubiquitous glucose meter for diabetes monitoring (using glucose oxidase or glucose dehydrogenase in a thin-film electrode layer strip), first reported in 1962 (Clark and Lyons, 1962). Novel sensor-based assays continue to be introduced, including assays for cholesterol and cardiac markers. Sensor-based disposable assays are primarily used in home-care applications.

Rapid strip assays (e.g., the lateral flow strip test described in Figure II.5.17.11) also continue to be developed, especially for countries other than the United States, where pregnancy testing remains the primary application. The tests have a wide-range of applications including HIV, troponin T, malaria, fertility, respiratory disease, and drugs of abuse.

Cartridge-based systems, such as the Abbott i-STAT® or the Abaxis Piccolo®, are also electrode-based, but generally comprise multiple reagents in liquid or gel form on a cartridge in addition to the actual electrode layer. Typically, each cartridge can detect multiple analytes in a single drop of blood on panels for blood gases, electrolytes, blood chemistries, coagulation, cardiac markers, and hematology. Cartridge-based systems are used primarily in stat laboratories and in physician's offices.

New assays and cartridges for additional analytes, comprising novel biomarker receptors in biomaterial layers, are developed routinely for the platforms described above. New, even more powerful platforms are being developed for POC applications, especially those that can either increase sensitivity and specificity (e.g., those using nucleic acid amplification techniques), perform very complex assay procedures (e.g., microfluidic cell-based assays) or are highly multiplexed assays (e.g., nanoparticle-based diagnostic platforms).

Disposables

An important discussion in the diagnostic development community centers on the amount of complexity that should reside in a disposable device versus an instrument. There are many extant instruments that use robotics and automated pipetting to transfer liquid reagents and samples, and perform all assay steps in a sequence of reservoirs, tubes, and cuvettes. Such devices require some maintenance and cleaning, but use relatively little in biomaterials or costly custom reagents. On the other hand, almost all reaction steps can be integrated using complex engineered biomaterials and cartridge-based disposables without the need for complex instrumentation. The field of microfluidics has essentially enabled any laboratory process to be performed on a laboratory chip. All that is needed to drive and read them are pumps and electrodes – and in some cases, only fingers and eyeballs.

The question that has emerged is one of trade-offs. Integrated assays are more complex and frequently costlier on a per-test basis than instrument-based assays, but they require a smaller investment for instrumentation and can be used in settings that do not have access to highly trained laboratory personnel.

Biomaterials may help diagnostics achieve the benefits of an integrated disposable without some of the mechanical complexity of microfluidic cartridges. Biomaterials can be engineered in adjacent layers and reagent pads, such that assays requiring multi-step reactions can be performed without externally driven fluid transport and transfer steps. As an example, biomaterials similar to ones used on lateral flow strips can be combined on paper strips to form so-called paper microfluidics assays (Fu et al., 2010; Martinez et al., 2009).

Direction of New Diagnostic Platforms

New diagnostic devices are developed using novel bioengineering and biomaterials-based advances to address many very different requirements, (Jain, 2010; Yager et al., 2008) including: (1) improved sensitivity and specificity to detect analytes at ever lower levels and in ever smaller samples that are less painful to obtain; (2) added biomarkers to address many more disease parameters; (3) improved multiplexing to test for multiple analytes in a single sample; (4) improved throughput to run many different samples in a short period; (5) reduced complexity to allow testing outside of laboratories by personnel with minimal training; (6) reduced cost to allow their use in low-resource settings that cannot afford most laboratory-based technologies common in the developed world; and (7) better communication among platforms, and between platforms and existing databases. This last advance will allow rapid interpretation of individual diagnostic results in the context of the overall health of a patient, and take into account medication and treatment, the epidemiological environment and, in the near future, predictive information such as fully sequenced individual genomes or genetic disease markers.

A few new technologies that usually comprise biomaterials stand out as examples of novel technologies that are currently finding their way into commercial diagnostic devices.

1. *Microfludics*: Microfluidics deals with the behavior, precise control, and manipulation of fluids that are geometrically constrained to a small (typically submillimeter) scale. Some of the advantages of microfluidic devices include less waste, lower cost of reagents, and smaller sample volumes for diagnostics, faster analysis and response times due to short diffusion distances, fast heating, high surface-to-volume ratios, small heat capacities, and compactness of the systems.

   Over the past 20 years, a variety of microfluidics-based diagnostic devices have been developed by various research groups. These devices hold promise for miniaturization of analysis equipment, improvement in response times, and simplification of analysis procedures. These systems, a few of which have entered commercial use, typically consist of a small microfluidic chip surrounded by a desktop-sized analysis instrument (Guia and Xu, 2005; Gulliksen et al., 2005; Mcmillan, 2002).

Further integration of electronic and fluidic components will allow additional miniaturization and simplification of the analysis instrument. Another means to further miniaturize instruments and integrate elements of micro total analysis systems (μTAS) is to eliminate as many power-consuming and otherwise complex elements as possible. These elements can be replaced with passive components that operate without external power by manipulating fluids using gravity, air pressure or simple manual actions (Weigl et al., 2008).

Current major barriers against widespread adoption of microfluidics in diagnostic devices include the lack of standardization, and of a large-scale, low-cost manufacturing method for microfluidic disposables.

2. *Micro- and nanoparticle diagnostics*: Biomaterials in the form of small particles have found widespread use in diagnostics as capture and detection tools. Because many different types can be combined in a single assay and individually coded with marker molecules, functionalized particles enable a high degree of multiplexing (i.e., assaying for more than one analyte from a single sample). In addition, the interaction between a biomaterial and analyte molecules contained in a sample volume can be accelerated if the biomaterial is provided in the form of particles rather than a surface – the diffusion distances between the reaction partners are shorter, and more biomaterial surface can be provided. This is especially important if the analytes are large molecules or even bio-particles themselves, such as cells. Biomaterial particles can be magnetic, and thus can be captured and concentrated after interaction with the analyte using a magnet. Different types of particles can also be labeled with different markers (e.g., fluorescent dyes) that allow parallel detection of several analytes in a single sample volume (Sekhon and Kamboj, 2010).

   Biomaterial particles are in commercial use in multiplexed diagnostic systems such as the Luminex (Dunbar, 2006), Illumina (Fan et al., 2006), and Nanostring (Geiss et al., 2008; Malkov et al., 2009) platforms.
3. *Microelectrical sensors (*Arruda et al., 2009) utilizing silica nanowire (Cui et al., 2001; Jain, 2007), impedimetric, surface acoustic wave, magnetic nanoparticle, and microantenna technologies are being developed that offer a significant advantage to conventional optical protein assays – the target molecules do not need to be labeled. Their ability to transduce protein-binding events into electrical signals, and their ability to be easily integrated with microarray, microfluidic, and telemetry technologies, makes them potentially cheaper and easier to use. However, common limitations associated with the microelectrical sensors, including problems with sensor fabrication and sensitivity, must first be resolved.
4. *Nucleic acid amplification assays*: PCR or the more recently developed isothermal amplification methodologies (Gill and Ghaemi, 2008; Morisset et al., 2008) have not traditionally included biomaterials. In recent efforts to develop nucleic acid assays for POC and low-resource diagnostic applications, however, reagents for those assays now frequently are provided as stabilized surface coatings in vials or microfluidic channels. Further, detection of amplified nucleic acid sequences has also been demonstrated with lateral flow strips, some of which are integrated in the cartridges to avoid work area contamination.
5. *Prognostic and predictive biomarkers*: As multiplexing capabilities become more widespread, and genomic- and proteomics-based surveys reveal ever more biomarkers that are associated with a wide variety of diseases, disease states, and genetic predispositions, the field of prognostic and predictive diagnostics is emerging as a significant field of research. While in some cases a single biomarker is predictive, in general most of these assays need to be multiplexed to have predictive value, as the underlying proteomic patterns are usually complex and influenced by many different conditions. IVD for metabolic diseases, as well as cancer and a variety of chronic and aging-related diseases, will soon become possible. Already, existing but expensive platforms such as the biomaterial particles are in commercial use in multiplexed diagnostic systems such as the Luminex (Dunbar, 2006), Illumina (Fan et al., 2006), and Nanostring (Geiss et al., 2008; Malkov et al., 2009) platforms will likely be joined soon by other much more cost-effective, massively multiplexed biomarker assays (Simon, 2010).
6. *Phone-based applications*: Cell phones are emerging as a unifying principle for POC diagnostics (Blobel et al., 2008). Cell phone cameras serve as imaging devices (Braun et al., 2005), and on-chip computing capability can provide some image processing; when additional computing is needed, images can be sent to larger imaging stations wirelessly (McLean et al., 2009). These imaging devices can be coupled to simple microscope attachments (Frean, 2007) that have been shown to be capable of detecting, for example, malaria parasites in erythrocytes (Breslauer et al., 2009) or differentiation between normal and abnormal pap smears. Further, cameras can simply capture and transmit visually readable results from devices such as lateral flow strip tests (Martinez et al., 2008).

   Beyond imaging, cell phones can provide a data link for any electronic diagnostic device to a laboratory information system or a telemedicine provider.

Combined with other innovations that allow multiplexing, predictive diagnostics, and simplification of complex assays, they may usher in a new era of telemedicine-enabled, prevention and cost-effectiveness focused and personalized medicine in which most routine diagnostic monitoring happens in people's homes.

Defining the Requirements for New Diagnostics

While diagnostic devices have evolved enormously over the last century, the diagnosis of a disease – and especially the prediction of its course – is still a very inexact science. Even in high-resource medical settings, for most syndromes only a few laboratory parameters are being measured – and those usually only once or twice. From recent work on genomics and proteomics, however, we know that thousands of proteins in many more configurations interact with myriad other biomolecules within and outside of cells. In principle, the presence, concentration, and change in any of these biomolecules can provide diagnostic information (Bolouri, 2010).

The tests developed so far are only scratching the surface of what may be possible in the future. At the moment, most diagnostic tests can detect only a few biomarkers with sufficient sensitivity to be useful. They can measure them only at a single point in time, as a bulk quantity, and in a bodily fluid that is not localized in individual organs or cells. Finally, the tests do not correlate the biomarkers to a patient's individual medical history or genetic make-up very well. Over time, new devices will be developed that can address these issues and provide a much more comprehensive and predictive picture of a person's health (Blair, 2010). Their costs promise to be significant, and barriers to access may be substantial. One hopes that these devices will benefit not only the select few, but patients in all settings, regardless of wealth.

ACKNOWLEDGMENTS

Support for this chapter was provided through funding by Award Number U54EB007949 from the National Institute of Biomedical Imaging and Bioengineering for the Center for Point-of-Care Diagnostics for Global Health. The content is solely the responsibility of the authors and does not necessarily represent the official views of the National Institute of Biomedical Imaging and Bioengineering or the National Institutes of Health.

BIBLIOGRAPHY

Arruda, D. L., Wilson, W. C., Nguyen, C., Yao, Q. W., Caiazzo, R. J., Talpasanu, I., Dow, D. E., & Liu, B. C. (2009). Microelectrical sensors as emerging platforms for protein biomarker detection in point-of-care diagnostics. *Expert Review of Molecular Diagnostics*, **9**(7), 749–755. PM:19817557.

Blair, E. D. (2010). Molecular diagnostics and personalized medicine: Value-assessed opportunities for multiple stakeholders. *Personalized Medicine*, **7**(2), 143–161. ISI:000276030800009.

Blobel, B., Pharow, P., & Nerlich, M. (2008). *eHealth: Combining Health Telematics, Telemedicine, Biomedical Engineering, and Bioinformatics to the Edge: Global Experts Summit Textbook*. Presented at: IOS Press, Amsterdam; Washington, DC.

Bolouri, H. (2010). *Personal Genomics and Personalized Medicine*. London; Singapore; Hackensack, NJ: Imperial College Press. Distributed by World Scientific Pub.

Braun, R. P., Vecchietti, J. L., Thomas, L., Prins, C., French, L. E., Gewirtzman, A. J., Saurat, J. H., & Salomon, D. (2005). Telemedical wound care using a new generation of mobile telephones: A feasibility study. *Archives of Dermatology*, **141**(2), 254–258. ISI:000226913500017.

Breslauer, D. N., Maamari, R. N., Switz, N. A., Lam, W. A., & Fletcher, D. A. (2009). Mobile phone based clinical microscopy for global health applications. *PLoS ONE*, **4**(7). ISI:000268260100006.

Christian, G. D. (2004). *Analytical Chemistry*. Hoboken, NJ: J. Wiley.

Clark, L. C. J., & Lyons, C. (1962). Electrode systems for continuous monitoring in cardiovascular surgery. *Annals of The New York Academy of Sciences*, **102**. MEDLINE:14021529.

Cui, Y., Wei, Q., Park, H., & Lieber, C. M. (2001). Nanowire nanosensors for highly sensitive and selective detection of biological and chemical species. *Science*, **293**(5533), 1289–1292. PM:11509722.

Dunbar, S. A. (2006). Applications of Luminex® xMAP™ technology for rapid, high-throughput multiplexed nucleic acid detection. *Clinica Chimica Acta*, **363**(1-2), 71–82. http://www.sciencedirect.com/science/article/B6T57-4GWC13H-1/2/5f9f50db6a3e08a014df523a6d7bd5d9.

Fan, J. B., Gunderson, K. L., Bibikova, M., Yeakley, J. M., Chen, J., Garcia, E. W., Lebruska, L. L., Laurent, M., Shen, R., & Barker, D. (2006). Illumina Universal Bead Arrays. *Methods in Enzymology*, **410**, 57–73.

Frean, J. (2007). Microscopic images transmitted by mobile cameraphone. *Transactions of the Royal Society of Tropical Medicine and Hygiene*, **101**(10), 1053–1054. ISI:000250027800020.

Fu, E., Lutz, B., Kauffman, P., & Yager, P. (2010). Controlled reagent transport in disposable 2D paper networks. *Lab on a Chip*, **10**(7). MEDLINE:20300678.

Geiss, G. K., Bumgarner, R. E., Birditt, B., Dahl, T., Dowidar, N., Dunaway, D. L., Fell, H. P., Ferree, S., George, R. D., Grogan, T., James, J. J., Maysuria, M., Mitton, J. D., Oliveri, P., Osborn, J. L., Peng, T., Ratcliffe, A. L., Webster, P. J., Davidson, E. H., & Hood, L. (2008). Direct multiplexed measurement of gene expression with color-coded probe pairs. *Nature Biotechnology*, **26**(3), 317–325. ISI:000254123400025.

Gill, P., & Ghaemi, A. (2008). Nucleic acid isothermal amplification technologies: A review. *Nucleosides, Nucleotides and Nucleic Acids*, **27**(3), 224–243. http://www.informaworld.com/10.1080/15257770701845204.

Guia, A., & Xu, J. (2005). Planar electrodes: The future of voltage clamp. In J. X. J. Yuan (Ed.), *Ion Channels in the Pulmonary Vasculature*. Boca Raton, FL: Taylor and Francis Group, 635–649.

Gulliksen, A., Solli, L. A., Drese, K. S., Sorensen, O., Karlsen, F., Rogne, H., Hovig, E., & Sirevag, R. (2005). Parallel nanoliter detection of cancer markers using polymer microchips. *Lab on a Chip*, **5**(4), 416–420. PM:15791339.

Hawkins, K. R. (2007). *Designing the Diffusion Immunoassay (DIA): How Properties of the Analyte Affect DIA Performance*. PhD Thesis: University of Washington.

Jain, K. K. (2007). Applications of nanobiotechnology in clinical diagnostics. *Clinical Chemistry*, **53**(11), 2002–2009. PM:17890442.

Jain, K. K. (2010). Innovative diagnostic technologies and their significance for personalized medicine. *Molecular Diagnosis and Therapy*, **14**(3), 141–147. PM:20560675.

Mabey, D., Peeling, R. W., Ustianowski, A., & Perkins, M. D. (2004). Diagnostics for the developing world. *Nature Reviews Microbiology*, **2**(3), 231–240. ISI:000220431800014.

Maciel, R. J. (1985). Standard curve fitting in immunodiagnostics: A primer. *Journal of Clinical Immunoassay*, **8**, 98–106.

Malkov, V. A., Serikawa, K. A., Balantac, N., Watters, J., Geiss, G., Mashadi-Hossein, A., & Fare, T. (2009). Multiplexed measurements of gene signatures in different analytes using the Nanostring nCounter Assay System. *BMC Research Notes*, **2**. MEDLINE:19426535.

Martinez, A. W., Phillips, S. T., Carrilho, E., Thomas, S. W., Sindi, H., & Whitesides, G. M. (2008). Simple telemedicine for developing regions: Camera phones and paper-based microfluidic devices for real-time, off-site diagnosis. *Analytical Chemistry*, **80**(10), 3699–3707. ISI:000255871500023.

Martinez, A. W., Phillips, S. T., Whitesides, G. M., & Carrilho, E. (2009). Diagnostics for the developing world: Microfluidic paper-based analytical devices. *Analytical Chemistry. American Chemical Society*, **82**(1), 3–10. http://dx.doi.org/10.1021/ac9013989.

McLean, R., Jury, C., Bazeos, A., & Lewis, S. M. (2009). Application of camera phones in telehaematology. *Journal of Telemedicine and Telecare*, **15**(7), 339–343. ISI:000271210100003.

Mcmillan, W. A. (2002). Rapid, realtime PCR with fully integrated specimen preparation. *Proceedings of the 8th International Symposium on Microbial Detection*, 1–13.

McPherson, M. J., & Müller, S. G. (2000). *PCR*. Oxford, New York: BIOS Scientific Publishers, Springer.

Meltzer, S. J. (1998). *PCR in Bioanalysis*. Totowa, N.J: Humana Press.

Morisset, D., Stebih, D., Cankar, K., Zel, J., & Gruden, K. (2008). Alternative DNA amplification methods to PCR and their application in GMO detection: A review. *European Food Research and Technology*, **227**(5), 1287–1297. WOS:000258276500001.

Sekhon, B. S., & Kamboj, S. R. (2010). Inorganic nanomedicine – part 1. *Nanomedicine*, **6**(4), 516–522. PM:20417313.

Simon, R. (2010). Clinical trial designs for evaluating the medical utility of prognostic and predictive biomarkers in oncology. *Personilized Medicine*, **7**(1), 33–47. PM:20383292.

Strobel, H. A., & Heineman, W. R. (1989). *Chemical Instrumentation: A Systematic Approach*. New York: Wiley.

Weigl, B. H., Domingo, G. J., LaBarre, P. D., & Gerlach, J. L. (2008). Towards non- and minimally instrumented, microfluidics-based diagnostic devices. *Lab on a Chip*, **8**(12), 1999–2014. PM:19023463.

Yager, P., Domingo, G. J., & Gerdes, J. C. (2008). Point-of-care diagnostics for global health. *Annual Review of Biomedical Engineering*, **10**, 107–144. PM:18358075.

CHAPTER II.5.18 MEDICAL APPLICATIONS OF SILICONES

Jim Curtis[1] *and André Colas*[2]

[1]Dow Corning Corporation, Midland, MI, USA

[2]Dow Corning Europe S.A., Parc Industriel, Seneffe, Belgium

MEDICAL APPLICATIONS

Silicones, with their unique material properties, have found widespread application in healthcare. Properties attributed to silicone include biocompatibility and biodurability, which can be expressed in terms of other material properties such as hydrophobicity, low surface tension, and chemical and thermal stability. These properties were the basis for the initial use of silicone in the medical field. For example, their hydrophobic (water-repellent) character caused silicones to be considered for blood coagulation prevention in the mid-1940s. Researchers from the Universities of Toronto and Manitoba obtained a methylchlorosilane from the Canadian General Electric Company and coated syringes, needles, and vials with the material. When rinsed with distilled water, the silane hydrolyzed, forming a silicone coating on the substrate. The researchers published results from their clotting time study in 1946, finding that the silicone treatment "on glassware and needles gives a surface which preserves blood from clotting for many hours" (Jaques et al., 1946) (see Chapter II.3.5). Researchers at the Mayo Clinic took notice of the work by their Canadian colleagues, indicating that silicone: "was the most practical of any known [substance] for coating needle, syringe and tube" (Margulies and Barker, 1949). They also demonstrated that leaving blood in silicone-coated syringes had no significant effect on the blood as measured by coagulation time after being dispensed from the syringe. Soon the use of silicone precoating of needles, syringes, and blood collection vials became commonplace. In addition to the blood-preserving quality of silicone, it was soon discovered that silicone-coated needles were less painful (Figure II.5.18.1). Today most hypodermic needles, syringes, and other blood-collecting apparatus are coated or lubricated with silicone.

In addition, silicone materials have found other uses in the medical laboratory. Silicone tubing is used for fluid transport between vessels. Another application is in dentistry, where silicone impression materials have been used since the 1950s (Starke, 1975). Modern silicone impression materials can be obtained using addition or condensation polymerization methods to cure the implants (Ciapetti et al., 1998).

Extracorporeal Equipment

Silicone tubing and membranes found application in numerous extracorporeal machines, due in large part to their hemocompatibility and gas permeability properties. Silicone has been used in kidney dialysis, blood oxygenators, and heart bypass machines. Blood compatibility was also a factor in the application of silicone in several mechanical heart valves (Figure II.5.18.2). The use of silicone in extracorporeal applications continues today. Hemocompatibility testing has suggested that platinum-cured silicone tubing may be superior to poly(vinyl chloride) (PVC) in several respects (Harmand and Briquet, 1999).

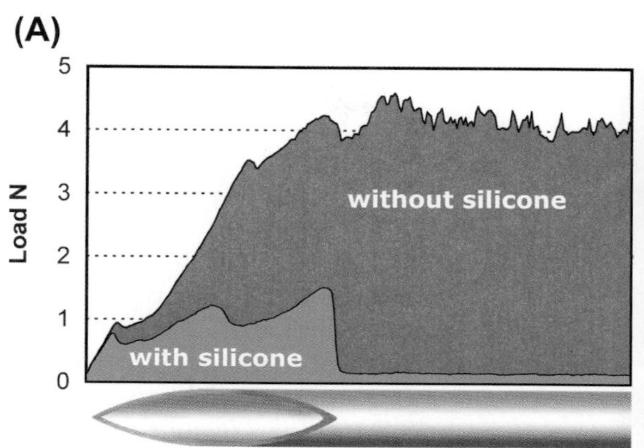

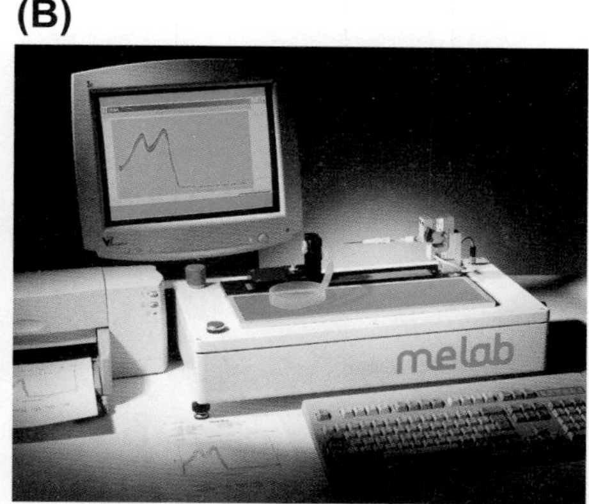

FIGURE II.5.18.1 (A) Penetration force of silicone coated and non-coated hypodermic needles as measured by (B) Melab equipment using DIN 13097. *(Photo courtesy of Melab GmbH © Dow Corning AV12291).*

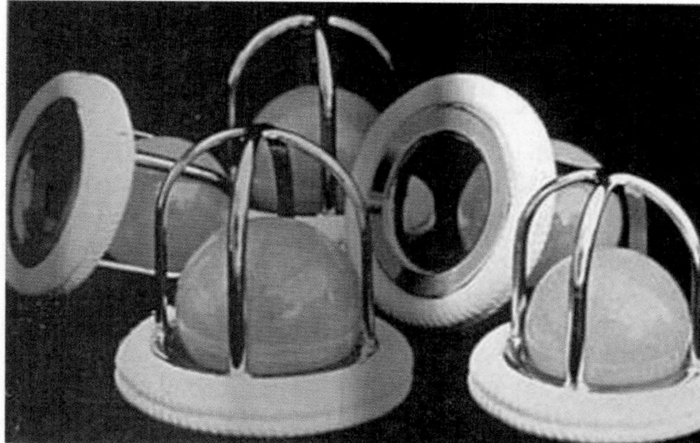

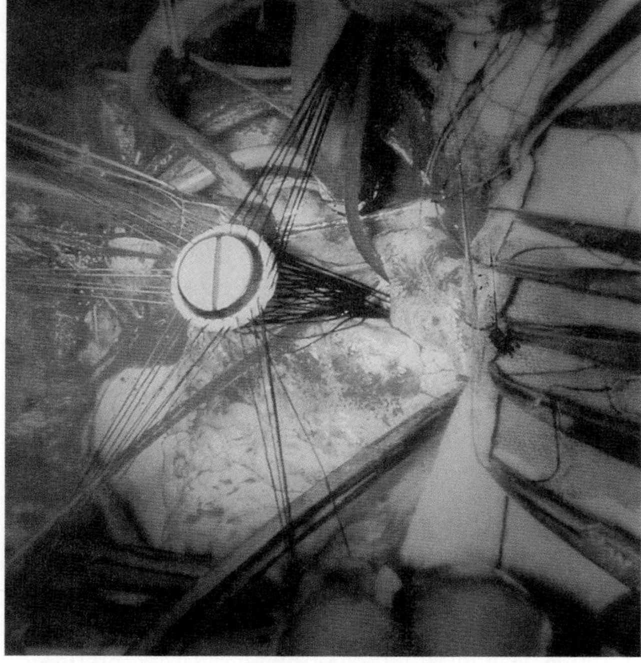

FIGURE II.5.18.2 Examples of early heart valves containing silicone elastomer. *(© Dow Corning, top: AV06433, bottom: AV06434)*

Catheters, Drains, and Shunts

The properties of silicone elastomers have also found application in numerous catheters, shunts, drains, and the like (Figure II.5.18.3). These include devices fabricated

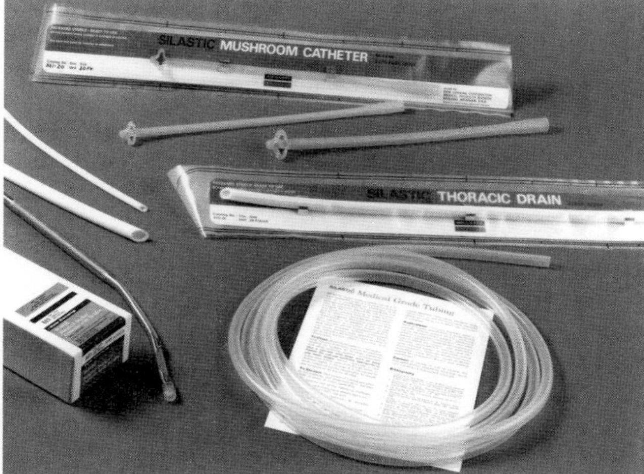

FIGURE II.5.18.3 Examples of silicone in catheters, drains, tubes, and cannulae. *(© Dow Corning AV06432)*

with silicone extrusions, as well as devices with non-silicone substrates that are silicone-coated to provide less host reaction. For example, although several all-silicone urology catheters are on the market, the Silastic® Foley is a latex catheter whose exterior and interior are coated with silicone elastomer (Figure II.5.18.4).

Figure II.5.18.5 shows the various components of the Cystocath® suprapubic drainage system (Baeham, 1973), which was used for bladder drainage after gynecological surgery that complicated or prevented normal urethral urination. The system included: (A) the catheter, a silicone tube whose non-wetting surface minimized encrustation; (B) the body seal made of flexible silicone elastomer that conformed easily to the body contour and allowed the patient freedom of movement; (C) pressure-sensitive silicone adhesive that adhered well to skin; and (D) the trocar needle used to pierce the bladder and overlying tissue. After vaginal surgery the bladder was inflated and located. Silastic® Medical Adhesive B was applied by brush to the clean abdomen over the bladder and the bottom of the body seal component. The pressure-sensitive adhesive (described in Chapter I.2.2.B) had excellent properties conducive to the application. It provided good adherence to dry or wet skin,

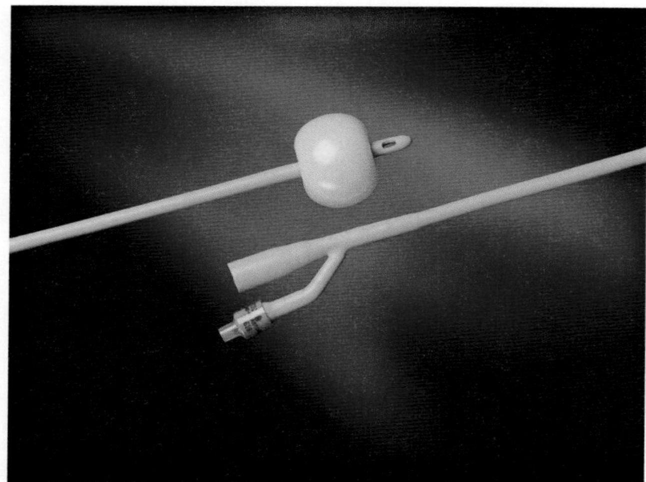

FIGURE II.5.18.4 Silastic® Foley catheter. *(Courtesy of C. R. Bard, Inc).*

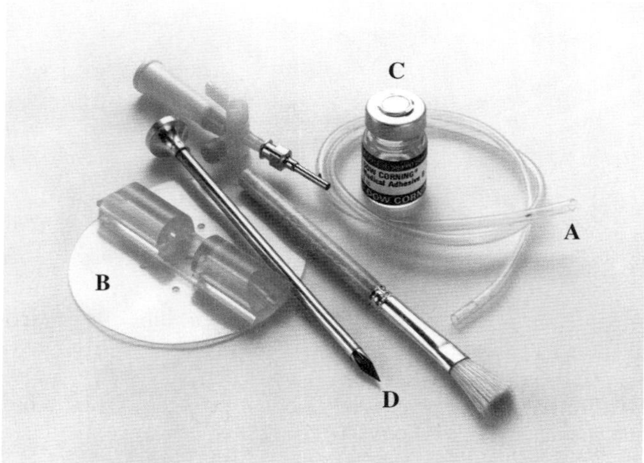

FIGURE II.5.18.5 Cystocath® suprapubic drainage system. (A)–(D): See text details. *(© Dow Corning AV06436)*

without causing irritation or sensitization, and good permeability to oxygen, carbon dioxide, and moisture vapor; it also formed a waterproof and urine-proof seal. After a short wait for the solvent to evaporate, allowing the adhesive to become tacky, the body seal was adhered to the abdomen. The trocar was advanced through the center of the body seal and pierced through the skin and bladder. The stylet was removed and the silicone catheter threaded through the needle and well into the bladder. The needle was withdrawn, leaving the catheter in place. The silicone tube was secured in the retention groove and the distal end attached to a siphon drainage system.

Long-Term Implants

The chemical stability and elastic nature of silicone are beneficial for many applications involving long-term implantation. The first published report of silicone elastomers being implanted in humans was in April 1946, when Dr. Frank H. Lahey told of his use of these materials for bile duct repair. He obtained the material, called "bouncing clay" at the time, from the experimental laboratory of the General Electric Company (GE). Citing its elastic properties, he reported: "It is flexible, it will stretch, it will bounce like rubber, and it can be cast in any shape" (Lahey, 1946).

In 1948, Dr. DeNicola implanted an artificial urethra fashioned from the same type of GE silicone tubing used previously by Lahey. The first apparently successful replacement of the human male urethra by artificial means was conducted under general anesthesia. The 3¾ inch (9.5 cm) long silicone tube was threaded over a narrow catheter whose distal end was in the bladder. Fourteen months after implantation, the artificial urethra "had been retained with normal genitourinary function. … There is no evidence at this time that the tube is acting as a foreign body irritant…" (DeNicola, 1950).

A particularly notable early silicone implant was the hydrocephalus shunt, which benefitted from the thermal stability of silicone. This application became quite celebrated when tenderly described in *Reader's Digest* (LaFay, 1957). Charles Case "Casey" Holter was born on the seventh of November 1955 with a neural tube defect called lumbosacral myelomeningocele. By December, the baby had contracted meningitis, and surgeons at the Children's Hospital of Philadelphia closed the defect. A few weeks later, hydrocephalus caused young Casey's head to swell as cerebrospinal fluid (CSF) collected in his brain. At the time, there were few treatment options, and this affliction was fatal for most children who contracted it. After infection concerns with the daily venting of CSF through the fontanels (spaces between cranial bones that have not completely fused), Dr. Eugene Spitz implanted a polyethylene shunt catheter in Casey to drain excess CSF from the brain into the atrium of the heart. A valve was needed that would allow the CSF to drain when pressure began to build in the brain, but would close to prevent backflow when the pressure equalized. Spitz did his neurosurgery residency at the University of Pennsylvania in 1952, gaining clinical experience with a ball and spring valve developed by the Johnson Foundation, an arm of Johnson & Johnson. So it was this valve that was first implanted in young Casey. Basically a scaled-down version of an automotive pressure relief valve, it frequently clogged with tissue. Casey's father John, a machinist by profession, asked Spitz about the valve, the CSF, and the pressures involved. Spitz confided in Holter that "a competent one-way valve" that would be stable in the human body was needed (Baru et al., 2001).

It is said that necessity is the mother of invention, and this is a poignant example. A desperately concerned father went home to his garage workshop that evening and constructed a prototype valve from two rubber condoms and flexible tubing. However, autoclaving caused the material to shrink a bit and the valve to leak. Holter discussed the shrinkage problem with a local rubber company, where the head of

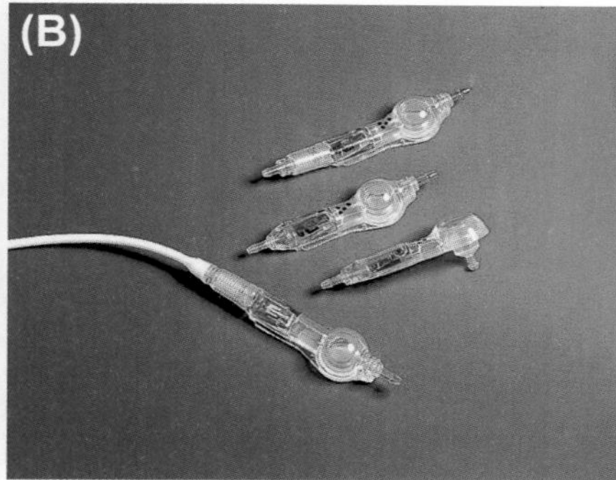

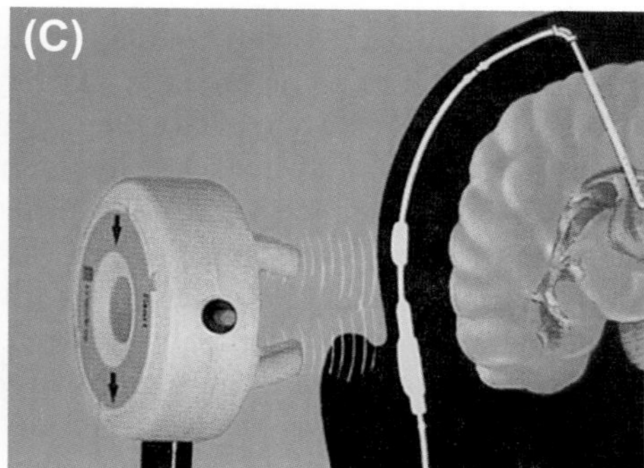

FIGURE II.5.18.6 (A) Original Silastic hydrocephalus shunt. (B, C) Modern Codman Hakim programmable valve shunt. *(Courtesy of Codman, a Johnson & Johnson company).*

research suggested he replace the natural rubber with a thermally stable material known as silicone. Holter obtained Silastic® brand silicone elastomer and tubing free of charge from Dow Corning. In March of 1956, Holter believed the valve that would come to bear his name was ready. At the time, Holter's son was too ill to undergo the surgery; however, Spitz saw promise in the valve design and successfully implanted the ventriculoatrial shunt in another hydrocephalic child. Casey was sufficiently stable for surgery by April, when Spitz implanted him with a Holter valve (Baru et al., 2001). The Holter valve was so successful that its production began that summer, and the valve is still being made in almost unchanged form today (Aschoff et al., 1999) (Figure II.5.18.6).

Early healthcare applications resulted in substantial interest in the emerging silicone materials and their promising properties. The two leading silicone suppliers, General Electric and Dow Corning, began receiving inquiries from the medical field at unprecedented rates. By 1959, Dow Corning was so inundated with requests for materials and information that the Dow Corning Center for Aid to Medical Research was established to act as a clearing house for all information on medical uses of silicone, and to supply medical scientists with research quantities of various silicone materials, all without cost to the researcher. The Center corresponded with more than 35,000 physicians and researchers from all over the world and in numerous areas of healthcare (Braley, 1973).

The upsurge of interest in silicones for healthcare applications continued in the early 1960s. Before the end of the decade, silicone materials were being employed or evaluated in numerous healthcare applications – in orthopedics, catheters, drains and shunts of numerous descriptions, as components in kidney dialysis, blood oxygenators, and heart bypass machines, heart valves, and aesthetic implants, to name just a few.

Silicone materials continue to find application in various parts of the body today. Including orthopedic applications, silicone implants are literally found from head to toe. In the eye, silicone foam and elastomer scleral bands and bucklers, silicone vitreous fluid replacement, and elastomer intraocular lenses help restore vision after retinal reattachment or cataract surgery. In addition, the incorporation of siloxane into soft contact lens material resulted in the development of the silicone hydrogel contact lens. First launched in 1999, this innovative lens material greatly increases the amount of oxygen that reaches the cornea, allowing the lenses to be worn for as long as one month of continuous wear. The use of these lenses has resulted in fewer complications, and much improvement in patient symptoms of dryness and discomfort compared with the previous soft lenses (Dillehay, 2007).

In the ear, examples include silicone elastomer tubes for otological ventilation, as well as electronic cochlear implants that are encapsulated and insulated with silicone. Silicone elastomers are used in voice prostheses placed in the throat between the trachea and esophagus after laryngectomy. This is a particularly challenging location for any elastomeric material, as yeast and bacterial biofilm colonization often develop with long-term use (Delank and Scheuermann, 2008).

Silicone elastomers find application in many devices implanted in the thoracic cavity. A key example is the cardiac pacemaker, where silicone is used to encapsulate and insulate. Two interesting examples involving the stomach are the popular gastric band implant (Lap-Band®) for weight loss, and the Angelchik antireflux device for management of gastro-esophageal reflux or hiatal hernia which have not been resolved by more conservative treatments (Timoney, et al. 1990). The latter has the distinction of being the first medical device containing silicone gel to be approved by the US Food and Drug Administration (FDA) (FDA approval P790006, 1979).

Silicone Orthopedic Implants

The most significant orthopedic applications of silicone are hand and foot joint implants. Dr. Alfred Swanson, with assistance from Dow Corning, developed silicone finger joint implants such as those shown in Figure II.5.18.7 (Swanson, 1968). Similar implants were developed for the other small joints of the foot and hand. In addition to double-stemmed finger joint implants in each of the metacarpophalangeal joints (large arrow), Figure II.5.18.7D also shows a single-stemmed Silastic ulnar head implant at the distal terminus of the ulna (small arrow). Nearly four decades later, silicone remains the most prevalent type of small joint implant.

Cement restrictors made of silicone elastomer with added barium sulfate for radiopacity (Figure II.5.18.8A) are used in joint replacement surgery involving cement. For example, one of these implants, also known as a bone plug, is inserted deeply into the prepared intramedullary canal of the femur during hip replacement surgery (Figure II.5.18.8B). It fits securely to restrict distal migration of poly(methyl)methacrylate bone cement intra- and postoperatively.

Another early orthopedic application of silicone was in 1969, when the French GUEPAR (Groupe d'Utilisation et d'Étude des Prothèses ARticulaires) posterior-offset hinged total knee implant was introduced. This design was constructed of the metal Vitallium, with a shock-absorbing silicone bumper that prevented impact of the anterior portions of the tibial and femoral components during knee extension (Mazas and GUEPAR, 1973).

Aesthetic Implants

Silicones have been used extensively in aesthetic and reconstructive plastic surgery for over 40 years. Silicone elastomer is used in implanted prosthetics of numerous descriptions. Silicone implants are widely used in the breast, scrotum, chin, nose, cheek, calf, and buttocks. Some of these devices may also employ a softer-feeling substance known as silicone gel. The gel is a lightly cross-linked silicone elastomer, without silica or other reinforcing filler, that is swollen with polydimethylsiloxane fluid. The gel is contained within a silicone elastomer shell in breast, testicular, and chin implants. Surgeons implant these medical devices for aesthetic reasons, to correct congenital deformity or during reconstructive surgery after trauma or cancer treatment.

With silicone materials and prototypes supplied free of charge from Dow Corning, Doctors Cronin and Gerow developed and tested their silicone gel-filled breast implant in 1961. They implanted the first pair in a woman in 1962 (Cronin and Gerow, 1963). Word of their success and the superiority of these silicone implants to the existing foam type led to the popularity of the silicone gel breast implant (Gerow, 1976).

> The most prominent of these aesthetic implants is the silicone breast implant. Breast enlargement by surgical means has been practiced for over a century. In 1895, Czerny reported transplanting a lipoma to a breast in order to correct a defect resulting from the removal of a fibroadenoma (Czerny, 1895). The insertion of glass balls into the breasts was described by Schwartzmann in 1930, and again by Thorek in 1942. The Ivalon sponge, introduced by Pangman in 1951, was the first augmentation prosthesis to be retained fairly consistently. This surgical sponge, formulated of poly(vinyl alcohol) cross-linked with formaldehyde, was at first hand-trimmed to the desired shape by the implanting surgeon, and later preformed by Clay Adams, Inc. There was some early recognition of the tendency for tissue growth into the open-cell foam, and in 1958 Pangman patented the concept of encapsulating the foam with an alloplastic (manmade) envelope. His patent also contemplated the use of other fill materials, such as silicone, in place of foam. The Polystan and Etheron polyurethane sponge implants began to be used as breast implants in 1959 and 1960, respectively. These sponge implants became popular in the early 1960s, later being supplanted by the Cronin-type silicone gel-filled breast implants.

Figure II.5.18.9 shows the appearance of these implants in 1964. The shells of Cronin-type implants were vacuum-molded with anterior and posterior elastomer pieces sealed together creating a seam around the perimeter of the base. The posterior shell had exposed loops of Dacron® mesh attached. The surgeons believed that prosthesis fixation to the chest wall was necessary to prevent implant migration.

Since this early 1964 design, Dow Corning and the numerous other companies that manufactured silicone breast implants made prosthesis design improvements, including elimination of the seam and realization that fixation is frequently unnecessary.

In the early 1990s, these popular devices became the subject of a torrent of contentious allegations regarding their safety. The controversy in the 1990s initially involved breast cancer, then evolved to autoimmune connective tissue disease, and continued to evolve to the frequency of local or surgical complications such as rupture, infection or capsular contracture. Epidemiology studies have consistently found no association between breast implants and breast cancer (McLaughlin et al., 1998, 2007; Park et al., 1998; Brinton et al., 2000; Mellemkjær et al., 2000). In fact, some studies suggest that women with implants may have decreased risk of breast cancer (Brinton et al., 1996; Deapen et al., 1997). Reports of cancer at sites other than the breast are inconsistent or attributed to lifestyle factors (Herdman and Fahey, 2001). The epidemiologic research on autoimmune or connective-tissue disease has also been remarkably uniform, and

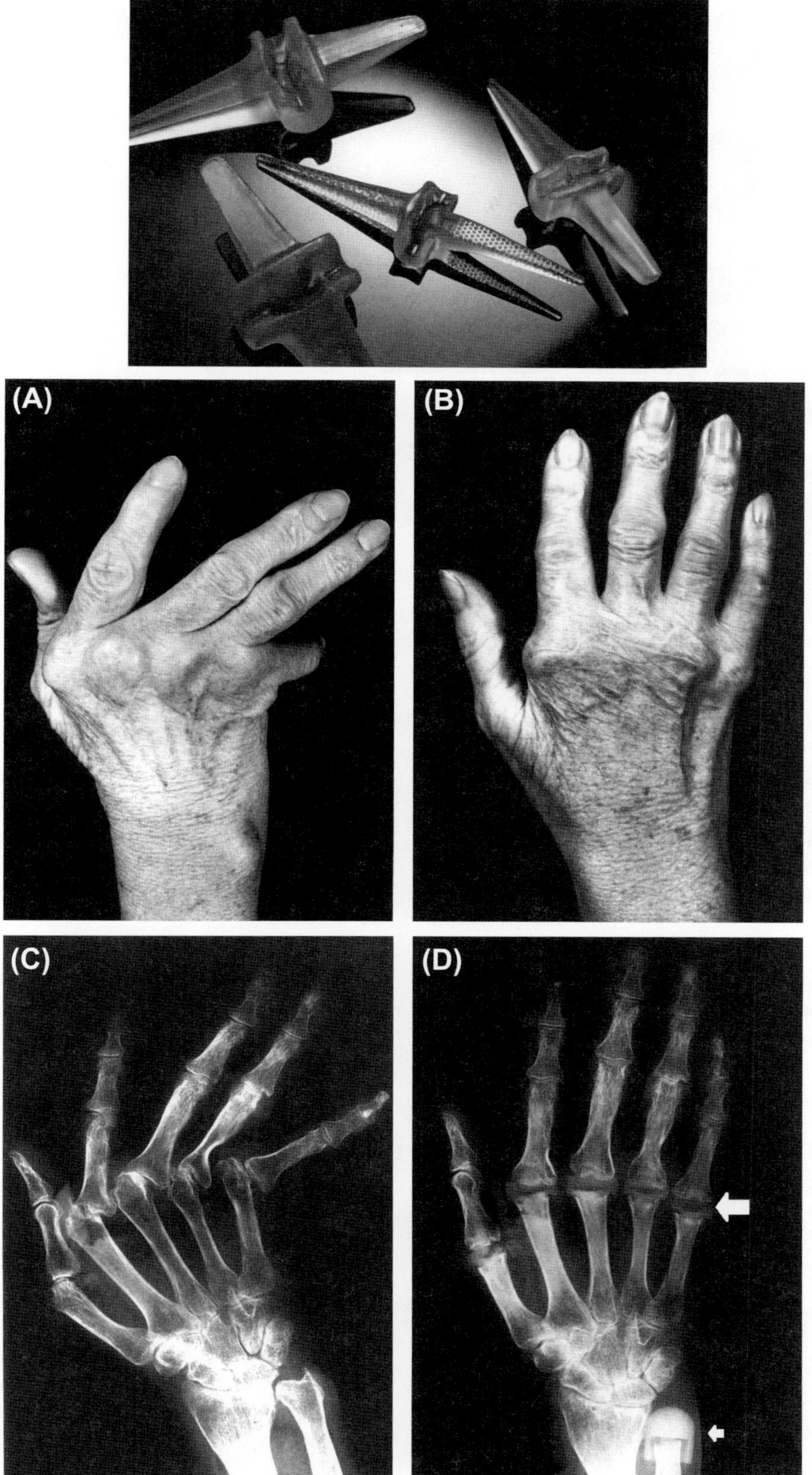

FIGURE II.5.18.7 Silicone elastomer finger joint implants: (A, C) Photograph and X-ray of arthritic right hand prior to restorative implantation surgery. (B, D) Postoperative photograph and X-ray view of the same hand. *(© Dow Corning, top: AV06438, A: AV06439, B: AV06441, C: AV06440, D: AV06442.)*

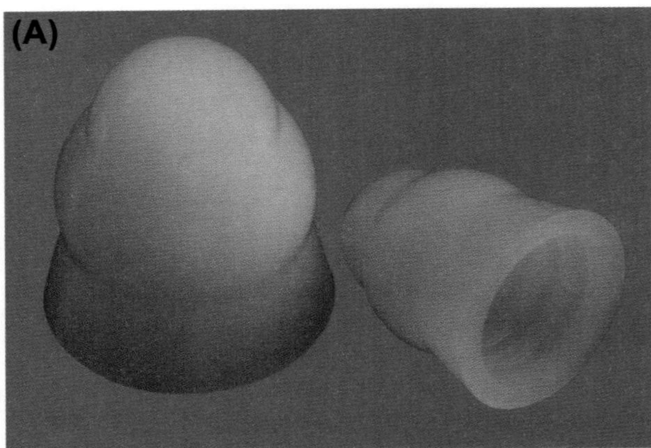

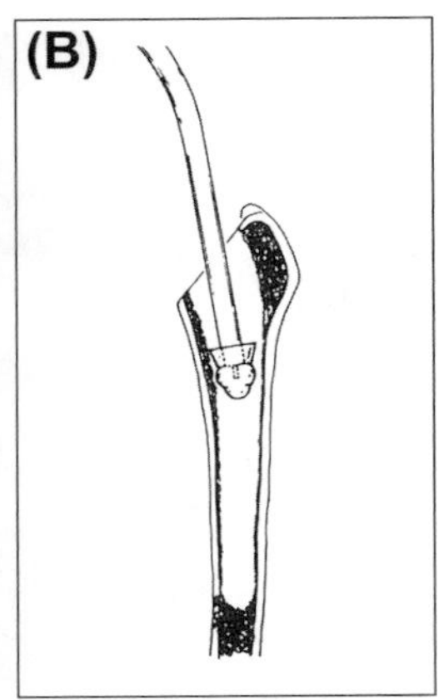

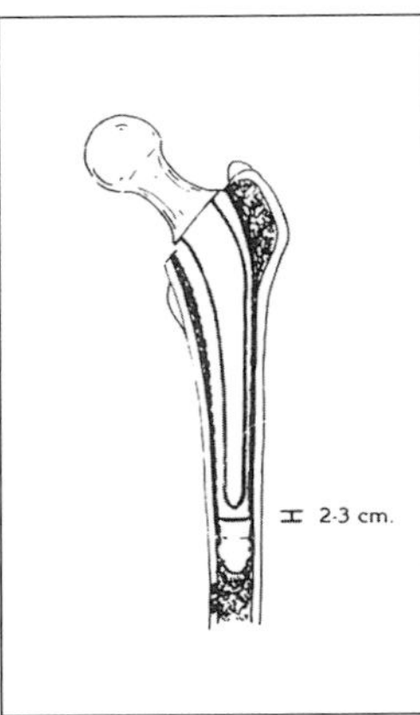

FIGURE II.5.18.8 (A) Silicone cement restrictors (bone plugs). (B) Sketch showing placement. *(© Dow Corning, A: AV12072, B: AV12071.)*

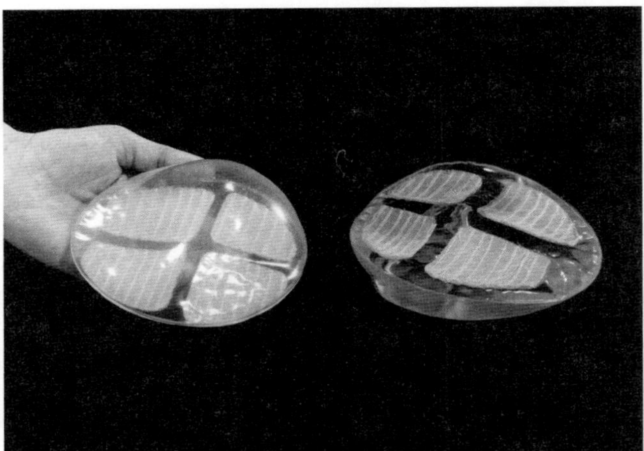

FIGURE II.5.18.9 Silastic mammary prosthesis, 1964. *(© Dow Corning, AV06444.)*

FIGURE II.5.18.10 Silicone testicular implants. (© *Dow Corning, AV06437.)*

concludes there is no causal association between breast implants and connective-tissue disease (Gabriel et al., 1994; Sánchez-Guerrero et al., 1995; Hennekens et al., 1996; Edworthy et al., 1998; Nyrén et al., 1998; Kjøller et al., 2001; McLaughlin et al., 2007).

> Numerous systematic reviews (UK IEAG, 1993; France ANDEM, 1996; US IOM, 1996; Australia TDEC, 1997; Germany BgVV, 1998; UK IRG, 1998; EU EQUAM, 2000; US NSP/MDL926, 2001; Spain/EU STOA, 2003) commissioned by various governments have repeatedly and consistently borne out that the evidence fails to support a cause-and-effect relationship between silicone breast implants and systemic diseases. Following FDA approval action on 17 November 2006 making silicone gel breast implants widely available in the United States again, the controversy has diminished.

Largely without any specific safety concern or allegation critical of it, another silicone gel-filled implant was swept up in the breast implant controversy. At the time, most testicular implants were constructed of the same materials as silicone gel breast implants. Silicone artificial testicles had been used nearly as long as the breast implants. Dow Corning, one of several companies that manufactured these implants, was producing them as early as 1964. These devices served to ameliorate psychological stress associated with testicle loss due to cancer, traumatic injury or those absent at birth. The Teflon® strips seen in Figure II.5.18.10 shield each implant shell during suturing through an elastomer loop at the superior pole to achieve fixation for proper anatomical orientation in the scrotum.

BIOCOMPATIBILITY

There has been much discussion regarding the various definitions of the term *biocompatibility*. We now take it to mean "the ability of a material to perform with an appropriate host response in a specific situation" (Black, 1992; Remes and Williams, 1992). Historically it has been tacitly understood that silicone materials are intrinsically biocompatible, since they have been used successfully in so many healthcare applications. However, given the modern definition of the term, no material can be assumed to be universally biocompatible, since this implies that it is suitable for every conceivable healthcare application involving contact with the host patient.

Numerous silicone materials have undergone biocompatibility testing. Many have passed every bioqualification test; however, others have not. Several factors can affect the results of such testing, including the composition of the material. As described in Chapter I.2.2.B, the basic polydimethylsiloxane (PDMS) polymer can be modified to replace methyl with other functional groups. In some cases, those groups may be responsible for untoward host response. There may be by-products from the preparation of silicone materials that might trigger tissue reaction. For example, these could come from the use of a peroxide initiator under inappropriate temperature and processing conditions.

Purity is another factor that can affect biotest results. Medical silicone materials, including fluids, gels, elastomers, and adhesives, are manufactured by several companies today. Some of these firms manufacture these medical materials following good manufacturing practice (GMP) principles in dedicated, registered, and inspected facilities. Others sell materials generated on their industrial production line into the healthcare market.

Selection of appropriate preclinical material bioqualification tests for their application is the responsibility of the medical device or pharmaceutical manufacturer. Several national, international, and governmental agencies have provided guidance or regulation. Several silicone manufacturers offer special grades of materials that have met these specific requirements. The buyer should carefully investigate the supplier's definition, since there are no universal special grade definitions. At Dow Corning, Silastic BioMedical Grade materials have been qualified to meet or exceed the requirements of ISO 10993-1, USP (*United States Pharmacopeia*) Class V Plastics tests (acute systemic toxicity and intracutaneous reactivity), hemolysis, cell culture, skin sensitization, mutagenicity, pyrogenicity, and 90 day implant testing. Other physio-chemical qualification tests have been conducted, such as certain tests from the *European Pharmacopoeia*. Specific information regarding material biotesting can be found in other chapters of this text. Testing of the device in finished form should follow material bioqualification tests such as those described above.

BIODURABILITY

Traditionally we have thought of *biocompatibility* as the situation in which the biomaterial has minimal adverse impact on the host. Conversely, *biodurability* is where the host has a minimal adverse effect on the biomaterial (see Chapter II.4.2). The material properties of silicone, such as hydrophobicity, have been related to biocompatibility properties such as hemolytic potential, and the relative purity, and high molecular weight polymeric nature and chemical structure of the material, provide a theoretical basis for its lack of toxicity. The biodurability of silicone in medical applications is probably related to its exceptional thermal and chemical stability properties.

Silicones are used in numerous applications requiring high temperature resistance (Noll, 1968; Stark et al., 1982). During thermogravimetric analysis and in the absence of impurities, poly-dimethylsiloxane degradation starts only at around 400°C. Thus, silicones remain essentially unaffected by repeated sterilization by autoclaving, and they can usually also be dry-heat sterilized. Other sterilization methods can be used, such as ethylene oxide exposure and gamma and e-beam irradiation – although care must be taken to ensure complete sterilant outgassing in the former, and that dosage does not affect performance properties in the latter.

Although silicones can be chemically degraded, particularly at elevated temperatures, by substances capable of acting as depolymerization catalysts (Stark et al., 1982), their hydrophobic nature limits the extent of their contact with many aqueous solutions. Typically, the biologic milieu does not present a particularly hostile chemical environment for silicone. A notable exception, however, is the stomach, which excretes large amounts of hydrochloric acid, capable of attacking PDMS if it remains there too long. Based on silicone elastomer performance in long-term implantation applications, its biodurability is generally considered excellent (Table II.5.18.1).

The chemical stability associated with silicones became so well-established that it has been formulated into other biomaterials, such as polyurethane, to enhance their biodurability (Pinchuk et al., 1988; Ward, 2000; Christenson et al., 2002; Ward et al., 2006a,b).

Notwithstanding the chemical stability of silicone, certain factors have been shown to affect its durability in terms of long-term *in vivo* performance. The hydrophobic elastomer is somewhat lipophilic, and can be swollen by lipids or other nonpolar agents. Early experience with *in vivo* failure of silicone-containing heart valves was traced to elastomer absorption of lipids from the blood that resulted in significant dimensional swelling (McHenry et al., 1970). In most cases the absorption was low and failures did not occur, but in a small percentage of cases, the silicone was absorbing quantities sufficient to render the valves variant. Researchers speculated that variations in silicone poppet manufacture, such as cure, might have been a factor (Carmen and Mutha, 1972). Absorption of lipids was a variable reported by Swanson and LeBeau (1974) and Langley and Swanson (1976). The work of Brandon et al. (2002, 2003) has shown that the shells of silicone gel-filled breast implants also absorb silicone fluid (from the gel), causing a minor diminution in mechanical properties, one that is reversed after extraction of the elastomer.

TABLE II.5.18.1 Biodurability Studies of Silicone Elastomer and Medical Implants

Year	Researcher	Synopsis
1960	Ames	Explant examination of a clinical silicone ventriculocisternostomy shunt used in the treatment of hydrocephalus showed "the silicone rubber tubing was unchanged by three years implantation in the tissues of the brain and in the cervical subarachnoid space." Similarly, after over three years' implantation of silicone tubing in the peritoneal cavity of dogs, Ames wrote: "The physical properties of the tubing itself are apparently unchanged by prolonged contact with tissues."
1963	Sanislow and Zuidema	Silastic® T-tubes were placed in the common ducts of dogs and explanted nine months later. They were found to be free of bile-salt precipitation and completely patent. Four were tested for tensile strength and compared with a control sample from the same lot of elastomer. "These tests suggested that little physical change occurred in the Silastic as a result of prolonged contact with animal bile." The tensile strength after nine months was reported as 1130 psi (7.8 MPa), the same value as reported for the non-implanted control.
1964	Leininger et al.	The Battelle Memorial Institute examined the biodurability of five plastics by implanting films in dogs for 17 months' duration. The materials tensile strength and elongation were measured and compared with non-implanted controls. Although sizeable changes were seen in the tensile properties of polyethylene, Teflon®, and nylon, the results for Mylar and Silastic® remained essentially the same.
1974	Swanson and LeBeau	"Dog-bone"-shaped specimens of medical grade silicone rubber were implanted subcutaneously in dogs. Tensile properties and lipid content were measured at six months and two years postimplantation. A slight, but statistically significant, decrease in measured ultimate tensile strength and elongation were observed, as well as a small weight increase attributed to lipid absorption.
1976	Langley and Swanson	Mechanical test specimens were implanted in dogs for two years. Tensile strength, elongation, and tear resistance showed no statistically significant changes. Lipid absorption into the elastomer ranged from 1.4 to 2.6%.
2000	Curtis et al.	Six silicone breast implants surgically excised after 13.8 to 19.3 years, and 10 similar non-implanted units were tested to determine shell tensile properties and molecular weight of silicone gel extracts. The "study observed only minor changes (less than the explant or implant lot-to-lot variation range) in the tensile strength of Dow Corning silicone breast implants after nearly 20 years of human implantation." The gel extract molecular weight was either unchanged by implantation or increased slightly.
2003	Brandon et al.	In the most comprehensive study of breast implant biodurability heretofore published, the authors reported their results of tensile, cross-link density, and percent extractable measurements made on 42 explants and 51 controls. The study included some of the oldest explants, with human implantation durations up to 32 years. The researchers also performed a literature search, and plotted all published explant tensile modulus data against implantation duration, finding no temporal relationship. Neither was a relationship with implant time seen for the cross-link density results, supporting the biodurability of the silicone elastomer utilized in the implant shells. The researchers concluded, "There was little or no degradation of the base polydimethylsiloxane during *in vivo* aging in any of the implants we examined."
2008	Taylor et al.	Silicone biodurability after long-term implantation was examined by a highly sensitive NMR spectroscopy technique, as well as NMR relaxometry measurements of explanted gel breast implants and matched non-implanted controls. No evidence of chemical degradation of the cross-linked silicone matrix was observed in specimens explanted after as many as 32 years *in vivo*, underscoring the biostability of the cross-linked silicone shell and gel.

CONCLUSION

A variety of silicone materials have been prepared, many possessing excellent properties including chemical and thermal stability, low surface tension, hydrophobicity, and gas permeability. These characteristics helped originate the use of silicones in the medical field and are key to the materials' reported biocompatibility and biodurability. Since the 1960s, silicones have enjoyed expanded medical application and today are one of the most thoroughly tested and important biomaterials.

ACKNOWLEDGMENTS

The authors thank Doctors S. Hoshaw and P. Klein, both from Dow Corning, for their contribution regarding breast implant epidemiology.

BIBLIOGRAPHY

Ames, R. H. (1960). Response to Silastic tubing. *Bull. Dow Corning Center Aid Med. Res.*, **2**(4), 1.

Aschoff, A., Kremer, P., Hashemi, B., & Kunze, S. (1999). The scientific history of hydrocephalus and its treatment. *Neurosurg. Rev.*, **22**, 67.

Australia TDEC (Therapeutic Devices Evaluation Committee, Therapeutic Goods Administration). (1997). *Australian Therapeutic Device Bulletin*, **33**, 3.

Baeham, K. A. (1973). Suprapubic bladder drainage in gynaecological surgery. *Austr. NZ J. Surg.*, **43**(1), 32–36.

Baru, J. S., Bloom, D. A., Muraszko, K., & Koop, C. E. (2001). John Holter's shunt. *J. Am. Coll. Surgeons*, **192**, 79.

Black, J. (1992). *Biological Performance of Materials: Fundamentals of Biocompatibility*. New York, NY: Marcel Dekker.

Braley, S. A. (1973). *Spare Parts for Your Body*. Midland, MI: Dow Corning Center for Aid to Medical Research.

Brandon, H. J., Jerina, K. L., Wolf, C. J., & Young, V. L. (2002). *In vivo* aging characteristics of silicone gel breast implants compared to lot-matched controls. *Plast. Reconstr. Surg.*, **109**(6), 1927.

Brandon, H. J., Jerina, K. L., Wolf, C. J., & Young, V. L. (2003). Biodurability of retrieved silicone gel breast implants. *Plast. Reconstr. Surg.*, **111**(7), 2295.

Brinton, L. A., Malone, K. E., Coates, R. J., Schoenberg, J. B., Swanson, C. A., et al. (1996). Breast enlargement and reduction: Results from a breast cancer case-control study. *Plast. Reconstr. Surg.*, **97**(2), 269.

Brinton, L. A., Lubin, J. H., Burich, M. C., Colton, T., Brown, S. L., et al. (2000). Breast cancer following augmentation mammoplasty (United States). *Cancer Causes Control*, **11**, 819.

Carmen, R., & Mutha, S. C. (1972). Lipid absorption by silicone rubber heart valve poppets: *In vivo* and *in vitro* results. *J. Biomed. Mater. Res.*, **6**, 327.

Christenson, E. M., Dadestan, M., Wiggins, M. J., Ebert, M., Ward, R., et al. (2002). The effect of silicone on the biostability of poly(ether urethane). *Soc. Biomater. 28th Ann. Meeting Trans.*, 111.

Ciapetti, G., Granchi, D., Stea, S., Savarino, L., Verri, E., et al. (1998). Cytotoxicity testing of materials with limited *in vivo* exposure is affected by the duration of cell–material contact. *J. Biomed. Mater. Res.*, **42**, 485–490.

Cronin, T. D., & Gerow, F. J. (1963). Augmentation mammaplasty: A new "natural feel" prosthesis. Transactions of the Third International Congress of Plastic Surgery. *Excerpta Medica., Int. Congr. Ser.*, **66**, 41.

Czerny, V. (1895). Plastischer ersatz der brustdruse durch ein lipoma. *Zentralbl. Chir.*, **17**, 72.

Curtis, J. M., Peters, Y. A., Swarthout, D. E., Kennan, J. J., & VanDyke, M. E. (2000). Mechanical and chemical analysis of retrieved breast implants demonstrate material durability. *Transactions of the Sixth World Biomaterials Congress*, 346.

Deapen, D. M., Bernstein, L., & Brody, G. S. (1997). Are breast implants anticarcinogenic? A 14-year follow-up of the Los Angeles Study. *Plast. Reconstr. Surg.*, **99**(5), 1346.

DeNicola, R. R. (1950). Permanent artificial (silicone) urethra. *J. Urol.*, **63**(1), 168–172.

Delank, K. W., & Scheuermann, K. (2008). Praktische Aspekte der prothetischen Stimmrehabilitation nach Laryngektomie. *Laryngorhinootologie*, **87**(3), 160–166.

Dillehay, S. M. (2007). Does the level of available oxygen impact comfort in contact lens wear? A review of the literature. *Eye Contact Lens*, **33**(3), 148–155.

Edworthy, S. M., Martin, L., Barr, S. G., Birdsell, D. C., Brant, R. F., et al. (1998). A clinical study of the relationship between silicone breast implants and connective tissue disease. *J. Rheumatol.*, **25**(2), 254.

European Union, European Committee on Quality Assurance & Medical Devices in Plastic Surgery (EQUAM). (2000). *Consensus Declaration of Breast Implants and Consensus Declaration of Advanced Technologies and Devices in Plastic Surgery.* 4th Consensus Conference, 22nd–24th June 2000 at Herzliya, Israel.

France Agence Nationale pour le Développement de l'Evaluation Médicale (ANDEM). (1996). *Silicone Gel-Filled Breast Implants.*

Gabriel, S. E., O'Fallon, W. M., Kurland, L. T., Beard, C. M., Woods, J. E., et al. (1994). Risk of connective-tissue diseases and other disorders after breast implantation. *New Engl. J. Med.*, **330**(24), 1697.

Germany, Bundesinstitut für gesundheitlichen Verbraucherschutz und Veterinärmedizin (BgVV). (1998).

Gerow, F. J. (1976). Breast Implants. In N. G. Georgiade (Ed.), *Reconstructive Breast Surgery*. St. Louis, PA: Mosby.

Harmand, M. F., & Briquet, F. (1999). *In vitro* comparative evaluation under static conditions of the hemocompatibility of four types of tubing for cardiopulmonary bypass. *Biomaterials*, **20**(17), 1561.

Hennekens, C. H., Lee, I. M., Cook, N. R., Hebert, P. R., Karlson, E. W., et al. (1996). Self-reported breast implants and connective-tissue diseases in female health professionals. A retrospective cohort study. *JAMA*, **275**(8), 616.

Herdman, R. C., & Fahey, T. J. (2001). Silicone breast implants and cancer. *Cancer Invest.*, **19**(8), 821.

Jaques, L. B., Fidlar, E., Feldsted, E. T., & MacDonald, A. G. (1946). Silicones and blood coagulation. *Can. Med. Assoc. J.*, **55**, 26.

Kjøller, K., Friis, S., Mellemkjær, L., McLaughlin, J. K., Winther, J. F., et al. (2001). Connective tissue disease and other rheumatic conditions following cosmetic breast implantation in Denmark. *Arch. Int. Med.*, **161**, 973.

LaFay, H. (1957). A father's last-chance invention saves his son. *Reader's Digest*, January, 29–32.

Lahey, F. H. (1946). Comments made following the speech "Results from using Vitallium tubes in biliary surgery," read by Pearse HE before the American Surgical Association, Hot Springs, VA. *Ann. Surg.*, **124**, 1027.

Langley, N. R., & Swanson, J. W. (1976). *Effects of Subcutaneous Implantation, through Two Years, on the Physical Properties of Medical Grade Tough Rubber (MDF-0198).* Internal Dow Corning report number 1976-I0030-4571, produced to FDA and in litigation discovery, MDL926 number P–000004408.

Leininger, R. I., Mirkovitch, V., Peters, A., & Hawks, W. A. (1964). Change in properties of plastics during implantation. *Trans. Am. Soc. Artif. Internal. Organs.* **X**, 320.

Margulies, H., & Barker, N. W. (1949). The coagulation time of blood in silicone tubes. *Am. J. Med. Sci.*, **218**, 42.

Mazas, F. B., & GUEPAR (1973). GUEPAR total knee prosthesis. *Clin. Orthop. Rel. Res.*, **94**, 211.

McHenry, M. M., Smeloff, E. A., Fong, W. Y., Miller, G. E., & Ryan, P. M. (1970). Critical obstruction of prosthetic heart valves due to lipid absorption by Silastic. *J. Thorac. Cardiovasc. Surg.*, **59**(3), 413.

McLaughlin, J. K., Nyrén, O., Blot, W. J., Yin, L., Josefsson, S., et al. (1998). Cancer risk among women with cosmetic breast implants: A population-based cohort study in Sweden. *J. Nat. Cancer Inst.*, **90**(2), 156.

McLaughlin, J. K., Lipworth, L., Murphy, D. K., & Walker, P. S. (2007). The safety of silicone gel-filled breast implants: A review of the epidemiologic evidence. *Ann. Plast. Surg.*, **59**(5), 569–580.

Mellemkjær, L., Kjøller, K., Friis, S., McLaughlin, J. K., Hogsted, C., et al. (2000). Cancer occurrence after cosmetic breast implantation in Denmark. *Int. J. Cancer*, **88**, 301.

Noll, W. (1968). *Chemistry and Technology of Silicones*. New York, NY: Academic Press.

Nyrén, O., Yin, L., Josefsson, S., McLaughlin, J. K., Blot, W. J., et al. (1998). Risk of connective tissue disease and related disorders among women with breast implants: A nation-wide retrospective cohort study in Sweden. *Br. Med. J.*, **316**(7129), 417.

Pangman, W. J. (1958). *U.S. Patent No. 2,842,775.*

Park, A. J., Chetty, U., & Watson, A. C.H. (1998). Silicone breast implants and breast cancer. *Breast*, **7**(1), 22.

Pinchuk, L., Martin, J. B., Esquivel, M. C., & MacGregor, D. C. (1988). The use of silicone/polyurethane graft polymers as a means of eliminating surface cracking of polyurethane prostheses. *J. Biomater. Appl.*, **3**(2), 260.

Remes, A., & Williams, D. F. (1992). Immune response in biocompatibility. *Biomaterials*, **13**(11), 731.

Sánchez-Guerrero, J., Colditz, G. A., Karlson, E. W., Hunter, D. J., Speizer, F. E., et al. (1995). Silicon breast implants and the risk of connective-tissue diseases and symptoms. *New Engl. J. Med.*, **332**(25), 1666.

Sanislow, C. A., & Zuidema, G. D. (1963). The use of silicone T-tubes in reconstructive biliary surgery in dogs. *J. Sur. Res.*, **III**(10), 497.

Schwartzmann, E. (1930). Die technik der mammaplastik. *Der. Chirurg.*, **2**(20), 932–945.

Spain/EU. European Parliament Scientific and Technological Options Assessment (STOA). (2003). *Update Health Risks Posed by Silicone Implants in General, with Special Attention to Breast Implants*.

Stark, F. O., Falenda, J. R., & Wright, A. P. (1982). Silicones. In G. Wilkinson, F. G. A. Sone, & E. W. Ebel (Eds.), *Comprehensive Organometallic Chemistry* (Vol. 2, pp. 305). Oxford, UK: Pergamon Press.

Starke, E. N., Jr. (1975). A historical review of complete denture impression materials. *JADA*, **91**, 1037–1041.

Swanson, A. B. (1968). Silicone rubber implants for replacement of arthritic or destroyed joints in the hand. *Surg. Clin. North Am.*, **48**, 1113.

Swanson, J. W., & LeBeau, J. E. (1974). The effect of implantation on the physical properties of silicone rubber. *J. Biomed. Mater. Res.*, **8**, 357.

Taylor, R. B., Eldred, D. E., Kim, G., Curtis, J. M., Brandon, H. J., et al. (2008). Assessment of silicone gel breast implant biodurability by NMR and EDS techniques. *J. Biomed. Mater. Res. A.*, **85**, 684–691.

Thorek, M. (1942). Amastia, hypomastia and inequality of the breasts. In *Plastic Surgery of the Breast and Abdominal Wall*. Springfield, IL: C. C. Thomas.

Timoney, A. G., Kelly, J. M., & Welfare, M. R. (1990). The Angelchik antireflux device: A 5-year experience. *Ann. R. Coll. Surg. Engl.*, **72**, 185–187.

UK Independent Expert Advisory Group (IEAG). (1993). *Evidence for an Association Between the Implantation of Silicones and Connective Tissue Disease*. UK Medical Devices Agency.

UK Independent Review Group (IRG). (1998). *Silicone Gel Breast Implants*. The Report of the Independent Review Group.

US Institute of Medicine (IOM). (2000). Safety of Silicone Breast Implants. In S. Bondurant, V. Ernster, & R. Herdman (Eds.), *Safety of Silicone Breast Implants*. Washington, DC: National Academy Press.

US National Science Panel (NSP). (2001). In *Silicone Gel Breast Implant Products Liability Litigation (MDL926) (2001) Rule 706 National Science Panel Report*. www.fjc.gov/BREIMLIT/SCIENCE/report.htm2001.

Ward, R. S. (2000). Thermoplastic silicone-urethane copolymers: A new class of biomedical elastomers. *Med. Dev. Diagnost. Ind.*, **22**(4), 68–77.

Ward, R., Anderson, J., McVenes, R., & Stokes, K. (2006a). *In vivo* biostability of polysiloxane polyether polyurethanes: Resistance to biologic oxidation and stress cracking. *J. Biomed. Mater. Res. A.*, 77(3), 580–589.

Ward, R., Anderson, J., Ebert, M., McVenes, R., & Stokes, K. (2006b). *In vivo* biostability of polysiloxane polyether polyurethanes: Resistance to metal ion oxidation. *J. Biomed. Mater. Res. A.*, 77(2). 380–289.

SECTION II.6

Applications of Biomaterials in Functional Tissue Engineering

SECTION

II.6

Applications of Biomaterials in Functional Tissue Engineering

CHAPTER II.6.1 INTRODUCTION: REBUILDING HUMANS USING BIOLOGY AND BIOMATERIALS

Frederick J. Schoen
Professor of Pathology and Health Sciences and Technology (HST), Harvard Medical School, Executive Vice Chairman, Department of Pathology, Brigham and Women's Hospital, Boston, MA, USA

Contemporary biomaterials investigation and development of important applications has been stimulated and informed by advances in materials science and engineering, an understanding of the interactions of materials with the physiological environment, and a logical evolution of the concepts and tools of cell and molecular biology. These developments have permitted the progression of concepts of and approaches to controlling tissue–biomaterials interactions, often exquisitely. In *tissue engineering* (also sometimes called regenerative therapeutics), the logical and rapidly developing extension of this technology, engineered surfaces, and architectures (at the bulk, micro-, and nanoscale) are intended to stimulate the replacement of dysfunctional or lacking tissue or stimulate other (patho)physiological interactions through highly precise reactions with proteins and cells at the molecular level. Biomaterials play a key role in tissue engineering and regenerative therapeutics.

Tissue engineering uses a set of tools at the interface of the biomedical and engineering sciences to support the growth of living cells or attract endogenous cells to aid tissue formation or regeneration to produce therapeutic or diagnostic benefit. In the most frequent paradigm, cells are seeded on a scaffold composed of synthetic polymer or natural material (collagen or chemically-treated tissue), a tissue is matured *in vitro*, and the construct is implanted in the appropriate anatomic location as a prosthesis (Langer and Vacanti, 1993). A typical scaffold is a bioresorbable polymer in a porous configuration in the desired geometry for the engineered tissue, often modified to be adhesive for cells, in some cases selective for a specific circulating cell population. Application-specific and differentiated or undifferentiated (stem) cells are used (Stocum, 2005; see Chapters II.1.7 and II.6.4). Recent emphasis is on micro- and nanoscale control of scaffold–cell interactions (Wheeldon et al., 2011).

The first phase is the *in vitro* formation of a tissue *construct*, by placing the chosen cells and scaffold in a metabolically and mechanically supportive environment with growth media (in a *bioreactor*), in which the cells proliferate and elaborate extracellular matrix. In the second phase, the construct is implanted in the appropriate anatomic location, where remodeling *in vivo* is intended to recapitulate the normal functional architecture of an organ or tissue (Orlando et al., 2011). The key processes occurring during the *in vitro* and *in vivo* phases of tissue formation and maturation are: (1) cell proliferation, sorting and differentiation; (2) extracellular matrix production and organization; (3) degradation of the scaffold; and (4) remodeling and potentially growth of the tissue. The general paradigm of tissue engineering is illustrated in Figure II.6.1.1. Biological and engineering challenges in tissue engineering are focused on the three principal components that comprise the "cell–scaffold–bioreactor system." Control of the various parameters in device fabrication (Table II.6.1.1) may have major impact on the ultimate result. Exciting new possibilities are opened by advances in stem cell technology, and the recent evidence that some multipotential cells possibly capable of tissue regeneration are released by the bone marrow and circulate systemically (Lakshmipathy and Verfaillie, 2005), while others may be resident in organs such as heart and the central nervous system formerly not considered capable of regeneration (Leri et al., 2011). Indeed, it may be possible to use the context of the *in vivo* physiological environment as a bioreactor, and thereby bypass the *in vitro* phase of initial tissue formation (Stevens et al., 2005; Mendelson and Schoen, 2006). The importance of developmental biological and biomimetic principles in tissue engineering is being realized (Gerecht-Nir et al., 2006; Ingber, 2010a,b), since a scaffold can contain specific chemical and structural information that controls tissue formation, in a manner analogous to cell–cell communication and patterning during embryological development.

In addition to the goal of developing implantable medical devices that use living cells (normal or genetically-manipulated), together with extracellular components

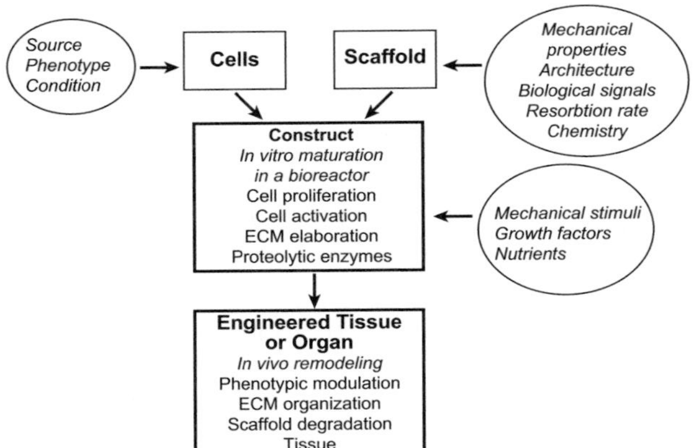

FIGURE II.6.1.1 Tissue engineering paradigm. In the first step of the typical tissue engineering approach, differentiated or undifferentiated *cells* are seeded on a bioresorbable *scaffold* and then the *construct* matured *in vitro* in a bioreactor. During maturation, the cells proliferate and elaborate extracellular matrix (ECM) to form a "new" tissue. In the second step, the construct is implanted in the appropriate anatomical position, where remodeling *in vivo* is intended to recapitulate the normal tissue/organ structure and function. The key variables in the principal components – cells, scaffold, and bioreactor – are indicated. (By permission from Rabkin, E. & Schoen, F. J. (2002). Cardiovascular tissue engineering. *Cardiovasc. Pathol.*, **11**, 305.)

TABLE II.6.1.1 Control of Structure and Function of an Engineered Tissue

Cells	Biodegradable Matrix/ Scaffold
Source	**Architecture/porosity/ chemistry**
Allogenic	Composition/charge
Xenogenic	Homogeneity/isotropy
Autologous	Stability/resorption rate
	Bioactive molecules/ligands
	Soluble factors
Type/phenotype	**Mechanical properties**
Single versus multiple types	Strength
Differentiated cells from primary or other tissue	Compliance
	Ease of manufacture
Adult bone marrow stem cells	
Pluripotent embryonic stem cells	
Density	
Viability	
Gene expression	
Genetic manipulation	
Bioreactor Conditions	
Nutrients/oxygen	
Growth factors	
Perfusion and flow conditions	
Electrical stimulation	
Mechanical Factors	
Pulsatile	
Hemodynamic shear stresses	
Tension/compression	

Modified from Rabkin, E. & Schoen, F. J. (2002). Cardiovascular tissue engineering. *Cardiovasc. Pathol.,* **11**, 305.

(either natural or synthetic), *tissue engineering* also seeks to: (1) understand structure–function relationships in normal and pathological tissues (particularly those related to embryological development and healing; Ballachandran et al., 2011); (2) to create models for human tissue that can be used for detection of toxins or infectious agents or for high-throughput drug testing (Khetani and Bhatia, 2008); and (3) to control cell and tissue responses to injury, physical stimuli, and biomaterials surfaces through chemical, pharmacological, mechanical, immunological, and genetic manipulation (Hubbell et al., 2009; Huebsch and Mooney, 2009; Lutolf et al., 2009; Reilly and Engler, 2010; Edalat et al., 2011). A key challenge in tissue engineering is to understand quantitatively how cells respond to molecular signals and integrate multiple inputs to generate a given response, so that cell responses specifically follow desired receptor–ligand interactions.

This section summarizes the major concepts and recent progress in translational tissue engineering. Sarkar et al (Chapter II.6.2) describes the essential overview concepts in the field, followed by chapters on tissue engineering scaffolds (Singh, Kasper and Mikos, Chapter II.6.3), cell sources (Caplan, Chapter II.6.4), micromechanical design criteria (Ghosh, Thodeti and Ingber, Chapter II.6.5), and bioreactors (Tandon et al., Chapter II.6.6). These general chapters are followed by descriptions of the major areas of progress and challenges for engineering tissues for specific applications, including bone (Brown, Kumbar and Laurencin et al., Chapter II.6.7), cartilage and ligament (Kuo, Li and Tuan, Chapter II.6.8), blood vessels, (Schutte and Nerem, Chapter II.6.9), heart valves (Schoen and Hoerstrup, Chapter II.6.10), cardiac muscle (Godier and Vunjak-Novakovic, Chapter II.6.11), skin (Mansbridge, Chapter II.6.12), esophageal and gastrointestinal tissue (Ratner, Chapter II.6.13), neuronal regeneration (Bellamkonda, Clements and Munson, Chapter II.6.14), immunoisolation (Rainbow and Lysaght, Chapter II.6.15), and engineering with decellularized tissues (Badylak, Brown and Gilbert, Chapter II.6.16).

The early phase of translation of tissue engineering technology to the clinic has begun. Tissue-engineered products for skin replacement are in clinical use (MacNeil, 2007). Further examples of previous and ongoing tissue engineering approaches used clinically include cartilage regeneration using autologous chondrocyte transplantation (Brittberg et al., 1994), a replacement thumb with bone composed of autologous periosteal cells and natural coral (hydroxyapatite) (Vacanti et al., 2001), bladder replacement (Atala et al., 2006), and blood vessels (McAllister et al., 2009). Biohybrid extracorporeal artificial organs using functional cells (Humes and Szcypka, 2004), tissue-engineered heart valves (Vesely, 2005; Sacks et al., 2009), blood vessels (Quint et al., 2012), ligaments (Laurencin and Freeman, 2005), and bone (Khan et al., 2008), are in active preclinical studies. Complex organs such as lung and heart (myocardium) are under development (Nichols et al., 2009; Badylak et al., 2011;

Vunjak-Novakovic et al., 2011). General problem areas include: (1) development of new cell lines and biomaterials; (2) evaluation of the optimal implant configuration, including a functional vasculature for three-dimensional tissues; (3) development of testing methodology appropriate to validate the concept; and (4) the reproducible manufacture and preservation of bioartificial devices until ready for use. A particularly interesting approach is the decellularization followed by recellularization of a complex organ (Badylak et al., 2011).

Cell-compatible biomaterial microarrays that allow rapid, microscale testing of biomaterial interactions with cells have recently been described, and may be extremely helpful in this regard (Anderson et al., 2005; Khademhosseini et al., 2006). Exquisite control of scaffold architecture and overall and regional surface chemistry is now also possible; these features may precisely regulate cell behavior (Lutolf et al., 2009). Another approach is *biomimetics*, in which natural structures and processes (e.g., development, wound healing) are mimicked. Modified native extracellular matrix (ECM) may have particularly advantageous properties in this regard (Badylak et al., 2009).

Future directions in biomaterials, tissue engineering, and regenerative therapeutics are indeed exciting, and priorities are under active discussion (Hacker and Mikos, 2006; Goldberg et al., 2007; Hellman and Nerem, 2007; Johnson et al., 2007; Daley and Scadden, 2008; Kohane and Langer, 2008).

BIBLIOGRAPHY

Anderson, D. G., Putnam, D., Lavik, E. B., Mahmood, T. A., & Langer, R. (2005). Biomaterial microarrays: Rapid, microscale screening of polymer–cell interaction. *Biomaterials*, **26**, 4892–4897.

Atala, A., Bauer, S. B., Soker, S., Yoo, J. J., & Retik, A. B. (2006). Tissue-engineered autologous bladders for patients needing cystoplasty. *Lancet*, **367**, 1241–1246.

Badylak, S. F., Freytes, D. O., & Gilbert, T. W. (2009). Extracellular matrix as a biological scaffold material: Structure and function. *Acta Biomater.*, **5**, 1–13.

Badylak, S. F., Taylor, D., & Uygun, K. (2011). Whole-organ tissue engineering: Decellularization and recellularization of three-dimensional matrix scaffolds. *Annu. Rev. Biomed. Eng.*, **13**, 27–53.

Balachandran, K., Alford, P. W., Wylie-Sears, J., Goss, J. A., Grosberg, A., et al. (2011). Cyclic strain induces dual-mode endothelial-mesenchymal transformation of the cardiac valve. *Proc. Natl. Acad. Sci. USA*, **108**, 19943–19948.

Brittberg, M., Lindahl, A., Nilsson, A., Ohlsson, C., Isaksson, O., et al. (1994). Treatment of deep cartilage defects in the knee with autologous chondrocyte transplantation. *New Engl. J. Med.*, **331**, 889–895.

Daley, G. Q., & Scadden, D. T. (2008). Prospects for stem cell therapy. *Cell*, **132**, 544–548.

Edalat, F., Bae, H., Manoucheri, S., Cha, J. M., & Khademhosseini, A. (2011). Engineering approaches toward deconstructing and controlling the stem cell environment. *Ann. Biomed. Eng.*, **40**, 1301–1315.

Gerecht-Nir, S., Radisic, M., Park, H., Cannizzaro, C., Boublik, J., et al. (2006). Biophysical regulation during cardiac development and application to tissue engineering. *Int. J. Dev. Biol.*, **50**, 233–243.

Goldberg, M., Langer, R., & Jia, X. (2007). Nanostructured materials for applications in drug delivery and tissue engineering. *J. Biomater Sci. Polym. Ed.*, **18**, 241–268.

Hacker, M. C., & Mikos, A. G. (2006). Trends in tissue engineering research. *Tiss. Eng.*, **12**, 2049–2057.

Hellman, K. B., & Nerem, R. M. (2007). Advancing tissue engineering and regenerative medicine. *Tiss. Eng.*, **13**, 2823–2824.

Hubbell, J. A., Thomas, S. N., & Swartz, M. A. (2009). Materials engineering for immunomodulation. *Nature*, **462**, 449–460.

Huebsch, N., & Mooney, D. J. (2009). Inspiration and application in the evolution of biomaterials. *Nature*, **462**, 426–432.

Humes, H. D., & Szczypka, M. S. (2004). Advances in cell therapy for renal failure. *Transpl. Immunol.*, **12**, 219–227.

Ingber, D. E. (2010a). Mechanical control of tissue and organ development. *Development*, **137**, 1407–1420.

Ingber, D. E. (2010b). From cellular mechanotransduction to biologically inspired engineering. *Ann. Biomed. Eng.*, **38**, 1148–1161.

Johnson, P. C., Mikos, A. G., Fisher, J. P., & Jansen, J. A. (2007). Strategic directions in tissue engineering. *Tiss. Eng.*, **12**, 2827–2937.

Khademhosseini, A., Langer, R., Borenstein, J., & Vacanti, P. (2006). Microscale technologies for tissue engineering and biology. *Proc. Natl. Acad. Sci. USA.*, **103**, 2480–2487.

Khan, Y., Yaszemski, M. J., Mikos, A. G., & Laurencin, C. T. (2008). Tissue engineering of bone: material and matrix considerations. *J. Bone Joint Surg. Am.*, **90**, 36.

Khetani, S. R., & Bhatia, S. N. (2008). Microscale culture of human liver cells for drug development. *Nat. Biotechnol.*, **26**, 120–126.

Kohane, D. S., & Langer, R. (2008). Polymeric materials in tissue engineering. *Pediatr. Res.*, **63**, 487–491.

Lakshmipathy, U., & Verfaillie, C. (2005). Stem cell plasticity. *Blood Rev.*, **19**, 29–38.

Langer, R., & Vacanti, J. P. (1993). Tissue engineering. *Science*, **260**, 920–926.

Laurencin, C. T., & Freeman, J. W. (2005). Ligament tissue engineering: An evolutionary materials science approach. *Biomaterials*, **26**, 7530–7536.

Leri, A., Kajstura, J., & Anversa, P. (2011). Role of cardiac stem cells in cardiac pathophysiology: A paradigm shift in human myocardial biology. *Circ. Res.*, **109**, 941–961.

Lutolf, M. P., Gilbert, P. M., & Blau, H. M. (2009). Designing materials to direct stem cell fate. *Nature*, **462**, 433–441.

MacNeil, S. (2007). Progress and opportunities for tissue-engineered skin. *Nature*, **445**, 874–880.

McAllister, T. N., Maruszewski, M., Garrido, S. A., Wystrychowski, W., Dusserre, N., et al. (2009). Effectiveness of haemodialysis access with an autologous tissue-engineered vascular graft: A multicentre cohort study. *Lancet*, **373**, 1440–1446.

Mendelson, K., & Schoen, F. J. (2006). Heart valve tissue engineering: Concepts, approaches, progress, and challenges. *Ann. Biomed. Engin.*, **34**, 1799–1819.

Nichols, J. E., Niles, J. A., & Cortiella, J. (2009). Design and development of tissue engineered ling: Progress and Challenges. *Organogenesis*, **5**, 57.

Orlando, G., Baptista, P., Birchall, M., De Coppi, P., Farney, A., et al. (2011). Regenerative medicine as applied to solid organ transplantation: Current status and future challenges. *Transpl. Int.*, **24**, 223–232.

Quint, C., Arief, M., Muto, A., Dardik, A., & Niklason, L. E. (2012). Allogeneic human tissue-engineered blood vessel. *J. Vasc. Surg.*, **55**, 790–798.

Rabkin, E., & Schoen, F. J. (2002). Cardiovascular tissue engineering. *Cardiovasc. Pathol.*, **11**, 305.

Reilly, G. C., & Engler, A. J. (2010). Intrinsic extracellular matrix properties regulate stem cell differentiation. *J. Biomech.*, **43**, 55–62.

Sacks, M. S., Schoen, F. J., & Mayer, J. E. (2009). Bioengineering challenges for heart valve tissue engineering. *Ann. Rev. Biomed. Engin.*, **11**, 289–313.

Stevens, M. M., Marini, R. P., Schaefer, D., Aronson, J., Langer, R., et al. (2005). *In vivo* engineering of organs: The bone bioreactor. *PNAS*, **102**, 1145–1455.

Stocum, D. L. (2005). Stem cells in CNS and cardiac regeneration. *Adv. Biochem. Eng. Biotechnol.*, **93**, 135–159.

Vacanti, C. A., Bonassar, L. J., Vacanti, M. P., & Shufflebarger, J. (2001). Replacement of an avulsed phalanx with tissue-engineered bone. *New. Engl. J. Med.*, **344**, 1511–1514.

Vesely, I. (2005). Heart valve tissue engineering. *Circ. Res.*, **97**, 743–755.

Vunjak-Novakovic, G., Lui, K. O., Tandon, N., & Chien, K. R. (2011). Bioengineering heart muscle: A paradigm for regenerative medicine. *Annu. Rev. Biomed. Eng.*, **13**, 245–267.

Wheeldon, I., Farhadi, A., Bick, A. G., Jabbari, E., & Khademhosseini, A. (2011). Nanoscale tissue engineering: Spatial control over cell-materials interactions. *Nanotechnology*, **22**, 212001.

Whitehead, K. A., Langer, R., & Anderson, D. G. (2009). Knocking down barriers: Advances in siRNA delivery. *Nat. Rev. Drug Discov.*, **8**, 129–138.

CHAPTER II.6.2 OVERVIEW OF TISSUE ENGINEERING CONCEPTS AND APPLICATIONS

Debanjan Sarkar[1,2,6], Weian Zhao[1,2,7], Sebastian Schaefer[1,2], James A. Ankrum[1,2], Grace S. L. Teo[1,2], Maria Nunes Pereira[1,2,4,5], Lino Ferreira[4,5], and Jeffrey M. Karp[1,2,3]

[1]Harvard-MIT Division of Health Sciences & Technology, USA

[2]Department of Medicine, Brigham and Women's Hospital, Harvard Medical School, USA

[3]Harvard Stem Cell Institute, USA

[4]Center of Neurosciences and Cell Biology, University of Coimbra, Portugal

[5]Biocant-Biotechnology Innovation Center, Cantanhede, Portugal

[6]Department of Biomedical Engineering, University at Buffalo, The State University of New York, USA

[7]Sue and Bill Gross Stem Cell Research Center, Chao Family Comprehensive Cancer Center, Department of Pharmaceutical Sciences, University of California, Irvine, USA

GENERAL INTRODUCTION

History of Tissue Engineering

The term "tissue engineering" as recognized today was first introduced at a panel meeting of the National Science Foundation in 1987, which led to the first tissue engineering meeting in early 1988 (Nerem, 2006; Vacanti, 2006). However, tissue engineering strategies date back to the seventies and eighties for developing skin substitutes (Vacanti, 2006). Despite these early approaches for replacement, repair, and regeneration of failing organs, the true emergence of tissue engineering as a medical field started in the early nineties when tissue engineering was defined as an interdisciplinary field that applies the principles of engineering and life sciences toward the development of biological substitutes that restore, maintain or improve tissue function (Langer and Vacanti, 1993). The field has since rapidly progressed worldwide, with nearly $4 billion invested in the field in the 1990s resulting in over 70 companies and several products on the market by the end of the millennium (Lysaght and Reyes, 2001). As of mid-2007, approximately 50 firms or business units with over 3000 employees offered commercial tissue-regenerative products or services with generally profitable annual sales in excess of $1.3 billion (Lysaght et al., 2008). Well over a million patients have been treated with these products. In addition, 110 development-stage companies with over 55 products in US Food and Drug Administration (FDA)-level clinical trials and other preclinical stages employed ~2500 scientists or support personnel, and spent $850 million development dollars in 2007. While early success has been challenging, technological advances alongside biological discoveries continue to propel the field of tissue engineering into exciting new frontiers.

GOALS OF TISSUE ENGINEERING AND CLASSIFICATION

Goals of Tissue Engineering

Tissue engineering aims to restore tissue and organ function by employing biological and engineering strategies to clinical problems. The functional failure of tissues and organs is a severe and costly healthcare problem, as their replacement is limited by the availability of compatible donors (Langer and Vacanti, 1993, 1999).

Artificial prostheses and mechanical devices save and improve the lives of millions of patients, but are not ideal since they are subject to mechanical failure upon long-term implantation. Furthermore, mechanical devices rarely integrate with host tissues, and can trigger a host immune response damaging healthy tissue around the implant. In addition, surgical reconstruction of organs and tissues are attempted where the organs or tissues are moved from their original location to replace a damaged tissue, e.g., saphenous vein as bypass graft, patella tendon for anterior cruciate ligament (ACL) repair. However, often this strategy fails to replace all the functions of the original tissue. Additionally, development of malignant tumors, surgical complications, and morbidity at the donor sites are major problems in surgical reconstruction of tissues. Thus, tissue engineering has emerged as another alternative for tissue or organ transplantation. The primary goal of tissue engineering is to provide a biological substitute to treat tissue/organ loss or failure by integrating multiple aspects of engineering, biology, and medicine. By recapitulating the normal tissue development process, tissue engineering represents a strategy to restore, maintain, and improve tissue function, which ultimately aims toward complete organ replacement.

Classification of Tissue Engineering Approaches

Traditional Tissue Engineering Approaches. Tissue engineering includes two main strategies: (1) transplantation of a tissue grown *in vitro* consisting of an artificial matrix with cells and growth factors; and (2) *in situ* regeneration of tissue utilizing a combination of an artificial matrix and growth factors as a guiding template to induce host cell regeneration of the tissue *in vivo* (Figure II.6.2.1).

In addition to traditional tissue engineering approaches, other methods used for tissue regeneration include local and systemic cell injection without a scaffold, and closed looped systems used as implantable or extracorporeal devices. While classically these methods are not regarded as tissue engineering, they have contributed significantly to tissue regeneration and are therefore briefly introduced here.

Cell Therapy. Cell therapy involves delivery of cells through systemic injection into the bloodstream or through direct transplantation into a local tissue (Mooney and Vandenburgh, 2008; Karp and Leng Teo, 2009). The major requirement for this strategy is to harvest the cells and grow them in large numbers for *in vivo* transplantation.

Direct injection of cells to a local site is a common strategy attempted to promote tissue regeneration. However, the survival of the delivered cells is typically low, often due to a lack of a rich nutrient and oxygen supply (Muschler et al., 2004). Alternatively, cells can be administrated via systemic injection, which relies on cells traveling through circulation to engraft in the target site. Cell transplants from bone marrow, peripheral blood or umbilical cord have been used to treat several blood-related diseases including leukemia, multiple myeloma, and immune deficiencies. The main goal of these strategies is to deliver hematopoietic (blood) stem cells to treat blood-related diseases (Thomas, 1987). Recently, mesenchymal stem cells, connective tissue progenitor cells, which repair or regenerate non-hematopoetic tissues, have been systemically injected to treat diseases including myocardial infarction (Barbash et al., 2003), bone diseases (Horwitz et al., 1999), and brain injury (Mahmood et al., 2003) in clinical trials. The main challenges for cell transplantation are: growing large number of cells without bacterial contamination; preservation of cell phenotype; and preventing accumulation of genetic mutations during culture expansion. Although cells have been successfully delivered to the heart to treat ischemic tissue following myocardial infarction, and into the joint to treat arthritis, irrespective of the delivery route, cell therapies face challenges due to widespread death of the transplanted cells, poor engraftment, and loss of control over the fate of the transplanted cells.

Closed Loop Methods. A variety of closed loop systems are used as extracorporeal or implantable devices which house the transplanted cells in a semipermeable membrane (Murua et al., 2008). The membrane permits diffusion of nutrients and excreted products, but prevents the movement of antibodies, pathogens or immunocompetent cells. There are several types of designs for such devices. Vascular type design uses a conduit structure around which the cells are transplanted in a chamber.

(A)

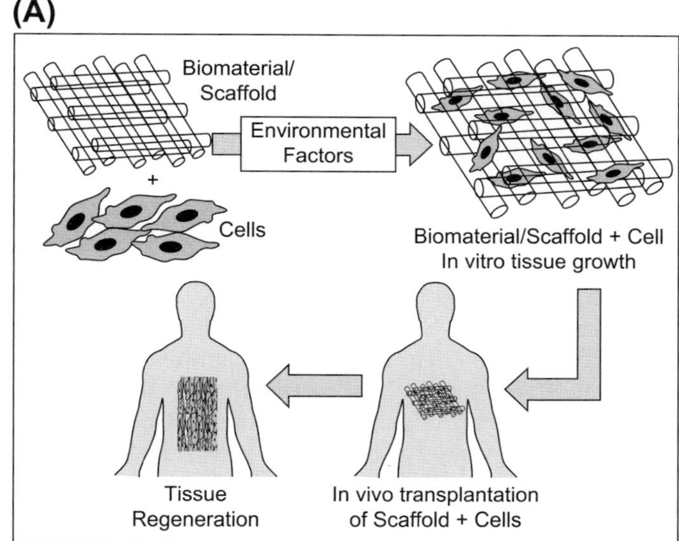

In vitro tissue engineering followed by transplantation

(B)

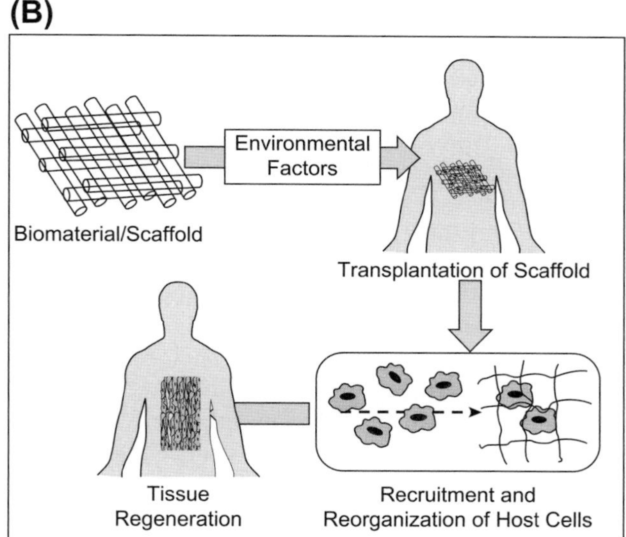

In vivo tissue engineering by transplantation of scaffold and recruitment and reorganization of host cells

FIGURE II.6.2.1 Tissue engineering approaches may be classified into two categories: (A) transplantation of *in vitro* grown tissues; and (B) promotion of tissue regeneration *in situ*. In both approaches the scaffolds or artificial matrices, often biodegradable polymers, are integrated with microenvironmental factors (such as cytokines, growth factors, mechanical forces, physico-chemical factors, spatial and temporal signals, and extracellular matrix molecules).

As blood flows through the conduit it provides nutrients to the transplanted cells, while cell-secreted substances diffuse into the bloodstream (Langer and Vacanti, 1993; Patzer, 2001). Additionally, micro/macrocapsule-based systems have been used as closed loop systems (Aebischer et al., 1991; Winn et al., 1991) where cells are encapsulated within hydrogel droplets. The encapsulated cells can then be cultured *in vitro* or transplanted *in vivo*, either to repopulate a defect site or to produce growth factors or other molecules that will have an effect on the targeted cell population. Closed loop extracoporeal devices have been used for the treatment of liver, pancreas, and kidney pathologies. Major problems associated with these types of devices include fouling, fibrous tissue overgrowth, restricted and hindered diffusion, and immunogenic response (Wiegand et al., 1993).

COMPONENTS OF TISSUE ENGINEERING

Tissue engineering strategies typically involve multiple components including cells, a physical template, and a combination of growth cues that promote tissue regeneration and integration of the construct into a functional and organized tissue.

The Cell

Cell Source. Cells are the building blocks of tissues, and play a critical role in promoting tissue healing and regeneration. Within tissue engineering strategies, cells may be a component of the *in vitro* construct or may be recruited *in vivo* with the aid of immobilized or soluble signals. Cell types utilized for tissue engineering are selected from a variety of sources which include autologous cells from the patient, allogeneic cells from another human, and xenogeneic cells from a different species. However, allogeneic and xenogeneic cells often suffer from immune rejection.

Cell Type. Common cell types include stem cells (capable of self-renewal and differentiation into multiple lineages), differentiated mature cells, or a mixture of differentiated cells. Stem cells (see also Chapter II.1.7 "Stem Cells: Key Concepts" and II.6.4 "Cell Sources for Tissue Engineering: Mesenchymal Stem Cells") may include embryonic stem cells and adult stem cells, such as mesenchymal stem cells and hematopoietic stem cells. Embryonic stem cells are an attractive cell source for tissue engineering as these cells can self-renew without differentiation, can be culture expanded, and most importantly can differentiate into any cell type. Since embryonic stem cells are isolated from an embryonic stage, they can develop into any of the three germ layers: endoderm (interior stomach lining, gastrointestinal tract, the lungs); mesoderm (muscle, bone, blood, urogential); or ectoderm (epidermal tissues and nervous system). Successful reprogramming of differentiated human somatic cells into pluripotent cells has led to the creation of induced pluripotent stem (iPS) cells (Takahashi and Yamanaka, 2006). These cells are functionally similar to embryonic stem cells, but do not require the destruction of an embryo and can be created from a patient's own cells, eliminating the risk of host rejection (Takahashi et al., 2007). Adult stem cells include, for example, mesenchymal (Pittenger et al., 1999), hematopoietic (Baum et al., 1992), neural (Snyder et al., 1997), and hepatic (Petersen et al., 1999) stem cells. In particular, hematopoietic stem cells have been used in clinics for a few decades for treating blood diseases (i.e., bone marrow transplantation). Mesenchymal stem cells (MSCs), which can be transplanted as an allogenic cell source to another patient without immunosuppressive drugs, are capable of differentiation into multiple lineages that may produce tissues including bone, cartilage, and muscle, and have been approved for use in multiple systems to treat bone defects. Mesenchymal stem cells can also modulate the host immune response through paracrine or endocrine mechanisms, and are currently being applied in clinical trials for treatment of immune diseases. Nevertheless, adult stem cells are rare, challenging to isolate and expand without altering cell phenotype, and limited in their differentiation potential.

Further investigation is needed to identify the best source of stem cells for each tissue engineering application. In the near future, it will be necessary to compare the performance of the different stem cell populations under the same testing conditions. Multiple variables should be addressed in the selection of the best stem cell population, including: (1) stem cell accessibility (e.g., the isolation of neural stem cells is invasive and relatively difficult compared to other stem cells); (2) number; (3) proliferation capacity; (4) differentiation profile; and (5) ethical issues.

In addition to stem cell populations, committed and/or differentiated cell types are also frequently used in tissue engineering approaches. For example, culture expanded chondrocytes have been used for over 15 years to treat cartilage defects (Brittberg et al., 1994). However, the potential of differentiated adult cells is often limited due to their low proliferation capability, loss of phenotype, and dedifferentiation in culture.

In addition to primary cells, the intrinsic biological potential and performance of a cell can be modified by transient or permanent alteration of specific genes (Hannallah et al., 2003). The introduction of new or altered genes is often accomplished with vectors created by modifying naturally-occurring viruses such as retrovirus, lentivirus, adenovirus or adeno-associated virus. There are several concerns for these approaches, such as transformation efficiency, safety of virus transfection, vector stability, and optimal function of the inserted genes. Non-viral transfection techniques have been developed to circumvent some of these issues; however, the long-term fate of these genetically modified cells still presents a potential risk.

The cell phenotype can also be regulated through manipulation of isolation and culture conditions. Although this may expand the available tools to manipulate cell characteristics, it can create more hurdles for consistently manufacturing high quality cells. In particular, care must be taken during the cell isolation process to maintain cell phenotype, purity, and differentiation state. It is critical to achieve a high purity of cells defined by rigorous characterization. In addition, it is important to consider that cell culture may "activate" cells, altering their phenotype from those found *in situ*. There are many decisions that must be made, often without suitable premise, such as the ideal number of cells to be transplanted, the maximum number of times cells can be passaged, the maximum length of time cells should be maintained in culture, the ideal differentiated state of the cells to produce a therapeutic effect, and the ideal storage conditions for the cells. These factors can significantly alter the outcome of a regenerative approach, and often need to be optimized on a case-by-case basis for specific animal models (Bueno et al., 2007).

MATERIALS

Biomaterials are used to develop scaffolds, which provide a template for cells to organize and restore structure and function to damaged or dysfunctional tissues. Guidance can be achieved through biophysical and biochemical cues that direct cell behavior, morphology, adhesion, and motility (Hubbell, 1995; Langer and Tirrell, 2004). Biomaterials can, at the same time, be used to supply nutrients, drugs, and bioactive factors that direct specific tissue growth. Accordingly, the material should be non-toxic and degradable within a clinically useful range into fully biocompatible products. Ideally, the material should also possess physico-mechanical and engineering properties suitable for the intended application, and be compatible for further functionalization with bioactive molecules. Biomaterial scaffolds for tissue engineering are discussed in more detail in Chapter II.6.3 "Tissue Engineering Scaffolds."

To accommodate these material requirements, the field of tissue engineering has witnessed tremendous development of new biomaterials over the past few decades. These materials are derived from both natural and synthetic sources, and possess a broad spectrum of structural and functional properties that make them suitable for many clinical applications.

Natural Materials

A wide range of natural-origin polymers, including proteins and polysaccharides, are used as carriers for cells and bioactive molecules (Malafaya et al., 2007). Natural materials are advantageous due to their inherent biological recognition through receptor–ligand interactions, cell-mediated proteolysis and remodeling, and low toxicity.

Protein-based natural polymers include collagen, gelatin, silk fibroin, fibrin, elastin, and soybean. Collagen is a major component of the extracellular matrix, the natural cell scaffold, which interacts with cells in all tissues, providing essential signals for the regulation of cell anchorage, migration, proliferation, differentiation, and survival (Yang et al., 2004). As a result, collagen has been studied for engineering artificial skin (collagen IV), bone (collagen I), and cartilage (collagen II), resulting in several tissue engineering products. Bilayered collagen gels seeded with human fibroblast and keratinocytes are used as a bioengineered artificial skin by Organogenesis, Inc. under the name of Apligraf®. Gelatin is a natural polymer derived from collagen that has significantly lower antigenicity in contrast to collagen and has been used for engineering bone, cartilage, and skin. Gelatin is also well-suited to serve as a carrier for bioactive molecules, including fibroblast growth factor and transforming growth factor (Ito et al., 2003). Silk fibroin has received significant attention as a versatile natural polymer due to its high strength-to-weight ratio and slow degradation (Sofia et al., 2001). Another commonly used natural material is fibrin, the structural component of blood clots that provides a transitory matrix for cell migration during wound healing. Fibrin has been used as a matrix for studying the regeneration of bone, cartilage, skin, nerve, and spinal cord (Ahmed et al., 2008).

Another class of natural polymers is polysaccharides that contain monomers (monosaccharides) linked together by O-glycosidic bonds to form linear and branched polymers (Malafaya et al., 2007). Several polysaccharides, including chitosan and alginates, have been studied for tissue engineering applications. Chitosan, the fully or partially deacetylated form of chitin found particularly in the shell of crustaceans, has attracted significant attention, due to its biocompatible properties, for regeneration of skin, bone, cartilage, and vascular grafts (Park et al., 2000; Mi et al., 2001; Kim et al., 2003). Alginates in the form of hydrogels, beads, and scaffolds are a widely utilized polysaccharide polymer used to promote angiogenesis and regenerate bone and cartilage (Marijnissen et al., 2002; Simmons et al., 2004; Tilakaratne et al., 2007). Other commonly employed polysaccharides include starch, hyaluronan, chrondoitin sulphate, cellulose, dextran, and polyhydroxyalkanoates.

Unfortunately, there are several limitations of natural materials which include purification, cost, immunogenic responses, and lack of mechanical properties and processability. Although some of these disadvantages have been avoided through recombinant protein expression technologies (Rodriguez-Cabello et al., 2005), synthetic biomaterials (see below) present a paradigm shift that overcomes many of these challenges while opening the door for custom-designed biomaterials.

Synthetic Materials

Many synthetic polymers have been designed and fabricated for tissue engineering purposes. Biodegradable synthetic polymers offer a number of advantages compared to natural materials, such as controlled mechanical properties and degradation kinetics, easy processability into custom shapes and structures, and easy modification of the material for specific applications.

The vast majority of synthetic biomaterial-based polymers belong to the polyester family which includes poly(glycolic acid), poly(lactic acid) and their copolymer poly(lactide-*co*-glycolide) (PLGA) (Shalaby, 1988). Polyesters have long been used in the clinic in the form of degradable sutures, and have more recently been studied to promote regeneration of bone, cartilage, bladder, skin, and vasculature. In particular, PLGA and its constituent polymers are used in a vast array of applications due to their ease of degradation by hydrolysis of ester bonds, controllable degradation rate, and minimal inflammatory response. Another class of polyester used in tissue engineering is polylactones, such as poly(caprolactone) (PCL), which are less crystalline than PLGA and degrade at a much slower rate. Interestingly, caprolactone-based materials have an added advantage given their shape memory properties which enables the polymer to change their shape after an increase in temperature (Lendlein et al., 2001). Other classes of polyesters that have been explored include poly(ortho esters) and poly(propylene fumarate) (Kharas et al., 1997). Recent development of elastomeric polyesters has shown significant promise as tissue engineering biomaterials that mimic the mechanical properties of soft tissues due to their elasticity compared to other polyesters, e.g., poly(glycerol sebacate) and poly(diol citrate) (Yang et al., 2006; Nijst et al., 2007).

Poly(anhydrides) are another class of synthetic polymers developed from the condensation of diacids or a mixture of diacids. These polymers are biocompatible and have well-defined degradation characteristics. Several poly (anhydride) homopolymers and copolymers have been synthesized to control their degradation rate and mechanical properties (Domb, 1987). Amino acid-based polymers, e.g., L-tyrosine based poly(carbonate), poly(imminocarbonate), and poly(phosphate) (Pulapura and Kohn, 1992; Gupta and Lopina, 2004) have been developed as a substitute for poly amino acids where the peptide linkages are partially replaced with non-peptide links. Biodegradable polyurethane elastomers have also been developed as synthetic biomaterials for regeneration of heart, cartilage, and skin (Guan et al., 2005). Polyurethanes with degradable linkages and biocompatible components have been developed from amino acids and peptides with a wide range of mechanical properties and degradation kinetics (Skarja and Woodhouse, 2001; Sarkar et al., 2009). Synthetic polymers with photopolymerizable groups are based on poly(ethylene glycol) and poly(acrylates). Upon exposure to ultraviolet radiation and cross-linking agents, these polymers can cross-link to form three-dimensional networks.

Due to complexities in the design and synthesis of biomaterials, recent techniques in combinatorial approaches to develop and screen synthetic biomaterials are rapidly gaining interest in this field (Anderson et al., 2004; Kohn, 2004; Yang et al., 2008). Such techniques can be used to rapidly synthesize a library of polymers by altering the structure and composition of polymers and to characterize their interactions with cells in a high-throughput manner.

Semi-Synthetic Materials

Combinations of natural and synthetic polymers have particular advantages compared to pure natural or synthetic counterparts for some applications. These biosynthetic hybrids are usually designed by incorporation of biologically active macromolecules onto the backbone of synthetic polymers to optimize both structure and function (Seliktar, 2005). From a tissue engineering perspective, the objective of using a biosynthetic hybrid material is to reproduce the intricate extracellular stimulation of the native cellular environment, yet exhibit enhanced control over the material properties (compared to purely natural materials). For example, semi-synthetic PEG(polyethylene glycol)-fibrinogen hydrogel has been developed to control cell migration and tissue regeneration where the synthetic PEG component controls density, stiffness, and biodegradability, and the fibrinogen component presents biofunctional domains for cell-mediated remodeling (Dikovsky et al., 2006).

SCAFFOLD DESIGN

Scaffolds act as the synthetic analog of the natural extracellular matrix. The role of scaffolds is to recapitulate the normal tissue development process by allowing cells to formulate their own microenvironment (Lee et al., 2008). The scaffold provides the necessary support for cells to attach, proliferate, and maintain their differentiated function and subsequent regeneration of new tissues. Ideally, a scaffold should have the following characteristics: (1) three-dimensional highly porous structure with an interconnected pore network to facilitate cell/tissue growth and diffusion of nutrients, metabolic waste, and paracrine factors; (2) biodegradable or bioresorbable with controllable degradation and resorption rates to match cell/tissue growth *in vitro* and *in vivo*; (3) suitable surface chemistry for cell attachment, proliferation, and differentiation; (4) mechanical properties to match those of the tissues at the site of implantation; and (5) be easily processed to form a variety of shapes and sizes.

Conventional Fabrication Methods

The formation of a porous structure is the main goal of scaffold fabrication. Most methods for fabricating porous scaffolds, including particulate leaching, freeze drying, gas infusion, and phase separation, create isotopically distributed interconnected pores. Porous structures are developed by introducing particles or bubbles when the scaffold is solidified, which are later removed leaving behind an interconnected network of pores. Although these techniques are relatively simple for developing a three-dimensional structure, they are limited by uncontrolled pore size and connectivity, poor mechanical strength, and residual solvent/porogens (Yang et al., 2001). Hydrocarbon templating is a combination of two distinct foam processes where water-insoluble hydrocarbon particulates are leached with the precipitation of the polymers. This process has several advantages, including control over scaffold thickness and pore structure, although use of organic solvents is typically a disadvantage for biological applications as it is difficult to remove completely.

Fiber-Based Scaffolds

Fibrous scaffold structures may be developed by electrospinning of polymers to generate continuous micro- or nanoscale diameter fibers. Additionally, the orientation of fibers can be controlled during electrospinning to develop random or aligned fibers. Electrospun scaffolds are advantageous due to their high surface-to-volume ratio and structural similarity to natural extracellular matrix. However, it is difficult to control the distance between fibers, an important factor that influences the migration of cells. Self-assembly of biopolymers, e.g., peptides and nucleic acids, using noncovalent interactions including H-bonding, hydrophobic, electrostatic interactions, and van der Waals forces into three-dimensional structures has also been used for scaffold development. The main advantage of such biopolymers is that their self-assembly relies on specific biorecognitions (e.g., DNA hybridization) which therefore make the formation of scaffold highly predictable and programmable.

Solid-Free-Form Methods

To enhance control over the three-dimensional organization of porous scaffolds, rapid prototyping techniques borrowed from electrical and mechanical engineering can be used. These techniques involve computer-assisted methods with solid-free-form (SFF) fabrication techniques (Hollister, 2005). Specifically, they involve acquiring a two-dimensional image of a target specimen by nondestructive imaging, then developing macroscale three-dimensional architecture with software, and finally fabrication of the three-dimensional matrix with highly precise and automated layer-by-layer SFF processes. There are several SFF processes, including laser-based, ink-jet type printing-based, and nozzle-based approaches. These methods offer precise control of the three-dimensional structure of the scaffolds; however, they are associated with higher costs and require more complex equipment compared to conventional salt leaching or gas foaming processes.

Hydrogel-Based Scaffold

Hydrogels are typically cross-linked networks swollen in water to suspend cells in three dimensions. Photocross-linkable hydrogels are frequently used to encapsulate cells (Ferreira et al., 2007a). Polymers with photocross-linkable moieties are polymerized into three-dimensional networks in the presence of cross-linkers under UV radiation. By controlling the structure with defined cross-linking density, mechanical properties, mass transport, and degradation characteristics, the gels can be tuned for a range of applications. Hydrogels are an appealing three-dimensional scaffold because they are structurally similar to the extracellular matrix of many tissues, can often be processed under relatively mild conditions, and may be delivered in a minimally invasive manner (Drury and Mooney, 2003). In addition, hydrogels are advantageous due to high water content, facile transport properties, and controlled degradation kinetics (Nuttelman et al., 2008). Furthermore, hydrogels can be chemically modified to improve the adhesion and proliferation of the cells on the gel matrices through inclusion of adhesion peptides, i.e., RGD(Arg-Gly-Asp tripeptide). In order to achieve more control over the cell placement within the hydrogels, three-dimensional patterned hydrogels are used. However, major problems with using hydrogels include structural instability and overall inferior mechanical properties for placement within dynamic environments.

DEVELOPMENT OF TISSUE ENGINEERING CONSTRUCTS

Strategies for Tissue Engineering

As outlined above, there are two main strategies to develop a tissue engineering construct. To facilitate tissue growth, scaffolds can be transplanted with or without cells. Each strategy's success largely depends on effective integration of different factors including cells, scaffolds, and environmental cues (e.g., growth factors, biological molecules).

Cells with Scaffold Transplantation. Scaffolds provide a substrate upon which cells attach, proliferate, and produce extracellular matrix within a predefined three-dimensional orientation. However, the main challenges for this strategy include: generating a complete functional tissue; preserving cell viability and function during transplantation; biological and mechanical integration with the surrounding tissue; and supplying oxygen and nutrients to the transplanted cells (Muschler et al., 2004). The transplanted cells within the core of large defects often die before contributing to the healing process. Moreover, the scaffolds are often filled with blood clots which present a harsh environment for the transplanted cells. Regardless, this strategy can be useful for transplantation of thin tissue grafts, such as corneal and skin grafts, where an immediate connection to the vascular system is not essential.

Scaffold-Only Transplantation. Another strategy involves *in situ* tissue regeneration by providing an instructive scaffold that can guide and control the regeneration process *in vivo* (Zhao and Karp, 2009). This process uses an artificial scaffold matrix that, once implanted, recruits host cells and relies on the body's innate regenerative ability. The scaffold recruits such cells through signal presentation (Richardson et al., 2001; Simmons et al., 2004; Lee et al., 2008). For example, transplantation of PEG-collagen matrices with enzyme cleavable matrix metalloproteinase (MMP) links that release bone morphogenic protein-2 (BMP2) can promote regeneration of critical size bone defects effectively through invasion of the host cells (Lutolf et al., 2003a). The absence of MMP sensitive linkages and/or BMP-2 hindered the regeneration process, indicating that transplantation of scaffolds with appropriate soluble and insoluble signals is critical. Thus, the success of scaffold transplantation-based approaches largely depends on the appropriate presentation of signals to mediate host cell mobilization and coordination of the subsequent behavior of the cells, e.g., adherence, migration, and proliferation.

Biological Factors in Tissue Growth

Type of Biological Factors. Biological factors including hormones, cytokines, growth factors, extracellular matrix molecules, cell surface molecules, and nucleic acids can significantly influence the function and behavior of cells in the scaffold. The temporal and spatial coordination of cellular processes is orchestrated by these signals from the extracellular environment. A large number of biomolecules have been explored to induce tissue regeneration, but the majority of these molecules can be categorized as follows: (1) small molecules, e.g., corticosteroids, hormones, are used for intercellular and intracellular signaling by binding to specific protein receptors; (2) proteins and peptides act on the cells as mitogens, morphogens, growth factors, and cytokines; and (3) oligonucleotides, either DNA or RNA, can either affect gene transcription and/or translation or can be incorporated into the cell's genome.

Delivery and Presentation of Biological Factors. The challenges and strategies to deliver these factors vary due to the differences in their physical and chemical properties. Presentation of factors as soluble cues is an important strategy to control the interaction of cells and their microenvironment. Strategies to embed factors within scaffold matrices or encapsulate them within micro- and nanoparticles have been used to control the release of the agents. Presentation of a single or multiple factors has been achieved by release of factors from polymeric scaffolds with controlled degradation characteristics or diffusion of the factors out of the scaffold. This allows delivery of multiple soluble cues to cells at distinct rates. This strategy has been demonstrated to release vascular endothelial growth factor (VEGF) and platelet derived growth factor (PDGF) in a PLGA scaffold to promote the growth of blood vessels *in situ* (Richardson et al., 2001). A similar strategy has been employed to regenerate bone with transplanted stromal cells in alginate scaffolds through dual delivery of BMP2 and transforming growth factor-β3 (TGF-β3) (Simmons et al., 2004). Additionally, presentation of factors in a spatio-temporally controlled manner is important for the development of spatially complex tissues. In particular, scaffolds developed by layer-by-layer deposition can control the releases of different growth factors and proteins. For example, by controlling spatial gradients of nerve growth factor (NGF) within the scaffold, axonal outgrowth can be achieved (Moore et al., 2006).

In addition to presenting growth factors in a soluble form, an alternative strategy is to bind factors to a surface, in either random or specific orientations. Non-covalent association of the matrix components, e.g., glycosaminoglycans (GAGs), can slowly release and potentiate binding to the cell membrane receptors (Sakiyama-Elbert and Hubbell, 2000). Alternatively, covalently bound growth factors can also influence cell behavior, i.e., covalently conjugated epidermal growth factor (EGF) enhances the survival of mesenchymal stem cells in a Matrigel™-based scaffold (Fan et al., 2007). The presence of protease sensitive peptide sequences within such growth factor proteins allows these molecules to be released on demand (Zisch et al., 2003). In addition, the adhesion of cells, which is a crucial step for cell survival and is mediated by cell-surface receptors with cell adhesion proteins, on the scaffold matrices can also be controlled by presentation of specific peptides and carbohydrates. In particular, a prototypical three amino acid sequence arginine-glycine-aspartic acid (RGD) is frequently found in many adhesion proteins and binds to many integrin receptors on cells. RGD peptide sequences have therefore been covalently immobilized on a synthetic material surface at a defined density and orientation to guide cell adhesion (Silva et al., 2008). To note, the graft density of RGD is essential to balance cell adhesion versus migration (Gobin and West, 2002). Finally, the covalent immobilization of protease sensitive cleavable linkages (i.e., MMP sensitive links) on synthetic matrices mimics the native extracellular matrix which can be used, in addition to soluble growth factors and other chemotactic cues, to guide migration of cells (Gobin and West, 2002; Lutolf et al., 2003b).

Mechano-Chemical Factors in Tissue Growth

In addition to soluble and immobilized biological factors, the physical and/or chemical nature of the scaffold is also known to play an important role in regulating cell fate. For example, recent studies suggest surface chemistries can regulate the differentiation of MSCs (Benoit et al., 2008). Similarly, the structure of scaffold can also influence the shape of stem cells, which impacts other cell functions such as differentiation. For instance, MSCs cultured in highly cross-linked hydrogels exhibit round morphology, but can be induced to spread by reducing the cross-linking density via photodegradation (Kloxin et al., 2009). Moreover, nanoscale geometry and matrix size can also influence cell adhesion, proliferation, and migration. By altering surface topography, cytoskeletal organization can be altered which directly influences molecular and biomechanical signals. Furthermore, mechanical forces exerted by scaffold matrix and the elasticity of the material also influence the cell fate. For instance, matrix stiffness can control the differentiation of MSCs (Engler et al., 2006; Winer et al., 2009). In conclusion, the physical environment, consisting of geometry, time-varying stress, strain, fluid flow, pressure, and potentially other biophysical parameters, e.g., osmotic pressure and electrical field, can regulate cell phenotype and tissue structure in a three-dimensional environment.

Integration of Multiple Factors

Successful integration of the appropriate cells, scaffolds, soluble cues, and mechano-chemical factors is a key in regulating cell fate and ultimately regenerating a functional tissue. Hence, combinatorial approaches in which the effect of different factors in combination is examined are useful (Flaim et al., 2005). Thus, an engineered "niche" or microenvironment is a useful approach for critical understanding of biology, as well as for engineering tissue regeneration (Lutolf and Blau, 2009). Static *in vitro* tissue engineering systems often fail to take into account the multiple factors that contribute to the healing and regeneration process. Thus, it is critical to understand the system as a whole through dynamic *in vitro* systems, such as bioreactors and relevant *in vivo* models.

MODELS FOR TISSUE ENGINEERING

Bioreactors

Importance of Bioreactors. The traditional tissue engineering approach of growing tissues on two-dimensional surfaces (i.e., petri dishes) or in three-dimensional scaffolds is limited by mass transfer, since the diffusion of metabolites, oxygen, and carbon dioxide under static conditions can only support a certain thickness of tissue (Lees et al., 1981). Consequently, cartilaginous tissue grown in petri dishes has been reported to reach only a maximum height of 0.5 mm (Martin et al., 1998). Multicellular spheroids also develop large, central necrotic regions with increasing diameter (Sutherland et al., 1986). Bioreactors can hence provide an enhanced environment for tissue growth. Their higher rates of mass transfer provide improved oxygen influx and waste disposal. Various cell seeding strategies (i.e., traditional static seeding through gravity, dynamic seeding which takes place in an agitated environment or seeding in a perfusion system) allows for desirable cell density and homogeneity, and mechanical conditioning (Martin and Vermette, 2005). By designing bioreactors it is possible to provide the desired conditions, i.e., mechanical signals, temperature, flow rate, oxygen, and carbon dioxide concentrations for simulating the *in vivo* environment required for regeneration of functional tissues (Burg et al., 2000; Freed et al., 2006). Bioreactors are discussed in more detail in Chapter II.6.6 "Bioreactors for Tissue Engineering."

Types of Bioreactors. Traditional industrial reactors such as the continuous stirred-tank reactor (CSTR) have been adapted as bioreactors in tissue engineering. In particular, the stirred-flask reactor improves reactor performance by agitating the media used for seeding scaffolds (Vunjak-Novakovic et al., 1996). Another system is the rotating-wall vessel, which aims to minimize the shear stress and turbulence of previous approaches. This system was developed for cell culture experiments in space, and thus is often referred to as a microgravity reactor (Unsworth and Lelkes, 1998). Another interesting bioreactor is the hollow fiber reactor that permits the cultivation of highly sensitive cells such as hepatocytes. Cells can either be situated on the exterior or in the lumen of the fibers. These reactors can also introduce two different solutions to the tissue culture system that are optimized to either provide metabolites or remove waste products. Bioartificial liver systems based on this reactor type are in clinical trials to treat patients with hepatic failure (Gerlach et al., 2008).

Another method of seeding and culturing cells is the direct perfusion reactor that can perform continuous perfusion of a cell suspension through a scaffold. This system can increase the efficiency and uniformity of cell seeding (Wendt et al., 2003). There are several other bioreactors designed to grow tissues that primarily serve a biomechanical function. Mechanical constraints can be induced through bioreactors for the successful regeneration of functional tissue, as demonstrated by the *in vitro* generation of a tissue-engineered heart valve (Karim et al., 2006) and a trachea (Martin et al., 2004; Macchiarini et al., 2008).

Limitations and Challenges. While bioreactors have been shown to enhance uniformity of cell seeding and perfusion of three-dimensional cultures, reactor conditions such as mass transfer properties, cell seeding, mechanical environment, and hydrodynamics must be uniquely optimized for each tissue. Fortunately, mathematical modeling (Mehta and Linderman, 2006) and computational fluid dynamics are available to help understand the complex interactions of these factors (Hutmacher and Singh, 2008). However, even within the best designed bioreactor, one limitation remains, i.e., the transport of nutrients inside the tissue. To overcome this barrier, controlled vascularization of the tissue must be fully understood and applied.

Role of *In Vivo* Models

While *in vitro* development of tissues is the starting point of a tissue-engineered product, an appropriate *in vivo* model is crucial to validate the tissues' function. Several factors cannot be assessed *in vitro*, for example the role of angiogenesis in a newly created tissue, host immune reactions to the graft, as well as functional considerations, such as rheological properties of engineered vessels and innervation of the graft (Fabian Schmidt, 2008). These factors must be tested to understand the efficacy of an engineered tissue or organ. Different animal models are used to investigate different tissue products, as all models have limitations that make them appropriate to assess some features but not others. For example, an immune-compromised mouse system is appropriate for initially evaluating graft integration with host tissue, but is not appropriate for evaluating host rejection. Thus, for any tissue-engineered product it is critical to analyze the outcomes of an *in vivo* study in the context of the model and its efficacy.

APPLICATIONS OF TISSUE ENGINEERING

The main goal of tissue engineering is to regenerate and replace human tissues and organs through a combination of biological, clinical, and engineering approaches.

Replacing/Regenerating Target Organs

Investigators have attempted to engineer almost every mammalian tissue. The following section describes some key developments in the field for several tissues:

- *Skin*: Tissue-engineered skin aims to restore barrier function to patients for whom this has been severely compromised, e.g., burn patients. Skin is a widely explored engineered tissue, and several commercial skin products are available, e.g., Epicel® by Genzyme is a product based on autologous cells grown to cover a wound; Apligraf® by Organogenesis is a dual layer skin equivalent with keratinocytes and fibroblasts on collagen gel; Dermagraft® by Advanced Tissue Science uses a similar approach with dermal fibroblast on resorbable polymer, among many others (MacNeil, 2007). These products treat burns, ulcers, deep wounds, and other injuries. Collagen, fibrin, hyaluronic acid, and poly(lactic glycolic acid) are mainly used in skin substitute matrices. Keratinocytes, melanocytes, and fibroblasts compose the majority of cells in skin tissue, thus expanding and transplanting these cells within biocompatible matrices is key to successful skin regeneration (Priya et al., 2008). Additionally, to achieve effective healing, tissue-engineered skin must attach to the wound bed and avoid rejection by the immune system. Despite great success in engineering skin tissue, pitfalls include scar tissue formation, wound contraction, incomplete healing of deep wounds, lack of full recovery of skin function, and imperfect regeneration of skin components such as glands and hair follicles (MacNeil, 2007). Thus, incorporating our understanding of the requirements of skin regeneration into new three-dimensional model systems is important to develop a fully functional skin replacement.
- *Liver*: Liver transplantation is end-stage treatment for many liver diseases. Several factors including drug use, alcohol abuse, and viruses like Hepatitis can cause acute liver failure. It is important to completely replace the damaged liver or support the patients that wait for donor organs or suffer from chronic liver diseases with tissue-engineered livers (Kulig and Vacanti, 2004). Intense efforts exist to develop a bridging device that can support a patient's liver function until a donor is available, e.g., dialysis, charcoal hemoperfusion, immobilized enzymes or exchange transfusion. Several extracorporeal systems use patients' own cells in a hollow-fiber, spouted-bed or flat-bed device, which reduce the chance of immune rejection. Tissue engineering approaches to transplant hepatocytes consist of several critical steps: culture and expansion of hepatocytes in a three-dimensional polymer substrate, maintaining the viability and differentiated state of the cells, engrafting a sufficient number of hepatocytes with vascularization for survival of the graft, and attaining the complex structural geometry of liver. Several bioartificial livers (BAL) have been developed and designed to flow patient's plasma through a bioreactor that houses/maintains hepatocytes sandwiched between artificial plates or capillaries. These BAL devices have different hepatocyte sources, treatment, and perfusate (i.e., blood or plasma). The liver tissue engineering field has developed two main strategies: (1) transplantation of suspended hepatocytes with extracellular matrix components; (2) use of biodegradable scaffolds to provide a platform for hepatocyte attachment. Recent advancement of microfabrication techniques has improved the three-dimensional design of artificial livers with microcapillary beds to mimic physiological conditions (Powers et al., 2002). However, challenges remain in liver tissue engineering due to the complex vascularized architecture of the liver, relatively high volumetric oxygen consumption rate of the tissue, and need for long-term culture for applications such as the evaluation of toxicity, efficacy, and infection (Griffith and Swartz, 2006).
- *Pancreas*: Diabetes is the fifth highest cause of death in the US, and diabetes-related expenses were estimated at more than $100 billion in 2007 (American Diabetes Association, 2008). One type of diabetes is type 1 diabetes mellitus (T1DM), an autoimmune disease which destroys the insulin-secreting cells of the pancreas. T1DM is relevant to tissue engineering, since

it can be treated by replacing the destroyed pancreatic islet cells. Techniques for pancreatic tissue engineering aim to release insulin from transplanted islets into the blood to restore normal blood glucose levels. Three main approaches are used: a tubular membrane that encapsulates islets and connects to blood vessels; hollow fibers containing islets embedded in a polymer matrix; and encapsulation of islets in microcapsules (Lacy et al., 1991; Sullivan et al., 1991; Lanza et al., 1995). The membranes used in the perfusion devices and coatings in microcapsules are developed from biocompatible polymers that allow insulin to diffuse into the bloodstream, while protecting the cells from destruction by immune cells. An insufficient source of islet cells presents a challenge, but recent advances with stem cells may overcome this.

- *Heart*: Cardiovascular disease is the leading cause of death in the US. Many patients are left with damaged or malfunctioning cardiac tissue that lead to arrythmias and diminished cardiac output. Tissue engineering is actively pursing treatments for myocardial infarction, congenital heart defects, and stenotic valves through regenerating cardiac tissues. Heart valves are developed by transplanting autologous cells onto a scaffold, growing and maturing the cell-seeded scaffold, and finally transplanting the valves into the patient. Scaffolds can be made from biomaterials, but decellularized heart valves from donors or animal and biomaterials are also used (Wilson et al., 1995). Decellularized heart valves consist of extracellular matrix which is repopulated with host cells, but they can potentially produce severe immune response (Simon et al., 2003). Alternatively, biomaterial-based heart valves are designed from various natural materials, e.g., collagen, fibrin, and synthetic polymers, e.g., PLGA, poly(hydroxy butyrate) (Hoerstrup et al., 2000). Biomaterial-based heart valves have advantageous characteristics, including malleability and improved mechanical strength. Cells for cardiovascular applications are usually obtained from donor tissues, e.g, peripheral arteries with mixed populations of myofibroblasts and endothelial cells, as well as established cell lines of myofibroblasts and endothelial cells. There are several successful applications of tissue-engineered heart valves for both *in vitro* and *in vivo* models. Additionally, efforts have been made to regenerate myocardium by scaffold-based approaches with natural and synthetic materials (Kofidis et al., 2002; Shimizu et al., 2003). Furthermore, recent devlopments have shown stem cells to be promising tools for the field of cardiovascular tissue engineering (Caspi et al., 2007). However, the complex metabolic, electrical, and mechanical nature of heart tissues continues to present a significant challenge to engineering myocardial tissues (Parker and Ingber, 2007).
- *Blood vessels*: Poly(tetrafluoro ethylene) PTFE and Dacron® grafts have traditionally been used as vascular grafts, particularly for large diameter vessels. However, these grafts are largely unsuccessful for small diameter blood vessels, due to thrombogenicity and compliance mismatch (Edelman, 1999). Tissue engineering strategies therefore provide great opportunities for development of blood vessels. Decellularized arteries with well-preserved extra cellular proteins have been repopulated with cells both *in vitro* and *in vivo* to generate blood vessels (Bader et al., 2000; Kaushal et al., 2001). Polymer-based scaffolds are often used as a template to guide the regeneration of blood vessels. By generating sufficient extracellular matrix and adequate mechanical responsiveness, scaffold guided blood vessels have shown significant promise (Niklason et al., 1999). Also, recently developed microfluidic and microfabrication techniques allow the complex architecture of a blood vessel with defined microstructures to be realized (Fidkowski et al., 2005). In an alternative approach, sheets of smooth muscle cells and extracellular matrix are generated under flow conditions in the absence of scaffold, and are subsequently rolled over a support to mature into a vessel structure (L'Heureux et al., 1998). An emerging approach focuses on combining endothelial cells with mesenchymal stem cells or smooth muscle cells to help stabilize the blood vessel and help aid in the vessel pruning process (Ferreira et al., 2007b). It is important to note that formation of new blood vessels is critical for engineering any tissue to supply nutrients and oxygen to cells. Unfortunately, vascularization of large organs is still a challenge for tissue engineering. Several approaches, including delivery of one or more growth factors from scaffold, have been used to stimulate angiogenesis in engineered tissues (Richardson et al., 2001).
- *Nervous system*: Generation of nerve tissues is a major focus in tissue engineering (Schmidt and Leach, 2003; Zhang et al., 2005). Injuries in the central nervous system (CNS) are often accompanied by permanent functional impairment, unlike peripheral nervous system (PNS) damage where the axons are able to re-extend and re-innervate, leading to functional recovery (Belkas et al., 2004). This is because the environment of the damaged site in CNS blocks the regeneration of neurons. Peripheral nerve grafts that include synthetic or biological substrates have been developed to act as a bridge to guide the nerve regeneration process. To restore the structures lost in the disorganization of axons during injury it is critical to build a bridge that spans the lesion gap with all the morphological, chemical, and biological cues that mimic normal tissue. Both natural and synthetic polymer bridges and conduits have been shown to aid nerve regeneration, and these graft materials may be

seeded with Schwann cells, for example, which aid the regeneration process (Bellamkonda and Aebischer, 1994; Zhang et al., 2005). Studies have also shown that controlled release of neurotrophic factors, neuronal adhesion molecules, and growth factors induces a local sprouting response, stimulates reinnervation, and remodels the growing axons (Schnell et al., 1994; Kapur and Shoichet, 2004; Slack et al., 2004). These methods of delivering cells and factors might also treat neurodegenerative disorders *in vivo* (i.e., encapsulated dopamine producing cells for treating Parkinson's disease) (Lindvall et al., 1990).

- *Bone*: There are numerous clinical applications that can benefit from bone regeneration therapies including spinal fusion to alleviate back pain, temporomandibular joint reconstruction to alleviate jaw pain, and restoration of contour and shape within reconstructed craniofacial bone. Significant progress in the development of bone regeneration therapies has been achieved, yet large critical-sized defects which do not heal remain a significant clinical challenge. Although several inorganic-, polymeric-, and hybrid-based scaffolds have been examined to engineer bone, it has been difficult to develop a material that displays optimal mechanical properties and degradation kinetics for bone repair (Muschler et al., 2004). It is clear that the scaffolds for bone regeneration should have an interconnected macroporosity to allow three-dimensional bone growth throughout the scaffold (Gao et al., 2001; Karp et al., 2003, 2004). Implantation of empty macroporous scaffolds that are devoid of cells typically do not improve the healing response (Grande et al., 1999; Louisia et al., 1999; Petite et al., 2000). Although combining biodegradable scaffolds with cells is a commonly explored strategy (Goshima et al., 1991; Holy et al., 2000; Ma et al., 2001; Hutmacher and Sittinger, 2003) these strategies have low success rates, as many of the cells may die due to a lack of vascularization (Petite et al., 2000). Since most tissue engineering scaffolds have heterogeneous pore sizes, and since the amount of blood that fills a scaffold depends on the geometry of the scaffold pores (Whang et al., 1999), it is likely that there is a high variability in the amount of blood that fills these scaffolds; a critical phenomena that is often neglected. It is important to consider that the diverse compositions of scaffolds with respect to materials, porosity, surface chemistry, morphology, degradation rate, pore sizes, and mechanical properties, employed in the field of bone engineering makes it challenging to compare results between studies, and this limits what can be learned and applied by others. Controlled systematic studies and increased understanding of the osteogenic microenvironment is required to provide rational design criteria for the next generation of bone engineering approaches.

In addition to the tissues described above, there are many other tissues which have been targeted for regeneration by tissue engineering strategies. These include cartilage, muscle, kidney, blood, cornea, gastrointestinal tissues, vocal chord, and many others.

Other Applications of Tissue Engineering

There are several non-traditional, yet useful and important applications of tissue engineering strategies (Griffith and Naughton 2002; Ingber et al., 2006). In particular, the field of drug screening is in need of advanced *in vitro* methods for the assessment of the activity and toxicity of drugs before clinical studies are initiated (Saltzman and Olbricht, 2002). Since most drugs are metabolized in the liver, microscale hepatocyte systems may allow high-throughput screening for liver toxicity (Bhadriraju and Chen, 2002; Khetani and Bhatia, 2008). Furthermore, humans or animals "on a chip" are in development. These chips are essentially several reactors with different cell types connected in series which can simulate the pathway of a chemical substance through several parts of the body, such as the liver, brain, and fat (Viravaidya et al., 2004; Khamsi, 2005). In the future, tissue engineering-based drug testing and toxicity assays can provide an alternate strategy to reduce the use of *in vivo* animal testing. Moreover, an artificially engineered tissue-engineered system can provide enhanced control of tissue microenvironments, both in physiological and pathological conditions (e.g., cancer tumor), contributing to the understanding of complex tissues and organs (Ali et al., 2009).

CURRENT CHALLENGES AND FUTURE DIRECTIONS

The past two decades have seen a dramatic increase in the exploration of tissue engineering as a promising approach to restore, maintain, and enhance tissues and organs. Tissue engineering concepts based on the application of a scaffold/cell construct have significant potential in the healthcare industry. Indeed, a number of engineered tissues have been approved by the FDA, and are used in the clinic for treatment of patients worldwide. Some examples of tissue engineering products and programs are included in Table II.6.2.1.

Challenges

Despite the excitement and early success, there are many hurdles to be addressed before tissue engineering reaches its eventual goal to treat millions of patients (Griffith and Naughton, 2002). In addition to efficacy, potential for facile scale-up, reliability, established regulatory routes, and societal acceptance issues, there are many technical

TABLE II.6.2.1 Tissue Engineering Products and Programs for Replacement or Restoration of Human Tissue Function

Application	Product (approved for application) and Program (currently under development)	Company
Skin	Apligraf®	Organogenesis
	TransCyte®, Dermagraft®	Advanced Tissue Sciences
	Epicel®	Genzyme
Cartilage	Carticel®	Genzyme
	Menaflex™	ReGen Biologics
	NeoCart®, VeriCart™	Histogenics
Bone	Osteocel®	Osiris Therapeutics
	Pura-matrix™	3DM
	OsteoScaf™	Tissue Regeneration Therapeutics
Bladder, Kidney	Neo-bladder	Tengion
	FortaPerm®	Organogenesis
Blood vessels	Lifeline™	Cytograft Tissue Engineering
	Omniflow® II	Bionova
Cardiac	Anginera™	Theregen
	CardioWrap®, CryoValve®,	Cryolife
Retinal	ECT Implant	Neurotech

challenges to overcome. Some of the major challenges are discussed below.

Cell Source

A cell source is one of the key aspects for an effective tissue engineering strategy. It is critical to access reliable cell sources that are adequate to repair the damaged tissue, and to understand at the molecular level how cells function. Specifically, it is important to comprehend how cells respond to molecular signals and integrate multiple inputs to generate a predictable response. Additionally, although local transplantation or injection of cells represents a potential approach, locally administered cells often die before significantly contributing to the healing response, due to diffusion limitations of nutrients and oxygen (Muschler et al., 2004). Cells need to be within ~200 μm of the nearest blood vessel, and it may take many weeks or months for vascularization to reach the cells, leading to cell and tissue death. This significantly reduces the capacity for an exogenous cell source to contribute to the regenerative process. In addition to issues with cell delivery, embryonic stem cells are fraught with several ethical concerns, but recent federal approval to expand research with these stem cells has invigorated the field of tissue engineering. More importantly, the recent development of iPS cells may replace embryo-derived embryonic stem cells, as iPS cells avoid the destruction of embryos, providing a cell source for drug screening, *in vitro* models, and future clinical applications. However, currently many existing techniques for creating iPS cells utilize viral transfection techniques that produce a very low yield. Consequently, there is currently a plethora of research ongoing to develop non-viral and highly efficient iPS cell techniques.

Vascularization

Vascularization, the growth of blood vessels, is a major engineering hurdle to overcome in creating artificial organs, particularly large-scale three-dimensional tissues (Soker et al., 2000). It is critical to have effective transport of oxygen, nutrients, and removal of cell-secreted waste for the survival of cells. This is the major reason why most successful engineered organs are restricted to tissues, such as skin, cartilage, and ligament. For these tissues, thin layers of cells (e.g., 1–2 cell layers in skin tissue) that are well-accessible to the blood vessels are sufficient and a proximal blood supply is not essential for survival (Rouwkema et al., 2008). For other tissues and organs, several potential strategies exist to address the issue of vascularization. Recent tissue studies have focused on prevascularizing the tissue constructs prior to implantation or delivering angiogenic growth factors. Although these methods have shown promising results, more critical issues remain as to methods of developing the complex vascular network using scalable technology and integrating the systems with host vasculature.

Material Design

It is a challenge to develop an ideal instructive biomaterial that can effectively induce the growth of tissues. In spite of significant progress in biomaterial development, it is not fully understood how the scaffold should be used to control cell–matrix interactions (Place et al., 2009). The type of signals and the mode of their presentation (e.g., density and organization of adhesion ligands, delivery of growth factors and cytokines) as soluble or insoluble cues can dramatically impact the functionality of the scaffold. Thus, it is critical to develop large-scale screening systems that can thoroughly and systematically analyze these effects. Specifically, new materials that control and manipulate transcription factors for regulating development and morphogen expression can be useful to control the formation of new tissues. It is important to consider how a cell acts at the molecular level, i.e., the cell's biochemical pathways and how these pathways are affected. Thus, the design of such materials is extremely important to understand the concerted effect of all the factors on the cell functions which are guided by their supporting scaffold, growth factor and cytokine profiles, and biomechanical forces. Additionally, immune rejection of engineered tissues and organs presents a serious problem

(Chan and Mooney, 2008). Strategies to evade the host immune system are clearly needed; particularly technologies that can alter or reduce the inflammatory response to increase tolerance will help to overcome this problem.

FUTURE PERSPECTIVES

Multiple challenges remain for translation of tissue-engineered products to the clinic. Cell type, source, and manipulation are critical parameters that need to be further studied and defined, in order to achieve the best clinical outcomes. Many approaches are too complex for scale-up to industrial level manufacture. Ideally, tissue-engineered products involving cells should be amenable to cryopreservation (Griffith and Naughton, 2002; Pancrazio et al., 2007). It is also critical to consider that *in vivo* animal models may not adequately represent the human condition. Furthermore, it is still not clear how the FDA will regulate combination products. This significantly increases the risk for new companies to develop multi-component systems.

Recent developments in the field of gene microarray analysis and imaging (e.g., magnetic resonance imaging (MRI)) provide valuable tools and strategies for advancing the field (Yamada et al., 2006; Pancrazio et al., 2007). More quantitative approaches such as computational modeling and systems biology will be useful to understand the mechanism of tissue development and regenerative processes. Although significant advances have been accomplished, most regenerative therapies are still in the developmental phase. Understanding the fundamental biology associated with normal tissue development is critical for the creation of highly integrated approaches to achieve controlled cell differentiation and tissue formation.

BIBLIOGRAPHY

Aebischer, P., Tresco, P. A., Winn, S. R., Greene, L. A., & Jaeger, C. B. (1991). Long-term cross-species brain transplantation of a polymer-encapsulated dopamine-secreting cell line. *Exp. Neurol.*, **111**(3), 269–275.

Ahmed, T. A., Dare, E. V., & Hincke, M. (2008). Fibrin: A versatile scaffold for tissue engineering applications. *Tissue Eng. Part B Rev.*, **14**(2), 199–215.

Ali, O. A., Huebsch, N., Cao, L., Granoff, G., & Mooney, D. J. (2009). Infection-mimicking materials to program dendritic cells *in situ*. *Nat. Mater.*, **8**(2), 151–158.

American Diabetes Association. (2008). Economic costs of diabetes in the U.S. in 2007. *Diabetes Care*, **31**(3), 596–615.

Anderson, D. G., Levenberg, S., & Langer, R. (2004). Nanoliter-scale synthesis of arrayed biomaterials and application to human embryonic stem cells. *Nat. Biotechnol.*, **22**(7), 863–866.

Bader, A., Steinhoff, G., Strobl, K., Schilling, T., Brandes, G., et al. (2000). Engineering of human vascular aortic tissue based on a xenogeneic starter matrix. *Transplantation*, **70**(1), 7–14.

Barbash, I. M., Chouraqui, P., Baron, J., Feinberg, M. S., Etzion, S., et al. (2003). Systemic delivery of bone marrow-derived mesenchymal stem cells to the infarcted myocardium: Feasibility, cell migration, and body distribution. *Circulation*, **108**(7), 863–868.

Baum, C. M., Weissman, I. L., Tsukamoto, A. S., Buckle, A. M., & Peault, B. (1992). Isolation of a candidate human hematopoietic stem-cell population. *Proc. Natl. Acad. Sci. USA*, **89**(7), 2804–2808.

Belkas, J. S., Shoichet, M. S., & Midha, R. (2004). Peripheral nerve regeneration through guidance tubes. *Neurol. Res.*, **26**(2), 151–160.

Bellamkonda, R., & Aebischer, P. (1994). Review: Tissue engineering in the nervous system. *Biotechnol. Bioeng.*, **43**(7), 543–554.

Benoit, D. S., Schwartz, M. P., Durney, A. R., & Anseth, K. S. (2008). Small functional groups for controlled differentiation of hydrogel-encapsulated human mesenchymal stem cells. *Nat. Mater.*, **7**(10), 816–823.

Bhadriraju, K., & Chen, C. S. (2002). Engineering cellular microenvironments to improve cell-based drug testing. *Drug Discov. Today*, **7**(11), 612–620.

Brittberg, M., Lindahl, A., Nilsson, A., Ohlsson, C., Isaksson, O., et al. (1994). Treatment of deep cartilage defects in the knee with autologous chondrocyte transplantation. *N. Engl. J. Med.*, **331**(14), 889–895.

Bueno, E. M., Laevsky, G., & Barabino, G. A. (2007). Enhancing cell seeding of scaffolds in tissue engineering through manipulation of hydrodynamic parameters. *J. Biotechnol.*, **129**(3), 516–531.

Burg, K. J., Holder, W. D., Jr., Culberson, C. R., Beiler, R. J., Greene, K. G., et al. (2000). Comparative study of seeding methods for three-dimensional polymeric scaffolds. *J. Biomed. Mater. Res.*, **52**(3), 576.

Caspi, O., Lesman, A., Basevitch, Y., Gepstein, A., Arbel, G., et al. (2007). Tissue engineering of vascularized cardiac muscle from human embryonic stem cells. *Circ. Res.*, **100**(2), 263–272.

Chan, G., & Mooney, D. J. (2008). New materials for tissue engineering: Towards greater control over the biological response. *Trends Biotechnol.*, **26**(7), 382–392.

Dikovsky, D., Bianco-Peled, H., & Seliktar, D. (2006). The effect of structural alterations of PEG-fibrinogen hydrogel scaffolds on 3-D cellular morphology and cellular migration. *Biomaterials*, **27**(8), 1496–1506.

Domb, L. R. (1987). Polyanhydrides I: Preparation of high molecular weight polyanhydrides. *J. Polym. Sci., Part A. Pol. Chem.*, **25**, 3373–3386.

Drury, J. L., & Mooney, D. J. (2003). Hydrogels for tissue engineering: Scaffold design variables and applications. *Biomaterials*, **24**(24), 4337–4351.

Edelman, E. R. (1999). Vascular tissue engineering: Designer arteries. *Circ. Res.*, **85**(12), 1115–1117.

Engler, A. J., Sen, S., Sweeney, H. L., & Discher, D. E. (2006). Matrix elasticity directs stem cell lineage specification. *Cell*, **126**(4), 677–689.

Fabian Schmidt, J. H. (2008). *In Vivo* Animal Models. In U. Meyer, J. Handschel, H. P. Wiesmann, & T. Meyer (Eds.), *Tissue Engineering Fundamentals of Tissue Engineering and Regenerative Medicine* (pp. 773–779). Berlin Heidelberg: Springer.

Fan, V. H., Tamama, K., Au, A., Littrell, R., Richardson, L. B., et al. (2007). Tethered epidermal growth factor provides a survival advantage to mesenchymal stem cells. *Stem Cells*, **25**(5), 1241–1251.

Ferreira, L. S., Gerecht, S., Fuller, J., Shieh, H. F., Vunjak-Novakovic, G., & Langer, R. (2007a). Bioactive hydrogel scaffolds for controllable vascular differentiation of human embryonic stem cells. *Biomaterials*, **28**(17), 2706–2717.

Ferreira, L. S., Gerecht, S., Shieh, H. F., Watson, N., Rupnick, M. A., et al. (2007b). Vascular progenitor cells isolated from human embryonic stem cells give rise to endothelial and smooth muscle like cells and form vascular networks *in vivo*. *Circ. Res.*, **101**(3), 286–294.

Fidkowski, C., Kaazempur-Mofrad, M. R., Borenstein, J., Vacanti, J. P., Langer, R., et al. (2005). Endothelialized microvasculature based on a biodegradable elastomer. *Tissue Eng.*, **11**(1–2), 302–309.

Flaim, C. J., Chien, S., & Bhatia, S. N. (2005). An extracellular matrix microarray for probing cellular differentiation. *Nat. Methods*, **2**(2), 119–125.

Freed, L. E., Guilak, F., Guo, X. E., Gray, M. L., Tranquillo, R., et al. (2006). Advanced tools for tissue engineering: Scaffolds, bioreactors, and signaling. *Tissue Eng.*, **12**(12), 3285–3305.

Gao, J., Dennis, J. E., Solchaga, L. A., Awadallah, A. S., Goldberg, V. M., et al. (2001). Tissue-engineered fabrication of an osteochondral composite graft using rat bone marrow-derived mesenchymal stem cells. *Tissue Eng.*, 7(4), 363–371.

Gerlach, J. C., Zellinger, K., & Patzer Li, J. F. (2008). Bioartificial Liver systems: Why? what? whither? *Regen Med.*, 3(4), 575–595.

Gobin, A. S., & West, J. L. (2002). Cell migration through defined, synthetic ECM analogs. *Faseb. J.*, **16**(7), 751–753.

Goshima, J., Goldberg, V. M., & Kaplan, A. I. (1991). The origin of bone formed in composite grafts of porous calcium phosphate ceramic loaded with marrow cells. *Clin. Orthop.*, **269**, 274–283.

Grande, D. A., Breitbart, A. S., Mason, J., Paulino, C., Laser, J., et al. (1999). Cartilage tissue engineering: Current limitations and solutions. *Clin. Orthop.*, Suppl. 367, S176–S185.

Griffith, L. G., & Naughton, G. (2002). Tissue engineering: Current challenges and expanding opportunities. *Science*, **295**(5557), 1009–1014.

Griffith, L. G., & Swartz, M. A. (2006). Capturing complex 3D tissue physiology *in vitro*. *Nat. Rev. Mol. Cell Biol.*, 7(3), 211–224.

Guan, J., Fujimoto, K. L., Sacks, M. S., & Wagner, W. R. (2005). Preparation and characterization of highly porous, biodegradable polyurethane scaffolds for soft tissue applications. *Biomaterials*, **26**(18), 3961–3971.

Gupta, A. S., & Lopina, S. T. (2004). Synthesis and characterization of -tyrosine based novel polyphosphates for potential biomaterial applications. *Polymer*, **45**(14), 4653–4662.

Hannallah, D., Peterson, B., Leiberman, J. R., Fu, F. H., & Huard. (2003). Gene therapy in orthopaedic surgery. *Instr. Course Lect.*, **52**, 753–768.

Hoerstrup, S. P., Sodian, R., Daebritz, S., Wang, J., Bacha, E. A., et al. (2000). Functional living trileaflet heart valves grown *in vitro*. *Circulation*, **102**(19 Suppl. 3), III44–III49.

Hollister, S. J. (2005). Porous scaffold design for tissue engineering. *Nat. Mater.*, **4**(7), 518–524.

Holy, C. E., Shoichet, M. S., & Davies, J. E. (2000). Engineering three-dimensional bone tissue in vitro using biodegradable scaffolds: Investigating initial cell-seeding density and culture period. *J. Biomed. Mater. Res.*, **51**(3), 376–382.

Horwitz, E. M., Prockop, D. J., Fitzpatrick, L. A., Koo, W. W., Gordon, P. L., et al. (1999). Transplantability and therapeutic effects of bone marrow-derived mesenchymal cells in children with osteogenesis imperfecta. *Nat. Med.*, 5(3), 309–313.

Hubbell, J. A. (1995). Biomaterials in tissue engineering. *Biotechnology (NY)*, **13**(6), 565–576.

Hutmacher, D. W., & Singh, H. (2008). Computational fluid dynamics for improved bioreactor design and 3D culture. *Trends Biotechnol.*, **26**(4), 166–172.

Hutmacher, D. W., & Sittinger, M. (2003). Periosteal cells in bone tissue engineering. *Tissue Eng.*, **9**(Suppl. 1), S45–S64.

Ingber, D. E., Mow, V. C., Butler, D., Niklason, L., Huard, J., et al. (2006). Tissue engineering and developmental biology: Going biomimetic. *Tissue Eng.*, **12**(12), 3265–3283.

Ito, A., Mase, A., Takizawa, Y., Shinkai, M., Honda, H., et al. (2003). Transglutaminase-mediated gelatin matrices incorporating cell adhesion factors as a biomaterial for tissue engineering. *J. Biosci. Bioeng.*, **95**(2), 196–199.

Kapur, T. A., & Shoichet, M. S. (2004). Immobilized concentration gradients of nerve growth factor guide neurite outgrowth. *J. Biomed. Mater. Res. A.*, **68A**(2), 235–243.

Karim, N., Golz, K., & Bader, A. (2006). The cardiovascular tissue-reactor: A novel device for the engineering of heart valves. *Artif. Organs*, **30**(10), 809–814.

Karp, J. M., & Leng Teo, G. S. (2009). Mesenchymal stem cell homing: The devil is in the details. *Cell Stem Cell*, **4**(3), 206–216.

Karp, J. M., Shoichet, M. S., & Davies, J. E. (2003). Bone formation on two-dimensional poly(DL-lactide-co-glycolide) (PLGA) films and three-dimensional PLGA tissue engineering scaffolds *in vitro*. *J. Biomed. Mater. Res. A.*, **64**(2), 388–396.

Karp, J. M., Sarraf, F., Shoichet, M. S., & Davies, J. E. (2004). Fibrin-filled scaffolds for bone-tissue engineering: An *in vivo* study. *J. Biomed. Mater. Res. A.*, **71**(1), 162–171.

Kaushal, S., Amiel, G. E., Guleserian, K. J., Shapira, O. M., Perry, T., et al. (2001). Functional small-diameter neovessels created using endothelial progenitor cells expanded *ex vivo*. *Nat. Med.*, 7(9), 1035–1040.

Khamsi, R. (2005). Labs on a chip: Meet the stripped down rat. *Nature*, **435**(7038), 12–13.

Kharas, G. B., Kamenetsky, M., Simantirakis, J., Beinlich, K. C., Rizzo, A. T., et al. (1997). Synthesis and characterization of fumarate-based polyesters for use in bioresorbable bone cement composites. *J. Appl. Polym. Sci.*, **66**(6), 1123–1137.

Khetani, S. R., & Bhatia, S. N. (2008). Microscale culture of human liver cells for drug development. *Nat. Biotechnol.*, **26**(1), 120–126.

Kim, S. E., Park, J. H., Cho, Y. W., Chung, H., Jeong, S. Y., et al. (2003). Porous chitosan scaffold containing microspheres loaded with transforming growth factor-beta1: Implications for cartilage tissue engineering. *J. Control Release*, **91**(3), 365–374.

Kloxin, A. M., Kasko, A. M., Salinas, C. N., & Anseth, K. S. (2009). Photodegradable hydrogels for dynamic tuning of physical and chemical properties. *Science*, **324**(5923), 59–63.

Kofidis, T., Akhyari, P., Boublik, J., Theodorou, P., Martin, U., et al. (2002). *In vitro* engineering of heart muscle: Artificial myocardial tissue. *J. Thorac. Cardiovasc. Surg.*, **124**(1), 63–69.

Kohn, J. (2004). New approaches to biomaterials design. *Nat. Mater.*, 3(11), 745–747.

Kulig, K. M., & Vacanti, J. P. (2004). Hepatic tissue engineering. *Transpl. Immunol.*, **12**(3–4), 303–310.

L'Heureux, N., Paquet, S., Labbe, R., Germain, L., & Auger, F. A. (1998). A completely biological tissue-engineered human blood vessel. *Faseb. J.*, **12**(1), 47–56.

Lacy, P. E., Hegre, O. D., Gerasimidi-Vazeou, A., Gentile, F. T., & Dionne, K. E. (1991). Maintenance of normoglycemia in diabetic mice by subcutaneous xenografts of encapsulated islets. *Science*, **254**(5039), 1782–1784.

Langer, R., & Tirrell, D. A. (2004). Designing materials for biology and medicine. *Nature*, **428**(6982), 487–492.

Langer, R., & Vacanti, J. P. (1993). Tissue engineering. *Science*, **260**(5110), 920–926.

Langer, R. S., & Vacanti, J. P. (1999). Tissue engineering: The challenges ahead. *Sci. Am.*, **280**(4), 86–89.

Lanza, R. P., Kuhtreiber, W. M., Ecker, D., Staruk, J. E., & Chick, W. L. (1995). Xenotransplantation of porcine and bovine islets without immunosuppression using uncoated alginate microspheres. *Transplantation*, **59**(10), 1377–1384.

Lee, J., Cuddihy, M. J., & Kotov, N. A. (2008). Three-dimensional cell culture matrices: State of the art. *Tissue Eng. Part B Rev.*, **14**(1), 61–86.

Lee, S. J., Van Dyke, M., Atala, A., & Yoo, J. J. (2008). Host cell mobilization for *in situ* tissue regeneration. *Rejuvenation Res.*, **11**(4), 747–756.

Lees, R. K., Sordat, B., & MacDonald, H. R. (1981). Multicellular tumor spheroids of human colon carcinoma origin. Kinetic analysis of infiltration and *in situ* destruction in a xenogeneic (murine) host. *Exp. Cell Biol.*, **49**(4), 207–219.

Lendlein, A., Schmidt, A. M., & Langer, R. (2001). AB-polymer networks based on oligo(epsilon-caprolactone) segments showing shape-memory properties. *Proc. Natl. Acad. Sci. USA*, **98**(3), 842–847.

Lindvall, O., Rehncrona, S., Brundin, P., Gustavii, B., Astedt, B., et al. (1990). Neural transplantation in Parkinson's disease: The Swedish experience. *Prog. Brain Res.*, **82**, 729–734.

Louisia, S., Stromboni, M., Meunier, A., Sedel, L., & Petite, H. (1999). Coral grafting supplemented with bone marrow. *J. Bone Joint Surg. Br.*, **81**(4), 719–724.

Lutolf, M. P., & Blau, H. M. (2009). Artificial stem cell niches. *Adv. Mater.*, **21**(32), 3255–3268.

Lutolf, M. P., Lauer-Fields, J. L., Smoekel, H. G., Metters, A. T., Weber, F. E., et al. (2003a). Synthetic matrix metalloproteinase-sensitive hydrogels for the conduction of tissue regeneration: Engineering cell-invasion characteristics. *Proc. Natl. Acad. Sci. USA*, **100**(9), 5413–5418.

Lutolf, M. P., Weber, F. E., Smoekel, H. G., Schense, J. C., Kohler, T., et al. (2003b). Repair of bone defects using synthetic mimetics of collagenous extracellular matrices. *Nat. Biotechnol.*, **21**(5), 513–518.

Lysaght, M. J., & Reyes, J. (2001). The growth of tissue engineering. *Tissue Eng.*, **7**(5), 485–493.

Lysaght, M. J., Jaklenec, A., & Deweerd, E. (2008). Great expectations: Private sector activity in tissue engineering, regenerative medicine, and stem cell therapeutics. *Tissue Eng. Part A*, **14**(2), 305–315.

Ma, P. X., Zhang, R., Xiao, G., & Franchesci, R. (2001). Engineering new bone tissue *in vitro* on highly porous poly(alpha-hydroxyl acids)/hydroxyapatite composite scaffolds. *J. Biomed. Mater. Res.*, **54**(2), 284–293.

Macchiarini, P., Jungebluth, P., Go, T., Asnagi, M. A., Rees, L. E., et al. (2008). Clinical transplantation of a tissue-engineered airway. *Lancet*, **372**(9655), 2023–2030.

MacNeil, S. (2007). Progress and opportunities for tissue-engineered skin. *Nature*, **445**(7130), 874–880.

Mahmood, A., Lu, D., Lu, M., & Chopp, M. (2003). Treatment of traumatic brain injury in adult rats with intravenous administration of human bone marrow stromal cells. *Neurosurgery*, **53**(3), 697–702; discussion 702–703.

Malafaya, P. B., Silva, G. A., & Reis, R. L. (2007). Natural-origin polymers as carriers and scaffolds for biomolecules and cell delivery in tissue engineering applications. *Adv. Drug Deliv. Rev.*, **59**(4–5), 207–233.

Marijnissen, W. J., van Osch, G. J., Aigner, J., van der Veen, S. W., Hollander, A. P., et al. (2002). Alginate as a chondrocyte-delivery substance in combination with a non-woven scaffold for cartilage tissue engineering. *Biomaterials*, **23**(6), 1511–1517.

Martin, I., Padera, R. F., Vunjak-Novakovic, G., & Freed, L. E. (1998). *In vitro* differentiation of chick embryo bone marrow stromal cells into cartilaginous and bone-like tissues. *J. Orthop. Res.*, **16**(2), 181–189.

Martin, I., Wendt, D., & Heberer, M. (2004). The role of bioreactors in tissue engineering. *Trends Biotechnol.*, **22**(2), 80–86.

Martin, Y., & Vermette, P. (2005). Bioreactors for tissue mass culture: Design, characterization, and recent advances. *Biomaterials*, **26**(35), 7481–7503.

Mehta, K., & Linderman, J. J. (2006). Model-based analysis and design of a microchannel reactor for tissue engineering. *Biotechnol. Bioeng.*, **94**(3), 596–609.

Mi, F. L., Shyu, S. S., Wu, Y. B., Lee, S. T., Shyong, J. Y., et al. (2001). Fabrication and characterization of a sponge-like asymmetric chitosan membrane as a wound dressing. *Biomaterials*, **22**(2), 165–173.

Mooney, D. J., & Vandenburgh, H. (2008). Cell delivery mechanisms for tissue repair. *Cell Stem Cell*, **2**(3), 205–213.

Moore, K., MacSween, M., & Soichet, M. (2006). Immobilized concentration gradients of neurotrophic factors guide neurite outgrowth of primary neurons in macroporous scaffolds. *Tissue Eng.*, **12**(2), 267–278.

Murua, A., Portero, A., Orive, G., Hernández, R. M., de Castro, M., et al. (2008). Cell microencapsulation technology: Towards clinical application. *J. Control Release*, **132**(2), 76–83.

Muschler, G. F., Nakamoto, C., & Griffith, L. G. (2004). Engineering principles of clinical cell-based tissue engineering. *J. Bone Joint Surg. Am.*, **86-A***(7)*, 1541–1558.

Nerem, R. M. (2006). Tissue engineering: The hope, the hype, and the future. *Tissue Eng.*, **12**(5), 1143–1150.

Nijst, C. L., Bruggeman, J. P., Karp, J. M., Ferreira, L., Zumbuehl, A., et al. (2007). Synthesis and characterization of photocurable elastomers from poly(glycerol-co-sebacate). *Biomacromolecules*, **8**(10), 3067–3073.

Niklason, L. E., Gao, J., Abbott, W. M., Hirschi, K. K., Houser, S., et al. (1999). Functional arteries grown *in vitro*. *Science*, **284**(5413), 489–493.

Nuttelman, C. R., Rice, M. A., Rydholm, A. E., Salinas, C. N., Shah, D. N., et al. (2008). Macromolecular monomers for the synthesis of hydrogel niches and their application in cell encapsulation and tissue engineering. *Prog. Polym. Sci.*, **33**(2), 167–179.

Pancrazio, J. J., Wang, F., & Kelley, C. A. (2007). Enabling tools for tissue engineering. *Biosens. Bioelectron.*, **22**(12), 2803–2811.

Park, Y. J., Lee, Y. M., Park, S. N., Sheen, S. Y., Chung, C. P., et al. (2000). Platelet derived growth factor releasing chitosan sponge for periodontal bone regeneration. *Biomaterials*, **21**(2), 153–159.

Parker, K. K., & Ingber, D. E. (2007). Extracellular matrix, mechanotransduction and structural hierarchies in heart tissue engineering. *Philos. Trans. R. Soc. Lond. B. Biol. Sci.*, **362**(1484), 1267–1279.

Patzer, J. F., 2nd (2001). Advances in bioartificial liver assist devices. *Ann. N. Y. Acad. Sci.*, **944**, 320–333.

Petersen, B. E., Bowen, W. C., Patrene, K. D., Mars, W. M., Sullivan, A. K., et al. (1999). Bone marrow as a potential source of hepatic oval cells. *Science*, **284**(5417), 1168–1170.

Petite, H., Viateau, V., Bensaïd, W., Meunier, A., de Pollak, C., et al. (2000). Tissue-engineered bone regeneration. *Nat. Biotechnol.*, **18**(9), 959–963.

Pittenger, M. F., Mackay, A. M., Beck, S. C., Jaiswal, R. K., Douglas, R., et al. (1999). Multilineage potential of adult human mesenchymal stem cells. *Science*, **284**(5411), 143–147.

Place, E. S., Evans, N. D., & Stevens, M. M. (2009). Complexity in biomaterials for tissue engineering. *Nat. Mater.*, **8**(6), 457–470.

Powers, M. J., Domansky, K., Kaazempur-Mofrad, M. R., Kalezi, A., Capitano, A., et al. (2002). A microfabricated array bioreactor for perfused 3D liver culture. *Biotechnol. Bioeng.*, **78**(3), 257–269.

Priya, S. G., Jungvid, H., & Kumar, A. (2008). Skin tissue engineering for tissue repair and regeneration. *Tissue Eng. Part B Rev.*, **14**(1), 105–118.

Pulapura, S., & Kohn, J. (1992). Tyrosine-derived polycarbonates: Backbone-modified pseudo-poly (amino acids) designed for biomedical applications. *Biopolymers*, **32**(4), 411–417.

Richardson, T. P., Peters, M. C., Ennett, A. B., & Mooney, D. J. (2001). Polymeric system for dual growth factor delivery. *Nat. Biotechnol.*, **19**(11), 1029–1034.

Rodriguez-Cabello, J. C., Reguera, J., Girotti, A., Alonso, M., & Testera, A. M. (2005). Developing functionality in elastin-like polymers by increasing their molecular complexity: The power of the genetic engineering approach. *Prog. Polym. Sci.*, **30**, 1119–1145.

Rouwkema, J., Rivron, N. C., & van Blitterswijk, C. A. (2008). Vascularization in tissue engineering. *Trends Biotechnol.*, **26**(8), 434–441.

Sakiyama-Elbert, S. E., & Hubbell, J. A. (2000). Development of fibrin derivatives for controlled release of heparin-binding growth factors. *J. Control Release*, **65**(3), 389–402.

Saltzman, W. M., & Olbricht, W. L. (2002). Building drug delivery into tissue engineering. *Nat. Rev. Drug Discov.*, **1**(3), 177–186.

Sarkar, D., Yang, J. C., Gupta, A. S., & Lopina, S. T. (2009). Synthesis and characterization of L-tyrosine based polyurethanes for biomaterial applications. *J. Biomed. Mater. Res. A.*, **90**(1), 263–271.

Schmidt, C. E., & Leach, J. B. (2003). Neural tissue engineering: Strategies for repair and regeneration. *Ann. Rev. Biomed. Eng.*, **5**, 293–347.

Schnell, L., Schneider, R., Kolbeck, R., Barde, Y. A., & Schwab, M. E. (1994). Neurotrophin-3 enhances sprouting of corticospinal tract during development and after adult spinal cord lesion. *Nature*, **367**(6459), 170–173.

Seliktar, D. (2005). Extracellular stimulation in tissue engineering. *Ann. NY. Acad. Sci.*, **1047**, 386–394.

Shalaby, S. (1988). Bioabsorbable Polymers. In J. Swarbrick (Ed.), *Encyclopedia of Pharmaceutical Technology* (pp. 465–476). Marcel Deker.

Shimizu, T., Yamato, M., Kikuchi, A., & Okano, T. (2003). Cell sheet engineering for myocardial tissue reconstruction. *Biomaterials*, **24**(13), 2309–2316.

Silva, E. A., Kim, E. S., Kong, H. J., & Mooney, D. J. (2008). Material-based deployment enhances efficacy of endothelial progenitor cells. *Proc. Natl. Acad. Sci. USA*, **105**(38), 14347–14352.

Simmons, C. A., Alsberg, E., Hsiong, S., Kim, W. J., & Mooney, D. J. (2004). Dual growth factor delivery and controlled scaffold degradation enhance *in vivo* bone formation by transplanted bone marrow stromal cells. *Bone*, **35**(2), 562–569.

Simon, P., Kasimir, M. T., Seebacher, G., Weigel, G., Ullrich, R., et al. (2003). Early failure of the tissue engineered porcine heart valve SYNERGRAFT in pediatric patients. *Eur. J. Cardiothorac. Surg.*, **23**(6), 1002–1006; discussion 1006.

Skarja, G. A., & Woodhouse, K. A. (2001). *In vitro* degradation and erosion of degradable, segmented polyurethanes containing an amino acid-based chain extender. *J. Biomat. Sci., Polym. Ed.*, **12**, 851–873.

Slack, S. E., Pezet, S., McMahon, S. B., Thompson, S. W., & Malcangio, M. (2004). Brain-derived neurotrophic factor induces NMDA receptor subunit one phosphorylation via ERK and PKC in the rat spinal cord. *Eur. J. Neurosci.*, **20**(7), 1769–1778.

Snyder, E. Y., Yoon, C., Flax, J. D., & Macklis, J. D. (1997). Multipotent neural precursors can differentiate toward replacement of neurons undergoing targeted apoptotic degeneration in adult mouse neocortex. *Proc. Natl. Acad. Sci. USA*, **94**(21), 11663–11668.

Sofia, S., McCarthy, M. B., Gronowicz, G., & Kaplan, D. L. (2001). Functionalized silk-based biomaterials for bone formation. *J. Biomed. Mater. Res.*, **54**(1), 139–148.

Soker, S., Machado, M., & Atala, A. (2000). Systems for therapeutic angiogenesis in tissue engineering. *World J. Urol.*, **18**(1), 10–18.

Sullivan, S. J., Maki, T., Borland, K. M., Mahoney, M. D., Solomon, B. A., et al. (1991). Biohybrid artificial pancreas: Long-term implantation studies in diabetic, pancreatectomized dogs. *Science*, **252**(5006), 718–721.

Sutherland, R. M., Sordat, B., Bamat, J., Gabbert, H., Bourrat, B., et al. (1986). Oxygenation and differentiation in multicellular spheroids of human colon carcinoma. *Cancer Res.*, **46**(10), 5320–5329.

Takahashi, K., & Yamanaka, S. (2006). Induction of pluripotent stem cells from mouse embryonic and adult fibroblast cultures by defined factors. *Cell*, **126**(4), 663–676.

Takahashi, K., Tanabe, K., Ohnuki, M., Narita, M., Ichisaka, T., et al. (2007). Induction of pluripotent stem cells from adult human fibroblasts by defined factors. *Cell*, **131**(5), 861–872.

Thomas, E. D. (1987). Bone marrow transplantation. *CA Cancer J. Clin.*, **37**(5), 291–301.

Tilakaratne, H. K., Hunter, S. K., Andracki, Me, E., Benda, J. A., & Rodgers, V. G. (2007). Characterizing short-term release and neovascularization potential of multi-protein growth supplement delivered via alginate hollow fiber devices. *Biomaterials*, **28**(1), 89–98.

Unsworth, B. R., & Lelkes, P. I. (1998). Growing tissues in microgravity. *Nat. Med.*, **4**(8), 901–907.

Vacanti, C. A. (2006). History of tissue engineering and a glimpse into its future. *Tissue Eng.*, **12**(5), 1137–1142.

Viravaidya, K., Sin, A., & Shuler, M. L. (2004). Development of a microscale cell culture analog to probe naphthalene toxicity. *Biotechnol. Prog.*, **20**(1), 316–323.

Vunjak-Novakovic, G., Freed, L. E., Biron, R. J., & Langer, R. (1996). Effects of mixing on the composition and morphology of tissue-engineered cartilage. *AIChE Journal*, **42**(3), 850–860.

Wendt, D., Marsano, A., Jakob, M., Heberer, M., & Martin, I. (2003). Oscillating perfusion of cell suspensions through three-dimensional scaffolds enhances cell seeding efficiency and uniformity. *Biotechnol. Bioeng.*, **84**(2), 205–214.

Whang, K., Healy, K. E., Elenz, D. R., Nam, E. K., Tsai, D. C., et al. (1999). Engineering bone regeneration with bioabsorbable scaffolds with novel microarchitecture. *Tissue Eng.*, **5**(1), 35–51.

Wiegand, F., Kroncke, K. D., & Kolb-BacHofen, V. (1993). Macrophage-generated nitric oxide as cytotoxic factor in destruction of alginate-encapsulated islets. Protection by arginine analogs and/or coencapsulated erythrocytes. *Transplantation*, **56**(5), 1206–1212.

Wilson, G. J., Courtman, D. W., Klement, P., Lee, J. M., & Yeger, H. (1995). Acellular matrix: A biomaterials approach for coronary artery bypass and heart valve replacement. *Ann. Thorac. Surg.*, **60**(Suppl. 2), S353–S358.

Winer, J. P., Janmey, P. A., McCormick, M. E., & Funaki, M. (2009). Bone marrow-derived human mesenchymal stem cells become quiescent on soft substrates but remain responsive to chemical or mechanical stimuli. *Tissue Eng. Part A*, **15**(1), 147–154.

Winn, S. R., Tresco, P. A., Zielinski, B., Greene, L. A., Jaeger, C. B., et al. (1991). Behavioral recovery following intrastriatal implantation of microencapsulated PC12 cells. *Exp. Neurol.*, **113**(3), 322–329.

Yamada, Y., Fujimoto, A., Ito, A., Yoshimi, R., & Ueda, M. (2006). Cluster analysis and gene expression profiles: A cDNA microarray system-based comparison between human dental pulp stem cells (hDPSCs) and human mesenchymal stem cells (hMSCs) for tissue engineering cell therapy. *Biomaterials*, **27**(20), 3766–3781.

Yang, C., Hillas, P. J., Baez, J. A., Nokelainen, M., Balan, J., et al. (2004). The application of recombinant human collagen in tissue engineering. *BioDrugs*, **18**(2), 103–119.

Yang, J., Webb, A. R., Pickerill, S. J., Hageman, G., & Ameer, G. A. (2006). Synthesis and evaluation of poly(diol citrate) biodegradable elastomers. *Biomaterials*, **27**(9), 1889–1898.

Yang, S., Leong, K. F., Du, Z., & Chua, C. K. (2001). The design of scaffolds for use in tissue engineering. Part I. Traditional factors. *Tissue Eng.*, **7**(6), 679–689.

Yang, Y., Bolikal, D., Becker, M. L., Kohn, J., Zeiger, D. N., et al. (2008). Combinatorial polymer scaffold libraries for screening cell-biomaterial interactions in 3D. *Adv. Mater.*, **20**(11), 2037–2043.

Zhang, N., Yan, H., & Wen, X. (2005). Tissue-engineering approaches for axonal guidance. *Brain Res. Brain Res. Rev.*, **49**(1), 48–64.

Zhao, W., & Karp, J. M. (2009). Controlling cell fate in vivo. *Chembiochem.*, **10**(14), 2308–2310.

Zisch, A. H., Lutolf, M. P., Ehrbar, M., Raeber, G. P., Rizzi, S. C., et al. (2003). Cell-demanded release of VEGF from synthetic, biointeractive cell ingrowth matrices for vascularized tissue growth. *Faseb. J.*, **17**(15), 2260–2262.

CHAPTER II.6.3 TISSUE ENGINEERING SCAFFOLDS

Milind Singh, F. Kurtis Kasper, and Antonios G. Mikos
Department of Bioengineering, Rice University, Houston, TX, USA

In the search for alternatives to conventional treatment strategies for the repair or replacement of missing or malfunctioning human tissues and organs, promising solutions have been explored through tissue engineering approaches (Langer and Vacanti, 1993). Biomaterials-based scaffolds have played a pivotal role in this quest. The fundamental purpose of a tissue engineering scaffold is to act as a three-dimensional template that may provide mechanical stability, deliver therapeutic agents, and facilitate processes critical in tissue repair, such as tissue induction, cell proliferation and differentiation, and/or guided tissue growth. To date, a wide variety of biomaterials have been explored for application as tissue engineering scaffolds, including metal-based implants, purified extracellular matrix (ECM) xenografts, ceramics, and natural and synthetic polymers. Bioresorbable synthetic polymers are particularly attractive for two main reasons. First, they degrade into products that can be safely metabolized and/or excreted, potentially leaving no residual foreign materials in the recipient, and second they are highly versatile with regard to the control over their physicochemical properties and ease of processability, allowing them to be tailored in an application-specific manner. This chapter provides an overview of bioresorbable synthetic polymeric scaffolds for tissue engineering applications.

SCAFFOLD DESIGN

When designing a polymeric scaffold, a combination of biological and engineering requisites is considered in an application-specific manner. One of the most essential design elements is the biocompatibility of the scaffolds, implying that the scaffold should not demonstrate immunogenicity or elicit an adverse inflammatory response (Babensee et al., 2000). In this regard, the bioresorbable scaffolds should also be sterilizable, and should degrade without significant cytotoxic, inflammatory or immunogenic degradation components. In addition, scaffold design considerations include suitable bulk properties, and the presentation of microenvironments to regulate cell adhesion, spreading, motility, survival, and differentiation.

For anchorage-dependent cells, cell-to-scaffold interactions need to be optimized. Cell-to-scaffold interactions have a direct impact on cell adhesion and morphology, which may in turn affect various cellular processes. These interactions are directly or indirectly influenced by physicochemical characteristics of the polymer surface, including features such as wettability (hydrophilicity), roughness, crystallinity, charge, and functionality (for reviews, see Ruardy et al., 1997; Singh et al., 2008a).

Synthetic polymer surfaces are usually devoid of specific ligands that cell surface receptors recognize, so modification with extracellular matrix (ECM) molecules is often applied to improve cell-to-scaffold interactions (Kraehenbuehl et al., 2008; Weber and Anseth, 2008). The ECM molecules are known to play important roles in integrin-mediated signaling and associated cellular functions (Howe et al., 1998). For this purpose, the presence of pendant functional groups on polymer chains is desirable, because these may be used to conjugate proteins or peptides. In some cases, specific surface functionalities may directly influence the desired cellular activity and function, precluding the need for ECM molecules. For example, it was recently shown that small molecules of specific functionalities (charged phosphate group and hydrophobic t-butyl group) tethered to a poly(ethylene glycol) (PEG) hydrogel can induce osteogenesis and adipogenesis of human mesenchymal stem cells in the absence of any cytokine (Benoit et al., 2008).

Substrate rigidity is another important factor that determines cell cytoskeletal shape and associated cellular function (see review by Discher et al., 2005). In general, increased cell adhesion and spreading leads to cell proliferation, while moderate cell adhesion and a rounded cell morphology corresponds to differentiated cellular function (Mooney et al., 1992), and substrate stiffness is known to affect the extent of cell spreading. For example, fibroblasts cultured on soft two-dimensional polyacrylamide substrates (substrate stiffness 14 kPa) display a rounded morphology compared to the cells cultured on stiffer substrates (substrate stiffness 30 kPa), which adopt a flat morphology (Lo et al., 2000). Usually, substrate stiffness similar to the native tissue yields a cellular phenotype similar to that tissue. Another polymer property that impacts cell–scaffold interactions is dimensionality and overall architecture. When compared to two-dimensional environments, a three-dimensional environment has been shown to lead to reduced cell adhesion to the substrate (Cukierman et al., 2001). In addition to impacting individual cell behavior, a three-dimensional environment also facilitates formation of a larger number of cell–cell contacts, allowing for cellular interactions that are often vital in tissue remodeling processes. A three-dimensional scaffold of the shape of native tissue is also desired, to define and guide the ultimate shape of the regenerated tissue.

Porosity, pore size, and interconnectivity of pores are morphological characteristics of scaffolds that have a direct influence on the cell-to-cell interactions, the surface area-to-volume ratio, and mass transport processes critical for cell survival. A high porosity is desirable to maximize the possible accommodation of cell mass and vascular infiltration; however, high porosity values often compromise the mechanical properties of the scaffold (Karageorgiou and Kaplan, 2005). Pore size is another important factor to consider, because it is known to influence the cellular infiltration, cell-to-cell interaction, and transport of nutrients and metabolites (Mikos et al., 1993c). For a given porosity, smaller pore sizes lead to an increased surface area-to-volume ratio, resulting in a larger area available for cell attachment. The minimum pore size that will allow cellular infiltration, either during initial seeding or through cellular proliferation and migration, is dependent on cell size (roughly 10 μm); however, optimal pore sizes are almost always larger, and are dependent on the topological features of the scaffold. Moreover, pore sizes can also be controlled in order to facilitate the process of vascularization, and to reduce fibrotic tissue formation (Ratner, 2007). Recently, it was shown that monodisperse pore sizes of ~35 μm in poly(2-hydroxyethyl methacrylate) (PHEMA) scaffolds resulted in an improvement of the vascularization of the implants and a reduction in fibrotic tissue formation compared to ~20 μm or ~70 μm pore sizes for soft tissue regeneration (Marshall et al., 2004a,b). In addition to optimization of pore size and porosity, the geometry of the pore network must also be considered. An interconnected pore network is desired in order to minimize inaccessible pore volume, and the tortuosity of this network is an important determinant of mass transport rates. Pore geometries may also critically affect cellular organization in the scaffolds (Ma and Zhang, 2001; Zmora et al., 2002). For example, pores in the form of tubular guidance channels have been widely utilized to promote directed neurite growth.

One of the major engineering considerations in tissue regeneration is the mechanical stability of bioresorbable scaffolds. Usually, the mechanical properties of the scaffolds are required to be similar to the native tissue. A weak structure may not be able to withstand the biomechanical forces encountered *in vivo*, whereas a stiffer scaffold may lead to stress shielding, a phenomenon where transfer of physical load to the scaffold leads to insufficient mechanical stimulation of tissues surrounding the implant. In addition, due to the degradation of bioresorbable polymeric scaffolds, their mechanical properties change as a function of time, so control over the degradation of tissue engineering scaffolds is a highly desired feature. Degradation of polymeric scaffolds is dependent on several parameters, as listed in Table II.6.3.1 (also see Chapters I.2.6 and II.4.3). Polymers also differ in the mechanism through which they degrade, e.g., surface erosion, bulk erosion, enzymatic action. Ideally, the tissue–polymer construct should provide sufficient mechanical stability consistently throughout the process of reconstruction. In addition, the degradation rate determines the available space for tissue growth and implant integration with the host tissue. In an optimal scenario, the degradation rate of the scaffolds and the rate of tissue formation should be matched to provide consistent structural and mechanical stability to the transplant.

Another important consideration when designing a scaffold for any tissue engineering application is the selection of a suitable material with desired characteristics, which is discussed in the following section.

TABLE II.6.3.1 Important Factors that Influence Scaffold Degradation

Polymer characteristics	
Polymer composition	Polydispersity index
Chemical structure and functionality	Hydrophilicity
Morphology	Chain motility and orientation
Molecular weight	Charge density
Crystallinity	Presence of additives or impurities
Scaffold architecture	
Pore size	Mass
Porosity	Density
Surface area-to-volume ratio	Shape
Surface roughness	Size
Processing effects (if any)	Sterilization (γ-irradiation)
***In vitro* factors**	
Medium composition	Incubation temperature
Medium refreshment frequency	Mechanical loads
pH	Cell type and density
Ionic strength	Enzyme concentrations
***In vivo* factors**	
Site of implantation	Tissue modeling and remodeling
Nutrient diffusion	Mechanical loads
Access to vasculature	Metabolism of degraded products
Dynamic pH	Enzyme concentrations

SCAFFOLD MATERIALS

Material selection for tissue engineering applications is based on several important factors including biocompatibility, surface characteristics, degradability, processability, and mechanical properties (see review by Thomson et al., 1995a). Synthetic bioresorbable polymers constitute a set of polymers that provide extreme versatility with regard to control over their physicochemical properties, and are generally easy to process into tissue engineering scaffolds. Synthesis of these polymers can be tailored to yield a specific molecular weight, chemical structure, end group chemistry, and composition (homopolymers, copolymers, and polymer blends).

A range of bioresorbable polymers have been explored for potential application in tissue engineering (for reviews, see Babensee et al., 1998; Gunatillake and Adhikari, 2003; Martina and Hutmacher, 2007) (Table II.6.3.2) (Chapter I.2.6). Among these, polyesters have been most widely investigated. Poly(α-hydroxy esters) (such as poly(lactic acid) (PLA), poly(glycolic acid) (PGA), and poly(lactic-*co*-glycolic acid) (PLGA)) have attracted extensive attention for a variety of biomedical applications. The ease of processability of these polymers into various shapes is a major advantage. However, there is a known concern that their degradation products may lead to a drop in local pH, creating an acidic environment that may harm cells and tissues. Moreover, the degradation products may also catalyze the degradation of these polymers, resulting in different degradation rates depending on the removal of these products (Li et al., 1990). Copolymers that combine poly(α-hydroxy esters) with amino acids, such as block copolymers of poly(lysine-co-lactic acid), have been produced that allow the addition of cell adhesion peptides to lysine groups (Cook et al., 1997). Another important class of polyesters are polylactones, among which poly(ε-caprolactone) (PCL) has been the most utilized as a tissue engineering scaffold. Blends and block copolymers of PCL with other poly(α-hydroxy esters) (such as poly(L-lactic acid-*co*-ε-caprolactone) (PLLACL) or poly(D,L-lactic acid-*co*-ε-caprolactone) (PDLLACL)) have been used to produce polymers with tailored properties, such as degradation rate. Polydioxanone is a bioresorbable poly(ether-ester), which has been used in fiber forms for fixation and other applications (Boland et al., 2005; Smith et al., 2008). Poly(propylene fumarate) (PPF), an unsaturated linear polyester based upon fumaric acid, has also been explored as part of an injectable formulation in tissue engineering applications (Timmer et al., 2003a,b). Cross-linking agents and initiation mechanisms (photoinitiation or thermal initiation) can be altered to form PPF networks of desired properties (Peter et al., 1999; He et al., 2000; Fisher et al., 2002a). Several composites and copolymers of PPF have been developed and investigated, mainly for bone tissue engineering and drug delivery applications.

TABLE II.6.3.2 Scaffold Materials, and Example Applications[a]

Materials	Example Areas of Applications
Polyesters	
Poly(glycolic acid) (PGA)	Bone, cartilage, liver, tendon, urothelium, and intestinal tissue engineering
Poly(L-lactic acid) (PLLA)	Bone, cartilage, ligament, and neural tissue engineering
Poly(D,L-lactic-*co*-glycolic acid) (PLGA)	Bone, cartilage, urothelium, and neural tissue engineering
Poly(lysine-*co*-lactic acid)	Bone, cartilage, and neural tissue engineering
Poly(ε-caprolactone) (PCL)	Bone and vessel tissue engineering
Poly(L-lactic acid-*co*-ε-caprolactone) (PLLACL)	Neural and meniscal tissue engineering
Poly(glycolic acid-*co*-ε-caprolactone) (PGACL)	Smooth muscle engineering
Poly(D,L-lactic acid-*co*-ε-caprolactone) (PDLLACL)	Vascular graft
Poly(dioxanone)	Bone tissue engineering, other orthopedic applications
Poly(propylene fumarate)	Bone tissue engineering, ocular and ophthalmic drug delivery
Poly(ethylene glycol) (PEG)-based polyesters	
PLLA blended with PEG	Soft tissue and tubular tissue engineering
PLLA–PEG block copolymer	Bone tissue engineering, drug delivery applications
PLGA–PEG block copolymer	Drug delivery applications
Poly(propylene fumarate-co-ethyelene glycol) [P(PF-*co*-EG)]	Bone and cardiovascular tissue engineering
Oligo(poly(ethylene glycol) fumarate)	Bone and cartilage tissue engineering
Pseudo-poly(amino acids)	
Tyrosine-derived polyiminocarbonates	Bone tissue engineering
Tyrosine-derived polycarbonate	
Tyrosine-derived polyacrylate	
Polyanhydrides	Orthopedic applications, drug delivery
Polyurethanes	Soft tissue engineering, biomedical applications
Polyorthoesters	Bone tissue engineering, drug delivery
Polyphosphates and polyphosphazenes	Cartilage, bone, liver, and neural tissue engineering

[a]Partially adapted and reproduced with permisson from Gunatillake and Adhikari (2003), and Martina and Hutmacher (2007).

To increase wettability, biocompatibility, and/or softness of bioresorbable polymers, blends and copolymers with non-degradable poly(ethylene glycol) (PEG) have been developed, such as block copolymers of PEG with PLLA, PLGA, and PCL (for reviews, see Hoffman, 2002; Baroli, 2007). Injectable, *in situ* forming hydrogels based on PEG, such as oligo[poly(ethylene glycol) fumarate] (OPF), have also been developed and investigated. An OPF-based hydrogel is biodegradable and demonstrates a high degree of swelling (Jo et al., 2001; Temenoff et al., 2002, 2003). These hydrogels have been recently investigated for bone, cartilage, and osteochondral tissue engineering applications (Shin et al., 2003; Holland et al., 2005; Kasper et al., 2006; Guo et al., 2009).

In addition to those derived from polyesters, a number of bioresorbable materials have been developed from various other polymer families. Many amorphous and soluble pseudo-poly(amino acids) (amino acids linked by both amide and non-amide bonds) have been processed into tissue engineering scaffolds, such as polycarbonates and polyacrylates. Polyanhydrides and poly(anhydrides-co-imides) have been vastly studied for drug delivery, and for use as hard tissue substitutes (Gunatillake and Adhikari, 2003). Degradable polyurethanes and their copolymers have been investigated for skin and other soft tissue engineering and replacement applications (Bruin et al., 1990; Spaans et al., 2000). Polyorthoesters have been used in bone tissue engineering and drug delivery (Andriano et al., 1999; Wang et al., 2004; Nguyen et al., 2008), and lastly, polyphosphates and polyphosphazenes have been mainly used in scaffolds for bone tissue engineering applications (Brown et al., 2008a; Nukavarapu et al., 2008; Deng et al., 2010).

To summarize, a wide variety of synthetic bioresorbable polymers have been investigated for their potential application in tissue engineering. Material selection is a critical factor of the scaffold design for an intended tissue engineering application. The following section briefly discusses a few established strategies for scaffold-assisted tissue regeneration.

APPLICATIONS OF SCAFFOLDS

Tissue Induction

After implantation, a porous acellular three-dimensional scaffold can act as a substrate to allow infiltration and ingrowth of the surrounding host tissue, a process known as tissue induction (Figure II.6.3.1A). For instance, an osteoinductive material has the property of promoting pre-osteoblast infiltration and bone induction. Tissue induction is commonly marked by events of host cell migration, proliferation, and differentiation, as well as vascularization. This scaffold-based approach has the potential of offering an "off-the-shelf" solution for defect repair in the form of implantable acellular devices. The selection of a material and scaffold design (such as pore size and porosity), however, are important factors that can affect the selectivity and extent of tissue induction. Regeneration of various tissues including skin, bone, ligament, and nerve has been investigated using this approach.

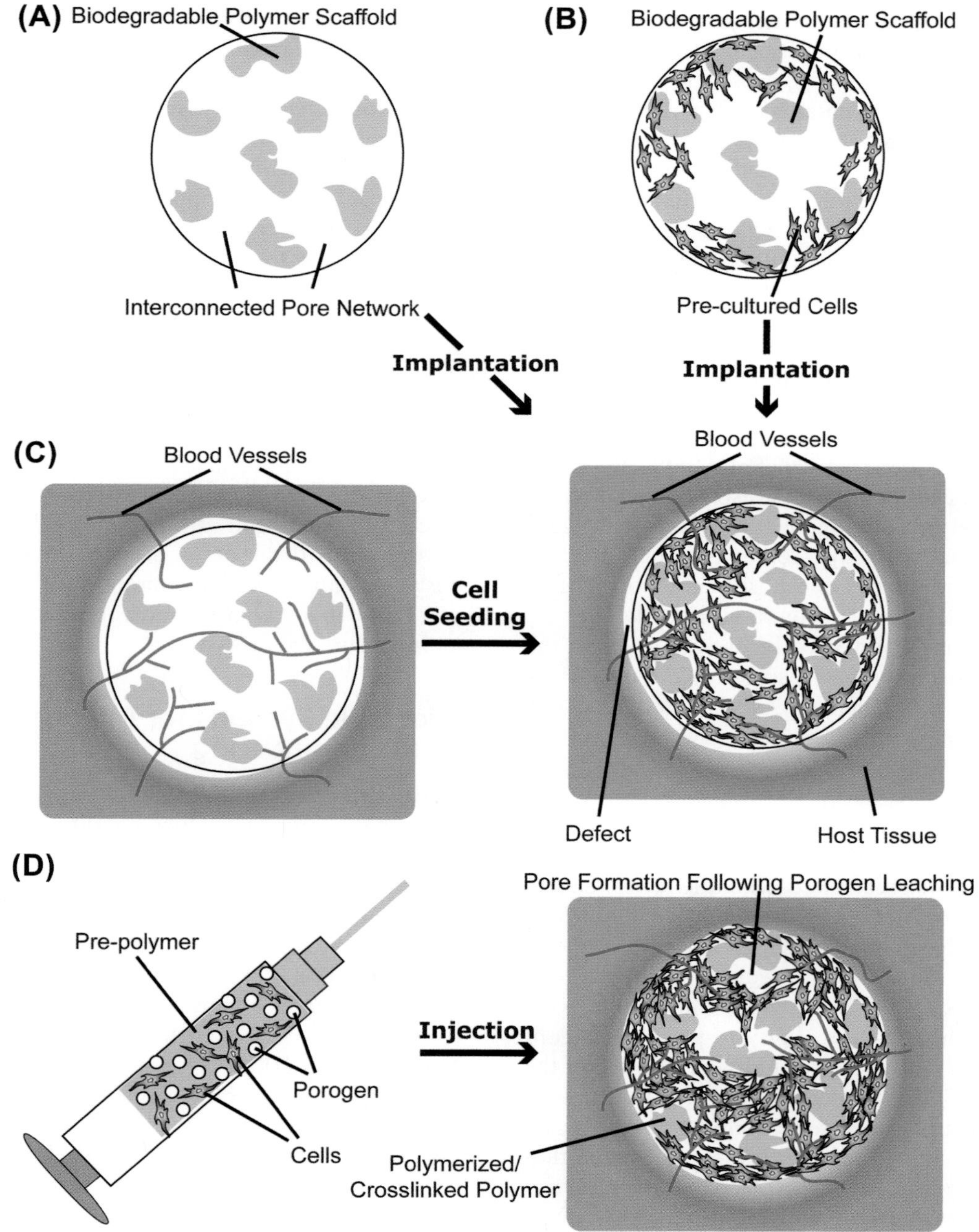

FIGURE II.6.3.1 Polymeric scaffolds in tissue engineering applied in prefabricated (A–C) or injectable form (D). Specific applications of prefabricated scaffolds: (A) tissue induction; (B) cell transplantation; and (C) prevascularization.

Cell Transplantation

Cell transplantation is a sub-class of cell-based therapies, where cells placed on a two-dimensional or in a three-dimensional polymeric scaffold facilitate the repair or regeneration of damaged tissue. In this approach, cells obtained from a donor site in a patient are harvested, expanded in culture, seeded onto an appropriate scaffold, and then transplanted to the defect site (Cima et al., 1991; Bancroft and Mikos, 2002) (Figure II.6.3.1B). In most practices, the cells are allowed to attach to the construct, and are cultured for a period of time to allow them to proliferate and/or differentiate before implantation. When using cell transplantation, an appropriate choice of material that favors the attachment and growth of seeded cells and integration of the transplant with the host tissue is critical. Using this approach, transplantation of a variety of differentiated cell types including osteoblasts, chondrocytes, fibroblasts, hepatocytes, and smooth muscle cells has been pursued. Due to the donor site morbidity often associated with the harvest of differentiated cells, transplantation of undifferentiated or pre-differentiated autologous adult stem cells (such as mesenchymal stem cells) has also been investigated. When engineering heterogeneous interfacial structures, a cell transplantation approach can be used to design multiphasic scaffolds containing more than one cell type regionalized spatially into the different phases of the scaffold along its axis (Cao et al., 2003; Schek et al., 2004; Chen et al., 2006; Spalazzi et al., 2008). Recently, a multilayered scaffold design has also been applied for pancreatic islet encapsulation in PEG-based hydrogels, where an additional outer layer serves as an immunoprotective barrier to minimize graft–host interaction (Weber et al., 2008). By using genetically modified cells that are programmed to produce desired bioactive factors at the site of interest, this approach also offers the possibility of simultaneous *in situ* delivery of cells and bioactive factors that may enhance the tissue regeneration process (Gilbert et al., 1993; Blum et al., 2003).

Prevascularization

Nutrient diffusion in a scaffold is generally restricted to a few hundred micrometers from the scaffold edge, which imposes one of the major challenges for the regeneration of large three-dimensional organs, such as the liver (Mooney and Mikos, 1999). Development of bioreactor technology has imparted the ability to grow large constructs *in vitro*, as discussed later. However, most of the cultured cells do not survive when implanted *in vivo* without their own blood vessels to supply nutrients and oxygen (Mooney and Mikos, 1999). Post-implantation scaffold vascularization may take place, but the rate may not be sufficient to prevent cell death and tissue necrosis, especially in the core of the scaffold. Bioactive factor delivery strategies (discussed later) have been utilized to expedite the vascularization process in polymeric scaffolds via the delivery of angiogenic factors such as vascular endothelial growth factor (VEGF) and basic fibroblast growth factor (bFGF) (Lee et al., 2002; Peattie et al., 2006; Patel et al., 2008), or by delivering gene expression vectors (e.g., plasmid DNA) that may allow the cells to produce such angiogenic factors *in situ* (Geiger et al., 2005). Co-culture of endothelial or progenitor cells with the cells of interest is another avenue of research to induce vascularization in the engineered tissue *in vitro* (see review by Lovett et al., 2009), which may help in the generation of a vascular network early after implantation. To promote sufficient vascularization of the implants, an alternate and effective strategy is to prevascularize a porous scaffold by first implanting it in a highly vascular site where the ingrowth of fibrovascular tissue or vascular tissue may take place by tissue infiltration. These prevascularized implants can then be seeded with cells and transplanted to the site of interest (Figure II.6.3.1C). Scaffold properties (such as pore size and morphology) are important factors that influence the extent of prevascularization, and determine the availability of space for cell seeding and tissue infiltration in the prevascularized implant (Mikos et al., 1993c). Apart from the regeneration of large organs, this strategy has also been applied for the treatment of critical-sized osseous defects. Prevascularized bone flaps were fabricated via ectopic bone formation by suturing an open chamber containing a mixture of a bioresorbable polymer and osteoinductive morcellized bone graft onto the cambium layer of the periosteum at a location remote from the defect site (Thomson et al., 1999), which may then be transplanted to the defect site and anastomosed to the host vasculature via microsurgery. An associated limitation of the prevascularization approach is that it requires multiple surgical procedures.

Injectable Systems for Minimally Invasive Tissue Engineering

In situ cross-linking biomaterials consist of space-filling injectable precursors, which solidify by cross-linking in the defect site (Figure II.6.3.1D). One method of classifying *in situ* cross-linking materials is based on the initiation mechanism of cross-linking, e.g., thermal initiation (such as PPF and OPF), photoinitiation (such as PEG-diacrylate (PEG-DA) or PPF), and ionic interaction (such as alginate or charged PLGA nanoparticulates) (for reviews see Hou et al., 2004; Kretlow et al., 2007; Van Tomme et al., 2008). Thermogelling polymers also fall under the category of injectable polymers; however, they undergo gelation by virtue of physical transition upon a sufficient temperature change (such as copolymers of poly(*N*-isopropylacrylamide) or poly(ethylene oxide-*b*-propylene oxide-*b*-ethylene oxide)) (Ruel-Gariepy and Leroux, 2004). One of the major advantages of *in situ* cross-linking materials is that they can be administered in a minimally invasive manner. Moreover, they also lend themselves to drug and cell delivery purposes, which can be achieved, under proper conditions of solidification, by simple mixing of cells and/or bioactive factors in the precursor solution (Kretlow et al., 2007). However, there are several additional design criteria that must be met, such as cytocompatibility of all the constituents, suitable rheological properties of the precursor solution, curing time, and polymerization conditions (e.g., temperature, pH, duration of exposure to the initiating wavelength of light, and heat release) that do not negatively impact the implanted cells and the surrounding tissue (Hou et al., 2004).

To facilitate cellular infiltration and guided tissue growth, an *in situ* cross-linking three-dimensional scaffold can be made highly porous via several methods, including particulate leaching or gas foaming (discussed later) (Peter et al., 1998; Behravesh et al., 2002). Since increased porosities usually lead to a reduction in mechanical properties, the mechanical properties of such materials can be tailored in an application-specific manner by either inclusion of micro- or nanophase materials that provide mechanical reinforcement or by varying the cross-linking mechanism, cross-linking agent, and/or cross-linking density (Timmer et al., 2003b; Shi et al., 2006). However, changes in the precursors or reaction/gelation conditions must be made so as to avoid any adverse reaction on the implanted cells or surrounding tissue. For example, an excessive concentration of the porogen salts, leachable components or cross-linking agents may not be suitable for implanted cells or surrounding host cells. In such cases, an alternate strategy of cell transplantation has been developed, where cells are encapsulated in a cytocompatible environment (in this case, gelatin microspheres) that is then included in the precursor solution to form a composite material (Payne et al., 2002a,b). Similar strategies have been widely applied for the delivery of bioactive factors, where bioresorbable micro- or nanoparticles loaded with bioactive agents can be included in the precursor solution to support sustained temporal release, as discussed in the following section.

Delivery of Bioactive Molecules

Bioactive molecules comprise many soluble molecules, including growth factors, angiogenic factors, cytokines, hormones, DNA, siRNA, and immunosuppressant drugs, which interact with and modulate the activity of a cell (Holland and Mikos, 2003; Kasper and Mikos, 2004). For instance, early in embryonic development, the fate of the uncommitted cells towards the formation of tissue patterns is governed by the spatiotemporal expression of specific signaling molecules, referred to as morphogens (Gurdon and Bourillot, 2001). Delivery of bioactive molecules is often desirable to induce tissue formation and/or selectively modulate various cellular activities, such as cell proliferation, differentiation or ECM production. For example, bone morphogenetic proteins (BMPs) constitute a family of proteins that have been exploited to enhance osteogenesis by bone cells (Cheng et al., 2003). *In vitro*, these bioactive factors are usually delivered in their soluble forms via the cell culture media. However, to administer such tissue inducing factors *in vivo* locally in a controlled manner, the concept of incorporating these factors into tissue engineering scaffolds has evolved. One approach is to deliver the bioactive factors directly from the supporting matrix (Figure II.6.3.2A). Another approach is to bind these molecules to the substrate surface at the time of scaffold processing (Figure II.6.3.2B). One of the most widely used examples is the use of cell adhesion molecules (such as RGD and YIGSR peptide sequences) covalently linked to the scaffold that provide attachment sites for anchorage dependent cells (see review by Hersel et al., 2003). Similarly, many growth and differentiation factors lead to an increase in desired cellular responses when they are linked covalently to the polymer substrate, as compared to the soluble forms (Hubbell, 2007). For example, recently it was shown that survival of mesenchymal stromal cells (MSCs), osteogenic colony formation of human bone marrow aspirates, and osteogenic differentiation of MSCs were all enhanced in the presence of epidermal growth factor

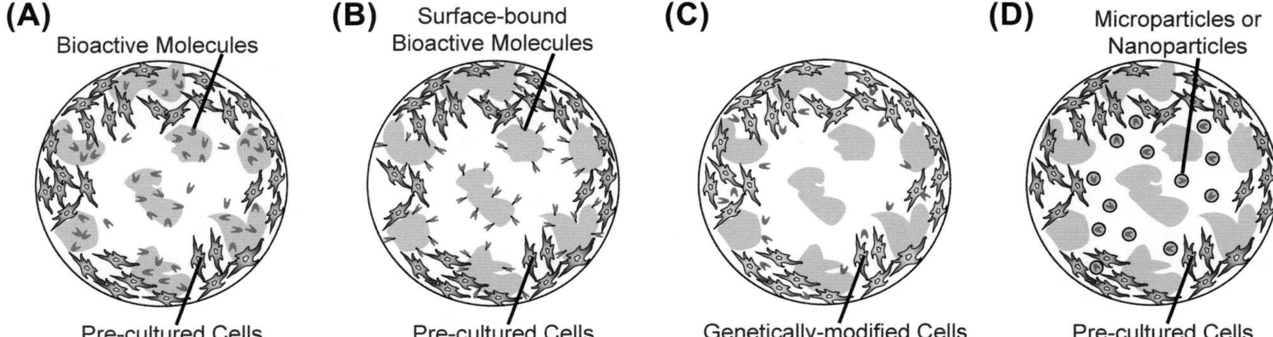

FIGURE II.6.3.2 Localized delivery of bioactive factors. (A) Bioactive factors released directly from a scaffold. (B) Bioactive factors chemically conjugated to the surface of a scaffold. (C) Localized production of bioactive agents by genetically modified cells seeded in a scaffold. (D) Controlled delivery of bioactive factors from a carrier (such as microparticles or nanoparticles) dispersed within the porosity of a scaffold.

(EGF) covalently attached to the substrate, as compared to the soluble EGF (Fan et al., 2007; Marcantonio et al., 2009; Platt et al., 2009). Alternatively, bioactive factors can be produced *in situ* using genetically modified cells seeded in the scaffold matrix that produce bioactive factors locally at the site of interest (Gilbert et al., 1993; Blum et al., 2003; Macdonald et al., 2007) (Figure II.6.3.2C). To achieve temporal control over the release of the bioactive factors (e.g., a sustained release over a desired period) in the scaffold, these molecules can be loaded in a carrier, such as bioresorbable micro- or nanoparticles (Figure II.6.3.2D) (Hedberg et al., 2002; Panyam and Labhasetwar, 2003; Kasper et al., 2005). More recently, lipid-based microtubular devices have also been explored as carriers for the delivery of plasmid DNA and growth factors (Meilander et al., 2003; Jain et al., 2006; Johnson et al., 2009). The release of bioactive factors from the carrier *in vivo* can be a function of many factors, such as molecular diffusion, degradation rate, and local enzyme concentration and activity (for polymers such as gelatin, a naturally derived polymer that degrades through enzymatic action) (Hedberg et al., 2002; Johnson et al., 2009). Depending on the application, the aforementioned delivery strategies also lend themselves to the delivery of multiple growth factors with desired release profiles (Richardson et al., 2001). For example, to explore the synergistic effects of simultaneous delivery of angiogenic and osteogenic proteins, dual growth factor delivery of VEGF and BMP-2 via a scaffold was achieved using acidic and basic gelatin microspheres as carriers, respectively (Patel et al., 2008). Spatial control over the expression of bioactive factors within a scaffold is desirable for the engineering of heterogeneous tissue structures (such as osteochondral tissue). For such applications, spatial patterning of bioactive factors at the time of scaffold processing is possible (Holland et al., 2005, 2007; Guo et al., 2009). Many such heterogeneous scaffold designs with biphasic, multiphasic or graded distribution of bioactive agents are currently under investigation, with the intention of controlling the spatiotemporal release of single or multiple growth factors.

In summary, the basic function of a tissue engineering scaffold is to provide a three-dimensional environment for tissue growth. Addition of other features to a scaffold (e.g., drug carrying ability or injectibility) imparts another degree of functionality to the scaffold that makes it more suitable for certain applications. In the following section, an overview of general scaffold fabrication techniques is provided.

SCAFFOLD PROCESSING TECHNIQUES

Various manufacturing methodologies have been developed to form porous three-dimensional substrates. In each of these techniques, one or more bioresorbable polymers of desired characteristics are processed uniquely to produce porous three-dimensional structures of particular shapes and morphologies (Hutmacher, 2000) (Table II.6.3.3). Polymer processing steps usually involve: (1) heat-assisted fusion/melting of the polymers above their glass transition or melting temperatures; (2) dissolution in organic solvents; (3) treatment with gases or supercritical fluids under high pressure; and/or (4) porogen (such as salt crystals, soluble microspheres or wax) leaching. Scaffold shape and morphology, mechanical

TABLE II.6.3.3 Fabrication Techniques for Three-Dimensional Scaffold Production, Required Polymer Characteristics, Geometry Produced, and Example Materials Processed[a]

Processing Technique	Required Polymer Characteristics	Structures	Examples
Fiber bonding	Thermoplastic or soluble	Thin foams or membranes	PGA fibers, PLA/PLGA coated PGA fibers
Solvent casting and particulate leaching	Soluble	Thin foams/mold-specific	PLA, PLGA, PPF foams
Superstructure engineering	Thermoplastic or soluble	Three-dimensional structures	PLA and PLGA membranes, sintered PLGA microspheres
Compression molding	Thermoplastic	Mold-specific	PLA, PLGA foams
Extrusion	Thermoplastic	Fixed cross-sectional profile	PLA, PLGA conduits
Freeze-drying	Soluble	Mold-specific	PLGA, PLGA/PPF foams
High internal phase emulsion templating	Soluble[b]	Mold-specific	PPF, poly(lactic acid-*co*-styrene), poly(ε-caprolactone-*co*-styrene)
Phase separation	Soluble	Mold-specific	PLA, PLGA foams
Gas foaming at near-critical conditions	Amorphous	Mold-specific	PLGA foams
Supercritical fluid processing	Amorphous	Mold-specific	PLA, PLGA foams
Solid freeform fabrication	Thermoplastic or soluble	Shape-specific	Complex three-dimensional structures of PLA, PLGA
Electrospinning	Soluble	Alligned or random fibers	PLA, PGA, PLGA, PCL fiber-based sheets and conduits

[a]Adapted with permission from Hutmacher (2010).
[b]Macromers in a liquid phase.

properties, and biocompatibility are greatly influenced by the choice of scaffold processing technique. A majority of these techniques allow for the incorporation of bioactive molecules into the polymer matrix. However, due to exposure to harsh chemical and thermal environments, retention of the bioactivity for sustained drug release has been a significant challenge. In the following sections, some established scaffold fabrication techniques are presented.

Fiber Bonding

Fiber-based technologies for scaffold fabrication have been very appealing, due to the large surface area-to-volume ratio that they provide, their early commercial availability, and the ease of fabrication of the fibers by industrial processes (such as hot drawing) (Freed et al., 1994). PGA fiber-based tassels and felts were some of the earliest constructs produced for organ regeneration purposes (Vacanti et al., 1991). However, the mechanical instability of such structures imposed a limitation on their *in vivo* application. To address this issue, a method was developed to fabricate bonded fiber networks of high porosities (up to 81%) (Mikos et al., 1993a). With this method, a non-bonded PGA fiber mesh is immersed in a PLA solution in methylene chloride. Following the evaporation of the organic solvent, the PLA–PGA composite is heated above the melting temperature of PGA. PGA fibers melt and bond with other PGA fibers at their contact points, while the molten PLA provides a structural casing around the fibers that prevents them from collapsing. PLA is finally removed by selective dissolution in methylene chloride, leaving the PGA fibers in an interlocked structure. Due to the specificity of the fiber bonding method with regard to the two polymers selected (they must be immiscible with appropriate relative melting temperatures) and to the solvent, the general use of this technique is precluded.

An alternate method of fiber bonding has also been developed, where PGA fibers are bonded by spray casting an atomized solution of PLA or PLGA in chloroform over the PGA fiber mesh placed on a rotating Teflon® mold (Mooney et al., 1996b). The PGA fibers are coated with a thin layer of sprayed polymer, which bonds the fibers at their cross-points. Solvent evaporation results in the formation of a thin tubular conduit made of bonded fibers. Such conduits were shown to support fibrovascular tissue in-growth following implantation (Mooney et al., 1996b). This fabrication technique, although useful for the preparation of thin fiber matrices, does not generally allow for the creation of complex three-dimensional structures. With the advent of computerized scaffold fabrication methodologies, complex fiber-based structures have now been created using solid freeform fabrication techniques, such as fused deposition modeling, as discussed later.

Solvent Casting/Particulate Leaching

Some of the drawbacks of fiber bonding techniques were addressed by the development of a new method, a solvent casting and particulate leaching (SC/PL) technique, which provides desired control over the porosity, pore size, surface area-to-volume ratio, and crystallinity of the prepared porous scaffolds (Mikos et al., 1994b). In an example of this method, sieved salt particles are dispersed in a solution of PLA dissolved in chloroform, which is used to cast a membrane on a platform (such as a glass petri dish). Following solvent evaporation, the dry polymer–salt composite is heated above the melting point of the polymer, then annealed or quenched at controlled cooling rates to produce a semicrystalline or amorphous polymer composite. The composites are then placed in water to leach out the salt particles, and subsequent drying yields polymer membranes that are highly porous. Regulation of the porogen-to-polymer weight ratio, porogen size, and cooling rate during the annealing/quenching step provides independent control over the production of scaffolds with desired porosity, pore sizes, and crystallinity, respectively. For example, PLA membranes fabricated using the SC/PL technique demonstrated porosities up to 93%, and median pore diameters up to 500 μm (Mikos et al., 1994b). Scaffolds produced using the SC/PL technique have been shown to support cell attachment and growth *in vitro*, as well as *in vivo* (Ishaug-Riley et al., 1997, 1998).

While the SC/PL technique is simple and inexpensive, extensive use of organic solvents during fabrication limits the possibility of using these matrices as carriers of specific bioactive agents. Moreover, the method is restricted to producing brittle thin membranes (up to 3 mm) (Wake et al., 1996). In one variation, blending of PEG with PLGA for the SC/PL technique resulted in an improvement in the pliability of the resulting membranes (Wake et al., 1996). In another method, thick scaffolds were constructed using particulate hydrocarbon wax as a porogen (Shastri et al., 2000). In this case, a mixture of viscous polymer solution and porogen is placed in a Teflon® mold. A hydrocarbon solvent (such as pentane or hexane) treatment is then used for selective extraction of the porogen from the mold, leaving a porous polymer foam matrix. PLA- and PLGA-based scaffolds fabricated using this method demonstrated porosities up to 87% and pore sizes above 100 μm, and this process supports the construction of thick shape-specific scaffolds.

In general, the SC/PL technique can be extended to any bioresorbable polymer that is soluble in an appropriate solvent (e.g., methylene chloride or chloroform), such as PLA or PLGA. In addition, a modified method of thermal cross-linking/particulate leaching has also been applied for cross-linkable polymers such as PPF. For example, by cross-linking the PPF in the presence of salt particles that are eventually leached in water, PPF scaffolds with porosities up to 75% and pore sizes in the range of 300–800 μm were prepared (Fisher et al., 2003). These scaffolds can be processed to allow for the incorporation of bioactive factors, and have been investigated for *in vivo* use in soft and hard tissue replacements (Fisher et al., 2002b; Hedberg et al., 2002, 2005). In addition, by incorporating another material phase (such as β-tricalcium phosphate or single-walled carbon nanotubes) in the scaffold design, reinforced porous PPF-based composite scaffolds can also be constructed to yield improved mechanical characteristics (Peter et al., 1997; Shi et al., 2005).

Superstructure Engineering

Superstructure refers to a three-dimensional structure made of superimposed two-dimensional structural elements such as fibers, pores or microspheres, which are ordered in a periodic, stochastic or fractal pattern (Wintermantel et al., 1996). Both individual elements and the pattern of organization determine the characteristics of the scaffold. Stable superstructures can be created from reproducible individual elements, the coherence of which determines the anisotropic structural behavior of the constructs (Wintermantel et al., 1996). An example of a superstructure is a three-dimensional structure produced by multiple stacked membranes fabricated using a membrane lamination process (Mikos et al., 1993b). In this method, individual membranes prepared by a SC/PL method are bonded together using chloroform on their contact surface to yield a shape-specific three-dimensional structure. Periodic arrangement of polymer microspheres is another simple example of a superstructure made of microspheres as the structural element. Recently, a variety of methods have been employed to produce polymer microsphere-based scaffolds by sintering the microspheres packed in a mold (Borden et al., 2004; Brown et al., 2008a; Jaklenec et al., 2008; Singh et al., 2008b, 2010). Here, the size of the microspheres and their packing configuration determines the coherence of the structure. For example, monodisperse microspheres in a cubic lattice yield a stable isotropic structure.

Compression Molding

Compression molding is a scaffold fabrication technique used with thermoplastic polymers. In an example of the compression molding technique, PLGA powder is mixed with gelatin microparticles and loaded in a Teflon® mold (Thomson et al., 1995b). The mixture is heated above the glass transition temperature of the amorphous polymer, while being compressed under a constant force. The composite is then removed from the mold, and the embedded gelatin is leached out in water. Porosity and pore sizes of the PLGA scaffolds thus produced can be varied selectively by altering the initial gelatin loading and gelatin microsphere size, respectively. This method avoids the use of any organic solvents, and may allow for the incorporation of bioactive factors in the polymer or porogen phase when performed at relatively low temperatures. By changing the mold geometry, it is also possible to prepare shape-specific scaffolds.

Several variations of compression molding exist. To extend the application of this process to bioresorbable semicrystalline polymers, such as PLA or PGA, a similar compression molding method has been use in which the mixture of polymer and porogen is heated above the melting temperature of the polymer. Melt-based compression molding has also been widely applied to create blends of polymers. Retention of the biological activity of bioactive factors at such elevated temperatures, however, is difficult. In another variation, compression molding can be applied together with the SC/PL technique to fabricate porous three-dimensional foams. In this method, solid pieces of polymer–salt composites are obtained by drying the cast polymer–salt solution or by coagulating it in an anti-solvent (Widmer et al., 1998; Hou et al., 2003). Particulate polymer–salt composites (<5 mm edge length) can then be compression molded, and subsequent salt leaching results in an open-cell foamed three-dimensional porous scaffold. This method has been shown to result in a relatively homogeneous pore morphology compared to the SC/PL method. Compared to the compression molding method, the SC/compression molding/PL method also offers the possibility of fabricating composites with another solid phase homogeneously distributed in the foamed three-dimensional structure. Using this method, foamed three-dimensional scaffolds reinforced homogeneously with osteoconductive hydroxyapatite microfibers were produced, and these demonstrated superior compressive strength compared to non-reinforced scaffolds of the same porosity within a certain range of polymer-to-fiber ratios (Thomson et al., 1998). Such a variation in scaffold fabrication is favorable for bone tissue engineering applications, because both incorporation of osteoconductive elements and improvement in mechanical properties are desired characteristics.

Extrusion

Extrusion is a well-known process applied to form objects of a predefined fixed cross-section; it has been used to process thermoplastic bioresorbable polymers for three-dimensional scaffold fabrication with macroscale cross-sectional areas (>1 mm). Processing of polymers using various extrusion methods, such as solid-state extrusion (die drawing), cylinder-piston (ram) extrusion or hydrostatic extrusion, generally changes the polymer chain orientation, leading to increases in their strength and modulus of elasticity (Ferguson et al., 1996). Using a mixture of polymer powders, extrusion methods can be readily applied to create solid blends of polymers. To create porous scaffolds, an SC/melt–extrusion/PL technique has been developed (Widmer et al., 1998). Dry polymer–salt composites produced by SC are cut into pieces (<5 mm edge length) and placed in a custom piston extrusion tool connected to a hydraulic press. The polymer is heated to the desired processing temperatures and equilibrated, then extruded by applying pressure using the hydraulic press to produce tubular constructs. Constructs of desired length are cut from the tube, and are then placed in water for salt leaching. Porous tubular conduits thus produced have been used for guided tissue engineering applications, including peripheral nerve regeneration (Evans et al., 2002). As with the SC/PL techniques, porogen-to-polymer weight ratio and porogen size can be selectively altered to produce scaffolds of desired porosity and pore sizes, respectively. In addition, processing temperature is another important variable that influences the pressure required for extrusion, and can influence the scaffold morphology and thermal degradation of the scaffold.

Freeze-Drying

Freeze-drying is a simple approach for polymer foam fabrication. The polymer is dissolved in a solvent such as benzene or glacial acetic acid, then frozen and lyophilized under high vacuum to remove the dispersed solvent (Hsu et al., 1997). Depending on the polymer–solvent system used, the resulting scaffolds demonstrate leaflet or capillary-like microfeatures. Several materials, including PLA, PLGA, and PLGA/PPF, have been used to fabricate polymer foam scaffolds using this method (Hsu et al., 1997). These scaffolds usually have a random pore structure and network. Due to their low density and pore connectivity, further processing, such as grinding and extrusion, is usually required to fabricate matrices useful for application.

An emulsion freeze-drying method is very similar to the freeze-drying approach; however, it involves the formation of an emulsion (Whang et al., 1995). The polymer is first dissolved in a solvent; then water is added to form a water-in-oil emulsion. The polymer is subsequently quenched in liquid nitrogen and freeze-dried to remove the dispersed water and solvent. Important process parameters include the polymer–solvent system, molecular weight of the polymer, polymer solution-to-water ratio, and emulsion viscosity. Using this technique, PLGA (methylene chloride as solvent) and PLA (dioxane as solvent) scaffolds with porosities greater than 90% have been prepared (Whang et al., 1995; Hu et al., 2002). Compared to the SC/PL technique, the emulsion freeze-drying method produces foams with a lower median pore diameter and higher specific pore surface area.

Another class of emulsion-derived porous polymeric foams known as polyHIPEs (Barby and Haq, 1982) are formed by a high internal phase emulsion templating method (see review by Cameron, 2005). A high internal phase emulsion (HIPE) consists of a monomer (external) phase and a droplet (internal) phase, and is defined by a characteristic internal phase volume fraction of at least 0.74 (Lissant, 1974). In this method, the continuous macromer phase is first polymerized around a template of the droplet phase, and subsequent removal of the droplet phase results in the formation of a porous polyHIPE. Many polyHIPEs have been recently investigated for their tissue engineering applications, including polyHIPEs based on PPF, poly(lactic acid-*co*-styrene), and poly(ε-caprolactone-*co*-styrene) (Busby et al., 2001, 2002; Christenson et al., 2007). Factors that govern pore size, pore interconnectivity, and porosity of the polyHIPEs include composition of the HIPE (e.g., composition and viscosity of the two phases, volume fraction of the droplet phase), the droplet radius, molecular weight of the macromer, and cross-linker density.

Phase Separation

In the thermally-induced liquid–liquid phase separation technique, the polymer is first dissolved in a solvent that has a low melting point and that can be easily sublimed (such as molten phenol, naphthalene or dioxane) (Lo et al., 1995; Nam and Park, 1999a,b). Lowering the solution temperature below the melting point of the solvent leads to liquid–liquid phase separation, and later quenching of the phase-separated solution creates a two-phase solid. Subsequent removal of the solvent via freeze-drying for several days results in a porous polymeric scaffold. There are several process variables that can affect the morphology of the scaffolds, including the polymer–solvent system, polymer concentration, molecular weight, and cooling rate. For example, this technique was used to prepare polymer foams with porosities greater than 90% and pore sizes close to 100 μm (Nam and Park, 1999a). This process also allows the incorporation of bioactive factors, which can be homogeneously mixed in the polymer solution before inducing phase separation (Nam and Park, 1999a). Depending on the choice of solvent, however, the bioactive factors may become denatured and lose some of their activity (Lo et al., 1995). More recently, phase separation has also been utilized to produce interconnected nanostructured networks (Chen and Ma, 2004; Yang et al., 2004). Using tetrahydrofuran as a solvent, PLA nanofibrous scaffolds were produced via phase separation with fibers in the range of 50–300 nm, and were shown to support neural stem cell differentiation and neurite outgrowth (Yang et al., 2004). It is, however, difficult to control the orientation of the fibers produced in such nanostructured networks (Murugan and Ramakrishna, 2007).

Gas Foaming and Supercritical Fluid Processing

Processing of many bioresorbable amorphous polymers with gases at ambient temperature and high (critical, near-critical or sub-critical) pressures has offered an alternative method for porous scaffold fabrication. Specifically, carbon dioxide (CO_2) has been most widely used, as it is inexpensive, non-toxic, non-flammable, recoverable, and reusable. The polymer is first saturated with CO_2 at high pressures in a vessel, which is followed by depressurization to ambient levels that results in nucleation of the gas, forming pores in the material. The amount of dissolved CO_2, the rate and type of gas nucleation, and the rate of gas diffusion to the pore nuclei are factors that primarily determine the pore structure and porosity of the resulting scaffold. Application of gas foaming (GF) at near-critical pressures (5.5 MPa) has been shown to produce highly porous scaffolds (>90% porosity and around 100 μm pore size) from preprocessed solid PLGA discs (Mooney et al., 1996a), and the technique has also been shown to allow the retention and delivery of active growth factors (Sheridan et al., 2000). In contrast to gas foaming at near-critical conditions, supercritical CO_2 (T_c = 304.1 K, P_c = 73.8 bar) absorbs into and liquifies many materials by lowering their glass transition temperatures (Jung and Perrut, 2001), and a supercritical fluid processing method has been used to incorporate bioactive factors and/or cells at the time of scaffold fabrication in a single step (Howdle et al., 2001; Ginty et al., 2006). Although these foaming techniques are advantageous over temperature-regulated or organic solvent-assisted scaffold fabrication for the incorporation of bioactive factors, an inherent limitation is the closed-cell structure and lack of pore interconnectivity. This problem was addressed by a modified technique of gas foaming/particulate leaching (GF/PL), where porogens (NaCl) included at the time of processing were leached out following gas foaming, resulting in an open pore structure of macropores (effected by porogen leaching), along with a non-interconnected microporous network (Harris et al., 1998; Murphy et al., 2000). The matrices produced demonstrate a relatively homogeneous pore structure, and higher mechanical strength than those fabricated with SC/PL.

In lieu of direct processing of polymers with CO_2, on-site production of CO_2 and foaming can be achieved using effervescent salts (Nam et al., 2000). For example, a combination of ascorbic acid, ammonium persulfate, and sodium bicarbonate was used to form highly porous hydrogels for bone tissue engineering (Behravesh et al., 2002), and these foams demonstrated excellent compatibility for marrow stromal cell differentiation and production of bone matrix *in vitro* (Behravesh and Mikos, 2003).

Solid Freeform Fabrication

Solid freeform fabrication (SFF), also known as rapid prototyping (RP), refers to a set of automated computer-assisted fabrication technologies, which have been utilized to produce porous three-dimensional scaffolds using bioresorbable materials. Examples of SFF techniques include three-dimensional printing (3DP), stereolithography, fused deposition modeling, selective laser sintering (SLS), and ballistic particle manufacturing. The underlying principle is analogous to a bottom-to-top approach, where mold-less creation of complex scaffolds is achieved by addition of materials to form cross-sectional levels in a layer-by-layer manner (obtained by computer-assisted design (CAD)), followed by layer fusion to yield a three-dimensional structure. These methods provide selective and precise spatial control over the scaffold morphology, including material composition, pore size, porosity, and microstructure. In addition, some technologies also offer the ability to pattern cells or bioactive molecules during the scaffold fabrication process.

SLS involves computerized laser scanning for the creation of patterned objects (Bartels et al., 1993). In SLS, a layer of polymer powder is first spread over a building platform. A desired two-dimensional profile is created by selectively scanning the polymer powder by an x–y controlled laser beam, which heats the polymer surface to or above its glass transition temperature. Upon cooling, the polymer particulates transition from a rubber to a glassy state, and fuse with the surrounding particulates. The layered scanning process is repeated with a fresh layer of polymer powder until the desired scaffold dimensions are met. Using this method, many bioresorbable materials have been processed into scaffolds, including PLLA, PCL, and PLGA, with or without a bioceramic component (Tan et al., 2003; Williams et al., 2005; Wiria et al., 2007; Simpson et al., 2008).

In stereolithography, computer-controlled movement of an ultraviolet laser is used to photopolymerize a liquid precursor solution in desired three-dimensional shapes. For example, a diethyl fumarate/PPF resin was used as a liquid base material in a custom-design stereolithography apparatus (Cooke et al., 2003). In a bottom-to-top manner, desired two-dimensional patterns of a fixed thickness (in this case, 100 μm) were created in the surface layer of the liquid resin over a building platform placed in a resin-filled tank. Sequential layers of 100 μm thickness were produced by lowering the stage in the tank by a fixed step-size of one layer, allowing the resin to cover the finished layer, and then exposing the fresh surface layer to UV. Both dimensions of the laser and resolution of the movement in the z-direction determine the overall resolution of the fabricated scaffolds. Using PEG-based photopolymerizable precursor solutions with cytocompatible photoinitiators, stereolithography has also been utilized to create fibroblast and Chinese hamster ovary (CHO) cell-loaded three-dimensional hydrogel constructs (Dhariwala et al., 2004; Arcaute et al., 2006). In some variations of the stereolithography technique, photopatterning of a precursor can be performed by selective exposure to UV via a designed photomask, which can also allow precise pore or cell positioning within the fabricated constructs (Albrecht et al., 2005; Bryant et al., 2007).

The 3DP technique is an automated inkjet printing-based technology (Sachs et al., 1993). In 3DP technique, a layer of polymer powder is first spread over a building platform. Using the print head of an inkjet printer, a desired two-dimensional profile is created by depositing a binder material (e.g., an organic solvent) over this layer to selectively join the polymer powder. The building platform is lowered and the process is repeated with a fresh layer of powder placed over the last finished layer until the desired part is obtained. The unbound powder is finally removed to yield the desired three-dimensional scaffold. Spatial resolution of this technique (~50–250 μm) is dependent on several factors, such as the size of the polymer particulate/powder being used, the size of the printed droplet, and the resolution of the printhead movement. 3DP is performed at ambient temperature, and allows the integration of controlled release features (Wu et al., 1996). Various direct and modified 3DP techniques have been utilized to create composite scaffolds for bone tissue engineering applications (Taboas et al., 2003; Lee et al., 2006; Ge et al., 2009).

Controlled dispensing methodologies have also been developed, where a polymer is deposited over a building platform in a layered fashion. The fused deposition modeling process is one such example, which involves controlled deposition of a melted/cast material through a computer-controlled nozzle in an automated layer-by-layer manner (Crump, 1992; Cao et al., 2003; Taboas et al., 2003).

In general, SFF techniques allow fabrication of precise complex shapes. In specific applications, some SFF techniques can also be used in combination with conventional scaffold fabrication techniques. For example, incorporation of porogens in the polymer powder for 3DP can be performed to introduce additional microfeatures in the scaffolds, which can improve the available surface area for cell attachment and mass transport, and enhance guided tissue regeneration (Kim et al., 1998b).

One limitation of the discussed techniques, with rare exceptions, is their inability to incorporate and pattern cells during the scaffold fabrication in a cytocompatible manner. Recently, development of new techniques that allow printing of cells during scaffold fabrication has been an area of great interest, the so-called organ printing or bioprinting technologies. In the automated process of bioprinting, controlled deposition of cell suspensions or cellular aggregates ("spheroids") is applied to create three-dimensional cellular constructs of desired geometries (see review by Mironov et al., 2008). Direct inkjet printing of cells on hydrogels has been used to produce two-dimensional or multi-layered structures (Boland et al., 2003; Wilson and Boland, 2003; Xu et al., 2005; Lee et al., 2009). Using modified controlled dispensing methodologies, direct printing of cells loaded in cross-linkable viscous hydrogel precursors has been accomplished via extrusion, microstereolithography or inkjet printing (Smith et al., 2004; Cohen et al., 2006; Lu et al., 2006; Wang et al., 2006; Smith et al., 2007; Moon et al., 2010). In addition, self-assembly of multicellular structures has also been investigated, where controlled dispensing of cellular aggregates is performed directly into a hydrogel (Mironov et al., 2009). The bioprinting techniques show promise to alleviate some major challenges associated with the traditional approaches, including cell seeding at high densities, user-defined positioning of multiple cell types, and creation of vascular beds within large three-dimensional structures (Boland et al., 2006; Mironov et al., 2009).

Electrospinning

Electrospinning is a versatile process with high production capability that has been used to form non-woven micro- and/or nanofibrous scaffolds using synthetic bioresorbable polymers (for reviews see Pham et al., 2006a; Murugan and Ramakrishna, 2007) (Figure II.6.3.3I). In a simple set-up, a dissolved solution of polymer is passed through a needle at a controlled rate. An electric field applied using a high voltage source across the needle and a grounded collector charges the surface of the polymer droplet held at the needle tip. When the forces of electrostatic repulsion within the solution overcome the surface tension, a thin polymer jet forms. Solvent evaporation occurs as the jet travels from the needle tip to the collector where polymer fibers are formed and deposited on the collector. Anisotropically distributed polymer fibers can be produced using a rotating mandrel or patterned electrodes, and these have been shown to influence the polarity of the cells (Figure II.6.3.3II) (Li et al., 2003; Yang et al., 2005). There are several variables (e.g., solution viscosity, polymer charge density, polymer molecular weight, surface tension, electric field strength, tip-to-collector distance, needle design, and the composition and design of the collector) that influence the process of electrospinning, and each can be varied selectively to obtain the desired fiber and scaffold morphology (Pham et al., 2006a). For example, by adjusting these variables, poly(ε-caprolactone) microfibers with a range of selected fiber diameters were produced (Figure II.6.3.3 III) (Pham et al., 2006b). In electrospun matrices, fiber size is the primary factor that defines the pore sizes and porosity of the scaffold. Recently, nanofibrous scaffolds have received special attention in tissue engineering, as the scaffolds are highly porous (>90% porosity) with a fiber architecture that resembles closely the nanofeatures of native ECM (Li et al., 2002). Due to the presence of nanofibers, a multi-layered PCL nanofiber/microfiber scaffold was shown to enhance cell attachment and spreading of rat marrow stromal cells, compared to scaffolds containing microfibers alone (Pham et al., 2006b). The electrospinning process also allows the possibility of controlled delivery of single or multiple bioactive factors. For example, using a coaxial needle design, core shell fibers with PEG as the core and PCL as the shell were produced that can be used for the controlled release

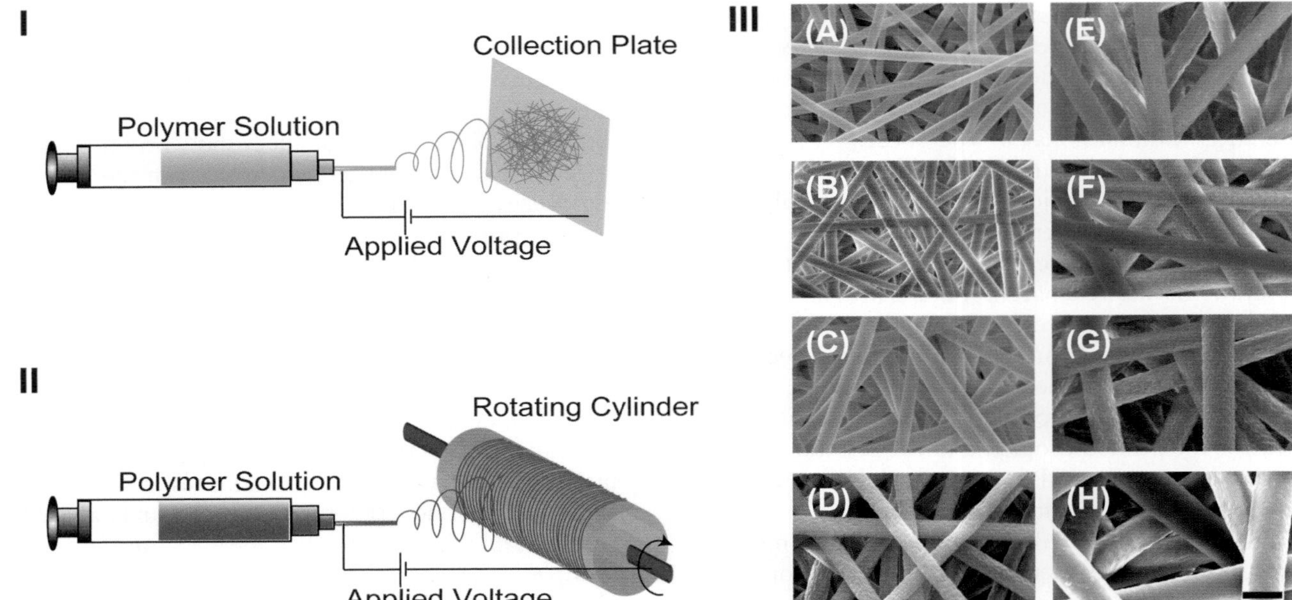

FIGURE II.6.3.3 Electrospinning process for the fabrication of fiber mesh scaffolds. (I, II) Schematic representations of electrospinning set-ups commonly used for the production of a non-woven fiber mesh (I), and an annulus of oriented fibers (II). (III) Scanning electron micrographs of electrospun non-woven poly(ε-caprolactone) microfibers of different fiber diameters: (A) 2 μm; (B) 3 μm; (C) 4 μm; (D) 5 μm; (E) 6 μm; (F) 7 μm; (G) 8 μm; and (H) 10 μm. (Scale bar: 10 μm) (Reproduced with permission from Pham et al., 2006a.)

of dual bioactive factors (Saraf et al., 2009, 2010). An issue that must often be addressed is that the nanodimensioned fibers have nanodimensioned interstices between them, hindering cell entry into the porous structure.

In summary, manufacturing methodologies described herein have been widely applied to process bioresorbable synthetic materials into porous three-dimensional scaffolds. The selection of the technique depends both on the envisioned application, and on the material under consideration. Post-fabrication, the attributes of the processed scaffolds are usually determined by characterizing the scaffolds, as discussed in the following section.

CHARACTERIZATION OF PROCESSED SCAFFOLDS

A number of techniques are available for the characterization of scaffolds (Table II.6.3.4). Surface chemistry of polymers (Chapter I.1.5) can be characterized by electron spectroscopy for chemical analysis (ESCA) and secondary ion mass spectroscopy (SIMS), among other techniques. Polarized infrared (IR) and near-edge X-ray absorption fine structure (NEXAFS) are employed to determine the orientation of the functional groups. Contact angle measurements provide information about the surface energy and wettability (hydrophilicity) of the polymer surface. Surface texture and roughness can be qualitatively or quantitatively assessed using scanning electron microscopy (SEM), scanning probe microscopy (SPM) or atomic force microscopy (AFM). AFM may also be employed to determine the surface rigidity.

TABLE II.6.3.4 Bioresorbable Polymeric Scaffold Characterization Techniques

Properties	Characterization Technique
Surface Characterization	
Chemistry	ESCA, SIMS
Texture	SEM, SPM, AFM
Orientation of functional groups	Polarized IR, NEXAFS
Surface energy and wettability	Contact angle measurement
Chemical Characterization	
Composition, sequence distribution of copolymers and crystal structure	NMR, XRD, FTIR, FTR
Mechanical Characterization	
Strength, stiffness, yield strength, etc.	Mechanical testing (compression, tension, shear, torsion)
Thermal Characterization	
Thermal properties (T_g, T_m, X_c)	DSC
Physical Characterization	
Molecular weight and distribution	GPC
Degradation	*In vitro*, *in vivo*
Mass loss	Gravimetry
Morphology	SEM, confocal microscopy
Pore-size and porosity	Mercury intrusion porosimetry, μ-CT
Interconnectivity	μ-CT

To determine the bulk properties (such as information regarding polymer composition, sequence distribution of co-polymers, and crystal structure), chemical characterization of the polymers can be performed using nuclear magnetic resonance (NMR) spectroscopy, Fourier transform IR (FTIR) spectroscopy, FT-Raman (FTR) spectroscopy, and X-ray diffraction (XRD). Bulk mechanical properties are usually characterized by uniaxial mechanical testing under different modes of loading (e.g., compression, tension, shear, and torsion), and provide valuable data regarding mechanical characteristics (such as strength and modulus). Thermal properties (glass transition temperature (T_g), melting temperature (T_m) and degree of crystallinity (X_c)) are determined by differential scanning calorimetry (DSC). Molecular weight and distribution are commonly determined by size exclusion chromatography methods, such as gel permeation chromatography (GPC). SEM and confocal microscopy can be employed to determine the microscopic structure and pore morphology qualitatively, while mercury intrusion porosimetry and μ-computed tomography (μ-CT) are useful techniques to determine the porosity and/or pore interconnectivity of the scaffolds quantitatively. Additionally, mass loss from the scaffold can be characterized by simple gravimetric evaluations. To characterize the *in vitro* degradation of the scaffolds, a combination of physical, chemical, and mechanical properties can be assessed as a function of time on scaffold samples placed under simulated physiological conditions. *In vivo* assessment is usually required, however, to assess "true" degradation rates, which are ideally coupled with the rate of tissue regeneration. Like *in vitro* degradation assessment, the *in vitro* release profile of any impregnated biomolecule from the scaffold can be quantitatively determined by using various techniques, such as radioactivity measurements of radiolabelled biomolecules.

CELL SEEDING AND CULTURE IN THREE-DIMENSIONAL SCAFFOLDS

Engineering of large tissues and organs *ex vivo* generally requires culture of cells seeded at high densities in bioresorbable scaffolds in a controlled culture environment. However, there are many practical challenges that need to be overcome, such as uniform cell seeding, sufficient nutrient supply, and waste removal. In addition, control over the culture environment, including temperature, oxygen tension, pH, biochemical and mechanical stimulation (such as compression and shear stress), is desired (Vunjak-Novakovic, 2003 and Chapter II.6.6). To gain control over the physicochemical and hydrodynamic environment in culture conditions, various strategies have been developed that are used for cell seeding and culture of polymer constructs, some of which are described here.

Static Seeding and Culture

Static conditions represent the simplest form of cell seeding and culture. A cell suspension is allowed to adsorb onto a porous polymeric scaffold, cells are permitted to attach, and the constructs are then cultured in the presence of culture media (Figure II.6.3.4A). A major drawback of this method is that the penetration of a majority of cells in the scaffolds is generally limited to depths of approximately a few hundred microns (Ishaug et al., 1997), leading to undesired heterogeneous tissue growth. To improve cell penetration, several variations of seeding techniques exist, such as cell seeding by injection, centrifugation, and application of vacuum (van Wachem et al., 1990; Mikos et al., 1994a; Kim et al., 1998a; Godbey et al., 2004; Roh et al., 2007), which have all been shown to improve the cell distribution. To improve the cell seeding density in hydrophobic polymeric materials, various methods have been developed. Pre-wetting a hydrophobic polymeric scaffold with ethanol and water helps to increase the available volume for cell penetration by the displacement of air from the pores (Mikos et al., 1994a). Similarly, hydrolysis of the polymer surface or infiltration with hydrophilic polymers has been shown to increase the cellularity of the constructs (Mooney et al., 1995; Gao et al., 1998). As molecular diffusion is the primary driving force for nutrient supply and waste removal under static conditions, poor cell survival in the central parts of the scaffold is usually seen due to the gradients of nutrients and metabolites generated across the scaffold (Goldstein et al., 2001; Vunjak-Novakovic, 2003). To gain uniform cellularity and continuous convection-driven mass

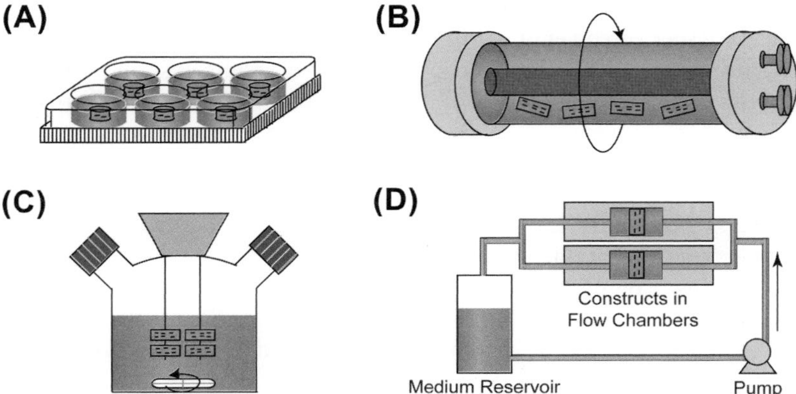

FIGURE II.6.3.4 Dynamic cell seeding and culture techniques used *in vitro*. (A) Static culture in a well plate. (B) Rotary vessel. (C) Spinner flask. (D) Flow perfusion system.

transport, various bioreactor devices have been built that are more suitable for the culture of large constructs and allow the possibility of multiple scaffold processing and scale-up, discussed next.

Spinner Flask Culture

In spinner flask culture, fluid flow driven by a magnetic stir bar creates a well-mixed cell suspension for dynamic seeding and culture of the scaffolds that are fixed inside the flask on needles or wires (Figure II.6.3.4B) (Sikavitsas et al., 2002). This technique has been applied to culture a variety of cells, including chondrocytes, hepatocytes, MSCs, and umbilical cord matrix stem cells (Yamada et al., 1998; Sikavitsas et al., 2002; Bailey et al., 2007). Optimization of stirring speed is usually desired, as turbulence and eddies generated at high speeds may improve cell penetration and mass transfer; however, these high speeds can also result in cell damage or ECM loss. Due to the velocity field created by the vortex of the stir bar and eddies generated on the rough surfaces of the scaffolds, convection is almost always higher at the exterior of the scaffold than in the interior (Goldstein et al., 2001; Vunjak-Novakovic, 2003). As a result, matrix formation in peripheral scaffold areas has been reported to be more prominent than in areas near the center (Vunjak-Novakovic et al., 2002; Meinel et al., 2004).

Rotary Vessel Culture

Rotary vessel culture usually involves the use of a rotating-wall cylindrical vessel (RWV) or two concentric vessels with the purpose of creating a microgravity environment for the constructs placed inside the vessel (Figure II.6.3.4C) (Freed and Vunjak-Novakovic, 1997). Centrifugal forces generated due to the rotation of the cylinder counterbalance the gravitational pull on the scaffolds, suspending the scaffolds in the culture medium in a low-shear laminar flow-dominated environment. By controlling the rotational speed, penetration of nutrients and oxygen into the scaffold can be adjusted. Rotary vessel culture is especially useful for applications where enhanced mass transport is needed in low shear environments. Rotary vessel culture has been applied for the engineering of musculoskeletal tissues (Freed et al., 1997; Goldstein et al., 2001; Sikavitsas et al., 2002; Detamore and Athanasiou, 2005), among other tissues.

Perfusion Culture

Direct flow perfusion culture systems have been developed for the *in vitro* reconstruction of large three-dimensional tissues and organs (Figure II.6.3.4D) (Griffith et al., 1997; Kim et al., 1998b; Bancroft et al., 2002; Grayson et al., 2008). In general, the culture medium continually circulates through the cell–polymer constructs. With a continuous flow of culture media through the porous structure of a scaffold, most mass transfer limitations are mitigated. In addition, flow-induced shear stresses can be modulated to upregulate desired cellular processes (such as cell differentiation or ECM production) in specific applications. For instance, increasing the shear stress resulted in an increased matrix deposition by MSCs cultured on three-dimensional scaffolds in a perfusion bioreactor (Sikavitsas et al., 2003). In another study, shear stress generated by direct perfusion was found to induce osteogenic differentiation of MSCs, even in the absence of other biochemical factors (Holtorf et al., 2005). Flow perfusion culture has also been shown to significantly outperform static culture, resulting in increased cellular proliferation, differentiation, and ECM deposition (van den Dolder et al., 2003; Datta et al., 2006). In addition to direct flow perfusion culture, other complex designs of perfusion bioreactors have also been developed, such as oscillatory flow perfusion bioreactors or pulsatile flow perfusion bioreactors (Brown et al., 2008b; Du et al., 2009).

Culture in Mechanically Stimulated Conditions

Mechanotransduction plays a key role in tissue development and maintenance (see reviews by Goodman and Aspenberg, 1993; Duncan and Turner, 1995; Sikavitsas et al., 2001; Ingber, 2006) (also see Chapter II.1.6). With the hypothesis that culture of cells in conditions that mimic the *in vivo* environment can optimize tissue regeneration *in vitro*, effects of biophysical forces on tissue growth *in vitro* have been widely studied, particularly for musculoskeletal and cardiovascular tissue engineering. The selection of mode of loading depends on the particular tissue engineering application, e.g., compression or shear for cartilage and bone tissue engineering, or tension for tendon tissue engineering. Level, loading profile (e.g., constant or cyclic), duration and frequency of mechanical stimulation are some of the variables that can influence cellular function.

Other Culture Conditions

As some cells respond to other particular biophysical cues, such as an electrical field (cardiac cells) or light intensity (retinal cells), inclusion of these as additional stimuli during the culture have also been studied (Hughes and Maffei, 1966; Radisic et al., 2004). Co-culture of different cell types is sometimes preferential to induce endogenous mutual signaling between the selected cell types, which has been used to facilitate *in vitro* organogenesis. Transwell culture systems have been used to selectively expose the apical and basal sides of the constructs to different media, desired in certain applications such as epithelial cell culture. Gradients of physical and chemical cues, which are known to influence cell motility, function, and survival, may also be introduced in the culture (Singh et al., 2008a). For instance, concentration gradients of growth factors have been widely applied for guided neurite growth in peripheral nerve regeneration. Moreover, the substrate on which cells are expanded in two dimensions is also important for specific applications. For example, engineering of continuous two-dimensional epithelial cell sheets is desired for corneal epithelial replacements (Hayashi et al., 2010). For this purpose, use of surfaces coated with temperature-responsive thermoreversible polymers (such as poly(*N*-isopropylacrylamide)) has been applied to engineer organized sheets of epithelial cells (Okano et al., 1995; Yamato et al., 2001) (Chapter I.2.11). The polymer is designed to support cell attachment at 37°C, whereas a reduction in temperature below its lower critical solution temperature swells the polymer, causing the cells to detach while remaining attached to their extracellular matrix and each other (Okano et al., 1995).

CONCLUSIONS

Continuous progress is being made towards the long-term goal of engineering human tissues and organs. As highlighted throughout the chapter, many challenges still remain that limit clinical success for several tissues. The integration of advances in biological sciences with modern material science will not only produce designed scaffolds, but will also generate new technologies for regenerative medicine applications.

BIBLIOGRAPHY

Albrecht, D. R., Tsang, V. L., Sah, R. L., & Bhatia, S. N. (2005). Photo- and electropatterning of hydrogel-encapsulated living cell arrays. *Lab. Chip.*, **5**, 111–118.

Andriano, K. P., Tabata, Y., Ikada, Y., & Heller, J. (1999). *In vitro* and *in vivo* comparison of bulk and surface hydrolysis in absorbable polymer scaffolds for tissue engineering. *J. Biomed. Mater. Res.*, **48**, 602–612.

Arcaute, K., Mann, B. K., & Wicker, R. B. (2006). Stereolithography of three-dimensional bioactive poly(ethylene glycol) constructs with encapsulated cells. *Ann. Biomed. Eng.*, **34**, 1429–1441.

Babensee, J. E., Anderson, J. M., McIntire, L. V., & Mikos, A. G. (1998). Host response to tissue engineered devices. *Adv. Drug. Deliv. Rev.*, **33**, 111–139.

Babensee, J. E., McIntire, L. V., & Mikos, A. G. (2000). Growth factor delivery for tissue engineering. *Pharm. Res.*, **17**, 497–504.

Bailey, M. M., Wang, L., Bode, C. J., Mitchell, K. E., & Detamore, M. S. (2007). A comparison of human umbilical cord matrix stem cells and temporomandibular joint condylar chondrocytes for tissue engineering temporomandibular joint condylar cartilage. *Tissue Eng.*, **13**, 2003–2010.

Bancroft, G. N., & Mikos, A. G. (2002). Bone tissue engineering by cell transplantation. In R. L. Reis, & D. Cohn (Eds.), *Polymer Based Systems on Tissue Engineering, Replacement and Regeneration* (pp. 251). Kluwer Academic Publishers.

Bancroft, G. N., Sikavitsas, V. I., van den Dolder, J., Sheffield, T. L., Ambrose, C. G., Jansen, J. A., & Mikos, A. G. (2002). Fluid flow increases mineralized matrix deposition in 3D perfusion culture of marrow stromal osteoblasts in a dose-dependent manner. *Proc. Natl. Acad. Sci. USA*, **99**, 12600–12605.

Barby, D., & Haq, Z. (1982). Low density porous cross-linked polymeric materials and their preparation. *Eur. Patent 0,060,138*.

Baroli, B. (2007). Hydrogels for tissue engineering and delivery of tissue-inducing substances. *J. Pharm. Sci.*, **96**, 2197–2223.

Bartels, K. A., Bovik, A. C., Crawford, R. C., Diller, K. R., & Aggarwal, S. J. (1993). Selective laser sintering for the creation of solid models from 3D microscopic images. *Biomed. Sci. Instrum.*, **29**, 243–250.

Behravesh, E., & Mikos, A. G. (2003). Three-dimensional culture of differentiating marrow stromal osteoblasts in biomimetic poly(propylene fumarate-co-ethylene glycol)-based macroporous hydrogels. *J. Biomed. Mater. Res. A.*, **66**, 698–706.

Behravesh, E., Jo, S., Zygourakis, K., & Mikos, A. G. (2002). Synthesis of *in situ* cross-linkable macroporous biodegradable poly(propylene fumarate-co-ethylene glycol) hydrogels. *Biomacromolecules*, **3**, 374–381.

Benoit, D. S., Schwartz, M. P., Durney, A. R., & Anseth, K. S. (2008). Small functional groups for controlled differentiation of hydrogel-encapsulated human mesenchymal stem cells. *Nat. Mater.*, **7**, 816–823.

Blum, J. S., Barry, M. A., & Mikos, A. G. (2003). Bone regeneration through transplantation of genetically modified cells. *Clin. Plast. Surg.*, **30**, 611–620.

Boland, T., Mironov, V., Gutowska, A., Roth, E. A., & Markwald, R. R. (2003). Cell and organ printing 2: Fusion of cell aggregates in three-dimensional gels. *Anat. Rec. A Discov. Mol. Cell Evol. Biol.*, **272**, 497–502.

Boland, E. D., Coleman, B. D., Barnes, C. P., Simpson, D. G., Wnek, G. E., & Bowlin, G. L. (2005). Electrospinning polydioxanone for biomedical applications. *Acta Biomater.*, **1**, 115–123.

Boland, T., Xu, T., Damon, B., & Cui, X. (2006). Application of inkjet printing to tissue engineering. *Biotechnol. J.*, **1**, 910–917.

Borden, M., Attawia, M., Khan, Y., El-Amin, S. F., & Laurencin, C. T. (2004). Tissue-engineered bone formation *in vivo* using a novel sintered polymeric microsphere matrix. *J. Bone Joint Surg. Br.*, **86**, 1200–1208.

Brown, J. L., Nair, L. S., & Laurencin, C. T. (2008a). Solvent/non-solvent sintering: A novel route to create porous microsphere scaffolds for tissue regeneration. *J. Biomed. Mater. Res. Part B Appl. Biomater.*, **86B**, 396–406.

Brown, M. A., Iyer, R. K., & Radisic, M. (2008b). Pulsatile perfusion bioreactor for cardiac tissue engineering. *Biotechnol. Prog.*, **24**, 907–920.

Bruin, P., Smedinga, J., Pennings, A., & Jonkman, M. (1990). Biodegradable lysine diisocyanate-based poly (glycolide-co-epsilon-caprolactone)-urethane network in artificial skin. *Biomaterials*, **11**, 291–295.

Bryant, S. J., Cuy, J. L., Hauch, K. D., & Ratner, B. D. (2007). Photo-patterning of porous hydrogels for tissue engineering. *Biomaterials*, **28**, 2978–2986.

Busby, W., Cameron, N. R., & Jahoda, C. A. (2001). Emulsion-derived foams (PolyHIPEs) containing poly(epsilon-caprolactone) as matrixes for tissue engineering. *Biomacromolecules*, **2**, 154–164.

Busby, W., Cameron, N. R., & Jahoda, C. A.B. (2002). Tissue engineering matrixes by emulsion templating. *Polym. Int.*, **51**, 871–881.

Cameron, N. R. (2005). High internal phase emulsion templating as a route to well-defined porous structures. *Polymer*, **46**, 1439–1449.

Cao, T., Ho, K. H., & Teoh, S. H. (2003). Scaffold design and *in vitro* study of osteochondral coculture in a three-dimensional porous polycaprolactone scaffold fabricated by fused deposition modeling. *Tissue Eng.*, 9(Suppl 1), S103–112.

Chen, G., Tanaka, J., & Tateishi, T. (2006). Osteochondral tissue engineering using a PLGA–collagen hybrid mesh. *Mater. Sci. Eng. C Biomim. Mater. Sens. Syst.*, **26**, 124–129.

Chen, V. J., & Ma, P. X. (2004). Nano-fibrous poly (L-lactic acid) scaffolds with interconnected spherical macropores. *Biomaterials*, **25**, 2065–2073.

Cheng, H., Jiang, W., Phillips, F. M., Haydon, R. C., Peng, Y., Zhou, L., Luu, H. H., An, N., Breyer, B., Vanichakarn, P., Szatkowski, J. P., Park, J. Y., & He, T. C. (2003). Osteogenic activity of the fourteen types of human bone morphogenetic proteins (BMPs). *J. Bone Joint Surg. Am.*, **85-A**, 1544–1552.

Christenson, E. M., Soofi, W., Holm, J. L., Cameron, N. R., & Mikos, A. G. (2007). Biodegradable fumarate-based polyHIPEs as tissue engineering scaffolds. *Biomacromolecules*, **8**, 3806–3814.

Cima, L. G., Vacanti, J. P., Vacanti, C., Ingber, D., Mooney, D., & Langer, R. (1991). Tissue engineering by cell transplantation using degradable polymer substrates. *J. Biomech. Eng.*, **113**, 143–151.

Cohen, D. L., Malone, E., Lipson, H., & Bonassar, L. J. (2006). Direct freeform fabrication of seeded hydrogels in arbitrary geometries. *Tissue Eng.*, **12**, 1325–1335.

Cook, A. D., Hrkach, J. S., Gao, N. N., Johnson, I. M., Pajvani, U. B., Cannizzaro, S. M., & Langer, R. (1997). Characterization and development of RGD-peptide-modified poly(lactic acid-co-lysine) as an interactive, resorbable biomaterial. *J. Biomed. Mater. Res.*, **35**, 513–523.

Cooke, M. N., Fisher, J. P., Dean, D., Rimnac, C., & Mikos, A. G. (2003). Use of stereolithography to manufacture critical-sized 3D biodegradable scaffolds for bone ingrowth. *J. Biomed. Mater. Res. Part B Appl. Biomater.*, **64**, 65–69.

Crump, S. S. (1992). Apparatus and method for creating three-dimensional objects. *US Patent 5121329*.

Cukierman, E., Pankov, R., Stevens, D. R., & Yamada, K. M. (2001). Taking cell–matrix adhesions to the third dimension. *Science*, **294**, 1708–1712.

Datta, N., Pham, Q. P., Sharma, U., Sikavitsas, V. I., Jansen, J. A., & Mikos, A. G. (2006). *In vitro* generated extracellular matrix and fluid shear stress synergistically enhance 3D osteoblastic differentiation. *Proc. Natl. Acad. Sci. USA*, **103**, 2488–2493.

Deng, M., Nair, L. S., Nukavarapu, S. P., Kumbar, S. G., Brown, J. L., Krogman, N. R., Weikel, A. L., Allcock, H. R., & Laurencin, C. T. (2010). Biomimetic, bioactive etheric polyphosphazene-poly(lactide-co-glycolide) blends for bone tissue engineering. *J. Biomed. Mater. Res. A.*, **92**(1), 114–125.

Detamore, M. S., & Athanasiou, K. A. (2005). Use of a rotating bioreactor toward tissue engineering the temporomandibular joint disc. *Tissue Eng.*, **11**, 1188–1197.

Dhariwala, B., Hunt, E., & Boland, T. (2004). Rapid prototyping of tissue-engineering constructs, using photopolymerizable hydrogels and stereolithography. *Tissue Eng.*, **10**, 1316–1322.

Discher, D. E., Janmey, P., & Wang, Y. L. (2005). Tissue cells feel and respond to the stiffness of their substrate. *Science*, **310**, 1139–1143.

Du, D., Furukawa, K. S., & Ushida, T. (2009). 3D culture of osteoblast-like cells by unidirectional or oscillatory flow for bone tissue engineering. *Biotechnol. Bioeng.*, **102**, 1670–1678.

Duncan, R. L., & Turner, C. H. (1995). Mechanotransduction and the functional response of bone to mechanical strain. *Calcif. Tissue Int.*, **57**, 344–358.

Evans, G. R.D., Brandt, K., Katz, S., Chauvin, P., Otto, L., Bogle, M., Wang, B., Meszlenyi, R. K., Lu, L., & Mikos, A. G. (2002). Bioactive poly(L-lactic acid) conduits seeded with Schwann cells for peripheral nerve regeneration. *Biomaterials*, **23**, 841–848.

Fan, V. H., Tamama, K., Au, A., Littrell, R., Richardson, L. B., Wright, J. W., Wells, A., & Griffith, L. G. (2007). Tethered epidermal growth factor provides a survival advantage to mesenchymal stem cells. *Stem Cells*, **25**, 1241–1251.

Ferguson, S., Wahl, D., & Gogolewski, S. (1996). Enhancement of the mechanical properties of polylactides by solid-state extrusion. II. Poly(L-lactide), poly(L/D-lactide), and poly(L/DL-lactide. *J. Biomed. Mater. Res.*, **30**, 543–551.

Fisher, J. P., Dean, D., & Mikos, A. G. (2002a). Photocrosslinking characteristics and mechanical properties of diethyl fumarate/poly(propylene fumarate) biomaterials. *Biomaterials*, **23**, 4333–4343.

Fisher, J. P., Vehof, J. W., Dean, D., van der Waerden, J. P., Holland, T. A., Mikos, A. G., & Jansen, J. A. (2002b). Soft and hard tissue response to photocross-linked poly(propylene fumarate) scaffolds in a rabbit model. *J. Biomed. Mater. Res.*, **59**, 547–556.

Fisher, J. P., Holland, T. A., Dean, D., & Mikos, A. G. (2003). Photoinitiated cross-linking of the biodegradable polyester poly(propylene fumarate). Part II. *In vitro* degradation. *Biomacromolecules*, **4**, 1335–1342.

Freed, L. E., & Vunjak-Novakovic, G. (1997). Microgravity tissue engineering. *In Vitro Cell. Dev. Biol. Anim.*, **33**, 381–385.

Freed, L. E., Vunjak-Novakovic, G., Biron, R. J., Eagles, D. B., Lesnoy, D. C., Barlow, S. K., & Langer, R. (1994). Biodegradable polymer scaffolds for tissue engineering. *Nat. Biotechnol.*, **12**, 689–693.

Freed, L. E., Langer, R., Martin, I., Pellis, N. R., & Vunjak-Novakovic, G. (1997). Tissue engineering of cartilage in space. *Proc. Natl. Acad. Sci. USA*, **94**, 13885–13890.

Gao, J., Niklason, L., & Langer, R. (1998). Surface hydrolysis of poly(glycolic acid) meshes increases the seeding density of vascular smooth muscle cells. *J. Biomed. Mater. Res.*, **42**, 417–424.

Ge, Z., Tian, X., Heng, B. C., Fan, V., Yeo, J. F., & Cao, T. (2009). Histological evaluation of osteogenesis of 3D-printed poly-lactic-co-glycolic acid (PLGA) scaffolds in a rabbit model. *Biomed. Mater.*, **4**, 21001.

Geiger, F., Bertram, H., Berger, I., Lorenz, H., Wall, O., Eckhardt, C., Simank, H. G., & Richter, W. (2005). Vascular endothelial growth factor gene-activated matrix (VEGF165-GAM) enhances osteogenesis and angiogenesis in large segmental bone defects. *J. Bone Miner. Res.*, **20**, 2028–2035.

Gilbert, J. C., Takada, T., Stein, J. E., Langer, R., & Vacanti, J. P. (1993). Cell transplantation of genetically altered cells on biodegradable polymer scaffolds in syngeneic rats. *Transplantation*, **56**, 423–427.

Ginty, P. J., Howard, D., Rose, F. R., Whitaker, M. J., Barry, J. J., Tighe, P., Mutch, S. R., Serhatkulu, G., Oreffo, R. O., Howdle, S. M., & Shakesheff, K. M. (2006). Mammalian cell survival and processing in supercritical $CO_{(2)}$. *Proc. Natl. Acad. Sci. USA*, **103**, 7426–7431.

Godbey, W. T., Hindy, S. B., Sherman, M. E., & Atala, A. (2004). A novel use of centrifugal force for cell seeding into porous scaffolds. *Biomaterials*, **25**, 2799–2805.

Goldstein, A. S., Juarez, T. M., Helmke, C. D., Gustin, M. C., & Mikos, A. G. (2001). Effect of convection on osteoblastic cell growth and function in biodegradable polymer foam scaffolds. *Biomaterials*, **22**, 1279–1288.

Goodman, S., & Aspenberg, P. (1993). Effects of mechanical stimulation on the differentiation of hard tissues. *Biomaterials*, **14**, 563–569.

Grayson, W. L., Bhumiratana, S., Cannizzaro, C., Chao, P. H., Lennon, D. P., Caplan, A. I., & Vunjak-Novakovic, G. (2008). Effects of initial seeding density and fluid perfusion rate on formation of tissue-engineered bone. *Tissue Eng. Part A*, **14**, 1809–1820.

Griffith, L. G., Wu, B., Cima, M. J., Powers, M. J., Chaignaud, B., & Vacanti, J. P. (1997). *In vitro* organogenesis of liver tissue. *Ann. N. Y. Acad. Sci.*, **831**, 382–397.

Gunatillake, P. A., & Adhikari, R. (2003). Biodegradable synthetic polymers for tissue engineering. *Eur. Cell Mater.*, **5**, 1–16. Discussion 16.

Guo, X., Park, H., Liu, G., Liu, W., Cao, Y., Tabata, Y., Kasper, F. K., & Mikos, A. G. (2009). *In vitro* generation of an osteochondral construct using injectable hydrogel composites encapsulating rabbit marrow mesenchymal stem cells. *Biomaterials*, **30**, 2741–2752.

Gurdon, J. B., & Bourillot, P. Y. (2001). Morphogen gradient interpretation. *Nature*, **413**, 797–803.

Harris, L. D., Kim, B. S., & Mooney, D. J. (1998). Open pore biodegradable matrices formed with gas foaming. *J. Biomed. Mater. Res.*, **42**, 396–402.

Hayashi, R., Yamato, M., Takayanagi, H., Oie, Y., Kubota, A., Hori, Y., Okano, T., & Nishida, K. (2010). Validation system of tissue engineered epithelial cell sheets for corneal regenerative medicine. *Tissue Eng. Part C Methods*, **16**(4), 553–560.

He, S., Yaszemski, M. J., Yasko, A. W., Engel, P. S., & Mikos, A. G. (2000). Injectable biodegradable polymer composites based on poly(propylene fumarate) cross-linked with poly(ethylene glycol)-dimethacrylate. *Biomaterials*, **21**, 2389–2394.

Hedberg, E. L., Tang, A., Crowther, R. S., Carney, D. H., & Mikos, A. G. (2002). Controlled release of an osteogenic peptide from injectable biodegradable polymeric composites. *J. Control. Release*, **84**, 137–150.

Hedberg, E. L., Kroese-Deutman, H. C., Shih, C. K., Lemoine, J. J., Liebschner, M. A., Miller, M. J., Yasko, A. W., Crowther, R. S., Carney, D. H., Mikos, A. G., & Jansen, J. A. (2005). Methods: A comparative analysis of radiography, microcomputed tomography, and histology for bone tissue engineering. *Tissue Eng.*, **11**, 1356–1367.

Hersel, U., Dahmen, C., & Kessler, H. (2003). RGD modified polymers: Biomaterials for stimulated cell adhesion and beyond. *Biomaterials*, **24**, 4385–4415.

Hoffman, A. S. (2002). Hydrogels for biomedical applications. *Adv. Drug. Deliv. Rev.*, **54**, 3–12.

Holland, T. A., & Mikos, A. G. (2003). Advances in drug delivery for articular cartilage. *J. Control. Release*, **86**, 1–14.

Holland, T. A., Bodde, E. W., Baggett, L. S., Tabata, Y., Mikos, A. G., & Jansen, J. A. (2005). Osteochondral repair in the rabbit model utilizing bilayered, degradable oligo(poly(ethylene glycol) fumarate) hydrogel scaffolds. *J. Biomed. Mater. Res. A.*, **75**, 156–167.

Holland, T. A., Bodde, E. W., Cuijpers, V. M., Baggett, L. S., Tabata, Y., Mikos, A. G., & Jansen, J. A. (2007). Degradable hydrogel scaffolds for *in vivo* delivery of single and dual growth factors in cartilage repair. *Osteoarthr. Cartilage*, **15**, 187–197.

Holtorf, H. L., Jansen, J. A., & Mikos, A. G. (2005). Flow perfusion culture induces the osteoblastic differentiation of marrow stroma cell-scaffold constructs in the absence of dexamethasone. *J. Biomed. Mater. Res. A.*, **72**, 326–334.

Hou, Q., Grijpma, D. W., & Feijen, J. (2003). Porous polymeric structures for tissue engineering prepared by a coagulation, compression moulding and salt leaching technique. *Biomaterials*, **24**, 1937–1947.

Hou, Q., Bank, P., & Shakesheff, K. (2004). Injectable scaffolds for tissue regeneration. *J. Mater. Chem.*, **14**, 1915–1923.

Howdle, S., Watson, M., Whitaker, M., Davies, M., Shakesheff, K., Popov, V., Mandel, F., & Wang, J. (2001). Supercritical fluid mixing: Preparation of thermally sensitive polymer composites containing bioactive materials. *Chem. Commun.*, **2001**, 109–110.

Howe, A., Aplin, A. E., Alahari, S. K., & Juliano, R. L. (1998). Integrin signaling and cell growth control. *Curr. Opin. Cell. Biol.*, **10**, 220–231.

Hsu, Y. Y., Gresser, J. D., Trantolo, D. J., Lyons, C. M., Gangadharam, P. R., & Wise, D. L. (1997). Effect of polymer foam morphology and density on kinetics of *in vitro* controlled release of isoniazid from compressed foam matrices. *J. Biomed. Mater. Res.*, **35**, 107–116.

Hu, Y., Grainger, D. W., Winn, S. R., & Hollinger, J. O. (2002). Fabrication of poly(alpha-hydroxy acid) foam scaffolds using multiple solvent systems. *J. Biomed. Mater. Res.*, **59**, 563–572.

Hubbell, J. A. (2007). Matrix-bound growth factors in tissue repair. *Swiss Med. Wkly.*, **137**(Suppl 155), 72S–76S.

Hughes, G. W., & Maffei, L. (1966). Retinal ganglion cell response to sinusoidal light stimulation. *J. Neurophysiol.*, **29**, 333–352.

Hutmacher, D. W. (2000). Scaffolds in tissue engineering bone and cartilage. *Biomaterials*, **21**, 2529–2543.

Ingber, D. E. (2006). Mechanical control of tissue morphogenesis during embryological development. *Int. J. Dev. Biol.*, **50**, 255–266.

Ishaug, S. L., Crane, G. M., Miller, M. J., Yasko, A. W., Yaszemski, M. J., & Mikos, A. G. (1997). Bone formation by three-dimensional stromal osteoblast culture in biodegradable polymer scaffolds. *J. Biomed. Mater. Res.*, **36**, 17–28.

Ishaug-Riley, S. L., Crane, G. M., Gurlek, A., Miller, M. J., Yasko, A. W., Yaszemski, M. J., & Mikos, A. G. (1997). Ectopic bone formation by marrow stromal osteoblast transplantation using poly(DL-lactic-co-glycolic acid) foams implanted into the rat mesentery. *J. Biomed. Mater. Res.*, **36**, 1–8.

Ishaug-Riley, S. L., Crane-Kruger, G. M., Yaszemski, M. J., & Mikos, A. G. (1998). Three-dimensional culture of rat calvarial osteoblasts in porous biodegradable polymers. *Biomaterials*, **19**, 1405–1412.

Jain, A., Kim, Y. T., McKeon, R. J., & Bellamkonda, R. V. (2006). *In situ* gelling hydrogels for conformal repair of spinal cord defects, and local delivery of BDNF after spinal cord injury. *Biomaterials*, **27**, 497–504.

Jaklenec, A., Wan, E., Murray, M. E., & Mathiowitz, E. (2008). Novel scaffolds fabricated from protein-loaded microspheres for tissue engineering. *Biomaterials*, **29**, 185–192.

Jo, S., Shin, H., & Mikos, A. G. (2001). Modification of oligo(poly(ethylene glycol) fumarate) macromer with a GRGD peptide for the preparation of functionalized polymer networks. *Biomacromolecules*, **2**, 255–261.

Johnson, M. R., Lee, H. J., Bellamkonda, R. V., & Guldberg, R. E. (2009). Sustained release of BMP-2 in a lipid-based microtube vehicle. *Acta Biomater.*, **5**, 23–28.

Jung, J., & Perrut, M. (2001). Particle design using supercritical fluids: Literature and patent survey. *J. Supercrit. Fluids*, **20**, 179–219.

Karageorgiou, V., & Kaplan, D. (2005). Porosity of 3D biomaterial scaffolds and osteogenesis. *Biomaterials*, **26**, 5474–5491.

Kasper, F. K., & Mikos, A. G. (2004). Biomaterials and gene therapy. In N. Peppas, & M. V. Sefton (Eds.), *Molecular and Cellular Foundations of Biomaterials* (pp. 131–163). Academic Press.

Kasper, F. K., Kushibiki, T., Kimura, Y., Mikos, A. G., & Tabata, Y. (2005). *In vivo* release of plasmid DNA from composites of oligo (poly(ethylene glycol) fumarate) and cationized gelatin microspheres. *J. Control. Release*, **107**, 547–561.

Kasper, F. K., Young, S., Tanahashi, K., Barry, M. A., Tabata, Y., Jansen, J. A., & Mikos, A. G. (2006). Evaluation of bone regeneration by DNA release from composites of oligo(poly(ethylene glycol) fumarate) and cationized gelatin microspheres in a critical-sized calvarial defect. *J. Biomed. Mater. Res. A.*, **78**, 335–342.

Kim, B. S., Putnam, A. J., Kulik, T. J., & Mooney, D. J. (1998a). Optimizing seeding and culture methods to engineer smooth muscle tissue on biodegradable polymer matrices. *Biotechnol. Bioeng.*, **57**, 46–54.

Kim, S. S., Utsunomiya, H., Koski, J. A., Wu, B. M., Cima, M. J., Sohn, J., Mukai, K., Griffith, L. G., & Vacanti, J. P. (1998b). Survival and function of hepatocytes on a novel three-dimensional synthetic biodegradable polymer scaffold with an intrinsic network of channels. *Ann. Surg.*, **228**, 8–13.

Kraehenbuehl, T. P., Zammaretti, P., Van der Vlies, A. J., Schoenmakers, R. G., Lutolf, M. P., Jaconi, M. E., & Hubbell, J. A. (2008). Three-dimensional extracellular matrix-directed cardioprogenitor differentiation: Systematic modulation of a synthetic cell-responsive PEG-hydrogel. *Biomaterials*, **29**, 2757–2766.

Kretlow, J. D., Klouda, L., & Mikos, A. G. (2007). Injectable matrices and scaffolds for drug delivery in tissue engineering. *Adv. Drug. Deliv. Rev.*, **59**, 263–273.

Langer, R., & Vacanti, J. P. (1993). Tissue engineering. *Science*, **260**, 920–926.

Lee, H., Cusick, R. A., Browne, F., Ho Kim, T., Ma, P. X., Utsunomiya, H., Langer, R., & Vacanti, J. P. (2002). Local delivery of basic fibroblast growth factor increases both angiogenesis and engraftment of hepatocytes in tissue-engineered polymer devices. *Transplantation*, **73**, 1589–1593.

Lee, K. W., Wang, S., Lu, L., Jabbari, E., Currier, B. L., & Yaszemski, M. J. (2006). Fabrication and characterization of poly(propylene fumarate) scaffolds with controlled pore structures using 3-dimensional printing and injection molding. *Tissue Eng.*, **12**, 2801–2811.

Lee, W., Pinckney, J., Lee, V., Lee, J. H., Fischer, K., Polio, S., Park, J. K., & Yoo, S. S. (2009). Three-dimensional bioprinting of rat embryonic neural cells. *Neuroreport*, **20**, 798–803.

Li, D., Wang, Y., & Xia, Y. (2003). Electrospinning of polymeric and ceramic nanofibers as uniaxially aligned arrays. *Nano Lett.*, **3**, 1167–1172.

Li, S., Garreau, H., & Vert, M. (1990). Structure-property relationships in the case of the degradation of massive aliphatic poly-(-hydroxy acids) in aqueous media. *J. Mater. Sci. Mater. Med.*, **1**, 123–130.

Li, W. J., Laurencin, C. T., Caterson, E. J., Tuan, R. S., & Ko, F. K. (2002). Electrospun nanofibrous structure: A novel scaffold for tissue engineering. *J. Biomed. Mater. Res.*, **60**, 613–621.

Lissant, K. J. (Ed.), (1974). *Emulsions and Emulsion Technology Part 1*. New York: Marcel Dekker Inc.

Lo, H., Ponticiello, M. S., & Leong, K. W. (1995). Fabrication of controlled release biodegradable foams by phase separation. *Tissue Eng.*, **1**, 15–28.

Lo, C. M., Wang, H. B., Dembo, M., & Wang, Y. L. (2000). Cell movement is guided by the rigidity of the substrate. *Biophys. J.*, **79**, 144–152.

Lovett, M., Lee, K., Edwards, A., & Kaplan, D. L. (2009). Vascularization strategies for tissue engineering. *Tissue Eng. Part B Rev.*, **15**, 353–370.

Lu, Y., Mapili, G., Suhali, G., Chen, S., & Roy, K. (2006). A digital micro-mirror device-based system for the microfabrication of complex, spatially patterned tissue engineering scaffolds. *J. Biomed. Mater. Res. A.*, **77**, 396–405.

Ma, P. X., & Zhang, R. (2001). Microtubular architecture of biodegradable polymer scaffolds. *J. Biomed. Mater. Res.*, **56**, 469–477.

Macdonald, K. K., Cheung, C. Y., & Anseth, K. S. (2007). Cellular delivery of TGFbeta1 promotes osteoinductive signalling for bone regeneration. *J. Tissue Eng. Regen. Med.*, **1**, 314–317.

Marcantonio, N. A., Boehm, C. A., Rozic, R. J., Au, A., Wells, A., Muschler, G. F., & Griffith, L. G. (2009). The influence of tethered epidermal growth factor on connective tissue progenitor colony formation. *Biomaterials*, **30**, 4629–4638.

Marshall, A., Barker, T., Sage, E., Hauch, K., & Ratner, B. (2004a). Pore size controls angiogenesis in subcutaneously implanted porous matrices. In *Transactions of the Seventh World Biomaterials Congress*. Sydney, Australia, pp. 710.

Marshall, A. J., Irvin, C. A., Barker, T., Sage, E. H., Hauch, K. D., & Ratner, B. D. (2004b). Biomaterials with tightly controlled pore size that promote vascular in-growth. *Polym. Preprints*, **45**, 100–101.

Martina, M., & Hutmacher, D. (2007). Biodegradable polymers applied in tissue engineering research: A review. *Polym. Int.*, **56**, 145–157.

Meilander, N. J., Pasumarthy, M. K., Kowalczyk, T. H., Cooper, M. J., & Bellamkonda, R. V. (2003). Sustained release of plasmid DNA using lipid microtubules and agarose hydrogel. *J. Control. Release*, **88**, 321–331.

Meinel, L., Karageorgiou, V., Fajardo, R., Snyder, B., Shinde-Patil, V., Zichner, L., Kaplan, D., Langer, R., & Vunjak-Novakovic, G. (2004). Bone tissue engineering using human mesenchymal stem cells: Effects of scaffold material and medium flow. *Ann. Biomed. Eng.*, **32**, 112–122.

Mikos, A. G., Bao, Y., Cima, L. G., Ingber, D. E., Vacanti, J. P., & Langer, R. (1993a). Preparation of poly(glycolic acid) bonded fiber structures for cell attachment and transplantation. *J. Biomed. Mater. Res.*, **27**, 183–189.

Mikos, A. G., Sarakinos, G., Leite, S. M., Vacanti, J. P., & Langer, R. (1993b). Laminated three-dimensional biodegradable foams for use in tissue engineering. *Biomaterials*, **14**, 323–330.

Mikos, A. G., Sarakinos, G., Lyman, M. D., Ingber, D. E., Vacanti, J. P., & Langer, R. (1993c). Prevascularization of porous biodegradable polymers. *Biotechnol. Bioeng.*, **42**, 716–723.

Mikos, A. G., Lyman, M. D., Freed, L. E., & Langer, R. (1994a). Wetting of poly(L-lactic acid) and poly(DL-lactic-co-glycolic acid) foams for tissue culture. *Biomaterials*, **15**, 55–58.

Mikos, A. G., Thorsen, A. J., Czerwonka, L. A., Bao, Y., Langer, R., Winslow, D. N., & Vacanti, J. P. (1994b). Preparation and characterization of poly(L-lactic acid) foams. *Polymer*, **35**, 1068–1077.

Mironov, V., Kasyanov, V., Drake, C., & Markwald, R. R. (2008). Organ printing: Promises and challenges. *Regen. Med.*, **3**, 93–103.

Mironov, V., Visconti, R. P., Kasyanov, V., Forgacs, G., Drake, C. J., & Markwald, R. R. (2009). Organ printing: Tissue spheroids as building blocks. *Biomaterials*, **30**, 2164–2174.

Moon, S., Hasan, S. K., Song, Y. S., Xu, F., Keles, H. O., Manzur, F., Mikkilineni, S., Hong, J. W., Nagatomi, J., Haeggstrom, E., Khademhosseini, A., & Demirci, U. (2010). Layer by layer 3d tissue epitaxy by cell laden hydrogel droplets. *Tissue Eng. Part C Methods*, **16**(1), 157–166.

Mooney, D. J., & Mikos, A. G. (1999). Growing new organs. *Sci. Am.*, **280**, 60–65.

Mooney, D., Hansen, L., Vacanti, J., Langer, R., Farmer, S., & Ingber, D. (1992). Switching from differentiation to growth in hepatocytes: Control by extracellular matrix. *J. Cell. Physiol.*, **151**, 497–505.

Mooney, D. J., Park, S., Kaufmann, P. M., Sano, K., McNamara, K., Vacanti, J. P., & Langer, R. (1995). Biodegradable sponges for hepatocyte transplantation. *J. Biomed. Mater. Res.*, **29**, 959–965.

Mooney, D. J., Baldwin, D. F., Suh, N. P., Vacanti, J. P., & Langer, R. (1996a). Novel approach to fabricate porous sponges of poly(D, L-lactic-co-glycolic acid) without the use of organic solvents. *Biomaterials*, **17**, 1417–1422.

Mooney, D. J., Mazzoni, C. L., Breuer, C., McNamara, K., Hern, D., Vacanti, J. P., & Langer, R. (1996b). Stabilized polyglycolic acid fibre-based tubes for tissue engineering. *Biomaterials*, **17**, 115–124.

Murphy, W. L., Peters, M. C., Kohn, D. H., & Mooney, D. J. (2000). Sustained release of vascular endothelial growth factor from mineralized poly(lactide-co-glycolide) scaffolds for tissue engineering. *Biomaterials*, **21**, 2521–2527.

Murugan, R., & Ramakrishna, S. (2007). Design strategies of tissue engineering scaffolds with controlled fiber orientation. *Tissue Eng.*, **13**, 1845–1866.

Nam, Y. S., & Park, T. G. (1999a). Porous biodegradable polymeric scaffolds prepared by thermally induced phase separation. *J. Biomed. Mater. Res.*, **47**, 8–17.

Nam, Y. S., & Park, T. G. (1999b). Biodegradable polymeric microcellular foams by modified thermally induced phase separation method. *Biomaterials*, **20**, 1783–1790.

Nam, Y. S., Yoon, J. J., & Park, T. G. (2000). A novel fabrication method of macroporous biodegradable polymer scaffolds using gas foaming salt as a porogen additive. *J. Biomed. Mater. Res.*, **53**, 1–7.

Nguyen, D. N., Raghavan, S. S., Tashima, L. M., Lin, E. C., Fredette, S. J., Langer, R. S., & Wang, C. (2008). Enhancement of poly(orthoester) microspheres for DNA vaccine delivery by blending with poly(ethylenimine). *Biomaterials*, **29**, 2783–2793.

Nukavarapu, S. P., Kumbar, S. G., Brown, J. L., Krogman, N. R., Weikel, A. L., Hindenlang, M. D., Nair, L. S., Allcock, H. R., & Laurencin, C. T. (2008). Polyphosphazene/nano-hydroxyapatite composite microsphere scaffolds for bone tissue engineering. *Biomacromolecules*, **9**, 1818–1825.

Okano, T., Yamada, N., Okuhara, M., Sakai, H., & Sakurai, Y. (1995). Mechanism of cell detachment from temperature-modulated, hydrophilic-hydrophobic polymer surfaces. *Biomaterials*, **16**, 297–303.

Panyam, J., & Labhasetwar, V. (2003). Biodegradable nanoparticles for drug and gene delivery to cells and tissue. *Adv. Drug Deliv. Rev.*, **55**, 329–347.

Patel, Z. S., Young, S., Tabata, Y., Jansen, J. A., Wong, M. E. K., & Mikos, A. G. (2008). Dual delivery of an angiogenic and an osteogenic growth factor for bone regeneration in a critical size defect model. *Bone*, **43**, 931–940.

Payne, R. G., McGonigle, J. S., Yaszemski, M. J., Yasko, A. W., & Mikos, A. G. (2002a). Development of an injectable, *in situ* cross-linkable, degradable polymeric carrier for osteogenic cell populations. Part 2. Viability of encapsulated marrow stromal osteoblasts cultured on cross-linking poly(propylene fumarate). *Biomaterials*, **23**, 4373–4380.

Payne, R. G., Yaszemski, M. J., Yasko, A. W., & Mikos, A. G. (2002b). Development of an injectable, *in situ* cross-linkable, degradable polymeric carrier for osteogenic cell populations. Part 1. Encapsulation of marrow stromal osteoblasts in surface cross-linked gelatin microparticles. *Biomaterials*, **23**, 4359–4371.

Peattie, R. A., Rieke, E. R., Hewett, E. M., Fisher, R. J., Shu, X. Z., & Prestwich, G. D. (2006). Dual growth factor-induced angiogenesis *in vivo* using hyaluronan hydrogel implants. *Biomaterials*, **27**, 1868–1875.

Peter, S. J., Nolley, J. A., Widmer, M. S., Merwin, J. E., Yaszemski, M. J., Yasko, A. W., Engel, P. S., & Mikos, A. G. (1997). *In vitro* degradation of a poly(propylene fumarate)/-tricalcium phosphate composite orthopaedic scaffold. *Tissue Eng.*, **3**, 207–215.

Peter, S. J., Miller, S. T., Zhu, G., Yasko, A. W., & Mikos, A. G. (1998). *In vivo* degradation of a poly(propylene fumarate)/beta-tricalcium phosphate injectable composite scaffold. *J. Biomed. Mater. Res.*, **41**, 1–7.

Peter, S. J., Kim, P., Yasko, A. W., Yaszemski, M. J., & Mikos, A. G. (1999). Cross-linking characteristics of an injectable poly(propylene fumarate)/beta-tricalcium phosphate paste and mechanical properties of the cross-linked composite for use as a biodegradable bone cement. *J. Biomed. Mater. Res.*, **44**, 314–321.

Pham, Q. P., Sharma, U., & Mikos, A. G. (2006a). Electrospinning of polymeric nanofibers for tissue engineering applications: A review. *Tissue Eng.*, **12**, 1197–1211.

Pham, Q. P., Sharma, U., & Mikos, A. G. (2006b). Electrospun poly(epsilon-caprolactone) microfiber and multilayer nanofiber/microfiber scaffolds: Characterization of scaffolds and measurement of cellular infiltration. *Biomacromolecules*, **7**, 2796–2805.

Platt, M. O., Roman, A. J., Wells, A., Lauffenburger, D. A., & Griffith, L. G. (2009). Sustained epidermal growth factor receptor levels and activation by tethered ligand binding enhances osteogenic differentiation of multi-potent marrow stromal cells. *J. Cell Physiol.*, **221**, 306–317.

Radisic, M., Park, H., Shing, H., Consi, T., Schoen, F. J., Langer, R., Freed, L. E., & Vunjak-Novakovic, G. (2004). Functional assembly of engineered myocardium by electrical stimulation of cardiac myocytes cultured on scaffolds. *Proc. Natl. Acad. Sci. USA*, **101**, 18129–18134.

Ratner, B. (2007). A paradigm shift: Biomaterials that heal. *Polym. Int.*, **56**, 1183–1185.

Richardson, T. P., Peters, M. C., Ennett, A. B., & Mooney, D. J. (2001). Polymeric system for dual growth factor delivery. *Nat. Biotechnol.*, **19**, 1029–1034.

Roh, J. D., Nelson, G. N., Udelsman, B. V., Brennan, M. P., Lockhart, B., Fong, P. M., Lopez-Soler, R. I., Saltzman, W. M., & Breuer, C. K. (2007). Centrifugal seeding increases seeding efficiency and cellular distribution of bone marrow stromal cells in porous biodegradable scaffolds. *Tissue Eng.*, **13**, 2743–2749.

Ruardy, T., Schakenraad, J. M., Van Der Mei, H. C., & Busscher, H. J. (1997). Preparation and characterization of chemical gradient surfaces and their application for the study of cellular interaction phenomena. *Surf. Sci. Rep.*, **29**, 3–30.

Ruel-Gariepy, E., & Leroux, J. C. (2004). *In situ*-forming hydrogels – review of temperature-sensitive systems. *Eur. J. Pharm. Biopharm.*, **58**, 409–426.

Sachs, E. M., Haggerty, J. S., Cima, M. J., & Williams, P. A. (1993). Three-dimensional printing techniques. *US Patent 5204055*.

Saraf, A., Lozier, G., Haesslein, A., Kasper, F. K., Raphael, R. M., Baggett, L. S., & Mikos, A. G. (2009). Fabrication of nonwoven coaxial fiber meshes by electrospinning. *Tissue Eng. Part C Methods*, **15**, 333–344.

Saraf, A., Baggett, L. S., Raphael, R. M., Kasper, F. K., & Mikos, A. G. (2010). Regulated non-viral gene delivery from coaxial electrospun fiber mesh scaffolds. *J. Control. Release*, **143**(1), 95–103.

Schek, R. M., Taboas, J. M., Segvich, S. J., Hollister, S. J., & Krebsbach, P. H. (2004). Engineered osteochondral grafts using biphasic composite solid free-form fabricated scaffolds. *Tissue Eng.*, **10**, 1376–1385.

Shastri, V. P., Martin, I., & Langer, R. (2000). Macroporous polymer foams by hydrocarbon templating. *Proc. Natl. Acad. Sci. USA*, **97**, 1970–1975.

Sheridan, M. H., Shea, L. D., Peters, M. C., & Mooney, D. J. (2000). Bioabsorbable polymer scaffolds for tissue engineering capable of sustained growth factor delivery. *J. Control. Release*, **64**, 91–102.

Shi, X., Hudson, J. L., Spicer, P. P., Tour, J. M., Krishnamoorti, R., & Mikos, A. G. (2005). Rheological behaviour and mechanical characterization of injectable poly(propylene fumarate)/single-walled carbon nanotube composites for bone tissue engineering. *Nanotechnology*, **16**, 531–538.

Shi, X., Hudson, J. L., Spicer, P. P., Tour, J. M., Krishnamoorti, R., & Mikos, A. G. (2006). Injectable nanocomposites of single-walled carbon nanotubes and biodegradable polymers for bone tissue engineering. *Biomacromolecules*, **7**, 2237–2242.

Shin, H., Quinten Ruhe, P., Mikos, A. G., & Jansen, J. A. (2003). *In vivo* bone and soft tissue response to injectable, biodegradable oligo(poly(ethylene glycol) fumarate) hydrogels. *Biomaterials*, **24**, 3201–3211.

Sikavitsas, V. I., Temenoff, J. S., & Mikos, A. G. (2001). Biomaterials and bone mechanotransduction. *Biomaterials*, **22**, 2581–2593.

Sikavitsas, V. I., Bancroft, G. N., & Mikos, A. G. (2002). Formation of three-dimensional cell/polymer constructs for bone tissue engineering in a spinner flask and a rotating wall vessel bioreactor. *J. Biomed. Mater. Res. A.*, **62**, 136–148.

Sikavitsas, V. I., Bancroft, G. N., Holtorf, H. L., Jansen, J. A., & Mikos, A. G. (2003). Mineralized matrix deposition by marrow stromal osteoblasts in 3D perfusion culture increases with increasing fluid shear forces. *Proc. Natl. Acad. Sci. USA*, **100**, 14683–14688.

Simpson, R. L., Wiria, F. E., Amis, A. A., Chua, C. K., Leong, K. F., Hansen, U. N., Chandrasekaran, M., & Lee, M. W. (2008). Development of a 95/5 poly(L-lactide-co-glycolide)/hydroxylapatite and beta-tricalcium phosphate scaffold as bone replacement material via selective laser sintering. *J. Biomed. Mater. Res. B Appl. Biomater.*, **84**, 17–25.

Singh, M., Berkland, C., & Detamore, M. S. (2008a). Strategies and applications for incorporating physical and chemical signal gradients in tissue engineering. *Tissue Eng. Part B Rev.*, **14**, 341–366.

Singh, M., Morris, C. P., Ellis, R. J., Detamore, M. S., & Berkland, C. (2008b). Microsphere-based seamless scaffolds containing macroscopic gradients of encapsulated factors for tissue engineering. *Tissue Eng. Part C Methods*, **14**, 299–309.

Singh, M., Sandhu, B., Scurto, A., Berkland, C., & Detamore, M. S. (2010). Microsphere-based scaffolds for cartilage tissue engineering: Using subcritical $CO_{(2)}$ as a sintering agent. *Acta Biomater.* **6**(1), 137–143.

Smith, C. M., Stone, A. L., Parkhill, R. L., Stewart, R. L., Simpkins, M. W., Kachurin, A. M., Warren, W. L., & Williams, S. K. (2004). Three-dimensional bioassembly tool for generating viable tissue-engineered constructs. *Tissue Eng.*, **10**, 1566–1576.

Smith, C. M., Christian, J. J., Warren, W. L., & Williams, S. K. (2007). Characterizing environmental factors that impact the viability of tissue-engineered constructs fabricated by a direct-write bioassembly tool. *Tissue Eng.*, **13**, 373–383.

Smith, M. J., McClure, M. J., Sell, S. A., Barnes, C. P., Walpoth, B. H., Simpson, D. G., & Bowlin, G. L. (2008). Suture-reinforced electrospun polydioxanone–elastin small-diameter tubes for use in vascular tissue engineering: A feasibility study. *Acta Biomater*, **4**(1), 58–66.

Spaans, C. J., Belgraver, V. W., Rienstra, O., de Groot, J. H., Veth, R. P., & Pennings, A. J. (2000). Solvent-free fabrication of micro-porous polyurethane amide and polyurethane-urea scaffolds for repair and replacement of the knee-joint meniscus. *Biomaterials*, **21**, 2453–2460.

Spalazzi, J. P., Dagher, E., Doty, S. B., Guo, X. E., Rodeo, S. A., & Lu, H. H. (2008). *In vivo* evaluation of a multiphased scaffold designed for orthopaedic interface tissue engineering and soft tissue-to-bone integration. *J. Biomed. Mater. Res. A*, **86**, 1–12.

Taboas, J. M., Maddox, R. D., Krebsbach, P. H., & Hollister, S. J. (2003). Indirect solid free form fabrication of local and global porous, biomimetic and composite 3D polymer-ceramic scaffolds. *Biomaterials*, **24**, 181–194.

Tan, K. H., Chua, C. K., Leong, K. F., Cheah, C. M., Cheang, P., Abu Bakar, M. S., & Cha, S. W. (2003). Scaffold development using selective laser sintering of polyetheretherketone-hydroxyapatite biocomposite blends. *Biomaterials*, **24**, 3115–3123.

Temenoff, J. S., Athanasiou, K. A., LeBaron, R. G., & Mikos, A. G. (2002). Effect of poly(ethylene glycol) molecular weight on tensile and swelling properties of oligo(poly(ethylene glycol) fumarate) hydrogels for cartilage tissue engineering. *J. Biomed. Mater. Res.*, **59**, 429–437.

Temenoff, J. S., Steinbis, E. S., & Mikos, A. G. (2003). Effect of drying history on swelling properties and cell attachment to oligo(poly(ethylene glycol) fumarate) hydrogels for guided tissue regeneration applications. *J. Biomater. Sci. Polym. Ed.*, **14**, 989–1004.

Thomson, R., Wake, M., Yaszemski, M., & Mikos, A. (1995a). Biodegradable polymer scaffolds to regenerate organs. *Adv. Polym. Sci.*, **122**, 245–274.

Thomson, R. C., Yaszemski, M. J., Powers, J. M., & Mikos, A. G. (1995b). Fabrication of biodegradable polymer scaffolds to engineer trabecular bone. *J. Biomater. Sci. Polym. Ed.*, **7**, 23–38.

Thomson, R. C., Yaszemski, M. J., Powers, J. M., & Mikos, A. G. (1998). Hydroxyapatite fiber reinforced poly(alpha-hydroxy ester) foams for bone regeneration. *Biomaterials*, **19**, 1935–1943.

Thomson, R. C., Mikos, A. G., Beahm, E., Lemon, J. C., Satterfield, W. C., Aufdemorte, T. B., & Miller, M. J. (1999). Guided tissue fabrication from periosteum using preformed biodegradable polymer scaffolds. *Biomaterials*, **20**, 2007–2018.

Timmer, M. D., Ambrose, C. G., & Mikos, A. G. (2003a). *In vitro* degradation of polymeric networks of poly(propylene fumarate) and the cross-linking macromer poly(propylene fumarate)-diacrylate. *Biomaterials*, **24**, 571–577.

Timmer, M. D., Ambrose, C. G., & Mikos, A. G. (2003b). Evaluation of thermal- and photo-cross-linked biodegradable poly(propylene fumarate)-based networks. *J. Biomed. Mater. Res. A.*, **66**, 811–818.

Vacanti, C. A., Langer, R., Schloo, B., & Vacanti, J. P. (1991). Synthetic polymers seeded with chondrocytes provide a template for new cartilage formation. *Plast. Reconstr. Surg.*, **88**, 753–759.

van den Dolder, J., Bancroft, G. N., Sikavitsas, V. I., Spauwen, P. H.M., Jansen, J. A., & Mikos, A. G. (2003). Flow perfusion culture of marrow stromal osteoblasts in titanium fiber mesh. *J. Biomed. Mater. Res. A.*, **64**, 235–241.

Van Tomme, S. R., Storm, G., & Hennink, W. E. (2008). *In situ* gelling hydrogels for pharmaceutical and biomedical applications. *Int. J. Pharm.*, **355**, 1–18.

van Wachem, P. B., Stronck, J. W., Koers-Zuideveld, R., Dijk, F., & Wildevuur, C. R. (1990). Vacuum cell seeding: A new method for the fast application of an evenly distributed cell layer on porous vascular grafts. *Biomaterials*, **11**, 602–606.

Vunjak-Novakovic, G. (2003). The fundamentals of tissue engineering: Scaffolds and bioreactors. *Novartis Found. Symp.*, **249**, 34–46. Discussion 46–51, 170–174, 239–141.

Vunjak-Novakovic, G., Obradovic, B., Martin, I., & Freed, L. E. (2002). Bioreactor studies of native and tissue engineered cartilage. *Biorheology*, **39**, 259–268.

Wake, M. C., Gupta, P. K., & Mikos, A. G. (1996). Fabrication of pliable biodegradable polymer foams to engineer soft tissues. *Cell Transplant.*, 5, 465–473.

Wang, C., Ge, Q., Ting, D., Nguyen, D., Shen, H. R., Chen, J., Eisen, H. N., Heller, J., Langer, R., & Putnam, D. (2004). Molecularly engineered poly(ortho ester) microspheres for enhanced delivery of DNA vaccines. *Nat. Mater.*, **3**, 190–196.

Wang, X., Yan, Y., Pan, Y., Xiong, Z., Liu, H., Cheng, J., Liu, F., Lin, F., Wu, R., Zhang, R., & Lu, Q. (2006). Generation of three-dimensional hepatocyte/gelatin structures with rapid prototyping system. *Tissue Eng.*, **12**, 83–90.

Weber, L. M., & Anseth, K. S. (2008). Hydrogel encapsulation environments functionalized with extracellular matrix interactions increase islet insulin secretion. *Matrix Biol.*, **27**, 667–673.

Weber, L. M., Cheung, C. Y., & Anseth, K. S. (2008). Multifunctional pancreatic islet encapsulation barriers achieved via multilayer PEG hydrogels. *Cell Transplant.*, **16**, 1049–1057.

Whang, K., Thomas, C. H., Healy, K. E., & Nuber, G. (1995). A novel method to fabricate bioabsorbable scaffolds. *Polymer*, **36**, 837–842.

Widmer, M. S., Gupta, P. K., Lu, L., Meszlenyi, R. K., Evans, G. R., Brandt, K., Savel, T., Gurlek, A., Patrick, C. W., Jr., & Mikos, A. G. (1998). Manufacture of porous biodegradable polymer conduits by an extrusion process for guided tissue regeneration. *Biomaterials*, **19**, 1945–1955.

Williams, J. M., Adewunmi, A., Schek, R. M., Flanagan, C. L., Krebsbach, P. H., Feinberg, S. E., Hollister, S. J., & Das, S. (2005). Bone tissue engineering using polycaprolactone scaffolds fabricated via selective laser sintering. *Biomaterials*, **26**, 4817–4827.

Wilson, W. C., Jr., & Boland, T. (2003). Cell and organ printing 1: Protein and cell printers. *Anat. Rec. A Discov. Mol. Cell Evol. Biol.*, **272**, 491–496.

Wintermantel, E., Mayer, J., Blum, J., Eckert, K. L., Lüscher, P., & Mathey, M. (1996). Tissue engineering scaffolds using superstructures. *Biomaterials*, **17**, 83–91.

Wiria, F. E., Leong, K. F., Chua, C. K., & Liu, Y. (2007). Poly-epsilon-caprolactone/hydroxyapatite for tissue engineering scaffold fabrication via selective laser sintering. *Acta Biomater.*, 3, 1–12.

Wu, B. M., Borland, S. W., Giordano, R. A., Cima, L. G., Sachs, E. M., & Cima, M. J. (1996). Solid free-form fabrication of drug delivery devices. *J. Control. Release*, **40**, 77–87.

Xu, T., Jin, J., Gregory, C., Hickman, J. J., & Boland, T. (2005). Inkjet printing of viable mammalian cells. *Biomaterials*, **26**, 93–99.

Yamada, K., Kamihira, M., Hamamoto, R., & Iijima, S. (1998). Efficient induction of hepatocyte spheroids in a suspension culture using a water-soluble synthetic polymer as an artificial matrix. *J. Biochem.*, **123**, 1017–1023.

Yamato, M., Utsumi, M., Kushida, A., Konno, C., Kikuchi, A., & Okano, T. (2001). Thermo-responsive culture dishes allow the intact harvest of multilayered keratinocyte sheets without dispase by reducing temperature. *Tissue Eng.*, 7, 473–480.

Yang, F., Murugan, R., Ramakrishna, S., Wang, X., Ma, Y. X., & Wang, S. (2004). Fabrication of nano-structured porous PLLA scaffold intended for nerve tissue engineering. *Biomaterials*, **25**, 1891–1900.

Yang, F., Murugan, R., Wang, S., & Ramakrishna, S. (2005). Electrospinning of nano/micro scale poly(L-lactic acid) aligned fibers and their potential in neural tissue engineering. *Biomaterials*, **26**, 2603–2610.

Zmora, S., Glicklis, R., & Cohen, S. (2002). Tailoring the pore architecture in 3-D alginate scaffolds by controlling the freezing regime during fabrication. *Biomaterials*, **23**, 4087–4094.

CHAPTER II.6.4 CELL SOURCES FOR TISSUE ENGINEERING: MESENCHYMAL STEM CELLS

Arnold I. Caplan

Professor of Biology, Director Skeletal Research Center, Case Western Reserve University, Cleveland, OH, USA

INTRODUCTION

Highly specialized, differentiated cells fabricate tissues and further differentiate to maintain these tissues. These specialized cells are derived from progenitor cells that are available to replace and regenerate the cells which expire or become injured. For example, during embryonic development, osteogenic progenitor cells give rise to osteoblasts in a multi-step, lineage sequence (Bruder and Caplan, 1990a). The osteoblasts fabricate osteoid that becomes calcified into bone; the osteoblasts organize as a monolayer sheet of electrically and functionally connected cells that are driven by vasculature at their backs to produce oriented sheets of collagen (osteoid) from their front surfaces (Caplan and Pechak, 1987). Some few osteoblasts further differentiate into osteocytes as they become encased in bone, and function to maintain bone (Bruder and Caplan, 1990b). Osteoblasts have 8–10 day half-lives in humans and the expired cells are replaced by cells derived from resident progenitors. Thus, to tissue engineer bone, the delivery of osteoblasts or their progenitors in a suitable delivery vehicle or scaffold would be required (Ohgushi and Caplan, 1999). Osteoblasts or their progenitors have been isolated from bone or from bone marrow, and this experience forms the basis for the logics involved in skeletal tissue engineering.

MESENCHYMAL STEM CELLS

The osteogenic potential of whole bone marrow has been known since the days of Roman or Greek medicine. More recently Friedenstein, Owen and their colleagues showed that osteoprogenitor cells could be isolated from marrow (Friedenstein et al., 1974; Friedenstein, 1980; Owen and Friedenstein, 1988). In the late 1980s, my colleagues and I formalized a protocol for isolating a class of progenitors from adult human and animal marrow that we called Mesenchymal Stem Cells, MSCs (Arias et al., 1991; Majumdar et al., 1998; Wakitani et al., 1995; Yoo et al., 1998; Young et al., 1998).

These marrow-derived MSCs are capable of differentiation into bone (Arias et al., 1991), cartilage (Yoo et al., 1998), muscle (Wakitani et al., 1995), marrow stroma (Majumdar et al., 1998), tendon/ligament (Young et al., 1998), fat and other connective tissue, as shown in Figure II.6.4.1. These MSCs could be culture expanded and their progeny shown to form cartilage, bone, tendon or fat in various preclinical animal models or in *in vitro* or *in vivo* test situations (Arias et al., 1991; Dennis et al., 1992; Majumdar et al., 1998; Wakitani et al., 1995, 1998; Yoo et al., 1998; Young et al., 1998; Solchaga et al., 1999). Indeed, MSCs could be attached to a scaffold and induced into a differentiation pathway in culture, and subsequently implanted into tissue defects in order to "jump-start" the cells to fabricate a specific tissue, such as bone (Kotobuki et al., 2004) or cartilage (Solchaga et al., 2006), in the *in vitro* setting.

It is now clear that conditions for the *in vitro* expansion and differentiation of human marrow-derived MSCs have been optimized (Lennon et al., 1969). Several laboratories have published details of such optimization, and a few have used autologous human MSCs in tissue-engineered implants in humans to regenerate bone (Augello et al., 2007) and human autologous chondrocytes for osteochondral defects in the knee (Brittberg et al., 1994; Dell'Accio et al., 2001; Pavesio et al., 2003; Saris et al., 2008). In these clinical circumstances, many millions of MSCs or chondrocytes have been obtained from culture expansion of adherent cells from marrow or cartilage biopsies. Importantly, the titers of MSCs in marrow sharply drops off as a function of age, as seen in the crude colony-forming assays depicted in Figure II.6.4.2.

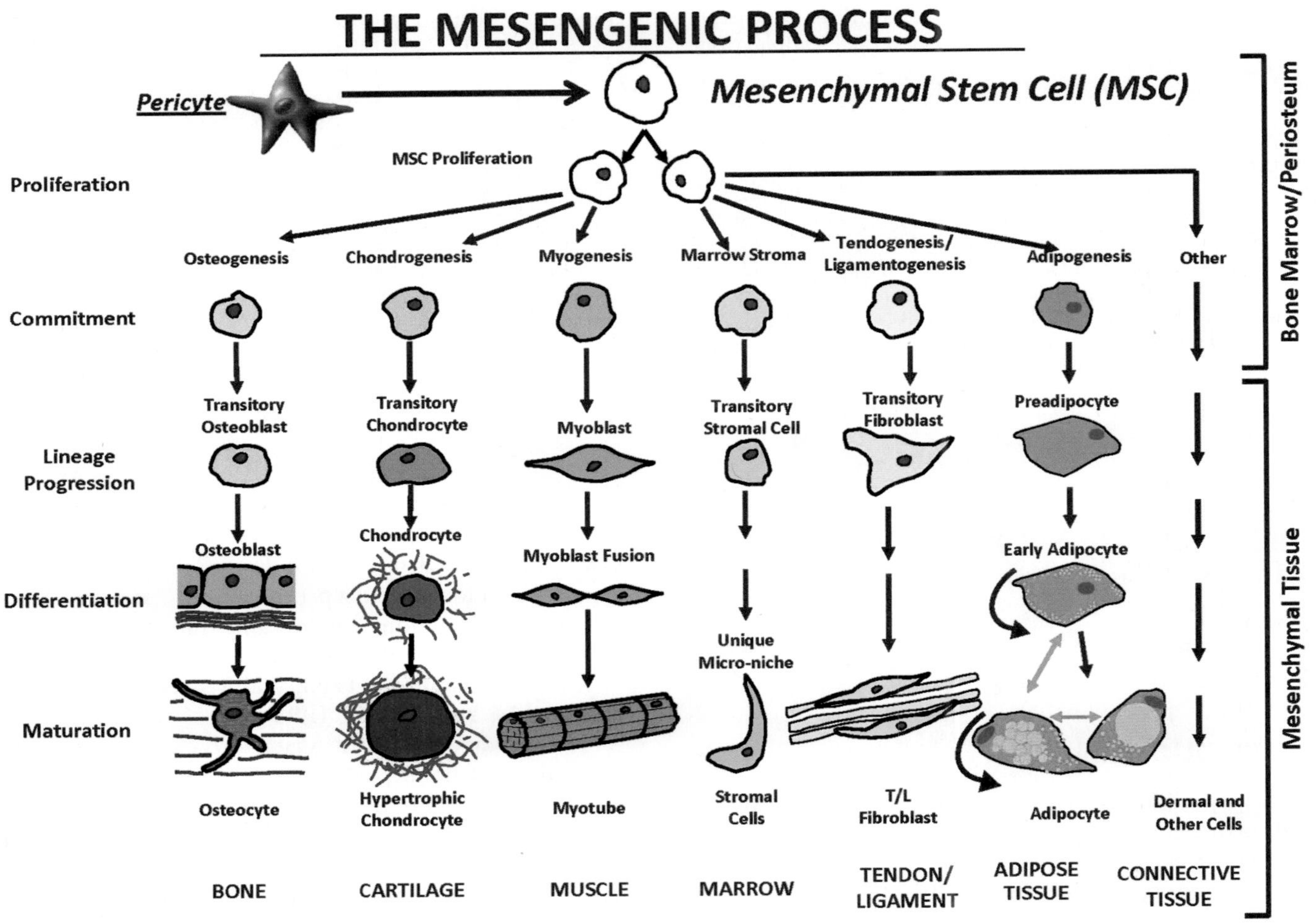

THE MESENGENIC PROCESS
Pericyte
Mesenchymal Stem Cell (MSC)
MSC Proliferation
Proliferation
Osteogenesis
Chondrogenesis
Myogenesis
Marrow Stroma
Tendogenesis/
Ligamentogenesis
Adipogenesis
Other
Commitment
Transitory
Osteoblast
Transitory
Chondrocyte
Myoblast
Transitory
Stromal Cell
Transitory
Fibroblast
Preadipocyte
Lineage
Progression
Osteoblast
Chondrocyte
Myoblast Fusion
Early Adipocyte
Differentiation
Unique
Micro-niche
Maturation
Osteocyte
Hypertrophic
Chondrocyte
Myotube
Stromal
Cells
T/L
Fibroblast
Adipocyte
Dermal and
Other Cells
BONE
CARTILAGE
MUSCLE
MARROW
TENDON/
LIGAMENT
ADIPOSE
TISSUE
CONNECTIVE
TISSUE
Bone Marrow/Periosteum
Mesenchymal Tissue

FIGURE II.6.4.1 The Mesengenic Pathway of differentiation lineages for adult marrow-derived Mesenchymal Stem Cells (MSCs).

Human MSC Numbers Decline With Age:
Caplan A. Clinics in Plastic Surgery 1994
1/10,000
1/100,000
1/250,000
1/400,000
1/2,000,000
Estimates
obtained by
CFU-F assay.
MSCs per Marrow Cells
Newborn
Teen
30
50
80
Age (Years)

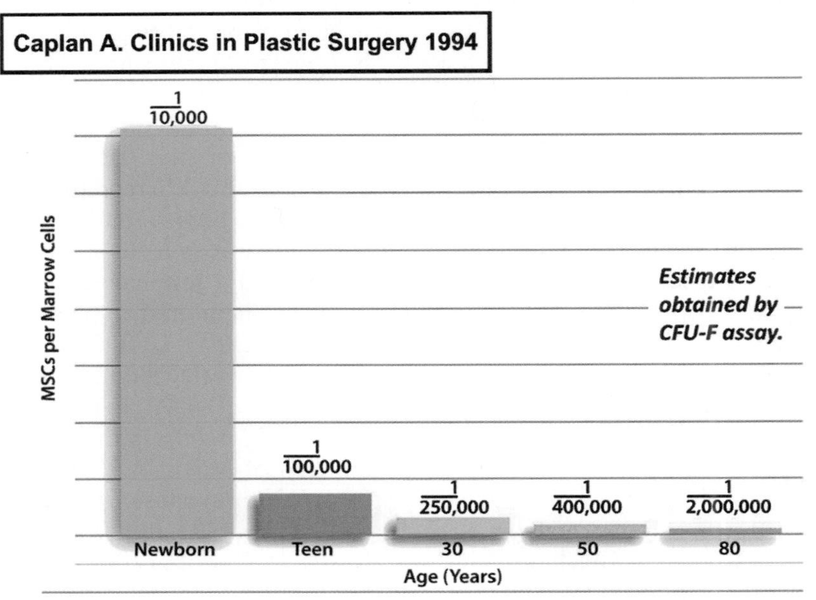

FIGURE II.6.4.2 Estimate of MSCs as CFU-f in whole marrow from individuals of different ages.

Although there are no precise quantitative assays for MSCs, the CFU-f assay has long been used as an estimate for progenitor cell titers from marrow (Friedenstein et al., 1974, 1980). Certainly, 5-, 50-, and 90-year-olds show dramatically different capacities to repair bone and cartilage, and this must be reflective of their MSC titers in the marrow.

MSC-NICHE

Recently, the earlier assertions that perivascular cells had osteochondrogenic potential (Brighton et al., 1992; Diaz-Flores et al., 1992) have been verified (Crisan et al., 2008) and stimulated me to suggest that "all" MSCs are pericytes (Caplan, 2008). The converse is not correct (i.e., some pericytes are not MSCs). Thus, Figure II.6.4.2 can be thought of as a plot of vascular density, and this decreased frequency of blood vessels would be expected to have a clear effect on the rate of healing of various tissues.

OTHER SOURCES OF MSCs

Several reports now clearly document that multipotent, mesenchymal progenitors can be isolated and expanded in culture from muscle, fat, skin, liver, brain, etc. (Gronthos et al., 2001; Nuk et al., 2001; Campagnoli et al., 2004; Gotherstrom et al., 2005; Kimelman et al., 2007). Retrospectively, this makes good sense if "all" MSCs are pericytes, since all of these tissues have vasculature. In addition, it is now clear that differentiated mesenchymal cells are plastic in their phenotypic characteristics (i.e., capable of transdifferentiation) (Herzog et al., 2003). Thus, we can take a pure population of human adipocytes and induce them with dexamethasone to become osteoblasts. This plasticity infers that the transdifferentiation pathways depicted in Figure II.6.4.3 are the result of transcriptive accessibility of promoters for the genes prominent in the newly-differentiated phenotypes. Conversely, the promoters for non-mesenchymal macromolecules are not accessible, unless other inductive or derepressive agents are introduced into the cells. The exact mechanism and fundamental details for mesenchymal plasticity have yet to be fully described. In addition, the basic unresolved issue is whether such *in vitro* plasticity has relevance to *in vivo* events. Indeed, the relative ease of such transdifferentiation challenges the use of such *in vitro* assays as definitions of MSCs, since a culture of human adipocytes and a culture of MSCs would both exhibit osteogenic progression when exposed to dexamethasone and ascorbate.

Multipotent mesenchymal progenitor cells have been isolated from a number of vascularized tissues. Indeed, a recent publication documented that if cell surface markers are used to sort out pericytes from several different

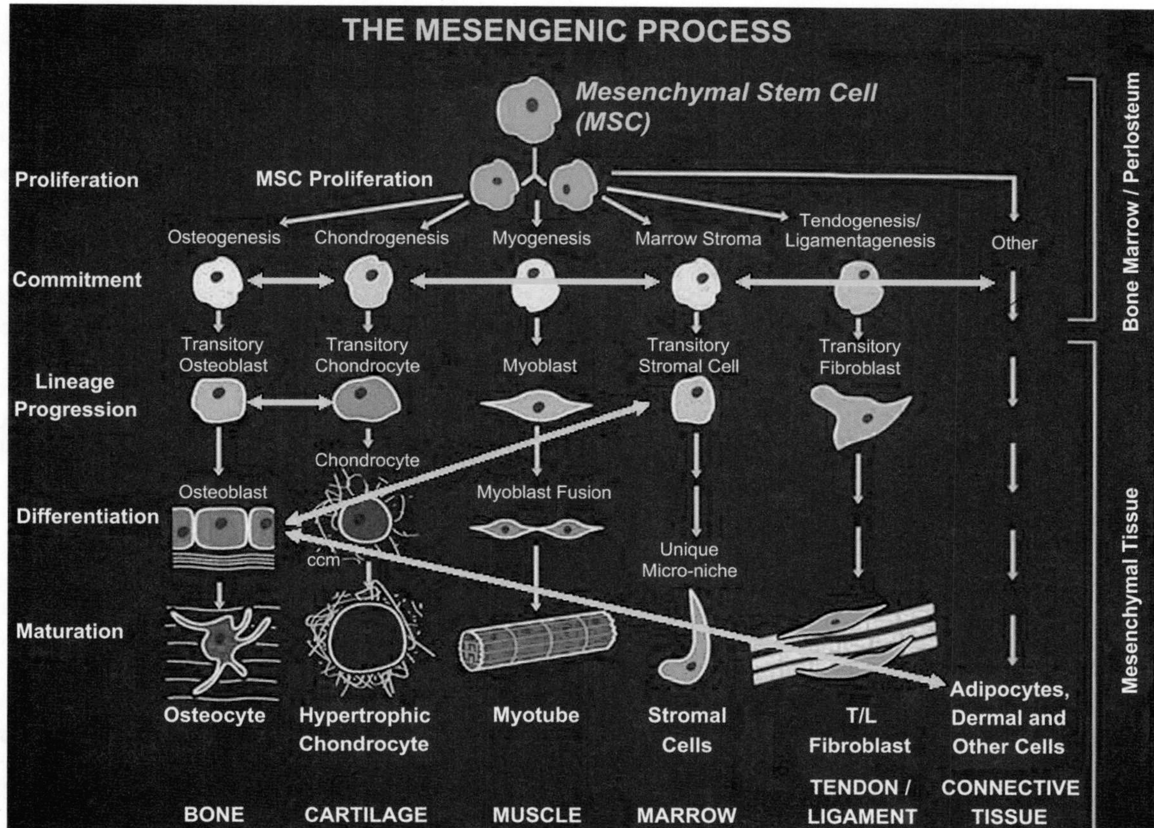

FIGURE II.6.4.3 The original Mesengenic Pathway diagram with non-vertical lines depicting the plasticity of MSCs and their descendents.

tissues; these cells also have MSC markers and are multipotent for several mesenchymal phenotypes (Crisan et al., 2008). Likewise, when cells are sorted for MSC markers, the sorted cells exhibit pericyte markers. We have recently shown that when labeled marrow-derived MSCs are added to cultures in which human vascular endothelial cells are forming capillary-like tubes in a human dermal fibroblast-produced extracellular matrix, the labeled human MSCs take up pericyte locations around the endothelial tubes (Sorrell et al., 2008). These same MSC preparations were multipotent in both *in vivo* and *in vitro* assays.

It is important to stress that the *in vitro* and *in vivo* conditions that are optimal for marrow-derived MSCs are not optimal for MSCs isolated from other tissues. Thus, the fetal calf serum used in the culture expansion of the adherent MSCs must be screened for each tissue source used to obtain those MSCs (Lennon et al., 1969). Furthermore, when optimally expanded MSCs from marrow and, for example, fat are tested for multi-potency, the test conditions must be separately optimized for each tissue-specific MSC. For example, as we have shown when human marrow MSCs are pelleted and the growth medium replaced with a defined medium containing TGF-β, the cells will differentiate into chondrocytes (Yoo et al., 1998). For this to occur for human fat-derived MSCs, BMP-6 must be added, in addition to TGF-β (Estes et al., 2006). This observation may be interpreted to show that both marrow- and fat-derived MSCs are chondrogenic, but that they are quite different in their response to TGF-β alone. Likewise, human marrow MSCs form osteogenic cells in culture when exposed to dexamethasone, while mouse marrow MSCs form adipocytes under these culture dexamethasone-supplemented conditions. Both mouse and human MSCs are multipotent, but optimal conditions must be separately used to isolate, expand, and assay them when derived from various different tissue sources. It follows that different scaffolds may prove to be necessary for MSCs from different tissue sources, and for different anatomic implantation sites.

It is important to stress that mature cell types can be used for tissue engineering applications. As mentioned above, autologous chondrocytes are culture expanded and used in specific scaffolds or in special surgical applications to repair both chondral and osteochondral defects (Brittberg et al., 1994; Dell'Accio et al., 2001; Pavesio et al., 2003; Saris et al., 2008). Likewise, endothelial cells can be isolated from autologous fat to be used in tissue-engineered vascular grafts (Arts et al., 2001). Last, both autologous and allogeneic tissue-engineered skin equivalents have been used in human clinical situations such as for burns or skin ulcers for more than 20 years (Supp and Boyce, 2002; Nakagawa et al., 2005). In all of these cases, optimal scaffolds and culture conditions for cell expansion have been developed for each cell type, and for each medical circumstance.

A NEW ERA OF MSC-BASED THERAPIES

Although not directly germane to materials science and tissue engineering, it is important to note that new information is available about the functioning of human MSCs that has resulted in their use for various clinical conditions as systematically introduced allogeneic MSCs (da Silva Meirelles et al., 2008). In brief, MSCs intrinsically secrete large amounts of bioactive molecules (Hyanesworth et al., 1996) that are immune-modulatory (Maitra et al., 2004; Beyth et al., 2005), are trophic (Caplan and Dennis, 2006), and support local tissue regeneration. The immuno-modulatory functions locally to inhibit T-cells, and to a lesser extent B-cells. Thus, allo-MSCs are not "seen" by the host's immune system. Indeed, human marrow-derived MSCs have dramatic effects in modulating the immune systems in rodents (Mahmood et al., 2003; Bai et al., 2007) and other animals. Clinically, rather spectacular clinical improvements of graft-versus-host-disease and Crohn's disease have been reported (see website: Osiris Therapeutics, Inc., www.osiris.com). Whether allo-MSCs can be used in tissue engineering constructs is yet to be thoroughly tested.

Allo-MSCs also have powerful trophic effects (Caplan and Dennis, 2006). These effects have been documented in animal models, and in some clinical trials of acute myocardial infarct, stroke, inflammatory bowel disease, acute kidney failure, tendonitis, spinal cord injury, and meniscus regeneration (Mahmood et al., 2003; Murphy et al., 2003; Maitra et al., 2004; Pittenger and Martin, 2004; Beyth et al., 2005; Lange et al., 2005; Caplan and Dennis, 2006; Bai et al., 2007; Brooke et al., 2007; Harman et al., 2007). These trophic effects involve four discrete activities: (1) anti-apoptosis especially in areas of ischemia; (2) anti-scarring; (3) angiogenesis; and (4) mitosis of tissue-intrinsic progenitors. In all cases, the allo-MSCs were delivered systemically, and they seem to home to sites of tissue injury. These newly discovered capacities of marrow MSCs have stimulated a new era of cell-based therapies. The use of the immuno-modulatory and trophic capacities of MSCs from other tissue sources is untested. From clinical trials (Brooke et al., 2007 and www.osiris.com), it is clear that systemically delivered marrow-derived human allo-MSCs will come to the healthcare marketplace long before tissue-engineered products.

LONG-TERM GOALS

Tissue engineering and cell-based therapies can be used to provide tissues that the body cannot naturally provide in cases of massive injury, disease, old age or surgical intervention. These technologies will allow us to better understand the complex dynamics of tissue repair and regeneration to overcome medically relevant deficiencies. Long-term, however, our goal must be to determine how to better manage the body's own intrinsic repair and regeneration potential so that less intrusive, more natural, protocols can be developed. The clearest example of this logic is distraction osteogenesis (Aronson, 1994; Rhee and Buchman, 2003) where broken or cut bones can be lengthened by mechanically manipulating the repair callus spanning the bone-break site. Although crude and time-inefficient, distraction osteogenesis manages the natural regenerative capacity to bring about a medically relevant solution. Other logics like this, including the targeting of systemically delivered MSCs, will change the course of healthcare delivery.

SYNOPSIS

The use of mature cell types or their progenitors to engineer tissue repair and/or tissue replacement strategies requires that adequate sources of cells be identified. Additionally, these cells are often expanded in number in plate cultures or bioreactors prior to the implantation of a cell-scaffold composite. Adult mesenchymal stem cells (MSCs) are the putative progenitors for a number of different skeletal and mesenchymal tissues. It appears that most or all MSCs reside as perivascular cells (i.e., pericytes) surrounding various blood vessels. Best-studied are the MSCs from bone marrow, but several other sources have been described, including fat and muscle. The optimization of isolation and *in vitro* expansion and assay parameters are essential for each tissue source, and for each species that serve as the source and recipient for cell-based therapies. New information about the intrinsic functioning of MSCs has changed the emphasis of their clinical delivery and utility.

ACKNOWLEDGMENTS

I thank my colleagues at the Skeletal Research Center for their hard work and support. Partially supported from funds from NIH and the David and Virginia Baldwin Fund.

BIBLIOGRAPHY

Arias, J. L., Fernandez, M. S., Laraia, V. J., Jr., Janicki, J., Heuer, A. H., et al. (1991). The avian eggshell as a model of biomineralization. In P. C. Rieke (Ed.), *Materials Synthesis Utilizing Biological Processes* (pp. 193–199). Pittsburgh, PA: Materials Research Society Symposium Proceedings.

Aronson, J. (1994). Experimental and clinical experience with distraction osteogenesis. *Cleft Palate Craniofac. J.*, **31**(6), 473–482.

Arts, C. H., Heijnen-Snyder, G. J., Joosten, P. P., Verhagen, H. J., Eikelboom, B. C., et al. (2001). A novel method for isolating pure microvascular endothelial cells from subcutaneous fat tissue ideal for direct cell seeding. *Lab. Invest.*, **81**, 1461–1465.

Augello, A., Tasso, R., Negrini, S. M., Cancedda, R., & Pennesi, G. (2007). Cell therapy using allogeneic bone marrow mesenchymal stem cells prevents tissue damage in collagen-induced arthritis. *Arthritis Rheum.*, **56**(4), 1175–1186.

Bai, L., Caplan, A. I., Lennon, D. P., & Miller, R. H. (2007). Human mesenchymal stem cells signals regulate neural stem cell fate. *Neurochem. Res.*, **32**, 353–362.

Beyth, S., Borovsky, Z., Mevorach, D., Liebergall, M., Bazit, Z., et al. (2005). Human mesenchymal stem cells alter antigen-presenting cell maturation and induce T cell unresponsiveness. *Blood*, **105**(5), 2214–2219.

Brighton, C. T., Lorich, D. G., Kupcha, R., Reilly, T. M., Jones, A. R., et al. (1992). The pericyte as a possible osteoblast progenitor cell. *Clin. Orthop. Relat. Res.*, **275**, 287–299.

Brittberg, M., Lindahl, A., Nilsson, A., Ohlsson, C., Isaksson, O., et al. (1994). Treatment of deep cartilage defects in the knee with autologous chondrocyte transplantation. *N. Engl. J. Med.*, **331**, 889–895.

Brooke, G., Cook, M., Blair, C., Han, R., Heazlewood, C., et al. (2007). Therapeutic applications of mesenchymal stromal cells. *Semin. Cell Dev. Biol.*, **18**, 846–858.

Bruder, S. P., & Caplan, A. I. (1990a). Osteogenic cell lineage analysis is facilitated by organ culture of embryonic chick periosteum. *Dev. Biol.*, **141**, 319–329.

Bruder, S. P., & Caplan, A. I. (1990b). Terminal differentiation of osteogenic cells in the embryonic chick tibia is revealed by a monoclonal antibody against osteocytes. *Bone*, **11**, 189–198.

Campagnoli, C., Roberts, I. A., Kumar, S., Bennett, P. R., Bellantuono, I., et al. (2004). Identification of mesenchymal stem/progenitor cells in human first-trimester fetal blood, liver, and bone marrow. *Arthritis Rheum.*, **50**, 817–827.

Caplan, A., (1994). The Mesengenic Process. *Clinics in Plastic Surgery*, **21**, 429–435.

Caplan, A. I. (2008). All MSCs are pericytes? *Cell Stem Cell*, **3**, 229–230.

Caplan, A. I., & Dennis, J. E. (2006). Mesenchymal stem cells as trophic mediators. *J. Cell Biochem.*, **98**, 1076–1084.

Caplan, A. I., & Pechak, D. G. (1987). The Cellular and Molecular Embryology of Bone Formation. In W. A. Peck (Ed.), *Bone and Mineral Research* (Vol. 5, pp. 117–184). New York, NY: Elsevier.

Crisan, M., Yap, S., Casteilla, L., Chen, C., Corselli, M., et al. (2008). A perivascular origin for mesenchymal stem cells in multiple human organs. *Cell Stem Cell*, **3**, 301–313.

da Silva Meirelles, L., Caplan, A. I., & Nardi, N. B. (2008). In search of the *in vivo* identity of mesenchymal stem cells. *Stem Cells* (published online in *Stem Cells Express*).

Dell'Accio, F., De Bari, C., & Luyten, F. P. (2001). Molecular markers predictive of the capacity of expanded human articular chondrocytes to form stable cartilage *in vivo*. *Arthritis Rheum.*, **44**, 1608–1619.

Dennis, J. E., Hyanesworth, S. E., Young, R. G., & Caplan, A. I. (1992). Osteogenesis in marrow-derived mesenchymal cell porous ceramic composites transplanted subcutaneously: Effect of fibronectin and laminin on cell retention and rate of osteogenic expression. *Cell Transpl.*, **1**, 23–32.

Diaz-Flores, L., Guitierrez, R., Lopez-Alonso, A., Gonzalez, R., & Varela, H. (1992). Pericytes as a supplementary source of osteoblasts in periosteal osteogenesis. *Clin. Orthop. Relat. Res.*, **275**, 280–286.

Estes, B. T., Wu, A. W., & Guilak, F. (2006). Potent induction of chondrocytic differentiation of human adipose-derived adult stem cells by bone morphogenetic protein 6. *Arthritis Rheum.*, **54**(4), 1222–1232.

Friedenstein, A. J. (1980). Stromal mechanisms of bone marrow: Cloning *in vitro* and retransplantation *in vivo*. *Hamatol Bluttransfus*, **25**, 19–29.

Friedenstein, A. J., Chailakhyan, R. K., Latsinik, N. V., Panasyuk, A. F., & Keiliss-Borok, I. V. (1974). Stromal cells responsible for transferring the microenvironment of the hemopoietic tissues. Cloning *in vitro* and retransplantation *in vivo*. *Transplantation*, **17**, 331–340.

Gotherstrom, C., West, A., Liden, J., Uzunel, M., Lahesmaa, R., et al. (2005). Difference in gene expression between human fetal liver and adult bone marrow mesenchymal stem cells. *Haematologica*, **90**, 1017–1026.

Gronthos, S., Franklin, D. M., Leddy, H. A., Robey, P. G., Storms, R. W., et al. (2001). Surface protein characterization of human adipose tissue-derived stromal cells. *J. Cell Physiol.*, **189**, 54–63.

Harman, R., Cowles, B., Orva, C., et al. (2007). A retrospective review of 52 cases of suspensory ligament injury in sport horses treated with adipose-derived stem and regenerative cell therapy. Presented at the Veterinary Orthopedic Society Convention Mar 3–10, 2007 in Sun Valley, ID.

Herzog, E. L., Chai, L., & Krause, D. S. (2003). Plasticity of marrow-derived stem cells. *Blood*, **102**, 3483–3493.

Hyanesworth, S. E., Baber, M. A., & Caplan, A. I. (1996). Cytokine expression by human marrow-derived mesenchymal progenitor cells *in vitro*: Effects of dexamethasone and IL-1α. *J. Cell Physiol.*, **166**, 585–592.

Kimelman, N., Pelled, G., Helm, G. A., Huard, J., Schwarz, E. M., et al. (2007). Gene- and stem cell-based therapeutics for bone regeneration and repair. *Tissue Eng.*, **13**, 1135–1150.

Kotobuki, N., Hirose, M., Funaoka, H., & Ohgushi, H. (2004). Enhancement of *in vitro* osteoblastic potential after selective sorting of osteoblasts with high alkaline phosphatase activity form human osteoblast-like cells. *Cell Transplant.*, **13**, 377–383.

Lange, C., Togel, F., Itirich, H., Clayton, F., Nolte-Ernsting, C., et al. (2005). Administered mesenchymal stem cells enhance recovery from ischemia/reperfusion-induced acute renal failure in rats. *Kidney Int.*, **68**, 1613–1617.

Lennon, D. P., Hyanesworth, S. E., Bruder, S. P., Jaiswall, N., & Caplan, A. I. (1969). Human and animal mesenchymal progenitor cells from bone marrow: Identification of serum for optimal selection and proliferation. *Vitro Cell Develop. Bio.*, **32**, 602–611.

Mahmood, A., Lu, D., & Chopp, M. (2003). Treatment of traumatic brain injury in adult rats with intravenous administration of human bone marrow stromal cells. *Neurosurgery*, **53**, 697–703.

Maitra, B., Szekely, E., Gjini, K., Laughlin, M. J., Dennis, J., et al. (2004). Human mesenchymal stem cells support unrelated donor hematopoietic stem cells suppress T-cell activation. *Bone Marrow Transpl.*, **33**, 597–604.

Majumdar, M. K., Thiede, M. A., Mosca, J. D., Moorman, M., & Gerson, S. L. (1998). Phenotypic and functional comparison of cultures of marrow-derived mesenchymal stem cells (MSCs) and stromal cells. *J. Cellular Physiol.*, **176**, 186–192.

Murphy, J. M., Fink, D. J., Hunziker, E. B., & Barry, F. P. (2003). Stem cell therapy in a caprine model of osteoarthritis. *Arthritis Rheum.*, **48**(12), 3464–3474.

Nakagawa, H., Akita, S., Fukui, M., Fujii, T., & Akino, K. (2005). Human mesenchymal stem cells successfully improve skin-substitute wound healing. *Br. J. Dermatol.*, **153**(1), 29–36.

Nuk, P. A., Zhu, M., Mizuno, H., Huang, J., Futrell, J. W., et al. (2001). Multilineage cells from human adipose tissue: Implications for cell-based therapies. *Tissue Eng.*, **7**, 211–228.

Ohgushi, H., & Caplan, A. I. (1999). Stem cell technology and bioceramics: From cell to gene. *Eng. J. Biomed. Mat. Res.*, **48**, 1–15.

Owen, M., & Friedenstein, A. J. (1988). Stromal stem cells: Marrow-derived osteogenic precursors. *Ciba Found Symp.*, **136**, 42–60.

Pavesio, A., Abatangelo, G., Borrione, A., Brocchetta, D., Hollander, A. P., et al. (2003). Hyaluronan-based scaffolds (Hyalograft® C) in the treatment of knee cartilage defects: Preliminary clinical findings. In Wiley (Ed.), *Tissue Engineering of Cartilage and Bone* (249, pp. 203–217). Novartis Foundation Symposium.

Pittenger, M. F., & Martin, B. J. (2004). Mesenchymal stem cells and their potential as cardiac therapeutics. *Cir. Res.*, **95**, 9–20.

Rhee, S. T., & Buchman, S. R. (2003). Pediatric mandibular distraction osteogenesis: The present and the future. *J. Craniofac. Surg.*, **14**, 803–808.

Saris, D. B.F., Vanlauwe, J., Victor, J., Haspl, M., Bohnsack, M., et al. (2008). Characterized chondrocyte implantation results in better structural repair when treating symptomatic cartilage defects of the knee in a randomized controlled trial versus microfracture. *Am. J. Sports Med.*, **36**(2), 235–246.

Solchaga, L. A., Dennis, J. E., Goldberg, V. M., & Caplan, A. I. (1999). Hyaluronic acid-based polymers as cell carriers for tissue engineered repair of bone and cartilage. *J. Orthop. Res.*, **17**, 205–213.

Solchaga, L. A., Tognana, E., Penick, K., Baskaran, H., Caplan, A. I., et al. (2006). A rapid vacuum-seeding technique for the assembly of large tissue-engineered cell/scaffold composites. *Tissue Eng.*, **12**(7), 1851–1863.

Sorrell, J. M., Baber, M. A., & Caplan, A. I. (2008). Influence of adult mesenchymal stem cells on *in vitro* vascular formation. *Tiss. Eng. Part A*, **15**(7), 1751–1761.

Supp, D. M., & Boyce, S. T. (2002). Overexpression of vascular endothelial growth factor accelerates early vascularization and improves healing of genetically modified cultured skin substitutes. *J. Burn Care Rehabil.*, **23**, 10–20.

Wakitani, S., Saito, T., & Caplan, A. I. (1995). Myogenic cells derived from rat bone marrow mesenchymal stem cells exposed to 5-azacytidine. *Muscle & Nerve*, **18**, 1417–1426.

Wakitani, S., Goto, T., Young, R. G., Mansour, J. M., Goldberg, V. M., et al. (1998). The repair of large full-thickness weight-bearing articular cartilage defects with allograft articular chondrocytes embedded in a collagen gel. *Tissue Eng.*, **4**, 429–442.

Yoo, J. U., Barthel, T. S., Nishimura, K., Solchaga, L. A., Caplan, A. I., et al. (1998). The chondrogenic potential of human bone-marrow-derived mesenchymal progenitor cells. *J. Bone and Joint Surg.*, **80**, 1745–1757.

Young, R. G., Butler, D. L., Weber, W., Gordon, S. L., Fink, D. J., et al. (1998). The use of mesenchymal stem cells in Achilles tendon repair. *J. Orthop. Res.*, **16**, 406–413.

CHAPTER II.6.5 MICROMECHANICAL DESIGN CRITERIA FOR TISSUE ENGINEERING BIOMATERIALS

Kaustabh Ghosh[1], *Charles K. Thodeti*[1], *and Donald E. Ingber*[1–3]

[1]Vascular Biology Program, Departments of Pathology and Surgery, Children's Hospital, Harvard Medical School, Boston, MA, USA

[2]Harvard School of Engineering and Applied Sciences

[3]Wyss Institute for Biologically Inspired Engineering at Harvard University, Boston, MA, USA

INTRODUCTION

The field of tissue engineering emerged from work striving to construct artificial tissues and organs *in vitro* by integrating the principles of cell biology, polymer chemistry, and materials science (Langer and Vacanti, 1993). In this approach, cells isolated from whole organs (e.g., liver, kidney) or multi-potential stem cells are cultured within three-dimensional scaffolding materials, which can be designed in myriad forms to provide the necessary structural and chemical cues to induce cell growth and tissue differentiation. This cell–scaffold construct is then implanted *in vivo* with the aim of promoting new tissue formation following injury or disease.

Skin was the first to be artificially constructed using natural scaffold materials in combination with patient-derived skin cells (Bell et al., 1981). Development of tissue-engineered skin is less challenging than constructing artificial internal organs, such as whole liver or kidney, because the relatively thin skin constructs can be implanted directly onto a pre-existing vascularized wound bed. In contrast, cells within larger three-dimensional cell–scaffold constructs from many solid organs (e.g., liver) die rapidly when implanted *in vivo* without prior vascularization, due to insufficient delivery of oxygen and nutrients. Interestingly, despite their simplicity, artificial skin grafts have not made a significant impact on the market. This is, in part, due to their failure to maintain optimal cell viability; however, the major difficulty has arisen from the tough regulatory and cost challenges involved in the development of medical products containing living cells. These limitations highlight the importance of exploring alternative materials-based tissue engineering approaches that rely on development of more sophisticated biomimetic scaffolds to promote angiogenesis, and induce recruitment and reprogramming of endogenous multipotent progenitor cells (e.g., either resident within remaining tissues or bone marrow-derived cells that circulate in the blood) when implanted *in vivo*. To do this effectively, however, we must fully-understand and recapitulate the dynamic and reciprocal interactions that occur within the cell–ECM microenvironment during normal tissue formation.

Biomaterials used in tissue engineering were initially viewed as inert carriers for cell delivery. However, modern strategies for regenerative medicine now employ methods to manipulate their biochemical and biophysical properties, in order to best promote orderly tissue growth and restore normal function. Materials that more closely mimic the inductive environment of normal developing tissues have been fabricated using natural ECM proteins (e.g., collagen, fibrin, hyaluronan, etc.), as well as synthetic polymers (PLGA, PEG, PVA, etc.) (Lutolf and Hubbell, 2005). To enhance cell adhesion, these polymeric scaffolds have also been covalently derivatized with proteins or peptides found within natural ECMs (Ghosh et al., 2006). The RGD (arginine-glycine-asparatic acid) tripeptide sequence is the most commonly used cell adhesion moiety (Hersel et al., 2003; Shu et al., 2004); however, other peptides such as the YIGSR and IKVAV peptides of laminin and VPGIG sequence of elastin have also been explored (Ranieri et al., 1995; Panitch et al., 1999; Lin et al., 2006). Importantly, these ECM-mimicking biomaterials support key cell functions, including cell spreading, migration, growth, and differentiation *in vitro*, as well as tissue repair *in vivo* (Shin et al., 2003; Lutolf and Hubbell, 2005).

To better mimic the fibrillar microarchitecture and anisotropy (alignment) of native ECMs, specialized fabrication techniques have been developed, such as electrospinning and three-dimensional weaving that produce fibrous biomaterial scaffolds with varying pore size and volume, ranging from macro- to nanoscale dimensions (Yang et al., 2001; Muschler et al., 2004; Tsang and Bhatia, 2007). These scaffolds may be rendered stable to hydrolytic or proteolytic degradation via inter- or intramolecular cross-linking, and cell recognition moieties can be chemically tethered to their polymer backbone to promote cell adhesion. To engineer mechanically active tissues that bear continuous cyclical hemodynamic and compressive loading (e.g., heart valve and cartilage, respectively), cell–biomaterial constructs are preconditioned prior to *in vivo* implantation by being cultured in specialized bioreactors that simulate the physical environment of native tissues (Waldman et al., 2003; McCulloch et al., 2004; Fehrenbacher et al., 2006; Isenberg et al., 2006). This mechanical force regimen significantly alters ECM structure, as well as cellular growth and function.

Despite efforts to mimic the properties of natural ECMs through various chemical and mechanical modifications, most tissue-engineered constructs fail to completely restore normal tissue form and function when implanted *in vivo*. Our inability to produce ideal tissue substitutes may be due to the failure to address the role of micromechanical forces in guiding cell and tissue development. These microscale forces result from the tension (contractility) generated in the cell cytoskeleton, which is exerted on the cells' adhesions to ECM and to neighboring cells. Cells sense these forces through transmembrane integrin and cadherin receptors that mediate cell–ECM anchorage and cell–cell adhesion, respectively (Chen et al., 2004). Mechanical stresses acting on these cell surface receptors are transmitted across the cell surface, and to internal cytoskeletal and nuclear scaffolds via specialized membrane adhesion complexes (e.g., focal adhesions, adherens junctions) (Wang et al., 2009). Mechanotransduction – the process by which mechanical signals are transduced into changes in cellular biochemistry and gene expression – can occur in these adhesion complexes or at many other locations along loadbearing structural elements in the cell where force induces molecular distortion (Ingber, 2006; Vogel and Sheetz, 2006). Importantly, these cell-generated forces influence various cell behaviors critical for developmental control, including growth, movement, differentiation, apoptosis, and stem cell fate switching (Chen et al., 1997; Engler et al., 2006; Ghosh et al., 2007).

In this chapter, we describe how micromechanical interactions between cells and ECM govern normal tissue morphogenesis and control cell fate switching. We also review various microscale design approaches to biomaterial fabrication that leverage this critical developmental paradigm. Finally, we explore how our growing knowledge of micromechanical control of cell function and biomaterials science can be integrated to develop improved biomimetic microenvironments for *in situ* tissue repair and regeneration.

MICROMECHANICAL CONTROL OF TISSUE FORM AND FUNCTION

Tissues and organs arise during embryogenesis from coordinated self assembly of large populations of cells that organize themselves into precise three-dimensional spatial patterns. Multicellular self-organization is facilitated by molecular self assembly of cell-derived proteins (e.g., collagen, laminin, fibronectin) into fibrous multimolecular ECM scaffolds that function as physical templates for orderly cell attachment and renewal (Vracko, 1974; Leivo, 1983). Cells attach to this anchoring ECM scaffold through transmembrane integrin receptors (Ruoslahti, 1991; Hynes, 1992), and they exert cytoskeleton-generated traction forces on these cell–ECM adhesions, as well as on their attachments to neighboring cells (Ingber, 1991; Chicurel et al., 1998a; Galbraith and Sheetz, 1998; Geiger et al., 2001; Ingber, 2003c). The resistance produced by cell–ECM and cell–cell linkages to cytoskeletal forces maintains the cell in a state of isometric tension or prestress that stabilizes cell and tissue form through a tensegrity mechanism (Ingber and Jamieson, 1985; Ingber, 1993, 2003b; Stamenovic and Ingber, 2009). The importance of ECM-dependent micromechanical forces in regulating tissue form is evident from observations of new epithelium formation in the early embryo and in specialized organs (e.g., kidney), which is always accompanied by concomitant formation of planar ECM scaffolds or "basement membranes," that promote consistent cell orientation and generation of epithelial form (Leivo, 1983; Yurchenco and Ruben, 1988).

ECM is also centrally involved in the morphogenetic patterning that occurs during embryonic organogenesis. For example, formation of lobular epithelial glands in salivary gland and lung, as well as branching capillary networks, appear to require establishment of local differentials in ECM remodeling (Bernfield and Banerjee, 1978; Moore et al., 2005). In general, regions of highest ECM turnover, which lead to basement membrane thinning, exhibit the highest rate of cell growth and tissue expansion, while cells in the neighboring areas (only several microns away) that experience increased ECM accumulation (due to low degradation relative to synthesis) remain quiescent. These microscale differentials in cell growth and ECM density drive the budding, folding, and branching of many epithelial and endothelial tissues (Ausprunk and Folkman, 1977; Bernfield and Banerjee, 1978).

These observations have formed the basis of a micromechanical model of tissue development (Figure II.6.5.1), which postulates that morphogenetic events such as epithelial budding and branching are controlled by local micromechanical variations that lead to spatial differentials in the cellular force balance (Ingber and Jamieson, 1985; Huang and Ingber, 1999). In this model, local enzymatic degradation of the prestressed (tensed) basement membrane causes it to stretch and thin in regions of high turnover, with concomitant distortion of the adherent epithelial cells. These stretched cells become more responsive to soluble mitogens and undergo proliferation (Singhvi et al., 1994; Chen et al., 1997; Folkman and Moscona, 1978) compared to their neighboring counterparts that remain quiescent in regions experiencing basement membrane accumulation. Reiteration of this micromechanical control scheme that couples ECM stretching and cell growth could then produce the complex fractal-like tissue patterns that characterize development of virtually all organs (Huang and Ingber, 1999; Moore et al., 2005). Stable tissue forms finally emerge when the stabilizing tissue prestress is restored by accumulating thickened ECMs and strengthened cell–cell adhesions that fully balance cell-generated tensional forces (Ingber and Jamieson, 1985; Hardin and Keller, 1988).

This micromechanical model of tissue morphogenesis is supported by the finding that changing the cell–ECM force balance within embryonic mouse lung rudiments alters branching morphogenesis (Figure II.6.5.1) (Moore et al., 2005). Specifically, suppressing cytoskeletal tension generation inhibits regional thinning of the basement membrane at the tips of expanding buds and interferes with epitheliogenesis, while increasing tension accelerates epithelial branching morphogenesis (Moore et al., 2005). These treatments also had similar stimulatory and inhibitory effects on formation of neighboring capillary networks, as well as on the growth of the entire organ. Importantly, this force-dependent spatial patterning was not mediated by alterations in the overall rate of

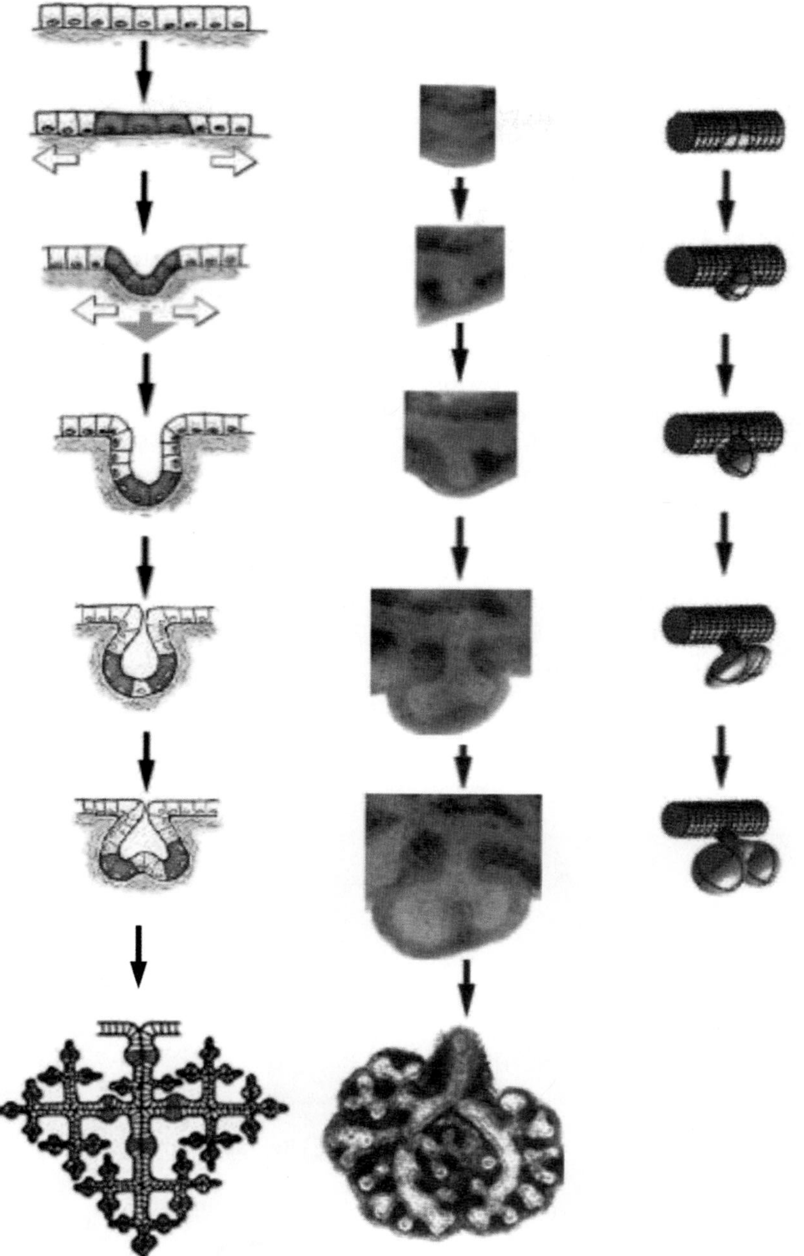

FIGURE II.6.5.1 Micromechanical control of tissue morphogenesis. (Left) Schematic diagrams of a micromechanical model of epithelial morphogenesis (Huang, S. & Ingber, D. (1999). Nat Cell Biol) showing progressive development of a simple planar epithelium into complex branching patterns (top to bottom). Local increases in basement membrane degradation (green), coupled with traction generated by overlying adherent cells (white arrows), result in local thinning and stretching of the basement membrane, as well as the overlying adherent epithelium (red), whereas neighboring cells only microns away remain quiescent (white). Cell distortion enhances cell sensitivity to soluble growth factors, which results in increased cell proliferation in regions of highest ECM turnover. Spreading and growing cells simultaneously deposit new ECM, which causes local basement extension and new bud formation. Reiteration of this process over time and space leads to the development of complex tissue architecture, with characteristic fractal-like patterns. (Center) Photomicrographs of individual epithelial buds in developing mouse lung epithelium corresponding to the stages shown at left (Moore, K. et al., (2005). Dev Dyn). (Right) Theoretical mechanical strain distributions within the basement membrane of the developing mouse lung epithelium at corresponding times during branching morphogenesis. Increased spacing between the strain field lines indicates regions where the basement membrane thins and experiences greatest mechanical strain (distortion). These regions of increased strain correlate precisely with regions of epithelial expansion and new bud formation (Moore, K. et al. (2005). Dev Dyn). (Reproduced from Ghosh, K. & Ingber, D., (2007). Adv. Drug Deliver. Rev. 59: 1306–1318, with permission from Elsevier ©2007).

cell proliferation within the mesenchyme and epithelium, but rather by establishment of local growth differentials within the forming gland (Moore et al., 2005). Thus, although tissue development is driven by soluble cytokines and changes in gene expression, it is also a highly mechanical process that is significantly influenced at the microscale by alterations in the mechanical force balance between cells and their surrounding adhesions to

ECM and neighboring cells. The tissue architecture generated by this mechanical patterning mechanism also feeds back to alter gradients of soluble developmental cues (Nelson et al., 2006), thereby establishing a vital feedback mechanism that further integrates changes in tissue form and function.

MICROSCALE DESIGN OF BIOMIMETIC SCAFFOLDS FOR TISSUE RECONSTRUCTION

Given the critical role of external physical cues and cell-generated forces in regulating tissue form and function, it is paramount that these factors be considered in the design of rational tissue engineering approaches aimed at restoring the normal structure and functionality of damaged tissues and organs. Tissue engineering scaffolds have been developed using various types of polymeric materials (natural, synthetic, and semi-synthetic) and techniques (gas foaming, phase separation, etc.) (Figure II.6.5.2A) (Agrawal and Ray, 2001; Lutolf and Hubbell, 2005). The design challenges focused on in the past lay largely in achieving appropriate macroscopic features, such as bulk stiffness and suitable degradation rate (to facilitate integration with host tissue), porosity (to promote nutrient exchange and tissue ingrowth), and adhesiveness (for cell adhesion and function). But more recent studies have shown that cell function within such bulk scaffolds can be more tightly regulated by introducing precise micro- or nanoscale structural features. For instance, microporous biopolymer scaffolds (pore size ~100–500 μm) induce cells to exhibit a flattened cell morphology similar to that seen when they are cultured on a two-dimensional tissue culture substrate, while nanofibrous scaffolds (Figure II.6.5.2B) (fiber diameter ~50–100 nm; pore size ~0.5–5 μm) cause these cells to assume a dendritic morphology similar to that seen in native tissues (Grinnell, 2003; Stevens and George, 2005; Ji et al., 2006). Further, creating nanoscale features within an otherwise microporous scaffold alone can markedly enhance cell spreading, growth, and expression of tissue-specific ECM components (Pattison et al., 2005). Such apparent differences in cell morphology likely arise from the differential ability of these topographies to induce distinct membrane protrusions (e.g., lamellipodia formation on flatter surfaces while filopodia form on more fibrous ones) (Cukierman et al., 2001; Curtis et al., 2004).

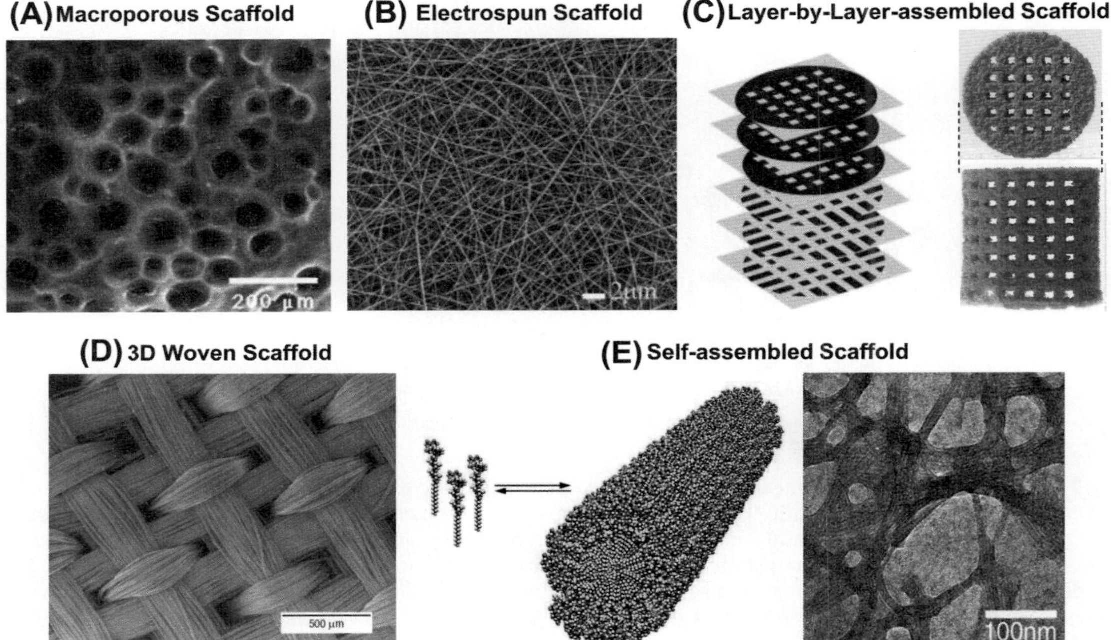

FIGURE II.6.5.2 Microscale design of biomimetic scaffolds for tissue reconstruction. Increasing knowledge of the influence of microscale ECM material properties on cell and tissue development has led to novel physical design strategies for tissue engineering biomaterials. (A) Initial attempts to engineer three-dimensional tissues used macroporous scaffolds. Their large surface area allows greater cell spreading, while the large pore size facilitates sufficient metabolic exchange (Eiselt, P. et al. (2000). Biomaterials 21: 1921–1927). (B) Electrospun scaffolds that better mimic the fibrous architecture of natural ECM are being increasingly used nowadays. Cells cultured on these scaffolds assume dendritic morphology similar to that exhibited by cells within native tissues (Ji, Y. et al. (2006). Biomaterials). (Reproduced with permission from Elsevier ©2006). (C) More sophisticated design strategies involving automated CAD/CAM programs with solid free-form fabrication approaches allow a bottom-up approach to scaffold construction, where the microscale material properties can be controlled in a precise spatiotemporal manner (Seitz, H. et al. (2005). J Biomed Mater Res). (D) Three-dimensional weaving technology has been recently used to produce fibrous scaffolds that mimic the anisotropy, porosity, and bulk mechanical properties of native ECM (Moutos, F. et al. (2007). Nat Mater). (E) Self-assembling molecules have been engineered that can spontaneously organize into complex three-dimensional structures in a manner reminiscent of the bottom-up molecular assembly of natural ECM materials (Hartgerink, J. et al. (2001). Science). (Reproduced with permission from AAAS ©2001.)

Because local variations in ECM structure and elasticity can trigger differential cell responses required for tissue morphogenesis and development (as seen during lung development), it would be desirable to engineer biomaterial scaffolds that exhibit such structural and mechanical variations across small length scales. One way to address this challenge is to use solid free-form fabrication (SFF) techniques (e.g., three-dimensional microprinting, soft lithography, etc.) that enable the development of biomaterials with complex three-dimensional microscale features using computer-assisted design and manufacturing (CAD/CAM) platforms (Figure II.6.5.2C). In this bottom-up approach, thin sequential layers of polymer are deposited one-at-a-time, with each layer having a prescribed physical structure and composition (Giordano et al., 1996; Seitz et al., 2005). By depositing polymer solutions at precise predetermined locations, and by varying the type and concentration of polymers used, the architecture and mechanical properties (stiffness, degradation rate, etc.) of the scaffold can be tightly controlled with microscale resolution. Such an approach can produce anisotropic scaffolds that better recapitulate the physical complexity of loadbearing tissues such as cartilage, which is composed of distinct tiers of cell and ECM organization. If so, this may lead to significant improvement in both scaffold integration at the defect site and tissue function. Such biomaterials have, in fact, been fabricated using a novel microscale three-dimensional weaving technology where the resultant porous scaffold closely resembles native cartilage, not only in terms of bulk (compressive, shear, and tensile) moduli, but also with respect to its physical anisotropy (tension-compression nonlinearity) (Figure II.6.5.2D) (Moutos et al., 2007). Specifically, small-pore scaffolds (pore dimensions 390 μm × 320 μm × 104 μm) produced by designing a biased woven structure that contained a higher fiber volume fraction in the weft direction than in the warp direction exhibited ~35% higher ultimate tensile stress when tested in the weft direction than in the warp direction, which is reminiscent of the mechanical anisotropy seen in native articular cartilage.

Native ECM is produced through complex self assembly reactions involving various natural biopolymers, such as collagen, elastin, and hyaluronic acid, among others. Due to this systematic bottom-up approach, the ECM displays a remarkable level of hierarchical organization that spans from nano- to mesoscale. Artificially replicating this self-assembly process may produce scaffolds that better mimic both the material and biological properties of native ECM. To this end, novel polypeptide systems have been developed that spontaneously self-assemble in physiological solvents to produce nanoscale, branched fibrous networks analogous to those seen in native ECM (Ryadnov and Woolfson, 2003). Peptides have also been modified to contain alternating hydrophilic and hydrophobic groups that enable them to interact with their complementary moieties to form β-sheet membranes, which further self-assemble to form higher-order structures (Zhang, 2003). Hydrogel scaffolds developed using such techniques can successfully encapsulate cells, as well as enhance cell viability and expression of tissue-specific ECM components (Kisiday et al., 2002). In another novel, and potentially valuable, application of this self-assembly process, amphiphilic molecules have been precisely engineered such that each nanofiber presents a template for hydroxyapatite mineralization, as well as osteoblast adhesion, which together enhance bone regeneration (Hartgerink et al., 2001) (Figure II.6.5.2E). An interesting feature of this technique is the ability of nanofibers to direct alignment of hydroxyapatite crystallization, which mimics the collagen fiber-induced hydroxyapatite alignment seen in native bone.

The apparent effects of scaffold structure on cell form and function may be due to differences in membrane receptor clustering and binding that result from alterations in the physical context in which adhesive cues are presented to cells. Varying the nanoscale (~50 nm) distribution of the integrin-binding RGD tripeptide on an otherwise non-adhesive two-dimensional surface reveals that cell spreading and motility on lower ligand density surfaces (~10^3–10^4 RGD molecules/μm^2) occur best when RGD is presented in clusters of nine ligands per cluster (Maheshwari et al., 2000). Notably, this influence of RGD clustering on cell function is markedly weakened when RGD is presented either at saturated levels (10^5 molecules/μm^2) or as individual peptides (Maheshwari et al., 2000). This functional difference correlates with the ability of RGD clusters to induce actin stress fiber formation (Maheshwari et al., 2000), confirming that adhesive ligands influence cell behavior by promoting optimal traction-mediated cell spreading and simultaneously activating integrin-dependent biochemical signaling inside the cell (Ingber, 2003c). Achieving this precise nanoscale control over adhesive signaling is desirable in biomaterial scaffolds, as that will allow the creation of well-defined three-dimensional cellular niches. To this end, poly(ethylene) glycol scaffolds containing separate RGD and PHSRN peptide sequences spaced precisely 4 nm apart have been developed that promote osteoblast growth and metabolism, while inhibiting ECM deposition, whereas different RGD-PHSRN spacings are less effective (Benoit and Anseth, 2005).

CELL AND ECM MECHANICS AS KEY REGULATORS OF TISSUE DEVELOPMENT

The mechanism of control of tissue development and related tissue engineering approaches described above are based on the concept that local changes in ECM structure and mechanics result in physical distortion of adjacent adherent cells. Cell distortion alters the level of cell prestress and deforms the cytoskeleton, which changes the cell's ability to respond to soluble growth factors and morphogens. To directly test this concept that cell shape distortion is critical for functional control, we applied microfabrication techniques first developed to make microchips for the computer industry to microengineer adhesive ECM islands with controlled size, shape, and position on the micrometer scale (Whitesides et al., 2001). Using this approach, cell shape can be controlled independently of the nature and density of immobilized ECM molecules or the presence or absence of soluble growth factors. Specifically, we created single cell-sized adhesive islands coated with a saturating density of ECM molecules (e.g., fibronectin) that were surrounded by non-adhesive regions. Cells spread by adhering to immobilized ECM ligands and exerting traction forces on these adhesions. Thus, these microengineered adhesive islands provided a simple method to control cell shape distortion, because cell spreading was limited only to the area of the ECM island (Singhvi et al., 1994; Chen et al., 1997).

When single capillary endothelial cells or primary hepatocytes were plated on each island, they spread and flattened, eventually taking on the precise size and shape (e.g., circle, square, hexagon, etc.) of the adhesive island. Most importantly, in the presence of optimal soluble mitogens, cells that distorted (spread) exhibited the highest growth rates (Singhvi et al., 1994; Chen et al., 1997), whereas round endothelial cells adherent to the same ECM coating underwent apoptosis (Figure II.6.5.3). Interestingly, cells that were cultured on intermediate size islands exhibited enhanced cytodifferentiation (secretion of blood proteins by liver epithelial cells) as well as histodifferentiation (e.g., tube formation by capillary cells) (Singhvi et al., 1994; Dike et al., 1999). In addition, when cells cultured on square-shaped ECM islands were stimulated with motility factors (e.g., PDGF, FGF), they preferentially extended lamellipodia at their corners, which are also the sites of maximal cytoskeletal tension and focal adhesion density (Parker et al., 2002).

Subsequent studies revealed that this form of cell shape-dependent developmental control results from cytoskeletal distortion and alterations in the level of cytoskeletal tension within the cell. For example, the Rho family GTPase, RhoA, mediates cell shape-dependent growth regulation by altering the balance of activities between its downstream effector proteins mDia1 and ROCK that are involved primarily in cytoskeletal remodeling (Mammoto et al., 2004). Dissipation of cytoskeletal tension using pharmacological or genetic methods inhibits cell distortion-dependent cell cycle progression, stimulates apoptosis, and suppresses lamellipodia formation (Parker et al., 2002; Numaguchi et al., 2003).

The mechanical properties of the ECM can also regulate cell shape, as well as these same behaviors including cell growth, motility, and differentiation, and again these effects are mediated at least in part through modulation of cytoskeletal tension. For example, cells plated on rigid ECM substrates generate higher levels of tension and proliferation, whereas softer substrates reverse this phenotype (Wang et al., 2000; Ghosh et al., 2007). ECM rigidity can also influence tissue-specific differentiation (Engler et al., 2004, 2006), as well as guide directional cell motility (durotaxis) by regulating cell-substrate mechanical interactions that lead to spatial differentials in cell traction forces and focal adhesion dynamics

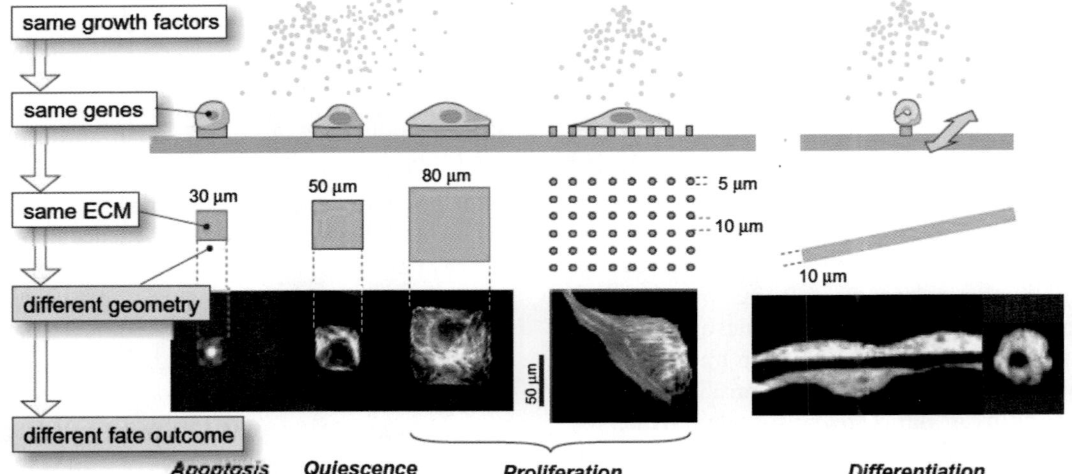

FIGURE II.6.5.3 ECM-mediated geometric control of cell shape regulates key developmental programs. Shown here is a schematic summary of experiments where microfabricated fibronectin islands (solid green) of desired shape and size were used to control cell shape and spreading. Under the influence of the same growth factors (light green dots) and ECM composition, round cells die (apoptosis) and spread cells proliferate, while those partially spread become quiescence (Huang, S. & Ingber, D. (1999). Nat Cell Biol; Chen, C. et al. (1997). Science). Note that cells plated on the small 5 μm dots in the center are in contact with approximately the same absolute amount of fibronectin as on the small 30 μm islands, yet they proliferate because they are allowed to extend over the ECM dots. The first four panels in the bottom row show fluorescence micrographs of phalloidin-stained actin "stress fibers." Allowing endothelial cells to extend on 10 μm-wide linear ECM arrays (right panel) stimulates tube formation, a manifestation of endothelial cell differentiation, as verified using confocal fluorescence microscopy of cytoplasm-stained cells (Dike, L. et al. (1999). In Vitro Cell Dev Biol Anim). (Reproduced from Huang, S. & Ingber, D. (2006). Breast Disease, 27–54, with permission from IOS Press ©2006, 2007.)

(Pelham and Wang, 1997; Lo et al., 2000). Importantly, matrix stiffness alone can control three-dimensional tissue morphogenesis, as normal mammary gland development commonly seen in compliant three-dimensional gels is disrupted when the gel stiffness is raised from 170 Pa to 1200 Pa (Paszek et al., 2005). Likewise, varying ECM elasticity from 700 Pa to 900 Pa can significantly alter three-dimensional endothelial capillary formation *in vivo*, with the intermediate stiffness (800 Pa) promoting the highest VEGF receptor expression and capillary ingrowth (Mammoto et al., 2009). Thus, it is critical to take into account the importance of mechanical features of the tissue microenvironment in future studies focused on engineering artificial inductive scaffolds.

BIOMATERIALS FOR STEM CELL DEVELOPMENT AND TISSUE REGENERATION

Engineered biomaterials that aim to promote complete regeneration of diseased or injured tissues and organs must recapitulate the developmental programs that drive endogenous tissue formation and function. This goal can be accomplished by developing biomimetic scaffolds that leverage the regenerative potential of stem and multipotent progenitor cells – the primitive cells that retain an ability to differentiate into mature cells of various tissue types. Since cell and ECM mechanics alone can regulate the development and function of mature tissue cells, it is conceivable that stem cell differentiation and functionality similarly depend on its physical microenvironment. Indeed, stem cells implanted in the absence of proper mechanostructural cues form teratomas (Wakitani et al., 2003; Hentze et al., 2007), which result from stem cells growing and differentiating into multiple different lineages in the absence of higher-order pattern controls. Thus, a major challenge in regenerative medicine will be to develop biomaterials that can provide appropriate physical cues required to precisely drive stem cell differentiation along a specific lineage, and stimulate regeneration in damaged tissues and organs.

In this regard, it is important to note that mechanical cues such as ECM topography and stiffness alone have been shown to direct mesenchymal stem cell (MSC) lineage specification. When human MSCs are grown on single cell-sized microfabricated ECM islands and exposed to a cocktail of soluble differentiation-inducing factors, adipogenic differentiation is favored on smaller (~1000 μm^2) ECM islands, while osteogenesis is induced in cells spread on larger (~10,000 μm^2) islands (McBeath et al., 2004). Notably, this shape-dependent lineage specification is mediated by cytoskeletal organization and tensile prestress as disruption of Rho activity and actin assembly favors adipogenic differentiation while inhibiting osteogenesis; moreover, overexpressing Rho has an opposite effect. Even more striking is the finding that this Rho-mediated lineage switching occurs independently of the appropriate differentiation-inducing medium (McBeath et al., 2004), suggesting that cell shape and tension alone can govern which cues are necessary and sufficient for stem cell commitment.

The nanotopography of biomaterial surfaces can similarly regulate stem cell differentiation. When human MSCs are cultured on titanium oxide (TiO_2) nanotubes, they adhere strongly but fail to differentiate when the nanotubes are small in size (~30 nm diameter), whereas they differentiate into bone on larger nanotubes (~70–100 nm diameter) that promote cytoskeletal stress-dependent changes in cell form and function (Oh et al., 2009). MSCs grown on ECM-coated flexible substrates can also be directed towards different lineages by tuning the substrate stiffness to match the elasticity of the whole living tissue. For example, neurogenesis is induced on softer substrates (0.1–1 kPa) that mimic brain's elasticity; stiffer substrates (8–17 kPa) that mimic muscle mechanical properties promote myogenic differentiation; and bone induction is observed on the stiffest substrates (25–40 kPa) that matched bone elasticity (Engler et al., 2006). Again, this elasticity-driven lineage switching is dependent on cytoskeletal tension, as disrupting non-muscle myosin II activity abolishes this effect (Engler et al., 2006) in a manner reminiscent of Rho-mediated cytoskeletal control of shape-dependent lineage specification (McBeath et al., 2004). Neural stem cell (NSC) differentiation can similarly be controlled in an ECM stiffness-dependent manner; compliant substrates (~100–500 Pa) favor neuron formation, whereas stiffer hydrogels drive glial cell formation (Saha et al., 2008). Thus, biomaterials developed for tissue engineering and regeneration must be tailored to provide the correct mechanostructural cues that promote desired tissue functionality while suppressing undesirable responses.

Embryonic stem (ES) cells, the pluripotent cells that can give rise to derivatives of all three embryonic germ layers (i.e., endoderm, mesoderm, and ectoderm), possess greater regenerative potential when compared with MSCs and, as such, they are being increasingly explored for cell-based therapies. Notably, human ES (hES) cell self-renewal and differentiation are remarkably sensitive to the physical dimensions of the culture microenvironment. Cell colonies generated from single hES cells initially cultured on micropatterned ECM islands of varying sizes revealed that smaller (200 μm diameter) hES cell colonies preferentially undergo endodermal differentiation, while larger (1200 μm) colonies differentiate into mesoderm (Bauwens et al., 2008; Lee et al., 2009). However, high mesodermal and cardiac induction is observed when the smaller endodermal-biased hES cell colonies are grown into embryoid bodies of large sizes, suggesting that the size of both the cell colony and embryoid body regulate hES cell fate determination (Bauwens et al., 2008). Taken together, these findings suggest that mechanical interactions at the cell–cell and cell–ECM interface guide tissue development not only by facilitating morphogenetic shape changes (as seen during lung branching), but also through precise spatiotemporal control of stem cell fate commitment.

To be effective, tissue and organ regeneration strategies must simultaneously promote the formation of a robust vascular network to maintain optimal oxygen delivery and metabolic exchange. But, despite years of research, achieving this single functionality still poses a huge challenge for tissue engineers. Past *in vitro* studies have shown that heterotypic interactions between endothelial cells and mesenchymal precursor cells, mediated by specific growth factors such as TGF-β and PDGF, cause significant enhancement in vessel stabilization through mesenchymal differentiation into mural cells (pericytes or smooth muscle cells) that are ultimately recruited to the perivascular niche (Hirschi et al., 1998; Ding et al., 2004). Indeed, when such co-cultures are grown in three-dimensional ECM gels and implanted in animals, long-lasting and robust vessels form that remain stable up to one year (Koike et al., 2004). hES cell-derived endothelial cells have also been co-cultured with mouse myoblasts and embryonic fibroblasts within three-dimensional porous polymeric scaffolds to obtain prevascularized tissue constructs *in vitro*, with improved vascular network and blood perfusion *in vivo* (Levenberg et al., 2005).

However, given the complexity of culturing multiple cell types in a controlled and reproducible manner, alternate approaches have been developed that use biomaterial scaffolds to release single or multiple angiogenic growth factors (e.g., VEGF, PDGF) in a spatiotemporally controlled manner to induce new vessel formation. These materials stimulate rapid and robust capillary ingrowth and maturation, as well as increased blood perfusion and improved tissue functionality (Richardson et al., 2001; Nillesen et al., 2007). More recent studies show that in addition to soluble cues (e.g., VEGF) scaffold elasticity can exert a significant effect on three-dimensional capillary formation by endothelial cells *in vivo* via transcriptional control of VEGF-receptor expression. In this study, ECMs with intermediate stiffness (800 Pa) promoted the highest VEGF-receptor expression and capillary ingrowth, compared with both softer (700 Pa) and stiffer (900 Pa) scaffolds *in vivo* (Mammoto et al., 2009). In addition to vascular density, the three-dimensional spatial patterning of the newly formed endothelial capillaries can also be controlled by providing appropriate mechanical cues. For instance, application of exogenous cyclic stretch to endothelial cell-seeded polymeric scaffolds results in three-dimensional sprouting in a direction perpendicular to the stretch axis (Matsumoto et al., 2007). Thus, formation of a robust vasculature within engineered tissues will benefit from the use of instructive biomimetic scaffolds that can effectively integrate and present both soluble and physical cues in a precisely controlled manner.

In addition to their pivotal role in cell-based regenerative approaches, biomaterials are also crucial for targeted delivery of therapeutic drugs and genes that, in turn, has direct implications for regenerative medicine. The physical properties of biomaterials are as important for optimizing drug and gene delivery, as they are for control of tissue development (Mitragotri and Lahann, 2009). Cellular uptake of injected microparticles, for instance, is greatly influenced by particle shape, where rod-like particles with an intermediate aspect ratio of 3 are preferably internalized relative to their spherical counterparts (Decuzzi and Ferrari, 2008; Gratton et al., 2008). Subsequent theoretical modeling and experimental work have revealed that particle internalization is greatly favored with increasing contact angle between the particle and the cell surface (which is high for cylindrical particles and low for spherical ones), which influences the development of a complex actin structure within the cell required for particle internalization (Champion and Mitragotri, 2006). Oblate particles are also expected to adhere more strongly to the vascular endothelium than spherical particles of the same volume (Decuzzi and Ferrari, 2006; Muro et al., 2008), a feature that has important implications for drug delivery to blood vessels. In addition, the degree of internalization depends on particle elasticity, as softer particles are less likely to be phagocytosed by macrophages, and thus have a longer lifetime in the circulation (Beningo and Wang, 2002; Geng et al., 2007). The efficiency of gene delivery to cells similarly depends on the elasticity of cell adhesive hydrogels, and the uptake and expression of plasmid DNA were shown to be higher on stiffer substrates (Kong et al., 2005).

MOLECULAR MECHANISMS OF CELLULAR MECHANOTRANSDUCTION

It is now clear that cells can actively sense and respond to physical cues from the ECM by modulating their level of contractility, which further alters ECM mechanics, thereby establishing a feedback loop that ultimately controls cell fate and function. This dynamic reciprocity of cell–ECM mechanical interactions is mediated by transmembrane integrin receptors that transmit mechanical forces across the cell surface and facilitate mechanochemical transduction events that control cell function and govern cell fate (Alenghat and Ingber, 2002). Integrins refer to a family of heterodimeric transmembrane proteins that bind various ECM molecules on the external surface of the cell (Hynes, 2002) and interact with actin-binding proteins (e.g., vinculin, paxillin, and zyxin) within focal adhesion anchoring complexes inside the cell (Zamir and Geiger, 2001; Ingber, 2003c) (Figure II.6.5.4A). Focal adhesions play a central role in mechanochemical transduction, because in addition to physically coupling the ECM to actin cytoskeleton and bearing high mechanical loads focused through integrins, they also contain many signaling molecules that alter their activity when mechanically stressed (Chicurel et al., 1998b; Alenghat and Ingber, 2002).

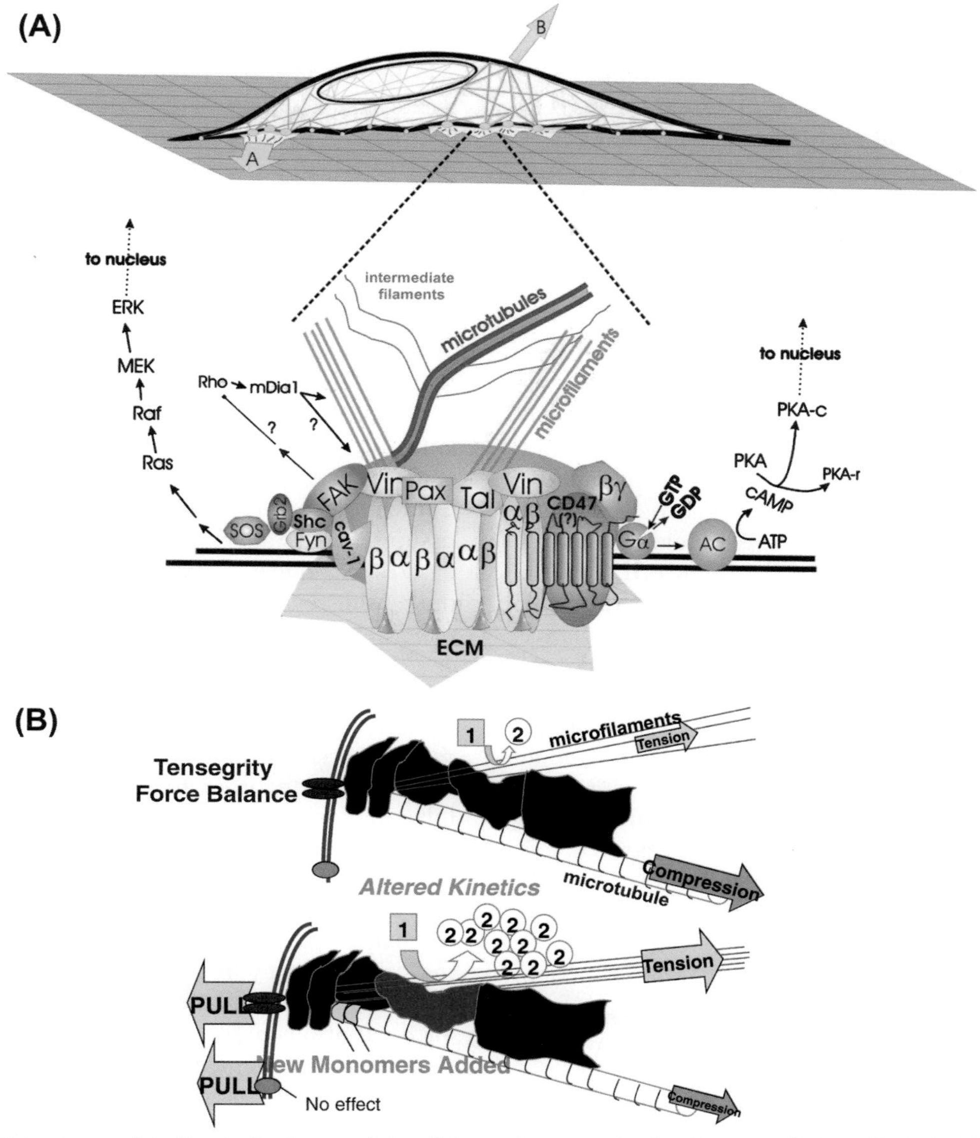

FIGURE II.6.5.4 (A) Integrin-associated focal adhesions mediate cellular mechanotransduction. Forces applied to cells via their basal ECM adhesions are transmitted across cell surface integrin receptors to their cytoplasmic domains down to the underlying focal adhesion scaffolds that link to the cytoskeleton. Forces such as fluid shear stresses, which act on the apical cell surface and produce generalized distortion can also result in increased tension on the cell's basal focal adhesions that are stiffened relative to the remainder of the cell and cytoskeleton. Internally-generated tension (contractility) and forces transmitted via intercellular junctions similarly reach focal adhesions through the cytoskeleton. Forces concentrated within the focal adhesion can stimulate clustering of dimeric (α,β) integrin-receptors, and induce recruitment of focal adhesion proteins (e.g., Vinculin (Vin), Paxillin (Pax), Talin (Tal)) that connect directly to actin microfilaments and indirectly to microtubules and intermediate filaments. Forces applied to this specialized focal adhesion complex activate various integrin-associated signaling cascades, including focal adhesion kinase (FAK), extracellular signal-regulated protein kinase (ERK), Rho, mDia1, heterotrimeric G-proteins, and protein kinase A (PKA), among others, which ultimately regulate gene expression and thereby cell behavior (Ingber, D. (2003c). J Cell Sci). (B) Tensegrity model as the architectural basis of cellular mechanotransduction. (Top) A schematic diagram of the tensegrity-based complementary force balance between tensed microfilaments, compressed microtubules, and transmembrane integrin-receptors (gray oval dimer) in living cells (intermediate filaments are not shown for simplicity). Black forms indicate regulatory proteins and enzymes that are physically immobilized on load-bearing cytoskeletal filaments; red oval represents a transmembrane protein that does not link to the internal cytoskeletal lattice. (Bottom) When force is applied to integrins, thermodynamic and kinetic parameters change locally for cytoskeleton-associated molecules that physically experience the mechanical load; when force is applied to non-adhesion receptors that do not link to the cytoskeleton, stress dissipates locally at the cell surface, and the biochemical response is muted. In this schematic, new tubulin monomers add onto the end of a microtubule (yellow symbols) when tension is applied to integrins, and the microtubule is decompressed as a result of a change in the critical concentration of tubulin. The blue form indicates a molecule that is physically distorted by stress transferred from integrins to the cytoskeleton and, as a result, changes its kinetics (increases its rate constant for chemical conversion of substrate 1 into product 2). In this manner, both cytoskeletal structure (architecture) and prestress (tension) in the cytoskeleton may modulate the cellular response to mechanical stress. (Ingber, D. (2003c). J Cell Sci; Stamenovic, D. & Ingber, D. E. (2009). Soft Matter 5: 1137–1145.)

Integrin-mediated force transfers occurs bidirectionally. In addition to sensing external mechanical forces, cells also exert cytoskeleton-generated contractile forces on these same integrin-ECM linkages (Ingber, 1997; Alenghat and Ingber, 2002; Ingber, 2006; Lele et al., 2006; Parker and Ingber, 2007). Moreover, since ECM is relatively rigid compared to the cell, differences in the ability of the ECM to physically resist cell-generated traction forces directly modify cytoskeletal prestress in a manner that is consistent with the "cellular tensegrity" hypothesis (Ingber et al., 1985; Ingber, 2006; Stamenovic and Ingber, 2009). In this model of cell structure, mechanical stability of the cell and cytoskeletal network results from inward-directed tensional forces borne by actin microfilaments and intermediate filaments that are balanced by microtubule struts and cell surface adhesions to ECM and neighboring cells (Stamenovic and Coughlin, 1999; Wang et al., 2001; Ingber, 2003b) (Figure II.6.5.4B). In addition to stabilizing cell shape and structure, the cytoskeletal tension-mediated cell–ECM force balance regulates cell sensitivity to external mechanical and chemical cues, thereby exerting a significant impact on overall cell function (Ingber, 2003b, 2006; Ghosh et al., 2008; Mammoto et al., 2009). Thus, although mechanical forces applied to integrins initiate a local response at the site of focal adhesion (Chicurel et al., 1998b; Geiger et al., 2001), the cell integrates this intracellular transduction response with other external cues in a tensegrity-dependent manner to elicit a global behavioral response (Ingber, 2003a).

IMPLICATIONS FOR FUTURE MATERIALS DESIGN FOR *IN SITU* TISSUE ENGINEERING

The availability of artificially-engineered skin and cartilage as commercial products serves as a testament to the significant progress that has been made in the field of tissue engineering over the past two decades. However, it must be noted that the success in artificially engineering these tissues is, to a large extent, due to the relative simplicity of their structure and composition, and the low vascularity of the cartilage matrix and skin epidermis. Yet, producing these "living" tissue constructs in a large-scale and reproducible manner has proved challenging, primarily owing to high cost and stringent quality control measures. Thus, to achieve the ultimate goal of regenerating more complex tissues and whole organs, we will need to develop novel *in situ* engineering approaches that can address these limitations by harnessing endogenous tissue and stem cells.

One way to accomplish this goal would be to design strategies for "targeted" delivery of biomaterials to specific injury or diseased sites in the body (Figure II.6.5.5). These injectable delivery systems could be used to target therapeutic agents (drugs, genes) or soluble factors to these critical locations, where they can then recruit endogenous tissue cells or bone marrow-derived stem cells required for tissue repair and regeneration. Peptides that selectively home to the vasculature of specific organs have been identified (Rajotte et al., 1998), and biomaterials derivatized with such targeting moieties could be used for tissue-specific therapies. These injectable, homing biomaterials will need to be programmed such that, upon reaching their target site, they self-assemble into stable, higher-order structures that integrate into the host tissue, and provide the correct adhesive and morphogenetic cues essential for orderly tissue ingrowth and development. Various approaches that facilitate self-assembly of nanoscale materials into supramolecular structures have already been developed. For instance, ampiphilic peptides have been engineered that can self-assemble into a three-dimensional hydrogel *in situ* and promote local tissue repair (Zhang, 2003). Self-assembled, tensegrity-based three-dimensional nanostructures have also been built using DNA that can potentially be used as templates for *in situ* cell and tissue engineering (Douglas et al., 2009).

Regardless of the chemical strategies, it will be advantageous to tune the microscale chemical and mechanical properties of three-dimensional scaffolds *in situ* and in real-time, in order to dynamically regulate the cellular microenvironment, as seen during different stages of endogenous tissue development. This concept has recently been explored using photosensitive PEGDA-based hydrogels, where three-dimensional chemical and mechanical micropatterns were created *in situ* by exposing cell-seeded hydrogels to light in a spatiotemporally defined manner (Hahn et al., 2006; Kloxin et al., 2009). Such variations in local scaffold properties produce distinct microenvironmental niches that permit precise spatiotemporal control over MSC differentiation (Kloxin et al., 2009). It may be possible to use similar approaches to pattern three-dimensional scaffolds in a way that can simultaneously establish differentials in cell growth and apoptosis, as seen during tissue morphogenesis. Such "smart" biomaterials offer great promise as scaffolds that can induce tissue and organ regeneration through recapitulation of the complex morphogenetic and differentiation events that underlie endogenous developmental processes.

CONCLUSION

Today, biomaterials are an indispensable tool in our efforts to repair and regenerate injured or diseased tissues and organs. They are no longer viewed as inert carriers for cell delivery; instead, biomaterials are now designed to actively interact with cells and promote tissue development, as well as restore its function. As cells adhere and spread on a substrate, they exert traction forces at their adhesion points that are balanced by the resistance (stiffness or elasticity) of the substrate. This

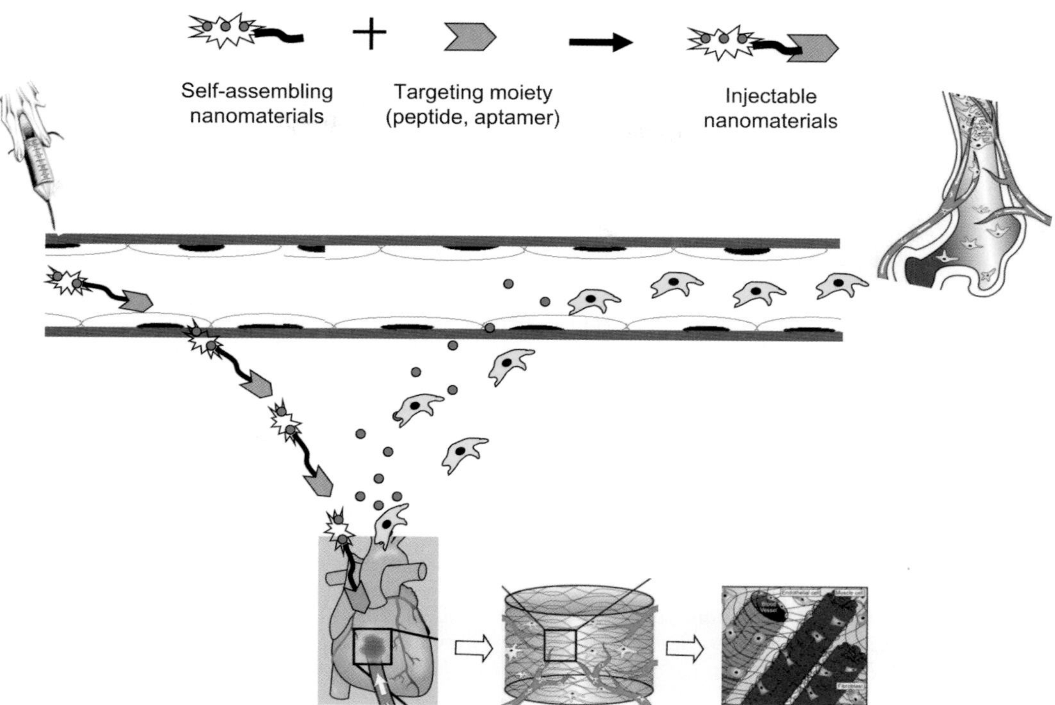

FIGURE II.6.5.5 Biomaterial design strategies for *in situ* tissue engineering. The schematic diagram depicts an example of how smarter approaches that leverage targeting nanomaterials and endogenous stem cells might be used to treat a diseased organ; myocardial infarction is used as an example. The first challenge will be to identify moieties (e.g., peptides, aptamers) that bind to site-specific ligands with high affinity. Nanomaterials derivatized with these targeting moieties (top) can be injected intravenously (middle left) and delivered selectively to sites of tissue damage (bottom left). These materials will need to be programmed to self-assemble into a three-dimensional scaffold that integrates into the host tissue (bottom center), but only when they reach to their target site. Once this occurs, they will be designed to release potent soluble bioactive factors that can increase expansion of critical stem cell populations within bone marrow (middle right), mobilize them into the circulation and recruit them to the infarct site (bottom left). Either the newly formed scaffold materials will need to release different factors over time that stimulate proliferation and differentiation of recruited stem cells and endothelial progenitor cells, or other materials with these inducing properties may be injected and targeted to these same sites at later times. In this manner, formation of vascular networks that can provide continued supplies of oxygen and nutrients will be stimulated in parallel with development of functional tissue structures (e.g., muscle bundles, nerves, connective tissue) (bottom right). This type of *in situ* tissue engineering approach may potentially lead to development of more effective and more cost-effective therapeutics for tissue and organ regeneration.

mechanical force balance drives cytoskeleton-dependent changes in cell shape, which can switch cells between different fates (e.g., growth versus differentiation) in both mature tissue cells and in various types of stem cells. Analysis of embryonic development reveals that micromechanical interactions between cells and the ECM are also critical for tissue morphogenesis, thus suggesting the existence of an overarching micromechanical control point for developmental control. However, the mechanical and structural features of developing tissues vary over space and time on the micrometer scale. Thus, biomaterials that allow precise spatiotemporal regulation of adhesive and micromechanical cues might have tremendous regenerative potential if they can better recapitulate the dynamic cellular microenvironment characteristic of endogenous developmental processes. The ultimate challenge, however, will be to promote tissue repair and regeneration *in situ* at the local site of tissue defect. This goal could be accomplished by developing injectable biomimetic nanomaterials that specifically home to injury or diseased sites, where they self-assemble into higher-order structures that provide the correct developmental and morphogenetic cues required for orderly tissue renewal.

ACKNOWLEDGEMENTS

This work was supported by grants from DoD and NIH. K.G., C.K.T., and D.E.I. are recipients of NIH U54 Interdisciplinary Research Training Grant, American Heart Association Scientist Development Award and DoD Breast Cancer Innovator Award, respectively. The authors thank Kristin Johnson for her assistance with the illustration.

BIBLIOGRAPHY

Agrawal, C. M., & Ray, R. B. (2001). Biodegradable polymeric scaffolds for musculoskeletal tissue engineering. *J. Biomed. Mater. Res.*, **55**, 141–150.

Alenghat, F. J., & Ingber, D. E. (2002). Mechanotransduction: All signals point to cytoskeleton, matrix, and integrins. *Sci. STKE*. 2002, 119, PE6.

Ausprunk, D. H., & Folkman, J. (1977). Migration and proliferation of endothelial cells in preformed and newly formed blood vessels during tumor angiogenesis. *Microvasc. Res.*, **14**, 53–65.

Bauwens, C. L., Peerani, R., Niebruegge, S., Woodhouse, K. A., Kumacheva, E., Husain, M., & Zandstra, P. W. (2008). Control of human embryonic stem cell colony and aggregate size heterogeneity influences differentiation trajectories. *Stem Cells*, **26**, 2300–2310.

Bell, E., Ehrlich, H. P., Sher, S., Merrill, C., Sarber, R., Hull, B., Nakatsuji, T., Church, D., & Buttle, D. J. (1981). Development and use of a living skin equivalent. *Plast. Reconstr. Surg.*, **67**, 386–392.

Beningo, K. A., & Wang, Y. L. (2002). Fc-receptor-mediated phagocytosis is regulated by mechanical properties of the target. *J. Cell Sci.*, **115**, 849–856.

Benoit, D. S., & Anseth, K. S. (2005). The effect on osteoblast function of colocalized RGD and PHSRN epitopes on PEG surfaces. *Biomaterials*, **26**, 5209–5220.

Bernfield, M. R., & Banerjee, S. D. (1978). *The basal lamina in epithelial mesenchymal interactions.* Biology and Chemistry of Basement Membranes, New York: Academic Press.

Champion, J. A., & Mitragotri, S. (2006). Role of target geometry in phagocytosis. *Proc. Natl. Acad. Sci. USA*, **103**, 4930–4934.

Chen, C. S., Mrksich, M., Huang, S., Whitesides, G. M., & Ingber, D. E. (1997). Geometric control of cell life and death. *Science*, **276**, 1425–1428.

Chen, C. S., Tan, J., & Tien, J. (2004). Mechanotransduction at cell–matrix and cell–cell contacts. *Annu. Rev. Biomed. Eng.*, **6**, 275–302.

Chicurel, M. E., Chen, C. S., & Ingber, D. E. (1998a). Cellular control lies in the balance of forces. *Curr. Opin. Cell Biol.*, **10**, 232–239.

Chicurel, M. E., Singer, R. H., Meyer, C. J., & Ingber, D. E. (1998b). Integrin binding and mechanical tension induce movement of mRNA and ribosomes to focal adhesions. *Nature*, **392**, 730–733.

Cukierman, E., Pankov, R., Stevens, D. R., & Yamada, K. M. (2001). Taking cell–matrix adhesions to the third dimension. *Science*, **294**, 1708–1712.

Curtis, A. S., Gadegaard, N., Dalby, M. J., Riehle, M. O., Wilkinson, C. D., & Aitchison, G. (2004). Cells react to nanoscale order and symmetry in their surroundings. *IEEE Trans. Nanobioscience*, **3**, 61–65.

Decuzzi, P., & Ferrari, M. (2006). The adhesive strength of non-spherical particles mediated by specific interactions. *Biomaterials*, **27**, 5307–5314.

Decuzzi, P., & Ferrari, M. (2008). The receptor-mediated endocytosis of nonspherical particles. *Biophys. J.*, **94**, 3790–3797.

Dike, L. E., Chen, C. S., Mrksich, M., Tien, J., Whitesides, G. M., & Ingber, D. E. (1999). Geometric control of switching between growth, apoptosis, and differentiation during angiogenesis using micropatterned substrates. *In Vitro Cell Dev. Biol. Anim.*, **35**, 441–448.

Ding, R., Darland, D. C., Parmacek, M. S., & D'Amore, P. A. (2004). Endothelial–mesenchymal interactions *in vitro* reveal molecular mechanisms of smooth muscle/pericyte differentiation. *Stem Cells Dev.*, **13**, 509–520.

Douglas, S. M., Dietz, H., Liedl, T., Hogberg, B., Graf, F., & Shih, W. M. (2009). Self-assembly of DNA into nanoscale three-dimensional shapes. *Nature*, **459**, 414–418.

Engler, A. J., Griffin, M. A., Sen, S., Bonnemann, C. G., Sweeney, H. L., & Discher, D. E. (2004). Myotubes differentiate optimally on substrates with tissue-like stiffness: Pathological implications for soft or stiff microenvironments. *J. Cell Biol.*, **166**, 877–887.

Engler, A. J., Sen, S., Sweeney, H. L., & Discher, D. E. (2006). Matrix elasticity directs stem cell lineage specification. *Cell*, **126**, 677–689.

Fehrenbacher, A., Steck, E., Roth, W., Pahmeier, A., & Richter, W. (2006). Long-term mechanical loading of chondrocyte-chitosan biocomposites *in vitro* enhanced their proteoglycan and collagen content. *Biorheology*, **43**, 709–720.

Folkman, J., & Moscona, A. (1978). Role of cell shape in growth control. *Nature*, **273**, 345–349.

Galbraith, C. G., & Sheetz, M. P. (1998). Forces on adhesive contacts affect cell function. *Curr. Opin. Cell Biol.*, **10**, 566–571.

Geiger, B., Bershadsky, A., Pankov, R., & Yamada, K. M. (2001). Transmembrane crosstalk between the extracellular matrix–cytoskeleton crosstalk. *Nat. Rev. Mol. Cell Biol.*, **2**, 793–805.

Geng, Y., Dalhaimer, P., Cai, S., Tsai, R., Tewari, M., Minko, T., & Discher, D. E. (2007). Shape effects of filaments versus spherical particles in flow and drug delivery. *Nat. Nanotechnol.*, **2**, 249–255.

Ghosh, K., Pan, Z., Guan, E., Ge, S., Liu, Y., Nakamura, T., Ren, X. D., Rafailovich, M., & Clark, R. A. (2007). Cell adaptation to a physiologically relevant ECM mimic with different viscoelastic properties. *Biomaterials*, **28**, 671–679.

Ghosh, K., Ren, X. D., Shu, X. Z., Prestwich, G. D., & Clark, R. A. (2006). Fibronectin functional domains coupled to hyaluronan stimulate adult human dermal fibroblast responses critical for wound healing. *Tissue Eng.*, **12**, 601–613.

Ghosh, K., Thodeti, C. K., Dudley, A. C., Mammoto, A., Klagsbrun, M., & Ingber, D. E. (2008). Tumor-derived endothelial cells exhibit aberrant Rho-mediated mechanosensing and abnormal angiogenesis *in vitro*. *Proc. Natl. Acad. Sci. USA*, **105**, 11305–11310.

Giordano, R. A., Wu, B. M., Borland, S. W., Cima, L. G., Sachs, E. M., & Cima, M. J. (1996). Mechanical properties of dense polylactic acid structures fabricated by three dimensional printing. *J. Biomater. Sci. Polym. Ed.*, **8**, 63–75.

Gratton, S. E., Ropp, P. A., Pohlhaus, P. D., Luft, J. C., Madden, V. J., Napier, M. E., & Desimone, J. M. (2008). The effect of particle design on cellular internalization pathways. *Proc. Natl. Acad. Sci. USA*, **105**, 11613–11618.

Grinnell, F. (2003). Fibroblast biology in three-dimensional collagen matrices. *Trends Cell Biol.*, **13**, 264–269.

Hahn, M. S., Miller, J. S., & Anseth, K. S. (2006). Three-dimensional biochemical and biomechanical patterning of hydrogels for guiding cell behavior. *Advanced Materials*, **18**, 2679–2684.

Hardin, J., & Keller, R. (1988). The behaviour and function of bottle cells during gastrulation of *Xenopus laevis*. *Development*, **103**, 211–230.

Hartgerink, J. D., Beniash, E., & Stupp, S. I. (2001). Self-assembly and mineralization of peptide-amphiphile nanofibers. *Science*, **294**, 1684–1688.

Hentze, H., Graichen, R., & Colman, A. (2007). Cell therapy and the safety of embryonic stem cell-derived grafts. *Trends Biotechnol.*, **25**, 24–32.

Hersel, U., Dahmen, C., & Kessler, H. (2003). RGD modified polymers: Biomaterials for stimulated cell adhesion and beyond. *Biomaterials*, **24**, 4385–4415.

Hirschi, K. K., Rohovsky, S. A., & D'Amore, P. A. (1998). PDGF, TGF-beta, and heterotypic cell–cell interactions mediate endothelial cell-induced recruitment of 10T1/2 cells and their differentiation to a smooth muscle fate. *J. Cell Biol.*, **141**, 805–814.

Huang, S., & Ingber, D. E. (1999). The structural and mechanical complexity of cell-growth control. *Nat. Cell Biol.*, **1**, E131–E138.

Hynes, R. O. (1992). Integrins: Versatility, modulation, and signaling in cell adhesion. *Cell*, **69**, 11–25.

Hynes, R. O. (2002). Integrins: Bidirectional, allosteric signaling machines. *Cell*, **110**, 673–687.

Ingber, D. (1991). Integrins as mechanochemical transducers. *Curr. Opin. Cell Biol.*, **3**, 841–848.

Ingber, D. E. (1993). Cellular tensegrity: Defining new rules of biological design that govern the cytoskeleton. *J. Cell Sci.*, **104**(Pt 3), 613–627.

Ingber, D. E. (1997). Integrins, tensegrity, and mechanotransduction. *Gravit Space Biol. Bull*, **10**, 49–55.

Ingber, D. E. (2003a). Mechanosensation through integrins: Cells act locally but think globally. *Proc. Natl. Acad. Sci. USA*, **100**, 1472–1474.

Ingber, D. E. (2003b). Tensegrity I. Cell structure and hierarchical systems biology. *J. Cell Sci.*, **116**, 1157–1173.

Ingber, D. E. (2003c). Tensegrity II. How structural networks influence cellular information processing networks. *J. Cell Sci.*, **116**, 1397–1408.

Ingber, D. E. (2006). Cellular mechanotransduction: Putting all the pieces together again. *FASEB J.*, **20**, 811–827.

Ingber, D. E., & Jamieson, J. D. (1985). Cells as tensegrity structures: Architectural regulation of histodifferentiation by physcial forces transduced over basement membrane. *Academic Press*, 13–32.

Ingber, D. E., Madri, J. A., & Jamieson, J. D. (1985). Neoplastic disorganization of pancreatic epithelial cell–cell relations. Role of basement membrane. *Am. J. Pathol.*, **121**, 248–260.

Isenberg, B. C., Williams, C., & Tranquillo, R. T. (2006). Small-diameter artificial arteries engineered *in vitro*. *Circ. Res.*, **98**, 25–35.

Ji, Y., Ghosh, K., Shu, X. Z., Li, B., Sokolov, J. C., Prestwich, G. D., Clark, R. A., & Rafailovich, M. H. (2006). Electrospun three-dimensional hyaluronic acid nanofibrous scaffolds. *Biomaterials*, **27**, 3782–3792.

Kisiday, J., Jin, M., Kurz, B., Hung, H., Semino, C., Zhang, S., & Grodzinsky, A. J. (2002). Self-assembling peptide hydrogel fosters chondrocyte extracellular matrix production and cell division: Implications for cartilage tissue repair. *Proc. Natl. Acad. Sci. USA*, **99**, 9996–10001.

Kloxin, A. M., Kasko, A. M., Salinas, C. N., & Anseth, K. S. (2009). Photodegradable hydrogels for dynamic tuning of physical and chemical properties. *Science*, **324**, 59–63.

Koike, N., Fukumura, D., Gralla, O., Au, P., Schechner, J. S., & Jain, R. K. (2004). Tissue engineering: Creation of long-lasting blood vessels. *Nature*, **428**, 138–139.

Kong, H. J., Liu, J., Riddle, K., Matsumoto, T., Leach, K., & Mooney, D. J. (2005). Non-viral gene delivery regulated by stiffness of cell adhesion substrates. *Nat. Mater.*, **4**, 460–464.

Langer, R., & Vacanti, J. P. (1993). Tissue engineering. *Science*, **260**, 920–926.

Lee, L. H., Peerani, R., Ungrin, M., Joshi, C., Kumacheva, E., & Zandstra, P. (2009). Micropatterning of human embryonic stem cells dissects the mesoderm and endoderm lineages. *Stem Cell Res.*, **2**, 155–162.

Leivo, I. (1983). Structure and composition of early basement membranes: Studies with early embryos and teratocarcinoma cells. *Med. Biol.*, **61**, 1–30.

Lele, T. P., Thodeti, C. K., & Ingber, D. E. (2006). Force meets chemistry: Analysis of mechanochemical conversion in focal adhesions using fluorescence recovery after photobleaching. *J. Cell Biochem.*, **97**, 1175–1183.

Levenberg, S., Rouwkema, J., Macdonald, M., Garfein, E. S., Kohane, D. S., Darland, D. C., Marini, R., Van Blitterswijk, C. A., Mulligan, R. C., D'Amore, P. A., & Langer, R. (2005). Engineering vascularized skeletal muscle tissue. *Nat. Biotechnol.*, **23**, 879–884.

Lin, X., Takahashi, K., Liu, Y., & Zamora, P. O. (2006). Enhancement of cell attachment and tissue integration by a IKVAV containing multi-domain peptide. *Biochim. Biophys. Acta.*, **1760**, 1403–1410.

Lo, C. M., Wang, H. B., Dembo, M., & Wang, Y. L. (2000). Cell movement is guided by the rigidity of the substrate. *Biophys. J.*, **79**, 144–152.

Lutolf, M. P., & Hubbell, J. A. (2005). Synthetic biomaterials as instructive extracellular microenvironments for morphogenesis in tissue engineering. *Nat. Biotechnol.*, **23**, 47–55.

Maheshwari, G., Brown, G., Lauffenburger, D. A., Wells, A., & Griffith, L. G. (2000). Cell adhesion and motility depend on nanoscale RGD clustering. *J. Cell Sci.*, **113**(Pt 10), 1677–1686.

Mammoto, A., Connor, K. M., Mammoto, T., Yung, C. W., Huh, D., Aderman, C. M., Mostoslavsky, G., Smith, L. E., & Ingber, D. E. (2009). A mechanosensitive transcriptional mechanism that controls angiogenesis. *Nature*, **457**, 1103–1108.

Mammoto, A., Huang, S., Moore, K., Oh, P., & Ingber, D. E. (2004). Role of RhoA, mDia, and ROCK in cell shape-dependent control of the Skp2-p27kip1 pathway and the G1/S transition. *J. Biol. Chem.*, **279**, 26323–26330.

Matsumoto, T., Yung, Y. C., Fischbach, C., Kong, H. J., Nakaoka, R., & Mooney, D. J. (2007). Mechanical strain regulates endothelial cell patterning *in vitro*. *Tissue Eng.*, **13**, 207–217.

McBeath, R., Pirone, D. M., Nelson, C. M., Bhadriraju, K., & Chen, C. S. (2004). Cell shape, cytoskeletal tension, and RhoA regulate stem cell lineage commitment. *Dev. Cell*, **6**, 483–495.

McCulloch, A. D., Harris, A. B., Sarraf, C. E., & Eastwood, M. (2004). New multi-cue bioreactor for tissue engineering of tubular cardiovascular samples under physiological conditions. *Tissue Eng.*, **10**, 565–573.

Mitragotri, S., & Lahann, J. (2009). Physical approaches to biomaterial design. *Nat. Mater.*, **8**, 15–23.

Moore, K. A., Polte, T., Huang, S., Shi, B., Alsberg, E., Sunday, M. E., & Ingber, D. E. (2005). Control of basement membrane remodeling and epithelial branching morphogenesis in embryonic lung by Rho and cytoskeletal tension. *Dev. Dyn.*, **232**, 268–281.

Moutos, F. T., Freed, L. E., & Guilak, F. (2007). A biomimetic three-dimensional woven composite scaffold for functional tissue engineering of cartilage. *Nat. Mater.*, **6**, 162–167.

Muro, S., Garnacho, C., Champion, J. A., Leferovich, J., Gajewski, C., Schuchman, E. H., Mitragotri, S., & Muzykantov, V. R. (2008). Control of endothelial targeting and intracellular delivery of therapeutic enzymes by modulating the size and shape of ICAM-1-targeted carriers. *Mol. Ther.*, **16**, 1450–1458.

Muschler, G. F., Nakamoto, C., & Griffith, L. G. (2004). Engineering principles of clinical cell-based tissue engineering. *J. Bone Joint Surg. Am.*, **86-A**, 1541–1558.

Nelson, C. M., Vanduijn, M. M., Inman, J. L., Fletcher, D. A., & Bissell, M. J. (2006). Tissue geometry determines sites of mammary branching morphogenesis in organotypic cultures. *Science*, **314**, 298–300.

Nillesen, S. T., Geutjes, P. J., Wismans, R., Schalkwijk, J., Daamen, W. F., & Van Kuppevelt, T. H. (2007). Increased angiogenesis and blood vessel maturation in acellular collagen-heparin scaffolds containing both FGF2 and VEGF. *Biomaterials*, **28**, 1123–1131.

Numaguchi, Y., Huang, S., Polte, T. R., Eichler, G. S., Wang, N., & Ingber, D. E. (2003). Caldesmon-dependent switching between capillary endothelial cell growth and apoptosis through modulation of cell shape and contractility. *Angiogenesis*, **6**, 55–64.

Oh, S., Brammer, K. S., Li, Y. S., Teng, D., Engler, A. J., Chien, S., & Jin, S. (2009). Stem cell fate dictated solely by altered nanotube dimension. *Proc. Natl. Acad. Sci. USA*, **106**, 2130–2135.

Panitch, A., Yamaoka, T., Fournier, M. J., Mason, T. L., & Tirrell, D. A. (1999). Design and biosynthesis of elastin-like artificial extracellular matrix proteins containing periodically spaced fibronectin CS5 domains. *Macromolecules*, **32**, 1701–1703.

Parker, K. K., Brock, A. L., Brangwynne, C., Mannix, R. J., Wang, N., Ostuni, E., Geisse, N. A., Adams, J. C., Whitesides, G. M., & Ingber, D. E. (2002). Directional control of lamellipodia extension by constraining cell shape and orienting cell tractional forces. *Faseb J.*, **16**, 1195–1204.

Parker, K. K., & Ingber, D. E. (2007). Extracellular matrix, mechanotransduction and structural hierarchies in heart tissue engineering. *Philos. Trans. R Soc. Lond. B Biol. Sci.*, **362**, 1267–1279.

Paszek, M. J., Zahir, N., Johnson, K. R., Lakins, J. N., Rozenberg, G. I., Gefen, A., Reinhart-King, C. A., Margulies, S. S., Dembo, M., Boettiger, D., Hammer, D. A., & Weaver, V. M. (2005). Tensional homeostasis and the malignant phenotype. *Cancer Cell*, **8**, 241–254.

Pattison, M. A., Wurster, S., Webster, T. J., & Haberstroh, K. M. (2005). Three-dimensional, nano-structured PLGA scaffolds for bladder tissue replacement applications. *Biomaterials*, **26**, 2491–2500.

Pelham, R. J., Jr., & Wang, Y. (1997). Cell locomotion and focal adhesions are regulated by substrate flexibility. *Proc. Natl. Acad. Sci. USA*, **94**, 13661–13665.

Rajotte, D., Arap, W., Hagedorn, M., Koivunen, E., Pasqualini, R., & Ruoslahti, E. (1998). Molecular heterogeneity of the vascular endothelium revealed by *in vivo* phage display. *J. Clin. Invest.*, **102**, 430–437.

Ranieri, J. P., Bellamkonda, R., Bekos, E. J., Vargo, T. G., Gardella, J. A., Jr., & Aebischer, P. (1995). Neuronal cell attachment to fluorinated ethylene propylene films with covalently immobilized laminin oligopeptides YIGSR and IKVAV.II. *J. Biomed. Mater. Res.*, **29**, 779–785.

Richardson, T. P., Peters, M. C., Ennett, A. B., & Mooney, D. J. (2001). Polymeric system for dual growth factor delivery. *Nat. Biotechnol.*, **19**, 1029–1034.

Ruoslahti, E. (1991). Integrins. *J. Clin. Invest.*, **87**, 1–5.

Ryadnov, M. G., & Woolfson, D. N. (2003). Engineering the morphology of a self-assembling protein fibre. *Nat. Mater.*, **2**, 329–332.

Saha, K., Keung, A. J., Irwin, E. F., Li, Y., Little, L., Schaffer, D. V., & Healy, K. E. (2008). Substrate modulus directs neural stem cell behavior. *Biophys. J.*, **95**, 4426–4438.

Seitz, H., Rieder, W., Irsen, S., Leukers, B., & Tille, C. (2005). Three-dimensional printing of porous ceramic scaffolds for bone tissue engineering. *J. Biomed. Mater. Res. B Appl. Biomater.*, **74**, 782–788.

Shin, H., Jo, S., & Mikos, A. G. (2003). Biomimetic materials for tissue engineering. *Biomaterials*, **24**, 4353–4364.

Shu, X. Z., Ghosh, K., Liu, Y., Palumbo, F. S., Luo, Y., Clark, R. A., & Prestwich, G. D. (2004). Attachment and spreading of fibroblasts on an RGD peptide-modified injectable hyaluronan hydrogel. *J. Biomed. Mater. Res. A.*, **68**, 365–375.

Singhvi, R., Kumar, A., Lopez, G. P., Stephanopoulos, G. N., Wang, D. I., Whitesides, G. M., & Ingber, D. E. (1994). Engineering cell shape and function. *Science*, **264**, 696–698.

Stamenovic, D., & Coughlin, M. F. (1999). The role of prestress and architecture of the cytoskeleton and deformability of cytoskeletal filaments in mechanics of adherent cells: A quantitative analysis. *J. Theor. Biol.*, **201**, 63–74.

Stamenovic, D., & Ingber, D. E. (2009). Tensegrity-guided self assembly: From molecules to living cells. *Soft Matter*, **5**, 1137–1145.

Stevens, M. M., & George, J. H. (2005). Exploring and engineering the cell surface interface. *Science*, **310**, 1135–1138.

Tsang, V. L., & Bhatia, S. N. (2007). Fabrication of three-dimensional tissues. *Adv. Biochem. Eng. Biotechnol.*, **103**, 189–205.

Vogel, V., & Sheetz, M. (2006). Local force and geometry sensing regulate cell functions. *Nat. Rev. Mol. Cell Biol.*, **7**, 265–275.

Vracko, R. (1974). Basal lamina scaffold-anatomy and significance for maintenance of orderly tissue structure. *Am. J. Pathol.*, **77**, 314–346.

Wakitani, S., Takaoka, K., Hattori, T., Miyazawa, N., Iwanaga, T., Takeda, S., Watanabe, T. K., & Tanigami, A. (2003). Embryonic stem cells injected into the mouse knee joint form teratomas and subsequently destroy the joint. *Rheumatology (Oxford)*, **42**, 162–165.

Waldman, S. D., Spiteri, C. G., Grynpas, M. D., Pilliar, R. M., Hong, J., & Kandel, R. A. (2003). Effect of biomechanical conditioning on cartilaginous tissue formation *in vitro*. *J. Bone Joint Surg. Am.*, **85-A**(Suppl. 2), 101–105.

Wang, H. B., Dembo, M., & Wang, Y. L. (2000). Substrate flexibility regulates growth and apoptosis of normal but not transformed cells. *Am. J. Physiol. Cell Physiol.*, **279**, C1345–C1350.

Wang, N., Naruse, K., Stamenovic, D., Fredberg, J. J., Mijailovich, S. M., Tolic-Norrelykke, I. M., Polte, T., Mannix, R., & Ingber, D. E. (2001). Mechanical behavior in living cells consistent with the tensegrity model. *Proc. Natl. Acad. Sci. USA*, **98**, 7765–7770.

Wang, N., Tytell, J. D., & Ingber, D. E. (2009). Mechanotransduction at a distance: Mechanically coupling the extracellular matrix with the nucleus. *Nat. Rev. Mol. Cell Biol.*, **10**, 75–82.

Whitesides, G. M., Ostuni, E., Takayama, S., Jiang, X., & Ingber, D. E. (2001). Soft lithography in biology and biochemistry. *Annu. Rev. Biomed. Eng.*, **3**, 335–373.

Yang, S., Leong, K. F., Du, Z., & Chua, C. K. (2001). The design of scaffolds for use in tissue engineering. Part I. Traditional factors. *Tissue Eng.*, **7**, 679–689.

Yurchenco, P. D., & Ruben, G. C. (1988). Type IV collagen lateral associations in the EHS tumor matrix. Comparison with amniotic and *in vitro* networks. *Am. J. Pathol.*, **132**, 278–291.

Zamir, E., & Geiger, B. (2001). Molecular complexity and dynamics of cell–matrix adhesions. *J. Cell Sci.*, **114**, 3583–3590.

Zhang, S. (2003). Fabrication of novel biomaterials through molecular self-assembly. *Nat. Biotechnol.*, **21**, 1171–1178.

CHAPTER II.6.6 BIOREACTORS FOR TISSUE ENGINEERING

Nina Tandon, Elisa Cimetta, Sarindr Bhumiratana, Amandine Godier-Furnemont, Robert Maidhof, and Gordana Vunjak-Novakovic

Department of Biomedical Engineering, Columbia University, New York, USA

INTRODUCTION

This chapter is a review of the principles of bioreactor design for tissue engineering. We describe the design and operation of tissue engineering bioreactors developed to direct the differentiation and functional assembly of cells cultured on three-dimensional biomaterial scaffolds. We first discuss the general design requirements for tissue engineering bioreactors, with a focus on mass transport considerations associated with environmental control, and biophysical signals necessary to modulate cell differentiation and the formation of engineered tissues. Next, we discuss the specifics of bioreactor design and operation using six examples of distinctly different tissue engineering systems: (1) cartilage tissue engineering with mechanical loading; (2) tissue engineering of anatomically shaped human bone; (3) cardiac tissue engineering with mechanical stretch; (4) cardiac tissue engineering with electrical stimulation and medium perfusion; (5) tissue engineering of heart valves with mechanical stimulation and perfusion; and (6) tissue engineering of blood vessels with pulsatile medium flow. Lastly, we address challenges in bioreactor design and operation in biological research, and translation into animal models and clinical treatment modalities.

BIOREACTOR DESIGN CONSIDERATIONS

Cells are the key architects of tissues and organs. The cells both respond to and remodel their microenvironment by taking specific differentiation paths, interacting with the neighboring cells, extracellular matrix, molecular and physical signals, and assembling functional tissue units (Kaplan et al., 2005). Both *in vivo* (during development/regeneration) and *in vitro* (in tissue engineering settings), the cues with which cells are presented are the principal determinants of the phenotypic nature of the resultant tissues. Notably, the same factors that determine cell fate and function *in vivo* also determine the progression of cell differentiation *in vitro*. Hence, for tissue engineering applications, major efforts have been invested into characterizing the native tissue environments, and describing these environments by parameters that may be recapitulated experimentally (Kirouac and Zandstra, 2008).

Cell function and the progression of tissue assembly depend on the entire context of cell environment: the availability of a scaffold for cell attachment and tissue formation; the maintenance of physiological conditions in the cell–tissue environment; the supply of nutrients, oxygen, metabolites, and growth factors; and the presence of physical regulatory factors. The regulatory factors of cell differentiation and tissue assembly can be utilized *in vitro* to engineer functional tissues by an integrated use of cells, scaffolds, and bioreactors, as depicted in Figure II.6.6.1. Bioreactors are the primary tools for mimicking the native environments, and providing tissue engineered constructs with physiologically relevant stimuli that can orchestrate the conversion of a "collection of cells" into a specific tissue (Freed and Vunjak-Novakovic, 1998; Bilodeau and Mantovani, 2006; Freed et al., 2006; Kretlow and Mikos, 2008).

Key Components

Bioreactors are composed of several building blocks: the culture chamber; medium reservoir; gas and medium exchange system; and instrumentation for monitoring and control (in many cases operated by computer

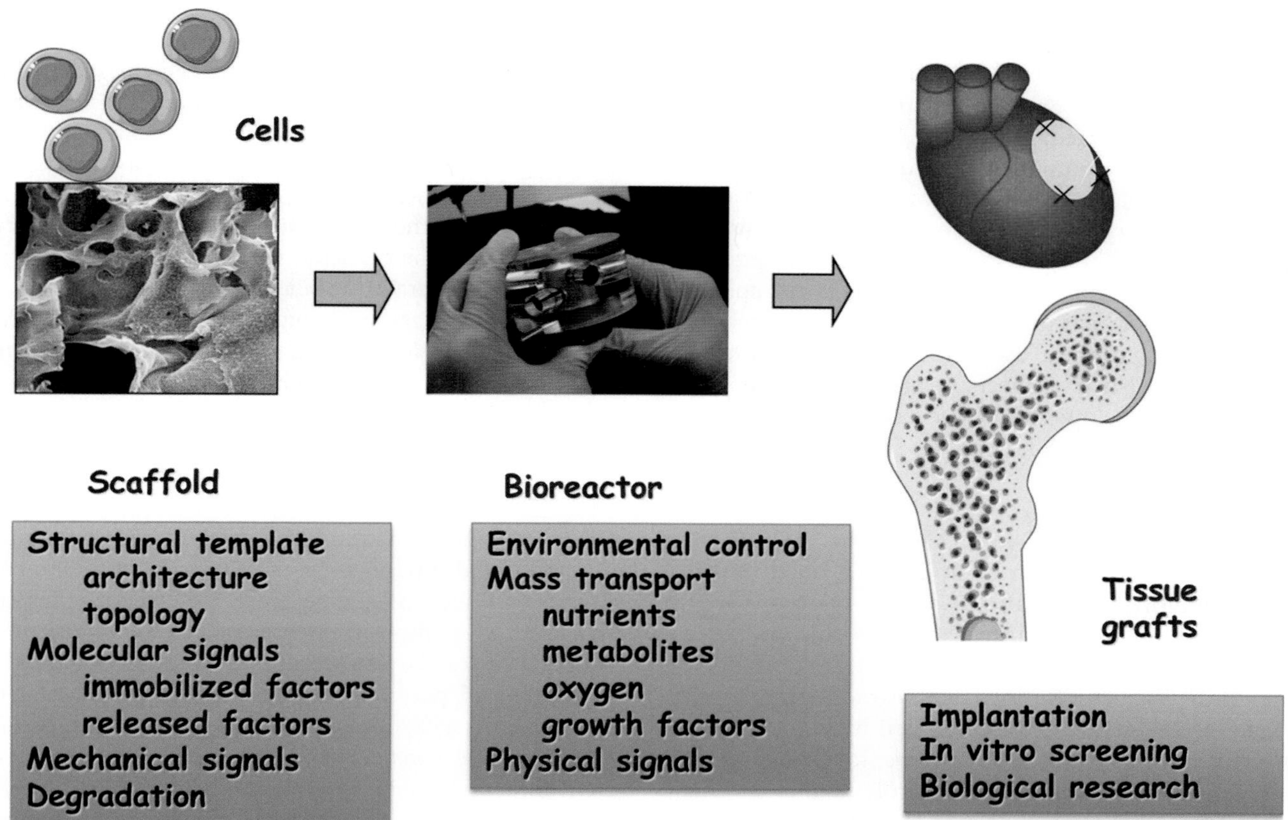

FIGURE II.6.6.1 The tissue engineering paradigm. The living cells (either differentiated or progenitor/stem cells) are seeded onto a biomaterial scaffold, and cultured in a bioreactor to carry out the process of tissue formation. The *scaffold* provides a template for cell attachment and tissue formation through its structural properties (overall architecture, surface properties), molecular factors (immobilized and released molecules and cell-receptor ligands), mechanical properties (stiffness and elasticity), and it degrades in parallel with the deposition of new tissue matrix. The *bioreactor* provides the entire milieu of environmental conditions necessary for regulating cell differentiation and functional assembly, through the control of cell environment (temperature, pH, medium composition) and application of physical stimuli (hydrodynamic shear, mechanical stretch, compression, electrical stimulation).

software). These building blocks are either modular (so that they can be used within several different tissue engineering setups) or integrated into a single device. In most cases, the bioreactor design is customized to meet the need of a specific application, although some of the components (as for example the gas exchange system) can be used within various bioreactor systems.

Technical Requirements

As described in the following sections, there is a great diversity in bioreactor designs that mirrors the range of environmental and regulatory signals that need to be provided to the cells to direct their differentiation into various lineages and assembly into engineered tissues. The environments needed to engineer bone or muscle or blood vessels are quite different from each other, in many respects. However, there are some universal requirements that need to be met by all tissue engineering bioreactors.

An essential prerequisite for all bioreactor components that come in contact with the cells and culture medium is their *biocompatibility*. In contrast to biomaterials used for clinical applications, which are typically designed to be bioactive in some way, all materials used for the bioreactor chamber, gas and medium exchange, and in-line sensors need to be as inert and neutral as possible, so that the cells and molecular factors in culture medium are not affected.

The bioreactor components need to be sterilizable (ideally by autoclaving, ethylene oxide or UV irradiation), and the *sterility* and *sterile containment* must be maintained during the entire culture period. This requirement extends to sampling of tissue constructs and culture medium (during which time the sterile envelope may be broken), imaging of the growing tissues or application of external stimuli (such as mechanical loading). The case studies discussed below show some examples of how this requirement is addressed for specific bioreactor designs.

Biomaterial Scaffold

An engineered tissue is, in most cases, formed by cell cultured on biomaterial scaffolds. In general, tissue engineering scaffolds are biodegradable. To this end, it is important that the scaffold degrades at a rate comparable to that of the tissue formation. Fast degradation not only disrupts the initial structure of the cell-biomaterial construct, but can also result in environmental changes (e.g., a decrease in pH associated with hydrolysis of polyesther scaffolds such as polyglycolic acid (PGA) and polylactic acid (PLA)), while slow degradation leaves too much scaffold material in the way of new tissue formation.

A scaffold serves as a structural template for cell attachment and tissue formation, and it is designed to define the shape, size, and structural properties of the forming tissues. The scaffold is also an informational template, by virtue of its mechanical properties, immobilized and released bioactive factors and cell receptor ligands, and overall architecture (including surface properties, pore size and connectiveness, orientation, and alignment). To mimic the anisotropy of native tissues (such as bone or heart), scaffolds are also made of materials that have different structural and mechanical properties in different directions (Peppas and Langer, 1994; Hubbel 1995; Langer and Tirrell, 2004). In all cases, the cells receive important regulatory signals – molecular, topological, and mechanical – from the scaffold. Hence, the bioreactor designs are largely determined by the specific scaffold properties, and the bioreactor chamber accommodating a flexible tubular scaffold used to engineer a blood vessel is distinctly different from the bioreactor chamber that accommodates an anatomically-shaped engineered bone construct.

Gas Exchange

A gas exchange unit is necessary to ensure the maintenance of physiological levels of gases inside the culture chambers: carbon dioxide to regulate the pH of the culture medium; and oxygen to ensure the correct metabolic activity of cells. In most cases, the bioreactor is placed into an incubator, and this function is provided by mass transport between the incubator gas and the culture medium circulating through a coil of silicone tubing or across a silicone membrane. The efficiency of gas exchange depends on the permeability coefficients of each gas through the materials used, the geometry of the system, temperature, composition of the gas atmosphere, and the flow rate that determines the residence time of the fluids during gas exchange. In general, thin silicone tubing and membranes are selected for gas exchange systems because of their high permeability for oxygen and carbon dioxide, high biocompatibility, and possibility for reuse as these materials can be autoclave-sterilized.

Environmental Control

One of the key functions of a bioreactor is to control, maintain, and modulate as necessary the environmental parameters such as temperature, pH, oxygen tension, and medium composition. Bioreactors allow the establishment of steady-state conditions thus maintaining, by definition, all parameters of interest at constant levels. On the contrary, standard culture systems such as Petri dishes can be considered batch bioreactors, where the environmental conditions are precisely defined only at time zero, and then continuously vary with time between two medium exchanges, often with diffusion-limited and unpredictable kinetics. It is important that bioreactors can be designed to introduce tightly controlled perturbations into the system (such as an episode of hypoxia), and investigate the dynamics of the resulting biological responses.

Mass Transport

Nutrients, oxygen, and regulatory molecules have to be efficiently transported from the bulk culture medium to the tissue surfaces (external mass transfer), and then through the tissue to the cells (internal mass transfer) to maintain the viability and metabolism of cells within tissue engineered constructs. At the same time, metabolites and CO_2 need to be removed from the cells, through the tissue matrix, and into the bulk medium. External mass transfer is determined by the flow conditions at construct surfaces, while the internal mass transport depends on molecular diffusion and convection, and the internal construct structure.

Transport of oxygen to the cells is by far the most critical among all molecules of interest, due to extremely low oxygen solubility in aqueous media. Fully oxygenated culture medium contains only 220 μM of oxygen at physiologic temperature. Oxygen concentration in blood plasma is even lower – only 130 μM. However, the total concentration of oxygen in oxyhemoglobin can reach 8600 μM, through reversible binding of oxygen to hemoglobin. Oxygen transport becomes critically inadequate in large non-perfused constructs (Muschler et al., 2004; Martin and Vermette, 2005), because the diffusional penetration depth of oxygen within native tissues is only 100–200 μm.

Transport phenomena occurring in bioreactors can be subjected to theoretical analysis and precise control once the geometries of the system, the transport distances, and fluid flow rates have been determined. The relative importance of the convective versus the diffusive component to the global mass transport can be calculated and analyzed in conjunction with the kinetics of uptake and release by the cultured cells.

Physical Signals

Tissues and organs in the body are subjected to complex biomechanical environments with dynamic stresses and strains, fluid flow, and electrical signals. Biophysical signals that play a role in cell physiology *in vivo* can also modulate the activity of cells within engineered tissues cultured *in vitro*. Innovative bioreactors have been developed to apply one or more regimes of controlled physical and/or electrical stimuli to three-dimensional engineered constructs, in an attempt to improve or accelerate the generation of a functional tissue (Freshney et al., 2007). *Shear stress* from the tangential forces applied to a surface (typically where cells reside) is important because of the profound effects of flow environment in biological systems (Wang et al., 2005). *Mechanical forces* play fundamental roles in determining the correct development and function of a multiplicity of organs and systems. The application of *mechanical stimulation* (compression, tension, torsion) can improve the engineering of tissues that normally provide mechanical function in the body (bone, muscle tissue, cartilage, tendon) (Stolberg and McCloskey, 2009). Likewise, *electrical stimulation* can improve the physiologic characteristics of tissues that are electrically excitable, such as myocardium (Tandon et al., 2009).

Scale

Tissue engineering bioreactors can have volumes ranging from hundreds of milliliters down to a few milliliters or in some cases only a few microliters. In recent years, micro-bioreactors with volumes of the order of micro- or even nano-liters are helping open new perspectives in studying cell biology and physiology, (Chang et al., 2007; Toner and Irimia, 2005) as they allow experiments conducted at the characteristic time and length scales of biological phenomena (Figallo et al., 2007). While macroscale observations of averaged cell–tissue properties are critical for engineering clinical-size grafts, they cannot be reliable in defining the cell-scale phenomena in biological systems, and should thus be supported by more precise microscale observations (Cimetta et al., 2009). In microscale systems, flow is always laminar (Reynolds numbers Re <1000), and transport is dominated by molecular diffusion or by convective regimes with defined hydrodynamic profiles. Reducing the characteristic dimensions to a microscale level allows working with very short transport distances, which are in turn associated with short time constants. As a result, biological responses are not limited any more by the slow kinetics of physical phenomena. Microscale systems also allow easier decoupling of the effects of mass transport phenomena (such as the generation of specific concentration patterns) from physical phenomena (such as the application of shear forces).

CARTILAGE TISSUE ENGINEERING WITH MECHANICAL LOADING

The avascular nature of articular cartilage predisposes it to a limited ability to regenerate once injured. Articular cartilage is a loadbearing tissue, and is exposed to cyclical stress, which induces fluid pressurization within the tissue and resultant fluid shear. In a 2000 paper, Hung, Ateshian, and colleagues (Mauck et al., 2000) developed a dynamic loading bioreactor to induce pressurization of the interstitial media within agarose scaffolds seeded with bovine articular chondrocytes. Agarose was selected as a scaffold material because of its open lattice structure that minimizes diffusional distances, while maintaining the three-dimensional structure necessary for the fixation of cells. Agarose is a clear polysaccharide hydrogel that has been used extensively for long-term chondrocyte culture, as it permits the application of deformational loading immediately upon cell encapsulation, and is suitable for the preparation of anatomically correct constructs. The spatial uniformity of cells after seeding (compared

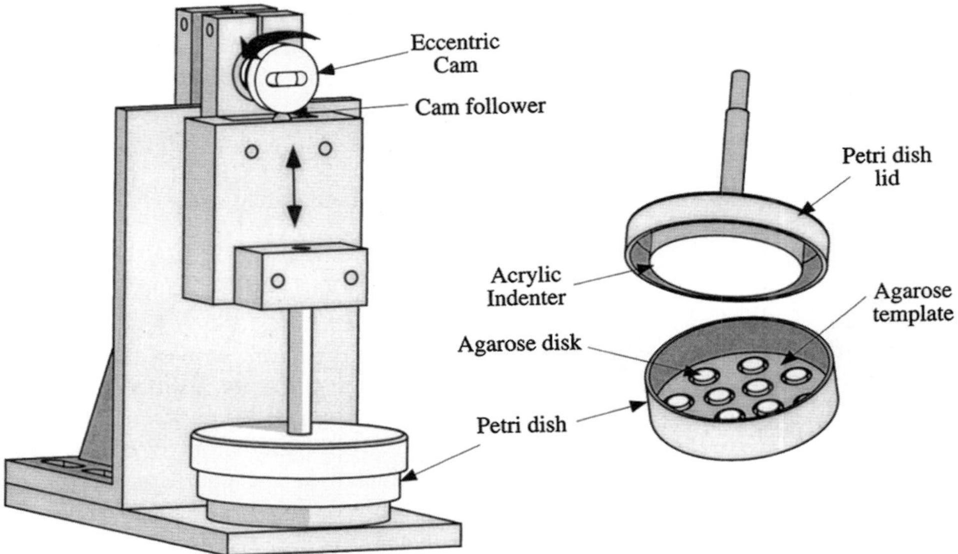

FIGURE II.6.6.2 Bioreactor with mechanical loading for tissue engineering of cartilage. Custom designed bioreactor demonstrating how scaffolds experience loading under sterile conditions. The culture well is fitted with a custom designed loading lid imposed by the cam follower system. (Reproduced by permission from Mauck et al., 2000.)

to seeded fibrous scaffolds) is considered an additional advantage. Cell encapsulation in agarose, combined with the staged application of growth factors, resulted in engineering of functional articular cartilage with mechanical properties far exceeding those achieved using culture systems without mechanical loading.

Hung, Ateshian, and colleagues developed a custom loading system with a cylindrical loading platen, designed to subject a set of cell-seeded agarose discs that were cultured under sterile conditions in a simple Petri dish covered with a lid to dynamic loading (Figure II.6.6.2). Impermeable platens were used for unconfined compression. The loading platen was connected to an eccentric circular cam driven by a motor. Optimal parameters were determined through empirical testing, using biomechanical properties of tissue constructs as the main readout. Following tare load and tare equilibrium, cyclical displacements were applied to agarose discs, with a saw tooth wave to determine lift-off limits, and recovery of the samples after each loading cycle. Dynamic axial compression was applied as a sinusoidal wave with 10% peak-to-peak compressive strain amplitude, at 1 Hz frequency, to mimic physiologic joint loading.

After four weeks of culture, with loading applied five days per week, with three "one-hour-on-one-hour-off" cycles per day, significant differences in mechanical properties of the engineered articular cartilage were observed as compared to free swelling controls. Specifically, the equilibrium aggregate modulus in the loaded group increased 21-fold over the time of cultivation, and reached a value that was 6-fold higher compared to the free swelling controls. As expected, the improved biomechanical properties correlated with increased amounts of cartilage proteoglycans (Mauck et al., 2000).

The increased nutrient supply interacted synergistically with the mechanical signals to advance tissue growth, as shown by combining dynamic deformational loading and the supplementation of TGF-β1 to the chondrocyte-laden agarose constructs. A particularly interesting result is that dynamic deformational loading applied in sequence with TGF-β (two weeks of growth factor supplementation followed by mechanical loading) yielded significantly increased overall mechanical properties. The equilibrium modulus reached 1306 ± 79 kPa and glycosaminoglycan levels reached 8.7 ± 1.6 % w.w. during the eight-week period of cultivation, and both of these values are similar to host cartilage (994 ± 280 kPa, 6.3 ± 0.9 % w.w.) (Lima et al., 2007).

TISSUE ENGINEERING OF ANATOMICALLY-SHAPED HUMAN BONE

Success of bone reconstructions is uniquely dependent on the viability, mechanical function, and precise geometry of the bone grafts. Traditionally, autologous or allogenic tissues are harvested, reshaped, and implanted, leading to suboptimal geometries, tissue quality, and morbidity. Tissue engineering offers the possibility of generating personalized tissues, tailored to the unique geometrical requirements of each patient and the defect being repaired, by using autologous or allogeneic cells, customized biomaterial scaffolds, and customized bioreactors. Critical to bone development are sufficient access to nutrients, regulated by mass transport, and exposure to shear that results from fluid flow and dynamic mechanical compression. Tissue engineering of bone requires the cultivation of spatially uniform seeded cells on mechanically strong scaffolds, with interstitial flow of culture

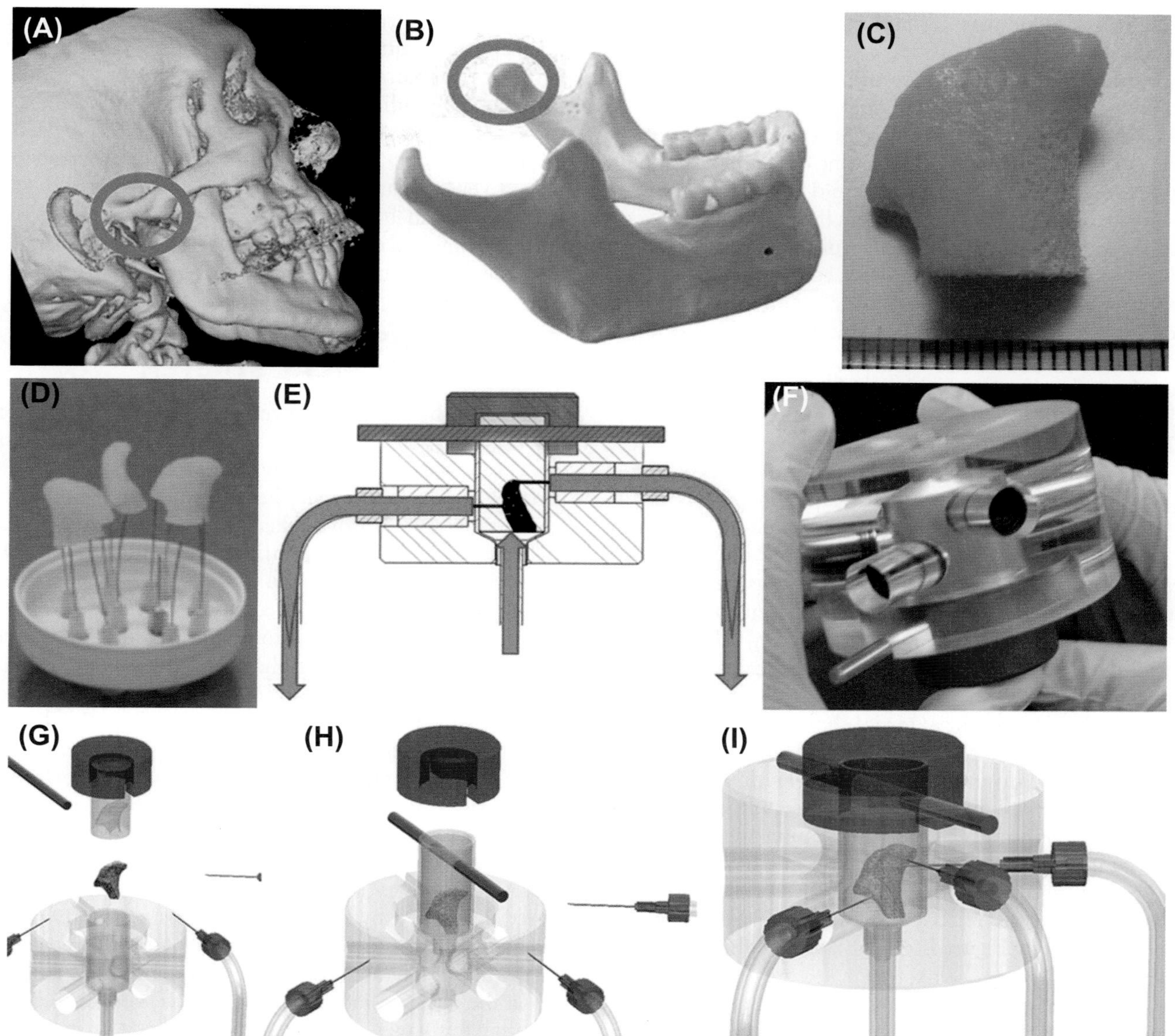

FIGURE II.6.6.3 Bioreactor for tissue engineering of anatomically shaped bone grafts. (A–C) Scaffold preparation. (A, B) Clinical micro-computed tomography (μCT) images were used to obtain high-resolution digital data for the reconstruction of exact geometry of human TMJ condyles. (C) These data were incorporated into MasterCAM software to machine TMJ-shaped scaffolds from fully decellularized trabecular bone. (D) A photograph illustrating the complex geometry of the final scaffolds that appear markedly different in each projection. (E) The scaffolds were seeded in stirred suspension of human mesenchymal stem cells, to three million cells per scaffold (~1 cm^3 volume), precultured statically for one week to allow cell attachment, and then the perfusion was applied for an additional four weeks. (F) A photograph of a perfusion bioreactor used to cultivate-anatomically shaped grafts *in vitro*. (G, H, I) Key steps in bioreactor assembly. See text for details. (Reproduced by permission from Grayson et al., 2009.)

medium to provide mass transport and hydrodynamic shear at levels necessary for proper bone formation.

The combined anatomical, biological, and functional requirements are probably most stringent for bones of the head and face. Craniofacial bone grafts with clinical utility require mechanical functionality, viability, capacity for integration with blood supply, and precise geometry, furthering the need for new methods to engineer anatomically correct yet functional tissues. Grayson et al. (2009) developed a bioreactor capable of meeting some of the above requirements. The group demonstrated the feasibility of engineering "designer bone grafts," in their report of engineered anatomically accurate temporomandibular joint (TMJ) condular bone (Figure II.6.6.3A–C), one of the most complex joints in the human body.

The group selected fully decellularized bone in the exact geometry of the TMJ (Figure II.6.6.3D) as the scaffold material. This way, the cells were cultured in a scaffold that had the structural, biochemical, and mechanical properties of native bone (through the use of fully decellularized bone), and the actual geometry of the final graft (through image-guided fabrication). The void volume of decellularized bone was >80%, as determined by microcomputer tomography (μCT), and the pore sizes were approximately 1 mm, as determined by scanning electron microscopy (SEM) and histological analysis.

The exact shape of the patient's TMJ was recreated from μCT images, and the resulting three-dimensional images were used to shape the scaffold using a computer-driven machine. Human mesenchymal stem cells (hMSCs) were seeded into the scaffold at a physiological density, and cultured in an "anatomical bioreactor" in which bone differentiation and bone formation were induced. The bioreactor itself provided controllable interstitial flow through the scaffold, by adjusting the input and output flow rates throughout the entire scaffold, based on the measurements and mathematical modeling of medium flow.

The bioreactor consisted of an external chamber machined from plastics (acrylic and polyetherimide) (Figure II.6.6.3F). The outer chamber housed a polydimethylsiloxane (PDMS) mold, which in turn held the scaffold. The PDMS mold was created in two steps: digital computer tomography images of a patient's TMJ condyle were used to generate a three-dimensional copolymer in the exact shape of the TMJ; and PDMS was poured over the model and cured to generate the mold. Once placed inside the mold, the scaffold was further compressed by the acrylic external chamber, which served to ensure perfusion through the scaffold rather than peripheral flow around the scaffold (Figure II.6.6.3G, H). Six ports were built into the acrylic chamber – three inlets and three outlets. A peristaltic pump was used to drive the medium through a common inlet, which interfaced to the three inlet ports (Figure II.6.6.3I). The exiting medium was recirculated through the scaffold via a medium reservoir (Figure II.6.6.3E), which also served as a bubble trap.

Modeling of interstitial flow through the bioreactor revealed flow patterns through the cell-seeded scaffold. The model incorporated physical properties of the scaffold, including its porosity. The fluid was assumed as incompressible and passing through a porous medium, with no-slip boundary conditions at scaffold boundaries. The inlet flow was assumed as fully developed, to achieve laminar flow with minimal shear at construct surfaces. The experimental flow results from the bioreactor substantiated the model. After seeding with hMSCs, tissue constructs were cultured in osteogenic medium for a total of five weeks, at a medium flow rate of 1.8 mL/min. Rapid, spatially uniform proliferation of hMSCs was facilitated by the high porosity and large pore sizes in decellularized bone scaffolds. Enhanced bone formation and mineralization were achieved, resulting in tissue constructs with physiological cell density, and bone-like architecture. The resulting bone grafts were centimeters in size, fully viable, with physiological cell density and the formation of mineralized bone matrix.

The use of this "anatomical" bioreactor system showed, for the first time, the feasibility of engineering anatomically correct pieces of three-dimensional human bone, with the ultimate aim of generating patient-specific bone grafts. Notably, the bioreactor system is modular, and can be readily modified to accommodate the cultivation of bone grafts with different shapes, by simply changing the PDMS mold and adjusting inlet port positions to ensure fluid will perfuse the entire scaffold. Overall, the system has significant potential for bone tissue engineering, in a domain where graft geometry, viability, and function define the success of bone repair.

CARDIAC TISSUE ENGINEERING WITH MECHANICAL STRETCH

Motivated by the positive results obtained using neonatal rat hearts to reconstitute tissue-like structures within an elastic scaffold, it was hypothesized that the application of passive and cyclic mechanical stretch within bioreactors would improve cell alignment and differentiation of cardiac cells (Zimmermann et al., 2002, 2006). Over the course of ten years, the Eschenhagen–Zimmermann group has established an effective approach to cardiac tissue engineering by cultivation of neonatal rat heart cells in collagen gel with the application of mechanical stretch. The addition of Matrigel (extracellular matrix from Engelbrecht Swarm tumors) has been essential for the differentiation and functional assembly of cardiac myocyte. This effect is most likely due to the high growth factor content in Matrigel, and was never reproduced using other scaffold materials. Interestingly, mixed populations of heart cells (myocytes, endothelial cells, and fibroblasts) could be cultured in type I collagen gels, but the contractile force developed by the engineered tissue and its responsiveness to calcium and isoprenaline of engineered cardiac constructs were markedly better if Matrigel was included.

In the first version of this approach, neonatal rat ventricular myocytes were suspended in a gel consisting of collagen type I and Matrigel that was cast into rectangular wells (11 × 17 × 4 mm) holding Velcro-coated silicone tubes (7 mm length, 3 mm outer, 2 mm inner diameter) kept at a fixed distance with a metal wire spacer (Figure II.6.6.4A1) (Fink et al., 2000). The mixture was allowed to gel before culture medium was added to the dish, and the resulting gel–cell constructs were cultured for four days before being exposed to unidirectional stretch, using a motorized device for six days (Figure II.6.6.4A2). In this device, tissues were connected at both ends to stretching bars, which pulled the matrix apart with a constant frequency of 1.5 Hz and +20% of original length. The entire arrangement was held in a CO_2 incubator at 37°C, and the culture medium was changed every other day. These early studies showed that the stretch improved organization of cardiac myocytes, increased the mitochondrial density, the length of myofilaments, and the force of contraction generated by the tissue construct.

In the next version of the approach, they strove to improve upon certain shortcomings, including the simplification of

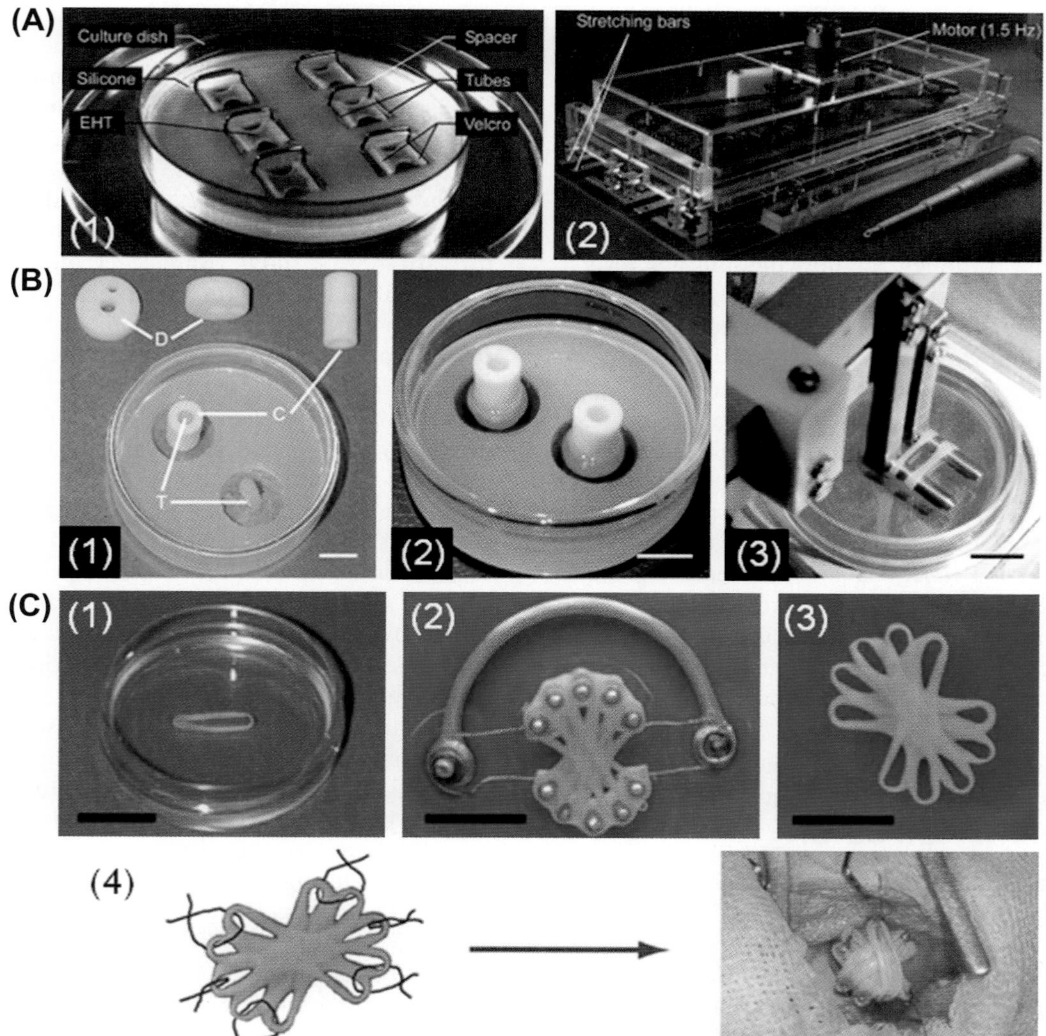

FIGURE II.6.6.4 Bioreactors for cardiac tissue engineering with mechanical stimulation. (A) Bioreactor for initial studies for stretching of rectangular engineered cardiac tissues. (1) Culture dish of six tissues before stretching. (2) Bioreactor incorporating motorized phasic stretching (1.5 Hz, +20% of original (spacer) length) (Reproduced by permission from Fink et al. 2000.) (B) Experimental set-up for circular cardiac tissue preparation and culture. (1) Casting mold assembly in which silicone tubing (T) was glued to the surface of glass culture dishes. Either Teflon™ disks (D) or cylinders (C) can be placed over silicone tubing to function as removable spacers during casting mold preparation and tissue culture, respectively. (2) Tissue condensation around the central Teflon™ cylinder in casting molds between culture days 1 to 4. (3) Tissues placed in bioreactor fitted with stretch apparatus to continue culture under unidirectional and cyclic stretch (10%, 2 Hz). Bars = 10 mm. (Reproduced by permission from Zimmermann et al., 2002.) (C) Tissues shown: (1) before; (2) during; and (3) after stacking. Scale bar = 10 mm. (4) implantation of cardiac tissues into rat model. (Reproduced by permission from Zimmermann et al., 2006.)

complicated casting techniques and improvement of inhomogeneous cell distribution. In the next generation set-up, neonatal rat cardiac cells were suspended in the collagen–Matrigel mix, and cast around cylindrical molds (Zimmermann et al., 2002). After seven days of culture, the rings of cardiac tissue were placed around two rods of a mechanical stretcher, and subjected to unidirectional cyclic stretch at 10% strain and 2 Hz (Figure II.6.6.4B1–3). With this approach, the level of structural and functional organization of the dissociated cells in hydrogel that was achieved within a relatively short time period was quite remarkable. Cells reconstituted cardiac-like tissue with interconnected, oriented cells displaying sarcomeres, adherens junctions, gap junctions, and desmosomes. Ultrastructurally, cardiac myocytes displayed a predominant orientation of sarcomeres along the longitudinal axis of the cell, and were composed of Z, I, A, and H bands, with the occasional presence of M bands. Importantly, capillary structures positive for CD31 were also noted. Functionally, the constructs exhibited contractile properties with a high ratio of twitch to resting tension, strong β-adrenegenic response, and action potentials characteristic of rat ventricular myocytes.

The size, homogeneity, and relative strength of the engineered tissues consisting of cardiac cells in hydrogel were further improved by fusing engineered cardiac tissue together (Zimmermann et al., 2006). A new bioreactor could accommodate five loop-shaped constructs as they fused to form a single synchronously contracting multi-loop construct ~15 mm in diameter and 1–4 mm thick (Figure II.6.6.4C 1–3). When these constructs were investigated for their ability to support or improve the contractile function of infarcted hearts in an immunosupressed rat model of cardiac ischemia (Figure II.6.6.4C4), grafted hearts were able to develop a greater active force than ungrafted, indicative of the engineered constructs playing a significant role in the regeneration of contractile activity in the infarcted hearts. Although the implants did not reverse the problems associated with infarction, they did significantly improve the diastolic and systolic heart function.

This elegant and effective approach has shown, in a rigorous and convincing way, that immature cardiac cell populations have a remarkable ability to assemble into cardiac constructs, if subjected to mechanical signals during cultivation. The loop-shaped constructs, assembled into larger structures, could not only slow down the progression of heart disease in an animal model of chronic infarction, but could also yield some functional improvements. The use of a hydrogel scaffold that contained growth factors necessary for cell differentiation, enabled encapsulation of the cells, and the application of mechanical stretch was critical for cell assembly and development of mechanical force.

CARDIAC TISSUE ENGINEERING WITH ELECTRICAL STIMULATION AND MEDIUM PERFUSION

In native heart, mechanical stretch is induced by electrical signals, and the orderly coupling between electrical pacing signals and mechanical contractions is crucial for the development and function of native myocardium. In an attempt to direct the differentiation and functional assembly of cardiac cells by factors guiding native development and function of the heart, the Vunjak-Novakovic group (Radisic et al., (2008); Radisic et al., (2004); Tandon (2009); Maidhof et al., (2010); Maidhof et al., (2011)) developed a "biomimetic" approach to cardiac tissue engineering. They engineered compact, synchronously contracting cardiac tissue constructs, by providing an environment combining two critical functions: (1) convective–diffusive oxygen transport (critical for cell survival and function) via perfusion bioreactors; and (2) excitation–contraction coupling (critical for cell differentiation and assembly) by using bioreactors with perfusion (interstitial flow) of culture medium and electrical stimulation.

Oxygen Supply by Perfusion

In native heart muscle, a capillary bed provides efficient exchange of oxygen, nutrients, and metabolites between the blood and tissue cells, while shielding cardiomyocytes from direct exposure to hydrodynamic shear. Therefore, the design of cardiac tissue engineering scaffolds and bioreactors should be based on the provision of efficient mass transport of nutrients (and most critically oxygen), along with the control of hydrodynamic shear. Both the transport rates and hydrodynamic shear increase with the increase in flow rate of culture medium; while low values of shear stress may induce phenotypic changes in cardiac cells, including elongation, higher values (e.g., >2.4 dyne cm^2) cause cell dedifferentiation, apoptosis, and death.

The group developed scaffolds and bioreactors that allow: (1) rapid cell inoculation into scaffolds using forward–reverse flow (Radisic et al., 2008); and (2) interstitial flow of culture medium through inoculated scaffolds. Cardiac cell populations isolated from neonatal rat heart ventricles and enriched for cardiac myocytes were suspended in Matrigel at concentrations of approximately 10^8 cells/mL (corresponding to physiological cell density in native myocardium), and loaded into a porous collagen scaffold (Ultrafoam™). By thermal gelation, cells were locked in place inside collagen scaffolds, and the perfusion of culture medium could be established. In these hybrid scaffolds, the gel phase provided cell encapsulation and growth factor supply, and the collagen scaffold provided the porous structure and biomechanical properties that supported the formation of contractile cardiac grafts.

The first-generation bioreactor consisted of perfusion loops comprised of tubing and cartridges (outfitted with silicone gaskets made from silicone tubing to hold scaffolds in place) that were maintained inside a cell culture incubator and connected to a peristaltic pump (Figure II.6.6.5A). During cell inoculation, forward–reverse flow was applied to increase the spatial uniformity of cell seeding, critical for engineering thick constructs with high and spatially uniform densities of viable cells. After cell seeding, the volume of culture medium was increased as required for longer-term cell culture, and the pump was switched to unidirectional flow.

Notably, the final cell viability in perfused constructs cultured for eight days was indistinguishable from the viability of the freshly isolated cells, and markedly higher than the cell viability in dish-grown constructs. Cells expressing cardiac markers (sarcomeric α-actin, sarcomeric tropomyosin, cardiac troponin I) were present throughout the perfused constructs. In response to electrical stimulation, perfused constructs contracted synchronously, had lower excitation thresholds, and readily recovered their baseline function levels after treatment with a gap junction blocker. However, most cells in perfused constructs were round and mononucleated, a situation that was likely due to the exposure of cardiac myocytes to hydrodynamic shear, in contrast to the native heart muscle where blood is confined within the capillary bed and is not in direct contact with cardiac myocytes.

Convective–Diffusive Oxygen Supply in Perfused Channeled Scaffolds

The observed effects of hydrodynamic shear on cardiomyocytes motivated the design of scaffolds with an

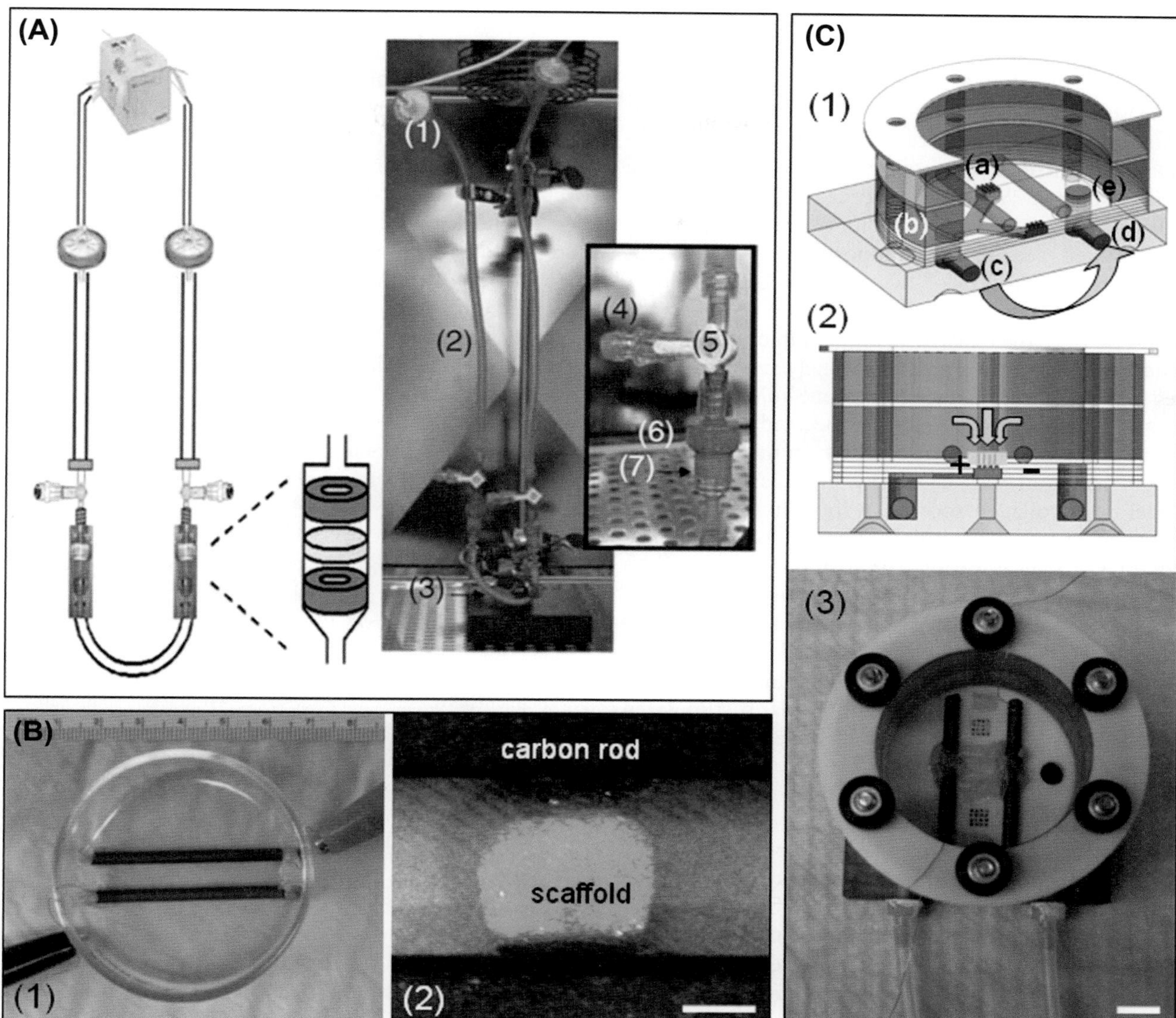

FIGURE II.6.6.5 Bioreactors for cardiac tissue engineering with perfusion and electrical stimulation. (A) Perfusion loop for cell seeding consisting of a multichannel peristaltic pump: (1) filters; (2,3) U-shaped tubing; (4) injection sites; (5) three-way stopcocks; (6) Apollo perfusion chambers with tissue construct. In the perfusion chamber, the construct was squeezed between (7) two silicone gaskets to force the flow through the 5-mm diameter scaffold core. (Reproduced by permission from Radisic et al., 2008.) (B) (1) photograph of an assembled electrical stimulation bioreactor; and (2) close-up view of scaffold positioned between electrodes. Scale bar = 3 mm. (Reproduced by permission from Tandon et al. 2009.) (C) Perfusion-stimulation bioreactor: (1) Alternating layers of custom machined Teflon™, silicone, and plastic are stacked to a "perfused dish" culture chamber where medium flows from the medium bath (a); into the "Y" shaped channel (b); out of the bioreactor (c); through a peristaltic pump, where it is pumped back to the bioreactor inlet (d); and back into the medium bath (e). Light blue arrows denote medium flow direction. (2) Cross-sectional view of the bioreactor showing flow path through the construct and carbon rod electrodes that provide electrical stimulation. (3) Final assembly of the bioreactor with spacers to keep construct loosely in place and platinum wires to connect electrodes to a cardiac stimulator. Scale bar = 1 cm. (Reproduced by permission from Maidhof et al., 2011.)

array of channels that provide a separate compartment for medium flow. In order to reduce the exposure of cardiac myocytes to hydrodynamic shear, the Vunjak-Novakovic group has designed porous elastomer scaffolds of poly(glycerol sebacate) with an array of parallel channels providing a separate compartment for medium flow (Maidhof et al., 2010, 2011). The developed a technique for spatial uniform seeding of these channel's scaffolds by stacking scaffolds two at a time, with channels not aligned to each other, so as to block medium flow through channels during seeding, and force the flow of cell suspension through the bulk phase of the scaffold. Following seeding, the scaffolds were separated (to reveal channels) and cultured individually. When cultured under these conditions, constructs were observed to contract synchronously in response to electrical stimulation after only three days of culture, while channels remained open. Improved construct properties were correlated with the enhanced supply of oxygen to the cells.

Electrical Stimulation of Contractile Constructs

In order to deliver electrical signals that can induce synchronous contractions of cultured constructs, the Vunjak-Novakovic group designed a custom bioreactor delivering signals mimicking those in native heart (Tandon et al., 2009). This bioreactor has several unique features, including the maintenance of a constant position of the scaffolds with respect to direction of the electrical field gradient, while neither restricting the contractions of cells, nor the ability to observe the constructs under a microscope. The bioreactor is fitted with a pair of parallel carbon rod electrodes that were held in place by silicone adhesive (Figure II.6.6.5B). The spacing between the electrodes accommodates the width of the constructs between them.

When neonatal rat cardiomyocytes suspended in Matrigel onto collagen sponges were cultured for three days subjected to trains of monophasic electrical pulses (5V/cm, 1 Hz, 2 ms duration) for an additional five days, they progressively developed into conductive and contractile cardiac constructs (Radisic et al., 2004). On a molecular level, electrical stimulation increased the expression of myosin heavy chain, Cx-43, creatine kinase-MM, and cardiac troponin-I. Morphologically, cells in stimulated constructs were more aligned and elongated, and contained abundant mitochondria, in addition to cells with clearly visible M and Z lines, and H, I, and A bands closely resembling those in native myocardium.

The effects of electrical stimulation depended strongly on the time of its initiation, indicating that cells required some time to reassemble their excitation–contraction coupling machinery before electrical stimulation could provide any reinforcement of its positive effects on cellular differentiation. Functionally, electrical field stimulation also induced cell alignment and coupling, increased the amplitude of synchronous contractions, and resulted in a remarkable level of ultrastructural organization of engineered myocardium, over only eight days of cultivation.

Bioreactors with Perfusion and Electrical Stimulation

Most recently, the group developed a bioreactor that simultaneously applies two conditioning stimuli – medium perfusion and electrical stimulation (Figure II.6.6.5C1–3) to scaffolds seeded using the perfusion loops (Figure II.6.6.5A) (Maidhof et al., 2011). The perfusion-stimulation bioreactor was machined from alternating layers of Teflon™, silicone, and plastic (polyetherimide), forming a "Y" shaped channel that routes culture medium through two constructs and into a single perfusion outlet. Constructs may freely contract, and are only loosely kept in place between silicone spacers and carbon rods (which also served as electrodes for stimulation) over an array of perforated holes through which culture medium was drawn. The upper bioreactor pieces form a medium bath above the constructs that serves for both gas exchange and as a trap for any bubbles generated during perfusion flow. A silicone tubing loop connects the bioreactor inlet and outlet and a peristaltic pump is used to flow culture medium through the system at a rate of 18 μL/minute.

The bioreactor was characterized by mathematical modeling, and in dye flow studies to show that the electrical field stimulus was linear, and that the flow was restricted to cultured tissues. Cardiac constructs were formed by seeding neonatal rat heart myocytes into highly porous poly(glycerol sebacate) scaffolds with an array of parallel channels (250 μm diameter, 500 μm wall-to-wall spacing) and subject to simultaneous perfusion and electrical stimulation (3 V/cm, 3 Hz, 2 ms duration monophasic square waves) for a period of eight days (three days of pre-culture and five days of stimulation).

Electrical Versus Mechanical Stimulation

Because of the high degree of electro-mechanical coupling, it is difficult to decipher whether there are indeed roles for the application of separate mechanical and/or electrical stimulation for engineering cardiac constructs. In constructs prevented from contracting by supplementation of verapamil, electrical stimulation maintained the gap junction protein connexin-43 (Cx-43) at levels comparable with stimulated drug-free controls, suggesting that electrical stimulation, even without contractile activity, aids in the establishment of functional gap junctions (Radisic et al., 2004). Similarly, mechanical stimulation resulted in functional coupling of cardiac cells (Zimmermann et al., 2002). Mechanical stretch and electrical stimulation each appeared to greatly enhance the differentiation of cardiomyocytes, and it may be speculated that a method incorporating the application of both factors could outperform a regime applying just one.

Constructs grown with simultaneous perfusion and electrical stimulation exhibited improved functional properties, including a significant increase in contraction amplitude, greater DNA content and a more uniform cell distribution throughout the scaffold thickness. Simultaneous perfusion and electrical stimulation also enhanced cell morphology, tissue organization, and cardiac protein expression, as shown by immunostaining and Western blot analysis. This novel bioreactor that enables the simultaneous application of culture medium perfusion and electrical conditioning may be appropriate for generating cardiac constructs with clinical size and organization, and may pave the way for further studies incorporating the delivery of combinations of biophysical forces in tissue engineering applications.

TISSUE ENGINEERING OF HEART VALVES WITH MECHANICAL STIMULATION AND PERFUSION

To repair heart valve defects, especially in children, it has been proposed to engineer replacement valves consisting of viable cells that could grow in sync with the developing heart. An autologous cell source offers the best chances of graft compatibility, but the method of obtaining differentiated heart valve cells (i.e., smooth muscle and endothelial cells) requires an invasive procedure. Therefore, research has been directed towards obtaining stem cells from less invasive sources (i.e., bone marrow-derived mesenchymal stem cells), and differentiating these cells during culture *in vitro*.

In the body the biophysical forces the heart valve cells are exposed to include: (1) mechanical stretch as the valves are forced open and closed; and (2) hydrodynamic shear as the blood flows over the valve surface. Each of these forces was independently shown to enhance the differentiation, organization, and mechanical properties of engineered heart tissue valves in *in vitro* bioreactor cultures. Therefore, it was proposed that a combination of these conditioning regimes applied during culture could yield further improvement of the engineered tissues.

To test this hypothesis, a flex-stretch-flow (FSF) bioreactor was developed to simultaneously apply mechanical stimulation and hydrodynamic shear to engineered heart valves (Figure II.6.6.6) (Engelmayr et al., 2008). The bioreactor consists of two stationary posts to which tissue specimens can be affixed, and cyclic flexure is applied by a third post attached to a linear actuator. Parameters for the amount of flexure of the engineered valves, and the flow rate for hydrodynamic shear, were selected based on measurements of typical valve opening and closing distances in the heart, and estimations of shear stress in the native heart. Culture medium is recirculated within the FSF bioreactor chamber by a magnetically coupled paddle-wheel driven by an immersible magnetic stirrer. The entire FSF bioreactor is constructed from sterilizable parts, and may be operated in an incubator for several weeks of culture.

Mesenchymal stem cells (MSCs) derived from sheep bone marrow aspirates were cultured on sheets of non-woven scaffold containing 50% fibers of polyglycolic acid and 50% fibers of poly-L-lactic acid that were assembled into a valved conduit by manual and machine needle punching, and cultured in the bioreactor. The combined application of mechanical stimulation and

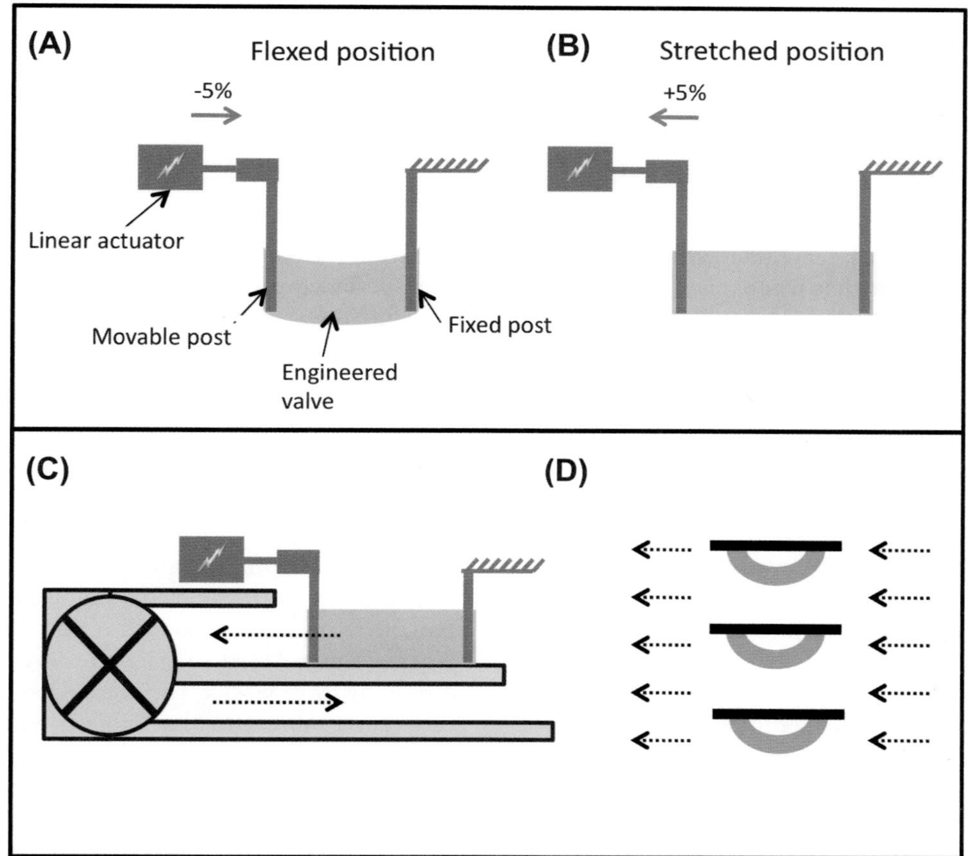

FIGURE II.6.6.6 Flex-stretch-flow bioreactor for engineering a heart valve. (A) Each scaffold is attached to a movable post coupled to a linear actuator and a stationary post. (Reproduced by permission from Engelmayr et al., 2008.) The scaffolds are then flexed (A) or stretched (B) via activation of the linear actuator along the direction of the blue arrow. Culture medium can be recirculated (C, dashed black arrows) within the bioreactor chamber via a magnetically coupled paddle-wheel to provide laminar flow and associated fluid shear stresses to scaffold specimens. In the no flex groups, scaffold specimens are maintained in the straight, undeformed configuration (solid black bars) with or without flow (D). In the flex groups, scaffold specimens are cycled at 1 Hz between the undeformed and flexed configurations (curved gray bars), with or without medium flow.

hydrodynamic shear resulted in increased collagen content and effective stiffness of engineered valves after three weeks of culture. In control groups that excluded either one or both of these conditions, collagen content and mechanical properties were lower. By the end of three weeks the combination of mechanical stimulation and hydrodynamic shear resulted in engineered valves with a modulus comparable to those generated starting from smooth-muscle cells in previous studies. Thus, the FSF bioreactor can be used to provide cyclic flexure and laminar flow, and can synergistically accelerate bone MSC-mediated tissue formation in engineered heart valves.

TISSUE ENGINEERING OF BLOOD VESSELS WITH PULSATILE MEDIUM FLOW

Atherosclerotic vascular disease is a major cause of mortality in the US. A commonly used treatment modality is bypass grafting with autologous veins or arteries, but adequate tissue for bypass vessels is lacking in many patients. Engineered blood vessel grafts could make up for this tissue shortage; however, issues including insufficient mechanical properties of the engineered vessels, and lack of a differentiated cell population, have limited clinical implementation. To help overcome these shortcomings, the design of a pulsatile perfusion bioreactor was proposed over a decade ago (Niklason et al., 1999).

During vasculogenesis and throughout life, vascular cells (endothelial and smooth muscle cells) are exposed to pulsatile physical forces generated with each heartbeat. Studies have shown that mechanical stretch is an important factor in extracellular protein synthesis and expression; specifically, cyclic stretch upregulates collagen and other extracellular matrix protein production in vascular cells. A biomimetic system to culture engineered vessels should replicate this cyclic stretch, with culture medium serving as the "blood" and a pulsatile pump providing the "heart beat."

A suspension of smooth muscle cells (SMCs) isolated from the medial layer of bovine aorta and expanded in culture was seeded onto tubular scaffolds made of fibrous polyglycolic acid (PGA). The surface of the PGA scaffolds was chemically modified with sodium hydroxide to cause ester hydrolysis on the surface of the fibers, and enhance hydrophilicity, adsorption of serum proteins, and cell attachment. The seeded scaffolds were fitted with silicone tubing, placed into the vessel lumen, and transferred into bioreactors and cultured with the application of pulsatile flow for eight weeks. At the end of cultivation, the silicone tubing was removed, a suspension of bovine endothelial cells was introduced, and the cells were allowed to adhere for three days to the lumen surfaces.

The bioreactor was designed to support the cultivation of tubular scaffolds seeded with cells and attached to a perfusion loop, so that culture medium was forced to flow through the lumen (Figure II.6.6.7). A pulsatile pump was used to vary the medium pressure inside the vessel over each cycle, causing the vessel to cyclically stretch in the radial direction. The cultivation conditions were designed to mimic those found during fetal development (~165 beats per minute and 5% radial distension of the vessel).

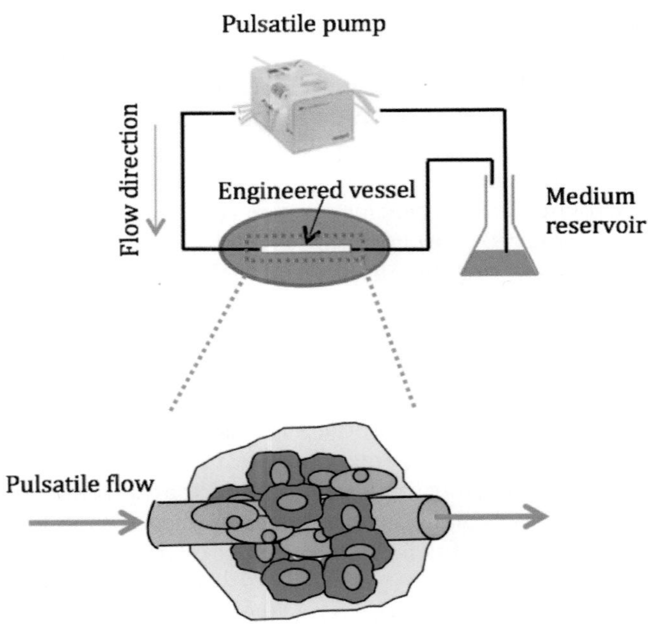

FIGURE II.6.6.7 Pulsatile flow bioreactor for engineering blood vessels. A pulsatile pump is used to flow culture medium through the lumen of an engineered blood vessel causing cyclic stretch of the tissue during culture. After exiting the vessel the medium is returned to a reservoir that provides oxygenation and allows for medium exchange throughout the culture period. (Niklason et al., 1999.)

After eight weeks of culture, the gross appearance of the engineered vessels was similar to that of native arteries. Bioreactor-cultured vessels developed significantly greater mechanical properties (rupture strength, suture retention) than those cultured without pulsatile flow, and matched and correlated with the increased collagen content. Upon implantation into the right saphenous vein of a miniature pig, vessels subjected to cyclic stretch remained open to blood flow for up to four weeks, whereas non-stretched vessels developed thrombosis after only three weeks. Taken together, these results indicate that a bioreactor that subjects engineered vessels to cyclic mechanical stretch can improve the differentiation of engineered vascular grafts prior to implantation.

CHALLENGES IN BIOREACTOR DESIGN

Part of the motivation behind bioreactor development has been to create greater correlation between *in vitro* studies and *in vivo* outcomes. The thought has been that advanced tissue culture platforms, recapitulating the *in vivo* environment in a controllable manner, would aid in the transition from cell studies to animal models, and eventually to human clinical trials and even the clinic. These engineered "biomimetic" environments allow cells to be measured and physically manipulated in unprecedented ways, and they are now replacing the simple but deficient environment of culture dishes. However, several important challenges remain to be resolved to establish conditions that are predictive of cell behavior *in vivo*, and provide bioreactors beyond the laboratory bench.

Producing Conditions More Predictive of Cell Behavior *In Vivo*

In terms of providing bioreactors that will better recapitulate the actual cell–tissue environment, and be more predictive of cellular responses *in vivo*, the main challenge is the difficulty of identifying and optimizing necessary biophysical and molecular cues. Although bioreactors outperform standard culture systems in their ability to provide matrix, molecular, and biophysical factors, current systems still lag behind nature's ability to deliver the highly coordinated sequences of spatial and temporal gradients of regulatory factors at the level of the cell which are necessary to regulate cell function in a developing and adult organism. This performance gap is likely responsible for at least some of tissue engineering's limited success in delivering on the promise of a "one stop shop" of engineered tissues for the entire body.

For reasons that are still little understood, certain tissues (such as cardiac and osteochondral) are more conducive to tissue assembly within bioreactors than others (such as pancreatic and hepatic). Perhaps some bioreactor limitations will be ameliorated through future research in developmental biology, which may shed more light on how tissues emerge from coordinated sequences of cell proliferation, differentiation, and functional assembly that are orchestrated by factors originating from the surrounding cells, matrix, and the external environment. Likewise, future research on regeneration of adult tissues may shed more light on how cells respond to the entire milieu of injury or disease, and help incorporate the most critical biophysical and/or molecular cues into the repertoire of current bioreactors. Some known biophysical cues, such as molecular gradients, although relatively facile to apply in microfluidic conditions, are difficult to produce in the larger-scale three-dimensional culture settings of the bioreactor, without applying significant shear stresses or using excessive amounts of expensive culture medium.

Furthermore, the large cell-number and media requirements for most bioreactors still preclude the performance of screening studies, which are key to optimizing culture conditions, let alone transitioning results from animal models to human stem cells. Overcoming this challenge may indeed involve the development of high-throughput microbioreactors that are equipped to provide certain well-known biophysical cues (e.g., transport, mechanical stimulation, electrical stimulation).

Providing Bioreactors Beyond the Laboratory Bench

In the current tissue engineering paradigm, a common approach is to culture cells on a biomaterial scaffold and regulate the environmental conditions via bioreactors toward reconstructing a functional tissue. With this approach, one can attempt to manipulate the cells' responses in predictable ways, by utilizing the same complex factors known to direct development and remodeling *in vivo*, with the eventual goal of producing a tissue with enough fidelity for implantation. Several challenges related to current bioreactor fabrication economics and logistics remain, before tissue engineering may deliver on this promise.

A bioreactor should ideally provide a native-like cell environment, and address the specific requirements for the tissue of interest. Therefore, bioreactors are often custom-designed to recapitulate the specific mechanisms of nutrient transfer and specific biophysical signals in each type of tissue. These custom-designed systems require an understanding not only of general cell culture principles, but also of the specific mechanisms of relevance to the tissue at hand. As cells are highly sensitive to changes in their environment, even minute changes in molecular and biophysical cues may have implications on the reproducibility of a bioreactor's regulation of cell survival, behavior, and differentiation capacity. Therefore, quality control and implementation of protocol transfer are of great importance to the tissue engineering community.

Further challenges to bioreactor design are also related to the complexity of regulatory issues, and the design constraints of surgical techniques and the operating theater. For tissue-engineered constructs to be viable and clinically useful, bioreactor design and protocols should be compatible with the transportation of biologically stable tissue engineered implants. Therefore, the application of engineered tissues *in vitro* – as experimental models of development and disease – will remain much easier than the cultivation of engineered tissues for implantation.

Interestingly, the best strategy for engineering functional tissue grafts is, in many cases, a simple one – not everything needs to be recapitulated, and one key factor provided within the right context can make the difference. The application of mechanical or electrical stimulation to high-density cultures of cells on scaffolds is an excellent example of this situation. The field has made remarkable strides with the application of limited sets of molecular and biophysical cues, and identified the next set of challenges that call for true interdisciplinary work of biomedical engineers, basic scientists, and clinicians.

WORKED EXAMPLES

Perfusion bioreactors have been used extensively in tissue engineering in order to provide adequate nutrient supply,

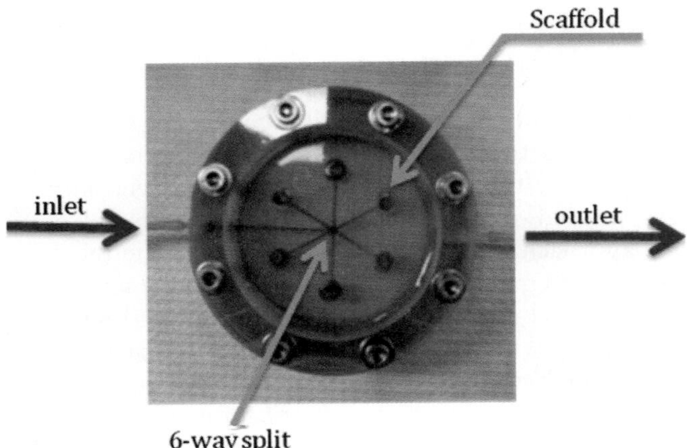

FIGURE II.6.6.8 Schematic of a simple bioreactor with direct perfusion of culture medium through cultured tissues.

as well as mechanical stimuli to the cells inside a scaffold. A round perfusion bioreactor (Figure II.6.6.8) consists of an inlet, an outlet, and a culture chamber, and allows the cultivation of six scaffolds simultaneously. The culture medium coming into the inlet splits into six channels, and flows through the scaffolds into the media chamber. One major function of a bioreactor is to mimic physiologic conditions *in vitro*. In previous studies, bone cells were exposed to physiologic shear stress ranging from 0.25 dynes/cm^2 to 25 dynes/cm^2. In an experiment designed to assess the effect of shear stress on bone tissue engineering, the medium flow rates corresponded to shear stresses of 1 dynes/cm^2 and 10 dynes/cm^2. Cylindrical scaffolds (5 mm in diameter and 5 mm long) can be modeled as 70% porous material consisting of an array of channels 400 μm in diameter (Figure II.6.6.9), with the cells lining the channel lumens. Estimate the flow rate required to provide the low and high shear stress, assuming fully developed laminar flow in the longitudinal direction, steady-state, incompressibility of culture medium, $\mu = 7.7 \times 10^{-4}$ Pa·s, and $\rho = 1$ g/ml.

Solution. The analysis can be done in the single channel within the scaffold.

Flow Rate in the Single Channel. The velocity profile of a fully developed laminar circular pipe flow can be derived from the Navier–Stokes equations:

$$v_z = \frac{1}{4\mu}\frac{\partial p}{\partial z}(r^2 - R^2)$$

The volumetric flow rate of single channel can be determined by:

$$Q_{\text{Channel}} = \int_0^{2\pi}\int_0^{R} V_z r dr d\theta$$

$$Q_{\text{Channel}} = \int_0^{2\pi}\int_0^{R} \frac{1}{4\mu}\frac{\partial p}{\partial z}(r^2 - R^2) r dr d\theta$$

$$Q_{\text{channel}} = -\frac{\pi R^4}{8\mu}\left(\frac{\partial P}{\partial z}\right)$$

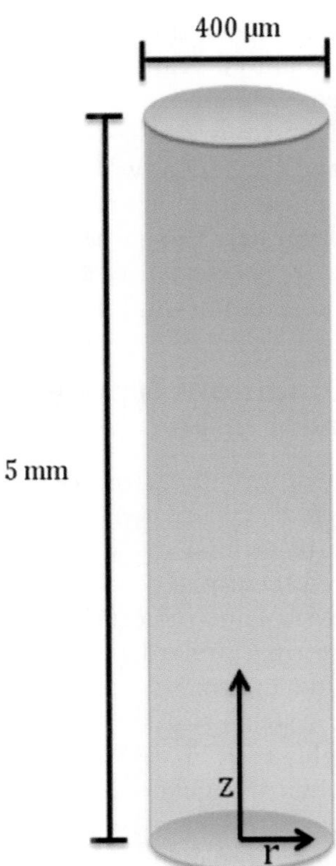

FIGURE II.6.6.9 Schematic of a single channel within a tissue construct perfused with culture medium.

The relationship can be expressed in terms of the pressure drop:

$$\frac{\Delta P}{L} = -\frac{\partial p}{\partial z}$$

Or:

$$Q_{\text{channel}} = \frac{\pi R^4 \Delta P}{8\mu L}$$

The average velocity is:

$$V = \frac{Q}{\pi R^2} = \frac{R^2 \Delta P}{8\mu L}$$

The maximum velocity occurs at the center of the tube:

$$v_{\max} = -\frac{R^2}{4\mu}\left(\frac{\partial p}{\partial z}\right) = \frac{R^2 \Delta P}{4\mu L}$$

Hence:

$$v_{\max} = 2V$$

The velocity profile can now be written as:

$$\frac{v_z}{v_{\max}} = 1 - \left(\frac{r}{R}\right)^2$$

and the shear stress equation becomes:

$$\tau_{\max Y} = -\mu \frac{dv_z}{dr}\Big|_{r=R}$$

$$\tau_{\max Y} = -\mu\left(-v_{\max}\frac{2r}{R^2}\right)\Big|_{r=R}$$

$$\tau_{maxY} = \frac{2\mu v_{max}}{R} = \frac{4\mu V}{R}$$

Now we can solve for the flow rate:

- Low shear stress

$$\tau_{maxY} = 1\,\text{dyne/cm}^2 = \frac{4(0.0077\text{dyne} \bullet \text{s/cm}^2)}{0.0200\,\text{cm}} V$$

$$V = 0.649\,\text{cm/s}$$

$$Q_{channel} = \pi R^2 V = 8.16 \times 10^{-4}\ \text{ml/s}$$

Checking the Reynolds number:

$$\text{Re} = \frac{\rho V(2R)}{\mu}$$

$$\text{Re} = \frac{(1\text{g/cm}^3)(0.649\,\text{cm/s})(2(0.02\,\text{cm}))}{(0.0077\text{dyne} \bullet \text{s/cm}^2)}$$

$$\text{Re} = 3.37 < 2100$$

we confirm that the laminar flow assumption was valid.

- High shear stress

$$\tau_{maxY} = 10\ \text{dyne/cm}^2 = \frac{4(0.0077\ \text{dyne} \bullet \text{s/cm}^2)}{0.0200\,\text{cm}} V$$

$$V = 6.49\ \text{cm/s}$$

$$Q_{channel} = \pi R^2 V = 8.16 \times 10^{-4}\text{ml/s}$$

Checking the Reynolds number:

$$\text{Re} = \frac{\rho V(2R)}{\mu}$$

$$\text{Re} = \frac{(1\,\text{g/cm}^3)(6.49\,\text{cm/s})(2(0.02\,\text{cm}))}{(0.0077\text{dyne} \bullet \text{s/cm}^2)}$$

$$\text{Re} = 33.7 < 2100$$

we confirm that the laminar flow assumption was valid. where v_z indicates the velocity in the z–direction (see Figure II.6.6.9), μ indicates the viscosity of the fluid, ρ indicates the pressure within the pipe, r indicates the radial position, and R indicates the radius of the pipe. $Q_{channel}$ indicates the flow within a channel, L indicates the length of the channel, V is average velocity, v_{max} is the maximum velocity, τ_{maxY} is the maximum shear in the Y-direction.

Scaffold Geometry. Cross-sectional area of the scaffold is:

$$A_{scaffold} = \pi(2.5\ \text{mm})^2 = 19.63\ \text{mm}^2$$

Void area of the scaffold is:

$$A_{void} = 0.7A_{scaffold} = 13.74\,\text{mm}^2$$

Cross-sectional area of single channel is:

$$A_{channel} = \pi(0.2\,\text{mm})^2 = 0.126\,\text{mm}^2$$

Therefore, each scaffold contains the following number of channels:

$$\frac{A_{void}}{A_{channel}} = \frac{13.74\ \text{mm}^2}{0.126\ \text{mm}^2} = 109$$

As there are six scaffolds per bioreactor, there are 654 channels in the scaffold.

Total Flow Rate

- Low shear stress:

$$Q_{total} = 654\ Q_{channel}$$

$$Q_{total} = 0.533\,\text{ml/s} = 32\,\text{ml/min}$$

- High shear stress:

$$Q_{total} = 654\ Q_{channel}$$

$$Q_{total} = 5.33\,\text{ml/s} = 320\,\text{ml/min}$$

ACKNOWLEDGMENT

We gratefully acknowledge the NIH support of the work described in this chapter (HL076485, DE016525, HL089913, EB002520, RR026244).

BIBLIOGRAPHY

Bilodeau, K., & Mantovani, D. (2006). Bioreactors for tissue engineering: Focus on mechanical constraints. A comparative review. *Tissue Engineering*, **12**(8), 2367–2383.

Cheng, X., Irimia, D., Dixon, M., Sekine, K., Demirci, U., et al. (2007). A microfluidic device for practical label-free CD4+ T-cell counting of HIV infected subjects. *Lab on a Chip*, **7**, 170–178.

Cheng, X., Gupta, A., Chen, C., Tompkins, R. A., Rodriguez, W., et al. (2009). Enhancing the performance of a point-of-care CD4+ T-cell counting microchip through monocyte depletion for HIV/AIDS diagnostics. *Lab on a Chip*, **9**, 1357–1364.

Cimetta, E., Figallo, E., Cannizzaro, C., Elvassore, N., & Vunjak-Novakovic, G. (2009). Micro-bioreactor arrays for controlling cellular environments: Design principles for human embryonic stem cell applications. *Methods*, **47**, 81–89.

Engelmayr, G. C., Soletti, L., Vigmostad, S. C., Budilarto, S. G., Federspiel, W. J., et al. (2008). A novel flex-stretch-flow bioreactor for the study of engineered heart valve tissue mechanobiology. *Annals of Biomedical Engineering*, **36**(5), 700–712.

Figallo, E., Cannizzaro, C., Gerecht, S., Burdick, J. A., Langer, R., et al. (2007). Micro-bioreactor array for controlling cellular microenvironments. *Lab on a Chip*, **7**, 710–719.

Fink, C., Ergün, S., Kralisch, D., Remmers, U., Weil, J., et al. (2000). Chronic stretch of engineered heart tissue induces hypertrophy and functional improvement. *The FASEB Journal: Official Publication of the Federation of American Societies for Experimental Biology*, **14**(5), 669–679.

Freed, L. E., & Vunjak-Novakovic, G. (1998). Culture of organized cell communities. *Advanced Drug Delivery Reviews*, **33**, 15–30.

Freed, L. E., Guilak, F., Guo, X. E., Gray, M. L., Tranquillo, R., et al. (2006). Advanced tools for tissue engineering: Scaffolds, bioreactors, and signaling. *Tissue Engineering*, **12**(12), 3285–3305.

Freshney, I., Obradovic, B., et al. (2007). Principles of tissue culture and bioreactor design. In R. P. Lanza, R. Langer, & J. Vacanti (Eds.), *Principles of Tissue Engineering* (pp. 155–184). San Diego, CA: Academic.

Grayson, W. L., Fröhlich, M., Yeager, K., Bhumiratana, S., Chan, M. E., et al. (2009). Regenerative medicine special feature: Engineering anatomically shaped human bone grafts. *Proceedings National Academy of Sciences USA*, **107**(8), 3299–3304.

Hubbel, J. A. (1995). Biomaterials in tissue engineering. *Biotechnology*, **13**, 565–575.

Kaplan, D., Moon, R. T., & Vunjak-Novakovic, G. (2005). It takes a village to grow a tissue. *Nature Biotechnology*, **23**(10), 1237–1239.

Kirouac, D. C., & Zandstra, P. W. (2008). The systematic production of cells for cell therapies. *Cell Stem Cell*, **3**, 369–381.

Kretlow, J. D., & Mikos, A. G. (2008). From material to tissue: Biomaterial development, scaffold fabrication, and tissue engineering. *AIChE Journal*, **54**(12), 3048–3067.

Langer, R., & Tirrell, D. A. (2004). Designing materials for biology and medicine. *Nature*, **428**, 487–492.

Lima, E. G., Bian, L., Ng, K. W., Mauck, R. L., Byers, B. A., et al. (2007). The beneficial effect of delayed compressive loading on tissue-engineered cartilage constructs cultured with TGF-beta3. *Osteoarthritis Cartilage*, **15**(9), 1025–1033.

Maidhof, R., Marsano, A., Lee, E. J., & Vunjak-Novakovic, G. (2010). Perfusion seeding of channeled elastomeric scaffolds with myocytes and endothelial cells for cardiac tissue engineering. *Biotechnology Progress*, **26**(2), 565–572.

Maidhof, R., Tandon, N., Lee, E. J., Luo, J., Duan, Y., Yeager, K., & Vunjak-Novakovic, G. (2011). Biomimetic perfusion and electrical stimulation applied in concert improved the assembly of engineered cardiac tissue. *Journal of Tissue Engineering and Regenerative Medicine*. DOI: 10.1002/term.525. [Epub ahead of print]

Martin, Y., & Vermette, P. (2005). Bioreactors for tissue mass culture: Design, characterization, and recent advances. *Biomaterials*, **26**(35), 7481–7503.

Mauck, R. L., Soltz, M. A., Wang, C. C., Wong, D. D., Chao, P. H., et al. (2000). Functional tissue engineering of articular cartilage through dynamic loading of chondrocyte-seeded agarose gels. *Journal of Biomechanical Engineering*, **122**(3), 252–260.

Muschler, G. F., Nakamoto, C., & Griffith, L. G. (2004). Engineering principles of clinical cell-based tissue engineering. *Journal of Bone and Joint Surgery. American Volume*, **86A**(7), 1541–1558.

Niklason, L. E., Gao, J., Abbott, W. M., Hirschi, K. K., Houser, S., et al. (1999). Functional arteries grown *in vitro*. *Science*, **284**(5413), 489–493.

Peppas, N. A., & Langer, R. (1994). New challenges in biomaterials. *Science*, **263**(5154), 1715–1720.

Radisic, M., Park, H., Shing, H., Consi, T., Schoen, F. J., et al. (2004). Functional assembly of engineered myocardium by electrical stimulation of cardiac myocytes cultured on scaffolds. *Proceedings National Academy of Sciences USA*, **101**(52), 18129–18134.

Radisic, M., Marsano, A., Maidhof, R., Wang, Y., & Vunjak-Novakovic, G. (2008). Cardiac tissue engineering using perfusion bioreactor systems. *Nature Protocols*, **3**(4), 719–738.

Stolberg, S., & McCloskey, K. E. (2009). Can shear stress direct stem cell fate? *Biotechnology Progress*, **25**(1), 10–19.

Tandon, N., Cannizzaro, C., Chao, P. H., Maidhof, R., Marsano, A., et al. (2009). Electrical stimulation systems for cardiac tissue engineering. *Nature Protocols*, **4**(2), 155–173.

Toner, M., & Irimia, D. (2005). Blood-on-a-chip. *Annu Rev Biomedical Engineering*, **7**, 77–103.

Wang, H., Riha, G. M., Yan, S., Li, M., Chai, H., et al. (2005). Shear stress induces endothelial differentiation from a murine embryonic mesenchymal progenitor cell line. *Arteriosclerosis Thrombosis and Vascular Biology*, **25**, 1817–1823.

Zimmermann, W. H., Schneiderbanger, K., Schubert, P., Didié, M., Münzel, F., et al. (2002). Tissue engineering of a differentiated cardiac muscle construct. *Circulation Research*, **90**(2), 223–230.

Zimmermann, W., Melnychenko, I., Wasmeier, G., Didié, M., Naito, H., et al. (2006). Engineered heart tissue grafts improve systolic and diastolic function in infarcted rat hearts. *Nature Medicine*, **12**(4), 452–458.

CHAPTER II.6.7 BONE TISSUE ENGINEERING

Justin L. Brown[1], *Sangamesh G. Kumbar*[2], *and Cato T. Laurencin*[3]

[1]Department of Bioengineering, The Pennsylvania State University, PA, USA

[2]Department of Orthopaedic Surgery, Department of Chemical, Materials and Biomolecular Engineering, University of Connecticut, CT, USA

[3]Connecticut Institute for Clinical and Translational Science Director, Institute for Regenerative Engineering, University of Connecticut, CT, USA

INTRODUCTION

The use of biocompatible polymeric materials for orthopedic applications, such as bone graft substitutes, has been under investigation since the 1940s (Blaine, 1946; Leveen and Barberio, 1949). By the 1970s the importance of attaining appropriate mechanical properties, an interconnected porosity, and a microstructure that promotes tissue ingrowth was realized (Hench et al., 1971; Weber et al., 1971). Recent research has highlighted the necessity for a subcellular dimension, or nanostructure, in synthetic bone grafts to promote the appropriate organization of bone cells in an effort to generate or regenerate bone tissue (Christenson et al., 2007; Horii et al., 2007; Hu et al., 2008).

This chapter begins with an overview of the biology of bone to provide a framework for what the application of biomaterials strives to recreate. The chapter then proceeds through a discussion of bone tissue engineering, beginning with the natural bone grafts, moving through bone graft substitutes, and finishing with a discussion of bioreactors used in bone tissue engineering.

BONE BIOLOGY

Bones are vascularized and innervated organs that are composed of bone tissue, bone marrow, and a surrounding connective tissue called periosteum. Bones serve a number of functions such as: support for muscles; protection of internal organs; production of blood; calcium homeostasis; acid/base buffering; and transmission of sound (Bilezikian et al., 2002; Rauschecker and Shannon, 2002). Bone tissue is the rigid calcified portion of the bone organ, and is critical for many of the functions.

Types of Bone Tissue

Bone tissue is classified as either cortical bone or trabecular bone. Cortical bone is dense and highly mineralized bone tissue that is found on the peripheral regions of bone. Cortical bone is 80–90% mineralized, and constitutes 80% by mass of the bone tissue in the body (Bilezikian et al., 2002). The high density of cortical bone makes it well-suited for the mechanical and structural properties of bone. The thickness and density of cortical bone is loosely correlated to mechanical loading; however, many other variables are also involved (Pearson and Lieberman, 2004). Trabecular bone is found on the interior of bones adjacent to the marrow cavity. It is approximately 80% porous, and exhibits less than 10% of the compressive strength and less than 5% of the compressive modulus of cortical bone (Bilezikian et al., 2002; Miyakoshi, 2004; Rezwan et al., 2006). However, trabecular bone exhibits higher surface area than cortical bone, and is considered more important for bone functions such as calcium homeostasis and acid/base regulation. Figure II.6.7.1 illustrates the hierarchical organization of bone, moving from the tissue level down to the subcellular level.

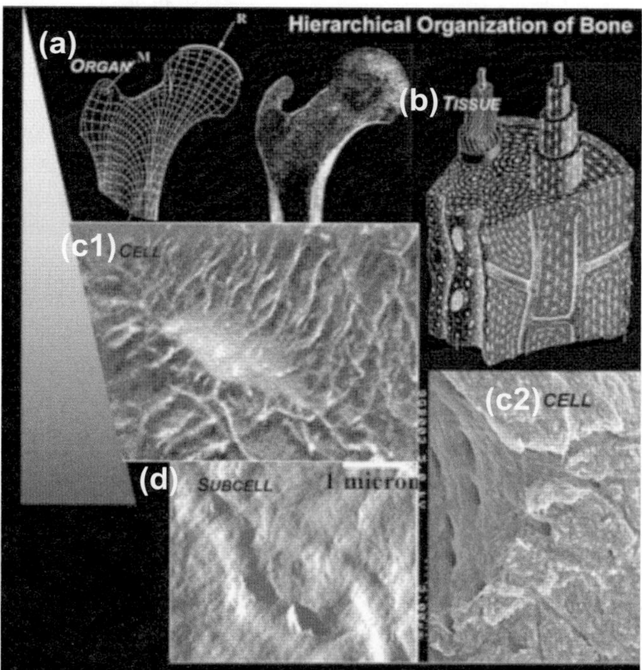

FIGURE II.6.7.1 Hierarchical organization of bone. This figure provides a view of the organization of bone beginning with organ level: (a) and moving to a depiction of the tissue level (b), which illustrates the network of osteocytes organized radially around Haversian canals. An individual osteocyte with the mineralized matrix removed is depicted in (c1) and clearly illustrates multiple processes extending away from the cell body; on the contrary, (c2) depicts the mineralized matrix without the cell, and illustrates on the left-hand side the interior of the lacuna where the osteocyte cell body would reside, and shows canals called canaliculi where the processes extending away from the osteocyte cell body ultimately connect with other osteocyte process. Finally, (d) depicts the interior of canaliculi and shows the presence of striations created by collagen fibers that formed the initial framework for the mineralized matrix (Knothe Tate, 2003).

CELLS INVOLVED

Osteoblasts

There are several distinct cell types involved in the formation and remodeling of bone tissue. These cells are osteoblasts, bone lining cells, osteocytes, and osteoclasts. Osteoblasts are the workhorses of bone formation. Osteoblasts are fully-differentiated cells derived from preosteoblasts or osteoprogenitor cells, which are progenitor cells derived from mesenchymal stem cells found in the bone marrow or the periosteum (Franz-Odendaal et al., 2006; Marie, 2008). The transition from a preosteoblast to an osteoblast occurs when the preosteoblast is stimulated to differentiate via soluble factors, such as bone morphogenetic proteins and wingless-int proteins (Franz-Odendaal et al., 2006; Zaidi, 2007). Once stimulated to differentiate, the preosteoblasts cease proliferation and begin to secrete proteins indicative of an osteoblast phenotype. The new osteoblasts are found at the surface of developing bone tissue, and exhibit a cuboidal morphology. They actively secrete a nonmineralized osteoid matrix at the location of newly forming bone. This osteoid matrix is the organic portion of bone extracellular matrix, and is composed primarily of collagen type I, which makes up approximately 90% of the matrix (Toole and Linsenmayer, 1977; Bilezikian et al., 2002). The remaining portion of the osteoid matrix is composed of proteoglycans and noncollagenous proteins, such as osteopontin, osteocalcin, and osteonectin.

The active production of the osteoid matrix, as well as the presence of the membrane protein alkaline phosphatase, distinguishes the osteoblast phenotype. The osteoid matrix around the osteoblast begins to calcify, and approximately 20% of the buried osteoblasts transition to osteocytes (Franz-Odendaal et al., 2006). The osteoblasts that do not transition to osteocytes undergo apoptosis.

Bone Lining Cells

Bone lining cells, much as their name suggests, are found lining the surface of bone. Unlike osteoblasts on the bone surface, bone lining cells have a long, slender, and flat morphology. Bone lining cells were initially considered to be preosteoblasts (Bilezikian et al., 2002); however, this is no longer thought to be the case. Instead, the current opinion is that osteoblasts that do not undergo apoptosis or differentiate to osteocytes become bone lining cells (Karsdal et al., 2002; Khosla et al., 2008; Matsuo and Irie, 2008). Two of the key phenotypic differences between the bone lining cells and osteoblasts are that bone lining cells express intercellular adhesion molecule 1, and they do not express osteocalcin (Everts et al., 2002). Recent research has shown that bone lining cells anchor hematopoietic stem cells, and provide these stem cells with appropriate signals to keep them in an undifferentiated state (Kollet et al., 2006). The bone lining cells then play a crucial role in the transitions involved with bone remodeling by communicating through gap junctions with osteocytes deep in the bone matrix, promoting differentiation of hematopoietic stem cells into osteoclasts (Kollet et al., 2006; Matsuo and Irie, 2008).

Additionally, the bone lining cells are responsible for preparing the surface of the bone by removing nonmineralized collagen fibrils through the use of matrix metalloproteinases. After remodeling, the bone lining cells deposit a smooth layer of collagen over the bone surface (Everts et al., 2002; Karsdal et al., 2002; Khosla et al., 2008; Matsuo and Irie, 2008).

Osteocytes

Osteocytes are terminally differentiated cells derived from mature osteoblasts that have become encased within a calcified matrix. In the transformation from an osteoblast to an osteocyte, the expression of many of the proteins that constitute an osteoblast phenotype, such as type I collagen, alkaline phosphatase, osteocalcin, and bone sialoprotein, are no longer produced (Franz-Odendaal et al., 2006; Zaidi, 2007). Additionally, osteocytes create a network among themselves by extending many long processes to adjacent osteocytes. This network, the lacunar–canalicular network, is used for nutrient and waste transfer, as well as for communication between the osteocytes via gap junctions (Franz-Odendaal et al., 2006).

The osteocyte cell body resides in the lacuna, and the osteocyte's processes extend out through the canaliculi to adjacent osteocytes and Haversian canals. The Haversian canals provide vasculature to suuply and remove nutrients. The concentric arrangement of the lacunar–canalicular network of osteocytes around a Haversian canal is referred to as an osteon (Bilezikian et al., 2002). The organization of the osteocytes within bone is illustrated in Figure II.6.7.1. Additionally, osteocytes have been implicated as the primary mechanosensors in bone. The mechanotransduction that occurs in osteocytes is believed to be initiated by fluid flux within the canaliculi, created by pressure gradients between lacunae when the bone is loaded. The fluid motion triggers depolarization of the osteocyte process, and is propagated to other osteocytes via gap junctions. The mechanotransduction in osteocytes contributes to the recruitment of osteoblasts or osteoclasts, depending on the loading condition (Wang et al., 2000; Goulet et al., 2008).

Osteoclasts

Osteoclasts, unlike the cells discussed thus far, are multinucleated cells derived from hematopoietic stem cells, as opposed to mesenchymal stem cells. It has been established that hematopoietic stem cells anchored to bone lining cells are induced to differentiate into osteoclasts in response to osteocyte–bone lining cell signaling (Kollet et al., 2006; Matsuo and Irie, 2008). The role of osteoclasts in bone metabolism is the resorption of bone. Osteoclasts exhibit a polarized plasma membrane. Osteoclasts involved in resorbing bone exhibit two distinct plasma membrane regions on the basal surface of the osteoclast; a ruffled portion of the plasma membrane which is where the resorption of the bone occurs, and a sealing region that binds the ruffled border to the bone extracellular matrix (Bilezikian et al., 2002; Vaananen and Laitala-Leinonen, 2008). The combination of the ruffled and sealing regions of the plasma membrane forms the resorption lacuna.

Initially, the osteoclast dissolves the mineralized portion of the bone matrix by secreting hydrochloric acid. After the mineral content is removed, the protein portion of the matrix is degraded by proteolytic enzymes (Kollet et al., 2006; Coxon and Taylor, 2008; Vaananen and Laitala-Leinonen, 2008). The resulting matrix fragments, and potentially the ions created from matrix dissolution, are transported through the osteoclast in vesicles that are emptied into the extracellular space on the basolateral side of the osteoclast. Much like the basal plasma membrane, the basolateral plasma membrane of the osteoclast is polarized and exhibits a functional secretory domain in the middle of the basolateral membrane. The functional secretory domain of the basolateral membrane is rich in microtubules, and is the site where the vesicles are emptied after the transcytotic transportation of the bone matrix degradation products (Kollet et al., 2006; Coxon and Taylor, 2008; Vaananen and Laitala-Leinonen, 2008).

BONE TISSUE DEVELOPMENT

Calcified bone tissue is formed by two distinct modes of ossification or calcification. These methods of ossification are classified as either intramembranous ossification or endochondral ossification (Bilezikian et al., 2002; Franz-Odendaal et al., 2006). The method of ossification depends on the type of bone being formed. Intramembranous ossification is involved in the formation of flat and irregularly shaped bones, such as the cranial bones. Endochondral ossification is involved in the formation of long bones (bones that are longer than they are wide), such as the femur, humerus, and metacarpal (Bilezikian et al., 2002).

Intramembranous Ossification

Intramembranous ossification begins without a preexisting cartilage model. Instead, mesenchymal stem cells form clusters. The mesenchymal stem cells then differentiate into osteoblasts, and the newly formed osteoblasts start to secrete an osteoid matrix. The osteoid matrix is calcified to form bone spicules. Osteoblasts trapped within the bone spicules either differentiate to osteocytes or undergo apoptosis. The bone spicules radiate outward from where the mesenchymal cluster originally formed. Eventually, spicules initiated by separate mesenchymal stem cell clusters join together to create a layer of calcified bone. Mesenchymal stem cells apical to the calcifying tissue differentiate to form the periosteum, whereas those basal to the calcifying tissue differentiate to osteoblasts which form subsequent layers of calcified tissue. The resulting bone tissue is classified as woven bone. Woven bone is formed quickly, and is characterized by randomly oriented collagen fibrils; however, it is not as mechanically viable as lamellar bone. Woven bone will be remodeled over time through resorption and deposition by osteoclasts and osteoblasts to form lamellar bone (Bilezikian et al., 2002; Franz-Odendaal et al., 2006; Shapiro, 2008).

Endochondral Ossification

Endochondral ossification occurs in several steps. Initially, endochondral ossification begins with a pre-existing cartilage template. The cartilage template begins to be calcified. As the cartilage template calcifies, the chondrocytes in the cartilage become hypertrophic and undergo apoptosis (Bilezikian et al., 2002; Tuan, 2004; Franz-Odendaal et al., 2006). Then, mesenchymal stem cells in the membrane surrounding the calcifying cartilage, periosteum, differentiate to osteoblasts. These osteoblasts lay down an osteoid matrix around the exterior of the cartilage template. At the same time, a bud of cells originating from the periosteum invades the interior of the partially calcified cartilage template. This periosteal bud leads to vascularization and innervation of the developing bone. The periosteal bud also supplies mesenchymal and hematopoietic stem cells to the center of the cartilage template. The mesenchymal stem cells differentiate to osteoblasts, and the hematopoietic stem cells differentiate to osteoclasts. These osteoblasts and osteoclasts remodel the partially calcified cartilage into woven bone, which is ultimately remodeled to become lamellar bone. Lamellar bone contains collagen fibrils that are arranged in parallel areas, and exhibits greater strength compared to woven bone (Shapiro, 2008). As the bone tissue created from cells originating from the periosteal bud increases, it radiates outward and eventually joins the bone tissue created by the osteoblasts on the surface of the cartilage template (Bilezikian et al., 2002; Franz-Odendaal et al., 2006; Shapiro, 2008).

Bone Tissue Engineering

Tissue engineering is the application of biological, chemical, and engineering principles toward the repair, restoration or regeneration of living tissue using biomaterials, cells, and factors alone or in combination (Laurencin et al., 1999). Strategies for tissue engineering often focus on one of the three elements; using biomaterials, cells or factors. For example, a common tissue engineering strategy involves fabricating biomaterials into porous scaffolds to facilitate cell growth and the eventual repair, restoration or regeneration of the tissue (Langer and Vacanti, 1993). These biomaterial scaffolds can be used without any further modification *in vivo*. The next iteration in the application of biomaterial-focused tissue engineering strategies involves culturing the biomaterial scaffold seeded with cells *in vitro*. The ultimate strategy for the *in vitro* use of a biomaterial scaffold involves seeding the scaffold and culturing *in vitro* to develop a replacement tissue that, on implantation, functions exactly as did the original host tissue (Langer, 2000).

Applying this tissue engineering strategy, a paradigm for a successful bone graft emerges. This paradigm is that the graft or construct should be osteoconductive, osteoinductive, osteogenic, resorbable or degradable, and possess mechanical properties near to that of the implant site. Osteoconduction refers to the ability of a scaffold or implant to promote attachment of osteoblastic cells on the surface and throughout the interior of the scaffold or implant. In an *in vitro* setting, osteoconduction is seen as an ability to promote the attachment, migration, and proliferation of osteoblasts (Kneser et al., 2002). Osteoinduction refers to the ability of a scaffold or implant to promote the differentiation of mesenchymal stem cells down an osteoblastic lineage, ultimately leading to the formation of mineralized tissue. Osteoinduction can also be viewed as an ability to promote phenotype progression of an osteoblast from an early osteoblast to a mature osteoblast, followed by differentiation to an osteocyte (Kneser et al., 2002).

Osteogenicity refers to the ability of a scaffold or implant to promote *de novo* bone formation, which would occur in the absence of host cell invasion. For a scaffold to be osteogenic, cells would need to be seeded on the scaffold prior to implantation (Kneser et al., 2002). The necessity of a bone tissue engineering construct to be degradable arises from the fact that bone is constantly remodeling. A non-resorbable or non-degradable implant would impede the natural remodeling process of bone, and extend the time it takes for the organ to return to natural function (Hutmacher, 2000). Finally, the graft, scaffold or implant should have mechanical properties that match that of the native bone tissue. The range for the mechanical properties depends on whether the bone tissue is cortical or trabecular. For trabecular bone the compressive strength varies from 4–12 Mpa, and the compressive modulus varies from 100–500 Mpa (Rezwan et al., 2006). Cortical bone exhibits a compressive strength from 130–180 Mpa, and a compressive modulus from 12–18 Gpa (Rezwan et al., 2006).

A problem with mechanical properties that exceed these ranges is stress shielding. Stress shielding results when the load on the bone is redistributed, with the scaffold or implant being the loadbearing region, and the surrounding bone being unloaded. The result of stress shielding is osteopenia of the bone surrounding the implant (Pitto et al., 2007). To solve the bone tissue engineering paradigm, there are presently a range of bone grafts and bone graft substitutes that fulfill all or some of the bone tissue engineering paradigm.

BONE GRAFTS

Autograft

Bone grafts are pieces of bone that are harvested from the patient, a donor or a cadaver, and placed at the desired site of bone repair, regeneration or restoration as needed. The long standing, and considered "gold standard," bone graft solution to the bone tissue engineering paradigm is an autograft (Greenwald et al., 2001; Kneser et al., 2002; Laurencin et al., 2006). Autografts are sections or fragments of bone removed from one site on the patient, typically the iliac crest, and implanted to another site based on need. Figure II.6.7.2 depicts an example of where two autografts were harvested from a patient, and shows the harvest site beginning to heal and reform bone after five months, with the help of a metal plate to serve as a template for the new bone growth.

Autografts harvested from the iliac crest are mostly trabecular bone with a thin shell of cortical bone. Since autografts originate within the patient, they are readily incorporated at the implant site and rarely elicit any immune responses, which allow autografts to have excellent wound healing properties. Autografts fulfill all four elements of the bone tissue engineering paradigm, primarily because they consist of native bone tissue moved from one region of a patient's body to another region. However, autografts have a few drawbacks; there is often donor site morbidity indicated by necrosis and infection at the location of autograft harvest that may cause the patient more pain from the harvest site than the implant site (Greenwald et al., 2001; Laurencin et al., 2006). Additionally, autografts are limited in availability to the amount of tissue that can be harvested from the donor site (Greenwald et al., 2001; Laurencin et al., 2006). The problem of autograft bone tissue availability increases in cases where the need is the highest; those that involve osteoporotic, pediatric or patients afflicted with bone cancer (Meister et al., 1990). It is these shortcomings plaguing the autograft that has increased the effort to find other bone graft substitutes.

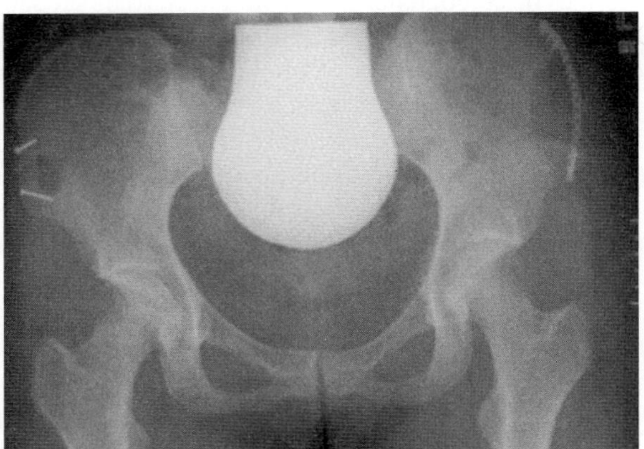

FIGURE II.6.7.2 X-ray of an individual that required an autograft harvest from both iliac crests. The right hip (left side of image) was harvested five months prior to the left hip. Both were reconstructed with a metal plate fixed with screws (most visible on the left hip). Notice how even five months post-harvest there persists a significant amount of bone that has not regrown (Huemer et al., 2004).

Allograft

The next bone graft solution to the bone tissue engineering paradigm is the allograft. Allografts are bone fragments obtained from donors or cadavers that are typically first frozen, irradiated and/or lyophilized. These steps are implemented to reduce the likelihood of disease transmission. Allografts are readily available in an array of shapes and sizes. Since allografts originate from a donor or cadaver there is no additional surgical site on the patient, which removes complications such as donor site morbidity (Greenwald et al., 2001; Kneser et al., 2002; Laurencin et al., 2006). However, the sterilization procedures performed on allograft bone tissue are not without consequence. In comparison to autografts, allografts are less osteoconductive, less osteoinductive, and are not osteogenic (Greenwald et al., 2001; Laurencin et al., 2006). Additionally, allografts that have been lyophilized exhibit much less mechanical integrity compared to autografts (Laurencin et al., 2006). Due to allografts being non-native tissue, they have been shown to occasionally produce an immune response, which requires allograft recipients to be placed on immunosuppressant drugs to prevent rejection of the allograft tissue (Paskert et al., 1987). The complications that arise from autografts and allografts have fueled the search for bone graft substitutes.

BONE GRAFT SUBSTITUTES

Bone graft substitutes can offer solutions to the bone tissue engineering paradigm, and are based on the tissue engineering concepts that arise from the definition of tissue engineering. These tissue engineering principles lead to bone graft substitutes that can be classified as those based on biomaterials, cells, factors or any combination of the three.

Allograft-Based Substitutes

Biomaterial-based bone graft substitutes can be further subdivided into allografts, natural polymers, synthetic polymers, and ceramics. Allograft-based bone graft substitutes use allograft bone tissue that has been thoroughly sterilized, decellularized, and demineralized. The methods by which manufacturers sterilize, decelluarize, and demineralize the bone graft substitutes are carefully controlled to create a product that retains the collagen, non-collagenous proteins, and some of the growth factors present in the original bone tissue (Gazdag et al., 1995). The result is demineralized bone matrix (DBM), which has been used in a variety of commercially available bone graft substitutes either as is or mixed with glycerol, hyaluronic acid or calcium phosphates to improve the handling and performance characteristics of the product (Martin et al., 1999; Schwartz et al., 2007). Osteotech, Inc.® has a line of DBM products under the trade name Grafton®. The Grafton® line of products contain materials that are simple DBM fragments, such as Grafton® DBM Crunch, as well as intact pieces of DBM that has been precut to a desired shape, such as Grafton® DBM Matrix PLF, DBM Matrix Plugs, and DBM Matrix Strips. Osteotech, Inc.® also has a number of commercially available DBM products that have been mixed with glycerol and other proprietary agents to create an injectable gel, Grafton® DBM Gel, moldable putty, Grafton® DBM Putty, and flexible strips, Grafton® DBM Flex and A-Flex™. Figure II.6.7.3 depicts a representative image of DBM formed into a simple rectangular solid, and also shows the presence of cells growing on the DBM. DBM-based bone graft substitutes exhibit a variable degree of osteoinduction based on the processing parameters (Kneser et al., 2006; Laurencin et al., 2006).

There is evidence that supports improved osteoinduction with demineralization due to exposure of soluble factors that would be occluded in mineralized bone (Gazdag et al., 1995; Peterson et al., 2004); however, DBM-based bone graft substitutes have limited osteoconductivity, no osteogenicity, and mechanical properties that are less than the desirable range (Gazdag et al., 1995; Peterson et al., 2004).

Natural Polymer-Based Substitutes

Natural polymers are gaining interest among the research community for bone tissue engineering applications, and there are also commercial bone graft substitutes derived from natural polymers available. One example of a commercially viable natural polymer product is Healos® from DePuy Orthopaedics, Inc. Healos® is a collagen microfiber matrix that has been coated with hydroxyapatite (Neen et al., 2006). The recommended use of Healos® involves coating it with bone marrow aspirate prior to implantation. By supplying the matrix with bone marrow aspirate there are progenitor cells present on the Healos® matrix, which makes it osteogenic in addition to being osteoconductive (Neen et al., 2006). There is no evidence that Healos® is osteoinductive, and it has poor mechanical properties (Neen et al., 2006). In addition to collagen, fibrin and chitosan are two other natural polymers that are being investigated for bone tissue engineering applications (Khan et al., 2008; Osathanon et al., 2008). The structures created by these polymers are typically fibers or foams (Wahl and Czernuszka, 2006; Osathanon et al., 2008; Song et al., 2008). These structures provide excellent osteoconduction; however, osteoinduction, osteogenicity, and mechanical properties are less than those provided by autograft tissue (Wahl and Czernuszka, 2006; Khan et al., 2008).

Synthetic Polymer-Based Substitutes

Synthetic polymer solutions for bone tissue engineering applications are varied and abundant in current research. The use of synthetic polymers provides control over the surface chemistry, degradation kinetics, and geometry in much finer detail than can be accomplished with natural polymers. Clinically, only a handful of synthetic polymers are US Food and Drug Administration (FDA) approved for use in non-life-threatening applications, such as bone graft substitutes. These are the following poly(α-hydroxy esters): poly(lactide); poly(glycolide); poly(lactide-*co*-glycolide); and poly(caprolactone). Figure II.6.7.4 illustrates a lattice structure made of poly(caprolactone) rods that has been implanted into a pig, and imaged three months post-implantation with μCT (micro-computed tomography) to demonstrate where new bone formation is occurring, which is not adjacent to existing bone tissue (Jones et al., 2004).

The mechanical properties of the construct are not documented, and the cracking and plastic deformation

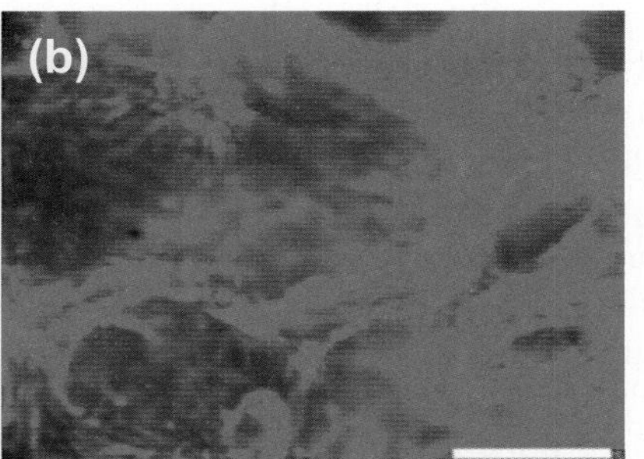

FIGURE II.6.7.3 Demineralized bone cut into a rectangular solid (a); and a high magnification fluorescent image showing the presence of viable cells on the demineralized bone matrix (b) (Ma et al., 2007).

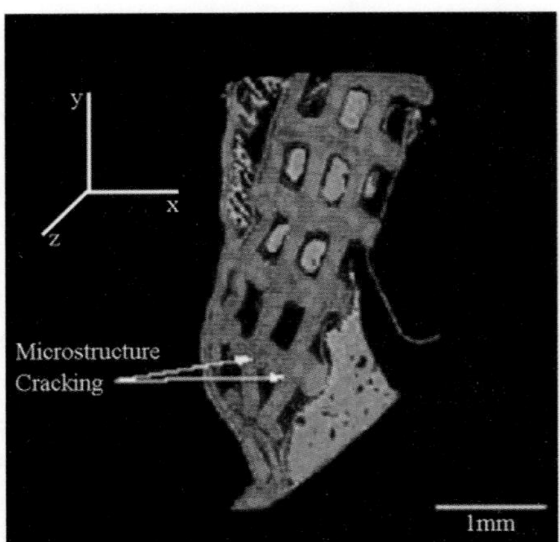

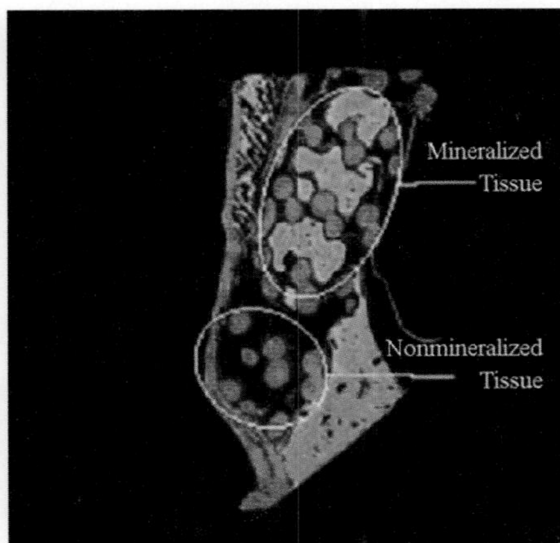

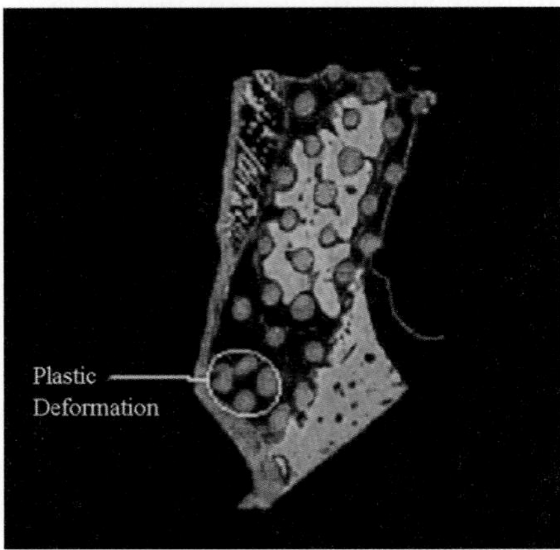

FIGURE II.6.7.4 Representation obtained of a synthetic bone graft after three months *in vivo* using μCT. The synthetic bone graft is a simple poly(caprolactone) (PCL) scaffold (red), with nonmineralized tissue (black), and mineralized tissue (blue green). The native bone is visible in the lower right corner of each image. The three images are different sections separated by 280 μm. The poly(caprolactone) scaffold was made by a rapid prototyping process, and consists of a simple three-dimensional lattice of connected rods. Also visible in the series of images are two defects caused by either degradation of the PCL, which the authors note that after three months has caused the diameters of the rods to shrink from 500 μm to 300 μm, or due to excessive loading on the scaffold compromising the material prior to degradation (Jones et al., 2004).

visible after three months indicates the material may not be mechanically viable. The first porous bone tissue engineering constructs fabricated from poly(α-hydroxy esters) were foams created by various porogen leaching techniques (Thomson et al., 1995). The porogen in these foams is typically a substance that readily dissolves in water, such as gelatin, salt or sugar (Thomson et al., 1995; Mooney et al., 1996). These scaffolds are created by dissolution of the poly(α-hydroxy esters) in an organic solvent. The dissolved polymer is then mixed with the porogen and cast. After the solvent has evaporated, the construct is immersed in water to remove the porogen (Thomson et al., 1995; Mooney et al., 1996; Rezwan et al., 2006). This technique can produce scaffolds that are highly porous.

The primary issue with porogen leaching is that the increase in porosity is directly correlated with a decrease in mechanical integrity. Additionally, at low porosities, which are more robust mechanically, the polymer surrounds each individual porogen, resulting in poor interconnectivity among the pores (Mooney et al., 1996; Rezwan et al., 2006). The sintered microsphere scaffold fabricated from poly(α-hydroxy esters) succeeded the scaffolds created with porogen leaching, and also resolved several of the problems with scaffolds created by porogen leaching. The maximum porosity achievable by the sintered microsphere scaffold is only around 45%, based on random packing of spheres; however, the interconnectivity of the porosity is 100%, which was a dramatic improvement over the scaffolds fabricated with porogen leaching (Devin et al., 1996; Mooney et al., 1996; Borden et al., 2002a,b; Rezwan et al., 2006). More recently, microscale scaffolds have been fabricated with a very specific architecture from poly(α-hydroxy esters) by using solid free-form fabrication techniques (Sachlos and Czernuszka, 2003; Ge et al., 2008). Solid free-form fabrication uses computer aided design to build structures layer by layer through techniques such as stereolithography, selective laser sintering, and three-dimensional printing (Hutmacher et al., 2004).

All of the scaffolds created with the above microscale technologies provide similar performance. The mechanical integrity of the scaffolds can be fabricated into a range that suits bone tissue applications, the scaffolds degrade in a controllable manner based on the selection of the poly(α-hydroxy esters), and they are all osteoconductive. However, for the preparation of microscale structures, each of the techniques is limited to a resolution an order of magnitude larger than a cell, and the resulting materials are not osteoinductive nor are they osteogenic without supplementing the structure with growth factors or osteoblastic cells (Rezwan et al., 2006). The research community has moved to examine scaffolds composed of poly(α-hydroxy esters) that exhibit a subcellular dimension. These scaffolds are almost invariably fiber based, and are made either with electrospinning, phase separation for crystalline poly(L-lactide), and precipitation in a non-solvent of a continuous fiber stream from a polymer solution (Yoshimoto et al., 2003; Smith and Ma, 2004; Tuzlakoglu et al., 2005). The diameter of these fiber scaffolds can vary from 50 nanometers to several micrometers (Yoshimoto et al., 2003; Smith and Ma, 2004). The subcellular dimension provided by nano- and microfibers has illustrated improved osteoconductivity as compared to the supercellular dimension of the microstructures covered previously. Additionally, there is some evidence that nanofibers may promote osteoinduction; however, the gains that nano- and microfibers make with osteoconduction and osteoinduction are compromised by the mechanical properties under compression of the nano- and microfiber structures (Patel et al., 2007; Hu et al., 2008).

Nano- and microfiber scaffolds exhibit near negligible compressive strength when compared to the microscale structures created by sintered microspheres, solid free-form fabrication, and porogen leaching (Kjelstrup-Hansen et al., 2006; Rezwan et al., 2006). The next degree of control being explored for bone tissue engineering constructs involves moving away from the FDA approved poly(α-hydroxy esters) and on to other biodegradable polymers that provide better degradation by-products or improve the osteoconductivity of polymers through the incorporation of integrin binding peptides, such as RGD (Arginine-Glycine-Aspartic Acid peptide) or, within the synthetic polymer structure. A noted drawback of poly(α-hydroxy esters) is that they degrade into carboxylic acids, which can be immunogenic (Bostman and Pihlajamaki, 1998; Mosier-Laclair et al., 2001).

An additional drawback to the usage of poly(α-hydroxy esters) is that they undergo bulk degradation, which results in sudden failure of the scaffold. Similar to the poly(α-hydroxy esters), poly(propylene fumarate) is a polyester that can be used to fabricate similar scaffolds to the poly(α-hydroxy esters), and also shares the same problems with bulk degradation and acidic degradation products (Lee et al., 2006; Rezwan et al., 2006). Poly[(amino acid ester)phosphazenes] address the negative degradation products found with poly(α-hydroxy esters), and have recently been investigated for bone tissue engineering applications. Poly[(amino acid ester)phosphazenes] degrade into amino acids, which are much easier for the body to metabolize, and a buffer solution consisting of ammonia and phosphate, which prevents any change in the pH potentially brought about by the increase in the concentration of amino acids. Poly[(amino acid ester) phosphazenes] have also exhibited a more favorable surface erosion degradation mechanism (Allcock et al., 1994; Ibim et al., 1997). Additionally, poly[(amino acid ester)phosphazenes] are suitable for many of the fabrication procedures discussed previously, such as sintered microsphere scaffolds, porogen leaching scaffolds, and electrospun nanofibers (Laurencin et al., 1996; Kumbar et al., 2006a; Brown, et al., 2008b).

The scaffolds fabricated from poly[(amino acid ester) phosphazenes] have illustrated osteoconduction, but studies investigating the osteoinductivity and mechanical properties have yet to be reported (Conconi et al., 2006; Kumbar et al., 2006a). Figure II.6.7.5 provides a representative image of sintered microsphere scaffolds, porogen leaching scaffolds, and electrospun nanofiber scaffolds created with degradable polyphosphazenes. Polyanhydrides are surface eroding polymers similar to polyphosphazenes; however, little research has been performed investigating porous polyanhydride scaffolds for bone tissue engineering (Muggli et al., 1999; Rezwan et al., 2006).

Ceramic-Based Substitutes

Ceramic-based biomaterials are prevalent and widespread as bone graft substitutes. These ceramic biomaterial bone graft substitutes are made primarily from calcium phosphates, calcium sulfate, and Bioglass®, which is a glass formulation containing lower amounts of silicon dioxide, and higher amounts of sodium oxide and calcium oxide compared to conventional glass. This specific glass formulation is bioactive, and undergoes dissolution in the body. Calcium phosphate bone graft substitutes are usually either tricalcium phosphate or hydroxyapatite, which is the primary mineral in bone. One example of a commercially available hydroxyapatite-based bone graft substitute is Pro-Osteon™ from Biomet® Osteobiologics. The Pro-Osteon™ bone graft substitute begins as natural coral, which is primarily calcium carbonate. The coral is then treated with a hydrothermal process in the presence of ammonium phosphate, to convert the calcium carbonate structure to hydroxyapatite (Ben-Nissan, 2003). The resulting structure exhibits a pore structure similar to trabecular bone. Figure II.6.7.6 is a depiction of the coralline hydroxyapatite resulting from the above hydrothermal process. Osteohealth® produces a product through the removal of the organic portion of bovine bone called Bio-Oss®. Bio-Oss® retains the structure of the hydroxyapatite formed in the bovine trabecular bone from which it is obtained.

A similar product to Bio-Oss® is Orthograf® which is manufactured by Dentsply, and which is also derived from bovine bone that has been treated to remove all

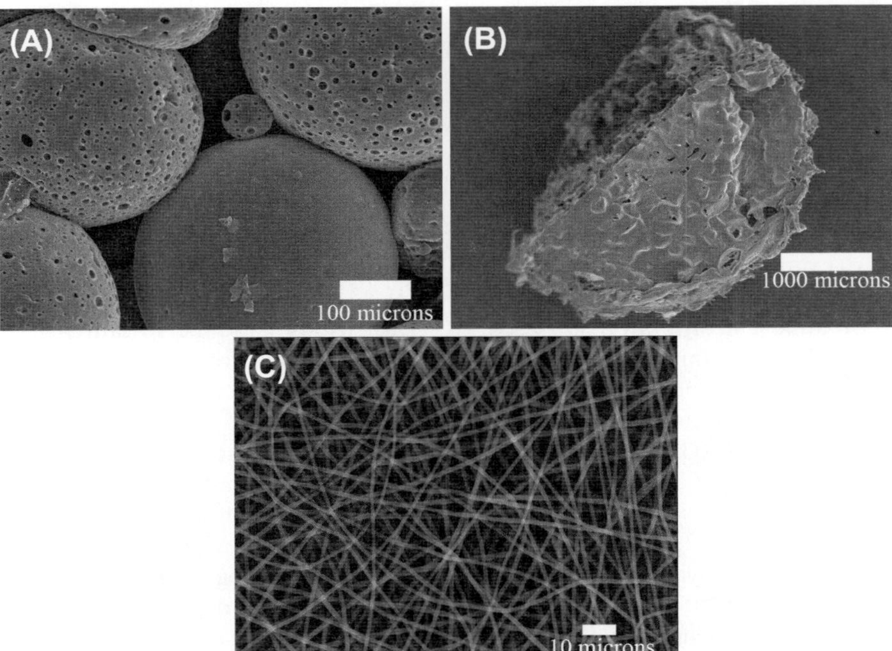

FIGURE II.6.7.5 Scanning electron microscope (SEM) images of three scaffold architectures fabricated from poly[(amino acid ester)phosphazenes]. (A) Sintered microsphere scaffold composed of poly[bis(ethyl alaninato)phosphazene]; (B) Scaffold made from leaching salt from poly[(ethyl alaninato-co-methylphenoxy)phosphazene]; (C) Nanofibers created by electrospinning poly[bis(methylphenoxy)phosphazene] (Nair et al., 2004). Images (A) and (B) are previously unpublished.

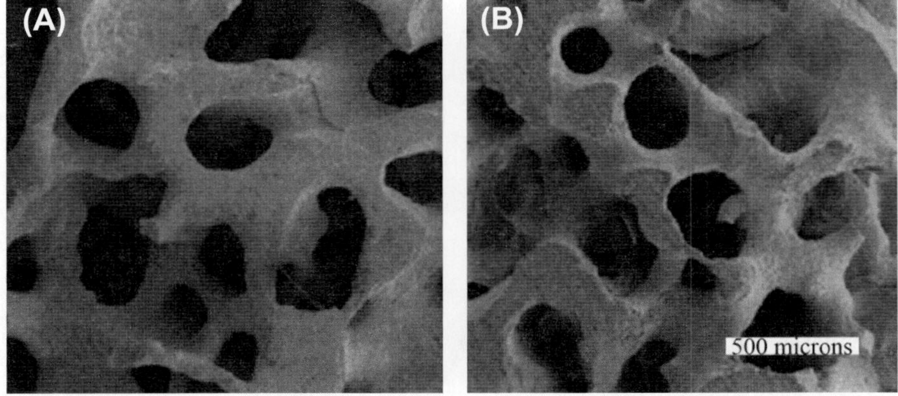

FIGURE II.6.7.6 SEM image of coral prior to conversion to hydroxyapatite (A); and after hydrothermal conversion to coralline hydroxyapatite (B) (Ben-Nissan, 2003).

organic components. All of the hydroxyapatite-based ceramic bone graft substitutes are slowly resorbed by the body as the bone surrounding the implant remodels. Hydroxyapatite-based ceramic bone graft substitutes are also osteoconductive; however, the porous hydroxyapatite bone graft substitutes typically have compressive strength and moduli that fall below the range of trabecular bone (Rezwan et al., 2006). Whether hydroxyapatite-based biomaterials are osteoinductive is a controversial topic. Osteoinduction appears to occur with hydroxyapatite and other calcium phosphate-based biomaterials.

Initially, the osteoinductivity of hydroxyapatite and other calcium phosphates was believed to be a property of the ceramic itself (Damien and Parsons, 1991). However, recent evidence suggests that it is a property of the microstructure of the material, which would be similar to results observed with polymer scaffolds (Yuan et al., 2002; Giannoudis et al., 2005; Li et al., 2008). Two calcium phosphate structures, one composed of hydroxyapatite and the other of a blend with 85% hydroxyapatite and 15% tricalcium phosphate, were fabricated into similar macrostructures; however, the hydroxyapatite/tricalcium phosphate structure also had a subcellular microstructure, and was found to be more osteoinductive than the hydroxyapatite with no subcellular microstructure (Yuan et al., 2002). Additionally, an examination of several distinct calcium phosphates with different chemical and crystal structures fabricated into similar macrostructures illustrated no significant difference in the expression of phenotype markers (Wang et al., 2004). Furthermore, examination of chemically identical types of hydroxyapatite formed into different geometries produced pronounced differences in the expression of phenotype markers, with the hydroxyapatite structure exhibiting a subcellular microstructure being more osteoinductive than the hydroxyapatite with no subcellular microstructure (Li et al., 2008).

An additional type of calcium salt is calcium sulfate. Wright Medical Technology produces a dense bone graft substitute from calcium sulfate called Osteoset®. Osteoset® has been shown to be resorbed nine months post-implantation, and has also been illustrated to be osteoconductive; however, conflicting evidence is found over whether Osteoset® and other calcium sulfate bone graft substitutes improve healing in bone (Clokie et al., 2002; Petruskevicius et al., 2002). Additionally, calcium sulfate bone graft substitutes have been found to be potentially immunogenic (Robinson et al., 1999; Lee et al., 2002).

Of the ceramic-based bone graft substitutes, those fabricated from Bioglass® are the most intriguing. Bioglass® was developed in the late 1960s and has been shown to fit the bone tissue engineering paradigm better than the other ceramics covered (Hench et al., 1971). Porous scaffolds have been fabricated from Bioglass® in a number of ways. One way, that produces a 90–95% porous structure, involves a foam replacement technique where a polyurethane foam is coated with a Bioglass® slurry. The coated foam is then heated, which burns the polyurethane out and processed through a heat-treatment schedule to fuse the Bioglass® particles and obtain the desired density and crystallinity (Vargas et al., 2009). Another technique sinters Bioglass® fibers to form porous mats of Bioglass® fibers. These Bioglass® fiber mats have been shown to be osteoconductive, and promote phenotype progression of preosteoblast cells, which suggests they may be osteoinductive. The compressive strength and modulus of these Bioglass® fiber rafts are in the range of trabecular bone, and they exhibit adequate porosity of around 44% (Brown et al., 2008a). Despite the advantages of Bioglass® over the other ceramics covered, there are only a few commercial applications. Perioglas® is a Bioglass® particulate produced by Novabone™ to serve as bone filler or graft extender. The other commercial Bioglass® product is Biogran®, which is also a particulate and is currently produced by Biomet® 3i.

Cell-Based Substitutes

Cell-based strategies for bone tissue engineering, similar to biomaterial-based strategies for bone tissue engineering, fall into several categories. These are transplantation of autogenous progenitor cells, transplantation of autogenous progenitor cells that have been expanded and/or differentiated in culture prior to implantation, transplantation of genetically modified cells, and transplantation of *ex vivo* generated tissue (Muschler et al., 2004). The simplest of these strategies is the transplantation of autogenous progenitor cells. Typically this is performed by aspirating bone marrow and placing the aspirate at the defect or surgical site, unmodified or centrifuged to remove red blood cells and hematopoietic cells, including monocytes, from the bone marrow. The transplantation of autogenous stem cells provides osteogenic potential to the defect, but does not provide osteoconductivity, osteoinductivity or mechanical strength (Connolly et al., 1989).

The second method involves culturing the progenitor cells extracted in the bone marrow in culture. This can increase the number of progenitor cells available to implant, but it also comes with risks. The added time in culture *in vitro* increases the likelihood of bacteria contaminating the progenitor cell population. Successful expansion of the progenitor cells *in vitro* can ultimately improve the healing at the defect or surgical site. Similar to the non-expanded cells, progenitor cells expanded *in vitro* prior to implantation are osteogenic, but not osteoconductive, osteoinductive or mechanically viable (Patterson et al., 2008).

The third strategy for cell-based bone tissue engineering is the use of genetically modified cells. This is similar to the use of *in vitro* expanded autogenous progenitor cells, except the progenitor cells have been treated with an adenovirus to express a protein of interest. For bone tissue, often the progenitor cells will be transfected to express bone morphogenetic protein 2, (BMP-2). BMP-2-expressing progenitor cells provide the cell-based construct to be osteoinductive as well as osteogenic; however, the construct would still not improve in regard to osteoinductivity or mechanical viability (Lieberman et al., 1999; Cui et al., 2006).

The final strategy for cell-based bone tissue engineering is the most ambitious and the least clinically relevant at present. This is the *in vitro* culture of autogenous or embryonic stem cells with the aid of a bioreactor to ultimately produce a viable piece of bone tissue *ex vivo*. The major limitation in this strategy is material transport in developing tissue, since the issue of developing a vascularized construct has not been solved (Muschler et al., 2004). Despite this limitation there is evidence that mesenchymal stem cells can be induced to form small spheroids and differentiate into osteoblasts (Kale et al., 2000). If successful, the *ex vivo* formation of bone tissue would meet all five criteria outlined in the bone tissue engineering paradigm.

Growth Factor-Based Substitutes

The next strategy for bone tissue engineering is based on the use of factors. Two factors that have been shown to be effective in a clinical setting and are approved by

TABLE II.6.7.1 Summary of the Properties Illustrated by the Different Types of Non-Composite Bone Graft Substitutes and How They Relate to the Desirable Characteristics of an Ideal Bone Graft Substitute

	Osteoconductive	Osteoinductive	Osteogenic	Mechanical Match
Allograft-based	Yes	Yes	No	No
Microscale biomaterials	Yes	No	No	Yes, within the range of trabecular bone
Nanoscale biomaterials	Yes	Potentially, supported by evidence with nanofibers made by self-assembly or phase separation	No	No
Ceramics	Yes	Potentially, supported by evidence for Bioglass®	No	Yes, range from less than trabecular bone to more than cortical bone depending on porosity
Cells	No	No	Yes	No
Growth factors	No	Yes	No	No

the FDA are recombinant human bone morphogenetic proteins 2 and 7, rhBMP-2 and rhBMP-7, respectively. Bone morphogenetic proteins are part of the transforming growth factor-β super family, and have been shown to exhibit osteoinductive potential (Fouletier-Dilling et al., 2007). Between the two rhBMPs, rhBMP-2 has been shown to be more osteoinductive than rhBMP-7 (Govender et al., 2002b; Ripamonti et al., 2007). Due to the soluble nature of both rhBMPs, they are typically packaged with some sort of carrier, often as simple as a collagen sponge (Gautschi et al., 2007). RhBMP-2 combined with a collagen sponge has been shown to perform better in spinal fusions than the "gold standard" autograft (Geiger et al., 2003).

In addition to spinal fusions, rhBMP-2 combined with a collagen sponge has illustrated effectiveness in fracture healing, where the rhBMP-2/collagen construct illustrated a significant improvement in fracture healing after 10 weeks (Govender et al., 2002a). RhBMP-2/collagen constructs have illustrated improved healing of critical size defects in animal models (Geiger et al., 2003). The use of rhBMP-7 in spinal fusions has been proven more effective than the autograft in high risk patients (Govender et al., 2002b; Gautschi et al., 2007). Medtronic®, Inc. developed a collagen sponge that is combined with rhBMP-2 to improve spinal fusion. This collagen sponge combined with rhBMP-2 and a spinal cage is marketed under the trade name of INFUSE®. The primary limitation to the usage of rhBMPs in bone tissue engineering applications is that they often cost a significant amount more than comparable procedures, and consequently the use of rhBMPs is primarily only in high risk cases where all other options have been exhausted (Gautschi et al., 2007).

Composite Substitutes

The final strategy for bone tissue engineering is based on composites which combine two or more of the elements detailed above. The goal of creating composites is to combine the benefits of each component. Table II.6.7.1 summarizes how the different bone tissue engineering strategies correlate to the bone tissue engineering paradigm. A common composite encountered consists of a biomaterial structure seeded with osteoblasts or osteoblastic cells. This is typically accomplished by either seeding of the biomaterial structure with osteoblast progenitor cells *ex vivo* just prior to implantation *in vivo*, or seeding of the biomaterial structure with osteoblasts or osteoblast progenitor cells followed by culture *in vitro* for several days to weeks prior to implantation.

An example of this technique that is presently used clinically is seeding Grafton® DBM with bone marrow aspirate from the patient *ex vivo*, followed by implantation of the Grafton® DBM/bone marrow aspirate composite at the desired surgical site. This composite combines the osteogenic potential of the progenitor cells in the bone marrow aspirate with the osteoinductive and osteoconductive properties of the Grafton® DBM, and has been shown to be comparable to an autograft in formation of bone at the surgical site (Russell, 2000; Lindsey et al., 2006).

Another example of a cell/biomaterial composite involves seeding a porous polymer scaffold created by selective laser sintering of poly(caprolactone) with fibroblasts transfected to express bone morphogenetic protein-7, BMP-7 (Williams et al., 2005). The selective laser sintered poly(caprolactone) scaffold provides a structure with millimeter scale features that is mechanically adequate and osteoconductive, and the incorporation of the BMP-7-expressing fibroblasts makes the composite construct osteoinductive. However, the fibroblasts are not expected to differentiate to osteoblasts, and consequently this composite is not osteogenic. A second example of composite scaffolds for bone tissue engineering involves the incorporation of a ceramic calcium phosphate with a polymer scaffold. This can be accomplished through several different methods, such as suspension of calcium phosphate particles with the polymer phase prior to processing to form a scaffold, spontaneous formation of calcium phosphates within the polymer phase during processing, and coating of calcium phosphates on the surface of a biomaterial scaffold through soaking in simulated body fluid. A composite created by incorporation of calcium phosphates with the polymer phase during processing is illustrated by electrospinning a slurry of β-tricalcium phosphate particles in dissolved poly(caprolactone) to create a scaffold consisting of nanofibers exhibiting β-tricalcium phosphate particles on the surface of the fibers (Erisken et al., 2008). The incorporation of the β-tricalcium phosphate particles improves the mechanical properties of the nanofiber scaffold, and may also improve the osteoinductivity of the scaffold; however, an improvement in the osteoinductivity of the nanofiber/β-tricalcium phosphate composite has not been illustrated (Erisken et al., 2008).

A composite created by spontaneous formation of calcium phosphates during processing is illustrated by a technique where microspheres are created by an emulsion technique where the organic phase contains the polymer, and the aqueous phase contains calcium and phosphate salts, which through careful control of temperature and pH precipitate onto the microsphere surface as amorphous hydroxyapatite (Khan et al., 2004). The microspheres are heat sintered to form a porous scaffold, and could yield promising results as a bone graft substitute; however, no published evaluation of these composite scaffolds' performance *in vitro* or *in vivo* is currently available. An example of a ceramic/polymer composite formed by precipitation of calcium phosphates on the surface of a scaffold is demonstrated by a technique where PLAGA (poly(lactic-co-glycolic acid)) microspheres are formed and placed in simulated body fluid, which cause nucleation of calcium phosphates on the surface of the microspheres. The microspheres are then compression molded into porous scaffolds (Davis et al., 2008). These scaffolds illustrate improved osteoconductivity *in vitro*, but do not show any improvement in osteoinductivity over the uncoated microsphere control (Davis et al., 2008).

A third example of a composite scaffold for bone tissue engineering involves the incorporation of growth factors with a polymer or ceramic scaffold. An advanced example of this concept involved the development of cross-linked poly(vinyl pyridine) microspheres containing either rhBMP-2 or rhBMP-7, which were then suspended in a PLAGA foam (Buket Basmanav et al., 2008). The concentration of the poly(vinyl pyridine) and the degree of cross-linking allowed for a staged release of rhBMP-2 and rhBMP-7. The rhBMP-2 was entrapped in the lower concentration and less cross-linked microspheres, and was released more rapidly than the rhBMP-7. The result of this rhBMP-microsphere-loaded PLAGA foam was an increase in the differentiation of bone marrow-derived mesenchymal stem cells *in vitro* as compared to the PLAGA foam loaded with control microspheres not containing any rhBMPs (Buket Basmanav et al., 2008). The results of this investigation are typical of what is expected by including BMPs into scaffolds, and provides a strategy for making an osteoconductive scaffold osteoinductive as well (Jeon et al., 2007; Buket Basmanav et al., 2008; Kempen et al., 2008). A critical issue with constructs containing growth factors, which is shared by growth factor strategies in general, is that they are prohibitively expensive for developing medical strategies to treat typically non-life-threatening injuries. The above examples of strategies for composite structures for bone tissue engineering represent only a small portion of the composite structures that have been investigated; however, the above examples do provide an accurate representation of the desired outcomes in preparing composite structures for bone tissue engineering.

POROSITY IN BONE GRAFT SUBSTITUTES

An important characteristic of a successful biomaterial-based bone graft substitute that promotes all the features of the bone tissue engineering paradigm involves the pore structure of the scaffold or construct. Without an adequate pore structure, migration into the scaffold is restricted, which subsequently limits the potential of the scaffold to be osteoconductive, osteoinductive, and osteogenic. The pore structure of the scaffold, both pore diameter and porosity, is a critical component in allowing cellular migration.

The concept of pore structure and cellular migration is similar across cell lines, and as such, this discussion focuses primarily on the characteristic dimension of the scaffold and the subsequent requirements to promote cell migration into the scaffold. For microscale scaffolds fabricated through techniques such as microsphere sintering, gas foaming, and particulate leaching the critical design aspects involve maintaining interconnected pores and pore diameters above 40 μm (Akay et al., 2004; Karageorgiou and Kaplan, 2005). An interconnected porosity is necessary to facilitate migration of cells throughout the scaffold, as well as to maintain a supply of nutrients and removal of waste from the cells on the scaffold interior. It was found that scaffolds with large, 100–300 μm, pore diameters facilitated faster migration throughout the scaffold (Borden et al., 2002b; Karageorgiou and Kaplan, 2005); however, despite the decreased rate of migration, the lower, 40 μm, pore diameter scaffolds achieved the same level of cellular penetration as those with larger pore diameters (Karageorgiou and Kaplan, 2005).

When the characteristic dimension of the scaffold shrinks to the nanoscale, these same concepts do not apply (Stevens and George, 2005). No longer does pore diameter appear to be as influential, since nanofiber scaffolds fabricated from self assembly and thermally-induced phase separation have illustrated cell invasion into and throughout the interior of the scaffold, despite pore diameters ranging from 300 nm–10 μm (Zhang, 2003; Semino et al., 2004; Silva et al., 2004; Chen et al., 2006; Horii et al., 2007). Two characteristics shared by

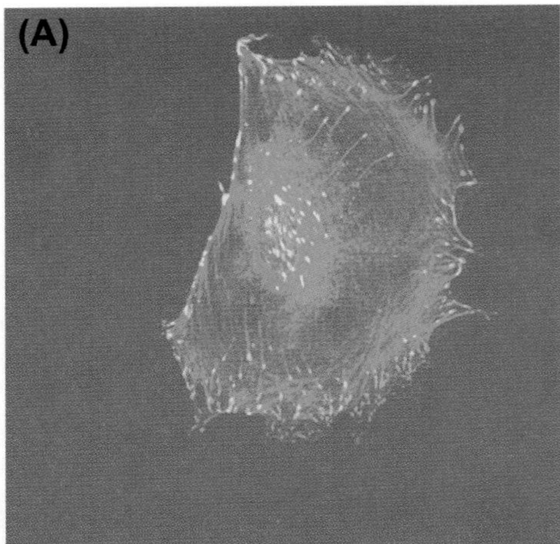

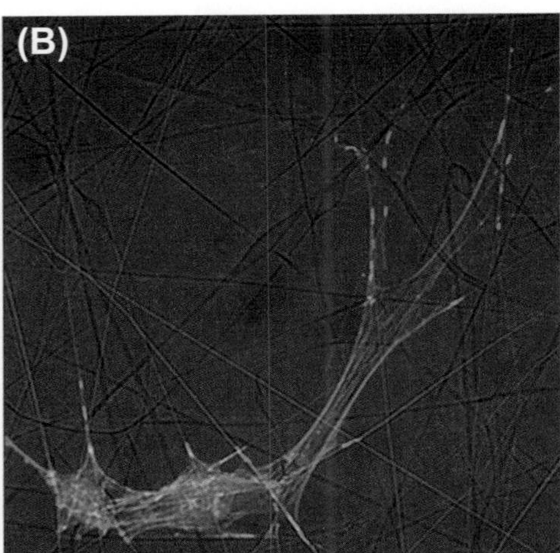

FIGURE II.6.7.7 Flourescent microscopy depicting preosteoblasts on either a flat poly(methyl methacrylate) surface (A); or poly(methyl methacrylate) nanofibers (B). The actin cytoskeleton of each cell is red, the protein vinculin, that is localized to points where the cell is adhered to the substrate, is green and the cell nucleus is stained blue. The image was taken 24 hours after the cells were seeded on the substrates. Of interest is how the preosteoblast on the flat surface spreads out radially, whereas the preosteoblasts on the nanofibers extend out along the fibers. What is not clear by the two-dimensional images provided is that the preosteoblast in (A) is much flatter in the z-direction than the preosteoblast in (B). Previously unpublished images.

both of these scaffold architectures are that they illustrate porosities greater than 95%, and that the fibers are randomly oriented in three dimensions (Ma and Zhang, 1999; Zhang, 2003; Horii et al., 2007). Likewise, electrospun nanofibers, which typically exhibit total porosities ranging from 40–75%, with isolated reports up to 90%, and pore diameters ranging from a few microns to 30 μm, have also exhibited cellular infiltration throughout the interior of the scaffold (Kim et al., 2003; Boland et al., 2004; Zhang et al., 2005; Venugopal et al., 2007).

These results illustrating cellular migration into the small pores contradict those previously found for microscale scaffolds, and consequently have raised questions regarding how cells are able to infiltrate nanoscale scaffolds. The ability of cell infiltration with a nanoscale scaffold starts with a fundamental difference in how cells respond to nanodimensional surfaces (Stevens and George, 2005). On nanostructures, cell morphology changes to exhibit pronounced pseudopodia-like processes which extend along individual fibers (Tan and Saltzman, 2004; Patel et al., 2007). Figure II.6.7.7 illustrates preosteoblasts seeded on a flat surface, which mimics a microstructure, and preosteoblasts seeded on nanofibers illustrating the aforementioned pseudopialike processes. These extensions on randomly orientated fibers produce an image appearing to be a well-spread cell; however, it is important to consider that most imaging techniques depict a two-dimensional field when there is a third dimension not necessarily seen. This implies that cells are extending down into the field of view as much as they are extending out within the field of view, and this theory has been corroborated through the use of confocal fluorescence imaging illustrating cells extending in three dimensions (Silva et al., 2004; Horii et al., 2007). These well-spread cells exhibiting pseudopodia-like processes have led researchers to conclude that cells utilize ameboidal migration to reach the interior of the scaffold (Friedl and Brocker, 2000a,b; Zhang et al., 2005). Presumably, this ameboidal migration is driven by soluble factors and proteins adsorbed to the surface of fibers below the cells, providing a gradient and driving infiltration to establish a uniform density of cells throughout the construct (Stevens and George, 2005; Patel et al., 2007). In conclusion, microscale scaffolds rely on pore diameter to allow cell migration; whereas nanoscale scaffolds rely on chemical gradients created by adsorbed factors and ameboidal motion to allow cell migration, which suggests that porosity is the critical property of nanoscale scaffolds.

DIMENSION IN BONE GRAFT SUBSTITUTES

In addition to porosity, the characteristic dimension of a polymer scaffold can have tremendous implications to the success of the scaffold as a bone graft substitute. Following are detailed descriptions of two scaffold types with dramatically different characteristic dimensions. The first, sintered microspheres, has a characteristic dimension an order of magnitude larger than that of a cell; while the second, nanofibers, demonstrate a characteristic dimension two orders of magnitude smaller than that of a cell.

Sintered Microspheres

Sintered biodegradable microsphere scaffolds were first developed by Cato Laurencin's laboratory in 1996, and were composed of PLAGA. These initial scaffolds were fabricated both with and without hydroxyapatite, and exhibited mechanical properties and porosity suitable for bone tissue engineering (Devin et al., 1996). Since that time the microsphere scaffold has undergone several compositional iterations. Microsphere scaffolds have been fabricated from other types of biodegradable polyesters, such as polylactide and poly(3-hydroxybutyrate-*co*-3-hydroxyvalerate) (Nof and Shea, 2002; Zhu et al., 2007). Sintered microsphere scaffolds have been fabricated from a polymer/ceramic composite microsphere composed of poly(lactide-*co*-glycolide) microspheres fabricated in such a way as to spontaneously induce calcium phosphate precipitation on the surface of the microspheres (Khan et al., 2004). Additionally, microsphere scaffolds have been combined with growth factors to provide differentiation cues to the seeded cells (Jaklenec et al., 2007).

A final recent iteration of the microsphere scaffold has dealt with the blend of poly(lactide-*co*-glycolide) with the biopolymer chitosan, as well as blends with other polyesters, such as poly(propylene fumarate) (Jiang et al., 2006; Kempen et al., 2006). The tremendous interest in the sintered microsphere scaffold is due to it being advantageous over other microscale scaffolds produced by such techniques as gas foaming and porogen leaching (Sachlos and Czernuzska, 2003; Rezwan et al., 2006). These advantages are reproducibility, scalability, and controllability over pore size, porosity, and mechanical properties. Microsphere scaffolds rely on the chemical properties of a polymer to sinter uniform spheres into scaffolds by elevating the polymer above the glass transition point, which causes the adjacent polymer chains to migrate and intertwine, forming a cohesive bond when the polymer is cooled (Borden et al., 2002a). Maintaining the temperature, size of microsphere, and time to sinter has proven to produce uniform scaffolds (Borden et al., 2002a). Gas foaming and porogen leaching do not have the same level of reproducibility, due to variation introduced by having multiple phases, either a heterogeneous mixture of a solid and liquid or a liquid and gas. These heterogeneous mixtures can undergo demixing, which introduces inconsistencies in the mixing of the phases and decreases reproducibility (Sachlos and Czernuzska, 2003; Rezwan et al., 2006).

Microsphere scaffolds maintain a very high level of interconnectivity among the pores, and allow control over the total interconnected porosity, as well as the average pore diameter (Borden et al., 2002a). Gas foaming and particle leaching provide control over the total porosity and pore size as well; however, they often have very poor interconnectivity (only 10–30% interconnectivity with gas foaming) of the pores, rendering a percentage of the pores inaccessible (Mooney et al., 1996). The inaccessible pores are formed when the initial heterogeneous mixture lacks contact between the solid or gaseous phases, which causes the resulting scaffold to contain pores surrounded by solid polymer. Another advantage of microsphere scaffolds, when compared to gas foaming and porogen leaching, is that they can be fabricated to exhibit mechanical properties that mimic that of trabecular bone (Borden et al., 2002a, 2004).

High mechanical properties are achievable with gas foaming and porogen leaching; however, to achieve this, the scaffold density must be increased, which causes the internal pore structure to be further compromised (Sachlos and Czernuzska, 2003; Rezwan et al., 2006). Microsphere scaffolds composed of polylactide and poly(lactide-*co*-glycolide) have been produced previously, and have yielded reasonable results in *in vitro* studies, as well as in *in vivo* studies examining critical size defect healing (Borden et al., 2002b, 2004; Botchwey et al., 2003; Yu et al., 2004; Jiang et al., 2006). *In vitro* studies have illustrated that rat calvarial cells demonstrated a phenotype with earlier expression of matrix proteins in a greater magnitude when cultured on microsphere scaffolds, compared with the same cells on tissue culture plastic (Borden et al., 2002b).

Nanofibers

Recent developments in tissue engineering have indicated that nanoscale structures are more advantageous for cellular phenotype expression and morphology when compared to microscale structures (Tuzlakoglu et al., 2005; Wan et al., 2005). A frequently employed type of nanoscale scaffold is based on polymeric nanofibers. Nanofibers can be fabricated through several different techniques. The initial and most common technique is through electrospinning, which has been used for over 70 years (Formhals and Schreiber-Gastell, 1934). Electrospun nanofibers are created by applying a voltage gradient between a target and a drop of polymer, either dissolved or melted (Formhals and Schreiber-Gastell, 1934; Nair et al., 2004; Smith and Ma, 2004). Once the voltage gradient overcomes the surface tension of the droplet, a polymer stream extends toward the target. As this stream travels, it is thinned out and eventually strikes the target as a nanofiber, with the accumulation of these fibers leading to the production of a nonwoven nanofiber mat (Nair et al., 2004; Smith and Ma, 2004).

Initially, electrospun nanofibers were made from non-degradable polymers for applications in filtration; however, recent developments have led to the use of electrospun nanofibers made of biodegradable polymers for tissue engineering applications (Nair et al., 2004). Nanofibers created by electrospinning typically have diameters ranging from 300–1200 nm, depending on the spinning conditions used (Kim et al., 2003; Yoshimoto et al., 2003; Nair et al., 2004; Smith and Ma, 2004; Tuzlakoglu et al., 2005; Kumbar et al., 2006b). Electrospun nanofibers are oriented lengthwise in only two dimensions, with the third dimension created from the stacking of the fibers on top of each other. This is why electrospun nanofiber scaffolds are typically referred to as mats. Nanofibers have also recently been developed using synthetic peptides designed to self-assemble into a three-dimensional nanofiber network. Self-assembly nanofibers often have a characteristic diameter that is within the range of 5–10 nm (Zhang, 2003; Semino et al., 2004; Silva et al., 2004; Smith and Ma, 2004; Horii et al., 2007). The very thin diameter of the fibers in the network, and the high porosity created with self-assembly, produces a scaffold poor in mechanical properties making them suitable primarily for hydrogel applications.

A final method of nanofiber fabrication is thermally-induced phase separation, which relies on the spinodal liquid–liquid phase separation of a polymer solution into a polymer poor phase and a polymer rich phase when the solution is rapidly cooled. Spinodal phase differs from binodal phase separation in that the two phases separate and exist continuously throughout the original mixture; whereas bimodal phase separation occurs via nucleation sites that build spherical particles. The selection of an appropriate polymer, one with a high degree of crystallinity, will allow the polymer rich phase to crystallize into nanofibers; whereas polymers with low degrees of crystallinity form microstructured foams (van de Witte et al., 1996; Chen et al., 2006). Nanofibers created with the thermally-induced phase separation technique exhibit a fiber diameter of 50–500 nm, which is similar to collagen, and a three-dimensional fiber structure making them very different to that of electrospun nanofibers (Smith and Ma, 2004). Evidence suggests that nanofibers exhibiting a three-dimensional spatial arrangement promote phenotype progression of osteoblasts, and may be osteoinductive (Hu et al., 2008). Producing a three-dimensional nanofibrous mechanically viable implant could be a tremendous leap forward in the field of bone tissue engineering.

Osteoinduction

Osteoinduction is the ability of a substance to cause stem cell differentiation down an osteoblastic lineage. Osteoinduction is known to occur when certain growth factors, such as bone morphogenetic proteins 2 and 7, and to a lesser extent vascular endothelial growth factor, are present (Urist, 1983; Geiger et al., 2003; Habibovic and de Groot, 2007; Dawson et al., 2009). What is less clear is whether a biomaterial can be osteoinductive in the absence of growth factor supplements. Current research has provided contradictory evidence as to whether or not a biomaterial is osteoinductive (Ye et al., 2007; Catros et al., 2009). No particular biomaterial has been conclusively shown to be osteoinductive, independent of structure and growth factors (Habibovic and de Groot, 2007; Habibovic et al., 2008).

This suggests that what may be more important than a particular biomaterial is the structure that the biomaterial is fabricated into. Research suggests that macroscale concavities (Graziano et al., 2008) in a surface, a microporosity (Habibovic et al., 2005), as well as subcellular structures (e.g., nanofibers) (Hu et al., 2008) may all promote osteoinduction. However, each of these elements has yet to be thoroughly challenged to definitively ascertain if they are indeed osteoinductive. For instance, the study investigating macroscale concavities included flat and convex surfaces as controls; however, all three surfaces were made from different materials, and the stem cells used were differentiated to osteoblast progenitors prior to seeding on the substrates. Another intriguing osteoinductive quality of substrates is that mesenchymal stem cell differentiation has demonstrated a dependence on the rigidity of a substrate (Engler et al., 2006; Khatiwala et al., 2007). These elements suggest that developing an osteoinductive bone graft substitute may depend on design and material considerations. Moving forward, research should first investigate how osteinduction occurs, and then deconstruct the independent elements of biomaterial chemistry, surface topography, and porosity to determine which, if any, element provides non-growth factor initiated osteoinduction.

IN VITRO CULTURE TECHNIQUES FOR BONE GRAFT SUBSTITUTES

A critical limitation of static culture conditions is that waste efflux and nutrient influx are governed by diffusion, which becomes exponentially more problematic with large three-dimensional constructs. As time passes, the accumulation of waste within a construct can lead to an acidic microenvironment that impedes calcification of the developing tissue, and compromises the viability of osteoblasts within a construct (Bushinsky et al., 1983; Han et al., 2009). To alleviate these issues the use of bioreactors becomes necessary when trying to achieve significant tissue growth *in vitro*. The bioreactor provides fluid flux that replenishes nutrients and removes waste; additionally, for tissues such as bone, bioreactors can provide mechanical stimulation to encourage the development of a mechanically viable tissue.

The earliest bioreactor that has been applied to bone tissue engineering involves dialysis membranes that are either gas or small molecule permeable (Figure II.6.7.8A). The bioreactor is divided into two compartments that are separated by small molecule permeable membrane, and each compartment also contains a gas permeable membrane (Vogler, 1989). The gas permeable membrane provides a liquid–air boundary to provide oxygen and CO_2, and the membrane between compartments serves as liquid–liquid boundary between the culture media surrounding the cells and a reservoir of fresh media. This boundary provides fresh nutrients while removing waste, but also allows the soluble factor gradients to remain intact during media changes. Recently, this bioreactor design has supported the development of simple osteoid tissue after culturing osteoprogenitor cells for 10 months (Mastro and Vogler, 2009). Despite the development of osteoid tissue, this bioreactor design still relies on passive diffusion to supply nutrients, and therefore limits the ultimate thickness of the construct to only a few cell layers (Mastro and Vogler, 2009).

The next iteration of the bioreactor involved rotation, either of the reactor itself or of the construct within the bioreactor (Figure II.6.7.8C). NASA developed rotating wall bioreactors to simulate microgravity. These bioreactors consist of a cylinder and a gas exchange membrane, either on one face of the cylinder

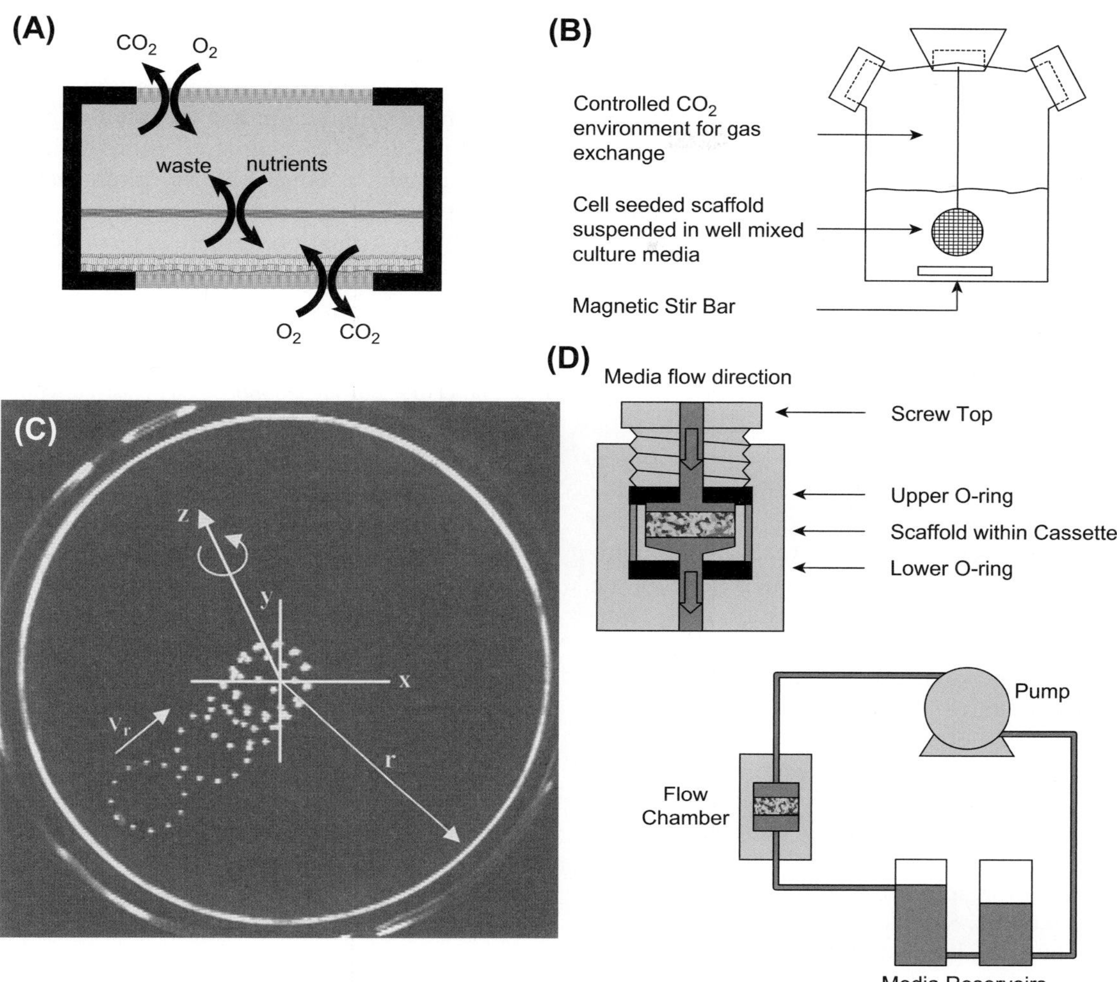

FIGURE II.6.7.8 Graphical depiction of four bioreactors. The simplest bioreactor that supports multi-layered growth of cells on the bottom surface and consists of two gas permeable membranes on the top and bottom and a dialysis membrane in the middle (A). A spinner flask bioreactor that utilizes a stir bar to force nutrient flux through the constructs, which remain in a fixed position (B) (Martin and Vermette, 2005). A high aspect ratio rotating vessel bioreactor demonstrating the trajectory of a lighter than water construct within the rotating wall bioreactor, the fluid shear on the scaffold is low; however, the back panel of the bioreactor is gas permeable membrane and the rotation maintains a large volume of well-mixed media to support nutrient supply (C) (Botchwey et al., 2004). Finally, a schematic illustrating a perfusion bioreactor that utilizes a pump to force media through a construct, which research suggests also promotes bone tissue formation through mechanical stimulation provided by fluid shear forces on the construct (D) (Bancroft et al., 2002). Copyright (2002) National Academies of Sciences, U.S.A.

or as a separate inner cylinder. In this design the shear force exerted by fluid rotation negates the effect of gravity on the constructs, and also provides a well-mixed volume of culture media. This strategy was employed by tissue engineers to culture constructs seeded with osteoblasts, in the hope that the rotating well-mixed volume of culture media would overcome diffusion limitations and allow tissue development throughout the thickness of the construct (Botchwey et al., 2004). Research suggests this is certainly the case, with osteoblasts found throughout the thickness of the construct (Yu et al., 2004); however, the simulation of microgravity seems to (not surprisingly due to a lack of mechanical stimulation) discourage bone tissue development; this is demonstrated both by lower levels of phenotype markers, and by lower mineralization as compared to static cultured constructs (Sikavitsas et al., 2002; Yu et al., 2004).

The next type of bioreactor that involves rotating fluid is the spinner flask (Figure II.6.7.8B). Spinner flask bioreactors suspend the constructs on thin needles that are extended into a stirred flask of media. This configuration not only provides a well mixed environment to avoid accumulation of waste in local environment of the construct, but also provides mechanical stimulation through shear forces on the constructs. Since the spinner flask also provides a well-mixed environment, it is not surprising to see that cells migrate throughout the construct suspended in the spinner flask, similar to those cultured in rotating wall bioreactors (Sikavitsas et al., 2002; Stiehler et al., 2008). Additionally, because the spinner flask provides much higher shear forces than the rotating wall bioreactor, there is a marked increase in the calcification of the construct and the differentiation of osteoblasts in the spinner flask (Sikavitsas et al., 2002; Stiehler et al., 2008; Wang et al., 2008).

The final bioreactor for bone tissue engineering is a perfusion bioreactor (Figure II.6.7.8D). The perfusion bioreactor works by forcing culture media through a construct with a pump. This system provides intricate control over the shear experienced by the cells in culture. Similar to the spinner flask, osteoblasts generated more calcium and exhibited increased expression of phenotype markers in the perfusion bioreactor (Gomes et al., 2003; Pham et al., 2008).

CONCLUSION

The concepts of tissue engineering have made profound advances in developing clinically relevant solutions for tissue such as skin, bladder, and to some extent bone; however, where the clinical strategies for skin and bladder exist as a straightforward solution based on the requirements of those tissues, bone presents a more complicated situation. Skin and bladder tissue are essentially a uniform two-dimensional sheet; whereas, bone often takes many unusual three-dimensional and non-uniform geometries (Bannasch et al., 2003; Vats et al., 2003; Atala et al., 2006; Bolland and Southgate, 2008).

The evolution of bone tissue engineering began with osteoblasts cultured on rudimentary polymer foams, and has progressed to include an array of bone graft substitutes cultured in bioreactors to drive the development of *de novo* bone tissue. Significant clinical achievements have occurred in using biodegradable scaffolds, with or without growth factors or cells, as synthetic bone grafts to heal large defects in bone tissue (Damron, 2007; Rosa et al., 2008; Wlodarski et al., 2008; Dawson et al., 2009; Gosain et al., 2009). However, despite the ground that has been covered thus far, there persist unmet goals and challenges still ahead. No research has yet demonstrated the capability to grow *de novo* bone in an *in vitro* setting; so far only rudimentary calcified cell masses approaching bone tissue have been developed. Moving forward, future researchers should consider the flexibility in design of the macroscale structure to accommodate the unusual architectures and mechanics necessary for bone graft substitutes, biodegradability of the structure such that natural healthy bone is ultimately all that persists, the osteoconductivity of the bone graft substitute to promote proliferation of progenitor cells and osteoblasts throughout the bone graft substitute, and the osteoinductivity of the bone graft substitute to promote the maturation of the progenitor cells and osteoblasts into organized bone tissue.

To date, the one aspect the earliest bioreactor got right has been lacking in all future designs; the ability to maintain gradients in soluble growth factors secreted by cells. The development of the bioreactor for bone tissue engineering has moved from a series of static chambers separated by dialysis membranes to reactors that provide mechanical stimulation; the future success of the bioreactor will depend on the ability to incorporate aspects of the existing bioreactors: nutrient influx; waste efflux; mechanical stimulation; and the establishment of soluble factor gradients. Providing all four of these bioreactor elements with a bone graft substitute may lead to the development of hierarchical bone tissue, and ultimately the *de novo* formation of bone *in vitro*.

BIBLIOGRAPHY

Akay, G., Birch, M. A., & Bokhari, M. A. (2004). Microcellular polyHIPE polymer supports osteoblast growth and bone formation *in vitro*. *Biomaterials*, **25**, 3991–4000.

Allcock, H. R., Pucher, S. R., & Scopelianos, A. G. (1994). Poly[(amino acid ester)phosphazenes]: Synthesis, crystallinity, and hydrolytic sensitivity in solution and the solid state. *Macromolecules*, **27**, 1071–1075.

Atala, A., Bauer, S. B., Soker, S., Yoo, J. J., & Retik, A. B. (2006). Tissue-engineered autologous bladders for patients needing cystoplasty. *Lancet*, **367**, 1241–1246.

Bancroft, G. N., Sikavitsas, V. I., van den Dolder, J., Sheffield, T. L., Ambrose, C. G., et al. (2002). Fluid flow increases mineralized matrix deposition in 3D perfusion culture of marrow stromal osteoblasts in a dose-dependent manner. *Proc. Natl. Acad. Sci. USA*, **99**, 12600–12605.

Bannasch, H., Föhn, M., Unterberg, T., Bach, A. D., Weyand, B., et al. (2003). Skin tissue engineering. *Clin. Plast. Surg.*, **30**, 573–579.

Ben-Nissan, B. (2003). Natural bioceramics: From coral to bone and beyond. *Curr. Opin. Solid State Mater. Sci.*, **7**, 283–288.

Bilezikian, J. P., Raisz, L. G., & Rodan, G. A. (Eds.), (2002). *Principles of Bone Biology* (2nd ed.). San Diego, CA: Academic Press.

Blaine, G. (1946). Experimental observations on the use of absorbable and non-absorbable plastics in bone surgery. *Br. J. Surg.*, **33**, 245–250.

Boland, E. D., Telemeco, T. A., Simpson, D. G., Wnek, G. E., & Bowlin, G. L. (2004). Utilizing acid pretreatment and electrospinning to improve biocompatibility of poly(glycolic acid) for tissue engineering. *J. Biomed. Mater. Res. B. Appl. Biomater.*, **71**, 144–152.

Bolland, F., & Southgate, J. (2008). Bio-engineering urothelial cells for bladder tissue transplant. *Expert. Opin. Biol. Ther.*, **8**, 1039–1049.

Borden, M., Attawia, M., Khan, Y., & Laurencin, C. T. (2002a). Tissue engineered microsphere-based matrices for bone repair: Design and evaluation. *Biomaterials*, **23**, 551–559.

Borden, M., Attawia, M., & Laurencin, C. T. (2002b). The sintered microsphere matrix for bone tissue engineering: *In vitro* osteoconductivity studies. *J. Biomed. Mater. Res.*, **61**, 421–429.

Borden, M., Attawia, M., Khan, Y., El-Amin, S. F., & Laurencin, C. T. (2004). Tissue-engineered bone formation *in vivo* using a novel sintered polymeric microsphere matrix. *J. Bone. Joint. Surg. Br.*, **86**, 1200–1208.

Bostman, O. M., & Pihlajamaki, H. K. (1998). Late foreign-body reaction to an intraosseous bioabsorbable polylactic acid screw. A case report. *J. Bone. Joint. Surg. Am.*, **80**, 1791–1794.

Botchwey, E. A., Dupree, M. A., Pollack, S. R., Levine, E. M., & Laurencin, C. T. (2003). Tissue engineered bone: Measurement of nutrient transport in three-dimensional matrices. *J. Biomed. Mater. Res. A.*, **67**, 357–367.

Botchwey, E. A., Pollack, S. R., Levine, E. M., Johnston, E. D., & Laurencin, C. T. (2004). Quantitative analysis of three-dimensional fluid flow in rotating bioreactors for tissue engineering. *J. Biomed. Mater. Res. A.*, **69**, 205–215.

Brown, R. F., Day, D. E., Day, T. E., Jung, S., Rahaman, M. N., et al. (2008a). Growth and differentiation of osteoblastic cells on 13–93 bioactive glass fibers and scaffolds. *Acta Biomater.*, **4**, 387–396.

Brown, J. L., Nair, L. S., & Laurencin, C. T. (2008b). Solvent/non-solvent sintering: A novel route to create porous microsphere scaffolds for tissue regeneration. *J. Biomed. Mater. Res. B. Appl. Biomater.*, **86B**, 396–406.

Buket Basmanav, F., Kose, G. T., & Hasirci, V. (2008). Sequential growth factor delivery from complexed microspheres for bone tissue engineering. *Biomaterials*, **29**, 4195–4204.

Bushinsky, D. A., Krieger, N. S., Geisser, D. I., Grossman, E. B., & Coe, F. L. (1983). Effects of pH on bone calcium and proton fluxes *in vitro*. *Am. J. Physiol.*, **245**, F204–F209.

Catros, S., Zwetyenga, N., Bareille, R., Brouillaud, B., Renard, M., et al. (2009). Subcutaneous-induced membranes have no osteoinductive effect on macroporous HA-TCP *in vivo*. *J. Orthop. Res.*, **27**, 155–161.

Chen, V. J., Smith, L. A., & Ma, P. X. (2006). Bone regeneration on computer-designed nano-fibrous scaffolds. *Biomaterials*, **27**, 3973–3979.

Christenson, E. M., Anseth, K. S., van den Beucken, J. J., Chan, C. K., Ercan, B., et al. (2007). Nanobiomaterial applications in orthopedics. *J. Orthop. Res.*, **25**, 11–22.

Clokie, C. M., Moghadam, H., Jackson, M. T., & Sandor, G. K. (2002). Closure of critical sized defects with allogenic and alloplastic bone substitutes. *J. Craniofac. Surg.*, **13**(111), 21; discussion, 122–123.

Conconi, M. T., Lora, S., Menti, A. M., Carampin, P., & Parnigotto, P. P. (2006). *In vitro* evaluation of poly[bis(ethyl alanato)phosphazene] as a scaffold for bone tissue engineering. *Tissue. Eng.*, **12**, 811–819.

Connolly, J., Guse, R., Lippiello, L., & Dehne, R. (1989). Development of an osteogenic bone-marrow preparation. *J. Bone. Joint. Surg. Am.*, **71**, 684–691.

Coxon, F. P., & Taylor, A. (2008). Vesicular trafficking in osteoclasts. *Semin Cell. Dev. Biol.*, **19**(5), 424–433.

Cui, Q., Xiao, Z., Li, X., Saleh, K. J., & Balian, G. (2006). Use of genetically engineered bone-marrow stem cells to treat femoral defects: An experimental study. *J. Bone. Joint. Surg. Am.*, **88**(Suppl. 3), 167–172.

Damien, C. J., & Parsons, J. R. (1991). Bone graft and bone graft substitutes: A review of current technology and applications. *J. Appl. Biomater.*, **2**, 187–208.

Damron, T. A. (2007). Use of 3D beta-tricalcium phosphate (Vitoss) scaffolds in repairing bone defects. *Nanomed.*, **2**, 763–775.

Davis, H. E., Rao, R. R., He, J., & Leach, J. K. (2008). Biomimetic scaffolds fabricated from apatite-coated polymer microspheres. *J. Biomed. Mater. Res. A.*, **90**(4), 1021–1031.

Dawson, E., Bae, H. W., Burkus, J. K., Stambough, J. L., & Glassman, S. D. (2009). Recombinant human bone morphogenetic protein-2 on an absorbable collagen sponge with an osteoconductive bulking agent in posterolateral arthrodesis with instrumentation. A prospective randomized trial. *J. Bone Joint Surg. Am.*, **91**, 1604–1613.

Devin, J. E., Attawia, M. A., & Laurencin, C. T. (1996). Three-dimensional degradable porous polymer-ceramic matrices for use in bone repair. *J. Biomater. Sci. Polym. Ed.*, **7**, 661–669.

Engler, A. J., Sen, S., Sweeney, H. L., & Discher, D. E. (2006). Matrix elasticity directs stem cell lineage specification. *Cell*, **126**, 677–689.

Erisken, C., Kalyon, D. M., & Wang, H. (2008). Functionally graded electrospun polycaprolactone and beta-tricalcium phosphate nanocomposites for tissue engineering applications. *Biomaterials*, **29**, 4065–4073.

Everts, V., Delaisse, J. M., Korper, W., Jansen, D. C. & Tigchelaar-Gutter, W. et al. (2002). The bone lining cell: Its role in cleaning Howship's lacunae and initiating bone formation. *J. Bone Miner. Res.*, **17**, 77–90.

Formhals, A. & Schreiber-Gastell, R. (1934). Anonymous Process and Apparatus for Preparing Artificial Threads. *Patent No. 1,975,504*.

Fouletier-Dilling, C. M., Gannon, F. H., Olmsted-Davis, E. A., Lazard, Z., Heggeness, M. H., et al. (2007). Efficient and rapid osteoinduction in an immune-competent host. *Hum. Gene Ther.*, **18**, 733–745.

Franz-Odendaal, T. A., Hall, B. K., & Witten, P. E. (2006). Buried alive: How osteoblasts become osteocytes. *Dev. Dyn.*, **235**, 176–190.

Friedl, P., & Brocker, E. B. (2000a). T cell migration in three-dimensional extracellular matrix: Guidance by polarity and sensations. *Dev. Immunol.*, **7**, 249–266.

Friedl, P., & Brocker, E. B. (2000b). The biology of cell locomotion within three-dimensional extracellular matrix. *Cell. Mol. Life Sci.*, **57**, 41–64.

Gautschi, O. P., Frey, S. P., & Zellweger, R. (2007). Bone morphogenetic proteins in clinical applications. *ANZ J. Surg.*, **77**, 626–631.

Gazdag, A. R., Lane, J. M., Glaser, D., & Forster, R. A. (1995). Alternatives to autogenous bone graft: Efficacy and indications. *J. Am. Acad. Orthop. Surg.*, **3**, 1–8.

Ge, Z., Jin, Z., & Cao, T. (2008). Manufacture of degradable polymeric scaffolds for bone regeneration. *Biomed. Mater.*, **3**(2), 022001.

Geiger, M., Li, R. H., & Friess, W. (2003). Collagen sponges for bone regeneration with rhBMP-2. *Adv. Drug. Deliv. Rev.*, **55**, 1613–1629.

Giannoudis, P. V., Dinopoulos, H., & Tsiridis, E. (2005). Bone substitutes: An update. *Injury*, **36**, S20–S27.

Gomes, M. E., Sikavitsas, V. I., Behravesh, E., Reis, R. L., & Mikos, A. G. (2003). Effect of flow perfusion on the osteogenic differentiation of bone marrow stromal cells cultured on starch-based three-dimensional scaffolds. *J. Biomed. Mater. Res. A.*, **67**, 87–95.

Gosain, A. K., Chim, H., & Arneja, J. S. (2009). Application-specific selection of biomaterials for pediatric craniofacial reconstruction: Developing a rational approach to guide clinical use. *Plast. Reconstr. Surg.*, **123**, 319–330.

Goulet, G. C., Cooper, D. M., Coombe, D., & Zernicke, R. F. (2008). Influence of cortical canal architecture on lacunocanalicular pore pressure and fluid flow. *Comput. Methods Biomech. Biomed. Engin.*, **11**, 379–387.

Govender, S., Csimma, C., Genant, H. K., Valentin-Opran, A., Amit, Y., et al. (2002a). Recombinant human bone morphogenetic protein-2 for treatment of open tibial fractures: A prospective, controlled, randomized study of four hundred and fifty patients. *J. Bone Joint Surg. Am.*, **84-A**, 2123–2134.

Govender, P. V., Rampersaud, Y. R., Rickards, L., & Fehlings, M. G. (2002b). Use of osteogenic protein-1 in spinal fusion: Literature review and preliminary results in a prospective series of high-risk cases. *Neurosurg. Focus*, **13**, e4.

Graziano, A., d'Aquino, R., Cusella-De Angelis, M. G., De Francesco, F., Giordano, A., et al. (2008). Scaffold's surface geometry significantly affects human stem cell bone tissue engineering. *J. Cell Physiol.*, **214**, 166–172.

Greenwald, A. S., Boden, S. D., Goldberg, V. M., Khan, Y., Laurencin, C. T., et al. (2001). Bone-graft substitutes: Facts, fictions, and applications. *J. Bone Joint Surg. Am.*, **83-A**(Suppl. 2 Pt 2), 98–103.

Habibovic, P., & de Groot, K. (2007). Osteoinductive biomaterials: Properties and relevance in bone repair. *J. Tissue Eng. Regen. Med.*, **1**, 25–32.

Habibovic, P., Yuan, H., van der Valk, C. M., Meijer, G., van Blitterswijk, C. A., et al. (2005). 3D microenvironment as essential element for osteoinduction by biomaterials. *Biomaterials*, **26**, 3565–3575.

Habibovic, P., Kruyt, M. C., Juhl, M. V., Clyens, S., Martinetti, R., et al. (2008). Comparative *in vivo* study of six hydroxyapatite-based bone graft substitutes. *J. Orthop. Res.*, **26**, 1363–1370.

Han, S. H., Chae, S. W., Choi, J. Y., Kim, E. C., Chae, H. J., et al. (2009). Acidic pH environments increase the expression of cathepsin B in osteoblasts: The significance of ER stress in bone physiology. *Immunopharmacol. Immunotoxicol.*, **31**(3), 428–431.

Hench, L. L., Splinter, R. J., Allen, W. C., & Greenlee, T. K. (1971). Bonding mechanisms at the interface of ceramic prosthetic materials. *J. Biomed. Mater. Res.*, **5**, 117–141.

Horii, A., Wang, X., Gelain, F., & Zhang, S. (2007). Biological designer self-assembling peptide nanofiber scaffolds significantly enhance osteoblast proliferation, differentiation and 3-D migration. *PLoS ONE*, **2**, e190.

Hu, J., Liu, X., & Ma, P. X. (2008). Induction of osteoblast differentiation phenotype on poly(L-lactic acid) nanofibrous matrix. *Biomaterials*, **29**, 3815–3821.

Huemer, G. M., Puelacher, W., & Schoeller, T. (2004). Improving the iliac crest donor site by plate insertion after harvesting vascularized bone. *J. Craniomaxillofac. Surg.*, **32**, 387–390.

Hutmacher, D. W. (2000). Scaffolds in tissue engineering bone and cartilage. *Biomaterials*, **21**, 2529–2543.

Hutmacher, D. W., Sittinger, M., & Risbud, M. V. (2004). Scaffold-based tissue engineering: Rationale for computer-aided design and solid free-form fabrication systems. *Trends in Biotechnology*, **22**, 354–362.

Ibim, S. E., Ambrosio, A. M., Kwon, M. S., El-Amin, S. F., Allcock, H. R., et al. (1997). Novel polyphosphazene/poly(lactide-co-glycolide) blends: Miscibility and degradation studies. *Biomaterials*, **18**, 1565–1569.

Jaklenec, A., Hinckfuss, A., Bilgen, B., Ciombor, D. M., Aaron, R., et al. (2007). Sequential release of bioactive IGF-I and TGF-beta(1) from PLGA microsphere-based scaffolds. *Biomaterials*, **29**(10), 1518–1525.

Jeon, O., Song, S. J., Kang, S. W., Putnam, A. J., & Kim, B. S. (2007). Enhancement of ectopic bone formation by bone morphogenetic protein-2 released from a heparin-conjugated poly(L-lactic-co-glycolic acid) scaffold. *Biomaterials*, **28**, 2763–2771.

Jiang, T., Abdel-Fattah, W., & Laurencin, C. T. (2006). *In vitro* evaluation of chitosan/poly(lactic acid-glycolic acid) sintered microsphere scaffolds for bone tissue engineering. *Biomaterials*, **27**, 4894–4903.

Jones, A. C., Milthorpe, B., Averdunk, H., Limaye, A., Senden, T. J., et al. (2004). Analysis of 3D bone ingrowth into polymer scaffolds via micro-computed tomography imaging. *Biomaterials*, **25**, 4947–4954.

Kale, S., Biermann, S., Edwards, C., Tarnowski, C., Morris, M., et al. (2000). Three-dimensional cellular development is essential for *ex vivo* formation of human bone. *Nat. Biotechnol.*, **18**, 954–958.

Karageorgiou, V., & Kaplan, D. (2005). Porosity of 3D biomaterial scaffolds and osteogenesis. *Biomaterials*, **26**, 5474–5491.

Karsdal, M. A., Larsen, L., Engsig, M. T., Lou, H., Ferreras, M., et al. (2002). Matrix metalloproteinase-dependent activation of latent transforming growth factor-beta controls the conversion of osteoblasts into osteocytes by blocking osteoblast apoptosis. *J. Biol. Chem.*, **277**, 44061–44067.

Kempen, D. H., Lu, L., Kim, C., Zhu, X., Dhert, W. J., et al. (2006). Controlled drug release from a novel injectable biodegradable microsphere/scaffold composite based on poly(propylene fumarate). *J. Biomed. Mater. Res. A.*, **77**, 103–111.

Kempen, D. H., Lu, L., Hefferan, T. E., Creemers, L. B., Maran, A., et al. (2008). Retention of *in vitro* and *in vivo* BMP-2 bioactivities in sustained delivery vehicles for bone tissue engineering. *Biomaterials*, **29**, 3245–3252.

Khan, Y., Yaszemski, M. J., Mikos, A. G., & Laurencin, C. T. (2008). Tissue engineering of bone: Material and matrix considerations. *J. Bone Joint Surg. Am.*, **90**(Suppl. 1), 36–42.

Khan, Y. M., Katti, D. S., & Laurencin, C. T. (2004). Novel polymer-synthesized ceramic composite-based system for bone repair: An *in vitro* evaluation. *J. Biomed. Mater. Res. A.*, **69**, 728–737.

Khatiwala, C. B., Peyton, S. R., Metzke, M., & Putnam, A. J. (2007). The regulation of osteogenesis by ECM rigidity in MC3T3-E1 cells requires MAPK activation. *J. Cell Physiol.*, **211**, 661–672.

Khosla, S., Westendorf, J. J., & Oursler, M. J. (2008). Building bone to reverse osteoporosis and repair fractures. *J. Clin. Invest.*, **118**, 421–428.

Kim, K., Yu, M., Zong, X., Chiu, J. & Fang, D. et al. (2003). Control of degradation rate and hydrophilicity in electrospun non-woven poly(D,L-lactide) nanofiber scaffolds for biomedical applications. *Biomaterials*, **24**, 4977–4985.

Kjelstrup-Hansen, J., Hansen, O., Rubahn, H. G., & Boggild, P. (2006). Mechanical properties of organic nanofibers. *Small*, **2**(5), 660–666.

Kneser, U., Schaefer, D. J., Munder, B., Klemt, C., Andree, C., et al. (2002). Tissue engineering of bone. *Minimally Invasive Therapy and Allied Technologies*, **11**, 107,116(10).

Kneser, U., Schaefer, D. J., Polykandriotis, E., & Horch, R. E. (2003). Tissue engineering of bone: The reconstructive surgeon's point of view. *J. Cell Mol.*, **36**, 1409–1424.

Kollet, O., Dar, A., Shivtiel, S., Kalinkovich, A., Lapid, K., et al. (2006). Osteoclasts degrade endosteal components and promote mobilization of hematopoietic progenitor cells. *Nat. Med.*, **12**, 657–664.

Kumbar, S. G., Bhattacharyya, S., Nukavarapu, S. P., Khan, Y. M., Nair, L. S., et al. (2006a). *In vitro* and *in vivo* characterization of biodegradable poly(organophosphazenes) for biomedical applications. *J. Inorg. Organomet. P. Mater.*, **16**, 365–385.

Kumbar, S. G., Nair, L. S., Bhattacharyya, S., & Laurencin, C. T. (2006b). Polymeric nanofibers as novel carriers for the delivery of therapeutic molecules. *J. Nanosci. Nanotechnol.*, **6**, 2591–2607.

Langer, R. (2000). Tissue engineering. *Mol. Ther.*, **1**, 12–15.

Langer, R., & Vacanti, J. P. (1993). Tissue engineering (methods of replacing or substituting for damaged or diseased tissues). *Science*, **260**, 920–926.

Laurencin, C. T., El-Amin, S. F., Ibim, S. E., Willoughby, D. A., Attawia, M., et al. (1996). A highly porous 3-dimensional polyphosphazene polymer matrix for skeletal tissue regeneration. *J. Biomed. Mater. Res.*, **30**, 133–138.

Laurencin, C. T., Ambrosio, A. M., Borden, M. D., & Cooper, J. A., Jr. (1999). Tissue engineering: Orthopedic applications. *Annu. Rev. Biomed. Eng.*, **1**, 19–46.

Laurencin, C. T., Khan, Y., & El-Amin, S. F. (2006). Bone graft substitutes. *Expert Rev. Med. Devices*, **3**, 49–57.

Lee, G. H., Khoury, J. G., Bell, J. E., & Buckwalter, J. A. (2002). Adverse reactions to OsteoSet bone graft substitute, the incidence in a consecutive series. *Iowa Orthop. J.*, **22**, 35–38.

Lee, K. W., Wang, S., Lu, L., Jabbari, E., Currier, B. L., et al. (2006). Fabrication and characterization of poly(propylene fumarate) scaffolds with controlled pore structures using 3-dimensional printing and injection molding. *Tissue Eng.*, **12**, 2801–2811.

Leveen, H. H., & Barberio, J. R. (1949). Tissue reaction to plastics used in surgery with special reference to Teflon. *Ann. Surg.*, **129**, 74–84.

Li, X., van Blitterswijk, C. A., Feng, Q., Cui, F., & Watari, F. (2008). The effect of calcium phosphate microstructure on bone-related cells *in vitro*. *Biomaterials*, **29**, 3306–3316.

Lieberman, J. R., Daluiski, A., Stevenson, S., Wu, L., McAllister, P., et al. (1999). The effect of regional gene therapy with bone morphogenetic protein-2-producing bone-marrow cells on the repair of segmental femoral defects in rats. *J. Bone Joint Surg. Am.*, **81**, 905–917.

Lindsey, R. W., Wood, G. W., Sadasivian, K. K., Stubbs, H. A., & Block, J. E. (2006). Grafting long bone fractures with demineralized bone matrix putty enriched with bone marrow: Pilot findings. *Orthopedics*, **29**, 939–941.

Ma, P. X., & Zhang, R. (1999). Synthetic nano-scale fibrous extracellular matrix. *J. Biomed. Mater. Res.*, **46**, 60–72.

Ma, S., Lin, H., Miao, Y., Liu, X., Wang, B., et al. (2007). The effect of three-dimensional demineralized bone matrix on *in vitro* cumulus-free oocyte maturation. *Biomaterials*, **28**, 3198–3207.

Marie, P. J. (2008). Transcription factors controlling osteoblastogenesis. *Archives of Biochemistry and Biophysics*, **473**, 98–105.

Martin, G. J., Jr., Boden, S. D., Titus, L., & Scarborough, N. L. (1999). New formulations of demineralized bone matrix as a more effective graft alternative in experimental posterolateral lumbar spine arthrodesis. *Spine*, **24**, 637–645.

Martin, Y., & Vermette, P. (2005). Bioreactors for tissue mass culture: Design, characterization, and recent advances. *Biomaterials*, **26**, 7481–7503.

Mastro, A. M., & Vogler, E. A. (2009). A three-dimensional osteogenic tissue model for the study of metastatic tumor cell interactions with bone. *Cancer Res.*, **69**, 4097–4100.

Matsuo, K., & Irie, N. (2008). Osteoclast–osteoblast communication. *Arch. Biochem. Biophys.*, **473**, 201–209.

Meister, K., Segal, D., & Whitelaw, G. P. (1990). The role of bone grafting in the treatment of delayed unions and nonunions of the tibia. *Orthop. Rev.*, **19**, 260–271.

Miyakoshi, N. (2004). Effects of parathyroid hormone on cancellous bone mass and structure in osteoporosis. *Curr. Pharm. Des.*, **10**, 2615–2627.

Mooney, D. J., Baldwin, D. F., Suh, N. P., Vacanti, J. P., & Langer, R. (1996). Novel approach to fabricate porous sponges of poly(D,L-lactic-co-glycolic acid) without the use of organic solvents. *Biomaterials*, **17**, 1417–1422.

Mosier-Laclair, S., Pike, H., & Pomeroy, G. (2001). Intraosseous bioabsorbable poly-L-lactic acid screw presenting as a late foreign-body reaction: A case report. *Foot Ankle Int.*, **22**, 247–251.

Muggli, D. S., Burkoth, A. K., & Anseth, K. S. (1999). Crosslinked polyanhydrides for use in orthopedic applications: Degradation behavior and mechanics. *J. Biomed. Mater. Res.*, **46**, 271–278.

Muschler, G. F., Nakamoto, C., & Griffith, L. G. (2004). Engineering principles of clinical cell-based tissue engineering. *J. Bone Joint Surg. Am.*, **86-A**, 1541–1558.

Nair, L. S., Bhattacharyya, S., Bender, J. D., Greish, Y. E., Brown, P. W., et al. (2004). Fabrication and optimization of methylphenoxy substituted polyphosphazene nanofibers for biomedical applications. *Biomacromolecules*, **5**, 2212–2220.

Neen, D., Noyes, D., Shaw, M., Gwilym, S., Fairlie, N., et al. (2006). Healos and bone marrow aspirate used for lumbar spine fusion: A case controlled study comparing healos with autograft. *Spine*, **31**, E636–E640.

Nof, M., & Shea, L. D. (2002). Drug-releasing scaffolds fabricated from drug-loaded microspheres. *J. Biomed. Mater. Res.*, **59**, 349–356.

Osathanon, T., Linnes, M. L., Rajachar, R. M., Ratner, B. D., Somerman, M. J., et al. (2008). Microporous nanofibrous fibrin-based scaffolds for bone tissue engineering. *Biomaterials*, **29**, 4091–4099.

Paskert, J. P., Yaremchuk, M. J., Randolph, M. A., & Weiland, A. J. (1987). The role of cyclosporin in prolonging survival in vascularized bone allografts. *Plast. Reconstr. Surg.*, **80**, 240–247.

Patel, S., Kurpinski, K., Quigley, R., Gao, H., Hsiao, B. S., et al. (2007). Bioactive nanofibers: Synergistic effects of nanotopography and chemical signaling on cell guidance. *Nano. Lett.*, **7**, 2122–2128.

Patterson, T. E., Kumagai, K., Griffith, L., & Muschler, G. F. (2008). Cellular strategies for enhancement of fracture repair. *J. Bone Joint Surg. Am.*, **90**(Suppl. 1), 111–119.

Pearson, O. M., & Lieberman, D. E. (2004). The aging of Wolff's "law:" Ontogeny and responses to mechanical loading in cortical bone. *Am. J. Phys. Anthropol., (Suppl. 39)*, 63–99.

Peterson, B., Whang, P. G., Iglesias, R., Wang, J. C., & Lieberman, J. R. (2004). Osteoinductivity of commercially available demineralized bone matrix. Preparations in a spine fusion model. *J. Bone Joint Surg. Am.*, **86-A**, 2243–2250.

Petruskevicius, J., Nielsen, S., Kaalund, S., Knudsen, P. R., & Overgaard, S. (2002). No effect of Osteoset, a bone graft substitute, on bone healing in humans: A prospective randomized double-blind study. *Acta Orthop. Scand.*, **73**, 575–578.

Pham, Q. P., Kurtis Kasper, F., Scott Baggett, L., Raphael, R. M., Jansen, J. A., et al. (2008). The influence of an *in vitro* generated bone-like extracellular matrix on osteoblastic gene expression of marrow stromal cells. *Biomaterials*, **29**, 2729–2739.

Pitto, R. P., Mueller, L. A., Reilly, K., Schmidt, R., & Munro, J. (2007). Quantitative computer-assisted osteodensitometry in total hip arthroplasty. *Int. Orthop.*, **31**, 431–438.

Rauschecker, J. P., & Shannon, R. V. (2002). Sending sound to the brain. *Science*, **295**, 1025–1029.

Rezwan, K., Chen, Q. Z., Blaker, J. J., & Boccaccini, A. R. (2006). Biodegradable and bioactive porous polymer/inorganic composite scaffolds for bone tissue engineering. *Biomaterials*, **27**, 3413–3431.

Ripamonti, U., Heliotis, M., & Ferretti, C. (2007). Bone morphogenetic proteins and the induction of bone formation: From laboratory to patients. *Oral Maxillofac. Surg. Clin. North Am.*, **19**, 575–589, vii.

Robinson, D., Alk, D., Sandbank, J., Farber, R., & Halperin, N. (1999). Inflammatory reactions associated with a calcium sulfate bone substitute. *Ann. Transplant*, **4**, 91–97.

Rosa, A. L., de Oliveira, P. T., & Beloti, M. M. (2008). Macroporous scaffolds associated with cells to construct a hybrid biomaterial for bone tissue engineering. *Expert Rev. Med. Devices*, **5**, 719–728.

Russell, J. L. (2000). Grafton demineralized bone matrix: Performance consistency, utility, and value. *Tissue Eng.*, **6**, 435–440.

Sachlos, E., & Czernuszka, J. T. (2003). Making tissue engineering scaffolds work. Review: The application of solid freeform fabrication technology to the production of tissue engineering scaffolds. *Eur. Cell Mater.*, **5**, 29–39; discussion 39–40.

Schwartz, Z., Goldstein, M., Raviv, E., Hirsch, A., Ranly, D. M., et al. (2007). Clinical evaluation of demineralized bone allograft in a hyaluronic acid carrier for sinus lift augmentation in humans: A computed tomography and histomorphometric study. *Clin. Oral Implants Res.*, **18**, 204–211.

Semino, C. E., Kasahara, J., Hayashi, Y., & Zhang, S. (2004). Entrapment of migrating hippocampal neural cells in three-dimensional peptide nanofiber scaffold. *Tissue Eng.*, **10**, 643–655.

Shapiro, F. (2008). Bone development and its relation to fracture repair. The role of mesenchymal osteoblasts and surface osteoblasts. *Eur. Cell. Mater.*, **15**, 53–76.

Sikavitsas, V. I., Bancroft, G. N., & Mikos, A. G. (2002). Formation of three-dimensional cell/polymer constructs for bone tissue engineering in a spinner flask and a rotating wall vessel bioreactor. *J. Biomed. Mater. Res.*, **62**, 136–148.

Silva, G. A., Czeisler, C., Niece, K. L., Beniash, E., Harrington, D. A., et al. (2004). Selective differentiation of neural progenitor cells by high-epitope density nanofibers. *Science*, **303**, 1352–1355.

Smith, L. A., & Ma, P. X. (2004). Nano-fibrous scaffolds for tissue engineering. *Colloids Surf. B. Biointerfaces*, **39**, 125–131.

Song, J. H., Kim, H. E., & Kim, H. W. (2008). Electrospun fibrous web of collagen-apatite precipitated nanocomposite for bone regeneration. *J. Mater. Sci. Mater. Med.*, **19**, 2925–2932.

Stevens, M. M., & George, J. H. (2005). Exploring and engineering the cell surface interface. *Science*, **310**, 1135–1138.

Stiehler, M., Bünger, C., Baatrup, A., Lind, M., Kassem, M., et al. (2008). Effect of dynamic 3-D culture on proliferation, distribution, and osteogenic differentiation of human mesenchymal stem cells. *J. Biomed. Mater. Res. Part A*, **89**(1), 96–107.

Tan, J., & Saltzman, W. M. (2004). Biomaterials with hierarchically defined micro- and nanoscale structure. *Biomaterials*, **25**, 3593–3601.

Thomson, R. C., Yaszemski, M. J., Powers, J. M., & Mikos, A. G. (1995). Fabrication of biodegradable polymer scaffolds to engineer trabecular bone. *J. Biomater. Sci. Polym. Ed.*, **7**, 23–38.

Toole, B. P., & Linsenmayer, T. F. (1977). Newer knowledge of skeletogenesis: Macromolecular transitions in the extracellular matrix. *Clin. Orthop. Relat. Res.*, **129**, 258–278.

Tuan, R. S. (2004). Biology of developmental and regenerative skeletogenesis. *Clin. Orthop. Relat. Res.*, **427**(Suppl.), S105–S117.

Tuzlakoglu, K., Bolgen, N., Salgado, A. J., Gomes, M. E., Piskin, E., et al. (2005). Nano- and micro-fiber combined scaffolds: A new architecture for bone tissue engineering. *J. Mater. Sci. Mater. Med.*, **16**, 1099–1104.

Urist, M. (1983). Bone cell differentiation and growth factors. *Science*, **220**, 680.

Vaananen, H. K., & Laitala-Leinonen, T. (2008). Osteoclast lineage and function. *Arch. Biochem. Biophys.*, **473**, 132–138.

van de Witte, P., Dijkstra, P. J., van den Berg, J. W. A., & Feijen, J. (1996). Phase separation processes in polymer solutions in relation to membrane formation. *J. Membrane Sci.*, **117**, 1–31.

Vargas, G. E., Mesones, R. V., Bretcanu, O., López, J. M. P., Boccaccini, A. R., et al. (2009). Biocompatibility and bone mineralization potential of 45S5 Bioglass®-derived glass–ceramic scaffolds in chick embryos. *Acta Biomater.*, **5**(1), 374–380.

Vats, A., Tolley, N. S., Polak, J. M., & Gough, J. E. (2003). Scaffolds and biomaterials for tissue engineering: A review of clinical applications. *Clin. Otolaryngol. Allied Sci.*, **28**, 165–172.

Venugopal, J., Vadgama, P., Kumar, T. S. S., & Ramakrishna, S. (2007). Biocomposite nanofibres and osteoblasts for bone tissue engineering. *Nanotechnology*, 055101.

Vogler, E. A. (1989). A compartmentalized device for the culture of animal cells. *Biomater. Artif. Cells Artif. Organs*, **17**, 597–610.

Wahl, D. A., & Czernuszka, J. T. (2006). Collagen-hydroxyapatite composites for hard tissue repair. *Eur. Cell Mater.*, **11**, 43–56.

Wan, Y., Wang, Y., Liu, Z., Qu, X., Han, B., et al. (2005). Adhesion and proliferation of OCT-1 osteoblast-like cells on micro- and nano-scale topography structured poly(l-lactide). *Biomaterials*, **26**, 4453–4459.

Wang, C., Duan, Y., Markovic, B., Barbara, J., Howlett, C. R., et al. (2004). Phenotypic expression of bone-related genes in osteoblasts grown on calcium phosphate ceramics with different phase compositions. *Biomaterials*, **25**, 2507–2514.

Wang, L., Cowin, S. C., Weinbaum, S., & Fritton, S. P. (2000). Modeling tracer transport in an osteon under cyclic loading. *Ann. Biomed. Eng.*, **28**, 1200–1209.

Wang, T., Wu, H., Wang, H., Lin, F., & Sun, J. (2008). Regulation of adult human mesenchymal stem cells into osteogenic and chondrogenic lineages by different bioreactor systems. *J. Biomed. Mater. Res. Part A*, **88**(4), 935–946.

Weber, J. N., White, E. W., & Lebiedzik, J. (1971). New porous biomaterials by replication of echinoderm skeletal microstructures. *Nature*, **233**, 337–339.

Williams, J. M., Adewunmi, A., Schek, R. M., Flanagan, C. L., Krebsbach, P. H., et al. (2005). Bone tissue engineering using polycaprolactone scaffolds fabricated via selective laser sintering. *Biomaterials*, **26**, 4817–4827.

Wlodarski, K. H., Wlodarski, P. K., & Galus, R. (2008). Bioactive composites for bone regeneration. *Review Ortop. Traumatol. Rehabil.*, **10**, 201–210.

Ye, F., Lu, X., Lu, B., Wang, J., Shi, Y., et al. (2007). A long-term evaluation of osteoinductive HA/beta-TCP ceramics *in vivo*: 4.5 years study in pigs. *J. Mater. Sci. Mater. Med.*, **18**, 2173–2178.

Yoshimoto, H., Shin, Y. M., Terai, H., & Vacanti, J. P. (2003). A biodegradable nanofiber scaffold by electrospinning and its potential for bone tissue engineering. *Biomaterials*, **24**, 2077–2082.

Yu, X., Botchwey, E. A., Levine, E. M., Pollack, S. R., & Laurencin, C. T. (2004). Bioreactor-based bone tissue engineering: The influence of dynamic flow on osteoblast phenotypic expression and matrix mineralization. *Proc. Natl. Acad. Sci. USA*, **101**, 11203–11208.

Yuan, H., Van Den Doel, M., Li, S., Van Blitterswijk, C. A., De Groot, K., et al. (2002). A comparison of the osteoinductive potential of two calcium phosphate ceramics implanted intramuscularly in goats. *J. Mater. Sci. Mater. Med.*, **13**, 1271–1275.

Zaidi, M. (2007). Skeletal remodeling in health and disease. *Nat. Med.*, **13**, 791–801.

Zhang, S. (2003). Fabrication of novel biomaterials through molecular self-assembly. *Nat. Biotechnol.*, **21**, 1171–1178.

Zhang, Y. Z., Venugopal, J., Huang, Z. M., Lim, C. T., & Ramakrishna, S. (2005). Characterization of the surface biocompatibility of the electrospun PCL-collagen nanofibers using fibroblasts. *Biomacromolecules*, **6**, 2583–2589.

Zhu, X. H., Gan, S. K., Wang, C. H., & Tong, Y. W. (2007). Proteins combination on PHBV microsphere scaffold to regulate Hep3B cells activity and functionality: A model of liver tissue engineering system. *J. Biomed. Mater. Res. A.*, **83**, 606–616.

CHAPTER II.6.8 CARTILAGE AND LIGAMENT TISSUE ENGINEERING: BIOMATERIALS, CELLULAR INTERACTIONS, AND REGENERATIVE STRATEGIES

Catherine K. Kuo[1], Wan-Ju Li[2], and Rocky S. Tuan[3]

[1]Department of Biomedical Engineering, Tufts University, Medford, MA, USA

[2]Departments of Orthopedics and Rehabilitation, and Biomedical Engineering, University of Wisconsin-Madison, Madison, WI, USA

[3]Department of Orthopaedic Surgery and Center for Cellular and Molecular Engineering, University of Pittsburgh School of Medicine, Pittsburgh, PA, USA

INTRODUCTION TO CARTILAGE AND LIGAMENT TISSUE ENGINEERING

The musculoskeletal system is responsible for complex movements that are performed many thousands of times over a lifetime. Two connective tissues of the musculoskeletal system, cartilage and ligament, protect the body from injuries during these movements, primarily by absorbing loads and maintaining joint stability, respectively. Relative to other musculoskeletal tissues, cartilage and ligament have low oxygen and nutrient requirements, low cell density, and poor regenerative capacity, yet they experience some of the highest mechanical loads in the body. When these loads exceed a critical threshold that causes permanent tissue damage, or if diseases cause severe tissue degeneration, these problems often result in a significant locomotive impairment. For the repair of both tissues, given their very low self-regenerative capacity, typically the only recourse is surgical intervention. Current surgical reparative techniques rely upon total joint replacement or grafting, and are often accompanied with further musculoskeletal problems; a more ideal solution would be to use a biological approach to repair the defects and fully restore the cartilage or ligament tissue to its pre-injured state. This is the promise of tissue engineering, a new field which generally aims to incorporate specific cell types into a biodegradable scaffold which, when implanted, will gradually regenerate into a tissue that closely resembles the original tissue and restores functionality. In this chapter, we outline the state of the art of cartilage and ligament tissue engineering, divided into two sections, with emphasis ranging from biomaterial selection and functionalization, cellular activities critical for tissue function, scaffold design to mimic native tissue composition, to recent outcomes of long-term implantation studies in animal models.

CARTILAGE TISSUE ENGINEERING

Introduction to Cartilage Tissue Engineering

Cartilage is a connective tissue that functions to absorb mechanical impact, assist joint motion, provide structural support, and connect soft and hard tissues. There are three types of human cartilage: hyaline, elastic, and fibrocartilage, located at different parts of the body, and each type of cartilage features a unique composition and structure specifically suitable for carrying out the aforementioned tissue functions. Like other musculoskeletal tissues primarily responsible for mechanical and structural functions, cartilage plays an important role in protecting the body from mechanical impacts during physical activities. Therefore, any damage significantly impeding cartilage function may also affect functional capabilities of the musculoskeletal system.

Clinical Relevance and Limitations of Current Repair Strategies. Cartilage damage is a common problem often seen in clinical orthopedics. The causes and mechanisms leading to cartilage damage can be physical, biological or both. Physical causes, such as trauma, often damage the tissue structure and injure cells, creating permanent lesions in cartilage. On the other hand, biological causes, such as those related to autoimmunity and inflammation, trigger catabolic cellular and biochemical activities that degrade cartilage. Among these, osteoarthritis (OA), an aging-related disease, is the leading cause of cartilage degeneration. While OA is a joint problem commonly found in the aged population, the current trend has shown a decrease in the average age of the OA population. Obesity is one of the major factors triggering the early development of OA, and is likely to involve biological and physical mechanisms. As one of the leading causes of disability among the US adult population, OA significantly not only affects the quality of an individual's life, but also brings enormous societal and financial burdens to the healthcare system. Annually, OA treatment incurs more than 185.5 billion US dollars of healthcare cost, and this number is expected to increase rapidly in the next few decades when the baby boomer generation ages (Kotlarz et al., 2009). By the year 2030, it is estimated that 25% or more of the population in the US would likely suffer from arthritis. These demographic and health-cost estimates clearly illustrate the disease burden of OA, and indicate the urgent need to develop an effective therapy to treat OA, as well as other cartilage-related health problems.

Cartilage is a hypo-cellular tissue that does not contain blood and lymphatic vessels to support circulation of repair cells and molecules, such as growth factors and cytokines, to initiate wound healing of lesions once injured. Therefore, cartilage lesions often progress with time in an almost irreversible manner. Current treatments to cartilage lesions include non-surgical therapies and surgical interventions (Hunziker, 2002). Non-surgical treatments, such as drug or physical therapy, are generally used for the early stage of OA or cartilage with a minor lesion, whereas surgical interventions are necessary for mid- and late-stage OA or cartilage with a severe lesion. The clinical gold standard to treat severe cartilage lesions is total joint replacement, an effective but final resort surgical procedure to restore joint functions. For patients with less severe cartilage lesions, other surgical procedures, such as arthroscopic abrasion and debridement, microfracture, osteochondral plug transplantation or mosaicplasty, and autologous chondrocyte implantation, are used to repair lesions. Clinical results have shown that although these surgical procedures are able to provide some improvements to the disease state, there are inherent complications, such as fibrocartilage formation and donor site tissue morbidity.

The Modern Cartilage Engineering Strategy. Cartilage tissue engineering is an emerging method for therapeutic repair and regeneration of cartilage. The first cartilage tissue engineering attempt may be traced back to the 1970s, when a pediatric surgeon, Dr. W. T. Green, cultured rabbit chondrocytes on decalcified bone and transplanted the cellular structure into cartilage defects (Green, 1977). Although the outcome of the study was not satisfactory, this pioneer study introduced the innovative concept of cartilage tissue engineering. In 1991, Vacanti et al. implanted chondrocyte-laden poly(lactide-co-glycolide) (PLGA) scaffolds subcutaneously in nude mice to regenerate cartilage (Vacanti et al., 1991). The research team successfully demonstrated the feasibility of using a tissue engineering approach to regenerate new cartilage. Two years later, Freed and co-workers used a bioreactor system to grow chondrocyte-laden biodegradable scaffolds *ex vivo*, and showed significantly enhanced production of cartilage matrix in the bioreactor-grown cartilage, compared to static culture-grown cartilage (Freed et al., 1993). In the past two decades, the introduction of stem cells, smart biomaterials, and bioreactors have continued to enhance cartilage tissue engineering.

The principal concept of cartilage tissue engineering is to seed chondrocytes or chondroprogenitor cells within a three-dimensional biomaterial scaffold which is then cultured in a bioreactor to produce functional cartilage *in vitro*, which is then implanted *in vivo*. Alternatively, a three-dimensional biomaterial scaffold is used as a carrier to deliver therapeutic cells to facilitate cartilage regeneration at a defect site, without being pre-cultured *in vitro*. Critical to the success of these cartilage tissue engineering strategies are the physical and chemical properties of biomaterial scaffolds, which must be able to provide mechanical strength, as well as to regulate biological activities of the seeded cells and their chondrocytic phenotype. In the past two decades, a great number of biomaterials and three-dimensional structures, including many novel materials and scaffolds, have been introduced into cartilage tissue engineering. By including cutting-edge nanotechnologies, peptide synthesis and controlled release, novel scaffolds with intelligent materials and structures are capable of changing their physical or chemical properties in response to physiological needs to enhance cell growth and tissue regeneration.

Macromolecular Composition and Structure of Native Cartilage

Cartilage Composition. Chondrocytes are the sole cell type in cartilage, and make up 5–10% of total tissue volume. Besides chondrocytes, other major components of cartilage are specific extracellular matrix (ECM) macromolecules, and water (Buckwalter and Mankin, 1998). The ECM constitutes 20–40% of the tissue volume, and the water content amounts to 70% of the total tissue weight. Despite a minor fraction of the total tissue volume occupied by chondrocytes, activities of chondrocytes are critical to the anabolism and catabolism of cartilage, i.e., chondrocytes continue to synthesize and break down ECM components to maintain healthy matrix turnover under the influence of autocrine or paracrine signals (Stockwell, 1979).

The ECM of cartilage consists of collagens, non-collagenous proteins, and proteoglycans (Buckwalter and Mankin, 1998). Collagenous ECM components include collagen types I, II, VI, IX, X, and XI (Eyre et al., 1992). In hyaline cartilage, a small quantity of collagen type I accumulates in the superficial zone, while collagen type II is ubiquitously present in cartilage and accounts for 90–95% of total collagen (Kuettner, 1992). Both collagen types I and II play a mechanical role in maintaining tissue structure integrity by providing strength to resist tensile loading. Collagen type VI is localized around chondrocytes, and has been postulated to play a role in mechanobiological regulation. Collagen type IX functions as a linker molecule that binds to collagen type II to stabilize collagen type II fibril assembly (Mendler et al., 1989). Collagen type X is produced by hypertrophic chondrocytes, and is mostly found in the calcified zone of the cartilage growth plate. Collagen type XI constitutes 5–10% of the total collagen content. Its function is not clear but likely associated with stabilization of the collagen network (Swoboda et al. 1989).

Non-collagenous proteins of the cartilage ECM are also involved in structuring the ECM and regulating chondrocyte activities (Heinegard and Pimentel, 1992). For example, the proteoglycan link protein serves to link aggrecan with hyaluronic acid (Ruoslahti, 1988), and fibronectin regulates chondrocyte adhesion to the ECM (Hayashi et al., 1996). Proteoglycans make up the bulk of the non-collagenous ECM components of cartilage, including large aggregating proteoglycans, such as aggrecan and versican, and small leucine-rich proteoglycans, such as decorin and biglycan (Prydz and Dalen, 2000). In hyaline articular cartilage, aggrecan is composed of a core protein, covalently modified with sulfated glycosaminoglycan (sGAG) molecules that make up about 90% of the total proteoglycan mass (Roughley and Lee, 1994). Sulfated GAGs, such as chondroitin sulfate, dermatan sulfate, heparan sulfate, and keratan sulfate, have a long polysaccharide chain that consists of repeating disaccharide units with sulfate groups. These sulfated polysaccharides carry a large amount of negative charge to attract and retain water between the matrix molecules, thus creating hydrostatic resistance to compressive loading (Mow and Wang, 1999).

Cartilage Structure. Although the articular cartilage of a human joint is only a few millimeters in thickness, it is capable of mediating smooth joint motion, and protecting subchondral bone from mechanical damage. This property of the articular cartilage is attributed to the existence of four regional zones in the tissue, with specific cell activities, and ECM structure and composition associated with each zone (Buckwalter and Mankin, 1998). The four zones from the top down are the superficial, middle, deep, and calcified zones (Farnworth, 2000). The superficial zone consists of a large amount of fibrillar collagen and water, but a small amount of proteoglycans (Venn, 1978). Collagen fibrils run parallel to each other and the joint surface, conferring high tensile and shear strength (Kempson et al., 1973). Chondrocytes in this zone are flattened, ellipsoid-shaped, and synthesize low molecular weight proteoglycans (Hasler et al., 1999). The middle zone lies beneath the superficial zone, and has several times the volume of the superficial zone. This zone features a lower cell number, the presence of spherical chondrocytes, and a large amount of proteoglycans (Venn, 1978). Chondrocytes produce fewer but larger randomly-oriented collagen fibrils in the middle zone, compared to those fibrils in the superficial zone. Located beneath the middle zone is the deep zone; this zone contains the largest collagen fibrils, the greatest amount of proteoglycans, and the lowest cell number, compared to the other three zones. Chondrocytes in this zone maintain a spherical shape, and the collagen fibrils are perpendicular to the joint surface (Hunziker et al., 1997). The deep zone is capable of providing substantial resistance to compressive loading because of the large amount of proteoglycans and the perpendicular collagen fibrils. The calcified cartilage zone is a transition zone located between the deep zone and subchondral bone. Chondrocytes in this zone are hypertrophic and exhibit low metabolic activities (Bullough and Jagannath, 1983).

Important Considerations in Designing and Fabricating a Cartilage Tissue Engineered Scaffold

Functionally, a tissue engineered scaffold represents an artificial ECM to accommodate, protect, and interact with cells residing in the structure during tissue regeneration. An ideal tissue engineered scaffold should be able to substitute for native ECM, and faithfully carry out ECM functions. In order to design an ideal scaffold for cartilage tissue engineering, it is essential to identify the critical cellular activities and material properties required for robust cartilage regeneration, and to select the biomaterial(s) that is best able to represent these characteristics.

Cellular Activity Requirements

Maintenance of the Phenotypic Spherical Shape of Chondrocytes. Cell morphology is an important aspect of the phenotype of a cell, and is critical in the regulation of cell activities. In native cartilage, chondrocytes are embedded in dense ECM, and exhibit a characteristic spherical morphology which is critical for maintaining the chondrocytic phenotype. Numerous *in vitro* studies have shown that after chondrocytes are isolated from cartilage and cultured as a monolayer surface, the cells change from a spherical to a spindle, elongated shape, and dedifferentiate into fibroblast-like cells. Unlike phenotypic chondrocytes that produce collagen type II and aggrecan, dedifferentiated cells produce collagen type I, decorin, and biglycan. Indeed, spindle shaped dedifferentiated chondrocytes show distinct cell activities, compared to spherical chondrocytes. However, the dedifferentiated cells can be redifferentiated back to phenotypic chondrocytes if they are encapsulated in a hydrogel or cultured on a low-adhesivity surface to re-establish a spherical morphology. Assumption and maintenance of a spherical cell shape is also a requirement for effective chondrogenesis of embryonic limb mesenchymal cells and adult mesenchymal stem cells (MSCs), as exemplified by a high cell-seeding density of these cells used in three-dimensional micromass or pellet culture for chondrogenic differentiation (see below).

Establishment of Biologically Favorable Three-dimensional Cell–Cell and Cell–Matrix Interactions. During embryonic chondrogenesis, mesenchymal cells undergo cell condensation, which is a critical and necessary step for chondrogenic differentiation (Fell, 1925). The cell condensation process is responsible for the initiation of specific, three-dimensional cell–cell interactions, such as N-cadherin mediated cell adhesion and gap junctional communication (Oberlender and Tuan, 1994), to activate chondrogenic signaling events for cartilage formation. For *ex vivo* chondrogenesis, the commonly used high-density cell pellet culture imitates the cell condensation process by densely packing cells via centrifugation to enhance three-dimensional cell–cell interactions. In addition to cell–cell interactions, cell–matrix interactions represent another critical mechanism that is involved in the regulation of chondrocyte activities in mature cartilage. After initial chondrogenesis, the resultant chondrocytes synthesize abundant ECM which is deposited around the cells, thus gradually separating cells from each other. In this manner, in mature cartilage, chondrocytes interact intimately with the pericellular ECM to maintain cartilage functions.

Material Property Requirements

Presentation of Biocompatible Surface Chemistry and Topography on Scaffolds. Surface chemical and physical properties of scaffolding materials must be biocompatible, and not elicit immunogenic or inflammatory responses from seeded cells or adjacent host tissues surrounding the scaffold. In addition, byproducts released from the scaffold as a result of biodegradation or bioresorption must be non-toxic, and not cause undesirable host reactions. Alternatively, any unwanted byproducts must be effectively removed from the local microenvironment. This characteristic is particularly important for cartilage tissue engineering because of the non-vascular nature of cartilage, which precludes vasculature as an efficient means of byproduct removal, with diffusion as the only mechanism available.

Construction of Scaffolding Structures with ECM-Mimetic Geometry. There is mounting evidence that physical properties of a biomaterial scaffold can instruct cell behavior, thus ultimately determining the fate of an engineered tissue. For example, scaffolds with nanosized or microsized structure create different patterns of cell–matrix interaction with seeded chondrocytes to modulate cell morphology and cytoskeletal organization, in turn resulting in different gene and protein expressions (Li et al., 2006a).

Mechanical Properties. In addition to functioning as structural templates to direct tissue growth and to determining the size and shape of regenerated cartilage, scaffolds must also provide protection for the neocartilage from mechanical damage. In the process of cartilage tissue engineering, mechanical challenges could occur during *in vitro* regeneration with mechanical loading stimulating tissue growth or post-implantation when the joints are subjected to natural weightbearing functions. Ideal scaffolds for cartilage tissue engineering must be able to protect cells from damage, and maintain structural integrity of tissue-engineered constructs to provide a mechanically stable environment to seeded cells during the tissue regeneration process.

Biomaterials in Cartilage Tissue Engineering

Synthetic Biomaterials for Cartilage Tissue Engineering. *Poly(α-hydroxy esters)* – Members of poly(α-hydroxy esters) are the most commonly used synthetic biodegradable polymer for cartilage tissue engineering. An advantage of using poly(α-hydroxy esters) as cartilage

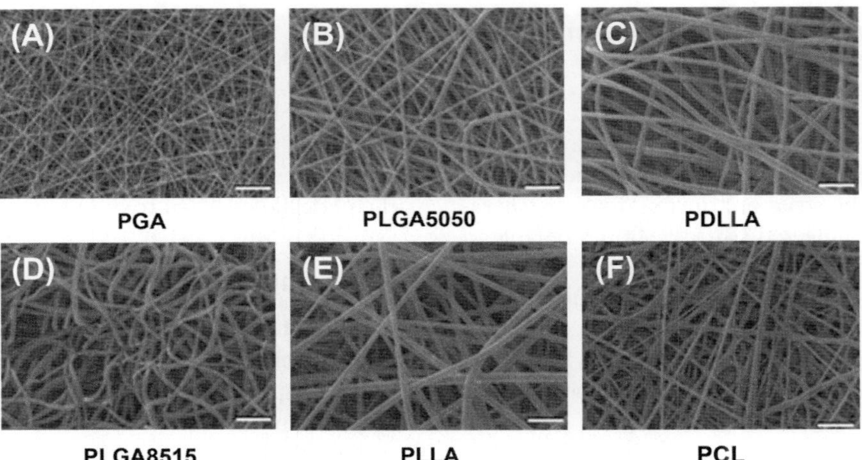

FIGURE II.6.8.1 Scanning electron microscopy (SEM) micrograph of electrospun poly(α-hydroxy ester) fibrous scaffolds composed of randomly oriented ultra-fine fibers. (A) PGA; (B) PLGA5050; (C) PDLLA; (D) PLGA8515; (E) PLLA; and (F) PCL. Bar: 10 μm. (Reproduced with permission from Elsevier B.V.)

tissue engineering scaffolds is their prior approval by the US Food and Drug Agency for clinical use as bioresorbable biomaterials. Among the poly(α-hydroxy ester) members, poly(glycolic acid) (PGA), poly(lactic acid) (PLA), PLGA, and poly (ε-caprolactone) (PCL) have been extensively investigated for the use in fabrication of cartilage tissue-engineered scaffolds. PGA is a semi-crystalline, hydrophilic polymer, and may rapidly degrade in an aqueous solution. PLA, with the addition of a methyl group, is a more hydrophobic, semi-crystalline polymer, and degrades more slowly than PGA (Middleton and Tipton, 2000). The three stereoisomers, D, L, and D and L mixture, referring to the different conformation of the methyl group, exhibit distinct properties. For example, poly(l-lactic acid) (PLLA) has a higher melting temperature and degrades more slowly than poly(d, l-lactic acid) (PDLLA). Furthermore, PGA and PLA can be copolymerized in different ratios to form a new polymer, PLGA, with properties different from either of the two constituent polymers. Poly (ε-caprolactone) (PCL) is another semi-crystalline biodegradable polymer. Compared to other poly(α-hydroxy ester) members, PCL is less commonly used in the fabrication of scaffolds, mainly because its slow degradation may interfere with deposition of newly synthesized ECM. Recently, PCL has been blended or copolymerized with other poly(α-hydroxy esters) to fabricate new polymers with controlled properties for meeting the needs of fabricating working scaffolds (Middleton and Tipton, 2000).

The use of PLA scaffolds to culture chondrocytes, followed by implantation of the cell-seeded scaffold in an animal model to regenerate cartilage was first reported in the early 1990s (Vacanti et al., 1991). After this pioneer study, PLA, PGA, and PLGA have been fabricated into various structured scaffolds to culture chondrocytes or MSCs to regenerate cartilage *in vitro* or *in vivo*. Studies have shown that effective *in vivo* cartilage regeneration can be achieved by seeding and culturing chondrocytes on PGA scaffolds *in vitro*, followed by *in vivo* implantation (Chu et al., 1995; Liu et al., 2002), suggesting that the process of *in vitro* cartilage tissue engineering might allow chondrocytes to produce sufficient ECM to stabilize the newly synthesized tissue structure while PGA undergoes degradation. Unlike natural biopolymers that are often prepared in a hydrogel form, the poly(α-hydroxy esters) are commonly fabricated into various pre-formed structures, such as fiber, foam, and sponge.

Developing nanostructural materials as scaffolds is an emerging trend in tissue engineering. Nanostructured scaffolds, such as nanofibrous scaffolds, can be fabricated using the electrospinning technique. Several poly(α-hydroxy esters) have been successfully fabricated into nanofibrous scaffolds (Figure II.6.8.1) (Li et al., 2006b). These different poly(α-hydroxy ester)-based nanofibrous scaffolds possess a range of mechanical properties and degradation profiles that render them as suitable scaffolds for different engineered tissues. For example, chondrocytes cultured in PLLA or PCL nanofibrous scaffolds maintained their chondrocytic phenotype (Li et al., 2003), and MSCs in the same structure were successfully induced into chondrocytes (Li et al., 2005). Furthermore, an *in vivo* study showed that the chondrocyte- or MSC-laden nanofibrous scaffold was able to repair full-thickness cartilage defects created in pig femoral condyles (Figure II.6.8.2) (Li et al., 2009).

Poly(ethylene glycol) (PEG) or polyethylene oxide (PEO) – PEG or PEO is a non-biodegradable synthetic polymer that has been extensively used for cartilage tissue engineering. The major advantage of using PEG or PEO for scaffold fabrication is that the polymer can be functionalized to conjugate with a peptide or protein of interest to enhance chondrocyte activities or induce chondrogenesis. In addition, the polymer can be chemically modified to carry reactive groups, such as acrylates, that can be activated by high-energy light sources, and cross-linked to improve mechanical and degradation

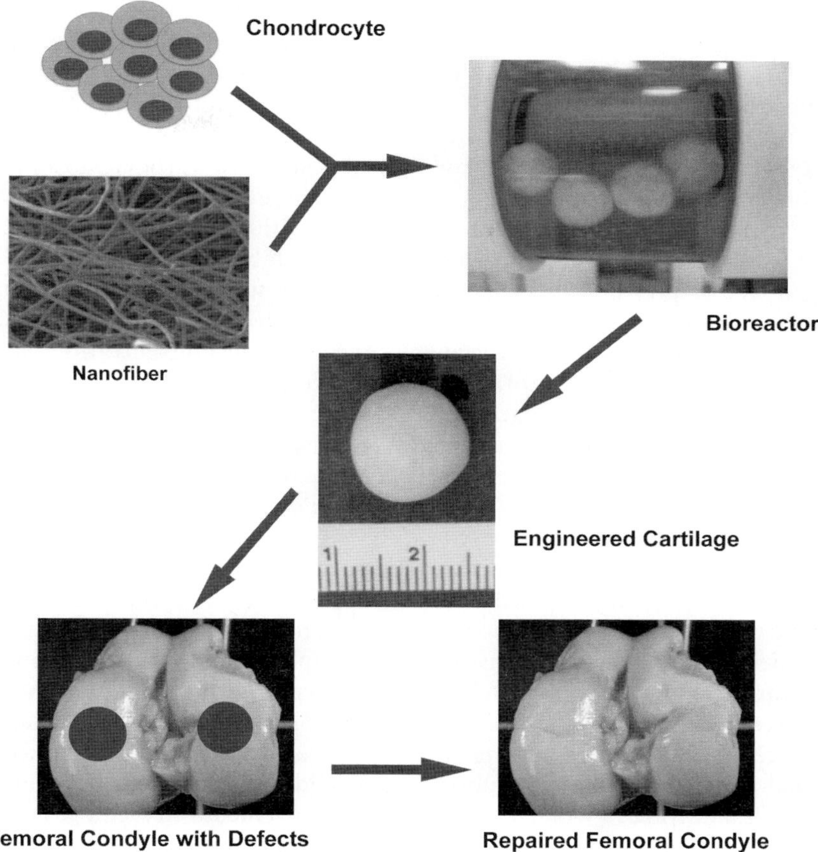

FIGURE II.6.8.2 *In vivo* cartilage repair using engineered cartilage developed from chondrocyte-laden nanofibrous scaffold. Cultured chondrocytes were seeded into poly(lactic acid) nanofibrous scaffolds and cultured in a rotating wall vessel bioreactor. With the enhanced cell culture medium circulation, the cellular scaffold turned into a mature, functional cartilage implant after 42 days of culture. Artificially created cartilage defects of swine knees were successfully repaired by the engineered cartilage six months after surgery.

properties. Another desirable feature is that a modified PEG or PEO hydrogel mixed with therapeutic cells can be directly injected to fill irregular-shaped cartilage defects using a minimally invasive procedure (Elisseeff et al., 1999a). The minimally invasive implantation significantly reduces the time needed for the healing process.

For cartilage tissue engineering, dimethacrylate-functionalized PEO hydrogels mixed with chondrocytes were injected subcutaneously into mice and photopolymerized *in situ* using UV light (Elisseeff et al., 1999b). The results showed that chondrocytes in the solidified PEO hydrogel survived and actively produced cartilage matrix, and the regenerated neocartilage histologically resembled native cartilage. In addition, for the regeneration of healthy functional cartilage, PEG scaffolds have been designed to degrade at a controlled rate to free part of the scaffold space for natural ECM deposition. Functional groups, such as esters, can be chemically linked to unmodified PEG to improve biodegradation of the PEG hydrogel, since the ester linkage can be hydrolytically cleaved in culture medium (Bryant and Anseth, 2003). Various natural and synthetic polymers have also been grafted with PEG (Metters et al., 1999; Temenoff et al., 2002) to enhance cartilage-specific ECM production (Bryant and Anseth, 2003) and cartilage regeneration (Lee et al., 2003a; Wang et al., 2003).

Natural Biopolymers for Cartilage Tissue Engineering. Natural biopolymers, including protein-based and carbohydrate-based polymers, are derived from animals or plants, and have a long history of being used as biomaterials. The advantage of using natural biopolymers is that they have biofunctional molecular domains that can act to enhance cellular activities, such as cell adhesion and migration. On the other hand natural biopolymers for scaffolds also have drawbacks. They may have varying properties between different batches and may elicit immune response.

Protein-Based Materials. *Collagen* – Collagen represents the major protein type in the body, and has been widely used in various medical applications. Among the 29 identified collagen types, collagen type I is the most commonly used collagen for culturing chondrocytes (Fuss et al., 2000) or MSCs (Wakitani et al., 1994) in both *in vitro* and *in vivo* cartilage tissue engineering. Clinically, collagen type I, CaRes®, and collagen type I/III bilayer membrane, such as Maix® or Chondro-Gide®, have been approved to treat cartilage defects. Collagen types I and II have been compared in terms of their capability as a scaffold for cartilage repair (Frenkel et al., 1997; Lee et al., 2003b), and the results suggest that collagen type II is a more suitable scaffold material to maintain chondrocytic phenotype (Nehrer et al., 1997). However, in comparison to collagen type I, isolation and preparation of collagen type II in a sufficient quantity for scaffold fabrication is more challenging, because collagen type II is present exclusively in hyaline cartilage, whereas abundant collagen type I is found in many other tissues. As a biomaterial, collagen can be prepared in the form of a hydrogel, dehydrated porous sponge or electrospun fibrous structure for scaffold applications. Regardless of the structure type, the principal hurdle of using collagen matrices for cartilage tissue engineering is the short-term durability of the matrix, which reduces the structural stability of cell-seeded constructs during cartilage regeneration.

Fibrin – Fibrin, a component of the extrinsic coagulation cascade, is found commonly in blood clots. This biopolymer is an insoluble fibrillar protein converted from a soluble fibrinogen through the enzymatic activity of thrombin, and has been widely used as a clinical fixative or biological glue for many years, including recently as a biomaterial for cartilage tissue engineering. Specifically, it has been used as a glue to secure other scaffolds at the repair site, as a three-dimensional scaffold to culture chondrocytes (Silverman et al., 1999) and MSCs (Worster et al., 2001) for tissue regeneration, or as a carrier to deliver growth factors (Fortier et al., 2002) to facilitate cartilage repair. Like collagen, a fibrin-based glue, Tissucol®, is also approved to repair cartilage defects clinically. A previous study with 12-month post-surgery results showed more effective cartilage repair in patients implanted with fibrin embedded with autologous chondrocytes, compared to patients treated with a cartilage abrasion technique. Although the outcome appears promising, the inherent poor mechanical property and the immunogenicity of fibrin may limit its use in cartilage tissue engineering.

Silk fibroin – Fibroin, the principal component of insect silk, is another natural biopolymer that has recently been used in cartilage tissue engineering. Natural silks from different species, for example the spider or the silkworm, have different material properties, because the constituent silk fibroin can be arranged differently to construct different molecular structures. Owing to its considerable mechanical properties, silk holds great promise as a tissue engineering scaffold, particularly in a highly mechanically demanding tissue environment, because as that in cartilage. Silk fibroin has been made into a sponge scaffold to culture chondrocytes for rabbit cartilage repair, or fabricated into foam, microfibrous, or electrospun nanofibrous scaffolds to culture MSCs for both *in vitro* and *in vivo* cartilage tissue engineering.

Polysaccharide-Based Materials. *Alginate* – Alginate is a brown algae-derived carbohydrate polymer consisting of repeating L-guluronic acid and D-mannuronic acid. Alginate can be gelled in the presence of calcium or other divalent cations, and reversibly re-solubilized upon removal of the cations. In addition, alginate is capable of encapsulating cells in the matrix to maintain their spherical shape, which makes alginate a desirable scaffold for studying redifferentiation of chondrocytes (Liu et al., 1998), or promoting chondrogenic differentiation of MSCs (Caterson et al., 2001). Alginate scaffolds with varying stiffness can be fabricated by using different concentrations of alginate solutions or by varying cation concentrations during cross-linking (Kuo et al., 2001; 2008). By varying alginate concentration, mechanical properties of the resultant hydrogel can be significantly improved. However, the major concern with the use of alginate in cartilage tissue engineering is that the slow degradation of alginate often causes inappropriate immune response *in vivo* (Pier et al., 1994).

Agarose – Another carbohydrate polymer, agarose, is purified from brown seaweed, and contains a repeated disaccharide sequence of L- and D-galactose. Similar to alginate, the stiffness of the agarose hydrogel can be manipulated by varying its concentration. Dependent on the type and molecular weight, agarose gels have different solidification temperatures. By altering the temperature, agarose can change between a liquid gel form and a solidified structure form, which is a preferred property for convenient cell seeding. For cartilage tissue engineering applications, agarose has been extensively used as a three-dimensional scaffold to redifferentiate chondrocytes (Benya and Shaffer, 1982) or differentiate MSCs into chondrocytes (Fukumoto et al., 2003; Huang et al., 2004) *ex vivo*. However, the use of agarose for *in vivo* applications is limited by its extremely slow degradation rate (Cook et al., 2003) that could elicit foreign body giant cell reaction (Rahfoth et al., 1998).

Hyaluronic acid – The most widely used carbohydrate-based biopolymer in cartilage tissue engineering is hyaluronic acid. The biopolymer is a polysaccharide composed of repeating glucuronic acid and N-acetylglucosamine. Because hyaluronic acid is one of the major ECM components of cartilage, its use as a scaffolding material is biologically compatible for the activities of chondrocytes. Nevertheless, the applicability of native, unmodified hyaluronic acid is limited because the highly hydrophilic molecule is easily degraded in culture medium, and the hyaluronic acid gel does not possess sufficient mechanical strength. Therefore, a number of chemical modifications, including esterification (Campoccia et al., 1998) and other chemical cross-linking (Vercruysse et al., 1997), have been developed to improve its properties. For example, Hyaff 11® is a commercially available, three-dimensional structure based on an esterified derivative of hyaluronic acid, which exhibits improved degradation and mechanical properties, and has been used as a tissue engineering scaffold to seed chondrocytes and MSCs for cartilage tissue engineering (Facchini et al., 2006). A three-year clinical study using the esterified hyaluronic acid scaffold with autologous chondrocytes has shown satisfactory results on cartilage repair.

Chitosan – Chitosan is composed of repeating glucosamine and N-acetylglucosamine disaccharides, and is derived from chitin, a type of polysaccharide commonly found in shellfish exoskeletons. Insoluble chitin becomes soluble chitosan after sufficient acetyl groups in a chitin polymer are removed. Because chitosan carries high cationic charges, it can interact with a variety of anionic polymers, such as chondroitin sulfate, to form a hydrogel (Denuziere et al., 1998). In addition, owing to its excellent biocompatibility and the ease of scaffold fabrication, chitosan is considered a promising biomaterial for cartilage tissue engineering. Chondrocytes cultured in a chondroitin sulfate-augmented chitosan hydrogel maintained their chondrocytic morphology and produced collagen type II (Lahiji et al., 2000; Sechriest et al., 2000). An *in vivo* study showed that chitosan injection effectively repaired defects of rat knee cartilage (Lu et al., 1999).

Chondroitin sulfate – Chondroitin sulfate is the major sulfated glycosaminoglycan of cartilage proteoglycan. Due to its excellent biocompatibility, and biological and functional roles in the cartilage ECM macrostructure (Bryant et al., 2004; Li et al., 2004), chondroitin sulfate has been shown to help wound healing and restore arthritic joint functions. Chondroitin sulfate monomers are commonly modified with methacrylate or aldehyde groups to produce reactive side-groups. With UV-activating these functionalized side-groups, a chondroitin sulfate gel can be solidified via photopolymerization. A number of *in vitro* studies have demonstrated the viability of encapsulated chondrocytes in the cross-linked chondroitin sulfate gel, suggesting potential applications for cell delivery to repair cartilage defects. To improve the stability of the chondroitin sulfate hydrogel in culture medium, nonbiodegradable PEG (Li et al., 2004) or polyvinyl alcohol (PVA) (Bryant et al., 2004) has been incorporated into chondroitin sulfate to produce a co-gel. Within the co-gel, the negatively charged chondroitin sulfate attracts water into the gel to create a swelling pressure that contributes to mechanical stiffness of the material, similar to that in native cartilage.

Advanced Scaffolds and Signaling Factors in Cartilage Tissue Engineering

Recent advances in the fields of material science, protein chemistry, control release, and nanotechnology have increasingly been applied to the development of cartilage tissue engineered scaffolds. These next generation biomaterial matrices, so-called "smart" scaffolds, are made of materials with a bio-inspired, functionalized surface, constructed of an ECM-mimetic structure, or capable of releasing biomolecules in a controlled release manner. The "smart" scaffold is designed to actively instruct cell behavior to facilitate cartilage regeneration.

"Smart" Scaffolds with Controlled Release Capability. One of the emerging research trends to enhance cartilage regeneration is the development of "smart" tissue engineered scaffolds with the capability of controlled release of growth factors, such as transforming growth factor-beta (TGF-β) and insulin-like growth factor-1 (IGF-1), that act to enhance matrix production of chondrocytes or induce chondrogenic differentiation of MSCs (Park et al., 2009). Delivery of growth factors can be achieved via biodegradation of scaffolding polymers, with the release kinetics of target molecules controlled as a function of the degradation properties of the polymer (Basmanav et al., 2008). Indeed, by using biodegradable polymers with differential degradation properties, it is possible to develop "smart" tissue engineered scaffolds capable of sequentially delivering different growth factors to enhance tissue regeneration (Yilgor et al., 2009). In addition to the capability of controlled release, the "smart" scaffolds should also function to protect delivered molecules from denaturation. Conventionally, growth factors are embedded in polymeric scaffolds by direct incorporation into a polymer solution during a fabrication process; however, scaffold fabrication usually involves organic solvents or chemical reactions that are likely to harm biological activities of to-be-delivered growth factors. Recently, several research groups have developed dual-structure scaffolds to protect bioactivities of delivered molecules by encapsulating the molecules with microparticles embedded in scaffolds (Holland et al., 2004).

Nanostructured Scaffolds with Biomimetic Architecture. In addition to biological signals, effective tissue engineering-based cartilage regeneration in *ex vivo* culture requires that scaffolds provide appropriate physical and chemical cues to guide cellular activities, such as cell differentiation and matrix production. Therefore, a biologically favorable culture environment must be used for effective cartilage regeneration. It is thus noteworthy that the size of the structural elements of a biomaterial scaffold have been proven to be crucial to cell activities. Recent studies (Webster et al., 2000; Dalby et al., 2002) have shown that cells cultured on substrates with varying topography exhibit different behaviors; a substrate composed of nanometer-scale components is biologically preferred. Among nanostructures, a nanofibrous structure composed of ultra-fine, continuous fibers feature the properties of high porosity, high surface area-to-volume ratio, and most importantly, morphological similarity to natural ECM, making it well-suited for tissue engineering applications (Li et al., 2002). To fabricate nanofibers, several fabrication approaches are available, and among these approaches, the electrospinning technique is considered a convenient and highly versatile approach. Electrospinning has been used for fiber production in the textile industry since 1934 (Formhals, 1934), and has recently been used for tissue engineering applications. Research evidence has shown that the unique physical features of electrospun nanofibers induce biologically favorable activities (Li et al., 2003), and nanofibrous scaffolds are ideal for tissue engineering and regenerative medicine applications (Li et al., 2009).

LIGAMENT TISSUE ENGINEERING

Introduction to Ligament Tissue Engineering

Ligaments are a dense connective tissue between bony segments with specific macromolecular composition and arrangement, enabling a defined response to the unique loading state at each ligament. Their strength and elasticity permit our rigid skeletal frame to rotate around specialized joints and produce complex movements, while retaining mechanical stability. The normal response to load is nonlinear, anisotropic, and viscoelastic, with significant stress relaxation (Kwan et al., 1993) to slowly dissipate elastic energy (Hukins et al., 1990). A tear or dislocation in the tissue (Kennedy et al., 1974) or prolonged ligament immobilization (Woo et al., 1987) will compromise ligament function and joint stability, and result in a reparative response with subsequent scar formation. Several studies with animal models have demonstrated that injured ligaments do not recover normal mechanical properties, even after one year of healing (reviewed by Woo et al., 2006).

Clinical Relevance and Limitations of Current Repair Strategies. The poor healing ability of ligaments is compounded by the fact that ligament injuries are not rare, as an estimated 100,000 anterior cruciate ligament (ACL) reconstructions are performed each year in the US (CDC, 1996), with injury rates among females between ages 14–17 increasing at the fastest rate (Csintalan et al., 2008). Thus there has been a tremendous effort among clinicians, scientists, and engineers to improve the rate and outcome of recovery after a ligament injury. There are three standard reconstructive treatments for serious ligament injury. For more than a decade, the most common surgical reconstruction technique for ligament injury has involved the transplantation of healthy autografted (self-donated) connective tissue, particularly patellar or semitendinosus-gracilis (hamstring) tendon tissue. This is not an ideal solution, since removal of local non-regenerative tendon tissue at the knee joint disrupts the normal biomechanical state, which has been found to require up to six months to recover to preoperative strength and endurance (Soon et al., 2004). Further, one study found that after seven years, knee flexibility, sensitivity, and laxity still remained significantly reduced (Lidén et al., 2007). Alternatives to autografting, such as allograft transplantation, for ligament reconstruction have been found to be more prone to rupture, and afford reduced functionality compared to autografts, mainly due to chemical treatments (Krych et al., 2008). Furthermore, concerns remain regarding potential immune reactions and disease transmission of untreated donated tissue. Finally, non-degradable synthetic polymer grafts made from materials such as polytetrafluorethylene or polypropylene have been found to suffer from mechanical problems, such as particle wear and failure rates on the order of 40–80%, primarily from fatigue rupture (Vunjak-Novakovic et al., 2004).

The Modern Ligament Engineering Strategy. Given these issues, there is a need for a treatment that could restore normal ligament function immediately, and retain or improve upon that level of function over the long term. We, and others, consider tissue engineering as an exciting alternative, offering the possibility of replacement with normal, living tissue. The modern ligament engineering strategy involves culturing cells of stem or fibroblastic phenotype onto a three-dimensional biodegradable scaffold, usually fibrous, and typically with a biochemical or physical stimulus to maintain or enhance ligament regeneration. Complex bioreactors have been developed that provide simultaneous dynamic fluid flow and cyclic tensile stimulation in heated and humidified environments. Central to the success of this strategy is the biomaterial which the cells are cultured on; the choice of biomaterial is important to the extent of cellular adhesion, initial scaffold strength and mechanical properties, scaffold degradation rate, and long-term implantation success. Tissue engineers generally obtain the best results when they design their regeneration strategy to mimic the tissue of interest, and thus it is critical to define and design for native ligament properties as summarized in the following section.

Native Ligament Macromolecular Composition and Structure

Ligament Composition. Ligaments derive their tensile strength and elasticity from the basic primary building block of their structure: collagen fibers. Collagen types I and III are primary constituents, along with elastin, and more minor proteins including fibronectin, decorin and tenascin. The mass ratios of these protein constituents to overall ligament mass depend upon location in the body, and vary especially with elastin content, which is most dense in ligaments that experience high tensile strains, such as spinal ligaments. The elastin affords greater flexibility (ductility) at the expense of strength, and also minimizes energy losses during elastic recovery, as exploited by some purely elastin ligaments found in cows and horses associated with grazing. Circumferentially oriented around the collagen fiber bundles are specialized elongated fibroblast cells, commonly referred to as tenocytes, which are responsible for maintaining the density and composition of the ligament matrix. The tenocytes adhere to the fiber bundles, and are surrounded by a glycosaminoglycan-rich matrix which enables fibers to slide along each other during bending, and also provides limited blood vessel and nerve supply. The neural innervations are primarily proprioceptive in nature, and act to provide feedback on the degree of joint flexion/extension.

Ligament Structure. Since ligaments are responsible for maintaining joint stability, their structure is designed to withstand high tensile loads. Beginning with a single triple-helical collagen molecule, the collagen molecules are organized in parallel nanometer-scale fibrils, and subsequently into micrometer scale fibers. The parallel collagen fiber matrix and surrounding proteoglycan-rich matrix reside in the middle zone, and resist tensile and sometimes simultaneous torsional loads, as seen in the anterior cruciate ligament (~90° twist). Within a few millimeters of bone, the ligament structure gradually becomes more bone-like, which is a critical region of normal function. The ligament insertion into bone, known as the enthesis, is typically described as a gradual, four-layer progression from non-calcified collagen fibers to fully calcified bone matrix. In-between are two layers of cartilaginous tissue containing fibrocartilagenous cells aligned with the collagen fibers, wherein the layer proximal to the bone contains calcium, and the layer proximal to the ligament does not.

Important Considerations of a Ligament Tissue Engineered Scaffold

Cellular Activity Requirements. One of the most important design considerations for ligament tissue engineering is that despite their white appearance, ligaments are not avascular tissues, and an active oxygen supply is critical for cell survival. This can be especially challenging, since ligaments must also be strong, stiff and elastic, often requiring densely packed natural or synthetic fibers. Additionally, tenocytes are natively distributed throughout the ligament both radially and longitudinally. The native long, spindle-like morphology of ligament cells is likely also important for normal genetic expression, and should be permitted by the scaffold geometry. The fabrication of three-dimensional scaffolds with fully distributed cells, and with adequate O_2 supply, remains a formidable challenge in ligament tissue engineering, especially with ligament thicknesses greater than the oft-cited oxygen diffusion limit of 200–300 μm. Several authors have been actively addressing these cellular activity requirements, which will be discussed in upcoming sections.

Material Property Requirements. In addition to cellular activity requirements, there are four principal material property requirements for engineering ligaments. First, the material must enable cellular adhesion and proliferation, and must not be cytotoxic or immunogenic to attached cells, local surrounding tissues or the overall body. Second, the scaffold material should have an architecture that enables full cellular distribution, additional tissue ingrowth (e.g. endothelial or connective), and is biomechanically capable of resisting major planes of stresses expected to be seen *in vivo*. Third, the material must possess adequate elastic modulus, strength, and ductility, especially under tension, and should also have similar viscoelastic properties to native ligament. Mechanical mismatches are particularly important since ligament is a loadbearing tissue; they can result in scaffold rupture (too weak), stress-shielding effects where local connective tissues begin to atrophy (too strong) or joint laxity and energy imbalances (potentially from incorrect viscoelasticity). Finally, the material should biodegrade without toxic byproducts at a rate that is equal to the rate of tissue formation, and ideally be completely eliminated from the body at a time coincident with full ligament regeneration.

While at present no material or system has met all of these requirements, they have been actively investigated through a number of different single and composite material approaches using synthetic and natural biopolymers, as reviewed in the following paragraphs.

Biomaterials in Ligament Tissue Engineering

Synthetic Biomaterials in Ligament Tissue Engineering. An ideal engineered ligament would contain enough starting material for immediate loadbearing post-implantation, and would degrade at a rate comparable with that of developing cellular and tissue ingrowth. After a period of several weeks or months the starting material, or scaffold, would be completely replaced by regenerating ligament cells and matrix. The requirement for biodegradability and fabrication into specific ligament-like geometries has generated interest in the use of synthetic polymer materials for ligament tissue engineering, dating back to at least 1999 (Lin et al., 1999).

Polyhydroxyesters degrade by hydrolysis, and thus have been very popular for ligament tissue engineering because the degradative molecule, H_2O, is naturally present in abundance *in vivo*. An early cell attachment study by Ouyang et al. (2002) found that on two-dimensional matrices anterior cruciate ligament (ACL)-derived fibroblasts adhered and proliferated on PCL, PCL/PLA (50:50), and PLGA (50:50) substrates at a rate that was not significantly less than tissue-culture plastic (O_2-plasma treated polystyrene). However, high hydrophilicity of certain polyhydroxyesters is thought to be detrimental to general cell adhesion (van Wachem et al., 1985), and hence it is regarded that a balance of hydrophobic and hydrophilic components should be considered for ligament tissue engineering. Several studies have utilized these findings, and extended ligament cell culture into three dimensions by fabricating ligament-like polyhydroxyester scaffolds, several of which are reviewed here.

Lu et al. (2005) fabricated three-dimensional braided scaffolds of PGA, PLGA, and PLLA filaments at two levels of bundling using a circular braiding loom designed to mimic the hierarchical structure of native ligament. To improve cellular adhesion, these investigators modified

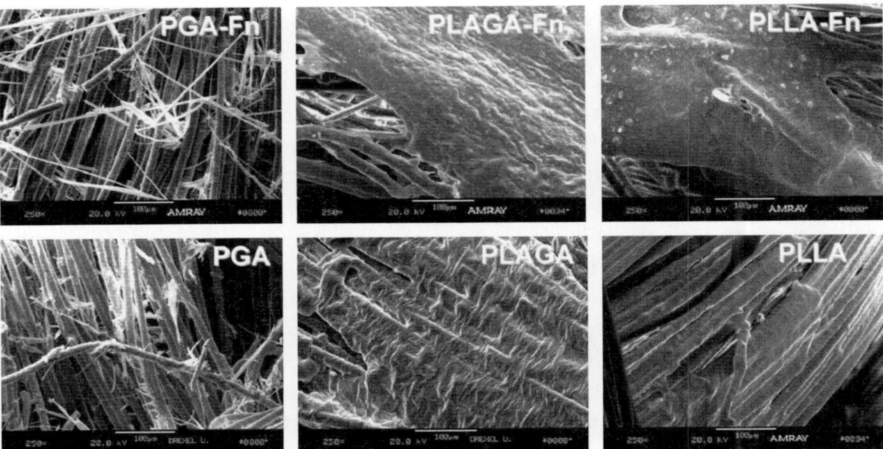

FIGURE II.6.8.3 ACL fibroblast growth and matrix formation on different synthetic braided scaffolds visualized with SEM (Lu et al., 2005). Images were taken after 14 days of *in vitro* culture in 10% fetal bovine serum. (Left) Culture with PGA resulted in substantial matrix degradation from acidic byproducts. (Middle)-co-glycolic acid (PLAGA) in an 82:18 mass ratio showed more sustainable matrix formation, particularly with the addition of fibronectin (Fn). (Right) PLLA scaffolds also displayed considerable matrix formation, which again was amplified by the addition of Fn. (Reproduced with permission from Elsevier B.V.)

the surfaces by immersing the scaffolds in a solution of human recombinant fibronectin (Fn). Ligament cells were isolated from rabbit ACL tissue by digestion in collagenase, and grown to confluence before being added to the scaffolds. After 14 days of culture, using scanning electron microscopic analysis it was found that cells seeded on PLLA-Fn and PLAGA-Fn scaffolds produced the most matrix, and that PGA was detrimental to matrix formation, owing to rapidly produced acidic byproducts (Figure II.6.8.3).

Furthermore, PGA-Fn had significantly reduced cell numbers compared to PLAGA-Fn and PLLA-Fn scaffolds, which were not significantly different from each other. The authors concluded that both the PLLA-Fn and PLAGA-Fn scaffolds would be effective for ligament tissue engineering, but acknowledged that the PLLA-Fn scaffolds may be the more appropriate choice because of their slower degradation rate. A study of the tensile properties of braided PLGA 10:90 scaffolds found ultimate tensile strengths in the range of 100–400 MPa, and with a circular braiding scheme a maximum load of over 900 N, both of which were considered practical and safe for initial implantation in a human ACL replacement surgery (Cooper et al., 2005). Recent work has also elucidated ideal braiding angles of these scaffolds to most accurately mimic the non-linear stress–strain relationship of native ligament (Cooper et al., 2007). Taken together, these studies show great promise for future use of braided synthetic scaffolds for ligament replacement, particularly if the degradation profiles are optimally balanced against tissue ingrowth using an animal model.

PLGA 10:90 scaffolds have also been fabricated with collagen type I to form porous, rolled microsponges for ligament tissue engineering (Chen et al., 2004). These hybrid scaffolds were considered advantageous because of the combined cellular binding affinity and mechanical strength afforded by collagen and PLGA, respectively. To fabricate the microsponges, the PLGA was immersed in a solution of bovine collagen type I, freeze-dried, and cross-linked with glutaraldehyde (unreacted aldehyde groups were blocked with glycine). In this study, ligament cells were isolated from beagle ACL samples with collagenase and cultured for 16 days *in vitro* before rolling and subcutaneous implantation into nude mice. Using paraffin sectioning with hematoxylin and eosin staining, coupled with SEM, the authors concluded that cells were viable in the scaffold even after 12 weeks of implantation, and that the hybrid scaffolds may be appropriate for ligament tissue engineering.

One of the advantages of synthetic polymers is that they can be processed relatively easily into fibers with nanometer-scale diameters – on the order of native collagen fibrils. As presented earlier, nanofibers are typically fabricated using the 70+ year old electrospinning process, which utilizes an electric field (typically kV potentials over ~5–20 cm) to overcome surface tension of a viscous polymer solution and form a chaotic nano-diameter polymer jet that is deposited onto a surface of interest (Pham et al., 2006). Lee et al. (2005) produced aligned polyurethane nanofiber scaffolds with average fiber diameter of 650 nm and 82% porosity using an electrospinning apparatus and a rotating collector target. Seeded human ligament cells were cultured for 48 hours then subjected to 5% uniaxial strain at 0.083 Hz using vacuum flexion on a silicone membrane for an additional 24 hours. It was demonstrated that cells seeded onto aligned scaffolds produced significantly more collagen mass per cell of DNA, compared to randomly oriented fiber controls, and that the aligned nanofibers enabled cell morphologies most comparable with *in vivo* morphology. This study was one of several demonstrating the potential applicability of aligned nanofibers for ligament tissue engineering. Composite nanofiber and microfiber scaffolds have also been fabricated for ligament tissue engineering (Sahoo et al., 2006). In this study, electrospun PLGA (65:35) nanofibers with 300–900 nm diameter range were deposited onto a mesh of 25 μm diameter PLGA (10:90) microfibers producing scaffolds with 0.8–1.3 mm thickness and 2–50 μm pores (evaluated with SEM). The rationale for this system was that the microfibers would provide mechanical strength and degradation resistance, and the nanofibers would provide hydrophilicity and a very high surface area for optimal cell attachment. The authors seeded porcine bone-marrow derived mesenchymal stem cells (BMSCs), and observed an average collagen production of 1.55 ng/cell after seven days, which was considered high relative to previous studies with other scaffold geometries. Furthermore, mRNA expression levels of critical ligament genes, including collagen type I, decorin, and biglycan relative to the GAPDH housekeeping gene, were all slightly upregulated (5–20%), suggesting partial differentiation into ligament lineages. The authors acknowledged, however, that the ultimate tensile strengths and failure loads were still far below those of native ligament, and that it was difficult to have confidence that the cells were directed into ligament lineages. Nevertheless, the high level of matrix production demonstrates the potential of the method, and a follow-up study (Sahoo et al., 2007) reports further optimization of these scaffolds using surface modifications including PCL and collagen type I deposition to enhance cell attachment and proliferation, and also a right-angle fiber weaving method to increase failure load by ~50% over knitted fabrication.

These studies demonstrate some of the advantages imparted by synthetic nanofiber technologies for ligament tissue engineering, which will continue to expand and evolve as more specific ligament cell–nanofiber interactions are elucidated. At present, it appears that most studies show results favoring aligned nanofibers over randomly oriented fibers, likely due to their resemblance to native ligament fibril orientation, but the ideal fiber materials, diameters, spacing, and angles remain unresolved for ligament tissue engineering. Currently, a single material is typically used, but the opportunity exists to incorporate fibers of different diameters and mechanical properties into a single aligned scaffold in an effort to mimic the multi-protein matrix of ligament, e.g., collagen type I, collagen type III, and elastin. Finally, the rapid degradation rate of synthetic nanofibers must be controlled, and perhaps decelerated, before *in vivo* implantation will be feasible. Overall, however, synthetic nanofiber scaffolds are cytocompatible and have tailorable diameters and degrees of alignment, and are one of the most exciting prospects for the design of engineered ligaments.

Natural Polymeric Biomaterials in Ligament Tissue Engineering. We have seen that synthetic polymeric biomaterials have reproducible mechanical and chemical properties, are easily fabricated into different shapes and sizes, degrade by hydrolysis, and are efficacious for ligament engineering research. However, they may lack functional chemical groups for cellular binding, and furthermore they may release acid byproducts or unnatural polyesters into the bloodstream during degradation. For these reasons there has been considerable interest in the application of natural, protein-based fiber materials as scaffolds for ligament tissue engineering. The most direct and obvious choice for this material is collagen type I, because of its prevalence in ligament tissues. Unfortunately, no method yet exists to organize and cross-link collagen fibers in a cytocompatible manner, and as such collagen gels remain exceptionally weak with typical elastic moduli between 10–30 kPa and ultimate tensile strengths between 5–10 kPa (Roeder et al., 2002). Thus, collagen-based scaffolds, while useful to investigate mechanisms of tendon differentiation and regeneration (Kuo and Tuan, 2008), are challenging to be used for ligament replacement.

Silk fibroin, rather than collagen, has been the most popular natural polymeric biomaterial for ligament tissue engineering. Silk fibroin is one of two proteins excreted by *Bombyx mori* silkworms during cocoon production, and is typically isolated from its sister protein sericin using sodium carbonate, urea, and/or soap detergents and near-boiling temperatures (Yamada et al., 2001). *B. Mori* silk fibroin is 70–80% by mass of the silk bicomplex (Kato et al., 1998), contains a heavy (350 kDa) and light chain unit (25 kDa), and is held together by the sticky sericin protein, which unfortunately is cytotoxic. The principal advantage of silk is its remarkable tensile strength and toughness (area under stress–strain curve) which is unmatched for natural proteins. Reported tensile mechanical properties of *B. Mori* silk fibroin range between 5–9 GPa for elastic modulus, 250–400 MPa for tensile strength, and 23–26% for failure strain (Jiang et al., 2006). The protein also displays surface amino acids for cell adhesion, remains structurally whole in aqueous solutions but slowly degrades (weeks–months) proteolytically *in vivo* (Minoura et al., 1990; Greenwald et al., 1994), and can be fabricated into gels, films, braided fibers or nanofibers. These characteristics make silk fibroin one of the best natural polymers for support of cellular and tissue ingrowth for tissue engineering.

Silk fibroin has been used for several decades as a suture material because of its strength and slow biodegradability, and since at least 2002 has been considered for ligament tissue engineering. Altman et al. (2002a) were one of the first to design a braided silk fibroin scaffold and test for adherence, proliferation, and genetic expression of seeded bone-marrow derived MSCs. A braided geometry with four levels of bundles twisted with 0.5 cm pitch was chosen to effectively reduce the linear stiffness of a single fiber, to better model native ACL stiffness and minimize stress-shielding effects. To model physiological performance a 400 N cyclic load was applied to a 3.67 mm diameter scaffold (designed to fit a human knee bone tunnel), and a subsequent fatigue analysis with linear extrapolation indicated a matrix life of 3.3 million cycles, which was expected to far outlast *in vivo* degradation. After 14 days of *in vitro* culture with human MSCs, the authors found that DNA content (and thus cell number) increased five-fold compared to 24 hours of culture; this was complemented by scanning electron microscopic analysis showing considerable matrix deposition over the culture period. Furthermore, mRNA expression levels of ligament-associated genes were comparable to those in native human ACLs, especially an average collagen type I to type III ratio of ~9, an absence of collagen type II and bone sialoprotein (which would indicate cartilage or bone differentiation, respectively), and baseline expression of the tendon/ligament marker tenascin-C. The success of this initial work was considerable and highlighted the potential applicability of silk fibroin that has been properly processed and organized for *in vivo* ligament replacement. It also led to several modifications, optimization studies, and *in vivo* evaluation studies for utilizing silk fibroin in ligament tissue engineering, several of which are reviewed below.

Scaffold geometries other than braided structures have also been fabricated with silk fibroin. Liu et al. (2008) surmised from previous studies of synthetic braided ligaments that such braided structures may have limited nutrient diffusion and tissue infiltration, particularly towards the center on the radial axis. To circumvent this potential issue, the authors fabricated a silk fibroin hybrid scaffold with geometries at two levels – a knitted scaffold and an interspersed microporous silk sponge. The knitted scaffold was fabricated using a 40-needle knitting machine, and the sponge was added by immersing the knitted scaffold in a low concentration silk solution, freeze-drying to form pores, and then immersing in a methanol solution to prevent resolubilization. Human MSCs were seeded by pipetting onto knitted-sponge or knitted-alone scaffolds, with unseeded scaffolds as a secondary control and a total culture time of 14 days. Compared with knitted scaffolds alone, the knitted-sponge scaffolds showed significantly higher biological responses with nearly every evaluation method, including cellular proliferation, GAG production, viable cell density, mRNA expression of collagen types I and III, and tenascin-C, and collagen-based matrix production, confirming the positive benefit of the microporous silk sponge. However, no significant differences in maximum tensile load or stiffness were recorded compared to unseeded scaffolds after 14 days, suggesting that the secreted matrix did not contribute towards scaffold strength. Also, the tensile strength and stiffness were far below (<20%) those of the adult human ACL (Woo et al., 1991). Nevertheless, the cytocompatability and rapid ligament-like matrix development is impressive, and further demonstrates the effectiveness of a cyto-friendly composite structure. With the design of a stronger starting matrix, the knitted-sponge scaffold may be an attractive candidate for *in vivo* ligament replacement, conferring rapid tissue regeneration and (importantly) initial tensile strength.

In addition to synthetic scaffolds, multi-material or composite natural scaffolds have also been successfully fabricated for ligament tissue engineering applications (Chen et al., 2008). In this work a knitted silk-fibroin base matrix was infiltrated with a freeze-dried collagen type I microsponge creating a similar composite structure to that used by Liu et al. (2008), but with a collagen type I sponge instead of a silk sponge. MSCs were seeded into the scaffolds which were then implanted into a rabbit medial cruciate ligament (MCL) transection model to evaluate *in vivo* repair potential over a total of 12 weeks. *In vitro*, the authors found that gene expression levels of ligament-associated genes by cells seeded in silk–collagen scaffolds compared to silk alone were considerably higher, with collagen type I elevated by 250%, and decorin elevated by over 500%. Histological analysis with H&E and Masson's trichrome staining of the repairing MCL found more tissue ingrowth compared to silk alone and untreated MCLs after 2, 4, and 12 weeks; Figure II.6.8.4 shows the repair results after 14 days of implantation.

In addition, the mean collagen diameter of the repaired region observed with transmission electron microscopy was closer to that of native ligament (~65 nm). Although the mechanical properties of the repair tissue were considerably less than those of native MCL, the results suggest that a silk and collagen sponge matrix may be an effective treatment for MCL transections, particularly because of its support of rapid tissue ingrowth. Future anlaysis comparing *in vitro* and *in vivo* repair between knitted silk scaffolds, with either a collagen type I sponge or silk-fibroin sponge, would provide additional information for the optimization of sponge pore size.

One of the most important evaluation tests in ligament engineering is the success of an engineered graft to be regenerated by the body into a functional ligament. Recent studies by Fan et al. have investigated silk scaffold ligament regeneration in rabbit (Fan et al., 2008) and pig models (Fan et al., 2009), providing new insights into the kinetics of ligament tissue ingrowth and scaffold degradation. First, a rabbit ACL reconstruction model was tested by transecting healthy rabbit ACLs and surgically implanting an MSC-seeded knitted silk scaffold with microporous silk sponge. *In vitro*, cells were seeded onto the silk grafts for an eight-hour adhesion time before implantation by anchoringinto bone tunnels of rabbit knees. A morphological and histological evaluation of implanted silk grafts replacing the rabbit ACL after eight weeks suggested substantial production of collagen type I, collagen type III, and tenascin-C, and that the ligament–bone attachment was stable (evaluated with micro-CT). Similar to other studies with knitted silk scaffolds, the tensile strength and stiffness were unfortunately below those of native human ACL.

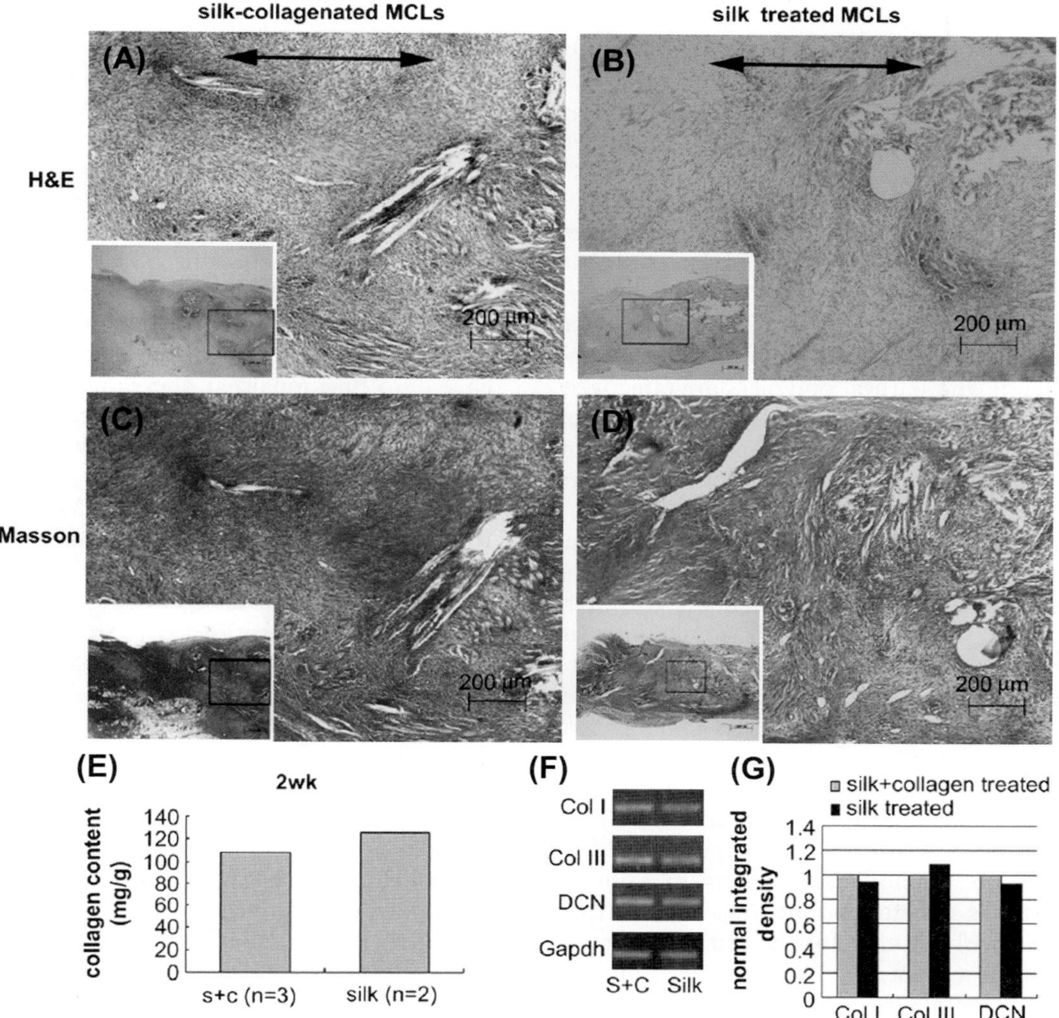

FIGURE II.6.8.4 Histological and gene expression analysis of the extent of *in vivo* repair after 14 days using silk and silk/collagen scaffolds in a rabbit model of MCL transection by Chen et al. (2008). (A,B) H&E staining showed relatively unorganized matrix formation but a higher cell density in silk/collagen scaffolds than silk alone. (C,D) Masson's trichrome staining revealed the beginnings of collagen fiber formation (indicated by blue areas), and more collagen deposition in the silk/collagen scaffold group. (E) Total collagen content was reduced in the silk/collagen group (S + C), but, as detected by RT-PCR (reverse transcription-polymerase chain reaction) shown in (F,G), this could have been due to a reduction in the expression of collagen type III, a less appropriate collagen type for the ligamentous phenotype than collagen type I and decorin, both of which are showing higher levels of expression in the silk/collagen scaffold group. (Reproduced with permission from Elsevier B.V.)

A subsequent study using the larger porcine model showed similar encouraging results. To compensate for the additional loadbearing, the knitted-sponge silk scaffold was rolled around a braided silk cord and again was seeded with MSCs for eight hours before implantation. After 24 weeks, the gross morphological and histological characteristics were evaluated and as before, collagen type I, collagen type III, and tenascin-C were all distinctly present, as seen by immunohistochemistry. However, with an average failure load of 398 N, the authors noted that regenerated scaffolds at 24 weeks could be effective for mild daily loadbearing, but would likely not survive trauma or vigorous exercise. Unfortunately, the scaffold strength at earlier time points was not measured; it is clear, however, that 24 weeks would likely be too long for most patients to wait before mild loadbearing would be feasible. Nevertheless, the clinical implications of this study are profound, and demonstrate the *in vivo* effectiveness of a multi-structural silk scaffold for ligament tissue engineering.

Because of their inherent biocompatibility, natural polymeric macromolecules will likely remain at the forefront of biomaterials for ligament tissue engineering. Although collagen type I gels may eventually be the scaffold material of choice, their ultimate tensile strength and elastic modulus are currently too low to act as a loadbearing material. In the interim, silk fibroin has emerged as an excellent natural biomaterial alternative to collagen, and has already been shown to regenerate ligament in large-scale animal models. Assuming the success of longer-term animal trials, it is anticipated that impending clinical trials with silk fibroin-based engineered ligaments will confirm the efficacy of silk to restore ligament function after injury, and offer exciting new options for ACL or general ligament repair.

Advanced Scaffolds and Signaling Factors in Ligament Tissue Engineering

Functionalized Ligament Scaffolds. One of the key factors in effective application of material scaffolds in tissue engineering is the optimization of cell–biomaterial interactions, particularly in terms of the ability of cells to adhere, proliferate, and secrete matrix onto the scaffold. Synthetic and natural polymers are effectively long chains of repeating chemical units, and it is thus possible to link small molecules covalently to their surfaces to enhance cell adhesion, proliferation, and matrix production. Such an approach has high potential, since biomaterials that natively lack chemical cell attachment groups can also be functionalized, thus expanding the range of implantable biomaterials. Cell–matrix interactions are typically mediated via cell surface integrin receptors, which are specialized transmembrane proteins that are connected cytoplasmically to the actin cytoskeleton. The most common example of an integrin-interacting matrix epitope is the peptide sequence, RGD (arginine–glycine–aspartic acid) (Orlando and Cheresh, 1991), which has been used to functionalize a number of biomaterials.

The practice of functionalizing grafts to improve engineered tissues is sometimes done passively by merging one scaffold with another for the purpose of combining mechanical properties with integrin-binding capability. An example of passive functionalization is a recent work by Garcia-Fuentes et al. (2009) who blended hyaluronan, a common native glycosaminoglycan, with silk fibroin and seeded MSCs for general regenerative applications. Matsumura et al. (2004) conducted a more direct functionalization study, and modified polyethylene-co-vinyl alcohol (PEVA) films with carboxyl groups (COOH) and subsequently covalently attached collagen type I, designed to enhance periodontal ligament adhesion to PEVA-coated titanium dental implants. Other general tissue engineering applications have used a variety of functionalizing groups, including phosphate (PO_4^{3-}), amide ($R_1(CO)NR_2R_3$) and sulfonate groups (RSO_2O^-).

There has been some interest in adding functional groups to non-degradable synthetic graft surfaces in the hope of enabling tissue growth and avoiding poor tissue integration, foreign body immune responses, and high failure rates. Zhou et al. (2007) functionalized polyethylene terephthalate (PET) grafts with poly(sodium styrene sulfonate) (PNaSS) functional groups and observed fibroblastic cell response. PET is used for an array of commercial applications including containers, water bottles, polar fleeces, and wind sails, and does not degrade *in vivo*. These investigators functionalized PET by first exposing PET fabrics to ozone gas (O_3), which is unstable and breaks into O_2 and O·, the latter of which transfers its free radical to the PET surface making it much more reactive. Under an inert argon atmosphere, the samples were then immersed in a bath of monomer 15% w/v sodium p-(styrene sulfonate) (NaSS) at 65–70°C, forming polyNaSS on the PET surface by radical polymerization. The human fibroblast McCoy cell line was seeded onto the functionalized PET surface and observed after four days of culture with calcein AM, a fluorescent label that is only retained inside living cell membranes. Captured images revealed considerably more cell adherence onto functionalized fibers than non-functionalized fibers. Additionally, dynamic fluid testing indicated that cells adhered to polyNaSS-PET scaffolds compared to PET alone required 12-fold more shear stress for 50% of the adhered cells to be removed (12 dyn/cm^2 compared to 1 dyn/cm^2). The authors attributed these profound results to two factors: the enhanced surface hydrophilicity enabling cell spreading; and the opportunity for fibronectin to bind to the PNaSS, increasing the number of focal adhesion contacts and the potential for organized cytoskeletal formation. It will be particularly interesting when these functionalized grafts are implanted *in vivo* to test if the foreign body capsulation response is still present, and if not, how they perform over long-term implantation.

Decellularized Ligament Scaffolds. It has been argued that the best replacement biomaterial for ligaments is those derived from the ligaments themselves (when they are available), because the tissues already have ideal mechanical properties and because the endogenous integrin-binding sites are abundant and present in the native ECM. However, some studies have suggested that allograft tissues can contain residual donor cells, even with strict sterilization and cleaning (Zheng et al., 2005), and may cause significant inflammatory response when implanted *in vivo* (Malcarney et al., 2005). Furthermore, there are concerns about the extent and efficiency of cellular infiltration, particularly to the dense center areas of the graft. Thus, the application of chemical treatments to yield a fully decellularized and more porous scaffold, as opposed to the use of minimally treated allografts (Musahl et al., 2004), is preferred for ligament tissue engineering. Whitlock et al. (2007) recently addressed this issue with a novel oxidative chemical treatment and a battery of *in vitro* cytocompatibility and tissue tests. The authors isolated adult chicken flexor digitorum profundus tendons and added 1.5% peracetic acid to act as an oxidizing agent (using OH radicals) to create pores in the tissue and remove loose DNA. Simultaneously, the detergent Triton™ X-100 (polyethylene glycol octylphenyl ether) was also added at 2% concentration to lyse cell membranes. When compared to non-treated controls, the oxidized scaffolds had no nuclei visibly present (H&E, DAPI staining), and at minimum 70% less total DNA, which was considered a very promising result. The scaffolds also appeared much more porous, as observed by SEM, and on average had 25% less elastic modulus and stiffness, which was not statistically significant. Results from a subcutaneous rat *in vivo* cell infiltration study showed that after three weeks, cell nuclei were present in the outside layers and some inner layers of the scaffold, and no inflammatory reaction or capsule formation was present. Future studies with human ligament allografts utilizing a combination of a lysing agent and oxidative agent are warranted, and these treatments may prove to be effective in minimizing the foreign body and capsulation responses found with standard allografts.

The Effects of Growth Factors on Natural and Engineered Ligaments. One of the most promising strategies to augment natural regeneration is the introduction of growth factors with specific activities on target tissues. Of the many growth factors in the body, the following five growth factors have received the most attention for ligament engineering, due to their notable upregulation during ligament healing (Molloy et al., 2003): insulin-like growth factor-I (IGF-I); transforming growth factor-β (TGF-β); vascular endothelial growth factor (VEGF); platelet-derived growth factor (PDGF); and basic fibroblastic growth factor (bFGF). Once considered impractical due to extreme costs, growth factors have recently been re-evaluated for their possible application for ligament tissue engineering, made possible by biotechnological advancements in the production and purification of recombinant proteins. Of these five growth factors, bFGF has shown particular efficacy and has been known as an effective promoter of ligament regeneration since early wound-healing studies of canine dental defects (Murakami et al., 1999), and is now beginning to be incorporated into full ligament tissue engineering strategies. Sahoo et al. (2009) blended bFGF with PLGA, and produced electrospun nanofibrous scaffolds capable of releasing 60% of the growth factor over the course of one week. When these scaffolds were seeded with bone marrow-derived rabbit MSCs and compared against scaffolds without bFGF, the authors found significant increases in cell proliferation, immunostaining of tenascin-C and collagen types I and III, and gene expression levels of collagen types I and III, fibronectin, and biglycan over 14 days, which are all promising indicators of enhanced ligament differentiation. Recently, bFGF was also employed *in vivo* in a rabbit ACL repair model via loading into gelatin hydrogels to form a three-part engineered ligament consisting of a braided fibronectin-coated PLLA core, a collagen membrane wrapping sheet, and the bFGF-loaded gelatin hydrogels (Kimura et al., 2008). Observation at eight weeks after implantation revealed that the addition of bFGF increased both the maximal strength and stiffness of the regenerated ACL by approximately 50%, increased collagen mass in the regenerated tissue by 2.5–4-fold, and produced histological cross-sections that more closely resembled native ligament. This study demonstrates that local and controlled release of growth factors such as bFGF can be potent accelerators of ligament regeneration *in vivo*.

In addition to bFGF, TGF-β has also been implicated as an important growth factor in tendon and ligament development, and is actively being investigated for applications in ligament tissue engineering. We characterized the spatiotemporal distribution of TGF-β in the developing chick tendon during embryonic days 13–16 (Figure II.6.8.5) (Kuo et al., 2008). Histologic results demonstrated rapid tissue organization and development during this time period. Immunohistochemical staining showed TGF-β2 and -β3, but not TGF-β1, were present within the tendon mid-substance on all days studied. TGF-β2 and -β3 exhibited similar distribution patterns, but differed in timing and intensity. Taken together, these findings strongly support the postulate that TGF-β2 and -β3 are involved in tendon development, and that these isoforms may have independent roles. In the near future, as production of purified growth factor continues to become more economical, more research on regenerative growth factors will be done, leading to information that will guide the use of growth factors as integral components of functional ligament replacement tissues.

The Effects of Cyclic Strain on Engineered Ligaments. It is well-known that locomotion induces tensile strain on ligament tissues, and for more than 100 years it has been theorized, and later demonstrated, that the density and remodeling of bone is related to its loading state. Thus, it should be no surprise that since nearly the inception of the field, tissue scientists have studied the effect of forces on fibroblasts (Chiquet-Ehrismann et al., 1994) and later on ACL cells (Toyoda et al., 1998). Subsequent to these preliminary experiments, more elaborate bioreactors with uniaxial or multiaxial applied forces were developed for musculoskeletal tissue engineering. Altman et al. (2002b) developed a novel cyclic strain bioreactor that simultaneously applied 10% tensile strain and 25% torsional strain to MSC-seeded collagen type I gels, designed to mimic the natural 90° twist on ACL collagen fibers during knee flexion/extension. The stimulation was applied at a rate of 0.0167 Hz for up to 21 days. Compared to static construct controls, the mechanically stimulated gels had significantly higher cross-sectional cell density and a 2.5-fold increase in cell alignment. The most striking finding was that the mRNA expression of collagen type I, collagen type III, and tenascin-C in mechanically stimulated gels were all significantly upregulated compared to static controls, and approached native ligament expression, although bone and cartilage markers were not upregulated. This was the first study to demonstrate that MSCs could begin to be differentiated into ligament-like lineages using mechanical stimulation alone. The application of cyclic strain in stem cell-based bioreactor systems to

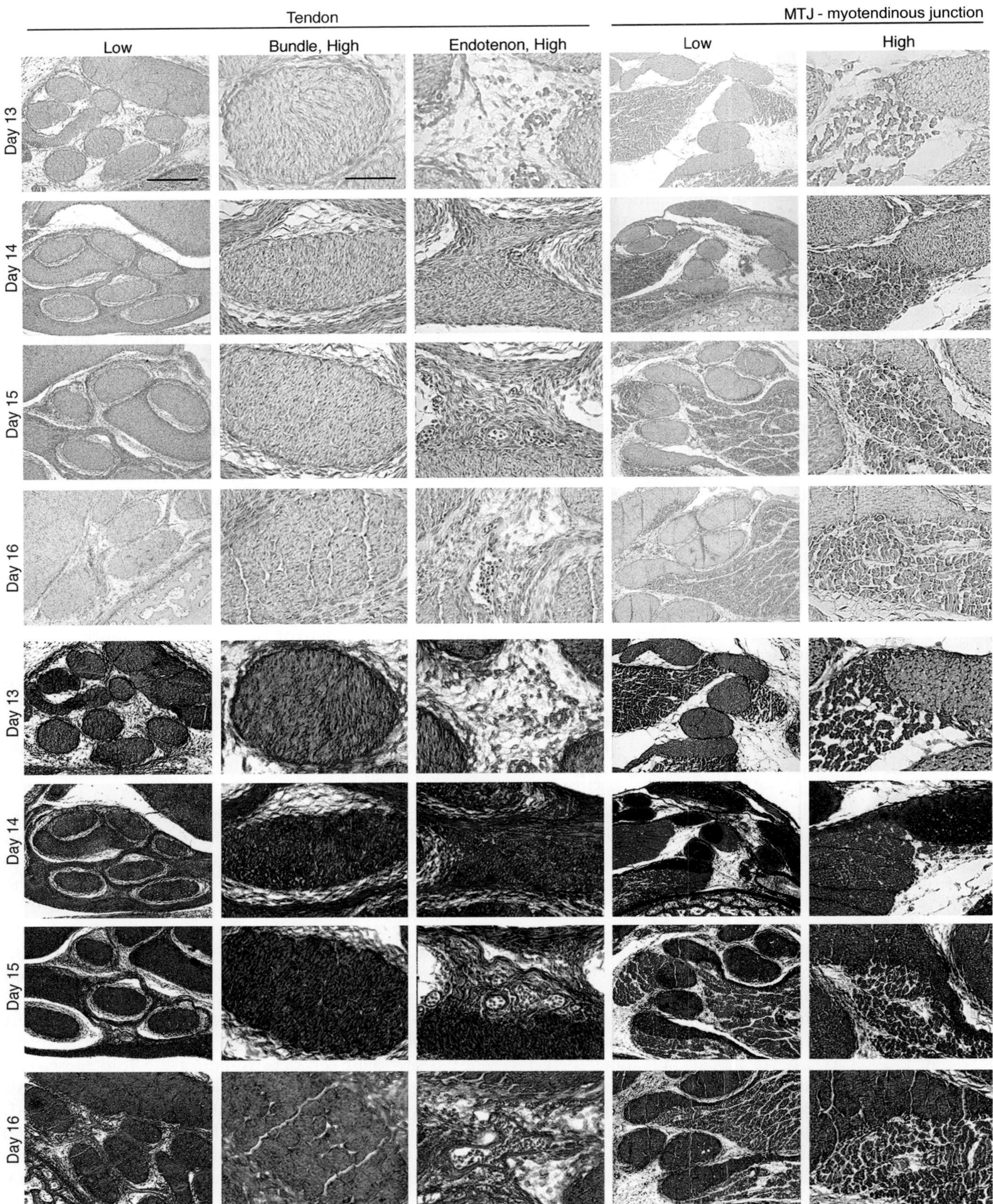

FIGURE II.6.8.5 Histology of intermediate tendon of chick embryos at developmental days 13, 14, 15, and 16. (Top) Hematoxylin-eosin staining. (Bottom) Mallory's trichrome staining. Low magnification, Bar = 200 μm; high magnification, Bar = 50 μm.

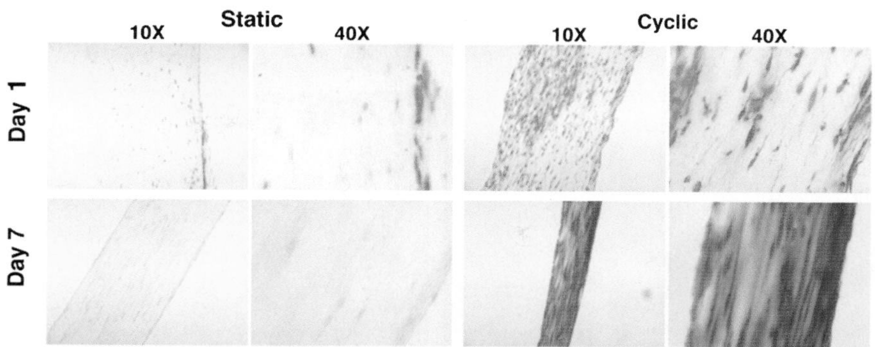

FIGURE II.6.8.6 Histology of MSC-collagen constructs harvested after 1 and 7 days of static and cyclic loading. Longitudinal 5 μm thick sections histologically stained with Mallory's trichrome. A higher level of matrix staining is seen consistently in constructs subjected to cyclic loading.

promote cell proliferation, cell alignment, and ligament-marker expression is now an accepted and widely utilized method in ligament tissue engineering. This and other studies have demonstrated enhanced matrix production by MSCs when cyclically loaded under uniaxial tension in long-term cultures, but have not elucidated the mechanisms for these results (Bhatt et al., 2007; Riboh et al., 2008; Zhang et al., 2008). To investigate potential mechanisms, we conducted short-term studies with human MSCs in a similar model system utilizing collagen type I scaffolds and uniaxial tensile loading, and investigated putative tendon marker expression (Kuo and Tuan, 2008). Results showed that while static uniaxial tensile loads upregulated scleraxis expression, cyclic loading significantly enhanced collagen matrix production (Figure II.6.8.6). Cyclic loading was necessary to sustain mRNA levels of scleraxis, a tendon-specific marker gene, and differentially regulated additional developmental and mature tendon marker molecules including collagens, Wnts, and MMPs. The results of this study supported the premise that dynamic mechanical loading enhances tenogenesis of hMSCs, and provided insights into the mechanisms of this process.

Synthetic biomaterials have also been implemented in cyclic strain bioreactors for ligament tissue engineering. Moe et al. (2005) seeded human dermal fibroblasts on PLGA 10:90 knitted scaffolds and applied 1.8% tensile strain for four hours daily over two weeks at either static, 0.1 Hz or 1 Hz strain rates. Using H&E staining, the authors found the most substantial cellular alignment with a 0.1 Hz applied strain rate; however, the scaffold stiffness of mechanically stimulated constructs was significantly less than that of static controls. Unfortunately, the stiffness was not normalized to cross-sectional area (i.e., elastic modulus), thus it is difficult to know if cell contraction and/or matrix formation was partially responsible for this result. Raif and Seedhom (2005) also seeded cells on a knitted scaffold, but used non-degradable PET fibers and bovine synovial cells, which appear to have the capacity to be dedifferentiated into multipotent cells (De Bari et al., 2001). The applied strain parameters varied considerably between 0.65–4.5% strain magnitude at 1 Hz and were applied for 1–4 hours per day for either 1 or 35 days. The authors found that cell proliferation during short-term application was reduced 1 hour post-stimulation, but was upregulated 22 hours later. Furthermore, cell proliferation increased as cyclic strain amplitude was increased, suggesting a higher affinity for differentiation or matrix production at lower strain amplitude, or simply less cell proliferation. The long-term (35 day) study did not find significant differences, except that higher strains tended to result in a higher scaffold cellular density.

In addition to mechanical stimulation, recent studies have investigated the combinatorial effect of mechanical stimulation and cellular alignment. The developing underlying theory of the influence of patterned scaffold structures is known as "contact guidance," which states that cellular response, especially alignment and proliferation, is dependent on the size and type of the channeling structure. Mechanically stimulating ACL cells seeded on polydimethylsiloxane (PDMS) micropatterned surfaces with an applied 8% uniaxial strain at 0.5 Hz for four hours a day over two days revealed the expression of several novel genes influenced by mechanical stimulation (Park et al., 2006). Specifically, microarray-based real-time PCR analysis showed that expression of the following genes decreased during normal culture, but increased after mechanical stimulation: MGP (matrix Gla-protein, 3.8-fold); GADD45A (growth arrest and DNA damage-inducible gene, 2.3-fold); UNC5B (unc-5 homolog B, 1.6-fold); TGFB1 (transforming growth factor-β1, 1.4-fold); COL4A1 (collagen type IV α1, 1.2-fold), and COL4A2 (collagen type IV α2, 1.2-fold); The authors noted that MGP is a small matrix protein that may be involved in cellular differentiation, and that GADD45A may influence cell cycle proteins and play a role in genomic stability. The exact functions of some of these genes in ligament biology are not well-understood, and certainly warrant further investigation, especially with other cell types. In a similar study, Jones et al. (2005) seeded rat MCL cells on a

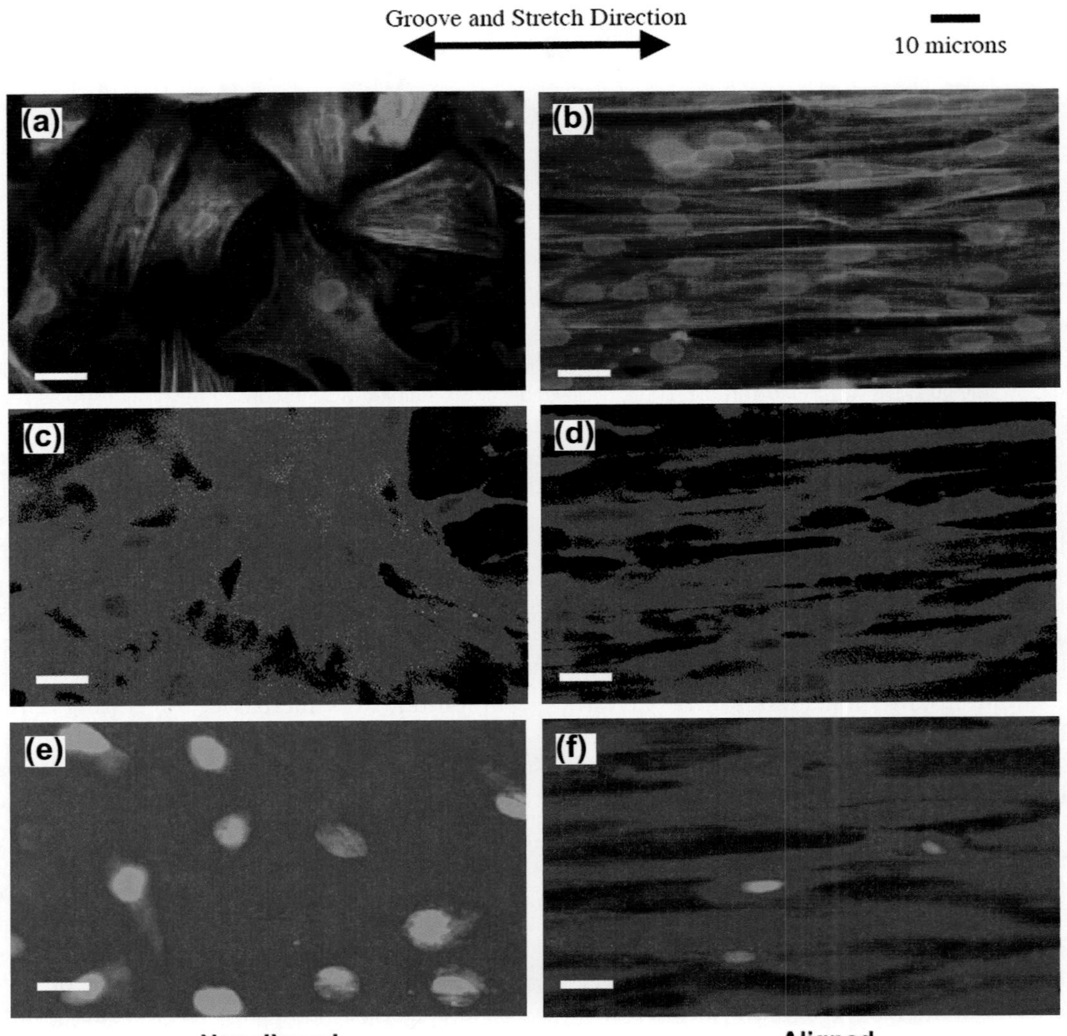

FIGURE II.6.8.7 Fluorescence images of actin filaments (red) and nuclei (blue) of MCL fibroblasts cultured on collagen-coated PDMS surfaces (Jones et al., 2005). (Left column) Cells cultured on smooth PDMS substrates showed randomly aligned cytoskeletal and nuclear morphology, even after 3.5% cyclic strain at 1 Hz for 2 hours (c). (Right column) Cells cultured in PDMS microgrooves with 10 μm width and 6 μm depth were substantially more aligned, based on cytoskeletal and nuclear morphology, compared with smooth PDMS, especially after being cyclically strained (d). (e,f) Control cells grown in the custom cyclic strain incubator without strain applied showed similar results to cells cultured in a standard incubator (a,b). (Reproduced with permission from Elsevier B.V.)

microgrooved PDMS substrate and applied 3.5% strain at 1 Hz for two hours. The authors found considerably more alignment in the groove and stretch direction, compared to cells grown on a smooth PDMS surface (Figure II.6.8.7), and that intercellular propagation of mechanically induced Ca^{2+} flux was significantly enhanced with the application of cyclic strain, but not when grooves alone were introduced. Taken together, these findings suggest that uniaxial mechanical stimulation, and not only forced cellular alignment, is necessary to produce positive benefits for ligament tissue engineering.

While mechanical stimulation is a relatively new practice in tissue engineering, it is actively being employed to develop a variety of musculoskeletal engineered tissues, including ligament, tendon, muscle, and bone constructs. In most, if not all, cases the addition of cyclic strain has conferred a positive benefit especially with cell proliferation, density, and differentiation of stem cells towards musculoskeletal lineages. However, one important admonition is that the ideal mechanical stimulation regime is far from being described, and specific biological models of mechanical stimulation to tissues are lacking. Thus, there is a need for further optimization of applied tissue strains with ligament tissue systems, and perhaps in the future applying an optimized mechanical stimulation regime to engineered ligaments will be commonplace.

Summary of the State of the Art of Ligament Tissue Engineering

The field of ligament tissue engineering is progressing at an incredibly rapid pace; in just 10 years engineered ligaments have advanced from concept to large animal studies. The potential healthcare implications of engineered ligaments are extensive, and with an aging population will become more important with time. Although ligament engineering has thus far focused on ACL and MCL regeneration, research may eventually expand to target all ligaments of the body.

We have seen that design factors for ligament engineering include native ligament anatomy, biomaterial mechanical properties, biomaterial degradation rate, cellular adherence/spreading on biomaterials, and matrix formation. From a biomaterials perspective, it is clear that matching biomaterial properties to the native ligament structure and function is a critical consideration. Yet what has also been important for the progression of the field has been the utilization of combinatorial approaches – researchers often merge two concepts into one to produce an improved outcome. Examples of this include merging braided scaffolds with sponges, or merging two materials into a single scaffold, or functionalizing a biodegradable surface, or adding mechanical stimulation to aligned cells. In all cases there is not one superior engineered ligament design, and thus far the ideal engineered ligament has yet to be created. Some of the important future milestones of ligament tissue engineering include improving the strength and biological integrity of the ligament–bone junction of implanted engineered ligaments, developing scaffolds and models that match the rate of scaffold degradation with the rate of tissue formation, matching native ligament elastic and viscoelastic mechanical properties, and developing ligament disease models through tissue engineering. Meeting the demands of these requirements will require concentrated interdisciplinary efforts from biologists, chemists, biomaterials scientists, tissue engineers, and clinicians which will eventually provide a new and improved option for repair of injured ligaments to thousands of patients in need of help.

ACKNOWLEDGMENT

The authors would like to acknowledge Joseph E. Marturano for his helpful contributions to this review.

BIBLIOGRAPHY

Altman, G. H., Horan, R. L., Lu, H. H., Moreau, J., Martin, I., et al. (2002a). Silk matrix for tissue engineered anterior cruciate ligaments. *Biomaterials*, **23**, 4131–4141.

Altman, G. H., Horan, R. L., Martin, I., Farhadi, J., Stark, P. R., et al. (2002b). Cell differentiation by mechanical stress. *FASEB J.*, **16**, 267–269.

Basmanav, F. B., Kose, G. T., & Hasirci, V. (2008). Sequential growth factor delivery from complexed microspheres for bone tissue engineering. *Biomaterials*, **29**, 4195–4204.

Benya, P. D., & Shaffer, J. D. (1982). Dedifferentiated chondrocytes re-express the differentiated collagen phenotype when cultured in agarose gels. *Cell*, **30**, 215–224.

Bhatt, K. A., Chang, E. I., Warren, S. M., Lin, S. -E., Bastidas, N., et al. (2007). Uniaxial mechanical strain: An *in vitro* correlate to distraction osteogenesis. *J. Surg. Res.*, **143**, 329–336.

Bryant, S. J., & Anseth, K. S. (2003). Controlling the spatial distribution of ECM components in degradable PEG hydrogels for tissue engineering cartilage. *J. Biomed. Mater. Res.*, **64A**, 70–79.

Bryant, S. J., Davis-Arehart, K. A., Luo, N., Shoemaker, R. K., Arthur, J. A., et al. (2004). Synthesis and characterization of photopolymerized multifunctional hydrogels: Water-soluble poly(vinyl alcohol) and chondroitin sulfate macromers for chondrocyte encapsulation. *Macromolecules*, **37**, 6726–6733.

Buckwalter, J. A., & Mankin, H. J. (1998). Articular cartilage: Tissue design and chondrocyte-matrix interactions. *Instr. Course Lect.*, **47**, 477–486.

Bullough, P. G., & Jagannath, A. (1983). The morphology of the calcification front in articular cartilage. Its significance in joint function. *J. Bone Joint Surg. Br.*, **65**, 72–78.

Campoccia, D., Doherty, P., Radice, M., Brun, P., Abatangelo, G., et al. (1998). Semisynthetic resorbable materials from hyaluronan esterification. *Biomaterials*, **19**, 2101–2127.

Caterson, E. J., Nesti, L. J., Li, W. J., Danielson, K. G., Albert, T. J., et al. (2001). Three-dimensional cartilage formation by bone marrow-derived cells seeded in polylactide/alginate amalgam. *J. Biomed. Mater. Res.*, **57**, 394–403.

CDC (Centers for Disease Control and Prevention) (1996). *National Hospital Discharge Survey*. Atlanta, GA: National Center for Health Statistics.

Chen, G., Sato, T., Sakane, M., Ohgushi, H., Ushida, T., et al. (2004). Application of PLGA-collagen hybrid mesh for three-dimensional culture of canine anterior cruciate ligament cells. *Mater. Sci. Eng. C*, **24**, 861–866.

Chen, X., Qi, Y. Y., Wang, L. L., Yin, Z., Yin, G. L., et al. (2008). Ligament regeneration using a knitted silk scaffold combined with collagen matrix. *Biomaterials*, **29**, 3683–3692.

Chiquet-Ehrismann, R., Tannheimer, M., Koch, M., Brunner, A., Spring, J., et al. (1994). Tenascin-C expression by fibroblasts is elevated in stressed collagen gels. *J. Cell Biol.*, **127**, 2093–2101.

Chu, C. R., Coutts, R. D., Yoshioka, M., Harwood, F. L., Monosov, A. Z., et al. (1995). Articular cartilage repair using allogeneic perichondrocyte-seeded biodegradable porous polylactic acid (PLA): A tissue-engineering study. *J. Biomed. Mater. Res.*, **29**, 1147–1154.

Cook, J. L., Williams, N., Kreeger, J. M., Peacock, J. T., & Tomlinson, J. L. (2003). Biocompatibility of three-dimensional chondrocyte grafts in large tibial defects of rabbits. *Am. J. Vet. Res.*, **64**, 12–20.

Cooper, J. A., Lu, H. H., Ko, F. K., Freeman, J. W., & Laurencin, C. T. (2005). Fiber-based tissue-engineered scaffold for ligament replacement: Design considerations and *in vitro* evaluation. *Biomaterials*, **26**, 1523–1532.

Cooper, J. W., Woods, M. D., & Laurencin, C. T. (2007). Tissue engineering of the anterior cruciate ligament using a braid-twist scaffold design. *J. Biomech.*, **40**, 2029–2036.

Csintalan, R. P., Inacio, M. C. S., & Funahashi, T. T. (2008). Incidence rate of anterior cruciate ligament reconstructions. *Perm. J.*, **12**, 17–21.

Dalby, M. J., Riehle, M. O., Johnstone, H. J., Affrossman, S., & Curtis, A. S. (2002). Polymer-demixed nanotopography: Control of fibroblast spreading and proliferation. *Tissue Eng.*, **8**, 1099–1108.

De Bari, C., Dell'Accio, F., Tylzanowski, P., & Luyten, P. (2001). Multipotent mesenchymal stem cells from human synovial membrane. *Arthritis Rheum.*, **44**, 1928–1942.

Denuziere, A., Ferrier, D., Damour, O., & Domard, A. (1998). Chitosan–chondroitin sulfate and chitosan–hyaluronate polyelectrolyte complexes: Biological properties. *Biomaterials*, **19**, 1275–1285.

Elisseeff, J., Anseth, K., Sims, D., McIntosh, W., Randolph, M., et al. (1999a). Transdermal photopolymerization for minimally invasive implantation. *Proc. Natl. Acad. Sci. USA*, **96**, 3104–3107.

Elisseeff, J., Anseth, K., Sims, D., McIntosh, W., Randolph, M., et al. (1999b). Transdermal photopolymerization of poly(ethylene oxide)-based injectable hydrogels for tissue-engineered cartilage. *Plast. Reconstr. Surg.*, **104**, 1014–1022.

Eyre, D. R., Wu, J. J., & Woods, P. (1992). Cartilage-Specific Collagens. Structural Studies. *In Articular Cartilage and Osteoarthritis: Workshop Conference Hoechst Werk Kalle-Albert, Wiesbaden, May 12–16, 1991. K. E. Kuettner*. New York, NY: Raven Press. xxxix, 757.

Facchini, A., Lisignoli, G., Cristino, S., Roseti, L., De Franceschi, L., et al. (2006). Human chondrocytes and mesenchymal stem cells grown onto engineered scaffold. *Biorheology*, **43**, 471–480.

Fan, H., Liu, H., Wong, E. J. W., Toh, S. L., & Goh, J. C. H. (2008). *In vivo* study of anterior cruciate ligament regeneration using mesenchymal stem cells and silk scaffold. *Biomaterials*, **29**, 3324–3337.

Fan, H., Liu, H., Toh, S. L., & Goh, J. C. H. (2009). Anterior cruciate ligament regeneration using mesenchymal stem cells and a silk scaffold in a large animal model. *Biomaterials*, **30**, 4967–4977.

Farnworth, L. (2000). Osteochondral defects of the knee. *Orthopedics*, **23**, 146–157; quiz 158–149.

Fell, H. B. (1925). The histogenesis of cartilage and bone in the long bones of the embryonic fowl. *J. Morphol.*, **40**, 417–451.

Formhals, A. (1934). *U.S. Patent 1,975,504*.

Fortier, L. A., Mohammed, H. O., Lust, G., & Nixon, A. J. (2002). Insulin-like growth factor-I enhances cell-based repair of articular cartilage. *J. Bone Joint Surg. Br.*, **84**, 276–288.

Freed, L. E., Vunjak-Novakovic, G., & Langer, R. (1993). Cultivation of cell-polymer cartilage implants in bioreactors. *J. Cell Biochem.*, **51**, 257–264.

Frenkel, S. R., Toolan, B., Menche, D., Pitman, M. I., & Pachence, J. M. (1997). Chondrocyte transplantation using a collagen bilayer matrix for cartilage repair. *J. Bone Joint Surg. Br.*, **79**, 831–836.

Fukumoto, T., Sperling, J. W., Sanyal, A., Fitzsimmons, J. S., Reinholz, G. G., et al. (2003). Combined effects of insulin-like growth factor-1 and transforming growth factor-beta1 on periosteal mesenchymal cells during chondrogenesis *in vitro*. *Osteoarthr. Cartilage*, **11**, 55–64.

Fuss, M., Ehlers, E. M., Russlies, M., Rohwedel, J., & Behrens, P. (2000). Characteristics of human chondrocytes, osteoblasts and fibroblasts seeded onto a type I/III collagen sponge under different culture conditions. A light, scanning and transmission electron microscopy study. *Anat. Anz.*, **182**, 303–310.

Garcia-Fuentes, M., Meinel, A. J., Hilbe, M., Meinel, L., & Merkle, H. P. (2009). Silk fibroin/hyaluronan scaffolds for human mesechymal stem cell culture in tissue engineering. *Biomaterials*, **30**, 5068–5076.

Green, W. T., Jr. (1977). Articular cartilage repair. Behavior of rabbit chondrocytes during tissue culture and subsequent allografting. *Clin. Orthop.*, 237–250.

Greenwald, D., Shumway, S., Albear, P., & Gottlieb, L. (1994). Mechanical comparison of 10 suture materials before and after *in vivo* incubation. *J. Surg. Res.*, **56**, 372–377.

Hasler, E. M., Herzog, W., Wu, J. Z., Muller, W., & Wyss, U. (1999). Articular cartilage biomechanics: Theoretical models, material properties, and biosynthetic response. *Crit. Rev. Biomed. Eng.*, **27**, 415–488.

Hayashi, T., Abe, E., & Jasin, H. E. (1996). Fibronectin synthesis in superficial and deep layers of normal articular cartilage. *Arthritis Rheum.*, **39**, 567–573.

Heinegard, D. K., & Pimentel, E. R. (1992). Articular cartilage matrix proteins. *In Articular Cartilage and Osteoarthritis: Workshop Conference Hoechst Werk Kalle-Albert, Wiesbaden, May 12–16, 1991. K. E. Kuettner*. New York, NY: Raven Press. xxxix, 757.

Holland, T. A., Tessmar, J. K., Tabata, Y., & Mikos, A. G. (2004). Transforming growth factor-beta1 release from oligo(poly(ethylene glycol) fumarate) hydrogels in conditions that model the cartilage wound healing environment. *J. Control Release*, **94**, 101–114.

Huang, C. Y., Reuben, P. M., D'Ippolito, G., Schiller, P. C., & Cheung, H. S. (2004). Chondrogenesis of human bone marrow-derived mesenchymal stem cells in agarose culture. *Anat. Rec.*, **278A**, 428–436.

Hukins, D. W.L., Kirby, M. C., Sikoryn, T. A., Aspden, R. M., & Cox, A. J. (1990). Comparison of structure, mechanical properties, and functions of lumbar spinal ligaments. *Spine*, **15**, 787–795.

Hunziker, E. B. (2002). Articular cartilage repair: Basic science and clinical progress. A review of the current status and prospects. *Osteoarthr. Cartilage*, **10**, 432–463.

Hunziker, E. B., Michel, M., & Studer, D. (1997). Ultrastructure of adult human articular cartilage matrix after cryotechnical processing. *Microsc. Res. Tech.*, **37**, 271–284.

Jiang, P., Liu, H., Wang, C., Wu, L., Huang, J., et al. (2006). Tensile behavior and morphology of differently degummed silkworm (*Bombyx mori*) cocoon silk fibres. *Mater. Lett.*, **60**, 919–925.

Jones, B. F., Wall, M. E., Carroll, R. L., Washburn, S., & Banes, A. J. (2005). Ligament cells stretch-adapted on a microgrooved substrate increase intercellular communication in response to a mechanical stimulus. *J. Biomech.*, **38**, 1653–1664.

Kato, N., Sato, S., Yamanaka, A., Yamada, H., Fuwa, N., et al. (1998). Silk protein, sericin, inhibits lipid peroxidation and tyrosinase activity. *Biosci. Biotech. Bioch.*, **62**, 145–147.

Kempson, G. E., Muir, H., Pollard, C., & Tuke, M. (1973). The tensile properties of the cartilage of human femoral condyles related to the content of collagen and glycosaminoglycans. *Biochim. Biophys. Acta.*, **297**, 456–472.

Kennedy, J. C., Weinberg, H., & Wilson, A. S. (1974). The anatomy and function of the anterior cruciate ligament: As determined by clinical and morphological studies. *J. Bone. Joint Surg.*, **56A**, 223–235.

Kimura, Y., Hokugo, A., Takamoto, T., Tabata, Y., & Kurosawa, H. (2008). Regeneration of anterior cruciate ligament by biodegradable scaffold combined with local controlled release of basic fibroblast growth factor and collagen wrapping. *Tissue Eng. Part C*, **14**, 47–57.

Kotlarz, H., Gunnarsson, C. L., Fang, H., & Rizzo, J. A. (2009). Insurer and out-of-pocket costs of osteoarthritis in the US: Evidence from national survey data. *Arthritis Rheum.* **60**, 3546–3553.

Krych, A. J., Jackson, J. D., Hoskin, T. L., & Dahm, D. L. (2008). A meta-analysis of patellar tendon autograft versus patellar tendon allograft in anterior cruciate ligament reconstruction. *Arthroscopy*, **24**, 292–298.

Kuettner, K. E. (1992). Biochemistry of articular cartilage in health and disease. *Clin. Biochem.*, **25**, 155–163.

Kuo, C. K., & Ma, P. X. (2001). Ionically crosslinked alginate hydrogels as scaffolds for tissue engineering: Part 1. Structure gelation rate and mechanical properties. *Biomaterials, 22*, 511–521.

Kuo, C. K., & Ma, P. X. (2008). Maintaining dimensions and mechanical properties of ionically crosslinked alginate hydrogel scaffolds in vitro. *J. Biomed. Mater. Res. A, 84*, 899–907.

Kuo, C. K., & Tuan, R. S. (2008). Mechanoactive tenogenic differentiation of human mesenchymal stem cells. *Tissue Eng. A.*, **14**, 1615–1627.

Kuo, C. K., Petersen, B. C., & Tuan, R. S. (2008). Spatiotemporal protein distribution of TGF-βs, their receptors, and extracellular matrix molecules during embryonic tendon development. *Dev. Dynam.*, **237**, 1477–1489.

Kwan, M. K., Lin, T. H.C., & Woo, S. L.Y. (1993). On the viscoelastic properties of the anteromedial bundle of the anterior cruciate ligament. *J. Biomech.*, **26**, 447–452.

Lahiji, A., Sohrabi, A., Hungerford, D. S., & Frondoza, C. G. (2000). Chitosan supports the expression of extracellular matrix proteins in human osteoblasts and chondrocytes. *J. Biomed. Mater. Res.*, **51**, 586–595.

Lee, C. H., Shin, H. J., Cho, I. H., Kang, Y., Kim, I. A., et al. (2005). Nanofibers alignment and direction of mechanical strain affect the ECM production of human ACL fibroblast. *Biomaterials*, **26**, 1261–1270.

Lee, W. K., Ichi, T., Ooya, T., Yamamoto, T., Katoh, M., et al. (2003a). Novel poly(ethylene glycol) scaffolds crosslinked by hydrolyzable polyrotaxane for cartilage tissue engineering. *J. Biomed. Mater. Res.*, **67A**, 1087–1092.

Lee, C. R., Grodzinsky, A. J., Hsu, H. P., & Spector, M. (2003b). Effects of a cultured autologous chondrocyte-seeded type II collagen scaffold on the healing of a chondral defect in a canine model. *J. Orthop. Res.*, **21**, 272–281.

Li, Q., Williams, C. G., Sun, D. D., Wang, J., Leong, K., et al. (2004). Photocrosslinkable polysaccharides based on chondroitin sulfate. *J. Biomed. Mater. Res. A.*, **68**, 28–33.

Li, W. J., Laurencin, C. T., Caterson, E. J., Tuan, R. S., & Ko, F. K. (2002). Electrospun nanofibrous structure: A novel scaffold for tissue engineering. *J. Biomed. Mater. Res.*, **60**, 613–621.

Li, W. J., Danielson, K. G., Alexander, P. G., & Tuan, R. S. (2003). Biological response of chondrocytes cultured in three-dimensional nanofibrous poly(epsilon-caprolactone) scaffolds. *J. Biomed. Mater. Res.*, **67A**, 1105–1114.

Li, W. J., Tuli, R., Okafor, C., Derfoul, A., Danielson, K. G., et al. (2005). A three-dimensional nanofibrous scaffold for cartilage tissue engineering using human mesenchymal stem cells. *Biomaterials*, **26**, 599–609.

Li, W. J., Jiang, Y. J., & Tuan, R. S. (2006a). Chondrocyte phenotype in engineered fibrous matrix is regulated by fiber size. *Tissue Eng.*, **12**, 1775–1785.

Li, W. J., Cooper, J. A., Jr., Mauck, R. L., & Tuan, R. S. (2006b). Fabrication and characterization of six electrospun poly(alpha-hydroxy ester)-based fibrous scaffolds for tissue engineering applications. *Acta. Biomater.*, **2**, 377–385.

Li, W. J., Chiang, H., Kuo, T. F., Lee, H. S., Jiang, C. C., et al. (2009). Evaluation of articular cartilage repair using biodegradable nanofibrous scaffolds in a swine model: A pilot study. *J. Tissue Eng. Regen. Med.*, **3**, 1–10.

Lidén, M., Ejerhed, L., Sernert, N., Laxdal, G., & Kartus, J. (2007). Patellar tendon or semitendinosus tendon autografts for anterior cruciate ligament reconstruction: A prospective, randomized study with a 7-year follow-up. *Am. J. Sports Med.*, **35**, 740–748.

Lin, V. S., Lee, M. C., O'Neal, S., McKean, J., & Sung, K. P. (1999). Ligament tissue engineering using synthetic biodegradable fiber scaffolds. *Tissue Eng.*, **5**(5), 443–451.

Liu, H., Lee, Y. W., & Dean, M. F. (1998). Re-expression of differentiated proteoglycan phenotype by dedifferentiated human chondrocytes during culture in alginate beads. *Biochim. Biophys. Acta.*, **1425**, 505–515.

Liu, H., Fan, H., Wang, Y., Toh, S. L., & Goh, J. C. H. (2008). The interaction between a combined knitted silk scaffold and microporous silk sponge with human mesenchymal stem cells for ligament tissue engineering. *Biomaterials*, **29**, 662–674.

Liu, Y., Chen, F., Liu, W., Cui, L., Shang, Q., et al. (2002). Repairing large porcine full-thickness defects of articular cartilage using autologous chondrocyte-engineered cartilage. *Tissue Eng.*, **8**, 709–721.

Lu, H. H., Cooper, J. A., Jr., Manuel, S., Freeman, J. W., Attawia, M. A., et al. (2005). Anterior cruciate ligament regeneration using braided biodegradable scaffolds: *In vitro* optimization studies. *Biomaterials*, **26**, 4805–4816.

Lu, J. X., Prudhommeaux, F., Meunier, A., Sedel, L., & Guillemin, G. (1999). Effects of chitosan on rat knee cartilages. *Biomaterials*, **20**, 1937–1944.

Malcarney, H. L., Bonar, F., & Murrell, G. A. C. (2005). Early inflammatory reaction after rotator cuff repair with a porcine small intestine submucosal implant. *Am. J. Sports Med.*, **33**, 907–911.

Matsumura, K., Hyon, S. H., Nakajima, N., Iwata, H., Watazu, A., et al. (2004). Surface modification of poly(ethylene-co-vinyl alcohol): Hydroxyapatite immobilization and control of periodontal ligament cells differentiation. *Biomaterials*, **25**, 4817–4824.

Mendler, M., Eich-Bender, S. G., Vaughan, L., Winterhalter, K. H., & Bruckner, P. (1989). Cartilage contains mixed fibrils of collagen types II, IX, and XI. *J. Cell Biol.*, **108**, 191–197.

Metters, A. T., Anseth, K. S., & Bowman, C. N. (1999). Fundamental studies of biodegradable hydrogels as cartilage replacement materials. *Biomed. Sci. Instrum.*, **35**, 33–38.

Middleton, J. C., & Tipton, A. J. (2000). Synthetic biodegradable polymers as orthopedic devices. *Biomaterials*, **21**, 2335–2346.

Minoura, N., Tsukada, M., & Nagura, M. (1990). Physico-chemical properties of silk fibroin membrane as a biomaterial. *Biomaterials*, **11**(6), 430–434.

Moe, K., Tay, T. E., Goh, J. C. H., Ouyang, H. W., & Toh, S. L. (2005). Cyclic uniaxial strains on fibroblasts-seeded PLGA scaffolds for tissue engineering of ligaments. *P. Spine. Int. Soc. Opt. Eng.*, **5852**, 665–670.

Molloy, T., Wang, Y., & Murrell, G. A. C. (2003). The roles of growth factors in tendon and ligament healing. *Sports Med.*, **33**, 381–394.

Mow, V. C., & Wang, C. C. (1999). Some bioengineering considerations for tissue engineering of articular cartilage. *Clin. Orthop.*, S204–223.

Murakami, S., Takayama, S., Ikezawa, K., Shimabukuro, Y., Kitamura, M., et al. (1999). Regeneration of periodontal tissues by basic fibroblast growth factor. *J. Peridontal Res.*, **34**, 425–430.

Musahl, V., Abramowitch, S. D., Gilbert, T. W., Tsuda, E., Wang, J. H. C., et al. (2004). The use of porcine small intestinal submucosa to enhance the healing of the medial collateral ligament – a functional tissue engineering study in rabbits. *J. Orthop. Res.*, **22**, 214–220.

Nehrer, S., Breinan, H. A., Ramappa, A., Shortkroff, S., Young, G., et al. (1997). Canine chondrocytes seeded in type I and type II collagen implants investigated *in vitro*. *J. Biomed. Mater. Res.*, **38**, 95–104.

Oberlender, S. A., & Tuan, R. S. (1994). Expression and functional involvement of N-cadherin in embryonic limb chondrogenesis. *Development*, **120**, 177–187.

Orlando, R. A., & Cheresh, D. A. (1991). Arginine-glycine-aspartic acid binding leading to molecular stabilization between integrin alpha v beta 3 and its ligand. *J. Bio. Chem.*, **266**, 19543–19550.

Ouyang, H. W., Goh, J. C.H., Mo, X. M., Teoh, S. H., & Lee, E. H. (2002). Characterization of anterior cruciate ligament cells and bone marrow stromal cells on various biodegradable polymeric films. *Mater. Sci. Eng. C*, **20**, 63–69.

Park, H., Temenoff, J. S., Tabata, Y., Caplan, A. I., Raphael, R. M., et al. (2009). Effect of dual growth factor delivery on chondrogenic differentiation of rabbit marrow mesenchymal stem cells encapsulated in injectable hydrogel composites. *J. Biomed. Mater. Res. A.*, **88**, 889–897.

Park, S. A., Kim, I. A., Lee, Y. J., Shin, J. W., Kim, C. R., et al. (2006). Biological responses of ligament fibroblasts and gene expression profiling on micropatterned silicone substrates subjected to mechanical stimuli. *J. Biosci. Bioeng.*, **102**, 402–412.

Pham, Q. P., Sharma, U., & Mikos, A. G. (2006). Electrospinning of polymeric nanofibers for tissue engineering applications: A review. *Tissue Eng.*, **12**, 1197–1211.

Pier, G. B., DesJardin, D., Grout, M., Garner, C., Bennett, S. E., et al. (1994). Human immune response to *Pseudomonas aeruginosa* mucoid exopolysaccharide (alginate) vaccine. *Infect Immun.*, **62**, 3972–3979.

Prydz, K., & Dalen, K. T. (2000). Synthesis and sorting of proteoglycans. *J. Cell Sci.*, **113**(Pt 2), 193–205.

Rahfoth, B., Weisser, J., Sternkopf, F., Aigner, T., von der Mark, K., et al. (1998). Transplantation of allograft chondrocytes embedded in agarose gel into cartilage defects of rabbits. *Osteoarthr. Cartilage*, **6**, 50–65.

Raif, E. M., & Seedhom, B. B. (2005). Effect of cyclic tensile strain on proliferation of synovial cells seeded onto synthetic ligament scaffolds: An *in vitro* simulation. *Bone*, **36**, 433–443.

Riboh, J., Chong, A. K. S., Pham, H., Longaker, M., Jacobs, C., et al. (2008). Optimization of flexor tendon tissue engineering with a cyclic strain bioreactor. *J. Hand Surg.*, **33**, 1388–1396.

Roeder, B. A., Kokini, K., Sturgis, J. E., Robinson, J. P., & Voytik-Harbin, S. L. (2002). Tensile mechanical properties of three-dimensional type I collagen extracellular matrices with varied microstructure. *J. Biomech. Eng.*, **124**, 214–222.

Roughley, P. J., & Lee, E. R. (1994). Cartilage proteoglycans: Structure and potential functions. *Microsc. Res. Tech.*, **28**, 385–397.

Ruoslahti, E. (1988). Structure and biology of proteoglycans. *Annu. Rev. Cell Biol.*, **4**, 229–255.

Sahoo, S., Ouyang, H., Goh, J. C.H., Tay, T. E., & Toh, S. L. (2006). Characterization of a novel polymeric scaffold for potential application in tendon/ligament tissue engineering. *Tissue Engineering*, **12**(1), 91–99.

Sahoo, S., Cho-Hong, J. G., & Siew-Lok, T. (2007). Development of hybrid polymer scaffolds for potential applications in ligament and tendon tissue engineering. *Biomed. Mater.*, **2**, 169–173.

Sahoo, S., Ang, L. -T., Goh, J. C. H., & Toh, S. L. (2009). Bioactive nanofibers for fibroblastic differentiation of mesenchymal precursor cells for ligament/tendon tissue engineering applications. *Differentiation*, **79**, 102–110.

Sechriest, V. F., Miao, Y. J., Niyibizi, C., Westerhausen-Larson, A., Matthew, H. W., et al. (2000). GAG-augmented polysaccharide hydrogel: A novel biocompatible and biodegradable material to support chondrogenesis. *J. Biomed. Mater. Res.*, **49**, 534–541.

Silverman, R. P., Passaretti, D., Huang, W., Randolph, M. A., & Yaremchuk, M. J. (1999). Injectable tissue-engineered cartilage using a fibrin glue polymer. *Plast. Reconstr. Surg.*, **103**, 1809–1818.

Soon, M., Chang, P., Neo, C. P. C., Mitra, A. K., & Tay, B. K. (2004). Morbidity following anterior cruciate ligament reconstruction using hamstring autograft. *Ann. Acad. Med. Singap.*, **33**, 214–219.

Stockwell, R. A. (1979). *Biology of Cartilage Cells*. Cambridge, NY: Cambridge University Press.

Swoboda, B., Holmdahl, R., Stob, H., & von der Mark, K. (1989). Cellular heterogeneity in cultured human chondrocytes identified by antibodies specific for 2(XI) collagen chains. *J. Cell Bio.*, *109*, 1363–1369.

Temenoff, J. S., Athanasiou, K. A., LeBaron, R. G., & Mikos, A. G. (2002). Effect of poly(ethylene glycol) molecular weight on tensile and swelling properties of oligo(poly(ethylene glycol) fumarate) hydrogels for cartilage tissue engineering. *J. Biomed. Mater. Res.*, **59**, 429–437.

Toyoda, T., Matsumoto, H., Fujikawa, K., Saito, S., & Inoue, K. (1998). Tensile load and the metabolism of anterior cruciate ligament cells. *Clini. Orthop. Relat. Res.*, **353**, 247–255.

Vacanti, C. A., Langer, R., Schloo, B., & Vacanti, J. P. (1991). Synthetic polymers seeded with chondrocytes provide a template for new cartilage formation. *Plast. Reconstr. Surg.*, **88**, 753–759.

van Wachem, P. B., Beugeling, T., Feijen, J., Bantjes, A., Detmers, J. P., et al. (1985). Interaction of cultured human endothelial cells with polymeric surfaces of different wettabilities. *Biomaterials*, **6**(6), 403–408.

Venn, M. F. (1978). Variation of chemical composition with age in human femoral head cartilage. *Ann. Rheum. Dis.*, **37**, 168–174.

Vercruysse, K. P., Marecak, D. M., Marecek, J. F., & Prestwich, G. D. (1997). Synthesis and *in vitro* degradation of new polyvalent hydrazide cross-linked hydrogels of hyaluronic acid. *Bioconjug. Chem.*, **8**, 686–694.

Vunjak-Novakovic, G., Altman, G., Horan, R., & Kaplan, D. L. (2004). Tissue engineering of ligaments. *Annu. Rev. Biomed. Eng.*, **6**, 131–156.

Wakitani, S., Goto, T., Pineda, S. J., Young, R. G., Mansour, J. M., et al. (1994). Mesenchymal cell-based repair of large, full-thickness defects of articular cartilage. *J. Bone Joint Surg. Am.*, **76**, 579–592.

Wang, D. A., Williams, C. G., Li, Q., Sharma, B., & Elisseeff, J. H. (2003). Synthesis and characterization of a novel degradable phosphate-containing hydrogel. *Biomaterials*, **24**, 3969–3980.

Webster, T. J., Ergun, C., Doremus, R. H., Siegel, R. W., & Bizios, R. (2000). Enhanced functions of osteoblasts on nanophase ceramics. *Biomaterials*, **21**, 1803–1810.

Whitlock, P. W., Smith, T. L., Poehling, G. G., Shilt, J. S., & Van Dyke, M. (2007). A naturally derived, cytocompatible, and architecturally optimized scaffold for tendon and ligament regeneration. *Biomaterials*, **28**, 4321–4329.

Woo, S. L.Y., Gomez, M. A., Sites, T. J., Newton, P. O., Orlando, C. A., et al. (1987). The biomechanical and morphological changes in the medial collateral ligament of the rabbit after immobilization and remobilization. *J. Bone Joint Surg.*, **69**, 1200–1211.

Woo, S. L.Y., Hollis, J. M., Adams, D. J., Lyon, R. M., & Takai, S. (1991). Tensile properties of human femur-anterior cruciate ligament-tibia complex: The effects of specimen age and orientation. *Ame. J. Sports Med.*, **19**, 217–225.

Woo, S. L.Y., Abramowitch, S. D., Kilger, R., & Liang, R. (2006). Biomechanics of knee ligaments: Injury, healing, and repair. *J. Biomech.*, **39**, 1–20.

Worster, A. A., Brower-Toland, B. D., Fortier, L. A., Bent, S. J., Williams, J., et al. (2001). Chondrocytic differentiation of mesenchymal stem cells sequentially exposed to transforming growth factor-beta1 in monolayer and insulin-like growth factor-I in a three-dimensional matrix. *J. Orthop. Res.*, **19**, 738–749.

Yamada, H., Nakao, H., Takasu, Y., & Tsubouchi, K. (2001). Preparation of undegraded native molecular fibroin solution from silkworm cocoons. *Mater. Sci. Eng. C*, **14**, 41–46.

Yilgor, P., Tuzlakoglu, K., Reis, R. L., Hasirci, N., & Hasirci, V. (2009). Incorporation of a sequential BMP-2/BMP-7 delivery system into chitosan-based scaffolds for bone tissue engineering. *Biomaterials*, **30**, 3551–3559.

Zhang, L., Kahn, C. J.F., Chen, H. -Q., Tran, N., & Wang, X. (2008). Effect of uniaxial stretching on rat bone mesenchymal stem cell: Orientation and expressions of collagen types I and III and tenascin-C. *Cell Biology International*, **32**, 344–352.

Zheng, M. H., Chen, J., Kirilak, Y., Willers, C., Xu, J., et al. (2005). Porcine small instestine submucosa (SIS) is not an acellular collagenous matrix and contains porcine DNA: Possible implications in human implantation. *J. Biomed. Mater. Res. B: Applied Biomaterials*, **73B**, 61–67.

Zhou, J., Ciobanu, M., Pavon-Djavid, G., Gueguen, V., & Migonney, V. (2007). Morphology and adhesion of human fibroblast cells cultured on bioactive polymer grafted ligament prosthesis. *Annu. Int. Conf. IEEE Eng. Med. Bio. Soc.*, **2007**, 5115–5118.

CHAPTER II.6.9 BLOOD VESSEL TISSUE ENGINEERING

Stacey C. Schutte[1] *and Robert M. Nerem*[2]
[1]Emory University School of Medicine, Woodruff Memorial Building, Atlanta, GA, USA
[2]Georgia Institute of Technology, Atlanta, GA, USA

INTRODUCTION

Cardiovascular disease (CVD) is a major clinical and economic issue, as well as a leading cause of death in developed nations. It affects 80 million Americans and was attributed to 35.3% of all deaths in the United States in 2005. Atherosclerosis is the most common form of CVD and leads to narrowing of the arteries, most commonly the coronary, carotid, and peripheral arteries, which can lead to reduced function or myocardial infarction, stroke or amputation of limbs. Coronary artery disease is the most prevalent CVD accounting for 52% of all CVD (American Heart Association, 2008; Rosamond et al., 2008). Coronary stenosis can be treated by coronary artery bypass grafting (CABG) or by percutaneous coronary intervention (PCI), commonly known as angioplasty, which is used today with endovascular stents (see Chapter II.5.3). While PCI has become a common alternative, CABG is still considered the "gold-standard" of treatment, is the preferred treatment option for patients with left main or triple-vessel coronary artery disease with reduced left ventricular function (Bravata et al., 2007; Taggart, 2007; Malvindi et al., 2008), and requires significantly fewer repeated procedures than PCI (Bravata et al., 2007; Dixon et al., 2008).

There are several choices of material for bypass graft conduits. Autologous venous grafts, most commonly from the saphenous vein, were the material of choice at the advent of CABG surgery. More recently autologous arterial grafts, especially from the internal mammary artery, have gained popularity in CABG procedures, due to the lower incidence of intimal thickening (Canver, 1995; Bahn et al., 1999; Hayward et al., 2008). Although preferred, not all patients have sufficient amounts of healthy tissue necessary for bypass grafting. There are several non-autologous tissue options, such as the use of allogeneic umbilical or saphenous vein or xenogeneic bovine internal thoracic artery, which are often made nonviable through use of glutaraldehyde fixation. The patency rates for these grafts are low, ranging from 65% for cryopreserved saphenous vein grafts to less than 16% for xenografts (Canver, 1995). Synthetic materials, such as expanded polytetrafluoroethylene (ePTFE) and Dacron® have been used successfully to replace larger diameter vessels (>6 mm); however, these materials have not been successful for small diameter (diameter ~3 mm) bypass procedures (such as coronary artery bypass grafts) due to occlusion (Edelman, 1999).

While there is a significant need for bypass graft material due to the prevalence of disease, there are two other areas where there is a need for tissue-engineered blood vessels: pediatrics and shunts for hemodialysis. Conduits of ePTFE are widely used for pediatric vascular repair (Petrossian et al., 2006), with Dacron® and homografts used to a lesser extent. One major limitation of these grafts is the lack of growth potential. Without a graft that can be remodeled and grow with the patient, repeated procedures are required to replace grafts that the patient has outgrown. A graft capable of repair is also ideal for hemodialysis shunts. Creation of a fistula is the preferred method for dialysis access; however, not all patients have vasculature such that a fistula can be created. Healing of the area can take several weeks, and the area where the artery and vein are joined is small. There is a risk of failure, as repeated punctures can result in aneurysm. An arteriovenous (AV) shunt can be used in place of the fistula, where a graft is placed between the artery and vein in the arm, allowing for greater access. Healing is more rapid allowing for earlier access; however, grafts are usually made from ePTFE which can become stenotic due to thrombosis, and the damage due to numerous punctures can cause graft failure.

Tissue engineering may provide a solution to these problems. An ideal graft would be non-immunogenic, non-thrombogenic, have mechanical properties similar to native vessel, be capable of being remodeled, and be vasoactive. "Off-the-shelf" availability is also desired, as many patients cannot wait for these life-saving treatments. This chapter will discuss the enabling technologies and current research into the creation of a tissue engineered vascular graft that could address these needs.

ENABLING TECHNOLOGIES

In the tissue engineering of a blood vessel substitute there are a number of critical issues that are being addressed by a wide variety of investigators. One of these relates to the scaffold that will be used in creating the assembly of cells into the architecture required for a blood vessel. Figure II.6.9.1 shows a cross-section of an artery that engineers are trying to reproduce. Here the enabling technology is that of biomaterials, and since this book is devoted to that topic, any further discussion will be left to the following section in this chapter where the use of biomaterials in different tissue engineering approaches will be outlined. Other critical issues include the sourcing of cells to be used, the achievement of appropriate mechanical properties, and the vascularization of the wall of an engineered blood vessel. It is these three issues that will be discussed in this section.

Cell sourcing is a critical issue for blood vessel tissue engineering, with each cell type having its own unique challenges. While there is some debate over the need for autologous cells for the smooth muscle cells of the medial layer and fibroblasts of the adventitia layer, the need for autologous endothelial cells is agreed upon. Both Advanced Tissue Sciences, Inc. and Organogenesis, Inc. conducted animal experiments using allogeneic smooth muscle cells that, over a period of up to a month, exhibited no immune-mediated rejection. Unfortunately, the results of these studies were never

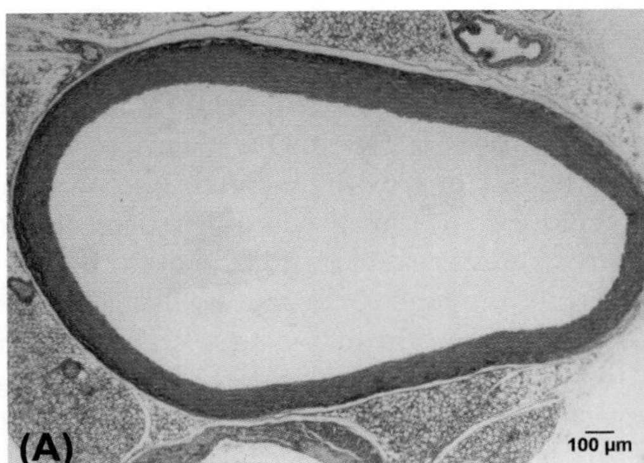

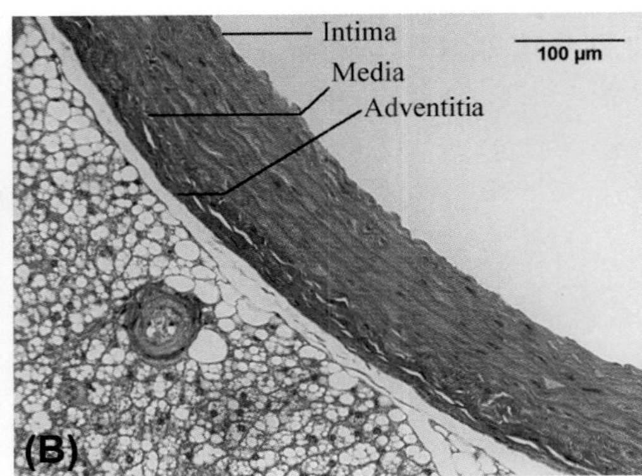

FIGURE II.6.9.1 Cross-section of a rat muscular artery. The sections are stained with Masson's trichrome which stains collagen blue, nuclei purple, and cytoplasm and muscle fiber red. The entire cross-section is shown (A). The different layers of the artery can be seen more clearly at higher magnification (B). The innermost layer is the intima, a confluent monolayer of endothelial cells sitting on a thin basement membrane. This layer provides a non-thrombogenic lining. The innermost layer is the medial layer; it consists primarily of collagen, elastin, and smooth muscle cells. The thickness and amount of elastin depends on the type of blood vessel. The medial layer provides the mechanical integrity and the vasoactive response of the blood vessel. The outermost layer is the adventitia; it is primarily collagen and fibroblasts, and provides anchoring and allows for ingrowth of nerves and the vasa vasorum.

published. These studies do suggest that at least the possibility of using allogeneic smooth muscle cells exists; however, it must be emphasized that there is a general consensus that autologous endothelial cells could only be employed if there was a strategy to create immune acceptance.

In an acellular graft, endothelial cells migrate no more than 1–2 cm into the graft (along the luminal surface), leaving the remainder of the graft exposed and prone to thrombosis; thrombosis is especially problematic in small diameter grafts where it often leads to significant stenosis or occlusion. There are several endothelial cell sources that have been explored that are summarized in Table II.6.9.1. Along with endothelialization, there are strategies for creating non-thrombogenic linings to minimize thrombosis, often with a short-term goal of prevention until endothelialization can occur. These strategies include use of heparin, thrombomodulin, and surfaces modified to release nitric oxide (Jordan and Chaikof, 2007).

Moreover, large numbers of smooth muscle cells are needed for blood vessel tissue engineering. Expanding smooth muscle cells *in vitro* is common practice, but can result in a modulation to a more synthetic smooth muscle cell phenotype where cells lose much of their contractile apparatus. Cell culture techniques designed to avoid this phenotypic shift result in very long culture times, and are only successful for a small number of passages. After prolonged culture, modulation back to a contractile phenotype is uncertain, but may be possible under some culture conditions such as serum starvation. The long culture times are also a problem in instances where "off-the-shelf" availability is desired. Recruitment of smooth muscle cells *in vivo*, use of allogeneic cells, and use of stem cells each could provide a solution.

Fibroblasts are not always used in every approach, but can be used in the creation of an adventitial layer or in place of the smooth muscle cells to create the medial layer. Fibroblasts secrete extracellular matrix (ECM) and have contractile apparatus, potentially allowing for the use of fibroblasts in place of smooth muscle cells. While the adventitia is often ignored, recent studies have shown that tissue-engineered blood vessels with an adventitia layer containing both fibroblasts and endothelial cells had improved vascularization when implanted subcutaneously (Guillemette et al., 2008).

TABLE II.6.9.1 Possible Endothelial Cell Sources

Endothelial Cell Source	Comment
Autologous endothelial cells (biopsy)	Requires *in vitro* expansion Not available "off-the-shelf" Used in early clinical trials
Allogeneic endothelial cells	Possible "off-the-shelf" availability Required immune acceptance strategy
Genetically engineered cells	Possible "off-the-shelf" availability Engineered to provide endothelial cell function
Adult progenitor cells or adult stem cells	Autologous cells or immunoprotected cells Could be recruited from bone marrow or blood Provide endothelial-like function
Embryonic stem cells	Unlimited supply, rapidly expanding Scientific and ethical hurdles Long-term fate of cells uncertain

Adapted from Nerem and Ensley, 2004.

Adult Stem Cells and Blood Vessel Tissue Engineering

While embryonic stem cells have been studied as a cell source in tissue-engineered blood vessels, these have been investigated primarily as an endothelial cell source. Adult stem cells have become a popular choice, due to the lack of ethical concerns and relative ease in harvesting. Mesenchymal stem cells (MSC) are a popular choice for tissue engineering. They have been shown to express smooth muscle cell phenotypic markers to different degrees under different culture conditions (Gong and Niklason, 2008); when MSCs are combined with endothelial progenitor cells (EPCs), pulmonary artery patches can remain patent *in vivo* (Mettler et al., 2008). MSCs have also been investigated as a potential endothelial cell source; genetic modification resulted in MSCs that produce nitric oxide (Zhang et al., 2006a). Other adult stem cells have also been studied and may be potential cell sources. Muscle-derived stem cells have been shown as a possible cell source when used with a biodegradable scaffold (Nieponice et al., 2008), and hair follicle cells have been shown to express smooth muscle cell markers, as well as generate a contractile force in engineered tissue (Liu et al., 2008a). Research will continue in these areas as many adult stem cells are immunoprotected and can provide necessary function, possibly eliminating the need for allogeneic cell sources.

Another critical issue in the tissue engineering of a blood vessel is achieving the appropriate mechanical properties. This includes having a strength that is comparable to a native artery, but also having similar compliance and visco-elastic properties. The mechanical stress environment within a blood vessel is quite complex. Blood vessels are subjected to shear stress, transmural pressure, and axial extension, each of which has an effect on the blood vessel wall, as depicted in Figure II.6.9.2. Flow through the lumen produces a shear stress on the endothelial cells lining the blood vessel wall. The wall shear stress influences not only endothelial cell alignment, but also behavioral characteristics of this inner lining of cells (Fisher et al., 2001). The direction and magnitude of the shear stress thus influences the health of the vessel wall. Areas of disturbed or stagnant flow are prone to forming atherosclerotic plaques, and

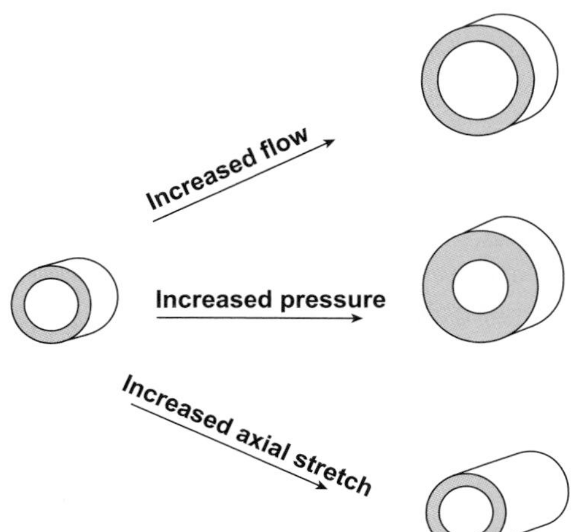

FIGURE II.6.9.2 Effects of hemodynamics on blood vessel. Increased blood flow results in an increase in the inner diameter of the blood vessel. Blood flow also influences endothelial cell alignment and is correlated with atherosclerosis. Areas of oscillatory or stagnant flow are prone to atherosclerotic plaques. Increased pressure results in an increase in wall thickness; this occurs during development and growth, as well as during prolonged hypertension. Axial stretch results in an increase in vessel length.

such atherosclerosis is a highly localized disease. Transmural pressure produces hoop stress in the vessel wall. This hoop stress influences the organization of collagen and elastin in the wall. Elevated stress, such as with long-term hypertension, will result in remodeling and thickening of the vessel wall. The blood vessels also see axial stresses that affect cell and matrix organization, and vessel length.

In order to mimic the physiological environment of the artery, including its mechanical nature, bioreactors are often employed. The most common form of mechanical stimulation is the application of cyclic strain; cyclic strain of the engineered tissue has been shown to increase tissue strength (Seliktar et al., 2000; Isenberg and Tranquillo, 2003), and to promote a more contractile cell phenotype (Stegemann and Nerem, 2003). Cyclic strain increases matrix metalloproteinase 2 (MMP-2) activity, which can have beneficial or detrimental effects depending on the duration and magnitude of the MMP activity (Seliktar et al., 2003). The ideal strain profiles are still being determined; strain amplitude and rate influences the mechanical properties of the engineered tissue, and recent studies show that a more complex profile incrementally increasing strain rate may yield better results (Syedain et al., 2008).

Early work looked only at the application of cyclic strain, and often required the use of an internal silicon sleeve for transference of strain due to insufficient mechanical properties of the engineered tissue. Since then, several groups have developed bioreactors combining cyclic strain with other stimuli. Cyclic strain has been combined with flow through the lumen (Hoerstrup et al., 2001; Hahn et al., 2007; Iwasaki et al., 2008), perfusion and shear (Sodian et al., 2002; Williams and Wick, 2004), electrical stimulation (Abilez et al., 2006; Mantero et al., 2007) or longitudinal extension (Mironov et al., 2003; McCulloch et al., 2004). Figure II.6.9.3 shows a tissue-engineered blood vessel in a bioreactor capable of imparting cyclic strain, shear, and longitudinal extension (Zaucha et al., 2009). While the use of bioreactors can provide a more physiological environment for culturing and studying tissue-engineered blood vessels, it may not be necessary. There are several approaches that will be discussed in the next section that avoid the use of bioreactors in order to minimize the culture time and try to achieve off-the-shelf availability.

Finally, a critical issue for any thick tissue engineered structure is that of vascularization. Without a vascular network, cell death will occur in tissues greater than 1–2 cm thick, and some experiments and the structure of natural aorta suggest that a tissue thicker than approximately 0.2 cm requires an intrinsic vasculature (called the *vasa vasorum*). Microfabrication techniques such as plasma etching and lithography can be used to create capillary-like networks while maintaining tolerances to ±0.1 μm (Borenstein et al., 2002). These techniques have been used to fabricate devices and networks in both natural and synthetic polymers, including silk fibroin

FIGURE II.6.9.3 Culture of an engineered blood vessel in a bioreactor capable of luminal flow, cyclic strain, and axial stretch. The bottom picture shows a collagen-based tissue-engineered blood vessel within the culture chamber. (As described in Zaucha et al., 2009.)

(Bettinger et al., 2007), poly(lactic acid) (PLA) (Richards Grayson et al., 2003), and poly(glycerol sebacate) (PGS) (Bettinger et al., 2006).

TISSUE ENGINEERING APPROACHES

The different approaches used in creating a tissue-engineered vascular substitute can be divided into three scaffolding approaches: two of which employ scaffolds that are either biologic or synthetic polymer in nature, and a third in which cells secrete ECM to form their own scaffold.

Since the time of Weinberg and Bell (Weinberg and Bell, 1986), type I collagen hydrogels have been explored for use in vascular tissue engineering. These are a popular choice as the hydrogels are easy to mold, provide a more natural environment, and are free of cytotoxic degradation by-products. Smooth muscle cells or fibroblasts can be easily encapsulated with the hydrogel, and will compact and remodel the gel; endothelial cells can be seeded and will form confluent monolayers on the lumen (L'Heureux et al., 1993). Figure II.6.9.4 shows the process used for creating a tubular cell seeded hydrogel. Due to the ease of fabrication, hydrogels have been created using fibrin (Grassl et al., 2003) and collagen–fibrin mixtures (Cummings et al., 2004; Rowe and Stegemann, 2006), resulting in improved strength and collagen production. While fibrin-based tissue has demonstrated substantial strength, ε-amino caproic acid (ACA) is needed to prevent proteolytic degradation; ACA concentration is an important factor as it also affects the fibrin gel compaction and strength (Rowe et al., 2007). The hydrogel scaffolds are quite weak, so several strategies have been used for improving mechanical properties. Support sleeves of poly(ethyleneterphthalate) (Dacron®) (Weinberg and Bell, 1986), and cross-linked collagen (Berglund et al., 2003) have all been used. Several non-cytotoxic cross-linking agents have been investigated to improve strength. Methylacrylamide photo-cross-linking (Brinkman et al., 2003), enzymatic cross-linking by transglutaminase (Orban et al., 2004), overexpressed lysyl oxidase (Elbjeirami et al., 2003), and non-enzymatic glycation (Girton et al., 1999) have also been employed to improve collagen gel strength. Biochemical stimulation such as with transforming growth factor-β (TGF-β), insulin, and aprotinin have all been shown to improve the mechanical properties of the tissues (Neidert et al., 2002; Yao et al., 2005; Williams et al., 2006).

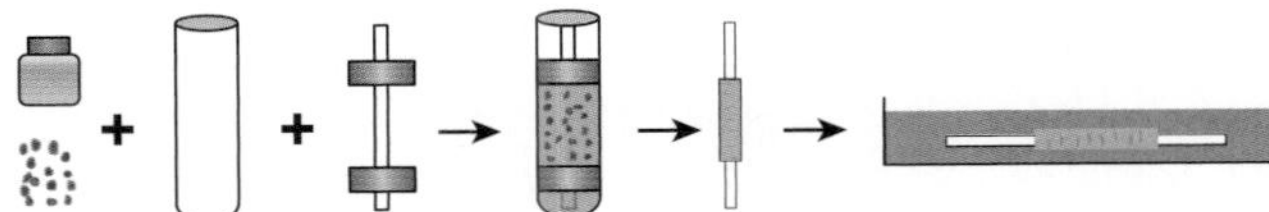

FIGURE II.6.9.4 Creation of a hydrogel-based engineered blood vessel. Smooth muscle cells are mixed with the solubolized ECM solution. The solution is placed into a tubular mold and incubated. After the solution gels, approximately 1 hour, the tissue is removed from the outer mold and placed into culture.

ECM proteins can also be formed into tubular structures using other techniques. Collagen has been nanopatterned, improving the strength of the tissue to levels close to that of native arteries (Zorlutuna et al., 2009). Silk fibroin has been electrospun into tubular forms and cell seeded (Zhang et al., 2009). Fibronectin–fibrinogen scaffolds have been fabricated using a dip coating/extrusion method (Harding et al., 2002). Woven chitosan has been coated with a chitosan/collagen layer (Zhang et al., 2006b). Esterified hyaluronic acid (HYAFF) has also been used as a scaffold for blood vessel tissue engineering. Tubular shapes have been made by suturing sheets into a tubular form (Remuzzi et al., 2004) or by coagulating a HYAFF coating in an ethanol bath and then air drying (Zavan et al., 2008). Grafts made by the latter method have been tested *in vivo* in a pig model. The HYAFF grafts promoted elastin synthesis and deposition, although several grafts were found to have partial or complete occlusion after harvesting.

TABLE II.6.9.2 Examples of Synthetic Materials Used in Vascular Tissue Engineering

Polymer	Comments	Reference
Poly(glycolic acid) (PGA)	Easily modified, can be synthesized in bulk; after dynamic culture provides strong, vasoactive tissue	Niklason et al., 1999
Poly(D, L-lactide)-7 co-(1,3-trimethylene carbonate)	Low cost microsponge; after dynamic culture mechanical properties were similar to human artery	Buttafoco et al., 2006b
Polydiaxonone (PDO) – Elastin	PDO has slower resorption rate with low inflammatory response and thrombogenicity. Elastin allows for cell adhesion and growth and improved elasticity	Sell et al., 2006
Poly(ethylene glycerol) (PEG)	Elastic with tunable mechanical properties; resists thrombosis; dissolves in aqueous solution, and can be photopolymerized in presence of cells	Hahn et al., 2007
Poly(1,8-octanediol citrate) (POC)	Elastomer with tunable compliance; can withstand physiological burst pressures	Webb et al., 2007
PGA/Polycaprolactone (PCL)	Two wrapped sheets; similar strength and compliance to native artery	Iwasaki et al., 2008
Poly(glycerol sebacate) (PGS)	Elastomer that promotes cell adhesion and proliferation; promotes elastin deposition in dynamic culture	Gao et al., 2008
Fibrin-Poly(lactide) (PLA)	Strong polymer for initial strength; fibrin for cell adhesion and promotion of ECM synthesis. Led to supraphysiological strength	Tschoeke et al., 2009
Poly(lactide-co-β-malic acid) with extended carboxyl arms (PLMA-ECA)	Improved hemocompatability and cell affinity compared to PLA	Wang et al., 2009

One limitation of many approaches is the lack of mature elastin. Most approaches appear to assume that mature elastin will be synthesized *in vivo*, although few approaches have demonstrated an ability for adult smooth muscle cells to produce, deposit, and appropriately assemble mature elastin in an engineered tissue. Several methods have been used to incorporate elastin into the engineered tissue prior to implantation. Intact elastin scaffolds can be isolated from arteries and incorporated into the tissue-engineered blood vessel (Berglund et al., 2004). Tubular scaffolds of collagen and elastin have been created by electrospinning (Buttafoco et al., 2006a).

To provide a more natural ECM environment, decellularized tissues have been used; decellularization methods are important and can affect the mechanical integrity of the tissue and the cell viability of seeded cells (Dahl et al., 2003; Jo et al., 2007). This approach also allows for the use of xenogeneic tissue that can be cellularized *in vitro* or could be populated with autologous cells *in vivo*. Use of a decellularized artery is a natural choice; it has similar mechanical properties as fresh artery (Liu et al., 2008b). Decellularized veins with venous valves have been explored, and have clinical application for grafting of peripheral veins such as the femoral vein (Teebken et al., 2009). Decellularized ureters have been explored as vascular grafts (Derham et al., 2008). Endothelialized decellularized ureter grafts have similar compliance as compared to the coronary artery, and remain patent in canine models (Narita et al., 2008). Small intestinal submucosa (SIS) has been tested for use as a vascular graft and has similar mechanical properties to native vessels (Roeder et al., 1999). The SIS grafts have been tested successfully *in vivo* in a growth model showing that they are not only able to arterialize, but also to adapt and grow in a pediatric model (Robotin-Johnson et al., 1998). The intestinal collagen layer of the small intestine was also used as a vascular graft. When treated with heparin, the small diameter grafts remained patent in a rabbit carotid (Huynh et al., 1999).

While use of ECM scaffolds is popular, synthetic polymer scaffolds do have some benefits and have been widely explored. The mechanical properties of synthetic scaffolds are tunable, and the resulting grafts can have "off-the-shelf" availability. Many of the materials can be made in bulk at low cost and are easily modified. Dacron® and ePTFE are popular choices for bypass grafting of larger vessels; however, they are not appropriate for small diameter blood vessel grafting or pediatric applications, since they do not remain patent and cannot be remodeled as needed for growth. For these applications biodegradable polymer scaffolds have been studied. The most widely used polymer is polyglycolic acid (PGA). Because it is rapidly resorbed, *in vitro* culture is required prior to implantation. There are many different single and combination polymers that have been tested for vascular tissue engineering applications. Table II.6.9.2 shows a variety of synthetic materials that are being used for vascular tissue engineering. Many are still in the development stages, while others listed in Table II.6.9.3 have progressed to the point of preclinical or clinical testing. There has been limited clinical success with grafts made with P(LA/CL) and either PGA or PLA which have been used in low pressure pediatric applications. There are concerns with synthetic materials though; cell adhesion and ingrowth are a challenge. To overcome this the polymers can be modified with adhesion motifs or combined

TABLE II.6.9.3 Tissue Engineering Approaches Tested *In Vivo*

Polymer	Comments	Reference
Tested *In Vivo*, Preclinical		
SIS	Used as carotic graft in canine model. After 2 months, all were patent. Remodeling led to changes in mechanical properties becoming close to that of the surrounding tissue	Roeder et al., 2001
PGA	Remained patent during 24-day study in porcine model	Niklason et al., 1999
PGA-Polyglactin	Remained patent for 24 weeks when implanted as a segment of ovine pulmonary artery	Shinoka et al., 1998
PGA-Polyhydroxyalkanoate (PHA)	Remained patent with no anyeurism for 5 months as ovine abdominal aorta segment	Shum-Tim et al., 1999
Autologous, cell self-assembled in peritoneal or pleural cavity	Used as femoral artery interposition graft in canine; 8 of 11 patent after 6.5 months	Chue et al., 2004
Decellularized umbilical artery coated with Poly(steryne sulfonate)/Poly(allylamine hydrochloride) (PSS/PAH)	Remained patent during 12-week study as carotid artery bypass in rabbit model	Kerdjoudj et al., 2008
PGA-Poly(4-hydroxybutyrate) (P4HB)	No stenosis or aneuryism for ovine pulmonary patch. Seeded with EPCs and mesenchymal stem cells	Mettler et al., 2008
PCL	Longer degradation time; 12 weeks in rat abdominal aorta showed cell in-growth with confluent endothelial cell monolayer by 12 weeks	Pektok et al., 2008
PGA/collagen-Poly(L-PLLA)	Patent for 12 months in porcine aorta	Torikai et al., 2008
Poly(ester-urea) urethane-Poly(2-meth-acryloyloxyethyl phosphorylcholine-co-methacryloxyethyl butylurethane) (PEUU-PMBU)	Addition of PMBU improved patency in rat aorta	Hong et al., 2009
PLLA coated with Poly(L-lactide-co-ε-caprolactone)(P(LA/CL))	Remained patent without anyeurism for 1 year as infrarenal aortic interposition graft in mice	Mirensky et al., 2009
PCL/collagen	Maintained structural integrity as rabbit aortoiliac bypass (1 month)	Tillman et al., 2009
Tested *In Vivo*, Clinical		
PGA or PLA with P(LA/CL)	Results reported for 23 patients as extracardiac total cavopulmonary connection. No graft related failures	Shinoka and Breuer, 2008
Autologous fibroblast cell secreted scaffold	60% patency in 8 patients after 6 months	McAllister et al., 2009

with other synthetic polymers or ECM proteins. Immunogenicity and thrombogenicity have been a problem for many of these materials, especially in the smaller diameter applications. Degradation by-products are a concern as well, as they may alter the local environment and lead to alterations in cell function or cell death (Athanasiou et al., 1996; Agrawal and Athanasiou, 1997). *In vitro* culture prior to implantation would allow for the material to degrade, but would require long culture times for complete degradation of the materials.

The third method has employed a cell secreted scaffolding material. The two notable approaches are the cell sheet approach, and the use of the body as a bioreactor. In the cell sheet approach, the smooth muscle cells or fibroblasts are cultured until they are post-confluent and have formed a tissue. The sheets are wrapped around an inner mandrel to create a tube, and cultured to promote adhesion of the layers. All three layers of the blood vessel can be created by wrapping the fibroblast layer around the smooth muscle layer and then seeding endothelial cells into the lumen (L'Heureux et al., 1998). In private communication this laboratory has discussed their modified approach using a single step to create the media and adventia layers. Using a compartmentalized tissue culture plate with a custom spacer, the smooth muscle cells and fibroblasts can form tissue. The spacer can then be removed to form a single layer of tissue with two distinct regions. This approach leads to a small reduction in culture time, and also results in better adhesion between the two layers. Use of a single rolling step also improves manufacturability. A similar cell sheet approach is being used by Cytograft Tissue Engineering Inc. to create autologous tissue. These tissue-engineered blood vessels use fibroblasts to create the medial layer; endothelial cells are seeded prior to implantation. Mechanical testing of an engineered blood vessel using fibroblast sheets only has been shown to have burst pressures comparable to artery, and compliance between that of an artery and vein (Konig et al., 2009). Figure II.6.9.5 shows the completely autologous tissue engineered AV shunt for dialysis prior to implantation.

In order to form a tissue-engineered blood vessel with autologous cells in a shorter amount of time, the use of the body as a bioreactor to create a blood-vessel-like structure was developed (Campbell et al., 1999). The body's foreign-body response forms granulation tissue around a silastic mold, resulting in a blood-vessel-like structure

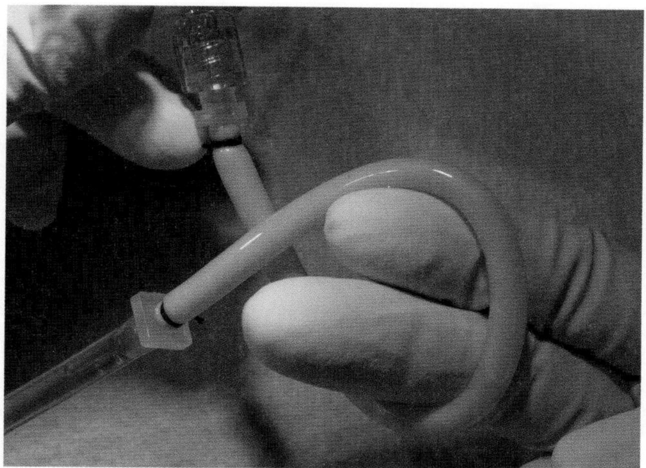

FIGURE II.6.9.5 Tissue engineered blood vessel created using cell sheet technology. Shown is the Cytograft tissue being tested as an AV shunt. (Reprinted with permission of Cytograft.)

with myofibroblasts and a monolayer of mesothelium. The tissue is removed from the tubing and everted so that the mesothelium is facing the lumen. In canines the tissue-engineered blood vessels cultured in the peritoneal cavity for three weeks resulted in a tissue with a burst strength over 2500 mmHg; when implanted as a graft in the same animals, 73% of the grafts remained patent (Chue et al., 2004). This approach is promising due to its short culture time and its use of autologous cells; however, off-the-shelf availability will likely be needed for some patients. Thus, even though this approach has relatively short culture times, it would not meet the needs of all patients.

CONCLUDING DISCUSSION

In spite of the progress reported in this chapter, there still is much that needs to be done, and there are many questions that investigators are trying to answer. As an example, what is the best source of cells? Should the approach be based on an allogeneic source so as to have off-the-shelf availability or an autologous source so as to bypass the immunogenicity issue? Or does one implant an acellular substitute and employ methods for recruiting the patient's own cells?

What is the best scaffold material to be used in fabricating the implant, whether it be an acellular one or one seeded with cells? Recognizing that the behavior of cells is orchestrated by a symphony of signals, a symphony that includes signaling from the extracellular matrix, does this make the biological scaffolds better or can this be incorporated into a synthetic scaffold by attaching appropriate ECM molecules to the surface of the synthetic material?

How can the appropriate physiological properties be achieved? This includes having mechanical properties that mimic those of native vessels, but what about vasoactivity? In the large vessels that have been the focus of efforts to tissue engineer a blood vessel and that are not resistance vessels, how important is it for this engineered vessel to be able to regulate its diameter?

Finally, what about vascularization? Will the wall of an engineered blood vessel be thick enough to require this? If so, what are the approaches that could lead to vascularization? And then the related question, what about neural control?

Thus there are many questions that remain, and that provide for the ongoing research that is taking place. Even so, there are some initial clinical studies, and these are starting to see some success in the arena of tissue-engineered blood vessels. The clinical studies to date have been very small in number. As the numbers grow and the applications increase, additional information and challenges will occur. Remodeling or host cell recruitment is key for most of the vascular tissue engineering approaches. Interpatient variability will have a significant impact on the rate of remodeling and the mechanical integrity of the tissue. Age, sex, health, and lifestyle of the patients will have significant effects. Degradation and synthesis of new ECM must be controlled so that the graft material does not weaken to the point of aneurysm or rupture.

Tissue-Engineered Blood Vessels in the Clinic

Advances in tissue engineering of blood vessels have resulted in recent clinical successes. For the pediatric population, synthetic vascular grafts of P(CL/LA) with PGA or PLA have been used successfully as grafts for extracardiac total cavopulmonary connection. This graft has "off-the-shelf" availability and a high success rate; only one of the 23 patients had any graft-related complications (Shinoka and Breuer, 2008). Despite its success, this material is not suited for all applications. The mechanical strength reported for these grafts is too low to be expected to survive in a high pressure environment, and significant stenosis has been found in conduits that are smaller than 18 mm.

Grafts using the cell sheet technique are in clinical studies as AV shunts (Cytograft Tissue Engineering, Inc., CA). There were promising results after six months in a population with a high risk of failure (McAllister et al., 2009). These living grafts use autologous fibroblasts that secrete their own matrix, resulting in tissue with burst strengths greater than 2000 mmHg. There were also several failures due to aneurysm in this high risk population. Additionally, the process to create the grafts is very long; the grafts in the published study took 6–9 months to create. The lack of "off-the-shelf" availability may limit the patient population that could benefit from the engineered tissue.

In closing, it is clear that considerable progress has been made in the last 10–15 years. There is even some initial clinical success. Even so, there is much still to do, and the tissue engineering of a blood vessel will remain a fertile area for research for years to come.

BIBLIOGRAPHY

Abilez, O., Benharash, P., Miyamoto, E., Gale, A., Xu, C., et al. (2006). P19 progenitor cells progress to organized contracting myocytes after chemical and electrical stimulation: Implications for vascular tissue engineering. *J. Endovasc. Ther.*, **13**, 377–388.

Agrawal, C. M., & Athanasiou, K. A. (1997). Technique to control pH in vicinity of biodegrading PLA-PGA implants. *J. Biomed. Mater. Res.*, **38**, 105–114.

American Heart Association (2008). *Heart Disease and Stroke Statistics–2008 Update*. Dallas, TX: American Heart Association.

Athanasiou, K. A., Niederauer, G. G., & Agrawal, C. M. (1996). Sterilization, toxicity, biocompatibility and clinical applications of polylactic acid/polyglycolic acid copolymers. *Biomaterials*, **17**, 93–102.

Bahn, A., Gupta, V., Kumar, S., Sharma, R., Singh, B., et al. (1999). Radial artery in CABG: Could the early results be comparable to internal mammary artery graft? *Ann. Thorac. Surg.*, **67**, 1631–1636.

Berglund, J. D., Mohseni, M. M., Nerem, R. M., & Sambanis, A. (2003). A biological hybrid model for collagen-based tissue engineered vascular constructs. *Biomaterials*, **24**, 1241–1254.

Berglund, J. D., Nerem, R. M., & Sambanis, A. (2004). Incorporation of intact elastin scaffolds in tissue-engineered collagen-based vascular grafts. *Tissue Eng.*, **10**, 1526–1535.

Bettinger, C. J., Orrick, B., Misra, A., Langer, R., & Borenstein, J. T. (2006). Microfabrication of poly(glycerol-sebacate) for contact guidance applications. *Biomaterials*, **27**, 2558–2565.

Bettinger, C. J., Cyr, K. M., Matsumoto, A., Langer, R., Borenstein, J. T., et al. (2007). Silk fibroin microfluidic devices. *Adv. Mater. Deerfield*, **19**, 2847–2850.

Borenstein, J. T., Terai, H., King, K. R., Weinberg, E. J., Kaazempur-Mofrad, M. R., et al. (2002). Microfabrication technology for vascularized tissue engineering. *Biomed. Microdevices*, **4**, 167–175.

Bravata, D. M., Gienger, A. L., McDonald, K. M., Sundaram, V., Perez, M. V., et al. (2007). Systematic review: The comparative effectiveness of percutaneous coronary interventions and coronary artery bypass graft surgery. *Ann. Intern. Med.*, **147**, 703–716.

Brinkman, W. T., Nagapudi, K., Thomas, B. S., & Chaikof, E. L. (2003). Photo-cross-linking of type I collagen gels in the presence of smooth muscle cells: Mechanical properties, cell viability, and function. *Biomacromolecules*, **4**, 890–895.

Buttafoco, L., Kolkman, N. G., Engbers-Buijtenhuijs, P., Poot, A. A., Dijkstra, P. J., et al. (2006a). Electrospinning of collagen and elastin for tissue engineering applications. *Biomaterials*, **27**, 724–734.

Buttafoco, L., Boks, N. P., Engbers-Buijtenhuijs, P., Grijpma, D. W., Poot, A. A., et al. (2006b). Porous hybrid structures based on P(DLLA-co-TMC) and collagen for tissue engineering of small-diameter blood vessels. *J. Biomed. Mater. Res. B. Appl. Biomater.*, **79**, 425–434.

Campbell, J. H., Efendy, J. L., & Campbell, G. R. (1999). Novel vascular graft grown within recipient's own peritoneal cavity. *Circ. Res.*, **85**, 1173–1178.

Canver, C. C. (1995). Conduit options in coronary artery bypass surgery. *Chest*, **108**, 1150–1155.

Chue, W. L., Campbell, G. R., Caplice, N., Muhammed, A., Berry, C. L., et al. (2004). Dog peritoneal and pleural cavities as bioreactors to grow autologous vascular grafts. *J. Vasc. Surg.*, **39**, 859–867.

Cummings, C. L., Gawlitta, D., Nerem, R. M., & Stegemann, J. P. (2004). Properties of engineered vascular constructs made from collagen, fibrin, and collagen-fibrin mixtures. *Biomaterials*, **25**, 3699–3706.

Dahl, S. L., Koh, J., Prabhakar, V., & Niklason, L. E. (2003). Decellularized native and engineered arterial scaffolds for transplantation. *Cell Transplant*, **12**, 659–666.

Derham, C., Yow, H., Ingram, J., Fisher, J., Ingham, E., et al. (2008). Tissue engineering small-diameter vascular grafts: Preparation of a biocompatible porcine ureteric scaffold. *Tissue Eng. Part A*, **14**, 1871–1882.

Dixon, S. R., Grines, C. L., & O'Neill, W. W. (2008). The year in interventional cardiology. *J. Am. Coll. Cardiol.*, **51**, 2355–2369.

Edelman, E. R. (1999). Vascular tissue engineering: Designer arteries. *Circ. Res.*, **85**, 1115–1117.

Elbjeirami, W. M., Yonter, E. O., Starcher, B. C., & West, J. L. (2003). Enhancing mechanical properties of tissue-engineered constructs via lysyl oxidase crosslinking activity. *J. Biomed. Mater. Res. A.*, **66**, 513–521.

Fisher, A. B., Chien, S., Barakat, A. I., & Nerem, R. M. (2001). Endothelial cellular response to altered shear stress. *Am. J. Physiol. Lung Cell Mol. Physiol.*, **281**, L529–L533.

Gao, J., Crapo, P., Nerem, R., & Wang, Y. (2008). Co-expression of elastin and collagen leads to highly compliant engineered blood vessels. *J. Biomed. Mater. Res. A.*, **85**, 1120–1128.

Girton, T. S., Oegema, T. R., & Tranquillo, R. T. (1999). Exploiting glycation to stiffen and strengthen tissue equivalents for tissue engineering. *J. Biomed. Mater. Res.*, **46**, 87–92.

Gong, Z., & Niklason, L. E. (2008). Small-diameter human vessel wall engineered from bone marrow-derived mesenchymal stem cells (hMSCs). *FASEB J.*, **22**, 1635–1648.

Grassl, E. D., Oegema, T. R., & Tranquillo, R. T. (2003). A fibrin-based arterial media equivalent. *J. Biomed. Mater. Res. A.*, **66**, 550–561.

Guillemette, M. D., Gauvin, R. J., Perron, C., Germain, L., & Auger, F. A. (2008). Tissue engineered vascular adventitia with vasa vasorum drastically improve implanted graft perfusion via inosculation. San Diego, CA: TERMIS-NA 2008 Annual Conference.

Hahn, M. S., McHale, M. K., Wang, E., Schmedlen, R. H., & West, J. L. (2007). Physiologic pulsatile flow bioreactor conditioning of poly(ethylene glycol)-based tissue engineered vascular grafts. *Ann. Biomed. Eng.*, **35**, 190–200.

Harding, S. I., Afoke, A., Brown, R. A., MacLeod, A., Shamlou, P. A., et al. (2002). Engineering and cell attachment properties of human fibronectin–fibrinogen scaffolds for use in tissue engineered blood vessels. *Bioprocess Biosyst. Eng.*, **25**, 53–59.

Hayward, P. A., Hare, D. L., Gordon, I., & Buxton, B. F. (2008). Effect of radial artery or saphenous vein conduit for the second graft on 6-year clinical outcome after coronary artery bypass grafting. Results of a randomised trial. *Eur. J. Cardiothorac. Surg.*, **34**, 113–117.

Hoerstrup, S. P., Zund, G., Sodian, R., Schnell, A. M., Grunenfelder, J., et al. (2001). Tissue engineering of small caliber vascular grafts. *Eur. J. Cardiothorac. Surg.*, **20**, 164–169.

Hong, Y., Ye, S. H., Nieponice, A., Soletti, L., Vorp, D. A., et al. (2009). A small diameter, fibrous vascular conduit generated from a poly(ester urethane)urea and phospholipid polymer blend. *Biomaterials*, **30**(13), 2457–2467.

Huynh, T., Abraham, G., Murray, J., Brockbank, K., Hagen, P. O., et al. (1999). Remodeling of an acellular collagen graft into a physiologically responsive neovessel. *Nat. Biotechnol.*, **17**, 1083–1086.

Isenberg, B. C., & Tranquillo, R. T. (2003). Long-term cyclic distention enhances the mechanical properties of collagen-based media-equivalents. *Ann. Biomed. Eng.*, **31**, 937–949.

Iwasaki, K., Kojima, K., Kodama, S., Paz, A. C., Chambers, M., et al. (2008). Bioengineered three-layered robust and elastic artery using hemodynamically-equivalent pulsatile bioreactor. *Circulation*, **118**, S52–S57.

Jo, W. M., Sohn, Y. S., Choi, Y. H., Kim, H. J., & Cho, H. D. (2007). Modified acellularization for successful vascular xenotransplantation. *J. Korean Med. Sci.*, **22**, 262–269.

Jordan, S. W., & Chaikof, E. L. (2007). Novel thromboresistant materials. *J. Vasc. Surg.*, **45**(Suppl. A), A104–A115.

Kerdjoudj, H., Berthelemy, N., Rinckenbach, S., Kearney-Schwartz, A., Montagne, K., et al. (2008). Small vessel replacement by human umbilical arteries with polyelectrolyte film-treated arteries: *In vivo* behavior. *J. Am. Coll. Cardiol.*, **52**, 1589–1597.

Konig, G., McAllister, T. N., Dusserre, N. A., Garrido, S. A., Iyican, C., et al. (2009). Mechanical properties of completely autologous human tissue engineered blood vessels compared to human saphenous vein and mammary artery. *Biomaterials*, **30**, 1542–1550.

L'Heureux, N., Germain, L., Labbe, R., & Auger, F. A. (1993). *In vitro* construction of a human blood vessel from cultured vascular cells: A morphologic study. *J. Vasc. Surg.*, **17**, 499–509.

L'Heureux, N., Paquet, S., Labbe, R., Germain, L., & Auger, F. A. (1998). A completely biological tissue-engineered human blood vessel. *FASEB J.*, **12**, 47–56.

Liu, J. Y., Peng, H. F., & Andreadis, S. T. (2008a). Contractile smooth muscle cells derived from hair-follicle stem cells. *Cardiovasc. Res.*, **79**, 24–33.

Liu, G. F., He, Z. J., Yang, D. P., Han, X. F., Guo, T. F., et al. (2008b). Decellularized aorta of fetal pigs as a potential scaffold for small diameter tissue engineered vascular graft. *Chin. Med. J. (Engl)*, **121**, 1398–1406.

Malvindi, P. G., Dunning, J., & Vitale, N. (2008). For which patients with left main stem disease is percutaneous intervention rather than coronary artery bypass grafting the better option? *Interact Cardiovasc. Thorac. Surg.*, **7**, 306–314.

Mantero, S., Sadr, N., Riboldi, S. A., Lorenzoni, S., & Montevecchi, F. M. (2007). A new electro-mechanical bioreactor for soft tissue engineering. *J. Applied Biomat. Biomech.*, **5**, 107–116.

McAllister, T. N., Maruszewski, M., Garrido, S. A., Wystrychowski, W., Dusserre, N., et al. (2009). Effectiveness of haemodialysis access with an autologous tissue-engineered vascular graft: A multicentre cohort study. *Lancet*, **373**, 1440–1446.

McCulloch, A. D., Harris, A. B., Sarraf, C. E., & Eastwood, M. (2004). New multi-cue bioreactor for tissue engineering of tubular cardiovascular samples under physiological conditions. *Tissue Eng.*, **10**, 565–573.

Mettler, B. A., Sales, V. L., Stucken, C. L., Anttila, V., Mendelson, K., et al. (2008). Stem cell-derived, tissue-engineered pulmonary artery augmentation patches *in vivo*. *Ann. Thorac. Surg.*, **86**, 132–140.

Mirensky, T. L., Nelson, G. N., Brennan, M. P., Roh, J. D., Hibino, N., et al. (2009). Tissue-engineered arterial grafts: Long-term results after implantation in a small animal model. *J. Pediatr. Surg.*, **44**, 1127–1132.

Mironov, V., Kasyanov, V., McAllister, K., Oliver, S., Sistino, J., et al. (2003). Perfusion bioreactor for vascular tissue engineering with capacities for longitudinal stretch. *J. Craniofac. Surg.*, **14**, 340–347.

Narita, Y., Kagami, H., Matsunuma, H., Murase, Y., Ueda, M., et al. (2008). Decellularized ureter for tissue-engineered small-caliber vascular graft. *J. Artif. Organs*, **11**, 91–99.

Neidert, M. R., Lee, E. S., Oegema, T. R., & Tranquillo, R. T. (2002). Enhanced fibrin remodeling *in vitro* with TGF-beta1, insulin and plasmin for improved tissue-equivalents. *Biomaterials*, **23**, 3717–3731.

Nerem, R. M., & Ensley, A. E. (2004). The tissue engineering of blood vessels and the heart. *Am. J. Transplant*, **4**(Suppl. 6), 36–42.

Nieponice, A., Soletti, L., Guan, J., Deasy, B. M., Huard, J., et al. (2008). Development of a tissue-engineered vascular graft combining a biodegradable scaffold, muscle-derived stem cells and a rotational vacuum seeding technique. *Biomaterials*, **29**, 825–833.

Niklason, L. E., Gao, J., Abbott, W. M., Hirschi, K. K., Houser, S., et al. (1999). Functional arteries grown *in vitro*. *Science*, **284**, 489–493.

Orban, J. M., Wilson, L. B., Kofroth, J. A., El-Kurdi, M. S., Maul, T. M., et al. (2004). Crosslinking of collagen gels by transglutaminase. *J. Biomed. Mater. Res. A.*, **68**, 756–762.

Pektok, E., Nottelet, B., Tille, J. C., Gurny, R., Kalangos, A., et al. (2008). Degradation and healing characteristics of small-diameter poly(epsilon-caprolactone) vascular grafts in the rat systemic arterial circulation. *Circulation*, **118**, 2563–2570.

Petrossian, E., Reddy, V. M., Collins, K. K., Culbertson, C. B., MacDonald, M. J., et al. (2006). The extracardiac conduit Fontan operation using minimal approach extracorporeal circulation: Early and midterm outcomes. *J. Thorac. Cardiovasc. Surg.*, **132**, 1054–1063.

Remuzzi, A., Mantero, S., Colombo, M., Morigi, M., Binda, E., et al. (2004). Vascular smooth muscle cells on hyaluronic acid: Culture and mechanical characterization of an engineered vascular construct. *Tissue Eng.*, **10**, 699–710.

Richards Grayson, A. C., Choi, I. S., Tyler, B. M., Wang, P. P., Brem, H., et al. (2003). Multi-pulse drug delivery from a resorbable polymeric microchip device. *Nat. Mater.*, **2**, 767–772.

Robotin-Johnson, M. C., Swanson, P. E., Johnson, D. C., Schuessler, R. B., & Cox, J. L. (1998). An experimental model of small intestinal submucosa as a growing vascular graft. *J. Thorac. Cardiovasc. Surg.*, **116**, 805–811.

Roeder, R., Wolfe, J., Lianakis, N., Hinson, T., Geddes, L. A., et al. (1999). Compliance, elastic modulus, and burst pressure of small-intestine submucosa (SIS), small-diameter vascular grafts. *J. Biomed. Mater Res.*, **47**, 65–70.

Roeder, R. A., Lantz, G. C., & Geddes, L. A. (2001). Mechanical remodeling of small-intestine submucosa small-diameter vascular grafts: A preliminary report. *Biomed. Instrum. Technol.*, **35**, 110–120.

Rosamond, W., Flegal, K., Furie, K., Go, A., Greenlund, K., et al. (2008). Heart disease and stroke statistics – 2008 update: A report from the American Heart Association Statistics Committee and Stroke Statistics Subcommittee. *Circulation*, **117**, e25–e146.

Rowe, S. L., & Stegemann, J. P. (2006). Interpenetrating collagen-fibrin composite matrices with varying protein contents and ratios. *Biomacromolecules*, **7**, 2942–2948.

Rowe, S. L., Lee, S., & Stegemann, J. P. (2007). Influence of thrombin concentration on the mechanical and morphological properties of cell-seeded fibrin hydrogels. *Acta. Biomater.*, **3**, 59–67.

Seliktar, D., Black, R. A., Vito, R. P., & Nerem, R. M. (2000). Dynamic mechanical conditioning of collagen-gel blood vessel constructs induces remodeling *in vitro*. *Ann. Biomed. Eng.*, **28**, 351–362.

Seliktar, D., Nerem, R. M., & Galis, Z. S. (2003). Mechanical strain-stimulated remodeling of tissue-engineered blood vessel constructs. *Tissue Eng.*, **9**, 657–666.

Sell, S. A., McClure, M. J., Barnes, C. P., Knapp, D. C., Walpoth, B. H., et al. (2006). Electrospun polydioxanone-elastin blends: Potential for bioresorbable vascular grafts. *Biomed. Mater.*, **1**, 72–80.

Shinoka, T., & Breuer, C. (2008). Tissue-engineered blood vessels in pediatric cardiac surgery. *Yale J. Biol. Med.*, **81**, 161–166.

Shinoka, T., Shum-Tim, D., Ma, P. X., Tanel, R. E., Isogai, N., et al. (1998). Creation of viable pulmonary artery autografts through tissue engineering. *J. Thorac. Cardiovasc. Surg.*, **115**, 536–545.

Shum-Tim, D., Stock, U., Hrkach, J., Shinoka, T., Lien, J., et al. (1999). Tissue engineering of autologous aorta using a new biodegradable polymer. *Ann. Thorac. Surg.*, **68**, 2298–2304.

Sodian, R., Lemke, T., Fritsche, C., Hoerstrup, S. P., Fu, P., et al. (2002). Tissue-engineering bioreactors: A new combined cell-seeding and perfusion system for vascular tissue engineering. *Tissue Eng.*, **8**, 863–870.

Stegemann, J. P., & Nerem, R. M. (2003). Phenotype modulation in vascular tissue engineering using biochemical and mechanical stimulation. *Ann. Biomed. Eng.*, **31**, 391–402.

Syedain, Z. H., Weinberg, J. S., & Tranquillo, R. T. (2008). Cyclic distension of fibrin-based tissue constructs: Evidence of adaptation during growth of engineered connective tissue. *Proc. Natl. Acad. Sci. USA*, **105**, 6537–6542.

Taggart, D. P. (2007). Coronary artery bypass graft vs. percutaneous coronary angioplasty: CABG on the rebound? *Curr. Opin. Cardiol.*, **22**, 517–523.

Teebken, O. E., Puschmann, C., Breitenbach, I., Rohde, B., Burgwitz, K., et al. (2009). Preclinical development of tissue-engineered vein valves and venous substitutes using re-endothelialised human vein matrix. *Eur. J. Vasc. Endovasc. Surg.*, **37**, 92–102.

Tillman, B. W., Yazdani, S. K., Lee, S. J., Geary, R. L., Atala, A., et al. (2009). The *in vivo* stability of electrospun polycaprolactone-collagen scaffolds in vascular reconstruction. *Biomaterials*, **30**, 583–588.

Torikai, K., Ichikawa, H., Hirakawa, K., Matsumiya, G., Kuratani, T., et al. (2008). A self-renewing, tissue-engineered vascular graft for arterial reconstruction. *J. Thorac. Cardiovasc. Surg.*, **136**, 37–45. 45 e1.

Tschoeke, B., Flanagan, T. C., Koch, S., Harwoko, M. S., Deichmann, T., et al. (2009). Tissue-engineered small-caliber vascular graft based on a novel biodegradable composite fibrin-polylactide scaffold. *Tissue Eng. Part A*, **15**, 1909–1918.

Wang, W., Liu, Y., Wang, J., Jia, X., Wang, L., et al. (2009). A novel copolymer poly(lactide-co-beta-malic acid) with extended carboxyl arms offering better cell affinity and hemacompatibility for blood vessel engineering. *Tissue Eng. Part A*, **15**, 65–73.

Webb, A. R., Macrie, B. D., Ray, A. S., Russo, J. E., Siegel, A. M., et al. (2007). *In vitro* characterization of a compliant biodegradable scaffold with a novel bioreactor system. *Ann. Biomed. Eng.*, **35**, 1357–1367.

Weinberg, C. B., & Bell, E. (1986). A blood vessel model constructed from collagen and cultured vascular cells. *Science*, **231**, 397–400.

Williams, C., & Wick, T. M. (2004). Perfusion bioreactor for small diameter tissue-engineered arteries. *Tissue Eng.*, **10**, 930–941.

Williams, C., Johnson, S. L., Robinson, P. S., & Tranquillo, R. T. (2006). Cell sourcing and culture conditions for fibrin-based valve constructs. *Tissue Eng.*, **12**, 1489–1502.

Yao, L., Swartz, D. D., Gugino, S. F., Russell, J. A., & Andreadis, S. T. (2005). Fibrin-based tissue-engineered blood vessels: Differential effects of biomaterial and culture parameters on mechanical strength and vascular reactivity. *Tissue Eng.*, **11**, 991–1003.

Zaucha, M. T., Raykin, J., Wan, W., Gauvin, R., Auger, F. A., et al. (2009). A novel cylindrical biaxial computer controlled bioreactor and biomechanical testing device for vascular tissue engineering. *Tissue Eng. Part A*, **15**(11), 3331–3340.

Zavan, B., Vindigni, V., Lepidi, S., Iacopetti, I., Avruscio, G., et al. (2008). Neoarteries grown *in vivo* using a tissue-engineered hyaluronan-based scaffold. *FASEB J.*, **22**, 2853–2861.

Zhang, J., Qi, H., Wang, H., Hu, P., Ou, L., et al. (2006a). Engineering of vascular grafts with genetically modified bone marrow mesenchymal stem cells on poly (propylene carbonate) graft. *Artif. Organs*, **30**, 898–905.

Zhang, L., Ao, Q., Wang, A., Lu, G., Kong, L., et al. (2006b). A sandwich tubular scaffold derived from chitosan for blood vessel tissue engineering. *J. Biomed. Mater. Res. A.*, **77**, 277–284.

Zhang, X., Wang, X., Keshav, V., Johanas, J. T., Leisk, G. G., et al. (2009). Dynamic culture conditions to generate silk-based tissue-engineered vascular grafts. *Biomaterials*, **30**, 3213–3223.

Zorlutuna, P., Elsheikh, A., & Hasirci, V. (2009). Nanopatterning of collagen scaffolds improve the mechanical properties of tissue engineered vascular grafts. *Biomacromolecules*, **10**, 814–821.

CHAPTER II.6.10 HEART VALVE TISSUE ENGINEERING

Frederick J. Schoen[1] and Simon P. Hoerstrup[2]

[1]Professor of Pathology and Health Sciences and Technology (HST), Harvard Medical School, Executive Vice Chairman, Department of Pathology, Brigham and Women's Hospital, Boston, MA, USA

[2]Cardiovascular Surgery Research University and University Hospital Zurich, Zurich, Switzerland

As summarized in Chapter II.5.3B of this book, diseased heart valves are usually replaced by mechanical prostheses or tissue-derived valves. Heart valve replacement is today a common procedure that is lifesaving and enhances quality of life in many patients (Rahimtoola, 2010). However, there remain key limitations to the uniform success of currently available substitute heart valve technology. In particular, mechanical prosthetic valves necessitate long-term anticoagulation to reduce thromboembolic complications; as a consequence, patients so treated assume a significant risk of hemorrhagic complications (Schoen, 2001). Moreover, tissue failure by cuspal mineralization and non-calcific structural damage frequently leads to failure of bioprostheses fabricated from glutaraldehyde fixed (and hence non-viable) porcine aortic valve or bovine pericardium (see Chapter II.4.5 of this book and Schoen and Levy, 2005). There is also evidence that the recently introduced bioprosthetic heart valves implanted through a catheter may suffer dysfunctional degeneration due to procedure-related mechanical damage as a consequence of prosthesis "crimping" (Noble et al., 2009; Zegdi et al., 2011 as well as the typical bioprosthesis-related complications). Aortic heart valve allografts derived from deceased humans used as transplants are severely limited in availability, and suffer durability limitations (Mitchell et al., 1998). Thus, currently available substitute valves have characteristic complications, and neither mechanical nor bioprosthetic devices, nor allograft valves permit growth of the prosthesis to accommodate the somatic growth of the recipient. In patients with congenital heart disease, this often necessitates multiple valve replacement operations during childhood and adolescence, to permit implantation of successively larger valves to match patient growth. Indeed, much of the motivation for the tissue engineering approach to heart valve replacement has derived from the particular need for heart valve replacement in children. Nearly 20,000 children are born with congenital heart defects each year in the United States; many congenital cardiac defects are either valve-related or require surgical intervention that replaces or repairs a valve. Examples are congenital aortic or pulmonary valve (AV or PV) stenosis, or complex congenital cardiac anomalies (such as tetralogy of Fallot), which include reconstruction of the right ventricular outflow tract with a valved conduit. These conduits often fail. Moreover, in this young patient population with congenital heart disease, the anticoagulation required with mechanical valves is particularly dangerous, and tissue valve substitutes undergo accelerated calcification (see Chapter II.5.3.B of this book).

Recent scientific and technological progress has stimulated the goal of generating a living valve replacement device that would not only obviate the limitations of conventional valve replacement, but would also have viable cells that would permit not only adaptation to changing environmental conditions in the recipient but also growth of the valve with a rapidly growing patient. Moreover, owing to the understanding that the mechanisms of bioprosthetic and other tissue valve calcification are potentiated by non-viable components (see Chapter II.4.5 of this book), a healthy, living valve would likely not suffer calcification. Innovative work toward a living valve fabricated from engineered tissue is active in many laboratories, and may eventually lead to clinical application.

> … much of the motivation for the tissue engineering approach to heart valve replacement has derived from the particular need for heart valve replacement in children. Recent scientific and technological progress has stimulated the goal of generating a living valve replacement device that would … have viable cells and thereby permit adaptation to changing environmental conditions in the recipient, and growth with a growing patient.

This chapter provides a contemporary review of key concepts of heart valve tissue engineering (HVTE). We describe the evolving understanding of heart valve function and biology (which provides key design criteria), the promise and difficulties demonstrated by *in vivo* studies done to date, and the critical challenges that will be encountered and hopefully surmounted in translating the promise of this potentially exciting therapeutic modality from the laboratory to the clinical realm. The reader is referred to comprehensive recent reviews of HVTE for additional information (Breuer et al., 2004; Vesely, 2005; Mendelson and Schoen, 2006; Schoen, 2008, 2011; Hjortnaes et al., 2009; Sacks et al., 2009a; Weber et al., 2011a).

DESIGN CRITERIA AND CHALLENGES IN TISSUE-ENGINEERED HEART VALVE (TEHV)

The design criteria and characteristics for engineered tissue valve substitutes were originally articulated for conventional replacement heart valves in 1962 by pioneering heart surgeon Dr. Dwight Harken (Harken et al., 1962), a modified formulation is shown in Table II.6.10.1. The realization of the HVTE principle requires a living, functional tissue whose cellular components can assume a homeostatic functional state, repair ongoing structural injury to cells and extracellular matrix (ECM), and accommodate somatic growth of the recipient.

Key Structure–Function Correlations in the Cardiac Valves

Heart valves are structural specializations of the cardiac tissues that ensure unidirectional blood flow through the heart; they open and close approximately 40 million times a year and three billion times over an average lifetime. Normal valves are free of obstruction in the open position, and without regurgitation in the closed position. The motions of the heart valves during opening and closing are driven by mechanical forces exerted by the surrounding blood and heart. Heart valve opening and closing induces repetitive and substantial changes in the shape, dimensions, and stress of the leaflets and supporting valvular structures (Figure II.6.10.1) (Sacks and Yoganathan, 2007; Schoen, 2008; Sacks et al., 2009b). Their ability to permit unobstructed forward flow depends on the mobility, pliability, and structural integrity of the moving parts of the valves; these are delicate flaps, that are generally called *leaflets* (in the tricuspid valve (TV) and mitral valve (MV)) and *cusps* (in the pulmonary valve (PV) and aortic valve (AV)). An engineered tissue valve and its components must not only accommodate those deformations, but must also have ongoing strength, flexibility, and durability, beginning at

TABLE II.6.10.1 Design Objectives For and Characteristics of Replacement Heart Valves

Feature to Optimize	Conventional (Mechanical, Bioprosthetic)	Tissue-Engineered
Closure of leaflets	Rapid and complete	Rapid and complete
Size of orifice area	Less than that of natural valves	Better
Mechanical properties	Initially suitable and stable	Initially suitable and stable
Surgical insertion	Easy and permanent	Easy and permanent
Risk of thrombosis	Yes, especially mechanical valves, which require anticoagulation, causing vulnerability to hemorrhage	No, endothelial surface to inhibit thrombogenesis
Risk of structural dysfunction	Degradation of synthetic materials rare with mechanical valves Tissue degradation and calcification of leaflets with bioprosthetic valves	Resistant to degradation, calcification
Risk of infection	Ever present	Resistant to infection
Cellular viability	No	Yes, able to repair injury, remodel, and potentially grow with patient

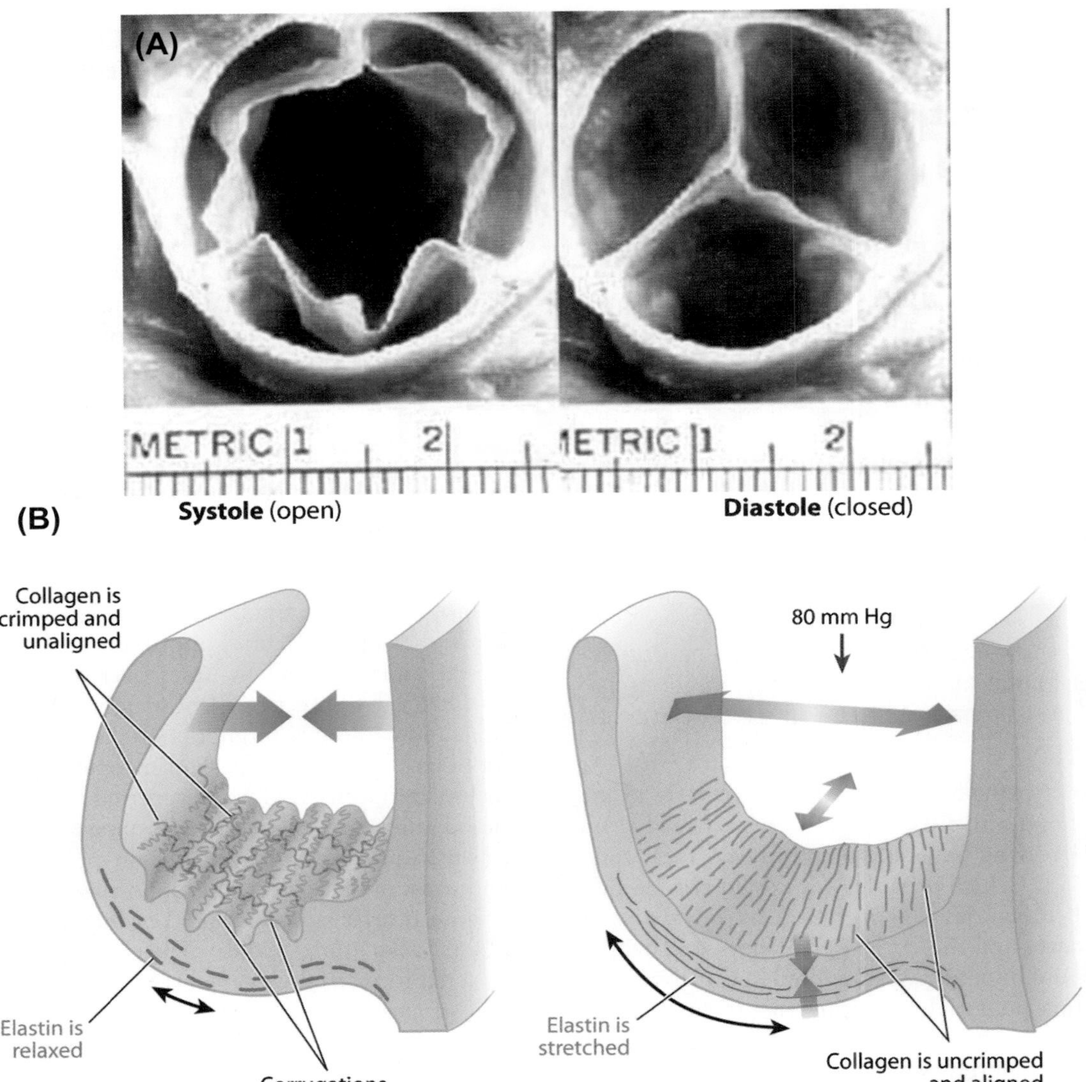

FIGURE II.6.10.1 Aortic valve functional structure at both macro- and microscopic levels. (A) Outflow aspect of aortic valve in open (left) and closed (right) configurations, corresponding to systole and diastole, respectively. (B) Schematic representation of architecture and configuration of collagen and elastin in systole and diastole. ((A) From Schoen, F. J. & Edwards, W. D. (2001). Valvular heart disease: General principles and stenosis. In: *Cardiovascular Pathology,* 3rd ed., Silver, M. D., Gotlieb, A. I. & Schoen, F. J. (Eds.). W. B. Saunders, pp. 402–442. (B) Reprinted, with permission, from Schoen FJ: Mechanisms of function and disease in natural and replacement heart valves. Ann Rev Mech Dis, 2012; 7:161-183, by Annual Reviews www.annualreviews.org.)

the instant of implantation and continuing indefinitely thereafter, despite potentially evolving tissue architecture owing to remodeling. This key requirement, coupled with the realization that the consequences of failure are likely to be serious, renders the heart valves a particularly challenging application for an engineered tissue.

Most frequently diseased, most frequently substituted (often from AV derived from animals or deceased humans), and most widely studied, the AV provides a paradigm for functional valvular architectural specialization, tissue dynamics, and biological interactions. Individual AV cusps (of which there are normally three) attach to the aortic wall in a crescentic (or semilunar) fashion, ascending to the commissures (where adjacent cusps come together at the aorta), and descending to the basal attachment of each cusp to the aortic wall. Behind the cusps are dilated pockets in the aortic root, called *sinuses of Valsalva*.

Moreover, the cusps and leaflets of the heart valves are vital and dynamic tissues that respond to changes in local mechanical forces and other environmental conditions (Schoen, 2008). The combination of substantial changes in the size and shape of the valve cusps during the second-to-second movements of the cardiac cycle and sufficient strength and durability to withstand repetitive and substantial mechanical stress and strain over billions of cycles is accomplished by a highly dynamic and complex internal microarchitecture, consisting of specialized cells and extracellular matrix (ECM). The four cardiac valves have a qualitatively similar, layered architectural

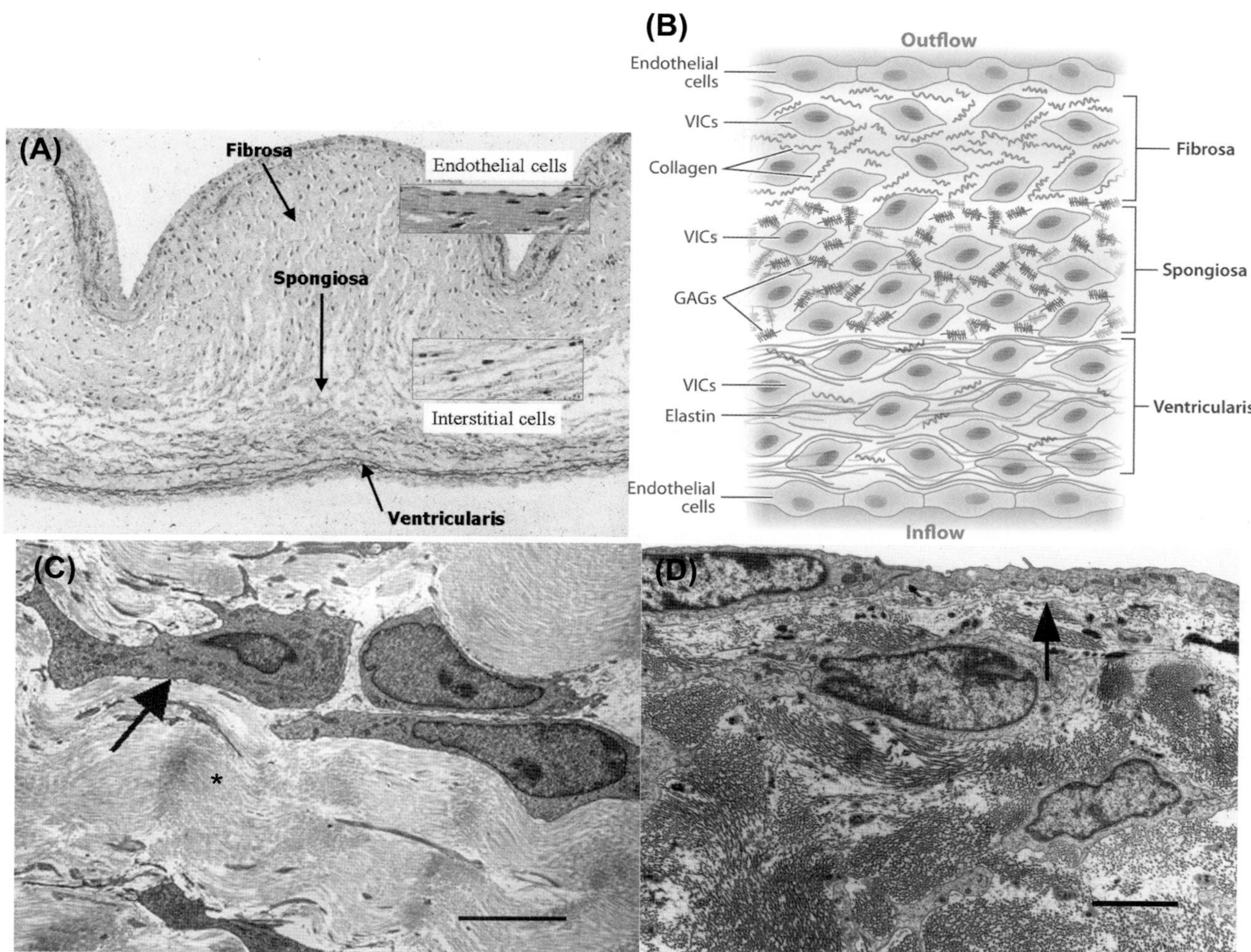

FIGURE II.6.10.2 Aortic valve histology and ultrastructure. (A) Aortic valve histology, shown as low-magnification photomicrograph of cross-section cuspal configuration in the nondistended state (corresponding to systole), emphasizing the three major layers: ventricularis, spongiosa, and fibrosa. (B) Schematic tissue architecture at the level of cells and extracellular matrix, highlighting layered structure and layer-specific cellular components and ECM. In (A) and (B) the outflow surface is at the top. (A) Original magnification 100 ×. Movat pentachrome stain (collagen is yellow; elastin is black). (C) Transmission electron photomicrograph of fresh porcine aortic valve at 0 mm Hg transvalvular pressure (characteristic of the systolic configuration), demonstrating the fibroblast morphology of valvular interstitial cells (VICs, arrow), the dense, surrounding closely apposed collagen (asterisk) with wavy crimp, and the potential for VIC–collagen and VIC–VIC interactions. (D) Transmission electron photomicrograph of the surface of the aortic valve demonstrating valvular endothelial cell (VEC; arrow) and proximity of deeper VICs, and potential for VEC–VIC interactions. In (C) and (D), bar = 5 microns. ((A) Reproduced by permission from Schoen, F. J. & Edwards, W. D. (2001). Valvular heart disease: General principles and stenosis. In: *Cardiovascular Pathology*, 3rd ed., Silver, M. D., Gotlieb, A. I. & Schoen, F. J. (Eds.). W. B. Saunders, pp. 402–442. (B) Reprinted, with permission, from Schoen FJ. Mechanisms of function and disease in natural and replacement heart valves. Ann Rev Path Mech Dis, 2012; 7:161–183 by Annual Reviews www.annualreviews.org. (C) Reproduced by permission from Schoen, F. J. (2008). Evolving concepts of cardiac valve dynamics: The continuum of development, functional structure, pathobiology and tissue engineering. *Circulation*, 118, 1864–1880. Copyright American Heart Association. (D) Reproduced by permission from Hilbert, S. L., Schoen, F. J. & Ferrans, V. J. (2004). Allograft heart valves: Morphologic, biomechanical and explant pathology studies. In: *Cardiac Reconstructions with Allograft Tissues*, 2nd ed., Hopkins, R. (Ed.). Springer-Verlag, Inc.: New York, NY, pp. 193–233.)

pattern composed of cells, including the valvular endothelial cells (VECs) at the blood-contacting surfaces and the deep valvular interstitial cell (VICs), and valvular extracellular matrix (VECM), including collagen, elastin, and amorphous ECM (predominately glycosaminoglycans (GAGs)) (Figure II.6.10.2 and Table II.6.10.2). Thus, the AV (and analogously the PV) has a dense collagenous layer close to the outflow surface and continuous with valvular supporting structures, which provides strength (the *fibrosa*); a central core of loose connective tissue (the *spongiosa*) rich in glycosoaminoglycans (GAGs); and a layer rich in elastin below the inflow surface (the *ventricularis*). PV structure is analogous to, but less robust than, that of the AV, consistent with the lower pressure environment (Joyce et al., 2009). Moreover, since heart valves are sufficiently thin (approximately 200–300 microns in thickness) to be nourished by diffusion from the blood bathing the valves, normal leaflets and cusps have only scant and inconsistent blood vessels.

The major stress-bearing component within the heart valves is collagen, particularly in the fibrosa, which undergoes realignment and crimping to accommodate

TABLE II.6.10.2 Key Cellular and Extracellular Matrix Components of the Aortic Valve

Component	Location	Putative Function	Comments – Key Questions
Endothelial cells	Lining inflow and outflow valve surfaces	Provide thromboresistance, mediate inflammation	Role in transducing shear and modulating VIC function; functional differences from vascular wall EC; differences in inflow side to outflow side functions/responses largely unknown
Interstitial cells	Deep to surface, throughout all layers	Synthesize and remodel matrix elements	Presently considered the major modulator of long-term valve durability and a key mediator of disease; regional heterogeneity; regulation of activation and functional role of contractile potential poorly understood
Elastin	Concentrated in *ventricularis* layer	Extend in diastole, recoil in systole	Potential mechanistic role in disease not defined
Glycosoamino-glycans	Concentrated in *spongiosa* layer	Absorb shear of relative movements and cushion shock between ventricularis and fibrosa during cyclical valve motion	Potential mechanistic role in disease not defined
Collagen	Concentrated in *fibrosa* layer	Provide strength and stiffness to maintain coaptation during diastole	Likely the most important structural element; crimp and orientation/alignment provide directional anisotropy of properties, and accommodate cyclical cuspal shape changes

Modified from Schoen, F. J. (1997). Aortic valve structure–function correlations: Role of elastic fibers no longer a stretch of the imagination (Editorial). *J. Heart Valve Dis*. **6**, 1–6.

the repetitive changes in the size and shape of the cusps during the cardiac cycle. In the closed phase of the AV (*diastole*), the back pressure (normally approximately 80 mm Hg) stretches the valve cusps as they come together to seal the orifice, to prevent backward leakage of blood. Moreover, the direction of collagen fibers in the cusps determines the directions in which the tissue has the greatest compliance (i.e., orthogonal to the collagen fiber orientation) or can withstand the greatest tensile stresses (i.e., parallel to the collagen fiber orientation). The strains during valve closure and the mechanical properties of the AV cusps are anisotropic (i.e., different in different directions). Since the need for compliance and stretching in the radial direction is greater than that in the circumferential direction, and the need for stiffness correspondingly greater circumferentially (i.e, toward the attachments) the collagen fiber orientation is predominantly circumferential. The second-to-second cyclical internal rearrangements in collagen (i.e., alignment of fibers and extension of microscopic crimp) are extremely sensitive to the instantaneous mechanical stresses, and are completed very early following closing. In contrast, during valve opening (*systole*), the tissue of the cusps that was stretched during diastole becomes relaxed, owing to recoil of the elongated, taut elastin (in the ventricularis). This elastic recoil restores the retracted configuration of the cusp, with a decreased cuspal surface area, a more random directional distribution, and crimp (i.e., folding/pleating) of collagen fibrils. The GAG-rich spongiosa is thought to facilitate the relative rearrangements of the collagenous and elastic layers during the cardiac cycle. Since the dynamics described above are crucial for normal valve function, the quantity, quality, and architecture of the valvular ECM, particularly collagen, elastin, and GAGs, are the major determinants of both the short-term function and long-term (lifetime) durability of a valve.

VICs, the most abundant cell type in the heart valves and distributed throughout all the valve layers, synthesize the valvular ECM and express matrix degrading enzymes (including matrix metalloproteinases (MMPs) and their inhibitors (TIMPs)) that mediate and regulate remodeling and repair of collagen and other matrix components. When stimulated by the mechanical environment, certain chemicals, and as a part of the response to injury, VICs are highly plastic and may transition from one phenotypic state to another during valvular homeostasis, adaptation and pathology (Liu et al., 2007; Schoen, 2008). Thus, VIC health is critical for normal valve function; likewise, VIC's are likely to be the key cell in tissue-engineered heart valves.

> The quantity, quality, and architecture of the valvular ECM, particularly collagen, elastin, and GAGs, are the major determinants of both the short-term function and long-term (lifetime) durability of a valve. Since VICs maintain and remodel the ECM, VIC health is critical for normal valve function. Likewise, VICs are likely to be the key cell in tissue-engineered heart valves.

The VECs that line blood contacting surfaces of the valves share some characteristics of circulatory endothelial cells generally. However, they are phenotypically different in some key features from vascular endothelial cells in the adjacent aorta and elsewhere in the circulation. For example, in response to fluid shear stress, porcine aortic VECs align perpendicular to flow (Butcher et al., 2004), whereas endothelial cells from the nearby aorta align parallel to flow, and the transcriptional gene expression profile of aortic wall and aortic VECs are different when the different cells are exposed to the same mechanical environment (Butcher et al., 2006).

TABLE II.6.10.3	Key Concepts of Heart Valve Function and Pathology That Inform Tissue Engineering
Valve function/durability depends on collagen and other ECM	
ECM quality depends on VIC viability and function	
VIC have a complex phenotype	
VIC elaborate and remodel valvular ECM	
Non-viable VIC are deleterious	
Valve degeneration is caused by calcification (in cells) and mechanical/chemical damage to collagen	
VICs are damaged during processing of tissue valves	
Thus, a key goal of an engineered tissue valve is an environment whereby viable VIC can repair structural injury, remodel ECM in the appropriate quantity spatial pattern and quality, and permit valve growth	

Adult heart valve VIC's *in situ* have characteristics of fibroblasts; they are quiescent, without synthetic or destructive activity for ECM. However, VICs are activated during intrauterine valvular maturation by abrupt changes in the valvular mechanical stress state, and in disease states. Additionally, valvular cells continuously repair the low level of injury to the valvular ECM that occurs during physiological function (Liu et al., 2007; Schoen, 2008). Since the macroscopic mechanical state of the valve is likely transmitted to the VICs through their interactions with the surrounding VECM, there is considerable interest in the effects of mechanical forces on VIC function, the mechanisms of response of VICs to their physical environment, and the mechanical properties of isolated VICs. Thus, in disease states and with bioprosthetic, allograft, and autograft valves, healthy, functional valve cells are beneficial; they mediate ongoing valve function and remodeling over a lifetime (Aikawa et al., 2006). In contrast, stressed or injured valve cells can potentiate significant pathology (as summarized in Table II.6.10.3) (Schoen, 2008, 2012).

Role of Heart Valve Developmental Biology

Some of the processes and regulatory pathways active in valvular development and maturation may be recapitulated in tissue-engineered valves; an understanding of developmental biology, physiology, and pathophysiology of heart valves will likely inform tissue engineering (Armstrong and Bischoff, 2004; Combs and Yutzey, 2009). During embryological development, the three germ layers – *ectoderm*, *endoderm*, and *mesoderm* – give rise to cells that differentiate to form the body's tissues and complex organs. The heart develops primarily from the embryonic layer called mesoderm, in which the initial commitment of precursor cells to a cardiac lineage depends on complex signaling pathways. Cardiac myocytes become organized into a linear heart tube that subsequently undergoes looping; further growth of the looped heart tube and development of septa leads to the multichambered heart.

Subsequently, the valve cusps and leaflets originate as outgrowths (known as *endocardial cushions*) from the mesodermal-derived connective tissue called *mesenchyme*. Endothelial cells lining the inside surface of the cushion forming area undergo an *epithelial-to-mesenchymal transition* (EMT) and migrate from the blood-contacting internal heart surface deep into the connective tissue of the subendocardium (called *cardiac jelly*) to become precursors of mature VIC (Figure II.6.10.3). EMT comprises a series of cell–cell and cell–matrix interactions that release epithelial cells from a surface, and confer the ability to move through three-dimensional ECM and synthesize ECM components (Armstrong and Bischoff, 2004). The newly formed mesenchymal cells remodel the cushions into leaflets and cusps. Evidence for EMT is provided by mesenchymal cellular expression of α-SMA, a marker that is not typically expressed by endothelial cells. Numerous signaling pathways, growth and transcriptional factors (including vascular endothelial growth factor (VEGF), nuclear factor in activated T cells (NFATc1), and Notch) regulate the process of heart valve formation.

TISSUE ENGINEERING APPROACHES TO HEART VALVES

Engineering of heart valve tissue is an active goal of many laboratories, and this technology may eventually achieve clinical application. As emphasized above, the long-term success of an engineered tissue heart valve will depend on the ability of its living cellular components (VECs and VICs) to function normally, to maintain homeostasis, and to repair structural injury to the ECM.

There is a spectrum of approaches to engineered tissue heart valves, utilizing processes with overlapping features: (1) formation of a tissue *in vitro* by seeding cells on a biodegradable scaffold and maturing a living tissue model *construct* (to be implanted *in vivo*) in a bioreactor; (2) cell-seeded natural biodegradable scaffolds; (3) guided tissue regeneration via implanted degradable tissue which is remodeled by endogenous cells; (4) implantation of decellularized valvular material; and (5) other approaches that use endogenous pathophysiologic processes to generate and/or functionally differentiate tissue *in vivo* (variations on these themes are summarized in Figure II.6.10.4).

The most widely-studied paradigm of tissue engineering to facilitate valve regeneration uses cells that are preseeded on a synthetic, biodegradable polymer scaffold fabricated in the shape of a trileaflet valve, and matured *in vitro* in a controlled metabolic and mechanical environment (in a *bioreactor*). The intent is for the cells to differentiate, proliferate, and produce ECM to form the construct. Subsequently, the construct is implanted orthotopically (i.e., in the appropriate valve site) as a valve prosthesis, and further remodeling *in vivo* is intended to recapitulate the normal functional architecture of a valve. The cells produce matrix materials (structural proteins and polysaccharides) and other bioactive

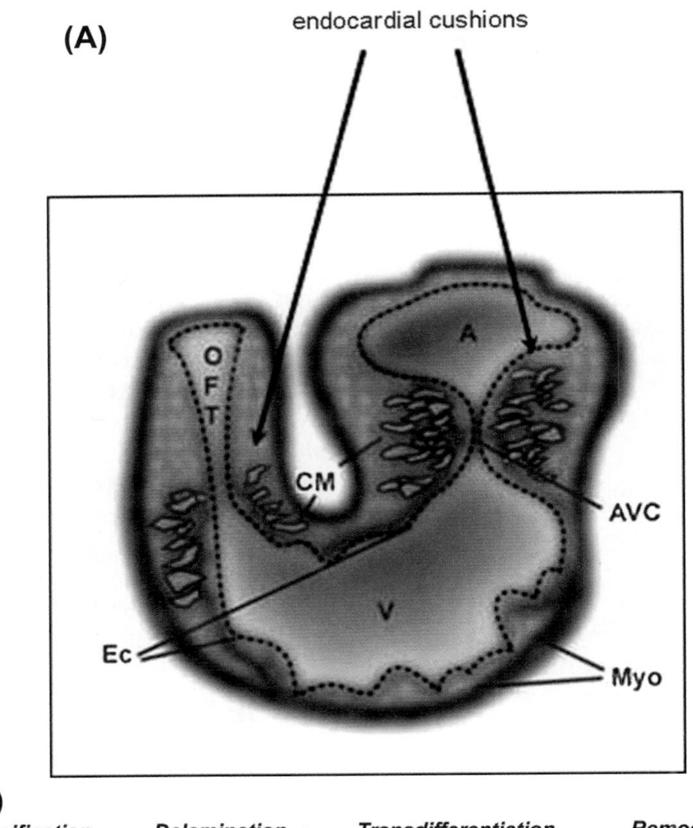

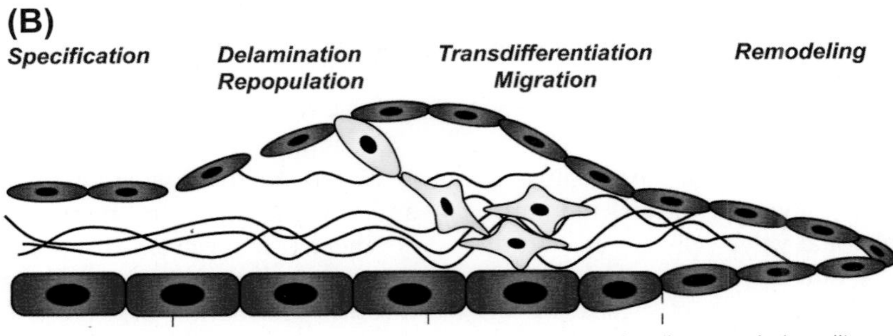

FIGURE II.6.10.3 Overview of heart valve development. (A): Looped heart tube showing locally expanded swellings of cardiac jelly and mesenchymal cells at cardiac cushions. (B): The developing heart tube contains an outer layer of myocardium and an inner lining of endothelial cells separated by an ECM referred to as the cardiac jelly. During heart valve formation, a subset of endothelial cells overlying the future valve site are specified to delaminate, differentiate, and migrate into the cardiac jelly, a process referred to as endothelial–mesenchymal transition, transformation or transdifferentiation (EMT). Locally expanded swellings of cardiac jelly and mesenchymal cells are referred to as cardiac cushions. In a poorly understood process, cardiac cushions undergo extensive remodeling from bulbous swellings to eventual thinly tapered heart valves. (Reproduced by permission from Armstrong, E. J. & Bischoff, J. (2004). Heart valve development: Endothelial cell signaling and differentiation. *Circ. Res.,* **95,** 459–470. Copyright American Heart Association.)

components (e.g., chemical signals, matrix metalloproteinases MMPs and inhibitors, etc.). The ECM provides the basic framework for the tissue and is primarily responsible for the tissue structural and mechanical properties. The polymer scaffold temporarily provides matrix on which cells can organize, grow, proliferate, and produce ECM, and the temporary biomechanical characteristics for the replacement "tissue" until the cells produce the ECM which will ultimately provide the structural integrity and biomechanical profile. The general principles involved in tissue engineering, including considerations for cells, scaffolds, and bioreactors, are summarized in chapters II.6.2, II.6.3, and II.6.6, respectively, of this book. Thus, key pathophysiologic processes occur during the *in vitro* and *in vivo* phases of tissue formation and maturation in TEHV, including: (1) cell proliferation, sorting and differentiation; (2) ECM production and organization; (3) degradation of the polymer scaffold; and (4) remodeling and, potentially, growth of the tissue commensurate with the growth of the individual.

> Tissue-engineered heart valves grown as valved conduits from autologous cells (derived from vascular wall or bone marrow) seeded on biodegradable synthetic polymers and matured *in vitro* have been explored by several groups working in the US and abroad.

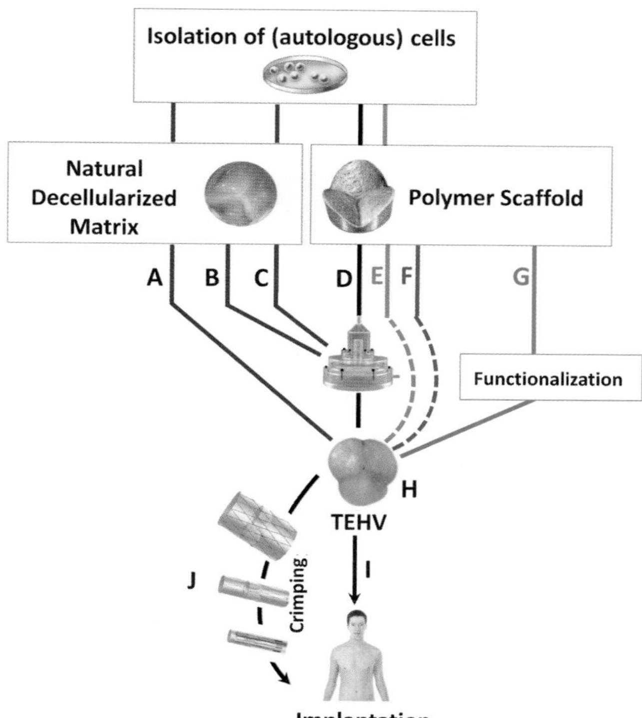

FIGURE II.6.10.4 Spectrum of approaches to heart valve tissue engineering, with overlapping features. The generation of natural decellularized matrix-based constructs involving cell seeded (A), unseeded (B), and *in vitro* conditioned constructs (C). The formation of a tissue *in vitro* by seeding autologous cells onto a biodegradable starter matrix and maturing a living tissue model, called the *construct*, in a bioreactor (D). The use of cell-seeded natural biodegradable scaffolds without *in vitro* biomimetic conditioning, also referred to as the *single step* or *in situ* approach (E); The concept of guided tissue regeneration via implanted degradable tissue, which is remodeled by endogenous cells (F); and other approaches that use endogenous pathophysiologic processes to generate "functionalized" starter matrices stimulating tissue maturation *in vivo* (G). After *in vitro* generation, the valvular replacement (H; TEHV) can be implanted via conventional open-heart surgery (I) or using novel cathether-based approaches involving valve crimping (J).

Tissue-engineered heart valves grown as valved conduits from autologous cells (derived from vascular wall or bone marrow) seeded on biodegradable synthetic polymers and matured *in vitro* have been explored by several groups working in the US and abroad. Cells used have varied across a wide spectrum, from mature vascular wall cells, to VICs, to bone marrow-derived mesenchymal stem cells (Schmidt et al., 2007). In this location, implanted constructs generated *in vitro* have evolved *in vivo* to a complex structure that, at the level of light microscopic histology, resembles that of the native aortic valve described earlier in this chapter. Moreover, the progression of the cell and ECM changes are analogous to those occurring during development and physiologic valve remodeling described earlier (Rabkin-Aikawa et al., 2004), suggesting that the dynamic, chemical, and mechanical environment *in vivo* provides signals that induce functional organization of the tissue construct to a heart valve. This approach has been used to produce pulmonary arterial wall (without a valve) replacements to repair complex congenital heart disease in children (Shin'oka et al., 2005). Pulmonary arterial wall replacements fabricated from vascular wall cells (predominantly vascular smooth muscle cells) seeded onto a biodegradable polymer and implanted into very young lambs enlarged proportionally to overall animal growth over a two year period (Hoerstrup et al., 2006). Although wall thickness remained constant over the observed growth period, it is not yet fully understood whether this "functional growth," as we have called it, is comparable to physiological growth with new tissue accretion, or is rather passive dilation with valve cuspal and conduit enlargement.

The scaffold must also have appropriate geometrical design, density, and mechanical characteristics (e.g., strength and compliance) that allow valve-like function during *in vivo* "maturation" of the TE construct, and have non-toxic original materials and degradation products, biocompatibility, and a controlled rate of resorption. Hoerstrup et al. (2000) implanted PGA/P4HB valve constructs seeded with vascular wall cells and incubated *in vitro* for three weeks in lambs. Valves so fabricated have functioned in the pulmonary circulation of growing lambs for up to five months (Figure II.6.10.5). Postoperative echocardiography demonstrated mobile functioning leaflets without stenosis, thrombus or aneurysm up to 20 weeks. Histology (16 and 20 weeks) showed uniform layered cuspal tissue with endothelium. Complete degradation of the polymers occurred by eight weeks, and extracellular matrix content (collagen, glycosaminoglycans, and elastin) and DNA content increased to levels of native tissue and higher at 20 weeks. Stock has also demonstrated formation of pulmonary artery tissue with histology very similar to native tissue (Stock, 2000). Sutherland et al. (2005) utilized mesenchymal stem cells isolated from ovine bone marrow seeded on a polyglycolic-lactic acid/polyglycolic acid (PGLA/PGA) biodegradable scaffold to create autologous semilunar heart valves, and implanted into the pulmonary position of sheep. The valves were evaluated by echocardiography at implantation and at four months. These valves demonstrated trivial or mild regurgitation at implantation, and minimal gradients. Histology showed disposition of ECM and distribution of cell phenotypes in the engineered valves somewhat reminiscent of that in native pulmonary valves.

Interestingly, these studies have shown the gradual development of a tri-layered structure, including variations in collagen, GAG, and elastin after 4–5 months. However, it remains unknown whether these layers are functionally equivalent to those of the native valve, and what mechanisms guide their formation *in vivo*. Measurements of the mechanical properties of TEHV leaflet tissue suggest that these compare favorably with those of the native PV (Sacks et al., 2009a).

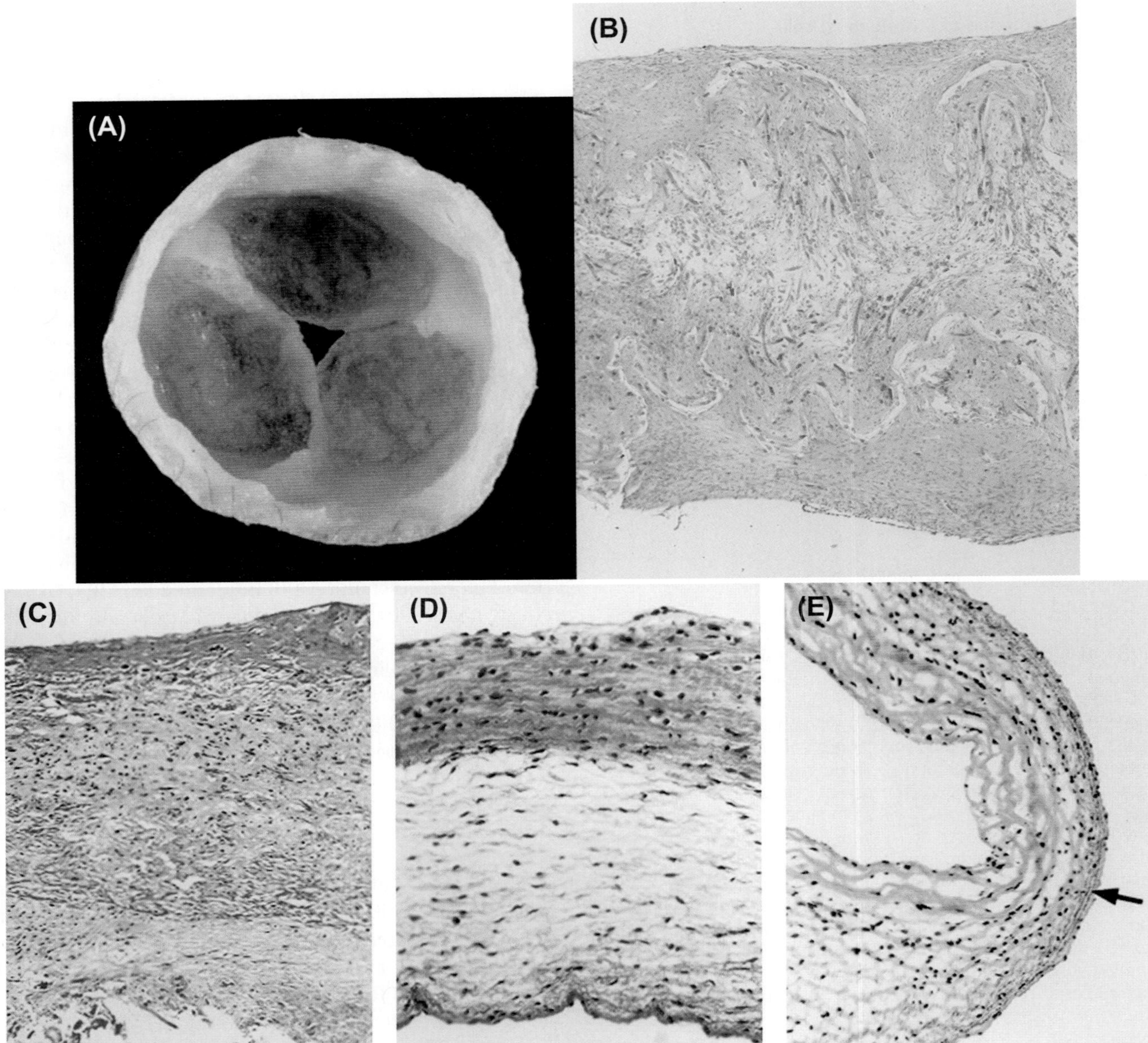

FIGURE II.6.10.5 Tissue-engineered heart valves *in vitro* and *in vivo*. (A): Gross photo of TE heart valve after 14 days of conditioning the cell seeded polymeric scaffold in a bioreactor (*in vitro*). (B), (C), (D) and (E): Photomicrographs of engineered heart valve cusps *in vitro* and *in vivo,* demonstrating evolution to near-normal structure after 16–20 weeks. (B): After 14 days of pulsatile flow, leaflet portion is composed of dense fibrous tissue near surface and loose central core (hematoxylin and eosin stain, magnification x 20). (C) At six weeks, there is early organization of tissue predominantly in outer (outflow) layer (top) (magnification 50 ×). (D): Cross-section of leaflet at 16 weeks shows layered cellular fibrous tissue, which is more dense near outflow surface (top) (original magnification 100 ×). (E) Cross-section of leaflet at 20 weeks demonstrates collagen (yellow), GAGs (blue), and elastin (arrow, inflow surface; original magnification 100 ×). (B), (C) and (D): Hematolylin and eosin. (E) Movat stain. (From Hoerstrup, S. P., Sodian, R., Daebritz, S., Wang, J., Bacha, E. A. et al. (2000). Functional living trileaflet heart valves grown *in vitro. Circulation,* **102,** III-44–49. Copyright American Heart Association)

In the second heart valve tissue engineering approach, fibroblast- or VIC-seeded natural degradable scaffolds under investigation include degradable scaffolds such as hyaluronan or fibrin gel (Masters et al., 2005; Robinson et al., 2008). Cell-free collagen constructs fabricated by directed collagen gel contraction are also being investigated (Shi et al., 2006).

The third, alternative tissue engineering strategy, called "guided tissue regeneration" (Brody and Pandit, 2007), uses an implanted scaffold of a naturally-derived biomaterial or decellularized valve designed to attract circulating endothelial and other precursor cells, and provide a fertile environment for their adherence, growth, and differentiation. In this approach, and in contrast to the preparation of conventional bioprosthetic heart valves, the materials are not aldehyde-fixed or otherwise chemically preserved, as are conventional bioprosthetic heart valves. Cell-free porcine small intestinal submucosa (SIS) has been investigated experimentally in a single valve cusp model (Matheny et al., 2000). In general, natural tissue-derived valve scaffolds possess desirable three-dimensional architecture, mechanical properties, and potential adhesion/migration sites capable of promoting cell attachment and ingrowth.

This concept is further employed in the fourth strategy, in which decellularized natural valve tissue is intended to become recellularized, and not to undergo early remodeling. Rather, the goal is to provide a durable and functional substrate. In some cases, *in vitro* "re-endothelialization" is performed prior to implantation, with the goal of catalyzing recellularization *in vivo*. Decellularized tissue scaffolds derived from valve (in some cases with *in vitro* "re-endothelialization" performed prior to implantation of the valve) have been used in clinical studies in the pulmonary position (Cebotari et al., 2006; Dohmen et al., 2007). However, decellularized porcine valves implanted in humans as aortic valve replacements elicited a strong inflammatory response and suffered structural failure, which has inhibited further use (Simon et al., 2003). The specific patient and implant variables accounting for the spectrum of outcomes are not yet understood, and the long-term fate of these implants, the role of endothelial cell seeding, and the extent of cellular ingrowth into decellularized tissue *in vivo* are not yet known.

Finally, other strategies have included the creation of a cellularized graft by maturation of tissue formed in association with either a microporus polyurethane valve assembly implanted into the subcutaneous space in rabbits (Hayashida et al., 2007, 2008) or a photooxidized bovine pericardial valve implanted intraperitoneally in sheep (de Vissscher et al., 2007). Additionally, in principle, a functional valve might be grown *de novo* in the valve location, as a synthetic valve replacement that undergoes transformation to valvular tissue over time via endogenous processes. Accumulating evidence suggests that circulating cells can be recruited *in vivo* to adhere to intravascular sites of injury or prosthetic material, via a pathway that likely mimics the adherence of inflammatory cells during physiological inflammation (Kumar et al., 2010). Thus, a potential strategy may be to coat a degradable polymer scaffold in the configuration of a valve with appropriate cell-signaling molecules (or use a biological matrix already containing such information as discussed above), in an effort to encourage and direct endothelial progenitor cell (EPC) and other cell adhesion and differentiation. An experiment utilizing decellularized porcine aortic valves containing fibronectin and hepatocyte growth factor suggested that the growth factor enhances early endothelial cell recruitment to, and coverage of, the grafts (Ota et al., 2005). Attempts to attract EPCs from peripheral blood onto grafts via antibodies directed at proposed EPC markers, such as anti-CD34 antibodies and kinase insert domain receptor (KDR), are under investigation (Aoki et al., 2005; Rotmans et al., 2005; Markway et al., 2008).

CHALLENGES FOR FUTURE TRANSLATION OF ENGINEERED TISSUE VALVES TO THE CLINIC

Translation of heart valve tissue engineering from the laboratory to the clinical realm has exciting potential, but also has formidable challenges and uncertainties. Although modest progress has been made toward the goal of a clinically useful tissue-engineered heart valve, ultimate human benefit will be dependent upon advances in biodegradable polymers and other scaffolds, cellular manipulation, strategies for rebuilding the extracellular matrix, and techniques to characterize and potentially non-invasively assess the speed and quality of tissue healing, remodeling, and potential complications. Specific hurdles include optimization and scale up of processing components and conditions *in vitro* to generate a highly functional and durable valve *in vivo*, selection and validation of suitable animal models, development of guidelines for characterization, and assurance of the quality of an *in vitro* fabricated tissue-engineered heart valve destined for human implantation, validation of the hypothesis that tissue evolution *in vivo* can be predicted and measured, and the development of strategies to understand, monitor, and potentially control patient variability in tissue remodeling *in vivo*. Additional technical (e.g., sterilization, storage, shelf-life, etc.), regulatory, and other challenges remain before this form of therapy can be validated as sufficiently safe and effective to warrant translation to clinical use. The major identifiable challenges and the research directions that will be necessary to overcome them are summarized in Table II.6.10.4.

> The possibility of therapeutic regeneration of the heart valves is indeed exciting; however, immense technical, regulatory, and other challenges remain before this form of therapy is validated as sufficiently safe and effective to warrant translation to clinical use.

A number of groups have attempted to develop engineered heart valve tissues *in vitro* using a variety of approaches described above, and although the goals of these studies were often to form functional tissue though a variety of cell types, the elucidation of predictive determinants (including cells, scaffold, bioreactor conditions) influencing reproducible tissue formation *in vitro* has proved difficult. Additionally, although many groups have focused much effort on controlling and quantifying EMC deposition, and emphasized that construct tissue properties are enhanced by dynamic conditioning using bioreactors with a flow regime that mimics that of the intended application (Cox et al., 2010; Ramaswamy et al., 2010), a rigorous evaluation of the *quality* of ECM produced is only beginning (Eckert et al., 2011). Though encouraging, current engineered tissue heart valve work is largely empirically-based, limited in scope, and has not yet clarified the time course of changes in structure

TABLE II.6.10.4 Major Challenges in Clinical Translation of Heart Valve Tissue Engineering

Challenges	Research Directions
TEHV components and their interactions are complex, heterogeneous, and dynamic	Define cell/scaffold/bioreactor combinations that optimize construct composition and properties (*in vitro*)
Correlation of *in vitro* generated construct structure and properties with *in vivo* outcomes has not been demonstrated	Determine and validate correlations between *in vitro* conditions, elements, structure, properties, and *in vivo* function
Quality control of construct structure and function is likely to be difficult	Develop guidelines, tools, and metrics for the pre-implantation characterization of TEHV structure, function, and quality
Animal models may not reliably predict human outcomes	Develop and validate animal models that will test key biological processes and correlate with human outcomes
TEHV structure is likely to be evolving *in vivo* and ongoing function may be less predictable than with conventional valve replacement technology	Develop guidelines, tools, and metrics for the *in vivo* characterization of dynamic TEHV structure, function, and quality
TEHV function will depend upon patient response to implantation, and integration with the recipient's tissues more than with conventional valve replacement, and individual patient responses may be highly variable	Identify/validate biomarkers that predict and assess patient variability in implant success/failure and are capable of non-invasive *in vivo* monitoring and potential control
Remodeling processes after implantation may release or change seeded cells and recruit host cells	Develop tools to monitor the fate of transplanted and endogenous cells (location, function, viability, phenotype)
Regulatory processes and approaches are not yet well-established	Create suitable regulatory approaches to engineered tissue valves that will ensure safety and efficacy

and mechanical behavior concurrent with tissue formation and remodeling, and, or demonstrated, the ability to measure tissue evolution.

A particularly difficult challenge is to develop appropriate methodologies to evaluate the evolving structural remodeling and functionality of valves implanted *in vivo*, especially in a non-invasive manner, so that a particular valve can be followed over time, without necessitating animal sacrifice (Hjortnaes et al., 2009). A schema for the interrelationships among and challenges in tissue characterization for clinical heart valve tissue engineering is summarized in Figure II.6.10.6. To establish such criteria, preclinical animal studies with long-term follow-up are required, based on derivation of predictive models that can be used for evaluating the outcome of implanting tissue-engineered heart valves in humans. Evaluation of valve durability, biocompatibility, biomechanical properties, and the capability to repair and remodel *in vivo* are the principal elements. These properties need to be evaluated against a gold standard before and after implantation. With respect to autologous tissue-engineered heart valves, the aim must be for a design that mimics functionally, if not structurally, the performance of a healthy native heart valve *in vivo*, or at least be of similar and preferably better, functional quality as the heart valve replacements currently in use. It is also imperative, with respect to the role of tissue-engineered heart valves in treating congenital heart disease in pediatric patients or young adults, to show that the tissue-engineered heart valve has the ability to grow and remodel (Figure II.6.10.7).

As a component of an agenda for translating the notion of TEHVs from an extraordinarily interesting research curiosity to a clinically useful surgical tool, it is necessary to understand mechanisms, define animal models, develop biomarkers, develop assays/tools, define surrogate and true endpoints – and the major clinical goals – "i.e., to characterize and assure quality tissue constructs, accommodate patient-to-patient heterogeneity in tissue remodeling, and predict outcomes as early as possible. While studies using animal models such as sheep suggest some level of efficacy and safety, further detailed studies will be needed in these models, other animal models, and in humans. There is considerable controversy over to what extent results from available animal models translate directly to humans, and the most suitable animal model for testing tissue-engineered valves has not yet been determined. Juvenile sheep, for example, have been used for preclinical testing of conventional cardiovascular implants by the FDA and other regulatory agencies, since they replicate some of the features of clinically observed calcific degeneration due to the increased calcium metabolism associated with rapid growth (Barnhart et al., 1982). However, sheep produce an exuberant fibrotic response to cardiovascular implants; valves implanted in sheep generally overgrow more rapidly with fibrotic tissue than they do in humans, and, in tissue engineering studies, may overestimate tissue remodeling relative to humans. Non-human primates and perhaps senescent models may be more appropriate to investigate remodeling phenomena of tissue-engineered heart valves (Weber et al., 2011b). In addition, owing to immunologic considerations, the choice of an animal model for preclinical testing for allogenic or xenogeneic cell-based therapies presents unique challenges.

Moreover, currently available heart valve replacements have predictable behavior in many recipients, whereas *in vivo* remodeling of tissue-engineered heart valves will likely display considerable variability among patients, owing to heterogeneity among individuals in physiological tissue remodeling potential. Principles

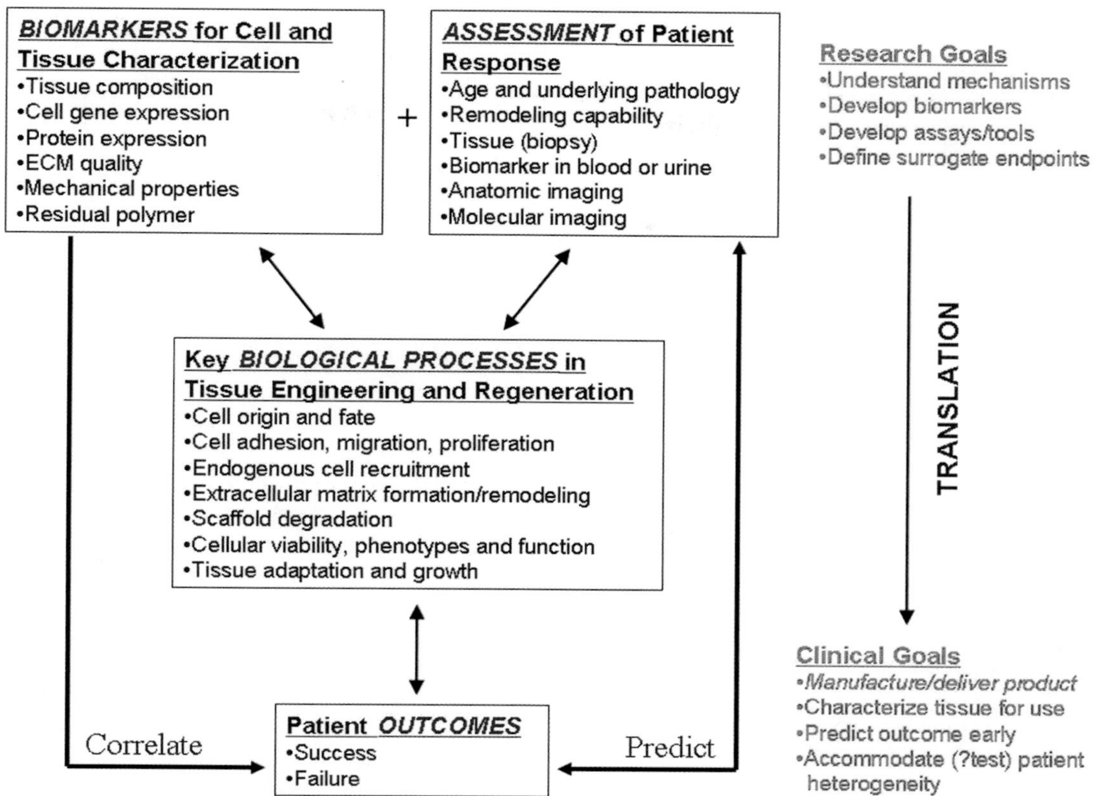

FIGURE II.6.10.6 Paradigm for translating research in heart valve tissue engineering from the laboratory to the clinic. Biomarkers for cell and tissue characterization in conjunction with structural, chemical, and molecular information obtained via *in vitro* and *in vivo* models are necessary for understanding key biological processes in tissue engineering and regenerative medicine. These concepts and data can be used to predict and measure patient success and failure. Data from clinical experience further informs the development of appropriate biomarkers, which may result in reassessment of the appropriate characterization parameters. (Reproduced by permission from Mendelson, K. & Schoen, F. J. (2006). Heart valve tissue engineering: Concepts, approaches, progress, and challenges. *Ann. Biomed. Engin.*, **34**, 1799–1819. Copyright Springer)

analogous to those of pharmacogenetics, a field which seeks to understand the role of genetics in inter-individual variation in drug metabolism, may be useful (Pereira and Weinshilboum, 2009). Some patients might remodel their tissue-engineered valves too slowly, and this could lead to failure (Figure II.6.10.8). Person-to-person variation could be a result of mutations or polymorphisms in key proteins central to ECM remodeling. Indeed, as implants have become more interactive and integrative with the host tissues, there has arisen a corresponding need to understand and potentially control human variation in different facets of biomaterial–tissue interaction and the healing process.

To understand, monitor, and potentially control patient-to-patient differences in wound healing and tissue remodeling capability *in vivo*, measurable biomarkers that predict implant outcomes must be identified. Specific molecular biomarkers may be identified and validated by assessing tissue healing and remodeling during *in vitro* and *in vivo* experiments; suitable biomarkers will need to be followed *in vivo*, possibly via chemical assays in the serum or urine or via molecular imaging. Key targets for characterizing tissue-engineered constructs include tissue composition, cellular gene expression and phenotype, ECM, key effectors of tissue remodeling, and tissue quality. These biomarkers could correlate directly with success and failure in order to generate surrogate endpoints, namely outcome measurements (such as laboratory assays or imaging results) that substitute for but reflect the mechanism of a significant clinical event or characteristic (such as valvular regurgitation, on stenosis, thromboembolism, calcification, infection, or death). Validated surrogate end points could be assessed longitudinally in an individual patient, in order to predict outcome as early as possible in the patient's course, and influence necessary changes in management.

Conventional and innovative invasive and/or non-invasive anatomic and functional imaging modalities will certainly be important tools to assess success and failure. Molecular imaging is particularly exciting in this regard; it requires the identification of a molecular target, selection of a ligand that binds the target, selection of an appropriate imaging system, and synthesis of a molecular imaging agent to detect the desired target. *In vivo* molecular imaging has been used to demonstrate key enzymatic and cellular events in atherosclerosis and thrombosis (Jaffer et al., 2009). Molecular imaging can probe polymorphisms of ECM gene expression *in vivo*

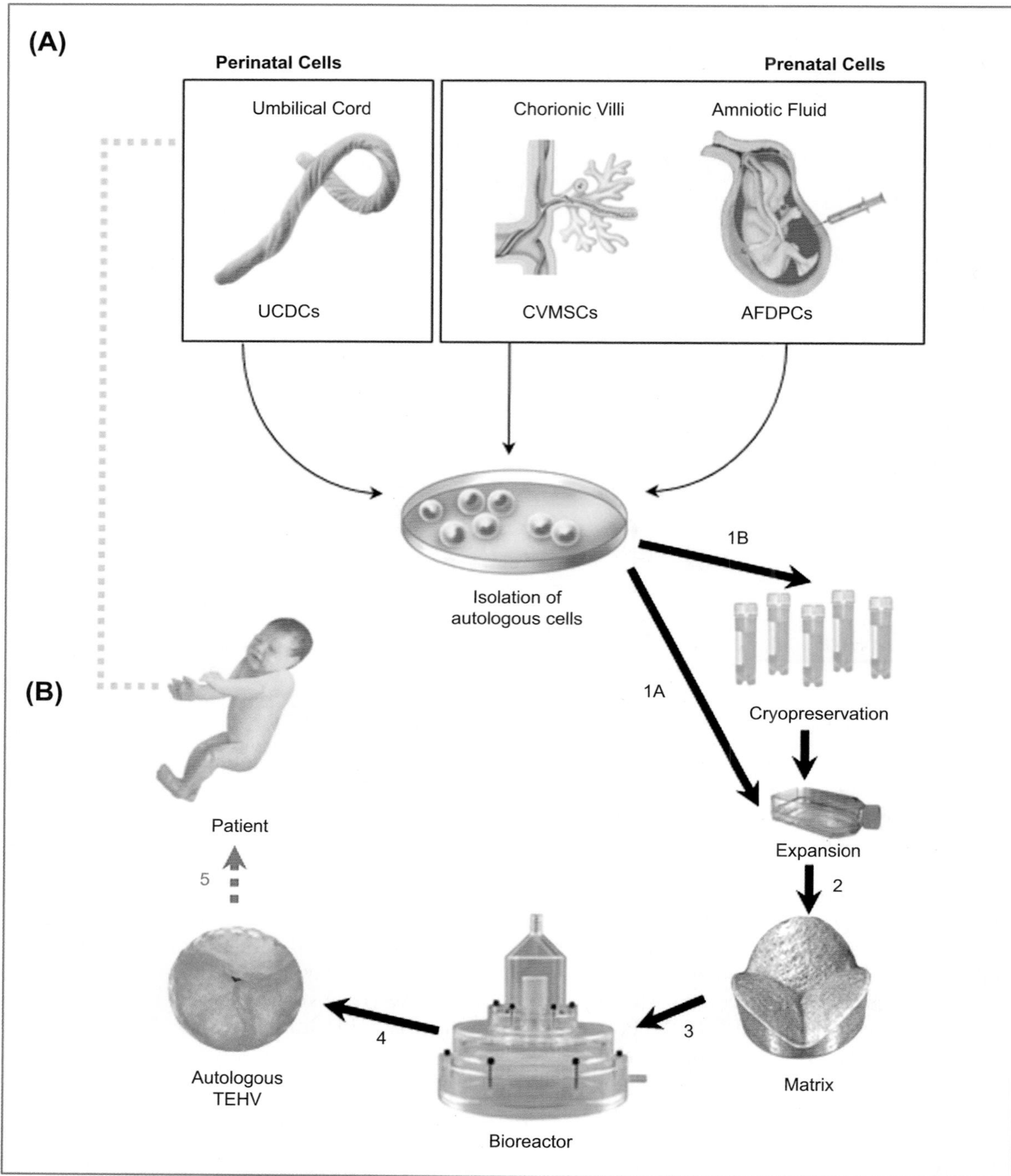

FIGURE II.6.10.7 Pediatric heart valve tissue engineering. The concept of pediatric heart valve tissue engineering uses peri- and pre-natally accessible cell sources (A), for the fabrication of autologous tissue-engineered constructs prior to or shortly after birth (B). For this purpose fetal cells are harvested prenatally or perinatally, expanded *in vitro,* and either directly further processed (1A) or cryopreserved for later use (1B) depending on the ideal time window for surgical correction. Relevant cell sources include umbilical cord- or cord-blood-derived cells (UCDCs), chorionic villi-derived multipotent stem cells (CVMSCs), and amniotic fluid-derived progenitor cells (AFDPCs). After expansion, cells are seeded onto a biodegradable scaffold (2). After a short static phase, the constructs are positioned in a bioreactor system (3), and conditioned *in vitro.* When optimal tissue formation is achieved, tissue-engineered constructs (4) are ready for implantation (5).

in models of wound healing and cardiovascular disease, and can potentially be translated to perform real-time *in vivo* characterization of scaffold matrices (either seeded or with the potential of attracting endogenous cells) implanted in animal models.

Another important laboratory consideration for seeded scaffolds relates to the origin and fate of the cells seeded *in vitro*, and whether the seeded cells remain viable and attached to the scaffold following *in vivo* implantation. Molecular imaging could be utilized to track the

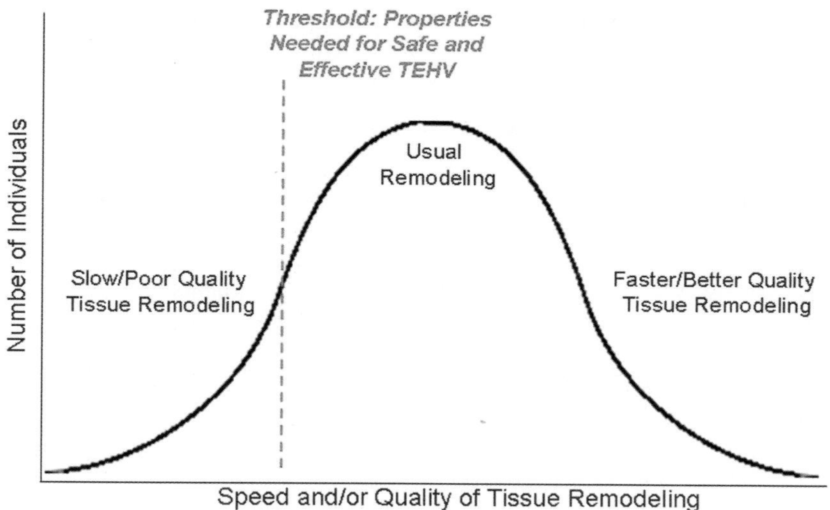

FIGURE II.6.10.8 An hypothesis for inter-individual variability in tissue remodeling. While most individuals will remodel tissue with a usual speed and quality of remodeling, some people will display slow and poor quality of remodeling while others will show fast and better quality of remodeling. Inadequate remodeling could lead to implant failure and its consequences for the patient. The threshold of properties needed for tissue-engineered heart valves and the means of conducting post-implantation surveillance of the patient and graft need to consider this variability. Success or failure may be followed and predicted non-invasively. (Reproduced by permission from Mendelson, K. & Schoen, F. J. (2006). Heart valve tissue engineering: Concepts, approaches, progress, and challenges. *Ann. Biomed. Engin.*, **34**, 1799–1819. Copyright Springer)

presence, migration, proliferation, and function of bone marrow-derived progenitor cells used to seed scaffolds both *in vitro* and *in vivo*. In future experiments, endothelial progenitor and mesenchymal stem cells might be labeled during the *in vitro* stage, and then analyzed *in vivo* using molecular imaging to ensure that they differentiate into appropriate cell lineages and that they remain functional and attached to the scaffold over time.

Since risk–benefit relationships of engineered tissue may be less predictable than those of accepted technology, and since contemporary heart valve replacements have considerable success in most situations (notwithstanding the limitations, and except in pediatrics), acceptance of tissue engineering by the surgical community may be slow. Finally, there may be practical hurdles and barriers that relate to commercialization that go beyond the "academic" considerations discussed above, including the ethics of use in children, especially where the long-term durability has not been proven (or been possible to prove), and the costs of this therapy are likely to be high. A hypothetical case study was developed for bioengineering students and is concerned with choosing between two devices used for development of a pediatric tissue-engineered heart valve (TEHV) (Merryman, 2008). The specific issues faced by start-up, development-phase, and established companies in the general field of tissue engineering/regenerative medicine in the product development, intellectual property, and regulatory arenas have recently been discussed (Johnson et al., 2010).

Several investigators have attempted to develop engineered tissue heart valves that could be deployed through a catheter (Metzner et al., 2010; Schmidt et al., 2010; Weber et al. 2011b), a mode of delivery that now uses valves and is a procedure presently clinically restricted to elderly and otherwise high-risk patients (Walther et al., 2007). The valve substitutes presently used for minimally invasive replacement procedures are non-viable bioprosthetics (e.g., glutaraldehyde-fixed, etc.), and therefore are potentially vulnerable to calcification and progressive dysfunctional degeneration similar to that which occurs in conventional (surgical) non-viable tissue valves. In addition, as stated at the beginning of this chapter, there is growing evidence that catheter-based bioprosthetic valves show mechanical damage, and even accelerated degeneration, as a result of the substantial deformations that occur during the crimping/deployment procedures (Noble et al., 2009; Zegdi et al., 2011). Thus, a living engineered valve with the capacity for self-repair might be a more durable treatment option and enable the application of minimally invasive treatment modalities, especially in younger patients (Fernandes et al., 2007). Conceptually, the combination of minimally invasive valve replacement procedures with tissue engineering is appealing. Moreover, when combined with minimally invasive cell harvest (e.g., autologous adult stem cells), a clinically relevant complete minimally invasive tissue-engineered valve is particularly exciting, and may emerge as a useful clinical approach (Weber et al., 2011b) that synergizes the benefits of two novel technologies.

CONCLUSIONS

The goal of heart valve tissue engineering is to regenerate a functional structure containing endothelial and interstitial cells capable of continuously remodeling the ECM that functions structurally and biomechanically as a valve leaflet. Despite an exciting potential for tissue-engineered heart valves, significant technical barriers must be overcome before widespread clinical application can be envisioned. Further success toward a clinically useful tissue-engineered heart valve will be dependent upon additional advances in biodegradable polymers, stem cell and other cell manipulation, strategies for recreating the ECM, understanding how to harvest the potential of endogenous recruitment of cells, and techniques to non-invasively assess the speed and quality of tissue healing and remodeling, both *in vitro* and *in vivo*. This need is likely to engender a host of novel testing strategies and methods, which will include *in vitro* safety studies, *ex vivo* performance characterization in functional testing devices akin to bioreactors, and *in vivo* preclinical studies. The key concepts that unify the dynamic pathobiology of heart valves and the mechanisms of heart valve disease may be used to improve biologic valve substitutes, and potentially enable heart valve regeneration.

BIBLIOGRAPHY

Aikawa, E., Whittaker, P., Farber, M., Mendelson, K., Padera, R. F., et al. (2006). Human semilunar cardiac valve remodeling by activated cells from fetus to adult: Implications for postnatal adaptation, pathology, and tissue engineering. *Circulation*, **113**, 1344–1352.

Aoki, J., Serruys, P. W., van Beusekom, H., Ong, A. T., McFadden, E. P., et al. (2005). Endothelial progenitor cell capture by stents coated with antibody against CD34: The HEALING-FIM (Healthy Endothelial Accelerated Lining Inhibits Neointimal Growth-First In Man) Registry. *J. Am. Coll. Cardiol.*, **45**, 1574–1579.

Armstrong, E. J., & Bischoff, J. (2004). Heart valve development: Endothelial cell signaling and differentiation. *Circ. Res.*, **95**, 459–470.

Barnhart, G. R., Jones, M., Ishihara, T., Rose, D. M., Chavez, A. M., et al. (1982). Degeneration and calcification of bioprosthetic cardiac valves. Bioprosthetic tricuspid valve implantation in sheep. *Am. J. Pathol.*, **106**, 136–139.

Breuer, C. K., Mettler, B. A., Anthony, T., Sales, V. L., Schoen, F. J., et al. (2004). Application of tissue-engineered principles toward the development of a semilunar heart valve substitute. *Tissue Eng.*, **10**, 1725–1736.

Brody, S., & Pandit, A. (2007). Approaches to heart valve tissue engineering scaffold design. *J. Biomed. Mater Res. Part B: Appl. Biomater.*, **83**, 16–43.

Butcher, J. T., Penrod, A. M., Garcia, A. J., & Nerem, R. (2004). Unique morphology and focal adhesion development of valvular endothelial cells in static and fluid flow environments. *Arterioscler. Thromb. Vasc. Biol.*, **24**, 1429–1434.

Butcher, J. T., Tressel, S., Johnson, T., Turner, D., Sorescu, G., et al. (2006). Transcriptional profiles of valvular and vascular endothelial cells reveal phenotypic differences: Influence of shear stress. *Arterioscler. Thromb. Vasc. Biol.*, **26**, 69–77.

Cebotari, S., Lichtenberg, A., Tudorache, I., Hilfiker, A., Mertsching, H., et al. (2006). Clinical application of tissue engineered human heart valves using autologous progenitor cells. *Circulation*, **114**, I132–I137.

Combs, M. D., & Yutzey, K. E. (2009). Heart valve development: Regulatory networks in development and disease. *Circ. Res.*, **105**, 408–421.

Cox, M. A. J., Kortsmit, J., Driessen, N., Bouten, C. V. C., & Baaijens, F. P. T. (2010). Tissue-engineered heart valves develop native-like collagen fiber architecture. *Tissue Engin. A.*, **16**, 1527–1537.

de Vissscher, G., Vranken, I., Lebacq, A., Van Kerrebroeck, C., Ganame, J., et al. (2007). *In vivo* cellularization of a cross-linked matrix by intraperitoneal implantation: A new tool in heart valve tissue engineering. *European Heart J.*, **28**, 1389–1396.

Dohmen, P. M., Lembcke, A., Holinski, S., Kivelitz, D., Braun, J. P., et al. (2007). Mid-term clinical results using a tissue-engineered pulmonary valve to reconstruct the right ventricular outflow tract during the Ross procedure. *Ann. Thorac. Surg.*, **84**, 729–736.

Eckert, C. E., Mikulis, B. T., Gottlieb, D., Gemeke, D., Legrice, I., et al. (2011). Three-dimensional quantitative micromorphology of pre- and post-implanted engineered heart valve tissues. *Ann. Biomed. Eng.*, **39**, 205–222.

Fernandes, S. M., Khairy, P., Sanders, S. P., & Colan, S. D. (2007). Bicuspid aortic valve morphology and interventions in the young. *J. Am. Coll. Cardiol.*, **4922**, 2211–2214.

Harken, D. F., Taylor, W. J., LeFemine, A. A., Lunzer, S., Low, H. B., et al. (1962). Aortic valve replacement with a caged ball valve. *Am. J. Cardiol.*, **9**, 292–299.

Hayashida, K., Kanda, K., Yaku, H., Ando, J., & Nadayama, Y. (2007). Development of an *in vivo* tissue-engineered, autologous heart valve (the biovalve): Preparation of a prototype model. *J. Thorac. Cardiovasc. Surg.*, **134**, 152–159.

Hayashida, K., Kanda, K., Oie, T., Okamoto, Y., Ishibashi-Ueda, H., et al. (2008). Architecture of an *in vivo* tissue engineered autologous conduit "biovalve." *J. Biomed. Mater Res. Part B: Appl. Biomater.*, **86B**, 1–8.

Hilbert, S. L., Schoen, F. J., & Ferrans, V. J. (2004). Allograft heart valves: Morphologic, biomechanical and explant pathology studies. In R. Hopkins (Ed.), *Cardiac Reconstructions with Allograft Tissues* (2nd ed., pp. 193–233). New York, NY: Springer-Verlag, Inc.

Hjortnaes, J., Bouten, C. V. C., Van Herwerden, L. A., Gründeman, P. F., & Kluin, J. (2009). Translating autologous heart valve tissue engineering from bench to bed. *Tissue Eng.*, **15**, 307–317.

Hoerstrup, S. P., Sodian, R., Daebritz, S., Wang, J., Bacha, E. A., et al. (2000). Functional living trileaflet heart valves grown *in vitro*. *Circulation*, **102**, III-44–49.

Hoerstrup, S. P., Cummings, I., Lachat, M., Schoen, F. J., Jenni, R., et al. (2006). Functional growth in tissue engineered living, vascular grafts: Follow-up to 100 weeks in a large animal model. *Circulation*, **114**, 159–166.

Jaffer, F. A., Libby, P., & Weissleder, R. (2009). Optical and multimodality molecular imaging: Insights into atherosclerosis. *Arterioscler. Thromb. Vasc. Biol.*, **29**, 1017–1024.

Johnson, P. C., Bertram, T. A., Tawil, B., & Hellman, K. B. (2010). Hurdles in tissue engineering/regenerative medicine product commercialization: A survey of North American academy and industry. *Tissue Eng. Part A*, **17**, 5–15.

Joyce, E. M., Liao, J., Schoen, F. J., Mayer, J. E., & Sacks, M. S. (2009). Functional collagen fiber architecture of the pulmonary heart valve cusp. *Ann. Thorac. Surg.*, **87**, 1240–1249.

Kumar, V., Abbas, A. K., & Fausto, N. (Eds.). (2010). *Robbins and Cotran Pathologic Basis of Disease* (7th ed., pp. 53–62). New York, NY: Elsevier-Saunders.

Liu, A. C., Joag, V. R., & Gotlieb, A. I. (2007). The emerging role of valve interstitial cell phenotypes in regulating heart valve pathology. *Am. J. Pathol.*, **171**, 1407–1418.

Markway, B. D., McCarty, O. J. T., Marzec, U. M., Courtman, D. W., Hanson, S. R., et al. (2008). Capture of flowing endothelial cells using surface-immobilized anti-kinase insert domain receptor antibody. *Tissue Eng: Part C*, **14**, 97–105.

Masters, K. S., Shah, D. N., Leinwand, L. A., & Anseth, K. S. (2005). Crosslinked hyaluronan scaffolds as a biologically active carrier for valvular interstitial cells. *Biomaterials*, **26**, 2517–2525.

Matheny, R. G., Hutchison, M. L., Dryden, P. E., Hiles, M. D., & Shaar, C. J. (2000). Porcine small intestine submucosa as a pulmonary: Valve leaflet substitute. *J. Heart Valve Dis.*, **9**, 769–775.

Mendelson, K., & Schoen, F. J. (2006). Heart valve tissue engineering: Concepts, approaches, progress, and challenges. *Ann. Biomed. Engin.*, **34**, 1799–1819.

Merryman, W. D. (2008). Development of a tissue engineered heart valve for pediatrics: A case study in bioengineering ethics. *Sci. Eng. Ethics.*, **14**, 93–101.

Metzner, A., Stock, U. A., Lino, K., Fischer, G., Huemme, T., et al. (2010). Percutaneous pulmonary valve replacement: Autologous tissue-engineered valved stents. *Cardiovasc. Res.*, **88**, 453–461.

Mitchell, R. N., Jonas, R. A., & Schoen, F. J. (1998). Pathology of explanted cryopreserved allograft heart valves: Comparison with aortic valves from orthotopic heart transplants. *J. Thorac. Cardiovasc. Surg.*, **115**, 118–127.

Noble, S., Asgar, A., Cartier, R., Virmani, R., & Bonan, R. (2009). Anatomo-pathological analysis after CoreValve revalving system implantation. *EuroIntervention*, **5**, 78–85.

Ota, T., Sawa, Y., Iwai, S., Kitajima, T., Ueda, Y., et al. (2005). Fibronectin-hepatocyte growth factor enhances reendothelialization in tissue-engineered heart valve. *Ann. Thorac. Surg.*, **80**, 1794–1801.

Pereira, N. L., & Weinshilboum, R. M. (2009). Cardiovascular pharmacogenomics and individualized drug therapy. *Nat. Rev. Cardiol.*, **6**, 632–638.

Rabkin-Aikawa, E., Farber, M., Aikawa, M., & Schoen, F. J. (2004). Dynamic and reversible changes of interstitial cell phenotype during remodeling of cardiac valves. *J. Heart Valve Dis.*, **13**, 841–847.

Rahimtoola, S. H. (2010). Choice of prosthetic heart valve in adults: An update. *J. Am. Coll. Cardiol.*, **55**, 2413–2426.

Ramaswamy, S., Gottlieb, D., Engelmayr, G. C., Aikawa, E., Schmidt, D. E., et al. (2010). The role of organ level conditioning on the promotion of engineered heart valve tissue development *in vitro* using mesenchymal stem cells. *Biomaterials*, **31**, 1114–1125.

Robinson, P. S., Johnson, S. L., Evans, M. C., Barocas, V. H., & Tranquillo, R. T. (2008). Functional tissue-engineered valves from cell-remodeled fibrin with commissural alignment of cell-produced collagen. *Tissue Eng. Part A*, **14**, 83–95.

Rotmans, J. I., Heyligers, J. M. M., Verhagen, H. J. M., Velema, E., Nagtegaal, M. M., et al. (2005). *In vivo* cell seeding with anti-CD34 antibodies successfully accelerates endothelialization but stimulates initial hyperplasia in porcine arteriovenous expanded polytetrafluoroethylene grafts. *Circulation*, **112**, 12–18.

Sacks, M. S., & Yoganathan, A. P. (2007). Heart valve function: A biomechanical perspective. *Philos. Trans. R. Soc. Lond. B. Biol. Sci.*, **362**, 1369–1391.

Sacks, M. S., Schoen, F. J., & Mayer, J. E., Jr. (2009a). Bioengineering challenges for heart valve tissue engineering. *Annu. Rev. Biomed. Eng.*, **11**, 289–313.

Sacks, M. S., Merryman, W. D., & Schmidt, D. E. (2009b). On the biomechanics of heart valve function. *J. Biomech.*, **42**, 1804–1824.

Schmidt, D., Mol, A., Kelm, J. M., & Hoerstrup, S. P. (2007). *In vitro* heart valve tissue engineering. *Methods Mol. Med.*, **140**, 319–330.

Schmidt, D., Dijkman, P. E., Driessen-Mol, A., Stenger, R., Mariani, C., et al. (2010). Minimally-invasive implantation of living tissue engineered heart valves. *J. Amer. Coll. Cardiol.*, **6**, 510–520.

Schoen, F. J. (1997). Aortic valve structure – function correlations: Role of elastic fibers no longer a stretch of the imagination (editorial). *J. Heart Valve Dis.*, **6**, 1–6.

Schoen, F. J. (2001). Pathology of heart valve substitution with mechanical and tissue prostheses. In M. D. Silver, A. L. Gotlieb, & F. J. Schoen (Eds.), *Cardiovascular Pathology* (3rd ed., pp. 629–677). Philadelphia, PA: W.B. Saunders.

Schoen, F. J. (2008). Evolving concepts of cardiac valve dynamics: The continuum of development, functional structure, pathobiology, and tissue engineering. *Circulation*, **118**, 1864–1880.

Schoen, F. J., & Edwards, W. D. (2001). Valvular heart disease: General principles and stenosis. In: M. D. Silver, A. I. Gotlieb, & F. J. Schoen (Eds.), *Cardiovascular Pathology* (3rd ed., pp. 402–442). Philadelphia, PA: W.B. Saunders.

Schoen, F. J., & Levy, R. J. (2005). Calcification of tissue heart valve substitutes: Progress toward understanding and prevention. *Ann. Thorac. Surg.*, **79**, 1072–1080.

Shi, Y., Rittman, L., & Vesely, I. (2006). Novel geometries for tissue-engineered tendonous collagen constructs. *Tissue Eng.*, **12**, 2601–2609.

Shin'oka, T., Matsumura, G., Hibino, N., Naito, Y., Watanabe, M., et al. (2005). Midterm clinical result of tissue-engineered vascular autografts seeded with autologous bone marrow cells. *J. Thorac. Cardiovasc. Surg.*, **129**, 1330–1338.

Simon, P., Kasimir, M. T., Seebacher, G., Weigel, G., Ullrich, R., et al. (2003). Early failure of the tissue engineered porcine heart valve SYNERGRAFT in pediatric patients. *Eur. J. Cardiothorac. Surg.*, **23**, 1002–1006.

Stock, U. A., Nagashima, M., Khalil, P. N., Nollert, G., Herden, T., et al. (2000). Tissue engineered valved conduits in the pulmonary circulation. *J. Thorac. Cardiovasc. Surg.*, **119**, 732–740.

Sutherland, F. W. H., Perry, T. E., Yu, Y., Sherwood, M. C., Rabkin, E., Masuda, Y., et al. (2005). From stem cells to viable autologous semilunar heart valve. *Circulation*, **111**, 2783–2791.

Vesely, I. (2005). Heart valve tissue engineering. *Circ. Res.*, **97**, 743–755.

Walther, T., Simon, P., Dewey, T., Wimmer-Greinecker, G., Falk, V., et al. (2007). Transapical minimally invasive aortic valve implantation: Multicenter experience. *Circulation*, **116**, I240–I245.

Weber, B., Emmert, M. Y., Schoenauer, R., Brokopp, C., Baumgartner, L., et al. (2011a). Tissue engineering on matrix: Future of autologous tissue replacement. *Semin. Immunopathol.*, **33**, 307–315.

Weber, B., Scherman, J., Emmert, M. Y., Gruenenfelder, J., Verbeek, R., et al. (2011b). Injectable living marrow stromal cell-based autologous tissue engineered heart valves: First experiences with a one-step intervention in primates. *Eur. Heart J.*, (Epub ahead of print).

Zegdi, R., Bruneval, P., Blanchard, D., & Fabiani, J. N. (2011). Evidence of leaflet injury during percutaneous aortic valve deployment. *Eur. J. Cardiothorac. Surg.*, **40**, 257–259.

CHAPTER II.6.11 CARDIAC MUSCLE TISSUE ENGINEERING

Amandine Godier-Furnemont and
Gordana Vunjak-Novakovic
Department of Biomedical Engineering, Columbia University, New York, USA

INTRODUCTION

The human heart is a hugely complex tissue composed of multiple cell types that interact in a highly ordered manner as a compact and densely vascularized synctium. In this chapter we will focus on tissue engineering of cardiac muscle (myocardium). The metabolic demands of cardiac tissue, as well as the unique electromechanical properties governing its function, make the heart an especially challenging tissue to grow *in vitro* and regenerate *in vivo*. Development of human sources of cardiac cells, and the biomaterial scaffolds and bioreactors to appropriately recapitulate the right environment for these cells, have been crucial to progress within the field of cardiac tissue engineering. In this chapter we discuss the key requirements for tissue engineering of functional heart muscle, and how these requirements can be met in a laboratory setting. We describe selection of biomaterials, signaling paradigms, and the evaluating criteria for engineered heart tissue. Next, we describe advances in cardiac tissue engineering, envisioned applications of the tissue for *in vivo* therapeutic use (regenerative medicine) and as *in vitro* models of disease. Lastly, we address the challenges in engineering heart tissue *in vitro* and subsequent translation to the clinical setting.

CARDIOVASCULAR DISEASE AND THE NEED FOR ENGINEERED MYOCARDIUM

Cardiovascular Disease

A healthy heart will beat about three billion times during an average lifespan, pumping blood through our vasculature. However, once damaged, the heart does not heal by itself. Cardiovascular disease (CVD) is responsible for almost 30% of deaths worldwide. In the US, CVD is responsible for almost 40% of all deaths, with about 70 million Americans living with heart disease (American Heart Association, 2008). For those living with cardiovascular disease, it can lead to significant disabilities and contribute significantly to escalating healthcare costs. In the US, the economic burden of cardiovascular disease is approximately $300 billion per year, and this figure is expected to rise. The standard therapy for end stage heart failure remains heart transplantation, and the shortage of donated organs contributes to serious loss of life while waiting for an organ, with no other viable means to treat the heart. With practically no ability of cardiac cells to regenerate, scientists are working to develop functional muscle replacements in the laboratory setting that may be used to prevent deterioration of heart function or even contribute to regenerate the muscle.

Characteristics of Native Myocardium

The myocardium is a densely packed, highly ordered, and anisotropic structure, whose function is governed by the propagation of electrical signals, which induce mechanical contraction of cardiac myocytes. A specialized group of cardiac cells, pacemaker cells, initiate an electrical signal that is propagated throughout the heart; intercalated discs connect cardiomyocytes, with gap junctions allowing for rapid signal propagation, and synchronized contraction of the heart. Electrical conduction induces calcium release, and causes mechanical contraction of the cell's myofibrils.

The myocardium consists of cardiomyocytes, cardiac fibroblasts, and endothelial cells that populate the dense, embedded vascular network. Cardiomyocytes comprise about 20–40% of the total cells by number, and about 80–90% of the total tissue by volume. Cardiomyocytes are multinucleated and elongated cells that orient themselves parallel to the heart wall. The aligned cells are interspaced by dense vascular bed, with typical inter-capillary distances being only 20 μm (Rakusan and Korecky, 1982), which is the cross-sectional width of a single cell. This extensive vascular network supports the high metabolic activity of the heart tissue through transport of nutrients and, most critically, oxygen. The exchange of oxygen and its metabolic product carbon dioxide between the cells and blood occurs via hemoglobin in the blood. As an oxygen carrier, hemoglobin allows for a 100-fold increase in the amount of oxygen that can be transported through the capillaries to tissue, significantly increasing the total tissue mass that can be sustained.

Mechanical contractions of cardiomyocytes are mediated by the heart's dynamic extracellular matrix (ECM). Collagen is the primary loadbearing protein that makes up the connective tissue, although it alone constitutes a minute amount of the heart's mass. Collagen transduces the force expended by the myocytes in systole, as well as providing a passive stiffness during diastole, preventing dilation and edema of the muscle over time. Collagen type I makes up 85% of the fibrillar collagen in the heart, and collagen types IV and VI constitute the basement membrane. The heart's extracellular matrix includes fibronectin, laminin, vitronectin, and elastin, all of which contribute to cell adhesion and the loadbearing capacity of the heart.

At the cellular level, various proteins are involved in the mechanotransduction underlying the macroscopic beating of the heart. While collagen is responsible for diastolic stiffness, preventing further extension of already elongated cardiomyocytes, intracellular titin is responsible for passive stiffness at short sarcomere lengths. Active muscle contraction is regulated at the cellular level, starting with cell membrane depolarization that activates intracellular calcium release from the Sarcoplasmic Reticulum (SR), and raising cytoplasmic calcium concentrations. The free cytoplasmic calcium binds to the troponin–tropomyosin complex, activating movement of the tropomyosin to unblock the attachment site for the myosin cross-bridge and allowing it to attach to the actin, resulting in contraction. The uptake of calcium into the SR reverses the process. The function of the heart is maintained by intercellular communication and junctions, which are in turn maintained by a functional extracellular matrix providing the necessary alignment, elasticity, and stiffness necessary to sustain the mechanical demands of the muscle.

Myocardial Infarction

Upon myocardial infarction, a patient can lose 50 grams of muscle mass. This substantial loss of functional muscle is due to hypoxia – lack of oxygen supply to the cells due to the interrupted blood flow – leading to the release of apoptotic factors that ultimately cause cell death. The inflammatory response elicited by myocardial infarction leads to rapid macrophage invasion, formation of granulation tissue, and ultimately remodeling of the myocardium into fibrous tissue.

The heart undergoes a three-step healing process that is characterized by the inflammatory, proliferative, and maturation phases. Within seconds of myocardial ischemia, hypoxia sets in in the myocardium, and within minutes, adenosine triphosphate (ATP) depletion of cardiomyocytes leads to an inability to contract, and eventually necrosis of the myocytes. During the inflammatory phase, dying cardiovascular cells release pro-inflammatory signals, leading to rapid infiltration of neutrophils that begin to clear the cell debris within the infarct. Within 24 hours, the myocardium is invaded by monocytes, which phagocytose apoptosed neutrophils, leading to the release of cytokines such as TGF-β that initiate fibrotic tissue remodeling during the proliferative phase (Frangogiannis, 2008). The first phase recruits Ly-$6C^{hi}$ monocytes via CCR2 chemokines, while the second phase recruits Ly-$6C^{lo}$ monocytes via CX_3CR1 chemokines. Ly-$6C^{lo}$ monocytes exhibit phagocytic, proteolytic, and inflammatory properties, while Ly-$6C^{lo}$ monocytes promote healing via myofibroblast accumulation, angiogenesis, and matrix deposition (Nahrendorf et al., 2007).

During phase I, monocytes are recruited to the site of injury, and begin to differentiate into macrophages. Initially, classical (M1) activation of monocytes and macrophages takes place, leading to the production of nitric oxide (NO) and pro-inflammatory cytokines such as IL-1β, IL-6, and TNF-α. M1 activation is typically associated with inflammation, tumor resistance, and graft rejection, and is followed by phase II with the recruitment of anti-inflammatory monocytes, and their activation into anti-inflammatory macrophages (alternative macrophage activation, M2).

M2 monocytes/macrophages exhibit a non-inflammatory profile with the expression of arginase 1 and 2, and TGF-β. M2 activation is associated with immunoregulation, matrix deposition, remodeling, and graft acceptance (Martinez et al., 2006; Nahrendorf et al., 2007; Brunelli and Rovere-Querini, 2008; Lambert et al., 2008). M2 macrophages become the dominant macrophage at the later stages of myocardial remodeling, and continue to turn over the ECM during scar formation, as well as activating resident and infiltrating endothelial and fibroblast cells promoting angiogenesis. The scar tissue alters the ECM content in the myocardium, reduces contractile function of the heart, and leads to wall thinning and dilatation, extracellular matrix remodeling, and ultimately heart failure.

CONSIDERATIONS FOR ENGINEERING CARDIAC MUSCLE

The limited ability of the heart to regenerate following infarction has spurred the current therapies for terminal heart failure that involve the injection of stem cells (experimental therapy, with variable results), and heart transplantation (clinical therapy, limited by the scarcity of donor hearts). Tissue engineering could radically change the way we treat heart disease, if some of the recent advances in the derivation of human cardiomyocytes and the development of technologies for cell delivery and tissue engineering are translated into clinical practice.

Engineered heart muscle must be able to support physiologic levels of diastolic loads, while producing sufficient systolic force to result in whole muscle contraction that is sufficient to create the pressure needed to pump blood through the body. The structure supporting the cardiac myocytes must not fatigue over time, or lead to dilation and thinning, but must rather be able to withstand the change in force and contraction of the muscle with sufficient compliance to sustain the heart cells. The graft must not illicit an immune response, present phenotypic stability of cells throughout its lifetime, and must be prepared to function and survive in the ischemic environment of failing heart tissue. Most importantly, engineered heart muscle intended for clinical use must have the ability to integrate both electrically and mechanically with the host muscle, and a graft will need to establish a vasculature *in vivo* that is sufficient to supply the graft with the oxygen and nutrients necessary to support long-term survival and function, and prevent necrosis of the tissue.

The myocardium is a very complex tissue, and the macroscopic function of the heart is made possible from specialized cell organization and control at every hierarchical level, from the subcellular to the macroscopic. We discussed how molecular interactions enable nanoscale movements of actin, resulting in microscopic contraction of individual cells – an action propagated throughout the heart and resulting in the whole organ contraction. At the macroscopic scale, tissue engineers need to develop constructs that will effect change in the ailing heart. To engineer a centimeter thick muscle (human myocardium is ~1.5 cm thick), we will need to overcome the current limitations of diffusional supply of nutrients and, most critically, oxygen necessary to sustain dense cell populations *in vitro*. At the mm scale, cell alignment and intercellular connections necessary for assembling the anisotropic structure of the heart can be facilitated by scaffolds with grooves or aligned fibers, and further facilitated by electrical and mechanical stimulation. To induce cell alignment and coupling, cells must be cultured at a high density (~10^8 cells/cm^3), and to sustain them, vascularization must be addressed, as will be discussed in detail in the section on Challenges and Future Applications.

At the μm and nm scale, the cell's cytoskeleton must be fully developed to enable the physical contraction of cells. Engineers must address cell adhesion kinetics and scaffold stiffness that enables cells to display the appropriate morphology to orient their scaffold in a way that permits physiological cell contraction. Several groups have demonstrated the importance of cell shape in determining cytoskeletal arrangement and resulting contractile capabilities. At the subcellular and nanometer scale, functional gap junctions and channels must develop in the cardiomyocytes to enable intracellular junctions capable of electrical impulse propagation – a conduction velocity of about 0.5 m/s through the ventricular mass. Additionally, cell–matrix adhesions dictate morphology and cell orientation, and proper ion channel kinetics to allow for ion transfer to induce physiologic cell contractions.

Tissue engineering offers the possibility of controlling the assembly of myocardium at each of these levels – from the subcellular to macroscopic tissue organization. The general approach of tissue engineers is shown in Figure II.6.11.1, and involves three principle components: (1) cells; (2) scaffold; and (3) bioreactor.

Cells act as the building blocks of tissue. Responding to the signaling imparted by the scaffold and bioreactor, cells adapt to their environment and themselves engineer new tissues, changing their individual phenotypes, forming intracellular junctions, and creating and interacting with the extracellular matrix. Engineered scaffolds provide the three-dimensional environment for cells to adhere to, and to interact with each other. Scaffolds tailored to a specific application can be engineered to be bioactive (interact with the cells), with specific porosities, mechanical properties, and degradation rates to complement tissue development. In tissue engineering applications for the heart, scaffolds are generally intended to provide instructional and temporary support to cells – inducing alignment, providing appropriate stiffness to induce cells to generate physiological force, but also to degrade over time, as the cells digest the scaffold and replace it with generated extracellular matrix proteins.

Lastly, bioreactors provide cells seeded on scaffolds, with environmental control that is biomimetic. Bioreactors are engineered to provide basic control at the level of carbon dioxide and oxygen control, up to the inclusion of perfusion loops for medium flow, and electrical and mechanical apparatus for stimulation of the constructs. The inclusion of such tools for the culture conditions, and the manner in which bioreactors are constructed,

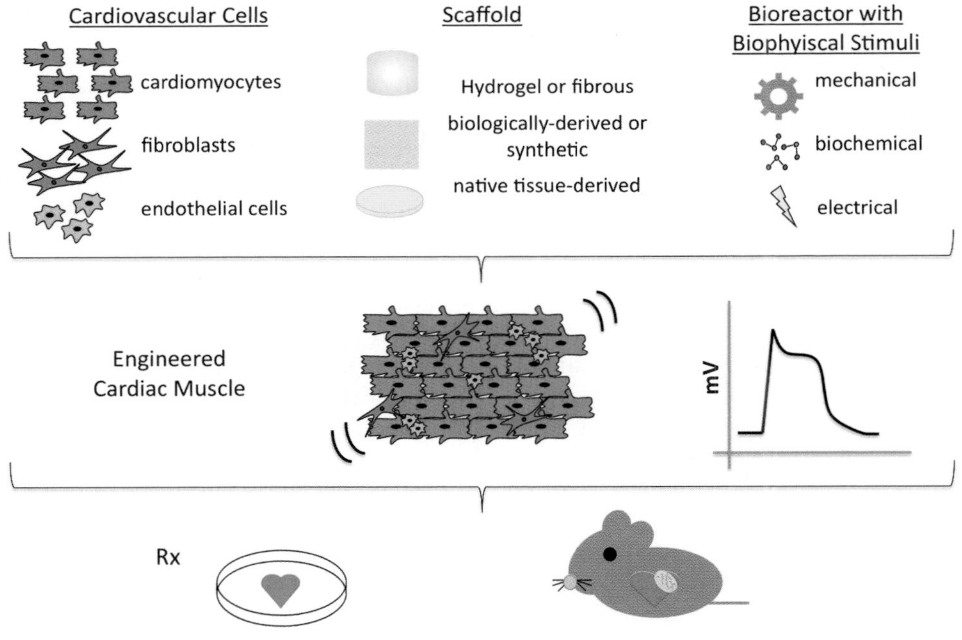

FIGURE II.6.11.1 Cardiac muscle tissue engineering paradigm. Cardiovascular cells are cultured on a biomaterial scaffold, in a bioreactor providing environmental control and biophysical stimulation. The resulting engineered constructs are studied *in vitro* (as models of disease, for drug testing), and *in vivo* (in the implantation model of cardiac ischemia).

enable scientists to provide dynamic spatiotemporal control over the culture of scaffolds. Overall, the use of cells seeded on scaffolds and cultured in advanced bioreactors can provide the overall culture conditions conducive to cardiac tissue formation *in vitro*. Each of these components will be discussed in further detail below.

The Heart Does Regenerate Itself … Sort of

It has been widely thought that the heart does not regenerate itself, with cardiomyocytes arrested in a post-mitotic state. In 2009, Bergmann et al. in their landmark paper in *Science* (Bergmann et al., 2009), reported that cardiomyocytes do in fact renew themselves at a rate of 1% per year during the first 25 years of life, and this is decreased to a rate of 0.45% per year by the age of 75. This means that less than half of the heart cells we are born with are changed during the average lifespan. How did they figure this out? Researchers took advantage of Cold War nuclear bomb tests that released Carbon-14 (prior to these tests mainly Carbon-11 was found in nature) into the atmosphere, which was subsequently incorporated into our DNA, to effectively carbon-date our cells.

Cardiovascular Cell Populations

The traditional cell model for cardiac tissue engineering has been neonatal rat cardiac myocytes. Neonatal rat cardiomyocytes are commonly used because they are relatively easy to obtain and culture, and exhibit phenotypic stability for up to 10 days in monolayer culture. Additionally, the neonatal cells provide a baseline from which maturation of the cells with biophysical signaling can be studied, and an excellent opportunity for comparing data between numerous studies from different research groups. With the use of neonatal rat myocytes, there is a mixed population of cardiovascular cells, although this can be enriched to select for cardiomyocytes, cardiac fibroblasts or endothelial cells.

It is becoming increasingly clear that a functional cardiac graft cannot uniquely contain cardiomyocytes, but rather should include various subpopulations, including vascular cells and cardiac fibroblasts. It has been shown that preculture of scaffolds with cardiac fibroblasts and/or endothelial cells may contribute to the stabilization of cardiomyocytes, and overall enhancement of the graft's function. This is likely due to the secretion of soluble factors and extracellular matrix components that increase adhesion and promote a functional phenotype in the co-cultured cardiac myocytes. Furthermore, endothelial cells have the unique capability to form stable vessel-like structures within the tissue, which may become functional vascular networks over time.

To make cardiac tissue engineering a clinical reality, human sources of cardiomyocytes and supporting vascular cells need to be established. Adult cardiomyocytes are terminally differentiated, and have an extremely low rate of self-renewal (Bergmann et al., 2009), necessitating the use of stem and progenitor cells. The presence of resident cardiac progenitor cells in the heart has been demonstrated (Laugwitz et al., 2005); however, the low frequency of these cells and very difficult harvesting do not make them a practical cell source. Adult stem cells (e.g., from bone marrow, blood or fat) have demonstrated little proliferative and myogenic potential that could be readily scaled to the necessary numbers needed for clinically relevant pieces of muscle. At this time, exogenous human stem cell sources – embryonic and induced pluripotent stem (iPS) cells – represent the two sources of cells with myogenic potential that are being actively explored. As with their application to other tissue types, the use of embryonic and iPS stem cells will need to be closely monitored for tumorigenic potential and genetic stability before they can be safely introduced into the clinical setting. Overall, the selection of specific cell populations will play an important role in the outcome of experiments, and each cell type will likely require optimization of the derivation and differentiation methods to achieve the formation of functional tissue grafts.

Choice of the Biomaterial Scaffold

Biomaterials for cardiac tissue engineering must have the appropriate porosity, structural arrangement, and stiffness and compliance characteristic of native myocardium. Several types of biomaterials have been explored for cardiac applications, and some of these categories are shown in Figure II.6.11.2. Broadly, these can be categorized as porous, fibrous or hydrogel based, and can be fabricated in a variety of ways to seed or encapsulate the cells within. For each of these categories, it is important to mimic the native architecture of the heart matrix at certain developmental stages. Numerous studies have been performed to achieve this goal using: (1) a porous matrix that allows for rapid inoculation of cells that will self-align; (2) a fibrous scaffold that provides the appropriate organization for alignment at the micrometer scale; or (3) a biologically derived hydrogel. Mechanical properties are dependent on the material and fabrication techniques, and are an important parameter, as substrate stiffness has been shown to affect stem cell lineage specification (Engler et al., 2006) and cardiomyocyte contraction.

Gel substrates used for cardiac tissue engineering include collagen, Matrigel, fibrin, and hyaluronic acid. Native tissue-derived porous scaffolds include decellularized myocardium (Ott et al., 2008; Wainwright et al., 2009) (discussed in the section Biophysical Stimuli: Biological, Electrical, and Mechanical Signals), which boast fully maintained ECM components, as well as intact mechanical properties and ultrastructure. Other ECM-derived products are commercially available as powders (that can be used to make hydrogel) or porous sheets. The biocompatibility, constituency and preserved bioactivity can make ECM a biomaterial of choice for *in vivo* cardiac applications.

Synthetic biomaterials including various polyesters (e.g., polyglycolic acid, PGA; polylactic acid, PLA) have

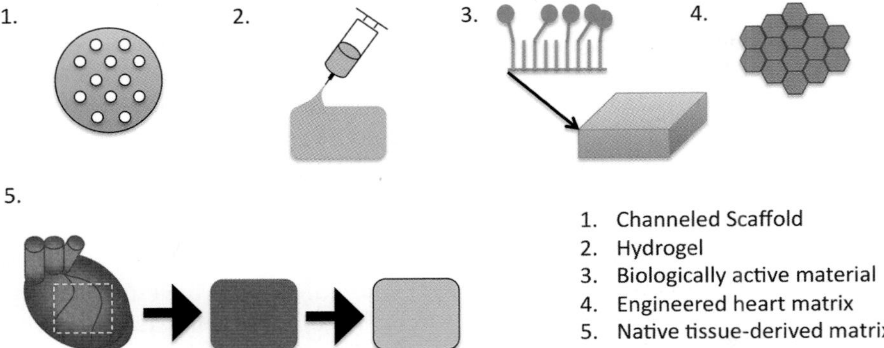

FIGURE II.6.11.2 Examples of biomaterials for cardiac muscle tissue engineering.

been explored (Akhyari et al., 2008), yet their lack of biological function makes these materials less amenable to tissue growth, and typically elicit an immune response *in vivo*. Various polymer structures have integrins and binding proteins tethered to them to increase cell adhesion and survival. Others incorporate microspheres loaded with various growth factors, with tailored kinetics such that local delivery of the growth factors to the cells within the polymer is time-dependent and can be used to induce differentiation or stabilize phenotype of the cell population. This custom tailoring of scaffolds for specific applications has enabled control of the physical environment of the cells, and – in a very precise manner – control of the temporal dynamics of the cellular microenvironment. A scaffold with bound targeted gene sequences has even been introduced as a means for ischemia-activated gene therapy for the heart with spatiotemporal control (Tang et al., 2005).

Structural scaffold design for the heart is particularly appealing given the anisotropy, high cell density, and high metabolic demand of the cells. Tailored to these three criteria, several groups have developed channeled and grooved scaffolds to foster cell alignment and elongation, and at the same time ensure sufficient cell infiltration in the scaffold and access to the oxygen and nutrients. Some examples include aligned woven or fibrous scaffolds (Boublik et al., 2005; Lunkenheimer et al., 2006), organ-printing methods with hydrogels (Mironov et al., 2003; Fedorovich et al., 2007), engineered honeycomb-like structures (Engelmayr et al., 2008), and use of passive or dynamic tension of hydrogels to induce cell alignment (Zimmermann et al., 2006a,b).

The combination of various biomaterials has led to results that may strike a balance between mechanical properties, biological function, and structural integrity. Using hydrogels to encapsulate cells and inoculate fibrous scaffolds is an approach that provides a way to infuse and retain cells within a stiff matrix, and have a biologically relevant microenvironment for the cell, while retaining the mechanical properties characteristic of the tissue (Radisic et al., 2004, 2007).

Biophysical Stimuli: Biological, Electrical, and Mechanical Signals

In recent years, there has been a general consensus that physiological stimuli need to be applied to the cultures of cells on scaffolds, to guide the functional assembly of tissues. The heart's complex function revolves around organized sequences of electrical and biological signals that induce mechanical contraction, and various groups have sought to apply one or more sequences of this paradigm in the *in vitro* cultivation of cardiac tissue. The biochemical, hydrodynamic, mechanical, and electrical signals have been explored both independently and in combinations, with a growing notion that a "biomimetic" approach, in which the cells are subjected to sequences of signals they encounter *in vivo*, is necessary to unlock the full biological potential of the cells.

Biochemical Cues. The inclusion of soluble factors in culture media is critical for inducing and maintaining the differentiation of stem cells into cardiac phenotypes. In grafts that include vascular cell populations, growth factors such as VEGF and PDGF are also needed to promote vascular network formation within the tissue, *in vitro* or *in vivo*. Finally, the supporting cells (e.g., pericytes) or the factors they release may be needed to stabilize and mature the newly formed vascular conduits.

A 2007 paper by Laflamme and Murry (Laflamme et al., 2007) described the use of a "prosurvival cocktail" (PSC) to protect against apoptosis of embryonic stem cells that are injected directly into the ischemic heart environment and exposed to harsh apoptotic and pro-inflammatory factors. The inclusion of such factors utilized in this cocktail with grafted engineered tissues may prove instrumental in the survival of the muscle in the *in vivo* setting.

One of the major constituents of the PSC includes Matrigel to act against anoikis (apoptosis due to an inability of a cell to form cell–matrix adhesions). Additional factors include IGF-1 (insulin-like growth factor 1) to activate the Akt pathway (involved in cell survival via anti-apoptotic processes); pinacidil (KATP channel opener which mimics ischemic preconditioning); cyclosporine A, to block cyclophilin D-dependent mitochondrial death pathways; the pan-caspase inhibitor z-VADfmk (anti-inflammatory and anti-apoptotic agent); and a cell-permeant TAT peptide from BclXL, to block mitochondrial apoptotic pathways. Interestingly, Matrigel, a mouse basement membrane preparation, was the only factor shown to induce significant protection when used alone, suggesting that anoikis due to cell detachment is an important path to death after cell transplantation. In addition to promoting cell adhesion through integrin receptors, Matrigel can facilitate cell retention.

Matrigel is a heterogenous solution made of collagen and laminin, rich in extracellular matrix components and growth factors, that promotes the survival and differentiation of various cell types. Matrigel is unique in its ability to induce endothelial cells to form tube-like structures, and has been used extensively in cardiac tissue engineering applications. Radisic et al. (2004) used it to seed Ultrafoam collagen sponges, and with electrical signaling induced functional tissue assembly of neonatal rat cardiomyocytes. Using the same cells, Zimmermann et al. (2002a, 2006a) used a collagen and Matrigel mixture to achieve similar results using mechanical stimulation, and further demonstrated successful engraftment in injured hearts of rats. Lastly, Laflamme et al. (2007) used Matrigel as a delivery vehicle for embryonic stem cells to the heart, and cited its value in guarding against cell anoikis in the infarct. The repeated use – and advantages – of Matrigel in cardiac tissue engineering is not yet fully understood, but it appears critical to the success of tissue assembly and successful engraftment *in vitro*.

Electrical Cues. The first report that electrical stimulation could be used to enhance functional tissue assembly by cardiac myocytes cultured on biomaterial scaffolds came in 2004 (Radisic et al., 2004) (the study is discussed in the section on Biomimetic Culture Systems for Cardiac Tissue Engineering). This system utilized carbon electrodes to deliver the current to the culture media, wherein an electric field was established. Since this landmark study, others have explored different methods and parameters for stimulating the cells, and several platforms for electrical stimulation of cultured cells and tissue constructs have been proposed (Tandon et al., 2009).

The same group later analyzed various candidate electrodes (nanoporous carbon, stainless steel, titanium, and titanium nitride) to determine the best electrode for cardiac tissue engineering applications (Tandon et al., 2006). By studying electrode characteristics and charge-transfer profiles, including Faradaic and non-Faradaic charge transfer, current injection, carbon was identified as the best electrode for such an application. This was further confirmed by the quality of engineered cardiac tissues that resulted from exposure to electrical fields, including assessment of the amplitude, and threshold of contraction and maximum capture rate.

In 2011, the Radisic group compared monophasic and biphasic stimulation of cardiac constructs with differing ratios of cardiovascular cell populations (Chiu et al., 2011). The biphasic stimulation group significantly differed from monophasic stimulation, and achieved higher cell density, upregulation of Cx-43, electrical excitability, three-dimensional organization and elongation of cells, and – most importantly – lower excitation thresholds than constructs subjected to monophasic stimulation. Additional studies have determined that direct currrent (DC) stimulation of cardiomyocytes, a regime relevant to early cardiac development, may also enhance cell elongation and migration. The electric field-mediated effects of galvanotaxis and galvanotropism may play a future role in three-dimensional cultures, as the directionality of cells plays an especially important role for the heart tissue.

The use of microarray bioreactors and bioreactors with interdigitated arrays of electrodes has allowed systematic studies of electrical stimulation of single cells, embryoid bodies formed from human embryonic stem cells, and cardiac tissue constructs (Marsano et al., 2008; Cimetta et al., 2009a; Serena et al., 2009; Tandon et al., 2009). These technologies have provided critical information regarding how an external electric field is processed by cardiomyocytes, and continues to contribute to our understanding of electrical stimulation on centimeter-scale tissues.

Mechanical Conditioning. As an electrical signal propagates through the gap junctions of cardiomyocytes in the form of ions each individual cardiac myocyte contracts, resulting in the synchronized contraction of the heart muscle. Fink et al. (2000) proposed that directly applying the mechanical stretch to embryonic chick or neonatal rat heart cells cultured in hydrogel would increase cell alignment and function. In that and later studies (Zimmermann et al., 2006a,b), unidirectional and cyclic stretch of the hydrogel constructs resulted in mature, adult like engineered heart tissue, and *in vivo*, led to stable muscle grafts that prevented deterioration of heart function and exhibited electromechanical integration with the host tissue. While the mechanism by which heart cells sense and generate force in response to exogenous factors is not totally understood, these research groups have demonstrated that their role is instrumental in forming functional cardiac tissues. Various mechanical stimulation studies are discussed in detail in the section on Culture in Hydrogel with Mechanical Stimulation.

Mass Transport

In the human body, oxygen transport through the blood and to the tissues relies on the combination of convective (bulk movement of oxygen in blood) and diffusive (passive movement to tissue driven by concentration gradients) components. Hemoglobin inside red blood cells enables high concentrations of oxygen to be transported to tissues. Oxygen has low solubility in blood (only ~7 mg/liter), but when bound to hemoglobin, the oxy-hemoglobin complex can be carried through blood, enabling a high rate of oxygen transport to provide sufficient oxygen supply to all of the cells in the body (Boron, 2002). The delivery of oxygen from red blood cells to tissues is facilitated by oxygen gradients. Oxygen must unbind from the hemoglobin and be transported across the blood vessel walls before reaching tissue. The oxygen gradients allow for primarily transcellular transport through capillary endothelium, as well as some convective-driven transport through the intercellular clefts in the capillary endothelium. Oxygen concentration gradients facilitate its movement through the depth of tissues, allowing a rapid and continuous supply of oxygen to reach tissues at the rate at which it is consumed.

To achieve sufficient mass transport throughout engineered tissue, researchers have used perfusion-based tissue culture, channeled scaffolds, and oxygen carriers, to enhance availability of oxygen and nutrient supply throughout the depth of the tissue (Radisic et al., 2005, 2006a,b, 2008). Early studies characterized the diffusion limits of cardiac tissue, demonstrating oxygen gradients prevent sustained cell viability in engineered tissue at >100 μm construct depth, whereas cardiac muscle is typically 1 cm in thickness. It was demonstrated that perfusion of culture medium supplemented with perfluorocarbons (PFCs) significantly enhanced cell viability, and penetration of oxygen and nutrients into greater depth of cultured constructs.

To enable the application of convective regimes of oxygen transport, specific scaffold architectures needed to be developed. While medium flow through porous scaffolds enables higher cell survival, perfusion of medium through the scaffolds resulted in non-physiologic shear stress on the cells. In native tissue, blood flows through the vasculature lined with endothelial cells that are exposed to the shear forces, while the muscle is shielded from these forces. Exposure of muscle cells to shear may affect their phenotypic stability and ability to form the cell–matrix interactions necessary for function.

To provide a native-like oxygen supply, without exposing cells to hydrodynamic shear, channeled scaffolds have been introduced as a means of mimicking vasculature (Radisic et al., 2008). While tailoring scaffolds to increase porosity was an early approach to minimizing diffusion limits to provide oxygen and nutrients to the cells, later advances in scaffold development used porous elastomers that were laser pierced to enable medium perfusion through the channels, rather than through the cell-seeded bulk phase of the scaffold. This is shown in the bottom left panel of Figure II.6.11.3. While the channels made thus far have been in the order of 200 μm in diameter, and spaced several hundred μm apart, *in vivo* capillaries are in the order of 10 μm, with intercapillary distances of about 20 μm, approximately one capillary between every two cardiomyocytes (Rakusan and

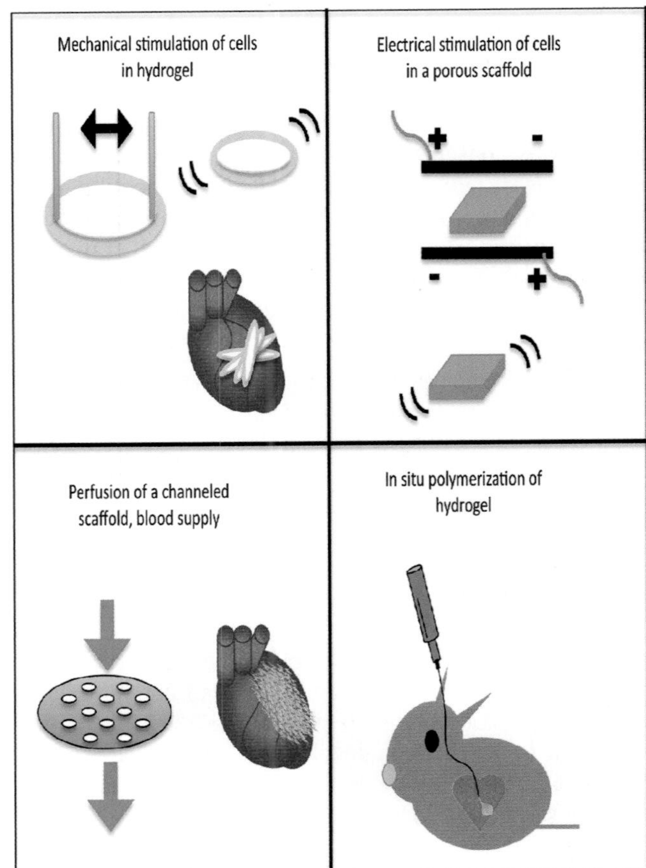

FIGURE II.6.11.3 Recent studies in cardiac muscle tissue engineering.

Korecky, 1982). Patterning of high resolution features to create vascular structures at physiological density and size, within engineered heart muscle, may be the next step forward in scaffold development to combat the issue of sustaining engineered heart muscle.

Bioreactor systems with medium perfusion are also very convenient for achieving spatially uniform, high-density seeding of cells on scaffolds with the maintenance of cell viability. Perfusion of media through non-channeled scaffolds exposed cardiac cells to non-physiologic hydrodynamic shear, as intramyocardial capillaries typically shield cardiomyocytes from shear stress while allowing for mass transport throughout the depth of the tissue. Channeled scaffolds were introduced into perfusion culture systems to mimic the capillary bed and reduce the amount of shear seen by muscle cells. Channels were strategically placed to ensure sufficient oxygen and nutrients would diffuse to the cells in thick constructs. Methods developed for direct perfusion (interstitial flow) cell seeding were later extended to seeding of channeled scaffolds to breach closer to the *in vivo* setting. The main technical problem that needed to be solved is to perfuse cell suspension through the scaffold pores, while keeping the channels open.

A perfusion culture system was developed for use with channeled, porous or fibrous scaffolds, and the PFC-supplemented media to provide *in vivo*-like oxygen supply (Radisic et al., 2005). Mathematical modeling was also used to predict interstitial flow through each scaffold type that would support low levels of hydrodynamic shear to ensure cell viability, and the ability for cells to elaborate their matrix was not compromised.

More recently, methods for perfusion seeding of both non-channeled and channeled scaffolds with cardiomyocytes have been developed (Radisic et al., 2008). Notably, these methods enable spatial specificity of the seeded cells, with cardiomyocytes seeded into the porous bulk phase, and endothelial cells seeded onto the channel walls.

The use of perfusion-based culture systems combined with other biophysical signaling paradigms in advanced bioreactors may further enable researchers to culture thick, functional cardiac tissues for large animal-scale studies in the future. One of the conditions for the effective use of these systems is the development of biomaterial scaffolds with porosity in the bulk phase (to allow cell attachment, and protect the cells from flow-induced hydrodynamic shear), and an array of parallel channels (to allow organized flow of culture medium, with short diffusional distances, and lumens for the attachment of endothelial cells).

Biomimetic Culture Systems for Cardiac Tissue Engineering

Several groups have undertaken development of bioreactors to recapitulate the *in vivo* milieu in a controllable, *in vitro* system (Bursac et al., 1999; Zimmermann et al., 2002b, 2006b; Radisic et al., 2004; Gerecht et al., 2007; Marsano et al., 2008; Tandon et al., 2008, 2009; Cimetta et al., 2009b,c; Serena et al., 2009). A combination of scaffolds and bioreactors allows for integration of environmental control (temperature, pH, osmolality, oxygen levels, nutrients, and metabolites) and biophysical stimuli (electrical, mechanical, perfusion, hydrodynamic shear) to guide cell differentiation and assembly into a syncytium. Environmental control, as well as modular systems for introducing mechanical and electrical signals, allows for these culture systems to be used as high-throughput screening platforms for drugs. In cardiac tissue engineering, the principal requirements include sufficient oxygen supply, mechanical and electrical signals, and the application of molecular factors.

Evaluation of Engineered Cardiac Tissue

Engineered cardiac constructs are evaluated with respect to their molecular composition (marker expression, composition), structure (cell morphology, ultrastructure, tissue organization, and anisotropy), and function (electromechanical coupling, signal propagation, force generation, responses to pharmaceutical agents) (Zimmermann et al., 2002a, 2006a; Radisic et al., 2004).

Both *in vitro* and *in vivo* assessments can be used to characterize the function and therapeutic potential of engineered cardiac muscle. Gross assessments include the seeding distribution, morphology, and survival of cells in a scaffold. Additional staining can be done for cardiac proteins (connexin-43, cardiac troponin I, α- and β-Myosin Heavy Chain), as well as cytoskeletal proteins (actin), and high-magnification imaging (TEM) to evaluate ultrastructure (A, I, M, Z bands, mitochondria, and cardiac-specific subcellular structures).

Functional assessments include quantifying the frequency and fractional area or volume change of an engineered construct and can be done optically, by taking videos and using an algorithm to quantify the contractions. Additionally, inclusion of a force transducer allows for estimation of the force of contraction generated by the cells. To assess electrophysiological parameters, constructs are paced with an electric field, to determine excitation threshold and maximum capture rate. Alternatively, electrodes and imaging with voltage sensitive dyes are used to measure signal propagation and conduction velocity.

Translational studies involve *in vivo* models of human disease, in which engineered heart tissues are tested for their ability to prevent deterioration of, or reverse, heart damage. Animals undergo a surgical ligation of their left anterior descending (LAD) artery to prevent blood flow to the left ventricle (induce a myocardial infarction). This is the most common model of myocardial infarction, and engineered muscle constructs are implanted over the injured site; in some studies, the damaged ventricle is excised and replaced with the engineered tissue. Following several days to weeks *in vivo*, animals are sacrificed, at which point hearts are excised for histological analysis. Typically, researchers look for signs of an immune response, integration of the implant with host tissue, cell migration, differentiation or change in phenotype, vascularization of the infarct or construct, and signs of apoptosis or necrosis in the graft.

Functional assessments *in vivo* include standard echocardiography measurements of the heart that can be taken at various stages, including before and after infarction of the heart, and at various time points following implantation of a graft. Echocardiography can provide critical information regarding the function of the heart. Key parameters include ejection fraction, cardiac output, end systolic diameter, end diastolic diameter, and fractional shortening. An imaging technique, SPECT (Single Photon Emitting Computed Tomography) can be used to determine perfusion in the heart and graft, and various cell-labeling techniques can be employed to track the movement and fate of cells *in vivo*.

SOME REPRESENTATIVE STUDIES IN CARDIAC TISSUE ENGINEERING

There are several approaches for engineering cardiac tissues designed to overcome diffusional limitations of mass transport, and to maximize the conductive and contractile functions in a different way, that use a remarkably diverse range of biomaterials. The most extensively used approaches include cell cultivation: (1) in stackable cell sheets; (2) anisotropic cardiac-like scaffolds; (3) on porous scaffolds derived from native heart tissue; (4) in hydrogels with mechanical stimulation; (5) in porous scaffolds with electrical stimulation; and (6) injectable hydrogels.

Scaffold-Free Methods: Cell Sheets and Discs

The simplest form of cardiac tissue engineering is scaffold-free cultivation of cell sheets. In their 2002 paper Tatsuya Shimizu and colleagues (Shimizu et al., 2002), reported the creation of scaffold-free cardiac grafts using a novel cell-sheet stacking technique. The group cultured neonatal rat cardiomyocytes on temperature-sensitive polymer surfaces that, with temperature change, allowed for detachment of intact monolayer cell sheets from the adhesive surface, leaving gap junctions, integrins, and extracellular matrix produced by the cells intact. The group stacked four monolayer sheets and demonstrated the formation of intracellular junctions between the layers, resulting in three-dimensional cardiac constructs approximately 80 μm thick. The study demonstrated the viability and contractile function of cardiac grafts for up to three months when implanted subcutaneously, stressing the utility of this approach. In this study, only four sheets were layered to form a graft, although it is conceivable that thicker cardiac tissue grafts could be generated in this manner. To address the diffusion constraints associated with multiple-layer stacking, the authors propose that endothelial monolayer sheets could be placed between layers to promote vascularization.

The approach of engineering heart tissue without the use of a scaffold has been more recently explored by the Murry group (Stevens et al., 2008), whose development of a scaffold-free cardiac patch with human embryonic stem cells demonstrated that hES matrix formation and cell aggregation alone were sufficient to generate synchronously contracting cardiac tissue. The group cultured embryonic stem cell-derived cardiomyocytes in a rotating orbital shaker over 11 days, where cell aggregates into macroscopic cell discs ranging from 1.9 to 10.7 mm in diameter, depending on cell number. Embryonic stem cells differentiate into a heterogeneous cell population, and can give rise to any of the three germ layers.

To characterize the cardiac patches, researchers used antibodies against epithelial, endodermal, and neuroectodermal markers, and observed minimal presence of non-cardiac cell populations by day 11 in culture. To complement this trend, an enriched cardiomyocyte population was observed over time, as non-cardiomyocyte populations appeared to diffuse out of the disc into the media, while cardiomyocyte populations within the graft matured. At day 2 of culture, cardiomyocytes exhibited high levels of Nkx 2.5 (an early stage embryonic cardiac marker), but low levels of ß-MHC (a marker of maturation), demonstrating the maturation of Nkx 2.5 expressing cells as they increasingly expressed ß-MHC. Calcium transients were imaged to show that the cardiomyocytes were electromechanically coupled, and that synchronous beating was associated with increases in intracellular calcium levels that are necessary to initiate cell contractions.

Reporting that greater than 75% of the cells within the graft were cardiomyocytes, with the remaining 25% most likely a mixture of cardiac fibroblasts and endothelial cells, the Murry group demonstrated a feasible way to generate synchronous cell aggregates *in vitro* (Stevens et al., 2008). Further studies would need to characterize the mechanical integrity of the grafts, three-dimensional architecture and organization of cells, and quantify their ability to generate sufficient force to adapt to the *in vivo* environment.

Both studies that we discuss here demonstrate the possibility to develop cardiac tissues without the use of biomaterial scaffolds. However, it remains to be seen whether the scaffold-less cardiac tissues withstand mechanical forces required to function as cardiac muscle, and imposed by engineered biomaterials specific for cardiac tissue engineering.

Cardiac-Specific Biomaterials

With the aim of modeling the anisotropy of native cardiac tissue, Lisa Freed's group at MIT developed an "accordion-like honeycomb" using microfabrication technologies and Poly (glycerol sebacate) (PGS) polymer (Engelmayr et al., 2008) (Figure II.6.11.2, panel 4). The goal was to mimic the anisotropy of the heart through varying alignments of pores, rather than use typical scaffolds that provide unidirectional alignment representative of only small regions of isotropy in the heart. PGS has been extensively used in cardiac tissue applications, because of its high and controllable elasticity and stiffness, as well as slow degradation via hydrolysis. Scaffolds were fabricated to have physiological mechanical properties, and extensive testing of the honeycomb-like scaffolds demonstrated remarkable stability over time (determined via fatigue testing over 24 hours to 3 weeks), and had anisotropic mechanical properties.

These scaffolds were cultured with neonatal rat heart cells, but without any external stimuli. The honeycombs successfully guided cells to orient in the preferred direction, and engineered grafts achieved DNA content to be in line with previously reported engineered tissues. Electrophysiological recordings demonstrated synchronous contractions of cells within scaffolds, and when comparing engineered accordion-like versus rectangular and square honeycomb tissues, a hierarchical dependence of mechanical properties on degree of anisotropy was found. Specifically, accordion-like honeycomb has a higher degree of anisotropy as compared to rectangular honeycomb, whereas square honeycomb is isotropic. Using accordion-like honeycomb structures, engineered tissues were found to have mechanical properties most closely resembling right ventricular myocardium of the rat. The study demonstrated the importance of microstructure and scaffold architecture on the ability for cells to electrically couple and achieve physiological mechanical properties, although the 250 μm thickness of the grafts necessitates further development for *in vivo* relevance, where vascularization of thick grafts remains crucial.

Native Myocardium as a Biomaterial

A 2008 study reported by Doris Taylor's group (Ott et al., 2008) revealed that decellularized rat hearts could be seeded with cardiomyocytes and endothelial cells, resulting in regions of contractile activity observed following four days in culture. Decellularization of tissues, pioneered by the Badylak laboratory (Robinson et al., 2005), removes all cellular elements from a tissue, leaving an intact, functional extracellular matrix (process demonstrated in Figure II.6.11.2, panel 5). The architecture and mechanical properties of the native tissue matrix provide a natural environment for cells to orient, integrate, and remodel their environment. Decellularized hearts were first perfusion seeded with endothelial cells to line the native vasculature, followed by seeding of cardiomyocytes to repopulate the acellular myocardial regions. Continuous perfusion of culture media through the heart's vasculature took advantage of the natural network that allows for sufficient nutrient and oxygen exchange throughout the depth of the tissue. Adaptations of this approach may be promising for creation of cardiac patches, using pieces of intact decellularized myocardium as a scaffold. Additionally, further work similar to that of the Badylak lab, where decellularized tissues are emulsified and reconstituted as an "extracellular-matrix" gel may prove beneficial, as it incorporates cardiac-specific extracellular matrix components in the appropriate ratios, and enables tailoring of mechanical and degradation scaffold properties.

Culture in Hydrogel with Mechanical Stimulation

The model system of Eschenhagen and Zimmermann used mechanical conditioning of neonatal rat cardiomyocytes encapsulated in ring-shaped hydrogel made of a mixture of collagen and Matrigel. Early studies using this system (Fink et al., 2000) used a mixture of collagen type I and Matrigel with encapsulated neonatal rat myocytes to form gels that were subjected to unidirectional and cyclic stretch over six days. The functional improvements in cell morphology, their increased mitochondrial density and force of contraction, were again demonstrated in their subsequent studies. Using an improved set-up for mechanical stretch, the same group compared the engineered heart tissue (EHT) to newborn and adult rats, concluding that mechanical stimulation drove the constructs to a more mature cardiac muscle structure (Zimmermann et al., 2002b). This was further evidenced by the presence of mature sarcomeric structure, membrane junctions (adherens, gap junctions, desmosomes), T-tubule system (transverse tubules are central to electromechanical coupling), and the presence of a basement membrane surrounding myocytes (suggesting they secrete ECM to enhance and stabilize their alignment). Force generation of the EHTs approached that of native heart muscle (0.4–0.8 mN twitch force, 0.1–0.3 mN restin tension), stable, typical resting membrane potentials (−66 to −78 mV) and characteristic waveform kinetics.

In their 2006 study Zimmermann and colleagues (Zimmermann et al., 2006a), implanted 1–4 mm thick, 15 mm diameter EHTs into immunosuppressed rats, and at 28 days postoperatively found significant (400 μm thick) engrafted muscle layers in the host heart tissue. A schematic of the culture and implantation is shown in Figure II.6.11.3, top left. The grafts demonstrated complete integration and functional coupling with the host tissue, as evidenced by undelayed signal propagation through the muscle, no evidence of arrythmias, and additionally the grafts delayed wall dilation, contributed to ventricular wall thickening, and demonstrated improved fractional area shortening (FAS) as compared to all control groups. Overall, the tissue engineering systems developed by this group have rigorously demonstrated the importance of mechanical forces in regulated tissue development from the subcellular to macroscopic level, with functional EHTs *in vitro*, and their success at delaying heart deterioration *in vivo*. Further understanding of the role of mechanical stimulation, and how conditioning regimes can be tailored to complement the mechanical properties of scaffolds, should provide great insight into regulating development of force by cells, preventing hypertrophy, and even affecting differentiation pathways of stem cells as they mature to cardiomyocytes.

Culture in Porous Scaffold with Electrical Stimulation

In the 2004 study by the Vunjak-Novakovic group (Radisic et al., 2004), electrical stimulation has been established as a method for aiding in the function assembly of myocardium in a three-dimensional culture system. Neonatal rat myocytes cultured in Matrigel on porous collagen sponges were precultured before being subject to electrical stimulation for 5 days. A schematic of this set-up is shown in the top right panel of Figure II.6.11.3. The electrical stimulation regimen was designed to induce synchronous contractions during cell culture, by selecting physiologically appropriate parameters: 1 Hz, 2 ms pulse duration, at 5 V/cm, monophasic, square wave.

In comparison to non-stimulated controls, this study demonstrated that electrical stimulation resulted in cell alignment, increased the frequency of gap junctions, and increased amplitude of synchronous construct contractions. The functional organization of the cardiac muscle was further evidenced by ultrastructural organization, with upregulation of cardiac troponin I, sarcomeric alpha-actin, Cx-43, alpha-MHC, and beta-MHC. The number of mitochondria present between myofibrils was also increased, which is another significant finding, as mitochondria function as calcium stores in mature cardiomyocytes, allowing for rapid contraction of cells when calcium is released and binds to TnC to initiate contraction.

The study also showed that the outcomes of electrical stimulation strongly depended on the time when electrical activity was initiated. Introducing the signals too early in culture deteriorated the cardiac phenotype rather than enhanced their contractile apparatus, as was seen when electrical stimulation began on day 5 of culture. This study introduced electrical stimulation as a key regulator of functional tissue assembly, and the differentiation and maturation of differentiated stem cells.

Additional studies have explored the importance of electrical signal type (AC versus DC, biphasic versus monophasic, square waveform versus others) as it relates to functional cardiac tissue assembly (Chiu et al., 2011). It may be that different stimulation paradigms play a definitive role at different time points in tissue development, as is true in the developing embryo, and further studies will need to elucidate which signals are appropriate for which culture regimen. Further understanding of electrical–ion interfaces, and signal transport through media and scaffolds to the cells, will help us fully understand what signals are being seen by the cells themselves. Development of conductive biomaterials to facilitate signal transduction to cells, and foster intercellular communication, may provide a means for cells to couple to each other at the culture phase (*in vitro*) and at the integration phase (*in vivo*).

Injectable Hydrogels

Currently, cell therapy in humans for cardiovascular diseases injects adult stem cells into the heart in a saline or physiologic solution. Localization and method of cell delivery, introducing cells into the ischemic environment of an injured heart, without the ability to form cell–matrix junctions, limits the survival and engraftment rate of delivered cells. This method is shown in the bottom right panel of Figure II.6.11.3.

To address this issue, Martens et al. (2009) introduced the idea of using hydrogels as a delivery vehicle to introduce and retain cells *in situ*. The advantages of such an approach include providing a cell-friendly microenvironment to engrafted cells to prevent anoikis and shield cells from the infarcted muscle, and the ability to deliver cells directly into the muscle rather than laying over a cardiac patch that must integrate with the muscle and vasculature from the surface. Hydrogels can be tailored to polymerize as a result of chemical- or temperature-dependent conditions, and can be delivered with cells by minimally invasive catheter-based procedures, with high accuracy of cell localization to the area of interest.

CHALLENGES AND FUTURE APPLICATIONS

Cardiac muscle engineering remains a relatively young and emerging field, and many of the key signals that regulate the tissue's assembly have already been identified. As researchers strive towards creation of full-thickness, functional heart muscle, they face many of the same challenges as other tissue engineering applications.

Some of the key challenges facing cardiac tissue revolve around the complexity of both biochemical and physical signaling that is involved in the tissue function. The high density and metabolic activity of the tissue demands that researchers find a way to vascularize the tissue even *in vitro*. Achieving the density of highly packed myocardium is an additional challenge, where both availability of cells for such studies, and the ability to successfully maintain such high densities in culture, make this difficult to achieve. Although millimeter-thick cardiac tissues may be sufficient for *in vitro* studies, clinically relevant pieces of tissue on the order of centimeters need to be developed to conduct appropriate translational studies.

The ability for engineered constructs to generate physiologic force remains a challenge, although initial results show that mechanical conditioning of constructs enhances the threshold of force generation. Evaluation of force generation, as well as contraction and electrical activity of the cardiac tissue, is further compounded by the heterogeneity of current constructs. Past studies using neonatal rat heart cells will contain cardiac myocytes, but also fibroblasts and endothelial cells which do not contain the appropriate contractile apparatus to participate in electromechanical activity. Without control of the cell type distribution, as well as non-uniform cell seeding, it is difficult to distinguish between signal propagation and force generation that is due to the contribution of pure cardiomyocytes. Thus, it may be useful to study pure cardiomyocytes to probe signal propagation and force generation of the cells, as well as develop better controlled distribution of cell types within the engineered constructs.

Generating Cardiac Grafts. Generation of replacement tissues for the human body is the ultimate goal of tissue engineering. For cardiac tissue engineering to become a clinical reality, scientists will have to select the appropriate cells, biomaterials, and regulatory factors. One possibility is to guide endogenous cells to repopulate an engineered biomaterial implanted in the body. Another possibility is to grow a fully functional piece of tissue in the laboratory. Stem cells remain the most promising source of cells, and various groups are exploring human adult, embryonic, and induced pluripotent stem cells.

Challenges in the translation of engineered cardiac grafts will require phenotypic stability of the cells and mediating the immune host response to the graft. Specifically, engineered cardiac grafts will have to: (1) be conditioned to tolerate the stiff, ischemic environment of injured myocardium; (2) generate enough force to enable heart function and prevent further remodeling of the heart; (3) electrically couple with host cells to prevent arrhythmias; (4) connect to host vasculature to ensure graft survival; and (5) be resilient to immune cells, inflammatory and apoptotic signals present in the heart following myocardial infarction.

Our inability to effectively vascularize tissue grafts, and most critically cardiac tissue grafts, remains the main factor limiting the size of grafts. This problem has been approached in various ways, including vascular cells in the construct to facilitate functional vessel formation, using channeled scaffolds seeded with endothelial cells to mimic native vasculature, and including pro-angiogenic factors to attract native vasculature to infiltrate into the scaffold once implanted *in vivo*. Yet, for many applications in tissue engineering, these approaches have not been sufficient to enable cell survival *in vitro*, and especially not *in vivo*. The critical challenge *in vivo* is that once an engineered tissue is removed from its *in vitro* culture system, where medium perfusion enables nutrient transport, it does not have the ability to immediately reconnect with the host vasculature. Until the graft is vascularized, insufficient supply of oxygen and nutrients renders it ischemic, and ultimate apoptosis and cell necrosis lead to a failed graft. Future approaches will need to address this critical period, from when a construct is removed from its *in vitro* culture and is implanted, until sufficient revascularization occurs. Solutions to enable immediate vascularization or sustained cell survival until that point is reached will greatly advance graft survival and function.

Models of Human Disease. Developing functional heart tissue *in vitro* has the added benefit of being used as a model of human disease. Scientists may use engineered heart tissues for drug screening assessments, and apply different physical strains on the heart to study the onset on various pathologies. Furthermore, the emergence of induced pluripotent stem cells will allow scientists to investigate the onset and progression of various diseases in an *in vitro* setting. Cells obtained from patients with a history of cardiovascular disease can be made into iPS cells and differentiated to cardiovascular cells. These populations may then be used to engineer tissues and study various pathologies in the *in vitro* models of human disease. Such studies may prove valuable in screening for new technologies and treatments, and provide an additional assessment between animal studies and clinical translation by predicting how various therapies may fare in the human body.

Scaffolds of the Future. This chapter focused on cardiac tissue engineering for various applications – regeneration of the ailing heart, models of studying human disease, and drug screening. The complexity of heart muscle demands the use of multiple physical and biochemical signaling, similar to those that govern cell differentiation during normal development. The scaffold remains a key component in the tissue engineering paradigm, as it determines interactions between the cells and their environment. To date, researchers have developed scaffolds that guide cell alignment and improve accessibility to oxygen and nutrients through networks of channels in conjunction with medium perfusion. The mechanical properties – stiffness, elasticity, degradation rates – have been tailored to aid the development of functional cardiomyocytes with physiologic force generation. In addition, the adhesion molecules facilitate cell–matrix interactions, and tethered growth factors promote cardiac development. When placed into bioreactors, the cells in scaffolds can be subjected to stretch, hydrodynamic shear, and electrical stimulation.

The next great advances in tissue engineering may require novel scaffold development, with the creation of scaffolds that are adaptable to their environment, facilitating the communication between cells and exogenous signals. Cardiac tissue engineering scaffolds will ultimately need to facilitate functional tissue assembly by inducing the cells to align, interconnect, and communicate in the anisotropic fashion of native heart muscle. The scaffold should ideally act as that basement membrane surrounding individual myocytes, yet allow for 100 million cells to be packed in 1 cm^3.

> Microtechnologies have been extensively used to study developmental processes of cardiac cells and tissue assembly, and to build *in vitro* cardiac models for study of disease processes. In their 2009 paper, Cimetta et al. (2009b) describe microbioreactors for cultivation of embryonic stem cells, demonstrating that controlling transport phenomena can direct regulatory signals in a spatiotemporal manner and enable us to recreate physiological states *in vitro*. Similarly, Khademhosseini et al. (2007) and Cimetta et al. (2009c) used micropatterning to create cardiac organoids that can be used as *in vitro* models for physiological studies of cardiac processes at the level of a single cardiac fiber.

WORKED EXAMPLE

The heart is a densely packed, metabolically demanding muscle, which is highly vascularized to ensure sufficient supply of oxygen and nutrients through the depth of the tissue. Metabolically demanding cardiomyocytes demonstrate poorer organization and weaker cardiac phenotype at oxygen concentrations of less than 80 μM. Channeled scaffolds have been proposed as a way to enhance nutrient supply throughout the thickness of engineered cardiac tissue by mimicking vasculature in cardiac muscle (Figure II.6.11.4).

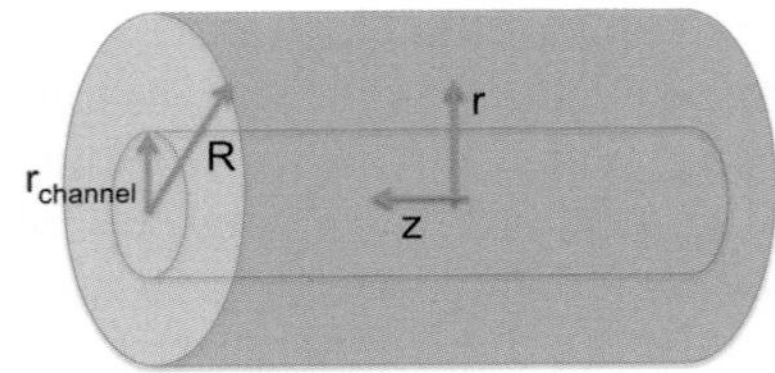

FIGURE II.6.11.4 Schematic presentation of engineered cardiac muscle tissue with channels for flow of culture medium within a scaffold populated with cardiac cells. Convective-diffusive transport of oxygen is analyzed in cylindrical coordinates.

Culture medium has an oxygen concentration of 220 μM, and the scaffold has 100 μm diameter channels, spaced 400 μm apart. Values for consumption (k) of resting, non-contractile cardiac tissue and the diffusion constant can be found below.

Assume: (1) an isotropic, constant diffusion term; (2) a steady-state system; (3) constant tissue consumption rate; and (4) axisymmetric, well-mixed media flowing through the channel.

Boundary Conditions: Media is well mixed, with constant oxygen concentration $C(r = a) = C_0$; no supply from neighboring vessels (no flux at $r = R$, so $\partial C/\partial r$ $(r = R) = 0$, where r and R are the radi of the channel and tissue cylinder as described in Figure II.6.11.4 and (a) is an arbitrary radial position with the tissue.

(a) Given the following equation and above assumptions, develop an expression to describe the oxygen concentration within tissue surrounding a channel, based on the radius of the channel, and dependent on the radius from the channel. Is the oxygen concentration at the midpoint between channels elevated enough to sustain the cardiomyocytes?
(b) How may channeled scaffolds contribute to increased tissue viability in standard culture systems?
(c) Name two additional components that can be added to a culture system to increase cell viability in cardiac constructs.

$\varrho_{tissue} = 1.06$ g/mL
$r_{channel} = 50$ μm
$d_{interchannel} = 400$ μm
$\kappa = 2.0$ mL O_2 /min*100 g tissue
$D = 2.0 * 10^{-5}$ cm²/s
$\varrho_{O2} = 1.308$ g/L
t(s) = time
C(r, t) = oxygen concentration

SOLUTION

(a)

$$\frac{\partial C}{\partial t} = div(D \nabla C) - \kappa$$

$\partial C/\partial t$ disappears based on Assumption 2.
∇D disappears based on Assumption 1.
$\partial^2 C/\partial \theta^2$ and $\partial^2 C/\partial z^2$ disappears from Assumption 4.
K = constant based on Assumption 3.

where:
z = axial distance (Fig. II.6.11.4)
theta = angle around the central axis

Apply assumptions to starting equation to arrive at an expression.

$$\frac{\partial C}{\partial t} = div(D \nabla C) - \kappa$$
$$\frac{\partial C}{\partial t} = \nabla D \nabla C + D \nabla^2 C - \kappa$$
$$D\nabla^2 C - \kappa = 0$$
$$D\left\{\frac{1}{r}\frac{\partial}{\partial r}\left(r\frac{\partial C}{\partial r}\right) + \frac{1}{r^2}\frac{\partial^2 C}{\partial \theta^2} + \frac{\partial^2 C}{\partial z^2}\right\} - \kappa$$
$$\frac{D}{r}\frac{\partial}{\partial r}\left(r\frac{\partial C}{\partial r}\right) = \kappa$$
$$\int \frac{\partial}{\partial r}\left(r\frac{\partial C}{\partial r}\right) dr = \int \frac{\kappa}{D} r dr$$
$$r\frac{\partial C}{\partial r} = \frac{\kappa}{D}\frac{r^2}{2} + A$$
$$\frac{\partial C}{\partial r} = \frac{\kappa}{D}\frac{r}{2} + \frac{A}{r}$$
$$\int \frac{\partial C}{\partial r} dr = \int \left(\frac{\kappa}{D}\frac{r}{2} + \frac{A}{r}\right) dr$$
$$C(r) = \frac{\kappa}{4D} r^2 + A \ln r + B$$

where:
A, B = integration constants

Apply first boundary condition: $C(r = a) = C_0$

$$\frac{\partial C}{\partial r} = 0$$
$$\frac{\partial C}{\partial r} = \frac{\kappa}{D}\frac{r}{2} + \frac{A}{r}$$
$$\frac{\partial C}{\partial r}(R) = \frac{\kappa}{D}\frac{R}{2} + \frac{A}{R} = 0$$
$$A = -\frac{R^2\kappa}{2D}$$

Apply second boundary condition: $\partial C/\partial r(r = R) = 0$

$$C(r = a) = C_0$$
$$C(r = a) = \frac{\kappa}{4D} a^2 + A \ln a + B$$
$$C(r = a) = \frac{\kappa}{4D} a^2 - \frac{R^2\kappa}{2D} \ln a + B = C_0$$
$$B = C_0 - \frac{\kappa}{4D} a^2 + \frac{R^2\kappa}{2D} \ln a$$

Substitute in Integration Constants, A and B, to achieve final expression:

$$C(r) = \frac{\kappa}{4D} r^2 - \frac{R^2\kappa}{2D} \ln r + C_0 - \frac{\kappa}{4D} a^2 + \frac{R^2\kappa}{2D} \ln a$$
$$C(r) = \frac{\kappa}{4D}\left(r^2 - a^2\right) + \frac{R^2\kappa}{2D} \ln\frac{a}{r} + C_0$$
$$C(r) = \frac{\kappa}{4D}\left(r^2 - a^2 - 2R^2 \ln\frac{r}{a}\right) + C_0$$

Now, given the distance between these channels and physiological parameters, what is the minimum concentration of oxygen that will be sustained throughout the engineered tissue (occurs at the midpoint between two channels)?

$C_{O_2,\text{culture medium}}$ = 220 μM or 0.00538 O_2/g after conversion

Channel radius: $r_{channel} = 50\ \mu m$

Inter-channel distance: $d = 2R = 400\ \mu m$

Oxygen consumption (κ) in resting, non-contractile heart: 2.0 mL O_2 per minute per 100 g tissue $= 3.53$ ml $O_2/cm^3 s$ after conversion

Diffusion constant: $D = 2.0 * 10^{-5}\ cm^2/s$ in tissue

Plug into master equation:

$$C(r) = \frac{\kappa}{4D}\left(r^2 - a^2 - 2R^2 \ln\frac{r}{a}\right) + C_0$$

Solve for $C(r = R)$.

$C(r = R) = 0.0021$ ml/g or ~85 μM O_2 at the midpoint between two channels. This is sufficient to sustain the cardiomyocytes.

(b) Channeled scaffolds decrease the diffusion distance required for oxygen and nutrients to penetrate the tissue at sufficient levels to sustain the cells.

(c) An oxygen carrier such as perfluorocarbons can substantially increase the amount of oxygen carried in culture media, as hemoglobin does in blood. The inclusion of flow through channeled constructs would eliminate diffusion as the sole source of mass transport in the tissue, and increase the amount of oxygen and nutrients that may be transported to the cells inside the scaffold.

ACKNOWLEDGMENT

The work described in this article has been supported by NIH (R01 HL076485, R21 HL089913, P41-EB002520).

BIBLIOGRAPHY

American Heart Association (2008). *Cardiovascular Disease Statistics*. 2008.

Akhyari, P., Kamiya, H., Haverich, A., Karck, M., & Lichtenberg, A. (2008). Myocardial tissue engineering: The extracellular matrix. *European Journal of Cardio-Thoracic Surgery: Official Journal of the European Association for Cardio-Thoracic Surgery*, **34**, 229–241.

Bergmann, O., Bhardwaj, R. D., Bernard, S., Zdunek, S., Barnabé-Heider, F., et al. (2009). Evidence for cardiomyocyte renewal in humans. *Science*, **324**, 98–102.

Boron, W. F. (2002). *Medical Physiology* (1st ed.). Philadelphia, PA: W. B. Saunders Company.

Boublik, J., Park, H., Radisic, M., Tognana, E., Chen, F., et al. (2005). Mechanical properties and remodeling of hybrid cardiac constructs made from heart cells, fibrin, and biodegradable, elastomeric knitted fabric. *Tissue Engineering*, **11**, 1122–1132.

Brunelli, S., & Rovere-Querini, P. (2008). The immune system and the repair of skeletal muscle. *Pharmacol. Res.*, **58**, 117–121.

Bursac, N., Papadaki, M., Cohen, R. J., Schoen, F. J., Eisenberg, S. R., et al. (1999). Cardiac muscle tissue engineering: Toward an *in vitro* model for electrophysiological studies. *The American Journal of Physiology*, **277**, H433–H444.

Chiu, L. L., Iyer, R. K., King, J. P., & Radisic, M. (2011). Biphasic electrical field stimulation aids in tissue engineering of multicell-type cardiac organoids. *Tissue Engineering Part A*, **17**, 1465–1477.

Cimetta, E., Figallo, E., Cannizzaro, C., Elvassore, N., & Vunjak-Novakovic, G. (2009a). Micro-bioreactor arrays for controlling cellular environments: Design principles for human embryonic stem cell applications. *Methods*, **47**, 81–89.

Cimetta, E., Cannizzaro, C., Elvassore, N., & Vunjak-Novakovic, G. (2009b). Microarray bioreactors for steady-state and transient studies of stem cells. *Methods*, **47**, 81–89.

Cimetta, E., Pizzato, S., Bollini, S., Serena, E., Coppi, P., et al. (2009c). Production of arrays of cardiac and skeletal muscle myofibers by micropatterning techniques on a soft substrate. *Biomedical Microdevices*, **11**, 389–400.

Engelmayr, G., Cheng, M., Bettinger, C., Borenstein, J., Langer, R., et al. (2008). Accordion-like honeycombs for tissue engineering of cardiac anisotropy. *Nature Materials*, **7**, 1003–1010.

Engler, A. J., Sen, S., Sweeney, H. L., & Discher, D. E. (2006). Matrix elasticity directs stem cell lineage specification. *Cell*, **126**, 677–689.

Fedorovich, N. E., Alblas, J., De Wijn, J. R., Hennink, W. E., Verbout, A. J., et al. (2007). Hydrogels as extracellular matrices for skeletal tissue engineering: State-of-the-art and novel application in organ printing. *Tissue Engineering*, **13**, 1905–1925.

Fink, C., Ergün, S., Kralisch, D., Remmers, U., Weil, J., et al. (2000). Chronic stretch of engineered heart tissue induces hypertrophy and functional improvement. *FASEB Journal: Official Publication of the Federation of American Societies for Experimental Biology*, **14**, 669–679.

Frangogiannis, N. G. (2008). The immune system and cardiac repair. *Pharmacol. Res.*, **58**, 88–111.

Gerecht, S., Burdick, J. A., Ferreira, L. S., Townsend, S. A., Langer, R., et al. (2007). Hyaluronic acid hydrogel for controlled self-renewal and differentiation of human embryonic stem cells. *Proc. Natl. Acad. Sci. USA*, **104**, 11298–11303.

Khademhosseini, A., Eng, G., Yeh, J., Kucharczyk, P., Langer, R., et al. (2007). Microfluidic patterning for fabrication of contractile cardiac organoids. *Biomedical Microdevices*, **9**, 149–157.

Laflamme, M., Chen, K., Naumova, A., Muskheli, V., Fugate, J., et al. (2007). Cardiomyocytes derived from human embryonic stem cells in pro-survival factors enhance function of infarcted rat hearts. *Nature Biotechnology*, **25**, 1015–1024.

Lambert, J. M., Lopez, E. F., & Lindsey, M. L. (2008). Macrophage roles following myocardial infarction. *Int. J. Cardiol.*, **130**, 147–158.

Laugwitz, K. L., Moretti, A., Lam, J., Gruber, P., Chen, Y., et al. (2005). Postnatal isl1+ cardioblasts enter fully differentiated cardiomyocyte lineages. *Nature*, **433**, 647–653.

Lunkenheimer, P. P., Redmann, K., Westermann, P., Rothaus, K., Cryer, C. W., et al. (2006). The myocardium and its fibrous matrix working in concert as a spatially netted mesh: A critical review of the purported tertiary structure of the ventricular mass. *European Journal of Cardio-Thoracic Surgery: Official Journal of the European Association for Cardio-Thoracic Surgery*, **29**(Suppl. 1), S41–S49.

Marsano, A., Maidhof, R., Tandon, N., Gao, J., Wang, Y., et al. (2008). Engineering of functional contractile cardiac tissues cultured in a perfusion system. *Conference Proceedings: Annual International Conference of the IEEE Engineering in Medicine and Biology Society*, 2008, 3590–3593.

Martens, T. P., Godier, A. F., Parks, J. J., Wan, L. Q., Koeckert, M. S., et al. (2009). Percutaneous cell delivery into the heart using hydrogels polymerizing *in situ*. *Cell Transplantation*, **18**, 297–304.

Martinez, F. O., Gordon, S., Locati, M., & Mantovani, A. (2006). Transcriptional profiling of the human monocyte-to-macrophage differentiation and polarization: New molecules and patterns of gene expression. *J. Immunol.*, **177**, 7303–7311.

Mironov, V., Boland, T., Trusk, T., Forgacs, G., & Markwald, R. R. (2003). Organ printing: Computer-aided jet-based 3D tissue engineering. *Trends in Biotechnology*, **21**, 157–161.

Nahrendorf, M., Swirski, F. K., Aikawa, E., Stangenberg, L., Wurdinger, T., et al. (2007). The healing myocardium sequentially mobilizes two monocyte subsets with divergent and complementary functions. *J. Exp. Med.*, **204**, 3037–3047.

Ott, H., Matthiesen, T., Goh, S., Black, L., Kren, S., et al. (2008). Perfusion-decellularized matrix: Using nature's platform to engineer a bioartificial heart. *Nature Medicine*, **14**, 213–221.
Radisic, M., Park, H., Shing, H., Consi, T., Schoen, F. J., et al. (2004). Functional assembly of engineered myocardium by electrical stimulation of cardiac myocytes cultured on scaffolds. *Proceedings of the National Academy of Sciences of the United States of America*, **101**, 18129–18134.
Radisic, M., Deen, W., Langer, R., & Vunjak-Novakovic, G. (2005). Mathematical model of oxygen distribution in engineered cardiac tissue with parallel channel array perfused with culture medium containing oxygen carriers. *American Journal of Physiology Heart and Circulatory Physiology*, **288**, H1278–H1289.
Radisic, M., Malda, J., Epping, E., Geng, W., Langer, R., et al. (2006a). Oxygen gradients correlate with cell density and cell viability in engineered cardiac tissue. *Biotechnology and Bioengineering*, **93**, 332–343.
Radisic, M., Park, H., Chen, F., Salazar-Lazzaro, J. E., Wang, Y., et al. (2006b). Biomimetic approach to cardiac tissue engineering: Oxygen carriers and channeled scaffolds. *Tissue Engineering*, **12**, 2077–2091.
Radisic, M., Park, H., Gerecht, S., Cannizzaro, C., Langer, R., et al. (2007). Biomimetic approach to cardiac tissue engineering. *Philosophical Transactions of the Royal Society of London Series B, Biological Sciences*, **362**, 1357–1368.
Radisic, M., Marsano, A., Maidhof, R., Wang, Y., & Vunjak-Novakovic, G. (2008). Cardiac tissue engineering using perfusion bioreactor systems. *Nature Protocols*, **3**, 719–738.
Rakusan, K., & Korecky, B. (1982). The effect of growth and aging on functional capillary supply of the rat heart. *Growth*, **46**, 275–281.
Robinson, K. A., Li, J., Mathison, M., Redkar, A., Cui, J., et al. (2005). Extracellular matrix scaffold for cardiac repair. *Circulation*, **112**, I135–I143.
Serena, E., Figallo, E., Tandon, N., Cannizzaro, C., Gerecht, S., et al. (2009). Electrical stimulation of human embryonic stem cells: Cardiac differentiation and the generation of reactive oxygen species. *Experimental Cell Research*, **315**, 3611–3619.
Shimizu, T., Yamato, M., Isoi, Y., Akutsu, T., Setomaru, T., et al. (2002). Fabrication of pulsatile cardiac tissue grafts using a novel 3-dimensional cell sheet manipulation technique and temperature-responsive cell culture surfaces. *Circ. Res.*, **90**, e40–e48.
Stevens, K. R., Pabon, L., Muskheli, V., & Murry, C. E. (2008). Scaffold-free human cardiac tissue patch created from embryonic stem cells. *Tissue Engineering Part A*, **15**, 1211–1222.
Tandon, N., Cannizzaro, C., Figallo, E., Voldman, J., & Vunjawk-Novakovic, G. (2006). Characterization of electrical stimulation electrodes for cardiac tissue engineering. *Conference Proceedings: Annual International Conference of the IEEE Engineering in Medicine and Biology Society*, **1**, 845–848.
Tandon, N., Marsano, A., Cannizzaro, C., Voldman, J., & Vunjak-Novakovic, G. (2008). Design of electrical stimulation bioreactors for cardiac tissue engineering. *Conference Proceedings: Annual International Conference of the IEEE Engineering in Medicine and Biology Society*, **2008**, 3594–3597.
Tandon, N., Cannizzaro, C., Chao, P. H., Maidhof, R., Marsano, A., et al. (2009). Electrical stimulation systems for cardiac tissue engineering. *Nature Protocols*, **4**, 155–173.
Tang, Y. L., Tang, Y., Zhang, Y. C., Agarwal, A., Kasahara, H., et al. (2005). A hypoxia-inducible vigilant vector system for activating therapeutic genes in ischemia. *Gene Therapy*, **12**, 1163–1170.
Wainwright, J. M., Czajka, C. A., Patel, U. B., Freytes, D. O., Tobita, K., et al. (2009). Preparation of cardiac extracellular matrix from an intact porcine heart. *Tissue Engineering Part C Methods*, **16**, 525–532.
Zimmermann, W. H., Schneiderbanger, K., Schubert, P., Didie, M., Munzel, F., et al. (2002a). Tissue engineering of a differentiated cardiac muscle construct. *Circulation Research*, **90**, 223–230.
Zimmermann, W. H., Schneiderbanger, K., Schubert, P., Didie, M., Munzel, F., et al. (2002b). Tissue engineering of a differentiated cardiac muscle construct. *Circulation Research*, **90**, 223–230.
Zimmermann, W., Melnychenko, I., Wasmeier, G., Didie, M., Naito, H., et al. (2006a). Engineered heart tissue grafts improve systolic and diastolic function in infarcted rat hearts. *Nature Medicine*, **12**, 452–458.
Zimmermann, W. -H., Melnychenko, I., Wasmeier, G., Didie, M., Naito, H., et al. (2006b). Engineered heart tissue grafts improve systolic and diastolic function in infarcted rat hearts. *Nat. Med.*, **12**, 452–458.

CHAPTER II.6.12 TISSUE-ENGINEERED SKIN SUBSTITUTES

J. N. Mansbridge
Histogen, Inc. San Diego, CA, USA

INTRODUCTION

Skin substitutes were among the earliest tissue-engineered devices, and much experience has been obtained at all levels, from bench models to therapeutics. The earliest systems were devised by Bell (Bell et al., 1979), in the late 1970s. He grew three-dimensional fibroblast cultures by casting fibroblasts in a collagen gel. Since then, this system has become the basis of testing and therapeutic skin substitutes. Several other scaffold systems have been developed, human-derived, animal-derived, and using synthetic polymer materials, to give a variety of products with a wide range of properties. This chapter will survey skin substitutes, their properties and applications, and will provide some insight into the factors affecting the design and implementation of a commercial system to produce a therapeutic skin substitute. I will provide an example in the form of the TransCyte® manufacturing process that illustrates many of the principles involved.

The basis of many tissue-engineered skin substitutes is the collagen gel system that was introduced by Bell (Bell et al., 1981). Collagen is soluble at low pH, but forms a gel if the pH is raised. This system has been much studied and has given rise to commercial laboratory test systems and also to therapeutics. When cast in a collagen gel, fibroblasts respond to the collagen environment by changing their gene expression pattern and modifying the arrangement of the collagen strands (Guidry et al., 1992). Strikingly, the gel contracts more than 90% and increases in strength, forming a structure with some resemblance to the dermis (Auger et al., 1995). While such a structure, if seeded with keratinocytes, has many features of skin, many other systems have been developed recently involving other three-dimensional culture techniques, that show a lesser or greater degree of complexity. The collagen gel technique has been applied very widely in university research and in research institutes. While academic studies may aim for systems that resemble human skin most closely, commercial production entails other considerations, such as issues of manufacture, production cost, and selling price.

TYPES OF SKIN SUBSTITUTES

Skin substitutes may be broadly divided into those intended for therapeutic application and those designed for testing products applied to the skin or for investigative experimental purposes. The issues related to the two types of product differ considerably. The differences in the types of skin function that they attempt to reproduce and the manufacturing processes are very different. However, at least one commercial product, Apligraf®, has been used for both therapeutic and experimental applications (Nouri et al., 2005).

The goal of a skin substitute is ultimately to generate material that mimics the properties of normal skin or that can be used to replace skin. Skin has many properties. Apart from providing a tough and self-maintaining surface to the body, it also provides initial contact between the outside world and the defensive systems of the body, and serves a variety of other physiological functions such as temperature control, neurological sensing, and so forth. It is also host to structures such as sweat glands and hair follicles. None of the skin substitutes achieve all these functions, and usually concentrate on a limited range.

Test materials are usually cell based, and use culture systems that are based on classical tissue culture techniques. While all are based on fibroblasts and keratinocytes, they may use other cells derived from different areas of the skin, such as the gingiva or the vagina, and may include other cells of the skin, such as Langerhans cells or melanocytes. Inclusion of these cells provides systems that may be used to investigate the responses of the skin immune system or its response to light.

Therapeutic products are designed to aid in the healing of wounds, either large, acute wounds, such as those following burns or chronic wounds such as diabetic foot ulcers and venous stasis ulcers. The production of therapeutic entities is highly regulated and will be considered in detail below.

Therapeutic skin substitutes have taken many forms from constructs made from non-biological polymers through to complex cellular structures involving the differentiation of cells *in vitro*. Representative samples are given in Table II.6.12.1.

Clearly, the amount of effort and expense involved in making therapeutic skin substitutes increases with the complexity of the product. In particular, cell culture is inherently expensive (cost of media, aseptic operation, etc.) and generates issues related to immunological response. Sterility and storage are also factors of increasing difficulty when using living cells. Thus, attempts have been made to produce acellular products. Indeed, some of the earliest products in this area were of this type (Yannas et al., 1980).

The simplest therapeutic skin substitute device consists of a degradable polymer, which may be textured to permit ingrowth of cells that can be manufactured by well-established methods then terminally sterilized. Storage of the finished product is not an issue. Such devices cause a foreign-body capsule to form which, if the structure is of suitable porosity, forms a dermis-like provisional matrix, which persists after loss of the scaffold (Yannas et al., 1989; Compton et al., 1998; Yannas, 1998). The cells used are autologous and the system uses well-optimized, physiological mechanisms of wound healing. Their products do not provide enhanced wound healing components, such as growth factors, that may be required in cases where healing has stalled, such as in chronic wounds.

The second type of dermal template uses biological molecules to form the scaffold. These are usually collagen with or without chondroitin sulphate (Boyce and Hansbrough, 1988). An optimal physical structure, such as porosity, is obtained by varying the composition and lyophilizing under conditions that cause ice crystals of the required size to form. Sterilization of the finished product is straightforward, and the material may be stored in alcohol or lyophilized. Materials of this kind have had a long and successful history in the treatment of acute wounds. Host fibroblasts will bind to collagen, but it is likely that other molecules, such as fibronectin and vitronectin, which also bind to the collagen, provide additional adhesion sites for the cells. A provisional matrix is formed, related to a foreign-body response, and remodeling of this and the scaffold forms a dermis-like material. These materials have also been seeded with fibroblasts and keratinocytes as the basis for cell-based therapy (Boyce et al., 1993; Still et al., 2003).

While the precise composition of the collagen preparations in terms of the proportions of types I, III, V, etc., and the modifications of the chondroitin sulphate may be uncertain, the systems discussed above are reasonably well-defined. The next stage of complexity involves other matrix proteins, which substantially increases the uncertainty of composition. These products include matrices derived from decellularized dermis, human (AlloDerm®, Graftjacket®) or animal (PrimiMatrix™), from pig intestinal submucosa (Oasis® and its derivatives), bladder or using human cells grown in culture (TransCyte®, currently discontinued). While the exact composition cannot be comprehensively ensured with these products, it is assumed that since they are generated by living cells, they will be cell-compatible and not toxic to cells. Since these materials are obtained directly or indirectly from living animals or humans, freedom from disease (HIV, hepatitis, transmissable animal pathogens, etc.) has to be ensured. The above products have not been found to cause any clinically important, immunological reaction. Sterilization has to be balanced between destruction of microorganisms and destruction of the material, but is comparatively straightforward using

TABLE II.6.12.1 Commercial Therapeutic Skin Substitutes

Type	Commercial Product	Scaffold	Cells	Differentiation	Reference
Non-biological polymer	Suprathel®	PLGA[1], PEG[2]	No	No	www.suprathel.de/
Biological polymer scaffold	Gelfoam®	Gelatin	No	No	www.pfizer.com/files/products/uspi_gelfoam_sponge.pdf
Dermal template	Integra®	Collagen, chondroitin sulphate	No	No	www.integra-ls.com
Dermal	AlloDerm®, Graftjacket®	No	No	No	www.wmt.com/Downloads/NEW 141-405 GJ Comprehensive Brochure.pdf
Dermal	PriMatrix™	No	No	No	teibio.com/Literature/PriMatrix/.../PN 607-999-001v02.pdf
Cell-derived extracellular matrix	TransCyte®	Cell-derived extracellular matrix on nylon template	No	No	www.ideo.com/work/print/transcyte
Live fibroblasts and extracellular matrix	Dermagraft®	Cell-derived extracellular matrix on PLGA template	Fibroblasts	No	www.advancedbiohealing.com/pdf/Dermagraft0802su[1].pdf
Bilayer, live keratinocyte, fibroblast construct	Apligraf®	Collagen	Fibroblasts, keratinocytes	Yes	www.apligraf.com/
Bilayer, live keratinocyte, fibroblast construct	OrCel®	Collagen	Fibroblasts, keratinocytes	Yes	www.forticellbioscience.com/technology.html
Epidermal sheets	EpiDex®	No	Hair follicle keratinocytes	Yes	www.sciencedaily.com/releases/2008/01/080104140344.htm
Bilayer, live keratinocyte, fibroblast construct	Xgene, AccuGraft™	No	Fibroblasts, keratinocytes	Yes	www.xgene.com/html/technologies/AccuGraft.htm
Hair follicles	No	Collagen	No	Yes	Zheng et al., 2005

[1]PLGA: Poly(lactic-co-glycolic acid).
[2]PEG: Polyethylene glycol.

ethylene oxide, UV, γ or electron beam irradiation. Preservation is straightforward using alcohol, lyophilization or cryopreservation. It is thought that these materials provide binding proteins and structures that promote healing, in addition to those present in a collagen-based construct. Claims are made for the presence of growth factors. Since most of these materials are derived from adult tissues that are usually quiescent and contain only very small quantities of growth factors and other molecules associated with healing (such as fibronectin, vitronectin, tenascin), their contribution may be very limited. An exception is TransCyte®, in which the matrix is laid down by rapidly growing cells that secrete substantial amounts of growth factors, some of which bind to the matrix. By contrast, materials derived from mature dermis or other tissues do provide structures, such as blood vessel tracks, that can be used to improve inosculation and colonization of the implant.

Inclusion of living cells in a product brings many complications related to the cost of materials, cell source, immunological considerations, complexity of manufacture, sterility, preservation to provide shelf-life, transport, regulatory issues, and so forth. These issues will be discussed in the remainder of this chapter. As a result, these products are inevitably expensive, and it is important that they show real benefit from the inclusion of cells. A few reasons for cell inclusion are the provision of growth factors and cytokines, provisional matrix proteins, active and non-senescent cells for tissue regeneration, and responsiveness to environmental conditions. These capabilities have been shown, in the case of skin substitutes, to improve the healing of chronic wounds and their cost is less than the cost of the sequelae of a non-healing wound such as amputation.

Among cellular three-dimensional skin substitutes, several strategies have been followed. At the simplest level, three-dimensional skin substitutes consist of just fibroblasts and their culture products. When seeded in a three-dimensional scaffold, fibroblasts grow rapidly and secrete extracellular matrix that consists of collagens types I, III, V, and VI, with some collagen IV as well as other collagens. Many non-collagenous proteins are also present, e.g., the tissue forms of fibronectin, tenascin, decorin, SPARC, etc. The material is a provisional matrix that resembles foreign-body capsule in many respects. As

TABLE II.6.12.2 Commercial Skin Substitutes for Testing

Product	Company	Scaffold	Cell type	Website
EpiDerm™, several designations[1]	MatTek	Collagen-coated plates	Keratinocytes	www.mattek.com
EpiDerm 201™	MatTek	Collagen	Fibroblasts, keratinocytes	www.mattek.com
MelanoDerm™	MatTek	Collagen-coated dishes	Keratinocytes, melanocytes	www.mattek.com
EpiOcular™	MatTek	Inserts	Keratinocytes	www.mattek.com
EpiAirway™	MatTek	Inserts	Tracheal/bronchial epithelial cells	www.mattek.com
EpiVaginal™	MatTek	Inserts	Ectocervical epithelial cells	www.mattek.com
EpiOral™	MatTek	Collagen-coated inserts	Oral keratinocytes	www.mattek.com
Episkin®, RHE	SkinEthic	Collagen	Keratinocytes	www.skinethic.com
RHPE	SkinEthic	Acellular collagen	Keratinocytes, melanocytes	www.skinethic.com
RHCE	SkinEthic	Acellular collagen	Immortalized corneal keratinocytes	www.skinethic.com
HOE	SkinEthic	Acellular collagen	Transformed buccal squamous cell carcinoma cells	www.skinethic.com
HGE	SkinEthic	Acellular collagen	Normal gingival epithelial cells	www.skinethic.com

[1]Several products at different levels of keratinocyte differentiation adapted to testing for corrosivity, phototoxicity, and differentiation markers.

discussed elsewhere in this volume, if material of non-host origin, such as a polymer fragment or certain bacteria, is placed in connective tissue, a foreign-body response ensues that consists of the formation of multinucleated giant cells and a fibrous capsule. It appears that capsule formation is an independent fibroblast function that may account for the deposition of extracellular matrix in the three-dimensional fibroblast culture. At the next level, a structure is formed with a fibroblast-populated dermal-like layer overlain with keratinocytes that may form a stratified epithelium as a bilayered structure. In general, the fibroblast layer is based on isolated collagen (usually bovine or rat-tail), either as a gel (Apligraf®) or as a sponge (Orcel®). In a collagen gel, fibroblasts respond to their environment by activating protein kinase Cζ (Xu and Clark, 1997), and switching off collagen synthesis (Mauch et al., 1988). In some test systems, the bilayered construct may be further augmented with other cells of the epidermis, such as Langerhans cells or melanocytes.

For therapeutic purposes, the keratinocyte layer is seeded as a submerged culture, and then raised to the air–liquid interface after a few days. It is then grown until it stratifies and forms multiple layers including a basal layer, a spinous layer, a granular layer, and a stratum corneum (Prunieras et al., 1983). For testing purposes it may be harvested at an earlier stage to make a model of corneal or a simple epithelium. The single keratinocyte layer corneal model is the basis of models to test corrosivity and irritancy to replace the controversial Draize test, in which responses are assessed by placing new products on a rabbit eye. Other epithelia, such as gingival or vaginal, may also be used with a similar fibroblast/collagen gel substratum. In some cases, the fibroblasts and epithelial cells may be obtained from the same biopsy.

A totally different approach to the generation of cellular skin substitutes is used by XGene in the production of AccuSkin™. Here, a slurry of cells, both fibroblasts and keratinocytes, is seeded. The fibroblasts and keratinocytes retain the capacity to sort themselves into a dermis and an epidermis without the use of any scaffold. Materials of this type are used for testing, as well as for therapeutic purposes.

Table II.6.12.2 lists a series of skin substitutes that are used as toxicological model systems for testing products that may come in contact with the skin. This includes a very wide range of cosmetics, household products, and other items. In general, the skin substitutes are used to evaluate the ability of the test material to kill the cells in the construct (corrosivity) or their ability to activate an immune response (irritancy). In addition, testing for phototoxicity, photoprotection, and other properties is also performed. Assays used to assess corrosivity generally include cell viability, frequently determined using MTT reduction, enzyme release (such as lactate dehydrogenase) or dye uptake (Triglia et al., 1991; Faller and Bracher, 2002). Irritancy is determined by the release of inflammatory cytokines, such as IL1 or prostaglandins (zur Muhlen et al., 2004) from the construct. There is a major movement in Europe and America towards replacing animal testing, which has been used traditionally, with *in vitro* testing (Edwards et al., 1994). Many of the systems kits used are commercial versions of laboratory tissue culture, but several involve multiple cell types and three-dimensional culture. While quality control of these products is critical, they do not entail the regulatory requirements of therapeutics, and are simpler to produce. Nonetheless, these systems all involve live cells, which entail additional considerations such as their limited shelf-life.

Three-dimensional systems have also been used extensively for investigating mechanisms in skin biology (Smola et al., 1993), e.g., growth factor interactions, wound healing, and as models of dermatological diseases. For instance, Asselineau has developed models of xeroderma pigmentosum using cells derived from patients of complementation group C showing specific morphological and DNA repair changes relative to normal, and the feasibility of correcting the defects by genetic modification (Bernerd et al., 2004).

COMMERCIAL PRODUCTION OF SKIN SUBSTITUTES

This discussion is primarily directed towards therapeutic products, as their regulatory requirements are generally more stringent than those used for testing. However, many of these general principles also apply to the manufacture of test systems.

Overall Considerations

Production of therapeutic skin substitutes is highly regulated by the US Food and Drugs Administration (FDA) to ensure safe and reliable products. This entails extensive quality assurance procedures. Of particular importance is detailed documentation of every aspect of the development and manufacturing processes including design, protocols, completion of manufacturing stages, deviation from protocols, correction of the deviations and measures to prevent deviations from protocols, adequacy and execution of quality control systems, traceability of components, environmental monitoring, facility maintenance, etc. Essentially all aspects of the operation of the process of making skin substitutes are covered by quality assurance operations, which represent a substantial cost. Examining records to ensure correct completion of all steps and correction of mistakes is frequently a rate-limiting step in the release of a product.

Cell Source

Sourcing of cells for tissue-engineered skin substitutes is comparatively easy, as this organ is on the surface of the body and biopsies can be taken easily. In addition, discarded skin is often available as the result of surgical procedures, such as circumcision, facelift, breast reduction, and abdomenoplasty. Indeed, keratinocytes may be obtained from plucked hairs (Vermorken et al., 1985). Many of the cells obtained in this way are derived from a region (the hair follicle bulge) that contains the stem cells for hair follicles and, under some circumstances, also for the interfollicular epidermis (Cotsarelis et al., 1990). The various skin sources provide cells that vary in age and sex, and thus may be optimal for different applications. In general, therapeutic products are made from neonatal tissues, but adult tissues may well be adequate for making products for testing purposes. In the case of testing, there may be further advantages in using a cell bank pooled from many sources. This reduces variation that might be found between individuals and may detect responses that may occur in a minority of the population. The use of stem cells, adult, embryonic or induced, would involve substantial additional expense and is unnecessary for skin substitutes.

The ease of obtaining skin cells has made it possible to compare allogeneic and autologous sources in therapeutic applications. Early applications employed autologous sources to avoid any complications from immunological rejection and this approach is still used. However, it has become evident that the use of autologous cells is unnecessary and imposes substantial practical burdens that make commercial manufacture difficult (see Box 1). Several companies have, therefore, explored allogeneic approaches and the experience has indicated that immunological barriers to such implants are, at worst, minor. No case has been recorded of immunological rejection of a skin substitute in the clinic, despite the implantation of several hundred thousand constructs (Mansbridge, 2007).

BOX 1 Immunological Rejection of Skin Substitutes

Medawar demonstrated that transplanted organs are almost invariably acutely rejected (Medawar, 1957) within a couple of weeks. This has become a dogma of transplant biology. However, extensive experience with allogeneic skin substitutes has shown no evidence for rejection. Not only has acute rejection not been observed clinically, but investigation of antibody formation and attempts to activate lymphocytes in response to constructs have failed (Briscoe et al., 1999; Laning et al., 2001). It appears that at least some tissues grown in culture fail to elicit an immune response. The probable explanation is that tissue-engineered constructs lack antigen presenting cells (APCs). Indeed, incorporation of APCs into a skin substitute causes it to become susceptible to immunological attack (Rouabhia, 1996). Skin normally contains endothelial cells, Langerhans cells, and other cells which are APCs, and which normally cause the rejection of unmatched skin grafts. These cells are lost during the culture expansion steps that are involved in generating skin substitutes. Similar types of immunity to acute immunological rejection are also seen in certain anatomical locations, such as the cornea, that have long been regarded as "privileged." These tissues may also lack the characteristic cell types required to cause rejection. The scope of cells showing this type of immunity has not been determined. It appears that fibroblasts, smooth muscle cells, and probably keratinocytes and chondrocytes show the same property. In addition, it has been found that fibroblasts are capable of inhibiting T-cell activation (Haniffa et al., 2007) and, on a second restimulation of the T-cells, directing them towards an IL-10, Th2 response. The inhibitory activity of the fibroblasts is stimulated by T-cell-derived γ-interferon. It may be noted that the response of fibroblast in another immunologically-related property, HLA Class II induction, is much reduced in some tissue-engineered constructs (Kern et al., 2002).

A second form of rejection, chronic rejection, also occurs. This takes place months or years after transplantation and, in the case of organs, causes loss of the transplant. This type of rejection has not been observed clinically with skin substitutes. However, transplanted cells are lost rapidly during the completion of wound repair, and then slowly during the ensuing months, and are presumably replaced by the host. Thus, in this case, chronic rejection may be masked by cell loss. However, in a transplant where continued presence of the cells is required, as in correction of a genetic defect, chronic rejection may become a problem. Indeed, where allogeneic skin substitutes have been used clinically to correct various forms of the genetic disease epidermolysis bullosa (Sibbald et al., 2005) the effects, although much more durable than other methods of treatment, ultimately require repeated application.

Supplementary Proteins and Genetic Modification

Two areas exist in which supplementary proteins or genetic modification might be important. These are the correction of genetic diseases and enhancement of wound healing. Considerable effort has been put into the first of these with some success in model systems, although few have been applied to human clinical studies. An example of this approach to the group of skin diseases comprising epidermolysis bullosa is given in Box 2. In general, the number of patients suffering from any individual genetic disease is very small and, even with relaxed regulatory requirements (such as orphan disease status), it is difficult to sustain such a product as a commercially viable operation.

Wound healing is clearly regulated by growth factors and over the last two decades many attempts to enhance wound healing by the addition of growth factors have occurred. Factors added in one way or another include PDGF (Pierce et al., 1991), FGF1 (Pandit et al., 1998), FGF2 (Pierce et al., 1992), FGF7 (Xia et al., 1999), FGF10 (Jimenez and Rampy, 1999), TGF-α (Schultz et al., 1991), EGF (Chen et al., 1993), and TGF-β (Jones et al., 1991). While many factors have been claimed to show some effect in animal experiments, only one gave rise to a commercial product. This is PDGF-B chain (Becaplermin, Regranex) that has shown clinical efficacy in treating chronic wounds (Steed, 1995). In clinical trials, the improvement in wound healing was about the same as that found with unmodified skin substitutes. This growth factor is normally released by platelets at the wound site, and is specialized for retention in wounds. PDGF-B chain has a matrix binding domain and will remain at the wound site. In general, other growth factors have been found to provide insignificant benefit in humans.

Efforts have been made to extend the persistence of added growth factors by supplying them in the form of DNA, either as plasmids to be taken up by host cells or by using transformed cells (Tyrone et al., 2000). Such measures have been successful in animal tests, but have not reached the market. In general, this approach has not generated results sufficiently striking with respect to issues of safety and cost to make it worthwhile to pursue them as commercial products.

It is difficult to choose agents that are likely to improve wound healing. Each cell in the wound carries a complete genome, including all the growth factors and cytokines, together with the control systems required to regulate the production and secretion of these proteins. These systems have been optimized over millennia, and it is unlikely that we can improve wound healing by adding factors. An exception to this may be the addition of an antimicrobial peptide, β-defensin, as suggested by Supp (Smiley et al., 2007) or cathelicidin (McFarland et al., 2008). The approach is to provide cells in a wound with a capability they would not otherwise have, such as antimicrobial activity. While keratinocytes and cells such as neutrophils and macrophages secrete such peptides, others, such as fibroblasts, do not. Further, it might be possible to use peptides from a non-human source. However, these peptides are extremely variable between species, and may be adapted to the skin flora of the host species. This makes it unlikely that peptides from other species, such as frog or moth peptides (magainins or cecropins), would have optimal activity against the contaminants of a chronic human wound (Zasloff, 2002).

BOX 2 Epidermolysis Bullosa – an Example of Genetic Disease

Epidermolysis bullosa is a group of severe blistering diseases of the skin caused by mutations in any of at least 10 genes. Mutations in keratins 5 and 14 cause epidermolysis bullosa simplex, which causes a break in the basal keratinocyte layer which, while severe in children, tends to ameliorate with age. Mutations in collagen VII cause the dystrophic form of the disease, in which a split between dermis and epidermis occurs below the basement membrane, together with syndactyly and susceptibility to squamous cell carcinoma. Mutations in any of the three subunits of laminin 332 cause junctional epidermolysis bullosa. It shows a split in the basement membrane and is frequently rapidly fatal. Other mutations causing related diseases occur in integrins a6 and b4, the bullous pemphigoid antigens, and in plectin. Details are outside the scope of this discussion, but may be found in many reviews such as Uitto and Richard (2005). While many of the attempts to treat these diseases have been based on gene therapy approaches (Robbins et al., 2001; Ortiz-Urda, 2002), skin substitutes using genetically normal cells have been used on several occasions with comparative success. Skin substitutes provide wound covering, relief of pain, and comparatively durable healing (Eisenberg and Llewelyn, 1998; Dellambra et al., 2000; Falabella et al., 2000; Fine, 2000; Jiang et al., 2002; Fivenson et al., 2003; Ozerdem et al., 2003; Dagregorio and Guillet, 2005; Sibbald et al., 2005).

Cell Expansion

In general, cell sources (biopsies, foreskins, surgical discards, differentiated stem cells) supply limited quantities of cells that need to be expanded to provide adequate numbers of cells for tissue engineering purposes. Several devices have been developed for performing this operation: roller bottles, large flasks, the cell cube, automated plate devices (Prenosil and Villeneuve, 1998), and bioreactors using beads or some other support. Expansion in roller bottles is a well-established and reliable technique, widely used, and works very well for cell lines (such as CHO and HeLa) that have undergone evolution *in vitro* and been adapted to culture conditions. However, roller bottles have disadvantages, particularly for recently isolated primary cell strains. Cell seeding, every medium change and harvesting the cells all require multiple aseptic operations for each bottle, raising the likelihood of adventitious contamination and requiring skilled operators. It is also very difficult to control the oxygen tension in roller bottles. Bottles could be gassed with the correct mixture of nitrogen, carbon dioxide, and reduced oxygen, to avoid problems associated with high (atmospheric) oxygen tension as described in Box 3. However, they must then be incubated under conditions to maintain the gas composition, requiring low oxygen incubators, which are cumbersome and expensive to run. Bioreactors using bead carriers avoid the large number of aseptic operations, and oxygen tension is much more easily controlled. Moreover, it is much easier to control the nutritional composition of the medium in bioreactors by continuous flow techniques.

Fibroblasts are traditionally grown in Dulbecco's Modified Eagle Medium (DMEM) that was developed for this purpose in the early days of tissue culture. It is routinely supplemented with serum, glutamine, nonessential amino acids and, in conventional tissue culture, with antimicrobials (penicillin, streptomycin, gentamycin, amphotericin, etc.). While antimicrobials may be used for the production of skin substitutes for testing purposes, they have to be omitted from medium used for therapeutic products. They may obscure contamination that could become evident when applied to a patient and lead to adverse consequences. All tissue culture for therapeutic skin implants is performed under antibiotic-free conditions.

The use of fetal calf serum in fibroblast tissue culture has become standard. The original media developed for the cells (fibroblasts) available at the time all required supplementation by fetal serum. Despite numerous attempts, a serum-free medium for fibroblasts that will sustain more than 6–8 passages (subcultures) has not been devised. This is insufficient cell expansion for tissue engineering where, for fibroblasts, at least 60 generations (more than 20 passages) are required from isolation to the final product. This includes a margin to be sure that the cells are not senescent when implanted into the patient, which is desirable. While some sort of serum appears essential for fibroblast replication, it has been found that fetal calf serum may be replaced with calf serum at much lower cost, with little or no loss in proliferation rate or growth potential.

Several systems have been developed for the expansion of keratinocytes. A widely used system was developed early by Rheinwald and Green that used non-proliferative (irradiated or mitomycin C-treated) mouse 3T3 cells as a feeder layer. Under these conditions, keratinocytes can be expanded for 30 or more generations. However, for therapeutic purposes, the contact with mouse cells leads to human keratinocytes being considered as xenogeneic from a regulatory point of view. Thus, cells expanded in this way are suitable for test systems, but not for implants. For implants, keratinocytes are expanded in serum-free medium in the presence of bovine pituitary extract. Under these conditions, the cells cannot be subcultured beyond about 6–8 passages, so the cell banks have a limited life.

BOX 3 Effects of Culture Conditions on Cellular Aging

The use of high glucose (0.4%) DMEM (Dulbecco's Modified Eagle Medium) arises from the view that living systems benefit from the provision of abundant supplies of nutrients. The concentration of glucose present in high glucose DMEM is comparable to levels reached by uncontrolled diabetic patients, where it is known to cause long-term deleterious effects, such as the formation of advanced glycation products (AGEs), which are inflammatory through the receptor for AGEs, RAGE (Vlassara, 1990). The use of low glucose DMEM has not been investigated, but may be superior, although it may entail more frequent changes of medium. Fibroblasts, when grown in three-dimensional culture in high glucose DMEM, metabolize glucose mostly by glycolysis over about 5–8 days, producing large amounts of lactate and alanine (using nitrogen from glutamine). Oxidation through mitochondria, which generates nine times as much ATP per glucose, increases subsequently. It is not clear whether the cells use glycolysis early to achieve a high glucose charge during cell proliferation, changing to higher yielding oxidation when collagen is being deposited, or just as a consequence of glucose availability.

Oxygen is another factor that is frequently regarded as beneficial when supplied in non-limiting quantities, but may be deleterious to cells at high concentrations. In traditional tissue culture, cells are grown in atmospheric oxygen tension (21% of atmospheric pressure), but the oxygen tension in tissues is rather less (equivalent to 5–6% of atmospheric pressure), and fibroblasts have been demonstrated to grow better at low partial pressures of oxygen (Falanga and Kirsner, 1993; Falanga et al., 2002). Oxygen is a di-free radical and very reactive, oxidizing proteins and lipids, and generating DNA adducts such as 8-oxoguanosine. Oxidation of proteins may lead to inactivation, oxidation of lipids to signal transduction related to inflammation and damage to DNA that invokes repair processes that block cell proliferation and other processes. The phenomenon of senescence in fibroblast cultures has been largely attributed to these processes, and is termed Stress Induced Premature Senescence (SIPS) (Toussaint et al., 2002). Culture of cells in reduced oxygen tension may be difficult to achieve in traditional tissue culture, but is facilitated by the use of bioreactors.

Construct Manufacture

The manufacturing process needs to be designed to fulfill multiple goals simultaneously. Its first basic function is to provide adequate nutrition and mass transport to allow the development of the tissue-engineered skin substitute (see Chapter II.6.6). In addition, the system needs to be scaleable, able to maintain sterility for long periods, to maximize automation, and to provide a product adapted to the users' needs. Several design concepts to achieve this have emerged.

The scaleability allows expansion without substantial additional resources, and thus economies of scale. Simple addition of production units does not provide this. This consideration is one of the major drawbacks of autologous tissue engineering, as complete tissue culture suites are required for each patient and no economy is achieved with expansion.

Secondly, tissue culture has to be performed under sterile conditions. It follows that the parts of the equipment in contact with the product (product contact parts) should be sterilizable, either portable so that the can be sterilized in a radiation (gamma or e-beam facility), treatable with ethylene oxide or able to be steam-treated. In general, radiation treatment of disposable systems is most convenient in small- to moderate-sized installations, although some commonly used scaffolds, such as PLGA, are radiation sensitive.

It is also important to minimize aseptic operations. Every aseptic operation is an opportunity for contamination, and has to be performed using skilled operators. This consideration and the impossibility of terminal sterilization of a living product are the reason that the bioreactor (either a Petri dish-like device or a closed bioreactor) has been incorporated in the final product in skin substitutes.

Thirdly, the manufacturing system should be designed to maximize automation. There are two reasons for this. The first is to minimize labor costs and to allow the use of trained but less highly educated personnel. By comparison, tissue culture systems require periodic assessment by highly skilled practitioners. The second and more important reason is that machine failures are less frequent than operator errors. While most of the operator errors are minor (e.g., not signing a document where and when required), they still require expensive quality assurance investigation.

A fourth consideration is that since the bioreactor in many current systems forms part of the final packaging, for reasons given above, the customer has to be able to apply the contained product easily. For this reason, the bioreactor should be designed to allow easy access to the skin substitute. This is less of a problem in the case of test products, but therapeutics may be used in an operating theatre and the requirements of entry into the sterile field should be built into the bioreactor design.

It is also important that the system should be reliable and consistent. Tissue-engineered systems for manufacturing skin substitutes are very consistent by comparison with classical tissue culture systems. For instance, Dermagraft® is held by specifications to less than ±25% in one of its most variable properties (MTT reductase activity).

Shelf-Life

Shelf-life is a valuable property to incorporate into a tissue-engineered product. It is less critical for skin substitutes used for testing, as scheduling delivery with the customer is straightforward and it is rare that a test system is required instantly on an emergency basis. Therapeutic skin substitutes also usually have some latitude, but quantities of a burn product might be required very quickly. The availability of product off-the-shelf may alleviate inventory control. However, the real value is ensuring consistent quality. Testing for sterility or mycoplasma takes from 2–4 weeks. If it is to be ensured that the product is sterile, a shelf-life of this duration must be incorporated. For non-living products, there is little problem. Most of the skin substitute products of this type are sterile and stable, and can be maintained at refrigeration temperatures for years. TransCyte® was originally stored at −80°C, which required special equipment. It was subsequently found to store at −20°C in a frost-free freezer for 18 months without loss of efficacy. Methods of stasis preservation are discussed in Box 4.

BOX 4 Long-Term Preservation of Skin Substitutes

Many techniques for maintaining viable cells *in vitro* have been studied. The simplest involve maintenance in the cold or at room temperature in the presence of suitable media containing sugars or gelling agents, such as agarose. Many cells will survive for several weeks under these conditions in a quiescent state. For storage periods of months, three methods have been investigated – cryopreservation, vitrification, and desiccation (anhydrobiosis).

Cryopreservation systems have been developed for both Apligraf® (Watson and Toner, 1999) and Dermagraft® (Applegate and Kim, 1998). Both methods are based on slow cooling in the presence of cryoprotectants (glycerol in the case of Apligraf, and dimethylsulphoxide in the case of Dermagraft). The techniques are based on those successfully developed and routinely used for suspended cells, sperm, ova, etc., and are well-understood on a theoretical basis (Karlsson and Toner, 1996). Development of systems that might be applied to three-dimensional tissues on the scale of tissue-engineered constructs has proved more difficult than to isolated cells. Much of the understanding of the process is related to mechanisms involved in slow cooling so as to avoid lethal ice formation. However, experiments with Dermagraft indicated a substantial range of cooling rates over which satisfactory survival could be obtained. The method requires storage below −70°C. This can be achieved using dry ice, and transport, while needing precautions, is possible. This approach has been realized on a commercial scale (Applegate et al., 1999).

Vitrification involves very rapid cooling of tissues in the presence of high concentrations of cryoprotectants that promote glass formation. In this process, water forms a glass rather than ice crystals as it cools, and this has been used successfully for tissues that are small enough to allow the high heat transfer rates that are required. The technique requires storage at temperatures below −120°C, usually using liquid nitrogen, which causes transport problems (Fujita et al., 2000; Elder et al., 2005).

Experience with Dermagraft and Apligraf indicated that thawing was at least as important as the freezing process, and should be optimized with the same care. In general, it should be performed very rapidly to minimize ice crystal growth as thawing occurs. In the case of Apligraf, additional prolonged incubation was required for removal of the cryoprotective agents and the process was only performed in-house. Dermagraft was shipped frozen to the final customer, who was responsible for the thawing and cryoprotectant removal process. This required careful training of customers.

Anhydrobiosis is a technique of achieving survival by dehydration; a method of stasis achieved by many organisms. It has been applied successfully to human platelets, and is promising in the case of human cells (Wolkers et al., 2002). Achievement of anhydrobiosis has tremendous potential in tissue engineering. Storage is at room temperature, so it does not require special freezing systems and product preserved in this way can be easily transported. At the time of writing, anhydrobiosis remains a dream, but its advantages make it a worthwhile goal to pursue.

Quality Control

Before release, the product is tested to ensure safety, in particular that it is free of pathogens, and that it achieves analytical properties that have been found to provide efficacy in a clinical trial. However, analytical testing may cover other aspects of product development, such as demonstration of equivalence to a predicate device (for a 510k submission) or a prototype. There is a marked contrast between release testing and equivalence testing. Equivalence testing is only performed a small number of times, and should include everything that might be important for the expected function of the device. Since testing of this kind is only performed a small number of times, expensive tests are practical if they provide the necessary information. By contrast, release testing is performed on every lot, and should include the minimum of testing sufficient to show that the product has reached the required standard. Since such tests are performed many times, cost becomes a major issue.

Specifications for testing should be established early. For equivalence, the range of values satisfying a criterion can be obtained from the predicate device, the prototype or by consideration of the expected function of the device (e.g., the minimal strength required for efficacy). Release criteria are generally set initially from the performance of the manufacturing system using a factor of three standard deviations either side of the mean. This range may be tightened subsequently, based on improved manufacturing performance or the results of a clinical trial.

THE TRANSCYTE® SYSTEM

Introduction

TransCyte® was designed as a temporary covering for severe burns. Following third degree burns, burn eschar (dead tissue) is removed as soon as the patient can be sufficiently stabilized. This process leaves a large open wound that must be covered. The traditional treatment, and still the gold standard, is to apply cadaveric skin. This provides an excellent covering for about two weeks, but is then rejected and subsequent applications of cadaveric skin are rejected more rapidly. TransCyte was intended to replace the cadaveric skin, the advantages being the reduced possibility of disease transmission, and the finding that it did not appear to be rejected.

The production of TransCyte illustrates several of the approaches, developed at Advanced Tissue Sciences, which are valuable in the commercial production of tissue-engineered constructs. These include considerations of efficient mass transport, scalability, minimal aseptic operations, maximum automation, and "user friendly" packaging, as described above. In addition, as much of the assembly of the bioreactor systems as possible was performed in clean, although not aseptic, conditions prior to sterilization, much of the process could be performed by operators without a college degree. The reasons that this device was discontinued are discussed at the end of this section.

Product Concept

The TransCyte bioreactor was of clamshell design (Figure II.6.12.1) that was a compromise between efficient manufacture and ease of use by the burn surgeon. After thawing, draining, and removal of a closure device, the bioreactor could be opened and presented to the surgeon like a book, from which the TransCyte could be lifted and applied. Care had to be exercised here as TransCyte had two different sides, so labeling to ensure correct application had to be devised.

The material was based on a known biocompatible wound dressing, Biobrane™, which was used as the scaffold. Biobrane consists of a knitted nylon fabric, backed by silicone and "porcine collagen peptides" (essentially denatured pig collagen). The silicone has laser-punched holes at intervals that allow drainage of excess fluid. This material was used as a scaffold for the three-dimensional culture of fibroblasts, which laid down a provisional matrix, consisting of collagen, fibronectin, tenascin, etc. (Table II.6.12.3). At the end of the manufacturing process, the material was frozen for storage, which killed the cells, leaving extracellular matrix. This material was colonized by fibroblasts from the host, endothelial cells,

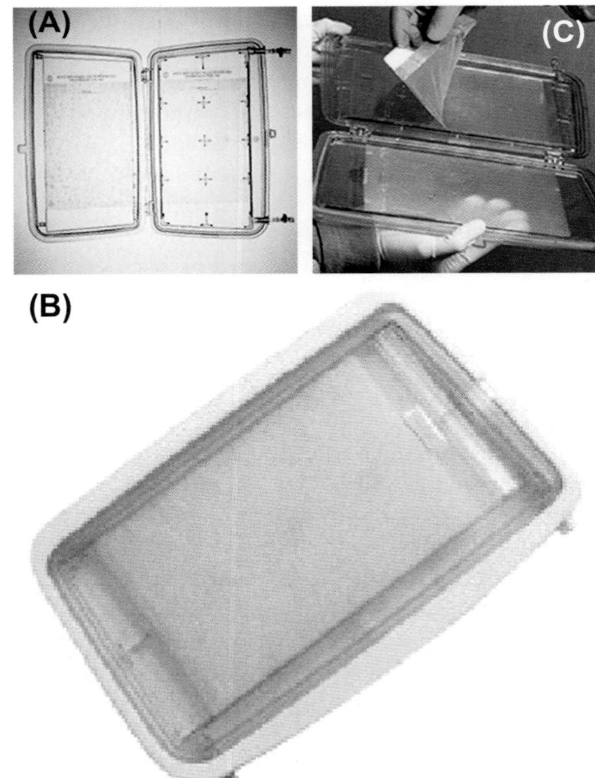

FIGURE II.6.12.1 (A, B): The TransCyte® bioreactor and packaging system; (C) after thawing, draining and removal of a closure device, the bioreactor could be opened and presented to the surgeon like a book.

TABLE II.6.12.3 Composition of the Extracellular Matrix of TransCyte®

Component	mRNA Copies/Cell	Immunohistology	Other Assays	ELISA* µg/cm²	Expression Array
Collagen 1α1	282 ± 25	✓	~0.8 mg/cm²		✓
Fibronectin	4.7 ± 0.6	✓		4	✓
Decorin	3.0 ± 0.8	✓			✓
Collagen III	3.1 ± 0.7	✓			✓
Tenascin	3.3 ± 0.2	✓		10	✓
Collagen V	0.33 ± 0.05	✓	✓ PAGE/MS**		✓
Collagen VII	0.36 ± 0.08				✓
Collagen VI					✓
Elastin	0.93 ± 0.16				✓
Lumican					✓
Biglycan					✓
Overall GAG			37 µg/cm²		
SPARC		✓			
Thrombospondin II		✓			

*EUSA: Enzyme-Linked Immunosorbent Assay.
**PAGE/MS: PolyAcrylamide Gel Electrophoresis/Mass Spectrometry.

etc., and provided a conductive substrate for keratinocyte migration. In clinical trials on third degree burns, TransCyte proved at least as good as cadaveric skin, and remained adherent for much longer (Hansbrough et al., 1997). It subsequently proved efficacious on severe second degree burns, reducing pain and, in some cases, obviating the need for further surgery.

Manufacturing System

During production of the construct in the bioreactor, fibroblasts were fed by a continuous flow system. The bioreactor, including the Biobrane sheets and the closure device, was assembled, sterilized, and attached to a system of tubes that recirculated medium. This system included all parts that came in contact with the medium, except for medium and waste bags, and could be sterilized as a unit by gamma radiation. Gamma radiation was the chosen sterilization method for its convenience and lack of residue. The scaffold material and other components were not affected by γ-radiation.

The flow rate (about 5 mL/min) resulted in complete medium replacement in the bioreactor in about eight minutes. Considerable care was taken to ensure that the fluid motion was laminar and uniform. This flow was maintained by peristaltic pumps, one channel per bioreactor. Other systems were investigated in early experiments in which the flow through manifolded bioreactors was controlled by capillary constrictions. However, this proved to be unreliable, and the product was unacceptably variable. Fibroblast growth under a flow regime was superior to static cultivation and, in retrospect, a higher flow rate might be an improvement. Care was taken to ensure that the flow was laminar, as turbulence might have led to uneven growth.

This system could be sterilized as a unit of 12 bioreactors, and connected to medium and waste with two or three aseptic steps. It then ran automatically for 2 to 4 days until the next medium change, which required another two to three aseptic steps. Thus, automation was maximized and aseptic steps minimized.

Harvest Time, ATP Estimation

In an ideal system, the manufacturing process is sufficiently reliable that times of harvest, feeding, and so forth are reliably repeatable. In practice, using cells, there is always some variability and a method is needed to determine the correct harvest date. In the case of TransCyte, this was performed by determining a quantity based on the amount of energy consumed by the culture. It was reasoned that the proliferation of the cells and deposition of extracellular matrix would consume energy at a reproducible rate. Since most glucose is consumed in generating energy (glucose consumed is about 20 times the weight of cells plus matrix) a number was derived, based on glucose used and lactate secreted, that was related to the amount of ATP used by the cells. Both glucose and lactate had to be determined to allow for the switch from glycolysis to full oxidation (Figure II.6.12.2.) Glucose and lactate were measured at each medium change and at harvest. The notional quantity of ATP consumed, accumulated over the three-dimensional culture period, was correlated with quality control specifications, such as the number of cells (from DNA determination) and collagen deposited (by Sirius Red binding), to provide a predictable window within which the process would result in acceptable product.

Quality Control

The quality control parameters used for the release of this product were estimates of DNA (determined by Hoechst 33258 fluorescence), which demonstrated that cells had

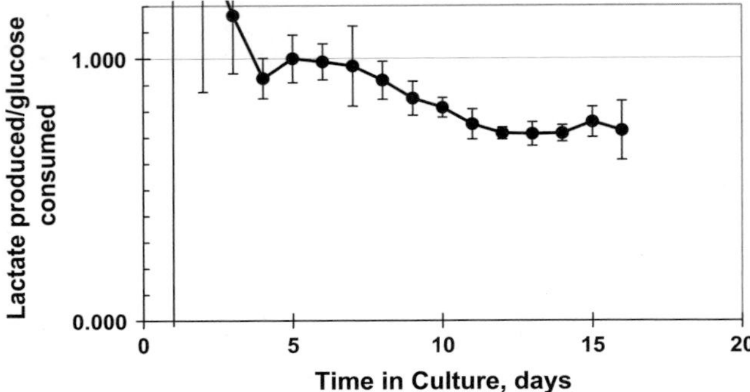

FIGURE II.6.12.2 Production of lactate by TransCyte® as a function of time in culture. A ratio of lactate production to glucose consumption of 1 represents 100% glycolysis. The declining ratio represents increasing oxidation of glucose.

actually grown during the production process, and collagen (determined by Direct Red binding), which demonstrated that extracellular matrix had been deposited.

In an alternative approach, it was reasoned that a satisfactory product had to adhere to the wound, and remain attached for an extended period of time (more than two weeks) and be easily removed. This was tested during the characterization phase of development by determining the force required to remove the implant using the apparatus developed by Richard Skalak and colleagues (Dong et al., 1993). It was, however, too complex to use as a release criterion.

Reason for Failure of TransCyte

TransCyte is, at present, discontinued as a commercial product. The major reason for this was the size of the market rather than the manufacturing process. The number of severe burns in the US is comparatively small, and while TransCyte had distinct advantages over cadaveric skin in terms of availability and duration of adhesion, its high price resulted in lack of acceptance and only a moderate revenue stream. Attempts were made to extend the market. TransCyte proved effective in second degree burns and, particularly, in severe second degree burns by preventing progression to third degree burns. However, this was insufficient to make its manufacture worthwhile. The conclusion of this analysis is that tissue-engineered or regenerative medicine products have to be either made in sufficient quantities or command a sufficiently high price, to provide enough incentive to establish the commercial manufacturing process. The same issues are important in developing treatment for genetic diseases, which are frequently rare and unable to sustain a commercial approach. As a result, when the TransCyte manufacturing process was shut down during a transition from one company to another, its restarting was delayed. Thus, while the manufacturing process was well adapted to commercial production and achieved consistently high yields, the market and effort involved in recommissioning the system prevented its resumption.

CONCLUSION

Skin substitutes are the most straightforward of the tissue-engineered organ constructs, and have been developed over many years for the testing of commercial products, therapeutics, and as models of skin biology and pathology. They are based on the ability of fibroblasts and keratinocytes to form a skin-like structure spontaneously, but have been extended by the inclusion of other cell types to provide models of the very wide range of properties displayed by intact skin. The production of these constructs on a commercial scale, particularly for therapeutic purposes, provides a series of challenges and room for ingenuity in the balancing of large-scale production of reliable product with the business requirements of a viable commercial enterprise.

BIBLIOGRAPHY

Applegate, D. R., & Kim, J. (1998). Cryopreservation process scaleup for a tissue engineered, metabolically active, human dermal replacement, Dermagraft. *Cryobiology*, **37**, 409.

Applegate, D. R., Liu, K., & Mansbridge, J. (1999). Practical considerations for large-scale cryopreservation of a tissue engineered human dermal replacement. *ASME* Advances in Heat and Mass Transfer in Biotechnology. *HTO-vol 363/BED vol 44*, 77–91.

Auger, F. A., Lopez Valle, C. A., Guignard, R., Tremblay, N., Noel, B., et al. (1995). Skin equivalent produced with human collagen. *In Vitro Cell Dev. Biol. Anim.*, **31**, 432–439.

Bell, E., Ivarsson, G., & Merrill, C. (1979). Production of a tissue-like structure by contraction of collagen lattices by human fibroblasts of different proliferative potential *in vitro*. *Proc. Natl. Acad. Sci. USA*, **76**, 1274–1278.

Bell, E., Ehrlich, H. P., Buttle, D. J., & Nakatsuji, T. (1981). Living tissue formed *in vitro* and accepted as skin-equivalent tissue of full thickness. *Science*, **211**, 1052–1054.

Bernerd, F., Asselineau, D., Frechet, M., Sarasin, A., & Magnaldo, T. (2004). Reconstruction of DNA-repair deficient xeroderma pigmentosum skin *in vitro*: A model to study hypersensitivity to UV light. *Photochem. Photobiol.*, **1**, 1.

Boyce, S. T., & Hansbrough, J. F. (1988). Biologic attachment, growth, and differentiation of cultured human epidermal keratinocytes on a graftable collagen and chondroitin-6-sulfate substrate. *Surgery*, **103**, 421–431.

Boyce, S. T., Greenhalgh, D. G., Kagan, R. J., Housinger, T., Sorrell, J. M., et al. (1993). Skin anatomy and antigen expression after burn wound closure with composite grafts of cultured skin cells and biopolymers. *Plast. Reconstr. Surg.*, **91**, 632–641.

Briscoe, D. M., Dharnidharka, V. R., Isaacs, C., Downing, G., Prosky, S., et al. (1999). The allogeneic response to cultured human skin equivalent in the hu-PBL-SCID mouse model of skin rejection. *Transplantation*, **67**, 1590–1599.

Chen, J. D., Kim, J. P., Zhang, K., Sarret, Y., Wynn, K. C., et al. (1993). Epidermal growth factor (EGF) promotes human keratinocyte locomotion on collagen by increasing the alpha 2 integrin subunit. *Exp. Cell. Res.*, **209**, 216–223.

Compton, C. C., Butler, C. E., Yannas, I. V., Warland, G., & Orgill, D. P. (1998). Organized skin structure is regenerated *in vivo* from collagen-GAG matrices seeded with autologous keratinocytes. *J. Invest. Dermatol.*, **110**, 908–916.

Cotsarelis, G., Sun, T. T., & Lavker, R. M. (1990). Label-retaining cells reside in the bulge area of pilosebaceous unit: Implications for follicular stem cells, hair cycle, and skin carcinogenesis. *Cell*, **61**, 1329–1337.

Dagregorio, G., & Guillet, G. (2005). Artificial skin as a valuable adjunct to surgical treatment of a large squamous cell carcinoma in a patient with epidermolysis bullosa. *Dermatol. Surg.*, **31**, 474–476.

Dellambra, E., Pellegrini, G., Guerra, L., Ferrari, G., Zambruno, G., et al. (2000). Toward epidermal stem cell-mediated *ex vivo* gene therapy of junctional epidermolysis bullosa. *Hum. Gene. Ther.*, **11**, 2283–2287.

Dong, C., Mead, E., Skalak, R., Fung, Y. C., Debed, J. C., et al. (1993). Development of a device for measuring adherence of skin grafts to the wound surface. *Ann. Biomed. Eng.*, **21**, 51–55.

Edwards, S. M., Donnelly, T. A., Sayre, R. M., Rheins, L. A., Spielmann, H., et al. (1994). Quantitative *in vitro* assessment of phototoxicity using a human skin model, Skin2. *Photodermatol. Photoimmunol. Photomed.*, **10**, 111–117.

Eisenberg, M., & Llewelyn, D. (1998). Effectiveness of Ortec's CCS treatment. *Medical Industry Week*, **33**, 16.

Elder, E., Chen, Z., Ensley, A., Nerem, R., Brockbank, K., et al. (2005). Enhanced tissue strength in cryopreserved, collagen-based blood vessel constructs. *Transplant Proc.*, **37**, 4625–4629.

Falabella, A. F., Valencia, I. C., Eaglstein, W. H., & Schachner, L. A. (2000). Tissue-engineered skin (Apligraf) in the healing of patients with epidermolysis bullosa wounds. *Arch. Dermatol.*, **136**, 1225–12230.

Falanga, V., & Kirsner, R. S. (1993). Low oxygen stimulates proliferation of fibroblasts seeded as single cells. *J. Cell Physiol.*, **154**, 506–510.

Falanga, V., Zhou, L., & Yufit, T. (2002). Low oxygen tension stimulates collagen synthesis and COL1A1 transcription through the action of TGF-beta1. *J. Cell Physiol.*, **191**, 42–50.

Faller, C., & Bracher, M. (2002). Reconstructed skin kits: Reproducibility of cutaneous irritancy testing. *Skin Pharmacol. Appl. Skin Physiol.*, **15**, 74–91.

Fine, J. D. (2000). Skin bioequivalents and their role in the treatment of inherited epidermolysis bullosa. *Arch. Dermatol.*, **136**, 1259–1260.

Fivenson, D. P., Scherschun, L., Choucair, M., Kukuruga, D., Young, J., et al. (2003). Graftskin therapy in epidermolysis bullosa. *J. Am. Acad. Dermatol.*, **48**, 886–892.

Fujita, T., Takami, Y., Ezoe, K., Saito, T., Sato, A. K., et al. (2000). Successful preservation of human skin by vitrification. *J. Burn Care Rehabil.*, **21**, 304–309.

Guidry, C., McFarland, R. J., Morris, R., Witherspoon, C. D., & Hook, M. (1992). Collagen gel contraction by cells associated with proliferative vitreoretinopathy. *Invest. Ophthalmol. Vis. Sci.*, **33**, 2429–2435.

Haniffa, M. A., Wang, X. N., Holtick, U., Rae, M., Isaacs, J. D., et al. (2007). Adult human fibroblasts are potent immunoregulatory cells and functionally equivalent to mesenchymal stem cells. *J. Immunol.*, **179**, 1595–1604.

Hansbrough, J. F., Mozingo, D. W., Kealey, G. P., Davis, M., Gidner, A., et al. (1997). Clinical trials of a biosynthetic temporary skin replacement, Dermagraft-Transitional Covering, compared with cryopreserved human cadaver skin for temporary coverage of excised burn wounds. *J. Burn Care Rehabil.*, **18**, 43–51.

Jiang, Q. J., Izakovic, J., Zenker, M., Fartasch, M., Meneguzzi, G., et al. (2002). Treatment of two patients with Herlitz junctional epidermolysis bullosa with artificial skin bioequivalents. *J. Pediatr.*, **141**, 553–559.

Jimenez, P. A., & Rampy, M. A. (1999). Keratinocyte growth factor-2 accelerates wound healing in incisional wounds. *J. Surg. Res.*, **81**, 238–242.

Jones, S. C., Curtsinger, L. J., Whalen, J. D., Pietsch, J. D., Ackerman, D., et al. (1991). Effect of topical recombinant TGF-beta on healing of partial thickness injuries. *J. Surg. Res.*, **51**, 344–352.

Karlsson, J. O., & Toner, M. (1996). Long-term storage of tissues by cryopreservation: Critical issues. *Biomaterials*, **17**, 243–256.

Kern, A., Liu, K., & Mansbridge, J. (2002). Modulation of interferon-gamma response by dermal fibroblast extracellular matrix. *Ann. N Y Acad. Sci.*, **961**, 364–367.

Laning, J. C., Deluca, J. E., Isaacs, C. M., & Hardin-Young, J. (2001). *In vitro* analysis of CD40-CD154 and CD28-CD80/86 interactions in the primary T-cell response to allogeneic "non-professional" antigen presenting cells. *Transplantation*, **71**, 1467–1474.

Mansbridge, J. (2007). *Transplantation of Engineered Cells and Tissues*. Cambridge UK: Woodhead.

Mauch, C., Hatamochi, A., Scharffetter, K., & Krieg, T. (1988). Regulation of collagen synthesis in fibroblasts within a three-dimensional collagen gel. *Exp. Cell Res.*, **178**, 493–503.

McFarland, K. L., Klingenberg, J. M., Boyce, S. T., & Supp, D. M. (2008). Expression of genes encoding antimicrobial proteins and members of the toll-like receptor/nuclear factor-kappaB pathways in engineered human skin. *Wound Repair Regen.*, **16**, 534–541.

Medawar, P. B. (1957). *The Uniqueness of the Individual*. New York, NY: Basic Books.

Nouri, K., Zhang, Y. P., Singer, L., Zhu, L., Huo, R., et al. (2005). Effect of the 1,450 nm diode non-ablative laser on collagen expression in an artificial skin model. *Lasers Surg. Med.*, **37**, 97–102.

Ortiz-Urda, S., Thyagarajan, B., Keene, D. R., Lin, Q., Fang, M., Calos, M. P., & Khavari, P. A. (2002). Stable nonviral genetic correction of inherited human skin disease. *Nat. Med.*, **8**, 1166–1170.

Ozerdem, O. R., Wolfe, S. A., & Marshall, D. (2003). Use of skin substitutes in pediatric patients. *J. Craniofac. Surg.*, **14**, 517–520.

Pandit, A., Ashar, R., Feldman, D., & Thompson, A. (1998). Investigation of acidic fibroblast growth factor delivered through a collagen scaffold for the treatment of full-thickness skin defects in a rabbit model. *Plast. Reconstr. Surg.*, **101**, 766–775.

Pierce, G. F., Mustoe, T. A., Altrock, B. W., Deuel, T. F., & Thomason, A. (1991). Role of platelet-derived growth factor in wound healing. *J. Cell. Biochem.*, **45**, 319–326.

Pierce, G. F., Tarpley, J. E., Yanagihara, D., Mustoe, T. A., Fox, G. M., et al. (1992). Platelet-derived growth factor (BB homodimer), transforming growth factor-beta 1, and basic fibroblast growth factor in dermal wound healing. Neovessel and matrix formation and cessation of repair. *Am. J. Pathol.*, **140**, 1375–1388.

Prenosil, J. E., & Villeneuve, P. E. (1998). Automated production of cultured epidermal autografts and sub-confluent epidermal autografts in a computer controlled bioreactor. *Biotechnol. Bioeng.*, **59**, 679–683.

Prunieras, M., Regnier, M., & Woodley, D. (1983). Methods for cultivation of keratinocytes with an air–liquid interface. *J. Invest. Dermatol.*, **81**, 28s–33s.

Rheinwald, J. G., & Gleen, H. (1975). Serial cultivation of strains of human epidermal keratinocytes: The formation of keratinizing colonies from single cells. *Cell*, **60**, 331–343.

Robbins, P. B., Lin, Q., Goodnough, J. B., Tian, H., Chen, X., et al. (2001). *In vivo* restoration of laminin 5 beta 3 expression and function in junctional epidermolysis bullosa. *Proc. Natl. Acad. Sci. USA*, **98**, 5193–5198.

Rouabhia, M. (1996). *In vitro* production and transplantation of immunologically active skin equivalents. *Lab. Invest.*, **75**, 503–517.

Schultz, G., Rotatori, D. S., & Clark, W. (1991). EGF and TGF-alpha in wound healing and repair. *J. Cell Biochem.*, **45**, 346–352.

Sibbald, R. G., Zuker, R., Coutts, P., Coelho, S., Williamson, D., et al. (2005). Using a dermal skin substitute in the treatment of chronic wounds secondary to recessive dystrophic epidermolysis bullosa: A case series. *Ostomy Wound Manage.*, **51**, 22–46.

Smiley, A. K., Gardner, J., Klingenberg, J. M., Neely, A. N., & Supp, D. M. (2007). Expression of human beta defensin 4 in genetically modified keratinocytes enhances antimicrobial activity. *J. Burn Care Res.*, **28**, 127–132.

Smola, H., Thiekotter, G., & Fusenig, N. E. (1993). Mutual induction of growth factor gene expression by epidermal-dermal cell interaction. *J. Cell. Biol.*, **122**, 417–429.

Steed, D. L. (1995). Clinical evaluation of recombinant human platelet-derived growth factor for the treatment of lower extremity diabetic ulcers. Diabetic Ulcer Study Group. *J. Vasc. Surg.*, **21**, 71–78; discussion 79–81.

Still, J., Glat, P., Silverstein, P., Griswold, J., & Mozingo, D. (2003). The use of a collagen sponge/living cell composite material to treat donor sites in burn patients. *Burns*, **29**, 837–841.

Toussaint, O., Royer, V., Salmon, M., & Remacle, J. (2002). Stress-induced premature senescence and tissue ageing. *Biochem. Pharmacol.*, **64**, 1007–1009.

Triglia, D., Braa, S., Yonan, C., & Naughton, G. (1991). Cytotoxicity testing using neutral red and MTT assays on a three-dimensional human skin substrate. *Toxicol. in Vitro*, **5**, 573–578.

Tyrone, J. W., Mogford, J. E., Chandler, L. A., Ma, C., Xia, Y., et al. (2000). Collagen-embedded platelet-derived growth factor DNA plasmid promotes wound healing in a dermal ulcer model. *J. Surg. Res.*, **93**, 230–236.

Uitto, J., & Richard, G. (2005). Progress in epidermolysis bullosa: From eponyms to molecular genetic classification. *Clin. Dermatol.*, **23**, 33–40.

Vermorken, A. J., Verhagen, H., Vermeesch-Markslag, A. M., Wirtz, P., Bernard, B. A., et al. (1985). Differentiation of keratinocytes *in vitro*: A new culture vessel mimicking the *in vivo* situation. *Mol. Biol. Rep.*, **10**, 205–213.

Vlassara, H. (1990). Advanced nonenzymatic tissue glycosylation: Cell-mediated interactions implicated in the complications associated with diabetes and aging. *Blood Purif.*, **8**, 223–232.

Watson, S., & Toner, M. (1999). *Cryopreservation of harvested skin and cultured skin or cornea equivalents by slow freezing.* USA: Organogenesis Inc.

Wolkers, W. F., Tablin, F., & Crowe, J. H. (2002). From anhydrobiosis to freeze-drying of eukaryotic cells. *Comp. Biochem. Physiol. A Mol. Integr. Physiol.*, **131**, 535–543.

Xia, Y. P., Zhao, Y., Marcus, J., Jimenez, P. A., Ruben, S. M., et al. (1999). Effects of keratinocyte growth factor-2 (KGF-2) on wound healing in an ischaemia-impaired rabbit ear model and on scar formation. *J. Pathol.*, **188**, 431–438.

Xu, J., & Clark, R. A. (1997). A three-dimensional collagen lattice induces protein kinase C-zeta activity: Role in alpha2 integrin and collagenase mRNA expression. *J. Cell. Biol.*, **136**, 473–483.

Yannas, I. V., Burke, J. F., Gordon, P. L., Huang, C., & Rubenstein, R. H. (1980). Design of an artificial skin. II. Control of chemical composition. *J. Biomed. Mater. Res.*, **14**, 107–132.

Yannas, I. V., Lee, E., Orgill, D. P., Skrabut, E. M., & Murphy, G. F. (1989). Synthesis and characterization of a model extracellular matrix that induces partial regeneration of adult mammalian skin. *Proc. Natl. Acad. Sci. USA*, **86**, 933–937.

Yannas, I. V. (1998). Studies on the biological activity of the dermal regeneration template. *Wound Repair Regen.*, **6**, 518–523.

Zasloff, M. (2002). Antimicrobial peptides of multicellular organisms. *Nature*, **415**, 389–395.

zur Muhlen, A., Klotz, A., Weimans, S., Veeger, M., Thorner, B., et al. (2004). Using skin models to assess the effects of a protection cream on skin barrier function. *Skin Pharmacol. Physiol.*, **17**, 167–175.

Zheng, Y., Du, X., Wang, W., Boucher, M., Parimoo, S., et al. (2005). Organogenesis from dissociated cells: Generation of mature cycling hair follicles from skin-derived cells. *J. Invest. Dermatol.*, **124**, 867–876.

CHAPTER II.6.13 ESOPHAGEAL AND GASTROINTESTINAL TISSUE ENGINEERING

Buddy D. Ratner
Professor, Bioengineering and Chemical Engineering,
Director of University of Washington Engineered Biomaterials (UWEB), Seattle, WA, USA

INTRODUCTION

The gastrointestinal (GI) tract, extending from the mouth to the anus, is over 5 meters long, and is responsible for taking in food and liquids, extracting useful dietary components from that input, excreting unneeded components of that input and also less well defined regulatory processes. A functional GI system is critical for a good quality of life, though life can be sustained bypassing the GI tract, for example with parenteral feeding through a catheter into a large, central vein. The GI system can be damaged due to cancer, trauma, other diseases, acid reflux (esophagus), use of certain drugs, accidental or intentional ingestion of toxins and excess alcohol consumption. Congenital defects such as esophageal atresia, tracheoesophageal fistulas and short bowel syndrome also comprise the functionality of the GI tract. In many cases, surgical repair of the GI tract may be called for. Largely, surgical repair associated with the GI tract has high morbidity rates from stricture, leakage, infection, scarring, ulceration, migration and overall problematic healing. To address these issues and return patients to a reasonable quality of life, tissue engineered replacement of parts of the GI tract may be a solution. Numerous publications have appeared in recent years on tissue engineering for repair of esophagus, stomach, colon and small intestine.

ANATOMY AND STRUCTURE

To understand issues in tissue engineering of GI tract components, an appreciation of the gross anatomy and the microscopic anatomy is valuable. Figure II.6.13.1 illustrates the major components of the GI tract. Figure II.6.13.2 is a schematic of the tissue structure throughout the GI tract. Figure II.6.13.3 illustrates the microscopic structures of intestinal tissue. Note that every portion of the GI tract as illustrated in Figure II.6.13.1 has a unique microscopic structure, often different from that seen in Figure II.6.13.3, but following the pattern suggested by Figure II.6.13.2. Within the many layers seen in figures II.6.13.2 and II.6.13.3 are structures comprised of epithelial cells, smooth muscle cells, striated muscle cells, fibroblasts, endothelium (blood vessels) and nerve.

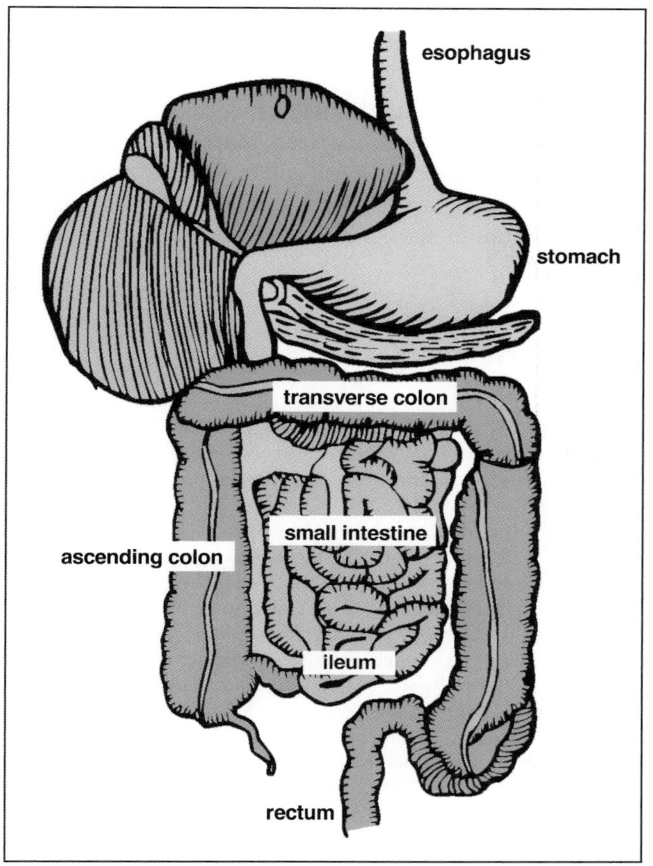

FIGURE II.6.13.1 The human gastrointestinal (GI) tract.

TISSUE ENGINEERING OF GI TISSUES

Tissue engineering replacement tissues for the GI tract require that each of the considerations below be addressed and emulated in the neo-tissue that is produced.

Biomechanics

Tissues of the GI tract are relatively strong (tough) and typically they are radially elastic and less elastic longitudinally. A tissue engineering construct must demonstrate biomechanics close to that of the natural tissue being replaced.

Muscular

GI tract tissues all contain smooth muscle and skeletal muscle responsible for swallowing and for peristalsis.

Mucosal Lining

The lining protects the digestive tract, secretes substances and, in some parts of the tract, absorbs products of digestion.

Innervation

Parasympathetic and sympathetic nerves are found throughout the GI tract. These nerves sense stimuli to the GI tract, control peristalsis and control digestion.

A Blood Vessel Network

The GI tract tissue is well vascularized to nourish and oxygenate cells and remove wastes.

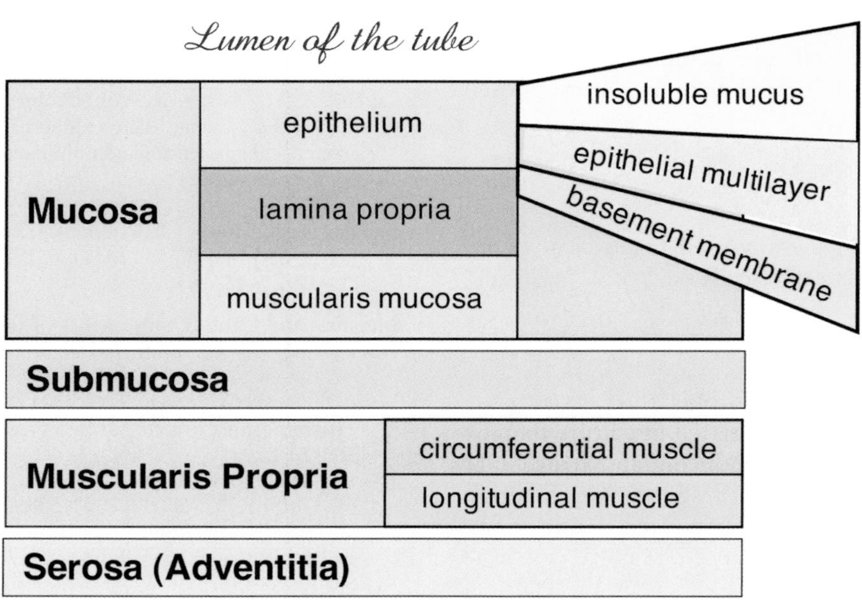

FIGURE II.6.13.2 General structure of the multilayer epithelial tissue comprising the GI tract.

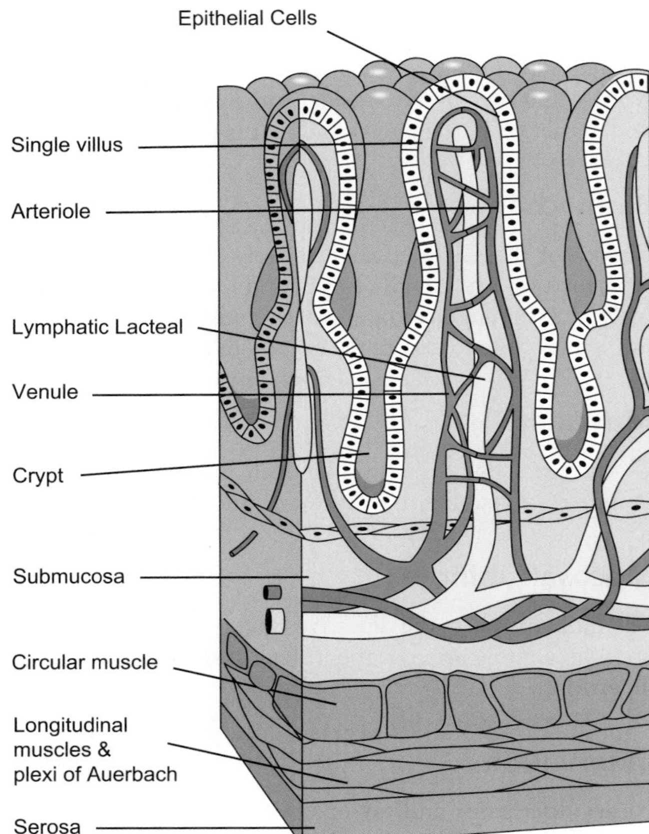

FIGURE II.6.13.3 An illustration showing the microscopic structure of the intestinal wall.

Cell Types

Multiple cell types comprise the GI tract tissues including epithelial cells, smooth muscle cells, striated muscle cells, fibroblasts, endothelium and nerve. The engineered tissue needs the correct cells that demonstrate appropriate functionality and are located were they are supposed to be located.

Scaffolds

A scaffold structure is generally used to physically support, orient and instruct cells (see II.6.3).

Bioreactor Issues

Where cells are preseeded on scaffolds, the desired tissue formation typically has to be initiated in culture thereby increasing total cell number, producing an extracellular matrix and, in many cases, permitting mechanical conditioning of cells prior to implantation (see II.6.6).

Surgical Issues

When the tissue is formed after bioreactor growth and conditioning, how will the surgeon implant the tissue? The integration of the tissue engineered construct can often be accelerated by placing it in contact with a highly vascularized tissue such as the omentum to speed blood vessel ingrowth.

Considering the above issues, a number of strategies have been applied to re-engineer damaged parts of the GI tract with living replacements or regeneration. Predominant among these approaches have been seeding cells on scaffolds and the use of decellularized extracellular matrices (decellularized matrices can be used unseeded, or pre-seeded with cells). Specific studies will not be discussed in this short summary, but a number of more detailed works are lists in the bibliography, below.

BIBLIOGRAPHY

Esophagus

Kajitani, M., Wadia, Y., Hinds, M. T., Teach, J., Schwartz, K. R., & Gregory, K. W. (2001). Successful repair of esophageal injury using an elastin based biomaterial patch. *ASAIO J.*, **47**, 342–345.

Badylak, S., Meurling, S., Chen, M., Spievack, A., & Simmons-Byrd, A. (2000). Resorbable bioscaffold for esophageal repair in a dog model. *J. Pediatr. Surg.*, **35**, 1097–1103.

Takimoto, Y., Nakamura, T., Yamamoto, Y., Kiyotani, T., Teramachi, M., & Shimizu, Y. (1998). The experimental replacement of a cervical esophageal segment with an artificial prosthesis with the use of collagen matrix and a silicone stent. *J. Thorac. Cardiovasc. Surg.*, **116**, 98–106.

Isch, J. A., Engum, S. A., Ruble, C. A., Davis, M. M., & Grosfeld, J. L. (2001). Patch esophagoplasty using AlloDerm as a tissue scaffold. *J. Pediatr. Surg.*, **36**, 266–268.

Grikscheit, T., Ochoa, E. R., Srinivasan, A., Gaissert, H., & Vacanti, J. P. (2003). Tissue-engineered Esophagus: Experimental substitution by onlay patch or interposition. *J. Thorac. Cardiovasc. Surg.*, **126**, 537–544.

Bhrany, A. D., Beckstead, B. L., Lang, T. C., Farwell, D. G., Giachelli, C. M., & Ratner, B. D. (2006). Development of an Esophagus Acellular Matrix (EAM) Tissue Scaffold. *Tissue Engineering*, **12**(2), 319–330.

Saxena, A. K., Kofler, K., Ainödhofer, H., & Höllwarth, M. E. (2009). Esophagus tissue engineering: Hybrid approach with esophageal epithelium and unidirectional smooth muscle tissue component generation in vitro. *J. Gastrointest. Surg.*, **13**(6), 1037–1043.

Badylak, S. F., Hoppo, T., Nieponice, A., Gilbert, T. W., Davison, J. M., & Jobe, B. A. (2011). *Tissue Engineering Part A*, **17** (11-12), 1643–1650.

Intestine and Other Components of the GI Tract

Day, R. M. (2006). Epithelial stem cells and tissue engineered intestine. *Current stem cell research & therapy*, **1**(1), 113–120.

Levin, D. E., & Grikscheit, T. C. (2012). Tissue-engineering of the gastrointestinal tract. *Current Opinion in Pediatrics*, **24**(3), 365–370.

Basu, J., Mihalko, K., Payne, R., Rivera, E., Knight, T., Genheimer, C., Guthrie, K., et al. (2011). Regeneration of rodent small intestine tissue following implantation of scaffolds seeded with a novel source of smooth muscle cells. *Regenerative Medicine*, **6**(6), 721–731.

Bitar, K. N., & Raghavan, S. (2012). Intestinal tissue engineering: Current concepts and future vision of regenerative medicine in the gut. *Neurogastroenterology & Motility*, **24**, 7–19.

CHAPTER II.6.14 NEURONAL TISSUE ENGINEERING

Ravi V. Bellamkonda, Isaac P. Clements, and Jennifer M. Munson
Department of Biomedical Engineering, Georgia Institute of Technology, Atlanta, GA, USA

The human nervous system gathers, transmits, processes, and stores information with remarkable speed and fidelity. These capabilities are made possible by networks of highly specialized neural cells, organized within an intricate three-dimensional framework of interconnected neural tissue. Only through a gradual and complex sequence of development does this neural tissue form and mature, in a process beginning soon after conception, and continuing long after the time of birth.

Due to its level of complexity and specialization, the nervous system is vulnerable to damage that exceeds its natural capacity to heal or regenerate. It should not be surprising, therefore, that neural structures originally formed and refined over the timespan of years are not always trivially restored after injury. Recovery is also hindered by a variety of other factors, including the limited capacity of neurons to proliferate, and a scarring response to injury that limits the extent of damage, but also interferes with healing.

Despite intrinsic challenges to the repair of neural tissue, effective therapeutic strategies are not beyond reach. An increasing understanding of the nervous system, coupled with advances in engineering and materials science, has formed the basis of the field of neural tissue engineering. Neural tissue engineering strategies use biomaterials to influence the physical and chemical environment at the site of neural injury or disease. By controlling the spatial and temporal distribution of various types of signaling cues, pro-regenerative environments can be engineered for the purpose of restoring the complex physical and biochemical architecture of neural tissue.

CHALLENGES TO NEURAL TISSUE HEALING AND THE ENABLING THERAPEUTIC ROLE OF BIOMATERIALS

The vertebrate nervous system can be subdivided into the central nervous system (CNS) and peripheral nervous system (PNS). The chief components of the CNS are the brain and spinal cord, while the PNS includes the peripheral nerves that branch out from the CNS. Within both of these divisions, neural tissue is composed primarily of neurons and supporting glial cells. However, basic cell types and tissue organization vary greatly between the brain, spinal cord, and peripheral nerves. Consequently, each of these divisions responds uniquely to injury, and therapeutic interventions must be specialized accordingly.

Peripheral Nervous System

Neurons of the PNS connect the CNS with sensory and motor targets. The cell body of each PNS neuron is located in or near the spinal cord or base of the brain. From this cell body, a long axon extends uninterrupted to the tissue it innervates (up to one meter or more). A single peripheral nerve can contain thousands of motor and sensory axons, enveloped along with other cell types within flexible tubes of collagen and other supporting extracellular matrix (ECM) components.

The glial cells of the PNS are Schwann cells, which wrap around the axons in concentric layers to form insulating sheaths rich in the protein myelin. These myelin sheaths enhance the speed of signal conduction down the length of the axon. Schwann cells also provide trophic support to the axons and play a major role in injury response.

Peripheral Nerve Injury. When a peripheral nerve is damaged such that axons within it are severed, the portions of the affected axons lying distal to the injury site are cut off from the centrally located cell body. These distal axon segments subsequently degrade, and the resulting debris is cleared away by macrophages and Schwann cells as part of a process termed Wallerian degeneration. The Schwann cells proliferate and take on a pro-regenerative phenotype, rearranging themselves into aligned tracts called "bands of Bungner" and secreting factors conducive to the in-growth of regenerating axons.

Meanwhile, the portions of the injured neurons proximal to the damage site prepare for regeneration. The cell body of each injured axon undergoes major metabolic changes and initiates a program of protein synthesis to support axonal regeneration. At the tip of each injured axon, a growth cone develops which leads the regenerating axon through the intact structure of the distal nerve segment and back toward its original target, at the approximate rate of 1–3 mm per day (Gutmann et al., 1942).

Treatment for Nerve Injury. The process of peripheral nerve regeneration occurs spontaneously, as long as the injury does not result in an extended gap in the nerve. In cases where a small gap exists, the two nerve stumps can be surgically re-apposed. However, when the nerve gap is longer, this procedure creates unacceptable tension on the nerve. In these cases, an autografted segment of nerve is the clinical gold standard for treatment. Healthy segments of nerve are harvested from elsewhere in the body, and used to bridge the gap across the injury site. Unfortunately, this technique comes with several drawbacks, including the need for an additional surgery and a loss-of-function at the site from which the donor nerves are taken. Furthermore, patient recovery after autograft treatment is less than ideal.

For these reasons, much research has been devoted to developing a biomaterials-based implant as an alternative to an autograft. The goal of this research has been to engineer a construct that can be implanted across an injury gap in place of a nerve graft to provide support and guidance to regenerating cells and axons. Such a construct would employ biomaterials to recreate a pro-regenerative environment, similar to that which is found in the distal segment of injured nerve. For example, the bands of Bungner found in the distal nerve segment provide a physically aligned surface topography containing oriented tracks of adhesive molecules. Diffusible biochemical trophic and signaling factors are also released by cells in the distal nerve stump. In the following sections, we will see how appropriately selected biomaterials can be employed to recreate these types of natural physical and chemical cues, for the purpose of guiding migrating cells and regenerating axons across the nerve gap, and into the distal nerve segment (Figure II.6.14.1).

Spinal Cord

In some ways, spinal cord architecture resembles that of the peripheral nerves. Both types of neural tissue contain extended cables of bundled axons that are myelinated and supported by associated glial cells. Spinal cord neurons, however, are not limited to connecting the neurons of the brain with those of peripheral nerves, but can also synapse with each other to form reflex loops and complex processing circuits. As part of the CNS, spinal cord glial cells include oligodendrocytes and astrocytes. The former myelinate and support axons, while the latter serve a variety of roles, including injury response and maintenance of the blood–brain barrier (BBB). This barrier is composed of tight cell junctions surrounding the vasculature, and restricts the passage of molecules from the bloodstream to the CNS. The CNS also contains microglia, scavenger-like cells that carry out similar roles to macrophages, and also play a major role in the response to injury.

Response of the Spinal Cord to Injury. In contrast to the injury response of peripheral nerves, the sequence of physiological events resulting from traumatic injury to the spinal cord serves to inhibit rather than to promote axonal regeneration. The spinal cord injury response can be divided into two major phases. The primary, or acute, phase of events results from direct damage to spinal cord tissue caused by mechanical forces. A fractured vertebra, for example, might crush a section of the spinal cord, destroying neural cells, severing axonal connections, and disrupting the local vasculature. Within minutes of the initial injury, the spinal cord swells, further disrupting blood flow, and uncontrolled waves of neural activity spread through the area. These events disrupt ionic concentrations that are normally tightly regulated, leading to a loss of local neural function and regulation of the remaining vasculature.

In response to this primary set of events, a cascade of secondary injury events ensues, expanding the zone of cell death and tissue damage beyond the site of initial injury. Edema and ischemic cell death continue, due to disruption of the vasculature. Excitatory neurotransmitters, normally used to transmit signals across synapses, are released in bulk from neurons during the initial injury. The resulting excitotoxicity triggers the production of damaging free radicals, and also mediates cell death by several mechanisms. In one chain of events, excitatory transmitters open gated membrane channels, allowing an influx of calcium that triggers programmed pathways of cell death (apoptosis).

The secondary cascade of events also contributes to establishing barriers to regeneration. Cell death, coupled with infiltration of cells and other factors through the compromised blood spinal cord barrier triggers a destructive inflammatory response. Resident astrocytes and microglia take on a reactive phenotype, and contribute to the formation of a glial scar around the zone of necrotic cell death. This scar tissue contains growth-inhibitory factors such as chondroitin sulfate proteoglycans (CSPGs), which impede axonal regeneration across the injury site. Other major growth-inhibitory components include various molecules exposed from within damaged myelin sheaths.

Treatment for Spinal Injury. Current treatments for spinal cord repair and regeneration are limited. Because acute injury events occur so quickly after the initial

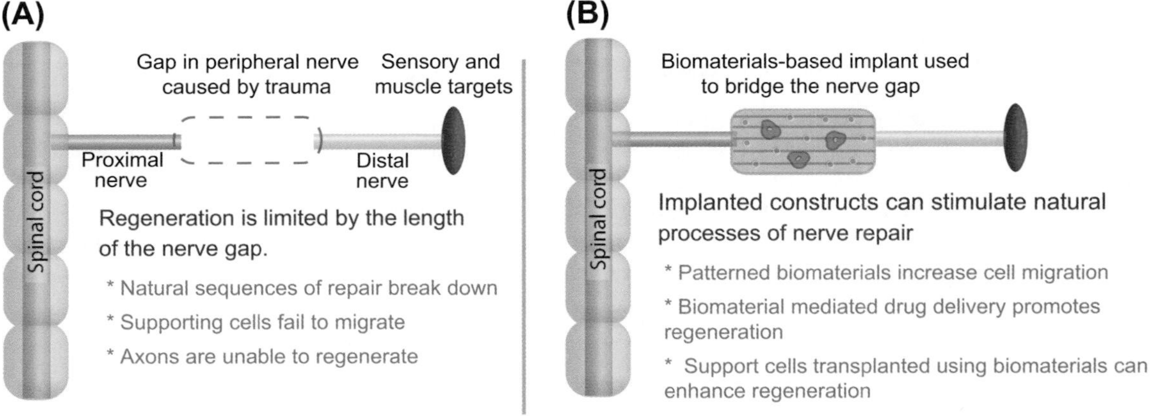

FIGURE II.6.14.1 (A) Summary of biological constraints to peripheral nerve regeneration. (B) Capacity of biomaterials-based constructs to alleviate biological constraints and promote regeneration.

damage, existing therapies are typically designed to limit the extent of secondary injury, often with the administration of high doses of steroids (Bracken et al., 1990). Even for this purpose, few options for clinical intervention exist.

Tissue-engineered therapies offer the potential to more effectively modulate the spinal cord's secondary response to injury. For example, biomaterials can be used to engineer advanced delivery systems for drugs able to influence secondary injury events. These types of systems can be designed to deliver biochemical factors with a level of control and precision not achievable with conventional approaches.

Another biomaterials-based treatment approach is to facilitate axonal regeneration across the injury site. Contrary to former and longstanding beliefs, axons in the CNS have an inherent capacity to regenerate. It is mainly due to the growth-inhibitory nature of the post-injury environment that regeneration is prevented from occurring. Two main strategies for promoting axonal regeneration include: (1) making the environment of the injury gap less inhibitory to axonal growth; and (2) providing an engineered nerve bridge as an alternative pathway for regeneration (similar to a PNS treatment approach).

As we'll see in the following sections, biomaterials offer the ability to pursue each of these strategies. By controlling the physical and chemical environment of the injury gap, axonal regeneration can be promoted (Figure II.6.14.2). Significantly, the task of inducing regenerating axons to re-enter spinal cord tracts distal to the injury site poses a second set of challenges.

Brain

The brain and spinal cord are both part of the CNS, and so they share many basic structural similarities, including glial cell types and the presence of the blood–brain barrier (BBB). For these reasons, the responses of the brain and spinal cord to injury bear much in common. As in the spinal cord, acute injury events involved in traumatic brain injury (TBI) are defined as those resulting from direct tissue damage. Again, the acute phase is followed by a sequence of secondary events that result in an expanded zone of cell death, inflammation, and glial scarring. Similar strategies for limiting secondary injury events are applicable in both tissues.

In other ways, the structure of the brain and its response to injury are unique. Even more so than in the spinal cord, neurons in the brain exhibit a wide variety of distinct morphologies. Some neuronal cell types branch extensively, synapsing with other neurons over large regions. Others span only small distances, transmitting localized information. This complex and distributed architecture of brain tissue complicates strategies for directing the functional repair of neural circuitry.

The brain is also unique, in that it is divided into distinct structures, many of which are composed of neurons specialized to perform specific tasks. Pathological conditions stemming from injury, disorders, and disease can selectively target certain regions of the brain, causing a characteristic result. For example, Parkinson's disease is a neurodegenerative disease that targets regions of the basal ganglia, destroying neurons responsible for producing the neurotransmitter dopamine.

A variety of tissue engineering-based treatment strategies are emerging for the treatment of neurodegenerative disease and TBI (Figure II.6.14.3). These approaches rely on biomaterials, which can be used to alter the local chemical and physical environment around the injury site, in order to encourage tissue repair and correct biochemical deficits. Significantly, biomaterials-based constructs also confer the ability to transplant cells as an additional therapeutic approach for treating neural injury and disease.

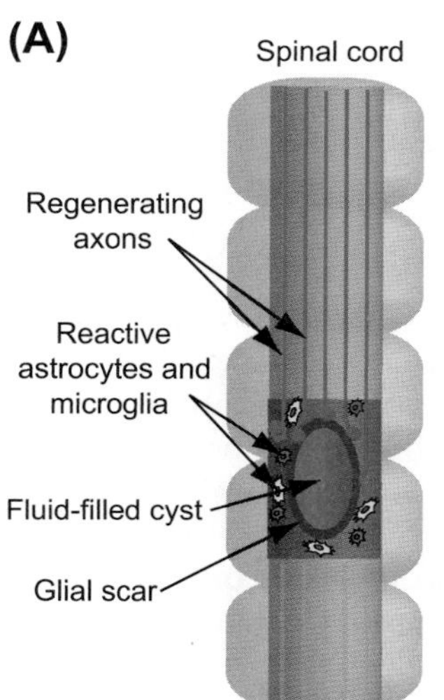

Repair and regeneration after spinal cord injury is limited.

* Secondary injury cascades cause additional reactivity and cell death
* Inflammatory response results in cells death and impairs regeneration
* Growth inhibitory molecules block regenerating axons
* CNS neurons do not proliferate to replace lost cells

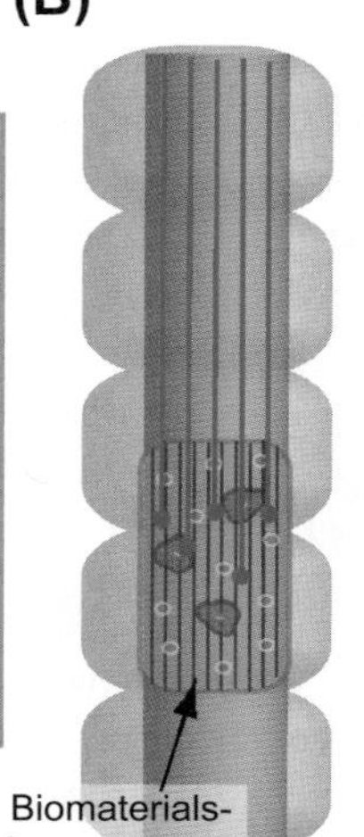

Biomaterials-based implants can limit secondary injury events and promote regeneration.

* Patterned biomaterials can increase cell migration and guide regenerating axons
* Released drugs can limit secondary injury, modulate growth inhibitory environment, stimulate cell migration, and promote axonal regeneration
* Transplanted cells can replace support cells and neurons to enhance repair and regeneration and limit secondary injury events.

FIGURE II.6.14.2 (A) This simplified diagram summarizes constraints to spinal cord regeneration. (B) An implanted biomaterials-based construct can reduce damaging post-injury events and contribute to neural repair and regeneration.

Traumatic brain injury and degenerative disease can result in neuronal loss and biochemical deficits

* As in the spinal cord, secondary injury cascades cause further damage and cell death
* Growth inhibitory molecules block regenerating axons, and CNS neurons cannot proliferate
* Injury and neurodegenerative diseases can cause imbalances of neurotransmitters resulting in severe loss of function

Biomaterials-based implants can restore balances of biochemical cues

* Patterned materials can increase cell migration and axonal growth
* Released drugs can limit secondary injury and promote regeneration
* Released drugs can restore biochemical imbalances
* Transplanted cells can replace support cells and neurons to limit secondary injury, enhance repair, and restore neural function.

FIGURE II.6.14.3 The use of biomaterials in treatment strategies for neurodegenerative disease and traumatic brain injury.

BIOMATERIALS-BASED CUES FOR THE TREATMENT OF NEURAL INJURY/DISEASE

Neural cells are highly responsive to natural cues present in their surrounding environment, especially during periods of growth and development. The course of a regenerating axon, for example, is guided by such factors as the physical topography and chemistry of the surface along which it grows, as well as by signaling molecules in the local environment. These physical, biochemical, and even electrical cues exert great influence on cellular viability, growth, and activity.

A fundamental strategy in tissue engineering-based treatment approaches is to artificially recreate environmental cues in such a way as to influence neural cell behavior. This task is often delicate, however, as natural cues are presented in complex combinations within the intricate terrain of neural tissue. Accordingly, neural cells are sensitive to the precise nature of a given type of cue, as well as to its precise spatial location and time course.

Significantly, the effective use of biomaterials within tissue-engineered constructs enables spatiotemporal control over the presentation of environmental cues, such that the repair and regeneration of neuronal deficits can be engineered. Tissue engineering techniques also employ biomaterials to transplant cells for the purpose of replacing or supplementing the function of injured or diseased neural cells. Cell transplantation constructs can present cues able to influence not only the host neural tissue of the implantation site, but also the transplanted cells contained within the constructs.

Structural and Topographic Cues Can Physically Support and Guide Neural Repair Mechanisms

Structural Support. At the most basic level, tissue-engineered scaffolds can promote neural repair by providing physical support through the mechanical structure of their component biomaterials. This macro-level physical support might serve, for example, to isolate an area of injured tissue to establish a local environment conducive to healing or regeneration. In some applications, the mechanical properties of a scaffold might also serve to provide physical support in the place of lost surrounding tissue.

The influence on neural tissue of scaffold structure can be illustrated by the case of engineered nerve guidance channels, which can be used in place of nerve grafts to promote nerve regeneration across extended gaps in peripheral nerves. In its most basic embodiment, a nerve guidance channel consists of an empty tube of inert material, such as smooth silicone. The ends of the injured nerve stumps can be affixed into the ends of the tube, which serves to provide an isolated pathway conducive to regeneration. The walls of the tube block the in-growth of scar tissue forming cells, holding open a space through which regenerating cells and axons can grow (Figure II.6.14.4).

In many neural repair applications, implanted biomaterials are required to serve only a temporary function. In these situations, it is often desirable to use biodegradable materials, which usually consist of natural or synthetic polymers. Common synthetic polymers used in such applications include poly(α-hydroxy-acids), such as poly(lactic acid) (PLA), poly(glycolic acid) (PGA), and the copolymer poly(lactic-*co*-glycolic acid) (PLGA) (see Chapter I.2.6). Each of these synthetic biodegradable polymers is broken down by hydrolysis of ester bonds to release nontoxic by-products. The use of the copolymer PLGA is especially useful, because adjusting the copolymer ratio can be used to tune the rate of biodegradability. This breakdown time can be tailored to range from less than a month to several years, depending on the specific application.

Other biodegradable polymers, including collagen, fibrin, chitosan, and alginate are derived and purified from natural sources. These natural polymers are typically degraded by intrinsic enzymes, and are attractive because of their inherent biocompatibility and low toxicity. A disadvantage of natural polymers is that their chemical composition is not as well-defined as that of synthetic materials.

Influence of Topographic Cues on Cellular Behavior. Biomaterials with patterned surfaces can provide fine topographic cues in addition to tissue-level structural support. Topographic cues are physical features of the appropriate geometry and size to influence individual cells. Cellular attachment, alignment, migration, growth,

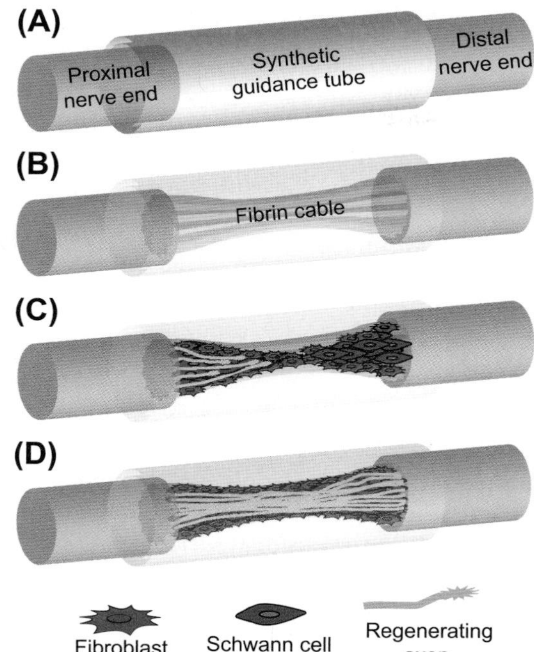

FIGURE II.6.14.4 Peripheral nerve regeneration through an empty nerve-guidance tube occurs in an ordered sequence. (A) The ends of a severed nerve are attached into a nerve-guidance tube. (B) The tube fills with plasma derived from the nerve stumps. Fibrin precursors from within the plasma coalesce into an oriented fibrin bridge that physically bridges the nerve gap. (C) Cells involved in regeneration, including fibroblasts and Schwann cells, migrate through this aligned fibrin cable and begin to proliferate and differentiate to prepare the way for regenerating axons. (D) Axons regenerate from the proximal stump along with the migrating Schwann cells, bridging the gap and re-entering the distal nerve stump.

and gene expression can all be regulated by controlling the nature and distribution of topographic cues provided by the component biomaterials of a tissue-engineered construct.

The capacity of cells to respond to artificial topographic cues stems from the sensitivity of cells to natural topographic cues in their normal biological surroundings. Cells are particularly sensitive to topography during periods of growth, development, and regeneration. For example, during neural tissue development, aligned extracellular matrix (ECM) plays a major role in guiding neural cell migration and differentiation (Rakic, 1971). By mimicking the natural structures along which neural cells develop and reside, engineered substrates with particular topographies can greatly influence cellular behavior.

The mechanisms by which topographic cues influence cellular behavior are complex, but involve interactions between the substrate topography and cellular structures that are sensitive to physical cues and able to effect cell-wide changes in behavior. For example, neural growth cones, located at the front of extending axons or neurites, are particularly sensitive to topographical features as they guide neuronal extension during periods of growth and development. Filipodial protrusions on the end of the growth cones continually advance or

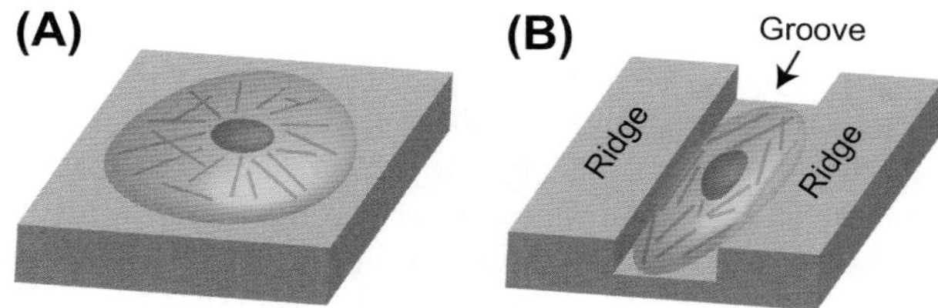

FIGURE II.6.14.5 The reorganization of cytoskeletal components can be constrained by the surface topography of the surrounding substrate, resulting in cellular alignment and directed migration. (A) A cell cultured on a flat surface contains cytoskeletal components radiating predominantly from the centralized nucleus. (B) When the same type of cell is cultured on a grooved surface, the orientation of the cytoskeletal components are constrained by the grooved surface topography. (Figure adapted from Wilkinson et al., 2002, with permission.)

retract in response to physical and chemical cues as they "explore" their surrounding environment. Ultimately, growth cones direct axon/neurite extension by mediating reorganization of the neuronal cytoskeleton.

This cytoskeletal reorganization is required for changes in cellular orientation as well as for growth and migration. The cytoskeleton is rearranged by depolymerization and polymerization of its component microtubules, intermediate filaments, and microfilaments (Wilkinson et al., 2002). It is hypothesized that these rod-like structures are limited in their capacity to bend, similar to a railroad track, and that their polymerization is thus dependent on the surrounding topography. This hypothesis might explain, for example, why cellular alignment and growth can be directionally constrained by certain types of topographic features, such as aligned grooves (Figure II.6.14.5).

Size Scales of Topography

The response of a cell to its surrounding topography strongly depends on the size of the topographic cues. Micron scale features might act at the cellular level, while sub-micron and nanoscale features act at subcellular levels (Seidlits et al., 2008). Significantly, biological structures such as growth cone filopodia lie within the nanometer range, and are highly sensitive to nanoscale topographic cues (Dalby et al., 2004). For example, basement membranes, across which axons typically grow, are composed of a dense meshwork of three-dimensional topography, featuring pores and fibers with dimensions ranging from tens to hundreds of nanometers (Flemming et al., 1999; Abrams et al., 2000).

Due to the sensitivity of neural cells to nanoscale cues, attention must be given to nanoscale topography in the process of tissue engineering design. Significantly, few choices of biomaterials are truly smooth at the nanoscale, so most implanted materials present nanoscale cues, whether intended or not (Curtis and Wilkinson, 1997).

Engineering Topographic Cues to Control Cell Behavior. One of the most basic ways of altering the topography of a biomaterial is to vary its roughness by randomly modulating its surface texture (Norman and

Desai, 2006) (see Chapter I.2.15). Biomaterials roughened to include surface features at the appropriate size scales have been shown to promote cellular adhesion and viability. For example, silicon substrates treated with chemical etchants or reactive ion etching techniques to include surface features down to the nanoscale have been shown to increase the adhesion and survival of both neurons and glia (Turner et al., 1997; Fan et al., 2002). It is proposed that nanoscale surface roughness promotes cellular adhesion by increasing cell membrane contact area as cells conform to fit the roughened surface (Fan et al., 2002).

More precisely defined topography can be used to influence cellular behavior in a more controlled manner. The ability to pattern arbitrary physical features allows for the fabrication of patterned surfaces that more closely mimic the surfaces of physiological structures and supporting cells. Cell culture experiments on topographically patterned substrates have explored the effects on cell behavior of a variety of patterned surface features including grooves, pits, ridges, steps, and waves. These experiments have demonstrated the sensitivity of cells to geometric features down to nanoscale dimensions, and have shed light on the mechanisms by which cells are influenced by their surrounding topographies. Neurite extension, for example, can be directly guided along grooved or ridged surfaces, in a manner resembling fasciculation, the naturally occurring process in which axons grow along other pre-existing axons (Rajnicek et al., 1997; Johansson et al., 2006). Other experiments have explored the effects of topographic cues presented in three dimensions. Matrices in the form of permissive gels or porous materials are useful for stimulating cell attachment, and for providing a substrate for physical support and growth. Hydrogels, for example, are composed of a three-dimensional structure of polymer cross-linkages that can be controlled to achieve desired properties such as stiffness and pore size. These parameters can be tailored to create biomaterial scaffolds with similar stiffness to soft-tissues and a low interfacial tension, favorable to cellular in-growth (Woerly et al., 1991; Bellamkonda et al., 1995).

Different types of oriented topographies can be created within three-dimensional substrates using a variety of techniques. For example, gels can be given an aligned orientation by flowing them during gelling. Externally applied magnetic fields have been used to align collagen gels (Ceballos et al., 1999), and a variety of techniques have been used to create aligned subluminal or microchannels within three-dimensional polymer substrates. Oriented three-dimensional substrates have been used to create nerve guidance channels, as well as guidance bridges for treating spinal cord injury gaps. Success has been met in both applications, although in the case of spinal cord repair, additional challenges must be addressed before clinically relevant results are achieved.

Indirect Alignment of Axons

Although topographic cues can be used to directly influence neuronal growth, a potentially more effective strategy is to provide indirect guidance, by targeting the influence of patterned topography towards non-neuronal cells. For example, oriented topography can be used to align glial cells, which in turn can stimulate and support directed axonal growth not only through physical guidance cues, but also through aligned pathways of biochemical guidance cues, such as secreted ECM molecules (Biran et al., 2003; Thompson and Buettner, 2004). Fibroblasts, meningeal cells, astrocytes, and Schwann cells, aligned on topographically defined substrates, have all been demonstrated to promote and direct axonal growth (Figure II.6.14.6) (Miller et al., 2001; Biran et al., 2003; Walsh et al., 2005; Kim et al., 2008).

Techniques for Topographic Patterning. Conventional photolithography techniques are most commonly used to pattern two-dimensional surfaces with microscale features. Silicon is often used as a substrate, due to established patterning techniques developed for the microelectronics industry. Polymeric substrates can also be lithographically patterned, and are typically more suitable due to their definable mechanical properties and/or biodegradability. Another advantage of polymeric substrates is their capacity for direct pattern transfer via casting or embossing (Curtis and Wilkinson, 1997). Polydimethylsiloxane (PDMS) and poly(methyl methacrylate) (PMMA) are two of the most commonly used materials for casting replicas of another patterned substrate.

While standard photolithography techniques can be used to pattern surfaces with microscale features, other techniques are required to pattern nanoscale features (100 nm and below) (Curtis and Wilkinson, 1997). For example, electron beam lithography (EBL) can be used to create arbitrary patterns with feature sizes as small as 3–5 nm (Vieu et al., 2000). However, processing with EBL techniques is time-consuming and requires expensive equipment (Norman and Desai, 2006). To increase speed and reduce cost, patterns initially produced by EBL techniques can be replicated with polymer materials

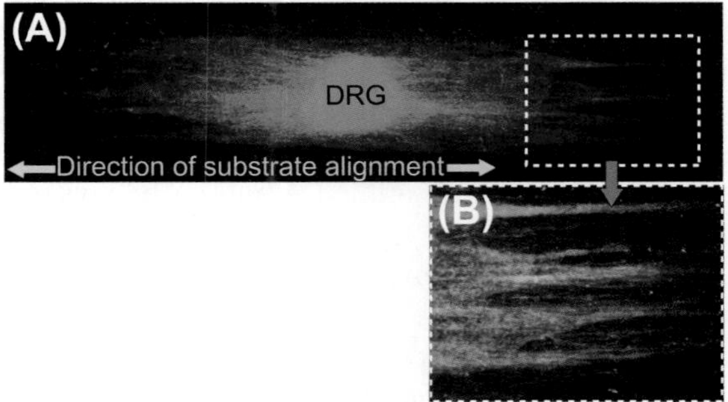

FIGURE II.6.14.6 Image of a rat dorsal root ganglion (DRG) cultured on a surface of aligned polymer fibers created using electrospinning techniques. (A) Schwann cells (green) migrate from the DRG, following the orientation of the submicron fibers. (B) In this magnified image, axons are additionally labeled (red), and can be seen to grow along the aligned tracts of Schwann cells.

such as PDMS, through techniques such as hot embossing and solvent casting (Figure II.6.14.7) (Dalby et al., 2004; Johansson et al., 2006; Norman and Desai, 2006).

While techniques such as EBL enable arbitrary control over nanoscale surface features, simpler and more inexpensive techniques are often preferable for patterning nanoscale topography when precise control over individual features is not required (Norman and Desai, 2006). Patterning is discussed in detail in Chapter I.2.13.

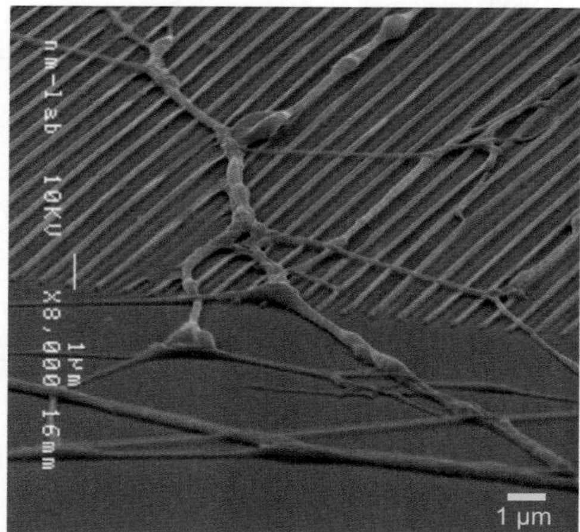

FIGURE II.6.14.7 Scanning electron microscope image of neurites extending along and across patterned ridges, but not within the grooves between the ridges. (Ridges are 100 nm in width and height.) This topography was created by coating a silicon surface with PMMA polymer, and then imprinting the polymer with a master stamp made using electron beam lithography techniques. (Figure reproduced from Johannson et al., 2006, with permission.)

Electrospinning techniques, for example, can be used to create polymer fibers with diameters down to the nanoscale range (Figure II.6.14.8A). Aligned nanofibers are thought to mimic fibrous extracellular matrix proteins such as collagen and fibrinogen which have similar dimensions and alignment (Norman and Desai, 2006). The aligned topography of nanofibers can be used to promote directed nerve regeneration by guiding glial migration and axonal growth (Yang et al., 2005; Kim et al., 2008; Clements, 2009). A variety of nerve regeneration scaffolds, employing electrospun nanofibers at their core, have been shown to effectively facilitate nerve regeneration *in vivo* (Figure II.6.14.8B).

Table II.6.14.1 summarizes the use of biomaterial-based topographic cues for neural tissue engineering applications.

Biochemical Cues Can Promote Neural Cell Activity Leading to Repair and Regeneration

Neural cells are surrounded by a complex sea of biomolecules, many of which the neural cells specifically recognize as biochemical cues. These biochemical cues come

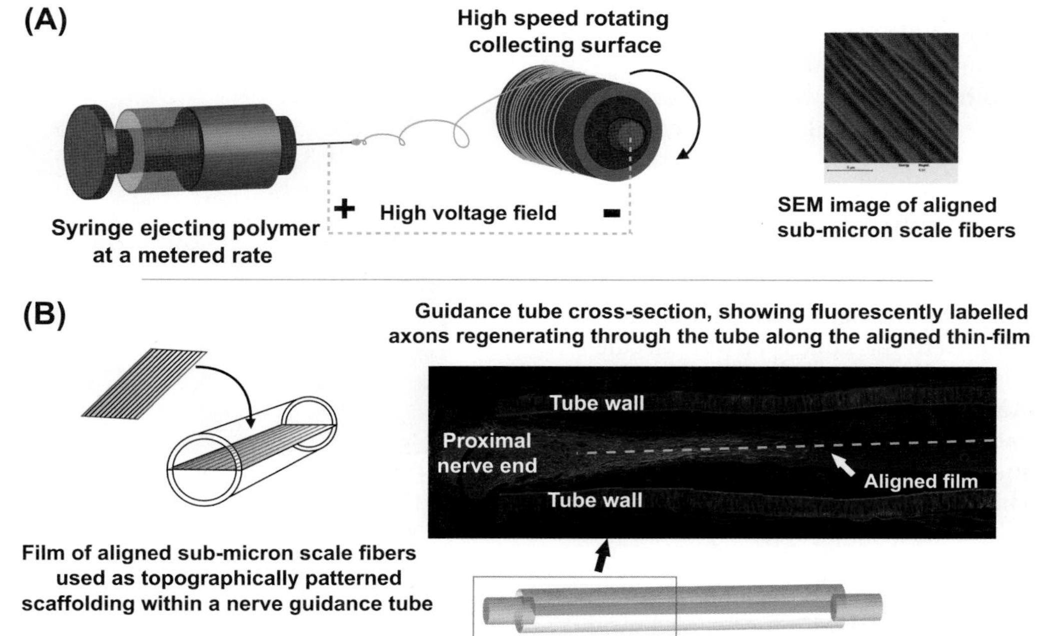

FIGURE II.6.14.8 (A) To electrospin fibers, a liquid polymer melt is introduced to a high voltage field as it is slowly ejected from a syringe. Charge build-up on the drop of polymer at the syringe tip creates repulsive forces that cause tiny streams of polymer to be ejected towards a metal collecting surface. By collecting the ejected fibers over extended periods of time, mats of fibers can be collected. Using a cylinder or disk rotating at high speeds as the collecting surface is one method to collect the fibers in aligned films or sheets. (B) When an empty unpatterned tube is used to repair a nerve gap, the length of the gap across which regeneration will occur is limited. Across long gaps the normal processes of regeneration break down, and glial cells and axons fail to cross the nerve gap. When the guidance tube is supplemented with internal scaffolding of aligned substrates, as in this example, the length across which regeneration will occur can be greatly extended. Aligned topographic cues provided by the scaffolding material enable regeneration across longer gaps, in part by promoting the alignment and directed migration of regenerating cells.

TABLE II.6.14.1 The Use of Biomaterial-Based Topographic Cues for Neural Tissue Engineering Applications

Type of Topography	Fabrication Technique	Biomaterials	Influence	Neural Applications
Roughened surfaces	Reactive ion etching, chemical etching	Rigid materials such as silicon and quartz. Pattern can be transferred onto polymers	Promotes cell adhesion, viability, and activation. Enhances implant/host tissue integration	*In vitro* models (Fan et al., 2002), implanted constructs
Patterned surfaces (e.g., grooves, ridges, steps, etc.)	Photolithography, electron beam lithography. Patterns can be transferred onto other polymers	Silicon, polymers	Promotes cell attachment, alignment, migration, and differentiation. Guidance of axonal/ neurite outgrowth	Influencing and studying cell behavior *in-vitro* and *in vivo* (Goldner et al., 2006)
Three-dimensional solid structures with random or patterned topography (pores, channels, sublumina, etc.)	Freeze drying, injection molding, templating, solvent casting, phase separation, three-dimensional printing, multiphoton excitation, magnetic alignment	Degradable (e.g., PLGA, chitosan) and non-degradable materials (PAN-PVC) can be used to create hydrogels, sponges, and tubes	Promotes cell attachment, alignment, migration, and differentiation. Provides patterned structural support	*In vitro* models, peripheral nerve guidance scaffolds (Chamberlain et al., 1998; Ceballos et al., 1999), spinal cord guidance scaffolds (Yu and Schoichet, 2005)
Random or aligned micro/nano scale diameter fibers	Electrospinning, wet spinning, phase separation, peptide self-assembly	Synthetic polymers (e.g., PLA, PCA), co-polymers (e.g., PLGA), natural polymers (e.g., collagen)	Promotes cell attachment, alignment, migration, and differentiation. Can simulate aligned ECM. Provides aligned structural support.	*In-vitro* cell differentiation models (Yang et al., 2004), peripheral nerve guidance tubes (Ngo et al. 2003; Kim et al., 2008)

PAN-PVC: poly(acrylonitrile)-poly(vinyl chloride)
PCA: poly(caprolactone)
PLA: poly(lactic acid)
PLGA: poly(lactic-co-glycolic acid)
ECM: extracellular matrix

in different forms, and influence cell behavior through a variety of mechanisms.

Surface-Bound Biomolecules Mediate Adhesion. Surface-bound adhesion molecules are glycoproteins found on cell membranes and within extracellular matrix that provide binding sites able to mediate cellular attachment and growth. Neural cells express several different families of membrane-bound receptors that are able to selectively bind adhesion molecules. In the case of homophilic binding, a given adhesion molecule on one cell binds to the same type of molecule on another cell. This is the case for neural cell adhesion molecules (NCAMs) and cadherins, both of which are important for mediating intercellular adhesion. Neural cell membranes also contain integrins, dual subunit receptors that bind adhesion molecules in the surrounding extracellular matrix, particularly laminin and fibronectin. Other types of bound molecules, including several types found in the terrain of injured CNS tissue, mediate repulsion and inhibition of neural growth (Figure II.6.14.9).

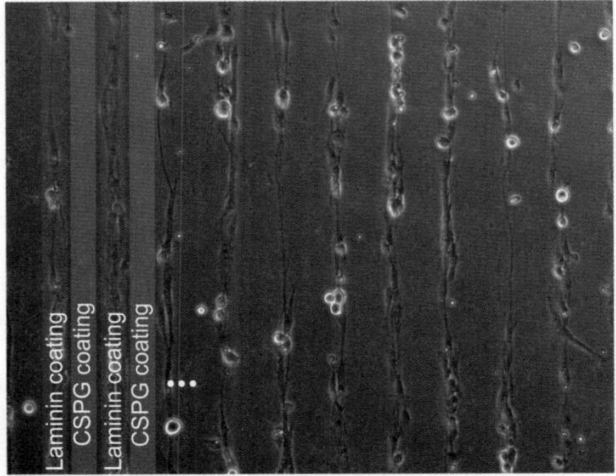

FIGURE II.6.14.9 Neurons from chick dorsal root ganglia were cultured on surfaces containing alternating stripes of adhesive laminin and chondroitin sulfate proteoglycans (CSPGs). (CSPGs are growth-inhibitory molecules present within the glial scar following spinal cord injury.) The neurons adhere within and extend neurite along the laminin lanes, but avoid contact with the CSPG-coated lanes.

ECM adhesion molecules are frequently used *in vitro* to promote neural cell attachment. Poly-L-lysine, collagen, laminin, and fibronectin are all commonly applied to cell culture surfaces for this purpose. Constructs designed for implantation into the nervous system can be similarly coated with surface bound molecules for the purpose of achieving a more seamless integration between the implant and host neural tissue. For example, anti-inflammatory agents such as dexamethasone can be coated on the surface of an implantable device to modulate scar tissue formation (Zhong and Bellamkonda, 2007).

Surface bound adhesion molecules influence neural cells in a fashion similar to topographic cues. In fact, it is difficult to separate the effects on cellular behavior of surface topography and chemistry, which actually work closely and synergistically together. For example, after peripheral nerve injury, aligned Schwann cells secrete laminin, creating a substrate that is not only topographically oriented, but also rich in growth-promoting adhesion molecules. Tissue-engineered implants are commonly designed with the same combinational strategy, containing topographically patterned surfaces coated with growth promoting biomolecules (Yu et al., 1999; Miller et al., 2002).

Diffusible Biomolecules Activate Intracellular Signaling Pathways. Other biomolecules are released in soluble form, and influence neural cells primarily by activating secondary messenger pathways. Neurotrophins are a family of neurotrophic factors critical in supporting neural cell development, growth, and maintenance. Nerve growth factor (NGF), for example, is the most studied of the neurotrophins and is found throughout the nervous system. NGF and other neurotrophins, including brain-derived neurotrophic factor (BDNF) and NT-3, bind to membrane-bound surface receptors, such as tyrosine kinase receptors, to activate intracellular signaling cascades.

Other biochemical cues mediate different types of effects. Tropic factors, for example, attract or repel extending axons, in a process termed chemotropism. Concentration gradients of chemotropic factors are particularly crucial during development in guiding directed axonal growth towards the appropriate target tissue. Other biochemical cues important in intercellular signaling include molecules known as cytokines. After CNS injury, reactive astrocytes and microglia release inflammatory cytokines, which elicit an increased local immune response that exacerbates the secondary injury process.

Anti-inflammatory drugs can be used to treat CNS injury by modulating the cascade of secondary injury events. Coatings of dexamethasone or α-melanocyte stimulating hormone (α-MSH), for example, have been shown to modulate the local response to neural implants. The anti-inflammatory steroid methylprednisolone (MP) has been shown to reduce inflammation after spinal cord injury (Bracken et al., 1990). In one study, sustained release of MP was achieved by encapsulating the drug within degradable PLGA-based nanoparticles, which were suspended with an agarose hydrogel able to conform to the injury site (Chvatal et al., 2008).

Biomaterials Enable the Controlled Delivery of Diffusible Biochemical Cues

Natural biochemical cues are presented in neural tissues according to a precise spatial distribution and time course. For example, tightly localized gradients of biochemicals guide neuronal growth and differentiation during neural development. When neural tissue is injured or diseased, however, the regulation of biochemical cues can be disrupted, resulting in cell death and the establishment of barriers to regeneration.

The controlled delivery of therapeutic drugs can be used to counteract these effects, restoring a balance of biochemical cues able to bring about repair and regeneration. Biomaterials are useful vehicles for mediating this control over drug delivery, because their versatile properties can be tailored to achieve precise spatiotemporal control over drug release. Whereas direct drug injections can be invasive, and injected drugs tend to diffuse away from the target site, biomaterials-based delivery systems can be used to non-invasively deliver drugs in a more controlled and fine-tuned manner. Chapter II.5.16 introduces many different drug delivery modalities.

Classes of Biomaterials-Based Delivery Systems

Drug delivery systems commonly used in neural tissue engineering applications are often polymer-based, and can consist of non-degradable or biodegradable substrates, depending on the requirements of the system and the desired release profiles. In implantable systems that do not degrade, drugs are released by outward diffusion through a porous polymeric substrate. The release profile in these non-degradable systems depends on the porosity and loading of the substrate. A disadvantage is that the polymer remains in the site indefinitely.

In biodegradable drug delivery systems, loaded drugs are released by the degradation of the polymeric matrix (often via hydrolysis), as well as by diffusion through pores. For hydrophobic molecules like steroids, the degradation of the matrix is the primary determinant of delivery time. Commonly used biodegradable polymers for drug delivery into neural tissue include PLLA, PGA, and copolymers such as PLGA. Another synthetic polymer that is widely used is polyethylene glycol (PEG). Naturally derived polymers, like collagen, agarose, and alginate are also available, although natural materials are difficult to modify with peptides or other chemical cues that may be desirable in a regenerating system.

Three major types of drug delivery vehicles used in neural specific applications include solid implantables, scaffolds, and hydrogels. However, as we will see in the following sections, there is significant overlap between these divisions.

Solid Implants for Controlled Presentation of Cues

In its most basic form, a solid implant for drug delivery consists of a biodegradable macrostructure, such as a disk or block. The structure is synthesized externally, and then implanted in a solid form to yield prolonged and localized release at the target site. Benefits of this type of system include ease of manufacturing, quick administration, and localization of drug release. Prefabricated implants, however, do not geometrically conform to the site of injury, and consequently do not deliver drugs evenly to the surrounding tissues. To illustrate, following surgical removal of brain tumors, a rigid PLGA disk is commonly placed within the site of resection, where it slowly releases chemotherapeutic drug designed to kill remaining tumor cells. However, this non-conformable wafer does not completely contact the surrounding cells in the cavity, and untreated cancer cells can cause tumor recurrence (Fleming, 2002).

Microscale diameter polymer spheres provide an alternative to solid macrostructures (Figure II.6.14.10). Together in high quantities, microspheres can encapsulate large amounts of drug, offering the benefits of a single macroscale implant, while allowing a higher degree of space filling of the target site. Microsphere-based delivery systems are popular in the treatment of degenerative brain diseases, where localized and time-controlled release profiles are critical. For example, microspheres have been used experimentally to deliver Levodopa (L-DOPA) to Parkinson's disease models, and acetylcholinesterase inhibitors for Alzheimer's disease (Arica, 2005; Gao et al., 2007). In these cases, implantable microspheres are delivered via injection to the target site, allowing bypass of the blood–brain barrier. The delivery of drugs using microspheres also protects active compounds, and minimizes diffusion of drug away from the target site.

Due to the small dimensions of individual microspheres, their delivery can be controlled using non-conventional techniques. For example, ferromagnetic alginate microspheres can be moved via external magnetic control to yield a microsphere distribution gradient, and hence a gradient of prolonged NGF release (Ciofani, 2009). Although still in the early stages of development, these types of advanced delivery techniques might one day enable finely-unable, patient-specific release profiles.

FIGURE II.6.14.10 These PCL (polycaprolactone) microspheres were fabricated using a double emulsion solvent extraction/evaporation method. The microspheres are biodegradable and porous, and can release loaded drugs over the timespan of days or weeks. (Figure adapted from Yang et al., 2001, with permission.)

A drawback to the use of freely injected microspheres is that they can move around in the implantation site, and thus do not evenly contact all surfaces of the target tissue. To better adapt to the geometry of the target site, microspheres can be incorporated within three-dimensional systems, such as the hydrogel-based delivery systems described in the next section.

Hydrogel-Based Delivery Systems

Hydrogels can be used as conformable three-dimensional drug delivery systems, able to release embedded drugs while maintaining close contact with the surfaces of the implantation site. Hydrogels can also be tailored to have mechanical properties similar to neural tissue, making them desirable for use in systems of variable geometries that might be sensitive to damage by the implantation of a solid matrix. To conform a hydrogel to a target site, it is first injected in liquid form to fill the site, and subsequently induced *in situ* to gel via cross-linking by a stimulus, such as chemical, temperature or light. Using these techniques, the hydrogel can be chemically controlled to swell to a desired size, and then to degrade at a given rate.

> Photopolymerizable systems cross-link in the presence of light and can include poly(ethylene glycol) (PEG)-based polymers, poly(vinyl alcohol)-based polymers, and polysaccharide derivatives (Nguyen, 2002). Finely-tuned biochemical gradients can be established using photopolymerizable systems (Figure II.6.14.11).

Many natural materials including collagen, agarose, and methylcellulose, serve as excellent hydrogels. These natural substrates can be triggered to gel based on temperature (agarose) and ionic concentrations (collagen). As an example, collagen has been used to successfully deliver the growth factors EGF and FGF to the damaged spinal cord, resulting in increased cell proliferation and reduced cavitation (Hamann, 2005).

Often hydrogels are combined with a drug release system such as microspheres to yield a more even spatial distribution of drug, in addition to a tunable release profile. As an example, lipid microtubules, made of chemically-defined phospholipid derivatives, have been loaded with BDNF and delivered within an agarose hydrogel to target injured spinal cord. The lipid microtubule/hydrogel mixture was delivered to the damaged spinal cord in liquid form, and then cooled *in situ* to conform to the shape of the injury site. This combined release system resulted in enhanced neurite outgrowth and a decreased inflammatory response (Jain, 2006).

Other Scaffold-Based Drug Delivery Systems

Aside from hydrogels, other scaffold-based drug delivery systems are also able to offer structural support and guidance while delivering biochemical cues. Drug delivery scaffolds have been commonly applied to the peripheral nervous system and the spinal cord, where a defined architecture is beneficial in enhancing the longitudinal regeneration of cells across extended injury gaps.

Scaffold-based systems are particularly well-suited for use in multifaceted repair strategies, incorporating combinations of structural, topographical, and biochemical cues. For example, drug-loaded microspheres, embedded within hydrogels, can be incorporated with a physically patterned scaffold to enable finely-tunable drug delivery and to allow for synergistic combinations of physical and biochemical cues. These types of holistic approaches, which seek to provide combinations of cues to influence multiple aspects of healing and regeneration, offer great therapeutic potential in the treatment of neural injury and disease. Table II.6.14.2 provides further examples of biomaterial-based drug delivery systems for neural tissue engineering applications.

Neural Cells are also Sensitive to Electrical Cues. In addition to topographic and biochemical cues, neural tissue is also responsive to other cue types as well. For example, natural electrical cues play a significant role within neural tissue, especially during development and regeneration. Extracellular matrix materials and other

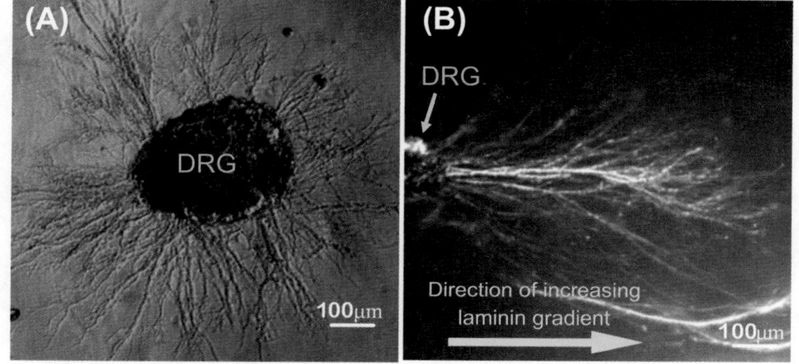

FIGURE II.6.14.11 Chick dorsal root ganglia (DRG) were cultured in agarose hydrogels. (A) This DRG was cultured in a plain agarose hydrogel that was photopolymerized with UV light. Neurites extended from the DRG evenly in all directions. (B) This DRG was cultured in a hydrogel photopolymerized to establish a directional gradient of the adhesive ECM molecule laminin 0.017 (μg/ml/mm). The extending neurites were sensitive to the gradient of biochemical cues, and grew directionally towards the higher concentrations of laminin. (Figure modified from Dodla et al., 2006.)

TABLE II.6.14.2 Examples of Biomaterials-Based Drug Delivery System for Neural Tissue Engineering Application

Biomaterial	Biomolecule	Types of Implant	Applications	Benefits
Nonbiodegradables (e.g., HEMA, EVA, polysulfones)	NGF, NT-3, BDNF, GDNF	Microspheres, scaffolds	PNS injury (Aebischer et al., 1989), Spinal cord injury (Kapur and Shoichet, 2004)	Provide long-term support and extended drug delivery, flexibility of shape/size
Polyesters (e.g., PLGA, PLLA)	GDNF, MP, BDNF, DOPA, AChE, NGF	Microspheres, scaffolds, implants	Neurodegenerative disease (McRae and Dahlstrom, 1994), CNS injury (Saltzman et al., 1999), PNS injury (Piotrowicz and Shoichet, 2006)	Biodegradable, degradation rate is tunable
Poly(ethylene glycol) (PEG)	NT-3, NGF	Hydrogel, scaffolds	Spinal cord injury (Piantino et al., 2006)	Limited immune response, easily modified for cell adhesion, tunable biodegradation
Agarose/alginate	NGF, laminin, CNTF	Scaffolds, hydrogels	PNS injury (Yu et al., 1999)	Can form to injection site (temperature triggered/Ca^{2+} triggered), can vary stiffness
Chitosan	Fibronectin, BDNF, CNTF	Hydrogels, implants, microspheres	CNS degeneration (Mittal et al., 1994)	Can form to injection site (pH triggered), functionalizable
Collagen/fibrin	PDGF, NGF, NT-3	Hydrogels, implants, scaffolds	Spinal cord injury (Houweling et al., 1998)	Promotes cell adhesion, covalently modifiable, reduced immune response

HEMA: 2-hydroxyethyl methacrylate
EVA: ethylene vinyl acetate

components within neural tissue contain charged molecules, and can even exhibit piezoelectric properties (i.e., generate transient electrical currents in response to mechanical deformation) (Valentini et al., 1992).

Influence of Electrical Cues on Cellular Behavior. Naturally occurring electrical cues can impact neuronal behavior in a variety of ways. For example, growth inhibitory CSPG molecules in the CNS are negatively charged, and serve to constrain neuronal growth and adhesion. Patterns of action potential signaling between neurons are critical for establishing and modulating the strength of synaptic connections within developing and neuronal networks (Fields et al., 1990).

A variety of cell culture experiments have been designed to probe the effects of electrical cues on neurite outgrowth. Neurons cultured within electric fields have been shown to alter the direction and rate of neurite extension, based on the parameters of the electrical stimulation. The precise means by which electrical cues influence neural growth are not fully known, but several mechanisms have been proposed. *In vitro* experiments have demonstrated the influence of electric charges and fields on the growth cones of developing neurites. This influence is possibly mediated in part by effects of electrical fields on the distribution of receptors on the growth cone membrane. Electrical cues can also influence voltage-sensitive ion channels on the growth cone membrane, modulating the influx of calcium. Intracellular calcium concentrations have been shown to exert potent effects on growth cone motility (Cohan et al., 1987).

As with other types of cues, electrical cues can also be used to indirectly influence neurons by targeting supporting cells. For example, electric fields have been used to orient astrocytes, which in turn enhanced and directed neurite extension by offering an aligned framework for growth (Alexander et al., 2006).

Biomaterials-Based Electrical Cues. Biomaterials can be used to provide different types of electrical cues for the purpose of influencing cellular growth and differentiation. The most straightforward technique for electrically stimulating neural tissue is to apply current with metallic electrodes. For this purpose, electrodes can be integrated within implantable devices and scaffolds to promote healing and regeneration. Even brief bouts of electrical stimulation applied to peripheral nerves at the time of injury and surgical repair can have long-term effects on the speed and accuracy of regeneration (Figure II.6.14.12) (Brushart et al., 2002; English et al., 2007).

An alternative to metal-based electrodes for transferring current are conducting polymers. Polypyrrole (PPy) and poly(3,4-ethylenedioxythiophene) (PEDOT) are two of the most commonly used conducting polymers for neural tissue engineering applications. Each of these polymers has a reversible electrochemistry, allowing the transfer of ions. Conducting polymers have many advantages over metals, including mechanical properties that are more similar to those of soft tissue.

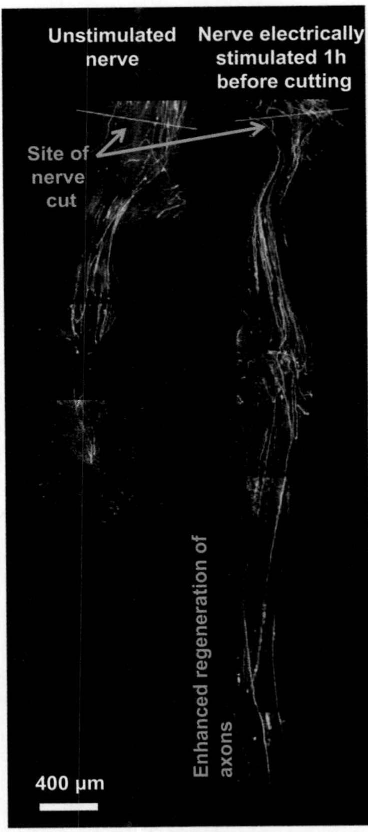

FIGURE II.6.14.12 These images demonstrate the effects of electrical stimulation on axonal regeneration from transected nerves. The nerve on the right was electrically stimulated for 1 hour at the time of injury. The nerve on the left received no stimulation. Enhanced patterns of regeneration and larger numbers of regenerated axon profiles were present in the stimulated nerve after two weeks. (Experiment detailed in English et al., 2007.)

Piezoelectric biomaterials convert mechanical strain into charge displacement, resulting in the generation of transient local currents. PVDF (polyvinylidene fluoride) is an example of a piezoelectric polymer able to transduce minute vibrational forces into electrical charges. PVDF culturing substrates have been shown to promote neurite outgrowth *in vitro*, and similar piezoelectric materials have been incorporated into nerve guidance channels to promote peripheral nerve regeneration (Valentini et al., 1992; Fine et al., 1991).

These types of studies demonstrate the therapeutic potential of biomaterials-based electrical cues in a clinical setting, although much remains to be learned about the influence of electrical cues on healing and regeneration.

Scaffold-Facilitated Cell Transplantation

A different strategy for treating neural injury and disease is to transplant neural cells to the afflicted site. Transplanted cells can interact spontaneously with host neural tissue in complex ways to restore biochemical deficits and replace lost tissue. For these reasons, cell-based therapies are particularly warranted in cases where the complexities of injury or disease are such that effective treatment with artificial cues is difficult to engineer.

Candidate cell types for transplantation include glial cells and stem or stem-like neural progenitor cells. Transplanted glial cells have the potential to stimulate neural repair through the natural production of factors such as ECM and biochemical signaling molecules. Stem or stem-like cells have the additional potential to differentiate into neuronal cells to replace lost or damaged tissue.

The conventional technique for transplanting cells is by injection into the injury site. Although this is the simplest technique to implement clinically, it is associated with several disadvantages, including poor survival of transplanted cells, and movement of cells away from the target site.

Biomaterials Allow for More Effective Cell Transplantation Strategies. Biomaterials-based cell transplantation scaffolds may be used to contain transplanted cells within the target site, as well as to control the physical and biochemical environment around the cells. Thus, in the design of a cell transplantation scaffold, consideration must be given not only to the interactions between the scaffold and host tissue, but also to the interactions between the scaffold and the transplanted cells it contains. Topographical and biochemical cues presented by the scaffold should be designed to promote cell viability and desired behavior.

Natural polymers like collagens, agarose, fibrin, and alginate have all been successfully used as substrates for transplanting cells, as have synthetic materials such as the biodegradable polymer PLGA. Component materials of a cell transplantation scaffold should by structured to provide a suitable substrate for cell seeding, and often must be shaped to physically fill gaps left by damaged tissue, and to encourage host cell adhesion to the construct.

Schwann cells, the glial cells of the PNS, can be readily purified and delivered into both PNS and CNS tissues for the purpose of contributing to a pro-regenerative environment (Morrissey et al., 1991). For example, nerve guidance channels can be loaded with Schwann cells suspended in ECM proteins. These cell-loaded guidance scaffolds have been shown to promote axonal outgrowth and myelination more effectively than scaffolds containing ECM proteins alone (Paino et al., 1994; Xu et al., 1997). In the spinal cord, transplanted Schwann cells are effective at promoting the in-growth of regenerating axons, although it is still a major challenge to encourage regenerating axons to then exit the scaffold and continue growth. Notably, combining Schwann cell transplantation with the delivery of anti-inflammatory agents has been shown to help mitigate this obstacle, through the beneficial effects of the drugs on host tissue surrounding the graft (Chen et al., 1996).

Other candidate cell types for encouraging axonal regeneration have been studied to a lesser extent, including CNS glial cells and macrophages. As you remember, astrocytes and microglia become activated after CNS injury, and contribute to the formation of a growth-inhibitory glial scar. Further studies are needed to fully-understand neuronal–glial interactions in the CNS, and how glial cells may aid in regeneration. Macrophages are able to act at the damage site to clear remaining debris. In this way, these cells may indirectly aid in the regeneration of axons by promoting angiogenesis, Schwann cell migration, and clearing of growth-inhibitory molecules (Franzen et al., 1998). Significantly, the future creation of multifunctional cell implant systems, containing multiple cell types within precisely defined scaffolds, may dramatically improve repair and regeneration.

Delivery of Stem Cells and Stem-Like Cells

The implantation of neural progenitor cells (NPCs) and stem cells (SCs) is a more recently developed technique. NPCs are multipotent cells that can develop into neural cell types, while SCs are pluripotent cells that have further differentiation capabilities. This approach is appealing, because these cell types have the potential to differentiate to replace lost or damaged cells. However, there are challenges associated with ensuring the survival, proliferation, and proper differentiation of these multi- or pluripotent cell types. NPCs (more commonly used than SCs because of availability and compatibility) are highly migratory cells that require proper adhesion and signaling interactions with the surrounding ECM, in addition to proper biochemical cues. Thus, the cell transplantation scaffold must be precisely designed in order to signal NPCs to differentiate into the appropriate cell types, while also contributing to cell growth and viability. As an example, exposure of neural progenitor cells to glial growth factor yields differentiation to Schwann cells in the PNS, while exposure to ciliary neurotrophic factor yields differentiation to astrocytes in the CNS. The presence of other growth factors, such as EGF and FGF, furthermore aid in the proliferation and survival of NPCs and SCs (McKay, 1997). These biochemical factors can all be incorporated into biodegradable materials to promote proper differentiation and longevity.

Newer biomimetic materials are tailored with precise control over the presentation of peptides, in order to influence the viability and differentiation of neural progenitor and stem cells. These materials incorporate specific chemical structures that closely mimic endogenous materials, such as collagen and laminin, while allowing finely-tunable degradation time and chemical structure for more reproducible and dependable delivery systems. For example, short peptides can be attached to polymer matrices in order to mimic specific integrin-binding domains. This technique allows the beneficial effects of laminin or fibronectin coatings to be exploited in a highly spatially controllable fashion – without requiring the use of a biological source (Potter et al., 2008; Tate et al., 2009). Similarly, cytokines can also be incorporated and conjugated to biomaterials for cell transplantation. By integrating these types of biochemical and topographical

TABLE II.6.14.3 Use of Transplanted Cells in Neural Tissue Engineering Applications

Cell Types	Benefits	Examples
Glial cells: Schwann cells, oligodendrocytes	Deliver growth factors, provide structural support, re-myelinate axons	Schwann cells were transplanted in nondegradable (PVA) scaffolds to enhance spinal cord repair (Xu et al., 1997). A PLGA foam scaffold was seeded with Schwann cells and implanted, yielding enhanced PNS regeneration (Hadlock et al., 2000)
Immune response cells: macrophages, microglia, astrocytes	Contribute to the clearing of dead cells and debris. Other roles not fully understood	Activated macrophages and microglia were transplanted on nitrocellulose membranes in spinal cord to promote regeneration of sensory axons (Prewitt et al., 1997)
Stem cells, neural progenitor cells (NPCs)	Replace damaged neurons and support cells, deliver growth factors, migrate to damaged sites	Neural stem cells were seeded within a PLGA polymer scaffold, which was implanted at the site of spinal cord injury. The treatment enhanced axonal regeneration and functional recovery (Teng et al., 2002)

cues with the control and precision enabled by biomaterials, biomimetic approaches seek to greatly enhance cell transplantation therapies.

In summary, cell-based therapies offer a new dimension to the treatment of neural injury. Transplanted cells offer the potential to deliver ECM-based and diffusible growth factors to indirectly support regenerating neurons, and to replace damaged cells. The use of appropriately selected and structured biomaterials is integral to the success of such scaffold-based cell transplantation therapies. Table II.6.14.3 summarizes the use of transplanted cells in neural tissue engineering applications, and provides some further examples.

CONCLUSIONS

Neural tissue engineering strategies exploit the versatile properties of biomaterials in order to influence cells and tissues of the nervous system, for the purpose of restoring lost function. By delivering topographic, biochemical, and other types of cues, implanted biomaterials can direct cellular growth and behavior in order to overcome biological constraints on neural repair and regeneration. In order to simulate the native architecture of the brain, biomaterials-based cues must be delivered with spatial precision and at controllable timescales. Furthermore, because neural tissues within the CNS and PNS vary widely in both structure and response to disease or injury, unique treatment strategies must be tailored for individual applications. Continuing advances in biomaterials-based neural tissue engineering techniques will enable effective therapies that are increasingly effective in the treatment of neural injury and disease.

BIBLIOGRAPHY

Abrams, G. A., Goodman, S. L., Nealey, P. F., Franco, M., & Murphy, C. J. (2000). Nanoscale topography of the basement membrane underlying the corneal epithelium of the rhesus macaque. *Cell and Tissue Research*, **299**, 39–46.

Aebischer, P., Salessiotis, A. N., & Winn, S. R. (1989). Basic fibroblast growth-factor released from synthetic guidance channels facilitates peripheral-nerve regeneration across long nerve gaps. *Journal of Neuroscience Research*, **23**, 282–289.

Alexander, J. K., Fuss, B., & Colello, R. J. (2006). Electric field-induced astrocyte alignment directs neurite outgrowth. *Neuron Glia Biology*, **2**, 93–103.

Arica, B., et al. (2005). Carbidopa/levodopa-loaded biodegradable microspheres: in vivo evaluation on experimental Parkinsonism in rats. *Journal of Controlled Release*, **102**, 689–697.

Bellamkonda, R., Ranieri, J. P., Bouche, N., & Aebischer, P. (1995). Hydrogel-based 3-dimensional matrix for neural cells. *Journal of Biomedical Materials Research*, **29**, 663–671.

Biran, R., Noble, M. D., & Tresco, P. A. (2003). Directed nerve outgrowth is enhanced by engineered glial substrates. *Experimental Neurology*, **184**, 141–152.

Bracken, M. B., Shepard, M. J., Collins, W. F., Holford, T. R., Young, W., Baskin, D. S., Eisenberg, H. M., Flamm, E., Leo-summers, L., Maroon, J., Marshall, L. F., Perot, P. L., Piepmeier, J., Sonntag, V. K.H., Wagner, F. C., Wilberger, J. E., & Winn, H. R. (1990). A randomized, controlled trial of methylprednisolone or naloxone in the treatment of acute spinal-cord injury – results of the 2nd national acute spinal-cord injury study. *New England Journal of Medicine*, **322**, 1405–1411.

Brushart, T. M., Hoffman, P. N., Royall, R. M., Murinson, B. B., Witzel, C., & Gordon, T. (2002). Electrical stimulation promotes motoneuron regeneration without increasing its speed or conditioning the neuron. *Journal of Neuroscience*, **22**, 6631–6638.

Ceballos, D., Navarro, X., Dubey, N., Wendelschafer-Crabb, G., Kennedy, W. R., & Tranquillo, R. T. (1999). Magnetically aligned collagen gel filling a collagen nerve guide improves peripheral nerve regeneration. *Experimental Neurology*, **158**, 290–300.

Chamberlain, L. J., Yannas, I. V., Hsu, H. P., Strichartz, G., & Spector, M. (1998). Collagen-GAG substrate enhances the quality of nerve regeneration through collagen tubes up to level of autograft. *Experimental Neurology*, **154**, 315–329.

Chen, A. Q., Xu, X. M., Kleitman, N., & Bunge, M. B. (1996). Methylprednisolone administration improves axonal regeneration into Schwann cell grafts in transected adult rat thoracic spinal cord. *Experimental Neurology*, **138**, 261–276.

Chvatal, S. A., Kim, Y. T., Bratt-Leal, A. M., Lee, H. J., & Bellamkonda, R. V. (2008). Spatial distribution and acute anti-inflammatory effects of Methylprednisolone after sustained local delivery to the contused spinal cord. *Biomaterials*, **29**, 1967–1975.

Ciofani, G., et al. (2009). Magnetic alginate microspheres: System for the position controlled delivery of nerve growth factor. *Biomedical Microdevices*, **11**, 517–527.

Clements, I. P., Kim, Y., English, A. E., Lu, X., Chung, H., & Bellamkonda, R. V. (2009). Thin-film enhanced nerve guidance channels for peripheral nerve repair. *Biomaterials*, **30**(23-24), 3834–3846.

Cohan, C. S., Connor, J. A., & Kater, S. B. (1987). Electrically and chemically mediated increases in intracellular calcium in neuronal growth cones. *Journal of Neuroscience*, **7**, 3588–3599.

Curtis, A., & Wilkinson, C. (1997). Topographical control of cells. *Biomaterials*, **18**, 1573–1583.

Dalby, M. J., Gadegaard, N., Riehle, M. O., Wilkinson, C. D.W., & Curtis, A. S. G. (2004). Investigating filopodia sensing using arrays of defined nano-pits down to 35 nm diameter in size. *International Journal of Biochemistry and Cell Biology*, **36**, 2005–2015.

Dodla, M. C., & Bellamkonda, R. V. (2006). Anisotropic scaffolds facilitate enhanced neurite extension in vitro. *Journal of Biomedical Materials Research Part A*, **78A**, 213–221.

English, A. W., et al. (2007). Electrical stimulation promotes peripheral axon regeneration by enhanced neuronal neurotrophin signaling. *Developmental Neurobiology*, **67**, 158–172.

Fan, Y. W., Cui, F. Z., Chen, L. N., Zhai, Y., Xu, Q. Y., & Lee, I. S. (2002). Adhesion of neural cells on silicon wafer with nano-topographic surface. *Applied Surface Science*, **187**, 313–318.

Fields, R. D., Neale, E. A., & Nelson, P. G. (1990). Effects of patterned electrical-activity on neurite outgrowth from mouse sensory neurons. *Journal of Neuroscience*, **10**, 2950–2964.

Fine, E. G., Valentini, R. F., Bellamkonda, R., & Aebischer, P. (1991). Improved nerve regeneration through piezoelectric vinylidenefluoride-trifluoroethylene copolymer guidance channels. *Biomaterials*, **12**, 775–780.

Fleming, A. B., & Saltzman, W. M. (2002). Pharmacokinetics of the carmustine implant. *Clin Pharmacokinet*, **41**, 403–419.

Flemming, R. G., Murphy, C. J., Abrams, G. A., Goodman, S. L., & Nealey, P. F. (1999). Effects of synthetic micro- and nano-structured surfaces on cell behavior. *Biomaterials*, **20**, 573–588.

Franzen, R., et al. (1998). Effects of macrophage transplantation in the injured adult rat spinal cord: A combined immunocytochemical and biochemical study,". *Journal of Neuroscience Research*, **51**, 316–327.

Gao, P., Xu, H., Ding, P. T., Gao, Q. Z., Sun, J. Y., & Chen, D. W. (2007). Controlled release of huperzine A from biodegradable microspheres: *In vitro* and *in vivo* studies. *International Journal of Pharmaceutics*, **330**, 1–5.

Goldner, J. S., et al. (2006). Neurite bridging across micropatterned grooves. *Biomaterials*, **27**, 460–472.

Gutmann, E., Guttmann, L., Medawar, P. B., & Young, J. Z. (1942). The rate of regeneration of nerve. *Journal of Experimental Biology*, **19**, 14–44.

Hadlock, T., Sundback, C., Hunter, D., Cheney, M., & Vacanti, J. P. (2000). A polymer foam conduit seeded with Schwann cells promotes guided peripheral nerve regeneration. *Tissue Engineering*, **6**, 119–127.

Hamann, M. C.J., et al. (2005). Injectable intrathecal delivery system for localized administration of EGF and FGF-2 to the injured rat spinal cord. *Experimental Neurology*, **194**, 106–119.

Houweling, D. A., Lankhorst, A. J., Gispen, W. H., Bar, P. R., & Joosten, E. A. J. (1998). Collagen containing neurotrophin-3 (NT-3) attracts regrowing injured corticospinal axons in the adult rat spinal cord and promotes partial functional recovery. *Experimental Neurology*, **153**, 49–59.

Jain, A., et al. (2006). In situ gelling hydrogels for conformal repair of spinal cord defects, and local delivery of BDNF after spinal cord injury. *Biomaterials*, **27**, 497–504.

Johansson, F., Carlberg, P., Danielsen, N., Montelius, L., & Kanje, M. (2006). Axonal outgrowth on nano-imprinted patterns. *Biomaterials*, **27**, 1251–1258.

Kapur, T. A., & Shoichet, M. S. (2004). Immobilized concentration gradients of nerve growth factor guide neurite outgrowth. *Journal of Biomedical Materials Research Part A*, **68A**, 235–243.

Kim, Y. T., Haftel, V. K., Kumar, S., & Bellamkonda, R. V. (2008). The role of aligned polymer fiber-based constructs in the bridging of long peripheral nerve gaps. *Biomaterials*, **29**, 3117–3127.

McKay, R. (1997). Stem cells in the central nervous system. *Science*, **276**, 66–71.

McRae, A., & Dahlstrom, A. (1994). Transmitter-loaded polymeric microspheres induce regrowth of dopaminergic nerve-terminals in striata of rats with 6-Oh-Da induced parkinsonism. *Neurochem. Int.*, **25**(1), 27–33.

Miller, C., Jeftinija, S., & Mallapragada, S. (2001). Micropatterned Schwann cell-seeded biodegradable polymer substrates significantly enhance neurite alignment and outgrowth. *Tissue Engineering*, **7**, 705–715.

Miller, C., Jeftinija, S., & Mallapragada, S. (2002). Synergistic effects of physical and chemical guidance cues on neurite alignment and outgrowth on biodegradable polymer substrates. *Tissue Engineering*, **8**, 367–378.

Mittal, S., Cohen, A., & Maysinger, D. (1994). *In-vitro* effects of brain-derived neurotrophic factor released from microspheres. *Neuroreport*, **5**, 2577–2582.

Morrissey, T. K., Kleitman, N., & Bunge, R. P. (1991). Isolation and functional-characterization of Schwann-cells derived from adult peripheral-nerve. *Journal of Neuroscience*, **11**, 2433–2442.

Ngo, T. T., Waggoner, P. J., Romero, A. A., Nelson, K. D., Eberhart, R. C., & Smith, G. M. (2003). Poly(L-lactide) microfilaments enhance peripheral nerve regeneration across extended nerve lesions. *Journal of Neuroscience Research*, **72**, 227–238.

Nguyen, K. T., & West, J. I. (2002). Photopolymerizable hydrogels for tissue engineering applications. *Biomaterials*, **23**, 4307–4314.

Norman, J., & Desai, T. (2006). Methods for fabrication of nanoscale topography for tissue engineering scaffolds. *Annals of Biomedical Engineering*, **34**, 89–101.

Paino, C. L., Fernandezvalle, C., Bates, M. L., & Bunge, M. B. (1994). Regrowth of axons in lesioned adult-rat spinal-cord – promotion by implants of cultured Schwann-cells. *Journal of Neurocytology*, **23**, 433–452.

Piantino, J., Burdick, J. A., Goldberg, D., Langer, R., & Benowitz, L. I. (2006). An injectable, biodegradable hydrogel for trophic factor delivery enhances axonal rewiring and improves performance after spinal cord injury. *Experimental Neurology*, **201**, 359–367.

Piotrowicz, A., & Shoichet, M. S. (2006). Nerve guidance channels as drug delivery vehicles. *Biomaterials*, **27**, 2018–2027.

Potter, W., Kalil, R. E., & Kao, W. J. (2008). Biomimetic material systems for neural progenitor cell-based therapy. *Frontiers in Bioscience*, **13**, 806–821.

Prewitt, C. M. F., Niesman, I. R., Kane, C. J. M., & Houle, J. D. (1997). Activated macrophage/microglial cells can promote the regeneration of sensory axons into the injured spinal cord. *Experimental Neurology*, **148**, 433–443.

Rajnicek, A. M., Britland, S., & McCaig, C. D. (1997). Contact guidance of CNS neurites on grooved quartz: Influence of groove dimensions, neuronal age and cell type. *Journal of Cell Science*, **110**, 2905–2913.

Rakic, P. (1971). Neuron-glia relationship during granule cell migration in developing cerebellar cortex – golgi and electron-microscopic study in macacus rhesus. *Journal of Comparative Neurology*, **141**, 283–312.

Saltzman, W. M., Mak, M. W., Mahoney, M. J., Duenas, E. T., & Cleland, J. L. (1999). Intracranial delivery of recombinant nerve growth factor: Release kinetics and protein distribution for three delivery systems. *Pharmaceutical Research*, **16**, 232–240.

Seidlits, S. K., Lee, J. Y., & Schmidt, C. E. (2008). Nanostructured scaffolds for neural applications. *Nanomedicine*, **3**, 183–199.

Tate, C. C., Shear, D. A., Tate, M. C., Archer, D. R., Stein, D. G., & LaPlaca, M. C. (2009). Laminin and fibronectin scaffolds enhance neural stem cell transplantation into the injured brain. *Journal of Tissue Engineering and Regenerative Medicine*, **3**, 208–217.

Teng, Y. D., Lavik, E. B., Qu, X., Park, K. I., Ourednik, J., Zurakowski, D., Langer, R., & Snyder, E. Y. (2002). Functional recovery following traumatic spinal cord injury mediated by a unique polymer scaffold seeded with neural stem cells. *Proceedings of the National Academy of Sciences*, **99**, 3024–3029.

Thompson, D. M., & Buettner, H. M. (2004). Oriented Schwann cell monolayers for directed neurite outgrowth. *Annals of Biomedical Engineering*, **32**, 1120–1130.

Turner, S., Kam, L., Isaacson, M., Craighead, H. G., Shain, W., & Turner, J. (1997). Cell attachment on silicon nanostructures. *Journal of Vacuum Science and Technology B*, **15**(6), 2248–2854.

Valentini, R. F., Vargo, T. G., Gardella, J. A., & Aebischer, P. (1992). Electrically charged polymeric substrates enhance nerve-fiber outgrowth *in-vitro*. *Biomaterials*, **13**, 183–190.

Vieu, C., Carcenac, F., Pepin, A., Chen, Y., Mejias, M., Lebib, A., Manin-Ferlazzo, L., Couraud, L., & Launois, H. (2000). Electron beam lithography: Resolution limits and applications. *Applied Surface Science*, **164**, 111–117.

Walsh, J. F., Manwaring, M. E., & Tresco, P. A. (2005). Directional neurite outgrowth is enhanced by engineered meningeal cell-coated substrates. *Tissue Engineering*, **11**(7-8), 1085–1094.

Wilkinson, C. D. W., Riehle, M., Wood, M., Gallagher, J., & Curtis, A. S. G. (2002). The use of materials patterned on a nano- and micro-metric scale in cellular engineering. *Materials Science and Engineering C: Biomimetic and Supramolecular Systems*, **19**, 263–269.

Woerly, S., Marchand, R., & Lavallee, C. (1991). Interactions of copolymeric poly(glyceryl methacrylate)-collagen hydrogels with neural tissue – effects of structure and polar groups. *Biomaterials*, **12**(2), 197–203.

Xu, X. M., Chen, A., Guenard, V., Kleitman, N., & Bunge, M. B. (1997). Bridging Schwann cell transplants promote axonal regeneration from both the rostral and caudal stumps of transected adult rat spinal cord. *Journal of Neurocytology*, **26**, 1–16.

Yang, F., et al. (2004). Fabrication of nano-structured porous PLLA scaffold intended for nerve tissue engineering. *Biomaterials*, **25**, 1891–1900.

Yang, F., Murugan, R., Wang, S., & Ramakrishna, S. (2005). Electrospinning of nano/micro scale poly(L-lactic acid) aligned fibers and their potential in neural tissue engineering. *Biomaterials*, **26**, 2603–2610.

Yang, Y. Y., et al. (2001). Morphology, drug distribution, and in vitro release profiles of biodegradable polymeric microspheres containing protein fabricated by double-emulsion solvent extraction/evaporation method. *Biomaterials*, **22**, 231–241.

Yu, T. T., & Shoichet, M. S. (2005). Guided cell adhesion and outgrowth in peptide-modified channels for neural tissue engineering. *Biomaterials*, **26**, 1507–1514.

Yu, X. J., Dillon, G. P., & Bellamkonda, R. V. (1999). A laminin and nerve growth factor-laden three-dimensional scaffold for enhanced neurite extension. *Tissue Engineering*, **5**, 291–304.

Zhong, Y. H., & Bellamkonda, R. V. (2007). Dexamethasone-coated neural probes elicit attenuated inflammatory response and neuronal loss compared to uncoated neural probes. *Brain Research*, **1148**, 15–27.

CHAPTER II.6.15 IMMUNOISOLATION

Roshni S. Rainbow[1] and Michael J. Lysaght†[2]

[1]Department of Anatomy and Cellular Biology, Tufts University School of Medicine, Boston, MA, USA

[2]Center for Biomedical Engineering, Brown University, Providence, RI, USA

"Good fences make good neighbors."

Robert Frost

INTRODUCTION

Synthetic membranes represent a unique and important class of biomaterials widely used in contemporary medical devices. They provide a means of selective transport, usually by allowing the passage of small molecules while blocking the transfer of larger counterparts. For example, such membranes are the principle enabling biomaterials for hemodialysis, the mainstay treatment for end-stage kidney failure. Synthetic membranes also play a critical enabling role in blood oxygenation during cardiac bypass surgery, as well as in a host of smaller therapies and special purpose devices. Membrane transport and membrane biocompatibility have become highly advanced subspecialties.

The potential role for membrane barriers in tissue engineering derives from their capacity for selective transport. Many of the molecules required by cells for survival, growth, and metabolism, as well as bioactive secretory products of such cells, have much lower molecular weights than the effector molecules of the immune system. Starting in the late 1970s, researchers recognized that foreign cells could survive and function in an immunocompetent host, provided that these non-self cells were contained within, or encapsulated by, a semi-permeable barrier with an appropriate molecular weight cut-off. Since that time, a large body of experimental data has been amassed to confirm that this approach is both feasible, and of potential therapeutic use in a variety of disease conditions. The literature of encapsulation is extensive and contains an estimated 25,000 peer-reviewed publications; many of these describe successful application in animal models for a variety of disease conditions. The technique has been applied experimentally in human clinical trials with a cumulative total of over 250 patients. Encapsulated cell therapy has yet to be approved by the FDA, but regulatory filings are currently under review by the agency.

There are various ways to conceptualize immunoisolation, and to appreciate its potential contributions to cell-based medicine. The first relates to cell sourcing, a major challenge for tissue engineering. Generally, cells are derived from one of three sources: *autologous* (from the patient themself), *allogeneic* (from another human donor) or *xenogeneic* (from another species). All three sources have major practical limitations. Autologous cells are rarely available in the quantity needed, and may be compromised by the patient's age or disease. In some cases, a small quantity of autologous cells can be harvested from the patient, expanded in tissue culture, and then returned to the body. This approach is technically satisfactory, but very expensive. Allogeneic cells can usually be found in sufficient quantity, but require the recipient to receive life-long pharmacologic immunosuppression. Unfortunately, the side-effects of long-term immunosuppression are severe, and this approach only makes sense in the case of life-threatening disease. Xenogeneic cells also require immunosuppression, and additionally raise a host of vexing safety issues. Encapsulation skirts some of these constraints by allowing allogeneic and xenogeneic cells to be deployed in a contained format, without immunosuppression and often in a retrievable fashion.

Alternatively, encapsulation can be thought of as an advantageous form of gene therapy. Functional genes, usually ones that produce a biologically important protein, can be added to human cells by the established "cut-and-paste" techniques of recombinant molecular biology. The newly designed cell can be expanded to large numbers under controlled conditions, tested carefully, encapsulated, and implanted into a patient. With suitable device design, the encapsulated cell can be recovered from the patient, should it be necessary or desirable. Many consider this approach safer than simply injecting cells or gene constructs into a patient. Finally, encapsulation is sometimes thought of as a method of overcoming the stifling restrictions on the number of human organs suitable and available for transplantation.

All three visions of encapsulation (a method to overcome the limitations on cell sourcing, an improved approach to gene therapy, and a way around the numerical confines of organ transplants) capture the versatility and promise of encapsulation. This chapter explains the mechanisms of immunoisolation in more detail, and provides an overview of the different developed encapsulation formats, along with their advantages and disadvantages. The current status of this technology, and the important challenges for future development, will also be discussed. Of special interest to the student will be the importance of small non-idealities in membrane selectivity, and the difficulties of bringing a therapy clearly effective in large animal models into human clinical practice.

PRINCIPLES OF IMMUNOISOLATION

Traditional cell transplantation strategies utilize pharmacologic immunosuppression to circumvent recipient immune system response to the transplanted allogeneic donor tissue or organ. In the absence of immunosuppression, a recipient's immune system would respond to the presence of foreign cells or proteins from the transplanted tissue or organs via the antibody-mediated (termed humoral) and cell-mediated response pathways. Humoral response involves the binding of host antibodies (also called immunoglobulins) to foreign substances, called antigens, which elicit antibody responses. Antibody binding triggers a host of complex biochemical pathways that ultimately result in destruction of the foreign cell or protein. Cell-mediated response begins with physical contact between host and graft, and ultimately relies on the activation of killer T-cells and macrophages to secrete antibodies, cytokines, and low molecular weight metabolites, such as free radicals and reactive oxygen and nitrogen intermediates, to attack the transplanted cells. Because the success of a transplant depends on the inhibition of these immune responses, immunosuppressants are a vital part of traditional whole organ transplant strategies. Immunosuppressants are powerful drugs with serious side-effects, and their use is generally limited to life-threatening organ failure. Immunosuppression should not be thought of as a therapy that "shuts down" the immune system. Dosing is complex and requires suppression levels that are neither too low nor too high. Dosages too low will lead to a compromised graft, while dosages too high can have serious or even fatal side-effects.

Immunoisolation does not employ immunosuppression, and instead utilizes selectively permeable membranes to protect allogeneic or xenogeneic transplanted cells and tissue from immunological rejection. As illustrated in Figure II.6.15.1, these membranes serve as a mechanical analog of pharmacological immunosuppression by surrounding donor cells with a membrane whose pores are suitably sized to inhibit the transport of immunologically active molecules and prevent cell–cell contact.

Membranes are often described by a parameter termed nominal molecular weight cut-off (nMWCO) that indicates the approximate maximum molecular weight

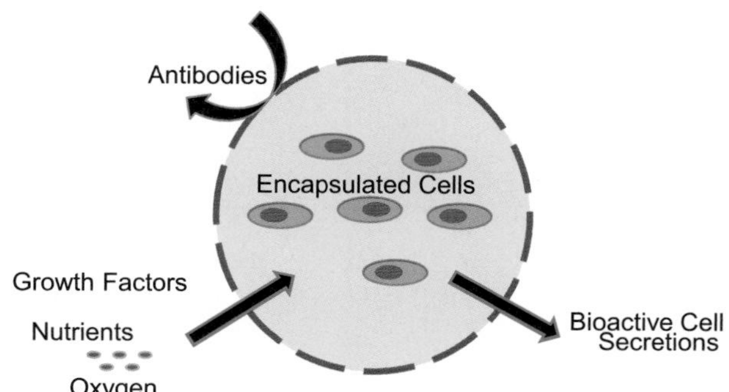

FIGURE II.6.15.1 Principles of Immunoisolation. Cells are supported on a synthetic matrix and surrounded by a selectively permeable membrane. The membrane contains pores suitably sized to limit physical contact between the cells within the intracapsular space and immunologically active pericapsular host proteins and cells. Encapsulated cells are able to survive and function since the membrane is permeable to smaller molecules, including dissolved oxygen, nutrients, waste products, and growth factors. In addition, many cell-derived bioactive therapeutic molecules are small enough to diffuse through the capsule wall and into the host.

substance which can pass through the membrane in a given transport mode (Dionne et al., 1996). The molecules needed for cell survival are less than 50,000 Daltons, while immune molecules are generally larger, with molecular weights beginning around 150,000–200,000 Daltons (such as Immunoglobulin E, IgE), and extending to just under 900,000 Daltons (such as Immunoglobulin M, IgM) (Colton 1995; Gentile et al., 1995). For immunoisolation, the ideal membrane should have an nMWCO around 50,000 Daltons, which allows for the passage of both the molecules necessary to sustain the transplanted cells and for any relevant transplant-derived therapeutics, but which prevents entry of lethal numbers of the recipient's immunomolecules.

Transport through immunoisolatory membranes is also described in terms of sieving coefficient, although strictly speaking this term applies to convective rather than diffusive transport. For a given membrane and solute, sieving coefficient represents the ratio of the concentrations of the solute on the filtrate and retentate side, with a value of 1 indicating total permeability of the solute through the membrane. Rejection, on the other hand, is equal to one minus the sieving coefficient, and is a more intuitive parameter. As illustrated in Figure II.6.15.2, an ideal membrane (blue curve) would only allow for the passage of solutes equal to or smaller than its nMWCO. Real membranes (red curve), however, do in fact allow for some transport of molecules above the nMWCO. Determining the tolerable levels of high MW species for a given device and cell–membrane combination is a major part of the development process for an application.

As in the case of immunosuppression, immunoisolation should not be thought of in absolute terms. Low molecular weight metabolites and cytokines generated by the host can still pass through the membranes and attack the grafted tissue. Additionally, because synthetic membranes typically have a Gaussian distribution of pore sizes, a few large proteins well above the nMWCO will be able to pass through the membrane. Together, there is a possibility that these fractional amounts will give rise to a host immune response and inflammation in the graft microenvironment (Hernández et al., 2010). A number of ways to circumvent these effects have been proposed, including the co-administration of anti-inflammatory drugs along with the encapsulated cells (Bunger et al., 2005), using more hypoimmunogeneic donor cells, such as human mesenchymal stem cells (Goren et al., 2010), or co-encapsulating immunoprivileged cells, such as Sertoli cells (Mital et al., 2010), to create a more immunoprotective microenvironment.

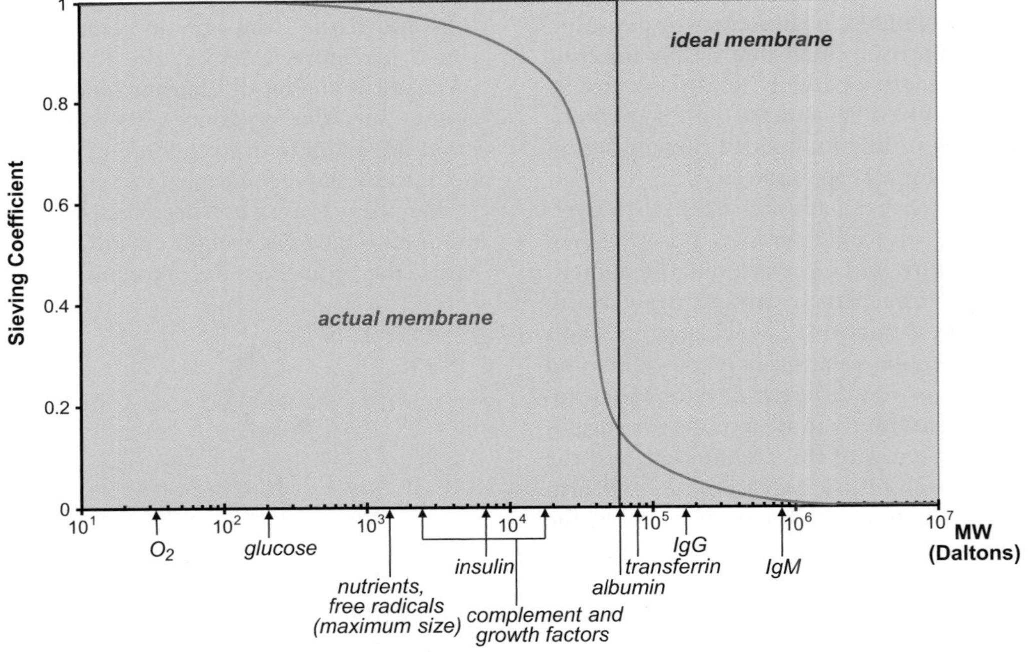

FIGURE II.6.15.2 Selective Permeability of Immunoisolatory Membranes. Immunoisolatory membranes are often characterized by two parameters: sieving coefficient (SC) and nominal molecular weight cut-off (nMWCO). Both are determined during filtration, even though most immunisolatory devices operate in the diffusive mode. SC is the ratio of solute concentration in the filtrate to that in the retentate, where a value of 1.0 indicates a membrane freely permeable to the solute in question, and a value of 0.0 indicates an impermeable membrane. The lines on the graph are plots of SC as a function of permeant molecular weight, ranging from 30 to 10^6 Daltons. The blue line signifies ideal membranes for which all solutes less than the nMWCO are passed and all larger are rejected. In practice, however, membranes have a distribution of pore size around a mean leading to the red sigmoid curve. The nMWCO can be thought of as the molecular weight of the largest solute capable of passing through the membrane's mean pore size. For immunisolation, the acceptable level of "leakage" of molecules above nMWCO must be determined empirically for a particular application. Once this is established, graphs like those in this figure are used to perform quality control on membrane production lots. Figure II.6.15.2 is an idealized plot; for actual data, refer to Table 1 and Figure 7 in Dionne et al. (1996). *Biomaterials,* **17**(3), 257–266.

Choice of biomaterial and site of implantation are important considerations during encapsulated cell therapy development.

Additionally, encapsulated cells may be subject to acute inflammation upon implantation. Depending on the biomaterial used, proteins from the host will begin to adsorb upon the capsule surface, which will lead to the activation of macrophages and fibroblasts *in situ* (Zielinski and Lysaght, 2007). After one week, this initial immune response will either result in fibrotic tissue growth around the capsule or will progress to a chronic inflammation and ultimate graft rejection. Either of these situations can result in inhibiting the diffusion through the encapsulating membrane, as a result of which the choice of encapsulating material and the site of implantation remain an important consideration in encapsulated cell therapy development.

DEVICES FOR IMMUNOISOLATION

Devices for immunoisolation vary in design configurations depending upon the intended application. Based on their size and shape, as illustrated in Figure II.6.15.3, devices may be implanted *in vivo* as microcapsules or macrocapsules, or configured as flow-through devices connected to the recipient's circulatory system (Lysaght and Aebischer, 1999). All encapsulation formats have one design constraint in common: cells must be no further than 250 microns from the external supply of nutrients and oxygen (Colton, 1995). Beyond this critical distance cells cannot survive, and this constraint has been demonstrated in numerous unsatisfactory device designs that exhibited central necrosis, a core of dead cells.

Microcapsules (Figure II.6.15.3A), the smallest type of immunoisolatory vehicles, are small spheres with diameters typically ranging from 200–600 μm. Although structurally fragile, microcapsules are easy to produce with readily available laboratory equipment. Microcapsules have a high surface area to volume ratio, a characteristic facilitating mass transfer. Capsules can contain up to a thousand cells, a cell number which typically fills only 1–2% of the capsule volume. Hundreds to several thousand microcapsules can easily be implanted, depending on the requirements for a particular application. Species scaling is not an issue, as the number of capsules can be increased or decreased to adjust therapeutic dose. However, the small size makes explanting microcapsules difficult, and the large number of capsules makes quality control difficult; both represent a significant hurdle to the regulatory approval process.

Macrocapsules (Figure II.6.15.3B) are much larger in size, with a typical length and inner diameter of 1–10 cm and 500 micron – 2 millimeter, respectively. Although flat sheet diffusion chambers have also been employed, these types of capsules are most commonly composed of single hollow fiber capillaries, similar to those used in

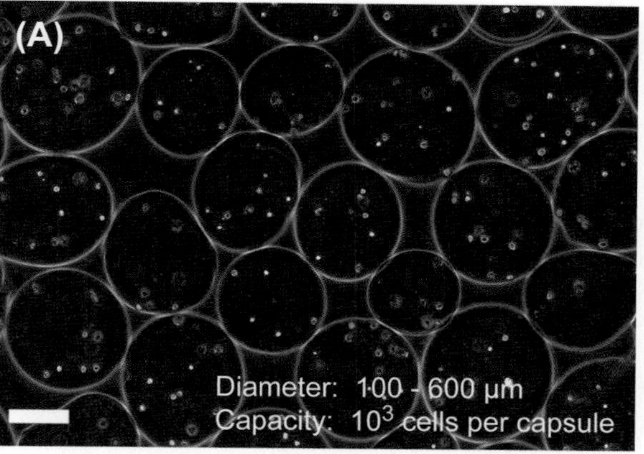

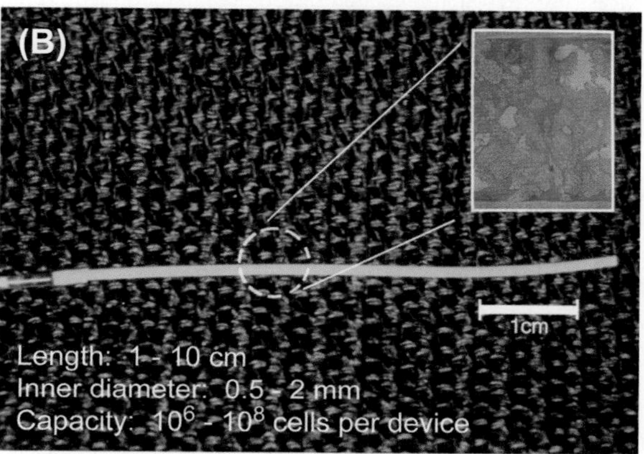

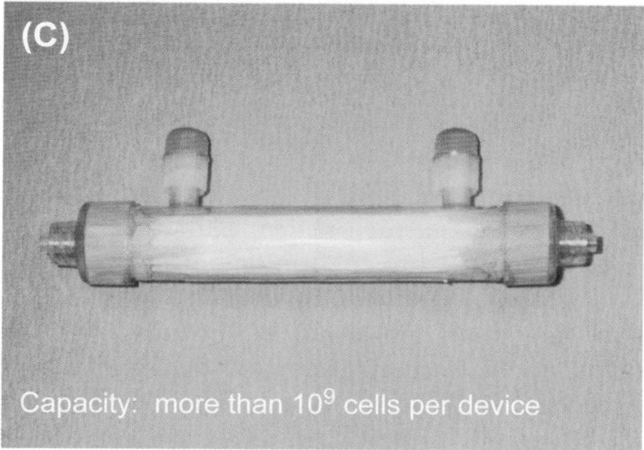

FIGURE II.6.15.3 Immunoisolatory Device Configurations. (A) is a phase contrast microscopy image of microcapsules containing human fibroblasts formed into clusters, which appear as white dots within the capsules. Scale bar is 200 microns. Microcapsules are not perfectly spherical, nor are they perfectly monodisperse. Capsules can contain a single islet or upwards of a thousand cells. A therapeutic dosage in mice is usually in the range of 500 microcapsules, and scales linearly with the weight of larger animals. (B) is a photograph of a single capillary macrocapsule device and contains an inset micrograph illustrating the cells contained within the device. The length and inner diameter of capillary of such devices typically ranges from 1–10 cm and 0.5–2.0 mm, respectively. Capacity is around a million cells. Unlike microcapsules, macrocapsules are usually easier to retrieve. (C) presents a typical flow-through device, similar to a conventional hemodialyzer, with a capacity of more than 10^9 cells per device. They are typically not implanted, but rather utilized in an intermittent extracorporeal circuit. The device shown is approximately 10 inches long, 1 inch in diameter, and contains several thousand capillaries.

BOX 1 | Species Scaling

After preliminary design is completed, most new medical devices and virtually all encapsulated cell therapy products are first tested in small rodents, usually mice or rats. Preliminary results may dictate further design iterations, but eventually the candidate products reach the stage where it is necessary to evaluate performance in a larger species, such as a canine, sheep, or nonhuman primate. For drugs and, in this chapter's context, microcapsules, scale-up is directly proportional to the increase in body weight. For example, if a 100-gram mouse is successfully treated with microcapsules containing 500 islets, then a 20-kilogram dog would require 100,000 islets.* Some adjustment may subsequently be required, but linear scaling provides a good starting point. Scale-up of mechanical devices (such as a spinal-fixation cage or joint replacement) is more complex. In the context of encapsulated cell products, scale-up of macrocapsules has proven particularly difficult. The problem arises because at least one dimension of the macrocapsule must remain 500 microns or less, in order to keep cells within reach of nutrient and oxygen supply (see text). In this case, 500 islets for a diabetic mouse would fit into a macrocapsule device smaller than the size of a paper match. However, scaled-up to accommodate a canine model, the device would be equivalent of a string over 15 feet long, and would not fare well after implantation. Because of these issues, to date, macrocapsules have been limited to applications requiring a relatively low dosage of administered therapeutic agent, and thus a small number of cells.

*Although this dosage is correct, the implant itself has proven to be cumbersome. Not only is it difficult to manufacture and control the quality of this many defect-free capsules, but the large volume of capsules needed (approximately 1.4 cm^3 assuming a capsule diameter of 300 μm and one islet per microcapsule) also make it challenging to implant the capsules safely.

hemodialyzers. Both types of structures are preformed, filled with cells, and then sealed. Elaborate procedures have been developed to confirm the integrity of the seal. Macrocapsules are strong, robust, and have been successfully implanted into a number of locations *in vivo*, including the intrathecal space of the spinal column, and the capsule space of the eye. Their design, validation, and quality control have passed the US Food and Drug Administration (FDA) regulatory scrutiny, at least for human clinical trials. Cell number per implant is based on the size of the device, and can range from 10^6 to 10^8 cells for tubular and flat-sheet diffusion chambers, respectively. Unlike microcapsules, species scaling is an issue with this larger device (see Box 1).

Similar in material properties to macrocapsules, flow-through devices (Figure II.6.15.3C) can be connected via an extracorporeal circuit to the recipient's vascular system, with the device's selectively permeable membrane separating the donor cells in the device from the recipient's blood (Lysaght and Aebischer, 1999). The most common designs employ thousands of hollow fiber membrane capillaries in a configuration similar to that of devices used in conventional hemodialysis and plasma filtration. Devices are much larger than micro- and macrocapsules, and allow for a capacity of more than 10^9 cells per device. Cells are located either in the lumen of the fiber or in the extra luminal spaces. These devices are intended for use in an extracorporeal circuit, and are typically used for 4–6 hours of treatment and then discarded. Prominent applications include fulminate liver failure or acute renal failure where the goal is to bridge patients to recovery, transplantation or maintenance therapy.

All encapsulation formats have one design constraint in common: cells must be no further than 250 mm from the external supply of nutrients and oxygen.

Virtually all immunoisolation devices, regardless of format, employ the same three components: a selectively permeable membrane barrier to isolate the cells; internal matrix scaffolding to physically support the cells within the device; and the living cells themselves (Li, 1998). In the following section, we outline these three components and some of the materials typically used.

Membranes

A wide variety of materials have been used in fabricating membranes for immunoisolation therapies, ranging from hydrogel-based polymers, to synthetic thermoplastics and polymer blends. Table II.6.15.1 lists some of the commonly used materials for immunoisolation.

Macrocapsules are generally composed of synthetic thermoplastics in a hollow-fiber design. Membranes are produced via phase inversion, a process by which polymer–solvent solutions are extruded as annular tubes surrounding a core fluid that is miscible with the solvent, but a precipitant for the polymer. Living cells are then loaded into the capsules, which are then sealed prior to implantation.

Most macrocapsules are formed from polysulfone, a material that is widely used in dialysis and which provides a convenient source of medical-grade polymer. Early on, devices were made from copolymers of polyacrylonitrile and polyvinyl chloride, which had previously been favored for the fabrication of specialty dialysis membranes.

Microcapsules are typically comprised of weak polyelectrolyte hydrogels, often coated with an external film formed from oppositely charged electrolytes. Capsules are prepared by suspending cells in dilute aqueous solutions containing a weak polyelectrolyte, and extruding the cell-in-polymer suspension into droplets that fall into a bath containing divalent electrolytes, which bind to and cross-link the charged polymer. Optionally, the capsules may then be overcoated with a layer of oppositely charged polyelectrolytes to provide an outer layer with a lower nMWCO than the uncoated capsule. Equipment for sterile extrusion of uniform droplets is not particularly expensive, and homemade extruders are also an option. The necessary techniques are easy to

TABLE II.6.15.1 Materials Commonly Utilized in Immunoisolatory Devices

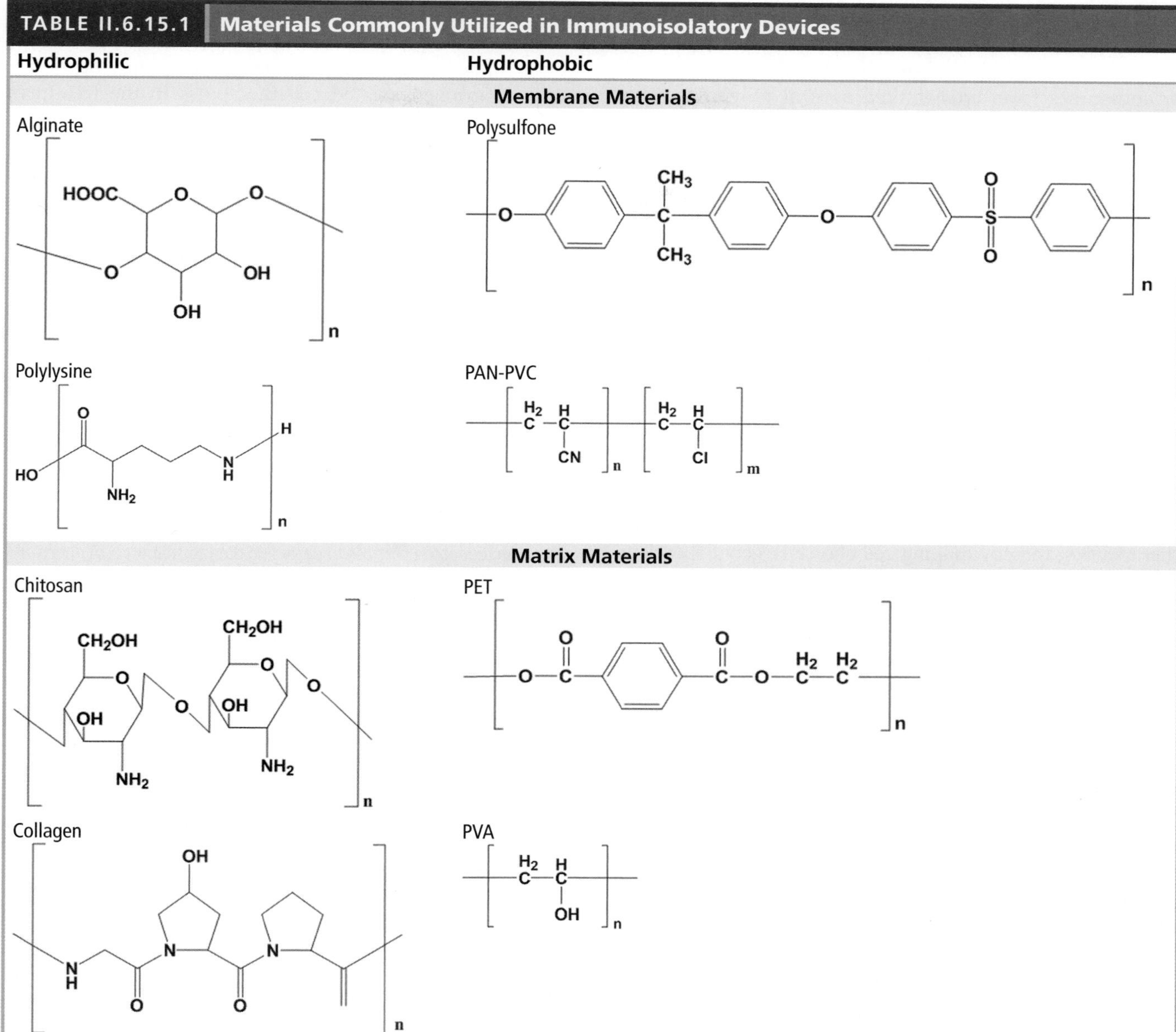

learn. Alginate is by far the most widely used encapsulant, and polylysine is the most common outer coating. These materials were among the first to be used, are readily available, and are easy to work with. Furthermore, they have built a successful track record in laboratory and animal experiments, and alternative materials have not shown any clear advantages. Microcapsules can also be fabricated by interfacial precipitation using water-insoluble polymers, such as polyacrylates.

Matrices

Immunoisolation devices typically utilize some internal matrix component that provides the encapsulated cells with a microenvironment similar to that which is found in their natural environment. Analogous to the extracellular matrix (ECM) found *in vivo*, the matrix provides cells with the appropriate biophysical stimuli that support cell viability and regulate cellular function. The matrix may also play a role in governing cellular proliferation, differentiation, and ECM production. The matrix provides a structural component within the device that ensures cells remain well dispersed throughout the device. Without some matrix, cells tend to aggregate and amass to a size which results in central necrosis.

Matrices can be divided into two categories: hydrogels and solid scaffolds. Hydrogels are typically prepared from a dilute solution of natural polymers (such as collagen, alginate or agarose) to which cells are added to form a cell–polymer suspension. In the case of microcapsules, the cells are extruded into spheres; for macrocapsules and flow-through devices, the suspension is injected into the preformed capsule and allowed to solidify. Solid matrices may be fabricated from synthetic polymers (such as polyvinyl alcohol or polyethylene terephthalate) or sometimes from cross-linked natural polymers, such as chitosan. These matrices are deployed in a very wide variety of geometries, ranging from solid beads to open cell foams. They tend to be utilized in macrocapsules, and are typically added to the device before the cells. Hydrogels tend to be better for immobilizing suspension cell cultures within a device, while solid scaffolds are ideal for the attachment of adherent cells. Different matrices have different advantages and disadvantages, and as a result, no consensus has emerged in the field as to a preferred matrix material.

Cells

Living cells, encapsulated within the matrix and membrane, make up the final component of an immunoisolatory device. These cells almost always function in a device by secreting therapeutics, thereby making cell choice largely dependent on the application of interest. Every cell has a specific set of metabolic and antigenicity requirements which must be considered during device formulation and implantation. The site of the transplant must also be considered. For example, cells with high metabolic activity will fare better in a nutrient-rich environment. A highly antigenic device may be better suited placed in an immunoprivileged site, such as the eye or behind the blood–brain barrier.

Two broad classes of cells are employed: primary and cell lines. The former are cells which have been isolated and purified from a donor subject, but which are only capable of a limited number of divisions. The number of permissible cell divisions for primary cells is termed "the Hayflick limit," and typically is in the range of 25 divisions, after which cells become senescent. Since cells generally have divided several times prior to harvesting from a donor subject, the extent to which they can be expanded *in vivo* is limited. In contrast, cell lines have gained the ability to proliferate indefinitely through naturally occurring or deliberate mutations. In principle, a single founder population can be used to seed a large number of devices. With appropriate cell banking technology, cell lines become homogeneous, allowing for the functional or safety tests performed on one cohort to be applied to the entire bank. Cell lines are clearly more convenient and cost effective than primary cells. However, not all cell types are available as cell lines; for example, there is no established line for pancreatic beta cells utilized in the bioartificial pancreas. In addition, indefinitely dividing cell lines raise a potential host of regulatory concerns.

Some cells naturally secrete the therapeutic protein of interest for a given application, such as in the case of diabetes treatment using pancreatic beta cells that naturally secrete insulin at a rate responsive to the local glucose concentration. In other instances, cells may be genetically modified using the by-now standard tools of recombinant molecular biology to produce a specific therapeutic. Such modifications are usually, though not always, performed on cell lines. One example where genetic modification is utilized in encapsulated cell therapy is in the treatment of ocular degeneration using recombinant ciliary neurotrophic factor (CNTF) (Tao, 2006). As noted in an earlier section, cells may be autologous, allogeneic or xenogeneic. Allogeneic cell lines or autologous primary cells have been most widely used in human clinical trials. Xenogeneic cells are often used for *in vitro* studies or in early animal studies.

Clinical Trials

Successful human clinical trials represent perhaps the most important step between early development of a new device or therapy format, and its final acceptance into medical practice. The minimum baseline for ethical conduct of a human clinical trial requires that three conditions be met: (1) the overall benefit of the trial must outweigh the risk; (2) the patient must give free and informed consent; and (3) the clinical protocol must be reviewed in advance by a duly constituted independent review board. The gold standards are trials conducted for the purpose of obtaining data to support FDA approval of the device. Whenever practical, such trials have a control group in addition to the study group receiving the new device. Patients are randomized to either the study arm or control arm, thereby eliminating selection bias. Regulatory level trials have a detailed protocol, which is usually discussed in advance with FDA personnel, and is preceded by careful animal trials and studies of material toxicity and biostability. Successful outcomes are defined in advance. Not surprisingly, such trials tend to be expensive, and the total cost may exceed \$50,000 to \$100,000 per patient. Regulatory level trials are almost always conducted, or at least sponsored, by industry. A second level of trial is that which is performed by an academic investigator. This type of clinical trial usually follows investigations in small and large animal models, and will have been reviewed by the appropriate review boards at the investigator's hospital or academic medical center. These trials are often credible, but frequently are conducted without a control group and without a prospective definition of successful outcomes. Benefits may be technique-dependent and achieved in a fashion that makes the study difficult to reproduce. If the trials are successful they may receive press attention, and further development may be pursued by industry groups who will conduct regulatory level trials. While rarely definitive, academic studies play a very important role in the development of new therapeutic technologies, and bring to bear the brilliance and skills of academic investigators. Less salubrious are "first in man" trials, conducted by the biotech industry and other groups motivated to establish the claim that their products or protocols "reached the stage of successful human clinical trials." Such studies are often conducted in parts of the world with lax regulatory requirements. Control arms are rare, and preclinical studies may or may not be appropriate. While these types of studies occasionally prove to be fruitful, they should generally be taken with a grain of salt.

ENCAPSULATED CELL THERAPY APPLICATIONS

Immunoisolation was first proposed as a therapy for the treatment of diabetes in the late 1970s. Since then, numerous peer-reviewed publications have shown the

TABLE II.6.15.2 Encapsulated Cell Therapies Evaluated in Humans

Target Disease	Therapy Format	Total Patient Exposure (Study Arm Participants and Controls)	Results	Principle Investigators
Acute Liver Failure	Extracorporeal flow-through device (bioartificial liver)	116 patients (85 controls)	Clear benefit in certain patient subsets	A. A. Demetriou B. Solomon
Acute Renal Failure	Extracorporeal flow-through device (bioartificial kidney)	50 patients (18 controls)	Inconsistent results with clear benefit in some, but not all, trials	H. D. Humes
Retinal Pigmentosa and Macular Degeneration	Intraocular implantable macrocapsule releasing human CNTF	166 patients (168 controls)	Study underway with initially promising results	R. A. Bush W. Tao
Chronic Pain	Intrathecal implantable macrocapsule releasing adrenal analgesics	85 patients total	Definitive benefits in some patients	M. Goddard N. de Tribolet
Huntington's Disease	Intracerebral implantable macrocapsule releasing human CNTF	6 patients (no controls)	Improvement in some patients	J. Bloch P. Aebischer
Amyotrophic Lateral Sclerosis	Intrathecal implantable macrocapsule releasing CNTF	6 patients (no controls)	No decline in disease progression	P. Aebischer E. E. Baetge
Diabetes	Macro- and microcapsule releasing insulin	22 patients (no controls)	No benefit demonstrated in any well-controlled long-term trial	Several

potential utility of immunoisolation in the treatment of a variety conditions, both *in vitro* and in animal models. Furthermore, immunoisolation is one of the few subfields within tissue engineering that has successfully provided a platform for evaluation in human clinical trials. Table II.6.15.2 summarizes some of the applications that have been explored in human clinical trials.

In some of these applications, such as liver or kidney failure, encapsulated cells serve as organ replacement therapy, usually as a temporary "bridge to transplant" for patients awaiting whole organ transplantations, as is the case with liver (Demetriou et al., 2004) or renal failure (Tumlin et al., 2008). In other applications, such as Parkinson's disease (Aebischer et al., 1994), ALS (Aebischer et al., 1996), Huntington's disease (Bloch et al., 2004), and chronic pain (Buchser et al., 1996), encapsulated cells represent destination therapies, intended to be permanently implanted to synthesize and deliver cell-derived therapeutics *in situ*. This is particularly relevant in cases where conventional drug delivery is inadequate or ineffective, because the therapeutic agents cannot cross the blood–brain and blood–retina barriers.

While the utility of immunoisolation has been explored in humans for each of the applications listed in Table II.6.15.2, none of these therapies have yet received FDA approval for routine clinical care. Several of the trials demonstrated significant benefit to many of the subjects, but the overall improvements did not rise to the level of statistical difference between the study and control arms as required by the FDA. The studies were relatively small, often with well under 100 patients in the study arm, and underpowered to demonstrate statistical benefit in complex polyfactorial disease states like liver failure or chronic pain. By way of comparison, even a straightforward drug trial often includes several thousand patients. Biotech start-up companies, which have headed the research in the field of immunoisolation, lack the funds to support studies of this size.

> Failure to demonstrate statistical significance with respect to therapeutic benefit is one of the major limitations in translating encapsulated cell therapies to the clinic.

To further illustrate the historical and current context of encapsulated cell therapy as a viable clinical treatment option, the rest of this chapter will expand upon the utility of immunoisolation in the treatment of two diseases, diabetes and adult-onset blindness caused by retinal and macular degeneration, with the former being the oldest and most extensively researched application in the field, and the latter being a promising therapy, currently (as of 2011) undergoing FDA-approved clinical trials.

Diabetes

Encapsulated cells in the format of a biohybrid pancreas were first proposed in 1975 by Chick and his colleagues for the treatment of diabetes (Chick et al., 1975). This group went on to demonstrate the reversal of diabetes in a rat model using encapsulated pancreatic islets sourced from healthy rats (Chick et al., 1977). The islets, containing pancreatic beta cells, were encapsulated within a hollow-fiber flow-through device, and connected to the vascular system of diabetic rats via an arteriovenous shunt. No immunosuppressive drugs were employed.

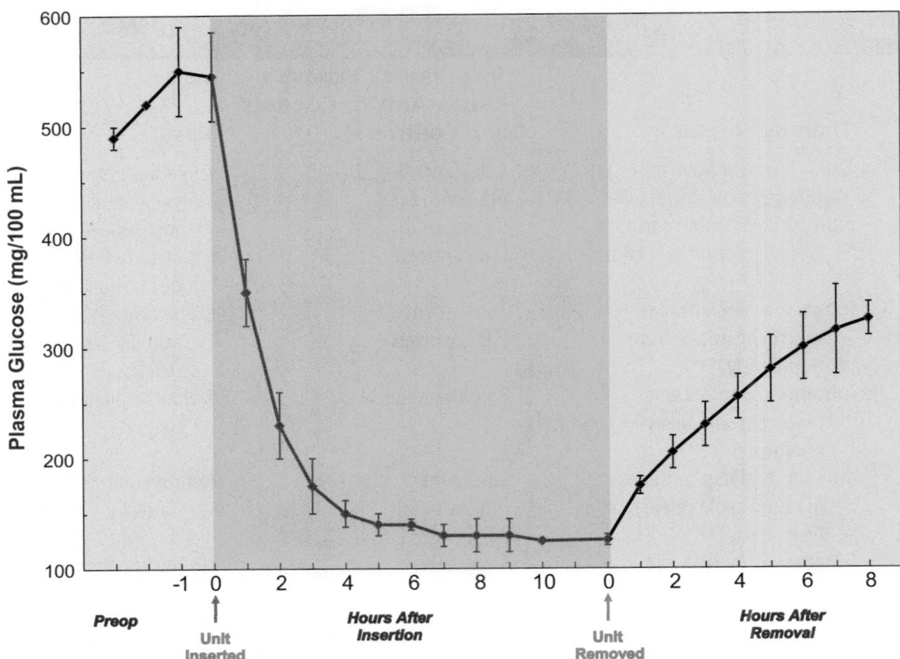

FIGURE II.6.15.4 First Report of a Bioartificial Pancreas. Pancreatic islet cells from healthy neonatal rats were encapsulated within a semi-permeable hollow fiber membrane device and connected to the circulatory system of alloxan-induced diabetic rats via an arteriovenous shunt. This "bioartificial pancreas" caused the plasma glucose concentration to drop steadily and stabilize to a normal non-diabetic range. Blood sugar did not continue to decrease below normal (hypoglycemia), suggesting a feedback loop between the encapsulated islets and the circulating plasma. Glucose returned to elevated levels when the device was removed. This experiment was reported in 1977, 35 years ago. Subsequent investigators have optimized the device and simplified its configuration, but have had very limited success in scaling the bioartifical pancreas up for application in dogs or primates. (Adapted from Chick et al. (1977). *Science,* **197**(4305), 780–782.) Reprinted with permission from AAAS.

Insulin was secreted from the encapsulated cells, diffused out of the capsules, and stabilized the diabetic host's plasma glucose concentration to a normal range, as shown in Figure II.6.15.4. This finding began and validated the field of immunoisolation by demonstrating the utility of encapsulated cells as a viable alternative to traditional transplantation and pharmaceutical treatment (Lim and Sun, 1980; Thanos and Elliott, 2009). The transplanted cells produced insulin that was delivered to the host systemically, and thereby eliminated the need for insulin administration orally or via injection. Just as in the case of insulin secreted by the pancreas, secretions from encapsulated cells were responsive to circulating blood level, a feature that has been thoroughly confirmed in later studies.

Since this first demonstration in 1977, diabetes reversal in rats has been reproduced and reported in dozens of peer-reviewed publications using transplantable macro- and microcapsules containing allo- and xenogeneic islets (Thanos and Elliott, 2009). Porcine islets are particularly popular because porcine insulin was the standard treatment in humans until recombinant human insulin became available in the early 1980s. However, reversal of diabetes in larger animal models, as opposed to mice and rats, has proven to be particularly difficult (Pierson et al., 2009). In canine models, only few groups report cell survival and short-term glucose control with allo- and xenogeneic islets with insulin response (Lanza et al., 1995; Wang et al., 2008). Even fewer reported the success of encapsulated islets in primates, with the majority of these studies assessing encapsulated cell viability in non-diabetic models (Elliott et al., 2005). The reversal and stabilization of glucose levels in these larger animals often required additional insulin administration to supplement the amount that was being delivered by the encapsulated cells.

Reproducing the success of rodent models in larger animals is difficult due to the complexity of species scaling (see Box 1). First, the device requires a large number of cells, making cell harvesting and processing more difficult than in small animal applications. Beta cells cluster into islets that contain about 3000 cells each. About 5000–10,000 islets per kilogram of body weight are believed to be required for successful pancreatic replacement. While the islets for rodent models can be isolated by hand, due to the small number required (about 500 islets), larger animal models require so many cells (300,000–400,000 islets) that automated processing and multiple donor sources are often needed. This results in lower reliability and consistency within the donor cell population. To date, attempts to produce a robust cell line releasing insulin in response to glucose have been unsuccessful. Additionally, the number of cells necessary poses a limitation on the size of the immunoisolatory device used, in that it must be able to house the required number of cells without adversely affecting the transport properties of the device. Finally, as discussed earlier, encapsulated cells will secrete small amounts of soluble antigens that can diffuse through the immunoisolatory membrane and elicit immune response in the host tissue. This phenomenon is dose-responsive, and so the large number of cells makes it a particular problem for pancreatic replacement in large animals.

Lack of success in reproducible long-term reversal of diabetes in large animal models has not deterred some researchers from exploring the utility of these implants in humans. For instance, in 1994, Soon-Shiong reported insulin independence and tight glycemic control in a single type 1 diabetic patient nine months after an allogeneic transplant of islets encapsulated within alginate microcapsules (Soon-Shiong et al., 1994). However, the patient was never completely taken off supplementary insulin treatment. Over the years, approximately half a dozen other human clinical trials have been conducted around the world (Thanos and Elliott, 2009), many in countries with lax regulation. Some, but not all, involve microcapsules, and have been criticized for the lack of capsule validation through pre-clinical and animal studies, lack of study arm controls, and failure to publish results in appropriate peer-review studies (Grose 2007; Sykes et al., 2007).

The artificial pancreas remains today where it has been for well over a decade. Efficacy has been abundantly demonstrated in small animal models, but despite extensive efforts and impressive incentives for success, this technology has yet to be translated to a clinically relevant device. Development of a practical bioartificial pancreas remains as one of the great bioengineering challenges of the early 21st century.

Ocular Degeneration

Treatment of retinal and macular degeneration, both of which can lead to adult onset vision loss or even blindness, represents an area of research currently (as of 2011) regarded as offering great promise for bringing encapsulated cells to the clinic. An ocular therapy, called NT-501, is currently under investigation in FDA-regulated Phase II human clinical trials for the treatment of retinitis pigmentosa (RP), and has completed a Phase II for the treatment of dry age-related macular degeneration (AMD) (Sieving et al., 2006; Tao, 2006; Emerich and Thanos, 2008). Formulated by the US-based biotechnology company Neurotech Pharmaceuticals, Inc. (Lincoln, RI, USA), NT-501 is an intraocular implantable macrocapsule containing a genetically modified human retinal pigment epithelial cell line that secretes ciliary neurotrophic growth factor (CNTF). The device consists of a polysulfone hollow-fiber membrane which houses the cells, and an internal poly(ethylene terephthalate) yarn scaffold matrix. The entire device is sealed and has a small titanium anchoring loop embedded in one end that is used to tether the device to the sclera after it is placed in the vitreous body of the eye.

CNTF has been shown to be very effective in delaying retinal degeneration in RP animal models, and thus is a promising therapeutic for the treatment of RP and AMD, both of which currently have no available cure or viable treatment options. By delivering CNTF directly *in situ* to the retina, NT-501 circumvents the blood–retina barrier that systemically restricts the retinal tissue, a circumstance which poses a challenge to traditional drug therapies.

NT-501 has fared well in the regulatory process, having been granted “Fast Track” status by the FDA. This status allows for accelerated regulatory review of a drug therapy that exhibits potential to meet the treatment needs of a serious or life-threatening condition that currently has no available treatment options. The randomized, multi-center RP trials (early- and late-stage) each have 60 patients enrolled, while the completed AMD trial had a total of 51 patients. Results from the Phase 2 RP trials, along with the results from the completed AMD trial, reveal a significant increase in retinal thickness with treatment. Additionally, the AMD trials have shown the stability of visual acuity with NT-501 treatment. Together, these results all suggest that the intraocular delivery of CNTF by NT-501 may in fact be effective in decelerating ocular degeneration and loss of vision.

The success of ocular implants in clinical trials, as compared to diabetes implants, is due to a number of factors including the site of implantation, the cell type, and the relatively small population of cells implanted. Unlike the majority of diabetes implants, NT-501 is placed in the eye, which is a more immune-privileged region *in vivo* than the peritoneal cavity. Additionally, NT-501 contains an allogeneic cell line that is easier to source and validate than the xenogeneic or primary human cells utilized in the diabetes implants. Finally, NT-501 contains half a million cells, while most encapsulated cell therapies for treatment of diabetes contain over a billion. The smaller number of cells provides a smaller immune challenge, and significantly simplifies the design and manufacture of the implant.

CONCLUDING PERSPECTIVES

Encapsulated cell therapy remains a promising approach to the treatment of many serious diseases which still lack satisfactory therapies. The technology, especially for macroencapsulation, is well-developed. Clinical trials have been encouraging, although not yet definitive. The remaining challenges relate to more extensive clinical trials, and to regulatory approval.

BIBLIOGRAPHY

Aebischer, P., Buchser, E., Joseph, J. M., Favre, J., de Tribolet, N., et al. (1994). Transplantation in humans of encapsulated xenogeneic cells without immunosuppression: A preliminary report. *Transplantation*, 58(11), 1275–1277.

Aebischer, P., Schluep, M., Deglon, N., Joseph, J. M., Hirt, L., et al. (1996). Intrathecal delivery of CNTF using encapsulated genetically modified xenogeneic cells in amyotrophic lateral sclerosis patients. *Nat. Med.*, **2**(6), 696–699.

Bloch, J., Bachoud-Levi, A. C., Deglon, N., Lefaucheur, J. P., Winkel, L., et al. (2004). Neuroprotective gene therapy for Huntington's disease, using polymer-encapsulated cells engineered to secrete human ciliary neurotrophic factor: Results of a phase I study. *Hum. Gene. Ther.*, **15**(10), 968–975.

Buchser, E., Goddard, M., Heyd, B., Joseph, J. M., Favre, J., et al. (1996). Immunoisolated xenogenic chromaffin cell therapy for chronic pain. Initial clinical experience. *Anesthesiology*, **85**(5), 1006–1012.

Bunger, C., Tiefenbach, B., Jahnke, A., Gerlach, C., Freier, T., et al. (2005). Delection of the tissue response against alginate-pll capsules by temporary release of co-encapsulated steroids. *Biomaterials*, **26**(15), 2346–2360.

Chick, W. L., Like, A. A., Lauris, V., Galletti, P. M., Richardson, P. D., et al. (1975). A hybrid artificial pancreas. *ASAIO*, **21**, 8–15.

Chick, W. L., Perna, J. J., Lauris, V., Low, D., Galletti, P. M., et al. (1977). Artificial pancreas using living beta cells: Effects on glucose homeostasis in diabetic rats. *Science*, **197**(4305), 780–782.

Colton, C. K. (1995). Implantable biohybrid artificial organs. *Cell Transplant*, **4**(4), 415–436.

Demetriou, A. A., Brown, R. S., Jr., Busuttil, R. W., Fair, J., McGuire, B. M., et al. (2004). Prospective, randomized, multicenter, controlled trial of a bioartificial liver in treating acute liver failure. *Ann. Surg.*, **239**(5), 660–670.

Dionne, K. E., Cain, B. M., Li, R. H., Bell, W. J., Doherty, E. J., et al. (1996). Transport characterization of membranes for immunoisolation. *Biomaterials*, **17**(3), 257–266.

Gentile, F. T., Doherty, E. J., Rein, D. H., Shoichet, M. S., & Winn, S. R. (1995). Polymer science for macroencapsulation of cells for central nervous system transplantation. *React. Poly.*, **25**, 207–227.

Goren, A., Dahan, N., Goren, E., Baruch, L., & Machluf, M. (2010). Encapuslated human mesenchymal stem cells: A unique hypoimmunogenic platform for long-term cellular therapy. *FASEB J.*, **24**(1), 22–31.

Grose, S. (2007). Critics slam Russian trial to test pig pancreas for diabetics. *Nat. Med.*, **13**(4), 390–391.

Elliott, R. B., Escobar, L., Calafiore, R., Basta, G., Garkavenko, O., et al. (2005). Transplantation of micro- and macroencapsulated piglet islets into mice and monkeys. *Transplant Proc.*, **37**(1), 466–469.

Emerich, D. F., & Thanos, C. G. (2008). NT-501: An ophthalmic implant of polymer-encapsulated ciliary neurotrophic factor-producing cells. *Curr. Opin. Mol. Ther.*, **10**(5), 506–515.

Hernández, R., Orive, G., Murua, A., & Pedraz, J. (2010). Microcapsules and microcarriers for *in situ* cell delivery. *Adv. Drug Deliv. Rev.*, **62**(7–8), 711–730.

Lanza, R. P., Kühtreiber, W. M., & Chick, W. L. (1995). Encapsulation technologies. *Tissue Eng.*, **1**(2), 181–196.

Li, R. H. (1998). Materials for immunoisolated cell transplantation. *Adv. Drug Deliv. Rev.*, **33**(1–2), 87–109.

Lim, F., & Sun, A. M. (1980). Microencapsulated islets as bioartificial endocrine pancreas. *Science*, **210**(4472), 908–910.

Lysaght, M. J., & Aebischer, P. (1999). Encapsulated cells as therapy. *Sci. Am.*, **280**(4), 76–82.

Mital, P., Kaur, G., & Dufour, J. (2010). Immunoprotective Sertoli cells: Making allogeneic and xenogeneic transplantation feasible. *Reproduction*, **139**(3), 495–504.

Pierson, R. N., Dorling, A., Ayares, D., Rees, M. A., Seebach, J. D., et al. (2009). Current status of xenotransplantation and prospects for clinical application. *Xenotransplantation*, **16**(5), 263–280.

Sieving, P. A., Caruso, R. C., Tao, W., Coleman, H. R., Thompson, D. J., et al. (2006). Ciliary neurotrophic factor (CNTF) for human retinal degeneration: Phase I trial of CNTF delivered by encapsulated cell intraocular implants. *Proc. Natl. Acad. Sci. USA*, **103**(10), 3896–3901.

Soon-Shiong, P., Heintz, R. E., Merideth, N., Yao, Q. X., Yao, Z., et al. (1994). Insulin independence in a type 1 diabetic patient after encapsulated islet transplantation. *Lancet*, **343**(8903), 950–951.

Sykes, M., Pierson, R. N., O'Connell, P., D'Apice, A., Cowan, P., et al. (2007). Reply to Critics slam Russian trial to test pig pancreas for diabetes. *Nat. Med.*, **13**(6), 662–663.

Tao, W. (2006). Application of encapsulated cell technology for retinal degenerative diseases. *Expert Opin. Biol. Ther.*, **6**(7), 717–726.

Thanos, C. G., & Elliott, R. B. (2009). Encapsulated porcine islet transplantation: An evolving therapy for the treatment of type I diabetes. *Expert Opin. Biol. Ther.*, **9**(1), 29–44.

Tumlin, J., Wali, R., Williams, W., Murray, P., Tolwani, A. J., et al. (2008). Efficacy and safety of renal tubule cell therapy for acute renal failure. *J. Am. Soc. Nephrol.*, **19**(5), 1034–1040.

Wang, T., Adcock, J., Kuhtreiber, W., Qiang, D., Salleng, K. J., et al. (2008). Successful allotransplantation of encapsulated islets in pancreatectomized canines for diabetic management without the use of immunosuppression. *Transplantation*, **85**(3), 331–337.

Zielinski, B. A., & Lysaght, M. J. (2007). Immunoisolation. In R. Lanza, R. Langer, & J. Vacanti (Eds.), *Principles of Tissue Engineering* (pp. 399–404). Burlington, MA: Elsevier.

CHAPTER II.6.16 TISSUE ENGINEERING WITH DECELLULARIZED TISSUES

Stephen F. Badylak, Bryan N. Brown, and Thomas W. Gilbert

McGowan Institute for Regenerative Medicine, University of Pittsburgh, Pittsburgh, PA, USA

INTRODUCTION

Materials for the repair or reconstruction of injured, missing or weakened tissues can be composed of either synthetic or naturally occurring components. The optimal material for each application should be selected based upon criteria such as the structural and mechanical requirements of the intended application, the type of response expected from the adjacent host tissue, and the ability of the material to support normal tissue growth and function over the long-term. Unlike conventional surgical repair procedures, in which a biomaterial mesh simply needs to have sufficient strength to hold adjacent tissues together and avoid an adverse host response, tissue engineering and regenerative medicine approaches utilize "scaffold" materials to facilitate the restoration of the normal structure and function of the tissue or organ of interest. The present chapter will focus exclusively upon the use of decellularized tissues (i.e., extracellular matrix) as scaffold materials in tissue engineering and regenerative medicine applications.

Every tissue and organ is composed of cells and extracellular matrix (ECM). The ECM consists of the secreted products, both structural and functional, of the resident cells of each tissue and organ. The composition and ultrastructure of the ECM is determined by factors that influence the phenotype of its resident cells, including mechanical forces, biochemical milieu, oxygen requirements, pH, and inherent gene expression patterns. In turn, the ECM influences the attachment, migration, proliferation, and three-dimensional organization of cells, as part of a process of dynamic reciprocity (Bissell et al., 1982; Boudreau et al., 1995; Ingber, 1991). The ECM is essentially an "information highway" between its embedded cells and serves a critical function in tissue and organ homeostasis, as well as response to injury.

For these same reasons, multiple forms of allogeneic and xenogeneic ECM have been isolated, processed into application-appropriate configurations, and investigated as scaffolds for tissue engineering and regenerative medicine purposes in multiple body systems. Tissue engineering and regenerative medicine approaches utilizing ECM scaffolds have been successfully used in both pre-clinical and clinical studies. Examples of ECM scaffold sources, configurations, and tissues that have been reconstructed using ECM scaffolds are listed in Table II.6.16.1, and examples of commercially available ECM scaffold materials, their source tissue, and configuration are provided in Table II.6.16.2.

The success of ECM scaffolds in tissue engineering and regenerative medicine applications may be attributed in large part to their ability to modulate the default mechanisms of tissue repair. In general, the response following implantation of an acellular non-chemically cross-linked ECM scaffold has been described as "constructive remodeling." That is, ECM scaffolds are capable of inducing the formation of new tissue structures that are arranged in a spatially appropriate pattern for the tissue of interest. For example, ECM scaffolds have been shown to be capable of inducing the formation of functionally innervated muscular tissue in a model of abdominal wall reconstruction (Agrawal et al., 2009; Valentin et al., 2010). This is in direct contrast to the default mechanism of mammalian response to injury that involves inflammation and scarring.

The objective of this chapter is to provide a rationale for the use of ECM as a scaffold for tissue engineering and regenerative medicine applications, an overview of the methods used in the preparation of ECM materials, a description of the composition and structure of such scaffold materials, and the mechanisms by which these biologic materials function as inductive scaffolds for tissue reconstruction.

RATIONALE FOR THE DECELLULARIZATION OF TISSUES AND ORGANS AND THE USE OF DECELLULARIZED TISSUES AS SCAFFOLDS IN TISSUE ENGINEERING AND REGENERATIVE MEDICINE

Individual ECM components such as collagen I, laminin, and fibronectin have been isolated, purified, and processed for use in cell culture, as biomaterial coatings, and as three-dimensional scaffolds for tissue engineering and regenerative medicine applications (Dejana et al., 1987; Macarak and Howard, 1983; Glowacki and Mizuno, 2008). These approaches have typically been successful for such applications. However, an advantage of utilizing ECM materials that have been derived through the decellularization of intact tissues or organs as substrates or scaffolds for cell growth and differentiation is the presence of tissue-specific structural and functional molecules (and their inhibitors) in the same relative amounts that exist in nature, and in their native three-dimensional ultrastructure. The ECM is capable of presenting these factors efficiently to resident or migrating cell surface receptors, protecting growth factors from degradation, and modulating the synthesis of new ECM molecules (Entwistle et al., 1995; Bonewald, 1999; Kagami et al., 1998; Roberts et al., 1988; Freise et al., 2009). If one considers the ECM as a substrate for *in vivo* cell growth, migration, and differentiation, it is reasonable to think of the ECM as a temporary (i.e., degradable) controlled release vehicle for naturally derived functional molecules. Many groups have attempted to create ECM analogs using individual ECM components and/or synthetic materials (Causa et al., 2007; Alini et al., 2003; Koh et al., 2008); however, the diversity and complex structure of the molecules that make up the ECM predict the difficulty of creating such a scaffold *in vitro*. It is for these reasons that the isolation of ECM through the decellularization of tissues and organs is an effective method for the production of materials to be used in tissue engineering and regenerative medicine approaches to tissue reconstruction.

METHODS OF DECELLULARIZATION

The tissues from which ECM scaffold materials are harvested, the species of origin, the decellularization method, and the methods by which the material is sterilized can vary widely. Antigenic epitopes associated with cell membranes and intracellular components of tissues and organs are the primary cause of the adverse immunologic response elicited by allogeneic and xenogeneic tissue transplants (Erdag and Morgan, 2004; Gock et al., 2004; Ross et al., 1993), while the molecules that constitute the extracellular matrix are generally conserved across species lines, and are well-tolerated even by xenogeneic recipients (Bernard MPC et al., 1983; Bernard MPM et al., 1983; Constantinou CDJ, 1991; Exposito JYDA et al., 1992). Therefore, a suitable ECM scaffold material must be effectively treated to remove cellular material or chemically cross-linked to mask the antigens prior to implantation. The most commonly used chemical cross-linking agents are gluteraldehyde, carbodiimide, genipin, and hexamethylene-diisocyanate (Bhrany et al., 2006; Billiar et al., 2001; Harper, 2001). Although a description of the specific effects of chemical cross-linking on ECM scaffold materials is not within the scope of this chapter, the choice to decellularize or chemically cross-link the

TABLE II.6.16.1 Partial List of ECM Scaffold Donor Sources, Configurations, and Applications

Donor Source	Configurations	Applications
	Sheet	Abdominal wall (Clarke et al., 1996; Prevel et al., 1995)
	Multilaminate sheet (Freytes et al., 2004)	Artery (Badylak et al., 1989; Lantz et al., 1990)
Allogeneic	Powder (Gilbert et al., 2005)	Bladder (Boruch et al., 2009; Piechota et al., 1998; Yoo et al., 1998)
Xenogeneic	Gel (Freytes, et al., 2008)	Bone (Suckow et al., 1999; Moore et al., 2004)
	3D construct (Badylak et al., 2000; Nieponice et al., 2009)	Cartilage (Welch et al., 2002; Peel et al., 1998)
Species	Intact decellularized organ (Wainwright et al., 2009; Ott et al., 2008; Soto-Gutierrez et al., 2011; Petersen et al., 2012)	Dura mater (Cobb et al., 1996, 1999)
		Esophagus (Nieponice et al., 2009; Badylak et al., 2005)
Bovine		Heart (Badylak et al., 2006; Uygun et al., 2010; Hammond et al., 2011)
Equine		Liver (Lin et al., 2004)
		Lung (Ott et al., 2008; Petersen et al., 2012; Manni et al., 2011)
Porcine		Nerve (Kim et al., 2004; Sondell et al., 1998)
Human		Skin (Prevel et al., 1995; Black et al., 1998; Bello et al., 2001; Mostow et al., 2005)
		Tendon (Gilbert et al., 2009; Zantop et al., 2006; Dejardin et al., 2001; Musahl et al., 2004)
Age		Thoracic wall (Gilbert et al., 2008)
		Trachea (Gilbert et al., 2008; Macchianni et al., 2008; Remlinger et al., 2010)
Fetal		Tympanic membrane (Parekh et al., 2009)
Adult		Vocal fold (Gilbert et al., 2009; Huber et al., 2003)
Organ or tissue		
Bladder (Piechota et al., 1998; Hodde et al., 2002; Chen et al., 1999)		
Heart (Wainwright et al., 2009; Bader et al., 1998)		
Liver (Lin et al., 2004; Uygun et al., 2010; Soto-Gutierrez et al., 2011)		
Lung (Nichols et al., 2009; Price et al., 2010; Petersen et al., 2012; Song et al., 2011)		
Muscle (Borschel et al., 2004)		
Nerve (Kim et al., 2004; Hudson et al., 2004)		
Skin (Buinewicz & Rosen, 2004; Armour et al., 2006)		
Small intestine (Badylak et al., 1989)		
Tendon (Tischer et al., 2007; Cartmell & Dunn, 2004)		

material has important effects on the host tissue response, which will be discussed later in the chapter.

Although it seems logical that the decellularization process will by definition affect the structure and composition of the extracellular matrix, the ultimate goal of any decellularization protocol is to remove all cellular material without adversely affecting the biochemical composition, mechanical behavior, topographical ligand landscape, and eventual biologic activity of the remaining scaffold material (Crapo et al., 2011). Decellularization processes generally begin with exposure of the tissue to ionic solutions to disrupt the cell membrane, followed by separation of cellular components from the ECM using enzymatic treatments (e.g., trypsin or nucleases), solubilization of cytoplasmic

TABLE II.6.16.2 Partial List of Commercially Available Scaffold Materials Composed of Extracellular Matrix

Product	Company	Material		Form	Use
AlloDerm®	Lifecell	Human skin	Cross-linked	Dry sheet	Abdominal wall, breast, ear, nose, and throat (ENT), head and neck reconstruction, grafting
Allopatch®	Muscolskeletal Transplant Foundation	Human fascia lata	Cross-linked	Dry sheet	Orthopedic applications
Axis™ dermis	Mentor	Human dermis	Natural	Dry sheet	Pelvic organ prolapse
CollaMend®	Bard	Porcine dermis	Cross-linked	Dry sheet	Soft tissue repair
CuffPatch™	Arthrotek	Porcine small intestinal submucosa (SIS)	Cross-linked	Hydrated sheet	Reinforcement of soft tissue
DurADAPT™	Pegasus Biologicals	Horse pericardium	Cross-linked	–	Repair dura mater after craniotomy
Dura-Guard®	Synovis Surgical	Bovine pericardium	–	Hydrated sheet	Spinal and cranial repair
Durasis®	Cook SIS	Porcine small intestinal submucosa (SIS)	Natural	Dry sheet	Repair dura mater
Durepair®	TEI Biosciences	Fetal bovine skin	Natural	Dry sheet	Repair of cranial or spinal dura
FasLata®	Bard	Cadaveric fascia lata	Natural	Dry sheet	Soft tissue repair
Graft Jacket®	Wright Medical Tech	Human skin	Cross-linked	Dry sheet	Foot ulcers
Matristem®	ACell	Porcine Urinary Bladder	Natural	Dry Sheet, Powder	Soft tissue repair and reinforcement, burns, gynecologic
Oasis®	Healthpoint	Porcine small intestinal submucosa (SIS)	Natural	Dry sheet	Partial and full thickness wounds; superficial and second degree burns
OrthADAPT™	Pegasus Biologicals	Horse pericardium	Cross-linked	–	Reinforcement, repair and reconstruction of soft tissue in orthopedics
Pelvicol®	Bard	Porcine dermis	Cross-linked	Dry sheet	Soft tissue repair
Peri-Guard®	Synovis Surgical	Bovine pericardium	–	–	Pericardial and soft tissue repair
Permacol™	Tissue Science Laboratories	Porcine skin	Cross-linked	Hydrated sheet	Soft connective tissue repair
PriMatrix™	TEI Biosciences	Fetal bovine skin	Natural	Dry sheet	Wound management
Restore™	DePuy	Porcine small intestinal submucosa (SIS)	Natural	Sheet	Reinforcement of soft tissues
Stratasis®	Cook SIS	Porcine small intestinal submucosa (SIS)	Natural	Dry sheet	Treatment of urinary incontinence
Surgimend™	TEI Biosciences	Fetal bovine skin	Natural	Dry sheet	Surgical repair of damaged or ruptured soft tissue membranes
Surgisis®	Cook SIS	Porcine small intestinal submucosa (SIS)	Natural	Dry sheet	Soft tissue repair and reinforcement
Suspend™	Mentor	Human fascia lata	Natural	Dry sheet	Urethral sling
TissueMend®	TEI Biosciences	Fetal bovine skin	Natural	Dry sheet	Surgical repair and reinforcement of soft tissue in rotator cuff
Vascu-Guard®	Synovis Surgical	Bovine pericardium	–	–	Reconstruction of blood vessels in neck, legs, and arms
Veritas®	Synovis Surgical	Bovine pericardium	–	Hydrated sheet	Soft tissue repair
Xelma™	Molnlycke	ECM protein, poly(glycolic acid) PGA, water	–	Gel	Venous leg ulcers
Xeniform™	TEI Biosciences	Fetal bovine skin	Natural	Dry sheet	Repair of colon, rectal, urethral, and vaginal prolapse; pelvic reconstruction; urethral sling
Zimmer Collagen Patch®	Tissue Science Laboratories	Porcine dermis	Cross-linked	Dry sheet	Orthopedic applications

and nuclear components using detergents (e.g., most commonly sodium dodecyl sulfate, sodium deoxycholate or Triton™ X-100), and ultimately removal of cellular debris from the ECM scaffold material. Alkaline and acidic solutions are also commonly used to remove nucleic acids from ECM scaffolds, and to disinfect the material prior to sterilization. All of these steps can be coupled with physical treatments (e.g., freezing, direct pressure, sonication, and agitation) to increase their effectiveness. Following decellularization, all residual chemicals must be removed by thorough rinsing to avoid an adverse host–tissue response to the decellularization agents.

Most commercially available biologic scaffold materials are manufactured by first processing the organ of interest into a sheet prior to decellularization. However, due to the density, mass, and three-dimensional architecture of many whole organs such as the heart, liver, and kidney, these approaches are not effective for removing cellular material from these tissues (Crapo et al., 2011). Recent reports have described methodology for decellularizing intact hearts by perfusion of the same reagents described above through the existing vascular network (Ott et al., 2008; Wainwright et al., 2010). Although additional work is still needed, these preliminary reports suggest that the native three-dimensional architecture and biochemical composition of the tissue can be largely preserved, thus providing a promising approach for complex organ engineering (Badylak et al., 2012).

Despite the goal of preserving the intact ECM after the decellularization process, each treatment has an effect on the tissue. Detergents have been shown to disrupt the collagen that is present within certain tissues, thereby decreasing the mechanical strength of the resulting ECM scaffold. The same detergent may, however, have no effect on the collagen present within another tissue (Cartmell and Dunn, 2004; Woods and Gratzer, 2005). This is likely due to the diversity of the types of collagen known to exist within tissues and organs, many of which serve tissue-specific functions (van der Rest and Garrone, 1991). Studies have also shown that removal of glycosaminoglycans (GAGs) from the scaffold can adversely affect scaffold remodeling and the viscoelastic behavior of the scaffold (Cartwright et al., 2006; Lovekamp et al., 2006). Therefore, decellularization methods require optimization for each tissue and organ to remove cellular material without compromising the structural and functional properties of the remaining ECM.

Despite efforts to fully decellularize tissues, most commonly used methods are insufficient to achieve complete decellularization, as most if not all ECM scaffold materials retain some amount of DNA (Derwin et al., 2006; Farhat et al., 2008; Gilbert et al., 2009). In addition, antigens such as the galactosyl alpha 1,3 galactose (i.e., gal-epitope) have been shown to be present in porcine ECM. Fortunately, the gal-epitope in ECM scaffolds does not activate complement or bind IgM antibody, presumably because of the small amount and widely scattered distribution of antigen (McPherson et al., 2000; Raeder et al., 2002). So, while the full removal of cellular content remains the goal of the decellularization process, it appears that there is a threshold below which the amount or integrity of the cellular material is insufficient to invoke immune-mediated rejection. The presence and effects of these molecules will be discussed more fully later in the chapter.

ECM CONFIGURATION

Following the decellularization of a tissue or organ, the resulting ECM may take on a variety of shapes and sizes which are dependant on the particular architecture of the decellularized organ of interest or the methods used in the decellularization process (Gilbert et al., 2006). As mentioned previously, many tissues and organs such as dermis, small intestine, and urinary bladder are typically processed into a sheet-like configuration prior to decellularization. The sheet form may be insufficient in its mechanical properties and/or three-dimensional morphology (i.e., shape and size) depending on the application of interest. Therefore, a number of methods have been utilized for the processing of ECM scaffolds into a variety of application-specific shapes and sizes. Intact ECM scaffolds have been molded and vacuum pressed into shapes that include tubes (Badylak et al., 2005), cones (Nieponice et al., 2006), and multilaminate sheets, among others (Freytes et al., 2004). These scaffolds have been utilized in applications ranging from esophageal repair (tubular) (Badylak et al., 2005) to gastro-esophageal junction repair (cone-shaped) (Nieponice et al., 2006), and orthopedic applications (multi-laminate sheets) (Dejardin et al., 2001). ECM materials have also been comminuted to create a powder form of the scaffold, which is of interest for injectable and space-filling applications (Gilbert et al., 2005). A hydrogel form of ECM has also been produced via enzymatic degradation of whole ECM scaffolds (Freytes et al., 2008). The ability of ECM scaffolds to be formed into varied shapes and sizes further add to their utility as scaffolds for tissue engineering and regenerative medicine applications.

Composition of Extracellular Matrix

Most of the component molecules of the ECM are well-recognized and form a complex, tissue-specific meshwork of proteins, glycosaminoglycans, glycoproteins, and small molecules (Laurie et al., 1989; Baldwin, 1996; Martins-Green and Bissel, 1995). The logical division of the ECM into structural and functional molecules is not possible, because many of the component molecules have both structural and functional roles in health and disease. For example, both collagen and fibronectin, molecules that were once considered to exist purely for their “structural” properties, are now known to have a variety of “functional” moieties, with properties ranging from cell adhesion and motility to angiogenesis or inhibition of angiogenesis. These "bimodal" or multifunctional molecules provide a hint of the diverse cryptic peptide sequences that exist within certain parent molecules and which, in themselves, have biologic effects that significantly affect the ECM scaffold remodeling process (Agrawal et al., 2010, 2011a,b). This section will highlight a number of the molecules that are known to exist within many of the ECM scaffolds described in this chapter. However, it should be noted that the ECM of each tissue and organ is unique, and the exact composition depends largely on the resident cell population and the function of the organ of interest. Chapter I.2.7 also discusses many natural biomolecules and their use as biomaterials.

Collagen. There are more than 20 distinct types of collagen, each with a unique biologic function. These proteins account for nearly 90% of the dry weight of the ECM of most tissues and organs (van der Rest and Garrone, 1991). Type I collagen is the major structural protein present in tissues, and is found ubiquitously throughout the plant and animal kingdoms. Type I collagen is particularly abundant in ligamentous, tendinous, and connective tissue structures, and provides the strength necessary to accommodate the uniaxial and multiaxial mechanical loading to which these tissues are commonly subject. These same tissues provide a convenient source of collagen for many medical device applications. For example, bovine type I collagen is harvested from Achilles tendon, and is perhaps the most commonly used xenogeneic ECM component intended for therapeutic applications.

Other collagen types also exist in the ECM of most tissues, but typically in much lower quantities. These alternative collagen types provide distinct, tissue-specific, physical and mechanical properties to the ECM, while simultaneously acting as ligands that interact with the resident cell populations. For example, type III collagen is found within the submucosal tissue of selected organs such as the urinary bladder; a location in which tissue flexibility and compliance are required for appropriate function, as opposed to the more rigid properties required of a tendon or ligament supplied by type I collagen (Piez, 1984). Type IV collagen is present within the basement membrane of most vascular structures, and within tissues that contain an epithelial cell component (Piez, 1984; Barnard and Gathercole, 1991; Yurchenco et al., 1994). The ligand affinity of type IV collagen for endothelial cells is the reason for its use as a biocompatible coating for medical devices intended to have a blood interface. Type VI collagen is a relatively small molecule that serves as a connecting unit between glycosaminoglycans and larger structural proteins such as type I collagen, thus providing a gel-like consistency to the ECM (Yurchenco et al., 1994), and has recently been shown to sequester latent forms of matrix metalloproteinases (Freise et al., 2009). Type VII collagen is found within the basement membrane of the epidermis, and functions as an anchoring fibril to protect the overlying keratinocytes from sheer stresses (Yurchenco et al., 1994). In nature, collagen is intimately associated with glycosylated proteins, growth factors, and other structural proteins such as elastin and laminin to provide unique tissue properties (Yurchenco et al., 1994). Each of these types of collagen exists within many of the ECM scaffolds used in tissue engineering and regenerative medicine applications.

Fibronectin. Fibronectin, second only to collagen in quantity within the ECM of most tissues, was the first primarily "structural" molecule identified to also possess a functional motif. Fibronectin is a dimeric molecule of 250,000 MW subunits, and exists both in tissue and soluble isoforms and possesses ligands for adhesion of many cell types (McPherson and Badylak, 1998; Miyamoto et al., 1998; Schwarzbauer, 1991, 1999). The ECM of submucosal structures, basement membranes, and interstitial tissues all contain abundant fibronectin (McPherson and Badylak, 1998; Schwarzbauer, 1999). The cell-friendly characteristics of this protein have made it an attractive substrate for *in vitro* cell culture. Fibronectin has also been used as a coating for synthetic scaffold materials to promote host biocompatibility. Fibronectin is rich in the Arg-Gly-Asp (RGD) subunit, a tripeptide that is important in cell adhesion via the $\alpha_5\beta_1$ integrin (Yurchenco et al., 1994), and is found at an early stage within the ECM of developing embryos. Fibronectin is critical for normal biologic development, especially the development of vascular structures. The importance of this molecule and its interactions with other matrix components cannot be overstated with regard to cell–matrix communication.

Laminin. Laminin is a complex adhesion protein found in the ECM, especially within the basement membrane (Schwarzbauer, 1999). This protein plays an important role in early embryonic development, and is perhaps the best studied of the ECM proteins found within embryonic bodies (Li et al., 2002). Laminin is a trimeric cross-linked polypeptide that exists in numerous forms dependent upon the particular mixture of peptide chains (e.g., $\alpha1$, $\beta1$, $\gamma1$) (Timpl, 1996; Timpl and Brown, 1996). The prominent role of laminin in the formation and maintenance of vascular structures is particularly noteworthy when considering the ECM as a scaffold for tissue reconstruction (Ponce et al., 1999; Werb et al., 1999). The crucial role of the beta-1 integrin chain in mediating hematopoietic stem cell interactions with fibronectin and laminin is well-established (Ponce et al., 1999; Werb et al., 1999). Loss of the beta-1 integrin receptors in mice results in intrapartum mortality. This protein appears to be among the first and most critical ECM factors in the process of cell and tissue differentiation. The specific role of laminin in tissue reconstruction when ECM is used as a scaffold for tissue engineering and regenerative medicine applications is unclear; however, its importance in normal development suggests that this molecule is essential for self-assembly of cell populations, and for organized functional tissue development, as opposed to scar tissue formation.

Glycosaminoglycans. A mixture of glycosaminoglycans (GAGs) is present within native ECM, with the amount and relative distribution dependent upon the tissue location of the ECM in the host, the age of the host, and the microenvironment. These GAGs promote water retention, bind growth factors and cytokines, and contribute to the gel properties of the ECM. The heparin-binding properties of numerous cell surface receptors and of many growth factors (e.g., fibroblast growth factor family, vascular endothelial cell growth factor) make the heparin-rich GAGs important components of naturally occurring substrates for cell growth. The glycosaminoglycans present in ECM include chondroitin sulfates A and B, heparin, heparan sulfate, and hyaluronic acid (Entwistle et al., 1995; Hodde et al., 1996). Hyaluronic acid has been extensively investigated as a scaffold material for tissue reconstruction and as a carrier for selected cell populations in therapeutic tissue engineering applications (Chapter I.2.7). The concentration of hyaluronic acid within ECM is highest in fetal and newborn tissues (greater than 20% of total GAGs), and has been associated with desirable healing properties. The specific role, if any, of hyaluronic acid upon progenitor cell proliferation and differentiation during adult wound healing is unknown.

Growth Factors. A characteristic of ECM scaffolds that clearly distinguishes them from other scaffolds for tissue reconstruction is the diversity of the structural and functional proteins that contribute to its composition. The bioactive molecules that reside within the ECM and their unique spatial distribution patterns provide a reservoir of biologic signals. Although the quantity of cytokines and growth factors present within ECM is very small, these molecules act as potent modulators of cell behavior. The list of growth factors found within ECM is extensive, and includes the fibroblast growth factor (FGF) family, vascular endothelial cell growth factor (VEGF), stromal-derived growth factor (SDF-1), transforming growth factor beta (TGF-beta), keratinocyte growth factor (KGF), epithelial cell growth factor (EGF), platelet-derived growth factor (PDGF), hepatocyte growth factor (HGF), and bone morphogenetic protein (BMP), among others (Bonewald, 1999; Kagami et al., 1998; Roberts et al., 1988). These factors exist in multiple isoforms, each with a unique biologic activity. Purified forms of growth factors have been investigated in recent years as therapeutic methods of encouraging blood vessel formation (e.g., VEGF), stimulating deposition of granulation tissue (PDGF), and bone (BMP), and encouraging epithelialization of wounds (KGF). However, this therapeutic approach has been disappointing, because of the difficulty in determining optimal dose and methods of delivery, the ability to sustain and localize the growth factor release at the desired site, and the inability to turn the factor "on" and "off" as needed during the course of tissue repair. An important function of the ECM is its role as a reservoir for latent forms of many growth factors and its ability to release these factors during the process of *in vivo* degradation (Freise et al., 2009).

MECHANISMS BY WHICH ECM SCAFFOLDS FUNCTION AS INDUCTIVE TEMPLATES FOR TISSUE RECONSTRUCTION

The mechanisms by which decellularized tissue (i.e., the ECM remaining following decellularization) scaffolds support a constructive remodeling process and the formation of new functional tissue when appropriately utilized in tissue engineering and regenerative medicine applications are not fully-understood. However, several biologically important events have been clearly associated with positive outcomes. Among these events are: (1) a robust innate immune response consisting of neutrophils at early time points, changing to a predominantly mononuclear cell/macrophage cell population during the 2–3 week period following implantation (Chapter II.2.2); (2) rapid and complete scaffold degradation and the release of bioactive matricryptic peptides from parent molecules within the ECM; (3) abundant angiogenesis which accompanies the deposition of new ECM by the cells that have infiltrated the scaffold; and (4) spatial and temporal remodeling which is clearly respondent to mechanical forces. Each of these events will be discussed separately, but it should be understood that they are occurring simultaneously and are interdependent upon each other for a constructive remodeling outcome following the implantation of an ECM scaffold.

Host Cell and Immune Responses to Implanted ECM Scaffolds

The mechanisms of the host cellular and humoral response to whole organ transplantation are reasonably well-understood. Xenogeneic and allogeneic cellular antigens are recognized by the host, elicit immune activation, and cause the production of pro-inflammatory mediators with downstream cytotoxicity and transplant tissue rejection. The mechanisms of the host immune response to acellular scaffolds derived from ECM, either allogeneic or xenogeneic, are neither as well-studied nor as well-understood as whole organ and tissue transplantation. The preparation of ECM scaffolds for tissue engineering and regenerative medicine applications involves the decellularization of the tissue or organ from which the ECM is to be harvested (Gilbert et al., 2006). The removal of the cellular component produces a different type of "tissue graft" than is typically presented with autogeneic, allogeneic or xenogeneic whole organ grafts. An ECM scaffold consists primarily of the ECM constituent molecules, many of which have been found to be conserved across species (van der Rest and Garrone, 1991), thus mitigating many adverse components of the host immune response (Allman et al., 2001).

Potential Immune Activating Molecules within ECM Scaffolds

Following the removal of resident cells from the tissue of interest, the majority of the components of the remaining ECM scaffold are conserved across species and are, therefore, largely non-immunogenic. However, many ECM scaffolds have been shown to contain a number of components that are thought to induce adverse host immune- and/or rejection-type responses when present in large quantities. These components include the α-Gal epitope and DNA. The α-Gal epitope is known to cause hyperacute rejection of organ transplants (Collins et al., 1994; Cooper et al., 1993; Galili et al., 1985; Oriol et al., 1993). However, studies of α-Gal-positive ECM scaffold implantation have not shown adverse responses that can be attributed to the α-Gal epitope (Raeder et al., 2002; Daly et al., 2009). A recent study investigated the effects of the presence of the α-Gal epitope upon the remodeling of ECM scaffolds in a non-human primate model (Daly et al., 2009). The study compared the host response to ECM derived from allogeneic, xenogeneic porcine, and xenogeneic α-Gal$^{-/-}$ porcine sources. The results of the study showed that although those animals implanted with an ECM scaffold containing the α-Gal epitope exhibited an increase in serum anti-Gal antibodies, there were no adverse effects of the α-Gal epitope upon the remodeling response. Several studies have shown the presence of DNA fragments remaining within ECM scaffolds following the decellularization and sterilization processes (Roberts et al., 1988; Derwin et al., 2006; Zheng et al., 2005). A recent study examined the presence of DNA within a number of commercially available ECM scaffolds (Gilbert et al., 2008), some of which are listed in Table II.6.16.2. The results of the study showed that, although all of the products tested contained small amounts of DNA, the remnants generally consisted of fragments of less than 300 bp. Despite the presence of small amounts of both the α-Gal epitope and DNA within ECM scaffolds, adverse clinical effects have not been observed. This is likely due to the minute amounts of these components present, and the rapid degradation of the ECM scaffold. It has been shown that the presence of large amounts of cellular material within an implanted ECM scaffold lead to scar tissue formation, as opposed to the modulation of the host response towards a constructive remodeling outcome (Brown et al., 2008). Therefore, it is probable that there is a threshold amount of these components required to induce adverse effects upon the remodeling response.

Innate Immune Response to ECM Scaffolds

In general, innate immune cells (neutrophils and macrophages) are the first cells to encounter and respond to implanted biomaterials (also see Chapter II.2.2). The immediate cellular response observed following the implantation of an ECM scaffold consists almost exclusively of neutrophils, as one might expect, but there is also a significant mononuclear cell component. In the absence of large amounts of cellular debris within the scaffold, chemical cross-linking or an excess of contaminants such as endotoxin, the neutrophil infiltrate diminishes almost entirely within 72 hours and is replaced by a mononuclear cell population. This type of response, characterized by a large infiltration of innate immune cells, has been conventionally interpreted as either acute or chronic inflammation with associated negative implications. However, the presence of these cells, especially mononuclear macrophages, has been shown to be essential to the formation of the type of constructive remodeling response that has been observed following the implantation of ECM scaffolds (Brown et al., 2008, 2012; Valentin et al., 2006, 2009; Badylak et al., 2008).

A histologically similar population of neutrophils and macrophages is observed following the implantation of ECM scaffolds which either have or have not been processed using chemical cross-linking agents such as glutaraldehyde or carbodiimide; however, the tissue remodeling outcome observed following the implantation of chemically cross-linked ECM scaffolds is distinctly different than that observed with the use of non-cross-linked scaffolds (Valentin et al., 2006). The host tissue response typically observed following implantation of an acellular ECM scaffold that has not been chemically cross-linked is characterized by a dense infiltration of neutrophils at early time-points, changing to primarily mononuclear cells thereafter. This infiltrate of innate immune cells is accompanied by rapid degradation of the ECM scaffold, and replacement with organized, site-specific, functional host tissue (Valentin et al., 2006; Badylak et al., 2002; Gilbert et al., 2007c). If the scaffold has been processed using chemical cross-linking agents such as glutaraldehyde or carbodiimide, the host response is characterized by a similar presence of a large number of neutrophils and macrophages, but results in a more typical pro-inflammatory response consisting of dense fibrous tissue encapsulation and the prolonged presence of a multinucleate cell population (Valentin et al., 2006). Although histologically similar populations of neutrophils and macrophages are present in the host response to either scaffold type, studies have linked the differences observed in remodeling outcomes, in part, to differences in the phenotype of the host innate immune cells which participate in the host response to implanted ECM scaffolds (Brown et al., 2008; Badylak et al., 2008).

Mononuclear macrophages are plastic innate immune cells that are capable of changing their phenotype in response to local stimuli during the process of wound healing. The macrophages participating in the host response following implantation of a biomaterial are exposed to multiple stimuli, including cytokines and effector molecules secreted by cells (including other macrophages) participating in the host response, microbial agents, epitopes associated with the implanted biomaterial, and the degradation products (if any) of the biomaterial, among others. Recently, macrophages have been characterized by differential receptor expression, cytokine, and effector molecule production, and function as having either an M1 or M2 phenotype (Mantovani et al., 2004; Mills et al., 2000). M1 or classically activated macrophages, produce pro-inflammatory cytokines and effector molecules, possess the ability to efficiently present antigen, and are inducer and effector cells in Th1-type inflammatory responses (Gordon and Taylor, 2005; Mosser, 2003; Verreck et al., 2006; Mantovani et al., 2005). M2 or alternatively activated macrophages, are generally characterized by minimal production of pro-inflammatory cytokines, their involvement in Th2-type responses, and their ability to facilitate tissue repair and regeneration. The overall M1/M2 profile of the host macrophage response to an implanted material will fall somewhere along a continuum between the M1 and M2 extremes (Mills et al., 2000). It should be noted that M1 and M2 macrophages are indistinguishable by the routine histologic methods that are generally used to evaluate the host response to a biomaterial. A full description of the activating signals, characteristic surface markers, gene expression, effector molecule production, and biologic activity associated with M1 and M2 macrophages is beyond the scope of this chapter; however, this subject has been reviewed extensively elsewhere (Martinez et al., 2008).

Recent studies of the M1/M2 profile of the macrophages responding to implanted ECM scaffolds have shown that acellular, non-cross-linked ECM scaffolds elicit an enhanced M2-type macrophage response, and result in constructive tissue remodeling (Brown et al., 2008, 2012; Badylak et al., 2008). Chemically cross-linked ECM scaffolds, however, elicit a predominantly M1-type macrophage response and result in a more typical foreign-body-type of response that includes the deposition of dense collagenous connective tissue, and a lack of constructive remodeling. Autograft controls also exhibited a predominance of the M1 phenotype, and resulted in scarring. The exact mechanisms by which acellular non-cross-linked ECM scaffolds are capable of modulating the default host macrophage response are, as of yet, unknown. However, it is increasingly clear that the M1/M2 polarization profile of the macrophages that participate in the host response to ECM scaffolds is related to the downstream outcome associated with their implantation (Brown et al., 2012). Further, characterization and control of the M1/M2 phenotype may provide a tool by which a constructive and functional tissue remodeling outcome can be predicted and/or promoted.

T-Cell-Mediated Immune Response to ECM Scaffolds

In addition to eliciting a robust, but friendly, host innate immune response, acellular non-cross-linked ECM scaffolds have consistently been shown to evoke a Th2-type T-cell response (Allman et al., 2001, 2002). The Th2 response is generally associated with transplant acceptance. One study utilized a mouse model of subcutaneous implantation to examine the T-cell response to xenogeneic muscle tissue, syngeneic muscle tissue, and an acellular ECM scaffold (Allman et al., 2001). Results showed that the xenogeneic tissue implant was associated with a response consistent with rejection. That is, the xenogeneic muscle implant showed signs of necrosis, granuloma formation, and encapsulation. The syngeneic tissue and the ECM scaffold elicited an acute inflammatory response that resolved with time, and resulted in an organized tissue morphology at the remodeling site. Tissue cytokine analysis revealed that the ECM group elicited expression of IL-4, and suppressed the expression of IFN-γ compared to the xenogeneic tissue implants. The ECM group elicited the production of an ECM-specific antibody response; however, it was restricted to the IgG1 isotype. Reimplantation of the mice with another ECM scaffold led to a secondary anti-ECM antibody response that was also restricted to the IgG1 isotype, and there was no evidence of the formation of a Th1-type response. Further investigation confirmed that the observed responses were in fact T-cell-dependent. Finally, it has been shown that, while both T- and B-cells respond to ECM scaffolds, they are not required for acceptance or constructive remodeling of an ECM implant (Allman et al., 2001). This further indicates the importance of the host innate immune response in driving/determining the downstream remodeling outcome following implantation of an ECM scaffold.

Degradation of ECM Scaffolds. ECM scaffolds are rapidly degraded *in vivo*. A recent study showed that 10 layer ^{14}C-labeled ECM scaffolds were 60% degraded at 30 days post-implantation and 100% at 90 days post-surgery in a model of canine Achilles tendon repair (Gilbert et al., 2007c). During this period, the scaffold was populated and degraded by host cells, and resulted in the formation of site-specific functional host tissue. The major mechanism of excretion of the degraded scaffold was found to be via hematogenous circulation and elimination by the kidneys, urine, and exhaled CO_2 (Record et al., 2001). The mechanisms of *in vivo* degradation of ECM scaffolds are complex, and include both cellular and enzymatic pathways. The process is mediated by inflammatory cells, such as macrophages, which produce oxidants as well as proteolytic enzymes that aid in the degradation of the matrix. Another study utilizing ^{14}C-labeled ECM scaffolds showed that peripheral blood monocytes are required for the early and rapid degradation of both ECM scaffolds, and that cross-linked ECM scaffolds are resistant to macrophage-mediated degradation (Valentin et al., 2009).

ECM scaffolds have also been degraded *in vitro* by chemical and physical methods. Recent findings suggest that the degradation products of ECM scaffolds are bioactive (Brennan et al., 2006, 2008; Haviv et al., 2005; Li et al., 2004; Reing et al., 2009; Sarikaya et al., 2002). Studies have shown antimicrobial activity associated with the degradation products of ECM scaffolds; however, in the absence of degradation, antimicrobial activity was not seen, suggesting that some of the bioactive properties of the ECM are derived from its degradation products, rather than from whole molecules present in the ECM (Holtom et al., 2004). Degradation products of ECM scaffolds have also been shown to be chemoattractants for progenitor and non-progenitor cell populations (Brennan et al., 2008; Haviv et al., 2005; Li et al., 2004; Reing et al., 2009). An ECM scaffold that cannot degrade (i.e., is chemically cross-linked) may not release bioactive degradation products, including those bioactive molecules that may be responsible for modulating the host response towards constructive remodeling. Furthermore, chemical cross-linking may alter ligand–receptor interactions important in determining cell–scaffold interactions.

One of the biologic effects of ECM degradation is the recruitment of host stem and progenitor cells to the site of degradation. A study of ECM scaffold remodeling in a mouse Achilles tendon model examined the ability of ECM scaffolds and autograft tissue to recruit bone marrow-derived cells (Zantop et al., 2006). Bone marrow-derived cells were observed in the sites of remodeling associated with both ECM scaffolds and autograft control tissue, among what appeared to be predominantly mononuclear cells at early time points (1 and 2 weeks) post-surgery. Both scaffold types remodeled into tissue resembling the native Achilles tendon; however, by 16 weeks the presence of bone marrow-derived cells was observed only in the ECM treated group. Another study, also utilizing a model of mouse Achilles tendon, examined the ability of ECM scaffold explants to cause the chemotaxis of progenitor cells after 3, 7, and 14 days of *in vivo* remodeling (Beattie et al., 2009). The results of the study showed greater migration of progenitor cells towards tendons repaired with ECM scaffolds, compared to tendons repaired with autologous tissue and uninjured normal contralateral tendon. The mechanisms by which ECM scaffolds recruit progenitor cell populations during *in vivo* remodeling and the specific phenotype of the cells recruited remains unknown, and the effects of the recruited cells upon the scaffold remodeling process is also not well-understood.

Angiogenesis and New ECM Deposition

Angiogenesis is implicit to the success of many tissue engineering and regenerative medicine strategies involving biomaterials. The in-growth of vessels into a tissue-engineered construct provides a means for nourishing the tissue growing on or within the implanted material. In the absence of angogenesis, many implanted biomaterials may fail to integrate with the surrounding tissue, and/or fail to support cell populations that have been seeded onto or migrate into the material. The process of angiogenesis within the native extracellular matrix of tissues and organs has been studied in-depth. The mechanisms by which ECM scaffolds facilitate angiogenesis are not clear. Angiogenesis has, however, been shown to be a prominent feature of ECM scaffold remodeling, and is commonly observed within the first 1–3 days following scaffold implantation. These vessels remain within the implantation site, and continue to provide a blood supply to the remodeling tissue throughout the remodeling process. It has been shown that certain ECM scaffolds contain bioactive VEGF (a potent angiogenic factor) and basic fibroblast growth factor (bFGF) (Chun et al., 2007). Factors such as these may account, in part, for the ability of ECM scaffolds to promote angiogenesis at early time points following implantation.

Response to Mechanical Stimuli

Several studies have shown that early site-appropriate mechanical loading facilitates the remodeling of ECM scaffold materials into site-specific tissue (Boruch et al., 2009; Hodde et al., 1997). In a recent pre-clinical study, partial cystectomies repaired with an ECM scaffold material were exposed to long-term catheterization and prevention of bladder filling, with an associated lack of cyclic distention and decrease in maximal bladder distention, compared to bladders that experienced an early return to normal micturition following ECM scaffold implantation (Boruch et al., 2009). The presence of physiologic amounts of mechanical loading in the early timeframe promoted remodeling of the ECM scaffold material into tissue that had a highly differentiated transitional urothelium, vasculature, innervation, and islands of smooth muscle cells. Delayed return of normal mechanical loading was insufficient to overcome the lack of early mechanical signals, and resulted in degradation of the ECM scaffold material without constructive remodeling.

Similar results were observed in a study of Achilles tendon repair in rabbits, with and without post-surgical immobilization (Hodde et al., 1997). With early mechanical loading, the SIS was rapidly infiltrated by mononuclear cells and new, highly aligned host collagenous connective tissue eventually replaced the ECM scaffold. In contrast, after a period of immobilization, cell infiltration into the scaffold was limited to the periphery of the graft, and little or no ECM remodeling was observed. The lack of cellular infiltration in the immobilized Achilles tendon differed from the abundant mononuclear cell population observed in the early catherization group in the study described above, possibly due to the inadvertent pressure on the bladder from surrounding tissues. It is possible that non-specific mechanical loading may have contributed to the cell infiltration, but site-appropriate mechanical loading is necessary for site-specific remodeling.

When ECM was used for vascular reconstruction, the host remodeling response was different in the arterial system than in the venous system (Lantz et al., 1993; Sandusky Jr. et al., 1992). In the arterial system, a smooth muscle layer developed in the adventitia of the remodeling ECM, while the remodeling response after repair of the superior vena cava led to the formation of primarily collagenous tissue (Lantz et al., 1993; Sandusky Jr. et al., 1992). These differences in the remodeling response are thought to be due to the different mechanical environments present in the high pressure arterial circulation and low pressure venous circulation.

The mechanisms by which ECM scaffolds promote site-specific remodeling in the presence of site-appropriate mechanical loading have been partially elucidated. Static and cyclic stretching of cells seeded on an ECM scaffold *in vitro* modulated the collagen expression by the cells, enhanced the cell and collagen alignment, and improved in the mechanical behavior of the scaffold (Almarza et al., 2008; Androjna et al., 2007; Gilbert et al., 2007b; Nguyen et al., 2009; Wallis et al., 2008). In addition, during *in vivo* remodeling, it is thought that the progenitor cells that are recruited to the site of ECM remodeling by matricryptic peptides formed during scaffold degradation will differentiate into site-appropriate cells in the presence of local mechanical cues (Reing et al., 2009; Zantop et al., 2006; Badylak et al., 2001; Beattie et al., 2009; Crisan et al., 2008). Several *in vitro* studies have shown that mechanical loading can induce progenitor cells to differentiate into fibroblasts, smooth muscle cells, and osteoblasts (Altman et al., 2002; Matziolis et al., 2006; Nieponice et al., 2007).

CONCLUSION

ECM scaffolds have been isolated from a number of tissues and organs for the purpose of tissue engineering and regenerative medicine. The methods by which these scaffolds are derived are highly dependent on the structure of the organ of interest, and may include both chemical and physical methods. Decellularization processes should be designed such that the majority of the antigenic cellular components are removed while attempting to maintain the highly complex, tissue-specific combination of structural and functional molecules present within the ECM. Following decellularization, ECM scaffolds can be processed into application-specific configurations. These scaffolds have been used successfully in a wide variety of clinical and preclinical settings. The success of ECM scaffolds in these applications is due, in large part, to their ability to modulate the default mammalian host response from scar tissue formation to constructive remodeling. The ability of an ECM scaffold to promote constructive remodeling has been related to a number of biologic events that occur during the remodeling process. Among these are the ability of the scaffold to elicit a friendly host innate immune response, degrade rapidly and fully while releasing bioactive peptides from ECM parent molecules, promote angiogenesis at early time-points following implantation, and respond to the mechanical microenvironment in which it is placed. These processes occur concurrently through the course of scaffold remodeling, and are likely interrelated. Failure of an ECM scaffold to undergo any one of these processes may result in degradation of the scaffold without constructive remodeling, fibrous encapsulation of the implanted material and a lack of scaffold degradation or the formation of scar tissue within the implant site.

BIBLIOGRAPHY

Agrawal, V., Brown, B. N., Beattie, A. J., Gilbert, T. W., & Badylak, S. F. (2009). Evidence of innervation following extracellular matrix scaffold-mediated remodelling of muscular tissues. *J. Tissue Eng. Regen. Med.*

Agrawal, V., Johnson, S. A., Reing, J., Zhang, L., Tottey, S., Wang, G., Hirschi, K. K., Braunhut, S., Gudas, L. J., & Badylak, S. F. (2010). Epimorphic regeneration approach to tissue replacement in adult mammals. *Proc. Natl. Acad. Sci. U S A*, **107**(8), 3351–3355.

Agrawal, V., Tottey, S., Johnson, S. A., Freund, J. M., Siu, B. F., & Badylak, S. F. (2011a). Recruitment of progenitor cells by an extracellular matrix cryptic peptide in a mouse model of digit amputation. *Tissue Eng. Part A*, **17**(19-20), 2435–2443.

Agrawal, V., Kelly, J., Tottey, S., Daly, K. A., Johnson, S. A., Siu, B. F., Reing, J., & Badylak, S. F. (2011b). An isolated cryptic peptide influences osteogenesis and bone remodeling in an adult mammalian model of digit amputation. *Tissue Eng. Part A*, **17**(23-24), 3033–3044.

Alini, M., Li, W., Markovic, P., Aebi, M., Spiro, R. C., & Roughley, P. J. (2003). The potential and limitations of a cell-seeded collagen/hyaluronan scaffold to engineer an intervertebral disc-like matrix. *Spine*, **28**(5), 446–454. Discussion 53.

Allman, A. J., McPherson, T. B., Badylak, S. F., Merrill, L. C., Kallakury, B., Sheehan, C., et al. (2001). Xenogeneic extracellular matrix grafts elicit a TH2-restricted immune response. *Transplantation*, **71**(11), 1631–1640.

Allman, A. J., McPherson, T. B., Merrill, L. C., Badylak, S. F., & Metzger, D. W. (2002). The Th2-restricted immune response to xenogeneic small intestinal submucosa does not influence systemic protective immunity to viral and bacterial pathogens. *Tissue Eng.*, **8**(1), 53–62.

Almarza, A. J., Yang, G., Woo, S. L., Nguyen, T., & Abramowitch, S. D. (2008). Positive changes in bone marrow-derived cells in response to culture on an aligned bioscaffold. *Tissue Eng. Part A.*, **14**(9), 1489–1495.

Altman, G. H., Horan, R. L., Martin, I., Farhadi, J., Stark, P. R., Volloch, V., et al. (2002). Cell differentiation by mechanical stress. *FASEB J.*, **16**(2), 270–272.

Androjna, C., Spragg, R. K., & Derwin, K. A. (2007). Mechanical conditioning of cell-seeded small intestine submucosa: A potential tissue-engineering strategy for tendon repair. *Tissue Eng.*, **13**(2), 233–243.

Armour, A. D., Fish, J. S., Woodhouse, K. A., & Semple, J. L. (2006). A comparison of human and porcine acellularized dermis: Interactions with human fibroblasts in vitro. *Plast. Reconstr. Surg.*, **117**(3), 845–856.

Bader, A., Schilling, T., Teebken, O. E., Brandes, G., Herden, T., Steinhoff, G., et al. (1998). Tissue engineering of heart valves–human endothelial cell seeding of detergent acellularized porcine valves. *Eur. J. Cardiothorac. Surg.*, **14**(3), 279–284.

Badylak, S. F., Lantz, G. C., Coffey, A., & Geddes, L. A. (1989). Small intestinal submucosa as a large diameter vascular graft in the dog. *J. Surg. Res.*, **47**(1), 74–80.

Badylak, S., Meurling, S., Chen, M., Spievack, A., & Simmons-Byrd, A. (2000). Resorbable bioscaffold for esophageal repair in a dog model. *J. Pediatr. Surg.*, **35**(7), 1097–1103.

Badylak, S. F., Park, K., Peppas, N., McCabe, G., & Yoder, M. (2001). Marrow-derived cells populate scaffolds composed of xenogeneic extracellular matrix. *Exp. Hematol.*, **29**(11), 1310–1318.

Badylak, S., Kokini, K., Tullius, B., Simmons-Byrd, A., & Morff, R. (2002). Morphologic study of small intestinal submucosa as a body wall repair device. *J. Surg. Res.*, **103**(2), 190–202.

Badylak, S. F., Vorp, D. A., Spievack, A. R., Simmons-Byrd, A., Hanke, J., Freytes, D. O., et al. (2005). Esophageal reconstruction with ECM and muscle tissue in a dog model. *J. Surg. Res.*, **128**(1), 87–97.

Badylak, S. F., Kochupura, P. V., Cohen, I. S., Doronin, S. V., Saltman, A. E., Gilbert, T. W., et al. (2006). The use of extracellular matrix as an inductive scaffold for the partial replacement of functional myocardium. *Cell Transplant.*, **15**(Suppl. 1), S29–S40.

Badylak, S. F., Valentin, J. E., Ravindra, A. K., McCabe, G. P., & Stewart-Akers, A. M. (2008). Macrophage phenotype as a determinant of biologic scaffold remodeling. *Tissue Eng.*, **14**(11), 1835–1842.

Badylak, S. F., Freytes, D. O., & Gilbert, T. W. (2009). Extracellular matrix as a biological scaffold material: Structure and function. *Acta Biomater.*, **5**(1), 1–13.

Badylak, S. F., Weiss, D. J., Caplan, A., & Macchiarini, P. (2012). Engineered whole organs and complex tissues. *Lancet*, **379**(9819), 943–952.

Baldwin, H. S. (1996). Early embryonic vascular development. *Cardiovasc. Res.*, 31 Spec. No: E34–45.

Barnard, K., & Gathercole, L. J. (1991). Short and long range order in basement membrane type IV collagen revealed by enzymic and chemical extraction. *Int. J. Biol. Macromol.*, **13**(6), 359–365.

Beattie, A. J., Gilbert, T. W., Guyot, J. P., Yates, A. J., & Badylak, S. F. (2009). Chemoattraction of progenitor cells by remodeling extracellular matrix scaffolds. *Tissue Eng. Part A.*, **15**(5), 1119–1125.

Bello, Y. M., Falabella, A. F., & Eaglstein, W. H. (2001). Tissue-engineered skin. Current status in wound healing. *Am. J. Clin. Dermatol.*, **2**(5), 305–313.

Bernard MPC, M. L., Myers, J. C., Ramirez, F., Eikenberry, E. F., & Prockop, D. J. (1983). Nucleotide sequences of complementary deoxyribonucleic acids for the pro alpha 1 chain of human type I procollagen. Statistical evaluation of structures that are conserved during evolution. *Biochemistry*, **22**, 5213–5223.

Bernard MPM, J. C., Chu, M. L., Ramirez, F., Eikenberry, E. F., & Prockop, D. J. (1983). Structure of a cDNA for the pro alpha 2 chain of human type I procollagen. Comparison with chick cDNA for pro alpha 2(I) identifies structurally conserved features of the protein and the gene. *Biochemistry*, **22**, 1139–1145.

Bhrany, A. D., Beckstead, B. L., Lang, T. C., Farwell, D. G., Giachelli, C. M., & Ratner, B. D. (2006). Development of an esophagus acellular matrix tissue scaffold. *Tissue Eng.*, **12**(2), 319–330.

Billiar, K., Murray, J., Laude, D., Abraham, G., & Bachrach, N. (2001). Effects of carbodiimide crosslinking conditions on the physical properties of laminated intestinal submucosa. *J. Biomed. Mater. Res.*, **56**(1), 101–108.

Bissell, M. J., Hall, H. G., & Parry, G. (1982). How does the extracellular matrix direct gene expression? *J. Theor. Biol.*, **99**(1), 31–68.

Black, A. F., Berthod, F., L'Heureux, N., Germain, L., & Auger, F. A. (1998). In vitro reconstruction of a human capillary-like network in a tissue-engineered skin equivalent. *Faseb J.*, **12**(13), 1331–1340.

Bonewald, L. F. (1999). Regulation and regulatory activities of transforming growth factor beta. *Crit. Rev. Eukaryot. Gene. Expr.*, **9**(1), 33–44.

Borschel, G. H., Dennis, R. G., & Kuzon, W. M., Jr. (2004). Contractile skeletal muscle tissue-engineered on an acellular scaffold. *Plast Reconstr. Surg.*, **113**(2), 595–602. Discussion 603–594.

Boruch, A. V., Nieponice, A., Qureshi, I. R., Gilbert, T. W., & Badylak, S. F. (2009). Constructive remodeling of biologic scaffolds is dependent on early exposure to physiologic bladder filling in a canine partial cystectomy model. *J. Surg. Res.*

Boudreau, N., Myers, C., & Bissell, M. J. (1995). From laminin to lamin: Regulation of tissue-specific gene expression by the ECM. *Trends Cell Biol.*, **5**(1), 1–4.

Brennan, E. P., Reing, J., Chew, D., Myers-Irvin, J. M., Young, E. J., & Badylak, S. F. (2006). Antibacterial activity within degradation products of biological scaffolds composed of extracellular matrix. *Tissue Eng.*, **12**(10), 2949–2955.

Brennan, E. P., Tang, X. H., Stewart-Akers, A. M., Gudas, L. J., & Badylak, S. F. (2008). Chemoattractant activity of degradation products of fetal and adult skin extracellular matrix for keratinocyte progenitor cells. *J. Tissue Eng. Regen. Med.*, **2**(8), 491–498.

Brown, B. N., Londono, R., Tottey, S., Zhang, L., Kukla, K. A., Wolf, M. T., Daly, K. A., Reing, J. E., & Badylak, S. F. (2012). Macrophage phenotype as a predictor of constructive remodeling following the implantation of biologically derived surgical mesh materials. *Acta. Biomater.*, **8**(3), 978–987.

Buinewicz, B., & Rosen, B. (2004). Acellular cadaveric dermis (AlloDerm): A new alternative for abdominal hernia repair. *Ann. Plast. Surg.*, **52**(2), 188–194.

Brown, B. N., Valentin, J. E., Stewart-Akers, A. M., McCabe, G. P., & Badylak, S. F. (2008). Macrophage phenotype and remodeling outcomes in response to biologic scaffolds with and without a cellular component. *Biomaterials*, **30**(8), 1482–1491.

Cartmell, J. S., & Dunn, M. G. (2004). Development of cell-seeded patellar tendon allografts for anterior cruciate ligament reconstruction. *Tissue Eng.*, **10**(7-8), 1065–1075.

Cartwright, L. M., Shou, Z., Yeger, H., & Farhat, W. A. (2006). Porcine bladder acellular matrix porosity: Impact of hyaluronic acid and lyophilization. *J. Biomed. Mater. Res. A.*, **77**(1), 180–184.

Causa, F., Netti, P. A., & Ambrosio, L. (2007). A multi-functional scaffold for tissue regeneration: The need to engineer a tissue analogue. *Biomaterials*, **28**(34), 5093–5099.

Chen, F., Yoo, J. J., & Atala, A. (1999). Acellular collagen matrix as a possible "off the shelf" biomaterial for urethral repair. *Urology*, **54**(3), 407–410.

Chun, S. Y., Lim, G. J., Kwon, T. G., Kwak, E. K., Kim, B. W., Atala, A., et al. (2007). Identification and characterization of bioactive factors in bladder submucosa matrix. *Biomaterials*, **28**(29), 4251–4256.

Clarke, K. M., Lantz, G. C., Salisbury, S. K., Badylak, S. F., Hiles, M. C., & Voytik, S. L. (1996). Intestine submucosa and polypropylene mesh for abdominal wall repair in dogs. *J. Surg. Res.*, **60**(1), 107–114.

Cobb, M. A., Badylak, S. F., Janas, W., & Boop, F. A. (1996). Histology after dural grafting with small intestinal submucosa. *Surg. Neurol.*, **46**(4), 389–393. Discussion 393–384.

Cobb, M. A., Badylak, S. F., Janas, W., Simmons-Byrd, A., & Boop, F. A. (1999). Porcine small intestinal submucosa as a dural substitute. *Surg. Neurol.*, **51**(1), 99–104.

Collins, B. H., Chari, R. S., Magee, J. C., Harland, R. C., Lindman, B. J., Logan, J. S., et al. (1994). Mechanisms of injury in porcine livers perfused with blood of patients with fulminant hepatic failure. *Transplantation*, **58**(11), 1162–1171.

Constantinou CDJ, S. A. (1991). Structure of cDNAs encoding the triple-helical domain of murine alpha 2 (VI) collagen chain and comparison to human and chick homologues. Use of polymerase chain reaction and partially degenerate oligonucleotide for generation of novel cDNA clones. *Matrix*, **11**, 1–9.

Cooper, D. K., Good, A. H., Koren, E., Oriol, R., Malcolm, A. J., Ippolito, R. M., et al. (1993). Identification of alpha-galactosyl and other carbohydrate epitopes that are bound by human anti-pig antibodies: Relevance to discordant xenografting in man. *Transpl. Immunol.*, **1**(3), 198–205.

Crapo, P. M., Gilbert, T. W., & Badylak, S. F. (2011). An overview of tissue and whole organ decellularization processes. *Biomaterials*, **32**(12), 3233–3243.

Crisan, M., Yap, S., Casteilla, L., Chen, C. W., Corselli, M., Park, T. S., et al. (2008). A perivascular origin for mesenchymal stem cells in multiple human organs. *Cell Stem Cell.*, **3**(3), 301–313.

Daly, K., Stewart-Akers, A., Hara, H., Ezzelarab, M., Long, C., Cordero, K., et al. (2009). Effect of the alphaGal epitope on the response to small intestinal submucosa extracellular matrix in a nonhuman primate model. *Tissue Eng. Part A.*, **15**(12), 3877–3888.

Dejana, E., Colella, S., Languino, L. R., Balconi, G., Corbascio, G. C., & Marchisio, P. C. (1987). Fibrinogen induces adhesion, spreading, and microfilament organization of human endothelial cells *in vitro*. *J. Cell Biol.*, **104**(5), 1403–1411.

Dejardin, L. M., Arnoczky, S. P., Ewers, B. J., Haut, R. C., & Clarke, R. B. (2001). Tissue-engineered rotator cuff tendon using porcine small intestine submucosa. Histologic and mechanical evaluation in dogs. *Am. J. Sports Med.*, **29**(2), 175–184.

Derwin, K. A., Baker, A. R., Spragg, R. K., Leigh, D. R., & Iannotti, J. P. (2006). Commercial extracellular matrix scaffolds for rotator cuff tendon repair. Biomechanical, biochemical, and cellular properties. *J. Bone Joint Surg. Am.*, **88**(12), 2665–2672.

Entwistle, J., Zhang, S., Yang, B., Wong, C., Li, Q., Hall, C. L., et al. (1995). Characterization of the murine gene encoding the hyaluronan receptor RHAMM. *Gene*, **163**(2), 233–238.

Erdag, G., & Morgan, J. R. (2004). Allogeneic versus xenogeneic immune reaction to bioengineered skin grafts. *Cell Transplant*, **13**(6), 701–712.

Exposito JYDA, M., Solursh, M., & Ramirez, F. (1992). Sea urchin collagen evolutionarily homologous to vertebrate pro-alpha 2(I) collagen. *J. Biol. Chem.*, **267**, 15559–15562.

Farhat, W. A., Chen, J., Haig, J., Antoon, R., Litman, J., Sherman, C., et al. (2008). Porcine bladder acellular matrix (ACM): Protein expression, mechanical properties. *Biomed. Mater.*, **3**(2), 25015.

Freise, C., Erben, U., Muche, M., Farndale, R., Zeitz, M., Somasundaram, R., et al. (2009). The alpha 2 chain of collagen type VI sequesters latent proforms of matrix-metalloproteinases and modulates their activation and activity. *Matrix Biol.*, **28**(8), 480–489.

Freytes, D. O., Badylak, S. F., Webster, T. J., Geddes, L. A., & Rundell, A. E. (2004). Biaxial strength of multilaminated extracellular matrix scaffolds. *Biomaterials*, **25**(12), 2353–2361.

Freytes, D. O., Martin, J., Velankar, S. S., Lee, A. S., & Badylak, S. F. (2008). Preparation and rheological characterization of a gel form of the porcine urinary bladder matrix. *Biomaterials*, **29**(11), 1630–1637.

Galili, U., Macher, B. A., Buehler, J., & Shohet, S. B. (1985). Human natural anti-alpha-galactosyl IgG. II. The specific recognition of alpha (1–3)-linked galactose residues. *J. Exp. Med.*, **162**(2), 573–582.

Gilbert, T. W., Stolz, D. B., Biancaniello, F., Simmons-Byrd, A., & Badylak, S. F. (2005). Production and characterization of ECM powder: Implications for tissue engineering applications. *Biomaterials*, **26**(12), 1431–1435.

Gilbert, T. W., Sellaro, T. L., & Badylak, S. F. (2006). Decellularization of tissues and organs. *Biomaterials*, **27**(19), 3675–3683.

Gilbert, T. W., Stewart-Akers, A. M., Simmons-Byrd, A., & Badylak, S. F. (2007). Degradation and remodeling of small intestinal submucosa in canine Achilles tendon repair. *J. Bone Joint Surg. Am.*, **89**(3), 621–630.

Gilbert, T. W., Stewart-Akers, A. M., & Badylak, S. F. (2007a). A quantitative method for evaluating the degradation of biologic scaffold materials. *Biomaterials*, **28**(2), 147–150.

Gilbert, T. W., Stewart-Akers, A. M., Sydeski, J., Nguyen, T. D., Badylak, S. F., & Woo SL-, Y. (2007b). Gene expression by fibroblasts seeded on small intestinal submucosa and subjected to cyclic stretching. *Tissue Eng.*, **13**(6), 1313–1323.

Gilbert, T. W., Stewart-Akers, A. M., Simmons-Byrd, A., & Badylak, S. F. (2007c). Degradation and remodeling of small intestinal submucosa in canine Achilles tendon repair. *J. Bone Joint Surg. Am.*, **89**(3), 621–630.

Gilbert, T. W., Gilbert, S., Madden, M., Reynolds, S. D., & Badylak, S. F. (2008). Morphologic assessment of extracellular matrix scaffolds for patch tracheoplasty in a canine model. *Ann. Thorac. Surg.*, **86**(3), 967–974. discussion 967–974.

Gilbert, T. W., Nieponice, A., Spievack, A. R., Holcomb, J., Gilbert, S., & Badylak, S. F. (2008). Repair of the thoracic wall with an extracellular matrix scaffold in a canine model. *J. Surg. Res.*, **147**(1), 61–67.

Gilbert, T. W., Agrawal, V., Gilbert, M. R., Povirk, K. M., Badylak, S. F., & Rosen, C. A. (2009). Liver-derived extracellular matrix as a biologic scaffold for acute vocal fold repair in a canine model. *Laryngoscope*, **119**(9), 1856–1863.

Gilbert, T. W., Freund, J. M., & Badylak, S. F. (2009). Quantification of DNA in biologic scaffold materials. *J. Surg. Res.*, **152**(1), 135–139.

Glowacki, J., & Mizuno, S. (2008). Collagen scaffolds for tissue engineering. *Biopolymers*, **89**(5), 338–344.

Gock, H., Murray-Segal, L., Salvaris, E., Cowan, P., & D'Apice, A. J. (2004). Allogeneic sensitization is more effective than xenogeneic sensitization in eliciting Gal-mediated skin graft rejection. *Transplantation*, **77**(5), 751–753.

Gordon, S., & Taylor, P. R. (2005). Monocyte and macrophage heterogeneity. *Nat. Rev. Immunol.*, **5**(12), 953–964.

Hammond, J. S., Gilbert, T. W., Howard, D., Zaitoun, A., Michalopoulos, G., Shakesheff, K. M., Beckingham, I. J., & Badylak, S. F. (2011). Scaffolds containing growth factors and extracellular matrix induce hepatocyte proliferation and cell migration in normal and regenerating rat liver. *J. Hepatol.*, **54**(2), 279–287.

Harper, C. (2001). Permacol: Clinical experience with a new biomaterial. *Hosp. Med.*, **62**(2), 90–95.

Haviv, F., Bradley, M. F., Kalvin, D. M., Schneider, A. J., Davidson, D. J., Majest, S. M., et al. (2005). Thrombospondin-1 mimetic peptide inhibitors of angiogenesis and tumor growth: Design, synthesis, and optimization of pharmacokinetics and biological activities. *J. Med. Chem.*, **48**(8), 2838–2846.

Hodde, J. P., Badylak, S. F., Brightman, A. O., & Voytik-Harbin, S. L. (1996). Glycosaminoglycan content of small intestinal submucosa: A bioscaffold for tissue replacement. *Tissue Eng.*, 2(3), 209–217.

Hodde, J. P., Badylak, S. F., & Shelbourne, K. D. (1997). The effect of range of motion on remodeling of small intestinal submucosa (SIS) when used as an Achilles tendon repair material in the rabbit. *Tissue Eng.*, 3(1), 27–37.

Hodde, J. P., Record, R. D., Tullius, R. S., & Badylak, S. F. (2002). Retention of endothelial cell adherence to porcine-derived extracellular matrix after disinfection and sterilization. *Tissue Eng.*, **8**(2), 225–234.

Holtom, P. D., Shinar, Z., Benna, J., & Patzakis, M. J. (2004). Porcine small intestine submucosa does not show antimicrobial properties. *Clin. Orthop. Relat. Res.* (427), 18–21.

Huber, J. E., Spievack, A., Simmons-Byrd, A., Ringel, R. L., & Badylak, S. (2003). Extracellular matrix as a scaffold for laryngeal reconstruction. *Ann. Otol. Rhinol. Laryngol.*, **112**(5), 428–433.

Hudson, T. W., Liu, S. Y., & Schmidt, C. E. (2004). Engineering an improved acellular nerve graft via optimized chemical processing. *Tissue Eng.*, **10**(9-10), 1346–1358.

Ingber, D. (1991). Extracellular matrix and cell shape: Potential control points for inhibition of angiogenesis. *J. Cell. Biochem.*, **47**(3), 236–241.

Kagami, S., Kondo, S., Loster, K., Reutter, W., Urushihara, M., Kitamura, A., et al. (1998). Collagen type I modulates the platelet-derived growth factor (PDGF) regulation of the growth and expression of beta1 integrins by rat mesangial cells. *Biochem. Biophys. Res. Commun.*, **252**(3), 728–732.

Kim, B. S., Yoo, J. J., & Atala, A. (2004). Peripheral nerve regeneration using acellular nerve grafts. *J. Biomed. Mater. Res. A.*, **68**(2), 201–209.

Koh, H. S., Yong, T., Chan, C. K., & Ramakrishna, S. (2008). Enhancement of neurite outgrowth using nano-structured scaffolds coupled with laminin. *Biomaterials*, **29**(26), 3574–3582.

Lantz, G. C., Badylak, S. F., Coffey, A. C., Geddes, L. A., & Blevins, W. E. (1990). Small intestinal submucosa as a small-diameter arterial graft in the dog. *J. Invest. Surg.*, **3**(3), 217–227.

Lantz, G. C., Badylak, S. F., Hiles, M. C., Coffey, A. C., Geddes, L. A., Kokini, K., et al. (1993). Small intestinal submucosa as a vascular graft: A review. *J. Invest Surg.*, **6**(3), 297–310.

Laurie, G. W., Horikoshi, S., Killen, P. D., Segui-Real, B., & Yamada, Y. (1989). *In situ* hybridization reveals temporal and spatial changes in cellular expression of mRNA for a laminin receptor, laminin, and basement membrane (type IV) collagen in the developing kidney. *J. Cell Biol.*, **109**(3), 1351–1362.

Li, F., Li, W., Johnson, S., Ingram, D., Yoder, M., & Badylak, S. (2004). Low-molecular-weight peptides derived from extracellular matrix as chemoattractants for primary endothelial cells. *Endothelium*, **11**(3-4), 199–206.

Li, S., Harrison, D., Carbonetto, S., Fassler, R., Smyth, N., Edgar, D., et al. (2002). Matrix assembly, regulation, and survival functions of laminin and its receptors in embryonic stem cell differentiation. *J. Cell Biol.*, **157**(7), 1279–1290.

Lin, P., Chan, W. C., Badylak, S. F., & Bhatia, S. N. (2004). Assessing porcine liver-derived biomatrix for hepatic tissue engineering. *Tissue Eng.*, **10**(7-8), 1046–1053.

Lin, P., Chan, W. C., Badylak, S. F., & Bhatia, S. N. (2004). Assessing porcine liver-derived biomatrix for hepatic tissue engineering. *Tissue Eng.*, **10**(7-8), 1046–1053.

Lovekamp, J. J., Simionescu, D. T., Mercuri, J. J., Zubiate, B., Sacks, M. S., & Vyavahare, N. R. (2006). Stability and function of glycosaminoglycans in porcine bioprosthetic heart valves. *Biomaterials*, **27**(8), 1507–1518.

Macarak, E. J., & Howard, P. S. (1983). Adhesion of endothelial cells to extracellular matrix proteins. *J. Cell Physiol.*, **116**(1), 76–86.

Macchiarini, P., Jungebluth, P., Go, T., Asnaghi, M. A., Rees, L. E., Cogan, T. A., et al. (2008). Clinical transplantation of a tissue-engineered airway. *Lancet*, **372**(9655), 2023–2030.

Manni, M. L., Czajka, C. A., Oury, T. D., & Gilbert, T. W. (2011). Extracellular matrix powder protects against bleomycin-induced pulmonary fibrosis. *Tissue Eng. Part A.*, **17**(21-22), 2795–2804.

Mantovani, A., Sica, A., Sozzani, S., Allavena, P., Vecchi, A., & Locati, M. (2004). The chemokine system in diverse forms of macrophage activation and polarization. *Trends Immunol.*, **25**(12), 677–686.

Mantovani, A., Sica, A., & Locati, M. (2005). Macrophage polarization comes of age. *Immunity*, **23**(4), 344–346.

Martinez, F. O., Sica, A., Mantovani, A., & Locati, M. (2008). Macrophage activation and polarization. *Front. Biosci.*, **13**, 453–461.

Martins-Green, M., & Bissel, M. F. (1995). Cell-extracellular matrix interactions in development. *Semin. Dev. Biol.*, **6**, 149–159.

Matziolis, G., Tuischer, J., Kasper, G., Thompson, M., Bartmeyer, B., Krocker, D., et al. (2006). Simulation of cell differentiation in fracture healing: Mechanically loaded composite scaffolds in a novel bioreactor system. *Tissue Eng.*, **12**(1), 201–208.

McPherson, T., & Badylak, S. F. (1998). Characterization of fibronectin derived from porcine small intestinal submucosa. *Tissue Eng.*, **4**, 75–83.

McPherson, T. B., Liang, H., Record, R. D., & Badylak, S. F. (2000). Galalpha(1,3)Gal epitope in porcine small intestinal submucosa. *Tissue Eng.*, **6**(3), 233–239.

Mills, C. D., Kincaid, K., Alt, J. M., Heilman, M. J., & Hill, A. M. (2000). M-1/M-2 macrophages and the Th1/Th2 paradigm. *J. Immunol.*, **164**(12), 6166–6173.

Miyamoto, S., Katz, B. Z., Lafrenie, R. M., & Yamada, K. M. (1998). Fibronectin and integrins in cell adhesion, signaling and morphogenesis. *Ann. N. Y. Acad. Sci.*, **857**, 119–129.

Moore, D. C., Pedrozo, H. A., Crisco, J. J., 3rd, & Ehrlich, M. G. (2004). Preformed grafts of porcine small intestine submucosa (SIS) for bridging segmental bone defects. *J. Biomed. Mater. Res. A.*, **69**(2), 259–266.

Mosser, D. M. (2003). The many faces of macrophage activation. *J. Leukocyte Biol.*, **73**(2), 209–212.

Mostow, E. N., Haraway, G. D., Dalsing, M., Hodde, J. P., & King, D. (2005). Effectiveness of an extracellular matrix graft (OASIS Wound Matrix) in the treatment of chronic leg ulcers: A randomized clinical trial. *J. Vasc. Surg.*, **41**(5), 837–843.

Nguyen, T. D., Liang, R., Woo, S. L.-Y., Burton, S. D., Wu, C., Almarza, A., et al. (2009). Effects of cell seeding and cyclic stretch on the fiber remodeling in an extracellular matrix-derived bioscaffold. *Tissue Eng. Part A.*, **15**(4), 957–963.

Nichols, J. E., Niles, J. A., & Cortiella, J. (2009). Design and development of tissue engineered lung: Progress and challenges. *Organogenesis*, **5**(2), 57–61.

Nieponice, A., Gilbert, T. W., & Badylak, S. F. (2006). Reinforcement of esophageal anastomoses with an extracellular matrix scaffold in a canine model. *Ann. Thorac. Surg.*, **82**(6), 2050–2058.

Nieponice, A., Maul, T. M., Cumer, J. M., Soletti, L., & Vorp, D. A. (2007). Mechanical stimulation induces morphological and phenotypic changes in bone marrow-derived progenitor cells within a three-dimensional fibrin matrix. *J. Biomed. Mater. Res. A.*, **81**(3), 523–530.

Nieponice, A., McGrath, K., Qureshi, I., Beckman, E. J., Luketich, J. D., Gilbert, T. W., et al. (2009). An extracellular matrix scaffold for esophageal stricture prevention after circumferential EMR. *Gastrointest Endosc.*, **69**(2), 289–296.

Oriol, R., Ye, Y., Koren, E., & Cooper, D. K. (1993). Carbohydrate antigens of pig tissues reacting with human natural antibodies as potential targets for hyperacute vascular rejection in pig-to-man organ xenotransplantation. *Transplantation*, **56**(6), 1433–1442.

Ott, H. C., Matthiesen, T. S., Goh, S. K., Black, L. D., Kren, S. M., Netoff, T. I., et al. (2008). Perfusion-decellularized matrix: Using nature's platform to engineer a bioartificial heart. *Nat. Med.*, **14**(2), 213–221.

Parekh, A., Mantle, B., Banks, J., Swarts, J. D., Badylak, S. F., Dohar, J. E., et al. (2009). Repair of the tympanic membrane with urinary bladder matrix. *Laryngoscope*, **119**(6), 1206–1213.

Peel, S. A. F., Chen, H., Renlund, R., Badylak, S. F., & Kandel, R. A. (1998). Formation of a SIS–cartilage composite graft in vitro and its use in the repair of articular cartilage defects. *Tissue Eng.*, **4**(2), 143–155.

Petersen, T. H., Calle, E. A., Colehour, M. B., & Niklason, L. E. (2012). Matrix composition and mechanics of decellularized lung scaffolds. *Cells Tissues Organs*, **195**(3), 222–231.

Piechota, H. J., Dahms, S. E., Probst, M., Gleason, C. A., Nunes, L. S., Dahiya, R., et al. (1998). Functional rat bladder regeneration through xenotransplantation of the bladder acellular matrix graft. *Br. J. Urol.*, **81**(4), 548–559.

Piez, K. A. (1984). *Molecular and Aggregate Structures of The Collagens*. New York: Elsevier.

Ponce, M. L., Nomizu, M., Delgado, M. C., Kuratomi, Y., Hoffman, M. P., Powell, S., et al. (1999). Identification of endothelial cell binding sites on the laminin gamma 1 chain. *Circ. Res.*, **84**(6), 688–694.

Prevel, C. D., Eppley, B. L., Summerlin, D. J., Jackson, J. R., McCarty, M., & Badylak, S. F. (1995). Small intestinal submucosa: Utilization for repair of rodent abdominal wall defects. *Ann. Plast. Surg.*, **35**(4), 374–380.

Prevel, C. D., Eppley, B. L., Summerlin, D. J., Sidner, R., Jackson, J. R., McCarty, M., et al. (1995). Small intestinal submucosa: utilization as a wound dressing in full-thickness rodent wounds. *Ann. Plast. Surg.*, **35**(4), 381–388.

Price, A. P., England, K. A., Matson, A. M., Blazar, B. R., & Panoskaltsis-Mortari, A. (2010). Development of a decellularized lung bioreactor system for bioengineering the lung: The matrix reloaded. *Tissue Eng. Part A.*, **16**(8), 2581–2591.

Raeder, R. H., Badylak, S. F., Sheehan, C., Kallakury, B., & Metzger, D. W. (2002). Natural anti-galactose alpha1,3 galactose antibodies delay, but do not prevent the acceptance of extracellular matrix xenografts. *Transpl. Immunol.*, **10**(1), 15–24.

Record, R. D., Hillegonds, D., Simmons, C., Tullius, R., Rickey, F. A., Elmore, D., & Badylak, S. F. (2001). In vivo degradation of 14C-labeled small intestinal submucosa (SIS) when used for urinary bladder repair. *Biomaterials*, **22**(19), 2653–2659.

Reing, J. E., Zhang, L., Myers-Irvin, J., Cordero, K. E., Freytes, D. O., Heber-Katz, E., et al. (2009). Degradation products of extracellular matrix affect cell migration and proliferation. *Tissue Eng. Part A.*, **15**(3), 605–614.

Remlinger, N. T., Czajka, C. A., Juhas, M. E., Vorp, D. A., Stolz, D. B., Badylak, S. F., Gilbert, S., & Gilbert, T. W. (2010). Hydrated xenogeneic decellularized tracheal matrix as a scaffold for tracheal reconstruction. *Biomaterials*, **31**(13), 3520–3526.

Roberts, R., Gallagher, J., Spooncer, E., Allen, T. D., Bloomfield, F., & Dexter, T. M. (1988). Heparan sulphate bound growth factors: A mechanism for stromal cell mediated haemopoiesis. *Nature*, **332**(6162), 376–378.

Ross, J. R., Kirk, A. D., Ibrahim, S. E., Howell, D. N., Baldwin, W. M., 3rd, & Sanfilippo, F. P. (1993). Characterization of human anti-porcine "natural antibodies" recovered from *ex vivo* perfused hearts – predominance of IgM and IgG2. *Transplantation*, **55**(5), 1144–1150.

Sandusky, G. E., Jr., Badylak, S. F., Morff, R. J., Johnson, W. D., & Lantz, G. (1992). Histologic findings after *in vivo* placement of small intestine submucosal vascular grafts and saphenous vein grafts in the carotid artery in dogs. *Am. J. Pathol.*, **140**(2), 317–324.

Sarikaya, A., Record, R., Wu, C. C., Tullius, B., Badylak, S., & Ladisch, M. (2002). Antimicrobial activity associated with extracellular matrices. *Tissue Eng.*, **8**(1), 63–71.

Schwarzbauer, J. E. (1991). Fibronectin: From gene to protein. *Curr. Opin. Cell Biol.*, **3**(5), 786–791.

Schwarzbauer, J. (1999). Basement membranes: Putting up the barriers. *Curr. Biol.*, **9**(7), R242–R244.

Sondell, M., Lundborg, G., & Kanje, M. (1998). Regeneration of the rat sciatic nerve into allografts made acellular through chemical extraction. *Brain Res.*, **795**(1-2), 44–54.

Song, J. J., Kim, S. S., Liu, Z., Madsen, J. C., Mathisen, D. J., Vacanti, J. P., & Ott, H. C. (2011). Enhanced in vivo function of bioartificial lungs in rats. *Ann. Thorac. Surg.*, **92**(3), 998–1005.

Soto-Gutierrez, A., Zhang, L., Medberry, C., Fukumitsu, K., Faulk, D., Jiang, H., Reing, J., Gramignoli, R., Komori, J., Ross, M., Nagaya, M., Legasse, E., Stolz, D., Strom, S. C., Fox, I. J., & Badylak, S. F. (2011). A whole-organ regenerative medicine approach for liver replacement. *Tissue Eng. Part C Methods*, **17**(6), 677–686.

Suckow, M. A., Voytik-Harbin, S. L., Terril, L. A., & Badylak, S. F. (1999). Enhanced bone regeneration using porcine small intestinal submucosa. *J. Invest Surg.*, **12**(5), 277–287.

Timpl, R. (1996). Macromolecular organization of basement membranes. *Curr. Opin. Cell Biol.*, **8**(5), 618–624.

Timpl, R., & Brown, J. C. (1996). Supramolecular assembly of basement membranes. *Bioessays*, **18**(2), 123–132.

Tischer, T., Vogt, S., Aryee, S., Steinhauser, E., Adamczyk, C., Milz, S., et al. (2007). Tissue engineering of the anterior cruciate ligament: A new method using acellularized tendon allografts and autologous fibroblasts. *Arch. Orthop. Trauma Surg.*, **127**(9), 735–741. Tissue Eng Regen Med 2008 Dec;2(8):491–498.

Uygun, B. E., Soto-Gutierrez, A., Yagi, H., Izamis, M. L., Guzzardi, M. A., Shulman, C., Milwid, J., Kobayashi, N., Tilles, A., Berthiaume, F., Hertl, M., Nahmias, Y., Yarmush, M. L., & Uygun, K. (2010). Organ reengineering through development of a transplantable recellularized liver graft using decellularized liver matrix. *Nat. Med.*, **16**(7), 814–820.

Valentin, J. E., Badylak, J. S., McCabe, G. P., & Badylak, S. F. (2006). Extracellular matrix bioscaffolds for orthopaedic applications. A comparative histologic study. *J. Bone Joint. Surg. Am.*, **88**(12), 2673–2686.

Valentin, J. E., Stewart-Akers, A. M., Gilbert, T. W., & Badylak, S. F. (2009). Macrophage participation in the degradation and remodeling of extracellular matrix scaffolds. *Tissue Eng. Part A.*, **15**(7), 1687–1694.

Valentin, J. E., Turner, N. J., Gilbert, T. W., & Badylak, S. F. (2010). Functional skeletal muscle formation with a biologic scaffold. *Biomaterials*, **31**(29), 7475–7484.

van der Rest, M., & Garrone, R. (1991). Collagen family of proteins. *Faseb J.*, **5**(13), 2814–2823.

Verreck, F. A., de Boer, T., Langenberg, D. M., van der Zanden, L., & Ottenhoff, T. H. (2006). Phenotypic and functional profiling of human proinflammatory type-1 and anti-inflammatory type-2 macrophages in response to microbial antigens and IFN-gamma- and CD40L-mediated costimulation. *J. Leukocyte Biol.*, **79**(2), 285–293.

Wainwright, J. M., Czajka, C. A., Patel, U. B., Freytes, D. O., Tobita, K., Gilbert, T. W., et al. (2010). Preparation of cardiac extracellular matrix from an intact porcine heart. *Tissue Eng. Part C: Methods*, **16**(3), 525–32.

Wallis, M. C., Yeger, H., Cartwright, L., Shou, Z., Radisic, M., Haig, J., et al. (2008). Feasibility study of a novel urinary bladder bioreactor. *Tissue Eng. Part A.*, **14**(3), 339–348.

Welch, J. A., Montgomery, R. D., Lenz, S. D., Plouhar, P., & Shelton, W. R. (2002). Evaluation of small-intestinal submucosa implants for repair of meniscal defects in dogs. *Am. J. Vet. Res.*, **63**(3), 427–431.

Werb, Z., Vu, T. H., Rinkenberger, J. L., & Coussens, L. M. (1999). Matrix-degrading proteases and angiogenesis during development and tumor formation. *Apmis*, **107**(1), 11–18.

Woods, T., & Gratzer, P. F. (2005). Effectiveness of three extraction techniques in the development of a decellularized bone-anterior cruciate ligament-bone graft. *Biomaterials*, **26**(35), 7339–7349.

Yoo, J. J., Meng, J., Oberpenning, F., & Atala, A. (1998). Bladder augmentation using allogenic bladder submucosa seeded with cells. *Urology*, **51**(2), 221–225.

Yurchenco, P., Birk, D. E., & Mecham, R. P. (1994). *Extracellular Matrix Assembly and Structure*. San Diego: Academic Press.

Zantop, T., Gilbert, T. W., Yoder, M. C., & Badylak, S. F. (2006). Extracellular matrix scaffolds are repopulated by bone marrow-derived cells in a mouse model of achilles tendon reconstruction. *J. Orthop. Res.*, **24**(6), 1299–1309.

Zheng, M. H., Chen, J., Kirilak, Y., Willers, C., Xu, J., & Wood, D. (2005). Porcine small intestine submucosa (SIS) is not an acellular collagenous matrix and contains porcine DNA: Possible implications in human implantation. *J. Biomed. Mater. Res. B. Appl. Biomater.*, **73**(1), 61–67.

PART III

Practical Aspects of Biomaterials

SECTION III.1

Implants, Devices, and Biomaterials: Special Considerations

SECTION

III.1

Implants, Devices, and Biomaterials: Special Considerations

CHAPTER III.1.1 INTRODUCTION: IMPLANTS, DEVICES, AND BIOMATERIALS: SPECIAL CONSIDERATIONS

Frederick J. Schoen
Professor of Pathology and Health Sciences and Technology (HST), Harvard Medical School, Executive Vice Chairman, Department of Pathology, Brigham and Women's Hospital, Boston, MA, USA

Section III.1 addresses some special concerns about the use of biomaterials in surgical implants and medical devices. The themes of this section are that implant retrieval and failure mode analysis provide information that improves the design and use of medical devices and thereby patient outcomes, and that the characteristics of biomaterials surfaces play a critical role in determining the tissue reactions to implants. Moreover, in this respect, sterilization of implants, as discussed by Lambert et al. in Chapter III.1.2, is not only critical to preventing infection, but this procedure can also alter the surface chemistry of biomaterials (and indeed potentially their bulk properties), and thereby inadvertently affect implant performance. A central concept of this section is that following contact of biomaterials with tissues, careful and detailed analysis of implants, the biomaterials that comprise them, and the surrounding tissues, can be a powerful tool in understanding the mechanisms and causes of tissue–biomaterials interactions, both desired and adverse, and in ensuring the efficacy and safety of medical devices.

Although most surgical implants serve their recipients well by perfoming their desired function (i.e., *functionality*), the outcome of biomaterials and medical device implantation is not always optimal, in large part owing to suboptimal tissue–biomaterials interactions (i.e., *biocompatibility*). Implants can be retrieved after *in vitro* studies or at either reoperation or necropsy/autopsy of animals or humans, and they may or may not be failed. In order to understand tissue reactions, described in Section II.2, the correlation of biomaterials surface chemistry with performance is a critical step (as by detailed by Ratner in Chapter III.1.3). The literature contains numerous instances where problem-oriented evaluation and analysis of medical implant research has yielded important insights into both deficiencies and complications limiting the success of implants, and understanding why certain implants are successful. The approaches to implant analysis (for both individual cases and for cohorts of implants of similar type) and numerous instances where such analyses have been useful are summarized in Chapter III.1.4 by Schoen, and Chapter III.1.5 by Anderson et al. Implant research has guided development of new and modified implant designs and materials, assisted in decisions of implant selection and management of patients, and permitted *in vivo* study of the mechanisms of biomaterials–tissue interactions, both local and distant from the device.

Preclinical studies of modified designs and materials are crucial to developmental advances. These investigations usually include *in vitro* functional testing (such as fatigue studies at accelerated rates) and implantation of functional devices in the intended location in an appropriate animal model, followed by noninvasive and invasive monitoring, specimen explantation, and detailed pathological and material analysis. Relative to clinical studies, animal investigation may permit more detailed monitoring of device function and enhanced observation of morphologic detail, as well as frequent assay of laboratory parameters, and allow *in situ* observation of fresh implants following elective sacrifice at desired intervals. In addition, specimens from experimental animals can minimize the artifacts that can occur inadvertently when specimens are harvested under clinical circumstances. Furthermore, advantageous technical adjuncts may be available in animal but not human investigations, such as injection of cells or radiolabeled imaging markers or fixation by pressure perfusion that maintains tissues and cells in their physiological configuration following removal. Animal studies often facilitate observation of specific complications in an accelerated timeframe, such as calcification of bioprosthetic valves, in which the equivalent of 5–10 years in humans is simulated in 4–6 months in juvenile sheep (Schoen and Levy, 2005). Moreover, in preclinical animal studies, experimental conditions can generally be held constant among groups of subjects, including nutrition, activity levels, and treatment conditions, and concurrent control implants are often possible. This is clearly not possible in clinical studies. However, animal models that faithfully duplicate the relevant human anatomy, physiology and pathology are not often available.

Nevertheless, clinicopathologic analysis of cohorts of patients who have received a new or modified prosthesis type evaluates device safety and efficacy to an extent beyond that obtainable by either *in vitro* tests of durability and biocompatibility or preclinical investigations of implant configurations in large animals. Moreover, through analysis of rates and modes of failure, as well as characterization of the morphology and mechanisms of specific failure modes, retrieval studies can contribute to the development of methods for enhanced clinical recognition of failures. The information gained should serve to guide future development of improved prosthetic devices, to eliminate complications, and to develop and validate diagnostic and therapeutic management strategies to reduce the frequency and clinical impact of complications. Furthermore, since patient-specific factors could impact performance, demonstration of an individual propensity toward certain complications (see below) could impact greatly on further management. Moreover, history has shown that some medical devices demonstrate important but rare complications only during clinical trials or postmarket surveillance (complications that were not predicted by animal investigations). Subsequent study of clinically important device failure mechanisms may yield prevention strategies; such strategies may then be screened *in vivo* in animal studies; favorable therapies are then tested in clinical trials. Thus, an important future goal is the effective integrated use of data derived from implant retrieval investigations (along with other clinical and experimental data) to inform regulatory decisions and device improvements in an ongoing, incremental and iterative fashion throughout the product lifecycle.

Clinical implant retrieval and evaluation can yield additional benefits. Implant retrieval studies have demonstrated that success of a material or design feature in one application may not necessarily translate to another. Although analysis of implants and medical devices has traditionally concentrated on failed devices, important data can accrue from implants serving the patient well until death or removal for unrelated causes. Indeed, detailed analyses of implant structural features following implantation can yield an understanding of not only predisposition to specific failure modes, but also structural correlates of favorable performance. Implant retrieval and evaluation may also provide specimens and data that can be used to educate patients, their families, physicians, residents, students, engineers, and biomaterials scientists, as well as the general public. As a basic research resource, the process of implant retrieval and evaluation yields data that can be used to develop and test hypotheses, and to improve protocols and techniques.

For investigation of bioactive materials/devices, and combination and potentially tissue-engineered medical devices in which the interactions between the implant and the surrounding tissue are complex, research based on implant retrieval and evaluation continues to be critical. In such instances, novel and innovative approaches must be used in the investigation of *in vivo* tissue compatibility. In such implant types, the scope of the concept of "biocompatibility" is much broader, and the approaches employed in implant retrieval and evaluation require identification of the phenotypes and functions of cells, and the architecture and remodeling of extracellular matrix (Mendelson and Schoen, 2006; Hjortnaes et al., 2009). These are circumstances in which individual patient characteristics (for example, genetic variation in molecules important in matrix remodeling) could have a profound influence on the outcome in some patients (Ye, 2000; Jones et al., 2003). This potentially yields a new area of study: "biomateriogenomics," conceptually analogous to the emerging area of pharmacogenomics (Wang et al., 2011). Indeed, individuals with genetic defects in coagulation proteins may be unusually susceptible to thrombosis of prosthetic heart valves (Gencbay et al., 1998). Thus, a critical role of implant retrieval will be the identification of patient and characteristics and post-implant tissue responses (as assessed by *biomarkers*) that will be predictive of (i.e., surrogates for) success and failure. A most exciting possibility is that such biomarkers may be used to non-invasively image/monitor the maturation/remodeling of tissue-engineered devices *in vivo* in individual patients (Sameni et al., 2003; Mendelson and Schoen, 2006; Hjortnaes et al., 2009).

BIBLIOGRAPHY

Gencbay, M., Turan, F., Degertekin, M., Eksi, N., Mutlu, B., et al. (1998). High prevalence of hypercoagulable states in patients with recurrent thrombosis of mechanical heart valves. *J. Heart Valve. Dis.*, **7**, 601–609.

Hjortnaes, J., Bouten, C. V., Van Herwerden, L. A., Gründeman, P. F., & Kluin, J. (2009). Translating autologous heart valve tissue engineering from bench to bed. *Tissue Eng. Part B Rev.*, **15**, 307–317.

Hjortnaes, J., Gottlieb, D., Figueiredo, J. L., Melero-Martin, J., Kohler, R. H., et al. (2010). Intravital molecular imaging of small-diameter tissue-engineered vascular grafts in mice: A feasibility study. *Tissue Eng. Part C. Methods*, **16**, 597–607.

Jones, G. T., Phillips, V. L., Harris, E. L., Rossaak, J. I., & van Rij, A. M. (2003). Functional matrix metalloproteinases-9 polymorphism (C-1562T) associated with abdominal aortic aneurysm. *Vasc. Surg.*, **38**, 1363–1367.

Mendelson, K. M., & Schoen, F. J. (2006). Heart valve tissue engineering: Concepts, approaches, progress, and challenges. *Ann. Biomed. Engin.*, **34**, 1799–1819.

Sameni, M., Dosescu, J., Moin, K., & Sloane, B. F. (2003). Functional imaging of proteolysis: Stromal and inflammatory cells increase tumor proteolysis. *Mol. Imaging*, **2**, 159–175.

Schoen, F. J., & Levy, R. J. (2005). Calcification of tissue heart valve substitutes: Progress toward understanding and prevention. *Ann. Thorac. Surg.*, **79**, 1072–1080.

Wang, L., McLeod, H. L., & Weinshilboum, R. M. (2011). Genomics and drug response. *N. Engl. J. Med.*, **364**, 1144–1153.

Ye, S. (2000). Polymorphism in matrix metalloproteinase gene promoters: Implication in regulation of gene expression and susceptibility of various diseases. *Matrix. Biol*, **19**, 623–629.

CHAPTER III.1.2 STERILIZATION OF IMPLANTS AND DEVICES

Byron Lambert[1] and Jeffrey Martin[2]
[1]Sterilization Science, Abbott Vascular, Temecula, CA, USA
[2]Director of Corporate Quality Technology, Alcon® a Novartis Company, Fort Worth, Texas, USA

INTRODUCTION

Successful sterilization of biomaterials used in implants and devices is a critical prerequisite for their successful clinical application. It requires knowledge of sterility concepts and sterilization technologies that render products sterile. Understanding the effect of sterilization processes on the biomaterials themselves is also increasingly important. These topics are the focus of this chapter.

Delivering a sterile product is the responsibility of a hospital, a manufacturer of an aseptically processed device or a manufacturer of a terminally sterilized device. This chapter is largely focused on the latter: industrial terminal sterilization. Equipping the biomaterials scientist to find a terminal sterilization solution for a product, if possible, and thereby avoiding aseptic processing of the product, is a desired outcome of the chapter.

The current regulatory expectation of the term "sterile" for blood-contacting medical devices and implants is to produce only one non-sterile device out of one million. If you think about it, this is an extreme target. Fortunately, it is normally attainable, and clearly a patient heading to the hospital appreciates this high safety target. The patient, however, would cease to be pleased if that sterile device failed during treatment because the sterilization cycle was so rigorous in killing microbes that it also damaged the product. The challenge for the biomaterials scientist responsible for defining a terminal sterilization process is an optimization problem – to determine and define a cost effective sterilization process window where sterility is achieved, and yet deleterious material effects are minimized. Many traditional medical devices are made with materials that are compatible with multiple sterilization modalities. In this scenario, finding a terminal sterilization solution is straightforward. However, the optimization problem is much more complex for biologics and combination devices, both of which are important and rapidly growing markets.

To navigate these challenges, it is useful for the biomaterials scientist to have a concrete understanding of common terminal sterilization technologies (see below, Sterilization Technologies), which are the basis of both achieving sterility and any product material effects. Material compatibility constraints often drive the choice of sterilization modality. The next section of the chapter, therefore, provides an overview of sterilization material compatibility challenges and guidance (see below, Material Compatibility). Robust sterilization validation methods embodied in national and international standards are the foundation for the strong patient safety record of terminal sterilization. However, some materials, especially biologics, may have undesirable responses to the techniques used, and then may require special sterilization methods. Sterility assurance concepts are, therefore, reviewed in sufficient detail to give the biomaterials scientist confidence in evaluating all options for finding a sterilization solution (see below, Sterility and Patient Safety). The chapter closes with a view toward exciting challenges on the terminal sterilization horizon. The combination device market offers compelling patient benefits, and equally compelling challenges for terminal sterilization. Awareness of these challenges and knowledge to meet them will serve the biomaterials scientist well (see below, Summary and Future Challenges).

STERILIZATION TECHNOLOGIES

How is microbial contamination on fully packaged devices reduced by 9, 12 or even greater orders of magnitude? This section provides a concrete understanding of industrial sterilization technologies that deliver these enormous microbial reduction levels. Two terminal sterilization modalities, radiation sterilization and ethylene oxide sterilization, dominate the industrial (non-hospital) terminal sterilization market. These robust workhorse terminal sterilization technologies utilize well-established validation processes (see below, Sterility and Patient Safety) that, in most instances, far exceed regulatory requirements for a one in a million sterility assurance level. Under well-controlled and validated conditions, neither of these modalities inflicts significant material degradation damage on many traditional devices. In addition, both are capable of processing high volumes of product at low cost. These important technologies are reviewed in some detail, along with a brief overview of a number of other methodologies that are less frequently utilized.

Radiation Sterilization

Safety First. Radiation sterilization doses are lethal. Terminal sterilization doses typically range from 8 to 35 kGy. An acute lethal dose to man is approximately 0.01 kGy, requiring an exposure time of only a fraction of a second in some processes. To ensure worker safety in these high radiation environments, significant shielding, robust interlocks, and the utmost care are required in radiation processing facilities. Another safety concern is exposure to ozone, a toxic gas, produced by radiation as it passes through air. Appropriate ventilation and ozone monitors are required to provide fresh air, and ensure the safety of personnel prior to entering a processing cell.

Technology Overview. Terminal sterilization by radiation sterilization is elegantly simple. Fully packaged medical devices are exposed to a validated dose from a radiation source that emits electrons or photons that penetrate through the final packaging and inactivate the device's microbial load. One parameter, radiation dose, correlates directly with microbial kill, and this is easily measured to provide process control. The microbial kill mechanism involves radiation-induced scission of DNA chains to stop microbial reproduction. While radiation destroys the ability of most microorganisms to reproduce, the resistance of viruses such as HIV-1 remains a concern (Smith et al., 2001).

International radiation sterilization standards (ISO 11137-1; see Table III.1.2.1) call out three radiation sterilization modalities: gamma, electron beam, and X-ray. Gamma and electron beam dominate the radiation sterilization market, and are described in enough detail below to provide a working knowledge of the technologies. Sources of the radiation, processing facility layouts, mechanism of radiation energy deposition into materials, and key processing parameters are described. Cycle times range from minutes to half a day. Technologies available for R&D, pilot, and low volume processing are also described. These small-scale systems will play an increasingly important role in qualifying materials and providing special processes to meet demanding material compatibility needs of combination devices.

Gamma Sterilization. Cobalt-60 gamma sterilization accounts for approximately 80% of the radiation sterilization market, 40% of the overall terminal sterilization industrial market. ^{60}Co is a radioactive element that undergoes nuclear decay producing useful gamma radiation. Note: while the gamma source itself is radioactive, the product sterilized by the exposure to the radiation does not become radioactive in the slightest measure.

High product volume ^{60}Co processing plants are relatively simple. Small ^{60}Co pellets are doubly encapsulated in rods, which are arranged into racks and stored in

TABLE III.1.2.1 Sterilization Standards

Sterilization Modality	Reference Standard or Guidance Document
Radiation sterilization	ANSI/AAMI/ISO 11137-1 Sterilization of health care products – Radiation – Part 1: Requirements for development, validation, and routine control of a sterilization process for medical devices
	AAMI TIR29: 2002 Guide for process control in radiation sterilization
	AAMI TIR33: 2006 Sterilization of health care products – Radiation sterilization – Substantiation of a selected sterilization dose – Method VD_{MAX}
	AAMI TIR37: 2007 Sterilization of health care products – Radiation – Guidance on sterilization of human tissue-based products
	AAMI TIR40: 2009 Sterilization of health care products – Radiation – Guidance on dose setting utilizing a Modified Method 2
	EU GMP Annex 12: Use of ionizing radiation in the manufacture of medicinal products
Ethylene oxide (EO)	ANSI/AAMI/ISO 11135-1 Sterilization of health care products – Ethylene oxide – Part 1: Requirements for the development, validation and routine control of a sterilization process for medical devices
	AAMI TIR14: 2009 Contract sterilization using ethylene oxide
	AAMI TIR16: 2009 Microbiological aspects of ethylene oxide sterilization
	AAMI TIR28: 2009 Product adoption and process equivalency for ethylene oxide sterilization
EO residuals	ANSI/AAMI ISO 10993-7: 2008 Biological evaluation of medical devices – Part 7: Ethylene oxide sterilization residuals
Moist heat (saturated steam)	ANSI/AAMI/ISO 17665-1 Sterilization of health care products – Moist heat – Part 1: Requirements for the development, validation, and routine control of a sterilization process for medical devices
Packaging	ANSI/AAMI/ISO 11607-1 Packaging for terminally sterilized medical devices – Part 1: Requirements for materials, sterile barrier systems, and packaging systems
	AAMI TIR22: 2007 Guidance for ANSI/AAMI/ISO 11607: Packaging for terminally sterilized medical devices – Part 1 and Part 2: 2006
	ASTM D4169-05 Standard practice for shipping containers and systems
General	EN ISO 14937 Sterilization of health care products – General requirements for characterization of a sterilizing agent and the development, validation and routine control of a sterilization process for medical devices
	EN 556-1 Sterilization of medical devices – Requirements for medical devices to be designated "STERILE" – Part 1: Requirements for terminally sterilized medical devices
	AAMI ST67 Sterilization of health care products – Requirements for products labeled "STERILE"
Material compatibility	AAMI TIR17: 2008 Compatibility of materials subject to sterilization
ISO 14160:2011?	Sterilization of health care products – Liquid chemical sterilizing agents for single-use medical devices utilizing animal tissues and their derivatives – Requirements for characterization, development, validation and routine control of a sterilization process for medical devices

AAMI: Association for the Advancement of Medical Instrumentation
ANSI: American National Standards Institute
EU: European Standard
EU GMP: European Union Good Manufacturing Practice
ISO: International Organization for Standardization

a pool of water, with some 20 feet of water acting as shielding to keep the constantly emitted gamma radiation away from the processing room when product is not being processed. When the room is cleared, interlocks satisfied, and the facility is ready to process product, a mechanical elevator system raises the radiation source from the pool into the room. A conveyor system moves many totes of fully packaged product into the room and around the racks of ^{60}Co, often passing by multiple times, as seen in Figure III.1.2.1.

^{60}Co gamma particles have no charge and penetrate uniformly through a very long distance of material. This is a significant processing benefit. Configurations of product within shipper boxes and processing load configurations are rarely a limitation due to the high gamma penetration. Product is then brought back out of the processing room. Product is released for distribution after dosimeters are read and documented to confirm that the product received the proper dose.

The beneficial sterilization effects of ^{60}Co gamma rays (along with any deleterious material effects) come as a result of the energy of the photons being deposited into the microbes and materials. By definition, the energy deposited per unit weight is the radiation dose (typical units are kJ/kg, i.e., kilogray or kGy). Energy deposition from a gamma ray into a material occurs through a Compton Scattering interaction that generates high energy electrons (0.5 MeV). These primary electrons travel through the material and deposit their energy in 60–100 electron volt bundles as they generate secondary electrons. It is these secondary electrons that cause all microbial kill. They do so by forming oxidizing free radicals. In addition to causing scission of DNA, the radicals can also cause scission or crosslinking of polymer chains (see Material Compatibility).

Electron Beam Sterilization. Electron beam (e-beam) sterilization accounts for approximately 20% of the radiation sterilization market, 10% of the overall terminal sterilization industrial market. High energy electrons for e-beam sterilization are generated by a variety of technologies that take 200 volt electrons from the power grid and accelerate them to 0.2 million to 10 million electron volts (0.2–10 MeV, typically 5–10 MeV). At these energy levels, no part of the e-beam processing plant (or the product irradiated) is or becomes radioactive; e-beam accelerators turn off like a light bulb. Before turning on the accelerator, however, the processing room must be cleared and safety interlocks satisfied to avoid human exposure to the high doses of radiation and ozone that are generated. Accelerated electrons for device sterilization are typically magnetically focused into a 1–5 cm diameter beam, and magnetically scanned at high frequency across the conveyor width, typically 12–48 inches.

The conveyor system transports product through the shielding and in front of the beam. Typically, one product, one shipper box or one thin tote of product at a time passes by the beam's scan horn. Electrons from accelerators do not penetrate nearly as far as photons from gamma sources. A rule of thumb is that the maximum penetration (in centimeters) from a two-sided e-beam process is 0.8 times the beam energy (in MeV) divided by the density of product (in g/cm^3). Resulting maximum penetration distances are only a fraction of a meter, with higher beam energies resulting in higher penetration. Limited penetration can lead to inefficient processing and/or large distributions of delivered dose. In practice, this requires significant dosimetry work to develop load configurations that ensure an appropriate range of e-beam doses. Due to the limited penetration

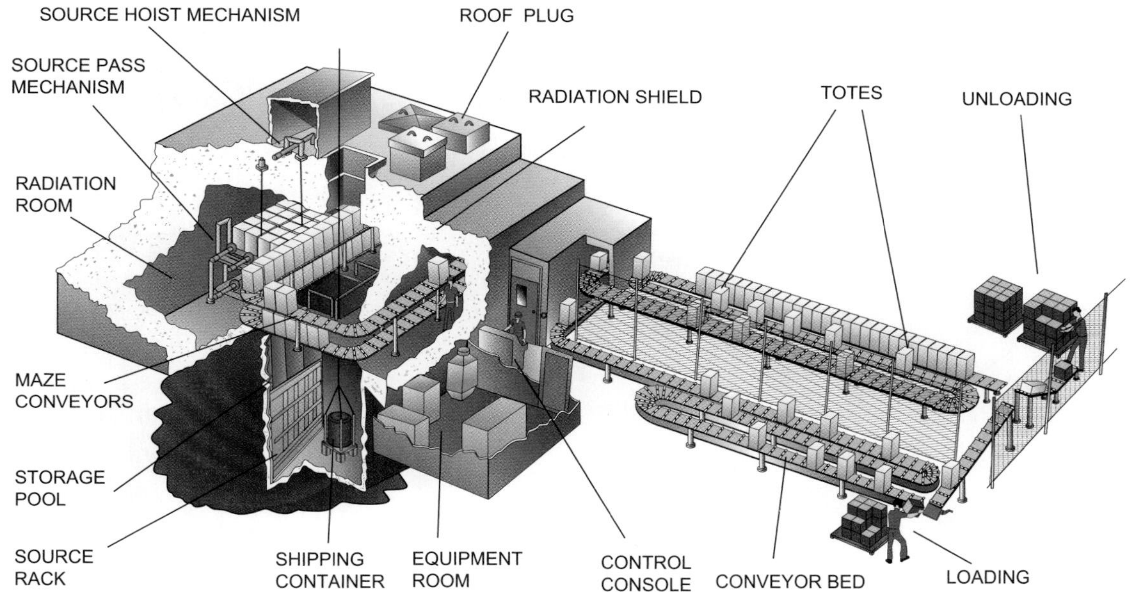

FIGURE III.1.2.1 High volume production gamma irradiation facility (courtesy of MDS Nordion, Canada).

of accelerated electrons, products need to be carefully "dose-mapped," as their size and shape provide shielding, scattering interfaces, and incident angles which impact dose distribution. These differences provide a less homogeneous dose profile than that of ^{60}Co.

Product is then brought back out of the processing room. Product is released for distribution after dosimeters are read and documented to confirm that the product received the proper dose. E-beam processing facilities can either be in a processing room as described above for a gamma facility or be of a self-shielded design, as shown in Figure III.1.2.2.

The mechanism of energy deposition from e-beams is that the high-energy electrons journey through the material and deposit their energy in 60–100 electron volt bundles as they generate secondary electrons. Like gamma processing, the resulting secondary electrons cause microbial kill, as well as all product material effects. The main advantage of e-beam relative to gamma in terms of material compatibility is that the dose is delivered very quickly. Products being treated in gamma sterilization processes often have dwell times of several hours, whereas products typically dwell in an e-beam process for only minutes. The short processing time limits oxidative degradation, since oxygen does not have time to diffuse into the product. The main disadvantage of e-beams versus ^{60}Co is that the negatively charged electrons do not penetrate deeply into a material compared with neutral gamma rays, which penetrate through the product being sterilized.

X-ray Sterilization. X-ray sterilization accounts for a small fraction of the radiation sterilization market. It is a hybrid between gamma sterilization and e-beam sterilization. Radiation is generated from high-energy electrons from accelerators, typically 5 MeV electrons, impinging on a high atomic number ("z") target, often tungsten, to generate X-rays. The X-ray photons behave nearly identically to photons from gamma sources in terms of energy deposition and high penetration capabilities. Like e-beam plants, no part of the X-ray processing plant is or becomes radioactive if X-rays are generated by electrons with energies no greater than 5 MeV.

Important Processing Parameters

Radiation sterilization processing parameters that are of practical importance to the biomaterials scientist are those that affect the maximum dose to the product, and those that affect the throughput of product through the process. The maximum dose to the product is dependent on:

- sterilization dose (minimum dose to the product);
- dose uniformity during the radiation sterilization process; and
- number of sterile cycles for which the device will be qualified.

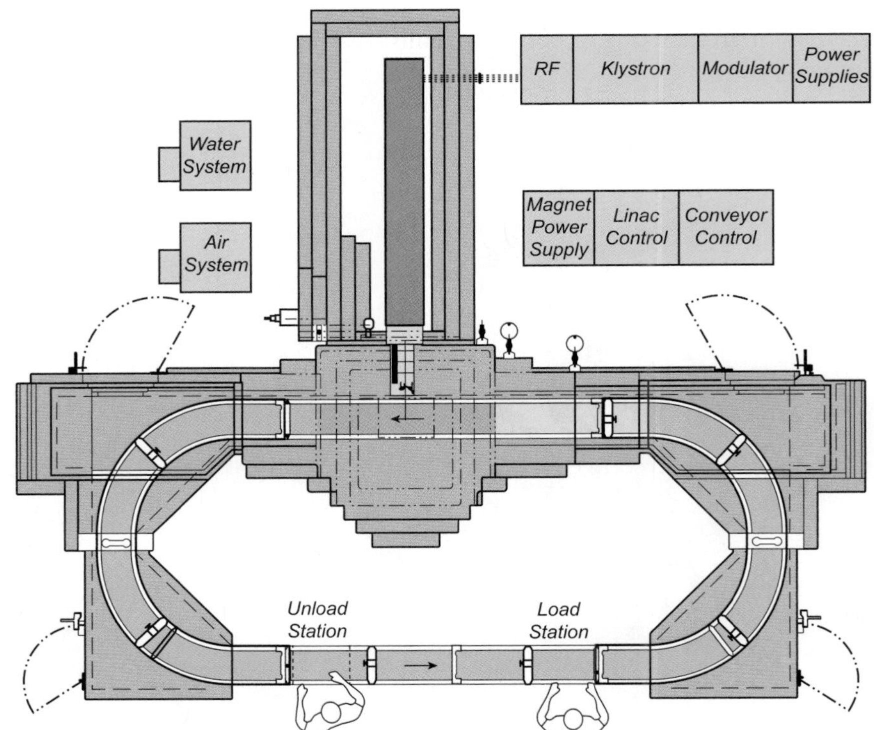

FIGURE III.1.2.2 Medium volume self-shielded production electron beam irradiation facility.

Dose uniformity ratio (DUR) is a measure of the range of doses delivered to product by a given radiation sterilization process. DUR is defined, for a given load configuration and density, as the ratio of the maximum to the minimum dose received by any one product. The DUR of a radiation sterilization process is always greater than 1.0; radiation sterilization does not provide a single dose, but a distribution of doses delivered throughout any one product. Dose uniformity is dependent on the radiation source (gamma is typically significantly more uniform than e-beam), product density, product shipper box or pallet size, sterilizer configuration, and for e-beam, energy of the electrons. High density products with large boxes, resulting in large distances through which the radiation must travel, result in a high distribution of doses, e.g., a DUR greater than 2.

Throughput for a radiation sterilization process is primarily dependent on the radiation source intensity, i.e., for gamma sterilizers the number of curies of ^{60}Co, and for e-beam and X-ray sterilizers the number of electrons, the current. Secondarily, throughput is a function of product density and quantity of product processed at a given time. For example, while it may be beneficial to reduce the quantity of product processed at a given time to reduce the dose distribution, and hence the overall dose to product, this change also reduces throughput efficiency.

R&D, Pilot, and Low Volume Technologies

Small radiation sources are very powerful tools for the biomaterial scientist and his/her team. An R&D gamma cell like that shown in Figure III.1.2.3 provides a narrow dose range, very close to a DUR of 1.0. Along with delivering precise doses for the sterilization validation process (see below, Sterility and Patient Safety), the biomaterials scientist will be well-served by these narrow dose ranges during material qualification exercises, especially if the source is easily accessible to provide fast iterations during the product development process.

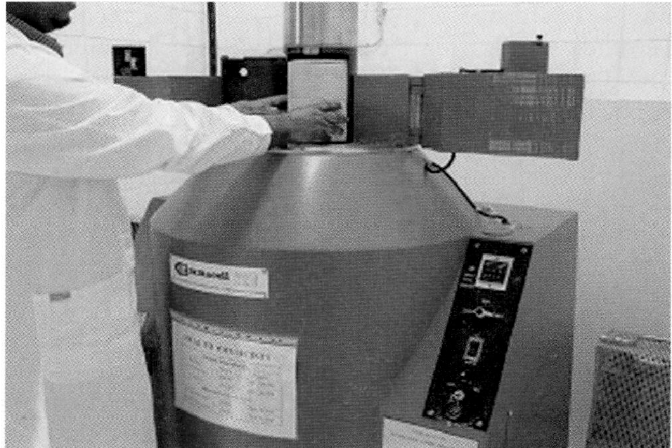

FIGURE III.1.2.3 Low volume self-contained R&D gamma cell manufactured by MDS Nordion (courtesy of IAEA, STI/PUB/1313).

In addition, small self-shielded low energy e-beam accelerators are available to process single units. The low energy of the beam, typically 0.5–2.0 MeV, may only penetrate a single product or a sample portion of a product at a time. This low energy keeps the accelerator to a manageable size and cost, and allows for rapid product development iterations.

Ethylene Oxide (EO) Technologies

Safety First. Ethylene oxide gas is toxic, carcinogenic, and very explosive. EO sterilization cycles must be designed to avoid conditions or parameters that may be explosive. Precautions must be taken to avoid any equipment, tools, etc., that may spark in the vicinity of EO chambers. Explosions can be horrific in magnitude, and extreme care must be taken. After an EO process is complete, employees must take appropriate precautions to avoid exposure to EO gas, e.g., use of respirators when working in the vicinity of product that has not been fully aerated to dissipate the gas. Finally, the manufacturer of EO sterilized product is responsible for following international standards to ensure that product is allowed to aerate in order to get EO residual levels below permissible limits for distribution (ISO 10993-7; see Table III.1.2.1).

Technology Overview. Ethylene oxide sterilization accounts for approximately 50% of the industrial terminal sterilization market, and is another conceptually simple terminal sterilization process. Fully packaged devices are placed into an ethylene oxide chamber and exposed to a validated combination of humidity, ethylene oxide gas, temperature, and time. Microbial-resistant sterile barrier packaging must be used, to allow ingress and egress of EO and water vapor within a defined load configuration. Following the sterilization process, EO levels are brought below acceptable levels through completion of a validated in-chamber vacuum purge process or a post-sterilization aeration process. Product is released for distribution following review and documentation of routine monitoring parameters and, in many instances, biologic indicator test results. Total cycle times range from six hours to several days.

The international ethylene oxide sterilization standard, ISO 11135-1 (see Table III.1.2.1) refers to the use of both ethylene oxide, and mixtures of ethylene oxide and a diluent. Diluents were commonly used as a deterrent to explosions, but are less common today since many common diluents are HCFCs and their use has been curtailed. Instead, nitrogen blanketing and other cycle design features are used to reduce the presence of air/oxygen and keep EO non-flammable. Diluents do not affect the principles of operation and validation of an EO cycle.

EO sterilization principles and practical aspects are described below in enough detail to provide a working knowledge of the technology. Chamber dynamics, processing plant layout, mechanism of microbial kill, and key processing parameters are described. Cycle times range from several hours to several days. Technologies available for R&D, pilot, and low volume processing are also described.

Ethylene oxide is a highly reactive cyclic ether with two carbons and one oxygen, CH_2CH_2O (Figure III.1.2.4). It is a gas at room temperature, with a boiling point of 11°C. It is pressurized and stored as a liquid for use in EO processing plants. The mechanism of microbial kill is alkylation of the amine groups of DNA. Moisture facilitates microbial kill; product has to be exposed to a humid environment before EO exposure. EO kill rate is a function of temperature and concentration of EO gas.

In practice, approximately three cubic meters of product is placed on pallets in well-defined configurations. Product is humidified, either in a dedicated preconditioning area (typically 6–24 hours at 50–80% relative humidity (RH) and mild temperatures) or within the EO processing chamber with dynamic vacuum/steam pulsing. One to 40 pallets of product are placed into industrial EO chambers. A three-pallet chamber is shown in Figure III.1.2.5.

Along with humidification, early phases often include evacuation to remove oxygen, and injection of nitrogen to keep the cycle non-explosive. Evacuation levels can be down to 2 torr in some cycles, with very fast evacuation rates. EO gas is injected into the chamber with the humidified product, and allowed to dwell for a validated period of time, EO concentration, temperature, and humidity. Typical in-chamber exposure times are 6–24 hours at processing temperatures of 40–65°C and RH of 30–90%. The EO is then purged, for example, with dynamic vacuum/steam pulsing followed by vacuum/nitrogen pulsing.

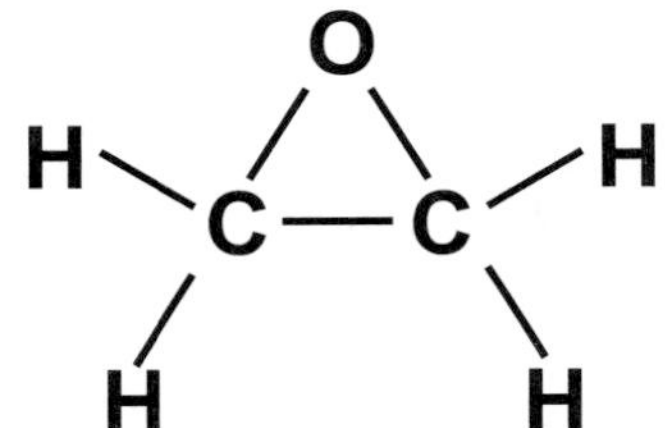

FIGURE III.1.2.4 Ethylene oxide molecule.

FIGURE III.1.2.5 Medium volume ethylene oxide production sterilization chamber manufactured by Getinge (Courtesy of Getinge Inc.).

If EO residual levels do not meet regulatory requirements for product release at the end of the cycle, product is placed into a dedicated forced air heated aeration room.

Important Processing Parameters

Ethylene oxide sterilization processing parameters that are of practical importance to the biomaterials scientist are those that can affect the product and package, and those that affect throughput. Deep vacuum levels, rate of vacuum cycles, chamber temperature and concentration, and time of exposure to humidity, EO gas, and any diluents utilized can all potentially affect packaging and/or device material components, especially pharmaceuticals and biologics. The size of the chamber, the length of the cycle, and any preconditioning and/or aeration steps affect cycle times and therefore product throughput.

R&D, Pilot, and Low Volume Technologies

Small ethylene oxide sterilizers are very powerful tools for the biomaterials scientist and his team. As well as providing a vehicle for well-controlled microbial kill challenge testing, they can be used to facilitate finding material compatibility solutions and qualifying product at most challenging parameters. Special low temperature and/or low humidity EO cycles can be developed and validated to avoid material compatibility concerns. A well-controlled eight cubic foot R&D EO chamber is shown in Figure III.1.2.6. Small EO chambers are also used extensively within the hospital setting.

Other Terminal Sterilization Technologies

Moist heat, i.e., steam, sterilization has long been a workhorse technology for reusable devices in hospitals and certain industrial applications. The high temperatures limit its use for most devices with plastic materials, including most single-use devices. Moist heat sterilization, like other major industrial and hospital processes, is well-characterized in national and international sterilization standards (see Table III.1.2.1). Gas phase hydrogen peroxide sterilization has, in the last decade, come into broad application in hospitals, but has limited industrial applications. This was the first technology to drive the generic ISO 14937 "General Criteria" sterilization standard (see Table III.1.2.1). A number of additional gas chemical sterilization technologies have been developed, but have not found significant industrial terminal sterilization application, including chlorine dioxide, ozone, nitrogen dioxide, supercritical carbon dioxide, and propylene oxide. In addition, dry heat sterilization has limited applications.

Liquid chemical sterilization is sometimes used for sterilization, in applications either with animal or human tissues that are not compatible with terminal sterilization options or as a high level disinfection for reuse of

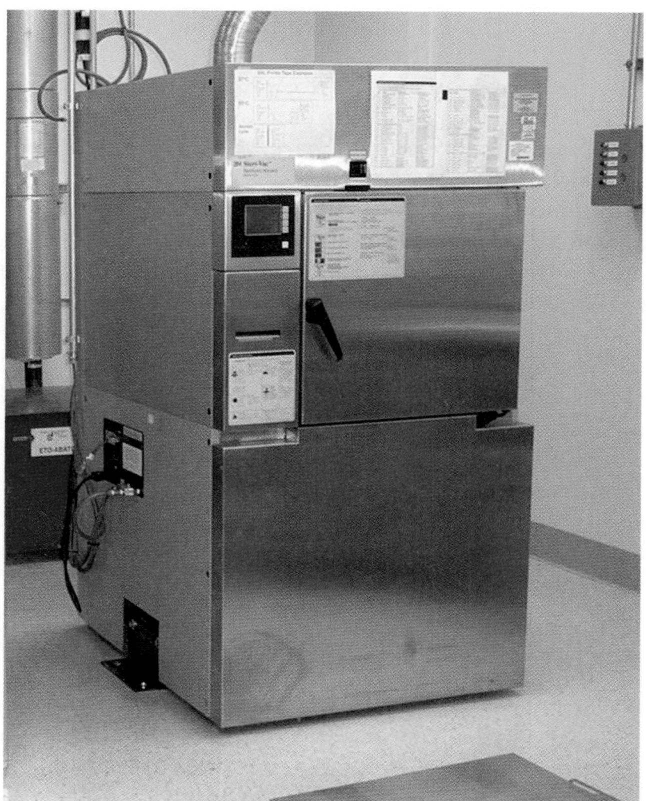

FIGURE III.1.2.6 Low volume ethylene oxide R&D sterilization chamber (Courtesy of Getinge Inc.).

certain devices, such as blood dializers. The process involves the immersion of the device into formulations of either an aldehyde or an oxidizing agent (e.g., gluteraldehyde, hydrogen peroxide or peracetic acid), sometimes with buffers, anti-corrosive agents, and detergents. The method does not provide the process control or sterility assurance levels of terminal sterilization processes (Chamberlain et al., 1999). Automation of the process, however, has led to significant success in providing safe tissue product.

MATERIAL COMPATIBILITY

The attention of this chapter will now shift to the response of the materials being terminally sterilized, especially the molecular effects which in turn can affect physical, chemical, and mechanical properties of the material. Specific guidance is provided for the two major industrial sterilization technologies: radiation and EO. Many of the principles and processes outlined, however, are applicable to all sterilization technologies. In addition, there are material compatibility resources available in the literature for some of the less utilized technologies, for example, for hydrogen peroxide and ozone (AAMI, 2008), nitrogen dioxide (Kulla et al., 2009), and supercritical carbon dioxide (White et al., 2006).

To reap the high patient safety benefits of robust terminal sterilization technologies, the biomaterials scientist needs to manage potential material compatibility concerns. This is necessary in order to ensure a high quality product, and to claim compliance to international sterilization standards that require assessment of material compatibility with terminal sterilization.

Requirements in Sterilization Standards to Assess Material Compatibility

International sterilization standards require the assessment of material compatibility under worst case sterilization conditions. For radiation sterilization, ISO 11137-1 (see Table III.1.2.1) requires the following:

> *8.1.1 When treated with **the maximum acceptable dose**, product shall meet its specified functional requirements throughout its defined lifetime.*

For ethylene oxide sterilization, ISO 11135-1: 2006 (see Table III.1.2.1) requires the following:

> *7.2.1 It shall be confirmed that the product and its packaging meet specified requirements for safety, quality, and performance following the application of the defined sterilization process at **the most challenging process parameters** for the product/package. The influence of the tolerances for the process parameters shall be taken into consideration.*

Effect of Terminal Sterilization Modalities on Materials

Knowledge of a sterilization modality's processing parameters and how they interact with materials (see above, Sterilization Technologies) is a key element in assessing the compatibility of materials with a given terminal sterilization process. A summary of material effects with six sterilization modalities is given in AAMI TIR17 (see Table III.1.2.1); radiation and ethylene oxide sterilization are reviewed below.

Radiation Sterilization. Radiation chemistry initiated by secondary electrons (see above, Sterilization Technologies) produces material effects during radiation sterilization. These effects include generation of free radicals that can cause polymer chain scission or cross-linking, as well as interaction with oxygen that can do further damage to the synthetic polymer or biologic product being sterilized. The radiation chemistry of polymeric materials and industrial applications have been studied and extensively reviewed (e.g., Forsythe and Hill, 2000; Clough, 2001; AAMI, 2008). Many device materials are essentially inert to the relatively low doses applied during radiation sterilization. Some of these materials are actually cross-linked by the radiation, e.g., polyethylene (PE), with a neutral or positive effect on functional properties. Some polymers of medical device interest, however, tend toward scission and are mechanically degraded to a significant extent following radiation sterilization, in particular polytetrafluoroethylene (PTFE), fluorinated ethylene propylene (FEP), polyacetals, and natural polypropylene (PP). In some cases, the presence of air can accelerate chain scission via a radiation-induced oxidation process. Degrading materials may be acceptable in certain applications, but must be evaluated carefully after exposure to a maximum dose to ensure clinically acceptable performance over the shelf-life of the device. In glassy polymers like polymethylmethacrylate (PMMA), low energy electrons may be trapped within the polymer, causing yellow or brownish discoloration, a challenge for radiation sterilization of contact lenses. Also, in cross-linking materials like PE (UHMWPE [Ultra High Molecular Weight Polyethylene] in orthopedic implants) it may be important to anneal out the trapped free radicals, to avoid undesirable, long-term free radical-induced reactions. Bioabsorbable polymers, e.g., polylactide (PLA) and PLGA or poly(lactic-co-glycolic acid), show significant molecular weight (MW) reduction as a function of sterilization dose. This change in initial MW of the polymer may affect *in vivo* polymer degradation dynamics, leading to a change in device performance, such as drug release kinetics from degradable PLGA microparticles. In special cases, it may be possible to account for this MW decrease by increasing the initial pre-sterile molecular weight. In such cases, radiation would be the sterilization modality of choice, since irradiations can be done in a refrigerated state to maintain structural integrity by keeping well below the glass transition temperature of the polymer.

Secondary effects in material may arise from or be avoided by managing the radiation sterilization process. The temperature of a device will be elevated during irradiation, for a few seconds for e-beam sterilization (typically to 50°C with polymeric materials) or for a few hours for gamma sterilization (typically 30–40°C). For low glass temperature bioabsorbable materials, this may be significant. If so, it can be managed by processing the device at a reduced temperature. If oxygen and humidity contribute to deleterious material interactions with radiation, the device can be packaged in an inert packaging environment to manage the problem.

Biologics and Human-Based Tissue: Compatibility with Radiation Sterilization. Progress has been made achieving compatibility of tissue products with radiation sterilization. Sterilization process concerns are addressed in AAMI TIR37: *Sterilization of health care products – Radiation – Guidance on sterilization of human tissue-based products* (see Table III.1.2.1). Low dose radiation sterilization validation methods (see Sterility and Patient Safety) offer promise of expanded use of radiation sterilization with tissues and other biologic products. There are also other interesting cases where biologics are terminally sterilized with radiation after special processing to produce sterile biologics, e.g., sterile liposomal vaccines and human insulin. Special processing may include lyophilization, cryoprotection, and addition of free radical scavengers or radio-protectors (e.g., the anti-oxidant, ascorbic acid) (Mohammed et al., 2006; Terryn et al., 2007). Processes have also been developed using radical scavengers and other tools to improve radiation compatibility of tissue products. These developments offer promise for the expanding market of tissue engineering scaffolds.

Ethylene Oxide Sterilization. Alkylation chemistry, the microbial kill mechanism for DNA, may produce other material effects, but in general the effects are small. Other EO process parameters, i.e., temperature, humidity, and evacuation cycles, can also affect device materials. Materials must be able to withstand the most challenging range of conditions used in a given EO cycle. As described previously, this may include humidity preconditioning before, and aeration cycles afterwards, for several days at 40°C; EO cycles for 6–24 hours at processing temperatures of 40–65°C; RH in the range of 30% to 90%; and evacuation levels down to 2 torr in some cycles with very fast evacuation rates.

Bioabsorbable polymers may be difficult to process under these conditions, in particular if structure integrity is required. Both the temperature and the humidity can degrade bioabsorable material properties. Packaging materials need to be able to withstand the evacuation rates and pressures. Materials with known deleterious responses to typical single EO sterilization cycles include polyacrylates, e.g., polymethylmethacrylate (PMMA), and some styrene resins, e.g., polystyrene (PS) and styrene acrylonitrile (SAN). These materials may be acceptable in certain applications, but must be evaluated carefully after exposure to worse case cycle ("most challenging parameters") to ensure clinically acceptable performance over the shelf-life of the device. It is not expected that EO sterilization will affect device performance over time.

Pharmaceuticals and Biologics: Compatibility with EO Sterilization. In the pharmaceutical/biologics and human-based tissue industries, the issue of sterilization can take on a variety of special conditions and concerns. In the pharmaceutical industry, the predominant use of EO (and radiation) sterilization is used for packaging components that enter the aseptic processes (bottles, plugs, caps). EO is also used to sterilize some dry pharmaceutical active components. Liquids and temperature-sensitive drugs will likely *not* be compatible. Few biologics are compatible with EO sterilization due to concerns regarding temperature and EO residues, as well as the ability for EO to alkylate chemically reactive species such as amines or proteins. This is analogous to the reaction of EO with DNA to sterilize product.

Other Sterilization Modalities. For heat sterilization modalities, material compatibility challenges arise from the temperatures to which the devices are exposed and, for moist heat, from hydrolysis reactions. For oxidative sterilization technologies, e.g., hydrogen peroxide, ozone, and chlorine dioxide, the oxidative mechanism of microbial reduction is also the mechanism by which devices are damaged.

Optimizing Chances for Finding Material Compatibility Solutions

Starting with the best candidate materials is the surest way to optimize material compatibility with terminal sterilization. Guidance on selecting material compatible with six terminal sterilization modalities is available (AAMI, 2008). The most challenging material selection scenario is the combination of a difficult-to-sterilize primary material, which is central to the clinical use of the device, with secondary materials that are also sensitive to sterilization, e.g., an active agent (pharmaceutical or biologic), a bioabsorbable material, and/or active electronics. The best approach is to select a sterilization modality that is compatible with the primary material, and then try to select secondary materials that are compatible with the selected sterilization modality.

It is also important to avoid material processing errors that lead to performance failures caused by sterilization. Both extrusion and molding processes can build in significant polymer stresses, leading to product testing failures post-sterilization. This can be identified by testing product pre-sterilization as a control to compare with post-sterilization testing.

Choosing or developing clinically relevant test methods is also important. It can be tempting to use easily available test methods to demonstrate material performance before and after sterilization. If the test method does not correlate well with clinical performance, the test may not be an appropriate indicator of material compatibility. Clinically-relevant device attributes and test methods vary enormously depending on the device. Physical, analytical, and microbiological tests need to be considered. To characterize drug release kinetics and other combination device properties, micro- and nano-characterization of drug–polymer matrices may be required (Ding et al., 2009). Material biocompatibility can change as a function of sterilization modality and process. Device endotoxin levels are typically not sensitive to the sterilization modality, and testing may often be performed before or after sterilization. The effect of sterilization on the surface of biomaterials should also be considered. Radiation typically has a minimal effect on biomaterial surfaces, while EO, oxidative gases, and liquid chemical sterilants can have either a positive or negative surface effect depending on the application of the biomaterial (Chamberlain et al., 1999).

Ensuring adequate device performance throughout its intended shelf-life is the final material compatibility challenge for the biomedical scientist. The most conservative route is to age the materials in the final product format at room temperature, and confirm the desired functionality throughout the desired shelf-life. If the product is designed to be stored in refrigerator or freezer conditions, this approach can require exceptionally long development timeframes, and can limit the availability of important technologies to the market. As a result, accelerated aging models have been developed to safely and conservatively estimate device performance over time. The cornerstone of all models, however, is confirmation of the model with aging at storage conditions.

Arrhenius (semi-logarithmic) temperature-based accelerated aging models have been used extensively in the medical device industry for product (AAMI, 2008) and packaging (ASTM, 2007). Practical methods have been developed that can accelerate aging by two-fold to greater than ten-fold. Basic conservative aging models that are simple and cost-effective have been shown to be appropriate for most materials (Lambert and Tang, 2000). With additional resources, more sophisticated correlations between accelerated and real-time conditions can be developed to provide more aggressive aging factors.

For devices combined with pharmaceutical materials, regulatory agencies expect to have accelerated stability protocols based on ICH (International Conference on Harmonisation) guidelines. Distinguishing between accelerated device-aging models and pharmaceutical stability requirements is both challenging and important.

STERILITY AND PATIENT SAFETY

For a device with significant sterilization material compatibility challenges, a robust understanding of sterility concepts and associated patient safety issues may be the key to discovering a cost-effective sterilization solution. At one end of the spectrum of sterilization options is overkill terminal sterilization. This approach has for decades served the industry well for devices with limited material compatibility concerns. The other end of the spectrum includes liquid chemical sterilization of biologics, and aseptic processing of combination devices, relatively costly options from many perspectives. Between these extremes are creative terminal sterilization validation methodologies, and terminal sterilization to non-traditional specifications. Without the ability to skillfully navigate this continuum of options, the biomaterials scientist may not be able to cost-effectively bring a product to market.

This section begins with a review of sterilization-related product and patient safety issues. This is followed by an overview of basic terminal sterilization principles related to logarithmic microbial reduction, validation of sterility, and maintaining sterility over time through sterile barrier packaging. The end of the section reviews efforts to provide creative sterilization validation methods to reduce material compatibility concerns, and differentiates terminal sterilization from sterilization achieved by aseptic processing.

PRODUCT AND PATIENT SAFETY ISSUES

Review of the US Food and Drug Administration (FDA) product recalls and the Center for Disease Control (CDC) publications indicates very few instances related to inadequate industry terminal sterilization practices (Favero, 2001). Therefore, it appears the existing standards and medical device manufacturers, and associated contractors, are providing safe and effective devices in terms of product sterility. In the following case studies we will take a look at where the industry safety concerns may lie by examining a few sterilization-related incidents with subsequent patient infections or injury.

Three case studies described here in the text illustrate how small changes in product design and/or the materials used, in combination with the sterilization protocols used, caused significant medical problems. The cases are examples which had significant patient safety issues. Case 1 involved a highly respected major device and pharmaceutical manufacturer utilizing a moist heat terminal sterilization process with cooling water. Case 2 is from a medical supply company totally unaware of the requirements related to the GMP (Good Manufacturing Practice) and aseptically processed products or drugs. Case 3 highlights an instance where a sterilizer and sterilization technology was sold to hospitals without regulatory approval. Sterilization technologies need to be evaluated by the manufacturers as being suitable for their devices and the materials used. Hospitals should not use sterilization technologies that are not indicated in the device manufacturer's directions for use. These instances, although significant, are relatively rare.

Case 1

From October 1970 until March 1, 1971, eight US hospitals encountered 150 cases of bacteremias associated with *Enterobacter cloacae* or *Erwinia*, a genus of *Enterobacteriaceae* which is gram-negative. All hospitals had used the same intraveneous (IV) fluids provided by a major manufacturer. The CDC studies for IV systems indicate that at least 6% of all IV tubing or bottles are contaminated after the system has been in use. A higher risk of contamination is attributable to systems that are not changed after 48 hours. However, in these cases the incidence was much higher. CDC sampling and culturing of fluids directly indicated no contamination. Bottles that were opened by unscrewing the bottle cap and closing again revealed contamination. After further investigation, it was determined that the manufacturer was in the process of implementing a new closure system and that the contamination was associated with the new cap design. The older cap design contained a red rubber disc and a Glisonite wafer which was pressed against the bottle opening. It was discovered that the red rubber had antibacterial properties against the contaminating organisms. The new cap design allowed these organisms to be drawn up into the lined cap while cooling after being autoclaved, but the red rubber had been eliminated and no longer provided an antibacterial effect. The organisms were prevalent in the manufacturing environment due to spillage of solutions during manufacturing. Recalls were initiated for all affected products. Ulitimately, the epidemiological studies estimated the total outbreak of bloodstream infections to be between 2000 and 8000 episodes caused by these contaminated IV fluids. Approximately 10% of the case patients in the hospitals studies died while bacteremic or shortly thereafter (CDC, 1997).

Case 2

From December 12, 2004 until February 15, 2005, local health departments and the CDC identified 36 cases of *Pseudomonas fluorescens* infections in patients that were previously administered a heparin-saline flush from prefilled syringes. The majority of the patients had central venous catheters and lines which required removal and treatment with antibiotics. Prefilled syringes cultured by a few of the hospitals involved all grew the contaminating organism. In almost all cases blood cultures from the patients grew *Pseudomonas fluorescens.* The investigation performed by the CDC indicated that seven of nine product lots tested were contaminated. The company filling the syringes had contracted a pharmacy to prepare a heparin concentrate which was returned to the manufacturer and then added to saline bags, and was subsequently used to fill the syringes. Although preparation of drug products to fill prescriptions is considered pharmaceutical compounding and not manufacturing, in this instance the supplier was acting like a manufacturer and subject to the FDA Good Manufacturing Practices. This manufacturer did not perform final product sterility testing of the finished products, which could have detected the contamination and prevented the product distribution (CDC, 2005).

Case 3

April 13, 1998, the FDA issued a safety alert regarding the use of the AbTox Plazlyte™ sterilization system. The Plazlyte™ sterilizers utilized a proprietary, low temperature gas plasma, along with vaporized peracetic acid. Although the system provided sufficient lethality, the FDA had not cleared the safety, performance or instructions for use of this sterilizer. The sterilizers were sold directly to hospitals seeking alternatives for ethylene oxide sterilizers. The warning was initiated due to serious eye injuries to corneal endothelial cells which resulted in corneal transplantation in some patients. The problem occurred when surgical instruments were sterilized, and copper and zinc salts formed on the surfaces of the sterile instruments. The copper compound residues were toxic to corneal endothelia and resulted in blindness (FDA, 1998). During the time period of January 8–14, 1998, six of eight patients undergoing intraocular surgery incurred corneal edema and opacification of the cornea (CDC, 1998).

The majority of patient safety issues do not come from non-sterile, aseptically-processed product, or from improperly sterilized products, regardless of the delivered Sterility Assurance Level (SAL). Patient safety issues develop from the introduction of foreign materials into the body during patient treatment at hospitals. There is a relatively high incidence of surgical site infections (SSI), estimated to be at 2% for all procedures. In 2002 there were approximately 1.7 million hospital-acquired infections (HAIs). Of these, 22% were urinary tract infections, 14% were bloodstream infections, 15% were pneumonia, 22% were surgical site infections, and 17% were related to other issues (Klevens et al., 2007).

As can be seen from these examples, patient safety goes well beyond the ability to provide a sterile product. Nonetheless, the validation of product sterility is an absolute must, and part of the regulatory submission and approval process. Principles of terminal sterilization validation are the next topics covered.

TERMINAL STERILIZATION VALIDATION PRINCIPLES

Logarithmic Microbial Reduction

An amazing feature of terminal sterilization processes is the ability to enormously reduce microbial contamination levels on fully packaged product, with exceptionally high process control to meet regulatory requirements for sterility. The word sterile is defined as "free from viable microorganisms." Sterilization is defined as a "validated process used to render product free from viable microorganisms" (ISO/TS 11139, 2006). These definitions are problematic, as they imply an absolute condition. Sterilization processes are based on microbial inactivation which is exponential in nature, in most cases, and follows first order kinetics. Therefore, the sterility of a product is expressed in terms of probability. While the probability can be reduced to a very low number, it can never be reduced to zero. The probability of a non-sterile unit associated with inactivation of a microbial population is quantified by the term Sterility Assurance Level (SAL).

Logarithmic microbial reduction to achieve various sterility assurance levels is illustrated in Figure III.1.2.7. The SAL required for regulatory purposes is 10^{-3} or 10^{-6}, the probability of one in one thousand units as non-sterile or one in one million units as non-sterile, respectively. The SAL of 10^{-6} is a lesser number, but provides a greater assurance of sterility.

Sterilization Validations and Consensus Standards

If at all possible, regulators require a terminal sterilization process over the risk associated with aseptic processing. The strong track record of terminally sterilized product in regards to patient safety owes much to robust sterilization validation methods. It is, therefore, important to understand the basic principles surrounding terminal sterilization validations. The terminology of international sterilization validation standards is provided in this section to equip the biomaterials scientist to interface effectively with regulatory and sterilization vendor personnel.

Sterilization is a special process where the results of the processing cannot be demonstrated by routine testing. For instance, in order to demonstrate an SAL of 10^{-6} it would require the sterility testing of approximately one million products with only one positive. This is neither possible nor practical. Therefore, sterilization processes are validated. EN 556-1 2001 indicates the following:

> *Evidence that a medical device is sterile comes from: (1) the initial validation of the sterilization process and subsequent revalidations that demonstrate the acceptability of the process; and (2) information gathered during routine control and monitoring which demonstrates that the validated process has been delivered in practice.*
> *The achievement of sterility is predicted from the bioburden level on products, the resistance of the micro-organisms comprising that bioburden, and the extent of treatment imposed during sterilization.*

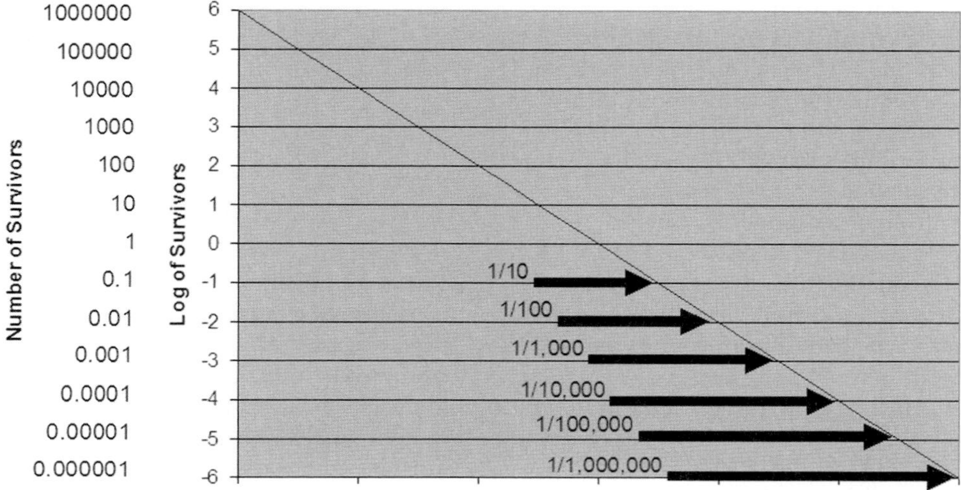

FIGURE III.1.2.7 Sterility Assurance Levels: Illustration of logarithmic microbial reduction.

How are these sterilization validations attained? The medical device and pharmaceutical industries, in conjunction with regulatory authorities, have cooperatively developed industry standards. Device manufacturers, contract sterilization providers, regulatory authorities, and academic resources are all part of the process of developing a consensus standard. These standards provide requirements and guidance for the performance of the sterilization validation. Table III.1.2.1 provides a listing of standards documents utilized for validating and controlling a particular sterilization process.

Sterilization equipment and the process must be qualified and validated. The following paragraphs provide the general steps outlined in the standards for the validation of the sterilization process. Terminology used in the standards is introduced as it is recognized by regulatory bodies around the world. The first considerations are termed "Product Definition" and "Process Definition." For a radiation process this includes defining product families, establishing the maximum acceptable dose (the highest dose the device can withstand and still function properly), and establishing the sterilization dose (the validated minimum dose the device can receive to ensure sterility). The considerations for an ethylene oxide process include defining product families, confirming that the product design allows EO and humidity to penetrate the product and package, defining the hardest to kill location within the product, and determining the microbial rate of inactivation.

Sterilization equipment and systems go through an Installation Qualification (IQ). The IQ entails the calibration of measurement equipment. Utilities and services are verified that they are provided and capable of supporting the equipment requirements. The system software and control hardware are tested and qualified as appropriate for the process. In the case of contract sterilization services, the IQ and Operational Qualification are generally performed by the contract facility, and must be reassessed on a regular basis for inadvertent changes and degradation of equipment performance.

The IQ is followed by an Operational Qualification (OQ). The OQ tests the installed equipment to determine that it operates and performs within predetermined specifications. Procedures for the operation of equipment are established. For example, the OQ for a gamma irradiator would entail dose distribution studies in varying density loads, for all possible modes of operation, and determining the impact of system restarts. The OQ for gaseous sterilization systems (e.g., EO, steam, dry heat) include empty chamber temperature distribution studies, rates and depth of evacuation, injection rates and temperatures of gases, etc.

The next element of the sterilization validation is the Performance Qualification (PQ). ISO 14937 (see Table III.1.2.1) defines the PQ as the: "process of obtaining and documenting evidence that the equipment, as installed and operated in accordance with operational procedures, consistently performs in accordance with predetermined criteria, and thereby yields product meeting its specification." For radiation processes this entails the demonstration of the dose distribution for the product-specific loading configuration, establishing the minimum and maximum dose location(s), and demonstrating that the product can be routinely processed within the range of the minimum and maximum allowable doses. In the case of ethylene oxide sterilization, the PQ would include a microbial qualification under subnominal conditions, and measurement of product temperature and humidity distributions. The reduction or dissipation of sterilant residuals in full cycle worse case conditions (highest gas concentration and longest exposure time) is also required.

Maintaining Sterility Over Time: Primary Package Validation

If the sterilization process is controlled to the extent that the standards require, e.g., such that the process delivers to the product a 9 to 12 or greater log reduction in microbial load, then what do we have to worry about? The package. The product packaging and sterile barrier system is critical to maintaining the safety and integrity of the sterilized product. Device packaging must maintain the product sterility throughout the expected shelf-life. As indicated previously in the section Material Compatibility, this would require accelerated and real-time aging studies. Packaging failures (i.e., open seals, pinholes in materials, material degradation, etc.) represent one of the most common reasons for product recalls. Compromised packaging jeopardizes the safety and integrity of the sterile product, and becomes a patient safety issue. The standards for packaging validation provide the industry with the requirements for validation of device packaging (ISO 11607-1; see Table III.1.2.1).

Packaging failures represent one of the most common reasons for product recalls. Examples of possible causes of such failures include pinholes worn into the sterile barrier from device protrusions, cracks in rigid packaging components, dislodged devices from the packaging fixtures, product damage or wear, and dust generated by device components rubbing against other materials.

Packaging materials used for the sterilization of gaseous sterilants must allow for the ingress of the sterilant to the device surfaces, while packaging for irradiated products can be a vapor barrier. A breathable package may be desirable for irradiated products, as some materials (polyethylene, PVC, and polyurethane used in tubing, drapes, and gowns) tend to generate gases during radiation sterilization. If the gases are trapped within the package, they may present potentially offensive odors when opened. Odor generation may be reduced incorporating antioxidants or by using higher molecular weight materials and breathable packaging materials (Hemmerich, 2000). For product material compatibility purposes, and to protect the product from the environment (i.e., water vapor, oxygen, etc.), one may have to provide packaging with a very low rate of gas transmission.

Product and packaging are subject to an accelerated aging program, and a simulation of the transportation environment. The shipping simulation can be performed according to one of several approaches, either ASTM D4169-05 *Standard practice for shipping containers and systems*, or by an appropriate International Safe Transit Association (ISTA) procedure such as ISTA 2A – Packaged-Products 150 lb (68 kg) or less. Following performance of the shipping simulation and aging, the packaging and products are inspected for packaging failures. Examples of possible failures include pinholes worn into the sterile barrier from device protrusions, cracks in rigid packaging components, dislodged devices from the packaging fixtures, product damage or wear, and dust generated by device components rubbing against other materials.

Alternative Sterilization Validation Methodologies and Specifications

Where does the biomaterials scientist go if standard terminal sterilization processes and validation methodologies do not meet the product need? Straight to liquid chemical sterilization or aseptic processing? These are relatively expensive options that may not provide as high a level of process control as terminal sterilization. It is, therefore, hoped that the biomaterials scientist will explore creative sterilization validation methods, and ensure that the sterility assurance specification being used is appropriate.

Validation Considerations for Minimizing Process Impact. In the majority of sterilization validations, the sterilization process delivers a significantly greater amount of lethality than is actually required. Typically, for devices composed of materials and packaging that are compatible with the sterilization process, the sterilization scientist uses a sterilization process that is easy to validate, such as substantiation of a standard sterilization dose of 25 kGy or an "overkill" approach. EO processes often use half-cycle overkill approaches that deliver SAL values of at least 10^{-12}, and often values of 10^{-20} to 10^{-40}, an overkill of enormous proportions. If product and or package functionality are a potential issue, one should explore options of minimizing the conditions of the terminal sterilization process in order to overcome material compatibility challenges.

EO sterilization process optimization can be achieved by exploring the possibilities of using a bioburden-based validation process and minimizing the conditions of exposure to the process (Annex A of ISO 11135-1). Figure III.1.2.8 illustrates the benefit of simply using actual product bioburden (orange line; assuming a

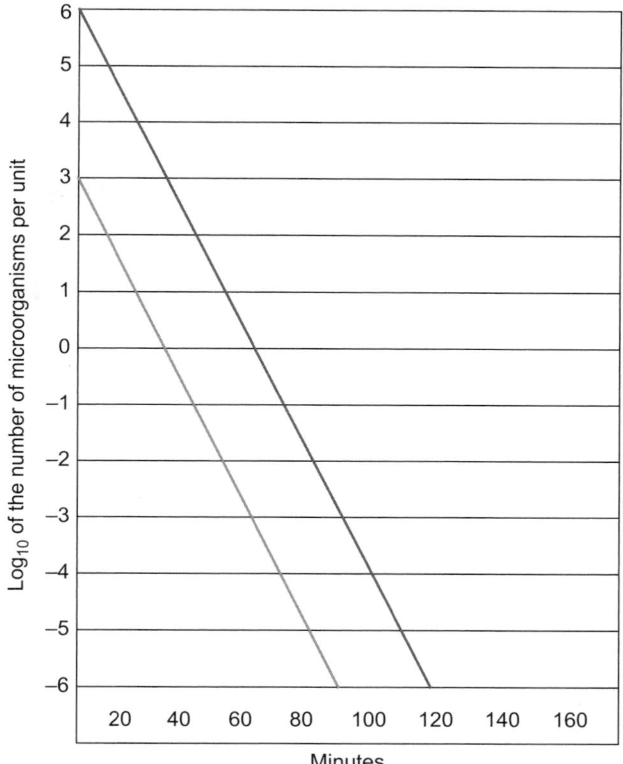

FIGURE III.1.2.8 Microbial contamination level versus EO processing time: High and very high initial contamination levels.

realistic but high bioburden of 1000) as opposed to using the overkill method where an initial bioburden of 1,000,000 is assumed (purple line). The times required to achieve the desired SAL are considerably less. Using the overkill assumption, a 12 log reduction down to an SAL of 10^{-6} is achieved at about 120 minutes, whereas with an initial population of 1000 a 9 log reduction to attain the 10^{-6} SAL is achieved in approximately 95 minutes. If the microbial resistance of the actual product bioburden is not as robust as the severe biological indicator challenge assumed in the overkill method, and the slope of the actual bioburden line (orange line) decreases, the time to achieve an SAL of 10^{-6} is reduced even further.

Even though radiation sterilization validation processes are based on natural product bioburden, as opposed to an overkill method, the sterilization validation may not be optimized to deliver the lowest dose, but an acceptable or convenient dose. New radiation sterilization validation methods have recently been developed (See Table III.1.2.1: Method VD_{MAX}, AAMI TIR33; Modified Method 2, AAMI TIR40) that allow for the lowest possible doses to be achieved for a given product. Along with optimizing the minimum sterilization dose, the biomaterials scientist may want to reduce the range of doses provided in the radiation process. An unnecessarily high distribution of doses throughout one product leads to high product doses and, therefore, significant overkill and potential damage to product. Achieving a low sterilization dose and a narrow distribution of doses may lower the overall product dose to allow for successful radiation sterilization of a sensitive product. In addition, as mentioned in the section Material Compatibility, cold processing and processing with an inert environment within the product packaging may also help reduce material degradation even further.

Application of Sterility Assurance Level (SAL)

The typical SAL for most devices distributed in Europe and the US is 10^{-6} (EN 556-1; see Table III.1.2.1). Historically the US permitted a greater SAL value of 10^{-3} for items such as surgical drapes and gowns (ST67; see Table III.1.2.1). However, with the requirement for CE marking at the start of 1998, and in recognition of certain Pharmacopeia requirements, the SAL value decreased to a greater assurance of sterility, 10^{-6}. This is not to say that a device cannot be submitted and approved by the regulatory authorities with an SAL value greater than 10^{-6}. Regulatory approval for these lower sterility assurance levels, such as 10^{-3}, may be granted if it is shown that the device functionality cannot be achieved with a SAL of 10^{-6}. Alternatively, if the biomaterials scientists, the microbiologists, and sterilization personnel cannot demonstrate that the device is compatible from a product performance perspective with any sterilization process, they may need to move the product to an aseptic process where contamination rates may not be greater than 0.1%. With either of these scenarios, it must also be shown that the benefits of the device for the target population outweigh the risks (potential harms and severity of these harms to the patient) associated with the use of the device. Another major perceived concern for a greater sterility assurance level such as 10^{-3} is the potential use of the device on elderly patients, immuno-compromised patients, those with HIV, transplant patients, burn patients, newborns, and children. This perception is countered by the fact that these high risk patients routinely are treated with aseptically processed products.

Terminal Sterilization Versus Aseptic Processing

In order to put terminal sterilization and sterility assurance levels in perspective, it is useful to compare terminal sterilization with aseptic processing. In aseptic processing there is no inactivation of the product bioburden. These products are not exposed to a sterilization process in their final packaging. Aseptic processing is defined as "handling of sterile product, containers, and/or devices in a controlled environment, in which the air supply, materials, equipment, and personnel are regulated to maintain sterility" (ISO 13408-1, 2008). Aseptic processing includes compounding, filtration, and filling. Through simulations or media fills of the aseptic process, the manufacturer must demonstrate the effectiveness of the controls over the risks of contamination associated with each step of the process. The control over potential contamination in an aseptic process results in a frequency of a non-sterile unit occurring of less than one in one thousand (0.1%). The sterility of the final product is based on the filter efficiency and integrity, the lack of available contamination in the surrounding environment, and controlling a series of operations in order to "maintain" sterility.

Since better process control and patient safety is achieved through terminal sterilization processes, the current regulatory environment requires medical device manufacturers to demonstrate that a product cannot be terminally sterilized before they can convert to an aseptic process. The extent of this demonstration is not well defined. To serve the needs of the industry when a terminal sterilization option is not achievable, there is current activity within the ISO TC198 Working Group 9 on development of a standard for solid medical devices (ISO/Committee Draft 13408-7, *Aseptic processing of health care products – Part 7: Aseptic qualification of solid medical devices and combination medical devices*). It is hoped that the creative terminal sterilization options provided in this chapter will permit some biomaterials scientists to avoid this option.

SUMMARY AND FUTURE CHALLENGES

The medical device industry has been well served by robust workhorse terminal sterilization technologies, radiation and ethylene oxide, in addition to many other technologies that meet lesser needs. The important processing parameters and mechanisms of microbial kill are well established. This information provides foundational insights into the interactions of the sterilization modality with a device's materials. This information, along with available guidance on the compatibility of materials subject to sterilization, allows for the biomaterials scientist to overcome material compatibility challenges in order to reap the benefits of terminal sterilization relative to aseptic processing. Patient safety is ensured by terminal sterilization, as a result of the very high assurance of sterility specifications utilized, strong process control, and robust sterilization validation methodologies that have been incorporated into national and international standards.

The challenge for the biomaterials scientist is to skillfully apply terminal sterilization technologies to as many products as possible, especially the more recent combination devices. There are many material compatibility challenges that will make this a formidable challenge. Examples include drug delivery systems with active biological agents, bioresorbable temperature-sensitive polymers, and radiation-sensitive electronics. The ability to find terminal sterilization solutions to avoid aseptic processing will allow the device industry to provide these innovative therapies at the lowest price, and with the highest associated process control and patient safety.

The solution to these challenges will be an effort from a combination of disciplines. Material scientists and engineers need to develop increasingly robust materials that can withstand terminal sterilization processes. Even so, biomaterials scientists will need to carefully choose a sterilization modality that is compatible with key primary materials; this may require the utilization or development of novel sterilization technologies. Once a technology is defined that is compatible with the most sensitive materials of the device, supporting materials must be selected, processed, and tested to optimize chances for success with the sterilization modality. Optimal accelerated aging techniques to ensure product functionality over time are also required. In addition, to optimize chances of success with sensitive materials, the sterilization community needs to continue to develop and publish creative and gentle sterilization validation methodologies (e.g., minimizing radiation dose; ethylene oxide concentration, humidity, and temperature). Finally, regulatory authorities, industry forums, and international bodies involved in writing standards need to continue to challenge the bases for sterility assurance levels to assure that the terminal sterilization requirement for combination devices is not being set to an arbitrarily high standard.

BIBLIOGRAPHY

See Table III.1.2.1 for additional Sterilization Standard references.

AAMI. (2008). *TIR17 Compatibility of materials subject to sterilization. AAMI.*

ASTM. (2007). *F1980–07 Standard guide for accelerated aging of sterile barrier systems for medical devices. ASTM.*

CDC, Center for Disease Control. (1997). Epidemiologic notes and reports, nosocomial bacteremias associated with intraveneous fluid therapy – USA. *MMWR*, December 26, **46**(51) 1227–1233.

CDC, Center for Disease Control. (1998). Corneal Decompensation after Intraocular Ophthalmic Surgery FDA Safety Alert: Warning Regarding the Use of the AbTox Plazlyte™ Sterilization System. *MMWR,* April 24, **47**(15), 306–308. http://www.fda.gov/MedicalDevices/Safety/AlertsandNotices/PublicHealthNotifications/ucm062297.htm?utm_source-fdaSearch_utm_medium_website_utm_term-AbTox_utm-content-1.

CDC, Center for Disease Control. (2005). Pseudomonas bloodstream infections associated with a heparin/saline flush – Missouri, New York, Texas, and Michigan, 2004–2005. *MMWR*, March 25, **54**(11), 269–272.

Chamberlain, V. C., Lambert, B. J., Tang, F. W., et al. (1999). Sterilization Effects. In A. F. von Recum (Ed.), *Handbook of Biomaterials Evaluation* (pp. 253–261). Taylor and Francis, Columbus, OH.

Clough, R. L. (2001). High-energy radiation and polymers: A review of commercial processes and emerging applications. *Nuclear Instruments and Methods in Physics Research Section B: Beam Interactions with Materials and Atoms*, **185**, 8–33.

Ding, N., Pacitti, S. D., Tang, F. W., et al. (2009). XIENCE V™ stent design and rationale. *Journal of Interventional Cardiology*, **22**(Suppl. 1), S18–S27.

EN556-1. (2001). *Sterilization of medical devices – Requirements for medical devices to be designated "STERILE" – Part 1: Requirements for terminally sterilized medical devices.* BSI: British Standards Institute.

Favero, M. S. (2001). Sterility Assurance: Concepts for Patient Safety. In W. A. Rutala (Ed.), *Disinfection, Sterilization and Antisepsis: Principles and Practices in Healthcare Facilities* (pp. 110–119). Association for Professionals in Infection Control and Epidemiology. NOTE: In addition to the summary of CDC data by decade in the Favero paper, much evidence exists for non-sterilization causes of hospital acquired infections, e.g., hip implant infection rates are 0.2% in high volume clinics versus 4% in low volume hospitals, with no causes related to initial device contamination (*Biomaterials Science*, 2nd ed, 4.8, 1st paragraph).

FDA, US Food and Drug Administration. (1998). *Safety alert: Warning regarding the use of the AnTox Plazlyte™ sterilization system*: http://www.fda.gov/MedicalDevices/Safety/AlertsandNotices/PublicHealthNotifications/UCM062297.

Forsythe, J. S., & Hill, D. J. T. (2000). The radiation chemistry of fluoropolymers. *Progress in Polymer Science*, **25**(1), 101–136.

Klevens, M., Edwards, J. R., Richards, C. L., Jr., Horan, T. C., Gaynes, R. P., et al. (2007). Estimating health care-associated infections and deaths in U.S. hospitals, 2002. *Public Health Reports*, **122**, March–April.

Hemmerich, K. J. (2000). Polymer materials selection for radiation-sterilized products. *Medical Device and Diagnostic Industry*. February.

ISO 13408-1 (2008). *Aseptic processing of health care products – Part 1: General requirements. ISO.*

ISO/TS 11139. (2006). *Sterilization of healthcare products – Vocabulary. ISO.*

Kulla, J., Reich, R., Bioedel, S. Jr., et al. (2009). Sterilising combination products using oxides of nitrogen. *Medical Device and Diagnostic Industry*, **31**(3), March. http://www.mddionline.com/article/sterilizing-combination-products-using-oxides-nitrogen. Accessed 24.05.10.

Lambert, B. J., & Tang, F. W. (2000). Rationale for practical medical device accelerated aging programs in AAMI TIR17. *Radiation Physics and Chemistry*, **57**, 349–353.

Mohammed, A. R., Bramwell, V. W., Coombes, A. G., & Perrie, Y. (2006). Lyophilisation and sterilisation of liposomal vaccines to produce stable and sterile products. *Methods*, **40**, 30–38.

Smith, R. A., Ingels, J., Lochemes, J. J., Dutkowsky, J. P., & Pifer, L. L. (2001). Gamma irradiation of HIV-1. *Journal of Orthopaedic Research*, **19**(5), 815–819.

Terryn, H., Maquille, A., Houée-Levin, C., & Tilquin, B. (2007). Irradiation of human insulin in aqueous solution, first step towards radiosterilization. *International Journal of Pharmaceutics*, **343**, 4–11.

White, A., Burns, D., Christensen, T. W., et al. (2006). Effective terminal sterilisation using supercritical carbon dioxide. *Journal of Biotechnology*, **123**(4), 504–515.

CHAPTER III.1.3 CORRELATION, MATERIALS PROPERTIES, STATISTICS AND BIOMATERIALS SCIENCE

Buddy D. Ratner
Professor, Bioengineering and Chemical Engineering,
Director of University of Washington Engineered Biomaterials (UWEB), Seattle, WA, USA

INTRODUCTION

In chapters prior to this one, this textbook has presented much on the characterization of biomaterials, assessment of biological reaction to biomaterials, and medical applications of biomaterials. What are the prospects for physical and chemical measurements for predicting the performance of new materials in complex medical applications? In other words, can we find relationships between the composition and structure of a biomaterial and its biological interactions, particularly in complex systems like living organisms? This chapter will address this question.

The reality is that physical or chemical measurements which can reliably predict *in vivo* biocompatibility are at this time unavailable for most biomaterials. It would be ideal, when considering a new material for medical applications, to use a spectroscopic technique to measure the properties and, from that physical measurement, predict how well the material will work in a particular application. Animal experiments are expensive, of questionable value for predicting performance in humans, and raise ethical issues (see Chapters II.3.7 and III.2.7). Human clinical trials are very expensive, and also raise ethical issues (Chapters III.2.7 and III.2.9). Can we predict or prescreen *in vivo* or *in vitro* performance from measurements of surface and other physical and chemical properties? This chapter, addressing correlation, will examine this question, and offer suggestions for future exploration.

Correlation implies *dependence* of one set of data on another. Correlations can take many forms. For biomaterials, some possibilities are:

1. Measure specific surface physicochemical properties (for example, contact angle), and correlate that with a biointeraction response such as a protein adsorption or blood cell interaction.
2. Measure protein adsorption or cell adhesion and predict *in vivo* performance.
3. Measure mechanical properties and predict the wear of an implanted hip prosthesis.
4. Measure an animal response to materials (for example, blood platelet consumption by the material), and predict clinical performance.

This chapter will primarily focus on those correlations that fit within number 1 above, but the comments made will have relevance to many correlations that can be envisioned for biomaterials.

BIOCOMPATIBILITY AND MEDICAL DEVICE PERFORMANCE

A re-examination of the definition of biocompatibility presented in the Introduction to this textbook and in Chapter II.3.2 is appropriate at this point.

> *Biocompatibility is the ability of a material to perform with an appropriate host response in a specific application (Williams, 1987).*

The "biocompatibility" of a medical device can be defined in terms of the success of that device in fulfilling its intended function. Thus, the blood filtration module of a hemodialysis system might be defined by its ability to appropriately fractionate soluble blood components, its robustness over its intended lifetime, and its non-damaging interaction with the patient's blood. Alternately, we can define a "blood compatibility" for the membrane, a "blood compatibility" for the silicone rubber header, a "blood compatibility" for the tubing, and a "blood and soft tissue compatibility" for the percutaneous connection between the apparatus and the patient's bloodstream. Similarly, for a hip joint prosthesis, we can discuss the fatigue resistance of the device, the corrosion resistance of the device, the distribution of stresses transferred to the bone by the device, the solid angle of mobility provided, and the overall success of the device in restoring a patient to an ambulatory state. Again, the hip joint prosthesis performance might also be couched in terms of the tissue reaction to the bone cement, the tissue reaction to an uncemented titanium prosthesis stem, and the tissue reaction to the acetabular cup. In both of these examples, two cases are offered: in the first case, a whole system (device) performance is assessed, and in the second case, the biological reaction to specific components of the device (the biomaterials) is examined. The difference between the consideration of the whole device and the materials that comprise it is a critically important point. In certain contexts, only the performance of the complete device can be labeled as biocompatible. This distinction can be inferred from the US Food and Drug Administration (FDA) policy that only complete devices, and never materials, receive "approval." Yet biomaterials scientists and engineers refer to the performance of the individual materials as "biocompatibility." The materials performance can also be described as a "bioreaction" or "bioresponse." This is a textbook on biomaterials, so it is appropriate to focus on the materials. "Bioreaction" is a much more straightforward descriptor than "biocompatible," and this term will largely be used here and defined by example.

Toxicology assays, at times imprecisely referred to as biocompatibility assessment, were presented in Chapter II.3.3. These assays deal with measurement of substances that leach from materials, most of which will induce some cell or tissue reaction. Such assays will be discussed here, because the published correlations are clear, interpretable, and measurable. In contrast, if we concentrate on bioreactions to implanted materials that *do not* leach substances (i.e., most "biocompatible" biomaterials) the surface properties immediately assume a high profile as the prime candidate to control bioreaction. However, hardness, porosity, shape, movement, and specific implant site are also important. These issues will be addressed in working through important concepts in correlating material properties to bioreaction.

Using words such as "correlation" and "dependence" requires some appreciation of data and statistics. This topic will be briefly addressed before the discussion of biomaterials and correlation.

DATA, INFORMATION, AND STATISTICS

Laboratory experiments generate data that comes from the measurement of the properties (biological and physical) of materials. Frequently we are given that data as a column of numbers (maybe a spreadsheet). Staring at these numbers is often unhelpful in understanding the system under study. What we really desire is not *data*, but *information*, about our system. This idea is illustrated in Figure III.1.3.1, proposing how clinical performance data and physicochemical measurement data might lead to the development of an improved implant device. Correlation is one way to process data so that it yields information.

Correlation is a statistical tool useful for analyzing data so as to appreciate its variance, significance, and interrelationships. A general introduction to statistics is not presented here, but every biomaterials research and engineer should be versed in these mathematical tools. A few useful, general books on statistics as applied to scientific problems are Bevington and Robinson (2002), Mosteller and Tukey (1977), Anderson and Sclove (1986), and Urdan (2010). A book that can be useful in interpreting the meaning of statistics is Huck (2000). A search of Amazon.com found over 10,000 textbooks on statistics and probability.

CORRELATION

Correlation is a relationship (dependency) between two or more variables. Correlation does not necessarily mean cause and effect. For example, most people walking with open umbrellas in the rain will have wet feet. But, in this situation, it is obvious that umbrellas do not cause wet feet. We have a high correlation, but it misses the controlling factor (causative factor) in this example – the

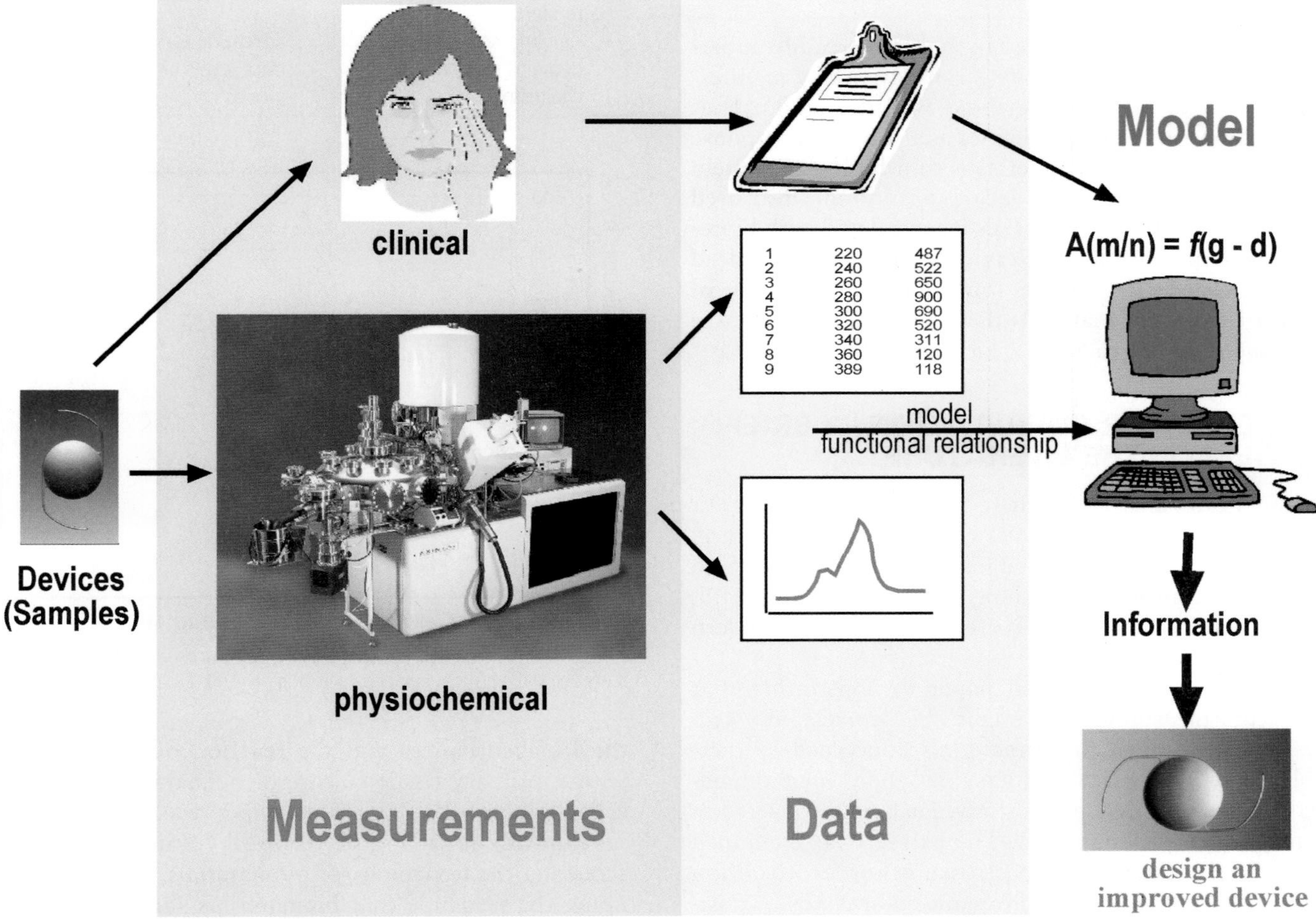

FIGURE III.1.3.1 Clinical and physicochemical data are converted to useful information to optimize the performance of an implant device.

rain. Thus, we can propose that A (the umbrella) causes B (wet feet), B causes A, they cause each other or they are both caused by C (the rain). We cannot prove causation, but it can be strongly suggested. Often, where correlations are observed, the causative factor is obscured, and so we have data but little useful information. Where relationships are established between a dependent variable and an independent (or explanatory) variable, this is referred to as regression analysis.

It may be more productive to look at this problem in terms of calibration and prediction. Calibration in a practical (e.g., analytical) sense has been defined by Martens and Naes as the use of empirical data and prior knowledge for determining how to predict unknown quantitative information Y from available measurements X, via some mathematical transfer function (Martens and Naes, 1989). This could be as simple as plotting Y versus X and using a least squares fit to deal with modest levels of random noise or as complex as a multivariate calibration model to accommodate noisy data, interfering agents, multiple causes, non-linearities, and outliers. Multivariate methods will be elaborated upon toward the end of this chapter.

ASPECTS OF THE BIOREACTION TO BIOMATERIALS

Bioreaction, a process related to, but more readily understood than "biocompatibility," can have many manifestations. Some of these are listed in Table III.1.3.1. A bioreaction is most simply defined as an observed response upon interaction of a material with a biological system or system containing biomolecules. Can simple measured physical properties of materials be correlated with bioreactions? There are many examples where this is indeed the case. Table III.1.3.2 lists some of the surface physical measurements for materials that one might hypothesize as influencing bioreaction or biocompatibility.

THE CASE FOR CORRELATION: A BRIEF REVIEW OF THE LITERATURE

In the 1970s and 1980s there was widespread interest in using correlation to predict biomaterial performance. The lessons learned provided insights, but also tempered the enthusiasm for establishing such relationships. Still, papers continue to explore correlation and a few modern efforts will be described here.

An early and influential paper demonstrating that physical measurements might be correlated with observed reactions to biomaterials concerned extractables from biomaterials (Homsy, 1970). Many biomaterials were examined in this study. Each was extracted in a pseudoextracellular fluid. The extract was examined by infrared absorbance spectroscopy for hydrocarbon bands that are indicative of organic compounds. A positive correlation was observed between the strength of the IR absorbances and the reaction of the materials with a primary tissue culture of neonatal mouse heart cells (Figure III.1.3.2). This paper was important in scientifically justifying *in vitro* cell culture analysis for screening the toxicology of biomaterials, and for developing the principle that biomaterials should not unintentionally leach substances.

TABLE III.1.3.1 Some Bioreactions

• protein adsorption	• macrophage adhesion
• protein retention	• neutrophil attachment
• lipid adsorption	• cell spreading
• bacterial adhesion	• phagocytosis
• platelet adhesion	• biodegradation
• hemolysis	• angiogenesis
• platelet activation	• apoptosis
• expression of new genes	• cell differentiation
• downregulation of genes	• mRNA production
• cytokine release	• fibrous encapsulation

Note: These reactions are commonly observed with implant materials; however, their relationship to "biocompatibility" is not direct.

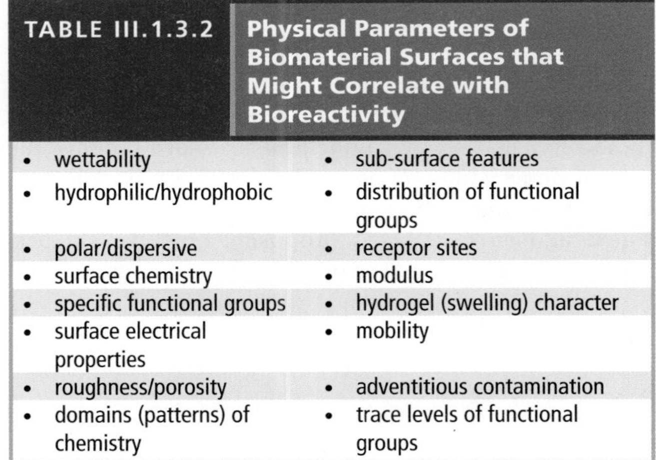

TABLE III.1.3.2 Physical Parameters of Biomaterial Surfaces that Might Correlate with Bioreactivity

• wettability	• sub-surface features
• hydrophilic/hydrophobic	• distribution of functional groups
• polar/dispersive	• receptor sites
• surface chemistry	• modulus
• specific functional groups	• hydrogel (swelling) character
• surface electrical properties	• mobility
• roughness/porosity	• adventitious contamination
• domains (patterns) of chemistry	• trace levels of functional groups

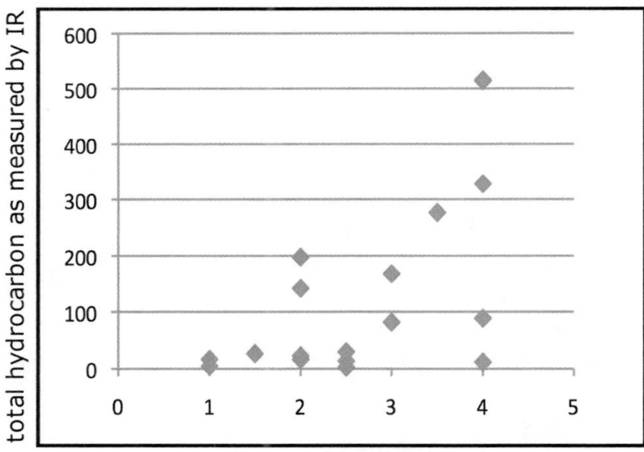

FIGURE III.1.3.2 A hypothesis for a minimum biointereaction for surfaces with critical surface tensions around 22 dynes/cm.

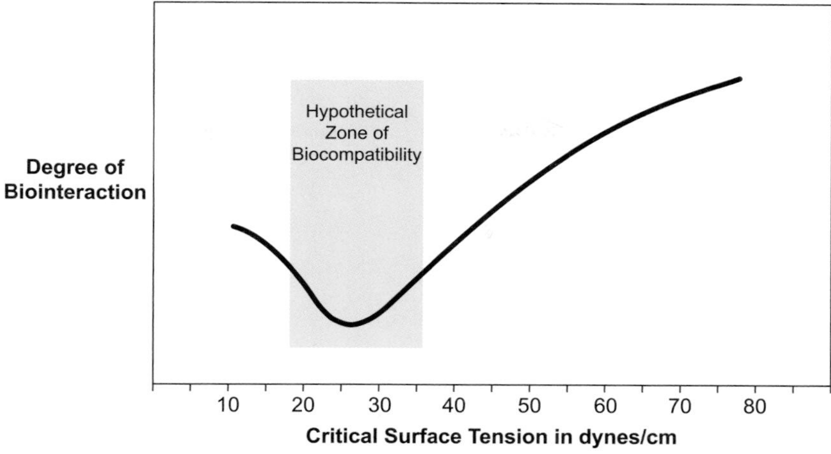

FIGURE III.1.3.3 A correlation between baboon platelet consumption as measured in an *ex vivo* shunt system and hydrogel water content. (From Hanson et al., 1980.)

In the late 1960s, Robert Baier and colleagues offered an intriguing hypothesis concerning surface properties and bioreaction that continues to stimulate new experiments (Baier, 1972). This hypothesis is based upon interfacial energetics of surfaces as approximated from contact angle measurements, and suggests that materials with critical surface tensions (see Chapter I.1.5) of approximately 22 dyne/cm would exhibit minimum bioreactivity (Figure III.1.3.3). Support for this hypothesis has been generated in a number of experiments spanning different types of biointeractions (Dexter, 1979; Baier et al., 1985). However, in a larger number of cases, this minimum has not been observed, raising questions about the generality of this concept (Lyman, 1970; Mohandas et al., 1974; Chang et al., 1977; Yasuda et al., 1978; Neumann, 1979).

Some of the clearest biointeraction correlations have been observed in simple, non-proteinaceous media. Linear trends of cell (mammalian and bacterial) adhesion versus various measures of surface energy have been noted (Mohandas et al., 1974; Chang et al., 1977; Yasuda et al., 1978; Neumann et al., 1979). For example, Chang and co-workers (1977) found that the adhesion of washed pig platelets to solid substrates increased with increasing water contact angle, a parameter that generally correlates well with solid surface tension. It is interesting that these simple linear trends often vanish or diminish if protein is present in the attachment medium (Chang et al., 1977; Neumann et al., 1979; van der Valk et al., 1983). More complex surface energetic parameters have also been explored to correlate bioreaction to surface properties (Kaelble and Moacanin, 1977).

Correlations between material properties and long-term events upon implantation are less frequently seen in the literature. However, some important examples have been published. The baboon A–V shunt model of arterial thrombosis has yielded a number of intriguing correlations (Harker and Hanson, 1979). Using an *ex vivo* femoral–femoral shunt, this model measures a first order rate constant of platelet destruction induced by the shunt material (the units are platelets destroyed/cm^2/day). The values for this surface reactivity parameter are independent of flow rate (after the flow rate is sufficiently high to ensure kinetically limited reaction), length of time that the reaction is observed, blood platelet count, and surface area of the material in contact with the blood. In one experiment, a series of hydrogels grafted to the luminal surfaces of 0.25 cm i.d. tubes was studied (Hanson et al., 1980). The platelet consumption (see Chapter II.3.5) was found to increase in a simple, linear fashion with the equilibrium water content of the hydrogels. This correlation, illustrated in Figure III.1.3.4, is particularly intriguing because the hydrogel materials studies were amide-, carboxylic acid-, and hydroxyl-based. The only clear, correlating parameter was equilibrium water content. In another study, the platelet consumption of a series of polyurethanes was observed to decrease in a linear fashion as the fraction of the polyurethane C1s ESCA surface spectrum that was indicative of hydrocarbon moieties increased (Hanson et al., 1982).

There have been many attempts to correlate specific protein adsorption with biological reaction. A 1975 study showed that the number of platelets adsorbed to polyurethanes correlated inversely with the amount of albumin adsorbed in competition with fibrinogen and IgG (Lyman et al., 1975). For a series of polyurethanes, platelet attachment was shown to correlate with the amount of fibrinogen adsorbed (Chinn et al., 1991). However, two surprising outlier points could not be explained – two materials that adsorbed high fibrinogen levels adhered low levels of platelets. Bailly et al. (1996) used a direct ELISA method to measure adsorbed fibrinogen on catheters, and found that *in vitro* platelet adhesion and *in vivo* catheter thrombogenicty correlated with the amount of adsorbed fibrinogen. Hu et al. (2001) have isolated specific domains within fibrinogen that help to explain their observation that the level of adsorbed fibrinogen correlates with the magnitude of the foreign-body reaction at short (~3 day) implantation times.

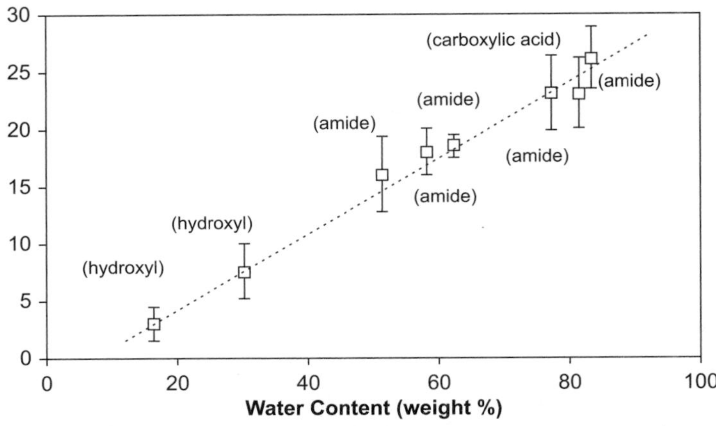

FIGURE III.1.3.4 Four very different surfaces with similar average roughness (R_a).

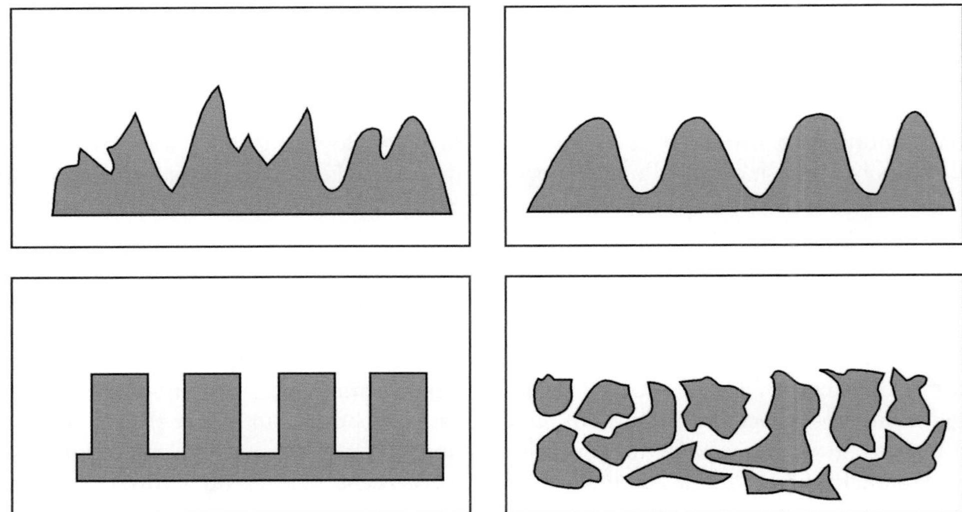

FIGURE III.1.3.5 Cross-sectional views of four surfaces with similar average roughness (R_a).

These studies all implicate levels of adsorbed fibrinogen with complex biological reactions.

A material parameter that lends itself to correlation is roughness or surface texture. Roughness is readily measured using a scanning electron microscope, a profilometer or an atomic force microscope (see Chapter I.1.5). Relationships between roughness and blood hemolysis (Wielogorski et al., 1976) or thrombogenicity (Hecker and Scandrett, 1985) were reported. Textures and roughness are also extremely important to the fixation of materials into hard tissue, and to the nature of the foreign-body response observed (Thomas and Cook, 1985; Schmidt and von Recum, 1991; Brauker et al., 1992) (see Chapter I.2.15). A complication in the use of the roughness parameters is differentiation between porosity and roughness, and also an appreciation of the difference between the average feature amplitude (often call R_a) and the nature of the roughness (e.g., are rolling hills and jagged rocks of the same height also of the same roughness; see Figure III.1.3.5)? Specific porosities, which might be measured as rougher or smoother surfaces, profoundly impact biological reactions (Brauker et al., 1992; Bryers et al., 2012).

Clinical results correlated with material properties are infrequent, in part because materials used in clinical studies are generally not characterized to measure the parameters appropriate to make such correlations, and also because of the complexity of interpreting data from humans. However, a few such studies have been published. In the 1970s, a contact angle measurement criterion was established as a quality control paramater for qualifying the clinical success of processed umbilical cord vascular grafts (Shapiro et al., 1978). In another example, rigid gas permeable contact lens wettability was correlated with subject discomfort, and a predictive trend was noted (Bourassa and Benjamin, 1989). Catheters that are used in humans were evaluated in a test system closely simulating clinical application (Wilner et al., 1978). Catheters were classified into three groups related to their probable success, but clear relationships to surface properties were not discerned. Studies by Bailly et al. (1996) were more successful at establishing relationships between catheter properties and *in vivo* thrombogenicity. The complications in performing control studies, the difficulties in assembling a sufficiently large experimental population, the complexity of the materials (devices), and the human biology make clinical correlation challenging.

ISSUES COMPLICATING SIMPLE CORRELATION

Correlations should allow us to take readily measured physical properties and use that information in the design of improved biomaterials. Although many studies suggesting intriguing possibilities are cited here, this textbook does not present these correlations as rules that can be used in biomaterials design. It should be clear by now that, although many correlations have been noted, there is often contradictory evidence about what the correlating factors are, and the nature of the correlations. Also, in many systems, no obvious correlations have been noted.

The most widely-used correlating factor has been surface energetics, possibly because contact angles can be readily measured in any laboratory (Chapter I.1.5). Surface energetic parameters relate back to the second law of thermodynamics, and it is well-established that the interactions of simple colloid particles can be modeled using thermodynamic and electrostatic arguments. If living cells are treated as simple colloid particles with fixed mass, charge density, polar forces, and hydrophobic interactions, thermodynamic (energetic) modeling may be appropriate (Gerson and Scheer, 1980; Fletcher and Pringle, 1985). However, living cells most often cannot be viewed as "hard, charged spheres." Living cells can change their surface characteristics in response to surfaces and other stimuli. Also, specific (e.g., receptor) interactions do not lend themselves to this simple thermodynamic modeling. For example, two surfaces with similar (but not the same) immobilized oligopeptides, and hence essentially the same surface energy, may interact very differently with cells, if one of the oligopeptides represents a minimal recognized sequence for the cell surface receptor. This was observed with fibroblast cell attachment where an immobilized peptide containing an RGD unit (arginine-glycine-aspartic acid) and a closely related immobilized peptide containing an RGE segment (where the E indicates glutamic acid) were compared (Massia and Hubbell, 1990). The RGD peptide was highly active in inducing cell spreading, while the RGE peptide was not (see Chapter I.2.17 for details on this experiment). Finally, the nature of the correlations may be multivariate rather than univariate. This concept will now be discussed.

MULTIVARIATE CORRELATION

Cause and effect observations are widely used by scientists and engineers to develop generalizations valuable for predicting and modeling phenomena. For example, if the temperature of a solution is increased (the cause), the reaction rate between two chemicals in the solution will increase in a well-defined manner (the effect). There is a simple univariate correlation between temperature and reaction rate. However, many systems, particularly multicomponent systems that are so often important to biomaterials science, have competing reactions that are dependent upon each other (e.g., the product of one reaction may influence the rate of another reaction). Thus, we do not see a simple relationship, but rather many changes occurring simultaneously. Our eye cannot discriminate the key trend(s) in this "stew" of changing numeric values. Multivariate statistics is a class of statistical methods that looks for trends, patterns, and relationships among many variables. Also, where contemporary analytical instrumentation produces large amounts of complex (e.g., spectral) data, multivariate statistics can assist in examining the data for similarities, differences, and trends. Where large data sets overload our ability to discern relationships, multivariate methods thrive on large amounts of data and, in fact, become more accurate and useful. This class of statistical methods has come into its own only with the introduction of powerful computers, since the methods are computation-intensive. Many general introductions to multivariate statistics are available (Sharaf et al., 1986; Massart et al., 1988; Martens and Naes, 1989; Wickens, 1995; Brereton, 2003; Manly, 2004). Multivariate statistics applied to problems involving chemistry are often referred to as "chemometrics."

An important general principle in multivariate analysis is dimensionality reduction. A plot of x versus y requires us to think in two dimensions. A "three-dimensional" plot of x, y, and z can still be easily visualized. Where we have w, x, y, and z as the axes, we lose the ability to absorb the information in graphical form and visualize trends to the data. However, if we take our three-dimensional example, we can visualize a projection (shadow) of the three-dimensional data cluster in two dimensions. We have reduced the dimensionality from 3 to 2. Similarly, our four-dimensional data set can be projected (by a computer) into a three-dimensional space. Thus, we have a data representation we can visualize in order to look for trends. This dimensionality reduction is readily performed by computers using linear algebra methods. The projection of a five-dimensional data set is a four-dimensional shadow, a structure in space our minds cannot visualize. But a computer has no such limitations in looking for relationships in complex multidimensional data sets. The number of dimensions that the computer can work with is, for all practical purposes, unlimited.

There are many multivariate statistical algorithms useful for analyzing data. They are sometimes divided into classification methods (also called cluster analysis methods) and factor analysis methods (Mellinger, 1987). Classification methods find similarities in groups of data points, and arrange them accordingly. Factor analysis methods take data and transform them into new "factors" which are linear combinations of the original data. In this way the dimensionality of the problem is reduced. Factor analysis methods useful for multivariate correlation with data sets such as are acquired in biomaterials research include principal component analysis (PCA) (Wold et al., 1987) and partial least squares (PLS) regression (Geladi and Kowalski, 1986). Review articles oriented to the issues in biomaterials science are Wagner et al. (2006) and Park et al. (2009). Two important points about these methods are that they do not require a hard model (rarely do we have such a quantitative model), and that they make use of all the data (i.e., we do not have choose which data we want to put into the correlation model). There are numerous examples of the application of these methods to biomaterials research (Wojciechowski and Brash, 1993; Perez-Luna and Ratner, 1994; Kempson et al., 2010; Barnes et al., 2012). The power of these statistical methods is being recognized, and they will become standard data analysis tools. This is because they make efficient use of all data, thrive on large amounts of data produced by modern instruments, are objective in that we do not have choose which data to use, are congruent with biomaterials studies that typically have many interrelated variables, and reduce the influence of noise and irrelevant variables, thereby effectively increasing the signal-to-noise ratio.

Multivariate statistical methods can be a great boon to data analysis, but they will not solve all problems in biomaterials science. They should be considered as powerful hypothesis generators. The correlations and trends noted using such analysis represent a new view of the significance of data that we cannot appreciate by staring at spectra or tables. New hypotheses about the importance of materials variables can be formulated, and then they must be tested. Multivariate statistical methods also provide powerful tools for experimental design, but these will not be discussed here.

CONCLUSIONS

Perhaps the reader expected this chapter to provide instruction on the importance of wettability, or roughness or carboxyl group concentration on biological reaction? Unfortunately, biomaterials science is not yet at the state where we can assemble a handbook of design data about the relationships between surface structures and biological reactions. We do not fully-understand the controlling variables from the biology or the materials science sufficiently well to generalize many of our observations. What this chapter does do is to suggest that such relationships probably exist, and to point out that there are powerful mathematical methods that have the potential to help us generalize our data into correlations and trends useful in biomaterials and medical device design.

BIBLIOGRAPHY

Anderson, T. W., & Sclove, S. L. (1986). *The Statistical Analysis of Data*. Palo Alto, CA: The Scientific Press.

Baier, R. E. (1972). The role of surface energy in thrombogenesis. *Bull. NY Acad. Med.*, **48**(2), 257–272.

Baier, R. E., DePalma, V. A., Goupil, D. W., & Cohen, E. (1985). Human platelet spreading on substrata of known surface chemistry. *J. Biomed. Mater. Res.*, **19**, 1157–1167.

Bailly, A. L., Laurent, A., Lu, H., Elalami, I., Jacob, P., Mundler, O., Merland, J. J., Lautier, A., Soria, J., & Soria, C. (1996). Fibrinogen binding and platelet retention: Relationship with the thrombogenicity of catheters. *J. Biomed. Mater. Res.*, **30**, 101–108.

Barnes, C. A., Brison, J., Robinson, M., Graham, D. J., Castner, D. G., & Ratner, B. D. (2012). Identifying individual cell types in heterogeneous cultures using secondary ion mass spectrometry imaging with C60 etching and multivariate analysis. *Analytical Chemistry*, **84**(2), 893–900.

Bevington, P. R., & Robinson, D. K. (2002). *Data Reduction and Error Analysis for the Physical Science* (3rd ed). New York, NY: McGraw-Hill.

Bourassa, S., & Benjamin, W. J. (1989). Clinical findings correlated with contact angles on rigid gas permeable contact lens surfaces *in vivo*. *J. Am. Optom. Assoc.*, **60**(8), 584–590.

Brauker, J., Martinson, L., Young, S., & Johnson, R. C. (1992). *Neovascularization at a membrane-tissue interface is dependent on microarchitecture*. Berlin: Transactions of the Fourth World Biomaterials Congress. April 24-28, 1992, p. 685.

Brereton, R. G. (2003). *Data Analysis for ther Laboratory and Chemical Plant*. New York, NY: John Wiley & Sons.

Bryers, J. D., Giachelli, C. M., & Ratner, B. D. (2012). Engineering biomaterials to integrate and heal: The biocompatibility paradigm shifts. *Biotechnology and Bioengineering*, **109**, 1898–1911.

Chang, S. K., Hum, O. S., Moscarello, M. A., Neumann, A. W., Zingg, W., Leutheusser, M. J., & Ruegsegger, B. (1977). Platelet adhesion to solid surfaces. The effect of plasma proteins and substrate wettability. *Med. Progr. Technol.*, **5**, 57–66.

Chang, S. K., Neumann, A. W., Moscarello, M. A., Zingg, W., & Hum, O. S. (1977). *Substrate wettability and in vitro platelet adhesion*. Grand Island, New York, NY: Abstracts of the 51st Colloid and Surface Science Symposium. June 19-22, 1977.

Chinn, J. A., Posso, S. E., Edelman, P. G., Ratner, B. D., & Horbett, T. A. (1991). Fibrinogen adsorption and platelet adhesion to novel polyurethanes. *Abstracts, American Chemical Society Spring Meeting*. April 14-19, 1991, Atlanta, GA.

Dexter, S. C. (1979). Influence of substratum critical surface tension on bacterial adhesion: *In situ* studies. *J. Coll. Interf. Sci.*, **70**(2), 346–354.

Fletcher, M., & Pringle, J. H. (1985). The effect of surface free energy and medium surface tension on bacterial attachment to solid surfaces. *J. Coll. Interf. Sci.*, **104**(1), 5–14.

Geladi, P., & Kowalski, B. R. (1986). Partial least-squares regression: A tutorial. *Anal. Chim. Acta*, **185**, 1–17.

Gerson, D. F., & Scheer, D. (1980). Cell surface energy, contact angles and phase partition III. Adhesion of bacterial cells to hydrophobic surfaces. *Biochim. Biophys. Acta*, **602**, 506–510.

Hanson, S. R., Harker, L. A., Ratner, B. D., & Hoffman, A. S. (1980). *In vivo* evaluation of artificial surfaces with a nonhuman primate model of arterial thrombosis. *J. Lab. Clin. Med.*, **95**(2), 289–304.

Hanson, S. R., Harker, L. A., Ratner, B. D., & Hoffman, A. S. (1982). Evaluation of artificial surfaces using baboon arteriovenous shunt model. In G. D. Winter, D. F. Gibbons, & H. Plenk, Jr (Eds.), *Biomaterials 1980, Advances in Biomaterials* (Vol. 3, pp. 519–530). Chichester, UK: John Wiley and Sons.

Harker, L. A., & Hanson, S. R. (1979). Experimental arterial thromboembolism in baboons. Mechanism, quantitation, and pharmacologic prevention. *J. Clin. Invest*, **64**, 559–569.

Hecker, J. F., & Scandrett, L. A. (1985). Roughness and thrombogenicity of the outer surfaces of intravascular catheters. *J. Biomed. Mater. Res.*, **19**, 381–395.

Homsy, C. A. (1970). Biocompatibility in selection of materials for implantation. *J. Biomed. Mater. Res.*, **4**, 341–356.

Hu, W. J., Eaton, J. W., & Tang, L. (2001). Molecular basis of biomaterial-mediated foreign body reactions. *Blood*, **98**(4), 1231–1238.

Huck, S. W. (2000). *Reading Statistics and Research* (3rd ed.). Reading, MA: Addison-Wesley.

Kaelble, D. H., & Moacanin, J. (1977). A surface energy analysis of bioadhesion. *Polymer*, **18**, 475–482.

Kempson, I. M., Martin, A. L., Denman, J. A., French, P. W., Prestidge, C. A., & Barnes, T. J. (2010). Detecting the presence of denatured human serum albumin in an adsorbed protein monolayer using TOF-SIMS. *Langmuir*, **26**(14), 12075–12080.

Lyman, D. J., Klein, K. G., Brash, J. L., & Fritzinger, B. K. (1970). The interaction of platelets with polymer surfaces. *Thrombos. Diathes. Haemorrh*, **23**, 120–128.

Lyman, D. J., Knutson, K., McNeill, B., & Shibatani, K. (1975). The effects of chemical structure and surface properties of synthetic polymers on the coagulation of blood. IV. The relation between polymer morphology and protein adsorption. *Trans. Am. Soc. Artif. Int. Organs*, **21**, 49–54.

Manly, B. F.J. (2004). *Multivariate Statistical Methods: A Primer* (3rd ed.). Boca Raton, FL: Chapman and Hall/CRC Press.

Martens, H., & Naes, T. (1989). *Multivariate Calibration*. New York, NY: John Wiley and Sons.

Massart, D. L., Vandeginste, B. G.M., Deming, S. N., Michotte, Y., & Kaufman, L. (1988). *Chemometrics: A Textbook*. Amsterdam: Elsevier Science Publishers.

Massia, S. P., & Hubbell, J. A. (1990). Covalent surface immobilization of Arg-Gly-Asp- and Tyr-Ile-Gly-Ser-Arg-containing peptides to obtain well-defined cell-adhesive substrates. *Anal. Biochem.*, **187**, 292–301.

Mellinger, M. (1987). Multivariate data analysis: Its methods. *Chemometrics and Intelligent Laboratory Systems*, **2**, 29–36.

Mohandas, N., Hochmuth, R. M., & Spaeth, E. E. (1974). Adhesion of red cell to foreign surfaces in the presence of flow. *J. Biomed. Mater. Res.*, **8**, 119–136.

Mosteller, F., & Tukey, J. W. (1977). *Data Analysis and Regression*. Reading, MA: Addison Wesley.

Neumann, A. W., Moscarello, M. A., Zingg, W., Hum, O. S., & Chang, S. K. (1979). Platelet adhesion from human blood to bare and protein-coated polymer surfaces. *J. Polym. Sci., Polym. Symp.*, **66**, 391–398.

Park, J. -W., Min, H., Kim, Y. -P., Kyong Shon, H., Kim, J., Moon, D. W., & Lee, T. G. (2009). Multivariate analysis of ToF-SIMS data for biological applications. *Surface and Interface Analysis*, **41**(8), 694–703.

Perez-Luna, V. H., Horbett, T. A., & Ratner, B. D. (1994). Developing correlations between fibrinogen adsorption and surface properties using multivariate statistics. *J. Biomed. Mater. Res.*, **28**, 1111–1126.

Schmidt, J. A., & von Recum, A. F. (1991). Texturing of polymer surfaces at the cellular level. *Biomaterials*, **12**, 385–389.

Shapiro, R. I., Cerra, F. B., Hoffman, J., & Baier, R. (1978). Surface chemical features and patency characteristics of chronic human umbilical vein arteriovenous fistulas. *Surg. Forum*, **29**, 229–231.

Sharaf, M. A., Illman, D. L., & Kowalski, B. R. (Eds.), (1986). *Chemometrics*. New York, NY: John Wiley & Sons.

Thomas, K. A., & Cook, S. D. (1985). An evaluation of variables influencing implant fixation by direct bone apposition. *J. Biomed. Mater. Res.*, **19**, 875–901.

Urdan, T. C. (2010). *Statistics in Plain English* (3rd ed.). New York, NY: Routledge.

van der Valk, P., van Pelt, A. W.J., Busscher, H. J., de Jong, H. P., Wildevuur, C. R.H., & Arends, J. (1983). Interaction of fibroblasts and polymer surfaces: Relationship between surface free energy and fibroblast spreading. *J. Biomed. Mater. Res.*, **17**, 807–817.

Wagner, M., Graham, D., & Castner, D. (2006). Simplifying the interpretation of ToF-SIMS spectra and images using careful application of multivariate analysis. *Applied Surface Science*, **252**(19), 6575–6581.

Wickens, T. D. (1995). *The Geometry of Multivariate Statistics*. Hillsdale, NJ: Lawrence Earlbaum Associates.

Wielogorski, J. W., Davy, T. J., & Regan, R. J. (1976). The influence of surface rugosity on haemolysis occurring in tubing. *Biomed. Eng.*, **11**, 91–94.

Williams, D. F. (1987). *Definitions in biomaterials. Proceedings of a Consensus Conference of the European Society for Biomaterials*. Chester, England, March 3–5 1986, Vol. 4, The Netherlands: Elsevier.

Wilner, G. D., Casarella, W. J., Baier, R., & Fenoglio, C. M. (1978). Thrombogenicity of angiographic catheters. *Circ. Res.*, **43**(3), 424–428.

Wojciechowski, P. W., & Brash, J. L. (1993). Fibrinogen and albumin adsorption from human blood plasma and from buffer onto chemically functionalized silica substrates. *Coll. Surf. B: Biointerfaces*, **1**, 107–117.

Wold, S., Esbensen, K., & Geladi, P. (1987). Principal component analysis. *Chemometrics and Intelligent Laboratory Systems*, **2**, 37–52.

Yasuda, H., Yamanashi, B. S., & Devito, D. P. (1978). The rate of adhesion of melanoma cells onto nonionic polymer surfaces. *J. Biomed. Mater. Res.*, **12**, 701–706.

CHAPTER III.1.4 DEVICE FAILURE MODE ANALYSIS

Frederick J. Schoen[1] *and Allan S. Hoffman*[2]

[1]Professor of Pathology and Health Sciences and Technology (HST), Harvard Medical School, Executive Vice Chairman, Department of Pathology, Brigham and Women's Hospital, MA, USA

[2]Professor of Bioengineering and Chemical Engineering, UWEB Investigator, University of Washington, Seattle, WA, USA

The design of implants, the biomaterials selected for them, their fabrication, and the protocols used to test them before use in humans, are all intended to minimize the possibility of device failure. Indeed, the majority of such devices serve their patients well, alleviating pain and disability, enhancing quality of life, and/or increasing survival. Nevertheless, some medical devices fail, often following extended intervals of satisfactory function. Problems involving medical device technology are often called "medical device error" (Goodman, 2002). Unraveling a cause of failure in an individual case usually requires systematic integration of clinical and laboratory information pertaining to the patient, and careful and protocol-driven pathological analyses. Analysis of a cohort of failed devices, often called "failure mode analysis", involves many such cases and sometimes additional investigation, such as review of corporate quality assurance and/or other documents. Irrespective of implant site or desired function of the device, the overwhelming majority of clinical complications produced by medical devices fall into several well-defined categories, which are summarized in Table III.1.4.1.

TABLE III.1.4.1 Patient–Device Interactions Causing Clinical Complications of Medical Devices

Thrombosis
• thrombotic occlusion
• thromboembolism
• anticoagulation-related hemorrhage (owing to the therapy to prevent thrombosis)
Infection
Inappropriate healing
• too little/incomplete
• excessive tissue overgrowth
Structural failure due to materials degeneration
• wear
• fracture/fatigue
• calcification
• tearing
Adverse local tissue interactions
• inflammation
• toxicity
• tumor formation
Migration
• whole device
• embolization or lymphatic spread of materials fragments
Systemic and miscellaneous effects
• allergy
• heart valve noise

Unraveling a cause of failure in an individual case usually requires systematic integration of clinical and laboratory information pertaining to the patient, and careful and protocol-driven pathological analyses. Analysis of a cohort of failed devices, often called "failure mode analysis", involves many such cases and sometimes additional investigation, such as review of corporate quality assurance and/or other documents.

Medical device failures have some features in common with medical errors in general, in that they result from an alignment of "windows of opportunity" in a system's defenses (Carthey et al., 2001). In general, demographic, structural, functional or physiologic factors relating to the patient (e.g., age, anatomy of the implant site, activity level, genetic predisposition to complications to thrombosis, allergies) and the implantation procedure (e.g., implant selection, technical aspects of the surgery, potential damage to the implant), are superimposed on the biomaterials and device design, which collectively induce vulnerabilities to particular failure modes (Figure III.1.4.1). Thus, we will see that failure of a device in the clinical setting often involves multiple contributing factors. For example, a material or design inadequacy that could potentially cause excessive wear or fracture in an orthopedic prosthesis (hip or knee joint) may only come

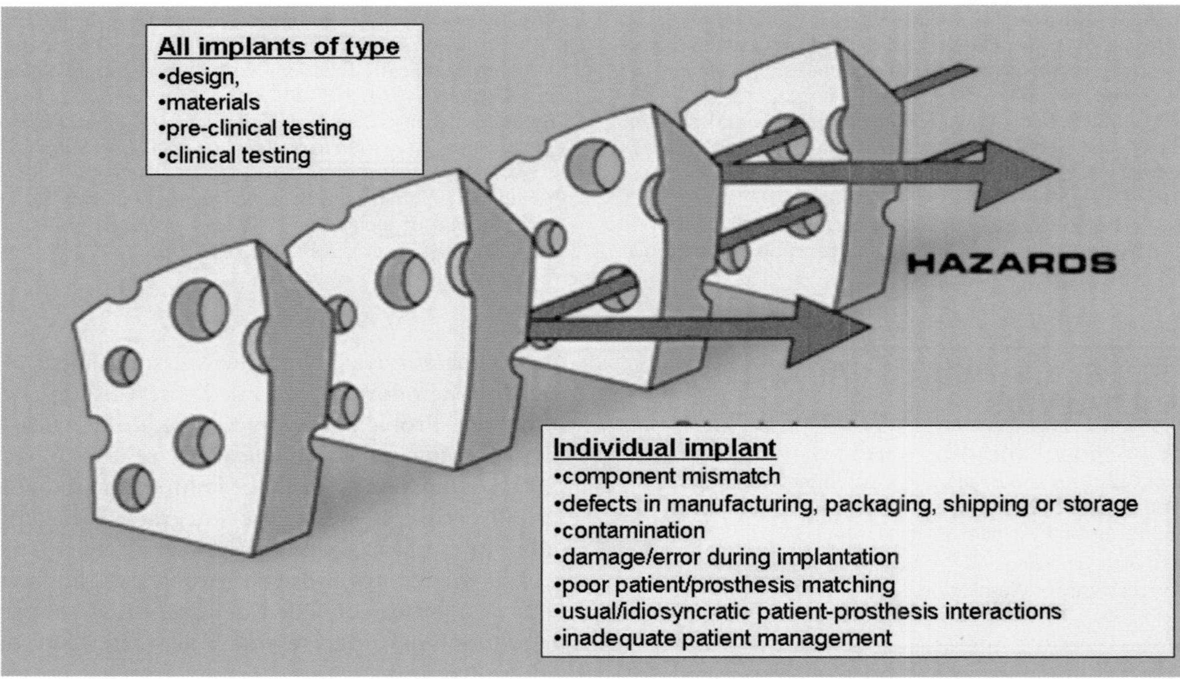

FIGURE III.1.4.1 Analysis of implant failure using systems approach to understanding multiple factors (including latent factors, i.e., those relating to any implant of a particular type, and active factors, i.e., conditions affecting a particular implant), useful in critical incident and near-miss reporting in other high technology industries. When potential risks "line up" a failure can occur. (Concepts modified from Carthey, J., de Leval, M. R. & Reason, J. T. (2001). The human factor in cardiac surgery: Errors and near misses in a high technology medical domain. Ann. Thorac. Surg. **72,** 300–305. Reproduced by permission from The Society for Thoracic Surgeons)

to light in a young, athletic patient. Moreover, a tendency toward thrombosis of a new design of prosthetic heart valve may only cause a problem in a patient whose anti-coagulation effectiveness drops below a critical point, who has a genetic hypercoagulability disorder or who has local blood stasis owing to atrial fibrillation. The determination of the cause(s) and contributory mechanism(s) of implant or other device failure is accomplished by a process called *implant retrieval and evaluation* (see Chapter II.1.5).

> In general, demographic, structural, functional or physiologic factors relating to the patient (e.g., age, anatomy of the implant site, activity level, genetic predisposition to complications to thrombosis, allergies) and the implantation procedure (e.g., implant selection, technical aspects of the surgery, potential damage to the implant), are superimposed on the biomaterials and device design, which collectively induce vulnerabilities to particular failure modes.

The fundamental objective of incident investigation is to identify the root cause of an incident, and eliminate or decrease the risk (based on the probability and severity) of recurrence (Baretich, 2007). The results of implant failure analysis have several potential implications for quality care in individual patients or cohorts of prior and potential recipients, and thus the US Food and Drug Administration (FDA) Medical Device Reporting regulations require investigation and reporting of certain device-related incidents (Shepherd, 1996). Analysis of an isolated failed implant or of multiple implants that have suffered a consistent failure mode can provide important information for individual patient care. For example, such information can:

- Impact on implant selection and patient–prosthesis matching.
- Mandate altered management of a specific patient, such as a change in the type or dose of drug therapy, such as anticoagulation, and/or closer monitoring of the patient with the at-risk type of prosthesis by echocardiography for a heart valve or X-ray or bone scan to monitor a hip joint.
- Reveal a vulnerability of a sub-group of patients (often called a *cohort*) with a specific prosthesis type to a particular mode or mechanism of device failure. This can both lead to closer scrutiny of that group of patients with such a device already implanted, and/or stimulate the need for a prospective change in design, materials selection, processing or fabrication, or potential action by regulatory agencies such as further testing or withdrawal from use.

The results of failure analysis can also influence product liability litigation, either in an individual case or in a class action proceeding involving multiple patients who have received a device with a specific real or potential failure mode.

The purpose of this chapter is to define a conceptual approach to the determination of the factors responsible for a device or implant failure. The aim is to provide a basis for using failures to guide patient management and enhance high-quality design, biomaterials development, selection, and implantation of implants.

> Analysis often reveals considerations which relate to the development, testing or manufacturing processes, and thereby potentially affect all devices in a cohort, or conditions germane primarily to a particular failed device and the patient who received it.

Questions to be addressed in a specific case might be:

- Was the device chosen the correct type and size for the patient?
- Was it optimally implanted or was there a technical error in placement?
- Was the implant damaged during fabrication or implantation by the surgeon or interventionalist?
- Did the patient have an idiosyncratic pathophysiologic response to the implant, such as allergy or an inherited heightened tendency toward blood clotting on the component biomaterials?
- Was there a design flaw in this prosthetic device type or a poor choice of materials in using an appropriate design?
- Did failure arise because the pre-clinical testing phases of the device did not (or were unable to) reveal some defect of design or materials which only became apparent following use of the device in a large number of patients or for longer duration following large-scale clinical utilization?

The array of potential causes is summarized in Table III.1.4.2 and conceptualized in Figure III.1.4.2. This chapter is not meant to describe the detailed mechanisms of either deterioration or biological responses of materials, although they may be contributory to complications (which are covered in Sections II.2 and II.4 of this book).

Multiple recent instances of medical device cohort failures have occurred in the last decade; particularly notable are the metal-on-metal hip replacements implicated for excessive wear (Browne et al., 2010), late thrombosis in drug-eluting endovascular stents (Holmes et al., 2010), and some implantable cardioverter-defibrillators with defective leads, making them unable to provide life-saving heart rhythm corrections (Becker, 2005; Maron and Hauser, 2007).

TABLE III.1.4.2 Potential Contributory Causes of Implant or Device Failure
Inadequate materials properties
Inadequate materials testing
Design flaw
Inadequate pre-clinical animal test models
Components missized/mismatched during fabrication
Defects introduced during manufacturing, packaging, shipping, and storage
Damage or contamination during sterilization
Damage or contamination during implantation
Technical error during implantation
Poor patient–prosthesis matching
Unavoidable, physiologic patient–prosthesis interactions
Inadequate patient management
Unusual/abnormal patient response

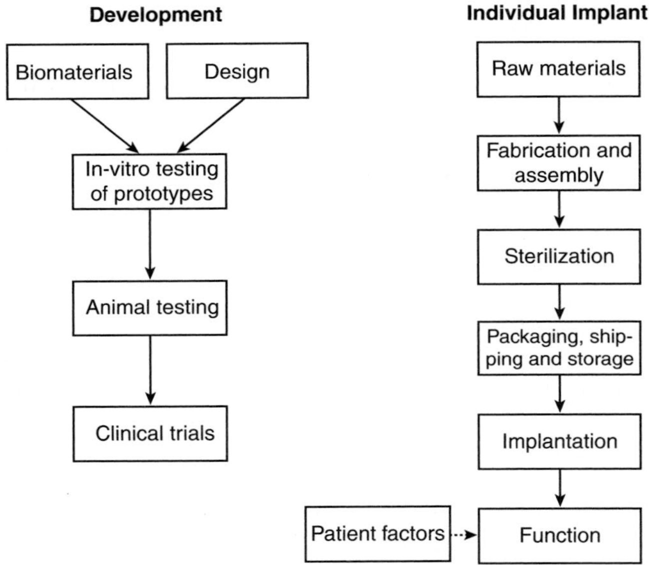

FIGURE III.1.4.2 Potential causes of implant or device failure.

ROLE OF BIOMATERIALS–TISSUE INTERACTIONS: EFFECT OF MATERIALS ON THE PATIENT AND THE EFFECT OF THE PATIENT ON THE MATERIALS

Biomaterials Selection

Mechanical failure of biomaterials and adverse biomaterials–tissue interactions are often implicated as the key factors in a device failure. The wide range of different material "breakdown" mechanisms which can cause an implant or device to fail is summarized in Table III.1.4.3. A device or implant can fail simply because the component biomaterials did not have, and/or were not tested for, the requisite physical, chemical or biological properties for the intended application. For example, the poor resistance of Teflon® (polytetrafluoroethylene) to abrasive wear, and its unsuitability for use in both hip joint prostheses (Charnley et al., 1969) and artificial heart valves (Silver and Wilson, 1977) was appreciated only following extensive clinical use. Similarly, the inadequate durability of a carbon-reinforced Teflon® in temporomandibular joint replacement was encountered in clinical usage (Trumpy et al., 1996). In addition, the earliest silicone ball poppets in the Starr–Edwards caged-ball heart valves frequently absorbed lipids from blood during function, and consequently swelled and became brittle, and sometimes fractured (Chin et al., 1971). "Ball variance," as this complication was known, caused valve dysfunction and/or downstream embolization of fragments. This problem stimulated development of an improved processing protocol for curing medical silicone elastomers that enhanced their mechanical properties, mitigated the lipid absorption, and thereby effectively eliminated ball variance. Conversely, a material used to form the femoral stem of a hip joint prosthesis which is too stiff can cause "stress-shielding," leading to structurally poor surrounding bone remodeling and consequent loosening of the prosthesis (Huiskes, 1998).

TABLE III.1.4.3 Some Mechanisms of Biomaterial "Breakdown"
Mechanical
• Creep
• Wear
• Stress cracking
• Fracture
Physiochemical
• Adsorption of biomolecules (such as proteins leading to irreversible fouling)
• Absorption of H_2O or lipids (leading to softening or hardening)
• Leaching of low molecular weight compounds (such as plasticizer) leading to weakening or embrittlement
• Dissolution (leading to disintegration)
Chemical/biochemical
• Oxidation or reduction (leading to bond scission or cross-linking)
• Hydrolysis (such as amide or ester bonds, leading to bond scission)
• Mineralization/calcification
Electrochemical
• Corrosion

Design

Device design is critical to performance. Design of the femoral stems of hip joint prostheses has evolved from clinical experience with fractured stems in human implants to computer-aided design and manufacturing techniques. The design of a heart valve affects the pattern of blood flow and associated platelet damage, and/or the presence of regions of blood stasis, both of which can lead to thrombosis (Yoganathan et al., 1978; Yoganathan et al., 1981).

Device design is critical to performance. Design of the femoral stems of hip joint prostheses has evolved from clinical experience with fractured stems in human implants to computer-aided design and manufacturing techniques. The design of a heart valve affects the pattern of blood flow and associated platelet damage, and/or presence of regions of blood stasis, both of which can lead to thrombosis.

However, redesigning a medical device to eliminate one complication can have unintended, potentially serious consequences. When a widely used tilting disk heart valve was redesigned to allow more complete opening, and thereby enhance its hemodynamic function as well as reduce the incidence of thrombosis, a large number of mechanical failures occurred (Walker et al., 1995). Nearly all failures had similar characteristics; they resulted from fracture of the welds anchoring the metallic struts (which confine and guide the motion of the disk) to the housing, with consequent separation of the strut and escape of the disk. Pathology studies revealed that the underlying problem was metal fatigue resulting from design-related high-velocity over-rotation of the disk during closure, excessively stressing the welds, potentially coupled with intrinsic flaws in the welded regions (Schoen et al., 1992). Interestingly, many such failures occurred in young patients during exercise, when cardiac forces and thereby valve closing velocities, are enhanced (Blot, 2005). Another example was a specific design of a bileaflet tilting disk heart valve which developed fractures that were initiated in various locations along the contacts of the disks with the housing (Baumgartner et al., 1997). In this case, the likely cause was a design flaw which led to the formation of cavitation bubbles during function as the disk moved away from the housing during the earliest phases of valve opening. The hypothesis was that implosion of these bubbles initiated microscopic cracks in the pyrolytic carbon components which then precipitated gross fractures.

TESTING OF BIOMATERIALS–DESIGN CONFIGURATIONS

In some cases, the pre-clinical *in vitro* or *in vivo* evaluation studies that were carried out on the material itself, or on the design and device prototype, either did not adequately simulate the range and/or nature of conditions which are encountered in actual clinical use or were "under powered" (did not have enough replicates) to reveal relatively low-frequency complications, especially where those complications are potentiated by infrequent patient-related factors. For example, some patients have genetic conditions that predispose to hypercoagulability; such individuals would be particularly vulnerable to clotting in cardiovascular devices (Valji and Linenberger, 2009). Individuals taking high therapeutic doses of (cortico) steroid medications for inflammatory/autoimmune disease may not heal implants properly. The inability to reveal certain complications in pre-clinical studies may be a result of the necessarily limited numbers of animals used, or differences in animal versus human anatomy or physiology. In the case of a new bileaflet tilting disk heart valve design which suffered clinical failures due to thrombosis initiated at the points where the hemidisks contacted and pivoted in the housing, the testing of implants in sheep failed to reveal a tendency toward thrombosis (Gross et al., 1996). Furthermore, appropriate tests are not available for some physiologic responses, especially but not exclusively where those problems potentially arise from immunological interactions of xenograft (animal) or allograft (from another person) tissues with humans. For example, the potential for a significant immunological response, such as has been suggested but not proven in some patients with silicone gel breast implants, and is always a possibility in tissue-based biomaterials, is difficult to evaluate adequately in pre-clinical studies. It is also possible that available tests were overlooked because the specific problem types that occurred later were not anticipated.

A particularly illustrative example points out not only the need for improved and comprehensive pre-clinical testing regimes, but also that a change in biomaterial implemented to alleviate one complication may precipitate another problem if corresponding design changes are not considered. Specifically, bioprosthetic heart valves were fabricated from photooxidized bovine pericardium to mitigate the well-known potentiating effect of glutaraldehyde on calcification. However, many of these valves failed by abrasion-induced tearing of the pericardial tissue against the cloth which covered a portion of the stent (Schoen, 1998). The issue in this case was that photooxidized pericardium has a higher compliance than the glutaraldehyde-preserved pericardium used, but the design was unchanged from that of a valve which had been clinically successful. Higher tissue compliance led to greater excursion of the cuspal material during valve closing; the greater tissue movement of the more compliant material caused the tissue to contact and thereby abrade against the cloth. However, this was not revealed during extensive pre-implantation *in vitro* bench fatigue testing, most likely owing to the markedly accelerated rate of *in vitro* durability testing which limits tissue movements used owing to a high rate of cycling (generally approximately 20 times or more actual). Moreover, animal testing done in sheep was not (and is typically not) extended more than approximately five months, except in a few specimens which in this case did not exhibit gross failure.

BIOLOGICAL TESTING OF IMPLANTS

After prototype devices have been designed and fabricated, biologic tests *in vitro* and in animals are normally used to gain more information for eventual regulatory approval and introduction of such a device into the clinic. However, as with the *in vitro* tests on the materials or designs, *in vivo* tests on the final device or implant could be poorly chosen, key tests could be overlooked, and/or the animal model selected or permitted may not be the most appropriate one available. For example, there have been continuing arguments over the past several decades among biomaterials scientists as to whether the dog is an appropriate animal model for evaluating blood–surface interactions, especially since dog platelets are considered to be much more adherent to foreign surfaces than human platelets. Furthermore, there is no adequate animal model to study the intense inflammatory reaction to wear debris and other particulates which can accumulate adjacent to a hip joint (Holt et al., 2007). Another case in point highlights a problem that arose owing to differences in healing in some animal models versus humans with a caged-ball mechanical prosthetic heart valve type which had a ball fabricated from silicone and cloth-covered cage struts (Schoen et al., 1984). This design was intended to encourage anchoring and organization of thrombus in the fabric, and thereby decrease thromboembolism. Pre-clinical studies of this valve concept utilized mitral valve implants in pigs, sheep, and calves; in such models the cloth-covered struts were rapidly healed by endothelium-coated fibrous tissue (Bull et al., 1969). However, subsequent clinical implantation of these valves was complicated by cloth wear and embolization before healing occurred. This vivid "case history" illustrates and reinforces the concept that human implantation may reveal important problems not predicted by pre-clinical animal testing. The more vigorous healing that occurred in the preclinical implants in this case is typical in animals compared to humans; this sometimes makes prediction of such problems very difficult. This reinforces the need for closely monitored clinical (human) trials during the introduction of a new implant type, and continued post-market surveillance following widespread general availability and implantation thereafter (Blackstone, 2005).

Animal models also must be used thoughtfully in the study of calcification of tissue valves (Schoen and Levy, 1999), another problem that results from an adverse biomaterial–tissue interaction (see Chapter II.4.5). Two animal models are generally used: (1) subcutaneous implants of valve materials in weanling (approximately three-week-old) rats, which achieve clinical levels of calcification in eight weeks or less; and (2) mitral valve replacements in juvenile sheep, which calcify extensively in 3–5 months. However, when older animals are used (in either model), calcification is vastly diminished. Use of inadequately severe models could lead to overestimation of the efficacy of an anticalcification strategy. This raises a more generic problem: that demonstration of the absence or prevention of specific pathologic features by a therapeutic option in a study requires proper controls to ensure that the model used is capable of provoking the pathology being investigated.

RAW MATERIALS, FABRICATION, AND STERILIZATION

In unusual circumstances, a batch of raw material may become defective or damaged. The fabrication process used to manufacture a device can introduce defects or contaminants which can lead to failure of a device, even though the biomaterials and design are well-matched and appropriate for the application and otherwise intact at the start of device assembly. The fracture in the weld of the tilting disk heart valve mentioned above illustrates well the introduction of defects and vulnerabilities during the fabrication process that can contribute to failure.

Incomplete sterilization of the implant (potentiating infection) or damage to the material by the sterilization process also can occur. Indeed, the use of certain materials places serious limitations on sterilization conditions. For example, PMMA intraocular lenses cannot be heat sterilized because the shape (and thereby optics) of the rigid poly(methyl methacrylate) (PMMA) lens would change above the glass transition temperature of the PMMA (100°C). Sterilization by ethylene oxide (EO) gas cannot be used with some biomaterials, because of EO solubility within and/or reactivity with the biomaterial.

These limitations are sometimes so severe that a sterilization protocol is used which is inadequate to sterilize the device. Incomplete sterilization, of course, can lead to infection, as exemplified by cohorts of porcine aortic valve or bovine pericardial bioprostheses contaminated with *Myobacterium chelonei*, an organism related to that causing human tuberculosis (Rumisek et al., 1985; Strabelli et al., 2010). The problem is that the antibacterial and antifungal efficacy of low concentrations of glutaraldehyde may be poor. The use of combined sterilization agents (e.g., alcoholic glutaraldehyde solutions) has been used to solve this problem without damaging the valve tissue.

PACKAGING, SHIPPING, AND STORAGE

Contamination or degradation can occur not only during fabrication and sterilization of devices, but also can be introduced during packaging and shipping. In all medical devices and implants, inadequate packaging or improper storage can contribute to limited shelf life. Transdermal drug delivery patches have to be carefully protected so that the drug does not leak out of the device. Tissue-derived heart valves and other devices may be degraded by excessive heat or freezing, and so are packaged with temperature indicators. Packaging and storage are also critical issues for condoms, made from natural rubber, which is sensitive to oxidation, a process that can degrade the rubber chains during storage on the shelf. Moreover, any drug delivery implant or device (e.g., skin patch, stent, degradable microparticles) could become saturated with the drug while the device is stored, and deliver a burst of drug until the steady-state concentration gradient is reached after contact with the body.

CLINICAL HANDLING AND SURGICAL PROCEDURE

Given an implant made of appropriate materials, properly designed, fabricated, sterilized, tested, manufactured, packaged, shipped, and stored, the "moment of truth" occurs at the instant when the package is opened, and the device is handled and implanted or contacted extracorporeally with the patient's "biosystem." Key issues here relate to the technical aspects of the surgery and the skill of the surgeon, the possibility of infection due to pre-implantation contamination or mechanical damage caused by improper handling by surgical personnel and/or their instruments. An example is the kinking of a tubular vascular graft, during implantation, thereby impeding its flow and potentially inducing thrombosis.

THE RECIPIENT

Although patients are generally well-matched to their prostheses, factors related to the particular recipient, but infrequent in the population at large, may cause or contribute to some failures. For example, since young patients exhibit accelerated calcification of tissue heart valves, such devices are generally avoided in children and adolescents, in whom calcification is accelerated (see Chapter II.4.5). Moreover, appropriateness of prosthesis selection (type or size) can play a role. For example, it would not be appropriate to use a mechanical valve (which requires anticoagulation therapy) in a patient with a known bleeding problem, such as a stomach ulcer. Moreover, some patients are allergic (i.e., exhibit hypersensitivity) to nickel or other metallic elements, and patients with implants containing nickel have had hypersensitive reactions that necessitated reoperation with removal. Thus, a prosthesis containing nickel should not be implanted in a patient with a known allergy to nickel. In addition, a prosthesis inappropriately sized can be deleterious, such as a heart valve with an insufficiently sized orifice (Rahimtoola, 1998). Finally, a patient can abuse or misuse an implant, or can exhibit an unexpected "abnormal" physiologic response. For example, a hip implant recipient who over-exercises before adequate healing has occurred could cause implant loosening.

Abnormal responses may be related to normal or disease-altered physiology. All individuals with implants are vulnerable to infection (see Chapter II.2.8), and they often receive prophylactic antibiotics when undergoing procedures that may introduce bacteria into the bloodstream. However, individuals with cancer or who have received organ transplants, who are often immunosuppressed, are particularly vulnerable to infection at any tumor, transplant or implant site; they may be at even higher risk to develop implant-related infections. Moreover, "abnormal" or "skewed" pathophysiologic responses may also occur in any large population of patients. For example, some patients have genetic or acquired abnormalities of coagulation which render them particularly vulnerable to thrombosis. They are at heightened thrombotic risk for cardiovascular implant or device therapies (De Stefano et al., 1996; Girling and de Swiet, 1997). This general issue of patient heterogeneity in biologic processes may become even more critical in the future with cell-based and other tissue engineered medical devices, in which biological processes play a critical role in both efficacy and safety (Mendelson, 2006).

Failure can also be related to use, independent of patient factors. Implanted interposition grafts (called fistulas) linking artery to vein in the arm are often used for the access to the vasculature required by hemodialysis. Use requires frequent puncture by needles. Repetitive puncture can lead to graft fragmentation, local hemorrhage or aneurysm formation (Georgiadis et al., 2008).

CONCLUSIONS

In this chapter, we have described the wide range and diversity of factors that can contribute to the failure of a medical device or implant, and an approach to thinking about them in a particular case. It is important to emphasize that inadequate properties or behavior of the biomaterials or poor design features are not always responsible for failure, and other factors may be superimposed on particular inherent vulnerabilities of the biomaterial and/or design. Moreover, a well-meaning and heavily tested modification of biomaterials and/or designs to solve an important complication can lead to unintended and unpredictable consequences. Indeed, abnormal physiological responses of a patient due to an implant or treatment with a therapeutic device can be expected in some, but is not always predictable in a random subgroup of patients. It is up to the biomaterial scientists and engineers to alert and educate the public and their representatives, as well as the corporate community, to these possibilities. In this way, the biomaterials scientist or engineer can play a major role in ensuring the success of medical devices and implants.

BIBLIOGRAPHY

Baretich, M. F. (2007). Medical device incident investigation. *Biomed. Sci. Instrum.*, **43**, 302–305.

Baumgartner, F. J., Munro, A. I., & Jamieson, W. R. (1997). Fracture embolization of a Duromedics mitral prosthesis. *Tex. Heart Inst. J.*, **24**, 122–124.

Becker, C. (2005). Stuck with the check? Guidant recall raises questions of medical tab. *Mod. Healthc.*, **35**, 20–22.

Blackstone, E. H. (2005). Could it happen again? The Björk–Shiley convexo-concave heart valve story. *Circulation*, **111**, 2717–2719.

Blot, W. J., Ibrahim, M. A., Ivery, T. D., Acheson, D. E., Brookmeyer, R., Weyman, A., Defauw, J., Smith, J. K., & Harrison, D. (2005). Twenty-five–year experience with the Björk-Shiley Convexoconcave Heart Valve – A continuing clinical concern. *Circulation*, **111**, 2850–2857.

Browne, J. A., Bechtold, C. D., Berry, D. J., Hanssen, A. D., & Lewallen, D. G. (2010). Failed metal-on-metal hip arthroplasties: A spectrum of clinical presentations and operative findings. *Clin. Orthop. Relat. Res.*, **468**, 2313–2320.

Bull, B., Fuchs, J. C., & Braunwald, N. S. (1969). Mechanism of formation of tissue layers on the fabric lattice covering intravascular prosthetic devices. *Surgery*, **65**, 640–648.

Carthey, J., de Leval, M. R., & Reason, J. T. (2001). The human factor in cardiac surgery: Errors and near misses in a high technology medical domain. *Ann. Thorac. Surg.*, **72**, 300–305.

Charnley, J., Kamangar, A., & Longfield, M. D. (1969). The optimum size of prosthetic heads in relation to the wear of plastic sockets in total replacement of the hip. *Med. & Biol. Eng.*, **7**, 31–39.

Chin, H. P., Harrison, E. C., Blankenhorn, D. H., & Moacanin, J. (1971). Lipids in silicone rubber valve prostheses after human implantation. *Circulation*, **43**, I-51–I-56.

De Stefano, V., Finazzi, G., & Mannucci, P. M. (1996). Inherited thrombophilia. Pathogenesis, clinical syndromes, and management. *Blood*, **87**, 3531–3544.

Georgiadis, G. S., Lazarides, M. K., Panagoutsos, S. A., Kantartzi, K. M., Lambidis, C. D., et al. (2008). Surgical revision of complicated false and true vascular access-related aneurysms. *J. Vasc. Surg.*, **47**, 1284–1291.

Girling, J., & de Swiet, M. (1997). Acquired thrombophilia. *Bailleres Clin. Ob. Gyn.*, **11**, 447–462.

Goodman, G. R. (2002). Medical device error. *Crit. Care Nurs. Clin. North Am.*, **14**, 407–416.

Gross, J. M., Shu, M. C. S., Dai, F. F., Ellis, J., & Yoganathan, A. P. (1996). A microstructural flow analysis within a bileaflet mechanical heart valve hinge. *J. Heart Valve Dis.*, **5**, 581–590.

Holmes, D. R., Jr., Kereiakes, D. J., Garg, S., Serruys, P. W., Dehmer, G. J., et al. (2010). Stent thrombosis. *J. Am. Coll. Cardiol.*, **56**(17), 1357–1365.

Holt, G., Murnaghan, C., Reilly, J., & Meek, R. M. (2007). The biology of aseptic osteolysis. *Clin. Orthop. Relat. Res.*, **460**, 240–252.

Huiskes, R. (1998). The causes of failure of hip and knee arthroplasties. *Neder Tijdsch Voor Geneesk*, **142**, 2035–2040.

Maron, B. J., & Hauser, R. G. (2007). Perspectives on the failure of pharmaceutical and medical device industries to fully protect public health interests. *Am. J. Cardiol.*, **100**, 147–151.

Mendelson, K. M., & Schoen, F. J. (2006). Heart valve tissue engineering: Concepts, approaches, progress, and challenges. *Ann. Biomed. Engin.*, **34**, 1799–1819.

Rahimtoola, S. H. (1998). Valve prosthesis–patient mismatch: An update. *J. Heart Valve Dis.*, **7**, 207–210.

Rumisek, J. D., Albus, R. A., & Clarke, J. S. (1985). Late *Myobacterium chelonei* bioprosthetic valve endocarditis: Activation of implanted contaminant? *Ann. Thorac. Surg.*, **39**, 277–279.

Schoen, F. J. (1998). Pathologic findings in explanted clinical bioprosthetic valves fabricated from photooxidized bovine pericardium. *J. Heart Valve. Dis.*, **7**, 174–179.

Schoen, F. J., & Levy, R. J. (1999). Tissue heart valves: Current challenges and future research perspectives. *J. Biomed. Mater. Res.*, **47**, 439–465.

Schoen, F. J., Goodenough, S. H., Ionescu, M. I., & Braunwald, N. S. (1984). Implications of late morphology of Braunwald–Cutter mitral valve prostheses. *J. Thorac. Cardiovasc. Surg.*, **88**, 208–216.

Schoen, F. J., Levy, R. J., & Piehler, H. R. (1992). Pathological considerations in replacement cardiac valves. *Cardiovasc. Pathol.*, **1**, 29–52.

Shepherd, M. (1996). SMDA '90 (Safe Medical Devices Act of 1990): User facility requirements of the final medical device reporting regulation. *J. Clin. Eng.*, **21**, 114–148.

Silver, M. D., & Wilson, C. J. (1977). The pathology of wear in the Beall model 104 heart valve prosthesis. *Circulation*, **56**, 617–622.

Strabelli, T. M., Siciliano, R. F., Castelli, J. B., Demarchi, L. M., Leão, S. C., et al. (2010). *Mycobacterium chelonae* valve endocarditis resulting from contaminated biological prostheses. *J. Infect.*, **60**(6), 467–473.

Trumpy, I. G., Roald, B., & Lyberg, T. (1996). Morphologic and immunohistochemical observation of explanted proplast-Teflon temporomandibular joint interpositional implants. *J. Oral. Maxillo. Surg.*, **54**, 63–68.

Valji, K., & Linenberger, M. (2009). Chasing clot: Thrombophilic states and the interventionalist. *J. Vasc. Interv. Radiol.*, **10**, 1403–1416.

Walker, A. M., Funch, D. P., Sulsky, S. I., & Dreyer, N. A. (1995). Patient factors associated with strut fracture in Bjork–Shiley 60° convexo-concave heart valves. *Circulation*, **92**, 3235–3239.

Yoganathan, A. P., Corcoran, W. H., Harrison, E. C., & Carl, J. R. (1978). The Bjork–Shiley aortic prosthesis: Flow characteristics, thrombus formation and tissue overgrowth. *Circulation*, **58**, 70–76.

Yoganathan, A. P., Reamer, H. H., Corcoran, W. H., Harrison, E. C., Shulman, I. A., et al. (1981). The Starr–Edwards aortic ball valve: Flow characteristics, thrombus formation, and tissue overgrowth. *Artif. Organs*, **5**, 6–17.

CHAPTER III.1.5 IMPLANT RETRIEVAL AND EVALUATION

James M. Anderson[1], Frederick J. Schoen[2], Stanley A. Brown (retired)[3] and Katharine Merrit (retired)[3]

[1]Department of Pathology, University of Cleveland, OH, USA

[2]Professor of Pathology and Health Sciences and Technology (HST), Harvard Medical School, Executive Chairman, Department of Pathology, Brigham and Women's Hospital, Boston, MA, USA

[3]17704 Stoneridge Dr., Gaithersburg MD, USA

INTRODUCTION

Implant retrieval and evaluation is a term used to describe a method of scientific analysis designed to determine the efficacy and safety or biocompatibility of biomaterials, prostheses, and medical devices *in vivo*. A key objective of this process is to determine if a medical device and its constituent biomaterials present potential harm to the patient. Implant retrieval and evaluation addresses the effects of both the recipient on the implant, and the implant on the recipient, especially biocompatibility, in the context of existing pathology. Implant retrieval and evaluation methodology is used to analyze an individual implant in the context of surgical removal or autopsy from a particular patient, or as an analysis of a collection of implants of a given generic type or a specific model of a device. When properly done, comprehensive analysis of cohorts of implants of similar types reveals paradigms of success as well as failure.

Implant retrieval and evaluation, both preclinical (animal) and clinical (humans), provides the opportunity to identify the modes, interactions, and mechanisms that lead to the success or failure of medical devices under end-use conditions. Appropriate implant retrieval and evaluation requires the development of a multilevel strategy to address device-specific parameters, the role of the patient, and device–patient interactions that lead to implant failure. Implant retrieval and evaluation is at its best highly structured and multidisciplinary, requiring the expertise and experience of scientists, engineers, and clinicians.

In this chapter, implant retrieval and evaluation are considered from the perspective of ultimate human clinical use. Therefore, this chapter draws upon many important perspectives presented in chapters dealing with materials science and engineering, biology, biochemistry, and medicine, host reactions and their evaluation, the testing of biomaterials, the degradation of materials in the biological environment, the application of materials in medicine and dentistry, the surgical perspective of implantation, the correlation of material surface properties with biological responses of biomaterials, and failure analysis (as discussed in Chapter III.5.3). Although this chapter focuses on implant retrieval and evaluation in the clinical environment, many of the goals and perspectives presented are important, and largely identical, to those used in preclinical (i.e., animal) investigation. Emphasis will be on the appropriate rationale and overall contributions of hypothesis-driven explant analysis applicable to failed as well as nonfailed and, in some situations, unimplanted devices. In this chapter, implants are considered to be composed of one or more biomaterials that are part of a specific design for a specific application. For these purposes, implants are prostheses, medical devices, or artificial organs.

GOALS

The general concept of implant retrieval in the context of clinical device use is presented in Figure III.1.5.1. Key factors determining the outcome of a medical device are the biomaterials and device design, patient-specific functional anatomy and pathology, and technical aspects of the implantation. An implant is placed in a patient by a surgeon. The term "surgeon" is broadly used here to include also "interventionalists" such as cardiologists or radiologists, etc., who implant devices through catheters or cannulas. Through device selection, implantation technique, and implant manipulation during the insertion procedure, the surgeon potentially has considerable impact on the success or failure of a device. This feature constitutes a key difference between devices and drugs (the efficacy and safety of the latter generally are not affected by the process of insertion *per se*). All implants have interactions of the constituent biomaterials with the surrounding tissues. These interactions may be local at the

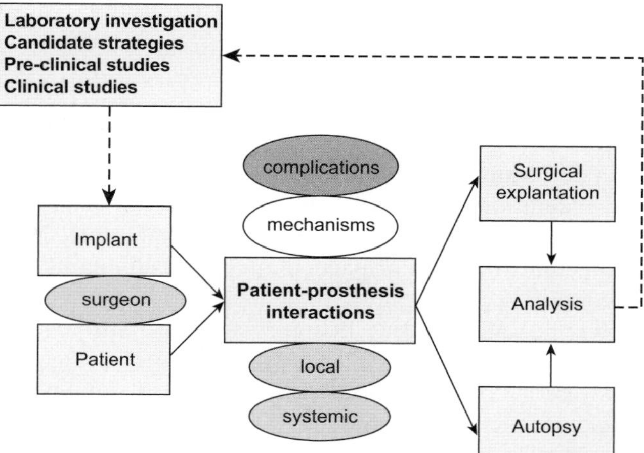

FIGURE III.1.5.1 Role of implant retrieval and evaluation in the development and use of clinical devices.

TABLE III.1.5.1 General Goals of Implant Retrieval and Evaluation
Determine rates, patterns, and mechanisms of implant failure
Identify effects of patient and prosthesis factors on performance
Establish factors that promote implant success
Determine dynamics, temporal variations, and mechanisms of tissue–materials and blood–materials interactions
Develop design criteria for future implants
Determine adequacy and appropriateness of animal models

site of implantation or distant from the device, and their mechanisms may be elucidated by careful and appropriate study. Clinically important deleterious biomaterials–tissue interactions manifest in a patient as *complications*. Implants are retrieved after successful function or failure by either surgical removal (at reoperation) or autopsy (by consent of the next-of-kin obtained after the patient's death). However, conventional autopsy permission and procedures generally do not permit destructive examination of the extremities or face. The information derived from implant retrieval and evaluation is frequently used to guide development of new, and modification of existing, implant designs and materials, to assist in decisions of implant selection, and to otherwise alter the management of patients (such as anticoagulation drug regimens for prosthetic heart valves, activity limitations for patients with prosthetic joints, or periodic imaging of an operative site).

The general goals of retrieved implant evaluation are presented in Table III.1.5.1. While many implant failures can be characterized as implant- or material-dependent or clinically- or biologically-dependent, many modes and mechanisms of failure are dependent on both implant and biological factors (Schoen and Levy, 2005; Goodman and Wright, 2007; Burke and Goodman, 2008; Revell, 2008a,b).

To appropriately appreciate the dynamics and temporal variations of tissue–materials and blood–materials interactions of implants, a fundamental understanding of these interactions is important. Appropriately conceived, structured and implemented, implant retrieval and evaluation should elucidate materials, design, and biological factors in implant performance, and enhance design criteria for future development. Finally, implant retrieval and evaluation should offer the opportunity to determine the adequacy and appropriateness of animal models used in preclinical testing of particular types of implants and biomaterials.

The goals of routine hospital surgical pathology or autopsy examination of a prosthetic device are generally restricted to essential documentation that a specific device has been removed at reoperation or that the patient died with the device, and diagnosis of a clinical abnormality that required the original therapeutic intervention or was a major complication of the device. Detailed correlation of morphologic features with clinical signs, symptoms, and dysfunctional physiology is usually not performed. Indeed, few pathologists have appropriate training, motivation and necessary resources to do complete and sophisticated analysis of retrieved implants. Nevertheless, directed and informed pathological examination of prostheses retrieved during preclinical animal studies or at reoperation or autopsy of human patients can provide highly valuable information. First, preclinical studies of modified designs and materials are crucial to developmental advances for technological innovation in biomaterials and medical devices. These investigations usually include implantation of functional devices in the intended location in an appropriate animal model, followed by noninvasive and invasive monitoring, followed by specimen explantation, and detailed pathological and material analysis. Second, for individual patients, demonstration of a propensity toward certain complications (e.g., a genetic propensity toward thrombosis) could have great impact on further management. Third, clinicopathologic analysis of cohorts of patients who have received a new or modified prosthesis type evaluates device safety and efficacy to an extent beyond that obtainable by either *in vitro* tests of durability and biocompatibility or preclinical investigations of implant configurations in large animals. Moreover, through analysis of rates and modes of failure, as well as characterization of the morphology and mechanisms of specific failure modes in patients with implanted medical devices, retrieval studies can contribute to the development of methods for enhanced clinical recognition of failures. The information gained should serve to guide both future development of improved prosthetic devices to eliminate complications, and diagnostic and therapeutic management strategies to reduce the frequency and clinical impact of complications. Emphasis is usually directed toward failed implants; however, careful and sophisticated analysis of removed prostheses that are functioning properly is also needed. Indeed, detailed analyses of implant structural features prior to, and their evolution following, implantation can yield an understanding of structural correlates of favorable performance, and identify predisposition to specific failure modes, which can be extremely valuable.

Device retrieval analysis has an important regulatory role, as specified in the US Safe Medical Devices Act of 1990 (PL101–629), the first major amendment to the Federal Food, Drug, and Cosmetics Act since the Medical Device Amendments of 1976. The user requirements of the 1990 legislation require health care personnel and hospitals to report (within 10 days) to the US Food and Drug Administration (FDA) or manufacturers or both (depending on the nature of the occurrence) all prosthesis-associated complications that cause death, serious illness or injury. Such incidents are often discovered during a pathologist's diagnostic evaluation of an implant in the autopsy suite or the surgical pathology laboratory.

Implant retrieval and evaluation programs may also provide specimens and data that serve as a teaching or research resource. The information gained can be used to educate patients, their families, physicians, residents, students, engineers, and materials scientists, as well as the general public. As a research resource, implant retrieval and evaluation yields data that can be used to develop and test hypotheses, and to improve protocols and techniques.

COMPONENTS OF IMPLANT RETRIEVAL AND EVALUATION

Implant retrieval and evaluation is a structured, multifaceted, and interdisciplinary effort by scientists with expertise in materials science, materials engineering, biomechanics, biology, pathology, microbiology, radiology, medicine, and surgery (Table III.1.5.2).

TABLE III.1.5.2 Important Components and Features of an Implant Retrieval and Evaluation Program
Entire activity is hypothesis-driven
Specimens are appropriately accessioned, cataloged, and identified
Known and potential failure modes are thoughtfully considered
Patient's medical history and laboratory results are reviewed
Data are collected on well-designed, study-specific forms
Careful gross examination, photography, and other basic analyses are always done
Advanced analytical techniques done by specialists are considered in selected cases
Analytical protocols and techniques for assessing host and implant responses are rigorously followed
Correlations and cause-and-effect relationships among material, design, mechanical, manufacturing, clinical, and biological variables are sought
Quantitative data are collected wherever possible and appropriate
Statistical and multivariant analyses are used

The first component is the appropriate accessioning, cataloging, and identification of retrieved implants. Patient anonymity* is required in any implant retrieval and evaluation program, and this can be achieved through appropriate accessioning, cataloging, coding of demographic information, and restriction of access. At most institutions, implant retrieval is carried out through the Surgical Pathology and Autopsy Services of the hospital.

As an example the Implant Retrieval and Evaluation Program at University Hospitals of Cleveland Case Medical Center uses the general accessioning and cataloging scheme for surgical pathology specimens. Gross and microscopic diagnoses in a standardized format are provided for patient's charts, for the appropriate clinicians and surgeons, and for the database for the Implant Retrieval and Evaluation Program. The Department of Surgery provides specimens to the Division of Surgical Pathology, with the patient's name and hospital identification number, clinical diagnoses, and notes on the patient and/or implant history.

An in-depth evaluation of a retrieved implant requires a review of the patient's medical history and radiographs, and other studies (e.g., echocardiograms, MRI scans), where pertinent. Table III.1.5.3 provides a partial list of anatomic and physiologic conditions that may influence the failure or success of orthopedic and cardiovascular implants. The identification of acute and chronic problems presented by the patient often will provide guidance on how the evaluation of a specific implant should be carried out. For example, a clinical diagnosis of infection necessitates that biohazard precautions, sterile handling and microbiologic cultures should be used in the evaluation. In addition, gross examination and photography play an important role, and must be carried out before specific techniques are used to evaluate implants.

It is critical to use a strategy that will allow optimal information yield using appropriate analytical protocols and techniques to assess host and implant responses. This strategy is directed toward developing correlative and cause-and-effect relationships among material, design, mechanical, manufacturing, clinical, and biological variables (Chapter III.1.3). Finally, whenever possible, analytical protocols and techniques should produce quantitative information that can be analyzed statistically. These analyses may also include clinical information.

*In the US The Privacy Rule provides federal protection for personal health information held by health care providers, and gives patients an array of rights with respect to that information. At the same time, the Privacy Rule is balanced so that it permits the disclosure of personal health information needed for patient care and other important purposes. The Privacy Rule was a component of The Health Insurance Portability and Accountability Act of 1996 (HIPAA), Public Law 104-191, which was enacted on August 21, 1996 (http://www.hhs.gov/ocr/privacy/hipaa/understanding/index.html).

TABLE III.1.5.3 Patient Conditions and Other Factors Influencing Implant Failure or Success

Orthopedic	Cardiovascular
Polyarthritis syndromes	Atherosclerosis
Connective-tissue disorders	Diabetes
Osteoarthritis	Infection
Trauma	Hypertension
Infection	Ventricular hypertrophy
Metabolic disease	Arrhythmias
Endocrine disease	Coagulation abnormalities
Tumor	Cardiac function
Primary joint disease	Recipient activity level
Osteonecrosis	
Recipient activity level	

Procedures used to evaluate devices and prostheses after function in animals and humans are largely the same. However, subject to humane treatment considerations, enumerated in institutional and federal guidelines and legislation, animal studies permit more detailed monitoring of device function and enhanced observation of morphologic detail (including blood–tissue–biomaterials interactions), as well as frequent assay of laboratory parameters (such as indices of platelet function or coagulation), and allow *in situ* observation of fresh implants following elective sacrifice at desired intervals. In addition, specimens from experimental animals can often be obtained rapidly, thereby minimizing the autolytic changes that occur when tissues are removed from their blood supply. Furthermore, advantageous technical adjuncts may be available in animal but not human investigations, including *in vivo* studies, such as injection of radiolabeled ligands for imaging platelet deposition, fixation by pressure perfusion that maintains tissues and cells in their physiological configuration following removal, and injection of various substances that serve as informative markers during analysis (such as indicators of endothelial barrier integrity). Animal studies often facilitate observation of specific complications in an accelerated time frame, such as calcification of bioprosthetic valves, in which the equivalent of 5–10 years in humans is simulated in 4–6 months in sheep (Schoen and Levy, 2005). Moreover, in preclinical studies, experimental conditions can be held constant among groups of subjects, including nutrition, activity levels, and treatment conditions. Consequently, concurrent control implants, in which only a single critical parameter varies, are often available in animal but not human studies.

Most implant retrieval involves examination of failed implants that have been surgically removed from patients or have been encountered at autopsy. Nevertheless, critical information can accrue from examination of "successful" implants removed after either fulfilling their function (e.g., a bone fixation device) or death of the patient resulting from an unrelated cause (e.g., heart valve prosthesis in a patient who dies of cancer). However, such studies may present difficulties, including but not limited to, access to the anatomic site, timing of the explantation following excessive postmortem interval, and family permission (Jacobs et al., 1999).

APPROACH TO ASSESSMENT OF HOST AND IMPLANT RESPONSES

Evaluation of the implant without attention to the tissue will produce an incomplete evaluation, and no understanding of the host response. Table III.1.5.4 provides a partial list of techniques for evaluating implants and tissues. We anticipate that novel implants and materials may dictate further techniques to be used in evaluating host–biomaterials interactions, and new analytical techniques will likely be developed.

Adequate and appropriate implant retrieval and evaluation require the quantitative assessment of host and implant responses using analytical protocols and non-destructive and destructive testing procedures. Techniques utilized for implant and tissue evaluation should be both device-designed and material-specific, permitting the collection of quantitative data that, in turn, can provide correlations and cause-and-effect relationships between material, design, mechanical, manufacturing, clinical, and biological variables. The evaluation techniques appropriate for a particular sample may be mutually exclusive, and the choice of a subset of techniques for evaluation may be dictated by the key clinical questions, and the availability and condition of the retrieved implant and tissue specimens. While a multilevel strategy is necessary for implant evaluation, analytical protocols and techniques are only utilized following a thorough understanding of the patient's medical history, laboratory results, and associated information, and known and potential failure modes and mechanisms are considered.

In general, analytical protocols and techniques for assessing host and implant responses can be divided into two categories: nondestructive and destructive testing procedures. Only after appropriate accessioning, cataloging, and identification, and a complete review of the patient's medical history and radiography, can the analytical protocols and techniques for implant evaluation be specified.

It should be noted that the techniques for implant evaluation are most commonly destructive techniques, that is, the implant or portions of the implant and the surrounding tissues must be destroyed or altered to obtain the desired information on the properties of the implant or material. The availability of the implant and tissue specimens will dictate the choice of technique.

Detailed analysis of the implant–tissue interface often necessitates that both a piece of tissue and a piece of the implant which it contacts be sectioned and examined in continuity (i.e., as one unit). This may necessitate special procedures, especially for polymer and ceramic implants in hard tissue (see below), which generally precludes any other analysis of that specimen. Similarly, chemical analysis of a piece of tissue for metal ion concentration requires acid digestion, and thus precludes morphological analysis. Thus, consideration must be given to either selection of multiple specimens or division of a specimen before processing, in order to permit part to be used for morphological and part for chemical analyses. Each of these portions may need to be further subdivided to be processed according to the requirements of specific tests. Moreover, since many of the procedures include cutting or sectioning a portion of the retrieved device, one must consider the legal aspects of device ownership, and destruction of what may become evidence in a court of law. Strict institutional guidelines need to be adhered to, and permission for use of destructive methods may be necessary. For clinicopathologic correlation studies that

TABLE III.1.5.4 Techniques for Implant Evaluation

Implant	Tissue
Atomic absorption spectrophotometry	Atomic absorption spectrophotometry
ATR-FTIR	Autoradiography
Burst strength	Biochemical analysis
Compliance studies	Cell culture
Contact angle measurements	Chemical analysis
Digestability	Enzyme histochemistry
EDAX	Gel electrophoresis
ESCA	Histology
Extractability	Immunocytochemistry
Fatigue studies	Immunofluorescence
Fracture analysis	Immunohistochemistry
FTIR	Immunoperoxidase
Gel permeation chromatography	*In situ* hybridization
Glass transition temperature	Microbiologic cultures
Hardness studies	Molecular analysis for cellular gene expression
Light microscopy	Radiography
Macrophotography	Morphometry
Melt temperature	Transmission electron microscopy
Metallographic examination	Scanning electron microscopy
Particulate analysis	Tissue culture
Polarized light microscopy	
Porosity analysis	
Radiography	
Scanning electron microscopy	
Shrink temperature	
SIMS	
Stereomicroscopy	
Stress analysis	
Tensile studies	
Transmission electron microscopy	
Topography analysis	

involve utilization of both patient data and pathological findings, approval of the local institutional review board (IRB) in the US is required. This is largely to ensure that the confidentiality and other interests of the patients from whom the implants were removed will be maintained, especially for purposes of publication or other dissemination of information.

Technical Problems of Metallic Implants, Bone, and Calcified Tissue

Evaluation of an implant and the surrounding host tissue when the implant was placed into bone, bone has formed around it, local tissue has calcified, or teeth presently pose unique problems. Dental implants and orthopedic implants are usually associated with hard tissue, but other implants may have contact with calcified tissue. Material containing calcified tissue, bone or teeth should be identified before the laboratory begins processing it or there may be harm to some of the equipment used for sectioning soft tissue. In addition some consensus standards, such as ASTM F561 and ISO/FDIA 12891-1, have been developed that address these special issues. Bone is a dynamic tissue, and analysis of bone formation and bone resorption may be a critical part of the study. Selection of techniques for evaluation must be undertaken carefully. Consideration needs to be given to what questions are being addressed before anything is cut or any harsh solutions are used.

If the implant and tissue are handled correctly, the techniques listed in Table III.1.5.4 are applicable to the study of hard tissue and associated implants. As stated earlier, some techniques are mutually exclusive. For example, histologic analysis of particle size and quantity, and chemical digestion and analysis for metal ions cannot be done on the same sample. Following routine fixation to preserve cellular detail, either of two generic approaches may be taken with implants adjacent to hard tissue: preparation of undecalcified sections or decalcification prior to embedding. Fully decalcified bone can be embedded and sectioned using methods common to soft-tissue pathology. To prepare undecalcified sections, hard tissue and the associated implant component or section are dehydrated and embedded in a hard plastic (usually methyl methacrylate) and thick (approximately 500 μm) sections are cut with a diamond saw and ground down to form the final thin section (approximately 50 μm). The advantage of undecalcified sections is that areas of mineralization and newly formed bone can be clearly distinguished from soft tissue. Specimen X-rays can also be taken to identify sites and extent of mineralization. Assistance with techniques for evaluation of specimens containing bone is available from the Society of Histotechnology and elsewhere (McNeil et al., 1997; Sanderson and Bachus, 1997; Callis and Strerchi, 1998).

Other special techniques may need to be developed and/or adapted to study particular issues. For example, stents composed of metallic wires are now commonly used in many clinical applications, including peripheral vascular disease intervention, biliary obstruction, endovascular repair of aneurysms, and percutaneous coronary interventions. In the examination of vascular stents, it is particularly important to determine if the stent is open or has become obstructed, and identify specific biomaterial–tissue interactions, without disturbing the key interfaces during dissection. A rapid and cost-effective method was recently described to dissolve most metallic stents, leaving the vascular and luminal tissues intact, in order to facilitate this type of analysis (Bradshaw et al., 2009).

Other Special Issues

Because implant retrieval programs may obtain data that will be used in regulatory or litigation procedures, it is important to document the disposition and appearance of the tissue, and of the implant, at each stage in the evaluation. Ample use of photography will greatly help in the description of the tissue and implant condition at each step of analysis. Careful labeling that is recorded during photography is helpful in later reviews of the evaluation procedure.

Concerns relative to the safety of laboratory personnel arise in the examination of retrieved materials and tissues, especially with respect to infection. If the device and/or tissues arrive in the laboratory without having been adequately disinfected or treated, special precautions need to be observed with the use of barrier clothing and gloves, and the use of biosafety cabinets. All safety regulations of the institution need to be known and followed. Procedures for disposal of hazardous chemicals also need to be carefully observed. If untreated devices or tissues are to be packaged and shipped from the clinical setting to the laboratory, care must be taken to follow the rules and regulations of the shipping agency and the Department of Transportation for transportation of biohazardous material. ASTM Committee F04 has a draft standard WK13292, Standard Guide for Shipping Possibly Infectious Materials, Tissues, and Fluids. This document contains guidelines and references for all the regulatory issues involved in shipping explant related materials.

An additional and very complex matter relates to implant ownership (Beyleveld et al., 1995). Several parties may have (and claim) rights to possession of a removed device under various circumstances: the patient, the surgeon, the pathologist, the hospital, an entity such as an insurance company, and attorneys representing both the plaintiff and the defendant in product liability or malpractice litigation. Moreover, ownership may not automatically entail the right to test. These issues are both controversial and unresolved.

The Multilevel Strategy to Implant Evaluation

Implant retrieval and analysis protocols often utilize a multilevel approach along the spectrum of essential documentation to the use of sophisticated research tools. Some investigators have advocated a basic two-level approach (Table III.1.5.5). Level I studies include routine evaluation modalities capable of being done in virtually any laboratory, and that characterize the overall safety and efficacy of a device. Level II studies comprise well-defined and meaningful test methods that are difficult, time-consuming or expensive to perform, require special expertise or yield more investigative or esoteric data. Since some level II analyses may be mutually exclusive, some material might routinely need to be accessioned, set aside (during level I analyses), and prepared for more specialized level II studies, in the event that they should be indicated later. Level II evaluation is usually undertaken with specific investigative objectives directed toward a focused research question or project. Prioritized, practical approaches using these guidelines have been described for heart valves, cardiac assist devices and artificial hearts, and other cardiovascular devices (Schoen et al., 1990; Borovetz et al., 1995; Schoen, 1995, 2001; Schwartz and Edelman, 2002, 2004).

Both ASTM and ISO implant retrieval standards (F561 and ISO 12891, respectively) have utilized a three-stage approach to implant analysis. Stage I analysis consists of routine device identification and description. Stage II is more detailed, time-consuming, and expensive, and includes photography and non-destructive failure analysis. Since stage I and II protocols are the same for all material types, they are combined in F561. Stage III protocols include destructive analytical techniques, many of which are specific to particular material types, and thus separate guidelines are provided for metallic, polymeric, and ceramic materials. Combinations of these protocols provide guidelines for analysis of the various components of composites and tissue engineered materials.

TABLE III.1.5.5 Study Prioritization
Level I Studies
Gross dissection
Photographic documentation
Microbiologic cultures
Radiography
Light microscopic histopathology
Level II Studies
Scanning electron microscopy
Transmission electron microscopy
Energy dispersive X-ray microanalysis (EDXA)
Analysis of adsorbed and absorbed proteins
Mechanical properties measurement
Materials surface analysis
Leukocyte immunophenotypic studies
Molecular analyses for cellular gene expression or extracellular matrix production

THE ROLE OF IMPLANT RETRIEVAL IN DEVICE DEVELOPMENT

For implants placed into many anatomic sites, prosthesis-associated pathology is a major determinant of quality of life and prognosis of patients. Implant retrieval and evaluation plays a critical role in the evolution of medical devices through development and clinical use (Figure III.1.5.1). Implants are examined as: (1) fabricated but unimplanted prototypes, to reveal changes in device components induced by the fabrication process that can predict (or lead to an understanding of) failure modes observed subsequently; (2) specimens that have been subjected to *in vitro* tests of biocompatibility or durability; (3) specimens removed from animal models following *in vivo* function, usually as actual devices; (4) specimens that accrue in carefully controlled clinical trials; and (5) failed or functioning specimens that are explanted in the course of ongoing surveillance of general clinical use of a device following regulatory approval. In each case, implant retrieval is concerned with the documentation of device safety and efficacy, and problems that dictate modification of design, materials or use of the implants in patients (e.g., patient–device matching or patient management). Devices are explanted and analyzed with attention directed toward all complications, but especially those that may be considered special vulnerabilities engendered by the device type. Efforts are made to retrieve and evaluate as many such implants as possible, both failed and nonfailed. Evaluation of retrieved animal implants plays a special role in the documentation for the FDA-Investigational Device Exemption (IDE) required for a clinical trial and for the FDA-Premarket Approval (PMA) required for general marketing. The literature contains numerous instances where problem-oriented implant evaluation studies have yielded important insights into the deficiencies and complications that have limited the success of various implants. Implant retrieval has also contributed to the assessment of the safety and efficacy of implant modifications intended to be improvements. The conceptual approach to the analysis of a failed implant is explored in Chapter III.1.3.

> Common failure modes of cardiovascular implants include thrombosis, embolism, calcification, and infection, whereas those of orthopedic and dental implants are loosening, wear, and infection. Implant retrieval studies have, and will continue to have, significant clinical, as well as research and development, utility. Specific implant designs can lead to specific types of complications and failure. Implant retrieval and evaluation with the identification of failure modes and mechanisms have played a significant role in the evolution of medical devices and will continue to be crucial for the evaluation of "next" generation devices.

WHAT CLINICALLY USEFUL INFORMATION HAS BEEN LEARNED FROM IMPLANT RETRIEVAL AND ANALYSIS?

Cardiovascular Implants

Cardiovascular implants commonly involve both blood and soft-tissue interactions with materials. The complications most commonly found with these implants, i.e., heart valve prostheses, vascular grafts, cardiac assist devices/artificial hearts, and vascular stents are listed in Tables III.1.5.6 and III.1.5.7. Perspectives, approaches, and techniques for evaluating cardiovascular implants have been described (Schoen et al., 1990; Borovetz et al., 1995; Schoen, 1995, 2001; Schwartz and Edelman, 2002, 2004). Specific examples from the field of cardiovascular prostheses are summarized next and in Table III.1.5.8, to illustrate the scope and utility of this activity.

TABLE III.1.5.6 Complications of Heart Valve Substitutes, Vascular Grafts, and Cardiac Assist/Replacement Devices

Heart Valve Prostheses	Vascular Grafts	Cardiac Assist/ Replacement Devices
Thrombosis	Thrombosis	Thrombosis
Embolism	Embolism	Embolism
Paravalvular leak	Infection	Endocarditis
Anticoagulation-related hemorrhage	Perigraft erosion	Extraluminal infection
Infective endocarditis	Perigraft seroma	Component fracture
Extrinsic dysfunction	False aneurysm	Hemolysis
Incomplete valve closure	Anastomotic hyperplasia	Calcification
Cloth wear	Disintegration/ degradation	
Hemolytic anemia		
Component fracture		
Tissue valves		
Cusp tearing		
Cusp calcification		

TABLE III.1.5.7 Complications of Vascular Stents

Thrombosis
Proliferative restenosis
Strut-related inflammation
Foreign-body reaction
Incomplete expansion
Overexpansion
Infection
Malposition

Experience with caged-ball mechanical prosthetic heart valves in the 1960s demonstrated both thrombus and degradation as complications. Degradation of silicone due to lipid absorption (called ball variance) was virtually eliminated by improved curing of silicone poppets (Hylen et al., 1972). Subsequently, a cloth-covered caged-ball mechanical prosthetic heart valve that had a ball fabricated from silicone and a metallic cage covered by polypropylene mesh was introduced. The cloth-covered cage struts were designed to encourage tissue ingrowth, and thereby decrease thromboembolism. Preclinical studies in pigs, sheep, and calves demonstrated rapid healing by fibrous tissue of the cloth-covered struts. However, subsequent clinical studies of these valves showed abundant cloth wear in some patients, sufficient to cause escape of the ball through the spaces between the cage struts, a circumstance that was usually fatal without emergency surgery (Schoen et al., 1984). This analysis demonstrated the general concept that valve prostheses used in humans may have important complications that were not predicted by animal investigations, in this case because the vigorous healing that occurred in animals, but not in humans, obscured the potential problem.

Modification of a tilting-disk valve design intended to permit enhanced opening and thereby reduce thromboembolism also led to a new complication – strut fracture with escape of the disk. Careful studies of retrieved valves made a critical contribution to understanding and managing this serious clinical problem (see Prosthetic Heart Valve Case Study). Multiple coordinated studies of retrieved experimental and clinical implants have characterized and facilitated therapeutic approaches toward elimination of calcification-induced failure modes in bioprosthetic tissue heart-valve replacement (Schoen and Levy, 1999, 2005).

Such studies have identified the causes and morphology of failure, patterns of mineral deposition, nature of the mineral phase, and early events in the mechanisms of calcification. Calcification of valve tissue in both circulatory and subcutaneous animal models exhibits morphological features similar to that observed in clinical specimens, but calcification is markedly accelerated in the experimental explants. Such studies have provided the means to test approaches to inhibit bioprosthetic valve calcification by modifying host, implant, and mechanical influences. These studies serve to emphasize that biological, as well as mechanical, failure mechanisms can be understood using specifically chosen and controlled animal model investigations, guided by and correlated with the results of careful studies of retrieved clinical specimens. Furthermore, they show how implant retrieval studies can be used as a critical component of a program to ensure efficacy and safety of potential therapeutic modifications.

An additional use of retrieval studies is exemplified by a case in which a change in biomaterials was shown

TABLE III.1.5.8 Clinical Utility of Retrieval Studies: Heart Valve Substitutes

Implant Type	Knowledge Gained/Lessons Learned
Caged ball valves	• Poppets fabricated from industrial silicone absorbed blood lipids, and became swollen and brittle
	• Fragments of degraded heart valve material may embolize to other organs
	• Thrombosis can occur at stasis points downstream of the ball
Cloth-covered caged ball valves	• Cloth wear can cause hemolysis and cloth emboli
	• Cloth wear accompanied by silicone poppet wear can precipitate poppet escape
	• Healing of fabric may be more vigorous in animals than in humans
	• Quantitation of data (e.g., polymeric poppet weight and dimensions) may facilitate the understanding of a failure mode
Caged disk valves	• Teflon® has poor wear resistance as a valve occluder
	• Poor design features may potentiate thrombosis
Tilting-disk valves	• Thrombosis may initiate downstream to the edge of a partially open disk at a region of stasis
	• A "minor" change in valve design can result in a new propensity toward failure
	• Animal implants instrumented with strain gauges can be used to test a hypothetical mechanical failure mechanism
	• Understanding a failure mode can lead to both new methods for noninvasive diagnosis (e.g., X-ray and acoustic), and modified patient management strategies (e.g., drugs to depress cardiac contractility)
Bileaflet tilting-disk valves	• Cavitation may cause critical materials damage in some valve designs
	• Thrombosis may be initiated in regions of microstasis at component junctions
	• Microscopic areas of stasis may be predicted by computer-assisted computation
	• Animal implant models may fail to predict vulnerability to thrombosis in humans
Bioprosthetic heart valves	• Tissue calcification is a major failure mode
	• Calcification is most pronounced in areas of leaflet flexion, where deformations are maximal
	• Calcification is accelerated in young recipients, especially children
	• Heart valve calcification can be studied outside of the circulation (e.g., subcutaneous implants in rats)
	• Calcification is initiated principally at cell remnants deep in the tissue
Cryopreserved allograft valves	• These valves are not viable and cannot grow
	• Failure is not immunologically mediated and, therefore, immunosuppression is inappropriate
Substitution of new materials	• Pyrolytic carbon has favorable clinical durability
	• Detailed examination of functional (not failed) prostheses may yield worthwhile data

Case Study — Prosthetic Heart Valve Case Study

Often cited as the prototypical device failure analysis that contributed important clinical information, the problem of the Björk-Shiley tilting disk mechanical valve prosthesis provides an instructive case history. The prosthesis of original design, initially in which a Delrin® polymer disk and later a pyrolytic carbon disk was held in place by a wire superstructure composed of inflow and outflow struts, was associated with an unacceptable late failure rate owing to thrombotic occlusion (Figure III.1.5.2). The inflow strut was an integral part of the valve base while the outflow strut was welded to the base ring.

In order to reduce the risk of thrombosis by improving valve flow, the company manufacturing the valve (Shiley, Inc.) redesigned the prosthesis to achieve opening angles of the disk of 60° or 70° from the plane of the valve base and rounded the surfaces of the disk ("convexo-concave" or "C-C"). However, clinical use of the redesigned Björk-Shiley C-C heart valve led to an unusually large cluster of cases in which the metallic outlet strut fractured in two places, leading to disk escape (Figure III.1.5.3). The complication was fatal in the majority of patients in whom it occurred. In 2005, strut fracture was reported in at least 633 of the 86,000 valves of this type implanted worldwide during 1978–1986. Clinical studies identified large valve size, mitral site, young recipient age, and valve manufacture date as risk factors for this failure mode (Blot et al., 2005). The mechanisms of this problem were elucidated through careful analysis using retrieved implant analysis, coupled with other methods.

Pathologic studies of retrieved Björk-Shiley valves, enhanced by scanning electron microscopy, demonstrated a pronounced wear facet at the tip of the outlet struts, and localized pyrolytic carbon wear deposits at sites where the closing disk contacted the strut (Schoen et al., 1992), providing a fatigue fracture mechanism that began on the inflow side of the outlet strut often initiated within a weld joint (Figure III.1.5.4). The analysis suggested that the first strut leg fracture typically initiated at or near the point of maximum bending stress, was traced to a site of weld shrinkage porosity and/or inclusion in most cases, and was followed by fracture of the second leg. Occasionally, valves with only a single strut fracture were encountered. This showed that the underlying problem in catastrophic failure was metal fatigue failure, probably initiated by over-rotation of the disk during valve closure in this design, leading to excessive bending stresses at or near the welds joining the outlet struts to the housing, and potentially coupled with intrinsic weld flaws.

Moreover, in subsequent animal studies in which Björk-Shiley 60–70° C-C valves were implanted in sheep and instrumented with strain gauges showed that impact forces varied greatly with cardiac activity, and that loads occurring during exercise were significantly elevated. This correlated with clinical data that showed that fracture often occurred during exertion.

The Björk-Shiley heart valve case demonstrates that elucidation of a failure mode by detailed materials failure analysis can have

Case Study **Prosthetic Heart Valve Case Study (*Continued*)**

potential impact on patient management. Understanding this mode of failure has enabled development of non-invasive testing modalities (via high-definition radiography (O'Neill et al., 1995) or acoustic characterization of strut status (Plemons and Hovenga, 1995)) to establish when one strut has fractured prior to the onset of clinical failure, to caution patients with these valves against vigorous exercise, and to consider re-replacement of properly functioning valves at high risk of fracture. Moreover, this case has reinforced certain principles of valve testing which should occur before widespread clinical use of a new or modified prosthesis design (Blackstone, 2005).

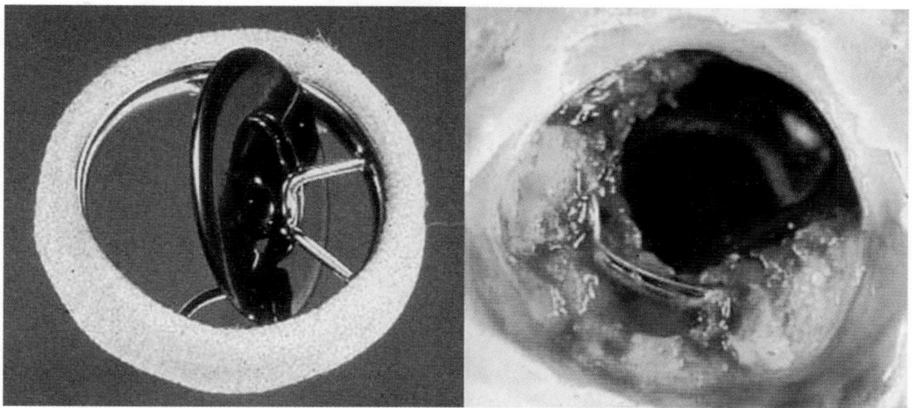

FIGURE III.1.5.2 Björk-Shiley mechanical heart valve prosthesis. Left: Original model with pyrolytic carbon disk, unimplanted. Right: thrombosed prosthesis. (B) from Schoen F.J., Levy, R.J., Piehler, H.R. (1992). Pathological Considerations in replacement heart valves. Cardiovascular Pathology 1:29–52.

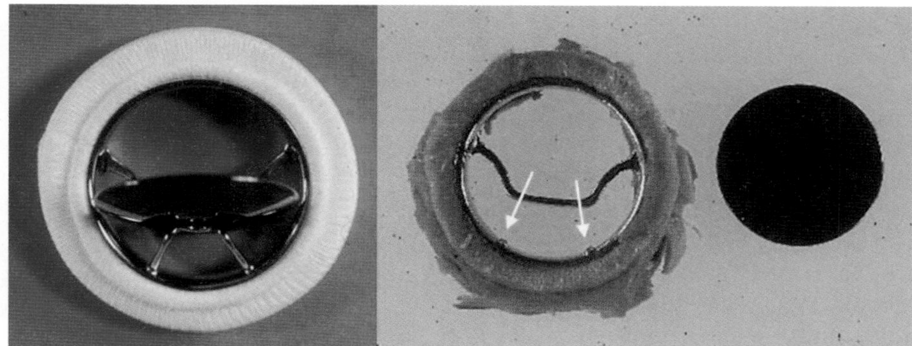

FIGURE III.1.5.3 Modified Björk-Shiley mechanical heart valve prosthesis with convexo-concave disk and 70° opening. Left: Unimplanted valve. Right: Explanted valve with outflow strut fracture at welds (arrows). Disk had escaped from housing and the outflow strut piece was not located. (B) from Schoen F.J., Levy, R.J., Piehler, H.R. (1992). Pathological considerations in replacement heart valves. Cardiovascular Pathology 1:29–52.

to be beneficial. Implant retrieval studies of a low-profile disk valve composed of a disk and cage fabricated from Teflon® demonstrated poor wear properties of the disk. After a new model of the valve with disk and struts fabricated from pyrolytic carbon was introduced, clinical experience suggested that the use of carbon contributed to a major advancement in the durability of prosthetic heart valves. Retrieval studies of carbon valves recovered at autopsy or surgery, and analyzed by surface scanning electron microscopy and surface profilometry indicated that, compared with valves composed of Teflon®, carbon valves exhibited minimal abrasive wear (Schoen et al., 1982). This study revealed that although analysis of implants and medical devices has traditionally concentrated on those devices that failed in service, important data can accrue from implants serving the patient until death or removal for unrelated causes. Thus, detailed examination of the properly functioning prostheses removed from patients after a long duration of implantation may yield worthwhile data, provided that a focused question is asked of the material.

Other retrieval studies have revealed: (1) the cause of excessive thrombosis of a new bileaflet tilting-disk design (Gross et al., 1996); (2) the cause of mechanical failure of a new pericardial bioprosthesis fabricated from photofixed bovine pericardium that had different mechanical properties than conventional glutaraldehyde-fixed tissue (Schoen, 1998a); and (3) the cause of failure and characteristic changes that occur in cryopreserved allograft aortic heart valves transplanted from one individual

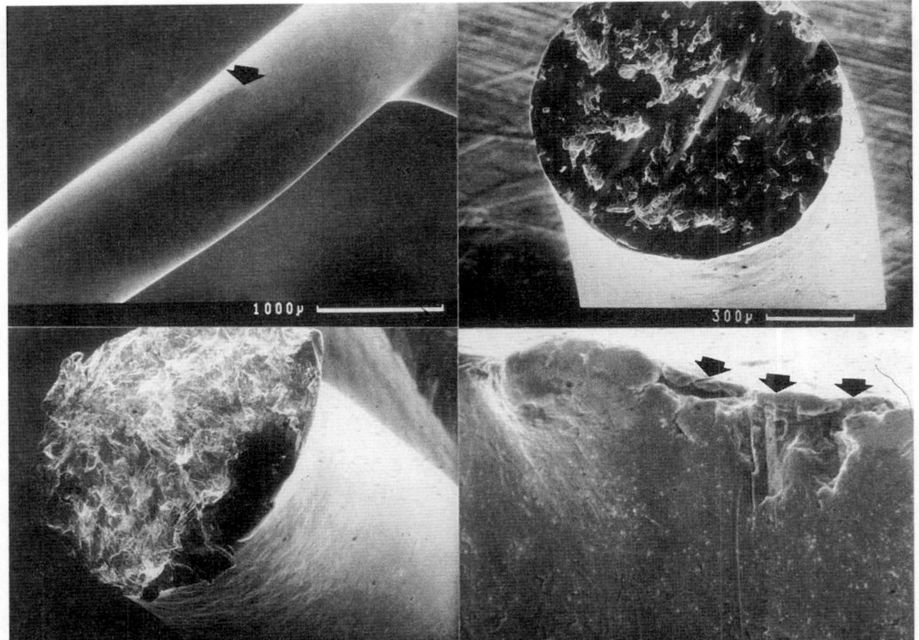

FIGURE III.1.5.4 Analysis of retrieved Björk-Shiley convexo-concave mechanical heart valve prosthesis. Upper left: Elliptical wear flat on tip of outlet strut resulting from contact with disk during closure. Upper right: Worn and abraded surface of first strut leg to fracture. Origin of the fatigue fracture is at bottom center. Lower left: Fracture surface of second leg to fracture; stigmata of fatigue fracture are absent. Lower right: Residual weld fracture at the base of valve showing weld porosity and secondary cracking (arrows). From Schoen F.J., Levy, R.J., Piehler, H.R. (1992). Pathological Considerations in replacement heart valves. Cardiovascular Pathology 1:29–52.

to another (Mitchell et al., 1998). Studies of ventricular assist devices have demonstrated the importance of valves or biomaterial junctions in initiating thrombosis (Fyfe and Schoen, 1993; Wagner et al., 1993), and several studies have described the failure modes of specific types of vascular grafts in different locations (Canizales et al., 1982; Downs et al., 1991; Guidoin et al., 1993), and the morphology of vascular stents (Anderson et al., 1992; van Beusekom et al., 1993; Farb et al., 2002).

Oxidative biodegradation and environmental stress cracking of polyether polyurethane pacemaker lead insulation have been identified as failure mechanisms of pacemaker leads (Wiggins et al., 2001). Reactive oxygen radicals, released by adherent macrophages and foreign-body giant cells in the foreign-body reaction oxidize the polyether component of the polyurethane, leading to oxidative chain cleavage of the polymer and subsequent failure. These findings have lead to the development of new polyurethanes that are oxidation-resistant or inhibit the adhesion and activation of adherent cells in the foreign-body reaction at the tissue–material interface (Anderson et al., 2007).

TABLE III.1.5.9 Complications of Orthopedic Implants

Bone resorption	Loosening
Corrosion	Mechanical mismatch
Fatigue	Fracture
Fibrosis	Motion and pain
Fixation failure	Particulate formation
Fracture	Surface wear
Incomplete osseous integration	Stress riser
Infection	
Interface separation	

TABLE III.1.5.10 Complications of Dental Implants

Adverse foreign-body reaction	Loosening
Biocorrosion	Foreign-body reaction
Electrochemical galvanic coupling	Corrosion
Fatigue	Particulate formation
Fixation failure	Wear
Fracture	
Infection	
Interface separation	
Loss of mechanical force transfer	

Orthopedic and Dental Implants

Complications most commonly found with orthopedic and dental implants are presented in Table III.1.5.9 and Table III.1.5.10, respectively. The clinical utility of hard-tissue (orthopedic and dental) implant retrieval and evaluation is summarized in Table III.1.5.11.

Early studies were concerned primarily with metallic devices used for internal fixation of fractures. Examination of some early designs showed failure at regions of poor metallurgy or weak areas due to poor design (Cahoon and Paxton, 1968). Other studies (Wright

TABLE III.1.5.11 Clinical Utility of Retrieval Studies: Hard-Tissue Implants

Implant Type	Knowledge Gained/Lessons Learned
Plates, screws, and rods used for fracture fixation	• Do not mix metal alloys in same device
	• Match the hardness (degree of cold-work) and stiffness with the application
	• Metallic implant wear and corrosion may lead to problems associated with allergic reactions
Tooth root implants	• Careful surgical technique avoids thermal tissue necrosis of the implant–tissue interface
	• Implant surface finish is a critical determinant of outcome
Femoral stems of total hip replacements (fracture analysis)	• Cobalt alloys require high-quality casting
	• High-strength superalloys may be advantageous
	• Welded regions may fail in tensile-loaded locations
	• Part numbers should not be etched in tensile-loaded locations
Femoral stem, modular interface	• Corrosion at an interface is dependent on metallurgical and mechanical design factors
Polymeric component to total joint replacements (excluding one-piece flex hinge joints, such as finger)	• Teflon® performs poorly in wear applications
	• Reinforcement of UHMWPE (ultra-high molecular weight polyethylene) with chopped carbon fiber is ineffective
	• Laboratory simulation should be done using clinically realistic implant wear motion and loading patterns
	• Mechanical designs that produce high localized stresses may cause delamination
	• Radiation sterilization to produce cross-linking is useful, but can lead to molecular-chain scission, oxidation, and aging
Analysis of tissues surrounding total joint replacements	• Wear particles derived from breakdown of the implant–bone interface or wear of the articulation (bearing) material cause inflammation and loosening
Titanium and titanium-alloy implants	• A good material in one implant application may not necessarily be good in another
	• Commercially pure titanium is appropriate for tooth root implant or fracture fixation devices
	• Titanium alloys may yield severe wear in some total joint applications

et al., 1982; Cook et al., 1985) reinforce the importance of material and design, such as the pitfall of stamping the manufacturer's name in the middle of the plate where stresses are high. Modern total hip replacements have also had fatigue fractures initiated at labels etched at high-stress regions (Woolson et al., 1997). Galvanic corrosion due to mixed metal alloys was also demonstrated in early retrieval studies (Scales et al., 1959). Analyses of fixation devices have shown a correlation between corrosion and biological reactions, such as allergic reactions to metallic elements in the implants (Brown and Merritt, 1981; Cook et al., 1988; Merritt and Rodrigo, 1996).

The development of the low-friction total hip arthroplasty by Charnley is a classical example of how implant retrieval can yield important clinical information which impacts implant materials and design (Charnley et al., 1969). The initial choice for the low-friction polymeric cup (replacing the acetabulum) was a Teflon®-like PTFE. As described above for heart valves, this relatively soft material had a very high wear rate, and within a few years many hip replacements had been removed due to severe wear and intense inflammatory reactions. The analysis of the retrievals has also demonstrated the relationships between large head diameter, and the wear debris and excessive wear that have served as the basis for our understanding of joint bearing design. Clinical and retrieval studies are now examining the performance of metal-on-metal total hip bearing couples (Kwon et al., 2010; Ebramzadeh et al., 2011).

Teflon® was replaced by ultra-high molecular weight polyethylene (UHMWPE) as the acetabular cup bearing surfaces, and retrieval studies have continuously stimulated changes in the methods of processing UHMWPE. Retrieval studies have demonstrated that reinforcing UHMWPE with carbon fibers did not sufficiently improve its strength (Wright et al., 1988). Heat pressing to improve surface hardness was not effective with some total knee designs (Bloebaum et al., 1991; Wright et al., 1992).

The physical and mechanical properties of UHMWPE are altered by ionizing radiation, due to the generation of free radicals. Oxidative degradation of UHMWPE components radiation sterilized in air-permeable packaging occurs during sterilization, during shelf-storage prior to implantation, and *in vivo*. Oxidation has a detrimental effect on UHMWPE, leading to a decrease in the elongation to failure, an increase in elastic modulus, and a decrease in fatigue crack propagation resistance (Kurtz, 2009), and has resulted in the premature failure of some UHMWPE components *in vivo*. Contemporary radiation sterilization practices (e.g., sterilization in an inert gas or vacuum) for UHMWPE have largely eliminated the potential for shelf aging; however, *in vivo* oxidation occurs and remains a concern (Kurtz et al., 2006).

Case Study **Orthopedic Hip Prosthesis Case Study**

One of the most dramatic examples of effective implant retrieval and analysis is that of the zirconia femoral ball head made under the trade name of Prozyr® by Saint-Gobain Ceramiques Desmarquest. By mid-1998 they had made thousands of these with very low failure rates. However, in 1998 they changed a processing step from batch furnace sintering to a continuous flow "tunnel furnace" (Clarke et al., 2003). In 2001 reports of catastrophic head fracture were being reported by the primarily European manufacturers. On August 10, 2001, the French Agency of Health Security of Health Products issued a recall on all TH batches from the tunnel furnace. On August 16, 2001, the British Therapeutic Goods Administration issued a similar recall. The major manufacturers involved were European and Australian divisions of Biomet, DePuy, Smith & Nephew, Zimmer, and Stryker Howmedica Osteonics.

What followed was an intense international study of device failure. The company organized a committee of experts. They met with the national regulatory authorities to discuss the issue. Failed and non-implanted devices were studied. As of April 2003 (http://www.prozyr.com/PAGES_UK/Biomedical/breakages.htm), Saint-Gobain was reporting on their web site that six TH batches had fracture problems; the two major failure batches had failure rates of 33% (227 out of 683) and 8.7% (60 out of 692).

So what was learned? Zirconia is a metastable ceramic which will undergo a phase transformation at room temperature (Chevalier et al., 2004) which is partially stabilized by the addition of Yttria. With proper sintering, the material obtains maximum density, and the result is a very strong material (Clarke et al., 2003; Chevalier et al., 2004) which was thus considered a better ceramic than its more brittle alternative, alumina. However, some of the batch furnace runs produced balls which were not fully densified in the center of the ball. This unstable material was exposed by boring out the center for taper fit to the metallic femoral stems. Very careful analysis of failures, non-implanted balls, and the "carrots" produced by the hollow core drills used to drill out the balls revealed that this material was subject to environmental degradation, phase transformation, and fracture (Clarke et al., 2003; Chevalier, 2006). While subsequent batches had better thermal controls and the carrots were fully dense, the global impact of such a large failure rate and recall resulted in the company's decision to pull Prozyr® off the market (Chevalier, 2006). Subsequent analysis of other Y-TZP products may well restore clinical confidence in this material.

Highly cross-linked UHMWPE formulations have been in clinical use for a decade as a means to improve wear resistance of acetabular hip components. Clinical studies of highly cross-linked acetabular hip components support reduced wear within the first decade of use (Muratoglu et al., 2004; Bragdon et al., 2007). However, cross-linking adversely affects uniaxial ductility, fracture toughness, and fatigue crack propagation resistance (Rimnac and Pruitt, 2008; Sobieraj and Rimnac, 2009). Due to the changes in mechanical properties of first-generation highly cross-linked UHMWPE materials, there has been concern regarding the potential propensity of *in vivo* fracture of highly cross-linked UHMWPE joint replacement components. In fact, *in vivo* fractures of highly cross-linked UHMWPE hip components have been reported, though the numbers are small (Tower et al., 2007; Furmanski et al., 2008). Also, when UHMWPE is cross-linked via ionizing radiation, free radicals are generated that can lead to oxidative degradation; thus, post-processing above (remelting) or below (annealing) the melt temperature is usually conducted. Remelting effectively eliminates free radicals, but also alters crystallinity, which affects mechanical properties. Annealing preserves the original crystal structure better, but is not as effective at reducing free radicals as remelting so that oxidation is possible with annealed cross-linked UHMWPE formulations (Rimnac and Pruitt, 2008). More recently, sequential radiation/annealing and incorporation of vitamin E (as a free-radical scavenger) have been introduced (the second-generation highly cross-linked UHMWPE materials) as methods by which to more effectively reduce or eliminate free radicals while preserving crystallinity (Oral et al., 2006; Sobieraj et al., 2008).

Implant retrieval and analysis has provided critical information toward understanding the anchoring of implants in bone. Branemark and associates demonstrated that direct bone–metal osseointegration of titanium tooth-root implants was possible with carefully controlled surgical technique and implant surface preparation (Albrektsson et al., 1982). Porous coatings may be applied to implants to facilitate bony ingrowth; relationships between pore size and bony ingrowth have come from retrieval studies (Cook et al., 1988).

Examination of the interface between the bone and the acrylic bone cement in long-term implant recipients examined at autopsy demonstrated the presence of inflammation with macrophages, thus providing the initial description of bone resorption due to particulate debris (Charnley, 1970). The relationship between the amount of particulate wear debris generated and bone resorption due to inflammation, i.e., osteolysis, was graphically demonstrated by the retrieval and tissue studies of Willert and Semlitsch (1977). Extensive research has been conducted over the years looking at osteolysis or "particle disease," and characterization of the polymeric debris, as correlated with implant type and design (Schmalzied et al., 1997). Studies on tissue reactions to particulates and wear debris require special histologic evaluation, as described by Wright and Goodman (1996), in the various consensus standards, in other references cited earlier, and in many other studies readily available in the orthopedic or histotechnology literature (see Goodman and Wright, 2007; Burke and Goodman, 2008; Purdue, 2008; Revell, 2008a,b).

Implant retrieval studies have demonstrated that the success of a material in one application may not necessarily translate to another. For example, titanium does well as a bone–implant interface, but it may be subject to severe wear as a bearing surface (Agins et al., 1988). An alternative approach in the hip was to use titanium alloy stems with modular press-fit heads made of a more wear-resistant cobalt alloy. However, some of these designs demonstrated significant corrosion, which was first attributed to use of mixed metals (Collier et al., 1991), and later to microgalvanic corrosion of the cast heads (Mathiesen et al., 1991), and to micromotion and fretting corrosion between components (Brown et al., 1995).

As with orthopedic devices, complications of dental implants may be related to the mechanical–biomechanical aspects of force transfer, or the chemical–biochemical aspects of elements transferred across biomaterial and tissue interfaces, or both. The complex synergism that exists between tissue and biomaterial responses presents a significant challenge in the identification of the failure mechanisms of dental implants. Since many dental devices are percutaneous, in that they contact bone but are also exposed to the oral cavity, the problems of infection are also significant (Moore, 1987; Sussman and Moss, 1993). Lemons (1988) and others have provided appropriate perspectives to be taken in the evaluation of dental implants.

CONCLUSIONS

Although the focus of this chapter has been on the retrieval and evaluation of implants in humans, the perspectives, approaches, techniques, and methods may also apply to the evaluation of new biomaterials, preclinical testing for biocompatibility, and premarket clinical evaluation. Each type of *in vivo* or clinical setting has its unique implant–host interactions, and therefore requires the development of a unique strategy for retrieval and evaluation.

Indeed, for investigation of bioactive materials and tissue engineered medical devices, in which interactions between the implant and the surrounding tissue are complex, novel and innovative approaches must be used in the investigation of *in vivo* tissue compatibility. In such cases, the scope of the concept of "biocompatibility" is much broader, and the approaches of implant retrieval and evaluation require measures of the phenotypes and functions of cells, and the architecture and remodeling of extracellular matrix and the vasculature (Mizuno et al., 2002; Peters et al., 2002; Rabkin et al., 2002; Rabkin and Schoen, 2002). Moreover, a critical role of implant retrieval will be the identification of tissue characteristics (biomarkers) that will be predictive of (i.e., surrogates for) success and failure. A most exciting possibility is that such biomarkers may be used to noninvasively image/monitor the maturation/remodeling of tissue engineered devices *in vivo* in individual patients (Mendelson and Schoen, 2006).

BIBLIOGRAPHY

Agins, H. J., Alcock, N. W., Bansal, M., Salvati, E. A., Wilson, P. D., et al. (1988). Metallic wear in failed titanium-alloy hip replacements. *J. Bone Joint Surg.*, **70A**, 347–356.

Albrektsson, T., Branemark, P.-I., Hansson, H.-A., & Lindstrom, J. (1982). Osseo-integrated titanium implants. *Acta. Orthop. Scand.*, **52**, 155–170.

Anderson, J. M. (1986). Procedures in the retrieval and evaluation of vascular grafts. In H. E. Kambic, A. Kantrowitz, & P. P. Sung (Eds.), *Vascular Graft Update: Safety and Performance* (pp. 156–165). Philadelphia, PA: ASTM STP 898 American Society for Testing and Materials.

Anderson, J. M. (1993). Cardiovascular device retrieval and evaluation. *Cardiovasc. Pathol.*, **2**(Suppl. 3), 199S–208S.

Anderson, J. M., Christenson, E. M., & Hiltner, A. (2007). Biodegradation mechanisms of polyurethane elastomers. *Corros. Eng. Sci. Techn.*, **42**(4), 312–323.

Anderson, P. G., Bajaj, R. K., Baxley, W. A., & Roubin, G. S. (1992). Vascular pathology of balloon-expandable flexible coil stent in humans. *J. Am. Coll. Cardiol.*, **19**, 372–381.

ASTM F561–05a (2005). Practice for retrieval and analysis of implanted medical devices, and associated tissues and fluids. *ASTM Annual Book of Standards, Vol. 13.01*, 2008.

Beyleveld, D., Howells, G. G., & Longley, D. (1995). Heart valve ownership: Legal, ethical and policy issues. *J. Heart Valve Dis.*, **4**, S2–S5.

Blackstone, E. H. (2005). Could it happen again? The Björk-Shiley convexo-concave heart valve story. *Circulation*, **111**, 2850–2857.

Bloebaum, R. D., Nelson, K., Dorr, L. D., Hoffman, A., & Lyman, D. J. (1991). Investigation of early surface delamination observed in retrieved hard-pressed tibial inserts. *Clin. Orthop.*, **269**, 120–127.

Blot, W. J., Ibrahim, M. A., Ivey, T. D., Acheson, D. E., Brookmeyer, R., et al. (2005). Twenty-five-year experience with Björk-Shiley convexoconcave heart valve: A continuing clinical concern. *Circulation*, **111**, 2850–2857.

Borovetz, H. S., Ramasamy, N., Zerbe, T. R., & Portner, P. M. (1995). Evaluation of an implantable ventricular assist system for humans with chronic refractory heart failure. Device explant protocol. *ASAIO J.*, **41**, 42–48.

Bradshaw, S. H., Kennedy, L., Dexter, D. F., & Veinot, J. P. (2009). A practical method to rapidly dissolve metallic stents. *Cardiovasc. Pathol.*, **18**, 127–133.

Bragdon, C. R., Kwon, Y. M., Geller, J. A., Greene, M. E., Freiberg, A. A., et al. (2007). Minimal 6-year followup of highly cross-linked polyethylene in THA. *Clin. Orthop. Rel. Res.*, **465**, 122–127.

Brown, S. A., & Merritt, K. (1981). Metal allergy and metallurgy. In A. Weinstein, D. Gibbons, S. Brown, & W. Ruff (Eds.), *Implant Retrieval: Material and Biological Analysis, NBS SP 601* (pp. 299–321). Washington, DC: National Bureau of Standards Special Pubication.

Brown, S. A., Flemming, C. A.C., Kawalec, J. S., Placko, H. E., Vassaux, C., et al. (1995). Fretting accelerated crevice corrosion of modular hip tapers. *J. Appl. Biomater.*, **6**, 19–26.

Burke, M., & Goodman, S. (2008). Failure mechanisms in joint replacement. In P. Revell (Ed.), *Joint Replacement Technology* (pp. 264–285). Cambridge, UK: Woodhead.

Cahoon, J. R., & Paxton, H. W. (1968). Metallurgical analysis of failed orthopaedic implants. *J. Biomed. Mater. Res.*, **2**, 1–22.

Callis, G., & Strerchi, D. (1998). Decalcification of bone: Literature review and practical study of various decalcifying agents. Methods, and their effects on bone histology. *J. Histotechnol.*, **21**, 49–58.

Canizales, S., Charara, J., Gill, F., Guidoin, R., Roy, P. E., et al. (1982). Expanded polytetrafluoroethylene prostheses as secondary blood access sites for hemodialysis: Pathological findings in 29 excised grafts. *Can. J. Surg.*, **154**, 17–26.

Charnley, J. (1970). The reaction of bone to self-curing acrylic cement. A long-term histological study in man. *J. Bone Joint Surg.*, **53B**, 340–353.

Charnley, J., Kamangar, A., & Longfield, M. D. (1969). The optimum size of prosthetic heads in relation to the wear of plastic sockets in total replacement of the hip. *Med. Biol. Eng.*, **7**, 31–39.

Chevalier, J. (2006). What future for zirconia as a biomaterial? *Biomaterials*, **27**, 535–543.

Chevalier, J., Deville, S., Munch, E., Julian, R., & Lair, F. (2004). Critical effect of cubic phase on aging of 3 mol% yttria-stabilized zirconia ceramics for hip replacement prostheses. *Biomaterials*, **25**, 5539–5545.

Clarke, I. C., Manaka, M., Green, D. D., Williams, P., Pezzotti, G., et al. (2003). Current status of zirconia used in total hip implants. *J. Bone Joint Surg. AM.*, **85A**, 73–84.

Collier, J. P., Suprenant, V. A., Jensen, R. E., & Mayor, M. B. (1991). Corrosion at the interface of cobalt-alloy heads on titanium alloy stems. *Clin. Orthop.*, **271**, 305–312.

Cook, S. D., Renz, E. A., Barrack, R. L., Thomas, K. A., Harding, A. F., et al. (1985). Clinical and metallurgical analysis of retrieved internal fixation devices. *Clin. Orthop.*, **184**, 236–247.

Cook, S. D., Barrack, R. L., Thomas, K. A., & Haddad, R. J., Jr. (1988). Quantitative analysis of tissue growth into human porous total hip components. *J. Arthroplasty*, **3**, 249–262.

Cook, S. D., Thomas, K. A., & Haddad, R. J., Jr. (1988). Histologic analysis of retrieved human porous-coated total joint components. *Clin. Orthop.*, **234**, 90–101.

Downs, A. R., Guzman, R., Formichi, M., Courbier, R., Jausseran, J. M., et al. (1991). Etiology of prosthetic anastomotic false aneurysms: Pathologic and structural evaluation in 26 cases. *Can. J. Surg.*, **34**, 53–58.

Ebramzadeh, E., Campbell, P. A., Takamura, K. M., Lu, Z., Sangiorgio, S. N., et al. (2011). Failure modes of 433 metal-on metal hip implants: How, why, and wear. *Orthop. Clin. N. Am.*, **42**, 241–250.

Farb, A., Sangiorgi, G., Carter, A. J., Walley, V. M., Edwards, W. D., et al. (2002). Pathology of acute and chronic coronary stenting in humans. *Circulation*, **105**, 2932–2933.

Furmanski, J., Atwood, S., Bal, B. Anderson, M. R. & Penenberg, B. L. et al. (2008). Fracture of highly cross-linked UHMWPE acetabular liners. *Proceedings of the 7th Annual AAOS, SE18.*

Fyfe, B., & Schoen, F. J. (1993). Pathologic analysis of 34 explanted Symbion ventricular assist devices and 10 explanted Jarvik-7 total artificial hearts. *Cardiovasc. Pathol.*, **2**, 187–197.

Goodman, S. B., & Wright, T. M. (2007). AAOS/NIH workshop – Osteolysis and implant wear: Biological, biomedical engineering and surgical principles. *J. Am. Acad. Orthop. Surg.*, **16**(Suppl. 1).

Gross, J. M., Shu, M. C.S., Dai, F. F., Ellis, J., & Yoganathan, A. P. (1996). A microstructural flow analysis within a bileaflet mechanical heart valve hinge. *J. Heart Valve Dis.*, **5**, 581–590.

Guidoin, R., Chakfe, N., Maurel, S., How, T., Batt, M., et al. (1993). Expanded polytetrafluoroethylene arterial prostheses in humans: Histopathological study of 298 surgically excised grafts. *Biomaterials*, **14**, 678–693.

http://www.prozyr.com/PAGES_UK/Biomedical/breakages.htm Accessed 7.04.03.

Hylen, J. C., Hodam, R. P., & Kloste, F. E. (1972). Changes in the durability of silicone rubber in ball-valve prostheses. *Ann. Thorac. Surg.*, **13**, 324.

ISO Soliclus/FDIA 12891-1. Retrieval and Analysis of Implantable Medical Devices.

Jacobs, J. J., Patterson, L. M., Skipor, A. K., Urban, R. M., Black, J., et al. (1999). Postmortem retrieval of total joint replacement components. *J. Biomed. Mater. Res. (Appl. Biomater)*, **48**, 385–391.

Kurtz, S. M. (2009). *The UHMWPE Handbook: Ultra-High Molecular Weight Polyethylene in Total Joint Replacement* (2nd ed.). New York, NY: Elsevier Academic Press.

Kurtz, S. M., Hozack, W. J., Purtill, J. J., Marcolongo, M., Kraay, M. J., et al. (2006). Significance of *in vivo* degradation for polyethylene in total hip arthroplasty. *Clin. Orthop. Rel. Res.*, **453**, 47–57.

Kwon, Y.-M., Glyn-Jones, S., Simpson, D. J., Kamali, A., McLardy-Smith, P., et al. (2010). Analysis of wear of retrieved metal-on metal hip resurfacing implants revised due to pseudotumors. *J. Bone Joint Surg.*, **92B**, 356–361.

Lemons, J. E. (1988). Dental implant retrieval analyses. *J. Dent. Educ.*, **52**, 748–756.

Mathiesen, E. B., Lindgren, J. U., Biomgren, G. G. A., & Reinholt, F. P. (1991). Corrosion of modular hip prostheses. *J. Bone Joint Surg.*, **73B**, 569–575.

McNeil, P. J., Durbridge, T. C., Parkinson, I. H., & Moore, R. J. (1997). Simple method for the simultaneous determination of formation and resorption in undecalcified bone embedded in methyl methacrylate. *J. Histotechnol.*, **20**, 307–311.

Mendelson, K. M., & Schoen, F. J. (2006). Heart valve tissue engineering: Concepts, approaches, progress, and challenges. *Ann. Biomed. Engin.*, **34**, 1799–1819.

Merritt, K., & Brown, S. A. (1981). Metal sensitivity reactions to orthopaedic implants. *Int. J. Dermatol.*, **20**, 89–94.

Merritt, K., & Rodrigo, J. J. (1996). Immune response to synthetic materials: Sensitization of patients receiving orthopaedic implants. *Clin. Orthop.*, **329S**, S233–S243.

Mitchell, R. N., Jonas, R. A., & Schoen, F. J. (1998). Pathology of explanted cryopreserved allograft heart valves: Comparison with aortic valves from orthotopic heart transplants. *J. Thorac. Cardiovasc. Surg.*, **115**, 118–127.

Mizuno, S., Tateishi, T., Ushida, T., & Glowacki, J. (2002). Hydrostatic fluid pressure enhances matrix synthesis and accumulation by bovine chondrocytes in three-dimensional culture. *J. Cell Biol.*, **193**, 319–327.

Moore, W. E.C. (1987). Microbiology of periodontal disease. *J. Periodontal. Res.*, **22**, 335–341.

Muratoglu, O. K., Greenbaum, E. S., Bragdon, C. R., Jasty, M., Freiberg, A. A., et al. (2004). Surface analysis of early retrieved acetabular polyethylene liners: A comparison of conventional and highly crosslinked polyethylenes. *J. Arthroplasty*, **19**(1), 68–77.

O'Neill, W. W., Chandler, J. G., Gordon, R. E., Bakalyar, D. M., Abolfathi, A. H., et al. (1995). Radiographic detection of strut separations in Björk-Shiley convexo-concave mitral valves. *N. Engl. J. Med.*, **333**, 414–419.

Oral, E., Christensen, S. D., Malhi, A. S., Wannomae, K. K., & Muratoglu, O. K. (2006). Wear resistance and mechanical properties of highly cross-linked, ultrahigh-molecular weight polyethylene doped with vitamin E. *J. Arthroplasty*, **21**, 580–591.

Peters, M. D., Polverini, P. J., & Mooney, D. J. (2002). Engineering vascular networks in porous polymer matrices. *J. Biomed. Mater. Res.*, **60**, 668–678.

Plemons, T. D., & Hovenga, M. (1995). Acoustic classification of the state of artificial heart valves. *J. Acoust. Soc. Am.*, **97**, 2326–2333.

Purdue, E. (2008). Alternative macrophage activation in periprosthetic osteolysis. *Autoimmunity*, **41**(3), 212–217.

Rabkin, E., & Schoen, F. J. (2002). Cardiovascular tissue engineering. *Cardiovasc. Pathol.*, **11**, 305.

Rabkin, E., Hoerstrup, S. P., Aikawa, M., Mayer, J. E., Jr., & Schoen, F. J. (2002). Evolution of cell phenotype and extracellular matrix in tissue-engineered heart valves during *in vitro* maturation and *in vivo* remodeling. *J. Heart Valve Dis.*, **11**, 308–314.

Revell, P. A. (2008a). Biological causes of prosthetic joint failure. In P. A. Revell (Ed.), *Joint Replacement Technology* (pp. 315–348). Cambridge, UK: Woodhead.

Revell, P. A. (2008b). The combined role of wear particles, macrophages, and lymphocytes in the loosening of total joint prostheses. *J. R. Soc. Interface*, **5**, 1263–1278.

Rimnac, C., & Pruitt, L. (2008). How do material properties influence wear and fracture mechanisms? *Journal of the AAOS*, **16**(Suppl. 1), S94–S100.

Sanderson, C., & Bachus, K. N. (1997). Staining technique to differentiate mineralized and demineralized bone in ground sections. *J. Histotechnol.*, **20**, 119–122.

Scales, J. T., Winter, G. D., & Shirley, H. T. (1959). Corrosion of orthopaedic implants. *J. Bone Joint Surg.*, **41B**, 810–820.

Schmalzied, T. P., Campbell, P., Schmitt, A. K., Brown, I. C., & Amstutz, H. C. (1997). Shapes and dimensional characteristics of polyethylene wear particles generated *in vivo* by total knee replacements compared to total hip replacements. *J. Biomed. Mater. Res. (Appl. Biomater.)*, **38**, 203–210.

Schoen, F. (1989). *Interventional and Surgical Cardiovascular Pathology*. Philadelphia, PA: W. B. Saunders.

Schoen, F. J. (1995). Approach to the analysis of cardiac valve prostheses as surgical pathology or autopsy specimens. *Cardiovasc. Pathol.*, **4**, 241–255.

Schoen, F. J. (1998a). Pathologic findings in explanted clinical bioprosthetic valves fabricated from photooxidized bovine pericardium. *J. Heart Valve Dis.*, **7**, 174–179.

Schoen, F. J. (1998b). Role of device retrieval and analysis in the evaluation of substitute heart valves. In K. B. Witkin (Ed.), *Clinical Evaluation of Medical Devices: Principles and Case Studies* (pp. 209–231). Totowa, NJ: Humana Press.

Schoen, F. J. (2001). Pathology of heart valve substitution with mechanical and tissue prostheses. In M. D. Silver, A. Gotlieb, & F. J. Schoen (Eds.), *Cardiovascular Pathology* (3rd ed.), (pp. 629–677). Philadelphia, PA: W. B. Saunders.

Schoen, F. J., & Levy, R. J. (1999). Tissue heart valves: Current challenges and future research perspectives. *J. Biomed. Mater. Res.*, **47**, 439–465.

Schoen, F. J., & Levy, R. J. (2005). Calcification of tissue heart valve substitutes: Progress toward understanding and prevention. *Ann. Thorac. Surg.*, **79**, 1072–1080.

Schoen, F. J., Titus, J. L., & Lawrie, G. M. (1982). Durability of pyrolytic carbon-containing heart valve prostheses. *J. Biomed. Mater. Res.*, **16**, 559–570.

Schoen, F. J., Goodenough, S. H., Ionescu, M. I., & Braunwald, N. S. (1984). Implications of late morphology of Braunwald–Cutter mitral heart valve prostheses. *J. Thorac. Cardiovasc. Surg.*, **88**, 208–216.

Schoen, F. J., Anderson, J. M., Didisheim, P., Dobbins, J. J., Gristina, A. G., et al. (1990). Ventricular assist device (VAD) pathology analyses: Guidelines for clinical study. *J. Appl. Biomater.*, **1**, 49–56.

Schoen, F. J., Levy, R. J., & Piehler, H. R. (1992). Pathological considerations in replacement cardiac valves. *Cardiovasc. Pathol.*, **1**, 29–52.

Schwartz, R. S., & Edelman, E. R. (2002). Drug-eluting stents in pre-clinical studies. Recommended evaluation from a consensus group. *Circulation*, **106**, 1867–1873.

Schwartz, R. S., & Edelman, E. R. (2004). Preclinical evaluation of drug-eluting stents for peripheral applications. Recommendations from an expert consensus group. *Circulation*, **110**, 2498–2505.

Sobieraj, M. C., & Rimnac, C. M. (2009). Ultra high molecular weight polyethylene: Mechanics, morphology, and clinical behavior. *J. Mechanical Behavior of Biomedical Materials*, **2**(5), 433–443.

Sobieraj, M. C., Kurtz, S. M., Wang, A., Manley, M. M., & Rimnac, C. M. (2008). Notched stress-strain behavior of a conventional and a sequentially annealed highly crosslinked UHMWPE. *Biomaterials*, **29**, 4575–4583.

Sussman, H. I., & Moss, S. S. (1993). Localized osteomyelitis secondary to endodontic implant pathosis. A case report. *J. Periodontol.*, **64**, 306–310.

Sutula, L. C., Collier, J. P., Saum, K. A., Currier, B. H., Sandford, W. M., et al. (1995). Impact of gamma sterilization on clinical performance of polyethylene in the hip. *Clin. Orthop.*, **319**, 28–40.

Tower, S. S., Currier, J. H., Currier, B. H., Lyford, K. A., Van Citters, D. W., et al. (2007). Rim cracking of the cross-linked longevity polyethylene acetabular liner after total hip arthroplasty. *J. Bone Joint Surg. Am.*, **89**, 2212–2217.

USPHS (1985). *U.S. Department of Health and Human Services, Public Health Service, National Institutes of Health. Guidelines for Blood–Material Interactions*. Bethesda, MD: NIH Publication No. 85–12185, National Institutes of Health.

van Beusekom, H. M. M., van der Giessen, W. J., van Suylen, R. J., Bos, E., Bosman, F. T., et al. (1993). Histology after stenting of human saphenous vein bypass grafts: Observations from surgically excised grafts 3 to 320 days after stent implantation. *J. Am. Coll. Cardiol.*, **21**, 45–54.

Wagner, W. R., Johnson, P. C., Kormos, R. L., & Griffith, B. P. (1993). Evaluation of bioprosthetic valve-associated thrombus in ventricular assist device patients. *Circulation*, **88**, 203–2029.

Walker, A. M., Funch, D. P., Sulsky, S. L., & Dreyer, N. A. (1995). Patient factors associated with strut fracture in Björk-Shiley 60° convexo-concave heart valves. *Circulation*, **92**, 3235–3239.

Weinstein, A., Gibbons, D., Brown, S., & Ruff, W. (Eds.), (1981). *Implant Retrieval: Material and Biological Analysis*. Rockville, MD: US Department of Commerce, National Bureau of Standards.

Wiggins, M. J., Wilkoff, B., Anderson, J. M., & Hiltner, A. (2001). Biodegradation of polyether polyurethane inner insulation in bipolar pacemaker leads. *J. Biomed. Mater Res.*, **58A**, 302–307.

Willert, H. G., & Semlitsch, M. (1977). Reaction of articular capsule to wear products of artificial joint prostheses. *J. Biomed. Mater Res.*, **11**, 157–164.

Woolson, S. T., Milbauer, J. P., Bobyn, J. D., Yue, S., & Maloney, W. J. (1997). Fatigue fracture of a forged cobalt–chromium–molybdenum femoral component inserted with cement. *J. Bone Joint Surg.*, **79A**, 1842–1848.

Wright, T. M., & Goodman, S. B. (Eds.), (1996). *Implant Wear: The Future of Joint Replacement*. Rosemont, IL: American Academy of Orthopedic Surgeons.

Wright, T. M., Hood, R. W., & Burstein, A. H. (1982). Analysis of material failures. *Orthop. Clin. N. Am.*, **13**, 33–44.

Wright, T. M., Rimnac, C. M., Faris, P. M., & Bansal, M. (1988). Analysis of surface damage in retrieved carbon fiber-reinforced and plain polyethylene tibial components from posterior stabilized total knee replacements. *J. Bone Joint Surg.*, **70-A**, 1312–1319.

Wright, T. M., Rimnac, C. M., Stulberg, S. D., Mintz, L., Tsao, A. K., et al. (1992). Wear of polyethylene in total joint replacements. Observations from retrieved PCA knee implants. *Clin. Orthop.*, **276**, 126–134.

GENERAL REFERENCES

Fraker, A. C., & Griffin, C. D. (Eds.), (1985). *Corrosion and Degradation of Implant Materials: Second Symposium*. Philadelphia, PA: ASTM Special Publication, STP 859.

Ratner, B. D. (Ed.), (1988). *Surface Characterization of Biomaterials*. Amsterdam: Elsevier.

Revell, P. A. (Ed.), (2008). *Joint Replacement Technology*. Cambridge, UK: Woodhead.

Shanbhag, A., Rubash, H. E., & Jacobs, J. J. (Eds.), (2006). *Joint Replacement and Bone Resorption. Pathology, Biomaterials, and Clinical Practice*. New York, NY: Taylor & Francis.

Transactions of the Society for Biomaterials Symposium on Retrieval and Analysis of Surgical Implants and Biomaterials. (1988). Snowbird, UT: August 12–14, pp. 1–67.

von Recum, A. F. (1999). *Handbook of Biomaterials Evaluation: Scientific, Technical, and Clinical Testing of Implant Materials* (2nd ed.). Washington, DC: Taylor & Francis.

Weinstein, A., Gibbons, D., Brown, S. & Ruff, W. (Eds.). (1981). *Implant Retrieval: Material and Biological Analysis*. US Dept of Commerce, NBS SP 601.

SECTION III.2

Voluntary Standards, Regulatory Compliance, and Non-Technical Issues

SECTION

III.2

Voluntary Standards, Regulatory Compliance, and Non-Technical Issues

CHAPTER III.2.1 INTRODUCTION: VOLUNTARY STANDARDS, REGULATORY COMPLIANCE, AND OTHER NON-TECHNICAL ISSUES

Frederick J. Schoen[1] *and Jack E. Lemons*[2]

[1]Professor of Pathology and Health Sciences and Technology (HST), Harvard Medical School, Executive Vice Chairman, Department of Pathology, Brigham and Women's Hospital, Boston, MA, USA

[2]University Professor, Schools of Dentistry, Medicine and Engineering, University of Alabama at Birmingham, Birmingham, AL, USA

This section highlights some of the key non-technical considerations in the development of medical devices that depend so importantly on biomaterials. Several principles underlie the approach presented in this section. The major driving force to technological innovation in biomaterials, implants, and medical devices is to improve the quality of life and survival of patients. The early phase of the process includes these steps: (1) identification of a clinical problem that needs solving ("clinical pull"); (2) determination that a viable market exists for it; and (3) the judicious and sophisticated application of existing, or development of, novel technology to solve this problem. In this sense, hospitals and clinics provide a "problem-rich" environment, and a laboratory for testing and translating the latest technologies and materials into clinical successes. An invention, by itself, is of little value; the value of the idea represented by an invention is through innovation, i.e., the validation and implementation of that idea or embodiment thereof to affect individuals and populations through its utilization and acceptance by others.

In addition, it is important to recognize that the future of a new biomaterial associated with a device- and/or implant-based treatment utilizing a biomaterial usually depends on the successful commercialization and implementation of a medical product (Chapter III.2.2). For any product, such as an implant or medical device, the needs of commercialization (involving development, design optimization, a viable fabrication process, financing, regulatory approval, reimbursement, manufacturing and scale-up, a suitable and profitable business model, marketing, etc.) are vital, and the scientific development, fabrication, clinical trials, and subsequent regulatory approval must occur in the context of, and concurrent with, business considerations (Figure III.2.1.1). Indeed, medical device innovation is complex; the generic process and the requisite skills in technological innovation in medicine, including identification, invention and implementation phases, have been recently described, both systematically and in detail (Zenios et al., 2010). Moreover, although demonstration of clinical effectiveness and safety are vital to clinical translation of any new medical product, today's economic pressures demand serious consideration of cost-benefit and cost-effectiveness, i.e., the relative gain in health from a measure (years of life, pain-free gained) and the relative costs of different ways to achieve an equivalent gain in health, respectively (Sculco, 2010; Lammers et al., 2011; Moreno et al., 2012). Finally, as embodied by the total collection of chapters in this book, progress in biomaterials and their implementation in clinically-useful medical devices requires multidisciplinary collaboration, thinking, and communication, where engineers, physical scientists, biologists, and clinicians are on an equal footing, and share a common language and objectives – an approach generally termed "convergence" (Sharp and Langer, 2011).

The chapters that comprise this section summarize the key principles enumerated above and provide a "toolkit" for medical device innovation. Specifically, the chapter by Lemons (Chapter III.2.3) describes consensus standards for biomaterials and devices. Consensus standards in general can be separated into categories of specifications, test methods, terminology (nomenclature and definitions), provisional, and guidances (Chapter III.2.3). Related to biomaterials, standards have been developed and repetitively tested for metallics, ceramics, polymerics, and combinations of these classifications. In broad terms for biomaterials of synthetic origin, biomaterials may also be classified in terms of biological responses such as bio-tolerant, -active, and -degradable.

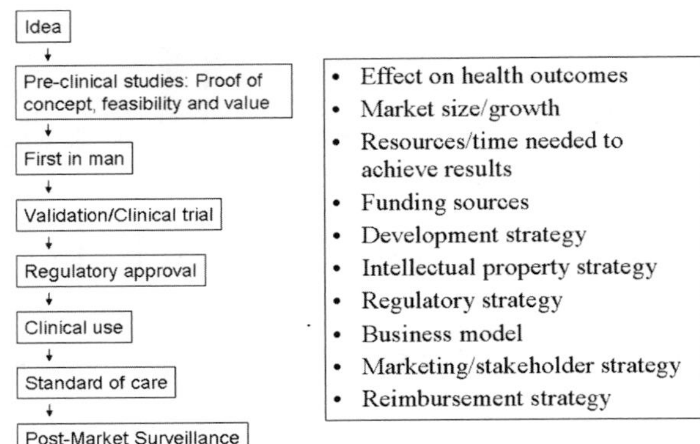

FIGURE III.2.1.1 Key steps in technical development (flow chart on left) and non-technical considerations (box on right) must be integrated to achieve successful innovation in biomaterials and medical devices.

Also, since regulatory processes consider devices (rather than biomaterials, per se, grouping can also be made in terms of the clinical application. Consensus standards have provided an opportunity to document biomaterial charateristics needed for biocompatibility, the methods for testing and presentation of results (precision and accuracy), terminology including basic definitions, and importantly, guidelines for evolving device systems and the properties of new or modified biomaterials. Over past decades, as the device disciplines have evolved from synthetic origin bio-tolerant biomaterials to surface bioactive biomaterials, naturally-derived biomaterials, and to biodegradable and biointeractive biomaterials, the focus has shifted to Tissue Engineered Medical Product Standards (TEMPS). This shift has resulted in reconsideration of standards needed for evaluations, including a move from *in vitro* to *in vivo* testing methods. Also, consensus standards have become more harmonized throughout developed countries, especially as device products have expanded throughout the world. Many consensus standards for medical devices have now been evaluated and accepted for regulatory processes and approvals. Thus, in many submissions device developers only need to list "meets standard x." This process has enhanced considerably the value of standards, especially when harmonized internationally.

Anders and Tolkoff (Chapter III.2.2) provide aspiring entrepreneurs a basic introduction to the considerations they can use to size up an idea in the context of clinical utility and commercial potential. Most medical and dental device applications are regulated by government agencies charged to ensure patient health and welfare. Thus, agencies evaluate safety and efficiency of devices and related treatments. In the USA, most devices constituted to include biomaterials are regulated by the US Food and Drug Administration (FDA) Center for Devices and Radiological Health (CDRH). As biomaterials, devices and treatments have evolved to combination products (comprising mixtures of materials, drugs and cells (and their products) many must be considered by combining expertise from the FDA centers regulating biologicals (including cells) and drugs. Duncan (Chapter III.2.4) summarizes the key concepts that govern the regulation of medical devices by the Food and Drug Administration in the USA; regulation abroad follows generally similar principles although not exactly the same approaches. Cahn and Erickson (Chapter III.2.5) discusses reimbursement, the process by which medical devices and any other medical services are paid for; this chapter emphasizes that reimbursement is often critical to market success and that those inventing, developing or marketing new medical devices have an understanding of how their products will be reimbursed. The integration of regulatory and reimbursement considerations is emphasized by several case histories of actual products presented by Baura (Chapter III.2.6). This is followed by related discussions of pertinent ethical and legal issues by Bianco et al. (Chapter III.2.7) and Mayesh and Vicari (Chapter III.2.8), respectively. Grunkemeier et al., in the following chapter in this section, addresses the special challenges associated with clinical trials of biomaterials and medical devices (Chapter III.2.9) that necessitate very different approaches than those used in the development and implementation of drugs. Langer and colleagues (Chapter III.2.10) write with considerable experience about the approach to creating small start-up companies to develop medical devices born in the university setting. In the last chapter in this section, Feigal (Chapter III.2.11) addresses the special regulatory aspects of the post-market period, when a product is in the marketplace, is no longer investigational and is available for commercial use.

BIBLIOGRAPHY

Lammers, M. J., Grolman, W., Smulders, Y. E., & Rovers, M. M. (2011). The cost-utility of bilateral cochlear implantation: A systematic review. *Laryngoscope*, **121**, 2604–2609.

Moreno, S. G., Novielli, N., & Cooper, N. J. (2012). Cost-effectiveness of the implantable HeartMate II left ventricular assist device for patients awaiting heart transplantation. *J. Heart Lung Transplant.*, in press.

Sculco, T. P. (2010). The economics of new age arthroplasty: Can we afford it? *Orthopedics*, 33, 628.

Sharp, P. A., & Langer, R. (2011). Promoting convergence in biomedical science. *Science*, 333, 527.

Zenios, S., Makower, J., & Yock, P. (Eds.), (2010). *Biodesign: The Process of Innovating Medical Technologies*. Cambridge, UK: Cambridge University Press.

CHAPTER III.2.2 COMMERCIALIZATION: WHAT IT TAKES TO GET A PRODUCT TO MARKET

Joshua Tolkoff[1] and Richard Anders[2]
[1]Accelerator Executive, CIMIT, Boston, MA, USA
[2]Senior Lecturer, Massachusetts Institute of Technology, Cambridge, MA, USA

INTRODUCTION

Previous chapters of this book cover technical and biological issues relating to biomaterials. Transforming a biomaterial into a commercial product with clinical utility is extraordinarily difficult. There are two major routes to commercializing a product: licensing it to an established company or creating a company to move it to market. In this chapter we focus on the latter pathway, but our discussion will also include the key elements involved in a successful license.

A business starts with an idea. And, like the mythical Pygmalion, those who come up with ideas have a tendency to fall in love with them. In the world of promising ideas, there are some which when brought to fruition make a significant difference, and those that have little impact. We urge the entrepreneur to test their idea against a list of important success factors to determine if it will have real impact. It is, after all, easy to find ideas that appear to have good business prospects, but ultimately very few have the potential for real market success.

A medical product must meet many criteria: it must be well-designed; fulfill complex regulatory requirements; obtain reimbursement; be appealing to one group of (often conservative) medical practitioners while able to withstand the attacks of other practitioners who may be working against its adoption. Failure of any of these is usually fatal to the product.

In this chapter we give aspiring entrepreneurs a basic introduction to the considerations they can use to size up an idea. But the subject is vast, and this introduction is merely a brief survey. Still, we hope that if you are seeking to commercialize a product, what follows will give you an entry point to the hard work you will need to do to honestly assess, modify or even discard your idea. We also hope that by giving you insight into how to increase your own chances of success, more clinically useful products will get to market, which will help us all.

Life science businesses fail for many reasons, including lack of financing and poor execution. But in our experience they most often fail because the idea driving the business premise does not provide sufficient value to the prospective customer or market. The chasm between an interesting idea and an idea with significant new value arises for many reasons, some of them quite predictable: an uncompelling need; the market size doesn't justify the cost to bring the product to market; the idea infringes on other's patents; there is no reimbursement. At the beginning of an enterprise, the team is setting out the blueprint for the next five-to-ten years of the company's work. They have an obligation to carefully evaluate every facet of the future company. Problems such as those listed above can be identified at this early planning stage and, ideally, avoided. Then, the team should again evaluate the key risks to business success. This continuing circle – considering risks, evaluating them in light of the best available data and opinions, strengthening the plan, and doing it again – is critical if the team is to hit key milestones and be successful. Elucidating shortcomings early is a useful (though painful) task. It can save the adept practitioner from literally years of work chasing unproductive blind alleys to their inevitable, fruitless, conclusion.

THE NEED YOU PROPOSE TO SOLVE

There are techniques for developing ideas for medical products. One, based on the so-called Biodesign Process, was evolved by Paul Yock, Josh Makower, and others at Stanford (Zenios et al., 2009). The subject of an excellent textbook, their technique starts with fieldwork – observing real-life clinical procedures, and using those observations to uncover important needs and formulate novel solutions. This process from the start is designed around finding needs to solve, not ideas to solve them.

But that's not what usually happens. More often, a scientist or inventor makes a discovery and starts thinking about how and where it can be applied. Many applications are considered, and chosen as much for whether the idea can be applied to them as whether they are worth solving. As a result, although it may seem too obvious to state, one must keep in mind that a new medical product must solve a clinical problem.

Developing something innovative and useful is not the same as solving an important need. For example, a new treatment for smallpox, while undoubtedly an achievement, is not the solution of an important clinical need (the possibility of bioterrorism notwithstanding), since the disease has largely been eradicated.

Although it is hard to give absolute rules, a solution to an important need might typically have some of the following characteristics:

- Provably decreases mortality or serious morbidity in an identifiable patient group;
- Substantially lowers the cost of treatment relative to the standard of care;
- Meaningfully improves an existing treatment by such things as filtering out patients who will be unresponsive or will have serious adverse effects;
- Makes an earlier diagnosis of an illness for which an earlier diagnosis means substantially better outcomes.

Less significant solutions often include the following:

- Early detection of conditions for which early detection confers no benefit or even worse, early detection of things that are not even provably connected with the condition;
- Improvement of patient comfort but not patient outcomes. Since payors typically will not pay for comfort, it will be hard to find a market;
- Small decreases in surgery time, especially when this comes at significantly increased cost.

Determining whether one solves an important need is difficult to do in a vacuum. It will typically require extensive research of the literature. One will likely wish to speak with clinicians to have them validate the idea's value. Depending on the degree of protection, this may need to be done under non-disclosure agreements (NDAs), but to maximize the chances someone will speak with you. NDAs should be avoided if at all possible. Your goal, when all this has been done, is to very succinctly – ideally in a paragraph or so – write down the need you proprose to solve, explain why its solution is important, and in what ways your idea solves it.

MARKET SIZE AND GROWTH

Although it is possible to develop a product for a small market, the size of the opportunity has an obvious impact on the amount of money that should be spent to develop it. Thus, it is essential to understand how large the potential market for your product/technology is, and how ready the clinical users will be to adopt it. This can be evaluated at the earliest stages of a project, and again when prototypes are available, when animal testing is done, at early clinical trials, etc. Important considerations include:

- The projected growth of the patient population (i.e., diseases of aging are likely to increase in aging populations);
- The number of procedures currently performed and the number of units sold;
- The current solutions, including how they are applied, their outcomes, their cost (especially reimbursed cost), length of time they take to use, and the tools and training they require;
- Whether your solution is incremental or a path-breaking replacement;
- The companies selling the existing solutions and whether your device complements their product line (i.e., appeals to the same markets without replacing their product) or competes with it;
- The path of innovation, and what the most likely new devices will be in one, five or ten years;
- The number of patients who COULD BE and are LIKELY to be served by your solution;
- The specialty that will use your solution, and whether this is a change from current practice.

Before an entrepreneur moves to exploring the numerous tactical business considerations, which we will outline in the section entitled "Business and Commercialization Issues," they should gain insight into one more aspect(s) of the market: how the various constituencies will align in support of or against the proposed new product.

EFFECT ON VARIOUS CONSTITUENCIES

New products displace old ones, and disruptive new products upset complex webs of relationships and interests. In a business as complex and conservative as the healthcare industry, it is important to understand that introducing solutions also creates problems. For example, the arrival of percutaneously introduced vascular stents caused a dramatic shift in power between cardiac surgeons and interventional cardiologists, with surgeons being less than completely enthusiastic advocates of the new technology. In developing a new medical product, it is – perhaps surprisingly – not enough to simply improve patient care. You should also pay attention to how the new product will affect the interests of each constituency. For example:

- *Patients*: who balance ultimate results against recuperation time, likelihood of success, risks, and many other factors;
- *Clinicians*: whose concerns include patient safety, difficulty of learning to use the new product, and whether they – or a different specialist – control the patient workup;
- *Insurers*: who worry about the additional cost of new technologies, and risks to their competitive position if they fail to reimburse the technologies.

CLINICAL TRIALS

It is never too early to begin to understand the clinical trials a company will need to perform. Here is a short list of some of the reasons:

- Regulatory approvals often require clinical data;
- Reimbursement often requires proof of economic performance that can only be obtained through clinical trials;
- Marketing claims can often only be made after clinical trials have substantiated them;
- Investors are usually much more interested in investing in companies that have proven their product works. Additionally, investors are loath to invest in companies where the expected clinical trials will be long, large, and extremely expensive;
- Clinicians are much more interested in products for which there is clinical validation;
- Some products that would undoubtedly be valuable simply cannot be brought to market because the cost of the trials would be too high.

Clinical trials are generally the costliest initiatives undertaken by a life science company. But because a clinical trial needs to accomplish so many disparate objectives (viz. the list above) it is extremely important to design them as efficiently as possible, and to understand their purpose as early as possible.

Intellectual Property

A start-up's most significant asset is usually its intellectual property (IP). Therefore, every start-up needs a strategy for developing, increasing, and protecting this key asset. Understanding competitive IP, and the strengths and weaknesses of your IP, are critical for fundraising and prioritizing all other expenditures.

Patents are expensive and slow to obtain. It can cost $10,000 or more to have a law firm file a US Patent. Even a provisional patent, which gives you the right to file an actual patent within a year, can run upwards of $2000. Initial responses ("office actions") from the United States Patent and Trademark Office can take two years or more, and there is a long road to a patent following that. As a result, there is often a substantial gap between the time one starts a company and the time patents are in hand. During that time, founders must convince savvy investors that the company's patent applications are likely to lead to issued, strong patents. The company must clearly understand the strength of their patent applications, and have an in-depth understanding of how they are superior to the competitive IP.

Selecting a strong patent attorney is crucial. Use advisers, investors, and the company board to get recommendations of at least three patent firms, and interview each to assess their expertise in the company's scientific domain, their ability to be prudent advisors to the company, and the personal chemistry you will have in working with them. A good patent attorney will be a source of strategic IP advice regarding patents, including IP budget priorities. In some cases, IP attorneys will defer their fees for start-ups until the company has obtained funding.

There are alternatives to using patent attorneys, for example, filing a provisional patent is often straightforward work for a technical person. Another source of expertise is the patent agent, a technical person with additional training analyzing and preparing patents. An economical and effective approach is to draft the initial application oneself or with the help of a patent agent, and then have an attorney do a final review prior to submission.

It is also important to understand the patents close to your IP to ensure you have "freedom to operate," meaning that you are able to build your device without infringing anyone else's patents. Again, patent lawyers and patent agents can assist you. Much of this work can be done by the company founders, as long as they have good advisors.

Regulatory Strategy

It is far beyond the scope of this chapter to provide a detailed view of the regulatory process. Our purpose instead is to provide a basic understanding of the regulatory complexities, and an appreciation of the need to plan the regulatory strategy early. There are few things more critical to a life science company than its regulatory strategy and choices, and the better you understand the regulatory framework, and the earlier you understand your strategic choices, the greater your chance of success.

In the United States, the Food and Drug Administration (FDA) is committed to ensuring a product is safe and effective for its intended purpose. The level of FDA review is determined by the nature of the product, its indication for use, and the manner in which it is to be used.

Broadly speaking, in the U.S. medical products for human use are regulated under one of three divisions: drugs (Center for Drug Evaluation and Research or CDER); devices (Center for Devices and Radiological Health or CDRH); and biologics (Center for Biologics Evaluation and Research or CBER). Most biomaterial-based products will fall within CDRH. The device pathway through CDRH (or at times, CBER) is usually simpler, cheaper, and faster than the drug pathway.

CDRH regulates how a device can be brought to market, and what controls are placed on the marketing and manufacturing of the device. The manner in which a device can be brought to market depends first on the device classification (I, II or III), and to a lesser extent on the CDRH medical specialty panel that will be reviewing it. This, and a great deal more about the regulatory framework, is explained in further detail at the FDA website, www.fda.gov.

Class I devices – such as bandages, adjustable beds, many surgical instruments, and manual stethoscopes – are deemed to pose minimal risk to patients and are therefore regulated most lightly. Such products can usually (count on exceptions in the complex regulatory world) be brought to market without any advance approvals or clearances from the FDA. Additionally, the company will be subject only to a set of general rules ("general controls") about how the product must be labeled, registered, and manufactured.

Class II devices – e.g., mercury dental amalgam, electronic stethoscopes, most catheters, needles, and monitoring devices – pose higher risks than Class I devices. They must usually be cleared in advance by the FDA before they can be brought to market. This is done by the 510(k) premarket notification process, the overwhelmingly dominant path to the US market for complex medical devices. A 510(k) clearance can often be completed in months, usually without clinical trials, and is relatively inexpensive.

To receive a 510(k) clearance, one must usually show that the device is "substantially equivalent" in safety and effectiveness to one or more "predicate" devices.

There is a great deal of misunderstanding about what are true predicates, and substantial art and skill in selecting appropriate predicates and preparing a submission under the 510(k) rules. Thus, this is usually done with the help of experts.

Class III devices – including devices that are life-sustaining, such as cardiac assist pumps and also most implants such as bone grafts, heart valves, orthopedic implants, arterial grafts, and pacemakers – are regulated most strictly. The FDA requires Class III devices to undergo a more elaborate approval before allowing them on the market. This so-called Pre-market Approval (PMA), requires a minimum of 180 days, and often a year or more, to issue after submission of significant amounts of data, and typically will require many millions of dollars in preclinical and clinical trials. One of the hurdles to surmount is that the device must typically be tested in a pivotal human trial of effectiveness, and this trial must be conducted under an Investigational Device Exemption (IDE).

Many other countries have their own regulatory approval bodies. In the European Economic Area (EC), a manufacturer can market a device when it has received a CE ("Conformité Européenne" or "European Conformity") mark, meaning that the manufacturer's product conforms to applicable EC directives. Many other countries, including China, piggyback on the United States regulatory system, and in such countries, US FDA approval is sufficient for a manufacturer to bring their product to market.

Entrepreneurs should keep the following regulatory considerations in mind:

- In the US, a company must carefully plan its regulatory strategy and be sure the FDA is comfortable with corporate marketing strategies. In particular, before approval do not be vague about product claims in your product manuals or literature. Instead be transparent in your intent, document your work meticulously, and work with the FDA to make them comfortable with your approach.
- Don't cut corners with the rules. For example, if you believe your device can be shoehorned into Class I as a wound cover, but your intent is to use it in brain surgery, you are likely at some point to face significant regulatory problems and delays, and your business may be shut down. It is far better to confront the issue head on and seek an advance determination that your product is, indeed, Class I.
- As elaborate as the regulatory infrastructure is, it is not algorithmic. Often regulatory consultants or even FDA insiders will disagree on how something is regulated. Thus, there can be significant value to managing your approach intelligently and carefully.
- Understand the regulatory framework around your product as completely as you can. Among many reasons for this, investors expect that entrepreneurs will have thought a great deal about this critical aspect of a product.
- FDA regulations are enabled by law, but evolve by agency action. They are in constant flux (currently the most notable example being the 510(k) regulations), so keep abreast of these changes through trade organizations, current FDA news, and by working with experienced consultants.

Because of the aforementioned complexities, companies are increasingly introducing their products outside the US where regulations are easier to comply with and approvals are faster.

Reimbursement

In the United States, most medical expenses are paid by healthcare insurance, and so insurance carriers are a crucial part of the medical market infrastructure. Business plans from companies in the 1990s rarely discussed reimbursement because it was assumed that a superior product desired by the clinician would be paid for. But this has changed dramatically as medical costs have escalated. While drugs, once approved, have a very favorable reimbursement profile, medical devices, even when approved, often face a complex and uncertain reimbursement process. Nevertheless it is never too early to obtain an understanding of the reimbursement landscape, because without adequate reimbursement a product will be doomed.

Very generally, reimbursement within the US falls into one of four bins:

- Unreimbursed "self-pay" procedures, such as cosmetic surgery, typically paid for by the patient;
- Physician services, described using CPT (or its synonym, HCPCS level I) codes and reimbursed by the insurer (Medicare or others) based on how they wish to pay for these services;
- Hospital admissions to treat a condition, typically reimbursed by Medicare (and often other insurers, though with different payment amounts) as part of a single fixed payment to the hospital (which of course in the complex world of reimbursement can change subject to various factors) under a diagnosis-related-group or DRG, and a separate CPT-based payment to the physician;
- An outpatient-based procedure performed in a hospital or ambulatory surgical center, reimbursed by Medicare (and often other insurers) using one or more ambulatory procedure codes (APCs) which are like DRGs with the notable exception that several APCs can be stacked together for a single patient visit.

Subject to insurance regulations and its own commitments to its clients, a private insurer has great latitude to set its own reimbursement levels for any particular product or to not reimburse it at all. However, broadly speaking, private insurers follow Medicare, and therefore obtaining Medicare reimbursement is critical for a new product.

There are several general business considerations to keep in mind which will make the reimbursement environment easier to negotiate.

- Some devices are used in a self-pay indication. In such a case, with no insurer, there is no reimbursement issue (although of course self-payers also care about cost).
- Products used as integral parts of existing DRGs or APCs – for example, an improved heart bypass pump for cardiac surgery – have narrowed reimbursement concerns. Perhaps the biggest risk is that the new device fundamentally changes the procedure, thereby bumping it into a different DRG or APC, or worse, transmuting it into something no longer covered under any DRG or APC. Assuming this is not the case, customers will care about the value the new product adds to the process – reduced procedure time, lowered co-morbidities or hospitalization time, perceived importance of the product by clinicians and even patients, and so forth – compared to its cost.
- For new services and procedures which require a CPT code, the code may already exist and be reimbursed. But keep in mind, a CPT code *is not a guarantee* your product will be reimbursed. For example, if a product dramatically changes how a procedure is applied, perhaps by substantially enlarging the pool of patients who will receive it or allowing a different specialist to perform it, insurers may reconsider reimbursement. That is one reason why experienced investors look carefully when device entrepreneurs claim their new product fits into an existing code.
- Each country has its own reimbursement process, and different payers within countries have their own processes.

When a product is unreimbursed, in many cases the entrepreneur may need to obtain a CPT or DRG code. This alone can be an extended multi-year process, often longer than the regulatory approval process. And unlike the regulatory process, with its established (mostly followed) timelines, there are no such schedules for the reimbursement process. In the current cost-constrained environment, there is negative pressure to add expensive new procedures. Some of the things required to obtain reimbursement include:

- Determining who makes reimbursement decisions at each payer;
- Describing the clinical and economic value of the proposed product/procedure, including key economic benefits and consideration for the proposed new reimbursement;
- Indicating how this procedure will affect total costs;
- Collecting clinician feedback on the proposed product/procedure;
- Performing clinical trials of the new product, including the economic issues as part of the study;
- Publishing the clinical results showing the economic value produced by the new product;
- Getting regulatory approval of the new product;
- Having the new product used by many practitioners in the use indicated;
- Getting support from clinicians, key opinion leaders, and medical societies;
- Obtaining relevant codes from the American Medical Association (AMA) or Medicare.

For these reasons, obtaining anything remotely close to national reimbursement (i.e., from Medicare and most insurers) for an unreimbursed product can take years. The process is quite expensive for two reasons: first, the actual cost of obtaining reimbursement; and second, the difficulty in convincing practitioners to use a product without an automatic reimbursement. Thus, it is critical for small companies to work with experienced reimbursement experts. And if a company needs to obtain new reimbursement codes, the experts may well need to be retained full-time on staff.

BUSINESS AND COMMERCIALIZATION ISSUES

Once you have confirmed a compelling unmet need in a large market, understood the regulatory and reimbursement issues, and considered the clinical trials your company will need to perform, you have done your basic due diligence. If your idea has passed muster, congratulations! But running a business brings a number of other issues to the fore which you should consider (Figure III.2.2.1).

Funding

Developing a new product employing a biomaterial technology is an expensive and time-consuming process. Table III.2.2.1 shows an example of a typical timeline for development of a PMA product within the United States. Many start-up companies do portions of this work outside the US to save time and money. Nevertheless, the steps, approximate times, and rough budgets are good guidelines for 2012.

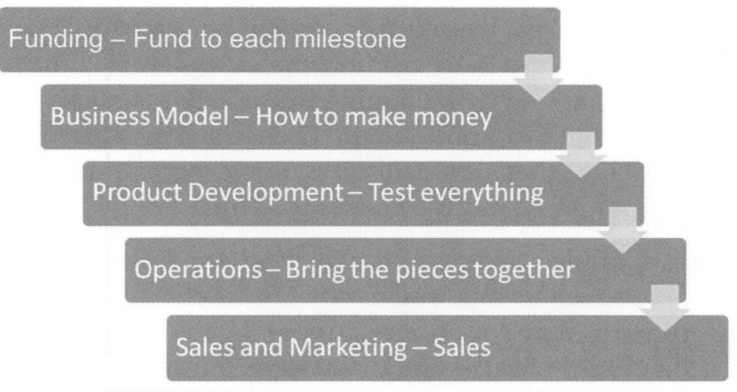

FIGURE III.2.2.1 Steps to commercialization.

TABLE III.2.2.1 Sample Timeline & Budget for Development of a PMA Product within the US

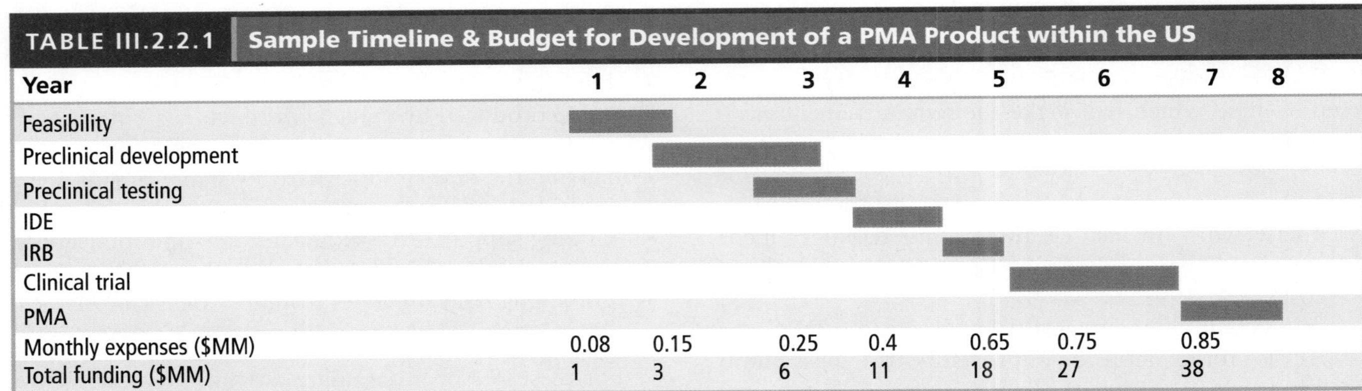

Year	1	2	3	4	5	6	7	8
Feasibility								
Preclinical development								
Preclinical testing								
IDE								
IRB								
Clinical trial								
PMA								
Monthly expenses ($MM)	0.08	0.15	0.25	0.4	0.65	0.75	0.85	
Total funding ($MM)	1	3	6	11	18	27	38	

IDE: Investigational Device Exemption.
IRB: Institutional Review Board.
PMA: Pre-market approval.

During the development stages of a company, moving quickly and meeting the key milestones of the project plan within budget are critical to having adequate resources to succeed. Since a medical business rarely has revenues during the development stages, raising capital is essential. Below are the three major sources of capital you are likely to encounter. Keep in mind that the earlier investors typically incur the highest risk, and therefore rightly expect the highest potential returns.

- Friends and family (sometimes jokingly referred to as "three f" financing: friends: family: and fools). This includes unpaid "sweat equity" put in by founders. Typical amounts are tens to hundreds of thousands of dollars;
- "Angel Investors," often wealthy individuals who have made money in business and enjoy the chance to make money or become involved in new start-ups. They will typically invest hundreds of thousands to low millions of dollars in a venture;
- Professional investors or Venture Capitalists (VCs) who manage large pools of money, and are often seeking companies of a stage and type able to use millions of dollars, which of course requires an eventual payback of tens to hundreds of millions of dollars.

A business without funding will not likely survive. Therefore, an entrepreneur should learn about the funding process early. This is an extremely intricate topic, and we offer one hopefully useful tip: call potential funders before you are actually looking for money to solicit their advice and get to know them. Then when you are actually looking, they can calibrate your progress, and also you are a known entity. Three more tips:

- If possible, do not cold call these funders. Network first to clinicians and scientists in the field. Many of them may know investors and can arrange introductions. Use the initial investor's introductions to obtain other introductions.
- Use care before approaching professional venture capitalists. If your idea has major holes or you lack experience, it would be better to hone your approach a bit first.
- Spend time understanding, as deeply as possible, the various factors described in this chapter. Savvy potential investors will care about all of them, and will be reassured when they meet a CEO who has really thought through the business.

Business Model

A business model refers to the method a company will use to make money from its product. Although it may seem straightforward, there are usually many ways to bring a product to market, and it usually takes a great deal of effort to determine the best one.

Considerations that can affect the model include: resources available to the company; relevant experiences of the team; buying patterns of customers; rate of product adoption in the marketplace; and the value of the product to each stakeholder. Some of the approaches used in medical device business models include the following.

Direct Sales. The company sells its products directly to end-users. Although this typically captures more value for a company than licensing, it is very expensive and can be slow to ramp-up to levels sufficient to support the company. Additionally, products must generally be expensive with high margins to warrant the cost of direct selling, as even a single sales call – which is often unsuccessful – can cost hundreds to thousands of dollars.

Recurring Revenue. In a recurring revenue model, the company has a continuing stream of revenue from a product. There are numerous ways to do this, including disposables, service contracts, leases, and tolls. Most innovative medical devices use some method of obtaining recurring revenues.

Disposables. Some products, such as surgical instruments, are purchased once and used multiple times. Others, such as hypodermic needles, are purchased, used, discarded, and replaced. Still others are purchased once and require additional components every time they are used. For example, many diagnostic instruments require reagents or special disposable sample holders. Such recurring revenue can be very attractive to a small company, amplifying the initial sale many times over a product's life.

Service Contracts. Another source of recurring revenue is to add a service component to expensive capital equipment. Such service costs can be in the range of 10–15% per year.

Leases and Tolls. One way of lessening a customer's cost for a product or converting it from a capital expenditure to an expense is to lease it or rent it on a per use basis or for a fixed number of prepaid uses. This model is especially suited for expensive imaging equipment, monitoring equipment, complex pumps, dialysis equipment, and the like.

Licenses. Companies that cannot or do not want to bring a product to market themselves can license it to another company for commercialization. A license agreement often includes an upfront licensing fee with recurring royalties for product sales. This can be a good way to create revenue, particularly in markets and geographies that the company will be unlikely to enter for an extended time. One danger is that a license necessarily involves a company giving up certain rights (such as exclusivity) to others. By doing so, it may diminish its value to potential acquirers or prospective partners who desire these foregone rights.

Development Strategy

Operations

"Operations" often refers to manufacturing, but also includes a wide variety of other tasks involved in running a business. These include purchasing materials, delivering products to customers, managing employees, and keeping track of product costs and capital expenses. For capital equipment product sales, "operations" generally includes managing installation, warranty, and ongoing service updates, and for software products it would include installation, assurance of compatibility, training, and ongoing support and upgrades. The comments below refer specifically to issues related to the manufacture of biomaterials products and implants.

Key Operations: Keep them in House or Outsource? One of the critical questions a start-up faces is whether to keep operations in house or outsource them. This can have day-to-day implications, but also larger strategic ones. For example, if a business expects to eventually be acquired by a larger company, generally the acquirer will have their own operations expertise, and therefore will not pay more for a start-up's operations. Developing operational capacity and expertise for biomaterials and implants requires significant effort, capital, and management attention. These are the reasons that most small companies find a partner to manufacture their products. However, in spite of these costs and complexities, start-ups with novel biomaterials often find it important to create unique processes for efficient manufacturing and repeatable product performance. Developing these finicky processes may take significant expertise and multiple iterations to assure a stable, repeatable, predictable, and affordable outcome, and the novel processes themselves then become part of the IP of the company, covered either by patents or trade secrets. In such cases, developing operational expertise internally builds a valuable asset of the company, and often is necessary for the success of the venture.

Even in companies that have outsourced the bulk of their operations, it is critical to have adequate and experienced oversight. Materials suppliers must be qualified, and contracts and oversight must be in place to ensure consistent quality, traceability, and delivery. Failure at any point in the supply chain with these partners will result in product problems for the start-up company. For a summary of these basic considerations, refer to Table III.2.2.2.

Sales and Marketing

"Sales" is the process of offering the product to a purchaser. This includes: finding the initial leads; setting up meetings; demonstrating the product; enlisting users to speak with potential purchasers; working with purchasing

TABLE III.2.2.2 In-house or Outsource: Basic Considerations

	Advantages	Disadvantages
Outsourced development	• More resources • Resources available • Infrastructure in place • Flexibility: can cut or add resources quickly • Acquirers don't value infrastructure • Novel processes available	• More expensive • Less motivated/slower • What they have may not be what you need • You may not be their top priority • Acquirer may need more capacity • No control over access to novel processes
Internal development	• Founders' dedication overcomes obstacles • Control is optimized • Technology may need unique infrastructure • Processes kept as trade secrets	• Alternatives may be overlooked • Formal systems may be limited • Small company systems may not scale-up • May miss better processes

departments and committees; and doing whatever else it takes to get the sale made! Marketing involves developing the value proposition for a product – that is, the economic and other benefits to be gained by using it – and communicating that inside and outside the company. The marketing department performs the product management function – identifying strategies for locating customers, preparing literature and other support materials for the sales department, and helping develop new product ideas based on the market feedback it obtains. One important thing to keep in mind is that a marketing department is very limited in the claims it can make about a regulated device, unless it can prove those claims are true. The best way to do that is by clinical trials; therefore, it is very important to involve the marketing department in clinical trial design early in the process to ensure they can make the claims they require.

Although sometimes people confuse sales with marketing, the function of marketing is to lay the groundwork of building tools to communicate what the product does and locating customers so the sales department can sell to them. Experience from large companies demonstrates the value of using dedicated salespeople to call on potential customers and build long-term relationships with them as a prelude to selling them products. But it is a rare start-up that can afford both a dedicated marketing department and full-fledged internal sales effort. Therefore, start-ups should consider a number of alternatives for selling.

The most obvious sales model uses employees as salespeople. Unfortunately, this can be very expensive, requiring management, salary, expenses, training, and additional sales resources. Additionally, the company must be prepared to wait for possibly a long period of time before it will see significant revenues from a salesforce.

Manufacturer's representatives (or "reps") are independent business people who sell products made by different manufacturers. A rep typically works in a defined geographical region and works on commission. Reps generally know their customers well, know the buying processes in the healthcare entities they visit, and get paid when products are sold. They provide affordable selling outreach and greatly expand the sales territory a company is able to cover. But reps, who need training and management in the product, have divided attentions because they usually sell multiple products from many companies. Thus, they rarely provide the "missionary" effort required for acceptance of a novel product, and can lose confidence if a product does not catch on quickly. Finally, in the event a company wishes to restructure a sales effort, rep contracts can sometimes make that difficult.

Stocking distributors purchase products directly from a company at a significant discount, typically 30–40% of the expected sale price. Distributors tend to specialize in particular medical areas, creating extensive catalogs of products which they sell. The distributor maintains an inventory of products, and handles the shipping and billing to the customer. The distributor will have its own sales staff, with each salesperson responsible for various accounts and/or various product lines. In the United States, a distributor will often have sales people in an entire state or region, giving them a strong sales footprint. Because distributors may carry even more products than sales reps, they can be even harder to motivate, and one complaint companies have about distributors is that they don't sell, but "merely" take sales orders from customers who call them.

Due to the increasingly difficult regulatory environment in the United States, companies are increasingly looking to introduce their products abroad. Because the cost and complexity of selling a product in another country can seem even more daunting to a start-up than selling in the US, international sales almost always begin with international distributors. There are substantial, experienced, and knowledgeable distributors in many countries. Even so, a start-up will likely need to dedicate a full-time experienced professional to finding and managing the international distributors.

Team

Founding a company is an intensive process that takes dedication, perseverance, and a dogged determination to overcome innumerable obstacles. A company includes a number of individuals, each with different skills. The first people to come on board are, by definition, founders. Usually founders of a biomaterials start-up will have either a technical or marketing background. This is because the key assets in a start-up are typically IP (technical), understanding of the unmet market need/value proposition (marketing), and the ability to inspire others to invest in the company (often marketing, sometimes technical).

The Chief Executive Officer (CEO) has the most critical job in a company. The CEO plans and explains the company's strategy, obtains (often through investors) the money and other resources the company will need, and hires and manages the team who implements the strategy. An experienced, charismatic CEO can plan on spending the equivalent of a full-time job raising money and working with the investors for the first two or three years of the company's active life. Unfortunately it is not uncommon for the first CEO to initially be, or eventually become, an inappropriate fit for the company. In part that is due to the fact that many CEOs are brought into a company by "being there" as a founder or an acquaintance of the founder. Such accidents of proximity by no means insure that such a person has the appropriate skills to run a company, especially as the company's needs evolve.

Assuming adequate financial resources, the CEOs key job is to hire the best team possible to implement the strategy. This will include consultants and employees. Early strategic experts in areas such as regulatory will likely be consultants. Working part-time they can develop the regulatory requirements for product development, bench and animal testing, and clinical trials. The first hires will be in key positions critical to the company's success, such as R&D. Later hires will include marketing, regulatory, clinical, reimbursement, and operations staff. The CEO is constantly balancing the budget with the timing for adding other personnel.

A board of directors is not only statutorily required for companies in most states, but is also required by many investors, who will often demand one or more seats on the board (and investors should strive to balance this with an equal number of independent directors). But a good board does more than just fulfill these obligations. It also provides a pool of connections and expertise for the CEO to draw on as needed, and provides needed oversight to ensure the company stays on track. Generally the best board members have deep experience in the field, and significant operational expertise as CEOs or general managers. Investing board members often receive no stock, but independent board members typically receive 1–4% of the company for their service.

A board of scientific or clinical advisors validates the company's science and product. For a company with a superb idea, it is not difficult to recruit a top-notch board. Often, clinicians and scientists are extremely approachable and interested in new products for patient care. These boards usually meet three or four times per year, are available for advice to the team at other times, and often are the first users of new devices in animal or clinical tests. Advisors will typically receive a small amount of stock in return for their services.

THE STAGES OF LIFE SCIENCE: FROM CONCEPT TO ADOPTION

As a company grows, it will need ever increasing amounts of capital, and for this to occur with minimal pain for all concerned, investors will expect to see the value of the company continue to climb as new investments are brought in. Generally this means the company must continually be reducing the chance of failure. To be clear: successful management of a medical device company is about progressively eliminating risk. There are various reasonably well-defined risk-reduction stages in a medical product's life cycle (Figure III.2.2.2), each with its own considerations. They are set out below.

Idea

In the idea stage, the entrepreneur is first and foremost dedicated to ensuring their idea solves a real problem.

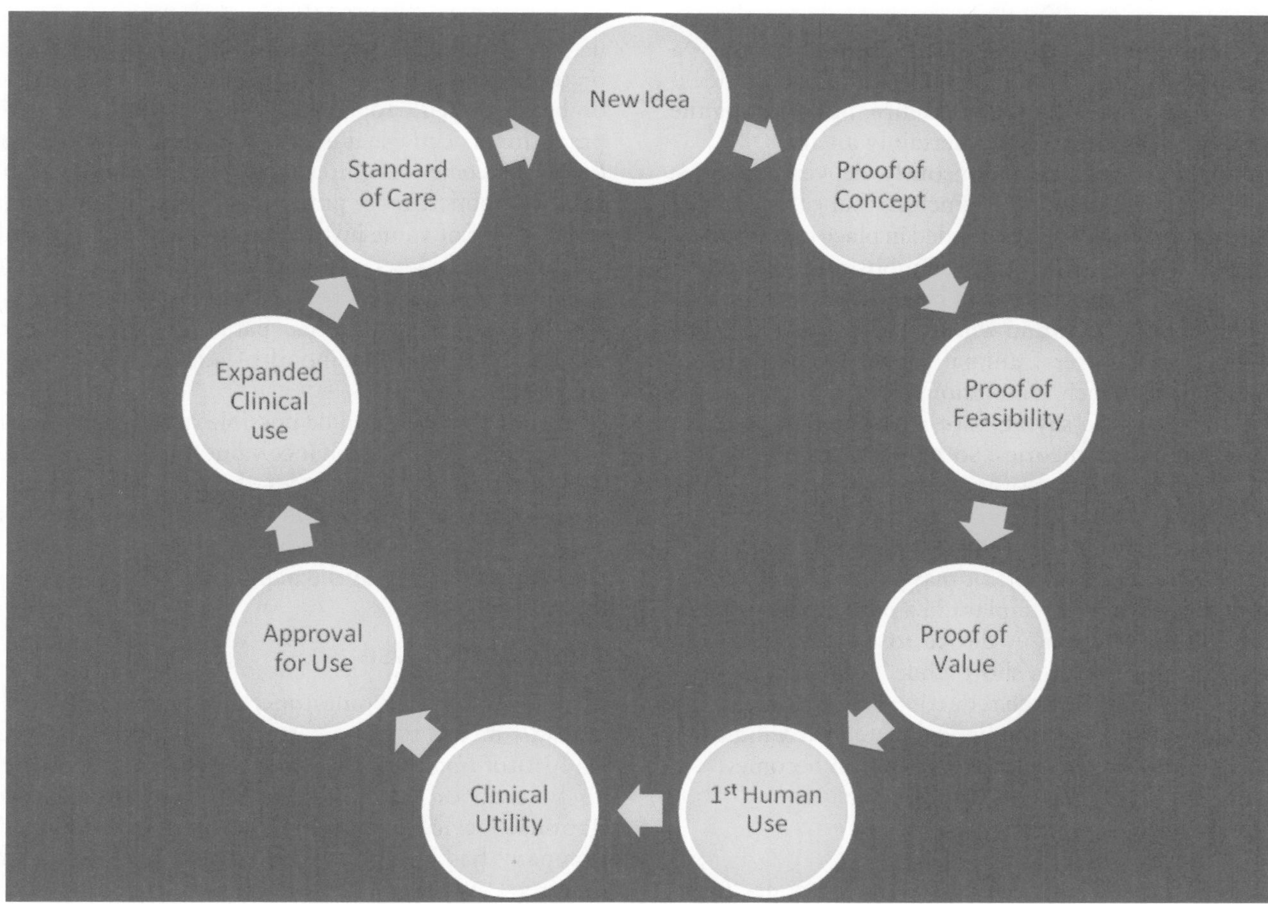

FIGURE III.2.2.2 The Circle of Life in a Life Science Company.

Assuming so, this is the stage to read and absorb the medical literature, learn about patients and treatments, and investigate the competitive products and the IP around those products. It is also a good time to start expanding on the idea to understand how it will: work in practice; actually be turned into a product; and provide clinical value. This is the time when the successful entrepreneur will be starting the never-ending process of stress-testing the idea against the considerations put forth in the beginning of this chapter: need; market; clinical trials; intellectual property; regulatory; reimbursement. Before in-vivo or even in-vitro testing, this is the time for the successful entrepreneur to start the "in-cerebro" work, that is, the never-ending process of stress-testing the idea against the considerations put forth in the beginning of this chapter: need; market; clinical trials; intellectual property; regulatory; reimbursement.

As the idea progresses to an invention, it is important to follow best-practices such as capturing the idea as clearly and completely as possible in writing and with sketches and three-dimensional computer models for use when filing, and perhaps defending, patents.

Proof of Concept

This milestone is achieved when the entrepreneur can demonstrate to themselves and their advisors that the invention can work. One should select the fastest, most cost effective, and elegant tests sufficient to convince a critical but friendly expert of the validity of the idea. The demonstration will involve science, but selecting the appropriate demonstrations is certainly an art.

For example, in a percutaneously delivered cardiac valve replacement, making a structure that can be folded sufficiently small and then expanded in place is doubtless a critical test. For anti-microbials, demonstrating that their presence in agar plates prevents bacterial growth may be compelling. Be creative and wide-ranging: consider crude prototypes in cadavers; animal parts purchased from butcher shops; bench simulations; mechanical testing; and so forth. As the saying goes, one dumb experiment is worth ten expert theories, so build things that can be tested. No matter how crude your prototypes, you will learn from these tests.

Sometimes one can develop very clever ideas for testing something. For example, it may be possible to try an early prototype product implant in a patient who is going to have an amputation. A device to treat morbid obesity could be placed for a short while in patients who are being prepared for stomach reduction surgery. In general, the better a concept is tested in simulated yet valid tests, the more compelling the value proposition becomes.

Proof of Feasibility

Proof of concept is achieved by convincing your "insiders." Proof of feasibility, by contrast, is achieved by convincing multiple, skeptical, outsiders. To do this, a product must be fabricated in small quantities and tested in a controlled biological model. Although the product should be reproducible, it will not likely be fabricated in any sort of production environment. The emphasis should not be on reducing cost or manufacturing complexities, or improving shelf-life or human factors. Instead, the point is to develop the most expeditious, inexpensive ways to make small quantities so preliminary clinical testing may begin.

Animal Models. If there is an available predictive animal model, it can be an excellent vehicle for feasibility testing. One should spend time learning the intricacies, strengths, and shortcomings of various models. For example, porcine (pig) models are often favored for cardiovascular testing due to the similar reactions pig and human blood vessels have with biomedical devices. Canine models may not be good for testing thrombogenicity since dogs have more aggressive clotting mechanisms than humans.

Short-Term Results. If there is no animal model suitable to test your invention, it may be possible, as described above, to perform short-term human testing to get valid data on device performance.

Proof of Value

This milestone ties the theoretical functioning of a product to the practical requirements for using it. A device or technology may work, but may take so long to use or be so difficult to deploy that only the most skilled practitioners can use it effectively. For example, if a new knee replacement requires more precise alignment in the receiving bones, most practitioners may be unwilling to use it. Proof of value involves an exploration of whether, all things considered, the product adds value. Considerations include time to perform the procedure, expertise required, practice patterns, patient comfort, and time for rehabilitation, likelihood of adverse outcomes, and so forth.

Although it may still be possible to fine-tune the product and procedure to enhance value and manage the risks and unintended consequences, this is the time for an honest, unflinching look by the team and their advisors, to determine whether the product is likely to be useful, and therefore successful in the market.

First Human Use

This is a really big milestone, with significant implications. It means the company has managed to persuade an Institutional Review Board (IRB) to allow the device to be used on patients, and has received an Investigational Device Exemption (IDE) from the FDA. The company has also persuaded one or more clinicians to use it and will soon see how it performs, in the target population, under the intended procedure.

First human use is often performed outside the United States, since other countries typically have a shorter review and approval cycle, lower costs, and faster patient recruitment. In addition, in the event a start-up must make modifications to its device or protocol after these trials, there will often be less adverse publicity than if the trials occurred in the US.

Validation of Clinical Utility

Validation of clinical utility occurs upon early statistical confirmation of what had heretofore been anecdotal evidence of product and procedural value.

Trial results from the initial human use site provide the first data used for this validation. This data – including description of device, protocol, clinical outcomes, economics, post-procedure recovery, and other parameters of the trial – is measured, captured, and made available for others to review. While early clinical adopters may be willing to use devices that have limited human use, the average practitioner will wait for test results from numerous users at several institutions. And the more serious the indication, the more conservative the average practitioner will be.

The difficulty is that some devices – especially biomaterials-based ones – can require extensive time and large numbers of patients for true validation. A new drug-eluting stent will take at least 6–12 months (perhaps more) of patient follow-up in hundreds or thousands of patients before most clinicians will have confidence in its performance. A new orthopedic implant may take many years before the true story is known.

Approval for Use

This occurs upon clear and unequivocal receipt of notice from a regulator that the company is cleared for commercial launch. For more information, see the "Regulatory Strategy" section above.

Expanded Clinical Use

The product is in general market release, in use by mainstream clinicians. It has been estimated that 5–10% of clinicians can be characterized as "early adopters" willing to use a product before it is widely accepted. But mainstream users need more. They rely on things such as testimonials from one or more early adopters, discussion at medical conferences, and journal articles. The product still requires user training and sales, but there is much less "missionary" selling; most potential users are aware of the new technology, and it requires just a little bit of persuasion to give them the confidence to try it.

Standard of Care

By this phase, the product is in widespread use by mainstream clinicians as part of their standard patient practice. By the time a product has achieved this market-leading milestone, it usually will have large competitors making knock-offs with incremental or substantial improvements. One sure indication that a device or technology has reached this milestone is when the device and related procedures are taught in medical schools and residency programs. But by then, the start-up is no longer a start-up. It will have typically sold the business or licensed the product to a large medical device company or in rare cases will be a large company itself.

The Next Phase

A new idea or invention! And the cycle begins anew.

BIBLIOGRAPHY

Zenios, S., Makower, J., Yock, P., Editors (2010). *Biodesign: The Process of Innovating Medical Technologies* (1st ed.). MA, USA: Cambridge University Press.

CHAPTER III.2.3 VOLUNTARY CONSENSUS STANDARDS

Jack E. Lemons
University Professor, Schools of Dentistry, Medicine and Engineering, University of Alabama at Birmingham, Birmingham, AL, USA

WHAT ARE STANDARDS?

Consensus standards are documents that have been developed by committees to represent a consensus opinion on test methods, materials, devices or procedures. Most standards organizations review their documents within five years to ensure that they are up-to-date. The mechanisms by which they are developed are described in subsequent sections. Development of standards is an ongoing process, and the latest publications should be consulted for new standards.

A test method standard describes the test specimen to be used, the conditions under which it is to be tested, how many specimens and what controls are to be tested, and how the data are to be analyzed. Many methods are validated by "round-robin testing," meaning that several laboratories have followed the method and their results are analyzed to determine the degree to which they agree to a specified degree of precision and accuracy. Once a test method has been standardized, it can be used in any other laboratory; the details are sufficient to ensure that different facilities will obtain similar results for the same samples. Stating that a test was "conducted in accordance with…" ensures that the results can be duplicated. Some representative test methods are listed in Table III.2.3.1.

A material or specification standard describes the chemical, mechanical, physical, and electrical properties of the material. Any test method standards cited are to be used to ensure that a significant sample meets the requirements of the standard. Some representative material standards are listed in Table III.2.3.2.

For implant materials, there is also a requirement that the materials meet general biocompatibility test criteria. There are two formats for the biocompatibility language in the material standards of the American Society for Testing and Materials (ASTM). For materials that can be well-characterized by chemical, mechanical, and physical tests, and have demonstrated a well-characterized biological response, reference to the published biological testing data and clinical experience is often sufficient. For materials that are not well-characterized, for example, the wide class of materials called epoxy resins, biological test methods are cited, and each particular formulation must be tested independently. This area is evolving for combination biological and synthetic (tissue-engineered) products.

A device standard describes the device and its laboratory-based performance. General design aspects, dimensions, and dimensional tolerances are given using schematic drawings. The materials to be used are described by reference to materials standards. Methods for testing the device are also cited. Since test methods only describe how to do a test, it is in the device-related standards that performance is addressed. For example, the fatigue life requirements of biocompatibility requirements of the device and its materials would be stated in a device standard. Some representative device standards are listed in Table III.2.3.3.

TABLE III.2.3.1 Some Representative ASTM Standard Test Methods
A. Mechanical testing standards
ASTM D412 Test methods for rubber properties in tension
ASTM D638 Test method for tensile properties of plastics
ASTM D695 Test method for compressive properties of rigid plastics
ASTM D790 Test methods for flexural properties of unreinforced and reinforced plastics and electrical insulating materials
B. Metallographic methods
ASTM E3 Preparation of metallographic specimens
ASTM E7 Terminology relating to metallography
ASTM E45 Determining the inclusion content of steel
ASTM E112 Determining the average grain size
C. Corrosion testing
ASTM G3 Conventions applicable to electrochemical measurements in corrosion testing
ASTM G5 Reference test method for making potentiostatic and potentiodynamic anodic polarization measurements
ASTM G59 Conducting potentiodynamic polarization resistance measurements
ASTM F746 Pitting and crevice corrosion of surgical alloys
ASTM F897 Fretting corrosion of osteosynthesis plates and screws
ASTM F1875 Practice for fretting corrosion testing of modular implant interfaces: Hip femoral head-bore and cone taper interface
ASTM F2129 Conducting cyclic potentiodynamic polarization measurements to determine the corrosion susceptibility of small implant devices
D. Polymer testing
ASTM D2238 Test methods for absorbance of polyethylene due to methyl groups at 1378 cm^{-14}
ASTM D3124 Test method for vinylidene unsaturation in polyethylene by infrared spectrophotometry

ASTM: American Society for Testing and Materials

TABLE III.2.3.2 Some Typical ASTM Materials Standards (4)
ASTM F75 Cast cobalt-chromium-molybdenum alloy for surgical implant applications
ASTM F139 Stainless steel sheet and strip for surgical implants (special quality)
ASTM F451 Acrylic bone cements
ASTM F603 High-purity dense aluminum oxide for surgical implant applications
ASTM F604 Silicone elastomers used in medical applications
ASTM F641 Implantable epoxy electronic encapsulants

ASTM: American Society for Testing and Materials

TABLE III.2.3.3 Some Representative AAMI and ASTM Device Standards
AAMI CVP3 Cardiac valve prostheses
AAMI VP20 Vascular graft prostheses
AAMI RD17 Hemodialyzer blood tubing
AAMI ST8 Hospital steam sterilizers
ASTM E667 Clinical thermometers (maximum self-registering, mercury-in-glass)
ASTM F367 Holes and slots with spherical contour for metric cortical bone screws
ASTM F703 Implantable breast prostheses
ASTM F623 Foley catheters

AAMI: Association for the Advancement of Medical Instrumentation
ASTM: American Society for Testing and Materials

TABLE III.2.3.4 Some Representative AAMI (1) and ASTM (4) Procedure Standards
AAMI ROH-1986 Reuse of hemodialyzers
AAMI ST19 Biological indicators for saturated steam sterilization process in healthcare facilities
AAMI ST21 Biological indicators for ethylene oxide sterilization processes in healthcare facilities
ASTM F86 Surface preparation and marking of metallic surgical implants
ASTM F561 Retrieval and analysis of implanted medical devices and associated tissues
ASTM F565 Care and handling of orthopedic implants and instruments
ASTM F983 Permanent marking of orthopedic implant components

AAMI: Association for the Advancement of Medical Instrumentation
ASTM: American Society for Testing and Materials

WHO USES STANDARDS?

The term "voluntary standards" implies that the documents are not mandatory; anyone can use them. This terminology also refers to the way that standards are developed. Standards are used by manufacturers, users, test laboratories and, in many instances, college professors and their students. One's use or compliance with a standard is voluntary. Using them is often to everyone's advantage. At this time, standards can also be utilized as a part of the regulatory (US Food and Drug Administration (FDA)) approval process.

Manufacturers often use standards as guidelines in making and testing their materials and devices. Manufacturers also cite standards in their sales literature as a concise way of describing their product. Stating that a device is made from cast cobalt-chromium-molybdenum alloy in accordance with ASTM F75 tells the user precisely what the material is. Conformance to standards is also a way to expedite device review by the FDA.

On a more personal level, for example, after purchasing a piece of plastic pipe at the hardware store labeled with "ASTM D1784," one could go to ASTM Volume 8.02 and find that this is a specification for rigid poly (vinyl chloride) compounds. If you have "DIN" stamped on the bottom of your ski boots, you know they conform to the standards of the Deutsches Institut für Normung, and the ski shop will have standards for adjusting your bindings.

As an example of why one would want device standards used for medical devices, consider screws for fixing bone fractures. There are device standards for bone screws, plates, taps, and screwdrivers. The physician can purchase a screw and a screwdriver, and be confident that the components will fit as intended. A surgeon about to remove a plate implanted at another hospital can evaluate radiographs and see that the device has 4.5 mm bone screws of a specific design. Knowing this, a standard 4.5 mm screwdriver can be used to remove the screws.

Standardized test methods should simplify life. For example, many who teach undergraduate and graduate biomedical engineering courses use standards. In a mechanical testing laboratory, several ASTM standard test methods for mechanical testing, such as D790, "Flexural properties of plastics and electrical insulating materials," might be used. This method describes the samples, test apparatus, test speeds, and equations used to calculate the results. During a laboratory session, the students follow the test directions. In writing the methods section of their reports, all they had to write is "the test was done according to D790."

WHO WRITES STANDARDS?

In the United States, voluntary consensus standards are developed by a number of organizations. In the medical electronics, sterilization, vascular prosthesis, and cardiac valve areas, most standards are developed by committees within the Association for the Advancement of Medical Instrumentation (AAMI). In the implant materials and implants area, most standards are set by ASTM Committee F04 on Medical and Surgical Materials and Devices. These documents may then be reviewed and accepted by the American National Standards Institute (ANSI). ANSI is the official US organization that interacts with other national organizations in developing international standards within the International Standards Organization (ISO), such as TC 150 on Medical Materials and Devices, and TC 194 on Biological Evaluation of Medical Devices. The USP provides information on minimal biocompatibility testing for materials intended to be used in medical devices.

Dental material standards are written by the American Dental Association (ADA). Similar committees exist in other countries: the Canadian Standards Association (CSA); the British Standards Institute (BSI); the Association Française de Normalisation (AFNOR) in France; and the Deutsches Institut für Normung (DIN) in Germany, which is a voluntary organization.

The initiation, development, and process for the completion of national ASTM F04 consensus standards for medical and surgical materials and devices has evolved significantly within recent years. In part, this has been a result of multiple interactions among those involved with the basic sciences, applied research and development, business marketing and sales, clinical applications, regulatory agencies, professional societies, legal and insurance professions, device recipients, and associated advocacy groups. To establish consensus opinions satisfactory to all of these interest groups is far from a simple process.

History and Current Structure of ASTM F04

The ASTM was organized in 1898, whereas Committee F04 on Medical Devices and Surgical Materials and Devices was founded in 1962. The committee has grown to include a current membership of approximately 600 individuals, representing a variety of disciplines and interests. The ASTM F04 Committee has more than 100 active standards, and is structured into more than 30 technical subcommittees. The overview structure includes five divisions divided according to responsibilities specific to organizational activities (process) or the development of specific types of standard document e.g., full consensus standards include six types: (1) classification; (2) guide; (3) practice; (4) specification; (5) terminology; and (6) test methods, plus a provisional status. The divisions are: (I) resources; (II) orthopedic devices; (III) medical/surgical devices; (IV) tissue engineered products; and (V) administration. Each division is subdivided into subcommittees and task groups according to areas of interest and activities. This structure is intended to be flexible and can be rearranged to suit new or more efficient operations at any scheduled meeting of the executive committee. The divisions, subcommittees, and task groups have an appointed chair and vice-chair, whereas the executive is nominated and elected on a two-year cycle.

Standards Development Process

After a request for a new standard is received by a member or group (task group or subcommittee) and accepted by the executive committee, a task group activity is initiated. An appropriate chair is recommended and approved by the administrative committees, and the process is started. Consensus standards development follows a reference document for content, form, and structure. An assignment of an appropriate document number is made for records purposes, and a staff manager confirms that the committee and subcommittee representation associated with this activity is classified and balanced with respect to producers and non-producers. A first draft of the proposed document is reviewed (usually three to five times) before the task force reaches a consensus.

Critical to these proceedings is the necessity that adequate information is available within the public domain in order to substantiate the requirements listed with the final standard. If data are limited or unknown, round-robin tests or new test methods must be developed, and then confirmed to be applicable and valid. Sometimes, documents are held until basic (necessary) information for the standard is developed. Again, a standard must be based on known results, and documents are not intended to represent areas of new research.

As a part of the process, once a draft document has been circulated within the task group and consensus is reached, the task group chair may recommend initial voting at the task group level. At this time, the opinions gained could lead to further improvements and no formal voting rules are required within the task group interactions. At the next stages, the subcommittees and main committees of ASTM F04 must ensure a balance among the various voting interests, with adequate representation from the general interest, user, consumer, and producer segments of the membership. The total of the user, consumer, and general interest votes must equal or exceed the number of producer votes. To prevent domination by any one interest group, only one vote per voting interest is permitted. The ASTM staff confirms the numerical status (adequate response and balance) for each formal vote, and all members are permitted to vote on any ballot within their committee and membership. If approved or approved with editorial (no substantive) changes at the subcommittee level, the document proceeds to main committee ballot and society review. If, however, a negative vote is received during formal voting, the task group and subcommittee chairs must resolve the negative to the satisfaction of the negative voter or must provide rational and justification to find the negative voter nonpersuasive. This opinion must be accepted by the task force and approved at the subcommittee and main committee levels by a formal recorded vote based on written documentation and associated discussions. The staff manager works with the committee to confirm the validity of the vote, and to document the action.

The general experience has been that it requires three to five years (six to ten meetings) to go from a first draft to a full standard accepted at the main committee level. Several procedural steps are also required in the process, including approval of a rationale statement, use of standardized units (SI) and terminology, and acceptance by the editorial and precision and bias subcommittees. At the stage of society acceptance, the final document is reviewed by the Committee on Standards prior to publication by ASTM.

After approval, a given standard may remain "active as published" for up to five years. At five years, that standard must be reaffirmed or revised to suit the information available at that time. For records purposes, the date of last formal approval is included as a part of the standards designation.

BIOCOMPATIBILITY STANDARDS

There is a wide range of tests that may be used to determine the biological response to materials. Short-term uses require only short-term tests. Long-term uses require tests applicable to the particular device and tissue type. Since not all tests are necessary for all applications, national and international standards organizations have developed matrix documents that indicate what methods are appropriate for specific applications. These documents can be used as guidelines in preparing a submission to the US Food and Drug Administration (FDA) and other national regulatory agencies for approval of a new material or device. Similar matrix documents have been standardized by the CSA, BSI, and ISO. Test method documents have also been developed by the National Institutes of Health (NIH) and the US Pharmacopeia (USP). Guidelines for dental materials have been developed by the ADA and ISO.

Much of the standards activity is now associated with the International Standards Organization (ISO) with biological evaluation of medical devices under the consideration of TC 194 and presented in the developing documents of ISO 10993. There are various parts to this document. Part 1 is definitions and the guidance on selection of evaluation test categories that should be done. The other parts of ISO 10993 give more discussion and detail on the selection of individual tests that should be done for a particular biological interaction or biological effort (e.g., contact with blood, systemic toxicity, genotoxicity). Often, the details of test methods are not given in the ISO documents, and reference is made to other documents such as ASTM and USP standards for procedures and methodology.

Material selection and evaluation of biological risk are integral components of the design process for medical devices being considered by TC 194. This evaluation is a component of the risk management plan in line with ISO/IEC 14971 – Application of Risk Management to Medical Devices, encompassing identification of all hazards and the estimation of their associated risks. Criteria to define the acceptable biological (toxicological) risk must be established at the start of the risk assessment and design management processes. The biological safety evaluation must be designed and performed to demonstrate the achievement of the specified criteria for safety. Adequate risk assessment requires characterization of toxicological hazards and exposures. Following the risk management structure described in ISO 14971, a major component in hazard identification is material characterization.

In the following section we review some of the steps taken to establish the biocompatibility of a new material for a specific application, in this case a long-term orthopedic implant. We use ASTM standard F748 "Practice for Selecting Generic Biological Test Methods for Materials and Devices" as a guideline. The standard test methods described are those used within ASTM.

In Vitro Tests

F619. Practice for Extraction of Medical Plastics. A method for extraction of medical plastics in liquids that simulate body fluids. The extraction vehicle is then used for chemical or biological tests. Extraction fluids include saline, vegetable oil (sesame or cottonseed), and water.

F813. Practice for Direct Contact Cell Culture Evaluation of Materials for Medical Devices. A cell culture test using American Type Culture Collection (ATCC) L929 mouse connective tissue cells. This method or this type of cell culture method can be used as the first stage of biological testing. It is also used for quality control in a production setting. There are other ASTM standard cell culture methods, and others not standardized by ASTM that could also be used.

F756. Assessment of Hemolytic Properties of Materials. An *in vitro* test to evaluate the hemolytic properties of materials intended for use in contact with blood. Procedure A is static; procedure B is done under dynamic conditions.

Short-Term *In Vivo* Testing

F719. Testing Biomaterials in Rabbits for Primary Skin Irritation. A procedure to assess the irritancy of a biomaterial in contact with intact or abraded skin. This test would be indicated for surgical glove material or skin dressing.

F720. Practice for Testing Guinea Pigs for Contact Allergens: Guinea Pig Maximization Test. A two-stage induction procedure employing Freund's complete adjuvant and sodium lauryl sulfate, followed two weeks later by a challenge with the extract material. Ten animals per test material.

F749. Practice for Evaluating Material Extract by Intracutaneous Injection in the Rabbit. Extraction vehicles (as per F619) of saline and vegetable oil are injected intracutaneously and the skin reaction graded for erythema, edema, and necrosis. Two rabbits per extraction vehicle.

F750. Practice for Evaluating Material Extracts by Systemic Injection in the Mouse. Intravenous injection of saline extracts and intraperitoneal injection of oil extracts. Animals are observed for evidence of toxicity. Five mice per extract and five mice per extract vehicle controls.

F763. Practice for Short-Term Screening of Implant Materials. This method provides for several implant types and sites for short-term screening *in vivo*. This method is essentially the first stage of animal testing of solid pieces of the implant material.

F1983. Assessment of Compatibility of Absorbable/Resorbable Biomaterials for Implant Applications. This material type presents unique features for tissue evaluation, in that the materials are not inert, and a chronic inflammatory reaction may be observed during the degradation period. The time periods at which reactions are examined are based on the anticipated rates of degradation of the test material.

Additional tests for special issues are also included in ASTM standards, such as examination and reactions to particles, immunotoxicity, and retrieval and analysis of implants and tissues.

These are additional *in vivo* tests that have not yet been standardized by ASTM:

Thrombogenicity. Tests for the propensity of materials to cause blood coagulation have not been standardized. Guidelines for such tests have been developed by the NIH Heart Lung and Blood Institute.

Mutagenicity. There are a number of *in vitro* and *in vivo* tests to determine if chemicals cause cell mutations. Although not specifically developed for implants, guidelines do exist as part of the OECD (Organisation for Economic Co-operation and Development) guidelines for testing of chemicals, and within ASTM, e.g., E1262, Guide for the Performance of the Chinese Hamster Ovary Cell/Hypoxanthine Guanine Phosphoribosyl Transferase Gene Mutation Assay; E1263, Guide for Conduct of Micronucleus Assays in Mammalian Bone Marrow Erythrocytes; E1280, Guide for Performing the Mouse Lymphoma Assay for Mammalian Cell Mutagenicity; E1397, Practices for the *In Vitro* Rat Hepatocyte DNA Repair Assay; and E1398, Practices for the *In Vivo* Rat Hepatocyte DNA Repair Assay, which are in Vol. 11.05 of the ASTM *Annual Book of Standards.*

Pyrogenicity. A pyrogen is a chemical that causes fever. The USP rabbit test is a standard *in vivo* test. One can also test for bacterial endotoxins, which are pyrogenic, using the Limulus amebocyte lysate (LAL) test.

The oxygen carrying cell (amebocyte) of the horseshoe crab, *Limulus polyphemus*, lyses when exposed to endotoxin.

Long-Term *In Vivo* Testing

There are two aspects to the long-term testing issue. One is the response of tissue to the material; the other is the response of the material (degradation) to implantation.

F981. Practice for Assessment of Compatibility of Biomaterials for Surgical Implants with Respect to Effects of Materials on Muscle and Bone. Long-term implantation of test materials in the muscle and bone of rats, rabbits, and dogs. Two species are recommended. For rabbit muscle implants the standard calls for four rabbits per sacrifice period, with one control and two test materials placed in the paravertebral muscles on each side of the spine. For bone implants in rabbits the standard calls for three implants per femur.

A general necropsy is performed at the time of sacrifice. Muscle and bone implant sites are removed at sacrifice and the implants left *in situ* until the tissue has been fixed in formalin. Implants may be removed prior to embedding and sectioning.

No standards have been established for any long-term testing of devices. However, for a device intended for a particular application, it is essential to conduct a functional device test. For a fracture fixation plate, it could be proposed to use plates to fix femoral osteotomies in dogs. This study would consider the effects of the implant on the tissues, as well as the effect of implantation on the properties of the device, i.e., material degradation.

The methodology for long-term carcinogenicity testing of implants also has not yet been standardized by the ASTM, although F1439 (Standard Guide for Performance of Lifetime Bioassay for the Tumorigenic Potential of Implant Materials) does provide guidelines for test selection. This is normally a life survival and tumor production test, typically in rats. ISO 10993-3 provides considerations for genotoxicity, carcinogenicity, and reproductive toxicity testing with reference to some test methods.

TISSUE-ENGINEERED MEDICAL PRODUCTS

A rapidly evolving area for standards has been tissue-engineered medical products (TEMPs). This area was identified early in the product development cycle, and focused initially on terminology and guidance standardization. Combination products, including synthetic and biologic origin biomaterials within bioactive and biodegradable device designs, required very different types of test for product safety and efficiency. Key within this activity was testing of biocompatibility and methods for sterilization of substances with biologic type properties. Examples of some standards from the ASTM F04 Division IV documents are provided below.

NANOTECHNOLOGY

The continued expansion of nanotechnology-based medical and dental devices has now evolved as an area in need of standardization. In this regard, higher resolution microscopy and spectroscopy instruments and methods are required, thereby resulting in a focus on test methods and round-robin testing to ensure precision and accuracy. It is anticipated that this area and the need for standardization will increase significantly over the next few years.

WORKSHOP AND SYMPOSIA

Planning for a new standard often results in identification of areas where data are needed prior to developing a consensus standard. As an example, the ASTM F04 annual meetings normally initiate with a workshop or symposium within a pre-identified focus topic. Workshops include invited short papers with abstracts, usually as half-day sessions where participants are asked to participate in the following subcommittee group meeting for standard development. Symposia often follow and require a two-year planning cycle. The meeting is based on peer-reviewed papers and the proceedings are published as a Standard Technical Publication (STP) as a component of the ASTM Journal (ASTMI) or a co-sponsoring professional society document.

Importantly, these meetings, which are open to all stakeholders, provide interactions for acceptance or not of basic data on existing or new surgical implant products. Examples of some previous and scheduled ASTM F04 and co-sponsored workshop and symposia titles are listed in Table III.2.3.5.

Biomaterials Biomolecules, Cells, and Tissue-Engineered Constructs for Tissue-Engineered Medical Products

Test Methods for

F2131-02 *In Vitro* Biological Activity of Recombinant Human Bone Morphogenetic Protein-2 (rhBMP-2) Using the W-20 Mouse Stromal Cell Line

F2260-03 Determining Degree of Deacetylation in Chitosan Salts by Proton Nuclear Magnetic Resonance (1H NRM) Spectroscopy

F2259-03 Determining the Chemical Composition and Sequence in Alginate by Proton Nuclear Magnetic Resonance (1H NMR) Spectroscopy

Guides for

F2450-04 Assessing Microstructure of Polymeric Scaffolds for Use in Tissue-Engineered Medical Products

F2064-00 Characterization and Testing of Alginate as Starting Materials Intended for Use in Biomedical and Tissue-Engineered Medical Products Application

F2150-02 Characterization and Testing of Biomaterial Scaffolds Used in Tissue-Engineered Medical Products
F2103-01 Characterization and Testing of Chitosan Salts as Starting Materials Intended for Use in Biomedical and Tissue-Engineered Medical Product Applications
F2347-03 Characterization and testing of Hyaluronan as Starting Materials Intended for Use in Biomedical and Tissue-Engineered Medical Product Applications
F2027-00 Characterization and Testing of Substrate Materials for Tissue-Engineered Medical Products

Test Methods for

F2149-01 Automated Analyses of Cells – the Electrical Sensing Zone Method of Enumerating and Sizing Single Cell Suspensions

Guides for

F2212-02 Characterization of Type I Collagen as Starting Material for Surgical Implants and Substrates for Tissue-Engineered Medical Products (TEMPs)
F2315-03 Immobilization or Encapsulation of Living Cells or Tissue in Alginate Gels

INTERNATIONALIZATION OF STANDARDS

The various USA-based standards organizations harmonize the consensus standards within the world communities through a number of different interactions. The American National Standards Institute (ANSI) represents many of the standards groups, such as ASTM and AAMI, for the International Standards Organization (ISO). The various committees within the USA act through Technical Advisory Groups (TAGs), and participate within the ISO committees as formal voting participants. Where applicable, considerable similarities exist between AAMI, ASTM, and ISO standards. Although processes are different within ISO, the outcome is published standard documents.

Organizations such as the ASTM also maintain international structures (ASTMI), where international participation results in an ASTMI standard and in some cases a memorandum of understanding (MoU) for utilization of the standard as applicable to specific products. In many situations, regulatory organizations such as the FDA seek and accept the most applicable standard(s) related to procedures for product acceptance.

TABLE III.2.3.5 Examples of ASTM and Collaborative Workshop and Symposia Held Using a Consensus Standard Format

1. Porous Implants for Hard Tissue Application, ASTM STP 953, 1987.
2. Calcium Phosphate Coatings for Implants, ASTM STP 1196, 1994.
3. UHMW Polyethylene, ASTM STP 1307, 1998.
4. Synthetic Bioabsorbable Polymers for Implants, ASTM STP 1396, 2000.
5. Bone Graft Substitutes, ASTMI/AAOS, 2003.
6. Spinal Implants: Are We Evaluating Them Appropriately? ASTMI STP 1431, 2003.
7. Cross-linked and Thermally Treated UHMWPE, ASTMI STP 1445, 2004
8. Titanium, Niobium, Zirconium and Tantalum for Surgical Application, ASTMI, 1471, 2006.
9. Osteolysis and Implant Wear: AAOS/NIH, Nov. 2007.
10. Proposed Regulator Strategy for Neurotoxicity Testing, ASTMI/FDA, Nov. 2009.
11. Fretting Fatigue of Metallic Medical Devices and Materials, ASTMI E8/F04, Nov. 2009.
12. Symposium on Mobile Bearing Total Knee (MBK) Replacement Devices, May, 2010.
13. Static and Dynamic Spinal Implants: Are We Evaluating Them Appropriately? 2010.
14. Workshop on Metal on Metal Total Hip Replacement Devices, ASTM F04/AAOS, May, 2011.

CHAPTER III.2.4 REGULATORY OVERVIEW FOR MEDICAL PRODUCTS USING BIOMATERIALS

Elaine Duncan
Paladin Medical, Inc. Stillwater, MN, USA

INTRODUCTION

Assessment of the safety and effectiveness of new (and the ongoing evaluation of approved) medical devices is challenging. The US Food, Drug & Cosmetic Act amendment in 1976 extended the powers of the US Food and Drug Administration (FDA) over medical devices. The key areas of the FDA that are responsible for the regulation of therapeutic medical products are the Center for Biologics Evaluation and Research, the Center for Devices and Radiological Health, and the Center for Drug Evaluation and Research. The FDA's Center for Devices and Radiological Health (CDRH) is responsible for regulating firms who manufacture, repackage, relabel, and/or import medical devices sold in the United States. In addition, CDRH regulates radiation-emitting electronic products (medical and non-medical) such as lasers, X-ray systems, ultrasound equipment, microwave ovens, and color televisions. Thus, CDRH plays an essential role in promoting and protecting the public health by ensuring that medical devices marketed in the United States provide a reasonable assurance of safety and effectiveness and confer a favorable risk–benefit profile for their intended use population. Many subsequent amendments have continued to refine and broaden the agency's control to include many other specialties, such as veterinary drugs, biologics, and combination products.

It is important to emphasize that the agency does not specifically regulate biomaterials, by choice. Rather the agency prefers to see regulatory constraints on the use of a biomaterial follow the intended use of the device (drug, biologic, or combination product, i.e., a product comprised of two or more regulated components, i.e., drug/device, biologic/device, drug/biologic or drug/device/biologic, that are physically, chemically or otherwise combined or mixed, and produced as a single entity) of which the biomaterial is a component. This section provides a broad regulatory overview with a few specific requirements incumbent upon an engineer developing a medical device incorporating a biomaterial. General references describing the details of the regulatory review process and strategies are available (Maisel, 2004; Muni and Zuckerman, 2007; Ergina et al., 2009; Zenios et al., 2010; Sweet et al., 2011).

GLOBAL REGULATORY STRATEGY ACCORDING TO INTENDED USE

In most countries the regulation of the use of a biomaterial depends primarily on the risk associated with the intended use of the product into which the material is incorporated. For example, a biomaterial may be the major component of an implanted artificial joint, a surface in contact with blood or tissues (such as an intravenous line) or a material within a free-standing device (such as a dialysis membrane). The inherent risk intrinsic to the intended use of the device establishes the requirements on the biomaterial and thereby the level of performance validation required.

This section will focus on how compliance with US FDA regulations (21 CFR Part 820: Quality Systems Regulations) and the commonly adopted quality system standard: ISO 13485:2003 standard for "Quality Management Systems – Medical Devices – System Requirements for Regulatory Purposes" defines the responsibilities of the developer of a medical device incorporating a biomaterial.

The European Union adopted the Medical Device Directives (MDD) and the Active Implantable Medical Device Directive (AIMDD), and also continues to evolve both the directives and supporting requirements. The general construction of the European directives builds on recognized international standards and Essential Requirements. The Essential Requirements specifically require that the chemical, physical, and biological properties guarantee characteristics and performance of the product. Demonstration of conformance to the Essential Requirements is directly tied to demonstration of meeting a recognized European or international-level standard. Although these directives have regulated the "safety" of a device, only in a few instances is the efficacy of a device brought under scrutiny. Changes, which took effect in 2010, update the Directives to now require that the Conformity Assessment process for active implantable medical devices as well as for class III and implantable medical devices include a clinical investigation unless it is duly justified to rely on existing data (section 1.2 of Annex 7 of directive 90/385/EEC and section 1.1a of Annex X of directive 93/42/EEC). Any such justification will have to be based on a proper clinical evaluation. This requirement further harmonizes the European requirements with the FDA regulations to demonstrate a device is not only safe but also the the device performs as intended.

As a result of global harmonization, it is generally fair to say device regulations are divided into two main groups: (1) development and manufacturing regulations; and (2) premarket entry requirements. Internationally or nationally recognized standards, as discussed previously in this chapter, play a major role in defining the requirements for conformity in both of these regulatory theaters. Premarket entry requirements vary more broadly between countries, but because of the harmonization of the quality standards, the development and manufacturing controls for devices using biomaterials enable most devices to be developed and manufactured for the global market.

Despite various premarket regulatory requirement differences, the strategies are fairly uniform in segmenting medical device regulatory thresholds around the perception of relative risk (and perception of benefit.) Thus a low-risk device generally carries the requirement to be manufactured in a proper environment and promoted only for the well-established intended use. A moderate-risk device typically requires some intermediary to review the technical documentation of the product prior to its reaching the marketplace, to ensure that the device is safe for the intended use. The highest risk category thus require the greatest scrutiny of the performance of the device prior to market, which may include extensive animal trials or human clinical trials to establish not only safety, but also efficacy. In the US, these classifications are Class I, Class II, and Class III, respectively.

A well-constructed Device Master Record will go a long way in satisfying the regulatory requirements of most countries, even though it may need to be presented in various regulatory formats. Documentation of conformity to international standards and essential requirements will become an increasingly important function for all medical device developers. Harmonization efforts may soon do away with the historic "me too" practice of the 510(k) submission, which requires the applicant to compare the properties and performance of the candidate product to a previously cleared device. Instead, perhaps an FDA submission for a low- or medium-risk device may certify conformity to standards or FDA published guidance documents, and equivalence analysis may be less crucial. As this textbook goes to press, the FDA continues to reassess the premarket clearance procedure and the implementation of other 30-year old regulations to improve the methods without endorsing a wholesale change in practice. An example of continuing updates of policy and practice are published through public guidance documents such as draft guidance "510(k) Device Modifications: Deciding When to Submit a 510(k) for a Change to an Existing Device": http://www.fda.gov/medicaldevices/deviceregulationandguidance/guidancedocuments/ucm265274.htm

The regulatory environment is volatile, so the premarket regulatory scrutiny of a new medical device may vary with changing perceptions of risk-to-benefit. Some commonly known products which enjoy high confidence of regulators may be "down-classified" (requiring less oversight), whereas a sudden outbreak of failures of a device with comparable risk factors may suddenly enjoy increased testing prior to market release or even post-market surveillance. These trends are a natural by-product of changes in our understanding of biomaterial interactions (systemically and locally), changes in expectations for device performance, and are possibly the result of devices functioning longer in a more diverse population. Unfortunately, a textbook cannot stay current with such variations in regulatory enforcement, so it is best to stay up to-date by viewing the latest information on www.fda.gov.

A device previously cleared by the FDA, but which must undergo a significant change, may also require regulatory review or approval prior to market introduction of the change. It is the responsibility of the manufacturer to document that the changes do not affect the safety and efficacy of the device. The nature of the change, and the potential for new risk due to the change, will determine the level of regulatory scrutiny required prior to introducing the modified product to market. At the present time, guidance documents which describe methodology to assess the safety and efficacy of such changes can be found on the web at: http://www.fda.gov/downloads/MedicalDevices/DeviceRegulationandGuidance/GuidanceDocuments/ucm080243.pdf or http://www.fda.gov/downloads/MedicalDevices/DeviceRegulationandGuidance/GuidanceDocuments/ucm089612.pdf (FDA website addresses are subject to change).

Regulations and design requirements for biomaterials in medical devices typically focus on concerns for the safety and efficacy of biomaterials when the medical device is used as intended. The potential harm derived by using an unsafe biomaterial is viewed in the context of the device application. Therefore, more stringent testing of biomaterial safety and efficacy is required for devices with longer exposure, those used in life-supporting or life-sustaining devices, or where repeated exposure to the biomaterial component could engender greater risk. Increasing concern for the safety of a device when it is mishandled or fails to meet user's expectations (or level of understanding) has seen increased attention on human factor engineering and usability testing of devices. Usability testing of device design and labeling can span the range of testing to ensure a surgical instrument's use does not damage a delicate implant during surgery, to the ability of a patient to "self-diagnose" adequately "over-the-counter" (OTC) biomaterial-based devices ranging from wound dressings to anti-snoring mouth guards.

In general, all regulations that govern medical devices place the burden of proof upon the manufacturer to document the quality of the biomaterials from which the medical devices may be fabricated. Biomaterial suitability for use determinations are strongly influenced by the FDA's guidance documents, which usually describe the level of testing the FDA expects to see in an application for market. The FDA also publishes a list of "recognized standards" which may also influence the material selection process, and the testing required. The voluntary standards and guidance documents, plus what the FDA calls "general control" create boundaries or expectations for performance, but are inadequate to ensure the biomaterials selected will perform as intended within the given device for the stated intended use for all circumstances.

DESIGN CONTROL AND RISK ANALYSIS

The FDA Modernization Act of 1990 introduced the requirement for preproduction design validation of medical devices. Finally, in October 1996, the FDA issued the resultant regulations that enhanced the good manufacturing practice regulations (GMP) to include, among other provisions, the requirement that a device's development would also be subject to written procedures. The new regulation adapted principles for design processes initiated in ISO 9001 known as "design controls." The FDA's regulations of design control are not voluntary for most Class II and Class III devices. Furthermore, design control requirements were excluded from the "exemption" in the Investigational Device Exemption regulations (21 CFR Part 812) which had previously exempted investigational devices from GMPs. After the effective date of the regulation, almost all new medical devices involving a biomaterial would be subject to design control and review according to Quality System Regulations (21 CFR Part 820).

Design Control and Review, as described in §820.30 and §820.40, lay out the fundamental requirements for development of a new medical device. Implementing these requirements as early as possible in the development of a new medical device ensures early documentation of critical "design inputs" that often heavily influence the success of a new medical device. Any new medical device involving a biomaterial has intrinsic "input requirements" which may prove difficult to elucidate. For example, it's obvious that a biomaterial would need to be "biocompatible," but design engineers must attempt to clarify the meaning of that requirement. Is the "input" that the material must be biocompatible for a few hours, a few days or will the device be a long-term implant? When the duration for biocompatibility is defined we have more information with which to establish performance requirements. Is "biocompatibility" to be judged as how well the material creates an interface with bone, how well it tolerates contact with another synthetic material which might induce a chemical interaction or is there long-term blood contact? These "input" details are critical information necessary to accurately describe the user's expectation for the service of the biomaterials within the device.

One of the first tasks to meet the Design Control regulations is the establishment of Design Inputs (§820.30(c)). Typically a robust project must define "user requirements," and then further refine this information to "engineering design requirements," because the "user requirements" may be too vaguely expressed to be sufficiently useful for materials selection and evaluation. Even this process of refining the input requirement can lead to design drift as the engineer attempts to interpret the user's statements into quantitative and measurable design parameters.

Design "outputs" are often an area of confusion for beginning medical device designers, because the regulation allows a broad interpretation of what constitutes an "output." It is generally agreed in the industry that "outputs" are the sum total of the finished design, inclusive of design documentation, design testing results, and the device itself. Nevertheless, the regulations require that any device subject to Design Control must demonstrate that the Design Outputs meet the Design Inputs (see §820.30(d)).

Each new medical device requires the systematic Design Review (§820.30(e)) of the device performance and documentation through each defined phase of development. Proof that the device design (outputs) meets the device inputs may require iterative reviews of test results conducted over a range of design phases. These phases or stages should be identified in a Design Plan (§820.30 (b)) which is reviewed repeatedly to confirm that design development is on track, and that requirements are being met in a systematic manner throughout the development project.

Setting up a Design Plan for a new biomaterial can be especially difficult, because the performance of the biomaterial within the device is intrinsically linked to the performance of the device. Qualifying a new biomaterial in a new, unproven medical device may require iterative qualification cycles, as the biomaterial, then the device, then the biomaterial is repeatedly tweaked and evaluated. Possibly the only way to evaluate and define performance is through repetitive animal and/or bench models, which are perhaps themselves not yet validated. Critical to success in these circumstances is accurate, thorough, and relentless documentation of changes as the development progresses, and written justification of compromises made to input requirements.

Certainly a more straightforward scenario is substituting a new biomaterial in an existing and well-established medical device where performance is well-characterized. In this situation the new biomaterial would be evaluated against the performance of the current biomaterial to determine if the new material performs as well, without compromising overall device performance. Even this can be challenging if the relationship of material properties to device performance was never elucidated.

Biomedical engineers often find that the prior biomaterial performance requirements were not well-characterized for the intended use when the material was originally adopted. Many older medical devices have undergone design or materials "creep," whereby the performance requirements of the components and/or biomaterials were never separately qualified for the specific role played within the device function, yet change after change has slowly moved the device performance along to keep up with customer requirements. Qualifying a substitute biomaterial in these circumstances can be time-consuming and expensive. Gradually as principles of Design Control and Review infiltrate the medical device industry, biomaterial characterization associating material properties more succinctly to intended use may streamline biomaterial substitutions.

Fundamental to establishing the suitability of the biomaterial within the medical device is the task of assessing risks and how to mitigate these risks. ISO 14971:2007 describes methods for conducting and documenting risk analysis, but this standard also describes the concept of Risk Management. Risk Management begins at the design level, but continues through the lifecycle by continuous risk assessment and management during development and marketing phases, including how to incorporate risk assessment into corrective and preventative action and management oversight. For this present discussion, Design Level Hazard Analysis and Risk Assessment are of immediate interest.

In the design level hazard analysis the exercise is to evaluate the potential hazard that may exist if the device (biomaterial) failed to meet the user requirement (or engineering requirement). A hazard is the "potential" to do harm (either to the patient, the user or even device performance). A difficult chore is assessment of risk posed by the potential hazard, because one must understand the "likelihood" or "potential" for the hypothetical hazard to actually occur in the defined circumstances. This exercise can lead to excessive speculation or contrarily, a denial that there is any risk at all. Despite the often awkward and painful process of listing and evaluating these potential hazards, the resultant laundry list of potential hazards and mitigations (usually testing, labeling, and defining release criteria) should help to reduce the potential for the identified risk to occur. This list ideally creates the road map for a "risk-based" approach to the verification and validation plan for the new product (§820.30(f) and (g)). From here designers can document that device outputs, including risk mitigations, have met user inputs.

Complex new medical devices, particularly those with innovative biomaterials, may require iterative testing, and protracted design phases and multiple reviews prior to final release of the device to manufacturing (Design Transfer §820.30(h)). Design review stages for such products may require bench and standardized testing on early prototypes, then animal testing on prototypes maximized for the test subject, then still more sophisticated and possibly different "prototypes" for limited clinical trials.

In each design phase the device, as embodied by the prototypes, may be fabricated differently, use different formulations of biomaterials or other components, and may have different physical shapes or sizes. Processes such as packaging and sterilization may also differ. As the design evolves there must be Design Control that documents the many Design Changes (§820.30 (h)). All of the design changes, the results of testing, and the myriad design specifications must be maintained in the Design History File (§820.30(j)). Early design documentation may begin in a laboratory notebook and evolve to state-of-the-art electronic documentation systems, but in every circumstance the development engineer is charged with the responsibility of providing documentation that the device is properly validated for intended use. Good documentation will prove invaluable if it should be necessary to justify why the test results using early prototypes are still valid as supporting evidence of the suitability of the final design. Even with such support, most medical device manufacturers will conduct a final design validation on devices produced from final production runs prior to full market release. For Premarket Approval (PMA)-regulated devices, this is necessary prior to FDA establishment that the device is safe and effective, and thereby can be released to market.

BIOCOMPATIBILITY ASSESSMENT FOR BIOMATERIALS IN MEDICAL DEVICES

Standardized biocompatibility assays are conducted prior to clinical evaluation of the device or for those not requiring a clinical trial, prior to clearance to market the product. Regulatory requirements based upon the risk of the device will typically dictate the extent of standardized biocompatibility testing. These recommendations, such as those mapped in ISO 10993: Part 1, reflect the perceived risk associated with the use of the product. Many of the standardized biocompatibility screens focus on the chemical toxicity of leachable components or degradation by-products. These exposure-based categories have for years been seen as a logical (although sometimes awkward) structure for determination of which testing protocols best evaluate the potential toxicity of the material/device. It is tempting to make the list of tests a "tick-list" without considering the value of the test and its appropriateness to the product. Ethical use of animals and considerations for cost-control compel the biomedical engineer to prudently and judiciously take advantage of published literature, internal corporate databases, and commercial databases offering technical information on materials. Biomaterials suppliers are more frequently establishing "Device Master Files" (MAF) which may serve as a reference library of prior testing, held confidential at the FDA for use by reviewers when a biomaterial is included in a new medical device. Such historic testing may not eliminate the need for additional testing in the specific application or due to differences in processing, but the device developer should attempt to take every advantage from well-characterized materials that are offered for use in medical devices, rather than incur excessive costs and potential setbacks due to selecting industrial-grade materials not intended for use in medical devices.

Despite the goal to conduct biocompatibility assays on finished devices, most test regimens cannot make use of the entire, finished device. It may be necessary to evaluate a biomaterial separate from the finished device, with consideration for the exposure of the materials in the finished device by including exposure to effects from manufacturing and sterilization processes. Such "environmental" stressors may be important in determining the potential that the material may have toxic and accessible process by-products. The standardized biocompatibility screens are not without limitations. For example, these screens are not designed to evaluate the long-term effects of any structural forces on the device which might affect the biomaterial once incorporated into an implant. As an example, osteolysis, due to the presence of wear particulates, has been identified as a cause for implant loosening, yet the materials responsible for osteolysis routinely pass all of the biocompatibility screens.

For new biomaterials or new formulations of historic materials, developers may wish to consider a risk-based approach to establishing allowable limits for leachable substances. ISO 10993-17 provides a method by which maximum tolerable levels can be calculated from available data on health risks that can be systemic or local, immediate or delayed, and range in severity from minor to life-threatening. The derivation of the allowable limits should be conducted by an individual knowledgeable and sufficiently experienced to form a qualified opinion based upon scientific data and knowledge of medical devices. Such an effort may be useful in establishing whether or not expensive and long-term chronic studies on animals to rule out carcinogenicity are warranted.

Medical device regulatory authorities recognize that many, if not most, of the standardized biocompatibility assays were not designed to assess performance of combination devices, tissue-derived products, and devices designed to be modified by the body after implantation. New standardized testing protocols are necessary, but are unlikely in the short-term. In the interim, biomaterials development companies must contend with the cost of these standardized assays, despite their lack of practical value to predict the performance of the new biomaterial in a new medical device application. Acceptance of a new biomaterial continues to rely heavily on the results of animal trials in each device application, with the hope that the biomaterial can be transitioned to additional new uses as the material gains a reputation. Unfortunately, this migration of a biomaterial from one use to the next has historically resulted in the material's eventual use in an application for which it is ill-suited, with the resultant whiplash against the material in all applications. The best immunization against fad-material adoption is training engineers in the requirements of Design Control and Review, but in particular, the necessity for defining the input requirements for the biomaterial in the device, in the intended use.

MANUFACTURING CONTROLS AND POST-MARKET OVERSIGHT

Most regulatory authorities require that medical device manufacturers demonstrate acceptable manufacturing practices, although the specific obligations and criteria required may vary from country to country. In the US, the law of the land is found in 21 CFR Part 820: Quality System Regulations. These regulations are more or less harmonized with ISO 13485:2003, but different details still keep manufacturers on their toes when different auditors come to call.

An emerging threat to the safety of medical devices is the potential for counterfeit or tainted materials to make their way into a device after initial design qualification. Medical device manufacturers must increase vigilance with vendors of raw materials and components to ensure purity of the material supply. It is no longer sufficient to call out a material by a "name brand" in a specification. Component and material purchasing must be accompanied by trustworthy integrity testing which gives evidence that the source has delivered materials to the agreed specification. "Certification" to a standard should only be trusted when there is an active and robust auditing program in place. "Certified" vendors may help secure the supply chain, but periodic testing of materials is still critical to ensure only qualified biomaterials are used in critical medical devices.

Post-marketing oversight of medical devices in the marketplace is increasing as the sophistication of medical devices increases. Conventional clinical trials designed to prove efficacy seldom establish the long-term performance of the product. More importantly, the interaction of the medical device with the patient in their normal environment, with other devices and drugs or as the patient may age, is not typically assessed during a clinical trial. The FDA has regulations concerning recalls, product removals, and corrections, and heavily stresses the importance of corrective and preventive action (CAPA) programs for marketed products. The FDA's enforcement and oversight will continue to increase in an effort to monitor device performance in the hands of the user, not just as-manufactured in the factory.

Post-approval limitations to the distribution of the medical device typically include restrictions to be placed on the claims for the product, and on the content of labeling. Restrictions on the content of labeling and advertising medical devices are increasing as the FDA seeks to avoid creating undue expectations by patients for the performance of a medical device. Some "off-label use" of a device by a physician has been tolerated in the past, but promotion by industry of new use or "performance claims" for a device requires a new application to support a different indication for use or claims of performance not previously reviewed by the FDA.

Most regulatory authorities around the world now require the reporting of adverse effects. The degree of hazard that must be reported to the governing agency will vary from country to country. Tracking of implantable or life-supporting devices is limited, but is still a common regulatory constraint for high-risk implantable devices, such as defibrillators.

PREMARKET CLEARANCE (510(K)), PREMARKET APPROVAL (PMA) OR "CE MARK"

For the European market, once a developer of a medical device has demonstrated conformance with the Essential Requirements of either the MDD or AIMDD, the company can often file a Certificate of Conformance or request that a Notified Body conduct an examination to ensure conformance. The route to conformity assessment is spelled out in the Directives, and is determined by the classification of the device and the applicable rule, which is primarily a risk-based stratagem. Higher risk devices require clinical data and review of a Design Dossier by a Notified Body. The CE Mark must be applied, based upon assessment of conformance, prior to the sale of any medical device in a country within the European Union, and in other countries that have voluntarily adopted the recognition of the CE Mark. Canada, Australia, and many Asian countries follow a similar pattern for device review and market approval, but their market entry requirements may differ significantly from those of the EU.

The US FDA requires that the manufacturer demonstrate safety and efficacy of the product, unless the device can be shown to be substantially equivalent to a predicate product that is legally marketed in the United States (via the 510(k)). For most new products, those without predicates, and certain high-risk products (typically Class III), a PMA is required. Certain Class I and exempted Class II medical devices are permitted to be marketed without the FDA's prior authorization, although manufacturing (Quality System Regulations) and/or documentation controls may still apply.

Premarket applications for medical devices containing biomaterials may take a variety of regulatory pathways, but for most new or innovative biomaterial-based, implantable devices, the PMA is required. The FDA has continued to permit some Class III devices to enter the market after clearance by 510(k), although the US Congress has now mandated that FDA down-classify any remaining "transitional" devices or show cause why the products should remain Class III and require a PMA. The FDA database can provide information for comparative analysis of predicate devices, and help to identify the device classification and regulations for new products, but finding the "keywords" and how to most effectively "mine" the database for information can take time. The best place to start is to review the information on: http://www.fda.gov/MedicalDevices/DeviceRegulationandGuidance/Databases/default.htm. (Zuckerman et al., 2011)

For devices subject to the PMA process, the culminating step in the lengthy and expensive regulatory review process may be the presentation to the FDA's clinical advisory panel (this is known as "panel track"). Panels of medical experts focused in a medical sub-specialty convene at the FDA's request to make recommendations on the acceptance of a new medical device. These sessions typically focus on whether, in the opinion of the panel, the clinical evaluation of the product has demonstrated efficacy. Proof of efficacy is highly dependent upon statistical analysis of prospective data collected under tightly controlled clinical evaluation protocols. These recommendations often impact the FDA's decision to allow a new device to be marketed, and limitations on labeling claims.

The FDA has sought to take some of the uncertainty and time out of the approval process by offering manufacturers collateral review steps. The modular-PMA is finding favor with medical device companies, because they can submit sections of testing as completed. This does not, however, eliminate the potential that the clinical panel may fail to recommend the product for approval based upon the lack of convincing performance data produced during the Investigational Device Exemption (IDE) studies.

The FDA has issued a revised draft guidance: The 510(k) program: Evaluating Substantial Equivalence in Premarket Notifications [510(k)] http://www.fda.gov/MedicalDevices/DeviceRegulationandGuidance/GuidanceDocuments/ucm282958.htm, which seeks to clarify how to compare a new product to an existing device and its associated technology for the purposes of filling an application for premarket clearence [510(k)]. The speed at which a medical device technology can become obsolete mandates that the FDA's review process for moderate and low-risk devices be equally swift. But Congress has expressed concerns that device safety may be compromised due to antiquated regulatory strategies which may allow inadequately tested devices to become broadly marketed or allow undue industry influence over the review process. It is obvious that requiring every new medical device to undergo full premarket approval would guarantee stagnation of the industry and escalating costs for new products, not to mention overwrought FDA reviewers. It is already common for new medical devices to reach the European market years before the same device is available in the USA. A delicate balancing act between the needs of the stakeholders will continue to affect the economic impact of medical device regulations, the availability of new technology for patients, and ultimately the cost of healthcare.

CLINICAL AND ANIMAL TRIALS OF UNAPPROVED DEVICES

Clinical evaluations for significant-risk devices containing biomaterials are cleared for study in human patients by the FDA under an IDE (Investigational Device Exemption regulations – 21 CFR Part 812.) A protocol (study plan) and informed consent form must be approved by an institutional review board (IRB). The patient must give prior consent before enrollment in the investigational study. Investigators must be trained in the use of the product, and sign an agreement to conduct the study according to regulations and according to the approved protocol. Investigators must now disclose certain levels of financial involvement with the study sponsor. Case report forms and data collection methods must be carefully crafted in order to document data required to support the claims for the product under study. Software used to capture or evaluate data must be validated, and any electronic-capture systems must be compliant with 21 CFR Part 11.

When the study constitutes a "significant risk," the sponsor must submit and wait for approval of an IDE application prior to commencing the trial. US clinical studies must now be posted on the internet at www.clinicaltrial.gov. This is an effort to make the trials more accessible to more potential patients. The public disclosure of the existence of the clinical trial also reduces the potential that trials with less than sterling results will go unreported to the FDA.

Clinical trials for medical devices containing biomaterials must develop objective evidence of safety of the device, and indirectly the safety of the biomaterial as applied in the device. Any safety problem encountered with the use of a biomaterial during the study is reported as a part of the adverse effects of the clinical trial. Any special performance claim attributed to the biomaterial, such as tissue in-growth, needs to be evaluated both as a part of a preclinical animal study, and if possible, in the clinical study. In the United States, only claims for a biomaterial supported with statistically valid evidence will be permitted in the device labeling.

Basic requirements for the conduct of human clinical trials continue to be harmonized internationally, although significant differences in trial monitoring and regulatory oversight remain. Clinical trials of US-developed medical devices outside of the US, prior to starting investigational trials in the US, are likely to remain the quickest and least expensive way to obtain clinical experience and market approval for a new medical device. The potential ethics of this practice are not challenged by the US FDA, and indeed seem to be encouraged. Ironically, ethical constraints on testing devices in animals in Europe often motivate European medical device manufacturers to come to the US for animal studies.

Specific clinical performance claims of a biomaterial must be supported by clinically significant objective evidence. Obviously it can be difficult on a clinical level to isolate biomaterial performance from device performance. This further complicates assessment of the isolated performance characteristics of a biomaterial for any future marketing benefits. Often, indirect or noninvasive clinical assays, such as X-rays, prove inadequate to demonstrate biomaterial performance. As a result, understanding long-term biomaterial performance usually depends upon high-quality animal studies to demonstrate interfacial characteristics between the tissues and biomaterial. Often these studies have contextual limitations and are not published, due to proprietary concerns. Even though animal study protocols strive to provide the basis for predicting behavior of the material in the human application, their objectives are typically tightly focused and not entirely relevant to applications beyond the specific intended use. Impediments to success include the differences in animal anatomy, animal metabolism, healing characteristics and behavior, when compared to humans.

Despite these limitations, pivotal animal studies are crucial to developing the confidence that a material is safe for human clinical trials. Pivotal trials supporting a device submission must be conducted under the Good Laboratory Practices Regulations (CFR Part 58). These regulations establish the quality standards for the research data that may be used in support of an FDA application, and help the sponsor of the study to have confidence that the study has been monitored for conduct according to the protocol.

STERILIZATION, SHELF-LIFE, AND AGING

Physical properties and durability of the biomaterial in its intended use are critical parameters to establish. If failures begin to occur in the field, differences from baseline parameters can provide crucial evidence to assist in establishing the root cause. Data demonstrating the shelf-life aging profile of medical device materials is required for an increasing number of new device applications to the FDA. The device manufacturer can expect increasing demands by the FDA for evidence that the device will perform to specification, at least to the expiration date placed on the labeling. Such data are already required if a medical device manufacturer incorporates drugs and/or biologics as a surface treatment or as enhancements to device performance. Most sterilization processing can affect biomaterial performance in a product to some degree, even though the effects may not be immediately evident. It is known that effects, such as those from irradiation, may not be immediately apparent but can accelerate over time. The exact parameters of a sterilization cycle must be defined and documented, and the information maintained for future reference. The FDA requires that validation of materials performance requirements and safety testing of the device is conducted after the same sterilization cycle that will be applied to the market-released product. Long-term shelf-life aging studies, which include a variety of sterilization processes, should begin on any new biomaterial as soon as possible so that potential effects of environmental stressors (heat, gas, radiation, oxidation, etc.) can be quantified as early as possible. Such materials aging studies should begin using standardized specimens and well-defined test methods as early as possible, to ensure the candidate material can remain stable within the device under development.

INNOVATIVE TECHNOLOGIES REQUIRE SPECIAL CONSIDERATIONS

Silo-style regulations of drugs, biologics, and devices have proven inadequate to the challenges from novel biomaterial technology. In October 2002, Congress passed the Medical Device User Fee and Modernization Act of 2002 that established the Office of Combination Products. Combination products use device, biologics or drugs in combination to enhance or deliver the desired treatment (or diagnostic). The premarket review jurisdiction of such combination products is predetermined by interagency agreement. The FDA has now created a Combination Products website (http://www.fda.gov/CombinationProducts/default.htm) to serve as a portal to various guidance documents and performance reports. As new biomaterials incorporate biologic functions, integrate within organ systems to replace lost function, and evolve engineered performance requirements as tissue scaffold devices, more biomaterials will likely fall into the category of combination products. Manufacturers may request the FDA to make a determination of the review authority if the Center with authority is uncertain. Early determination of the regulatory pathway for a combination biomaterial application can help developers avoid setbacks due to inadequate scientific assessments. Determining the "mode of action" dictates the regulatory pathway for a combination product. For combination products, joint review between two FDA Centers is commonly required, and more than one type of premarket application may be necessary.

Continued concerns for the safety of biomaterials derived from human or animal products have increased the regulations associated with these medical products. The FDA has issued guidance for the use of animal materials used in medical devices. The guidance titled "Medical Devices

Containing Materials Derived from Animal Sources (Except for *in Vitro* Diagnostic Devices") has provided details on the FDA's concerns, and how the FDA expects manufacturers will control the sources of the materials.

Tissue-engineered products designed to replace major organ function are in clinical research, thus the FDA is exploring methods for predictably characterizing the materials that support these novel technologies. The FDA sponsors studies that help to understand the processes surrounding cell maturation, cell signaling, differentiation and formation of organs, and functional tissue. This research will help to contribute to methods for predicting behavior and perhaps quality control testing.

The FDA has formed a taskforce to evaluate how nanotechnology may impact device safety. The taskforce released a report (Nanotechnology Task Force Report released July 2007) which made the case to continue to develop innovative and safe FDA-regulated products, such as "tiny sensors to detect disease markers in the body" to imaging tools, and even packaging that can extend shelf-life. While recognizing the potential for the technology to impact all aspects of medical product assessment, the FDA recognizes that nanoscale biomaterials can have chemical or physical properties that are different from those of the bulk material, including altered magnetic, electrical or optical activity. Nanobiomaterials could have altered structural integrity, and differences in biological activity. Because of their special properties, nanobiomaterials may pose different safety issues, and conventional testing methods may not identify the potential risks.

SUMMARY

Development engineers must document conformance of the product to requirements, and be able to show that the established specifications result in a product that meets user needs. By following the basic steps for Design Control and Review, biomaterials which serve as device components can be demonstrated safe and effective for sale in the global medical device market. Medical device regulations and best practices for evaluating biomaterials in new medical products change faster than book chapters can be updated. This is a broad outline of the regulatory structures. Consult the FDA website and numerous published resources for the most up-to-date regulations.

BIBLIOGRAPHY

Challoner, D. R., & Vodra, W. W. (2011). Medical devices and health-creating a new regulatory framework for moderate-risk devices. *N. Engl. J. Med.*, **365**, 977–979.

Curfman, G. D., & Redberg, R. F. (2011). Medical devices: Balancing regulation and innovation. *N. Engl. J. Med.*, **365**, 975–977.

Ergina, P. L., Cook, J. A., Blazeby, J. M., Boutron, I., Clavien, P. A., et al. (2009). Challenges in evaluating surgical innovation. *Lancet*, **374**, 1097–1104.

Maisel, W. H. (2004). Medical device regulation: An introduction for the practicing physician. *Ann. Intern. Med.*, **140**, 296–302.

Muni, N. I., & Zuckerman, B. D. (2007). The process of regulatory review for new cardiovascular devices. In E. M. Antman (Ed.), (3rd ed.) *Cardiovascular Therapeutics – A Companion to Braunwald's Heart Disease*, (pp. 67–75). Philadelphia, PA: Saunders.

Sweet, B. V., Schwemm, A. K., & Parsons, D. M. (2011). Review of the processes for FDA oversight of drugs, medical devices, and combination products. *J. Man. Care Pharm.*, **17**, 40–50.

Zenios, S., Makower, J., & Yock, P. (2010). *Biodesign: The Process of Innovating Medical Technologies* (pp. 1–742). Cambridge, UK: Cambridge University Press.

Zuckerman, D. M., Brown, P., & Nissen, S. E. (2011). Medical device recalls and the FDA approval process. *Arch. Intern. Med.*, **171**, 1006–1011.

FDA. (Revised as of April 1, 2011). Regulation CFR Part 58: Good Laboratory Practices Regulations.

FDA. (Revised as of April 1, 2011). Regulation 21 CFR Part 11: Electronic Records, Electronic signatures.

FDA. (Revised as of April 1, 2011). Regulation 21 CFR Part 812: Investigational Device Exemptons.

FDA. (Revised as of April 1, 2011). Regulation 21 CFR Part 820: Quality Systems Regulations.

FDA (November 6, 1998). Guidance: Medical Devices Containing Materials Derived from animal Sources (Except for *in Vitro* Diagonictic Devices).

ISO 10993–17:2002 Biological evaluation of medical devices–part 17: Establishment of allowable limits for leachable substances.

ISO (2003). ISO 13485:2003: Quality Management Systems – Medical Devices – System Requirements for Regularity Purposes.

ISO (2001). ISO 14971:2007: Medical devices – Application of risk management to medical devices.

CHAPTER III.2.5 PRINCIPLES OF REIMBURSEMENT FOR MEDICAL DEVICES

Fred Cahn

BioMedical Strategies, La Jolla, CA, USA

SIGNIFICANCE OF REIMBURSEMENT

"Reimbursement" is a term used to describe the process by which medical products are paid for. In most cases medical devices are not chosen and purchased by the patient who receives them; instead, they are purchased by a hospital or physician (the "provider") as a necessary component of a medical service that is provided to the patient. Also, in most cases the patient does not pay either the hospital or physician directly for that service; instead, the provider is "reimbursed" for providing the service by a health insurance company or a Government program, such as Medicare in the United States (the "payer"). Thus, if the provider is paid adequately by the payer for providing a medical service, the provider will be willing and able to purchase from the manufacturer the medical devices that are required to perform the service.

There is considerable complexity in the reimbursement processes, and there are many considerations that can go into reimbursement policies for existing and new products. It is important that those responsible for inventing, developing or marketing new medical devices have an understanding of how their products will be reimbursed, because reimbursement is often critical to market success. Thus, reimbursement should be treated as an input to be addressed repeatedly during the design control procedures of a new medical product, beginning with marketing feasibility in the initial design plan. Reimbursement considerations should be addressed in the regulatory strategy, and in the clinical study planning, since these are the stages at which the foundation of a successful reimbursement plan is laid.

BACKGROUND

Health Plans

In the United States, at this time, there is no universal healthcare plan or policy. Patients usually fall into one of the following categories:

- Private insurance (usually provided by, and mostly paid by, employers for their employees and their families). Private insurance plans generally fall into two basic types:
  - Fee for service, in which the provider is paid on a per patient visit basis;
  - Health maintenance organization (HMO) in which the provider is paid on a "capitated" basis (a fixed payment per patient per month), and the provider is responsible for providing all necessary care to the patients in the plan.
- Medicaid administered by state governments on a fee for service or HMO basis, for residents who are classified as poor or who fall into a special needs category.
- Medicare, a US Government program for the elderly and certain others. Medicare has both fee for service ("Part A" and "Part B") and HMO (Medicare Advantage) options.
- Veterans and Military.
- "Self-pay," representing patients who are not covered by any health plan, and who are expected to pay for services themselves. Self-pay also includes patients in a healthcare plan who wish to obtain services not included ("covered") in the plan (e.g., cosmetic surgery).

Usually, a manufacturer trying to understand the likely reimbursement policies for their product will initially be concerned with trying to understand the policies of Medicare fee for service, which is not only one of the largest plans, but also one in which the policies are relatively transparent because they are defined by public law and by published administrative policies. Many of the policies of large private fee for service plans are also published and easily obtained from the Internet, and frequently they are based on Medicare policies. These fees for service plans usually represent the majority of the market for medical devices, at present.

Because Medicaid policies are different in each state, it is usually too difficult to assemble comprehensive information on them and on veterans or military, unless the product is of special interest to these groups.

Self-pay is a special case that can affect the marketability of certain products. While self-pay may also be practical for routine medical care by some "middle income" patients, very few of these patients have the financial resources to pay for hospitalization. In this case, in the United States, providers can be required to provide emergency care, and they attempt to collect their "charges" (retail prices) from the patient. Most self-pay patients cannot afford to pay all or most of the charges, and their care is subsidized by the provider and by government subsidies to the providers. The result of this situation is that there are certain types of care for which hospital providers frequently lose money, for example, emergency trauma care, which affects younger and often less responsible patients who are frequently not covered by health plans. Hospitals tend to resist the purchase of products that are disproportionately used for these patients.

On the other hand, there are a few products specifically supplied to patients who are willing and able to pay for them, and reimbursement is not an issue (e.g., for cosmetic surgery).

POINTS OF SERVICE

In fee for service plans, reimbursement methods for medical devices depend on the "point of service" where they are used:

- Physicians' offices
- Inpatient hospital
- Outpatient hospital
- Ambulatory Surgical Center (ASC)
- Laboratories (for diagnostic devices)
- Other institutions (skilled nursing facilities, long-term care, etc.)
- Durable Medical Equipment ("DME," for home use, by prescription)
- Pharmacy, by prescription (e.g., diabetic supplies)
- Over-the-counter.

Over-the-counter products are not reimbursed by health plans. Pharmacy fees for service reimbursement are covered under Medicare "Part D" or private pharmacy benefit plans, but very few medical devices are paid for in this way. The "other institutions" are usually not significant users of advanced medical devices, with the exception of chronic wound care. Thus, for the reimbursement of most medical devices, we are concerned primarily in the reimbursement of physicians, hospitals, laboratories, and DME.

THE "HOSPITAL BILL" AND THE "CLAIM"

In the US, physicians and hospitals are paid separately. Internally, a hospital maintains a "chargemaster," which contains the (retail) prices of all services, goods, and procedures for which a separate charge exists. It is used to generate an itemized patient bill. Similarly, physicians also bill patients for each service they perform. These are the bills that would be presented to a self-pay patient. However, for reimbursement from health plans, the reimbursement is based on negotiated contracts (private plans) or Government rules (Medicare), and the reimbursement process begins with claims from the providers.

There are two basic types of claim, an institutional (hospital) claim and a professional claim (physicians, also used for DME and laboratory). When a physician provides services in his or her office, there will be only a professional claim, but for a service provided in a hospital there will be two separate claims for the same services, one from the physician for his professional services, and one from the institution for use of facilities and supplies. While these claims will include summaries of the hospital or physician charges, these are not the amounts that will be paid. Instead, the procedures and diagnoses are identified on the claims by combinations of codes, which form the basis of the reimbursement system. The format and rules for coding the claims are mandated by the Government for Medicare claims, but in general all payers use the Medicare formats and rules.

CLAIMS PROCESSING

Processing of claims can be divided into three basic components: coding; coverage; and payment.

Coding

All physician and hospital claims include one or more diagnosis codes (Table III.2.5.1). Currently, the ICD-9-CM diagnosis coding system is used, but it is scheduled to be superseded by the ICD-10-CM system in October, 2012.

Hospital inpatient claims also include one or more procedure codes that represent the services performed. Currently, the ICD-9-CM procedure coding system is used, but it too will be superseded soon by the ICD-10-CM procedure coding system. Hospital outpatient and physician claims do not use the ICD-9-CM procedure codes; instead they use two other methods to code procedures and related services:

- CPT procedure codes (published by the American Medical Association, AMA) are five digit numeric codes that identify procedures performed on the physician's claim at all points of service (inpatient, outpatient, ASC, office). CPT codes also identify these procedures on claims by hospital outpatient departments and ASCs.
- HCPCS codes, which are alphanumeric codes issued by the Centers for Medicare and Medicaid Services (CMS (Centers for Medicare and Medicaid Services), the Government agency that administers the Medicare Program), extend the CPT system to identify services not included in the CPT codes, including certain drugs, biologics, and devices.

Another type of code appears on institutional claims: the "revenue code." These codes are used to summarize hospital charges by department; Medicare uses this charge information for statistical purposes to determine payment rates in subsequent years.

Coding Implications for New Devices and Technologies

Codes are the only way the payer can recognize the performance of a procedure or the use of a device in the procedure. Thus, the proper use of existing codes or the creation of new codes can be essential for the use of the device. For physician procedures, the AMA CPT Editorial Panel is the sole determiner of whether existing codes can be used with new technology, as well as whether new codes will be created. The members of the editorial panel are physicians who generally represent their professional societies. Manufacturers do not directly participate in these decisions.

The AMA, following the advice of specialty medical societies that participate in the CPT editorial process, makes these coding decisions; criteria include clinical coherence and the amount of physician time and skill required to perform

TABLE III.2.5.1 Procedure Coding

Point of Service	Physician Claim	Institutional Claim	DME Claim	Laboratory Claim
Physician office	CPT and HCPCS	N/A	HCPCS	CPT and HCPCS
Hospital inpatient	CPT	ICD-9-CM Procedure codes	N/A	N/A
Hospital outpatient or ASC	CPT	CPT and HCPCS	HCPCS	CPT and HCPCS

ASC: Ambulatory Surgical Center.
CPT: Current Procedural Terminology.
HCPCS: Healthcare Common Procedure Coding System.

the procedure (measured by relative value units (RVUs), discussed below), whether the procedure is commonly performed, and whether it provides equivalent or superior clinical results. AMA has a procedure for requesting new codes and for confirming the correct use of existing ones.

For a new product that will be used in a medical procedure, there are several possibilities:

- AMA can decide that it can be used under an existing CPT code. For example, a device improves patient outcomes, but still meets the code description and requires the same amount of skill and work.
- The description of an existing code could be modified or a new CPT code could be created. Generally, creation of a new CPT code takes at least two years.
- Physicians can use a "miscellaneous code;" however, payment is unlikely from Medicare and other payers.
- The AMA can assign a "Level III" code to track its use. No RVUs are assigned; Medicare or other payers nearly always use a level III code as proof that the procedure is "experimental," and thus ineligible for payment.

Resistance to the creation of a new code or to the use of an existing one can arise when a new technology enables a less skilled physician or surgeon, or one in a different specialty, to carry out a procedure.

PAYMENT

Medicare's payments are established by law and Federal regulations. Medicare has different payment systems, depending on the type of claim and the point of service.

Physician Payments for Office Procedures

Physician professional payments are based on "relative value units" (RVUs) that the AMA assigns to each CPT code. The RVU consists of three components: a Work RVU, representing an estimate of the physician skill and time required to carry out the procedure; a "practice expense" RVU; and a malpractice RVU. In the case of Medicare, each of these RVU values is subject to a geographical adjustment, and the sum of the adjusted RVU is multiplied by an annual conversion factor. For private plans, the same logic is frequently used, but the conversion factor would be negotiated.

Most medical supplies, drugs, and devices used in an office procedure are "bundled" into the practice expense. Some relatively expensive drugs and devices are paid for separately, in addition to the RVU-based payment. These are identified by the HCPCS codes, which are reported along with CPT codes on the physician's claim.

Physician Payments for Hospital Procedures

If the procedure is carried out in an institution, a different, lower, practice expense RVU is used, reflecting the lower expenses a physician experiences by carrying out the procedure in an institution. Also, there would be no payment to the physician for HCPCS codes, since the institution purchases those items.

Hospital Payments for Outpatient Procedures

Hospital outpatient departments perform procedures for which the patient is not expected to stay overnight. Under Medicare, outpatient visits are paid by a system called the Hospital Outpatient Prospective Payment System. (Procedures carried out in Ambulatory Surgical Centers (ASC), which perform similar procedures but are private facilities that are independent of hospitals, are paid under a different, but similar system.)

CPT and HCPCS codes identify procedures on outpatient claims. The principle of the outpatient payment system is to group all of the procedures into a single payment amount, called the Ambulatory Payment Classification (APC). The procedure codes, and the diagnosis codes, are used to group procedures that are considered clinically similar, and also similar in terms of the cost to the hospital of the resources (e.g., operating room type and time, staff, supplies, drugs) required, into an APC. Each APC has a payment rate; both the grouping into "similar" resource requirements and the payment amounts are determined by statistical analysis of the codes and charges from prior years' claims. Depending on the services provided, hospitals may be paid for more than one APC, and they may also be paid separately for certain "expensive" drugs and devices (identified by HCPCS codes). Some services, such as clinical laboratory, are not bundled into the APCs; they are paid separately under the laboratory fee schedule.

It is important that companies developing new technology that will be used in hospital outpatient departments have a thorough understanding of the outpatient payment system. Some procedures are disadvantaged because they are relatively expensive compared with the rest of the group, and products used in those procedures may be uncompetitive. There are provisions with rigorous requirements for CMS to recognize and classify entirely new procedures, to pay for certain devices separately, and to provide transitional add-on payments for new technologies. Critical to obtaining favorable classifications or add-on payments is data that shows substantial clinical benefits for the technology, as well as economic data proving the expense to the hospital of employing a new, relatively expensive technology. On the other hand, cost-saving technologies can often be marketed successfully to hospitals.

Hospital Payment for Inpatient Procedures

Hospitals are paid for inpatient discharges according to the Inpatient Prospective Payment System. This system follows the principle of bundling services more rigorously than the outpatient system. The payment for all of the hospital resources used between admission and discharge, including laboratory, are grouped into a single MS-DRG ("modified severity adjusted diagnosis related group") by a computer algorithm known as the "grouper;" each MS-DRG has a payment amount associated with it. Adjustments are made for geographical and other hospital-specific factors, and there is an additional payment for "outlier" cases, but there are no separate payments for drugs or devices.

An exception is the "new technology transitional add-on payment." This payment pays a portion of the additional cost of using a new technology in an existing MS-DRG. New technologies can qualify for this payment for two to three years; after that, the historical statistical data exists for properly classifying the procedure into the existing MS-DRGs. Requirements for the add-on payment are FDA premarket approval (or rarely, 510[k] clearance), an expected increase in the charges associated with one or more MS-DRGs above a threshold, and demonstration of "substantial clinical improvement." To demonstrate that the charges exceed the threshold, claims data is required. Sponsors can collect these claims during clinical studies, and/or historical charge data can be used to support their claim. If the technology is considered to be eligible for the add-on payment, a new ICD-9-CM procedure code is created by CMS so that the use of the technology can be identified on claims.

As with devices used in hospital outpatient departments, manufacturers developing products that will be used in hospital inpatient departments should have a thorough understanding of how the inpatient payment system will affect the economics of using their product, and the criteria and procedures for obtaining add-on payments.

Private payers pay for inpatient stays in various ways, as specified in contracts between payer and provider. These are not publicly available. They could be a multiple of Medicare's payment, or they may be based on *per diem* rates or a percentage of charges.

Payment for Outpatient Clinical Laboratory Tests

Laboratory tests are identified by one or more CPT codes and paid for by Medicare by a fee schedule. Private payers usually pay laboratories a multiple of the Medicare fee. In many cases, multiple CPT codes identify a single test, and a payment is made for each; for example there could be a code for sample preparation, codes for each analyte tested.

The existence of codes representing components of a test often means that a new test can be coded and paid for immediately under the existing fee schedule. This ability to fit new tests into the existing structure is advantageous for new tests that have similar cost structures to existing ones. However, the introduction of more expensive technologies can be complicated by the time and uncertainty for obtaining a new CPT code, and by the "gap filling" and "cross walking" methodology that CMS uses to price new tests, which restricts the price that a manufacturer can charge.

Payment for Durable Medical Equipment (DME)

DME represents equipment that can withstand repeated use, primarily serves a medical purpose, and generally is not useful to a person without an illness or injury. DME also includes prosthetics, orthotics, and some medications considered necessary to use the equipment. DME is identified on claims by HCPCS codes. DME is paid under a fee schedule for each state. Medicare is in the process of implementing a competitive bidding process for DME.

COVERAGE

Coverage determines whether a provider will be paid at all. Even if there are codes to identify the procedure, and payment amounts in the fee schedules or prospective payment systems, the combination of procedure and diagnosis must meet the payers' criteria for being "reasonable and necessary," which lacks a common definition.

Usually, coverage is not an issue for inpatient procedures, since the assignment of a procedure code to an MS-DRG includes recognition of its medical necessity. However, physician and hospital outpatient payments identified by CPT and HCPCS codes are usually covered only under limited circumstances. Many are considered to be bundled into the physician or hospital payments for procedures which are covered. Many have no coverage under any circumstance.

Each payer has their own coverage policies and, within Medicare, each regional contractor has its own coverage policy; however, some procedures are covered under National coverage policies. For common procedures, coverage policies of Medicare and most large payers are published on the Internet.

Coverage decisions are made by medical directors, who are usually physicians employed by the payers. The medical directors claim that their coverage decisions are based on principles of "evidence-based medicine" which, however, is loosely defined and inconsistently applied. In practice, evidence-based medicine represents a compilation of peer-reviewed publications, but only studies from controlled randomized multicenter trials are given serious consideration.

It is common for coverage decisions to consider new CPT or HCPCS codes to represent "experimental" procedures (thus not reasonable and necessary), and to demand published peer-reviewed clinical data from large-scale multicenter clinical studies conducted under "real world" conditions that are not practical to conduct on new technologies. Furthermore, essentially all studies on new technology are sponsored by the company developing it, which can bias the results in favor of the technology.

However, publications based on the clinical data submitted to obtain premarket approval from the FDA can be convincing to some or most medical directors.

Frequently, coverage decisions use reports from independent "technology assessment" organizations that compile and assess the published literature and sell these assessments to health plans. However, these reports are of varying quality and may be biased against genuinely new technologies; furthermore, these organizations do not have the resources for thorough and impartial investigations.

A recent development is government financing of "comparative effectiveness" studies sponsored by organizations such as the Agency for Healthcare Research and Quality. Comparative effectiveness studies are directed towards major public health problems, those which affect large patient populations with significant mortality or morbidity, and to treatments that are well-established. Such studies are believed to avoid company bias, and they are able to compare the products of different companies. Company sponsored clinical studies rarely compare their products with the most effective alternatives. (Clinical studies designed to demonstrate "safety and effectiveness" to the FDA frequently compare products with placebo or a "standard of care," which may not be the most effective alternative.)

It can be important for manufacturers introducing new technologies to be aware of the concerns of medical directors. Technologies that add to the payers' expense and also may be used by a large patient population receive special, critical attention. Technologies that result in dramatically better clinical outcomes may be favorably received, even if they are more expensive.

CLAIMS DATABASES

The reimbursement system creates a large body of data that can be of use in understanding the size of markets for medical devices, and for their likely impact on the economics of payers and providers.

Detailed data on clinical procedures and the usage of products in them is available to qualified researchers for purchase from CMS, subject to data use agreements intended to protect the privacy of patients. Data is available for physician claims, hospital outpatient claims, ASC claims, and hospital inpatient claims on a per claim basis, as well as aggregated data, some of which is available for download from the Internet. An analysis of this claims data can be essential in supporting new technology add-on applications for inpatient and outpatient procedures, and can be used to demonstrate to CMS that a particular procedure would fit better into a higher paying APC or MS-DRG.

Medicare data is heavily biased to an older patient population. "All-payer" databases are also available from the National Center for Health Statistics and from the Agency for Healthcare Research and Quality, and from some states. There are also private companies that collect and analyze this data commercially.

CHAPTER III.2.6 CORPORATE CONSIDERATIONS ON BIOMATERIALS AND MEDICAL DEVICES: CASE STUDIES IN REGULATION AND REIMBURSEMENT

Gail D. Baura
Keck Graduate Institute of Applied Life Sciences, Claremont, CA, USA

As previous chapters have demonstrated, developing a new medical device for the US market is a complex process. Certainly, most Class II (moderate risk) and Class III (high risk) devices require US Food and Drug Administration Center for Devices and Radiologic Health (FDA CDRH, often just called "FDA") notification or premarket approval. The premarket submission requires that Good Manufacturing Processes (GMPs), including design controls, be followed. For Class III devices, significant manufacturing controls must also be applied. The manufacturing facility of the new device must be registered, and the device listed. The FDA often inspects the manufacturing facility as part of the approval process. Depending on its market, a new medical device often needs medical and private health insurance reimbursement, so that potential end users can afford to purchase the device. Finally, the US Patent Office (USPTO) requires that claims for a new invention be filed within a year of first public disclosure (often, the offer for sale), in order to protect the invention.

In this chapter we discuss these important issues, within the context of biomaterials for medical devices. Many of these issues are also covered in other chapters, so we focus here on case studies of design controls, described in more detail in Chapter III.2.3, which comprise the fundamental requirements for development of a new medical device. Case studies exemplifying the different considerations for an electrocardiogram (ECG) electrode, a tissue heart valve, and a permanent skin substitute material are highlighted.

REGULATORY STRATEGY

When a product development project is first initiated, it is prudent to determine the regulatory strategy as early as possible, in order to facilitate project planning. Class II devices require less time and resources for a 510(k) submission than do Class III devices for a premarket approval (PMA) submission. A combination product (device + drug, device + biologic) may not even be regulated as a device, but rather as a drug or biologic. Often, a Request for Designation is made to the FDA Office of Combination Products (OCP). If possible, a manufacturer tries to have a specific combination product designated as a medical device, because regulatory requirements are less substantial compared to those for a drug or biologic.

After the determination has been made that CDRH is the appropriate regulatory center, the company's regulatory department must decide if the device is Class II or Class III. As of 2009, for a Class II device, such as an ECG electrode, one predicate device or a group of predicate devices, whose lineages can be traced to substantially equivalent devices released in the US before May 28, 1976, must be determined. These predicate devices must not currently be considered Class III devices. Most implantable devices have been designated as Class III. A truly new type of device is classified as Class III by default.

Cypher® Stent Regulatory Strategy

Johnson and Johnson's Cordis division developed the first drug-eluting stent that received FDA approval. Each Cypher® stent (Figure III.2.6.1) consists of a stainless steel mesh tube, treated with parylene C, and covered with a basecoat and topcoat. The basecoat is formulated from sirolimus and two polymers: 67% polyethylene-co-vinyl acetate (PEVA) and 33% poly n-butyl methacrylate (PBMA). A drug-free topcoat of PBMA enables controlled release of sirolimus over 90 days. Based on *in vivo* study data, sirolimus inhibits smooth muscle cell migration andproliferation, which aids in the prevention of restenosis (Cordis, 2003).

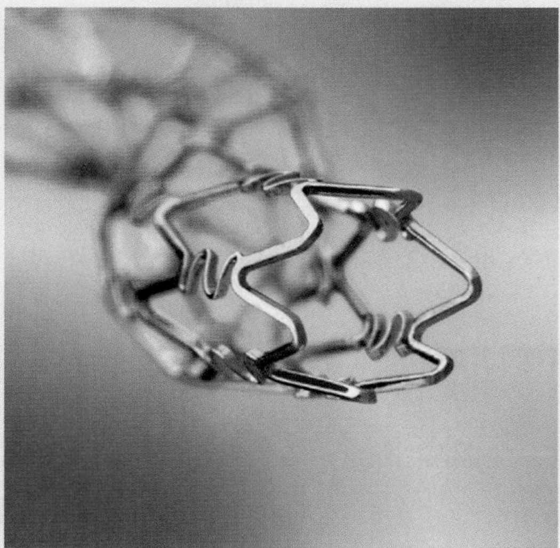

FIGURE III.2.6.1 Cypher® Coronary Stent (Cordis Corporation, Bridgewater, NJ, USA).

In its Request for Designation to the FDA OCP in 2000, Cordis stated that the primary mode of action for a drug-eluting stent is to provide: "a mechanical buttress that resists mechanical compression," while a coating on the stent serves "the ancillary purpose of retarding formation of intimal hyperplasia." The FDA agreed with this request, and designated CDRH as having primary jurisdiction (OCP, 2000). Cordis filed its Cypher® PMA submission, P020026, on June 28, 2002, and quickly responded to 24 separate requests for additional data by filing corresponding amendments. It received premarket approval for the Cypher® stent on April 24, 2003 (CDRH, 2003a), a total of 300 days.

Whenever possible, Class II status is preferable. Compared to a PMA submission, a 510(k) submission does not require clinical data or manufacturing data. Further, a 510(k) is less expensive to file, and less time is required until the FDA issues its premarket clearance decision. For the fiscal year 2009, the 510(k) and original PMA review fees were $3695 and $200,725, respectively. For fiscal year 2006, the most recent year with available statistics, 510(k) submissions required an average of 95 days between filing and notification. Similarly, original PMA submissions that were approved required an average of 335 days between filing and approval (ODE, 2008). It should be noted that the original PMA approval time for a new type of device, with which the FDA has no previous experience, may easily extend from the 335 day average by additional months, if not years. Approval may require the majority vote of an FDA Advisory Panel. For example, Gensia's GenESA device, which was the closed loop infusion device for the new drug GenESA, required 44 months until CDRH approval (CDRH, 1997a).

MEDICARE REIMBURSEMENT

Equally important to the regulatory strategy is the product's Medicare reimbursement strategy, assuming that reimbursement is critical for product success. Overwhelmingly, Medicare's national reimbursement decisions are adopted by private insurance carriers. Moreover, more clinical studies may be required for reimbursement than for FDA approval. Therefore, it is prudent to begin planning for all these studies, potentially with study overlap, as soon as possible.

The national Medicare coverage process for a new device that is not currently reimbursed generally takes one year. The manufacturer of this new device meets preliminarily with the Centers for Medicare and Medicaid (CMS), and then submits a formal request for National Coverage Determination (NCD). The formal request must include a statement of desired benefit categories, as well as supporting documentation such as the PMA summary of safety and effectiveness data, any 510(k) predicate devices, and a description of any other clinical trials in progress. Within 90 days of receipt of the formal request, CMS determines whether or not to accept the request. If the request is accepted, another 90 days passes, during which CMS accepts 30 days of public comment on the device and, if necessary, generates an internal health technology assessment. The technology assessment may extend this 90 day timeframe. At the end of this 90 day period, a proposed decision memorandum is issued.

During the first 30 days after the proposed memorandum is issued, the public may comment. Within 90 days of when the proposed decision was issued, a final decision memorandum is issued. If the decision is made to provide coverage for the device, CMS makes this payment change effective within 180 days of the issuance of the proposed decision memorandum. A final decision may be reconsidered or appealed by the manufacturer or any other interested party, such as an insurance carrier or individual. CMS's illustration of the National Coverage Determination process is shown in Figure III.2.6.2.

It is also possible to apply for local Medicare coverage with each of 10 Medicare regions.

Cypher® Stent Medicare Reimbursement Strategy

Separate from the National Coverage Process, Cordis requested a new payment code for reimbursement of the insertion of drug-eluting coronary artery stents. Cordis argued that: "due to the absence of restenosis in patients treated with the drug-eluting stents based on the preliminary trial results, bypass surgery may no longer be the preferred treatment for many patients ... Lower payments due to the decline in Medicare bypass surgeries will offset the higher payments associated with assigning all cases receiving the drug-eluting stent." The request was made in the early 2002 timeframe (CMS, 2002b), before FDA approval of the Cypher® stent. CMS solicited comments about the requested procedure code in the Federal Register on May 9, 2002, and approved a new code (effective October 1, 2002) on August 1, 2002 (CMS, 2002a).

DESIGN CONTROLS

During the course of a product development project, design controls are mandated by the FDA. As shown in the illustration of the Waterfall Design Process (Figure III.2.6.3), a product is developed in response to user needs.

The process includes design inputs, the design process, and design outputs. During the design input phase, requirement specifications and design specifications are written. The inclusion of any FDA-recognized consensus standard in a requirement specification is not mandatory. "Conformance with recognized consensus standards is strictly voluntary" (CDRH, 2007). Regardless of requirement origin, subsequent demonstration that requirements are met must also include sufficient evidence of medical device safety and effectiveness.

Risk analysis spans both the design input and design output phases. Initial risk analysis, during which potential device hazards are assessed, is clearly part of design input. Based on this analysis, the device design is modified to minimize the occurrence of specific hazards. However, in its guidance document, the FDA states that "the results of risk analysis" are an example of design output (CDRH, 1997b). For the purposes of design control, we consider risk analysis implementation and sign-off to be design output. These days, under ISO 14971: Medical

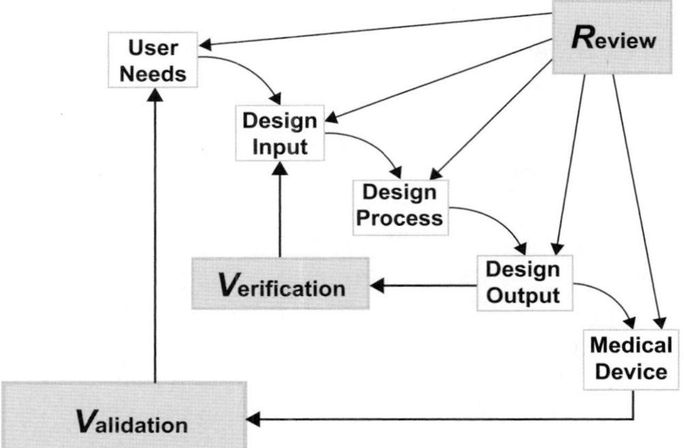

FIGURE III.2.6.3 Waterfall Design Process (CDRH, 1997b).

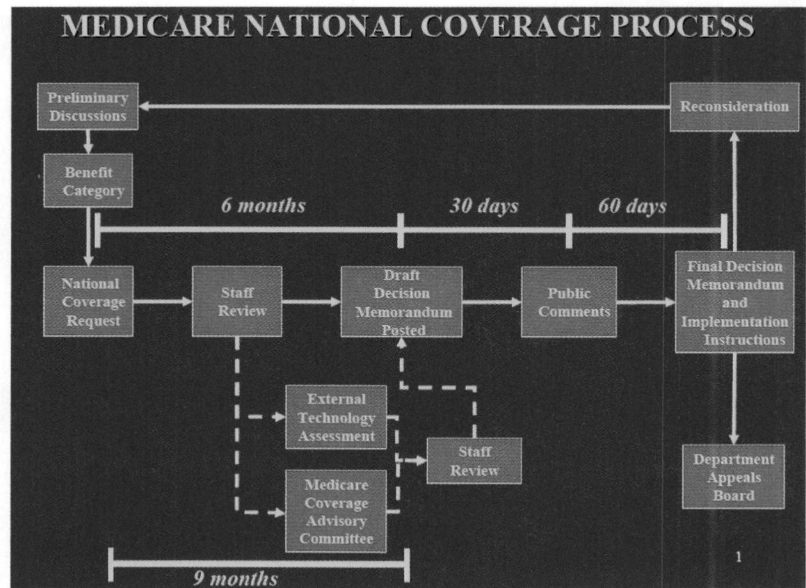

FIGURE III.2.6.2 Medicare National Coverage Process (CMS, 2009).

devices – Application of risk management to medical devices (ISO, 2007), risk management is considered a continuous process which spans well beyond design inputs and design outputs.

During the design output phase, each specific requirement from the design input phase is verified. As part of the design output phase, the new medical device is validated to ensure that user needs and clinical performance requirements are met. During each phase, a design review takes place before moving to the next phase. After validation, the design is transferred to manufacturing. Device documents are maintained in a design history file. While not mandatory for design control, it is helpful to initiate the product development process with a product requirements specification, which is not under rigid document revision control. In this way, initial marketing needs are documented.

Biomaterials Design Controls Examples

In this subsection, we describe specific examples of how biomaterials are handled by design inputs and outputs. The class of the device under development predetermines the testing complexity.

ECG Electrode. For an ECG electrode (Figure III.2.6.4), which is a Class II device, biomaterials specification and testing are relatively straightforward.

A manufacturer typically follows the national standard ANSI/AAMI EC12:2000/(R)2005 Disposable ECG electrodes. In this standard for ECG electrodes, electrical performance and biologic response requirements are defined (AAMI, 2005a). The biologic response requirements for cytotoxicity, sensitization, and irritation point to the national and international standard ANSI/AAMI/ISO 10993-1:2003 Biological evaluation of medical devices – Part 1: Evaluation and testing (AAMI, 2003). In the cytotoxicity tests, the lysis and growth inhibition of mouse fibroblast cells are determined, after exposure to an extract dilution of the electrode adhesive or gel. In the sensitization tests, erythema and edema of human skin are assessed after two applications of electrode adhesive or gel. In the irritation tests, erythema and edema of rabbit skin are assessed after electrode adhesive or gel application. In addition to these verification tests, the manufacturer may wish to provide validation clinical data demonstrating that human heart rates determined with the ECG electrode under test are insignificantly different from heart rates determined with a predicate ECG electrode. However, clinical data are not required for a 510(k) submission. This type of simple 510(k) submission is typically cleared by the FDA within 90 days.

Tissue Heart Valve. For a porcine tissue heart valve, which is a Class III device, biomaterials specification and testing becomes more complicated. Indeed, to demonstrate safety and effectiveness of the Carpentier-Edwards S.A.V. Bioprosthesis Model 2650 aortic valve (Figure III.2.6.5), ISO 10993 testing, animal studies, and clinical trials were performed.

This valve consists of a porcine aortic valve mounted on a flexible frame. The recovered porcine valve tissue is treated with ethanol and polysorbate-80, and packaged and terminally sterilized in glutaraldehyde.

As with the ECG electrode, cytotoxicity, sensitivity, and irritation tests were performed, using samples of materials involved in the valve manufacture. Some of these materials were silk suture thread, silicone rubber, and polyethylene terephthlate cloth and film. Additionally, other ISO 10993 tests were performed to assess genotoxicity, hemocompatibility, systemic toxicity, and implantation.

As a precursor to clinical trials, *in vivo* animal studies were conducted. Since these studies were conducted in the 2001 timeframe, they followed the FDA Draft Replacement Heart Valve Guidance of 1994 (CDRH, 1994). However, the animal studies are consistent with the current national and international standard ANSI/AAMI/ISO 5840:2005 Cardiovascular implants – Cardiac valve

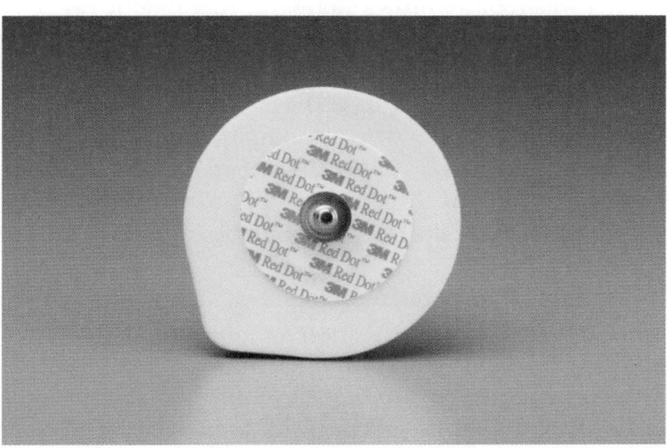

FIGURE III.2.6.4 Red Dot™ Foam ECG Monitoring Electrode (3M, St. Paul, MN, USA).

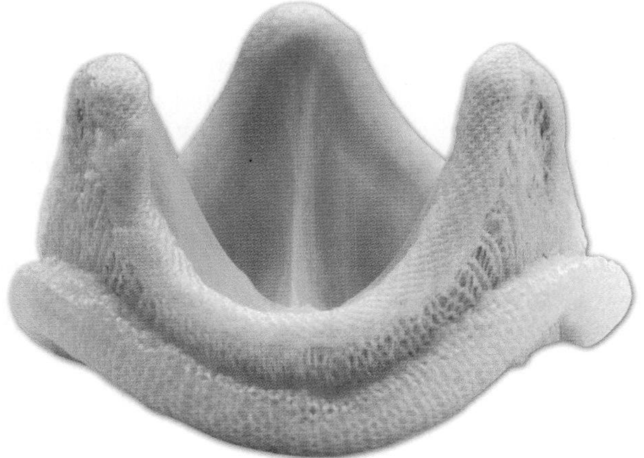

FIGURE III.2.6.5 Carpentier-Edwards S.A.V. Bioprosthesis Model 2650 Aortic Valve (Edwards Lifesciences, Irvine, CA, USA).

prostheses (AAMI, 2005b). Eleven juvenile sheep were implanted with the valve in the aortic or mitral position for five months. Similarly, 12 weanling rats and 12 juvenile rabbits were implanted subcutaneously with porcine aortic valve tissue for 90 days. In both studies, histopathology and calcification were assessed. In the sheep study, hemodynamic performance was also assessed.

Finally, before PMA submission in 2001, clinical trials were conducted. Again, the clinical trials followed the FDA Draft Replacement Heart Valve Guidance of 1994. These studies do not meet the current requirements of ISO 5840:2005 in terms of minimum number of enrolled patients. In one prospective study, the adverse event rates were recorded at one year for valve-related mortality, explants, reoperation, bleeding, endocarditis, hemolysis, nonstructural dysfunction, perivalvular leak, structural valve deterioration, thromboembolism, and valve thrombosis (Edwards, 2002). The time interval between premarket submission and FDA approval for this aortic tissue valve was 11 months. Eight amendments were required during the approval process (CDRH, 2002).

In the current version of ISO 5840:2005, a minimum of 150 tissue valve patients are followed for at least five years each. The complication rates for structural valve deterioration, thromboembolism, valve thrombosis, anticoagulant-related hemorrhage, endocarditis, valve dysfunction/paravalvular leak, and re-operation are recorded at one year. The complication rates for similar criteria are recorded at five years. In both cases, the observed complication rates must be within acceptable complication limits (AAMI, 2005b).

Permanent Skin Substitute. Design controls for Class III devices with which the FDA has little or no previous experience are the most difficult. As already discussed, requirements are to be determined by the manufacturer, regardless of accepted consensus standards. A device that meets these requirements must demonstrate safety and effectiveness in its FDA premarket submission. The threshold for this demonstration is made solely by the FDA.

When Advanced Tissue Sciences (ATS) submitted its request to market its Dermagraft® permanent skin substitute in December, 2000, it did not realize that 57 months would elapse before FDA approval. Dermagraft® is a cryopreserved dermal substitute composed of human fibroblasts, extracellular matrix, and a bioabsorbable scaffold (Figure III.2.6.6).

No FDA consensus standard existed in 2000, upon which permanent skin substitute submission requirements and testing could be based. In order to obtain the Indication for Use of "treatment of full-thickness diabetic foot ulcers greater than six weeks duration," ATS worked with the FDA on study protocols whose positive results would demonstrate safety and effectiveness. When ATS met with the FDA General and Plastic Surgery Devices Panel in January, 1998, its biocompatibility

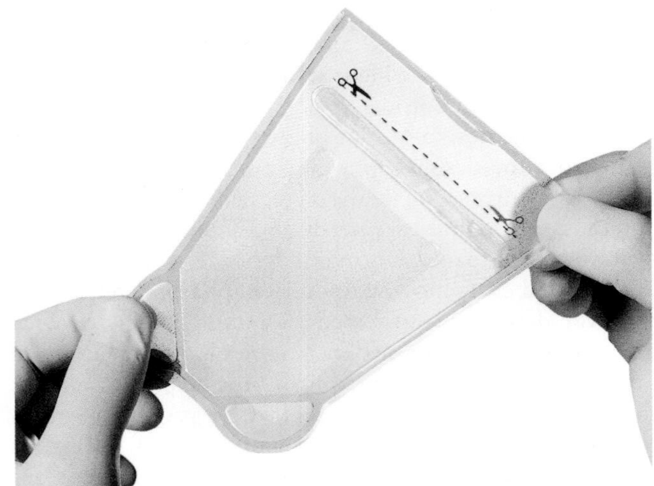

FIGURE III.2.6.6 Dermagraft® Skin Substitute (Advanced BioHealing, Westport, CT, USA).

results were accepted. The cytotoxicity, irritation, systemic toxicity, genotoxcity, and stability/shipping data were deemed "adequate." It was not stated in the Panel meeting transcript if any of these tests were based on ISO 10993 specifications.

The Panel voted 7-to-2 in favor of approval with a postmarket study, in order to resolve questions about data from two clinical trials. Specifically, only a subset of patients in the pivotal randomized clinical trial met the intent-to-treat hypothesis. While this subset was treated with grafts within a retrospectively determined narrower therapeutic range of cell viability, the therapeutic range was not statistically validated (CDRH, 1998a). After the Panel meeting, the FDA continued to raise concerns with ATS about the narrower therapeutic range issue (ATS, 1998). Moreover, the premarket approval manufacturing inspection for Dermagraft® resulted in a warning letter citing issues related to contaminants, and corrective and preventive action procedures (CDRH, 1998b). On June 10, 1998, premarket approval was denied (ATS, 1998).

ATS was required to validate this therapeutic range of cell viability with a second pivotal randomized clinical trial. In its second PMA submission, ATS demonstrated in 314 patients that 91% of Dermagraft® patients had wound closure at 12 weeks, compared to 78% of the control patients ($p = 0.044$). Further, it demonstrated that the ulcers of Dermagraft® patients significantly closed faster than ulcers in the control group ($p = 0.040$) (ATS, 2001).

Because of the extended three year delay between the Panel meeting and FDA approval in September, 2001 (CDRH, 2001), ATS accumulated more debt before shipping Dermagraft® to market. ATS filed for bankruptcy in 2002, and sold its assets to its joint venture partner Smith & Nephew. After Dermagraft® failed to gain FDA approval for an additional indication for venous leg ulcers in 2005, Smith & Nephew sold the technology to Advanced BioHealing in 2006 (Stuart, 2008).

From a general business perspective, being first-to-market is a competitive advantage for achieving the dominant market share. However, from a medical device industry perspective, being first-to-market with a Class III device means being first-to-deal with the FDA, which is an unpredictable process.

MANUFACTURING CONTROLS

Manufacturing controls partner with design controls to ensure product quality and safety. These controls oversee the manufacturing process and manufacturing facilities. At the front end of the manufacture of a specific device, purchasing controls are required for evaluating suppliers. Production and process controls are required to eliminate environmental and contaminant concerns. Inspection, measuring, and test equipment must be regularly calibrated and maintained. Manufacturing processes must be validated. Procedures must be in place for receiving acceptance activities, final acceptance activities, and nonconforming products. Corrective and preventive action, complaint file handling, installation, and servicing (where appropriate) procedures are also required (CDRH, 2003b).

For a PMA submission for a Class III device, a complete description of implemented manufacturing controls is required. While a manufacturing controls description is not required for a 510(k) submission, it is assumed that manufacturing controls are in place for the Class II device under review.

REGISTRATION, LISTING, AND INSPECTION

A manufacturer must register each of its facilities annually with the FDA, and list devices that are specified or manufactured at each facility. As of the fiscal year 2008, a fee must be paid during annual registration. For fiscal year 2009, this fee is $1851.

Registration, listing, design controls, and manufacturing controls are verified during an FDA inspection. Generally, a manufacturer is notified one week before the inspection occurs.

An initial inspection of a manufacturer's facility is conducted before its first product is market released. For Class III devices only, an inspection is generally conducted during the PMA submission process. (This is the inspection for which ATS received a warning letter.) However, if two PMAs are being reviewed during the same year period for a mature company, a second inspection may not occur. For both Class III and Class II devices, an inspector also returns to review a device recall, correction, removal or reported medical device complication. An inspector may also visit to follow up on observations from a previous inspection, which had been documented in a Form 483 warning letter.

INTELLECTUAL PROPERTY

While a few companies still patent *all* their new technologies, these days it is more cost effective to patent only technologies that create a barrier to market entry for competitors. A US patent can easily cost $25,000 to $30,000 from initial filing to issue. This cost includes attorney fees and US Patent Office (USPTO) fees. For the fiscal year 2009, the basic USPTO fees for filing a utility application, assuming at most 20 claims, total $1310. If other countries are chosen for protection, then the cost for issued patents escalates exponentially.

Three types of US patents exist: design; plant; and utility patents. A design patent protects the look or ornamentation of an article of manufacture. A plant patent protects the discovery and asexual reproduction of any distinct and new variety of plant. A utility patent protects a process, machine, article of manufacture or composition of matter.

In order for an invention to be protected as a utility patent, it must be utile, novel, and not obvious. Utility is not a high threshold to cross. An invention may be broadly interpreted to be useful, as long as it is not a law of nature or abstract idea. For an invention to be novel, it cannot be known or used by others anywhere in the world, more than one year before the inventor files his/her US patent application. Therefore, any previous public disclosure of the invention by another person, such as in an obscure foreign abstract or in an abandoned US patent, acts as prior art. If the inventor has publicly disclosed the invention, they must file for a US patent within one year of disclosure. Any similar prior art must not be obvious. In other words, a person having ordinary skill in the art of the invention must not consider the invention to be apparent. An example of obviousness would be the combination of two previous inventions.

The patent attorney representing the inventor writes broad claims for the invention, and negotiates with USPTO for allowance of these claims. For an issued patent, the invention is generally protected for 20 years from the date of US filing.

An issued patent does not guarantee that another company will not infringe on the patented invention. When this occurs, the company of the inventor is often forced to sue for infringement of specific patent claims. In anticipation of defending patent claims in the future, an inventor provides a written description, enabling disclosure, and best mode of the invention in the patent specification (the patent application text that excludes the claims) at the time of patent filing. The enabling disclosure refers to the requirement that a person having ordinary skill in the art of the invention would be taught by the patent how to make the invention. The best mode refers the best design of the invention being claimed at the time the patent application was filed. When an infringement lawsuit goes to trial in a US District Court, a jury decides the case. Either party may appeal the decision to the Court of Appeals for the Federal Circuit.

The Cypher® Stent US Patents

In 2005, in the US District Court of Delaware, Johnson and Johnson's Cordis's Cypher® stent (Figure III.2.6.1) was found to infringe on Boston Scientific's drug-eluting stent patent US 6,120,536. Claim 1 in patent '536 describes a drug-eluting stent with a coating containing a non-thrombogenic surface. Cordis appealed to the Court of Appeals for the Federal Circuit, stating that the Boston Scientific claim was obvious, in light of similar claims from Medtronic patent 5,545,208 (Figure III.2.6.7). In January, 2009, the Court of Appeals reversed the original judgment: "because the court erred as a matter of law in failing to hold the '536 patent to have been obvious" (United States Court of Appeals for the Federal Circuit, 2009).

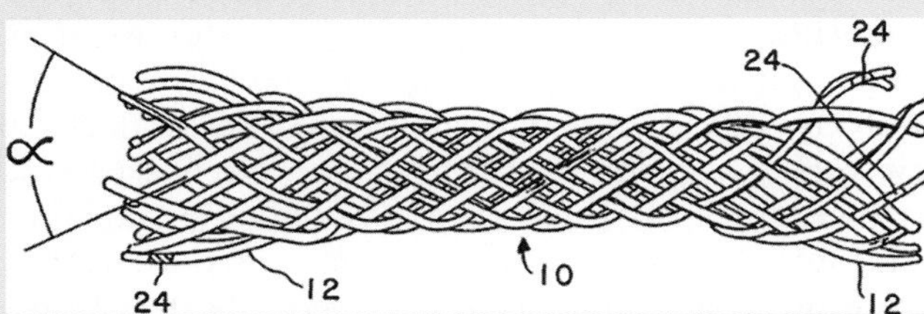

FIGURE III.2.6.7 An illustration of the stent covered by Medtronic's US patent 5,545,208 (Wolff and Hull, 1996).

Developing a medical device with key biomaterial components for the US market is not for the faint of heart. Especially with these devices, it becomes an art to navigate the regulatory roads towards FDA approval and Medicare reimbursement. In this chapter, we have discussed issues which are important for releasing a device to the US market: regulatory strategy; Medicare reimbursement; design controls; manufacturing controls; registration, listing, and inspection; and intellectual property. We leave it to others to discuss these changes and issues for market released products, such as US postmarket surveillance, medical device reports, corrective action and preventative action, and recalls (Baura, 2012). As this chapter goes to press in 2012, the FDA is considering major changes to the regulatory process.

FURTHER READING

For more detailed information on medical devices and US medical device regulation, please see:

Baura, G. D. (2002). *System Theory and Practical Applications of Biomedical Signals*. Hoboken, NJ: Wiley-IEEE Press.

Baura, G. D. (2008). *A Biosystems Approach to Industrial Patient Monitoring and Diagnostic Devices*. San Rafael, CA: Morgan Claypool.

Baura, G. D. (2011). *Medical Device Technologies: A Systems Based Overview Using Engineering Standards*. Waltham, MA: Elsevier Academic Press.

Baura, G. D. (2012). US Medical Device Regulation: An Introduction to Biomedical Product Development, Market Release, and Postmarket Surveillance. *Manuscript in preparation*.

BIBLIOGRAPHY

AAMI. (2003). *ANSI/AAMI/ISO 10993-1:2003 Biological Evaluation of Medical Devices – Part 1: Evaluation and Testing*.

AAMI. (2005a). *ANSI/AAMI EC12:2000/(R)2005 Disposable ECG Electrodes*. Arlington, VA.

AAMI. (2005b). *ANSI/AAMI/ISO 5840:2005 Cardiovascular Implants – Cardiac Valve Prostheses*.

ATS. (1998). *June 11 Press Release: FDA Responds to Premarket Approval Application*. San Diego, CA.

ATS. (2001). *Summary of Safety and Effectiveness Data: Dermagraft, P000036*. Rockville, FDA.

Baura, G. D. (2012). US Medical Device Regulation: An Introduction to Biomedical Product Development, Market Release, and Postmarket Surveillance. *Manuscript in preparation*.

CDRH. (1994). *Draft Replacement Heart Valve Guidance*. Rockville, FDA.

CDRH. (1997a). *Approval Order P940001*. Rockville, FDA.

CDRH. (1997b). *Design Control Guidance for Medical Device Manufacturers*. Rockville, FDA.

CDRH. (1998a). *General and Plastic Surgery Devices Panel Transcript: January 29, 1998*. Rockville, FDA.

CDRH. (1998b). *Advanced Tissue Sciences Warning Letter WL-21-8*. Rockville, FDA.

CDRH. (2001). *Approval Order P000036*. Rockville, FDA.

CDRH. (2002). *Approval Order P010041*. Rockville, FDA.

CDRH. (2003a). *Approval Order P020026*. Rockville, FDA.

CDRH. (2003b). *Quality System Information for Certain Premarket Application Reviews: Guidance for Industry and FDA Staff*. Rockville, FDA.

CDRH. (2007). *Guidance for Industry and FDA Staff: Recognition and Use of Consensus Standards*. Rockville, FDA.

CMS. (2002a). *Medicare Program; Changes to the Hospital Inpatient Prospective Payment Systems and Fiscal Year 2003 Rates. Final Rule, August 1, 2002*: Washington, DC, US Federal Register.

CMS. (2002b). *Medicare Program; Changes to the Hospital Inpatient Prospective Payment Systems and Fiscal Year 2003 Rates; Proposed Rule, May 9, 2002*. Washington, DC, US Federal Register.

CMS. (2009). *Medicare National Coverage Process*. Washington, DC, HHS.

Cordis. (2003). *Summary of Safety and Effectiveness Data: Cypher Sirolimus-eluting coronary stent: P020026*. Rockville, FDA.

Edwards. (2002). *Summary of Safety and Effectiveness Data: Carpentier-Edwards S.A.V. Bioprothesis, Model 2650 (Aortic), P010041*. Rockville, FDA.

ISO. (2007). *ISO 14971:2007 Medical Devices – Application of Risk Management to Medical Devices*. Geneva.

OCP. (2000). *Request For Designation RFD 0.006*: Rockville, *FDA*.

ODE. (2008). *Annual Report Fiscal Year 2006 and Fiscal Year 2007*: Rockville, FDA.

Stuart, M. (2008). The Rebirth of Dermagraft. *Start-up, 13*.

United States Court of Appeals for the Federal Circuit. (2009). Boston Scientific Scimed, Inc. and Boston Scientific Corporation v. Cordis Corporation and Johnson & Johnson, Inc. 2008–1073.

Wolff, R. G., & Hull, V. W. (1996). *Intralumenal Drug Eluting Prosthesis*. U.S. 5,545,208.

CHAPTER III.2.7 ETHICAL ISSUES IN BIOMATERIALS AND MEDICAL DEVICES

Taufiek Rajab[1], Andrew L. Rivard[2], Karen R. Wasiluk[2], Robert P. Gallegos[1], and Richard W. Bianco[2]
[1]Brigham and Women's Hospital, Boston, MA, USA
[2]Department of Surgery, University of Minnesota, Minneapolis, MN, USA

INTRODUCTION

Ethics is the branch of philosophy that addresses questions of morality. Biomedical ethics, therefore, narrows this focus onto the morality judgments that must be made in the field of medicine, and its allied subjects including biomaterial science. Biomaterial science has evolved at a tremendous pace, with innovations that have empowered clinicians and scientists beyond the imagination of their predecessors. This empowerment has resulted in new sets of ethical questions that were not posed previously. Questions of morality have direct relevance to the practical work of scientists and clinicians in the field as the influence of Institutional Review Boards and ethics committees continues to grow. Furthermore, there is also increasing public scrutiny of biomedical science in general. Therefore, it is important that engineers, scientists, and physicians become familiar with the relevant ethical issues, and actively participate in the debate that guides professional standards in the field of biomaterial science.

Medical device innovation often originates with an observation by a creative individual who encounters a problem in clinical practice and postulates a possible solution. To reach this solution, a linear pathway from the original idea through bench research, preclinical testing in an appropriate animal model, followed by clinical trials and eventual human use is intuitive. However, this linear pathway does not reflect the complexity of innovation. In reality, biomaterial development is recursive, with multiple iterations that may return to *in vitro* and *in vivo* testing multiple times, due to the highly regulated environment that is necessary to ensure safety for human patients. Each phase of this recursive process, including ideation, bench-work, animal research, and clinical research, has its own unique ethical requirements, which manifest in regulatory guidelines. This chapter will focus on the ethical principles that have been established and applied to the field of biomedical research and development. At the conclusion of this chapter, we will describe a pathway for the development of a new prosthetic heart valve which will be used to illustrate the application of biomedical ethics as it applies today.

PROTECTION OF PATIENTS

The paramount ethical principle in biomaterial research is the protection of patients. With this goal in mind, the US Food and Drug Administration (FDA) has established a set of regulatory guidelines to standardize studies related to medical devices. As a result, scientific investigation into the development of medical devices must be performed in accordance with defined best practice standards, which carry the weight of law. Preclinical studies must use Good Laboratory Practice (GLP), the manufacture of devices must follow Good Manufacturing Practice (GMP), and clinical trials must be in compliance with Good Clinical Practice (GCP). These Good Practice guidelines are quality systems concerned with organizational processes that are founded upon ethical principles and ensure accountability at every level from the designers, the manufacturers, and the investigators.

GOOD LABORATORY PRACTICE

Preclinical studies must be conducted according to good laboratory practice (GLP). GLP is outlined by the Organization for Economic Cooperation and Development (OECD), an international organization which countries may apply to become members. Therefore, data collected under GLP standards is accepted much more readily, as it is considered to be of exceptional quality as opposed to data obtained without adherence to GLP. Within the Code of Federal Regulations, the FDA has outlined rules for GLP. All preclinical studies in the United States that use animals must be performed using GLP prior to clinical studies in human patients. Results, either from preclinical studies in the United States that are not conducted in compliance with the FDA rules, or from studies performed in countries outside the United States that do not conform to GLP guidelines, may be inadmissible in support of an Investigational Device Exemption for use in human subjects. In a preclinical study conducted using GLP each involved entity has defined responsibilities.

The research sponsor is responsible for informing the contract laboratory that the preclinical study must be performed using GLP, and for monitoring contracted studies to make sure they are compliant. Management responsibilities include designating study directors, overseeing adequate quality assurance, ensuring that all articles are appropriately tested and that all personnel understand their roles. They also approve standard operating procedures and verify that resources are available. Management should also determine that the facility itself is GLP-compliant, document approval of the study plan, and verify that the study plan is available for inspection and quality assurance. The monitor may serve as the liaison between the sponsor and the study director, assess scientific and technical capabilities of sites to meet the requirements of the study, and be responsible for providing characterized test materials to the laboratory where the GLP study is to be performed. Study directors have overall responsibility for the performance of a study, including the interpretation, analysis, documentation, and reporting of the results. As such, the study director represents the single point of control for a study, and ensures that a defined study plan is followed. Raw data should be fully documented, and it must be accessible for inspection by the monitor. Documentation of studies performed using GLP should be complete enough to allow experiments to be accurately reproduced by another laboratory. Study directors must also be prepared to document any deviation from the study protocol, as well as to note unforeseen circumstances affecting the integrity and quality of the data, and must take corrective action if necessary. Principal investigators may be assigned to oversee parts of a study performed at a test site, and to ensure that all phases of the study are conducted in accordance with GLP and the study plan. They are also obliged to ensure that experimental data are recorded accurately, and that deviations from the study plan or operating procedures are communicated promptly to the study director. Principal investigators should also sign a compliance statement regarding the phase of the study conducted at their test site for the study director's final report. Study personnel have the responsibility of following all protocols, reporting any deviations to the study director, keep their training updated, and informing the study director of any unusual responses or unforeseen circumstances. Finally, the quality assurance unit reports to management separately from the study director and all other study personnel and is responsible for monitoring each study to make sure that it is compliant with GLP. They are responsible for inspecting the study at intervals that ensure the integrity of the data and maintaining signed records of each periodic inspection with any problems and corrective actions taken. The quality assurance unit should also review the final report to ensure that all reported methods and results accurately reflect the raw data.

In summary, GLP is a quality system that ensures the integrity and quality of data supporting the approval and eventual manufacture of a regulated medical device. It includes compliant organization and management, a defined study plan, defined standard operating procedures, suitable facilities and material, documentation and archiving of results, as well as an independent quality assurance program. Personnel at every level of the study must understand how their role in the research is conducted within GLP regulations.

GOOD MANUFACTURING PRACTICE

Good manufacturing practice (GMP), also known as current GMP, constitutes a series of quality system requirements that govern the methods, the facilities, and the controls used by manufacturers, processors, and packagers of medical devices intended for human use. GMP ensures that all finished products are safe and in compliance with the US Federal Food, Drug, and Cosmetic Act as mandated by the FDA. GMP regulations address all facets of manufacture of medical devices, but most GMP requirements allow manufacturers the flexibility to individually determine how they will comply with the required controls. Design controls ensure that specified design requirements are met. These controls include design validation, as well as an archived design history of all changes to the design. Document controls require manufacturers to designate an individual or individuals to review and approve all documents and subsequent changes. Identification and traceability controls mandate that each unit, lot or batch of finished devices must be identified with a control number. Production and process controls ensure that devices conform to their specifications by establishment of check-points where deviations from device specifications could occur as the result of the manufacturing process. Acceptance controls include inspections, tests, and other verifications to ensure that only accepted products are used, installed or distributed. Procedures for controlling nonconforming product must be established, including the documentation of corrective and preventative actions. Labeling and packaging controls ensure appropriate handling, storage, and distribution in order to prevent mix-ups, damage, contamination or other adverse effects on the products. All records need to be maintained in a location accessible to manufacturing officials and FDA inspectors. These include the device design, the design history records, the quality system records, and all complaint files. By mandating these controls, the GMP regulations aim to protect patients from defective medical devices, and ensure that high-quality products are manufactured.

GOOD CLINICAL PRACTICE

Once pre-clinical assessment has indicated that an investigational biomaterial or medical device is suitable for use in human patients, clinical trials must be designed according to good clinical practice (GCP) guidelines. GCP protects the rights of human subjects participating in clinical trials consistent with ethical principles, and ensures the integrity of clinical research data. Therefore, GCP defines a standard for clinical practices which encompasses the design, conduct, monitoring, termination, audit, analysis, documentation, and reporting of the studies and which ensures that the studies are scientifically and ethically sound, and that the clinical properties of the product or device under investigation are properly established (World Health Organization, 1995). All clinical trials should be conducted in accordance with the following ethical principles, based upon the Declaration of Helsinki. A clinical trial should not be initiated unless the anticipated benefits outweigh the risks. Furthermore, the rights, safety, and well-being of trial subjects take precedence over the interests of science and society. Clinical trials should be scientifically sound and described in a detailed protocol that is subject to review by an institutional review board or independent ethics committee. Informed consent must be obtained without coercion from each subject prior to participation. Each individual providing medical care to the trial subjects should be qualified to do so. Confidentiality of records that could identify subjects should be protected in accordance with applicable regulatory requirements. Investigational products should be manufactured, handled, and stored according to GMP, and used in accordance with the approved protocol. Quality assurance systems for every aspect of the clinical trial should be implemented. By conducting clinical research in accordance with these ethical principles, GCP provides a basis for ensuring that participants in clinical trials are not exposed to undue risk, while guaranteeing that data generated from the research are valid, accurate, and fully documented.

PROTECTION OF RESEARCH SUBJECTS

In addition to their moral obligation towards patients, scientists and clinicians involved in biomaterial research have an obligation to protect their research subjects. This includes animals as well as human subjects.

Animal Use and Care

Although biomaterials undergo rigorous procedures during *in vitro* testing, this process is limited in that the assessment cannot provide definitive evidence of human safety and efficacy. Animal models are designed for anatomical, physiological, and pathological similarity to humans. *In vivo* models can provide information about the performance of a biomaterial within the context of complex interrelated living systems which can be extrapolated to human patients. Therefore, the results of animal studies have direct implications for patient safety, as well as unforeseen catastrophic events. However, animals have a moral standing even with respect to human beings, as discussed specifically in Chapter II.3.6 of this volume). Consequently, research scientists have a moral obligation not only to human patients, but also to their animal subjects. When using animals for biomaterial research, humane care and ethical treatment of the animals is essential. Guidelines with standards for treatment of animals as research subjects were first published in 1963 by the Institute of Laboratory Animal Resources. Two basic ethical principles relate to the use of animals in biomaterial research. Firstly, experimentation on living animals should be reduced to the minimum necessary to obtain the necessary scientific information. Secondly, pain, distress, and other harm to the animals used should be reduced to the minimum necessary to obtain valid scientific data. Only biomaterials and medical devices exhibiting the most promise with regard to biocompatibility, function, and efficacy through rigorous *in vitro* laboratory testing should proceed to a preclinical *in vivo* assessment using an animal model. Preclinical assessment is required for FDA approval for a device to proceed into a clinical trial. The primary purpose of *in vivo* testing is to provide a critical assessment of safety in humans, with a secondary purpose of determining efficacy in the context of comparable technology. Given the growing public sensitivity to the use of animals in research, extra care must be taken to recognize and comply with the current definitions of humane animal use and care.

Activities involving live animals should be overseen and guided by the Animal Care and Use Committee of the sponsoring institution. Investigators also need to design studies in compliance with local laws and regulations. In the United States, FDA guidelines assume compliance with the Animal Welfare Act, which governs the use of animals in research (see discussion in Chapter II.3.6 of this volume), and also related guidelines of AAALAC (Association for Assessment and Accreditation of Laboratory Animal Care), and the National Institutes of Health (NIH) (Public Health Service (PHS) Policy on Humane Care and Use of Laboratory Animals, 2002). These guidelines are enforced by the Animal and Plant Health Inspection Service (APHIS) of the US Department of Agriculture. Specific species are covered by this Act (including non-human primates, dogs, cats, rabbits, and guinea pigs), while other species are specifically exempt (mice, rats, and birds specifically bred for research use, as well as horses and other livestock species used in agricultural research). The Act also provides specifications for the procurement of animals from licensed suppliers, as well as husbandry and veterinary care. Institutions that use species covered by these regulations must be registered by APHIS and submit annual reports that contain a categorization of all animals used by the institution in the previous year with respect to the level of discomfort the animals were subject to during the course of research procedures. Regulations also require all research protocols involving animals to be reviewed and approved by an Institutional Animal Care and Use Committee prior to the initiation of the proposed study. In addition to a detailed plan of the investigation, submitted research protocols must include a discussion of alternative approaches that are less harmful to the animals, as well as a justification for the number of animals required to address the purpose of the study without being unnecessarily duplicative. In addition to the federal guidelines, some state governments have mandated additional regulations. For example, at some state-supported institutions Institutional Animal Care and Use Committee reports and deliberations must be conducted in public (e.g., Florida, Massachusetts, North Carolina, and Washington). Different regulations may apply in other countries and jurisdictions.

Human Research Subjects

Origins of human research subjects' protection are found in the Nuremberg Code, which outlined standards developed for the Nuremberg Military Tribunal against which the human experimentation conducted by Nazi Germany was judged. The Nuremberg Code outlines many of the guiding principles inherent in ethical conduct of research involving human subjects. These include freely given informed consent without coercion, as well as the option for the human subject to withdraw at any time from the study. In 1964, the 18th World Medical Assembly in Helsinki, Finland adopted the Declaration of Helsinki, which made recommendations similar to those found in the Nuremberg Code. In the United States, the NIH issued Policies for the Protection of Human Subjects in 1966 that were based on the Declaration of Helsinki. In 1974, the National Commission for the Protection of Human Subjects of Biomedical and Behavioral Research was established and issued recommendations four years later that were titled the Belmont Report. This report identifies both basic ethical principles of research involving human subjects, as well as guidelines for conducting research in accordance with these principles. The main tenets defined by the Belmont Report are firstly respect for persons (recognition of autonomy and personal dignity of individuals, as well as special protection of those persons with diminished autonomy), secondly beneficence (the obligation to maximize anticipated benefits), thirdly non-maleficience (minimizing possible risks of harm), and finally justice (fair distribution of benefits and burdens of research). These tenets continue to form the basis of all acceptable conduct of research involving human subjects. Based on the Belmont Report, the Department of Health and Human Services (DHHS) codified regulations relating to protection of human subjects, and in 1991 the Federal Policy for the Protection of Human Subjects (or "Common Rule") was adopted. This policy was designed to unify the protection system for human subjects in all relevant federal agencies that conduct, support or otherwise regulate human subjects' research. Regulatory compliance is monitored through routine site visits and audits conducted by federal officials, as well as through the establishment of Institutional Review Boards (IRB).

Any biomaterial investigator involved in human subject research will be regulated by an IRB. The overarching purpose of an IRB is to ensure that all research is conducted with appropriate safeguards for human subjects, as mandated by the federal regulations. An IRB is a group made up of at least five individuals with diverse experience and expertise to professionally qualify the group to adequately review and monitor research activities involving human subjects that are commonly conducted by the institution. The IRB has the legal authority to approve, disprove or require modifications in the experimental design of research activities involving human subjects at the institution with which it is affiliated. Although each institution may have additional committees that review proposed research involving human subjects, no research may be initiated that has not been approved by the IRB. While the IRB overall must possess the scientific expertise to review a specific research design, the membership of the IRB must include at least one member with interests that are primarily scientific, and one member with interests that are primarily nonscientific. Members of an IRB should make every effort to avoid gender or race bias, and the member from the community at large should be a suitable representative. The number of members of an IRB can exceed the proscribed number of five, but it should not become so large that it can no longer function effectively. Investigators may be members of an IRB, but they may not participate in the review and approval of any project in which a potential conflict of interest could arise. The basic IRB review of a submitted protocol focuses on the following components: a risk/benefit analysis; adequacy of informed consent; appropriate selection of subjects; ongoing monitoring of subjects; mechanisms to ensure confidentiality; examination of additional safeguards; evaluation of incentives for participation; and plans for continuing review. The risk–benefit analysis aims to determine whether the risks to the subject are reasonable in relation to the benefits to the subject. Risk analysis is a formal procedure that includes hazard identification, evaluation of failure modes, risk estimation, risk evaluation, risk control, and continuous risk review. The risk analysis is based on the evaluated biomaterial or medical device, as well as the manufacturer's claims attributed to it. It also takes into account whether the device is novel or just an incremental modification of an existing device. The risks associated with the research are also distinguished from the risks of therapies that a subject would face, even if not participating in the research. It is important to determine that the probability and degree of harm associated with the research has been minimized as much as possible. When reviewing protocols involving medical devices, both the risks of the device and the risks associated with the procedure for using the device (e.g., the surgery to implant a heart valve) must be taken into consideration. Sponsors should make the initial risk assessment along with the study proposal. If the device study presents significant risk, the sponsor must submit an Investigational Device Exemption (IDE) to the FDA for approval. The sponsor must communicate the results of the IDE to the IRB. If the IRB finds that a device study presents non-significant risks only, the study may proceed without submission of an IDE. Adequacy of informed consent is another important consideration for the IRB. Human research subjects also need to be provided with an accurate and fair description of the anticipated risks, benefits, and possible discomforts. Federal regulations also require that the following information be provided to each subject as part of informed consent: a statement that the study involves research; explanations of the purposes; expected duration; descriptions of any planned procedures (including identification of procedures that are experimental); reasonably anticipated risks or discomforts; benefits to the subjects or to others; a disclosure of alternative treatments or procedures if they are advantageous to the subject; a statement describing confidentiality of records; contact information; and a statement that participation is voluntary and that refusal to participate will not result in the loss of benefits to which a subject is entitled. For research involving more than minimal risk there should be explanations of compensation or medical treatments available if injury occurs as a result of the research. The IRB also needs to ensure that the informed consent document presents the information to prospective subjects in language that can be easily understood, even by those with no medical background. Sometimes it is also appropriate that subjects be re-educated and consented periodically. The criteria for selection of subjects should take into consideration the requirements of the scientific design of the study, the susceptibility to risk, the potential benefits, and whether the selection of subjects from a proposed subject population is equitable (i.e., depending on the benefit of the study to the population as a whole, rather than disproportionately favoring one segment of the population). IRBs should also evaluate whether adequate precautions are taken to safeguard the privacy of information linked to individuals that will be recorded as part of the study. Subjects should be informed that federal officials have the right to inspect research records as part of their regulatory oversight. The initial IRB review of a protocol also includes an assessment of how often a research project should be re-evaluated by the IRB. Repeated monitoring is crucial, as preliminary data may indicate that the research design or the information presented to subjects must be changed, or even that the study should be terminated before the scheduled end date. Only after research has begun can the preliminary data be used to estimate the actual risk–benefit ratio.

Although institutions are ultimately responsible for ensuring that all regulatory requirements relating to human subject research are met, IRBs and investigators also bear part of that responsibility, and can be held accountable. At the level of the investigator, the most likely sources of noncompliance include failure to submit protocols or changes in approved protocols in a timely fashion to the IRB, and problems with obtaining informed consent. Often the IRB can resolve these deficiencies without jeopardizing the safety of the research subjects. However, research involving human subjects conducted by an investigator who has avoided or ignored an IRB cannot be allowed to proceed. Once discovered, the IRB and the institution must halt the research and take measures to correct any regulatory infractions. The fitness of the investigator to engage in research involving human subjects should also be evaluated. Noncompliance with regulations can also occur at the level of the IRB. This may arise from inadequate review of research protocols, not conducting a continuous review of research with a frequency that is commensurate with the degree of risk to subjects, failing to maintain adequate records, and consistently holding meetings without the majority of members present. Failure of an IRB to perform their responsibilities in accordance with DHHS regulations can be grounds for suspension or withdrawal of the institutional assurance. Finally, noncompliance at the institutional level is usually the result of a more systemic failure to meet their responsibilities. Institutions must provide appropriate staffing and support of an IRB so that it can function in accordance with the regulations, as well as ensure that investigators meet their obligations to the IRB as an integral part of their research using human subjects.

CONFLICTS OF INTEREST

In the process of developing a biomaterial, investigators may become involved with the business aspects of the technology. Examples include patents, company stock, royalties, a contract or other financial compensation from the medical device company, the academic institution or both. It is desirable to achieve a balance such that these rewards do not unduly influence data collection and analysis. Since research data are vulnerable to alteration by unethical participants, it is vital that principal investigators, companies, and academic institutions manage both perceived and real conflicts of interest to ensure accountability at all levels. Such potential conflicts of interest are common, and should ideally be identified and managed by the academic institution. A simple disclosure of financial interests may be enough to manage a conflict. However, in other circumstances the perceived conflict of interest may be so great that it is prudent to remove the conflicted individual from the research setting.

CONCLUSION

The emphasis of the above sections has been that compliance to ethical standards is crucial in facilitating good biomaterials and medical device research. Ethical investigators must conduct good science in the setting of GLP, GCP, and GMP standards. The rights of patients, as well as research subjects, need to be protected. Conflicts of interest should be disclosed and managed appropriately. These fundamental elements can then come together to form medical advances, which ultimately improve patient safety and healthcare outcomes. It is the full responsibility of all those contributing to the biomedical and device industry, and those advancing clinical healthcare, to abide by these widely accepted ethical standards.

A PRACTICAL EXAMPLE: THE DESIGN OF A NOVEL HEART VALVE

In this section, we will frame our discussion of the ethical principles in biomaterial research using the example of the process of designing a new heart valve. The development and ultimate production of a new heart valve begins with the incorporation of novel technology. This may either be a significant change in design or the inclusion of a new biomaterial that may improve the performance of the valve. There are many stages of assessment through which a device must be characterized before it is approved for use in human patients. In each stage there may be test results which prompt a return to further *in vitro* or *in vivo* testing, disrupting the linear progression of development. These recursive iterations are vital to ensuring maximum patient safety and efficacy of the valve. Following the development of a prototype valve, the first *in vitro* assessments are performed. This subjects the valve to conditions that exceed physiologic limits to determine durability, wear, and hemodynamic properties. All *in vitro* experiments must follow GLP. Such experiments should be thoroughly documented, such that an independent laboratory could reproduce the experiments and obtain results suitable for a direct comparison.

Although a device undergoes rigorous procedures during its *in vitro* testing this process is limited, in that the assessment cannot provide definitive evidence of safety and efficacy in a complex interrelated living system. The primary purpose of *in vivo* animal testing is to protect patients by providing a critical assessment of predicted safety in humans. A secondary purpose of *in vivo* animal testing is determining efficacy with regard to the claims of the developer in the context of comparable technology. In this way, *in vivo* animal testing serves to protect human patients from inferior medical devices. However, only devices exhibiting the most promise with regards to biocompatibility, function, and efficacy through rigorous *in vitro* laboratory testing should proceed to preclinical animal model testing. In this way, experimentation on living animals is reduced to the minimum necessary to obtain the necessary scientific information. Animal experimentation cannot be avoided, because preclinical assessment is required for FDA approval necessary for a device to proceed into a clinical trial. The preclinical assessment of a valve incorporates multivariable analysis with respect to hemodynamic performance, *in vivo* pathological effects, ease of handling, and results in the best approximation of use of the new heart valve within a clinical setting prior to proceeding into a Phase I clinical trial. There are two organizations in the United States that dictate regulatory requirements for preclinical assessments of cardiac devices: the United States Food and Drug Administration (FDA), and the International Standards Organization (ISO), in conjunction with the American National Standards Institute, Inc. (ANSI), and the Association for the Advancement of Medical Instrumentation (AAMI). ISO 5840 has defined regulatory standards governing preclinical studies, which are specific to cardiovascular valve prostheses (ANSI/AAMI/ISO 5840:2005). Although the current FDA guidelines are in draft form, they describe nonbinding recommendations for the manufacturing, preclinical *in vitro* and *in vivo* studies, clinical investigations, and labeling for heart valves that differ, augment, and complement the regulations defined by ISO 5840:2005. The previous FDA guidance was issued in 1994, before the FDA Good Guidance Practices were implemented in 2000, and before ISO 5840:2005 was published. The 1994 draft was withdrawn in 2005 (70 FR 824, January 5, 2005), and the FDA is in the process of drafting a new guidance. The FDA has recognized only the 2005 version of ISO 5840 (71 FR 16313, March 31, 2006). ISO standards are reviewed every five years and, if necessary, revised to reflect changes in the industry for which the standards define regulatory compliance. One major difference in the current ISO 5840:2005 guidelines from its previous version is the emphasis that is now placed on the use of risk analysis, a formal procedure by which preclinical studies are designed to mitigate to the extent possible the probability and degree of harm caused by the use of a device. Specific steps are outlined in ISO 5840:2005 to perform a risk analysis and include: hazard identification; associated failure modes; risk estimation; risk evaluation; risk control; and risk review. Risk analyses must be updated throughout the characterization of a heart valve through all phases of development. This encourages the manufacturer to strive for continued improvements in design, as well as to ensure safety and efficacy with less reliance on years of clinical assessment for verification of effectiveness. Within the regulatory framework mandated by the FDA and the ISO 5840, there are criteria that must be incorporated into the design and assessment of a new heart valve as follows:

(1) All studies must be completed using Good Laboratory Practice guidelines.
(2) Valves to be assessed must be of clinical quality.
(3) Studies must be conducted using a control valve that is of similar design and has been approved for clinical use. The only exception to this is a proof-of-concept study in which a new heart valve is so different from existing valves that have been approved for clinical use that a suitable control valve cannot be identified.
(4) *In vivo* studies must be designed to incorporate site-specific implantation of the new heart valve in both mitral and aortic positions. Of note, it is important to standardize as many of the variables as possible, including the surgeon, the device size, and the animal demographics.
(5) Ideally, the limitation of the number of implants should not be determined by the minimum regulatory requirements, but should be reflected by the study length.
(6) There should be a complete pathological examination of all animals within a particular protocol, with two independent assessments, including one performed by a veterinary pathologist, to distinguish complications caused by the model from those caused by the new heart valve.

One regulation not mandated by ISO 5840 is the choice of a definitive *in vivo* model used for preclinical assessment of heart valves, as there are a variety of *in vivo* models that are relevant. These include, but are not limited to, sheep, pigs, calves, dogs, and non-human primates. Currently, there is no ideal model for preclinical assessment of heart valves, but existing models can be improved upon by incorporating additional outcomes, for example, including site-specific testing. Other potential improvements include the increased use of focused screening models, such as isolated hearts, the use of risk analysis for determining not only the end points, but also the choice of controls and power of the study, and increased communication with regards to negative findings. Compliant preclinical assessment provides not only an accurate correlation of *in vitro* and *in vivo* performance of a device, but also provides a basis for clinical trials.

Once preclinical assessment has indicated that the new valve is suitable for use in human patients, clinical trials may be designed to evaluate the safety and efficacy with regard to toxicity, efficacy, and field conditions. Complete documentation must cover therapeutic indications, contraindications, safety precautions, and safety information. Given the broad public health significance, it is critical for the clinical trial process and resultant data to conform to rigorous ethical standards. Therefore, investigators need to submit a study protocol for IRB review. This protocol will include a risk–benefit analysis, adequacy of informed consent, appropriate selection of subjects, ongoing monitoring of subjects, mechanisms to ensure confidentiality, examination of additional safeguards, evaluation of incentives for participation, and plans for continuing review. The sponsor must also submit an Investigational Device Exemption (IDE) to the FDA for approval. Subsequently, the clinical trial may be conducted according to the general principles described above. Often, during the development process, preclinical assessment, and clinical research, most cardiac valves will have undergone only short-term safety and efficacy assessment in a limited number of carefully selected human subjects before being approved for marketing.

In contrast, post-marketing safety surveillance is utilized in an effort to detect rare or long-term adverse effects not easily recognized during preclinical testing by examining a much larger patient population over a prolonged period of time. Although such events may be rare, their potentially catastrophic consequences cannot be ignored. With better documentation of such problems, future device iterations can be redesigned to attempt to eliminate failures.

BIBLIOGRAPHY

Code of Federal Regulations. (June 23, 2005). Title 45 Public Welfare, Part 46 Protection of Human subjects. Revised.

Guidelines for the Conduct of Research Involving Human Subjects at the National Institutes of Health. (August 2004). 5th printing.

International Conference on Harmonization. (1996). Good Clinical Practice: Consolidated Guidance. *ICH Guidelines E6*, 8–9.

The National Commission for the Protection of Human Subjects of Biomedical and Behavioral Research. (1979). *The Belmont Report on Ethical Principles and Guidelines for the Protection of Human Subjects of Research, April 18*.

UNDP/World Bank/WHO/TDR. (2000). *Good Laboratory Practice (GLP) Handbook: Quality practices for regulated non-clinical research and development*.

US Government Printing Office. (1949). *Trials of War Criminals before the Nuremberg Military Tribunals under Control Council Law No. 10*. (Vol. 2). Washington, DC: US Government Printing Office.

World Health Organization (1995). Guidelines for Good Clinical Practice (GCP) for Trials on Pharmaceutical Products. *WHO Technical Report Series*. No. 850, Annex 3.

World Medical Association (1964). *World Medical Association (WMA) Declaration of Helsinki on Ethical Principles for Medical Research Involving Human Subjects*. Helsinki, Finland: Adopted by the 18th WMA General Assembly. June 1964.

CHAPTER III.2.8 LEGAL ASPECTS OF BIOMATERIALS

Jay P. Mayesh and Angela R. Vicari
Kaye Scholer, New York, USA

INTRODUCTION

Students of biomaterials engineering know that no product lasts forever, and that implantable medical devices have unwanted side-effects. In today's litigious society, these factors often transform patients and device manufacturers into warring parties in always unwelcome and sometimes financially disastrous products liability litigation over the safety of medical devices. Products liability law imposes legal responsibility on manufacturers of products (ladders, cars, and medical devices, to name a few), as well as other companies involved in the "stream of commerce," such as wholesalers, distributors, and retailers, for injuries incurred by the consumer.

Products liability falls within the area of civil law called *torts*. A tort is simply a wrongful act that may give rise to a lawsuit, for example an auto negligence lawsuit or professional malpractice claim against a doctor or lawyer. If the plaintiff is successful, a tort lawsuit results in an award of money to the plaintiff called *damages*.

Products liability plaintiffs typically rely on four theories of liability. First, they claim that the manufacturer was *negligent*, meaning that the manufacturer failed to use reasonable care in designing and manufacturing the product. Second, plaintiffs claim that the manufacturer breached legally enforceable promises, called *warranties*, because the product did not meet recognized performance expectations or have the qualities expected of products of its type. Third, plaintiffs sue under *strict liability*, where a manufacturer is held responsible for a product that was unreasonably dangerous to the consumer or carried inadequate warnings, regardless of the degree of care exercised or any promises made by the manufacturer. Strict liability rests on two assumptions about law and economics – first, that imposing liability on a manufacturer, even without a showing of carelessness on the manufacturer's part, is fair because the manufacturer is best able to discover and correct defects in products before they cause harm, and second, that in the event of personal injury attributable to a product, the manufacturer can afford to compensate the injured party, add the cost of injury to the product and, if necessary, raise the price of the product to recover the cost. Finally, plaintiffs sue device manufacturers for *consumer fraud*, which involves a wide range of improper practices with respect to the advertising, marketing, and sale of medical devices. Many consumer fraud claims allege that a device is not performing in the way it was represented or advertised to perform (because of false advertising, product defects or other reasons). To avoid liability under each of these legal theories, device manufacturers must design, manufacture, and sell products that are reasonably safe, they must disclose written warnings to physicians (and sometimes to patients) about risks associated with the products, and they must advertise, market, and sell products in a truthful manner.

This chapter describes "mass tort" products liability litigation involving implantable medical devices – that is, personal injury litigation implicating thousands of individuals making a claim for an injury as a result of an alleged failure of a medical device. Although the individual injuries vary, most if not all claims arise out of the same mode of failure, insufficient warning or a false claim. Faced with numbers of cases involving hundreds of millions of dollars, the parties often attempt to consolidate mass tort claims in order to streamline pretrial procedures, either by class action or by consolidated litigation, in which all cases are assigned to one judge. In the federal courts, such consolidated litigation is termed "multidistrict litigation" or "MDL." Many states have their own analogs to MDL.

Although medical device mass torts raise complicated scientific and medical issues, in the United States legal system lay jurors, untrained in medical science, are the arbiters of reasonable safety, adequate disclosure, and causation. The vagaries of the jury system always create a degree of risk and uncertainty for litigants. That risk is compounded in medical device litigation when jurors are asked to discern which of many competing medical and scientific theories is more credible. Outcomes become even more unpredictable in cases involving unsettled or evolving areas of science.

To help illustrate the subtle and complex world of mass tort litigation, the six largest subjects of medical device litigation are described below, followed by a discussion of some of the obstacles faced by litigants in prosecuting and defending these claims, including the problems posed by science in the courtroom.

INTRAUTERINE DEVICES

The first mass tort litigation involving a medical device arose out of injuries to women who received intrauterine devices ("IUDs"). In the early 1970s, the IUD was presented to the public as a safe, effective alternative to oral contraceptives. In the following decades, serious reproductive health risks associated with IUDs ensued, prompting women to sue the manufacturers.

A. H. Robins, the manufacturer of the Dalkon Shield, bore the brunt of IUD litigation. Because the Dalkon Shield predated the Medical Device Amendments of 1976 (see later discussion), no requirements for premarket testing were in place. As described in the case *Tetuan v. A. H. Robins Co.* (241 Kan. 441 [1987]), A. H. Robins was anxious to make inroads in the market and stave off competitors, and thus began marketing the device in 1971 having performed few safety and efficacy studies. While internal corporate documents reflected the company's concern with the paucity of information on the product, A. H. Robins nevertheless publicly touted the benefits of the Dalkon Shield, distributing product cards that claimed the Dalkon Shield was superior to other forms of contraception, and placing advertisements in popular magazines. These efforts earned the Dalkon Shield a dominant position in the IUD market.

As further detailed in *Tetuan v. A. H. Robins Co*, shortly after the device came onto the market the manufacturer began to receive adverse incident reports from doctors about health problems believed to have been induced by the Dalkon Shield, including septic abortions and an increased incidence of pelvic inflammatory disease ("PID"). The Dalkon Shield, like other IUDs, has a string that descends through the cervix from the uterus to allow the user to ensure the device is in place. Before marketing the Dalkon Shield, the manufacturer had knowledge that the string had a "wicking" tendency, meaning it could transport fluid by capillary action into the uterus and introduce bacteria into an otherwise sterile environment. The string was surrounded by a nylon-6 sheath, but the sheath biodegraded in the moist body environment, permitting wicking. The wicking was alleged to be the cause of the comparatively high rate of infections and injuries in women who used the Dalkon Shield.

Between 1970 and 1974, approximately 2.2 million women in the United States used the Dalkon Shield (*In re Northern Dist. of Cal., Dalkon Shield IUD Prods. Liability Litigation*, 693 F.2d 847, 848 [9th Cir. 1982]). The device had been linked to 16 deaths and 25 miscarriages by the middle of 1975 (H.R. Rep. No. 94-853, p. 8 [1976]). By the end of 1975, the lawsuits against A. H. Robins were sufficiently numerous that they were consolidated before a district court in Kansas for pretrial purposes (*In re A. H. Robins Co., Inc.*, "*Dalkon Shield Products Liability Litigation*," 406 F. Supp. 540 [J.P.M.L. 1975]), and the number of lawsuits continued to multiply. By early 1976, more than 500 lawsuits seeking compensatory and punitive damages totaling more than $400 million had been filed (H.R. Rep. No. 94-853, p. 8 [1976]). Although A. H. Robins was successful in early suits, once certain damaging corporate documents became public the tide turned, and plaintiffs began to win large verdicts. In May of 1985, in the pivotal *Tetuan* case, a Kansas jury awarded $9.25 million in damages. Shortly after the *Tetuan* verdict, A. H. Robins filed for bankruptcy reorganization (Vairo, 1992). Under the reorganization plan a trust fund of $2.475 billion was established to pay outstanding claims against the manufacturer. The bulk of the fund came from another healthcare company that agreed to make the contribution in order to acquire A. H. Robins (Vairo, 1992).

IUD plaintiffs have claimed injuries such as uterine perforations, infections, ectopic pregnancies, spontaneous abortions, fetal injuries and birth defects, sterility, and hysterectomies (*In re Northern Dist. of Cal., Dalkon Shield IUD Prods. Liability Litigation*, 693 F.2d 847 [9th Cir. 1982]). The theories asserted included failure to warn, unsafe design, breach of warranty, and fraud. Plaintiffs prevailed at trial and were able to negotiate large settlements because the defenses that manufacturers typically rely on in medical device litigation were less likely to avail the IUD defendants. Sometimes medical device manufacturers argue that their device cannot be linked with the type of injuries suffered by the plaintiffs. In IUD cases, causation had been a less defensible issue as compared to other medical device litigation because medical experts generally agree on the type of injuries IUDs can produce (Vairo, 1992). Although an IUD manufacturer may successfully argue that the plaintiff's injury had another cause, such as a sexually transmitted disease, the general causative correlation between IUDs and certain injuries is not in question.

In light of evidence that they were not sufficiently candid about health problems associated with their devices, IUD manufacturers also had difficulty proving they satisfied their duty to warn physicians of the risks of IUD use. For instance, in *Nelson v. Dalkon Shield Claimants Trust* (1994 WL 255392 [D.N.H. June 8, 1994]) the court refused to dismiss the plaintiff's claims because the evidence showed that A. H. Robins failed to issue public warnings about the risks associated with the Dalkon Shield's string. Statutes of limitation, which provide an injured party with a fixed amount of time to file suit after an injury occurs, frequently have provided IUD manufacturers with their strongest defense. Injuries caused by IUDs may not manifest themselves for many years, and an even greater amount of time may pass before the user becomes aware that the IUD may have caused the injury.

The Dalkon Shield was ultimately linked to "thousands of serious injuries to otherwise healthy women" (Vladeck, 2005). By October 1984, the manufacturer had settled or litigated approximately 7,700 Dalkon Shield cases (Sobol, 1991).

PEDICLE SCREWS

Pedicle screws are bone screws that are implanted in the pedicles of the spine and are used to anchor a variety of stabilizing hardware. Litigation involving pedicle screws was sparked by a December 1993 exposé on the ABC news program, 20/20. The story publicized allegations that orthopedic screws, which had received FDA approval only for use in the long bones of the arms and legs, were being implanted in the pedicles, a procedure associated with a high rate of complications (Hoyle and Madeira, 2001). The story spawned a multitude of lawsuits, which were consolidated before a Pennsylvania district court for all pretrial procedures (*In re Orthopedic Bone Screw Products Liability Litigation*, 1998 WL 118060 [E.D. Pa. Jan. 12, 1988]).

Plaintiffs alleged several variants of device failure. They complained that bone screws implanted in their spines broke, loosened, corroded or were malpositioned, causing, among other injuries, pseudarthrosis, neurogenic bladder, and arachnoiditis. In addition to device failure, plaintiffs also alleged that manufacturers committed fraud-on-the-FDA by seeking FDA approval for orthopedic screw use only in the long bones of the arms and legs, but then promoting their use in spinal surgery. Ultimately, the United States Supreme Court rejected this theory of liability in a case called *Buckman v. Plaintiffs' Legal Committee* (531 U.S. 341 [2001]), in which it held that such fraud-on-the-FDA claims are not actionable by plaintiffs in product liability cases. The Supreme Court reasoned that if courts were permitted to find insufficient disclosures made by a manufacturer to the FDA, even though those same disclosures were deemed appropriate by the FDA, the FDAs ability to protect the public would be frustrated. Plaintiffs also attempted, unsuccessfully, to impose liability on doctors and medical associations, alleging they conspired with manufacturers to commit fraud by conducting medical seminars that promoted bone screws for unapproved uses.

In addition, plaintiffs argued that manufacturers should be liable for failing to warn of the risks inherent in pedicle screws. Many failure-to-warn claims involving pedicle screws were dismissed by courts under the "learned intermediary" doctrine, which provides that, in the case of most medical devices, a manufacturer has discharged its duty to warn once it has informed the medical community (not the patient) of the risks associated with the device. Dismissals under this doctrine occur when the implanting surgeon testifies that he was aware of the risks through sources entirely independent of the manufacturer's product literature.

In addition, plaintiffs claimed, usually without success, that the pedicle screws were defective. In *Toll v. Smith & Nephew Richards, Inc.*, 1998 WL 398062 [E.D. La. July 14, 1998]), for instance, the court dismissed the case because the plaintiff could not make the required showing that an alternative design capable of preventing the alleged injury existed. Courts have found that lack of FDA approval for use of orthopedic screws in the spine does not strengthen plaintiffs' claims of defect, and that "off-label" use of a medical device (that is, use for a purpose other than which it is FDA-approved) is not prohibited. Finally, since spinal surgery can have complications regardless of the instrumentation used, plaintiffs' claims have frequently failed because they could not show their injuries were not complications of surgery unrelated to the use of the screws.

As the discussion indicates, many claims against pedicle screw manufacturers were dismissed by courts prior to trial. However, a handful of cases reached juries, with mixed results. A Louisiana jury rendered a verdict against one manufacturer in the amount of $318,000 after finding the pedicle screw was unreasonably dangerous (Anonymous, 1995a), and a Texas jury awarded the plaintiff $451,000 against one manufacturer for misrepresenting the safe and effective use of its pedicle screws (Anonymous, 1998b). On the other hand, a Pennsylvania jury rendered a verdict for the defendant manufacturer (Anonymous, 1995b), and a Tennessee trial ended with a hung jury (Anonymous, 1998c).

The financial strain of the massive litigation prompted one manufacturer, AcroMed, to settle and agree to contribute $100 million and the proceeds of its insurance policies to a settlement fund. In exchange, the manufacturer, as well as its distributors and physicians and hospitals that used its products, were released from suit under any products liability theory (*In re Orthopedic Bone Screw Products Liability Litigation*, 176 F.R.D. 158 [E.D. Pa. 1997]). Approval of this $100 million settlement was based, in part, on an expert economist's opinion that AcroMed had an estimated value of about $104 million, and that based on AcroMed's financial condition, limited insurance coverage, defense costs, and adverse verdict potential, the $100 million settlement was "at the outer boundary of what AcroMed can afford to pay" (*In re Orthopedic Bone Screw Products Liability Litigation*, 176 F.R.D. at 170). Interestingly, within a few weeks after the settlement was declared final, AcroMed was sold for $325 million, more than three times the expert's estimated value of $104 million.

The death knell of the pedicle screw litigation came in July 1998 when the FDA approved bone screws as safe and effective for spinal surgery.

SILICONE BREAST IMPLANTS

Since the early 1960s, an estimated two million silicone breast implants have been implanted in women, for both breast augmentation and reconstruction. Silicone breast implants are made out of a poly(dimethyl siloxane) ("PDMS") elastomer shell, to which fumed amorphous silica is added, encasing PDMS gel or saline (see Chapters I.2.2.B and II.5.18). Some implants have been manufactured with a thin layer of polyurethane foam covering the elastomer shell; this was thought to decrease the formation of capsular contracture (hardening of the scar capsule surrounding the implant with resultant disfigurement), one of the common local complications of breast implants.

Physicians have long recognized that silicone breast implants (including those filled with saline) occasionally cause local complications such as capsular contracture, and that implants can rupture, often necessitating surgery to remove and replace the implants. However, it was not until the early 1990s, spurred on by a television exposé and the publication of several case reports of women with implants who developed autoimmune diseases, that breast implants exploded into mass tort litigation. The FDA, having classified silicone breast implants as Class III devices that "present a potential unreasonable risk of illness or injury" in 1988, enforced the requirement that manufacturers collect and provide safety data on the devices. At this point there were no controlled epidemiological studies exploring the relationship between breast implants and systemic disease. Citing the lack of safety data, the FDA imposed a moratorium on the use of silicone breast implants, except in clinical studies.

Meanwhile, tens of thousands of implant recipients sued the implant and raw material manufacturers, claiming that they had developed autoimmune diseases, such as lupus, scleroderma, and fibromyalgia. To establish that the implants caused illness, plaintiffs' lawyers relied on uncontrolled case reports of autoimmune disease in women with implants, and on anecdotal testimony by treating physicians to persuade juries that the implants caused illness. They also relied on early case reports of autoimmune disease following direct injections of silicone liquid and paraffin into the breasts of Japanese women. Although none of these sources of evidence scientifically established that breast implants cause autoimmune disease, several early plaintiffs won large verdicts.

After publication of the first well-controlled epidemiological studies refuting an association between implants and recognized autoimmune diseases, many plaintiffs modified their injury claims. They alleged that they had developed "atypical" autoimmune diseases with signs and symptoms that would not have been looked for in the studies that tracked the classic autoimmune diseases.

Plaintiffs premised liability on assertions that manufacturers defectively designed the implants and failed to warn physicians and patients of health risks. A central allegation in plaintiffs' lawsuits was that silicone has myriad ill effects on the immune system. Plaintiffs claimed that, as a result of rupture of the implants and "gel bleed" of low molecular weight PDMS through the elastomer shell, silicone microdroplets migrate to remote organs where the silicone causes a chronic inflammatory response and the development of silicone granulomas (Plaintiffs' Submission and Proposed Findings, 1997).

Moreover, they claimed that silicone is an antigen capable of eliciting an immunologic response in the body, and that silicone acts as an adjuvant, heightening the body's immunologic response to other substances. Plaintiffs claimed that silicone, through these mechanisms, exacerbates existing autoimmune diseases, and causes classic and atypical autoimmune diseases in exposed women.

In addition to citing gel bleed as a factor in causation, plaintiffs offered several theories that implicate biodegradation of silicone. They alleged that silicone, migrating throughout the body, is picked up by phagocytes and transformed into crystalline silica which, they claimed, was immunogenic and caused connective-tissue disorders. Moreover, they alleged that the silicone biodegrades into silanols, relying on *in vivo* NMR spectroscopy studies that purported to identify silicone and its metabolic byproducts in the blood and livers of exposed women. Other scientists were unable to replicate these NMR findings (Macdonald et al., 1995; Mayesh and O'Hea, 1997).

In a medical science atmosphere of anecdotal case reports and uncertain elaborate scientific theories, even though no epidemiological studies supported causation, the industry was prepared to spend approximately $4 billion to settle claims globally. The plaintiffs demanded more money, and settlement fell apart. Faced with potential liability and enormous defense costs, Dow Corning, the largest manufacturer of breast implants, was forced into bankruptcy.

Over time, scientific evidence began to mount against causation. Each epidemiology study from leading institutions such as the Mayo Clinic, Harvard Medical School, Johns Hopkins University, the University of Michigan, and the University of California failed to show any association between silicone breast implants and any classic or atypical autoimmune disease. Nonetheless, occasionally manufacturers were still subjected to substantial jury verdicts.

In 1996, the coordinating MDL judge, before whom all lawsuits filed in federal court were consolidated for pretrial proceeding, appointed a National Science Panel of four impartial scientific experts. An immunologist, epidemiologist, toxicologist, and rheumatologist were charged to evaluate the scientific data in relation to connective tissue diseases and immunologic dysfunction (National Science Panel, 1998). These experts were instructed to "review and critique the scientific literature pertaining to the possibility of a causal association between silicone breast implants and connective tissue diseases, related signs and symptoms, and immune system dysfunction."

On three occasions in 1996 and 1997, the panel of four court-appointed scientists heard testimony from expert witnesses chosen by the lawyers for plaintiffs and for the manufacturers. The panel received more than 2000 documents from counsel, and the panel members performed their own literature searches. The panel concluded in November 1998 that available data did not support a connection between silicone breast implants and any defined connective tissue disease or other autoimmune or rheumatic condition (National Science Panel, 1998).

IMPLANTABLE CARDIAC DEFIBRILLATORS

Implantable cardiac defibrillators ("ICD") are small, battery-powered devices used to treat a variety of potentially fatal cardiac arrhythmias by delivering electrical therapy to the heart. They do so through small wires called "leads" that on one end are attached to the ICD, and on the other end are attached directly to the patient's heart muscle through a coronary vein. If electrodes on the leads detect that the patient's heart is out of rhythm, the ICD sends an electric shock to the heart muscle through the leads in order to correct the problem (see Chapter II.5.3.C).

Since the first human use of an ICD was reported in 1980, use of these devices has expanded dramatically, and their design has undergone remarkable change. Modern ICDs are implanted transvenously, provide physiologic pacing in up to three cardiac chambers, perform a number of automated functions and self-checks, and can deliver pacing or high voltage therapies to treat life-threatening arrhythmias. A pacemaker-dependent ICD patient will receive more than 100 million paced beats over the device's lifetime.

As with many evolving technologies, advances brought with them new safety risks. The risks spawned two major products liability litigations involving ICDs, both of which centered on the manufacturer's failure to warn of known problems with the product.

In February 2002, Guidant Corporation identified certain electrical flaws with one of its ICD products. As a result, the company made manufacturing changes to reduce the short circuit risk. However, the company did not notify physicians or patients of these failures or of the manufacturing changes. It continued to sell ICDs manufactured using the older process while it phased in the newer version. The company justified its decision to continue selling older devices and not notifying physicians of the new modifications on the grounds that the overall failure rate of the device, even including these events, was extremely low and because the device was performing better than its design specifications. It also concluded that the risk of explanting the device was greater than the risk of device malfunction.

The company eventually announced that there were problems with its ICD after the death of a 21-year-old college student with a genetic heart disease. Guidant acknowledged that the student's ICD short-circuited, and subsequently told his doctors that it was aware of 25 other cases in which the ICD had been affected by the same flaw. Following this disclosure to the student's physicians, Guidant was contacted about a *New York Times* article that was set to run entitled "Maker of Heart Devices Kept Flaws from Doctors" (Meier, 2005). The company's first public announcement came hours before the *New York Times* published its article – more than three years after it had allegedly become aware of the problem. By June 2005, there had been two deaths reported to the FDA suspected to be associated with this malfunction. By October 2005, there were 21 clinical failures, including three patient deaths, worldwide.

Over the next several months, Guidant announced several additional physician notifications covering a significant number of similar products. The company also began issuing a series of recalls for what amounted to over 100,000 devices in the United States. The company had different problems with different devices, some more serious than others, involving electrical shorts, as well as other defects, including memory errors and premature battery depletion.

A spate of personal injury lawsuits followed, and a multidistrict litigation commenced in November 2005. Unlike many of the other large mass torts discussed in this chapter, the parties settled the litigation soon after. In July 2007, the parties entered into an initial $195 million proposed settlement. After more individual settlement claims than anticipated were submitted, the parties commenced renegotiation, resulting in an increase in the settlement fund to $240 million. The agreement covered 8550 patient claims, including all of those that had been consolidated in the multidistrict litigation, as well as other filed and unfiled claims throughout the United States.

Guidant was not the only ICD manufacturer to face lawsuits. In 2006, patients with ICDs utilizing Medtronic's Sprint Fidelis leads as a component began to suffer painful shocks. An investigation by a physician at the Minneapolis Heart Institute concluded that the shocks were caused by fractures in the leads, and that the leads were failing at a significantly higher rate than other leads, a sentiment echoed by another physician at Cornell University Medical Center in New York (*In re Medtronic, Inc. Sprint Fidelis Leads Products Liability Litigation*, 592 F. Supp. 2d 1147 [D. Minn. 2009]). Plaintiffs claim that upon being confronted with this information, Medtronic undertook a campaign to defend its leads, even though it knew that they were unsafe. As part of that campaign, Medtronic purportedly delayed filing "adverse event reports" with the FDA concerning failures of the leads.

In May 2007, Medtronic filed a supplemental application to the FDA containing design and manufacturing changes to the Sprint Fidelis leads. According to plaintiffs, Medtronic filed this supplement in order to correct defects endemic to the leads. Yet, Medtronic did not advise the FDA that it was filing the premarket approval ("PMA") supplement because of the lead failures, instead merely informing it that the proposed changes were intended to make the leads more "robust." The FDA approved Medtronic's PMA supplement in July 2007, but previously manufactured (and allegedly defective) Sprint Fidelis leads continued to be shipped to hospitals and implanted into patients.

On September 10, 2007, Medtronic filed more than 120 adverse event reports concerning Sprint Fidelis leads. On October 7, 2007, Medtronic suspended sales of the leads but did not inform the FDA, doctors or the public of its decision, and they continued to be implanted into patients. On October 15, 2007, Medtronic recalled the Sprint Fidelis leads, and the FDA issued a "Class I" recall shortly thereafter, the most serious type of medical device recall. At the time of the recall, approximately 257,000 Sprint Fidelis leads remained implanted in patients.

Following the recall, plaintiffs across the country began to file actions against Medtronic alleging (among other things) claims for negligence, strict products liability, fraud, and breach of express and implied warranties. According to the plaintiffs, the leads were not adequately tested prior to seeking FDA approval, the method of manufacturing, which involved direct resistance spot welding of two different metals, was prone to damaging them, and Medtronic knew (but failed to disclose to the FDA) that this welding technique was likely to result in the leads failing. Finally, plaintiffs claimed that Medtronic failed to take adequate steps to ensure that the Sprint Fidelis leads were not damaged during production, including failing to perform adequate testing on the leads' components, and failing to take corrective action to prevent lead failures.

Although a multidistrict litigation was commenced on February 21, 2008, in the United States District Court for the District of Minnesota, many of the claims were dismissed on the ground of preemption (*In re Medtronic, Inc. Sprint Fidelis Leads Products Liability Litigation*, 592 F. Supp. 2d 1147 [D. Minn. 2009]). As discussed further below, preemption is a powerful defense available to manufacturers in cases involving devices that were approved through the FDAs premarket approval process.

ARTIFICIAL HEART VALVES

Artificial heart valves are composed primarily of metal or carbon alloys, and are classified according to their structure as caged-ball, single tilting-disk or bileaflet tilting-disk valves (Vongpatanasin et al., 1996) (see Chapter II.5.3.A). All three types have been the subject of product liability litigation.

In the late 1970s, 15 suits were brought against a manufacturer of a caged-ball valve. Plaintiffs claimed that defects in the valve caused it to wear out prematurely, resulting in major embolic complications, premature open heart surgery to replace the valve, and catastrophic popet-ball escape from the valve cage. The theories of liability included negligence in the design, manufacture, and testing of the valve, breach of express or implied warranties, and strict liability (*In re Cutter Labs, Inc. "Braunwald-Cutter" Aortic Valve Products Liability Litigation*, 465 F. Supp. 1295 [J.P.M.L. 1979]). The manufacturer incurred significant legal expenses, yet was only partially successful in defending the lawsuits (*Lindsay v. Cutter Laboratories, Inc.*, 536 F. Supp. 799 [W.D. Wis. 1982]).

Beginning in the mid-1980s, a manufacturer of a tilting-disk valve, implanted in 50,000 to 100,000 patients between 1979 and 1986 (*Bowling v. Pfizer*, 143 F.R.D. 141 [S.D. Ohio 1992]), became the object of numerous lawsuits due to the valve's potential to fracture. Plaintiffs alleged that even before the valve was marketed, the first instance of valve failure due to "strut fracture" occurred in clinical trials. The valve consists of a disk located inside a metal ring covered by a Teflon sewing ring, which is sutured to the heart. The disk opens and closes rhythmically, allowing blood to pass through the heart. The disk is held in place by two wire holders, the inflow and outflow struts. When the overflow strut fractures, the disk escapes from the ring, causing uncontrolled blood flow through the heart, usually resulting in death. According to a Congressional report on the valve, based in part on examination of the manufacturer's internal documents, strut fracture was most likely caused or exacerbated by deficiencies in quality control procedures.

The number of lawsuits mounted, and eventually the manufacturer entered into a settlement with a class of plaintiffs implanted with its valve who had not yet experienced fracture. Under the terms of the settlement, the manufacturer agreed to establish a $75 million Patient Benefit Fund for research and heart valve replacement surgery, and a Medical Compensation Fund of between $80 million and $130 million to provide cash payments to valve recipients. The settlement also guaranteed immediate cash payments in the event of a fracture, and provided for contribution of another $10 million for spouses of class members (*Bowling v. Pfizer*, 143 F.R.D. 141 [S.D. Ohio 1992]).

The issue of whether a plaintiff whose heart valve has not failed can make a legally valid claim against a manufacturer has been addressed repeatedly. Generally, courts have demonstrated little willingness to entertain lawsuits based on a valve recipient's fear of failure because either physical injury or an actual device failure is a prerequisite to imposing liability on a manufacturer. However, exceptions to the general trend can be found. For instance, in a California case, *Kahn v. Shiley*, 217 Cal. App. 3d 848 [1990], the court allowed a plaintiff to proceed with her fraud claim because the court reasoned fraud does not challenge the safety or efficacy of the medical device. Because fraud allegations focus exclusively on the defendant's conduct, the fact that the plaintiff's valve had not failed was immaterial. In another case, *Michael v. Shiley, Inc.* (46 F.3d 1316 [2d Cir. 1995]), a court allowed a plaintiff to proceed to trial because, unlike most plaintiffs asserting fear of future valve failure, she suffered a tangible injury because she underwent surgery to replace her heart valve.

The most recent round of litigation concerning artificial heart valves involves the Silzone heart valve, which has a patented silver elemental coating on the sewing cuff, the part of the valve that is sewn to the patient's body. The purpose of the coating is to use silver's antimicrobial properties to help combat endocarditis, an infection of heart tissues and a common cause of prosthetic valve failure. Although the silver-coated Silzone valve was approved for commercial distribution in 1998, the FDA prohibited the company from claiming that the silver coating would reduce the risk of endocarditis.

Following FDA approval, the device manufacturer enrolled 792 patients in a multi-national clinical trial designed to study whether the silver-coated heart valve reduced the incidence of endocarditis in humans. The study was never completed, however, because the monitoring board found that recipients of the Silzone valve were more likely than controls to experience a complication called "paravalvular leak," requiring the valve to be removed and replaced.

With these results, the manufacturer voluntarily recalled all of its unimplanted Silzone products. As part of the recall, the manufacturer immediately notified hospitals and physicians, instructing them not to use the valves. The company also sent letters regarding the care and management of patients with its implanted valves, and established a reimbursement program to pay for uninsured medical costs associated with the detection, diagnosis, and treatment of paravalvular leak.

These remedial measures, however, did not stop a tide of tort litigation by the patients who received the silver-coated Silzone valves. Some of these claimants alleged bodily injuries as a result of an explant or other complications, which they attributed to the Silzone product. Others, who had not had their valve explanted, and who were asymptomatic with no apparent clinical injury to date, sought compensation for past and future costs of medical monitoring.

A number of individual lawsuits were filed and consolidated in a multidistrict litigation in a Minnesota federal court, which then consolidated various class action complaints consolidated into one class action. A class action is a form of lawsuit where a large group of people collectively bring their claims. The proposed class must consist of a group of individuals that have suffered a common injury, and in many cases, the party seeking certification must show that common issues between the class and the defendants will predominate the proceedings, as opposed to individual fact-specific conflicts between class members and the defendants.

Class actions alleging physical injury from defective medical devices often fail to meet these requirements, because such cases inevitably involve individual issues. In addition, nationwide classes are often not certified because there is no national body of product liability law. Plaintiffs must sue under the law of a particular state, and differences in state law make class resolution of personal injury claims impracticable.

The Silzone consolidated class action was brought by five named plaintiffs on behalf of themselves and over 11,000 other Silzone valve recipients. The litigation presents a sort of class action rollercoaster ride, as the case bounced between the district court and the Eighth Circuit Court of Appeals for five years. In 2004, the district court certified a nationwide class of plaintiffs who had been implanted with the Silzone valve seeking injunctive relief in the form of medical monitoring, as well as a consumer protection class under Minnesota's consumer protection law. This decision was reversed by the Eighth Circuit Court of Appeals in 2005. The appellate court rejected the medical monitoring class because the class presented "a myriad of individual issues making class certification improper" (*In re St. Jude Med., Inc.*, 425 F.3d 1116, 1122 [8th Cir. 2005]). However, it remanded the case to the district court for it to reconsider the consumer class decision, because the lower court did not properly analyze whether the law of Minnesota should apply to the claims of all of the potential members of the proposed nationwide class.

On remand, the district court determined that Minnesota law should apply to all claims in the nationwide class, and recertified the consumer protection class (*In re St. Jude Medical, Inc.*, 2006 WL 2943154 [D. Minn. 2006]). The manufacturer appealed again and won. In 2008, the Eighth Circuit held that the district court erred in certifying the consumer protection class, because there were material variations in the representations made by the manufacturer concerning the valve, as well as material differences in the kind and degree of reliance on those representations by the potential class members. The Eighth Circuit recognized that trial would require physician-by-physician inquiry into each doctor's sources of information about the valve, and held that the case was not suited for trial as a class action (*In re St. Jude Medical, Inc.*, 522 F.3d 836 [8th Cir. 2008]).

Following the Eighth Circuit's rejection of the consumer protection class action the plaintiffs continued to pursue their claims by trying to substitute "omissions" for affirmative misrepresentations in their allegations, but in light of the appellate court's prior rulings, the district court refused to certify the class (*In re St. Jude Medical, Inc.*, MDL No. 01-1396 (JRT/FLN), 2009 WL 1789376 [D. Minn. June 23, 2009]).

As of April 24, 2009, there were only three individual Silzone cases pending in federal court, and seven individual state court suits concerning Silzone. The plaintiffs in these cases were requesting damages between $10,000 to $100,000 and, in some cases, an unspecified amount (St. Jude Medical Form 10-Q [May 12, 2009]).

HIP/KNEE PROSTHESIS IMPLANTS

Prosthetic hip implants used in total hip replacements are made up of two components – a socket-like shell that is surgically inserted into the acetabulum of the pelvis, and a ball-like device to replace the femoral head (see Chapter II.5.6). The Inter-Op Shell manufactured by Sulzer Orthopedics was one such implant. It was constructed of an outer, porous titanium covered cup and an inner polyethylene cup, and was designed to bond with the natural bone.

In September 2000, Sulzer began to receive adverse reports from surgeons complaining that patients were experiencing symptoms post-operatively that suggested a defect in the hip implant. There were reports of severe pain, loose implants, second surgeries, inability to bear weight or walk, and other health complications. Sulzer initiated an investigation, and after eight weeks, they identified the cause of the problem. During the manufacturing process, a small amount of a mineral-based oil lubricant leaked into the machine coolant. Residual amounts of the lubricant remained on the surfaces of the shell, interfering with the bonding process between bone and shell. On December 8, 2000, Sulzer recalled many of its Inter-Op acetabular shells. According to the company, residue remained on the surface of the shells following the machining process which could prevent the implant from properly bonding with the bone.

Initially, Sulzer offered to pay for revision surgery, which cost between $20,000 and $70,000 per patient. Many patients, however, were frustrated with the fact that Sulzer did not offer to pay for pain and suffering associated with revision surgery and rehabilitation. Also, many of the patients' revision surgeries would be covered by Medicare, and so patients felt the offer was not genuine compensation for their surgery or for their pain. In 2001, patients began filing lawsuits against Sulzer, basing their claims on theories of strict liability for manufacturing a defective product, breach of warranty, negligence, and strict liability for failure to warn. Damages sought against Sulzer included: all pain and suffering associated with the recall implant; all past and potential future medical expenses associated with the defective hip implant; and all loss of earnings and/or earning capacity associated with the surgery. Many patients also sought punitive damages.

Around the same time, in early 2001, Sulzer began receiving adverse incident reports regarding its Natural Knee II Tibial Baseplate, an orthopedic knee implant. The Tibial Baseplate is one component of a system used for complete knee replacements. The Baseplate is inserted into the tibia or shin bone. While some implants are cemented into the bone, the Natural Knee II, like the Inter-Op Shell, was designed to allow the bone to grow into and around the implant. Again, surgeons reported that patients were experiencing poor recovery progress and ambulatory pain. X-rays showed radiolucencies under the baseplate, suggesting that the implant had separated from the bone. Sulzer discovered that the same manufacturing problem that caused the hip implants to fail was causing the knee implants to fail to bond. In May 2001 Sulzer announced a recall of its knee implants, and informed surgeons who had implanted its Tibial Baseplates of the unanticipated adverse clinical outcomes.

As knee implant patients began to file suits, and hip implant patients continued to file suits, Sulzer realized that it would have a difficult time trying a case in which it had voluntarily discovered and already admitted responsibility for the cause of the defect. In attempts to settle the matter without going bankrupt, Sulzer opened its books up to plaintiffs' attorneys and financial advisors who engaged in extensive discovery regarding the company's liability insurance and financial condition. A $1 billion settlement was approved on May 8, 2002.

PREEMPTION

Prior to 1976, medical device manufacturers were not required to seek FDA approval for a new medical device. In that year, a new regulatory scheme was implemented under the Medical Device Amendments of 1976 ("MDA"). The new regulatory regime established various levels of oversight for medical devices, depending on the risks that they presented. Class I, which includes such devices as elastic bandages and examination gloves, is subject to the lowest level of oversight. Class II, which includes, for example, powered wheelchairs and surgical drapes, is subject to certain performance standards and postmarket surveillance measures. Most medical devices are considered Class II devices. The devices receiving the most federal oversight are those in Class III. These devices usually sustain or support life, are implanted or present potential serious risk of illness or injury.

The MDA also includes a "preemption" provision, Section 360k(a), which states that: "no State … may establish … any requirement (1) which is different from, or in addition to, any requirement applicable under this chapter, and (2) which relates to the safety and effectiveness of the device …" Device manufacturers frequently argue that the MDAs preemption provision precludes plaintiffs from bringing lawsuits under state tort law. The preemption defense succeeds if a court finds that Congress intended that the federal regulatory scheme govern the question of device safety to the exclusion of state tort law.

The extent to which state tort law claims are preempted is determined by the manner in which a medical device is approved by the FDA. In 2008, in a case called *Riegel v. Medtronic Inc.* (128 S. Ct. 1999), the Supreme Court addressed preemption in the context of Class III medical devices. New Class III medical devices, such as the balloon catheter at issue in the *Riegel* case, must receive premarket approval ("PMA") before they can be legally marketed. The PMA process is the "rigorous" process through which the FDA weighs any probable benefit to health from the use of the device against any probable risk of injury or illness. A manufacturer must submit what is typically a multivolume application including, among other things, full reports of all studies and investigations of the device's safety and effectiveness, a full statement of the device's components, ingredients, and properties, and of the principle or principles of operation, a full description of the methods used in, and the facilities and controls used for, the manufacture, processing, and a specimen of the proposed labeling. The FDA then spends an average of 1200 hours reviewing each application, and grants premarket approval only if it finds there is a reasonable assurance of the device's safety and effectiveness. In *Riegel*, the Supreme Court held that PMA approval imposes "specific federal requirements" applicable to a particular device, and that state law tort claims impose requirements that "are different from, or in addition to" the requirements imposed by federal law. Thus, the Court found that the plaintiff's claims against Medtronic under New York law based on strict liability, breach of implied warranty, and negligence in the design, inspection, distribution, labeling, marketing, and sale of the catheter were preempted by the MDA.

However, most medical devices are not approved through the PMA process; rather, they are marketed after receiving FDA approval through the less rigorous "premarket notification" process. Premarket notification facilitates the marketing of medical devices that are new to a company, but "substantially equivalent" to a device already on the market (known as a "predicate" device). Substantial equivalence means that the new device has the same intended use as the predicate, and is at least as safe and effective as the predicate device. In 2005, the FDA authorized the marketing of 3148 devices through the premarket notification process, and granted PMA approval to just 32 devices.

Unlike PMA-approved Class III medical devices, there is only a limited preemption defense available in litigation involving devices that were approved through the premarket notification process. Whether state law tort claims are preempted depends on whether FDA regulation of these devices constitutes a specific "requirement" applicable to the particular device. Courts have held that certain non-PMA-approved medical devices, such as tampons (*Papike v. Tambrands Inc.*, 107 F.3d 737 [9th Cir. 1997]), and contact lens solutions (*Tuttle v. CIBA Vision Corp.*, 2007 WL 677134 [D. Utah March 1, 2007]), meet this threshold, because specific FDA regulations govern the labeling of such products.

Interestingly, Congress has chosen to subject medical devices and prescription drugs to different regulatory schemes. Despite its enactment of an express preemption provision for medical devices in the MDA, Congress has not enacted such a provision for prescription drugs. Thus, in 2009, in a case called *Wyeth v. Levine* (129 S. Ct. 1187 [2009]), the Court held that the FDAs approval of prescription *drug* labeling does not preempt a plaintiff's state law tort suit, alleging deficiencies in the approved label.

The disparity between the medical device and prescription drug regulatory schemes creates possible tension regarding the scope of a preemption defense in a case involving a "combination" product that combines drugs, devices, and/or biological products, such as the Silzone artificial heart valve discussed above, a drug-eluting stent or a tapered metallic spinal fusion cage. At least one court has held that where the FDA regulates a combination product as a medical device, express preemption under the MDA applies (*Riley v. Cordis Corp.*, 2009 WL 1606650 [D. Minn. June 5, 2009]). However, in the Silzone heart valve MDL, the court questioned, although it did not decide, whether a combination product would be subject to the MDAs express preemption provision (*In re St. Jude Medical, Inc. Silzone Heart Valves Prods. Liab. Litig.*, 2004 WL 45503 [D. Minn. 2004]).

SCIENCE IN THE COURTROOM

When medical device lawsuits reach the trial phase, the expert witness assumes a critical role. Whether an allegedly defective medical device caused the plaintiff's injury is an issue deemed beyond the ken of the average juror, and courts therefore require the litigants to present scientific evidence in the form of expert opinions. The courtroom, however, is a forum ill-suited to discussion of scientific principles, and the scientific evidence communicated to jurors often fails to meet the standards of reliability our judicial system envisions.

Laypersons who lack schooling in the fundamentals of scientific inquiry are prone to biases. For instance, a layperson would be more inclined to accept coincidence as proof. As Marcia Angell, the former editor of the *New England Journal of Medicine*, noted, many people might find a reasonable proposition that mere temporal relationship, i.e., that health complications followed breast implantation, is sufficient proof that implants caused the injuries (Angell, 1996). A scientist, on the other hand, is trained to understand that association is not causation, and anecdotal reports are no substitute for scientific data. Studies also show that laypersons are likely to ignore epidemiological evidence, and that they have a so-called "hindsight bias," i.e., they tend to favor facts that are more consistent with the ultimate outcome (Haskel, 2007).

The gap between a layperson's and a scientist's understanding of cause and effect yawns even wider in the courtroom, because both the goals and the methods of science and litigation are at odds (Mayesh and Ried, 1986). Litigation and science employ disparate standards of proof, as well as disparate measures of causation. Science examines causative correlations in the population at large, whereas litigation asks whether a particular device caused a particular plaintiff's injury. Moreover, litigation and scientific inquiry demand quantitatively different standards of proof. Scientific inquiry seeks to establish causal relationships to a 95% degree of certainty, whereas judicial inquiry requires only a 51% probability of correctness. Under these circumstances, it is easily imaginable that a jury could determine that a systemic disease was caused by exposure to an implanted device, while a scientist would find the same evidence merely sufficient to suggest an interesting hypothesis (Mayesh and Ried, 1986b).

As a practical matter, the judicial system's goal of timely conflict resolution would not be well-served by imposing the standards of the scientific community on the courtroom. Lawsuits must be resolved in a few years. Questions of science typically take decades of testing and data accumulation before repeatability can be achieved and the question thereby answered. To resolve the tension, the judicial system has fashioned a compromise by allowing juries to hear only that scientific evidence deemed by the judge to be reliable and generally accepted within the scientific community. In modern federal practice, this is called the *Daubert* standard after the US Supreme Court decision, *Daubert v. Merrell Dow* (509 U.S. 579 [1993]). Federal trial judges are given wide discretion to admit or preclude evidence under this standard (Mayesh and Ramallo, 2007).

The saga of DNA identification is an example of how judges serve as the gatekeepers of whether scientific evidence is allowed before the jury. Until DNA typing became generally accepted as reliable, it was not allowed into the courtroom. Now that it has been established as accepted methodology, DNA proof may be put before the jury although, of course, subject to rigorous dispute over correct methodology. This model, however, does not always prove satisfactory.

Because litigants face admissibility hurdles over scientific evidence, litigation-driven science has become a growing phenomenon. Fortunately, the few courts that have confronted litigation science to date have generally been skeptical of the value of these studies. For example, in *In re Breast Implant Litigation* (11 F. Supp. 2d 1217 [D. Colo. 1998]), the court excluded the testimony and opinions of a plaintiff's rheumatology expert, in part, because his patients were litigants who had been referred to him. The court also excluded the testimony of the plaintiff's biomaterials expert, in part, because the vast majority of his business came from plaintiffs involved in breast implant litigation.

Because scientific inquiry is typically characterized by a degree of uncertainty, deft litigants may succeed in casting doubt upon strong scientific proof. On the other side, experts for hire are sometimes allowed to spin unproven theories in order to strengthen the inclination of jurors to assume that anecdotes are the equal of data, and association the correlate of cause. At the conclusion of the evidence, 12 laypersons, who have listened to experts express conflicting opinions, vote on which is more persuasive. The opinions themselves need not be held with more than a "reasonable degree" of scientific certainty, which the giver of the opinion may define as "more probable than not." Although the jurors usually need to be unanimous, their measure of confidence need only reach "a fair preponderance of the credible evidence," meaning 51%. Little wonder, then, that there is a disconnect between real science and courtroom science. As the great jurist, Learned Hand, observed in the beginning of the last century: "But how can the jury judge between two statements each founded upon an experience confessedly foreign in kind to their own? It is just because they are incompetent for such a task that the expert is necessary at all" (Hand, 1902).

BIOMATERIALS ACCESS INSURANCE ACT

Plaintiffs suing the manufacturer of an allegedly defective device sometimes also join as defendants the suppliers of its raw materials and component parts. The incentive to sue a raw material supplier is particularly strong if the supplier has significant financial resources and the medical device manufacturer has limited assets. For instance, DuPont, a supplier of Teflon used in TMJ implants, spent $26 million litigating more than 650 lawsuits over implants used to treat TMJ after the small manufacturer of the implants went bankrupt (Murphy, 2000). Although bulk suppliers have consistently succeeded in having suits against them dismissed, the cost of litigation can nonetheless be substantial. After weighing the risk of becoming embroiled in expensive litigation against tiny profits derived from the medical device market, bulk suppliers began to deny manufacturers access to their products, and the United States faced a serious biomaterials shortage in the 1990s.

To address the dwindling supply of biomaterials essential for the manufacture of implantable devices in 1998 Congress passed the Biomaterials Access Assurance Act ("BAAA"). The act shields suppliers of raw materials and component parts from liability, unless the supplier is also the manufacturer or seller of the device or furnished materials or components that did not comply with contractual requirements or certain other specifications. Under the act, a supplier named in a lawsuit is entitled to move for dismissal immediately. Once a motion to dismiss is filed, the supplier is excused from participation in any "discovery," the expensive and time-consuming process by which litigants seek information from each other before trial. Limited discovery is allowed only to determine whether a supplier failed to comply with contractual requirements or specifications.

After dismissal from the lawsuit, the supplier may still face liability in certain, limited circumstances. After a verdict against the device manufacturer, the plaintiff or manufacturer can require the supplier to pay part of the judgment if the court determines that the supplier's negligent or intentional conduct was a cause of the plaintiff's injury. The plaintiff may utilize this procedure if the full amount of damages cannot be recovered from the manufacturer.

Although the BAAA was enacted in 1998, there appears to be only one reported decision concerning the act or its interpretation. In 2008, the United States District Court for the District of Colorado dismissed claims brought against a company that manufactured the femoral hip head component of a hip replacement system, because the company was not the manufacturer of the implant, was not the seller of the implant, and did not furnish raw materials or component parts that failed to meet applicable contractual requirements or specifications (*Whaley v. Morgan Advanced Ceramics, Ltd.*, 2008 WL 901523 [D. Colo. March 31, 2008]).

LIABILITY OF THE DESIGN ENGINEER

For several reasons, an individual design engineer is an unlikely defendant in medical device cases. First, plaintiffs know that manufacturers are more likely to have the financial resources to compensate them for their injuries. Second, a plaintiff gains a "David and Goliath" tactical advantage by suing only the manufacturer, and not the design engineer. A jury will be more inclined to sympathize with a plaintiff if the case is viewed as a confrontation between a single individual and a corporate giant.

If sued, a design engineer is theoretically not immune from liability. However, the theories available to plaintiffs are less expansive when suing a design engineer as opposed to a manufacturer. For example, courts have recognized that the policy objectives underlying strict liability would not be furthered by applying the theory of strict liability to design engineers. Courts have reasoned that design engineers provide a professional service, and do not occupy the same superior position that allows manufacturers to discover defects and spread economic losses. Although the law of strict liability is subject to some flux and uncertainty, the great weight of judicial precedent provides a good deal of assurance that a design engineer who develops a medical device will not be held liable.

That being said, design engineers may be liable under a theory of professional malpractice, which asks whether the design engineer has exercised the degree of care reasonably expected in the profession. Liability will only be imposed if the design engineer is found to be at fault for failing to live up to that standard.

In addition, a design engineer can be held criminally liable if he or she is also a corporate or compliance officer. For example, in *U.S. v. Caputo* (517 F.3d 935 [7th Cir. 2008]), an appeals court upheld the criminal convictions of a device designer and his assistant. They were alleged to have marketed an unapproved medical device, known as an autoclave, which left a harmful blue-green residue on brass instruments employed for procedures in the eye. Their convictions were upheld because they wore multiple hats: the device designer was also the President and CEO of the manufacturer, and his assistant was also the Vice President of Regulatory Affairs and Chief Compliance Officer. As a practical matter, however, a design engineer is very rarely named as a defendant.

TABLE III.2.8.1 Department of Justice Settlements 2003–2007

Company	Product	Date	Settlement Amount
Serono, S.A.	Serostim	October 2005	$704 million
The Purdue Frederick Company	OxyContin	May 2007	$634 million
Bristol-Myers Squibb	Abilify	September 2007	$500 million
Schering-Plough Corporation	Temodar and Intron A	August 2006	$435 million
Pfizer	Neurontin	May 2004	$430 million
InterMune, Inc.	Actimmune	October 2006	$36.9 million
Eli Lilly and Company	Evista	December 2005	$36 million
Pfizer, Inc.	Genotropin	April 2007	$35 million
Jazz Pharmaceuticals, Inc.	Xyrem	September 2007	$20 million
Cell Therapeutics, Inc.	Trisenox	April 2007	$10.5 million
Medicis	Loprox	May 2007	$9.8 million

DEVICE MARKETING AND PROMOTION

With increasing frequency, plaintiffs suing medical device manufacturers allege that the manufacturer promoted the product in an improper manner. One type of improper promotion is known as "off-label" promotion, whereby a manufacturer promotes a medical device for some other indication than that for which it has been approved by the FDA. Once the FDA has approved a medical device for marketing, a physician may prescribe it for any use, even uses not approved by the FDA. Often, such off-label use of a medical device becomes the standard of care. Nonetheless, although off-label *use* of medical devices by physicians is accepted and occurs frequently, device manufacturers are prohibited from engaging in off-label *promotion*. A medical device that is labeled or promoted for a use that has not been approved or cleared by the FDA is considered to be misbranded by the FDA.

Government scrutiny of off-label promotion is on the rise. During calendar years 2003 through 2007, the FDA issued 42 regulatory letters in response to off-label promotions requesting companies to stop dissemination of violative promotions. During that same time, the Department of Justice settled 11 civil and criminal cases that involved, at least partially, off-label promotion. The settlements in these cases were for staggering amounts (Table III.2.8.1). And, although they involved pharmaceutical companies, medical device manufacturers are certainly not immune to such investigation.

DEFENSIVE MANUFACTURING AND MARKETING

Although little can be done to prevent a plaintiff from initiating a lawsuit, the best defense to a products liability suit is having manufactured a safe, well-designed product with adequate warnings. In virtually every lawsuit alleging injury caused by a medical device, the plaintiff – and eventually jurors – will have access to the manufacturer's internal corporate documents. These include lot histories, manufacturing specifications, results of toxicology and safety tests, quality assurance documents, FDA submissions and compliance reviews, adverse event incident reports, and intracorporate memos discussing all of these, whether paper or email. The majority of these documents will have been prepared many years prior to the lawsuit, often by personnel who are no longer employed by the manufacturer at the time of trial.

If the manufacturer makes the proper investment up front in designing, manufacturing, and selling the product, these documents can be the best proof that the manufacturer performed all necessary safety testing, that the product conformed with all government and industry standards, that no manufacturing defects occurred, that warnings and instructions for use were legally appropriate, and that the manufacturer complied with all regulatory requirements for device approval. Moreover, the documents should establish that the manufacturer took appropriate post-market safety surveillance, and had adequate quality assurance systems. This includes documentation that it was responsive to complaints from doctors and patients, had a method of tracking such complaints as well as resolving them, and when appropriate, took post-market action, such as product recall and issuing revised warnings.

CONCLUSION

Very few things are certain in either law or science. However, when it comes to products liability, scientists and lawyers alike can safely bet that scientists will always test and expand the limits of current technology, creative plaintiffs' counsel will constantly be developing novel theories of liability, and defendants will be equally creative in trying to rebut those claims.

BIBLIOGRAPHY

Anderson, W., Parsons, B., & Rennie, D. (2001). Daubert's backwash: Litigation-generated science. *U. Mich. J. L. Ref.*, **34**(4), 619–682.

Anonymous. (1995a). Federal jury finds acromed device unreasonably dangerous, awards $318,000 in Reeves. *Mealey's Litig. Rep. Pedicle Screws l(17).*

Anonymous. (1998b). Jury awards $451,000 in first pedicle screw lawsuit to reach a verdict. (1998). *Med.-Leg. Aspects Breast Implants*, **6**(7), 1.

Anonymous. (1995b). Pedicle screw case ends in defense verdict. *Penn. Law Weekly*. 19 June.

Anonymous. (1998c). Tennessee pedicle trial ends with hung jury. *Mealey's Litig. Rep. Pedicle Screws*, **4**(4).

Biomaterials Access Assurance Act. (1998). *U.S. Code*, **Vol. 21**, secs. 1601–1606.

Hand, L. (1902). Historical and practical considerations regarding expert testimony. *Harv. L. Rev.*, **15**, 40–58.

Hoyle, L., & Madeira, E. (2001). The Philadelphia story: Mass torts in the city of brotherly love. *Sedona Conf. J.*, **2**, 119.

Macdonald, Plavac, N., Peters, W., Lugowski, S., & Smith, D. (1995). Failure of Si NMR to detect increased blood silicone levels in silicone gel breast implant recipients. *Anal. Chem.*, **67**(2), 3799–3801.

Mayesh, J. P., & O'Hea, J. A. (1997). Can plaintiffs win second generation claims? *Leader's Prod. Liabil. L. Strat.*, **15**(7), 5, 6.

Mayesh, J. P., & Ramallo, O. (2007). Nine years of *Joiner*: A review of appellate cases applying the abuse of discretion standard to *Daubert appeals. LJN's Prod. Liabil. L. Strat.*, **25**(12), 3–4, 8.

Mayesh, J. P., & Ried, W. M. (1986a). The problems caused by "junk science." *Leader's Prod. Liabil. Newslett.*, **5**(3), 1, 6.

Medical Device Amendments of 1976. (1976). *U.S. Code*, **Vol. 21**. secs. 360c–360k.

Meier, B. (2005). Maker of heart device kept flaw from doctors. *The New York Times*. May 24, 2005.

Murphy, A. (2000). The Biomaterials Access Assurance Act of 1998 and supplier liability: Who you gonna sue? *Del. J. Corp. L.*, **25**, 715–7339.

National Science Panel. (1998). *Silicone breast implants in relation to connective tissue diseases and immunologic dysfunction. Report to the Honorable Sam C. Pointer, Jr., Coordinating Judge for the Federal Breast Implant Multidistrict Litigation (17 November 1998).*

Plaintiff's Submission and Proposed Findings to the National Science Panel on Silicone Gel Breast Implants. (6 October 1997).

United States Government Accountability Office. (2008). *Report to the Ranking Member, Committee on Finance, U.S.* Senate, FDA's Oversight of the Promotion of Drugs for Off-Label Uses, http://www.gao.gov/new.items/d08835.pdf. Date accessed 07/08/2009.

Sobol, R. (1991). *Bending the law: the story of the Dalkon Shield bankruptcy.*

St. Jude Medical Form 10-Q (May 12, 2009).

Sulzer Implant Settlement Class Member and Attorney Guide, http://www.sulzerimplantsettlement.com, Date accessed 07/08/2009.

Vairo, G. M. (1992). The Dalkon Shield claimants trust: Paradigm lost (or found)? *Fordham L. Rev.*, **61**, 617–660.

Vladeck, D. (2005). Preemption and regulatory failure. *Pepperdine L. Rev.*, **33**, 95–132.

Vongpatanasin, W., Hillis, L. D., & Lange, R. A. (1996). Medical progress: Prosthetic heart valves. *N. Eng. J. Med.*, **335**, 407–416.

CHAPTER III.2.9 CLINICAL TRIALS FOR MEDICAL DEVICES

Gary L. Grunkemeier,[1] Ruyun Jin,[1] Lian Wang,[1] and Albert Starr[2]

[1]Medical Data Research Center, Providence Health & Services, Portland, Oregon, USA

[2]Oregon Health & Science University, Portland, Oregon, USA

INTRODUCTION

The designation "Clinical Trial" is usually interpreted to mean "Randomized Clinical Trial" (RCT), and an RCT is usually considered to be the gold standard for biomedical studies. We take the position that randomized controls are not necessary or even the best option for many medical device studies, and that the maligned historical controls, if used in carefully designed prospective observational studies, will yield as good or better information regarding the safety and efficacy of a new medical device, with advantages in the areas of generalizability (external validity), speed, cost, and ethical considerations.

Our experience with clinical trials for valve replacement, the first life-supporting device in heart surgery, runs the gamut from first-in-man, to the rapid development and implantation with significant refinements, starting in small numbers of patients hoping for improved performance, and finally to large-scale distribution and long-term follow-up of all patients receiving the device – all before the US Food and Drug Administration (FDA) became involved in 1974, when the Bureau of Medical Devices and Diagnostic Products was created. The FDA had the benefit of looking back at this heart valve development experience for developing guidelines for future medical devices. Also, early on before the FDA or major medical societies provided guidelines, our group had experience with randomized clinical trials of coronary artery bypass surgery (CABG). This has given us insights into the process that provides a clear perspective for the evaluation of various clinical trial methodologies. From the very beginning, surgeons and cardiologists of our team at the Oregon Health and Sciences University were allied with in-house mathematicians and statisticians, providing a powerful tool in pushing forward the frontiers of cardiac surgery.

Since our experience is primarily with heart valve replacement devices, we will begin by giving a brief history of heart valve development. Then we will critically assess the pinnacle position customarily assigned to *randomized* studies in the so-called "hierarchy of evidence," discuss the pragmatic view being adopted by the FDA, and the consequent alternatives to RCTs that have been advocated and used successfully for medical devices (especially heart valves). Then we mention some methods that have been proposed to improve observational studies, and to make comparative studies more adaptive, and conclude the heart valve story with the latest technological breakthrough – transcatheter implantation.

THE FIRST SUCCESSFUL HEART VALVE PROSTHESIS

Our perspective on clinical trials for medical devices was shaped by our work in the clinical investigation of heart valve prostheses. Creating them and deploying them in patients required original thinking in many areas of science, medicine, and ethics. A comprehensive review of heart valve function and dysfunction, and of the various devices available for repairing and replacing them, is given elsewhere in this volume (Schoen and Padera, 2010). Here, we give a brief discussion of some of the considerations involving the first human valve implants.

The first Starr-Edwards caged-ball heart valves were produced in a cabin workshop (Figure III.2.9.1) by retired engineer M. Lowell Edwards (Figure III.2.9.2), and implanted into black Labrador dogs. In 1960, after many dogs had been implanted, the decision was made to begin a clinical series. Although many dogs were dying, because the valves were being occluded by blood clots, humans have a less intense clotting mechanism, plus they would be given anticoagulant medicine which the dogs, for practical reasons, were not given. Interestingly, the cardiologists pushed the surgeon into acting sooner than he had intended, because their patients were dying of end-stage heart failure. The technique of randomization was certainly known by this time, it had been developed for agricultural experiments in the 1920s (Fisher, 1926) and extended to clinical applications by 1951 (Hill, 1951). But the laboratory dogs had not been randomized, and no thought was given to doing so with the patients. Instead, a dying patient was presented with a possibility of living.

FIGURE III.2.9.1 The cabin in which Lowell Edwards produced the first Starr-Edwards heart valves.

FIGURE III.2.9.2 Lowell Edwards inside his workshop cabin.

The first successful implant was in 1960 (Starr and Edwards, 1961). Continual bench testing was being performed on the valve configuration and, as might be expected with any mechanical solution to a mechanical problem, incremental improvements in design were introduced. Between 1960 and 1966 there were nine mitral and five aortic valve "sub-models" (Figure III.2.9.3), each with slight improvements to certain mechanical parameters. (Such development would have been impossible if an RCT were required at each incremental step.) By 1964–1966 the development cycle had resulted in a stable design that was then unchanged until the valves were finally discontinued in 2007. Now (2010) in its 52nd year, the company that Lowell Edwards founded is still the leading manufacturer of heart valves (Figure III.2.9.4), and has distributed more than 1.5 million devices to date – none of which relied on RCTs to gain regulatory approval.

CRITIQUE OF MEDICAL DEVICE RCTs

Some studies must be observational, since it is not always possible to perform an RCT. An RCT may not be optimal or possible in public health research (Victora et al., 2004) or, for example, to study myocardial preservation techniques in open heart surgery (Liu et al., 1998). In situations where an RCT is possible, it may not be necessary to perform one. Obvious mechanical solutions to mechanical problems, such as seat belts or airbags in automobiles, parachutes in airplanes (Smith and Pell, 2003), and many other situations (Glasziou et al., 2007) do not require randomization. In fact, the frequency of surgical RCTs was only 9% in 1993 and 8% in 2006 (Barkun et al., 2009). Drugs, as opposed to medical devices, usually require an RCT, but the beneficial effects of aspirin and penicillin were discovered without one. It has been estimated that fewer than 20% of all clinical policies are based on randomized studies (Hornberger and Wrone, 1997).

We will discuss how heart valve prostheses have been evaluated clinically, including their regulatory requirements. But first, we present a general critique of randomized studies which have problems that are often not appreciated. Then we discuss why they are particularly ill-suited for the evaluation of many medical devices.

Hierarchy of Evidence: Fact or Myth?

It is widely, though not universally (Concato, 2004), assumed that there is a spectrum of evidential value for

ENGINEERING DEVELOPMENT CHARTS

STARR-EDWARDS MITRAL VALVE. SEPTEMBER 1960 – MARCH 1966

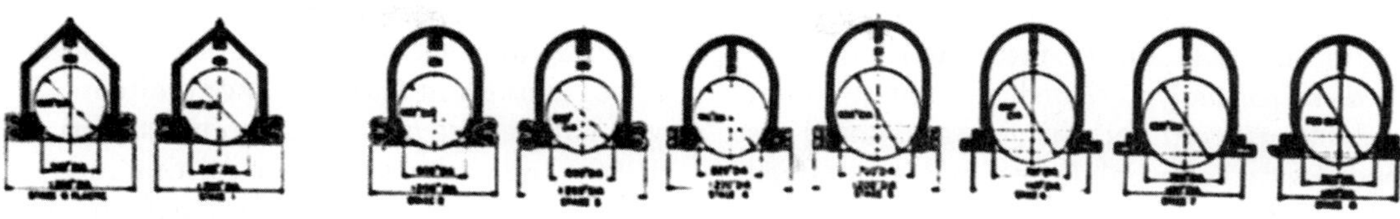

CLINICAL EXPERIMENTAL

STANDARD PRODUCTION

STARR-EDWARDS AORTIC VALVE. FEBRUARY 1962 – JANUARY 1964

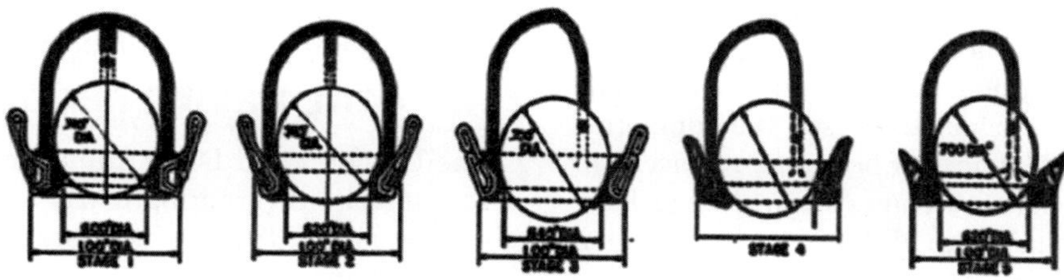

FIGURE III.2.9.3 Early incremental development cycles of Starr-Edwards mitral (upper panel) and aortic (lower panel) valve prosthesis.

FIGURE III.2.9.4 Current (2010) headquarters of Edwards Lifesciences (compare to Figure III.2.9.1). (Reproduced with permission from Edwards Lifescience)

medical studies. At the lowest end is the anecdotal evidence of a case report, at the highest end is the "unassailable" evidence of the RCT. But, we (Grunkemeier and Starr, 1992; Grunkemeier et al., 2006), and others (Black, 1996; Concato et al., 2000; Padkin et al., 2001; McCulloch et al., 2002; Victora et al., 2004; Grapow et al., 2006; Hordijk-Trion et al., 2006; Steg et al., 2007; Vandenbroucke, 2008) believe that there is a valid and important role in the advancement of medical knowledge for the large, well-conducted, real-world prospective observational study, using historical controls. In fact, a recent review article (Treasure, 2009) suggested that the usual hierarchy of evidence for medicines should be inverted for the purpose of evaluating interventions, and an associated editorial (Moghissi, 2009) concluded that: "Logically, therefore, the benchmark which applies to the evaluation of drugs needs to be modified for the interventional procedures." This inversion of the hierarchy of evidence had been proposed earlier (Vandenbroucke, 2008).

A classic 1988 study demonstrated that analysis of an existing cardiac database could reproduce the results of randomized studies comparing surgery to medical management of coronary artery disease (Hlatky et al., 1988). A 1999 paper comparing surgery to coronary angioplasty showed no differences between the findings of an RCT and database analyses (Lewsey et al., 1999). Many subsequent studies have agreed that the value of observational (non-randomized) studies is not less than that of RCTs. A paper in the *British Medical Journal* on the relative merits of observational and randomized studies summarized an exhaustive review of the topic from a previous Health Technology Assessment study (Britton et al., 1998), and concluded that: "Treatment effects obtained from randomized and non-randomized studies may differ, but one method does not give a consistently greater effect than the other" (McKee et al., 1999).

Two reviews in the 2000 *New England Journal of Medicine* showed that the results of RCTs and large observational studies in many areas of research reached the same conclusions with regard to treatment effects. One paper reviewed 136 reports concerning 19 diverse treatments (Benson and Hartz, 2000), and the other evaluated 99 reports dealing with five clinical topics (Concato et al., 2000). A critical, but balanced, editorial accompanying these two papers expressed fear that these publications might result in fewer RCTs being done, and cited three situations where RCTs disagreed with observational studies, but added that: "Detection of serious but rare side-effects requires very large numbers of patients, and can be achieved only through analysis of records from routine clinical practice" (Pocock and Elbourne, 2000). A balanced editorial discussion of these two papers in the *British Medical Journal* presented an analysis using the combined data sets from both studies, and noted the "striking concordance between the estimates obtained with the two research designs" (Ioannidis et al., 2001). A 2006 editorial in the *Journal of Thoracic and Cardiovascular Surgery* (Grapow et al., 2006) discussing studies of the relative benefits of CABG and percutaneous coronary intervention (PCI) concludes that: "only by considering both results from RCTs and large cohort studies can real-life clinical situations be reflected to a degree capable of providing applicable treatment guidelines useful in a daily clinical setting." A 2004 paper discussed the conflicting conclusions regarding the effect of hormone replacement therapy on acute coronary outcomes from a large RCT that found evidence contrary to all known observational studies (Garbe and Suissa, 2004); this should not be too surprising, considering a published indictment of "most published research findings" (Ioannidis, 2005).

Carefully planned, prospective observational studies have several advantages over randomized studies, especially with regard to the evaluation of a medical device. Results in "real-world" patients are obtained faster and cheaper. Randomized studies are expensive, ponderous, and slow. Once a study question arises, it may take a few years before the RCT is planned and implemented with patient recruitment, and a few more years after that before the results are available. Often, especially when the study involves a medical device, the results of the trial are obsolete by the time the trial results are available, because the device has been superseded. In the meantime, an ongoing observational study following the same patients, and observing the same outcomes and trends can provide an immediate answer to the same study question.

Internal Validity at the Expense of External Validity

Participants in RCTs are highly selected, and do not represent the patients to whom the winning therapy will eventually be applied. The providers (hospitals and physicians) that administer the RCT are highly selected, as are the patients who consent to this rather unnatural method of treatment selection. The patients must satisfy strict inclusion/exclusion criteria, resulting in less risky patients without common comorbidities. But, once the device is approved, it is used in the "real-world" on a much wider range of patients, and at a much wider range of centers (Britton et al., 1999; Hordijk-Trion et al., 2006; Steg et al., 2007; D'agostino and D'agostino, 2007). The result of these selection criteria severely limits the generalizability of the results to patients outside of the RCT (McKee et al., 1999). Moreover, RCTs have their own "internal" problems. The randomization itself is not always done or done correctly (Altman and Bland, 1999). Blinding is not always done effectively (Ney, 1989; Berger and Exner, 1999), and can lead to masking an effect or even reversing the direction of an effect (Kunz and Oxman, 1998). Intention-to-treat analysis is not always done correctly (Hollis and Campbell, 1999), and such problems are accentuated for RCTs that evaluate surgical techniques (McCulloch et al., 2002; Anyanwu and Treasure, 2004). And: "the very act of random allocation may reduce the effectiveness of the intervention" (Black, 1996). Practical clinical trials have been advocated as a means to restore the goal of external validity (Tunis et al., 2003; Glasgow et al., 2005).

MEDICAL DEVICE REGULATORY TRIALS

The US Food and Drug Administration (FDA) has been responsible for granting marketing approval of medical devices in the US since the passage of the Medical Device Amendments (MDA) in 1976 (Holden, 1976). The MDA lists three device categories: Class 1 are common, low-risk devices, subject to "general controls" only; Class 2 devices are higher risk, and subject to "special controls;" Class 3 devices are the highest risk, and require a pre-market approval (PMA) application, involving a comprehensive clinical study. In approving such devices, the FDA is mandated to "rely upon valid scientific evidence to determine whether there is reasonable assurance that the device is safe and effective." Valid scientific evidence is broadly defined to include "evidence from well-controlled investigations, partially controlled studies, studies and objective trials without matched controls, well-documented case histories conducted by qualified experts, and reports of significant human experience with a marketed device, from which it can fairly and responsibly be concluded by qualified experts that there is reasonable assurance of the safety and effectiveness of a device under its conditions of use" (Office of the Federal Register, 2009). The FDA insists on RCTs for drugs, but is more equivocal on the preferred testing paradigm for devices (Chen et al., 2006), partly because of the fundamental differences between drugs and devices.

Medical Devices Differ from Drugs

The RCT methodology was developed for agricultural experiments (Fisher, 1926), and has been successfully adapted to the clinical testing of drugs. But medical devices are not drugs. Drugs solve a biochemical problem, their action is systemic, cannot be seen directly, and is often not well-understood. In contrast, medical devices solve a mechanical problem, their action is local, and the mechanism can be directly or indirectly observed. A heart valve can be seen to open, allow forward blood flow, and then close, preventing appreciable retrograde blood flow (Figure III.2.9.5), and it can do this repeatedly

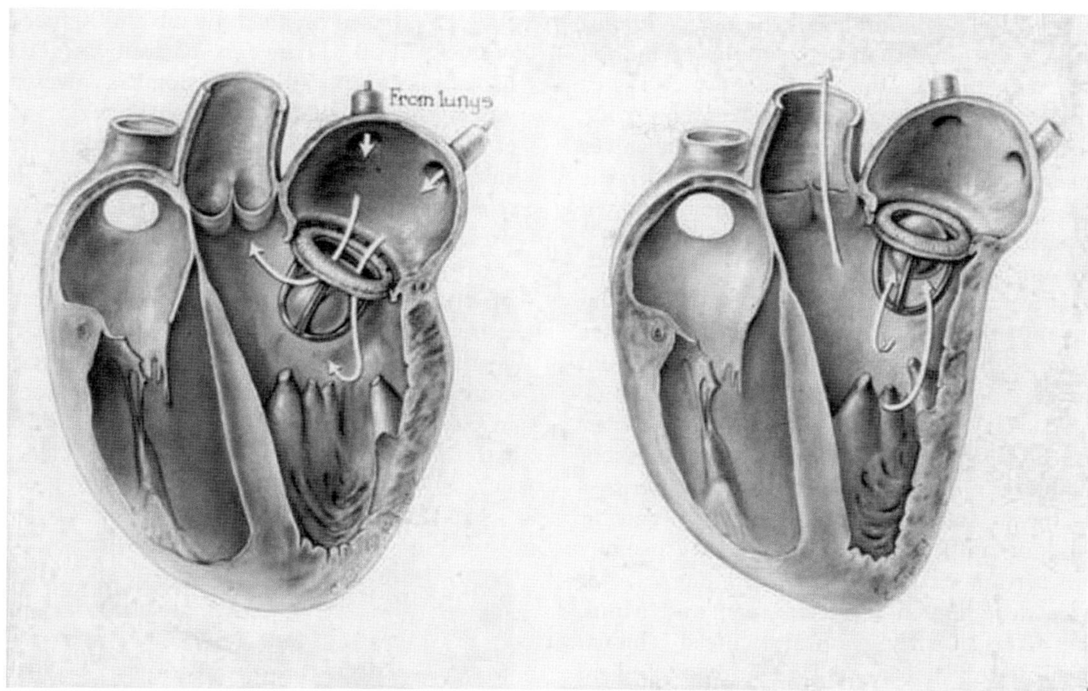

FIGURE III.2.9.5 Schematic of the mechanical function of the caged-ball mitral valve.

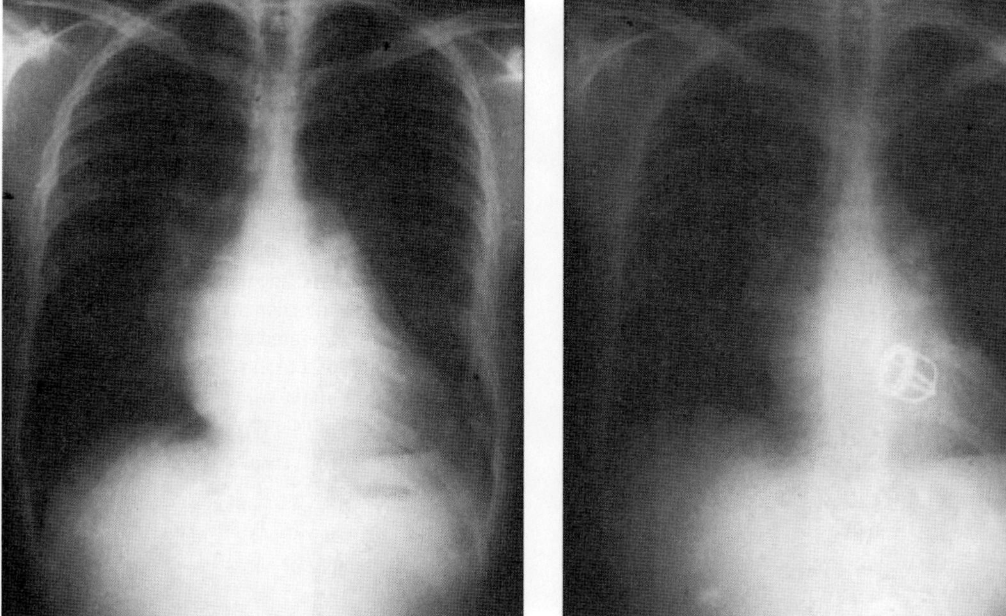

FIGURE III.2.9.6 Effectiveness of the valve prosthesis in reducing cardiac enlargement: before (left); and after (right) mitral valve replacement.

for many years. Several Starr-Edwards heart valves have been implanted for more than 45 years, and hence have cycled approximately 1.8 billion times (Gao et al., 2004). The clinical efficacy can be measured objectively, for instance as reduced heart size and pulmonary congestion on a chest X-ray (Figure III.2.9.6). Also, drugs are basically temporary; they can be re-dosed or discontinued with ease, whereas implanted devices are more difficult to change. And, unlike a drug regimen, noncompliance with an implanted medical device is not an option. Incremental improvements in the original Starr-Edwards heart valves each modified a small aspect of the design (Figure III.2.9.3), with a resultant small change in function. With a drug, a slightly different formulation might have a greatly different, and unpredictable, effect (Neaton et al., 2007; Pennello and Thompson, 2008).

An almost universal aspect of medical devices is their ongoing evolutionary improvement, so that by the time a large, long-term RCT is completed, the device manufacturer may have introduced another, improved model. This happened with the two large heart valve RCTs of the 1980s (Hammermeister et al., 2000; Oxenham et al., 2003); by the time these trials were completed, the valves used were no longer being manufactured. Also, by the time they were completed, the results that they obtained had already been reported by numerous observational studies (Grunkemeier et al., 2006).

The effect of a drug is independent of the prescribing physician, but the effect of an implanted medical device depends on the skill of the implanting physician and surgical team (Ergina et al., 2009). Finally, for drugs, multi-center trials can recruit thousands of patients over a relatively short time, but for devices, the number of patients available is much smaller. Recent FDA-written publications have acknowledged these differences: "However, use of an RCT study design is especially challenging in certain circumstances, including cardiovascular device development" (Zuckerman and Muni, 2005), and: "Consequently, nonrandomized clinical studies play a substantial role in the evaluation of medical devices" (Li and Yue, 2008).

FDA Heart Valve Guidance

The original FDA requirement for approval of a new heart valve specified the collection of certain data; completeness and accuracy of the collected data were important, but the decision for marketing approval was based on the subjective opinion of the reviewers (Sapirstein, 2001). In 1993 the FDA, in consultation with a committee of experts from industry and academia, produced a Guidance Document for heart valve prostheses (Division of Cardiovascular, Respiratory, and Neurological Devices; Center for Devices and Radiological Health; US Food and Drug Administration, 1994). This document requires a well-designed, adequately powered, prospective, single-arm study. The results of this study are measured against target values called Objective Performance Criteria (OPC), which is really a euphemism for historical controls. The OPC values were gleaned from a large literature on FDA-approved heart valves (Grunkemeier et al., 1992; Wheatley, 2004), and a sample size of 800 valve-years was shown to be adequate to evaluate the performance of a heart valve candidate for FDA approval (Grunkemeier et al., 1994).

Why would the FDA permit historical controls for pre-market approval of new heart valves? First, the existing valves worked very well, and had solved the big initial problems of insecure attachment, rejection, infection, etc. The OPC concept was based on the fact that heart valves have well-defined performance profiles, and that the expected complications are well-known (Table III.2.9.1) and well-defined (Aitkin et al., 1983; Akins et al., 2008b,c). Also, the complication rates for approved valves tend to cluster around certain norms (Figure III.2.9.7), based on a review of 37,000 mechanicalvalves with 187,000 valve-years of follow-up from 95 published series, and 24,000 biological valves with 132,000 valve-years from 70 published series (Grunkemeier et al., 2000). The FDA felt justified for making this decision since: "... the use of a prospective concurrent controlled clinical trial, the gold standard of study design, presented insurmountable problems for the testing of a cardiac valve prosthesis – problems such as the size of study samples necessary for the statistical analysis of extremely low adverse event rates. A compromise was adopted, whereby rates for the most frequently occurring complications for approved valve prostheses, as reported in peer-reviewed literature, were to be applied as comparators to those complications experienced with any new valve design" (Sapirstein, 2001).

TABLE III.2.9.1 Complications for Evaluating Clinical Performance of Mechanical and Biological Heart Valves, and Objective Performance Criteria (OPC) Values for Complication Rates (Percent/Year)*

Definitions from the Guidelines for Reporting Mortality and Morbidity after Cardiac Valve Interventions	OPC Values	
	Mechanical Valves	**Biological Valves**
Embolism: any embolic event that occurs in the absence of infection after the immediate perioperative period. May be manifested by a neurologic event or a noncerebral embolic event.	3.0	2.5
Valve thrombosis: any thrombus not caused by infection attached to or near an operated valve that occludes part of the blood flow path, interferes with valve function or is sufficiently large to warrant treatment.	0.8	0.2
Bleeding: any episode of major internal or external bleeding that causes death, hospitalization or permanent injury, or necessitates transfusion.	3.5 (1.5)	1.4 (0.9)
Parivalvular leak: resulting in dysfunction of an operated valve, as diagnosed by reoperation, autopsy or clinical investigation.	1.2 (0.6)	1.2 (0.6)
Endocarditis: any infection involving an operated valve.	1.2	1.2

*The complications and definitions are from the Guidelines for Reporting Mortality and Morbidity after Cardiac Valve Interventions (Akins et al., 2008a,b,c). The OPC values are taken from Appendix K of the Draft Replacement Heart Valve Guidance (Division of Cardiovascular, Respiratory, and Neurological Devices; Center for Devices and Radiological Health; US Food and Drug Administration, 1994). Numbers in parentheses are the OPC for major events.

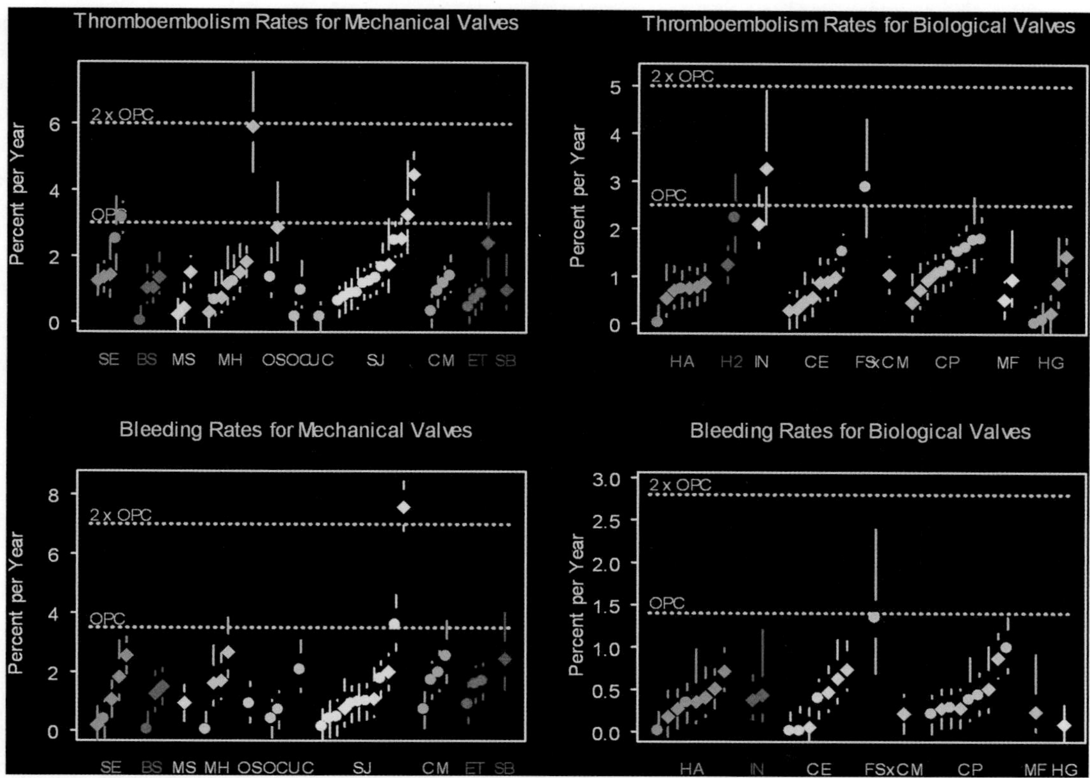

FIGURE III.2.9.7 Figures demonstrating the well-established rates and ranges of thromboembolism and bleeding for mechanical and biological valves. The height of each symbol represents the rate from one published series, with 90% confidence intervals shown by the vertical bars. The OPC (Objective Performance Criteria) are shown by horizontal lines. The FDA requirement, that the upper confidence limit be less than twice the OPC, is achieved by all but two of these valve series.

Support for the FDA Position

An endorsement of the FDA-mandated 800 valve-year criterion was recently reported. A huge randomized four-year study, designed to include 4400 patients from 17 centers, was undertaken to evaluate a new valve sewing ring designed to reduce endocarditis (Schaff et al., 1999). Since endocarditis is so rare, the study required over 10,000 valve-years of follow-up, but was prematurely terminated by the DSMB (Data Safety Monitoring Board) because of an unanticipated, statistically significant increase in re-operation for paravalvular leak (Schaff et al., 2002). However, a single-arm study of only 800 valve-years would have shown that the new sewing ring had a significantly greater incidence of major paravalvular leak than the OPC allowed (Grunkemeier et al., 2006).

Further endorsement of this FDA decision can be inferred from a randomized comparison of the earliest, caged-ball Starr-Edwards valve, and the most widely-used, modern, bileaflet St. Jude Medical valve. After five years of follow-up, the study found no difference in rates of complication or of symptomatic improvement between the two valves in either the aortic or mitral position (Murday et al., 1998). Note that even though these two valve models look nothing alike (Figure III.2.9.8), they perform the same basic mechanical functions (opening and closing) equally well.

The OPC approach is outlined in a recent paper written by members of the CDRH (Center for Devices and Radiological Health) (Chen et al., 2006), which states the FDA position: "The FDA is willing to accept alternative designs and controls, such as OPC, when based on sound arguments that they will provide sufficiently robust data ... Since their introduction in 1994, the FDA has approved 15 cardiac valve prostheses with data derived from single-arm studies ... a study design with OPC that has served well for open surgical cardiac valve replacement with established technology." At a Circulatory System Device Panel meeting, the FDA suggested using an OPC for survival to transplant for left ventricular assist devices used as a bridge to transplant (Neaton et al., 2007). An enhanced version of the OPC concept has also been recommended for evaluation of coronary artery stents (O'Malley et al., 2003).

Current FDA Draft Heart Valve Guidance

In January 2010, the FDA issued a Draft Guidance for heart valve IDE (Investigational Device Exemption) and PMA (US Department of Health and Human Services, 2010a). The section on *Clinical Investigations* follows the theme of the previous guidance document for conducting a clinical trial for heart valves. The FDA continues to endorse the use of

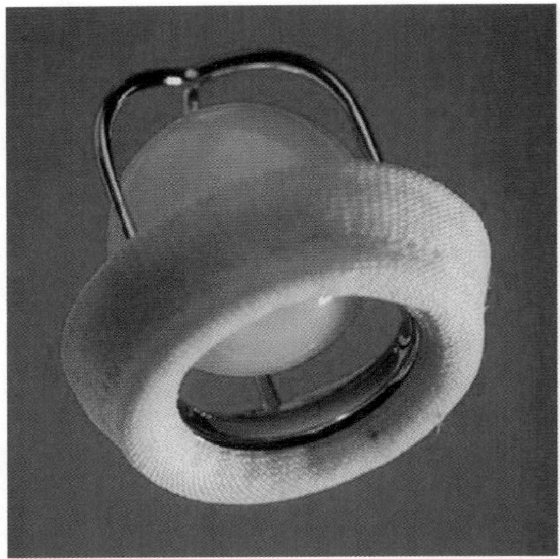
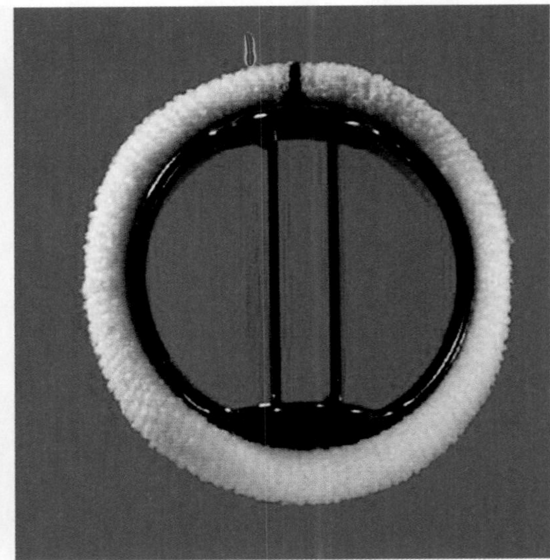

FIGURE III.2.9.8 Starr-Edwards (left); and St. Jude Medical (right) mechanical heart valves. (Reproduced with permission from St. Jude Medical)

historical controls (OPC): "Based on our long history in the evaluation of heart valves and the extensive reporting in the literature, FDA believes that a single-arm study with literature-based controls is the least burdensome approach to clinical evaluation of a replacement heart valve. Your clinical study should include appropriate controls including literature-based objective performance criteria (OPCs) for safety data, and literature articles and reports for both safety and effectiveness data."

IMPROVING OBSERVATIONAL STUDIES

Two 2004 papers in *The Lancet* (Lawlor et al., 2004; Vandenbroucke, 2004), and an accompanying commentary (Concato and Horwitz, 2004), describe the strengths of observational studies, and suggest ways of improving them. A recent series of three articles in the *British Medical Journal* provides a "critical appraisal" of observational (cohort) studies (Mamdani et al., 2005; Normand et al., 2005; Rochon et al., 2005). Their premise is that there is a place for observational studies, since RCTs: "face important ethical and logistical constraints, and have been criticized for focusing on highly selected populations and outcomes." They offer ways to improve cohort studies, but go on to say: "However, the promise of cohort studies as a useful source of evidence needs to be balanced against concerns about the validity of that evidence." And the series concerns itself with best methods for producing valid evidence, in the design (Rochon et al., 2005), assessment (Mamdani et al., 2005), and analysis (Normand et al., 2005) of such studies.

A 2007 article in the *Journal of the American Medical Association* (*JAMA*) compared three methods for removing the effects of selection bias in observational studies: multivariable model risk adjustment; propensity score analyses; and instrumental variable analysis (Stukel et al., 2007). An accompanying editorial (D'agostino and D'agostino, 2007) concluded: "RCTs should not always be considered as the only source of valid scientific information. The data collected from such studies are strong only if it can be shown that in fact a truly random sample of eligible patients participate and complete the protocol as designed." A 2007 letter in *JAMA* (Vandenbroucke, 2007) makes an interesting point about safety rather than efficacy: "When restricted to negative unexpected effects, observational research is equally reliable as randomized trials, which was recently shown by a comparison of randomized versus observational studies of adverse effects of treatment" (Papanikolaou et al., 2006; Vandenbroucke, 2006).

Propensity Score Analysis

The propensity score (PS) method was introduced in 1983 (Rosenbaum and Rubin, 1983), but has only become popular in recent years. A PubMed search (March 2010) for papers with "propensity score" in the title produced the following numbers of citations for the years 2001 to 2009, respectively (of course many studies used the PS method without including it in the title of the paper): 4, 5, 12, 16, 23, 25, 34, 43, and 43. One of the co-introducers of the PS method published a very readable description, including an intuitive appreciation for the validity of this method (Rubin, 1997). The FDA has suggested that: "For nonrandomized controlled trials, the use of propensity score analysis may allow a more appropriate comparison of control to treatment data" (Chen et al., 2006). An informative FDA perspective on the use of propensity score (PS) methods in medical device trials was recently published (Yue, 2007), along with seven discussion papers and a rejoinder.

Guidelines for Reporting (STROBE)

The well-known CONSORT (CONsolidated Standards Of Reporting Trials) statement was developed to improve the reporting of RCTs (Begg et al., 1996). Similarly, the recently developed STROBE (STrengthening the Reporting of OBservational studies in Epidemiology) statement contains recommendations on what should be included in an accurate and complete report of an observational study. The STROBE recommendations were simultaneously published in many journals (Von Elm et al., 2007a,b,c,d,e,f), and a supplementary elaboration was also simultaneously published in several journals (Vandenbroucke et al., 2007a,b,c). The STROBE checklist for cohort studies (Table III.2.9.2), and similar lists for case-control and cross-sectional studies are available at the STROBE website (Strobe, 2007). This website currently (2012) lists 110 journals that refer to the STROBE statement in their Instructions to Authors.

ADAPTIVE AND BAYESIAN TRIALS

In a traditional trial, patients are assigned equally to two treatments, and only at the end of the trial when the predetermined number of patients have been treated are we allowed to look at the data and decide which treatment is more effective. We begin by assuming that the therapies are equivalent (clinical equipoise), but if we were permitted to look at the accumulating data, we might discover that one of the treatments has already shown itself to be superior, and we could stop the trial and offer the superior treatment to the remaining patients. Another possibility is that the treatments are so similar that the trial should be stopped for futility or more patients than originally planned should be recruited. In an *adaptive*, also called *response-adaptive* trial (Rosenberger and Lachin, 1993), patient outcomes can be used as soon as they become available. This confers an ethical advantage, since fewer patients will be subjected to the inferior treatment during the trial.

Adaptive designs have been around for a long time. So-called *multi-armed bandit* designs originated in 1952 (Robbins, 1952), named after the *one-armed bandit* nickname for a slot machine gambling device. When used in a clinical trial, a bandit design aims to minimize the number of patients who get the inferior treatment. (In fact, the term *treatment arm* comes from the analogy to the bandit problem, where the "arms" represent different slot machines.)

An interesting form of adaptive trial is called *play the winner* (Zelen, 1969). The assignment of the current patient is decided by the result of the previous patient. If the previous patient's treatment was successful, it is used again on the next patient; if not, the other treatment is used. This requires that the treatment outcome is known quickly (e.g., procedural success). As an extreme example, in a trial of parachute versus no parachute, there would be only two possible outcomes: if the first patient is assigned "parachute," and the outcome is survival, then all future patients would be assigned parachute, and no patients would die. The worst case would be that the first patient is assigned "no parachute," and then there would be only one death in the trial.

Bayesian Analysis

Bayesian statistical methods are better for many applications, including clinical trials, than the more commonly used "frequentist" methods (reminds one of the putative superiority of Apple Macintosh computers in a world dominated by Windows-based computers). Bayesian designs allow prior knowledge to formally influence the results derived from the current study data, to apply knowledge about similar devices to formally augment the data generated by the present study device, just as one tends to do in everyday life. For example if 0 out of 5 new devices fail, the point estimate for failure is 0%, with an extremely wide *confidence interval*. But if this result comes from a device that is very similar to a previous device, about whose failure rate a substantial amount of (historical) information is already available, that information can be used to augment the results from a small sample of new devices (Connor and Berry, 2005). Moreover, the Bayesian summarization of the results is in terms of the probability of an effect being in a certain interval (*credible interval*). This is a natural, understandable summary statistic – the answer that most clinicians want, and that many erroneously think they are getting from a *confidence interval*, the frequentist counterpart. Medical inference is naturally deductive, as is the Bayesian approach, whereas traditional statistical methods are inductive (Goodman, 1999). Thus, most clinicians are Bayesians at heart. The Bayesian approach lends itself rather ideally to medical device trials (Hobbs and Carlin, 2008), especially when there is an incremental change to a new model, and information is available for previous models that are very similar to the new device (Figure III.2.9.3). Bayesian trials can often be shorter, with fewer patients involved, than a frequentist trial, because of the contribution of the prior information to the study.

The FDA is actively supporting the use of Bayesian methods for new device applications. An FDA workshop in April 2008 emphasized Bayesian analysis and adaptive designs in the evaluation of medical devices.

- In February 2010 the FDA published a *Guidance for the Use of Bayesian Statistics in Medical Device Clinical Trials* (US Department of Health and Human Services, 2010b).
- A recent talk on Bayesian Statistics, "Bayesian Statistics at the FDA: The Pioneering Experience with Medical Devices," was given by Greg Campbell, Director, Division of Biostatistics, of the CDRH (Center for Devices and Radiological Health) (Campbell, 2009).

TABLE III.2.9.2 STROBE Checklist of Items That Should be Included in Reports of Cohort Studies

	Item No	Recommendation
Title and abstract	1	(a) Indicate the study's design with a commonly used term in the title or the abstract (b) Provide in the abstract an informative and balanced summary of what was done and what was found
Introduction		
Background/ rationale	2	Explain the scientific background and rationale for the investigation being reported
Objectives	3	State specific objectives, including any prespecified hypotheses
Methods		
Study design	4	Present key elements of study design early in the paper
Setting	5	Describe the setting, locations, and relevant dates, including periods of recruitment, exposure, follow-up, and data collection
Participants	6	(a) Give the eligibility criteria, and the sources and methods of selection of participants. Describe methods of follow-up (b) For matched studies, give matching criteria and number of exposed and unexposed
Variables	7	Clearly define all outcomes, exposures, predictors, potential confounders, and effect modifiers. Give diagnostic criteria, if applicable
Data sources/ measurement	8*	For each variable of interest, give sources of data and details of methods of assessment (measurement). Describe comparability of assessment methods if there is more than one group
Bias	9	Describe any efforts to address potential sources of bias
Study size	10	Explain how the study size was arrived at
Quantitative variables	11	Explain how quantitative variables were handled in the analyses. If applicable, describe which groupings were chosen and why
Statistical methods	12	(a) Describe all statistical methods, including those used to control for confounding (b) Describe any methods used to examine subgroups and interactions (c) Explain how missing data were addressed (d) If applicable, explain how loss to follow-up was addressed (e) Describe any sensitivity analyses
Results		
Participants	13*	(a) Report numbers of individuals at each stage of study, e.g., numbers potentially eligible, examined for eligibility, confirmed eligible, included in the study, completing follow-up, and analyzed (b) Give reasons for non-participation at each stage (c) Consider use of a flow diagram
Descriptive data	14*	(a) Give characteristics of study participants (e.g., demographic, clinical, social) and information on exposures and potential confounders (b) Indicate number of participants with missing data for each variable of interest (c) Summarize follow-up time (e.g., average and total amount)
Outcome data	15*	Report numbers of outcome events or summary measures over time
Main results	16	(a) Give unadjusted estimates and, if applicable, confounder-adjusted estimates and their precision (e.g., 95% confidence interval). Make clear which confounders were adjusted for and why they were included (b) Report category boundaries when continuous variables were categorized (c) If relevant, consider translating estimates of relative risk into absolute risk for a meaningful time period
Other analyses	17	Report other analyses done, e.g., analyses of subgroups and interactions, and sensitivity analyses
Discussion		
Key results	18	Summarize key results with reference to study objectives
Limitations	19	Discuss limitations of the study, taking into account sources of potential bias or imprecision. Discuss both direction and magnitude of any potential bias
Interpretation	20	Give a cautious overall interpretation of results considering objectives, limitations, multiplicity of analyses, results from similar studies, and other relevant evidence
Generalizability	21	Discuss the generalizability (external validity) of the study results
Other Information		
Funding	22	Give the source of funding and the role of the funders for the present study and, if applicable, for the original study on which the present article is based

*Give information separately for exposed and unexposed groups.

- A Special Issue on Medical Device Clinical Studies appeared in the *Journal of Biopharmaceutical Statistics* (Vol. **18**(1), 2008), with Guest Editor Lilly Yue of CDRH/FDA. Many of the papers are written by FDA staff, and Bayesian statistics are prominently featured.
- The FDA and Johns Hopkins University co-sponsored a workshop, "Can Bayesian Approaches to Studying New Treatments Improve Regulatory Decision-Making?" The proceedings were published in the August 2005 issue of *Clinical Trials* (Alderson, 2005).

Bayesian Adaptive Trials

Bayesian methods are particularly well-suited to adaptive clinical trials, as was pointed out in a recent FDA guidance document that lists the advantages of a Bayesian adaptive trial, including modifying the sample size (to stop or to continue patient recruitment), and possibly stopping the trial early either for success, futility or harm (US Department of Health and Human Services, 2010b). A primary advantage of adaptive trials, and particularly Bayesian adaptive trials, stems from ethical considerations. These designs are most ethical, because they minimize the number of patients who ultimately get the inferior treatment, within the clinical trial itself, as opposed to the usual fixed sample size trial (Berry, 2004, 2006). This chapter is not intended to be a cookbook on how to design a Bayesian adaptive trial, but excellent cookbooks do exist (Spiegelhalter et al., 2004).

CLINICAL TRIALS FOR TRANSCATHETER VALVES

We began this chapter with the story of the first successful heart valve. The latest technology and current clinical study challenge for this type of medical device is the percutaneous or transcatheter replacement heart valve. In the US, the first FDA-approved (2010) transcatheter valve is for the pulmonary position (Medtronic Melody® Transcatheter Pulmonary Valve), intended to treat desperate congenital situations under the Humanitarian Device Exemption, in a relatively small percentage of patients. The overwhelming clinical impact will be from devices developed to treat the growing population of adult patients with severe aortic stenosis.

Note on terminology: Both *percutaneous* and *transcatheter* have been used to describe this type of heart valve intervention. Since these valves can sometimes be implanted via an incision in the left ventricular apex, as well as by the more common femoral artery, using catheterization techniques, *transcatheter* is the more-encompassing adjective. Also, since the native (or failed bioprosthetic) valve is not explanted, but compressed against the aortic root, the term "implantation" is preferable to "replacement." Thus, although PAVR (Percutaneous Aortic Valve Replacement) is still being used (Chiam and Ruiz, 2009), TAV (Transcatheter Aortic Valve) and TAVI (TAV Implantation) are the preferred acronyms.

Historical Perspective

The first CABG surgery was performed in 1967 (Favaloro et al., 1967). Then, just 10 years later, cardiologists discovered a way to repair clogged coronary arteries using catheters, at first just dilating them with a catheter-placed balloon (Gruentzig, 1982), then inserting a (bare metal) stent as well to keep the artery expanded (the Palmaz-Schatz stent was FDA-approved in 1994), and more recently using drug-eluting stents (the Cypher® stent was FDA-approved in 2003; the Taxus® stent was FDA-approved in 2004) to attenuate restenosis of the stent. Contrast that 10-year lag time with how long it took to discover a way to perform heart valve insertion via catheter; the first transcatheter aortic valve implantation was performed more than 40 years (Cribier et al., 2002) after Starr's surgical replacement (Starr and Edwards, 1961). However late arriving, this new cardiologic/surgical partnership paradigm will be transformational in the treatment of heart valve disease in the future.

In the early 1990s, there was another pending paradigm shift in heart valve prostheses, marked by the introduction of *stentless* bioprosthetic valves. The original tissue or bioprosthetic heart valves, developed by Carpentier and first marketed by Hancock, a company later bought by Medtronic, were porcine valves with a frame-like structure called a *stent* that supported the leaflets and attached to the sewing ring. Then, to obtain anticipated improvements in hemodynamics and durability, motivated by the perceived advantages of human homograft transplant valves, several valve manufacturers started producing porcine valves *without* a stent. At that time, we argued for the use of single-arm studies for evaluation of these valves: "Moreover, for evaluating a new heart valve, the primary purpose should be estimation rather than comparison with a current device. This can be accomplished faster, in a more generalizable way, and with more concern for patient care by using non-randomized comparison groups" (Grunkemeier and Starr, 1992). In fact, several stentless valves were FDA-approved by the late 1990s, based on prospective, single-arm, historically-controlled (using OPC) observational studies. As we have seen, implantable devices, especially innovative devices such as TAVs, and their implantation are adaptive in nature, and thus eminently well-suited to adaptive study designs and the use of historical controls.

Clinical Trials: The Societies Weigh In

In 2005, the "Clinical Development of Percutaneous Heart Valve Technology," a joint position statement from the Society of Thoracic Surgeons, the American Association for Thoracic Surgery, the American College of Cardiology, the Society for Cardiovascular Angiography and Interventions, the US Food and Drug Administration, the Centers for Medicare and Medicaid Services, and numerous representatives from industry, recommended that RCTs should be used to evaluate TAVI (Vassiliades et al., 2005a,b,c,d).

However, a more recent (2008) position paper gives another perspective from the Europeans, who are already quite experienced with TAVI. Both the Edwards Lifesciences SAPIEN and Medtronic CoreValve® devices have already received their CE (Conformité Européenne) Mark approvals in 2007, and are being marketed in Europe, with about 50,000 implants to date (2012). This position paper is jointly from the European Association of Cardio-Thoracic Surgery and the European Society of Cardiology, in collaboration with the European Association of Percutaneous Cardiovascular Interventions (Vahanian et al., 2008a,b,c), and states that: "The Committee believes that randomized trials are highly desirable once greater experience has been acquired, and only small modifications in the technology used are to be expected." This statement agrees with our previous arguments: "Although randomization is not practicable for establishing the adequacy of a new heart valve, it is an optimal strategy for comparing subtle differences in an established valve" (Grunkemeier et al., 2000). "Marketing approval can be based on an absolute comparison with qualifying standards through the use of historical controls. However, short-term studies (randomized or not) cannot detect increased rates of rare or late events. Long-term post-market studies, including RCTs where appropriate, can be used to detect rare complications and differences with regard to late-occurring complications among otherwise comparable valves" (Grunkemeier, 1998).

This revolutionary TAV therapy was mentioned as the motivation for the FDA rethinking the design of device trials (Chen et al., 2006). A recent overview points out the benefits of the Bayesian approach and of propensity score analysis in clinical trials, and discusses how best to assess results and outcomes: "which may require a paradigm shift in mindset. Apart from the randomized controlled trial, some of the more novel concepts in trial design which may be more suitable in this area are also explored" (Chiam and Ruiz, 2008).

The Devices and Their Clinical Trials

To date there are at least 17 TAV contenders, seven of which have already been used clinically (Chiam and Ruiz, 2009). Two of these are currently the most well-developed: the Edwards Lifesciences SAPIEN valve (Figure III.2.9.9, left); and the Medtronic CoreValve® (Figure III.2.9.9, right). Technical details of these valves are given elsewhere in this volume (Chapter II.5.3); below we focus on their clinical studies.

Edwards Lifesciences SAPIEN Valve. Edwards Lifesciences, the company started by Lowell Edwards as chronicled at the beginning of this chapter, is producing a balloon-expandable bovine pericardial TAV. The valve is based on the first-ever-in-man TAV implanted in 2002 (Cribier et al., 2002). Edwards has designed an RCT, called the PARTNER (Placement of AoRTic traNscathetER valves) Trial to seek FDA approval (US National Institutes of Health, 2007). This is two studies in one: in one study (PARTNER A), the SAPIEN valve is randomized against a conventional, surgically implanted valve; in the other study (PARTNER B), SAPIEN is randomized against medical management in patients who were not considered suitable for surgery. In November 2011 Edwards received FDA approval to use the SPAIEN valve in these latter, "unoperable", patients. The PARTNER Trial design specifies 22 exclusion criteria, so deployment will not be to the "real-world" of aortic valve patients, and the problem of internal versus external validity in this study must be addressed. The SAPIEN leaflet material has already changed from untreated equine to treated bovine pericardium, the

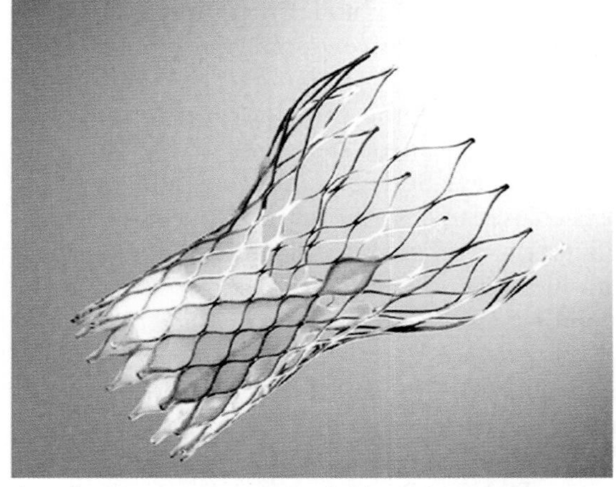

FIGURE III.2.9.9 Transcatheter heart valve implantation devices: Edwards Lifesciences SAPIEN (left); and Medtronic CoreValve® (right). Reproduced with permission from Edwards LifeSciences and CoreValve.

catheter sizes have changed, a next-generation Edwards SAPIEN XT valve has already received the CE Mark. Not only are the devices continually evolving, but the operator skills and patient selection (Chiam and Ruiz, 2009) are also improving with time, and consequently the mortality and major complications are continuing to decrease. "In the initial 15 patients there were five deaths for an early mortality of 33%. After this initial learning curve, seven of the remaining 56 patients died, for an early mortality of 12.5%" (Ye et al., 2010). These changes are almost entirely on the SAPIEN arms of the study, not the control arms. In this setting, a fixed, randomized study seems to violate what was said in a short, but important, FDA publication (Sapirstein, 2001) which mentions that the OPC method was chosen for heart valves because an RCT: "presented insurmountable problems for the testing of a cardiac valve prosthesis ..." And, although this: "approval process ... has the limitations characteristic of any nonrandomized or non-concurrent controlled study. It was accepted as necessary in part to accommodate timely availability of technologic advances for the fabrication of prosthetic cardiac valves" (Sapirstein, 2001).

In September 2010, the PARTNER Trial reported promising early results for the study in 358 patients with severe, symptomatic aortic stenosis, who were deemed inoperable. They concluded that the SAPIEN valve: "... as compared with standard therapy, significantly reduced the rates of death from any cause, the composite end point of death from any cause or repeat hospitalization, and cardiac symptoms, despite the higher incidence of major strokes and major vascular events" (Leon et al., 2010).

Medtronic CoreValve®. CoreValve, Inc. was founded in 2001, and performed its first-in-man transcatheter valve procedure in 2004. Its ReValving System is comprised of a porcine (originally bovine) pericardial tissue valve, mounted on a self-expanding frame. As of November 2010 it had been implanted in more than 20,000 patients worldwide. Medtronic acquired CoreValve in 2009, and in 2010 began the CoreValve Advance International Post Market Study. This is an observational study in the "real-world" patient population of an estimated 1000 patients in 70 cardiac centers in 13 countries outside the US (US National Institutes of Health, 2010). Yet, to get approval for marketing in the US, an RCT of about 1200 patients will soon be launched (Medtronic, Inc., 2010).

SUMMARY

We discuss clinical trials for medical devices using heart valves as an example. During the past 50 years, many excellent heart valve prostheses have been developed and successfully introduced into clinical use. Since 1976, the FDA has overseen the market approval process in the US. Perhaps surprisingly, the FDA requirement for the definitive clinical trial is based on historical controls, called Objective Performance Criteria (OPC), rather than on randomized clinical trials (RCT). This method was introduced in a 1993 guidance document, and again upheld in a 2010 draft guidance document. Although RCT is the preferred method of clinical evaluation for drugs, for heart valves in particular, and for most medical devices in general, a well-designed and carefully monitored single-armed prospective study using historical controls is sufficient, and more efficient in terms of cost and duration of the study. For comparative trials, the FDA has also advocated adaptive designs, and in particular Bayesian adaptive designs. These result in more humane trials, because the fewest number of patients in the trial are subjected to what turns out to be the inferior therapy. The most recent development in this arena is heart valves intended to be implanted using a catheter. Although not yet approved for marketing in the US, these revolutionary valves promise to transform the treatment of aortic stenosis, especially in the elderly. Two different companies are mounting clinical trials of these devices, both using RCTs. It will be interesting to see whether any new clinical information about the performance of these valves comes from the RCTs that is not already known, or will become known, from the nonrandomized observational studies.

BIBLIOGRAPHY

Aitkin, M., Laird, N., & Francis, B. (1983). Reanalysis of the Stanford Heart Transplant Data. *JASA*, **78**, 264–274.

Akins, C. W., Miller, D. C., Turina, M. I., Kouchoukos, N. T., Blackstone, E. H., et al. (2008a). Guidelines for reporting mortality and morbidity after cardiac valve interventions. *Eur. J. Cardiothorac. Surg.*, **33**, 523–528.

Akins, C. W., Miller, D. C., Turina, M. I., Kouchoukos, N. T., Blackstone, E. H., et al. (2008b). Guidelines for reporting mortality and morbidity after cardiac valve interventions. *J. Thorac. Cardiovasc. Surg.*, **135**, 732–738.

Akins, C. W., Miller, D. C., Turina, M. I., Kouchoukos, N. T., Blackstone, E. H., et al. (2008c). Guidelines for reporting mortality and morbidity after cardiac valve interventions. *Ann. Thorac. Surg.*, **85**, 1490–1495.

Alderson, N. E. (2005). Editorial. Clinical Trials, **2**, 271–272.

Altman, D. G., & Bland, J. M. (1999). Statistics notes. Treatment allocation in controlled trials: Why randomise? *BMJ*, **318**, 1209.

Anyanwu, A. C., & Treasure, T. (2004). Surgical research revisited: Clinical trials in the cardiothoracic surgical literature. *Eur. J Cardiothorac. Surg.*, **25**, 299–303.

Barkun, J. S., Aronson, J. K., Feldman, L. S., Maddern, G. J., Strasberg, S. M., et al. (2009). Evaluation and stages of surgical innovations. *Lancet*, **374**, 1089–1096.

Begg, C., Cho, M., Eastwood, S., Horton, R., Moher, D., et al. (1996). Improving the quality of reporting of randomized controlled trials. The CONSORT statement. *JAMA*, **276**, 637–639.

Benson, K., & Hartz, A. J. (2000). A comparison of observational studies and randomized, controlled trials. *N. Engl. J. Med.*, **342**, 1878–1886.

Berger, V. W., & Exner, D. V. (1999). Detecting selection bias in randomized clinical trials. *Control Clin. Trials*, **20**, 319–327.

Berry, D. A. (2004). Bayesian statistics and the efficiency and ethics of clinical trials. *Stat. Sci.*, **19**, 175–187.

Berry, D. A. (2006). Bayesian clinical trials. *Nat. Rev. Drug Discov.*, **5**, 27–36.

Black, N. (1996). Why we need observational studies to evaluate the effectiveness of health care. *BMJ*, **312**, 1215–1218.

Britton, A., McKee, M., Black, N., McPherson, K., Sanderson, C., et al. (1998). Choosing between randomised and non-randomised studies: A systematic review. *Health Technol. Assess.* 2(i–iv), 1–124.

Britton, A., McKee, M., Black, N., McPherson, K., Sanderson, C., et al. (1999). Threats to applicability of randomised trials: Exclusions and selective participation. *J. Health Serv. Res. Policy*, **4**, 112–121.

Campbell, G. (2009). *Guidance for the use of Bayesian statistics in medical device clinical trials*. http://www.stat.fsu.edu/50th/Campbell.ppt. Accessed 5.3.2010.

Chen, E., Sapirstein, W., Ahn, C., Swain, J., & Zuckerman, B. (2006). FDA perspective on clinical trial design for cardiovascular devices. *Ann. Thorac. Surg.*, **82**, 773–775.

Chiam, P. T., & Ruiz, C. E. (2008). Percutaneous transcatheter aortic valve implantation: Assessing results, judging outcomes, and planning trials. The interventionalist perspective. *JACC Cardiovasc. Interv.*, **1**, 341–350.

Chiam, P. T. L., & Ruiz, C. E. (2009). Percutaneous transcatheter aortic valve implantation: Evolution of the technology. *Am. Heart J.*, **157**, 229–242.

Concato, J. (2004). Observational versus experimental studies: What's the evidence for a hierarchy? *NeuroRx*, **1**, 341–347.

Concato, J., & Horwitz, R. I. (2004). Beyond randomised versus observational studies. *Lancet*, **363**, 1660–1661.

Concato, J., Shah, N., & Horwitz, R. I. (2000). Randomized, controlled trials, observational studies, and the hierarchy of research designs. *N. Engl. J. Med.*, **342**, 1887–1892.

Connor, J. T., & Berry, S. M. (2005). Bayesian analysis for medical device trials. *Am. J. Gastroenterol.*, **100**, 1732–1735.

Cribier, A., Eltchaninoff, H., Bash, A., Borenstein, N., Tron, C., et al. (2002). Percutaneous transcatheter implantation of an aortic valve prosthesis for calcific aortic stenosis: First human case description. *Circulation*, **106**, 3006–3008.

D'agostino, R. B., Jr., & D'agostino, R. B., Sr. (2007). Estimating treatment effects using observational data. *JAMA*, **297**, 314–316.

Division of Cardiovascular, Respiratory, and Neurological Devices; Center for Devices and Radiological Health; US Food and Drug Administration (1994). Draft Replacement Heart Valve Guidance. Version 4.1.

Ergina, P. L., Cook, J. A., Blazeby, J. M., Boutron, I., Clavien, P. A., et al. (2009). Challenges in evaluating surgical innovation. *Lancet*, **374**, 1097–1104.

Favaloro, R. G., Effler, D. B., Groves, L. K., Sones, F. M., Jr., & Fergusson, D. J. (1967). Myocardial revascularization by internal mammary artery implant procedures. Clinical experience. *J. Thorac. Cardiovasc. Surg.*, **54**, 359–370.

Fisher, R. A. (1926). The arrangement of field experiments. *J. Min. Agric. G. Br.*, **33**, 503–513.

Gao, G., Wu, Y., Grunkemeier, G. L., Furnary, A. P., & Starr, A. (2004). Forty-year survival with the Starr-Edwards heart valve prosthesis. *J. Heart Valve Dis.*, **13**, 91–96. Discussion 96.

Garbe, E., & Suissa, S. (2004). Hormone replacement therapy and acute coronary outcomes: Methodological issues between randomized and observational studies. *Hum. Reprod.*, **19**, 8–13.

Glasgow, R. E., Magid, D. J., Beck, A., Ritzwoller, D., & Estabrooks, P. A. (2005). Practical clinical trials for translating research to practice: Design and measurement recommendations. *Med. Care.*, **43**, 551–557.

Glasziou, P., Chalmers, I., Rawlins, M., & McCulloch, P. (2007). When are randomised trials unnecessary? Picking signal from noise. *BMJ*, **334**, 349–351.

Goodman, S. N. (1999). Toward evidence-based medical statistics. 1: The P value fallacy. *Ann. Intern. Med.*, **130**, 995–1004.

Grapow, M. T., Von Wattenwyl, R., Guller, U., Beyersdorf, F., & Zerkowski, H. R. (2006). Randomized controlled trials do not reflect reality: Real-world analyses are critical for treatment guidelines! *J. Thorac. Cardiovasc. Surg.*, **132**, 5–7.

Gruentzig, A. (1982). Results from coronary angioplasty and implications for the future. *Am. Heart. J.*, **103**, 779–783.

Grunkemeier, G. L. (1998). Clinical studies of prosthetic heart valves using historical controls. In K. M. Becker, & J. J. Whyte (Eds.), *Clinical Evaluation of Medical Devices: Principles and Case Studies* (pp. 83–102). New York, NY: Humana Press Inc.

Grunkemeier, G. L., & Starr, A. (1992). Alternatives to randomization in surgical studies. *J. Heart Valve Dis.*, **1**, 142–151.

Grunkemeier, G. L., Starr, A., & Rahimtoola, S. H. (1992). Prosthetic heart valve performance: long-term follow-up. *Curr. Probl. Cardiol.*, **17**, 329–406.

Grunkemeier, G. L., Johnson, D. M., & Naftel, D. C. (1994). Sample size requirements for evaluating heart valves with constant risk events. *J. Heart Valve Dis.*, **3**, 53–58.

Grunkemeier, G. L., Li, H. H., Naftel, D. C., Starr, A., & Rahimtoola, S. H. (2000). Long-term performance of heart valve prostheses. *Curr. Probl. Cardiol.*, **25**, 73–154.

Grunkemeier, G. L., Jin, R., & Starr, A. (2006). Prosthetic heart valves: Objective Performance Criteria versus randomized clinical trial. *Ann. Thorac. Surg.*, **82**, 776–780.

Hammermeister, K., Sethi, G. K., Henderson, W. G., Grover, F. L., Oprian, C., et al. (2000). Outcomes 15 years after valve replacement with a mechanical versus a bioprosthetic valve: Final report of the Veterans Affairs randomized trial. *J. Am. Coll. Cardiol.*, **36**, 1152–1158.

Hill, A. B. (1951). The clinical trial. *Br. Med. Bull.*, **7**, 278–282.

Hlatky, M. A., Califf, R. M., Harrell, F. E., Jr., Lee, K. L., Mark, D. B., et al. (1988). Comparison of predictions based on observational data with the results of randomized controlled clinical trials of coronary artery bypass surgery. *J. Am. Coll. Cardiol.*, **11**, 237–245.

Hobbs, B. P., & Carlin, B. P. (2008). Practical Bayesian design and analysis for drug and device clinical trials. *J. Biopharm. Stat.*, **18**, 54–80.

Holden, C. (1976). Medical devices law is on the books at last. *Science*, **192**, 1216.

Hollis, S., & Campbell, F. (1999). What is meant by intention to treat analysis? Survey of published randomised controlled trials. *BMJ*, **319**, 670–674.

Hordijk-Trion, M., Lenzen, M., Wijns, W., De Jaegere, P., Simoons, M. L., et al. (2006). Patients enrolled in coronary intervention trials are not representative of patients in clinical practice: Results from the Euro Heart Survey on Coronary Revascularization. *Eur. Heart J.*, **27**, 671–678.

Hornberger, J., & Wrone, E. (1997). When to base clinical policies on observational versus randomized trial data. *Ann. Intern. Med.*, **127**, 697–703.

Ioannidis, J. P. (2005). Why most published research findings are false. *PLoS Med.*, **2**, e124.

Ioannidis, J. P., Haidich, A. B., & Lau, J. (2001). Any casualties in the clash of randomised and observational evidence? *BMJ*, **322**, 879–880.

Kunz, R., & Oxman, A. D. (1998). The unpredictability paradox: Review of empirical comparisons of randomised and non-randomised clinical trials. *BMJ*, **317**, 1185–1190.

Lawlor, D. A., Davey Smith, G., Kundu, D., Bruckdorfer, K. R., & Ebrahim, S. (2004). Those confounded vitamins: What can we learn from the differences between observational versus randomised trial evidence? *Lancet*, **363**, 1724–1727.

Leon, M. B., Smith, C. R., Mack, M., Miller, D. C., Moses, J. W., et al. (2010). Transcatheter aortic-valve implantation for aortic stenosis in patients who cannot undergo surgery. *N. Engl. J. Med.*, **363**, 1579–1607.

Lewsey, J. D., Murray, G. D., Leyland, A. H., & Boddy, F. A. (1999). Comparing outcomes of percutaneous transluminal coronary angioplasty with coronary artery bypass grafting; can routine health service data complement and enhance randomized controlled trials? *Eur. Heart J.*, **20**, 1731–1735.

Li, H., & Yue, L. Q. (2008). Statistical and regulatory issues in nonrandomized medical device clinical studies. *J. Biopharm. Stat.*, **18**, 20–30.

Liu, Z., Valencia, O., Treasure, T., & Murday, A. J. (1998). Cold blood cardioplegia or intermittent cross-clamping in coronary artery bypass grafting? *Ann. Thorac. Surg.*, **66**, 462–465.

Mamdani, M., Sykora, K., Li, P., Normand, S. L., Streiner, D. L., et al. (2005). Reader's guide to critical appraisal of cohort studies: 2. Assessing potential for confounding. *BMJ*, **330**, 960–962.

McCulloch, P., Taylor, I., Sasako, M., Lovett, B., Griffin, D., et al. (2002). Randomised trials in surgery: Problems and possible solutions. When are randomised trials unnecessary? Picking signal from noise. *BMJ*, **324**, 1448–1451.

McKee, M., Britton, A., Black, N., McPherson, K., Sanderson, C., et al. (1999). Methods in health services research. Interpreting the evidence: Choosing between randomised and non-randomised studies. *BMJ*, **319**, 312–315.

Medtronic, Inc. (2010). *Medtronic Clinical Trial Receives FDA Approval to Evaluate New CoreValve® System for Aortic Valve Implantation.* http://wwwp.medtronic.com/Newsroom/NewsReleaseDetails.do?itemId=1287081500254&lang=en_US. Accessed 19.10.2010.

Moghissi, K. (2009). Editorial comment: Are randomised trials needed in the era of rapidly evolving technologies? *Eur. J. Cardiothorac. Surg.*, **35**, 478–479.

Murday, A., Miles, J., Taylor, B., & Treasure, T. (1998). A prospective randomised trial of St. Jude versus Starr-Edwards aortic and mitral valve prostheses. 1998 EACTS meeting.

Neaton, J. D., Normand, S. L., Gelijns, A., Starling, R. C., Mann, D. L., et al. (2007). Designs for mechanical circulatory support device studies. *J. Card. Fail.* **13**, 63–74.

Ney, P. G. (1989). Double-blinding in clinical trials. *CMAJ*, **140**, 15.

Normand, S. L., Sykora, K., Li, P., Mamdani, M., Rochon, P. A., et al. (2005). Readers guide to critical appraisal of cohort studies: 3. Analytical strategies to reduce confounding. *BMJ*, **330**, 1021–1023.

O'Malley, A. J., Normand, S. L., & Kuntz, R. E. (2003). Application of models for multivariate mixed outcomes to medical device trials: Coronary artery stenting. *Stat. Med.*, **22**, 313–336.

Office of the Federal Register. (2009). *Determination of safety and effectiveness.* The Code of Federal Regulations (Title 21, Volume 8, Section 860.7).

Oxenham, H., Bloomfield, P., Wheatley, D. J., Lee, R. J., Cunningham, J., et al. (2003). Twenty year comparison of a Bjork-Shiley mechanical heart valve with porcine bioprostheses. *Heart*, **89**, 715–721.

Padkin, A., Rowan, K., & Black, N. (2001). Using high quality clinical databases to complement the results of randomised controlled trials: The case of recombinant human activated protein C. *BMJ*, **323**, 923–926.

Papanikolaou, P. N., Christidi, G. D., & Ioannidis, J. P. (2006). Comparison of evidence on harms of medical interventions in randomized and nonrandomized studies. *CMAJ*, **174**, 635–641.

Pennello, G., & Thompson, L. (2008). Experience with reviewing Bayesian medical device trials. *J. Biopharm. Stat.*, **18**, 81–115.

Pocock, S. J., & Elbourne, D. R. (2000). Randomized trials or observational tribulations? *N. Engl. J. Med.*, **342**, 1907–1909.

Robbins, H. (1952). Some aspects of the sequential design of experiments. *Bull. Amer. Math. Soc.*, **58**, 527–535.

Rochon, P. A., Gurwitz, J. H., Sykora, K., Mamdani, M., Streiner, D. L., et al. (2005). Reader's guide to critical appraisal of cohort studies: 1. Role and design. *BMJ*, **330**, 895–897.

Rosenbaum, P. R., & Rubin, D. B. (1983). The central role of the propensity score in observational studies for causal effects. *Biometrika*, **70**, 41–55.

Rosenberger, W. F., & Lachin, J. M. (1993). The use of response-adaptive designs in clinical trials. *Control Clin. Trials*, **14**, 471–484.

Rubin, D. B. (1997). Estimating causal effects from large data sets using propensity scores. *Ann. Intern. Med.*, **127**, 757–763.

Sapirstein, W. (2001). Designing trials for testing prosthetic cardiac valves: A Food and Drug Administration perspective. *Am. Heart J.*, **141**, 861–863.

Schaff, H., Carrel, T., Steckelberg, J. M., Grunkemeier, G. L., & Holubkov, R. (1999). Artificial Valve Endocarditis Reduction Trial (AVERT): Protocol of a multicenter randomized trial. *J. Heart Valve Dis.*, **8**, 131–139.

Schaff, H. V., Carrel, T. P., Jamieson, W. R., Jones, K. W., Rufilanchas, J. J., et al. (2002). Paravalvular leak and other events in silzone-coated mechanical heart valves: A report from AVERT. *Ann. Thorac. Surg.*, **73**, 785–792.

Schoen, F. J., & Padera, R. F. (2010). Cardiovascular Medical Devices 1: Substitute Heart Valves. In B. D. Ratner, A. S. Hoffman, F. J. Schoen, & J. E. Lemons (Eds.), *Biomaterials Science: An Introduction to Materials in Medicine.* New York, NY: Elsevier.

Smith, G. C., & Pell, J. P. (2003). Parachute use to prevent death and major trauma related to gravitational challenge: Systematic review of randomised controlled trials. *BMJ*, **327**, 1459–1461.

Spiegelhalter, D. J., Abrams, K. R., & Myles, J. P. (2004). *Bayesian Approaches to Clinical Trials and Health-Care Evaluation (Statistics in Practice).* Chichester, UK: Wiley.

Starr, A., & Edwards, M. (1961). Mitral replacement: Clinical experience with a ball valve prosthesis. *Ann. Surg.*, **154**, 726–740.

Steg, P. G., Lopez-Sendon, J., Lopez De Sa, E., Goodman, S. G., Gore, J. M., et al. (2007). External validity of clinical trials in acute myocardial infarction. *Arch. Intern. Med.*, **167**, 68–73.

Strobe (Strengthening the Reporting of Observational Studies in Epidemiology). (2007). *STROBE checklists*: http://www.strobe-statement.org/index.php?id=available-checklists. Accessed 3.5.2010.

Stukel, T. A., Fisher, E. S., Wennberg, D. E., Alter, D. A., Gottlieb, D. J., et al. (2007). Analysis of observational studies in the presence of treatment selection bias: Effects of invasive cardiac management on AMI survival using propensity score and instrumental variable methods. *JAMA*, **297**, 278–285.

Treasure, T. (2009). Are randomised trials needed in the era of rapidly evolving technologies? *Eur. J. Cardiothorac. Surg.*, **35**, 474–478. Discussion 478–479.

Tunis, S. R., Stryer, D. B., & Clancy, C. M. (2003). Practical clinical trials: Increasing the value of clinical research for decision making in clinical and health policy. *JAMA*, **290**, 1624–1632.

US Department of Health and Human Services. (2010a). *Draft Guidance for Industry and FDA Staff: Heart Valves – Investigational Device Exemption (IDE) and Premarket Approval (PMA) Applications.* http://www.fda.gov/MedicalDevices/DeviceRegulationandGuidance/GuidanceDocuments/ucm193096.htm. Date accessed 3/5/2010.

US Department of Health and Human Services. (2010b). *Guidance for the Use of Bayesian Statistics in Medical Device Clinical Trials.* http://www.fda.gov/MedicalDevices/DeviceRegulationandGuidance/GuidanceDocuments/ucm071072. Accessed 3.5.2010.

US National Institutes of Health. (2007). *THE PARTNER TRIAL: Placement of AoRTic TraNscathetER Valve Trial.* http://clinicaltrials.gov/ct2/show/NCT00530894. Accessed 3.5.2010.

US National Institutes of Health. (2010). *CoreValve Advance International Post Market Study.* http://clinicaltrials.gov/ct2/show/NCT01074658?term=corevalve&rank=1. Accessed 19.10.2010.

Vahanian, A., Alfieri, O., Al-Attar, N., Antunes, M., Bax, J., et al. (2008a). Transcatheter valve implantation for patients with aortic stenosis: A position statement from the European Association of Cardio-Thoracic Surgery (EACTS) and the European Society of Cardiology (ESC), in collaboration with the European Association of Percutaneous Cardiovascular Interventions (EAPCI). *EuroIntervention*, **4**, 193–199.

Vahanian, A., Alfieri, O., Al-Attar, N., Antunes, M., Bax, J., et al. (2008b). Transcatheter valve implantation for patients with aortic stenosis: A position statement from the European Association of Cardio-Thoracic Surgery (EACTS) and the European Society of Cardiology (ESC), in collaboration with the European Association of Percutaneous Cardiovascular Interventions (EAPCI). *Eur. Heart J.*, **29**, 1463–1470.

Vahanian, A., Alfieri, O. R., Al-Attar, N., Antunes, M. J., Bax, J., et al. (2008c). Transcatheter valve implantation for patients with aortic stenosis: A position statement from the European Association of Cardio-Thoracic Surgery (EACTS) and the European Society of Cardiology (ESC), in collaboration with the European Association of Percutaneous Cardiovascular Interventions (EAPCI). *Eur. J. Cardiothorac. Surg.*, **34**, 1–8.

Vandenbroucke, J. P. (2004). When are observational studies as credible as randomised trials? *Lancet*, **363**, 1728–1731.

Vandenbroucke, J. P. (2006). What is the best evidence for determining harms of medical treatment? *CMAJ*, **174**, 645–646.

Vandenbroucke, J. P. (2007). Analytic approaches to observational studies with treatment selection bias. *JAMA*, **297**, 2077–2078. author reply 2078.

Vandenbroucke, J. P. (2008). Observational research, randomised trials, and two views of medical science. *PLoS Med.*, **5**, e67.

Vandenbroucke, J. P., Von Elm, E., Altman, D. G., Gotzsche, P. C., Mulrow, C. D., et al. (2007a). Strengthening the Reporting of Observational Studies in Epidemiology (STROBE): Explanation and elaboration. *PLoS Med.*, **4**, e297.

Vandenbroucke, J. P., Von Elm, E., Altman, D. G., Gotzsche, P. C., Mulrow, C. D., et al. (2007b). Strengthening the Reporting of Observational Studies in Epidemiology (STROBE): Explanation and elaboration. *Ann. Intern. Med.*, **147**, W163–194.

Vandenbroucke, J. P., Von Elm, E., Altman, D. G., Gotzsche, P. C., Mulrow, C. D., et al. (2007c). Strengthening the Reporting of Observational Studies in Epidemiology (STROBE): Explanation and elaboration. *Epidemiology*, **18**, 805–835.

Vassiliades, T. A., Jr., Block, P. C., Cohn, L. H., Adams, D. H., Borer, J. S., et al. (2005a). The clinical development of percutaneous heart valve technology: A position statement of the Society of Thoracic Surgeons (STS), the American Association for Thoracic Surgery (AATS), and the Society for Cardiovascular Angiography and Interventions (SCAI). *Ann. Thorac. Surg.*, **79**, 1812–1818.

Vassiliades, T. A., Jr., Block, P. C., Cohn, L. H., Adams, D. H., Borer, J. S., et al. (2005b). The clinical development of percutaneous heart valve technology: A position statement of the Society of Thoracic Surgeons (STS), the American Association for Thoracic Surgery (AATS), and the Society for Cardiovascular Angiography and Interventions (SCAI) Endorsed by the American College of Cardiology Foundation (ACCF) and the American Heart Association (AHA). *J. Am. Coll. Cardiol.*, **45**, 1554–1560.

Vassiliades, T. A., Jr., Block, P. C., Cohn, L. H., Adams, D. H., Borer, J. S., et al. (2005c). The clinical development of percutaneous heart valve technology: A position statement of the Society of Thoracic Surgeons (STS), the American Association for Thoracic Surgery (AATS), and the Society of Cardiovascular Angiography and Intervention (SCAI). *Catheter. Cardiovasc. Interv.*, **65**, 73–79.

Vassiliades, T. A., Jr., Block, P. C., Cohn, L. H., Adams, D. H., Borer, J. S., et al. (2005d). The clinical development of percutaneous heart valve technology: A position statement of the Society of Thoracic Surgeons (STS), the American Association for Thoracic Surgery (AATS), and the Society of Cardiovascular Angiography and Intervention (SCAI). *J. Thorac. Cardiovasc. Surg.*, **129**, 970–976.

Victora, C. G., Habicht, J. P., & Bryce, J. (2004). Evidence-based public health: Moving beyond randomized trials. *Am. J. Public Health*, **94**, 400–405.

Von Elm, E., Altman, D. G., Egger, M., Pocock, S. J., Gotzsche, P. C., et al. (2007a). The Strengthening the Reporting of Observational Studies in Epidemiology (STROBE) statement: Guidelines for reporting observational studies. *Lancet*, **370**, 1453–1457.

Von Elm, E., Altman, D. G., Egger, M., Pocock, S. J., Gotzsche, P. C., et al. (2007b). The Strengthening the Reporting of Observational Studies in Epidemiology (STROBE) statement: Guidelines for reporting observational studies. *Ann. Intern. Med.*, **147**, 573–577.

Von Elm, E., Altman, D. G., Egger, M., Pocock, S. J., Gotzsche, P. C., et al. (2007c). Strengthening the Reporting of Observational Studies in Epidemiology (STROBE) statement: Guidelines for reporting observational studies. *BMJ*, **335**, 806–808.

Von Elm, E., Altman, D. G., Egger, M., Pocock, S. J., Gotzsche, P. C., et al. (2007d). The Strengthening the Reporting of Observational Studies in Epidemiology (STROBE) statement: Guidelines for reporting observational studies. *Prev. Med.*, **45**, 247–251.

Von Elm, E., Altman, D. G., Egger, M., Pocock, S. J., Gotzsche, P. C., et al. (2007e). The Strengthening the Reporting of Observational Studies in Epidemiology (STROBE) statement: Guidelines for reporting observational studies. *Bull. World Health Organ.*, **85**, 867–872.

Von Elm, E., Altman, D. G., Egger, M., Pocock, S. J., Gotzsche, P. C., et al. (2007f). The Strengthening the Reporting of Observational Studies in Epidemiology (STROBE) statement: Guidelines for reporting observational studies. *Epidemiology*, **18**, 800–804.

Wheatley, D. J. (2004). Clinical evaluation: Statistical considerations and how to meet them in clinical practice. *J. Heart Valve Dis.*, **13**(Suppl. 1), S11–13.

Ye, J., Cheung, A., Lichtenstein, S. V., Nietlispach, F., Albugami, S., et al. (2010). Transapical transcatheter aortic valve implantation: Follow-up to 3 years. *J. Thorac. Cardiovasc. Surg.*, **139**, 1107–1113. 1113 e1.

Yue, L. Q. (2007). Statistical and regulatory issues with the application of propensity score analysis to nonrandomized medical device clinical studies. *J. Biopharm. Stat.*, **17**, 1–13. Discussion 15–17, 19–21, 23–27 passim.

Yue, L. Q. (Ed.) (2008). Special Issue on Medical Device Clinical Studies. *J. Biopharm. Stat.*, **18**(1), 1–3.

Zelen, M. (1969). Play the winner rule and the controlled clinical trial. *J. Am. Stat. Assoc.*, **64**, 131–146.

Zuckerman, B. D., & Muni, N. I. (2005). Cardiovascular device development: An FDA perspective. *Am. J. Ther.*, **12**, 176–178.

CHAPTER III.2.10 ENTREPRENEURSHIP IN BIOMATERIALS

Robert Langer,[1] Jason Fuller[2] and Mark Levin[2]
[1]Department of Chemical Engineering, Massachusetts Institute of Technology Cambridge, MA, USA
[2]Third Rock Ventures, Boston, MA, USA

INTRODUCTION

Start-up companies are an integral part of the biomedical product development landscape. Although most medical devices are manufactured, marketed, and sold by large companies, their fundamental technologies are often born in the university setting, and incubated in small start-up companies (Kahn, 1991).

Starting a company can be a great opportunity to raise the awareness of academic work, and also to have a significant impact on the industry and on patients' lives. It is a complex process of bringing together people, ideas, money, and business. The start-up environment is a fast-paced, dynamic, and collaborative home for a project, and with the right vision, team, technology, and plan, a start-up company can be highly efficient at creating significant development progress and creating considerable value.

However, we want the reader to understand the risks inherent in starting a company: a relatively small percentage of start-up companies get a product to market (Gompers, 2008), and the effort required can be substantial. Therefore, the decision to spin out one's research into a company should not be taken lightly.

Many paths can bring a technology or product to market, and many of the functional challenges of patent protection, product design, regulatory approval, reimbursement, and ethical issues are described earlier in this book. These challenges are faced both in established companies and start-ups. However, the culture, pace, politics, risk profile, and potential reward differ between start-up and large company. It is recommended that the biomaterials specialist consider entrepreneurship as one of many potential avenues for developing a successful product.

This chapter will step into the shoes of an entrepreneur planning to start a company. First, we will survey the entrepreneurial ecosystem with a focus on the interconnections between its members and the transfer of resources into the company. After an overview of the start-up process, we will cover the criteria entrepreneurs use to decide which company ideas are worth spinning out, and which ones are better suited to academic exercises or large company projects. We will finish with several tips on entrepreneurial decision making, and how venture capital firms decide which companies to fund.

The fundamental message of this chapter is that academics, entrepreneurs, and investors should carefully choose which companies to start, join or fund. In this chapter, we will develop five criteria for starting a new company. As you read the various sections, keep these criteria in mind, and we will revisit them at the end of the chapter.

Suggested criteria for choosing research projects to spin into a company include the following (Kent Bowen et al., 2004):

1. Is this a high impact area?
   a. Clinically high impact (will it help patients?)
   b. High impact to potential commercial partners (will it create shareholder value?)
   c. Scientifically high impact (is the technology new and important?)
2. Have you published a seminal paper?
3. Have you obtained a protecting patent?
4. Have you demonstrated efficacy data in animals?
5. Can you build a great team?

THE ENTREPRENEURIAL ECOSYSTEM AND THE START-UP PROCESS

The creation of a start-up company assumes that a small, focused, and highly motivated team can efficiently develop a new product and that the company can be the appropriate investment vehicle for those involved. In a start-up, inventors of intellectual property, shareholders, employees, and other key players acknowledge that the start-up is the best vehicle for commercializing the technology, as opposed to licensing to a company, for example. This advantage may exist for a number of reasons: for example, universities may lack the facilities or infrastructure to take a project to the next step or large companies may be constrained by their organizational bureaucracy. In other cases, the choice of spinning out a project into a new company is purely financial (Chatterji, 2009).

Although start-up companies lack some of the structural challenges common in universities and large companies, they have their own risks. Start-up companies have a narrow focus, sometimes on a single product or narrow platform, they cannot diversify risks in the ways large companies can. Therefore, if a start-up encounters an enormous setback, with no way to compensate, it may trigger the company's demise. Start-up companies do fail for reasons beyond anyone's control, and all parties should understand that although the risks are high, the rewards are great if the company succeeds. The challenge for the start-up company is then to share these risks and rewards in an equitable way between all of its members. Bringing together all of these partners is the first challenge of building a company.

Figure III.2.10.1 summarizes the process of transferring these people and resources into the start-up company, NewCo. In the figure, the four main components of a start-up are color-coded as described below.

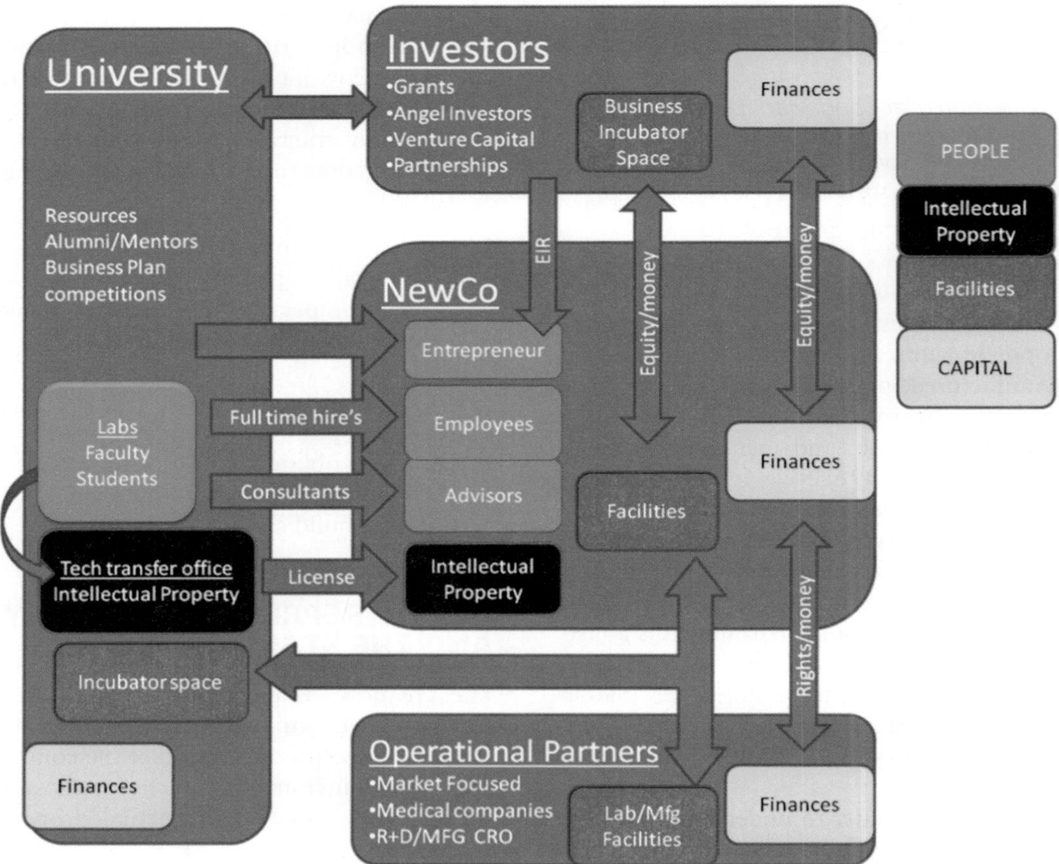

FIGURE III.2.10.1 The entrepreneurial ecosystem. A representation of the complex interplay between investors, universities and existing companies in the industry.

The essential components of a biomedical start-up include:

1. People (green);
2. Technology (intellectual property) (blue);
3. Facilities (red);
4. Capital (yellow).

It is the combination of these components that leads to successful companies. When brought together, these raw materials lead to the creation of high value products, as shown in Figure III.2.10.2. The process of bringing them together is a complex and high-risk exercise.

Each of the pieces is typically legally "transferred" from other organizations to the start-up. The typical organizations in the entrepreneurial ecosystem include:

1. The university (or hospital);
2. The start-up;
3. Funding agencies;
4. Corporate partners.

Figure III.2.10.1 shows the connections between these organizations. At the center of the ecosystem, the start-up must build relationships with the university, the funding agencies, and its corporate partners. As the start-up builds these relationships and negotiates to bring in

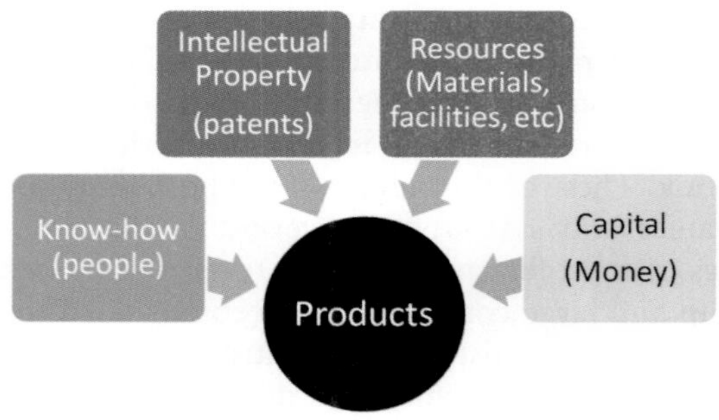

FIGURE III.2.10.2 Inputs to generating innovative life science products.

people, IP, facilities, and capital, a number of decisions are required.

These decisions are made by the shareholders in the company. When a company is incorporated, it becomes a discrete legal entity. Shares of the company, each of which represents ownership of a fraction of the company, are distributed between its owners. The shares have value because the shareholders are entitled to their fraction of the start-up's future profit. Capital investment (cash), intellectual property rights, and services can all be exchanged for shares. Since the start-up company is private and there is no market where shareholders can sell their shares, it typically sells chunks of shares to equity investors in exchange for a cash investment. This equity investment sets the price of the shares, and gives the company money to begin operating. As the pieces come together, the relationship with each partner is formalized with a contract outlining the specific terms of the relationship. The company exchanges money or shares in exchange for people (labor), technology (IP rights), and facilities.

The following subsections review each of these relationships within the start-up ecosystem. We will first discuss the university (or research hospital), a common source of technologies and scientists. Our discussion of the university will consider the research lab environment, the technology transfer office, and other university resources. Next, we will cover the start-up company itself and its components. Funding agencies will be covered next, as there are many options for funding a start-up company. Finally, we will cover corporate partners and their role in helping build new, innovative companies.

The University (or Hospital)

Universities and/or research hospitals are a site of innovation. New technology, new biomaterials, new devices, and new biological concepts can all provide the motivation and genesis for a new company. As Figure III.2.10.1 shows, several connections exist between the university and the start-up company. Universities play a central role in connecting the other organizations in the ecosystem, and in disseminating knowledge (Rothaermel and Ku, 2007). The primary interface for the start-up and the university is between academic experts who may become founders or advisers to the company and through the licensing of intellectual property. The university's fundamental strengths are innovation, cutting-edge expertise, and IP generation. Its research labs are often staffed by the brightest innovators in the world who invent intellectual property ("IP"), which the university owns and then transfers to the start-up company. As Figure III.2.10.1 shows, thought leaders interface directly with the start-up company, as well as technology transfer officers.

Innovation Engine: The Research Lab. The research lab is the university's innovation engine. Many aspects of the university research lab make it the ideal site for fundamental discovery. The university's research goal is to push the frontiers of scientific and technological discovery. Therefore, researchers are not directly focused on inventing a specific product, but instead can think very creatively about new technologies or the application of technologies to new problems. They can take high risks as part of the learning process. This culture contrasts with the culture inside an early-stage start-up, where specific milestones (set at the beginning) determine the company's success. The academic lab must create publications, whereas the start-up company must meet very specific stated goals on time.

Culture The academic culture tends to reward people who enjoy and excel at being the first to communicate a new idea or discovery (Stalcup, 2006). Although the multidisciplinary nature of modern technology and its application to medicine require teams of researchers from diverse backgrounds to work together, ultimately, all academics are required to be the contributing author of numerous top publications. This creates a healthy competition between academics, which speeds academic progress, since multiple labs want to be the first to publish.

Another feature of the academic culture is the unlimited pursuit of intellectual inquiry. There is no limit to how risky a research project could be, in terms of leading to a tangible outcome, as long as it has a high likelihood of leading to academic publications. Because academic research is understood to be a learning process, and fundamental research is high-risk in nature, it is very difficult to put definitive timelines on results in academic research. This is in contrast with development work done in a company setting, which typically has very specific timelines and milestones for completion, but creativity is also very important in small companies.

The culture inside a start-up company is focused on working together to meet milestones. If the whole company does not succeed from one milestone to the next, the company could be shut down. Most key employees in the start-up company are shareholders, and therefore their compensation is tied to the company's success. The competitor of a start-up is not the person across the lab bench; instead, it's the clock. In a start-up company, creativity is important, but reaching the next phase of development, and with it the next value milestone, is required. Figure III.2.10.3 illustrates some of the factors that induce the change in culture as a company progress through the development process.

What Makes a Lab Conducive to Entrepreneurship? Specific cultural and practical aspects tend to appear in entrepreneurial academic labs. If you are a faculty member running a research lab, you have significant control over the culture in your lab. If you are a student interested in joining an entrepreneurial lab, you might evaluate the lab on its history of the aspects listed below.

1. Hiring researchers with an interest in starting companies, and encouraging them to act on those interests.
2. Solving problems that have impact and could potentially be translated into products, especially in areas where you are willing go against the conventional wisdom.
3. Being willing to work on high-risk problems. The solutions to the problems may take a long time or have a low probability of success; however, if the solution to the problem is very important, it's worth trying to solve. This can lead to the fundamental discovery and patent, which can be taken to the next step (product development) in a start-up.

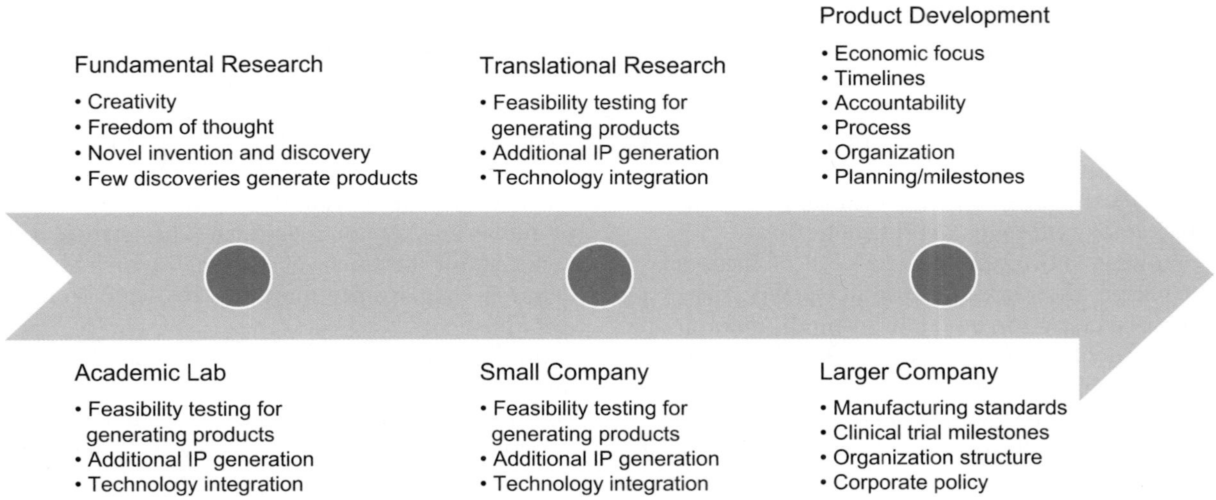

FIGURE III.2.10.3 Activities and cultures through company growth.

4. Encouraging a collaborative environment. The academic process can induce competition. There are other ways to promote a culture of collaboration that will help people think like entrepreneurs. For example, broad projects that require technologies from multiple backgrounds induce people to work together. Focusing diverse teams on important problems is a critical focus.
5. Paying attention to IP early and focusing work in areas where there is room for significant IP filings. Finding areas were a new discovery can lead to a seminal patent will help lead to fundable spin-off companies.
6. Building a good relationship with the technology licensing office at your university (see Figure III.2.10.1).
7. Building relationships with venture capital firms and corporate partners. It is important to understand what problems they are facing, so that when you have an idea that is relevant to those problems, you will know where to turn (see Figure III.2.10.1).
8. Helping to promote entrepreneurship at your university. Many schools encourage faculty and students to participate in entrepreneurship programs, and some offer grants and other support for small translational research projects.

How to Decide when it Makes Sense to Move a Company Out of the Academic Lab Many factors contribute to the decision of when to move a research project from the academic research lab into a company. As mentioned above, it is typically after fundamental IP has been filed and significant proof of concept has been completed that development work starts in a new company. Before the development work begins, the researcher should ask a number of questions, including the following:

Do you have enough data and confidence?

You will take the plunge into the company either as a full-time employee or as a significant adviser to the company. You will have to be involved in getting funding, recruiting employees and senior management, and making the company move forward. It is your time and energy, your career, and your reputation that you will invest in building this company. It is in no-one's interest for you to prematurely try to start a company. Therefore, you should wait until you feel confident in taking the leap into the start-up world. This may require extensive interaction with advisers in IP, commercial and/or clinical development, and entrepreneurs, but you should have a strong positive feeling about your venture. To succeed, a company concept takes more than great lab work and great data: it takes a diverse team of great people to build a successful company. It also takes a diverse team of great people to know which companies to build.

Have you created enough value to generate interest from investors?

Equity investors exchange cash for equity in the company. Pre-money valuation is the company's value immediately before the transaction, and post-money valuation is its value after the transaction. Pre- and post-money valuations are related through the following equation:

$$\text{Pre-money valuation} + \text{venture investment} = \text{post-money valuation}$$

The relationship between share price, valuation, and number of shares is:

$$\text{Share price} = \text{pre-money valuation} / \text{number of pre-money shares}$$

Sometimes (but not always) share price is set to \$1/share at the first venture funding, and therefore pre-money valuation is equal to the number of pre-money shares. Share price can change after each round of venture funding, and at each round of funding the following relationship is valid:

$$\text{Share price} \times \text{Number of new shares issued} = \text{Venture investment}$$

The company's value increases as the company makes more progress and generates more data. Therefore, it can be to the founders' advantage to generate as much data as possible using grants before taking an equity investment. There are several benefits to having more data: you will be more confident, possibly making it easier or quicker to raise money, and you will retain more ownership in the company after funding.

What is your IP protection?

NewCo will license IP from the university. Usually, the company must be formed before the license agreement is executed. Sometimes the university technology licensing office will pressure founders to execute a license agreement in an attempt to quickly recoup the cost of the patent application. This is just one of the considerations in the decision of when to incorporate the venture. This is not necessarily the right time to start a company, but it is possible that the technology transfer office may license the IP to another company (having a great relationship with them helps).

What are the funding opportunities for the company?

Without significant funding, it may be pointless to create a virtual company (described in the sidebar), especially if there is significant funding to continue the research in the lab.

Definition of Terms

Founders are generally the thought leaders involved in assembling the vision the company is founded around. Founding a company is a process by which the company becomes legally incorporated. Corporations have limited liability, which protects their shareholders and employees' assets from legal liability related to the company's operations.

A **virtual company** is a company that has no physical assets, facilities, lab space or offices. All of the company's employees work from home or from their own remote or virtual locations.

Pre-money valuation is a company's value before it has taken any cash investment. Post-money valuation is a company's value after a dilutive equity investment. Post-money = pre-money + amount invested. In a dilutive equity investment, investors exchange money in exchange for shares in the company. As the company issues more shares (at a given valuation), the current owners are diluted or own less of the company. Non-dilutive funding comes from sources that do not exchange equity for money, but rather give money for free.

Licensing terms are agreed upon reimbursement for the rights to operate a business in an area covered by the patents a university holds. The terms of an agreement typically include cash payment, equity, and/or royalties.

Due diligence is the investigation of an opportunity as a potential investment. Due diligence is done by investors, entrepreneurs, large companies, and others to determine an investment's merit. The process usually involves learning about the expected challenges and timeframe for generating a valuable enterprise, the costs associated with the venture, and the expected future value of the investment.

What is your academic grant funding situation?

Because pre-money valuations increase, and the probability of getting venture funding increases with every positive experimental result, it can make sense to continue making progress in the academic lab so long as funding allows and so long as there is no significant pressure to license the IP from the university. There are clear examples of start-up companies that have been started in a virtual mode. This is one potential way to start a company early with a very small investment, and avoid much of the challenge of being diluted by a large investment at a low early-stage valuation. It allows the research scientist to leverage the research dollars raised from grants. One potential drawback to this approach is that universities often issue policies restricting research sponsorship from flowing into labs where the principal investor is a founder or shareholder of the sponsoring company.

Technology Transfer Office. In the previous section, we discussed that the purpose of the university lab is to innovate and to publish and patent its innovation. The technology transfer office is the central clearinghouse for licensing IP from the university.

Chapter III.2.4 in this book reviews the patent process itself. The transfer of IP rights to a new company is an important step, and is outlined in Figure III.2.10.4. University employees and faculty are usually required to assign the IP rights to the university for any invention where they used university resources. They are required to document an invention through a standard disclosure (Jensen et al., 2003), which is a legal document that describes the invention and becomes the basis of a provisional patent. The submission of a disclosure triggers an internal process by which the technology transfer office evaluates the invention to determine whether it merits investing in patent protection. There is a fairly extensive literature on evaluating patents (Allison et al., 2004). Case agents are usually assigned to a particular project and/or faculty member, and when a decision is made to move forward with an application, they will help manage the process of hiring the appropriate legal counsel to draft the application.

The same patent agent will typically be heavily involved in the process of licensing the IP to a start-up company that will be commercializing the technology, as shown in Figure III.2.10.4. Sometimes there is a licensing agent (a different person form the patent agent) who manages the licensing process. In some cases, technology transfer offices also have other resources for entrepreneurs. These can include investing venture or seed funds, business plan competitions, translational research grants for researchers, and so forth.

Patents As described earlier in this book, the Bayh-Dole Act of 1980 gave universities the opportunity to retain IP for government grant funded research. As owners of the IP, universities decide to whom they will license this IP, and under what terms (Lin and Kulatilaka, 2006; Macho-Stadler et al., 2007). When a university licenses a patent to a company, the company and the university then enter a partnership where the university typically retains the ownership of the patent, but grants the company permission to develop and

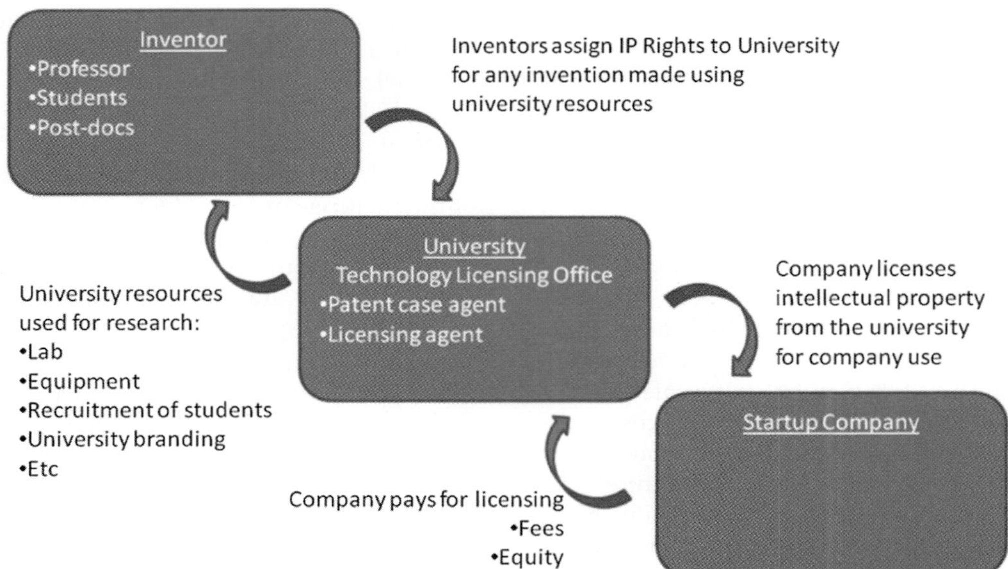

FIGURE III.2.10.4 The flow of resources and intellectual property between inventors, universities and startup companies.

sell product(s) covered by the particular patents being licensed. The company and or the university can usually enforce the IP protection rights against another organization if that organization is infringing on its patent's rights by developing or selling a product covered under the patent.

Sometimes competitive situations arise where more than one group is interested in licensing a particular patent. Large and small companies can both have an interest in the same IP, so entrepreneurs can find themselves competing with large, established companies for IP rights. Some universities consider the opinions of the inventors, and rely on them to help determine which organization would be the best partner for the university. Often, the small company interested in licensing the technology is being founded by faculty or students involved in inventing the technology, which adds an additional layer to the university's decision-making. One factor that favors the academic founder is that many universities have identified the process of entrepreneurship as an important part of their mission, as well as being very lucrative (Nelson, 2001; Owen-Smith and Powell, 2003). Sometimes universities believe that some technologies are better suited to licensing to large companies, and although generating revenue is not the university's mission, licensing agreements with large companies can bring in large amounts of cash quickly.

Licensing terms for a given industry or a given university are not standardized (Thursby et al., 2001). When dealing with a new start-up company, the university will typically propose being compensated with equity in the company (in addition to cash). This introduces a component of risk on the university's part, since the equity will only have any real value if the company succeeds. However, if the company is very successful, the university will share in the rewards.

The university is not interested in owning a controlling stake in the company; rather, it wants to recoup the financial cost of pursuing patent applications for a number of technologies. Successful IP licenses pay for a large number of patents that are not lucrative; therefore, the university needs to generate enough revenue to cover the costs of ongoing IP submissions. The university may also seek royalties as a percentage of revenue (or profit) from selling the product. This ensures that the university will also share in the proceeds if the company succeeds from a revenue standpoint.

In addition to IP, many university faculty and students are involved with start-up companies either as employees, entrepreneurs or advisers. These individuals can benefit significantly from their activities with the start-up company. Although entrepreneurial and industrial experience is rarely seen as supporting an associate professor's case for tenure, many universities look favorably on faculty who are actively involved in building great companies. These faculties are often limited by a conflict-of-interest agreement that they must sign when they initiate their employment at the university. The inherent conflict-of-interest results from academic investigators potentially becoming biased in their scientific work, because they have a financial interest in the results. This is particularly relevant in the case of human clinical trials, where a single researcher could directly influence (even if subconsciously) the selection of individuals. For this reason, academics who are company shareholders (and therefore have a financial interest in the company) are sometimes not allowed to have active operational collaborations between their academic labs and the company (Bagley and Dauchy, 2002).

Venture Groups/Arms Many universities facilitate company creation and technology transfer by operating an entrepreneurship center, venture group or a special center as part of the technology transfer. In any case, this office advises faculty and the technology transfer officers on the array of options for licensing the technology and many start-up issues, like hiring a team, finding lab space, raising money, and the like. In some cases, this group has the ability to invest a small amount of money in the new company on behalf of the university. This process has the effect of financially aligning the university with both IP ownership and equity ownership.

Alumni Networks Most universities keep records of their alumni for fundraising purposes. Often, the most successful alumni are the most likely to give large gifts, as well as the most likely to be interested in providing mentorship to new companies spinning out of the university. Your alumni office may be willing to share that information with budding entrepreneurs. This can be a great way to find experts with complementary backgrounds to your own, especially those with experience building companies.

Some universities have official mentorship programs designed to provide specialized advice to entrepreneurs at the university. These groups try to retain successful entrepreneurs who typically have good intentions. Within the context of the university system, mentors typically have noble intentions; however, there can be an occasional experience with a "mentor" who is completely self-interested and trying to hijack a new company concept. This should not deter you from working with the alumni from your institution; however, you should use good judgment when interacting with potential mentors, regardless of when or how you meet them.

Business Plan Competitions Business plan competitions have enormously increased the awareness and interest in student entrepreneurship at many universities. These competitions help motivate and mobilize student teams around exciting new company concepts. These business plan competitions provide prize money and support for moving a company concept forward. Perhaps most importantly, they provide a forum for company builders to meet with other entrepreneurs and to bounce their ideas off other teams in a very non-threatening and supportive environment. Often, the prize money these organizations award is not significant enough to build a company; however, there are many other benefits to competing in a business plan competition, such as great publicity, visibility to potential investors, and free mentorship. If your venture would benefit from publicity, business plan competitions typically have great relationships with the media, and often set up interviews for print, radio or video news with their finalists' teams. Additionally, the finalists and winners of a business plan competition are often local celebrities on campus, which can help significantly in the recruiting process, and also in raising visibility within the university system and the technology transfer office.

Classes Many schools of engineering, science, and management offer classes in entrepreneurship, business plan writing, entrepreneurial finance, and venture capital. All of these classes can be extremely useful in helping students understand and set expectations for working with investors and entrepreneurs.

The Start-Up Company

The company is the central entity in the start-up process. Referring back to Figure III.2.10.1, there are many components to a successful start-up company. All of these pieces are necessary, and they must all work together. The founders must not underestimate the challenges of building the company: they must overcome significant hurdles in the process of putting the company together. Again, from Figure III.2.10.1, the main components of a start-up company are people, IP, capital, and physical assets.

People. People are the most important component of the start-up company. Many investors and academics will jokingly say, "I'd rather back an A team with a B idea than a B team with an A idea." A great entrepreneurial team will unselfishly work hard, work together, and be completely honest with itself about where the company is going. A great team will adjust the strategy and vision of the company as needed. Every start-up will encounter unexpected hurdles, and it will be up to the team to creatively overcome them.

Entrepreneurs Entrepreneurs are the people who pull together the company concept in the beginning. As depicted in Figure III.2.10.1, entrepreneurs can come from universities, from venture capital firms or be unaffiliated. The entrepreneur is a project manager, cheerleader, and generally the company's leader. As you can imagine, the goal of the entrepreneur is to create value and to sell that value to investors, employees, and collaborators. Being a good salesperson is a crucial skill for the entrepreneur.

Although entrepreneurs can come from anywhere, typically they are not the inventors of a technology; rather, they meet the inventors through their networks and work closely with them to generate companies. Some academics have friends who are entrepreneurs who they work with over and over again, and other serial entrepreneurs work with different scientific founders. Serial entrepreneurs are professionals who start company after company. They generally have a better track record than first-time entrepreneurs (Gompers, 2008). Many research studies have analyzed factors that make entrepreneurs successful. Some of the most important characteristics include social skills (Baron and Markman, 2000), personal networking (Witt, 2004), and motivation (Buttner and Moore, 1997).

To evaluate a new company idea, an entrepreneur will undertake a "due diligence" process of analyzing the opportunity. The entrepreneur is trying to convince himself or herself that the opportunity is worth taking, and that it is worth investing time into building the company. This process typically involves asking experts to help evaluate the technology, market opportunity, appetite for funding, and general excitement about the company. This process is somewhat similar to the due diligence process a venture capitalist would use to evaluate an opportunity for funding. The entrepreneur evaluates the opportunity for its ability to raise funding and succeed.

If you are an academic or a student who is interested in playing the role of the entrepreneur, you should certainly take an objective view of the project and ask yourself whether you would be passionate about the idea if it were not your own. When you have invested so much in an idea already, it can be difficult to objectively analyze the opportunity.

Building a successful company takes hard work on the part of a number of people. The team is very important, and typically includes experts from business, technology, and medicine. One of the key aspects to a successful start-up is the seamless collaboration between business and technical leaders. The general culture in the company has a great effect on the organization's efficiency. Generating great cultures is a non-trivial task, and many entrepreneurs fail because they do not have the appropriate temperament. Many books have been written about how managers can and should interact with each other to form great cultures (Kotter and Heskett, 1992; Collins, 2001; Flamholtz, 2001; Johnson and Phillips, 2003).

Founders Founders are individuals who are involved in the company at its inception. They can be academic experts who lend technical credibility, help recruit talented individuals to the company or invent the technology, or who are entrepreneurs. Ultimately, these experts typically will join the company's scientific advisory board, and in some cases the board of directors. Sometimes, they can be full-time company employees, but they typically maintain an academic or clinical appointment to maintain their professional credibility.

Management Team The management team is the group that makes strategic operational decisions. The typical management team is comprised of a Chief Executive Officer ("CEO"), a Chief Medical Officer ("CMO"), a Chief Scientific Officer ("CSO"), a Chief Operating Officer ("COO"), and others. In the early days of a start-up company, all of these positions may not be filled. Management teams are flexible, and are designed to cover all the needed skill-sets. Collectively, this team should be an expert in every relevant area. Often the recruitment of the full-time management team will be coordinated with the venture group backing the company, and will not be the responsibility of the founding academics. The entrepreneur may or may not become a part of the management team as the company moves forward.

Board of Directors The board of directors ("BOD") typically is responsible for choosing the company's strategic direction. This group of people has specific legal responsibilities, and is required to provide governance to the company (Bagley and Dauchy, 2002). In addition to their official legal requirements, the BOD usually represents the company's investors and advises the management team on how to move forward in the investors' best interest. The board usually includes the CEO and/or president of the company, as well as representatives of the investors, and often a representative of the founders and an independent representative who is typically an expert in the industry and well-connected with other groups, especially with established companies that may become interested in acquiring the new start-up.

Employees Employees are perhaps the most important component of the start-up. They perform the company's operations and support the executive management team in executing the plan developed with the board. Typically, employees can come from labs that invented the key technology or collaborating labs. Having world-renowned founders and a top management team immensely helps in recruiting great employees. There are several resources on recruiting great employees and building effective cultures (Kotter and Heskett, 1992; Collins, 2001).

Advisers Start-up companies typically will hire any number of industry experts to help them operate. One of the challenges with an early-stage company is that there are limitations to the number of full-time employees the company can afford; however, a large number of job functions need to be completed. Many of the job functions required for operating an early-stage company only require a fraction of the resources of a full-time employee. In particular, some of the technology-agnostic functions like accounting or financial governance do not require a full-time employee. In this case, it is common to have a part-time employee serve in a flexible capacity on an hourly basis.

Ideas and Intellectual Property. Intellectual property is very important for the new company. It is often the only "property" the NewCo owns. Most start-ups don't employ in-house IP attorneys at a very early stage. Typically, their IP advice comes from an external counsel, who may be compensated with options or shares of equity in the company instead of with cash. This is especially common for early-stage companies before they have raised capital. Detailed analysis of IP is covered in an earlier chapter in this book, as well as in the section on technology transfer.

Although the IP licensed from the university is typically the basis for the company, additional IP is generated inside the company. Developing a strategy for IP development is an important activity of the management team. IP is especially important in medical device companies because there are many theoretical ways to solve most medical problems. Having a strong IP portfolio can be key to protecting the start-up's market. The selection of IP experts is key to understanding the competitive landscape, as well as pushing the company's own patent applications.

Physical Resources. Physical resources are required for the company to operate. These can be in the form of offices, lab space, assets, equipment, and the like. As mentioned earlier in this section, early-stage companies can often operate in a virtual fashion and collaborate with the inventor's labs within the university. Although this process can be a delicate balancing act of avoiding and managing potential conflicts of interest, it is often more appropriate to minimize the requirements for physical resources by operating the company out of a home office, for example. However, most universities have strict policies on IP ownership, and if a member of the university generates an invention, idea or another form of IP using any university resources, then the university has a claim to (at least partial) ownership of that IP.

Capital. There are many sources of capital, and raising capital is a very complex process. The main considerations are how much capital to raise, where to raise it from, and at what valuation. Multiple sources of capital are described in detail in the next section, but in this section we will cover the strategic decisions a company must make to meet its fundraising needs.

Early-stage capital investment is very important for the company, and the management team must constantly operate the company in the most capital-efficient way. The company must raise enough money to operate and grow comfortably, but it must not raise or spend more money than required, since this leads to unneeded dilution.

There is a clear trade-off between raising a large and a small amount of money. If the company intends to grow rapidly, it needs to raise a significant amount of money. Many finance textbooks and courses cover the basics of corporate/entrepreneurial finance (Smith and Smith, 2000; Denis, 2004; Brealey et al., 2006). Ultimately, there is no specific answer to how fast a company should grow, and it is a strategic decision for the management team and the BOD. From the perspective of the founders, dilution is a major trade-off.

The next strategic question is determining the source from which the start-up should raise money. The main sources of funding are grants, venture capital, and corporate partnerships.

Grants Grants are a great source of capital for the company, and a number of government and non-government grant agencies are described in the "Sources of Capital" section. Because grant funding is not exchanged for equity, it is non-dilutive and is therefore very attractive.

Grant writing can be a very time-consuming process. The time and effort put into grant writing can pay off if the grant is awarded. However, grants are not awarded to the majority of grant applicants. Therefore, when a small company is thinking strategically about raising money, it must also consider that the time, energy, and resources spent preparing applications for the grant process may not pay off in the long run. The company must find the right balance between risk-taking and stability.

An additional feature of grant funding is that the oversight is largely only financial in nature. Grant money is typically not managed by a person representing the interests of the grant who sits on the BOD. This is completely unique when compared to capital raised through the sale of equity to investors. This stems from the fact that private investors have purchased equity for the main purpose of generating return on capital during a given duration.

Venture Capital Venture capital is the most common source of funding for start-ups. When choosing whether to work with a venture capital firm, you should consider a number of things. The potential pros of building a company with a venture capital firm include:

1. The probability of the company's success increases (experienced company builders get involved and bring extensive networks to the company).
2. Things will move faster in the direction of financial return.
3. It is typically easier to recruit senior management to the company after the validation of significant venture funding, especially from a firm with a good track record.

But the potential cons of building a company with a venture capital firm include:

1. There will be substantial dilution for the founder(s) and entrepreneur(s).
2. Founders will give up substantial control in corporate decision-making.
3. Investors are likely to identify new senior management.

Ultimately, the entrepreneur(s) must see the venture capitalists as true partners in the business, and this is easiest when they have a strong relationship. Venture capital firms and angels almost always contribute to building the company. This can be through their position on the company's board or, in some cases, by contributing entrepreneurial or operational capabilities. More experienced investors with better track records have a higher probability of taking a start-up company to a successful exit (Gompers, 2008).

Talking to Venture Capital Firms Venture capital firms have a significant preference for companies founded or recommended by someone in their network. Therefore, networking is very important. If you are interested in potentially starting a company and raising venture money in the future, it makes sense to build relationships with several venture capital firms in your area (both of expertise and geography). Firms are always in need of expertise for evaluating new investment opportunities, and often hire experts in various fields to help them understand opportunities. If you are actively looking for venture funding but do not have an existing network in the venture industry, it likely makes sense to find an entrepreneur (either an academic or a business professional) who has a successful track record of building companies and raising venture capital. Ideally, you'll find an entrepreneur with expertise in building and selling successful companies.

Important Differentiating Factors between Venture Firms Many factors define the venture firms you may consider working with.

SIZE Large firms tend to want to do bigger deals, and may be interested in investing larger amounts over a longer period of the company's lifetime. Their downside is that their partners may be very busy or focused on other later-stage companies in their portfolio and have less time to work with early-stage investment.

VALUATION Valuation is the often the largest point of contention or negotiation between start-up and venture capitalists. The entrepreneur should remember that valuation is important, and stock ownership is only relevant if the company is ultimately successful. Building a good working relationship with your investors and board of advisers is more important from this standpoint than the actual valuation.

The structure of the deal and terms on the term sheet are just as important as the actual valuation (Wilmerding, 2003; Gold et al., 2005). Many venture capital firms attempt to add terms that give them preferential status over the entrepreneurs and employees of the company.

EXPERIENCE Experienced investors bring more to a deal than money. They have years of experience building companies, and they will likely have a better context for setting the direction of the company than inexperienced investors. Many successful entrepreneurs start venture capital firms, and individuals as well as firms have great reputations. Venture capital funding offers are three times more likely to be accepted when they come from a firm with a strong reputation. In one study these respected firms often get a 10 to 14% discount on equity (Hsu, 2004).

WILLINGNESS TO HELP BUILD THE COMPANY Some venture capital firms will help identify business leaders and entrepreneurs to join the company or join the company themselves.

When you begin speaking with venture capital investors, don't be alarmed if things move slowly. If and when you have developed a relationship with venture capitalists and you would like to discuss funding opportunities, your first challenge will be to piqué their interest to convince them to initiate the due diligence process (Dooley and Dooley, 2003). This process will take place over a timeframe that is mutually agreed upon by you and the venture capitalist. The venture capitalist will ask you a number of questions over the course of weeks or months in a process somewhat analogous to an extended job interview. At the end of the process, you may get an "offer" in the form of a term sheet that outlines the valuation and terms for potential investment or the venture capitalist may decide to pass on the investment. In either case, you will have learned a lot about what it takes to build a company, and you can apply this experience to the next venture capitalist interaction you have.

Sources of Capital

The previous section reviewed the funding scenario from the perspective of the start-up's strategic decision-making process. This section will review more specific information about each of the sources of capital: grants; venture capital; and corporate partnerships. Figure III.2.10.1 shows the various types of funding that can flow into a start-up company. Each of these sources has its own unique challenges and opportunities. Below is a review of each of these sources.

Grants. Grants, by definition, are money that is given to an organization for a specific project related to an interest of the grant agency. Often grant money comes from government sources, and in addition to academic funding, many government agencies have grant programs available for early start-up research efforts inside companies. The purpose of these grants is to improve the competitiveness of the local or national economy, to support the generation of new and better medical technologies, and to develop new companies.

Perhaps the most common entrepreneurial grants issued by the government are the US Small Business Administration's Small Business Innovation Research (SBIR) grant and the Small Business Technology Transfer Program (STTR). These grants are typically awarded to small businesses developing technology in an area that is important for national health, national security or the like. These grants are awarded in escalating phases. Phase one feasibility grants are typically intended for the initial phases of projects, and are awarded for a 12-month period in an amount up to $100,000. These grants are intended to allow the company to demonstrate a proof of principle,and generate initial data on their concept. Phase two grants allow the company to expand on the data generated in the phase one grant, and they are awarded for up to 24 months and in an amount ranging from $500,000 to $750,000. The final phase three grants are typically funded outside the SBIR, and involve the commercialization of the technology. Solicitations for these SBIR grants are posted on several government and other websites available through http://www.sbir.gov. These grants are offered by the Department of Defense, Department of Energy, National Institutes of Health, National Science Foundation, and National Cancer Institute, among other government organizations.

Several firms offer consulting services to help write and/or advise start-ups in the strategic process of applying for SBIR grants. If the entrepreneur has limited experience writing grant applications, it may be advantageous to hire an experienced leader to supervise the grant writing process. Studies of SBIR recipients have concluded that SBIR awardees are more successful when the academic principal investors are involved in the company (Toole and Czarnitzki, 2007).

In some cases, non-government organizations offer grants. Examples of philanthropic organizations include the Bill & Melinda Gates Foundation, the Carnegie Foundation for the Advancement of Teaching, the Kauffman Foundation, the Ford Foundation, and others. Many larger corporations have extensive corporate foundations that donate money to causes in society's best interests.

Venture Capital Firms and Business Angels. In 2008, there were 382 venture capital investments in medical devices and equipment companies, which totaled $3.4 billion. Therefore, the average venture investment in 2008 was $9 million (Money Tree, 2009). The collective investment of business angels is harder to track, but a recent report predicts that angel investing in US medical devices and equipment was $3.1 billion in 2007 (Sohl, 2008).

Venture capital firms are usually partnerships that raise an investment fund from a variety of investors called limited partners. Sources of funding for venture firms are wealthy individuals, institutional investors, endowments, and the general partners (individuals) in the fund themselves. Venture capital firms raise money with the intent of returning the investment plus a substantial return to their investors. Financial incentives are used to ensure motivation, and usually there are incentives to build companies that have a high probability of "exit." A successful exit is the sale of the company for a significant multiple of the amount invested, and this usually means the company must be put together quickly and meet a number of milestones that increase its value moving toward acquisition or going public. This approach may (or may not) be in the entrepreneur's best interest. Often, when start-up companies take a significant equity investment, the company's founders lose financial control of the company through dilution. This may be the best option for the entrepreneur, but some early-stage company builders do not like to relinquish control, since they have some personal connection to the company they have built.

Corporate Partners

Many large companies are involved in start-ups, and this is becoming a more popular, particularly for medical device and pharmaceutical companies who are struggling to gain access to new drugs or devices as their pipelines are not full. Public markets follow economic trends, and at the time of press were mostly out of the picture, acquisitions are currently the primary exit possibility for start-up companies. Therefore, building a relationship with large companies is of utmost importance for the successful start-up company, which should focus on making itself an attractive acquisition target.

Medical device, pharmaceutical, and biotech companies have developed sponsored research programs with universities to help academics fund translational research (Business Wire, 2009). This approach usually involves research money flowing into the university similar to a grant, but it is usually accompanied with a blanket license agreement that grants a license in any resulting IP to the sponsoring company. Therefore, in some ways, this approach is not compatible with starting a new company.

Another model that has also become interesting to venture investors is early-stage collaborations with large companies. This model theoretically allows the technology to spin into a start-up company, yet creates a relationship with a potential acquirer early to maximize the probability of acquisition by signing an option agreement on the company. Admittedly, this strategy can limit upside for the start-up company if it is wildly successful, but it can also help keep management focused on the company's operation, since it will not have to focus on finding an acquisition partner. In addition, it helps diversify the financial risk, as well as leverage the large company's scientific resources.

The start-up company benefits from working with a large company because publically announced collaborations can bring money, credibility, and expertise into the company. These aspects of doing a collaboration deal lead to value creation and ultimately liquidity.

As a start-up company is courting potential buyout partners, there is a trade-off in timing analogous to the beginning of the company. As the start-up operates further it generates higher value, which should be reflected in the buyout price. However, as more capital is used, current shareholders take on significantly more dilution.

Even in the absence of an official collaboration with a large company, close interactions and discussion about technologies can help early-stage companies. At a minimum, close interactions with large companies can help any start-up management team build the company with the maximum probability of acquisition.

BUILDING A SUCCESSFUL START-UP

As mentioned throughout the text, many things need to fall in place for a start-up company to move forward. Ultimately, only you can define what success means for your start-up venture. However, the authors have assembled some general guidelines based on past experience of building companies that have received funding.

This section reviews some of our thinking on the key questions of why, how, when, and which company to start. Many potential entrepreneurs blindly fall in love with an idea or stick to an idea just because it was their idea, pushing it forward passionately. But in the end, the company's projects do not reach the market. In the pharmaceutical world, there is always the risk of not clearing clinical trials, which will typically signal the end is near for the company. On the other hand, many company concepts fail for other reasons. Some common stumbling blocks for early-stage companies include:

1. Wrong management (CEO, CSO or others make poor operational decisions).
2. Running out of funding.
3. Missing technical milestones.
4. Participating in the wrong clinical trial.
5. Burning too much money.

Many of these mistakes could have been avoided, or at the very least, red flags should have alerted the start-up team that there were substantial problems with the concept that should have come up during the due diligence process.

The following sections provide a guideline for the potential academic entrepreneur to do his own due diligence, and to be honest about the opportunity before he or she jumps in blindly. Indeed, great company concepts are not that common, and the process for self-evaluation should be just as rigorous as the process the venture capitalist carries out.

The potential entrepreneur should spend significant time reflecting on himself or herself, as well as the questions posed here. Often, many entrepreneurs gloss over this part of the process. They suffer from being biased toward their own idea, sometimes so much that they are blinded to the rest of the landscape. In fact, sometimes other opportunities are there but go unnoticed, because the entrepreneur is too focused on his or her own work.

Why Start a Company?

Starting a new company is not something that should be taken lightly. It is a lot of hard work. It takes a lot from the people involved, and there is significant risk involved with pushing forward in a given entrepreneurial area. As an entrepreneur or an employee of an early-stage company, you will need to roll up your sleeves and get your hands dirty. You will most likely do everything from sweep the floor to cashing the big checks from time to time. The start-up mentality is very different from that in the corporate or academic world. You should ask yourself whether you are truly passionate about the idea you are considering building a company around, and then ask yourself why you are passionate about it.

Choosing the Right Company Concept

Many academic scientists are interested in building companies. The reader might ask, what will make the difference between a successful start-up and one that fails? No formula can guarantee success in a start-up company. However, the authors have developed a list of criteria that can form a general guideline for research projects that have high potential to raise money and ultimately become successful companies (Kent Bowen et al., 2004).

1. Choose a high impact area?
   a. Clinically high impact (will it help patients?)
   b. High impact to potential commercial partners (will it create shareholder value?)
   c. Scientifically high impact (is the technology new and important?)
2. Publish a seminal paper?
3. Obtain a protecting patent?
4. Demonstrate efficacy in animals?
5. Build a great team?

Doing Your Own Due Diligence to Prepare for the Venture Capitalist Pitch

The due diligence process has been defined over and over again by the venture capital industry. Generally speaking, this is the process of getting comfortable contributing a significant investment. In the case of the entrepreneur, the investment is his or her own blood, sweat, and tears. In determining whether and what to invest, venture capital firms use several criteria, and there are several books, websites, and publications on venture capital due diligence (Camp, 2002; Gladstone and Gladstone, 2002). Here, we will focus on the issues of specific relevance to the technical evaluation of an early-stage biomedical company concept.

This list of potential due diligence questions will sum up most of the content of this chapter in a simple table that the reader can easily refer back to in the future. The answers to these questions are typically summarized in a presentation that is delivered to the venture capital group to spark its interest in doing its own due diligence.

1. Team: Is the academic founding team outstanding? When thinking about all the potential top academic founders in the world, are a substantial number of very top people involved as founders of this company? This is important because it gives the company confidence that this will be the leading company in the field. Has the entrepreneur or have the founders done a good job of assembling the top scientific team in the area? Have you done background checks on the other founders?

   Is there an extraordinary management team in place or can one be recruited? One of the largest stumbling blocks for an early-stage company is an inexperienced management team that demands retention of control of the company. If the company's intention is to obtain significant funding, it will need to have seasoned management with a serious track record of success.
2. Technology: This includes the existing technology and the technology plan that could be laid out for the company. Is there a high probability of success? Has

the company thought through a research and development path for the next several years? What are the milestones along the way? How much will it cost to reach each of those milestones? How long will it take, and how much will it cost to get to clinical validation of the product? Are there any simple, cheap experiments that could significantly improve the probability of success? Is the technology in an area that is of critical importance to large companies that may acquire the company in the future?

3. Financial: How will this project appear to a potential investor? What is a reasonable pre-money valuation? Given the pre-money valuation, funding requirements, and expected exit options, is it possible to generate appropriate venture capital returns?
4. Market: What is the overall market for the product or concept? What are the commercial issues with respect to this market? Who are the competitors, and how will the new product be positioned relative to the existing products? Who are likely acquirers, and have they done similar deals in the past? What has the value of these deals been?
5. IP: Is the IP strong enough to build a company? Have you asked an independent IP attorney to review the IP? Is there freedom to operate given the other IP in the space?

CONCLUSIONS

We hope to have made a positive impact on our readers by giving them a better sense of the basic mechanics of starting a company and what some of the challenges would be. We started the chapter by giving a basic overview of the company-building process, and explained where the pieces of a start-up company come from and how they are transferred into the new company. Next, we analyzed how to distinguish between an opportunity that may represent a successful company and one that may be less attractive.

Together with the other information in this book, and perhaps some personal interaction with business and legal advisers, readers can better evaluate the start-up process as something that makes sense to pursue for their technology.

Case Study The Drug Eluting Stent Industry

Innovation in the stent industry exemplifies many of the concepts explored in this chapter. The market forces around the stent market combined with innovation in many start-ups led to an explosion in the market and a revolution in the way patients were treated.

The Clinical Need

Coronary artery disease occurs when atherosclerotic plaques clog the arteries that supply the heart. It affects about 13 million Americans, and this number has grown significantly due to an aging population, obesity, lack of exercise, and other factors. In the 1980s the surgical option for severe coronary artery disease was bypass surgery, with the non-surgical option of balloon angioplasty. Balloon angioplasty has grown in popularity, since it is much cheaper and less invasive. However, balloon angioplasty is often accompanied by restenosis and reclosing of the artery in 30–40% of cases within six months of the procedure.

Bare Metal Stents

To address the high recurrence of arterial narrowing and blockage, bare metal stents were introduced. Deployed with balloon angioplasty, stents physically limit an artery's ability to reclose after the procedure. However, even with a stent in place many arteries developed scar tissue and reconstrict. With the market demanding solutions offering better quality of life, local delivery, and more cost efficiency, the first use of the coronary stent (Palmaz) started in the early 1990s. Others joined the market, and by the early 2000s the number of stent procedures had reached one million. This was a revolution in the way coronary artery disease was treated. However, there were still over 150,000 patients in the US requiring reprocedure due to restenosis.

Drug Eluting Stents

The first commercially available drug eluting stent was the Cypher stent from Johnson and Johnson. These stents combined the mechanical support of keeping the artery open plus the ability to locally deliver restenosis-preventing drugs released directly from the stent. The use of drug eluting stents decreased restenosis significantly to below 10%. Because outcomes were increased significantly, drug eluting stents were quickly adopted and used in hospitals.

Development of Technology in Small Companies

Cordis, which was founded in 1957 in a garage in Miami, had retained the culture of a small and innovative company. Several technologies came out of this company including angioplasty catheters, pacemakers, and ultimately stents. It was the combination of several product/technologies developed in Cordis, Biosense, and other start-ups that enabled the image guided drug eluting stent market to grow so quickly. This is a common practice in the medical device world; many new technologies are developed at small innovative companies, and then acquired by large companies to fill their product pipelines.

Reimbursement and Adoption

With any new medical device, reimbursement is very important. Medical procedures are reimbursed by Medicare and/or private healthcare providers (insurance companies). In the case of drug eluting stents, cost was very high. Cypher drug eluting stents sold for $3195 at their introduction. Medicare, for the first time in history, announced its reimbursement level for drug eluting stents before FDA approval was complete. The cost increase for using a drug eluting stent instead of a bare metal stent is approximately $2800. This cost is associated with the increased cost of the stent, and the cost of additional blood thinning drugs (Plavix for six months). Medicare decided to increase the reimbursement for the procedure by only $1800; therefore, hospitals were forced to absorb the extra cost of the procedure. Several studies have found the stents to be useful in terms of outcome, but their cost effectiveness is still debated.

Additional reading on the development of the drug eluting stent market, as well as statistics from this case study can be found in Currie and Arundine, 2005; Denend and Zenios, 2006; Ofek and Wickersham, 2008.

BIBLIOGRAPHY

Allison, J. R., Lemley, M. A., Moore, K. A., & Trunkey, R. D. (2004). Valuable Patents. *Georgetown Law Journal*, **92**, 435. SSRN eLibrary. Available at: http://papers.ssrn.com/sol3/papers.cfm?abstract_id=426020. (Accessed July 13, 2009).

Bagley, C. E., & Dauchy, C. E. (2002). *The Entrepreneur's Guide to Business Law*. Mason, OH: Thomson South-Western.

Baron, R. A., & Markman, G. D. (2000). Beyond social capital: How social skills can enhance entrepreneurs' success. *The Academy of Management Executive (1993–2005)*, 106–116.

Brealey, R. A., Myers, S. C., & Allen, F. (2006). *Corporate Finance* (8th ed.): McGraw-Hill Irwin.

Business Wire. (2009). Harvard announces pioneering collaboration with Merck & Co., Inc. to advance osteoporosis research. *Reuters*.

Buttner, E. H., & Moore, D. P. (1997). Women's organizational exodus to entrepreneurship: Self-reported motivations and correlates with success. *Journal of Small Business Management*, **35**(1).

Camp, J. J. (2002). *Venture capital due diligence*. New York, NY: John Wiley and Sons.

Chatterji, A. K. (2009). Spawned with a silver spoon? Entrepreneurial performance and innovation in the medical device industry. *Strategic Management Journal*, **30**(2), 185–206.

Collins, J. C. (2001). *Good to Great*. New York, NY: HarperBusiness.

Currie, D. M., & Arundine, M. (2005). Medicare and drug-eluting stents. In J. S. Rakich, B. B. Longest, & K. Darr (Eds.), *Cases in Health Services Management* (pp. 63–74). Baltimore, MD: Health Professions Press.

Denend, L., & Zenios, S. (2006). Drug eluting stents: A paradigm shift in the medical device industry. Standford Graduate School of Business. *Case Study*, OIT-50, 02/13/06.

Denis, D. J. (2004). Entrepreneurial finance: An overview of the issues and evidence. *Journal of Corporate Finance*, **10**(2), 301–326.

Dooley, J. F., & Dooley, J. F. (2003). Convincing a venture capitalist to invest in your idea. *Nature Biotechnology*, **21**(7). Supp.

Flamholtz, E. (2001). Corporate culture and the bottom line. *European Management Journal*, **19**(3), 268–275.

Gladstone, D., & Gladstone, L. (2002). *Venture Capital Handbook*. Upper Saddle River, NJ: Prentice Hall/FT Press.

Gold, L. A., et al. (2005). *Private Equity and Venture Capital Financing*. Boston, MA: MCLE.

Gompers, P. A., Kovner, A., Lerner, J., & Scharfstein, D. S. (2008). Performance persistence in entrepreneurship. *Harvard Business School Working Knowledge*. Available at: http://hbswk.hbs.edu/item/6045.html. (Accessed July 13, 2009).

Hsu, D. H. (2004). What do entrepreneurs pay for venture capital affiliation? *Journal of Finance*, **59**(4), 1805–1844.

Jensen, R. A., Thursby, J. G., & Thursby, M. C. (2003). Disclosure and licensing of university inventions: The best we can do with the s**t we get to work with. *International Journal of Industrial Organization*, **21**(9), 1271–1300.

Johnson, L., & Phillips, B. (2003). *Absolute Honesty*. New York, NY: AMACOM Div American Mgmt Assn.

Kahn, A. (1991). The dynamics of medical device innovation: An innovator's perspective. *The Changing Economics of Medical Technology*, **89**.

Kent Bowen, H., Kazaka, A., Muir-Harmony, A., & LaPierre, B. (2004). *The Langer Lab: Commercializing Science*. Boston, MA: Harvard Business School.

Kotter, J. P., & Heskett, J. L. (1992). *Corporate Culture and Performance*. New York, NY: Maxwell Macmillan International.

Lin, L., & Kulatilaka, N. (2006). Network effects and technology licensing with fixed fee, royalty, and hybrid contracts. *Journal of Management Information Systems*, **23**(2), 91–118.

Macho-Stadler, I., Pérez-Castrillo, D., & Veugelers, R. (2007). Licensing of university inventions: The role of a technology transfer office. *International Journal of Industrial Organization*, **25**(3), 483–510.

Money Tree. (2009). *PricewaterhouseCoopers: Global: Insights & Solutions: MoneyTree™ Survey Report*. Available at:https://www.pwcmoneytree.com/MTPublic/ns/index.jsp. (Accessed July 13, 2009).

Nelson, R. R. (2001). Observations on the Post-Bayh-Dole rise of patenting at American Universities. *The Journal of Technology Transfer*, **26**(1), 13–19.

Ofek, E., & Wickersham, P. (2008). Examining the adoption of drug-eluting stents. *Harvard Business School Case*. 509–028.

Owen-Smith, J., & Powell, W. W. (2003). The expanding role of university patenting in the life sciences: Assessing the importance of experience and connectivity. *Research Policy*, **32**(9), 1695–1711.

Rothaermel, F. T., & Ku, D. (2008). Inter-cluster innovation differentials: The role of research universities. *IEEE Transactions on Engineering Management*, **55**(1), 9–22.

Smith, R. L., & Smith, J. K. (2000). *Entrepreneurial Finance*. New York, NY: Wiley.

Sohl, J. (2008). The angel investor market in 2008: A down year in investment dollars but not in deals. *Center for Venture Research*, **2**.

Stalcup, A. (2006). The mechanics of getting tenure. *Analytical and Bioanalytical Chemistry*, **385**(1), 1–5.

Thursby, J. G., Jensen, R., & Thursby, M. C. (2001). Objectives, characteristics and outcomes of university licensing: A survey of major US universities. *The Journal of Technology Transfer*, **26**(1), 59–72.

Toole, A. A., & Czarnitzki, D. (2007). Biomedical academic entrepreneurship through the SBIR program. *Journal of Economic Behavior and Organization*, **63**(4), 716–738.

Wilmerding, A. (2003). *Deal Terms*. Eagan, MN: Aspatore Books.

Witt, P. (2004). Entrepreneurs' networks and the success of start-ups. *Entrepreneurship & Regional Development*, **16**(5), 391–412.

CHAPTER III.2.11 POSTMARKET CONSIDERATIONS IN BIOMATERIALS AND MEDICAL DEVICES

David W. Feigal, Jr

Adjunct Faculty, Sandra Day O'Connor School of Law, Arizona State University, Santa Rosa Valley, CA, USA

The postmarket period is defined as the time when a product is in the marketplace, that is, it is no longer investigational and is available for commercial use. Consumer protection laws, such as the US Food Drug and Cosmetic Act (FDCA), create regulations, with the force of law, that determine both the manufacturers' requirements after placing products into commerce, and the authorities of regulatory agencies, such as the US Food and Drug Administration (FDA) to enforce these laws. From its inception over 100 years ago, the FDA has been concerned with products in the marketplace, and was given authorities for postmarketing regulation long before it had the premarket authority to grant marketing authorization.

The first consumer protection laws for medical products date back to 1902 and 1906, when Congress enacted the first Federal laws regulating biologics and drugs. While biologics were required to meet standards of safety, purity, and potency, the drug consumer protections focused on "truth in advertising" (misbranding) and adulterated products, and provided the US Food and Drug Administration (FDA) with the authority to remove such products from interstate commerce (Hutt, 1989). Devices would not be regulated by the FDA until laws were passed in 1933 and 1935 defining devices and extending some of the FDA's drug authority to these products. Emphasis remained on fraudulent claims and unsafe products. The FDA did not have authority to report adverse effects in the postmarket period for drugs until 1973, and it was not mandatory for medical devices until 1984 (Gross, 2007), eight years after the more comprehensive medical device regulation in the United States.

Many forces lead to the Medical Device amendments of 1976 which expanded the FDA's authority to premarket and postmarket regulation of medical devices. Arguably, one of the most salient factors was public concern about device safety problems which had been identified in the postmarketing period. Problems with the Shiley heart valve, the Dalcon Shield, orthopedic and cosmetic implants were not caused by misbranding or adulteration. Design and manufacturing problems were unknown until products were on the market. The need for postmarket safety shaped the authorities given to the FDA in the 1976 and 1984 device amendments to the FDCA and Safe Medical Device Act that followed in 1990. The FDA was given the power to recall medical devices, and require tracking, postmarketing studies, and postmarketing surveillance reporting – not only of manufacturers, but also of health facilities 30 years before all the same authorities were granted to the FDA for drugs.

How, then, do these authorities apply to biomaterials? When the biomaterial for all intents and purposes *is* the device, such as latex gloves, silicone implants, and ceramic orthopedic implants, the reporting of problems for the biomaterial will be intrinsically tied in with reports of problems with the devices, even though not all problems will be caused by the biomaterial (Bright, 2007). When the biomaterial is a small component, coating or is only part of a drug delivery container, biomaterial problems may be more difficult to identify, and may not come to the attention of the biomaterial supplier. The Quality Systems (or "device good manufacturing practices") require device manufacturers to have Corrective and Preventive Action (CAPA) systems in place to identify problems. Reporting medical device problems to the FDA or other regulatory authorities is the first line of defense to find and prevent unsafe medical devices, and to ensure that new or unanticipated problems are quickly identified and promptly remedied.

THE FDA's POSTMARKETING PROGRAMS

Postmarketing programs can be broadly grouped by whether they study exposure or adverse events, and whether the data collection is active or passive. A clinical trial is an example of an active exposure study; a registry of all recipients of a specific device using healthcare billing records is a passive exposure study; a spontaneous adverse event reporting program is a passive event driven program, while programs that actively train participating hospitals to report "near misses" and adverse events is an example of an active event driven program.

The FDA has the authority to require manufacturers, user facilities, and importers to participate in some of these programs, and can require clinical trials, registries, and tracking as part of the conditions for approval. Some postmarketing programs, such as clinical trials and active registries, require informed consent, Institutional Review Board (IRB) approval, and must comply in the US and other countries with health record privacy laws. Other programs, like MedWatch or tracking programs, are not considered clinical studies, *per se*, and although they must protect the confidentiality of the patients and reporters, do not require the patient's consent to provide medical information to the device manufacturer or regulatory authority.

PASSIVE EVENT SURVEILLANCE: SPONTANEOUS REPORTS

MedWatch

In 1993 the FDA reorganized its postmarketing adverse experience reporting systems into the MedWatch program (Kessler, 1993) to provide a single reporting system for drugs, biologics, and medical devices. Millions of reports have been collected, including over 200,000 medical device reports (MDRs) each year. Adverse event reporting is not limited to events where the casual relationship between the medical product and the adverse event is known.

Definitions

The FDA's postmarketing system relies on several important definitions, quoted below from the regulations (US Government Printing Office, 2009):

Medical Device Report (*MDR*) reportable event (or reportable event) means: (1) an event that user facilities become aware of that reasonably suggests that a device has or may have caused or contributed to a death or serious injury; or (2) an event that manufacturers or importers become aware of that reasonably suggests that one of their marketed devices: (1) may have caused or contributed to a death or serious injury, or (2) has malfunctioned and that the device or a similar device marketed by the manufacturer or importer would be likely to cause or contribute to a death or serious injury if the malfunction were to recur. (*The regulations presume that the malfunction will recur.*)

Caused or *contributed* means that a death or serious injury was or may have been attributed to a medical device, or that a medical device was or may have been a factor in a death or serious injury, including events occurring as a result of: (1) failure; (2) malfunction; (3) improper or inadequate design; (4) manufacture; (5) labeling or (6) user error.

Malfunction means the failure of a device to meet its performance specifications or otherwise perform as intended. Performance specifications include all claims made in the labeling for the device. The intended performance of a device refers to the intended use for which the device is labeled or marketed.

Serious injury means an injury or illness that: (1) is life-threatening; (2) results in permanent impairment of a body function or permanent damage to a body structure or (3) necessitates medical or surgical intervention to preclude permanent impairment of a body function or permanent damage to a body structure. *Permanent* means irreversible impairment or damage to a body structure or function, excluding trivial impairment or damage.

All manufacturers selling finished medical devices or ready for use components are required to monitor and identify significant adverse events involving medical devices in order to detect and correct problems in a timely manner. Events which must be reported include device-related deaths, serious injuries, and reportable malfunctions. The FDA requires a report within 30 calendar days from manufacturers, except when the event requires remedial action to prevent an unreasonable risk of substantial harm to public health, in which case the report must be made within five days. User facilities are required to report deaths and serious injuries to the manufacturer or the FDA within 10 calendar days.

Serious adverse events (SAEs) are of particular importance, and even single reports or small clusters of reports have allowed identification of important problems that have allowed prompt actions to mitigate risk. An important objective of medical device reports is to identify signals of potential new problems that can be evaluated further with appropriate methods. Manufacturers' quality systems have corrective and preventive action (CAPA) systems which usually include training and reporting programs, so that anyone within the company who learns of an SAE can initiate the reporting process. In the US, criminal convictions and substantial fines have resulted from failure to report MDRs (Justice Department, 2003).

While the FDA uses a single form for all medical products, there are specific information fields and types of problems that are unique to medical devices. Medical devices reports indicate whether the device malfunctioned or whether there was a "user error," i.e., if the device functioned properly but an adverse event was caused by an error in how the product was used. For example, if there was an injury from a device from an unskilled operator a MedWatch report should be filed.

MDR Limitations

It is recognized that although manufacturers are required to report all known reportable adverse reactions, only a fraction of adverse events associated with medical devices are reported in the MDR system. Some events are more likely to be reported than others: unusual events; problems with a newly implanted device; clusters of events (which often occur by chance); and publicized device problems. Unfortunately, although it is known that the degree of under-reporting varies widely, it is not known what the degree of reporting is, even for a single device, making it difficult to interpret trends.

MDR information is often incomplete, and may not even include enough data to identify the model of the medical device or even with certainty the manufacturer, and reports may be missing key details that would help assess the relationship between a device malfunction and injury. MDR reports themselves have a disclaimer that they are not intended to establish a causal relationship, but by casting a broad net, they hope not to miss signals of new problems, even if initially they are not recognized as caused by the device.

BIOMATERIALS AND MEDWATCH

For some devices most adverse experiences associated with medical devices have nothing directly to do with the biomaterial. However, when the biomaterial provides essential properties that determine the use of the device it may be hard to separate biomaterial failure from device failure. For example, the biomaterial may be the problem in the case of adverse experiences associated with rupture of the silicone shell of breast implants or with metal fatigue of a dental implant, or the wear debris from the coating of an orthopedic implant. If the final device manufacturer has also developed the biomaterial then it is straightforward for the manufacturer to consider the role of the biomaterial to the other features of the finished device. It is more difficult when the biomaterial is supplied by a separate developer who may have incomplete knowledge about the final product, but may get a complaint about an SAE that requires reporting. In general, the FDA will hold the final manufacturers responsible, but there are examples when they have used their enforcement tools to deal with a problem that was caused by a common source supplier of a component or biomaterial.

ACTIVE EVENT SURVEILLANCE

Active event surveillance systems typically train a group of hospitals and clinicians in reporting methods, and provide more accurate estimates of trends for relatively common adverse events. Cooperative efforts to capture adverse events have also been organized to assess specific technologies by specialty healthcare professional groups.

In countries with national health systems, the Ministry of Health may be able to create active surveillance at a national level. In the US, with larger passive surveillance systems most reporting comes from manufacturers, while in the UK most reporting comes from the hospitals and facilities of the National Health Service.

Active surveillance has several advantages over passive surveillance. First, the data collection is more standardized. Any monitoring system for events must be based on a precise common vocabulary for both events and outcomes. For example, device failure modes for implantable devices can be organized into predefined categories, rather than the somewhat idiosyncratic clinical descriptions found in MedWatch reports.

Second, much more relevant detail is available to characterize patients. Clinical descriptions are usually often more consistent and completely documented in an active surveillance system. The real-world use active surveillance programs can better identify whether there are subpopulations particularly vulnerable to specific device failures. Third, the medical devices can be better characterized. Not only can the make and model number be more accurately captured in active surveillance, but there is also an opportunity to evaluate other device characteristics, such as the age of the device.

The FDA's MedSun Program

In 1998 the FDA began to recruit acute care hospitals and long-term care facilities to participate in the MedSun active surveillance network (Rich, 2002). Healthcare facilities, as part of their own Quality Assurance programs, usually have internal incident reporting systems to identify potential risks and injuries to patients from problems relating to patient falls, medication errors, equipment malfunctions, and other misadventures. The FDA has obtained ongoing funding to establish a network of several hundred institutions that would anonymously share not only adverse events, but also "near misses." The analysis of "near misses" in the aviation industry had been adapted to improve anesthesia equipment decades earlier (Barach and Small, 2000), which produced anesthesia safety changes that resulted in improvements, largely from better human factor engineering. The hope was that an active surveillance program in hospitals could provide an early warning about device design and manufacturing, and use problems for a broader group of medical devices.

While the strength of spontaneous reporting was identifying low frequency, and events of unusual severity or in unusual situations, the MedSun program is designed to provide a better insight into some of the more common problems, and to train participating facilities to take corrective actions to decrease risk. The program has organized subnetworks with special interests in medical devices for cardiovascular, pediatric, ophthalmologic, diagnostic, and home use.

FDA TRACKING AND CLINICAL TRIAL REQUIREMENTS

Mandated Clinical Trials

At the time of approval of a Class III device, premarket approval (PMA) application companies must agree to "conditions of approval" requirements which can include commitments to conduct clinical trials or registries to assess safety issues. These active studies are more likely to represent "real world" use and a much larger experience than the studies conducted prior to approval, starting with the use or exposure to the device, and prospectively collecting adverse experience and other outcomes.

The FDA also has an infrequently used authority from the Food and Drug Modernization Act in section 522 of the Act (Food Drug and Cosmetic Act, Sec. 522 Postmarket Surveillance). (Hence, these are referred to as 522 studies.) When the device is an implant to be in place for more than a year or is a life-sustaining device, or when failure of the device could lead to a serious adverse event, the FDA can mandate a 522 clinical study. Since the FDA already has mechanisms under conditions of approval to require Class III studies, the 522 studies have been applied to Class II products cleared under 510(k), where there are no authorities to impose conditions for approval. In practice, perhaps in part because most high risk products are Class III products, the FDA has required few 522 studies.

Tracking. Products meeting the same criteria as a 522 study can also be required to have a tracking system for product traceabililty. The purpose of the tracking is not to collect safety data or data of any kind, beyond information on how to locate a patient with a tracked device. The tracking is required for devices where there is some likelihood that patients with those devices might need to be contacted if there was a product recall or other safety program, such as reprogramming a pacemaker or early replacement of a battery. Active implants are the most common products to have tracking requirements.

NEW FDA DEVICE SAFETY INITIATIVES

The Center for Devices has recently announced efforts to improve the MDR system. The five main areas they highlighted are:

- Working toward an electronic reporting system for adverse medical device events;
- Unique ways to identify medical devices, including standardized and globally accepted names;
- Ways to improve device information in patient records;
- Improved internal collaboration on postmarket safety issues; and
- Identifying opportunities to improve the safety of medical devices through collaborative efforts with professional organizations and the medical device industry.

The first is designed to improve both the timely acquisition of reports at the FDA, and to allow better cross-manufacturer evaluation of emerging signals. As companies automate they may improve their internal capacity to evaluate new problems. The FDA is not only interested in an improved MDR system, but is also interested in making company annual reports more useful. The fourth initiative is an internal FDA effort, although it may change the way that the premarket staff evaluate applications for new products. Manufacturers may need to systematically describe their postmarketing experience with older approved products that use some of the same technology. Finally, the last bullet point emphasizes the need to solve many problems across a whole group of manufacturers, even though the tendency historically has been to focus on the company where a problem is first identified.

Postmarket Surveillance Outside the United States

The medical device regulatory framework is much more varied from country to country than pharmaceutical regulation, in part reflecting the shorter history of device regulation compared to drugs. While the premarket approval and risk classification in Europe and some other countries is an activity delegated to non-government third parties, Notified Bodies, as described in the previous chapter, and results in market authorization across the entire European Union (EU), postmarket safety systems are a country-by-country responsibility, and are not delegated to third parties.

Passive spontaneous reporting systems, like the US FDA's MedWatch program, have their parallels in most countries. The medical device quality systems which include many of the elements that collect and respond to reports of adverse events and complaints are also nearly universal. The EU, US, Canada, Australia, and Japan participate in the Global Harmonization Task Force (GHTF) which, like the International Conference on Harmonization (ICH) for drugs and biologics, develops harmonized guidance documents. The GHTF Study Group 2 develops postmarket surveillance/vigilance guidances which can be found at www.ghtf.org/sg2.

SUMMARY: WHAT DOES POSTMARKETING SURVEILLANCE TEACH US?

First, postmarketing surveillance is one of the best ways to identify low frequency events which are usually not detected in premarket evaluation, even when clinical trials are required to market a new medical device. The type of event may be unsuspected or it may be an "event of interest" based on known problems with similar devices, or suspected from failure mode assessments from the design of the device. Postmarket surveillance has an unusual breadth, since the FDA and other competent authorities will receive reports from all products in their countries and, in this increasingly harmonized world, reports from other countries. Early detection provides manufacturers with opportunities for early mitigation, with redesign or better training in the use of their products.

Second, changes in the frequency of reports over time can detect problems or suggest mitigation of problems when a design change is made that would be difficult to assess in clinical testing. Assessing the temporal trends in the rate of lead failure of pacemakers, for example, has been useful to ensure safe use or identify problems that may be the result of design or biomaterial changes.

Third, problems that develop only after prolonged use or exposure to a biomaterial may only be available from postmarket surveillance. Late failures of implants, pacemaker leads or orthopedic implants are all difficult to determine from premarket testing or clinical studies.

Fourth, postmarket surveillance is a "real world" test of medical devices and biomaterials in patient subgroups, with off-label uses, with concomitant use of other products that are not all possible to test before marketing.

In summary, while postmarket surveillance is necessarily more observational than the controlled experiments more typical of premarket evaluation, it provides an important safety net and is the only practical method to detect medical device problems that occur at too low a frequency or only after prolonged use, or are unlikely to be observed until a product has real-world use.

BIBLIOGRAPHY

Barach, P., & Small, S. D. (2000). Reporting and preventing medical mishaps: Lessons from non-medical near miss reporting systems. *BMJ*, **320**(7237), 759–763.

Bright, R. A. (2007). Surveillance and epidemiology as tools for evaluating the materials used in medical devices. In S. L. Brown, R. A. Bright, & D. R. Tavris (Eds.), *Medical Device Epidemiology and Surveillance* (pp. 273–290). Chichester, West Sussex, UK; Hoboken, NJ: John Wiley & Sons.

Food Drug and Cosmetic Act, (1938) Sec. 522 Postmarket Surveillance, in 21 USC 360.

Gross, T. P. (2007). Medical device regulation in the USA. In S. L. Brown, R. A. Bright, & D. R. Tavris (Eds.), *Medical Device Epidemiology and Surveillance* (pp. 5–19). Chichester, West Sussex, UK; Hoboken, NJ: John Wiley & Sons.

Hutt, P. B. (1989). A history of government regulation of adulteration and misbranding of medical devices. *Food Drug. Cosm. Law. J.*, **44**, 99.

Justice Department. (2003). *Civil fraud recoveries total $2.1 billion for FY 2003*. Available from: http://www.justice.gov/opa/pr/2003/November/03_civ_613.htm.

Kessler, D. A. (1993). Introducing MEDWatch. A new approach to reporting medication and device adverse effects and product problems. *JAMA*, **269**(21), 2765–2768.

Rich, S. (2002). MedSun: User facility reporting for the new millennium. *Int. J. Trauma Nurs.*, **8**(2), 57–58.

US Government Printing Office. (2009). Part 803 Medical Device Reporting, Subpart A_General Provisions Sec. 803.3 How does FDA define the terms used in this part? *21 CFR*. April 1, 2009.

Appendices

APPENDIX A: PROPERTIES OF BIOLOGICAL FLUIDS

*Steven M. Slack**

This appendix represents a compilation of information relevant to biomaterials scientists regarding the properties and composition of several body fluids, i.e., blood, plasma (serum), cerebrospinal fluid, synovial fluid, saliva, tear fluid, and lymph. Where possible, ranges of values are provided, but the reader should recognize that significant variations are possible, particularly in states of disease. Further, the data reported here reflect adult measurements and may be substantially different in a pediatric population. Values for cerebrospinal fluid refer to the lumbar region, those for synovial fluid refer to the knee joint, and those for lymph refer to the thoracic duct, unless otherwise specified. Table A1 lists the physico-chemical properties of these fluids. Table A2 provides the typical cellular composition of human blood. Table A3 shows the normal volumes of these fluids in males and females, and presents equations whereby such volumes can be estimated from the mass of the individual. Next, Table A4 lists the approximate concentrations of the major proteins present in various biological fluids. Tables A5 and A6 present the concentrations of inorganic and organic species, respectively, in these fluids. Table A7 provides data on the major plasma proteins, i.e., concentration, molecular weight, isoelectric point (p*I*), sedimentation coefficient (*S*), diffusion coefficient (*D*), extinction coefficient (E_{280}), partial specific volume (V_{20}), carbohydrate content (*C*), and half-life. Finally, Tables A8 and A9 present information on the proteins involved in the complement pathway and blood coagulation pathway, respectively. Some of the information contained in this appendix have been previously published in Black, J., & Hastings, G. (eds.) (1998). *Handbook of Biomaterial Properties*. Chapman & Hall: New York, NY, pp. 114–124.

*Dr. Steven M. Slack** died on April 29, 2010. He was Professor in the Department of Biomedical Engineering and Associate Dean of Graduate Studies in the Herff College of Engineering, at the University of Memphis in Tennessee. He will be sorely missed by the four editors of this textbook, by his colleagues around the world, and especially by his many students at the University of Memphis.

TABLE A1 Physico-chemical Properties of Several Biological Fluids

Property	Whole Blood	Plasma (Serum)	Cerebrospinal Fluid	Synovial Fluid	Saliva	Tear Fluid
Freezing-point depression[a]	0.557–0.577	0.512–0.568	0.540–0.603	—	0.07–0.34	0.572–0.642
Osmolality[b]	—	275–295	290–324	292–300	—	309–347
pH[c]	7.35–7.45	7.35–7.43	7.35–7.70	7.29–7.45	5.0–7.1	7.3–7.7
Refractive index[d]	16.2–18.5	1.3485–1.3513	1.3349–1.3351	—	—	1.3361–1.3379
Relative viscosity[e]	2.18–3.59	1.18–1.59	1.020–1.027	>300	—	1.26–1.32
Specific conductivity[f]	—	0.0117–0.0123	0.0119	—	—	—
Specific gravity[g]	1.052–1.061	1.024–1.027	1.006–1.008	1.008–1.015	1.002–1.012	1.004–1.005
Specific heat[h]	0.87	0.94	—	—	—	—
Surface tension[i]	55.5–61.2	56.2	60.0–63.0	—	15.2–26.0	40–50

[a]Units are °C.
[b]Units are mosm/kg H_2O. Calculated from freezing-point depression.
[c]pH measured from arterial blood and plasma, and from cisternal portion of CSF.
[d]Measured at 20°C.
[e]Measured *in vitro* at 37°C for whole blood, plasma, and synovial fluid, and at 38°C for cerebrospinal fluid. The viscosity of serum is slightly less than plasma due to the absence of fibrinogen.
[f]Units are S/cm. Measured at 25°C for plasma, 18°C for CSF.
[g]Relative to water at 20°C.
[h]Units are cal/g °C.
[i]Units are dyn/cm. Measured at 20°C.

TABLE A2 Cellular Composition of Blood

Cell Type	Cells/μl	Half-Life in Circulation
Erythrocytes	$4.6–6.2 \times 10^6$ (M) $4.2–5.2 \times 10^6$ (F)	25 ± 2 days
Leukocytes		
Neutrophils	3000–5800	6–8 hours
Eosinophils	50–250	8–12 hours
Basophils	15–50	?
Monocytes	300–500	1–3 days
Lymphocytes	1500–3000	Variable
Platelets	$1.5–3.5 \times 10^5$	3.2–5.2 days
Reticulocytes	$2.3–9.3 \times 10^4$	—

TABLE A3 Volumes of Various Biological Fluids[a]

Parameter	Whole Blood	Erythrocytes	Plasma	CS Fluid	Tear Fluid
Volume (ml)	4490 (M) 3600 (F)	2030 (M) 1470 (F)	2460 (M) 2130 (F)	100–160	4.0–13

[a]The following equations can be used to estimate blood volume (BV), erythrocyte volume (EV), and plasma volume (PV) from the known body mass (*b*, kg) with a coefficient of variation of approximately 10%:

Males (M)	Females (F)
$BV = 41.0 \times b + 1530$	$BV = 47.16 \times b + 864$
$PV = 19.6 \times b + 1050$	$PV = 28.89 \times b + 455$
$EV = 21.4 \times b + 490$	$EV = 18.26 \times b + 409$

TABLE A4 Protein Concentrations (mg/dl) in Various Biological Fluids

Protein	Plasma (Serum)	Cerebrospinal Fluid	Synovial Fluid	Saliva	Tear Fluid	Lymph
Total	6000–8000	20–40	500–1800	140–640	430–1220	2910–7330
Albumin	4000–5500	11.5–19.5	400–1000	0.2–1.2	400	1500–2670
α_1-Acid glycoprotein	50–115	0.1–0.25	—	—	—	260
αA?????	—	—	—	6–70	—	—
α_1-Antitrypsin	85–185	0.4–1.0	45–110	—	1.5	—
Ceruloplasmin	15–60	0.07–1.0	1–7.5	—	4	—
Fibrinogen	200–400	0.065	—	—	—	—
Fibronectin (μg/ml)	150–300	1–3	150	<1	3–9	—
Haptoglobin	70–140	0.075–0.4	—	9	—	—
Hemopexin	50–120	—	—	—	—	—
IgA	100–400	0.1–0.3	60–115	2.2–15	4–80	—
IgG	650–1600	0.7–2.0	150–46	0.3–1.8	4–60	780
IgM	30–120	—	9–20	0.1–1.2	trace	—
Lysozyme	0.3–0.8	—	—	13–66	100–280	—
α_2-Macroglobulin	150–450	0.3–0.65	10–50	—	—	—
Transferrin	200–320	0.5–1.2	—	—	10	—

TABLE A5 Concentrations of Major Inorganic Substances (mmol/L) in Various Biological Fluids

Electrolyte	Whole Blood	Plasma (Serum)	Cerebrospinal Fluid	Synovial Fluid	Saliva	Tear Fluid	Lymph
Bicarbonate	19–23	21–30	21.3–25.9	—	2–13	20–40	—
Calcium	2.42	2.1–2.6	1.02–1.34	1.2–2.4	0.69–2.46	0.35–0.77	1.7–2.8
Chloride	77–86	98–109	122–132	87–138	6.5–42.9	110–135	87–103
Magnesium	1.48–1.85	0.80–1.05	0.55–1.23	>Serum	0.065–0.38	—	—
Total phosphorus	10.1–14.3	2.5–4.8	0.442–0.694	>Serum	3.9–9.3	0.11–10.3	2.0–3.6
Potassium	40–60	3.5–5.6	2.62–3.3	3.5–4.5	14–41	31–36	3.9–5.6
Sodium	79–91	125–145	137–153	133–139	5.2–24.4	126–166	118–132
Sulfate	0.1–0.2	0.31–0.58	—	Same as serum	—	—	—

TABLE A6 Concentrations of Organic Compounds (mg/dl) in Various Biological Fluids

Species	Whole Blood	Plasma (Serum)	Cerebrospinal Fluid	Synovial Fluid	Saliva	Tear Fluid	Lymph
Amino acids	4.8–7.4	3.6–7.0	1.0–1.5	—	—	5.0	—
Bilirubin	0.3–1.1	0.2–0.8	<0.01	—	—	—	0.8
Cholesterol	115–225	120–200	0.16–0.77	0.3–1.0	—	10.6–24.4	34–106
Creatine	2.9–4.9	0.13–0.77	0.46–1.9	—	—	—	—
Creatinine	1–2	0.6–1.2	0.65–1.05	—	0.5–2	—	0.8–8.9
Fatty acids	250–390	150–500	—	—	—	—	—
Glucose	80–100	85–110	50–80	—	10–30	10	140
Hyaluronic acid	—	—	—	250–365	—	—	—
Lipids, total	445–610	400–850	0.77–1.7	—	—	—	—
Phospholipid	225–285	150–300	0.2–0.8	13–15	—	—	—
Urea	20–40	20–30	13.8–36.4	—	14–75	20–30	—
Uric acid	0.6–4.9	2.0–6.0	0.5–2.6	7–8	0.5–2.9	—	1.7–10.8
Water (g)	81–86	93–95	94–96	97–99	99.4	98.2	81–86

TABLE A7 Properties of the Major Plasma Proteins

Protein	Plasma Concentration (mg/ml)	Molecular Weight (Da)	p*I*	*S*[a]	*D*[b]	E_{280}^{c}	V_{20}^{d}	*C*[e]	Half-life (days)
Prealbumin	0.12–0.39	54,980	4.7	4.2	—	14.1	0.74	—	1.9
Albumin	40–55	66,500	4.9	4.6	6.1	5.8	0.733	0	17–23
α_1-Acid glycoprotein	0.5–1.15	44,000	2.7	3.1	5.3	8.9	0.675	41.4	5.2
α_1-Antitrypsin	0.85–1.85	54,000	4.0	3.5	5.2	5.3	0.646	12.2	3.9
C1q	0.05–0.1	459,000	—	11.1	—	6.82	—	8	—
C3	1.5–1.7	185,000	6.1–6.8	9.5	4.5	—	0.736	—	—
C4	0.3–0.6	200,000	—	10.0	—	—	—	—	—
Ceruloplasmin	0.15–0.60	160,000	4.4	7.08	3.76	14.9	0.713	8	4.3
Fibrinogen	2.0–4.0	340,000	5.5	7.6	1.97	13.6	0.723	2.5	3.1–3.4
Fibronectin	0.15–0.2	450,000	—	13–13.6	2.1–2.3	13.5	0.72	4–9	0.33
α_2-Haptoglobin									
Type 1.1	1.0–2.2	100,000	4.1	4.4	4.7	12.0	0.766	19.3	2–4
Type 2.1	1.6–3.0	200,000	4.1	4.3–6.5	—	12.2	—	—	
Type 2.2	1.2–2.6	400,000	—	7.5	—	—	—	—	
Hemopexin	0.5–1.2	57,000	5.8	4.8	—	19.7	0.702	23.0	9.5
IgA (monomer)	1.0–4.0	162,000	—	7.0	3.4	13.4	0.725	7.5	5–6.5
IgG	6.5–16.5	150,000	6.3–7.3	6.5–7.0	4.0	13.8	0.739	2.9	20–21
IgM	0.3–1.2	950,000	—	18–20	2.6	13.3	0.724	12	5.1
Lysozyme	0.003–0.008	14,400	10.5	—	—	—	—	—	—
α_2 – Macroglobulin	1.5–4.5	725,000	5.4	19.6	2.4	8.1	0.735	8.4	7.8
Transferrin	2.0–3.2	76,500	5.9	5.5	5.0	11.2	0.758	5.9	7–10

[a]Sedimentation constant in water at 20°C, expressed in Svedberg units.
[b]Diffusion coefficient in water at 20°C, expressed in 10^{-7} cm^2/sec.
[c]Extinction coefficient for light of wavelength 280 nm traveling 1 cm through a 10 mg/ml protein solution.
[d]Partial specific volume of the protein at 20°C, expressed as mlg^{-1}.
[e]Carbohydrate content of the protein, expressed as the percentage by mass.

TABLE A8 Properties of Proteins Involved in the Complement System

Protein	Serum Concentration (mg/L)	Relative Molecular Weight M_r (Da)	Sedimentation Constant S_{20w} (10^{-13} sec)
C1q	50–100	459,000	11.1
C1r	35–40	83,000	7.5
C1s	32–40	83,000	4.5
C2	20–35	108,000	4.5
C3	1500–1700	185,000	9.5
C4	300–600	200,000	10.0
C5	120–180	185,000	8.7
C6	42–60	128,000	5.5
C7	4–60	121,000	6.0
C8	35–50	151,000	8.0
C9	45–70	71,000	4.5
Factor B	220–330	92,000	5–6
Factor D	Trace	24,000	3.0
Properdin	25–35	220,000	5.4
C1 inhibitor	145–170	100,000	—
Factor H	475–575	150,000	6.0
Factor I	30–45	88,000	5.5

TABLE A9 Properties of Proteins Involved in Blood Coagulation

Protein	Plasma Concentration (μg/ml)	Relative Molecular Weight M_r (Da)	Biological Half-Life $t_{1/2}$ (Hours)
Fibrinogen	2000–4000	340,000	72–120
Prothrombin	70–140	71,600	48–72
Factor III (tissue factor)	—	45,000	—
Factor V	4–14	330,000	12–15
Factor VII	Trace	50	2–5
Factor VIII	0.2	330,000	8–12
Factor IX	5.0	57,000	24
Factor X	12	58,800	24–40
Factor XI	2.0–7.0	160,000	48–84
Factor XII	15–47	80,000	50–60
Factor XIII	10	320,000	216–240
Protein C	4.0	62,000	10
Protein S	22	77,000	—
Protein Z	3.0	62,000	60
Prekallikrein	35–50	85,000	—
High molecular weight kininogen	70–90	120,000	—
Antithrombin III	210–250	58,000	67

APPENDIX B: PROPERTIES OF SOFT MATERIALS

M. Cristina L. Martins
Universidade do Porto, Portugal

TABLE B1 Mechanical and Physical Properties of Common Polymers Used as Biomaterials and the Respective Applications

Polymer	Tensile Strength (MPa)	Tensile Modulus (GPa)	Elongation (%)	Tg (°C)	Tm (°C)	Water Absorption (%)	Water Contact Angle (°)	Biomedical Applications
Polyethylene:								Tubing[1,4,5]; shunts[1]; catheters[3–5]
Low density polyethylene – LDPE[1]	4–16	0.1–0.3	90–800	−20	95–115	<0.01	93–95	
High density polyethylene –HDPE[1]	21–38	0.4–1.2	20–1000	−125	135–138	<0.01	91	Plastic surgery implants
Ultra high molecular weight polyethylene – UHMWPE[2]	39–48	0.8–1.6	350–525	−160	125–138	–	–	Acetabulum in total hip prostheses[4]; artificial knee prostheses[4]
Polypropylene[1]	30–38	1.1–1.6	200–700	0	165	<0.01	104	Heart valve structures[1]; oxygenator[3,4] and plasmapheresis[1] membranes; unabsorbable sutures[1,3,4]; disposable syringes[1,3,4]
Polyvinylchloride – PVC (rigid)[1] (plasticized)[1]	35–62 10–24	2.4–4.1 –	2–40 200–450	87–90	212	0.1–0.4	80	Catheters[3]; maxillofacial prostheses[6]; blood bags[1,3,5,7]; tubings[1,5,7]; plasmapheresis membrane[1]
Poly(tetrafluoroethylene) – PTFE[1]	14–35	0.4	200–400	−10	327	0	110	Oxygenator membrane[1]; vascular graft[1,5]; catheter coatings[1,5]; facial prosthesis[1]; heart valve structures[4]; stapes prosthesis (in tympanoplasty)[4,5]
Poly (dimethylsiloxane)[1]	2–10	–	100–600	(−)120–(−)123	–	0.02	101–109	Oxygenator membrane[1,4]; tubing[1,5,7]; shunts[1,4]; breast, joint, tracheal, bladder, and maxillofacial prostheses[4]; heart pacemaker leads[4]; heart valve structures[4]; burn dressings[4,5]
Poly (ethylene terephthalate) – PET or Dacron[1]	59–72	2.8–4.1	50–300	69–82	265–270	0.1–0.2	73–78	Vascular grafts[1,4]; heart valve structures[3]; shunts[1]; laryngeal, esophageal, and bladder prostheses[5,4]; nonabsorbable sutures[3,4,6]; tendon reconstruction[6]
Polyamides (nylons)[1]	62–68	1.2–2.9	60–300	45–75	200–270	1.5	–	Hemodialysis membrane[1]; nonabsorbable sutures[1–3]
Poly (ether urethane) (e.g., Pellethane)[8]	35–48	<0.01	350–600	(−)43– (−)60	188–204	–	62–107	Percutaneous leads[1]; catheters[1,7]; tubings[1,3]; intra-aortic balloons[1]; wound dressings[1]
Poly (ether urethane urea) (e.g., Biomer)[8]	31–41	<0.01	600–800	(−)53–(−)67	120–150	0.02	63–69	Artificial heart components[1]; heart valves[1]; tubing[4]; vascular graft prostheses[4]
Polystyrene[1]	35–83	2.8–4.1	1–3	116	137	0.10	45[b]	Tissue culture dish[1,3]
Polycarbonate[1]	55–66	2.4	100–130	150	267	0.2	62[b]	Connectors[4]; oxygenator, hemodialysis, and plasmapheresis membranes[4]
Polysulfones[1]	70.3	2.5	50–100	190–285	–	0.22	–	Artificial heart components[1,7]; heart valve structures[1,7]; oxygenator, hemodialysis, and plasmapheresis membrane[4]

(Continued)

TABLE B1 Mechanical and Physical Properties of Common Polymers Used as Biomaterials and the Respective Applications *(Continued)*

Polymer	Tensile Strength (MPa)	Tensile Modulus (GPa)	Elongation (%)	Tg (°C)	Tm (°C)	Water Absorption (%)	Water Contact Angle (°)	Biomedical Applications
Poly(aryl-ether-ether-ketone): PEEK[9] (OPTIMA LT1)	93	4[c]	30–40	~143	~343	–	–	Orthopedic and spinal implants
Poly (methyl methacrylate) – PMMA[1]	48–76	3.1	2–10	110	160	0.3–0.4	62	Dentures[1,4,6]; plasmapheresis[1] and hemodialysis membranes[3,4]; bone cement[1,3–6]; intraocular lens[1,3–5]; middle ear prosthesis[1,3]
Poly (2-hydroxyethylmethacrylate) – PHEMA[1]	0.3	0.8	50	115 (dry)	–	>1	–	Catheter coating[1,7]; drug delivery devices[1]; soft contact lenses[1,6]; vascular prosthesis coatings[1]; burn dressings[1,6]
Poly(vinyl alcohol)[4] (98–99% hydrolyzed) (87–89% hydrolyzed)	67–110 24–79		0–300	85 58	230 180	–	–	Drug delivery devices[4,6]
Poly (ε-caprolactone) (MW: 44,000)[10]	16	0.4	80	−62	57	<0.2	–	Drug delivery devices[7]
Poly (glycolic acid)-PGA[11] (fibers)	57 750	6.5 13.4	–	35 –	225 233	–	–	Drug delivery devices; absorbable sutures[6,7]
Poly (lactic acid) – PLA[10] L-PLA (M_w: 50,000–300,000) D,L-PLA (M_w: 107,000–550,000)	28–48 29–35	1.2–3.0 1.9–2.4	6–2 6–5	54–59 51–53	170–178 –	–	–	Drug delivery devices[6]
Poly(ortho ester)[d] (M_w: 99,700)	20	0.82	220	55	–	0.2–1	–	Drug delivery devices[1,7]
Poly(β-hydroxybutyrate) – PHB[10] (M_w: 370,000) P(HB-co-22 mol% hydroxyvaleric acid)[10] (M_w: 227,000)	36 16	2.5 0.62	2.5 36	1 −5	171 137	– –	– –	Drug release devices; sutures; artificial skin[10]; cardiovascular patches; orthopedic pins; nerve guides; wound dressings; bone marrow scaffolds[12]
Cellulose acetate[4]	13.1–58.6	0.6–1.8	6–50	–	306	2.0–6.5	–	Hemodialysis membrane[6,7]
Collagen[13,e] (uncross-linked fibers) (cross-linked fibers)	0.70–1.12 29.7–63.9	(1.5–2.2) × 10^{-3} 0.271–0.495	61.1–74.8 12.9–18.3	–	–	–	–	Hemostatic agent; vascular prosthesis; heart valves; tendons and ligaments; wound and burn dressings; absorbable sutures[6,14]
Silk[15,16,f] (with sericin) (without sericin)	500 610–690	5–12 15–17	19 4–16	–	–	–	–	Nonabsorbable sutures[6,7]; wound healing dressings; scaffolds for regeneration of bone, cartilage, tendon, and ligament tissues[12,16]
Chitosan[17,g] (M_v: 8.73 (±0.55) × 10^5) (Deacetylation (%) = 98.17 ± 0.35)	56.2–67.6	2.93–3.30	26.9–33.3	50.8–53.2	–	–	–	Surgical sutures; dental implants; artificial skin; rebuilding of bone; corneal contact lenses; drug release devices[18]

[a]Applications are dependent of the polymer morphologic form.
[b]values are averages of advancing and receding contact angles.
[c]Flexural modulus.
[d]Transcyclohexane dimethanol (t-CDM): 1,6-hexanediol (1,6-HD) = 35:65.
[e]Acid-soluble collagen obtained from rat tail tendons. Collagen fibers coextruded using NaCl (coagent).
[f]*Bombyx mori* single-brin silkworm silk.
[g]Chitosan films.

BIBLIOGRAPHY

1. Marchant, R. E., & Wang, I. (1994). Physical and chemical aspects of biomaterials used in humans. In R. S. Greco (Ed.), *Implantation Biology – The host Response and Biomedical Devices* (pp. 13–38). Boca Raton, Fl: CRC Press.
2. Steven, M. K. (2004). *The UHMWPE Handbook: Ultra-high Molecular Weight Polyethylene in Total Joint Replacement.* Academic Press.
3. Lee, H. B., Kim, S. S., & Khang, G. (1995). Polymeric Biomaterials. In J. D. Bronzino (Ed.), *The Biomedical Engineering Handbook* (pp. 581–597). Boca Raton, Fl: CRC Press.
4. Kroschwitz, J. I. (1990). *Concise Encyclopedia of Polymer Science and Engineering.* New York, NY: John Wiley and Sons Inc.
5. Dumitriu, S., & Dimitriu-Medvichi, C. (1994). Hydrogel and general properties of biomaterials. In S. Dumitriu (Ed.), *Polymeric Biomaterials* (pp. 3–97). New York, NY: Marcel Dekker Inc.
6. Williams, D. (1990). Concise Encyclopedia of Medical Dental *Materials*. UK: Pergamon Press.
7. Helmus, M. N., & Hubbell, J. A. (1993). Materials selection. *Cardiovascular Pathology*, 2(suppl.), 53S–71S.
8. Lamba, N. M.K., Woodhouse, K. A., & Cooper, S. L. (1998). *Polyurethanes in Biomedical Applications*. Boca Raton, Fl: CRC Press.
9. Kurtz, S. M., & Devine, J. N. (2007). PEEK biomaterials in trauma, orthopedic and spinal implants. *Biomaterials*, **28**, 4845–4869.
10. Engelberg, I., & Kohn, J. (1991). Physico-mechanical properties of degradable polymers used in medical applications: A comparative study. *Biomaterials*, **12**, 292–304.
11. Fambri, L., Migliaresi, C., Kesenci, K., & Piskin, E. (2002). Biodegradable polymers. In R. Barbucci (Ed.), *Integrated Biomaterials Science* (pp. 119–170). New York, NY: Springer.
12. Chen, G. Q., & Wu, Q. (2005). The application of hydroxyalkanoates as tissue engineering materials. *Biomaterials*, **26**, 6565–6578.
13. Pins, G. D., Huang, E. K., Christiansen, D. L., & Silver, F. H. (1997). Effect of static axial strain on the tensile properties and failure mechanisms of self-assembled collagen fibers. *Journal of Applied Polymer Science*, **63**, 1429–1440.
14. Li, S. T. (1995). Biological Biomaterials: Tissue-Derived Biomaterials (Collagen). In J. D. Bronzino (Ed.), *The Biomedical Engineering Handbook* (pp. 627–647). Boca Raton, Fl: CRC Press.
15. Perez-Rigueiro, J., Viney, C., Llorca, J., & Elices, M. (2000). Mechanical properties of single-brin silkworm silk. *J Applied Polymer Science*, **75**, 1270–1277.
16. Vepari, C., & Kaplan, D. L. (2007). Silk as a biomaterial. *Progress in Polymer Science*, **32**, 991–1007.
17. Lim, L. Y., Khor, E., & Koo, O. (1998). γ Irradiation of chitosan. *Journal of Biomedical Material Research (Applied Biomaterials)*, **43**, 282–290.
18. Rinaudo, M. (2006). Chitin and chitosan: Properties and applications. *Progress in Polymer Science*, **31**, 603–632.

APPENDIX C: CHEMICAL COMPOSITION OF METALS USED FOR IMPLANTS

Jack E. Lemons
University Professor, Schools of Dentistry, Medicine and Engineering, University of Alabama at Birmingham, Birmingham, AL, USA

Biomaterial	Nominal Composition (Elements)	Tensile Strength (MPa), (Min)	Elongation (%), (Min)	Comments
Iron (Fe) Alloys	Fe-18Cr-2.5Mo	480	30	Cast and Wrought Forms
	Fe-18Cr-14Ni-2.5Mo	150–1350	10–40	
	Fe-Zr-10Ni-3 Mn	74–1100	10–35	
	Fe-15Ni-2.5Mo	490–860	4–40	
	Fe-22Cr-13Ni-5 Mn-2.5Mo	690–1035	12–35	
	Fe-23 Mn-21Cr-1Mo	827–1379	12–30	
Cobalt (Co) Alloys	35Co-35Ni-20Cr-10Mo	241–1580	3–60	Cast and Wrought Forms
	Co-28Cr-6Mo	655–1192	8–20	
	40Co-20Cr-16Fe-15Ni-7Mo	1240–2275	1–65	
	Co-20Cr-15 W-IONi	190–896	20–45	
Titanium (Ti) and Alloys	99Ti (Grade I-IV)	240–550	4–30	Cast and Wrought Forms
	Ti-6Al-4 V	860	8	
	Ti-6Al-4 V (ELI)	825–860	8–10	
	Ti-6AL-7Nb	800	10	
	Ti-13Nb-13Zr	560–860	8–15	
	Ti-12Mo-6Zr-2Fe	931	12	
	Ti-15Mo	690–724	12–20	
	Ti-3Al-2.5 V	621–862	10–15	
Tantalum (Ta)	99Ta	172–517	12–30	Deposit and Wrought Forms
Zirconium (Zi) Alloy	Zr-2.5Nb	450	15	Wrought Forms
Nickel (Ni) Titanium (Ti) Alloy	45-57Ni-Ti	551	10-May	Wrought Forms

*Data from ASTMF04, Volume 13.01, Medical and Surgical Materials and Devices, 2011.

APPENDIX D: THE BIOMATERIALS LITERATURE

Buddy D. Ratner
Professor, Bioengineering and Chemical Engineering, Director of University of Washington Engineered Biomaterials (UWEB), Seattle, WA, USA

In the last ten years, we have seen the launch of many new biomaterials journals, and the publication of new books, paralleling the growth of the field and the increasing globalization of biomaterials and medical devices. Though this list of journals and books cannot be complete, it does illustrate the breadth of the field and may provide some guidance in working with this complex, multi-disciplinary literature.

As of the 2010 listing of impact factors by the Institute for Scientific Information, the following are some of the lead journals in the field classified as "biomaterials:"

European Cells & Materials (eCM) (9.650)
Biomaterials (7.882)
Lab on a Chip (6.26)
Nanomedicine (6.202)
Biomacromolecules (5.325)
Acta Biomaterialia (4.822)
Tissue Engineering (4.636)
Soft Matter (4.457)
Journal of Biomedical Materials Research Part A (3.044)

This list is particularly interesting because many of the most influential journals impacting biomaterials are not "classical" biomaterials journals.

Biomaterials Journals (or Journals with Significant Biomaterials Content)

Acta Biomaterialia (Elsevier)
Advanced Drug Delivery Reviews (Elsevier)
Advanced Healthcare Materials (WILEY-VCH Verlag GmbH & Co.)
American Journal of Drug Delivery (Wolters Kluwer)
American Society of Artificial Internal Organs Transactions
Annals of Biomedical Engineering (Springer – Official Publication of the Biomedical Engineering Society)
Annual Reviews of Biomedical Engineering
Artificial Cells, Blood Substitutes, and Biotechnology (Informa Healthcare) (T. M. S. Chang, editor)
Artificial Organs (Raven Press)
Artificial Organs Today (T. Agishi, editor; VSP Publishers)
Bio-Medical Materials and Engineering (T. Yokobori, editor; Pergamon Press)
Biofabrication (Institute of Physics)
Biofouling (Harwood Academic Publishers)
Bioinspiration and Biomimetics (Institute of Physics)
Biointerphases (American Institute of Physics)
Biomacromolecules (American Chemical Society)
Biomaterial–Living System Interactions (BioMir; V. I. Sevastianov, editor)
Biomaterials (including Clinical Materials) (Elsevier)
Biomaterials Forum (Society For Biomaterials)
Biomaterials: Processing, Testing and Manufacturing Technology (Butterworth)
Biomatter (Landes Bioscience)
Biomedical Engineering OnLine (Electronic – http://www.biomedical-engineering-online.com/start.asp)
Biomedical Materials: Materials for Tissue Engineering and Regenerative Medicine (Institute of Physics)
Biomedical Microdevices (Kluwer)
Biosensors and Bioelectronics (Elsevier)
Cell Stem Cell (Cell Press – Official Journal of the International Society for Stem Cell Research (ISSCR)) (Elsevier)
Cell Transplantation – The Regenerative Medicine Journal (Cognizant Communication Corporation)
Cells Tissues Organs (Karger)
Clinical Biomechanics (Elsevier)
Colloids and Surfaces B: Biointerfaces (Elsevier)
Dental Materials (Elsevier, Official Publication of the Academy of Dental Materials)
Drug Delivery Systems & Sciences (Euromed Scientific)
Drug Delivery Technology (Montville, NJ)
Drug Targeting and Delivery (Academic Press)
e-biomed: *The Journal of Regenerative Medicine* (Mary Ann Liebert, Inc.) *discontinued*
European Cells and Materials (eCM) (Electronic – http://www.eurocellmat.org.uk)
Expert Review of Medical Devices (Expert Reviews Ltd, London)
Frontiers of Medical and Biological Engineering (Y. Sakurai, editor; VSP Publishers)
IEEE Transactions on Biomedical Engineering
International Journal of Artificial Organs (Wichtig Editore)
International Journal of Biomaterials (Hindawi Publishing Corp.).Openaccess:http://www.hindawi.com/journals/ijbm/contents.html
International Journal of Nanoscience (World Scientific Journals, Imperial College Press, London)
Journal of Applied Biomaterials and Biomechanics (Wichtig Editore)
Journal of Bioactive and Compatible Polymers (Sage Publications)
Journal of Biochips & Tissue Chips (OMICS Publishing Group)
Journal of Biomaterials Applications (Sage Publications)
Journal of Biomaterials Science: Polymer Edition (VSP Publishers)
Journal of Biomedical Materials Research A (Wiley-VCH – Official Publication of the Society for Biomaterials)
Journal of Biomedical Materials Research B: Applied Biomaterials (Wiley-VCH – Official Publication of the Society for Biomaterials)
Journal of Biomimetics, Biomaterials, and Tissue Engineering (Transtech Publications, Inc)
Journal of Controlled Release (Elsevier)
Journal of Developmental Biology and Tissue Engineering (academicjournals.org)

Journal of Drug Targeting (Harwood Academic Publishers)
Journal of Engineering in Medicine (Institution of Mechanical Engineers)
Journal of Long Term Effects of Medical Implants (CRC Press)
Journal of Materials Science: Materials in Medicine (Kluwer Academic Publishers – Official Journal of the European Society for Biomaterials)
Journal of Medical Devices (ASME)
Journal of Mechanical Behavior of Biomedical Materials (Elsevier)
Journal of Nanobiotechnology (BioMed Central, Springer)
Journal of Neural Engineering (Institute of Physics)
Journal of Oral Tissue Engineering (Japanese Association of Regenerative Dentistry)
Journal of Tissue Engineering (Wiley Interscience)
Lab on a Chip (RSC Publishing)
Langmuir (American Chemical Society)
Macromolecular Bioscience (Wiley-VCH)
Materials in Medicine (Springer – Official Publication of the European Society for Biomaterials)
Medical Device and Diagnostics Industry (Canon Publications)
Medical Device Research Report (AAMI)
Medical Device Technology (Astor Publishing Corporation)
Medical Plastics and Biomaterials (Canon Communications, Inc.; not publishing)
Nano (American Chemical Society)
Nano LIFE (World Scientific Journals, Imperial College Press, London)
Nanobiology (Carfax Publishing Co.)
Nanobiotechnology (Springer Publishers)
Nanomedicine (Future Medicine)
Nanotechnology (Institute of Physics Publishing Co.)
Nature Biotechnology (Nature Publishing Group)
Nature Materials (Nature Publishing Group)
Small (Wiley-VCH)
Soft Materials (Taylor & Francis Group, London)
Soft Matter (RSC Publishing, London)
Stem Cells (AlphaMed Press)
The Open Biomaterials Journal (Bentham Publishers – http://www.bentham.org/open/tobiomtj/EBM.htm)
Tissue Engineering (Mary Ann Liebert, Inc.)
Trends in Biomaterials & Artificial Organs (Society For Biomaterials And Artificial Organs – India)
Wound Repair and Regeneration (Blackwell Publishing)

Biomaterials Books

Plastics, Rubber and Health, G. Akovali, Smithers RAPRA Publishing: Akron, OH (2006).

Integrated Biomaterials Science, R. Barbucci, Springer: New York, NY (2002).

Advanced Biomaterials: Fundamentals, Processing, and Applications, B. Basu, D. S. Katti, A. Kumar, Wiley-American Ceramic Society: New York, NY (2009).

Biological Performance of Materials: Fundamentals of Biocompatibility, 4th ed., J. Black, CRC Press: Boca Raton, F. (2005).

Tissue Engineering Using Ceramics and Polymers, A.R. Boccaccini and J. Gough, (Eds.). Woodhead Publishing Ltd: Cambridge, UK (2007).

Biomaterials, A Textbook (Kindle Edition), T. Boland, Amazon Digital Services.

Contemporary Biomaterials – Material and Host Response, Clinical Applications, New Technology and Legal Aspects, J. W. Boretos and M. Eden, (Eds.). Noyes Publ.: Park Ridge, NJ (1984).

Titanium in Medicine: Material Science, Surface Science, Engineering, Biological Responses and Medical Applications, D. M. Brunette, P. Tengvall, and M. Textor, (Eds.). Springer-Verlag: Berlin, Heidelberg, New York (2001).

Biomaterials Fabrication and Processing Handbook, P. K. Chu, and X. Liu, (Eds.). CRC Press: Boca Raton, FL (2008).

An Introduction to Tissue–Biomaterial Interactions, Kay C. Dee, David A. Puleo, and Rena Bizios. Wiley-Liss: New York, NY (2002).

Biomaterials and Tissue Engineering. In: *Urology*, J. Denstedt, and A. Atala, (Eds.). Woodhead Publishing Ltd: Cambridge, UK (2009).

Applications of Biomaterials in Facial Plastic Surgery, A. I. Glasgold, and F. H. Silver. CRC Press: Boca Raton, FL (1991).

Cells and Biomaterials for Intervertebral Disc Regeneration, S. Grad, M. Alini, D. Eglin, D. Sakai, J. Mochida, S. Mahor, E. Collin, B. Dash, and A. Pandit. Morgan & Claypool Publishers: San Francisco, CA (2010).

Osseo-Integrated Implants, G. Heimke. CRC Press: Boca Raton, FL (1990).

Biomaterials: An Interfacial Approach, L. L. Hench, and E. C. Ethridge. Academic Press: New York, NY (1982).

An Introduction to Biomaterials, 2nd edn, J. O. Hollinger. CRC Press: Boca Raton, FL (2011).

Printed Biomaterials: Novel Processing and Modeling Techniques for Medicine and Surgery, R. Narayan, T. Boland, and Y-S. Lee, (Eds.). Springer: New York, NY (2009).

Biomaterials: An Introduction, 3rd edn, J. B. Park. Springer: New York, NY (2007).

Artificial Cells, Cell Engineering and Therapy, S. Prakash, (Ed.). Woodhead Publishing: Cambridge, UK (2007).

Mechanics of Biomaterials: Fundamental Principles for Implant Design, L. A. Pruitt, and A. M. Chakravartula. Cambridge University Press: Cambridge, UK, (2011).

Biomaterials: A Nano Approach, S. Ramakrishna, M. Ramalingam, T. S. Sampath Kumar, and W. O. Soboyejo. CRC Press: Boca Raton, FL (2010).

Handbook of Biomaterials Evaluation, 1st edn, A. F. Von Recum, (Ed.). Macmillan: New York, NY (1986).

Interventional and Surgical Cardiovascular Pathology: Clinic Correlations and Basic Principles, F. J. Schoen. W. B. Saunders: Philadelphia, PA (1989).

Biocompatibility: Interactions of Biological and Implanted Materials,Vol. 1 – Polymers, F. H. Silver, and C. Doillon. VCH: New York, NY (1989).

Biomaterials Science and Biocompatibility, F. H. Silver, and D. L. Christiansen. Springer: New York, NY (1999).

Biomaterials: The Intersection of Biology and Materials Science, J. S. Temenoff, and A. G. Mikos. Prentice Hall: Upper Saddle River, NJ (2008).

Surfaces and Interfaces for Biomaterials, P. Vadgama. CRC Press: Boca Raton, FL. (2005).

Concise Encyclopedia of Medical and Dental Materials, 1st edn., D. Williams (Ed.). Pergamon Press: Oxford, UK (1990).

Definitions in Biomaterials (Proceedings of a Consensus Conference of the European Society for Biomaterials, March 3–5, 1986, Chester, UK). D. F. Williams (Ed.). Elsevier Science Ltd: Amsterdam. (1987).

Williams Dictionary of Biomaterials, D. F. Williams (Ed.). Liverpool University Press: Liverpool, UK (1999).

Biomaterials, J. Y. Wong, and J. D. Bronzino. CRC Press: Boca Raton, FL (2007).

A Laboratory Course in Biomaterials, W. Xian. CRC Press: Boca Raton, FL (2011).

CRC Handbook of Bioactive Ceramics, T. Yamamuro, L. L. Hench, and J. Wilson. CRC Press: Boca Raton, FL (1990).

(Note: A more extensive listing of biomaterials books can be found on www.biomat.net.)

Index

Note: Page numbers with "b" denote boxes; "f" figures; "t" tables.